P9-CAT-104

GENIUM'S

HANDBOOK OF

Safety, Health, and Environmental Data

for common hazardous substances

GENIUM'S HANDBOOK OF Safety, Health, and Environmental Data for common hazardous substances

Genium Publishing Corporation
One Genium Plaza
Schenectady, N.Y. 12304-4690

McGRAW-HILL
New York San Francisco Washington, D.C. Auckland Bogotá
Caracas Lisbon London Madrid Mexico City Milan
Montreal New Delhi San Juan Singapore
Sydney Tokyo Toronto

McGraw-Hill 

A Division of The McGraw-Hill Companies

3 4 5 6 7 8 9 0 KGP/KGP 9 0 3 2 1 0 9

P/N 134337-7
PART OF
ISBN 0-07-024577-0 or 0-07-913614-1

The sponsoring editor for this book was Robert Esposito.

Printed and bound by Quebecor/Kingsport.

Contents

Volume I

Volume II

Volume III

Introduction

As a long-time developer of Material Safety Data Sheets, Genium Publishing needs to gather comprehensive and accurate data about hazardous materials. This is the same information that you, as a safety, health, or environmental professional, may need for your work. From our own experiences searching for data, we know that finding the information that you want can be both frustrating and time-consuming. We have applied this search-for-data process to thousands of hazardous materials, used rules and judgement with some of the data to derive items such as the pictograms, organized the data into chemical profiles, and present these profiles in this publication, ***The Handbook of Safety, Health, and Environmental Data*** (HSHE). HSHE is a compilation of data from a variety of sources that acts as a quick, yet comprehensive, source of information about hazardous materials.

Profiles included in this publication were selected from materials regulated by OSHA, EPA, DOT, and other federal agencies; materials listed in the Hazardous Substances Data Bank, developed and maintained by the National Library of Medicine; and other commonly used materials. If only limited information was found for a material, then the profile was not included in this collection. The key to locating data was the CAS number. Only those materials identified by a CAS number have been included, and only sources which identify materials by CAS number were used. Finally, most data was obtained from electronic sources, using the computer to extract the information. Note that an important data source was the Genium MSDS Collection because the MSDSs in this Collection contain the results of many Genium-conducted data searches.

Based upon Genium's data needs, and the data needs of other health, safety, and environmental professionals, over 100 data fields were identified and categorized into the following five sections:

Material Identification: Not only is there a need to locate information about a material via many different identifications such as name, CAS number, and EINECS number, but once you've located information about a material you may need to verify or search for additional information using identifiers such as other material names, RTECS number, DOT number, or the molecular formula.

Physical Properties: Sources were searched for data in 21 fields of information that are generally of interest to HSE professionals.

Toxicity Data: For many materials, there is so much toxicity data that it's difficult to find the most important. Criteria were established to select only the studies of greatest interest to HSE professionals.

Information about Hazards to Humans and How to Avoid These Hazards: From easy to recognize and understand pictograms, to health, fire and reactivity overviews, to respirator selection guidelines, the material hazards are identified, classified , and presented in an easy-to-understand manner.

Information about the Material in the Environment: Data include methods for identifying the presence of the material in water, soil, and air; environmental regulations; environmental physical properties data; and summaries for ecotoxicity, environmental fate, and cleanup/disposal.

HSHE contains data collected from respected sources such as HSDB and RTECS. Though the data have been standardized for consistent presentation, no attempt was made to evaluate the accuracy of the data. From past experience, the sources of information that were used have proven to be accurate and reliable. However, occasionally there are inconsistencies among these sources. Depending upon your specific use for these data, professional judgement should always be applied to this type of information. We recommend quickly scanning the following introduction to get a basic understanding of the data presented, and then using this introduction as a reference for further explicating the data when necessary.

1. Material identification: The information in the Material Identification section gives multiple fields–from common names to identification numbers–for cross-referencing and identifying chemical compounds.

1.1 Index Code: The unique material identifier developed by Genium consisting of the first three letters of the material name (unless there are less than three letters in the material name), followed by four digits.

1.2 CAS #: This number, assigned by the American Chemical Society's Chemical Abstracts Service, is a unique index which allows cross-referencing and identification of a chemical even when various synonyms have been used. All data was collected by using the CAS number as the unique identifier in the data sources.

1.3 Material Name: Each material is referenced under what Genium has determined to be the most commonly used name.

1.4 RTECS: A unique number assigned by NIOSH in the ***Registry of Toxic Effects of Chemical Substances***.

1.5 DOT: A number, published by the Department of Transportation, used to identify materials for regulation of their transportation.

1.6 EINECS Number: An identification number published as part of the European Inventory of Existing Chemical Substances. It identifies chemical substances marketed in the European community between 1/1/71 and 9/18/81. Also included in this field are ELINICS numbers which are a supplemental list to the EINICS list, consisting of additional materials marketed in the European community between 9/18/81 and 6/30/90.

1.7 Molecular Formula: Simplified molecular formulas have been presented. The format for this simplified molecular formula is as follows: If a material contains carbon or hydrogen, the symbol for these elements (along with the number of atoms of these elements in the material) are listed first and second, respectively. The symbols for the remaining elements within the material follow in alphabetical order. If a material does not contain carbon or hydrogen, the element symbols and the number of atoms are listed in alphabetical order.

1.8 Structured MF: The molecular formula found in the literature. In some sources this was specifically identified as a structured molecular formula, while in other sources it was just given as the molecular formula, but based on the way the data was presented some semblance of structure was indicated. A standard for presenting the structured molecular formula was not developed. What is presented is what was found in the literature. Therefor the kind of structural representations for some materials may differ from the kind of structural representations for other materials. In some cases the number of atoms of each element shown in the structured molecular formula may differ from the simplified molecular formula due to such things as a variation in the number of water molecules, or ions that may be present in complex molecules.

1.9 Formula Weight: When possible, the molecular weight reported in the literature is given. When necessary, the molecular weight was calculated from the simplified molecular formula.

1.10 Chemical Structure: A two-dimensional diagram of the material's molecular structure, taken from the literature, is included.

1.11 Synonyms: The list of material names, listed alphabetically and printed in all caps, includes all known chemical names, tradenames, foreign names, and common names. Exact duplicates have been eliminated, however, differences in punctuation were judged to be a different name and are included.

1.12 Description: The information found in the data sources was organized into a color, physical state and odor format. Note that some materials may have more than one physical form or color, so the user cannot use the description as a unique identifier of the chemical compound.

1.13 Use: This section contains a brief synopsis of common uses of the chemical as found in the literature. These uses have not been verified, and it is possible that the chemical is no longer used for the purposes given in this statement (in fact, there may be cases where the material is no longer used at all!).

2. **Physical Properties:** The Physical Properties section consists of data (facts) that assist HSE professionals in the determination of material hazards and the possible effects on people and the environment.

2.1 **Boiling Point:** The temperature at which a liquid's vapor pressure equals the surrounding atmospheric pressure so that the liquid rapidly vaporizes. Materials with low boiling points (such as butane) may be fire hazards. The boiling point is listed in both Celsius and Fahrenheit, and may appear as a range. If the literature has noted specific conditions, such as pressure conditions, or whether the compound decomposes at this temperature, we have included that information here.

2.2 **Freezing Point:** The temperature at which a material changes from a liquid to a solid state upon cooling (or from a solid to a liquid upon heating, generally referred to as the melting point). When materials freeze, they may expand enough to burst their container, or the type or degree of hazard may change. The freezing point (melting point) is also listed in both Celsius and Fahrenheit, and like the boiling point, may appear as a range. If the literature has noted specific conditions, such as pressure conditions, or whether the compound decomposes or sublimates at this temperature, we have included that information here.

2.3 **Specific Gravity:** The specific gravity, which may appear as a range, is the ratio of the density of the material to the density of a reference material, at a specific temperature. A common reference material for solids and liquids is water, which has a density of 1 kg/l, or 1 g/ml, or 1 g/cm^3 at 4 degrees C. A common reference material for gases is air, which has a density of 1.29 g/l at 0 degrees C, 760 mm Hg. Specific gravity is a dimensionless number. Occasionally the literature will report a specific gravity value with respect to an uncommon material, and/or at an uncommon temperature or pressure. If a solid or liquid has a specific gravity less than 1, then it will float on water; a value greater than 1 and it will sink in water. If a gas has a specific gravity less than 1 it will float to the ceiling or off into the atmosphere; greater than 1 and it will sink to the floor or ground.

2.4 **Vapor Density:** The ratio of the formula mass of the material to the average formula mass of the gasses which constitute air (29 grams per mole). The vapor density may be reported as a range, or may be described as >, calculated, or estimated. The literature may also report the temperature at which the measurement was made. This ratio is commonly reported in the literature, but it is only correct for a pure gas at room temperature. A liquid cannot liberate vapors more concentrated than its saturated vapor concentration. A more accurate parameter is the Saturated Vapor Density, which Genium calculated and is described below.

2.5 **Saturated Vapor Density:** The saturated vapor concentration of a liquid is the ratio of its vapor pressure at a given temperature to the atmospheric pressure. Using this ratio, the % of the compound in air and the remaining % of air at saturation (for example, 19.7% hexane and 80.3% air) can be calculated. The saturated vapor density is then determined by multiplying the decimal equivalent of the % of the compound in air by its formula mass (FM) and the decimal equivalent of the % of air by its FM; adding this air/liquid vapor mixture at saturation; and dividing the sum by 29 and multiplying by the density of pure air (1.2 kg/m^3, 0.075 lbs/ft^3). Saturated air/liquid vapor mixtures may be heavier than air, but not as heavy as formula mass ratios indicate. Temperature differences and turbulence create density differences between volumes of air and often have a greater influence on the movement of contaminated air than the actual saturated vapor density of the chemical.

2.6 **Density:** The ratio of weight (mass) to volume of a material, usually reported in grams per cubic centimeter (g/cc) or grams per milliliter (g/ml), and sometimes reported as a range. A reference temperature is frequently given with this information.

2.7 Bulk Density: The mass (weight) per unit volume of a solid particulate material as it is normally packed, with voids between particulates containing air. This property is not commonly given, but when it is, it is usually reported in pounds per cubic foot (lb/ft^3) or pounds per gallon (lb/gal), and may appear with a reference temperature.

2.8 Vapor Pressure: The pressure a saturated vapor exerts above its own liquid in a closed container. Materials with high vapor pressures will quickly evaporate into the atmosphere, hence, knowing vapor pressures and evaporation rates can help predict how quickly a compound will become airborne and thus how quickly workers will be exposed to the chemical. Vapor pressure is usually measured in millimeters of mercury or kPa at a designated temperature (this temperature is listed when it is reported in the literature). Vapor pressure information may be reported subjectively using terms such as "negligible" or "extremely low."

2.9 Water Solubility: The extent to which a material will dissolve in water is a very important fact for HSE professionals. Numeric values are found with units such as percentage by weight in a solution, grams of solute per liter of solution, grams of solute dissolved in 100g of water, or other units. In some cases, the temperature at which these measurements were taken is given. Water solubility may be a range of values or it may be qualified by terms such as "greater than", "less than", "approximately", etc.. The literature also reports water solubility in more subjective terms such as "negligible", "slight", "moderate", "appreciable", "complete", "soluble in all proportions", etc..

2.10 Other Solubilities: This section contains descriptions (usually not numeric values) of the solubility in other common solvents such as alcohols.

2.11 Surface Tension: The force acting on the surface of a liquid, tending to minimize the area of the surface. Quantitatively it is the tension across a unit length of a line on the surface of a liquid. The surface tension of a liquid depends on the temperature; it diminishes as temperature increases and becomes zero at the critical temperature.

2.12 Odor Threshold: The values for odor threshold vary widely because this parameter is the point at which an individual can detect an odor, and each individual's ability varies widely and can be subjective. The odor threshold is commonly expressed as, milligrams per liter (mg/l), milligrams per cubic meter (mg/m^3), or may be described as a range.

2.13 pH: A 14 point scale which measures the hydrogen ion concentration in a solution. Low values indicate acidity, while high values indicate alkalinity. The mid-point value of 7 is neutral. pH can be reported as a range, and sometimes all that is reported in the literature is that the solution is acid or alkaline.

2.14 Refractive Index: The change in direction (apparent bending) of a light ray passing from one medium to another of different density. The ratio of the sine of the angle of incidence to the sine of the angle of refraction is the index of refraction of the second medium. Index of refraction of a substance may also be expressed as the ratio of the velocity of light in a vacuum to its velocity in the substance. It varies with the wavelength of the incident light, temperature, and pressure. The usual light source is the D line of sodium. The refractive index can appear as a range, and is usually accompanied by the temperature at which the index was measured.

2.15 Evaporation Rate: This is the rate at which a material vaporizes from the liquid or solid state in comparison with a known material's vaporization rate (usually normal butyl acetate). This rate is useful for evaluating the health or fire hazard potential of a material.

2.16 Critical Temperature: The temperature above which a gas cannot be liquefied by pressure. All values are expressed in degrees Celsius, and may appear as a range or with a descriptor.

2.17 Critical Pressure: The pressure required to liquefy a gas at its critical temperature. Values are generally expressed in atm (atmospheres), Pa (Pascals), or psia (pounds per square inch atmospheric). Values may also appear as a range or with a descriptor.

2.18 Ionization Potential: Ionization is a chemical change by which ions are formed from a neutral molecule of an inorganic solid, liquid, or gas. The ionization potential is the energy per unit charge, for a particular kind of atom, necessary to remove an electron from the atom to infinite distance. The ionization potential is usually expressed in volts and is numerically equivalent to the work done in removing the electron from the atom, expressed in electron-volts.

2.19 Flash Point: The flash point is the lowest temperature at which a flammable liquid gives off sufficient vapor to form an ignitable mixture with air near its surface or within a vessel. This point is usually determined in laboratory experiments in a cup (open, closed, Tag open cup, Pensky-Martin are some descriptors of cups used to determine flash point). It is usually expressed in degrees Celsius, but can appear as a range, or be described as "combustible", "does not burn", "burns with difficulty", "flammable", or "noncombustible."

2.20 Autoignition Temperature: The minimum temperature at which a substance ignites without application of flame or spark. Along with the flash point, the autoignition temperature gives an indication of relative flammability. It is usually expressed in degrees Celsius and can be given as a range. There may also be a descriptor such as ">," "<," "estimated," or "ignites spontaneously in air."

2.21 LEL: The Lower Explosive Limit refers to the lowest concentration of gas or vapor that burns or explodes if an ignition source is present at ambient temperatures. It is usually expressed in % v/v, but can appear as a textual description.

2.22 UEL: The Upper Explosive Limit refers to the highest concentration of a material in air that produces an explosion or fire, or that ignites when it contacts an ignition source. Concentrations above the UEL are too rich to be ignited. The UEL is usually expressed as a % v/v, but can appear as a textual description, and may have a reference temperature.

3. RTECS Toxicity Data: The *Registry of Toxic Effects of Chemical Substances* published by NIOSH reports the toxicity of materials as reported in a variety of studies. The objective of all of these studies is to try and determine the potential effects of a material on humans. RTECS studies are organized into the following categories:

Acute Oral
Acute Inhalation
Acute Dermal
Chronic (Multiple Dose) Oral
Chronic (Multiple Dose) Inhalation
Chronic (Multiple Dose) Dermal
Irritation–Eye
Irritation–Skin
Reproductive/Teratogenic
Mutagenic
Tumorigenic

For many materials the number of studies reported in RTECS can be overwhelming, making it difficult for the HSE professional to find the results of greatest interest. For each RTECS category we have established a selection process that results in listing only those studies that are generally of greatest interest to HSE professionals. The selection process for each category is described below.

3.1 Acute Oral: Studies were selected in the following order:

1. All studies involving human subjects;
2. All studies involving monkeys;

3. All Rat LD_{50} studies; these results are widely used as an indicator of the potential danger to humans from ingesting this material. LD_{50} stands for "lethal dose 50" and is the amount of material given to a population in a single dose that results in the death of 50% of the population.

If the above sequence did not result in at least two studies being selected, then additional studies were reviewed until two studies were selected, or there were no additional acute oral studies listed in RTECS. The selection sequence first looks for studies on "higher order" species and then for study type (LD_{50}, LD_{lo}, LD, others).

3.2 Acute Inhalation: Studies were selected in the following order:

1. All studies involving human subjects;
2. All studies involving monkeys;
3. All Rat LC_{50} studies; these results are widely used as an indicator of the potential danger to humans from inhaling airborne amounts of the material. LC_{50} stands "for lethal concentration 50" and is the amount of material given to a population in a single exposure that results in the death of 50% of the population within one hour (usually).

If the above sequence did not result in at least two studies being selected, then additional studies were reviewed until two studies were selected, or there were no additional acute inhalation studies listed in RTECS. The selection sequence first looks for studies on "higher order" species and then for study type (LC_{50}, LC_{lo}, LC, others)

3.3 Acute Dermal: Studies were selected in the following order:

1. All studies involving human subjects;
2. All studies involving monkeys;
3. All Rabbit LD_{50} studies; these results are widely used as an indicator of the potential danger to humans from absorbing the material through the skin. LD_{50} stands for "lethal dose 50" and is the amount of material given to a population in a single exposure that results in the death of 50% of the population.

If the above sequence did not result in at least two studies being selected, then additional studies were reviewed until two studies were selected, or there were no additional acute dermal studies listed in RTECS. The selection sequence first looks for studies on "higher order" species and then for study type (LD_{50}, LD_{lo}, LD, others)

3.4 Chronic (Multiple Dose) Oral: The purpose of these studies is to try and predict the dangers to humans from ingesting the material over a long period of time. Studies were selected in the following order:

1. All studies involving human subjects;
2. All studies involving monkeys;
3. All studies on rats.

If the above sequence did not result in at least two studies being selected, then additional studies were reviewed until two studies were selected, or there were no additional multiple dose oral studies listed in RTECS. The selection sequence proceeds by looking for studies on "higher order" species first.

3.5 Chronic (Multiple Dose) Inhalation: The purpose of these studies is to try and predict the dangers to humans from inhaling airborne amounts of the material over a long period of time. Studies were selected in the following order:

1. All studies involving human subjects;
2. All studies involving monkeys;
3. All studies on rats.

If the above sequence did not result in at least two studies being selected, then additional studies were reviewed until two studies were selected, or there were no additional multiple dose inhalation studies listed in RTECS. The selection sequence proceeds by looking for studies on "higher order" species first.

3.6 Chronic (Multiple Dose) Dermal: The purpose of these studies is to try and predict the dangers to humans from absorbing the material through the skin over a long period of time. Studies were selected in the following order:

1. All studies involving human subjects;
2. All studies involving monkeys;
3. All studies on rabbits.

If the above sequence did not result in at least two studies being selected, then additional studies were reviewed until two studies were selected, or there were no additional multiple dose dermal studies listed in RTECS. The selection sequence proceeds by looking for studies on "higher order" species first.

3.7 Irritation–Eye: The first step in the selection process was to disregard all studies where the reaction was listed as "not reported." Then studies were selected in the following order:

1. All studies involving human subjects;
2. All studies involving monkeys;
3. All Rat Standard Draize studies; these results are widely used as an indicator of the potential danger to humans from having the material come in contact with the eye.

If the above sequence did not result in at least two studies being selected, then additional studies were reviewed until two studies were selected, or there were no additional eye irritation studies listed in RTECS. The selection sequence first looks for studies on "higher order" species, then for study type (first Standard Draize, then nonstandard types), then for "Reaction" (severe, followed by moderate, then mild, then any other result).

3.8 Irritation–Skin: The first step in the selection process was to disregard all studies where the reaction was listed as "not reported." Then studies were selected in the following order:

1. All studies involving human subjects;
2. All studies involving monkeys;
3. All Rabbit Standard Draize studies; these results are widely used as an indicator of the potential danger to humans from having the material come in contact with the skin.

If the above sequence did not result in at least two studies being selected, then additional studies were reviewed until two studies were selected, or there were no additional skin irritation studies listed in RTECS. The selection sequence first looks for studies on "higher order" species, then for study type (first Standard Draize, followed by Open Draize, then Immersion Draize), then for "Reaction" (severe, followed by moderate, then mild, then any other result).

3.9 Reproductive/Teratogenic: The first step in the selection process was to disregard all studies where the route of entry was listed as "unreported." Then studies were selected in the following order:

1. All studies involving human subjects;
2. All studies involving monkeys;
3. All rat inhalation studies;
4. All rat oral studies.

If the above sequence did not result in at least two studies being selected, then additional studies were reviewed until two studies were selected, or there were no additional reproductive/teratogenic studies listed in RTECS. The selection sequence first looks for studies on "higher order" species, then for routes of entry in the following order: inhalation, oral, skin, subcutaneous, others.

3.10 Mutagenic: Studies were selected in the following order:

1. All studies involving human subjects;

2. All studies involving monkeys.

If the above sequence did not result in at least two studies being selected, then additional studies were reviewed until two studies were selected, or there were no additional mutagenic studies listed in RTECS. The selection sequence first looks for studies on "higher order" species, then for study type in the following order:

DNA damage
DNA repair
unscheduled DNA synthesis
DNA inhibition
DNA adduct
mutations in microorganisms
mutations in mammalian somat
cytogenetic analysis
sister chromatid exchange
micronucleous test
dominant lethal test
body fluid assay
host-mediated assay
specific locus test
sex chromosome loss
morphological transformation
sperm morphology
other mutation test systems

3.11 Tumorigenic: The first step in the selection process was to disregard all studies where the route was listed as "unreported." Then studies were selected in the following order:

1. All studies involving human subjects;
2. All studies involving monkeys;
3. All rat inhalation studies;
4. All rat oral studies.

If the above sequence did not result in at least two studies being selected, then additional studies were reviewed until two studies were selected, or there were no additional tumorigenic studies listed in RTECS. The selection sequence first looks for studies on "higher order" species, then for routes of entry in the following order: inhalation, oral, skin, subcutaneous, others.

4. Hazard Overviews: The Hazard Overview is comprised of six subsections: hazard pictograms; a fire diamond; descriptive overviews for health, fire, and reactivity; target organ pictograms; exposure limits; and respirator recommendations. Each of these subsections is described below.

4.1 Physical Hazard Pictograms: In these chemical profiles, Genium has made considerable effort to compile data from a variety of sources and present this (sometimes conflicting) data in one convenient summary. Given the serious and immediate danger of some chemicals, however, Genium has defined six physical hazard pictograms that appear on the profile if very specific criteria are met. Pictograms are available for the following physical hazards:

Poison
Corrosive
Flammable
Explosive

Compressed Gas
Radioactive

The criteria for determining classification as a hazard for these six areas are given below in the order in which they would appear within the chemical profile.

4.1.1 Poison: If the material is a poison, the skull-and-crossbones pictogram and the word "poison" appear. A material is classified as a poison if it is a highly toxic material. There are four categories of highly toxic chemicals:

4.1.1.1 A chemical that has a median lethal dose (LD_{50}) of 50 milligrams or less per kilogram of body weight when administered orally to albino rats weighing between 200 and 300 grams each.

4.1.1.2 A chemical that has a median lethal dose (LD_{50}) of 200 milligrams or less per kilogram of body weight when administered by continuous contact for 24 hours (or less, if death occurs within 24 hours) with the bare skin of albino rabbits weighing between 2 and 3 kilograms each.

4.1.1.3 A chemical that has a median lethal concentration (LC_{50}) of gas or vapor in air of 200 parts per million (ppm) or less by volume, or 2 milligrams per liter or less of mist, fume, or dust, when administered by continuous inhalation for 1 hour (or less, if death occurs within 1 hour) to albino rates weighing between 200 and 300 grams each, provided such concentration or condition, or both, are likely to be encountered by humans when the chemical is used in any reasonably foreseeable manner.

4.1.1.4 A chemical that is a liquid having a saturated vapor concentration (ppm) at 68 °F (20 °C) equal to or greater than its LC_{50} (vapor) value (ppm), if the LC_{50} value is 3000 ppm or less when administered by continuous inhalation for 1 hour (or less, if death occurs within 1 hour) to albino rats weighing between 200 and 300 grams each, provided such concentration or condition, or both, are likely to be encountered by humans when the chemical is used in any reasonably foreseeable manner.

4.1.2 Corrosive: The corrosive pictogram, a hand and test tube with the word "corrosive" beneath, appears when a material causes visible destruction of or irreversible alterations to living tissue by chemical action at the site of contact.

4.1.3 Explosive: The explosive pictogram, a starburst with the word "Explosive" underneath, is used when a material exhibits any of the following characteristics:

- Explosive when heated;
- Explosive when shocked;
- Forms potentially explosive mixtures with water;
- May polymerize explosively;
- May form explosive dust-air mixtures;
- Is capable, in itself, of detonation, explosive decomposition, or explosive reaction at normal temperature and pressure, or does not require strong initiating source or heating under confinement.

4.1.4 Flammable: The flammable pictogram, a picture of a flame with the word 'Flammable" underneath, is used to classify a liquid with a flash point less that or equal to 141 °F; a gas that at atmospheric pressure/temperature has a Lower Flammability Limit (LFL) of 13% or less, or the difference between the LFL and the Upper Flammability Limit (UFL) is wider than 12%; or projects a flame more than 18 inches beyond the ignition source; or is a solid that ignites readily or is liable to cause a fire.

4.1.5 Compressed Gas: The compressed gas hazard pictogram, a picture of a gas cylinder with the words "Compressed Gas" beneath, denotes any material or mixture having in the container either an absolute pressure greater than 276 kPa (40 lb/in^2) at 21 °C (70 °F), or an absolute pressure greater than 717 kPa (104 lb/in^2) at 54 °C (129 °F) or both, or any liquid flammable material having a Reid vapor pressure greater than 276 kPa (40 lb/in^2) at 38 °C (100 °F).

4.1.6 Radioactive: The radioactive pictogram, the standard radiation tri-foil symbol with the word "Radioactive" underneath, is used if a material emits ionizing radiation.

4.2 Fire Diamond: In accordance with the National Fire Protection Association (NFPA) Publication 704, fire diamonds were developed for the Genium MSDS Collection and are reprinted in these chemical profiles. The fire diamond is a well-recognized rating system which indicates the dangers associated with a material during a fire. Explanations of the four rating categories on the diamond are:

H (far left position) is the health hazard rating, and indicates the short-term degree of hazard:
- 0 = ordinary combustible hazards in a fire
- 1 = slightly hazardous
- 2 = hazardous
- 3 = extreme danger
- 4 = deadly

F (top position) is the flammability rating, and indicates the susceptibility to burning:
- 0 = will not burn
- 1 = will ignite if preheated
- 2 = will ignite if moderately heated
- 3 = will ignite at normal room temperature; will burn with extreme rapidity
- 4 = will burn easily and rapidly at normal room temperature/pressure; or ignite spontaneously when exposed to air

R (far right position) is the instability rating, and indicates the energy released if the material is burned, decomposed, or mixed:
- 0 = stable and not reactive with water
- 1 = unstable if heated; changes or decomposes on exposure to air, light, or moisture
- 2 = violent chemical change; reacts violently with water or forms potentially explosive mixtures with water
- 3 = shock and heat may detonate; reacts explosively with water without heating or confinement
- 4 = may detonate

S (bottom position) contains special hazard symbols:
- OX = Oxidizer–an oxidizer may not burn itself, but may ignite and will intensify the burning of combustible materials
- ~~W~~ = Use no water/material reacts with water–may become explosive, may produce a flammable or poisonous material, may produce excessive heat, etc.

4.3 Textual Hazard Overviews: In addition to the pictograms and fire diamond, data concerning the hazardous nature of each chemical have been compiled from a variety of (sometimes contradictory) sources. This section is designed to give you an overview of data reported in the literature. Data have been divided into the following categories: Health, Fire, and Reactivity.

4.3.1 Health Overview: This section has been structured to list information in a specific order and, as much as possible, use standardized words or phrases. The structure of this section is as follows:

4.3.1.1 Listed first is the corrosiveness/irritability of the chemical. Effects caused to the eyes, skin, and respiratory tract are reported. Chemicals may be described as corrosive, highly irritating, irritating or may cause irritation.

4.3.1.2 Listed second is the material's classification as Poison, Toxic, or Harmful. Criteria for determining this classification are:

Poison (highly toxic). The four categories are:

A) A chemical that has a median lethal dose (LD_{50}) of 50 milligrams or less per kilogram of body weight when administered orally to albino rats weighing between 200 and 300 grams each.

B) A chemical that has a median lethal dose (LD_{50}) of 200 milligrams or less per kilogram of body weight when administered by continuous contact for 24 hours (or less, if death occurs within 24 hours) with the bare skin of albino rabbits weighing between 2 and 3 kilograms each.

C) A chemical that has a median lethal concentration (LC_{50}) of gas or vapor in air of 200 pats per million (ppm) or less by volume, or 2 milligrams per liter or less of mist, fume, or dust, when administered by continuous inhalation for 1 hour (or less, if death occurs within 1 hour) to albino rats weighing between 200 and 300 grams each, provided such concentration or condition, or both, are likely to be encountered by humans when the chemical is used in any reasonably foreseeable manner.

D) A chemical that is a liquid having a saturated vapor concentration (ppm) at 68 °F (20 °C) equal to or greater than its LC_{50} (vapor) value (ppm), if the LC_{50} value is 3000 ppm or less when administered by continuous inhalation for 1 hour (or less, if death occurs within 1 hour) to albino rats weighing between 200 and 300 grams each, provided such concentration or condition, or both, are likely to be encountered by humans when the chemical is used in any reasonably foreseeable manner.

Toxic: A material is toxic if:

A) It has a median lethal dose (LD_{50}) of more than 50 milligrams per kilogram, but no more than 500 milligrams per kilogram of body weight, when administered orally to albino rats weighing between 200 and 300 grams each;

B) It has a median lethal dose (LD_{50}) of more than 200 milligrams per kilogram, but no more that 1000 milligrams per kilogram of body weight, when administered by continuous contact for 24 hours (or less, if death occurs within 24 hours) with the bare skin of albino rabbits weighing between 2 and 3 kilograms each;

C) It has a median lethal dose (LD_{50}) in air of more that 200 parts per million (ppm), but no more than 2000 ppm of gas or vapor by volume, or more than 2 milligrams per liter, but no more than 20 milligrams per liter, or mist, fume, or dust, when administered by continuous inhalation for 1 hour (or less, if death occurs within 1 hour) to albino rats weighing between 200 and 300 grams each, provided such concentration or condition, or both, are likely to be encountered by humans when the chemical is used in any reasonably foreseeable manner;

D) It is a liquid having a saturated vapor concentration (ppm) at 68 °F (20 °C) of more than one-fifth its LC_{50} (vapor) value (ppm), if the LC_{50} value is not more than 5000 ml/m^3 (ppm) when administered by continuous inhalation for 1 hour

(or less, if death occurs within 1 hour) to albino rats weighing between 200 and 300 grams each, and the criteria for "highly toxic material" is not met, provided such concentration, or condition, or both, are likely to be encountered by humans when the chemical is used in any reasonably foreseeable manner.

Harmful: A material is classified as harmful if:

A) It has a median lethal dose (LD_{50}) of more than 500 milligrams per kilogram, but no more that 2000 milligrams per kilogram of body weight, when administered orally to albino rats weighing between 200 and 300 grams each;
B) It has a median lethal dose (LD_{50}) of more than 1000 milligrams per kilogram, but no more than 2000 milligrams per kilogram of body weight, when administered by continuous contact for 24 hours (or less, if death occurs within 24 hours) with the bare skin of albino rabbits weighing between 2 and 3 kilograms each.

4.3.1.3 Appearing third is a list of the acute effects, which are classified as adverse effects developing rapidly after exposure (usually in high concentrations) of short duration.

4.3.1.4 Appearing fourth is a list of chronic effects, which are defined as adverse effects with symptoms that develop slowly over time and persist and/or recur frequently. Special emphasis was placed on chronic effects such as carcinogenicity, reproductive effects, and effects noted in humans. If effects have only been found in laboratory animals, this is also noted.

4.3.1.5 Appearing last is a list of target organs that may be effected by exposure to this chemical.

4.3.2 Fire Overview: This section has been structured to list information in a specific order and, as much as possible, use standardized words or phrases. The structure of this section is as follows:

4.3.2.1 If data is available, the material is classified as flammable, combustible, will burn, or not combustible. Criteria for determining this classification are:

A) Flammable: A material is flammable if it is a liquid with a flash point less than or equal to 141 °F; a gas that at atmospheric pressure/temperature has a Lower Flammability Limit (LFL) of 13% or less or the difference between the LFL and the Upper Flammability Limit (UFL) is wider that 12%; or projects a flame more than 18 inches beyond the ignition source; or is a solid that ignites readily or is liable to cause fire.
B) Combustible: A material is combustible if the flash point is greater than 141 °F, but less than 200 °F.
C) Will Burn: If the criteria for Flammable, Combustible, or Non-Combustible are not met, the material is usually classified as Will Burn. This means that the material will burn, but not in accordance with the criteria that would be used to classify the material in the other three categories.
D) Non-Combustible: Materials that will not burn. This usually includes any material that will not burn in air when exposed to a temperature of 1500 °F (815.5 °C) for a period of 5 minutes.

4.3.2.2 Next, when available, appears a listing of the significant hazards in a fire situation. Items addressed in this area include the explosiveness of the compound, formation of dangerous products of combustion, and reaction with water.

4.3.2.3 Next appear recommended agents for combating a fire. This list is not a complete description, and is not intended as a guide for combating a fire. Size, location, and other factors may determine the best agent for combating a fire.

4.3.2.4 Finally, when available, special precautions associated with a fire involving the material are reported.

4.3.3 Reactivity Overview: This section has been structured to list information in a specific order and, as much as possible, use standardized words or phrases. The structure of this section is as follows:

4.3.3.1 Data regarding the material's stability is listed first. A material is either stable or unstable. If unstable, a brief description of the instability is included.

4.3.3.2 If polymerization occurs, a reaction wherein one or more small molecules combine to form larger molecules, it is noted. Of special concern are hazardous polymerizations that can be harmful to humans, including polymerization that releases large amounts of energy that can cause fires or explosions or burst containers, or that produce toxic materials.

4.3.3.3 Next appears conditions to avoid which may lead to the possibility of hazardous reactions.

4.3.3.4 Chemical incompatibilities are reported next. This is a listing of materials that should not come in contact with the material. This can include contact which may occur within a laboratory situation, or more general hazards. No attempt has been made to distinguish between incompatibilities that may be significant only in a laboratory setting and incompatibilities that may be significant in other situations.

4.3.3.5 Last, is a listing of hazardous decomposition products.

4.4 Carcinogenicity: Seven agencies have developed lists of materials that are carcinogenic. For each agency the material is either "listed" or "not listed" If the material is listed, then the rating assigned to the material by that agency is also given. Agencies whose carcinogen lists were looked at are:

IARC - International Agency for Research on Cancer
NIOSH - National Institute for Occupational Safety and Health
NTP - National Toxicology Program
ACGIH - American Conference of Governmental Industrial Hygienist
OSHA - Occupational Safety and Health Administration
EPA - Environmental Protection Agency
DFG - Deutsche Forschungsgemeinschaft (Germany)

4.5 Target Organ Pictograms: Like the Physical Hazard Pictograms, the Target Organ Pictograms offer the reader a quick visual indication of the harmful effects of a material. These pictograms indicate which organs or parts of the body are most likely to be affected by the material. The eyes, skin and respiratory system are considered to be target organs if the material is corrosive, severely irritating, or irritating to these systems. For the remaining possible target organs, the designation was made only if the words "target organ" were found associated with the organ in the literature. Up to six target organ pictograms may be shown from the following fifteen target organs.

Eyes
Skin
Respiratory System
Mucous Membranes
Gastrointestinal System
Nervous System
Liver
Kidneys
Cardiovascular System
Blood
Bone

Teeth
Lymphatic System
Reproductive System
Glandular System

If more than six target organs were listed in the literature, then pictograms are given for only the six most significant. However, all target organs are listed in the Health Hazard Overview.

4.6 Exposure Limits: This section reports the concentration of a material in workplace air that is deemed the maximum acceptable level by various agencies. Some agencies report limits as a Time-Weighted Average (TWA), a Short-term Exposure Limit (STEL) or a Ceiling limit (C). The agencies and levels are reported in the following order:

4.6.1 OSHA PEL: The Occupational Safety and Health Administration has established Permissible Exposure Limits (PEL) which must be legally adhered to by employers. These may be expressed as a time-weighted average (TWA) limit, a short-term exposure limit (STEL), or a ceiling exposure limit (a limit that must never be exceeded even instantaneously). Some materials are specifically listed along with their exposure limits, but in some instances exposure limits are given for classes of materials (inorganic lead, for instance) or elements in a material (such as arsenic). In these cases, the molecular formula was analyzed to see if the material had an exposure limit. If the material fell into more than one class or contained more than one element with an exposure limit, then the more stringent exposure limits were listed. Exposure limits may be specified in parts per million (ppm), or milligrams per cubic meter (mg/m^3).

4.6.2 OSHA PEL vacated 1989 limits: These limits were issued in 1989, but were subsequently held to be unenforceable by the U.S. Court of Appeals. In many cases they were more stringent than previous limits, and included previously unregulated substances. These limits reflect OSHA's ideal limits (lowest value) for worker exposure to the material.

4.6.3 ACGIH TLV: The American Conference of Governmental Industrial Hygienists specifically identifies materials and their exposure limits, or lists an element, such as aluminum, and suggests exposure limits. To determine if there is an element exposure limit, the molecular formula was analyzed. In cases where more than one element with an exposure limit was part of a material, the more stringent limit is listed. ACGIH uses the threshold limit value (TLV) to express the maximum airborne concentration of a material to which most healthy workers can be exposed during a normal daily and weekly work schedule without adverse effects. ACGIH reports TLV limits using time-weighted average (TWA), the allowable concentration for a normal 8-hour day or 40-hour work week; short-term exposure limits (STEL), the maximum concentration for a continuous exposure period of 15 minutes with a maximum of four exposure periods per day with at least 60 minutes between each exposure; and ceiling limits (C), a concentration which cannot be exceeded at any time. Limits can be expressed in parts per million (ppm), or in milligrams per cubic meter (mg/m^3).

4.6.4 NIOSH REL: The National Institute for Occupational Safety and Health certifies respiratory and air-sampling devices, and makes exposure limit recommendations to OSHA. NIOSH specifically identifies materials and their exposure limits, or identifies an element, such as cadmium, and suggests exposure limits. To determine if there is an element exposure limit, the molecular formula was analyzed. In cases where more than one element with an exposure limit was part of a material, the more stringent limit is listed. The NIOSH Recommended Exposure Limit (REL) is the highest allowable airborne concentration that is not expected to injure a worker, and is usually based on a 10-hour work day. It may be expressed at either a time-weighted average (TWA), or a ceiling limit (C), and may be expressed in either parts per million (ppm) or milligrams per cubic meter (mg/m^3).

4.6.5 NIOSH IDLH: NIOSH also reports an IDLH limit (immediately dangerous to life and health). This limit, which can be expressed in either parts per million (ppm) or milligrams per cubic meter (mg/m^3), is the maximum concentration to which one could be exposed for thirty minutes without impairing the ability to escape or having irreversible health effects.

4.6.6 DFG MAK: In Germany, the Commission for the Investigation of Health Hazards of Chemical Compounds in the Work Area establishes maximum concentration values (MAK) for substances found in the workplace. MAKs can be expressed as time-weighted averages (TWA) or peak exposures. MAK limits either specifically identify materials and their exposure limits, or list elements, such as barium, and provide suggested exposure limits. To determine if there is an element exposure limit, the molecular formula was analyzed. In cases where more than one element with an exposure limit was part of a material, the more stringent limit is listed. MAKs may be expressed as either parts per million (ppm) or milligrams per cubic meter (mg/m^3).

4.6.7 AIHA WEEL: The American Industrial Hygiene Association (AIHA) identifies hazardous substances, commonly found in the workplace, which do not have established exposure limits (such as TLVs). For these substances, AIHA establishes a Workplace Environmental Exposure Level (WEEL). Values can be expressed as time-weighted averages (TWA) or ceiling values. WEELs may be expressed as either parts per million (ppm) or milligrams per cubic meter (mg/m^3).

4.7 Respirator Recommendations: This section of the chemical profile was developed by an expert on the selection and use of respirators for use in dealing with hazardous materials in the workplace. Information for this section was gathered primarily from OSHA, ACGIH, and NIOSH. Where available, OSHA PELs were used, since these are the standards that will be enforced by OSHA personnel. It is extremely important to note, however, that the standards governing protective gear, and the design of protective equipment are in constant flux.

For each material for which there are established airborne exposure limits, up to five respirator categories have been established. A category consists of an exposure range and the suggested respirator to use when exposed to airborne material concentrations within that range. If the respirator suggestions include a cartridge respirator, then the cartridge color is also listed. Special precautions or concerns are given as "Notes." The listed respirators and equipment are basic types, common to most manufacturers of such equipment. However there are many variations in these common configurations, as well as many specialized pieces of equipment that may be chosen in lieu of these suggestions. Respirator recommendations have been made using ANSI Z88-1992 guidelines. However, appropriate technical personnel should always be consulted before making decisions on the appropriate respirator to use in specific exposure situations.

5. Environmental Overview: Data collected for this section is associated with the effects that the material may have on the environment.

5.1 Ecotoxicity: The purpose of this section is to indicate the possible effects that exposure to the material will have on life forms other than humans. Like the RTECS data presented earlier (which relates to human exposure), this section presents studies and results. These studies are designed to provide an indication of the effects to be expected on non-human life forms if the material is released to the environment. For many materials the number of studies reported in the literature can be overwhelming, making it difficult for the HSE professional to find the results of greatest interest or importance. A selection process was implemented that results in listing only those studies that are generally of greatest interest to HSE professionals. The selection process is described below.

A) If two or less studies were found in the literature, all are reported.

B) Early Life Stage studies were segregated. These were grouped by species (for instance marine fish, freshwater fish, etc.) and test type. For identical test types and species, the test results with the lowest toxic dose are reported.

C) All remaining studies were classified by species (for instance marine fish, freshwater fish, etc.) and test type. Studies that could not be classified by species or test type were eliminated. For identical test types and species, the test results with the lowest toxic dose are reported.

5.2 Environmental Fate: The environmental fate is a summary describing the transport and ultimate environmental effects of a chemical substance when released to air, soil, or water. This summary is based on the substance's chemical behavior and affinities. The information reported in this section was compiled from data found in the literature. The largest source of information for these summaries is the ***Hazardous Substances Data Bank*** (HSDB).

5.3 Clean-up/Disposal: This section is the specific emergency action guideline found in the ***Emergency Response Guide Book.***, published by the U.S. Department of Transportation.

5.4 Environmental Physical Data: The data presented here give some indication of how the chemical will dissipate when discharged to the environment, how it will effect plants and animals, and how it will eventually decompose.

5.4.1 Henry's Law Constant: Henry's Law Constant is the equilibrium ratio of concentration in air (expressed as a partial pressure) and concentration in water. Materials with a high Henry's Law Constant are more volatile when dissolved in water. Data in this field is numerical (typically with units of atm•m^3/mol), and can include descriptors such as "calculated" or "estimated." In some cases the values are given as a range..

5.4.2 Octanol/Water Partition Coefficient: Data in this field are numerical, and can include descriptors such as "calculated," "measured," or "estimated." In some cases the values are given as a range. Generally the data are reported as the log of the actual value, but the actual value may be inadvertently listed in a few cases.

5.4.3 Sorption Partition Coefficient: Data in this section is numerical, and can include descriptors such as "calculated" or "estimated," In some cases the values are given as a range. Data is reported both as the actual value and the log of the actual value (many times with no specification as to which is being reported). When doubt existed as to which parameter was being reported, the size of the value was used as the discriminator.

5.4.4 BCF: The Bioconcentration Factor (BCF) indicates the extent to which a chemical is passed through the food chain from soil to plants and animals where it may accumulate and ultimately be passed to humans. Data in this field can include descriptors such as "calculated" or "estimated." In some cases the values are given as a range. Additionally, statements such as "no food chain concentration potential," "none," and "none likely" may appear.

5.4.5 BOD: Biochemical Oxygen Demand (BOD) indicates the dissolved oxygen required to decompose organic matter in water, and is used to estimate the degree of contamination in water supplies. "None" or "not pertinent" may appear, or the BOD could be described as "theoretical." In some cases the values are given as a range. When reported, the BOD will usually include a percentile and time period, or a fraction and percentile.

5.5 Regulations: The Environmental Protection Agency (EPA) is responsible for enforcing a number of regulations that relate to chemicals in the environment. This section states whether or not a chemical is specifically covered by one or more of these regulations and, as applicable, the minimum quantity of the chemical that must be present before the user is required to follow certain requirements of the regulation.

5.5.1 RCRA 40 CFR: The Resource Conservation and Recovery Act, enacted in 1976 and subsequently amended, controls solid-waste disposal and encourages recycling. This section indicates whether a substance is listed under this regulation, or not listed. If listed, the Hazardous Waste number and waste characterization assigned by RCRA is included.

5.5.2 CERCLA: Enacted in 1980 and amended thereafter, the Comprehensive Environmental Response, Compensation, and Liability Act provides for identification and cleanup of hazardous materials released on land, into the air, waterways, and groundwater. It covers areas affected by newly released materials and older leaking or abandoned dump sites. This field states whether a material is listed or not listed in CERCLA Table 302.4. If listed, this field includes the Listing Code and Reportable Quantity (RQ).

5.5.3 SARA: Signed into law in 1986, the Superfund Amendments and Reauthorization Act (SARA) is an extension of CERCLA, and is intended to encourage and support local and state emergency planning efforts. SARA provides citizens and local governments with information about potential chemical hazards, and calls for facilities that store hazardous materials to provide officials and citizens with data on the type and amount on hand at specific locations. This field states whether a material is listed or not listed in section 372.65 of SARA. If listed, the Threshold Planning Quantity (TPQ) for the material is included.

5.5.4 SARA EHS: This field states if a material is listed or not listed in Appendix B to part 355, the SARA Extremely Hazardous Substances (EHS) section. If listed, the Threshold Planning Quantity (TPQ) is also noted.

5.5.5 TSCA: The Toxic Substances Control Act (TSCA) controls the exposure to and use of raw industrial chemicals not subject to other laws. This section indicates whether the chemical is listed or not listed by TSCA.

5.6 Analytical Methods: The Analytical Methods sections gives an overview of methods listed in the Environmental Monitoring Methods Index produced by the Environmental Protection Agency. These are the primary methods used by analytical laboratories to detect the presence of the chemical in various media. The information is divided into nine media categories: Air, Drinking Water, Food, Indoor Air/Expired Air, Plasma, Soil, Urine, Water/Waste Water, and Other. All listings include the abbreviation of the organization that issued the testing method followed by the method number(s). Organizational abbreviations used in this section are:

ACOE	Army Corps of Engineers
ACS	American Chemical Society
AOAC	Association of Official Analytical Chemists
APHA	American Public Health Association
ASDWA	Association of State Drinking Water Administrators
ASTM	American Society for Testing Materials
CEM	CEM Corporation
CLP	EPA Contract Laboratory Program
DOE	Department of Energy
ENVCA	Environment Canada
EPA	Environmental Protection Agency (various groups)
FDA	Food and Drug Administration
FISON	Applied Research Laboratories Fison Instruments
HACH	Hach Chemical Company
ISO	International Organization for Standardization
ISWSD	Illinois State Water Survey
NCASI	National Council of the Paper Industry for Air and Stream
NIOSH	National Institute for Occupational Safety and Health
NIST	National Institute for Standards and Technology
NOAA	National Oceanographic and Atmospheric Administration
NTIS	National Technical Information Service
OSHA	Occupational Safety and Health Administration
SCARB	State of California Air Resources Board
USGS	Text Products Section Distribution Branch

ABA5000 CAS #: 71751-41-2

ABAMECTIN

RTECS: CL1203000
Molecular Formula: Mixture
Formula Weight: N/A
Synonyms: AFFIRM; AGRIMEK; AVID EC; AVOMEC; MK 936; VERTIMEC; ZEPHYR

Physical Properties

Freezing Point: 150 °C (302 °F) to 155 °C (311 °F)
Specific Gravity: 1.16
Water Solubility: 7.8 ppb
Other Solubilities: 350g/l toluene; 100g/l acetone; 70g/l isopropanol; 25g/l chloroform; 20g/l ethanol; 19.5g/l methanol

RTECS Toxicity Data

Acute Oral: Rat LD_{50} Dose: 10 mg/kg. Monkey LD_{50} Dose: 17 mg/kg.
Acute Inhalation: Rat LC_{50} Dose: 1100 mg/m^3/4hr.
Acute Dermal: Rabbit LD_{50} Route: Skin; Dose: >2 gm/kg.

Hazard Overviews

Carcinogenicity: IARC - Not listed; NIOSH - Not listed; NTP - Not listed; ACGIH - Not listed; OSHA - Not listed; EPA - Not listed; MAK - Not listed

Environmental

Regulations

RCRA 40CFR: Not listed
CERCLA: 40CFR 302.4: Not listed
SARA 40CFR 372.65: Listed
SARA EHS 40CFR 355: Not listed
TSCA: Not listed

ABI5000 CAS #: 514-10-3

ABIETIC ACID

RTECS: TP8580000
EINECS Number: 208-178-3
Molecular Formula: $C_{20}H_{30}O_2$
Structured MF: $(CH_3)_2CH(O_{14}H_{16})(CH_3)_2COOH$
Formula Weight: 302.46

O
OH

Chemical Structure

Synonyms: ABIETINIC ACID; 13-ISOPROPYLPODOCARPA-7,13-DIEN-15-OIC ACID; KYSELINA ABIETOVA; PODOCARPA-7,13-DIEN-15-OIC ACID,13-ISOPROPYL-; SYLVIC ACID
Description: yellow powder
Use: manufacture of esters (ester gums), e.g., methyl ester, vinyl and glyceryl esters for use in lacquers and varnishes; manufacture of "metal resinates", soaps, plastics, and paper sizes; assists growth of lactic and butyric acid bacteria

Physical Properties

Boiling Point: 250 °C (482 °F)
Freezing Point: 173 °C (343.4 °F)
Water Solubility: Insoluble in Water
Other Solubilities: Acetone: Soluble; Alcohol: Soluble; Benzene: Very Soluble; CS_2: Soluble; Chloroform: Soluble; dilute NaOH solution: Soluble; Ether: Very Soluble; MeOH: Soluble.
Flash Point: Combustible

Hazard Overviews

Explosive

Fire Diamond

Health: Irritating to eyes/skin/respiratory tract. Also Causes: CNS disturbances.
Fire: Combustible and explosive. Use dry chemical, carbon dioxide, water spray, or foam.
Reactivity: Stable. Hazardous polymerization cannot occur. Hazardous decomposition products: acrid smoke; irritating fumes.
Carcinogenicity: IARC - Not listed; NIOSH - Not listed; NTP - Not listed; ACGIH - Not listed; OSHA - Not listed; EPA - Not listed; MAK - Not listed

Primary Target Organs:

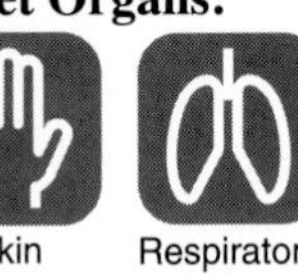

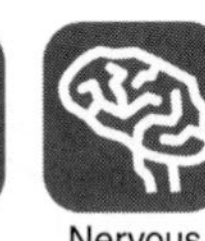
Eyes Skin Respiratory System Mucous Membranes Nervous System

Environmental

Ecotoxicity: Fishes: Salmo gairdneri: 96h LC_{50},S 0.7 mg/l

Regulations

RCRA 40CFR: Not listed
CERCLA: 40CFR 302.4: Not listed
SARA 40CFR 372.65: Not listed
SARA EHS 40CFR 355: Not listed
TSCA: Listed

ACA5000 CAS #: 18181-80-1

ACAROL

RTECS: DD2100000
EINECS Number: 242-070-7
Molecular Formula: $C_{17}H_{16}Br_2O_3$
Formula Weight: 428.14
Synonyms: ASCAROL 2E; BENZENEACETIC ACID,4-BROMO-ALPHA-(4-BROMOPHENYL)-ALPHA-HYDROXY-,1-METHYLETHYL ESTER; BENZILIC ACID,4,4'-DIBROMO-,ISOPROPYL ESTER; 4-BROMO-ALPHA-(4-BROMOPHENYL)-ALPHA-HYDROXYBENZENEACETICACID 1-METHYL ETHYL ESTER; BROMOPROPYLATE; CIBA-GEIGY GS 19851; 4,4'-DIBROMOBENZILIC ACID ISOPROPYL ESTER; ENT 27552; FOLBEX VA; GEIGY 19851; GEIGY GS-19851; GS-19851; GS 19851; ISOPROPYL BROMOBENZILATE; ISOPROPYL 4,4'-DIBROMOBENZILATE; ISOPROPYL DIBROMOBENZILATE; ISOPROPYL 4,4'-DIBROMOBENZYLATE; 1-METHYLETHYL 4-BROMO-ALPHA-(4-BROMOPHENYL)-ALPHA-HYDROXYBENZENEACETATE; 1-METHYLETHYL4-BROMO-ALPHA-(4-BROMOPHENYL)-ALPHA-HYDROXYBENZENEACETATE; NEORON; NSC 195087; PHENISOBROMOLATE
Description: crystalline solid
Use: acaricide

Physical Properties

Freezing Point: 77 °C (170.6 °F)
Specific Gravity: 1.59
Saturated Vapor Density: 1.200000001 kg/m^3
Vapor Pressure: 5.1 $x10^{-8}$ mm Hg at 20 °C
Water Solubility: < 0.5 mg/L at 20 °C
Other Solubilities: readily Soluble in most organic solvents

RTECS Toxicity Data

Acute Oral: Rat LD_{50} Dose: 5 gm/kg. Mouse LD_{50} Dose: 8 gm/kg.
Acute Dermal: Rabbit LD_{50} Route: Skin; Dose: 10200 mg/kg. Rat LD_{50} Route: Skin; Dose: >4 gm/kg.
Irritation Eye: Rabbit Standard Draize Test Dose: 600 ug; Reaction: mild.
Irritation Skin: Rabbit Open Draize Test Dose: 121 mg open; Reaction: moderate.
Mutagenic: Rat Morphological Transformation; Route: Oral; Dose: 4620 mg/kg/77D-C.

Hazard Overviews

Carcinogenicity: IARC - Not listed; NIOSH - Not listed; NTP - Not listed; ACGIH - Not listed; OSHA - Not listed; EPA - Not listed; MAK - Not listed

Environmental

Ecotoxicity: LD_{50} mallard ducks oral greater than 2000 mg/kg, 3 mo old females /sample purity 97.5

Regulations

RCRA 40CFR: Not listed
CERCLA: 40CFR 302.4: Not listed
SARA 40CFR 372.65: Not listed
SARA EHS 40CFR 355: Not listed
TSCA: Not listed

Analytical Methods

Food: FDA 212.1, 232.1, 232.4, 242.1
Plasma: EPA 001; FDA 211.1, 231.1, 252

ACE1000 CAS #: 37517-30-9

ACEBUTOLOL

RTECS: EJ3500500
EINECS Number: 253-539-0
Molecular Formula: $C_{18}H_{28}N_2O_4$
Formula Weight: 336.43
Synonyms: (+-)-ACEBUTOLOL; 1-(2-ACETYL-4-N-BUTYRAMIDOPHENOXY)-2-HYDROXY-3-ISOPROPYLAMINOPROPANE; 3'-ACETYL-4'-(2-HYDROXY-3-(ISOPROPYLAMINO)PROPOXY)BUTYRANILIDE; N-(3-ACETYL-4-(2-HYDROXY-3-((1-METHYLETHYL)AMINO)PROPOXY)PHENYL) BUTANAMIDE; (+-)-N-(3-ACETYL-4-(2-HYDROXY-3-((1-METHYLETHYL)AMINO)PROPOXY)PHENYL)BUTANAMIDE; 5'-BUTYRAMIDO-2'-(2-HYDROXY-3-ISOPROPYLAMINOPROPOXY)ACETOPHENONE; DL-ACEBUTOLOL; PRENT
Description: crystals
Use: medication: beta adrenergic blocking agent; medication: antihypertensive, antianginal, antiarrhythmic

Physical Properties

Freezing Point: 119 °C (246.2 °F) to 123 °C (253.4 °F)
Water Solubility: 200 mg/mL Water
Other Solubilities: Solubilities of 70 mg/ml in Alcohol at room temp (approx 25 °C)

RTECS Toxicity Data

Acute Oral: Man TD_{Lo} Dose: 137 mg/kg/24D-I; Toxic Effects: Gastrointestinal - Nausea or vomiting; Kidney, Ureter, and Bladder - Other changes in urine composition; Biochemical - Transaminases. Woman TD_{Lo} Dose: 152 mg/kg; Toxic Effects: Cardiac - Change in conduction velocity; Vascular - BP lowering not characterized in autonomic section. Woman TD_{Lo} Dose: 109 mg/kg/22D-I; Toxic Effects: Gastrointestinal

- Nausea or vomiting; Kidney, Ureter, and Bladder - Other changes in urine composition; Biochemical - Transaminases.

Acute Dermal: Mouse LD_{50} Route: Subcutaneous Dose: 125 mg/kg.

Hazard Overviews

Carcinogenicity: IARC - Not listed; NIOSH - Not listed; NTP - Not listed; ACGIH - Not listed; OSHA - Not listed; EPA - Not listed; MAK - Not listed

Environmental

Regulations

RCRA 40CFR: Not listed
CERCLA: 40CFR 302.4: Not listed
SARA 40CFR 372.65: Not listed
SARA EHS 40CFR 355: Not listed
TSCA: Not listed

ACE1150 CAS #: 83-32-9

ACENAPHTHENE

RTECS: AB1000000
EINECS Number: 201-469-6
Molecular Formula: $C_{12}H_{10}$
Structured MF: $C_{10}H_6(CH_2)_2$
Formula Weight: 154.21

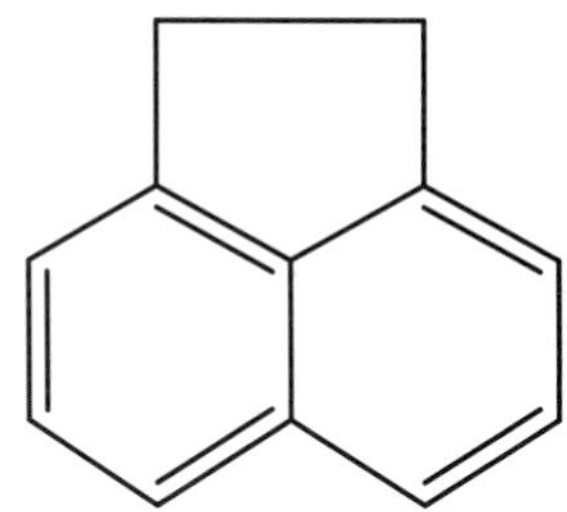

Chemical Structure

Synonyms: ACENAPHTHYLENE,1,2-DIHYDRO-; 1,8-DIHYDROACENAPHTHALENE; 1,2-DIHYDROACENAPHTHYLENE; 1,8-DIHYDROACENAPHTHYLENE; 1,8-ETHYLENENAPHTHALENE; ETHYLENENAPHTHALENE; NAPHTHYLENEETHYLENE; PERI-ETHYLENE NAPHTHALENE; PERIETHYLENENAPHTHALENE

Description: white crystals

Use: dye intermediate; pharmaceuticals; insecticide; fungicide; herbicide and manufacture of plastics

Physical Properties

Boiling Point: 279 °C (534 °F) at 760 mm Hg
Freezing Point: 95 °C (203 °F)
Specific Gravity: 1.0242 at 90 °C/4 °C
Vapor Density: 5.32 Air=1
Density: 1.02 g/mL at 25 °C
Vapor Pressure: 10 mm Hg at 131.2 °C
Water Solubility: < 1 mg/mL at 20 C
Other Solubilities: 95% Ethanol: 1-5 mg/ml at 20 °C; Acetone: 50-100 mg/ml at 20 °C; Benzene: 1 g/5 mL; Chloroform: 1 g/2.5 mL; DMSO: 10-50 mg/ml at 20 °C; Ether: Soluble; Glacial Acetic Acid: 1 g/100 mL; Toluene: 1 g/5 mL;Methanol: 1 g/56 mL; Propanol: 1 g/25 mL.
Odor Threshold: 0.5048 mg/m^3
Refraction Index: 1.6048 at 95 °C/D
Ionization Potential (eV): 7.68 +/-0.2
Flash Point: Not available; probably combustible
LEL: 0.6% v/v

RTECS Toxicity Data

Chronic (Multiple Dose) Inhalation: Rat Dose: 12 mg/m^3/4H/22W-I; Toxic Effects: Lungs, Thorax, or Respiration - Fibrosis, focal (pneumoconiosis).

Mutagenic: Other Microorganisms Mutations in Microorganisms; Dose: 3 mg (-S9).

Hazard Overviews

Fire Diamond

Health: Irritating to eyes/skin/respiratory tract. Also Causes: vomiting.

Fire: Combustible. Use dry chemical, carbon dioxide, water spray, fog, or foam. Carbon oxide(s) are possible hazardous combustion products.

Reactivity: Stable. Hazardous polymerization cannot occur. Avoid: exposure to heat; ignition sources. Incompatible with: molecular oxygen in the presence of alkali-earth metal bromides; ozone in the presence of alkali-earth metal hydroxides; aromatic alcohols and ketones by reaction with transition metal catalysts. Hazardous decomposition products: carbon oxides and thick, acrid smoke.

Carcinogenicity: IARC - Not listed; NIOSH - Not listed; NTP - Not listed; ACGIH - Not listed; OSHA - Not listed; EPA - Not listed; MAK - Not listed

Primary Target Organs:

Eyes

Skin

Respiratory System

Mucous Membranes

Environmental

Ecotoxicity: LC_{50} Cyprinodon variegatus (sheepshead minnow) 2,230 ug/l/96 hr in a static bioassay LC_{50} Salmo trutta (brown trout) 580 ug/l/96 hr at 12.0 °C, wt 0.16 g, 10.3 mg/l dissolved oxygen, water hardness EDTA 45.8 mg/l as calcium carbonate ($CaCO_3$), total alkalinity 44.1 mg/l as calcium carbonate, pH 7.2-7.4, flow-through bioassay LC_{50} Mysidopsis bahia (mysid shrimp) 970 ug/l/96 hr in a static bioassay LC_{50} Aplexa hynorum (snail) adult > 2040 ug/l/96 hr at 22.9 °C, 7.2 mg/l dissolved oxygen, water hardness EDTA 43.3 mg/l as calcium carbonate ($CaCO_3$), total alkalinity 41.9 mg/l as calcium carbonate, pH 7.5-7.6, flow-through bioassay

Environmental Fate: Should biodegrade rapidly in the environment. The reported biodegradation half-lives in aerobic soil and surface waters range from 10 to 60 days and

1 to 25 days, respectively. However, it may persist under anaerobic conditions or at high concentration due to toxicity to micro-organisms. It is not expected to hydrolyze or bioconcentrate in the environment; yet, it should undergo direct photolysis in sunlit environmental media. A calculated K_{oc} range of 2065 to 3230 indicates it will be slightly mobile in soil. In aquatic systems, it can partition from the water column to organic matter contained in sediments and suspended solids. A Henry's Law constant of 1.55 $x10^{-4}$ atm-cu m/mole at 25 °C suggests volatilization from environmental waters may be important. The volatilization half-lives from a model river and a model pond, the latter considers the effect of adsorption, have been estimated to be 11 hr and 39 days, respectively. It is expected to exist entirely in the vapor-phase in ambient air. In the atmosphere, the reaction with photochemically produced hydroxyl radicals (half-life of 7.2 hr) is likely to be an important fate process.

Environmental Physical Data

Henry's Law Constant: 1.55 $x10^{-4}$
Octanol/Water Partition Coefficient: log K_{ow} = 3.92
Sorption Partition Coefficient: K_{oc} = estimated at 2065 to 3230
BCF: bluegill concentration factor 387

Regulations

RCRA 40CFR: Not listed
CERCLA: 40CFR 302.4: Listed per CWA Section 307(a) RQ: 100 lb (45.35 kg)
SARA 40CFR 372.65: Not listed
SARA EHS 40CFR 355: Not listed
TSCA: Listed

Analytical Methods

Air: EPA TO-13; California 429
Soil: CLP LC_SV, MC_SVOA, OHC; EPA 16, 1625, PAH-005, PAH-007, PAH-011, PAH-012, S-004-1; SW846 3630B, 3640A, 8100, 8250A, 8270B, 8270C, 8275A, 8310, 8410; DOE OS050
Water / Groundwater: EPA PAH-002, PAH-006, S-002-1, 1625, 610, 625, 625-S, 6; APHA 6040-B, 6410-B, 6440-B, 6440-C; ASTM D4657, D4763; USGS O3113, O3118
Drinking Water: EPA 550, 550.1
Indoor / Expired Air: NIOSH 5506, 5515; EPA IP-7-A, IP-7-B
Plasma: EPA 29
Other: EPA PAH-009

ACE1300 CAS #: 208-96-8

ACENAPHTHYLENE

RTECS: AB1254000
EINECS Number: 205-917-1
Molecular Formula: $C_{12}H_8$
Formula Weight: 152.20

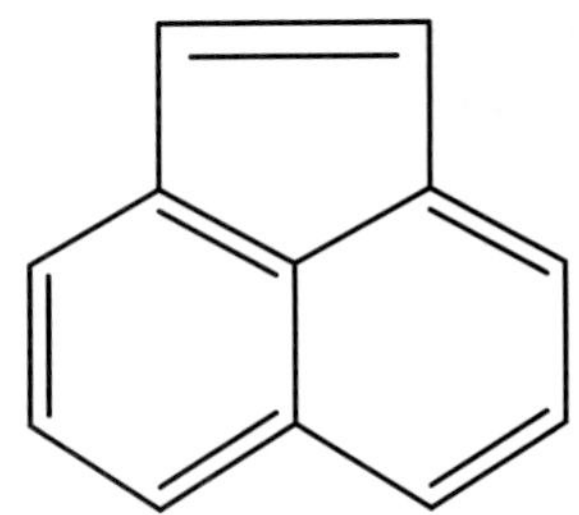

Chemical Structure

Synonyms: CYCLOPENTA(DE)NAPHTHALENE
Description: prisms or plates

Physical Properties

Boiling Point: 265 °C (509 °F) to 275 °C (527 °F)
Freezing Point: 92 °C (197.6 °F) to 93 °C (199.4 °F)
Specific Gravity: 0.8988 at 16 °C/2 °C
Saturated Vapor Density: 1.200050325 kg/m^3
Vapor Pressure: 0.001 kPa at 25 °C
Water Solubility: 3.93 mg/L distilled Water at 25 °C
Other Solubilities: 95% Ethanol: Very Soluble; Benzene: Very Soluble; Ether: Very Soluble.
Ionization Potential (eV): 8.22 +/-0.2

RTECS Toxicity Data

Acute Oral: Mouse LD_{50} Dose: 1760 mg/kg; Toxic Effects: Automatic Nervous System - Other (direct) parasympathomimetic; Lungs, Thorax, or Respiration - Respiratory depression; Blood - Hemorrhage.
Chronic (Multiple Dose) Inhalation: Rat Dose: 500 ug/m^3/4H/17W-I; Toxic Effects: Lungs, Thorax, or Respiration - Structural or functional change in trachea or bronchi; Lungs, Thorax, or Respiration - Bronchiolar dilation; Nutritional and gross metabolic - Weight loss or decreased weight gain.
Mutagenic: Bacteria - S Typhimurium Mutations in Microorganisms; Dose: 1 mmol/L/2H (-S9).

Hazard Overviews

Health: Irritating to eyes/skin/respiratory tract. Other Acute Effects: may be harmful by inhalation, ingestion, or skin absorption.
Fire: Hazards: emits toxic fumes. Extinguishing agents: water spray; carbon dioxide, dry chemical powder or appropriate foam. Precautions: combustible liquid.
Reactivity: Incompatible with: strong oxidizing agents. Hazardous decomposition products: toxic fumes of: carbon monoxide, carbon dioxide.
Carcinogenicity: IARC - Not listed; NIOSH - Not listed; NTP - Not listed; ACGIH - Not listed; OSHA - Not listed; EPA - Not listed; MAK - Not listed

Primary Target Organs:

Eyes Skin Respiratory System

Environmental

Environmental Fate: Should biodegrade in the environment. The reported biodegradation half-lives in aerobic soil range from 12 to 121 days. It is not expected to hydrolyze or bioconcentrate in the environment; yet, may undergo direct photolysis in sunlit environmental media. A calculated K_{oc} range of 950 to 3315 indicates it will have a low to slight mobility class in soil. In aquatic systems, it may partition from the water column to organic matter contained in sediments and suspended solids. A Henry's Law constant of 1.13×10^{-5} atm-cu m/mole at 25 °C suggests volatilization from environmental waters may be important. The volatilization half-lives from a model river and a model pond, the later considers the effect of adsorption, have been estimated to be 4 and 184 days, respectively. It is expected to exist entirely in the vapor-phase in ambient air. In the atmosphere, reactions with photochemically produced hydroxyl radicals and ozone (respective estimated half-lives of 5 and 1 hr) are likely to be important fate processes.

Environmental Physical Data

Henry's Law Constant: calculated at 1.13×10^{-5}
Octanol/Water Partition Coefficient: log K_{ow} = 3.94
Sorption Partition Coefficient: K_{OC} = estimated at 950 to 3315
BCF: estimated at 2.11

Regulations

RCRA 40CFR: Not listed
CERCLA: 40CFR 302.4: Listed per CWA Section 307(a) RQ: 5000 lb (2268 kg)
SARA 40CFR 372.65: Not listed
SARA EHS 40CFR 355: Not listed
TSCA: Listed

Analytical Methods

Air: EPA TO-13
Soil: CLP LC_SV, MC_SVOA, OHC; EPA 16, 1625, PAH-005, PAH-007, PAH-011, PAH-012, S-004-1; SW846 1311, 3630B, 3640A, 8100, 8250A, 8270B, 8270C, 8275A, 8310, 8410; DOE OS050
Water / Groundwater: EPA PAH-002, PAH-006, S-002-1, 1625, 610, 625, 625-S, 6; APHA 6410-B, 6440-B, 6440-C; ASTM D4657; USGS O3118
Drinking Water: EPA 525.1, 525.2, 550, 550.1
Indoor / Expired Air: NIOSH 5506, 5515; EPA IP-7-A, IP-7-B
Plasma: EPA 29
Other: EPA PAH-009

ACE1450 CAS #: 82-86-0

1,2-ACENAPHTHYLENEDIONE

RTECS: AB1024500
EINECS Number: 201-441-3
Molecular Formula: $C_{12}H_6O_2$
Formula Weight: 182.18

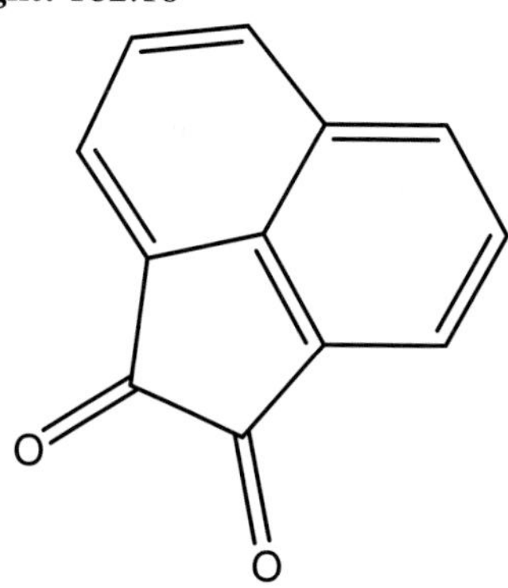

Chemical Structure

Synonyms: ACENAPHTHAQUINONE; 1,2-ACENAPHTHENEDIONE; 1,2-ACENAPHTHENEQUINONE; ACENAPHTHOQUINONE
Description: yellow needles
Use: fungicide; dye synthesis

Physical Properties

Boiling Point: 280 °C (536 °F)
Freezing Point: 261 °C (501.8 °F)
Specific Gravity: 1.48
Water Solubility: Insoluble in Water
Other Solubilities: Slightly Soluble in Acetic Acid; Slightly Soluble in Alcohol, hot Benzene, hot Petroleum Ether
Ionization Potential (eV): 8.6

Hazard Overviews

Health: Irritating to eyes/skin/respiratory tract. Other Acute Effects: may be harmful by inhalation, ingestion, or skin absorption.
Fire: Extinguishing agents: water spray; carbon dioxide, dry chemical powder or appropriate foam. Precautions: combustible liquid.
Reactivity: Incompatible with: strong oxidizing agents. Hazardous decomposition products: toxic fumes of: carbon monoxide, carbon dioxide.
Carcinogenicity: IARC - Not listed; NIOSH - Not listed; NTP - Not listed; ACGIH - Not listed; OSHA - Not listed; EPA - Not listed; MAK - Not listed

Primary Target Organs:

Environmental

Regulations

RCRA 40CFR: Not listed
CERCLA: 40CFR 302.4: Not listed
SARA 40CFR 372.65: Not listed
SARA EHS 40CFR 355: Not listed
TSCA: Listed

ACE1600 CAS #: 152-72-7

ACENOCOUMAROL

RTECS: GN4900000
EINECS Number: 205-807-3
Molecular Formula: $C_{19}H_{15}NO_6$
Formula Weight: 353.32
Synonyms: ACENOCOUMARIN; ACENOCUMAROL; ACENOKUMARIN; 3-(ALPHA-ACETONYL-4-NITROBENZYL)-4-HYDROXYCOUMARIN; 3-(ALPHA-ACETONYL-P-NITROBENZYL)-4-HYDROXY-COUMARIN; ASCUMAR; 2H-1-BENZOPYRAN-2-ONE,4-HYDROXY-3-(1-(4-NITROPHENYL)-3-OXOBUTYL)-; COUMARIN,3-(ALPHA-ACETONYL-P-NITROBENZYL)-4-HYDROXY-; G-23,350; G-23350; G23350; 4-HYDROXY-3-(1-(4-NITROPHENYL)-3-OXOBUTYL)-2H-1-BENZOPYRAN-2-ONE; NICOUMALONE; 3-(ALPHA-(P-NITROPHENOL)-BETA-ACETYLETHYL)-4-HYDROXYCOUMARIN; 3-(ALPHA-(4'-NITROPHENYL)-BETA-ACETYLETHYL)-4-HYDROXYCOUMARIN; 3-(ALPHA-P-NITROPHENYL-BETA-ACETYLETHYL)-4-HYDROXYCOUMARIN; NITROPHENYLACETYLETHYL-4-HYDROXYCOUMARINE; NITROVARFARIAN; NITROWARFARIN; SINCOUMAR; SINKUMAR; SINTHROM; SINTHROME; SINTROM; SINTROMA; SYNCOUMAR; SYNCUMAR; SYNTROM; ZOTIL
Description: off-white to light-tan powder or crystals; odorless
Use: medication: anticoagulant

Physical Properties

Freezing Point: 197 °C (386.6 °F)
Water Solubility: Almost Insoluble in Water
Other Solubilities: Soluble in solution of alkali hydroxides.

RTECS Toxicity Data

Acute Oral: Rat LD_{50} Dose: 513 mg/kg. Mouse LD_{50} Dose: 1470 mg/kg.

Hazard Overviews

Carcinogenicity: IARC - Not listed; NIOSH - Not listed; NTP - Not listed; ACGIH - Not listed; OSHA - Not listed; EPA - Not listed; MAK - Not listed

Environmental

Regulations

RCRA 40CFR: Not listed
CERCLA: 40CFR 302.4: Not listed
SARA 40CFR 372.65: Not listed
SARA EHS 40CFR 355: Not listed
TSCA: Not listed

ACE1750 CAS #: 33665-90-6

ACESULFAME

EINECS Number: 251-622-6
Molecular Formula: $C_4H_5NO_4S$
Formula Weight: 163.15
Synonyms: ACETOSULFAM; 6-METHYL-3,4-DIHYDRO-1,2,3-OXATHIAZIN-4-ONE 2,2-DIOXIDE; 6-METHYL-1,2,3-OXATHIAZIN-4(3H)-ONE 2,2-DIOXIDE; 1,2,3-OXATHIAZIN-4(3H)-ONE,6-METHYL-,2,2-DIOXIDE
Description: needles
Use: in prepn of compositions to be used as carriers for edible compounds; sweetener in edible compositions of oral pharmaceuticals

Physical Properties

Freezing Point: 123.5 °C (254.3 °F)
Density: 1.83 g/cu cm
Water Solubility: 30 g/100 mL Water at 20 °C
Other Solubilities: solubility in g/100 ml at 20 °C: 1.0 in Methanol, 0.08 in Acetone, about 13 in Acetic Acid; Soluble in Glycerin water.

Hazard Overviews

Carcinogenicity: IARC - Not listed; NIOSH - Not listed; NTP - Not listed; ACGIH - Not listed; OSHA - Not listed; EPA - Not listed; MAK - Not listed

Environmental

Regulations

RCRA 40CFR: Not listed
CERCLA: 40CFR 302.4: Not listed
SARA 40CFR 372.65: Not listed
SARA EHS 40CFR 355: Not listed
TSCA: Not listed

ACE1900 CAS #: 105-57-7

ACETAL

RTECS: AB2800000
DOT: UN1088; IMO3.2
EINECS Number: 203-310-6
Molecular Formula: $C_6H_{14}O_2$
Structured MF: $(C_2H_5O)_2{\bullet}CHCH_3$
Formula Weight: 118.17

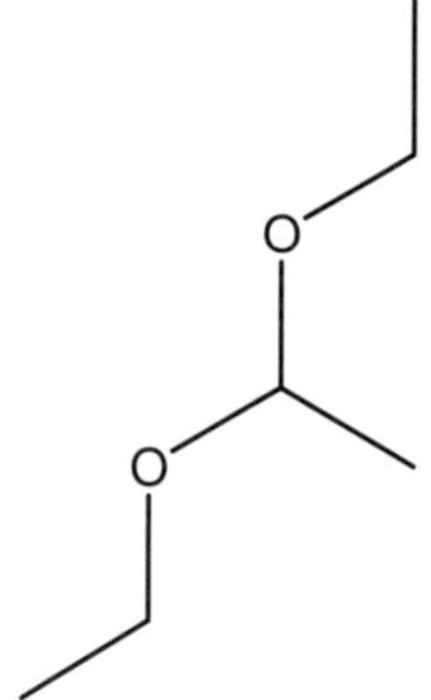

Chemical Structure

Synonyms: ACETAAL; ACETAL DIETHYLIQUE; ACETALDEHYDE ETHYL ACETAL; ACETALDEHYDE,DIETHYL ACETAL; ACETALE; ACETOL; 1,1-DIAETHOXY-AETHAN; DIAETHYLACETAL; 1,1-DIETHOXY-ETHAAN; 1,1-DIETHOXYETHANE; DIETHYL ACETAL; DIETHYLACETAL; 1,1-DIETOSSIETANO; ETHANE,1,1-DIETHOXY-; ETHYLIDENE DIETHYL ETHER; ETHYLIDENEDIETHYL ETHER

Description: colorless liquid; fruity odor

Use: solvent; in organic syntheses; formerly hypnotic; solvent in perfumes; synthetic flavoring ingredient; in cosmetics

Physical Properties

Boiling Point: 102.7 °C (217 °F) at 760 mm Hg
Freezing Point: -100 °C (-148 °F)
Specific Gravity: 0.8254 at 20 °C/4 °C
Vapor Density: 4.1 Air=1
Vapor Pressure: 20 mm Hg at 19.6 °C
Water Solubility: 5 g/100 g Water
Other Solubilities: Soluble in Chloroform, Acetone.
Surface Tension: 21.65 dynes/cm at 20 °C
Refraction Index: 1.38193 at 20 °C/D
Ionization Potential (eV): 9.2 +/-0.6
Flash Point: -21 °C Closed Cup
Autoignition Temperature: 230 °C
LEL: 1.6% v/v
UEL: 10.4% v/v

RTECS Toxicity Data

Acute Oral: Rat LD_{50} Dose: 4600 mg/kg. Mouse LD_{50} Dose: 3500 mg/kg.

Acute Inhalation: Rat LC_{Lo} Dose: 4000 ppm/4hr.

Irritation Eye: Rabbit Standard Draize Test Dose: 500 mg/24H; Reaction: mild.

Irritation Skin: Rabbit Standard Draize Test Dose: 500 mg/24H; Reaction: mild. Rabbit Open Draize Test Dose: 10 mg/24H open; Reaction: mild.

Hazard Overviews

Flammable

Fire Diamond

Health: Irritating to eyes/skin/respiratory tract. Also Causes: by inhalation: narcosis, CNS depression; by ingestion: drowsiness, analgesia.

Fire: Flammable. Forms explosive mixtures in air. Use dry chemical, alcohol foam, or carbon dioxide. Use a water spray to cool fire-exposed tanks or containers. Water or foam may cause frothing; water may be ineffective in extinguishing acetal fires.

Reactivity: Stable. Hazardous polymerization can occur upon storage. Avoid: heat; sparks; open flame; direct contact with dilute acids. Incompatible with: strong oxidizers. Hazardous decomposition products: carbon monoxide; carbon dioxide.

Carcinogenicity: IARC - Not listed; NIOSH - Not listed; NTP - Not listed; ACGIH - Not listed; OSHA - Not listed; EPA - Not listed; MAK - Not listed

Primary Target Organs:

Eyes

Skin

Respiratory System

Nervous System

Environmental

Environmental Fate: If released to soil, it will have high mobility. Volatilization may be important from moist and dry soil surfaces. Insufficient data are available to determine the rate or importance of biodegradation in soil or water. If released to water, it would not adsorb to suspended solids and sediment, based on an estimated K_{oc} value of 68. Would volatilize from water surfaces with estimated half-lives for a model river and model lake of 13 hours and 7.3 days, respectively. An estimated BCF value of 2.6 suggests that it will not bioconcentrate in aquatic organisms. It is easily cleaved by dilute acids but extremely resistant to hydrolysis by bases. Therefore, hydrolysis is expected to be an important fate process in water. If released to the atmosphere, it will exist in the vapor phase. Vapor-phase is degraded in the atmosphere by reaction with photochemically produced hydroxyl radicals with an estimated half-life of about 19.6 hours.

Cleanup/Disposal: Guide No. 127: Eliminate all ignition sources (no smoking, flares, sparks or flames in immediate area). All equipment used when handling the product must be grounded. Do not touch or walk through spilled material. Stop leak if you can do it without risk. Prevent entry into waterways, sewers, basements or confined areas. A vapor suppressing foam may be used to reduce vapors. Absorb or cover with dry earth, sand or other non-combustible material and transfer to containers. Use clean non-sparking tools to collect absorbed material. Large Spills: Dike far ahead of liquid spill for later disposal. Water spray may reduce vapor; but may not prevent ignition in closed spaces.

Environmental Physical Data

Henry's Law Constant: estimated at 9.75×10^{-5}
Octanol/Water Partition Coefficient: log K_{ow} = 0.84
Sorption Partition Coefficient: K_{oc} = estimated at 68
BCF: estimated at 2.6

Regulations

RCRA 40CFR: Not listed
CERCLA: 40CFR 302.4: Not listed
SARA 40CFR 372.65: Not listed
SARA EHS 40CFR 355: Not listed
TSCA: Listed

ACE2050 CAS #: 75-07-0

ACETALDEHYDE

RTECS: AB1925000
DOT: UN1089; IMO3.1
EINECS Number: 200-836-8
Molecular Formula: C_2H_4O
Structured MF: CH_3CHO
Formula Weight: 44.05

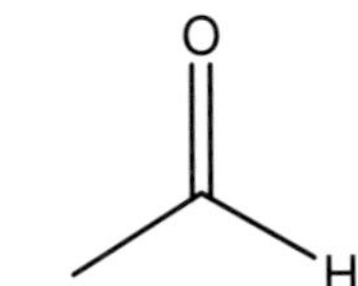

Chemical Structure

Synonyms: ACETALDEHYD; ACETIC ALDEHYDE; ACETYLALDEHYDE; ALDEHYDE ACETIQUE; ALDEIDE ACETICA; ETHANAL; ETHYL ALDEHYDE; OCTOWY ALDEHYD
Description: colorless liquid; fruity to suffocating odor
Use: manufacture of pentaerythritol, peracetic acid, pyridines, paraldehyde, acetic acid, acetic anhydride, 2-ethylhexanol, aldol, chloral, 1,3-butylene glycol, trimethylolpropane, butanol, perfumes, aniline dyes, plastics and synthetic rubber; in silvering mirrors and in hardening gelatin fibers; as a chemical intermediate and synthetic flavoring substance and adjuvant

Physical Properties

Boiling Point: 21 °C (70 °F)
Freezing Point: -123.5 °C (-190.3 °F)
Specific Gravity: 0.788 at 16 °C/4 °C
Vapor Density: 1.52 Air=1
Saturated Vapor Density: 1.806370236 kg/m^3
Vapor Pressure: 740 mm Hg at 20 °C
Water Solubility: Miscible with Water
Other Solubilities: miscible with Ether, Benzene; miscible with Gasoline, solvent Naphtha, Toluene, Xylene, Turpentine, & Acetone.
Surface Tension: 21.2 mN/m at 20 °C
Odor Threshold: 0.0002 to 4.14 mg/m^3
Refraction Index: 1.3316 at 20 °C/D
Evaporation Rate: 3.0 Ethylether=1
Critical Temperature: 193 °C
Critical Pressure: 5.5500 x10^6 Pa
Ionization Potential (eV): 10.22
Flash Point: -38.89 °C Closed Cup
Autoignition Temperature: 175 °C
LEL: 4.0% v/v
UEL: 60% v/v

RTECS Toxicity Data

Acute Oral: Rat LD_{50} Dose: 661 mg/kg; Toxic Effects: Peripheral nerve and sensation - Spastic parapysis with or without sensory change; Behavioral - Altered sleep time (including change in righting reflex); Lungs, Thorax, or Respiration - Dyspnea. Mouse LD_{50} Dose: 900 mg/kg.
Acute Inhalation: Rat LC_{50} Dose: 13300 ppm/4hr; Toxic Effects: Behavioral - Excitment; Lungs, Thorax, or Respiration - Dyspnea. Human TC_{Lo} Dose: 134 ppm/30M; Toxic Effects: Lungs, Thorax, or Respiration - Other changes.
Acute Dermal: Rabbit LD_{50} Route: Skin; Dose: 3540 mg/kg. Rabbit LD_{Lo} Route: Subcutaneous Dose: 1200 mg/kg; Toxic Effects: Behavioral - Convulsions or effect on seizure threshold; Lungs, Thorax, or Respiration - Other changes.
Chronic (Multiple Dose) Oral: Rat Dose: 18900 mg/kg/4W-C; Toxic Effects: Behavioral - Fluid intake; Blood - Changes in serum composition; Blood - Other changes.
Chronic (Multiple Dose) Inhalation: Rat Dose: 2217 ppm/6H/4W-I; Toxic Effects: Sense organs and special senses - Tumors; Kidney, Ureter, and Bladder - Urine volume decreased; Nutritional and gross metabolic - Weight loss or decreased weight gain. Hamster Dose: 4560 ppm/6H/90D-I; Toxic Effects: Lungs, Thorax, or Respiration - Structural or functional change in trachea or bronchi; Kidney, Ureter, and Bladder - Other changes; Nutritional and gross metabolic - Weight loss or decreased weight gain.
Irritation Eye: Rabbit Standard Draize Test Dose: 40 mg; Reaction: severe.
Irritation Skin: Rabbit Open Draize Test Dose: 500 mg open; Reaction: mild.
Reproductive/Teratogenic: Rat Route: Oral; Dose: 4800 mg/kg; Duration: female 1-20D of pregnancy; Effects on Embryo or Fetus - Fetotoxicity; Specific Developmental Abnormalities - Respiratory system; Hepatobiliary system. Rat Route: Oral; Dose: 5040 mg/kg; Duration: female 1-21D of pregnancy; Specific Developmental Abnormalities - Central nervous system; Endocrine system; Urogenital system. Rat Route: Oral; Dose: 5040 mg/kg; Duration: female 1-21D of pregnancy; Effects on Newborn - Growth statistics.
Mutagenic: Human DNA Damage; Cell Type: lymphocyte; Dose: 1560 umol/L. Human DNA Damage; Cell Type: other cell types; Dose: 3 mmol/L. Human DNA Inhibition; Cell Type: other cell types; Dose: 30 mmol/L. Human DNA Inhibition; Cell Type: HeLa cell; Dose: 10 mmol/L. Human Mutations in Mammalian Somatic Cells; Cell Type: fibroblast; Dose: 5 mmol/L. Human Cytogenetic Analysis; Cell Type: leukocyte; Dose: 1000 ppm/72H-C. Human Sister Chromatid Exchenge; Cell Type: lymphocyte; Dose: 1200 umol/L. Human Sister Chromatid Exchenge; Cell Type: fibroblast; Dose: 40 umol/L. Human Other Mutation Test Systems; Cell Type: other cell types; Dose: 30 mmol/L.
Tumorigenic: Rat Route: Inhalation; Dose: 735 ppm/6H/2Y-I; Toxic Effects: Tumorigenic - Carcinogenic by RTECS criteria; Sense organs and special senses - Tumors. Rat Route: Inhalation; Dose: 1410 ppm/6H/65W-I; Toxic Effects:

Tumorigenic - Equivocal tumorigenic agent by RTECS criteria; Sense organs and special senses - Tumors.

Hazard Overviews

Explosive Flammable

Fire Diamond

Health: Irritating to eyes/skin/respiratory tract. Also Causes: drowsiness, dizziness, possible liver/kidney damage. Chronic Effects: dermatitis, possible cancer hazard, may cause birth defects.

Fire: Flammable. Forms explosive peroxides upon exposure to air. Can form explosive mixtures in air. For small fires use dry chemical, carbon dioxide, water spray, or alcohol-resistant foam. For large fires use water spray, fog, or alcohol-resistant foam.

Reactivity: Unstable, oxidizes readily in air. Polymerization can occur when contaminated with acids, alkalies, and trace metals. Avoid: exposure to heat and ignition sources. Incompatible with: acetic acid; acid anhydrides; alcohols; ammonia (anhydrous); halogens; ketones; phenols; phosphorus isocyanate; sodium hydroxide; cobalt acetate and oxygen; hydrogen peroxide; dessicants; dinitrogen pentaoxide; halocarbons; mercury (II) oxosalts; trace metals; other acids and alkalies. Hazardous decomposition products: toxic carbon monoxide gas.

Carcinogenicity: IARC - Group 2B, Possibly carcinogenic to humans; NIOSH - Listed as carcinogen; NTP - Class 2B, Reasonably anticipated to be a carcinogen, sufficient evidence of carcinogenicity from studies in experimental animals; ACGIH - Class A3, Animal carcinogen; OSHA - Not listed; EPA - Class B2, Probable human carcinogen based on animal studies; MAK - Class B, Justifiably suspected of having carcinogenic potential

Primary Target Organs:

Eyes

Skin

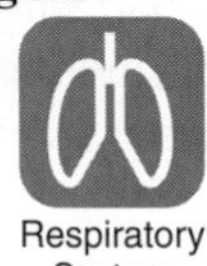
Respiratory System

Nervous System

Liver

Kidneys

Exposure Limits

OSHA PEL: TWA: 200 ppm; 360 mg/m^3.

OSHA PEL Vacated 1989 Limits: TWA: 100 ppm; 180 mg/m^3; STEL: 150 ppm; 270 mg/m^3.

ACGIH TLV: STEL: 25 ppm; 45 mg/m^3; Ceiling.

NIOSH IDLH: 2000 ppm.

DFG MAK: TWA: 50 ppm; 90 mg/m^3.

Respirator Recommendation

Exposure Range: >200 to 1000 ppm Air Purifying, Negative Pressure, Half Mask

Exposure Range: >1000 to 2000 ppm Supplied Air, Constant Flow/Pressure Demand, Full Face

Exposure Range: >2000 to unlimited ppm Self-contained Breathing Apparatus, Pressure Demand, Full Face

Cartridge Color: black

Environmental

Ecotoxicity: EC_{50} Pimephales promelas (fathead minnow) 30.8 mg/l/96 hr (confidence limit 28.0-34.0 mg/l), flow-through bioassay with measured concentrations, 23.9 °C, dissolved oxygen 7.2 mg/l, hardness 53.0 mg/l calcium carbonate, alkalinity 43.2 mg/l calcium carbonate

Environmental Fate: If released into water it will rapidly biodegrade and volatilize (half-life 3 hours for a typical river). If spilled on land it will also rapidly evaporate and leach into the ground where it will biodegrade. In the atmosphere it will degrade in a matter of hours by reaction with hydroxyl radicals and photolysis.

Cleanup/Disposal: Guide No. 129: Eliminate all ignition sources (no smoking, flares, sparks or flames in immediate area). All equipment used when handling the product must be grounded. Do not touch or walk through spilled material. Stop leak if you can do it without risk. Prevent entry into waterways, sewers, basements or confined areas. A vapor suppressing foam may be used to reduce vapors. Absorb or cover with dry earth, sand or other non-combustible material and transfer to containers. Use clean non-sparking tools to collect absorbed material. Large Spills: Dike far ahead of liquid spill for later disposal. Water spray may reduce vapor; but may not prevent ignition in closed spaces.

Environmental Physical Data

Henry's Law Constant: 7.89 x10^{-5}

Octanol/Water Partition Coefficient: log K_{ow} = estimated at 0.43

BCF: none likely

BOD: 93 to 127%, 5 days

Regulations

RCRA 40CFR: Listed Hazardous Waste No. U001 Ignitable Waste

CERCLA: 40CFR 302.4: Listed per CWA Section 311(b)(4) per RCRA Section 3001 RQ: 1000 lb (453.5 kg)

SARA 40CFR 372.65: Listed

SARA EHS 40CFR 355: Not listed

TSCA: Listed

Analytical Methods

Air: EPA TO-11, TO-5, 0100; ASTM D4490

Soil: SW846 8315A

Water / Groundwater: ASTM D3695

Drinking Water: EPA 554

Indoor / Expired Air: NIOSH 2538, 2539, 3507; EPA IP-6A, IP-6B, IP-6C, 0100

ACE2200 CAS #: 107-29-9

ACETALDEHYDE OXIME

RTECS: AB2975000

DOT: UN2332; IMO3.2

EINECS Number: 203-479-6

Molecular Formula: C_2H_5NO

Formula Weight: 59.07

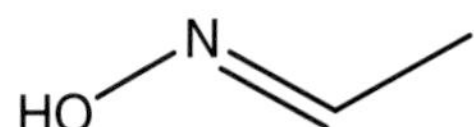

Chemical Structure

Synonyms: ACETALDEHYDE,OXIME; ACETALDOXIME; ALDOXIME; ETHANAL OXIME; ETHYLIDENEHYDROXYLAMINE; HYDROIMINOETHANE

Description: needles

Use: intermediate in chemical synthesis

Physical Properties

Boiling Point: 114.5 °C (238 °F)
Freezing Point: 47 °C (116.6 °F)
Specific Gravity: 0.9656 at 20 °C/4 °C
Water Solubility: Very Soluble in Water
Other Solubilities: Very Soluble in Alcohol, Ether
Refraction Index: 1.415 at 20 °C/D
Ionization Potential (eV): 10.0 +/-2.0
Flash Point: < 22 °C

RTECS Toxicity Data

Mutagenic: Mouse Mutations in Microorganisms; Cell Type: lymphocyte; Dose: 230 mg/L (+S9). Mouse Mutations in Mammalian Somatic Cells; Cell Type: lymphocyte; Dose: 15 gm/L.

Hazard Overviews

Flammable

Health: Severely irritating to eyes; irritating to skin/respiratory tract. Harmful. Other Acute Effects: harmful if inhaled or swallowed; target organ: blood.

Fire: Flammable. Hazards: emits toxic fumes. Extinguishing agents: carbon dioxide, dry chemical powder or appropriate foam. Precautions: combustible liquid.

Reactivity: Incompatible with: strong oxidizing agents. Hazardous decomposition products: toxic fumes of: carbon monoxide, carbon dioxide, nitrogen oxides.

Carcinogenicity: IARC - Not listed; NIOSH - Not listed; NTP - Not listed; ACGIH - Not listed; OSHA - Not listed; EPA - Not listed; MAK - Not listed

Primary Target Organs:

  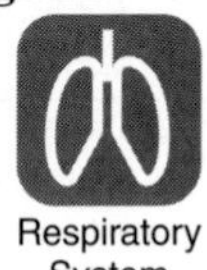 

Eyes, Skin, Respiratory System, Blood

Environmental

Environmental Fate: If released to the atmosphere, it should degrade by reaction with photochemically produced hydroxyl radicals which has an estimated half-life of approximately 7.3 days in average air. If released to soil, significant leaching may occur. Evaporation from dry surfaces may also occur. If released to water, volatilization is not expected to be rapid, although significant volatilization may occur from shallow, rapidly moving rivers. The volatilization half-lives from a model river (1 meter deep) and from an environmental pond have been estimated to be 3.6 and 39.5 days, respectively. No data are available pertaining to the chemical degradation in soil or water. A single biological screening study has noted that two microbes isolated from soil are able to grow on it as a sole carbon source.

Cleanup/Disposal: Guide No. 129: Eliminate all ignition sources (no smoking, flares, sparks or flames in immediate area). All equipment used when handling the product must be grounded. Do not touch or walk through spilled material. Stop leak if you can do it without risk. Prevent entry into waterways, sewers, basements or confined areas. A vapor suppressing foam may be used to reduce vapors. Absorb or cover with dry earth, sand or other non-combustible material and transfer to containers. Use clean non-sparking tools to collect absorbed material. Large Spills: Dike far ahead of liquid spill for later disposal. Water spray may reduce vapor; but may not prevent ignition in closed spaces.

Environmental Physical Data

Henry's Law Constant: estimated at 7.94 x10^{-6}
Octanol/Water Partition Coefficient: log K_{ow} = -0.13
Sorption Partition Coefficient: K_{oc} = estimated at 5 to 20
BCF: estimated at 0.5

Regulations

RCRA 40CFR: Not listed
CERCLA: 40CFR 302.4: Not listed
SARA 40CFR 372.65: Not listed
SARA EHS 40CFR 355: Not listed
TSCA: Listed

ACE2350 CAS #: 107-89-1

ACETALDOL

RTECS: ES3150000
EINECS Number: 203-530-2
Molecular Formula: $C_4H_8O_2$
Formula Weight: 88.10

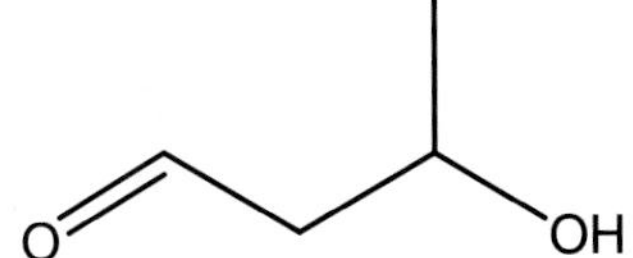

Chemical Structure

Synonyms: ALDOL; BUTANAL,3-HYDROXY-; 3-BUTANOLAL; BUTYRALDEHYDE,3-HYDROXY-; 3-HYDROXYBUTANAL; 3-HYDROXYBUTYRALDEHYDE; BETA-HYDROXYBUTYRALDEHYDE; OXYBUTANAL; OXYBUTYRIC ALDEHYDE

Description: clear, white to yellow syrupy liquid

Use: mfr rubber vulcanizers, accelerators & age resisters; in perfumes; in ore flotation; medication: former hypnotic & sedative agent; in engraving; solvent; in solvent mixt for cellulose acetate; in fungicides, org synthesis, cadmium plating; in dyes, drugs; dyeing assistant; in synthetic polymers, printer's rollers

Physical Properties

Boiling Point: 83 °C (181 °F) at 20 mm Hg
Freezing Point: Decomposes at 85 °C (185 °F)
Specific Gravity: 1.103 at 20 °C/4 °C
Water Solubility: Soluble in all Proportions
Other Solubilities: miscible with organic solvents.
Refraction Index: 1.4238 at 20 °C/D
Flash Point: 65.556 °C Open Cup
Autoignition Temperature: 250 °C

RTECS Toxicity Data

Acute Oral: Rat LD_{50} Dose: 2180 mg/kg.

Acute Dermal: Rabbit LD_{50} Route: Skin; Dose: 140 mg/kg.

Irritation Eye: Rabbit Standard Draize Test Dose: 100 mg/24H; Reaction: moderate. Rabbit Standard Draize Test Dose: 100 mg; Reaction: mild.

Irritation Skin: Rabbit Standard Draize Test Dose: 500 mg/24H; Reaction: mild. Rabbit Open Draize Test Dose: 10 mg/24H open; Reaction: mild.

Hazard Overviews

Fire: Combustible.

Carcinogenicity: IARC - Not listed; NIOSH - Not listed; NTP - Not listed; ACGIH - Not listed; OSHA - Not listed; EPA - Not listed; MAK - Not listed

Environmental

Regulations

RCRA 40CFR: Not listed
CERCLA: 40CFR 302.4: Not listed
SARA 40CFR 372.65: Not listed
SARA EHS 40CFR 355: Not listed
TSCA: Listed

ACE2500 CAS #: 60-35-5

ACETAMIDE

RTECS: AB4025000
EINECS Number: 200-473-5
Molecular Formula: C_2H_5NO
Structured MF: CH_3CONH_2
Formula Weight: 59.07

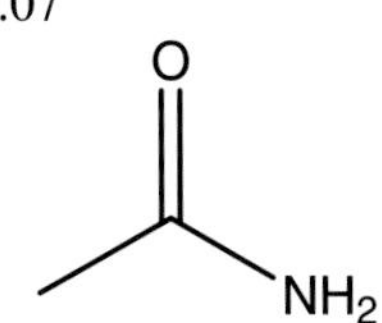

Chemical Structure

Synonyms: ACETIC ACID AMIDE; ACETIMIDIC ACID; AMID KYSELINY OCTOVE; ETHANAMIDE; METHANECARBOXAMIDE

Description: hexagonal crystals; odorless

Use: in organic synthesis as a reactant, solvent and peroxide stabilizer; use in lacquers, explosives, soldering flux; as a hygroscopic agent, wetting agent, penetrating agent and pharmaceuticals

Physical Properties

Boiling Point: 222 °C (432 °F) at 760 mm Hg
Freezing Point: 81 °C (177.8 °F)
Specific Gravity: 1.159 at 20 °C/4 °C
Density: 1.16 g/mL
Vapor Pressure: 10 mm Hg at 105 °C
Water Solubility: 1 g dissolves in 0.5 ml Water
Other Solubilities: Insoluble in Ether.
Odor Threshold: Recognition 140 to 160 mg/m^3
Refraction Index: 1.4274 at 78 °C/D
Ionization Potential (eV): 9.65 +/-0.2
Flash Point: Not available; probably combustible

RTECS Toxicity Data

Acute Oral: Rat LD_{50} Dose: 7 gm/kg. Mouse LD_{50} Dose: 12900 mg/kg; Toxic Effects: Peripheral nerve and sensation - Spastic parapysis with or without sensory change; Sense organs and special senses - Lacrimation.

Acute Dermal: Rat LD_{50} Route: Subcutaneous Dose: 10 gm/kg; Toxic Effects: Peripheral Nerve and sensation - Flaccid paralysis without anesthesia; Automatic Nervous System - Smooth muscle relaxant (mechanism undefined, spasmolytic); Cardiac - Change in rate. Mouse LD_{50} Route: Subcutaneous Dose: 8300 mg/kg; Toxic Effects: Peripheral Nerve and sensation - Flaccid paralysis without anesthesia; Automatic Nervous System - Smooth muscle relaxant (mechanism undefined, spasmolytic); Cardiac - Change in rate.

Reproductive/Teratogenic: Rabbit Route: Oral; Dose: 39 gm/kg; Duration: female 6-18D of pregnancy; Effects on Fertility - Post-implantation mortality; Specific Developmental Abnormalities - Musculoskeletal system.

Rabbit Route: Oral; Dose: 13 gm/kg; Duration: female 6-18D of pregnancy; Effects on Embryo or Fetus - Fetotoxicity.

Mutagenic: Rat Morphological Transformation; Cell Type: embryo; Dose: 5 mg/L. Mouse Micronucleus Test; Route: Oral; Dose: 3390 umol/kg.

Tumorigenic: Rat Route: Oral; Dose: 431 gm/kg/1Y-C; Toxic Effects: Tumorigenic - Carcinogenic by RTECS criteria; Liver - Tumors; Nutritional and gross metabolic - Weight loss or decreased weight gain. Rat Route: Oral; Dose: 546 gm/kg/52W-C; Toxic Effects: Tumorigenic - Neoplastic by RTECS criteria; Liver - Tumors.

Hazard Overviews

Fire Diamond

Health: Irritating to skin. Chronic Effects: possible cancer hazard.

Fire: Will burn. Use water fog, dry chemical, alcohol foam, or carbon dioxide to fight fires. Use a water spray to cool tanks or containers exposed to the fire. Do not use a solid stream of water because this can scatter and spread the fire.

Reactivity: Stable. Hazardous polymerization cannot occur. Avoid: ignition sources (heat; sparks; open flame; lighted tobacco products). Incompatible with: strong oxidizing agents. Hazardous decomposition products: carbon monoxide; oxides of nitrogen.

Carcinogenicity: IARC - Group 2B, Possibly carcinogenic to humans; NIOSH - Not listed; NTP - Not listed; ACGIH - Not listed; OSHA - Not listed; EPA - Not listed; MAK - Class B, Justifiably suspected of having carcinogenic potential

Primary Target Organs:

Skin

Environmental

Ecotoxicity: Fishes: Gambusia affinis: LC_{50} (72 hr): 15,500-20,000 mg/l

Environmental Fate: If released to soil, it would leach through soil and probably biodegrade. It would also probably biodegrade if released into water. Bioconcentration in fish and adsorption to sediment would not be significant. In the atmosphere, it will primarily exist as an aerosol and be subject to gravitational settling and washout by rain.

Environmental Physical Data

Henry's Law Constant: estimated at 5.52×10^{-9}

Octanol/Water Partition Coefficient: log K_{ow} = estimated at -1.26

BCF: calculated at 0.06

Regulations

RCRA 40CFR: Not listed

CERCLA: 40CFR 302.4: Listed per CAA Section 112 RQ: 100 lb (45.35 kg)

SARA 40CFR 372.65: Listed

SARA EHS 40CFR 355: Not listed

TSCA: Listed

ACE2650 CAS #: 1068-90-2

ACETAMIDOMALONIC ACID DIETHYL ESTER

RTECS: OO0360000

EINECS Number: 213-952-9

Molecular Formula: $C_9H_{15}NO_5$

Formula Weight: 217.22

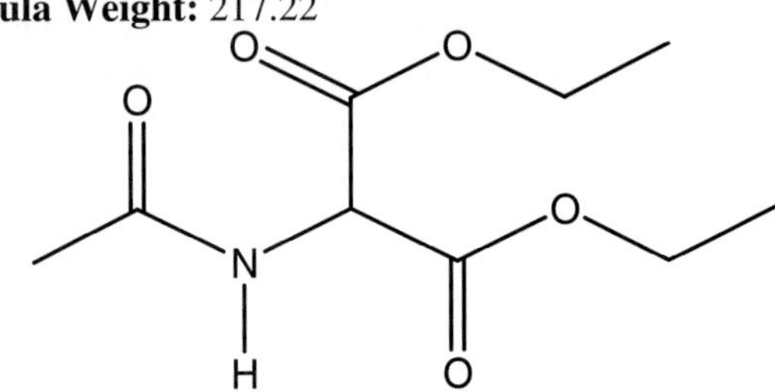

Chemical Structure

Synonyms: DIETHYL 2-ACETAMIDOMALONATE; DIETHYL ACETAMIDOMALONATE; DIETHYL ACETYLAMINOMALONATE; DIETHYLESTER KYSELINY ACETYLAMINOMALONOVE; MALONIC ACID,ACETAMIDO-,DIETHYL ESTER; PROPANEDIOIC ACID,(ACETYLAMINO)-,DIETHYL ESTER

Description: crystals

Physical Properties

Boiling Point: 185 °C (365 °F) at 20 mm Hg

Freezing Point: 95 °C (203 °F) to 96 °C (204.8 °F)

Water Solubility: Slightly Soluble in Hot Water

Other Solubilities: Slightly Soluble in Ether; Soluble in hot Alcohol

RTECS Toxicity Data

Irritation Eye: Rabbit Standard Draize Test Dose: 500 mg/24H; Reaction: mild.

Hazard Overviews

Carcinogenicity: IARC - Not listed; NIOSH - Not listed; NTP - Not listed; ACGIH - Not listed; OSHA - Not listed; EPA - Not listed; MAK - Not listed

Environmental

Regulations

RCRA 40CFR: Not listed

CERCLA: 40CFR 302.4: Not listed

SARA 40CFR 372.65: Not listed

SARA EHS 40CFR 355: Not listed

TSCA: Listed

ACE2800 **CAS #: 103-90-2**

ACETAMINOPHEN

RTECS: AE4200000
EINECS Number: 203-157-5
Molecular Formula: $C_8H_9NO_2$
Structured MF: $CH_3CONHC_6H_4OH$
Formula Weight: 151.16

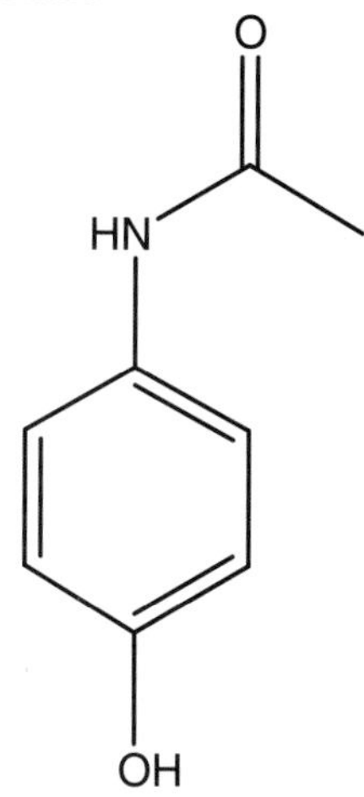

Chemical Structure

Synonyms: ABENSANIL; ACAMOL; ACETAGESIC; ACETALGIN; ACETAMIDE,N-(4-HYDROXYPHENYL)-; ACETAMIDE,N-(P-HYDROXYPHENYL)-; 4-ACETAMIDOPHENOL; P-ACETAMIDOPHENOL; ACETAMINOFEN; P-ACETAMINOPHENOL; ACETANILIDE,4'-HYDROXY-; N-ACETYL-4-AMINOPHENOL; N-ACETYL-P-AMINOPHENOL; P-ACETYLAMINOPHENOL; ALGOTROPYL; ALPINYL; ALVEDON; AMADIL; ANAFLON; ANELIX; ANHIBA; APADON; APAMID; APAMIDE; APAP; BEN-U-RON; BICKIE-MOL; CALPOL; CETADOL; CLIXODYNE; DATRIL; DIAL-A-GESIC; DIROX; DOLIPRANE; DYMADON; ENELFA; ENERIL; EXCEDRIN; EXDOL; FEBRILIX; FEBRO-GESIC; FEBROLIN; FENDON; FINIMAL; G 1; GELOCATIL; HEDEX; HOMOOLAN; 4'-HYDROXYACETANILIDE; 4-HYDROXYACETANILIDE; P-HYDROXYACETANILIDE; 4-HYDROXYANILID KYSELINY OCTOVE; N-(4-HYDROXYPHENYL)ACETAMIDE; N-(P-HYDROXYPHENYL)ACETAMIDE; JANUPAP; KORUM; LESTEMP; LIQUAGESIC; LONARID; LYTECA; LYTECA SYRUP; MOMENTUM; MULTIN; NAPA; NAPAFEN; NAPAP; NAPRINOL; NOBEDON; PACEMO; PALDESIC; PANADOL; PANALEVE; PANASORB; PANETS; PANEX; PANOFEN; PARACETAMOL; PARACETAMOLE; PARACETAMOLO; PARACETANOL; PARAPAN; PARASPEN; PARELAN; PARMOL; PASOLIND; PEDRIC; PHENDON; PHENOL,P-ACETAMIDO-; PYRINAZINE; SALZONE; SK-APAP; TABALGIN; TAPAR; TEMLO; TEMPANAL; TEMPRA; TRALGON; TUSSAPAP; TYLENOL; VALADOL; VALGESIC

Description: large monoclinic prisms or needles; odorless

Use: as an analgesic and antipyretic; in the treatment of a wide variety of arthritic and rheumatic conditions involving musculoskeletal pain, as well as in other painful disorders such as headache, dysmenorrhea, myalgias and neuralgias; as an analgesic and antipyretic in diseases accompanied by discomfort and fever, such as the common cold and other viral infections; the manufacture of azo dyes and photographic chemicals, as an intermediate for pharmaceutical and as a stabilizer for hydrogen peroxide

Physical Properties

Freezing Point: 169 °C (336.2 °F)
Specific Gravity: 1.293 at 21 °C/4 °C
Water Solubility: Very Slightly Soluble in Cold Water
Other Solubilities: DMSO: >=100 mg/ml at 22 °C; Acetone: 50-100 mg/ml at 22 °C; Ethyl Acetate: Soluble; Petroleum Ether: Practically Insoluble; Ether: Insoluble; Benzene: Insoluble; Alcohol: 1 g/7 mL; Chloroform: 1 g/50 mL; Solutions of alkali hydroxides: Soluble.
pH: Saturated aqueous solution 5.5 to 6.5
Flash Point: Not available; probably combustible

RTECS Toxicity Data

Acute Oral: Man TD_{Lo} Dose: 714 mg/kg; Toxic Effects: Cardiac - EKG changes not diagnostic of above. Woman TD_{Lo} Dose: 4962 ug/kg; Toxic Effects: Gastrointestinal - Changes in structure or function of exocrine pancreas; Liver - Liver function tests impaired; Blood - Other changes. Woman TD_{Lo} Dose: 490 mg/kg; Toxic Effects: Behavioral - Somnolence (general depressed activity); Gastrointestinal - Other changes; Kidney, Ureter, and Bladder - Other changes. Infant TD_{Lo} Dose: 1440 mg/kg/6D; Toxic Effects: Behavioral - Irritability; Gastrointestinal - Hypermotility, diarrhea; Nutritional and gross metabolic - Body temperature increase. Child TD_{Lo} Dose: 591 mg/kg/2D-I; Toxic Effects: Liver - Liver function tests impaired; Kidney, Ureter, and Bladder - Chgs in tubules (inc acute renal failure, acute tubular necrosis; Blood - Aplastic anemia. Child TD_{Lo} Dose: 801 mg/kg; Toxic Effects: Behavioral - General anesthetic; Gastrointestinal - Nausea or vomiting; Liver - Other changes. Human LD_{Lo} Dose: 143 mg/kg; Toxic Effects: Behavioral - General anesthetic. Human LD_{Lo} Dose: 357 mg/kg; Toxic Effects: Behavioral - Anorexia (human); Behavioral - Coma; Gastrointestinal - Nausea or vomiting. Man LD_{Lo} Dose: 714 mg/kg; Toxic Effects: Liver - Other changes. Man LD_{Lo} Dose: 143 mg/kg/24H-I; Toxic Effects: Behavioral - Anorexia (human); Liver - Hepatitis (hepatocellular necrosis), zonal; Liver - Jaundice, other or unclassified. Woman LD_{Lo} Dose: 260 mg/kg; Toxic Effects: Behavioral - Coma; Gastrointestinal - Nausea or vomiting; Kidney, Ureter, and Bladder - Chgs in tubules (inc acute renal failure, acute tubular necrosis. Woman LD_{Lo} Dose: 650 mg/kg; Toxic Effects: Vascular - BP lowering not characterized in autonomic section; Vascular - Other changes; Nutritional and gross metabolic - Metabolic acidosis. Child LD_{Lo} Dose: 360 mg/kg/2D; Toxic Effects: Gastrointestinal - Nausea or vomiting; Liver - Other changes; Skin and appendages - Dermatitis, other. Child LD_{Lo} Dose: 50 mg/kg; Toxic Effects: Cardiac - Other changes; Lungs, Thorax, or Respiration - Acute pulmonary edema; Kidney, Ureter, and Bladder - Chgs in tubules (inc acute renal failure, acute tubular necrosis. Rat LD_{50} Dose: 2404 mg/kg.

Acute Dermal: Mouse LD_{50} Route: Subcutaneous Dose: 310 mg/kg. Frog LD_{Lo} Route: Subcutaneous Dose: 50 mg/kg; Toxic Effects: Behavioral - Altered sleep time (including change in righting reflex); Behavioral - Ataxia; Lungs, Thorax, or Respiration - Other changes.

Chronic (Multiple Dose) Oral: Rat Dose: 105 gm/kg/35D-C; Toxic Effects: Liver - Other changes; Nutritional and gross metabolic - Weight loss or decreased weight gain; DEATH. Rat Dose: 68 gm/kg/13W-C; Toxic Effects: Liver - Other changes; Kidney, Ureter, and Bladder - Changes in kidney weight; Nutritional and gross metabolic - Weight loss or decreased weight gain. Rat Dose: 6080 mg/kg/19D-I; Toxic Effects: Gastrointestinal - Other changes; Liver - Changes in liver weight; Blood - Changes in leukocyte (WBC) cell count.

Reproductive/Teratogenic: Woman Route: Oral; Dose: 650 mg/kg; Duration: female 29W of pregnancy; Effects on Newborn - Apgor score (human only); Other neonatal measures or effects; Other postnatal measures or effects. Woman Route: Oral; Dose: 417 mg/kg; Duration: female 20W of pregnancy; Specific Developmental Abnormalities - Craniofacial (including nose and tongue); Hepatobiliary system. Woman Route: Oral; Dose: 1300 mg/kg; Duration: female 31-32W of pregnancy Maternal Effects - Other effects on females; Effects on Embryo or Fetus - Other effects to embryo or fetus; Effects on Newborn - Other neonatal measures or effects. Rat Route: Oral; Dose: 1500 mg/kg; Duration: female 8-19D of pregnancy; Effects on Fertility - Post-implantation mortality; Effects on Embryo or Fetus - Fetotoxicity. Rat Route: Oral; Dose: 12500 mg/kg; Duration: female 14D prior to mating Effects on Embryo or Fetus - Cytological changes (inc. somatic cell genetic material). Rat Route: Oral; Dose: 13 gm/kg; Duration: female 14D prior to mating Effects on Embryo or Fetus - Fetotoxicity. Rat Route: Oral; Dose: 35 gm/kg; Duration: male 70D prior to mating; Paternal Effects - Testes, epididymis, sperm duct; Other effects on male.

Mutagenic: Human DNA Inhibition; Cell Type: lymphocyte; Dose: 300 umol/L. Human Cytogenetic Analysis; Cell Type: lymphocyte; Dose: 200 mg/L. Human Cytogenetic Analysis; Route: Oral; Dose: 42860 ug/kg. Human Sister Chromatid Exchenge; Route: Oral; Dose: 42860 ug/kg. Human Sister Chromatid Exchenge; Cell Type: lymphocyte; Dose: 1 mmol/L. Human Micronucleus Test; Route: Oral; Dose: 42857 ug/kg/8H-I. Human Other Mutation Test Systems; Cell Type: lymphocyte; Dose: 200 mg/L.

Tumorigenic: Rat Route: Oral; Dose: 164 gm/kg/78W-C; Toxic Effects: Tumorigenic - Carcinogenic by RTECS criteria; Kidney, Ureter, and Bladder - Tumors. Rat Route: Oral; Dose: 329 gm/kg/78W-C; Toxic Effects: Tumorigenic - Carcinogenic by RTECS criteria; Liver - Tumors.

Hazard Overviews

Health: Irritating to eyes/skin/respiratory tract. Other Acute Effects: may be harmful by inhalation, ingestion, or skin absorption; prolonged or repeated exposure may cause allergic reactions in certain sensitive individuals; target organs: liver, kidneys. Chronic Effects: Possible human carcinogen.

Fire: Will burn. Hazards: emits toxic fumes. Extinguishing agents: water spray; carbon dioxide, dry chemical powder or appropriate foam. Precautions: combustible liquid.

Reactivity: Incompatible with: strong oxidizing agents. Hazardous decomposition products: toxic fumes of: carbon monoxide, carbon dioxide, nitrogen oxides.

Carcinogenicity: IARC - Group 3, Not classifiable as to carcinogenicity to humans; NIOSH - Not listed; NTP - Not listed; ACGIH - Not listed; OSHA - Not listed; EPA - Not listed; MAK - Not listed

Primary Target Organs:

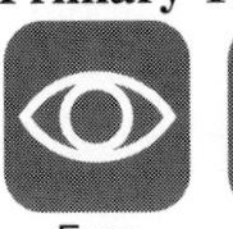
Eyes

Skin

Respiratory System

Liver

Kidneys

Environmental

Regulations

RCRA 40CFR: Not listed
CERCLA: 40CFR 302.4: Not listed
SARA 40CFR 372.65: Not listed
SARA EHS 40CFR 355: Not listed
TSCA: Listed

ACE2950 CAS #: 103-84-4

ACETANILIDE

RTECS: AD7350000
EINECS Number: 203-150-7
Molecular Formula: C_8H_9NO
Structured MF: $NHCOCH_3$
Formula Weight: 135.16

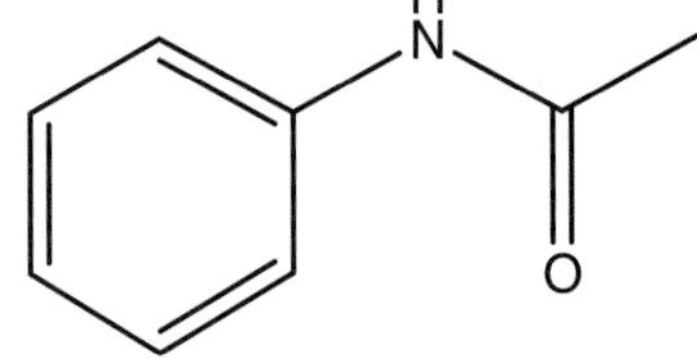

Chemical Structure

Synonyms: ACETAMIDE,N-PHENYL-; ACETAMIDOBENZENE; ACETANIL; ACETANILID; ACETIC ACID ANILIDE; ACETIC ACID,AMIDE,N-PHENYL-; ACETOANILIDE; ACETYLAMINOBENZENE; ACETYLANILINE; N-ACETYLANILINE; AN; ANILINE,N-ACETYL-; ANTIFEBRIN; BENZENAMINE,N-ACETYL-; PHENALGENE; PHENALGIN; N-PHENYLACETAMIDE

Description: white shining crystalline scales or plates; odorless

Use: manufacture of medicinals and dyes; stabilizer for hydrogen peroxide solutions; additive for cellulose ester varnishes; antipyretic, analgesic

Physical Properties

Boiling Point: 304 °C (579 °F) at 760 mm Hg
Freezing Point: 114.3 °C (237.74 °F)
Specific Gravity: 1.219 at 15 °C
Vapor Density: 4.65

Density: 1.219 g/mL at 15 °C
Vapor Pressure: 1 mm Hg at 114.0 °C
Water Solubility: 1 g Soluble in 185 ml Water
Other Solubilities: Soluble in Toluene; Very Soluble in hot Toluene, in Carbon Tetrachloride.
Odor Threshold: 270 mg/m^3
Refraction Index: Alpha 1.515
Ionization Potential (eV): 8.30 +/-0.2
Flash Point: 169.444 °C Open Cup
Autoignition Temperature: 524 to 535 °C

RTECS Toxicity Data

Acute Oral: Human TD_{Lo} Dose: 14 mg/kg/D; Toxic Effects: Lungs, Thorax, or Respiration - Cyanosis; Kidney, Ureter, and Bladder - Chgs in tubules (inc acute renal failure, acute tubular necrosis; Blood - Methemoglobinemia-Carboxyhemoglobin. Man TD_{Lo} Dose: 405 mg/kg; Toxic Effects: Behavioral - Sleep; Lungs, Thorax, or Respiration - Cyanosis; Lungs, Thorax, or Respiration - Respiratory stimulation. Man LD_{Lo} Dose: 56 mg/kg/H-I; Toxic Effects: Behavioral - Hallucinations, distorted perceptions; Gastrointestinal - Decreased motility or constipation; Nutritional and gross metabolic - Body temperature decrease. Rat LD_{50} Dose: 800 mg/kg.
Mutagenic: Mouse Micronucleus Test; Route: Intraperitoneal; Dose: 50 mg/kg.

Hazard Overviews

Health: Irritating to eyes/skin/respiratory tract. Toxic. Other Acute Effects: harmful if swallowed, inhaled, or absorbed through skin; may cause cyanosis (blue-gray coloring of skin and lips caused by lack of oxygen).
Fire: Will burn. Hazards: emits toxic fumes. Extinguishing agents: water spray; carbon dioxide, dry chemical powder or appropriate foam. Precautions: combustible liquid.
Reactivity: Incompatible with: strong oxidizing agents, strong bases. Hazardous decomposition products: toxic fumes of: carbon monoxide, carbon dioxide, nitrogen oxides.
Carcinogenicity: IARC - Not listed; NIOSH - Not listed; NTP - Not listed; ACGIH - Not listed; OSHA - Not listed; EPA - Not listed; MAK - Not listed
Primary Target Organs:

Eyes

Skin

Respiratory System

Environmental

Ecotoxicity: Fishes: Lepomis macrochirus: static bioassay in freshwater at 23 °C, mild aeration applied after 24 hr. % survival after material added, 320 mg/l 24 hr (40%); Fishes: Lepomis macrochirus: 24h LC_{50} 5,000 mg/l (1294(145)); Insects: Culex sp. larvae: 24h LC_{50} 7,500 mg/l 48h LC_{50} 7,425 mg/l

Environmental Physical Data

Octanol/Water Partition Coefficient: log K_{ow} = 1.16
BOD: acclimated > 1 lb/lb, 10 days

Regulations

RCRA 40CFR: Not listed
CERCLA: 40CFR 302.4: Not listed
SARA 40CFR 372.65: Not listed
SARA EHS 40CFR 355: Not listed
TSCA: Listed

ACE3100 CAS #: 59-66-5

ACETAZOLAMIDE

RTECS: AC8225000
EINECS Number: 200-440-5
Molecular Formula: $C_4H_6N_4O_3S_2$
Formula Weight: 222.25

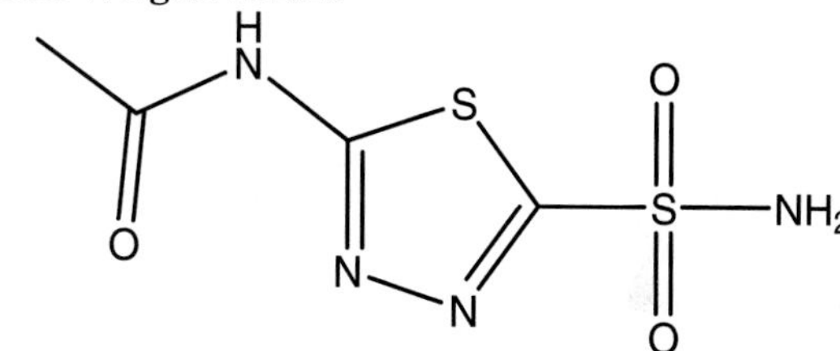

Chemical Structure

Synonyms: #6063; 6063; ACETAMIDE; ACETAMIDE,N-(5-(AMINOSULFONYL)-1,3,4-THIADIAZOL-2-YL)-; ACETAMIDE,N-(5-SULFAMOYL-1,3,4-THIADIAZOL-2-YL)-; 5-ACETAMIDE-1,3,4-THIADIAZOLE-2-SULFONAMIDE; 2-ACETAMIDO-5-SULFONAMIDO-1,3,4-THIADIAZOLE; ACETAMIDOTHIADIAZOLESULFONAMIDE; ACETAMOX; ACETAZIDE; ACETAZOLAMID; ACETAZOLEAMIDE; ACETOZALAMIDE; 2-ACETYLAMINO-1,3,4-THIADIAZOLE-5-SULFONAMIDE; N-(5-(AMINOSULFONYL)-1,3,4-THIADIAZOL-2-YL)ACETAMIDE; ATENAZOL; CARBONIC ANHYDRASE INHIBITOR NO 6063; CARBONIC ANHYDRASE INHIBITOR NO. 6063; CIDAMEX; DEFILTRAN; DEHYDRATIN; DIACARB; DIAKARB; 4-DIAMOX; DIAMOX; DIDOC; DILURAN; DIURAMID; DIURETICUM-HOLZINGER; DIURIWAS; DIUTAZOL; DONMOX; DUIRAMID; EDEMOX; EUMICTON; FONURIT; GLAUPAX; GLUPAX; NATRIONEX; NEPHRAMID; NEPHRAMIDE; PHONURIT; N-(5-SULFAMOYL-1,3,4-THIADIAZOL-2-YL)ACETAMIDE; 1,3,4-THIADIAZOLE-2-SULFONAMIDE,5-ACETAMIDO; 1,3,4-THIADIAZOLE-2-SULFONAMIDE,5-ACETAMIDO-; VETAMOX
Description: white to faintly yellowish white, crystalline, powder; odorless
Use: formerly used as a diuretic but is now used to reduce intraocular pressure in the treatment of glaucoma and as a carbonic anhydras inhibitor

Physical Properties

Freezing Point: 258 °C (496.4 °F) to 259 °C (498.2 °F)
Water Solubility: Sparingly Soluble in Cold Water
Other Solubilities: Slightly Soluble in Alcohol; Insoluble in Chloroform, Diethyl Ether, Carbon Tetrachloride; Slightly Soluble in Acetone.
Flash Point: Not available; probably combustible

RTECS Toxicity Data

Acute Oral: Man TD_{Lo} Dose: 54 mg/kg/5D-I; Toxic Effects: Lungs, Thorax, or Respiration - Dyspnea. Mouse LD_{50} Dose: 4300 mg/kg.

Acute Dermal: Mouse LD_{50} Route: Subcutaneous Dose: >3 gm/kg. Guinea Pig LD_{50} Route: Subcutaneous Dose: >1500 mg/kg.

Reproductive/Teratogenic: Rat Route: Oral; Dose: 6600 mg/kg; Duration: female 1-22D of pregnancy; Specific Developmental Abnormalities - Musculoskeletal system. Rat Route: Oral; Dose: 240 mg/kg; Duration: female 9-10D of pregnancy; Specific Developmental Abnormalities - Musculoskeletal system. Rat Route: Oral; Dose: 3300 mg/kg; Duration: female 1-22D of pregnancy; Effects on Fertility - Post-implantation mortality; Effects on Embryo or Fetus - Fetal death. Rat Route: Oral; Dose: 1500 mg/kg; Duration: female 5-7D of pregnancy; Effects on Embryo or Fetus - Fetal death; Effects on Newborn - Sex ratio. Rat Route: Oral; Dose: 1 gm/kg; Duration: female 10D of pregnancy; Effects on Embryo or Fetus - Extra embryonic structures.

Hazard Overviews

Health: Irritating to eyes/skin/respiratory tract. Toxic. Other Acute Effects: may be harmful by inhalation, ingestion, or skin absorption; prolonged or repeated exposure may cause allergic reactions in certain sensitive individuals; exposure can cause headache; fatiguel anorexial gastrointestinal disturbancesl drowsiness. Chronic Effects: may cause congenital malformation in the fetus; may cause harm to the unborn child; target organs: kidneys; eyes; blood.

Fire: Will burn. Hazards: emits toxic fumes. Extinguishing agents: water spray; carbon dioxide, dry chemical powder or appropriate foam. Precautions: combustible liquid.

Reactivity: Stable. Hazardous polymerization will not occur. Incompatible with: strong oxidizing agents. Hazardous decomposition products: toxic fumes of: carbon monoxide, carbon dioxide, nitrogen oxides, sulfur oxides.

Carcinogenicity: IARC - Not listed; NIOSH - Not listed; NTP - Not listed; ACGIH - Not listed; OSHA - Not listed; EPA - Not listed; MAK - Not listed

Primary Target Organs:

Eyes

Skin

Respiratory System

Kidneys

Blood

Environmental

Regulations

RCRA 40CFR: Not listed
CERCLA: 40CFR 302.4: Not listed
SARA 40CFR 372.65: Not listed
SARA EHS 40CFR 355: Not listed
TSCA: Listed

ACE3250 CAS #: 919-54-0

ACETHION

RTECS: AI6825000
Molecular Formula: $C_8H_{17}O_4PS_2$
Formula Weight: 272.34
Synonyms: ACETHIONE; ACETIC ACID,((DIETHOXYPHOSPHINOTHIOYL)THIO)-,ETHYL ESTER; ACETIC ACID,((DIETHOXYPHOSPHINOTHIOYL)THIO)-,ETHYL ESTER(9CI); ACETIC ACID,MERCAPTO-,ETHYL ESTER,S-ESTER WITHO,O-DIETHYL PHOSPHORODITHIOATE; AZETHION; O,O-DIETHYL S-CARBOETHOXYMETHYL DITHIOPHOSPHATE; O,O-DIETHYL S-CARBOETHOXYMETHYL PHOSPHORODITHIOATE; ENT 25650; ETHOXYPHAS; ETHOXYPHOS; ETHYL ((DIETHOXYPHOSPHINOTHIOYL)THIO)ACETATE; HERCULES 4580; PHOSPHORODITHIOIC ACID,O,O-DIETHYL ESTER,S-ESTER WITHETHYL MERCAPTOACETATE; PROTHION
Use: insecticide

RTECS Toxicity Data

Acute Oral: Rat LD_{50} Dose: 1100 mg/kg. Mouse LD_{50} Dose: 1200 mg/kg.

Hazard Overviews

Carcinogenicity: IARC - Not listed; NIOSH - Not listed; NTP - Not listed; ACGIH - Not listed; OSHA - Not listed; EPA - Not listed; MAK - Not listed

Environmental

Regulations

RCRA 40CFR: Not listed
CERCLA: 40CFR 302.4: Not listed
SARA 40CFR 372.65: Not listed
SARA EHS 40CFR 355: Not listed
TSCA: Not listed

ACE3400 CAS #: 64-19-7

ACETIC ACID

RTECS: AF1225000
DOT: UN2789; UN2790; IMO8.0
EINECS Number: 200-580-7
Molecular Formula: $C_2H_4O_2$
Structured MF: CH_3CO_2H
Formula Weight: 60.05

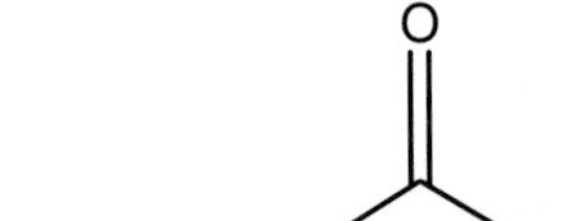

Chemical Structure

Synonyms: ACETIC ACID (AQUEOUS); ACETIC ACID,GLACIAL; ACIDE ACETIQUE; ACIDO ACETICO; AZIJNZUUR; ESSIGSAEURE; ETHANOIC ACID; ETHYLIC ACID; GLACIAL ACETIC ACID;

GLACIAL ACETIC ACID (PURE COMPOUND); KYSELINA OCTOVA; METHANECARBOXYLIC ACID; OCTOWY KWAS; PYROLIGNEOUS ACID; VINEGAR ACID; VOSOL

Description: colorless liquid, crystals; pungent vinegar odor

Use: to prepare dilute acetic acids and strong ammonium acetate solution; has been used for the destruction of warts; manufacture of acetic anhydride, cellulose acetate, vinyl acetate, chloroacetic acid; production of plastics, pharmaceuticals, dyes, insecticides and photographic chemicals; latex coagulant; oil-well acidifier; textile printing; manufacture of vitamins, antibiotics and hormones; astringent, antimicrobial agent; used in eardrops; cosmetic ingredient; solvent in production of terephthalic acid; mild expectorant; in liniments; manufacture of acetyl compounds, acetate rayon a rubber; laundry sour; printing calico and dyeing silk; as acidulant and preservative in foods; solvent for gums, resins, volatile oils and many other substances; widely used in commercial organic synthesis; in veterinary medicine as a vesicant, caustic and in the destruction of warts; metabolic inter mediate; and food preservative

Physical Properties

Boiling Point: 118 °C (244 °F)
Freezing Point: 16.6 °C (61.88 °F)
Specific Gravity: 1.0492 at 20 °C/4 °C
Vapor Density: 2.1 Air=1
Saturated Vapor Density: 1.227049002 kg/m^3
Density: 1.053 at 16.67 °C (liquid)
Vapor Pressure: 16 mm Hg at 25 °C
Water Solubility: Miscible with Water
Other Solubilities: miscible with Acetone, Benzene; Soluble in Alcohol; miscible with Glycerine.
Surface Tension: 28.8 dynes/cm at 10 °C
Odor Threshold: 0.21 to 1.0 ppm
pH: Aqueous solution 1.0 molar 2.4
Refraction Index: 1.3718 at 20 °C/D
Evaporation Rate: Evaporation rate at 25 °C, wind speed 4 0.24 g/sq m/sec
Critical Temperature: 322 °C
Critical Pressure: 839 psia
Ionization Potential (eV): 10.66
Flash Point: 39 °C Closed Cup
Autoignition Temperature: 426 °C
LEL: 4% v/v
UEL: 16% v/v

RTECS Toxicity Data

Acute Oral: Human TD_{Lo} Dose: 1470 ug/kg; Toxic Effects: Gastrointestinal - Changes in structure or function of esophagus; Gastrointestinal - Ulceration or bleeding from small intestine; Gastrointestinal - Ulceration or bleeding from large intestine. Rat LD_{50} Dose: 3310 mg/kg.

Acute Inhalation: Rat LC_{Lo} Dose: 16000 ppm/4hr. Human TC_{Lo} Dose: 816 ppm/3M; Toxic Effects: Sense organs and special senses - Other; Sense organs and special senses - Other; Lungs, Thorax, or Respiration - Other changes.

Acute Dermal: Rabbit LD_{50} Route: Skin; Dose: 1060 uL/kg. Rabbit LD_{Lo} Route: Subcutaneous Dose: 600 mg/kg.

Chronic (Multiple Dose) Oral: Rat Dose: 22680 mg/kg/9W-C; Toxic Effects: Behavioral - Food intake (animal); Nutritional and gross metabolic - Weight loss or decreased weight gain.

Chronic (Multiple Dose) Inhalation: Rat Dose: 5070 ug/m^3/24H/95D-C; Toxic Effects: Kidney, Ureter, and Bladder - Other changes in urine composition; Blood - Changes in leukocyte (WBC) cell count; Biochemical - True cholinesterase.

Irritation Eye: Rabbit Standard Draize Test Dose: 50 ug open; Reaction: severe. Rabbit Nonstandard Exposure Dose: 5 mg/30S rinse; Reaction: mild.

Irritation Skin: Human Standard Draize Test Dose: 50 mg/24H; Reaction: mild. Rabbit Standard Draize Test Dose: 20 mg/24H; Reaction: moderate. Rabbit Standard Draize Test Dose: 50 mg/24H; Reaction: mild.

Reproductive/Teratogenic: Rat Route: Oral; Dose: 700 mg/kg; Duration: female 18D after birth; Effects on Newborn - Behavioral. Rat Route: Intratesticular; Dose: 400 mg/kg; Duration: male 1D prior to mating; Effects on Fertility - Male fertility index.

Mutagenic: Human Sister Chromatid Exchenge; Cell Type: lymphocyte; Dose: 5 mmol/L. Mouse Unscheduled DNA Synthesis; Route: Skin; Dose: 79279 ug/kg.

Hazard Overviews

Corrosive

Flammable

Fire Diamond

Health: Corrosive to eyes/skin/respiratory tract. Chronic Effects: skin may become darkened and thick, tooth enamel erosion, digestive disorders.

Fire: Concentrations greater than 50% are flammable. Concentrations below 50% are nonflammable. Can form explosive mixtures in the air. Use water spray, dry chemical, carbon dioxide, or alcohol-resistant foam.

Reactivity: Stable. Hazardous polymerization cannot occur. Avoid: heat; ignition sources; carbonates; hydroxides; oxides; phosphates. Incompatible with: some forms of plastics, rubber, and coatings; acetaldehyde; 5-azidotetrazole; 2-aminoethanol; ammonium nitrate; bromine trifluoride; chromic acid; chlorine trifluoride; chlorosulfonic acid; diallymethyl carbinol and ozone; ethylenediamine; hydrogen peroxide; sodium peroxide; potassium hydroxide; sodium hydrox-ide; potassium permanganate; nitric acid and acetone; oleum; perchloric acid; phosphorus trichloride; potassium t-butoxide; phosphorus isocyanate; n-xylene. Hazardous decomposition products: carbon dioxide; carbon monoxide; toxic, irritating vapors.

Carcinogenicity: IARC - Not listed; NIOSH - Listed as carcinogen; NTP - Not listed; ACGIH - Not listed; OSHA - Not listed; EPA - Not listed; MAK - Not listed

Primary Target Organs:

    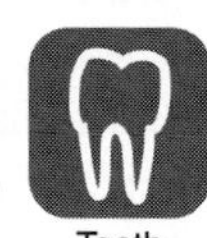

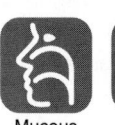

Eyes Skin Respiratory System Mucous Membranes Gastro-intestinal Teeth

Exposure Limits

OSHA PEL: TWA: 10 ppm; 25 mg/m^3.

ACGIH TLV: TWA: 10 ppm; 25 mg/m^3; STEL: 15 ppm; 37 mg/m^3.

NIOSH IDLH: 50 ppm.

DFG MAK: TWA: 10 ppm; 25 mg/m^3.

Respirator Recommendation

Exposure Range: >10 to <50 ppm Air Purifying, Negative Pressure, Half Mask

Exposure Range: 50 to unlimited ppm Self-contained Breathing Apparatus, Pressure Demand, Full Face

Cartridge Color: black

Environmental

Ecotoxicity: TLm Culex (larvae) 1,500 mg/l/24-48 hr /Conditions of bioassay not specified LC_{50} Fathead minnows 175 mg/l/1 hr; 106 mg/l/24 hr; 106 mg/l/48 hr; 79 mg/l/72 hr; 79 mg/l/96 hr (static bioassay in reconstituted water at 18-22 °C, pH < 5.9) EC_{50} Alfalfa fumigation 7.8 mg/cu m/2 hr, effect: leaf injury TLm Mosquito fish 251 mg/l/24-96 hr /Conditions of bioassay not specified Microcystis aeruginosa (algae) 90 mg/l toxic effect: cell multiplication inhibition Entosiphon sulcatum (protozoa) 78 mg/l toxic effect: cell multiplication inhibition LD_0 Creek chub 100 mg/l/24 hr; Detroit river /Conditions of bioassay not specified TLm Sunfish 75 mg/l/96 hr 18-20 °C, soft water TLm Brine shrimp 22 mg/l/48 hr /Conditions of bioassay not specified

Environmental Fate: If released to the atmosphere, it is degraded in the vapor-phase by reaction with photochemically produced hydroxyl radicals (estimated typical half-life of 26.7 days). It occurs in atmospheric particulate matter in acetate form and physical removal from air can occur via wet and dry deposition. If released to water, it will biodegrade readily. If released to soil, it will biodegrade readily. Evaporation from dry surfaces is likely to occur.

Cleanup/Disposal: Guide No. 132: Fully encapsulating, vapor protective clothing should be worn for spills and leaks with no fire. Eliminate all ignition sources (no smoking, flares, sparks or flames in immediate area). All equipment used when handling the product must be grounded. Do not touch or walk through spilled material. Stop leak if you can do it without risk. Prevent entry into waterways, sewers, basements or confined areas. A vapor suppressing foam may be used to reduce vapors. Absorb with earth, sand or other non-combustible material and transfer to containers (except for Hydrazine). Use clean non-sparking tools to collect absorbed material. Large Spills: Dike far ahead of liquid spill for later disposal. Water spray may reduce vapor; but may not prevent ignition in closed spaces. Guide No. 153: Eliminate all ignition sources (no smoking, flares, sparks or flames in immediate area). Do not touch damaged containers or spilled material unless wearing appropriate protective clothing. Stop leak if you can do it without risk. Prevent entry into waterways, sewers, basements or confined areas. Absorb or cover with dry earth, sand or other non-combustible material and transfer to containers. Do not get water inside containers.

Environmental Physical Data

Henry's Law Constant: 1 x10^{-7}

Octanol/Water Partition Coefficient: log K_{ow} = -0.17

BCF: none

BOD: 52 to 62%, 5 days

Regulations

RCRA 40CFR: Not listed

CERCLA: 40CFR 302.4: Listed per CWA Section 311(b)(4) RQ: 5000 lb (2268 kg)

SARA 40CFR 372.65: Not listed

SARA EHS 40CFR 355: Not listed

TSCA: Listed

Analytical Methods

Air: ASTM D4490

Soil: DOE OM500R

Water / Groundwater: EPA 023

Indoor / Expired Air: NIOSH 1603

ACE3550 CAS #: 591-87-7

ACETIC ACID, ALLYL ESTER

RTECS: AF1750000

EINECS Number: 209-734-8

Molecular Formula: $C_5H_8O_2$

Formula Weight: 100.13

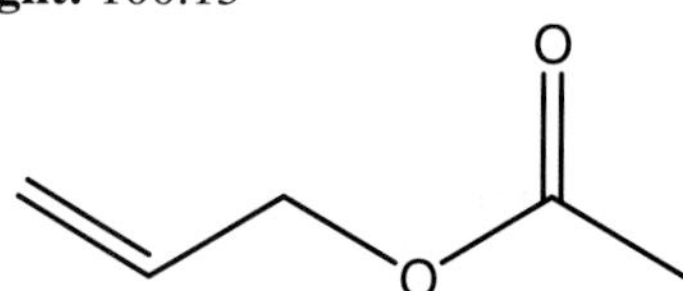

Chemical Structure

Synonyms: ACETIC ACID,2-PROPENYL ESTER; ACETIC ACID,2-PROPENYL ESTER (9CI); 3-ACETOXY-1-PROPENE; 3-ACETOXYPROPENE; ALKYL ACETIC ACID; ALLYL ACETATE; ALLYL ACETIC ACID; 2-PROPENYL METHANOATE

Description: colorless liquid; acrid odor at high levels

Use: synthetic flavor useful in cheese, butter, fruit; cockroach repellent

Physical Properties

Boiling Point: 103.5 °C (218 °F) at 760 mm Hg

Freezing Point: -22 °C (-7.6 °F)

Specific Gravity: 0.9276 at 20 °C/4 °C

Vapor Density: 3.45 Air=1

Water Solubility: Slightly Soluble in Water

Other Solubilities: Soluble in Acetone; miscible with Alcohol & Ether

Refraction Index: 1.4049 at 20 °C/D

Flash Point: 21 °C

Autoignition Temperature: 374 °C

RTECS Toxicity Data

Acute Oral: Rat LD_{50} Dose: 130 mg/kg. Mouse LD_{50} Dose: 170 mg/kg; Toxic Effects: Behavioral - Somnolence (general depressed activity).
Acute Inhalation: Rat LC_{50} Dose: 1000 ppm/1hr.
Acute Dermal: Rabbit LD_{50} Route: Skin; Dose: 1021 mg/kg.
Irritation Eye: Rabbit Standard Draize Test Dose: 100 mg/24H; Reaction: moderate.
Irritation Skin: Rabbit Standard Draize Test Dose: 500 mg/24H; Reaction: mild. Rabbit Open Draize Test Dose: 10 mg/24H open; Reaction: mild.

Hazard Overviews

Poison

Flammable

Health: Irritating to eyes/skin/respiratory tract. Poison. Other Acute Effects: may be fatal if inhaled, swallowed, or absorbed through skin; prolonged exposure can cause damage to the lungs.
Fire: Flammable. Hazards: vapor may travel considerable distance to source of ignition and flash back; container explosion may occur; may undergo autopolymerization. Extinguishing agents: carbon dioxide, dry chemical powder or appropriate foam; water may be effective for cooling, but may not effect extinguishment. Precautions: combustible liquid.
Reactivity: Incompatible with: oxidizing agents, bases, peroxides, may undergo autopolymerization. Hazardous decomposition products: toxic fumes of: carbon monoxide, carbon dioxide.
Carcinogenicity: IARC - Not listed; NIOSH - Not listed; NTP - Not listed; ACGIH - Not listed; OSHA - Not listed; EPA - Not listed; MAK - Not listed
Primary Target Organs:

Eyes

Skin

Respiratory System

Environmental

Regulations

RCRA 40CFR: Not listed
CERCLA: 40CFR 302.4: Not listed
SARA 40CFR 372.65: Not listed
SARA EHS 40CFR 355: Not listed
TSCA: Listed

ACE3700 CAS #: 142-92-7

ACETIC ACID, HEXYL ESTER

RTECS: AI0875000
EINECS Number: 205-572-7
Molecular Formula: $C_8H_{16}O_2$
Structured MF: $CH_3CO_2C_6H_{13}$
Formula Weight: 144.24

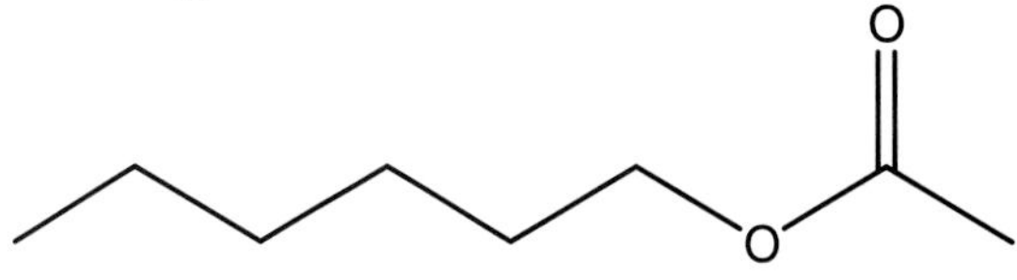

Chemical Structure

Synonyms: 1-HEXYL ACETATE; HEXYL ACETATE; L-HEXYL ACETATE; N-HEXYL ACETATE; HEXYL ALCOHOL,ACETATE; HEXYL ETHANOATE; HEXYLESTER KYSELINY OCTOVE
Description: colorless, oily liquid; pleasant fruity sweet ester odor
Use: solvent for cellulose esters & other resins; spray base; flavor ingredient; fragrance for perfumes, soaps, detergents & creams; flavoring agent for candy, baked goods & ice creams; solvent for cellulose, styrene & phenolic resins

Physical Properties

Boiling Point: 171.5 °C (341 °F) at 760 mm Hg
Freezing Point: -80.9 °C (-113.62 °F)
Specific Gravity: 0.8779 at 15 °C/4 °C
Vapor Density: 4.97 Air=1
Saturated Vapor Density: 1.208708609 kg/m^3
Vapor Pressure: 0.185 kPa at 25 °C
Water Solubility: Insoluble in Water
Other Solubilities: Very Soluble in Alcohol or Ether
Refraction Index: 1.4092 at 20 °C/D
Flash Point: 43.333 to 45 °C

RTECS Toxicity Data

Acute Oral: Rat LD_{50} Dose: 41500 uL/kg.
Acute Dermal: Rabbit LD_{50} Route: Skin; Dose: >5 gm/kg.
Irritation Eye: Rabbit Standard Draize Test Dose: 500 mg/24H; Reaction: mild.
Irritation Skin: Rabbit Standard Draize Test Dose: 500 mg/24H; Reaction: mild.

Hazard Overviews

Flammable

Health: Irritating. Other Acute Effects: may be harmful by inhalation, ingestion, or skin absorption.
Fire: Flammable. Hazards: vapor may travel considerable distance to source of ignition and flash back; container explosion may occur; forms explosive mixtures in air.

Extinguishing agents: carbon dioxide, dry chemical powder or appropriate foam; water may be effective for cooling, but may not effect extinguishment. Precautions: combustible liquid.

Reactivity: Stable. Incompatible with: oxidizing agents, strong acids, strong bases, reducing agents. Hazardous decomposition products: toxic fumes of: carbon monoxide, carbon dioxide.

Carcinogenicity: IARC - Not listed; NIOSH - Not listed; NTP - Not listed; ACGIH - Not listed; OSHA - Not listed; EPA - Not listed; MAK - Not listed

Primary Target Organs:

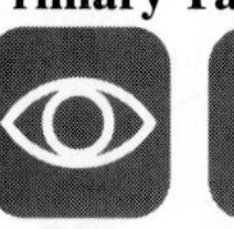
Eyes

Skin

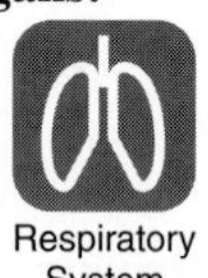
Respiratory System

Environmental

Cleanup/Disposal: Guide No. 127: Eliminate all ignition sources (no smoking, flares, sparks or flames in immediate area). All equipment used when handling the product must be grounded. Do not touch or walk through spilled material. Stop leak if you can do it without risk. Prevent entry into waterways, sewers, basements or confined areas. A vapor suppressing foam may be used to reduce vapors. Absorb or cover with dry earth, sand or other non-combustible material and transfer to containers. Use clean non-sparking tools to collect absorbed material. Large Spills: Dike far ahead of liquid spill for later disposal. Water spray may reduce vapor; but may not prevent ignition in closed spaces.

Regulations

RCRA 40CFR: Not listed
CERCLA: 40CFR 302.4: Not listed
SARA 40CFR 372.65: Not listed
SARA EHS 40CFR 355: Not listed
TSCA: Listed

ACE3850 CAS #: 108-24-7

ACETIC ANHYDRIDE

RTECS: AK1925000
DOT: UN1715; IMO8.0
EINECS Number: 203-564-8
Molecular Formula: $C_4H_6O_3$
Structured MF: $(CH_3CO)_2O$
Formula Weight: 102.10

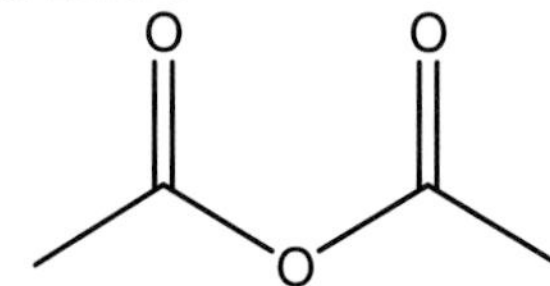

Chemical Structure

Synonyms: ACETANHYDRIDE; ACETIC ACID ANHYDRIDE; ACETIC ACID,ANHYDRIDE; ACETIC ACID,ANHYDRIDE (9CI); ACETIC OXIDE; ACETIC OXIDE ANHYDRIDE; ACETYL ANHYDRIDE; ACETYL ETHER; ACETYL OXIDE; ANHYDRID KYSELINY OCTOVE; ANHYDRIDE ACETIQUE; ANIDRIDE ACETICA; AZIJNZUURANHYDRIDE; ESSIGSAEUREANHYDRID; ETHANOIC ANHYDRATE; ETHANOIC ANHYDRIDE; OCTOWY BEZWODNIK

Description: colorless liquid; acetic acid-like odor

Use: in manufacture of acetyl compounds and cellulose acetates; as acetualizer and solvent in examining wool fat, glycerol, fatty and volatile oils and resins; in organic syntheses as dehydrating agent in nitrations, sulfonations and other reactions where removal of water is necessary

Physical Properties

Boiling Point: 139 °C (282 °F)
Freezing Point: -73 °C (-99.4 °F)
Specific Gravity: 1.08 at 15 °C/4 °C
Vapor Density: 3.52 Air=1
Density: 1.08 g/mL
Vapor Pressure: 4 mm Hg
Water Solubility: Slowly Soluble in Water
Other Solubilities: 95% Ethanol: Very Soluble (>=10mg/ml); Acetic Acid: Very Soluble; Benzene: Soluble; Chloroform: Soluble; Ether: Very Soluble; Ethyl Acetate: Very Soluble.
Surface Tension: 32.7 dynes/cm at 20 °C
Odor Threshold: 0.56 to 1.44 mg/m^3
pH: < 7
Refraction Index: 1.3904 at 20 °C/D
Evaporation Rate: 0.46 Butyl Acetate=1
Critical Temperature: 296 °C
Critical Pressure: 679 psia
Ionization Potential (eV): 10.00
Flash Point: 49 °C Closed Cup
Autoignition Temperature: 316 °C
LEL: 2.9% v/v
UEL: 10.3% v/v

RTECS Toxicity Data

Acute Oral: Rat LD_{50} Dose: 1780 mg/kg.
Acute Inhalation: Rat LC_{50} Dose: 1000 ppm/4hr.
Acute Dermal: Rabbit LD_{50} Route: Skin; Dose: 4 mL/kg.
Chronic (Multiple Dose) Inhalation: Rat Dose: 2470 ug/m^3/24H/95D-C; Toxic Effects: Kidney, Ureter, and Bladder - Other changes in urine composition; Blood - Changes in leukocyte (WBC) cell count; Biochemical - True cholinesterase.
Irritation Eye: Rabbit Standard Draize Test Dose: 250 ug open; Reaction: severe.
Irritation Skin: Rabbit Open Draize Test Dose: 10 mg/24H open; Reaction: mild. Rabbit Open Draize Test Dose: 540 mg open; Reaction: mild.

Hazard Overviews

Corrosive

Flammable

Fire Diamond

Health: Corrosive to eyes/skin/respiratory tract. Also Causes: severe burns of the mouth, throat, stomach. Chronic Effects: Skin irritation.
Fire: Flammable. Can form explosive mixtures in the air. Use carbon dioxide, dry chemical or alcohol foam. Do not use water. Vapors may flow along surfaces to distant ignition sources and flash back. Abundant water spray can be used to dilute spills to noncombustible mixtures, extinguish fire and to cool fire-exposed closed containers. Vapors may flow along surfaces to distant ignition sources and flash back.
Reactivity: Stable. Hazardous polymerization cannot occur. Avoid: heat; ignition sources. Incompatible with: oxidizing materials; water; alcohols; alkaline materials; aniline; 2-aminoethanol; ethylenediamine; chlorosulfonic acid; chromium trioxide and acetic acid; ethyleneimine; glycerol; oleum; hydrofluoric acid; sodium hydroxide; permanganates; hydrosulfuric acid; glycerol and phosphoryl chloride; boric acid; barium peroxide; 1,3-diphenyltriazene; chromium trioxide; hydrochloric acid and water; nitric acid; hypochlorous acid; perchloric acid and water; peroxyacetic acid; potassium permanganate; tetrafluoroboric acid; acetic acid and water; 4-toluenesulfonic acid and water; sodium hydrogen sulfate and ethanol; hydrogen peroxide; ammonium nitrate and hexamethylenetetrammonium acetate and nitric acid; metal nitrates; chromic acid; glycerol and phosphoryl chloride; N-tert-butylphthalimic acid and tetrafluoroboric acid; iron; steel; certain other metals. Hazardous decomposition products: acetic acid; carbon monoxide.
Carcinogenicity: IARC - Not listed; NIOSH - Listed as carcinogen; NTP - Not listed; ACGIH - Not listed; OSHA - Not listed; EPA - Not listed; MAK - Not listed
Primary Target Organs:

Eyes

Skin

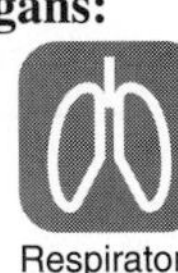
Respiratory System

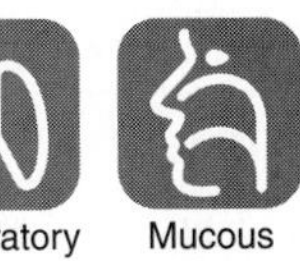
Mucous Membranes

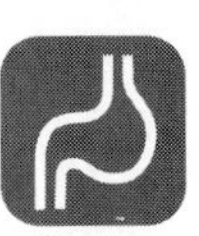
Gastro-intestinal

Exposure Limits
OSHA PEL: TWA: 5 ppm; 20 mg/m^3.
OSHA PEL Vacated 1989 Limits: STEL: 5 ppm; 20 mg/m^3; Ceiling.
ACGIH TLV: TWA: 5 ppm; 21 mg/m^3.
NIOSH REL: STEL: 5 ppm; 20 mg/m^3.
NIOSH IDLH: 200 ppm.
DFG MAK: TWA: 5 ppm; 20 mg/m^3.
Respirator Recommendation
Exposure Range: >5 to 50 ppm Air Purifying, Negative Pressure, Half Mask
Exposure Range: >50 to <200 ppm Air Purifying, Negative Pressure, Full Face
Exposure Range: 200 to unlimited ppm Self-contained Breathing Apparatus, Pressure Demand, Full Face
Cartridge Color: black

Environmental

Ecotoxicity: Aquatic toxicity: 75 ppm/96 hr/bluegill/TLm/fresh water 100-300 ppm/48 hr/shrimp/LC_{50}/salt water
Cleanup/Disposal: Guide No. 137: Fully encapsulating, vapor protective clothing should be worn for spills and leaks with no fire. Do not touch damaged containers or spilled material unless wearing appropriate protective clothing. Stop leak if you can do it without risk. Use water spray to reduce vapors; do not put water directly on leak, spill area or inside container. Keep combustibles (wood, paper, oil, etc.) away from spilled material. Small Spills: Cover with dry earth, dry sand, or other non-combustible material followed with plastic sheet to minimize spreading or contact with rain. Use clean non-sparking tools to collect material and place it into loosely covered plastic containers for later disposal. Prevent entry into waterways, sewers, basements or confined areas.

Environmental Physical Data

Octanol/Water Partition Coefficient: log K_{ow} = measured at -0.2
BCF: none
BOD: 53%, 1-5 days

Regulations

RCRA 40CFR: Not listed
CERCLA: 40CFR 302.4: Listed per CWA Section 311(b)(4) RQ: 5000 lb (2268 kg)
SARA 40CFR 372.65: Not listed
SARA EHS 40CFR 355: Not listed
TSCA: Listed

Analytical Methods

Air: ASTM D4490
Indoor / Expired Air: NIOSH 3506
Plasma: EPA 001

ACE4000 CAS #: 26446-35-5

ACETIN

RTECS: AK3595000
EINECS Number: 247-704-6
Molecular Formula: $C_5H_{10}O_4$
Formula Weight: 134.13

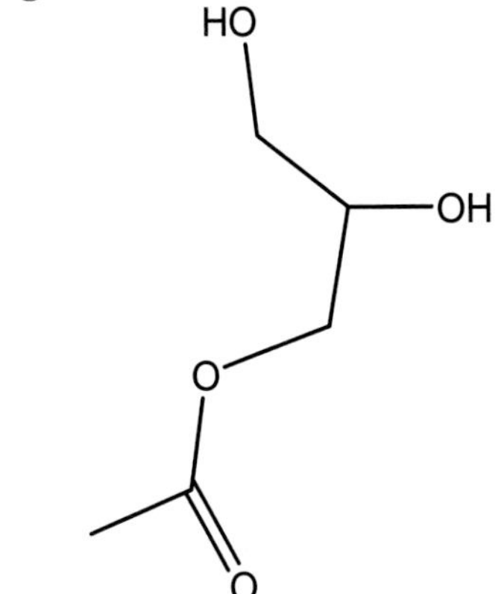

Chemical Structure

Synonyms: ACETIC ACID,MONOGLYCERALDEHYDE; ACETIN,MONO-; ACETOGLYCERIDE; ACETYL MONOGLYCERIDE;

GLYCERIN MONOACETATE; GLYCEROL ACETATE; GLYCEROL ALPHA-MONOACETATE; GLYCEROL MONOACETATE; GLYCEROL,1-ACETATE; GLYCERYL ACETATE; GLYCERYL MONOACETATE; MONACETIN; 1-MONO-ACETIN; 1-MONOACETIN; ALPHA-MONOACETIN; MONO-ACETIN; MONOACETIN; MONOACETYL GLYCERINE; 1,2,3-PROPANETRIOL,MONOACETATE

Description: colorless liquid, commercial product is pale yellow liquid; characteristic odor

Use: manufacture of smokeless powder and dynamite; solvent for basic dyes; tanning leather; food additive

Physical Properties

Boiling Point: 129 °C (264 °F) to 131 °C (268 °F) at 3 mm Hg
Freezing Point: -78 °C (-108.4 °F)
Specific Gravity: 1.206 at 20 °C/4 °C
Density: 1.21 g/mL
Vapor Pressure: 3 mm Hg at 130 °C
Water Solubility: Soluble in Water
Other Solubilities: 95% Ethanol: >=100 mg/ml at 18 °C; Acetone: >=100 mg/ml at 18 °C; Benzene: Insoluble; DMSO: >=100 mg/ml at 18 °C; Ether: Slightly Soluble; Toluene: Insoluble.
Refraction Index: 1.4157
Flash Point: > 94.0 °C

RTECS Toxicity Data

Mutagenic: Hamster Sister Chromatid Exchenge; Cell Type: ovary; Dose: 1500 mg/L. Insects - D Melanogaster Sex Chromosome Loss; Route: Oral; Dose: 5 pph.

Hazard Overviews

Health: Irritating to eyes/skin/respiratory tract. Other Acute Effects: may be harmful by inhalation, ingestion, or skin absorption.

Fire: Will burn. Hazards: emits toxic fumes. Extinguishing agents: water spray; carbon dioxide, dry chemical powder or appropriate foam. Precautions: combustible liquid.

Reactivity: Stable. Hazardous polymerization will not occur. Hazardous decomposition products: toxic fumes of: carbon monoxide, carbon dioxide.

Carcinogenicity: IARC - Not listed; NIOSH - Not listed; NTP - Not listed; ACGIH - Not listed; OSHA - Not listed; EPA - Not listed; MAK - Not listed

Primary Target Organs:

Eyes

Skin

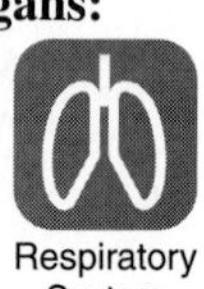
Respiratory System

Environmental

Regulations

RCRA 40CFR: Not listed
CERCLA: 40CFR 302.4: Not listed
SARA 40CFR 372.65: Not listed
SARA EHS 40CFR 355: Not listed
TSCA: Listed

ACE4150 CAS #: 102-01-2

ACETOACETANILIDE

RTECS: AK4200000
EINECS Number: 202-996-4
Molecular Formula: $C_{10}H_{11}NO_2$
Formula Weight: 177.20

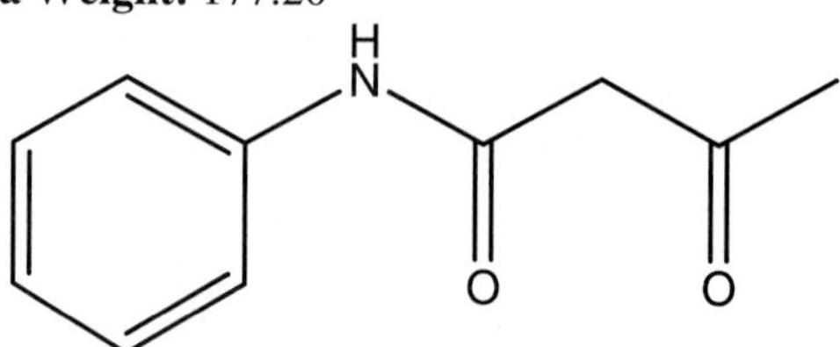

Chemical Structure

Synonyms: AAN; ACETANILIDE,2-ACETYL-; ACETOACETAMIDOBENZENE; ACETOACETANILID; ACETOACETIC ACID ANILIDE; ACETOACETIC ANILIDE; ((ACETOACETYL)AMINO)BENZENE; ACETOACETYLANILINE; ACETYLACETANILIDE; ALPHA-ACETYLACETANILIDE; N-(ACETYLACETYL)ANILINE; ALPHA-ACETYL-N-PHENYLACETAMIDE; ANILID KYSELINY ACETOCTOVE; BUTANAMIDE,3-OXO-N-PHENYL-; BUTANAMIDE,3-OXO-N-PHENYL-(9CI); BUTANOIC ACID,3-OXO-,AMIDE,N-PHENYL-; BETA-KETOBUTYRANILIDE; 3-OXO-N-PHENYLBUTANAMIDE; N-PHENYLACETOACETAMIDE; 1-(PHENYLCARBAMOYL)-2-PROPANONE; N-PHENYL-3-OXOBUTANAMIDE

Description: white crystalline solid, plates or needles

Use: manufacture of yellow dyes such as Hansa and benzidine yellow; in rubber compounding and organic synthesis

Physical Properties

Boiling Point: Decomposes
Freezing Point: 86 °C (186.8 °F)
Specific Gravity: 1.1 at Melting Point
Saturated Vapor Density: 1.20008069 kg/m^3
Density: 1.26 g/mL at 20 °C
Vapor Pressure: 0.01 mm Hg at 20 °C
Water Solubility: < 1 mg/mL at 21 C
Other Solubilities: 95% Ethanol: >=100 mg/ml at 21 °C; Acetone: >=100 mg/ml at 21 °C; Acids: Soluble; Alkali hydroxide solutions: Soluble; Benzene: Soluble in hot; Chloroform: Soluble; DMSO: >=100 mg/ml at 21 °C; Ether: Soluble; Petroleum Ether: Soluble in hot.
Flash Point: 185 °C Open Cup

RTECS Toxicity Data

Acute Oral: Rat LD_{50} Dose: 2450 mg/kg. Mouse LD_{50} Dose: 3400 mg/kg.

Acute Dermal: Rat LD_{50} Route: Subcutaneous Dose: 7 gm/kg. Guinea Pig LD_{50} Route: Skin; Dose: >1 gm/kg.

Chronic (Multiple Dose) Oral: Rat Dose: 14024 mg/kg/14D-C; Toxic Effects: Sense organs and special senses - Other; Blood - Changes in erythrocite (RBC) cell count; Nutritional and gross metabolic - Weight loss or decreased weight gain.

Hazard Overviews

Health: May cause irritation to eyes/skin. Harmful. Other Acute Effects: harmful if swallowed or absorbed through skin; absorption into the body leads to the formation of methemoglobin which in sufficient concentration causes cyanosis; onset may be delayed 2 to 4 hours or longer; target organ: blood.

Fire: Will burn. Hazards: emits toxic fumes. Extinguishing agents: water spray; carbon dioxide, dry chemical powder or appropriate foam. Precautions: combustible liquid.

Reactivity: Incompatible with: strong oxidizing agents. Hazardous decomposition products: toxic fumes of: carbon monoxide, carbon dioxide, nitrogen oxides.

Carcinogenicity: IARC - Not listed; NIOSH - Not listed; NTP - Not listed; ACGIH - Not listed; OSHA - Not listed; EPA - Not listed; MAK - Not listed

Primary Target Organs:

Blood

Environmental

Regulations

RCRA 40CFR: Not listed
CERCLA: 40CFR 302.4: Not listed
SARA 40CFR 372.65: Not listed
SARA EHS 40CFR 355: Not listed
TSCA: Listed

ACE4300 CAS #: 34256-82-1

ACETOCHLOR

RTECS: AB5457000
EINECS Number: 251-899-3
Molecular Formula: $C_{14}H_{20}ClNO_2$
Formula Weight: 269.8
Synonyms: ACENIT; ACETAL; O-ACETOTOLUIDIDE,2-CHLORO-N-(ETHOXYMETHYL)-6'-ETHYL-(8CI); AZETOCHLOR; 2-CHLORO-N-(ETHOXYMETHYL)-6'-ETHYL-O-ACETOTOLUIDIDE; 2-CHLORO-N-(ETHOXYMETHYL)-N-(2-ETHYL-6-METHYLPHENYL)ACETAMIDE; 2-CHLORO-2'-METHYL-6-ETHYL-N-ETHOXYMETHYLACETANILIDE; ERUNIT; HARNESS; MG 02; MON 097; NEVIREX
Description: light amber to purple oily liquid
Use: control of most annual grasses and certain broadleaf weeds and yellow nutsedge tolerant crops include corn (all types), soybeans, peanuts, sugarcane, cotton, potatoes, rape and sunflower

Physical Properties

Freezing Point: <
Specific Gravity: 1.1 at 30 °C
Saturated Vapor Density: 1.2 kg/m^3
Density: 1.1358 g/cu cm at 20 °C
Vapor Pressure: 3.4 $x10^{-8}$ at 25 °C
Water Solubility: 223 ppm at 25 °C
Other Solubilities: Soluble in Alcohol, Acetone, Toluene, Carbon Tetrachloride.
Flash Point: > 93 °C Tag Closed Cup

RTECS Toxicity Data

Acute Oral: Rat LD_{50} Dose: 763 mg/kg. Mouse LD_{50} Dose: 1550 mg/kg; Toxic Effects: Behavioral - Somnolence (general depressed activity); Behavioral - Convulsions or effect on seizure threshold; Lungs, Thorax, or Respiration - Dyspnea.

Hazard Overviews

Fire: Will burn.

Carcinogenicity: IARC - Not listed; NIOSH - Not listed; NTP - Not listed; ACGIH - Not listed; OSHA - Not listed; EPA - Not listed; MAK - Not listed

Environmental

Regulations

RCRA 40CFR: Not listed
CERCLA: 40CFR 302.4: Not listed
SARA 40CFR 372.65: Not listed
SARA EHS 40CFR 355: Not listed
TSCA: Not listed

Analytical Methods

Food: FDA 212.2, 232.4, 242.1

ACE4450 CAS #: 968-81-0

ACETOHEXAMIDE

RTECS: YR7350000
EINECS Number: 213-530-4
Molecular Formula: $C_{15}H_{20}N_2O_4S$
Formula Weight: 324.42
Synonyms: ACETOHEXAMID; 1-(4-ACETYLBENZENESULFONYL)-3-CYCLOHEXYLUREA; 1-(P-ACETYLBENZENESULFONYL)-3-CYCLOHEXYLUREA; N-(4-ACETYLBENZENESULFONYL)-N'-CYCLOHEXYLUREA; N-(P-ACETYLBENZENESULFONYL)-N'-CYCLOHEXYLUREA; 4-ACETYL-N-((CYCLOHEXYLAMINO)CARBONYL)BENZENESULFONAMIDE; 1-((P-ACETYLPHENYL)SULFONYL)-3-CYCLOHEXYLUREA; N-(P-ACETYLPHENYLSULFONYL)-N'-CYCLOHEXYLUREA; BENZENESULFONAMIDE; BENZENESULFONAMIDE,4-ACETYL-N-((CYCLOHEXYLAMINO)CARBONYL)-; CYCLAMIDE; 3-CYCLOHEXYL-1-(P-ACETYLPHENYLSULFONYL)UREA; DIMELIN; DIMELIN (ANTIDIABETIC); DIMELOR; DYMELOR; GAMADIABER; GAMADIABET; HYPOGLICIL; METAGLUCINA; MINORAL; ORDIMEL; TSIKLAMID; U 14812; U-14812; UREA,1-((P-ACETYLPHENYL)SULFONYL)-3-CYCLOHEXYL-
Description: white crystalline powder; practically odorless
Use: a hypoglycemic and oral antidiabetic drug

Physical Properties

Freezing Point: 188 °C (370.4 °F) to 190 °C (374 °F)
Water Solubility: Practically Insoluble in Water

Other Solubilities: 95% Ethanol: 1-5 mg/ml at 18 °C; Acetone: 10-50 mg/ml at 18 °C; Alcohol: 1 in 230; Chloroform: 1 in 210; DMSO: >=100 mg/ml at 18 °C; Dilute solutions of alkali hydroxides: Soluble; Ether: Insoluble; Pyridine: Soluble.
Flash Point: Not available; probably combustible

RTECS Toxicity Data

Acute Oral: Rat LD_{50} Dose: 5 gm/kg. Mouse LD_{50} Dose: >2500 mg/kg.

Hazard Overviews

Fire: Will burn.
Carcinogenicity: IARC - Not listed; NIOSH - Not listed; NTP - Not listed; ACGIH - Not listed; OSHA - Not listed; EPA - Not listed; MAK - Not listed

Environmental

Regulations

RCRA 40CFR: Not listed
CERCLA: 40CFR 302.4: Not listed
SARA 40CFR 372.65: Not listed
SARA EHS 40CFR 355: Not listed
TSCA: Listed

ACE4600 CAS #: 513-86-0

ACETOIN

RTECS: EL8790000
EINECS Number: 208-174-1
Molecular Formula: $C_4H_8O_2$
Formula Weight: 88.10

Chemical Structure

Synonyms: ACETYL METHYL CARBINOL; ACETYLMETHYLCARBINOL; 2,3-BUTANOLONE; 2-BUTANOL-3-ONE; 2-BUTANONE,3-HYDROXY-; DIMETHYLKETOL; 2-HYDROXY-3-BUTANONE; 3-HYDROXY-2-BUTANONE; 1-HYDROXYETHYL METHYL KETONE; GAMMA-HYDROXY-BETA-OXOBUTANE; METHANOL,ACETYLMETHYL-
Description: slightly yellow liquid or crystalline solid; bland, woody, yogurt odor
Use: synthetic flavoring in foods and non-alcoholic beverages; aroma carrier; prepn of flavors & essences

Physical Properties

Boiling Point: 148 °C (298 °F) at 760 mm Hg
Freezing Point: 15 °C (59 °F)
Density: DL 1.0062 at 20 °C
Water Solubility: Miscible with Water
Other Solubilities: Slightly Soluble in Alcohol; Soluble in Acetone.
Refraction Index: 1.4171 at 20 °C
Flash Point: Combustible

RTECS Toxicity Data

Acute Oral: Rat LD_{50} Dose: >5 gm/kg.
Acute Dermal: Rabbit LD_{50} Route: Skin; Dose: >5 gm/kg. Rat LD_{Lo} Route: Subcutaneous Dose: 14 gm/kg; Toxic Effects: Peripheral Nerve and sensation - Flaccid paralysis without anesthesia; Behavioral - Convulsions or effect on seizure threshold.
Chronic (Multiple Dose) Oral: Rat Dose: 66 gm/kg/13W-C; Toxic Effects: Liver - Changes in liver weight; Blood - Normocytic anemia; Nutritional and gross metabolic - Weight loss or decreased weight gain.
Irritation Skin: Rabbit Standard Draize Test Dose: 500 mg/24H; Reaction: moderate.
Reproductive/Teratogenic: Rat Route: Oral; Dose: 12600 mg/kg; Duration: male 42D prior to mating; Paternal Effects - Testes, epididymis, sperm duct.

Hazard Overviews

Corrosive

Health: Corrosive to eyes/skin/respiratory tract. Harmful. Other Acute Effects: harmful if swallowed, inhaled, or absorbed through skin; material is extremely destructive to tissue of the mucous membranes and upper respiratory tract, eyes and skin; inhalation may result in spasm, inflammation and edema of the larynx and bronchi, chemical pneumonitis and pulmonary edema; symptoms of exposure may include burning sensation, coughing, wheezing, laryngitis, shortness of breath, headache, nausea and vomiting; target organs: liver, kidneys.
Fire: Combustible. Hazards: emits toxic fumes; vapor may travel considerable distance to source of ignition and flash back. Extinguishing agents: carbon dioxide, dry chemical powder or appropriate foam; water spray. Precautions: combustible liquid.
Reactivity: Incompatible with: strong oxidizing agents, strong bases, store away from heat and direct sunlight. Hazardous decomposition products: toxic fumes of: carbon monoxide, carbon dioxide.
Carcinogenicity: IARC - Not listed; NIOSH - Not listed; NTP - Not listed; ACGIH - Not listed; OSHA - Not listed; EPA - Not listed; MAK - Not listed

Primary Target Organs:

Environmental

Regulations

RCRA 40CFR: Not listed
CERCLA: 40CFR 302.4: Not listed
SARA 40CFR 372.65: Not listed
SARA EHS 40CFR 355: Not listed
TSCA: Listed

ACE4750 CAS #: 67-64-1

ACETONE

RTECS: AL3150000
DOT: UN1090; IMO3.1
EINECS Number: 200-662-2
Molecular Formula: C_3H_6O
Structured MF: CH_3COCH_3
Formula Weight: 58.09

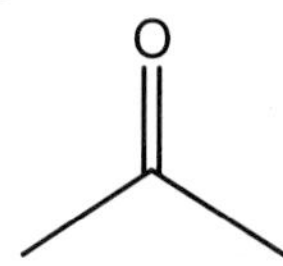

Chemical Structure

Synonyms: ACETON; CHEVRON ACETONE; DIMETHYL KETONE; DIMETHYLFORMALDEHYDE; DIMETHYLKETAL; EPA PESTICIDE CHEMICAL CODE 004101; KETONE PROPANE; KETONE,DIMETHYL; BETA-KETOPROPANE; METHYL KETONE; 2-PROPANONE; PROPANONE; PYROACETIC ACID; PYROACETIC ETHER
Description: colorless liquid; sweetish odor
Use: in the manufacture of smokeless powder, paints, varnishes, lacquers, organic chemicals, pharmaceuticals, sealants, adhesives, methyl isobutyl ketone, mesityl oxide, acetic acid (ketene process), diacetone alcohol, chloroform, iodoform, bromoform, explosives, airplane dopes, rayon, photographic films, isoprene, methyl isobutyl carbinol, methyl methacrylate and bisphenol A; as solvent for cellulose acetate, nitrocellulose, acetylene, fats, oils, waxes, resins, rubber, plastics, rubber cements, pharmaceuticals, potassium iodide and permanganate; in storing acetylene gas, in purifying paraffin, in hardening and dehydrating tissues an to clean and dry parts of precision equipment; in extraction of various principles from animal and plant substances, as a delusterant for cellulose acetate fibers, in specification testing of vulcanized rubber products, as a cosmetic ingredient, as a dye intermediate and in paint and varnish removers

Physical Properties

Boiling Point: 56.2 °C (133 °F) at 760 mm Hg
Freezing Point: -95.35 °C (-139.63 °F)
Specific Gravity: 0.7899 at 20 °C/4 °C
Vapor Density: 2 Air=1
Saturated Vapor Density: 1.565868784 kg/m^3
Density: 0.786 gm/cc at 22.7 °C
Vapor Pressure: 231 mm Hg at 25 °C
Water Solubility: Miscible
Other Solubilities: 95% Ethanol: >=100 mg/ml at 22 °C; Acetone: >=100 mg/ml at 22 °C; Benzene: Soluble; Chloroform: miscible; DMSO: >=100 mg/ml at 22 °C; Ether: miscible; Methanol: miscible; Most oils: miscible; Organic solvents: miscible.
Surface Tension: 26.2 nM/M at 0 °C
Odor Threshold: 47.5 to 1613.9 mg/m^3
Refraction Index: 1.3588 at 20 °C/D
Evaporation Rate: < 1 Butyl Acetate=1
Critical Temperature: 235 °C
Critical Pressure: 46.4 atm
Ionization Potential (eV): 9.69
Flash Point: -20 °C
Autoignition Temperature: 465 °C
LEL: 2.15% v/v
UEL: 13% v/v

RTECS Toxicity Data

Acute Oral: Man TD_{Lo} Dose: 2857 mg/kg; Toxic Effects: Behavioral - Coma; Kidney, Ureter, and Bladder - Other changes. Man TD_{Lo} Dose: 2857 mg/kg; Toxic Effects: Behavioral - Coma; Biochemical - Other. Rat LD_{50} Dose: 5800 mg/kg; Toxic Effects: Behavioral - Altered sleep time (including change in righting reflex); Behavioral - Tremor.
Acute Inhalation: Rat LC_{50} Dose: 50100 mg/m^3/8hr. Human TC_{Lo} Dose: 500 ppm; Toxic Effects: Sense organs and special senses - Other; Sense organs and special senses - Conjunctive irritation; Lungs, Thorax, or Respiration - Other changes. Man TC_{Lo} Dose: 440 ug/m^3/6M; Toxic Effects: Brain and coverings - Recordings from specific areas of CNS. Man TC_{Lo} Dose: 10 mg/m^3/6hr; Toxic Effects: Biochemical - Other carbohydrates. Man TC_{Lo} Dose: 12000 ppm/4hr; Toxic Effects: Gastrointestinal - Nausea or vomiting; Behavioral - Muscle weakness.
Acute Dermal: Rabbit LD_{Lo} Route: Skin; Dose: 20 mL/kg. Guinea Pig LD_{50} Route: Skin; Dose: >9400 mg/kg.
Chronic (Multiple Dose) Oral: Rat Dose: 273 gm/kg/13W-C; Toxic Effects: Liver - Changes in liver weight; Kidney, Ureter, and Bladder - Changes in kidney weight; Blood - Normocytic anemia. Mouse Dose: 546 gm/kg/13W-C; Toxic Effects: Liver - Changes in liver weight; Endocrine - Changes in spleen weight.
Chronic (Multiple Dose) Inhalation: Rat Dose: 19000 ppm/3H/8W-I; Toxic Effects: Brain and coverings - Changes in brain weight. Rat Dose: 199 mg/m^3/8H/45D-I; Toxic Effects: Behavioral - Muscle contraction or spasticity.
Irritation Eye: Rabbit Standard Draize Test Dose: 20 mg/24H; Reaction: moderate. Rabbit Standard Draize Test Dose: 20 mg; Reaction: severe.
Irritation Skin: Rabbit Standard Draize Test Dose: 500 mg/24H; Reaction: mild. Rabbit Open Draize Test Dose: 395 mg open; Reaction: mild.

Reproductive/Teratogenic: Rat Route: Oral; Dose: 273 gm/kg; Duration: male 13W prior to mating; Paternal Effects - Spermatogenesis. Mammal Route: Inhalation; Dose: 31500 ug/m^3/ Duration: female 1-13D of pregnancy; Effects on Fertility - Post-implantation mortality.

Mutagenic: Hamster Cytogenetic Analysis; Cell Type: fibroblast; Dose: 40 gm/L. Yeast - S Cerevisiae Sex Chromosome Loss; Dose: 47600 ppm.

Hazard Overviews

Flammable

Fire Diamond

Health: Irritating to eyes/skin/respiratory tract. Also Causes: muscle weakness, mental confusion, coma (high concentrations) ingestion: GI irritation, kidney and liver damage, metabolic changes, coma. Chronic Effects: dermatitis.

Fire: Flammable. Can form explosive mixtures in the air. Do not extinguish fire unless flow can be stopped. For small fires, use dry chemical, carbon dioxide, water spray or alcohol-resistant foam. For large fires, use water spray, fog, or alcohol-resistant foam. Use water in flooding quantities as fog because solid streams may be ineffective.

Reactivity: Stable. Hazardous polymerization cannot occur. Avoid: contact with plastic eyeglass frames, jewelry, pens, pencils, and rayon garments. Incompatible with: hydrogen peroxide; acetic acid; nitric acid; nitric acid and sulfuric acid; chromic anhydride; chromyl chloride; nitrosyl chloride; hexachloromelamine; nitrosyl perchlorate; nitryl perchlorate; permonosulfuric acid; thiodiglycol and hydrogen peroxide; oxidizing materials; activated carbon; chromium trioxide; dioxygen difluoride and carbon dioxide; potassium-tert-butoxide; air; bromoform; bromine; chloroform and alkalies; trichloromelamine; sulfur dichloride. Hazardous decomposition products: carbon dioxide; carbon monoxide.

Carcinogenicity: IARC - Not listed; NIOSH - Not listed; NTP - Not listed; ACGIH - Not listed; OSHA - Not listed; EPA - Class D, Not classifiable as to human carcinogenicity; MAK - Not listed

Primary Target Organs:

Eyes

Skin

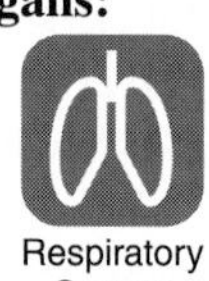

Respiratory System

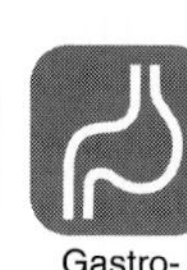

Gastro-intestinal

Nervous System

Exposure Limits

OSHA PEL: TWA: 1000 ppm; 2400 mg/m^3.

OSHA PEL Vacated 1989 Limits: TWA: 750 ppm; 1800 mg/m^3; STEL: 1000 ppm; 2400 mg/m^3.

ACGIH TLV: TWA: 750 ppm; 1780 mg/m^3; STEL: 1000 ppm; 2380 mg/m^3.

NIOSH REL: TWA: 250 ppm; 590 mg/m^3.

NIOSH IDLH: 2500 ppm; LEL.

DFG MAK: TWA: 500 ppm; 1200 mg/m^3.

Respirator Recommendation

Exposure Range: >1000 to <2500 ppm Supplied Air, Constant Flow/Pressure Demand, Full Face

Exposure Range: 2500 to unlimited ppm Self-contained Breathing Apparatus, Pressure Demand, Full Face

Note: use ov (black) cartridge for nusiance(<1000)

Environmental

Ecotoxicity: LD_{100} Asellus aquaticus 3 ml/l (within 3 days of exposure) /Conditions of bioassay not specified LC_{50} Mexican axolotl 20.0 mg/l/48 hr (3-4 weeks after hatching) /Conditions of bioassay not specified TLm Mosquito fish 13,000 mg/l/24, 48, 96 hr /Conditions of bioassay not specified LD_{100} Gammarus fossarum 10 ml/l (within 48 hr) /Conditions of bioassay not specified LC_{50} Poecilia reticulata (guppy) 7,032 ppm/14 days /Conditions of bioassay not specified LC_{50} Ring-necked pheasant oral greater than 40,000 ppm, in diet, age 10 days, (no mortality to 40,000 ppm) LC_{50} Salmo gairdneri (Rainbow trout) 5,540 mg/l/96 hr at 12 °C (95% confidence limit 4,740-6,330 mg/l), wt 1.0 g /static bioassay LC_{50} Clawed toad 24.0 mg/l/48 hr (3-4 weeks after hatching) /Conditions of bioassay not specified TLm Daphnia magna 10 mg/l/24, 48 hr /Conditions of bioassay not specified

Environmental Fate: If released on soil, it will both volatilize and leach into the ground and probably biodegrade. If released into water, it will probably biodegrade. It will also be lost due to volatilization (estimated half-life 20 hr from a model river). Bioconcentration in aquatic organisms and adsorption to sediment should not be significant. In the atmosphere, it will be lost by photolysis and reaction with photochemically produced hydroxyl radicals. Half-life estimates from these combined processes average 22 days and are shorter in summer and longer in winter. It will also be washed out by rain.

Cleanup/Disposal: Guide No. 127: Eliminate all ignition sources (no smoking, flares, sparks or flames in immediate area). All equipment used when handling the product must be grounded. Do not touch or walk through spilled material. Stop leak if you can do it without risk. Prevent entry into waterways, sewers, basements or confined areas. A vapor suppressing foam may be used to reduce vapors. Absorb or cover with dry earth, sand or other non-combustible material and transfer to containers. Use clean non-sparking tools to collect absorbed material. Large Spills: Dike far ahead of liquid spill for later disposal. Water spray may reduce vapor; but may not prevent ignition in closed spaces.

Environmental Physical Data

Henry's Law Constant: 3.97 x10^{-5}

Octanol/Water Partition Coefficient: log K_{ow} = -0.24

BCF: negligible

BOD: theoretical 122%, 5 days

Regulations

RCRA 40CFR: Listed Hazardous Waste No. U002 Ignitable Waste

CERCLA: 40CFR 302.4: Listed per RCRA Section 3001 RQ: 5000 lb (2268 kg)

SARA 40CFR 372.65: Not listed
SARA EHS 40CFR 355: Not listed
TSCA: Listed

Analytical Methods

Air: EPA 0100, OA-002-1, VA-005-1, VG-006-1, TO-11, TO-5; ASTM D3686, D3687, D4490
Soil: CLP LC_VOA, MC_VOA, OHC; EPA 1624; SW846 1311, 5031, 5032, 5041, 5041A, 8015B, 8240B, 8260A, 8260B, 8315, 8315A; DOE OG015R
Water / Groundwater: EPA VW-008-1; ASTM D3695, D4763
Drinking Water: EPA 524.2
Food: AOAC 975.06
Indoor / Expired Air: NIOSH 1300; EPA IP-6A, IP-6B, IP-6C, 0100
Plasma: EPA 29
Other: EPA VS-006-1

ACE4900 CAS #: 1752-30-3

ACETONE THIOSEMICARBAZIDE

RTECS: AL7350000
EINECS Number: 217-137-9
Molecular Formula: $C_4H_9N_3S$
Formula Weight: 131.22

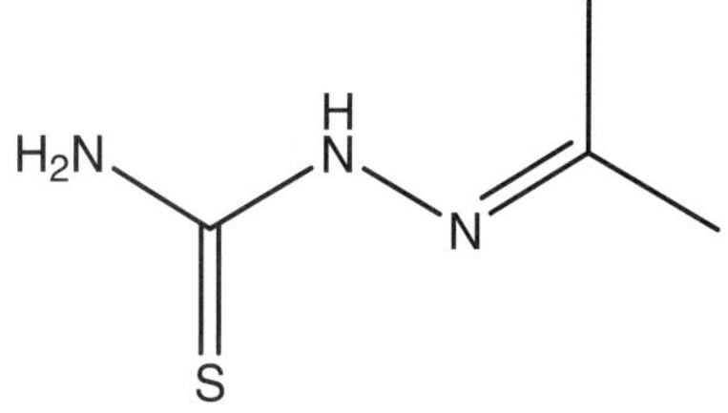

Chemical Structure

Synonyms: ACETONE THIOSEMICARBAZONE; ACETONTHIOSEMIKARBAZON; HYDRAZINECARBOTHIOAMIDE,2-(1-METHYLETHYLIDENE)-; HYDRAZINECARBOTHIOAMIDE,2-(1-METHYLETHYLIDENE)-(9CI); 2-(1-METHYLETHYLIDENE)HYDRAZINECARBOTHIOAMIDE; THIOSEMICARBAZONE ACETONE
Description: white, needle-shaped crystalline powder; odorless

RTECS Toxicity Data

Acute Oral: Rat LD_{Lo} Dose: 10 mg/kg.

Hazard Overviews

Carcinogenicity: IARC - Not listed; NIOSH - Not listed; NTP - Not listed; ACGIH - Not listed; OSHA - Not listed; EPA - Not listed; MAK - Not listed

Environmental

Regulations

RCRA 40CFR: Not listed
CERCLA: 40CFR 302.4: Not listed
SARA 40CFR 372.65: Not listed TPQ: 1000/10000 lb
SARA EHS 40CFR 355: Listed TPQ: 1000 lb
TSCA: Listed

ACE5050 CAS #: 75-05-8

ACETONITRILE

RTECS: AL7700000
DOT: UN1648; IMO3.2
EINECS Number: 200-835-2
Molecular Formula: C_2H_3N
Structured MF: CH_3CN
Formula Weight: 41.05

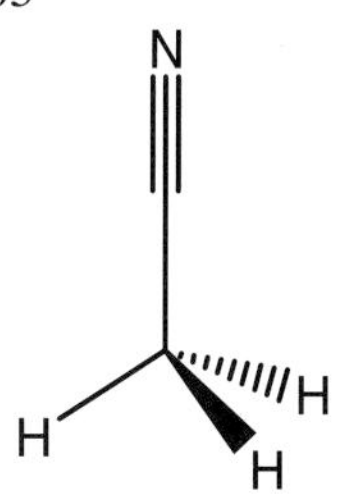

Chemical Structure

Synonyms: ACETONITRIL; CYANOMETHANE; CYANURE DE METHYL; ETHANENITRILE; ETHYL NITRILE; METHANECARBONITRILE; METHANE,CYANO-; METHYL CYANIDE; METHYLKYANID
Description: clear, colorless liquid; aromatic odor
Use: chemical intermediate in the synthesis of acetophenone, 1-naphthaleneacetic acid, thiamine and acetamidine, in pesticide manufacture, as an extractant for animal and vegetable oils, as a pharmaceutical solvent, as a solvent for inorganic salts and in organic synthesis, as a polymer solvent and in acrylic fibers; for separation of butadiene by extractive distillation, in perfumes, in nitrile rubber, in ABS resins, as a solvent in hydrocarbon extraction processes, as a specialty solvent, as a catalyst, to remove tars, phenols and coloring matter from petroleum hydrocarbons which are not soluble in acetonitrile, to recrystallize steroids, as an indifferent medium in physicochemical investigations, as a medium for promoting reactions involving ionizations and as a solvent in non-aqueous titrations

Physical Properties

Boiling Point: 81.6 °C (179 °F) at 760 mm Hg
Freezing Point: -45 °C (-49 °F)
Specific Gravity: 0.78745 at 15 °C/4 °C
Vapor Density: 1.42 Air=1
Saturated Vapor Density: 1.257078947 kg/m^3
Density: 0.7857 g/mL at 20 °C
Vapor Pressure: 87 mm Hg at 24 °C
Water Solubility: Miscible with Water
Other Solubilities: Equal wt of acetonitrile and the following materials are miscible at room temp: formic acid, Acetic Acid, Methanol, cellosolve solvent, formaldehyde,

acetaldehyde, di-n-butyl amine, acetic anhydride, Pyridine, nitrobenzene, aniline, Xylene.
Surface Tension: 29.04 dynes/cm at 20 °C
Odor Threshold: 70.0 mg/m^3
Refraction Index: 1.33934 at 30 °C/D
Evaporation Rate: 5.79 Butyl Acetate=1
Critical Temperature: 275 °C
Critical Pressure: 701 psia
Ionization Potential (eV): 12.20
Flash Point: 6 °C Open Cup
Autoignition Temperature: 524 °C
LEL: 4.4% v/v
UEL: 16% v/v

RTECS Toxicity Data

Acute Oral: Man TD_{Lo} Dose: 571 mg/kg; Toxic Effects: Behavioral - Convulsions or effect on seizure threshold; Gastrointestinal - Nausea or vomiting; Nutritional and gross metabolic - Metabolic acidosis. Man TD_{Lo} Dose: 64 mg/kg; Toxic Effects: Behavioral - Excitement. Child TD_{Lo} Dose: 800 mg/kg; Toxic Effects: Behavioral - Hallucinations, distorted perceptions; Behavioral - Convulsions or effect on seizure threshold; Gastrointestinal - Nausea or vomiting. Rat LD_{50} Dose: 2460 mg/kg.

Acute Inhalation: Rat LC_{50} Dose: 7551 ppm/8hr; Toxic Effects: Behavioral - Altered sleep time (including change in righting reflex); Behavioral - Convulsions or effect on seizure threshold; Blood - Hemorrhage. Human TC_{Lo} Dose: 160 ppm/4hr; Toxic Effects: Lungs, Thorax, or Respiration - Other changes.

Acute Dermal: Rabbit LD_{50} Route: Skin; Dose: 1250 uL/kg. Rabbit LD_{Lo} Route: Subcutaneous Dose: 105 mg/kg.

Chronic (Multiple Dose) Inhalation: Rat Dose: 655 ppm/7H/90D-I; Toxic Effects: Lungs, Thorax, or Respiration - Chronic pulmonary edema; Liver - Other changes; Kidney, Ureter, and Bladder - Chgs in tubules (inc acute renal failure, acute tubular necrosis. Monkey Dose: 350 ppm/7H/91D-I; Toxic Effects: Brain and coverings - Changes in circulation; Lungs, Thorax, or Respiration - Emphysema; Blood - Changes in erythrocite (RBC) cell count.

Irritation Eye: Rabbit Standard Draize Test Dose: 20 mg open; Reaction: severe.

Irritation Skin: Rabbit Open Draize Test Dose: 500 mg open; Reaction: mild.

Reproductive/Teratogenic: Rat Route: Inhalation; Dose: 1800 ppm/6H; Duration: female 6-20D of pregnancy; Effects on Fertility - Post-implantation mortality. Hamster Route: Inhalation; Dose: 5000 ppm/1H; Duration: female 8D of pregnancy; Effects on Fertility - Post-implantation mortality; Specific Developmental Abnormalities - Central nervous system. Hamster Route: Inhalation; Dose: 8000 ppm/1H; Duration: female 8D of pregnancy; Effects on Embryo or Fetus - Fetotoxicity; Specific Developmental Abnormalities - Musculoskeletal system.

Mutagenic: Hamster Sister Chromatid Exchenge; Cell Type: ovary; Dose: 5 gm/L. Insects - D Melanogaster Sex Chromosome Loss; Route: Inhalation; Dose: 131 ppm.

Hazard Overviews

Poison

Flammable

Fire Diamond

Health: Poison. Also Causes: headache, nausea, vomiting, weakness, lethargy, stupor, chest pain, respiratory depression, convulsions, death.

Fire: Flammable. Can form explosive mixtures in air. Use carbon dioxide, dry chemical, or foams to put out fire. Never direct solid streams of water into burning pools of liquid since this can scatter and spread flames. Use water sprays to cool fire-exposed containers, to disperse vapor, to dilute spills to nonflammable mixtures, and to reduce the fire's intensity.

Reactivity: Stable. Hazardous polymerization cannot occur. Avoid: heat and ignition sources (open flames; lighted cigarettes or pipes; uninsulated heating elements; sparks). Incompatible with: chlorosulfonic acid; erbium perchlorate; oleum; sulfuric acid. Hazardous decomposition products: carbon dioxide; toxic oxides of nitrogen; carbon monoxide; hydrogen cyanide.

Carcinogenicity: IARC - Not listed; NIOSH - Not listed; NTP - Not listed; ACGIH - Class A4, Not classifiable as a human carcinogen; OSHA - Not listed; EPA - Not listed; MAK - Not listed

Primary Target Organs:

Eyes

Skin

Respiratory System

Nervous System

Liver

Kidneys

Exposure Limits
OSHA PEL: TWA: 40 ppm; 70 mg/m^3.
OSHA PEL Vacated 1989 Limits: TWA: 40 ppm; 70 mg/m^3; STEL: 60 ppm; 105 mg/m^3.
NIOSH IDLH: 500 ppm.
DFG MAK: TWA: 40 ppm; 70 mg/m^3.
Respirator Recommendation
Exposure Range: >40 to <500 ppm Supplied Air, Constant Flow/Pressure Demand, Half Mask
Exposure Range: 500 to unlimited ppm Self-contained Breathing Apparatus, Pressure Demand, Full Face
Note: poor warning properties

Environmental

Ecotoxicity: TLm Pimephales promelas (fathead minnow) 1000 mg/l/96 hr (soft water) /Conditions of bioassay not specified Toxicity Threshold (Cell Multiplication Inhibition Test) Pseudomonas putida (bacteria) 680 mg/l Toxicity Threshold (Cell Multiplication Inhibition Test) Microcystis aeruginosa (algae) 520 mg/l /Conditions of bioassay not specified

Environmental Fate: If released to soil, aerobic biodegradation is likely to occur. It is expected to be mobile in soil and may evaporate from soil surfaces. Biodegradation is expected to be a major loss process in water. Acclimatization increases the biodegradation rate substantially. Volatilization may become competitive with other loss processes particularly at shallow water depths. Hydrolysis, photolysis, adsorption to suspended particles and sediments and bioconcentration in aquatic organisms are not likely to be important fate mechanisms. It is likely to be unreactive towards direct photolysis in air and the half-lives for its reaction with OH radicals and ozone have been estimated to be 535 days and 860 days, respectively. Therefore, it will persist in the troposphere for a long time and may be transported a long distance from its source of emission. Wet deposition may remove some of the atmospheric material.

Cleanup/Disposal: Guide No. 131: Fully encapsulating, vapor protective clothing should be worn for spills and leaks with no fire. Eliminate all ignition sources (no smoking, flares, sparks or flames in immediate area). All equipment used when handling the product must be grounded. Do not touch or walk through spilled material. Stop leak if you can do it without risk. Prevent entry into waterways, sewers, basements or confined areas. A vapor suppressing foam may be used to reduce vapors. Small Spills: Absorb with earth, sand or other non-combustible material and transfer to containers for later disposal. Use clean non-sparking tools to collect absorbed material. Large Spills: Dike far ahead of liquid spill for later disposal. Water spray may reduce vapor; but may not prevent ignition in closed spaces.

Environmental Physical Data

Henry's Law Constant: 2.93 $x10^{-5}$
Octanol/Water Partition Coefficient: log K_{ow} = -0.34
Sorption Partition Coefficient: K_{oc} = estimated at 16
BCF: none
BOD: 17%, 5 days

Regulations

RCRA 40CFR: Listed Hazardous Waste No. U003 Toxic Waste Ignitable Waste
CERCLA: 40CFR 302.4: Listed per RCRA Section 3001 RQ: 5000 lb (2268 kg)
SARA 40CFR 372.65: Listed
SARA EHS 40CFR 355: Not listed
TSCA: Listed

Analytical Methods

Air: ASTM D3686, D3687, D4490
Soil: SW846 8015B, 8033, 8240B, 8260A, 8260B
Water / Groundwater: ASTM D3371, D3695
Indoor / Expired Air: NIOSH 1606
Plasma: EPA 29
Other: EPA 1666, 1671

ACE5200 CAS #: 98-86-2

ACETOPHENONE

RTECS: AM5250000
EINECS Number: 202-708-7
Molecular Formula: C_8H_8O
Structured MF: $C_6H_5COCH_3$
Formula Weight: 120.16

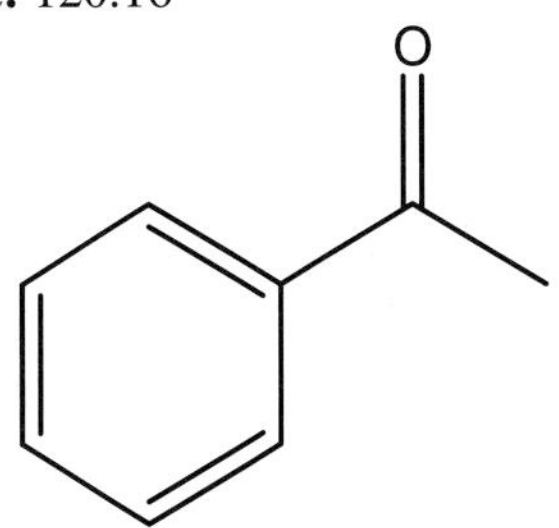

Chemical Structure

Synonyms: ACETOFENON; ACETOPHENON; ACETYLBENZENE; BENZENE,ACETYL-; BENZOYL METHIDE; ETHANONE,1-PHENYL-; ETHANONE,1-PHENYL-(9CI); HYPNON; HYPNONE; KETONE,METHYL PHENYL; METHYL PHENYL KETONE; PHENYL METHYL KETONE; 1-PHENYLETHANONE

Description: colorless slightly oily liquid; sweet, pungent odor

Use: in perfumery to impart orange-blossom-like odor; catalyst for polymerization of olefins; in org synthesis, esp photosensitizer; specialty solvent for plastics & resins; intermed for odorant, ethyl methyl phenylglycidate, riot control agent, 2-chloroacetophenone, 2-bromoacetophenone (for dyes), 3-nitroacetophenone; flavoring agent in non-alcoholic beverages, ice cream, candy, etc; fragrance in soaps, detergents, creams, lotions, perfumes; flavorant in tobacco.

Physical Properties

Boiling Point: 202 °C (396 °F)
Freezing Point: 20.5 °C (68.9 °F)
Specific Gravity: 1.033 at 15 °C/15 °C
Vapor Density: 4.14
Saturated Vapor Density: 1.202183869 kg/m^3
Bulk Density: 8.56 lbs/gal at 20 °C
Vapor Pressure: 0.44 mm Hg at 25 °C
Water Solubility: Insoluble in Water
Other Solubilities: Soluble in Acetone & Benzene; 1:5 in 50% Ethanol.
Surface Tension: 12 dynes/cm at 30 °C
Odor Threshold: 0.8347 to 2.9460 mg/m^3
Refraction Index: 1.5339 at 20 °C/D
Evaporation Rate: 0.06 Ether=1
Critical Temperature: 428 °C
Critical Pressure: 560 psia
Ionization Potential (eV): 9.28 +/-0.4
Flash Point: 82.222 °C Open Cup
Autoignition Temperature: 570 °C

RTECS Toxicity Data

Acute Oral: Rat LD_{50} Dose: 815 mg/kg. Mouse LD_{50} Dose: 740 mg/kg.
Acute Inhalation: Rat LC; Dose: >210 ppm/8hr. Mammal LC_{50} Dose: 1200 mg/m^3.
Acute Dermal: Rabbit LD_{50} Route: Skin; Dose: 15900 uL/kg. Mouse LD_{Lo} Route: Subcutaneous Dose: 330 mg/kg; Toxic Effects: Behavioral - Somnolence (general depressed activity).

Irritation Eye: Rabbit Standard Draize Test Dose: 750 ug; Reaction: severe.
Irritation Skin: Rabbit Open Draize Test Dose: 515 mg open; Reaction: mild.
Mutagenic: Hamster Cytogenetic Analysis; Cell Type: lung; Dose: 600 mg/L.

Hazard Overviews

Fire Diamond

Health: Irritating to eyes/skin. Also Causes: CNS depression and narcosis, transient corneal injury; anesthetic effects.
Fire: Combustible. Use dry chemical, carbon dioxide, water spray, fog, or regular foam.
Reactivity: Stable. Hazardous polymerization cannot occur. Avoid: heat; ignition sources. Incompatible with: oxidizers. Hazardous decomposition products: acrid smoke; carbon oxide gases.
Carcinogenicity: IARC - Not listed; NIOSH - Not listed; NTP - Not listed; ACGIH - Not listed; OSHA - Not listed; EPA - Class D, Not classifiable as to human carcinogenicity; MAK - Not listed
Primary Target Organs:

Eyes

Skin

Nervous System

Exposure Limits
ACGIH TLV: TWA: 10 ppm; 49 mg/m^3.
AIHA WEEL: TWA: 10 ppm.
Respirator Recommendation
Exposure Range: >10 to 100 ppm Air Purifying, Negative Pressure, Half Mask
Exposure Range: >100 to 1000 ppm Air Purifying, Negative Pressure, Full Face
Exposure Range: >1000 to 10,000 ppm Supplied Air, Constant Flow/Pressure Demand, Full Face
Exposure Range: >10,000 to unlimited ppm Self-contained Breathing Apparatus, Pressure Demand, Full Face
Cartridge Color: black
Note: use dust/mist filter if mist present

Environmental

Ecotoxicity: LC_{50} Pimephales promelas (fathead minnow) >200 mg/l/1 hr, static bioassay in Lake Superior water at 18-22 °C
Environmental Fate: If released to soil, microbial degradation is likely to be the major degradation pathway. It is expected to be moderately to highly mobile in soil and may evaporate from dry soil surfaces. Biodegradation and volatilization are expected to be the major loss processes in water. The estimated biodegradation half-lives in groundwater, river water and lake water samples were 32 days, 8 days and 4.5 days, respectively. The volatilization half-life from a river 1 m deep flowing at 1 m/sec with a wind speed of 3 m/sec is estimated to be 3.8 days. Hydrolysis, oxidation and adsorption to suspended particles and sediments and bioconcentration in aquatic organisms are not likely to be important fate processes. Oxidation by hydroxyl radicals in air has an estimated half-life of 2.2 days. Other oxidants (e.g., ozone) and photolysis do not appear to be important loss mechanism of this compound in air. Wet deposition may be important for the removal of atmospheric material.
Cleanup/Disposal: Guide No. 156: Eliminate all ignition sources (no smoking, flares, sparks or flames in immediate area). All equipment used when handling the product must be grounded. Do not touch damaged containers or spilled material unless wearing appropriate protective clothing. Stop leak if you can do it without risk. A vapor suppressing foam may be used to reduce vapors. For chlorosilanes, use AFFF alcohol-resistant medium expansion foam to reduce vapors. Do not get water on spilled substance or inside containers. Use water spray to reduce vapors or divert vapor cloud drift. Prevent entry into waterways, sewers, basements or confined areas. Small Spills: Cover with dry earth, dry sand, or other non-combustible material followed with plastic sheet to minimize spreading or contact with rain. Use clean non-sparking tools to collect material and place it into loosely covered plastic containers for later disposal.

Environmental Physical Data

Henry's Law Constant: 1.07 $x10^{-5}$
Octanol/Water Partition Coefficient: log K_{ow} = 1.58
Sorption Partition Coefficient: K_{oc} = sediments 1.34 to 2.43
BCF: none noted

Regulations

RCRA 40CFR: Listed Hazardous Waste No. U004 Toxic Waste
CERCLA: 40CFR 302.4: Listed per RCRA Section 3001 RQ: 5000 lb (2268 kg)
SARA 40CFR 372.65: Listed
SARA EHS 40CFR 355: Not listed
TSCA: Listed

Analytical Methods

Soil: SW846 3640A, 8250A, 8270B, 8270C
Water / Groundwater: EPA 1625; ASTM D3695
Indoor / Expired Air: EPA IP-1B
Plasma: EPA 29

ACE5350 CAS #: 88-15-3

2-ACETOTHIOPHENE

RTECS: OB6300000
EINECS Number: 201-804-6
Molecular Formula: C_6H_6OS
Formula Weight: 126.18

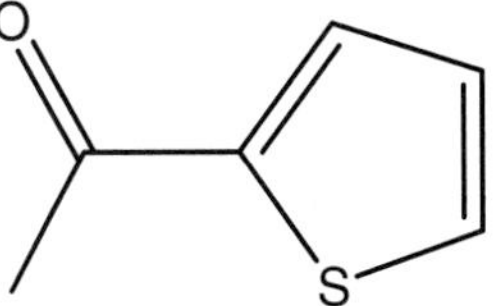

Chemical Structure

Synonyms: 2-ACETOTHIENONE; 2-ACETYLTHIOPHENE

Physical Properties

Boiling Point: 214 °C (417.2 °F)
Freezing Point: 10 °C (50 °F) to 11 °C (51.8 °F)
Specific Gravity: 1.16790
Water Solubility: Slightly Soluble
Other Solubilities: Ethanol: Miscible; Ether: Miscible; Carbon Tetrachloride: Soluble
Refraction Index: 1.5667
Ionization Potential (eV): 9.20 +/-0.05

Hazard Overviews

Health: May cause irritation. Harmful. Other Acute Effects: may be harmful by inhalation, ingestion, or skin absorption; exposure can cause nausea, headache and vomiting.
Fire: Hazards: emits toxic fumes. Extinguishing agents: water spray; carbon dioxide, dry chemical powder or appropriate foam. Precautions: combustible liquid.
Reactivity: Incompatible with: strong bases, strong oxidizing agents, strong reducing agents. Hazardous decomposition products: toxic fumes of: carbon monoxide, carbon dioxide, sulfur oxides, hydrogen sulfide gas.
Carcinogenicity: IARC - Not listed; NIOSH - Not listed; NTP - Not listed; ACGIH - Not listed; OSHA - Not listed; EPA - Not listed; MAK - Not listed

Environmental

Regulations

RCRA 40CFR: Not listed
CERCLA: 40CFR 302.4: Not listed
SARA 40CFR 372.65: Not listed
SARA EHS 40CFR 355: Not listed
TSCA: Listed

ACE5500 CAS #: 2425-25-4

ACETOXON

RTECS: AI7000000
DOT: UN2783; IMO6.1
Molecular Formula: $C_8H_{17}O_5PS$
Formula Weight: 256.28
Synonyms: ACETAPHOS; ACETIC ACID,((DIETHOXYPHOSPHINYL)THIO)-,ETHYL ESTER; ACETIC ACID,((DIETHOXYPHOSPHINYL)THIO)-,ETHYL ESTER (9CI); ACETIC ACID,MERCAPTO-,ETHYL ESTER,S-ESTER WITHO,O-DIETHYL PHOSPHOROTHIOATE; ACETOFOS; ACETOPHOS; O,O-DIETHYL S-(CARBETHOXY)METHYL PHOSPHOROTHIOLATE; O,O-DIETHYL S-CARBETHOXYMETHYL THIOPHOSPHATE; O,O-DIETHYL S-CARBOETHOXYMETHYL PHOSPHOROTHIOATE; O,O-DIETHYL S-CARBOETHOXYMETHYL THIOPHOSPHATE; ETHYL ((DIETHOXYPHOSPHINYL)THIO)ACETATE; PHOSPHOROTHIOIC ACID,O,O-DIETHYL ESTER,S-ESTER WITHETHYL MERCAPTOACETATE
Use: insecticide

RTECS Toxicity Data

Acute Oral: Rat LD_{50} Dose: 45 mg/kg. Mouse LD_{50} Dose: 210 mg/kg.

Hazard Overviews

Carcinogenicity: IARC - Not listed; NIOSH - Not listed; NTP - Not listed; ACGIH - Not listed; OSHA - Not listed; EPA - Not listed; MAK - Not listed

Environmental

Cleanup/Disposal: Guide No. 152: Do not touch damaged containers or spilled material unless wearing appropriate protective clothing. Stop leak if you can do it without risk. Prevent entry into waterways, sewers, basements or confined areas. Cover with plastic sheet to prevent spreading. Absorb or cover with dry earth, sand or other non-combustible material and transfer to containers. Do not get water inside containers.

Regulations

RCRA 40CFR: Not listed
CERCLA: 40CFR 302.4: Not listed
SARA 40CFR 372.65: Not listed
SARA EHS 40CFR 355: Not listed
TSCA: Not listed

ACE5800 CAS #: 644-31-5

ACETYL BENZOYL PEROXIDE

RTECS: SD7860000
DOT: UN2081; IMO5.2
EINECS Number: 211-412-7
Molecular Formula: $C_9H_8O_4$
Formula Weight: 180.15
Synonyms: ACETOZONE; ACETYL BENZOYL PEROXIDE,SOLID,OR >40% IN SOLUTION; BENZOYL ACETYL PEROXIDE; BENZOZONE; PEROXIDE,ACETYL BENZOYL; PEROXIDE,ACETYL BENZOYL (SOLID)
Description: white crystals or needles
Use: strong oxidizing agent; bleaching agent for food oils; active germicide; disinfectant; bleaching flour

Physical Properties

Boiling Point: 130 °C (266 °F) at 19 mm Hg
Freezing Point: 36 °C (96.8 °F) to 37 °C (98.6 °F)
Water Solubility: Decomposed by Water
Other Solubilities: Decomposed by alkaloids, organic matter, and some organic solvents.

Hazard Overviews

Carcinogenicity: IARC - Not listed; NIOSH - Not listed; NTP - Not listed; ACGIH - Not listed; OSHA - Not listed; EPA - Not listed; MAK - Not listed

Environmental

Cleanup/Disposal: Guide No. 147: Eliminate all ignition sources (no smoking, flares, sparks or flames in immediate area). Keep combustibles (wood, paper, oil, etc.) away from spilled material. Do not touch damaged containers or spilled material unless wearing appropriate protective clothing. Keep substance wet using water spray. Stop leak if you can do it without risk. Small Spills: Take up with inert, damp, noncombustible material using clean non-sparking tools and place into loosely covered plastic containers for later disposal. Large Spills: Wet down with water and dike for later disposal. Prevent entry into waterways, sewers, basements or confined areas. Do not clean-up or dispose of, except under supervision of a specialist.

Regulations

RCRA 40CFR: Not listed
CERCLA: 40CFR 302.4: Not listed
SARA 40CFR 372.65: Not listed
SARA EHS 40CFR 355: Not listed
TSCA: Not listed

ACE5950 CAS #: 506-96-7

ACETYL BROMIDE

RTECS: AO5955000
DOT: UN1716; IMO8.2
EINECS Number: 208-061-7
Molecular Formula: C_2H_3BrO
Structured MF: CH_3COBr
Formula Weight: 122.96

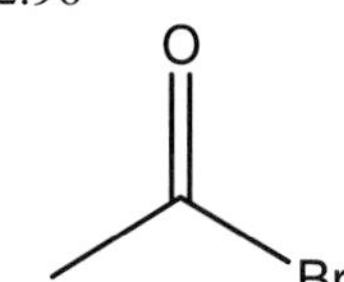

Chemical Structure

Synonyms: ACETIC ACID,BROMIDE; ETHANOYL BROMIDE
Description: colorless, yellow upon exposure to air liquid; sharp, pungent odor
Use: chemical intermediate for dyes; acetylating agent in organic synthesis

Physical Properties

Boiling Point: 76 °C (169 °F)
Freezing Point: -96 °C (-140.8 °F)
Specific Gravity: 1.52 at 9 °C
Vapor Density: 4.24 Air=1
Saturated Vapor Density: 1.821772952 kg/m^3
Vapor Pressure: 16.2 kPa at 25 °C
Water Solubility: Decomposes
Other Solubilities: Soluble in Acetone.
Odor Threshold: Chemically pure gas 5.00 x10^{-4} mg/m^3
pH: Acid
Refraction Index: 1.45376 at 16 °C/D
Ionization Potential (eV): 10.24 +/-0.1
Flash Point: Not flammable

RTECS Toxicity Data

Acute Inhalation: Mammal LC_{50} Dose: 48 gm/m^3.

Hazard Overviews

Corrosive

Fire Diamond

Health: Corrosive to eyes/skin/respiratory tract. Also Causes: ocular damage, esophageal perforation, skin blackening, thickening, bronchitis, tooth enamel erosion, pharyngitis, conjunctivitis.
Fire: Will burn. Use dry chemical, carbon dioxide, fog, or regular foam. Do not use water to extinguish fire. Reacts violently with water.
Reactivity: Unstable; reacts violently with water, alcohol. Hazardous polymerization cannot occur. Avoid: heat; water; alcohol. Incompatible with: water; alcohol; hydroxylic compounds; protic organic solvents; aprotic solvents; dimethyl formamide; dimethyl sulfoxide; ethers. Hazardous decomposition products: carbon dioxide; carbonyl bromide; bromine.
Carcinogenicity: IARC - Not listed; NIOSH - Not listed; NTP - Not listed; ACGIH - Not listed; OSHA - Not listed; EPA - Not listed; MAK - Not listed
Primary Target Organs:

Eyes Skin Respiratory System Mucous Membranes Teeth

Environmental

Cleanup/Disposal: Guide No. 156: Eliminate all ignition sources (no smoking, flares, sparks or flames in immediate area). All equipment used when handling the product must be grounded. Do not touch damaged containers or spilled material unless wearing appropriate protective clothing. Stop leak if you can do it without risk. A vapor suppressing foam may be used to reduce vapors. For chlorosilanes, use AFFF alcohol-resistant medium expansion foam to reduce vapors. Do not get water on spilled substance or inside containers. Use water spray to reduce vapors or divert vapor cloud drift. Prevent entry into waterways, sewers, basements or confined areas. Small Spills: Cover with dry earth, dry sand, or other non-combustible material followed with plastic sheet to minimize spreading or contact with rain. Use clean non-sparking tools to collect material and place it into loosely covered plastic containers for later disposal.

Environmental Physical Data

BCF: no food chain concentration potential

Regulations

RCRA 40CFR: Not listed
CERCLA: 40CFR 302.4: Listed per CWA Section 311(b)(4) RQ: 5000 lb (2268 kg)
SARA 40CFR 372.65: Not listed
SARA EHS 40CFR 355: Not listed
TSCA: Listed

ACE6100 **CAS #: 75-36-5**

ACETYL CHLORIDE

RTECS: AO6390000
DOT: UN1717; IMO3.2
EINECS Number: 200-865-6
Molecular Formula: C_2H_3ClO
Structured MF: CH_3COCl
Formula Weight: 78.50

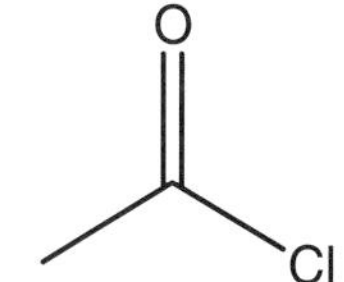

Chemical Structure

Synonyms: ACETIC ACID,CHLORIDE; ACETIC CHLORIDE; ETHANOYL CHLORIDE
Description: colorless to pale yellow liquid; pungent, unpleasant odor
Use: acetylating agent; in testing for cholesterol; determination of water in organic liquids; dye stuffs; pharmaceuticals

Physical Properties

Boiling Point: 52 °C (126 °F)
Freezing Point: -112 °C (-169.6 °F)
Specific Gravity: 1.1051 at 20 °C/4 °C
Vapor Density: 2.7 Air=1
Density: 1.104 g/mL
Vapor Pressure: 135 mm Hg at 7.5 °C
Water Solubility: Decomposes
Other Solubilities: 95% Ethanol: Reaction; Acetone: >=100 mg/ml at 21 °C; Benzene: Soluble; Chloroform: Soluble; DMSO: Reaction; Ether: Soluble; Glacial Acetic Acid: Soluble; Petroleum Ether: Soluble.
Surface Tension: 26 dynes/cm at 20 °C
Odor Threshold: Acetic acid (breakdown product) 1 ppm
pH: Acid
Refraction Index: 1.3898 at 20 °C/D
Evaporation Rate: > 1 Butyl Acetate=1
Critical Temperature: 246 °C
Critical Pressure: 845 psia
Ionization Potential (eV): 10.78
Flash Point: 4.4 °C Closed Cup
Autoignition Temperature: 390 °C
LEL: 7.3% v/v
UEL: 19% v/v

RTECS Toxicity Data

Acute Oral: Rat LD_{50} Dose: 910 mg/kg; Toxic Effects: Peripheral nerve and sensation - Spastic parapysis with or without sensory change; Behavioral - Excitment; Lungs, Thorax, or Respiration - Other changes.
Acute Inhalation: Human TC_{Lo} Dose: 2 ppm/1M; Toxic Effects: Sense organs and special senses - Other; Lungs, Thorax, or Respiration - Other changes.

Hazard Overviews

Corrosive

Flammable

Fire Diamond

Health: Corrosive to eyes/skin/respiratory tract. Also Causes: coughing, difficult breathing, chest pain, headache, dizziness, choking, death.
Fire: Flammable. Use carbon dioxide or dry chemical to put out acetyl chloride fires. Never use water or foams since they can react violently with the acetyl chloride. Be especially careful to keep water or foams from acetyl chloride containers since the container can then violently rupture.
Reactivity: Stable. Hazardous polymerization cannot occur. Avoid: exposure to heat and ignition sources (lighted cigarettes or pipes; open flames; uninsulated heating elements). Incompatible with: dimethyl sulfoxide; alcohols; amines; ethanol; alkaline materials; water. Hazardous decomposition products: toxic hydrogen chloride and phosgene gases.
Carcinogenicity: IARC - Not listed; NIOSH - Not listed; NTP - Not listed; ACGIH - Not listed; OSHA - Not listed; EPA - Class D, Not classifiable as to human carcinogenicity; MAK - Not listed
Primary Target Organs:

Eyes

Skin

Respiratory System

Environmental

Ecotoxicity: Aquatic toxicity: 10-100 ppm (est.)
Environmental Fate: If released in the atmosphere it will react with atmospheric moisture and its half-life will depend on the humidity of the air. No half-lives for air with different water contents could be found; however, it is known to fume in moist air.
Cleanup/Disposal: Guide No. 132: Fully encapsulating, vapor protective clothing should be worn for spills and leaks with no fire. Eliminate all ignition sources (no smoking, flares, sparks or flames in immediate area). All equipment used when handling the product must be grounded. Do not touch or walk through spilled material. Stop leak if you can do it without risk. Prevent entry into waterways, sewers, basements or confined areas. A vapor suppressing foam may be used to

reduce vapors. Absorb with earth, sand or other non-combustible material and transfer to containers (except for Hydrazine). Use clean non-sparking tools to collect absorbed material. Large Spills: Dike far ahead of liquid spill for later disposal. Water spray may reduce vapor; but may not prevent ignition in closed spaces.

Environmental Physical Data

BCF: no food chain concentration potential

Regulations

RCRA 40CFR: Listed Hazardous Waste No. U006 Toxic Waste Reactive Waste Corrosive Waste
CERCLA: 40CFR 302.4: Listed per CWA Section 311(b)(4) per RCRA Section 3001 RQ: 5000 lb (2268 kg)
SARA 40CFR 372.65: Not listed
SARA EHS 40CFR 355: Not listed
TSCA: Listed

ACE6400 CAS #: 674-82-8

ACETYL KETENE

RTECS: RQ8225000
DOT: UN2521; IMO3.3
EINECS Number: 211-617-1
Molecular Formula: $C_4H_4O_2$
Formula Weight: 84.04

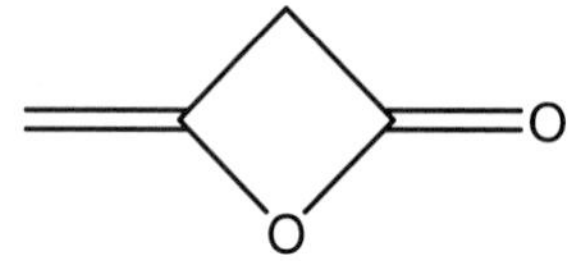

Chemical Structure

Synonyms: 3-BUTENOIC ACID,3-HYDROXY-,BETA-LACTONE; 3-BUTENO-BETA-LACTONE; DIKETENE; DIKETENE,INHIBITED; ETHENONE,DIMER; KETENE,DIMER; 4-METHYLENE-2-OXETANONE; 2-OXETANONE,4-METHYLENE-; VINYLACETO-BETA-LACTONE
Description: colorless, clear liquid when pure; pungent odor
Use: chem int for acetoacetic esters, acetoacetanilides, n,n-dialkylacetoacetamides, dyes, pharmaceuticals, for pigments & toners, food preservatives, insecticides, dicrotophos, & the fungicide dehydroacetic acid

Physical Properties

Boiling Point: 127.4 °C (261 °F)
Freezing Point: -6.5 °C (20.3 °F)
Specific Gravity: 1.0897
Vapor Density: 2.9 Air=1
Saturated Vapor Density: 1.224056697 kg/m^3
Vapor Pressure: 1.07 kPa at 20 °C
Water Solubility: Soluble in Water
Refraction Index: 1.4379
Critical Temperature: 310 °C
Critical Pressure: 5.47 mPa
Ionization Potential (eV): 9.6 +/-0.1
Flash Point: 33.889 °C Open Cup

RTECS Toxicity Data

Acute Oral: Rat LD_{50} Dose: 560 uL/kg. Mouse LD_{Lo} Dose: 800 mg/kg.
Acute Inhalation: Guinea Pig LC_{50} Dose: 3 gm/m^3/2hr. Rat LC_{Lo} Dose: 20000 ppm/1hr.
Acute Dermal: Rabbit LD_{50} Route: Skin; Dose: 2830 uL/kg.
Irritation Eye: Rabbit Standard Draize Test Dose: 25 mg; Reaction: severe. Rabbit Standard Draize Test Dose: 50 ug/24H; Reaction: severe.
Irritation Skin: Rabbit Standard Draize Test Dose: 20 mg/24H; Reaction: moderate. Rabbit Open Draize Test Dose: 500 mg open; Reaction: mild.

Hazard Overviews

Corrosive

Flammable

Fire Diamond

Health: Corrosive to eyes/skin/respiratory tract. Toxic. Other Acute Effects: toxic if inhaled; harmful if swallowed; may be harmful if absorbed through the skin; severe lachrymator; inhalation may result in spasm, inflammation and edema of the larynx and bronchi, chemical pneumonitis and pulmonary edema; symptoms of exposure may include burning sensation; coughing; wheezing; laryngitis; shortness of breath; headache; nausea; vomiting. Corrosive to eyes/skin/respiratory tract. Toxic. Other Acute Effects: toxic if inhaled; harmful if swallowed; may be harmful if absorbed through the skin; severe lachrymator; inhalation may result in spasm, inflammation and edema of the larynx and bronchi, chemical pneumonitis and pulmonary edema; symptoms of exposure may include burning sensation; coughing; wheezing; laryngitis; shortness of breath; headache; nausea; vomiting.
Fire: Flammable. Hazards: vapor may travel considerable distance to source of ignition and flash back; container explosion may occur; forms explosive mixtures in air. Extinguishing agents: carbon dioxide; water may be effective for cooling, but may not effect extinguishment; do not use dry chemical powder extinguisher on this material. Precautions: combustible liquid.
Reactivity: Stable. May undergo autopolymerization. Incompatible with: oxidizing agents, acids, bases, amines, glass may undergo autopolymerization. Hazardous decomposition products: toxic fumes of: carbon monoxide, carbon dioxide.
Carcinogenicity: IARC - Not listed; NIOSH - Not listed; NTP - Not listed; ACGIH - Not listed; OSHA - Not listed; EPA - Not listed; MAK - Not listed

Primary Target Organs:

Environmental

Environmental Fate: If released to water, volatilization will not be rapid but possibly significant with estimated half-lives of 4 hours and 4 days from a model environmental river and a model lake, respectively. Adsorption to sediment and bioconcentration are not likely fate processes. It is expected to biodegrade rapidly in both soil and aquatic conditions. If released to the atmosphere, it will exist primarily in the vapor phase. Vapor-phase will degrade in the atmosphere by reaction with photochemically produced hydroxyl radicals with an estimated half-life of approximately 7.5 hours and by reaction with atmospheric ozone with an estimated half- life of 24 hours. Removal of atmospheric compound may occur through wet deposition. If released to soil, it is expected to have very high mobility based on estimated K_{oc} values of 3.1 to 6.3. Volatilization is expected from dry soils, and may also be important from moist soils.

Cleanup/Disposal: Guide No. 131: Fully encapsulating, vapor protective clothing should be worn for spills and leaks with no fire. Eliminate all ignition sources (no smoking, flares, sparks or flames in immediate area). All equipment used when handling the product must be grounded. Do not touch or walk through spilled material. Stop leak if you can do it without risk. Prevent entry into waterways, sewers, basements or confined areas. A vapor suppressing foam may be used to reduce vapors. Small Spills: Absorb with earth, sand or other non-combustible material and transfer to containers for later disposal. Use clean non-sparking tools to collect absorbed material. Large Spills: Dike far ahead of liquid spill for later disposal. Water spray may reduce vapor; but may not prevent ignition in closed spaces.

Environmental Physical Data

Henry's Law Constant: estimated at 6.07 x10^{-4}

Sorption Partition Coefficient: K_{OC} = estimated at 6.3

BCF: estimated at 0.365

Regulations

RCRA 40CFR: Not listed

CERCLA: 40CFR 302.4: Not listed

SARA 40CFR 372.65: Not listed

SARA EHS 40CFR 355: Not listed

TSCA: Listed

ACE6700 CAS #: 77-90-7

ACETYL TRIBUTYL CITRATE

RTECS: TZ8330000

EINECS Number: 201-067-0

Molecular Formula: $C_{20}H_{34}O_8$

Formula Weight: 402.88

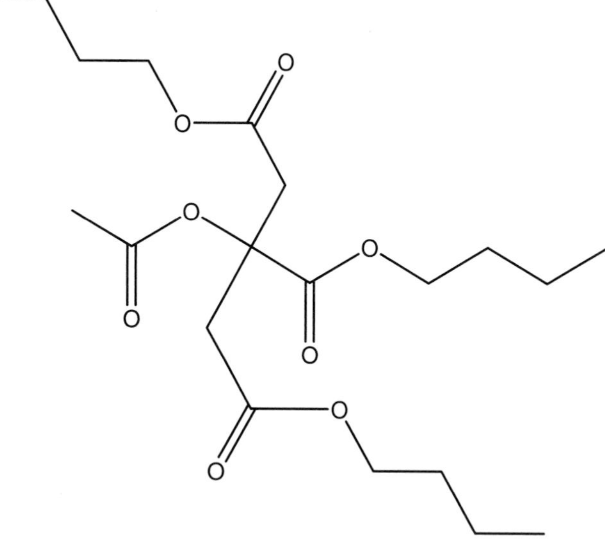

Chemical Structure

Synonyms: 2-ACETOXY-1,2,3-PROPANETRICARBOXYLIC ACID TRIBUTYL ESTER; ACETYL BUTYL CITRATE; ACETYLCITRIC ACID TRIBUTYL ESTER; ACETYLCITRIC ACID,TRIBUTYL ESTER; BLO-TROL; CITRIC ACID,TRIBUTYL ESTER,ACETATE; CITROFLEX A; CITROFLEX A 4; 1,2,3-PROPANETRICARBOXYLIC ACID,2-(ACETYLOXY)-,TRIBUTYLESTER; TRIBUTYL 2-ACETOXY-1,2,3-PROPANETRICARBOXYLATE; TRIBUTYL ACETYLCITRATE; TRIBUTYL O-ACETYLCITRATE; TRIBUTYL 2-(ACETYLOXY)-1,2,3-PROPANETRICARBOXYLATE; TRIBUTYL 2-(ACETYLOXY)-1,2,3-PROPANETRICARBOXYLIC ACID; TRIBUTYL CITRATE ACETATE

Description: colorless liquid; very faint sweet, herbaceous odor

Use: flavor ingredient; plasticizer for cellulose acetate resins; for cellulose acetate butyrate resins; for cellulose nitrate resins; for ethyl cellulose resins; for cellulose acetate propionate resins

Physical Properties

Boiling Point: 172 °C (342 °F) to 174 °C (345 °F) at 1 mm Hg

Freezing Point: -80 °C (-112 °F)

Specific Gravity: 1.048 at 25 °C/25 °C

Vapor Pressure: 1 mm Hg at 173 °C

Water Solubility: Insoluble in Water

Other Solubilities: Soluble in Alcohol, Ether

Flash Point: 204 °C

Hazard Overviews

Health: May be irritating to eyes/skin. Other Acute Effects: may be harmful by inhalation, ingestion, or skin absorption.

Fire: Will burn. Hazards: emits toxic fumes. Extinguishing agents: water spray; carbon dioxide, dry chemical powder or appropriate foam. Precautions: combustible liquid.

Reactivity: Hazardous polymerization will not occur. Incompatible with: strong oxidizing agents, protect from moisture. Hazardous decomposition products: toxic fumes of: carbon monoxide, carbon dioxide.

Carcinogenicity: IARC - Not listed; NIOSH - Not listed; NTP - Not listed; ACGIH - Not listed; OSHA - Not listed; EPA - Not listed; MAK - Not listed

Environmental

Regulations
RCRA 40CFR: Not listed
CERCLA: 40CFR 302.4: Not listed
SARA 40CFR 372.65: Not listed
SARA EHS 40CFR 355: Not listed
TSCA: Listed

Analytical Methods
Plasma: EPA 001

ACE6850 CAS #: 121-60-8

4-(ACETYLAMINO)BENZENESUL-FONYL CHLORIDE

RTECS: DB8837500
EINECS Number: 204-485-1
Molecular Formula: $C_8H_8ClNO_3S$
Formula Weight: 233.68

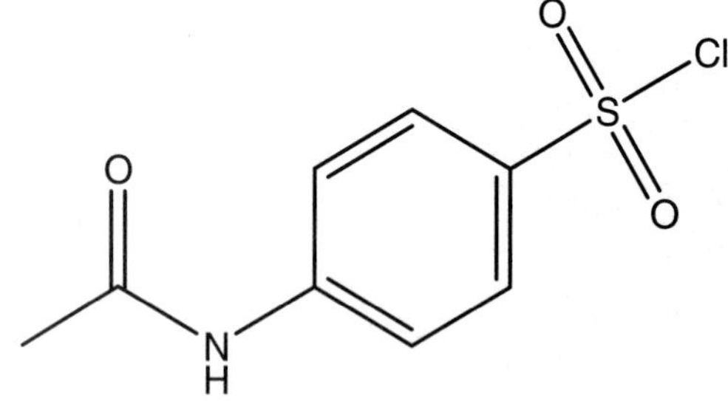

Chemical Structure

Synonyms: 4-ACETAMIDOBENZENESULFONYL CHLORIDE; P-ACETAMIDOBENZENESULFONYL CHLORIDE; 4-ACETAMIDOPHENYLSULFONYL CHLORIDE; P-ACETAMIDOPHENYLSULFONYL CHLORIDE; P-ACETAMINOBENZENESULFONYL CHLORIDE; ACETANILIDE-P-SULFONYL CHLORIDE; P-ACETYLAMINOBENZENESULFOCHLORIDE; P-ACETYLAMINOBENZENESULFONYL CHLORIDE; ACETYLSULFANILYL CHLORIDE; N(4)-ACETYLSULFANILYL CHLORIDE; N(SUP 4)-ACETYLSULFANILYL CHLORIDE; N-ACETYLSULFANILYL CHLORIDE; ASC; BENZENESULFONIC ACID,4-ACETAMIDO-,CHLORIDE; BENZENESULFONYL CHLORIDE,4-(ACETYLAMINO)-; 4'-(CHLOROSULFONYL)ACETANILIDE; 4-CHLOROSULFONYLACETANILIDE; P-(CHLOROSULFONYL)ACETANILIDE; DAGENAN CHLORIDE; SULFANILYL CHLORIDE,N-ACETYL-; SULFANILYL CHLORIDE,N-ACETYL-(6CI,7CI,8CI)
Description: light tan to brownish powder or fine crystals or prisms; slight odor of acetic acid
Use: intermediate in prepn of sulfanilamide & its derivatives; chem int for sulfonamide pharmaceuticals-eg, sulfathiazole

Physical Properties
Freezing Point: 149 °C (300.2 °F)
Water Solubility: Decomposes in Water
Other Solubilities: Very Soluble in Ether, Alcohol. Soluble in hot Chloroform. Soluble in hot Benzene

RTECS Toxicity Data
Acute Oral: Rat LD_{50} Dose: >3200 mg/kg; Toxic Effects: Behavioral - Muscle weakness; Behavioral - Ataxia. Mouse LD_{50} Dose: 1600 mg/kg; Toxic Effects: Behavioral - Muscle weakness; Behavioral - Ataxia.

Hazard Overviews

Corrosive

Health: Corrosive to eyes/skin/respiratory tract. Harmful. Other Acute Effects: harmful if swallowed, inhaled, or absorbed through skin; inhalation may result in: spasm; inflammation and edema of the larynx and bronchi; chemical pneumonitis; pulmonary edema; exposure may cause burning sensation; coughing; wheezing; laryngitis; shortness of breath; headache; nausea; vomiting.
Fire: Hazards: emits toxic fumes; water hydrolyzes material liberating acidic gas which in contact with metal surfaces can generate flammable and/or explosive hydrogen gas. Extinguishing agents: do not use water; carbon dioxide, dry chemical powder or appropriate foam. Precautions: combustible liquid.
Reactivity: Incompatible with: strong oxidizing agents, strong bases, may decompose on exposure to moist air or water. Hazardous decomposition products: carbon monoxide, carbon dioxide, sulfur oxides, hydrogen chloride gas.
Carcinogenicity: IARC - Not listed; NIOSH - Not listed; NTP - Not listed; ACGIH - Not listed; OSHA - Not listed; EPA - Not listed; MAK - Not listed
Primary Target Organs:

Eyes

Skin

Respiratory System

Environmental

Regulations
RCRA 40CFR: Not listed
CERCLA: 40CFR 302.4: Not listed
SARA 40CFR 372.65: Not listed
SARA EHS 40CFR 355: Not listed
TSCA: Listed

ACE7000 CAS #: 4075-79-0

4-ACETYLAMINOBIPHENYL

RTECS: AE6125000
Molecular Formula: $C_{14}H_{13}NO$
Formula Weight: 211.27

Synonyms: ACETAMIDE,N-(1,1'-BIPHENYL)-4-YL-; ACETAMIDE,N-(1,1'-BIPHENYL)-4-YL-(9CI); 4-ACETAMIDOBIPHENYL; ACETANILIDE,4'-PHENYL-; N-ACETYL-4-AMINOBIPHENYL; N-ACETYLXENYLAMIN; 4-BIPHENYLACETAMIDE; N-4-BIPHENYLACETAMIDE; BIPHENYL,4-ACETAMIDO; N-(1,1'-BIPHENYL)-4-YLACETAMIDE; N-(4-BIPHENYLYL)ACETAMIDE; 4'-FENYLACETANILID; 4'-PHENYLACETANILIDE; 4-PHENYLACETANILIDE; P-PHENYLACETANILIDE
Description: crystals
Use: no evidence of current use in US

Physical Properties

Freezing Point: 172 °C (341.6 °F)
Water Solubility: Insoluble in Water
Other Solubilities: Very Soluble in Alcohol, Acetone, Methanol

RTECS Toxicity Data

Mutagenic: Bacteria - S Typhimurium Mutations in Microorganisms; Dose: 10 ug/plate (-S9).
Tumorigenic: Rat Route: Oral; Dose: 2770 mg/kg/21W-C; Toxic Effects: Tumorigenic - Carcinogenic by RTECS criteria; Sense organs and special senses - Tumors; Skin and appendages - Tumors. Rat Route: Oral; Dose: 4070 mg/kg/34W-C; Toxic Effects: Tumorigenic - Carcinogenic by RTECS criteria; Sense organs and special senses - Tumors; Skin and appendages - Tumors. Rat Route: Oral; Dose: 3425 mg/kg/26W-C; Toxic Effects: Tumorigenic - Carcinogenic by RTECS criteria; Skin and appendages - Tumors. Rat Route: Oral; Dose: 2240 mg/kg/16W-C; Toxic Effects: Tumorigenic - Equivocal tumorigenic agent by RTECS criteria; Sense organs and special senses - Tumors; Skin and appendages - Tumors.

Hazard Overviews

Carcinogenicity: IARC - Not listed; NIOSH - Not listed; NTP - Not listed; ACGIH - Not listed; OSHA - Not listed; EPA - Not listed; MAK - Not listed

Environmental

Regulations

RCRA 40CFR: Not listed
CERCLA: 40CFR 302.4: Not listed
SARA 40CFR 372.65: Not listed
SARA EHS 40CFR 355: Not listed
TSCA: Listed

ACE7150 **CAS #: 53-96-3**

2-ACETYLAMINOFLUORENE

RTECS: AB9450000
EINECS Number: 200-188-6
Molecular Formula: $C_{15}H_{13}NO$
Formula Weight: 223.26

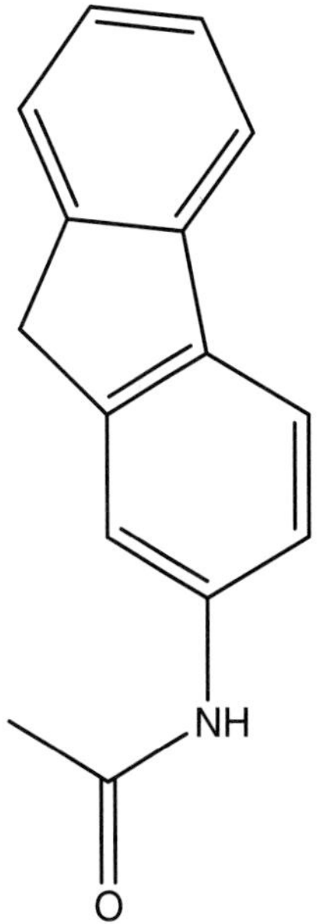

Chemical Structure

Synonyms: 2-AAF; AAF; ACETAMIDE,N-9H-FLUOREN-2-YL-; ACETAMIDE,N-FLUOREN-2-YL-; ACETAMIDE,N-9H-FLUOREN-2-YL-(9CI); 2-ACETAMIDOFLUORENE; 2-ACETAMINOFLUORENE; ACETOAMINOFLUORENE; 2-ACETYLAMINO-FLUOREN; 2-(ACETYLAMINO)FLUORENE; ACETYLAMINOFLUORENE; N-ACETYL-2-AMINOFLUORENE; AZETYLAMINOFLUOREN; 2-FAA; FAA; FLUORENE,2-ACETAMIDO-; 2-FLUORENYLACETAMIDE; N-2-FLUORENYLACETAMIDE; N-9H-FLUOREN-2-YLACETAMIDE; N-FLUOREN-2-YL-ACETAMIDE; N-FLUOREN-2-YLACETAMIDE; N-FLUORENYL-2-ACETAMIDE
Description: tan, crystalline powder
Use: in carcinogenesis studies, particularly for tumor induction as a positive control, and for the mechanism of hydroxylation by the liver; in the study of mutagenicity of aromatic amines as a positive control

Physical Properties

Freezing Point: 194 °C (381.2 °F)
Vapor Pressure: 0.00023 mm Hg at 100 °C
Water Solubility: Insoluble in Water
Other Solubilities: 95% Ethanol: 10-50 mg/ml at 20.5 °C; Acetone: 33-50 mg/ml at 20.5 °C; Acetic Acid: Soluble; Alcohols: Soluble; DMSO: >=100 mg/ml at 20.5 °C; Ether: Soluble; Fat solvents: Soluble; Glycols: Soluble; Methanol: 10-50 mg/ml at 19 °C.
Flash Point: Not available; probably combustible

RTECS Toxicity Data

Acute Oral: Mouse LD_{50} Dose: 810 mg/kg; Toxic Effects: Peripheral nerve and sensation - Spastic parapysis with or without sensory change; Behavioral - Altered sleep time (including change in righting reflex); Kidney, Ureter, and Bladder - Other changes.
Reproductive/Teratogenic: Rat Route: Oral; Dose: 300 mg/kg; Duration: female 9D of pregnancy; Specific Developmental Abnormalities - Central nervous system. Mouse Route: Intraperitoneal; Dose: 100 mg/kg; Duration:

female 10D of pregnancy; Effects on Embryo or Fetus - Fetotoxicity; Specific Developmental Abnormalities - Musculoskeletal system. Mouse Route: Intraperitoneal; Dose: 100 mg/kg; Duration: female 8D of pregnancy; Effects on Fertility - Post-implantation mortality. Mouse Route: Intraperitoneal; Dose: 625 mg/kg; Duration: male 5D prior to mating; Paternal Effects - Spermatogenesis. Mouse Route: Intraperitoneal; Dose: 2500 mg/kg; Duration: male 5D prior to mating; Paternal Effects - Testes, epididymis, sperm duct.

Mutagenic: Human DNA Damage; Cell Type: fibroblast; Dose: 1 mmol/L. Human Unscheduled DNA Synthesis; Cell Type: fibroblast; Dose: 100 umol/L/5H. Human Unscheduled DNA Synthesis; Cell Type: HeLa cell; Dose: 10 umol/L. Human Unscheduled DNA Synthesis; Cell Type: liver; Dose: 1 umol/L. Human DNA Adduct; Cell Type: HeLa cell; Dose: 5 umol/L. Human DNA Adduct; Cell Type: other cell types; Dose: 1200 nmol/L. Human Mutations in Mammalian Somatic Cells; Cell Type: lymphocyte; Dose: 25 umol/L. Human Sister Chromatid Exchenge; Cell Type: lymphocyte; Dose: 4400 ug/L.

Tumorigenic: Rat Route: Oral; Dose: 4368 mg/kg/2Y-C; Toxic Effects: Tumorigenic - Carcinogenic by RTECS criteria; Liver - Tumors; Kidney, Ureter, and Bladder - Tumors. Rat Route: Oral; Dose: 672 mg/kg/8W-C; Toxic Effects: Tumorigenic - Carcinogenic by RTECS criteria; Liver - Tumors. Rat Route: Oral; Dose: 2100 mg/kg/25W-C; Toxic Effects: Tumorigenic - Carcinogenic by RTECS criteria; Liver - Tumors. Rat Route: Oral; Dose: 1220 mg/kg/19W-C; Toxic Effects: Tumorigenic - Carcinogenic by RTECS criteria; Liver - Tumors; Skin and appendages - Tumors. Rat Route: Oral; Dose: 2100 mg/kg/30W-C; Toxic Effects: Tumorigenic - Carcinogenic by RTECS criteria; Skin and appendages - Tumors. Rat Route: Oral; Dose: 2610 mg/kg/23W-C; Toxic Effects: Tumorigenic - Carcinogenic by RTECS criteria; Liver - Tumors; Skin and appendages - Tumors. Rat Route: Oral; Dose: 1344 mg/kg/16W-C; Toxic Effects: Tumorigenic - Carcinogenic by RTECS criteria; Liver - Tumors; Liver - Hepatitis, fibrous (cirrhosis, post-necrotic scarring). Rat Route: Oral; Dose: 2940 mg/kg/35W-C; Toxic Effects: Tumorigenic - Carcinogenic by RTECS criteria; Liver - Tumors; Nutritional and gross metabolic - Weight loss or decreased weight gain.

Hazard Overviews

Health: May cause irritation. Toxic. Other Acute Effects: may be fatal if inhaled, swallowed, or absorbed through skin; absorption into the body leads to the formation of methemoglobin which in sufficient concentration causes cyanosis; onset may be delayed 2 to 4 hours or longer. Chronic Effects: may cause heritable genetic damage; may alter genetic material; target organ: liver. Carcinogen.

Fire: Will burn. Hazards: emits toxic fumes. Extinguishing agents: carbon dioxide, dry chemical powder or appropriate foam. Precautions: combustible liquid.

Reactivity: Incompatible with: acids, acid anhydrides, oxidizing agents. Hazardous decomposition products: carbon monoxide, carbon dioxide, nitrogen oxides.

Carcinogenicity: IARC - Not listed; NIOSH - Listed as carcinogen; NTP - Class 2B, Reasonably anticipated to be a carcinogen, sufficient evidence of carcinogenicity from studies in experimental animals; ACGIH - Not listed; OSHA - Listed as a carcinogen; EPA - Not listed; MAK - Not listed

Primary Target Organs:

Liver

Respirator Recommendation

Exposure Range: unlimited Self-contained Breathing Apparatus, Pressure Demand, Full Face

Note: no established TLV

Environmental

Environmental Fate: If released to soil, it is expected to have low mobility. Chemical hydrolysis, oxidation and volatilization are not expected to be significant. If released to water, it may undergo direct photolysis and is expected to strongly adsorb to suspended solids and sediments Chemical hydrolysis, oxidation, volatilization, and bioaccumulation are not expected to be significant. If released to the atmosphere it may undergo vapor phase adsorption to air borne particulate matter, it may react with photochemically generated hydroxyl radicals (estimated vapor phase half-life = 5.92 hr) or it may undergo direct photolysis.

Cleanup/Disposal: Guide No. 171: Do not touch or walk through spilled material. Stop leak if you can do it without risk. Prevent dust cloud. Avoid inhalation of asbestos dust. Small Dry Spills: With clean shovel place material into clean, dry container and cover loosely; move containers from spill area. Small Spills: Take up with sand or other noncombustible absorbent material and place into containers for later disposal. Large Spills: Dike far ahead of liquid spill for later disposal. Cover powder spill with plastic sheet or tarp to minimize spreading. Prevent entry into waterways, sewers, basements or confined areas.

Environmental Physical Data

Octanol/Water Partition Coefficient: log K_{ow} = 3.22

Sorption Partition Coefficient: K_{oc} = estimated at 1020

BCF: estimated at 165

Regulations

RCRA 40CFR: Listed Hazardous Waste No. U005 Toxic Waste

CERCLA: 40CFR 302.4: Listed per RCRA Section 3001 RQ: 1 lb (0.454 kg)

SARA 40CFR 372.65: Listed

SARA EHS 40CFR 355: Not listed

TSCA: Listed

Analytical Methods

Soil: SW846 8270B, 8270C

Plasma: EPA 29

ACE7300 CAS #: 28322-02-3

4-ACETYLAMINOFLUORENE

RTECS: AB9530000
EINECS Number: 248-964-3
Molecular Formula: $C_{15}H_{13}NO$
Formula Weight: 223.29

Chemical Structure

Synonyms: 4-AAF; ACETAMIDE,N-9H-FLUOREN-4-YL; ACETAMIDE,N-FLUOREN-4-YL; ACETAMIDE,N-9H-FLUOREN-4-YL-(9CI); 4-ACETAMIDOFLUORENE; 4-ACETYLAMINOFLUOREN; 4-FLUORENYLACETAMIDE; N-4-FLUORENYLACETAMIDE; N-9H-FLUOREN-4-YLACETAMIDE; N-FLUOREN-4-YLACETAMIDE
Description: light beige powder
Use: research chemical

Physical Properties

Freezing Point: 199 °C (390.2 °F) to 200 °C (392 °F)
Water Solubility: < 1 mg/mL at 20 C
Other Solubilities: 95% Ethanol: <1 mg/ml at 20 °C; Acetone: <1 mg/ml at 20 °C; DMSO: 10-50 mg/ml at 20 °C; Toluene: <1 mg/ml at 22 °C.
Flash Point: Not available; probably combustible

RTECS Toxicity Data

Mutagenic: Rat Unscheduled DNA Synthesis; Cell Type: liver; Dose: 100 umol/L/2H. Rat Morphological Transformation; Cell Type: embryo; Dose: 63600 ug/L.
Tumorigenic: Rat Route: Oral; Dose: 4175 mg/kg/17W-C; Toxic Effects: Tumorigenic - Equivocal tumorigenic agent by RTECS criteria; Gastrointestinal - Tumors. Rat Route: Oral; Dose: 5240 mg/kg/57W-C; Toxic Effects: Tumorigenic - Equivocal tumorigenic agent by RTECS criteria; Skin and appendages - Tumors; Tumorigenic effects - Uterine tumors. Rat Route: Oral; Dose: 6300 mg/kg/30W-C; Toxic Effects: Tumorigenic - Equivocal tumorigenic agent by RTECS criteria; Gastrointestinal - Tumors; Endocrine - Tumors.

Hazard Overviews

Fire: Will burn.

Carcinogenicity: IARC - Not listed; NIOSH - Not listed; NTP - Not listed; ACGIH - Not listed; OSHA - Not listed; EPA - Not listed; MAK - Not listed

Environmental

Regulations

RCRA 40CFR: Not listed
CERCLA: 40CFR 302.4: Not listed
SARA 40CFR 372.65: Not listed
SARA EHS 40CFR 355: Not listed
TSCA: Not listed

ACE7450 CAS #: 99-92-3

4-ACETYLANILINE

RTECS: AM5500000
EINECS Number: 202-801-2
Molecular Formula: C_8H_9NO
Formula Weight: 135.17

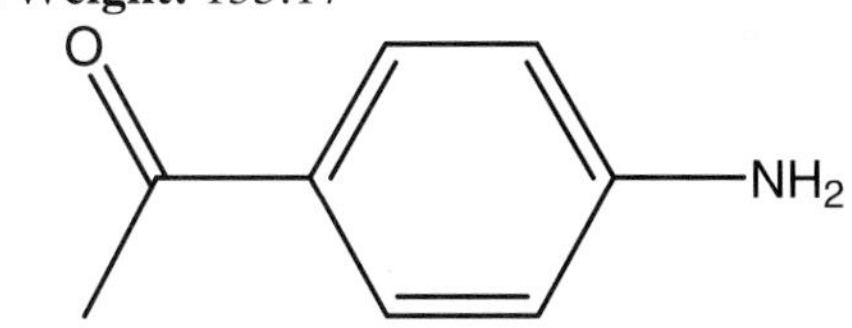

Chemical Structure

Synonyms: ACETOPHENONE,4'-AMINO-; ACETOPHENONE,4-AMINO-; ACETOPHENONE,P-AMINO-; ACETOPHENONE,4'-AMINO-(8CI); P-ACETYLANILINE; P-AMINO ACETOPHENONE; P-AMINOACETOFENONU; 4'-AMINOACETOPHENONE; P-AMINOACETOPHENONE; P-AMINOACETYLBENZENE; 1-(4-AMINOPHENYL)ETHANONE; ETHANONE,1-(4-AMINOPHENYL)-; ETHANONE,1-(4-AMINOPHENYL)-(9CI)
Description: yellow monoclinic prisms or needles; pleasant, characteristic odor
Use: chemical intermediate for flavaniline (basic dye), acetohexamide, and an oral hypoglycemic drug

Physical Properties

Boiling Point: 293 °C (559 °F) to 295 °C (563 °F)
Freezing Point: 106 °C (222.8 °F)
Saturated Vapor Density: 1.200000461 kg/m^3
Vapor Pressure: 7.98 $x10^{-5}$ mm Hg at 25 °C
Water Solubility: Slightly Soluble in Water
Other Solubilities: Soluble in Hydrochloric Acid; Soluble in Alcohol and Ether.
Ionization Potential (eV): 7.8 +/-0.1

RTECS Toxicity Data

Acute Oral: Rat LD_{50} Dose: 381 mg/kg. Mouse LD_{50} Dose: 596 mg/kg.

Hazard Overviews

Health: Irritating to eyes/skin/respiratory tract. Toxic. Other Acute Effects: harmful if swallowed, inhaled, or absorbed through skin; absorption into the body leads to the formation of methemoglobin which in sufficient concentration causes cyanosis.

Fire: Hazards: emits toxic fumes. Extinguishing agents: water spray; carbon dioxide, dry chemical powder or appropriate foam. Precautions: combustible liquid.

Carcinogenicity: IARC - Not listed; NIOSH - Not listed; NTP - Not listed; ACGIH - Not listed; OSHA - Not listed; EPA - Not listed; MAK - Not listed

Primary Target Organs:

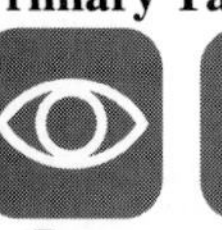
Eyes

Skin

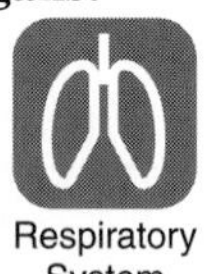
Respiratory System

Environmental

Environmental Fate: If released to the atmosphere, it is expected to degrade rapidly (estimated half-life of 3.6 hr) by reaction with photochemically produced hydroxyl radicals. If released to soil, it may undergo a covalent chemical bonding with humic materials which can result in its chemical alteration to a latent form and prevent leaching. In the absence of covalent bonding, it is expected to be highly mobile in soil. If released to water, covalent bounding with humic materials in the water column and sediments may result in partitioning from the water column to sediments. By analogy to similar aromatic amines, in the water column it may be susceptible to photooxidation (via hydroxyl and peroxy radicals). Since it absorbs UV light in the environmental range, direct photolysis in the environment is possible. Insufficient data are available to assess the relative importance of biodegradation in soil or water

Environmental Physical Data

Henry's Law Constant: estimated at 3.47×10^{-9}

Octanol/Water Partition Coefficient: log K_{ow} = 0.83

Sorption Partition Coefficient: K_{oc} = estimated at 4 to 67

BCF: estimated at 2.5

Regulations

RCRA 40CFR: Not listed

CERCLA: 40CFR 302.4: Not listed

SARA 40CFR 372.65: Not listed

SARA EHS 40CFR 355: Not listed

TSCA: Listed

ACE7600 CAS #: 616-91-1

ACETYLCYSTEINE

RTECS: HA1660000

EINECS Number: 210-498-3

Molecular Formula: $C_5H_9NO_3S$

Formula Weight: 163.20

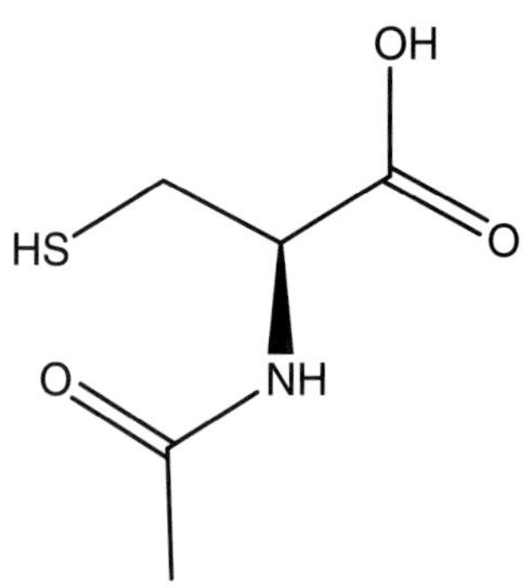

Chemical Structure

Synonyms: L-ALPHA-ACETAMIDO-BETA-MERCAPTOPROPIONIC ACID; ACETEIN; N-ACETYL-L-CYSTEINE; N-ACETYL-N-CYSTEINE; N-ACETYLCYSTEINE; N-ACETYL-3-MERCAPTOALANINE; AIRBRON; BRONCHOLYSIN; CYSTEINE,N-ACETYL-,L-; L-CYSTEINE,N-ACETYL-; L-CYSTEINE,N-ACETYL-(9CI); FLUIMICIL INFANTIL; FLUIMUCETIN; FLUIMUCIL; INSPIR; MERCAPTURIC ACID; MERCAPTURIC ACID,-; MUCOFILIN; MUCOLYTICUM LAPPE; MUCOLYTICUM-LAPPE; MUCOLYTIKUM LAPPE; MUCOMYST; MUCOSOLVIN; NAC; NSC 111180; PARVOLEX; RESPAIRE

Description: white, crystalline powder; slight acetic odor

Use: mucolytic agent (adjuvant) for bronchopulmonary disorders; medicine, biochemical research; mucolytic; corneal vilnerary; antidote to acetaminophen poisoning; secretolytic agent

Physical Properties

Freezing Point: 109.5 °C (229.1 °F)

Water Solubility: 1 g in 5 ml Water, 4 ml alcohol

Other Solubilities: Soluble in Alcohol, hot isopropyl Alcohol, methyl Acetate, and Ethyl Acetate.

pH: 1 in 100 ml 2 to 2.8

RTECS Toxicity Data

Acute Oral: Rat LD_{50} Dose: 5050 mg/kg. Mouse LD_{50} Dose: 4400 mg/kg.

Hazard Overviews

Health: Acute Effects: may be harmful by inhalation, ingestion, or skin absorption; exposure can cause bronchospasm; nausea; vomiting; stomatitis; nasal discharge.

Fire: Hazards: emits toxic fumes. Extinguishing agents: water spray; carbon dioxide, dry chemical powder or appropriate foam. Precautions: combustible liquid.

Reactivity: Stable. Hazardous polymerization will not occur. Incompatible with: heavy metals, heavy metal salts, rubber, oxygen, oxidizing agents. Hazardous decomposition products: toxic fumes of: carbon monoxide, carbon dioxide, nitrogen oxides, sulfur oxides.

Carcinogenicity: IARC - Not listed; NIOSH - Not listed; NTP - Not listed; ACGIH - Not listed; OSHA - Not listed; EPA - Not listed; MAK - Not listed

Environmental

Regulations

RCRA 40CFR: Not listed
CERCLA: 40CFR 302.4: Not listed
SARA 40CFR 372.65: Not listed
SARA EHS 40CFR 355: Not listed
TSCA: Listed

ACE7750 CAS #: 74-86-2

ACETYLENE

RTECS: AO9600000
DOT: UN1001; IMO2.1
EINECS Number: 200-816-9
Molecular Formula: C_2H_2
Structured MF: HC= =CH
Formula Weight: 26.02

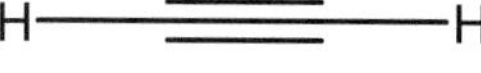

Chemical Structure

Synonyms: ACETYLEN; ACETYLENE (LIQUEFIED); ETHINE; ETHYNE; NARCYLEN; WELDING GAS
Description: colorless gas; slight, pleasant ethereal odor
Use: illuminant; oxyacetylene welding, cutting & soldering metals; signalling; precipitating metals; fuel for motor boats; mfr of acetic acid, vinyl chloride, acrylates, acrylonitrile, acetaldehyde, perchloroethylene, trichloroethylene, carbon black, etc; intermed for acrylic acid, tetrahydrofuran, chlorinated solvents, etc; for brazing, metallizing, hardening, flame scarfing, & local heating in metallurgy; flame in glass ind; acetylene in mfr of synthetic rubber, butyrolactone, vinyl alkyl ether, pyrrolidone.

Physical Properties

Boiling Point: -84 °C (-119 °F) at 760 mm Hg
Freezing Point: -80.8 °C (-113.44 °F)
Specific Gravity: 0.6208 at -82 °C/4 °C
Vapor Density: 0.91 Air=1 at 1 Atm
Vapor Pressure: 40 atm at 16.8 °C
Water Solubility: 2% by weight
Other Solubilities: Soluble in many organic materials; Soluble in Benzene, Chloroform; Soluble in Ether; Acetone dissolves 25 vol at 15 °C, 760 mm Hg; Acetone dissolves 300 vol at 12 atm; Slightly Soluble in Carbon Disulfide.
Odor Threshold: Detection 240 mg/m^3
Refraction Index: 1.00051 at 0 °C/D
Critical Temperature: 35.2 °C
Critical Pressure: 6190 kPa
Ionization Potential (eV): 11.40
Flash Point: -17.778 °C Closed Cup
Autoignition Temperature: 305 °C
LEL: 2.5% v/v
UEL: 100% v/v

RTECS Toxicity Data

Acute Inhalation: Human LC_{Lo} Dose: 50 pph/5M. Human TC_{Lo} Dose: 20 pph; Toxic Effects: Behavioral - Headache; Lungs, Thorax, or Respiration - Dyspnea.

Hazard Overviews

Explosive

Flammable

Compressed Gas

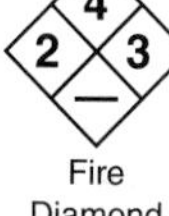

Fire Diamond

Health: Stored as a compressed gas which can cause frostbite. A simple asphyxiant which can displace available oxygen needed for breathing.
Fire: Highly flammable. Forms explosive mixtures in air. May cause flash fire. Stop gas flow before extinguishment. Use water spray to extinguish and cool fire-exposed containers. If leak cannot be stopped, allow to burn and vacate premises.
Reactivity: Unstable, under pressure, in excess of 15 psig, can decompose violently to form carbon and hydrogen if heated or shocked (no oxygen required). Hazardous polymerization can occur under pressure. Avoid: ignition sources (open flame; lighted tobacco; electrical or mechanical sparks; uninsulated heating elements); accidental or uncontrollably rapid release of acetylene gas from high-pressure cylinders, tank cars, or pipelines. Incompatible with: silver; mercury; copper; chlorine gas; oxygen; chlorine; fluorine; brass; bromine; cesium hydride; cobalt; copper salts; iodine; nitric acid; potassium; rubidium hydride. Hazardous decomposition products: carbon dioxide; carbon monoxide.
Carcinogenicity: IARC - Not listed; NIOSH - Listed as carcinogen; NTP - Not listed; ACGIH - Not listed; OSHA - Not listed; EPA - Not listed; MAK - Not listed
Primary Target Organs:

Eyes

Skin

Nervous System

Respirator Recommendation
Exposure Range: >2500 to 125000 ppm Supplied Air, Constant Flow/Pressure Demand, Half Mask
Exposure Range: >125,000 to 790,000 ppm Supplied Air, Constant Flow/Pressure Demand, Full Face
Exposure Range: 790,000 to unlimited ppm Self-contained Breathing Apparatus, Pressure Demand, Full Face
Note: simple asphyxiant

Environmental

Ecotoxicity: TLm River trout 200 mg/l/33 hr /Conditions of bioassay not specified
Cleanup/Disposal: Guide No. 116: Eliminate all ignition sources (no smoking, flares, sparks or flames in immediate area). All equipment used when handling the product must be grounded. Stop leak if you can do it without risk. Do not touch or walk through spilled material. Do not direct water at spill or source of leak. Use water spray to reduce vapors or divert vapor cloud drift. If possible, turn leaking containers so

that gas escapes rather than liquid. Prevent entry into waterways, sewers, basements or confined areas. Isolate area until gas has dispersed.

Environmental Physical Data

BCF: not pertinent
BOD: not pertinent

Regulations

RCRA 40CFR: Not listed
CERCLA: 40CFR 302.4: Not listed
SARA 40CFR 372.65: Not listed
SARA EHS 40CFR 355: Not listed
TSCA: Listed

Analytical Methods
Air: ASTM D2820, D4490

ACE7900 CAS #: 142-26-7

ACETYLETHANOLAMINE

RTECS: AC3120000
EINECS Number: 205-530-8
Molecular Formula: $C_4H_9NO_2$
Formula Weight: 103.12

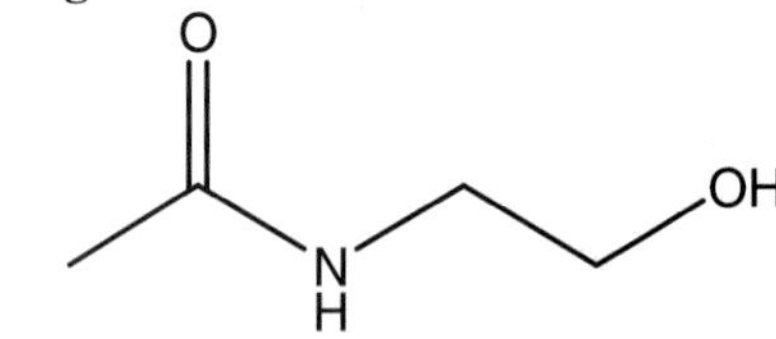

Chemical Structure

Synonyms: 2-ACETAMIDOETHANOL; ACETIC ACID,AMIDE,N(2-HYDROXYETHYL)-; N-ACETYL ETHANOLAMINE; 2-ACETYLAMINOETHANOL; ACETYLCOLAMINE; N-ACETYLETHANOLAMINE; N-ETHANOLACETAMIDE; HYDROXYETHYL ACETAMIDE; BETA-HYDROXYETHYLACETAMIDE; N-(2-HYDROXYETHYL)ACETAMIDE; N-(BETA-HYDROXYETHYL)ACETAMIDE; N-BETA-HYDROXYETHYLACETAMIDE
Description: needles; brown, viscous liquid

Physical Properties

Boiling Point: 166 °C (331 °F) to 167 °C (333 °F) at 8 mm Hg
Freezing Point: 63 °C (145.4 °F) to 65 °C (149 °F)
Specific Gravity: 1.1079 at 25 °C/4 °C
Water Solubility: Soluble in all Proportions
Other Solubilities: Soluble in hot Acetone. Slightly Soluble in Benzene, Petroleum Ether
Refraction Index: 1.4674 at 20 °C/D
Flash Point: 176.667 °C Open Cup
Autoignition Temperature: 460 °C

RTECS Toxicity Data

Acute Oral: Rat LD_{50} Dose: 26950 mg/kg.
Acute Dermal: Rabbit LD_{50} Route: Skin; Dose: >20 mL/kg.
Irritation Eye: Rabbit Standard Draize Test Dose: 500 mg; Reaction: severe.
Irritation Skin: Rabbit Open Draize Test Dose: 500 mg open; Reaction: mild.

Hazard Overviews

Health: Severely irritating to eyes/skin/respiratory tract. Other Acute Effects: harmful if swallowed, inhaled, or absorbed through skin; high concentrations are extremely destructive to tissues of the mucous membranes and upper respiratory tract, eyes and skin; symptoms of exposure may include burning sensation; coughing; wheezing; laryngitis; shortness of breath; headache; nausea and vomiting.
Fire: Will burn. Hazards: emits toxic fumes. Extinguishing agents: water spray; carbon dioxide, dry chemical powder or appropriate foam. Precautions: combustible liquid.
Reactivity: Incompatible with: strong oxidizing agents. Hazardous decomposition products: toxic fumes of: carbon monoxide, carbon dioxide, nitrogen oxides.
Carcinogenicity: IARC - Not listed; NIOSH - Not listed; NTP - Not listed; ACGIH - Not listed; OSHA - Not listed; EPA - Not listed; MAK - Not listed
Primary Target Organs:

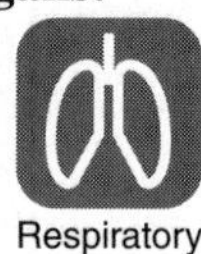

Eyes Skin Respiratory System

Environmental

Regulations

RCRA 40CFR: Not listed
CERCLA: 40CFR 302.4: Not listed
SARA 40CFR 372.65: Not listed
SARA EHS 40CFR 355: Not listed
TSCA: Listed

ACE8050 CAS #: 1001-55-4

(ACETYLOXY)ACETONITRILE

RTECS: MC7700000
Molecular Formula: $C_4H_5NO_2$
Formula Weight: 99.10
Synonyms: ACETIC ACID CYANOMETHYL ESTER; ACETONITRILE,(ACETYLOXY)-; 2-ACETOXYACETONITRILE; ACETOXYACETONITRILE; ACETYLOXYACETONITRILE; CYANOMETHYL ACETATE; GLYCOLONITRILE,ACETATE; GLYCOLONITRILE,ACETATE (ESTER); KYANMETHYLESTER KYSELINY OCTOVE
Use: mfr intermediates in prodn of pharmaceuticals; component of synthetic resins

RTECS Toxicity Data

Acute Oral: Rat LD_{50} Dose: 32 mg/kg.
Acute Inhalation: Rat LC_{Lo} Dose: 16 ppm/4hr.
Acute Dermal: Rabbit LD_{50} Route: Skin; Dose: 43 mg/kg.

Irritation Eye: Rabbit Standard Draize Test Dose: 20 mg open; Reaction: severe.

Hazard Overviews

Carcinogenicity: IARC - Not listed; NIOSH - Not listed; NTP - Not listed; ACGIH - Not listed; OSHA - Not listed; EPA - Not listed; MAK - Not listed

Environmental

Regulations

RCRA 40CFR: Not listed
CERCLA: 40CFR 302.4: Not listed
SARA 40CFR 372.65: Not listed
SARA EHS 40CFR 355: Not listed
TSCA: Not listed

ACE8200 CAS #: 114-83-0

1-ACETYL-2-PHENYLHYDRAZINE

RTECS: AJ2900000
EINECS Number: 204-055-3
Molecular Formula: $C_8H_{10}N_2O$
Formula Weight: 150.20

Chemical Structure

Synonyms: ACETIC ACID PHENYLHDRAZONE; ACETIC ACID PHENYLHYDRAZONE; ACETYLPHENYLHYDRAZINE; BETA-ACETYLPHENYLHYDRAZINE; N-ACETYL-N'-PHENYLHYDRAZINE; APH; FENYLHYDRAZID KYSELINY OCTOVE; HYDRACETIN; N'-PHENYLACETHYDRAZIDE; 2-PHENYLHYDRAZIDE ACETIC ACID; PYRODIN; PYRODINE
Description: colorless to white prisms or solid
Use: antipyretic

Physical Properties

Freezing Point: 129 °C (264.2 °F) to 132 °C (269.6 °F)
Water Solubility: < 1 mg/mL at 19 C
Other Solubilities: 95% Ethanol: 10-50 mg/ml at 23 °C; Acetone: 50-100 mg/ml at 23 °C; Benzene: Soluble; Chloroform: Soluble; DMSO: >=100 mg/ml at 23 °C; Ether: Slightly Soluble.
Flash Point: Not available; probably combustible

RTECS Toxicity Data

Acute Oral: Mouse LD_{50} Dose: 270 mg/kg.
Chronic (Multiple Dose) Oral: Dog Dose: 112 mg/kg/4W-I; Toxic Effects: Blood - Methemoglobinemia-Carboxyhemoglobin; Blood - Changes in erythrocite (RBC) cell count; Biochemical - Dehydrogenases.
Mutagenic: Bacteria - S Typhimurium Mutations in Microorganisms; Dose: 333 ug/plate (+S9), 3333 ug/plate (-S9).
Tumorigenic: Mouse Route: Oral; Dose: 31 gm/kg/79W-I; Toxic Effects: Tumorigenic - Neoplastic by RTECS criteria; Vascular - Tumors; Lungs, Thorax, or Respiration - Tumors.

Hazard Overviews

Health: Irritating to eyes/skin/respiratory tract. Toxic. Other Acute Effects: harmful if swallowed, inhaled, or absorbed through skin; may cause allergic skin reaction; ; may cause sensitization by skin contact; possible risk of irreversible effects. Chronic Effects: damage to the liver and blood. Possible carcinogen.
Fire: Will burn. Hazards: emits toxic fumes. Extinguishing agents: water spray; carbon dioxide, dry chemical powder or appropriate foam. Precautions: combustible liquid.
Reactivity: Incompatible with: strong oxidizing agents, strong acids, strong bases. Hazardous decomposition products: toxic fumes of: carbon monoxide, carbon dioxide, nitrogen oxides.
Carcinogenicity: IARC - Not listed; NIOSH - Not listed; NTP - Not listed; ACGIH - Not listed; OSHA - Not listed; EPA - Not listed; MAK - Not listed
Primary Target Organs:

Eyes

Skin

Respiratory System

Environmental

Regulations

RCRA 40CFR: Not listed
CERCLA: 40CFR 302.4: Not listed
SARA 40CFR 372.65: Not listed
SARA EHS 40CFR 355: Not listed
TSCA: Listed

ACE8350 CAS #: 30560-19-1

ACETYLPHOSPHORAMIDOTHIOIC ACID O,S-DIMETHYL ESTER

DOT: UN2783; UN2784; UN3017; UN3018; IMO3.2; IMO6.1
EINECS Number: 250-241-2
Molecular Formula: $C_4H_{10}NO_3PS$
Formula Weight: 183.16
Synonyms: CHEVRON ORTHENE; O,S-DIMETHYL ACETYL PHOSPHORAMIDOTHIOATE; ORTHENE; ORTHO 124120; ORTRAN
Description: colorless to white solid or crystals
Use: contact & systemic insecticide; effective against alfalfa looper, aphids, armyworms, webworms, & whitefly; for peppers, brussel sprouts, cotton, cranberries; cock roach

control in buildings; insect control in forests, tobacco, & on ornamentals; golf course spray.

Physical Properties

Freezing Point: 82 °C (179.6 °F) to 89 °C (192.2 °F)
Specific Gravity: 1.35
Saturated Vapor Density: 1.200000014 kg/m^3
Vapor Pressure: 1.7 x10^{-6} mm Hg at 25 °C
Water Solubility: About 650 g/L at 0 °C
Other Solubilities: moderately Soluble in Acetone, Alcohol; low solubility in aromatic solvents.

Hazard Overviews

Carcinogenicity: IARC - Not listed; NIOSH - Not listed; NTP - Not listed; ACGIH - Not listed; OSHA - Not listed; EPA - Class C, Possible human carcinogen; MAK - Not listed

Environmental

Ecotoxicity: LC_{50} Coturnix 3,275 ppm/5 days (95% confidence interval: 2691-3986 ppm) LC_{50} Gammarus pseudolimnaeus > 50 ug/l/96 hr at 12 °C, mature. Static bioassay without aeration, pH 7.2-7.5, water hardness 40-50 mg/l as calcium carbonate and alkalinity of 30-35 mg/l
Environmental Fate: If released to soil or water, it will degrade by microbial degradation and aqueous hydrolysis. The rate of hydrolysis increases with increasing pH; therefore, degradation can occur more rapidly in alkaline media than in acidic media. Estimated K_{oc} values of 2-8 suggest that it will leach readily in soil. A review of available literature has determined that the average soil half-life is about 3 days. If released to the atmosphere, it can exist in both the vapor and particulate-phases; vapor-phase degrades rapidly by reaction with photochemically produced hydroxyl radicals (estimated half-life of 6 hours). Physical removal from the atmosphere may occur through wet and dry deposition.
Cleanup/Disposal: Guide No. 131: Fully encapsulating, vapor protective clothing should be worn for spills and leaks with no fire. Eliminate all ignition sources (no smoking, flares, sparks or flames in immediate area). All equipment used when handling the product must be grounded. Do not touch or walk through spilled material. Stop leak if you can do it without risk. Prevent entry into waterways, sewers, basements or confined areas. A vapor suppressing foam may be used to reduce vapors. Small Spills: Absorb with earth, sand or other non-combustible material and transfer to containers for later disposal. Use clean non-sparking tools to collect absorbed material. Large Spills: Dike far ahead of liquid spill for later disposal. Water spray may reduce vapor; but may not prevent ignition in closed spaces. Guide No. 152: Do not touch damaged containers or spilled material unless wearing appropriate protective clothing. Stop leak if you can do it without risk. Prevent entry into waterways, sewers, basements or confined areas. Cover with plastic sheet to prevent spreading. Absorb or cover with dry earth, sand or other non-combustible material and transfer to containers. Do not get water inside containers.

Environmental Physical Data

Henry's Law Constant: estimated at 5.0 x10^{-13}
Octanol/Water Partition Coefficient: log K_{ow} = -0.85
Sorption Partition Coefficient: K_{oc} = estimated at 2 to 8
BCF: estimated at 0.1 to 0.03

Regulations

RCRA 40CFR: Not listed
CERCLA: 40CFR 302.4: Not listed
SARA 40CFR 372.65: Listed
SARA EHS 40CFR 355: Not listed
TSCA: Not listed

Analytical Methods

Soil: EPA PMD-TLC
Water / Groundwater: EPA 1657
Food: FDA 232.3, 232.4, 242.1; AOAC 985.22
Other: EPA 1656

ACE8500 CAS #: 350-03-8

3-ACETYLPYRIDINE

RTECS: OB5425000
EINECS Number: 206-496-7
Molecular Formula: C_7H_7NO
Formula Weight: 121.15

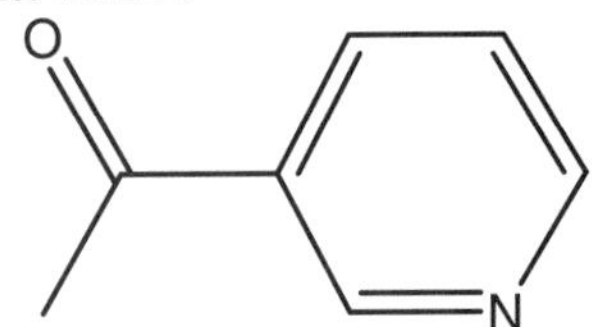

Chemical Structure

Synonyms: 3-ACETOPYRIDINE; BETA-ACETYLPYRIDINE; METHYL 3-PYRIDYL KETONE; METHYL BETA-PYRIDYL KETONE; METHYL PYRIDYL KETONE; 1-(3-PYRIDENYL)ETHANONE; PYRIDINE,3-ACETYL-

Physical Properties

Boiling Point: 220 °C (428 °F)
Freezing Point: 14 °C (57.2 °F)
Specific Gravity: 1.102
Water Solubility: Soluble
Other Solubilities: Ethanol: Soluble; Ether: Soluble; acid: Soluble
Refraction Index: 1.5341

RTECS Toxicity Data

Acute Oral: Rat LD_{50} Dose: 46 uL/kg; Toxic Effects: Behavioral - Somnolence (general depressed activity); Behavioral - Convulsions or effect on seizure threshold; Musculoskelital - Other changes. Quail LD_{50} Dose: 422 mg/kg.
Mutagenic: Yeast - S Cerevisiae Sex Chromosome Loss; Dose: 5000 ppm.

Hazard Overviews

Poison

Health: Irritating to eyes/skin/respiratory tract. Poison. Other Acute Effects: may be fatal if inhaled, swallowed, or absorbed through skin. Chronic Effects: overexposure may cause reproductive disorder(s) based on tests with laboratory animals; possible risk of irreversible effects; possible teratogen; target organs: central nervous system.
Fire: Hazards: emits toxic fumes. Extinguishing agents: water spray; carbon dioxide, dry chemical powder or appropriate foam. Precautions: combustible liquid.
Reactivity: Incompatible with: strong oxidizing agents, strong acids, strong reducing agents. Hazardous decomposition products: toxic fumes of: carbon monoxide, carbon dioxide, nitrogen oxides.
Carcinogenicity: IARC - Not listed; NIOSH - Not listed; NTP - Not listed; ACGIH - Not listed; OSHA - Not listed; EPA - Not listed; MAK - Not listed
Primary Target Organs:

Eyes

Skin

Respiratory System

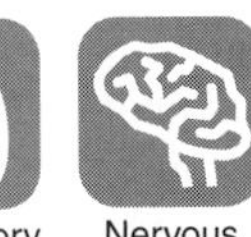
Nervous System

Environmental

Regulations

RCRA 40CFR: Not listed
CERCLA: 40CFR 302.4: Not listed
SARA 40CFR 372.65: Not listed
SARA EHS 40CFR 355: Not listed
TSCA: Listed

ACE8650 CAS #: 1122-54-9

4-ACETYLPYRIDINE

RTECS: OB5426000
EINECS Number: 214-350-9
Molecular Formula: C_7H_7NO
Structured MF: $CH_3COC_5H_4N$
Formula Weight: 121.15

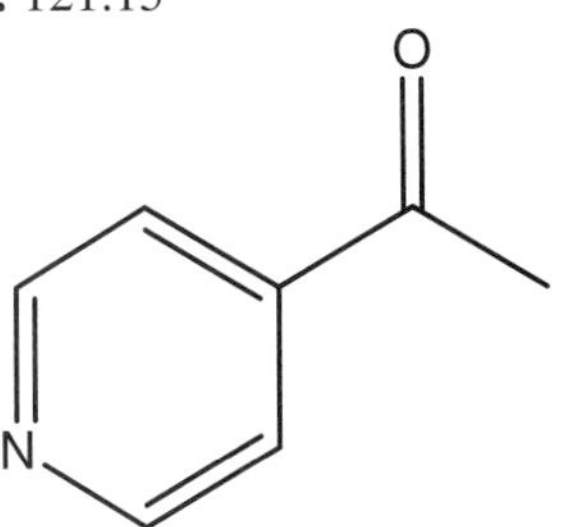

Chemical Structure

Synonyms: GAMMA-ACETYLPYRIDINE; ETHANONE,1-(4-PYRIDINYL)-; KETONE,METHYL 4-PYRIDYL; METHYL 4-PYRIDYL KETONE; PYRIDINE,4-ACETYL-; 1-(4-PYRIDINYL)ETHANONE; 4-PYRIDYL METHYL KETONE
Description: dark amber liquid
Use: as an acetylcholine antagonist and an organoleptic

Physical Properties

Boiling Point: 212 °C (414 °F)
Freezing Point: 16 °C (60.8 °F)
Saturated Vapor Density: 1.224584483 kg/m^3
Density: 1.111 g/cu cm at 23.4 °C
Vapor Pressure: 4.9 mm Hg at 25.0 °C
Water Solubility: >= 100 mg/mL at 19 C
Other Solubilities: 95% Ethanol: >=100 mg/ml at 19 °C; Acetone: >=100 mg/ml at 19 °C; DMSO: >=100 mg/ml at 19 °C.
Refraction Index: 1.535
Ionization Potential (eV): 9.3
Flash Point: > 93.3 °C

RTECS Toxicity Data

Mutagenic: Yeast - S Cerevisiae Gene Conversion; Dose: 9900 ppm. Yeast - S Cerevisiae Sex Chromosome Loss; Dose: 6200 ppm.

Hazard Overviews

Health: Irritating to eyes/skin/respiratory tract. Harmful. Other Acute Effects: harmful if swallowed, inhaled, or absorbed through skin.
Fire: Will burn. Extinguishing agents: water spray; carbon dioxide, dry chemical powder or appropriate foam. Precautions: combustible liquid.
Reactivity: Incompatible with: strong oxidizing agents, strong acids, strong reducing agents. Hazardous decomposition products: toxic fumes of: carbon monoxide, carbon dioxide, nitrogen oxides.
Carcinogenicity: IARC - Not listed; NIOSH - Not listed; NTP - Not listed; ACGIH - Not listed; OSHA - Not listed; EPA - Not listed; MAK - Not listed
Primary Target Organs:

Eyes

Skin

Respiratory System

Environmental

Regulations

RCRA 40CFR: Not listed
CERCLA: 40CFR 302.4: Not listed
SARA 40CFR 372.65: Not listed
SARA EHS 40CFR 355: Not listed
TSCA: Listed

ACE8800 **CAS #: 50-78-2**

ACETYLSALICYLIC ACID

RTECS: VO0700000
EINECS Number: 200-064-1
Molecular Formula: $C_9H_8O_4$
Structured MF: $CH_3COOC_6H_4COOH$
Formula Weight: 180.15

Chemical Structure

Synonyms: A.S.A; A.S.A. EMPIRIN; AC 5230; ACENTERINE; ACESAL; ACETAL; ACETICYL; ACETILSALICILICO; ACETILUM ACIDULATUM; ACETISAL; ACETOL; ACETONYL; ACETOPHEN; ACETOSAL; ACETOSALIC ACID; ACETOSALIN; 2-ACETOXYBENZOIC ACID; O-ACETOXYBENZOIC ACID; ACETYLIN; 2-(ACETYLOXY)BENZOIC ACID; ACETYLSAL; ACETYLSALICYLSAURE; ACIDE ACETYLSALICYLIQUE; ACIDO O-ACETIL-BENZOICO; ACIDO ACETILSALICILICO; ACIDUM ACETYLSALICYLICUM; ACIMETTEN; ACISAL; ACYLPYRIN; ASA; ASAGRAN; ASATARD; ASPALON; ASPERGUM; ASPIRDROPS; ASPIRIN; ASPIRINE; ASPRO; ASTERIC; BENASPIR; BIALPIRINIA; CAPRIN; O-CARBOXYPHENYL ACETATE; COLFARIT; CONTRHEUMA RETARD; CRYSTAR; DELGESIC; DOLEAN PH 8; DURAMAX; ECM; ECOTRIN; EMPIRIN; ENDYDOL; ENTERICIN; ENTEROPHEN; ENTEROSARINE; ENTROPHEN; EXTREN; GLOBOID; HELICON; IDRAGIN; KYSELINA 2-ACETOXYBENZOOVA; KYSELINA ACETYLSALICYLOVA; MEASURIN; NEURONIKA; NOVID; POLOPIRYNA; RHEUMIN TABLETTEN; RHODINE; RONAL; SALACETIN; SALCETOGEN; SALETIN; SALICYLIC ACID ACETATE; SOLPYRON; SP 189; XAXA; YASTA

Description: white crystals; odorless in dry air; hydrolyzes in moist air to produce an acetic acid odor

Use: in analgesics, anti-inflammatories, antipyretics, anticoagulants and antirheumatics; in food, animal feed, drug and cosmetic manufacturing; in the treatment of acute and chronic rheumatic states and for relief of the less severe types of pain such as headache, neuritis, myalgia and toothache; in veterinary medicine as an analgesic and an antipyretic

Physical Properties

Boiling Point: Decomposes at 140 °C (284 °F)
Freezing Point: 135 °C (275 °F)
Specific Gravity: 1.4
Saturated Vapor Density: 1.200000207 kg/m^3
Density: 1.35 g/mL
Vapor Pressure: 2.52 x10^{-5} mm Hg at 25 °C
Water Solubility: 1 g Soluble in: 300 ml Water at 25 °C
Refraction Index: 1.4842 to 1.4936 at 25 °C/D
Flash Point: 250 °C
LEL: dust 40 g/m^3

RTECS Toxicity Data

Acute Oral: Human TD_{Lo} Dose: 669 mg/kg/11D; Toxic Effects: Liver - Liver function tests impaired. Human TD_{Lo} Dose: 2880 mg/kg/8W; Toxic Effects: Sense organs and special senses - Tinnitus; Gastrointestinal - Nausea or vomiting; Gastrointestinal - Decreased motility or constipation. Human TD_{Lo} Dose: 480 mg/kg/7D-I; Toxic Effects: Sense organs and special senses - Tinnitus; Behavioral - Somnolence (general depressed activity); Gastrointestinal - Other changes. Man TD_{Lo} Dose: 857 mg/kg; Toxic Effects: Behavioral - Coma; Lungs, Thorax, or Respiration - Respiratory stimulation. Man TD_{Lo} Dose: 1625 mg/kg; Toxic Effects: Behavioral - Coma; Nutritional and gross metabolic - Body temperature increase. Woman TD_{Lo} Dose: 525 mg/kg/5D-I; Toxic Effects: Liver - Hepatitis (hepatocellular necrosis), diffuse. Woman TD_{Lo} Dose: 480 mg/kg/5D-I; Toxic Effects: Kidney, Ureter, and Bladder - Chgs in tubules (inc acute renal failure, acute tubular necrosis; Biochemical - Other. Woman TD_{Lo} Dose: 800 mg/kg; Toxic Effects: Kidney, Ureter, and Bladder - Chgs in tubules (inc acute renal failure, acute tubular necrosis; Musculoskelital - Other changes. Infant TD_{Lo} Dose: 120 mg/kg; Toxic Effects: Lungs, Thorax, or Respiration - Respiratory stimulation; Kidney, Ureter, and Bladder - Hematuria; Nutritional and gross metabolic - Dehydration. Child TD_{Lo} Dose: 10 mg/kg/1D-I; Toxic Effects: Lungs, Thorax, or Respiration - Acute pulmonary edema; Kidney, Ureter, and Bladder - Chgs in tubules (inc acute renal failure, acute tubular necrosis. Child TD_{Lo} Dose: 39 mg/kg/13D-I; Toxic Effects: Liver - Hepatitis (hepatocellular necrosis), diffuse. Child LD_{Lo} Dose: 104 mg/kg; Toxic Effects: Lungs, Thorax, or Respiration - Acute pulmonary edema; Gastrointestinal - Nausea or vomiting; Blood - Hemorrhage. Rat LD_{50} Dose: 200 mg/kg.

Acute Dermal: Mouse LD_{50} Route: Subcutaneous Dose: 1020 mg/kg.

Chronic (Multiple Dose) Oral: Rat Dose: 200 mg/kg/4D-I; Toxic Effects: Gastrointestinal - Ulceration or bleeding from stomach. Rat Dose: 8127 mg/kg/43W-C; Toxic Effects: Kidney, Ureter, and Bladder - Other changes in urine composition. Rat Dose: 9500 mg/kg/3W-I; Toxic Effects: DEATH.

Chronic (Multiple Dose) Inhalation: Rat Dose: 25 mg/m^3/4H/17W-I; Toxic Effects: Brain and coverings - Recordings from specific areas of CNS; Blood - Change in clotting factors; Blood - Changes in serum composition.

Reproductive/Teratogenic: Woman Route: Oral; Dose: 700 mg/kg; Duration: female 35-36W of pregnancy Specific Developmental Abnormalities - Central nervous system; Cardiovascular (circulatory) system; Effects on Newborn - Biochemical and metabolic. Woman Route: Oral; Dose: 546 mg/kg; Duration: female 37-39W of pregnancy Effects on Newborn - Other postnatal measures or effects. Woman Route: Oral; Dose: 546 mg/kg; Duration: female 37-39W of pregnancy Specific Developmental Abnormalities - Central nervous system; Craniofacial (including nose and tongue); Other developmental abnormalities. Woman Route: Oral; Dose: 17550 mg/kg; Duration: female 12-39W of pregnancy Maternal Effects - Parturition. Woman Route: Oral; Dose: 100 mg/kg; Duration: female 37W of pregnancy; Effects on Newborn - Other neonatal measures or effects. Woman Route: Oral; Dose: 17280 mg/kg; Duration: female 1-39W of pregnancy; Specific Developmental Abnormalities - Cardiovascular (circulatory) system; Respiratory system; Effects on Newborn - Apgor score (human only). Woman Route: Oral; Dose: 189 mg/kg; Duration: female 12-39W of pregnancy Maternal Effects - Parturition; Effects on Embryo or Fetus - Fetotoxicity; Specific Developmental Abnormalities - Blood and lymphatic systems (including spleen and marrow). Rat Route: Oral; Dose: 1 gm/kg; Duration: female 12D of pregnancy; Effects on Fertility - Post-implantation mortality; Effects on Embryo or Fetus - Fetal death. Rat Route: Oral; Dose: 2100 mg/kg; Duration: male 14D prior to mating; Paternal Effects - Testes, epididymis, sperm duct. Rat Route: Oral; Dose: 500 mg/kg; Duration: female 9D of pregnancy; Effects on Embryo or Fetus - Fetal death. Rat Route: Oral; Dose: 200 mg/kg; Duration: female 9D of pregnancy; Effects on Embryo or Fetus - Fetotoxicity. Rat Route: Oral; Dose: 10 mg/kg; Duration: female 22D of pregnancy; Maternal Effects - Parturition; Effects on Newborn - Stillbirth; Live birth index. Rat Route: Oral; Dose: 500 mg/kg; Duration: female 9D of pregnancy; Specific Developmental Abnormalities - Central nervous system; Eye, ear; Musculoskeletal system. Rat Route: Oral; Dose: 125 mg/kg; Duration: female 12D of pregnancy; Specific Developmental Abnormalities - Musculoskeletal system.

Mutagenic: Human DNA Inhibition; Cell Type: lymphocyte; Dose: 100 umol/L. Human Cytogenetic Analysis; Cell Type: fibroblast; Dose: 100 mg/L. Human Cytogenetic Analysis; Cell Type: leukocyte; Dose: 100 ug/L. Human Cytogenetic Analysis; Cell Type: lymphocyte; Dose: 10 mg/L. Human Other Mutation Test Systems; Cell Type: lymphocyte; Dose: 75 mg/L.

Hazard Overviews

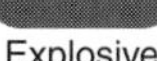
Explosive

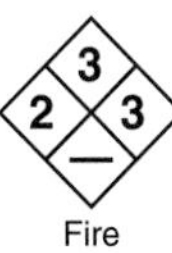

Fire Diamond

Health: Irritating to eyes/skin. Also Causes: stimulation and depression of CNS, local gum bleeding, vomiting, hypoglycemia, hyperpnea, headache, tinnitus, confusion, mania, convulsions, sweating, skin eruptions, gastrointestinal hemorrhages. Since aspirin inhibits platelet aggregation and irritates gastric mucosa, there is a tendency to bleeding. Death is usually due to respiratory failure or cardiovascular collapse while in a coma. Chronic Effects: salicylism, gastrointestinal irritation, ulcers, hemorrhagic strokes.

Fire: Combustible. Use dry chemical, water spray, or regular foam. Acetylsalicylic acid dust is a serious explosion hazard when dispersed in air.

Reactivity: Stable in dry air. Hazardous polymerization cannot occur. Avoid: contact with heat or flames. Hazardous decomposition products: oxides of carbon..

Carcinogenicity: IARC - Not listed; NIOSH - Listed as carcinogen; NTP - Not listed; ACGIH - Not listed; OSHA - Not listed; EPA - Not listed; MAK - Not listed

Primary Target Organs:

Eyes

Skin

Gastro-intestinal

Nervous System

Cardio-vascular

Blood

Exposure Limits

OSHA PEL Vacated 1989 Limits: TWA: 5 mg/m^3.

ACGIH TLV: TWA: 5 mg/m^3.

NIOSH REL: TWA: 5 mg/m^3.

Respirator Recommendation

Exposure Range: >5 to 50 mg/m^3 Air Purifying, Negative Pressure, Half Mask

Exposure Range: >50 to 500 mg/m^3 Air Purifying, Negative Pressure, Full Face

Exposure Range: >500 to <5000 mg/m^3 Supplied Air, Constant Flow/Pressure Demand, Full Face

Exposure Range: 5000 to unlimited mg/m^3 Self-contained Breathing Apparatus, Pressure Demand, Full Face

Cartridge Color: magenta (P100)

Environmental

Ecotoxicity: Marine tests MicrotoxTM (Photobacterium) test 5min EC_{50} 26 mg/l Artoxkit M (Artemia salina) test 24h LC_{50}381 mg/l; Freshwater tests Streptoxkit F (Streptocephalus proboscideus) test24h LC 50 178 mg/l

Environmental Fate: With a pKa of 3.49, the ion, will occur in varying proportions that are pH dependent. Above pH 5.5, virtually all will exist as the ion. Anions generally do not volatilize or adsorb to particulate matter as strongly as their neutral counterparts and therefore, volatilization, adsorption and bioconcentration are not expected to be important environmental fate processes. Limited aqueous screening test data suggests that it should biodegrade upon acclimation under anaerobic conditions. Hydrolysis is expected to be important (half-lives ranged from 12.5 days to 1.2 hours). Based on data for salicylic acid, it may also undergo photochemical degradation in sunlit environmental media. In the atmosphere, it is expected to exist in both the vapor and particulate phase. For gaseous material, reactions with photochemically produced hydroxyl radicals may be important (estimated half-life of 19.8 days). Removal by precipitation may occur.

Cleanup/Disposal: Guide No. 154: Eliminate all ignition sources (no smoking, flares, sparks or flames in immediate area). Do not touch damaged containers or spilled material unless wearing appropriate protective clothing. Stop leak if you can do it without risk. Prevent entry into waterways, sewers, basements or confined areas. Absorb or cover with dry earth, sand or other non-combustible material and transfer to containers. Do not get water inside containers.

Environmental Physical Data

Henry's Law Constant: calculated at 1.30 x10^{-9}
Octanol/Water Partition Coefficient: log K_{ow} = 1.19
Sorption Partition Coefficient: K_{oc} = estimated at 42
BCF: estimated at 0.67 to 0.73

Regulations

RCRA 40CFR: Not listed
CERCLA: 40CFR 302.4: Not listed
SARA 40CFR 372.65: Not listed
SARA EHS 40CFR 355: Not listed
TSCA: Listed

ACE8950 CAS #: 120-66-1

N-ACETYL-O-TOLUIDINE

RTECS: AN2900000
EINECS Number: 204-414-4
Molecular Formula: $C_9H_{11}NO$
Formula Weight: 149.21

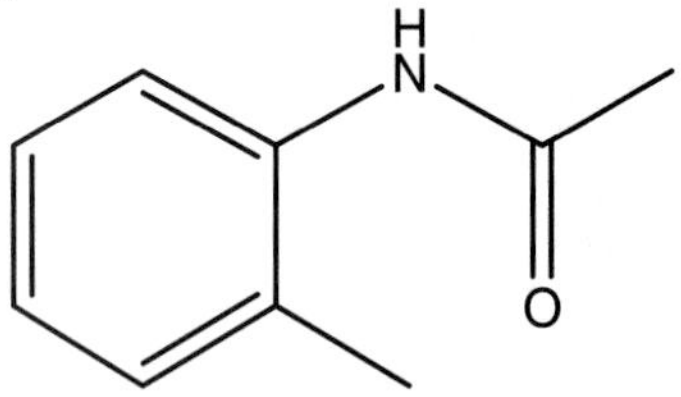

Chemical Structure

Synonyms: ACETAMIDE,N-(2-METHYLPHENYL)-; ACETAMIDE,N-(2-METHYLPHENYL)-(9CI); O-ACETOTOLUIDE; O-ACETOTOLUIDIDE; N-ACETYL-O-TOLUIDIDE; ACETYL-O-TOLUIDINE; 2'-METHYLACETANILIDE; 2-METHYLACETANILIDE; O-METHYLACETANILIDE; N-(2-METHYLPHENYL)-ACETAMIDE; N-(2-METHYLPHENYL)ACETAMIDE; N-O-TOLYLACETAMIDE
Description: colorless crystals
Use: chem int for dyes, eg, CI sulfur yellow 12

Physical Properties

Boiling Point: 296 °C (565 °F)
Freezing Point: 110 °C (230 °F)
Specific Gravity: 1.168 at 15 °C
Density: 1.168 g/mL
Water Solubility: Slightly Soluble in Water
Other Solubilities: 95% Ethanol: Soluble; Acetone: Soluble; Benzene: Soluble in hot solvent; Chloroform: Soluble; Ether: Soluble; Toluene: hot solvent.

RTECS Toxicity Data

Acute Oral: Mouse LD_{50} Dose: 1450 mg/kg.
Mutagenic: Bacteria - S Typhimurium Mutations in Microorganisms; Dose: 1 mg/plate (+S9). Bacteria - S Typhimurium Mutations in Microorganisms; Dose: 47 nmol/plate (-S9).

Hazard Overviews

Carcinogenicity: IARC - Not listed; NIOSH - Not listed; NTP - Not listed; ACGIH - Not listed; OSHA - Not listed; EPA - Not listed; MAK - Not listed

Environmental

Regulations

RCRA 40CFR: Not listed
CERCLA: 40CFR 302.4: Not listed
SARA 40CFR 372.65: Not listed
SARA EHS 40CFR 355: Not listed
TSCA: Listed

ACI3000 CAS #: 5610-64-0

ACID BLACK 52

EINECS Number: 227-029-3
Molecular Formula: $C_{40}H_{21}CrN_6Na_2O_{14}S_2$
Formula Weight: 887.69

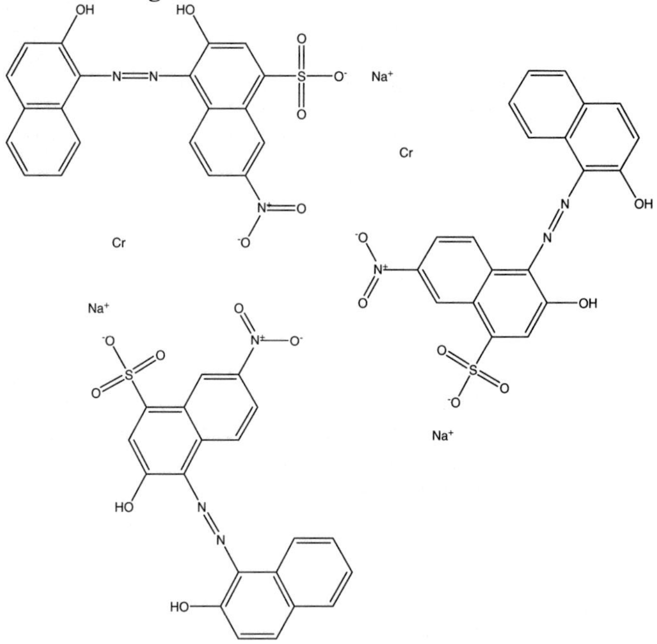

Chemical Structure

Synonyms: CHROMATE(3-),BIS(3-HYDROXY-4-((2-HYDROXY-1-NAPHTHALENYL)AZO)-7-NITRO-1-NAPHTHALENESULFONATO(3-))-,DISODIUM HYDROGEN
Use: dye for nylon, leather; anodized aluminum; aminoplasts; coloring & printing wool & silk

Physical Properties

Water Solubility: Soluble in Water
Other Solubilities: slightly Soluble in Ethanol and Acetone

Hazard Overviews

Health: Irritating to eyes/skin/respiratory tract. Toxic. Other Acute Effects: harmful if swallowed, inhaled, or absorbed through skin. Chronic Effects: Possible carcinogen.
Fire: Hazards: emits toxic fumes. Extinguishing agents: water spray; carbon dioxide, dry chemical powder or appropriate foam. Precautions: combustible liquid.
Reactivity: Incompatible with: strong oxidizing agents. Hazardous decomposition products: toxic fumes of: carbon monoxide, carbon dioxide, nitrogen oxides, sulfur oxides.
Carcinogenicity: IARC - Group 3, Not classifiable as to carcinogenicity to humans; NIOSH - Not listed; NTP - Not listed; ACGIH - Class A4, Not classifiable as a human carcinogen; OSHA - Not listed; EPA - Not listed; MAK - Class A2, Unmistakably carcinogenic in animal experimentation only
Primary Target Organs:

Eyes

Skin

Respiratory System

Exposure Limits
OSHA PEL: TWA: 0.5 mg/m^3; as Cr, CR-II,CR-III. Other Values: 0.1 mg/m^3; ceiling Cr-VI as CrO_3.
NIOSH REL: TWA: 0.5 mg/m^3; as Cr;Cr-II;Cr-III;Cr(VI) =.001.

Environmental

Ecotoxicity: TL_{50} Pimephales promelas (fathead minnow) 7.0 mg/l/24 hr; 6.2 mg/l/48 and 96 hr

Regulations

RCRA 40CFR: Not listed
CERCLA: 40CFR 302.4: Listed as Compound per CWA Section 307(a) per CAA Section 112
SARA 40CFR 372.65: Listed as Compound
SARA EHS 40CFR 355: Not listed
TSCA: Listed

ACI6000 CAS #: 62476-59-9

ACIFLUORFEN, SODIUM SALT

RTECS: DG5643200
EINECS Number: 263-560-7
Molecular Formula: $C_{14}H_6ClF_3NNaO_5$
Formula Weight: 383.65
Synonyms: ACIFLUORFEN SODIUM; BLAZER; BLAZER 2S; 5-(2-CHLORO-4-(TRIFLUOROMETHYL)PHENOXY)-2-NITROBENZOICACID SODIUM SALT; LS 80.1213; MC 10978; RH 6201; SCIFLUORFEN; SODIUM 5-(2-CHLORO-4-(TRIFLUOROMETHYL)PHENOXY)-2-NITROBENZOATE; SODIUM SALT OF ACIFLUORFEN

Physical Properties

Freezing Point: Powder 124 °C (255.2 °F) to 125 °C (257 °F)
Water Solubility: > 25 g/100g

RTECS Toxicity Data

Acute Oral: Rat LD_{50} Dose: 1300 mg/kg. Duck LD_{50} Dose: 2821 mg/kg.
Acute Inhalation: Rat LC_{50} Dose: >6900 mg/m^3/4hr.
Acute Dermal: Rabbit LD_{50} Route: Skin; Dose: >2 gm/kg.
Mutagenic: Insects - D Melanogaster Specific Locus Test; Route: Multiple routes; Dose: 10 ppb.

Hazard Overviews

Carcinogenicity: IARC - Not listed; NIOSH - Not listed; NTP - Not listed; ACGIH - Not listed; OSHA - Not listed; EPA - Not listed; MAK - Not listed

Environmental

Regulations

RCRA 40CFR: Not listed
CERCLA: 40CFR 302.4: Not listed
SARA 40CFR 372.65: Listed
SARA EHS 40CFR 355: Not listed
TSCA: Listed

ACO5000 CAS #: 499-12-7

ACONITIC ACID

RTECS: UD2380000
EINECS Number: 207-877-0
Molecular Formula: $C_6H_6O_6$
Formula Weight: 174.12

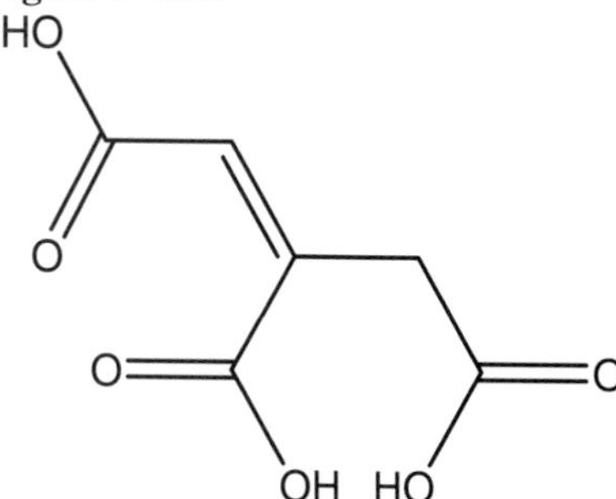

Chemical Structure

Synonyms: ACHILLEAIC ACID; ACHILLEIC ACID; CITRIDIC ACID; CITRIDINIC ACID; EQUISETIC ACID; GLUTACONIC ACID,3-CARBOXY-; 2-PENTENEDIOIC ACID,3-CARBOXY-; 1,2,3-PROPENE TRICARBOXYLIC ACID; 1-PROPENE-1,2,3-TRICARBOXYLIC ACID; PROPENE-1,2,3-TRICARBOXYLIC ACID; PYROCITRIC ACID
Description: white crystalline powder
Use: mfr itaconic acid; as plasticizer for buna rubber & plastics; synthetic flavoring substance & adjuvant; prepn of plasticizers & wetting agents; antioxidant; org syntheses
Boiling Point: Decomposes at 209 °C (408 °F)
Freezing Point: Decomposes at 194 °C (381.2 °F) to 195 °C (383 °F)

Physical Properties

Water Solubility: 1 g dissolves in 55 ml Water at 13 °C

Other Solubilities: Slightly Soluble in Ether; Soluble in Alcohol; Slightly Soluble in Ether.

Hazard Overviews

Fire Diamond

Health: Irritating to eyes/skin.
Fire: Will burn. Use dry chemical, carbon dioxide, water spray, or foam.
Reactivity: Stable. Hazardous polymerization cannot occur. Hazardous decomposition products: acrid smoke; fumes.
Carcinogenicity: IARC - Not listed; NIOSH - Not listed; NTP - Not listed; ACGIH - Not listed; OSHA - Not listed; EPA - Not listed; MAK - Not listed
Primary Target Organs:

Eyes

Skin

Environmental

Regulations

RCRA 40CFR: Not listed
CERCLA: 40CFR 302.4: Not listed
SARA 40CFR 372.65: Not listed
SARA EHS 40CFR 355: Not listed
TSCA: Listed

ACR1000 CAS #: 260-94-6

ACRIDINE

RTECS: AR7175000
EINECS Number: 205-971-6
Molecular Formula: $C_{13}H_9N$
Structured MF: $C_{13}H_9N$
Formula Weight: 179.21

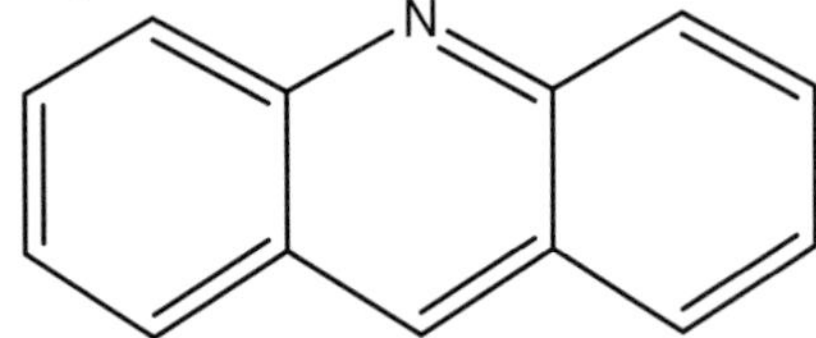

Chemical Structure

Synonyms: AKRIDIN; 9-ASAANTHRACENE; 10-AZAANTHRACENE; 9-AZAANTHRACENE; 2,3-BENZOQUINOLINE; BENZO(B)QUINOLINE; COAL TAR PITCH VOLATILES:ACRIDINE; 2,3,5,6-DIBENZOPYRIDINE; DIBENZO(B,C)PYRIDINE; DIBENZO(B,E)PYRIDINE
Description: colorless or faintly yellow crystals, needles, or prisms; weak irritating odor
Use: manufacture of dyes and intermediates; some dyes derived from it

Physical Properties

Boiling Point: 346 °C (655 °F) at 760 mm Hg
Freezing Point: 111 °C (231.8 °F)
Specific Gravity: 1.005 at 20 °C/4 °C
Vapor Pressure: 1 mm Hg at 129.4 °C
Water Solubility: Slightly Soluble in Hot Water
Other Solubilities: Benzene: Very Soluble; Carbon Disulfide: Very Soluble; Ether: Soluble; Ethanol: Soluble; Hydrocarbons: Soluble.
pH: Weak base
Ionization Potential (eV): 7.8 +/-0.2
Flash Point: Not pertinent (combustible solid)

RTECS Toxicity Data

Acute Oral: Rat LD_{50} Dose: 2 gm/kg; Toxic Effects: Behavioral - Somnolence (general depressed activity); Gastrointestinal - Necrotic changes; Blood - Changes in erythrocite (RBC) cell count. Mouse LD_{50} Dose: 500 mg/kg; Toxic Effects: Sense organs and special senses - Lacrimation; Behavioral - Somnolence (general depressed activity); Behavioral - Excitment.
Acute Dermal: Mouse LD_{50} Route: Subcutaneous Dose: 400 mg/kg.
Chronic (Multiple Dose) Oral: Rat Dose: 4500 mg/kg/45D-I; Toxic Effects: Kidney, Ureter, and Bladder - Other changes in urine composition; Blood - Pigmented or nucleated red Blood cells; Blood - Changes in other cell count.
Mutagenic: Mammal DNA Adduct; Cell Type: lymphocyte; Dose: 100 umol/L. Chicken DNA Adduct; Cell Type: leukocyte; Dose: 100 umol/L.

Hazard Overviews

Health: Severely irritating to eyes/skin/respiratory tract. Harmful. Other Acute Effects: harmful if swallowed, inhaled, or absorbed through skin; symptoms of exposure may include burning sensation; coughing; wheezing; laryngitis; shortness of breath; headache; nausea; vomiting; sneezing; photosensitivity; exposure to light can result in allergic reactions resulting in dermatologic lesions, which can vary from sunburnlike responses to edematous, vesiculated lesions or bullae; absorption into the body leads to the formation of methemoglobin which in sufficient concentration causes cyanosis; onset may be delayed 2 to 4 hours or longer; exposure can cause damage to the eyes; target organs: nerves, blood, eyes.
Fire: Combustible. Hazards: emits toxic fumes. Extinguishing agents: carbon dioxide, dry chemical powder or appropriate foam; water spray. Precautions: combustible liquid.
Reactivity: Incompatible with: strong oxidizing agents. Hazardous decomposition products: toxic fumes of: carbon monoxide, carbon dioxide, nitrogen oxides.
Carcinogenicity: IARC - Not listed; NIOSH - Not listed; NTP - Not listed; ACGIH - Not listed; OSHA - Not listed; EPA - Not listed; MAK - Not listed

Primary Target Organs:

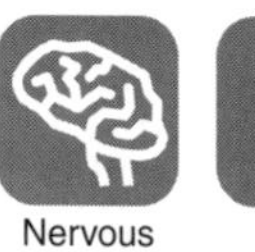

Eyes Skin Respiratory System Nervous System Blood

Exposure Limits
OSHA PEL: TWA: 0.2 mg/m^3.

Environmental

Ecotoxicity: Aquatic toxicity: 0.7 ppm/*/perch/kill/fresh water *Time period not specified

Cleanup/Disposal: Guide No. 153: Eliminate all ignition sources (no smoking, flares, sparks or flames in immediate area). Do not touch damaged containers or spilled material unless wearing appropriate protective clothing. Stop leak if you can do it without risk. Prevent entry into waterways, sewers, basements or confined areas. Absorb or cover with dry earth, sand or other non-combustible material and transfer to containers. Do not get water inside containers.

Environmental Physical Data

Octanol/Water Partition Coefficient: log K_{ow} = 3.4 to 3.6
BCF: no food chain concentration potential

Regulations

RCRA 40CFR: Not listed
CERCLA: 40CFR 302.4: Not listed
SARA 40CFR 372.65: Not listed
SARA EHS 40CFR 355: Not listed
TSCA: Listed

Analytical Methods

Water / Groundwater: ASTM D4763

ACR3000 CAS #: 2465-29-4

ACRIDINE RED

RTECS: ZD5500000
EINECS Number: 219-568-8
Molecular Formula: $C_{15}H_{15}ClN_2O$
Formula Weight: 274.77
Synonyms: ACRIDIN RED; ACRIDINE RED 3B; ACRIDINE RED,HYDROCHLORIDE; C.I. 45000; DIMETHYLDIAMINOXANTHENYL CHLORIDE; 3H-XANTHEN-6-AMINE,N-METHYL-3-(METHYLIMINO)-,MONOHYDROCHLORIDE
Use: basic dye; biological stain

RTECS Toxicity Data

Mutagenic: Insects - D Melanogaster Sex Chromosome Loss; Route: Oral; Dose: 1000 ppm.

Tumorigenic: Rat Route: Subcutaneous; Dose: 1215 mg/kg/59W-I; Toxic Effects: Tumorigenic - Equivocal tumorigenic agent by RTECS criteria; Tumorigenic - Tumors at site of application.

Hazard Overviews

Carcinogenicity: IARC - Not listed; NIOSH - Not listed; NTP - Not listed; ACGIH - Not listed; OSHA - Not listed; EPA - Not listed; MAK - Not listed

Environmental

Regulations

RCRA 40CFR: Not listed
CERCLA: 40CFR 302.4: Not listed
SARA 40CFR 372.65: Not listed
SARA EHS 40CFR 355: Not listed
TSCA: Listed

ACR5000 CAS #: 107-02-8

ACROLEIN

RTECS: AS1050000
DOT: UN1092; IMO3.1
EINECS Number: 203-453-4
Molecular Formula: C_3H_4O
Structured MF: CH_2=CHCHO
Formula Weight: 56.06

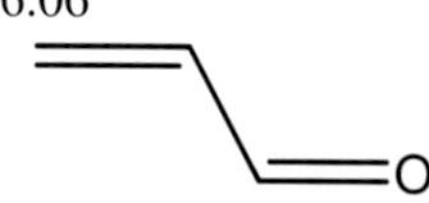

Chemical Structure

Synonyms: ACQUINITE; ACRALDEHYDE; ACRALDEHYDEACROLEINA; TRANS-ACROLEIN; ACROLEINA; ACROLEINE; ACROLEIN,INHIBITED; ACRYLALDEHYD; ACRYLALDEHYDE; ACRYLIC ALDEHYDE; AKROLEIN; AKROLEINA; ALDEHYDE ACRYLIQUE; ALDEIDE ACRILICA; ALLYL ALDEHYDE; AQUALIN; AQUALINE; BIOCIDE; CROLEAN; EPA PESTICIDE CHEMICAL CODE 000701; ETHYLENE ALDEHYDE; MAGNACIDE; MAGNACIDE H; NSC 8819; PAPITE; 2-PROPENAL; PROP-2-EN-1-AL; PROP-2-ENAL; PROPENAL; 2-PROPEN-1-ONE; PROPYLENE ALDEHYDE; SLIMICIDE

Description: colorless to yellow liquid; penetrating, pungent odor

Use: as a lacrimogenic warning agent in methyl chloride refrigerant, as a component of military poison gases, as a synthetic reagent in the manufacture of methionine, glycerol and glutaraldehyde; as an aquatic herbicide and as an algaecide for water treatment; as an intermediate for polyurethane and polyester resins, in pharmaceuticals, as an herbicide, as a biocide, in the manufacture of colloidal forms of metals, in making plastics and perfumes, to modify food starch, in the manufacture of 1,3,6-hexanetriol, as a fungicide and bactericide, as a liquid fuel, as an antimicrobial agent and as a slimicide in paper manufacture; an intermediate for acrylic acid and its esters and is used in the manufacture of 2hydroxyadipaldehyde, quinoline, pentaerythritol, cycloaliphatic epoxy resins, oil-well additives and water-treatment formulae

Physical Properties

Boiling Point: 52.5 °C (127 °F) at 760 mm Hg
Freezing Point: -88 °C (-126.4 °F)
Specific Gravity: 0.8389 at 20 °C
Vapor Density: 1.94 Air=1
Density: 0.8389 g/mL at 20 °C
Bulk Density: 7.03 lbs/gal at 20 °C
Vapor Pressure: 200 mm Hg at 17.5 °C
Water Solubility: 40% by weight
Other Solubilities: 95% Ethanol: >=100 mg/ml at 21 °C; Acetone: >=100 mg/ml at 21 °C; Alcohol: Soluble; DMSO: >=100 mg/ml at 21 °C; Ethanol: >10%; Ether: Soluble.
Surface Tension: 24 dynes/cm at 20 °C
Odor Threshold: 0.0525 to 37.5 mg/m^3
pH: 10% solution in water at 25 °C maximum 6
Refraction Index: 1.4022 at 19 °C/D
Critical Temperature: 254 °C
Critical Pressure: 737 psia
Ionization Potential (eV): 10.13
Flash Point: -26 °C Closed Cup
Autoignition Temperature: Unstable at 220 °C
LEL: 2.8% v/v
UEL: 31% v/v

RTECS Toxicity Data

Acute Oral: Rat LD_{50} Dose: 26 mg/kg. Mouse LD_{50} Dose: 13900 ug/kg; Toxic Effects: Behavioral - Somnolence (general depressed activity); Skin and appendages - Hair; Nutritional and gross metabolic - Weight loss or decreased weight gain.
Acute Inhalation: Rat LC_{50} Dose: 18 mg/m^3/4hr. Human LC_{Lo} Dose: 5500 ppb. Human LC_{Lo} Dose: 153 ppm/10M. Man TC_{Lo} Dose: 1 ppm; Toxic Effects: Sense organs and special senses - Lacrimation. Child TC_{Lo} Dose: 300 ppb/2hr; Toxic Effects: Lungs, Thorax, or Respiration - Structural or functional change in trachea or bronchi; Lungs, Thorax, or Respiration - Respiratory obstruction; Lungs, Thorax, or Respiration - Other changes.
Acute Dermal: Rabbit LD_{50} Route: Skin; Dose: 200 mg/kg. Rabbit LD_{Lo} Route: Subcutaneous Dose: 250 mg/kg; Toxic Effects: Lungs, Thorax, or Respiration - Dyspnea.
Chronic (Multiple Dose) Inhalation: Rat Dose: 4 ppm/6H/62D-I; Toxic Effects: Lungs, Thorax, or Respiration - Other changes; Nutritional and gross metabolic - Weight loss or decreased weight gain; DEATH. Rat Dose: 4900 ppb/6H/13W-I; Toxic Effects: Sense organs and special senses - Other; Endocrine - Changes in adrenal weight; DEATH. Rat Dose: 510 ug/m^3/24H/9W-C; Toxic Effects: Kidney, Ureter, and Bladder - Other changes in urine composition; Blood - Changes in leukocyte (WBC) cell count; Nutritional and gross metabolic - Weight loss or decreased weight gain. Rat Dose: 3 ppm/6H/3W-I; Toxic Effects: Sense organs and special senses - Other; Endocrine - Changes in spleen weight. Monkey Dose: 3700 ppb/8H/6W-I; Toxic Effects: Lungs, Thorax, or Respiration - Structural or functional change in trachea or bronchi; DEATH.
Irritation Eye: Rabbit Standard Draize Test Dose: 1 mg; Reaction: severe. Rabbit Standard Draize Test Dose: 50 ug/24H; Reaction: severe.
Irritation Skin: Rabbit Standard Draize Test Dose: 2 mg/24H; Reaction: severe. Rabbit Open Draize Test Dose: 5 mg open; Reaction: severe.
Reproductive/Teratogenic: Rat Route: Oral; Dose: 840 mg/kg; Duration: female multigeneration; Effects on Newborn - Growth statistics. Rabbit Route: Intravenous; Dose: 6 mg/kg; Duration: female 9D of pregnancy; Effects on Fertility - Post-implantation mortality.
Mutagenic: Human DNA Damage; Cell Type: other cell types; Dose: 30 umol/L. Human DNA Adduct; Cell Type: fibroblast; Dose: 100 umol/L. Human Mutations in Mammalian Somatic Cells; Cell Type: fibroblast; Dose: 200 nmol/L. Human Sister Chromatid Exchenge; Cell Type: lymphocyte; Dose: 5 umol/L.

Hazard Overviews

Poison

Corrosive

Explosive

Flammable

Fire Diamond

Health: Severe irritation to eyes/skin/respiratory tract. Also Causes: watering eyes, difficulty breathing, chest tightness, nausea, vomiting, diarrhea, pulmonary edema, high blood pressure, and unconsciousness, corneal damage, palpebral damage, pus-like discharge from eyes. Severe irritation of the mouth, gastrointestinal tract. Chronic Effects: asthmatic response.
Fire: Flammable. Can form explosive mixtures in the air. May polymerize explosively. Forms peroxides. Use dry chemical, carbon dioxide, alcohol-resistant foam, or flooding quantities of water. Fight fire from maximum distance.
Reactivity: Unstable. Hazardous polymerization can occur from exposure to heat (approx 200 °C), sunlight, acid, or alkalis; formation of peroxides over time can also lead to polymerization; reaction may be violent. Avoid: heat; sunlight. Incompatible with: acids; amines; alkalis; sulfur dioxide; metal salts; oxidants; thiourea; dimethylamine; weak acid conditions. Hazardous decomposition products: carbon monoxide; peroxides.
Carcinogenicity: IARC - Group 3, Not classifiable as to carcinogenicity to humans; NIOSH - Not listed; NTP - Not listed; ACGIH - Not listed; OSHA - Not listed; EPA - Class C, Possible human carcinogen; MAK - Not listed
Primary Target Organs:

Eyes

Skin

Respiratory System

Nervous System
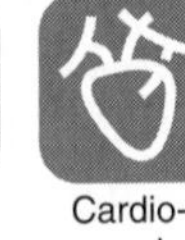
Cardio-vascular

Exposure Limits
OSHA PEL: TWA: 0.1 ppm; 0.25 mg/m^3.
OSHA PEL Vacated 1989 Limits: TWA: 0.1 ppm; 0.25 mg/m^3; STEL: 0.3 ppm; 0.8 mg/m^3.
ACGIH TLV: TWA: 0.1 ppm; 0.23 mg/m^3; STEL: 0.3 ppm; 0.69 mg/m^3.

NIOSH REL: TWA: 0.1 ppm; 0.25 mg/m^3. STEL: 0.3 ppm; 0.8 mg/m^3.
NIOSH IDLH: 2 ppm.
DFG MAK: TWA: 0.1 ppm; 0.25 mg/m^3.
Respirator Recommendation
Exposure Range: >0.1 to <2 ppm Supplied Air, Constant Flow/Pressure Demand, Full Face
Exposure Range: 2 to unlimited ppm Self-contained Breathing Apparatus, Pressure Demand, Full Face
Note: poor warning properties

Environmental

Ecotoxicity: LC_{50} Daphnia magna 0.23 mg/l/24 hr; 0.083 mg/l/48 hr; No discernible effect conc= 0.034 mg/l. /Conditions of bioassay not specified LD_{50} Mallard Duck (male, 3-5 mo old) oral 9.11 mg/kg (95% confidence limit 6.32 mg/kg) LC_{50} Pimephales promelas (fathead minnow) 14.0 ug/1/96 hr (confidence limit not reliable), flow-through bioassay with measured concentrations, 17.4 °C, dissolved oxygen 9.3 mg/l, hardness 45.2 mg/l calcium carbonate, alkalinity 42.9 mg/l Inhibition of cell multiplication starts at 0.04 mg/l in algae (Microcystis aeruginosa) Carp & Thread-fin shad are particularly sensitive, being killed at 1 to 2 ppm. Black bass, Blue gill, & Lamprey eel larvae appear to tolerate up to 5 ppm. /conditions of bioassay not specified Inhibition of cell multiplication starts at 0.21 mg/l in bacteria (Pseudomonas putida) The lowest observed avoidance concentration in insects was above 0.1 mg/l for Ephemerella walkeri (mayfly nymphs)
Environmental Fate: If released to moist soil, it is expected to be susceptible to extensive leaching. Biodegradation under aerobic conditions may be an important fate process. It is predicted to volatilize rapidly from dry soil surfaces. If released to water, it may biodegrade under aerobic conditions, volatilize (half-life of 7 hours from a model river), or undergo reversible hydration to beta-hydroxypropionaldehyde (half life of 21 days). The overall half-life in water is reported to range between 2 to 6 days. Bioaccumulation in aquatic organisms, adsorption to suspended solids and sediments, reaction with singlet oxygen or alkylperoxy radicals, and photolysis are not expected to be important fate processes in water. If released to the atmosphere, the dominant removal mechanism is expected to be reaction of vapor with photochemically generated hydroxyl radicals (half-life of 10-13 hours). Products of this reaction include: carbon dioxide, formaldehyde, and glycolaldehyde, and in the presence of nitrogen oxides include: peroxynitrate and nitric acid. Small amounts of this compound may be removed from the atmosphere by wet deposition. Reaction with ozone and direct photolysis are not expected to be important fate processes in the atmosphere.
Cleanup/Disposal: Guide No. 131: Fully encapsulating, vapor protective clothing should be worn for spills and leaks with no fire. Eliminate all ignition sources (no smoking, flares, sparks or flames in immediate area). All equipment used when handling the product must be grounded. Do not touch or walk through spilled material. Stop leak if you can do it without risk. Prevent entry into waterways, sewers, basements or confined areas. A vapor suppressing foam may be used to reduce vapors. Small Spills: Absorb with earth, sand or other non-combustible material and transfer to containers for later disposal. Use clean non-sparking tools to collect absorbed material. Large Spills: Dike far ahead of liquid spill for later disposal. Water spray may reduce vapor; but may not prevent ignition in closed spaces.

Environmental Physical Data

Henry's Law Constant: 1.36 x10^{-4}
Octanol/Water Partition Coefficient: log K_{ow} = -0.090
Sorption Partition Coefficient: K_{oc} = calculated at 5.0
BCF: bluegill 344
BOD: 33%, 10 days

Regulations

RCRA 40CFR: Listed Hazardous Waste No. P003 Toxic Waste
CERCLA: 40CFR 302.4: Listed per CWA Section 311(b)(4) per RCRA Section 3001 per CWA Section 307(a) RQ: 1 lb (0.454 kg)
SARA 40CFR 372.65: Listed TPQ: 500 lb
SARA EHS 40CFR 355: Listed TPQ: 1 lb
TSCA: Listed

Analytical Methods

Air: EPA TO-11, TO-5, 0100; ASTM D4490
Soil: SW846 5031, 5032, 8015B, 8030A, 8240B, 8260A, 8260B, 8315, 8315A, 8316; EPA 7
Water / Groundwater: EPA 603, 624, 624-S, 626; ASTM D3695
Food: EPA 5
Indoor / Expired Air: NIOSH 2501, 2539; EPA IP-6A, IP-6B, IP-6C, 0100
Plasma: EPA 29

ACR7000 CAS #: 107-13-1

ACRYLONITRILE

RTECS: AT5250000
DOT: UN1093; IMO3.2
EINECS Number: 203-466-5
Molecular Formula: C_3H_3N
Structured MF: CH_2=CHCN
Formula Weight: 53.06

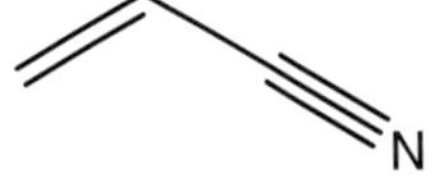

Chemical Structure

Synonyms: ACRITET; ACRYLNITRIL; ACRYLON; ACRYLONITRILE MONOMER; ACRYLONITRILE,INHIBITED; AKRYLONITRIL; AKRYLONITRYL; AN; CARBACRYL; CIANURO DI VINILE; CYANOETHYLENE; CYANURE DE VINYLE; ENT 54; FUMIGRAIN; MILLER'S FUMIGRAIN; NITRILE ACRILICO; NITRILE ACRYLIQUE; 2-PROPENENITRILE; PROPENENITRILE; PROPENONITRILE; TL 314; VCN; VENTOX; VINYL CYANIDE; VINYLCYANIDE; VINYLKYANID
Description: colorless liquid; pungent odor

Use: manufacture of acrylic and modacrylic fiber high-strength whiskers, acrylostyrene plastics, acrylonitrile-butadiene-styrene plastics, synthetic rubber, nitrile rubber, chemicals, adhesives and surface coatings; a chemical intermediate in the synthesis of antioxidant pharmaceuticals, dyes, textile fibers, surface-active agents and adiponitrile etc.; in organic synthesis to introduce a cyanoethyl group; as a modifier for natural polymers and in the cyanoethylation of cotton; in synthetic soil blocks (acrylonitrile polymerized in wood pulp) as a grain fumigant and a pesticide; a monomer for a semiconductive polymer that can be used like inorganic oxide catalysts in dehydrogenation of tertbutanol to isobutylene and water

Physical Properties

Boiling Point: 77.3 °C (171 °F) at 760 mm Hg
Freezing Point: -82 °C (-115.6 °F)
Specific Gravity: 0.8004 at 25 °C/4 °C
Vapor Density: 1.83 Air=1
Saturated Vapor Density: 1.330998185 kg/m^3
Density: 0.8075 g/mL at 20 °C
Vapor Pressure: 100 mm Hg at 22.8 °C
Water Solubility: 7% by weight
Other Solubilities: 95% Ethanol: >=100 mg/ml at 21.6 °C; Acetone: >=100 mg/ml at 21.6 °C; Benzene: Soluble; DMSO: >=100 mg/ml at 21.6 °C; Ether: Soluble; Most organic solvents: miscible.
Surface Tension: 27.3 dynes/cm at 24 °C
Odor Threshold: 8.1000 to 78.7500 mg/m^3
pH: 5% aqueous solution 5.5 to 7.5
Refraction Index: 1.3888 at 25 °C/D
Evaporation Rate: 4.54 Butyl Acetate=1
Critical Temperature: 263 °C
Critical Pressure: 660 psia
Ionization Potential (eV): 10.91
Flash Point: -1 °C Closed Cup
Autoignition Temperature: 481 °C
LEL: 3% v/v
UEL: 17% v/v

RTECS Toxicity Data

Acute Oral: Rat LD_{50} Dose: 78 mg/kg; Toxic Effects: Behavioral - Convulsions or effect on seizure threshold; Lungs, Thorax, or Respiration - Dyspnea; Gastrointestinal - Changes in structure or function of salivary glands. Mouse LD_{50} Dose: 27 mg/kg; Toxic Effects: Behavioral - Convulsions or effect on seizure threshold; Lungs, Thorax, or Respiration - Dyspnea; Gastrointestinal - Changes in structure or function of salivary glands.

Acute Inhalation: Monkey LC; Dose: >90 ppm/4hr. Rat LC_{50} Dose: 425 ppm/4hr. Man LC_{Lo} Dose: 1 gm/m^3/1hr; Toxic Effects: Behavioral - Somnolence (general depressed activity); Gastrointestinal - Hypermotility, diarrhea; Gastrointestinal - Nausea or vomiting. Human TC_{Lo} Dose: 16 ppm/20M; Toxic Effects: Sense organs and special senses - Other; Sense organs and special senses - Conjunctive irritation; Lungs, Thorax, or Respiration - Other changes.

Acute Dermal: Child LD_{Lo} Route: Skin; Dose: 2015 mg/kg; Toxic Effects: Behavioral - General anesthetic; Lungs, Thorax, or Respiration - Cyanosis; Gastrointestinal - Nausea or vomiting. Rabbit LD_{50} Route: Skin; Dose: 250 mg/kg.

Chronic (Multiple Dose) Oral: Rat Dose: 10920 mg/kg/1Y-C; Toxic Effects: Behavioral - Food intake (animal); Kidney, Ureter, and Bladder - Other changes in urine composition; Biochemical - Phosphatases. Rat Dose: 2100 mg/kg/21D-C; Toxic Effects: Endocrine - Other changes; Nutritional and gross metabolic - Changes in sodium. Rat Dose: 120 mg/kg/60D-I; Toxic Effects: Endocrine - Adrenal cortex hyperplasia; Endocrine - Changes in adrenal weight; Blood - Other changes.

Chronic (Multiple Dose) Inhalation: Rat Dose: 1500 ug/m^3/5H/26W-I; Toxic Effects: Liver - Other changes; Kidney, Ureter, and Bladder - Chgs in tubules (inc acute renal failure, acute tubular necrosis; Biochemical - Dehydrogenases. Rat Dose: 330 mg/m^3/4H/8W-I; Toxic Effects: Sense organs and special senses - Other; Nutritional and gross metabolic - Weight loss or decreased weight gain; DEATH. Monkey Dose: 330 mg/m^3/4H/8W-I; Toxic Effects: Behavioral - Muscle weakness; DEATH.

Chronic (Multiple Dose) Dermal: Rat Route: Skin; Dose: 3751 mg/kg/19W-I; Toxic Effects: Kidney, Ureter, and Bladder - Other changes in urine composition; Biochemical - True cholinesterase; Biochemical - Cytochrome oxidases (including oxidative phosphorylation).

Irritation Eye: Rabbit Standard Draize Test Dose: 100 mg; Reaction: moderate.

Irritation Skin: Rabbit Standard Draize Test Dose: 500 mg; Reaction: severe.

Reproductive/Teratogenic: Rat Route: Oral; Dose: 650 mg/kg; Duration: female 6-15D of pregnancy; Effects on Fertility - Female fertility index; Effects on Embryo or Fetus - Fetotoxicity; Specific Developmental Abnormalities - Musculoskeletal system. Rat Route: Oral; Dose: 650 mg/kg; Duration: female 6-15D of pregnancy; Specific Developmental Abnormalities - Musculoskeletal system; Cardiovascular (circulatory) system. Rat Route: Oral; Dose: 644 mg/kg; Duration: male 2W prior to mating; Paternal Effects - Spermatogenesis; Testes, epididymis, sperm duct. Rat Route: Inhalation; Dose: 40 ppm/6H; Duration: female 6-15D of pregnancy; Maternal Effects - Other effects on females; Nutritional and gross metabolic - Weight loss or decreased weight gain. Rat Route: Inhalation; Dose: 80 ppm/6H; Duration: female 6-15D of pregnancy; Specific Developmental Abnormalities - Musculoskeletal system. Rat Route: Inhalation; Dose: 25 ppm/6H; Duration: female 6-20D of pregnancy; Effects on Embryo or Fetus - Fetotoxicity.

Mutagenic: Human DNA Damage; Cell Type: other cell types; Dose: 200 mg/L. Human Mutations in Microorganisms; Cell Type: lymphocyte; Dose: 40 mg/L (+S9). Human Mutations in Mammalian Somatic Cells; Cell Type: lymphocyte; Dose: 25 mg/L. Human Sister Chromatid Exchenge; Cell Type: other cell types; Dose: 150 mg/L.

Tumorigenic: Rat Route: Oral; Dose: 18200 mg/kg/52W-C; Toxic Effects: Tumorigenic - Carcinogenic by RTECS criteria; Brain and coverings - Tumors. Rat Route: Inhalation;

Dose: 5 ppm/52W-I; Toxic Effects: Tumorigenic - Equivocal tumorigenic agent by RTECS criteria; Skin and appendages - Tumors. Rat Route: Inhalation; Dose: 20 ppm/4H/52W-I; Toxic Effects: Tumorigenic - Equivocal tumorigenic agent by RTECS criteria; Brain and coverings - Tumors. Rat Route: Inhalation; Dose: 40 ppm/4H/52W-I; Toxic Effects: Tumorigenic - Equivocal tumorigenic agent by RTECS criteria; Brain and coverings - Tumors. Rat Route: Oral; Dose: 3640 mg/kg/52W-C; Toxic Effects: Tumorigenic - Neoplastic by RTECS criteria; Sense organs and special senses - Tumors; Gastrointestinal - Tumors.

Hazard Overviews

Explosive

Flammable

Fire Diamond

Health: Severely irritating to eyes; irritating to skin. Toxic. Also Causes: nausea, headache, light-headedness, skin vesiculation, dermatitis, weakness, asphyxia. Suspect cancer hazard.

Fire: Explosive and flammable. Use carbon dioxide, alcohol foam, or dry chemical. Use water spray to dilute spills. Vapors can travel to ignition sources and flash back. In case of an advanced or massive fire, fight fire from a remote, protected, or shielded location.

Reactivity: Stable. Hazardous polymerization can occur in the absence of oxygen or upon exposure to light. Avoid: improper storage; ignition sources. Incompatible with: concentrated alkalies; strong oxidizing agents; bromine; ammonia; amines; copper; copper alloys; sulfuric acid; chlorosulfonic acid; nitric acid; oleum; potassium hydroxide; sodium hydroxide; 1,2,3,4-tetrahydrocarbazole. Hazardous decomposition products: hydrogen cyanide; oxides of nitrogen; carbon monoxide.

Carcinogenicity: IARC - Group 2A, Probably carcinogenic to humans; NIOSH - Listed as carcinogen; NTP - Class 2A, Reasonably anticipated to be a carcinogen, limited evidence of carcinogenicity from studies in humans; ACGIH - Class A2, Suspected human carcinogen; OSHA - Listed as a carcinogen; EPA - Class B1, Probable human carcinogen based on epidemiologic studies; MAK - Class A2, Unmistakably carcinogenic in animal experimentation only

Primary Target Organs:

Eyes

Skin

Nervous System

Liver

Kidneys

Cardio-vascular

Exposure Limits

ACGIH TLV: TWA: 2 ppm; 4.3 mg/m^3.

NIOSH REL: TWA: 1 ppm. STEL: 10 ppm; 15 mg/m^3; 15-minute, skin.

NIOSH IDLH: 85 ppm.

Respirator Recommendation

Exposure Range: >2 to 20 ppm Air Purifying, Negative Pressure, Half Mask

Exposure Range: >20 to <85 ppm Air Purifying, Negative Pressure, Full Face

Exposure Range: 85 to unlimited ppm Self-contained Breathing Apparatus, Pressure Demand, Full Face

Cartridge Color: black

Note: must change cartridges before each new shift

Environmental

Ecotoxicity: LC_{50} Pimephales promelas (fathead minnow) 10,100 ug/l/96 hr /Flow-through bioassay LC_{50} Pimephales promelas (fathead minnow) 2600 ug/l/30 days /Static bioassay LC_{50} Daphnia magna (water flea) 13 mg/l/24 hr; 7.6 mg/l/48 hr; no discernible effect = 0.78 mg/l. /Conditions of bioassay not specified

Environmental Fate: When released to the atmosphere, it will degrade primarily by reacting with photochemically produced hydroxyl radicals. The half-life for this process will be 3.5 sunlit days under relatively clean atmospheric conditions to somewhat over a day with smog. Therefore there would be opportunity for dispersal from source areas. If released in wastewater, it will slowly evaporate (half-life 1-6 days) and also biodegrade (complete degradation in approx. 1 week in receiving water in which microorganisms would be acclimated). If spilled on land, it will volatilize fairly rapidly due to its relatively high Henry's Law constant and low adsorption to soil. Some would leach into the ground where its fate is unknown.

Cleanup/Disposal: Guide No. 131: Fully encapsulating, vapor protective clothing should be worn for spills and leaks with no fire. Eliminate all ignition sources (no smoking, flares, sparks or flames in immediate area). All equipment used when handling the product must be grounded. Do not touch or walk through spilled material. Stop leak if you can do it without risk. Prevent entry into waterways, sewers, basements or confined areas. A vapor suppressing foam may be used to reduce vapors. Small Spills: Absorb with earth, sand or other non-combustible material and transfer to containers for later disposal. Use clean non-sparking tools to collect absorbed material. Large Spills: Dike far ahead of liquid spill for later disposal. Water spray may reduce vapor; but may not prevent ignition in closed spaces.

Environmental Physical Data

Henry's Law Constant: 0.063

Octanol/Water Partition Coefficient: log K_{ow} = 0.25

Sorption Partition Coefficient: K_{oc} = estimated at 9

BCF: bluegill 48

BOD: 70%, 5 days

Regulations

RCRA 40CFR: Listed Hazardous Waste No. U009 Toxic Waste

CERCLA: 40CFR 302.4: Listed per CWA Section 311(b)(4) per RCRA Section 3001 per CWA Section 307(a) RQ: 100 lb (45.35 kg)

SARA 40CFR 372.65: Listed TPQ: 10000 lb

SARA EHS 40CFR 355: Listed TPQ: 100 lb

TSCA: Listed

Analytical Methods

Air: EPA TO-2, 0031; ASTM D4490
Soil: SW846 5031, 5032, 5041, 5041A, 8015B, 8030A, 8031, 8240B, 8260A, 8260B, 8316; EPA 7
Water / Groundwater: EPA 603, 624, 624-S, 626; ASTM D3371, D3695
Drinking Water: EPA 524.2
Food: EPA 5
Indoor / Expired Air: NIOSH 1604
Plasma: EPA 29

ACR9000 CAS #: 814-68-6

ACRYLYL CHLORIDE

RTECS: AT7350000
DOT: NA9188
EINECS Number: 212-399-0
Molecular Formula: C_3H_3ClO
Formula Weight: 90.51

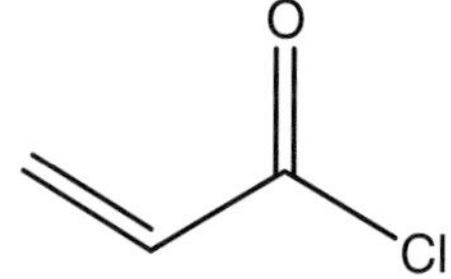

Chemical Structure

Synonyms: ACRYLIC ACID CHLORIDE; ACRYLOYL CHLORIDE; CHLORID KYSELINY AKRYLOVE; 2-PROPENOYL CHLORIDE; PROPENOYL CHLORIDE; 2-PROPENOYL CHLORIDE (9CI)
Description: clear, pale pink to pale yellow liquid
Use: monomer; intermed; med: spherical microcapsules for sustained drug release; for broader resistance to physical & chem attack than acrylic polymers; to increase fluorocarbon content in acrylic resins, without interfering with acrylate properties.

Physical Properties

Boiling Point: 75 °C (167 °F) to 76 °C (169 °F)
Specific Gravity: 1.1136 at 20 °C/4 °C
Water Solubility: Decomposes in Water
Other Solubilities: Soluble in chloride
Refraction Index: 1.4343
Flash Point: 16 °C

RTECS Toxicity Data

Acute Inhalation: Mouse LC_{50} Dose: 92 mg/m^3/2hr. Rat LC_{Lo} Dose: 25 ppm/4hr; Toxic Effects: Sense organs and special senses - Iritis; Behavioral - Ataxia; Lungs, Thorax, or Respiration - Emphysema.

Hazard Overviews

Poison

Corrosive

Flammable
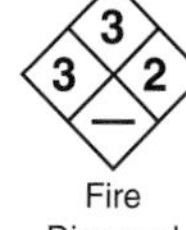

Fire Diamond

Health: Corrosive, causes severe burns of the eyes/skin/respiratory tract. Poison. Also Causes: wheezing, headache, nausea.
Fire: Flammable. Polymerizes. Use dry chemical, carbon dioxide, or regular foam. Use water spray only if prepared for the liberation of acidic gas which forms by reaction of acrylyl chloride and water. Container may explode in heat of fire.
Reactivity: Unstable, polymerizes. Hazardous polymerization can occur at temperatures above 39 °F (4 °C). Incompatible with: water. Hazardous decomposition products: toxic chlorine gas.
Carcinogenicity: IARC - Not listed; NIOSH - Not listed; NTP - Not listed; ACGIH - Not listed; OSHA - Not listed; EPA - Not listed; MAK - Not listed
Primary Target Organs:

Eyes

Skin

Respiratory System

Environmental

Cleanup/Disposal: Guide No. 171: Do not touch or walk through spilled material. Stop leak if you can do it without risk. Prevent dust cloud. Avoid inhalation of asbestos dust. Small Dry Spills: With clean shovel place material into clean, dry container and cover loosely; move containers from spill area. Small Spills: Take up with sand or other noncombustible absorbent material and place into containers for later disposal. Large Spills: Dike far ahead of liquid spill for later disposal. Cover powder spill with plastic sheet or tarp to minimize spreading. Prevent entry into waterways, sewers, basements or confined areas.

Regulations

RCRA 40CFR: Not listed
CERCLA: 40CFR 302.4: Not listed
SARA 40CFR 372.65: Not listed TPQ: 100 lb
SARA EHS 40CFR 355: Listed TPQ: 100 lb
TSCA: Listed

ACY5000 CAS #: 59277-89-3

ACYCLOVIR

RTECS: UP0791400
EINECS Number: 261-685-1
Molecular Formula: $C_8H_{11}N_5O_3$
Formula Weight: 225.21

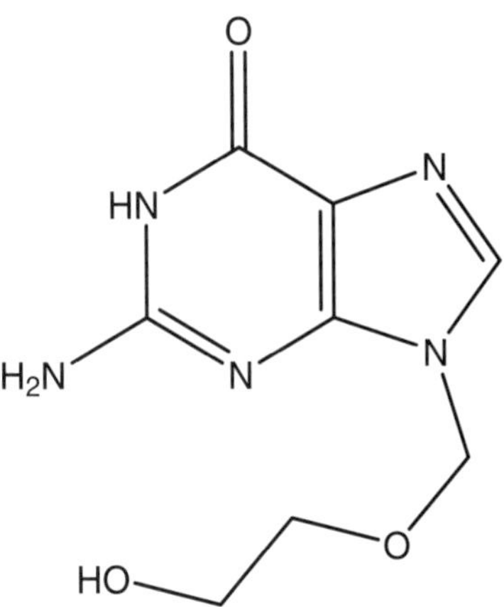

Chemical Structure

Synonyms: ACICLOVIR; ACICLOVIRUM; ACLOVIR; ACYCLOGUANOSINE; 2-AMINO-1,9-DIHYDRO-9-((2-HYDROXYETHOXY)METHYL)-6H-PURIN-6-ONE; BW 248U; 9-((2-HYDROXYETHOXY)METHYL)GUANINE; 9-(2-HYDROXYETHOXYMETHYL)GUANINE; 6H-PURIN-6-ONE,2-AMINO-1,9-DIHYDRO-9-((2-HYDROXYETHOXY)-METHYL)-; VIPRAL; VIRORAX; W-248-U; WELLCOME-248U; ZOVIRAX

Description: crystals

Use: medication as antiviral

Physical Properties

Freezing Point: 256.5 °C (493.7 °F) to 257 °C (494.6 °F)

Water Solubility: 0.2 mg/ml in alcohol

Other Solubilities: 0.2 mg/ml in Alcohol

RTECS Toxicity Data

Acute Oral: Woman TD_{Lo} Dose: 28 mg/kg/2D-I; Toxic Effects: Brain and coverings - Changes in surface EEG; Behavioral - Hallucinations, distorted perceptions; Lungs, Thorax, or Respiration - Sputum. Woman TD_{Lo} Dose: 12 mg/kg/1D-I; Toxic Effects: Brain and coverings - Meningeal changes; Behavioral - Somnolence (general depressed activity).

Mutagenic: Human Cytogenetic Analysis; Cell Type: lymphocyte; Dose: 250 mg/L/48H. Rabbit DNA Inhibition; Cell Type: kidney; Dose: 6800 ug/L.

Hazard Overviews

Health: Irritating. Other Acute Effects: may be harmful by inhalation, ingestion, or skin absorption.

Fire: Extinguishing agents: water spray; carbon dioxide, dry chemical powder or appropriate foam. Precautions: combustible liquid.

Reactivity: Stable. Hazardous polymerization will not occur. Hazardous decomposition products: toxic fumes of: carbon monoxide, carbon dioxide, nitrogen oxides.

Carcinogenicity: IARC - Not listed; NIOSH - Not listed; NTP - Not listed; ACGIH - Not listed; OSHA - Not listed; EPA - Not listed; MAK - Not listed

Primary Target Organs:

Eyes

Skin

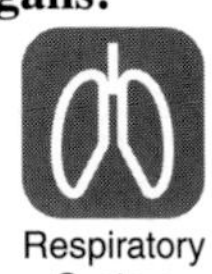
Respiratory System

Environmental

Regulations

RCRA 40CFR: Not listed

CERCLA: 40CFR 302.4: Not listed

SARA 40CFR 372.65: Not listed

SARA EHS 40CFR 355: Not listed

TSCA: Not listed

ADE5000 CAS #: 61-19-8

ADENOSINE 5'-PHOSPHATE

RTECS: AU7480500

EINECS Number: 200-500-0

Molecular Formula: $C_{10}H_{14}N_5O_7P$

Formula Weight: 347.23

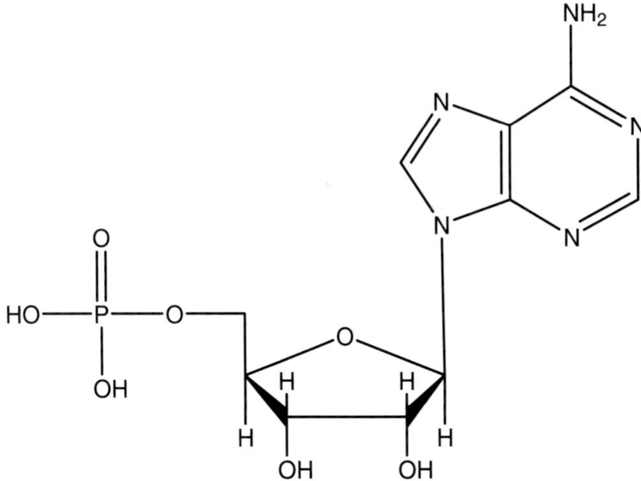

Chemical Structure

Synonyms: A5MP; ADENINE RIBOSIDE; ADENOSINE 5'-MONOPHOSPHATE; ADENOSINE PHOSPHATE; ADENOSINE 5'-PHOSPHORIC ACID; ADENOSINE,MONO(DIHYDROGEN PHOSPHATE) (ESTER); ADENOSINE-5'-MONOPHOSPHORIC ACID; ADENOSINE-5-MONOPHOSPHORIC ACID; ADENOVITE; ADENYL; T-ADENYLIC; 5'-ADENYLIC ACID; ADENYLIC ACID; TERT-ADENYLIC ACID; 5'-AMP; 5-AMP; AMP; AMP (NUCLEOTIDE); CARDIOMONE; ERGADENYLIC ACID; LYCEDAN; MUSCLE ADENYLIC ACID; MY-B-DEN; MYOSTON; NSC-20264; PHOSADEN; PHOSPHADEN; PHOSPHENTASIDE

Description: crystals, powder, or needles

Use: in biochemical research

Physical Properties

Boiling Point: Decomposes at 208 °C (406 °F)

Freezing Point: 196 °C (384.8 °F) to 200 °C (392 °F)

Water Solubility: Readily Soluble in Boiling Water

Other Solubilities: Soluble in 10% HCL; Insoluble in Alcohol.

RTECS Toxicity Data

Reproductive/Teratogenic: Rat Route: Intraperitoneal; Dose: 2800 mg/kg; Duration: female 7-13D of pregnancy; Effects on Embryo or Fetus - Fetotoxicity. Mouse Route: Intraperitoneal; Dose: 2800 mg/kg; Duration: female 7-13D of pregnancy; Effects on Embryo or Fetus - Fetotoxicity.
Mutagenic: Human Other Mutation Test Systems; Cell Type: other cell types; Dose: 100 umol/L. Rat DNA Inhibition; Cell Type: liver; Dose: 10 mmol/L.

Hazard Overviews

Carcinogenicity: IARC - Not listed; NIOSH - Not listed; NTP - Not listed; ACGIH - Not listed; OSHA - Not listed; EPA - Not listed; MAK - Not listed

Environmental

Regulations

RCRA 40CFR: Not listed
CERCLA: 40CFR 302.4: Not listed
SARA 40CFR 372.65: Not listed
SARA EHS 40CFR 355: Not listed
TSCA: Listed

ADI1000 CAS #: 64-95-9

ADIPHENINE

RTECS: AH2624000
EINECS Number: 200-599-0
Molecular Formula: $C_{20}H_{25}NO_2$
Formula Weight: 311.46
Synonyms: ACETIC ACID,DIPHENYL-,2-(DIETHYLAMINO)ESTER; ADIPHENIN; BENZENEACETIC ACID,ALPHA-PHENYL-,2-(DIETHYLAMINO)ETHYLESTER; BENZENEACETIC ACID,ALPHA-PHENYL-,2-(DIETHYLAMINO)ETHYLESTER,(9CI); 2-DIETHYLAMINOETHYL DIPHENYLACETATE; 2-DIETHYLAMINOETHYLESTER KYSELINY DIFENYLOCTOVE; DIFACIL; DIPHACIL; DIPHACYL; DIPHENYLACETIC ACID DIETHYLAMINOETHYL ESTER; DIPHENYLACETIC ACID,2-(DIETHYLAMINO)ETHYL ESTER; DIPHENYLACETYLDIETHYLAMINOETHANOL; ESTER DWUETYLOAMINOETYLOWY KWASU DWUFENYLOOCTOWEGO; PATROVINE; SKF 962A; SPASMOLYTIN; TRANSENTINE; TRANZETIL; TRASENTIN; TRASENTINE; TRAZENTYNA; VEGANTINE; WEGANTYNA
Description: crystals
Use: anticholinergic hydrochloride; (vet): anticholinergic, antispasmodic

Physical Properties

Freezing Point: 113 °C (235.4 °F) to 114 °C (237.2 °F)
Water Solubility: Crystals; Freely Soluble in Water
Other Solubilities: Very Sparingly Soluble in Alcohol, Ether
pH: Aqueous solution is neutral

RTECS Toxicity Data

Acute Oral: Mouse LD_{50} Dose: 600 mg/kg.
Acute Dermal: Mouse LD_{50} Route: Subcutaneous Dose: 400 mg/kg; Toxic Effects: Behavioral - Tremor; Behavioral - Convulsions or effect on seizure threshold; Behavioral - Muscle weakness.

Hazard Overviews

Carcinogenicity: IARC - Not listed; NIOSH - Not listed; NTP - Not listed; ACGIH - Not listed; OSHA - Not listed; EPA - Not listed; MAK - Not listed

Environmental

Regulations

RCRA 40CFR: Not listed
CERCLA: 40CFR 302.4: Not listed
SARA 40CFR 372.65: Not listed
SARA EHS 40CFR 355: Not listed
TSCA: Not listed

ADI5000 CAS #: 124-04-9

ADIPIC ACID

RTECS: AU8400000
DOT: NA9077
EINECS Number: 204-673-3
Molecular Formula: $C_6H_{10}O_4$
Structured MF: $COOH(CH_2)_4COOH$
Formula Weight: 146.14

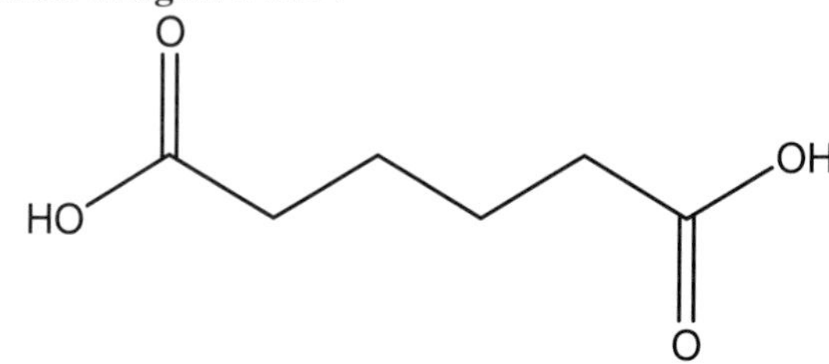

Chemical Structure

Synonyms: ACIFLOCTIN; ACINETTEN; ADILACTETTEN; ADIPATE; ADIPINIC ACID; 1,4-BUTANEDICARBOXYLIC ACID; 1,6-HEXANEDIOIC ACID; HEXANEDIOIC ACID; KYSELINA ADIPOVA; MOLTEN ADIPIC ACID
Description: white crystals, needles; odorless
Use: in food ind: acidulant & additive (in beverages, gelatins, powdered food, baking powders, to inhibit browning of fruits & other foodstuffs, buffer & neutralizing agent; for improving melting characteristics & texture of processed cheese; sequestrant in edible oils; in preservation of meats; in prodn of ester lubricants; in formation of putrescine; in prepn of tetraacylates; in alkyds for plasticizers; in polyester resins; for pharmaceuticals (eg, throat lozenges); perfume fixative.

Physical Properties

Boiling Point: 337.5 °C (640 °F) at 760 mm Hg

Freezing Point: 152 °C (305.6 °F)
Specific Gravity: 1.36 at 25 °C/4 °C
Vapor Density: 5.04 Air=1
Bulk Density: 40.5 lbs/cu ft
Vapor Pressure: 1 mm Hg at 159.5 °C
Water Solubility: Practically non-Hygroscopic
Other Solubilities: Insoluble in Acetic Acid; 0.633/100 parts (wt/wt) Ether at 19 °C; Practically Insoluble in petroleum benzin.
pH: Saturated aqueous solution at 25 °C 2.7
Flash Point: 196 °C Closed Cup
Autoignition Temperature: 420 °C
LEL: In air (dust) 10 mg/l
UEL: 15 mg/l

RTECS Toxicity Data

Acute Oral: Rat LD_{50} Dose: >11 gm/kg. Mouse LD_{50} Dose: 1900 mg/kg; Toxic Effects: Gastrointestinal - Other changes.
Irritation Eye: Rabbit Standard Draize Test Dose: 20 mg/24H; Reaction: moderate.

Hazard Overviews

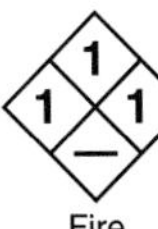

Fire Diamond

Health: Irritating to eyes/skin. Also Causes: irritation around the wrists, ankles, and neck; and to the mucous membranes of the nose, and throat.
Fire: Will burn. May form explosive dust-air mixtures. Use water, carbon dioxide, dry chemical, or alcohol foam. Cool fire exposed containers with a direct water spray.
Reactivity: Stable. Hazardous polymerization cannot occur. Avoid: excessive dust; ignition sources. Incompatible with: steel. Hazardous decomposition products: cyclopentanone; n-valeric acid.
Carcinogenicity: IARC - Not listed; NIOSH - Not listed; NTP - Not listed; ACGIH - Not listed; OSHA - Not listed; EPA - Not listed; MAK - Not listed
Primary Target Organs:

Eyes

Mucous Membranes

Exposure Limits
ACGIH TLV: TWA: 5 mg/m^3.
Respirator Recommendation
Exposure Range: >5 to 50 mg/m^3 Air Purifying, Negative Pressure, Half Mask
Exposure Range: >50 to <500 mg/m^3 Air Purifying, Negative Pressure, Full Face
Exposure Range: 500 to <5000 mg/m^3 Supplied Air, Constant Flow/Pressure Demand, Full Face Self-contained Breathing Apparatus, Pressure Demand, Full Face
Cartridge Color: magenta (P100)

Environmental

Ecotoxicity: LC_{50} Pimephales promelas (fathead minnow) 97 mg/l/72 hr /Static bioassay in Lake Superior Water at 18-22 °C
Environmental Fate: If released on land, it will leach into the ground and probably biodegrade. If released into water, it will readily biodegrade (half-life 3.5 days). Adsorption to sediment, bioconcentration in aquatic organisms, and volatilization should not be significant. It will be primarily associated with aerosols in the atmosphere and be subject to gravitational settling and degradation by reaction with photochemically produced hydroxyl radicals (vapor phase half-life 4.4 days).
Cleanup/Disposal: Guide No. 153: Eliminate all ignition sources (no smoking, flares, sparks or flames in immediate area). Do not touch damaged containers or spilled material unless wearing appropriate protective clothing. Stop leak if you can do it without risk. Prevent entry into waterways, sewers, basements or confined areas. Absorb or cover with dry earth, sand or other non-combustible material and transfer to containers. Do not get water inside containers.

Environmental Physical Data

Henry's Law Constant: calculated at 9.4 x10^{-7}
Octanol/Water Partition Coefficient: log K_{ow} = 0.08
BCF: negligible in fish at 2
BOD: theoretical 1.3%, 0.5 days

Regulations

RCRA 40CFR: Not listed
CERCLA: 40CFR 302.4: Listed per CWA Section 311(b)(4) RQ: 5000 lb (2268 kg)
SARA 40CFR 372.65: Not listed
SARA EHS 40CFR 355: Not listed
TSCA: Listed

ADI9000 CAS #: 111-69-3

ADIPONITRILE

RTECS: AV2625000
DOT: UN2205; IMO6.1
EINECS Number: 203-896-3
Molecular Formula: $C_6H_8N_2$
Structured MF: $NC(CH_2)_4CN$
Formula Weight: 108.16

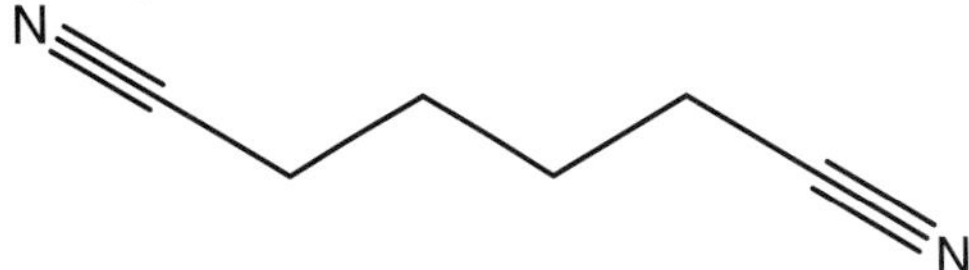

Chemical Structure

Synonyms: ADIPIC ACID DINITRILE; ADIPIC ACID NITRILE; ADIPINSAEUREDINITRIL; ADIPODINITRILE; ADIPONITRIL; ADIPYLDINITRILE; 1,4-DICYANOBUTANE; HEXANEDINITRILE; HEXANEDIOIC ACID,DINITRILE; NITRILE ADIPICO;

TETRAMETHYL CYANIDE; TETRAMETHYLENE CYANIDE; TETRAMETHYLENE DICYANIDE

Description: white to light yellow liquid; odorless

Use: intermediate in the manufacture of nylon and is used in organic synthesis

Physical Properties

Boiling Point: 295 °C (563 °F) at 760 mm Hg
Freezing Point: 1 °C (33.8 °F)
Specific Gravity: 0.962 at 20/4 °C
Vapor Density: 3.73 Air=1
Density: 0.9676 g/mL at 20 °C
Vapor Pressure: 2 mm Hg at 119 °C
Water Solubility: 4.5% by weight
Other Solubilities: 95% Ethanol: >=100 mg/ml at 23 °C; Acetone: >=100 mg/ml at 23 °C; Alcohol: Soluble; Chloroform: Soluble; DMSO: >=100 mg/ml at 23 °C.
Refraction Index: 1.4380 at 20 °C/D
Flash Point: 93.333 °C Open Cup
Autoignition Temperature: 550 °C
LEL: 1.0% v/v

RTECS Toxicity Data

Acute Oral: Rat LD_{50} Dose: 155 mg/kg. Mouse LD_{50} Dose: 172 mg/kg.

Acute Inhalation: Rat LC_{50} Dose: 1710 mg/m^3/4hr; Toxic Effects: Sense organs and special senses - Change in acuity; Lungs, Thorax, or Respiration - Other changes.

Acute Dermal: Rabbit LD_{Lo} Route: Skin; Dose: 1 mg/kg. Guinea Pig LD_{50} Route: Subcutaneous Dose: 50 mg/kg.

Chronic (Multiple Dose) Inhalation: Rat Dose: 493 mg/m^3/6H/4W-I; Toxic Effects: Blood - Pigmented or nucleated red Blood cells; Nutritional and gross metabolic - Weight loss or decreased weight gain; DEATH. Rat Dose: 99 mg/m^3/6H/13W-I; Toxic Effects: Blood - Pigmented or nucleated red Blood cells; Blood - Changes in other cell count; Blood - Changes in erythrocite (RBC) cell count. Rat Dose: 3028 ug/m^3/5H/26W-I; Toxic Effects: Blood - Other changes; Biochemical - True cholinesterase.

Reproductive/Teratogenic: Rat Route: Oral; Dose: 1120 mg/kg; Duration: female 6-19D of pregnancy; Effects on Embryo or Fetus - Fetotoxicity.

Hazard Overviews

Fire Diamond

Health: Irritating to eyes/skin/respiratory tract. Toxic. Also Causes: headache, convulsions, circulatory problems, cyanosis, weakness, mental confusion, rapid respirations, tachycardia, mydriasis, vertigo.

Fire: Combustible. Fight fire with dry chemical, water spray, or regular foam. Do not scatter material with high pressure water streams because adiponitrile floats on water and may spread fire. Releases highly toxic hydrogen cyanide when heated.

Reactivity: Stable. Hazardous polymerization cannot occur. Avoid: heat; ignition sources. Incompatible with: strong acids; oxidizing materials. Hazardous decomposition products: toxic hydrogen cyanide.

Carcinogenicity: IARC - Not listed; NIOSH - Not listed; NTP - Not listed; ACGIH - Not listed; OSHA - Not listed; EPA - Class D, Not classifiable as to human carcinogenicity; MAK - Not listed

Primary Target Organs:

Eyes

Skin

Respiratory System

Nervous System

Cardio-vascular

Exposure Limits
ACGIH TLV: TWA: 2 ppm; 8.8 mg/m^3.
NIOSH REL: TWA: 4 ppm; 18 mg/m^3.

Environmental

Ecotoxicity: TLm Guppies 775 mg/l/96 hr (soft water) /Conditions of bioassay not specified

Environmental Fate: If released to soil, aerobic biodegradation may be an important removal mechanism. Although it has the potential to undergo extensive leaching, biodegradation should limit movement through soil. Volatilization from soil surfaces is not expected to be significant. If released to water, aerobic biodegradation may be an important removal process (half-life in unacclimated river water at 20 °C is about 1 week). Chemical hydrolysis, adsorption to suspended solids and sediments, bioaccumulation in aquatic organisms and volatilization are not expected to be important fate processes in water. If released to the atmosphere, it is expected to exist almost entirely in the vapor phase. Reaction of vapor with photochemically generated hydroxyl radicals and wet deposition are expected to be an important fate processes (hydroxyl radical half-life under typical conditions 11.6 days). Some loss by wet deposition may also occur.

Cleanup/Disposal: Guide No. 153: Eliminate all ignition sources (no smoking, flares, sparks or flames in immediate area). Do not touch damaged containers or spilled material unless wearing appropriate protective clothing. Stop leak if you can do it without risk. Prevent entry into waterways, sewers, basements or confined areas. Absorb or cover with dry earth, sand or other non-combustible material and transfer to containers. Do not get water inside containers.

Environmental Physical Data

Henry's Law Constant: estimated at 7×10^{-9}
Octanol/Water Partition Coefficient: log K_{ow} = -0.32
Sorption Partition Coefficient: K_{oc} = estimated at 9 to 16
BCF: estimated at 1
BOD: 40%, 5 days

Regulations

RCRA 40CFR: Not listed
CERCLA: 40CFR 302.4: Not listed
SARA 40CFR 372.65: Not listed TPQ: 1000 lb
SARA EHS 40CFR 355: Listed TPQ: 1000 lb
TSCA: Listed

ADR5000 CAS #: 23214-92-8

ADRIAMYCIN

RTECS: AV9800000
EINECS Number: 245-495-6
Molecular Formula: $C_{27}H_{29}NO_{11}$
Formula Weight: 543.54
Synonyms: ADM; ADRIAMYCIN SEMIQUINONE; ADRIBLASTIN; ADRIBLASTINA; (85-CIS)-10-[(3-AMINO-2,3,6-TRIDEOXY-ALPHA-L-LYXOHEXAPYRANOSYL)OXY]-7,8,9,10-TETRAHYDRO-6,8,11-TRIHYDROXY-8-(HYDROXYACETYL)-1-METHOXY-5,12-NAPHTHACENEDIONE; 10-[(3-AMINO-2,3,6-TRIDEOXY-D-LYXOHEXOPYRANOSYL)OXY]-8-GLYCOLCYL-7,8,9,10-TETRAHYDRO-6,8,11-TRIHYDROXY-1-METHOXY-5,12-NAPHTHACENEDIONE; DOXORUBICIN; DX; F.I 106; FI 106; 1,2,3,4,6,11-HEXAHYDRO-4BETA,5,12-TRIHYDROXY-4-(HYDROXYACETYL)-10-METHOXY-6,11-DIOXONAPHTHACEN-1BETA-YL-3-AMINO-2,3,6-TRIDEOXY-ALPHA-L-LYXOHEXOPYRANOSIDE; 14'-HYDROXYDAUNOMYCIN; 14-HYDROXYDAUNOMYCIN; 14-HYDROXYDAUNORUBICINE; KW-125; 5,12-NAPHTHACENEDIONE,10-((3-AMINO-2,3,6-TRIDEOXY-ALPHA-L-LYXO-HEXOPYRANOSYL)OXY)-7,8,9,10-TETRAHYDRO-6,8,11-TRIHYDROXY-8-(HYDROXYACETYL)-1-METHOXY-, (8S-CIS)-; 5,12-NAPHTHACENEDIONE,10-((3-AMINO-2,3,6-TRIDEOXY-ALPHA-L-LYXO-HEXOPYRANOSYL)OXY)-7,8,9,10-TETRAHYDRO-6,8,11-TRIHYDROXY-1-METHOXY-5,12-NAPHTHACENEDIONE; NDC 38242-874; NSC-123127; NSC 123127
Description: red, crystalline solid; almost odorless
Use: medication: antineoplastic agent

Physical Properties

Freezing Point: 204 °C (399.2 °F)
Water Solubility: 2% Soluble in Water
Other Solubilities: Soluble in Methanol, aqueous Alcohols; Practically Insoluble in Acetone, Benzene.

RTECS Toxicity Data

Acute Oral: Mouse LD_{50} Dose: 570 mg/kg.
Reproductive/Teratogenic: Rat Route: Parenteral; Dose: 61336 ug/kg; Duration: male 56D prior to mating; Paternal Effects - Spermatogenesis. Rat Route: Intravenous; Dose: 500 ug/kg; Duration: male 1D prior to mating; Paternal Effects - Spermatogenesis. Rat Route: Intravenous; Dose: 10 mg/kg; Duration: male 1D prior to mating; Paternal Effects - Testes, epididymis, sperm duct. Rat Route: Intravenous; Dose: 3 mg/kg; Duration: male 1D prior to mating; Paternal Effects - Spermatogenesis; Testes, epididymis, sperm duct. Rat Route: Intravenous; Dose: 8 mg/kg; Duration: male 1D prior to mating; Paternal Effects - Spermatogenesis.
Mutagenic: Human DNA Damage; Cell Type: ovary; Dose: 10 umol/L. Human DNA Damage; Cell Type: other cell types; Dose: 1250 ug/L. Human DNA Damage; Cell Type: other cell types; Dose: 250 ug/L. Human DNA Damage; Cell Type: other cell types; Dose: 500 nmol/L. Human DNA Damage; Cell Type: lymphocyte; Dose: 50 ug/L. Human DNA Damage; Cell Type: other cell types; Dose: 700 nmol/L. Human DNA Damage; Cell Type: fibroblast; Dose: 100 umol/L. Human DNA Damage; Cell Type: other cell types; Dose: 1 umol/L. Human Unscheduled DNA Synthesis; Cell Type: lymphocyte; Dose: 100 ug/L. Human DNA Inhibition; Cell Type: fibroblast; Dose: 500 ug/L. Human DNA Inhibition; Cell Type: leukocyte; Dose: 10 umol/L. Human DNA Inhibition; Cell Type: HeLa cell; Dose: 1 umol/L. Human DNA Inhibition; Cell Type: other cell types; Dose: 10 nmol/L/2H. Human DNA Inhibition; Cell Type: other cell types; Dose: 30 nmol/L. Human DNA Inhibition; Cell Type: other cell types; Dose: 150 ug/L. Human DNA Inhibition; Cell Type: other cell types; Dose: 400 ug/L. Human Cytogenetic Analysis; Cell Type: other cell types; Dose: 1 umol/L. Human Cytogenetic Analysis; Cell Type: leukocyte; Dose: 20 ug/L. Human Cytogenetic Analysis; Cell Type: lymphocyte; Dose: 10 ug/L. Human Cytogenetic Analysis; Cell Type: fibroblast; Dose: 10 ug/L/1H. Human Sister Chromatid Exchenge; Cell Type: lymphocyte; Dose: 5 ng/L/48H. Human Micronucleus Test; Cell Type: lymphocyte; Dose: 20 ug/L. Human Other Mutation Test Systems; Cell Type: lymphocyte; Dose: 1 mg/L. Human Other Mutation Test Systems; Cell Type: lymphocyte; Dose: 10 umol/L. Human Other Mutation Test Systems; Cell Type: other cell types; Dose: 10 nmol/L/2H. Human Other Mutation Test Systems; Cell Type: lymphocyte; Dose: 500 ug/L. Human Other Mutation Test Systems; Cell Type: other cell types; Dose: 150 ug/L. Human Other Mutation Test Systems; Cell Type: other cell types; Dose: 50 nmol/L. Human Other Mutation Test Systems; Cell Type: other cell types; Dose: 800 ug/L. Human Other Mutation Test Systems; Cell Type: other cell types; Dose: 50 nmol/L.
Tumorigenic: Rat Route: Intravenous; Dose: 5 mg/kg; Toxic Effects: Tumorigenic - Carcinogenic by RTECS criteria; Skin and appendages - Tumors. Rat Route: Intravenous; Dose: 8 mg/kg; Toxic Effects: Tumorigenic - Neoplastic by RTECS criteria; Skin and appendages - Tumors. Rat Route: Intravenous; Dose: 10 mg/kg; Toxic Effects: Tumorigenic - Carcinogenic by RTECS criteria; Lungs, Thorax, or Respiration - Tumors; Skin and appendages - Tumors. Rat Route: Intravenous; Dose: 5 mg/kg; Toxic Effects: Tumorigenic - Equivocal tumorigenic agent by RTECS criteria; Skin and appendages - Tumors. Rat Route: Intravenous; Dose: 8 mg/kg; Toxic Effects: Tumorigenic - Carcinogenic by RTECS criteria; Skin and appendages - Tumors. Rat Route: Intravenous; Dose: 10 mg/kg; Toxic Effects: Tumorigenic - Carcinogenic by RTECS criteria; Skin and appendages - Tumors. Rat Route: Intravenous; Dose: 2 mg/kg; Toxic Effects: Tumorigenic - Neoplastic by RTECS criteria; Skin and appendages - Tumors.

Hazard Overviews

Health: Severely irritating to eyes/skin/respiratory tract. Toxic. Other Acute Effects: harmful if swallowed; may be harmful if inhaled; may be harmful if absorbed through the skin; symptoms of exposure may include burning sensation; coughing; wheezing; laryngitis; shortness of breath; headache; nausea; vomiting; vesicant; causes blisters on contact with skin; target organs: heart, bone marrow, blood, liver. Chronic Effects: may alter genetic material; may cause congenital malformation in the fetus. Probable carcinogen.

Fire: Hazards: emits toxic fumes. Extinguishing agents: carbon dioxide, dry chemical powder or appropriate foam. Precautions: combustible liquid.

Reactivity: Incompatible with: strong oxidizing agents. Hazardous decomposition products: toxic fumes of: carbon monoxide, carbon dioxide, hydrogen chloride gas, nitrogen oxides.

Carcinogenicity: IARC - Group 2A, Probably carcinogenic to humans; NIOSH - Not listed; NTP - Not listed; ACGIH - Not listed; OSHA - Not listed; EPA - Not listed; MAK - Not listed

Primary Target Organs:

Eyes

Skin

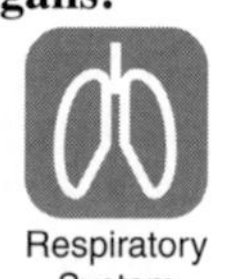
Respiratory System

Liver

Cardio-vascular

Bone

Environmental

Environmental Physical Data

Octanol/Water Partition Coefficient: log K_{ow} = 1.27

Regulations

RCRA 40CFR: Not listed
CERCLA: 40CFR 302.4: Not listed
SARA 40CFR 372.65: Not listed
SARA EHS 40CFR 355: Not listed
TSCA: Not listed

AFL1000 CAS #: 1162-65-8

AFLATOXIN B1

RTECS: GY1925000
EINECS Number: 214-603-3
Molecular Formula: $C_{17}H_{12}O_6$
Formula Weight: 312.06

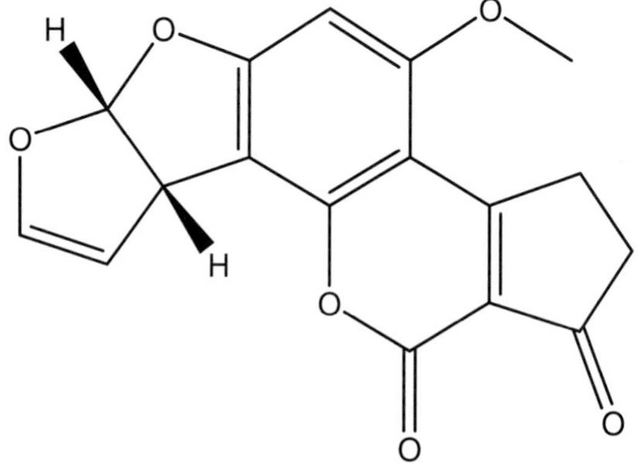

Chemical Structure

Synonyms: AFB1; AFBI; AFLATOXIN B; CYCLOPENTA(C)FURO(3',2':4,5)FURO(2,3-H)(1)BENZOPYRAN-1,11-DIONE,2,3,6AALPHA,9AALPHA-TETRAHYDRO-4-METHOXY-; CYCLOPENTA(C)FURO(3',2':4,5)FURO(2,3-H)(1)BENZOPYRAN-1,11-DIONE,2,3,6A,9A-TETRAHYDRO-4-METHOXY-; CYCLOPENTA(C)FURO(3',2':4,5)FURO(2,3-H)(1)BENZOPYRAN-1,11-DIONE,2,3,6A,9A-TETRAHYDRO-4-METHOXY-,(6AR-CIS)-; 6-METHOXYDIFUROCOUMARONE; PYRAN-1,11-DIONE; 2,3,6A ALPHA,9A ALPHA-TETRAHYDRO-4-METHOXYCYCLOPENTA(C)FURO(3',2':4,5)FURO (2,3-H)(1)BENZOPYRAN-1,11-DIONE; 2,3,6A,9A-TETRAHYDRO-4-METHOXYCYCLOPENTA(C)FURO(3',2':4,5)FURO(2,3-H)(1)BENZ

Description: crystals which exhibit blue fluorescence

Use: as positive standards for contaminated food and iron assays; in studies of metabolism and carcinogenesis

Physical Properties

Freezing Point: 268 °C (514.4 °F) to 269 °C (516.2 °F)
Water Solubility: < 1 mg/mL at 22 C
Other Solubilities: DMSO: 1-5 mg/ml at 22 °C; Acetone: 1-5 mg/ml at 22 °C; Alcohol: Soluble; Hydrocarbon solvents: Slightly Soluble; Chloroform: Soluble; Non-polar solvents: Insoluble; Moderately polar organic solvents: Freely Soluble.
Flash Point: Not available; probably combustible

RTECS Toxicity Data

Acute Oral: Rat LD_{50} Dose: 4800 ug/kg. Mouse LD_{50} Dose: 9 mg/kg.

Mutagenic: Human DNA Damage; Cell Type: liver; Dose: 2 ppm/3H. Human DNA Damage; Cell Type: HeLa cell; Dose: 32 mg/L/24H. Human DNA Damage; Cell Type: fibroblast; Dose: 300 umol/L. Human Unscheduled DNA Synthesis; Cell Type: fibroblast; Dose: 10 umol/L/2H. Human Unscheduled DNA Synthesis; Cell Type: HeLa cell; Dose: 10 nmol/L. Human Unscheduled DNA Synthesis; Cell Type: other cell types; Dose: 100 umol/L. Human Unscheduled DNA Synthesis; Cell Type: liver; Dose: 1 umol/L. Human Unscheduled DNA Synthesis; Cell Type: mammary gland; Dose: 10 umol/L. Monkey Unscheduled DNA Synthesis; Cell Type: liver; Dose: 50 ug/L. Monkey Unscheduled DNA Synthesis; Cell Type: kidney; Dose: 10 ug/L. Human DNA Inhibition; Cell Type: HeLa cell; Dose: 50 nmol/L/1H-C. Human DNA Inhibition; Cell Type: liver; Dose: 2 mg/L. Human DNA Inhibition; Cell Type: lung; Dose: 1 ppm. Human DNA Inhibition; Cell Type: fibroblast; Dose: 40 mg/L. Monkey DNA Inhibition; Cell Type: kidney; Dose: 2 mg/L. Human DNA Adduct; Cell Type: other cell types; Dose: 1 umol/L. Human DNA Adduct; Cell Type: other cell types; Dose: 1 umol/L. Human DNA Adduct; Cell Type: lymphocyte; Dose: 100 ug/L. Monkey DNA Adduct; Cell Type: other cell types; Dose: 1 umol/L. Human Mutations in Microorganisms; Cell Type: lymphocyte; Dose: 100 ug/L (+S9). Human Mutations in Mammalian Somatic Cells; Cell Type: lymphocyte; Dose: 3 nmol/L. Human Cytogenetic Analysis; Cell Type: lymphocyte; Dose: 100 nmol/L. Human Cytogenetic Analysis; Cell Type: leukocyte; Dose: 1 mg/L. Human Cytogenetic Analysis; Cell Type: HeLa cell; Dose: 100 nmol/L. Human Cytogenetic Analysis; Cell Type: lung; Dose: 1 mg/L. Human Cytogenetic Analysis; Cell Type: fibroblast; Dose: 60 umol/L. Human Sister Chromatid Exchenge; Cell Type: lymphocyte; Dose: 100 umol/L. Human Sister Chromatid Exchenge; Cell Type: fibroblast; Dose: 3 mg/L. Human Micronucleus Test; Cell Type: lymphocyte; Dose: 100 nmol/L. Human Morphological Transformation; Cell Type: other cell types; Dose: 2500 mg/L. Human Morphological Transformation; Cell Type: other cell types; Dose: 10 mg/L. Monkey Morphological Transformation; Route: Oral; Dose: 700 ug/kg. Human Other Mutation Test Systems; Cell Type: liver; Dose: 2 mg/L. Monkey Other Mutation Test Systems; Cell Type: kidney; Dose: 500 ug/L. Monkey Other Mutation Test Systems; Cell Type: kidney; Dose: 1 mg/L. Human DNA Adduct; Cell Type: other cell types; Dose: 1 umol/L. Human DNA Adduct; Cell Type: other cell types; Dose: 1 umol/L.

Hazard Overviews

Poison

Health: Irritating. Poison. Other Acute Effects: may be fatal if inhaled, swallowed, or absorbed through skin. Chronic Effects: may alter genetic material; overexposure may cause reproductive disorder(s) based on tests with laboratory animals; possible teratogen; target organ: liver. Carcinogen.
Fire: Will burn. Hazards: emits toxic fumes. Extinguishing agents: water spray; carbon dioxide, dry chemical powder or appropriate foam. Precautions: combustible liquid.
Reactivity: Incompatible with: strong oxidizing agents, light sensitive, air sensitive. Hazardous decomposition products: toxic fumes of: carbon monoxide, carbon dioxide, nitrogen oxides.
Carcinogenicity: IARC - Group 1, Carcinogenic to humans; NIOSH - Not listed; NTP - Not listed; ACGIH - Not listed; OSHA - Not listed; EPA - Not listed; MAK - Not listed
Primary Target Organs:

Eyes

Skin

Respiratory System

Liver

Environmental

Ecotoxicity: LD_{50} Duckling (one day old) oral 18.2 ug/50 g body weight
Cleanup/Disposal: Guide No. 154: Eliminate all ignition sources (no smoking, flares, sparks or flames in immediate area). Do not touch damaged containers or spilled material unless wearing appropriate protective clothing. Stop leak if you can do it without risk. Prevent entry into waterways, sewers, basements or confined areas. Absorb or cover with dry earth, sand or other non-combustible material and transfer to containers. Do not get water inside containers.

Regulations

RCRA 40CFR: Not listed
CERCLA: 40CFR 302.4: Not listed
SARA 40CFR 372.65: Not listed
SARA EHS 40CFR 355: Not listed
TSCA: Not listed

AFL3000 CAS #: 7220-81-7

AFLATOXIN B2

RTECS: GY1722000
EINECS Number: 230-618-8
Molecular Formula: $C_{17}H_{14}O_6$
Formula Weight: 314.08

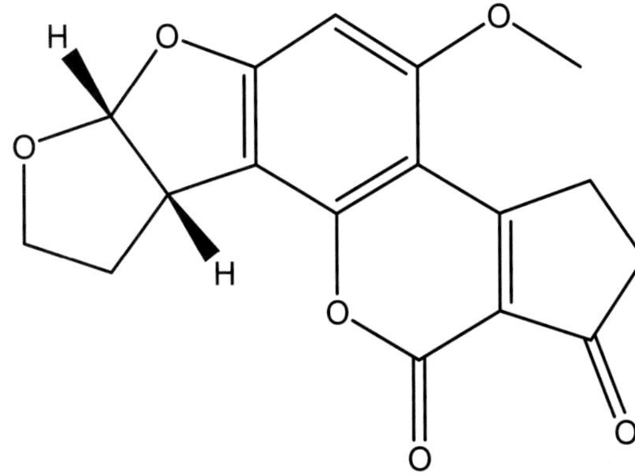

Chemical Structure

Synonyms: CYCLOPENTA(C)FURO(3',2':4,5)FURO(2,3-H)(1)BENZOPYRAN-1,11-DIONE,2,3,6AALPHA,8,9,9AALPHA-HEXAHYDRO-4-METHOXY-; CYCLOPENTA(C)FURO(3',2':4,5)FURO(2,3-H)(1)BENZOPYRAN-1,11-DIONE,2,3,6A,8,9,9A-HEXAHYDRO-4-METHOXY-,(6AR-CIS)-; DIHYDROAFLATOXIN B1; DIHYDROAFLATOXINE B1; 2,3,6A ALPHA,8,9,9A ALPHA-HEXAHYDRO-4-METHOXYCYCLOPENTA(C)FURO(3',2':4,5)FURO(2,3-H)(1)BENZOPYRAN-1,11-DIONE
Description: colorless to pale yellow crystals with blue fluorescence

Physical Properties

Freezing Point: 286 °C (546.8 °F) to 289 °C (552.2 °F)

RTECS Toxicity Data

Acute Oral: Duck LD_{50} Dose: 1700 ug/kg.
Mutagenic: Human DNA Inhibition; Cell Type: fibroblast; Dose: 100 umol/L. Rat Unscheduled DNA Synthesis; Cell Type: liver; Dose: 10 umol/L/1H.

Hazard Overviews

Poison

Health: Irritating. Poison. Other Acute Effects: may be fatal if inhaled, swallowed, or absorbed through skin. Chronic Effects: may alter genetic material; target organ: liver. Carcinogen.
Fire: Extinguishing agents: water spray; carbon dioxide, dry chemical powder or appropriate foam. Precautions: combustible liquid.

Reactivity: Incompatible with: strong oxidizing agents, strong acids, strong bases, light sensitive, air sensitive. Hazardous decomposition products: toxic fumes of: carbon monoxide, carbon dioxide.
Carcinogenicity: IARC - Not listed; NIOSH - Not listed; NTP - Not listed; ACGIH - Not listed; OSHA - Not listed; EPA - Not listed; MAK - Not listed
Primary Target Organs:

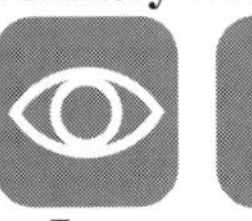
Eyes

Skin

Respiratory System

Liver

Environmental

Ecotoxicity: LD_{50} Duck oral 1700 ug/kg LD_{50} Duckling (day old) oral 84.8 ug/50 g body weight

Regulations

RCRA 40CFR: Not listed
CERCLA: 40CFR 302.4: Not listed
SARA 40CFR 372.65: Not listed
SARA EHS 40CFR 355: Not listed
TSCA: Not listed

AFL5000 CAS #: 1165-39-5

AFLATOXIN G1

RTECS: LV1720000
EINECS Number: 214-615-9
Molecular Formula: $C_{17}H_{12}O_7$
Formula Weight: 328.06

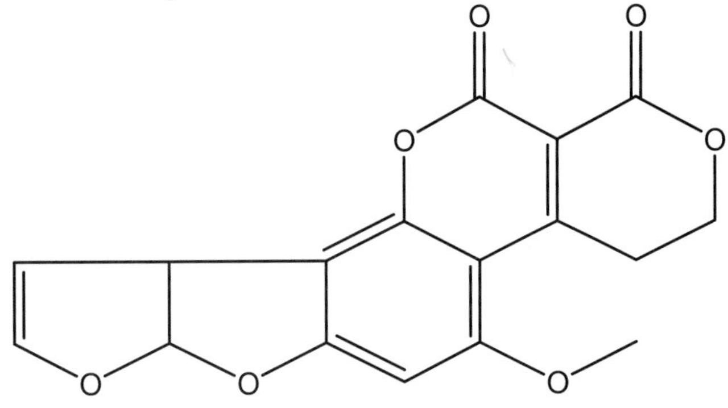

Chemical Structure

Synonyms: 1H,12H-FURO(3',2':4,5)FURO(2,3-H)PYRANO(3,4-C)(1)BENZOPYRAN-1,12-DIONE,3,4,7A,10A-TETRAHYDRO-5-METHOXY-; 1H,12H-FURO(3',2':4,5)FURO(2,3-H)PYRANO(3,4-C)(1)BENZOPYRAN-1,12-DIONE,3,4,7A,10A-TETRAHYDRO-5-METHOXY-,(7AR-CIS)-
Description: crystals which exhibit green fluorescene

Physical Properties

Freezing Point: 244 °C (471.2 °F) to 246 °C (474.8 °F)

RTECS Toxicity Data

Acute Oral: Duck LD_{50} Dose: 785 ug/kg.
Mutagenic: Human Unscheduled DNA Synthesis; Cell Type: fibroblast; Dose: 10 umol/L/30M. Human Unscheduled DNA Synthesis; Cell Type: HeLa cell; Dose: 10 umol/L. Monkey DNA Inhibition; Cell Type: kidney; Dose: 2 mg/L. Monkey Other Mutation Test Systems; Cell Type: kidney; Dose: 2 mg/L.

Hazard Overviews

Poison

Health: Irritating. Poison. Other Acute Effects: may be fatal if inhaled, swallowed, or absorbed through skin. Chronic Effects: may alter genetic material; target organ: liver. Carcinogen.
Fire: Extinguishing agents: water spray; carbon dioxide, dry chemical powder or appropriate foam. Precautions: combustible liquid.
Reactivity: Incompatible with: strong oxidizing agents, strong acids, strong bases, light sensitive, air sensitive. Hazardous decomposition products: toxic fumes of: carbon monoxide, carbon dioxide.
Carcinogenicity: IARC - Not listed; NIOSH - Not listed; NTP - Not listed; ACGIH - Not listed; OSHA - Not listed; EPA - Not listed; MAK - Not listed
Primary Target Organs:

Eyes

Skin

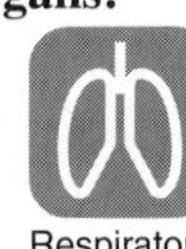
Respiratory System

Liver

Environmental

Ecotoxicity: LD_{50} rainbow trout1.90 mg/kg LD_{50} Duckling (day old) oral 39.2 ug/50 g body LD_{50} Duck oral 785 ug/kg

Regulations

RCRA 40CFR: Not listed
CERCLA: 40CFR 302.4: Not listed
SARA 40CFR 372.65: Not listed
SARA EHS 40CFR 355: Not listed
TSCA: Not listed

AFL7000 CAS #: 7241-98-7

AFLATOXIN G2

RTECS: LV1700000
EINECS Number: 230-643-4
Molecular Formula: $C_{17}H_{14}O_7$
Formula Weight: 330.07

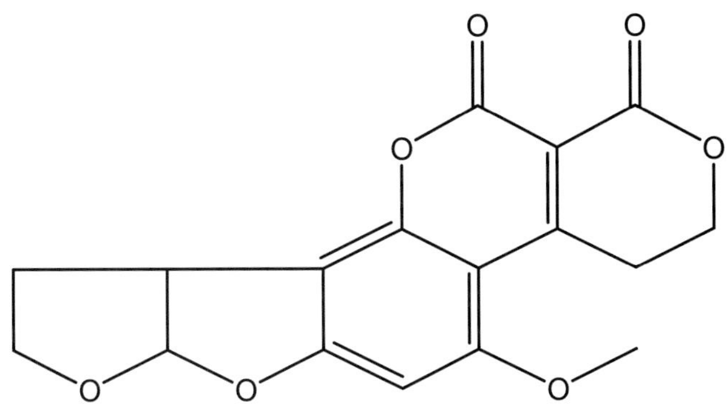

Chemical Structure

Synonyms: DIHYDROAFLATOXIN G1; 1H,12H-FURO(3',2':4,5)FURO(2,3-H)PYRANO(3,4-C)(1)BENZOPYRAN-1,12-DIONE,; 1H,12H-FURO(3',2':4,5)FURO(2,3-H)PYRANO(3,4-C)(1)BENZOPYRAN-1,12-DIONE,3,4,7AALPHA,9,10,10AALPHA-HEXAHYDRO-5-METHOXY-; 1H,12H-FURO(3',2':4,5)FURO(2,3-H)PYRANO(3,4-C)(1)BENZOPYRAN-1,12-DIONE,3,4,7A,9,10,10A-HEXAHYDRO-5-METHOXY-,(7AR-CIS)-; 1H,12H-FURO(3',2':4,5)FURO(2,3-H)PYRANO(3,4-C)(1)BENZOPYRAN-1,12-DIONE,3,4,7A-ALPHA,9,10,10A-ALPHA-HEXAHYDRO-5-METHOXY-; 3,4,7A-ALPHA,9,10,10A-ALPHA-HEXAHYDRO-5-METHOXY-; 3,4,7A ALPHA,9,10,10A ALPHA-HEXAHYDRO-5-METHOXY-1H,12H-FURO(3',2':4,5)FURO(2,3-H)PYRANO(3,4-C)(1)-BENZOPYRAN-1,12-DIONE

Description: crystals which exhibit green-blue fluorescence

Physical Properties

Freezing Point: 237 °C (458.6 °F) to 240 °C (464 °F)
Water Solubility: Slightly Soluble
Other Solubilities: Acetone: Soluble; Chloroform: Soluble; Hydrocarbon solvents: Slighty Soluble; Methanol: Soluble.
Flash Point: Not available; probably combustible

RTECS Toxicity Data

Acute Oral: Duck LD_{50} Dose: 2450 ug/kg.
Mutagenic: Rat Unscheduled DNA Synthesis; Cell Type: liver; Dose: 10 umol/L/1H. Hamster Unscheduled DNA Synthesis; Cell Type: liver; Dose: 100 nmol/L.

Hazard Overviews

Poison

Health: Irritating. Poison. Other Acute Effects: may be fatal if inhaled, swallowed, or absorbed through skin. Chronic Effects: laboratory experiments have shown mutagenic effects; target organ: liver. Possible human carcinogen.
Fire: Will burn. Extinguishing agents: water spray; carbon dioxide, dry chemical powder or appropriate foam. Precautions: combustible liquid.
Reactivity: Incompatible with: strong oxidizing agents, strong acids, strong bases. Hazardous decomposition products: toxic fumes of: carbon monoxide, carbon dioxide.

Carcinogenicity: IARC - Not listed; NIOSH - Not listed; NTP - Not listed; ACGIH - Not listed; OSHA - Not listed; EPA - Not listed; MAK - Not listed
Primary Target Organs:

Eyes

Skin

Respiratory System

Liver

Environmental

Ecotoxicity: LD_{50} Duck oral 2450 ug/kg LD_{50} Duckling (day old) oral 172.5 ug/50 g body wt
Cleanup/Disposal: Guide No. 154: Eliminate all ignition sources (no smoking, flares, sparks or flames in immediate area). Do not touch damaged containers or spilled material unless wearing appropriate protective clothing. Stop leak if you can do it without risk. Prevent entry into waterways, sewers, basements or confined areas. Absorb or cover with dry earth, sand or other non-combustible material and transfer to containers. Do not get water inside containers.

Regulations

RCRA 40CFR: Not listed
CERCLA: 40CFR 302.4: Not listed
SARA 40CFR 372.65: Not listed
SARA EHS 40CFR 355: Not listed
TSCA: Not listed

AFL9000 **CAS #: 1402-68-2**

AFLATOXINS

RTECS: AW5950000
Molecular Formula: Unspecified or Variable
Synonyms: AFLATOXIN
Use: for experimental purposes only

Physical Properties

Water Solubility: Slightly Soluble in Water
Other Solubilities: Soluble in Methanol, Acetone, and Chloroform

RTECS Toxicity Data

Acute Oral: Human LD_{Lo} Dose: 229 ug/kg/8W; Toxic Effects: Behavioral - Anorexia (human); Gastrointestinal - Ulceration or bleeding from small intestine; Liver - Jaundice, other or unclassified. Monkey LD_{50} Dose: 1750 ug/kg; Toxic Effects: Liver - Hepatitis (hepatocellular necrosis), zonal; Biochemical - Effect on inflammation or mediation of inflammation.
Chronic (Multiple Dose) Oral: Rat Dose: 900 mg/kg/60D-C; Toxic Effects: Blood - Pigmented or nucleated red Blood cells; Blood - Changes in other cell count; Blood - Changes in erythrocite (RBC) cell count. Rat Dose: 1 gm/kg/10W-I; Toxic Effects: Blood - Change in clotting factors; Blood - Changes in serum composition; Biochemical - Phosphatases. Rat Dose: 56 mg/kg/28D-C; Toxic Effects: Blood - Change in

clotting factors; Blood - Changes in serum composition; Nutritional and gross metabolic - Changes in iron. Rat Dose: 2100 ug/kg/3W-C; Toxic Effects: Blood - Other changes; Biochemical - Phosphatases; Biochemical - Dehydrogenases. Rat Dose: 2925 ug/kg/39D-I; Toxic Effects: Liver - Jaundice, other or unclassified; Blood - Hemorrhage; DEATH. Rat Dose: 1680 ug/kg/21D-C; Toxic Effects: Blood - Changes in erythrocite (RBC) cell count; Nutritional and gross metabolic - Weight loss or decreased weight gain; Biochemical - Transaminases. Rat Dose: 20 mg/kg/10D-I; Toxic Effects: Kidney, Ureter, and Bladder - Other changes in urine composition; Kidney, Ureter, and Bladder - Changes in kidney weight; Nutritional and gross metabolic - Changes in calcium. Rat Dose: 105 mg/kg/3W-C; Toxic Effects: Nutritional and gross metabolic - Weight loss or decreased weight gain.

Reproductive/Teratogenic: Hamster Route: Parenteral; Dose: 4 mg/kg; Duration: female 8D of pregnancy; Effects on Fertility - Litter size; Effects on Embryo or Fetus - Fetotoxicity; Specific Developmental Abnormalities - Musculoskeletal system. Hamster Route: Parenteral; Dose: 6 mg/kg; Duration: female 8D of pregnancy; Specific Developmental Abnormalities - Urogenital system.

Mutagenic: Human Cytogenetic Analysis; Cell Type: leukocyte; Dose: 50 mg/L. Rat Micronucleus Test; Route: Oral; Dose: 8500 ug/kg.

Tumorigenic: Rat Route: Oral; Dose: 7788 ug/kg/13W-C; Toxic Effects: Tumorigenic - Equivocal tumorigenic agent by RTECS criteria; Liver - Tumors. Rat Route: Oral; Dose: 2250 ug/kg (10-21D preg); Toxic Effects: Tumorigenic - Equivocal tumorigenic agent by RTECS criteria; Tumorigenic effects - Transplacental tumorigenesis; Liver - Tumors. Rat Route: Oral; Dose: 87 mg/kg/52W-C; Toxic Effects: Tumorigenic - Equivocal tumorigenic agent by RTECS criteria; Liver - Tumors; Gastrointestinal - Tumors. Rat Route: Oral; Dose: 66 mg/kg/52W-C; Toxic Effects: Tumorigenic - Equivocal tumorigenic agent by RTECS criteria; Gastrointestinal - Tumors; Liver - Tumors. Rat Route: Oral; Dose: 62 mg/kg/37W-C; Toxic Effects: Tumorigenic - Equivocal tumorigenic agent by RTECS criteria; Liver - Tumors; Kidney, Ureter, and Bladder - Kidney tumors. Monkey Route: Multiple routes; Dose: 71 mg/kg/6Y-I; Toxic Effects: Tumorigenic - Equivocal tumorigenic agent by RTECS criteria; Liver - Jaundice, other or unclassified; Biochemical - Transaminases.

Hazard Overviews

Carcinogenicity: IARC - Group 1, Carcinogenic to humans; NIOSH - Not listed; NTP - Listed; ACGIH - Not listed; OSHA - Not listed; EPA - Not listed; MAK - Not listed

Environmental

Regulations

RCRA 40CFR: Not listed
CERCLA: 40CFR 302.4: Not listed
SARA 40CFR 372.65: Not listed
SARA EHS 40CFR 355: Not listed
TSCA: Not listed

AGA3000 CAS #: 9002-18-0

AGAR

RTECS: AW7950000
EINECS Number: 232-658-1
Molecular Formula: Unknown
Synonyms: AGAR AGAR FLAKE; AGAR-AGAR; AGAR-AGAR FLAKE; AGAR-AGAR GUM; AGAROPECTIN,MIXT WITH AGAROSE; AGAROSE,MIXT WITH AGAROPECTIN; BENGAL; BENGAL GELATIN; BENGAL ISINGLASS; CEYLON; CEYLON ISINGLASS; CHINESE GELATIN; CHINESE ISINGLASS; DIGENEA SIMPLEX MUCILAGE; GELATIN; GELOSE; JAPAN AGAR; JAPAN ISINGLASS; JAPANESE GELATIN; LAYOR CARANG; MACASSAR GELATIN; MACASSAR GUM; VEGETABLE GELATIN
Description: pale buff, powder or thin pieces; odorless
Use: substitute for gelatin, etc.; meat canning; medicinal encapsulations and ointments; dental impressions; corrosion inhibitor; sizing, dyeing and printing fabrics and textiles; adhesives and culture media

Physical Properties

Water Solubility: Slowly Soluble in Hot Water
Other Solubilities: 95% Ethanol: <1 mg/ml at 17 °C; Acetone: <1 mg/ml at 17 °C; DMSO: <1 mg/ml at 17 °C; Most organic solvents: Insoluble.
Flash Point: Not available; probably combustible

RTECS Toxicity Data

Acute Oral: Rat LD_{50} Dose: 11 gm/kg. Mouse LD_{50} Dose: 16 gm/kg.

Hazard Overviews

Health: May be irritating to eyes/skin. Other Acute Effects: may be harmful by inhalation, ingestion, or skin absorption.
Fire: Will burn. Hazards: emits toxic fumes. Extinguishing agents: water spray; carbon dioxide, dry chemical powder or appropriate foam. Precautions: combustible liquid.
Reactivity: Incompatible with: strong oxidizing agents, protect from moisture. Hazardous decomposition products: toxic fumes of: carbon monoxide, carbon dioxide.
Carcinogenicity: IARC - Not listed; NIOSH - Not listed; NTP - Not listed; ACGIH - Not listed; OSHA - Not listed; EPA - Not listed; MAK - Not listed

Environmental

Regulations

RCRA 40CFR: Not listed
CERCLA: 40CFR 302.4: Not listed
SARA 40CFR 372.65: Not listed
SARA EHS 40CFR 355: Not listed
TSCA: Listed

AGA6000 CAS #: 2757-90-6

AGARITINE

RTECS: MA1284000
Molecular Formula: $C_{12}H_{17}N_3O_4$
Formula Weight: 267.28
Synonyms: L-GLUTAMIC ACID 5-(2-(4-(HYDROXYMETHYL)PHENYL)HYDRAZIDE); L-GLUTAMIC ACID,5-(2-(4-(HYDROXYMETHYL)PHENYL)HYDRAZIDE); L-GLUTAMIC ACID,5-(2-(4-(HYDROXYMETHYL)PHENYL)HYDRAZIDE)(9CI); GLUTAMIC ACID,5-(2-(ALPHA-HYDROXY-P-TOLYL)HYDRAZIDE),L-; L-GLUTAMIC ACID,5-(2-(ALPHA-HYDROXY-PARA-TOLYL)HYDRAZIDE); BETA-N-(GAMMA-L(+)-GLUTAMYL)-4-HYDROXYMETHYLPHENYLHYDRAZIDE; BETA-N-(GAMMA-L(+)GLUTAMYL)4-HYDROXYMETHYLPHENYLHYDRAZINE; BETA-N-[GAMMA-L(+)-GLUTAMYL]-4-HYDROXYMETHYLPHENYLHYDRAZINE
Description: glistening crystals
Use: pharmaceuticals

Physical Properties

Boiling Point: Decomposes at 205 °C (401 °F) to 209 °C (408.2 °F)
Freezing Point: Decomposes at 205 °C (401 °F) to 209 °C (408.2 °F)
Water Solubility: Very Freely Soluble in Water
Other Solubilities: Soluble in anhydrous organic solvents.
Flash Point: Not available; probably combustible

RTECS Toxicity Data

Mutagenic: Bacteria - S Typhimurium Mutations in Microorganisms; Dose: 2 umol/plate (+S9), 20 umol/plate (-S9).

Hazard Overviews

Fire: Will burn.
Carcinogenicity: IARC - Group 3, Not classifiable as to carcinogenicity to humans; NIOSH - Not listed; NTP - Not listed; ACGIH - Not listed; OSHA - Not listed; EPA - Not listed; MAK - Not listed

Environmental

Regulations

RCRA 40CFR: Not listed
CERCLA: 40CFR 302.4: Not listed
SARA 40CFR 372.65: Not listed
SARA EHS 40CFR 355: Not listed
TSCA: Not listed

ALA3000 CAS #: 15972-60-8

ALACHLOR

RTECS: AE1225000
EINECS Number: 240-110-8
Molecular Formula: $C_{14}H_{20}ClNO_2$
Formula Weight: 269.77
Synonyms: ACETAMIDE,2-CHLORO-N-(2,6-DIETHYLPHENYL)-N-(METHOXYMETHYL)-; ACETAMIDE,2-CHLORO-N-(2,6-DIETHYLPHENYL)-N-(METHOXYMETHYL)-(9CI); ACETANILIDE,2-CHLORO-2',6'-DIETHYL-N-(METHOXYMETHYL)-; ALACHLORE; ALANEX; ALANOX; ALATOX 480; ALOCHLOR; CHIMICLOR; CHLORESSIGSAEURE-N-(METHOXYMETHYL)-2,6-DIAETHYLANILID; 2-CHLORO-2',6'-DIETHYL-N-(METHOXYMETHYL)ACETANILIDE; 2-CHLORO-2',6'-DIETHYL-N-METHOXYMETHYLACETANILIDE; ALPHA-CHLORO-2',6'-DIETHYL-N-METHOXYMETHYLACETANILIDE; 2-CHLORO-N-(2,6-DIETHYL)PHENYL-N-METHOXYMETHYLACETAMIDE; 2-CHLORO-N-(2,6-DIETHYLPHENYL)-N-(METHOXYMETHYL)ACETAMIDE; CP 50144; LASAGRIN; LASSAGRIN; LASSO; LASSO MICRO-TECH; LAZO; METACHLOR; METHACHLOR; N-(METHOXYMETHYL)-2,6-DIETHYLCHLOROACETANILIDE; PILLARZO
Description: colorless to yellow crystals; odorless
Use: selective systemic pre & early post emergence herbicide for annual grasses & broadleaf weeds; in corn, sorghum, soybeans, peanuts, cotton, vegetables, forage crops, maize, groundnuts, lima beans, oilseed rape, brassicas, radish, oil radish, sunflowers, sugar cane, potatoes, peas, tobacco, some ornamentals.

Physical Properties

Boiling Point: 100 °C (212 °F) at 0.02 mm Hg
Freezing Point: 40 °C (104 °F) to 41 °C (105.8 °F)
Specific Gravity: 1.133 at 25 °C/15.6 °C
Saturated Vapor Density: 1.200000288 kg/m^3
Vapor Pressure: 2.2 x10^{-5} mm Hg at 25 °C
Water Solubility: 240 ppm at 24 °C
Other Solubilities: Soluble in Chloroform; Slightly Soluble in Heptane.

RTECS Toxicity Data

Acute Oral: Rat LD_{50} Dose: 930 mg/kg. Mouse LD_{50} Dose: 462 mg/kg.
Acute Inhalation: Rat LC_{50} Dose: >23400 mg/m^3/6hr. Rabbit LC_{50} Dose: >5100 mg/m^3.
Acute Dermal: Rabbit LD_{50} Route: Skin; Dose: 3500 mg/kg. Rat LD_{50} Route: Skin; Dose: 1200 mg/kg.
Mutagenic: Human DNA Damage; Cell Type: lymphocyte; Dose: 5 mg/L. Human Cytogenetic Analysis; Cell Type: lymphocyte; Dose: 1 mg/L.
Tumorigenic: Mouse Route: Oral; Dose: 142 gm/kg/78W-C; Toxic Effects: Tumorigenic - Carcinogenic by RTECS criteria; Lungs, Thorax, or Respiration - Tumors.

Hazard Overviews

Carcinogenicity: IARC - Not listed; NIOSH - Not listed; NTP - Not listed; ACGIH - Not listed; OSHA - Not listed; EPA - Not listed; MAK - Not listed

Environmental

Ecotoxicity: LC_{50} Carp 1 yr old 4.60 mg/l/96 hr at 18-20 °C, pH 6.9, wt 10 g LD_{50} Ictalurus (Catfish) 6.5 ppm in 96 hr /Conditions of bioassay not specified LD_{50} Procambarus acutus (Crayfish) 19.5 ppm in 96 hr LC_{50} Salmo gairdneri (Rainbow trout) 1.4 mg/l/96 hr at 12 °C (95% confidence interval 1.1-1.8 mg/l), wt 0.8 g. Static bioassay without aeration, pH 7.2-7.5, water hardness 40-50 mg/l as calcium carbonate and alkalinity of 30-35 mg/l LD_{50} Anas platyrhynchos (Mallard) male oral >2000 mg/kg (3 mo old)

Environmental Fate: In soil, it is transformed to its metabolites primarily by biodegradation. The half-life of disappearance from soil is about 15 days, although very little mineralization has been observed. It is highly to moderately mobile in soil and the mobilization decreases with an increase in organic carbon and clay content in soil. In water, both photolysis and biodegradation are important for loss, although the role of photolysis becomes important in shallow clean water, particularly in the presence of sensitizers. Bioconcentration in aquatic organisms is not important. Half-life due to reaction with hydroxyl radicals in the atmosphere has been estimated to be 2.1 hours. Partial removal will also occur as a result of dry and wet deposition.

Environmental Physical Data

Henry's Law Constant: estimated at 1.2 $x10^{-10}$
Octanol/Water Partition Coefficient: log K_{ow} = 2.63 to 3.53
Sorption Partition Coefficient: K_{oc} = 2.08 to 2.28
BCF: fathead minnow 6

Regulations

RCRA 40CFR: Not listed
CERCLA: 40CFR 302.4: Not listed
SARA 40CFR 372.65: Listed
SARA EHS 40CFR 355: Not listed
TSCA: Not listed

Analytical Methods

Soil: EPA PMD-TLC; SW846 8081
Water / Groundwater: EPA 645; USGS O3106
Drinking Water: EPA 505, 507, 508.1, 525.1, 525.2; AOAC 991.07; ASTM D5175, D5475
Food: FDA 212.1, 212.2, 232.1, 232.4, 242.1; AOAC 985.04, 988.03
Indoor / Expired Air: ASTM D4861
Plasma: EPA 001; FDA 211.1, 231.1, 252
Other: EPA 1656

ALA6000 CAS #: 56-41-7

(L)-ALANINE

RTECS: AY2990000
EINECS Number: 200-273-8
Molecular Formula: $C_3H_7NO_2$
Formula Weight: 89.09

O
OH
NH2

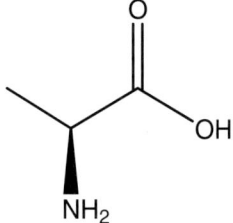

Chemical Structure

Synonyms: (S)-ALANINE; ALANINE; ALPHA-ALANINE; L-(+)-ALANINE; L-ALANINE; L-ALPHA-ALANINE; (S)-2-AMINOPROPANOIC ACID; L-2-AMINOPROPANOIC ACID; ALPHA-AMINOPROPIONIC ACID; L-ALPHA-AMINOPROPIONIC ACID; L-S-AMINOPROPIONIC ACID; PROPANOIC ACID,2-AMINO-,(S)-
Description: colorless crystals
Use: microbiological & biochemical research; dietary supplement; in pharmaceutical preparations for injection or infusion; dietary supplement and flavor compounds in maillard reaction products; stimulant of glucagon secretion

Physical Properties

Boiling Point: Sublimes at 250 °C (482 °F)
Freezing Point: Decomposes at 300 °C (572 °F)
Specific Gravity: 1.432 at 22 °C
Water Solubility: 127.3 g Soluble in 1 L Water at 0 °C
Other Solubilities: Insoluble in Acetone.
Ionization Potential (eV): 8.88 +/-1.0

RTECS Toxicity Data

Mutagenic: Human Sister Chromatid Exchenge; Cell Type: lymphocyte; Dose: 50 mg/L.

Hazard Overviews

Health: May be irritating to eyes/skin/respiratory tract. Other Acute Effects: may be harmful by inhalation, ingestion, or skin absorption.
Fire: Extinguishing agents: water spray; carbon dioxide, dry chemical powder or appropriate foam. Precautions: combustible liquid.
Reactivity: Incompatible with: strong oxidizing agents. Hazardous decomposition products: toxic fumes of: carbon monoxide, carbon dioxide, nitrogen oxides.
Carcinogenicity: IARC - Not listed; NIOSH - Not listed; NTP - Not listed; ACGIH - Not listed; OSHA - Not listed; EPA - Not listed; MAK - Not listed

Environmental

Environmental Physical Data

Octanol/Water Partition Coefficient: log K_{ow} = -2.96

Regulations

RCRA 40CFR: Not listed
CERCLA: 40CFR 302.4: Not listed
SARA 40CFR 372.65: Not listed
SARA EHS 40CFR 355: Not listed
TSCA: Listed

ALD1000 CAS #: 116-06-3

ALDICARB

RTECS: UE2275000
DOT: UN2757; UN2758; UN2991; UN2992; IMO3.2; IMO6.1
EINECS Number: 204-123-2
Molecular Formula: $C_7H_{14}N_2O_2S$
Structured MF: $CH_3SC(CH_3)_2CH=NOC(O)NHCH_3$

Formula Weight: 190.25

Synonyms: ALDECARB; ALDICARBE; AMBUSH; CARBAMIC ACID,METHYL-,0-((2-METHYL-2-(METHYLTHIO)PROPYLIDENE)AMINO) DERIV; CARBAMIC ACID,METHYL-,O-((2-METHYL-2-(METHYLTHIO)PROPYLIDENE)AMINO) DERIV; CARBAMYL; ENT 27,093; 2-METHYL-2-(METHYLTHIO)PROPANAL,O-((METHYLAMINO)CARBONYL))OXIME; 2-METHYL-2-(METHYLTHIO)PROPANAL,O-((METHYLAMINO)CARBONYL)OXIME; 2-METHYL-2-(METHYLTHIO)PROPIONALDEHYDE O-(METHYLCARBAMOYL)OXIME; 2-METHYL-2-(METHYLTHIO)PROPIONALDEHYDEO-(METHYLCARBAMOYL)OXIME; 2-METHYL-2-METHYLTHIO-PROPIONALDEHYD-O-(N-METHYL-CARBAMOYL)-OXIM; 2-METIL-2-TIOMETIL-PROPIONALDEID-O-(N-METIL-CARBAMOIL)-OSSIMA; OMS-771; OMS 771; PROPANAL,2-METHYL-2-(METHYLTHIO)-,O-((METHYLAMINO)CARBONYL)OXIME; PROPIONALDEHYDE,2-METHYL-2-(METHYLTHIO)-,O-(METHYLCARBAMOYL)OXIME; SULFONE ALDOXYCARB; TEMIC; TEMIK; TEMIK 10 G; TEMIK G10; UC-21149; UC 21149; UNION CARBIDE 21149; UNION CARBIDE UC-21149

Description: white, crystalline solid; slightly sulfurous odor

Use: soil-applied insecticide, acaricide and nematocide for use on cotton, sugar beets, potatoes, peanuts, ornamentals, yams, oranges, pecans, dry beans, soybeans and sugarcane; a systemic nematocide and insecticide used against millipedes, eelworms and many insect pests of root crops

Physical Properties

Boiling Point: Decomposes

Freezing Point: 99 °C (210.2 °F) to 100 °C (212 °F)

Specific Gravity: 1.195 at 25 °C/20 °C

Saturated Vapor Density: 1.200008779 kg/m^3

Vapor Pressure: 0.001 mm Hg at 25 °C

Water Solubility: 0.6% at 25 °C

Other Solubilities: 20% in ethyl Ether; 10% in Toluene; 15% in Chlorobenzene; 5% in Carbon Tetrachloride, 24% in methyl isobutyl ketone (each at 30 °C); Soluble in Ethanol; freely Soluble in Chloroform; Slightly Soluble in Petroleum Ether; Insoluble in Heptane.

Flash Point: Nonflammable

RTECS Toxicity Data

Acute Oral: Rat LD_{50} Dose: 650 ug/kg. Mouse LD_{50} Dose: 300 ug/kg.

Acute Inhalation: Rat LC_{50} Dose: 200 mg/m^3/5hr.

Acute Dermal: Rabbit LD_{50} Route: Skin; Dose: 1400 mg/kg. Rat LD_{50} Route: Skin; Dose: 2500 ug/kg. Rat LD_{50} Route: Subcutaneous Dose: 666 ug/kg.

Chronic (Multiple Dose) Oral: Domestic Animal Dose: 4 mg/kg/40D-I; Toxic Effects: Lungs, Thorax, or Respiration - Chronic pulmonary edema; Gastrointestinal - Hypermotility, diarrhea; DEATH.

Mutagenic: Human Mutations in Mammalian Somatic Cells; Cell Type: lymphocyte; Dose: 2 gm/L. Human Cytogenetic Analysis; Cell Type: lymphocyte; Dose: 350 mg/L. Human Sister Chromatid Exchenge; Cell Type: lymphocyte; Dose: 10 mg/L.

Hazard Overviews

Fire: Noncombustible.

Carcinogenicity: IARC - Group 3, Not classifiable as to carcinogenicity to humans; NIOSH - Not listed; NTP - Not listed; ACGIH - Not listed; OSHA - Not listed; EPA - Class D, Not classifiable as to human carcinogenicity; MAK - Not listed

Exposure Limits

AIHA WEEL: TWA: 0.07 mg/m^{3}skin.

Environmental

Ecotoxicity: LD_{50} Callipepla californica (California quail) oral 2.58 mg/kg (95% confidence limit 1.96-3.40 mg/kg), 10 mo old males LC_{50} Coturnix japonica (Japanese quail) oral 381 ppm in 5-day diet ad libitum (95% confidence limit 317-453 ppm), age 14 days LC_{50} Pimephales promelas (Fathead minnow) 1370 ug/l/96 hr /Flow-through bioassay LC_{50} Paramecium multimicronucleatum 145 ppm/9 hr; 122 ppm/13 hr; 104 ppm/17 hr; 93 ppm/24 hr /Static bioassay

Environmental Fate: If released to the soil it should not bind to the soil. It will be susceptible to chemical and possibly biological oxidation to form the sulfoxide and sulfone. Hydrolysis is both acid and base catalyzed with examples of hydrolysis half-lives in soil at 15 °C of 9.9 days at pH 6.34 and 7.0, 23 days at pH 7.2, and 3240 days at pH 5.4. Half-lives in soil have been reported to be 7 days in loam soil under field conditions, a few days in green house soil; a general range of persistence in soil of 1-15 days has been reported. It degraded faster in soil which had been previously treated with carbofuran. It does not degrade in groundwater under aerobic conditions unless relatively high pH (pH 8.5) exists; reported half-lives in groundwater under anaerobic conditions at pH 7.7-8.3 were 62-1300 days. It has been shown to be formed from aldicarb sulfoxide in groundwater under aerobic conditions and under anaerobic conditions in groundwater to which glucose had been added. It may volatilize from soil with the rate of its evaporation increasing with the rate of evaporation for water. It may leach to the groundwater in some soils where the rates of hydrolysis and oxidation are relatively slow, as in the slow hydrolysis reported at pH's around 5.4. If released to water it should not adsorb to sediments or bioconcentrate in aquatic organisms. It will be subject to hydrolysis which is both acid and base catalyzed with examples of half-lives of 131 days at pH 3.95 and 6 days at pH 8.85 at 20 °C, and 3240 days at pH 5.5 and 15 °C. No information on biodegradation in natural waters was found. It is susceptible to photolysis when irradiated at 254 nm, but may not be photolyzed by light >290 nm. Volatilization from water should not be an important fate process. Half-life is 5 days in lake and pond water. If released to the atmosphere it will be subject to reaction with hydroxyl radicals with an estimated vapor phase half-life of 3.49 days. No information on photolysis at environmentally significant wavelengths was found.

Cleanup/Disposal: Guide No. 131: Fully encapsulating, vapor protective clothing should be worn for spills and leaks with no fire. Eliminate all ignition sources (no smoking, flares,

sparks or flames in immediate area). All equipment used when handling the product must be grounded. Do not touch or walk through spilled material. Stop leak if you can do it without risk. Prevent entry into waterways, sewers, basements or confined areas. A vapor suppressing foam may be used to reduce vapors. Small Spills: Absorb with earth, sand or other non-combustible material and transfer to containers for later disposal. Use clean non-sparking tools to collect absorbed material. Large Spills: Dike far ahead of liquid spill for later disposal. Water spray may reduce vapor; but may not prevent ignition in closed spaces. Guide No. 151: Do not touch damaged containers or spilled material unless wearing appropriate protective clothing. Stop leak if you can do it without risk. Prevent entry into waterways, sewers, basements or confined areas. Cover with plastic sheet to prevent spreading. Absorb or cover with dry earth, sand or other non-combustible material and transfer to containers. Do not get water inside containers.

Environmental Physical Data

Henry's Law Constant: calculated at 1.5 x10^{-9}
Octanol/Water Partition Coefficient: log K_{OW} = 1.13
Sorption Partition Coefficient: K_{OC} = soils 8.2
BCF: fish 42

Regulations

RCRA 40CFR: Listed Hazardous Waste No. P070 Toxic Waste
CERCLA: 40CFR 302.4: Listed per RCRA Section 3001 RQ: 1 lb (0.454 kg)
SARA 40CFR 372.65: Listed TPQ: 100/10000 lb
SARA EHS 40CFR 355: Listed TPQ: 1 lb
TSCA: Listed

Analytical Methods

Air: EPA TO-5
Soil: SW846 8321A
Water / Groundwater: USGS O3107
Drinking Water: EPA 531.1; AOAC 991.06; ASTM D5315
Food: AOAC 974.04, 985.23
Plasma: FDA 242.2

ALD3000 CAS #: 1646-75-9

ALDICARB OXIME

RTECS: UE2286000
EINECS Number: 216-709-5
Molecular Formula: $C_5H_{11}NOS$
Structured MF: $(CH_3)_2C(SCH_3)CH{=}NOH$
Formula Weight: 133.20
Synonyms: 2-METHYL-2-(METHYLTHIO)PROPANAL OXIME; 2-METHYL-2-(METHYLTHIO)PROPANALOXIME; 2-METHYL-2-(METHYLTHIO)PROPIONALDEHYDE OXIME; 2-METHYL-2-(METHYLTHIO)PROPIONALDOXIME; 2-(METHYLTHIO)-2-METHYLPROPIONALDEHYDE OXIME; PROPANAL,2-METHYL-2-(METHYLTHIO)-,OXIME; PROPIONALDEHYDE,2-METHYL-2-(METHYLTHIO)-,OXIME; TEMIK OXIME
Description: colorless, clear viscous liquid; unpleasant, sulfurous or musty odor
Use: produced as a precursor in the production of the carbamate pesticide aldicarb

Physical Properties

Boiling Point: Partially decomposes at 210 °C (410 °F)
Freezing Point: 21 °C (69.8 °F)
Specific Gravity: 1.05
Vapor Density: Estimate 4.6 Air=1
Saturated Vapor Density: < 1.200567332 kg/m^3
Vapor Pressure: < 0.1 mm Hg at 20 °C
Water Solubility: 10 to 50 mg/mL at 20 °C
Other Solubilities: 95% Ethanol: >=100 mg/ml at 20 °C; Acetone: >=100 mg/ml at 20 °C; DMSO: >=100 mg/ml at 20 °C.
pH: About 7
Evaporation Rate: > 1 Butyl Acetate=1
Flash Point: 118 °C
Autoignition Temperature: 285 °C

RTECS Toxicity Data

Acute Oral: Rat LD_{50} Dose: 742 mg/kg; Toxic Effects: Sense organs and special senses - Lacrimation; Sense organs and special senses - Chromidracryorrhea; Behavioral - Food intake (animal).
Acute Inhalation: Rat LC_{50} Dose: 1230 mg/m^3/4hr; Toxic Effects: Sense organs and special senses - Lacrimation; Lungs, Thorax, or Respiration - Dyspnea.
Acute Dermal: Rabbit LD_{50} Route: Skin; Dose: 1900 mg/kg.
Mutagenic: Mouse Mutations in Mammalian Somatic Cells; Cell Type: lymphocyte; Dose: 1600 mg/L.

Hazard Overviews

Fire: Will burn.
Carcinogenicity: IARC - Not listed; NIOSH - Not listed; NTP - Not listed; ACGIH - Not listed; OSHA - Not listed; EPA - Not listed; MAK - Not listed

Environmental

Regulations

RCRA 40CFR: Not listed
CERCLA: 40CFR 302.4: Not listed
SARA 40CFR 372.65: Not listed
SARA EHS 40CFR 355: Not listed
TSCA: Listed

ALD6000 CAS #: 1646-88-4

ALDICARB SULFONE

RTECS: UE2080000
EINECS Number: 216-710-0
Molecular Formula: $C_7H_{14}N_2O_4S$
Formula Weight: 222.29
Synonyms: ALDOXYCARB; ALDOXYCARBE; ENT 4.9; ENT AI3-29261; 2-MESYL-2-METHYLPROPIONALDEHYDE O-METHYLCARBAMOYLOXIME; 2-METHYL-2-(METHYLSULFONYL)PROPANAL O-

((METHYLAMINO)CARBONYL)OXIME; 2-METHYL-2-(METHYLSULFONYL)PROPIONALDEHYDE O-(METHYLCARBAMOYL)OXIME; STANDAK; SULFOCARB; UC-21865

Physical Properties

Freezing Point: 140 °C (284 °F) to 142 °C (287.6 °F)
Water Solubility: 8000 mg/l at 20 °C

RTECS Toxicity Data

Acute Oral: Rat LD_{50} Dose: 20 mg/kg. Duck LD_{50} Dose: 33500 ug/kg.
Acute Inhalation: Rat LC_{50} Dose: 140 mg/m^3/4hr.
Acute Dermal: Rabbit LD_{50} Route: Skin; Dose: 200 mg/kg. Rat LD_{50} Route: Skin; Dose: 1 gm/kg.

Hazard Overviews

Carcinogenicity: IARC - Not listed; NIOSH - Not listed; NTP - Not listed; ACGIH - Not listed; OSHA - Not listed; EPA - Not listed; MAK - Not listed

Environmental

Ecotoxicity: Fishes: trout 96h LC_{50} 40 mg/l bluegill 96h LC_{50} 55 mg/l

Regulations

RCRA 40CFR: Listed Hazardous Waste No. P203 Toxic Waste
CERCLA: 40CFR 302.4: Listed per RCRA Section 3001 RQ: 1 lb (0.454 kg)
SARA 40CFR 372.65: Not listed
SARA EHS 40CFR 355: Not listed
TSCA: Not listed

Analytical Methods

Soil: SW846 8321A
Drinking Water: EPA 531.1; AOAC 991.06; ASTM D5315
Food: AOAC 985.23
Plasma: FDA 242.2

ALD9000 CAS #: 309-00-2

ALDRIN

RTECS: IO2100000
DOT: UN2761;UN2995;UN2996; NA2761; IMO6.1
EINECS Number: 206-215-8
Molecular Formula: $C_{12}H_8Cl_6$
Formula Weight: 364.93
Synonyms: ALDOCIT; ALDREC; ALDREX; ALDREX 30 E.C; ALDREX 30; ALDREX 40; ALDRINE; ALDRITE; ALDRON; ALDROSOL; ALGRAN; ALTOX; COMPOUND 118; 1,4:5,8-DIMETHANONAPHTHALENE,1,2,3,4,10,10-HEXACHLORO-1,4,4A,5,8,8A-HEXAHYDRO-,(1ALPHA,4ALPHA,4ABETA,5ALPHA,8ALPHA,8ABETA)-; 1,4:5,8-DIMETHANONAPHTHALENE,1,2,3,4,10,10-HEXACHLORO-1,4,4A,5,8,8A-HEXAHYDRO-,ENDO,EXO-; DRINOX; ENDO,EXO-1,2,3,4,10,10-HEXACHLORO-1,4,4A,5,8,8A-HEXAHYDRO-1,4:5,8-DIMETHANONAPHTHALENE; ENT 15,949; 1,2,3,4,10,10-HEXACHLORO-1ALPHA,4ALPHA,4ABETA,5ALPHA,8ALPHA,8ABETA-HEXAHYDRO-1,4:5,8-DIMETHANONAPHTHALENE; (1R,4S,4AS,5S,8R,8AR)-1,2,3,4,10,10-HEXACHLORO-1,4,4A,5,8,8A-HEXAHYDRO-1,4:5,8-DI METHANONAPHTHALENE,NOT LESS THAN 95%; (1ALPHA,4ALPHA,4A BETA,5ALPHA,8ALPHA,8A BETA)-1,2,3,4,10,10-HEXACHLORO-1,4,4A,5,8,8A-HEXAHYDRO-1,4:5,8-DIMETHANONAPHTHALENE; (1R,4S,5S,8R)-1,2,3,4,10,10-HEXACHLORO-1,4,4A,5,8,8A-HEXAHYDRO-1,4:5,8-DIMETHANONAPHTHALENE; 1,2,3,4,10,10-HEXACHLORO-1,4,4A,5,8,8A-HEXAHYDRO-1,4,5,8-DIMETHANONAPHTHALENE; 1,2,3,4,10,10-HEXACHLORO-1,4,4A,5,8,8A-HEXAHYDRO-1,4:5,8-DIMETHANONAPHTHALENE; 1,2,3,4,10,10-HEXACHLORO-1,4,4A,5,8,8A-HEXAHYDRO-1,4,5,8-DIMETHANONAPTHALENE; 1,2,3,4,10,10-HEXACHLORO-1,4,4A,5,8,8A-HEXAHYDRO-1,4-ENDO,EXO-5,8-DIMETHAN ONAPHTHALENE; 1,2,3,4,10,10-HEXACHLORO-1,4,4A,5,8,8A-HEXAHYDRO-1,4-ENDO-EXO-5,8-DIMETHANONAPHTHALENE; 1,2,3,4,10,10-HEXACHLORO-1,4,4A,5,8,8A-HEXAHYDRO-ENDO-1,4-EXO-5,8-DIMETHANONAPHTHALENE; HEXACHLOROHEXAHYDRO-ENDO-EXO-DIMETHANONAPHTHALENE; HEXACHLOROHEXAHYDRO-ENDO-EXO-DIMETHANONAPTHALENE; 1,2,3,4,10,10-HEXACHLORO-1,4,4A,5,8,8A-HEXAHYDRO-EXO-1,4-ENDO-5,8-DIMETHANONAPHTHALENE; HHDN; KORTOFIN; LATKA 118; OCTALENE; OCTALENE COMPOUND 118; SD 2794; SEEDRIN; SOILGRIN; TATUZINHO; TIPULA
Use: an insecticide

Physical Properties

Boiling Point: 145 °C (293 °F) at 2 mm Hg
Freezing Point: 104 °C (219.2 °F)
Specific Gravity: 1.6 at 20 °C/4 °C (solid)
Saturated Vapor Density: 1.200000421 kg/m^3
Density: 1.7 at 20 °C
Vapor Pressure: 0.000023 mm Hg at 20 °C
Water Solubility: 0.027 mg/L in Water at 27 °C
Other Solubilities: Soluble in Aromatics, Esters, Ketones, Paraffins and halogenated solvents; 0.20 mg/l at 25 °C; Soluble in Alcohol, Ether; Moderately Soluble in petroleum oils.
Flash Point: ~ 65.556 °C Closed Cup

RTECS Toxicity Data

Acute Oral: Human TD_{Lo} Dose: 14 mg/kg; Toxic Effects: Behavioral - Tremor; Behavioral - Excitment; Gastrointestinal - Nausea or vomiting. Child LD_{Lo} Dose: 1250 ug/kg. Rat LD_{50} Dose: 39 mg/kg.
Acute Inhalation: Rat LC_{Lo} Dose: 5800 ug/m^3/4hr.
Acute Dermal: Rabbit LD_{50} Route: Skin; Dose: 15 mg/kg; Toxic Effects: Sense organs and special senses - Other; Behavioral - Convulsions or effect on seizure threshold; Behavioral - Excitment. Rabbit LD_{Lo} Route: Subcutaneous Dose: 100 mg/kg.
Chronic (Multiple Dose) Oral: Rat Dose: 9100 ug/kg/26W-I; Toxic Effects: Endocrine - Other changes; Blood - Other changes. Rat Dose: 109 mg/kg/2Y-C; Toxic Effects: Liver - Changes in liver weight. Rat Dose: 348 mg/kg/58D-C; Toxic Effects: Biochemical - Other esterases.
Reproductive/Teratogenic: Rat Route: Subcutaneous; Dose: 10 mg/kg; Duration: female 2D prior to mating; Maternal Effects - Uterus, cervix, vagina. Rat Route: Intraperitoneal; Dose: 1950 ug/kg; Duration: male 13D prior to mating; Paternal Effects - Other effects on male; Endocrine - Androgenic. Rat Route: Intraperitoneal; Dose: 3900 ug/kg;

Duration: male 26D prior to mating; Paternal Effects - Prostate, seminal vessicle, Cowper's gland, accessory glands.

Mutagenic: Human Unscheduled DNA Synthesis; Cell Type: fibroblast; Dose: 1 umol/L. Human DNA Inhibition; Cell Type: lymphocyte; Dose: 100 mg/L. Human Cytogenetic Analysis; Cell Type: lymphocyte; Dose: 19125 ug/L. Human Cytogenetic Analysis; Cell Type: leukocyte; Dose: 19125 ug/L.

Tumorigenic: Rat Route: Oral; Dose: 200 mg/kg/2Y-C; Toxic Effects: Tumorigenic - Neoplastic by RTECS criteria; Lungs, Thorax, or Respiration - Tumors; Skin and appendages - Tumors. Rat Route: Oral; Dose: 188 mg/kg/2Y-C; Toxic Effects: Tumorigenic - Equivocal tumorigenic agent by RTECS criteria; Liver - Multiple effects; Lungs, Thorax, or Respiration - Other changes.

Hazard Overviews

Reactivity: Stable. Hazardous polymerization cannot occur. Avoid: elevated temperatures. Incompatible with: concentrated acids and phenols in the presence of oxidizers; may corrode some metals. Hazardous decomposition products: carbon oxide(s); hydrogen chloride; chlorine gas.

Carcinogenicity: IARC - Group 3, Not classifiable as to carcinogenicity to humans; NIOSH - Listed as carcinogen; NTP - Not listed; ACGIH - Class A3, Animal carcinogen; OSHA - Not listed; EPA - Class B2, Probable human carcinogen based on animal studies; MAK - Not listed

Exposure Limits

OSHA PEL: TWA: 0.25 mg/m^3; skin.

ACGIH TLV: TWA: 0.25 mg/m^3.

NIOSH REL: TWA: 0.25 mg/m^3.

NIOSH IDLH: 25 mg/m^3.

DFG MAK: TWA: 0.25 mg/m^3.

Respirator Recommendation

Exposure Range: >0.25 to 12.5 mg/m^3 Supplied Air, Constant Flow/Pressure Demand, Half Mask

Exposure Range: >12.5 to <25 mg/m^3 Supplied Air, Constant Flow/Pressure Demand, Full Face

Exposure Range: 25 to unlimited mg/m^3 Self-contained Breathing Apparatus, Pressure Demand, Full Face

Note: warning properties unknown

Environmental

Ecotoxicity: LC_{50} Leiostomus xanthurus (spot) 3.2 ug/l/2 days /Conditions of bioassay not specified LC_{50} Anguilla rostrata (American eel) 5 ppb/96 hr /Static bioassay LC_{50} Sphaeroides maculatus (Northern puffer) 36 ppb/96 hr /Static bioassay LC_{50} Pteronarcys californica (Stonefly) 1.3 ug/l/96 hr, second year class, at 15 °C (95% confidence limit 0.8-2.2 ug/l) static bioassay without aeration, pH 7.2-7.5, water hardness 40-50 mg/l as calcium carbonate and alkalinity of 30-35 mg/l LC_{50} Salmo gairdneri (Rainbow trout) 2.6 ug/l/96 hr, wt 0.6 g, at 13 °C (95% confidence limit 2.3-2.9 ug/l) static bioassay without aeration, pH 7.2-7.5, water hardness 40-50 mg/l as calcium carbonate and alkalinity of 30-35 mg/l EC_{50} Simocephalus serrulatus (Daphnid) 23 ug/l/48 hr, first instar, at 15 °C (95% confidence limit 17-30 ug/l) static bioassay without aeration, pH 7.2-7.5, water hardness 40-50 mg/l as calcium carbonate and alkalinity of 30-35 mg/l EC_{50} Cypridopsis vidua (Seed shrimp) 18 ug/l/48 hr, mature, at 21 °C (95% confidence limit 15-21 ug/l) static bioassay without aeration, pH 7.2-7.5, water hardness 40-50 mg/l as calcium carbonate and alkalinity of 30-35 mg/l LC_{50} Mugil cephalus (striped mullet) 2.0 ug/l/2 days /Conditions of bioassay not specified LC_{50} Aeroneuria pacifica (stonefly) 22 ug/l/30 days /Conditions of bioassay not specified TLm Gasterosteus aculeatus (Threespine stickleback) 27.4 ppb/96 hr /Static bioassay LD_{50} Colinis virginianus (Bobwhite) female oral 6.59 mg/kg (95% confidence limit) 3-4 mo old

Environmental Fate: Residues in soil and plants will volatilize from soil surfaces or be slowly transformed to dieldrin in soil. Biodegradation is expected to be slow and aldrin is not expected to leach. It was classified as moderately persistent meaning its half-life in soil ranged from 20-100 days. Residues in water will volatilize from the water surface and photooxidation is expected to be significant. Photolysis has been observed in water, although the absorption characteristics indicate it should not extensively directly photolyze in the environment. Bioconcentration will be significant. Adsorption to sediments is expected and biodegradation is expected to be slow. Vapor phase expected to be slow. Vapor phase residues in the atmosphere are expected to react with photochemically generated hydroxyl radicals with an estimated half-life of 35.46 min. It is expected to be adsorbed to particulate matter and no rate can be given for the reaction of adsorbed aldrin with hydroxyl radicals. Direct hydrolysis may also occur, in spite of the low absorption at >290 nm. it, is expected to be a slow process relative to reaction with radicals.

Cleanup/Disposal: Guide No. 131: Fully encapsulating, vapor protective clothing should be worn for spills and leaks with no fire. Eliminate all ignition sources (no smoking, flares, sparks or flames in immediate area). All equipment used when handling the product must be grounded. Do not touch or walk through spilled material. Stop leak if you can do it without risk. Prevent entry into waterways, sewers, basements or confined areas. A vapor suppressing foam may be used to reduce vapors. Small Spills: Absorb with earth, sand or other non-combustible material and transfer to containers for later disposal. Use clean non-sparking tools to collect absorbed material. Large Spills: Dike far ahead of liquid spill for later disposal. Water spray may reduce vapor; but may not prevent ignition in closed spaces. Guide No. 151: Do not touch damaged containers or spilled material unless wearing appropriate protective clothing. Stop leak if you can do it without risk. Prevent entry into waterways, sewers, basements or confined areas. Cover with plastic sheet to prevent spreading. Absorb or cover with dry earth, sand or other non-combustible material and transfer to containers. Do not get water inside containers.

Environmental Physical Data

Henry's Law Constant: estimated at 1.27×10^{-5}

Octanol/Water Partition Coefficient: log K_{ow} = 3.01

Sorption Partition Coefficient: K_{oc} = 2.61 to 4.45

BCF: mullusks 4571

BOD: not pertinent

Regulations

RCRA 40CFR: Listed Hazardous Waste No. P004 Toxic Waste
CERCLA: 40CFR 302.4: Listed per CWA Section 311(b)(4) per RCRA Section 3001 per CWA Section 307(a) RQ: 1 lb (0.454 kg)
SARA 40CFR 372.65: Listed TPQ: 500/10000 lb
SARA EHS 40CFR 355: Listed TPQ: 1 lb
TSCA: Not listed

Analytical Methods

Air: EPA TO-10, TO-4, 016
Soil: CLP LC_PEST, MC_PEST, OHC; EPA PMD-TLC, 16, 3, 024, 025, P-002-1, P-011-1; SW846 3630B, 3640A, 8080A, 8081, 8081A, 8250A, 8270B, 8270C, 8275; USGS O5104, O7104
Water / Groundwater: EPA P-003-1, P-004-1, 608, 617, 625, 625-S, 680, 022; APHA 6410-B, 6630-B, 6630-C, 6630-D; ASTM D3086; USGS O3104
Drinking Water: EPA 505, 508, 508.1, 525.1, 525.2; AOAC 990.06; ASTM D5175
Food: FDA 212.1, 212.2, 232.1, 232.4, 242.1; EPA 4; AOAC 961.05, 970.52, 972.05; USGS O9104
Indoor / Expired Air: NIOSH 5502; EPA IP-8; ASTM D4861
Plasma: EPA 001, 003, 004, 027, 028; FDA 211.1, 231.1, 251.1, 252, 253
Other: EPA P-009-1, 1656

ALF5000 CAS #: 71195-58-9

ALFENTANIL

RTECS: TX1452480
Molecular Formula: $C_{21}H_{32}N_6O_3$
Formula Weight: 416.52
Synonyms: ALFENTANILUM; ALFENTANYL
Description: crystals
Use: general anesthetic; medication; medication for minor surgical procedures

Physical Properties

Freezing Point: Crystals at 140.8 °C (285.44 °F)
Water Solubility: Soluble in Water

RTECS Toxicity Data

Reproductive/Teratogenic: Monkey Route: Intravenous; Dose: 100 ug/kg; Duration: female 24W of pregnancy; Effects on Newborn - Behavioral.

Hazard Overviews

Carcinogenicity: IARC - Not listed; NIOSH - Not listed; NTP - Not listed; ACGIH - Not listed; OSHA - Not listed; EPA - Not listed; MAK - Not listed

Environmental

Regulations

RCRA 40CFR: Not listed
CERCLA: 40CFR 302.4: Not listed
SARA 40CFR 372.65: Not listed
SARA EHS 40CFR 355: Not listed
TSCA: Not listed

ALG3000 CAS #: 9005-38-3

ALGIN

RTECS: AZ5820000
Molecular Formula: Unknown
Formula Weight: 32,000 - 200,000
Synonyms: L'-ALGILINE; ALGIN (POLYSACCHARIDE); ALGINATE KMF; ALGINIC ACID,SODIUM SALT; ALGIPON L-1168; AMNUCOL; ANTIMIGRANT C 45; CECALGINE TBV; COHASAL-IH; DARID QH; DARILOID QH; DUCKALGIN; HALLTEX; KELCO GEL LV; KELCOSOL; KELGIN; KELGIN F; KELGIN HV; KELGIN LV; KELGIN XL; KELGUM; KELSET; KELSIZE; KELTEX; KELTONE; LAMITEX; MANUCOL; MANUCOL DM; MANUCOL KMF; MANUCOL SS/LD 2; MANUCOL SS/LD2; MANUGEL F 331; MANUTEX; MANUTEX RS; MANUTEX RS-5; MANUTEX RS 1; MANUTEX RS 5; MANUTEX F; MANUTEX RS1; MANUTEX SA/KP; MANUTEX SH/LH; MEYPRALGIN R/LV; MINUS; MOSANON; NOURALGINE; OG 1; PECTALGINE; PROCTIN; PROTACELL 8; PROTANAL; PROTATEK; SNOW ALGIN H; SNOW ALGIN L; SNOW ALGIN M; SODIUM ALGINATE; SODIUM POLYMANNURONATE; STIPINE; TAGAT; TRAGAYA
Description: yellowish white, filamentous or granular solid or powder; almost odorless
Use: fining agent in brewing industry; suspending agent in soft drinks; pharmaceutic aid (suspending & tablet binding); flocculant for water treatment; sizing agent; thickener; foam & emulsion stabilizer in mayonnaise, dairy products, condiments, canned meat, cosmetics, & dye solns; bodying agent for bakery products; cement compositions; paper coating; water-base paints; appetite curbing prepn; antiperspirant; in blood-clotting materials; in boiler feedwater; for dental impression material; agent in silver recovery; in drilling muds.

Physical Properties

Water Solubility: Soluble in Water
Other Solubilities: Insoluble in Alcohol, Chloroform, Ether, aqueous acid solution below pH 3, & in hydro-Alcoholic solution greater than 30% wt/wt.
Flash Point: Combustible

RTECS Toxicity Data

Acute Oral: Rat LD_{50} Dose: >5 gm/kg.

Hazard Overviews

Health: May cause irritation to eyes/skin. Other Acute Effects: may be harmful by inhalation, ingestion, or skin absorption.
Fire: Combustible. Hazards: under fire conditions, material may decompose to form flammable and/or explosive mixtures

in air; in powder form capable of creating a dust explosion. Extinguishing agents: water spray; carbon dioxide, dry chemical powder or appropriate foam. Precautions: combustible liquid.

Carcinogenicity: IARC - Not listed; NIOSH - Not listed; NTP - Not listed; ACGIH - Not listed; OSHA - Not listed; EPA - Not listed; MAK - Not listed

Environmental

Regulations

RCRA 40CFR: Not listed
CERCLA: 40CFR 302.4: Not listed
SARA 40CFR 372.65: Not listed
SARA EHS 40CFR 355: Not listed
TSCA: Listed

ALG6000 CAS #: 9005-32-7

ALGINIC ACID

RTECS: AZ5775000
EINECS Number: 232-680-1
Molecular Formula: Unknown

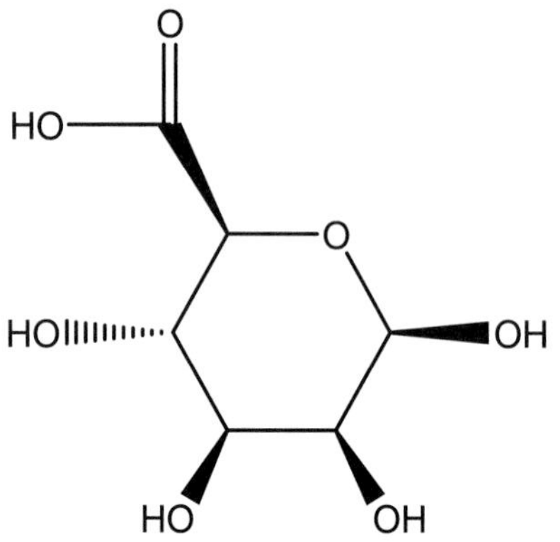

Chemical Structure

Synonyms: KELACID; LANDALGINE; NORGINE; POLYMANNURONIC ACID; SAZZIO

Description: white to yellowish white, fibrous powder; odorless

Use: sizing paper & textiles; binder for briquettes; mfr artificial horn, ivory, celluloid; emulsionizing mineral oils; mucilage; pharmaceutic aid (tablet binder & emulsifying agent); in food industry as thickener & emulsifier; protective colloid; tooth paste; cosmetics; textile sizing; coatings; waterproofing agent for concrete; boiler water treatment; oil-well drilling muds; storage of gasoline; dental impression material

Physical Properties

Water Solubility: Very Slightly Soluble in Water
Other Solubilities: Insoluble in organic solvents.
pH: 3 in 100 suspension in water 2 to 3.4

RTECS Toxicity Data

Acute Oral: Rat LD_{50} Dose: >5 gm/kg.

Hazard Overviews

Health: May cause irritation to eyes/skin. Other Acute Effects: may be harmful by inhalation, ingestion, or skin absorption.

Fire: Hazards: emits toxic fumes. Extinguishing agents: water spray; carbon dioxide, dry chemical powder or appropriate foam. Precautions: combustible liquid.

Carcinogenicity: IARC - Not listed; NIOSH - Not listed; NTP - Not listed; ACGIH - Not listed; OSHA - Not listed; EPA - Not listed; MAK - Not listed

Environmental

Regulations

RCRA 40CFR: Not listed
CERCLA: 40CFR 302.4: Not listed
SARA 40CFR 372.65: Not listed
SARA EHS 40CFR 355: Not listed
TSCA: Listed

ALI5000 CAS #: 81-48-1

ALIZUROL PURPLE

RTECS: CB7700000
EINECS Number: 201-353-5
Molecular Formula: $C_{21}H_{15}NO_3$
Formula Weight: 329.37

Chemical Structure

Synonyms: AHCOQUINONE BLUE IR BASE; ALIZARINE IRISOL R BASE; ALIZARINE VIOLET 3B BASE; 9,10-ANTHRACENEDIONE,1-HYDROXY-4-((4-METHYLPHENYL)AMINO)-; ANTHRAQUINONE,1-HYDROXY-4-P-TOLUIDINO-; C.I. 60725; C.I. DISPERSE BLUE 72; C.I. SOLVENT VIOLET 13; D AND C VIOLET NO 2; D+C VIOLET NO 2; D+C VIOLET NO. 2; DISPERSE BLUE 72; DISPERSOL VIOLET B-G; DURANOL BRILLIANT VIOLET 7G; N-(4-HYDROXY-1-ANTHRAQUINONYL)-4-METHYLANILINE; N-(4-HYDROXY-1-ANTHRAQUINONYL)-P-TOLUIDINE; 1-HYDROXY-4-(P-TOLUIDINO)ANTHRAQUINONE; IRISOL BASE; MODR DISPERZNI 72; OIL VIOLET IRS; OIL VIOLET ZIRS; RESIREN BLUE TR; RESOLIN BLUE RRL; RESORIN BLUE RRL; SOLVENT VIOLET 13;

SUMIKARON VIOLET B; N-(P-TOLYL)-4-HYDROXY-1-ANTHRAQUINONYLAMINE; TRANSETILE VIOLET P 3B; 11092 VIOLET; VIOLET ROZPOUSTEDLOVA 13; WAXOLINE PURPLE A

Use: dye for cosmetics (eg, hair & skin care products); dye for plastics & resins (eg, styrene); dye for synthetic textile fibers; dye for hydrocarbon solvents & gasoline; as dyes in home & industry

Physical Properties

Water Solubility: Insoluble

RTECS Toxicity Data

Mutagenic: Bacteria - S Typhimurium Mutations in Microorganisms; Dose: 25 ug/plate (+S9).

Hazard Overviews

Health: May cause irritation. Other Acute Effects: may be harmful by inhalation, ingestion, or skin absorption. The toxicological properties have not been thoroughly investigated.

Fire: Hazards: emits toxic fumes. Extinguishing agents: carbon dioxide, dry chemical powder or appropriate foam. Precautions: combustible liquid.

Reactivity: Stable. Hazardous polymerization will not occur. Hazardous decomposition products: thermal decomposition may produce carbon monoxide, carbon dioxide, and nitrogen oxides.

Carcinogenicity: IARC - Not listed; NIOSH - Not listed; NTP - Not listed; ACGIH - Not listed; OSHA - Not listed; EPA - Not listed; MAK - Not listed

Environmental

Regulations

RCRA 40CFR: Not listed
CERCLA: 40CFR 302.4: Not listed
SARA 40CFR 372.65: Not listed
SARA EHS 40CFR 355: Not listed
TSCA: Listed

ALK2330 CAS #: 2461-15-6

ALKYL GLYCIDYL ETHER

RTECS: TZ3300000
EINECS Number: 219-553-6
Molecular Formula: $C_{11}H_{22}O_2$
Formula Weight: 186.33

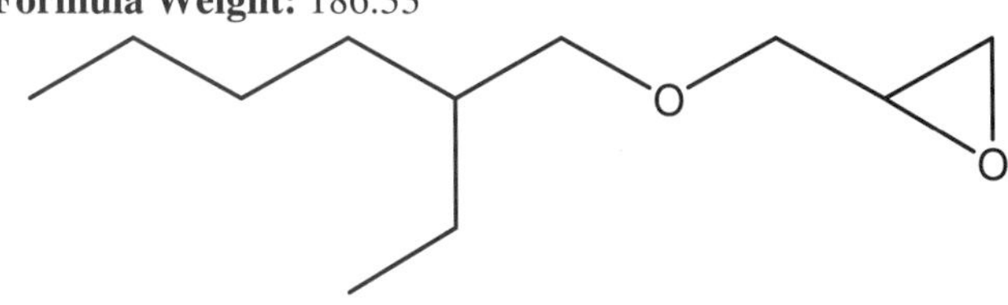

Chemical Structure

Synonyms: 1,2-EPOXY-3-((2-ETHYLHEXYL)OXY)PROPANE; 2-ETHYLHEXYL GLYCIDYL ETHER; (((2-ETHYLHEXYL)OXY)METHYL)OXIRANE; GLYCIDYL 2-ETHYLHEXYL ETHER; 1-GLYCIDYLOXY-2-ETHYLHEXANE; OXIRANE,(((2-ETHYLHEXYL)OXY)METHYL)-(9CI)

Description: colorless, clear liquid

Physical Properties

Boiling Point: 243 °C (469 °F)
Specific Gravity: 0.91
Vapor Density: > 1 Air=1
Density: 0.891 g/mL
Vapor Pressure: ~ 13
Water Solubility: < 1 mg/mL at 19 C
Other Solubilities: 95% Ethanol: >=100 mg/ml at 19 °C; Acetone: >=100 mg/ml at 19 °C; DMSO: >=100 mg/ml at 19 °C.
Refraction Index: 1.434
Evaporation Rate: < 1 Butyl Acetate=1
Flash Point: 96 °C

RTECS Toxicity Data

Acute Oral: Rat LD_{50} Dose: 7800 mg/kg.
Mutagenic: Bacteria - S Typhimurium Mutations in Microorganisms; Dose: 100 ug/plate (-S9).

Hazard Overviews

Health: Irritating to eyes/skin/respiratory tract. Other Acute Effects: may be harmful by inhalation, ingestion, or skin absorption.

Fire: Will burn. Hazards: emits toxic fumes. Extinguishing agents: water spray; carbon dioxide, dry chemical powder or appropriate foam. Precautions: combustible liquid.

Reactivity: Incompatible with: acids, bases, oxidizing agents. Hazardous decomposition products: carbon monoxide, carbon dioxide.

Carcinogenicity: IARC - Not listed; NIOSH - Not listed; NTP - Not listed; ACGIH - Not listed; OSHA - Not listed; EPA - Not listed; MAK - Not listed

Primary Target Organs:

Eyes

Skin

Respiratory System

Environmental

Regulations

RCRA 40CFR: Not listed
CERCLA: 40CFR 302.4: Not listed
SARA 40CFR 372.65: Not listed
SARA EHS 40CFR 355: Not listed
TSCA: Listed

ALK3660 **CAS #: 26027-38-3**

ALKYL-ARYL ETHER

RTECS: MD0906000
Molecular Formula: $C_{(2x+15)}H_{(4x+24)}O_{(x+1)}$

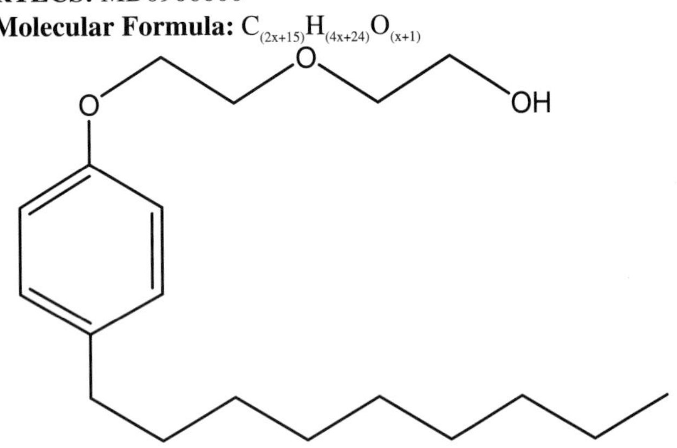

Chemical Structure

Synonyms: N-9; NONOXYNOL-9; NONYLPHENOXYPOLY(ETHYLENEOXY)ETHANOL; NP-9; PHENOL,P-NONYL-,MONOETHER WITH POLYETHYLENE GLYCOL

Physical Properties

Boiling Point: 250 °C (482 °F)
Freezing Point: 6 °C (42.8 °F)
Specific Gravity: 1.06

RTECS Toxicity Data

Reproductive/Teratogenic: Rat Route: Intravaginal; Dose: 50 mg/kg; Duration: female 1D prior to mating; Maternal Effects - Uterus, cervix, vagina; Other effects on females. Rat Route: Intravaginal; Dose: 50 mg/kg; Duration: female 3D of pregnancy; Effects on Fertility - Pre-implantation mortality.
Mutagenic: Mouse Morphological Transformation; Cell Type: fibroblast; Dose: 10 ppm.

Hazard Overviews

Health: Severely irritating to eyes; irritating to skin/respiratory tract. Other Acute Effects: may be harmful by inhalation, ingestion, or skin absorption; may cause allergic skin reaction.
Fire: Hazards: emits toxic fumes. Extinguishing agents: water spray; carbon dioxide, dry chemical powder or appropriate foam. Precautions: combustible liquid.
Reactivity: Incompatible with: strong oxidizing agents. Hazardous decomposition products: toxic fumes of: carbon monoxide, carbon dioxide.
Carcinogenicity: IARC - Not listed; NIOSH - Not listed; NTP - Not listed; ACGIH - Not listed; OSHA - Not listed; EPA - Not listed; MAK - Not listed

Primary Target Organs:

Eyes

Skin

Respiratory System

Environmental

Regulations

RCRA 40CFR: Not listed
CERCLA: 40CFR 302.4: Not listed
SARA 40CFR 372.65: Not listed
SARA EHS 40CFR 355: Not listed
TSCA: Not listed

ALK7650 **CAS #: 9036-19-5**

ALKYLPHENYLPOLY(OXYETH-YLENE)GLYCOL

RTECS: MD0907600
Molecular Formula: $C_{(2x+14)}H_{(4x+22)}O_{(x+1)}$
Formula Weight: N/A
Synonyms: CHARGER E; ETHOXYLATED OCTYL PHENOL; ETHYLAN CP; IGEPAL CA; IGEPAL CA 520; NEUTRONYX 622; NEUTRONYX 675; NONIDET P40; NONION HS 206; NONION HS 208; NP-40; TERT-OCTYLPHENOXY POLY(OXYETHYLENE)ETHANOL; OCTYLPHENOXYPOLY(ETHOXYETHANOL); TERT-OCTYLPHENOXYPOLY(ETHOXYETHANOL); OCTYLPHENOXYPOLY(ETHYLENEOXY)ETHANOL; OP 1062; POLYETHYLENE GLYCOL MONO(OCTYLPHENYL) ETHER; POLYETHYLENE GLYCOL OCTYLPHENYL ETHER; POLY(ETHYLENE OXIDE)OCTYLPHENYL ETHER; POLY(OXY-1,2-ETHANEDIYL),ALPHA-((1,1,3,3-TETRAMETHYLBUTYL)PHENYL)-OMEGA-HYDROXY-(9CI); POLYOXYETHYLENE MONOOCTYLPHENYL ETHER; POLY(OXYETHYLENE)OCTYLPHENOL ETHER; POLY(OXYETHYLENE)OCTYLPHENYL ETHER; SECOPAL OP 20; SYNPERONIC OP; SYNPERONIC OP 10; T 45 (POLYGLYCOL); T45; TRITON X 114; TRITON X 15; TRITON X 207

RTECS Toxicity Data

Acute Oral: Rat LD_{50} Dose: 4190 mg/kg.
Mutagenic: Human DNA Inhibition; Cell Type: lymphocyte; Dose: 5 ppm. Rat Other Mutation Test Systems; Route: Intraperitoneal; Dose: 870 mg/kg. Rat Other Mutation Test Systems; Route: Oral; Dose: 10200 mg/kg.

Hazard Overviews

Health: Irritating to eyes/skin/respiratory tract. Other Acute Effects: may be harmful by inhalation, ingestion, or skin absorption. Chronic Effects: the following effects have been observed in laboratory studies with this material, birth defects; fetotoxicity; embryolethality; anemia; bone marrow damage; hemolysis; immuno- suppression; and damage to the male reproductive tissues; target organs: female reproductive system, male reproductive system.

Fire: Hazards: emits toxic fumes. Extinguishing agents: carbon dioxide, dry chemical powder or appropriate foam; water spray. Precautions: combustible liquid.

Carcinogenicity: IARC - Not listed; NIOSH - Not listed; NTP - Not listed; ACGIH - Not listed; OSHA - Not listed; EPA - Not listed; MAK - Not listed

Primary Target Organs:

Eyes

Skin

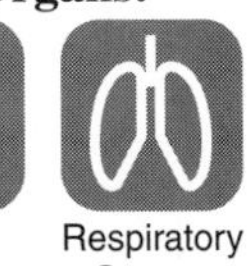
Respiratory System

Reproductive

Environmental

Regulations

RCRA 40CFR: Not listed
CERCLA: 40CFR 302.4: Not listed
SARA 40CFR 372.65: Not listed
SARA EHS 40CFR 355: Not listed
TSCA: Not listed

ALL1000 CAS #: 584-79-2

ALLETHRIN

RTECS: GZ1925000

DOT: UN2588; UN2902; UN2903; UN3021; NA2588; NA2902; IMO3.2; IMO6.1

EINECS Number: 209-542-4

Molecular Formula: $C_{19}H_{26}O_3$

Formula Weight: 302.45

Synonyms: (+)-ALLELRETHONYL (+)-CIS,TRANS-CHRYSANTHEMATE; D-ALLETHRIN; D-TRANS ALLETHRIN; ALLETHRIN I; ALLETHROLONE ESTER OF CHRYSANTHEMUMMONOCARBOXYLIC ACID; ALLEVIATE; ALLYL CINERIN; ALLYL CINERIN I; ALLYL HOMOLOG OF CINERIN I; 2-ALLYL-4-HYDROXY-3-METHYL-2-CYCLOPENTEN-1-ONE ESTER OFCHRYSANTHEMUMMONO-CARBOXYLIC ACID; D,L-2-ALLYL-4-HYDROXY-3-METHYL-2-CYCLOPENTEN-1-ONE-D,L-CHRYSANTHEMUM MONOCARBOXYLATE; 3-ALLYL-4-KETO-2-METHYLCYCLOPENTENYLCHRYSANTHEMUMMONOCARBOXYLATE; 3-ALLYL-2-METHYL-4-OXO-2-CYCLOPENTEN-1-YL CHRYSANTHEMATE; ALLYLRETHRONYL DL-CIS-TRANS-CHRYSANTHEMATE; BINAMIN FORTE; BIOALLETHRIN; BIOALTRINA; CINERIN I ALLYL HOMOLOG; CYCLOPROPANECARBOXYLIC ACID,2,2-DIMETHYL-3-(2-METHYLPROPENYL)-,ESTER WITH2-ALLYL-4-HYDROXY-3-METHYL-2-CYCLOPENTEN-1-ONE; CYCLOPROPANECARBOXYLIC ACID,2,2-DIMETHYL-3-(2-METHYL-1-PROPENYL)-,2-METHYL-4-OXO-3-(2-PROPENYL)-2-CYCLOPENTEN-1-YL ESTER; DEPALLETHRIN; 2,2-DIMETHYL-3-(2-METHYL-1-PROPENYL)CYCLOPROPANECARBOXYLICACID 2-METHYL-4-OXO-3-(2-PROPENYL)-2-CYCLOPENTEN-1-YLESTER; DL-2-ALLYL-4-HYDROXY-3-METHYL-2-CYCLOPENTEN-1-ONE ESTER OFDL CIS/TRANS2,2-DIMETHYL-3-(2-METHYLPROPENYL)-CYCLOPROPANECARBOXYLICACID; DL-3-ALLYL-2-METHYL-4-OXOCYCLOPENT-2-ENYL DL-CIS TRANSCHRYSANTHEMATE; ENT 16275; ENT 17,510; EXTHRIN; FDA 1446; FMC 249; NECARBOXYLIC ACID; NIA 249; OMS 468; PALLETHRINE; PYNAMIN; PYNAMIN FORTE; PYNAMIN-FORTE; PYRESIN; PYRESYN; PYROCIDE; SYNTHETIC PYRETHRINS; WASP STOPPER CF

Description: pale yellow oil

Use: insecticide for control of flies, mosquitoes, ants, & other household & public health insect pests, chewing & sucking insects on ornamentals, vegetables, etc; in animal houses; ectoparasiticide for control of human head lice; as effective as pyrethrins against house flies, less effective against roaches & other insects;. in combination with piperonyl butoxide or other synergists.

Physical Properties

Boiling Point: 140 °C (284 °F) at 0.1 mm Hg
Freezing Point: About 4 °C (39.2 °F)
Specific Gravity: 1.005 to 1.015 at 25 °C/4 °C
Vapor Pressure: 16 mPa at 30 °C
Water Solubility: Insoluble in Water
Other Solubilities: At room temp: > 1 kg/kg hexane, Methanol, Xylene.
Refraction Index: 1.5070 at 21 °C/D
Flash Point: > 60 °C

RTECS Toxicity Data

Acute Oral: Rat LD_{50} Dose: 685 mg/kg. Mouse LD_{50} Dose: 370 mg/kg.

Acute Inhalation: Mouse LC_{50} Dose: >2 gm/m^3. Rat LC_{Lo} Dose: 13800 mg/m^3/4hr; Toxic Effects: Behavioral - Tremor; Behavioral - Excitement.

Acute Dermal: Rabbit LD_{50} Route: Skin; Dose: 11332 mg/kg; Toxic Effects: Behavioral - Tremor; Behavioral - Excitment. Rat LD_{50} Route: Skin; Dose: 2500 mg/kg.

Chronic (Multiple Dose) Oral: Rat Dose: 21 gm/kg/12W-I; Toxic Effects: Liver - Changes in liver weight; Biochemical - Phosphatases; Biochemical - Other proteins.

Mutagenic: Hamster Cytogenetic Analysis; Cell Type: lung; Dose: 1900 ug/L. Bacteria - S Typhimurium Mutations in Microorganisms; Dose: 500 ug/plate (-S9).

Hazard Overviews

Fire: Combustible.

Carcinogenicity: IARC - Not listed; NIOSH - Not listed; NTP - Not listed; ACGIH - Not listed; OSHA - Not listed; EPA - Not listed; MAK - Not listed

Environmental

Ecotoxicity: LC_{50} Pteronarcys californica (stoneflies) 5.6 ug/l/96 hr at 15 °C (95% confidence limit 4.9-6.4 ug/l), second yr class /static bioassay LD_{50} Mallard duck oral less than 2000 mg/kg, male, 3-4 months old LC_{50} Salmo gairdneri (rainbow trout) 19.0 ug/l/96 hr at 13 °C, wt 0.9 g /static bioassay

Environmental Fate: If released to the atmosphere, it will degrade rapidly in the vapor phase by reaction with photochemically produced hydroxyl radicals (estimated half-life of 1.7 hr). If released to soil or water, it may be readily degraded in field conditions by sunlight and by most living organisms including soil microorganisms. Photodegradation studies using sunlamps or natural sunlight have found photodegradation rates of 90% in 8 hr to 11.1% in 15 min as either thin-films on glass plates or as aqueous suspensions.

Volatilization from dry surfaces is expected to be an important transport process (in the absence of strong surface adsorption). Estimated log K_{oc} values of 3.5 to 3.9 indicate that it may partition from the water column to sediment and suspended matter in the aquatic environment; also, it is not expected to leach in soil.

Cleanup/Disposal: Guide No. 131: Fully encapsulating, vapor protective clothing should be worn for spills and leaks with no fire. Eliminate all ignition sources (no smoking, flares, sparks or flames in immediate area). All equipment used when handling the product must be grounded. Do not touch or walk through spilled material. Stop leak if you can do it without risk. Prevent entry into waterways, sewers, basements or confined areas. A vapor suppressing foam may be used to reduce vapors. Small Spills: Absorb with earth, sand or other non-combustible material and transfer to containers for later disposal. Use clean non-sparking tools to collect absorbed material. Large Spills: Dike far ahead of liquid spill for later disposal. Water spray may reduce vapor; but may not prevent ignition in closed spaces. Guide No. 151: Do not touch damaged containers or spilled material unless wearing appropriate protective clothing. Stop leak if you can do it without risk. Prevent entry into waterways, sewers, basements or confined areas. Cover with plastic sheet to prevent spreading. Absorb or cover with dry earth, sand or other non-combustible material and transfer to containers. Do not get water inside containers.

Environmental Physical Data

Henry's Law Constant: estimated at 6.1×10^{-7}
Octanol/Water Partition Coefficient: log K_{OW} = 4.78
Sorption Partition Coefficient: log K_{OC} = estimated at 3.9
BCF: estimated at 2500

Regulations

RCRA 40CFR: Not listed
CERCLA: 40CFR 302.4: Not listed
SARA 40CFR 372.65: Not listed
SARA EHS 40CFR 355: Not listed
TSCA: Listed

Analytical Methods

Water / Groundwater: EPA 1660
Food: FDA 212.1, 212.2, 232.1; AOAC 953.05
Indoor / Expired Air: ASTM D4861
Plasma: FDA 211.1, 231.1, 252

ALL1420 CAS #: 28057-48-9

D-TRANS-ALLETHRIN

Hazard Overviews

Carcinogenicity: IARC - Not listed; NIOSH - Not listed; NTP - Not listed; ACGIH - Not listed; OSHA - Not listed; EPA - Not listed; MAK - Not listed

Environmental

Regulations

RCRA 40CFR: Not listed
CERCLA: 40CFR 302.4: Not listed
SARA 40CFR 372.65: Listed
SARA EHS 40CFR 355: Not listed
TSCA: Not listed

ALL1840 CAS #: 315-30-0

ALLOPURINOL

RTECS: UR0785000
EINECS Number: 206-250-9
Molecular Formula: $C_5H_4N_4O$
Formula Weight: 136.13

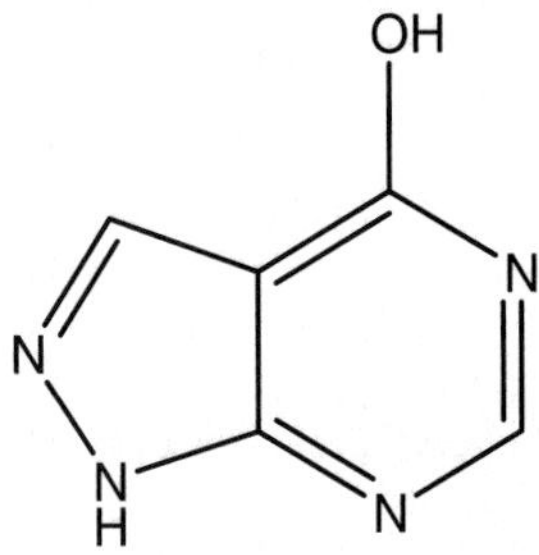

Chemical Structure

Synonyms: ADENOCK; AL-100; ALLOPUR; ALLO-PUREN; ALLOPURINOL(I); ALLOZYM; ALLURAL; ALORAL; ALOSITOL; ALULINE; ANOPROLIN; ANZIEF; APULONGA; APURIN; APUROL; ATISURIL; BLEMINOL; BLOXANTH; BW 56-158; BW 56158; CAPLENAL; CELLIDRIN; COSURIC; DABROSIN; DABROSON; 1,5-DIHYDRO-4H-PYRAZOLO(3,4-D)PYRIMIDINE-4-ONE; 1,5-DIHYDRO-4H-PYRAZOLO(3,4-D)PYRIMIDIN-4-ONE; DURA AL; EMBARIN; EPIDROPAL; EPURIC; FOLIGAN; GEAPUR; GICHTEX; GOTAX; HAMARIN; HEXANURET; HPP; 4'-HYDROXYPYRAZOLOL(3,4-D)PYRIMIDINE; 4-HYDROXY-1H-PYRAZOLO(3,4-D)PYRIMIDINE; 4-HYDROXY-3,4-PYRAZOLOPYRIMIDINE; 4-HYDROXYPYRAZOLO(3,4-D)PYRIMIDINE; 4-HYDROXYPYRAZOLOPYRIMIDINE; 4-HYDROXYPYRAZOLYL(3,4-D)PYRIMIDINE; KETANRIFT; KETOBUN-A; LEDOPUR; LOPURIN; LYSURON; MILURIT; MINIPLANOR; MONARCH; NEKTROHAN; NSC-1390; 1H-PYRAZOLO(3,4-D)PYRIMDIN-4-OL; 1H-PYRAZOLO(3,4-D)PYRIMIDIN-4-OL; 4H-PYRAZOLO(3,4-D)PYRIMIDIN-4-ONE; 4H-PYRAZOLO(3,4-D)PYRIMIDIN-4-ONE,1,5-DIHYDRO-; REMID; RIBALL; SUSPENDOL; TAKANARUMIN; URBOL; URICEMIL; URIPRIM; URIPURINOL; URITAS; UROBENYL; UROSIN; URTIAS 100; XANTURAT; ZYLOPRIM; ZYLORIC

Description: fluffy white to off-white powder or crystals; slight odor

Use: in the therapy of hematological disorders, hyperuricemia of gout, leukemia, lymphoma, malignancies which are receiving cancer therapy, calcium oxalate calculi, psoriasis, renal calculi, polycythemia, ver and myeloid metaplasia and other blood dyscrasias; in the prevention of uric acid nephropathy, damage to the kidneys and elevation of plasma uric acid levels in thiazide-treated patients; to lower hi plasma concentration of uric acid in patients with the Lesch-Nyhan syndrome; inhibits the action of xanthine oxidase and reduces incidence of urinary calculi

Physical Properties

Freezing Point: > 350 °C (662 °F)
Water Solubility: Very Slightly Soluble in Water
Other Solubilities: Soluble in mg/ml at 25 °C: n-octanol < 0.01; Chloroform 0.60; Ethanol 0.30; dimethyl sulfoxide 4.6.
Flash Point: Not available; probably combustible

RTECS Toxicity Data

Acute Oral: Man TD_{Lo} Dose: 120 mg/kg/4W-I; Toxic Effects: Blood - Other changes; Skin and appendages - Dermatitis, allergic; Skin and appendages - Dermatitis, other. Man TD_{Lo} Dose: 21429 ug/kg/5D-I; Toxic Effects: Behavioral - Muscle weakness; Liver - Jaundice, other or unclassified; Blood - Thrombocytopenia. Woman TD_{Lo} Dose: 42 mg/kg/7D-I; Toxic Effects: Kidney, Ureter, and Bladder - Other changes; Skin and appendages - Dermatitis, other. Woman LD_{Lo} Dose: 88 mg/kg/22D-I; Toxic Effects: Blood - Leukopenia.
Acute Dermal: Mouse LD_{50} Route: Subcutaneous Dose: 298 mg/kg.
Chronic (Multiple Dose) Dermal: Rat Route: Subcutaneous; Dose: 300 mg/kg/3D-I; Toxic Effects: Kidney, Ureter, and Bladder - Other changes; Kidney, Ureter, and Bladder - Changes in kidney weight; Biochemical - Xanthine, purine, or nucleotides including urate.
Reproductive/Teratogenic: Mouse Route: Intraperitoneal; Dose: 50 mg/kg; Duration: female 10D of pregnancy; Effects on Embryo or Fetus - Fetotoxicity; Specific Developmental Abnormalities - Craniofacial (including nose and tongue); Musculoskeletal system. Mouse Route: Intraperitoneal; Dose: 100 mg/kg; Duration: female 10D of pregnancy; Effects on Embryo or Fetus - Fetal death.

Hazard Overviews

Health: Irritating to eyes/skin/respiratory tract. Toxic. Other Acute Effects: harmful if swallowed, inhaled, or absorbed through skin; may cause allergic skin reaction; target organ: liver. Chronic Effects: skin rash or other signs which may indicate an allergic reaction; skin rash, followed by more severe hypersensitivity symptoms such as exfoliative, urticarial or purpuric lesions, with Stevens-Johnson syndrome, irreversible hepatotoxicity and on rare occasions death; rare occurences of transient leukopenia or leukocytosis and eosinophilia have been reported; isolated cases of peripheral neuritis, bone marrow depression and cataracts have been documented; renal failure has resulted from eosinophilia with epidermal necrolysis; other side effects include headache; drowsiness; nausea; vomiting; vertigo; diarrhea; gastric irritation.
Fire: Will burn. Hazards: emits toxic fumes. Extinguishing agents: carbon dioxide, dry chemical powder or appropriate foam. Precautions: combustible liquid.
Reactivity: Incompatible with: strong oxidizing agents. Hazardous decomposition products: toxic fumes of: carbon monoxide, carbon dioxide, nitrogen oxides.

Carcinogenicity: IARC - Not listed; NIOSH - Not listed; NTP - Not listed; ACGIH - Not listed; OSHA - Not listed; EPA - Not listed; MAK - Not listed
Primary Target Organs:

Eyes

Skin
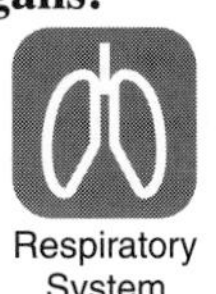
Respiratory System

Liver

Environmental

Regulations

RCRA 40CFR: Not listed
CERCLA: 40CFR 302.4: Not listed
SARA 40CFR 372.65: Not listed
SARA EHS 40CFR 355: Not listed
TSCA: Listed

ALL2260 CAS #: 76-24-4

ALLOXANTIN

EINECS Number: 200-947-1
Molecular Formula: $C_8H_6N_4O_8$
Formula Weight: 286.16

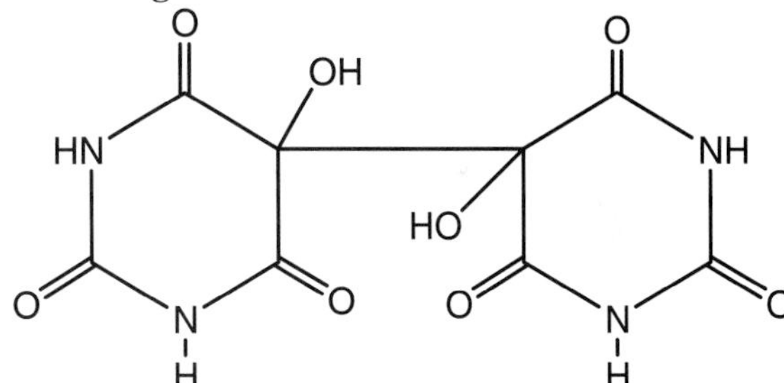

Chemical Structure

Synonyms: 5,5'-BIBARBITURIC ACID,5,5'-DIHYDROXY-; (5,5'-BIPYRIMIDINE)-2,2',4,4',6,6'(1H,1'H,3H,3'H,5H,5'H)-HEXONE,5,5'-DIHYDROXY; UROXIN; UROXINE
Description: rhombic prisms

Physical Properties

Boiling Point: Decomposes at 253 °C (487 °F) to 255 °C (491 °F)
Freezing Point: Decomposes at 253 °C (487.4 °F) to 255 °C (491 °F)
Water Solubility: Sparingly Soluble in Cold Water
Other Solubilities: Slightly Soluble in Alcohol, Ether.
pH: Aqueous solution is acid

Hazard Overviews

Carcinogenicity: IARC - Not listed; NIOSH - Not listed; NTP - Not listed; ACGIH - Not listed; OSHA - Not listed; EPA - Not listed; MAK - Not listed

Environmental

Regulations

RCRA 40CFR: Not listed
CERCLA: 40CFR 302.4: Not listed
SARA 40CFR 372.65: Not listed
SARA EHS 40CFR 355: Not listed
TSCA: Listed

ALL2680 CAS #: 107-18-6

ALLYL ALCOHOL

RTECS: BA5075000
DOT: UN1098; IMO3.2
EINECS Number: 203-470-7
Molecular Formula: C_3H_6O
Structured MF: $CH_2=CHCH_2OH$
Formula Weight: 58.08

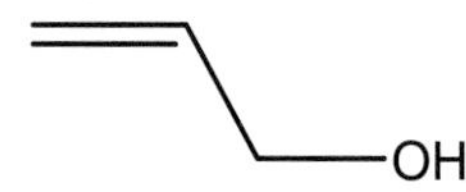

Chemical Structure

Synonyms: AA; ALCOOL ALLILCO; ALCOOL ALLYLIQUE; ALLILOWY ALKOHOL; ALLYL AL; ALLYLALKOHOL; ALLYLIC ALCOHOL; 3-HYDROXY PROPENE; 3-HYDROXYPROPENE; ORVINYLCARBINOL; 1-PROPENE-3-OL; 2-PROPENE-1-OL; 1-PROPEN-3-OL; 1-PROPENOL-3; 2-PROPEN-1-OL; 2-PROPENOL; PROPEN-1-OL-3; PROPENOL; 2-PROPENYL ALCOHOL; PROPENYL ALCOHOL; SHELL UNKRAUTTED A; SHELL UNKRAUTTOD A; VINYL CARBINOL; VINYLCARBINOL; WEED DRENCH
Description: colorless liquid; pungent mustard odor
Use: manufacture of allyl compounds, war gas, resins and plasticizers

Physical Properties

Boiling Point: 96 °C (205 °F) to 97 °C (207 °F)
Freezing Point: -129 °C (-200.2 °F)
Specific Gravity: 0.854 at 20 °C/4 °C
Vapor Density: 2 Air=1
Bulk Density: 7.11 lbs/gal at 20 °C
Vapor Pressure: 17 mm Hg
Water Solubility: > 10% in Water
Other Solubilities: Alcohol: miscible; Ether: Soluble.
Surface Tension: 25.8 dynes/cm
Odor Threshold: 1.9500 to 5.0000 mg/m^3
Refraction Index: 1.4135 at 20 °C/D
Critical Temperature: 272 °C
Critical Pressure: 55.5 atm
Ionization Potential (eV): 9.63
Flash Point: 21 °C Closed Cup
Autoignition Temperature: In air 443 °C
LEL: 2.5% v/v
UEL: 18% v/v

RTECS Toxicity Data

Acute Oral: Rat LD_{50} Dose: 64 mg/kg. Mouse LD_{50} Dose: 96 mg/kg; Toxic Effects: Lungs, Thorax, or Respiration - Acute pulmonary edema; Behavioral - Ataxia.
Acute Inhalation: Rat LC_{50} Dose: 76 ppm/8hr; Toxic Effects: Lungs, Thorax, or Respiration - Acute pulmonary edema. Monkey LC_{Lo} Dose: 1000 ppm/4hr; Toxic Effects: Gastrointestinal - Hypermotility, diarrhea; Blood - Hemorrhage; Biochemical - Effect on inflammation or mediation of inflammation.
Acute Dermal: Rabbit LD_{50} Route: Skin; Dose: 45 mg/kg. Rabbit LD_{Lo} Route: Subcutaneous Dose: 100 mg/kg.
Chronic (Multiple Dose) Oral: Rat Dose: 4200 mg/kg/15W-C; Toxic Effects: Behavioral - Food intake (animal); Kidney, Ureter, and Bladder - Changes in kidney weight; Nutritional and gross metabolic - Weight loss or decreased weight gain.
Irritation Eye: Human Standard Draize Test Dose: 25 ppm; Reaction: severe. Rabbit Standard Draize Test Dose: 20 mg; Reaction: severe.
Mutagenic: Hamster Mutations in Mammalian Somatic Cells; Cell Type: lung; Dose: 1 umol/L. Bacteria - S Typhimurium Mutations in Microorganisms; Dose: 100 umol/L (+S9). Bacteria - S Typhimurium Mutations in Microorganisms; Dose: 50 ug/plate (-S9).

Hazard Overviews

Corrosive

Flammable

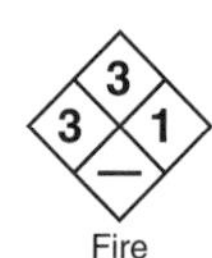

Fire Diamond

Health: Severe irritation to eyes/skin/respiratory tract. Toxic. Also Causes: mucous membrane irritation, hemoptysis, headache, nausea, vomiting, pulmonary edema, tears, photophobia, blurred vision, burns resulting in blindness; deep-seated pain, aching, blisters, burns serious systemic liver, kidney, and pancreatic injury, muscle spasms, severe gastrointestinal irritation, nausea, bloody vomit and feces.
Fire: Flammable. Can form explosive mixtures in the air. Use water spray, dry chemical, alcohol resistant foam, or carbon dioxide. Water may be ineffective. Use water spray to cool fire-exposed containers. Fight fire from maximum distance.
Reactivity: Stable. Hazardous polymerization can occur by elevated temperatures, oxidizers, and peroxides. Upon storage for several years, allyl alcohol polymerizes and forms a thick syrup. Avoid: heat; ignition sources. Incompatible with: sulfuric acid; 2,4,6-tris(bromoamino)-1,3,5-triazine; alkali and 2,4,6-trichloro-1,3,5-triazine; nitric acid; chlorosulfonic acid; oleum; sodium hydroxide; potassium trichloride; diallyl phosphite; tri-n-bromomelamine; carbon tetrachloride. Hazardous decomposition products: carbon oxides.
Carcinogenicity: IARC - Not listed; NIOSH - Not listed; NTP - Not listed; ACGIH - Not listed; OSHA - Not listed; EPA - Not listed; MAK - Not listed

Primary Target Organs:

Eyes Skin Respiratory System Mucous Membranes Liver Kidneys

Exposure Limits
OSHA PEL: TWA: 2 ppm; 5 mg/m^3; skin.
OSHA PEL Vacated 1989 Limits: TWA: 2 ppm; 5 mg/m^3; STEL: 4 ppm; 10 mg/m^3.
ACGIH TLV: TWA: 2 ppm; 4.8 mg/m^3; STEL: 4 ppm; 9.5 mg/m^3.
NIOSH REL: TWA: 2 ppm; 5 mg/m^3. STEL: 4 ppm; 10 mg/m^3; skin.
NIOSH IDLH: 20 ppm.
DFG MAK: TWA: 2 ppm; 5 mg/m^3.
Respirator Recommendation
Exposure Range: >2 to <20 ppm Air Purifying, Negative Pressure, Full Face
Exposure Range: 20 to unlimited ppm Self-contained Breathing Apparatus, Pressure Demand, Full Face
Cartridge Color: black

Environmental

Ecotoxicity: LD_{50} Goldfish 1 mg/l/24 hr modified ASTM 1345
Environmental Fate: When released to water it is not expected to volatilize, photooxidize or directly photolyze. Biodegradation is expected to be the predominant fate in water. Release to soil is expected to result in biodegradation and possible migration to groundwater. Volatilization from wet soil, direct photolysis, and bioconcentration are not expected to be significant. Volatilization from dry surfaces or soil should be significant. Release to the atmosphere is expected to result mainly in reaction with photochemically generated hydroxyl radicals with estimated half-lives of 6.03-14.7 hr. Direct photolysis is not expected to be significant. Due to the high water solubility, rainout may also occur.
Cleanup/Disposal: Guide No. 131: Fully encapsulating, vapor protective clothing should be worn for spills and leaks with no fire. Eliminate all ignition sources (no smoking, flares, sparks or flames in immediate area). All equipment used when handling the product must be grounded. Do not touch or walk through spilled material. Stop leak if you can do it without risk. Prevent entry into waterways, sewers, basements or confined areas. A vapor suppressing foam may be used to reduce vapors. Small Spills: Absorb with earth, sand or other non-combustible material and transfer to containers for later disposal. Use clean non-sparking tools to collect absorbed material. Large Spills: Dike far ahead of liquid spill for later disposal. Water spray may reduce vapor; but may not prevent ignition in closed spaces.

Environmental Physical Data

Henry's Law Constant: 4.9 x10^{-6}
Octanol/Water Partition Coefficient: log K_{ow} = 0.17
BCF: none expected
BOD: 57%, 10 days

Regulations

RCRA 40CFR: Listed Hazardous Waste No. P005 Toxic Waste
CERCLA: 40CFR 302.4: Listed per CWA Section 311(b)(4) per RCRA Section 3001 RQ: 100 lb (45.35 kg)
SARA 40CFR 372.65: Listed TPQ: 1000 lb
SARA EHS 40CFR 355: Listed TPQ: 100 lb
TSCA: Listed

Analytical Methods

Air: ASTM D3686, D3687
Soil: EPA 1624; SW846 5031, 8015B, 8240B, 8260A, 8260B
Indoor / Expired Air: NIOSH 1402
Plasma: EPA 29

ALL3100 **CAS #: 106-95-6**

ALLYL BROMIDE

RTECS: UC7090000
DOT: UN1099; IMO3.2
EINECS Number: 203-446-6
Molecular Formula: C_3H_5Br
Structured MF: CH_2=CH•CH_2Br
Formula Weight: 120.99

Chemical Structure

Synonyms: BROMALLYLENE; 1-BROMO-2-PROPENE; 3-BROMO-1-PROPENE; 3-BROMOPROPENE; 3-BROMOPROPYLENE; 1-PROPENE,3-BROMO-; PROPENE,3-BROMO-; 2-PROPENYL BROMIDE
Description: colorless to light yellow liquid; unpleasant, pungent odor
Use: mfr of other allyl compounds; insecticidal fumigant; chem int in organic synthesis, for resins and fragrances; contact poison

Physical Properties

Boiling Point: 71.3 °C (160 °F) at 760 mm Hg
Freezing Point: -119 °C (-182.2 °F)
Specific Gravity: 1.398 at 20 °C/4 °C
Vapor Density: 4.2 Air=1
Saturated Vapor Density: 1.898919835 kg/m^3
Vapor Pressure: 18.6 kPa at 25 °C
Water Solubility: Slightly Soluble in Water
Other Solubilities: miscible with Alcohol, Chloroform, Ether, Carbon Disulfide, Carbon Tetrachloride
Surface Tension: Estimated at 40 dynes/cm
Refraction Index: 1.46545 at 20 °C/D
Ionization Potential (eV): 9.96
Flash Point: -1 °C Closed Cup
Autoignition Temperature: 295 °C
LEL: 4.4% v/v
UEL: 7.3% v/v

RTECS Toxicity Data

Acute Oral: Rat LD_{50} Dose: 120 mg/kg.
Acute Inhalation: Rat LC_{50} Dose: 10 gm/m^3/30M. Mammal LC_{50} Dose: 4110 mg/m^3.
Mutagenic: Human Unscheduled DNA Synthesis; Cell Type: HeLa cell; Dose: 500 umol/L. Bacteria - S Typhimurium Mutations in Microorganisms; Dose: 1 umol/plate (+S9).

Hazard Overviews

Poison

Flammable

Health: Severely irritating to eyes; irritating to skin/respiratory tract. Poison. Other Acute Effects: may be fatal if inhaled, swallowed, or absorbed through skin; symptoms of exposure may include burning sensation, coughing, wheezing, laryngitis, shortness of breath, headache, nausea and vomiting; depending on the intensity and duration of exposure, effects may vary from mild irritation to severe destruction of tissue. Chronic Effects: damage to the liver; damage to the kidneys.
Fire: Flammable. Hazards: vapor may travel considerable distance to source of ignition and flash back; container explosion may occur; forms explosive mixtures in air; emits toxic fumes. Extinguishing agents: carbon dioxide, dry chemical powder or appropriate foam; water may be effective for cooling, but may not effect extinguishment. Precautions: combustible liquid.
Reactivity: Incompatible with: strong oxidizing agents, strong bases. Hazardous decomposition products: toxic fumes of: carbon monoxide, carbon dioxide, hydrogen bromide gas.
Carcinogenicity: IARC - Not listed; NIOSH - Not listed; NTP - Not listed; ACGIH - Not listed; OSHA - Not listed; EPA - Not listed; MAK - Not listed
Primary Target Organs:

Eyes

Skin

Respiratory System

Environmental

Environmental Fate: Should have very high mobility in soil. Volatilization is expected from both moist and dry soils. In water, it is expected to volatilize rapidly with estimated half-lives of 3.3 hours and 4.38 days from a model river and a model lake, respectively. Hydrolysis is expected to be an important fate process with an experimental half-life of 12 hours. Water solubility of 3835 mg/l indicates that bioconcentration and adsorption to sediment are not expected to be important in aquatic systems. Biodegradation may be an important fate process in both soil and aquatic conditions based on a BOD dilution test. It will exist in the vapor phase in the ambient atmosphere. If released to the atmosphere, it will degrade by reaction with photochemically produced hydroxyl radicals with an estimated half-life of approximately 19 hours. It will also react with atmospheric ozone with estimated half-life of 7.28 days. Removal from the atmosphere can occur though wet deposition.
Cleanup/Disposal: Guide No. 131: Fully encapsulating, vapor protective clothing should be worn for spills and leaks with no fire. Eliminate all ignition sources (no smoking, flares, sparks or flames in immediate area). All equipment used when handling the product must be grounded. Do not touch or walk through spilled material. Stop leak if you can do it without risk. Prevent entry into waterways, sewers, basements or confined areas. A vapor suppressing foam may be used to reduce vapors. Small Spills: Absorb with earth, sand or other non-combustible material and transfer to containers for later disposal. Use clean non-sparking tools to collect absorbed material. Large Spills: Dike far ahead of liquid spill for later disposal. Water spray may reduce vapor; but may not prevent ignition in closed spaces.

Environmental Physical Data

Henry's Law Constant: calculated at 1.11 x10^{-2}
Octanol/Water Partition Coefficient: log K_{ow} = 1.79
Sorption Partition Coefficient: K_{oc} = estimated at 44
BCF: estimated at 6

Regulations

RCRA 40CFR: Not listed
CERCLA: 40CFR 302.4: Not listed
SARA 40CFR 372.65: Not listed
SARA EHS 40CFR 355: Not listed
TSCA: Listed

ALL3520 **CAS #: 107-05-1**

ALLYL CHLORIDE

RTECS: UC7350000
DOT: UN1100; IMO3.1
EINECS Number: 203-457-6
Molecular Formula: C_3H_5Cl
Structured MF: $ClCH_2CH{=}CH_2$
Formula Weight: 76.53

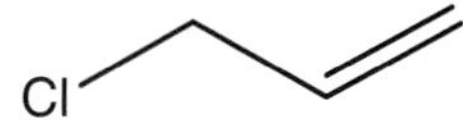

Chemical Structure

Synonyms: ALLILE (CLORURO DI); ALLYLCHLORID; ALLYLE (CHLORURE D'); P-AMINOPROPIOFENON; CHLORALLYLENE; 1-CHLORO PROPENE-2; CHLOROALLYLENE; 3-CHLOROPRENE; 1-CHLORO-2-PROPENE; 3-CHLORO-1-PROPENE; 3-CHLOROPROPENE; 3-CHLOROPROPENE-1; CHLORO-2-PROPENE; 3-CHLORO-1-PROPYLENE; 3-CHLOROPROPYLENE; ALPHA-CHLOROPROPYLENE; 3-CHLORPROPEN; 1-PROPENE,3-CHLORO-; PROPENE,3-CHLORO-; 2-PROPENYL CHLORIDE
Description: colorless, yellow, brown, purple liquid; pungent odor similar to garlic
Use: manufacture of glycerol, epichlorohydrin, allylamines, polymers, allyl alcohol, allyl silanes, allyl ethers of starch, 1,2,3-trichloropropane, sodium allylsulphonate, quaternary

ammonium salts, allyl ethers of a variety of alcohols, phenols, polyols and mono-, di and triallylamine; in the manufacture of allyl isothiocyanate, eugenol, 1,2-dibromo-3-chloropropane, glycerol chlorohydrins, cyclopropane an amines containing other alkyl groups (e.g., diallylmethylamine); thermosetting resins for varnishes, plastics, adhesives, synthesis of pharmaceuticals and insecticides; in a number of barbiturate hypnotic agents such as aprobarbital, butalbital, methohexital sodium, secobarbital, talbital and thiamylal sodium

Physical Properties

Boiling Point: 44 °C (111 °F) to 45 °C (113 °F)
Freezing Point: -134.5 °C (-210.1 °F)
Specific Gravity: 0.938 at 20 °C/4 °C
Vapor Density: 2.64 Air=1
Saturated Vapor Density: 2.079865699 kg/m^3
Density: 0.9392 g/mL at 20 °C
Vapor Pressure: 340 mm Hg at 20 °C
Water Solubility: 0.4% by weight
Other Solubilities: 95% Ethanol: 1-10 mg/ml at 19 °C; Acetone: >=100 mg/ml at 19 °C; Alcohol: miscible; Benzene: Soluble; Chloroform: miscible; DMSO: >=100 mg/ml at 19 °C; Ether: miscible; Ligroin: Soluble; Organic solvents: miscible; Petroleum Ether: miscible.
Surface Tension: 28.9 dynes/cm
Odor Threshold: Recognition 4.70 x10^{-1} ppm
Refraction Index: 1.4157 at 20 °C/D
Evaporation Rate: ~ 7 Butyl Acetate=1
Critical Temperature: 241 °C
Critical Pressure: 690 psia
Ionization Potential (eV): 10.05
Flash Point: -28.9 °C Open Cup
Autoignition Temperature: 392 °C
LEL: 3.3% v/v
UEL: 11.1% v/v

RTECS Toxicity Data

Acute Oral: Rat LD_{50} Dose: 460 mg/kg; Toxic Effects: Behavioral - Tremor; Behavioral - Convulsions or effect on seizure threshold; Lungs, Thorax, or Respiration - Dyspnea. Mouse LD_{50} Dose: 425 mg/kg;#Toxic Effects: Behavioral - Somnolence (general depressed activity); Behavioral - Tremor; Lungs, Thorax, or Respiration - Dyspnea;
Acute Inhalation: Rat LC_{50} Dose: 11 gm/m^3/2hr; Toxic Effects: Sense organs and special senses - Lacrimation; Gastrointestinal - Ulceration or bleeding from duodenum; Lungs, Thorax, or Respiration - Respiratory depression. Human LC_{Lo} Dose: 3000 ppm.
Acute Dermal: Rabbit LD_{50} Route: Skin; Dose: 2066 mg/kg.
Chronic (Multiple Dose) Oral: Mouse Dose: 15300 mg/kg/17W-I; Toxic Effects: Spinal Cord - Other degenerative changes; Peripheral Nerve and sensation - Structural change in nerve or sheath.
Chronic (Multiple Dose) Inhalation: Rat Dose: 8 ppm/7H/35D-I; Toxic Effects: Liver - Other changes; Kidney, Ureter, and Bladder - Other changes; Endocrine - Changes in spleen weight. Rabbit Dose: 206 mg/m^3/6H/13W-I; Toxic Effects: Peripheral Nerve and sensation - Flaccid paralysis without anesthesia; Peripheral Nerve and sensation - Structural change in nerve or sheath; Behavioral - Muscle weakness. Rabbit Dose: 8 ppm/7H/35D-I; Toxic Effects: Liver - Other changes; Kidney, Ureter, and Bladder - Other changes. Rabbit Dose: 206 mg/m^3/6H/13W-I; Toxic Effects: Spinal Cord - Demyelination; Peripheral Nerve and sensation - Flaccid paralysis without anesthesia; Behavioral - Muscle weakness.
Chronic (Multiple Dose) Dermal: Rat Route: Subcutaneous; Dose: 9949 mg/kg/13W-I; Toxic Effects: Spinal Cord - Other degenerative changes; Peripheral Nerve and sensation - Sensory change involving peripheral nerve; Peripheral Nerve and sensation - Recording from peripheral motor nerve. Rat Route: Subcutaneous; Dose: 1350 mg/kg/9D-I; Toxic Effects: Peripheral Nerve and sensation - Structural change in nerve or sheath; Blood - Changes in serum composition; Nutritional and gross metabolic - Changes in calcium.
Irritation Eye: Rabbit Standard Draize Test Dose: 500 mg; Reaction: moderate.
Reproductive/Teratogenic: Rat Route: Inhalation; Dose: 300 ppm/7H; Duration: female 6-15D of pregnancy; Specific Developmental Abnormalities - Musculoskeletal system. Rat Route: Intraperitoneal; Dose: 1200 mg/kg; Duration: female 1-15D of pregnancy; Effects on Fertility - Post-implantation mortality; Effects on Embryo or Fetus - Fetotoxicity; Specific Developmental Abnormalities - Other developmental abnormalities.
Mutagenic: Human Unscheduled DNA Synthesis; Cell Type: HeLa cell; Dose: 1 mmol/L. Bacteria - E Coli DNA Repair; Dose: 9380 ug/plate.
Tumorigenic: Mouse Route: Oral; Dose: 50 gm/kg/78W-I; Toxic Effects: Tumorigenic - Equivocal tumorigenic agent by RTECS criteria; Gastrointestinal - Tumors. Mouse Route: Oral; Dose: 78 gm/kg/78W-I; Toxic Effects: Tumorigenic - Equivocal tumorigenic agent by RTECS criteria; Gastrointestinal - Tumors.

Hazard Overviews

Corrosive

Flammable

Fire Diamond

Health: Corrosive to eyes/skin/respiratory tract. Toxic. Also Causes: headache, dizziness, unconsciousness, visual impairment, blindness, local blood vessel constriction, numbness, nausea and vomiting. Chronic Effects: liver/kidney damage, polyneuropathy, paresthesia, cramping pain, or numbness.
Fire: Flammable. Use dry chemical, carbon dioxide, regular foam, or water spray. Do not use solid streams of water because it may spread fire; use as a spray only. Remove cylinder from fire to prevent rupture due to heat.
Reactivity: Unstable, decomposes upon storage to release hydrogen chloride gas. Hazardous polymerization can occur at elevated temperatures and contact with acid catalysts such as Lewis acids, Ziegler acids, sulfuric acids, ferric chloride,

and aluminum chloride. Avoid: heat; ignition sources. Incompatible with: alkyl aluminum chlorides; Lewis acids; metals; aromatic hydrocarbons; nitric acid; ethylene imine; ethylene diamine; chlorosulfonic acid; oleum; sodium hydroxide; strong oxidizers. Hazardous decomposition products: carbon oxide(s); hydrogen chloride; phosgene gas.

Carcinogenicity: IARC - Group 3, Not classifiable as to carcinogenicity to humans; NIOSH - Not listed; NTP - Not listed; ACGIH - Class A3, Animal carcinogen; OSHA - Not listed; EPA - Class C, Possible human carcinogen; MAK - Class B, Justifiably suspected of having carcinogenic potential

Primary Target Organs:

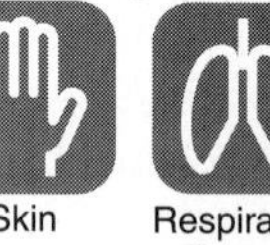
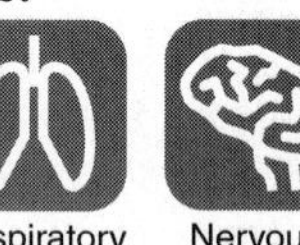

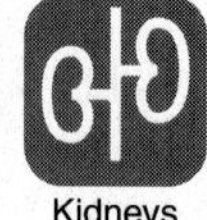

Eyes Skin Respiratory System Nervous System Liver Kidneys

Exposure Limits

OSHA PEL: TWA: 1 ppm; 3 mg/m^3.

OSHA PEL Vacated 1989 Limits: TWA: 1 ppm; 3 mg/m^3; STEL: 2 ppm; 6 mg/m^3.

ACGIH TLV: TWA: 1 ppm; 32 mg/m^3; STEL: 2 ppm; 6 mg/m^3.

NIOSH REL: TWA: 1 ppm; 3 mg/m^3. STEL: 2 ppm; 6 mg/m^3.

NIOSH IDLH: 250 ppm.

DFG MAK: TWA: 1 ppm; 3 mg/m^3.

Respirator Recommendation

Exposure Range: >1 to 50 ppm Supplied Air, Constant Flow/Pressure Demand, Half Mask

Exposure Range: >50 to <250 ppm Supplied Air, Constant Flow/Pressure Demand, Full Face

Exposure Range: 250 to unlimited ppm Self-contained Breathing Apparatus, Pressure Demand, Full Face

Note: poor warning properties

Environmental

Ecotoxicity: Aquatic toxicity: 48 ppm/96 hr/guppy/TLm/fresh water

Environmental Fate: It will rapidly degrade in the atmosphere by reaction with photochemically produced hydroxyl radicals (91% loss/day) as well as with ozone. Since the concentrations of hydroxyl radicals and ozone are greater in polluted atmospheres, its rate of loss will be greater under these situations. It will volatilize rapidly from water (half-life 2.6 hr in a typical river) and soil. While it may leach into the ground, it will not persist in ground water because it hydrolyzes (half-life 7 days).

Cleanup/Disposal: Guide No. 131: Fully encapsulating, vapor protective clothing should be worn for spills and leaks with no fire. Eliminate all ignition sources (no smoking, flares, sparks or flames in immediate area). All equipment used when handling the product must be grounded. Do not touch or walk through spilled material. Stop leak if you can do it without risk. Prevent entry into waterways, sewers, basements or confined areas. A vapor suppressing foam may be used to reduce vapors. Small Spills: Absorb with earth, sand or other non-combustible material and transfer to containers for later disposal. Use clean non-sparking tools to collect absorbed material. Large Spills: Dike far ahead of liquid spill for later disposal. Water spray may reduce vapor; but may not prevent ignition in closed spaces.

Environmental Physical Data

Henry's Law Constant: 0.44

Octanol/Water Partition Coefficient: log K_{ow} = -0.24

BCF: none noted

BOD: 14% BODT, 5 days

Regulations

RCRA 40CFR: Not listed

CERCLA: 40CFR 302.4: Listed per CWA Section 311(b)(4) RQ: 1000 lb (453.5 kg)

SARA 40CFR 372.65: Listed

SARA EHS 40CFR 355: Not listed

TSCA: Listed

Analytical Methods

Air: EPA TO-2; ASTM D4490

Soil: SW846 8010B, 8240B, 8260A, 8260B

Drinking Water: EPA 524.2

Indoor / Expired Air: NIOSH 1000; EPA IP-1B

Plasma: EPA 29

ALL3940 **CAS #: 2937-50-0**

ALLYL CHLOROFORMATE

RTECS: LQ5775000

DOT: UN1722; IMO8.1

EINECS Number: 220-916-6

Molecular Formula: $C_4H_5ClO_2$

Structured MF: CH_2=CH•CH_2•O•COCl

Formula Weight: 120.54

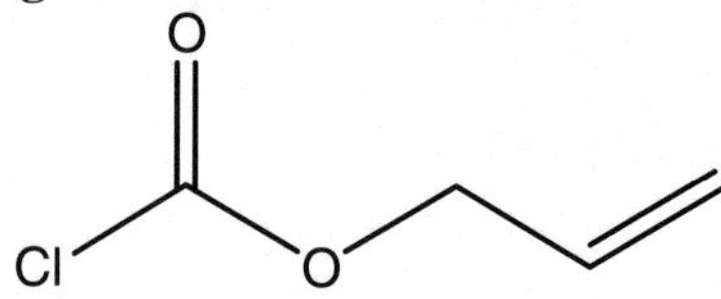

Chemical Structure

Synonyms: ALLYL CHLOROCARBONATE; ALLYLESTER KYSELINY CHLORMRAVENCI; CARBONOCHLORIDIC ACID,2-PROPENYL ESTER; CHLOROFORMIC ACID ALLYL ESTER; FORMIC ACID,CHLORO-,ALLYL ESTER; 2-PROPENYL CHLOROFORMATE

Description: colorless liquid; pungent odor

Use: intermediate in synthesis of numerous compounds; monomer in the manufacture of break-resistant optical lenses

Physical Properties

Boiling Point: 110 °C (230 °F) at 760 mm Hg

Freezing Point: -80 °C (-112 °F)

Specific Gravity: 1.1394 at 20 °C/4 °C

Vapor Density: 4.2 Air=1

Saturated Vapor Density: 1.299680581 kg/m^3

Vapor Pressure: 20 mm Hg at 25 °C

Water Solubility: Insoluble in Water

Other Solubilities: Can react with oxidizing materials
Surface Tension: Estimated at 25 dynes/cm
Odor Threshold: 1.4 ppm
pH: Acid
Refraction Index: 1.4223 at 20 °C/D
Flash Point: 31 °C Closed Cup

RTECS Toxicity Data

Acute Oral: Rat LD_{50} Dose: 244 mg/kg; Toxic Effects: Lungs, Thorax, or Respiration - Dyspnea. Mouse LD_{50} Dose: 210 mg/kg; Toxic Effects: Lungs, Thorax, or Respiration - Dyspnea.
Acute Inhalation: Rat LC_{50} Dose: 32400 ug/m^3; Toxic Effects: Lungs, Thorax, or Respiration - Dyspnea. Mouse LC_{50} Dose: 23100 ug/m^3; Toxic Effects: Lungs, Thorax, or Respiration - Dyspnea.

Hazard Overviews

Poison

Corrosive

Flammable
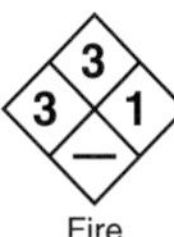

Fire Diamond

Health: Corrosive to eyes/skin/respiratory tract. Poison. Other Acute Effects: may be fatal if inhaled, swallowed, or absorbed through skin; lachrymator; inhalation may result in spasm, inflammation and edema of the larynx and bronchi, chemical pneumonitis and pulmonary edema; symptoms of exposure may include burning sensation; coughing; wheezing; laryngitis; shortness of breath; headache; nausea; vomiting.
Fire: Flammable. Hazards: vapor may travel considerable distance to source of ignition and flash back; container explosion may occur; forms explosive mixtures in air. Extinguishing agents: carbon dioxide, dry chemical powder or appropriate foam; do not use water. Precautions: combustible liquid.
Reactivity: Incompatible with: acids, may decompose on exposure to moist air or water, strong bases, alcohols, amines. Hazardous decomposition products: hydrogen chloride gas, carbon monoxide, carbon dioxide, phosgene gas.
Carcinogenicity: IARC - Not listed; NIOSH - Not listed; NTP - Not listed; ACGIH - Not listed; OSHA - Not listed; EPA - Not listed; MAK - Not listed
Primary Target Organs:

Eyes

Skin
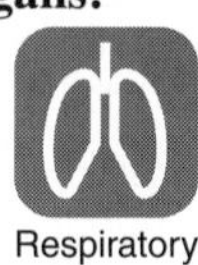
Respiratory System

Environmental

Cleanup/Disposal: Guide No. 155: Eliminate all ignition sources (no smoking, flares, sparks or flames in immediate area). All equipment used when handling the product must be grounded. Do not touch damaged containers or spilled material unless wearing appropriate protective clothing. Stop leak if you can do it without risk. A vapor suppressing foam may be used to reduce vapors. For chlorosilanes, use AFFF alcohol-resistant medium expansion foam to reduce vapors. Do not get water on spilled substance or inside containers. Use water spray to reduce vapors or divert vapor cloud drift. Prevent entry into waterways, sewers, basements or confined areas. Small Spills: Cover with dry earth, dry sand, or other non-combustible material followed with plastic sheet to minimize spreading or contact with rain. Use clean non-sparking tools to collect material and place it into loosely covered plastic containers for later disposal.

Environmental Physical Data

BCF: no food chain concentration potential

Regulations

RCRA 40CFR: Not listed
CERCLA: 40CFR 302.4: Not listed
SARA 40CFR 372.65: Not listed
SARA EHS 40CFR 355: Not listed
TSCA: Listed

ALL4360 **CAS #: 109-75-1**

ALLYL CYANIDE

RTECS: EM8050000
EINECS Number: 203-701-1
Molecular Formula: C_4H_5N
Formula Weight: 67.10

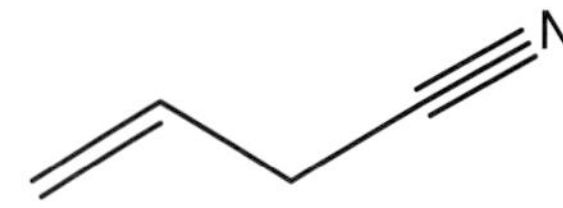

Chemical Structure

Synonyms: ACETONITRILE,VINYL-; ALLYLKYANID; ALLYLNITRILE; 1-BUTENE-4-NITRILE; BETA-BUTENONITRILE; TL 350; VINYLACETONITRILE

Physical Properties

Boiling Point: 119 °C (246.2 °F)
Freezing Point: -87 °C (-124.6 °F)
Specific Gravity: 0.83410
Vapor Pressure: 0.552 kPa at 0 °C
Water Solubility: Slightly Soluble
Other Solubilities: Ethanol: Miscible; Ether: Miscible
Refraction Index: 1.4060
Ionization Potential (eV): 10.22 +/-0.5

RTECS Toxicity Data

Acute Oral: Rat LD_{50} Dose: 115 mg/kg.
Acute Inhalation: Guinea Pig LC_{50} Dose: 2500 mg/m^3/4hr; Toxic Effects: Behavioral - Somnolence (general depressed activity); Behavioral - Convulsions or effect on seizure threshold; Lungs, Thorax, or Respiration - Dyspnea. Rat LC_{Lo} Dose: 500 ppm/4hr.
Acute Dermal: Rabbit LD_{50} Route: Skin; Dose: 1410 uL/kg. Rat LD_{50} Route: Subcutaneous Dose: 150 mg/kg.

Chronic (Multiple Dose) Inhalation: Rat Dose: 134 mg/m^3/2H/8W-I; Toxic Effects: Endocrine - Changes in thyroid weight; Nutritional and gross metabolic - Weight loss or decreased weight gain. Mouse Dose: 134 mg/m^3/2H/8W-I; Toxic Effects: Kidney, Ureter, and Bladder - Changes in kidney weight; Nutritional and gross metabolic - Weight loss or decreased weight gain.

Irritation Eye: Rabbit Standard Draize Test Dose: 500 mg/24H; Reaction: mild.

Irritation Skin: Rabbit Standard Draize Test Dose: 500 mg/24H; Reaction: mild. Rabbit Open Draize Test Dose: 10 mg/24H open; Reaction: mild.

Reproductive/Teratogenic: Rat Route: Inhalation; Dose: 50 ppm/6H; Duration: female 6-20D of pregnancy; Effects on Fertility - Post-implantation mortality; Effects on Embryo or Fetus - Fetotoxicity. Rat Route: Inhalation; Dose: 50 ppm/6H; Duration: female 6-20D of pregnancy; Specific Developmental Abnormalities - Craniofacial (including nose and tongue); Musculoskeletal system.

Hazard Overviews

Health: Irritating. Toxic. Other Acute Effects: harmful if swallowed, inhaled, or absorbed through skin.

Fire: Hazards: vapor may travel considerable distance to source of ignition and flash back; container explosion may occur; forms explosive mixtures in air; emits toxic fumes. Extinguishing agents: carbon dioxide, dry chemical powder or appropriate foam; water may be effective for cooling, but may not effect extinguishment. Precautions: combustible liquid.

Reactivity: Incompatible with: oxidizing agents, strong acids, strong bases, strong reducing agents. Hazardous decomposition products: toxic fumes of: carbon monoxide, carbon dioxide, nitrogen oxides, hydrogen cyanide.

Carcinogenicity: IARC - Not listed; NIOSH - Not listed; NTP - Not listed; ACGIH - Not listed; OSHA - Not listed; EPA - Not listed; MAK - Not listed

Primary Target Organs:

Eyes

Skin

Respiratory System

Environmental

Ecotoxicity: Fishes: Pimephales promelas 24h LC_{50} 210 mg/l 96h LC_{50} 182 mg/l

Environmental Physical Data

Octanol/Water Partition Coefficient: log K_{ow} = 0.12

Regulations

RCRA 40CFR: Not listed

CERCLA: 40CFR 302.4: Listed as Compound per CWA Section 307(a) per CAA Section 112

SARA 40CFR 372.65: Not listed

SARA EHS 40CFR 355: Not listed

TSCA: Listed

ALL4780 CAS #: 106-92-3

ALLYL GLYCIDYL ETHER

RTECS: RR0875000

DOT: UN2219; IMO3.3

EINECS Number: 203-442-4

Molecular Formula: $C_6H_{10}O_2$

Formula Weight: 114.06

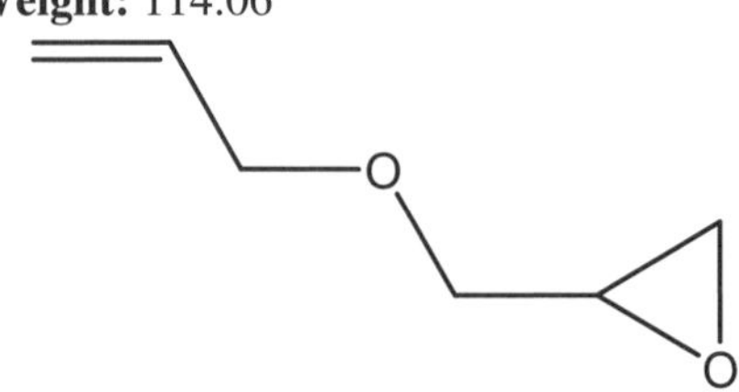

Chemical Structure

Synonyms: AGE; ALLIL-GLICIDIL-ETERE; 1-ALLILOSSI-2,3 EPOSSIPROPANO; ALLYL 2,3-EPOXYPROPYL ETHER; 1-ALLYL-2,3-EPOXYPROPANE; ALLYLGLYCIDAETHER; 1-ALLYLOXY-2,3-EPOXY-PROPAAN; 1-ALLYLOXY-2,3-EPOXYPROPAN; 1-(ALLYLOXY)-2,3-EPOXYPROPANE; 1-ALLYLOXY-2,3-EPOXYPROPANE; 1,2-EPOXY-3-ALLYLOXYPROPANE; ETHER,ALLYL 2,3-EPOXYPROPYL; GLYCIDYL ALLYL ETHER; OXIRANE,((2-PROPENYLOXY)METHYL)-; OXYDE D'ALLYLE ET DE GLYCIDYLE; PROPANE,1-(ALLYLOXY)-2,3-EPOXY-; ((2-PROPENYLOXY)METHYL) OXIRANE; ((2-PROPENYLOXY)METHYL)OXIRANE

Description: colorless liquid; strong, sweet odor

Use: a component of epoxy resin systems, resin intermediate and stabilizer of chlorinated compounds, vinyl resins and rubber

Physical Properties

Boiling Point: 153.9 °C (309 °F) at 760 mm Hg

Freezing Point: Forms Glass at -100 °C (-148 °F)

Specific Gravity: 0.9698 at 20 °C/40 °C

Vapor Density: 3.94 Air=1

Saturated Vapor Density: 1.216672377 kg/m^3

Density: 0.962 g/mL

Vapor Pressure: 3.6 mm Hg at 20 °C

Water Solubility: 14.1% in Water

Other Solubilities: 95% Ethanol: >=100 mg/ml at 18.5 °C; Acetone: >=100 mg/ml at 18.5 °C; DMSO: >=100 mg/ml at 18.5 °C; Toluene: miscible.

Odor Threshold: 44 mg/m^3

Refraction Index: 1.4348 at 20 °C

Flash Point: 57.222 °C Open Cup

RTECS Toxicity Data

Acute Oral: Rat LD_{50} Dose: 1600 mg/kg; Toxic Effects: Brain and coverings - Recordings from specific areas of CNS; Behavioral - Change in motor activity (specific assay); Behavioral - Ataxia. Mouse LD_{50} Dose: 390 mg/kg; Toxic Effects: Brain and coverings - Recordings from specific areas of CNS; Behavioral - Somnolence (general depressed activity); Behavioral - Ataxia.

Acute Inhalation: Rat LC_{50} Dose: 670 ppm/8hr; Toxic Effects: Sense organs and special senses - Corneal damage; Lungs, Thorax, or Respiration - Acute pulmonary edema; Gastrointestinal - Changes in structure or function of salivary glands. Mouse LC_{50} Dose: 270 ppm/4hr; Toxic Effects: Sense organs and special senses - Lacrimation; Lungs, Thorax, or Respiration - Dyspnea; Gastrointestinal - Changes in structure or function of salivary glands.
Acute Dermal: Rabbit LD_{50} Route: Skin; Dose: 2550 mg/kg.
Chronic (Multiple Dose) Inhalation: Rat Dose: 400 ppm/7H/10W-I; Toxic Effects: Lungs, Thorax, or Respiration - Emphysema; Kidney, Ureter, and Bladder - Changes in kidney weight; Nutritional and gross metabolic - Weight loss or decreased weight gain. Rat Dose: 200 ppm/6H/14D-I; Toxic Effects: Sense organs and special senses - Other; Lungs, Thorax, or Respiration - Structural or functional change in trachea or bronchi; Lungs, Thorax, or Respiration - Other changes.
Irritation Eye: Rabbit Standard Draize Test Dose: 750 ug/24H; Reaction: severe. Rabbit Standard Draize Test Dose: 97 mg; Reaction: severe.
Irritation Skin: Rabbit Standard Draize Test Dose: 2 mg/24H; Reaction: severe. Rabbit Standard Draize Test Dose: 485 mg/3D; Reaction: moderate.
Mutagenic: Hamster Cytogenetic Analysis; Cell Type: ovary; Dose: 64800 ug/L. Hamster Sister Chromatid Exchenge; Cell Type: ovary; Dose: 3300 ug/L. Hamster Sister Chromatid Exchenge; Cell Type: lung; Dose: 625 umol/L.
Tumorigenic: Rat Route: Inhalation; Dose: 10 ppm/6H/2Y-I; Toxic Effects: Tumorigenic - Equivocal tumorigenic agent by RTECS criteria; Sense organs and special senses - Tumors. Mouse Route: Inhalation; Dose: 10 ppm/6H/2Y-I; Toxic Effects: Tumorigenic - Neoplastic by RTECS criteria; Sense organs and special senses - Tumors.

Hazard Overviews

Flammable

Fire Diamond

Health: Severe irritation to eyes/skin/respiratory tract. Also Causes: pulmonary edema, dizziness, light-headedness, narcosis, dermatitis, skin sensitization, pain with swallowing, abdominal pain, diarrhea, shock possible burns of the eye. Chronic Effects: dermatitis, liver/kidney damage (animal data).
Fire: Flammable. For small fires use dry chemical, carbon dioxide, water spray, or regular foam. For large fires use water spray, fog, or regular foam.
Reactivity: Stable. Hazardous polymerization cannot occur. Avoid: heat; ignition sources. Incompatible with: strong oxidizers. Hazardous decomposition products: carbon oxide(s).
Carcinogenicity: IARC - Not listed; NIOSH - Not listed; NTP - Not listed; ACGIH - Not listed; OSHA - Not listed; EPA - Not listed; MAK - Class A2, Unmistakably carcinogenic in animal experimentation only

Primary Target Organs:

Eyes Skin Respiratory System Nervous System

Exposure Limits
OSHA PEL: STEL: 10 ppm; 45 mg/m^3.
OSHA PEL Vacated 1989 Limits: TWA: 5 ppm; 22 mg/m^3; STEL: 10 ppm; 44 mg/m^3.
ACGIH TLV: TWA: 5 ppm; 23 mg/m^3; STEL: 10 ppm; 47 mg/m^3.
NIOSH REL: TWA: 5 ppm; 22 mg/m^3. STEL: 10 ppm; 44 mg/m^3; skin.
NIOSH IDLH: 50 ppm.
Respirator Recommendation
Exposure Range: >10 to <50 ppm Supplied Air, Constant Flow/Pressure Demand, Full Face
Exposure Range: 50 to unlimited ppm Self-contained Breathing Apparatus, Pressure Demand, Full Face
Note: warning properties unknown

Environmental

Ecotoxicity: LD_{50} Goldfish 30 mg/l/96 hr /Conditions of bioassay not specified
Cleanup/Disposal: Guide No. 129: Eliminate all ignition sources (no smoking, flares, sparks or flames in immediate area). All equipment used when handling the product must be grounded. Do not touch or walk through spilled material. Stop leak if you can do it without risk. Prevent entry into waterways, sewers, basements or confined areas. A vapor suppressing foam may be used to reduce vapors. Absorb or cover with dry earth, sand or other non-combustible material and transfer to containers. Use clean non-sparking tools to collect absorbed material. Large Spills: Dike far ahead of liquid spill for later disposal. Water spray may reduce vapor; but may not prevent ignition in closed spaces.

Regulations

RCRA 40CFR: Not listed
CERCLA: 40CFR 302.4: Not listed
SARA 40CFR 372.65: Not listed
SARA EHS 40CFR 355: Not listed
TSCA: Listed

Analytical Methods

Indoor / Expired Air: NIOSH 2545

ALL5200 **CAS #: 556-56-9**

ALLYL IODIDE

RTECS: UD0450000
EINECS Number: 209-130-4
Molecular Formula: C_3H_5I
Formula Weight: 167.98

Chemical Structure

Synonyms: 3-IODO-1-PROPENE; 3-IODOPROPENE; 3-IODOPROPYLENE; 1-PROPENE,3-IODO-(9CI)

Description: yellowish, darkens upon exposure to light/air liquid; pungent odor

Physical Properties

Boiling Point: 103.1 °C (218 °F)
Freezing Point: -99 °C (-146.2 °F)
Vapor Density: 5.8
Density: 1.825 at 20 °C
Vapor Pressure: 95.8 kPa at 100 °C
Water Solubility: Insoluble
Other Solubilities: miscible with Alcohol, Chloroform and Ether
Refraction Index: 1.5540
Ionization Potential (eV): 9.298

RTECS Toxicity Data

Acute Oral: Rat LD_{50} Dose: 10 mg/kg;Toxic Effects: Behavioral - General anesthetic; Lungs, Thorax, or Respiration - Cyanosis; Gastrointestinal - Changes in structure or function of salivary glands.

Mutagenic: Human Unscheduled DNA Synthesis; Cell Type: HeLa cell; Dose: 50 umol/L. Bacteria - S Typhimurium Mutations in Microorganisms; Dose: 1 umol/plate (+S9).

Hazard Overviews

Flammable

Fire Diamond

Health: Severely irritating to eyes/respiratory tract; mildly irritating to skin. Also Causes: injury to liver, kidneys. Chronic Effects: lung/liver/kidney injury, decrease in systolic BP, tonicity of brain blood vessels

Fire: Flammable. For small fires use dry chemical, carbon dioxide, or regular foam. For large fires use fog, or regular foam. Use water only if other extinguishing agents are unavailable, do not spray directly on, or allow water to get inside containers. Containers may explode in fire.

Reactivity: Stable. Hazardous polymerization cannot occur. Avoid: sources of ignition; oxidizers. Incompatible with: light and strong oxidizers. Hazardous decomposition products: highly toxic fumes of iodine.

Carcinogenicity: IARC - Not listed; NIOSH - Not listed; NTP - Not listed; ACGIH - Not listed; OSHA - Not listed; EPA - Not listed; MAK - Not listed

Primary Target Organs:

Eyes

Skin

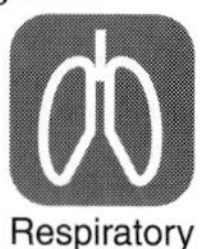

Respiratory System

Liver

Kidneys

Environmental

Regulations

RCRA 40CFR: Not listed
CERCLA: 40CFR 302.4: Not listed
SARA 40CFR 372.65: Not listed
SARA EHS 40CFR 355: Not listed
TSCA: Listed

ALL5620 CAS #: 57-06-7

ALLYL ISOTHIOCYANATE

RTECS: NX8225000
DOT: UN1545; IMO6.1
EINECS Number: 200-309-2
Molecular Formula: C_4H_5NS
Structured MF: $CH_2=CHCH_2N=C=S$
Formula Weight: 99.15

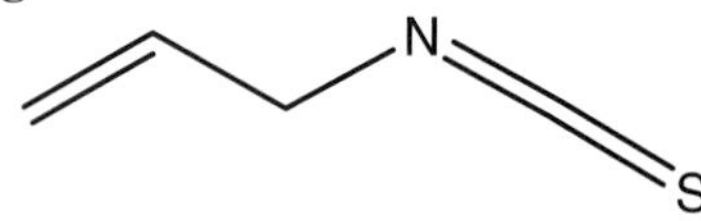

Chemical Structure

Synonyms: AITC; AITK; ALLYL ISORHODANIDE; ALLYL ISOSULFOCYANATE; ALLYL ISOSULPHOCYANATE; ALLYL ISOTHIOCYANATE,STABILIZED; ALLYL MUSTARD OIL; ALLYL SEVENOLUM; ALLYL THIOCARBONIMIDE; ALLYLISOTHIOKYANAT; ALLYLSENEVOL; ALLYLSENFOEL; ARTIFICIAL MUSTARD OIL; ARTIFICIAL OIL OF MUSTARD; CARBOSPOL; ISOTHIOCYANATE D'ALLYLE; 3-ISOTHIOCYANATO-1-PROPENE; 3-ISOTHIOCYANATOPROPENE; ISOTHIOCYANIC ACID ALLYL ESTER; ISOTHIOCYANIC ACID,ALLYL ESTER; MUSTARD OIL; OIL OF MUSTARD BPC 1949; OIL OF MUSTARD,ARTIFICIAL; OLEUM SINAPIS; OLEUM SINAPIS VOLATILE; 1-PROPENE,3-ISOTHIOCYANATO-; PROPENE,3-ISOTHIOCYANATO-; 2-PROPENYL ISOTHIOCYANATE; REDSKIN; SENF OEL; SENFOEL; SYNTHETIC MUSTARD OIL; VOLATILE MUSTARD OIL; VOLATILE OIL OF MUSTARD

Description: colorless to pale yellow, oily liquid; very pungent mustard odor

Use: as a flavoring agent, in military poison gas, in medicine as a rubefacient (counterirritant), as a fumigant, in ointments, in mustard plasters, as an adjuvant, as a fungicide, as a repellent for cats and dogs, as a denaturant for ethanol, as an ingredient in certain plastic glues cements to deter glue sniffers and as a preservative in animal feed

Physical Properties

Boiling Point: 148 °C (298 °F) to 154 °C (309 °F)
Freezing Point: -80 °C (-112 °F)
Specific Gravity: 1.013 to 1.02

Vapor Density: 3.41 Air=1
Saturated Vapor Density: 1.21412695 kg/m^3
Density: 1.013 to 1.016 g/mL at 25 °C
Vapor Pressure: 0.493 kPa at 20 °C
Water Solubility: Slightly Soluble in Water
Other Solubilities: 70% Aqueous Ethanol: Slightly Soluble; 95% Ethanol: 50-100 mg/ml at 23 °C; Acetone: >=100 mg/ml at 23 °C; Benzene: Soluble; Carbon Disulfide: Soluble; DMSO: >=100 mg/ml at 23 °C; Ether: Soluble; Most organic solvents: miscible.
Odor Threshold: 8 x10^{-3} ppm
Refraction Index: 1.5268 to 1.5820 at 20 °C/D
Flash Point: 46 °C Closed Cup

RTECS Toxicity Data

Acute Oral: Rat LD_{50} Dose: 112 mg/kg; Toxic Effects: Lungs, Thorax, or Respiration - Other changes; Gastrointestinal - Other changes. Mouse LD_{50} Dose: 308 mg/kg.
Acute Dermal: Rabbit LD_{50} Route: Skin; Dose: 88 mg/kg. Rat LD_{50} Route: Subcutaneous Dose: 92 mg/kg; Toxic Effects: Behavioral - Coma; Skin and appendages - Dermatitis, other.
Chronic (Multiple Dose) Oral: Rat Dose: 5600 mg/kg/14D-C; Toxic Effects: Behavioral - Somnolence (general depressed activity); Skin and appendages - Hair; DEATH. Rat Dose: 300 mg/kg/3D-I; Toxic Effects: Liver - Changes in liver weight; Blood - Changes in serum composition; Biochemical - Hepatic microsomal mixed oxidase(dealkylation, hyroxylation,etc).
Reproductive/Teratogenic: Rat Route: Subcutaneous; Dose: 100 mg/kg; Duration: female 8-9D of pregnancy; Effects on Embryo or Fetus - Fetotoxicity. Rat Route: Subcutaneous; Dose: 200 mg/kg; Duration: female 8-9D of pregnancy; Effects on Fertility - Post-implantation mortality.
Mutagenic: Mouse Mutations in Mammalian Somatic Cells; Cell Type: lymphocyte; Dose: 400 ug/L. Mammal Other Mutation Test Systems; Route: Skin; Dose: 800 ug/L.
Tumorigenic: Rat Route: Oral; Dose: 12875 mg/kg/2Y-I; Toxic Effects: Tumorigenic - Neoplastic by RTECS criteria; Kidney, Ureter, and Bladder - Tumors. Mouse Route: Skin; Dose: 12 gm/kg/12W-I; Toxic Effects: Tumorigenic - Equivocal tumorigenic agent by RTECS criteria; Tumorigenic - Increased incidence of tumors in susceptible strains; Tumorigenic - Facilitates action of known carcinogens.

Hazard Overviews

Poison

Corrosive

Flammable

Health: Corrosive to eyes/skin/respiratory tract. Poison. Other Acute Effects: may be fatal if inhaled, swallowed, or absorbed through skin; extremely destructive to tissue of the mucous membranes and upper respiratory tract, eyes and skin; inhalation may result in spasm, inflammation and edema of the larynx and bronchi, chemical pneumonitis and pulmonary edema; symptoms of exposure may include burning sensation, coughing, wheezing, laryngitis, shortness of breath, headache, nausea and vomiting; may cause allergic respiratory and skin reactions; repeated exposure may cause asthma; vesicant; causes blisters on contact with skin. Chronic Effects: may cause heritable genetic damage; may alter genetic material. Carcinogen.
Fire: Flammable. Hazards: vapor may travel considerable distance to source of ignition and flash back; emits toxic fumes. Extinguishing agents: carbon dioxide; dry chemical powder. Precautions: combustible liquid.
Reactivity: Incompatible with: water, alcohols, strong bases, amines, acids, strong oxidizing agents, heat. Hazardous decomposition products: thermal decomposition may produce carbon monoxide, carbon dioxide, and nitrogen oxides; hydrogen cyanide, sulfur oxides.
Carcinogenicity: IARC - Group 3, Not classifiable as to carcinogenicity to humans; NIOSH - Not listed; NTP - Not listed; ACGIH - Not listed; OSHA - Not listed; EPA - Not listed; MAK - Not listed
Primary Target Organs:

Eyes

Skin

Respiratory System

Exposure Limits
AIHA WEEL: STEL: 1 ppm 15-min skin.
Respirator Recommendation
Exposure Range: >1 to 50 ppm Supplied Air, Constant Flow/Pressure Demand, Half Mask
Exposure Range: >50 to 1000 ppm Supplied Air, Constant Flow/Pressure Demand, Full Face
Exposure Range: >1000 to unlimited ppm Self-contained Breathing Apparatus, Pressure Demand, Full Face

Environmental

Ecotoxicity: LC_{50} Pimephales promelas (fathead minnow) 85.6 ug/l/96 hr, flow-through bioassay with measured concentrations, 25.1 °C, dissolved oxygen 6.9 mg/l, and pH 7.8
Cleanup/Disposal: Guide No. 155: Eliminate all ignition sources (no smoking, flares, sparks or flames in immediate area). All equipment used when handling the product must be grounded. Do not touch damaged containers or spilled material unless wearing appropriate protective clothing. Stop leak if you can do it without risk. A vapor suppressing foam may be used to reduce vapors. For chlorosilanes, use AFFF alcohol-resistant medium expansion foam to reduce vapors. Do not get water on spilled substance or inside containers. Use water spray to reduce vapors or divert vapor cloud drift. Prevent entry into waterways, sewers, basements or confined areas. Small Spills: Cover with dry earth, dry sand, or other non-combustible material followed with plastic sheet to minimize spreading or contact with rain. Use clean non-sparking tools to collect material and place it into loosely covered plastic containers for later disposal.

Environmental Physical Data

Octanol/Water Partition Coefficient: log K_{ow} = calculated at 2.11

Regulations

RCRA 40CFR: Not listed
CERCLA: 40CFR 302.4: Listed as Compound per CWA Section 307(a) per CAA Section 112
SARA 40CFR 372.65: Not listed
SARA EHS 40CFR 355: Not listed
TSCA: Listed

ALL6040 CAS #: 2835-39-4

ALLYL ISOVALERATE

RTECS: NY1412000
EINECS Number: 220-609-7
Molecular Formula: $C_8H_{14}O_2$
Structured MF: $(CH_3)_2CHCH_2COOCH_2CH=CH_2$
Formula Weight: 142.19
Synonyms: ALLYL ISOVALERIANATE; ALLYL 3-METHYLBUTYRATE; BUTANOIC ACID,3-METHYL-,2-PROPENYL ESTER; BUTYRIC ACID,3-METHYL-,ALLYL ESTER; ISOVALERIC ACID,ALLYL ESTER; 3-METHYLBUTANOIC ACID,ALLYL ESTER; 3-METHYLBUTANOIC ACID,2-PROPENYL ESTER; 3-METHYLBUTYRIC ACID,ALLYL ESTER; 2-PROPENYL ISOVALERATE; 2-PROPENYL 3-METHYLBUTANOATE
Description: liquid; pleasant over-ripe fruit odor
Use: raw material to impart a fruit-like (apple, cherry) aroma; in soaps, detergents, creams, lotions and perfumes and as a synthetic flavoring agent and adjuvant in foods

Physical Properties

Boiling Point: 89 °C (192 °F) to 90 °C (194 °F)
Specific Gravity: 0.882 at 24/22 °C
Saturated Vapor Density: 1.280116334 kg/m^3
Density: 0.882 g/cu cm at 23.3 °C
Vapor Pressure: 13.0 mm Hg at 22.0 °C
Water Solubility: < 0.1 mg/mL at 165 C
Other Solubilities: 95% Ethanol: >=100 mg/ml at 21 °C; Acetone: >=100 mg/ml at 21 °C; DMSO: >=100 mg/ml at 21 °C.
Refraction Index: 1.4162 at 21 °C
Flash Point: 39.7 °C

RTECS Toxicity Data

Acute Oral: Rat LD_{50} Dose: 230 mg/kg.
Acute Dermal: Rabbit LD_{50} Route: Skin; Dose: 560 mg/kg.
Irritation Skin: Rabbit Standard Draize Test Dose: 500 mg/24H; Reaction: moderate.
Mutagenic: Mouse Mutations in Mammalian Somatic Cells; Cell Type: lymphocyte; Dose: 100 mg/L. Hamster Cytogenetic Analysis; Cell Type: ovary; Dose: 300 mg/L.
Tumorigenic: Rat Route: Oral; Dose: 31930 mg/kg/2Y-I; Toxic Effects: Tumorigenic - Carcinogenic by RTECS criteria; Blood - Leukemia; Blood - Lymphomax including Hodgkin's disease. Rat Route: Oral; Dose: 15965 mg/kg/2Y-I; Toxic Effects: Tumorigenic - Equivocal tumorigenic agent by RTECS criteria; Endocrine - Tumors; Blood - Leukemia.

Hazard Overviews

Flammable

Fire: Flammable.
Carcinogenicity: IARC - Group 3, Not classifiable as to carcinogenicity to humans; NIOSH - Not listed; NTP - Not listed; ACGIH - Not listed; OSHA - Not listed; EPA - Not listed; MAK - Not listed

Environmental

Cleanup/Disposal: Guide No. 131: Fully encapsulating, vapor protective clothing should be worn for spills and leaks with no fire. Eliminate all ignition sources (no smoking, flares, sparks or flames in immediate area). All equipment used when handling the product must be grounded. Do not touch or walk through spilled material. Stop leak if you can do it without risk. Prevent entry into waterways, sewers, basements or confined areas. A vapor suppressing foam may be used to reduce vapors. Small Spills: Absorb with earth, sand or other non-combustible material and transfer to containers for later disposal. Use clean non-sparking tools to collect absorbed material. Large Spills: Dike far ahead of liquid spill for later disposal. Water spray may reduce vapor; but may not prevent ignition in closed spaces.

Regulations

RCRA 40CFR: Not listed
CERCLA: 40CFR 302.4: Not listed
SARA 40CFR 372.65: Not listed
SARA EHS 40CFR 355: Not listed
TSCA: Listed

ALL6460 CAS #: 96-05-9

ALLYL METHACRYLATE

RTECS: UD3483000
EINECS Number: 202-473-0
Molecular Formula: $C_7H_{10}O_2$
Formula Weight: 126.16

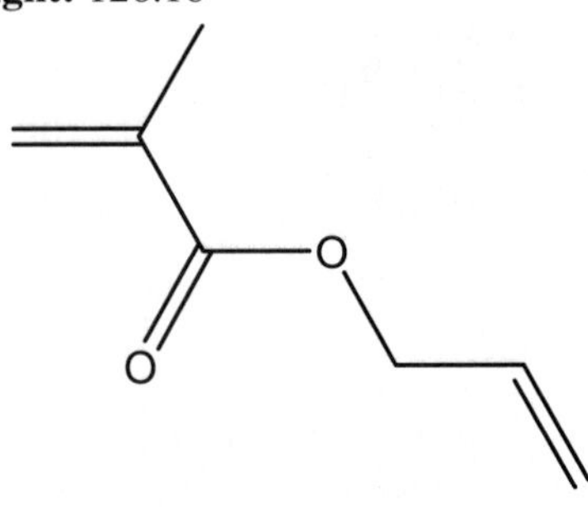

Chemical Structure

Synonyms: AGEFLEX AMA; ALLYLESTER KYSELINY METHAKRYLOVE; METHACRYLIC ACID,ALLYL ESTER; 2-PROPENOIC ACID,2-METHYL-,2-PROPENYL ESTER
Use: cross-linking agent; copolymerized with methyl methacrylate in such applications as dental plastics, lenses, adhesives, fiber reinforced plastics, acrylic ester rubbers, printing processes and coatings; in contact lenses

Physical Properties

Boiling Point: 80 °C (176 °F)
Specific Gravity: 0.929 at 20 °C/20 °C
Vapor Density: 4.2 Air=1
Water Solubility: Insoluble
Refraction Index: 1.4360 at 20 °C
Evaporation Rate: < 1 Butyl Acetate=1

RTECS Toxicity Data

Acute Oral: Rat LD_{50} Dose: 70 mg/kg. Mouse LD_{50} Dose: 57 mg/kg.
Acute Inhalation: Rat LC_{50} Dose: 1800 mg/m^3. Mouse LC_{50} Dose: 5500 mg/m^3.
Acute Dermal: Rabbit LD_{50} Route: Skin; Dose: 500 uL/kg.
Irritation Eye: Rabbit Standard Draize Test Dose: 500 mg/24H; Reaction: mild.
Irritation Skin: Rabbit Standard Draize Test Dose: 20 mg/24H; Reaction: moderate.

Hazard Overviews

Health: Severely irritating to eyes/respiratory tract. Toxic. Other Acute Effects: harmful by inhalation, in contact with skin and if swallowed; readily absorbed through skin; high concentrations are extremely destructive to tissues of the respiratory tract, eyes and skin; lachrymator; symptoms of exposure may include burning sensation, coughing, wheezing, laryngitis, shortness of breath, headache, nausea and vomiting; may cause allergic respiratory and skin reactions.
Fire: Hazards: vapor may travel considerable distance to source of ignition and flash back; may undergo autopolymerization; container explosion may occur. Extinguishing agents: carbon dioxide, dry chemical powder or appropriate foam. Precautions: combustible liquid.
Carcinogenicity: IARC - Not listed; NIOSH - Not listed; NTP - Not listed; ACGIH - Not listed; OSHA - Not listed; EPA - Not listed; MAK - Not listed
Primary Target Organs:

Eyes

Skin

Respiratory System

Environmental

Ecotoxicity: LC_{50} Pimephales promelas (fathead minnows) 0.99 g/l/96 hr (95% confidence limit 0.90-1.1 g/l); age 30 days old, water hardness 45.6 mg/l ($CaCO_3$), temp 24.9 °C, pH 7.66, dissolved oxygen 7.1 mg/l, alkalinity 44.4 mg/l ($CaCO_3$), Tank vol: 2.0 l, additions: 18 vol/day (flow-through bioassay)

Regulations

RCRA 40CFR: Not listed
CERCLA: 40CFR 302.4: Not listed
SARA 40CFR 372.65: Not listed
SARA EHS 40CFR 355: Not listed
TSCA: Listed

ALL6880 CAS #: 2179-59-1

ALLYL PROPYL DISULFIDE

RTECS: JO0350000
EINECS Number: 218-550-7
Molecular Formula: $C_6H_{12}S_2$
Structured MF: $H_2C{=}CHCH_2S_2CH_2CH_2CH_3$
Formula Weight: 148.30
Synonyms: DISULFIDE,ALLYL PROPYL; DISULFIDE,2-PROPENYL PROPYL; 4,5-DITHIA-1-OCTENE; ONION OIL; 2-PROPENYL PROPYL DISULFIDE; PROPENYL PROPYL DISULFIDE; PROPYL ALLYL DISULFIDE
Description: pale-yellow liquid; pungent, irritating odor odor like that of cooked onions
Use: chief volatile constituent of onion oil

Physical Properties

Boiling Point: 78 °C (172 °F) to 80 °C (176 °F) at 13 mm Hg
Freezing Point: -15 °C (5 °F)
Specific Gravity: 0.93 at 15 °C
Water Solubility: Insoluble
Other Solubilities: 95% Ethanol: >= 100 mg/ml at 20 °C; Acetone: >= 100 mg/ml at 20 °C; Carbon Disulfide: Soluble; Chloroform: Soluble; DMSO: >= 100 mg/ml at 20 °C; Ether: Soluble.
Refraction Index: 1.5219 at 20 °C/D
Flash Point: 56.3 °C

Hazard Overviews

Flammable

Fire: Flammable.
Carcinogenicity: IARC - Not listed; NIOSH - Listed as carcinogen; NTP - Not listed; ACGIH - Not listed; OSHA - Not listed; EPA - Not listed; MAK - Not listed
Exposure Limits
OSHA PEL: TWA: 2 ppm; 12 mg/m^3.
OSHA PEL Vacated 1989 Limits: TWA: 2 ppm; 12 mg/m^3; STEL: 3 ppm; 18 mg/m^3.
ACGIH TLV: TWA: 2 ppm; 13 mg/m^3; STEL: 3 ppm; 18 mg/m^3.
NIOSH REL: TWA: 2 ppm; 12 mg/m^3. STEL: 3 ppm; 18 mg/m^3.
DFG MAK: TWA: 2 ppm; 12 mg/m^3.

Respirator Recommendation
Exposure Range: >2 to 100 ppm Supplied Air, Constant Flow/Pressure Demand, Half Mask
Exposure Range: >100 to 2000 ppm Supplied Air, Constant Flow/Pressure Demand, Full Face
Exposure Range: >2000 to unlimited ppm Self-contained Breathing Apparatus, Pressure Demand, Full Face
Note: warning properties unknown

Environmental

Cleanup/Disposal: Guide No. 171: Do not touch or walk through spilled material. Stop leak if you can do it without risk. Prevent dust cloud. Avoid inhalation of asbestos dust. Small Dry Spills: With clean shovel place material into clean, dry container and cover loosely; move containers from spill area. Small Spills: Take up with sand or other noncombustible absorbent material and place into containers for later disposal. Large Spills: Dike far ahead of liquid spill for later disposal. Cover powder spill with plastic sheet or tarp to minimize spreading. Prevent entry into waterways, sewers, basements or confined areas.

Regulations

RCRA 40CFR: Not listed
CERCLA: 40CFR 302.4: Not listed
SARA 40CFR 372.65: Not listed
SARA EHS 40CFR 355: Not listed
TSCA: Not listed

ALL7300 **CAS #: 107-37-9**

ALLYL TRICHLOROSILANE

RTECS: VV1530000
DOT: UN1724
EINECS Number: 203-485-9
Molecular Formula: $C_3H_5Cl_3Si$
Structured MF: CH_2=CH•CH_2•$SiCl_3$
Formula Weight: 175.52

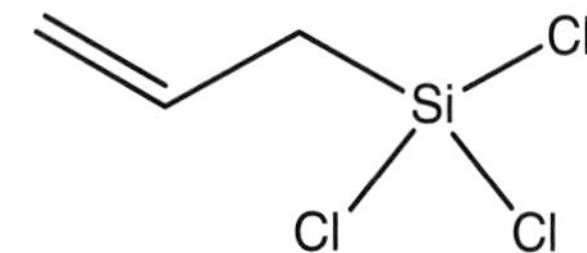

Chemical Structure

Synonyms: ALLYLSILICONE TRICHLORIDE; ALLYLTRICHLOROSILANE,STABILIZED; PROPEN-3-YLTRICHLOROSILANE; SILANE,ALLYLTRICHLORO-; SILANE,TRICHLOROALLYL-; SILANE,TRICHLORO-2-PROPENYL-; TRICHLOROALLYLSILANE
Description: colorless liquid; pungent irritating odor
Use: int for silicones; glass fiber finishes

Physical Properties

Boiling Point: 117.5 °C (244 °F)
Freezing Point: 35 °C (95 °F)
Specific Gravity: 1.217 at 27 °C
Vapor Density: 6 Air=1
Vapor Pressure: 8.71 kPa at 50 °C
Water Solubility: Insoluble
Surface Tension: Estimated at 20 dynes/cm
pH: Acid
Refraction Index: 1.487 at 20 °C/D
Flash Point: 35 °C Closed Cup

Hazard Overviews

Corrosive

Flammable

Fire Diamond

Health: Corrosive to eyes/skin/respiratory tract.
Fire: Flammable. Do not use water as an extinguishing agent since it reacts violently with allyl trichlorosilane to emit toxic hydrogen chloride gas. For small fires use dry chemical, carbon dioxide, or regular foam. For large fires use fog or regular foam.
Reactivity: Unstable. Hazardous polymerization can occur. Avoid: heat; acids; water; metal. Incompatible with: water; metal; acid; acid gases. Hazardous decomposition products: oxides of carbon; toxic chloride; hydrogen chloride; phosgene gas.
Carcinogenicity: IARC - Not listed; NIOSH - Not listed; NTP - Not listed; ACGIH - Not listed; OSHA - Not listed; EPA - Not listed; MAK - Not listed
Primary Target Organs:

Eyes

Skin

Respiratory System

Mucous Membranes

Environmental

Cleanup/Disposal: Guide No. 155: Eliminate all ignition sources (no smoking, flares, sparks or flames in immediate area). All equipment used when handling the product must be grounded. Do not touch damaged containers or spilled material unless wearing appropriate protective clothing. Stop leak if you can do it without risk. A vapor suppressing foam may be used to reduce vapors. For chlorosilanes, use AFFF alcohol-resistant medium expansion foam to reduce vapors. Do not get water on spilled substance or inside containers. Use water spray to reduce vapors or divert vapor cloud drift. Prevent entry into waterways, sewers, basements or confined areas. Small Spills: Cover with dry earth, dry sand, or other non-combustible material followed with plastic sheet to minimize spreading or contact with rain. Use clean non-sparking tools to collect material and place it into loosely covered plastic containers for later disposal.

Environmental Physical Data

BCF: no food chain concentration potential

Regulations

RCRA 40CFR: Not listed
CERCLA: 40CFR 302.4: Not listed
SARA 40CFR 372.65: Not listed

SARA EHS 40CFR 355: Not listed
TSCA: Listed

ALL7720 CAS #: 107-11-9

ALLYLAMINE

RTECS: BA5425000
DOT: UN2334; IMO6.1
EINECS Number: 203-463-9
Molecular Formula: C_3H_7N
Structured MF: CH_2=$CHCH_2NH_2$
Formula Weight: 57.09

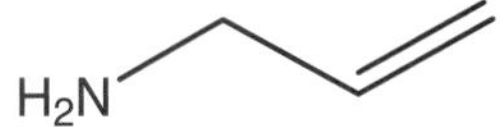

Chemical Structure

Synonyms: ALLYL AMINE; 3-AMINO-1-PROPENE; 3-AMINOPROPENE; 3-AMINOPROPYLENE; MONOALLYLAMINE; 2-PROPEN-1-AMINE; 2-PROPENAMINE; 2-PROPEN-1-AMINE (9CI); 2-PROPENYLAMINE
Description: colorless, light yellow liquid; ammonia odor
Use: manufacture of mercurial diuretics; organic synthesis

Physical Properties

Boiling Point: 55 °C (131 °F) to 58 °C (136 °F)
Freezing Point: -88 °C (-126.4 °F)
Specific Gravity: 0.76 at 20 °C/20 °C
Vapor Density: 2 Air=1
Density: 0.765 g/mL at 20 °C
Vapor Pressure: 9.77 kPa at 0 °C
Water Solubility: Miscible with Water
Other Solubilities: 95% Ethanol: Very Soluble; Benzene: Soluble; DMSO: Soluble; Ether: Very Soluble; Most organics: Soluble.
Surface Tension: 3.9 $x10^{-2}$ N/m at melting point
Refraction Index: 1.4205 at 20 °C/D
Evaporation Rate: > 1 Butyl Acetate=1
Ionization Potential (eV): 8.76
Flash Point: -29 °C Closed Cup
Autoignition Temperature: 374 °C
LEL: 2.2% v/v
UEL: 22% v/v

RTECS Toxicity Data

Acute Oral: Rat LD_{50} Dose: 102 mg/kg. Mouse LD_{50} Dose: 57 mg/kg.
Acute Inhalation: Rat LC_{50} Dose: 177 ppm/8hr; Toxic Effects: Sense organs and special senses - Other; Sense organs and special senses - Lacrimation; Lungs, Thorax, or Respiration - Dyspnea. Man TC_{Lo} Dose: 2500 ppb/5M; Toxic Effects: Sense organs and special senses - Lacrimation; Lungs, Thorax, or Respiration - Structural or functional change in trachea or bronchi.
Acute Dermal: Rabbit LD_{50} Route: Skin; Dose: 35 mg/kg; Toxic Effects: Skin and appendages - Primary irritation; Lungs, Thorax, or Respiration - Chronic pulmonary edema.
Chronic (Multiple Dose) Oral: Rat Dose: 1 gm/kg/11D-I; Toxic Effects: Cardiac - Changes in coronary arteries; Cardiac - Other changes; Vascular - Structural changes in vessels. Rat Dose: 6240 mg/kg/15W-C; Toxic Effects: Cardiac - Cardiomyopathy including infarction; Cardiac - Changes in heart weight; DEATH.
Chronic (Multiple Dose) Inhalation: Rat Dose: 20 ppm/7H/50D-I; Toxic Effects: Cardiac - Other changes; Liver - Changes in Liver weight; Nutritional and gross metabolic - Weight loss or decreased weight gain.
Irritation Eye: Rabbit Nonstandard Exposure Dose: 50 mg/20S rinse; Reaction: severe.
Irritation Skin: Rabbit Standard Draize Test Dose: 500 mg/24H; Reaction: severe.
Mutagenic: Rat Cytogenetic Analysis; Route: Oral; Dose: 2500 ng/kg.

Hazard Overviews

Corrosive

Flammable

Fire Diamond

Health: Toxic. Corrosive, causes severe burns to eyes/skin/respiratory tract. Also Causes: nausea, convulsions, pulmonary edema (high concentration), corneal edema, conjunctivitis. Chronic Effects: chemical pneumonia, liver/kidney damage (in animals).
Fire: Flammable. Can form explosive mixtures in air. Polymerizes. For small fires use dry chemical, carbon dioxide, water spray, or alcohol foam. For large fires use water spray, fog, or standard foam. Use water spray to disperse vapors and to dilute to nonflammable mixtures. Containers may explode in heat of fire
Reactivity: Stable. Hazardous polymerization can occur by elevated temperature, oxidizers, and peroxides. Avoid: heat; ignition sources. Incompatible with: oxidizing materials; acids; chlorine; hypochlorite; halogenated compounds; active metals; reactive organic compounds; rubber; cork. Hazardous decomposition products: nitrogen oxides; carbon dioxide; carbon monoxide; hydrocarbons; amine vapors.
Carcinogenicity: IARC - Not listed; NIOSH - Not listed; NTP - Not listed; ACGIH - Not listed; OSHA - Not listed; EPA - Not listed; MAK - Not listed
Primary Target Organs:

Eyes

Skin

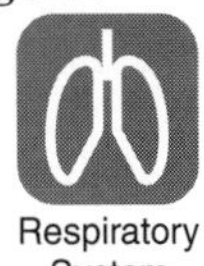
Respiratory System

Environmental

Ecotoxicity: bacteria (Pseudomonas putida): 16h EC_0 700 mg/l algae (Microcystis aeruginosa): 8d EC_0 0.35 mg/l; Amphibian: Mexican axolotl (3-4 wk after hatching): 48h

LC_{50}: 1.8 mg/l clawed toad (3-4 wk after hatching): 48h LC_{50}: 5.0 mg/l

Environmental Fate: If released to water, volatilization will be slow with estimated half-lives of 2.9 days and 23 days from a model environmental river and a model lake, respectively. Adsorption to sediment and bioconcentration are not likely fate processes. It is expected to biodegrade quickly in soil and aquatic conditions. If released to the atmosphere, it will exist primarily in the vapor phase. Vapor-phase will degrade in the atmosphere by reaction with photochemically produced hydroxyl radicals with an estimated half-life of approximately 6.9 hours and by reaction with atmospheric ozone with an estimated half-life of 23 hours. Removal of atmospheric material may occur through wet deposition. If released to soil, it is expected to have very high mobility based on an estimated K_{oc} value of 32. Volatilization is expected from dry soils, and may also be important from moist soils.

Cleanup/Disposal: Guide No. 131: Fully encapsulating, vapor protective clothing should be worn for spills and leaks with no fire. Eliminate all ignition sources (no smoking, flares, sparks or flames in immediate area). All equipment used when handling the product must be grounded. Do not touch or walk through spilled material. Stop leak if you can do it without risk. Prevent entry into waterways, sewers, basements or confined areas. A vapor suppressing foam may be used to reduce vapors. Small Spills: Absorb with earth, sand or other non-combustible material and transfer to containers for later disposal. Use clean non-sparking tools to collect absorbed material. Large Spills: Dike far ahead of liquid spill for later disposal. Water spray may reduce vapor; but may not prevent ignition in closed spaces.

Environmental Physical Data

Henry's Law Constant: estimated at 9.95×10^{-6}
Sorption Partition Coefficient: K_{OC} = estimated at 33
BCF: not significant
BOD: 35.4% BODT, 5 days

Regulations

RCRA 40CFR: Not listed
CERCLA: 40CFR 302.4: Not listed
SARA 40CFR 372.65: Listed TPQ: 500 lb
SARA EHS 40CFR 355: Listed TPQ: 500 lb
TSCA: Listed

ALL8140 CAS #: 1745-81-9

2-ALLYLPHENOL

RTECS: SJ3850000
EINECS Number: 217-119-0
Molecular Formula: $C_9H_{10}O$
Formula Weight: 134.19

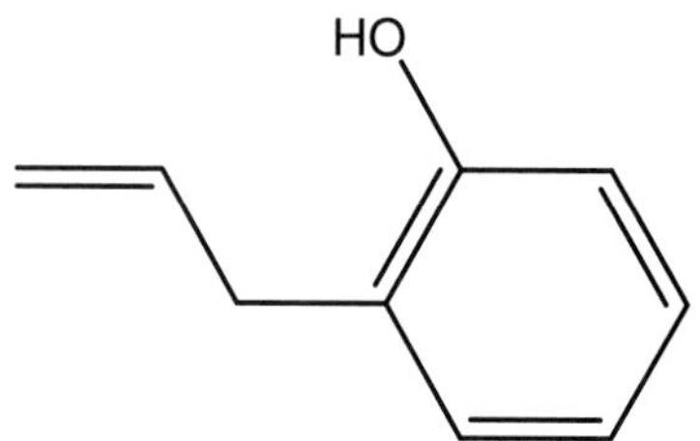

Chemical Structure

Physical Properties

Boiling Point: 220 °C (428 °F)
Freezing Point: -6 °C (21.2 °F)
Specific Gravity: 1.028
Water Solubility: Soluble
Other Solubilities: Ether: Very Soluble
Refraction Index: 1.5181

RTECS Toxicity Data

Tumorigenic: Mouse Route: Skin; Dose: 8400 mg/kg/30W-I; Toxic Effects: Tumorigenic - Carcinogenic by RTECS criteria; Skin and appendages - Tumors. Mouse Route: Skin; Dose: 3360 mg/kg/12W-I; Toxic Effects: Tumorigenic - Neoplastic by RTECS criteria; Skin and appendages - Tumors.

Hazard Overviews

Health: Irritating to eyes/skin/respiratory tract. Other Acute Effects: harmful if inhaled or swallowed.

Fire: Extinguishing agents: water spray; carbon dioxide, dry chemical powder or appropriate foam. Precautions: combustible liquid.

Reactivity: Incompatible with: strong oxidizing agents, acid chlorides, acid anhydrides. Hazardous decomposition products: toxic fumes of: carbon monoxide, carbon dioxide.

Carcinogenicity: IARC - Not listed; NIOSH - Not listed; NTP - Not listed; ACGIH - Not listed; OSHA - Not listed; EPA - Not listed; MAK - Not listed

Primary Target Organs:

Eyes

Skin

Respiratory System

Environmental

Regulations

RCRA 40CFR: Not listed
CERCLA: 40CFR 302.4: Not listed
SARA 40CFR 372.65: Not listed
SARA EHS 40CFR 355: Not listed
TSCA: Listed

ALL8560 CAS #: 1516-27-4

ALLYLTRIMETHYLAMMONIUM CHLORIDE

RTECS: BO4120000
EINECS Number: 216-164-3
Molecular Formula: $C_6H_{14}ClN$
Formula Weight: 135.63
Synonyms: AMMONIUM,ALLYLTRIMETHYL-,CHLORIDE; HOMONEURINE CHLORIDE; 2-PROPEN-1-AMINIUM,N,N,N-TRIMETHYL-,CHLORIDE
Use: no evidence of current commercial use in US

RTECS Toxicity Data

Acute Dermal: Mouse LD_{Lo} Route: Subcutaneous Dose: 180 mg/kg.

Hazard Overviews

Carcinogenicity: IARC - Not listed; NIOSH - Not listed; NTP - Not listed; ACGIH - Not listed; OSHA - Not listed; EPA - Not listed; MAK - Not listed

Environmental

Regulations

RCRA 40CFR: Not listed
CERCLA: 40CFR 302.4: Not listed
SARA 40CFR 372.65: Not listed
SARA EHS 40CFR 355: Not listed
TSCA: Listed

ALL8980 CAS #: 2114-11-6

ALLYLURETHANE

RTECS: EY8512000
EINECS Number: 218-309-6
Molecular Formula: $C_4H_7NO_2$
Formula Weight: 101.12

Chemical Structure

Synonyms: ALLYL CARBAMATE; CARBAMIC ACID,ALLYL ESTER; CARBAMIC ACID,2-PROPENYL ESTER

RTECS Toxicity Data

Mutagenic: Mouse Sister Chromatid Exchenge; Route: Intraperitoneal; Dose: 1100 umol/kg.

Tumorigenic: Mouse Route: Intraperitoneal; Dose: 279 mg/kg/4W-I; Toxic Effects: Tumorigenic - Neoplastic by RTECS criteria; Lungs, Thorax, or Respiration - Tumors.

Hazard Overviews

Carcinogenicity: IARC - Not listed; NIOSH - Not listed; NTP - Not listed; ACGIH - Not listed; OSHA - Not listed; EPA - Not listed; MAK - Not listed

Environmental

Regulations

RCRA 40CFR: Not listed
CERCLA: 40CFR 302.4: Not listed
SARA 40CFR 372.65: Not listed
SARA EHS 40CFR 355: Not listed
TSCA: Listed

ALP3000 CAS #: 67375-30-8

ALPHACYPERMETHRIN

RTECS: GZ1251400
Molecular Formula: $C_{22}H_{19}Cl_2NO_3$
Formula Weight: 416.32
Synonyms: ALFAMETHRIN; ALFOXYLATE; (+-)-ALPHAMETHRIN; ALPHAMETHRIN; (1R CIS S) AND (1S CIS R) ENANTIOMERIC ISOMER PAIR OFALPHA-CYANO-3-PHENOXYBENZYL-3-(2,2-DICHLOROVINYL)-2,2-DIMETHYLCYCLOPROPANE CARBOXYLATE; BESTOX; CONCORD; [1ALPHA(S*),3ALPHA]-(+ -)-CYANO(3-PHENOXYPHENYL)METHYL-3-(2,2-DICHLOROETHENYL)-2,2-DIMETHYLCYCLOPROPANECARBOXYLATE; DOMINEX; FASTAC; FASTAC 10 EC; FENDONA; WL 85871
Description: yellowish brown semisolid mass or colorless crystals
Use: control of wide range of chewing & sucking insects (particularly lepidotera, coleoptera, & hemiptera) in fruit (incl citrus), vegetables, vines, cereals, forestry, etc; against cockroaches, mosquitoes, flies, & other insect pests in public health, & in animal houses; against cotton leaf perforator, boll weevil, etc, in cotton; in coffee, maize, cruciferous crops, pasture; row crops; med (vet) animal ectoparasiticide.

Physical Properties

Boiling Point: 200 °C (392 °F) at 0.07 mm Hg
Freezing Point: 80.5 °C (176.9 °F)
Vapor Pressure: 170 nPa at 20 °C
Water Solubility: 0.01 to 0.2 mg/L at 20 °C
Other Solubilities: In Acetone 620, dichloromethane 550, cyclohexanone 515, Ethyl Acetate 440, chloroBenzene 420, acetophenone 390, o-Xylene 315, n-hexane 7 (all in g/l at 25 °C).
Flash Point: Burns with difficulty

RTECS Toxicity Data

Acute Oral: Rat LD_{50} Dose: 79 mg/kg.
Acute Dermal: Rabbit LD_{50} Route: Skin; Dose: >2 gm/kg. Rat LD_{50} Route: Skin; Dose: 500 mg/kg.

Mutagenic: Non-mammalian species Micronucleus Test; Route: Multiple routes; Dose: 10 ppb. Non-mammalian species Micronucleus Test; Route: Multiple routes; Dose: 5 ppb.

Hazard Overviews

Fire: Will burn.
Carcinogenicity: IARC - Not listed; NIOSH - Not listed; NTP - Not listed; ACGIH - Not listed; OSHA - Not listed; EPA - Not listed; MAK - Not listed

Environmental

Ecotoxicity: LC_{50} Rainbow trout 0.0028 mg/l/96 hr /Conditions of bioassay not specified LD_{50} Bee 0.059 ug/bee (24 hr)

Regulations

RCRA 40CFR: Not listed
CERCLA: 40CFR 302.4: Not listed
SARA 40CFR 372.65: Not listed
SARA EHS 40CFR 355: Not listed
TSCA: Not listed

Analytical Methods

Food: FDA 232.4, 242.1

ALP6000 CAS #: 77-20-3

ALPHAPRODINE

EINECS Number: 201-011-5
Molecular Formula: $C_{16}H_{23}NO_2$
Formula Weight: 261.35
Synonyms: ALPHAPRODIN; DEA CODE 9010; 1,3-DIMETHYL-4-PHENYL-4-PIPERIDINOL PROPANOATE; ALPHA-1,3-DIMETHYL-4-PHENYL-4-PIPERIDINYL PROPIONATE; 1,3-DIMETHYL-4-PHENYL-4-PROPIONOXYPIPERIDINE; ALPHA-1,3-DIMETHYL-4-PHENYL-4-PROPIONOXYPIPERIDINE; NISENTIL; NISINTIL; 4-PIPERIDINOL,1,3-DIMETHYL-4-PHENYL-,PROPANOATE (ESTER),CIS-; 4-PIPERIDINOL,1,3-DIMETHYL-4-PHENYL-,PROPIONATE; 4-PIPERIDINOL,1,3-DIMETHYL-4-PHENYL-,PROPIONATE (ESTER); PRISILIDENE; ALPHA-PRODINE; ALPHA-PRODINOL; PROPIONIC ACID,ALPHA-1,3-DIMETHYL-4-PHENYL-4-PIPERIDYLESTER
Description: white, crystalline powder; has slight odor
Use: medication: narcotic analgesic

Physical Properties

Freezing Point: 218 °C (424.4 °F) to 220 °C (428 °F)
Refraction Index: Alpha 1.499

Hazard Overviews

Carcinogenicity: IARC - Not listed; NIOSH - Not listed; NTP - Not listed; ACGIH - Not listed; OSHA - Not listed; EPA - Not listed; MAK - Not listed

Environmental

Regulations

RCRA 40CFR: Not listed
CERCLA: 40CFR 302.4: Not listed
SARA 40CFR 372.65: Not listed
SARA EHS 40CFR 355: Not listed
TSCA: Not listed

ALU1000 CAS #: 10043-67-1

ALUM, POTASSIUM

RTECS: WS5650000
EINECS Number: 233-141-3
Molecular Formula: $AlH_4KO_8S_2$
Formula Weight: 258.20
Synonyms: ALUM; ALUM,N.F; ALUM POTASSIUM; ALUMINUM POTASSIUM ALUM; ALUMINUM POTASSIUM DISULFATE; ALUMINUM POTASSIUM SULFATE; ALUMINUM POTASSIUM SULFATE (ALK(SO4)2); ALUMINUM POTASSIUM SULFATE ALUM; ALUMINUM POTASSIUM SULFATE,ALUM; ALUMINUM POTASSIUM SULFATE,ANHYDROUS; BURNT ALUM; BURNT POTASSIUM ALUM; DIALUMINUM DIPOTASSIUM SULFATE; EXSICCATED ALUM; POTASH ALUM; POTASSIUM ALUM; POTASSIUM ALUMINUM SULFATE; POTASSIUM ALUMINUM SULFATE (1:1:2); SULFURIC ACID,ALUMINUM POTASSIUM SALT (2:1:1)
Description: white powder; odorless
Use: in dyeing, printing fabrics; in mfr dyes, lakes, paper, vegetable glue, marble cements, explosives, artificial stones, statuary, pigments, matches paints; in tanning; in purifying water; hardening plaster; electrolytic copperplating; catalyst in synthesis of ammonia; mordant in staining; hardening agent in microscopy; waterproofing agent; aluminum salts; food additive (hardening gelatin, baking powders, firming agent); clarifying sugar; water correcting agent in brewing ind; in taxidermy; med: astringent, for gingivitis, warts, moles, etc, (vet) astringent; antiseptic, antimycotic; for fungal infections & eczematoid dermatitis; in leg tighteners; for stomatitis & vaginal & intrauterine therapy.

Physical Properties

Specific Gravity: 1.725
Water Solubility: 1 g dissolves in about 20 ml Cold Water
Other Solubilities: freely Soluble in Glycerol; Insoluble in Alcohol; Insoluble in Acetone; Soluble in dilute acid.
pH: 0.2 molar aqueous solution 3.3
Refraction Index: 1.454 to 1.4564

RTECS Toxicity Data

Mutagenic: Rat Other Mutation Test Systems; Route: Oral; Dose: 824 mg/kg/7D-C. Other Microorganisms Mutations in Microorganisms; Dose: 50 umol/L (-S9).

Hazard Overviews

Corrosive

Fire: Hazards: hazardous polymerization will not occur. Extinguishing agents: water spray; carbon dioxide, dry

chemical powder or appropriate foam. Precautions: combustible liquid.
Reactivity: Hazardous polymerization will not occur.
Carcinogenicity: IARC - Not listed; NIOSH - Not listed; NTP - Not listed; ACGIH - Not listed; OSHA - Not listed; EPA - Not listed; MAK - Not listed
Exposure Limits
OSHA PEL Vacated 1989 Limits: TWA: 2 mg/m^3; as Al soluble.
ACGIH TLV: TWA: 2 mg/m^3; as Al.
NIOSH REL: TWA: 2 mg/m^3; as Al soluble salts, alkyls.

Environmental

Regulations
RCRA 40CFR: Not listed
CERCLA: 40CFR 302.4: Not listed
SARA 40CFR 372.65: Not listed
SARA EHS 40CFR 355: Not listed
TSCA: Listed

ALU1280 CAS #: 10102-71-3
ALUMIINUM SODIUM SULFATE

EINECS Number: 233-277-3
Molecular Formula: $AlH_4NaO_8S_2$
Formula Weight: 242.10
Synonyms: ALUMINUM SODIUM SULFATE; SODA ALUM; SODIUM ALUM; SODIUM ALUMINUM SULFATE; SULFURIC ACID,ALUMINUM SODIUM SALT (2:1:1)
Description: colorless to white crystals, granules, or powder
Use: textiles (mordant, waterproofing); dry colors; ceramics; tanning; paper size precipitant; matches; inks; sugar refining; water purification; medicine; confectionery; baking powder; food additive; engraving; as firming agent in pickles and relishes

Physical Properties
Freezing Point: 61 °C (141.8 °F)
Specific Gravity: 1.675 at 20 °C
Water Solubility: Soluble in 1 part Water
Other Solubilities: Practically Insoluble in Alcohol.

Hazard Overviews

Corrosive

Carcinogenicity: IARC - Not listed; NIOSH - Not listed; NTP - Not listed; ACGIH - Not listed; OSHA - Not listed; EPA - Not listed; MAK - Not listed
Exposure Limits
OSHA PEL Vacated 1989 Limits: TWA: 2 mg/m^3; as Al soluble.
ACGIH TLV: TWA: 2 mg/m^3; as Al.
NIOSH REL: TWA: 2 mg/m^3; as Al soluble salts, alkyls.

Environmental

Regulations
RCRA 40CFR: Not listed
CERCLA: 40CFR 302.4: Not listed
SARA 40CFR 372.65: Not listed
SARA EHS 40CFR 355: Not listed
TSCA: Listed

ALU1560 CAS #: 7429-90-5
ALUMINUM

RTECS: BD0330000
DOT: UN1309; UN1383; UN1396; IMO4.1; IMO4.2; IMO4.3
EINECS Number: 231-072-3
Molecular Formula: Al
Structured MF: Al
Formula Weight: 26.98

Al

Chemical Structure

Synonyms: A 00; A 95; A 99; A 995; A 999; A999; A999V; AA 1099; AA1193; AA1199; AD 1; AD1M; ADO; ADOM; AE; ALAUN; ALLBRI ALUMINUM PASTE AND POWDER; ALUMINA FIBRE; ALUMINIUM; ALUMINIUM BRONZE; ALUMINIUM FLAKE; ALUMINUM-27; ALUMINUM 27; ALUMINUM A00; ALUMINUM DEHYDRATED; ALUMINUM METAL; ALUMINUM POWDER; AO A1; AO AL; AR2; AV00; AV000; C-PIGMENT 1; C.I. 77000; ELEMENTAL ALUMINUM; EMANAY ATOMIZED ALUMINUM POWDER; EPA PESTICIDE CHEMICAL CODE 000111; JISC 3108; JISC 3110; L16; METANA; METANA ALUMINUM PASTE; NORAL ALUMINIUM; NORAL ALUMINUM; NORAL EXTRA FINE LINING GRADE; NORAL INK GRADE ALUMINIUM; NORAL INK GRADE ALUMINUM; NORAL NON-LEAFING GRADE; PAP-1
Description: silvery-white metallic powder
Use: in mfr of printing inks, steel, alloys & paints, ceramic-faced armor for defense indust, permanent magnets, explosives & incendiaries; in dental alloys; in determining nitrates & nitrites, gold, arsenic, mercury; precipitating copper, arsenic, antimony; flashlight in photography; chem equipment; for photoengraving plates; cryogenic technology; foamed concrete; vacuum metallizing & coating; decorative stamping; insulation of liq fuels; structural material in building, canning, automobile, & aviation industries.

Physical Properties
Boiling Point: 2327 °C (4221 °F)
Freezing Point: 660 °C (1220 °F)
Specific Gravity: 2.7
Vapor Pressure: 1 mm Hg at 1284 °C
Water Solubility: Insoluble
Other Solubilities: Soluble in alkalies, Hydrochloric Acid, Sulfuric acid. Insoluble in concentrated Nitric acid, hot Acetic Acid.
Ionization Potential (eV): 5.98577
Flash Point: Combustible Solid
Autoignition Temperature: 760 °C

Hazard Overviews

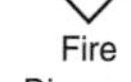

Explosive Flammable Fire Diamond

Health: Irritating to eyes/skin/respiratory tract. Also Causes: dyspnea, cough, lethargy, anorexia, increased respiration rate. Chronic Effects: pulmonary fibrosis, asthma, emphysema, dyspnea, cough, chronic obstructive lung disease.

Fire: Flammable. Explosive in air. Reacts with water. Isolate and permit large fires to burn; control smaller fires with sand, talc, or sodium chloride. Use nonsparking tools to ring small fires. Do not use water, carbon tetrachloride, or halon. A mixture of aluminum powder and water slowly forms hydrogen that can be hazardous if confined.

Reactivity: Stable. Hazardous polymerization cannot occur. Incompatible with: perchlorate/nitrate/water mixtures; powdered silver chloride; ammonium peroxodisulfate and water; peroxides; halocarbons; halogens; acids; hydrogen chloride gas; molten silicon steels; phosphorus; sulfur; selenium; interhalogens; oxidants; perchlorate salts; chlorates; carbon tetrachloride and chloroform amidinium nitrate; sodium acetylide; metal oxides; oxosalts (nitrates; sulfates); sulfides; hot copper oxide worked with an iron or steel tool; antimony; arsenic; anitmony trichloride vapor; iron powder and water; sodium hydroxide; diborane; alcohols; arseneic trioxide and sodium arsenate and sodium hydroxide. Hazardous decomposition products: metallic oxide smoke.

Carcinogenicity: IARC - Not listed; NIOSH - Listed as carcinogen; NTP - Not listed; ACGIH - Not listed; OSHA - Not listed; EPA - Not listed; MAK - Not listed

Primary Target Organs:

Eyes Skin Respiratory System

Exposure Limits

OSHA PEL: TWA: 15 mg/m^3; as Al, total.

OSHA PEL Vacated 1989 Limits: TWA: 2 mg/m^3; as Al soluble.

ACGIH TLV: TWA: 10 mg/m^3.

NIOSH REL: TWA: 10 mg/m^3; as Al soluble salts, alkyls.

DFG MAK: TWA: 6 mg/m^3.

Respirator Recommendation

Exposure Range: >5 to 50 mg/m^3 Air Purifying, Negative Pressure, Half Mask

Exposure Range: >50 to 500 mg/m^3 Air Purifying, Negative Pressure, Full Face

Exposure Range: >500 to 5000 mg/m^3 Supplied Air, Constant Flow/Pressure Demand, Full Face

Exposure Range: >5000 to unlimited mg/m^3 Self-contained Breathing Apparatus, Pressure Demand, Full Face

Cartridge Color: dust/mist filter (use P100 or consult supervisor for appropriate dust/mist filter)

Environmental

Cleanup/Disposal: Guide No. 135: Fully encapsulating, vapor protective clothing should be worn for spills and leak with no fire. Eliminate all ignition sources (no smoking, flares, sparks or flames in immediate area). Do not touch or walk through spilled material. Stop leak if you can do it without risk. Small Spills: Cover with dry earth, dry sand, or other non-combustible material followed with plastic sheet to minimize spreading or contact with rain. Use clean non-sparking tools to collect material and place it into loosely covered plastic containers for later disposal. Prevent entry into waterways, sewers, basements or confined areas. Guide No. 138: Eliminate all ignition sources (no smoking, flares, sparks or flames in immediate area). Do not touch or walk through spilled material. Stop leak if you can do it without risk. Use water spray to reduce vapors or divert vapor cloud drift. Do not get water on spilled substance or inside containers. Small Spills: Cover with dry earth, dry sand, or other non-combustible material followed with plastic sheet to minimize spreading or contact with rain. Dike for later disposal; do not apply water unless directed to do so. Powder Spills: Cover powder spill with plastic sheet or tarp to minimize spreading and keep powder dry. Do not clean-up or dispose of, except under supervision of a specialist. Guide No. 170: Eliminate all ignition sources (no smoking, flares, sparks or flames in immediate area). Do not touch or walk through spilled material. Stop leak if you can do it without risk. Prevent entry into waterways, sewers, basements or confined areas.

Regulations

RCRA 40CFR: Not listed

CERCLA: 40CFR 302.4: Not listed

SARA 40CFR 372.65: Listed

SARA EHS 40CFR 355: Not listed

TSCA: Listed

Analytical Methods

Soil: CLP 200.10_M, 200.62, 200.7_M, 202.1_M, 202.2, 202.2_M, 202.62, 6020_M, ICP-AES; EPA 200.7, 200.8, 202.2; SW846 3005A, 3010A, 3015, 3050A, 3050B, 3051, 3052, 6010A, 6010B, 6020, 7000A, 7020; ASTM D3974, D4698; USGS I5051, I5473, I5474

Water / Groundwater: EPA 200.0, 200.15, 200.7, 200.9; APHA 3111-A, 3111-D, 3111-E, 3113-B, 3120, 3500-AL; ASTM D1976, D2332, D4190, D857; USGS E-SPEC, I1051, I1052, I1054, I3051, I3052, I3054, I7051, I7052, I7054; CEM RD42; FISON AES-0029

Drinking Water: AOAC 920.196, 920.198, 993.14

Indoor / Expired Air: NIOSH 7013, 7300; ASTM D4185

Plasma: EPA 200.11

Urine: NIOSH 8310

Other: EPA 1620, 202.1; AOAC 990.08

ALU1840 CAS #: 7784-25-0

ALUMINUM AMMONIUM SULFATE

EINECS Number: 232-055-3
Molecular Formula: $AlH_7NO_8S_2$
Formula Weight: 237.14

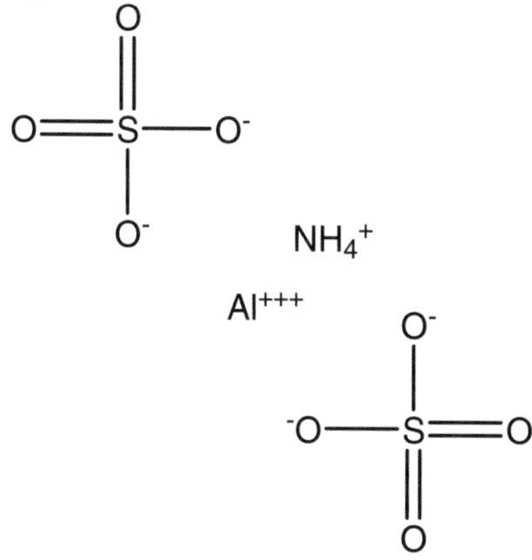

Chemical Structure

Synonyms: ALUM AMMONIUM; ALUMINUM AMMONIUM ALUM; ALUMINUM AMMONIUM DISULFATE (AL(NH4)(SO4)2); AMMONIUM ALUM; AMMONIUM ALUMINUM ALUM; BURNT AMMONIUM ALUM; CURB; EXSICCATED AMMONIUM ALUM; MONOAMMONIUM MONOALUMINUM SULFATE; SULFURIC ACID,ALUMINUM AMMONIUM SALT (2:1:1)
Description: white powder or colorless hexagonal crystals; odorless
Use: dyeing & printing fabrics; mfr pigments, lakes, artificial gems, paper, vegetable glue, marble & porcelain cements; medication: astringent, styptic; mordant in dyeing, water & sewage purification, sizing paper, retanning leather, clarifying agent, food additive, manufacture of lakes & pigments, & fur treatment

Physical Properties

Freezing Point: 280 °C (536 °F)
Specific Gravity: 2.45 at 20 °C
Water Solubility: Practically Insoluble in Water
Other Solubilities: Soluble in Glycerin; Soluble in dilute acid, Insoluble in Alcohol.

Hazard Overviews

Corrosive

Carcinogenicity: IARC - Not listed; NIOSH - Not listed; NTP - Not listed; ACGIH - Not listed; OSHA - Not listed; EPA - Not listed; MAK - Not listed
Exposure Limits
OSHA PEL Vacated 1989 Limits: TWA: 2 mg/m^3; as Al soluble.
ACGIH TLV: TWA: 2 mg/m^3; as Al.
NIOSH REL: TWA: 2 mg/m^3; as Al soluble salts, alkyls.

Environmental

Regulations

RCRA 40CFR: Not listed
CERCLA: 40CFR 302.4: Not listed
SARA 40CFR 372.65: Not listed
SARA EHS 40CFR 355: Not listed
TSCA: Listed

ALU2120 CAS #: 16962-07-5

ALUMINUM BOROHYDRIDE

RTECS: ED3200000
Molecular Formula: AlB_3H_{12}
Formula Weight: 71.53
Synonyms: ALUMINUM BOROHYDRIDE IN DEVICES; ALUMINUM HYDROBORATE; ALUMINUM TETRAHYDROBORATE; BORATE(1-),TETRAHYDRO-,ALUMINUM (3:1) (9CI)
Description: colorless liquid

Physical Properties

Boiling Point: 44.5 °C (112 °F)
Freezing Point: -64.5 °C (-84.1 °F)
Vapor Pressure: 400 mm Hg at 28.1 °C
Water Solubility: Violent Reaction
Other Solubilities: Reacts vigorously with H_2O and HCl evolving H_2
Flash Point: Very flammable
Autoignition Temperature: Ignites spontaneously in air

Hazard Overviews

Explosive

Flammable

Fire Diamond

Health: Causes: burns on all exposed tissues (skin, eyes, and the lining of the respiratory tract).
Fire: Explosive and flammable. It ignites spontaneously in air. Use dry chemical, or carbon dioxide. Never use water or foams, they will react violently. Extinguished material re-exposed directly to air ignites spontaneously.
Reactivity: Unstable, reactive reducing agent when exposed to air with spontaneous ignition. Hazardous polymerization are unlikely. Avoid: mixing with another material without first establishing chemical compatibility. Incompatible with: water; strong oxidizing materials; air; oxygen. Hazardous decomposition products: aluminum metal fumes; oxides of boron; hydrogen gas.
Carcinogenicity: IARC - Not listed; NIOSH - Not listed; NTP - Not listed; ACGIH - Not listed; OSHA - Not listed; EPA - Not listed; MAK - Not listed

Primary Target Organs:

Eyes

Skin

Respiratory System

Exposure Limits
OSHA PEL Vacated 1989 Limits: TWA: 2 mg/m^3; as Al soluble.
ACGIH TLV: TWA: 2 mg/m^3; as Al.
NIOSH REL: TWA: 2 mg/m^3; as Al soluble salts, alkyls.

Environmental

Regulations
RCRA 40CFR: Not listed
CERCLA: 40CFR 302.4: Not listed
SARA 40CFR 372.65: Not listed
SARA EHS 40CFR 355: Not listed
TSCA: Not listed

ALU2400 CAS #: 7727-15-3

ALUMINUM BROMIDE

RTECS: BD0350000
DOT: UN1725; UN2580; IMO8.0
EINECS Number: 231-779-7
Molecular Formula: $AlBr_3$
Structured MF: $AlBr_3$
Formula Weight: 266.72

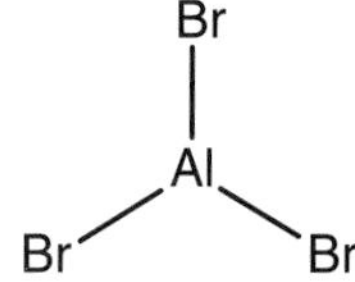

Chemical Structure

Synonyms: ALUMINUM BROMIDE (ANHYDROUS); ALUMINUM TRIBROMIDE; TRIBROMOALUMINUM
Description: colorless, rhombic plates or white to yellowish-red lumps
Use: acid catalyst in organic synthesis; bromination, alkylation, & isomerization catalyst in organic synthesis

Physical Properties
Boiling Point: 263.3 °C (506 °F) at 747 mm Hg
Freezing Point: 97.5 °C (207.5 °F)
Specific Gravity: 2.64 at 10 °C (fused)
Vapor Pressure: 1 mm Hg at 81.3 °C
Water Solubility: Soluble in Water
Other Solubilities: Soluble in Benzene, Nitrobenzene, Toluene, Xylene, simple hydrocarbons; Soluble in Ether.

RTECS Toxicity Data
Acute Oral: Rat LD_{50} Dose: 1598 mg/kg. Mouse LD_{50} Dose: 1623 mg/kg.

Hazard Overviews

Corrosive

Health: Corrosive to eyes/skin/respiratory tract. Toxic. Other Acute Effects: harmful if swallowed, inhaled, or absorbed through skin; inhalation may result in spasm, inflammation and edema of the larynx and bronchi, chemical pneumonitis and pulmonary edema; symptoms of exposure may include burning sensation; coughing; wheezing; laryngitis; shortness of breath; headache; nausea; vomiting; target organs: liver, nerves, blood, kidneys.
Fire: Hazards: emits toxic fumes; water hydrolyzes material liberating acidic gas which in contact with metal surfaces can generate flammable and/or explosive hydrogen gas. Extinguishing agents: carbon dioxide, dry chemical powder or appropriate foam; do not use water. Precautions: combustible liquid.
Reactivity: Stable. Hazardous polymerization will not occur. Incompatible with: strong oxidizing agents, do not allow water to enter container because of violent reaction. Hazardous decomposition products: toxic fumes of: carbon monoxide, carbon dioxide, hydrogen bromide gas, aluminum oxide.
Carcinogenicity: IARC - Not listed; NIOSH - Not listed; NTP - Not listed; ACGIH - Not listed; OSHA - Not listed; EPA - Not listed; MAK - Not listed
Primary Target Organs:

Eyes

Skin

Respiratory System

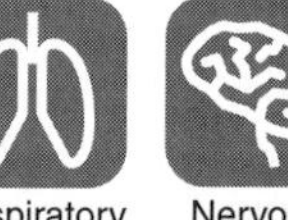
Nervous System

Liver

Kidneys

Exposure Limits
OSHA PEL Vacated 1989 Limits: TWA: 2 mg/m^3; as Al soluble.
ACGIH TLV: TWA: 2 mg/m^3; as Al.
NIOSH REL: TWA: 2 mg/m^3; as Al soluble salts, alkyls.

Environmental
Cleanup/Disposal: Guide No. 137: Fully encapsulating, vapor protective clothing should be worn for spills and leaks with no fire. Do not touch damaged containers or spilled material unless wearing appropriate protective clothing. Stop leak if you can do it without risk. Use water spray to reduce vapors; do not put water directly on leak, spill area or inside container. Keep combustibles (wood, paper, oil, etc.) away from spilled material. Small Spills: Cover with dry earth, dry sand, or other non-combustible material followed with plastic sheet to minimize spreading or contact with rain. Use clean non-sparking tools to collect material and place it into loosely covered plastic containers for later disposal. Prevent entry

into waterways, sewers, basements or confined areas. Guide No. 154: Eliminate all ignition sources (no smoking, flares, sparks or flames in immediate area). Do not touch damaged containers or spilled material unless wearing appropriate protective clothing. Stop leak if you can do it without risk. Prevent entry into waterways, sewers, basements or confined areas. Absorb or cover with dry earth, sand or other non-combustible material and transfer to containers. Do not get water inside containers.

Regulations

RCRA 40CFR: Not listed
CERCLA: 40CFR 302.4: Not listed
SARA 40CFR 372.65: Not listed
SARA EHS 40CFR 355: Not listed
TSCA: Listed

ALU2680 **CAS #: 16941-10-9**

ALUMINUM CALCIUM HYDRIDE

Molecular Formula: Al_2CaH_8
Formula Weight: 102.1
Description: slate-gray solid; odorless

Physical Properties

Water Solubility: Violent Reaction
Autoignition Temperature: Ignites spontaneously in air

Hazard Overviews

Explosive

Flammable

Fire Diamond

Health: Irritating to eyes/skin/respiratory tract. Also Causes: irritation and burns due to the reaction of the tissues' moisture with the aluminum calcium hydride.
Fire: Explosive and flammable. Use dry graphite or ground dolomite. Do not use water, carbon dioxide, dry chemical, or halogenated extinguishing agents to extinguish fire. Use a smothering technique. Extinguish small fires by applying a metal cover.
Reactivity: Stable inclosed, airtight, moisture-proof containers. Hazardous polymerization cannot occur. Incompatible with: water; moist air; oxygen. Hazardous decomposition products: fumes of metallic aluminum and metallic calcium; hydrogen gas.
Carcinogenicity: IARC - Not listed; NIOSH - Not listed; NTP - Not listed; ACGIH - Not listed; OSHA - Not listed; EPA - Not listed; MAK - Not listed

Primary Target Organs:

Eyes

Skin

Respiratory System

Mucous Membranes

Exposure Limits
OSHA PEL Vacated 1989 Limits: TWA: 2 mg/m^3; as Al soluble.
ACGIH TLV: TWA: 2 mg/m^3; as Al.
NIOSH REL: TWA: 2 mg/m^3; as Al soluble salts, alkyls.

Environmental

Regulations

RCRA 40CFR: Not listed
CERCLA: 40CFR 302.4: Not listed
SARA 40CFR 372.65: Not listed
SARA EHS 40CFR 355: Not listed
TSCA: Not listed

ALU2960 **CAS #: 1344-01-0**

ALUMINUM CALCIUM SODIUM SILICATE

EINECS Number: 215-685-3
Molecular Formula: Unknown
Synonyms: ALUMINOSILICIC ACID (UNSPECIFIED),CALCIUM SODIUM SALT,HYDRATE; ALUMINOSILICIC ACID,CALCIUM SODIUM SALT; ALUMINOSILICIC ACID,CALCIUM SODIUM SALT,HYDRATE; CALCIUM SODIUM ALUMINOSILICATE; CALCIUM SODIUM ALUMINOSILICATE HYDRATE; SODIUM CALCIUM ALUMINOSILICATE; SODIUM CALCIUM ALUMINOSILICATE,HYDRATED; SODIUM CALCIUM SILICOALUMINATE

Hazard Overviews

Corrosive

Carcinogenicity: IARC - Not listed; NIOSH - Not listed; NTP - Not listed; ACGIH - Not listed; OSHA - Not listed; EPA - Not listed; MAK - Not listed

Environmental

Regulations

RCRA 40CFR: Not listed
CERCLA: 40CFR 302.4: Not listed
SARA 40CFR 372.65: Not listed
SARA EHS 40CFR 355: Not listed
TSCA: Listed

ALU4360 **CAS #: 1299-86-1**

ALUMINUM CARBIDE (A)

EINECS Number: 215-076-2
Molecular Formula: C_3Al_4
Formula Weight: 143.96

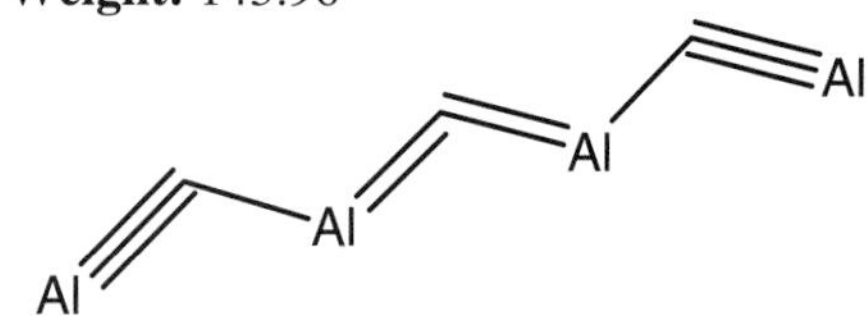

Chemical Structure

Description: yellow crystals, powder; odorless

Physical Properties

Boiling Point: Decomposes at 2200 °C (3992 °F)
Freezing Point: 2100 °C (3812 °F)
Specific Gravity: 2.36
Water Solubility: Dangerous Reaction

Hazard Overviews

Flammable

Fire Diamond

Health: Irritating to eyes/skin/respiratory tract. Also Causes: irritation and dryness of all directly exposed tissues, burns.
Fire: Flammable. Use carbon dioxide or dry chemicals. Do not use water sprays or foams. Forms methane gas on contact with water. If unavoidable contact with water occurs and large volumes of methane gas are generated, then direct fire-fighting techniques at controlling the methane gas and preventing a harmful methane gas-air explosion.
Reactivity: Stable. Hazardous polymerization cannot occur. Incompatible with: lead peroxide; potassium permanganate; water. Hazardous decomposition products: methane gas.
Carcinogenicity: IARC - Not listed; NIOSH - Not listed; NTP - Not listed; ACGIH - Not listed; OSHA - Not listed; EPA - Not listed; MAK - Not listed
Primary Target Organs:

Eyes

Skin

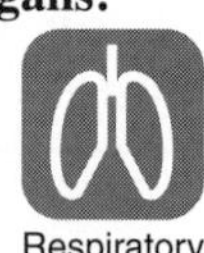
Respiratory System

Mucous Membranes

Exposure Limits
OSHA PEL Vacated 1989 Limits: TWA: 2 mg/m^3; as Al soluble.
ACGIH TLV: TWA: 2 mg/m^3; as Al.
NIOSH REL: TWA: 2 mg/m^3; as Al soluble salts, alkyls.

Environmental

Regulations

RCRA 40CFR: Not listed
CERCLA: 40CFR 302.4: Not listed
SARA 40CFR 372.65: Not listed
SARA EHS 40CFR 355: Not listed
TSCA: Listed

ALU4640 **CAS #: 7446-70-0**

ALUMINUM CHLORIDE, ANHYDROUS

RTECS: BD0525000
DOT: UN1726; UN2581; IMO8.0
EINECS Number: 231-208-1
Molecular Formula: $AlCl_3$
Structured MF: $AlCl_3$
Formula Weight: 133.34

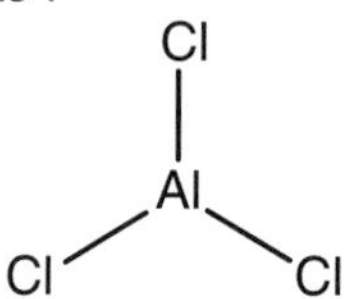

Chemical Structure

Synonyms: ALLUMINIO(CLORURO DI); ALUMINIUMCHLORID; ALUMINUM CHLORIDE (1:3); ALUMINUM TRICHLORIDE; ALUMINUM,(CHLORURE D'); CHLORURE D'ALUMINIUM; EPA PESTICIDE CHEMICAL CODE 013901; PEARSALL; TRICHLOROALUMINUM
Description: white when pure, grey, yellow, green crystalline solid; sharp acidic (hydrogen chloride) odor
Use: acid catalyst, especially in friedel-crafts type reactions; in cracking of petro; in mfr of rubbers, lubricants; intermed for aluminum compounds; in electrolytic prodn of aluminium; in preserving wood; disinfecting stables, slaughterhouses, etc; in deodorants & antiperspirant prepn; refining crude oil; dyeing fabrics; mfr parchment paper; med: astringent; pharmaceuticals & cosmetics, pigments, roof granules, special papers, photography, textiles (wool).

Physical Properties

Boiling Point: 182.7 °C (361 °F) at 752 mm Hg
Freezing Point: 190 °C (374 °F) at 2.5 atm
Specific Gravity: 2.44 at 25 °C
Vapor Pressure: 1 mm Hg at 100.0 °C
Water Solubility: 1 g/.9 g Water
Other Solubilities: freely Soluble in Carbon Tetrachloride; freely Soluble in many organic solvents; in absolute Alcohol 100 g/100 cc at 12.5 °C; in Chloroform 0.072 g/100 cc at 25 °C; Soluble in Ether; Slightly Soluble in Benzene.
Odor Threshold: Hydrogen Chloride 1 to 5 ppm
Flash Point: Nonflammable

RTECS Toxicity Data

Acute Oral: Rat LD_{50} Dose: 3450 mg/kg; Toxic Effects: Brain and coverings - Other degenerative changes; Gastrointestinal - Other changes; Kidney, Ureter, and Bladder - Hematuria. Mouse LD_{50} Dose: 1130 mg/kg.
Acute Dermal: Rabbit LD_{50} Route: Skin; Dose: >2 gm/kg.

Reproductive/Teratogenic: Rat Route: Oral; Dose: 11512 mg/kg; Duration: female 8-22D of pregnancy; Effects on Newborn - Weaning or lactation index; Growth statistics; Behavioral. Rat Route: Oral; Dose: 5723 mg/kg; Duration: female 1-21D of pregnancy; Effects on Newborn - Other neonatal measures or effects; Behavioral. Rat Route: Oral; Dose: 900 mg/kg; Duration: female 15D of pregnancy; Effects on Newborn - Growth statistics; Behavioral; Delayed effects.
Mutagenic: Rat DNA Damage; Cell Type: Ascites tumor; Dose: 500 umol/L. Mouse Cytogenetic Analysis; Route: Intraperitoneal; Dose: 444 mg/kg.

Hazard Overviews

Corrosive

Fire Diamond

Health: Corrosive to eyes/skin/respiratory tract.
Fire: Noncombustible. Use agents suitable for surrounding fire (except water). Reacts violently with water. Avoid using a direct water stream on this material. Use water spray only to cool surrounding combustibles.
Reactivity: Stable. Hazardous polymerization cannot occur. Avoid: exposure to water. Incompatible with: water; alkenes (especially isobutene); aluminum oxide and carbon oxide; aluminum and sodium peroxide; benzene and carbon tetrachloride; benzoyl chloride and naphthalene; ethylenimine and substituted anilines; ethylene oxide; nitrobenzene; nitrobenzene and phenol; nitromethane and organic matter; oxygen difluoride; phenyl azide; perchloryl benzene; sodium tetrahydroborate; ethylene; perchloryl fluoride and benzene. Hazardous decomposition products: aluminum fumes; chloride fumes.
Carcinogenicity: IARC - Not listed; NIOSH - Not listed; NTP - Not listed; ACGIH - Not listed; OSHA - Not listed; EPA - Not listed; MAK - Not listed
Primary Target Organs:

Eyes

Skin

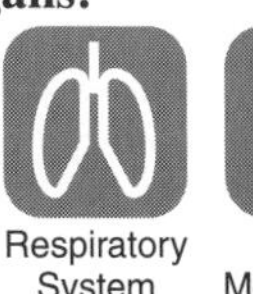
Respiratory System

Mucous Membranes

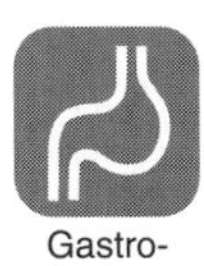
Gastro-intestinal

Exposure Limits
OSHA PEL Vacated 1989 Limits: TWA: 2 mg/m^3; as Al soluble.
ACGIH TLV: TWA: 2 mg/m^3; as Al.
NIOSH REL: TWA: 2 mg/m^3; as Al soluble salts, alkyls.

Environmental

Cleanup/Disposal: Guide No. 137: Fully encapsulating, vapor protective clothing should be worn for spills and leaks with no fire. Do not touch damaged containers or spilled material unless wearing appropriate protective clothing. Stop leak if you can do it without risk. Use water spray to reduce vapors; do not put water directly on leak, spill area or inside container. Keep combustibles (wood, paper, oil, etc.) away from spilled material. Small Spills: Cover with dry earth, dry sand, or other non-combustible material followed with plastic sheet to minimize spreading or contact with rain. Use clean non-sparking tools to collect material and place it into loosely covered plastic containers for later disposal. Prevent entry into waterways, sewers, basements or confined areas. Guide No. 154: Eliminate all ignition sources (no smoking, flares, sparks or flames in immediate area). Do not touch damaged containers or spilled material unless wearing appropriate protective clothing. Stop leak if you can do it without risk. Prevent entry into waterways, sewers, basements or confined areas. Absorb or cover with dry earth, sand or other non-combustible material and transfer to containers. Do not get water inside containers.

Environmental Physical Data

BCF: not pertinent
BOD: not pertinent

Regulations

RCRA 40CFR: Not listed
CERCLA: 40CFR 302.4: Not listed
SARA 40CFR 372.65: Not listed
SARA EHS 40CFR 355: Not listed
TSCA: Listed

ALU4920 **CAS #: 300-92-5**

ALUMINUM DISTEARATE

RTECS: BD0962000
DOT: UN1396
EINECS Number: 206-101-8
Molecular Formula: $C_{36}H_{71}AlO_5$
Formula Weight: 610.94
Synonyms: ALUMINUM HYDROXIDE DISTEARATE; ALUMINUM HYDROXYDISTEARATE; ALUMINUM,HYDROXYBIS(OCTADECANOATO-O)-; ALUMINUM,HYDROXYBIS(OCTADECANOATO-O)-(9CI); ALUMINUM,HYDROXYBIS(STEARATO)-; SPECIAL M
Description: white powder
Use: thickener in paints, inks, & greases; water repellent; lubricant in plastics & ropes; in cement production; stabilizer for packaging used for food (fda-approved); pigment suspending agent in paints & inks; used as a water repellent soap for surface treatment generally supplied as solutions in organic solvents; in the cosmetic industry in the preparation of clear cosmetic gels and pomdes; employed as a gasoline gelling agent

Physical Properties

Freezing Point: 145 °C (293 °F)
Specific Gravity: 1.009
Water Solubility: Insoluble in Water
Other Solubilities: Insoluble in Alcohol, Ether
Evaporation Rate: Negligible Butyl Acetate=1

Hazard Overviews

Carcinogenicity: IARC - Not listed; NIOSH - Not listed; NTP - Not listed; ACGIH - Not listed; OSHA - Not listed; EPA - Not listed; MAK - Not listed

Exposure Limits

OSHA PEL Vacated 1989 Limits: TWA: 2 mg/m^3; as Al soluble.

ACGIH TLV: TWA: 2 mg/m^3; as Al.

NIOSH REL: TWA: 2 mg/m^3; as Al soluble salts, alkyls.

Environmental

Cleanup/Disposal: Guide No. 138: Eliminate all ignition sources (no smoking, flares, sparks or flames in immediate area). Do not touch or walk through spilled material. Stop leak if you can do it without risk. Use water spray to reduce vapors or divert vapor cloud drift. Do not get water on spilled substance or inside containers. Small Spills: Cover with dry earth, dry sand, or other non-combustible material followed with plastic sheet to minimize spreading or contact with rain. Dike for later disposal; do not apply water unless directed to do so. Powder Spills: Cover powder spill with plastic sheet or tarp to minimize spreading and keep powder dry. Do not clean-up or dispose of, except under supervision of a specialist.

Regulations

RCRA 40CFR: Not listed
CERCLA: 40CFR 302.4: Not listed
SARA 40CFR 372.65: Not listed
SARA EHS 40CFR 355: Not listed
TSCA: Listed

ALU5200 CAS #: 555-75-9

ALUMINUM ETHYLATE

EINECS Number: 209-105-8
Molecular Formula: $C_6H_{18}AlO_3$
Formula Weight: 162.15

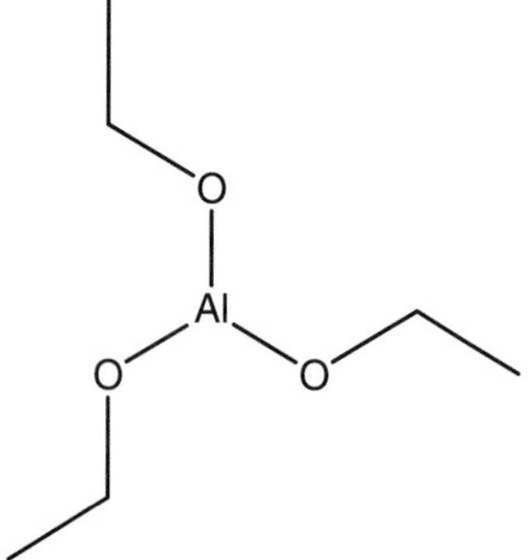

Chemical Structure

Synonyms: ALUMINUM ETHOXIDE; ALUMINUM TRIETHOXIDE; ETHANOL,ALUMINUM SALT; ETHYL ALCOHOL,ALUMINUM SALT; TRIETHOXYALUMINUM

Description: white crystals
Use: in reduction of aldehydes & ketones; catalyst for polymerizations; catalyst for preprn of esters from aldehydes and alcoholysis of diketones

Physical Properties

Boiling Point: 200 °C (392 °F) at 6-8 mm Hg
Freezing Point: 140 °C (284 °F)
Specific Gravity: 1.142 at 20 °C/ 0 °C
Water Solubility: Decomposed by Water
Other Solubilities: Insoluble in Alcohol; Slightly Soluble in Benzene, Ether.

Hazard Overviews

Corrosive

Health: Corrosive to eyes/skin/respiratory tract. Other Acute Effects: harmful if swallowed, inhaled, or absorbed through skin; inhalation may result in spasm, inflammation and edema of the larynx and bronchi, chemical pneumonitis and pulmonary edema; symptoms of exposure may include burning sensation; coughing; wheezing; laryngitis; shortness of breath; headache; nausea; vomiting.

Fire: Hazards: in powder form capable of creating a dust explosion. Extinguishing agents: carbon dioxide, dry chemical powder or appropriate foam; do not use water. Precautions: combustible liquid.

Reactivity: Incompatible with: air sensitive, may decompose on exposure to moist air or water, acids. Hazardous decomposition products: toxic fumes of: carbon monoxide, carbon dioxide, aluminum oxide.

Carcinogenicity: IARC - Not listed; NIOSH - Not listed; NTP - Not listed; ACGIH - Not listed; OSHA - Not listed; EPA - Not listed; MAK - Not listed

Primary Target Organs:

Eyes

Skin

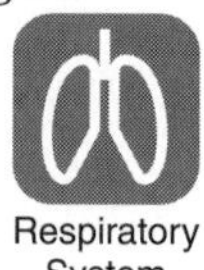

Respiratory System

Exposure Limits

OSHA PEL Vacated 1989 Limits: TWA: 2 mg/m^3; as Al soluble.

ACGIH TLV: TWA: 2 mg/m^3; as Al.

NIOSH REL: TWA: 2 mg/m^3; as Al soluble salts, alkyls.

Environmental

Regulations

RCRA 40CFR: Not listed
CERCLA: 40CFR 302.4: Not listed
SARA 40CFR 372.65: Not listed
SARA EHS 40CFR 355: Not listed
TSCA: Listed

ALU5480 CAS #: 7784-18-1

ALUMINUM FLUORIDE

RTECS: BD0725000
EINECS Number: 232-051-1
Molecular Formula: AlF_3
Structured MF: AlF_3
Formula Weight: 83.98

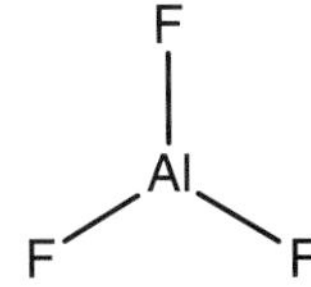

Chemical Structure

Synonyms: ALUMINIUM FLUORURE; ALUMINUM FLUORIDE (ALF3); ALUMINUM TRIFLUORIDE; FLUORID HLINITY
Description: white crystals; odorless
Use: in ceramics; as flux in metallurgy; in aluminum mfr; inhibitor of fermentation; catalyst in org reactions; production of aluminum; flux in ceramic glazes & enamels; manufacture of aluminum silicate; catalyst

Physical Properties

Boiling Point: Sublimes at 1272 °C (2322 °F)
Freezing Point: 1291 °C (2355.8 °F)
Specific Gravity: 2.882 at 25 °C/4 °C
Vapor Pressure: 1 mm Hg at 1238 °C
Water Solubility: 0.559 g/100 ml at 25 °C
Other Solubilities: Sparingly Soluble in acids & alkalies; Insoluble in Alcohol & Acetone.
Ionization Potential (eV): =< 15.45
Flash Point: Nonflammable

RTECS Toxicity Data

Acute Oral: Mouse LD_{50} Dose: 103 mg/kg.
Acute Dermal: Frog LD_{Lo} Route: Subcutaneous Dose: 1680 mg/kg.
Irritation Eye: Rabbit Standard Draize Test Dose: 500 mg/24H; Reaction: mild.

Hazard Overviews

Fire Diamond

Health: Irritating to eyes/skin/respiratory tract. Chronic Effects: chronic bone changes from fluoride poisoning, liver enlargement and inflammation. Nonreversible hyperactivity with nocturnal asthma has been documented.
Fire: Use water sprays, dry chemical, carbon dioxide, or foams.
Reactivity: Stable. Hazardous polymerization cannot occur. Incompatible with: potassium; sodium. Hazardous decomposition products: fumes of aluminum; oxides of fluorine.
Carcinogenicity: IARC - Not listed; NIOSH - Not listed; NTP - Not listed; ACGIH - Class A4, Not classifiable as a human carcinogen; OSHA - Not listed; EPA - Not listed; MAK - Not listed
Primary Target Organs:

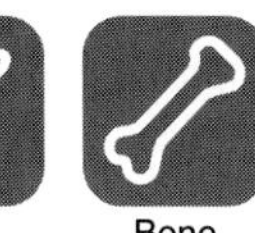

Eyes Skin Respiratory System Liver Bone

Exposure Limits
OSHA PEL: TWA: 2.5 mg/m^3; as F.
OSHA PEL Vacated 1989 Limits: TWA: 2 mg/m^3; as Al soluble.
ACGIH TLV: TWA: 2.5 mg/m^3; as F.
NIOSH REL: TWA: 2 mg/m^3; as Al soluble salts, alkyls.
DFG MAK: TWA: 2.5 mg/m^3; as F.

Environmental

Ecotoxicity: Aquatic toxicity: 60 ppm/*/fish/lethal/fresh water *Time period not specified
Cleanup/Disposal: Guide No. 157: Eliminate all ignition sources (no smoking, flares, sparks or flames in immediate area). All equipment used when handling the product must be grounded. Do not touch damaged containers or spilled material unless wearing appropriate protective clothing. Stop leak if you can do it without risk. A vapor suppressing foam may be used to reduce vapors. Do not get water inside containers. Use water spray to reduce vapors or divert vapor cloud drift. Prevent entry into waterways, sewers, basements or confined areas. Small Spills: Cover with dry earth, dry sand, or other non-combustible material followed with plastic sheet to minimize spreading or contact with rain. Use clean non-sparking tools to collect material and place it into loosely covered plastic containers for later disposal.

Environmental Physical Data

BCF: none noted
BOD: not pertinent

Regulations

RCRA 40CFR: Not listed
CERCLA: 40CFR 302.4: Not listed
SARA 40CFR 372.65: Not listed
SARA EHS 40CFR 355: Not listed
TSCA: Listed

ALU5760 CAS #: 21645-51-2

ALUMINUM HYDROXIDE

RTECS: BD0940000
EINECS Number: 244-492-7
Molecular Formula: AlH_3O_3
Structured MF: $Al(OH)_3$
Formula Weight: 77.99

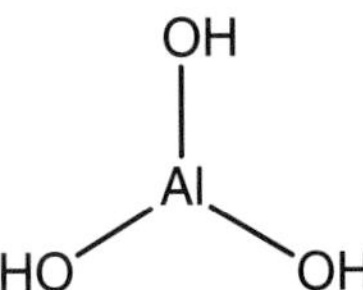

Chemical Structure

Synonyms: AF 260; ALCOA 331; ALCOA C 330; ALCOA C 333; ALCOA C 30BF; ALUGEL; ALUMIGEL; ALUMINA HYDRATE; ALUMINA HYDRATED; ALPHA-ALUMINA TRIHYDRATE; ALUMINA TRIHYDRATE; ALUMINIC ACID; ALUMINIUM HYDROXIDE; ALUMINUM HYDRATE; ALUMINUM HYDROXIDE GEL; ALUMINUM OXIDE HYDRATE; ALUMINUM OXIDE TRIHYDRATE; ALUMINUM OXIDE-3H2O; ALUMINUM OXIDE,TRIHYDRATE; ALUMINUM TRIHYDRAT; ALUMINUM TRIHYDRATE; ALUMINUM TRIHYDROXIDE; ALUMINUM(III) HYDROXIDE; ALUSAL; AMBEROL ST 140F; AMPHOJEL; BACO AF 260; BRITISH ALUMINUM AF 260; C 31C; C 31F; C 4D; C 31; C 33; C-31-F; C.I. 77002; CALMOGASTRIN; GHA 331; GHA 332; GHA 431; H 46; HIGILITE; HIGILITE H 31S; HIGILITE H 32; HIGILITE H 42; HYCHOL 705; HYDRAFIL; HYDRAL 705; HYDRAL 710; HYDRATED ALUMINA; HYDRATED ALUMINUM OXIDE; LIQUIGEL; MARTINAL; MARTINAL A/S; MARTINAL F-A; MARTINAL A; P 30BF; PGA; REHEIS F 1000; TRIHYDRATED ALUMINA; TRIHYDROXYALUMINUM

Description: white amorphous powder; odorless

Use: desiccant powder; in packaging materials; filler in paper, plastics, etc; mild abrasive; glass additive; smoke suppressant for plastics & latex foams; base for org lakes, flame retardants, mattress batting; finely divided form for rubber reinforcing agent, paper coating, filler, cosmetics; adsorbent; emulsifier; ion-exchanger, in chromatography; mordant in dyeing; filtering medium; mfr activated alumina, glass, paper, pottery, printing inks, etc; waterproofing fabrics; in antiperspirants, dentifrices, pharmacy gel.

Physical Properties

Freezing Point: 300 °C (572 °F)

Specific Gravity: 2.42

Water Solubility: Insoluble in Water

Other Solubilities: Soluble in Hydrochloric Acid, Sulfuric acid, alkaline aqueous solution, in strong acids in presence of water.

RTECS Toxicity Data

Acute Oral: Child TD_{Lo} Dose: 79 gm/kg/2Y-I; Toxic Effects: Behavioral - Change in motor activity (specific assay); Behavioral - Muscle contraction or spasticity; Musculoskelital - Osteomalacia. Child TD_{Lo} Dose: 122 gm/kg/4D; Toxic Effects: Gastrointestinal - Other changes; Nutritional and gross metabolic - Body temperature increase.

Chronic (Multiple Dose) Oral: Rat Dose: 8040 mg/kg/67D-C; Toxic Effects: Blood - Changes in serum composition; Nutritional and gross metabolic - Changes in phosphorus.

Hazard Overviews

Fire Diamond

Health: Irritating to eyes/skin. Also Causes: minor mechanical irritation, basic eye burns are possible.

Fire: Noncombustible. Use agent suitable for surrounding fire.

Reactivity: Stable. Hazardous polymerization cannot occur. Avoid: contact with strong acids and strong oxidizing materials. Incompatible with: bismuth. Hazardous decomposition products: oxides of aluminum.

Carcinogenicity: IARC - Not listed; NIOSH - Not listed; NTP - Not listed; ACGIH - Not listed; OSHA - Not listed; EPA - Not listed; MAK - Not listed

Primary Target Organs:

Eyes

Skin

Exposure Limits

OSHA PEL Vacated 1989 Limits: TWA: 2 mg/m^3; as Al soluble.

ACGIH TLV: TWA: 2 mg/m^3; as Al.

NIOSH REL: TWA: 2 mg/m^3; as Al soluble salts, alkyls.

DFG MAK: TWA: 6 mg/m^3.

Environmental

Regulations

RCRA 40CFR: Not listed

CERCLA: 40CFR 302.4: Not listed

SARA 40CFR 372.65: Not listed

SARA EHS 40CFR 355: Not listed

TSCA: Listed

ALU6040 CAS #: 555-31-7

ALUMINUM ISOPROPYLATE

RTECS: BD0975000

EINECS Number: 209-090-8

Molecular Formula: $C_9H_{24}AlO_3$

Formula Weight: 204.23

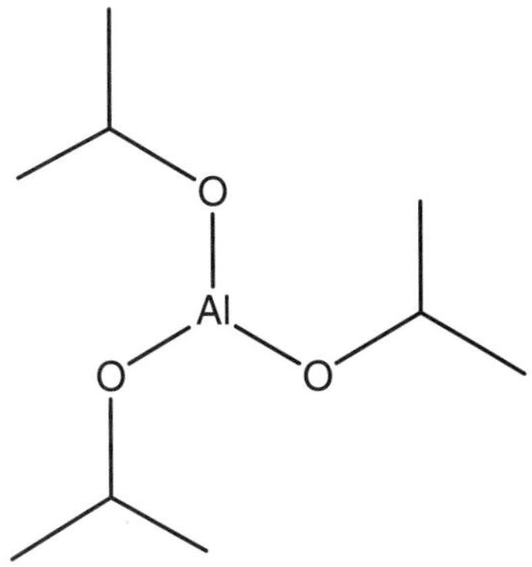

Chemical Structure

Synonyms: ALISO; ALUMINIUM ISOPROPOXIDE; ALUMINUM ISOPROPANOLATE; ALUMINUM ISOPROPOXIDE; ALUMINUM SEC-PROPANOLATE; ALUMINUM TRIISOPROPOXIDE; ALUMINUM TRIISOPROPYLATE; ALUMINUM TRIS(SEC-PROPOXIDE); ALUMINUM(II) ISOPROPYLATE; ISOPROPANOL ALUMINUM SALT; ISOPROPYL ALCOHOL,ALUMINUM SALT; ISOPROPYL ALCOHOL,ALUMINUM SALT (8CI); 2-PROPANOL,ALUMINUM SALT; 2-PROPANOL,ALUMINUM SALT (9CI); TRIISOPROPOXYALUMINUM; TRIISOPROPYLOXYALUMINUM; TRIS(ISOPROPOXY)ALUMINUM

Description: white solid or crystals

Use: in meerwein-ponndorf reactions; in alcoholysis and ester exchange; synthesis of higher alkoxides, chelates; synthesis of acylates; formation of aluminum soaps; formulation of paints; waterproofing finishes for textiles; dehydrating agent

Physical Properties

Boiling Point: 135 °C (275 °F) at 10 mm Hg
Freezing Point: 119 °C (246.2 °F)
Specific Gravity: 1.0346 at 20 °C/ 0 °C
Water Solubility: Decomposed by Water
Other Solubilities: Soluble in Ethanol, Isopropanol, Benzene, Toluene, Chloroform, Carbon Tetrachloride, petroleum hydrocarbons

RTECS Toxicity Data

Acute Oral: Rat LD_{50} Dose: 11300 mg/kg.

Hazard Overviews

Corrosive

Health: Corrosive to eyes/skin/respiratory tract. Other Acute Effects: harmful if swallowed, inhaled, or absorbed through skin; inhalation may result in spasm, inflammation and edema of the larynx and bronchi, chemical pneumonitis and pulmonary edema; symptoms of exposure may include burning sensation; coughing; wheezing; laryngitis; shortness of breath; headache; nausea; vomiting.

Fire: Hazards: emits toxic fumes; in powder form capable of creating a dust explosion. Extinguishing agents: carbon dioxide, dry chemical powder or appropriate foam. Precautions: combustible liquid.

Reactivity: Incompatible with: strong oxidizing agents, acids, may decompose on exposure to moist air or water. Hazardous decomposition products: toxic fumes of: aluminum oxide carbon monoxide, carbon dioxide.

Carcinogenicity: IARC - Not listed; NIOSH - Not listed; NTP - Not listed; ACGIH - Not listed; OSHA - Not listed; EPA - Not listed; MAK - Not listed

Primary Target Organs:

Eyes

Skin

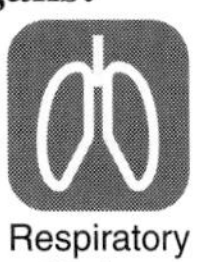
Respiratory System

Exposure Limits
OSHA PEL Vacated 1989 Limits: TWA: 2 mg/m^3; as Al soluble.
ACGIH TLV: TWA: 2 mg/m^3; as Al.
NIOSH REL: TWA: 2 mg/m^3; as Al soluble salts, alkyls.

Environmental

Regulations

RCRA 40CFR: Not listed
CERCLA: 40CFR 302.4: Not listed
SARA 40CFR 372.65: Not listed
SARA EHS 40CFR 355: Not listed
TSCA: Listed

ALU6320 CAS #: 13473-90-0

ALUMINUM NITRATE

RTECS: BD1040000
DOT: UN1438; IMO5.1
EINECS Number: 236-751-8
Molecular Formula: $AlH_3N_3O_9$
Structured MF: $Al(NO_3)_3 \cdot 9H_2O$
Formula Weight: 213.00

Synonyms: ALUMINUM TRINITRATE; ALUMINUM(III) NITRATE (1:3); NITRATO DE ALUMINIO; NITRIC ACID,ALUMINUM SALT; NITRIC ACID,ALUMINUM(3+) SALT

Description: white, colorless crystals; odorless

Use: tanning leather; antiperspirant; corrosion inhibitor; extraction of uranium; nitrating agent; textiles (mordant), mfr of incandescent filaments; catalyst in petroleum refining; nucleonics nonahydrate; salting-out agent in the extraction of actinides nonahydrate; preparation of insulating papers, on transformer core lamintes; in cathode ray tube heating elements

Physical Properties

Boiling Point: Decomposes at 100 °C (212 °F)
Freezing Point: Nonohydrate 73 °C (163.4 °F)
Specific Gravity: > 1 at 20 °C (solid)
Water Solubility: Soluble in Cold Water
Other Solubilities: Very Soluble in Alcohol; Very Slightly Soluble in Acetone;almost Insoluble in Ethyl Acetate & Pyridine

Refraction Index: 1.54
Flash Point: Nonflammable

RTECS Toxicity Data

Acute Oral: Rat LD_{50} Dose: 3654 mg/kg.
Irritation Eye: Rabbit Standard Draize Test Dose: 100 mg; Reaction: severe. Rabbit Nonstandard Exposure Dose: 100 mg/4S rinse; Reaction: mild.
Irritation Skin: Rabbit Standard Draize Test Dose: 500 mg; Reaction: mild.

Hazard Overviews

Fire Diamond

Health: Mildly irritating to eyes/skin/respiratory tract.
Fire: Noncombustible. Use agent suitable for surrounding fire. Strong oxidizer capable of igniting combustibles.
Reactivity: Stable. Hazardous polymerization cannot occur. Incompatible with: aluminum; boron phosphide; cyanides; esters; phospham; phosphorus; sodium cyanide; sodium hypophosphite; stannous chloride; thiocyanates. Hazardous decomposition products: oxides of nitrogen; oxides of aluminum; metallic aluminum fumes.
Carcinogenicity: IARC - Not listed; NIOSH - Not listed; NTP - Not listed; ACGIH - Not listed; OSHA - Not listed; EPA - Not listed; MAK - Not listed
Primary Target Organs:

Eyes

Skin

Respiratory System

Exposure Limits
OSHA PEL Vacated 1989 Limits: TWA: 2 mg/m^3; as Al soluble.
ACGIH TLV: TWA: 2 mg/m^3; as Al.
NIOSH REL: TWA: 2 mg/m^3; as Al soluble salts, alkyls.

Environmental

Ecotoxicity: Aquatic toxicity: 0.07 ppm/10 days/stickleback/killed/fresh water
Cleanup/Disposal: Guide No. 140: Keep combustibles (wood, paper, oil, etc.) away from spilled material. Do not touch damaged containers or spilled material unless wearing appropriate protective clothing. Stop leak if you can do it without risk. Do not get water inside containers. Small Dry Spills: With clean shovel place material into clean, dry container and cover loosely; move containers from spill area. Small Liquid Spills: Use a non-combustible material like vermiculite, sand or earth to soak up the product and place into a container for later disposal. Large Spills: Dike far ahead of liquid spill for later disposal. Following product recovery, flush area with water.

Environmental Physical Data

BOD: none

Regulations

RCRA 40CFR: Not listed
CERCLA: 40CFR 302.4: Not listed
SARA 40CFR 372.65: Not listed
SARA EHS 40CFR 355: Not listed
TSCA: Listed

ALU6600 CAS #: 688-37-9

ALUMINUM OLEATE

EINECS Number: 211-702-3
Molecular Formula: $C_{54}H_{102}AlO_4$
Formula Weight: 871.36
Synonyms: ALUMINUM TRIOLEATE; 9-OCTADECENOIC ACID (Z)-,ALUMINUM SALT; OLEIC ACID ALUMINUM SALT; OLEIC ACID,ALUMINUM SALT
Description: yellowish white, viscous mass
Use: in oil or turpentine soln as lacquer for metals; as size, waterproofing agent, drier for paints; in high pressure & high temp greases for thickening lubricating oils; medicine; lubricant for plastics; food additive; coating of pills for oral delivery in ruminants

Physical Properties

Freezing Point: 120 °C (248 °F)
Specific Gravity: 1.01
Water Solubility: Soluble in Hot Water
Other Solubilities: Insoluble in Alcohol.

Hazard Overviews

Carcinogenicity: IARC - Not listed; NIOSH - Not listed; NTP - Not listed; ACGIH - Not listed; OSHA - Not listed; EPA - Not listed; MAK - Not listed
Exposure Limits
OSHA PEL Vacated 1989 Limits: TWA: 2 mg/m^3; as Al soluble.
ACGIH TLV: TWA: 2 mg/m^3; as Al.
NIOSH REL: TWA: 2 mg/m^3; as Al soluble salts, alkyls.

Environmental

Regulations

RCRA 40CFR: Not listed
CERCLA: 40CFR 302.4: Not listed
SARA 40CFR 372.65: Not listed
SARA EHS 40CFR 355: Not listed
TSCA: Not listed

ALU6880 CAS #: 1344-28-1

ALUMINUM OXIDE

RTECS: BD1200000
EINECS Number: 215-691-6
Molecular Formula: Al_2O_3
Structured MF: Al_2O_3

Formula Weight: 101.94

O^{--} Al^{+++}

O^{--}

Al^{+++} O^{--}

Chemical Structure

Synonyms: A 1 (SORBENT); A1-0104 T 3/16"; A1-0109 P; A1-1401 P(MS); A1-1404 T 3/16"; A1-3438 T 1/8"; A1-3916 P; A1-3945 E 1/16"; A1-3970 P; A1-3980 T 5/32"; A1-4028 T 3/16"; A1-4126 E 1/16"; ABRAMANT; ABRAMAX; ABRAREX; ABRASIT; ACTIVATED ALUMINUM OXIDE; ACTIVATED ALUMINUM OXIDE (FORM); ALCOA F 1; ALMITE; ALON; ALON C; ALOXITE; ALPHA-ALUMINA; ALUMINA; BETA-ALUMINA; GAMMA-ALUMINA; ALUMINITE 37; ALUMINIUM OXIDE; ALPHA-ALUMINUM OXIDE; BETA-ALUMINUM OXIDE; GAMMA-ALUMINUM OXIDE; ALUMINUM OXIDE (BROCKMANN) (FORM); ALUMINUM SESQUIOXIDE; ALUMINUM TRIOXIDE; ALUMITE; ALUMITE (OXIDE); ALUNDUM; ALUNDUM 600; BROCKMANN,ALUMINUM OXIDE; CAB-O-GRIP; CATAPAL 5; CATAPAL S; CATAPAL SB ALUMINA; COMPALOX; CONOPAL; CORUNDUM,EMERY,ALUMINUM TRIOXIDE; DIADUR; DIALUMINUM TRIOXIDE; DISPAL; DISPAL ALUMINA; DISPAL M; DOTMENT 324; DOTMENT 358; DURAL; ETA-ALUMINA; EXOLON; EXOLON XW 60; F 360 (ALUMINA); FASERTON; FASERTONERDE; FIBER FP; G 0 (OXIDE); G 2 (OXIDE); GK (OXIDE); HYPALOX II; JUBENON R; KA 101; KETJEN B; KHP 2; LA 6; LUCALOX; LUDOX CL; MARTOXIN; MICROGRIT WCA; NEOBEAD C; PORAMINAR; PS 1; PS 1 (ALUMINA); Q-LOID A 30; RC 172DBM; SAFFIE

Description: white powder; odorless

Use: prodn of aluminum, mfr of abrasives, refractories, ceramics, electrical insulators, catalyst & catalyst supports, paper, spark plugs, crucibles; adsorbent for gases & water vapors, in petro ind, in chromatographic analysis, fluxes, light bulbs; in artificial gems, heat resistant fibers, food additive (dispersing agent); in hollow-fiber membrane units in water desalination, industrial ultrafiltration, hemodialysis; to reinforce aluminum; for unidirectional high temperature composites.

Physical Properties

Boiling Point: 2980 °C (5396 °F)
Freezing Point: About 2000 °C (3632 °F)
Specific Gravity: 4 at 20 °C/4 °C
Vapor Pressure: 1 mm Hg at 2158 °C
Water Solubility: Practically Insoluble in Water
Other Solubilities: Very Slightly Soluble in acid, alkali.
Refraction Index: 1.768
Flash Point: Noncombustible

RTECS Toxicity Data

Tumorigenic: Rat Route: Intrapleural; Dose: 90 mg/kg; Toxic Effects: Tumorigenic - Equivocal tumorigenic agent by RTECS criteria; Lungs, Thorax, or Respiration - Tumors. Rat Route: Implant; Dose: 200 mg/kg; Toxic Effects: Tumorigenic - Neoplastic by RTECS criteria; Tumorigenic - Tumors at site of application. Rat Route: Implant; Dose: 200 mg/kg; Toxic Effects: Tumorigenic - Equivocal tumorigenic agent by RTECS criteria; Tumorigenic - Tumors at site of application.

Hazard Overviews

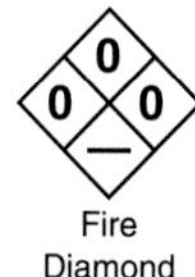

Fire Diamond

Health: Mildly irritating to respiratory tract.
Fire: Noncombustible. Use agent suitable for surrounding fire.
Reactivity: Stable. Hazardous polymerization cannot occur. Incompatible with: chlorine trifluoride; ethylene oxide.
Carcinogenicity: IARC - Not listed; NIOSH - Not listed; NTP - Not listed; ACGIH - Class A4, Not classifiable as a human carcinogen; OSHA - Not listed; EPA - Not listed; MAK - Not listed
Primary Target Organs:

Respiratory System

Exposure Limits
OSHA PEL: TWA: 15 mg/m^3; total dust.
OSHA PEL Vacated 1989 Limits: TWA: 10 mg/m^3. Other Values: respirable mg/m^3; 5.
ACGIH TLV: TWA: 10 mg/m^3.
NIOSH REL: TWA: 2 mg/m^3; as Al soluble salts, Alkyls.
DFG MAK: TWA: 6 mg/m^3.
Respirator Recommendation
Exposure Range: >5 to 50 mg/m^3 Air Purifying, Negative Pressure, Half Mask
Exposure Range: >50 to 500 mg/m^3 Air Purifying, Negative Pressure, Full Face
Exposure Range: >500 to 5000 mg/m^3 Supplied Air, Constant Flow/Pressure Demand, Full Face
Exposure Range: >5000 to unlimited mg/m^3 Self-contained Breathing Apparatus, Pressure Demand, Full Face
Cartridge Color: dust/mist filter (use P100 or consult supervisor for appropriate dust/mist filter)

Environmental

Regulations

RCRA 40CFR: Not listed
CERCLA: 40CFR 302.4: Not listed
SARA 40CFR 372.65: Listed
SARA EHS 40CFR 355: Not listed
TSCA: Listed

Analytical Methods

Indoor / Expired Air: NIOSH 7013

ALU7160 CAS #: 555-35-1

ALUMINUM PALMITATE

EINECS Number: 209-093-4
Molecular Formula: $C_{48}H_{96}AlO_6$
Formula Weight: 793.25

Synonyms: HEXADECANOIC ACID,ALUMINUM SALT; PALMITIC ACID ALUMINUM SALT; PALMITIC ACID,ALUMINUM SALT
Description: white to yellow mass or powder
Use: in packaging materials; waterproofing leather, paper, textiles, thickening for lubricating oils, thickening or suspending agent in paints and inks, production of high gross on leather and paper, ingredient of varnishes, lubricant for plastics, food additive; gelling agent in the preparation of high-energy rocket propellants; formulation of ointment bases; to improve the dyeability of polyolefins

Physical Properties

Freezing Point: 200 °C (392 °F)
Specific Gravity: 1.072
Water Solubility: Practically Insoluble in Water
Other Solubilities: Insoluble in Acetone; Soluble in Petroleum Ether or Oil Turpentine when fresh.

Hazard Overviews

Carcinogenicity: IARC - Not listed; NIOSH - Not listed; NTP - Not listed; ACGIH - Not listed; OSHA - Not listed; EPA - Not listed; MAK - Not listed
Exposure Limits
OSHA PEL Vacated 1989 Limits: TWA: 2 mg/m^3; as Al soluble.
ACGIH TLV: TWA: 2 mg/m^3; as Al.
NIOSH REL: TWA: 2 mg/m^3; as Al soluble salts, alkyls.

Environmental

Regulations

RCRA 40CFR: Not listed
CERCLA: 40CFR 302.4: Not listed
SARA 40CFR 372.65: Not listed
SARA EHS 40CFR 355: Not listed
TSCA: Listed

ALU7440 CAS #: 7784-30-7

ALUMINUM PHOSPHATE SOLUTION

RTECS: TB6450000
DOT: NA1760
EINECS Number: 232-056-9
Molecular Formula: AlH_3O_4P
Formula Weight: 121.95

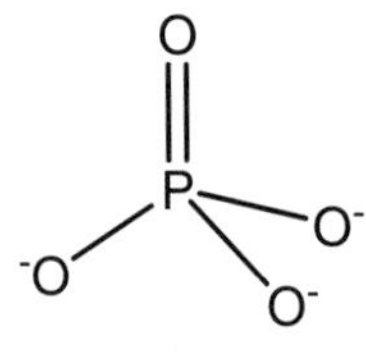

Al^{+++}

Chemical Structure

Synonyms: ALUMINOPHOSPHORIC ACID; ALUMINUM ACID PHOSPHATE; ALUMINUM MONOPHOSPHATE; ALUMINUM PHOSPHATE; ALUMINUM PHOSPHATE (1:1); ALUPHOS; FFB 32; MONOALUMINUM PHOSPHATE; PHOSPHALUGEL
Description: colorless liquid; little or no odor

Physical Properties

Boiling Point: > 100 °C (212 °F)
Freezing Point: 1850 °C (3362 °F)
Specific Gravity: 2.56
Vapor Pressure: 0.029 torr at 20 °C
Water Solubility: Completely
Other Solubilities: Very slightly soluble in HCl, HNO_3
pH: < 7
Evaporation Rate: 1.47
Flash Point: Will not burn

RTECS Toxicity Data

Acute Oral: Rat LD_{Lo} Dose: 4640 mg/kg; Toxic Effects: Behavioral - Somnolence (general depressed activity); Gastrointestinal - Other changes.
Acute Dermal: Rabbit LD_{50} Route: Skin; Dose: >4640 mg/kg; Toxic Effects: Skin and appendages - Dermatitis, irritative; Skin and appendages - Dermatitis, other.

Hazard Overviews

Corrosive

Fire Diamond

Health: Corrosive, causes severe burns to eyes/skin/respiratory tract.
Fire: Noncombustible. Use extinguishing agents suitable for surrounding fire.
Reactivity: Stable. Hazardous polymerization cannot occur. Avoid: elevated temperatures. Incompatible with: metals; carbonates; alkalis; powdered metals.
Carcinogenicity: IARC - Not listed; NIOSH - Not listed; NTP - Not listed; ACGIH - Not listed; OSHA - Not listed; EPA - Not listed; MAK - Not listed
Primary Target Organs:

Eyes

Skin

Respiratory System

Exposure Limits
OSHA PEL Vacated 1989 Limits: TWA: 2 mg/m^3; as Al soluble.
ACGIH TLV: TWA: 2 mg/m^3; as Al.
NIOSH REL: TWA: 2 mg/m^3; as Al soluble salts, alkyls.

Environmental

Cleanup/Disposal: Guide No. 154: Eliminate all ignition sources (no smoking, flares, sparks or flames in immediate area). Do not touch damaged containers or spilled material unless wearing appropriate protective clothing. Stop leak if you can do it without risk. Prevent entry into waterways, sewers, basements or confined areas. Absorb or cover with dry earth, sand or other non-combustible material and transfer to containers. Do not get water inside containers.

Regulations

RCRA 40CFR: Not listed
CERCLA: 40CFR 302.4: Not listed
SARA 40CFR 372.65: Not listed
SARA EHS 40CFR 355: Not listed
TSCA: Listed

ALU7720 CAS #: 20859-73-8

ALUMINUM PHOSPHIDE

RTECS: BD1400000
DOT: UN1397; UN3048; IMO4.3; IMO6.1
EINECS Number: 244-088-0
Molecular Formula: AlP
Structured MF: AlP
Formula Weight: 57.95
Synonyms: AIP; ALP; AL-PHOS; ALUMINIUM FOSFIDE; ALUMINIUM PHOSPHIDE; ALUMINIUM PHOSPHIDE (ALP); ALUMINUM MONOPHOSPHIDE; ALUMINUM PHOSPHIDE (ALP); CELPHIDE; CELPHINE; CELPHOS; DELICIA; DELICIA GASTOXIN; DETIA; DETIA GAS EX-B; DETIA-EX-B; FOSFURI DI ALLUMINIO; FUMITOXIN; GASTION; PHOSPHURES D'ALUMIUM; PHOSTOXIN; PHOSTOXIN-A; QUICKPHOS
Description: dark gray or dark yellow crystals; garlic odor
Use: insecticide; fumigant; source of phosphine; in semiconductor research; for animal feed, bulk grain, cottonseed, peanuts, processed food, leaf tobacco stores; space fumigant for flour mills, railcars, warehouses; to control burrowing rodents; with an igniting agent in sea flares; grain fumigant

Physical Properties

Freezing Point: > 1000 °C (1832 °F)
Specific Gravity: 2.85 at 15 °C/4 °C
Water Solubility: Decomposes
Ionization Potential (eV): 8.4 +/-0.4
Flash Point: Not pertinent (reacts with moisture)

RTECS Toxicity Data

Acute Oral: Man TD_{Lo} Dose: 21 mg/kg; Toxic Effects: Vascular - Other changes; Gastrointestinal - Gastritis. Woman TD_{Lo} Dose: 60 mg/kg; Toxic Effects: Vascular - Other changes; Lungs, Thorax, or Respiration - Dyspnea; Gastrointestinal - Gastritis. Man LD_{Lo} Dose: 86 mg/kg; Toxic Effects: Behavioral - Coma; Vascular - Other changes; Gastrointestinal - Nausea or vomiting. Woman LD_{Lo} Dose: 180 mg/kg; Toxic Effects: Cardiac - Pulse rate increased without fall in BP; Liver - Jaundice, other or unclassified; Kidney, Ureter, and Bladder - Chgs in tubules (inc acute renal failure, acute tubular necrosis.
Acute Inhalation: Man LC_{Lo} Dose: 2800 mg/m^3.

Hazard Overviews

Corrosive

Carcinogenicity: IARC - Not listed; NIOSH - Not listed; NTP - Not listed; ACGIH - Not listed; OSHA - Not listed; EPA - Not listed; MAK - Not listed
Exposure Limits
OSHA PEL Vacated 1989 Limits: TWA: 2 mg/m^3; as Al soluble.
ACGIH TLV: TWA: 2 mg/m^3; as Al.
NIOSH REL: TWA: 2 mg/m^3; as Al soluble salts, alkyls.

Environmental

Cleanup/Disposal: Guide No. 139: Fully encapsulating, vapor protective clothing should be worn for spills and leaks with no fire. Eliminate all ignition sources (no smoking, flares, sparks or flames in immediate area). Do not touch or walk through spilled material. Stop leak if you can do it without risk. Do not get water on spilled substance or inside containers. Use water spray to reduce vapors or divert vapor cloud drift. For chlorosilanes, use AFFF alcohol-resistant medium expansion foam to reduce vapors. Small Spills: Cover with dry earth, dry sand, or other non-combustible material followed with plastic sheet to minimize spreading or contact with rain. Dike for later disposal; do not apply water unless directed to do so. Powder Spills: Cover powder spill with plastic sheet or tarp to minimize spreading and keep powder dry. Do not clean-up or dispose of, except under supervision of a specialist. Guide No. 157: Eliminate all ignition sources (no smoking, flares, sparks or flames in immediate area). All equipment used when handling the product must be grounded. Do not touch damaged containers or spilled material unless wearing appropriate protective clothing. Stop leak if you can do it without risk. A vapor suppressing foam may be used to reduce vapors. Do not get water inside containers. Use water spray to reduce vapors or divert vapor cloud drift. Prevent entry into waterways, sewers, basements or confined areas. Small Spills: Cover with dry earth, dry sand, or other non-combustible material followed with plastic sheet to minimize spreading or contact with rain. Use clean non-sparking tools to collect material and place it into loosely covered plastic containers for later disposal.

Environmental Physical Data

BCF: not pertinent
BOD: not pertinent

Regulations

RCRA 40CFR: Listed Hazardous Waste No. P006 Toxic Waste Reactive Waste
CERCLA: 40CFR 302.4: Listed per RCRA Section 3001 RQ: 100 lb (45.35 kg)
SARA 40CFR 372.65: Listed TPQ: 500 lb
SARA EHS 40CFR 355: Listed TPQ: 100 lb
TSCA: Listed

ALU8280 **CAS #: 11138-49-1**

ALUMINUM SODIUM OXIDE

RTECS: BD1600000
EINECS Number: 234-391-6
Molecular Formula: Al_2O_3
Formula Weight: 82.0
Synonyms: BETA"-ALUMINA; BETA-ALUMINA; J 242; MAXIFLOC 8010; MONOFRAX H; NALCO 680; SODIUM ALUMINATE; SODIUM ALUMINUM OXIDE; SODIUM POLYALUMINATE
Description: white, colorless powder, liquid

Physical Properties

Freezing Point: 1650 °C (3002 °F)
Specific Gravity: 3.24
Water Solubility: Very Soluble
pH: Strongly alkaline
Flash Point: Nonflammable

Hazard Overviews

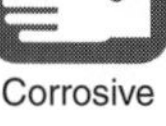

Corrosive

Fire Diamond

Health: Corrosive to eyes/skin/respiratory tract.
Fire: Noncombustible. Use agent suitable for surrounding fire.
Reactivity: Stable. Hazardous polymerization cannot occur. Avoid: heat. Incompatible with: acids. Hazardous decomposition products: fumes of sodium oxide.
Carcinogenicity: IARC - Not listed; NIOSH - Not listed; NTP - Not listed; ACGIH - Not listed; OSHA - Not listed; EPA - Not listed; MAK - Not listed
Primary Target Organs:

Eyes Skin Respiratory System Mucous Membranes

Exposure Limits
OSHA PEL Vacated 1989 Limits: TWA: 2 mg/m^3; as Al soluble.
ACGIH TLV: TWA: 2 mg/m^3; as Al.
NIOSH REL: TWA: 2 mg/m^3; as Al soluble salts, alkyls.

Environmental

Regulations

RCRA 40CFR: Not listed
CERCLA: 40CFR 302.4: Not listed
SARA 40CFR 372.65: Not listed
SARA EHS 40CFR 355: Not listed
TSCA: Not listed

ALU8560 **CAS #: 10043-01-3**

ALUMINUM SULFATE

RTECS: BD1700000
DOT: NA1760; NA9078
EINECS Number: 233-135-0
Molecular Formula: $Al_2H_6O_{12}S_3$
Structured MF: $Al_2(SO_4)_3 \cdot 18H_2O$
Formula Weight: 342.14

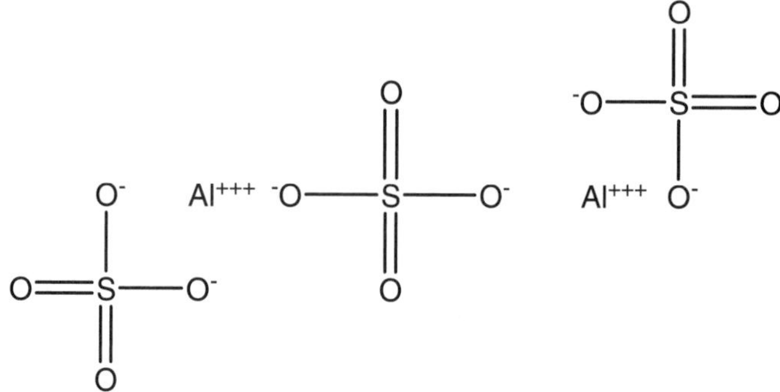

Chemical Structure

Synonyms: ALUM; ALUMINUM ALUM; ALUMINUM SULFATE (2:3); ALUMINUM SULFATE (AL2(SO4)3); ALUMINUM SULPHATE; ALUMINUM TRISULFATE; CAKE ALUM; DIALUMINUM SULFATE; DIALUMINUM SULPHATE; DIALUMINUM TRISULFATE; FILTER ALUM; PAPERMAKER'S ALUM; PEARL ALUM; PICKLE ALUM; SULFURIC ACID ALUMINUM(3+) SALT (3:2); SULFURIC ACID,ALUMINUM SALT (3:2)
Description: white, grey powder, crystals; odorless
Use: tanning leather, sizing paper, mordant in dyeing; fireproofing & waterproofing cloth; clarifying oils & fats; waterproofing concrete; deodorizing & decolorizing petro; antiperspirants; agricultural pesticides; mfr of aluminum salts, resinate; catalyst in mfr of ethane; lubricating compositions; food additive; applied to soils, to make less alkaline for rhododendrons, etc; med: anti-infective; water & sewage treatment (flocculant removal of trihalomethanes, trace metals, org matter, viruses); for decontamination of radiocontaminated surface; in cosmetics & soap mfr.

Physical Properties

Boiling Point: Anhydrous > 1600 °C (2912 °F)
Freezing Point: Decomposes at 770 °C (1418 °F)
Specific Gravity: 2.71 at 25 °C
Vapor Density: 2.7 Air=1
Vapor Pressure: Essentially 0 mm Hg
Water Solubility: 1 parts in 1 part Water
Other Solubilities: Soluble in dilute acid.
pH: Aqueous solution 1 g/1 ml water 2.9
Refraction Index: 1.47
Flash Point: Nonflammable

RTECS Toxicity Data

Acute Oral: Mouse LD_{50} Dose: 6207 mg/kg.

Reproductive/Teratogenic: Rat Route: Intratesticular; Dose: 27371 ug/kg; Duration: male 1D prior to mating; Paternal Effects - Spermatogenesis; Testes, epididymis, sperm duct. Mouse Route: Subcutaneous; Dose: 27371 ug/kg; Duration: male 30D prior to mating; Paternal Effects - Spermatogenesis; Testes, epididymis, sperm duct.
Mutagenic: Human Cytogenetic Analysis; Cell Type: lymphocyte; Dose: 20 mg/L. Human Sister Chromatid Exchenge; Cell Type: lymphocyte; Dose: 20 mg/L. Human Micronucleus Test; Cell Type: lymphocyte; Dose: 20 mg/L. Human Other Mutation Test Systems; Cell Type: lymphocyte; Dose: 20 mg/L.

Hazard Overviews

Fire Diamond

Health: Severe irritation to eyes/skin/respiratory tract. Chronic Effects: dermatitis.
Fire: Noncombustible. Use extinguishing agents suitable for surrounding fire.
Reactivity: Stable. Hazardous polymerization cannot occur. Avoid: exposure to moisture. Incompatible with: alkalis; metals (corrodes in the presence of moisture). Hazardous decomposition products: sulfur oxide(s).
Carcinogenicity: IARC - Not listed; NIOSH - Not listed; NTP - Not listed; ACGIH - Not listed; OSHA - Not listed; EPA - Not listed; MAK - Not listed
Primary Target Organs:

Eyes

Skin

Respiratory System

Mucous Membranes
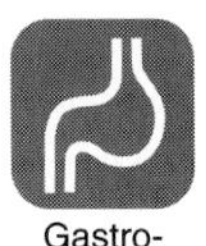
Gastro-intestinal

Exposure Limits
OSHA PEL Vacated 1989 Limits: TWA: 2 mg/m^3; as Al soluble.
ACGIH TLV: TWA: 2 mg/m^3; as Al.
NIOSH REL: TWA: 2 mg/m^3; as Al soluble salts, alkyls.

Environmental

Ecotoxicity: Aquatic toxicity: 14ppm/36 hr/fundulus/fatal/fresh water 240ppm/48 hr/mosquitofish/TLm/* *Water type not specified
Cleanup/Disposal: Guide No. 154: Eliminate all ignition sources (no smoking, flares, sparks or flames in immediate area). Do not touch damaged containers or spilled material unless wearing appropriate protective clothing. Stop leak if you can do it without risk. Prevent entry into waterways, sewers, basements or confined areas. Absorb or cover with dry earth, sand or other non-combustible material and transfer to containers. Do not get water inside containers. Guide No. 171: Do not touch or walk through spilled material. Stop leak if you can do it without risk. Prevent dust cloud. Avoid inhalation of asbestos dust. Small Dry Spills: With clean shovel place material into clean, dry container and cover loosely; move containers from spill area. Small Spills: Take up with sand or other noncombustible absorbent material and place into containers for later disposal. Large Spills: Dike far ahead of liquid spill for later disposal. Cover powder spill with plastic sheet or tarp to minimize spreading. Prevent entry into waterways, sewers, basements or confined areas.

Environmental Physical Data

BCF: no food chain concentration potential
BOD: none

Regulations

RCRA 40CFR: Not listed
CERCLA: 40CFR 302.4: Listed per CWA Section 311(b)(4) RQ: 5000 lb (2268 kg)
SARA 40CFR 372.65: Not listed
SARA EHS 40CFR 355: Not listed
TSCA: Listed

ALU8840 CAS #: 637-12-7

ALUMINUM TRISTEARATE

RTECS: WI2820000
EINECS Number: 211-279-5
Molecular Formula: $C_{54}H_{108}AlO_6$
Formula Weight: 877.35

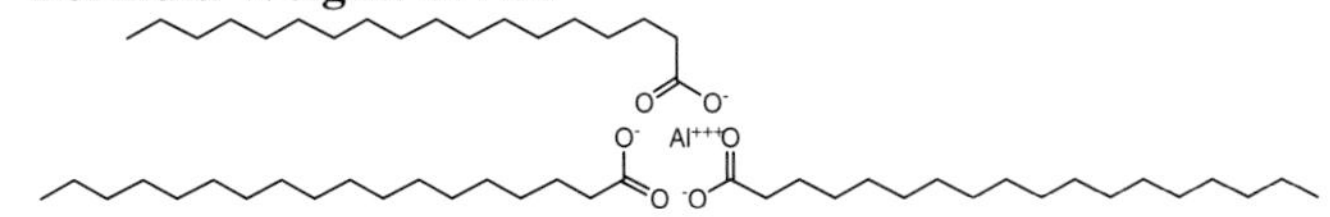

Chemical Structure

Synonyms: ALUGEL 34TN; ALUMINIUM STEARATE; ALUMINUM STEARATE; ALUMINUM STEARATE (1:3); METASAP XX; MONOALUMINUM STEARATE; OCTADECANOIC ACID,ALUMINUM SALT; ROFOB 3; SA 1500; STEARIC ACID,ALUMINUM SALT; TRIBASIC ALUMINUM STEARATE
Description: white powder
Use: waterproofing fabrics, ropes, cements; gelling agents & driers in paint & varnish; thickening lubricating oils; in light-sensitive photographic compositions; defoamer for oil drilling fluids, in beet sugar & yeast processing; retarder for polysulfide dental impression materials; greases, lubricants, cutting compounds, flatting agent, cosmetics & pharmaceuticals; water-repellent soap for surfaces, eg, natural stone; aluminum stearate; to gel petro fractions, in prepn of high-energy rocket fuels & plastic explosives.

Physical Properties

Freezing Point: 117 °C (242.6 °F) to 120 °C (248 °F)
Specific Gravity: 1.07
Water Solubility: Practically Insoluble in Water
Other Solubilities: when freshly made, Soluble in Alcohol, Benzene, Oil Turpentine, mineral oils

Hazard Overviews

Carcinogenicity: IARC - Not listed; NIOSH - Not listed; NTP - Not listed; ACGIH - Not listed; OSHA - Not listed; EPA - Not listed; MAK - Not listed

Exposure Limits

OSHA PEL Vacated 1989 Limits: TWA: 2 mg/m^3; as Al soluble.

ACGIH TLV: TWA: 2 mg/m^3; as Al.

NIOSH REL: TWA: 2 mg/m^3; as Al soluble salts, alkyls.

Environmental

Regulations

RCRA 40CFR: Not listed
CERCLA: 40CFR 302.4: Not listed
SARA 40CFR 372.65: Not listed
SARA EHS 40CFR 355: Not listed
TSCA: Listed

ALV5000 CAS #: 150-59-4

ALVERINE

EINECS Number: 205-763-5
Molecular Formula: $C_{20}H_{27}N$
Formula Weight: 281.43

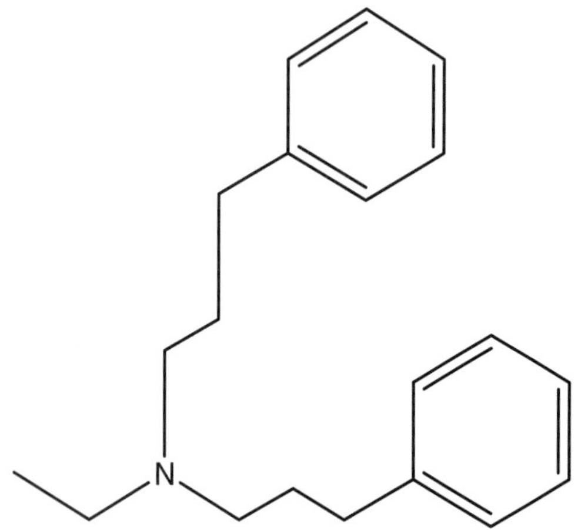

Chemical Structure

Synonyms: BENZENEPROPANAMINE,N-ETHYL-N-(3-PHENYLPROPYL)-; BIS(GAMMA-PHENYLPROPYL)ETHYLAMINE; DI(PHENYLPROPYL)ETHYLAMINE; DIPROPYLAMINE,N-ETHYL-3,3'-DIPHENYL-; DIPROPYLIN; DIPROPYLINE; N-ETHYL-3,3'-DIPHENYLDIPROPYLAMINE; N-ETHYL-N-(3-PHENYLPROPYL)BENZENEPROPANAMINE; PHENOPROPAMINE; PHENPROPAMINE; PROFENIL; SESTRON; SESTRON BASE; SPASMAVERINE

Description: liquid; sweet odor

Use: medication: anticholinergic; medication: antispasmatic and smooth muscle relaxant

Physical Properties

Boiling Point: 165 °C (329 °F) to 168 °C (334 °F) at 0.3 mm Hg

Water Solubility: Soluble in Water

Other Solubilities: Slightly Soluble in Chloroform; Soluble in dilute acids; Sparingly Soluble in Alcohol; Very Slightly Soluble in Ether.

pH: 10% aqueous solution is neutral

Hazard Overviews

Health: Irritating to eyes/skin/respiratory tract. Other Acute Effects: may be harmful by inhalation, ingestion, or skin absorption.

Fire: Hazards: emits toxic fumes. Extinguishing agents: water spray; carbon dioxide, dry chemical powder or appropriate foam. Precautions: combustible liquid.

Reactivity: Incompatible with: strong oxidizing agents. Hazardous decomposition products: toxic fumes of: carbon monoxide, carbon dioxide, nitrogen oxides.

Carcinogenicity: IARC - Not listed; NIOSH - Not listed; NTP - Not listed; ACGIH - Not listed; OSHA - Not listed; EPA - Not listed; MAK - Not listed

Primary Target Organs:

Eyes

Skin

Respiratory System

Environmental

Regulations

RCRA 40CFR: Not listed
CERCLA: 40CFR 302.4: Not listed
SARA 40CFR 372.65: Not listed
SARA EHS 40CFR 355: Not listed
TSCA: Not listed

AMA1000 CAS #: 21150-21-0

AMANIN

RTECS: NJ8326000
Molecular Formula: $C_{39}H_{53}N_9O_{14}S$
Formula Weight: 904.07

Synonyms: AMANINE; ALPHA-AMANITIN,1-L-ASPARTIC ACID-4-(2-MERCAPTO-L-TRYPTOPHAN)-; ALPHA-AMANITIN,1-L-ASPARTICACID-4-(2-MERCAPTO-L-TRYPTOPHAN)-

Hazard Overviews

Carcinogenicity: IARC - Not listed; NIOSH - Not listed; NTP - Not listed; ACGIH - Not listed; OSHA - Not listed; EPA - Not listed; MAK - Not listed

Environmental

Regulations

RCRA 40CFR: Not listed

CERCLA: 40CFR 302.4: Not listed
SARA 40CFR 372.65: Not listed
SARA EHS 40CFR 355: Not listed
TSCA: Not listed

AMA2600 CAS #: 23109-05-9

A-AMANITIN

RTECS: BD6195000
EINECS Number: 245-432-2
Molecular Formula: $C_{39}H_{54}N_{10}O_{14}S$
Formula Weight: 919.09

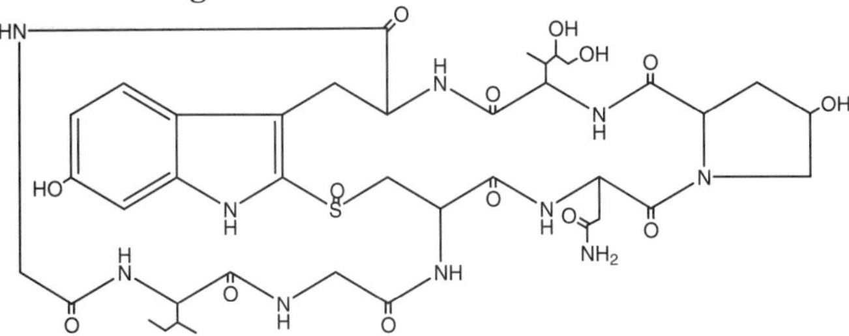

Chemical Structure

Synonyms: ALPHA-AMANITIN (8CI,9CI); ALPHA-AMANITINE; ALPHA-AMATOXIN; CYCLIC(L-ASPARAGINYL-4-HYDROXY-L-PROLYL-(R)-4,5-DIHYDROXY-L-ISOLEUCYL-6-HYDROXY-2-MERCAPTO-L-TRYPTOPHYLGLYCYL-L-ISOLEUCYLGLYCYL-L-CYSTEINYL), CYCLIC (4-8)-SULFIDE,(R)-S-OXIDE
Description: needles

Physical Properties

Freezing Point: 254 °C (489.2 °F) to 255 °C (491 °F)

RTECS Toxicity Data

Mutagenic: Rat DNA Inhibition; Cell Type: liver; Dose: 100 nmol/L. Rat Cytogenetic Analysis; Route: Intraperitoneal; Dose: 500 ug/kg.

Hazard Overviews

Poison

Health: Poison. Other Acute Effects: may be fatal if inhaled, swallowed, or absorbed through skin; symptoms may be delayed for as long as 6-24 hours; the manifestations of poisoning are severe nausea; vomiting; diarrhea; bloody stools; painful tenderness and enlarged liver; oliguria or anuria; jaundice; pulmonary edema; headache; mental confusion and depression; hypoglycemia; signs of cerebral injury with coma or convulsions; this may be followed by a day of apparent recovery and within one or two days the symptoms may reappear followed by death; target organs: liver, kidneys, g.i. system, central nervous system, heart.
Fire: Extinguishing agents: carbon dioxide, dry chemical powder or appropriate foam. Precautions: combustible liquid.
Reactivity: Stable. Hazardous polymerization will not occur. Hazardous decomposition products: toxic fumes of: carbon monoxide, carbon dioxide, nitrogen oxides, sulfur oxides.
Carcinogenicity: IARC - Not listed; NIOSH - Not listed; NTP - Not listed; ACGIH - Not listed; OSHA - Not listed; EPA - Not listed; MAK - Not listed
Primary Target Organs:

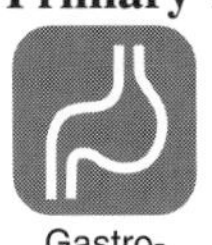
Gastro-intestinal

Nervous System

Liver

Kidneys

Cardio-vascular

Environmental

Regulations

RCRA 40CFR: Not listed
CERCLA: 40CFR 302.4: Not listed
SARA 40CFR 372.65: Not listed
SARA EHS 40CFR 355: Not listed
TSCA: Not listed

AMA4200 CAS #: 13567-07-2

B-AMANITIN

RTECS: NJ8324000
Molecular Formula: $C_{39}H_{53}N_{9}O_{14}S$
Formula Weight: 904.07
Synonyms: BETA-AMANITIN; ALPHA-AMANITIN,1-L-ASPARTIC ACID-(9CI); BETA-AMANITINE; BETA-AMATOXIN

Hazard Overviews

Carcinogenicity: IARC - Not listed; NIOSH - Not listed; NTP - Not listed; ACGIH - Not listed; OSHA - Not listed; EPA - Not listed; MAK - Not listed

Environmental

Regulations

RCRA 40CFR: Not listed
CERCLA: 40CFR 302.4: Not listed
SARA 40CFR 372.65: Not listed
SARA EHS 40CFR 355: Not listed
TSCA: Not listed

AMA5800 CAS #: 13567-11-8

GAMMA-AMANITIN

RTECS: BD6195200
EINECS Number: 236-970-9
Molecular Formula: $C_{39}H_{54}N_{10}O_{12}S$
Formula Weight: 886.99
Synonyms: GAMMA-AMANITINE; GAMMA-AMATOXIN

Hazard Overviews

Carcinogenicity: IARC - Not listed; NIOSH - Not listed; NTP - Not listed; ACGIH - Not listed; OSHA - Not listed; EPA - Not listed; MAK - Not listed

Environmental

Regulations

RCRA 40CFR: Not listed
CERCLA: 40CFR 302.4: Not listed
SARA 40CFR 372.65: Not listed
SARA EHS 40CFR 355: Not listed
TSCA: Not listed

AMA7400 CAS #: 768-94-5

AMANTADINE

RTECS: YD1925000
EINECS Number: 212-201-2
Molecular Formula: $C_{10}H_{17}N$
Formula Weight: 151.26

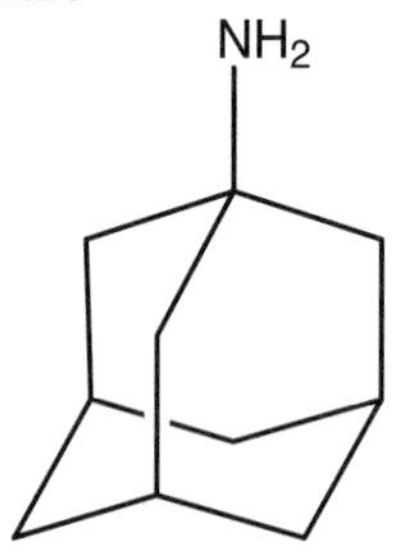

Chemical Structure

Synonyms: 1-ADAMANTAMINE; ADAMANTAMINE; 1-ADAMANTANAMINE; ADAMANTANAMINE; 1-ADAMANTYLAMINE; ADAMANTYLAMINE; 1-AMINOADAMANTANE; AMINOADAMANTANE; 1-AMINOADAMATANE; 1-AMINODIAMANTANE; 1-AMINOTRICYCLO(3.3.1.1(SUP 3,7))DECANE; 1-AMINOTRICYCLO[3.3.1.1(SUP 3,7)]DECANE; EXP-105-1; PK-MERZ; SYMMETREL; TRICYCLO(3.3.1.1(SUP 3.7))DECAN-1-AMINE; TRICYCLO(3.3.1.1(3,7))DECAN-1-AMINE
Description: crystals; odorless
Use: medication: antiviral, treatment of Parkinsonism; treatment of drug-induced extrapyramidal reactions; medication (vet): antiviral agent

Physical Properties

Boiling Point: Decomposes at 360 °C (680 °F)
Freezing Point: 160 °C (320 °F) to 190 °C (374 °F)
Water Solubility: Sparingly Soluble in Water
Other Solubilities: Practically Insoluble in Ether
pH: 1 in 5 solution 3.0 to 5.5

RTECS Toxicity Data

Acute Oral: Rat LD_{50} Dose: 900 mg/kg; Toxic Effects: Behavioral - Tremor; Behavioral - Convulsions or effect on seizure threshold. Mouse LD_{50} Dose: 900 mg/kg; Toxic Effects: Behavioral - Tremor; Behavioral - Ataxia.
Mutagenic: Bacteria - E Coli DNA Adduct; Dose: 10 umol/L.

Hazard Overviews

Health: Irritating to eyes/skin/respiratory tract. Toxic. Other Acute Effects: harmful if swallowed, inhaled, or absorbed through skin; adverse reactions include nausea; dizziness; insomnia; depression; anxiety and irritability; hallucinations; confusion; anorexia; dry mouth and constipation; ataxia; livedo reticularis; peripheral edema; headache; orthostatic hypotension; fatigue; skin rash; vomiting; dyspnea; congestive heart failure. Chronic Effects: possible teratogen; overexposure may cause reproductive disorder(s) based on tests with laboratory animals; the hydrochloride salt of this material is embryotoxic and teratogenic in rats.
Fire: Hazards: emits toxic fumes. Extinguishing agents: carbon dioxide, dry chemical powder or appropriate foam. Precautions: combustible liquid.
Reactivity: Incompatible with: strong oxidizing agents. Hazardous decomposition products: toxic fumes of: carbon monoxide, carbon dioxide, nitrogen oxides.
Carcinogenicity: IARC - Not listed; NIOSH - Not listed; NTP - Not listed; ACGIH - Not listed; OSHA - Not listed; EPA - Not listed; MAK - Not listed
Primary Target Organs:

Eyes Skin Respiratory System

Environmental

Regulations

RCRA 40CFR: Not listed
CERCLA: 40CFR 302.4: Not listed
SARA 40CFR 372.65: Not listed
SARA EHS 40CFR 355: Not listed
TSCA: Listed

AMA9000 CAS #: 915-67-3

AMARANTH

RTECS: QJ6550000
EINECS Number: 213-022-2
Molecular Formula: $C_{20}H_{11}N_2Na_3O_{10}S_3$
Formula Weight: 604.49

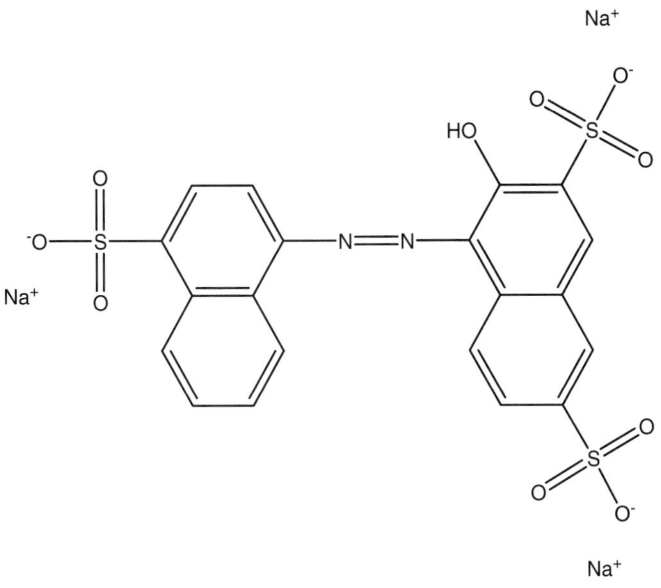

Chemical Structure

Synonyms: ACETACID RED 2BR; ACID AMARANTH; ACID AMARANTH I; ACID AMARANTH N; ACID LEATHER RED 12BW; ACID LEATHER RUBINE S; ACID RED 37; ACILAN RED SE; AIZEN AMARANTH; AMACID; AMACID AMARANTH; AMARANT; AMARANTH A; AMARANTH B; AMARANTH BPC; AMARANTH EXTRA; AMARANTH LAKE; AMARANTH S; AMARANTH S SPECIALLY PURE; AMARANTH USP; AMARANTH WD; AMARANTHE; AMARANTHE USP (BIOLOGICAL STAIN); AZO RED R; S-AZO RUBINE; AZO RUBINE S.FQ; AZO RUBINE S; AZO RUBINE SF; AZO RUBY S; AZORUBIN S; AZORUBINE S; BORDEAUX; BORDEAUX S; BORDEAUX S EXTRA CONC A EXPORT; BORDEAUX S EXTRA CONC. A.EXPORT; BORDEAUX S EXTRA PURE A; C.I. 16185; C.I. 184; C.I. ACID RED 27; C.I. ACID RED 27,TRISODIUM SALT; C.I. FOOD RED 9; CALCOCID AMARANTH; CANACERT AMARANTH; CERTICOL AMARANTH S; CERVEN KYSELA 27; CERVEN POTRAVINARSKA 9; CILEFA RUBINE 2B; D AND C RED 2; DAISHIKI AMARANTH; DOLKWAL AMARANTH; DYE FDC RED 2; DYE RED RASPBERRY; E 123; EDICOL AMARANTH; EDICOL SUPRA AMARANTH A; EEC NO. 123; EUROCERT AMARANTH; FAST RED; FD & C RED NO 2; FD AND C RED NO. 2; FD AND C RED NO. 2-ALUMINIUM LAKE; FD&C RED NO. 2 -ALUMINIUM LAKE; FOOD RED 2; FOOD RED 9; FRUIT RED A GEIGY; HD AMARANTH B; HD AMARANTH SUPRA; HEXACERT RED NO. 2; HEXACOL AMARANTH B EXTRA; HIDACID AMARANTH; HISPACID RED AM; 2-HYDROXY-1,1'-AZONAPHTHALENE-3,6,4'-TRISULFONIC ACID TRISODIUM SALT; 2-HYDROXY-1,1'-AZONAPHTHALENE-3,6,4'-TRISULFONIC ACIDTRISODIUM SALT; 3-HYDROXY-4-[(4-SULFO-1-NAPHTHALENYL)AZO]-2,7-NAPHTHALENEDISULFONIC ACID TRISODIUM SALT; 3-HYDROXY-4-((4-SULFO-1-NAPHTHALENYL)AZO)-2,7-NAPHTHLENEDISULFONIC ACID,; 3-HYDROXY-4-((4-SULFO-1-NAPHTHALENYL)AZO)-2,7-NAPHTHLENEDISULFONIC ACID,TRISODIUM SALT; 3-HYDROXY-4-((4-SULFO-1-NAPHTHYL)AZO)-2,7-NAPHTHALENEDISULFONIC ACID,; 3-HYDROXY-4-((4-SULFO-1-NAPHTHYL)AZO)-2,7-NAPHTHALENEDISULFONIC ACID,TRISODIUM SALT; 3-HYDROXY-4-((4-SULPHO-1-NAPHTHALENYL)AZO)-2,7-NAPHTHALENEDISULPHONIC ACID,; 3-HYDROXY-4-((4-SULPHO-1-NAPHTHALENYL)AZO)-2,7-NAPHTHALENEDISULPHONIC ACID,TRISODIUM SALT; 3-HYDROXY-4-((4-SULPHO-1-NAPHTHYL)AZO)-2,7-NAPHTHALENEDISULPHONIC ACID,; 3-HYDROXY-4-((4-SULPHO-1-NAPHTHYL)AZO)-2,7-NAPHTHALENEDISULPHONIC ACID,TRISODIUM SALT; JAVA AMARANTH; KAYAKU AMARANTH; KAYAKU FOOD COLOUR RED NO. 2; KCA FOODCOL AMARANTH A; KITON RUBINE S; L-RED 3; L-ROT 3; LISSAMINE; LISSAMINE AMARANTH AC; MAPLE AMARANTH; 2,7-NAPHTHALENEDISULFONIC ACID,3-HYDROXY-4-((4-SULFO-1-NAPHTHYL)AZO)-,TRISODIUM SALT; NAPHTHOL RED B; NAPHTHOL RED C; NAPHTHOL RED LZS; NAPHTHOL RED O; NAPHTHOL RED S; NAPHTHOL RED S CONC. SPECIALLY PURE; NAPHTHOL RED S SPECIALLY PURE; NAPHTHOL RED SI; NAPTHOLROT S; NEKLACID RED A; RAKUTO AMARANTH; RASPBERRY RED FOR JELLIES; 1302 RED; 1508 RED; RED DYE NO. 2; RED NO 2; RED NO. 2; SAN-EI AMARANTH; SCHULTZ NR. 212; SHIKISO AMARANTH; SOLAR RED O; 1-(4-SULFO-1-NAPHTHYLAZO)-2-NAPHTHOL-3,6-DISULFONIC ACID TRISODIUM SALT; 1-(4-SULFO-1-NAPHTHYLAZO)-2-NAPHTHOL-3,6-DISULFONIC ACIDTRISODIUM SALT; 1-(4-SULPHO-1-NAPHTHYLAZO)-2-NAPHTHOL-3,6-DISULPHONIC ACID,TRISODIUM SALT; 1-(4-SULPHO-1-NAPHTHYLAZO)-2-NAPHTHOL-3,6-DISULPHONICACID,TRISODIUM SALT; TAKAOKA AMARANTH; TERTRACID RED A; TETRACID RED A; TOYO AMARANTH; TRISODIUM SALT; TRISODIUM SALT OF 1-(4-SULFO-1-NAPHTHYLAZO)-2-NAPHTHOL-3,6-DISULFONIC ACID; TRISODIUM SALT OF 1-(4-SULPHO-1-NAPHTHYLAZO)-2-NAPHTHOL-3,6-DISULPHONIC ACID; TRISODIUM SALT OF1-(4-SULFO-1-NAPHTHYLAZO)-2-NAPHTHOL-3,6-DISULFONIC ACID; USACERT RED NO. 2; VICTORIA RUBINE O; VICTORIA RUBINE O FOR FOOD; WHORTLEBERRY RED; WOOL BORDEAUX 6RK; WOOL RED; WOOL RED 40F

Description: dark, reddish-brown powder; almost odorless

Use: in dyeing wool, silk, textiles, paper, phenolformaldehyde resins and leather; in color photography, pigments (as barium salt), drugs, canned goods, jams, mineral waters and as an indicator in hydrazine titrations; as a colorant of gelatin, maraschin cherries, sausage casings, frozen desserts, carbonated beverages, dry drink powders, sweets and confectionery products not containing oils and fats, bake products and cereals, puddings, aqueous drug solutions, tablets, capsules, mouthwashes, bath salts, hair rinses and custard powders

Physical Properties

Freezing Point: > 300 °C (572 °F)
Specific Gravity: 1.5
Density: 1.5 g/mL
Water Solubility: 1 g Soluble in 15 ml Water, 72 g/100 ml Water
Other Solubilities: DMSO: 10-50 mg/ml at 23 °C; Acetone: <1 mg/ml at 23 °C; Alcohol: Very slightly Soluble; Glycerine: Soluble; Propylene Glycol: Soluble; Most organic solvents: Insoluble; Vegetable oil: Insoluble.
pH: 1% solution in water approximately 10.8
Flash Point: Not available; probably combustible

RTECS Toxicity Data

Mutagenic: Hamster Cytogenetic Analysis; Cell Type: fibroblast; Dose: 1 gm/L/48H. Hamster Cytogenetic Analysis; Cell Type: lung; Dose: 2 gm/L.

Hazard Overviews

Health: Irritating to eyes/skin/respiratory tract. Harmful. Other Acute Effects: may be harmful by inhalation, ingestion, or skin absorption. Irritating to eyes/skin/respiratory tract. Harmful. Other Acute Effects: may be harmful by inhalation, ingestion, or skin absorption.

Fire: Will burn. Hazards: emits toxic fumes. Extinguishing agents: water spray; carbon dioxide, dry chemical powder or appropriate foam. Precautions: combustible liquid.
Carcinogenicity: IARC - Group 3, Not classifiable as to carcinogenicity to humans; NIOSH - Not listed; NTP - Not listed; ACGIH - Not listed; OSHA - Not listed; EPA - Not listed; MAK - Not listed
Primary Target Organs:

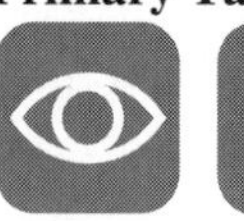
Eyes

Skin

Respiratory System

Environmental

Regulations

RCRA 40CFR: Not listed
CERCLA: 40CFR 302.4: Not listed
SARA 40CFR 372.65: Not listed
SARA EHS 40CFR 355: Not listed
TSCA: Listed

AMD5000 CAS #: 67485-29-4

AMDRO

RTECS: UW7583000
Molecular Formula: $C_{25}H_{24}F_6N_4$
Formula Weight: 494.53
Synonyms: AC 217300; AMDRO; CL 217300; COMBAT; HYDRAMETHYLNON; MATOX; MAXFORCE; WIPEOUT

Physical Properties

Freezing Point: 185 °C (365 °F) to 190 °C (374 °F)
Water Solubility: ppb
Other Solubilities: 360g/l acetone; 72g/l ethanol; 230g/l methanol

RTECS Toxicity Data

Acute Oral: Rat LD_{50} Dose: 1131 mg/kg. Duck LD_{50} Dose: >2510 mg/kg.
Acute Inhalation: Rat LC_{50} Dose: >5 gm/m^3/4hr.
Acute Dermal: Rabbit LD_{50} Route: Skin; Dose: >5 gm/kg.

Hazard Overviews

Carcinogenicity: IARC - Not listed; NIOSH - Not listed; NTP - Not listed; ACGIH - Not listed; OSHA - Not listed; EPA - Not listed; MAK - Not listed

Environmental

Regulations

RCRA 40CFR: Not listed
CERCLA: 40CFR 302.4: Not listed
SARA 40CFR 372.65: Listed
SARA EHS 40CFR 355: Not listed
TSCA: Not listed

AME5000 CAS #: 834-12-8

AMETRYNE

RTECS: XY9100000
DOT: UN2763
EINECS Number: 212-634-7
Molecular Formula: $C_9H_{17}N_5S$
Formula Weight: 227.35
Synonyms: A 1093; AMEPHYT; AMETREX; AMETRYN; CEMERIN; CRISATRINE; DORUPLANT; EPA PESTICIDE CODE 080801; 2-ETHYLAMINO-4-ISOPROPYLAMINO-6-METHYLMERCAPTO-S-TRIAZINE; 2-(ETHYLAMINO)-4-(ISOPROPYLAMINO)-6-(METHYLTHIO)-S-TRIAZINE; 2-ETHYLAMINO-4-ISOPROPYLAMINO-6-METHYLTHIO-1,3,5-TRIAZINE; 2-ETHYLAMINO-4-ISOPROPYLAMINO-6-METHYLTHIO-S-TRIAZINE; N-ETHYL-N-ISOPROPYL-6-METHYLTHIO-1,3,5-TRIAZINE-2,4-DIAMINE; N-ETHYL-N'-ISOPROPYL-6-METHYLTHIO-1,3,5-TRIAZINE-2,4-DIYLDIAMINE; N-ETHYL-N'-(1-METHYLETHYL)-6-(METHYLTHIO)-1,3,5-TRIAZINE-2,4-DIAMINE; EVIK; G 34162; G-34162; GARDOPAX; GESAPAX; GESTENE; 2-METHYLMERCAPTO-4-ETHYLAMINO-6-ISOPROPYLAMINO-S-TRIAZINE; 2-METHYLMERCAPTO-4-ISOPROPYLAMINO-6-ETHYLAMINO-S-TRIAZINE; 2-METHYLTHIO-4-ETHYLAMINO-6-ISOPROPYLAMINO-S-TRIAZINE; PRIMATOL Z 80; TOPAZOL; 1,3,5-TRIAZINE-2,4-DIAMINE,N-ETHYL-N'-(1-METHYLETHYL)-6-(METHYLTHIO)-; S-TRIAZINE,2-(ETHYLAMINO)-4-(ISOPROPYLAMINO)-6-(METHYLTHIO)-; S-TRIAZINE,2-ETHYLAMINO-4-ISOPROPYLAMINO-6-METHYLTHIO-; TRINATOX D
Description: colorless to white crystalline powder
Use: a selective herbicide used pre- & post-emergence to control broad leaved & grass weeds in bananas, citrus, cocoa, coffee, maize, oil palms, pineapples, sugarcane, tea, in potatoes (as vine desiccant) & in noncrop area

Physical Properties

Freezing Point: 88 °C (190.4 °F) to 89 °C (192.2 °F)
Saturated Vapor Density: 1.200000009 kg/m^3
Density: 1.19 g/cu cm at 20 °C
Vapor Pressure: 8.4 x10^{-7} mm Hg at 20 °C
Water Solubility: 185 ppm in Water at 20 °C
Other Solubilities: 500 g/l Acetone at 20 °C; 600 g/l dichloromethane at 20 °C; 14 g/l hexane at 20 °C; 450 g/l Methanol at 20 °C; 200 g/l octan-1-ol at 20 °C; 400 g/l Toluene at 20 °C; Soluble in alkali or acid.
Flash Point: Nonflammable

RTECS Toxicity Data

Acute Oral: Rat LD_{50} Dose: 508 mg/kg. Mouse LD_{50} Dose: 965 mg/kg.
Acute Inhalation: Rat LC_{50} Dose: >2200 mg/m^3/4hr.
Acute Dermal: Rabbit LD_{50} Route: Skin; Dose: 8160 mg/kg. Rat LD_{50} Route: Skin; Dose: >3 gm/kg.
Chronic (Multiple Dose) Oral: Rat Dose: 21840 mg/kg/2Y-C; Toxic Effects: Behavioral - Food intake (animal); Nutritional and gross metabolic - Weight loss or decreased weight gain. Rat Dose: 18200 mg/kg/52W-C; Toxic Effects: Behavioral - Food intake (animal); Blood - Changes in leukocyte (WBC) cell count; Nutritional and gross metabolic - Weight loss or decreased weight gain.

Irritation Eye: Rabbit Standard Draize Test Dose: 76 mg; Reaction: mild.
Reproductive/Teratogenic: Rat Route: Oral; Dose: 2500 mg/kg; Duration: female 6-15D of pregnancy; Maternal Effects - Other effects on females; Effects on Embryo or Fetus - Other effects to embryo or fetus. Rat Route: Oral; Dose: 336600 ug/kg; Duration: female 5-15D of pregnancy; Effects on Fertility - Post-implantation mortality; Effects on Embryo or Fetus - Fetotoxicity; Other effects to embryo or fetus.

Hazard Overviews

Fire: Noncombustible.
Carcinogenicity: IARC - Not listed; NIOSH - Not listed; NTP - Not listed; ACGIH - Not listed; OSHA - Not listed; EPA - Not listed; MAK - Not listed

Environmental

Ecotoxicity: LC_{50} Anas platyrhynchos (Mallard duck) oral 23000 ppm in 8 day diet LC_{50} Mugil cephalus (Mullet) 0.68 mg/l/96 hr /Conditions of bioassay not specified LC_{50} Salmo gairdneri (Rainbow trout) 3.2 mg/l/96 hr at 13 °C, wt 0.5 g /Static bioassay LC_{50} Oysters > 1.0 ppm/96 hr /Technical; Conditions of bioassay not specified
Environmental Fate: It is a weak base with a pKa of 3.12, indicating that it is almost entirely undissociated at environmental pHs. When applied to soil, it will adsorb moderately to the soil (average K_{oc} = 388). It is fairly persistent in soil; its half-life in soil is reported to be 6.0 months at 15 °C and 4.5 months at 30 °C). Degradation is more rapid in acidic soils than in neutral ones. An important mechanism by which it may be lost from soil is by volatilization. Chemicals with low Henry's Law constants may rise to the soil surface with evaporating water and as its concentration increases at the soil-air interface, so will the amount of chemical volatilizing. If released into water, it will partially adsorb to sediment and particulate matter in the water column. It should undergo photolysis in surface layers of water (half-life ca. 10.2 hr); the rate of photodegradation will be higher at low pHs and in the presence of some sensitizers. No biodegradation rates in water are available for natural water; biodegradation would be expected to be slow. Volatilization and bioconcentration in fish should not be significant. It would generally be released to the atmosphere as an aerosol while spraying and will be removed by gravitational settling. In the atmosphere, vapor-phase would degrade by reaction with photochemically produced hydroxyl radicals; its estimated half-life is 2.5 hr.
Cleanup/Disposal: Guide No. 151: Do not touch damaged containers or spilled material unless wearing appropriate protective clothing. Stop leak if you can do it without risk. Prevent entry into waterways, sewers, basements or confined areas. Cover with plastic sheet to prevent spreading. Absorb or cover with dry earth, sand or other non-combustible material and transfer to containers. Do not get water inside containers.

Environmental Physical Data

Henry's Law Constant: calculated at 1.2×10^{-9}
Octanol/Water Partition Coefficient: log K_{ow} = calculated at 250
Sorption Partition Coefficient: K_{oc} = soils 388.4
BCF: estimated at 185

Regulations

RCRA 40CFR: Not listed
CERCLA: 40CFR 302.4: Not listed
SARA 40CFR 372.65: Listed
SARA EHS 40CFR 355: Not listed
TSCA: Not listed

Analytical Methods

Water / Groundwater: EPA 619; USGS O3106
Drinking Water: EPA 507, 525.2; AOAC 991.07; ASTM D5475
Food: FDA 232.4, 242.1; AOAC 971.08

AMI1000 CAS #: 551-93-9

O-AMINOACETOPHENONE

EINECS Number: 209-002-8
Molecular Formula: C_8H_9NO
Formula Weight: 135.16

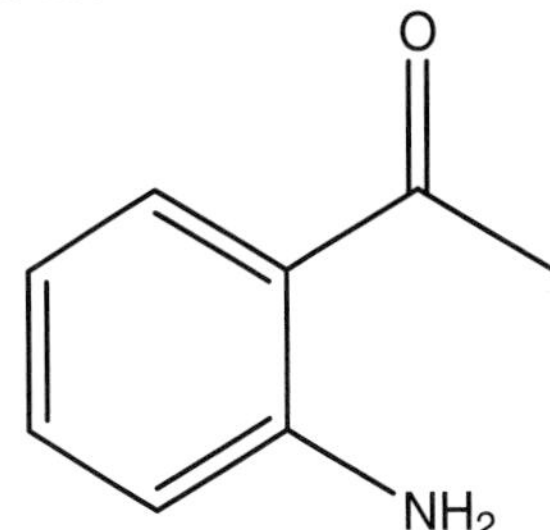

Chemical Structure

Synonyms: ACETOPHENONE,2'-AMINO-; 1-ACETYL-2-AMINOBENZENE; 2-ACETYLANILINE; O-ACETYLANILINE; 2'-AMINOACETOPHENONE; O-AMINOACETYLBENZENE; ETHANONE,1-(2-AMINOPHENYL)-
Description: yellow oily liquid or yellow crystals

Physical Properties

Boiling Point: 250 °C (482 °F) to 252 °C (486 °F) at 760 mm Hg
Freezing Point: 20 °C (68 °F)
Water Solubility: Practically Insoluble in Water
Other Solubilities: Soluble in Ether.
Refraction Index: 1.6160 at 20 °C/D

Hazard Overviews

Health: Irritating to eyes/skin/respiratory tract. Harmful. Other Acute Effects: harmful if swallowed, inhaled, or absorbed through skin; absorption into the body leads to the formation of methemoglobin which in sufficient concentration causes cyanosis; onset may be delayed 2 to 4 hours or longer.

Fire: Hazards: emits toxic fumes. Extinguishing agents: water spray; carbon dioxide, dry chemical powder or appropriate foam. Precautions: combustible liquid.

Reactivity: Incompatible with: acids, acid chlorides, acid anhydrides, chloroformates, strong oxidizing agents. Hazardous decomposition products: toxic fumes of: carbon monoxide, carbon dioxide, nitrogen oxides.

Carcinogenicity: IARC - Not listed; NIOSH - Not listed; NTP - Not listed; ACGIH - Not listed; OSHA - Not listed; EPA - Not listed; MAK - Not listed

Primary Target Organs:

Eyes

Skin

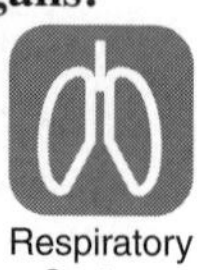
Respiratory System

Environmental

Regulations

RCRA 40CFR: Not listed
CERCLA: 40CFR 302.4: Not listed
SARA 40CFR 372.65: Not listed
SARA EHS 40CFR 355: Not listed
TSCA: Listed

AMI1080 CAS #: 134-50-9

9-AMINOACRIDINE HYDROCHLORIDE

RTECS: AR7350000
EINECS Number: 205-145-5
Molecular Formula: $C_{13}H_{11}ClN_2$
Formula Weight: 230.71

Synonyms: ACRAMINE YELLOW; 9-ACRIDINAMINE,MONOHYDROCHLORIDE; 9-ACRIDINAMINE,MONOHYDROCHLORIDE (9CI); ACRIDINE,9-AMINO,HYDROCHLORIDE; ACRIDINE,9-AMINO-,MONOHYDROCHLORIDE; AMINACRINE HYDROCHLORIDE; 5-AMINOACRIDINE HYDROCHLORIDE; AMINOACRIDINE HYDROCHLORIDE; 9-AMINOACRIDINE MONOHYDROCHLORIDE; MONACRIN; MONACRIN HYDROCHLORIDE; NSC-7571

Description: pale yellow crystals; odorless
Use: anti-bacterial, anti-infective

Physical Properties

Water Solubility: 1 g Soluble in 300 ml Water
Other Solubilities: 95% Ethanol: <1 mg/ml at 22 °C; Acetone: <1 mg/ml at 22 °C; DMSO: 5-10 mg/ml at 22 °C.
pH: 0.2% solution 5 to 6.5
Flash Point: Not available; probably combustible

RTECS Toxicity Data

Acute Oral: Mouse LD_{50} Dose: 78 mg/kg.

Mutagenic: Mammal DNA Adduct; Cell Type: lymphocyte; Dose: 10 pph. Hamster Cytogenetic Analysis; Cell Type: lung; Dose: 600 ug/L.

Hazard Overviews

Fire: Will burn.

Carcinogenicity: IARC - Not listed; NIOSH - Not listed; NTP - Not listed; ACGIH - Not listed; OSHA - Not listed; EPA - Not listed; MAK - Not listed

Environmental

Regulations

RCRA 40CFR: Not listed
CERCLA: 40CFR 302.4: Not listed
SARA 40CFR 372.65: Not listed
SARA EHS 40CFR 355: Not listed
TSCA: Listed

AMI1160 CAS #: 82-45-1

1-AMINOANTHRAQUINONE

RTECS: CB5075000
EINECS Number: 201-423-5
Molecular Formula: $C_{14}H_9NO_2$
Formula Weight: 223.24

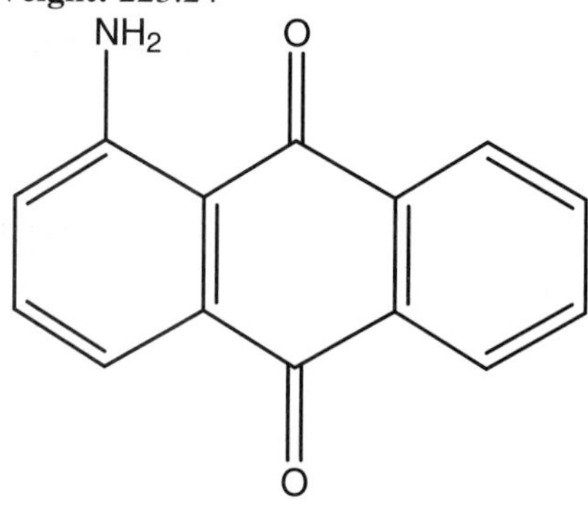

Chemical Structure

Synonyms: 1-AMINO-9,10-ANTHRACENEDIONE; 1-AMINOANTHRACHINON; 1-AMINO-9,10-ANTHRAQUINONE; ALPHA-AMINOANTHRAQUINONE; 9,10-ANTHRACENEDIONE,1-AMINO-; ALPHA-ANTHRAQUINONYLAMINE; C.I. 37275; DIAZO FAST RED AL

Physical Properties

Freezing Point: 254 °C (489.2 °F)
Water Solubility: Insoluble
Other Solubilities: Acetone: Very Soluble; Benzene: Very Soluble; Ethanol: Very Soluble; Chloroform: Very Soluble

RTECS Toxicity Data

Acute Oral: Mouse LD; Dose: >10 gm/kg.

Chronic (Multiple Dose) Oral: Rat Dose: 14 gm/kg/2W-I; Toxic Effects: Behavioral - Excitment; Kidney, Ureter, and Bladder - Proteinuria; Nutritional and gross metabolic - Weight loss or decreased weight gain. Guinea Pig Dose: 14 gm/kg/2W-I; Toxic Effects: Blood - Normocytic anemia; Blood - Changes in erythrocite (RBC) cell count; Nutritional and gross metabolic - Weight loss or decreased weight gain.

Chronic (Multiple Dose) Inhalation: Rat Dose: 10 mg/m^3/17W-I; Toxic Effects: Behavioral - Excitment; Blood - Changes in serum composition; Nutritional and gross metabolic - Weight loss or decreased weight gain.
Irritation Eye: Rabbit Standard Draize Test Dose: 500 mg/24H; Reaction: mild.
Mutagenic: Mouse DNA Damage; Route: Intraperitoneal; Dose: 250 mg/kg.
Tumorigenic: Rat Route: Oral; Dose: 2400 mg/kg/60W-I; Toxic Effects: Tumorigenic - Equivocal tumorigenic agent by RTECS criteria; Gastrointestinal - Tumors; Peripheral Nerve and sensation - Structural change in nerve or sheath. Rat Route: Oral; Dose: 3000 mg/kg/60W-I; Toxic Effects: Tumorigenic - Equivocal tumorigenic agent by RTECS criteria; Skin and appendages - Tumors.

Hazard Overviews

Health: Irritating to eyes/skin/respiratory tract. Other Acute Effects: may be harmful by inhalation, ingestion, or skin absorption.
Fire: Hazards: emits toxic fumes. Extinguishing agents: water spray; carbon dioxide, dry chemical powder or appropriate foam. Precautions: combustible liquid.
Reactivity: Incompatible with: acids, acid chlorides, acid anhydrides, chloroformates, strong oxidizing agents. Hazardous decomposition products: toxic fumes of: carbon monoxide, carbon dioxide, nitrogen oxides.
Carcinogenicity: IARC - Not listed; NIOSH - Not listed; NTP - Not listed; ACGIH - Not listed; OSHA - Not listed; EPA - Not listed; MAK - Not listed
Primary Target Organs:

Eyes

Skin

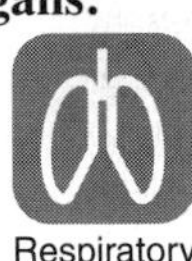
Respiratory System

Environmental

Regulations

RCRA 40CFR: Not listed
CERCLA: 40CFR 302.4: Not listed
SARA 40CFR 372.65: Not listed
SARA EHS 40CFR 355: Not listed
TSCA: Listed

AMI1240 ***CAS #: 117-79-3***

2-AMINOANTHRAQUINONE

RTECS: CB5120000
EINECS Number: 204-208-4
Molecular Formula: $C_{14}H_9NO_2$
Formula Weight: 223.24

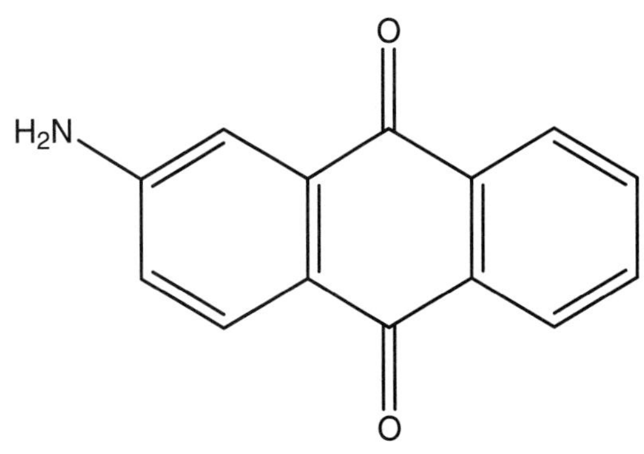

Chemical Structure

Synonyms: AAQ; 2-AMINO-9,10-ANTHRACEMEDIONE; 2-AMINO-9,10-ANTHRACENEDIONE; 2-AMINO-9,10-ANTHRAQUINONE; AMINOANTHRAQUINONE; BETA-AMINOANTHRAQUINONE; 9,10-ANTHRACENEDIONE,2-AMINO-; 9,10-ANTHRACENEDIONE,2-AMINO-(9CI); ANTHRAQUINONE,2-AMINO-; BETA-ANTHRAQUINONYLAMINE
Description: red or orange-brown needles
Use: dye and pharmaceutical intermediate

Physical Properties

Boiling Point: Sublimes
Freezing Point: 302 °C (575.6 °F)
Water Solubility: Insoluble in Water
Other Solubilities: 95% Ethanol: <1 mg/ml at 20 °C; Acetone: <1 mg/ml at 20 °C; Alcohol: Soluble; Benzene: Soluble; Chloroform: Soluble; DMSO: 10-50 mg/ml at 20 °C; Ether: Insoluble; P Toluene: <1 mg/ml at 20 °C; P Methanol: <1 mg/ml at 18 °C.
Flash Point: Not available; probably combustible

RTECS Toxicity Data

Chronic (Multiple Dose) Oral: Rat Dose: 336 gm/kg/8W-C; Toxic Effects: Kidney, Ureter, and Bladder - Urine volume increased; Kidney, Ureter, and Bladder - Proteinuria. Rat Dose: 77 gm/kg/48W-I; Toxic Effects: Kidney, Ureter, and Bladder - Chgs in tubules (inc acute renal failure, acute tubular necrosis; Endocrine - Hyperglycemia; Blood - Pigmented or nucleated red Blood cells.
Mutagenic: Mouse Micronucleus Test; Cell Type: other cell types; Dose: 6 mg/L. Bacteria - E Coli Mutations in Microorganisms; Dose: 1 mg/plate (+S9). Bacteria - S Typhimurium Mutations in Microorganisms; Dose: 1 mg/plate (+S9), 1 ug/plate (-S9).
Tumorigenic: Rat Route: Oral; Dose: 115 gm/kg/78W-C; Toxic Effects: Tumorigenic - Carcinogenic by RTECS criteria; Liver - Tumors. Rat Route: Oral; Dose: 225 gm/kg/78W-C; Toxic Effects: Tumorigenic - Neoplastic by RTECS criteria; Liver - Tumors. Rat Route: Oral; Dose: 32 gm/kg/77W-C; Toxic Effects: Tumorigenic - Neoplastic by RTECS criteria; Liver - Tumors. Rat Route: Oral; Dose: 1890 gm/kg/78W-C; Toxic Effects: Tumorigenic - Carcinogenic by RTECS criteria; Liver - Tumors. Rat Route: Oral; Dose: 3780 gm/kg/78W-C; Toxic Effects: Tumorigenic - Carcinogenic by RTECS criteria; Liver - Tumors.

Hazard Overviews

Health: Irritating to eyes/skin/respiratory tract. Toxic. Other Acute Effects: harmful if swallowed, inhaled, or absorbed through skin. Chronic Effects: may alter genetic material. Carcinogen.

Fire: Will burn. Hazards: emits toxic fumes. Extinguishing agents: water spray; carbon dioxide, dry chemical powder or appropriate foam. Precautions: combustible liquid.

Reactivity: Incompatible with: strong oxidizing agents. Hazardous decomposition products: toxic fumes of: carbon monoxide, carbon dioxide, nitrogen oxides.

Carcinogenicity: IARC - Group 3, Not classifiable as to carcinogenicity to humans; NIOSH - Not listed; NTP - Listed; ACGIH - Not listed; OSHA - Not listed; EPA - Not listed; MAK - Not listed

Primary Target Organs:

Eyes

Skin

Respiratory System

Environmental

Regulations

RCRA 40CFR: Not listed
CERCLA: 40CFR 302.4: Not listed
SARA 40CFR 372.65: Listed
SARA EHS 40CFR 355: Not listed
TSCA: Listed

Analytical Methods

Soil: SW846 8270C
Plasma: EPA 29

AMI1400 CAS #: 60-09-3

P-AMINOAZOBENZENE

RTECS: BY8225000
EINECS Number: 200-453-6
Molecular Formula: $C_{12}H_{11}N_3$
Structured MF: $H_2NC_6H_4N{=}NC_6H_5$
Formula Weight: 197.23

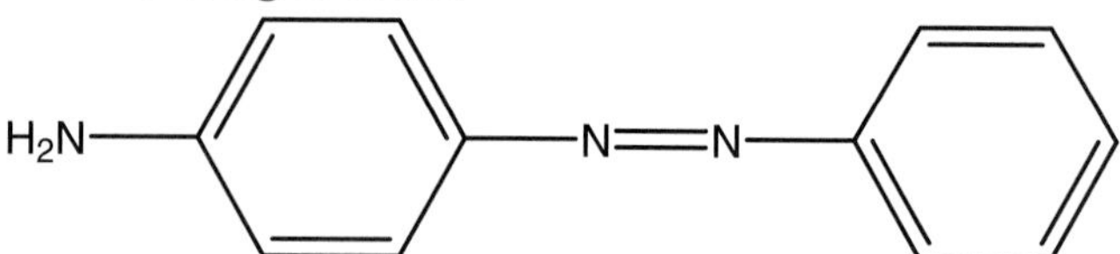

Chemical Structure

Synonyms: AAB; 4-AMINO-1,1'-AZOBENZENE; 4-AMINOAZOBENZENE; AMINOAZOBENZENE; AMINOAZOBENZENE (INDICATOR); 4-AMINOAZOBENZOL; P-AMINOAZOBENZOL; P-AMINODIPHENYLIMIDE; ANILINE YELLOW; ANILINE,P-(PHENYLAZO)-; AZOBENZENE,4-AMINO-; BENZENAMINE,4-(PHENYLAZO)-; BENZENAMINE,4-(PHENYLAZO)-(9CI); 4-BENZENEAZOANILINE; BRASILAZINA OIL YELLOW G; C.I. 11000; C.I. SOLVENT BLUE 7; C.I. SOLVENT YELLOW 1; CELLITAZOL R; CERES YELLOW R; FAST SPIRIT YELLOW; FAST SPIRIT YELLOW AAB; FAT YELLOW AAB; INDULINE R; OIL SOLUBLE ANILINE YELLOW; OIL YELLOW 2G; OIL YELLOW AAB; OIL YELLOW AB; OIL YELLOW AN; OIL YELLOW B; OIL YELLOW R; OIL-SOL. ANILINE YELLOW; ORGANOL YELLOW; ORGANOL YELLOW 2A; PARAPHENOLAZO ANILINE; 4-(PHENYLAZO)ANILINE; P-(PHENYLAZO)ANILINE; 4-(PHENYLAZO)BENZENAMINE; P-PHENYLAZONANILINE; P-PHENYLAZOPHENYLAMINE; SOLVENT YELLOW 1; SOMALIA YELLOW 2G; STEARIX BROWN 4R; SUDAN YELLOW R; SUDAN YELLOW RA; ZLUT ANILINOVA; ZLUT ROZPOUSTEDLOVA 1

Description: brownish-yellow needles with bluish cast; odorless

Use: as a dye for lacquer, varnish, wax products, oil stains and styrene resins; in insecticides; as an intermediate in the manufacture of acid yellow, diazo dyes and indulines

Physical Properties

Boiling Point: > 360 °C (680 °F)
Freezing Point: 128 °C (262.4 °F)
Specific Gravity: 1.05
Water Solubility: Slightly Soluble in Water
Other Solubilities: 95% Ethanol: 1-10 mg/ml at 19 °C; Acetone: 10-50 mg/ml at 19 °C; Alcohol: Soluble; Benzene: Soluble; Chloroform: Soluble; DMSO: 10-50 mg/ml at 19 °C; Ether: Soluble.
Flash Point: Not available; probably combustible

RTECS Toxicity Data

Mutagenic: Rat DNA Damage; Route: Intraperitoneal; Dose: 36 mg/kg. Rat Unscheduled DNA Synthesis; Cell Type: liver; Dose: 5 umol/L.

Tumorigenic: Rat Route: Oral; Dose: 89 gm/kg/57W-C; Toxic Effects: Tumorigenic - Equivocal tumorigenic agent by RTECS criteria; Liver - Tumors. Rat Route: Skin; Dose: 1965 mg/kg/2Y-I; Toxic Effects: Tumorigenic - Neoplastic by RTECS criteria; Skin and appendages - Tumors.

Hazard Overviews

Health: Irritating to eyes/skin/respiratory tract. Toxic. Other Acute Effects: harmful if swallowed, inhaled, or absorbed through skin. Chronic Effects: carcinogen; may alter genetic material; target organ: liver.

Fire: Will burn. Hazards: emits toxic fumes. Extinguishing agents: water spray; carbon dioxide, dry chemical powder or appropriate foam. Precautions: combustible liquid.

Reactivity: Incompatible with: strong oxidizing agents. Hazardous decomposition products: toxic fumes of: carbon monoxide, carbon dioxide, nitrogen oxides.

Carcinogenicity: IARC - Group 2B, Possibly carcinogenic to humans; NIOSH - Not listed; NTP - Not listed; ACGIH - Not listed; OSHA - Not listed; EPA - Not listed; MAK - Not listed

Primary Target Organs:

Eyes

Skin

Respiratory System

Liver

Environmental

Environmental Fate: If released on land, it should bind strongly to soil and undergo soil- or clay-catalyzed oxidation and possibly biodegrade. If released in water, it should bind strongly to sediment and undergo soil- or clay-catalyzed oxidation, photolyze, or biodegrade. It should not bioconcentrate appreciably in aquatic organisms. However, because of the lack of experimental data, the fate in natural waters is unknown. In the atmosphere, it should exist primarily adsorbed to particulate matter and in aerosols and be subject to gravitational settling. Vapor phase would degrade by reacting with photochemically produced hydroxyl radicals (half-life 5.8 hr).

Environmental Physical Data

Henry's Law Constant: estimated at 8.68×10^{-11}
Octanol/Water Partition Coefficient: log K_{ow} = estimated at 2.623
Sorption Partition Coefficient: K_{oc} = 624
BCF: calculated at 58

Regulations

RCRA 40CFR: Not listed
CERCLA: 40CFR 302.4: Not listed
SARA 40CFR 372.65: Listed
SARA EHS 40CFR 355: Not listed
TSCA: Listed

Analytical Methods

Soil: SW846 8270C
Plasma: EPA 29

AMI1480 CAS #: 97-56-3

O-AMINOAZOTOLUENE

RTECS: XU8800000
EINECS Number: 202-591-2
Molecular Formula: $C_{14}H_{15}N_3$
Structured MF: $CH_3C_6H_4N{=}NC_6H_3(NH_2)CH_3$
Formula Weight: 225.28

N=N NH2

Chemical Structure

Synonyms: AAT; O-AAT; O-AMIDOAZOTOLUOL; 2-AMINO-5-AZOTOLUENE; 4'-AMINO-2,3'-AZOTOLUENE; 4'-AMINO-2:3'-AZOTOLUENE; AMINOAZOTOLUENE; AMINOAZOTOLUENE (INDICATOR); O-AMINOAZOTOLUENO; O-AMINOAZOTOLUOL; 4'-AMINO-2,3'-DIMETHYLAZOBENZENE; 4-AMINO-2',3-DIMETHYLAZOBENZENE; O-AT; BENZENAMINE,2-METHYL-4-((2-METHYLPHENYL)AZO)-; BRASILAZINA OIL YELLOW R; BUTTER YELLOW; C.I. 11160B; C.I. 11160; C.I. SOLVENT YELLOW 3; 2',3-DIMETHYL-4-AMINOAZOBENZENE; FAST GARNET GBC BASE; FAST OIL YELLOW; FAST YELLOW AT; FAT YELLOW B; HIDACO OIL YELLOW; 2-METHYL-4-((2-METHYLPHENYL)AZO)BENZENAMINE; 2-METHYL-4-((O-TOLYL)AZO)ANILINE; OAAT; OIL YELLOW; OIL YELLOW 2R; OIL YELLOW 21; OIL YELLOW 2681; OIL YELLOW A; OIL YELLOW AT; OIL YELLOW C; OIL YELLOW I; OIL YELLOW T; ORGANOL YELLOW 2T; ORTHOAMINOASOTOLUOL; SOLVENT YELLOW 3; SOMALIA YELLOW R; SUDAN YELLOW RRA; TOLUAZOTOLUIDINE; O-TOLUENEAZO-O-TOLUIDINE; ORTHO-TOLUENEAZO-ORTHO-TOLUIDINE; O-TOLUOL-AZO-O-TOLUIDIN; ORTHO-TOLUOL-AZO-ORTHO-TOLUIDIN; 5-(O-TOLYLAZO)-2-AMINOTOLUENE; 4-(O-TOLYLAZO)-O-TOLUIDINE; TULABASE FAST GARNET GB; TULABASE FAST GARNET GBC; WAXAKOL YELLOW NL; ZLUT ROZPOUSTEDLOVA 3
Description: yellow to reddish-brown crystals
Use: dyes; used to color oils, fats, waxes, greases, solvents, shoe polishes and other wax polishes; medicine; and immunosuppressant

Physical Properties

Boiling Point: Sublimes at > 150 °C (302 °F)
Freezing Point: 101 °C (213.8 °F) to 102 °C (215.6 °F)
Specific Gravity: 0.57
Saturated Vapor Density: 1.200000008 kg/m^3
Vapor Pressure: 7.5×10^{-7} mm Hg at 25 °C
Water Solubility: < 1 mg/mL at 21 C
Other Solubilities: Soluble in oils and fats; Soluble in Acetone, Cellosolve & Toluene.
Flash Point: Not available; probably combustible

RTECS Toxicity Data

Acute Oral: Rat LD_{Lo} Dose: 1500 mg/kg. Mouse LD_{Lo} Dose: 800 mg/kg.
Acute Dermal: Mouse LD_{Lo} Route: Subcutaneous Dose: 1200 mg/kg.
Reproductive/Teratogenic: Mouse Route: Oral; Dose: 480 mg/kg; Duration: female multigeneration; Specific Developmental Abnormalities - Hepatobiliary system; Tumorigenic Effects - Transplacental tumorigenesis.
Mutagenic: Rat Unscheduled DNA Synthesis; Cell Type: liver; Dose: 200 nmol/L. Rat Unscheduled DNA Synthesis; Route: Oral; Dose: 200 mg/kg.
Tumorigenic: Rat Route: Oral; Dose: 15 gm/kg/57W-C; Toxic Effects: Tumorigenic - Neoplastic by RTECS criteria; Liver - Tumors. Rat Route: Oral; Dose: 31250 mg/kg/35W-C; Toxic Effects: Tumorigenic - Equivocal tumorigenic agent by RTECS criteria; Liver - Tumors.

Hazard Overviews

Fire Diamond

Health: Irritating to eyes/skin/respiratory tract. Also Causes: allergic reactions resulting in eczema, generally of hands/arms. Chronic Effects: liver cancer has occurred in animals.
Fire: Noncombustible. Use agent suitable for surrounding fire.
Reactivity: Stable. Hazardous polymerization cannot occur.
Carcinogenicity: IARC - Group 2B, Possibly carcinogenic to humans; NIOSH - Not listed; NTP - Class 2B, Reasonably anticipated to be a carcinogen, sufficient evidence of carcinogenicity from studies in experimental animals; ACGIH - Not listed; OSHA - Not listed; EPA - Not listed; MAK - Class A2, Unmistakably carcinogenic in animal experimentation only
Primary Target Organs:

Eyes

Skin

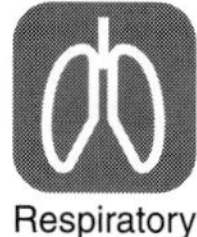
Respiratory System

Environmental

Environmental Fate: If released to soil, it is expected to display only limited mobility. Its amino ground may bind covalently with active sites in soil further limiting its mobility. It may only slowly volatilize from both moist and dry soils to the atmosphere. If released to water, it may bioconcentration in fish and aquatic organisms and it is expected to adsorb to sediment and suspended organic matter. It is not expected to volatilize from water to the atmosphere. The estimated half-life for volatilization from a model river is 1888 days. It absorbs UV light and it may undergo direct photochemical degradation in the upper layers of clear water. If released to the atmosphere, it may undergo a rapid gas-phase reaction with photochemically produced hydroxyl radicals with an estimated half-life of 2.7 hours. It absorbs UV light indicating that it has the potential to undergo direct photochemical degradation in the atmosphere.

Environmental Physical Data

Henry's Law Constant: estimated at 2.91×10^{-8}
Octanol/Water Partition Coefficient: log K_{ow} = 3.921
Sorption Partition Coefficient: K_{oc} = 1426 to 3236
BCF: calculated at 196 to 562

Regulations

RCRA 40CFR: Not listed
CERCLA: 40CFR 302.4: Not listed
SARA 40CFR 372.65: Listed
SARA EHS 40CFR 355: Not listed
TSCA: Listed

AMI1560 CAS #: 88-68-6

2-AMINOBENZAMIDE

RTECS: CU8993000
EINECS Number: 201-851-2
Molecular Formula: $C_7H_8N_2O$
Formula Weight: 136.2

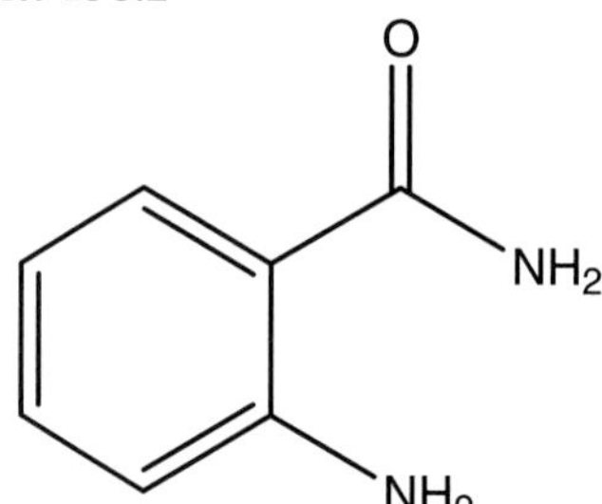

Chemical Structure

Synonyms: O-AMINOBENZAMIDE; ANTHRANILAMIDE; ANTHRANILIMIDIC ACID; BENZAMIDE,2-AMINO-; BENZAMIDE,O-AMINO-; BENZAMIDE,2-AMINO-(9CI); BENZOIC ACID,2-AMINO-,AMIDE; 2-CARBAMOYLANILINE
Description: leaflets

Physical Properties

Boiling Point: 300 °C (572 °F)
Freezing Point: 110 °C (230 °F) to 115 °C (239 °F)
Water Solubility: Soluble in Hot Water
Other Solubilities: Slightly Soluble in Benzene; Very Soluble in Ethyl Acetate.

Hazard Overviews

Health: Irritating to eyes/skin/respiratory tract. Other Acute Effects: may be harmful by inhalation, ingestion, or skin absorption.
Fire: Hazards: emits toxic fumes. Extinguishing agents: water spray; carbon dioxide, dry chemical powder or appropriate foam. Precautions: combustible liquid.
Reactivity: Incompatible with: strong oxidizing agents. Hazardous decomposition products: toxic fumes of: carbon monoxide, carbon dioxide, nitrogen oxides.
Carcinogenicity: IARC - Not listed; NIOSH - Not listed; NTP - Not listed; ACGIH - Not listed; OSHA - Not listed; EPA - Not listed; MAK - Not listed
Primary Target Organs:

Eyes

Skin

Respiratory System

Environmental

Regulations

RCRA 40CFR: Not listed
CERCLA: 40CFR 302.4: Not listed
SARA 40CFR 372.65: Not listed

SARA EHS 40CFR 355: Not listed
TSCA: Listed

AMI1640 CAS #: 137-07-5

2-AMINOBENZENETHIOL

RTECS: DC0600000
EINECS Number: 205-277-3
Molecular Formula: C_6H_7NS
Formula Weight: 125.20

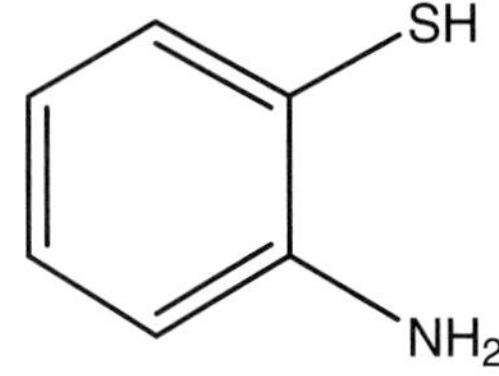

Chemical Structure

Synonyms: 2-AMINOTHIOPHENOL; O-AMINOTHIOPHENOL; O-MERCAPTOANILINE

Physical Properties

Boiling Point: 234 °C (453.2 °F)
Freezing Point: 19 °C (66.2 °F)
Specific Gravity: 1.168
Water Solubility: Insoluble
Other Solubilities: Ethanol: Soluble; Ether: Soluble
Refraction Index: 1.4606
Ionization Potential (eV): 7.6

RTECS Toxicity Data

Acute Oral: Rat LD_{Lo} Dose: 500 mg/kg; Toxic Effects: Blood - Methemoglobinemia-Carboxyhemoglobin.

Hazard Overviews

Corrosive

Health: Corrosive to eyes/skin/respiratory tract. Harmful. Other Acute Effects: harmful if swallowed, inhaled, or absorbed through skin; material is also extremely destructive to tissue of the mucous membranes; inhalation may result in spasm, inflammation and edema of the larynx and bronchi; chemical pneumonitis; pulmonary edema; exposure may cause burning sensation; coughing; wheezing; laryngitis; shortness of breath; headache; nausea; vomiting; absorption into the body leads to the formation of methemoglobin which in sufficient concentration causes cyanosis; onset may be delayed 2 to 4 hours or longer. Chronic Effects: coughing, chest pains; difficulty in breathing; lung irritation pulmonary edema; effects may be delayed.

Fire: Hazards: emits toxic fumes. Extinguishing agents: water spray; carbon dioxide, dry chemical powder or appropriate foam. Precautions: combustible liquid.

Reactivity: Incompatible with: acids, acid chlorides, acid anhydrides, chloroformates, strong oxidizing agents. Hazardous decomposition products: toxic fumes of: carbon monoxide, carbon dioxide, nitrogen oxides, sulfur oxides.

Carcinogenicity: IARC - Not listed; NIOSH - Not listed; NTP - Not listed; ACGIH - Not listed; OSHA - Not listed; EPA - Not listed; MAK - Not listed

Primary Target Organs:

Eyes

Skin

Respiratory System

Environmental

Regulations

RCRA 40CFR: Not listed
CERCLA: 40CFR 302.4: Not listed
SARA 40CFR 372.65: Not listed
SARA EHS 40CFR 355: Not listed
TSCA: Listed

AMI1720 CAS #: 934-32-7

2-AMINOBENZIMIDAZOLE

RTECS: DD5775000
EINECS Number: 213-280-6
Molecular Formula: $C_7H_7N_3$
Formula Weight: 133.17

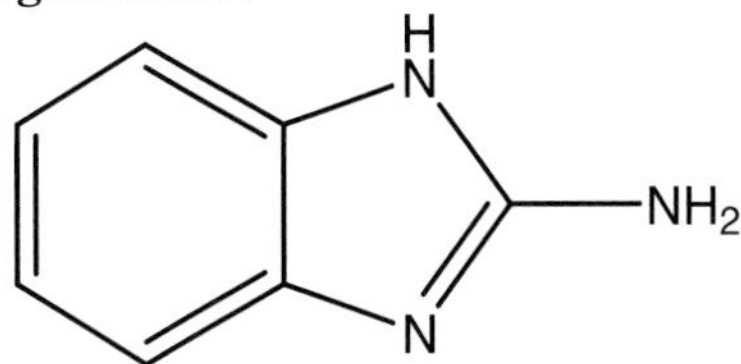

Chemical Structure

Synonyms: 2-AB; 2-AMINO-1H-BENZIMIDAZOLE; 1H-BENZIMIDAZOL-2-AMINE; 2-BENZIMIDAZOLAMINE; BENZIMIDAZOLE,2-AMINO-; 2-IMINOBENZIMIDAZOLINE
Description: white powder or plates
Use: in the photography industry as an antifoggant

Physical Properties

Freezing Point: 224 °C (435.2 °F)
Water Solubility: < 1 mg/mL at 20 C
Other Solubilities: 95% Ethanol: >=100 mg/ml at 16.5 °C; Acetone: 10-50 mg/ml at 16.5 °C; Alcohol: Soluble; Alkalies: Soluble; DMSO: >=100 mg/ml at 16.5 °C; Ether: Very slightly Soluble.
Flash Point: Not available; probably combustible

RTECS Toxicity Data

Acute Oral: Rat LD_{Lo} Dose: 500 mg/kg. Mouse LD_{40} Dose: 600 mg/kg.
Reproductive/Teratogenic: Rat Route: Oral; Dose: 426 mg/kg; Duration: female 8-15D of pregnancy; Effects on Embryo or Fetus - Fetal death.
Mutagenic: Bacteria - E Coli DNA Repair; Dose: 2 mg/disc. Bacteria - S Typhimurium Mutations in Microorganisms; Dose: 100 ug/plate (+S9). Bacteria - S Typhimurium Mutations in Microorganisms; Dose: 710 umol/L (-S9).

Hazard Overviews

Health: Irritating to eyes/skin/respiratory tract. Harmful. Other Acute Effects: harmful if swallowed; may be harmful if inhaled; may be harmful if absorbed through the skin.
Fire: Will burn. Hazards: emits toxic fumes. Extinguishing agents: water spray; carbon dioxide, dry chemical powder or appropriate foam. Precautions: combustible liquid.
Carcinogenicity: IARC - Not listed; NIOSH - Not listed; NTP - Not listed; ACGIH - Not listed; OSHA - Not listed; EPA - Not listed; MAK - Not listed
Primary Target Organs:

Eyes

Skin

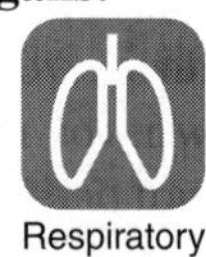
Respiratory System

Environmental

Regulations

RCRA 40CFR: Not listed
CERCLA: 40CFR 302.4: Not listed
SARA 40CFR 372.65: Not listed
SARA EHS 40CFR 355: Not listed
TSCA: Listed

AMI1800 CAS #: 150-13-0

4-AMINOBENZOIC ACID

RTECS: DG1400000
EINECS Number: 205-753-0
Molecular Formula: $C_7H_7NO_2$
Formula Weight: 137.13

Chemical Structure

Synonyms: ACIDO P-AMINOBENZOICO; AMBEN; AMINOBENZOIC ACID; GAMMA-AMINOBENZOIC ACID; P-AMINOBENZOIC ACID; PARA-AMINOBENZOIC ACID; 1-AMINO-4-CARBOXYBENZENE; ANTICANITIC VITAMIN; ANTI-CHROMOTRICHIA FACTOR; ANTICHROMOTRICHIA FACTOR; BACTERIAL VITAMIN H1; BENZOIC ACID,4-AMINO-; 4-CARBOXYANILINE; P-CARBOXYANILINE; P-CARBOXYPHENYLAMINE; CHROMOTRICHIA FACTOR; KYSELINA P-AMINOBENZOOVA; PAB; PABA; PABACYD; PABAFILM; PABANOL; PARAMINOL; PARANATE; POTABA; ROMAVIT; RVPABA; SUNBRELLA; TRICHOCHROMOGENIC FACTOR; TRICHROMOGENIC FACTOR; VITAMIN H'; VITAMIN BX; VITAMIN H
Description: white or slightly yellow crystals or crystalline powder; odorless
Use: manufacture of various esters(local anesthetics), folic acid and azo dyes; in sunburn preventatives; used in laboratories as sulfonamide antagonist; antirickettsial

Physical Properties

Freezing Point: 187 °C (368.6 °F) to 187.5 °C (369.5 °F)
Density: 1.37 g/mL
Water Solubility: 1 g dissolves in 170 ml Water at 25 °C
Other Solubilities: Benzene: Insoluble; Chloroform: Insoluble; Very Soluble in hot solvent; Ether: Soluble; Ethyl Acetate: Soluble; Glacial Acetic Acid: Soluble; Petroleum Ether: Insoluble.
pH: 0.5% solution 3.5
Ionization Potential (eV): 7.8 +/-1.0

RTECS Toxicity Data

Acute Oral: Rat LD_{50} Dose: >6 gm/kg. Mouse LD_{50} Dose: 2850 mg/kg; Toxic Effects: Behavioral - Somnolence (general depressed activity); Behavioral - Muscle weakness.
Chronic (Multiple Dose) Oral: Mammal Dose: 182 mg/kg/26W-I; Toxic Effects: Liver - Other changes; Blood - Changes in erythrocite (RBC) cell count; Blood - Changes in leukocyte (WBC) cell count.
Reproductive/Teratogenic: Rat Route: Oral; Dose: 2500 mg/kg; Duration: female 1-22D of pregnancy; Effects on Fertility - Post-implantation mortality.
Mutagenic: Mouse DNA Damage; Route: Intraperitoneal; Dose: 1 gm/kg.

Hazard Overviews

Health: Irritating to eyes/skin/respiratory tract. Harmful. Other Acute Effects: harmful if swallowed; may be harmful if inhaled, or absorbed through the skin; may cause allergic skin reaction; sensitization by skin contact.
Fire: Hazards: emits toxic fumes. Extinguishing agents: water spray; carbon dioxide, dry chemical powder or appropriate foam. Precautions: combustible liquid.
Reactivity: Incompatible with: strong oxidizing agents, may discolor on exposure to light, sensitive to air, sensitive to light. Hazardous decomposition products: toxic fumes of: carbon monoxide, carbon dioxide, nitrogen oxides.
Carcinogenicity: IARC - Group 3, Not classifiable as to carcinogenicity to humans; NIOSH - Not listed; NTP - Not listed; ACGIH - Not listed; OSHA - Not listed; EPA - Not listed; MAK - Not listed

Primary Target Organs:

Eyes

Skin

Respiratory System

Exposure Limits

AIHA WEEL: TWA: 5 mg/m^3.

Respirator Recommendation

Exposure Range: >5 to 50 mg/m^3 Air Purifying, Negative Pressure, Half Mask

Exposure Range: >50 to 500 mg/m^3 Air Purifying, Negative Pressure, Full Face

Exposure Range: >500 to 5000 mg/m^3 Supplied Air, Constant Flow/Pressure Demand, Full Face

Exposure Range: >5000 to unlimited mg/m^3 Self-contained Breathing Apparatus, Pressure Demand, Full Face

Cartridge Color: dust/mist filter (use P100 or consult supervisor for appropriate dust/mist filter)

Environmental

Environmental Physical Data

Octanol/Water Partition Coefficient: log K_{ow} = 0.68

Regulations

RCRA 40CFR: Not listed
CERCLA: 40CFR 302.4: Not listed
SARA 40CFR 372.65: Not listed
SARA EHS 40CFR 355: Not listed
TSCA: Listed

AMI1960 CAS #: 94-12-2

P-AMINOBENZOIC ACID, PROPYL ESTER

RTECS: DG3090000
EINECS Number: 202-306-1
Molecular Formula: $C_{10}H_{13}NO_2$
Formula Weight: 179.21

Chemical Structure

Synonyms: P-AMINOBENZOIC ACID PROPYL ESTER; BENZOIC ACID,4-AMINO-,PROPYL ESTER; BENZOIC ACID,P-AMINO-,PROPYL ESTER; KELOFORM P; NSC-23516; PROPAESIN; PROPAZYL; PROPESIN; PROPESINE; 4-(PROPOXYCARBONYL)ANILINE; PROPYL 4-AMINOBENZOATE; PROPYL P-AMINOBENZOATE; PROPYLCAIN; RAYTHESIN; RISOCAINE

Description: crystals or prisms

Use: medication: as local anesthetic, antipruritic agent

Physical Properties

Freezing Point: 75 °C (167 °F)
Water Solubility: 1.67 mmoles/L
Other Solubilities: freely Soluble in Alcohol, Benzene, Chloroform, Ether. about 7% Soluble in oils.

Hazard Overviews

Carcinogenicity: IARC - Not listed; NIOSH - Not listed; NTP - Not listed; ACGIH - Not listed; OSHA - Not listed; EPA - Not listed; MAK - Not listed

Environmental

Regulations

RCRA 40CFR: Not listed
CERCLA: 40CFR 302.4: Not listed
SARA 40CFR 372.65: Not listed
SARA EHS 40CFR 355: Not listed
TSCA: Not listed

AMI2040 CAS #: 1885-29-6

2-AMINOBENZONITRILE

RTECS: CB4575000
EINECS Number: 217-549-9
Molecular Formula: $C_7H_6N_2$
Formula Weight: 118.15

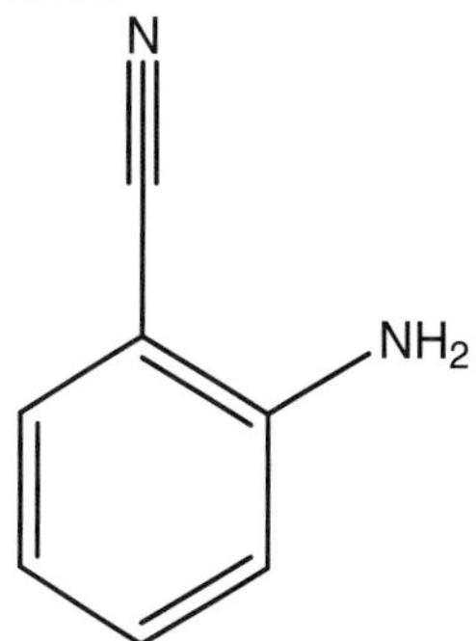

Chemical Structure

Synonyms: O-AMINOBENZONITRILE; BENZONITRILE,2-AMINO-(9CI); 2-CYANOANILINE; O-CYANOANILINE

Physical Properties

Boiling Point: 268 °C (514.4 °F)
Freezing Point: 48 °C (118.4 °F)
Water Solubility: Slightly Soluble
Other Solubilities: Ethanol: Very Soluble; Ether: Very Soluble; Acetone: Very Soluble; Benzene: Very Soluble; Chloroform: Very Soluble; Pyridine: Very Soluble

Hazard Overviews

Health: Irritating to eyes/skin/respiratory tract. Harmful. Other Acute Effects: may be harmful by inhalation, ingestion, or skin absorption.

Fire: Hazards: emits toxic fumes. Extinguishing agents: water spray; carbon dioxide, dry chemical powder or appropriate foam. Precautions: combustible liquid.

Reactivity: Incompatible with: strong acids, strong bases, strong oxidizing agents, strong reducing agents. Hazardous decomposition products: thermal decomposition may produce carbon monoxide, carbon dioxide, and nitrogen oxides.

Carcinogenicity: IARC - Not listed; NIOSH - Not listed; NTP - Not listed; ACGIH - Not listed; OSHA - Not listed; EPA - Not listed; MAK - Not listed

Primary Target Organs:

Eyes

Skin

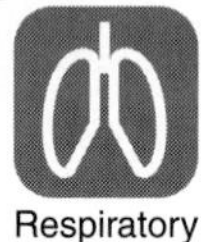
Respiratory System

Environmental

Regulations

RCRA 40CFR: Not listed
CERCLA: 40CFR 302.4: Not listed
SARA 40CFR 372.65: Not listed
SARA EHS 40CFR 355: Not listed
TSCA: Listed

AMI2120 CAS #: 2237-30-1

3-AMINOBENZONITRILE

RTECS: DI2454000
EINECS Number: 218-800-5
Molecular Formula: $C_7H_6N_2$
Formula Weight: 118.15

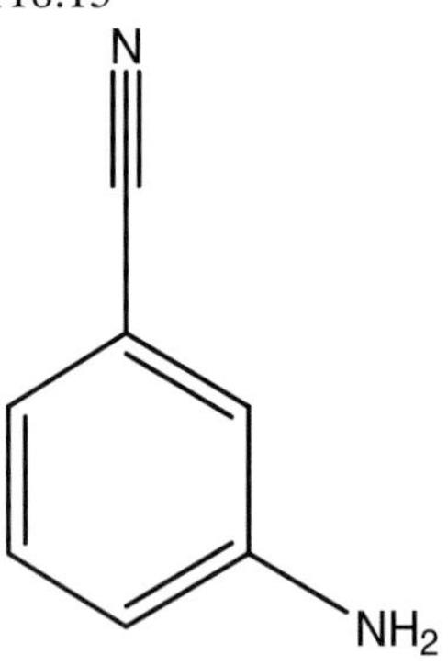

Chemical Structure

Synonyms: M-AMINOBENZONITRILE; M-ANTHRANILONITRILE; BENZONITRILE,3-AMINO-(9CI); 3-CYANOANILINE; M-CYANOANILINE

Physical Properties

Boiling Point: 289 °C (552.2 °F)
Freezing Point: 52 °C (125.6 °F)
Water Solubility: Slightly Soluble
Other Solubilities: Ethanol: Very Soluble; Ether: Very Soluble; Acetone: Very Soluble; Chloroform: Very Soluble; CS_2: Soluble
Ionization Potential (eV): 8.61 +/-0.05

RTECS Toxicity Data

Acute Oral: Quail LD_{50} Dose: 562 mg/kg.
Mutagenic: Bacteria - S Typhimurium Mutations in Microorganisms; Dose: 100 mg/L (-S9).

Hazard Overviews

Poison

Health: Irritating to eyes/skin/respiratory tract. Poison. Other Acute Effects: may be fatal if inhaled, swallowed, or absorbed through skin; absorption into the body leads to the formation of methemoglobin which in sufficient concentration causes cyanosis; onset may be delayed 2 to 4 hours or longer; target organ: blood.

Fire: Hazards: emits toxic fumes. Extinguishing agents: water spray; carbon dioxide, dry chemical powder or appropriate foam. Precautions: combustible liquid.

Reactivity: Incompatible with: acids, acid chlorides, acid anhydrides, chloroformates, strong oxidizing agents. Hazardous decomposition products: toxic fumes of: carbon monoxide, carbon dioxide, nitrogen oxides.

Carcinogenicity: IARC - Not listed; NIOSH - Not listed; NTP - Not listed; ACGIH - Not listed; OSHA - Not listed; EPA - Not listed; MAK - Not listed

Primary Target Organs:

Eyes

Skin

Respiratory System

Blood

Environmental

Regulations

RCRA 40CFR: Not listed
CERCLA: 40CFR 302.4: Not listed
SARA 40CFR 372.65: Not listed
SARA EHS 40CFR 355: Not listed
TSCA: Not listed

AMI2280 CAS #: 136-95-8

2-AMINOBENZOTHIAZOLE

RTECS: DL1050000
EINECS Number: 205-268-4
Molecular Formula: $C_7H_6N_2S$
Structured MF: $C_6H_4NC(NH_2)S$
Formula Weight: 150.20

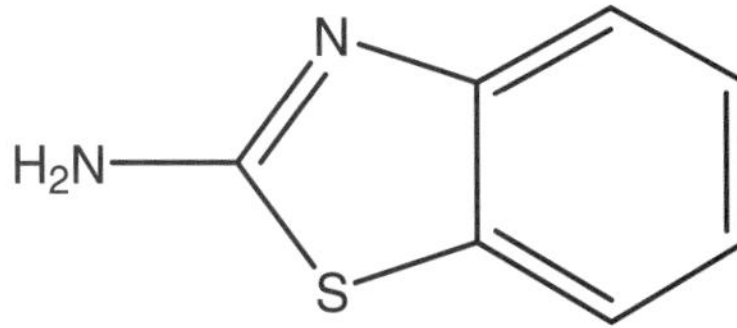

Chemical Structure

Synonyms: 2-AMINOBENZTHIAZOLE; MU-AMINOBENZTHIAZOLE; 2-BENZOTHIAZOLAMINE; BENZOTHIAZOLE,2-AMINO-; 2(3H)-BENZOTHIAZOLIMINE; 2-BENZOTHIAZOLYLAMINE; 2-IMINOBENZOTHIAZOLINE
Description: plates or leaflets; odorless
Use: as an azo dye intermediate and in photographic chemicals

Physical Properties

Boiling Point: Decomposes
Freezing Point: 132 °C (269.6 °F)
Specific Gravity: 0.5
Water Solubility: Slightly Soluble in Water
Other Solubilities: 95% Ethanol: >=100 mg/ml at 19 °C; Acetone: >=100 mg/ml at 19 °C; Alcohol: Very Soluble; Chloroform: Very Soluble; Concentrated acids: Soluble; DMSO: >=100 mg/ml at 19 °C; Ether: Very Soluble.
Flash Point: Not available; probably combustible

RTECS Toxicity Data

Acute Oral: Mouse LD_{50} Dose: >1 gm/kg.
Mutagenic: Bacteria - S Typhimurium Mutations in Microorganisms; Dose: 2500 ug/plate (-S9).

Hazard Overviews

Health: Irritating eyes/skin/respiratory tract. Other Acute Effects: may be harmful by inhalation, ingestion, or skin absorption.
Fire: Will burn. Hazards: emits toxic fumes. Extinguishing agents: water spray; carbon dioxide, dry chemical powder or appropriate foam. Precautions: combustible liquid.
Reactivity: Incompatible with: strong oxidizing agents. Hazardous decomposition products: toxic fumes of: carbon monoxide, carbon dioxide, nitrogen oxides, sulfur oxides.
Carcinogenicity: IARC - Not listed; NIOSH - Not listed; NTP - Not listed; ACGIH - Not listed; OSHA - Not listed; EPA - Not listed; MAK - Not listed

Primary Target Organs:

Eyes

Skin

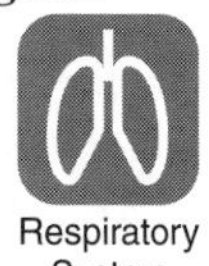
Respiratory System

Environmental

Regulations
RCRA 40CFR: Not listed
CERCLA: 40CFR 302.4: Not listed
SARA 40CFR 372.65: Not listed
SARA EHS 40CFR 355: Not listed
TSCA: Listed

AMI2360 CAS #: 92-67-1

4-AMINOBIPHENYL

RTECS: DU8925000
EINECS Number: 202-177-1
Molecular Formula: $C_{12}H_{11}N$
Structured MF: $C_6H_5C_6H_4NH_2$
Formula Weight: 169.24

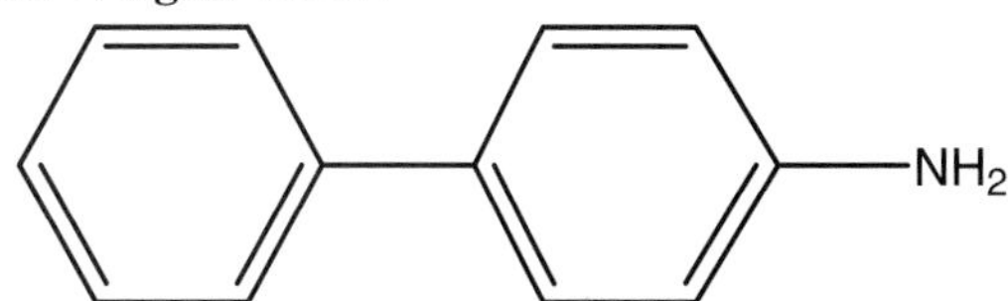

Chemical Structure

Synonyms: 4-ADP; 4-AMINOBIFENYL; P-AMINOBIPHENYL; 4-AMINODIFENIL; 4-AMINODIPHENYL; P-AMINODIPHENYL; ANILINOBENZENE; 4-BIFENYLAMIN; (1,1'-BIPHENYL)-4-AMINE; 4-BIPHENYLAMINE; BIPHENYLAMINE; P-BIPHENYLAMINE; BPA; P-DIPHENYLAMINE; PARAAMINODIPHENYL; 4-PHENYLANILINE; P-PHENYLANILINE; XENYLAMIN; P-XENYLAMINE; XENYLAMINE
Description: colorless, purple crystals; floral odor
Use: in organic research, in the detection of sulfates and as a carcinogen in cancer research; formerly used as a dyestuffs intermediate and as a raw material in the manufacture of rubber antioxidants

Physical Properties

Boiling Point: 302 °C (576 °F)
Freezing Point: 53 °C (127.4 °F)
Specific Gravity: 1.16 at 20 °C/20 °C
Vapor Density: 5.8 Air=1 at boiling point
Saturated Vapor Density: 1.200000458 kg/m^3
Vapor Pressure: 6×10^{-5} torr
Water Solubility: Slight
Other Solubilities: 95% Ethanol: 1-5 mg/ml at 22 °C; Acetone: 50-100 mg/ml at 22 °C; Alcohol: Soluble; Chloroform: Soluble; DMSO: 10-50 mg/ml at 22 °C; Ether: Soluble; Lipids: Soluble; Non-polar solvents: Soluble.
Flash Point: > 112 °C
Autoignition Temperature: 450 °C

RTECS Toxicity Data

Acute Oral: Rat LD_{50} Dose: 500 mg/kg; Toxic Effects: Behavioral - Coma; Lungs, Thorax, or Respiration - Dyspnea; Nutritional and gross metabolic - Weight loss or decreased weight gain. Mouse LD_{50} Dose: 205 mg/kg.

Mutagenic: Human Unscheduled DNA Synthesis; Cell Type: liver; Dose: 100 ug/L. Human Unscheduled DNA Synthesis; Cell Type: fibroblast; Dose: 800 ug/L. Human Mutations in Mammalian Somatic Cells; Cell Type: fibroblast; Dose: 60 mg/L.

Tumorigenic: Rat Route: Oral; Dose: 5 gm/kg/52W-C; Toxic Effects: Tumorigenic - Equivocal tumorigenic agent by RTECS criteria; Skin and appendages - Tumors. Rat Route: Subcutaneous; Dose: 4560 mg/kg/44W-I; Toxic Effects: Tumorigenic - Equivocal tumorigenic agent by RTECS criteria; Liver - Tumors; Kidney, Ureter, and Bladder - Kidney tumors. Rat Route: Subcutaneous; Dose: 5000 mg/kg/W-I; Toxic Effects: Tumorigenic - Equivocal tumorigenic agent by RTECS criteria; Gastrointestinal - Tumors; Liver - Tumors.

Hazard Overviews

Fire Diamond

Health: Toxic. Also Causes: headaches, lethargy, cyanosis, urinary burning, hematuria, methemoglobinemia (upon ingestion). Chronic Effects: known human bladder carcinogen.

Fire: Will burn. Use water spray, mist, or dry chemical.

Reactivity: Stable. Hazardous polymerization cannot occur. Hazardous decomposition products: nitrogen oxides.

Carcinogenicity: IARC - Group 1, Carcinogenic to humans; NIOSH - Listed as carcinogen; NTP - Class 1, Known to be a carcinogen; ACGIH - Class A1, Confirmed human carcinogen; OSHA - Listed as a carcinogen; EPA - Not listed; MAK - Class A1, Capable of inducing malignant tumors as shown by experience with humans

Primary Target Organs:

Skin

Gastro-intestinal

Environmental

Environmental Fate: It is easily oxidizable and probably also undergoes photolysis but there is little actual data on these processes. If released on land it will adsorb moderately to soil, probably binding to humic materials and undergoing redox reactions. In water it will adsorb to sediment, and probably undergo photolysis and oxidation. Oxidation by alkoxy radicals which are photochemically produced in eutrophic waters has an estimated half-life of 14 days. It is biodegradable and biodegradation may well occur in both soil and water but there are no rates available for soil or natural waters. It has a low potential for bioconcentration. In the atmosphere, degradation should occur due to direct photolysis, oxidation by ambient oxygen, and also photochemically produced hydroxyl radicals (estimated half-life 6.9 hr in the vapor phase).

Environmental Physical Data

Octanol/Water Partition Coefficient: log K_{ow} = 2.80
Sorption Partition Coefficient: K_{oc} = 417
BCF: calculated at 1.90

Regulations

RCRA 40CFR: Not listed
CERCLA: 40CFR 302.4: Listed per CAA Section 112 RQ: 1 lb (0.454 kg)
SARA 40CFR 372.65: Listed
SARA EHS 40CFR 355: Not listed
TSCA: Listed

Analytical Methods

Soil: SW846 3640A, 8250A, 8270B, 8270C
Water / Groundwater: EPA 1625
Plasma: EPA 29

AMI2520 CAS #: 96-20-8

2-AMINO-1-BUTANOL

RTECS: EK9625000
EINECS Number: 202-488-2
Molecular Formula: $C_4H_{11}NO$
Formula Weight: 89.16

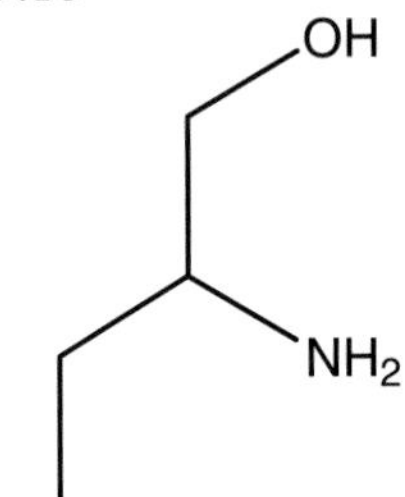

Chemical Structure

Synonyms: 2-AMINOBUTAN-1-OL; 2-AMINO-N-BUTYL ALCOHOL; BUTANOL-2-AMINE

Physical Properties

Boiling Point: 176 °C (348.8 °F)
Specific Gravity: 0.943

RTECS Toxicity Data

Acute Oral: Mouse LD_{50} Dose: 2300 mg/kg; Toxic Effects: Behavioral - Tetany.

Hazard Overviews

Corrosive

Health: Corrosive to eyes/skin/respiratory tract. Other Acute Effects: harmful if swallowed, inhaled, or absorbed through skin; extremely destructive to tissue of the mucous membranes and upper respiratory tract, eyes and skin; inhalation may result in spasm, inflammation and edema of the larynx and bronchi, chemical pneumonitis and pulmonary edema; symptoms of exposure may include burning sensation, coughing, wheezing, laryngitis, shortness of breath, headache, nausea and vomiting.
Fire: Hazards: emits toxic fumes. Extinguishing agents: carbon dioxide, dry chemical powder or appropriate foam. Precautions: combustible liquid.
Carcinogenicity: IARC - Not listed; NIOSH - Not listed; NTP - Not listed; ACGIH - Not listed; OSHA - Not listed; EPA - Not listed; MAK - Not listed
Primary Target Organs:

Eyes

Skin

Respiratory System

Environmental

Regulations

RCRA 40CFR: Not listed
CERCLA: 40CFR 302.4: Not listed
SARA 40CFR 372.65: Not listed
SARA EHS 40CFR 355: Not listed
TSCA: Listed

AMI2600 CAS #: 60-32-2

6-AMINOCAPROIC ACID

RTECS: MO6300000
EINECS Number: 200-469-3
Molecular Formula: $C_6H_{13}NO_2$
Formula Weight: 131.18

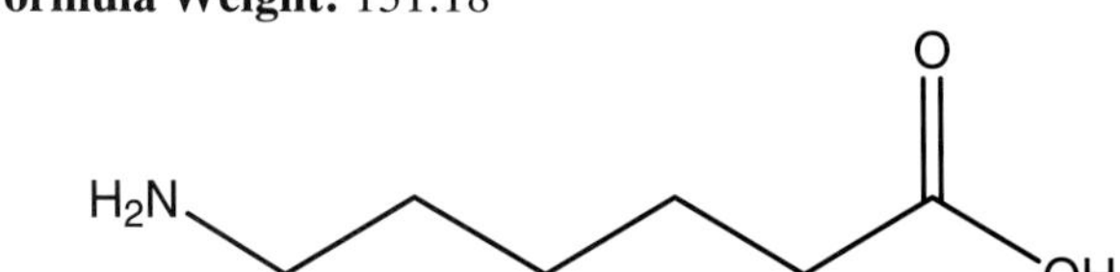

Chemical Structure

Synonyms: 177 J.D; ACEPRAMIN; ACEPRAMINE; ACS; AFIBRIN; AMICAR; AMIKAR; AMINOCAPROIC ACID; OMEGA-AMINOCAPROIC ACID; 6-AMINOHEXANOIC ACID; OMEGA-AMINOHEXANOIC ACID; AMINOKAPRON; CAPLAMIN; CAPRALENSE; CAPRAMOL; CAPROCID; CAPROLISIN; CL 10304; CY 116; EACA; EACA KABI; EACS; EPSAMON; EPSICAPRON; EPSIKAPRON; EPSILCAPRAMIN; EPSILON S; EPSILON-AMINOCAPROIC ACID; EPSILON-AMINOHEXANOIC ACID; EPSILON-LEUCINE; EPSILON-NORLEUCINE; HEMOCAPROL; HEMOPAR; HEPIN; HEXANOIC ACID,6-AMINO-; IPSILON; 177 JD; KYSELINA OMEGA-AMINOKAPRONOVA; NSC-26154; RESPRAMIN
Description: fine, white, crystalline powder or leaves; odorless, or nearly so
Use: antidote for overdosage in fibrinolytic therapy; captive chem int in production of nylon 6; medication (vet): antifibrinolytic agent

Physical Properties

Freezing Point: 202 °C (395.6 °F) to 203 °C (397.4 °F)
Water Solubility: 1 g in 3 ml of Water
Other Solubilities: Slightly Soluble in Alcohol. Practically Insoluble in Chloroform & Ether

RTECS Toxicity Data

Acute Oral: Man TD_{Lo} Dose: 1778 mg/kg/8D-I; Toxic Effects: Kidney, Ureter, and Bladder - Chgs in tubules (inc acute renal failure, acute tubular necrosis; Kidney, Ureter, and Bladder - Hematuria; Nutritional and gross metabolic - Body temperature increase. Monkey LD_{50} Dose: >7 gm/kg.
Chronic (Multiple Dose) Oral: Rat Dose: 32500 mg/kg/13W-I; Toxic Effects: Cardiac - Changes in heart weight; Blood - Changes in erythrocite (RBC) cell count; Blood - Changes in leukocyte (WBC) cell count. Rat Dose: 337 gm/kg/30D-C; Toxic Effects: Behavioral - Excitment; Behavioral - Food intake (animal); Nutritional and gross metabolic - Weight loss or decreased weight gain.
Irritation Eye: Rabbit Standard Draize Test Dose: 500 mg/24H; Reaction: mild.
Reproductive/Teratogenic: Rat Route: Oral; Dose: 153 gm/kg; Duration: female 40D prior to mating Effects on Fertility - Female fertility index. Rat Route: Oral; Dose: 190 gm/kg; Duration: male 38D prior to mating; Effects on Fertility - Male fertility index.

Hazard Overviews

Health: Irritating to eyes/skin/respiratory tract. Other Acute Effects: may be harmful by inhalation, ingestion, or skin absorption; exposure may cause itching; erythema; skin rash; diuresis; heartburn; nausea; diarrhea. Chronic Effects: target organ: blood; kidneys.
Fire: Extinguishing agents: water spray; carbon dioxide, dry chemical powder or appropriate foam. Precautions: combustible liquid.
Reactivity: Incompatible with: strong oxidizing agents. Hazardous decomposition products: toxic fumes of: carbon monoxide, carbon dioxide, nitrogen oxides.
Carcinogenicity: IARC - Not listed; NIOSH - Not listed; NTP - Not listed; ACGIH - Not listed; OSHA - Not listed; EPA - Not listed; MAK - Not listed

Primary Target Organs:

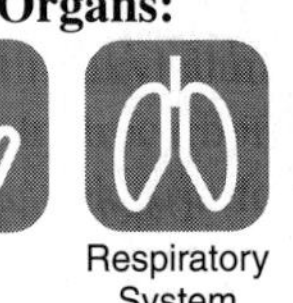

Environmental

Regulations

RCRA 40CFR: Not listed
CERCLA: 40CFR 302.4: Not listed
SARA 40CFR 372.65: Not listed
SARA EHS 40CFR 355: Not listed
TSCA: Listed

AMI2680 CAS #: 2032-59-9

AMINOCARB

RTECS: FC0175000
DOT: UN2757; UN2758; UN2991; UN2992; IMO3.2; IMO6.1
EINECS Number: 217-990-7
Molecular Formula: $C_{11}H_{16}N_2O_2$
Formula Weight: 208.26
Synonyms: A 363; AMINOCARBE; BAY 44646; BAYER 44646; BAYER 5080; CARBAMIC ACID,METHYL-,4-(DIMETHYLAMINO)-3-METHYLPHENYLESTER; CARBAMIC ACID,METHYL-,4-(DIMETHYLAMINO)-M-TOLYL ESTER; M-CRESOL,4-(DIMETHYLAMINO)-,METHYLCARBAMATE (ESTER); 4-DIMETHYLAMINE M-CRESYL METHYLCARBAMATE; 4-DIMETHYLAMINO-3-CRESYL METHYLCARBAMATE; 4-(DIMETHYLAMINO)-3-METHYLPHENOL METHYL CARBAMATE (ESTER); 4-(DIMETHYLAMINO)-3-METHYLPHENOL METHYLCARBAMATE (ESTER); (4-DIMETHYLAMINO-3-METHYL-PHENYL)N-METHYL-CARBAMAAT; (4-DIMETHYLAMINO-3-METHYL-PHENYL)N-METHYL-CARBAMAT; (4-DIMETHYLAMINO-3-METHYL-PHENYL)N-METHYL-CARBAMATE; 4-(DIMETHYLAMINO)-M-TOLYL METHYLCARBAMATE; (4-DIMETILAMINO-3-METIL-FENIL)-N-METIL-CARBAMMATO; ENT 25,784; ENT 25784; MATACIL; MATACIL 180D; N-METHYLCARBAMATE DE 4-DIMETHYLAMINO 3-METHYL PHENYLE; METHYLCARBAMIC ACID 4-(DIMETHYLAMINO)-M-TOLYL ESTER; 3-METHYL-4-DIMETHYLAMINOPHENYL METHYLCARBAMATE; MITACIL; PHENOL,4-(DIMETHYLAMINO)-3-METHYL-,METHYLCARBAMATE(ESTER); PHENOL,4-(DIMETHYLAMINO)-3-METHYL-,METHYLCARBAMATE(ESTER) (9CI)
Description: white to tan crystalline solid or crystals
Use: insecticide in control of forest pests incl spruce webworm & jack pine budworm (former use); control of lepidopterous larvae and other chewing insects in cotton, field crops, and in forestry (former use)

Physical Properties

Freezing Point: 93 °C (199.4 °F) to 94 °C (201.2 °F)
Vapor Pressure: Estimated 1.88 x10^{-6} mm Hg
Water Solubility: Slightly Soluble in Water
Other Solubilities: Soluble in polar organic solvents; moderately Soluble in aromatic solvents

RTECS Toxicity Data

Acute Oral: Rat LD_{50} Dose: 30 mg/kg. Mouse LD_{Lo} Dose: 94 mg/kg.
Acute Dermal: Rat LD_{50} Route: Skin; Dose: 275 mg/kg. Mouse LD_{50} Route: Skin; Dose: 31 mg/kg. Mouse LD_{50} Route: Subcutaneous Dose: 6900 ug/kg.
Mutagenic: Hamster Cytogenetic Analysis; Cell Type: ovary; Dose: 5 mmol/L. Bacteria - S Typhimurium Mutations in Microorganisms; Dose: 5 mmol/L (-S9).

Hazard Overviews

Carcinogenicity: IARC - Not listed; NIOSH - Not listed; NTP - Not listed; ACGIH - Not listed; OSHA - Not listed; EPA - Not listed; MAK - Not listed

Environmental

Ecotoxicity: LC_{50} Chironomus riparius 376.6 ppb/24 hr LD_{50} Mule deer oral 7.5-15.0 mg/kg males 13-15 mo old, acute LD_{50} Mallard oral 22.5 mg/kg (95% confidence limit 17.8-28.3) males 8 mo old, acute IC_{50} Selenastrum capricornutum (green algae) 0.1 x10^{-3} ug/l/14 days LC_{50} Coturnix japonica (Japanese quail) oral, 14 days old, (5 day ad libitum in diet) 2325 ppm (confidence intervals 1947-3020 ppm) LC_{50} Gammarus fasciatus (scud) 12 ug/l/96 hr at 21 °C, mature stage, (95% confidence limit 8.2-18 ug/l). Static bioassay without aeration, pH 7.2-7.5, water hardness 40-50 mg/l as calcium carbonate and alkalinity of 30-35 mg/l EC_{50} Daphnia magna (daphnid), immobilization, 10-100 ug/l/48 hr at 21 °C, first instar stage. Static bioassay without aeration, pH 7.2-7.5, water hardness 40-50 mg/l as calcium carbonate and alkalinity of 30-35 mg/l EC_{50} Chironomus plumosus (midge), immobilization, 270 ug/l/48 hr at 20 °C, fourth instar stage, (95% confidence limit 187-389 ug/l). Static bioassay without aeration, pH 7.2-7.5, water hardness 40-50 mg/l as calcium carbonate LC_{50} Salmo gairdneri (rainbow trout) 130 ug/l/96 hr at 10 °C, wt 1.5 g, (95% confidence limit 103-164 ug/l). Static bioassay without aeration, pH 7.2-7.5, water hardness 40-50 mg/l as calcium carbonate and alkalinity of 30-35 mg/l IC_{50} Chlamydomonas variabilis (green algae) 0.8 x10^{-3} ug/l/24 hr IC_{50} Daphnia magna (daphnid) 0.2 ug/l/24 hr LC_{50} Salmo gairdneri (rainbow trout) fingerlings 0.36 mg/l/96 hr, Matacil 1.8 d /Conditions of bioassay not specified LC_{50} Salmo salar (Atlantic salmon) 3.5 mg/l/96 hr /EC formulation/ /Conditions of bioassay not specified
Environmental Fate: If released to soil, it is expected to have a moderate high mobility in soils. Loss will occur primarily by hydrolysis and biodegradation. Loss from soil due to volatilization may not be important. Persistence in a forest soil was less than 8 days. In water, it appears to degrade as a result of hydrolysis, biodegradation and photolysis. The half-life in a creek water with a pH of 6.52 was 11.5 days. It should not bioconcentrate in aquatic organisms. It should not volatilize from soil or water. In the atmosphere, reaction with photochemically produced hydroxyl radicals will be an important loss processes for vapor phase in the atmosphere. The half-life due to this reaction has been estimated to be less

than 1 hr. Partial removal will also occur as a result of dry and wet deposition.

Cleanup/Disposal: Guide No. 131: Fully encapsulating, vapor protective clothing should be worn for spills and leaks with no fire. Eliminate all ignition sources (no smoking, flares, sparks or flames in immediate area). All equipment used when handling the product must be grounded. Do not touch or walk through spilled material. Stop leak if you can do it without risk. Prevent entry into waterways, sewers, basements or confined areas. A vapor suppressing foam may be used to reduce vapors. Small Spills: Absorb with earth, sand or other non-combustible material and transfer to containers for later disposal. Use clean non-sparking tools to collect absorbed material. Large Spills: Dike far ahead of liquid spill for later disposal. Water spray may reduce vapor; but may not prevent ignition in closed spaces. Guide No. 151: Do not touch damaged containers or spilled material unless wearing appropriate protective clothing. Stop leak if you can do it without risk. Prevent entry into waterways, sewers, basements or confined areas. Cover with plastic sheet to prevent spreading. Absorb or cover with dry earth, sand or other non-combustible material and transfer to containers. Do not get water inside containers.

Environmental Physical Data

Henry's Law Constant: estimated at 5.64×10^{-10}
Octanol/Water Partition Coefficient: log K_{ow} = 1.734
Sorption Partition Coefficient: K_{oc} = 2.01 to 2.32
BCF: brown bullhead 2.6 to 3.6

Regulations

RCRA 40CFR: Not listed
CERCLA: 40CFR 302.4: Not listed
SARA 40CFR 372.65: Not listed
SARA EHS 40CFR 355: Not listed
TSCA: Not listed

Analytical Methods

Soil: SW846 8321A
Water / Groundwater: EPA 632, 022
Food: FDA 232.4, 242.1; AOAC 985.02

AMI2760 **CAS #: 117-11-3**

1-AMINO-5-CHLOROANTHRAQUINONE

RTECS: CB5435000
EINECS Number: 204-174-0
Molecular Formula: $C_{14}H_8ClNO_2$
Formula Weight: 257.68

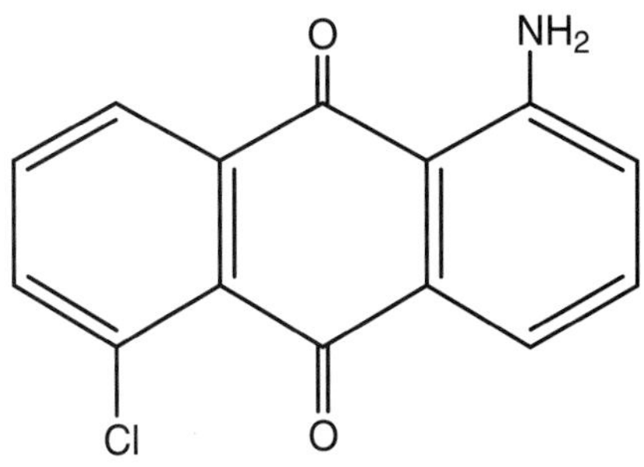

Chemical Structure

Synonyms: 9,10-ANTHRACENEDIONE,1-AMINO-5-CHLORO-; ANTHRAQUINONE,1-AMINO-5-CHLORO; 1-CHLOR-5-AMINOANTHRACHINON; 1-CHLORO-5-AMINOANTHRAQUINONE; 5-CHLORO-1-AMINOANTHRAQUINONE
Description:
Use: chemical intermediate for vat dyes

RTECS Toxicity Data

Irritation Eye: Rabbit Standard Draize Test Dose: 500 mg/24H; Reaction: mild.

Hazard Overviews

Carcinogenicity: IARC - Not listed; NIOSH - Not listed; NTP - Not listed; ACGIH - Not listed; OSHA - Not listed; EPA - Not listed; MAK - Not listed

Environmental

Regulations

RCRA 40CFR: Not listed
CERCLA: 40CFR 302.4: Not listed
SARA 40CFR 372.65: Not listed
SARA EHS 40CFR 355: Not listed
TSCA: Listed

AMI3080 **CAS #: 88-53-9**

2-AMINO-5-CHLORO-4-METHYLBENZENESULFONIC ACID

EINECS Number: 201-839-7
Molecular Formula: $C_7H_8ClNO_3S$
Formula Weight: 221.67

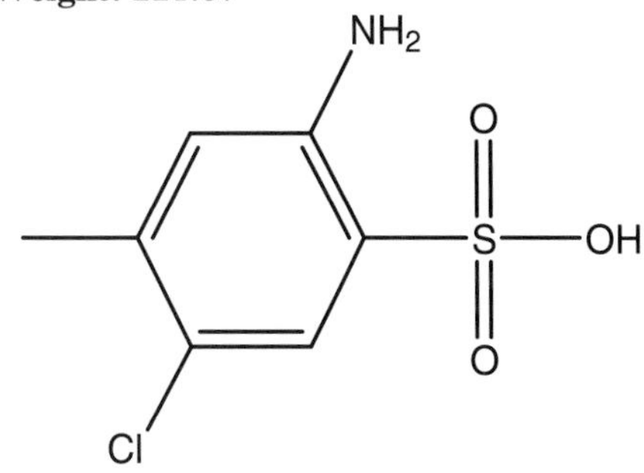

Chemical Structure

Synonyms: BENZENESULFONIC ACID,2-AMINO-5-CHLORO-4-METHYL-; LAKE RED C AMINE; RED LAKE CAMINE; P-TOLUENESULFONIC ACID,2-AMINO-5-CHLORO-

Hazard Overviews

Carcinogenicity: IARC - Not listed; NIOSH - Not listed; NTP - Not listed; ACGIH - Not listed; OSHA - Not listed; EPA - Not listed; MAK - Not listed

Environmental

Regulations

RCRA 40CFR: Not listed
CERCLA: 40CFR 302.4: Listed as Compound per CWA Section 307(a)
SARA 40CFR 372.65: Not listed
SARA EHS 40CFR 355: Not listed
TSCA: Listed

AMI3160 CAS #: 6358-07-2

2-AMINO-4-CHLORO-5-NITRO PHENOL

RTECS: SJ5736000
EINECS Number: 228-760-0
Molecular Formula: $C_6H_5ClN_2O_3$
Formula Weight: 188.58

Chemical Structure

Synonyms: 2-AMINO-4-CHLORO-5-NITROPHENOL; 2-AMINO-5-NITRO-4-CHLOROPHENOL
Description: orange powder
Use: azo dye intermediate

Physical Properties

Freezing Point: Decomposes at 225 °C (437 °F)
Water Solubility: < 0.1 mg/mL at 22 C
Other Solubilities: 95% Ethanol: <1 mg/ml at 20.5 °C; Acetone: <1 mg/ml at 20.5 °C; DMSO: 5-10 mg/ml at 20.5 °C.
Flash Point: Not available; probably combustible

RTECS Toxicity Data

Mutagenic: Bacteria - S Typhimurium Mutations in Microorganisms; Dose: 10 ug/plate (+S9), 33 ug/plate (-S9).

Hazard Overviews

Health: Irritating to eyes/skin/respiratory tract. Harmful. Other Acute Effects: may be harmful by inhalation, ingestion, or skin absorption.
Fire: Will burn. Hazards: emits toxic fumes. Extinguishing agents: water spray; carbon dioxide, dry chemical powder or appropriate foam. Precautions: combustible liquid.
Reactivity: Incompatible with: strong oxidizing agents, strong bases. Hazardous decomposition products: toxic fumes of: carbon monoxide, carbon dioxide, nitrogen oxides, hydrogen chloride gas.
Carcinogenicity: IARC - Not listed; NIOSH - Not listed; NTP - Not listed; ACGIH - Not listed; OSHA - Not listed; EPA - Not listed; MAK - Not listed
Primary Target Organs:

Eyes

Skin

Respiratory System

Environmental

Regulations

RCRA 40CFR: Not listed
CERCLA: 40CFR 302.4: Listed as Compound per CWA Section 307(a)
SARA 40CFR 372.65: Not listed
SARA EHS 40CFR 355: Not listed
TSCA: Listed

AMI3240 CAS #: 121-87-9

1-AMINO-2-CHLORO-4-NITROBENZENE

RTECS: BX1400000
EINECS Number: 204-502-2
Molecular Formula: $C_6H_5ClN_2O_2$
Formula Weight: 172.57

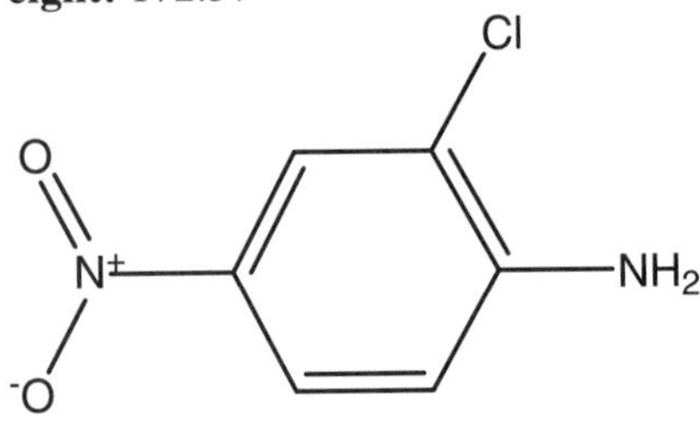

Chemical Structure

Synonyms: ANILINE,2-CHLORO-4-NITRO-; BENZENAMINE,2-CHLORO-4-NITRO-; 2-CHLORO-4-NITROANILINE; O-CHLORO-P-NITROANILINE; ORTHO-CHLORO-PARA-NITROANILINE; 2-CHLORO-4-NITROBENZENAMINE; 4-NITRO-2-CHLOROANILINE; OCPN; OCPNA
Description: yellow needles

Use: intermediate in mfr of dyes; intermediate in the production of n-(2-chloro-4-nitrophenol)-5-chlorosalicylanilide (niclosamid), a molluscicide

Physical Properties

Boiling Point: > 200 °C (392 °F)
Freezing Point: 108 °C (226.4 °F)
Water Solubility: Miscible with Water
Other Solubilities: Soluble in Benzene and Ether; Slightly Soluble in strong acids.
Flash Point: 205 °C

RTECS Toxicity Data

Acute Oral: Rat LD_{50} Dose: 6430 mg/kg. Mouse LD_{50} Dose: 1250 mg/kg.
Mutagenic: Bacteria - S Typhimurium Mutations in Microorganisms; Dose: 1 mg/plate (+/-S9).

Hazard Overviews

Health: Irritating to eyes/skin/respiratory tract. Other Acute Effects: may be harmful by inhalation, ingestion, or skin absorption; absorption into the body leads to the formation of methemoglobin which in sufficient concentration causes cyanosis; onset may be delayed 2 to 4 hours or longer; can cause CNS depression; exposure can cause headache; dizziness; weakness; incoordination; target organ: blood. Chronic Effects: target organs: spleen, bone marrow, liver, central nervous system, cardiovascular system, blood.
Fire: Will burn. Hazards: emits toxic fumes. Extinguishing agents: water spray; carbon dioxide, dry chemical powder or appropriate foam. Precautions: combustible liquid.
Reactivity: Stable. Hazardous polymerization will not occur. Incompatible with: strong oxidizing agents, strong bases, strong acids. Hazardous decomposition products: toxic fumes of: carbon monoxide, carbon dioxide, nitrogen oxides, hydrogen chloride gas, chlorine.
Carcinogenicity: IARC - Not listed; NIOSH - Not listed; NTP - Not listed; ACGIH - Not listed; OSHA - Not listed; EPA - Not listed; MAK - Not listed
Primary Target Organs:

Eyes

Skin
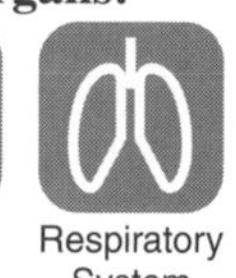
Respiratory System

Nervous System

Liver

Blood

Environmental

Ecotoxicity: Algae: Tetrahymena pyriformis 48h EC_{50}, growth 31 mg/l; Crustaceans: Daphnia magna 48h EC_{50} 1.8 mg/l; Fishes: Pimephales promelas 30-35d LC_{50} 222 mg/l
Environmental Fate: If released to the atmosphere, it will exist primarily in the vapor phase where it will be degraded by hydroxyl radicals with an estimated half-life of 4 days. In soil it is expected to have moderate mobility based on an estimated K_{oc} of 339; however, the aromatic amino group may react with organic matter in soil resulting in strong binding in soil. Under aerobic conditions it is resistant to biodegradation; a half-life of much greater that 4 weeks was measured using both non-acclimated and acclimated inocula in the OECD and Pitter tests. A 20% loss was measured after 17 days, using a mixed inoculum containing microorganisms from both activated sludge and river sediment. Showed only a 12% loss over 8 days in a water-based medium due to abiotic factors. It is not expected to volatilize from water surfaces (estimated Henry's Law constant of 1.54 $x10^{-8}$ atm-cu m/mole) or to bioconcentrate in aquatic organisms (estimated BCF of 24). It may bind to organic matter in the water column.
Cleanup/Disposal: Guide No. 153: Eliminate all ignition sources (no smoking, flares, sparks or flames in immediate area). Do not touch damaged containers or spilled material unless wearing appropriate protective clothing. Stop leak if you can do it without risk. Prevent entry into waterways, sewers, basements or confined areas. Absorb or cover with dry earth, sand or other non-combustible material and transfer to containers. Do not get water inside containers.

Environmental Physical Data

Henry's Law Constant: estimated at 1.54 $x10^{-8}$
Octanol/Water Partition Coefficient: log K_{ow} = 2.12
Sorption Partition Coefficient: K_{oc} = 339
BCF: estimated at 24

Regulations

RCRA 40CFR: Not listed
CERCLA: 40CFR 302.4: Listed as Compound per CWA Section 307(a)
SARA 40CFR 372.65: Not listed
SARA EHS 40CFR 355: Not listed
TSCA: Listed

Analytical Methods

Soil: SW846 8131

AMI3320 **CAS #: 81-49-2**

1-AMINO-2,4-DIBROMOANTHRAQUINONE

RTECS: CB5500000
EINECS Number: 201-354-0
Molecular Formula: $C_{14}H_7Br_2NO_2$
Formula Weight: 381.0

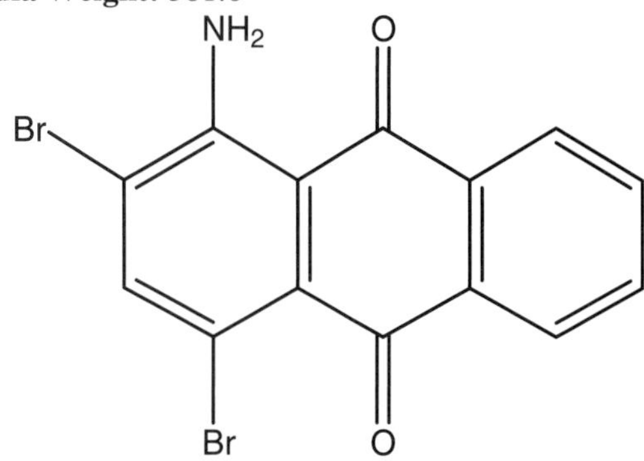

Chemical Structure

Synonyms: 1-AMINO-2,4-DIBROMANTHRACHINON; 1-AMINO-2,4-DIBROMO-9,10-ANTHRACENEDIONE; 9,10-ANTHRACENEDIONE,1-AMINO-2,4-DIBROMO-; ANTHRAQUINONE,1-AMINO-2,4-DIBROMO-; 2,4-DIBROMO-1-ANTHRAQUINONYLAMINE
Description: red powder; odorless
Use: dye and dye intermediate

Physical Properties

Freezing Point: 221 °C (429.8 °F)
Water Solubility: < 1 mg/mL at 23 C
Other Solubilities: DMSO: 1-10 mg/ml at 23 °C; 95% Ethanol: <1 mg/ml at 23 °C; Acetone: <1 mg/ml at 23 °C; Ether: Sparingly Soluble; Benzene: Sparingly Soluble; Chloroform: Soluble; Acetic Acid: Soluble; Concentrated sulfuric acid: Soluble.
Flash Point: > 200 °C

RTECS Toxicity Data

Chronic (Multiple Dose) Oral: Rat Dose: 130500 mg/kg/90D-C; Toxic Effects: Liver - Changes in liver weight; Blood - Changes in serum composition; Blood - Changes in erythrocite (RBC) cell count. Mouse Dose: 42075 mg/kg/90D-C; Toxic Effects: Liver - Changes in liver weight; Blood - Changes in serum composition; Nutritional and gross metabolic - Changes in chlorine.
Irritation Eye: Rabbit Standard Draize Test Dose: 500 mg/24H; Reaction: mild.
Mutagenic: Rat Morphological Transformation; Route: Oral; Dose: 329 gm/kg/39W-C. Bacteria - S Typhimurium Mutations in Microorganisms; Dose: 333 ug/plate (+S9). Bacteria - S Typhimurium Mutations in Microorganisms; Dose: 333 ug/plate (-S9).

Hazard Overviews

Fire: Will burn.
Carcinogenicity: IARC - Not listed; NIOSH - Not listed; NTP - Not listed; ACGIH - Not listed; OSHA - Not listed; EPA - Not listed; MAK - Not listed

Environmental

Environmental Fate: If released to the atmosphere, it should exist solely in the particulate phase based on an estimated vapor pressure of 1.44 $x10^{-9}$ mm Hg. Particulate-phase may be physically removed from the atmosphere by dry deposition. An estimated K_{oc} of 18000 suggests that it will be immobile in soil. Its amino group may bind covalently with active sites in soil further limiting its mobility. It is not expected to volatilize from moist soil surfaces given an estimated Henry's Law constant of 1.78 $x10^{-13}$ atm-cu m/mol. In water, it is expected to strongly adsorb to organic matter and particulates in the water column based on its K_{oc} value. It should not volatilize from water surfaces given its Henry's Law constant. It is expected to bioconcentrate in aquatic organisms; an estimated BCF value of 6400 was calculated using the estimated log Kow value for this compound.

Environmental Physical Data

Henry's Law Constant: estimated at 1.8 $x10^{-13}$
Octanol/Water Partition Coefficient: log K_{ow} = 5.31
Sorption Partition Coefficient: K_{oc} = 1.8 $x10^{4}$
BCF: estimated at 6400

Regulations

RCRA 40CFR: Not listed
CERCLA: 40CFR 302.4: Not listed
SARA 40CFR 372.65: Not listed
SARA EHS 40CFR 355: Not listed
TSCA: Listed

AMI3480 CAS #: 101-54-2

P-AMINODIPHENYLAMINE

RTECS: ST3150000
EINECS Number: 202-951-9
Molecular Formula: $C_{12}H_{12}N_2$
Structured MF: $C_6H_5NHC_6H_4NH_2$
Formula Weight: 184.26

HN— —NH2

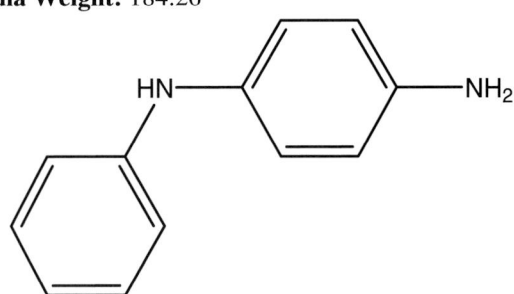

Chemical Structure

Synonyms: ACNA BLACK DF BASE; P-AMINODIFENYLAMIN; 4-AMINODIPHENYLAMINE; P-ANILINOANILINE; AZOSALT R; 1,4-BENZENEDIAMINE,N-PHENYL-; 1,4-BENZENEDIAMINE,N-PHENYL-(9CI); N,4'-BIANILINE; BLACK BASE P; C.I. 37240; C.I. 76085; C.I. AZOIC DIAZO COMPONENT 22; C.I. DEVELOPER 15; C.I. OXIDATION BASE 2; DIPHENYL BLACK; DIPHENYL BLACK BASE P; DIPHENYLAMINE,4-AMINO-; DIPHENYLAMINE,P-AMINO-; FAST BLUE R SALT; N-FENYL-P-FENYLENDIAMIN; LUXAN BLACK R; NAPHTHOELAN NAVY BLUE; OXY ACID BLACK BASE; PELTOL BR; PELTOL BR II; PHENYL 4-AMINOPHENYL AMINE; N-PHENYL-P-AMINOANILINE; N-PHENYL-1,4-BENZENEDIAMINE; P-PHENYLENEDIAMINE,N-PHENYL-; N-PHENYL-1,4-PHENYLENEDIAMINE; N-PHENYL-P-PHENYLENEDIAMINE; RODOL GRAY B BASE; SEMIDIN; P-SEMIDINE; SEMIDINE; VARIAMINE BLUE RT; VARIAMINE BLUE SALT RT
Description: purple powder, needles or crystals; odorless
Use: as a dye intermediate and constituent of hair dye; in calico printing, dyeing fur, pharmaceuticals and photographic chemicals

Physical Properties

Boiling Point: 354 °C (669 °F)
Freezing Point: 75 °C (167 °F)
Specific Gravity: 1.15
Density: 1.09 g/mL at 100 °C
Water Solubility: Slightly Soluble in Water

Other Solubilities: 95% Ethanol: >=100 mg/ml at 19 °C; Acetone: >=100 mg/ml at 19 °C; DMSO: >=100 mg/ml at 19 °C.
Flash Point: 193 °C

RTECS Toxicity Data

Acute Oral: Rat LD_{50} Dose: 464 mg/kg. Mouse LD_{50} Dose: 244 mg/kg; Toxic Effects: Blood - Hemorrhage; Nutritional and gross metabolic - Body temperature decrease.
Acute Dermal: Rabbit LD_{50} Route: Skin; Dose: >5 gm/kg; Toxic Effects: Sense organs and special senses - Other; Sense organs and special senses - Other; Behavioral - Food intake (animal).
Irritation Eye: Rabbit Standard Draize Test Dose: 100 mg/24H; Reaction: moderate.
Mutagenic: Bacteria - S Typhimurium Mutations in Microorganisms; Dose: 150 ug/plate (-S9).

Hazard Overviews

Poison

Health: Irritating to eyes/skin/respiratory tract. Poison. Other Acute Effects: harmful if swallowed, inhaled, or absorbed through skin; prolonged or repeated exposure may cause allergic reactions in certain sensitive individuals.
Fire: Will burn. Extinguishing agents: water spray; carbon dioxide, dry chemical powder or appropriate foam. Precautions: combustible liquid.
Reactivity: Incompatible with: strong oxidizing agents, strong acids. Hazardous decomposition products: toxic fumes of: carbon monoxide, carbon dioxide, nitrogen oxides.
Carcinogenicity: IARC - Not listed; NIOSH - Not listed; NTP - Not listed; ACGIH - Not listed; OSHA - Not listed; EPA - Not listed; MAK - Not listed
Primary Target Organs:

Eyes

Skin

Respiratory System

Environmental

Cleanup/Disposal: Guide No. 171: Do not touch or walk through spilled material. Stop leak if you can do it without risk. Prevent dust cloud. Avoid inhalation of asbestos dust. Small Dry Spills: With clean shovel place material into clean, dry container and cover loosely; move containers from spill area. Small Spills: Take up with sand or other noncombustible absorbent material and place into containers for later disposal. Large Spills: Dike far ahead of liquid spill for later disposal. Cover powder spill with plastic sheet or tarp to minimize spreading. Prevent entry into waterways, sewers, basements or confined areas.

Regulations

RCRA 40CFR: Not listed
CERCLA: 40CFR 302.4: Not listed
SARA 40CFR 372.65: Not listed
SARA EHS 40CFR 355: Not listed
TSCA: Listed

AMI3560 CAS #: 60-23-1

2-AMINOETHANETHIOL

RTECS: KJ0175000
EINECS Number: 200-463-0
Molecular Formula: C_2H_7NS
Formula Weight: 77.16

H_2N SH

Chemical Structure

Synonyms: 2-AMINOETHYL MERCAPTAN; BECAPTAN; CISTEAMINA; CYSTEAMIDE; CYSTEAMIN; CYSTEAMINE; CYSTEINAMINE; DECARBOXYCYSTEINE; LAMBRATEN; MEA; MECRAMINE; MERCAMINE; MERCAPTAMINE; (2-MERCAPTOETHYL)AMINE; BETA-MERCAPTOETHYLAMINE; THIOETHANOLAMINE

Physical Properties

Freezing Point: 100 °C (212 °F)
Water Solubility: Very Soluble
Other Solubilities: Ethanol: Very Soluble

RTECS Toxicity Data

Acute Oral: Mouse LD_{50} Dose: 625 mg/kg.
Acute Dermal: Rat LD_{50} Route: Subcutaneous Dose: 84 mg/kg; Toxic Effects: Peripheral Nerve and sensation - Spastic parapysis with or without sensory change; Behavioral - Convulsions or effect on seizure threshold; Skin and appendages - Hair. Mouse LD_{50} Route: Subcutaneous Dose: 84 mg/kg; Toxic Effects: Peripheral Nerve and sensation - Spastic parapysis with or without sensory change; Behavioral - Convulsions or effect on seizure threshold; Skin and appendages - Hair.
Reproductive/Teratogenic: Rat Route: Oral; Dose: 42 gm/kg; Duration: female 70D prior to mating Effects on Fertility - Litter size; Other measures of fertility.
Mutagenic: Human Unscheduled DNA Synthesis; Cell Type: fibroblast; Dose: 1 mmol/L. Hamster Cytogenetic Analysis; Cell Type: ovary; Dose: 1 mmol/L.

Hazard Overviews

Health: Irritating to eyes/skin/respiratory tract. Harmful. Other Acute Effects: harmful if swallowed; may be harmful if inhaled or absorbed through the skin; exposure can cause: nausea; headache; vomiting.
Fire: Hazards: emits toxic fumes. Extinguishing agents: water spray; carbon dioxide, dry chemical powder or appropriate foam. Precautions: combustible liquid.
Reactivity: Incompatible with: strong oxidizing agents, may decompose on exposure to air. Hazardous decomposition

products: toxic fumes of: carbon monoxide, carbon dioxide, nitrogen oxides, sulfur oxides.

Carcinogenicity: IARC - Not listed; NIOSH - Not listed; NTP - Not listed; ACGIH - Not listed; OSHA - Not listed; EPA - Not listed; MAK - Not listed

Primary Target Organs:

Eyes

Skin

Respiratory System

Environmental

Regulations

RCRA 40CFR: Not listed
CERCLA: 40CFR 302.4: Not listed
SARA 40CFR 372.65: Not listed
SARA EHS 40CFR 355: Not listed
TSCA: Not listed

AMI3720 CAS #: 929-06-6

2-(2-AMINOETHOXY)ETHANOL

RTECS: KJ6125000
DOT: NA1760
EINECS Number: 213-195-4
Molecular Formula: $C_4H_{11}NO_2$
Structured MF: $HOCH_2CH_2NHCH_2CH_2NH_2$
Formula Weight: 105.16

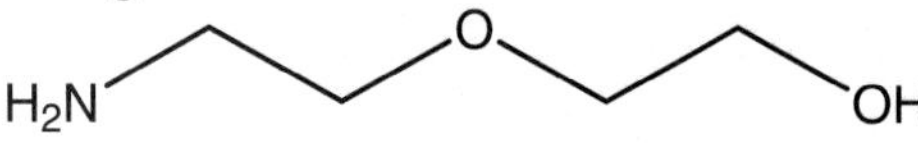

Chemical Structure

Synonyms: 2-AMINOETHOXYETHANOL; DGA; DIETHYLENE GLYCOL AMINE; DIGLYCOLAMINE; ETHANOL,2-(2-AMINOETHOXY)-; 2-(HYDROXYETHOXY)ETHYLAMINE; BETA-(BETA-HYDROXYETHOXY)ETHYLAMINE

Description: colorless liquid; mild amine odor

Use: removal of acid components from gases, especially carbon dioxide & hydrogen sulfide from natural gas; intermediate

Physical Properties

Boiling Point: 221 °C (430 °F)
Freezing Point: -12.5 °C (9.5 °F)
Specific Gravity: 1.0572 at 20 °C/20 °C
Vapor Density: 3.59 Air=1
Vapor Pressure: ~ 0.001 mm Hg
Water Solubility: Miscible with Water
Other Solubilities: miscible with Alcohols
Flash Point: 126.6 °C
Autoignition Temperature: 368 °C
LEL: 15% v/v
UEL: 27% v/v

RTECS Toxicity Data

Acute Oral: Rat LD_{50} Dose: 3 gm/kg; Toxic Effects: Behavioral - General anesthetic; Behavioral - Somnolence (general depressed activity). Mouse LD_{50} Dose: 2825 mg/kg; Toxic Effects: Behavioral - General anesthetic; Behavioral - Somnolence (general depressed activity).

Acute Dermal: Rabbit LD_{50} Route: Skin; Dose: 1190 uL/kg.

Irritation Eye: Rabbit Standard Draize Test Dose: 250 ug open; Reaction: severe. Rabbit Standard Draize Test Dose: 50 ug/24H; Reaction: severe.

Irritation Skin: Rabbit Standard Draize Test Dose: 5 mg/24H; Reaction: severe. Rabbit Open Draize Test Dose: 10 mg/24H open; Reaction: severe.

Hazard Overviews

Corrosive

Health: Corrosive to eyes/skin/respiratory tract. Other Acute Effects: harmful if swallowed, inhaled, or absorbed through skin; inhalation may result in spasm, inflammation and edema of the larynx and bronchi, chemical pneumonitis and pulmonary edema; symptoms of exposure may include burning sensation; coughing; wheezing; laryngitis; shortness of breath; headache; nausea; vomiting.

Fire: Will burn. Hazards: emits toxic fumes. Extinguishing agents: carbon dioxide, dry chemical powder or appropriate foam. Precautions: combustible liquid.

Carcinogenicity: IARC - Not listed; NIOSH - Not listed; NTP - Not listed; ACGIH - Not listed; OSHA - Not listed; EPA - Not listed; MAK - Not listed

Primary Target Organs:

Eyes

Skin

Respiratory System

Environmental

Cleanup/Disposal: Guide No. 154: Eliminate all ignition sources (no smoking, flares, sparks or flames in immediate area). Do not touch damaged containers or spilled material unless wearing appropriate protective clothing. Stop leak if you can do it without risk. Prevent entry into waterways, sewers, basements or confined areas. Absorb or cover with dry earth, sand or other non-combustible material and transfer to containers. Do not get water inside containers.

Regulations

RCRA 40CFR: Not listed
CERCLA: 40CFR 302.4: Not listed
SARA 40CFR 372.65: Not listed
SARA EHS 40CFR 355: Not listed
TSCA: Listed

AMI3800 CAS #: 17026-81-2

N-(3-AMINO-4-ETHOXYPHENYL)ACETAMIDE

RTECS: AD8575000
EINECS Number: 241-100-6
Molecular Formula: $C_{10}H_{14}N_2O_2$
Structured MF: $CH_3CONHC_6H_3(NH_2)OCH_2CH_3$
Formula Weight: 194.26

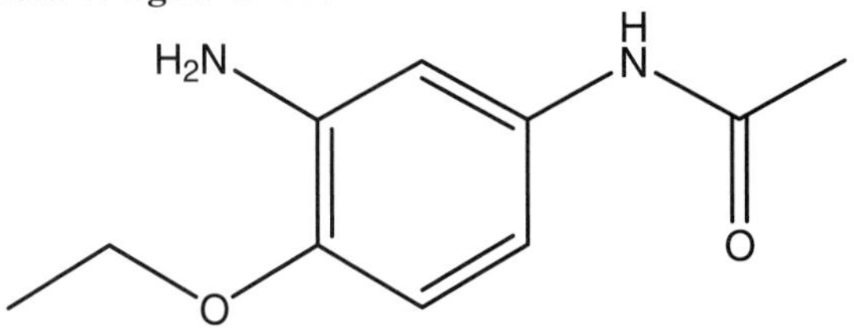

Chemical Structure

Synonyms: ACETAMIDE,N-(3-AMINO-4-ETHOXYPHENYL)-; ACETAMIDE,N-(3-AMINO-4-ETHOXYPHENYL)-(9CI); ACETANILIDE,3'-AMINO-4'-ETHOXY-; ACETANILIDE,3-AMINO-4-ETHOXY; ACETANILIDE,3-AMINO-4-ETHOXY-; P-ACETOPHENETIDIDE,3'-AMINO-; 5-(ACETYLAMINO)-2-ETHOXYANILINE; 2-AMINO-4-ACETAMINIFENETOL; 3'-AMINO-4'-ETHOXYACETANILIDE; 3-AMINO-4-ETHOXYACETANILIDE; 3-AMINO-4-ETHOXYANILID KYSELINY OCTOVE
Description: brown powder
Use: chem int for azo dyes (possible use)

Physical Properties

Freezing Point: 125 °C (257 °F) to 127 °C (260.6 °F)
Water Solubility: < 0.1 mg/mL at 21 C
Other Solubilities: 95% Ethanol: 10-50 mg/ml at 21.5 °C; Acetone: 1-5 mg/ml at 21.5 °C; DMSO: >=100 mg/ml at 21.5 °C.
Flash Point: Not available; probably combustible

RTECS Toxicity Data

Irritation Eye: Rabbit Standard Draize Test Dose: 500 mg/24H; Reaction: mild.
Mutagenic: Bacteria - E Coli Mutations in Microorganisms; Dose: 333 ug/plate (+S9). Bacteria - S Typhimurium Mutations in Microorganisms; Dose: 1 mg/plate (+S9), 33300 ng/plate (-S9).
Tumorigenic: Rat Route: Oral; Dose: 130 gm/kg/78W-C; Toxic Effects: Tumorigenic - Equivocal tumorigenic agent by RTECS criteria; Endocrine - Thyroid tumors. Mouse Route: Oral; Dose: 524 gm/kg/78W-C; Toxic Effects: Tumorigenic - Carcinogenic by RTECS criteria; Endocrine - Thyroid tumors. Mouse Route: Oral; Dose: 260 gm/kg/78W-C; Toxic Effects: Tumorigenic - Equivocal tumorigenic agent by RTECS criteria; Endocrine - Thyroid tumors.

Hazard Overviews

Fire: Will burn.
Carcinogenicity: IARC - Not listed; NIOSH - Not listed; NTP - Not listed; ACGIH - Not listed; OSHA - Not listed; EPA - Not listed; MAK - Not listed

Environmental

Regulations

RCRA 40CFR: Not listed
CERCLA: 40CFR 302.4: Not listed
SARA 40CFR 372.65: Not listed
SARA EHS 40CFR 355: Not listed
TSCA: Listed

AMI3880 CAS #: 109-58-0

(2-AMINOETHYL)CARBAMIC ACID

EINECS Number: 203-684-0
Molecular Formula: $C_3H_8N_2O_2$
Formula Weight: 104.11
Synonyms: CARBAMIC ACID,(2-AMINOETHYL)-; ETHYLENEDIAMINE CARBAMATE
Use: cross-linking agent in rubber processing

Hazard Overviews

Carcinogenicity: IARC - Not listed; NIOSH - Not listed; NTP - Not listed; ACGIH - Not listed; OSHA - Not listed; EPA - Not listed; MAK - Not listed

Environmental

Regulations

RCRA 40CFR: Not listed
CERCLA: 40CFR 302.4: Not listed
SARA 40CFR 372.65: Not listed
SARA EHS 40CFR 355: Not listed
TSCA: Listed

AMI3960 CAS #: 132-32-1

3-AMINO-9-ETHYLCARBAZOLE

RTECS: FE3590000
EINECS Number: 205-057-7
Molecular Formula: $C_{14}H_{14}N_2$
Formula Weight: 210.3

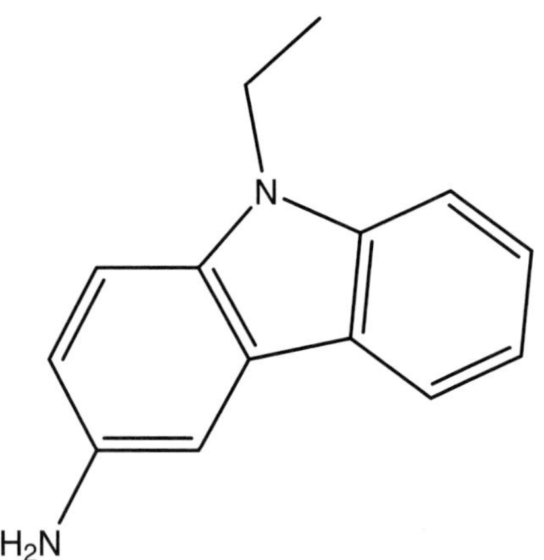

Chemical Structure

Synonyms: 3-AMINO-N-ETHYLCARBAZOL; 3-AMINO-N-ETHYLCARBAZOLE; 9H-CARBAZOL-3-AMINE,9-ETHYL-; CARBAZOLE,3-AMINO-9-ETHYL-

Description: crystalline compound

Use: manufacture of dyes and as an indicator for peroxidase activity

Physical Properties

Freezing Point: 127 °C (260.6 °F)

Water Solubility: < 1 mg/mL at 20 C

Other Solubilities: 95% Ethanol: 1-5 mg/ml at 20 °C; Acetone: >100 mg/ml at 20 °C; DMSO: >100 mg/ml at 20 °C.

RTECS Toxicity Data

Tumorigenic: Rat Route: Oral; Dose: 33 gm/kg/78W-C; Toxic Effects: Tumorigenic - Carcinogenic by RTECS criteria; Liver - Tumors; Skin and appendages - Tumors. Mouse Route: Oral; Dose: 87 gm/kg/78W-C; Toxic Effects: Tumorigenic - Carcinogenic by RTECS criteria; Liver - Tumors.

Hazard Overviews

Health: Irritating to eyes/skin/respiratory tract. Harmful. Other Acute Effects: harmful if swallowed, inhaled, or absorbed through skin; can cause CNS depression; exposure can cause; narcotic effect; damage to the heart; nausea; headache; vomiting; coughing; chest pains; difficulty in breathing; possible risk of irreversible effects. Chronic Effects: laboratory experiments have shown mutagenic effects; target organs: nerves, liver, kidneys. Possible carcinogen.

Fire: Hazards: emits toxic fumes. Extinguishing agents: water spray; carbon dioxide, dry chemical powder or appropriate foam. Precautions: combustible liquid.

Reactivity: Incompatible with: strong oxidizing agents, acids. Hazardous decomposition products: toxic fumes of: carbon monoxide, carbon dioxide, nitrogen oxides.

Carcinogenicity: IARC - Not listed; NIOSH - Not listed; NTP - Not listed; ACGIH - Not listed; OSHA - Not listed; EPA - Not listed; MAK - Class B, Justifiably suspected of having carcinogenic potential

Primary Target Organs:

Eyes

Skin

Respiratory System

Nervous System

Liver

Kidneys

Environmental

Cleanup/Disposal: Guide No. 154: Eliminate all ignition sources (no smoking, flares, sparks or flames in immediate area). Do not touch damaged containers or spilled material unless wearing appropriate protective clothing. Stop leak if you can do it without risk. Prevent entry into waterways, sewers, basements or confined areas. Absorb or cover with dry earth, sand or other non-combustible material and transfer to containers. Do not get water inside containers.

Regulations

RCRA 40CFR: Not listed
CERCLA: 40CFR 302.4: Not listed
SARA 40CFR 372.65: Not listed
SARA EHS 40CFR 355: Not listed
TSCA: Listed

Analytical Methods

Soil: SW846 8270C

AMI4040 CAS #: 111-41-1

AMINOETHYLETHANOLAMINE

RTECS: KJ6300000
EINECS Number: 203-867-5
Molecular Formula: $C_4H_{12}N_2O$
Structured MF: $NH_2(CH_2)_2NH_2CH_2CH_2OH$
Formula Weight: 104.15

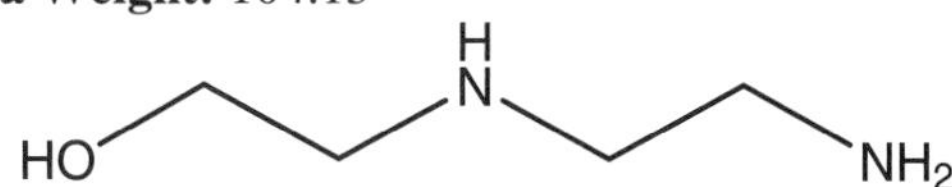

Chemical Structure

Synonyms: AMINOETHYL ETHANOLAMINE; 2-((2-AMINOETHYL)AMINO)ETHANOL; 2-((AMINOETHYL)AMINO)ETHANOL; 2-(2-AMINOETHYL)AMINOETHANOL; N-(BETA-AMINOETHYL)ETHANOLAMINE; N-AMINOETHYLETHANOLAMINE; 2-AMINO-2'-HYDROXYDIETHYLAMINE; ETHANOLETHYLENE DIAMINE; HYDROXYETHYL ETHYLENEDIAMINE; N-(2-HYDROXYETHYL)-1,2-ETHANEDIAMINE; N-HYDROXYETHYL-1,2-ETHANEDIAMINE; (2-HYDROXYETHYL)ETHYLENE DIAMINE; (2-HYDROXYETHYL)ETHYLENEDIAMINE; (BETA-HYDROXYETHYL)ETHYLENEDIAMINE; N-(2-HYDROXYETHYL)ETHYLENEDIAMINE; N-(BETA-HYDROXYETHYL)ETHYLENEDIAMINE; N-(HYDROXYETHYL)ETHYLENEDIAMINE; MONOETHANOLETHYLENEDIAMINE

Description: colorless liquid; mild ammonia-like odor

Use: textile finishing compounds(antifoaming agents, dyestuffs, cationic surfactant resins, rubber products, insecticides, and certain medicinals; sequestering agent for metal ions

Physical Properties

Boiling Point: 238 °C (460 °F) to 240 °C (464 °F)
Specific Gravity: 1.0254 at 25 °C
Vapor Density: 3.59 Air=1
Saturated Vapor Density: < 1.200040917 kg/m^3
Density: 1.025 g/mL
Bulk Density: 8.6 lbs/gal at 20 °C
Vapor Pressure: < 0.01 mm Hg at 20 °C
Water Solubility: Very Soluble in Water
Other Solubilities: 95% Ethanol: >=100 mg/ml at 21 °C; Acetone: >=100 mg/ml at 21 °C; Benzene: Slightly Soluble; Chloroform: miscible; DMSO: >=100 mg/ml at 21 °C; Ligroin: Very Soluble.
Refraction Index: 1.4861 at 20 °C/D
Flash Point: 135 °C Closed Cup
Autoignition Temperature: 368 °C
LEL: 1% v/v
UEL: Calculated at 8% v/v

RTECS Toxicity Data

Acute Oral: Rat LD_{50} Dose: 3 gm/kg. Mouse LD_{50} Dose: 3550 mg/kg.
Acute Dermal: Rabbit LD_{50} Route: Skin; Dose: 3560 uL/kg. Rat LD_{50} Route: Skin; Dose: 2250 mg/kg.
Irritation Eye: Rabbit Standard Draize Test Dose: 50 mg; Reaction: severe.
Irritation Skin: Rabbit Open Draize Test Dose: 445 mg open; Reaction: mild.
Mutagenic: Bacteria - S Typhimurium Mutations in Microorganisms; Dose: 2800 ug/plate (-S9).

Hazard Overviews

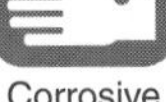
Corrosive

Fire Diamond: 1 / 3 / 0 / —

Health: Corrosive, causes severe burns to eyes, moderate burns to the skin/respiratory tract. Also Causes: permanent eye damage. Chronic Effects: allergic contact dermatitis to the fingertips, or secondary nail dystrophy.
Fire: Will burn. Use dry chemical, carbon dioxide, fog, or alcohol-resistant foam. Water spray may cause frothing.
Reactivity: Stable. Hazardous polymerization cannot occur. Avoid: ignition sources. Incompatible with: cellulose nitrate; acids; oxidizing materials. Hazardous decomposition products: carbon dioxide; nitrogen oxide gas.
Carcinogenicity: IARC - Not listed; NIOSH - Not listed; NTP - Not listed; ACGIH - Not listed; OSHA - Not listed; EPA - Not listed; MAK - Not listed

Primary Target Organs:

Eyes

Skin

Respiratory System

Environmental

Environmental Fate: If released to the atmosphere, it will exist primarily in the vapor-phase where it will degrade rapidly by reaction with photochemically produced hydroxyl radicals (estimated half-life of 3.3 hours). Physical removal from the atmosphere may be possible through wet deposition. If released to soil or water, it will probably biodegrade. The results of one biodegradation screening study have indicated that it biodegrades readily. It can exist in both the neutral and protonated forms in environmental media; the neutral form may leach readily; the mobility of the protonated form is not known.
Cleanup/Disposal: Guide No. 171: Do not touch or walk through spilled material. Stop leak if you can do it without risk. Prevent dust cloud. Avoid inhalation of asbestos dust. Small Dry Spills: With clean shovel place material into clean, dry container and cover loosely; move containers from spill area. Small Spills: Take up with sand or other noncombustible absorbent material and place into containers for later disposal. Large Spills: Dike far ahead of liquid spill for later disposal. Cover powder spill with plastic sheet or tarp to minimize spreading. Prevent entry into waterways, sewers, basements or confined areas.

Environmental Physical Data

Henry's Law Constant: estimated at 1.10×10^{-13}
Octanol/Water Partition Coefficient: **log K_{ow}** = -1.39
Sorption Partition Coefficient: K_{oc} = estimated at 4.2
BCF: estimated at 0.05

Regulations

RCRA 40CFR: Not listed
CERCLA: 40CFR 302.4: Not listed
SARA 40CFR 372.65: Not listed
SARA EHS 40CFR 355: Not listed
TSCA: Listed

AMI4120 CAS #: 140-31-8

1-AMINOETHYLPIPERAZINE

RTECS: TK8050000
DOT: UN2815
EINECS Number: 205-411-0
Molecular Formula: $C_6H_{15}N_3$
Structured MF: $C_6H_{15}N_3$
Formula Weight: 129.24

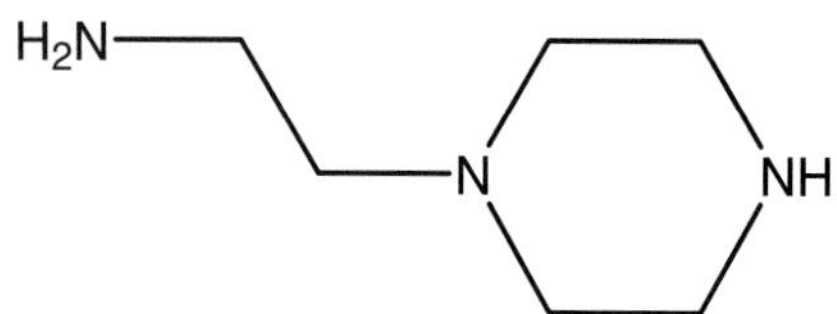

Chemical Structure

Synonyms: 1-(2-AMINOETHYL)PIPERAZINE; AMINOETHYLPIPERAZINE; N-(2-AMINOETHYL)PIPERAZINE; N-(AMINOETHYL)PIPERAZINE; N-(BETA-AMINOETHYL)PIPERAZINE; N-AMINOETHYLPIPERAZINE; PIPERAZINE,1-(2-AMINOETHYL)-; 1-PIPERAZINEETHANAMINE; 1-PIPERAZINEETHYLAMINE

Description: water-white, light-colored liquid; ammonia-like odor

Use: epoxy curing agent; intermediate for pharmaceuticals; anthelmintics, surface-active agents; synthetic fibers

Physical Properties

Boiling Point: 222 °C (432 °F)
Freezing Point: 17.6 °C (63.68 °F)
Specific Gravity: 0.9837
Vapor Density: 4.4 Air=1
Vapor Pressure: 0.075 mm Hg
Water Solubility: Soluble in Water
pH: 11.5
Refraction Index: 1.4983
Flash Point: 93 °C Open Cup
Autoignition Temperature: > 299 °C
LEL: 1.6% v/v
UEL: 6.5% v/v

RTECS Toxicity Data

Acute Oral: Rat LD_{50} Dose: 2140 uL/kg.

Acute Dermal: Rabbit LD_{50} Route: Skin; Dose: 880 uL/kg.

Irritation Eye: Rabbit Standard Draize Test Dose: 20 mg/24H; Reaction: moderate.

Irritation Skin: Rabbit Standard Draize Test Dose: 5 mg/24H; Reaction: severe.

Reproductive/Teratogenic: Rat Route: Oral; Dose: 1680 mg/kg; Duration: male 28D prior to mating; Paternal Effects - Spermatogenesis.

Mutagenic: Mouse Morphological Transformation; Cell Type: lymphocyte; Dose: 1 uL/L. Hamster Mutations in Mammalian Somatic Cells; Cell Type: ovary; Dose: 500 ug/L.

Hazard Overviews

Corrosive

Fire Diamond

Health: Corrosive to eyes/skin/respiratory tract. Also Causes: bronchitis, shortness of breath, pulmonary edema, burns to the lips, tongue, oral mucosa, esophagus, stomach, resulting in pain, drooling, swallowing difficulty, abdominal cramping. Chronic Effects: repeated skin contact can cause rash and sensitization.

Fire: Will burn. For small fires use carbon dioxide or dry chemical; for large fires use alcohol-type foam and water spray.

Reactivity: Stable. Hazardous polymerization cannot occur. Avoid: excess heating; nitrates. Incompatible with: strong acids; carbon monoxide; nitrites. Hazardous decomposition products: toxic vapors; gases

Carcinogenicity: IARC - Not listed; NIOSH - Not listed; NTP - Not listed; ACGIH - Not listed; OSHA - Not listed; EPA - Not listed; MAK - Not listed

Primary Target Organs:

Eyes

Skin

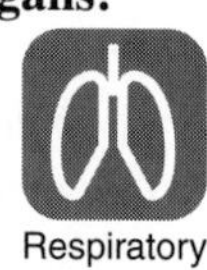

Respiratory System

Mucous Membranes

Environmental

Cleanup/Disposal: Guide No. 153: Eliminate all ignition sources (no smoking, flares, sparks or flames in immediate area). Do not touch damaged containers or spilled material unless wearing appropriate protective clothing. Stop leak if you can do it without risk. Prevent entry into waterways, sewers, basements or confined areas. Absorb or cover with dry earth, sand or other non-combustible material and transfer to containers. Do not get water inside containers.

Regulations

RCRA 40CFR: Not listed
CERCLA: 40CFR 302.4: Not listed
SARA 40CFR 372.65: Not listed
SARA EHS 40CFR 355: Not listed
TSCA: Listed

AMI4200 CAS #: 1760-24-3

AMINOETHYL-PROPYLTRIETHOXYSILANE

RTECS: KV7400000
EINECS Number: 217-164-6
Molecular Formula: $C_8H_{22}N_2O_3Si$
Formula Weight: 222.41

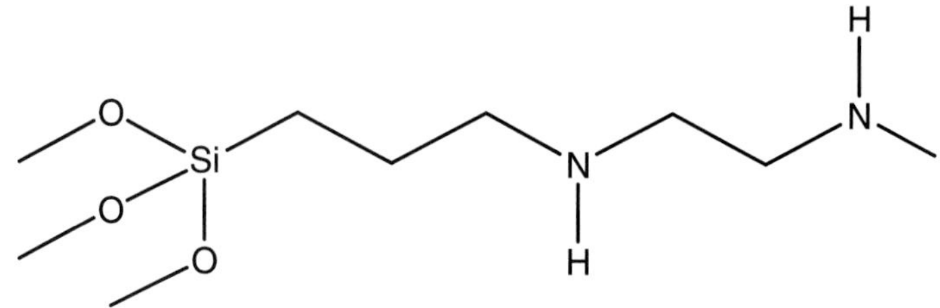

Chemical Structure

Synonyms: A 0700; AAS-M; AP 132; DOW CORNING Z-6020 SILANE; GF 91; KBM 603; NUCA 1120; PROSIL 3128; SH 6020; SILANE,(3-(2-AMINOETHYL)AMINOPROPYL)TRIMETHOXY-; SILICONE A-1120; N-(3-TRIMETHOXYSILYLPROPYL)-ETHYLENEDIAMINE; Z 6020

Physical Properties

Boiling Point: 146 °C (294.8 °F) at 15 mm Hg
Specific Gravity: 1.01000
Water Solubility: Reacts
Refraction Index: 1.447

RTECS Toxicity Data

Acute Oral: Rat LD_{50} Dose: 7460 uL/kg.
Acute Dermal: Rabbit LD_{Lo} Route: Skin; Dose: 16 mL/kg.
Irritation Eye: Rabbit Standard Draize Test Dose: 15 mg; Reaction: severe.
Irritation Skin: Rabbit Open Draize Test Dose: 500 mg open; Reaction: mild.

Hazard Overviews

Corrosive

Health: Corrosive to eyes/skin/respiratory tract. Other Acute Effects: harmful if swallowed, inhaled, or absorbed through skin; inhalation may result in spasm, inflammation and edema of the larynx and bronchi, chemical pneumonitis and pulmonary edema; symptoms of exposure may include burning sensation; coughing; wheezing; laryngitis; shortness of breath; headache; nausea; vomiting; may cause allergic respiratory and skin reactions; target organs: eyes, nerves, kidneys, liver.
Fire: Hazards: under fire conditions, material may decompose to form flammable and/or explosive mixtures in air; emits toxic fumes. Extinguishing agents: carbon dioxide; dry chemical powder; appropriate foam; do not use water. Precautions: combustible liquid.
Reactivity: Stable. Hazardous polymerization will not occur. Incompatible with: oxidizing agents, acids, may decompose on exposure to moist air or water, water and acids, reacts with material to liberate methanol. Hazardous decomposition products: toxic fumes of: carbon monoxide, carbon dioxide, silicon oxide, nitrogen oxides, formaldehyde.
Carcinogenicity: IARC - Not listed; NIOSH - Not listed; NTP - Not listed; ACGIH - Not listed; OSHA - Not listed; EPA - Not listed; MAK - Not listed
Primary Target Organs:

Eyes

Skin
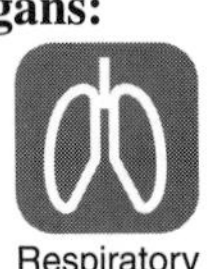
Respiratory System

Nervous System

Liver

Kidneys

Environmental

Regulations

RCRA 40CFR: Not listed
CERCLA: 40CFR 302.4: Not listed
SARA 40CFR 372.65: Not listed
SARA EHS 40CFR 355: Not listed
TSCA: Listed

AMI4280 CAS #: 153-78-6

2-AMINOFLUORENE

RTECS: LL5075000
EINECS Number: 205-817-8
Molecular Formula: $C_{13}H_{11}N$
Formula Weight: 181.23

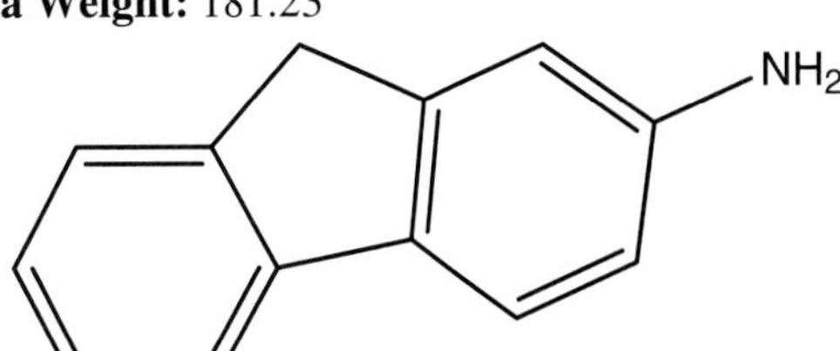

Chemical Structure

Synonyms: AMINOFLUOREN; 2-AMINO-FLUORENE; 2-FLUORENAMINE; 9H-FLUOREN-2-AMINE; FLUOREN-2-AMINE; 2-FLUORENEAMINE; FLUORENE,2-AMINO; FLUORENE,2-AMINO-; 2-FLUROENYLAMINE
Description: needles or long plates
Use: chemical research; experimental carcinogen

Physical Properties

Freezing Point: 131 °C (267.8 °F) to 132 °C (269.6 °F)
Water Solubility: < 1 mg/mL at 19.5 C
Other Solubilities: 95% Ethanol: 5-10 mg/ml at 19.5 °C; Acetone: >=100 mg/ml at 19.5 °C; DMSO: >=100 mg/ml at 19.5 °C; Ether: Soluble.
Flash Point: Not available; probably combustible

RTECS Toxicity Data

Mutagenic: Human Unscheduled DNA Synthesis; Cell Type: liver; Dose: 100 nmol/L. Human DNA Inhibition; Cell Type: HeLa cell; Dose: 225 umol/L. Human DNA Adduct; Cell Type: lymphocyte; Dose: 30 umol/L. Human Sister Chromatid Exchenge; Cell Type: lymphocyte; Dose: 25 umol/L.
Tumorigenic: Rat Route: Oral; Dose: 3600 mg/kg/32W-C; Toxic Effects: Tumorigenic - Carcinogenic by RTECS criteria; Sense organs and special senses - Tumors; Skin and appendages - Tumors. Rat Route: Oral; Dose: 4000 mg/kg/23W-C; Toxic Effects: Tumorigenic - Equivocal tumorigenic agent by RTECS criteria; Skin and appendages - Tumors. Rat Route: Oral; Dose: 3200 mg/kg/58W-C; Toxic Effects: Tumorigenic - Equivocal tumorigenic agent by RTECS criteria; Liver - Tumors; Kidney, Ureter, and Bladder - Tumors. Rat Route: Oral; Dose: 2420 mg/kg/23W-C; Toxic Effects: Tumorigenic - Neoplastic by RTECS criteria; Liver - Tumors; Skin and appendages - Tumors.

Hazard Overviews

Health: Irritating to eyes/skin/respiratory tract. Harmful. Other Acute Effects: harmful if swallowed, inhaled, or absorbed through skin. Chronic Effects: laboratory experiments have shown mutagenic effects. Possible human carcinogen.

Fire: Will burn. Hazards: emits toxic fumes. Extinguishing agents: water spray; carbon dioxide, dry chemical powder or appropriate foam. Precautions: combustible liquid.

Reactivity: Incompatible with: strong oxidizing agents. Hazardous decomposition products: toxic fumes of: carbon monoxide, carbon dioxide, nitrogen oxides.

Carcinogenicity: IARC - Not listed; NIOSH - Not listed; NTP - Not listed; ACGIH - Not listed; OSHA - Not listed; EPA - Not listed; MAK - Not listed

Primary Target Organs:

Eyes

Skin

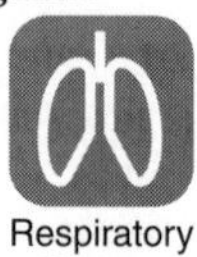
Respiratory System

Environmental

Environmental Fate: It is expected to biodegrade very slowly and is not expected to hydrolyze in the environment. A calculated K_{oc} range of 640 to 1100, indicates it is characterized by a low mobility class in soil. In aquatic systems, it may partition from the water column to organic matter contained in sediments and suspended solids. It also has a low potential to bioconcentrate in aquatic systems. A Henry's Law constant of 3.62 $x10^{-8}$ atm-cu m/mole at 25 °C, suggests volatilization from environmental waters will not be important. It is expected to exist in both the particulate and vapor phases in ambient air. For vapor phase in the atmosphere, the vapor phase reaction with photochemically produced hydroxyl radicals (half-life of 2 hr) is likely to be an important fate process. Otherwise, washout by precipitation and gravitational settling may be an important removal mechanism in atmospheric particulate matter.

Environmental Physical Data

Henry's Law Constant: estimated at 3.62 $x10^{-8}$

Octanol/Water Partition Coefficient: log K_{ow} = 3.00

Sorption Partition Coefficient: K_{oc} = estimated at 650 to 1100

BCF: calculated at 1.93 to 2.05

Regulations

RCRA 40CFR: Not listed

CERCLA: 40CFR 302.4: Not listed

SARA 40CFR 372.65: Not listed

SARA EHS 40CFR 355: Not listed

TSCA: Not listed

AMI4360 CAS #: 2432-74-8

6-AMINOHEXANENITRILE

EINECS Number: 219-409-2

Molecular Formula: $C_6H_{12}N_2$

Formula Weight: 112.17

Synonyms: 6-AMINOCAPRONITRILE; OMEGA-AMINOCAPRONITRILE; HEXANENITRILE,6-AMINO-

Use: no evidence of commercial use in us

Hazard Overviews

Health: Irritating to eyes/skin/respiratory tract. Harmful. Other Acute Effects: may be harmful by inhalation, ingestion, or skin absorption.

Fire: Hazards: emits toxic fumes. Extinguishing agents: water spray; carbon dioxide, dry chemical powder or appropriate foam. Precautions: combustible liquid.

Reactivity: Incompatible with: strong oxidizing agents. Hazardous decomposition products: toxic fumes of: carbon monoxide, carbon dioxide, nitrogen oxides.

Carcinogenicity: IARC - Not listed; NIOSH - Not listed; NTP - Not listed; ACGIH - Not listed; OSHA - Not listed; EPA - Not listed; MAK - Not listed

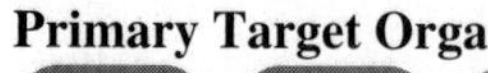

Primary Target Organs:

Eyes

Skin

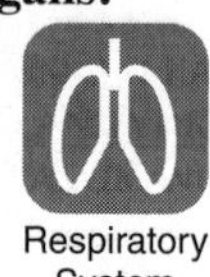
Respiratory System

Environmental

Regulations

RCRA 40CFR: Not listed

CERCLA: 40CFR 302.4: Not listed

SARA 40CFR 372.65: Not listed

SARA EHS 40CFR 355: Not listed

TSCA: Listed

AMI4440 CAS #: 143-23-7

N-(6-AMINOHEXYL)-1,6-HEXANEDIAMINE

EINECS Number: 205-593-1

Molecular Formula: $C_{12}H_{29}N_3$

Formula Weight: 215.38

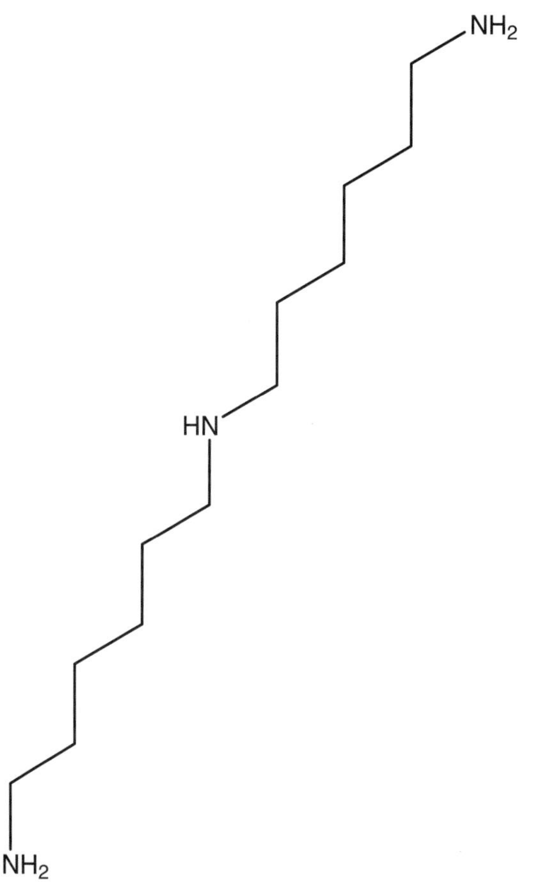

Chemical Structure

Synonyms: BIS(6-AMINOHEXYL)AMINE; BIS(HEXAMETHYLENE)TRIAMINE; 1,13-DIAMINO-7-AZATRIDECANE; DIHEXYLAMINE,6,6'-DIAMINO-; DIHEXYLENETRIAMINE; 1,6-HEXANEDIAMINE,N-(6-AMINOHEXYL)-

Hazard Overviews

Poison

Corrosive

Health: Corrosive to eyes/skin/respiratory tract. Poison. Other Acute Effects: may be fatal if inhaled, swallowed, or absorbed through skin; inhalation may result in spasm; inflammation and edema of the larynx and bronchi; chemical pneumonitis and pulmonary edema; burning sensation; coughing; wheezing; laryngitis; shortness of breath; headache; nausea; vomiting.

Fire: Hazards: emits toxic fumes. Extinguishing agents: water spray; carbon dioxide, dry chemical powder or appropriate foam. Precautions: combustible liquid.

Reactivity: Stable. Hazardous polymerization will not occur. Incompatible with: strong oxidizing agents, strong acids, protect from moisture. Hazardous decomposition products: toxic fumes of: carbon monoxide, carbon dioxide, nitrogen oxides.

Carcinogenicity: IARC - Not listed; NIOSH - Not listed; NTP - Not listed; ACGIH - Not listed; OSHA - Not listed; EPA - Not listed; MAK - Not listed

Primary Target Organs:

Eyes

Skin

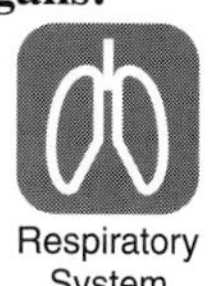
Respiratory System

Environmental

Regulations

RCRA 40CFR: Not listed
CERCLA: 40CFR 302.4: Not listed
SARA 40CFR 372.65: Not listed
SARA EHS 40CFR 355: Not listed
TSCA: Listed

AMI4520 **CAS #: 61-78-9**

P-AMINOHIPPURIC ACID

EINECS Number: 200-518-9
Molecular Formula: $C_9H_{10}N_2O_3$
Formula Weight: 194.19

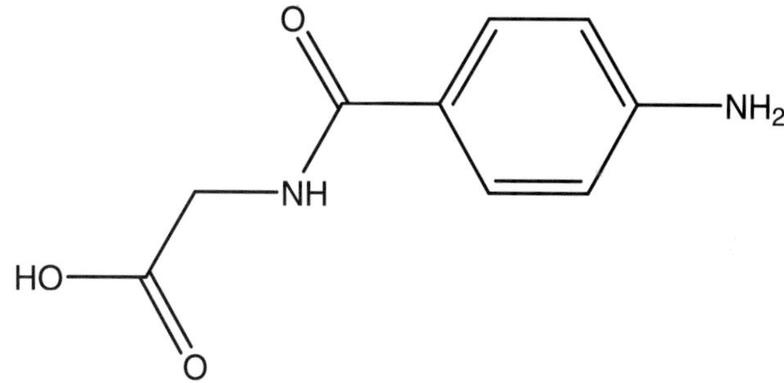

Chemical Structure

Synonyms: N-(P-AMINOBENZOYL)AMINOACETIC ACID; N-(4-AMINOBENZOYL)GLYCINE; N-(P-AMINOBENZOYL)GLYCINE; N-(PARA-AMINOBENZOYL)GLYCINE; 4-AMINOHIPPURIC ACID; PARA-AMINOHIPPURIC ACID; GLYCINE,N-(4-AMINOBENZOYL)-; HIPPURIC ACID,P-AMINO-; PAH; PAHA

Description: white, crystalline powder, needles, or prisms

Use: medication: diagnostic aid (renal function determination); intermediate

Physical Properties

Freezing Point: 198 °C (388.4 °F) to 199 °C (390.2 °F)
Water Solubility: 1 g Soluble in 45 ml Water
Other Solubilities: Soluble in Chloroform; Benzene; Acetone; Practically Insoluble in Ether; Carbon Tetrachloride; 1 g Soluble in 5 ml diluted Hydrochloric Acid.

Hazard Overviews

Health: May cause irritation to eyes/skin. Other Acute Effects: may be harmful by inhalation, ingestion, or skin absorption.

Fire: Hazards: emits toxic fumes. Extinguishing agents: carbon dioxide, dry chemical powder or appropriate foam; water spray. Precautions: combustible liquid.
Reactivity: Incompatible with: strong oxidizing agents. Hazardous decomposition products: carbon monoxide, carbon dioxide, nitrogen oxides.
Carcinogenicity: IARC - Not listed; NIOSH - Not listed; NTP - Not listed; ACGIH - Not listed; OSHA - Not listed; EPA - Not listed; MAK - Not listed

Environmental

Regulations

RCRA 40CFR: Not listed
CERCLA: 40CFR 302.4: Not listed
SARA 40CFR 372.65: Not listed
SARA EHS 40CFR 355: Not listed
TSCA: Listed

AMI4600 CAS #: 548-93-6

2-AMINO-3-HYDROXY-BENZOIC ACID

RTECS: DG2625000
EINECS Number: 208-962-5
Molecular Formula: $C_7H_7NO_3$
Formula Weight: 153.14

H_2N OH HO O

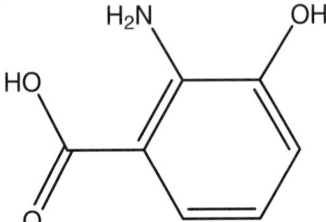

Chemical Structure

Synonyms: 2-AMINO-3-HYDROXYBENZOIC ACID; ANTHRANILIC ACID,3-HYDROXY-; 3-HYDROXYANTHRANILIC ACID; 3-HYDROXY-ANTHRANILSAEURE; 3-OHAA; 3-OXYANTHRANILIC ACID
Description: leaflets
Use: biological intermediate in the conversion of tryptophan to acetyl-coenzyme a; biological intermediate in the biosynthesis of niacin from tryptophan in niacin-deficient animals

Physical Properties

Freezing Point: 164 °C (327.2 °F)
Water Solubility: Slightly Soluble in Water
Other Solubilities: Soluble in Alcohol, Ether, Chloroform.

RTECS Toxicity Data

Mutagenic: Human Cytogenetic Analysis; Cell Type: embryo; Dose: 30 mg/L. Human Cytogenetic Analysis; Cell Type: leukocyte; Dose: 100 mg/L. Monkey Cytogenetic Analysis; Route: Subcutaneous; Dose: 1485 mg/kg/57D-I. Monkey Cytogenetic Analysis; Cell Type: kidney; Dose: 100 mg/L.
Tumorigenic: Mouse Route: Subcutaneous; Dose: 1600 mg/kg/8W-I; Toxic Effects: Tumorigenic - Carcinogenic by RTECS criteria; Kidney, Ureter, and Bladder - Tumors. Mouse Route: Subcutaneous; Dose: 185 mg/kg (13-17D preg); Toxic Effects: Tumorigenic - Equivocal tumorigenic agent by RTECS criteria; Tumorigenic effects - Transplacental tumorigenesis; Lungs, Thorax, or Respiration - Bromchiogenic carcinoma. Mouse Route: Subcutaneous; Dose: 8000 mg/kg/19W-I; Toxic Effects: Tumorigenic - Equivocal tumorigenic agent by RTECS criteria; Blood - Leukemia. Mouse Route: Subcutaneous; Dose: 2000 mg/kg/7W-I; Toxic Effects: Tumorigenic - Carcinogenic by RTECS criteria; Blood - Leukemia. Mouse Route: Subcutaneous; Dose: 2040 mg/kg/59W-C; Toxic Effects: Tumorigenic - Equivocal tumorigenic agent by RTECS criteria; Blood - Leukemia. Mouse Route: Subcutaneous; Dose: 121 mg/kg/21D-I; Toxic Effects: Tumorigenic - Equivocal tumorigenic agent by RTECS criteria; Lungs, Thorax, or Respiration - Bromchiogenic carcinoma; Liver - Tumors.

Hazard Overviews

Health: Irritating to eyes/skin/respiratory tract. Harmful. Other Acute Effects: harmful if swallowed, inhaled, or absorbed through skin. Chronic Effects: possible carcinogen.
Fire: Hazards: emits toxic fumes. Extinguishing agents: water spray; carbon dioxide, dry chemical powder or appropriate foam. Precautions: combustible liquid.
Reactivity: Incompatible with: strong oxidizing agents, strong acids, strong bases. Hazardous decomposition products: toxic fumes of: carbon monoxide, carbon dioxide, nitrogen oxides.
Carcinogenicity: IARC - Not listed; NIOSH - Not listed; NTP - Not listed; ACGIH - Not listed; OSHA - Not listed; EPA - Not listed; MAK - Not listed
Primary Target Organs:

Eyes

Skin

Respiratory System

Environmental

Regulations

RCRA 40CFR: Not listed
CERCLA: 40CFR 302.4: Not listed
SARA 40CFR 372.65: Not listed
SARA EHS 40CFR 355: Not listed
TSCA: Not listed

AMI4680 CAS #: 62-57-7

2-AMINOISOBUTYRIC ACID

RTECS: AY7000000
EINECS Number: 200-544-0
Molecular Formula: $C_4H_9NO_2$
Formula Weight: 103.14

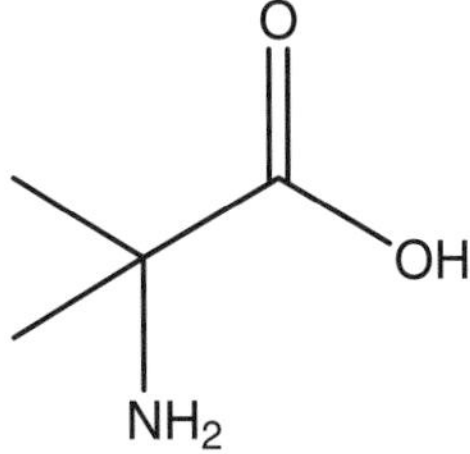

Chemical Structure

Synonyms: AIB; ALPHA-AMINOISOBUTANOIC ACID; ALPHA-AMINOISOBUTYRIC ACID; 2-AMINO-2-METHYLPROPANOIC ACID; CB 1637; ALPHA,ALPHA-DIMETHYLGLYCINE; 2-METHYLALANINE; ALPHA-METHYLALANINE

Physical Properties

Freezing Point: 335 °C (635 °F)
Water Solubility: Freely Soluble
Other Solubilities: Ethanol: Slightly soluble; Ether: Insoluble

Hazard Overviews

Health: May cause irritation. Other Acute Effects: may be harmful by inhalation, ingestion, or skin absorption. The toxicological properties have not been thoroughly investigated.
Fire: Hazards: emits toxic fumes. Extinguishing agents: water spray. Precautions: combustible liquid.
Reactivity: Stable. Hazardous polymerization will not occur. Hazardous decomposition products: thermal decomposition may produce carbon monoxide, carbon dioxide, and nitrogen oxides.
Carcinogenicity: IARC - Not listed; NIOSH - Not listed; NTP - Not listed; ACGIH - Not listed; OSHA - Not listed; EPA - Not listed; MAK - Not listed

Environmental

Regulations

RCRA 40CFR: Not listed
CERCLA: 40CFR 302.4: Not listed
SARA 40CFR 372.65: Not listed
SARA EHS 40CFR 355: Not listed
TSCA: Listed

AMI4840 CAS #: 2840-26-8

3-AMINO-4-METHOXY BENZOIC ACID

RTECS: DG2872000
EINECS Number: 220-634-3
Molecular Formula: $C_8H_9NO_3$
Formula Weight: 167.18

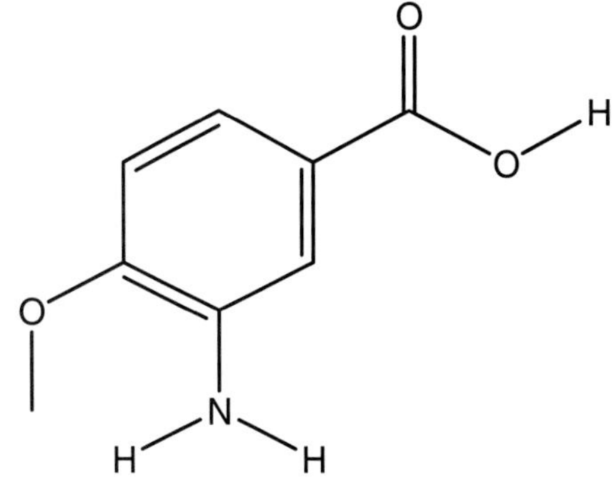

Chemical Structure

Synonyms: 3-AMINO-P-ANISIC ACID; 3-AMINO-4-METHOXYBENZOIC ACID; P-ANISIC ACID,3-AMINO-; KYSELINA 3-AMINO-4-METHOXYBENZOOVA

Physical Properties

Freezing Point: 239 °C (462.2 °F) to 241 °C (465.8 °F)

RTECS Toxicity Data

Irritation Eye: Rabbit Standard Draize Test Dose: 500 mg/24H; Reaction: mild.

Hazard Overviews

Health: Irritating to eyes/skin/respiratory tract. Other Acute Effects: may be harmful by inhalation, ingestion, or skin absorption.
Fire: Hazards: emits toxic fumes. Extinguishing agents: water spray; carbon dioxide, dry chemical powder or appropriate foam. Precautions: combustible liquid.
Reactivity: Incompatible with: strong oxidizing agents. Hazardous decomposition products: toxic fumes of: carbon monoxide, carbon dioxide, nitrogen oxides.
Carcinogenicity: IARC - Not listed; NIOSH - Not listed; NTP - Not listed; ACGIH - Not listed; OSHA - Not listed; EPA - Not listed; MAK - Not listed
Primary Target Organs:

Eyes

Skin

Respiratory System

Environmental

Regulations

RCRA 40CFR: Not listed
CERCLA: 40CFR 302.4: Not listed
SARA 40CFR 372.65: Not listed
SARA EHS 40CFR 355: Not listed
TSCA: Not listed

AMI4920 **CAS #: 543-82-8**

2-AMINO-6-METHOXY BENZOTHIAZOLE

RTECS: MQ4840000
EINECS Number: 208-851-1
Molecular Formula: $C_8H_{19}N$
Formula Weight: 129.28

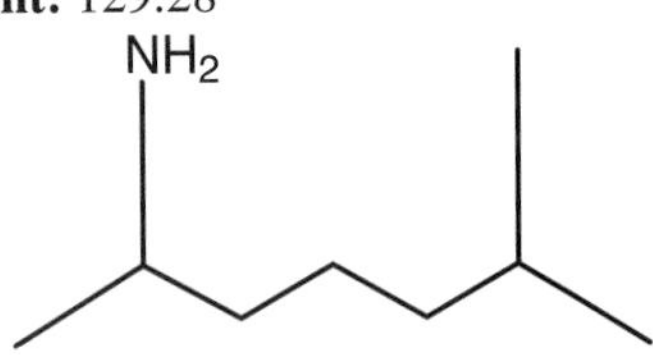

Chemical Structure

Synonyms: AMIDRINE; 2-AMINO-6-METHYLHEPTANE; 6-AMINO-2-METHYLHEPTANE; 1,5-DIMETHYLHEXYLAMINE; ALPHA,EPSILON-DIMETHYLHEXYLAMINE; 2-HEPTYLAMINE,6-METHYL-; 2-ISOOCTYL AMINE; 2-METHYL-6-AMINOHEPTANE; 6-METHYL-2-HEPTYLAMINE; 2-METIL-6-AMINO-EPTANO; OCTODRINE; SKF 51; VAPORPAC

RTECS Toxicity Data

Acute Oral: Rat LD_{50} Dose: 538 mg/kg; Toxic Effects: Behavioral - Somnolence (general depressed activity); Behavioral - Convulsions or effect on seizure threshold; Behavioral - Ataxia.
Acute Dermal: Mouse LD_{Lo} Route: Subcutaneous Dose: 100 mg/kg.

Hazard Overviews

Health: Irritating to eyes/skin/respiratory tract. Harmful. Acute Effects: harmful if swallowed; may be harmful if inhaled; may be harmful if absorbed through the skin.
Fire: Hazards: emits toxic fumes. Extinguishing agents: carbon dioxide, dry chemical powder or appropriate foam; water spray. Precautions: combustible liquid.
Reactivity: Incompatible with: acids, acid chlorides, acid anhydrides, strong oxidizing agents, carbon dioxide. Hazardous decomposition products: thermal decomposition may produce carbon monoxide, carbon dioxide, and nitrogen oxides.
Carcinogenicity: IARC - Not listed; NIOSH - Not listed; NTP - Not listed; ACGIH - Not listed; OSHA - Not listed; EPA - Not listed; MAK - Not listed
Primary Target Organs:

Eyes

Skin

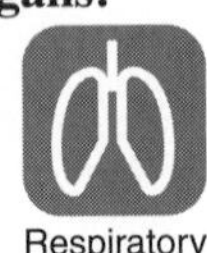
Respiratory System

Environmental

Regulations

RCRA 40CFR: Not listed
CERCLA: 40CFR 302.4: Not listed
SARA 40CFR 372.65: Not listed
SARA EHS 40CFR 355: Not listed
TSCA: Listed

AMI5000 **CAS #: 3731-51-9**

2-AMINOMETHYL PYRIDINE

RTECS: US1840000
EINECS Number: 223-090-5
Molecular Formula: $C_6H_8N_2$
Formula Weight: 108.16

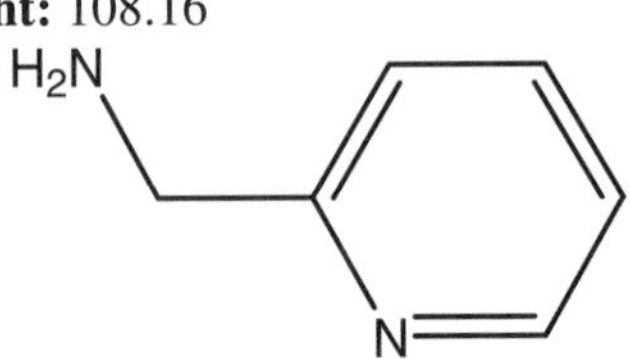

Chemical Structure

Synonyms: 2-AMINOMETHYLPYRIDINE; 2-PICOLINAMINE; 2-PICOLYLAMINE; 2-PYRIDINEMETHYLAMINE; (2-PYRIDYLMETHYL)AMINE

Physical Properties

Boiling Point: 197 °C (386.6 °F)
Specific Gravity: 1.05250
Water Solubility: Soluble
Refraction Index: 1.5431

RTECS Toxicity Data

Acute Oral: Quail LD_{50} Dose: 750 mg/kg.

Hazard Overviews

Corrosive

Health: Corrosive to eyes/skin/respiratory tract. Other Acute Effects: harmful if swallowed, inhaled, or absorbed through skin; inhalation may result in spasm, inflammation and edema of the larynx and bronchi, chemical pneumonitis and pulmonary edema; symptoms of exposure may include burning sensation; coughing; wheezing; laryngitis; shortness of breath; headache; nausea; vomiting.
Fire: Hazards: emits toxic fumes. Extinguishing agents: carbon dioxide, dry chemical powder or appropriate foam. Precautions: combustible liquid.
Reactivity: Incompatible with: strong oxidizing agents. Hazardous decomposition products: toxic fumes of: carbon monoxide, carbon dioxide, nitrogen oxides.
Carcinogenicity: IARC - Not listed; NIOSH - Not listed; NTP - Not listed; ACGIH - Not listed; OSHA - Not listed; EPA - Not listed; MAK - Not listed

Primary Target Organs:

Eyes

Skin

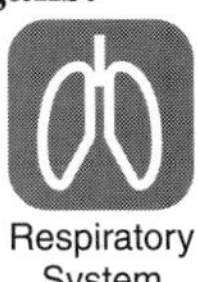
Respiratory System

Environmental

Regulations

RCRA 40CFR: Not listed
CERCLA: 40CFR 302.4: Not listed
SARA 40CFR 372.65: Not listed
SARA EHS 40CFR 355: Not listed
TSCA: Listed

AMI5080 **CAS #: 82-28-0**

1-AMINO-2-METHYLANTHRAQUINONE

RTECS: CB5740000
EINECS Number: 201-408-3
Molecular Formula: $C_{15}H_{11}NO_2$
Formula Weight: 237.3

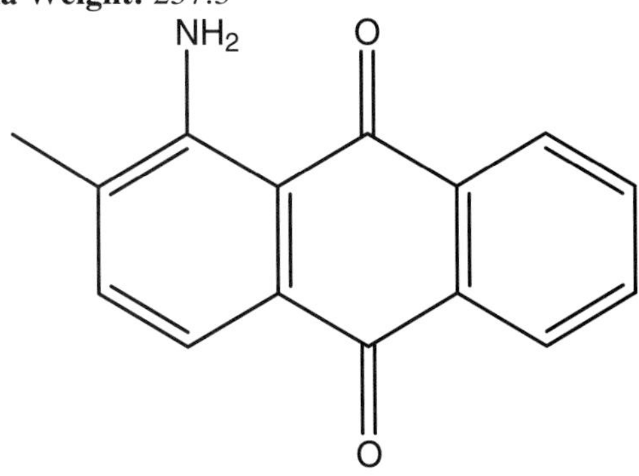

Chemical Structure

Synonyms: ACETATE FAST ORANGE R; ACETOQUINONE LIGHT ORANGE JL; 1-AMINO-2-METHYL-9,10-ANTHRACENEDIONE; 1-AMINO-2-METHYL-9,10-ANTHRAQUINONE; 9,10-ANTHRACENEDIONE,1-AMINO-2-METHYL-; 9,10-ANTHRACENEDIONE,1-AMINO-2-METHYL-(9CI); ANTHRAQUINONE,1-AMINO-2-METHYL-; ARTISIL ORANGE 3RP; C.I. 60700; C.I. DISPERSE ORANGE 11; CELLITON ORANGE R; CILLA ORANGE R; DISPERSE ORANGE; DISPERSE ORANGE (ANTHRAQUINONE DYE); DURANOL ORANGE G; 2-METHYL-1-ANTHRAQUINONYLAMINE; MICROSETILE ORANGE RA; NYLOQUINONE ORANGE JR; ORANZ DISPERZNI 11; PERLITON ORANGE 3R; SERISOL ORANGE YL; SUPRACET ORANGE R
Description: red powder
Use: in the dyeing of wool sheepskins and furs and in surface dyeing of thermoplastics

Physical Properties

Freezing Point: 205 °C (401 °F) to 206 °C (402.8 °F)
Water Solubility: < 1 mg/mL at 20 C
Other Solubilities: DMSO: 10-50 mg/ml at 20 °C; Acetone: 1-5 mg/ml at 20 °C; Toluene: <1 mg/ml at 22 °C; Carbon Tetrachloride: Slightly Soluble; Linseed oil: Soluble; Benzene: Soluble; Ether: Soluble; Chloroform: Soluble; Acetic Acid: Soluble.
Flash Point: Not available; probably combustible

RTECS Toxicity Data

Mutagenic: Bacteria - S Typhimurium Mutations in Microorganisms; Dose: 33 ug/plate (-S9).
Tumorigenic: Rat Route: Oral; Dose: 30 gm/kg/78W-C; Toxic Effects: Tumorigenic - Carcinogenic by RTECS criteria; Liver - Tumors; Kidney, Ureter, and Bladder - Kidney tumors. Rat Route: Oral; Dose: 60 gm/kg/78W-C; Toxic Effects: Tumorigenic - Carcinogenic by RTECS criteria; Liver - Tumors; Kidney, Ureter, and Bladder - Kidney tumors. Rat Route: Oral; Dose: 39 gm/kg/77W-C; Toxic Effects: Tumorigenic - Neoplastic by RTECS criteria; Liver - Tumors; Kidney, Ureter, and Bladder - Kidney tumors. Rat Route: Oral; Dose: 557 gm/kg/79W-C; Toxic Effects: Tumorigenic - Equivocal tumorigenic agent by RTECS criteria; Liver - Tumors; Kidney, Ureter, and Bladder - Kidney tumors. Rat Route: Oral; Dose: 1113 gm/kg/79W-C; Toxic Effects: Tumorigenic - Carcinogenic by RTECS criteria; Liver - Tumors; Kidney, Ureter, and Bladder - Kidney tumors.

Hazard Overviews

Health: May cause irritation. Toxic. Other Acute Effects: harmful if swallowed, inhaled, or absorbed through skin. Chronic Effects: may cause heritable genetic damage; may alter genetic material; target organs: liver, kidneys. Possible carcinogen.
Fire: Will burn. Hazards: emits toxic fumes. Extinguishing agents: water spray; carbon dioxide, dry chemical powder or appropriate foam. Precautions: combustible liquid.
Reactivity: Incompatible with: strong oxidizing agents. Hazardous decomposition products: toxic fumes of: carbon monoxide, carbon dioxide, nitrogen oxides.
Carcinogenicity: IARC - Group 3, Not classifiable as to carcinogenicity to humans; NIOSH - Not listed; NTP - Listed; ACGIH - Not listed; OSHA - Not listed; EPA - Not listed; MAK - Not listed
Primary Target Organs:

Liver

Kidneys

Environmental

Regulations

RCRA 40CFR: Not listed
CERCLA: 40CFR 302.4: Not listed
SARA 40CFR 372.65: Listed
SARA EHS 40CFR 355: Not listed
TSCA: Listed

AMI5240 **CAS #: 124-68-5**

2-AMINO-2-METHYLPROPANOL

RTECS: UA5950000
EINECS Number: 204-709-8
Molecular Formula: $C_4H_{11}NO$
Structured MF: $CH_3CCH_3NH_2CH_2OH$
Formula Weight: 89.14

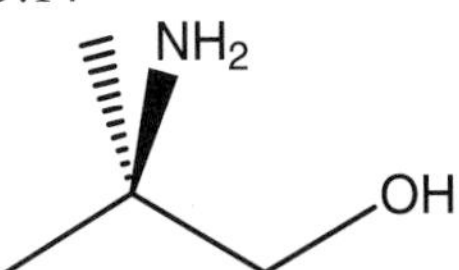

Chemical Structure

Synonyms: 2-AMINO-2,2-DIMETHYLETHANOL; 2-AMINODIMETHYLETHANOL; 2-AMINOISOBUTANOL; BETA-AMINOISOBUTANOL; 2-AMINO-2-METHYL-1-PROPANOL; 2-AMINO-2-METHYLPROPAN-1-OL; AMP; AMP-95; AMP REGULAR; 1,1-DIMETHYL-2-HYDROXYETHYLAMINE; HYDROXY-TERT-BUTYLAMINE; 2-HYDROXYMETHYL-2-PROPYLAMINE; ISOBUTANOL-2-AMINE; ISOBUTANOLAMINE; 2-METHYL-2-AMINOPROPANOL; 2-METHYL-2-AMINOPROPANOL-1; 1-PROPANOL,2-AMINO-2-METHYL-

Description: crystalline mass

Use: in synth of surface-active agents, vulcanization accelerators, pharmaceuticals; absorbent for acidic gases; emulsifying agent for cosmetic creams and lotions, mineral oil & paraffin wax emulsions, leather dressings, textile specialties, polishes, cleaning compounds; ingredient of diuretic pamabrom

Physical Properties

Boiling Point: 165 °C (329 °F) at 760 mm Hg
Freezing Point: 31 °C (87.8 °F)
Specific Gravity: 0.934 at 20 °C/20 °C
Vapor Density: 3 Air=1
Water Solubility: Miscible with Water
Other Solubilities: Soluble in Alcohols
pH: 0.1 aqueous solution 11.3
Refraction Index: 1.449 at 20 °C/D
Flash Point: 67 °C Closed Cup

RTECS Toxicity Data

Acute Oral: Rat LD_{50} Dose: 2900 mg/kg. Mouse LD_{50} Dose: 2150 mg/kg.

Chronic (Multiple Dose) Oral: Rat Dose: 44800 mg/kg/8W-C; Toxic Effects: Liver - Other changes; Skin and appendages - Hair; DEATH. Dog Dose: 11340 mg/kg/28D-C; Toxic Effects: Behavioral - Food intake (animal); Nutritional and gross metabolic - Weight loss or decreased weight gain; Biochemical - Phosphatases.

Chronic (Multiple Dose) Inhalation: Rat Dose: 230 ug/m^3/4H/13W-I; Toxic Effects: Cardiac - Changes in heart weight; Lungs, Thorax, or Respiration - Changes in lung weight; DEATH - Changes in uterine weight. Monkey Dose: 6 mg/m^3/89D-I; Toxic Effects: Lungs, Thorax, or Respiration - Respiratory depression; Nutritional and gross metabolic - Weight loss or decreased weight gain.

Hazard Overviews

Health: Severely irritating to eyes/skin/respiratory tract. Other Acute Effects: harmful if swallowed, inhaled, or absorbed through skin; inhalation may result in spasm, inflammation and edema of the larynx and bronchi, chemical pneumonitis and pulmonary edema; may cause burning sensation; coughing; wheezing; laryngitis; shortness of breath; headache; nausea; vomiting.

Fire: Combustible. Hazards: emits toxic fumes. Extinguishing agents: carbon dioxide, dry chemical powder or appropriate foam. Precautions: combustible liquid.

Reactivity: Stable. Hazardous polymerization will not occur. Incompatible with: oxidizing agents, strong acids, absorbs carbon dioxide from air. Hazardous decomposition products: thermal decomposition may produce carbon monoxide, carbon dioxide, and nitrogen oxides.

Carcinogenicity: IARC - Not listed; NIOSH - Not listed; NTP - Not listed; ACGIH - Not listed; OSHA - Not listed; EPA - Not listed; MAK - Not listed

Primary Target Organs:

Eyes

Skin

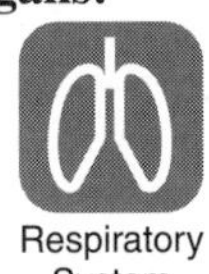
Respiratory System

Environmental

Cleanup/Disposal: Guide No. 128: Eliminate all ignition sources (no smoking, flares, sparks or flames in immediate area). All equipment used when handling the product must be grounded. Do not touch or walk through spilled material. Stop leak if you can do it without risk. Prevent entry into waterways, sewers, basements or confined areas. A vapor suppressing foam may be used to reduce vapors. Absorb or cover with dry earth, sand or other non-combustible material and transfer to containers. Use clean non-sparking tools to collect absorbed material. Large Spills: Dike far ahead of liquid spill for later disposal. Water spray may reduce vapor; but may not prevent ignition in closed spaces.

Regulations

RCRA 40CFR: Not listed
CERCLA: 40CFR 302.4: Not listed
SARA 40CFR 372.65: Not listed
SARA EHS 40CFR 355: Not listed
TSCA: Listed

AMI5320 **CAS #: 81-16-3**

2-AMINO-1-NAPHTHALENESULFONIC ACID

RTECS: QK1250000
EINECS Number: 201-331-5

Molecular Formula: $C_{10}H_9NO_3S$
Formula Weight: 223.25

Chemical Structure

Synonyms: KYSELINA 2-NAFTYLAMIN-1-SULFONOVA; KYSELINA TOBIASOVA; 1-NAPHTHALENESULFONIC ACID,2-AMINO-; 2-NAPHTHYLAMINE-1-SULFONIC ACID; TOBIAS ACID
Description: white needles
Use: intermediate for pigments; optical brightener

Physical Properties

Water Solubility: Slightly Soluble in Cold Water
Other Solubilities: Very Slightly Soluble in Alcohol, Ether

RTECS Toxicity Data

Acute Oral: Rat LD_{50} Dose: 19400 mg/kg.
Irritation Eye: Rabbit Standard Draize Test Dose: 500 mg/24H; Reaction: mild.

Hazard Overviews

Health: Irritating to eyes/skin/respiratory tract. Other Acute Effects: may be harmful by inhalation, ingestion, or skin absorption.
Fire: Hazards: emits toxic fumes. Extinguishing agents: water spray; carbon dioxide, dry chemical powder or appropriate foam. Precautions: combustible liquid.
Carcinogenicity: IARC - Not listed; NIOSH - Not listed; NTP - Not listed; ACGIH - Not listed; OSHA - Not listed; EPA - Not listed; MAK - Not listed
Primary Target Organs:

Eyes

Skin

Respiratory System

Environmental

Environmental Fate: Should biodegrade slowly in the environment. However, it may persist at high concentration due to toxicity to micro-organisms. It is not expected to undergo hydrolysis or bioconcentrate in the environment. It is expected to be highly mobile in soil. In aquatic systems, it should not partition from the water column to organic matter contained in sediments and suspended solids. Data regarding volatilization were unavailable. Therefore, the volatility in the environment and the partitioning in ambient air could not be predicted. However, if released to the atmosphere, the vapor phase reaction with photochemically produced hydroxyl radicals (estimated half-life of 2 hr) is likely to be important.

Environmental Physical Data

Octanol/Water Partition Coefficient: log K_{ow} = -1.16
Sorption Partition Coefficient: K_{oc} = 5 to 45
BCF: calculated at -0.11 to 0.75

Regulations

RCRA 40CFR: Not listed
CERCLA: 40CFR 302.4: Not listed
SARA 40CFR 372.65: Not listed
SARA EHS 40CFR 355: Not listed
TSCA: Listed

AMI5560 **CAS #: 99-57-0**

2-AMINO-4-NITROPHENOL

RTECS: SJ6300000
EINECS Number: 202-767-9
Molecular Formula: $C_6H_6N_2O_3$
Structured MF: $H_2NC_6H_3(NO_2)OH$
Formula Weight: 154.13

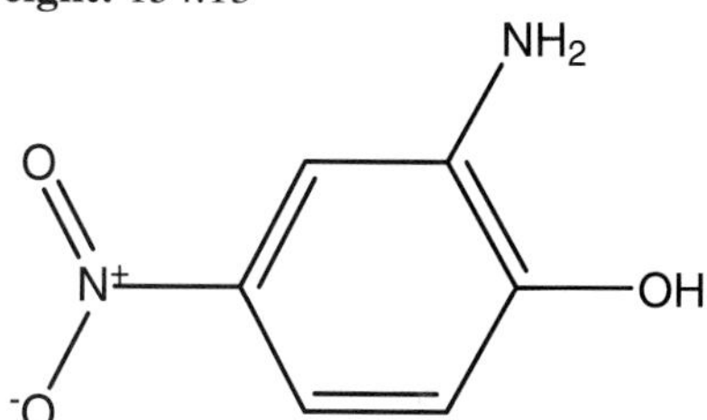

Chemical Structure

Synonyms: 1-AMINO-2-HYDROXY-5-NITROBENZENE; 3-AMINO-4-HYDROXYNITROBENZENE; 2-AMINO-4-NITROFENOL; C.I. 76530; 1-HYDROXY-2-AMINO-4-NITROBENZENE; 2-HYDROXY-5-NITROANILINE; 4-NITRO-2-AMINOFENOL; P-NITROAMINOFENOL; 4-NITRO-2-AMINOPHENOL; P-NITRO-O-AMINOPHENOL; PHENOL,2-AMINO-4-NITRO-; RODOL 42; URSOL 4GL
Description: orange prisms or yellow-brown leaflets; odorless
Use: dyes

Physical Properties

Freezing Point: 145 °C (293 °F) to 147 °C (296.6 °F)
Water Solubility: Slightly Soluble in Water
Other Solubilities: 95% Ethanol: 50-100 mg/ml at 20 °C; Acetone: >=100 mg/ml at 20 °C; Acetic Acid: Soluble; Acid: Soluble; DMSO: >=100 mg/ml at 20 °C; Ether: Soluble.
Flash Point: Not available; probably combustible

RTECS Toxicity Data

Acute Oral: Rat LD_{50} Dose: 2400 mg/kg; Toxic Effects: Behavioral - Somnolence (general depressed activity); Behavioral - Ataxia; Skin and appendages - Hair. Mouse LD_{50} Dose: 850 mg/kg.
Chronic (Multiple Dose) Oral: Rat Dose: 13750 mg/kg/15D-I; Toxic Effects: DEATH. Rat Dose: 32500 mg/kg/13W-I;

Toxic Effects: Liver - Changes in liver weight; Kidney, Ureter, and Bladder - Chgs in tubules (inc acute renal failure, acute tubular necrosis; DEATH.
Irritation Eye: Rabbit Standard Draize Test Dose: 100 mg/24H; Reaction: moderate.
Mutagenic: Hamster Cytogenetic Analysis; Cell Type: ovary; Dose: 249 mg/L. Hamster Sister Chromatid Exchenge; Cell Type: ovary; Dose: 5 mg/L.
Tumorigenic: Rat Route: Oral; Dose: 64375 mg/kg/2Y-C; Toxic Effects: Tumorigenic - Neoplastic by RTECS criteria; Endocrine - Tumors.

Hazard Overviews

Health: Irritating to eyes/skin/respiratory tract. Harmful. Other Acute Effects: harmful by inhalation, in contact with skin and if swallowed. Chronic Effects: possible risk of irreversible effects. Possible carcinogen. Laboratory experiments have shown mutagenic effects.
Fire: Will burn. Hazards: emits toxic fumes. Extinguishing agents: water spray; carbon dioxide, dry chemical powder or appropriate foam. Precautions: combustible liquid.
Carcinogenicity: IARC - Group 3, Not classifiable as to carcinogenicity to humans; NIOSH - Not listed; NTP - Not listed; ACGIH - Not listed; OSHA - Not listed; EPA - Not listed; MAK - Not listed
Primary Target Organs:

Eyes

Skin

Respiratory System

Environmental

Environmental Fate: If released to the atmosphere, it will exist in both the vapor phase and the particulate phase. In the vapor phase, it will react with hydroxyl radicals with an estimated half-life of 2 days. It may also react with nitrate radicals. Particulate phase may be removed from the atmosphere by wet and dry deposition. In soil, it may bind strongly to pure mineral soils from pH 3-7. This compound may also bind strongly in the presence of organic matter in the soil, as the aromatic amine should be very reactive. According to a K_{oc} of 98, based on the log Kow, however, it may be very mobile under high pH conditions where little organic matter is present. Incomplete information is available to determine its biodegradability but it was not used as a carbon source by Nocardia v. Direct photolysis may occur on soil surfaces as this compound has UV peaks at 308 and 373. In water, it is not expected to volatilize from water surfaces or to bioconcentrate in aquatic organisms. It may bind strongly to either pure mineral surfaces depending on the pH or to organic material present in the water column.

Environmental Physical Data

Henry's Law Constant: estimated at 2.2×10^{-12}
Octanol/Water Partition Coefficient: log K_{ow} = 1.13
Sorption Partition Coefficient: K_{oc} = estimated at 98
BCF: estimated at 4

Regulations

RCRA 40CFR: Not listed
CERCLA: 40CFR 302.4: Listed as Compound per CWA Section 307(a)
SARA 40CFR 372.65: Not listed
SARA EHS 40CFR 355: Not listed
TSCA: Listed

AMI5640 CAS #: 121-88-0

2-AMINO-5-NITROPHENOL

RTECS: SJ6302500
EINECS Number: 204-503-8
Molecular Formula: $C_6H_6N_2O_3$
Structured MF: $H_2NC_6H_3(NO_2)OH$
Formula Weight: 154.13

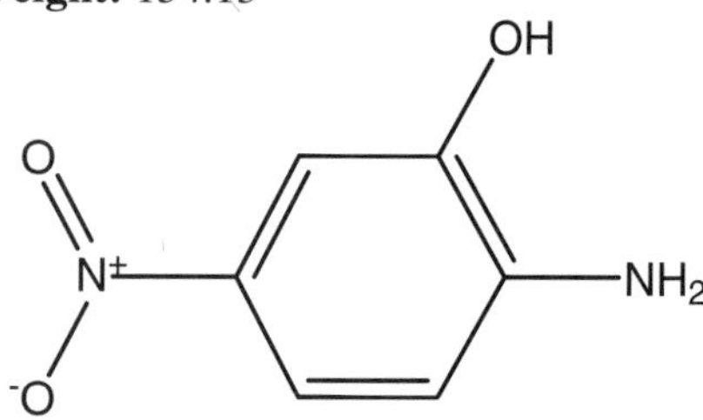

Chemical Structure

Synonyms: C.I. 76535; 3-HYDROXY-4-AMINONITROBENZENE; 2-HYDROXY-4-NITROANILINE; 3-NITRO-6-AMINOPHENOL; 5-NITRO-2-AMINOPHENOL; PHENOL,2-AMINO-5-NITRO-; RODOL YBA; URSOL YELLOW BROWN A
Description: olive-brown, brown to orange crystalline solid
Use: dye intermediate

Physical Properties

Freezing Point: 207 °C (404.6 °F) to 208 °C (406.4 °F)
Water Solubility: Slightly Soluble in Water
Other Solubilities: 95% Ethanol: 10-50 mg/ml at 20 °C; Acetone: 10-50 mg/ml at 20 °C; Benzene: Soluble; DMSO: >=100 mg/ml at 20 °C; Ether: Soluble.
Flash Point: Not available; probably combustible

RTECS Toxicity Data

Acute Oral: Rat LD_{50} Dose: >4 gm/kg.
Chronic (Multiple Dose) Oral: Rat Dose: 3756 mg/kg/16D-I; Toxic Effects: DEATH. Rat Dose: 52 gm/kg/13W-I; Toxic Effects: Gastrointestinal - Ulceration or bleeding from large intestine; Liver - Changes in liver weight; DEATH.
Mutagenic: Mouse Mutations in Mammalian Somatic Cells; Cell Type: lymphocyte; Dose: 25 mg/L. Hamster Cytogenetic Analysis; Cell Type: ovary; Dose: 905 mg/L. Hamster Cytogenetic Analysis; Cell Type: lung; Dose: 1 mg/L.
Tumorigenic: Rat Route: Oral; Dose: 51500 mg/kg/2Y-C; Toxic Effects: Tumorigenic - Equivocal tumorigenic agent by RTECS criteria; Gastrointestinal - Tumors.

Hazard Overviews

Health: Irritating to eyes/skin/respiratory tract. Toxic. Other Acute Effects: harmful if swallowed, inhaled, or absorbed through skin. Chronic Effects: may alter genetic material; may cause heritable genetic damage. Possible carcinogen.
Fire: Will burn. Hazards: emits toxic fumes. Extinguishing agents: water spray; carbon dioxide, dry chemical powder or appropriate foam. Precautions: combustible liquid.
Reactivity: Incompatible with: strong oxidizing agents. Hazardous decomposition products: toxic fumes of: carbon monoxide, carbon dioxide, nitrogen oxides.
Carcinogenicity: IARC - Group 3, Not classifiable as to carcinogenicity to humans; NIOSH - Not listed; NTP - Not listed; ACGIH - Not listed; OSHA - Not listed; EPA - Not listed; MAK - Not listed
Primary Target Organs:

Eyes

Skin

Respiratory System

Environmental

Regulations

RCRA 40CFR: Not listed
CERCLA: 40CFR 302.4: Listed as Compound per CWA Section 307(a)
SARA 40CFR 372.65: Not listed
SARA EHS 40CFR 355: Not listed
TSCA: Listed

AMI5720 CAS #: 119-34-6

4-AMINO-2-NITROPHENOL

RTECS: SJ6303000
EINECS Number: 204-316-1
Molecular Formula: $C_6H_6N_2O_3$
Structured MF: $H_2NC_6H_3(NO_2)OH$
Formula Weight: 154.13

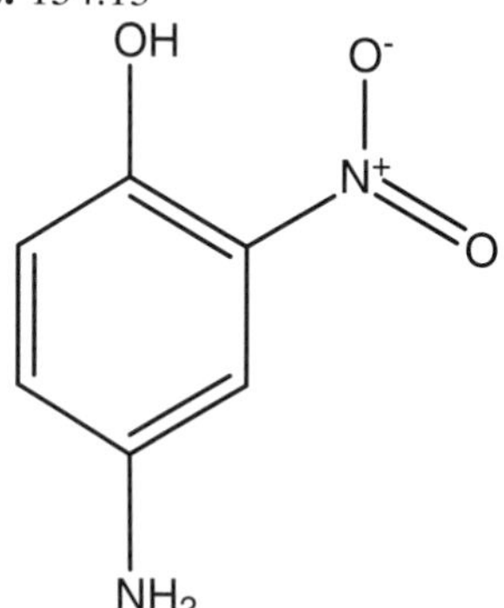

Chemical Structure

Synonyms: 4-AMINO-2-NITROFENOL; P-AMINO-O-NITROPHENOL; C.I. 76555; C.I. OXIDATION BASE 25; FOURRINE 57; FOURRINE BROWN PR; FOURRINE BROWN PROPYL; 4-HYDROXY-3-NITROANILINE; 2-NITRO-4-AMINOPHENOL; O-NITRO-P-AMINOPHENOL; ORTHO-NITRO-PARA-AMINOPHENOL; 3-NITRO-4-HYDROXYANILINE; OXIDATION BASE 25; PHENOL,4-AMINO-2-NITRO-
Description: dark red plates or needles
Use: dye for furs and hair

Physical Properties

Boiling Point: 110 °C (230 °F)
Freezing Point: 131 °C (267.8 °F)
Water Solubility: Soluble in Hot Water
Other Solubilities: 95% Ethanol: 10-50 mg/ml at 21.5 °C; Acetone: 10-50 mg/ml at 21.5 °C; DMSO: >=100 mg/ml at 21.5 °C; Ether: Soluble.
Flash Point: Not available; probably combustible

RTECS Toxicity Data

Acute Oral: Rat LD_{50} Dose: 1470 mg/kg. Mouse LD_{50} Dose: 1470 mg/kg.
Irritation Eye: Rabbit Standard Draize Test Dose: 100 mg/24H; Reaction: severe.
Mutagenic: Rat Morphological Transformation; Cell Type: embryo; Dose: 11 ug/plate. Mouse Mutations in Microorganisms; Cell Type: lymphocyte; Dose: 2500 ug/L (+S9).
Tumorigenic: Rat Route: Oral; Dose: 108 gm/kg/2Y-C; Toxic Effects: Tumorigenic - Carcinogenic by RTECS criteria; Kidney, Ureter, and Bladder - Tumors.

Hazard Overviews

Health: Severely irritating to eyes; irritating to skin/respiratory tract. Toxic. Other Acute Effects: harmful if swallowed, inhaled, or absorbed through skin. Chronic Effects: may alter genetic material; may cause heritable genetic damage. Carcinogen.
Fire: Will burn. Hazards: emits toxic fumes. Extinguishing agents: water spray; carbon dioxide, dry chemical powder or appropriate foam. Precautions: combustible liquid.
Reactivity: Incompatible with: strong oxidizing agents, strong bases, acid chlorides, acid anhydrides. Hazardous decomposition products: toxic fumes of: carbon monoxide, carbon dioxide, nitrogen oxides.
Carcinogenicity: IARC - Group 3, Not classifiable as to carcinogenicity to humans; NIOSH - Not listed; NTP - Not listed; ACGIH - Not listed; OSHA - Not listed; EPA - Not listed; MAK - Class B, Justifiably suspected of having carcinogenic potential
Primary Target Organs:

Eyes

Skin
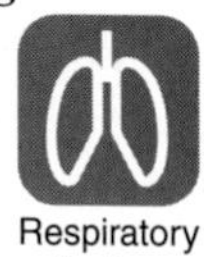
Respiratory System

Environmental

Ecotoxicity: LC_{50} Pimephales promelas (fathead minnow) 34.3 (29.9-39.4) mg/l 96 hr, wt 150 mg, flow-through bioassay, dissolved oxygen 7.4 (4.6-8.8) mg/l, water hardness 44.9

(42.4-46.6) mg/l as $CaCO_3$, pH 6.9-7.7, alkalinity 42.9 (39.6-61.4) mg/l $CaCO_3$, temp: 26.4 + or - 1.4 °C, Purity 97%

Environmental Fate: If released to the atmosphere, it will exist in both the vapor-phase and the particulate phase. In the vapor-phase, it will react with hydroxyl radicals with an estimated half-life of 2 days. It may also react with nitrate radicals. Particulate phase may be removed from the atmosphere by wet and dry deposition. Based on an estimated K_{oc} of 79, it should be highly mobile in soil; some leaching may be possible based on its high water solubility. This compound however, may bind strongly in the presence of organic matter in the soil as the aromatic amine should be very reactive. Incomplete information is available to determine biodegradability but it was only slightly used as a carbon source by Nocardia v. In water, it is not expected to volatilize from water surfaces or to bioconcentrate in aquatic organisms. It may bind strongly to organic material present in the water column.

Cleanup/Disposal: Guide No. 171: Do not touch or walk through spilled material. Stop leak if you can do it without risk. Prevent dust cloud. Avoid inhalation of asbestos dust. Small Dry Spills: With clean shovel place material into clean, dry container and cover loosely; move containers from spill area. Small Spills: Take up with sand or other noncombustible absorbent material and place into containers for later disposal. Large Spills: Dike far ahead of liquid spill for later disposal. Cover powder spill with plastic sheet or tarp to minimize spreading. Prevent entry into waterways, sewers, basements or confined areas.

Environmental Physical Data

Henry's Law Constant: estimated at 2.2×10^{-12}

Octanol/Water Partition Coefficient: log K_{ow} = 0.96

Sorption Partition Coefficient: K_{oc} = 79

BCF: estimated at 3

Regulations

RCRA 40CFR: Not listed

CERCLA: 40CFR 302.4: Listed as Compound per CWA Section 307(a)

SARA 40CFR 372.65: Not listed

SARA EHS 40CFR 355: Not listed

TSCA: Listed

AMI5800 CAS #: 121-66-4

2-AMINO-5-NITROTHIAZOLE

RTECS: XJ2800000

EINECS Number: 204-490-9

Molecular Formula: $C_3H_3N_3O_2S$

Formula Weight: 145.15

O, N+, -O, S, NH2, N

Chemical Structure

Synonyms: AMINONITROTHIAZOLE; AMINONITROTHIAZOLUM; AMINZOL SOLUBLE; AMNIZOL SOLUBLE; ENHEPTIN; ENHEPTIN-T; ENHEPTIN PREMIX; ENHEPTIN T; ENTRAMIN; NITRAMIN; NITRAMIN IDO; NITRAMINE; 5-NITRO-2-AMINOTHIAZOLE; NITROMIN IDO; 5-NITRO-2-THIAZOLAMINE; 5-NITRO-2-THIAZOLYLAMINE; 2-THIAZOLAMINE,5-NITRO-; 2-THIAZOLAMINE,5-NITRO-(9CI); THIAZOLE,2-AMINO-5-NITRO-

Description: greenish-yellow to orange-yellow fluffy powder

Use: antibacterial agent; used for the treatment of blackhead disease (histomoniasis, a protozoan enterohepatitis) in turkeys and chickens; used to treat trichomoniasis in pigeons; in the synthesis of another antiprotozoal agent, nithiazide; and used to make the antischistosomal drug nindazole

Physical Properties

Boiling Point: Decomposes at 202 °C (396 °F)

Freezing Point: Decomposes at 202 °C (395.6 °F)

Water Solubility: Very Sparingly Soluble in Water

Other Solubilities: 95% Ethanol: <1 mg/ml at 20 °C; Acetone: <1 mg/ml at 20 °C; Chloroform: Almost Insoluble; DMSO: 10-50 mg/ml at 20 °C; Dilute mineral acids: Soluble; Ether: 1 g/250 g; Toluene: <1 mg/ml at 16 °C; Methanol: <1 mg/ml at 16 °C.

Flash Point: Not available; probably combustible

RTECS Toxicity Data

Reproductive/Teratogenic: Rat Route: Oral; Dose: 600 mg/kg; Duration: male 10D prior to mating; Paternal Effects - Prostate, seminal vessicle, Cowper's gland, accessory glands. Rat Route: Oral; Dose: 700 mg/kg; Duration: male 14D prior to mating; Maternal Effects - Menstrual cycle changes or disorders.

Mutagenic: Rat Unscheduled DNA Synthesis; Cell Type: liver; Dose: 50 mg/L. Rat Morphological Transformation; Cell Type: embryo; Dose: 14500 ng/plate.

Tumorigenic: Rat Route: Oral; Dose: 28 gm/kg/2Y-C; Toxic Effects: Tumorigenic - Carcinogenic by RTECS criteria; Blood - Leukemia; Blood - Lymphomax including Hodgkin's disease. Rat Route: Oral; Dose: 23 gm/kg/46W-C; Toxic Effects: Tumorigenic - Neoplastic by RTECS criteria; Kidney, Ureter, and Bladder - Kidney tumors; Skin and appendages - Tumors. Rat Route: Oral; Dose: 12 gm/kg/2Y-C; Toxic Effects: Tumorigenic - Equivocal tumorigenic agent by RTECS criteria; Tumorigenic effects - Uterine tumors.

Hazard Overviews

Health: May cause irritation to eyes/skin/respiratory tract. Toxic. Other Acute Effects: harmful if swallowed, inhaled, or absorbed through skin. Chronic Effects: target organ: liver. Carcinogen.

Fire: Will burn. Hazards: emits toxic fumes. Extinguishing agents: water spray; carbon dioxide, dry chemical powder or appropriate foam. Precautions: combustible liquid.
Reactivity: Incompatible with: strong oxidizing agents, strong acids, acid chlorides, acid anhydrides. Hazardous decomposition products: toxic fumes of: carbon monoxide, carbon dioxide, nitrogen oxides, sulfur oxides.
Carcinogenicity: IARC - Group 3, Not classifiable as to carcinogenicity to humans; NIOSH - Not listed; NTP - Not listed; ACGIH - Not listed; OSHA - Not listed; EPA - Not listed; MAK - Not listed
Primary Target Organs:

Liver

Environmental

Regulations

RCRA 40CFR: Not listed
CERCLA: 40CFR 302.4: Not listed
SARA 40CFR 372.65: Not listed
SARA EHS 40CFR 355: Not listed
TSCA: Listed

AMI5880 CAS #: 95-55-6

2-AMINOPHENOL

RTECS: SJ4950000
EINECS Number: 202-431-1
Molecular Formula: C_6H_7NO
Formula Weight: 109.12

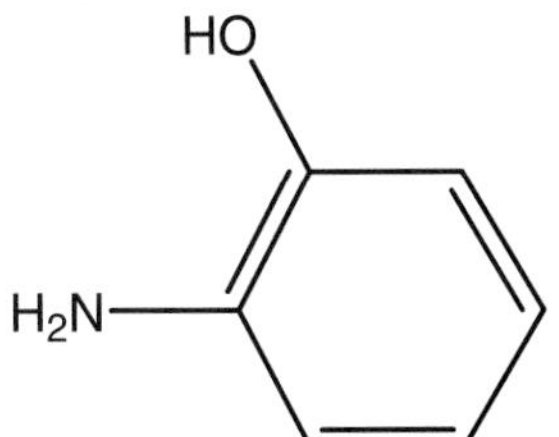

Chemical Structure

Synonyms: 1-AMINO-2-HYDROXYBENZENE; 2-AMINO-1-HYDROXYBENZENE; O-AMINOPHENOL; BASF URSOL 3GA; BENZOFUR GG; C.I. 76520; C.I. OXIDATION BASE 17; FOURAMINE OP; 1-HYDROXY-2-AMINOBENZENE; 2-HYDROXYANALINE; 2-HYDROXYANILINE; O-HYDROXYANILINE; O-HYDROXYPHENYLAMINE; NAKO YELLOW 3GA; PARADONE OLIVE GREEN B; PELAGOL 3GA; PELAGOL GREY GG; PHENOL,2-AMINO-; PHENOL,O-AMINO-; QUESTIOMYCIN B; ZOBA 3GA
Description: colorless rhombic crystals, needles or plates, rapidly becoming brown
Use: manufacture of azo and sulfur dyes; dyeing furs and hair; hydrochloride used in dyeing hair, fur, and leather

Physical Properties

Boiling Point: Sublimes at 153 °C (307 °F)
Freezing Point: 170 °C (338 °F) to 174 °C (345.2 °F)
Specific Gravity: 1.328
Density: 1.328 g/mL
Water Solubility: 1 g dissolves in 50 ml Cold Water
Other Solubilities: 95% Ethanol: Very Soluble; Benzene: Slightly Soluble; Ether: Very Soluble.
Flash Point: Low

RTECS Toxicity Data

Acute Oral: Rat LD_{50} Dose: 1300 mg/kg. Mouse LD_{50} Dose: 1250 mg/kg.
Acute Dermal: Rat LD_{50} Route: Subcutaneous Dose: 37 mg/kg. Cat LD_{Lo} Route: Subcutaneous Dose: 37 mg/kg; Toxic Effects: Behavioral - General anesthetic; Lungs, Thorax, or Respiration - Dyspnea; Lungs, Thorax, or Respiration - Cyanosis.
Irritation Eye: Rabbit Standard Draize Test Dose: 100 mg; Reaction: mild.
Reproductive/Teratogenic: Hamster Route: Intraperitoneal; Dose: 150 mg/kg; Duration: female 8D of pregnancy; Effects on Fertility - Post-implantation mortality; Specific Developmental Abnormalities - Central nervous system; Eye, ear. Hamster Route: Intraperitoneal; Dose: 150 mg/kg; Duration: female 8D of pregnancy; Specific Developmental Abnormalities - Body wall; Specific Developmental Abnormalities - Musculoskeletal system.
Mutagenic: Hamster DNA Inhibition; Cell Type: lung; Dose: 19 umol/L. Bacteria - E Coli DNA Repair; Dose: 20 ug/well.

Hazard Overviews

Health: Irritating to eyes/skin/respiratory tract. Harmful. Other Acute Effects: harmful if swallowed; may be harmful if inhaled or absorbed through the skin.
Fire: Will burn. Hazards: emits toxic fumes. Extinguishing agents: water spray; carbon dioxide, dry chemical powder or appropriate foam. Precautions: combustible liquid.
Reactivity: Incompatible with: strong oxidizing agents, acids, acid anhydrides, acid chlorides, chloroformates. Hazardous decomposition products: toxic fumes of: carbon monoxide, carbon dioxide, nitrogen oxides.
Carcinogenicity: IARC - Not listed; NIOSH - Not listed; NTP - Not listed; ACGIH - Not listed; OSHA - Not listed; EPA - Not listed; MAK - Not listed
Primary Target Organs:

Eyes

Skin

Respiratory System

Environmental

Ecotoxicity: Algae: Chlorella pyrenoidosa: 47 mg/l: toxic
Cleanup/Disposal: Guide No. 152: Do not touch damaged containers or spilled material unless wearing appropriate protective clothing. Stop leak if you can do it without risk. Prevent entry into waterways, sewers, basements or confined

areas. Cover with plastic sheet to prevent spreading. Absorb or cover with dry earth, sand or other non-combustible material and transfer to containers. Do not get water inside containers.

Environmental Physical Data

Octanol/Water Partition Coefficient: log K_{ow} = 0.52 to 0.62

Regulations

RCRA 40CFR: Not listed
CERCLA: 40CFR 302.4: Not listed
SARA 40CFR 372.65: Not listed
SARA EHS 40CFR 355: Not listed
TSCA: Listed

AMI5960 CAS #: 591-27-5

3-AMINOPHENOL

RTECS: SJ4900000
DOT: UN2512
EINECS Number: 209-711-2
Molecular Formula: C_6H_7NO
Formula Weight: 109.12

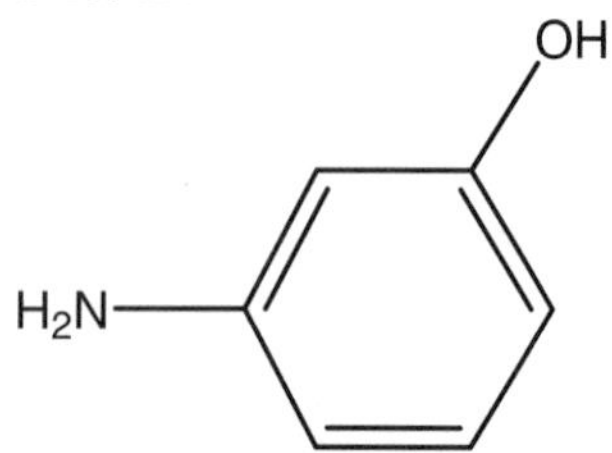

Chemical Structure

Synonyms: M-AMINOFENOL; 1-AMINO-3-HYDROXYBENZENE; 3-AMINO-1-HYDROXYBENZENE; M-AMINOPHENOL; META-AMINOPHENOL; BASF URSOL BG; BASF URSOL EG; C.I. 76545; C.I. OXIDATION BASE 7; FOURAMINE EG; FOURRINE 65; FOURRINE EG; FURRO EG; FUTRAMINE EG; M-HYDROXYAMINOBENZENE; 3-HYDROXYANILINE; M-HYDROXYANILINE; M-HYDROXYPHENYLAMINE; NAKO TEG; PELAGAL EG; PELAGOL EG; PHENOL,3-AMINO-; PHENOL,M-AMINO-; RENAL EG; TERTRAL EG; TETRAL EG; URSOL EG; ZOBA EG
Description: white prisms or crystals
Use: manufacture of dyes and of p-Aminosalicylic acid

Physical Properties

Boiling Point: 164 °C (327 °F) at 11 mm Hg
Freezing Point: 122 °C (251.6 °F) to 123 °C (253.4 °F)
Vapor Pressure: Low volatility
Water Solubility: Freely Soluble in Hot Water
Other Solubilities: 95% Ethanol: >=100 mg/ml at 24 °C; Acetone: >=100 mg/ml at 24 °C; Amyl Alcohol: Freely Soluble; Benzene: Slightly Soluble; DMSO: >=100 mg/ml at 24 °C; Ether: Soluble; Petroleum Ether: Very slightly Soluble.
Flash Point: Not available; probably combustible

RTECS Toxicity Data

Acute Oral: Rat LD_{50} Dose: 924 mg/kg; Toxic Effects: Peripheral nerve and sensation - Spastic parapysis with or without sensory change; Behavioral - Excitment; Lungs, Thorax, or Respiration - Other changes. Mouse LD_{50} Dose: 401 mg/kg; Toxic Effects: Peripheral nerve and sensation - Spastic parapysis with or without sensory change; Behavioral - Excitment; Lungs, Thorax, or Respiration - Other changes.
Acute Inhalation: Rat LC_{50} Dose: 1162 mg/m^3; Toxic Effects: Peripheral Nerve and sensation - Spastic parapysis with or without sensory change; Behavioral - Excitment; Lungs, Thorax, or Respiration - Other changes.
Chronic (Multiple Dose) Inhalation: Rat Dose: 24 mg/m^3/6H/17W-I; Toxic Effects: Behavioral - Muscle weakness.
Irritation Eye: Rabbit Standard Draize Test Dose: 100 mg/24H; Reaction: moderate.
Irritation Skin: Rabbit Standard Draize Test Dose: 12500 ug/24H; Reaction: mild.
Mutagenic: Rat Cytogenetic Analysis; Route: Inhalation; Dose: 28300 ug/m^3/24H. Hamster DNA Inhibition; Cell Type: lung; Dose: 190 umol/L.

Hazard Overviews

Health: Irritating to eyes/skin/respiratory tract. Toxic. Other Acute Effects: harmful if swallowed, inhaled, or absorbed through skin; depending on the intensity and duration of exposure, effects may vary from mild irritation to severe destruction of tissue; prolonged contact can cause damage to the eyes severe irritation or burns.
Fire: Will burn. Hazards: emits toxic fumes. Extinguishing agents: water spray; carbon dioxide, dry chemical powder or appropriate foam. Precautions: combustible liquid.
Reactivity: Incompatible with: acids, acid chlorides, acid anhydrides, chloroformates, strong oxidizing agents. Hazardous decomposition products: toxic fumes of: carbon monoxide, carbon dioxide, nitrogen oxides.
Carcinogenicity: IARC - Not listed; NIOSH - Not listed; NTP - Not listed; ACGIH - Not listed; OSHA - Not listed; EPA - Not listed; MAK - Not listed
Primary Target Organs:

Eyes

Skin

Respiratory System

Environmental

Ecotoxicity: Algae: Chlorella pyrenoidosa: toxic at 140 mg/l
Cleanup/Disposal: Guide No. 152: Do not touch damaged containers or spilled material unless wearing appropriate protective clothing. Stop leak if you can do it without risk. Prevent entry into waterways, sewers, basements or confined areas. Cover with plastic sheet to prevent spreading. Absorb or cover with dry earth, sand or other non-combustible material and transfer to containers. Do not get water inside containers.

Environmental Physical Data

Octanol/Water Partition Coefficient: log K_{ow} = 0.17

Regulations

RCRA 40CFR: Not listed
CERCLA: 40CFR 302.4: Not listed
SARA 40CFR 372.65: Not listed
SARA EHS 40CFR 355: Not listed
TSCA: Listed

AMI6040 **CAS #: 123-30-8**

P-AMINOPHENOL

RTECS: SJ5075000
EINECS Number: 204-616-2
Molecular Formula: C_6H_7NO
Formula Weight: 109.13

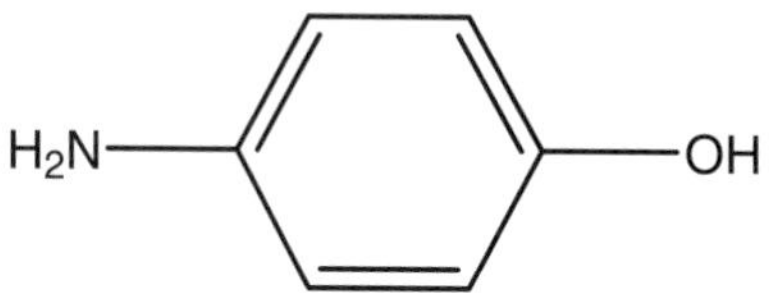

Chemical Structure

Synonyms: ACTIVOL; P-AMINOFENOL; 1-AMINO-4-HYDROXYBENZENE; 4-AMINO-1-HYDROXYBENZENE; 4-AMINOPHENOL; PARA-AMINOPHENOL; AZOL; BASF URSOL P BASE; BENZOFUR P; C.I. 76550; C.I. OXIDATION BASE 6A; CERTINAL; CITOL; DURAFUR BROWN RB; FOURAMINE P; FOURRINE 84; FOURRINE P BASE; FURRO P BASE; 4-HYDROXYANILINE; P-HYDROXYANILINE; NAKO BROWN R; PAP; PARANOL; PELAGOL GREY P BASE; PELAGOL P BASE; RENAL AC; RODINAL; TERTRAL P BASE; UNAL; URSOL P; URSOL P BASE; ZOBA BROWN P BASE
Description: colorless to white to reddish yellow crystals or plates
Use: photographic developer; intermediate in the manufacture of sulfur and azo dyes; dyeing textiles, hair, fur and feathers; pharmaceuticals; antioxidants oil additives

Physical Properties

Boiling Point: 284 °C (543 °F) at 760 mm Hg
Freezing Point: 189.6 °C (373.28 °F) to 190.2 °C (374.36 °F)
Water Solubility: 0.65% at 24 °C
Other Solubilities: 95% Ethanol: 1-10 mg/ml at 23 °C; Acetone: 10-50 mg/ml at 23 °C; Benzene: Insoluble; Chloroform: Insoluble; DMSO: >=100 mg/ml at 23 °C; Ether: Slightly Soluble; Ethyl methyl ketone: 9.3% at 58.5 °C.
Flash Point: Not available; probably combustible

RTECS Toxicity Data

Acute Oral: Rat LD_{50} Dose: 375 mg/kg; Toxic Effects: Behavioral - Muscle weakness; Lungs, Thorax, or Respiration - Cyanosis; Nutritional and gross metabolic - Body temperature decrease. Rabbit LD_{50} Dose: 10 gm/kg.
Acute Inhalation: Rat LC_{50} Dose: >5 mg/kg/1hr.
Acute Dermal: Rabbit LD_{50} Route: Skin; Dose: >10 gm/kg. Mouse LD_{Lo} Route: Subcutaneous Dose: 470 mg/kg.
Chronic (Multiple Dose) Oral: Rat Dose: 1326 mg/kg/4W-I; Toxic Effects: Blood - Pigmented or nucleated red Blood cells; Blood - Changes in erythrocite (RBC) cell count; Nutritional and gross metabolic - Weight loss or decreased weight gain.
Irritation Eye: Rabbit Standard Draize Test Dose: 100 mg; Reaction: mild.
Irritation Skin: Rabbit Standard Draize Test Dose: 12500 ug/24H; Reaction: mild.
Reproductive/Teratogenic: Rat Route: Oral; Dose: 2500 mg/kg; Duration: female 6-15D of pregnancy; Specific Developmental Abnormalities - Other developmental abnormalities. Rat Route: Oral; Dose: 667 mg/kg; Duration: female 11D of pregnancy; Effects on Fertility - Post-implantation mortality; Effects on Newborn - Growth statistics. Rat Route: Oral; Dose: 38850 mg/kg; Duration: female 13W prior to mating Effects on Fertility - Post-implantation mortality; Effects on Embryo or Fetus - Fetotoxicity; Fetal death. Rat Route: Oral; Dose: 11100 mg/kg; Duration: female 13W prior to mating Specific Developmental Abnormalities - Musculoskeletal system.
Mutagenic: Human DNA Damage; Cell Type: lymphocyte; Dose: 250 umol/L. Human DNA Inhibition; Cell Type: lymphocyte; Dose: 250 umol/L.

Hazard Overviews

Health: Irritating to eyes/skin/respiratory tract. Toxic. Other Acute Effects: harmful if swallowed, inhaled, or absorbed through skin; absorption into the body leads to the formation of methemoglobin which in sufficient concentration causes cyanosis; onset may be delayed 2 to 4 hours or longer; may cause sensitization by inhalation and skin contact; possible risk of irreversible effects; may cause allergic reaction. Chronic Effects: possible mutagen; possible teratogen; overexposure may cause reproductive disorders based on tests with laboratory animals; laboratory experiments have shown mutagenic effects; target organs: blood, kidneys, central nervous system.
Fire: Will burn. Hazards: emits toxic fumes. Extinguishing agents: water spray; carbon dioxide, dry chemical powder or appropriate foam. Precautions: combustible liquid.
Reactivity: Incompatible with: acids, may discolor on exposure to light, may discolor on exposure to air, chloroformates, strong oxidizing agents. Hazardous decomposition products: toxic fumes of: carbon monoxide, carbon dioxide, nitrogen oxides.
Carcinogenicity: IARC - Not listed; NIOSH - Not listed; NTP - Not listed; ACGIH - Not listed; OSHA - Not listed; EPA - Not listed; MAK - Not listed

Primary Target Organs:

Eyes

Skin
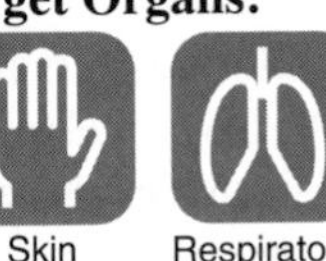
Respiratory System
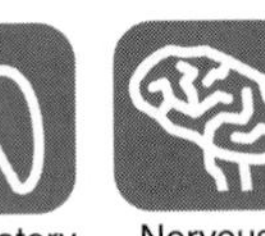
Nervous System

Kidneys

Blood

Environmental

Ecotoxicity: Algae: Chlorella pyrenoidosa: toxic at 140 mg/l Scenedesmus: toxic at 6 mg/l; Fishes: goldfish: approx. fatal conc.: 2.0 mg/l; 48 hr

Cleanup/Disposal: Guide No. 152: Do not touch damaged containers or spilled material unless wearing appropriate protective clothing. Stop leak if you can do it without risk. Prevent entry into waterways, sewers, basements or confined areas. Cover with plastic sheet to prevent spreading. Absorb or cover with dry earth, sand or other non-combustible material and transfer to containers. Do not get water inside containers.

Environmental Physical Data

Octanol/Water Partition Coefficient: log K_{ow} = 0.04

Regulations

RCRA 40CFR: Not listed
CERCLA: 40CFR 302.4: Not listed
SARA 40CFR 372.65: Not listed
SARA EHS 40CFR 355: Not listed
TSCA: Listed

AMI6120 CAS #: 51-78-5

P-AMINOPHENOL HYDROCHLORIDE

RTECS: SJ6070000
EINECS Number: 200-122-6
Molecular Formula: C_6H_8ClNO
Formula Weight: 145.60

Chemical Structure

Synonyms: 4-AMINOPHENOL HYDROCHLORIDE; C.I. 76551; C.I. OXIDATION BASE 6A; DURAFUR BROWN R; FOURAMINE CP; FOURRINE 83; FOURRINE P; FURRO P; FUTRAMINE P; 4-HYDROXYANILINIUM CHLORIDE; PELAGOL CP; PELAGOL GREY CP; PELTOL P; PHENOL,4-AMINO-,HYDROCHLORIDE; PHENOL,P-AMINO-,HYDROCHLORIDE

Description: crystalline powder

Use: photographic developer; int in mfr of sulfur and azo dyes; in dyeing furs and feathers

Physical Properties

Boiling Point: Decomposes ~ 306 °C (583 °F)
Freezing Point: Decomposes at ~ 306 °C (582.8 °F)
Water Solubility: Very Soluble in Water
Other Solubilities: Soluble in Alcohol

RTECS Toxicity Data

Mutagenic: Insects - D Melanogaster Specific Locus Test; Route: Oral; Dose: 20 mmol/L.

Hazard Overviews

Health: Irritating to eyes/skin/respiratory tract. Toxic. Other Acute Effects: harmful if swallowed, inhaled, or absorbed through skin; skin sensitization; dermatitis; asthma; methemoglobinemia; absorption into the body leads to the formation of methemoglobin which in sufficient concentration causes cyanosis; onset may be delayed 2 to 4 hours or longer. Chronic Effects possible risk of irreversible effects; possible teratogen; overexposure may cause reproductive disorder(s) based on tests with laboratory animals; target organs: blood, kidneys, central nervous system.

Fire: Hazards: emits toxic fumes. Extinguishing agents: water spray; carbon dioxide, dry chemical powder or appropriate foam. Precautions: combustible liquid.

Reactivity: Incompatible with: strong oxidizing agents. Hazardous decomposition products: toxic fumes of: carbon monoxide, carbon dioxide, nitrogen oxides, hydrogen chloride gas.

Carcinogenicity: IARC - Not listed; NIOSH - Not listed; NTP - Not listed; ACGIH - Not listed; OSHA - Not listed; EPA - Not listed; MAK - Not listed

Primary Target Organs:

Eyes

Skin

Respiratory System

Nervous System

Kidneys

Blood

Environmental

Regulations

RCRA 40CFR: Not listed
CERCLA: 40CFR 302.4: Not listed
SARA 40CFR 372.65: Not listed
SARA EHS 40CFR 355: Not listed
TSCA: Listed

AMI6200 CAS #: 317-34-0

AMINOPHYLLINE

RTECS: XH5600000
EINECS Number: 206-264-5
Molecular Formula: $C_{16}H_{24}N_{10}O_4$
Formula Weight: 420.44

Chemical Structure

Synonyms: AMINOCARDOL; AMINODUR; AMINOFILINA; AMINOPHYLLIN; AMMOPHYLLIN; CARDIOFILINA; CARDIOMIN; CARDOPHYLIN; CARDOPHYLLIN; CARENA; CARIOMIN; DIAPHYLLINE; 3,7-DIHYDRO-1,3-DIMETHYL-1H-PURINE-2,6-DIONE COMPD. WITH1,2-ETHANEDIAMINE (2:1); DIOPHLLIN; DIOPHYLLIN; DIUXANTHINE; DOBO; DURA-TAB S.M. AMINOPHYLLINE; DURA-TAB SM AMINOPHYLLINE; ETHOPHYLLINE; ETHYLENEDIAMINE,COMPD. WITH THEOPHYLLINE (1:2); ETILEN-XANTISAN TABL; EUFILINA; EUPHYLLIN; EUPHYLLINE; EURPHYLLIN; EUUFILIN; EUUFILLIN; GENOPHYLLIN; GRIFOMIN; INOPHYLLINE; LASODEX; LINAMPHETA; METAPHYLLIN; METAPHYLLINE; METHOPHYLLINE; MINAPHIL; MIOFILIN; NEOPHYILINE; NEOPHYLLINE; NOROFILINA; PETERPHYLLIN; PHYLCARDIN; PHYLLINDON; PHYLLOCONTIN; 1H-PURINE-2,6-DIONE,3,7-DIHYDRO-1,3-DIMETHYL-,COMPD WITH1,2-ETHANEDIAMINE (2:1); 1H-PURINE-2,6-DIONE,3,7-DIHYDRO-1,3-DIMETHYL-,COMPD.WITH 1,2-ETHANEDIAMINE (2:1) (9CI); RECTALAD-AMINOPHYLLINE; SOMOPHYLLIN; SOMOPHYLLIN O; STENOVASAN; TAD; TEFAMIN; TH/100; THEODROX; THEOLAMINE; THEOLONE; THEOMIN; THEOPHYLDINE; THEOPHYLLAMINE; THEOPHYLLAMINIUM; THEOPHYLLAMINUM; THEOPHYLLIN AETHYLENDIAMIN; THEOPHYLLIN ETHYLENEDIAMINE; THEOPHYLLINE ETHYLENEDIAMINE; THEOPHYLLINE,COMPD WITH ETHYLENEDIAMINE (2:1); THEPHYLDINE; VASOFILINA

Description: white or slightly yellowish granules or powder; slight ammoniacal odor

Use: med: respiratory smooth muscle relaxant, formerly diuretic, (vet) for heaves in horses; diuretic in dogs with congestive heart failure; bronchi dilator for cats; smooth muscle relaxant, myocardial stimulant; to treat hearing failure, & bronchial asthma (eg, in dogs); for reversible bronchospasm associated with chronic bronchitis & emphysema; diuretic agent.

Physical Properties

Water Solubility: 1 g dissolves in about 5 ml Water
Other Solubilities: Insoluble in Alcohol, Ether

RTECS Toxicity Data

Acute Oral: Man TD_{Lo} Dose: 19 mg/kg/3D; Toxic Effects: Behavioral - Withdrawal; Gastrointestinal - Nausea or vomiting. Woman TD_{Lo} Dose: 120 mg/kg; Toxic Effects: Endocrine - Hyperglycemia; Nutritional and gross metabolic - Changes in potassium; Nutritional and gross metabolic - Metabolic acidosis. Infant TD_{Lo} Dose: 7595 mg/kg/5D-I; Toxic Effects: Behavioral - Convulsions or effect on seizure threshold; Gastrointestinal - Hypermotility, diarrhea. Rat LD_{50} Dose: 243 mg/kg.

Acute Dermal: Rat LD_{50} Route: Subcutaneous Dose: 176 mg/kg. Mouse LD_{50} Route: Subcutaneous Dose: 186 mg/kg.

Chronic (Multiple Dose) Oral: Rat Dose: 2400 mg/kg/12D-I; Toxic Effects: Nutritional and gross metabolic - Weight loss or decreased weight gain; DEATH. Mouse Dose: 200 mg/kg/8D-I; Toxic Effects: Blood - Changes in serum composition; Immunological including allergic - Increase in humoral immune response. Mouse Dose: 175 mg/kg/7D-I.

Mutagenic: Mouse Cytogenetic Analysis; Cell Type: mammary gland; Dose: 500 mg/L/24H-C.

Hazard Overviews

Health: Irritating. Toxic. Other Acute Effects: harmful if swallowed, inhaled, or absorbed through skin; target organs: central nervous system, heart, lungs, kidneys. Chronic Effects: nausea; vomiting; diarrhea; anorexia; dyspepsia; epigastric pain; hematemesis; dizziness; vertigo; headache; light-headedness; nervousness; insomnia; hyperreflexia; fasciculations; clonic and tonic convulsions; palpitation tachycardia; extrasystoles; hypotension; circulatory failure; increased respiratory rates; urticaria.

Fire: Hazards: emits toxic fumes. Extinguishing agents: water spray; carbon dioxide, dry chemical powder or appropriate foam. Precautions: combustible liquid.

Reactivity: Incompatible with: strong oxidizing agents, may discolor on exposure to air. Hazardous decomposition products: toxic fumes of: carbon monoxide, carbon dioxide, nitrogen oxides.

Carcinogenicity: IARC - Not listed; NIOSH - Not listed; NTP - Not listed; ACGIH - Not listed; OSHA - Not listed; EPA - Not listed; MAK - Not listed

Primary Target Organs:

Eyes

Skin

Respiratory System

Nervous System

Kidneys

Cardio-vascular

Environmental

Regulations

RCRA 40CFR: Not listed
CERCLA: 40CFR 302.4: Not listed
SARA 40CFR 372.65: Not listed
SARA EHS 40CFR 355: Not listed
TSCA: Listed

AMI6280 **CAS #: 107-10-8**

1-AMINOPROPANE

RTECS: UH9100000
DOT: UN1277; IMO3.1
EINECS Number: 203-462-3
Molecular Formula: C_3H_9N

Structured MF: $CH_3(CH_2)_2NH_2$
Formula Weight: 59.11

Chemical Structure

Synonyms: MONO-N-PROPYLAMINE; MONOPROPYLAMINE; MONOPROPYLMINE; 1-PROPANAMINE; PROPANAMINE; 1-PROPYLAMINE; N-PROPYLAMINE; PROPYLAMINE
Description: colorless liquid; strong ammonia odor
Use: chemical intermediate for propyl isocyanate, rubber chemicals, dyes, textile resins, drugs (eg, prilocaine hydrochloride), pesticides (eg, profluralin), & petroleum additives

Physical Properties

Boiling Point: 48 °C (118 °F) to 49 °C (120 °F)
Freezing Point: -83 °C (-117.4 °F)
Specific Gravity: 0.719 at 20 °C/20 °C
Vapor Density: 2.04 Air=1
Vapor Pressure: 200 mm Hg at 15 °C
Water Solubility: Miscible with Water
Other Solubilities: Soluble in Acetone, Benzene & Chloroform.
Surface Tension: 57.72 dynes/cm at 20 °C
Odor Threshold: Water 9.01 x10^1 ppm
pH: Alkaline
Refraction Index: 1.389 at 20 °C/D
Critical Temperature: 223.8 °C
Critical Pressure: 46.8 atm
Ionization Potential (eV): 8.78 +/-0.02
Flash Point: < -37 °C Open Cup
Autoignition Temperature: 318 °C
LEL: 2.0% v/v
UEL: 10.4% v/v

RTECS Toxicity Data

Acute Oral: Rat LD_{50} Dose: 570 mg/kg.
Acute Inhalation: Rat LC_{50} Dose: 2310 ppm/4hr; Toxic Effects: Lungs, Thorax, or Respiration - Dyspnea; Liver - Hepatitis (hepatocellular necrosis), diffuse. Mouse LC_{50} Dose: 2500 mg/m^3/2hr.
Acute Dermal: Rabbit LD_{50} Route: Skin; Dose: 560 mg/kg; Toxic Effects: Skin and appendages - Primary irritation.
Chronic (Multiple Dose) Inhalation: Rat Dose: 400 ppm/7H/50D-I; Toxic Effects: Nutritional and gross metabolic - Weight loss or decreased weight gain; DEATH.
Irritation Eye: Rabbit Standard Draize Test Dose: 720 ug; Reaction: severe.

Hazard Overviews

Corrosive

Flammable

Health: Corrosive to eyes/skin/respiratory tract. Other Acute Effects: harmful if swallowed, inhaled, or absorbed through skin; material is extremely destructive to tissue of the mucous membranes and upper respiratory tract, eyes and skin; inhalation may result in spasm, inflammation and edema of the larynx and bronchi, chemical pneumonitis and pulmonary edema.
Fire: Flammable. Hazards: vapor may travel considerable distance to source of ignition and flash back; container explosion may occur; emits toxic fumes; forms explosive mixtures in air. Extinguishing agents: carbon dioxide, dry chemical powder or appropriate foam; water may be effective for cooling, but may not effect extinguishment. Precautions: combustible liquid.
Reactivity: Incompatible with: acids, acid chlorides, acid anhydrides, strong oxidizing agents. Hazardous decomposition products: thermal decomposition may produce carbon monoxide, carbon dioxide, and nitrogen oxides.
Carcinogenicity: IARC - Not listed; NIOSH - Not listed; NTP - Not listed; ACGIH - Not listed; OSHA - Not listed; EPA - Not listed; MAK - Not listed
Primary Target Organs:

Eyes

Skin

Respiratory System

Environmental

Ecotoxicity: Aquatic toxicity: Finfish toxicity critical concentration = 20 mg/l
Environmental Fate: If released to the atmosphere, it is rapidly degraded (estimated half-life of 12 hr) by reaction with photochemically produced hydroxyl radicals. If released to water, it is physically removed by volatilization. Volatilization half-lives of 2.44 and 26.5 days have been estimated for a shallow (1 m deep) model river and an environmental pond, respectively. Aquatic bioconcentration and adsorption to sediment are not expected to be important. If released to soil, it is expected to be very mobile and easily leached based upon estimated K_{oc} values of less than 50. Evaporation from dry soil is likely to occur. Several screening studies have demonstrated that it is readily biodegraded by activated and non-activated sludges.
Cleanup/Disposal: Guide No. 132: Fully encapsulating, vapor protective clothing should be worn for spills and leaks with no fire. Eliminate all ignition sources (no smoking, flares, sparks or flames in immediate area). All equipment used when handling the product must be grounded. Do not touch or walk through spilled material. Stop leak if you can do it without risk. Prevent entry into waterways, sewers, basements or confined areas. A vapor suppressing foam may be used to reduce vapors. Absorb with earth, sand or other non-combustible material and transfer to containers (except for Hydrazine). Use clean non-sparking tools to collect absorbed material. Large Spills: Dike far ahead of liquid spill for later disposal. Water spray may reduce vapor; but may not prevent ignition in closed spaces.

Environmental Physical Data

Henry's Law Constant: 1.2×10^{-5}
Octanol/Water Partition Coefficient: log K_{ow} = 0.48
Sorption Partition Coefficient: K_{oc} = estimated at 50
BCF: estimated at 0.13
BOD: 102% BODT, 13 days

Regulations

RCRA 40CFR: Listed Hazardous Waste No. U194 Toxic Waste Ignitable Waste
CERCLA: 40CFR 302.4: Listed per RCRA Section 3001 RQ: 5000 lb (2268 kg)
SARA 40CFR 372.65: Not listed
SARA EHS 40CFR 355: Not listed
TSCA: Listed

Analytical Methods

Soil: SW846 8260A, 8260B
Plasma: EPA 29

AMI6440 **CAS #: 78-96-6**

1-AMINO-2-PROPANOL

RTECS: UA5775000
EINECS Number: 201-162-7
Molecular Formula: C_3H_9NO
Structured MF: $CH_3CH(OH)CH_2NH_2$
Formula Weight: 75.11

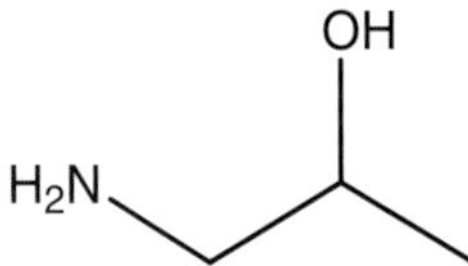

Chemical Structure

Synonyms: 1-AMINO-2-HYDROXYPROPANE; ALPHA-AMINOISOPROPYL ALCOHOL; 2-AMINO-1-METHYLETHANOL; 1-AMINOPROPAN-2-OL; 2-HYDROXYPROPANAMINE; 2-HYDROXY-1-PROPYLAMINE; 2-HYDROXYPROPYLAMINE; ISOPROPANOL AMINE; ISOPROPANOLAMINE; 1-METHYL-2-AMINOETHANOL; MIPA; MONO-ISO-PROPANOLAMINE; MONOISOPROPANOLAMINE; 2-PROPANOL,1-AMINO-; THREAMINE
Description: liquid; slight ammonia odor
Use: emulsifying agent, dry cleaning soaps, soluble textile oils, wax removers, metal cutting oils, cosmetics, emulsion paints, plasticizers, insecticides

Physical Properties

Boiling Point: 159.46 °C (319 °F) at 760 mm Hg
Freezing Point: 1.74 °C (35.132 °F)
Specific Gravity: 0.9611 at 20 °C/4 °C
Vapor Density: 2.6 Air=1
Saturated Vapor Density: 1.201179947 kg/m^3
Density: 0.97 g/mL at 18 °C
Vapor Pressure: 0.47 mm Hg at 25 °C
Water Solubility: Soluble in all Proportions
Other Solubilities: 95% Ethanol: >=100 mg/ml at 19.8 °C; Acetone: >=100 mg/ml at 19.8 °C; Benzene: Very Soluble; DMSO: >=100 mg/ml at 19.8 °C; Ether: Very Soluble.
Refraction Index: 1.4479 at 20 °C/D
Critical Temperature: 328 °C
Critical Pressure: 850 psia
Flash Point: 74 °C Open Cup
Autoignition Temperature: 374 °C
LEL: 2.2% v/v
UEL: Estimated at 12% v/v

RTECS Toxicity Data

Acute Oral: Rat LD_{50} Dose: 1715 mg/kg; Toxic Effects: Behavioral - Somnolence (general depressed activity); Gastrointestinal - Hypermotility, diarrhea; Nutritional and gross metabolic - Body temperature decrease.
Acute Dermal: Rabbit LD_{50} Route: Skin; Dose: 1640 uL/kg. Mammal LD_{50} Route: Skin; Dose: >1 gm/kg.
Irritation Eye: Rabbit Standard Draize Test Dose: 250 ug/24H; Reaction: severe. Rabbit Standard Draize Test Dose: 970 ug; Reaction: severe.
Irritation Skin: Rabbit Standard Draize Test Dose: 500 mg/24H; Reaction: mild. Rabbit Open Draize Test Dose: 485 mg open; Reaction: moderate.

Hazard Overviews

Corrosive

Health: Corrosive to eyes/skin/respiratory tract. Harmful. Other Acute Effects: harmful if swallowed, inhaled, or absorbed through skin; extremely destructive to tissue of the mucous membranes and upper respiratory tract, eyes and skin; inhalation may result in spasm, inflammation and edema of the larynx and bronchi, chemical pneumonitis and pulmonary edema; symptoms of exposure may include burning sensation, coughing, wheezing, laryngitis, shortness of breath, headache, nausea and vomiting.
Fire: Combustible. Hazards: emits toxic fumes. Extinguishing agents: carbon dioxide, dry chemical powder or appropriate foam. Precautions: combustible liquid.
Reactivity: Incompatible with: strong oxidizing agents. Hazardous decomposition products: toxic fumes of: carbon monoxide, carbon dioxide, nitrogen oxides.
Carcinogenicity: IARC - Not listed; NIOSH - Not listed; NTP - Not listed; ACGIH - Not listed; OSHA - Not listed; EPA - Not listed; MAK - Not listed
Primary Target Organs:

Eyes

Skin

Respiratory System

Environmental

Environmental Fate: If released to the atmosphere, it will degrade relatively rapidly by reaction with photochemically produced hydroxyl radicals (estimated half-life of 9.9 hr). If released to soil or water, it is expected to degrade via biodegradation. Several biodegradation studies have demonstrated that it is biodegradable. It may leach readily in soils; however, concurrent biodegradation may lessen the importance of leaching.

Cleanup/Disposal: Guide No. 171: Do not touch or walk through spilled material. Stop leak if you can do it without risk. Prevent dust cloud. Avoid inhalation of asbestos dust. Small Dry Spills: With clean shovel place material into clean, dry container and cover loosely; move containers from spill area. Small Spills: Take up with sand or other noncombustible absorbent material and place into containers for later disposal. Large Spills: Dike far ahead of liquid spill for later disposal. Cover powder spill with plastic sheet or tarp to minimize spreading. Prevent entry into waterways, sewers, basements or confined areas.

Environmental Physical Data

Henry's Law Constant: estimated at 2.34 $x10^{-10}$
Octanol/Water Partition Coefficient: log K_{ow} = -0.96
Sorption Partition Coefficient: K_{oc} = estimated at 7.1
BCF: estimated at 0.11
BOD: theoretical 5.1%, 5 days

Regulations

RCRA 40CFR: Not listed
CERCLA: 40CFR 302.4: Not listed
SARA 40CFR 372.65: Not listed
SARA EHS 40CFR 355: Not listed
TSCA: Listed

AMI6520 CAS #: 151-18-8

B-AMINOPROPIONITRILE

RTECS: UG0350000
EINECS Number: 205-786-0
Molecular Formula: $C_3H_6N_2$
Formula Weight: 70.09

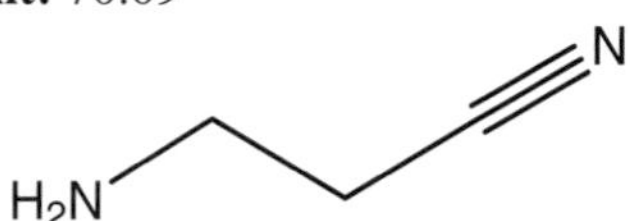

Chemical Structure

Synonyms: BETA-ALAMINENITRILE; BETA-ALANINENITRILE; 3-AMINOPROPANENITRILE; 3-AMINOPROPIONITRILE; AMINOPROPIONITRILE; BETA-AMINOPROPIONITRILE; BAPN; 2-CYANOETHYLAMINE; BETA-CYANOETHYLAMINE; PROPANENITRILE,3-AMINO-; PROPIONITRILE,3-AMINO-
Description: liquid
Use: chem int for beta-alanine & pantothenic acid

Physical Properties

Boiling Point: 185 °C (365 °F) at 760 mm Hg
Specific Gravity: 0.95840
Refraction Index: 1.4396 at 20 °C/D

RTECS Toxicity Data

Acute Dermal: Mouse LD_{Lo} Route: Skin; Dose: 12800 mg/kg; Toxic Effects: Behavioral - Somnolence (general depressed activity); Lungs, Thorax, or Respiration - Cyanosis.
Chronic (Multiple Dose) Oral: Rat Dose: 10 gm/kg/11D-I; Toxic Effects: Nutritional and gross metabolic - Weight loss or decreased weight gain; DEATH.
Reproductive/Teratogenic: Rat Route: Oral; Dose: 5 gm/kg; Duration: female 14-15D of pregnancy Effects on Fertility - Post-implantation mortality; Specific Developmental Abnormalities - Cardiovascular (circulatory) system. Mouse Route: Intraperitoneal; Dose: 1250 mg/kg; Duration: female 10D of pregnancy; Effects on Embryo or Fetus - Cytological changes (inc. somatic cell genetic material).
Mutagenic: Mouse Sister Chromatid Exchenge; Cell Type: embryo; Dose: 1250 mg/kg.

Hazard Overviews

Carcinogenicity: IARC - Not listed; NIOSH - Not listed; NTP - Not listed; ACGIH - Not listed; OSHA - Not listed; EPA - Not listed; MAK - Not listed

Environmental

Regulations

RCRA 40CFR: Not listed
CERCLA: 40CFR 302.4: Not listed
SARA 40CFR 372.65: Not listed
SARA EHS 40CFR 355: Not listed
TSCA: Listed

AMI6600 CAS #: 70-69-9

4-AMINOPROPIOPHENONE

RTECS: UG7350000
EINECS Number: 200-742-7
Molecular Formula: $C_9H_{11}NO$
Formula Weight: 149.21
Synonyms: 1-(4-AMINOPHENYL)-1-PROPANONE; P-AMINOPROPIOPHENONE; ETHYL P-AMINOPHENYL KETONE; PAPP; PARAMINOPROPIOPHENONE
Description: yellow crystalline needles

Physical Properties

Freezing Point: 140 °C (284 °F)
Water Solubility: Soluble
Other Solubilities: Alcohol: Soluble

RTECS Toxicity Data

Acute Oral: Rat LD_{50} Dose: 177 mg/kg; Toxic Effects: Gastrointestinal - Nausea or vomiting. Mouse LD_{50} Dose: 168 mg/kg.

Hazard Overviews

Carcinogenicity: IARC - Not listed; NIOSH - Not listed; NTP - Not listed; ACGIH - Not listed; OSHA - Not listed; EPA - Not listed; MAK - Not listed

Environmental

Cleanup/Disposal: Guide No. 154: Eliminate all ignition sources (no smoking, flares, sparks or flames in immediate area). Do not touch damaged containers or spilled material unless wearing appropriate protective clothing. Stop leak if you can do it without risk. Prevent entry into waterways, sewers, basements or confined areas. Absorb or cover with dry earth, sand or other non-combustible material and transfer to containers. Do not get water inside containers.

Regulations

RCRA 40CFR: Not listed
CERCLA: 40CFR 302.4: Not listed
SARA 40CFR 372.65: Not listed TPQ: 100/10000 lb
SARA EHS 40CFR 355: Listed TPQ: 100 lb
TSCA: Listed

AMI6680 CAS #: 156-87-6

3-AMINOPROPYLALCOHOL

RTECS: UA5600000
EINECS Number: 205-864-4
Molecular Formula: C_3H_9NO
Structured MF: $H_2NCH_2CH_2CH_2OH$
Formula Weight: 75.13

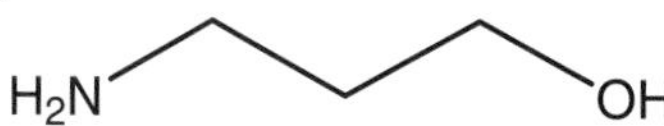

Chemical Structure

Synonyms: BETA-ALANINOL; 1-AMINO-3-HYDROXYPROPANE; 1-AMINO-3-PROPANOL; 3-AMINO-1-PROPANOL; 3-AMINOPROPANOL; GAMMA-AMINOPROPANOL; 3-AMINOPROPYL ALCOHOL; 3-HYDROXY-1-PROPYLAMINE; 3-HYDROXYPROPYLAMINE; GAMMA-HYDROXY-1-PROPYLAMINE; 1,3-PROPANOLAMINE; 3-PROPANOLAMINE; PROPANOLAMINE; 1-PROPANOL,3-AMINO-
Description: colorless to pale-yellow liquid; fishy odor
Use: chem int for dexpanthenol, a gastrointestinal drug; organic intermediate

Physical Properties

Boiling Point: 187 °C (369 °F) to 188 °C (370 °F) at 756 mm Hg
Freezing Point: 12.4 °C (54.32 °F)
Specific Gravity: 0.9824 at 26 °C/4 °C
Vapor Density: Estimate 2.6 Air=1
Vapor Pressure: 2.1 mm Hg
Water Solubility: Soluble in Water
Other Solubilities: miscible with Acetone & Chloroform.
pH: Acid
Refraction Index: 1.4617 at 20 °C
Ionization Potential (eV): 9.0
Flash Point: 79.444 °C

RTECS Toxicity Data

Acute Oral: Rat LD_{50} Dose: 2830 mg/kg.
Acute Dermal: Rabbit LD_{50} Route: Skin; Dose: 1250 mg/kg; Toxic Effects: Skin and appendages - Primary irritation.
Irritation Eye: Rabbit Standard Draize Test Dose: 250 ug/24H; Reaction: severe.
Irritation Skin: Rabbit Standard Draize Test Dose: 20 mg/24H; Reaction: moderate. Rabbit Open Draize Test Dose: 10 mg/24H open; Reaction: severe.
Reproductive/Teratogenic: Mouse Route: Intraperitoneal; Dose: 5700 ug/kg; Duration: female 11D of pregnancy; Specific Developmental Abnormalities - Musculoskeletal system.

Hazard Overviews

Corrosive

Fire Diamond

Health: Corrosive to eyes/skin/respiratory tract. Also Causes: severe gastrointestinal tract irritation, hemorrhaging may occur.
Fire: Combustible. Use dry chemical, carbon dioxide, water spray, or alcohol-resistant foam.
Reactivity: Stable. Hazardous polymerization cannot occur. Avoid: heat; ignition sources. Incompatible with: acids; strong oxidizers. Hazardous decomposition products: carbon; nitrogen oxides.
Carcinogenicity: IARC - Not listed; NIOSH - Not listed; NTP - Not listed; ACGIH - Not listed; OSHA - Not listed; EPA - Not listed; MAK - Not listed
Primary Target Organs:

Eyes

Skin

Respiratory System

Environmental

Cleanup/Disposal: Guide No. 153: Eliminate all ignition sources (no smoking, flares, sparks or flames in immediate area). Do not touch damaged containers or spilled material unless wearing appropriate protective clothing. Stop leak if you can do it without risk. Prevent entry into waterways, sewers, basements or confined areas. Absorb or cover with dry earth, sand or other non-combustible material and transfer to containers. Do not get water inside containers.

Regulations

RCRA 40CFR: Not listed
CERCLA: 40CFR 302.4: Not listed
SARA 40CFR 372.65: Not listed
SARA EHS 40CFR 355: Not listed
TSCA: Listed

AMI6840 CAS #: 13822-56-5

AMINOPROPYLTRIMETHOXYSILANE

EINECS Number: 237-511-5
Molecular Formula: $C_6H_{17}NO_3Si$
Formula Weight: 179.29

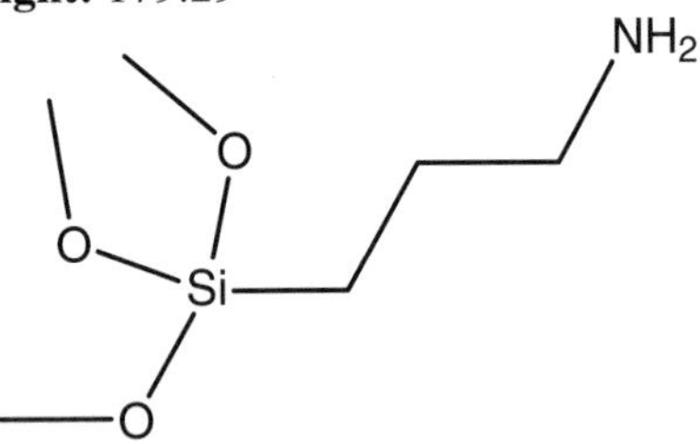

Chemical Structure

Physical Properties

Boiling Point: 210 °C (410 °F)
Freezing Point: -60 °C (-76 °F)
Specific Gravity: 1.014
Vapor Density: > 1 Air=1
Vapor Pressure: < 1
Water Solubility: Reacts
Other Solubilities: Soluble in Toluene
Refraction Index: 1.4255
Evaporation Rate: < 1 Butyl Acetate=1

Hazard Overviews

Corrosive

Health: Corrosive to eyes/skin/respiratory tract. Other Acute Effects: harmful if swallowed, inhaled, or absorbed through skin; inhalation may result in spasm, inflammation and edema of the larynx and bronchi, chemical pneumonitis and pulmonary edema; symptoms of exposure may include burning sensation; coughing; wheezing; laryngitis; shortness of breath; headache; nausea; vomiting.
Fire: Hazards: emits toxic fumes. Extinguishing agents: carbon dioxide, dry chemical powder or appropriate foam; water spray. Precautions: combustible liquid.
Reactivity: Incompatible with: strong oxidizing agents, may decompose on exposure to moist air or water, strong acids. Hazardous decomposition products: toxic fumes of: carbon monoxide, carbon dioxide, nitrogen oxides, silicon oxide.
Carcinogenicity: IARC - Not listed; NIOSH - Not listed; NTP - Not listed; ACGIH - Not listed; OSHA - Not listed; EPA - Not listed; MAK - Not listed

Primary Target Organs:

Eyes Skin Respiratory System

Environmental

Regulations

RCRA 40CFR: Not listed
CERCLA: 40CFR 302.4: Not listed
SARA 40CFR 372.65: Not listed
SARA EHS 40CFR 355: Not listed
TSCA: Not listed

AMI6920 CAS #: 54-62-6

AMINOPTERIN

RTECS: MA1050000
EINECS Number: 200-209-9
Molecular Formula: $C_{19}H_{20}N_8O_5$
Formula Weight: 440.47

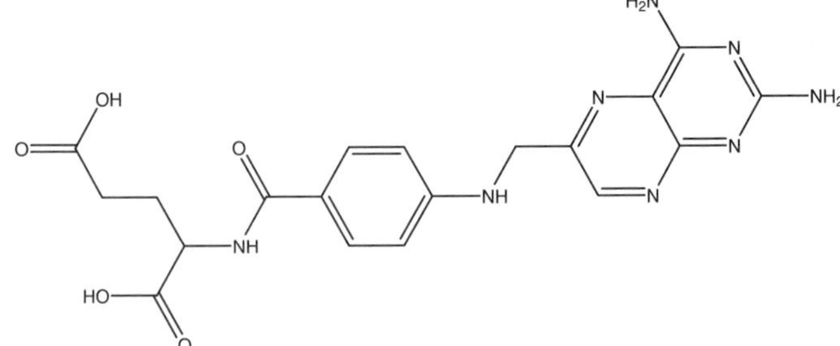

Chemical Structure

Synonyms: 4-AMINO-4-DEOXYPTEROYLGLUTAMATE; 4-AMINOFOLIC ACID; 4-AMINO-PGA; AMINOPTERIDINE; AMINOPTERINUM; 4-AMINOPTEROYLGLUTAMIC ACID; APGA; N-(4-(((2,4-DIAMINO-6-PTERIDINYL)METHYL)AMINO)BENZOYL)-L-GLUTAMIC ACID; N-(4-((2,4-DIAMINO-6-PTERIDYL)-METHYL)AMINO)BENZOYL)-L-GLUTAMICACID; ENT-26079; FOLIC ACID,4-AMINO-; GLUTAMIC ACID,N-(P-(((2,4-DIAMINO-6-PTERIDINYL)METHYL)AMINO)BENZOYL)-,L-; L-GLUTAMIC ACID,N-(4-(((2,4-DIAMINO-6-PTERIDINYL)METHYL)AMINO)BENZOYL)-; KYSELINA 4-AMINOLISTOVA; KYSELINA 4-AMINOPTEROYLGLUTAMOVA; KYSELINA N-(P-((2,4-DIAMINO-6-PTERIDINYLMETHYL)AMINO)BENZOYL)-L(+)-GLUTAMOVA; KYSELINAN-(P-((2,4-DIAMINO-6-PTERIDINYLMETHYL)BENZOYL)-L(+)-GLUTAMOVA; A-NINOPTERIN; NSC 739; PTERAMINA
Description: cluster of yellow needles or orange-yellow powder
Use: a folic acid antagonist; medication: formerly used for the treatment of leukemia in children; rodenticide

Physical Properties

Saturated Vapor Density: 1.200005825 kg/m^3
Water Solubility: 1 g/800 mL Water
Other Solubilities: Soluble in aqueous sodium hydroxide solutions
pH: Reconstituted solution 6.5 to 8.5
Flash Point: Will depend upon solvent; Generally < 38 °C

RTECS Toxicity Data

Acute Oral: Woman TD_{Lo} Dose: 120 ug/kg; Toxic Effects: Behavioral - Anorexia (human); Gastrointestinal - Other changes; Skin and appendages - Hair. Rat LD_{Lo} Dose: 2500 ug/kg; Toxic Effects: Gastrointestinal - Hypermotility, diarrhea; Nutritional and gross metabolic - Weight loss or decreased weight gain.
Chronic (Multiple Dose) Dermal: Rabbit Route: Subcutaneous; Dose: 24300 ug/kg/9D-I; Toxic Effects: Gastrointestinal - Hypermotility, diarrhea; Blood - Changes in bone marrow not included above; DEATH. Guinea Pig Route: Subcutaneous; Dose: 22200 ug/kg/20D-I; Toxic Effects: Gastrointestinal - Hypermotility, diarrhea; Blood - Changes in bone marrow not included above; DEATH.
Reproductive/Teratogenic: Woman Route: Oral; Dose: 240 ug/kg; Duration: female 55-66D of pregnancy Specific Developmental Abnormalities - Eye, ear; Craniofacial (including nose and tongue); Musculoskeletal system. Woman Route: Oral; Dose: 580 ug/kg; Duration: female 6-8W of pregnancy; Specific Developmental Abnormalities - Central nervous system; Craniofacial (including nose and tongue); Musculoskeletal system. Woman Route: Oral; Dose: 2880 ug/kg; Duration: female 56-67D of pregnancy Specific Developmental Abnormalities - Craniofacial (including nose and tongue); Musculoskeletal system; Effects on Newborn - Apgor score (human only). Woman Route: Oral; Dose: 165 ug/kg; Duration: female 40-44D of pregnancy Specific Developmental Abnormalities - Central nervous system; Other developmental abnormalities. Woman Route: Oral; Dose: 188 ug/kg; Duration: female 54-55D of pregnancy Specific Developmental Abnormalities - Craniofacial (including nose and tongue). Woman Route: Oral; Dose: 151 ug/kg; Duration: female 35-36D of pregnancy Effects on Fertility - Abortion. Woman Route: Oral; Dose: 240 ug/kg; Duration: female 1-30D of pregnancy; Specific Developmental Abnormalities - Central nervous system. Rat Route: Oral; Dose: 15 ug/kg; Duration: female 1-3D of pregnancy; Effects on Embryo or Fetus - Fetotoxicity. Rat Route: Oral; Dose: 150 ug/kg; Duration: female 9-11D of pregnancy; Effects on Embryo or Fetus - Fetal death.
Mutagenic: Mouse Micronucleus Test; Route: Intraperitoneal; Dose: 5 mg/kg/5D-C. Mouse Sperm Morphology; Route: Intraperitoneal; Dose: 2 mg/kg/5D.
Tumorigenic: Rat Route: Intramuscular; Dose: 1200 ug/kg/17W-C; Toxic Effects: Tumorigenic - Equivocal tumorigenic agent by RTECS criteria; Blood - Leukemia.

Hazard Overviews

Poison

Flammable

Health: Irritating to eyes/skin/respiratory tract. Poison. Other Acute Effects: may be fatal if inhaled, swallowed, or absorbed through skin. Chronic Effects: may cause congenital malformation in the fetus; laboratory experiments have shown mutagenic effects; blood effects.
Fire: Flammable. Hazards: emits toxic fumes. Extinguishing agents: water spray; carbon dioxide, dry chemical powder or appropriate foam. Precautions: combustible liquid.
Reactivity: Incompatible with: strong oxidizing agents, sensitive to light. Hazardous decomposition products: toxic fumes of: carbon monoxide, carbon dioxide, nitrogen oxides.
Carcinogenicity: IARC - Not listed; NIOSH - Not listed; NTP - Not listed; ACGIH - Not listed; OSHA - Not listed; EPA - Not listed; MAK - Not listed
Primary Target Organs:

Eyes

Skin

Respiratory System

Environmental

Cleanup/Disposal: Guide No. 151: Do not touch damaged containers or spilled material unless wearing appropriate protective clothing. Stop leak if you can do it without risk. Prevent entry into waterways, sewers, basements or confined areas. Cover with plastic sheet to prevent spreading. Absorb or cover with dry earth, sand or other non-combustible material and transfer to containers. Do not get water inside containers.

Regulations

RCRA 40CFR: Not listed
CERCLA: 40CFR 302.4: Not listed
SARA 40CFR 372.65: Not listed TPQ: 500/10000 lb
SARA EHS 40CFR 355: Listed TPQ: 500 lb
TSCA: Listed

AMI7000 **CAS #: 504-29-0**

2-AMINOPYRIDINE

RTECS: US1575000
DOT: UN2671; IMO6.1
EINECS Number: 207-988-4
Molecular Formula: $C_5H_6N_2$
Structured MF: $NH_2C_5H_4N$
Formula Weight: 94.11

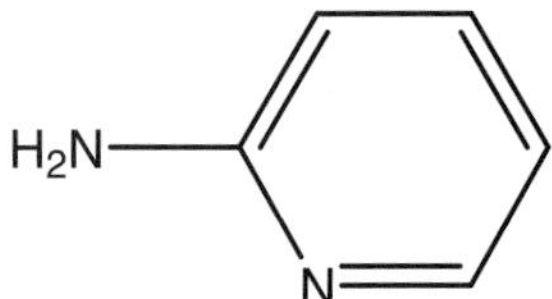

Chemical Structure

Synonyms: AMINO-2 PYRIDINE; ALPHA-AMINOPYRIDINE; AMINO-2-PYRIDINE; O-AMINOPYRIDINE; 2AP; 1,2-DIHYDRO-2-IMINOPYRIDINE; 2-PYRIDINAMINE; ALPHA-PYRIDINAMINE; 2-PYRIDYLAMINE; ALPHA-PYRIDYLAMINE

Description: white, colorless crystals; characteristic odor

Use: manufacture of pharmaceuticals esp. antihistamines; manufacture of medicines; organic synthesis

Physical Properties

Boiling Point: 104 °C (219 °F) at 20 mm Hg

Freezing Point: 58.1 °C (136.58 °F)

Saturated Vapor Density: 1.202836007 kg/m^3

Vapor Pressure: 0.8 mm Hg at 77 °F

Water Solubility: > 100% by weight

Other Solubilities: 95% Ethanol: >=100 mg/ml at 19 °C; Acetone: >=100 mg/ml at 19 °C; Benzene: Soluble; DMSO: 50-100 mg/ml at 19 °C; Ether: Soluble; Organic solvents: Soluble; Petroleum Ether: Soluble in hot.

pH: Basic

Ionization Potential (eV): 8.00

Flash Point: 68 °C Closed Cup

RTECS Toxicity Data

Acute Oral: Rat LD_{50} Dose: 200 mg/kg. Mouse LD_{50} Dose: 145 mg/kg.

Acute Inhalation: Human TC_{Lo} Dose: 5 ppm/5hr; Toxic Effects: Behavioral - Somnolence (general depressed activity); Behavioral - Convulsions or effect on seizure threshold; Behavioral - Antipsychotic.

Acute Dermal: Mouse LD_{50} Route: Subcutaneous Dose: 70 mg/kg. Guinea Pig LD_{50} Route: Skin; Dose: 500 mg/kg.

Hazard Overviews

Fire Diamond

Health: Toxic via skin absorption. Causes: dizziness, headache, weakness, nausea, respiratory distress, convulsions, elevated blood pressure, respiratory failure.

Fire: Combustible. Use dry chemical, water spray, or regular foam. Container may explode in heat of fire.

Reactivity: Stable. Hazardous polymerization cannot occur. Avoid: heat; ignition sources. Incompatible with: strong oxidizers. Hazardous decomposition products: nitrogen oxides.

Carcinogenicity: IARC - Not listed; NIOSH - Listed as carcinogen; NTP - Not listed; ACGIH - Not listed; OSHA - Not listed; EPA - Not listed; MAK - Not listed

Primary Target Organs:

Respiratory System Nervous System

Exposure Limits

OSHA PEL: TWA: 0.5 ppm; 2 mg/m^3.

ACGIH TLV: TWA: 0.5 ppm; 2 mg/m^3.

NIOSH REL: TWA: 0.5 ppm; 2 mg/m^3.

NIOSH IDLH: 5 ppm.

DFG MAK: TWA: 0.5 ppm; 2 mg/m^3.

Respirator Recommendation

Exposure Range: >0.5 to <5 ppm Supplied Air, Constant Flow/Pressure Demand, Half Mask

Exposure Range: 5 to unlimited ppm Supplied Air, Constant Flow/Pressure Demand, Full Face Self-contained Breathing Apparatus, Pressure Demand, Full Face

Note: warning properties unknown

Environmental

Environmental Fate: With a pKa of 6.86, it will exist in environmental media in varying proportions that are pH dependent. Ions generally do not volatilize. A Henry's Law constant of 2.81 $x10^{-9}$ atm-cu m/mole at 25 °C indicates that volatilization from environmental waters and moist soil should not be an important fate process. Yet, it should evaporate from dry surfaces, especially when present in high concentration such as in spill situations. In aquatic systems, it is not expected to bioconcentrate. A low K_{oc} indicates it should be highly mobile in soil. However, it may undergo a covalent chemical bonding the humic materials which can result in its chemical alteration to a latent form and prevent leaching. If released to water, covalent bonding with humic materials in the water column and sediments may result in significant partitioning from the water column. Biodegradation is expected to be slow in both aerobic and anaerobic soil and water. In the atmosphere, it is expected to exist almost entirely in the vapor phase and reactions with photochemically produced hydroxyl radicals should be important (estimated half-life of 8 hours). It may be physically removed from air by (rainfall and dissolution in clouds, etc.); however, the short atmospheric residence time suggests that wet deposition is of limited importance.

Cleanup/Disposal: Guide No. 153: Eliminate all ignition sources (no smoking, flares, sparks or flames in immediate area). Do not touch damaged containers or spilled material unless wearing appropriate protective clothing. Stop leak if you can do it without risk. Prevent entry into waterways, sewers, basements or confined areas. Absorb or cover with dry earth, sand or other non-combustible material and transfer to containers. Do not get water inside containers.

Environmental Physical Data

Henry's Law Constant: estimated at 2.81 $x10^{-9}$

Octanol/Water Partition Coefficient: log K_{ow} = 0.49

Sorption Partition Coefficient: K_{oc} = 44

BCF: estimated at 0.14

Regulations

RCRA 40CFR: Not listed
CERCLA: 40CFR 302.4: Not listed
SARA 40CFR 372.65: Not listed
SARA EHS 40CFR 355: Not listed
TSCA: Listed

AMI7080 CAS #: 504-24-5

4-AMINOPYRIDINE

RTECS: US1750000
DOT: UN2671; IMO6.1
EINECS Number: 207-987-9
Molecular Formula: $C_5H_6N_2$
Structured MF: $C_5NH_4NH_2$
Formula Weight: 94.13

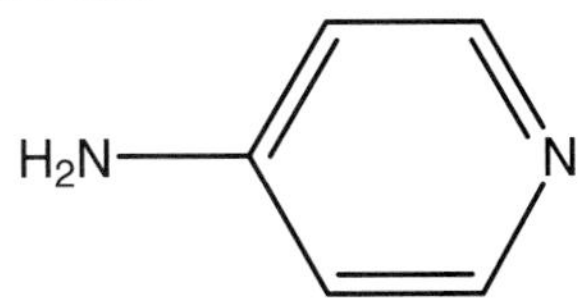

Chemical Structure

Synonyms: AMINO-4 PYRIDINE; GAMMA-AMINOPYRIDINE; P-AMINOPYRIDINE; 4-AP; AVITROL; AVITROL 200; MI-W-3; PHILIPS 1861; PHILLIPS 1861; 4-PYRIDINAMINE; PYRIDINE,4-AMINO-; 4-PYRIDYLAMINE; VMI 10-3
Description: white crystalline material; odorless
Use: chemical intermediate

Physical Properties

Boiling Point: 273.5 °C (524 °F)
Freezing Point: 158.9 °C (318.02 °F)
Specific Gravity: Solid 1.2607 at 25.3 °C
Water Solubility: 8796 mg/L at 25 °C
Other Solubilities: Soluble in Alcohol, Benzene and Ether.
Ionization Potential (eV): =< 8.77 +/-0.2
Flash Point: 164 °C Open Cup

RTECS Toxicity Data

Acute Oral: Woman TD_{Lo} Dose: 120 ug/kg; Toxic Effects: Behavioral - Convulsions or effect on seizure threshold. Man LD_{Lo} Dose: 590 ug/kg; Toxic Effects: Behavioral - Hallucinations, distorted perceptions; Lungs, Thorax, or Respiration - Dyspnea; Gastrointestinal - Nausea or vomiting. Rat LD_{50} Dose: 21 mg/kg.
Acute Dermal: Rat LD_{50} Route: Subcutaneous Dose: 19 mg/kg. Mouse LD_{Lo} Route: Subcutaneous Dose: 5 mg/kg.

Hazard Overviews

Poison

Health: Irritating to eyes/skin/respiratory tract. Poison. Other Acute Effects: may be fatal if inhaled, swallowed, or absorbed through skin; target organ: nerves.
Fire: Will burn. Hazards: emits toxic fumes. Extinguishing agents: water spray; carbon dioxide, dry chemical powder or appropriate foam. Precautions: combustible liquid.
Reactivity: Incompatible with: strong oxidizing agents, strong acids, acid chlorides, acid anhydrides. Hazardous decomposition products: toxic fumes of: carbon monoxide, carbon dioxide, nitrogen oxides.
Carcinogenicity: IARC - Not listed; NIOSH - Not listed; NTP - Not listed; ACGIH - Not listed; OSHA - Not listed; EPA - Class D, Not classifiable as to human carcinogenicity; MAK - Not listed
Primary Target Organs:

Eyes

Skin

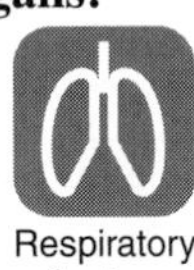
Respiratory System

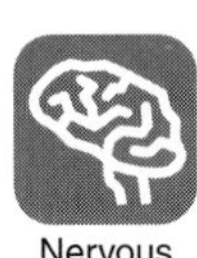
Nervous System

Environmental

Ecotoxicity: TLm Channel catfish 2.43 - 5.80 mg/l at 12 - 22 °C 96 hr /static bioassay LD_{50} Mallard, male oral (3-4 mos) 4.36 (3.36 - 5.66) mg/kg (95% confidence limits)
Environmental Fate: With a pKa of 9.17, it will exist in alkaline environmental media in varying proportions that are pH dependent. The conjugate acid should predominate under acidic conditions. Cations generally do not volatilize. A Henry's Law constant of 2.81 $x10^{-9}$ atm-cu m/mole at 25 °C indicates that volatilization from environmental waters and moist soil should not be an important fate process. Yet, it may evaporate from dry surfaces, especially when present in high concentration such as in spill situations. In aquatic systems, it is not expected to bioconcentrate. A low K_{oc} indicates it should be highly mobile in soil. However, it may undergo a covalent chemical bonding with humic materials which can result in its chemical alteration to a latent form and prevent leaching. If released to water, covalent bonding with humic materials in the water column and sediments may result in significant partitioning from the water column. Biodegradation is expected to be slow in both aerobic and anaerobic soil and water. Reported biodegradation half-lives in soil range from 90 to 960 days. In the atmosphere, it is expected to exist almost entirely in the vapor phase and reactions with photochemically produced hydroxyl radicals should be important (estimated half-life of 8 hours). It may be physically removed from air by (rainfall and dissolution in clouds, etc.); however, the short atmospheric residence time suggests that wet deposition is of limited importance.
Cleanup/Disposal: Guide No. 153: Eliminate all ignition sources (no smoking, flares, sparks or flames in immediate area). Do not touch damaged containers or spilled material unless wearing appropriate protective clothing. Stop leak if you can do it without risk. Prevent entry into waterways, sewers, basements or confined areas. Absorb or cover with dry earth, sand or other non-combustible material and transfer to containers. Do not get water inside containers.

Environmental Physical Data

Henry's Law Constant: estimated at 2.81×10^{-9}
Octanol/Water Partition Coefficient: log K_{ow} = 0.26
Sorption Partition Coefficient: K_{oc} = 33
BCF: calculated at 0.03

Regulations

RCRA 40CFR: Listed Hazardous Waste No. P008 Toxic Waste
CERCLA: 40CFR 302.4: Listed per RCRA Section 3001 RQ: 1000 lb (453.5 kg)
SARA 40CFR 372.65: Not listed TPQ: 500/10000 lb
SARA EHS 40CFR 355: Listed TPQ: 1,000 lb
TSCA: Listed

Analytical Methods

Soil: EPA PMD-AMN

AMI7160 CAS #: 58-15-1

AMINOPYRINE

RTECS: CD2625000
EINECS Number: 200-365-8
Molecular Formula: $C_{13}H_{17}N_3O$
Formula Weight: 231.29

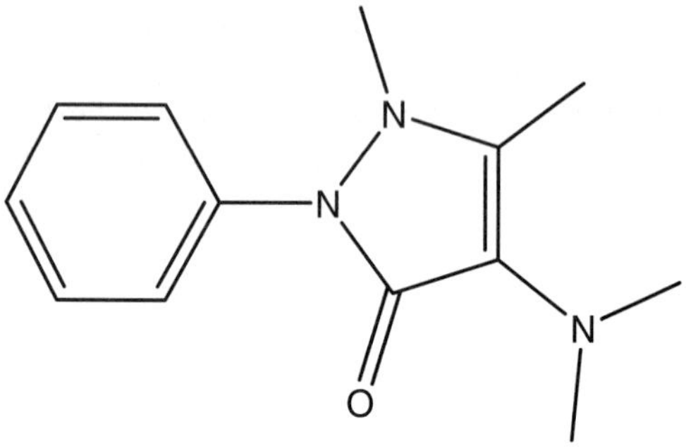

Chemical Structure

Synonyms: AMIDAZOFEN; AMIDAZOPHEN; AMIDAZOPHENE; AMIDOFEBRIN; AMIDOFEN; AMIDOPHEN; AMIDOPHENAZONE; AMIDOPYRAZOLINE; AMIDOPYRIN; AMIDOPYRINE; AMINOFENAZONE; AMINOPHENAZON; AMINOPHENAZONE; ANAFEBRINA; BRUFANEUXOL; DAP; DEREUMA; DIMAPYRIN; DIMETHYLAMINO-ANALGESINE; DIMETHYLAMINOANALGESINE; 4-(DIMETHYLAMINO)ANTIPYRINE; 4-DIMETHYLAMINOANTIPYRINE; DIMETHYLAMINOANTIPYRINE; P-DIMETHYLAMINOANTIPYRINE; DIMETHYLAMINOAZOPHENE; 4-(DIMETHYLAMINO)-1,2-DIHYDRO-1,5-DIMETHYL-2-PHENYL-3H-PYRAZOL-3-ONE; 4-DIMETHYLAMINO-2,3-DIMETHYL-1-PHENYL-3-PYRAZOLIN-5-ONE; 4-DIMETHYLAMINO-2,3-DIMETHYL-1-PHENYL-5-PYRAZOLONE; DIMETHYLAMINOPHENAZON; 4-DIMETHYLAMINOPHENAZONE; DIMETHYLAMINOPHENAZONE; DIMETHYLAMINOPHENYLDIMETHYLPYRAZOLIN; 4-DIMETHYLAMINO-1-PHENYL-2,3-DIMETHYLPYRAZOLONE; DIMETHYLAMINOPHENYLDIMETHYLPYRAZOLONE; 1,5-DIMETHYL-4-DIMETHYLAMINO-2-PHENYL-3-PYRAZOLONE; 2,3-DIMETHYL-4-DIMETHYLAMINO-1-PHENYL-5-PYRAZOLONE; DIPIRIN; DIPYRIN; DIPYRINE; FEBRININA; FEBRON; ITAMIDONE; 3-KETO-1,5-DIMETHYL-4-DIMETHYLAMINO-2-PHENYL-2,3-DIHYDROPYRAZOLE; MAMALLET-A; NETSUSARIN; NOVAMIDON; 1-PHENYL-2,3-DIMETHYL-4-DIMETHYLAMINOPYRAZOL-5-ONE; 1-PHENYL-2,3-DIMETHYL-4-DIMETHYLAMINOPYRAZOLONE-5; PIRAMIDON; PIRIDOL; PIROMIDINA; POLINALIN; PYRADONE; PYRAMIDON; PYRAMIDONE; 3-PYRAZOLIN-5-ONE,4-(DIMETHYLAMINO)-2,3-DIMETHYL-1-PHENYL-; 3H-PYRAZOL-3-ONE,4-(DIMETHYLAMINO)-1,2-DIHYDRO-1,5-DIMETHYL-2-PHENYL-; 3H-PYRAZOL-3-ONE,4-(DIMETHYLAMINO)-1,2-DIHYDRO-1,5-DIMETHYL-2-PHENYL-(9CI)
Description: colorless leaflets, prisms, or platelets; odorless
Use: in human and veterinary medicine as an antipyretic analgesic and anti-inflammatory agent; sometimes used in drug metabolism; as a tranquilizer

Physical Properties

Freezing Point: 107 °C (224.6 °F) to 109 °C (228.2 °F)
Water Solubility: 1 g / 18 ml Water
Other Solubilities: 95% Ethanol: >=100 mg/ml at 22 °C; Acetone: >=100 mg/ml at 22 °C; Alcohol: 1 g/1.5 mL; Benzene: 1 g/12 mL; Chloroform: 1 g/1 mL; DMSO: >=100 mg/ml at 22 °C; Ether: 1 g/13 mL.
pH: 5% water solution 7.5 to 9
Refraction Index: Alpha 1.520
Flash Point: Not available; probably combustible

RTECS Toxicity Data

Acute Oral: Rat LD_{50} Dose: 285 mg/kg. Mouse LD_{50} Dose: 350 mg/kg.
Acute Inhalation: Mouse LC_{50} Dose: 20410 mg/m^3.
Acute Dermal: Rabbit LD_{Lo} Route: Subcutaneous Dose: 250 mg/kg. Rat LD_{50} Route: Subcutaneous Dose: 295 mg/kg; Toxic Effects: Behavioral - Somnolence (general depressed activity).
Chronic (Multiple Dose) Inhalation: Rat Dose: 191 mg/m^3/4H/30D-I; Toxic Effects: Behavioral - Alteration of classical conditioning; Liver - Liver function tests impaired. Rat Dose: 20 mg/m^3/5H/17W-I; Toxic Effects: Blood - Changes in other cell count; Endocrine - Changes in spleen weight; Nutritional and gross metabolic - Weight loss or decreased weight gain.
Reproductive/Teratogenic: Rat Route: Oral; Dose: 60 mg/kg; Duration: female 9-14D of pregnancy; Specific Developmental Abnormalities - Musculoskeletal system; Effects on Newborn - Weaning or lactation index; Growth statistics. Rat Route: Oral; Dose: 900 mg/kg; Duration: female 9-14D of pregnancy; Effects on Embryo or Fetus - Fetotoxicity; Specific Developmental Abnormalities - Urogenital system.
Mutagenic: Rat Morphological Transformation; Route: Oral; Dose: 2520 mg/kg/6W. Mouse DNA Inhibition; Cell Type: other cell types; Dose: 100 mg/L.
Tumorigenic: Rat Route: Oral; Dose: 4200 mg/kg/2.5Y-C; Toxic Effects: Tumorigenic - Neoplastic by RTECS criteria; Skin and appendages - Tumors.

Hazard Overviews

Health: May cause irritation. Toxic. Other Acute Effects: harmful if swallowed, inhaled, or absorbed through skin; target organs: blood; bone marrow. Chronic Effects:

agranulocytosis may occur; can produce potentially fatal bone-marrow toxicity.
Fire: Will burn. Hazards: emits toxic fumes. Extinguishing agents: water spray; carbon dioxide, dry chemical powder or appropriate foam. Precautions: combustible liquid.
Reactivity: Incompatible with: strong oxidizing agents, strong acids, strong bases, sensitive to light. Hazardous decomposition products: toxic fumes of: carbon monoxide, carbon dioxide, nitrogen oxides.
Carcinogenicity: IARC - Not listed; NIOSH - Not listed; NTP - Not listed; ACGIH - Not listed; OSHA - Not listed; EPA - Not listed; MAK - Not listed
Primary Target Organs:

Blood

Bone

Environmental

Regulations
RCRA 40CFR: Not listed
CERCLA: 40CFR 302.4: Not listed
SARA 40CFR 372.65: Not listed
SARA EHS 40CFR 355: Not listed
TSCA: Listed

AMI7480 CAS #: 65-49-6

P-AMINOSALICYLIC ACID

RTECS: VO1225000
EINECS Number: 200-613-5
Molecular Formula: $C_7H_7NO_3$
Formula Weight: 153.13

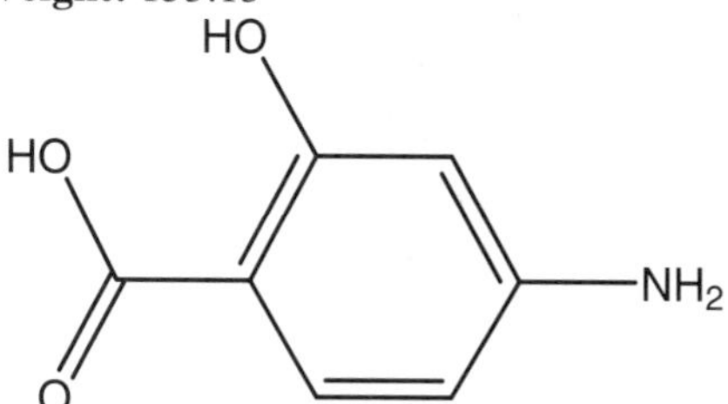

Chemical Structure

Synonyms: PARA-AMINO SALICYLIC ACID; 4-AMINO-2-HYDROXYBENZOIC ACID; AMINOPAR; 4-AMINOSALICYLIC ACID; AMINOSALICYLIC ACID; AMINOX; APACIL; APAS; BENZOIC ACID,4-AMINO-2-HYDROXY-; DEAPASIL; ENTEPAS; GABBROPAS; HELIPIDYL; HELLIPIDYL; 2-HYDROXY-4-AMINOBENZOIC ACID; 3-HYDROXY-4-CARBOXYANILINE; KYSELINA P-AMINOSALICYLOVA; NSC 2083; OSACYL; PAMACYL; PAMISYL; PARAMYCIN; PARASAL; PARASALICIL; PARASALINDON; PARA-PAS; PAS; PAS-C; PASA; PASALON; PASARA; PASCORBIC; PASEM; PASK; PASMED; PASNODIA; PASOLAC; PROPASA; REZIPAS; SALICYLIC ACID,4-AMINO-; SANIPIROL-4; SANIPRIOL-4
Description: white, or nearly white, bulky powder, minute crystals, needles, or plates; odorless, or has slight acetous odor
Use: medication: antitubercular

Physical Properties
Boiling Point: Decomposes at 224 °C (435 °F)
Freezing Point: 150 °C (302 °F) to 151 °C (303.8 °F)
Water Solubility: 1 1 g Soluble in About 7 mL Water
Other Solubilities: Insoluble in Chloroform; 1 g Soluble in 21 ml Alcohol; Slightly Soluble in Ether; Practically Insoluble in Benzene.
pH: Saturated aqueous solution 3 to 3.7

RTECS Toxicity Data
Acute Oral: Mouse LD_{50} Dose: 4 gm/kg. Rabbit LD_{50} Dose: 3650 mg/kg.
Acute Dermal: Mouse LD_{50} Route: Subcutaneous Dose: 4 gm/kg.
Irritation Eye: Rabbit Standard Draize Test Dose: 100 mg/24H; Reaction: moderate.
Mutagenic: Mouse Cytogenetic Analysis; Route: Oral; Dose: 50 mg/kg. Mouse Cytogenetic Analysis; Cell Type: mammary gland; Dose: 2 mmol/L/24H-C.

Hazard Overviews
Health: Irritating to eyes/skin/respiratory tract. Other Acute Effects: may be harmful by inhalation, ingestion, or skin absorption; target organs: g.i. system; liver.
Fire: Hazards: emits toxic fumes. Extinguishing agents: water spray; carbon dioxide, dry chemical powder or appropriate foam. Precautions: combustible liquid.
Reactivity: Incompatible with: strong oxidizing agents, sensitive to air, sensitive to light. Hazardous decomposition products: toxic fumes of: carbon monoxide, carbon dioxide, nitrogen oxides.
Carcinogenicity: IARC - Not listed; NIOSH - Not listed; NTP - Not listed; ACGIH - Not listed; OSHA - Not listed; EPA - Not listed; MAK - Not listed
Primary Target Organs:

Eyes

Skin

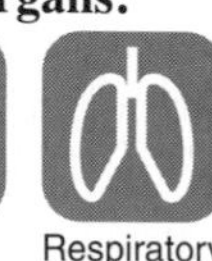
Respiratory System

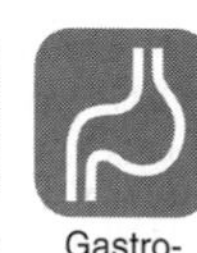
Gastro-intestinal

Liver

Environmental

Regulations
RCRA 40CFR: Not listed
CERCLA: 40CFR 302.4: Not listed
SARA 40CFR 372.65: Not listed
SARA EHS 40CFR 355: Not listed
TSCA: Listed

AMI7560 CAS #: 4005-51-0

2-AMINO-1,3,4-THIADIAZOLE

RTECS: XI3000000
EINECS Number: 223-657-7

Molecular Formula: $C_2H_3N_3S$
Formula Weight: 101.14

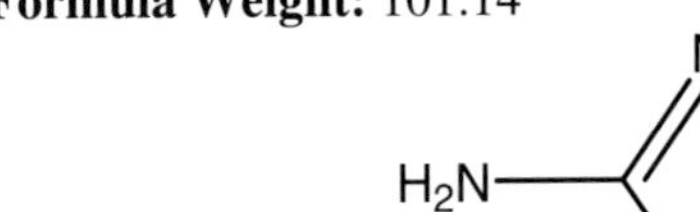

Chemical Structure

Synonyms: AMINOTHIADIAZOLE; ATDA; 1,3,4-THIADIAZOL-2-AMINE

Physical Properties

Water Solubility: 20 mg/ml

RTECS Toxicity Data

Acute Dermal: Rat LD_{50} Route: Subcutaneous Dose: 200 mg/kg.
Reproductive/Teratogenic: Rat Route: Oral; Dose: 20 mg/kg; Duration: female 10D of pregnancy; Effects on Fertility - Post-implantation mortality; Litter size. Rat Route: Oral; Dose: 5 mg/kg; Duration: female 10D of pregnancy; Specific Developmental Abnormalities - Musculoskeletal system; Other developmental abnormalities. Rat Route: Oral; Dose: 20 mg/kg; Duration: female 10D of pregnancy; Effects on Embryo or Fetus - Fetal death.

Hazard Overviews

Health: Irritating to eyes/skin/respiratory tract. Harmful. Other Acute Effects: harmful if swallowed, inhaled, or absorbed through skin. Chronic Effects: laboratory experiments have shown mutagenic effects.
Fire: Hazards: emits toxic fumes. Extinguishing agents: water spray; carbon dioxide, dry chemical powder or appropriate foam. Precautions: combustible liquid.
Reactivity: Incompatible with: strong oxidizing agents. Hazardous decomposition products: toxic fumes of: carbon monoxide, carbon dioxide, nitrogen oxides, sulfur oxides.
Carcinogenicity: IARC - Not listed; NIOSH - Not listed; NTP - Not listed; ACGIH - Not listed; OSHA - Not listed; EPA - Not listed; MAK - Not listed
Primary Target Organs:

Eyes

Skin

Respiratory System

Environmental

Regulations

RCRA 40CFR: Not listed
CERCLA: 40CFR 302.4: Not listed
SARA 40CFR 372.65: Not listed
SARA EHS 40CFR 355: Not listed
TSCA: Not listed

AMI7640 CAS #: 96-50-4

2-AMINOTHIAZOLE

RTECS: XJ2100000
EINECS Number: 202-511-6
Molecular Formula: $C_3H_4N_2S$
Formula Weight: 100.15

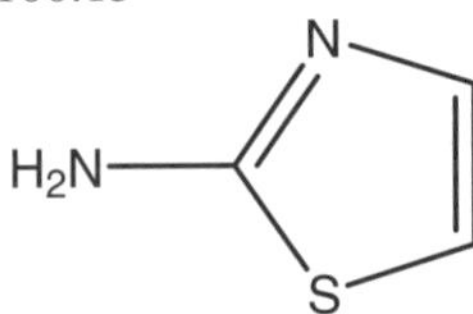

Chemical Structure

Synonyms: ABADOL; ABADOLE; AMINOTHIAZOLE; BASEDOL; 2-THIAZOLAMINE; 2-THIAZOLYLAMINE; 2-THIAZYLAMINE
Description: light brown to brown crystals or granular solid
Use: intermediate in the synthesis of sulfathiazole; in medicine as a thyroid inhibitor; and in the preparation of azo dyes

Physical Properties

Boiling Point: Decomposes
Freezing Point: 90 °C (194 °F)
Water Solubility: < 1 mg/mL at 20 C
Other Solubilities: DMSO: >=100 mg/ml at 20 °C; Acetone: <1 mg/ml at 20 °C; Dilute Hydrochloric Acid: Freely Soluble; Sulfuric acid (20%): Freely Soluble; Ether: Slightly Soluble; Hot Alcohol: Soluble.
Flash Point: Not available; probably combustible
Autoignition Temperature: 100 °C

RTECS Toxicity Data

Acute Oral: Rat LD_{50} Dose: 480 mg/kg; Toxic Effects: Behavioral - Tremor; Behavioral - Coma; Lungs, Thorax, or Respiration - Dyspnea. Rabbit LD_{50} Dose: 370 mg/kg; Toxic Effects: Behavioral - Tremor; Behavioral - Coma; Lungs, Thorax, or Respiration - Dyspnea.
Mutagenic: Mouse Mutations in Microorganisms; Cell Type: lymphocyte; Dose: 1214 mg/L (+S9). Mouse Mutations in Mammalian Somatic Cells; Cell Type: lymphocyte; Dose: 557 mg/L.

Hazard Overviews

Health: May cause irritation. Toxic. Other Acute Effects: harmful by inhalation, in contact with skin and if swallowed; target organ: thyroid.
Fire: Will burn. Extinguishing agents: water spray; carbon dioxide, dry chemical powder or appropriate foam. Precautions: combustible liquid.
Carcinogenicity: IARC - Not listed; NIOSH - Not listed; NTP - Not listed; ACGIH - Not listed; OSHA - Not listed; EPA - Not listed; MAK - Not listed

Primary Target Organs:

Glandular System

Environmental

Regulations

RCRA 40CFR: Not listed
CERCLA: 40CFR 302.4: Not listed
SARA 40CFR 372.65: Not listed
SARA EHS 40CFR 355: Not listed
TSCA: Listed

AMI7720 CAS #: 79-19-6

1-AMINO-2-THIOUREA

RTECS: VT4200000
EINECS Number: 201-184-7
Molecular Formula: CH_5N_3S
Formula Weight: 91.13

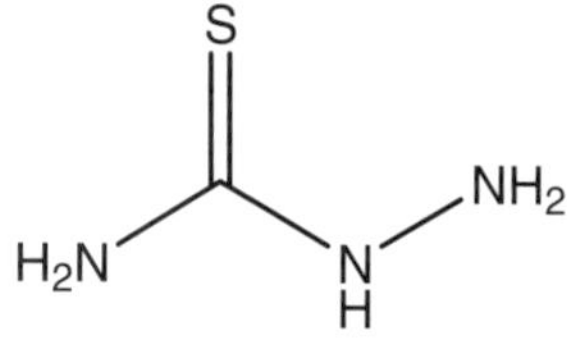

Chemical Structure

Synonyms: N-AMINO THIOUREA; 1-AMINOTHIOUREA; AMINOTHIO-UREA; AMINOTHIOUREA; N-AMINOTHIOUREA; HYDRAZINECARBOTHIOAMIDE; ISOTHIOSEMICARBAZIDE; SEMICARBAZIDE,3-THIO-; SEMICARBAZIDE,THIO-; SEMICARBIZIDE,3-THIO; THIOCARBAMOYLHYDRAZINE; THIOCARBAMYLHYDRAZINE; 2-THIOSEMICARBAZIDE; THIOSEMICARBAZIDE; TSC
Description: white crystalline powder or long needles; odorless
Use: reagent for detection of metals

Physical Properties

Freezing Point: 183 °C (361.4 °F)
Saturated Vapor Density: 1.785217241 kg/m^3
Water Solubility: Soluble in Water
Other Solubilities: Alcohol: Soluble.

RTECS Toxicity Data

Acute Oral: Rat LD_{50} Dose: 9160 ug/kg. Mouse LD_{Lo} Dose: 94 mg/kg.
Acute Dermal: Rabbit LD_{Lo} Route: Subcutaneous Dose: 14 mg/kg; Toxic Effects: Behavioral - Excitment; Endocrine - Hyperglycemia. Mouse LD_{50} Route: Subcutaneous Dose: 16407 ug/kg.
Mutagenic: Human DNA Damage; Cell Type: HeLa cell; Dose: 20 umol/L.
Tumorigenic: Rat Route: Oral; Dose: 1024 mg/kg/78W-C; Toxic Effects: Tumorigenic - Equivocal tumorigenic agent by RTECS criteria; Skin and appendages - Tumors. Rat Route: Oral; Dose: 2048 mg/kg/78W-C; Toxic Effects: Tumorigenic - Equivocal tumorigenic agent by RTECS criteria; Skin and appendages - Tumors.

Hazard Overviews

Poison

Health: May cause irritation. Poison. Other Acute Effects: may be fatal if inhaled, swallowed, or absorbed through skin.
Fire: Will burn. Extinguishing agents: water spray; carbon dioxide, dry chemical powder or appropriate foam. Precautions: combustible liquid.
Reactivity: Incompatible with: strong oxidizing agents, strong acids, strong bases. Hazardous decomposition products: toxic fumes of: carbon monoxide, carbon dioxide, sulfur oxides, nitrogen oxides.
Carcinogenicity: IARC - Not listed; NIOSH - Not listed; NTP - Not listed; ACGIH - Not listed; OSHA - Not listed; EPA - Not listed; MAK - Not listed

Environmental

Environmental Fate: No experimental data are available pertaining to environmental; therefore, the limited estimations of its environmental fate are based on its chemical structure and reported water solubility. If released to the atmosphere, it should degrade rapidly in the vapor-phase (half-life of 3.8 hr estimated from chemical structure) by reaction with photochemically produced hydroxyl radicals. In addition to the vapor-phase, it may exist in air adsorbed to particulates which way considerably affect its reactivity. If released to water, aquatic volatilization, bioconcentration and adsorption to sediment are not expected to be important. If released to soil, it may be susceptible to significant leaching. No data are available regarding biodegradation or chemical degradation processes in soil or water.
Cleanup/Disposal: Guide No. 154: Eliminate all ignition sources (no smoking, flares, sparks or flames in immediate area). Do not touch damaged containers or spilled material unless wearing appropriate protective clothing. Stop leak if you can do it without risk. Prevent entry into waterways, sewers, basements or confined areas. Absorb or cover with dry earth, sand or other non-combustible material and transfer to containers. Do not get water inside containers.

Environmental Physical Data

Henry's Law Constant: estimated at 1 x10^{-7}
BCF: not significant

Regulations

RCRA 40CFR: Listed Hazardous Waste No. P116 Toxic Waste
CERCLA: 40CFR 302.4: Listed per RCRA Section 3001 RQ: 100 lb (45.35 kg)

SARA 40CFR 372.65: Listed TPQ: 100/10000 lb
SARA EHS 40CFR 355: Listed TPQ: 100 lb
TSCA: Listed

Analytical Methods

Soil: SW846 3640A

AMI7800 CAS #: 591-08-2

N-(AMINOTHIOXOMETHYL)ACET-AMIDE

RTECS: YR7700000
EINECS Number: 209-699-9
Molecular Formula: $C_3H_6N_2OS$
Formula Weight: 118.15

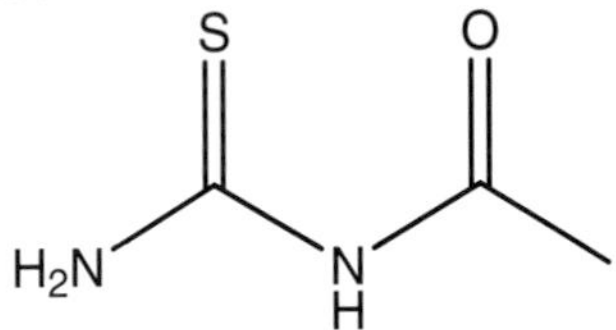

Chemical Structure

Synonyms: ACETAMIDE,N-(AMINOTHIOXOMETHYL)-; ACETOTHIOUREA; ACETYL THIOUREA; N-ACETYLTHIOCARBAMIDE; 1-ACETYL-2-THIOUREA; 1-ACETYLTHIOUREA; ACETYLTHIOUREA; N-ACETYL-2-THIOUREA; N-ACETYLTHIOUREA; N(THIOCARBAMYL)ACETAMIDE; UREA,1-ACETYL-2-THIO-
Description: prisms or crystals

Physical Properties

Freezing Point: 165 °C (329 °F)
Water Solubility: Soluble in Hot Water
Other Solubilities: Alcohol: Soluble; Dilute NaOH: Soluble; Ether: Slightly Soluble.

RTECS Toxicity Data

Acute Oral: Rat LD_{50} Dose: 50 mg/kg. Mouse LD_{Lo} Dose: 94 mg/kg.

Hazard Overviews

Poison

Health: Irritating. Poison. Other Acute Effects: may be fatal if inhaled, swallowed, or absorbed through skin.
Fire: Hazards: emits toxic fumes. Extinguishing agents: water spray; carbon dioxide, dry chemical powder or appropriate foam. Precautions: combustible liquid.
Reactivity: Incompatible with: strong oxidizing agents, strong acids, strong bases. Hazardous decomposition products: toxic fumes of: carbon monoxide, carbon dioxide, nitrogen oxides, sulfur oxides.

Carcinogenicity: IARC - Not listed; NIOSH - Not listed; NTP - Not listed; ACGIH - Not listed; OSHA - Not listed; EPA - Not listed; MAK - Not listed
Primary Target Organs:

Eyes

Skin

Respiratory System

Environmental

Environmental Fate: Only limited experimental data are available to predict environmental fate. A single aqueous hydrolysis study has reported a rate constant corresponding to a half-life of 2.7 hr at pH 9.65. Sufficient data are not available to estimate hydrolysis rates at lower pHs, although the rate at pH 7 is likely to be much slower. It absorbs light relatively weakly in the environmental region suggesting a potential for direct photolysis. If released to the atmosphere, it should degrade rapidly in the vapor-phase (half-life of 4.8 hr estimated from chemical structure) by reaction with photochemically produced hydroxyl radicals. It may additionally exist in air in the adsorbed-particulate phase. If released to water, hydrolysis may be important. If released to soil, it may be susceptible to leaching. Aqueous hydrolysis may be important, particularly in alkaline soils. No data are available regarding biodegradation in soil or water.
Cleanup/Disposal: Guide No. 154: Eliminate all ignition sources (no smoking, flares, sparks or flames in immediate area). Do not touch damaged containers or spilled material unless wearing appropriate protective clothing. Stop leak if you can do it without risk. Prevent entry into waterways, sewers, basements or confined areas. Absorb or cover with dry earth, sand or other non-combustible material and transfer to containers. Do not get water inside containers.

Environmental Physical Data

Henry's Law Constant: $< 1 \times 10^{-7}$
Octanol/Water Partition Coefficient: $\log K_{ow} = 0.22$
BCF: estimated at 0.9

Regulations

RCRA 40CFR: Listed Hazardous Waste No. P002 Toxic Waste
CERCLA: 40CFR 302.4: Listed per RCRA Section 3001 RQ: 1000 lb (453.5 kg)
SARA 40CFR 372.65: Not listed
SARA EHS 40CFR 355: Not listed
TSCA: Listed

Analytical Methods

Soil: SW846 8270C
Plasma: EPA 29

AMI7880 **CAS #: 95-53-4**

2-AMINOTOLUENE

RTECS: XU2975000
DOT: UN1708; IMO6.1
EINECS Number: 202-429-0
Molecular Formula: C_7H_9N
Structured MF: $CH_3C_6H_4NH_2$
Formula Weight: 107.15

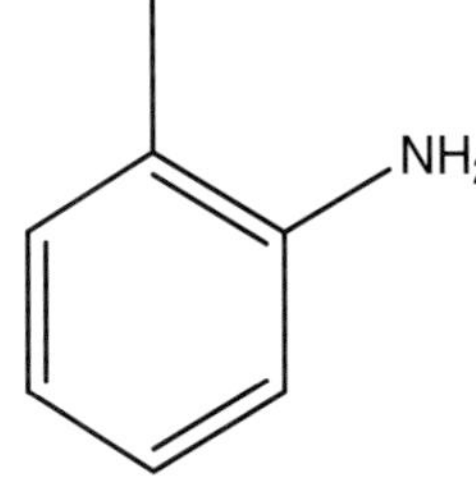

Chemical Structure

Synonyms: 1-AMINO-2-METHYLBENZENE; 2-AMINO-1-METHYLBENZENE; O-AMINOTOLUENE; ANILINE,2-METHYL-; BENZENAMINE,2-METHYL-; BENZENAMINE,2-METHYL-(9CI); C.I. 37077; 1-METHYL-2-AMINOBENZENE; 2-METHYL-1-AMINOBENZENE; 2-METHYLANILINE; O-METHYLANILINE; 2-METHYLBENZAMINE; 2-METHYLBENZENAMINE; O-METHYLBENZENAMINE; O-TOLUIDIN; 2-TOLUIDINE; O-TOLUIDINE; ORTHO-TOLUIDINE; O-TOLUIDYNA; O-TOLYLAMINE
Description: clear to yellowish liquid; amine odor
Use: textile printing dyes; vulcanization accelerator; organic synthesis; dye intermediate; antioxidant in the manufacture of rubber; ingredient in a clinical laboratory reagent used to test blood samples for glucose; making colors fast to acids; intermediate in pharmaceutical manufacture; chemicals production; and intermediate in the manufacture of pesticides

Physical Properties

Boiling Point: 200.2 °C (392 °F)
Freezing Point: -14.7 °C (5.54 °F)
Specific Gravity: 1.008 at 20 °C/20 °C
Vapor Density: 3.69
Density: 1.008 g/mL at 20 °C
Vapor Pressure: 0.32 torr
Water Solubility: Slightly Soluble in Water
Other Solubilities: 95% Ethanol: >=100 mg/ml at 15 °C; Acetone: >=100 mg/ml at 15 °C; Carbon Tetrachloride: miscible; DMSO: >=100 mg/ml at 15 °C; Dilute acids: Slightly Soluble; Ether: Soluble.
Surface Tension: 43.55 dynes/cm at 20 °C
Odor Threshold: 20 ppm
pH: 8
Refraction Index: 1.5688 at 20 °C/D
Critical Temperature: 421 °C
Critical Pressure: 544 psia
Ionization Potential (eV): 7.44
Flash Point: 85 °C Closed Cup
Autoignition Temperature: 482 °C
LEL: 1.5% v/v

RTECS Toxicity Data

Acute Oral: Rat LD_{50} Dose: 670 mg/kg; Toxic Effects: Blood - Normocytic anemia; Blood - Pigmented or nucleated red Blood cells; Blood - Methemoglobinemia-Carboxyhemoglobin. Mouse LD_{50} Dose: 520 mg/kg; Toxic Effects: Blood - Normocytic anemia; Blood - Pigmented or nucleated red Blood cells; Blood - Methemoglobinemia-Carboxyhemoglobin.
Acute Inhalation: Rat LC_{50} Dose: 862 ppm/4hr; Toxic Effects: Behavioral - Somnolence (general depressed activity); Behavioral - Tremor; Lungs, Thorax, or Respiration - Cyanosis. Man TC_{Lo} Dose: 25 mg/m^3; Toxic Effects: Kidney, ureter, and Bladder - Urine volume increased; Kidney, ureter, and Bladder - Hematuria; Blood - Methemoglobinemia-Carboxyhemoglobin.
Acute Dermal: Rabbit LD_{50} Route: Skin; Dose: 3250 mg/kg; Toxic Effects: Skin and appendages - Primary irritation.
Chronic (Multiple Dose) Oral: Rat Dose: 1125 mg/kg/5D-C; Toxic Effects: Blood - Changes in spleen; Nutritional and gross metabolic - Weight loss or decreased weight gain; DEATH. Rat Dose: 10200 mg/kg/30D-I; Toxic Effects: Kidney, Ureter, and Bladder - Proteinuria; Blood - Normocytic anemia; Nutritional and gross metabolic - Weight loss or decreased weight gain.
Irritation Eye: Rabbit Standard Draize Test Dose: 750 ug/24H; Reaction: severe.
Irritation Skin: Rabbit Standard Draize Test Dose: 500 mg/24H; Reaction: mild. Rabbit Open Draize Test Dose: 10 mg/24H open; Reaction: mild.
Reproductive/Teratogenic: Rat Route: Skin; Dose: 9520 mg/kg; Duration: male 17W prior to mating; Paternal Effects - Spermatogenesis; Effects on Newborn - Physical. Rat Route: Skin; Dose: 9520 mg/kg; Duration: female 17W prior to mating Maternal Effects - Menstrual cycle changes or disorders; Effects on Newborn - Growth statistics; Physical. Rat Route: Skin; Dose: 952 mg/kg; Duration: female 17W prior to mating Maternal Effects - Ovaries, fallopian tubes; Menstrual cycle changes or disorders.
Mutagenic: Human Unscheduled DNA Synthesis; Cell Type: HeLa cell; Dose: 50 uL/L. Human DNA Inhibition; Cell Type: HeLa cell; Dose: 4 mmol/L. Human Mutations in Microorganisms; Cell Type: lymphocyte; Dose: 300 mg/L (+S9). Human Mutations in Mammalian Somatic Cells; Cell Type: lymphocyte; Dose: 450 mg/L. Human Sister Chromatid Exchenge; Cell Type: lymphocyte; Dose: 200 umol/L. Human Micronucleus Test; Cell Type: lymphocyte; Dose: 2 mmol/L.
Tumorigenic: Rat Route: Oral; Dose: 109 gm/kg/2Y-C; Toxic Effects: Tumorigenic - Neoplastic by RTECS criteria; Blood - Tumors. Rat Route: Oral; Dose: 7250 mg/kg/23W-C; Toxic Effects: Tumorigenic - Equivocal tumorigenic agent by RTECS criteria; Kidney, Ureter, and Bladder - Tumors.

Hazard Overviews

Fire Diamond

Health: Severely irritating to eyes/skin. Also Causes: headache, cyanosis, unconsciousness, death, kidney/bladder irritation, drowsiness, nausea/vomiting, hematuria. Chronic Effects: dermatitis. Possible cancer hazard.

Fire: Combustible. Use dry chemical, carbon dioxide, or regular foam. Water may be ineffective because material is lighter than and floats on water which may only spread fire.

Reactivity: Stable. Hazardous polymerization cannot occur. Avoid: heat; ignition sources. Incompatible with: oxidizers; nitric acid; other acids and bases. Hazardous decomposition products: carbon oxides; nitrogen oxides.

Carcinogenicity: IARC - Group 2B, Possibly carcinogenic to humans; NIOSH - Listed as carcinogen; NTP - Class 2B, Reasonably anticipated to be a carcinogen, sufficient evidence of carcinogenicity from studies in experimental animals; ACGIH - Class A3, Animal carcinogen; OSHA - Not listed; EPA - Not listed; MAK - Class A2, Unmistakably carcinogenic in animal experimentation only

Primary Target Organs:

Eyes

Skin

Nervous System

Liver

Kidneys

Blood

Exposure Limits

OSHA PEL: TWA: 5 ppm; 22 mg/m^3.

ACGIH TLV: TWA: 2 ppm; 8.8 mg/m^3.

NIOSH IDLH: 50 ppm.

Respirator Recommendation

Exposure Range: >2 to <50 ppm Supplied Air, Constant Flow/Pressure Demand, Half Mask

Exposure Range: 50 to unlimited ppm Self-contained Breathing Apparatus, Pressure Demand, Full Face

Note: poor warning properties

Environmental

Ecotoxicity: Aquatic toxicity: 100 ppm/*/fish/lethal/fresh water *Time period not specified

Environmental Fate: If released on land, it will be lost by a combination of biodegradation, oxidation, and chemical binding to soil components. If released into water, it will also be primarily lost by biodegradation, oxidation and photooxidation. There will also be some adsorption to sediment. Bioconcentration in fish should not be an important fate process. In the atmosphere, it will photodegrade (estimated half-life 2.4 hr).

Cleanup/Disposal: Guide No. 153: Eliminate all ignition sources (no smoking, flares, sparks or flames in immediate area). Do not touch damaged containers or spilled material unless wearing appropriate protective clothing. Stop leak if you can do it without risk. Prevent entry into waterways, sewers, basements or confined areas. Absorb or cover with dry earth, sand or other non-combustible material and transfer to containers. Do not get water inside containers.

Environmental Physical Data

Henry's Law Constant: estimated at 2.72 x10^{-6}

Octanol/Water Partition Coefficient: log K_{ow} = 1.32

Sorption Partition Coefficient: K_{oc} = 21

BCF: calculated at 5.9

BOD: 143%, 5 days

Regulations

RCRA 40CFR: Listed Hazardous Waste No. U328 Toxic Waste

CERCLA: 40CFR 302.4: Listed per RCRA Section 3001 RQ: 100 lb (45.35 kg)

SARA 40CFR 372.65: Listed

SARA EHS 40CFR 355: Not listed

TSCA: Listed

Analytical Methods

Soil: SW846 3640A, 5031, 8015B, 8260B, 8270B, 8270C

Water / Groundwater: EPA 1625

Indoor / Expired Air: NIOSH 2002

Plasma: EPA 29

AMI7960 CAS #: 108-44-1

3-AMINOTOLUENE

RTECS: XU2800000

DOT: UN1708; IMO6.1

EINECS Number: 203-583-1

Molecular Formula: C_7H_9N

Structured MF: $CH_3C_6H_4NH_2$

Formula Weight: 107.15

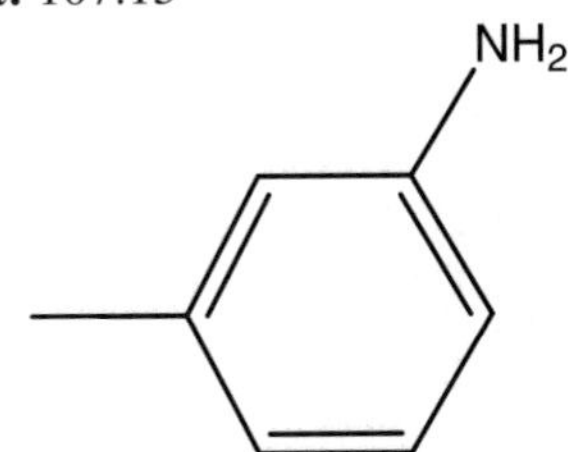

Chemical Structure

Synonyms: 1-AMINO-3-METHYLBENZENE; 3-AMINO-1-METHYLBENZENE; 1-AMINOPHENYLMETHANE; 3-AMINOPHENYLMETHANE; 3-AMINOTOLUEN; M-AMINOTOLUENE; ANILINE,3-METHYL-; BENZENAMINE,3-METHYL-; 3-METHYLANILINE; M-METHYLANILINE; 3-METHYLBENZENAMINE; M-METHYLBENZENAMINE; M-TOLUIDIN; 3-TOLUIDINE; M-TOLUIDINE; M-TOLUIDYNA; M-TOLYLAMINE

Description: colorless to light yellow liquid; aromatic, amine-like odor

Use: mfr dyes & other organic chemicals

Physical Properties

Boiling Point: 203 °C (397 °F) to 204 °C (399 °F)
Freezing Point: -30.4 °C (-22.72 °F)
Specific Gravity: 0.99 at 25 °C/25 °C
Vapor Density: Estimate 3.9 Air=1
Vapor Pressure: 1 mm Hg at 106 °F
Water Solubility: Slightly Soluble in Water
Other Solubilities: Infinitely Soluble in Carbon Tetrachloride, Heptane; Soluble in Acetone & Benzene.
Surface Tension: 36.9 dynes/cm at 20 °C
Odor Threshold: 0.46 to 6.0 ppm
Refraction Index: 1.5711 at 22 °C/D
Ionization Potential (eV): 7.50
Flash Point: 86.1 °C
Autoignition Temperature: 482 °C
LEL: 1.1% v/v
UEL: 6.6% v/v

RTECS Toxicity Data

Acute Oral: Rat LD_{50} Dose: 450 mg/kg. Mouse LD_{50} Dose: 740 mg/kg.
Chronic (Multiple Dose) Oral: Rat Dose: 8400 mg/kg/30D-I; Toxic Effects: Kidney, Ureter, and Bladder - Proteinuria; Blood - Normocytic anemia; Nutritional and gross metabolic - Weight loss or decreased weight gain.
Irritation Eye: Rabbit Standard Draize Test Dose: 20 mg/24H; Reaction: moderate.
Irritation Skin: Rabbit Standard Draize Test Dose: 500 mg/24H; Reaction: mild.

Hazard Overviews

Health: Irritating to eyes/skin/respiratory tract. Toxic. Other Acute Effects: harmful if swallowed, inhaled, or absorbed through skin; absorption into the body leads to the formation of methemoglobin which in sufficient concentration causes cyanosis; onset may be delayed 2 to 4 hours or longer. Chronic Effects: may alter genetic material; target organs: liver, blood, kidneys, bladder. Probable human carcinogen.
Fire: Combustible. Hazards: emits toxic fumes. Extinguishing agents: water spray; carbon dioxide, dry chemical powder or appropriate foam. Precautions: combustible liquid.
Reactivity: Hazardous polymerization will not occur. Incompatible with: acids, acid chlorides, acid anhydrides, chloroformates, strong oxidizing agents, air sensitive, light sensitive. Hazardous decomposition products: toxic fumes of: carbon monoxide, carbon dioxide, nitrogen oxides.
Carcinogenicity: IARC - Not listed; NIOSH - Not listed; NTP - Not listed; ACGIH - Class A4, Not classifiable as a human carcinogen; OSHA - Not listed; EPA - Not listed; MAK - Not listed
Primary Target Organs:

Eyes

Skin

Respiratory System

Liver

Kidneys

Blood

Exposure Limits
OSHA PEL Vacated 1989 Limits: TWA: 2 ppm; 9 mg/m^3.
ACGIH TLV: TWA: 2 ppm; 8.8 mg/m^3.
Respirator Recommendation
Exposure Range: >2 to 100 ppm Supplied Air, Constant Flow/Pressure Demand, Half Mask
Exposure Range: >100 to 2000 ppm Supplied Air, Constant Flow/Pressure Demand, Full Face
Exposure Range: >2000 to unlimited ppm Self-contained Breathing Apparatus, Pressure Demand, Full Face
Note: poor warning properties

Environmental

Environmental Fate: If released on land, it will be lost by a combination of biodegradation, oxidation, and chemical binding to soil components. If released into water, it will also be primarily lost by biodegradation and oxidation and in surface waters, photooxidation. There will also be some adsorption to sediment. Bioconcentration in fish should not be an important fate process. In the atmosphere, it will photodegrade (estimated half-life 2.4 hr).
Cleanup/Disposal: Guide No. 153: Eliminate all ignition sources (no smoking, flares, sparks or flames in immediate area). Do not touch damaged containers or spilled material unless wearing appropriate protective clothing. Stop leak if you can do it without risk. Prevent entry into waterways, sewers, basements or confined areas. Absorb or cover with dry earth, sand or other non-combustible material and transfer to containers. Do not get water inside containers.

Environmental Physical Data

Henry's Law Constant: estimated at 2.96 x10^{-6}
Octanol/Water Partition Coefficient: **log K_{ow}** = 1.40
Sorption Partition Coefficient: **K_{oc}** = 4 silt loam soils 44
BCF: calculated at 6.8
BOD: 74% BODT, 8 days

Regulations

RCRA 40CFR: Not listed
CERCLA: 40CFR 302.4: Not listed
SARA 40CFR 372.65: Not listed
SARA EHS 40CFR 355: Not listed
TSCA: Listed

AMI8040 **CAS #: 106-49-0**

4-AMINOTOLUENE

RTECS: XU3150000
DOT: UN1708; IMO6.1
EINECS Number: 203-403-1
Molecular Formula: C_7H_9N
Structured MF: $CH_3C_6H_4NH_2$
Formula Weight: 107.15

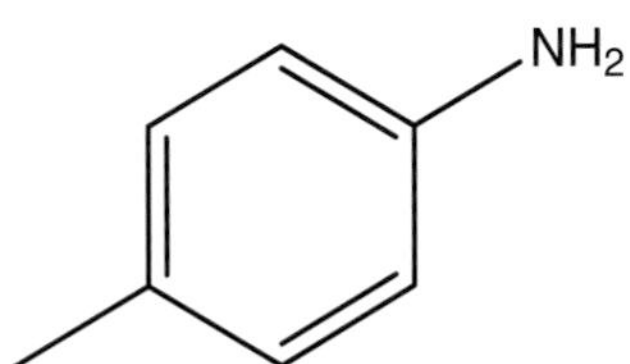

Chemical Structure

Synonyms: 1-AMINO-4-METHYLBENZENE; 4-AMINO-1-METHYLBENZENE; 4-AMINOTOLUEN; P-AMINOTOLUENE; ANILINE,P-METHYL-; BENZENAMINE,4-METHYL-; C.I. 37107; C.I. AZOIC COUPLING COMPONENT 107; 4-METHYLANILINE; P-METHYLANILINE; 4-METHYLBENZENAMINE; P-METHYLBENZENAMINE; NAPHTOL AS-KG; NAPHTOL AS-KGLL; P-TOLUIDIN; 4-TOLUIDINE; P-TOLUIDINE; P-TOLUIDYNA; TOLYAMINE; P-TOLYLAMINE; TOLYLAMINE

Description: lustrous plates or colorless leaflets; white solid; winelike aromatic odor

Use: chem int for many dyes; test reagent for lignin, nitrite, phloroglucinol; agent in preparation of ion exchange resins

Physical Properties

Boiling Point: 200.5 °C (393 °F)
Freezing Point: 44 °C (111.2 °F) to 45 °C (113 °F)
Specific Gravity: 1.046 at 20 °C/4 °C
Vapor Density: 3.9 Air=1
Vapor Pressure: 1 mm Hg at 108 °F
Water Solubility: 1 parts in about 135 parts Water
Other Solubilities: Soluble in Pyrimidine.
Surface Tension: 34.6 dynes/cm at 50 °C
Refraction Index: 1.5636 at 20 °C/D
Ionization Potential (eV): 7.50
Flash Point: 86.667 °C Closed Cup
Autoignition Temperature: 482 °C
LEL: 1.1% v/v
UEL: 6.6% v/v

RTECS Toxicity Data

Acute Oral: Rat LD_{50} Dose: 336 mg/kg; Toxic Effects: Peripheral nerve and sensation - Spastic parapysis with or without sensory change; Behavioral - Somnolence (general depressed activity); Behavioral - Convulsions or effect on seizure threshold. Mouse LD_{50} Dose: 330 mg/kg.

Acute Inhalation: Rat LC_{50} Dose: >640 mg/m^3/1hr.

Acute Dermal: Rabbit LD_{50} Route: Skin; Dose: 890 mg/kg.

Chronic (Multiple Dose) Oral: Rat Dose: 4560 mg/kg/30D-I; Toxic Effects: Kidney, Ureter, and Bladder - Proteinuria; Blood - Normocytic anemia; Nutritional and gross metabolic - Weight loss or decreased weight gain.

Irritation Eye: Rabbit Standard Draize Test Dose: 100 mg; Reaction: severe. Rabbit Standard Draize Test Dose: 20 mg/24H; Reaction: moderate.

Irritation Skin: Rabbit Standard Draize Test Dose: 500 mg/24H; Reaction: mild. Rabbit Standard Draize Test Dose: 500 mg/24H; Reaction: severe.

Mutagenic: Rat Unscheduled DNA Synthesis; Cell Type: liver; Dose: 100 umol/L. Mouse DNA Damage; Route: Intraperitoneal; Dose: 35 mg/kg.

Hazard Overviews

Health: Severely irritating to eyes/skin/respiratory tract. Toxic. Other Acute Effects: harmful if swallowed, inhaled, or absorbed through skin; high concentrations are extremely destructive to tissues of the mucous membranes and upper respiratory tract, eyes and skin; symptoms of exposure may include burning sensation, coughing, wheezing, laryngitis, shortness of breath, headache, nausea and vomiting; absorption into the body leads to the formation of methemoglobin which in sufficient concentration causes cyanosis; onset may be delayed 2 to 4 hours or longer. Chronic Effects: may alter genetic material; target organs: liver, blood. Probable carcinogen.

Fire: Combustible. Hazards: in powder form capable of creating a dust explosion; emits toxic fumes. Extinguishing agents: water spray; carbon dioxide, dry chemical powder or appropriate foam. Precautions: combustible liquid.

Reactivity: Stable. Hazardous polymerization will not occur. Incompatible with: acids, acid chlorides, acid anhydrides, chloroformates, strong oxidizing agents. Hazardous decomposition products: toxic fumes of: carbon monoxide, carbon dioxide, nitrogen oxides.

Carcinogenicity: IARC - Not listed; NIOSH - Listed as carcinogen; NTP - Not listed; ACGIH - Class A3, Animal carcinogen; OSHA - Not listed; EPA - Not listed; MAK - Class B, Justifiably suspected of having carcinogenic potential

Primary Target Organs:

Eyes

Skin

Respiratory System

Liver

Blood

Exposure Limits

OSHA PEL Vacated 1989 Limits: TWA: 2 ppm; 9 mg/m^3.

ACGIH TLV: TWA: 2 ppm; 8.8 mg/m^3.

Respirator Recommendation

Exposure Range: >2 to 100 ppm Supplied Air, Constant Flow/Pressure Demand, Half Mask

Exposure Range: >100 to 2000 ppm Supplied Air, Constant Flow/Pressure Demand, Full Face

Exposure Range: >2000 to unlimited ppm Self-contained Breathing Apparatus, Pressure Demand, Full Face

Note: poor warning properties

Environmental

Ecotoxicity: LD_{50} Starling oral 42.2 mg/kg

Environmental Fate: If released on land, it will be lost by a combination of biodegradation, oxidation, and chemical binding to soil components. If released into water, it will also be primarily lost by biodegradation, oxidation and photooxidation. There will also be some adsorption to sediment. Bioconcentration in fish should not be an important fate process. In the atmosphere, it will react with photochemically produced hydroxyl radical (estimated half-life 2.4 hr

Cleanup/Disposal: Guide No. 153: Eliminate all ignition sources (no smoking, flares, sparks or flames in immediate

area). Do not touch damaged containers or spilled material unless wearing appropriate protective clothing. Stop leak if you can do it without risk. Prevent entry into waterways, sewers, basements or confined areas. Absorb or cover with dry earth, sand or other non-combustible material and transfer to containers. Do not get water inside containers.

Environmental Physical Data

Henry's Law Constant: estimated at 7.22 x10^{-6}
Octanol/Water Partition Coefficient: log K_{ow} = 1.39
Sorption Partition Coefficient: K_{oc} = 4-aminotoluene 79
BCF: calculated at 6.7
BOD: 64% BODT, 5 days

Regulations

RCRA 40CFR: Listed Hazardous Waste No. U353 Toxic Waste
CERCLA: 40CFR 302.4: Listed per RCRA Section 3001 RQ: 100 lb (45.35 kg)
SARA 40CFR 372.65: Not listed
SARA EHS 40CFR 355: Not listed
TSCA: Listed

Analytical Methods

Soil: SW846 3640A
Water / Groundwater: ASTM D4763

AMI8120 CAS #: 2432-99-7

11-AMINOUNDECANOIC ACID

RTECS: YQ2293000
EINECS Number: 219-417-6
Molecular Formula: $C_{11}H_{23}NO_2$
Structured MF: $H_2N(CH_2)_{10}COOH$
Formula Weight: 201.35

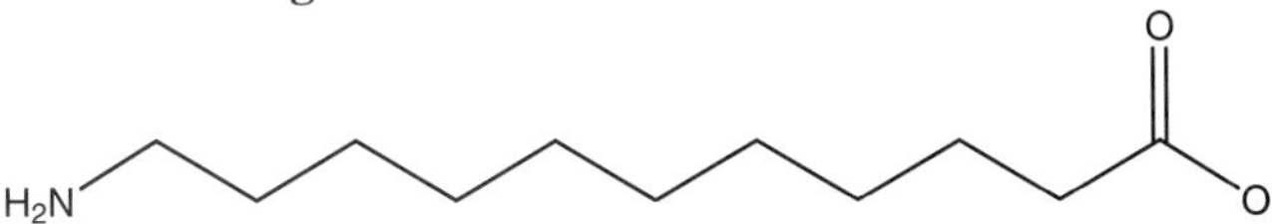

Chemical Structure

Synonyms: AMINOUNDECANOIC ACID; OMEGA-AMINOUNDECANOIC ACID; 11-AMINOUNDECYCLIC ACID; 11-AMINOUNDECYLIC ACID; UNDECANOIC ACID,11-AMINO-
Description: white crystalline solid or powder
Use: manufacture of Nylon 11 polymers (which are used in the automotive industry, for rollers and bearings on conveyors, as covering for braided cable, in fabrics such as filter cloth and netting, for flexible fishing line and for brush bristles); for dry or powder coatings and it can be machined into flat or tubular film, which is used as a packaging material

Physical Properties

Freezing Point: 190 °C (374 °F) to 192 °C (377.6 °F)
Water Solubility: < 0.1 mg/mL at 21 C
Other Solubilities: 5% Acetic Acid: 10-50 mg/ml at 16 °C; 95% Ethanol: <1 mg/ml at 20 °C; Acetone: <1 mg/ml at 21 °C; DMSO: <1 mg/ml at 20 °C; Methanol: <1 mg/ml at 22 °C; Toluene: <1 mg/ml at 21 °C.
Flash Point: Not available; probably combustible

RTECS Toxicity Data

Acute Oral: Rat LD_{Lo} Dose: 14700 ug/kg; Toxic Effects: Nutritional and gross metabolic - Weight loss or decreased weight gain.
Mutagenic: Hamster Sister Chromatid Exchenge; Cell Type: ovary; Dose: 500 mg/L. Hamster Morphological Transformation; Cell Type: embryo; Dose: 2500 mmol/L.
Tumorigenic: Rat Route: Oral; Dose: 655 gm/kg/2Y-C; Toxic Effects: Tumorigenic - Carcinogenic by RTECS criteria; Kidney, Ureter, and Bladder - Tumors. Rat Route: Oral; Dose: 328 gm/kg/2Y-C; Toxic Effects: Tumorigenic - Neoplastic by RTECS criteria; Skin and appendages - Tumors.

Hazard Overviews

Health: Irritating. Toxic. Other Acute Effects: harmful if swallowed, inhaled, or absorbed through skin. Chronic Effects: carcinogen; target organs: bladder, kidneys, liver.
Fire: Will burn. Hazards: emits toxic fumes. Extinguishing agents: water spray; carbon dioxide, dry chemical powder or appropriate foam. Precautions: combustible liquid.
Reactivity: Incompatible with: strong oxidizing agents, strong acids, strong bases. Hazardous decomposition products: toxic fumes of: carbon monoxide, carbon dioxide, nitrogen oxides.
Carcinogenicity: IARC - Group 3, Not classifiable as to carcinogenicity to humans; NIOSH - Not listed; NTP - Not listed; ACGIH - Not listed; OSHA - Not listed; EPA - Not listed; MAK - Not listed
Primary Target Organs:

Eyes

Skin

Respiratory System

Liver

Kidneys

Environmental

Regulations

RCRA 40CFR: Not listed
CERCLA: 40CFR 302.4: Not listed
SARA 40CFR 372.65: Not listed
SARA EHS 40CFR 355: Not listed
TSCA: Listed

Analytical Methods

Water / Groundwater: USGS O3117

AMI8200 CAS #: 6106-81-6

AMINOXYSCOPOLAMINE HYDROBROMIDE

EINECS Number: 228-066-8
Molecular Formula: $C_{17}H_{22}BrNO_5$

Formula Weight: 400.28

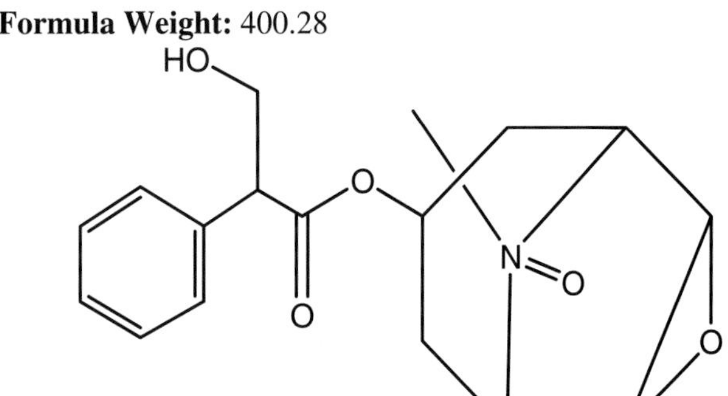

Chemical Structure

Synonyms: BENZENEACETIC ACID, ALPHA-(HYDROXYMETHYL)-,9-METHYL-3-OXA-9-AZATRICYCLO(3.3.1.0(2,4))NON-7-YL ESTER,N-OXIDE, HYDROBROMIDE,(7(S)-(1ALPHA,2BETA,4BETA,5ALPHA,7BETA))-; SCOPOLAMINE AMINOXIDE HYDROBROMIDE; SCOPOLAMINE,N-OXIDE,HYDROBROMIDE; 1ALPHA H,5ALPHA H-TROPAN-3ALPHA-OL,6BETA,7BETA-EPOXY-,(-)-TROPATE (ESTER),8-OXIDE,HYDROBROMIDE

Description: prisms

Use: medication: anticholinergic agent; anticholinergic agent; sedative for nonprescription sleeping aids (former use)

Physical Properties

Freezing Point: 135 °C (275 °F) to 138 °C (280.4 °F)

Water Solubility: 10 g/100ml at 0 °C

Other Solubilities: Slightly Soluble in Alcohol, Acetone

pH: 3% aqueous solution about 3.2

Hazard Overviews

Health: Irritating. Toxic. Other Acute Effects: harmful if swallowed, inhaled, or absorbed through skin; exposure can cause change in pupil size drowsiness; dry mouth; blurred vision; dizziness; confusion; delirium; hallucinations; coma; target organ: nerves.

Fire: Hazards: emits toxic fumes. Extinguishing agents: carbon dioxide, dry chemical powder or appropriate foam. Precautions: combustible liquid.

Reactivity: Stable. Hazardous polymerization will not occur. Hazardous decomposition products: toxic fumes of: carbon monoxide, carbon dioxide, nitrogen oxides.

Carcinogenicity: IARC - Not listed; NIOSH - Not listed; NTP - Not listed; ACGIH - Not listed; OSHA - Not listed; EPA - Not listed; MAK - Not listed

Primary Target Organs:

Eyes

Skin

Respiratory System

Nervous System

Environmental

Regulations

RCRA 40CFR: Not listed

CERCLA: 40CFR 302.4: Not listed

SARA 40CFR 372.65: Not listed

SARA EHS 40CFR 355: Not listed

TSCA: Listed

AMI8280 CAS #: 1951-25-3

AMIODARONE

RTECS: OB1360000

EINECS Number: 217-772-1

Molecular Formula: $C_{25}H_{29}I_2NO_3$

Formula Weight: 645.32

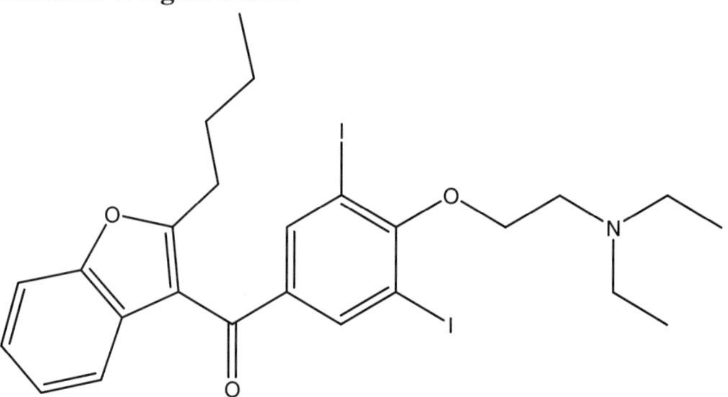

Chemical Structure

Synonyms: 2-BUTYL-3-BENZOFURANYL P-((2-DIETHYLAMINO)ETHOXY)-M,M-DIIODOPHENYL KETONE; 2-BUTYL-3-BENZOFURANYL4-(2-(DIETHYLAMINO)ETHOXY)-3,5-DIIODOPHENYL KETONE; (2-BUTYL-3-BENZOFURANYL)(4-(2-(DIETHYLAMINO)ETHOXY)-3,5-DIIDOPHENYL)METHANONE; 2-BUTYL-3-(4'-BETA-N-DIETHYLAMINOETHOXY-3',5'-DIIODOBENZOYL)BENZOFURAN; 2-BUTYL-3-(3,5-DIIODO-4-(2-DIETHYLAMINOETHOXY)BENZOYL)BENZOFURAN; 2-BUTYL-3-(3,5-DIIODO-4-(BETA-DIETHYLAMINOETHOXY)-BENZOYL)BENZOFURAN; 2-BUTYL-3-(3,5-DIIODO-4-(BETA-DIETHYLAMINOETHOXY)BENZOYL)BENZOFURAN; 2-N-BUTYL-3',5'-DIIODO-4'-N-DIETHYLAMINOETHOXY-3-BENZOYLBENZOFURAN; METHANONE,(2-BUTYL-3-BENZOFURANYL)(4-(2-(DIETHYLAMINO)ETHOXY)-3,5-DIIODOPHENYL)-(9CI)

Description: white to cream colored, crystalline powder

Use: medication: anti-arrhythmic; anti-anginal

Physical Properties

Freezing Point: 156 °C (312.8 °F)

Water Solubility: 1 mg/mL

Other Solubilities: solubilities of approx 12.8 mg/ml in Alcohol at 25 °C.

RTECS Toxicity Data

Acute Oral: Man TD_{Lo} Dose: 1714 mg/kg/21W-I; Toxic Effects: Lungs, Thorax, or Respiration - Cough; Lungs, Thorax, or Respiration - Cyanosis; Nutritional and gross metabolic - Weight loss or decreased weight gain. Man TD_{Lo} Dose: 133 mg/kg/23D-I; Toxic Effects: Skin and appendages - Photosensitivity. Man TD_{Lo} Dose: 3651 mg/kg/2.3Y-I; Toxic Effects: Liver - Hepatitis, fibrous (cirrhosis, post-necrotic scarring); Liver - Liver function tests impaired; Liver - Multiple effects.

Hazard Overviews

Health: Irritating. Harmful. Other Acute Effects: harmful if swallowed, inhaled, or absorbed through skin; photosensitizer; prolonged exposure can cause damage to the lungs and liver; significant heart block or sinus bradycardia; target organs: lungs, liver.

Fire: Extinguishing agents: water spray; carbon dioxide, dry chemical powder or appropriate foam. Precautions: combustible liquid.

Reactivity: Stable. Hazardous polymerization will not occur. Incompatible with: strong oxidizing agents. Hazardous decomposition products: emits toxic fumes under fire conditions;.

Carcinogenicity: IARC - Not listed; NIOSH - Not listed; NTP - Not listed; ACGIH - Not listed; OSHA - Not listed; EPA - Not listed; MAK - Not listed

Primary Target Organs:

Eyes

Skin

Respiratory System

Liver

Environmental

Regulations

RCRA 40CFR: Not listed
CERCLA: 40CFR 302.4: Not listed
SARA 40CFR 372.65: Not listed
SARA EHS 40CFR 355: Not listed
TSCA: Not listed

AMI8360 CAS #: 78-53-5

AMITON

RTECS: TF0525000
Molecular Formula: $C_{10}H_{24}NO_3PS$
Formula Weight: 269.38
Synonyms: CHIPMAN 6200; DIETHYL S-2-DIETHYLAMINOETHYL PHOSPHOROTHIOATE; O,O-DIETHYL S-2-DIETHYLAMINOETHYL PHOSPHOROTHIOATE; O,O-DIETHYL S-(BETA-DIETHYLAMINO)ETHYL PHOSPHOROTHIOLATE; O,O-DIETHYL S-2-DIETHYLAMINOETHYL PHOSPHOROTHIOLATE; O,O-DIETHYL S-DIETHYLAMINOETHYL PHOSPHOROTHIOLATE; O,O-DIETHYL S-(2-DIETHYLAMINOETHYL) THIOPHOSPHATE; S-(DIETHYLAMINOETHYL) O,O-DIETHYL PHOSPHOROTHIOATE; S-(2-(DIETHYLAMINO)ETHYL) O,O-DIETHYLPHOSPHOROTHIOATE; (2-DIETHYLAMINO)ETHYLPHOSPHOROTHIOIC ACID O,O-DIETHYL ESTER; S-(2-(DIETHYLAMINO)ETHYL)PHOSPHOROTHIOIC ACID O,O-DIETHYLESTER; O,O-DIETHYL-S-(BETA-DIETHYLAMINO)ETHYL PHOSPHOROTHIOATE; O,O-DIETHYL-S-2-(DIETHYLAMINO)ETHYLESTER KYSELINYTHIOFOSFORECNE; DSDP; ENT 24,980-X; EPA PESTICIDE CHEMICAL CODE 057302; INFERNO; METRAM; METRAMAC; PHOSPHOROTHIOIC ACID,S-(2-(DIETHYLAMINO)ETHYL)O,O-DIETHYL ESTER; R-5,158; RHODIA-6200; TETRAM

Description: colorless liquid

Use: contact insecticide, miticide; acaricide; a systemic insecticide to control aphids and mites on various species of plants acid oxalate form

Physical Properties

Boiling Point: 110 °C (230 °F) at 0.2 mm Hg
Freezing Point: Crystals at 98 °C (208.4 °F) to 99 °C (210.2 °F)
Vapor Pressure: 0.01 mm Hg at 80 °C
Water Solubility: Highly Soluble in Water
Other Solubilities: Highly Soluble in most organic solvents
Refraction Index: 1.4655 at 27 °C/D

RTECS Toxicity Data

Acute Oral: Rat LD_{50} Dose: 3300 ug/kg.

Acute Dermal: Rabbit LD_{50} Route: Subcutaneous Dose: 125 ug/kg. Rat LD_{50} Route: Subcutaneous Dose: 150 ug/kg.

Hazard Overviews

Carcinogenicity: IARC - Not listed; NIOSH - Not listed; NTP - Not listed; ACGIH - Not listed; OSHA - Not listed; EPA - Not listed; MAK - Not listed

Environmental

Regulations

RCRA 40CFR: Not listed
CERCLA: 40CFR 302.4: Not listed
SARA 40CFR 372.65: Not listed TPQ: 500 lb
SARA EHS 40CFR 355: Listed TPQ: 500 lb
TSCA: Not listed

AMI8440 CAS #: 3734-97-2

AMITON OXALATE

RTECS: TF1400000
EINECS Number: 223-100-8
Molecular Formula: $C_{12}H_{26}NO_7PS$
Formula Weight: 359.42
Synonyms: CHIPMAN R-6,199; CITRAM; O,O-DIETHYL S-(BETA-DIETHYLAMINO)ETHYL PHOSPHOROTHIOLATEHYDROGEN OXALATE; O,O-DIETHYL S-(2-ETHYL-N,N-DIETHYLAMINO)PHOSPHOROTHIOATEHYDROGEN OXALATE; S-(2-DIETHYLAMINOETHYL) O,O-DIETHYLPHOSPHOROTHIOATEHYDROGENOXALATE; O,O-DIETHYL-S-(2-

DIETHYLAMINO)ETHYLPHOSPHOROTHIOATEHYDROGEN OXALATE; ENT 20,993; HYDROGEN OXALATE OF AMITON; TETRAM 75; TETRAM MONOOXALATE; TETRAM,ACID OXALATE

Description: crystals

Physical Properties

Freezing Point: 98 °C (208 °F) to 99 °C (210 °F)
Vapor Pressure: Low
Water Solubility: Undergoes hydrolysis in Water

RTECS Toxicity Data

Acute Oral: Rat LD_{50} Dose: 3 mg/kg.
Acute Dermal: Rat LD_{50} Route: Skin; Dose: 2 mg/kg.

Hazard Overviews

Carcinogenicity: IARC - Not listed; NIOSH - Not listed; NTP - Not listed; ACGIH - Not listed; OSHA - Not listed; EPA - Not listed; MAK - Not listed

Environmental

Regulations

RCRA 40CFR: Not listed
CERCLA: 40CFR 302.4: Not listed
SARA 40CFR 372.65: Not listed TPQ: 100/10000 lb
SARA EHS 40CFR 355: Listed TPQ: 100 lb
TSCA: Not listed

AMI8520 CAS #: 33089-61-1

AMITRAZ

RTECS: ZF0480000
EINECS Number: 251-375-4
Molecular Formula: $C_{19}H_{23}N_3$
Formula Weight: 293.45
Synonyms: ACARAC; AMITRAZ ESTRELLA; AMITRAZE; AZADIENO; AZAFORM; BAAM; N,N-BIS(2,4-XYLYLIMINOMETHYL)METHYLAMINE; BOOTS BTS 27419; BTS 27,419; 1,5-DI(2,4-DIMETHYLPHENYL)-3-METHYL-1,3,5-TRIAZAPENTA-1,4-DIENE; N'-(2,4-DIMETHYLPHENYL)-N-(((2,4-DIMETHYLPHENYL)IMINO)METHYL)-N-METHYLMETHANIMIDAMIDE; N,N-DI-(2,4-XYLYLIMINOMETHYL)METHYLAMINE; ECTODEX; EDRIZAR; ENT 27967; FUMILAT A; METHANIMIDAMIDE,N'-(2,4-DIMETHYLPHENYL)-N-(((2,4-DIMETHYLPHENYL)IMINO)METHYL)-N-METHYL-; N-METHYL-BIS(2,4-XYLYLIMINOMETHYL)AMINE; 2-METHYL-1,3-DI(2,4-XYLYLIMINO)-2-AZAPROPANE; N,N'-((METHYLIMINO)DIMETHYLIDYNE)DI-2,4-XYLIDINE; N-METHYL-N'-2,4-XYLYL-N-(N-2,4-XYLYLFORMIMIDOYL)FORMAMIDINE; MITABAN; MITAC; R.D. 27419; TAKTIC; TRIATIX; TRIATOX; U-36059; UPJOHN U-36059

Physical Properties

Freezing Point: 86 °C (186.8 °F) to 87 °C (188.6 °F)
Specific Gravity: 1.12800
Water Solubility: 1 mg/l
Other Solubilities: >300 g/l acetone; toluene, xylene

RTECS Toxicity Data

Acute Oral: Woman LD_{Lo} Dose: 200 uL/kg; Toxic Effects: Brain and coverings - Other degenerative changes; Cardiac - Other changes; Kidney, Ureter, and Bladder - Chgs in tubules (inc acute renal failure, acute tubular necrosis. Rat LD_{50} Dose: 400 mg/kg.
Acute Inhalation: Rat LC_{50} Dose: 65 gm/m^3/6hr.
Acute Dermal: Rabbit LD_{50} Route: Skin; Dose: >200 mg/kg. Rat LD_{50} Route: Skin; Dose: >1600 mg/kg.
Reproductive/Teratogenic: Rat Route: Oral; Dose: 140 mg/kg; Duration: female 7D prior to mating; Effects on Newborn - Delayed effects.

Hazard Overviews

Carcinogenicity: IARC - Not listed; NIOSH - Not listed; NTP - Not listed; ACGIH - Not listed; OSHA - Not listed; EPA - Not listed; MAK - Not listed

Environmental

Regulations

RCRA 40CFR: Not listed
CERCLA: 40CFR 302.4: Not listed
SARA 40CFR 372.65: Listed
SARA EHS 40CFR 355: Not listed
TSCA: Not listed

Analytical Methods

Food: FDA 232.3, 232.4, 242.1

AMI8600 CAS #: 50-48-6

AMITRIPTYLINE

RTECS: HO9275000
EINECS Number: 200-041-6
Molecular Formula: $C_{20}H_{23}N$
Formula Weight: 277.39
Synonyms: ADEPRESS; ADEPRIL; AMITRIPTILINA; AMITRIPTYLIN; AMITRYPTYLINE; DAMILAN; DAMILEN; DAMITRIPTYLINE; 5H-DIBENZO(A,D)CYCLOHEPTENE-DELTA(5,GAMMA)-PROPYLAMINE,10,11-DIHYDRO-N,N-DIMETHYL-; 5H-DIBENZO(A,D)CYCLOPHEPTENE-DELTA(SUP5),GAMMA-PROPYLAMINE,10,11-DIHYDRO-N,N-DIMETHYL-; 3-(10,11-DIHYDRO-5H-DIBENZO(A,D)CYCLOHEPTEN-5-YLIDENE)-N,N-DIMETHYL-1-PROPANAMINE; 10,11-DIHYDRO-5-(GAMMA-DIMETHYLAMINOPROPYLIDENE)-5H-DIBENZO(A,D)CYCLOHEPTENE; 10,11-DIHYDRO-N,N-DIMETHYL-5H-DIBENZO(A,D)HEPTALENE-DELTA(SUP 5),GAMMA-PROPYLAMINE; 5-(3'-DIMETHYLAMINOPROPYLIDENE)-DIBENZO-(A,D)(1,4)-CYCLOHEPTADIENE; 5-(GAMMA-DIMETHYLAMINOPROPYLIDENE)-5H-DIBENZO(A,D)-10,11-DIHYDROCYCLOHEPTENE; 5-(3-DIMETHYLAMINOPROPYLIDENE)-10,11-DIHYDRO-5H-DIBENZO(A,D)CYCLOHEPTATRIENE; 5-(3-DIMETHYLAMINOPROPYLIDENE)-10,11-DIHYDRO-5H-DIBENZO(A,D)CYCLOHEPTENE; 5-(GAMMA-DIMETHYLAMINOPROPYLIDENE)-10,11-DIHYDRO-5H-DIBENZO(A,D)CYCLOHEPTENE; 5-(3-DIMETHYLAMINOPROPYLIDENE)-10,11-DIHYRO-5H-DIBENZO(A,D)CYCLOHEPTENE; 5-(GAMMA-DIMETHYLAMINOPROPYLIDINE)-5H-DIBENZO(A,D)(1,4)CYCLOHEPTADIENE; 5-(3-

DIMETHYLPROPYLIDENE)DIBENZO(A,D)(1,4)CYCLOHEPTADIENE; ELANI; ELAVIL; FLAVYL; LANTROL; LANTRON; LAROXIL; LAROXYL; LENTIZOL; MK-230; MK 230; N 750; PROHEPTADIENE; 1-PROPANAMINE,3-(10,11-DIHYDRO-5H-DIBENZO(A,D)CYCLOHEPTEN-5-YLIDENE)-N,N-DIMETHYL-; REDOMEX; RO 4-1575; SAROTEX; SEROTEN; TRIPTANOL; TRIPTISOL; TRYPTANOL; TRYPTIZOL

Description: minute crystals

Use: antidepressant for endogenous depression; medication: antidepressant

Physical Properties

Freezing Point: 196 °C (384.8 °F) to 197 °C (386.6 °F)
Water Solubility: Freely Soluble in Water
Other Solubilities: Freely Soluble in Hydrochloride.

RTECS Toxicity Data

Acute Oral: Human TD_{Lo} Dose: 14 mg/kg; Toxic Effects: Cardiac - Other changes. Woman TD_{Lo} Dose: 84 mg/kg; Toxic Effects: Cardiac - EKG changes not diagnostic of above. Woman TD_{Lo} Dose: 60 mg/kg; Toxic Effects: Cardiac - Arrythmias (including changes in conduction); Cardiac - Change in rate; Vascular - BP lowering not characterized in autonomic section. Woman TD_{Lo} Dose: 16800 ug/kg/2W-I; Toxic Effects: Peripheral nerve and sensation - Pareshtesia; Behavioral - Headache. Woman TD_{Lo} Dose: 13500 ug/kg/3D-C; Toxic Effects: Behavioral - Convulsions or effect on seizure threshold; Behavioral - Excitement. Infant TD_{Lo} Dose: 50 mg/kg; Toxic Effects: Behavioral - Convulsions or effect on seizure threshold; Cardiac - Change in rate; Behavioral - Muscle contraction or spasticity. Child TD_{Lo} Dose: 4500 ug/kg; Toxic Effects: Behavioral - Sleep. Man LD_{Lo} Dose: 29 mg/kg. Rat LD_{50} Dose: 320 mg/kg; Toxic Effects: Behavioral - Altered sleep time (including change in righting reflex); Behavioral - Ataxia; Gastrointestinal - Changes in structure or function of salivary glands.

Acute Dermal: Mouse LD_{50} Route: Subcutaneous Dose: 140 mg/kg. Dog LD_{Lo} Route: Subcutaneous Dose: 50 mg/kg; Toxic Effects: Peripheral Nerve and sensation - Spastic parapysis with or without sensory change; Behavioral - Muscle contraction or spasticity; Lungs, Thorax, or Respiration - Other changes.

Chronic (Multiple Dose) Oral: Rat Dose: 5760 mg/kg/22W-I; Toxic Effects: Cardiac - EKG changes not diagnostic of above; Nutritional and gross metabolic - Weight loss or decreased weight gain; DEATH.

Reproductive/Teratogenic: Rat Route: Oral; Dose: 75 mg/kg; Duration: female 4-18D of pregnancy; Specific Developmental Abnormalities - Musculoskeletal system. Rabbit Route: Oral; Dose: 540 mg/kg; Duration: female 8-16D of pregnancy; Specific Developmental Abnormalities - Other developmental abnormalities; Effects on Newborn - Growth statistics.

Hazard Overviews

Carcinogenicity: IARC - Not listed; NIOSH - Not listed; NTP - Not listed; ACGIH - Not listed; OSHA - Not listed; EPA - Not listed; MAK - Not listed

Environmental

Regulations

RCRA 40CFR: Not listed
CERCLA: 40CFR 302.4: Not listed
SARA 40CFR 372.65: Not listed
SARA EHS 40CFR 355: Not listed
TSCA: Not listed

AMI8680 CAS #: 61-82-5

AMITROLE

RTECS: XZ3850000
EINECS Number: 200-521-5
Molecular Formula: $C_2H_4N_4$
Formula Weight: 84.08

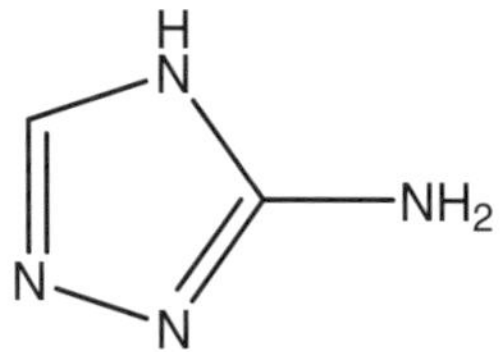

Chemical Structure

Synonyms: 3,A-T; AMEROL; AMINO TRIAZOLE WEEDKILLER 90; 2-AMINO-1,3,4-TRIAZOLE; 2-AMINOTRIAZOLE; 3-AMINO-1,2,4-TRIAZOLE; 3-AMINO-1H-1,2,4-TRIAZOLE; 3-AMINO-S-TRIAZOLE; 3-AMINOTRIAZOLE; AMINOTRIAZOLE; AMINOTRIAZOLE (PLANT REGULATOR); AMINOTRIAZOL-SPRITZPULVER; AMITOL; AMITRIL; AMITRIL T.L; AMITRIL TL; AMITROL; AMITROL-T; AMITROL 90; AMIZOL; AMIZOL D; AMIZOL DP NAU; AMIZOL F; AT; AT-90; AT LIQUID; ATA; AZAPLANT; AZAPLANT KOMBI; AZOLAN; AZOLE; CAMPAPRIM A 1544; CYTROL; CYTROL AMITROLE-T; CYTROLE; DIUROL; DIUROL 5030; DOMATOL; DOMATOL 88; ELMASIL; EMISOL; EMISOL 50; EMISOL F; ENT 25445; FENAMINE; FENAVAR; HERBIDAL TOTAL; HERBIZOLE; KLEER-LOT; ORGA-414; RADOXONE TL; RAMIZOL; SIMAZOL; SOLUTION CONCENTREE T271; 1H-1,2,4-TRIAZOL-3-AMINE; TRIAZOLAMINE; S-TRIAZOLE,3-AMINO-; VOROX; VOROX AS; VOROX AA; WEEDAR ADS; WEEDAR AT; WEEDAZIN; WEEDAZIN ARGINIT; WEEDAZOL; WEEDAZOL GP2; WEEDAZOL SUPER; WEEDAZOL T; WEEDAZOL TL; WEEDEX GRANULAT; WEEDOCLOR; X-ALL LIQUID

Description: white cystalline powder; odorless

Use: as a defoliant, a herbicide, a reagent in photography and a plant growth regulator; in non-selective weed control

Physical Properties

Freezing Point: 159 °C (318.2 °F)
Saturated Vapor Density: < 1.200000022 kg/m^3
Density: 1.138 mg/mL at 20 °C
Vapor Pressure: < 7.5 x10^{-6} mm Hg at 20 °C
Water Solubility: >= 100 mg/mL at 17.5 C
Other Solubilities: 95% Ethanol: 5-10 mg/ml at 17.5 °C; Acetone: 1-5 mg/ml at 17.5 °C; Alcohol: Soluble; Chloroform: Soluble; DMSO: >=100 mg/ml at 17.5 °C; Ether: Insoluble; Ethyl Acetate: Sparingly Soluble; Methanol: Soluble; Non-polar solvents: Insoluble.
pH: 10% solution 6.5 to 7.5

Flash Point: Noncombustible

RTECS Toxicity Data

Acute Oral: Rat LD_{50} Dose: 1100 mg/kg. Mouse LD_{50} Dose: 14700 mg/kg.

Acute Dermal: Rat LD_{50} Route: Skin; Dose: >10 gm/kg.

Chronic (Multiple Dose) Oral: Rat Dose: 313 mg/kg/90D-C; Toxic Effects: Endocrine - Thyroid weight (goiter). Hamster Dose: 4704 mg/kg/28W-C; Toxic Effects: Endocrine - Other changes; Blood - Changes in serum composition.

Reproductive/Teratogenic: Rat Route: Oral; Dose: 700 ug/kg; Duration: female 22D of pregnancy; Specific Developmental Abnormalities - Endocrine system. Mouse Route: Oral; Dose: 2600 mg/kg; Duration: female 6-18D of pregnancy; Effects on Embryo or Fetus - Fetotoxicity. Mouse Route: Oral; Dose: 1935 mg/kg; Duration: female 6-14D of pregnancy; Effects on Embryo or Fetus - Fetotoxicity; Fetal death.

Mutagenic: Human DNA Inhibition; Cell Type: fibroblast; Dose: 100 mmol/L. Rat Morphological Transformation; Cell Type: embryo; Dose: 80 mg/L.

Tumorigenic: Rat Route: Oral; Dose: 4595 mg/kg/2.5Y-C; Toxic Effects: Tumorigenic - Carcinogenic by RTECS criteria; Endocrine - Tumors; Endocrine - Thyroid tumors. Rat Route: Oral; Dose: 3670 mg/kg/2Y-C; Toxic Effects: Tumorigenic - Neoplastic by RTECS criteria; Endocrine - Thyroid tumors. Rat Route: Oral; Dose: 122 gm/kg/70W-C; Toxic Effects: Tumorigenic - Carcinogenic by RTECS criteria; Endocrine - Thyroid tumors. Rat Route: Oral; Dose: 105 gm/kg/60W-C; Toxic Effects: Tumorigenic - Carcinogenic by RTECS criteria; Endocrine - Thyroid tumors.

Hazard Overviews

Fire Diamond

Health: Mildly irritating to eyes/skin. Also Causes: intoxication, skin sensitization. Chronic Effects: possible cancer hazard.

Fire: Noncombustible. Use agent suitable for surrounding fire.

Reactivity: Stable, but decomposes in the presence of light. Hazardous polymerization cannot occur. Incompatible with: iron; aluminum; copper; copper alloys. Hazardous decomposition products: fumes of nitrogen oxides

Carcinogenicity: IARC - Group 2B, Possibly carcinogenic to humans; NIOSH - Listed as carcinogen; NTP - Class 2B, Reasonably anticipated to be a carcinogen, sufficient evidence of carcinogenicity from studies in experimental animals; ACGIH - Class A3, Animal carcinogen; OSHA - Not listed; EPA - Not listed; MAK - Not listed

Primary Target Organs:

Eyes

Skin

Exposure Limits

OSHA PEL Vacated 1989 Limits: TWA: 0.2 mg/m^3.

ACGIH TLV: TWA: 0.2 mg/m^3.

NIOSH REL: TWA: 0.2 mg/m^3.

DFG MAK: TWA: 0.2 mg/m^3.

Respirator Recommendation

Exposure Range: >0.2 to 2 mg/m^3 Air Purifying, Negative Pressure, Half Mask

Exposure Range: >2 to 20 mg/m^3 Air Purifying, Negative Pressure, Full Face

Exposure Range: >20 to 200 mg/m^3 Supplied Air, Constant Flow/Pressure Demand, Full Face

Exposure Range: >200 to unlimited mg/m^3 Self-contained Breathing Apparatus, Pressure Demand, Full Face

Cartridge Color: dust/mist filter (use P100 or consult supervisor for appropriate dust/mist filter)

Environmental

Ecotoxicity: LC_{50} ring-necked pheasant greater than 5000 ppm (no mortality to 5000 ppm), age 10 days LC_{50} Gammarus fasciatus (scud) greater than 10 mg/l/96 hr at 18 °C. /static conditions without aeration LC_{50} Daphnia Magna 30,000 ug/l/48 hr /Conditions of bioassay not specified LC_{50} Japanese quail oral greater than 5000 ppm (no mortality to 5000 ppm), age 12 days LC_{50} Bluegill 100 ppm/48 hr /Conditions of bioassay not specified LC_{50} Oncorhyncus kisutch 325,000 ug/l/48 hr /Conditions of bioassay not specified

Environmental Fate: If released to soil, it will degrade microbially and possibly chemically with a resultant average persistence of 2 to 4 weeks at recommended herbicidal concentrations. The degree of leaching in soil (which may be extensive according to estimation methods) may depend upon the chemical and organic content of an individual soil. Loss from soil by volatilization or photodegradation is minor. If released to the aquatic environment, it is not expected to hydrolyze, directly photolyze, volatilize or bioconcentrate in aquatic organisms significantly. Degradation in natural waters may be possible by oxidation with photochemically produced peroxy radicals or by photosensitized photolysis; biodegradation has not been shown to be a rapid removal process in water. Adsorption to hydrosoil may be an important transport mechanism. An initial maximum half-life of 68 days was observed when applied to an outdoor pond with persistence exceeding 200 days. If released to the atmosphere, vapor-phase will react rapidly with photochemically produced hydroxyl radicals (estimated half-life of 3.8 days at 25 °C), but will not react with ozone or directly photolyze.

Environmental Physical Data

Henry's Law Constant: $< 3.0 \times 10^{-12}$

Octanol/Water Partition Coefficient: log K_{OW} = -0.65

Sorption Partition Coefficient: K_{OC} = 110

BCF: estimated at 0.5

Regulations

RCRA 40CFR: Listed Hazardous Waste No. U011 Toxic Waste
CERCLA: 40CFR 302.4: Listed per RCRA Section 3001 RQ: 10 lb (4.535 kg)
SARA 40CFR 372.65: Listed
SARA EHS 40CFR 355: Not listed
TSCA: Listed

Analytical Methods

Soil: EPA PMD-TLC
Food: AOAC 967.06

AMM1000 CAS #: 7664-41-7

AMMONIA

RTECS: BO0875000
DOT: UN1005; IMO2.3
EINECS Number: 231-635-3
Molecular Formula: H_3N
Structured MF: NH_3
Formula Weight: 17.03

Chemical Structure

Synonyms: AM-FOL; AMMONIA ANHYDROUS; AMMONIA GAS; AMMONIA,ANHYDROUS; AMMONIA,ANHYDROUS,LIQUEFIED; AMMONIAC; AMMONIACA; AMMONIAK; AMONIAK; ANHYDROUS AMMONIA; AQUA AMMONIA; AQUEOUS AMMONIA; LIQUID AMMONIA; NITRO-SIL; R 717; SPIRIT OF HARTSHORN
Description: colorless liquid, gas; strong, pungent, and irritating odor
Use: in mfr of nitric acid, explosives, hydrazine, pesticides & detergents; in refrigeration & chem ind; cotton defoliant; intermed for urea, ammonium nitrate, ammonium salts, nylon, acrylonitrile & isocyanate; fertilizer; in metal treating operations, eg, nitriding, bright annealing, etc; for hydrogenation of fats & oils; source of pure nitrogen; in petro ind (neutralizing oil acidity, protecting equipment from corrosion); in rubber ind for stabilization of raw latex; catalyst to make synthetic resin; in water treatment, against tastes & odors; on citrus fruit to control fungal growth; med: (vet).

Physical Properties

Boiling Point: -33.35 °C (-28 °F)
Freezing Point: -77.7 °C (-107.86 °F)
Vapor Density: 0.59 Air=1
Density: 0.771 g/L at 760 mm Hg (gas)
Vapor Pressure: 1 mm Hg at -109.1 °C
Water Solubility: 47% in Water at 0 °C
Other Solubilities: 15% in Alcohol at 20 °C; 11% in absolute Alcohol at 30 °C; 20% in Ethanol at 0 °C; 10% in Ethanol at 25 °C; 16% in Methanol at 25 °C; Soluble in Chloroform & Ether.
Surface Tension: 23.4 dynes/cm at 4.1 °C
Odor Threshold: 46.8 ppm
pH: 1.0 N aqueous solution 11.6
Refraction Index: 1.325 at 16.5 °C/D
Critical Temperature: 132.4 °C
Critical Pressure: 111.5 atm
Ionization Potential (eV): 10.18
Flash Point: Indefinite < 0 °C
Autoignition Temperature: 651 °C
LEL: 16% v/v
UEL: 25% v/v

RTECS Toxicity Data

Acute Oral: Man TD_{Lo} Dose: 15 uL/kg; Toxic Effects: Gastrointestinal - Changes in structure or function of esophagus.
Acute Inhalation: Rat LC_{50} Dose: 2000 ppm/4hr. Human LC_{Lo} Dose: 5000 ppm/5M. Human TC_{Lo} Dose: 20 ppm; Toxic Effects: Sense organs and special senses - Ulcerated nasal septum; Sense organs and special senses - Conjunctive irritation; Lungs, Thorax, or Respiration - Structural or functional change in trachea.
Mutagenic: Rat Cytogenetic Analysis; Route: Inhalation; Dose: 19800 ug/m^3/16W. Bacteria - E Coli Mutations in Microorganisms; Dose: 1500 ppm/3H (-S9).
Tumorigenic: Rat Route: Oral; Dose: 1680 mg/kg/24W-C; Toxic Effects: Tumorigenic - Carcinogenic by RTECS criteria; Gastrointestinal - Tumors.

Hazard Overviews

Corrosive

Compressed Gas

Fire Diamond

Health: Corrosive causes burns to eyes/skin/respiratory tract. Also Causes: blindness; exposure to high levels may be fatal.
Fire: Will burn. Stop flow of gas. Use carbon dioxide or dry chemical to extinguish flame at gas valve. Use water spray to protect personnel shutting off gas. Remove cylinders from fire.
Reactivity: Stable. Hazardous polymerization cannot occur. Incompatible with: acids; interhalogens; boron halides; 1,2-dichloroethane; ethylene oxide; chloroformamidnium nitrate; oxygen and platinum; magnesium perchlorate; nitrogen trichloride; strong oxidants; heavy metals and their compounds; chlorine azide; bromine; iodine; iodine and potassium; tellurium halides; pentaborane; silver oxide; silver chloride; silver nitrate; silver azide; hypochlorites; chlorine or chlorine bleach; air and hydrocarbons; germanium derivatives; stibine; 1-chloro-2,4-dinitroben-zene; ethanol and silver nitrate; 2-, or 4-chloronitrobenzene (above 160 °C/30 bar); acetaldehyde; acrolein; boron; chlorosilane; hexachloromelamine; sulfur; hydrazine and alkali metals; potassium ferricyanide; potassium mercuric cyanide; nitrogen dioxide; phosphorus pentoxide; tetramethylammonium

amide. Hazardous decomposition products: fumes of ammonia; nitrogen oxides.
Carcinogenicity: IARC - Not listed; NIOSH - Not listed; NTP - Not listed; ACGIH - Not listed; OSHA - Not listed; EPA - Not listed; MAK - Not listed
Primary Target Organs:

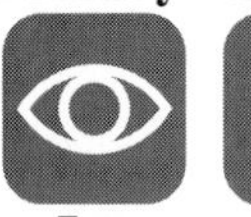

Eyes Skin Respiratory System

Exposure Limits
OSHA PEL: TWA: 50 ppm; 35 mg/m^3.
OSHA PEL Vacated 1989 Limits: STEL: 35 ppm; 27 mg/m^3.
ACGIH TLV: TWA: 25 ppm; 17 mg/m^3; STEL: 35 ppm; 24 mg/m^3.
NIOSH IDLH: 300 ppm.
DFG MAK: TWA: 20 ppm; 14 mg/m^3.
Respirator Recommendation
Exposure Range: >50 to <300 ppm Supplied Air, Constant Flow/Pressure Demand, Full Face
Exposure Range: 300 to unlimited ppm Self-contained Breathing Apparatus, Pressure Demand, Full Face
Cartridge Color: green

Environmental

Cleanup/Disposal: Guide No. 125: Fully encapsulating, vapor protective clothing should be worn for spills and leaks with no fire. Do not touch or walk through spilled material. Stop leak if you can do it without risk. If possible, turn leaking containers so that gas escapes rather than liquid. Prevent entry into waterways, sewers, basements or confined areas. Do not direct water at spill or source of leak. Use water spray to reduce vapors or divert vapor cloud drift. Isolate area until gas has dispersed.

Environmental Physical Data

Henry's Law Constant: 0.76
Octanol/Water Partition Coefficient: log K_{ow} = measured at -1.14

Regulations

RCRA 40CFR: Not listed
CERCLA: 40CFR 302.4: Listed per CWA Section 311(b)(4) RQ: 100 lb (45.35 kg)
SARA 40CFR 372.65: Listed
SARA EHS 40CFR 355: Listed TPQ: 500 lb
TSCA: Listed

Analytical Methods

Air: ASTM D4490
Soil: USGS I5553, I6522, I6523, I6552
Water / Groundwater: EPA 350.1, 350.2, 350.3; APHA 4500-NH3; ASTM D1426; NOAA NITRO-1, NITRO-17, NITRO-18, NITRO-2, NITRO-23, NITRO-24, NITRO-3, NITRO-4; USGS I1520, I1524, I1550, I2521, I2522, I2523, I2552, I2558, I3524, I4521, I4522, I4523, I4552, I7552; HACH 8038
Drinking Water: AOAC 973.49
Indoor / Expired Air: NIOSH 6015; EPA IP-9

AMM1160 CAS #: 631-61-8

AMMONIUM ACETATE

RTECS: AF3675000
EINECS Number: 211-162-9
Molecular Formula: $C_2H_7NO_2$
Structured MF: CH_3COONH_4
Formula Weight: 77.08

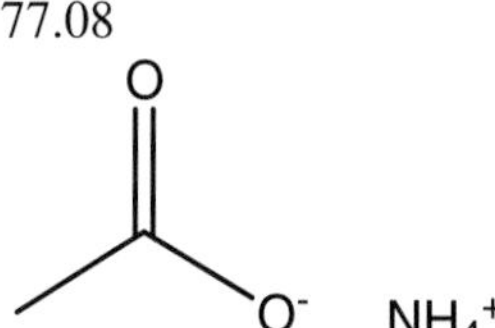

Chemical Structure

Synonyms: ACETIC ACID AMMONIUM SALT; ACETIC ACID,AMMONIUM SALT
Description: white crystals; slightly acetic odor
Use: medication (vet): as nutrient; explosives; foam rubbers; vinyl plastics; drugs; preserving meats; dyeing; stripping; as a reagent in analytical chemistry for determining lead and iron; separating lead sulfate from other sulfates; medication: diuretic; medication (vet): formerly a diuretic

Physical Properties

Boiling Point: Decomposes at 1 atm
Freezing Point: 114 °C (237.2 °F)
Specific Gravity: 1.17 at 20 °C/4 °C
Water Solubility: 1 parts in less than 1 part Water
Other Solubilities: 7.89 g/100 cc Methanol at 15 °C.
pH: Very concentrated solution is slightly acid 7
Flash Point: Nonflammable

Hazard Overviews

Fire Diamond

Health: Irritating to eyes/skin/respiratory tract. Also Causes: by ingestion: digestive tract irritation, systemic ammonia poisoning, diarrhea, diuresis, tremor. Chronic Effects: liver damage.
Fire: Will burn but doesn't ignite readily. Fight fire with dry chemical, carbon dioxide, water spray, fog, or foam.
Reactivity: Stable. Hazardous polymerization cannot occur. Avoid: excessive dust generation; contact with strong oxidizers. Incompatible with: strong oxidizers; sodium hypochlorite. Hazardous decomposition products: toxic nitric; ammonia fumes.
Carcinogenicity: IARC - Not listed; NIOSH - Not listed; NTP - Not listed; ACGIH - Not listed; OSHA - Not listed; EPA - Class D, Not classifiable as to human carcinogenicity; MAK - Not listed

Primary Target Organs:

Eyes

Skin

Respiratory System

Liver

Environmental

Ecotoxicity: TLm Mosquito fish 238 ppm/24 hr fresh water. /Conditions of bioassay not specified

Cleanup/Disposal: Guide No. 129: Eliminate all ignition sources (no smoking, flares, sparks or flames in immediate area). All equipment used when handling the product must be grounded. Do not touch or walk through spilled material. Stop leak if you can do it without risk. Prevent entry into waterways, sewers, basements or confined areas. A vapor suppressing foam may be used to reduce vapors. Absorb or cover with dry earth, sand or other non-combustible material and transfer to containers. Use clean non-sparking tools to collect absorbed material. Large Spills: Dike far ahead of liquid spill for later disposal. Water spray may reduce vapor; but may not prevent ignition in closed spaces.

Environmental Physical Data

BCF: no food chain concentration potential
BOD: 79%, 1-5 days

Regulations

RCRA 40CFR: Not listed
CERCLA: 40CFR 302.4: Listed per CWA Section 311(b)(4) RQ: 5000 lb (2268 kg)
SARA 40CFR 372.65: Not listed
SARA EHS 40CFR 355: Not listed
TSCA: Listed

AMM1320 CAS #: 9005-34-9

AMMONIUM ALGINATE

Molecular Formula: Unknown
Synonyms: ALGINIC ACID,AMMONIUM SALT; AMMONIUM POLYMANNURATE; ANALGINE; CALLATEX; COLLATEX ARM EXTRA; DIGAMON; PROTOMON; SUPERLOID
Description: colorless or slightly yellow, filamentous, grainy, granules, or powder; slight characteristic odor
Use: food additive; suspending agent in soft drinks; stabilizer for ice cream, ice milk, frozen custard, chocolate milk, confections, beverages, bakery products, canned meats, etc; to improve textures of dressings; creaming agent for rubber latex; protective colloid in boiler feedwater compounds; emulsion stabilizer; dental impression prepn; in drilling muds; in coatings; in flocculation of solids in water treatment; sizing agent; thickener.

Physical Properties

Water Solubility: Slowly Soluble in Water
Other Solubilities: Insoluble in Alcohol

Hazard Overviews

Carcinogenicity: IARC - Not listed; NIOSH - Not listed; NTP - Not listed; ACGIH - Not listed; OSHA - Not listed; EPA - Not listed; MAK - Not listed

Environmental

Regulations

RCRA 40CFR: Not listed
CERCLA: 40CFR 302.4: Not listed
SARA 40CFR 372.65: Not listed
SARA EHS 40CFR 355: Not listed
TSCA: Listed

AMM1480 CAS #: 1863-63-4

AMMONIUM BENZOATE

RTECS: DG3378000
EINECS Number: 217-468-9
Molecular Formula: $C_7H_9NO_2$
Structured MF: $C_6H_5COONH_4$
Formula Weight: 139.16

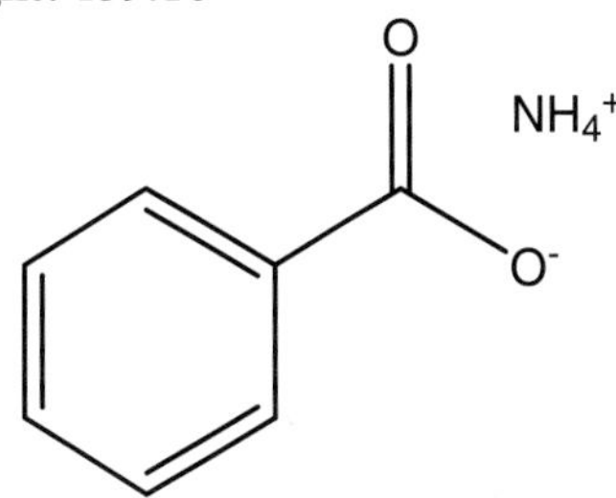

Chemical Structure

Synonyms: BENZOIC ACID,AMMONIUM SALT; VULNOC AB
Description: colorless rhombic crystals or crystalline powder; odorless or faint benzoic acid odor
Use: industrial preservative for paper wrappers; agent for reducing curing time in vulcanization of rubber; expectorant for chronic bronchitis; analytical reagent for various elements; to preserve glue and latex; medication: urinary anti-infective

Physical Properties

Boiling Point: Sublimes at 160 °C (320 °F)
Freezing Point: Decomposes at 198 °C (388.4 °F) to 198 °C (388.4 °F)
Specific Gravity: 1.26
Water Solubility: 19.6 g/100 cc at 145 °C
Other Solubilities: 1 g dissolves in 8 ml Glycerol.
pH: Aqueous solution is slightly acid
Flash Point: Not pertinent (combustible solid)

RTECS Toxicity Data

Acute Oral: Rat LD_{50} Dose: 825 mg/kg; Toxic Effects: Peripheral nerve and sensation - Spastic parapysis with or without sensory change; Behavioral - Altered sleep time (including change in righting reflex). Mouse LD_{50} Dose: 235 mg/kg; Toxic Effects: Peripheral nerve and sensation - Spastic parapysis with or without sensory change; Behavioral - Altered sleep time (including change in righting reflex).

Hazard Overviews

Health: Irritating to eyes/skin/respiratory tract. Toxic. Other Acute Effects: harmful if swallowed, inhaled, or absorbed through skin.
Fire: Combustible. Hazards: emits toxic fumes. Extinguishing agents: water spray; carbon dioxide, dry chemical powder or appropriate foam. Precautions: combustible liquid.
Reactivity: Incompatible with: strong oxidizing agents, strong acids, protect from moisture. Hazardous decomposition products: toxic fumes of: carbon monoxide, carbon dioxide, nitrogen oxides.
Carcinogenicity: IARC - Not listed; NIOSH - Not listed; NTP - Not listed; ACGIH - Not listed; OSHA - Not listed; EPA - Not listed; MAK - Not listed
Primary Target Organs:

Eyes

Skin

Respiratory System

Environmental

Environmental Physical Data

BCF: no food chain concentration potential
BOD: sewage 1.4 lb/lb, 71 days

Regulations

RCRA 40CFR: Not listed
CERCLA: 40CFR 302.4: Listed per CWA Section 311(b)(4) RQ: 5000 lb (2268 kg)
SARA 40CFR 372.65: Not listed
SARA EHS 40CFR 355: Not listed
TSCA: Listed

AMM1640 **CAS #: 1066-33-7**

AMMONIUM BICARBONATE

RTECS: BO8600000
EINECS Number: 213-911-5
Molecular Formula: CH_5NO_3
Structured MF: NH_4HCO_3
Formula Weight: 79.06
Synonyms: ACID AMMONIUM CARBONATE; AMMONIUM CARBONATE; AMMONIUM HYDROGEN CARBONATE; CARBONIC ACID,MONOAMMONIUM SALT; MONOAMMONIUM CARBONATE
Description: shiny, colorless, white hard prisms, crystalline solid; faint ammonia odor
Use: in cooling baths; in fire extinguishers; mfr porous plastics, ceramics, dyes, pigments; in compost heaps to accelerate decomp; fertilizer; defatting textiles; in cold wave soln; in chrome leather tanning; to remove gypsum from heat exchanges, etc; blowing agent for foam rubber; prodn of ammonium salts; leavening agent in prodn of baked foods; scale-removing cmpd; in smelling salts; med: (vet) expectorant, in bloat, colic.

Physical Properties

Boiling Point: Decomposes at 1 atm
Freezing Point: 107.5 °C (225.5 °F)
Specific Gravity: 1.57 at 20 °C/4 °C
Water Solubility: 11.9 g/100 cc at 0 °C
Other Solubilities: Glycerol (pharmaceutical grade): 1 g/10 ml; Insoluble in Alcohol, Acetone.
pH: 0.1 N solution at 25 °C 7.8
Refraction Index: 1.423
Flash Point: Nonflammable
Autoignition Temperature: 222 °C

Hazard Overviews

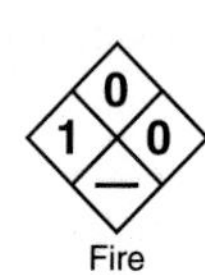

Fire Diamond

Health: Mildly irritating to eyes/skin/respiratory tract. Chronic Effects: slight irritation, dryness, or discomfort of the skin.
Fire: Noncombustible. Use agent suitable for surrounding fire.
Reactivity: Stable. Hazardous polymerization cannot occur. Avoid: dusty conditions. Incompatible with: acids; caustic alkalies; strong oxidizing agents. Hazardous decomposition products: carbon monoxide; carbon dioxide; oxides of nitrogen; ammonia.
Carcinogenicity: IARC - Not listed; NIOSH - Not listed; NTP - Not listed; ACGIH - Not listed; OSHA - Not listed; EPA - Not listed; MAK - Not listed
Primary Target Organs:

Eyes

Skin

Respiratory System

Mucous Membranes

Environmental

Cleanup/Disposal: Guide No. 154: Eliminate all ignition sources (no smoking, flares, sparks or flames in immediate area). Do not touch damaged containers or spilled material unless wearing appropriate protective clothing. Stop leak if you can do it without risk. Prevent entry into waterways, sewers, basements or confined areas. Absorb or cover with dry earth, sand or other non-combustible material and transfer to containers. Do not get water inside containers.

Environmental Physical Data

BCF: no food chain concentration potential

Regulations

RCRA 40CFR: Not listed
CERCLA: 40CFR 302.4: Listed per CWA Section 311(b)(4) RQ: 5000 lb (2268 kg)
SARA 40CFR 372.65: Not listed
SARA EHS 40CFR 355: Not listed

TSCA: Listed

AMM1800 **CAS #: 7789-09-5**

AMMONIUM BICHROMATE

RTECS: HX7650000
DOT: UN1439; IMO5.1
EINECS Number: 232-143-1
Molecular Formula: $Cr_2H_8N_2O_7$
Structured MF: $(NH_4)_2Cr_2O_7$
Formula Weight: 252.06

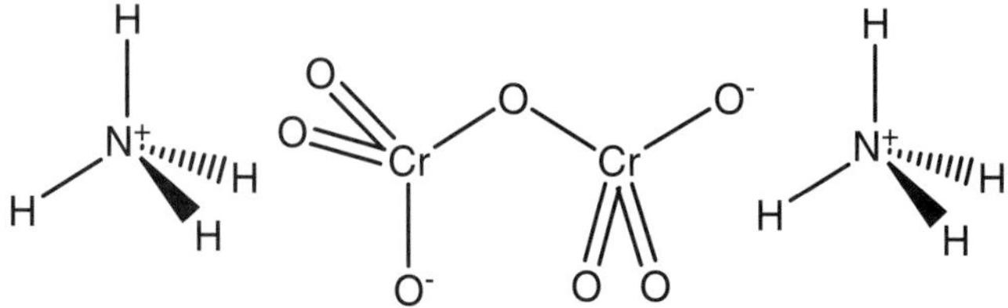

Chemical Structure

Synonyms: AMMONIO (BICROMATO DI); AMMONIO (DICROMATO DI); AMMONIUM CHROMATE ((NH4)2CR2O7); AMMONIUM (DICHROMATE D'); AMMONIUM DICHROMATE; AMMONIUM DICHROMATE (VI); AMMONIUMBICHROMAAT; AMMONIUMDICHROMAAT; AMMONIUMDICHROMAT; BICHROMATE D'AMMONIUM; DIAMMONIUM DICHROMATE; DIAMMONIUM SALT

Description: orange to red crystals, needles; odorless

Use: source of pure nitrogen; catalyst; dye mordant; pickling agent; in lithography & photography, pyrotechnics; intermed for chrome pigments for artists' colors, jointing pastes, etc; mfr of alizarin; chrome alum; oil purification; leather tanning; synthetic perfumes; finishing of porcelain; in metal finishing, chrome plating, anodizing, etc; for corrosion resistance in radiator coolants, etc; in dyeing of fur, leather, fabrics, etc; oxidizing of dyes; in mfr of glue; fungicides; in wood preservatives & fire retardants; in mfr of batterites, matches & explosives; chem rgnt, oxidizing agent, indicator, in bleaching of fats, oils, & waxes, in chem synthesis, & in anal chemistry.

Physical Properties

Boiling Point: Decomposes
Freezing Point: 170 °C (338 °F)
Specific Gravity: 2.155 at 25 °C/4 °C
Bulk Density: 82 lbs/cu ft
Water Solubility: 15.5% at 0 °C
Other Solubilities: Soluble in Alcohol; Insoluble in Acetone.
pH: 1% solution 3.95
Flash Point: Flammable solid
Autoignition Temperature: Ammomium Dichromate 190 °C

RTECS Toxicity Data

Acute Oral: Child LD_{Lo} Dose: 99 mg/kg; Toxic Effects: Gastrointestinal - Other changes; Kidney, Ureter, and Bladder - Chgs in tubules (inc acute renal failure, acute tubular necrosis; Blood - Change in clotting factors.

Acute Dermal: Guinea Pig LD_{Lo} Route: Subcutaneous Dose: 25 mg/kg.

Mutagenic: Bacteria - B Subtilis DNA Repair; Dose: 50 mmol/L. Bacteria - S Typhimurium DNA Repair; Dose: 50 mmol/L.

Hazard Overviews

Fire Diamond

Health: Severely irritating to eyes/skin/respiratory tract. Also Causes: skin and nasal passages ulcers; pulmonary edema. Chronic Effects: nose bleeds, asthma, tooth enamel erosion, dermatitis; liver and kidney damage.

Fire: Combustible as well as a strong oxidizer capable of igniting other materials. Fight fire with flooding amounts of water only. Do not use carbon dioxide, dry chemical, or foam.

Reactivity: Unstable, strong oxidizer which decomposes vigorously when heated above 190 °C and is mildly explosive in confined areas. Hazardous polymerization can occur. Avoid: excess heat; contact with water; combustibles; reducing materials. Incompatible with: ethylene glycol; reducing materials; combustibles; acids; alcohols; carbide; hydrazine; water. Hazardous decomposition products: nitrogen oxide; greenish chromic oxide smoke.

Carcinogenicity: IARC - Group 3, Not classifiable as to carcinogenicity to humans; NIOSH - Not listed; NTP - Not listed; ACGIH - Class A4, Not classifiable as a human carcinogen; OSHA - Not listed; EPA - Not listed; MAK - Class A2, Unmistakably carcinogenic in animal experimentation only

Primary Target Organs:

Eyes

Skin

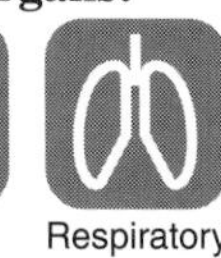
Respiratory System

Mucous Membranes

Liver

Kidneys

Exposure Limits

OSHA PEL: STEL: 0.1 mg/m^3; as CrO_3; Other Values: 0.1 mg/m^3; Clg Cr-VI as CrO_3.

NIOSH REL: TWA: 0.5 mg/m^3; as Cr;Cr-II;Cr-III;Cr(VI)=.001.

Respirator Recommendation

Exposure Range: >0.1 to 1 mg/m^3 Air Purifying, Negative Pressure, Half Mask

Exposure Range: >1 to 10 mg/m^3 Air Purifying, Negative Pressure, Full Face

Exposure Range: >10 to <15 mg/m^3 Supplied Air, Constant Flow/Pressure Demand, Full Face

Exposure Range: 15 to unlimited mg/m^3 Self-contained Breathing Apparatus, Pressure Demand, Full Face

Cartridge Color: magenta (P100)

Note: as chromium VI compounds

Environmental

Ecotoxicity: IL_{50} Eurasian Watermilfoil 1.9, 2.6, 8.0 and 9.5 ppm chromate for root weight, stem weight, root length, and stem length, respectively TLm Mosquito fish 136 and 212 ppm for 96 and 48 hr, respectively; Turbid

Cleanup/Disposal: Guide No. 141: Keep combustibles (wood, paper, oil, etc.) away from spilled material. Do not touch damaged containers or spilled material unless wearing appropriate protective clothing. Stop leak if you can do it without risk. Small Dry Spills: With clean shovel place material into clean, dry container and cover loosely; move containers from spill area. Large Spills: Dike far ahead of spill for later disposal.

Environmental Physical Data

BCF: brown algae 100 to 500
BOD: none

Regulations

RCRA 40CFR: Not listed
CERCLA: 40CFR 302.4: Listed per CWA Section 311(b)(4) RQ: 10 lb (4.535 kg)
SARA 40CFR 372.65: Listed as Compound
SARA EHS 40CFR 355: Not listed
TSCA: Listed

AMM1960 CAS #: 1341-49-7

AMMONIUM BIFLUORIDE

RTECS: BQ9200000
DOT: UN1727; UN2817; IMO8.0
EINECS Number: 215-676-4
Molecular Formula: F_2H_5N
Structured MF: NH_4F•HF
Formula Weight: 57.05

F^-

NH_4^+

H—F

Chemical Structure

Synonyms: ACID AMMONIUM FLUORIDE; AMMONIUM ACID FLUORIDE; AMMONIUM DIFLUORIDE; AMMONIUM FLUORIDE COMP WITH HYDROGEN FLUORIDE (1:1); AMMONIUM FLUORIDE COMP. WITH HYDROGEN FLUORIDE (1:1); AMMONIUM FLUORIDE ((NH4)(HF2)); AMMONIUM HYDROFLUORIDE; AMMONIUM HYDROGEN BIFLUORIDE; AMMONIUM HYDROGEN DIFLUORIDE; AMMONIUM HYDROGEN FLUORIDE; FLUORURE ACIDE D'AMMONIUM

Description: white crystals; odorless

Use: mfr of magnesium & magnesium alloys; mordant, & for brightening & anodizing aluminum; in acid dips for steel & for activation of metals before nickel plating; for corrosion resistance on magnesium & alloys; sterilizing food equipment; glass & porcelain industries; in laundering cloth; prodn of hydrogen fluoride; etching glass (white acid); electroplating; processing of beryllium; oil well acidizing; industrial chem cleaning; fungicides; in magnesium fluoride synthesis.

Physical Properties

Boiling Point: 239.5 °C (463 °F)
Freezing Point: 125.6 °C (258.08 °F)
Specific Gravity: 1.5
Water Solubility: 5.83 x10^5 ppm at 20 °C
Other Solubilities: 1.73 wt% in 9% Ethanol at 25 °C.
pH: 5% solution 3.5
Refraction Index: 1.390
Flash Point: Nonflammable

Hazard Overviews

Poison

Corrosive

Fire Diamond

Health: Corrosive, causes severe irritation or burns to the eyes/skin/respiratory tract. Poison. Also Causes: by ingestion: abdominal pain, muscle weakness, spasms, fatigue, convulsions. Chronic Effects: fluorosis with brittle bones, anemia; kidney damage.

Fire: Noncombustible. Use extinguishing agents suitable for surrounding fire. Be aware that contact with water forms hydrofluoric acid which is capable of etching glass, cement, and many metals.

Reactivity: Stable. Hazardous polymerization cannot occur. Incompatible with: water; acid; oxidizers. Hazardous decomposition products: fluorine; nitrogen oxide; ammonia gases.

Carcinogenicity: IARC - Not listed; NIOSH - Not listed; NTP - Not listed; ACGIH - Class A4, Not classifiable as a human carcinogen; OSHA - Not listed; EPA - Not listed; MAK - Not listed

Primary Target Organs:

Eyes

Skin

Respiratory System

Nervous System

Kidneys

Bone

Exposure Limits
OSHA PEL: TWA: 2.5 mg/m^3; as F.
ACGIH TLV: TWA: 2.5 mg/m^3; as F.
NIOSH REL: TWA: 2.5 mg/m^3; as F (Flourides).
DFG MAK: TWA: 2.5 mg/m^3; as F.

Environmental

Cleanup/Disposal: Guide No. 154: Eliminate all ignition sources (no smoking, flares, sparks or flames in immediate area). Do not touch damaged containers or spilled material unless wearing appropriate protective clothing. Stop leak if you can do it without risk. Prevent entry into waterways, sewers, basements or confined areas. Absorb or cover with

dry earth, sand or other non-combustible material and transfer to containers. Do not get water inside containers.

Environmental Physical Data

BCF: no food chain concentration potential

Regulations

RCRA 40CFR: Not listed
CERCLA: 40CFR 302.4: Listed per CWA Section 311(b)(4) RQ: 100 lb (45.35 kg)
SARA 40CFR 372.65: Not listed
SARA EHS 40CFR 355: Not listed
TSCA: Listed

AMM2120 CAS #: 10192-30-0

AMMONIUM BISULFITE

RTECS: WT3595000
DOT: UN2693; NA2693; IMO8.0
EINECS Number: 233-469-7
Molecular Formula: H_5NO_3S
Structured MF: NH_4HSO_3
Formula Weight: 99.12
Synonyms: AMMONIUM ACID SULFITE; AMMONIUM HYDROGEN SULFITE; AMMONIUM HYDROSULFITE; AMMONIUM MONOSULFITE; AMMONIUM SULFITE,HYDROGEN; MONOAMMONIUM SULFITE; SULFUROUS ACID,MONOAMMONIUM SALT
Description: colorless to yellow, rhombic prisms
Use: source of sulfur in fluid fertilizers; liq antimicrobial treatment for storage grain; in manufacture of industrial explosives & blasting agents; in sulfite pulping processes; corrosion control in drilling fluids; preservative

Physical Properties

Boiling Point: Sublimes at 150 °C (302 °F)
Freezing Point: Sublimes at 150 °C (302 °F)
Specific Gravity: 2.03
Vapor Pressure: 395 mm Hg
Water Solubility: 267 g/100 ml at 10 °C
Other Solubilities: 1.4 g/L in ortho-dichlorobenzene; 0.72 g/L in chlorobenzene; 0.57 g/L in benzene; 0.37 g/L in chloroform; 0.11 g/L in acetone; 0.05 g/L in n-octanol; 0.01 g/L in methanol; 0.04 g/L in lard oil; water solubility: 19.3 ng/L
Flash Point: Nonflammable

Hazard Overviews

Fire: Noncombustible.
Carcinogenicity: IARC - Not listed; NIOSH - Not listed; NTP - Not listed; ACGIH - Not listed; OSHA - Not listed; EPA - Not listed; MAK - Not listed

Environmental

Cleanup/Disposal: Guide No. 154: Eliminate all ignition sources (no smoking, flares, sparks or flames in immediate area). Do not touch damaged containers or spilled material unless wearing appropriate protective clothing. Stop leak if you can do it without risk. Prevent entry into waterways, sewers, basements or confined areas. Absorb or cover with dry earth, sand or other non-combustible material and transfer to containers. Do not get water inside containers.

Environmental Physical Data

Octanol/Water Partition Coefficient: log K_{ow} = 7.02
BCF: no food chain concentration potential

Regulations

RCRA 40CFR: Not listed
CERCLA: 40CFR 302.4: Listed per CWA Section 311(b)(4) RQ: 5000 lb (2268 kg)
SARA 40CFR 372.65: Not listed
SARA EHS 40CFR 355: Not listed
TSCA: Listed

AMM2280 CAS #: 12124-97-9

AMMONIUM BROMIDE

RTECS: BO9155000
EINECS Number: 235-183-8
Molecular Formula: BrH_4N
Structured MF: NH_4Br
Formula Weight: 97.96

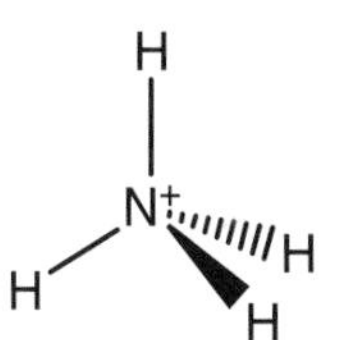

Chemical Structure

Synonyms: HYDROBROMIC ACID MONOAMMONIATE
Description: colorless to white to yellowish-white, crystals or granules; odorless
Use: former use sedative; process engraving & lithography; fireproofing of wood; corrosion inhibitor; analytical chemistry; textile finishing; fire retardant; mfr of photographic emulsions; metal treatment; catalyst in prodn of diphenylamine

Physical Properties

Boiling Point: 235 °C (455 °F) Vacuum
Freezing Point: Sublimes at 452 °C (845.6 °F)
Specific Gravity: 2.429 at 25 °C
Vapor Pressure: 1 mm Hg at 198.3 °C
Water Solubility: 97 g/100 cc at 25 °C
Other Solubilities: Practically Insoluble in Ethyl Acetate.
pH: Slightly acid
Refraction Index: 1.712 at 25 °C
Flash Point: Noncombustible

RTECS Toxicity Data

Acute Oral: Rat LD_{50} Dose: 2700 mg/kg; Toxic Effects: Sense organs and special senses - Other; Behavioral - Somnolence (general depressed activity); Lungs, Thorax, or Respiration - Other changes. Mouse LD_{50} Dose: 2860 mg/kg; Toxic Effects: Sense organs and special senses - Other; Behavioral - Somnolence (general depressed activity); Lungs, Thorax, or Respiration - Other changes.

Hazard Overviews

Health: Irritating to eyes/skin/respiratory tract. Harmful. Other Acute Effects: harmful if swallowed; may be harmful if inhaled; may be harmful if absorbed through the skin; prolonged contact with moist skin can produce severe irritation or burns; prolonged inhalation of dust can produce bronchitis; ingestion of large quantities can cause irritability; confusion; tremors; acne-like skin, erruptions; memory loss; headache; slurred speech; anorexia; target organ: central nervous system.

Fire: Noncombustible. Hazards: emits toxic fumes. Extinguishing agents: carbon dioxide, dry chemical powder or appropriate foam. Precautions: combustible liquid.

Reactivity: Stable. Hazardous polymerization will not occur. Incompatible with: strong oxidizing agents, strong acids, strong bases, heavy metals, silver salts, potassium, may discolor on exposure to air, protect from moisture. Hazardous decomposition products: toxic fumes of: ammonia hydrogen bromide gas.

Carcinogenicity: IARC - Not listed; NIOSH - Not listed; NTP - Not listed; ACGIH - Not listed; OSHA - Not listed; EPA - Not listed; MAK - Not listed

Primary Target Organs:

Eyes

Skin

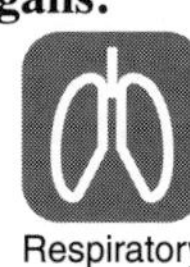
Respiratory System

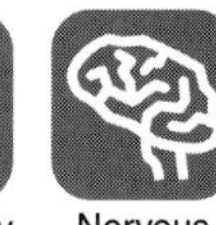
Nervous System

Environmental

Regulations

RCRA 40CFR: Not listed
CERCLA: 40CFR 302.4: Not listed
SARA 40CFR 372.65: Not listed
SARA EHS 40CFR 355: Not listed
TSCA: Listed

AMM2440 CAS #: 1111-78-0

AMMONIUM CARBAMATE

RTECS: EY8575000
DOT: NA9083
EINECS Number: 214-185-2
Molecular Formula: $CH_6N_2O_2$
Structured MF: $NH_4CO_2NH_2$
Formula Weight: 78.09

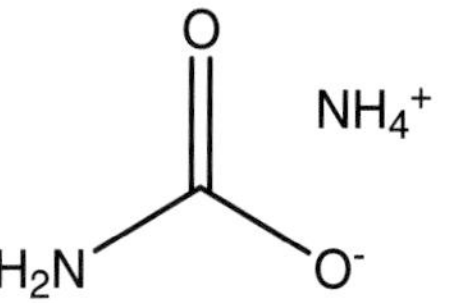

Chemical Structure

Synonyms: AMMONIUM AMINOFORMATE; CARBAMIC ACID,AMMONIUM SALT; CARBAMIC ACID,MONOAMMONIUM SALT

Description: colorless to white, crystalline rhombic powder; ammonia odor

Use: ammoniating agent; in fertilizers

Physical Properties

Boiling Point: 60 °C (140 °F)
Freezing Point: Sublimes at 60 °C (140 °F)
Vapor Pressure: 31 kPa
Water Solubility: Very Soluble in Cold Water
Other Solubilities: Very Soluble in Ammonium Hydroxide; Slightly Soluble in Alcohol; Insoluble in Acetone.
Odor Threshold: as NH3 < 5 ppm
Flash Point: Nonflammable

RTECS Toxicity Data

Acute Oral: Rat LD_{50} Dose: >681 mg/kg; Toxic Effects: Behavioral - Somnolence (general depressed activity); Lungs, Thorax, or Respiration - Dyspnea; Skin and appendages - Hair.

Hazard Overviews

Corrosive

Health: Corrosive to eyes/skin/respiratory tract. Other Acute Effects: harmful if swallowed, inhaled, or absorbed through skin; inhalation may result in spasm, inflammation and edema of the larynx and bronchi, chemical pneumonitis and pulmonary edema; symptoms of exposure may include burning sensation; coughing; wheezing; laryngitis; shortness of breath; headache; nausea; vomiting.

Fire: Noncombustible. Hazards: emits toxic fumes. Extinguishing agents: dry chemical powder. Precautions: combustible liquid.

Reactivity: Incompatible with: strong bases, strong acids, strong oxidizing agents, protect from moisture, store away from heat and direct sunlight. Hazardous decomposition products: carbon monoxide, carbon dioxide, nitrogen oxides, ammonia.

Carcinogenicity: IARC - Not listed; NIOSH - Not listed; NTP - Not listed; ACGIH - Not listed; OSHA - Not listed; EPA - Not listed; MAK - Not listed

Primary Target Organs:

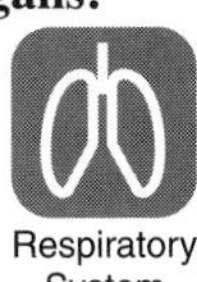

Eyes Skin Respiratory System

Environmental

Ecotoxicity: Aquatic toxicity: Decomposes to NH_42CO_3 and NH_3 5.5 to 7.0 mg/l NH_42CO_3 - lethal to fish. Toxicity threshold for freshwater fish appears to be between 30 to 40 ppm NH_42CO_3

Cleanup/Disposal: Guide No. 154: Eliminate all ignition sources (no smoking, flares, sparks or flames in immediate area). Do not touch damaged containers or spilled material unless wearing appropriate protective clothing. Stop leak if you can do it without risk. Prevent entry into waterways, sewers, basements or confined areas. Absorb or cover with dry earth, sand or other non-combustible material and transfer to containers. Do not get water inside containers.

Environmental Physical Data

BCF: no food chain concentration potential

Regulations

RCRA 40CFR: Not listed
CERCLA: 40CFR 302.4: Listed per CWA Section 311(b)(4) RQ: 5000 lb (2268 kg)
SARA 40CFR 372.65: Not listed
SARA EHS 40CFR 355: Not listed
TSCA: Listed

AMM2600 CAS #: 506-87-6

AMMONIUM CARBONATE

RTECS: BP1925000
DOT: NA9084
EINECS Number: 208-058-0
Molecular Formula: $CH_8N_2O_3$
Structured MF: $(NH_4)_2CO_3H_2O$
Formula Weight: 96.11
Synonyms: AMMONIA SESQUICARBONATE; AMMONIUMCARBONAT; CARBONATE D'AMMONIAQUE; CARBONIC ACID,AMMONIUM SALT; CARBONIC ACID,DIAMMONIUM SALT; CARBONIC ACID,DIAMMONIUM SALT (8CI,9CI); CRYSTAL AMMONIA; DIAMMONIUM CARBONATE; HARTSHORN
Description: colorless, white crystals, powder, cubes; strong ammonia odor
Use: in baking powders, for washing and defatting woolens; tanning; as mordant in dyeing; manufacture of rubber articles; casein glue, casein colors; in fire extinguishers; for separating cocas constituents; reagent in analytical chemistry; medication: expectorant; medication (vet): expectorant, carminative and stomachic; medication: smelling salts

Physical Properties

Boiling Point: Decomposes at 1 atm
Freezing Point: 58 °C (136.4 °F)
Specific Gravity: 1.5 at 20 °C/4 °C
Water Solubility: Soluble in Water
Odor Threshold: as ammonia gas < 5 ppm
Flash Point: Nonflammable

RTECS Toxicity Data

Acute Dermal: Rabbit LD_{Lo} Route: Subcutaneous Dose: 900 mg/kg. Frog LD_{Lo} Route: Subcutaneous Dose: 250 mg/kg.

Hazard Overviews

Fire Diamond

Health: Irritating to eyes/skin/respiratory tract. Acute Effects: difficulty in breathing; bronchitis; stomach cramps; burning and tingling sensation. Chronic Effects: None reported.
Fire: Noncombustible. Generation of ammonia gas may be an explosion hazard. Use agent suitable for surrounding fire.
Reactivity: Stable. Hazardous polymerization cannot occur. Avoid: air; water; acids; alkaloids. Incompatible with: acids; acid salts; alkaloids; alum; colomel; sodium hypochlorite; iron and zinc salts; tartar emetic. Hazardous decomposition products: oxides of carbon; nitrogen oxides; ammonia.
Carcinogenicity: IARC - Not listed; NIOSH - Not listed; NTP - Not listed; ACGIH - Not listed; OSHA - Not listed; EPA - Not listed; MAK - Not listed
Primary Target Organs:

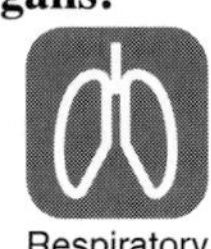

Eyes Skin Respiratory System Mucous Membranes

Environmental

Ecotoxicity: Aquatic toxicity: 24 ppm/3.5 hr/goldfish/killed/fresh water 10 ppm/>100 hr/goldfish/tolerated/fresh water

Cleanup/Disposal: Guide No. 154: Eliminate all ignition sources (no smoking, flares, sparks or flames in immediate area). Do not touch damaged containers or spilled material unless wearing appropriate protective clothing. Stop leak if you can do it without risk. Prevent entry into waterways, sewers, basements or confined areas. Absorb or cover with dry earth, sand or other non-combustible material and transfer to containers. Do not get water inside containers.

Environmental Physical Data

BCF: no food chain concentration potential

Regulations

RCRA 40CFR: Not listed
CERCLA: 40CFR 302.4: Listed per CWA Section 311(b)(4) RQ: 5000 lb (2268 kg)
SARA 40CFR 372.65: Not listed
SARA EHS 40CFR 355: Not listed
TSCA: Listed

AMM2760 **CAS #: 12125-02-9**

AMMONIUM CHLORIDE

RTECS: BP4550000
EINECS Number: 235-186-4
Molecular Formula: ClH_4N
Structured MF: NH_4Cl
Formula Weight: 53.50

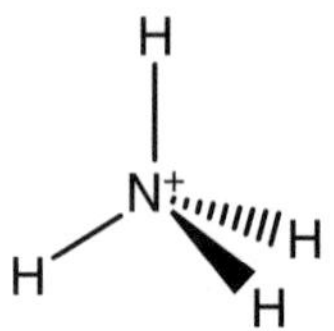

Chemical Structure

Synonyms: AMCHLOR; AMMON CHLOR; AMMONERIC; AMMONII CHLORIDUM; AMMONIUM CHLORATUM; AMMONIUM CHLORIDE INJECTION; AMMONIUM CHLORIDE TABLETS; AMMONIUM MURIATE; AMMONIUM MURIATE FUME; AMMONIUMCHLORID; CHLORAMMONIC; CHLORID AMMONIA; CHLORID AMONNY; CLORURO DE AMONIO; (COMPONENT OF) PV TUSSIN SYRUP; DARAMMON; GEN-DIUR; MURIATE OF AMMONIA; SAL AMMONIA; SAL AMMONIAC; SAL AMMONIAC FUME; SALAMMONITE; SALMIAC
Description: colorless, white crystals, powder; odorless
Use: in metallurgy: flux for coating iron with zinc; tinning; activator in chromizing; electroplating; in textile ind: lustering & dessicating cotton, mordant (dyeing & printing); tanning; mfr dyes; in batteries; freezing mixtures; explosives; washing powders; pipe cement; snow treatment; for drilling fluids; in salt substitutes; protectant for cell cultures against toxins; mfr of ammonia cmpd, resins, etc; fertilizer; defoliant; fire suppressant; med: systemic acidifier; (vet) expectorant, diaphoretic; acidifying diuretic, etc.

Physical Properties

Boiling Point: 520 °C (968 °F)
Freezing Point: Decomposes at 338 °C (640.4 °F)
Specific Gravity: 1.5274 at 25 °C
Vapor Pressure: 1 mm Hg at 160.4 °C
Water Solubility: 37% by weight
Other Solubilities: Soluble in liquid Ammonia; 0.6 g/100 ml Ethanol at 19 °C; 1 in 8 of Glycerol.
pH: Aqueous solution at 25 °C 5.5
Refraction Index: 1.642
Flash Point: Nonflammable

RTECS Toxicity Data

Acute Oral: Infant LD_{Lo} Dose: 2 gm/kg; Toxic Effects: Cardiac - Other changes. Rat LD_{50} Dose: 1650 mg/kg.
Acute Dermal: Rabbit LD_{Lo} Route: Subcutaneous Dose: 200 mg/kg. Mouse LD_{Lo} Route: Subcutaneous Dose: 500 mg/kg.
Irritation Eye: Rabbit Standard Draize Test Dose: 100 mg; Reaction: severe. Rabbit Standard Draize Test Dose: 500 mg/24H; Reaction: mild.
Mutagenic: Hamster Cytogenetic Analysis; Cell Type: fibroblast; Dose: 400 mg/L.

Hazard Overviews

Fire Diamond

Health: Irritating to eyes/skin/respiratory tract. Also Causes: ingestion can cause headache, hyperventilation, drowsiness, nausea, vomiting, diuresis, and possibly coma.
Fire: Noncombustible. Use extinguishing agents suitable for surrounding fire. Firefighters should protect against hydrogen chloride and ammonia gas produced in fire.
Reactivity: Stable, may volatilize and condense on cool surfaces, concentrated solutions may crystallize when exposed to low temperatures. Hazardous polymerization cannot occur. Avoid: excessive heat. Incompatible with: acids; alkalies and their carbonates; lead; silver salts; potassium chlorate; bromine trifluoride; bromide pentafluoride; ammonium compounds; nitrates; iodine heptafluoride; hydrogen cyanide. Hazardous decomposition products: ammonia; hydrochloric acid fumes.
Carcinogenicity: IARC - Not listed; NIOSH - Listed as carcinogen; NTP - Not listed; ACGIH - Not listed; OSHA - Not listed; EPA - Not listed; MAK - Not listed
Primary Target Organs:

Eyes

Skin

Respiratory System

Mucous Membranes
Gastro-intestinal
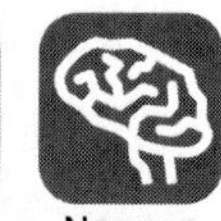
Nervous System

Exposure Limits
OSHA PEL Vacated 1989 Limits: TWA: 10 mg/m^3; STEL: 20 mg/m^3.
ACGIH TLV: TWA: 10 mg/m^3.
NIOSH REL: TWA: 10 mg/m^3. STEL: 20 mg/m^3.
Respirator Recommendation
Exposure Range: >10 to 100 mg/m^3 Air Purifying, Negative Pressure, Half Mask
Exposure Range: >100 to 1000 mg/m^3 Air Purifying, Negative Pressure, Full Face
Exposure Range: >1000 to 10,000 mg/m^3 Supplied Air, Constant Flow/Pressure Demand, Full Face
Exposure Range: >10,000 to unlimited mg/m^3 Self-contained Breathing Apparatus, Pressure Demand, Full Face
Cartridge Color: green with magenta (P100)

Environmental

Ecotoxicity: Aquatic toxicity: 6 ppm/96 hr/sunfish TLm/fresh water
Cleanup/Disposal: Guide No. 132: Fully encapsulating, vapor protective clothing should be worn for spills and leaks with no fire. Eliminate all ignition sources (no smoking, flares, sparks or flames in immediate area). All equipment used

when handling the product must be grounded. Do not touch or walk through spilled material. Stop leak if you can do it without risk. Prevent entry into waterways, sewers, basements or confined areas. A vapor suppressing foam may be used to reduce vapors. Absorb with earth, sand or other non-combustible material and transfer to containers (except for Hydrazine). Use clean non-sparking tools to collect absorbed material. Large Spills: Dike far ahead of liquid spill for later disposal. Water spray may reduce vapor; but may not prevent ignition in closed spaces.

Environmental Physical Data

BCF: no food chain concentration potential

Regulations

RCRA 40CFR: Not listed
CERCLA: 40CFR 302.4: Listed per CWA Section 311(b)(4) RQ: 5000 lb (2268 kg)
SARA 40CFR 372.65: Not listed
SARA EHS 40CFR 355: Not listed
TSCA: Listed

AMM2920 CAS #: 16919-58-7

AMMONIUM CHLOROPLATINATE

RTECS: BP5425000
EINECS Number: 240-973-0
Molecular Formula: $Cl_6H_8N_2Pt$
Formula Weight: 443.87

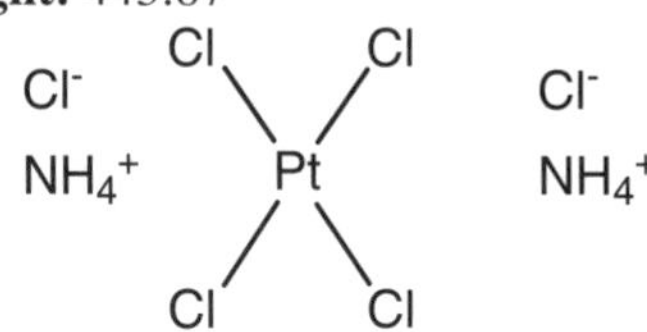

Chemical Structure

Synonyms: AMMONIUM HEXACHLOROPLATINATE; AMMONIUM HEXACHLOROPLATINATE (IV); AMMONIUM PLATINIC CHLORIDE; DIAMMONIUM HEXACHLOROPLATINATE(2-); PLATINATE(2-),HEXACHLORO-,DIAMMONIUM,(OC-6-11)-; PLATINATE(2-),HEXACHLORO-,DIAMMONIUM,(OC-6-11)-(9CI); PLATINIC AMMONIUM CHLORIDE; PLATINUM AMMONIUM CHLORIDE; QUATERNIUM-17
Description: orange, red, yellow crystals, powder
Use: platinum plating; mfr spongy platinum

Physical Properties

Freezing Point: Decomposes
Specific Gravity: 3.065
Water Solubility: 0.7 g/100 cc of Water at 15 °C
Other Solubilities: 0.005 g/100 cc of Alcohol; Insoluble in Ether & concentrated Hydrochloric Acid
Refraction Index: 1.8

RTECS Toxicity Data

Acute Oral: Rat LD_{50} Dose: 195 mg/kg.
Acute Inhalation: Rat LC; Dose: >565 mg/m^3/8hr. Human TC_{Lo} Dose: 900 ng/m^3; Toxic Effects: Sense organs and special senses - Other; Lungs, Thorax, or Respiration - Cough; Lungs, Thorax, or Respiration - Cyanosis.

Hazard Overviews

Explosive

Fire Diamond

Health: Irritating to eyes/skin/respiratory tract. Toxic. Also Causes: coughing, bluish face, pulmonary edema. Chronic Effects: platinosis, lymphocytosis, anaphylactic reactions, allergic contact skin sensitivity.
Fire: Explosive! Material is sensitive to impact and friction. Use dry chemical, carbon dioxide, water spray, fog, or regular foam.
Reactivity: Unstable, upon impact with other objects or certain chemicals. Hazardous polymerization cannot occur. Avoid: dust generation; friction. Incompatible with: hydrogen peroxide; lithium; oxygen difluoride; acids or acid fumes; potassium hydroxide. Hazardous decomposition products: toxic fumes of ammonia, chlorides, and nitrous oxides.
Carcinogenicity: IARC - Not listed; NIOSH - Not listed; NTP - Not listed; ACGIH - Not listed; OSHA - Not listed; EPA - Not listed; MAK - Not listed
Primary Target Organs:

Eyes

Skin

Respiratory System

Lymphatic System

Exposure Limits
OSHA PEL: TWA: 0.002 mg/m^3; as Pt.
ACGIH TLV: TWA: 0.002 mg/m^3; soluble as Pt.
NIOSH REL: TWA: 0.002 mg/m^3; as Pt soluble salts.
DFG MAK: TWA: 0.002 mg/m^3; Chloroplat.
Respirator Recommendation
Exposure Range: >0.002 to 0.02 mg/m^3 Air Purifying, Negative Pressure, Half Mask
Exposure Range: >0.02 to 0.2 mg/m^3 Air Purifying, Negative Pressure, Full Face
Exposure Range: >0.2 to 2 mg/m^3 Supplied Air, Constant Flow/Pressure Demand, Full Face
Exposure Range: >2 to unlimited mg/m^3 Self-contained Breathing Apparatus, Pressure Demand, Full Face
Cartridge Color: magenta (P100)
Note: as platinum salts, soluble

Environmental

Regulations

RCRA 40CFR: Not listed
CERCLA: 40CFR 302.4: Not listed
SARA 40CFR 372.65: Not listed
SARA EHS 40CFR 355: Not listed
TSCA: Listed

AMM3080 **CAS #: 7788-98-9**

AMMONIUM CHROMATE

RTECS: GB2880000
DOT: NA9086
EINECS Number: 232-138-4
Molecular Formula: $CrH_8N_2O_4$
Structured MF: $(NH_4)_2CrO_4$
Formula Weight: 152.07

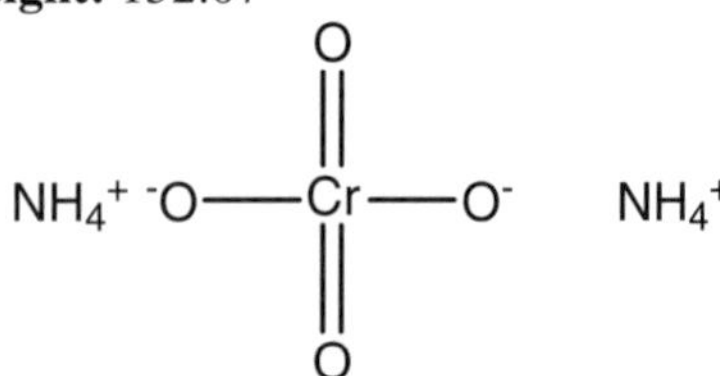

Chemical Structure

Synonyms: AMMONIUM CHROMATE(VI); DIAMMONIUM CHROMATE; DIAMMONIUM SALT; NEUTRAL AMMONIUM CHROMATE

Description: yellow crystals; ammonia odor

Use: in textile ind (printing pastes, fixing chromate dyes, dyeing of fur, fabrics, etc); in metallurgy (corrosion inhibitor; metal finishing, in plating, anodizing, coatings, & for corrosion resistance); rgnt in anal chemistry; catalyst; leather finishing; corrosion inhibitors in radiator coolants, engines, etc; photoreproduction processes, sensitizing agents for photoengraving, etc; in pigments, artists' colors, jointing pastes, etc; fungicides; wood preservatives; indicator, in bleaching of fats, oils, & waxes; in mfr of cement, glue, batteries, matches & explosives.

Physical Properties

Boiling Point: Decomposes at 180 °C (356 °F)
Freezing Point: Decomposes at 180 °C (356 °F)
Specific Gravity: 1.91 at 12 °C
Water Solubility: 40.5 g/100 cc at 30 °C
Other Solubilities: Sparingly Soluble in liquid Ammonia, Acetone; Slightly Soluble in Methanol; Practically Insoluble in Ethanol.
pH: Aqueous solution is alkaline
Flash Point: Nonflammable

RTECS Toxicity Data

Mutagenic: Bacteria - E Coli DNA Repair; Dose: 25 ug/well. Bacteria - S Typhimurium Mutations in Microorganisms; Dose: 35 ug/plate (+S9).

Hazard Overviews

Corrosive

Explosive

Fire Diamond: 1 / 3 / 3 / OX

Health: Corrosive to eyes/skin/respiratory tract. Also Causes: cough, wheezing, shortness of breath; pulmonary edema; dermatitis (rash), conjunctivitis, loss of sight, permanent blindness; poisoning; kidney failure, liver damage, hemolysis; shock; death. Chronic Effects: epistaxis, ulceration or perforation of the nasal septum, rhinitis, pharyngitis, nasal congestion, chest pain, asthma (via allergic sensitization), bronchitis, or respiratory tract cancer; hepatic; renal damage.

Fire: May burn, but it is not readily ignited. Explosive when exposed to shock or heated and an oxidizer. Use dry chemical, carbon dioxide, water spray, fog, or foam.

Reactivity: Stable. Hazardous polymerization cannot occur. Avoid: contact with hydrazine; reducing agents. Incompatible with: hydrazine; reducing agents (organic material; paper; wood; sulfur; aluminum; plastics). Hazardous decomposition products: fumes of ammonia; nitrogen oxide.

Carcinogenicity: IARC - Group 3, Not classifiable as to carcinogenicity to humans; NIOSH - Not listed; NTP - Not listed; ACGIH - Class A4, Not classifiable as a human carcinogen; OSHA - Not listed; EPA - Not listed; MAK - Class A2, Unmistakably carcinogenic in animal experimentation only

Primary Target Organs:

Eyes

Skin

Respiratory System

Liver

Kidneys

Blood

Exposure Limits

OSHA PEL: STEL: 0.1 mg/m^3; as CrO_3; Other Values: 0.1 mg/m^3; Clg Cr-VI as CrO_3.

NIOSH REL: TWA: 0.5 mg/m^3; as Cr;Cr-II;Cr-III;Cr(VI)=.001.

Respirator Recommendation

Exposure Range: >0.1 to 1 mg/m^3 Air Purifying, Negative Pressure, Half Mask

Exposure Range: >1 to 10 mg/m^3 Air Purifying, Negative Pressure, Full Face

Exposure Range: >10 to <15 mg/m^3 Supplied Air, Constant Flow/Pressure Demand, Full Face

Exposure Range: 15 to unlimited mg/m^3 Self-contained Breathing Apparatus, Pressure Demand, Full Face

Cartridge Color: magenta (P100)

Note: as chromium VI compounds

Environmental

Ecotoxicity: TLm Mosquitofish 270 and 240 ppm for 48 and 96 hr, respectively; Highly turbid water IL50 Eurasian watermilfoil 1.9, 2.6, 8.0 and 9.5 ppm as chromate for root weight, stem weight, root length, and stem length, respectively

Cleanup/Disposal: Guide No. 143: Keep combustibles (wood, paper, oil, etc.) away from spilled material. Do not touch damaged containers or spilled material unless wearing appropriate protective clothing. Use water spray to reduce vapors or divert vapor cloud drift. Prevent entry into waterways, sewers, basements or confined areas. Small Spills: Flush area with flooding quantities of water. Large Spills: Do not clean-up or dispose of, except under supervision of a specialist.

Environmental Physical Data
BCF: brown algae 100 to 500

Regulations
RCRA 40CFR: Not listed
CERCLA: 40CFR 302.4: Listed per CWA Section 311(b)(4) RQ: 10 lb (4.535 kg)
SARA 40CFR 372.65: Listed as Compound
SARA EHS 40CFR 355: Not listed
TSCA: Listed

AMM3240 CAS #: 513-74-6

AMMONIUM DITHIOCARBAMATE

EINECS Number: 208-166-8
Molecular Formula: $CH_6N_2S_2$
Formula Weight: 110.19
Synonyms: AMMONIUM SULFOCARBAMATE; CARBAMIC ACID,DITHIO-,MONOAMMONIUM SALT; CARBAMODITHIOIC ACID,MONOAMMONIUM SALT; DITHIOCARBAMIC ACID MONOAMMONIUM SALT
Description: yellow lustrous orthorhombic crystals when fresh; almost odorless when fresh, acquires odor of hydrogen sulfide after decomposition in air
Use: instead of hydrogen sulfide or ammonium sulfide for pptn of metals in chem analysis; synth of heterocyclic compd; chem int for rhodanine (analytical reagent)

Physical Properties
Freezing Point: 99 °C (210.2 °F)
Specific Gravity: 1.451 at 20 °C/4 °C
Water Solubility: Soluble in Water

Hazard Overviews
Carcinogenicity: IARC - Not listed; NIOSH - Not listed; NTP - Not listed; ACGIH - Not listed; OSHA - Not listed; EPA - Not listed; MAK - Not listed

Environmental

Regulations
RCRA 40CFR: Not listed
CERCLA: 40CFR 302.4: Not listed
SARA 40CFR 372.65: Not listed
SARA EHS 40CFR 355: Not listed
TSCA: Listed

AMM3400 CAS #: 14481-29-9

AMMONIUM FERROCYANIDE

EINECS Number: 238-476-9
Molecular Formula: $C_6H_{16}FeN_{10}$
Structured MF: $(NH_4)_4Fe(CN)_6$
Formula Weight: 284.12

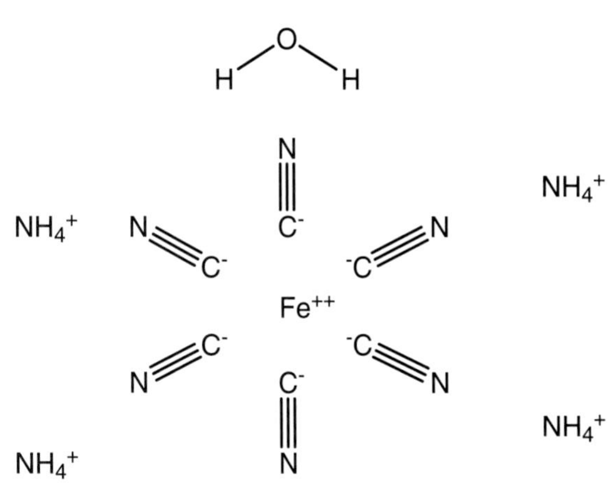

Chemical Structure

Synonyms: AMMONIUM HEXACYANOFERRATE (II); FERRATE(4-),HEXACYANO-,TETRAAMMONIUM; FERRATE(4-),HEXAKIS(CYANO-C)-,TETRAAMMONIUM,(OC-6-11)-; TETRAAMMONIUM HEXACYANOFERRATE; TETRAAMMONIUM HEXACYANOFERRATE(4-); TRIAMMONIUM HEXAKIS-(CYANO-C)FERRATE(4-)
Description: yellow crystals, turns blue in air

Physical Properties
Water Solubility: Freely Soluble in Water
Other Solubilities: Practically Insoluble in Alcohol
Odor Threshold: 0.5 ppm
Flash Point: Nonflammable

Hazard Overviews
Health: Irritating. Harmful. Other Acute Effects: may be harmful by inhalation, ingestion, or skin absorption.
Fire: Noncombustible. Hazards: emits toxic fumes. Extinguishing agents: water spray; carbon dioxide, dry chemical powder or appropriate foam. Precautions: combustible liquid.
Reactivity: Incompatible with: strong oxidizing agents, strong acids, light sensitive, air sensitive. Hazardous decomposition products: toxic fumes of: carbon monoxide, carbon dioxide, nitrogen oxides, hydrogen cyanide.
Carcinogenicity: IARC - Not listed; NIOSH - Not listed; NTP - Not listed; ACGIH - Not listed; OSHA - Not listed; EPA - Not listed; MAK - Not listed
Primary Target Organs:

Eyes

Skin

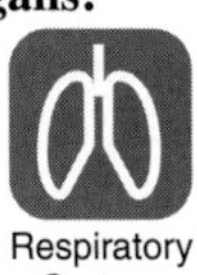
Respiratory System

Exposure Limits
OSHA PEL Vacated 1989 Limits: STEL: 1 mg/m^3; as Fe salts.
ACGIH TLV: TWA: 1 mg/m^3; as Fe Salt.
NIOSH REL: TWA: 1 mg/m^3; as Fe salts.
Respirator Recommendation
Exposure Range: >1 to 10 mg/m^3 Air Purifying, Negative Pressure, Half Mask

Exposure Range: >10 to 100 mg/m^3 Air Purifying, Negative Pressure, Full Face
Exposure Range: >100 to 1000 mg/m^3 Supplied Air, Constant Flow/Pressure Demand, Full Face
Exposure Range: >1000 to unlimited mg/m^3 Self-contained Breathing Apparatus, Pressure Demand, Full Face
Cartridge Color: dust/mist filter (use P100 or consult supervisor for appropriate dust/mist filter)
Note: as iron salts, soluble

Environmental

Regulations

RCRA 40CFR: Not listed
CERCLA: 40CFR 302.4: Listed as Compound per CWA Section 307(a) per CAA Section 112
SARA 40CFR 372.65: Not listed
SARA EHS 40CFR 355: Not listed
TSCA: Listed

AMM3560 CAS #: 13826-83-0

AMMONIUM FLUOBORATE

RTECS: BQ6100000
DOT: NA9088; IMO8.3
EINECS Number: 237-531-4
Molecular Formula: BF_4H_4N
Structured MF: NH_4BF_4
Formula Weight: 104.84

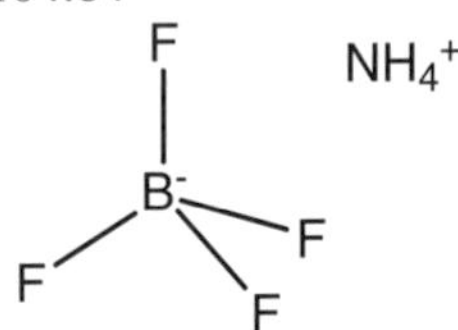

Chemical Structure

Synonyms: AMMONIUM BOROFLUORIDE; AMMONIUM FLUOROBORATE; AMMONIUM TETRAFLUOROBORATE; AMMONIUM TETRAFLUOROBORATE (1-); AMMONIUM TETRAFLUOROBORATE(1-); BORATE(1-),TETRAFLUORO-,AMMONIUM
Description: white rhombic crystals; odorless
Use: for the hi-temp fluxing action required by the metals industry; catalyst; in flame retardants; acts as a solid lubricant in cutting-oil emulsions for aluminum rolling and forming

Physical Properties

Boiling Point: Sublimes at 460 °C (860 °F)
Freezing Point: 230 °C (446 °F)
Specific Gravity: 1.871 at 15 °C
Vapor Pressure: 0.001 lb/sq inch at 100 °F
Water Solubility: 2.5 x10^5 ppm at 16 °C
Other Solubilities: Soluble in ammonium hydroxide
Flash Point: Nonflammable

Hazard Overviews

Corrosive

Fire Diamond

Health: Corrosive to eyes/skin/respiratory tract. Other Acute Effects: harmful if swallowed, inhaled, or absorbed through skin; inhalation may result in spasm, inflammation and edema of the larynx and bronchi, chemical pneumonitis and pulmonary edema; symptoms of exposure may include burning sensation; coughing; wheezing; laryngitis; shortness of breath; headache; nausea; vomiting.
Fire: Noncombustible. Hazards: emits toxic fumes. Extinguishing agents: use extinguishing media appropriate to surrounding fire conditions. Precautions: combustible liquid.
Reactivity: Incompatible with: strong acids, strong bases, may decompose on exposure to moist air or water, glass. Hazardous decomposition products: nitrogen oxides, ammonia, hydrogen fluoride.
Carcinogenicity: IARC - Not listed; NIOSH - Not listed; NTP - Not listed; ACGIH - Not listed; OSHA - Not listed; EPA - Not listed; MAK - Not listed
Primary Target Organs:

Eyes

Skin

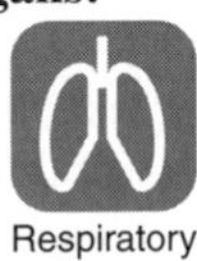

Respiratory System

Exposure Limits
OSHA PEL: TWA: 2.5 mg/m^3; as F.

Environmental

Cleanup/Disposal: Guide No. 154: Eliminate all ignition sources (no smoking, flares, sparks or flames in immediate area). Do not touch damaged containers or spilled material unless wearing appropriate protective clothing. Stop leak if you can do it without risk. Prevent entry into waterways, sewers, basements or confined areas. Absorb or cover with dry earth, sand or other non-combustible material and transfer to containers. Do not get water inside containers.

Environmental Physical Data

BOD: 87 ppm 50%

Regulations

RCRA 40CFR: Not listed
CERCLA: 40CFR 302.4: Listed per CWA Section 311(b)(4) RQ: 5000 lb (2268 kg)
SARA 40CFR 372.65: Not listed
SARA EHS 40CFR 355: Not listed
TSCA: Listed

AMM3720 **CAS #: 12125-01-8**

AMMONIUM FLUORIDE

RTECS: BQ6300000
DOT: UN2505; IMO6.1
EINECS Number: 235-185-9
Molecular Formula: FH_4N
Structured MF: NH_4F
Formula Weight: 37.04

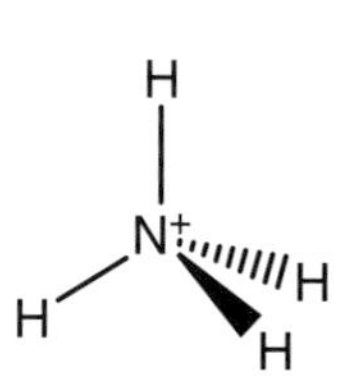

Chemical Structure

Synonyms: AMMONIUM FLUORIDE ((NH4)F); AMMONIUM FLUORURE; FLUOREK AMONOWY; FLUORURE D'AMMONIUM; FLUORURO AMONICO; NEUTRAL AMMONIUM FLUORIDE
Description: white powder, crystals; odorless
Use: etching and frosting glass, as antiseptic in brewing beer; preserving wood; in printing and dyeing textiles; as mothproofing agent; fluorides; analytical chemistry

Physical Properties

Boiling Point: Decomposes at 1 atm
Freezing Point: 125.6 °C (258.08 °F)
Specific Gravity: 1.015
Water Solubility: 100 g/100 ml of Water at 0 °C
Other Solubilities: Slightly Soluble in Alcohol; Insoluble in Ammonia.
Flash Point: Nonflammable

RTECS Toxicity Data

Acute Dermal: Frog LD_{Lo} Route: Subcutaneous Dose: 280 mg/kg.

Hazard Overviews

Poison

Corrosive

Fire Diamond

Health: Corrosive. Irritating to the eyes,skin, and respiratory tract. Poison. Also Causes: coughing, labored breathing, chills, fever, tremors, convulsions, kidney damage. Chronic Effects: mottled teeth, calcification of tendons/ligaments.
Fire: Noncombustible. Use agents suitable for surrounding fire.
Reactivity: Stable, but deliquescent. Hazardous polymerization cannot occur. Avoid: contact with heat and ignition sources. Incompatible with: quinine salts; chlorine trifluoride; acids; alkalis; soluble calcium salts; cement; glass; most metals. Hazardous decomposition products: fumes of fluorides; nitrogen oxides; ammonia; hydrogen fluoride.
Carcinogenicity: IARC - Not listed; NIOSH - Not listed; NTP - Not listed; ACGIH - Class A4, Not classifiable as a human carcinogen; OSHA - Not listed; EPA - Not listed; MAK - Not listed
Primary Target Organs:

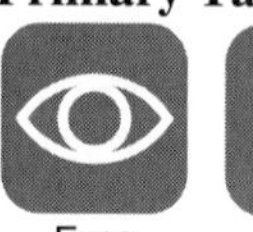
Eyes

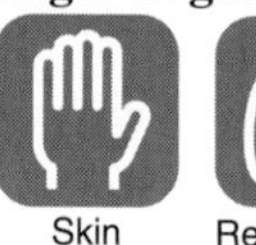
Skin

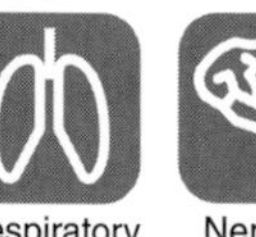
Respiratory System

Nervous System

Kidneys

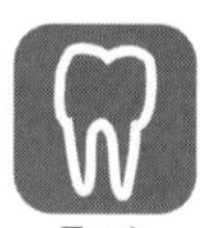
Teeth

Exposure Limits
OSHA PEL: TWA: 2.5 mg/m^3; as F.
ACGIH TLV: TWA: 2.5 mg/m^3; as F.
NIOSH REL: TWA: 2.5 mg/m^3; as F (Flourides).
DFG MAK: TWA: 2.5 mg/m^3; as F.

Environmental

Cleanup/Disposal: Guide No. 154: Eliminate all ignition sources (no smoking, flares, sparks or flames in immediate area). Do not touch damaged containers or spilled material unless wearing appropriate protective clothing. Stop leak if you can do it without risk. Prevent entry into waterways, sewers, basements or confined areas. Absorb or cover with dry earth, sand or other non-combustible material and transfer to containers. Do not get water inside containers.

Environmental Physical Data

BCF: no food chain concentration potential

Regulations

RCRA 40CFR: Not listed
CERCLA: 40CFR 302.4: Listed per CWA Section 311(b)(4) RQ: 100 lb (45.35 kg)
SARA 40CFR 372.65: Not listed
SARA EHS 40CFR 355: Not listed
TSCA: Listed

AMM3880 **CAS #: 540-69-2**

AMMONIUM FORMATE

RTECS: BQ6650000
EINECS Number: 208-753-9
Molecular Formula: CH_5NO_2
Structured MF: $HCOONH_4$
Formula Weight: 63.06

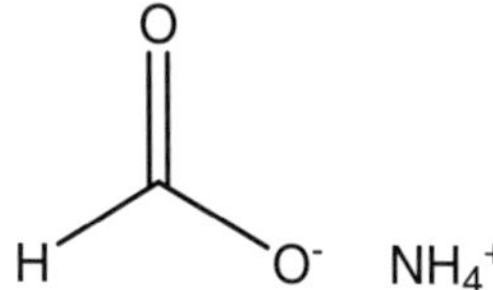

Chemical Structure

Synonyms: FORMIC ACID AMMONIUM SALT; FORMIC ACID,AMMONIUM SALT; MRAVENCAN AMONNY
Description: white monoclinic crystals; weak ammonia odor

Use: chem anal to ppt base metals from salts of noble metals; buffer

Physical Properties

Boiling Point: 180 °C (356 °F)
Freezing Point: 116 °C (240.8 °F)
Specific Gravity: 1.28
Water Solubility: 102 g/100 cc at 0 °C
Other Solubilities: Soluble in Ether, Ammonia
Flash Point: Nonflammable

RTECS Toxicity Data

Acute Oral: Mouse LD_{50} Dose: 2250 mg/kg.

Hazard Overviews

Health: Irritating to eyes/skin/respiratory tract. Other Acute Effects: may be harmful by inhalation, ingestion, or skin absorption.
Fire: Noncombustible. Hazards: emits toxic fumes. Extinguishing agents: water spray; carbon dioxide, dry chemical powder or appropriate foam. Precautions: combustible liquid.
Reactivity: Incompatible with: strong oxidizing agents, protect from moisture, strong acids. Hazardous decomposition products: toxic fumes of: carbon monoxide, carbon dioxide, nitrogen oxides.
Carcinogenicity: IARC - Not listed; NIOSH - Not listed; NTP - Not listed; ACGIH - Not listed; OSHA - Not listed; EPA - Not listed; MAK - Not listed
Primary Target Organs:

Eyes

Skin

Respiratory System

Environmental

Cleanup/Disposal: Guide No. 171: Do not touch or walk through spilled material. Stop leak if you can do it without risk. Prevent dust cloud. Avoid inhalation of asbestos dust. Small Dry Spills: With clean shovel place material into clean, dry container and cover loosely; move containers from spill area. Small Spills: Take up with sand or other noncombustible absorbent material and place into containers for later disposal. Large Spills: Dike far ahead of liquid spill for later disposal. Cover powder spill with plastic sheet or tarp to minimize spreading. Prevent entry into waterways, sewers, basements or confined areas.

Environmental Physical Data

BCF: no food chain concentration potential

Regulations

RCRA 40CFR: Not listed
CERCLA: 40CFR 302.4: Not listed
SARA 40CFR 372.65: Not listed
SARA EHS 40CFR 355: Not listed
TSCA: Listed

AMM4040 **CAS #: 10361-31-6**

AMMONIUM GLUCONATE

EINECS Number: 233-787-6
Molecular Formula: $C_6H_{(3x+12)}N_xO_7$
Formula Weight: Varies
Synonyms: D-GLUCONIC ACID,AMMONIUM SALT; GLUCONIC ACID,AMMONIUM SALT,D-
Description: white powder or needles
Use: as latent acid catalyst in textile printing; industrial chelating agent; emulsifying agent for cheese and salad dressings; component of disinfectant compd; componenet of skin care products

Physical Properties

Freezing Point: Decomposes at 154 °C (309.2 °F)
Water Solubility: 31.6 g/100 ml at 25 °C
Other Solubilities: Slightly Soluble in Alcohol; Practically Insoluble in most organic solvents

Hazard Overviews

Carcinogenicity: IARC - Not listed; NIOSH - Not listed; NTP - Not listed; ACGIH - Not listed; OSHA - Not listed; EPA - Not listed; MAK - Not listed

Environmental

Regulations

RCRA 40CFR: Not listed
CERCLA: 40CFR 302.4: Not listed
SARA 40CFR 372.65: Not listed
SARA EHS 40CFR 355: Not listed
TSCA: Listed

AMM4200 **CAS #: 1336-21-6**

AMMONIUM HYDROXIDE

RTECS: BQ9625000
DOT: UN1005; UN2073; UN2672; IMO2.2; IMO8.3
EINECS Number: 215-647-6
Molecular Formula: H_5NO
Structured MF: NH_4OH
Formula Weight: 35.05
Synonyms: AMMONIA AQUEOUS; AMMONIA SOLUTION; AMMONIA WATER 29%; AMMONIA,MONOHYDRATE; AMMONIUM HYDRATE; AQUA AMMONIA
Description: colorless liquid; suffocating odor
Use: in textiles, mfr of rayon, rubber; aniline dyes, ink; condensation polymerization; ceramics; photography; soaps; lubricants; fireproofing wood; saponifying fats & oils; org synth; detergent; household cleanser; food additive; removing stains, bleaching, printing, extracting plant colors & alkaloids; intermed for explosives & fertilizers; plastics; elastomers; fibers; in livestock feed; solvent in pulp & paper ind; refrigerant; electronic chem; pharmaceutical alkalizer; in

water treatment; for extracting metals from ores; in coating of paper; med: (vet) respiratory stimulant; carminative, insect bites; on bruises, sprains; inhalant; antacid.

Physical Properties

Freezing Point: -77 °C (-106.6 °F)
Specific Gravity: 0.9 at 25 °C/25 °C
Vapor Density: 0.6 Air=1
Water Solubility: Miscible with Cold Water
Odor Threshold: 47.0 ppm
Flash Point: Nonflammable
LEL: 16% v/v
UEL: 25% v/v

RTECS Toxicity Data

Acute Oral: Human LD_{Lo} Dose: 43 mg/kg. Rat LD_{50} Dose: 350 mg/kg; Toxic Effects: Gastrointestinal - Other changes; Liver - Other changes; Kidney, Ureter, and Bladder - Other changes.
Acute Inhalation: Human LC_{Lo} Dose: 5000 ppm. Human TC_{Lo} Dose: 408 ppm; Toxic Effects: Lungs, Thorax, or Respiration - Fibrosis, focal (pneumoconiosis); Lungs, Thorax, or Respiration - Acute pulmonary edema.
Acute Dermal: Rabbit LD_{Lo} Route: Subcutaneous Dose: 200 mg/kg. Mouse LD_{Lo} Route: Subcutaneous Dose: 160 mg/kg.
Irritation Eye: Rabbit Standard Draize Test Dose: 250 ug; Reaction: severe. Rabbit Standard Draize Test Dose: 44 ug; Reaction: severe.
Mutagenic: Bacteria - E Coli Mutations in Microorganisms; Dose: 10 mg/disc (-S9).

Hazard Overviews

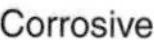
Corrosive

Fire Diamond

Health: Corrosive, causes severe burns to eyes/skin/respiratory tract. Toxic. Also Causes: blindness; exposure to high levels may be fatal.
Fire: Vapor may burn. Extinguish with dry chemical, water spray, carbon dioxide, or regular foam.
Reactivity: Stable. Hazardous polymerization cannot occur. Avoid: heat. Incompatible with: copper; zinc; galvinized surfaces; sulfuric; hydrochloric; heavy metals; halide salts; acrolein; acrylic acid; chlorosulfonic acid; dimethyl sulfate; fluorine; gold; aqua regia; oleum; beta-propiolactone; propylene oxide; silver oxide; silver nitrate; silver oxide and ethyl alcohol; nitromethane; silver permanganate; halogens. Hazardous decomposition products: carbon dioxide; toxic ammonia; nitrogen oxides.
Carcinogenicity: IARC - Not listed; NIOSH - Not listed; NTP - Not listed; ACGIH - Not listed; OSHA - Not listed; EPA - Not listed; MAK - Not listed

Primary Target Organs:

Eyes

Skin

Respiratory System

Mucous Membranes
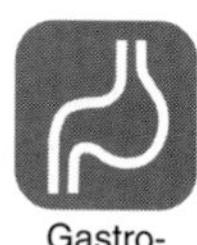
Gastro-intestinal

Environmental

Ecotoxicity: LC_{50} Daphnia magna 0.66 mg/l/48 hr 22 °C /Conditions of bioassay not specified LC_{50} Perch 0.29 mg/l/7 days /Un-ionized NH3/ /Conditions of bioassay not specified LC_{50} Salmo gairdnerii 8 ug/ml NH3/24 hr /Conditions of bioassay not specified LC_{50} Rutilus rutilus (roach) 0.42 mg/l/24 hr /Un-ionized NH3 LC_{50} Atlantic salmon smolt 0.02 mg NH3/l/24 hr (dissolved O2 of 10 mg/l, fresh water). /Conditions of bioassay not specified TLm Snail 90 mg/l/96 hr (soft water, 20 °C) /Conditions of bioassay not specified TLm Striped bass 0.97 ug/l/96 hr (15 °C); 0.73 ug/l/96 hr (23 °C) /Unionized NH3/ /Conditions of bioassay not specified
Cleanup/Disposal: Guide No. 125: Fully encapsulating, vapor protective clothing should be worn for spills and leaks with no fire. Do not touch or walk through spilled material. Stop leak if you can do it without risk. If possible, turn leaking containers so that gas escapes rather than liquid. Prevent entry into waterways, sewers, basements or confined areas. Do not direct water at spill or source of leak. Use water spray to reduce vapors or divert vapor cloud drift. Isolate area until gas has dispersed. Guide No. 154: Eliminate all ignition sources (no smoking, flares, sparks or flames in immediate area). Do not touch damaged containers or spilled material unless wearing appropriate protective clothing. Stop leak if you can do it without risk. Prevent entry into waterways, sewers, basements or confined areas. Absorb or cover with dry earth, sand or other non-combustible material and transfer to containers. Do not get water inside containers.

Environmental Physical Data

BCF: no food chain concentration potential

Regulations

RCRA 40CFR: Not listed
CERCLA: 40CFR 302.4: Listed per CWA Section 311(b)(4) RQ: 1000 lb (453.5 kg)
SARA 40CFR 372.65: Not listed
SARA EHS 40CFR 355: Not listed
TSCA: Listed

AMM4360 **CAS #: 7803-65-8**

AMMONIUM HYPOPHOSPHITE

EINECS Number: 232-266-0
Molecular Formula: H_6NO_2P
Formula Weight: 83.04

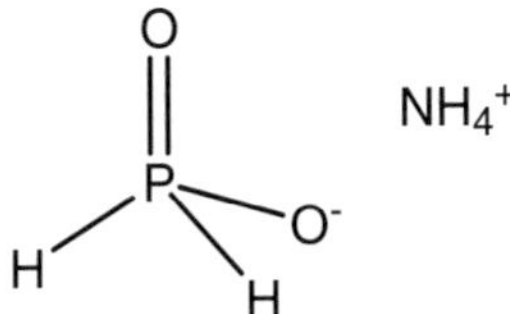

Chemical Structure

Synonyms: HYPOPHOSPHOROUS ACID,AMMONIUM SALT; PHOSPHINIC ACID,AMMONIUM SALT
Description: white tablets, crystals or granules
Use: catalyst in polyamide mfr

Physical Properties

Boiling Point: 240 °C (464 °F)
Freezing Point: 200 °C (392 °F)
Specific Gravity: 1.634
Water Solubility: 1 g/1 ml
Other Solubilities: Soluble in Ammonia, Acetone.
pH: Aqueous solution is practically neutral

Hazard Overviews

Carcinogenicity: IARC - Not listed; NIOSH - Not listed; NTP - Not listed; ACGIH - Not listed; OSHA - Not listed; EPA - Not listed; MAK - Not listed

Environmental

Regulations

RCRA 40CFR: Not listed
CERCLA: 40CFR 302.4: Not listed
SARA 40CFR 372.65: Not listed
SARA EHS 40CFR 355: Not listed
TSCA: Listed

AMM4520 CAS #: 12027-06-4

AMMONIUM IODIDE

EINECS Number: 234-717-7
Molecular Formula: H_4IN
Structured MF: NH_4I
Formula Weight: 144.96

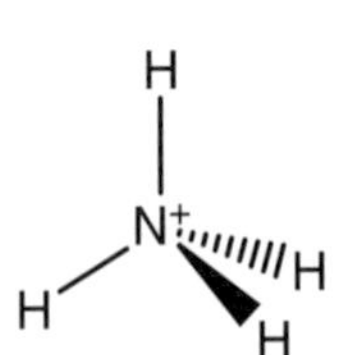

Chemical Structure

Description: colorless to white crystals or granular powder; odorless
Use: in photographic chemicals; medication: expectorant

Physical Properties

Boiling Point: 220 °C (428 °F) Vacuum
Freezing Point: Sublimes at 551 °C (1023.8 °F)
Specific Gravity: 2.514 at 25 °C
Vapor Pressure: 1 mm Hg at 210.9 °C
Water Solubility: 154.2 g/100 cc at 0 °C
Other Solubilities: Glycerol: 1 g/1.5 ml; Alcohol: 1 g/3.7 ml; Methanol: 1 g/2.5 ml; Very Soluble in Acetone, Ammonia; Slightly Soluble in Ether.
pH: 0.1 molar solution 4.6
Refraction Index: 1.7031
Flash Point: Nonflammable

Hazard Overviews

Health: Irritating to eyes/skin/respiratory tract. Other Acute Effects: may be harmful by inhalation, ingestion, or skin absorption.
Fire: Noncombustible. Hazards: emits toxic fumes. Extinguishing agents: carbon dioxide, dry chemical powder or appropriate foam. Precautions: combustible liquid.
Reactivity: Incompatible with: strong bases, strong acids, air sensitive, moisture sensitive, may decompose on exposure to light. Hazardous decomposition products: hydrogen iodide, ammonia nitrogen oxides.
Carcinogenicity: IARC - Not listed; NIOSH - Not listed; NTP - Not listed; ACGIH - Not listed; OSHA - Not listed; EPA - Not listed; MAK - Not listed
Primary Target Organs:

Eyes

Skin

Respiratory System

Environmental

Cleanup/Disposal: Guide No. 171: Do not touch or walk through spilled material. Stop leak if you can do it without risk. Prevent dust cloud. Avoid inhalation of asbestos dust. Small Dry Spills: With clean shovel place material into clean, dry container and cover loosely; move containers from spill area. Small Spills: Take up with sand or other noncombustible absorbent material and place into containers for later disposal. Large Spills: Dike far ahead of liquid spill for later disposal. Cover powder spill with plastic sheet or tarp to minimize spreading. Prevent entry into waterways, sewers, basements or confined areas.

Regulations

RCRA 40CFR: Not listed
CERCLA: 40CFR 302.4: Not listed
SARA 40CFR 372.65: Not listed
SARA EHS 40CFR 355: Not listed
TSCA: Listed

AMM4680 CAS #: 32612-48-9

AMMONIUM LAURYL ETHER SULFATE SOLUTION

RTECS: MD0100000
Molecular Formula: $C_{(2x+12)}H_{(4x+29)}NO_{(x+4)}S$
Formula Weight: Varies
Synonyms: AMMONIUM LAURETH SULFATE; AMMONIUM LAURYL ETHER SULFATE
Description: clear to light yellow liquid; faint odor

Physical Properties

Boiling Point: Initial < 100 °C (212 °F)
Specific Gravity: About 1.05
Water Solubility: Soluble
pH: 6.5 to 7.5
Flash Point: 23.3 to 27.8 °C Closed Cup
LEL: 3.3% v/v

RTECS Toxicity Data

Acute Oral: Rat LD_{50} Dose: 630 mg/kg.

Hazard Overviews

Flammable

Fire Diamond

Health: Irritating to eyes/skin. Also Causes: upon ingestion: diarrhea and spontaneous vomiting which may result in oral and esophageal burns. Chronic Effects: occupational dermatitis.
Fire: Flammable. Use dry chemical, carbon dioxide, and water spray as extinguishing media. Fight fire from maximum distance.
Reactivity: Stable. Hazardous polymerization cannot occur. Avoid: heat; ignition sources; oxidizing agents. Incompatible with: strong oxidizing agents. Hazardous decomposition products: toxic nitrogen oxides; carbon; sulfur and hydrocarbon partial oxidation products.
Carcinogenicity: IARC - Not listed; NIOSH - Not listed; NTP - Not listed; ACGIH - Not listed; OSHA - Not listed; EPA - Not listed; MAK - Not listed
Primary Target Organs:

Eyes

Skin

Environmental

Regulations

RCRA 40CFR: Not listed
CERCLA: 40CFR 302.4: Not listed
SARA 40CFR 372.65: Not listed
SARA EHS 40CFR 355: Not listed
TSCA: Listed

AMM4840 CAS #: 2235-54-3

AMMONIUM LAURYL SULFATE

RTECS: WT0825000
EINECS Number: 218-793-9
Molecular Formula: $C_{12}H_{29}NO_4S$
Formula Weight: 283.48
Synonyms: AKYPOSAL ALS 33; AMMONIUM DODECYL SULFATE; AMMONIUM N-DODECYL SULFATE; CONCO SULFATE A; DODECYL AMMONIUM SULFATE; LAURYL AMMONIUM SULFATE; LAURYL SULFATE AMMONIUM SALT; MAPROFIX NH; MONTOPOL LA 20; NEOPON LAM; PRESULIN; RICHONOL AM; SINOPON; SIPON LA 30; SIPROL 422; SIPROL L22; STERLING AM; SULFURIC ACID,LAURYL ESTER,AMMONIUM SALT; SULFURIC ACID,MONODODECYL ESTER,AMMONIUM SALT; TEXA PON A 400; TEXAPON A 400; TEXAPON SPECIAL
Description: clear liquid
Use: surface-active ingredient of hair shampoos, rug shampoos; anionic detergent

Hazard Overviews

Carcinogenicity: IARC - Not listed; NIOSH - Not listed; NTP - Not listed; ACGIH - Not listed; OSHA - Not listed; EPA - Not listed; MAK - Not listed

Environmental

Regulations

RCRA 40CFR: Not listed
CERCLA: 40CFR 302.4: Not listed
SARA 40CFR 372.65: Not listed
SARA EHS 40CFR 355: Not listed
TSCA: Listed

AMM5000 CAS #: 7803-55-6

AMMONIUM METAVANADATE

RTECS: YW0875000
DOT: UN2859; IMO6.1
EINECS Number: 232-261-3
Molecular Formula: H_4NO_3V
Formula Weight: 116.99

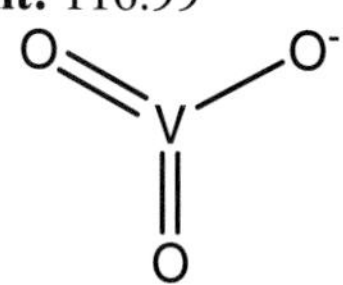

Chemical Structure

Synonyms: AMMONIUM VANADATE; AMMONIUM VANADATE(V); VANADATE; VANADIC ACID (HVO3),AMMONIUM SALT

Description: colorless, white, yellow crystals
Use: dyes; varnishes; indelible inks; drier for paints and inks; as a photographic developer; reagent in analytical chemistry; medication

Physical Properties

Freezing Point: Decomposes at 200 °C (392 °F)
Density: 2.326 g/cu m
Water Solubility: 1 parts in 165 parts Water
Other Solubilities: Insoluble in saturated ammonium chloride solution; Insoluble in Alcohol, Ether.
Flash Point: Nonflammable

RTECS Toxicity Data

Acute Oral: Rat LD_{50} Dose: 58100 ug/kg; Toxic Effects: Behavioral - Somnolence (general depressed activity); Gastrointestinal - Hypermotility, diarrhea; Nutritional and gross metabolic - Body temperature decrease. Mouse LD_{50} Dose: 25 mg/kg.
Acute Inhalation: Rat LC_{50} Dose: 7800 ug/m^3/4hr; Toxic Effects: Behavioral - Somnolence (general depressed activity); Gastrointestinal - Hypermotility, diarrhea; Nutritional and gross metabolic - Body temperature decrease.
Acute Dermal: Rabbit LD_{50} Route: Subcutaneous Dose: 13 mg/kg; Toxic Effects: Behavioral - Convulsions or effect on seizure threshold; Behavioral - Food intake (animal); Gastrointestinal - Hypermotility, diarrhea. Rat LD_{50} Route: Skin; Dose: 2102 mg/kg; Toxic Effects: Behavioral - Somnolence (general depressed activity); Gastrointestinal - Hypermotility, diarrhea; Nutritional and gross metabolic - Body temperature decrease. Rat LD_{50} Route: Subcutaneous Dose: 23 mg/kg; Toxic Effects: Behavioral - Coma.
Reproductive/Teratogenic: Hamster Route: Intraperitoneal; Dose: 11280 ug/kg; Duration: female 5-10D of pregnancy; Effects on Fertility - Post-implantation mortality; Effects on Embryo or Fetus - Fetal death. Hamster Route: Intraperitoneal; Dose: 22500 ug/kg; Duration: female 5-10D of pregnancy; Effects on Newborn - Sex ratio. Hamster Route: Intraperitoneal; Dose: 2820 ug/kg; Duration: female 5-10D of pregnancy; Specific Developmental Abnormalities - Musculoskeletal system.
Mutagenic: Human DNA Damage; Cell Type: lymphocyte; Dose: 200 umol/L. Human DNA Damage; Cell Type: ovary; Dose: 200 umol/L. Human Sister Chromatid Exchenge; Cell Type: lymphocyte; Dose: 40 umol/L. Human Micronucleus Test; Cell Type: lymphocyte; Dose: 10 umol/L. Human Sex Chromosome Loss; Cell Type: lymphocyte; Dose: 40 umol/L.

Hazard Overviews

Fire Diamond

Health: Irritating to eyes/skin/respiratory tract. Also Causes: sore throat, nose bleeds, cough, bronchitis, chest pain, pulmonary edema (fluid in lungs). Chronic Effects: greenish tongue, chronic bronchitis, dermatitis.
Fire: Noncombustible. Use extinguishing agents suitable for surrounding fire.
Reactivity: Stable. Hazardous polymerization cannot occur. Avoid: excessive dust generation. Hazardous decomposition products: ammonia gas; nitrogen oxide gas; vanadium oxide gas.
Carcinogenicity: IARC - Not listed; NIOSH - Not listed; NTP - Not listed; ACGIH - Not listed; OSHA - Not listed; EPA - Not listed; MAK - Not listed
Primary Target Organs:

Eyes

Skin

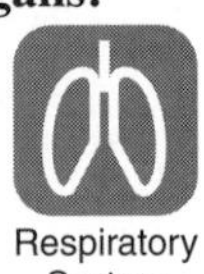
Respiratory System

Mucous Membranes

Environmental

Cleanup/Disposal: Guide No. 154: Eliminate all ignition sources (no smoking, flares, sparks or flames in immediate area). Do not touch damaged containers or spilled material unless wearing appropriate protective clothing. Stop leak if you can do it without risk. Prevent entry into waterways, sewers, basements or confined areas. Absorb or cover with dry earth, sand or other non-combustible material and transfer to containers. Do not get water inside containers.

Regulations

RCRA 40CFR: Listed Hazardous Waste No. P119 Toxic Waste
CERCLA: 40CFR 302.4: Listed per RCRA Section 3001 RQ: 1000 lb (453.5 kg)
SARA 40CFR 372.65: Not listed
SARA EHS 40CFR 355: Not listed
TSCA: Listed

Analytical Methods

Indoor / Expired Air: NIOSH 7504

AMM5160 **CAS #: 12027-67-7**

AMMONIUM MOLYBDATE (VI)

RTECS: QA5076000
EINECS Number: 234-722-4
Molecular Formula: $H_{24}Mo_7N_6O_{24}$
Formula Weight: 1163.89

Chemical Structure

Synonyms: AMMONIUM HEPTAMOLYBDATE; AMMONIUM MOLYBDATE; AMMONIUM MOLYBDATE ((NH4)6(MO7O24)); AMMONIUM MOLYBDATE (VI) ((NH4)6MO7O24); AMMONIUM PARAMOLYBDATE; AMMONIUM PARAMOLYBDATE ((NH4)6MO7O24); HEXAMMONIUM TETRACOSAOXOHEPTAMOLYBDATE; MOLYBDATE (MO7O24),HEXAMMONIUM; MOLYBDATE,HEXAAMMONIUM (9CI); MOLYBDIC ACID (H6MO7O24),HEXAAMMONIUM SALT

Description: colorless to slightly greenish to yellowish crystals

Use: in photography & for decorating ceramics; anal rgnt for detecting & determining phosphates, arsenates, lead; rgnt for alkaloids, etc; catalyst for dehydrogenation & desulfurization in petro & coal technology, prodn of molybdenum metal, source of molybdate ions; intermed for molybdenum pigments, phosphomolybdic acid, cobalt-molybdenum-alumina & molybdenum trioxide catalysts; catalyst for phthalocyanine dyes & pigments; in fertilizers.

Physical Properties

Freezing Point: 90 °C (194 °F)

Density: 2.498

Water Solubility: Tetrahydrate 43 g/100cc

Other Solubilities: Practically Insoluble in Alcohol; Soluble in acids & alkali

RTECS Toxicity Data

Mutagenic: Bacteria - B Subtilis DNA Repair; Dose: 50 mmol/L. Bacteria - E Coli Mutations in Microorganisms; Dose: 2 mmol/L (-S9).

Hazard Overviews

Health: Irritating to eyes/skin/respiratory tract. Harmful. Other Acute Effects: harmful if swallowed, inhaled, or absorbed through skin; possible risk of irreversible effects. Chronic Effects: laboratory experiments have shown mutagenic effects.

Fire: Hazards: emits toxic fumes. Extinguishing agents: noncombustible; use extinguishing media appropriate to surrounding fire conditions. Precautions: combustible liquid.

Reactivity: Incompatible with: strong oxidizing agents, strong acids. Hazardous decomposition products: ammonia, nitrogen oxides.

Carcinogenicity: IARC - Not listed; NIOSH - Not listed; NTP - Not listed; ACGIH - Not listed; OSHA - Not listed; EPA - Not listed; MAK - Not listed

Primary Target Organs:

Eyes

Skin

Respiratory System

Exposure Limits

OSHA PEL: STEL: 5 mg/m^3; as Mo soluble. Other Values: 15 mg/m^3; as Mo dust, insoluble.

OSHA PEL Vacated 1989 Limits: TWA: 10 mg/m^3; insoluble compound.

ACGIH TLV: TWA: 5 mg/m^3; soluble is 10.

DFG MAK: TWA: 5 mg/m^3; soluble; insoluble 1.

Respirator Recommendation

Exposure Range: >5 to 50 mg/m^3 Air Purifying, Negative Pressure, Half Mask

Exposure Range: >50 to 500 mg/m^3 Air Purifying, Negative Pressure, Full Face

Exposure Range: >500 to <1000 mg/m^3 Supplied Air, Constant Flow/Pressure Demand, Full Face

Exposure Range: 1000 to unlimited mg/m^3 Self-contained Breathing Apparatus, Pressure Demand, Full Face

Cartridge Color: dust/mist filter (use P100 or consult supervisor for appropriate dust/mist filter)

Note: as molybdenum, soluble

Environmental

Regulations

RCRA 40CFR: Not listed

CERCLA: 40CFR 302.4: Not listed

SARA 40CFR 372.65: Not listed

SARA EHS 40CFR 355: Not listed

TSCA: Listed

AMM5320 **CAS #: 12135-76-1**

AMMONIUM MONOSULFIDE

DOT: UN2683; IMO8.0

EINECS Number: 235-223-4

Molecular Formula: H_8N_2S

Structured MF: $(NH_4)_2S$

Formula Weight: 68.15

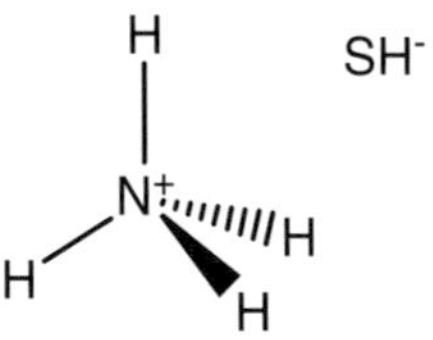

Chemical Structure

Synonyms: AMMONIUM SULFIDE ((NH4)2S); AMMONIUM SULFIDE (SOLUTION); DIAMMONIUM SULFIDE; TRUE AMMONIUM SULFIDE

Description: yellow, colorless- crystals; strong odor of rotten eggs and ammonia

Use: to apply patina to bronze; trace metal analysis; in photographic developers, metallurgy, textile & in certain cold wave sets; iron control in soda ash production; production of ammonium thiosulfite

Physical Properties

Boiling Point: 40 °C (104 °F)
Specific Gravity: 1.2
Water Solubility: Very Soluble in Cold Water
Other Solubilities: Very Soluble in Ammonia; Soluble in Ethanol
Flash Point: 72 °C Closed Cup
LEL: 4% v/v
UEL: 46% v/v

Hazard Overviews

Corrosive

Health: Corrosive to eyes/skin/respiratory tract. Toxic. Other Acute Effects: harmful if swallowed, inhaled, or absorbed through skin; readily absorbed through skin; inhalation may result in spasm, inflammation and edema of the larynx and bronchi, chemical pneumonitis and pulmonary edema; symptoms of exposure may include burning sensation; coughing; wheezing; laryngitis; shortness of breath; headache; nausea; vomiting.

Fire: Combustible. Hazards: vapor may travel considerable distance to source of ignition and flash back; container explosion may occur; emits toxic fumes; generates flammable and/or explosive hydrogen gas in contact with: strong oxidizing agents acid. Extinguishing agents: water spray; appropriate foam; dry chemical powder; use water spray to cool fire-exposed containers; do not use carbon dioxide extinguisher on this material. Precautions: combustible liquid.

Reactivity: Stable. Hazardous polymerization will not occur. Incompatible with: strong oxidizing agents, acids, strong bases, aluminum, zinc, copper. Hazardous decomposition products: toxic fumes of: hydrogen sulfide gas, sulfur oxides, ammonia.

Carcinogenicity: IARC - Not listed; NIOSH - Not listed; NTP - Not listed; ACGIH - Not listed; OSHA - Not listed; EPA - Not listed; MAK - Not listed

Primary Target Organs:

Eyes

Skin

Respiratory System

Environmental

Ecotoxicity: TLm Mosquito fish 248 ppm/48 hr fresh water. /Conditions of bioassay not specified

Cleanup/Disposal: Guide No. 132: Fully encapsulating, vapor protective clothing should be worn for spills and leaks with no fire. Eliminate all ignition sources (no smoking, flares, sparks or flames in immediate area). All equipment used when handling the product must be grounded. Do not touch or walk through spilled material. Stop leak if you can do it without risk. Prevent entry into waterways, sewers, basements or confined areas. A vapor suppressing foam may be used to reduce vapors. Absorb with earth, sand or other non-combustible material and transfer to containers (except for Hydrazine). Use clean non-sparking tools to collect absorbed material. Large Spills: Dike far ahead of liquid spill for later disposal. Water spray may reduce vapor; but may not prevent ignition in closed spaces.

Environmental Physical Data

BCF: no food chain concentration potential

Regulations

RCRA 40CFR: Not listed
CERCLA: 40CFR 302.4: Listed per CWA Section 311(b)(4) RQ: 100 lb (45.35 kg)
SARA 40CFR 372.65: Not listed
SARA EHS 40CFR 355: Not listed
TSCA: Listed

AMM5480 CAS #: 6484-52-2

AMMONIUM NITRATE

RTECS: BR9050000
DOT: UN0206; UN0223; UN1942; UN2067; UN2068; UN2426; IMO1.1; IMO5.1; IMO9.0
EINECS Number: 229-347-8
Molecular Formula: $H_4N_2O_3$
Structured MF: NH_4NO_3
Formula Weight: 80.06

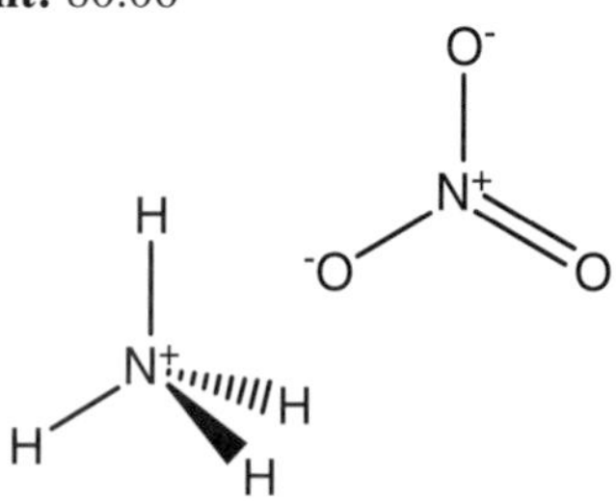

Chemical Structure

Synonyms: AMMONIUM (I) NITRATE (1:1); AMMONIUM SALTPETER; EPA PESTICIDE CHEMICAL CODE 076101; GERMAN SALTPETER; HERCO PRILLS; MERCO PRILLS; NITRAM; NITRATE D'AMMONIUM; NITRATE OF AMMONIA; NITRATO AMONICO; NITRIC ACID AMMONIUM SALT; NITRIC ACID,AMMONIUM SALT; NORWAY SALTPETER; VARIOFORM; VARIOFORM I

Description: colorless crystalline solid; odorless

Use: in freezing mixtures; mfr of cosmetic, matches, pyrotechnics, indust explosives & blasting agents; direct application fertilizer; intermed for nitrous oxide; herbicides & insecticides; absorbent for nitrogen oxides; oxidizer in rocket propellants; nutrient for antibiotics & yeast; catalyst; desiccant for cotton; in slow burning propellants & where smokeless exhaust required; in fuel oil & water based commercial explosives; in hair dyes, colorings; in military explosives eg, amatols, etc.

Physical Properties

Boiling Point: 210 °C (410 °F) at 11 mm Hg
Freezing Point: 169.6 °C (337.28 °F)
Specific Gravity: 1.725 at 25 °C
Vapor Density: 2.8 Air=1
Water Solubility: 118.3 g/100 cc of Water at 0 °C
Other Solubilities: Soluble in alkalies.
pH: 0.1 molar solution in water 5.43
Flash Point: Nonflammable

RTECS Toxicity Data

Acute Oral: Rat LD_{50} Dose: 2217 mg/kg.

Hazard Overviews

Fire Diamond

Health: Irritating to eyes/skin/respiratory tract. Also Causes: difficulty breathing, acidic urine, systemic acidosis, and abnormal hemoglobin.
Fire: Noncombustible. However, it is a strong oxidizing agent capable of igniting combustibles.
Reactivity: Stable. Hazardous polymerization cannot occur. Avoid: contaminating ammonium nitrate with oil, charcoal, or other organic substance; heat. Incompatible with: strong alkalies; reducing materials; ammonium dichromate; potassium dichromate; potassium chromate; chromium (VI) salts; barium chloride; sodium chloride; potassium nitrate; hot water; urea; sawdust; barium nitrate; copper iron (II) sulfide; acetic anhydride and nitric acid; ammonium chloride and water and zinc; ammonium sulfate and potassium; powdered metals (aluminum; antimony; bismuth; cadmium; chro-mium; cobalt; copper; iron; lead; magnesium; manganese; nickel; tin; zinc; brass); hydrocarbon oils; nonmetals; organic fuels; sugar; potassium permanganate; sulfur; trinitroanisole; acetic acid; aluminum and calcium nitrate and formamide; chloride salts; charcoal and metal oxides. Hazardous decomposition products: gases of nitrogen oxides.
Carcinogenicity: IARC - Not listed; NIOSH - Not listed; NTP - Not listed; ACGIH - Not listed; OSHA - Not listed; EPA - Not listed; MAK - Not listed
Primary Target Organs:

Eyes

Skin

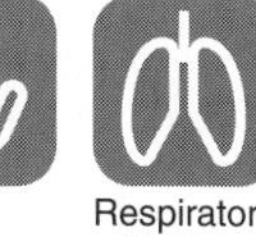
Respiratory System

Mucous Membranes

Blood

Environmental

Ecotoxicity: LD_{50} Aspergillus niger (fungus) 15 mg/l/40 hr (36 °C
Cleanup/Disposal: Guide No. 140: Keep combustibles (wood, paper, oil, etc.) away from spilled material. Do not touch damaged containers or spilled material unless wearing appropriate protective clothing. Stop leak if you can do it without risk. Do not get water inside containers. Small Dry Spills: With clean shovel place material into clean, dry container and cover loosely; move containers from spill area. Small Liquid Spills: Use a non-combustible material like vermiculite, sand or earth to soak up the product and place into a container for later disposal. Large Spills: Dike far ahead of liquid spill for later disposal. Following product recovery, flush area with water.

Regulations

RCRA 40CFR: Not listed
CERCLA: 40CFR 302.4: Not listed
SARA 40CFR 372.65: Listed
SARA EHS 40CFR 355: Not listed
TSCA: Listed

AMM5640 **CAS #: 9051-57-4**

AMMONIUM NONOXYNOL-4-SULFATE

RTECS: TR1583600
Molecular Formula: $C_{(2x+15)}H_{(4x+27)}NO_{(x+4)}S$
Formula Weight: Varies
Synonyms: ALIPAL CO 436; ALIPAL EP; ALIPAL EP 110; ALIPAL EP 120; CO 436; FENOPON CO 436; FENOPON EP 110; FENOPON EP 120; HITENOL N 093; NEWCOL 560SF; NIKKOL SNP; POLYETHYLENE GLYCOL NONYLPHENYL ETHER AMMONIUM BISULFATE; POLYETHYLENE GLYCOL NONYLPHENYL ETHER AMMONIUM SULFATE; POLYETHYLENE GLYCOL NONYLPHENYL ETHER SULFATE AMMONIUMSALT; POLY(OXY-1,2-ETHANEDIYL),ALPHA-SULFO-OMEGA-(NONYLPHENOXY)-,AMMONIUM SALT; POLY(OXYETHYLENE) NONYLPHENYL ETHER AMMONIUM SULFATE
Use: surfactant; surfactant for detergents & soaps; emulsifier for emulsion polymerization of vinyl acetate; antistatic agent for plastics & synthetic fibers

RTECS Toxicity Data

Acute Oral: Rat LD_{50} Dose: 8 gm/kg.

Hazard Overviews

Carcinogenicity: IARC - Not listed; NIOSH - Not listed; NTP - Not listed; ACGIH - Not listed; OSHA - Not listed; EPA - Not listed; MAK - Not listed

Environmental

Regulations

RCRA 40CFR: Not listed
CERCLA: 40CFR 302.4: Not listed

SARA 40CFR 372.65: Not listed
SARA EHS 40CFR 355: Not listed
TSCA: Listed

AMM5960 **CAS #: 1113-38-8**

AMMONIUM OXALATE

RTECS: RO2750000
DOT: NA2449
EINECS Number: 214-202-3
Molecular Formula: $C_2H_8N_2O_4$
Formula Weight: 124.10

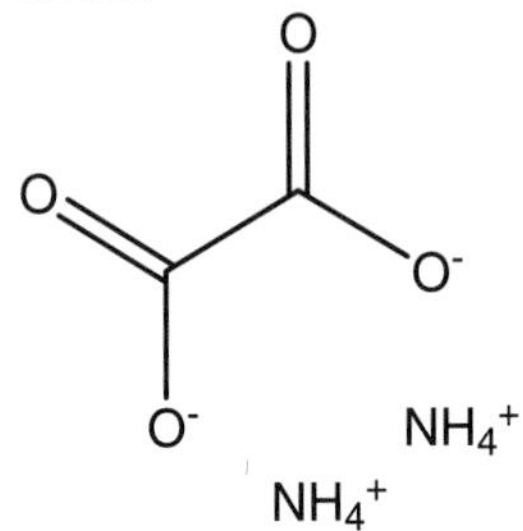

Chemical Structure

Synonyms: DIAMMONIUM OXALATE; ETHANEDIOIC ACID DIAMMONIUM SALT; ETHANEDIOIC ACID,DIAMMONIUM SALT; OXALIC ACID,DIAMMONIUM SALT
Description: colorless orthorhombic crystals, granules; odorless
Use: mfr explosives; electrolytic detinning of iron; in dyeing, metal polishes; for detection & determination of calcium, lead, & rare earth metals; analytical chemistry; mfr of oxalates; rust & scale removal

Physical Properties

Freezing Point: Decomposes
Specific Gravity: 1.502
Water Solubility: 3 g/100 cc
Other Solubilities: Slightly Soluble in Alcohol; Insoluble in Ammonia.
pH: 0.1 molar solution 6.4
Refraction Index: 1.439 to 1.594

Hazard Overviews

Poison

Fire Diamond

Health: Severely irritating to eyes/skin/respiratory tract. Poison. Also Causes: respiratory distress, intense burning in the throat/esophagus/stomach, severe vomiting (with evidence of blood), hypotension, weak/irregular pulse, cardiovascular collapse, headache, muscle cramps, convulsions, stupor, coma, oliguria (low urine output), hematuria (blood in urine), albuminuria (protein in urine). Excessive exposure can be fatal.

Fire: Will burn, but does not readily ignite. Use dry chemical, carbon dioxide, water spray or regular foam. Toxic nitrogen oxides and ammonia may be produced in a fire. Avoid breathing smoke and fumes.
Reactivity: Stable. Hazardous polymerization cannot occur. Avoid: heat; ignition sources. Incompatible with: sodium hypochlorite and ammonium acetate. Hazardous decomposition products: toxic fumes of nitrogen oxide and ammonia.
Carcinogenicity: IARC - Not listed; NIOSH - Not listed; NTP - Not listed; ACGIH - Not listed; OSHA - Not listed; EPA - Not listed; MAK - Not listed
Primary Target Organs:

Eyes

Skin

Respiratory System

Nervous System

Kidneys

Environmental

Cleanup/Disposal: Guide No. 154: Eliminate all ignition sources (no smoking, flares, sparks or flames in immediate area). Do not touch damaged containers or spilled material unless wearing appropriate protective clothing. Stop leak if you can do it without risk. Prevent entry into waterways, sewers, basements or confined areas. Absorb or cover with dry earth, sand or other non-combustible material and transfer to containers. Do not get water inside containers.

Regulations

RCRA 40CFR: Not listed
CERCLA: 40CFR 302.4: Not listed
SARA 40CFR 372.65: Not listed
SARA EHS 40CFR 355: Not listed
TSCA: Listed

AMM6600 **CAS #: 12007-89-5**

AMMONIUM PENTABORATE

EINECS Number: 234-521-1
Molecular Formula: $B_5H_4NO_8$
Structured MF: $NH_4B_5O_8$
Formula Weight: 272.14
Synonyms: AMMONIUM BORATE; AMMONIUM BORATE ((NH4)B5O8); AMMONIUM PENTABORATE ((NH4)B5O8); BORIC ACID (HB5O8),AMMONIUM SALT
Description: white powder; odorless
Use: intermediate for boron chemicals; as a power-level control in atomic submarines; as mild antiseptic or bacteriostat in eyewashes, mouthwashes, burn dressings, and diaper rash powders

Physical Properties

Specific Gravity: 1.58 at 15 °C
Water Solubility: 7.03 x10^4 ppm at 18 °C
Flash Point: Not flammable

Hazard Overviews

Health: Irritating to eyes/skin/respiratory tract. Other Acute Effects: may be harmful by inhalation, ingestion, or skin absorption.

Fire: Noncombustible. Hazards: emits toxic fumes. Extinguishing agents: water spray; carbon dioxide, dry chemical powder or appropriate foam. Precautions: combustible liquid.

Reactivity: Incompatible with: strong oxidizing agents. Hazardous decomposition products: toxic fumes of: carbon monoxide, carbon dioxide, nitrogen oxides, boron oxides; ammonia.

Carcinogenicity: IARC - Not listed; NIOSH - Not listed; NTP - Not listed; ACGIH - Not listed; OSHA - Not listed; EPA - Not listed; MAK - Not listed

Primary Target Organs:

Eyes

Skin

Respiratory System

Environmental

Regulations

RCRA 40CFR: Not listed
CERCLA: 40CFR 302.4: Not listed
SARA 40CFR 372.65: Not listed
SARA EHS 40CFR 355: Not listed
TSCA: Listed

AMM6760 CAS #: 7790-98-9

AMMONIUM PERCHLORATE

RTECS: SC7520000
DOT: UN1442
EINECS Number: 232-235-1
Molecular Formula: ClH_4NO_4
Structured MF: NH_4ClO_4
Formula Weight: 117.49

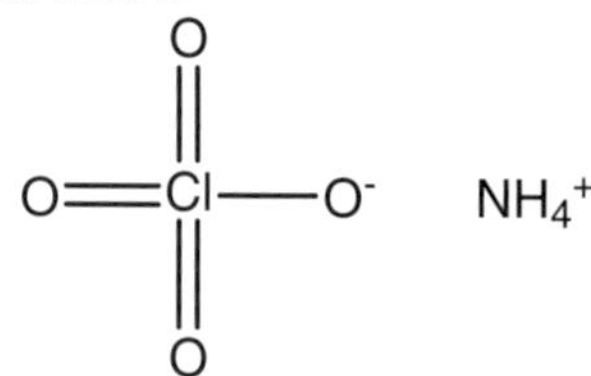

Chemical Structure

Synonyms: PERCHLORIC ACID,AMMONIUM SALT; PKHA
Description: white orthorhombic crystals; odorless
Use: analytical chemistry; etching agent; animal fattening agent; oxidizing agent in solid rocket propellants; component in explosive mixtures & pyrotechnics; chem int for alkali & alkaline metal perchlorates; engraving agent

Physical Properties

Freezing Point: Decomposes
Specific Gravity: 1.95
Water Solubility: 10.74 g/100 cc at 0 °C
Other Solubilities: Slightly Soluble in Ethanol; Soluble in Methanol; almost Insoluble in Ether; Ethyl Acetate.
Refraction Index: 1.482
Flash Point: Nonflammable

RTECS Toxicity Data

Acute Oral: Rat LD_{50} Dose: 4200 mg/kg; Toxic Effects: Behavioral - Convulsions or effect on seizure threshold; Behavioral - Ataxia; Behavioral - Coma. Mouse LD_{50} Dose: 1900 mg/kg; Toxic Effects: Behavioral - Convulsions or effect on seizure threshold; Behavioral - Ataxia; Behavioral - Coma.

Hazard Overviews

Explosive

Fire Diamond

Health: Severely irritating to eyes/skin/respiratory tract. Also Causes: pulmonary edema.

Fire: Explosive! Strong oxidizer. Fight fire with water only! Do not use dry chemical, carbon dioxide, or Halon. If material is a fine powder or if it is dry then it is a Class A Explosive and area should be evacuated if fire reaches cargo.

Reactivity: Unstable, shock-sensitive and self-reactive (if < 15 microns in diameter or a larger size if dry). Hazardous polymerization cannot occur. Avoid: exposure to heat; ignition sources; combustibles. Incompatible with: dicyclopentadienyliron (ferrocene); organic matter; sulfur; aluminum and other metals; nitrophenol and formaldehyde polymer; magnesium or lithium perchlorate (ignites at 240 °C and explodes at 290 °C); carbon (exothermic reaction at < 240 °C and mild explosion at > 240 °C); ethylene dinitrate (ignited at 60 °C after being mixed for 7 days); copper pipes; nitryl perchlorate; potassium periodate. Hazardous decomposition products: ammonia; hydrogen chloride; nitrogen oxide gases.

Carcinogenicity: IARC - Not listed; NIOSH - Not listed; NTP - Not listed; ACGIH - Not listed; OSHA - Not listed; EPA - Not listed; MAK - Not listed

Primary Target Organs:

Eyes

Skin

Respiratory System

Liver

Kidneys

Blood

Environmental

Cleanup/Disposal: Guide No. 143: Keep combustibles (wood, paper, oil, etc.) away from spilled material. Do not touch damaged containers or spilled material unless wearing appropriate protective clothing. Use water spray to reduce vapors or divert vapor cloud drift. Prevent entry into waterways, sewers, basements or confined areas. Small Spills: Flush area with flooding quantities of water. Large Spills: Do not clean-up or dispose of, except under supervision of a specialist.

Environmental Physical Data

BCF: no food chain concentration potential
BOD: none

Regulations

RCRA 40CFR: Not listed
CERCLA: 40CFR 302.4: Not listed
SARA 40CFR 372.65: Not listed
SARA EHS 40CFR 355: Not listed
TSCA: Listed

AMM6920 — CAS #: 7727-54-0

AMMONIUM PERSULFATE

RTECS: SE0350000
DOT: UN1444
EINECS Number: 231-786-5
Molecular Formula: $H_8N_2O_8S_2$
Structured MF: $(NH_4)_2S_2O_8$
Formula Weight: 228.20

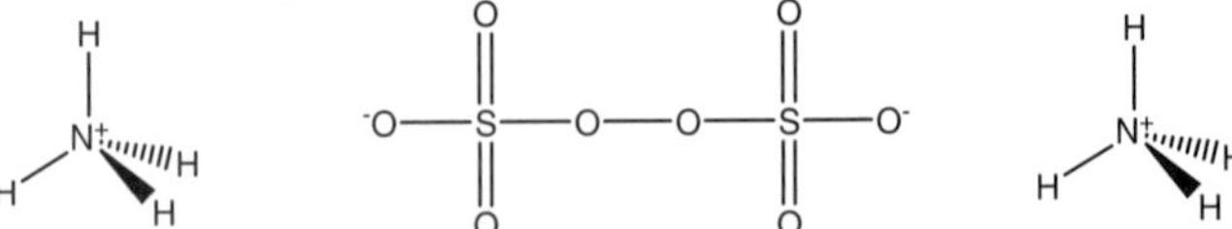

Chemical Structure

Synonyms: AMMONIUM PEROXYDISULFATE; DIAMMONIUM PEROXYDISULFATE; DIAMMONIUM PEROXYDISULPHATE; DIAMMONIUM PERSULFATE; PERSULFATE D'AMMONIUM
Description: colorless, light straw crystals, powder

Physical Properties

Boiling Point: Decomposes at 120 °C (248 °F) at 1 atm
Freezing Point: 120 °C (248 °F)
Specific Gravity: 1.98
Water Solubility: Soluble
pH: Acid
Flash Point: Nonflammable

RTECS Toxicity Data

Acute Oral: Rat LD_{50} Dose: 689 mg/kg.

Hazard Overviews

Fire Diamond

Health: Irritating to eyes/skin/respiratory tract. Also Causes: possible allergic reaction with tearing, difficulty breathing and life-threatening shock.

Fire: Noncombustible. However, it is a strong oxidizer capable of igniting combustible materials. Use flooding quantities of water to fight fire.
Reactivity: Stable, when pure and dry. Hazardous polymerization cannot occur. Incompatible with: sodium peroxide; iron; solutions of ammonia and silver salts; powdered aluminum and water; zinc and ammonia; sodium oxide; hydrogen. Hazardous decomposition products: nitrogen oxides; sulfur oxides; ammonia; sulfuric acid fumes.
Carcinogenicity: IARC - Not listed; NIOSH - Not listed; NTP - Not listed; ACGIH - Not listed; OSHA - Not listed; EPA - Not listed; MAK - Not listed
Primary Target Organs:

Eyes

Skin

Respiratory System

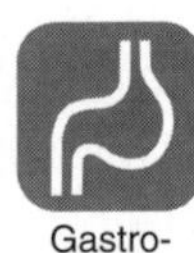
Gastro-intestinal

Exposure Limits
ACGIH TLV: TWA: 0.1 mg/m^3.

Environmental

Ecotoxicity: Aquatic toxicity: 120 ppm/48 hr/daphnia/TLm/fresh water
Cleanup/Disposal: Guide No. 140: Keep combustibles (wood, paper, oil, etc.) away from spilled material. Do not touch damaged containers or spilled material unless wearing appropriate protective clothing. Stop leak if you can do it without risk. Do not get water inside containers. Small Dry Spills: With clean shovel place material into clean, dry container and cover loosely; move containers from spill area. Small Liquid Spills: Use a non-combustible material like vermiculite, sand or earth to soak up the product and place into a container for later disposal. Large Spills: Dike far ahead of liquid spill for later disposal. Following product recovery, flush area with water.

Environmental Physical Data

BCF: no food chain concentration potential
BOD: none

Regulations

RCRA 40CFR: Not listed
CERCLA: 40CFR 302.4: Not listed
SARA 40CFR 372.65: Not listed
SARA EHS 40CFR 355: Not listed
TSCA: Listed

AMM7080 — CAS #: 7722-76-1

AMMONIUM PHOSPHATE

EINECS Number: 231-764-5
Molecular Formula: H_6NO_4P
Formula Weight: 115.03

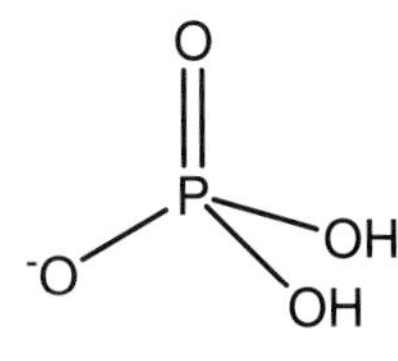

Chemical Structure

Synonyms: AMMONIUM ACID PHOSPHATE; AMMONIUM BIPHOSPHATE; AMMONIUM DIACID PHOSPHATE; AMMONIUM DIHYDROGEN ORTHOPHOSPHATE; AMMONIUM DIHYDROGEN PHOSPHATE ((NH4)H2PO4); AMMONIUM DIHYDROPHOSPHATE; AMMONIUM MONOBASIC PHOSPHATE; AMMONIUM ORTHOPHOSPHATE DIHYDROGEN; AMMONIUM PHOSPHATE ((NH4)H2PO4); AMMONIUM PHOSPHATE,MONOBASIC; AMMONIUM PRIMARY PHOSPHATE; DIHYDROGEN AMMONIUM PHOSPHATE; MONOAMMONIUM ACID PHOSPHATE; MONOAMMONIUM DIHYDROGEN ORTHOPHOSPHATE; MONOAMMONIUM DIHYDROGEN PHOSPHATE; MONOAMMONIUM DIHYDROGEN PHOSPHATE ((NH4)H2PO4); MONOAMMONIUM HYDROGEN PHOSPHATE; MONOAMMONIUM ORTHOPHOSPHATE; MONOBASIC AMMONIUM PHOSPHATE; PHOSPHORIC ACID,MONOAMMONIUM SALT; PRIMARY AMMONIUM PHOSPHATE

Description: white powder, crystals; weak ammonia odor

Use: as baking powder with sodium bicarbonate; in fermentations (yeast cultures, etc); fireproofing of paper, wood, fiberboard; fertilizer; to prevent afterglow in matches; plant nutrient soln; mfr of bread improvers; analytical chemistry; to protect pesticides in spray mixtures prepared with alkaline waters

Physical Properties

Freezing Point: 190 °C (374 °F)
Specific Gravity: 1.803 at 19 °C
Water Solubility: 22.7 g/100 cc at 0 °C
Other Solubilities: Slightly Soluble in Alcohol; Practically Insoluble in Acetone.
pH: 0.2 molar aqueous solution 4.2
Refraction Index: 1.525

Hazard Overviews

Fire Diamond

Health: Irritating to eyes/skin/respiratory tract. Also Causes: nausea, diarrhea, mild dermal irritation.
Fire: Noncombustible. Use agent suitable for surrounding fire.
Reactivity: Stable. Hazardous polymerization cannot occur. Incompatible with: sodium hypochlorite. Hazardous decomposition products: oxides of phosphorus fumes; nitrogen oxide fumes; ammonia.
Carcinogenicity: IARC - Not listed; NIOSH - Not listed; NTP - Not listed; ACGIH - Not listed; OSHA - Not listed; EPA - Not listed; MAK - Not listed

Primary Target Organs:

Eyes

Skin

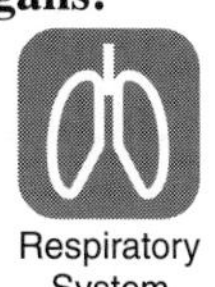
Respiratory System

Environmental

Regulations

RCRA 40CFR: Not listed
CERCLA: 40CFR 302.4: Not listed
SARA 40CFR 372.65: Not listed
SARA EHS 40CFR 355: Not listed
TSCA: Listed

AMM7240 CAS #: 131-74-8

AMMONIUM PICRATE

RTECS: BS3855000
DOT: UN0004; UN1310; IMO1.1; IMO4.1
EINECS Number: 205-038-3
Molecular Formula: $C_6H_6N_4O_7$
Structured MF: $NH_4C_6H_2N_3O_7$
Formula Weight: 246.14
Synonyms: AMMONIUM CARBAZOATE; AMMONIUM PICRONITRATE; EXPLOSIVE D; OBELINE PICRATE; PHENOL,2,4,6-TRINITRO-,AMMONIUM SALT; PHENOL,2,4,6-TRINITRO-,AMMONIUM SALT (9CI); PICRATOL; PICRIC ACID,AMMONIUM SALT; 2,4,6-TRINITROPHENOL AMMONIUM SALT
Description: bright yellow scales or orthorhombic crystals
Use: explosives; in fireworks, rocket propellants; pyrotechnics, explosive compositions

Physical Properties

Boiling Point: Decomposes
Freezing Point: Decomposes
Specific Gravity: 1.72
Water Solubility: 1 g/100 ml at20 °C
Other Solubilities: Slightly Soluble in Alcohol
Flash Point: Flammable solid

Hazard Overviews

Flammable

Explosive

Fire: Flammable.
Carcinogenicity: IARC - Not listed; NIOSH - Not listed; NTP - Not listed; ACGIH - Not listed; OSHA - Not listed; EPA - Not listed; MAK - Not listed

Environmental

Ecotoxicity: LC_{50} Lepomis macrochirus 220 ppm /96 hr static bioassay in fresh water at 23 °C, mild aeration applied after 24 hr LC_{50} Menidia beryllina 66 ppm /96 hr/ static bioassay in synthetic seawater at 23 °C, mild aeration applied after 24 hr

Cleanup/Disposal: Guide No. 113: Eliminate all ignition sources (no smoking, flares, sparks or flames in immediate area). All equipment used when handling the product must be grounded. Do not touch or walk through spilled material. Small Spills: Flush area with flooding quantities of water. Large Spills: Wet down with water and dike for later disposal. Keep wetted product wet by slowly adding flooding quantities of water.

Regulations

RCRA 40CFR: Listed Hazardous Waste No. P009 Reactive Waste
CERCLA: 40CFR 302.4: Listed per RCRA Section 3001 RQ: 10 lb (4.535 kg)
SARA 40CFR 372.65: Not listed
SARA EHS 40CFR 355: Not listed
TSCA: Listed

AMM7560 CAS #: 16919-19-0

AMMONIUM SILICOFLUORIDE

RTECS: VV7800000
DOT: UN2854; IMO6.1
EINECS Number: 240-968-3
Molecular Formula: $F_6H_8N_2Si$
Structured MF: $(NH_4)_2SIF_6$
Formula Weight: 178.15

Chemical Structure

Synonyms: AMMONIUM FLUOROSILICATE; AMMONIUM FLUOSILICATE; AMMONIUM HEXAFLUOROSILICATE; AMMONIUM SILICON FLUORIDE; BYE BUGS; (COMPONENT OF) DRIANONE; (COMPONENT OF) DRI-DIE; DIAMMONIUM FLUOROSILICATE; DIAMMONIUM FLUOSILICATE; DIAMMONIUM FLUOSILICATE ((NH4)2SIF6); DIAMMONIUM HEXAFLUOROSILICATE; DIAMMONIUM HEXAFLUOROSILICATE(2-); DIAMMONIUM SILICON HEXAFLUORIDE; EPA PESTICIDE CHEMICAL CODE 075301; FLUOROSILICIC ACID,AMMONIUM SALT; LAIDLAW U-SAN-O MOTH PROOFING SPRAY; SILICATE(2-),HEXAFLUORO-,DIAMMONIUM; SUPERIOR DRI-DIE
Description: white, crystalline powder; odorless
Use: in pesticides; formerly insecticide dri-die; wood preservative; in prophylactic dental prepn; in light metal casting; repellent or feeding depressant for carpet beetles & clothes moths; insecticide & miticide (eg, for lice & fleas on dogs & cats), silverfish, ants, german cockroaches, oriental cockroaches, & drywood termites (bye bugs); in soldering flux.

Physical Properties

Freezing Point: Decomposes
Specific Gravity: 2 at 20 °C (solid)
Water Solubility: 1.86×10^5 ppm at 17 °C
Other Solubilities: Slightly Soluble in Alcohol; Insoluble in Acetone
Refraction Index: Alpha 1.3696
Flash Point: Nonflammable

RTECS Toxicity Data

Acute Oral: Rat LD_{Lo} Dose: 100 mg/kg. Mouse LD_{50} Dose: 70 mg/kg; Toxic Effects: Peripheral nerve and sensation - Flaccid paralysis without anesthesia; Behavioral - Ataxia; Behavioral - Muscle contraction or spasticity.
Acute Dermal: Frog LD_{Lo} Route: Subcutaneous Dose: 224 mg/kg.

Hazard Overviews

Corrosive

Health: Corrosive to eyes/skin/respiratory tract. Toxic. Other Acute Effects: harmful if swallowed, inhaled, or absorbed through skin; inhalation may result in spasm, inflammation and edema of the larynx and bronchi, chemical pneumonitis and pulmonary edema; symptoms of exposure may include burning sensation; coughing; wheezing; laryngitis; shortness of breath; headache; nausea; vomiting.
Fire: Noncombustible. Hazards: emits toxic fumes. Extinguishing agents: water spray; carbon dioxide, dry chemical powder or appropriate foam. Precautions: combustible liquid.
Reactivity: Incompatible with: strong oxidizing agents, strong acids, do not store in glass. Hazardous decomposition products: nitrogen oxides, ammonia, hydrogen fluoride, silicon oxide.
Carcinogenicity: IARC - Not listed; NIOSH - Not listed; NTP - Not listed; ACGIH - Class A4, Not classifiable as a human carcinogen; OSHA - Not listed; EPA - Not listed; MAK - Not listed
Primary Target Organs:

Eyes

Skin

Respiratory System

Exposure Limits
OSHA PEL: TWA: 2.5 mg/m^3; as F.
ACGIH TLV: TWA: 2.5 mg/m^3; as F.
NIOSH REL: TWA: 2.5 mg/m^3; as F (Flourides).
DFG MAK: TWA: 2.5 mg/m^3; as F.

Environmental

Ecotoxicity: LD_{50} Silkworm larvae oral greater than 10 ppm
Cleanup/Disposal: Guide No. 151: Do not touch damaged containers or spilled material unless wearing appropriate

protective clothing. Stop leak if you can do it without risk. Prevent entry into waterways, sewers, basements or confined areas. Cover with plastic sheet to prevent spreading. Absorb or cover with dry earth, sand or other non-combustible material and transfer to containers. Do not get water inside containers.

Environmental Physical Data

BCF: no food chain concentration potential

Regulations

RCRA 40CFR: Not listed
CERCLA: 40CFR 302.4: Listed per CWA Section 311(b)(4) RQ: 1000 lb (453.5 kg)
SARA 40CFR 372.65: Not listed
SARA EHS 40CFR 355: Not listed
TSCA: Listed

AMM7720 CAS #: 7773-06-0

AMMONIUM SULFAMATE

RTECS: WO6125000
DOT: NA9089
EINECS Number: 231-871-7
Molecular Formula: $H_6N_2O_3S$
Structured MF: $NH_4NH_2SO_3$
Formula Weight: 114.13

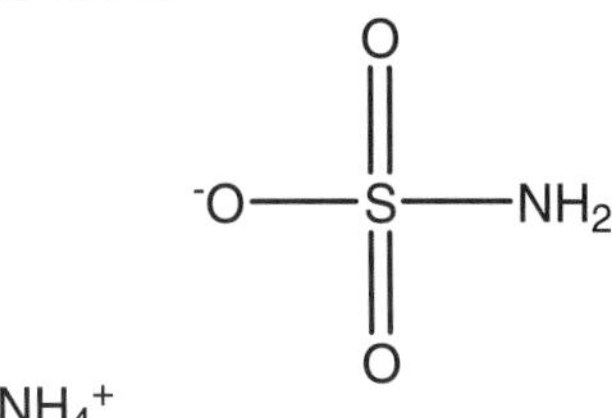

Chemical Structure

Synonyms: AMICIDE; AMMAT; AMMATE; AMMATE HERBICIDE; AMMATE X; AMMONIUM AMIDOSULFATE; AMMONIUM AMIDOSULFONATE; AMMONIUM AMIDOSULPHATE; AMMONIUM AMINOSULFONATE; AMMONIUM SULPHAMATE; AMMONIUM SULPHAMIDATE; AMMONIUMSALZ DER AMIDOSULFONSAURE; AMS; FELIDERM K; FYRAN J 3; FYRAN 200 K; IKURIN; MONOAMMONIUM SALT OF SULFAMIC ACID; MONOAMMONIUM SULFAMATE; SULFAMATE; SULFAMATE D'AMMONIUM; SULFAMIC ACID MONOAMMONIUM SALT; SULFAMIC ACID,MONOAMMONIUM SALT; SULFAMINSAURE
Description: colorless, white crystalline substance; odorless
Use: contact & translocated herbicide for woody plants, eg, alder, ash, herbaceous perennials eg, leafy spurge, bitter dock; in areas adjacent to cotton, & other plants susceptible to phenoxy cmpd; for weed control on power line rights-of-ways; noncrop areas, eg, driveways; soil treatment for weed control on fruit & ornamental trees; for flame-proofing textiles, wood & paper products; for generation of nitrous oxide gas; in electroplating soln; catalyst for acetylation of cellulose; scale removing & chem cleaning; to reduce tumor hazard of tobacco smoke;.

Physical Properties

Boiling Point: Decomposes at 160 °C (320 °F)
Freezing Point: 131 °C (267.8 °F)
Specific Gravity: Solid > 1 at 20 °C
Vapor Pressure: ~ 0 mm Hg at 20 °C
Water Solubility: 166.6 g/100 ml Water at 10 °C
Other Solubilities: Slightly Soluble in Ethanol; moderately Soluble in Glycerol, Glycol, & Formamide
pH: 0.2 molar solution in water 4.9
Flash Point: Nonflammable

RTECS Toxicity Data

Acute Oral: Rat LD_{50} Dose: 2 gm/kg. Mouse LD_{50} Dose: 3100 mg/kg; Toxic Effects: Behavioral - General anesthetic; Behavioral - Convulsions or effect on seizure threshold; Lungs, Thorax, or Respiration - Dyspnea.

Hazard Overviews

Fire Diamond

Health: Mildly irritating to eyes/respiratory tract.
Fire: Noncombustible. However, it is a strong oxidizer capable of igniting combustibles. Use agent suitable for surrounding fire.
Reactivity: Stable. Hazardous polymerization cannot occur. Avoid: heat. Incompatible with: hot acid solutions; corrosive to metals. Hazardous decomposition products: ammonia; oxides of sulfur and nitrogen.
Carcinogenicity: IARC - Not listed; NIOSH - Not listed; NTP - Not listed; ACGIH - Not listed; OSHA - Not listed; EPA - Not listed; MAK - Not listed
Primary Target Organs:

Eyes

Skin

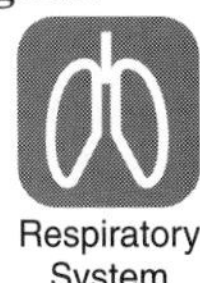
Respiratory System

Exposure Limits
OSHA PEL: TWA: 15 mg/m^3; total dust.
OSHA PEL Vacated 1989 Limits: TWA: 10 mg/m^3. Other Values: respirable mg/m^3; 5.
ACGIH TLV: TWA: 10 mg/m^3.
NIOSH REL: TWA: 10 mg/m^3.
NIOSH IDLH: 1500 mg/m^3.
DFG MAK: TWA: 15 mg/m^3.
Respirator Recommendation
Exposure Range: >5 to 50 mg/m^3 Air Purifying, Negative Pressure, Half Mask
Exposure Range: >50 to 500 mg/m^3 Air Purifying, Negative Pressure, Full Face
Exposure Range: >500 to 5000 mg/m^3 Supplied Air, Constant Flow/Pressure Demand, Full Face
Exposure Range: >5000 to unlimited mg/m^3 Self-contained Breathing Apparatus, Pressure Demand, Full Face

Cartridge Color: dust/mist filter (use P100 or consult supervisor for appropriate dust/mist filter)

Environmental

Ecotoxicity: Aquatic toxicity: 259 ppm/24 hr/catfish/LC_{50}/fresh water

Cleanup/Disposal: Guide No. 171: Do not touch or walk through spilled material. Stop leak if you can do it without risk. Prevent dust cloud. Avoid inhalation of asbestos dust. Small Dry Spills: With clean shovel place material into clean, dry container and cover loosely; move containers from spill area. Small Spills: Take up with sand or other noncombustible absorbent material and place into containers for later disposal. Large Spills: Dike far ahead of liquid spill for later disposal. Cover powder spill with plastic sheet or tarp to minimize spreading. Prevent entry into waterways, sewers, basements or confined areas.

Environmental Physical Data

BCF: none

Regulations

RCRA 40CFR: Not listed

CERCLA: 40CFR 302.4: Listed per CWA Section 311(b)(4) RQ: 5000 lb (2268 kg)

SARA 40CFR 372.65: Not listed

SARA EHS 40CFR 355: Not listed

TSCA: Listed

Analytical Methods

Soil: EPA PMD-AM-S

AMM7880 CAS #: 7783-20-2

AMMONIUM SULFATE

RTECS: BS4500000

DOT: UN2506; IMO8.0

EINECS Number: 231-984-1

Molecular Formula: $H_8N_2O_4S$

Structured MF: $(NH_4)_2SO_4$

Formula Weight: 132.14

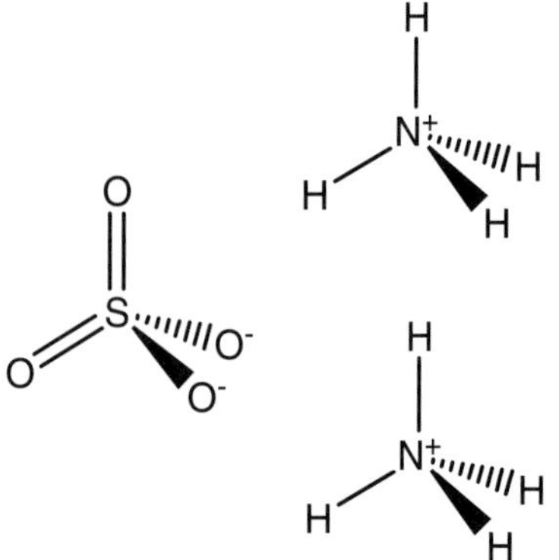

Chemical Structure

Synonyms: ACTAMASTER; AMMONIUM HYDROGEN SULFATE; AMMONIUM SULFATE (2:1); AMMONIUM SULFATE (SOLUTION); AMMONIUM SULPHATE; DIAMMONIUM SULFATE; DIAMMONIUM SULPHATE; DOLAMIN; EPA PESTICIDE CHEMICAL CODE 005601; MASCAGNITE; NSC 77671; SULFATOM AMMONIYA; SULFURIC ACID,DIAMMONIUM SALT

Description: white to brownish-gray crystals; odorless

Use: in mfr of ammonia alum, hydrogen sulfide, viscose silk; in anal chemistry; freezing mixtures; flameproofing fabrics & paper; tanning; galvanizing iron; in fractionation of proteins; in water treatment; food additive; nitrogen source in fertilizers; dry chem fire extinguisher agent; for shale stabilization drilling fluids; additive to supply nitrogen in fermentation process.

Physical Properties

Boiling Point: Decomposes > 280 °C (536 °F)

Freezing Point: Decomposes at > 280 °C (536 °F)

Specific Gravity: 1.769 at 50 °C

Water Solubility: 41.2 g/100g at 0 °C

Other Solubilities: Insoluble in Acetone, Alcohol, Ammonia.

Odor Threshold: 46.8 ppm

pH: 0.1 molar aqueous solution 5.5

Refraction Index: 1.521

Evaporation Rate: < 1 Butyl Acetate=1

Flash Point: Nonflammable

RTECS Toxicity Data

Acute Oral: Man TD_{Lo} Dose: 1500 mg/kg; Toxic Effects: Gastrointestinal - Hypermotility, diarrhea; Gastrointestinal - Nausea or vomiting; Gastrointestinal - Other changes. Rat LD_{50} Dose: 2840 mg/kg; Toxic Effects: Behavioral - Somnolence (general depressed activity); Behavioral - Muscle contraction or spasticity; Lungs, Thorax, or Respiration - Other changes.

Hazard Overviews

Fire Diamond

Health: Irritating to eyes/respiratory tract. Also Causes: nausea, vomiting, diarrhea, or increased urination.

Fire: Noncombustible. Use agent suitable for surrounding fire.

Reactivity: Stable. Hazardous polymerization cannot occur. Avoid: dispersion of particulates/dusts into air. Incompatible with: chlorate and heat; sodium hypochlorite; potassium and ammonium nitrate; potassium nitrite; sodium-potassium alloy and ammonium nitrate; strong oxidizers. Hazardous decomposition products: nitrogen oxide(s); sulfur oxide(s); ammonia.

Carcinogenicity: IARC - Not listed; NIOSH - Not listed; NTP - Not listed; ACGIH - Not listed; OSHA - Not listed; EPA - Not listed; MAK - Not listed

Primary Target Organs:

Environmental

Ecotoxicity: TLm Daphnia magna 423 mg/l/24 hr /Conditions of bioassay not specified

Cleanup/Disposal: Guide No. 154: Eliminate all ignition sources (no smoking, flares, sparks or flames in immediate area). Do not touch damaged containers or spilled material unless wearing appropriate protective clothing. Stop leak if you can do it without risk. Prevent entry into waterways, sewers, basements or confined areas. Absorb or cover with dry earth, sand or other non-combustible material and transfer to containers. Do not get water inside containers.

Environmental Physical Data

Octanol/Water Partition Coefficient: log K_{ow} = measured at -5.1

BCF: no food chain concentration potential

BOD: none

Regulations

RCRA 40CFR: Not listed

CERCLA: 40CFR 302.4: Not listed

SARA 40CFR 372.65: Listed

SARA EHS 40CFR 355: Not listed

TSCA: Listed

AMM8040 **CAS #: 12124-99-1**

AMMONIUM SULFIDE

RTECS: BS4900000

EINECS Number: 235-184-3

Molecular Formula: H_5NS

Structured MF: $(NH_4)_2SNH_4SH \cdot H_2O$

Formula Weight: 51.12

Synonyms: AMMONIUM BISULFIDE; AMMONIUM HYDROGEN SULFIDE; AMMONIUM HYDROSULFIDE; AMMONIUM MERCAPTAN; AMMONIUM SULFHYDRATE; MONOAMMONIUM SULFIDE; SIRNIK AMONNY; TRUE AMMONIUM SULFIDE

Description: yellow hygroscopic crystals; slight rotten egg odor

Physical Properties

Boiling Point: Decomposes

Freezing Point: 118 °C (244.4 °F)

Specific Gravity: 1.17

Vapor Pressure: 395 mm Hg

Water Solubility: Reacts

Other Solubilities: Insoluble in ether, benzene

pH: > 7

Flash Point: 22 °C Closed Cup

LEL: 4% v/v

UEL: Hydrogen Sulfide 46% v/v

RTECS Toxicity Data

Acute Oral: Rat LD_{50} Dose: 168 mg/kg. Mouse LD_{Lo} Dose: 80 mg/kg; Toxic Effects: Sense organs and special senses - Mydriasis (pupilliary dilation); Lungs, Thorax, or Respiration - Dyspnea; Lungs, Thorax, or Respiration - Respiratory stimulation.

Acute Dermal: Rabbit LD_{50} Route: Skin; Dose: 1682 mg/kg. Rabbit LD_{Lo} Route: Subcutaneous Dose: 7500 ug/kg; Toxic Effects: Sense organs and special senses - Mydriasis (pupilliary dilation); Lungs, Thorax, or Respiration - Dyspnea; Lungs, Thorax, or Respiration - Respiratory stimulation.

Hazard Overviews

Health: Irritating to eyes/skin/respiratory tract. Toxic. Also Causes: cyanosis, respiratory depression, vomiting, diarrhea. Forms poisonous hydrogen sulfide gas on contact with water.

Fire: Combustible. Use carbon dioxide, dry chemical, or foams to put out ammonium sulfide fires. Never use water since this can generate poisonous hydrogen sulfide gas.

Reactivity: Stable. Hazardous polymerization cannot occur. Avoid: exposure to any other substance unless chemical compatibility has been conclusively established. Incompatible with: acids or acidic vapors. Hazardous decomposition products: oxides of sulfur; ammonia; oxides of nitrogen; sulfides (hydrogen sulfide; polysulfides).

Carcinogenicity: IARC - Not listed; NIOSH - Not listed; NTP - Not listed; ACGIH - Not listed; OSHA - Not listed; EPA - Not listed; MAK - Not listed

Primary Target Organs:

Environmental

Ecotoxicity: Aquatic toxicity: 100 ppm/72 hr/goldfish/killed/fresh water; 248 ppm/48 hr/mosquito fish/TLm/fresh water

Cleanup/Disposal: Guide No. 132: Fully encapsulating, vapor protective clothing should be worn for spills and leaks with no fire. Eliminate all ignition sources (no smoking, flares, sparks or flames in immediate area). All equipment used when handling the product must be grounded. Do not touch or walk through spilled material. Stop leak if you can do it without risk. Prevent entry into waterways, sewers, basements or confined areas. A vapor suppressing foam may be used to reduce vapors. Absorb with earth, sand or other non-combustible material and transfer to containers (except for Hydrazine). Use clean non-sparking tools to collect absorbed material. Large Spills: Dike far ahead of liquid spill for later disposal. Water spray may reduce vapor; but may not prevent ignition in closed spaces.

Regulations

RCRA 40CFR: Not listed
CERCLA: 40CFR 302.4: Not listed
SARA 40CFR 372.65: Not listed
SARA EHS 40CFR 355: Not listed
TSCA: Listed

AMM8200 CAS #: 10196-04-0

AMMONIUM SULFITE

EINECS Number: 233-484-9
Molecular Formula: $H_8N_2O_3S$
Structured MF: $(NH_4)_2SO_3$
Formula Weight: 116.14
Synonyms: DIAMMONIUM SULFITE; SULFUROUS ACID,DIAMMONIUM SALT
Description: white, solid; odorless
Use: in prodn of caramel; in photography; as reducing agent; in bricks for blast furnace linings; in lubricants for metal cold-working; medicine; permanent wave solns; production of ammonium thiosulfite & ammonium chloride; wood pulping

Physical Properties

Freezing Point: Sublimes at 150 °C (302 °F)
Specific Gravity: 1.41
Water Solubility: Soluble in Water
Other Solubilities: Slightly Soluble in Alcohol; Insoluble in Acetone.
pH: Aqueous solution is alkaline
Refraction Index: 1.515
Flash Point: Nonflammable

Hazard Overviews

Fire: Noncombustible.
Carcinogenicity: IARC - Not listed; NIOSH - Not listed; NTP - Not listed; ACGIH - Not listed; OSHA - Not listed; EPA - Not listed; MAK - Not listed

Environmental

Ecotoxicity: TLm Gambusia affinis (mosquito fish) 240 ppm/48 hr fresh water. /Conditions of bioassay not specified TLm Daphnia magna (water flea) 203 mg/l/100 hr. /Conditions of bioassay not specified
Cleanup/Disposal: Guide No. 157: Eliminate all ignition sources (no smoking, flares, sparks or flames in immediate area). All equipment used when handling the product must be grounded. Do not touch damaged containers or spilled material unless wearing appropriate protective clothing. Stop leak if you can do it without risk. A vapor suppressing foam may be used to reduce vapors. Do not get water inside containers. Use water spray to reduce vapors or divert vapor cloud drift. Prevent entry into waterways, sewers, basements or confined areas. Small Spills: Cover with dry earth, dry sand, or other non-combustible material followed with plastic sheet to minimize spreading or contact with rain. Use clean non-sparking tools to collect material and place it into loosely covered plastic containers for later disposal.

Environmental Physical Data

BCF: no food chain concentration potential

Regulations

RCRA 40CFR: Not listed
CERCLA: 40CFR 302.4: Listed per CWA Section 311(b)(4) RQ: 5000 lb (2268 kg)
SARA 40CFR 372.65: Not listed
SARA EHS 40CFR 355: Not listed
TSCA: Listed

AMM8360 CAS #: 14307-43-8

AMMONIUM TARTRATE

EINECS Number: 238-245-2
Molecular Formula: $C_4H_{12}N_2O_6$
Structured MF: $C_4H_{12}N_2O_6$
Formula Weight: 184
Description: white solid; odorless

Physical Properties

Boiling Point: Decomposes at 1 atm
Specific Gravity: 1.601
Water Solubility: 5.81 x10^5 ppm at 15 °C
Flash Point: Not pertinent (combustible solid)

Hazard Overviews

Fire: Combustible.
Carcinogenicity: IARC - Not listed; NIOSH - Not listed; NTP - Not listed; ACGIH - Not listed; OSHA - Not listed; EPA - Not listed; MAK - Not listed

Environmental

Cleanup/Disposal: Guide No. 154: Eliminate all ignition sources (no smoking, flares, sparks or flames in immediate area). Do not touch damaged containers or spilled material unless wearing appropriate protective clothing. Stop leak if you can do it without risk. Prevent entry into waterways, sewers, basements or confined areas. Absorb or cover with dry earth, sand or other non-combustible material and transfer to containers. Do not get water inside containers.

Environmental Physical Data

BCF: no food chain concentration potential
BOD: sewage 0.30 lb/lb, 5 days

Regulations

RCRA 40CFR: Not listed
CERCLA: 40CFR 302.4: Not listed
SARA 40CFR 372.65: Not listed
SARA EHS 40CFR 355: Not listed
TSCA: Not listed

AMM8680 **CAS #: 1762-95-4**

AMMONIUM THIOCYANATE

RTECS: XK7875000
DOT: NA9092
EINECS Number: 217-175-6
Molecular Formula: CH_4N_2S
Structured MF: NH_4SCN
Formula Weight: 76.12

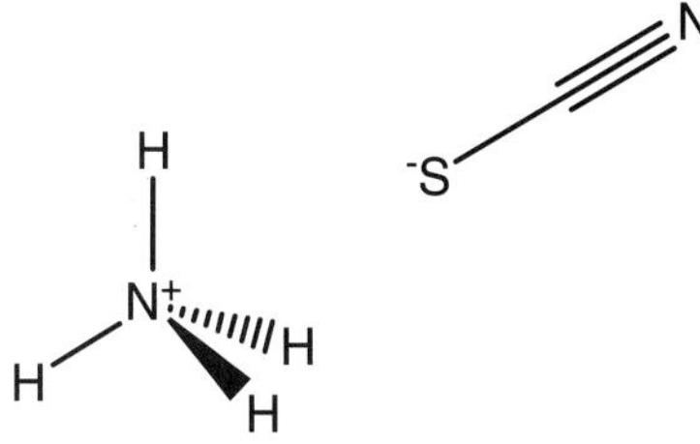

Chemical Structure

Synonyms: TRANS-AID; AMMONIUM RHODANATE; AMMONIUM RHODANIDE; AMMONIUM SULFOCYANATE; AMMONIUM SULFOCYANIDE; AMMONIUMTHIOCYANATE; AMTHIO; RHODANID; RHODANIDE; THIOCYANIC ACID,AMMONIUM SALT; WEEDAZOL TL
Description: colorless deliquescent crystals; odorless
Use: in matches; double-dyeing fabrics; photography; incr strength of weighted silks; in producing grayish-black coating on zinc; mfr transparent artificial resins, thiourea; in pesticides; in anal detection of metals, fluorine, iodoform, food preservatives; fertilizers; of freezing soln, esp liq rocket propellants; adhesives; in curing resins; pickling iron & steel; electroplating; temporary soil sterilizer; polymerization catalyst; weed killer & defoliant; separator of zirconium & hafnium, of gold & iron; stabilizer in glue formations; tracer in oil fields; adjuvant in textile dyeing & printing; in antibiotic fermentations.

Physical Properties

Boiling Point: Decomposes at 170 °C (338 °F)
Freezing Point: 149.6 °C (301.28 °F)
Density: 1.3057 g/mL
Water Solubility: 128 g/100 cc Water at 0 °C
Other Solubilities: Practically Insoluble in Chloroform, Ethyl Acetate.
Flash Point: Solid may be combustible; solution is not

RTECS Toxicity Data

Acute Oral: Human TD_{Lo} Dose: 430 mg/kg; Toxic Effects: Behavioral - Hallucinations, distorted perceptions; Gastrointestinal - Nausea or vomiting; Gastrointestinal - Other changes. Rat LD_{50} Dose: 750 mg/kg.
Acute Inhalation: Mammal LC; Dose: >100 mg/m^3.

Hazard Overviews

Fire Diamond

Health: Irritating to eyes/skin/respiratory tract. Also Causes: vomiting, agitation, delerium, convulsions, muscle spasms, death, loss of kidney function. Chronic Effects: runny eyes, rashes, weakness, nausea, diarrhea, confusion.
Fire: Use carbon dioxide, dry chemical, foams, or water spray to put out ammonium thiocyanate fires. Fire conditions may decompose the ammonium thiocyanate to form toxic gases more hazardous than ammonium thiocyanate itself.
Reactivity: Stable. Hazardous polymerization cannot occur. Incompatible with: lead nitrate; chlorates; nitrates; nitric acid; organic peroxides; oxidizing agents; peroxides; potassium chlorate; sodium chlorate. Hazardous decomposition products: ammonia; oxides of sulfur; oxides of nitrogen; thiocyanates; cyanide compounds.
Carcinogenicity: IARC - Not listed; NIOSH - Not listed; NTP - Not listed; ACGIH - Not listed; OSHA - Not listed; EPA - Not listed; MAK - Not listed
Primary Target Organs:

Eyes

Skin

Respiratory System

Nervous System

Environmental

Ecotoxicity: Aquatic toxicity: Sunfish 280-300 ppm/1 hr killed fresh water /Conditions of bioassay not specified

Environmental Physical Data

BCF: none
BOD: < 0.010 lb/lb, 5 days

Regulations

RCRA 40CFR: Not listed
CERCLA: 40CFR 302.4: Listed per CWA Section 311(b)(4) RQ: 5000 lb (2268 kg)
SARA 40CFR 372.65: Not listed
SARA EHS 40CFR 355: Not listed
TSCA: Listed

AMM8840 **CAS #: 7783-18-8**

AMMONIUM THIOSULFATE

RTECS: XN6465000
EINECS Number: 231-982-0
Molecular Formula: $H_8N_2O_3S_2$
Formula Weight: 148.20

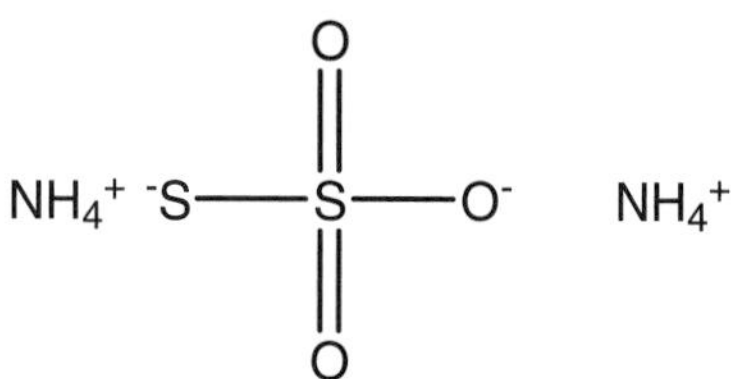

Chemical Structure

Synonyms: AMMO HYPO; AMMONIUM HYPOSULFITE; AMMONIUM THIOSULFATE ((NH4)2S2O3); AMTHIO; DIAMMONIUM SALT; DIAMMONIUM THIOSULFATE; THIO-SUL; THIOSULFURIC ACID,DIAMMONIUM SALT

Description: colorless to white, crystals; ammonia odor

Use: to clean "white" metal; photographic fixing agent especially for rapid development; in lubricants for metal cold-working; analytical reagent; fungicide; reducing agent; brightener in silver plating baths; cleaning compounds for zinc-base die-cast metals; hair waving preparations; fog screens; fertilizers; in desiccants and defoliants; flue-gas desulfurization; removal of nitrogen oxides and sulfur dioxide from flue gases; converting sulfur in hydrocarbons to a water-soluble form; and converting cellulose to hydrocarbons

Physical Properties

Boiling Point: Decomposes
Freezing Point: Decomposes at 150 °C (302 °F)
Specific Gravity: 1.679
Vapor Density: < 1 Air=1
Density: 1.679 g/cu cm
Vapor Pressure: 0.00 mm Hg
Water Solubility: Very Soluble in Cold Water
Other Solubilities: 95% Ethanol: Insoluble; Acetone: Slightly Soluble; Ether: Insoluble.
pH: 60% solution 6.5 to 7
Flash Point: Nonflammable

RTECS Toxicity Data

Acute Oral: Rat LD_{50} Dose: 2890 mg/kg; Toxic Effects: Lungs, Thorax, or Respiration - Emphysema; Kidney, Ureter, and Bladder - Chgs in tubules (inc acute renal failure, acute tubular necrosis; Blood - Hemorrhage. Mouse LD_{50} Dose: 2100 mg/kg; Toxic Effects: Behavioral - Somnolence (general depressed activity); Behavioral - Convulsions or effect on seizure threshold; Behavioral - Ataxia.

Acute Inhalation: Rat LC; Dose: >2260 mg/m^3/4hr. Mouse LC; Dose: >1800 mg/m^3/4hr.

Hazard Overviews

Health: Irritating to eyes/skin/respiratory tract. Harmful. Other Acute Effects: may be harmful by inhalation, ingestion, or skin absorption.

Fire: Noncombustible. Hazards: emits toxic fumes. Extinguishing agents: carbon dioxide, dry chemical powder or appropriate foam. Precautions: combustible liquid.

Reactivity: Incompatible with: strong oxidizing agents, strong acids, magnesium, aluminum, sensitive to heat. Hazardous decomposition products: sulfur oxides, ammonia, nitrogen oxides.

Carcinogenicity: IARC - Not listed; NIOSH - Not listed; NTP - Not listed; ACGIH - Not listed; OSHA - Not listed; EPA - Not listed; MAK - Not listed

Primary Target Organs:

Eyes

Skin

Respiratory System

Environmental

Cleanup/Disposal: Guide No. 171: Do not touch or walk through spilled material. Stop leak if you can do it without risk. Prevent dust cloud. Avoid inhalation of asbestos dust. Small Dry Spills: With clean shovel place material into clean, dry container and cover loosely; move containers from spill area. Small Spills: Take up with sand or other noncombustible absorbent material and place into containers for later disposal. Large Spills: Dike far ahead of liquid spill for later disposal. Cover powder spill with plastic sheet or tarp to minimize spreading. Prevent entry into waterways, sewers, basements or confined areas.

Environmental Physical Data

BCF: no food chain concentration potential
BOD: 0.62 lb/lb, 5 days

Regulations

RCRA 40CFR: Not listed
CERCLA: 40CFR 302.4: Not listed
SARA 40CFR 372.65: Not listed
SARA EHS 40CFR 355: Not listed
TSCA: Listed

AMO1000 CAS #: 57-43-2

AMOBARBITAL

RTECS: CQ5075000
EINECS Number: 200-330-7
Molecular Formula: $C_{11}H_{18}N_2O_3$
Formula Weight: 226.27

Synonyms: AMAL; AMASUST; AMITAL; AMOBARBITONE; AMOSPAN; AMYBAL; AMYLBARBITONE; AMYLOBARBITAL; AMYLOBARBITONE; AMYTAL; BARBAMIL; BARBAMYL; BARBAMYL ACID; BARBITURIC ACID,5-ETHYL-5-ISOPENTYL; BINOCTAL; DORLOTYN; DORMYTAL; 5-ETHYL-5-ISOAMYLBARBITURIC ACID; 5-ETHYL-5-ISOAMYLMALONYL UREA; 5-ETHYL-5-ISOPENTYLBARBITURIC ACID; ETHYLISOPENTYLBARBITURIC ACID; 5-ETHYL-5-(3-METHYLBUTYL)BARBITURIC ACID; 5-ETHYL-5-(3-METHYLBUTYL)-2,4,6(1H,3H,5H)-PYRIMIDINETRIONE; 5-ETHYL-5-(3-METHYLBUTYL)-2,4,6-(1H,3H,5H)-PYRIMIDINETRIONE; EUNOCTAL; EUROCTAL; ISOAMYETHYLBARBITURIC ACID; 5-ISOAMYL-5-ETHYLBARBITURIC ACID; ISOAMYLETHYLBARBITURIC ACID; ISOMYL; ISOMYTAL; MYLODORM; NSC 10815; PENTYMAL; PENTYMALUM; 2,4,6(1H,3H,5H)-PYRIMIDINETRIONE,5-ETHYL-5-(3-METHYLBUTYL)-,; 2,4,6(1H,3H,5H)-PYRIMIDINETRIONE,5-ETHYL-

5-(3-METHYLBUTYL)-(9CI); ROBARB; SCHIWANOX; SEDNOTIC; SOMNAL; STADADORM; STATADORM; SUMITAL; TALAMO

Description: white, crystalline powder; odorless

Use: a sedative, a narcotic and has been used as a "truth drug"

Physical Properties

Freezing Point: 156 °C (312.8 °F) to 158 °C (316.4 °F)

Water Solubility: 1 g dissolves in 1300 ml Water

Other Solubilities: 95% Ethanol: >=100 mg/ml at 16 °C; Acetone: >=100 mg/ml at 16 °C; Aliphatic hydrocarbons: Insoluble; Alkalies: Soluble; Benzene: Soluble; Chloroform: Soluble; DMSO: >=100 mg/ml at 16 °C; Petroleum Ether: Insoluble.

pH: Saturated aqueous solution is acid

Flash Point: Not available; probably combustible

RTECS Toxicity Data

Acute Oral: Rat LD_{50} Dose: 250 mg/kg; Toxic Effects: Behavioral - Altered sleep time (including change in righting reflex); Behavioral - Convulsions or effect on seizure threshold; Lungs, Thorax, or Respiration - Respiratory depression. Mouse LD_{50} Dose: 345 mg/kg.

Acute Dermal: Rabbit LD_{Lo} Route: Subcutaneous Dose: 170 mg/kg. Rat LD_{50} Route: Subcutaneous Dose: 190 mg/kg.

Hazard Overviews

Health: Toxic. Other Acute Effects: harmful if swallowed, inhaled, or absorbed through skin; overexposure can cause poor judgment, emotional instability, and at times a toxic psychosis; neurological signs may include nystagmus, dysarthria and ataxia; target organ: nerves. Chronic Effects: prolonged or repeated exposure can lead to habituation or addiction. The toxicolog ical properties have not been thoroughly investigated.

Fire: Will burn. Hazards: emits toxic fumes. Extinguishing agents: carbon dioxide, dry chemical powder or appropriate foam; water spray. Precautions: combustible liquid.

Reactivity: Stable. Hazardous polymerization will not occur. Hazardous decomposition products: toxic fumes of: carbon monoxide, carbon dioxide, nitrogen oxides.

Carcinogenicity: IARC - Not listed; NIOSH - Not listed; NTP - Not listed; ACGIH - Not listed; OSHA - Not listed; EPA - Not listed; MAK - Not listed

Primary Target Organs:

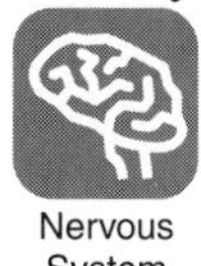

Nervous System

Environmental

Environmental Physical Data

Octanol/Water Partition Coefficient: log K_{ow} = 2.07

Regulations

RCRA 40CFR: Not listed

CERCLA: 40CFR 302.4: Not listed

SARA 40CFR 372.65: Not listed

SARA EHS 40CFR 355: Not listed

TSCA: Not listed

AMO5000 CAS #: 12172-73-5

AMOSITE

RTECS: CI6477000

Molecular Formula: Unspecified or Variable

Synonyms: AMOSITE ASBESTOS; ASBESTOS; BROWN ASBESTOS; MYSORITE

Use: asbestos cement building products; pressure, sewage & drainage pipes; fire resistant insulation boards; insulation products, incl sprays; manufacture of floor tile; in decorative insulations to replace timber in passenger ships; construction material for asbestos cement pipe & sheet

Physical Properties

Freezing Point: Decomposes

Specific Gravity: 3.1 to 3.25

Water Solubility: Insoluble

RTECS Toxicity Data

Mutagenic: Hamster DNA Inhibition; Cell Type: lung; Dose: 125 mg/L. Hamster Mutations in Mammalian Somatic Cells; Cell Type: lung; Dose: 10 mg/L.

Tumorigenic: Rat Route: Inhalation; Dose: 11 mg/m^3/2Y-I; Toxic Effects: Tumorigenic - Carcinogenic by RTECS criteria; Lungs, Thorax, or Respiration - Tumors. Rat Route: Inhalation; Dose: 12 mg/m^3/13W-I; Toxic Effects: Tumorigenic - Neoplastic by RTECS criteria; Lungs, Thorax, or Respiration - Tumors. Rat Route: Inhalation; Dose: 10 mg/m^3/52W-C; Toxic Effects: Tumorigenic - Equivocal tumorigenic agent by RTECS criteria; Lungs, Thorax, or Respiration - Tumors.

Hazard Overviews

Reactivity: Stable. Hazardous polymerization cannot occur. Incompatible with: strong acids; glacial acetic acid; hot water.

Carcinogenicity: IARC - Group 1, Carcinogenic to humans; NIOSH - Listed as carcinogen; NTP - Class 1, Known to be a carcinogen; ACGIH - Class A1, Confirmed human carcinogen; OSHA - Listed as a carcinogen; EPA - Class A, Human carcinogen; MAK - Class A1, Capable of inducing malignant tumors as shown by experience with humans

Exposure Limits

ACGIH TLV: TWA: 0.5 mg/m^3; f/cc.

Respirator Recommendation

Exposure Range: >0.1 to 1 f/cc Air Purifying, Negative Pressure, Half Mask

Exposure Range: >1 to 5 f/cc Air Purifying, Negative Pressure, Full Face

Exposure Range: >5 to 10 f/cc Powered Air Purifying Respirator, Half or Full Facepiece or Hood

Exposure Range: >10 to 100 f/cc Supplied Air, Constant Flow/Pressure Demand, Full Face

Exposure Range: >100 to unlimited f/cc Supplied Air, Pressure Demand, with Auxiliary SCBA

Cartridge Color: magenta (P100)

Environmental

Regulations

RCRA 40CFR: Not listed
CERCLA: 40CFR 302.4: Not listed
SARA 40CFR 372.65: Not listed
SARA EHS 40CFR 355: Not listed
TSCA: Not listed

AMO9000 CAS #: 26787-78-0

AMOXICILLIN

RTECS: XH8300000
EINECS Number: 248-003-8
Molecular Formula: $C_{16}H_{19}N_3O_5S$
Formula Weight: 365.41

OH O O N S NH NH2 O OH

Chemical Structure

Synonyms: D-(-)-ALPHA-AMINO-P-HYDROXYBENZYL PENICILLIN; ALPHA-AMINO-P-HYDROXYBENZYLPENICILLIN; [D-(-)-6-[2-AMINO-2-(P-HYDROXYPHENYL)ACETAMIDO]-3,3-DIMETHYL-7-OXO-4-THIA-1-AZABICYCLO[3.2.0]HEPTANE-2-CARBOXYLIC ACID TRIHYDRATE; (-)-6-[2-AMINO-2-(P-HYDROXYPHENYL)ACETAMIDO]-3,3-DIMETHYL-7-OXO-4-THIA-1-AZABICYCLO-[3.2.0]HEPTANE-2-CARBOXYLICACID; D-2-AMINO-2-(4-HYDROXYPHENYL)ACETAMIDOPENICILLANIC ACID; 6-(D-(-)-ALPHA-AMINO-P-HYDROXYPHENYLACETAMIDO)PENICILLANICACID; 6-[D(-)-ALPHA-AMINO-P-HYDROXYPHENYLACETAMIDO]PENICILLANICACID; 6-[[AMINO(4-HYDROXYPHENYL)ACETYL]-AMINO]-3,3-DIMETHYL-7-OXO-4-THIA-1-AZABICYCLO[3.2.0]HEPTANE-2-CARBOXYLIC ACID; AMOLIN; AMOPENIXIN; AMOXI; AMOXIL; AMOXIPEN; AMOXYCILLIN; AMPC; ANEMOLIN; BLP 1410; BRISTAMOX; BRL 2333; CLAMOXYL; DELACILLIN; EFPENIX; HICONCIL; HISTOCILLIN; 6-(D-(-)-P-HYDROXY-ALPHA-AMINOBENZYL)PENICILLIN; 6-(P-HYDROXY-ALPHA-AMINOPHENYLACETAMIDO)PENICILLANIC ACID; P-HYDROXYAMPICILLIN; IBIAMOX; PIRAMOX; SUMOX; 4-THIA-1-AZABICYCLO(3.2.0)HEPTANE-2-CARBOXYLIC ACID,6-(2-AMINO-2-(P-HYDROXYPHENYL)ACETAMIDO)-3,3-DIMETHYL-7-OXO-,D-; 4-THIA-1-AZABICYCLO(3.2.0)HEPTANE-2-CARBOXYLIC ACID,6-((AMINO(4-HYDROXYPHENYL)ACETYL)AMINO)-3,3-DIMETHYL-7-OXO-, (2S-(2ALPHA, 5ALPHA,6BETA(S*)))-
Description: ; penicillin-type odor
Use: medication: antibacterial agent

Physical Properties

Freezing Point: 194 °C (381.2 °F)
Water Solubility: 1 g Soluble in about 370 ml Water
Other Solubilities: 1 g Soluble in about 2000 ml Alcohol, about 290 ml Phosphate buffer (1%, pH 7), about 330 ml Methanol.

RTECS Toxicity Data

Acute Oral: Man TD_{Lo} Dose: 40 mg/kg/4D; Toxic Effects: Gastrointestinal - Hypermotility, diarrhea; Gastrointestinal - Other changes.

Hazard Overviews

Health: Harmful. Other Acute Effects: may be harmful by inhalation, ingestion, or skin absorption; may cause allergic reaction; shows cross-allergenicity with penicillin. The toxicological properties have not been thoroughly investigated.
Fire: Extinguishing agents: water spray; carbon dioxide, dry chemical powder or appropriate foam. Precautions: combustible liquid.
Reactivity: Stable. Hazardous polymerization will not occur. Hazardous decomposition products: toxic fumes of: carbon monoxide, carbon dioxide, nitrogen oxides, sulfur oxides.
Carcinogenicity: IARC - Not listed; NIOSH - Not listed; NTP - Not listed; ACGIH - Not listed; OSHA - Not listed; EPA - Not listed; MAK - Not listed

Environmental

Regulations

RCRA 40CFR: Not listed
CERCLA: 40CFR 302.4: Not listed
SARA 40CFR 372.65: Not listed
SARA EHS 40CFR 355: Not listed
TSCA: Not listed

AMP1000 CAS #: 300-62-9

AMPHETAMINE

RTECS: SH9450000
EINECS Number: 206-096-2
Molecular Formula: $C_9H_{13}N$
Formula Weight: 135.20
Synonyms: ACTEDRON; ADIPAN; ALLODENE; BETA-AMINOPROPYLBENZENE; ANOREXIDE; ANOREXINE; BENZEBAR; (+-)-BENZEDRINE; BENZEDRINE; BENZENEETHANAMINE,ALPHA-METHYL-,(+-)-; BENZOLONE; (+-)-DESOXYNOREPHEDRINE; DESOXYNOREPHEDRINE; DL-AMPHETAMINE; DL-BENZEDRINE; DL-ALPHA-METHYLPHENETHYLAMINE; DL-1-PHENYL-2-AMINOPROPANE; ELASTONON; FENYLO-IZOPROPYLAMINYL;

FINAM; ISOAMYCIN; ISOAMYNE; ISOMYN; MECODRIN; (+-)-ALPHA-METHYLBENZENEETHANAMINE; ALPHA-METHYLBENZENEETHANEAMINE; (+-)-ALPHA-METHYLPHENETHYLAMINE; ALPHA-METHYLPHENETHYLAMINE; (+-)-ALPHA-METHYLPHENYLETHYLAMINE; NOREPHEDRANE; NOREPHEDRINE,DEOXY-; NOVYDRINE; OKTEDRIN; ORTEDRINE; PERCOMON; PHENEDRINE; PHENETHYLAMINE,ALPHA-METHYL-,(+-)-; 1-PHENYL-2-AMINO-PROPAN; 1-PHENYL-2-AMINOPROPANE; BETA-PHENYLISOPROPYLAMIN; (+-)-BETA-PHENYLISOPROPYLAMINE; (PHENYLISOPROPYL)AMINE; BETA-PHENYLISOPROPYLAMINE; PHENYLISOPROPYLAMINE; PROFAMINA; PROPISAMINE; PSYCHEDRINE; RACEMIC DESOXY-NOR-EPHEDRINE; RACEMIC-DESOXYNOR-EPHEDRINE; RAPHETAMINE; RHINALATOR; SIMPATEDRIN; SIMPATINA; SYMPAMINE; SYMPATEDRINE; WECKAMINE

Description: liquid; amine odor

Use: drug of abuse; medication: central stimulant; medication (vet): CNS stimulant, in narcotic poisoning, anesthetic collapse, in depression from encephalitis

Physical Properties

Boiling Point: 200 °C (392 °F) to 203 °C (397 °F) at 760 mm Hg

Freezing Point: Crystals at > 300 °C (572 °F)

Specific Gravity: 0.913 at 25 °C/4 °C

Vapor Pressure: 1.30 kPa at 75 °C

Water Solubility: Sparingly Soluble in Water 1 part in 50 parts Water

Other Solubilities: Soluble in Chloroform; Soluble in Diethyl Ether, Ethanol; Soluble in Alcohol, Ether; readily Soluble in acids.

pH: Aqueous solution is alkaline

Refraction Index: 1.518 at 26 °C/D

Ionization Potential (eV): 8.5

Flash Point: 26.6 °C

RTECS Toxicity Data

Acute Oral: Rat LD_{50} Dose: 30 mg/kg. Mouse LD_{50} Dose: 21 mg/kg.

Acute Dermal: Monkey LD_{Lo} Route: Subcutaneous Dose: 20 mg/kg. Rabbit LD_{Lo} Route: Subcutaneous Dose: 20 mg/kg.

Reproductive/Teratogenic: Rat Route: Subcutaneous; Dose: 11 mg/kg; Duration: female 1-22D of pregnancy; Effects on Newborn - Behavioral.

Mutagenic: Mouse Micronucleus Test; Route: Oral; Dose: 25 mg/kg. Bacteria - E Coli DNA Adduct; Dose: 40 umol/L.

Hazard Overviews

Flammable

Fire: Flammable.

Carcinogenicity: IARC - Not listed; NIOSH - Not listed; NTP - Not listed; ACGIH - Not listed; OSHA - Not listed; EPA - Not listed; MAK - Not listed

Environmental

Environmental Physical Data

Octanol/Water Partition Coefficient: log K_{ow} = 1.76

Regulations

RCRA 40CFR: Not listed

CERCLA: 40CFR 302.4: Not listed

SARA 40CFR 372.65: Not listed TPQ: 1000 lb

SARA EHS 40CFR 355: Listed TPQ: 1000 lb

TSCA: Listed

AMP5000 CAS #: 1397-89-3

AMPHOTERICIN B

RTECS: BU2625000

EINECS Number: 215-742-2

Molecular Formula: $C_{47}H_{73}NO_{17}$

Formula Weight: 924.11

Chemical Structure

Synonyms: AMPHO-MORONAL; AMPHOMORONAL; AMPHOTERACIN B; AMPHOTERICIN BETA; AMPHOTERICINE B; AMPHOZONE; FUNGILIN; FUNGISONE; FUNGIZONE; IAB; IODOACETAMIDE; MYSTECLIN-F; NSC 527017; TEGOPEN

Description: deep yellow to orange powder, prisms or needles; odorless or practically so

Use: antifungal antibiotic; used to treat meningitis

Physical Properties

Boiling Point: Decomposes > 170 °C (338 °F)

Freezing Point: 170 °C (338 °F)

Water Solubility: About 0.1 mg/ml at pH=2 or 11

Other Solubilities: Insoluble in anhydrous Alcohol, Ether, Benzene & Toluene; Slightly Soluble in Methanol.

Flash Point: Not available; probably combustible

RTECS Toxicity Data

Acute Oral: Rat LD_{50} Dose: >5 gm/kg. Mouse LD_{50} Dose: >8 gm/kg.

Reproductive/Teratogenic: Rat Route: Intravenous; Dose: 5500 ug/kg; Duration: female 6-16D of pregnancy; Effects on Embryo or Fetus - Fetotoxicity. Rabbit Route: Intravenous; Dose: 20 mg/kg; Duration: male 10D prior to mating; Paternal Effects - Spermatogenesis.

Mutagenic: Human DNA Inhibition; Cell Type: lymphocyte; Dose: 10 umol/L. Human Other Mutation Test Systems; Cell Type: other cell types; Dose: 100 mg/L.

Hazard Overviews

Health: Irritating to eyes/skin/respiratory tract. Harmful. Other Acute Effects: harmful if inhaled or absorbed through skin; may be harmful if swallowed; exposure can cause nausea; headache; vomiting convulsions; chills; fever; malaise; muscle and joint pain; rash; anorexia; diarrhea; gastrointestinal cramps; hypertension; hypotension; cardiac arrhythmias; ventricular fibrillation; cardiac arrest; blurred vision; tinnitus; vertigo; anaemia and hypokalemia; prolonged or repeated exposure may cause allergic reactions in certain sensitive individuals; target organ: kidneys. Chronic Effects: possible mutagen.

Fire: Will burn. Extinguishing agents: carbon dioxide, dry chemical powder or appropriate foam. Precautions: combustible liquid.

Reactivity: Stable. Hazardous polymerization will not occur. Hazardous decomposition products: thermal decomposition may produce carbon monoxide, carbon dioxide, and nitrogen oxides.

Carcinogenicity: IARC - Not listed; NIOSH - Not listed; NTP - Not listed; ACGIH - Not listed; OSHA - Not listed; EPA - Not listed; MAK - Not listed

Primary Target Organs:

Eyes

Skin

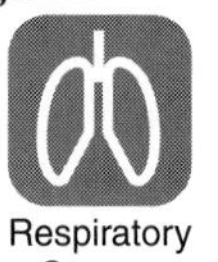
Respiratory System

Kidneys

Environmental

Regulations

RCRA 40CFR: Not listed
CERCLA: 40CFR 302.4: Not listed
SARA 40CFR 372.65: Not listed
SARA EHS 40CFR 355: Not listed
TSCA: Not listed

AMP9000 **CAS #: 69-53-4**

AMPICILLIN

RTECS: XH8350000
EINECS Number: 200-709-7
Molecular Formula: $C_{16}H_{19}N_3O_4S$
Formula Weight: 349.42

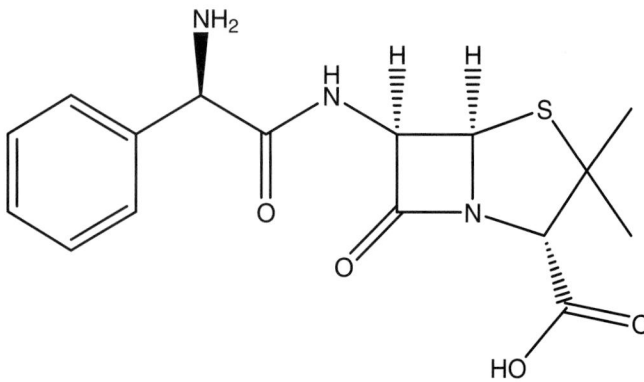

Chemical Structure

Synonyms: AB-PC; ADOBACILLIN; ALPEN; AMBLOSIN; AMFIPEN; AMINOBENZYLPENICILLIN; D(-)-ALPHA-AMINOBENZYLPENICILLIN; D-(-)-ALPHA-AMINOBENZYLPENICILLIN; D-(-)-ALPHA-AMINOPENICILLIN; 6-[D-(2-AMINO-2-PHENYLACETAMIDO)]-3,3-DIMETHYL-7-OXO-4-THIA-1-AZABICYCLO[3.2.0]HEPTANE-2-CARBOXYLIC ACID; 6-(D(-)-ALPHA-AMINOPHENYLACETAMIDO)PENICILLANIC ACID; 6-[D(-)-ALPHA-AMINOPHENYLACETAMIDO]PENICILLANIC ACID; AMIPENIX S; AMPI-BOL; D-(-)-AMPICILLIN; D-AMPICILLIN; AMPICILLIN A; AMPICILLIN ACID; AMPICILLIN ANHYDRATE; AMPICIN; AMPIKEL; AMPIMED; AMPIPENIN; AMPLISOM; AMPLITAL; AMPY-PENYL; AUSTRAPEN; AY-6108; BINOTAL; BONAPICILLIN; BRITACIL; BRL 1341; COPHARCILIN; DOKTACILLIN; GRAMPENIL; GUICITRINA; NSC-528986; NUVAPEN; OLIN KID; OMNIPEN; P-50; PENBRISTOL; PENBRITIN; PENBRITIN PAEDIATRIC; PENBRITIN SYRUP; PENBROCK; PENICILLIN,(AMINOPHENYLMETHYL)-; PENICLINE; PENTREX; PENTREXL; PENTREXYL; PFIZERPEN A; POLYCILLIN; PONECIL; PRINCIPEN; QIDAMP; ROSCILLIN; SEMICILLIN; SK-AMPICILLIN; SYNPENIN; 4-THIA-1-AZABICYCLO(3.2.0)HEPTANE-2-CARBOXYLIC ACID,6-(2-AMINO-2-PHENYLACETAMIDO)-3,3-DIMETHYL-7-OXO-,D-(-)-; TOKIOCILLIN; TOLOMOL; TOTACILLIN; TOTALCICLINA; TOTAPEN; ULTRABION; ULTRABRON; VICCILLIN; VICCILLIN S; VICILLIN S; WY-5103

Description: white, crystalline powder, or white, needle-like crystals; odorless or has a faint odor characteristic of penicillins

Use: antibiotic for a variety of infections, including those of the respiratory and urinary tracts, gonorrhoea, meningitis, septicaemia and enteric infections

Physical Properties

Freezing Point: Decomposes at 202 °C (395.6 °F)
Water Solubility: 1 gm dissolves in about 90 ml of Water
Other Solubilities: Soluble in Dimethyl Sulfoxide (anhydrous form); It is not stable in aqueous solutions of alkali hydroxides and carbonates, and is decomposed by dilute solutions of mineral acids.
pH: 10 g/ml aqueous solution 3.5 to 6.0

RTECS Toxicity Data

Acute Oral: Man TD_{Lo} Dose: 400 mg/kg/4W-I; Toxic Effects: Blood - Agranulocytosis; Blood - Other changes; Nutritional and gross metabolic - Body temperature increase. Woman TD_{Lo} Dose: 160 mg/kg/4D-I; Toxic Effects: Blood - Agranulocytosis; Blood - Thrombocytopenia.

Acute Dermal: Rat LD_{Lo} Route: Subcutaneous Dose: >5 gm/kg. Mouse LD_{Lo} Route: Subcutaneous Dose: >5 gm/kg.
Reproductive/Teratogenic: Rat Route: Oral; Dose: 2500 mg/kg; Duration: female 4-13D of pregnancy; Effects on Embryo or Fetus - Extra embryonic structures; Fetotoxicity; Fetal death.
Mutagenic: Human Cytogenetic Analysis; Cell Type: lymphocyte; Dose: 28 mg/L. Bacteria - E Coli DNA Damage; Dose: 100 ug/L.

Hazard Overviews

Health: Irritating to eyes/skin/respiratory tract. Harmful. Other Acute Effects: harmful if swallowed, inhaled, or absorbed through skin; possible allergic reaction. Chronic Effects: serious and sometimes fatal hypersensitivity (anaphylactic) reactions, in people receiving pencillin therapy by the parenteral route although also via oral administration, usually in individuals who have a history of sensitivity to multiple allergens, asthma, hay fever or urticaria; nausea; vomiting; diarrhea; glossitis; stomatitis; enterocolitis; pseudomembranous colitis; hypersensitivity reactions: erythematous maculopapular rashes, erythema multiforme, urticaria and exfoliative dermatitis; sensitization by inhalation and skin contact; possible risk of irreversible effects. Possible human carcinogen.
Fire: Hazards: emits toxic fumes. Extinguishing agents: water spray; carbon dioxide, dry chemical powder or appropriate foam. Precautions: combustible liquid.
Reactivity: Incompatible with: strong oxidizing agents, protect from moisture. Hazardous decomposition products: toxic fumes of: carbon monoxide, carbon dioxide, nitrogen oxides, sulfur oxides.
Carcinogenicity: IARC - Group 3, Not classifiable as to carcinogenicity to humans; NIOSH - Not listed; NTP - Not listed; ACGIH - Not listed; OSHA - Not listed; EPA - Not listed; MAK - Not listed
Primary Target Organs:

Eyes Skin Respiratory System

Environmental

Regulations

RCRA 40CFR: Not listed
CERCLA: 40CFR 302.4: Not listed
SARA 40CFR 372.65: Not listed
SARA EHS 40CFR 355: Not listed
TSCA: Not listed

AMT5000 CAS #: 90-44-8

AMTHRONE

RTECS: CB8925500
EINECS Number: 201-994-0
Molecular Formula: $C_{14}H_{10}O$
Formula Weight: 194.22

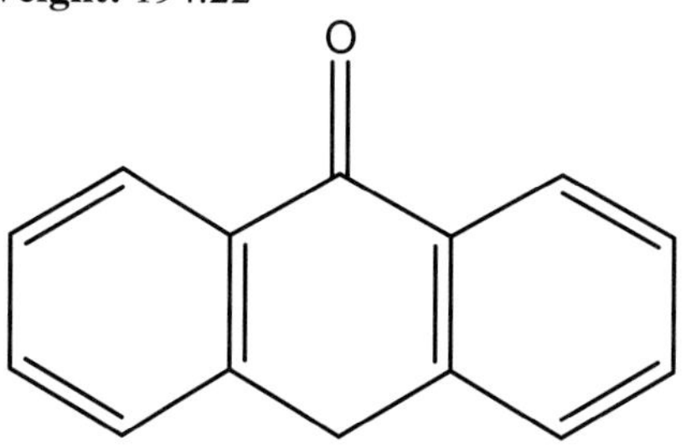

Chemical Structure

Synonyms: ANTHRACENE,9,10-DIHYDRO-9-OXO-; 9(10H)-ANTHRACENONE; ANTHRANONE; CARBOTHRONE; 9,10-DIHYDRO-9-OXOANTHRACENE
Description: colorless needles
Use: in colorimetric determination of sugar and animal starch in body fluids; in org synthesis; general reagent for carbohydrates; determination of animal starch in liver tissue; derivatives: melitracen

Physical Properties

Freezing Point: 155 °C (311 °F)
Other Solubilities: Soluble in Acetone, hot Benzene, concentrated Sulfuric acid, hot dilute alkali.
Ionization Potential (eV): 8.83 +/-0.03

RTECS Toxicity Data

Mutagenic: Bacteria - S Typhimurium Mutations in Microorganisms; Dose: 60 ug/plate (+S9).

Hazard Overviews

Health: Irritating to eyes/skin/respiratory tract. Other Acute Effects: may be harmful by inhalation, ingestion, or skin absorption; causes photosensitivity; exposure to light can result in allergic reactions resulting in dermatologic lesions, which can vary from sunburnlike responses to edematous, vesiculated lesions or bullae.
Fire: Hazards: emits toxic fumes. Extinguishing agents: water spray; carbon dioxide, dry chemical powder or appropriate foam. Precautions: combustible liquid.
Carcinogenicity: IARC - Not listed; NIOSH - Not listed; NTP - Not listed; ACGIH - Not listed; OSHA - Not listed; EPA - Not listed; MAK - Not listed
Primary Target Organs:

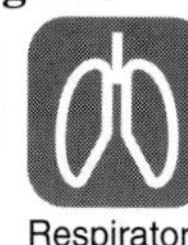

Eyes Skin Respiratory System

Environmental

Environmental Fate: If released to the atmosphere, it will exist in both the vapor and particulate phases in the ambient atmosphere based on an estimated vapor pressure of 1.8×10^{-5} mm Hg at 25 °C. Vapor-phase is degraded in the atmosphere by reaction with photochemically produced hydroxyl radicals with an estimated half-life of about 40 hours. An estimated

K_{oc} of 2300 suggests that it will have only slight mobility in soil. Volatilization from dry and moist soil surfaces should not be a major fate process for this compound. Based on limited data, this compound may biodegrade in both soil and water. In water, it is expected to adsorb to sediment and suspended matter based on its K_{oc} value. It may volatilize slowly from water surfaces given an estimated Henry's Law constant of 7.9 x10^{-7} atm-cu m/mole. Estimated half-lives for a model river and model lake are 65 and 480 days, respectively. Bioconcentration in aquatic organisms may occur based on an estimated BCF value of 360.

Environmental Physical Data

Henry's Law Constant: estimated at 7.9 x10^{-7}
Octanol/Water Partition Coefficient: log K_{ow} = 3.66
Sorption Partition Coefficient: K_{oc} = estimated at 2300
BCF: estimated at 360

Regulations

RCRA 40CFR: Not listed
CERCLA: 40CFR 302.4: Not listed
SARA 40CFR 372.65: Not listed
SARA EHS 40CFR 355: Not listed
TSCA: Listed

AMY1000 CAS #: 29883-15-6

AMYGDALIN

RTECS: OO8450000
EINECS Number: 249-925-3
Molecular Formula: $C_{20}H_{27}NO_{11}$
Formula Weight: 457.48

Chemical Structure

Synonyms: -AMYGDALIN; D-AMYGDALIN; R-AMYGDALIN; AMYGDALOSIDE; BENZENEACETONITRILE,ALPHA-((6-O-BETA-D-GLUCOPYRANOSYL-BETA-D-GLUCOPYRANOSYL)OXY)-,-; -ALPHA-((6-O-BETA-D-GLUCOPYRANOSYL-BETA-D-GLUCOPYRANOSYL)OXY)BENZENEACETONITR ILE; -ALPHA-((6-O-BETA-D-GLUCOPYRANOSYL-BETA-D-GLUCOPYRANOSYL)OXY)BENZENEACETONITRILE; D(-)-MANDELONITRILE-BETA-D-GENTIOBIOSIDE; MANDELONITRILE-BETA-GENTIOBIOSIDE; D-MANDELONITRILE-BETA-D-GLUCOSIDO-6-BETA-D-GLUCOSIDE; NSC-15780; NSC 15780

Physical Properties

Freezing Point: 223 °C (433.4 °F) to 226 °C (438.8 °F)
Water Solubility: Very Soluble in Hot Water
Other Solubilities: Slightly Soluble in Alcohol; Insoluble in Ether, Chloroform; Soluble in hot Alcohol

RTECS Toxicity Data

Acute Oral: Infant LD_{Lo} Dose: 50 mg/kg; Toxic Effects: Behavioral - Coma; Lungs, Thorax, or Respiration - Dyspnea; Gastrointestinal - Nausea or vomiting. Monkey LD_{Lo} Dose: 167 mg/kg; Toxic Effects: Behavioral - Somnolence (general depressed activity); Lungs, Thorax, or Respiration - Dyspnea; Nutritional and gross metabolic - Weight loss or decreased weight gain. Rat LD_{50} Dose: 405 mg/kg.
Chronic (Multiple Dose) Oral: Rat Dose: 6 gm/kg/30D-I; Toxic Effects: Blood - Pigmented or nucleated red Blood cells; Blood - Changes in erythrocite (RBC) cell count; Blood - Changes in leukocyte (WBC) cell count.
Reproductive/Teratogenic: Hamster Route: Oral; Dose: 300 mg/kg; Duration: female 8D of pregnancy; Specific Developmental Abnormalities - Central nervous system; Musculoskeletal system.
Mutagenic: Mouse Host-mediated Assay; Indicator Organism: Bacteria - S Typhimurium; Dose: 250 mg/kg.

Hazard Overviews

Health: Irritating to eyes/skin. Toxic. Other Acute Effects: harmful if swallowed; may be harmful if inhaled; may be harmful if absorbed through the skin; target organs: central nervous system, blood; heart.
Fire: Hazards: emits toxic fumes. Extinguishing agents: water spray; carbon dioxide, dry chemical powder or appropriate foam. Precautions: combustible liquid.
Reactivity: Incompatible with: strong oxidizing agents, strong acids, strong bases. Hazardous decomposition products: toxic fumes of: carbon monoxide, carbon dioxide, nitrogen oxides.
Carcinogenicity: IARC - Not listed; NIOSH - Not listed; NTP - Not listed; ACGIH - Not listed; OSHA - Not listed; EPA - Not listed; MAK - Not listed
Primary Target Organs:

Eyes

Skin

Nervous System

Cardio-vascular

Blood

Environmental

Regulations

RCRA 40CFR: Not listed
CERCLA: 40CFR 302.4: Not listed
SARA 40CFR 372.65: Not listed
SARA EHS 40CFR 355: Not listed

TSCA: Not listed

AMY1800 CAS #: 628-63-7

N-AMYL ACETATE

RTECS: AJ1925000
DOT: UN1104; IMO3.2; IMO3.3
EINECS Number: 211-047-3
Molecular Formula: $C_7H_{14}O_2$
Structured MF: $CH_3COO(CH_2)_4CH_3$
Formula Weight: 130.19

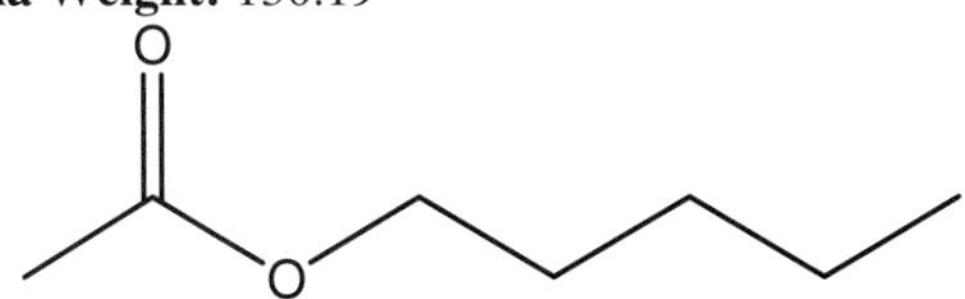

Chemical Structure

Synonyms: ACETATE D'AMYLE; ACETIC ACID,AMYL ESTER; ACETIC ACID,PENTYL ESTER; AMYL ACETATE; AMYL ACETIC ESTER; AMYL ACETIC ETHER; AMYLAZETAT; AMYLESTER KYSELINY OCTOVE; BANANA OIL; BIRNENOEL; CHLORDANTOIN; DYMON SWH WASP & HORNET SPRAY; EPA PESTICIDE CHEMICAL CODE 000169; HOLIDAY PET REPELLENT; HOLIDAY REPELLENT DUST; OCTAN AMYLU; PEAR OIL; PENT-ACETATE; PENT-ACETATE 28; 1-PENTANOL ACETATE; 1-PENTANOL,ACETATE; 1-PENTYL ACETATE; N-PENTYL ACETATE; PENTYL ACETATE; PENTYL ESTER OF ACETIC ACID; PRIM-AMYL ACETATE; PRIMARY AMYL ACETATE
Description: clear, colorless liquid; pear- or banana-like odor
Use: solvent for lacquers & paints, phosphors in fluorescent lamps; for extraction of penicillin; photographic film; leather & nail polish; warning odor; printing & finishing fabrics; in dry cleaning ind; flavoring agent in mfr of fruit flavor; vehicle for stiffening agent for straw hats; dog & cat repellent; insecticide & miticide to combat wasps, tc; med: anti-inflammatory agent.

Physical Properties

Boiling Point: 149.25 °C (301 °F) at 760 mm Hg
Freezing Point: -70.8 °C (-95.44 °F)
Specific Gravity: 0.8756 at 20 °C/4 °C
Vapor Density: 4.5 Air=1
Saturated Vapor Density: 1.227547187 kg/m^3
Vapor Pressure: 5 mm Hg at 25 °C
Water Solubility: 0.2% by weight
Other Solubilities: Soluble in acid.
Surface Tension: 12 dynes/cm at 30 °C
Odor Threshold: 0.0265 to 37.1 mg/m^3
Refraction Index: 1.4023 at 20 °C
Flash Point: 16 °C Closed Cup
Autoignition Temperature: 360 °C
LEL: 1.1% v/v
UEL: 7.5% v/v

RTECS Toxicity Data

Acute Oral: Rabbit LD_{50} Dose: 7400 mg/kg.
Acute Inhalation: Human TC_{Lo} Dose: 5000 mg/m^3/30M; Toxic Effects: Sense organs and special senses - Conjunctive irritation; Behavioral - Somnolence (general depressed activity); Lungs, Thorax, or Respiration - Other changes.

Hazard Overviews

Flammable

Fire Diamond

Health: Irritating to eyes/skin/respiratory tract. Also Causes: cough, chest pain, shortness of breath, headache, dizziness, nausea, fatigue, staggering, narcosis, anesthesia, or heart failure. Chronic Effects: dermatitis, neurological impairment.
Fire: Flammable. Can form explosive mixtures in air. For small fires use dry chemical, carbon dioxide, water spray, or alcohol-resistant foam. For large fires use water spray, fog, or alcohol-resistant foam. Material may float and re-ignite on water surface. Water spray can be used to cool fire-exposed closed containers and to disperse vapors.
Reactivity: Stable. Hazardous polymerization cannot occur. Avoid: heat; ignition sources. Incompatible with: nitrates; strong oxidizers; alkalis; acids. Hazardous decomposition products: carbon monoxide; toxic vapors/gases.
Carcinogenicity: IARC - Not listed; NIOSH - Listed as carcinogen; NTP - Not listed; ACGIH - Not listed; OSHA - Not listed; EPA - Not listed; MAK - Not listed
Primary Target Organs:

Eyes

Skin

Respiratory System

Nervous System

Exposure Limits
OSHA PEL: TWA: 100 ppm; 525 mg/m^3.
ACGIH TLV: TWA: 100 ppm; 532 mg/m^3.
NIOSH REL: TWA: 100 ppm; 525 mg/m^3.
NIOSH IDLH: 1000 ppm.
DFG MAK: TWA: 100 ppm; 525 mg/m^3.
Respirator Recommendation
Exposure Range: >100 to <1000 ppm Air Purifying, Negative Pressure, Full Face
Exposure Range: 1000 to unlimited ppm Self-contained Breathing Apparatus, Pressure Demand, Full Face
Cartridge Color: black
Note: dust/mist prefilter required if mist is present

Environmental

Ecotoxicity: Aquatic toxicity: 120 ppm/48 hr/daphnia/TLm/turbid water 180 ppm/96 hr/scenedesmus/TLm/fresh water
Environmental Fate: If released on land or in water, volatilization would be important (half-life 5.9 hr in a typical river) and biodegradation, should be a dominant degradative process. Adsorption to soil or sediment would not occur to any significant extent, so leaching into groundwater may occur. Some chemical hydrolysis may occur but only under fairly alkaline conditions. It would not be expected to

bioconcentrate in aquatic organism. In air, it will be scavenged by rain and degrade by reaction with photochemically produced hydroxyl radicals (estimated half-life 4.5 days).

Cleanup/Disposal: Guide No. 129: Eliminate all ignition sources (no smoking, flares, sparks or flames in immediate area). All equipment used when handling the product must be grounded. Do not touch or walk through spilled material. Stop leak if you can do it without risk. Prevent entry into waterways, sewers, basements or confined areas. A vapor suppressing foam may be used to reduce vapors. Absorb or cover with dry earth, sand or other non-combustible material and transfer to containers. Use clean non-sparking tools to collect absorbed material. Large Spills: Dike far ahead of liquid spill for later disposal. Water spray may reduce vapor; but may not prevent ignition in closed spaces.

Environmental Physical Data

Henry's Law Constant: 0.16
Octanol/Water Partition Coefficient: log K_{ow} = 2.18
Sorption Partition Coefficient: K_{oc} = 73
BCF: not significant
BOD: 0.3 to 0.8 lb/lb, 5 days

Regulations

RCRA 40CFR: Not listed
CERCLA: 40CFR 302.4: Listed per CWA Section 311(b)(4) RQ: 5000 lb (2268 kg)
SARA 40CFR 372.65: Not listed
SARA EHS 40CFR 355: Not listed
TSCA: Listed

Analytical Methods

Air: ASTM D3686, D3687, D4490
Water / Groundwater: ASTM D3695
Indoor / Expired Air: NIOSH 1450
Other: EPA 1666

AMY2600 CAS #: 626-38-0

SEC-AMYL ACETATE

RTECS: AJ2100000
DOT: UN1104; IMO3.3
EINECS Number: 210-946-8
Molecular Formula: $C_7H_{14}O_2$
Structured MF: $CH_3COOCH(CH_3)C_3H_7$
Formula Weight: 130.19
Synonyms: 2-ACETOXYPENTANE; 2-AMYLESTER KYSELINY OCTOVE; 1-METHYLBUTYL ACETATE; 2-PENTANOL ACETATE; 2-PENTANOL,ACETATE; 2-PENTANOL,ACETATE (8CI,9CI); 2-PENTYL ACETATE; 2-PENTYL ESTER OF ACETIC ACID; SEK.AMYLESTER KYSELINY OCTOVE
Description: colorless liquid; mild odor
Use: solvent for nitrocellulose & ethyl cellulose, artificial leather, celluloid products, cements, coated paper, lacquers, leather finishes, linoleum, nail enamels, plastic wood, textile sizing & printing compounds, washable wallpaper, & chlorinated rubber; metallic paints; fine perfumes; pearlescent coatings on artificial pearls

Physical Properties

Boiling Point: 130 °C (266 °F) to 131 °C (268 °F)
Freezing Point: -100 °C (-148 °F)
Specific Gravity: 0.862 to 0.866 at 20 °C/20 °C
Vapor Density: 4.5 Air=1
Saturated Vapor Density: 1.238566062 kg/m^3
Vapor Pressure: 7 mm Hg at 20 °C
Water Solubility: 0.2 g/100 g Water at 20 °C
Other Solubilities: Soluble in Alcohol, Ether.
Surface Tension: 28.9 dynes/cm at 20 °C
Odor Threshold: 0.08 ppm
Evaporation Rate: 0.9 Butyl Acetate=1
Critical Temperature: 326.1 °C
Critical Pressure: 28.0 atm
Flash Point: 32 °C Closed Cup
Autoignition Temperature: 360 to 379 °C
LEL: 1.0% v/v
UEL: 7.5% v/v

RTECS Toxicity Data

Acute Inhalation: Guinea Pig LC_{Lo} Dose: 10000 ppm/5hr; Toxic Effects: Sense organs and special senses - Other; Sense organs and special senses - Lacrimation; Behavioral - Somnolence (general depressed activity). Human TC_{Lo} Dose: 200 ppm; Toxic Effects: Sense organs and special senses - Other; Sense organs and special senses - Conjunctive irritation; Lungs, Thorax, or Respiration - Other changes.

Hazard Overviews

Flammable

Fire Diamond

Health: Irritating to eyes/skin/respiratory tract. Also Causes: (high concentrations): dizziness, drowsiness, heart failure, narcosis. Chronic Effects: dermatitis, hepatotoxicity.
Fire: Flammable. Can form explosive mixtures in the air. Use alcohol foam, carbon dioxide, or dry chemical. Apply water in flooding quantities as fog and from as far a distance as possible.
Reactivity: Stable. Hazardous polymerization cannot occur. Avoid: heat; ignition sources. Incompatible with: strong oxidizers; strong alkalies; nitrates; strong acids. Hazardous decomposition products: carbon dioxide; toxic gases and vapors.
Carcinogenicity: IARC - Not listed; NIOSH - Listed as carcinogen; NTP - Not listed; ACGIH - Not listed; OSHA - Not listed; EPA - Not listed; MAK - Not listed
Primary Target Organs:

Eyes

Skin

Respiratory System

Liver

Exposure Limits
OSHA PEL: TWA: 125 ppm; 650 mg/m^3.
ACGIH TLV: TWA: 125 ppm; 665 mg/m^3.
NIOSH REL: TWA: 125 ppm; 650 mg/m^3.
NIOSH IDLH: 1000 ppm.
DFG MAK: TWA: 100 ppm; 525 mg/m^3.
Respirator Recommendation
Exposure Range: >125 to <1000 ppm Air Purifying, Negative Pressure, Full Face
Exposure Range: 1000 to unlimited ppm Self-contained Breathing Apparatus, Pressure Demand, Full Face
Cartridge Color: black

Environmental

Ecotoxicity: Aquatic toxicity: 65 ppm/96 hr/Mosquito fish/TLm/turbid water (mixed isomers); 53 ppm/24 hr/brine shrimp/TLm 62%, 10 days/80%, 20 days
Cleanup/Disposal: Guide No. 129: Eliminate all ignition sources (no smoking, flares, sparks or flames in immediate area). All equipment used when handling the product must be grounded. Do not touch or walk through spilled material. Stop leak if you can do it without risk. Prevent entry into waterways, sewers, basements or confined areas. A vapor suppressing foam may be used to reduce vapors. Absorb or cover with dry earth, sand or other non-combustible material and transfer to containers. Use clean non-sparking tools to collect absorbed material. Large Spills: Dike far ahead of liquid spill for later disposal. Water spray may reduce vapor; but may not prevent ignition in closed spaces.

Environmental Physical Data

BCF: no food chain concentration potential
BOD: 53%, 5 days

Regulations

RCRA 40CFR: Not listed
CERCLA: 40CFR 302.4: Listed per CWA Section 311(b)(4) RQ: 5000 lb (2268 kg)
SARA 40CFR 372.65: Not listed
SARA EHS 40CFR 355: Not listed
TSCA: Listed

Analytical Methods

Air: ASTM D3686, D3687
Indoor / Expired Air: NIOSH 1450

AMY3400 CAS #: 625-16-1

TERT-AMYL ACETATE

Molecular Formula: $C_7H_{14}O_2$
Structured MF: $CH_3COOC(CH_3)_2C_2H_5$
Formula Weight: 130.18

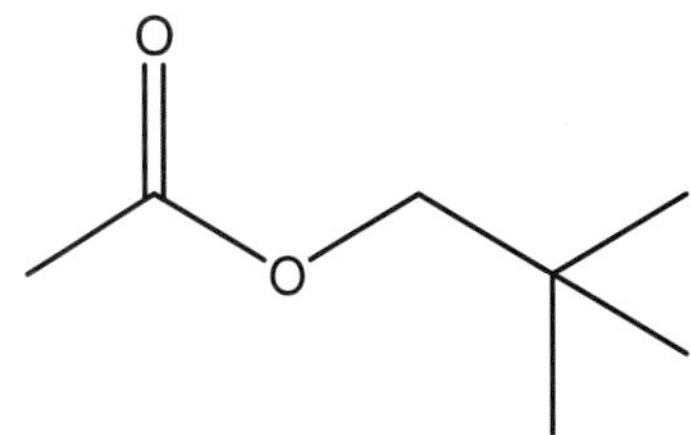

Chemical Structure

Description: colorless to yellow, watery liquid; banana odor

Physical Properties

Boiling Point: 124.5 °C (256 °F)
Freezing Point: > -100 °C (-148 °F)
Specific Gravity: 0.87
Vapor Density: 4.5 Air=1
Water Solubility: Very Slightly Soluble in Water
Surface Tension: Estimated at 29.2 dynes/cm
Odor Threshold: 0.0017 ppm
Critical Temperature: Estimated at 320.7 °C
Critical Pressure: 395 psia
Flash Point: Estimated at 26.11 °C Closed Cup
Autoignition Temperature: Estimated at 379 °C
LEL: 1.00% v/v
UEL: 7.5% v/v

Hazard Overviews

Flammable

Fire: Flammable.
Carcinogenicity: IARC - Not listed; NIOSH - Not listed; NTP - Not listed; ACGIH - Not listed; OSHA - Not listed; EPA - Not listed; MAK - Not listed

Environmental

Ecotoxicity: Aquatic toxicity: 65 ppm/96 hr/mosquito fish/TLm/turbid water. 120 ppm/48 hr/daphnia/TLm/24 °C; 53 ppm/24 hr/brine shripm/TLm. 70%, 15 days 80%, 20 days
Cleanup/Disposal: Guide No. 129: Eliminate all ignition sources (no smoking, flares, sparks or flames in immediate area). All equipment used when handling the product must be grounded. Do not touch or walk through spilled material. Stop leak if you can do it without risk. Prevent entry into waterways, sewers, basements or confined areas. A vapor suppressing foam may be used to reduce vapors. Absorb or cover with dry earth, sand or other non-combustible material and transfer to containers. Use clean non-sparking tools to collect absorbed material. Large Spills: Dike far ahead of liquid spill for later disposal. Water spray may reduce vapor; but may not prevent ignition in closed spaces.

Environmental Physical Data

BCF: no food chain concentration potential
BOD: 53%, 5 days

Regulations

RCRA 40CFR: Not listed
CERCLA: 40CFR 302.4: Not listed
SARA 40CFR 372.65: Not listed
SARA EHS 40CFR 355: Not listed
TSCA: Not listed

AMY4200 CAS #: 71-41-0

N-AMYL ALCOHOL

RTECS: SB9800000
DOT: UN1105; IMO3.2; IMO3.3
EINECS Number: 200-752-1
Molecular Formula: $C_5H_{12}O$
Structured MF: $CH_3(CH_2)_3CH_2OH$
Formula Weight: 88.15

Chemical Structure

Synonyms: ALCOOL AMYLIQUE; AMYL ALCOHOL; AMYL ALCOHOL,NORMAL; N-AMYL ALCOHOL,PRIMARY; N-AMYLALKOHOL; AMYLOL; BUTYL CARBINOL; N-BUTYLCARBINOL; 1-PENTANOL; N-PENTAN-1-OL; N-PENTANOL; PENTAN-1-OL; PENTANOL; PENTANOL-1; PENTASOL; 1-PENTYL ALCOHOL; N-PENTYL ALCOHOL; PENTYL ALCOHOL; PETAN-1-OL; PRIMARY AMYL ALCOHOL; PRIMARY-N-AMYL ALCOHOL
Description: colorless liquid; mild fusel oil
Use: in organic synthesis, as a solvent, in the manufacture of petroleum additives, in urea-formaldehyde plastics processing and in pharmaceuticals

Physical Properties

Boiling Point: 137.5 °C (280 °F)
Freezing Point: -79 °C (-110.2 °F)
Specific Gravity: 0.8146 at 20 °C/4 °C
Vapor Density: 3 Air=1
Saturated Vapor Density: 1.209017423 kg/m^3
Density: 0.815 g/cc
Bulk Density: 6.9 lbs/gal at 20 °C
Vapor Pressure: 2.8 mm Hg at 20 °C
Water Solubility: 2.7 g/100 ml at 22 °C
Other Solubilities: 95% Ethanol: >=100 mg/ml at 17 °C; Acetone: >=100 mg/ml at 17 °C; Alcohol: miscible; Benzene: miscible; DMSO: >=100 mg/ml at 17 °C; Ether: miscible.
Surface Tension: 25.60 dynes/cm
Odor Threshold: 0.4332 to 72.2 mg/m^3
Refraction Index: 1.4103 at 20 °C/D
Critical Temperature: 313 °C
Ionization Potential (eV): 10.04 +/-0.03
Flash Point: 33 °C Closed Cup
Autoignition Temperature: 300 °C
LEL: 1.2% v/v
UEL: 10% v/v

RTECS Toxicity Data

Acute Oral: Rat LD_{50} Dose: 2200 mg/kg. Mouse LD_{50} Dose: 200 mg/kg; Toxic Effects: Behavioral - General anesthetic.
Acute Inhalation: Rat LC_{Lo} Dose: 14000 mg/m^3/6hr; Toxic Effects: Behavioral - Somnolence (general depressed activity); Lungs, Thorax, or Respiration - Acute pulmonary edema; Kidney, ureter, and Bladder - Chgs in tubules (inc acute renal failure). Mouse LC_{Lo} Dose: 14000 mg/m^3/6hr; Toxic Effects: Behavioral - Somnolence (general depressed activity); Lungs, Thorax, or Respiration - Acute pulmonary edema; Kidney, ureter, and Bladder - Chgs in tubules (inc acute renal failure).
Acute Dermal: Rabbit LD_{50} Route: Skin; Dose: >3200 mg/kg; Toxic Effects: Brain and coverings - Recordings from specific areas of CNS; Behavioral - Ataxia; Lungs, Thorax, or Respiration - Dyspnea. Dog LD_{Lo} Route: Subcutaneous Dose: 1800 mg/kg; Toxic Effects: Gastrointestinal - Other changes; Blood - Hemorrhage.
Irritation Eye: Rabbit Standard Draize Test Dose: 20 mg/24H; Reaction: moderate. Rabbit Standard Draize Test Dose: 81 mg; Reaction: severe.
Irritation Skin: Rabbit Standard Draize Test Dose: 20 mg/24H; Reaction: moderate. Rabbit Standard Draize Test Dose: 3200 mg/kg/24H; Reaction: severe.
Mutagenic: Hamster Sex Chromosome Loss; Cell Type: lung; Dose: 25 mmol/L. Bacteria - E Coli Mutations in Microorganisms; Dose: 7000 ppm (-S9).

Hazard Overviews

Flammable

Fire Diamond

Health: Irritating to eyes/skin/respiratory tract. Also Causes: cough, difficulty breathing, CNS depression, dizziness, double vision, delirium, deafness, death, kidney damage, tearing, stinging, hyperemia of the conjunctiva, damage to color vision, nausea, vomiting, diarrhea.
Fire: For small fires, use dry chemical, carbon dioxide, water spray, or alcohol-resistant foam. For large fires, use water spray, fog, or alcohol-resistant foam.
Reactivity: Stable. Hazardous polymerization cannot occur. Avoid: exposure to heat and ignition sources. Incompatible with: strong oxidizers; hydrogen trisulfide (explosive). Hazardous decomposition products: carbon oxides.
Carcinogenicity: IARC - Not listed; NIOSH - Not listed; NTP - Not listed; ACGIH - Not listed; OSHA - Not listed; EPA - Not listed; MAK - Not listed
Primary Target Organs:

Eyes

Skin

Respiratory System

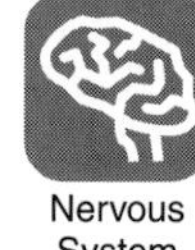
Nervous System

Environmental

Ecotoxicity: TLm Rainbow trout 370 to 490 mg/l/96 hr at 10 °C (static bioassay)

Environmental Fate: Both photolysis and hydrolysis should not be important in the environment. The from soil and aquatic media may occur by aerobic and anaerobic biodegradation, but the rates for these processes are not known. The estimated log K_{oc} value indicates it may be very mobile in soil. From the Henry's Law constant for this compound, the volatilization half-life from a model river has been estimated to be 2.8 days. Neither sorption to sediment and suspended solid in water nor bioconcentration in aquatic organisms are expected to be important in water. The half-life for the reaction with photochemically generated hydroxyl radicals in the atmosphere has been estimated to be 1.5 days. The high water solubility suggests partial removal from the atmosphere should occur by wet deposition.

Cleanup/Disposal: Guide No. 129: Eliminate all ignition sources (no smoking, flares, sparks or flames in immediate area). All equipment used when handling the product must be grounded. Do not touch or walk through spilled material. Stop leak if you can do it without risk. Prevent entry into waterways, sewers, basements or confined areas. A vapor suppressing foam may be used to reduce vapors. Absorb or cover with dry earth, sand or other non-combustible material and transfer to containers. Use clean non-sparking tools to collect absorbed material. Large Spills: Dike far ahead of liquid spill for later disposal. Water spray may reduce vapor; but may not prevent ignition in closed spaces.

Environmental Physical Data

Henry's Law Constant: estimated at 1.254 $x10^{-5}$
Octanol/Water Partition Coefficient: log K_{ow} = 1.42
Sorption Partition Coefficient: log K_{oc} = 1.95 to 2.15
BCF: not significant
BOD: 155%, 5 days

Regulations

RCRA 40CFR: Not listed
CERCLA: 40CFR 302.4: Not listed
SARA 40CFR 372.65: Not listed
SARA EHS 40CFR 355: Not listed
TSCA: Listed

Analytical Methods

Air: ASTM D4490
Water / Groundwater: ASTM D3695
Other: EPA 1666

AMY5000 **CAS #: 75-85-4**

TERT-AMYL ALCOHOL

RTECS: SC0175000
EINECS Number: 200-908-9
Molecular Formula: $C_5H_{12}O$
Structured MF: $(CH_3)_2(OH)CC{=}{=}CH$
Formula Weight: 88.15

Chemical Structure

Synonyms: T-AMYL ALCOHOL; AMYLENE HYDRATE; 2-BUTANOL,2-METHYL-; DIMETHYL ETHYL CARBINOL; DIMETHYLETHYLCARBINOL; 1,1-DIMETHYL-1-PROPANOL; ETHYL DIMETHYL CARBINOL; ETHYLDIMETHYLCARBINOL; 2-METHYL BUTANOL-2; 2-METHYL-2-BUTANOL; 3-METHYL-BUTANOL-(3); 3-METHYLBUTAN-3-OL; TERT-PENTANOL; T-PENTYL ALCOHOL; TERT-PENTYL ALCOHOL

Description: colorless liquid; camphor odor

Use: org synthesis; pharmaceutical solvent for tribromoethanol soln; frothing agent for ore flotation processes; med: sedative-hypnotic; intermed for synthesis of derivatives; in formulations for stabilizing, for degreasing metals, rosin flux removal & dry cleaning; solvent for resins & gums, eg, epoxy-containing novolak resins with low chloride content; in mixtures with surfactants; in petro recovery.

Physical Properties

Boiling Point: 102.4 °C (216 °F) at 760 mm Hg
Freezing Point: -8.8 °C (16.16 °F)
Specific Gravity: 0.8096 at 20 °C/4 °C
Vapor Density: 2.9. Air=1
Saturated Vapor Density: 1.254104537 kg/m^3
Vapor Pressure: 16.8 mm Hg at 25 °C
Water Solubility: 1 parts in 8 parts Water
Other Solubilities: Very Soluble in Acetone.
Surface Tension: 23.8 dynes/cm
pH: Solution is neutral
Refraction Index: 1.4052 at 20 °C/4 °C
Critical Temperature: 272.0 °C
Critical Pressure: 3880 kPa
Ionization Potential (eV): 9.8
Flash Point: 67 °C Closed Cup
Autoignition Temperature: 819 °C

RTECS Toxicity Data

Acute Oral: Rat LD_{50} Dose: 1 gm/kg; Toxic Effects: Behavioral - Ataxia. Mouse LD_{Lo} Dose: 2500 mg/kg.

Acute Dermal: Rat LD_{Lo} Route: Subcutaneous Dose: 1400 mg/kg. Mouse LD_{50} Route: Subcutaneous Dose: 2100 mg/kg.

Hazard Overviews

Flammable

Fire Diamond

Health: Irritating to eyes/skin/respiratory tract. Also Causes: headache, dizziness, difficulty breathing, cough, nausea, diarrhea, double vision, deafness, delirium, stinging, watering, redness, GI irritation.

Fire: Flammable. Can form explosive mixtures in the air. For small fires use dry chemical, carbon dioxide, water spray, or alcohol-resistant foam. For large fires use water spray, fog, or alcohol-resistant foam. Vapors may travel to an ignition source and flash back. Containers may explode in heat of fire.

Reactivity: Stable. Hazardous polymerization cannot occur. Avoid: exposure to heat; ignition sources; light. Incompatible with: oxidizers. Hazardous decomposition products: carbon oxides; acrid smoke.

Carcinogenicity: IARC - Not listed; NIOSH - Not listed; NTP - Not listed; ACGIH - Not listed; OSHA - Not listed; EPA - Not listed; MAK - Not listed

Primary Target Organs:

Eyes

Skin

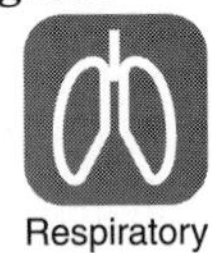
Respiratory System

Nervous System

Environmental

Ecotoxicity: Toxicity threshold (cell multiplication inhibition test): Algae: (Microcystis aeruginosa) 7d EC_0 105 mg/l; Fishes: creek chub 24h LD_0 1,300 mg/l 24h LD_{100} 2,000 mg/l

Environmental Fate: Has a low adsorptivity to soil and if released on soil, may leach. It is resistant to biodegradation in screening tests and was not biodegraded in soil. If released in water, it will be lost by volatilization. Its volatilization half-lives in a model river and model lake is estimated to be 3.9 hr days and 4.1 day, respectively. It would not be expected to biodegrade. Bioconcentration in aquatic organisms should not be important. In the atmosphere, It will react with photochemically-produced hydroxyl radicals resulting in an estimated half-life of 3.3 day. It may also be washed out of the atmosphere by rain.

Cleanup/Disposal: Guide No. 128: Eliminate all ignition sources (no smoking, flares, sparks or flames in immediate area). All equipment used when handling the product must be grounded. Do not touch or walk through spilled material. Stop leak if you can do it without risk. Prevent entry into waterways, sewers, basements or confined areas. A vapor suppressing foam may be used to reduce vapors. Absorb or cover with dry earth, sand or other non-combustible material and transfer to containers. Use clean non-sparking tools to collect absorbed material. Large Spills: Dike far ahead of liquid spill for later disposal. Water spray may reduce vapor; but may not prevent ignition in closed spaces.

Environmental Physical Data

Henry's Law Constant: estimated at 7.3 $x10^{-4}$

Octanol/Water Partition Coefficient: log K_{OW} = 0.89

Sorption Partition Coefficient: log K_{OC} = estimated at 2.9

BCF: calculated at 3

Regulations

RCRA 40CFR: Not listed

CERCLA: 40CFR 302.4: Not listed

SARA 40CFR 372.65: Not listed

SARA EHS 40CFR 355: Not listed

TSCA: Listed

AMY5800 CAS #: 110-58-7

AMYL AMINE

RTECS: SC0300000

EINECS Number: 203-780-2

Molecular Formula: $C_5H_{13}N$

Formula Weight: 87.19

Chemical Structure

Synonyms: 1-AMINO PENTANE; 1-AMINOPENTANE; AMYLAMINE; N-AMYLAMINE; MONOAMYLAMINE; NORLEUCAMINE; 1-PENTANAMINE; PENTYL AMINE; 1-PENTYLAMINE; N-PENTYLAMINE; PENTYLAMINE

Description: colorless liquid

Use: chemical intermediate; dyestuffs; rubber chemicals; insecticides; synthetic detergents; flotation agents; corrosion inhibitors; solvent; gasoline additive; pharmaceuticals

Physical Properties

Boiling Point: 104.4 °C (220 °F)

Freezing Point: -55 °C (-67 °F) to -50 °C (-58 °F)

Specific Gravity: 0.7547

Saturated Vapor Density: 1.295078463 kg/m^3

Density: 3.01

Vapor Pressure: 4.00 kPa at 25 °C

Water Solubility: Very Soluble

Other Solubilities: 95% Ethanol: Very Soluble; Acetone: Very Soluble; Benzene: Very Soluble; Ether: Very Soluble.

Refraction Index: 1.448

Flash Point: 7.2 °C

Hazard Overviews

Flammable

Health: Severely irritating to eyes/skin/respiratory tract. Toxic. Other Acute Effects: harmful if swallowed, inhaled, or absorbed through skin; symptoms of exposure may include burning sensation; coughing; wheezing; laryngitis; shortness of breath; headache; nausea; vomiting.

Fire: Flammable. Hazards: vapor may travel considerable distance to source of ignition and flash back; container explosion may occur; emits toxic fumes; forms explosive mixtures in air. Extinguishing agents: carbon dioxide, dry chemical powder or appropriate foam; water may be effective for cooling, but may not effect extinguishment. Precautions: combustible liquid.

Reactivity: Incompatible with: acids, acid chlorides, acid anhydrides, strong oxidizing agents, carbon dioxide. Hazardous decomposition products: thermal decomposition may produce carbon monoxide, carbon dioxide, and nitrogen oxides.

Carcinogenicity: IARC - Not listed; NIOSH - Not listed; NTP - Not listed; ACGIH - Not listed; OSHA - Not listed; EPA - Not listed; MAK - Not listed

Primary Target Organs:

Eyes

Skin

Respiratory System

Environmental

Ecotoxicity: Fishes: creek chub: 24h LD_0 30 mg/l 24h LD_{100} 50 mg/l

Cleanup/Disposal: Guide No. 132: Fully encapsulating, vapor protective clothing should be worn for spills and leaks with no fire. Eliminate all ignition sources (no smoking, flares, sparks or flames in immediate area). All equipment used when handling the product must be grounded. Do not touch or walk through spilled material. Stop leak if you can do it without risk. Prevent entry into waterways, sewers, basements or confined areas. A vapor suppressing foam may be used to reduce vapors. Absorb with earth, sand or other non-combustible material and transfer to containers (except for Hydrazine). Use clean non-sparking tools to collect absorbed material. Large Spills: Dike far ahead of liquid spill for later disposal. Water spray may reduce vapor; but may not prevent ignition in closed spaces.

Environmental Physical Data

Octanol/Water Partition Coefficient: log K_{ow} = calculated at 1.05

Regulations

RCRA 40CFR: Not listed
CERCLA: 40CFR 302.4: Not listed
SARA 40CFR 372.65: Not listed
SARA EHS 40CFR 355: Not listed
TSCA: Listed

AMY6600 CAS #: 543-59-9

AMYL CHLORIDE

DOT: UN1107; IMO3.2
EINECS Number: 208-846-4
Molecular Formula: $C_5H_{11}Cl$
Structured MF: $CH_3CH_2CH_2CH_2CH_2Cl$
Formula Weight: 106.60

Chemical Structure

Synonyms: N-AMYL CHLORIDE; N-BUTYLCARBONYL CHLORIDE; PENTANE,1-CHLORO-; N-PENTYL CHLORIDE; PENTYL CHLORIDE
Description: water-white liquid; sweet odor
Use: chemical intermediate

Physical Properties

Boiling Point: 107.8 °C (226 °F) at 760 mm Hg
Freezing Point: -99 °C (-146.2 °F)
Specific Gravity: 0.8818 at 20 °C/4 °C
Vapor Density: 3.7 Air=1
Saturated Vapor Density: 1.338204446 kg/m^3
Vapor Pressure: 4.36 kPa at 25 °C
Water Solubility: Insoluble in Water
Other Solubilities: Soluble in Alcohol, Ether, Benzene, Chloroform.
Surface Tension: 24.9 dynes/cm at 20 °C
Refraction Index: 1.41280 at 20 °C/D
Flash Point: 13 °C Open Cup
Autoignition Temperature: 260 °C
LEL: 1.6% v/v
UEL: 8.6% v/v

Hazard Overviews

Flammable

Health: Irritating to eyes/skin/respiratory tract. Other Acute Effects: may be harmful by inhalation, ingestion, or skin absorption.

Fire: Flammable. Hazards: vapor may travel considerable distance to source of ignition and flash back; container explosion may occur; forms explosive mixtures in air; emits toxic fumes. Extinguishing agents: carbon dioxide, dry chemical powder or appropriate foam; water may be effective for cooling, but may not effect extinguishment. Precautions: combustible liquid.

Reactivity: Incompatible with: strong oxidizing agents, strong bases. Hazardous decomposition products: toxic fumes of: carbon monoxide, carbon dioxide, hydrogen chloride gas, phosgene gas.

Carcinogenicity: IARC - Not listed; NIOSH - Not listed; NTP - Not listed; ACGIH - Not listed; OSHA - Not listed; EPA - Not listed; MAK - Not listed

Primary Target Organs:

Eyes

Skin

Respiratory System

Environmental

Cleanup/Disposal: Guide No. 129: Eliminate all ignition sources (no smoking, flares, sparks or flames in immediate area). All equipment used when handling the product must be grounded. Do not touch or walk through spilled material. Stop leak if you can do it without risk. Prevent entry into waterways, sewers, basements or confined areas. A vapor suppressing foam may be used to reduce vapors. Absorb or cover with dry earth, sand or other non-combustible material and transfer to containers. Use clean non-sparking tools to collect absorbed material. Large Spills: Dike far ahead of

liquid spill for later disposal. Water spray may reduce vapor; but may not prevent ignition in closed spaces.

Environmental Physical Data

BCF: no food chain concentration potential

Regulations

RCRA 40CFR: Not listed
CERCLA: 40CFR 302.4: Not listed
SARA 40CFR 372.65: Not listed
SARA EHS 40CFR 355: Not listed
TSCA: Listed

AMY7400 CAS #: 1002-16-0

AMYL NITRATE

RTECS: QV0600000
EINECS Number: 213-684-2
Molecular Formula: $C_5H_{11}NO_3$
Structured MF: $C_5H_{11}ONO_2$
Formula Weight: 133.15
Synonyms: AMYLESTER KYSELINY DUSICNE; NITRATE D'AMYLE
Description: clear, slightly yellow liquid; sickly sweet odor

Physical Properties

Boiling Point: 150 °C (302 °F) to 155 °C (311 °F)
Freezing Point: -123 °C (-189.4 °F)
Specific Gravity: 0.997 at 20 °C
Vapor Density: 4.59 Air=1
Vapor Pressure: 5 mm Hg at 28.8 °C
Water Solubility: Not Soluble in Water
Other Solubilities: miscible with hydrocarbons
Flash Point: 49 °C Open Cup

RTECS Toxicity Data

Acute Inhalation: Mouse LC_{Lo} Dose: 1807 ppm/7hr; Toxic Effects: Behavioral - Altered sleep time (including change in righting reflex); Behavioral - Convulsions or effect on seizure threshold; Behavioral - Ataxia. Rabbit LC_{Lo} Dose: 1807 ppm/7hr; Toxic Effects: Behavioral - Convulsions or effect on seizure threshold; Lungs, Thorax, or Respiration - Cyanosis.

Hazard Overviews

Flammable

Fire Diamond

Health: Causes: hypotension, tachycardia, nausea, headache, methemoglobinemia, blurred vision, diarrhea, abdominal pain, hemolysis.
Fire: Flammable and a strong oxidizer. For small fires use dry chemical, carbon dioxide, water spray, or alcohol-resistant foam. For large fires use water spray, fog, or alcohol-resistant foam. Vapors may travel to an ignition source and flash back. Containers may explode in heat of fire.
Reactivity: Stable. Hazardous polymerization cannot occur. Avoid: ignition sources; vapor generation. Incompatible with: organic and readily oxidized materials. Hazardous decomposition products: toxic fumes of nitrogen oxides.
Carcinogenicity: IARC - Not listed; NIOSH - Not listed; NTP - Not listed; ACGIH - Not listed; OSHA - Not listed; EPA - Not listed; MAK - Not listed
Primary Target Organs:

Cardio-vascular

Blood

Environmental

Cleanup/Disposal: Guide No. 140: Keep combustibles (wood, paper, oil, etc.) away from spilled material. Do not touch damaged containers or spilled material unless wearing appropriate protective clothing. Stop leak if you can do it without risk. Do not get water inside containers. Small Dry Spills: With clean shovel place material into clean, dry container and cover loosely; move containers from spill area. Small Liquid Spills: Use a non-combustible material like vermiculite, sand or earth to soak up the product and place into a container for later disposal. Large Spills: Dike far ahead of liquid spill for later disposal. Following product recovery, flush area with water.

Regulations

RCRA 40CFR: Not listed
CERCLA: 40CFR 302.4: Not listed
SARA 40CFR 372.65: Not listed
SARA EHS 40CFR 355: Not listed
TSCA: Not listed

AMY9000 CAS #: 513-35-9

AMYLENE

DOT: UN2460
EINECS Number: 208-156-3
Molecular Formula: C_5H_{10}
Formula Weight: 70.13

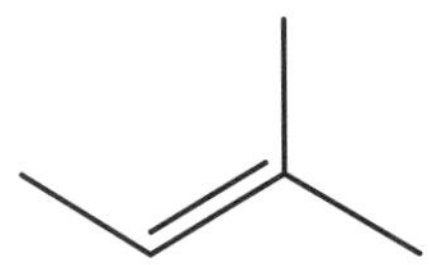

Chemical Structure

Synonyms: 2-BUTENE,2-METHYL-; ETHYLENE,TRIMETHYL-; BETA-ISO-AMYLENE; BETA-ISOAMYLENE; 2-METHYL-2-BUTENE; 2-METHYLBUT-2-ENE; 3-METHYL-2-BUTENE; 1,1,2-TRIMETHYLETHYLENE; TRIMETHYLETHYLENE
Description: colorless liquid; disagreeable odor
Use: principly chem int for isoprene; additive in high octane fuel mfr; chem int for the pharmaceutical solvent, tertiary pentyl alcohol; organic synthesis; hydrogenation,

halogenation, alkylation, condensation reactions; has been used for surgical anesthesia

Physical Properties

Boiling Point: 37.5 °C (100 °F) to 38.5 °C (101 °F)
Freezing Point: -133.8 °C (-208.84 °F)
Specific Gravity: 0.66 at 15 °C/4 °C
Vapor Pressure: 6.01 kPa at -25 °C
Water Solubility: Practically Insoluble in Water
Other Solubilities: Soluble in Benzene; Very Soluble in Petroleum Ether.
Odor Threshold: 0.25 ppm
Refraction Index: 1.3874 at 20 °C/D
Ionization Potential (eV): 8.68 +/-0.01
Flash Point: -17.7 °C

Hazard Overviews

Flammable

Health: Irritating to eyes/skin/respiratory tract. Harmful. Other Acute Effects: harmful if swallowed; may be harmful if inhaled, or absorbed through the skin; prolonged or repeated exposure to skin causes defatting and dermatitis; exposure can cause CNS depression; chemical pneumonitis; narcotic effect; coughing; chest pains; difficulty in breathing; nausea; dizziness; headache; unconsciousness. Chronic Effects: target organs: central nervous system, liver, kidneys.
Fire: Flammable. Hazards: emits toxic fumes; vapor may travel considerable distance to source of ignition and flash back; container explosion may occur; forms explosive mixtures in air. Extinguishing agents: carbon dioxide, dry chemical powder or appropriate foam; water may be effective for cooling, but may not effect extinguishment. Precautions: combustible liquid.
Reactivity: Incompatible with: acids, oxidizing agents. Hazardous decomposition products: toxic fumes of: carbon monoxide, carbon dioxide.
Carcinogenicity: IARC - Not listed; NIOSH - Not listed; NTP - Not listed; ACGIH - Not listed; OSHA - Not listed; EPA - Not listed; MAK - Not listed
Primary Target Organs:

Eyes

Skin

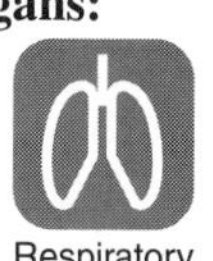
Respiratory System

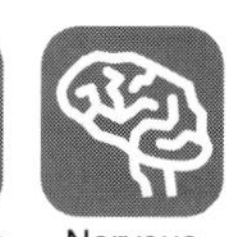
Nervous System

Liver

Kidneys

Environmental

Cleanup/Disposal: Guide No. 127: Eliminate all ignition sources (no smoking, flares, sparks or flames in immediate area). All equipment used when handling the product must be grounded. Do not touch or walk through spilled material. Stop leak if you can do it without risk. Prevent entry into waterways, sewers, basements or confined areas. A vapor suppressing foam may be used to reduce vapors. Absorb or cover with dry earth, sand or other non-combustible material and transfer to containers. Use clean non-sparking tools to collect absorbed material. Large Spills: Dike far ahead of liquid spill for later disposal. Water spray may reduce vapor; but may not prevent ignition in closed spaces.

Regulations

RCRA 40CFR: Not listed
CERCLA: 40CFR 302.4: Not listed
SARA 40CFR 372.65: Not listed
SARA EHS 40CFR 355: Not listed
TSCA: Listed

ANA5000 CAS #: 494-52-0

ANABASINE

RTECS: BV4375000
EINECS Number: 207-791-3
Molecular Formula: $C_{10}H_{14}N_2$
Formula Weight: 162.24
Synonyms: (-)-ANABASIN; ANABASIN; (-)-ANABASINE; S-(-)-ANABASINE; ANABAZIN; NEONICOTINE; NEONIKOTIN; PIPERIDINE,2-(3-PYRIDYL)-; (S)-3-(2-PIPERIDINYL)PYRIDINE; 3-(2-PIPERIDINYL)PYRIDINE; L-3-(2'-PIPERIDYL)PYRIDINE; PYRIDINE,3-(2-PIPERIDINYL)-,(S)-; PYRIDINE,3-(2-PIPERIDYL)-; PYRIDINE,3-(2-PYRIDYL) PIPERIDINE; 2-(3'-PYRIDYL) PIPERIDINE; (-)-2-(3'-PYRIDYL)PIPERIDINE
Description: liquid
Use: insecticide

Physical Properties

Boiling Point: 270 °C (518 °F) to 272 °C (522 °F)
Freezing Point: 9 °C (48.2 °F)
Specific Gravity: 1.0455 at 20 °C/4 °C
Water Solubility: Soluble in Water
Other Solubilities: Soluble in most organic solvents
Refraction Index: 1.5430 at 20 °C/D

RTECS Toxicity Data

Acute Oral: Dog LD_{Lo} Dose: 50 mg/kg; Toxic Effects: Behavioral - Convulsions or effect on seizure threshold; Behavioral - Ataxia; Lungs, Thorax, or Respiration - Dyspnea.
Acute Dermal: Guinea Pig LD_{50} Route: Subcutaneous Dose: 22 mg/kg. Guinea Pig LD_{Lo} Route: Skin; Dose: 100 mg/kg; Toxic Effects: Behavioral - Convulsions or effect on seizure threshold; Behavioral - Ataxia; Lungs, Thorax, or Respiration - Dyspnea.
Reproductive/Teratogenic: Pig Route: Oral; Dose: 20800 ug/kg; Duration: female 30-37D of pregnancy Specific Developmental Abnormalities - Craniofacial (including nose and tongue); Musculoskeletal system.

Hazard Overviews

Carcinogenicity: IARC - Not listed; NIOSH - Not listed; NTP - Not listed; ACGIH - Not listed; OSHA - Not listed; EPA - Not listed; MAK - Not listed

Environmental

Regulations

RCRA 40CFR: Not listed
CERCLA: 40CFR 302.4: Not listed
SARA 40CFR 372.65: Not listed
SARA EHS 40CFR 355: Not listed
TSCA: Not listed

ANE5000 CAS #: 104-46-1

ANETHOLE

RTECS: BZ8925000
EINECS Number: 203-205-5
Molecular Formula: $C_{10}H_{12}O$
Formula Weight: 148.20

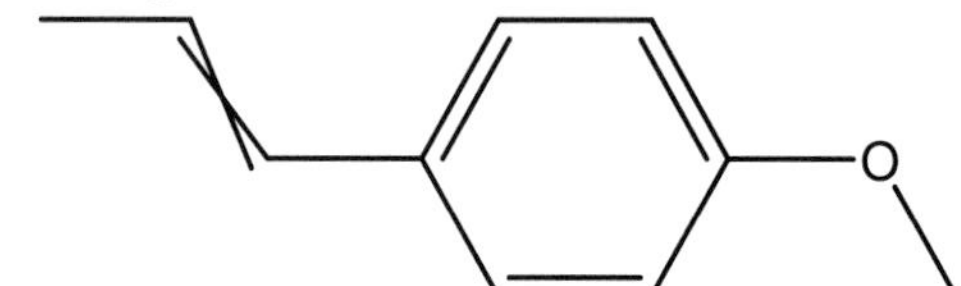

Chemical Structure

Synonyms: ACINTENE O; ANETHOL; P-ANETHOLE; ANISE CAMPHOR; ANISOLE,P-PROPENYL-; BENZENE,1-METHOXY-4-(1-PROPENYL)-; ISOESTRAGOLE; P-METHOXY-BETA-METHYLSTYRENE; 1-(P-METHOXYPHENYL)PROPENE; 1-METHOXY-4-(1-PROPENYL)BENZENE; 1-METHOXY-4-PROPENYLBENZENE; 4-METHOXYPROPENYLBENZENE; MONASIRUP; NAULI; OIL OF ANISEED; 1-PROPENE,1-(4-METHOXYPHENYL)-; PROPENE,1-(P-METHOXYPHENYL)-; P-PROPENYL ANISOLE; 4-PROPENYLANISOLE; P-1-PROPENYLANISOLE; P-PROPENYLANISOLE; P-PROPENYLMETHOXYBENZENE; P-PROPENYLPHENYL METHYL ETHER

Description: white crystals; anise oil odor

Use: manufacture of anisaldehyde; flavoring agent; in perfumery; particularly for soap and dentifrices; sensitizer in bleaching colors in color photography; as an imbedding material in microscopy; licorice candies

Physical Properties

Boiling Point: 234 °C (453 °F)
Freezing Point: 21.3 °C (70.34 °F)
Specific Gravity: 0.9878 at 20 °C/4 °C
Density: 0.9882 g/mL
Water Solubility: Slightly Soluble
Other Solubilities: 95% Ethanol: Very Soluble; Acetone: Very Soluble; Benzene: Very Soluble; Carbon Disulfide: Soluble; Chloroform: Very Soluble; Ether: Very Soluble; Ethyl Acetate: Soluble; organic solvents: Very Soluble; Petroleum Ether: Soluble.
Refraction Index: 1.5615
Flash Point: 90 °C

RTECS Toxicity Data

Acute Oral: Rat LD_{50} Dose: 2090 mg/kg; Toxic Effects: Behavioral - Somnolence (general depressed activity); Behavioral - Coma. Mouse LD_{50} Dose: 3050 mg/kg; Toxic Effects: Behavioral - Somnolence (general depressed activity); Behavioral - Coma.

Acute Dermal: Rabbit LD_{50} Route: Skin; Dose: >5 gm/kg.

Tumorigenic: Mouse Route: Intraperitoneal; Dose: 2400 mg/kg/8W-I; Toxic Effects: Tumorigenic - Equivocal tumorigenic agent by RTECS criteria; Lungs, Thorax, or Respiration - Tumors.

Hazard Overviews

Health: Severely irritating to eyes/skin/respiratory tract. Other Acute Effects: may be harmful by inhalation, ingestion, or skin absorption.

Fire: Combustible. Hazards: emits toxic fumes. Extinguishing agents: carbon dioxide, dry chemical powder or appropriate foam; water spray. Precautions: combustible liquid.

Reactivity: Incompatible with: strong oxidizing agents, sensitive to light. Hazardous decomposition products: toxic fumes of: carbon monoxide, carbon dioxide.

Carcinogenicity: IARC - Not listed; NIOSH - Not listed; NTP - Not listed; ACGIH - Not listed; OSHA - Not listed; EPA - Not listed; MAK - Not listed

Primary Target Organs:

Eyes

Skin

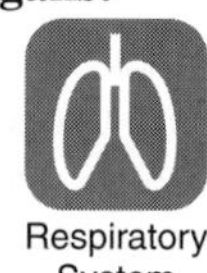
Respiratory System

Environmental

Environmental Fate: Insufficient data are available to predict the relative importance or rate of biodegradation in soil or water. It is expected to have medium to low mobility in soil. In water, it is expected to volatilize and have some adsorption to suspended solids and sediment. It will exist in the vapor phase in the ambient atmosphere. If released to the atmosphere, it will degrade by reaction with photochemically produced radicals with estimated half-lives of approximately 4.9 and 4.5 hours for the cis and trans isomers, respectively. It will also degrade by reaction with atmospheric ozone with estimated half-lives for the cis and trans isomers of 4.03 and 2.02 hours, respectively.

Environmental Physical Data

Henry's Law Constant: estimated at 7.18 $x10^{-5}$
Sorption Partition Coefficient: K_{OC} = estimated at 680
BCF: estimated at 43.4

Regulations

RCRA 40CFR: Not listed
CERCLA: 40CFR 302.4: Not listed
SARA 40CFR 372.65: Not listed
SARA EHS 40CFR 355: Not listed
TSCA: Listed

ANI1000 CAS #: 144-14-9

ANILERIDINE

Molecular Formula: $C_{22}H_{28}N_2O_2$
Formula Weight: 352.46
Synonyms: ADOPOL; ALIDINE; 1-(P-AMINOPHENETHYL)-4-PHENYLISONIPECOTIC ACID ETHYL ESTER; 1-(P-AMINOPHENETHYL)-4-PHENYLPIPERIDINE-4-CARBOXYLIC ACIDETHYL ESTER; N-BETA-(P-AMINOPHENYL)ETHYLNORMEPERIDINE; N-(BETA-(P-AMINOPHENYL)ETHYL)-4-PHENYL-4-CARBETHOXYPIPERIDINE; 1-[2-(4-AMINOPHENYL)ETHYL]-4-PHENYL-4-PIPERIDINECARBOXYLICACID ETHYL ESTER; APODOL; ETHYL 1-(4-AMINOPHENETHYL)-4-PHENYLISONIPECOTATE; ETHYL 1-(P-AMINOPHENETHYL)-4-PHENYLISONIPECOTATE; ISONIPECOTIC ACID,1-(P-AMINOPHENETHYL)-4-PHENYL-,ETHYLESTER; LERITIN; LERITINE; NIPECOTAN; 4-PIPERIDINECARBOXYLIC ACID,1-(2-(4-AMINOPHENYL)ETHYL)-4-PHENYL-,ETHYL ESTER
Description: white to yellowish white crystalline powder; odorless to practically odorless
Use: medication: narcotic analgesic

Physical Properties

Freezing Point: 83 °C (181.4 °F)
Water Solubility: Very Slightly Soluble in Water
Other Solubilities: Soluble in Ether; Soluble in 1 in 2 parts of Alcohol and 1 in 1 of Chloroform.
pH: Aqueous solution 2.0 to 2.5

Hazard Overviews

Carcinogenicity: IARC - Not listed; NIOSH - Not listed; NTP - Not listed; ACGIH - Not listed; OSHA - Not listed; EPA - Not listed; MAK - Not listed

Environmental

Regulations

RCRA 40CFR: Not listed
CERCLA: 40CFR 302.4: Not listed
SARA 40CFR 372.65: Not listed
SARA EHS 40CFR 355: Not listed
TSCA: Not listed

ANI1670 CAS #: 62-53-3

ANILINE

RTECS: BW6650000
DOT: UN1547; IMO6.1
EINECS Number: 200-539-3
Molecular Formula: C_6H_7N
Structured MF: $C_6H_5NH_2$
Formula Weight: 93.12

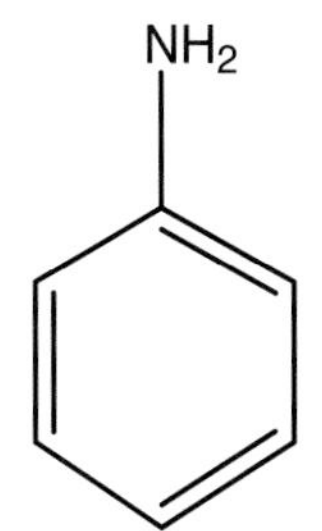

Chemical Structure

Synonyms: AMINOBENZENE; AMINOPHEN; ANILIN; ANILINA; ANILINE AND HOMOLOGS; ANILINE OIL; ANILINE OIL,LIQUID; ANYVIM; ARYLAMINE; BENZENAMINE; BENZENE,AMINO; BENZENE,AMINO-; BENZIDAM; BLUE OIL; C.I. 76000; C.I. OXIDATION BASE 1; CYANOL; EPA PESTICIDE CHEMICAL CODE 251400; HUILE D'ANILINE; KRYSTALLIN; KYANOL; PHENYLAMINE
Description: colorless, blue, brownish oily liquid; amine-like odor
Use: in the manufacture of rubber chemicals, agriculture chemicals and dyestuffs and in the production of MDI group isocyanates used in polyurethane; the parent substance for many dyes and drugs; in rubber accelerators, antioxidants, photographic chemicals explosives, petroleum refining, diphenylamine, phenolics, herbicides and fungicides; in marking inks, tetryl, optical whitening agents, resins, varnishes, perfumes, shoe polishes and many organic chemicals

Physical Properties

Boiling Point: 184 °C (363 °F) to 186 °C (367 °F)
Freezing Point: -6.3 °C (20.66 °F)
Specific Gravity: 1.022 at 20 °C/20 °C
Vapor Density: 3.22 Air=1
Saturated Vapor Density: 1.201710642 kg/m^3
Density: 1.0217 g/mL at 20 °C
Vapor Pressure: 4.9 x10^{-1} mm Hg at 25 °C
Water Solubility: 4% by weight
Other Solubilities: miscible in Carbon Tetrachloride, Acetone; miscible with lipids; Soluble in dilute Hydrochloric Acid; miscible with vegetable oils, essential oils; > 10% in Benzene; > 10% in Ethyl Ether; > 10% in Ethyl Alcohol; > 10% in Petroleum Ether.
Surface Tension: 45.5 dynes/cm
Odor Threshold: Recognition 1 ppm
pH: 0.2 molar aqueous solution 8.1
Refraction Index: 1.5863 at 20 °C/D
Evaporation Rate: < 1 Butyl Acetate=1
Critical Temperature: 426 °C
Critical Pressure: 10 mm Hg
Ionization Potential (eV): 7.70
Flash Point: 70 °C Closed Cup
Autoignition Temperature: 615 °C
LEL: 1.3% v/v
UEL: 25% v/v

RTECS Toxicity Data

Acute Oral: Child TD_{Lo} Dose: 3125 mg/kg; Toxic Effects: Lungs, Thorax, or Respiration - Cyanosis. Rat LD_{50} Dose: 250 mg/kg.

Acute Inhalation: Mouse LC_{50} Dose: 175 ppm/7hr. Rat LC_{Lo} Dose: 250 ppm/4hr.

Acute Dermal: Rabbit LD_{50} Route: Skin; Dose: 820 mg/kg. Rabbit LD_{Lo} Route: Subcutaneous Dose: 1 gm/kg; Toxic Effects: Lungs, Thorax, or Respiration - Other changes.

Chronic (Multiple Dose) Oral: Rat Dose: 913 mg/kg/2W-I; Toxic Effects: Endocrine - Changes in spleen weight; Nutritional and gross metabolic - Changes in iron; Biochemical - Transaminases. Rat Dose: 550 mg/kg/5D-C; Toxic Effects: Blood - Changes in spleen; Nutritional and gross metabolic - Weight loss or decreased weight gain; DEATH. Rat Dose: 210 mg/kg/12W-C; Toxic Effects: Blood - Pigmented or nucleated red Blood cells; Blood - Methemoglobinemia-Carboxyhemoglobin; Biochemical - Other esterases.

Chronic (Multiple Dose) Inhalation: Rat Dose: 3 mg/m^3/22W-I; Toxic Effects: Blood - Changes in serum composition; Blood - Other changes. Rat Dose: 87 ppm/6H/2W-I; Toxic Effects: Liver - Changes in Liver weight; Endocrine - Changes in spleen weight; Blood - Changes in erythrocite (RBC) cell count. Rat Dose: 300 ug/m^3/24H/80D-C; Toxic Effects: Behavioral - Muscle contraction or spasticity. Rat Dose: 5 mg/m^3/24H/21D-C; Toxic Effects: Blood - Methemoglobinemia-Carboxyhemoglobin.

Chronic (Multiple Dose) Dermal: Rat Route: Subcutaneous; Dose: 900 mg/kg/6D-I; Toxic Effects: Liver - Other changes; Biochemical - Hepatic microsomal mixed oxidase(dealkylation, hyroxylation,etc); Biochemical - Dehydrogenases.

Irritation Eye: Rabbit Standard Draize Test Dose: 102 mg; Reaction: severe. Rabbit Standard Draize Test Dose: 20 mg/24H; Reaction: moderate.

Irritation Skin: Rabbit Standard Draize Test Dose: 20 mg/24H; Reaction: moderate.

Reproductive/Teratogenic: Mouse Route: Oral; Dose: 4480 mg/kg; Duration: female 6-13D of pregnancy; Effects on Newborn - Growth statistics.

Mutagenic: Rat DNA Damage; Route: Intraperitoneal; Dose: 105 mg/kg. Rat Sister Chromatid Exchenge; Cell Type: liver; Dose: 200 umol/L.

Tumorigenic: Rat Route: Oral; Dose: 11 gm/kg/29W-C; Toxic Effects: Tumorigenic - Neoplastic by RTECS criteria; Kidney, Ureter, and Bladder - Tumors. Rat Route: Oral; Dose: 72800 mg/kg/2Y-C; Toxic Effects: Tumorigenic - Neoplastic by RTECS criteria; Blood - Tumors.

Hazard Overviews

Fire Diamond

Health: Irritating to eyes/skin/respiratory tract. Toxic. Also Causes: headache, ear ringing, confusion, faintness, disorientation, incoordination, lethargy, fatigue, drowsiness, nausea, vomiting, extremity paresthesias, muscle pain, photophobia, speech distur-bances, weak vision, sluggish pupillary reaction, unconsciousness, coma, death. Chronic Effects: anemia, energy loss, headache, and digestive disturbances. Combustible.

Fire: Combustible. Can form explosive mixtures in the air. For small fires use dry chemical, water spray (do not scatter with a high-pressure stream), carbon dioxide, or regular foam. For large fires use water spray, fog, or regular foam.

Reactivity: Stable. Hazardous polymerization cannot occur. Avoid: exposure to light; heat; ignition sources. Incompatible with: steam; oxidizers; acids; alkali or alkaline earth metals; dibenzoyl peroxide; benzenediazonium-2-carboxylate; fluorine nitrate; nitrosyl perchlorate; red fuming nitric acid; peroxodisulfuric acid; tetranitromethane; boron trichloride; peroxyformic acid; diisopropyl peroxydicarbonate; fluorine; acetic anhydride; trichloronitromethane (145 °C); chlorosulfonic acid; hexachloromelamine; nitric acid and nitrogen tetraoxide and sulfuric acid; nitrobenzene and glycerin; formaldehyde and perchloric acid; perchromates; beta-propiolactone; oleum; silver perchlorate; sulfuric acid; trichloromelamine; n-haloimides; peroxydisulfuric acid; perchloryl fluoride; diisopropyl peroxy-dicarbonate; trichloronitromethane; anilinium chloride; nitromethane; hydrogen peroxide; 1-chloro-2,3-epoxy-propane; peroxomonosulfuric acid; ozone; perchloryl; fluoride; perchloric acid; sodium peroxide and water. Hazardous decomposition products: carbon and nitrogen oxides.

Carcinogenicity: IARC - Group 3, Not classifiable as to carcinogenicity to humans; NIOSH - Listed as carcinogen; NTP - Not listed; ACGIH - Class A3, Animal carcinogen; OSHA - Not listed; EPA - Class B2, Probable human carcinogen based on animal studies; MAK - Class B, Justifiably suspected of having carcinogenic potential

Primary Target Organs:

Eyes

Skin

Nervous System

Liver

Kidneys

Blood

Exposure Limits

OSHA PEL: TWA: 5 ppm; 19 mg/m^3.

OSHA PEL Vacated 1989 Limits: TWA: 2 ppm; 8 mg/m^3.

ACGIH TLV: TWA: 2 ppm; 7.6 mg/m^3.

NIOSH IDLH: 100 ppm.

DFG MAK: TWA: 2 ppm; 8 mg/m^3.

Respirator Recommendation

Exposure Range: >5 to <100 ppm Supplied Air, Constant Flow/Pressure Demand, Full Face

Exposure Range: 100 to unlimited ppm Self-contained Breathing Apparatus, Pressure Demand, Full Face

Environmental

Ecotoxicity: LC_{50} Daphnia pulex 0.10 mg/l/48 hr. /Conditions of bioassay not specified LC_{50} Bass (8 exposure days beyond hatching) 4.4 mg/l in water hardness of $CaCO_3$ 200 mg/l. /Conditions of bioassay not specified LC_{50} Ambystoma mexicanum (mexican axolotl); (3-4 weeks after hatching) 440 mg/l/48 hr. /Conditions of bioassay not specified LC_{50} Leuciscus idus melanotus (golden orfe) 51-92 mg/l/48 hr. /Conditions of bioassay not specified LC_{100} Tetrahymena pyriformis (protozoa: ciliate) 21.5 mmol/l/24 hr. /Conditions of bioassay not specified

Environmental Fate: If released into water it will primarily be lost due to biodegradation and in surface waters, photooxidation (half-life of the order of days). It will not bioconcentrate in fish. If spilled on land it will be lost by a combination of biodegradation, oxidation and chemical binding to components of soil. If released into air, it will photodegrade (estimated half-life 3.3 hr).

Cleanup/Disposal: Guide No. 153: Eliminate all ignition sources (no smoking, flares, sparks or flames in immediate area). Do not touch damaged containers or spilled material unless wearing appropriate protective clothing. Stop leak if you can do it without risk. Prevent entry into waterways, sewers, basements or confined areas. Absorb or cover with dry earth, sand or other non-combustible material and transfer to containers. Do not get water inside containers.

Environmental Physical Data

Henry's Law Constant: calculated at 1.2 $x10^{-4}$

Octanol/Water Partition Coefficient: log K_{ow} = 0.90

Sorption Partition Coefficient: K_{oc} = 130 to 410

BCF: fish 0.78

BOD: 150%, 5 days

Regulations

RCRA 40CFR: Listed Hazardous Waste No. U012 Toxic Waste Ignitable Waste

CERCLA: 40CFR 302.4: Listed per CWA Section 311(b)(4) per RCRA Section 3001 RQ: 5000 lb (2268 kg)

SARA 40CFR 372.65: Listed TPQ: 1000 lb

SARA EHS 40CFR 355: Listed TPQ: 5,000 lb

TSCA: Listed

Analytical Methods

Air: ASTM D4490

Soil: SW846 3640A, 8131, 8250A, 8270B, 8270C

Water / Groundwater: EPA 1625; ASTM D4763

Indoor / Expired Air: NIOSH 2002

Plasma: EPA 29

Other: EPA 1665

ANI2340 CAS #: 142-04-1

ANILINE HYDROCHLORIDE

RTECS: CY0875000

EINECS Number: 205-519-8

Molecular Formula: C_6H_8ClN

Formula Weight: 129.60

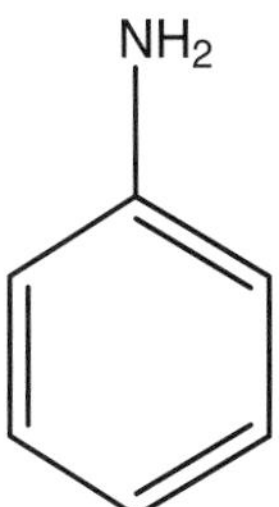

Chemical Structure

Synonyms: ANILINE CHLORIDE; ANILINE SALT; ANILINIUM CHLORIDE; BENZENAMINE,HYDROCHLORIDE; C.I. 76001; CHLORHYDRATE D'ANILINE; CHLORID ANILINU; HYDROCHLORIDE BENZENAMIDE; PHENYLAMINE HYDROCHLORIDE; SUL ANILINOVA

Description: white crystals or black granular solid

Use: intermediates; dyeing and printing; Aniline black

Physical Properties

Boiling Point: 245 °C (473 °F)

Freezing Point: 198 °C (388.4 °F)

Vapor Density: 4.46 Air=1

Density: 1.2215 g/mL

Water Solubility: < 1 mg/mL at 18 C

Other Solubilities: 95% Ethanol: <1 mg/ml at 18 °C; Acetone: <1 mg/ml at 18 °C; Chloroform: Insoluble; DMSO: <1 mg/ml at 18 °C; Ether: Insoluble.

Flash Point: 190 °C

RTECS Toxicity Data

Acute Oral: Rat LD_{50} Dose: 840 mg/kg. Mouse LD_{50} Dose: 841 mg/kg.

Chronic (Multiple Dose) Oral: Rat Dose: 1800 mg/kg/90D-C; Toxic Effects: Endocrine - Changes in spleen weight; Blood - Methemoglobinemia-Carboxyhemoglobin; Blood - Changes in erythrocite (RBC) cell count.

Irritation Eye: Rabbit Standard Draize Test Dose: 20 mg/24H; Reaction: moderate.

Irritation Skin: Rabbit Standard Draize Test Dose: 500 mg/24H; Reaction: moderate.

Reproductive/Teratogenic: Rat Route: Oral; Dose: 1400 mg/kg; Duration: female 7-20D of pregnancy; Specific Developmental Abnormalities - Blood and lymphatic systems (including spleen and marrow). Rat Route: Subcutaneous; Dose: 780 mg/kg; Duration: female 19-21D of pregnancy Maternal Effects - Other effects on females; Effects on Embryo or Fetus - Fetotoxicity.

Mutagenic: Human Sister Chromatid Exchenge; Cell Type: lymphocyte; Dose: 50 umol/L. Rat Morphological Transformation; Cell Type: embryo; Dose: 79500 ng/plate.

Tumorigenic: Rat Route: Oral; Dose: 130 gm/kg/2Y-C; Toxic Effects: Tumorigenic - Carcinogenic by RTECS criteria; Vascular - Tumors; Blood - Tumors. Rat Route: Oral; Dose: 238 gm/kg/2Y-C; Toxic Effects: Tumorigenic - Carcinogenic by RTECS criteria; Vascular - Tumors; Blood - Tumors. Rat

Route: Oral; Dose: 137 gm/kg/60W-C; Toxic Effects: Tumorigenic - Equivocal tumorigenic agent by RTECS criteria; Liver - Tumors; Blood - Tumors. Rat Route: Oral; Dose: 2163 gm/kg/2Y-C; Toxic Effects: Tumorigenic - Carcinogenic by RTECS criteria; Blood - Tumors. Rat Route: Oral; Dose: 4326 gm/kg/2Y-C; Toxic Effects: Tumorigenic - Carcinogenic by RTECS criteria; Blood - Tumors.

Hazard Overviews

Poison

Health: Irritating to eyes/skin/respiratory tract. Poison. Other Acute Effects: may be fatal if inhaled; swallowed; or absorbed through skin; symptoms of exposure may include burning sensation; coughing; wheezing; laryngitis; shortness of breath; headache; nausea; vomiting; absorption into the body leads to the formation of methemoglobin which in sufficient concentration causes cyanosis; onset may be delayed 2 to 4 hours or longer. Chronic Effects: may cause allergic skin reaction; laboratory experiments have shown mutagenic effects; target organs: blood; central nervous system; bladder. Possible human carcinogen.

Fire: Will burn. Hazards: emits toxic fumes. Extinguishing agents: water spray; carbon dioxide, dry chemical powder or appropriate foam. Precautions: combustible liquid.

Reactivity: Incompatible with: strong oxidizing agents. Hazardous decomposition products: toxic fumes of: carbon monoxide, carbon dioxide, nitrogen oxides, hydrogen chloride gas.

Carcinogenicity: IARC - Not listed; NIOSH - Not listed; NTP - Not listed; ACGIH - Not listed; OSHA - Not listed; EPA - Not listed; MAK - Not listed

Primary Target Organs:

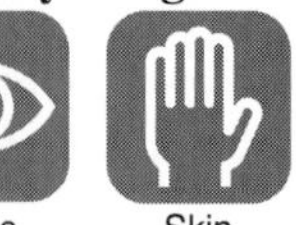
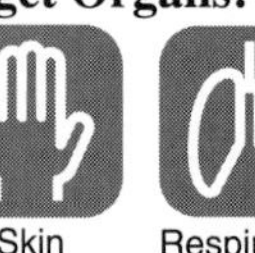
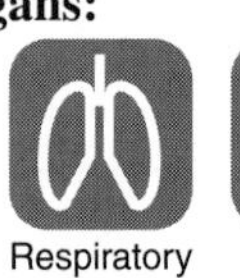

Eyes Skin Respiratory System Nervous System Blood

Environmental

Ecotoxicity: Fishes: 48h LC_{50} 5.5 mg/l

Cleanup/Disposal: Guide No. 153: Eliminate all ignition sources (no smoking, flares, sparks or flames in immediate area). Do not touch damaged containers or spilled material unless wearing appropriate protective clothing. Stop leak if you can do it without risk. Prevent entry into waterways, sewers, basements or confined areas. Absorb or cover with dry earth, sand or other non-combustible material and transfer to containers. Do not get water inside containers.

Regulations

RCRA 40CFR: Not listed
CERCLA: 40CFR 302.4: Not listed
SARA 40CFR 372.65: Not listed
SARA EHS 40CFR 355: Not listed
TSCA: Listed

ANI3010 CAS #: 123-11-5

P-ANISALDEHYDE

RTECS: BZ2625000
EINECS Number: 204-602-6
Molecular Formula: $C_8H_8O_2$
Formula Weight: 136.16

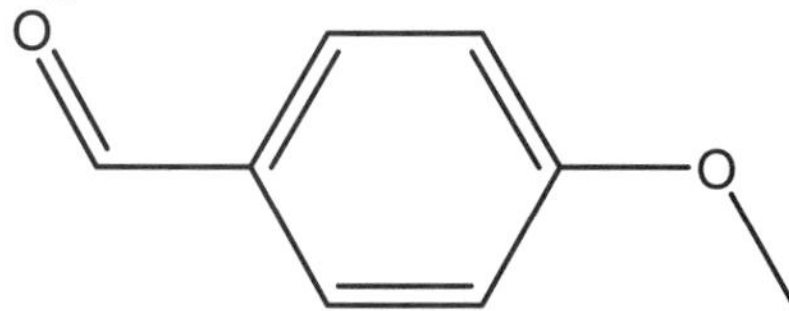

Chemical Structure

Synonyms: 4-ANISALDEHYDE; ANISIC ALDEHYDE; P-ANISIC ALDEHYDE; AUBEPINE; BENZALDEHYDE,4-METHOXY-; CRATEGINE; P-FORMYLANISOLE; 4-METHOXYBENZALDEHYDE; P-METHOXYBENZALDEHYDE; OBEPIN

Description: colorless to slightly yellow ,oily liquid; characteristic hawthorn odor

Use: in org synthesis; in prepn of insect attractants especially effective against cockroaches; fragrance chem used in soaps, detergents, creams, perfume; in flavor compositions such as anise, caramel, chocolate, strawberry & vanilla; chem int for antihistamines; in electroplating

Physical Properties

Boiling Point: 249.5 °C (481 °F) at 760 mm Hg
Freezing Point: -1 °C (30.2 °F)
Specific Gravity: 1.1191 at 15 °C/4 °C
Vapor Pressure: 0.161 kPa at 75 °C
Water Solubility: Insoluble in Water
Other Solubilities: Soluble in most organic solvents; at least 99% Soluble in Sodium Bisulfite solution.
Refraction Index: 1.5730 at 20 °C/D
Ionization Potential (eV): 8.43 +/-1.0
Flash Point: 121 °C

RTECS Toxicity Data

Acute Oral: Rat LD_{50} Dose: 1510 mg/kg; Toxic Effects: Behavioral - Somnolence (general depressed activity). Mouse LD_{50} Dose: 1859 mg/kg; Toxic Effects: Gastrointestinal - Necrotic changes; Gastrointestinal - Other changes; Liver - Fatty Liver degeneration.

Acute Dermal: Rabbit LD_{50} Route: Skin; Dose: >5 gm/kg.

Chronic (Multiple Dose) Oral: Rat Dose: 55200 ug/kg/30D-I; Toxic Effects: Cardiac - Changes in heart weight; Kidney, Ureter, and Bladder - Changes in kidney weight; Blood - Changes in other cell count.

Irritation Skin: Rabbit Standard Draize Test Dose: 500 mg/24H; Reaction: moderate.

Mutagenic: Human Sister Chromatid Exchenge; Cell Type: lymphocyte; Dose: 1 mmol/L. Mouse DNA Damage; Cell Type: lymphocyte; Dose: 7020 umol/L.

Hazard Overviews

Health: Irritating to eyes/skin/respiratory tract. Harmful. Other Acute Effects: harmful if swallowed; may be harmful if inhaled or absorbed through the skin.

Fire: Will burn. Hazards: emits toxic fumes. Extinguishing agents: water spray; carbon dioxide, dry chemical powder or appropriate foam. Precautions: combustible liquid.

Reactivity: Incompatible with: strong bases, strong oxidizing agents, strong reducing agents. Hazardous decomposition products: toxic fumes of: carbon monoxide, carbon dioxide.

Carcinogenicity: IARC - Not listed; NIOSH - Not listed; NTP - Not listed; ACGIH - Not listed; OSHA - Not listed; EPA - Not listed; MAK - Not listed

Primary Target Organs:

Eyes

Skin

Respiratory System

Environmental

Regulations

RCRA 40CFR: Not listed
CERCLA: 40CFR 302.4: Not listed
SARA 40CFR 372.65: Not listed
SARA EHS 40CFR 355: Not listed
TSCA: Listed

ANI4350 CAS #: 90-04-0

O-ANISIDINE

RTECS: BZ5410000
DOT: UN2431; IMO6.1
EINECS Number: 201-963-1
Molecular Formula: C_7H_9NO
Structured MF: $CH_3OC_6H_4NH_2$
Formula Weight: 123.2

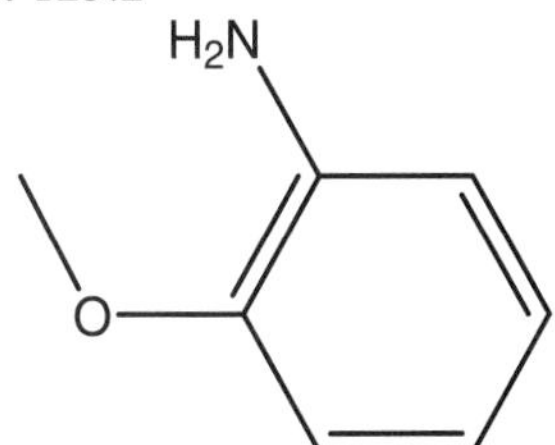

Chemical Structure

Synonyms: 2-AMINOANISOLE; O-AMINOANISOLE; ORTHO-AMINOANISOLE; 1-AMINO-2-METHOXYBENZENE; O-AMINOPHENOL METHYL ETHER; 2-ANISIDINE; O-ANISYLAMINE; ORTHO-ANISYLAMINE; BENZENAMINE,2-METHOXY-; BENZENAMINE,2-METHOXY-(9CI); 2-METHOXY-1-AMINOBENZENE; 2-METHOXYANILINE; O-METHOXYANILINE; ORTHO-METHOXYANILINE; 2-METHOXYBENZENAMINE; O-METHOXYPHENYLAMINE; ORTHO-METHOXYPHENYLAMINE; NSC 3122

Description: pale yellowish liquid or reddish or yellowish colored oil; colorless to pink liquid; red to yellow, oily liquid, solid below 41 deg F; characteristic amine odor

Use: in the manufacture of azo or triphenylmethane dyes and intermediates, in the preparation of organic compounds, in the synthesis guaicol, in the synthesis of hair dyes, as a corrosion inhibitor for steel storage and as an antioxidant for some polymercaptan resins; in the production of pharmaceuticals and textile-processing chemicals

Physical Properties

Boiling Point: 225 °C (437 °F)
Freezing Point: 5 °C (41 °F)
Specific Gravity: 1.098 at 15 °C/15 °C
Vapor Density: 4.25 Air=1
Saturated Vapor Density: Estimated 1.200071804 kg/m^3
Density: 1.097 g/mL at 20 °C
Vapor Pressure: Estimated 0.014 mm Hg at 25 °C
Water Solubility: < 0.1 mg/mL at 19 C
Other Solubilities: 95% Ethanol: >=100 mg/ml at 23 °C; Acetone: >=100 mg/ml at 23 °C; Benzene: miscible; DMSO: >=100 mg/ml at 23 °C; Dilute mineral acid: Soluble; Ether: miscible.
Refraction Index: 1.5715 at 10 °C/D
Ionization Potential (eV): 7.44
Flash Point: 118 °C Open Cup

RTECS Toxicity Data

Acute Oral: Rat LD_{50} Dose: 1150 mg/kg. Mouse LD_{50} Dose: 1400 mg/kg; Toxic Effects: Kidney, Ureter, and Bladder - Other changes; Blood - Normocytic anemia; Blood - Other changes.

Mutagenic: Mouse DNA Damage; Cell Type: lymphocyte; Dose: 1250 umol/L. Mouse DNA Inhibition; Route: Oral; Dose: 200 mg/kg.

Hazard Overviews

Health: Irritating to eyes/skin/respiratory tract. Toxic. Other Acute Effects: harmful if swallowed, inhaled, or absorbed through skin; may cause cyanosis (blue-gray coloring of skin and lips caused by lack of oxygen); may cause allergic respiratory and skin reactions; readily absorbed through skin. Chronic Effects: Probable carcinogen.

Fire: Will burn. Hazards: emits toxic fumes. Extinguishing agents: water spray; carbon dioxide, dry chemical powder or appropriate foam. Precautions: combustible liquid.

Carcinogenicity: IARC - Group 2B, Possibly carcinogenic to humans; NIOSH - Listed as carcinogen; NTP - Not listed; ACGIH - Class A3, Animal carcinogen; OSHA - Not listed; EPA - Not listed; MAK - Class A2, Unmistakably carcinogenic in animal experimentation only

Primary Target Organs:

Eyes

Skin

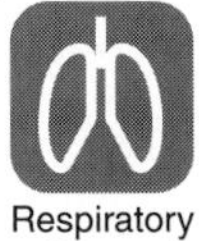
Respiratory System

Exposure Limits
OSHA PEL: TWA: 0.5 mg/m^3; skin.
ACGIH TLV: TWA: 0.1 ppm; 0.5 mg/m^3.
NIOSH IDLH: 50 mg/m^3.
Respirator Recommendation
Exposure Range: >0.5 to 25 mg/m^3 Supplied Air, Constant Flow/Pressure Demand, Half Mask
Exposure Range: >25 to <50 mg/m^3 Supplied Air, Constant Flow/Pressure Demand, Full Face
Exposure Range: 50 to unlimited mg/m^3 Self-contained Breathing Apparatus, Pressure Demand, Full Face
Note: poor warning properties

Environmental

Ecotoxicity: Bacteria: Pseudomonas fluorescens 24h EC 0 5,000 mg/l E. coli 24h EC 0 5,000 mg/l; Crustaceans: Daphnia magna 48h EC_{50} 6.8 mg/l Fishes: Leuciscus idus 96h LC_0 80 mg/l
Environmental Fate: If released to soil, it is expected to be immobilized by strong, irreversible covalent bonding with any humic materials. If released to water, it may react with either photochemically generated hydroxyl radicals or alkylperoxy radicals found in sunlit water (half-life 19-30 hours) or it may be strongly bound with humic materials found in suspended solids and sediments. Chemical hydrolysis, aerobic biodegradation, bioaccumulation in aquatic organisms, and volatilization are not expected to be important fate processes. If released to the atmosphere, it appears that reaction with photochemically generated hydroxyl radicals would be the dominant fate process (half-life 3 hours).
Cleanup/Disposal: Guide No. 153: Eliminate all ignition sources (no smoking, flares, sparks or flames in immediate area). Do not touch damaged containers or spilled material unless wearing appropriate protective clothing. Stop leak if you can do it without risk. Prevent entry into waterways, sewers, basements or confined areas. Absorb or cover with dry earth, sand or other non-combustible material and transfer to containers. Do not get water inside containers.

Environmental Physical Data

Henry's Law Constant: estimated at 1.4 $x10^{-6}$
Octanol/Water Partition Coefficient: log K_{ow} = 1.18
BCF: estimated at 3 to 5
BOD: 69.1% BODT, 14 days

Regulations

RCRA 40CFR: Not listed
CERCLA: 40CFR 302.4: Listed per CAA Section 112 RQ: 100 lb (45.35 kg)
SARA 40CFR 372.65: Listed
SARA EHS 40CFR 355: Not listed
TSCA: Listed

Analytical Methods

Soil: SW846 8270B, 8270C
Water / Groundwater: EPA 1625
Indoor / Expired Air: NIOSH 2514
Plasma: EPA 29

ANI5020 **CAS #: 104-94-9**

P-ANISIDINE

RTECS: BZ5450000
DOT: UN2431; IMO6.1
EINECS Number: 203-254-2
Molecular Formula: C_7H_9NO
Structured MF: $CH_3OC_6H_4NH_2$
Formula Weight: 123.15

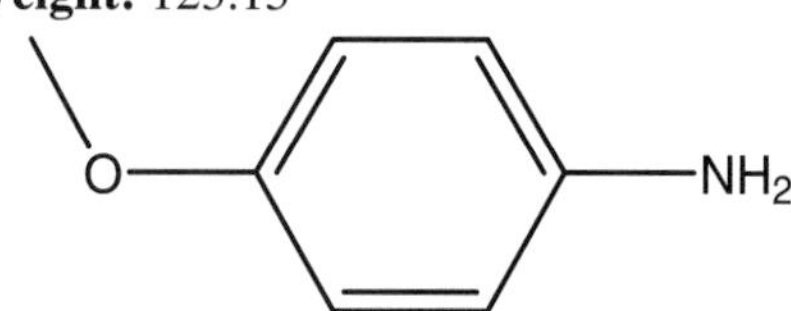

Chemical Structure

Synonyms: 4-AMINOANISOLE; P-AMINOANISOLE; PARA-AMINOANISOLE; 1-AMINO-4-METHOXYBENZENE; ANILINE,4-METHOXY-; ANILINE,P-METHOXY-; 4-ANISIDINE; ANISOLE,P-AMINO; ANISOLE,P-AMINO-; P-ANISYLAMINE; BENZENAMINE,4-METHOXY-; BENZENAMINE,4-METHOXY-(9CI); CCRIS 917; 4-METHOXY-1-AMINOBENZENE; 4-METHOXYANILINE; P-METHOXYANILINE; 4-METHOXYBENZENAMINE; 4-METHOXYBENZENEAMINE; P-METHOXYPHENYLAMINE; NSC 7921
Description: white crystalline fused solid
Use: in azo dyestuffs, as a dye intermediate, in hair dyes, as a corrosion inhibitor in steel storage and as an antioxidant for some polymercaptan resins

Physical Properties

Boiling Point: 246 °C (475 °F)
Freezing Point: 57 °C (134.6 °F)
Specific Gravity: 1.071 at 57 °C/4 °C
Vapor Density: 4.25 Air=1
Saturated Vapor Density: < 1.200512613 kg/m^3
Density: 1.071 g/cu cm at 57 °C
Vapor Pressure: < 0.1 mm Hg at 20 °C
Water Solubility: Sparingly Soluble in Water
Other Solubilities: 95% Ethanol: 10-50 mg/ml at 18 °C; Acetone: >=100 mg/ml at 18 °C; Alcohol: Soluble; Benzene: Soluble; DMSO: 50-100 mg/ml at 18 °C; Ether: Soluble.
Refraction Index: 1.5559
Ionization Potential (eV): 7.44
Flash Point: 5 °C Closed Cup

RTECS Toxicity Data

Acute Oral: Rat LD_{50} Dose: 1320 mg/kg. Mouse LD_{50} Dose: 1410 mg/kg; Toxic Effects: Sense organs and special senses - Other; Behavioral - Altered sleep time (including change in righting reflex); Behavioral - Ataxia.
Acute Dermal: Rat LD_{50} Route: Skin; Dose: 3200 mg/kg.
Mutagenic: Mouse DNA Damage; Cell Type: lymphocyte; Dose: 202 umol/L. Mouse DNA Inhibition; Route: Oral; Dose: 200 mg/kg.

Hazard Overviews

Fire Diamond

Health: Irritation of the respiratory tract. Also Causes: methemoglobinemia, cyanosis, headache, confusion, weakness, disorientation, drowsiness, unconsciousness, redness, swelling, stinging, tearing, skin sensitization, contact dermatitis. Chronic Effects: headache, dizziness, hemolytic anemia, liver and kidney damage.

Fire: Will burn. Use dry chemical, carbon dioxide, water spray, fog, or regular foam.

Reactivity: Stable. Hazardous polymerization cannot occur. Avoid: heat; ignition sources; light; moisture. Incompatible with: strong oxidizers; acids; acid chlorides; acid anhydrides; chloroformates; plastic; rubber; coatings. Hazardous decomposition products: carbon oxides; nitrogen oxides.

Carcinogenicity: IARC - Group 3, Not classifiable as to carcinogenicity to humans; NIOSH - Not listed; NTP - Not listed; ACGIH - Class A4, Not classifiable as a human carcinogen; OSHA - Not listed; EPA - Not listed; MAK - Not listed

Primary Target Organs:

Eyes

Skin

Respiratory System

Liver

Kidneys

Blood

Exposure Limits

OSHA PEL: TWA: 0.5 mg/m^3; skin.

ACGIH TLV: TWA: 0.1 ppm; 0.5 mg/m^3.

NIOSH REL: TWA: 0.5 mg/m^3.

NIOSH IDLH: 50 mg/m^3.

DFG MAK: TWA: 0.1 ppm; 0.5 mg/m^3.

Respirator Recommendation

Exposure Range: >0.5 to 25 mg/m^3 Supplied Air, Constant Flow/Pressure Demand, Half Mask

Exposure Range: >25 to <50 mg/m^3 Supplied Air, Constant Flow/Pressure Demand, Full Face

Exposure Range: 50 to unlimited mg/m^3 Self-contained Breathing Apparatus, Pressure Demand, Full Face

Note: poor warning properties

Environmental

Ecotoxicity: Algae: Scenedesmus pannonicusshort-term EC 50 14 mg /l; Crustaceans: Daphnia magna 24h EC 50 150 mg /l; Fishes: Poecilia reticulata 14d LC_{50} 190 mg /l

Environmental Fate: If released to soil, it is expected to be immobilized by strong, irreversible covalent bonding with any humic materials present. If released to water, it may react with either photochemically generated hydroxyl radicals or alkylperoxy radicals found in sunlit water (half-life 19-30 hours) or it may be strongly bound with humic materials found in suspended solids and sediments. Chemical hydrolysis, aerobic biodegradation, bioaccumulation in aquatic organisms, and volatilization are not expected to be important fate processes. If released to the atmosphere, it appears that reaction with photochemically generated hydroxyl radicals would be the dominant fate process (half-life 3 hours).

Cleanup/Disposal: Guide No. 153: Eliminate all ignition sources (no smoking, flares, sparks or flames in immediate area). Do not touch damaged containers or spilled material unless wearing appropriate protective clothing. Stop leak if you can do it without risk. Prevent entry into waterways, sewers, basements or confined areas. Absorb or cover with dry earth, sand or other non-combustible material and transfer to containers. Do not get water inside containers.

Environmental Physical Data

Henry's Law Constant: estimated at 1.4 $x10^{-6}$

Octanol/Water Partition Coefficient: log K_{ow} = 0.95

BCF: estimated at 3

BOD: 65.3% BODT, 14 days

Regulations

RCRA 40CFR: Not listed

CERCLA: 40CFR 302.4: Not listed

SARA 40CFR 372.65: Listed

SARA EHS 40CFR 355: Not listed

TSCA: Listed

Analytical Methods

Indoor / Expired Air: NIOSH 2514

ANI5690 **CAS #: 134-29-2**

O-ANISIDINE HYDROCHLORIDE

RTECS: BZ6500000

Molecular Formula: $C_6H_{10}ClNO$

Structured MF: $NH_2C_6H_4OCH_3$•HCl

Formula Weight: 159.63

Synonyms: 2-AMINOANISOLE HYDROCHLORIDE; O-AMINOANISOLE HYDROCHLORIDE; O-ANISIDINE HCL; 2-ANISIDINE HYDROCHLORIDE; O-ANISYLAMINE HYDROCHLORIDE; BENZENAMINE,2-METHOXY-,HYDROCHLORIDE (9CI); C.I. 37115; FAST RED BB BASE; 2-METHOXY-1-AMINOBENZENE HYDROCHLORIDE; 2-METHOXYANILINE HCL; 2-METHOXYANILINE HYDROCHLORIDE; O-METHOXYANILINE HYDROCHLORIDE; 2-METHOXYBENZENAMINE HYDROCHLORIDE; 2-METHOXYBENZENEAMINE HYDROCHLORIDE; O-METHOXYPHENYLAMINE HYDROCHLORIDE

Description: gray-black crystalline solid or light gray powder

Use: manufacture of dyes and as a starting material in the synthesis of guaiacol (o-methoxyphenol); a possible ingredient in permanent oxidation of hair dyes; however, as of 1977, it had not been used in the U.S

Physical Properties

Freezing Point: 225 °C (437 °F) to 227 °C (440.6 °F)

Water Solubility: 10 to 50 mg/mL at 21 °C

Other Solubilities: 95% Ethanol: >=100 mg/ml at 21 °C; Acetone: <1 mg/ml at 21 °C; DMSO: >=100 mg/ml at 21 °C.

Flash Point: Not available; probably combustible

RTECS Toxicity Data

Mutagenic: Rat Morphological Transformation; Cell Type: embryo; Dose: 195 ug/plate. Mouse Mutations in Mammalian Somatic Cells; Route: Oral; Dose: 2250 mg/kg/3D-C.

Tumorigenic: Rat Route: Oral; Dose: 180 gm/kg/2Y-C; Toxic Effects: Tumorigenic - Carcinogenic by RTECS criteria; Kidney, Ureter, and Bladder - Tumors; Kidney, Ureter, and Bladder - Kidney tumors. Rat Route: Oral; Dose: 360 gm/kg/78W-C; Toxic Effects: Tumorigenic - Carcinogenic by RTECS criteria; Kidney, Ureter, and Bladder - Tumors; Kidney, Ureter, and Bladder - Kidney tumors. Rat Route: Oral; Dose: 2905 gm/kg/83W-C; Toxic Effects: Tumorigenic - Carcinogenic by RTECS criteria; Kidney, Ureter, and Bladder - Tumors; Nutritional and gross metabolic - Weight loss or decreased weight gain. Rat Route: Oral; Dose: 5810 gm/kg/83W-C; Toxic Effects: Tumorigenic - Carcinogenic by RTECS criteria; Kidney, Ureter, and Bladder - Tumors; Nutritional and gross metabolic - Weight loss or decreased weight gain.

Hazard Overviews

Fire: Will burn.

Carcinogenicity: IARC - Group 2B, Possibly carcinogenic to humans; NIOSH - Not listed; NTP - Listed; ACGIH - Not listed; OSHA - Not listed; EPA - Not listed; MAK - Not listed

Environmental

Regulations

RCRA 40CFR: Not listed
CERCLA: 40CFR 302.4: Not listed
SARA 40CFR 372.65: Listed
SARA EHS 40CFR 355: Not listed
TSCA: Not listed

ANI6360 CAS #: 93-13-0

O-ANISIDYL-N-METHANESULFONIC ACID

EINECS Number: 202-220-4
Molecular Formula: $C_8H_{11}NO_4S$
Formula Weight: 217.2
Synonyms: O-ANISIDINOMETHANESULFONIC ACID; METHANESULFONIC ACID,O-ANISIDINO-; METHANESULFONIC ACID,((2-METHOXYPHENYL)AMINO)-; ((2-METHOXYPHENYL)AMINO)METHANESULFONIC ACID

Hazard Overviews

Carcinogenicity: IARC - Not listed; NIOSH - Not listed; NTP - Not listed; ACGIH - Not listed; OSHA - Not listed; EPA - Not listed; MAK - Not listed

Environmental

Regulations

RCRA 40CFR: Not listed
CERCLA: 40CFR 302.4: Not listed
SARA 40CFR 372.65: Not listed
SARA EHS 40CFR 355: Not listed
TSCA: Listed

ANI7030 CAS #: 117-37-3

ANISINDIONE

RTECS: NK5775300
EINECS Number: 204-186-6
Molecular Formula: $C_{16}H_{12}O_3$
Formula Weight: 252.26
Synonyms: ANDION; ANISIN INDANDIONE; 2-P-ANISYL-1,3-INDANDIONE; 2-PARA-ANISYL-1,3-INDANDIONE; 1,3-INDANDIONE,2-(P-METHOXYPHENYL)-; 1,3-INDANEDIONE,2-(4-METHOXYPHENYL)-; 1H-INDENE-1,3(2H)-DIONE,2-(4-METHOXYPHENYL)-; 2-(4-METHOXYPHENYL)INDAN-1,3-DIONE; 2-(P-METHOXYPHENYL)-1,3-INDANDIONE; 2-(P-METHOXYPHENYL)INDANE-1,3-DIONE; 2-(4-METHOXYPHENYL)-1H-INDENE-1,3(2H)-DIONE; MIRADON; SPE 2792; UNIDIONE; UNIDONE
Description: pale yellow crystals or white to cream-white fine crystalline powder; odorless or has slightly sweet odor
Use: medication: oral anticoagulant

Physical Properties

Freezing Point: 156 °C (312.8 °F) to 157 °C (314.6 °F)
Water Solubility: Practically Insoluble in Water
Other Solubilities: Slightly Soluble in Ether, Methanol, Hydrochloric Acid; Soluble in Methylene Chloride.

RTECS Toxicity Data

Acute Oral: Mouse LD_{50} Dose: 300 mg/kg.

Hazard Overviews

Carcinogenicity: IARC - Not listed; NIOSH - Not listed; NTP - Not listed; ACGIH - Not listed; OSHA - Not listed; EPA - Not listed; MAK - Not listed

Environmental

Regulations

RCRA 40CFR: Not listed
CERCLA: 40CFR 302.4: Not listed
SARA 40CFR 372.65: Not listed
SARA EHS 40CFR 355: Not listed
TSCA: Not listed

ANI7700 **CAS #: 100-66-3**

ANISOLE

RTECS: BZ8050000
DOT: UN2222; IMO3.3
EINECS Number: 202-876-1
Molecular Formula: C_7H_8O
Formula Weight: 108.13

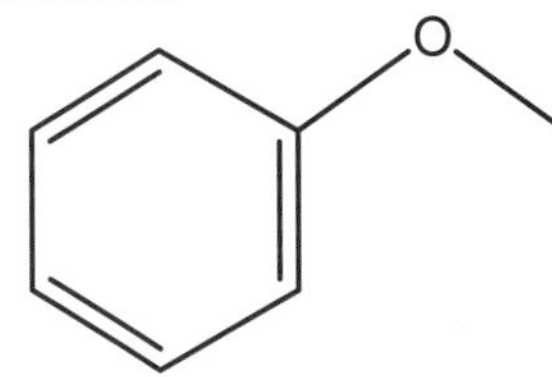

Chemical Structure

Synonyms: ANIZOL; BENZENE,METHOXY; BENZENE,METHOXY-; ETHER,METHYL PHENYL; METHOXYBENZENE; METHYL PHENYL ETHER; PHENYL METHYL ETHER
Description: clear straw color, mobile liquid; sweet anise-like spicy-sweet odor
Use: in perfumery; in organic syntheses; vermicide; flavoring; intermediate in the manufacture of organic compounds, for example fragrances and pharmaceuticals; solvent and heat transfer medium

Physical Properties

Boiling Point: 155.5 °C (312 °F) at 760 mm Hg
Freezing Point: -37.3 °C (-35.14 °F)
Specific Gravity: 0.9956 at 18 °C/4 °C
Vapor Density: 3.72 Air=1
Saturated Vapor Density: 1.215256572 kg/m^3
Vapor Pressure: 0.472 kPa at 25 °C
Water Solubility: Insoluble
Other Solubilities: Very Soluble in Acetone.
Surface Tension: 36.18 dyne/cm at 15 °C
Refraction Index: 1.51791 at 20 °C/D
Critical Temperature: 372.4 °C
Critical Pressure: 41.9 atm
Ionization Potential (eV): 8.20 +/-0.05
Flash Point: 52 °C Open Cup
Autoignition Temperature: 475 °C

RTECS Toxicity Data

Acute Oral: Rat LD_{50} Dose: 3700 mg/kg; Toxic Effects: Behavioral - Somnolence (general depressed activity); Gastrointestinal - Changes in structure or function of salivary glands; Kidney, Ureter, and Bladder - Hematuria. Mouse LD_{50} Dose: 2800 mg/kg; Toxic Effects: Behavioral - Tremor; Behavioral - Convulsions or effect on seizure threshold; Behavioral - Excitement.
Acute Inhalation: Rat LC_{50} Dose: >5000 mg/m^3. Mouse LC_{50} Dose: 3021 mg/m^3/2hr.
Irritation Skin: Rabbit Standard Draize Test Dose: 500 mg/24H; Reaction: moderate.
Mutagenic: Human DNA Inhibition; Cell Type: lymphocyte; Dose: 25 umol/L.

Hazard Overviews

Flammable

Health: Irritating to eyes/skin/respiratory tract. Other Acute Effects: may be harmful by inhalation, ingestion, or skin absorption.
Fire: Flammable. Extinguishing agents: water spray; carbon dioxide, dry chemical powder or appropriate foam. Precautions: combustible liquid.
Reactivity: Incompatible with: strong oxidizing agents, strong acids. Hazardous decomposition products: toxic fumes of: carbon monoxide, carbon dioxide.
Carcinogenicity: IARC - Not listed; NIOSH - Not listed; NTP - Not listed; ACGIH - Not listed; OSHA - Not listed; EPA - Not listed; MAK - Not listed
Primary Target Organs:

Eyes

Skin

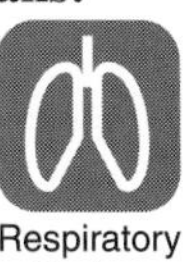

Respiratory System

Environmental

Environmental Fate: Has a low adsorptivity to soil and if released on soil, may leach. It has a moderately high Henry's Law constant and vapor pressure and would be expected to volatilize from both moist and dry soil surfaces. It is readily biodegradable in screening tests and may therefore biodegrade in soil. If released in water, it will be lost by volatilization. Its volatilization half-life in a model river and model lake is estimated to be 3.2 hr and 4.2 days, respectively. It would also be expected to biodegrade. Experiments performed in a model aquatic ecosystem demonstrated that it was metabolized in aquatic organisms and did not bioconcentrate. Over a period of 24 hr, the concentration was decreased by approximately 93% and degradation products were found in the water. In the atmosphere, it will react with photochemically-produced hydroxyl radicals resulting in an estimated half-life of 22 hr
Cleanup/Disposal: Guide No. 127: Eliminate all ignition sources (no smoking, flares, sparks or flames in immediate area). All equipment used when handling the product must be grounded. Do not touch or walk through spilled material. Stop leak if you can do it without risk. Prevent entry into waterways, sewers, basements or confined areas. A vapor suppressing foam may be used to reduce vapors. Absorb or cover with dry earth, sand or other non-combustible material and transfer to containers. Use clean non-sparking tools to collect absorbed material. Large Spills: Dike far ahead of liquid spill for later disposal. Water spray may reduce vapor; but may not prevent ignition in closed spaces.

Environmental Physical Data
Henry's Law Constant: estimated at 4.35 x10^{-3}
Octanol/Water Partition Coefficient: log K_{ow} = 2.11
Sorption Partition Coefficient: K_{oc} = 35
BCF: calculated at 24

Regulations
RCRA 40CFR: Not listed
CERCLA: 40CFR 302.4: Not listed
SARA 40CFR 372.65: Not listed
SARA EHS 40CFR 355: Not listed
TSCA: Listed

ANI8370 CAS #: 80-50-2

ANISOTROPINE METHYLBROMIDE

RTECS: YM3710000
EINECS Number: 201-285-6
Molecular Formula: $C_{17}H_{32}BrNO_2$
Formula Weight: 362.37

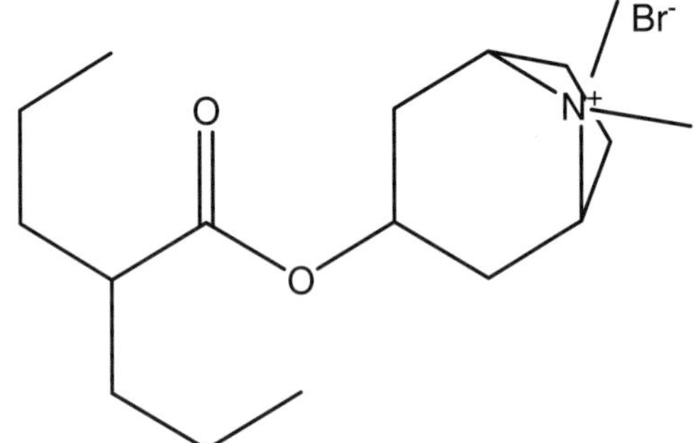

Chemical Structure

Synonyms: ANISOTROPINE METHOBROMIDE; 8-AZONIABICYCLO(3.2.1)OCTANE,8,8-DIMETHYL-3-((1-OXO-2-PROPYLPENTYL)OXY)-,BROMIDE,ENDO-; ENDO-8,8-DIMETHYL-3-[(1-OXO-2-PROPYLPENTYL)OXY]-8-AZONIABICYCLO[3.2.1]OCTANE BROMIDE; 3ALPHA-HYDROXY-8-METHYL-1ALPHAH,5ALPHAH-TROPANIUM BROMIDE2-PROPYLVALERATE; 3-ALPHA-HYDROXY-8-METHYL-1-ALPHA-H,5-ALPHA-H-TROPANIUMBROMIDE 2-PROPYLVALERATE; LYTISPASM; 8-METHYL-3-(2-PROPYLPENTANOYLOXY)TROPINIUM BROMIDE; 8-METHYLTROPINIUM BROMIDE 2-PROPYLPENTANOATE; 8-METHYLTROPINIUM BROMIDE 2-PROPYLVALERATE; OCTATROPINE METHYLBROMIDE; 2-PROPYLPENTANOYLTROPINIUM METHYLBROMIDE; 1ALPHA H,5ALPHA H-TROPANIUM,3ALPHA-HYDROXY-8-METHYL-,BROMIDE,2-PROPYLVALERATE; VALPIN; VAPIN
Description: white, glistening powder, crystals, or plates; probably odorless
Use: medication: antimuscarinic drug; medication: in mucous colitis & irritable colon, spastic colitis, splenic flexure syndrome, & biliary dyskinesia; medication: in treatment of cholelithiasis, pylorospasm, gastritis, duodenitis, enterocolitis, & peptic ulcer; anticholinergic as adjunctive therapy for peptic ulcers; anticholinergic for spastic colitis

Physical Properties
Freezing Point: 329 °C (624.2 °F)
Water Solubility: Soluble in Water
Other Solubilities: Soluble in Ethanol; Insoluble in Diethyl Ether.

RTECS Toxicity Data
Acute Oral: Rat LD_{50} Dose: 705 mg/kg. Mouse LD_{50} Dose: 850 mg/kg.
Acute Dermal: Mouse LD_{50} Route: Subcutaneous Dose: 133 mg/kg.

Hazard Overviews
Health: May cause irritation. Harmful. Other Acute Effects: harmful if swallowed, inhaled, or absorbed through skin; exposure can cause nausea; dizziness; headache; target organ: autonomic nervous system.
Fire: Hazards: emits toxic fumes. Extinguishing agents: water spray; carbon dioxide, dry chemical powder or appropriate foam. Precautions: combustible liquid.
Carcinogenicity: IARC - Not listed; NIOSH - Not listed; NTP - Not listed; ACGIH - Not listed; OSHA - Not listed; EPA - Not listed; MAK - Not listed
Primary Target Organs:

Nervous System

Environmental

Regulations
RCRA 40CFR: Not listed
CERCLA: 40CFR 302.4: Not listed
SARA 40CFR 372.65: Not listed
SARA EHS 40CFR 355: Not listed
TSCA: Not listed

ANI9040 CAS #: 100-07-2

P-ANISOYLCHLORIDE

RTECS: CA0270000
EINECS Number: 202-816-4
Molecular Formula: $C_8H_7ClO_2$
Formula Weight: 170.60

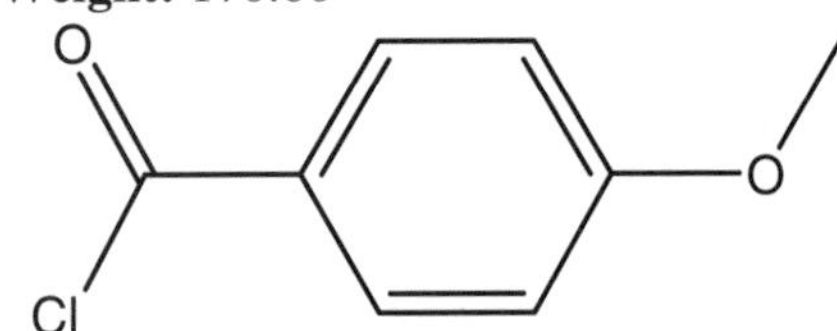

Chemical Structure

Synonyms: 4-ANISOYL CHLORIDE; ANISOYL CHLORIDE; P-ANISYL-CHLORIDE; BENZOIC ACID,4-METHOXY-,CHLORIDE; BENZOYL CHLORIDE,4-METHOXY-; BENZOYL CHLORIDE,METHOXY-; BENZOYL CHLORIDE,METHOXY-(9CI); 4-

METHOXYBENZOIC ACID CHLORIDE; P-METHOXYBENZOIC ACID CHLORIDE; 4-METHOXYBENZOYL CHLORIDE; METHOXYBENZOYL CHLORIDE; P-METHOXYBENZOYL CHLORIDE

Description: clear crystals or needles, or amber liquid

Use: intermediate for dyes & medicines

Physical Properties

Boiling Point: 262 °C (504 °F) to 263 °C (505 °F)

Freezing Point: 24 °C (75.2 °F) to 25 °C (77 °F)

Specific Gravity: 1.261 at 20 °C/4 °C

Water Solubility: Decomposed by Water

Other Solubilities: Soluble in Ether.

Surface Tension: Estimated at 25 dyne/cm at 20 °C

Refraction Index: 1.580 at 20 °C/D

Hazard Overviews

Corrosive

Fire Diamond

Health: Corrosive to eyes/skin/respiratory tract. Other Acute Effects: harmful if swallowed, inhaled, or absorbed through skin; inhalation may result in spasm, inflammation and edema of the larynx and bronchi, chemical pneumonitis and pulmonary edema; symptoms of exposure may include burning sensation, coughing, wheezing, laryngitis, shortness of breath, headache, nausea and vomiting.

Fire: Hazards: emits toxic fumes. Extinguishing agents: carbon dioxide, dry chemical powder or appropriate foam. Precautions: combustible liquid.

Reactivity: Incompatible with: strong oxidizing agents, may decompose on exposure to moist air or water, strong bases. Hazardous decomposition products: toxic fumes of: carbon monoxide, carbon dioxide, hydrogen chloride gas.

Carcinogenicity: IARC - Not listed; NIOSH - Not listed; NTP - Not listed; ACGIH - Not listed; OSHA - Not listed; EPA - Not listed; MAK - Not listed

Primary Target Organs:

Eyes

Skin

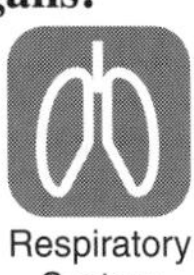
Respiratory System

Environmental

Regulations

RCRA 40CFR: Not listed

CERCLA: 40CFR 302.4: Not listed

SARA 40CFR 372.65: Not listed

SARA EHS 40CFR 355: Not listed

TSCA: Listed

ANT1000 CAS #: 91-75-8

ANTAZOLINE

RTECS: NJ2000000

EINECS Number: 202-094-0

Molecular Formula: $C_{17}H_{19}N_3$

Formula Weight: 265.35

Synonyms: 5512-M; ANTASTAN; ANTASTEN; ANTAZOLIN; ANTAZOLINA; ANTAZOLINUM; ANTIHISTAL; ANTISTIN; ANTISTINE; AZALONE; BEN-A-HIST; 2-(N-BENZYLANILINOMETHYL)-2-IMIDAZOLINE; 4,5-DIHYDRO-N-PHENYL-N-(PHENYLMETHYL)-1H=IMIDAZOLE-2-METHANAMINE; 4,5-DIHYDRO-N-PHENYL-N-PHENYLMETHYL-1H-IMIDAZOLE-2-METHANAMINE; HISTOSTAB; IMIDAMINE; PHENAZOLINE; 2-(N-PHENYL-N-BENZYLAMINOMETHYL) IMIDAZOLINE; 2-PHENYL-BENZYL-AMINO-METHYLIMIDAZOLIN; 2-(N-PHENYL-N-BENZYLAMINOMETHYL)IMIDAZOLINE

Description: white, crystalline powder; odorless

Use: medication (vet): antihistaminic; medication: antihistaminic

Physical Properties

Freezing Point: 120 °C (248 °F) to 122 °C (251.6 °F)

Water Solubility: 1 g/40 ml water

Other Solubilities: one gram dissolves in 25 ml Alcohol

pH: 1% aqueous solution 6

RTECS Toxicity Data

Acute Oral: Mouse LD_{50} Dose: 398 mg/kg.

Acute Dermal: Mouse LD_{50} Route: Subcutaneous Dose: 135 mg/kg.

Hazard Overviews

Health: Irritating to eyes/skin/respiratory tract. Toxic. Other Acute Effects: harmful if swallowed, inhaled, or absorbed through skin; overexposure may cause drowsiness and gastrointestinal disturbances; transient dizziness and anorexia have followed intravenous administration, and cardiac arrest has occurred. The toxicological properties have not been thoroughly investigated.

Fire: Hazards: emits toxic fumes. Extinguishing agents: carbon dioxide, dry chemical powder or appropriate foam. Precautions: combustible liquid.

Carcinogenicity: IARC - Not listed; NIOSH - Not listed; NTP - Not listed; ACGIH - Not listed; OSHA - Not listed; EPA - Not listed; MAK - Not listed

Primary Target Organs:

Eyes

Skin

Respiratory System

Environmental

Regulations

RCRA 40CFR: Not listed

CERCLA: 40CFR 302.4: Not listed

SARA 40CFR 372.65: Not listed
SARA EHS 40CFR 355: Not listed
TSCA: Not listed

ANT1500 CAS #: 613-13-8

2-ANTHRACENAMINE

RTECS: CA9275000
EINECS Number: 210-330-9
Molecular Formula: $C_{14}H_{11}N$
Formula Weight: 193.25

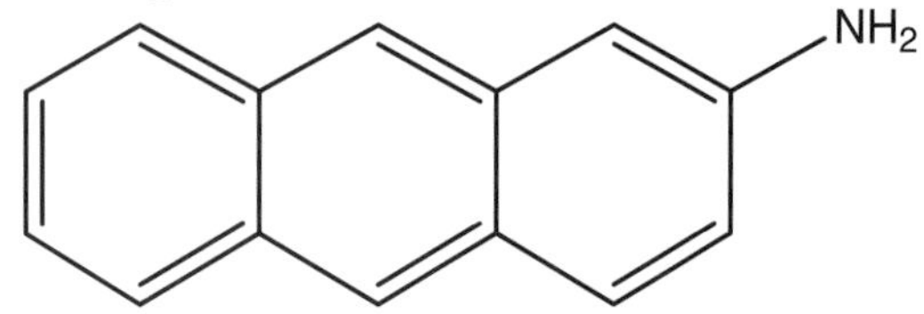

Chemical Structure

Synonyms: 2-AMINOANTHRACENE; BETA-AMINOANTHRACENE; 2-ANTHRACENAMIDE; ANTHRACENE AMINE; 2-ANTHRACINE AMIDE; 2-ANTHRACYLAMINE; 2-ANTHRAMINE; 2-ANTHRYLAMINE
Description: yellow leaflets
Use: organic synthesis

Physical Properties

Boiling Point: Sublimes at 93 °C (199 °F)
Freezing Point: 238 °C (460.4 °F) to 241 °C (465.8 °F)
Water Solubility: < 0.1 mg/mL at 22 C
Other Solubilities: 95% Ethanol: 1-10 mg/ml at 21 °C; Acetone: 10-50 mg/ml at 22 °C; Concentrated sulfuric acid: Insoluble; DMSO: 50-100 mg/ml at 22 °C; Ether: Slightly Soluble; Organic solvents: Slightly Soluble.
Flash Point: Not available; probably combustible

RTECS Toxicity Data

Reproductive/Teratogenic: Mouse Route: Oral; Dose: 210 mg/kg; Duration: female 1-21D of pregnancy; Effects on Newborn - Germ cell effects.
Mutagenic: Human Unscheduled DNA Synthesis; Cell Type: other cell types; Dose: 1 mg/L. Human DNA Inhibition; Cell Type: HeLa cell; Dose: 2 umol/L. Human DNA Adduct; Cell Type: lymphocyte; Dose: 30 umol/L.
Tumorigenic: Rat Route: Oral; Dose: 45 mg/kg/30D-I; Toxic Effects: Tumorigenic - Carcinogenic by RTECS criteria; Skin and appendages - Tumors. Rat Route: Oral; Dose: 100 mg/kg; Toxic Effects: Tumorigenic - Carcinogenic by RTECS criteria; Gastrointestinal - Tumors; Skin and appendages - Tumors.

Hazard Overviews

Health: Irritating to eyes/skin/respiratory tract. Harmful. Other Acute Effects: harmful if swallowed, inhaled, or absorbed through skin. Chronic Effects: laboratory experiments have shown mutagenic effects. Possible human carcinogen.
Fire: Will burn. Hazards: emits toxic fumes. Extinguishing agents: water spray; carbon dioxide, dry chemical powder or appropriate foam. Precautions: combustible liquid.
Reactivity: Incompatible with: strong oxidizing agents. Hazardous decomposition products: toxic fumes of: carbon monoxide, carbon dioxide, nitrogen oxides.
Carcinogenicity: IARC - Not listed; NIOSH - Not listed; NTP - Not listed; ACGIH - Not listed; OSHA - Not listed; EPA - Not listed; MAK - Not listed
Primary Target Organs:

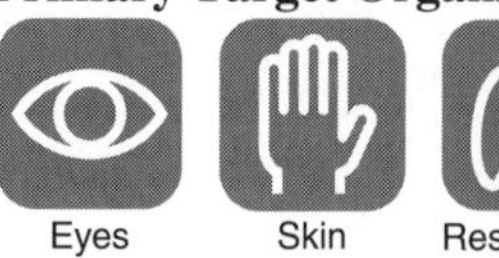

Environmental

Environmental Fate: If released to soil, it is expected to be immobilized by strong adsorption to the soil and it would probably photolyze on soil surfaces. Volatilization and hydrolysis are not expected to be environmentally relevant fate processes in either soil or water. Insufficient data are available to predict the significance of biodegradation. If released to water, it may adsorb strongly to suspended solids and sediments, bioaccumulate significantly in aquatic organisms, or photolyze in near-surface waters (estimated midday, midsummer half-life less than 4.5 days). If released to the atmosphere, it is expected to exist in both vapor and particulate form. This compound may react with photochemically generated hydroxyl radicals (estimated vapor phase half-life 1.8 hours), it may undergo direct photolysis, or it may be subject to dry deposition. Adsorption onto particulate matter may retard direct photolysis and reaction with hydroxyl radicals.

Environmental Physical Data

Henry's Law Constant: estimated at 3.0×10^{-7}
Sorption Partition Coefficient: K_{oc} = 94276 to 1.5904×10^{5}
BCF: estimated at 3141

Regulations

RCRA 40CFR: Not listed
CERCLA: 40CFR 302.4: Not listed
SARA 40CFR 372.65: Not listed
SARA EHS 40CFR 355: Not listed
TSCA: Not listed

ANT2000 CAS #: 120-12-7

ANTHRACENE

RTECS: CA9350000
DOT: UN1136; IMO3.2; IMO3.3
EINECS Number: 204-371-1
Molecular Formula: $C_{14}H_{10}$
Structured MF: $C_{14}H_{10}$
Formula Weight: 178.22

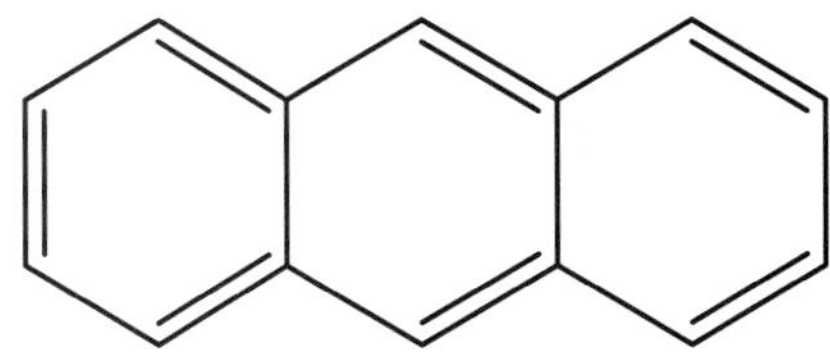

Chemical Structure

Synonyms: ANTHRACEN; ANTHRACENE OIL; ANTHRACIN; COAL TAR PITCH VOLATILES:ANTHRACENE; GREEN OIL; P-NAPHTHALENE; PARANAPHTHALENE; PARANAPTHALENE; TETRA OLIVE N2G

Description: colorless when pure, yellow crystals

Use: source of dyestuffs; in the manufacture of anthraquinone, alizarin dyes, insecticides and wood preservatives

Physical Properties

Boiling Point: 342 °C (648 °F)
Freezing Point: 218 °C (424.4 °F)
Specific Gravity: 1.25 at 27 °C/4 °C
Vapor Density: 6.15 Air=1
Density: 1.28 g/mL
Vapor Pressure: 1 mm Hg at 145 °C
Water Solubility: < 1 mg/mL at 20 C
Other Solubilities: 95% Ethanol: <1 mg/ml at 20 °C; Acetone: <1 mg/ml at 20 °C; Benzene: 1g/62 mL; Carbon Disulfide: 1 g/31 mL; Carbon Tetrachloride: 1 g/86 mL; Chloroform: 1 g/85 mL; DMSO: <1 mg/ml at 20 °C; Ether: 1 g/200 mL; Toluene: 1 g/125 mL.
Ionization Potential (eV): 7.4392 +/-0.2
Flash Point: 121 °C Closed Cup
Autoignition Temperature: 540 °C
LEL: 0.6% v/v

RTECS Toxicity Data

Acute Oral: Mouse LD; Dose: >17 gm/kg; Toxic Effects: Liver - Fatty Liver degeneration.

Irritation Skin: Mouse Standard Draize Test Dose: 118 ug; Reaction: mild.

Mutagenic: Rat DNA Damage; Cell Type: liver; Dose: 300 umol/L. Rat Morphological Transformation; Cell Type: embryo; Dose: 108 ug/plate.

Tumorigenic: Rat Route: Oral; Dose: 20 gm/kg/79W-I; Toxic Effects: Tumorigenic - Equivocal tumorigenic agent by RTECS criteria; Liver - Tumors. Rat Route: Subcutaneous; Dose: 3300 mg/kg/33W-I; Toxic Effects: Tumorigenic - Neoplastic by RTECS criteria; Tumorigenic - Tumors at site of application. Rat Route: Subcutaneous; Dose: 660 mg/kg/33W-I; Toxic Effects: Tumorigenic - Equivocal tumorigenic agent by RTECS criteria; Tumorigenic - Tumors at site of application.

Hazard Overviews

Fire Diamond

Health: Irritating to eyes/skin/respiratory tract. Also Causes: headache, nausea, loss of appetite, slowed reactions, adynamia, burning, itching, watering, edema, GI irritation. Chronic Effects: pigmentation of skin with cornification of surface layers, telangioectasis, sensitization.

Fire: Will burn. Use water spray, carbon dioxide, dry chemical, or foam. May explode in air.

Reactivity: Unstable, darkens upon exposure to sunlight. Hazardous polymerization cannot occur. Avoid: heat; ignition sources; sunlight. Incompatible with: calcium hypochlorite; fluorine; chromic acid; calcium oxychloride. Hazardous decomposition products: carbon oxide(s); acrid, irritating smoke.

Carcinogenicity: IARC - Group 3, Not classifiable as to carcinogenicity to humans; NIOSH - Not listed; NTP - Not listed; ACGIH - Not listed; OSHA - Not listed; EPA - Class D, Not classifiable as to human carcinogenicity; MAK - Not listed

Primary Target Organs:

Eyes

Skin

Respiratory System

Exposure Limits
OSHA PEL: TWA: 0.2 mg/m^3.

Environmental

Ecotoxicity: LC_{50} Rana pipiens (Leopard frog) 0.065 ppm/30 min; 0.025 ppm/5 hr /Both toxicity values based on phototoxicity study; Conditions of bioassay not specified LC_{50} Culicid mosquito larvae 26.8 ug/l/24 hr /Phototoxicity study; Conditions of bioassay not specified LC_{50} Leponis macrochirus (Bluegill sunfish, juvenile) 11.9 ug/l/96 hr /Phototoxicity study; Conditions of bioassay not specified

Environmental Fate: If released to soil it will be expected to adsorb very strongly to the soil and will not be expected to leach appreciably to groundwater. It will not hydrolyze but may be subject to biodegradation in soils with reported half-lives of 3.3-139 days. It may be subject to evaporation from the soil and other surfaces. If released to water it will strongly adsorb to sediment and particulate matter, but will not hydrolyze. It may bioconcentrate in species which lack microsomal oxidase, the presence of which allows organisms to rapidly metabolize polyaromatic hydrocarbons. It will be subject to direct photolysis near the surface of waters and may be subject to significant biodegradation. It may be subjected to significant evaporation with an estimated range of half-lives of 4.3-5.9 days predicted for evaporation from a river 1 m deep, flowing at 1 m/sec with a wind velocity of 3 m/sec. If released to the atmosphere, it will be subject to direct photolysis and the estimated vapor phase half-life in the atmosphere is 1.67 days as a result of reaction with photochemically produced hydroxyl radicals. Adsorption may retard the evaporation, biodegradation, bioconcentration, and photolysis processes.

Cleanup/Disposal: Guide No. 128: Eliminate all ignition sources (no smoking, flares, sparks or flames in immediate

area). All equipment used when handling the product must be grounded. Do not touch or walk through spilled material. Stop leak if you can do it without risk. Prevent entry into waterways, sewers, basements or confined areas. A vapor suppressing foam may be used to reduce vapors. Absorb or cover with dry earth, sand or other non-combustible material and transfer to containers. Use clean non-sparking tools to collect absorbed material. Large Spills: Dike far ahead of liquid spill for later disposal. Water spray may reduce vapor; but may not prevent ignition in closed spaces.

Environmental Physical Data

Henry's Law Constant: calculated at 2.72×10^{-3}
Octanol/Water Partition Coefficient: log K_{ow} = calculated at 4.45
Sorption Partition Coefficient: K_{oc} = 2.6×10^{4}
BCF: rainbow trout 4400 to 9200

Regulations

RCRA 40CFR: Not listed
CERCLA: 40CFR 302.4: Listed per CWA Section 307(a) RQ: 5000 lb (2268 kg)
SARA 40CFR 372.65: Listed
SARA EHS 40CFR 355: Not listed
TSCA: Listed

Analytical Methods

Air: EPA TO-13; California 429
Soil: CLP LC_SV, MC_SVOA, OHC; EPA 16, 1625, PAH-005, PAH-007, PAH-011, PAH-012, S-004-1; SW846 1311, 3630B, 3640A, 8100, 8250A, 8270B, 8270C, 8275A, 8310, 8410; DOE OS050
Water / Groundwater: EPA PAH-002, PAH-006, S-002-1, 1625, 610, 625, 625-S, 6; APHA 6040-B, 6410-B, 6440-B, 6440-C; ASTM D4657, D4763; USGS O3113, O3118
Drinking Water: EPA 525.1, 525.2, 550, 550.1
Indoor / Expired Air: NIOSH 5506, 5515; EPA IP-7-A, IP-7-B
Plasma: EPA 29
Other: EPA PAH-009

ANT2500 CAS #: 118-92-3

ANTHRANILIC ACID

RTECS: CB2450000
EINECS Number: 204-287-5
Molecular Formula: $C_7H_7NO_2$
Structured MF: $H_2NC_6H_4COOH$
Formula Weight: 137.14

O OH NH2

Chemical Structure

Synonyms: ORTHO-AMIDOBENZOIC ACID; 2-AMINOBENZOIC ACID; O-AMINOBENZOIC ACID; ORTHO-AMINOBENZOIC ACID; 1-AMINO-2-CARBOXYBENZENE; O-ANTHRANILIC ACID; BENZOIC ACID,O-AMINO-; 2-CARBOXYANILINE; CARBOXYANILINE; O-CARBOXYANILINE; KYSELINA O-AMINOBENZOOVA; KYSELINA ANTHRANILOVA; VITAMIN L; VITAMIN L1
Description: white to yellowish needle-like crystals
Use: in the production of dyes and pigments and in the manufacture of saccharin; in drugs, perfumes and pharmaceuticals; an important intermediate in the synthesis of many compounds; as a corrosion inhibitor for metals and as a mold inhibitor in soya sauce

Physical Properties

Boiling Point: Sublimes
Freezing Point: 144 °C (291.2 °F) to 146 °C (294.8 °F)
Specific Gravity: 1.412 at 20 °C
Vapor Density: Calculated 4.7 Air=1
Density: 1.412 g/mL at 20 °C
Vapor Pressure: Negligible
Water Solubility: Freely Soluble in Hot Water
Other Solubilities: 90% Alcohol: 10.7/100 at 10 °C; 95% Ethanol: 50-100 mg/ml at 22 °C; Acetone: 50-100 mg/ml at 22 °C; Alcohol: Freely Soluble; Benzene: Slightly Soluble; Chloroform: >10%; DMSO: >=100 mg/ml at 22 °C; Ethanol: >10%; Ether: Freely Soluble.
Evaporation Rate: 1 n-butyl acetate=1
Ionization Potential (eV): 7.6 +/-0.5
Flash Point: 171 °C
LEL: 0.03 oz/ft^3

RTECS Toxicity Data

Acute Oral: Rat LD_{50} Dose: 5410 mg/kg; Toxic Effects: Behavioral - Somnolence (general depressed activity); Behavioral - Excitment; Behavioral - Ataxia. Mouse LD_{50} Dose: 1400 mg/kg.
Reproductive/Teratogenic: Mouse Route: Oral; Dose: 34800 mg/kg; Duration: male 8D prior to mating; Effects on Fertility - Female fertility index.
Mutagenic: Human Mutations in Mammalian Somatic Cells; Cell Type: lymphocyte; Dose: 1667 mg/L. Mouse Mutations in Microorganisms; Cell Type: lymphocyte; Dose: 250 mg/L (+S9).
Tumorigenic: Rat Route: Oral; Dose: 16 gm/kg/25W-C; Toxic Effects: Tumorigenic - Equivocal tumorigenic agent by RTECS criteria; Kidney, Ureter, and Bladder - Tumors.

Mouse Route: Subcutaneous; Dose: 1345 mg/kg/21D-I; Toxic Effects: Tumorigenic - Equivocal tumorigenic agent by RTECS criteria; Lungs, Thorax, or Respiration - Bromchiogenic carcinoma; Liver - Tumors. Mouse Route: Subcutaneous; Dose: 2040 mg/kg (13-17D preg); Toxic Effects: Tumorigenic - Equivocal tumorigenic agent by RTECS criteria; Tumorigenic effects - Transplacental tumorigenesis; Liver - Tumors.

Hazard Overviews

Fire Diamond

Health: Irritating to eyes/skin. Also Causes: nausea, vomiting, acidosis, fever, methemoglobinemia, possible hepatitis.
Fire: Will burn. Use dry chemical, carbon dioxide, water spray, or regular foam. If sufficient quantities of finely divided particles are present, contact with an ignition source can cause an explosion.
Reactivity: Unstable, photo-oxidizes when exposed to light. Hazardous polymerization cannot occur. Avoid: heat; ignition sources; strong oxidizers. Incompatible with: strong oxidizers. Hazardous decomposition products: carbon; nitrogen oxides.
Carcinogenicity: IARC - Group 3, Not classifiable as to carcinogenicity to humans; NIOSH - Not listed; NTP - Not listed; ACGIH - Not listed; OSHA - Not listed; EPA - Not listed; MAK - Not listed
Primary Target Organs:

Eyes

Skin

Liver

Blood

Environmental

Ecotoxicity: Bacteria: Pseudomonas putida 17h EC_{10} 71 mg/l; Crustaceans: Daphnia magna Straus 24h EC_0 62 mg/l 24h EC_{50} 102 mg/l; Fishes: Leuciscus idus 1h LC_{50} 215-1,000 mg/l

Environmental Physical Data

Octanol/Water Partition Coefficient: log K_{ow} = 1.21

Regulations

RCRA 40CFR: Not listed
CERCLA: 40CFR 302.4: Not listed
SARA 40CFR 372.65: Not listed
SARA EHS 40CFR 355: Not listed
TSCA: Listed

ANT3000 CAS #: 84-65-1

ANTHRAQUINONE

RTECS: CB4725000
EINECS Number: 201-549-0
Molecular Formula: $C_{14}H_8O_2$
Formula Weight: 208.20

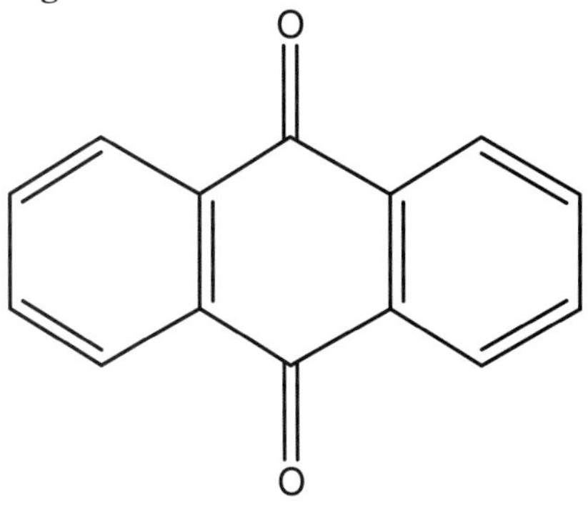

Chemical Structure

Synonyms: ANTHRACENE,9,10-DIHYDRO-9,10-DIOXO-; 9,10-ANTHRACENEDIONE; 9,10-ANTHRACHINON; ANTHRADIONE; 9,10-ANTHRAQUINONE; CORBIT; 9,10-DIHYDRO-9,10-DIOXOANTHRACENE; 9,10-DIOXOANTHRACENE; HOELITE; MORKIT
Description: almost colorless to light yellow to yellow-green crystals, needles, or prisms
Use: as a starting material for the manufacturing of vat dyes, organic inhibitor and to make seeds distasteful to birds

Physical Properties

Boiling Point: 377 °C (711 °F) at 760 mm Hg
Freezing Point: 286 °C (546.8 °F)
Specific Gravity: 1.42 to 1.44 at 20 °C/4 °C
Vapor Density: 7.16 Air=1
Saturated Vapor Density: 1.200000001 kg/m^3
Density: 1.438 g/mL
Vapor Pressure: 1.16 $x10^{-7}$ mm Hg at 25 °C
Water Solubility: Insoluble in Water
Other Solubilities: Soluble in concentrated Sulfuric acid; Soluble in Acetone.
Ionization Potential (eV): 9.25 +/-1.6
Flash Point: 185 °C Closed Cup

RTECS Toxicity Data

Acute Oral: Rat LD_{Lo} Dose: 15 gm/kg. Mouse LD_{50} Dose: >5 gm/kg.
Acute Inhalation: Rat LC_{50} Dose: >1300 mg/m^3/4hr.
Acute Dermal: Rat LD_{50} Route: Skin; Dose: >1 gm/kg.
Chronic (Multiple Dose) Oral: Rat Dose: 164 mg/kg/13W-C; Toxic Effects: Liver - Changes in liver weight; Kidney, Ureter, and Bladder - Changes in kidney weight; Nutritional and gross metabolic - Weight loss or decreased weight gain. Mouse Dose: 328 mg/kg/13W-C; Toxic Effects: Liver - Other changes; Liver - Changes in liver weight; Kidney, Ureter, and Bladder - Other changes.
Mutagenic: Mouse DNA Damage; Route: Intraperitoneal; Dose: 250 mg/kg. Bacteria - S Typhimurium Mutations in Microorganisms; Dose: 2 ug/plate (+S9). Bacteria - S Typhimurium Mutations in Microorganisms; Dose: 333 ug/plate (-S9).

Hazard Overviews

Health: Irritating to eyes/skin/respiratory tract. Harmful. Other Acute Effects: harmful if swallowed, inhaled, or absorbed through skin; may cause allergic skin reaction.

Fire: Will burn. Extinguishing agents: water spray; carbon dioxide, dry chemical powder or appropriate foam. Precautions: combustible liquid.

Reactivity: Incompatible with: strong oxidizing agents. Hazardous decomposition products: toxic fumes of: carbon monoxide, carbon dioxide.

Carcinogenicity: IARC - Not listed; NIOSH - Not listed; NTP - Not listed; ACGIH - Not listed; OSHA - Not listed; EPA - Not listed; MAK - Not listed

Primary Target Organs:

Eyes

Skin

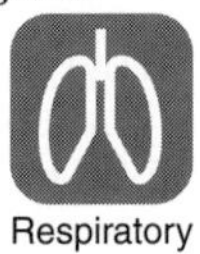
Respiratory System

Environmental

Ecotoxicity: Bacteria: Pseudomonas fluorescens 24h EC 0 5,000 mg /l; Crustaceans: Daphnia magna 30d EC_0,S 1 mg /l; Fishes: Pimephales promelas 96h LC_{50} 2,650 mg /l

Environmental Fate: If released to the atmosphere, it will exist in both the vapor-phase and the particulate phase based on an experimental vapor pressure of 1.16 $x10^{-7}$ mm Hg. In the vapor-phase, it should react with hydroxyl radicals with an estimated half-life of 11 days. Particulate phase may be physically removed from air by wet and dry deposition. It should have slight to low mobility in soil based on estimated K_{oc} values of 1664 and 3702. This compound is expected to biodegrade fairly rapidly with 68% degradation reported in 12 weeks by a mixed soil population. In water, it is expected to adsorb to particulate matter and sediment in the water column based on its K_{oc} values. Biodegradation is a major fate process in water; over a three day period, 82% (at 10 mg/L) was degraded by river water, 91% was degraded by sea water. It is also readily biodegraded by natural bacterial populations in groundwater and by activated sludge. A photolysis half-life of 2.8 hours was measured when exposed to incident light (295-500 nm). It may bioconcentrate in aquatic organisms based on estimated BCF values of 222 and 522.

Environmental Physical Data

Henry's Law Constant: 2.35 $x10^{-8}$
Octanol/Water Partition Coefficient: log K_{OW} = 3.39
Sorption Partition Coefficient: K_{OC} = 1664 to 3702
BCF: estimated at 222 to 522

Regulations

RCRA 40CFR: Not listed
CERCLA: 40CFR 302.4: Not listed
SARA 40CFR 372.65: Not listed
SARA EHS 40CFR 355: Not listed
TSCA: Listed

ANT3500 CAS #: 7440-36-0

ANTIMONY

RTECS: CC4025000
DOT: UN2871; IMO6.1
EINECS Number: 231-146-5
Molecular Formula: Sb
Structured MF: Sb
Formula Weight: 121.75

Sb

Chemical Structure

Synonyms: ANTIMONY BLACK; ANTIMONY METAL; ANTIMONY POWDER; ANTIMONY,REGULUS; ANTYMON; C.I. 77050; REGULUS OF ANTIMONY; STIBIUM

Description: blue-white, dark gray crystalline powder

Use: manufacture of alloys, such as Britannia or Babbitt metal, hard lead, white metal, type, bullets and bearing metal; in fireworks; for thermoelectric piles, blackening iron, coating metals, etc.

Physical Properties

Boiling Point: 1635 °C (2975 °F)
Freezing Point: 630 °C (1166 °F)
Specific Gravity: 6.684 at 25 °C
Vapor Pressure: 1 mm Hg at 886 °C
Water Solubility: Insoluble in Water
Other Solubilities: Sulfric acid: Soluble in hot.
Ionization Potential (eV): 8.64
Flash Point: Noncombustible Solid

RTECS Toxicity Data

Acute Oral: Rat LD_{50} Dose: 7 gm/kg.

Tumorigenic: Rat Route: Inhalation; Dose: 50 mg/m^3/7H/52W-I; Toxic Effects: Tumorigenic - Carcinogenic by RTECS criteria; Lungs, Thorax, or Respiration - Tumors.

Hazard Overviews

Explosive

Flammable

Fire Diamond

Health: Irritating to eyes/skin/respiratory tract. Also Causes: ingestion: violent vomiting, low blood pressure and shallow breathing. Chronic Effects: dizziness, dry throat, sleeplessness, anorexia, dermatitis.

Fire: Flammable and a moderate explosion hazard when exposed to heat and ignition sources. Fight fire with dry chemical.

Reactivity: Stable, in dry air. Hazardous polymerization cannot occur. Avoid: antimony or its alloys, magnesium, or zinc (reacts to form antimony trihydride). Incompatible with: cold, dilute acids; aqua regia; hot, concentrated sulfuric acid; hot, concentrated hydrochloric acid; oxidizing agents (nitrate salts; halogens; nitric acid; perchloric acids; chlorine trifluoride; potassium permanganate; ammonium nitrate; bromine trinitride; bromine trifluoride; chlorine monoxide; chlorine trifluoride; potassium nitrate; sodium nitrate; potassium oxide. Hazardous decomposition products: antimony hydroxide fumes.

Carcinogenicity: IARC - Not listed; NIOSH - Not listed; NTP - Not listed; ACGIH - Not listed; OSHA - Not listed; EPA - Not listed; MAK - Not listed

Primary Target Organs:

Eyes

Skin

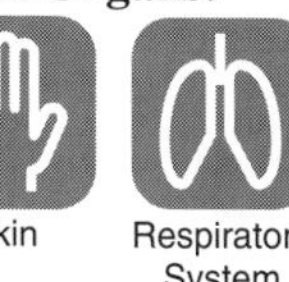
Respiratory System

Mucous Membranes

Nervous System

Exposure Limits

OSHA PEL: TWA: 0.5 mg/m^3; as Sb.

ACGIH TLV: TWA: 0.5 mg/m^3.

NIOSH REL: TWA: 0.5 mg/m^3; as Sb.

NIOSH IDLH: 50 mg/m^3; as Sb.

DFG MAK: TWA: 0.5 mg/m^3.

Respirator Recommendation

Exposure Range: >0.5 to 5 mg/m^3 Air Purifying, Negative Pressure, Half Mask

Exposure Range: >5 to <50 mg/m^3 Air Purifying, Negative Pressure, Full Face

Exposure Range: 50 to unlimited mg/m^3 Self-contained Breathing Apparatus, Pressure Demand, Full Face

Cartridge Color: magenta (P100)

Environmental

Cleanup/Disposal: Guide No. 170: Eliminate all ignition sources (no smoking, flares, sparks or flames in immediate area). Do not touch or walk through spilled material. Stop leak if you can do it without risk. Prevent entry into waterways, sewers, basements or confined areas.

Regulations

RCRA 40CFR: Not listed

CERCLA: 40CFR 302.4: Listed per CWA Section 307(a) RQ: 5000 lb (2268 kg)

SARA 40CFR 372.65: Listed

SARA EHS 40CFR 355: Not listed

TSCA: Listed

Analytical Methods

Air: EPA 29, 0060, 1638, 1639, 1669, ITM-001; ASTM D4490

Soil: CLP 200.10_M, 200.62, 200.7_M, 202.62, 204.1_M, 204.2_M, 206.3_M, 6020_M, ICP-AES; EPA 13, 200.7, 200.8; SW846 3005A, 3015, 3031, 3040, 3040A, 3051, 3052, 6010A, 6010B, 6020, 7000A, 7040, 7041, 7062, OSW-A; ASTM D1971; USGS I5055, I5475

Water / Groundwater: EPA 200.0, 200.15, 200.7, 200.9; APHA 3111-A, 3111-B, 3113-B, 3120, 3500-SB; ASTM D1976, D3697; USGS E-SPEC, I1055, I3055, I7055; CEM RD42

Drinking Water: AOAC 993.14

Food: EPA 14

Plasma: EPA 200.11; NIOSH 8005

Other: EPA 1620, 204.1, 204.2; AOAC 990.08

ANT4500 CAS #: 7647-18-9

ANTIMONY PENTACHLORIDE

RTECS: CC5075000

DOT: UN1730; UN1731; IMO8.0

EINECS Number: 231-601-8

Molecular Formula: Cl_5Sb

Structured MF: $SbCl_5$

Formula Weight: 299.05

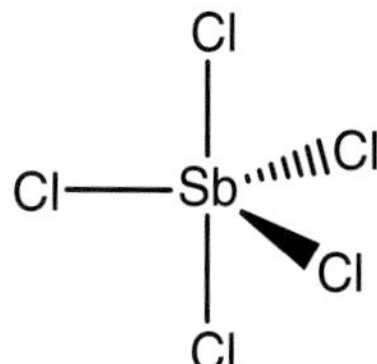

Chemical Structure

Synonyms: ANTIMOINE (PENTACHLORURE D'); ANTIMONIO (PENTACLORURO DI); ANTIMONPENTACHLORID; ANTIMONY CHLORIDE; ANTIMONY(V) CHLORIDE; ANTIMONY PERCHLORIDE; ANTIMOONPENTACHLORIDE; BUTTER OF ANTIMONY; PENTACHLOROANTIMONY; PENTACHLORURE D'ANTIMOINE; PERCHLORURE D'ANTIMOINE

Description: colorless to reddish-yellow, oily liquid; pungent, offensive odor

Use: catalyst when replacing a fluorine; substituent with chlorine in organic compounds; analysis (testing for alkaloids and cesium); dyeing intermediate; chlorine carrier in organic chlorinations; catalyst in production of org halogen cmpd

Physical Properties

Boiling Point: 79 °C (174 °F) at 22 mm Hg

Freezing Point: 2.8 °C (37.04 °F)

Specific Gravity: 2.336 at 20 °C/4 °C (liquid)

Vapor Density: 10.2 Air=1

Saturated Vapor Density: 1.214703267 kg/m^3

Vapor Pressure: 1 mm Hg at 22.7 °C

Water Solubility: Decomposes in Water

Other Solubilities: Soluble in Carbon Tetrachloride; Soluble in Chloroform.

Surface Tension: Estimated at 15 dynes/cm

pH: Acid

Refraction Index: 1.601 at 14 °C

Flash Point: Nonflammable

RTECS Toxicity Data

Acute Oral: Rat LD_{50} Dose: 1115 mg/kg; Toxic Effects: Behavioral - Muscle weakness; Liver - Hepatitis (hepatocellular necrosis), zonal; Kidney, Ureter, and Bladder - Other changes. Guinea Pig LD_{50} Dose: 900 mg/kg; Toxic Effects: Peripheral nerve and sensation - Spastic parapysis with or without sensory change; Gastrointestinal - Peritonitis; Gastrointestinal - Necrotic changes.

Acute Inhalation: Rat LC_{50} Dose: 720 mg/m^3/2hr. Mouse LC_{50} Dose: 620 mg/m^3; Toxic Effects: Lungs, Thorax, or Respiration - Acute pulmonary edema.
Mutagenic: Bacteria - B Subtilis DNA Repair; Dose: 65 ug/disc. Bacteria - B Subtilis Gene Conversion; Dose: 30 uL/disc.

Hazard Overviews

Poison

Corrosive

Fire Diamond

Health: Corrosive to eyes/skin/respiratory tract. Poison. Other Acute Effects: may be fatal if inhaled, swallowed, or absorbed through skin; inhalation may result in spasm, inflammation and edema of the larynx and bronchi, chemical pneumonitis and pulmonary edema; symptoms of exposure may include burning sensation; coughing; wheezing; laryngitis; shortness of breath; headache; nausea; vomiting; may cause nervous system disturbances. Chronic Effects: laboratory experiments have shown mutagenic effects. Possible carcinogen.
Fire: Noncombustible. Hazards: water hydrolyzes material liberating acidic gas which in contact with metal surfaces can generate flammable and/or explosive hydrogen gas; emits toxic fumes. Extinguishing agents: carbon dioxide, dry chemical powder or appropriate foam. Precautions: combustible liquid.
Reactivity: Incompatible with: strong oxidizing agents, may decompose on exposure to moist air or water. Hazardous decomposition products: toxic fumes of: carbon monoxide, carbon dioxide, hydrogen chloride gas, antimony/antimony oxides.
Carcinogenicity: IARC - Not listed; NIOSH - Not listed; NTP - Not listed; ACGIH - Not listed; OSHA - Not listed; EPA - Not listed; MAK - Not listed
Primary Target Organs:

Eyes

Skin

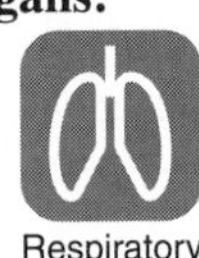
Respiratory System

Exposure Limits
OSHA PEL: TWA: 0.5 mg/m^3; as Sb.
ACGIH TLV: TWA: 0.5 mg/m^3; as Sb.
NIOSH REL: TWA: 0.5 mg/m^3; as Sb.

Environmental

Cleanup/Disposal: Guide No. 157: Eliminate all ignition sources (no smoking, flares, sparks or flames in immediate area). All equipment used when handling the product must be grounded. Do not touch damaged containers or spilled material unless wearing appropriate protective clothing. Stop leak if you can do it without risk. A vapor suppressing foam may be used to reduce vapors. Do not get water inside containers. Use water spray to reduce vapors or divert vapor cloud drift. Prevent entry into waterways, sewers, basements or confined areas. Small Spills: Cover with dry earth, dry sand, or other non-combustible material followed with plastic sheet to minimize spreading or contact with rain. Use clean non-sparking tools to collect material and place it into loosely covered plastic containers for later disposal.

Environmental Physical Data

BCF: no food chain concentration potential
BOD: none

Regulations

RCRA 40CFR: Not listed
CERCLA: 40CFR 302.4: Listed per CWA Section 311(b)(4) RQ: 1000 lb (453.5 kg)
SARA 40CFR 372.65: Listed as Compound
SARA EHS 40CFR 355: Not listed
TSCA: Listed

ANT5000 CAS #: 7783-70-2

ANTIMONY PENTAFLUORIDE

RTECS: CC5800000
DOT: UN1732; IMO8.0
EINECS Number: 232-021-8
Molecular Formula: F_5Sb
Structured MF: SbF_5
Formula Weight: 216.76

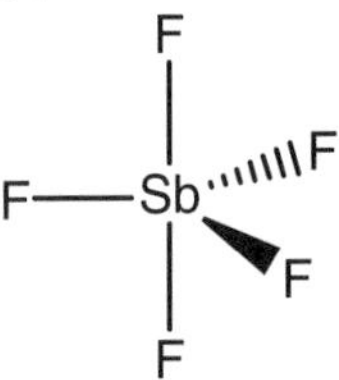

Chemical Structure

Synonyms: ANTIMONY (V) FLUORIDE; ANTIMONY FLUORIDE; ANTIMONY(V) FLUORIDE; ANTIMONY(V) PENTAFLUORIDE; PENTAFLUORIANTIMONY; PENTAFLUOROANTIMONY
Description: colorless oily liquid; sharp odor
Use: fluorination of organic cmpd; catalyst

Physical Properties

Boiling Point: 141 °C (286 °F)
Freezing Point: 8.3 °C (46.94 °F)
Specific Gravity: 3.097 at 25.8 °C
Vapor Pressure: 10 mm Hg
Water Solubility: Reacts with Water
Other Solubilities: Soluble in liquid sulfur dioxide.
Surface Tension: 20 dynes/cm
pH: Strong Lewis acid
Flash Point: Nonflammable

RTECS Toxicity Data

Acute Inhalation: Mouse LC_{50} Dose: 270 mg/m^3; Toxic Effects: Lungs, Thorax, or Respiration - Other changes; Liver - Other changes; Kidney, ureter, and Bladder - Other changes.

Hazard Overviews

Corrosive

Health: Corrosive to eyes/skin/respiratory tract. Toxic. Other Acute Effects: harmful if swallowed, inhaled, or absorbed through skin; inhalation may result in spasm, inflammation and edema of the larynx and bronchi, chemical pneumonitis and pulmonary edema; symptoms of exposure may include burning sensation; coughing; wheezing; laryngitis; shortness of breath; headache; nausea; vomiting. Chronic Effects: calcification of the bones, ligaments and tendons.

Fire: Noncombustible. Hazards: water hydrolyzes material liberating acidic gas which in contact with metal surfaces can generate flammable and/or explosive hydrogen gas; emits toxic fumes. Extinguishing agents: noncombustible; use extinguishing media appropriate to surrounding fire conditions; do not use water. Precautions: combustible liquid.

Reactivity: Incompatible with: acids, heat, sensitive to moisture. Hazardous decomposition products: toxic fumes of: carbon monoxide, carbon dioxide, hydrogen fluoride, antimony/antimony oxides.

Carcinogenicity: IARC - Not listed; NIOSH - Not listed; NTP - Not listed; ACGIH - Class A4, Not classifiable as a human carcinogen; OSHA - Not listed; EPA - Not listed; MAK - Not listed

Primary Target Organs:

Eyes

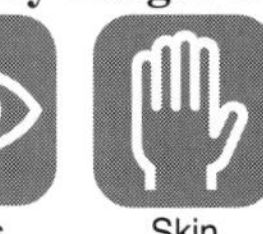

Skin

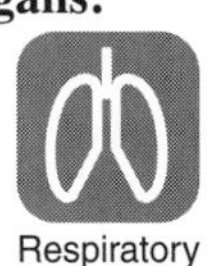

Respiratory System

Exposure Limits

OSHA PEL: TWA: 0.5 mg/m^3; as Sb.

ACGIH TLV: TWA: 0.5 mg/m^3; as Sb.

NIOSH REL: TWA: 0.5 mg/m^3; as Sb.

Environmental

Cleanup/Disposal: Guide No. 157: Eliminate all ignition sources (no smoking, flares, sparks or flames in immediate area). All equipment used when handling the product must be grounded. Do not touch damaged containers or spilled material unless wearing appropriate protective clothing. Stop leak if you can do it without risk. A vapor suppressing foam may be used to reduce vapors. Do not get water inside containers. Use water spray to reduce vapors or divert vapor cloud drift. Prevent entry into waterways, sewers, basements or confined areas. Small Spills: Cover with dry earth, dry sand, or other non-combustible material followed with plastic sheet to minimize spreading or contact with rain. Use clean non-sparking tools to collect material and place it into loosely covered plastic containers for later disposal.

Environmental Physical Data

BCF: no food chain concentration potential

BOD: none

Regulations

RCRA 40CFR: Not listed

CERCLA: 40CFR 302.4: Listed as Compound per CWA Section 307(a) per CAA Section 112

SARA 40CFR 372.65: Not listed TPQ: 500 lb

SARA EHS 40CFR 355: Listed TPQ: 500 lb

TSCA: Listed

ANT5500 CAS #: 28300-74-5

ANTIMONY POTASSIUM TARTRATE

RTECS: CC6825000

DOT: UN1551; IMO6.1

Molecular Formula: $C_4H_4KO_7Sb$

Structured MF: $C_4H_4O_7Sb{\bullet}K$

Formula Weight: 324.92

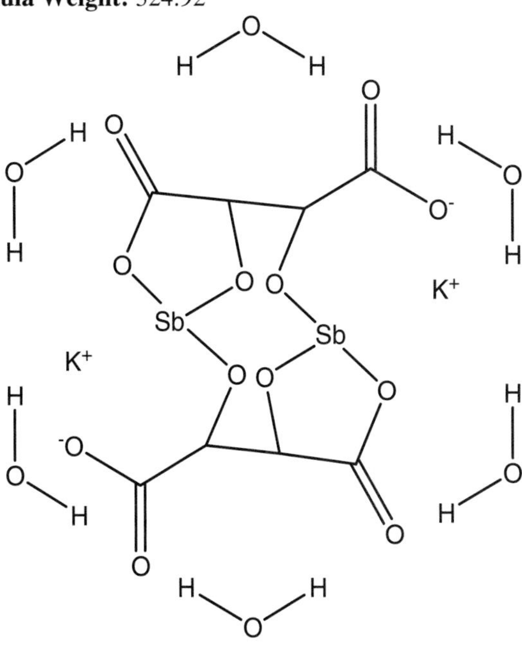

Chemical Structure

Synonyms: ANTIMONATE(2)-,BIS(MU-TARTRATO(4-))DI-,DIPOTASSIUM,TRIHYDRATE; ANTIMONY POTASSIUM TATRATE SOLID; ANTIMONYL POTASSIUM TARTRATE; EMETIQUE; ENT 50,434; EPA PESTICIDE CHEMICAL CODE 006201; POTASSIUM ANTIMONY TARTRATE; POTASSIUM ANTIMONYL D-TARTRATE; POTASSIUM ANTIMONYL TARTRATE; POTASSIUM ANTIMONYLTARTRATE; TARTAR EMETIC; TARTARIC ACID,ANTIMONY POTASSIUM SALT; TARTARIZED ANTIMONY; TARTOX; TARTRATE ANTIMONIO-POTASSIQUE; TARTRATE ANTIMONIOPOTASSIQUE; TARTRATED ANTIMONY; TASTOX

Description: ; odorless

Use: a mordant in the textile and leather industry, antischistosomal, parasiticide, expectorant, cuminatoric and insecticide

Physical Properties

Freezing Point: 100 °C (212 °F)

Specific Gravity: 2.6

Water Solubility: 10 to 50 mg/mL at 21 °C

Other Solubilities: 95% Ethanol: <1 mg/ml at 21 °C; Acetone: <1 mg/ml at 21 °C; Chloroform: Insoluble; DMSO: <1 mg/ml at 21 °C; Ether: Insoluble; Glycerol: 1 g/15 mL.
pH: Aqueous solution is slightly acid
Flash Point: Nonflammable

RTECS Toxicity Data

Acute Oral: Human LD_{Lo} Dose: 2 mg/kg. Rat LD_{50} Dose: 115 mg/kg.
Acute Dermal: Mouse LD_{50} Route: Subcutaneous Dose: 55 mg/kg; Toxic Effects: Behavioral - Muscle weakness; Lungs, Thorax, or Respiration - Dyspnea.
Chronic (Multiple Dose) Oral: Mouse Dose: 5698 mg/kg/14D-I; Toxic Effects: Liver - Other changes; Nutritional and gross metabolic - Weight loss or decreased weight gain; DEATH.
Mutagenic: Human Cytogenetic Analysis; Cell Type: fibroblast; Dose: 100 umol/L. Rat Cytogenetic Analysis; Route: Intraperitoneal; Dose: 2 mg/kg.

Hazard Overviews

Health: Irritating to eyes/skin/respiratory tract. Toxic. Other Acute Effects: harmful if swallowed, inhaled, or absorbed through skin; the most serious adverse effects are on the heart and liver; along with coughing; chest and abdominal pain; vomiting; fainting; collapse; less immediate adverse effects include gastrointestinal disturbances; headache; dizziness; weakness; damage to the kidneys; target organs: liver, kidneys, heart.
Fire: Noncombustible. Extinguishing agents: water spray; carbon dioxide, dry chemical powder or appropriate foam. Precautions: combustible liquid.
Reactivity: Stable. Hazardous polymerization will not occur. Incompatible with: strong oxidizing agents. Hazardous decomposition products: toxic fumes of: carbon monoxide, carbon dioxide, antimony/antimony oxides, potassium oxides.
Carcinogenicity: IARC - Not listed; NIOSH - Not listed; NTP - Not listed; ACGIH - Not listed; OSHA - Not listed; EPA - Not listed; MAK - Not listed
Primary Target Organs:

Eyes

Skin

Respiratory System

Liver

Kidneys

Cardio-vascular

Exposure Limits
OSHA PEL: TWA: 0.5 mg/m^3; as Sb.
ACGIH TLV: TWA: 0.5 mg/m^3; as Sb.
NIOSH REL: TWA: 0.5 mg/m^3; as Sb.

Environmental

Ecotoxicity: TLm Pimephales promelas (fathead minnows) 12 ppm as antimony/96 hr /Conditions of bioassay not specified
Cleanup/Disposal: Guide No. 151: Do not touch damaged containers or spilled material unless wearing appropriate protective clothing. Stop leak if you can do it without risk. Prevent entry into waterways, sewers, basements or confined areas. Cover with plastic sheet to prevent spreading. Absorb or cover with dry earth, sand or other non-combustible material and transfer to containers. Do not get water inside containers.

Environmental Physical Data

BCF: high

Regulations

RCRA 40CFR: Not listed
CERCLA: 40CFR 302.4: Listed per CWA Section 311(b)(4) RQ: 100 lb (45.35 kg)
SARA 40CFR 372.65: Listed as Compound
SARA EHS 40CFR 355: Not listed
TSCA: Not listed

ANT6000 CAS #: 7789-61-9

ANTIMONY TRIBROMIDE

RTECS: CC4400000
DOT: NA1549; IMO8.0
EINECS Number: 232-179-8
Molecular Formula: Br_3Sb
Structured MF: $SbBr_3$
Formula Weight: 361.51

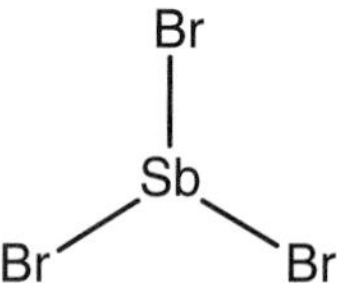

Chemical Structure

Synonyms: ANTIMONOUS BROMIDE; ANTIMONY BROMIDE; ANTIMONY TRIBROMIDE,SOLID OR SOLUTION; ANTIMONY(III) BROMIDE; STIBINE,TRIBROMO-; TRIBROMOSTIBINE
Description: colorless to yellow crystals, needles, or crystalline mass
Use: in analytical chem; mordant; in manufacture of antimony salts

Physical Properties

Boiling Point: 288 °C (550 °F) at 749 mm Hg
Freezing Point: 96 °C (204.8 °F)
Specific Gravity: 4.148 at 23 °C/23 °C
Vapor Pressure: 1 mm Hg at 93.9 °C
Water Solubility: Decomposed by Water
Other Solubilities: Soluble in Ammonia, Alcohol.
Refraction Index: 1.74
Critical Temperature: 904.5 °C
Critical Pressure: 56 atm
Flash Point: Not combustible

Hazard Overviews

Corrosive

Health: Corrosive to eyes/skin/respiratory tract. Other Acute Effects: harmful if swallowed, inhaled, or absorbed through skin; inhalation may result in spasm; inflammation and edema of the larynx and bronchi; chemical pneumonitis; pulmonary edema; exposure may cause burning sensation; coughing; wheezing; laryngitis; shortness of breath; headache; nausea; vomiting; anemia; cyanosis; dizziness; weakness; target organs: heart, blood. The toxicological properties have not been thoroughly investigated.

Fire: Noncombustible. Hazards: water hydrolyzes material liberating acidic gas which in contact with metal surfaces can generate flammable and/or explosive hydrogen gas; emits toxic fumes. Extinguishing agents: carbon dioxide, dry chemical powder or appropriate foam; do not use water. Precautions: combustible liquid.

Reactivity: Incompatible with: acids, strong bases, sensitive to light, sensitive to heat, reacts violently with water/sodium/potassium. Hazardous decomposition products: hydrogen bromide gas, antimony/antimony oxides.

Carcinogenicity: IARC - Not listed; NIOSH - Not listed; NTP - Not listed; ACGIH - Not listed; OSHA - Not listed; EPA - Not listed; MAK - Not listed

Primary Target Organs:

Eyes

Skin

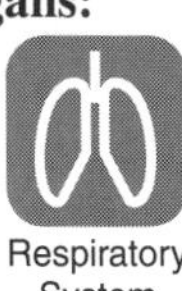
Respiratory System

Cardio-vascular

Blood

Exposure Limits

OSHA PEL: TWA: 0.5 mg/m^3; as Sb.

ACGIH TLV: TWA: 0.5 mg/m^3; as Sb.

NIOSH REL: TWA: 0.5 mg/m^3; as Sb.

Environmental

Ecotoxicity: Aquatic toxicity: 10 to 100 ppm/96 hr/*/TLm/* (as HBr reaction product with water) effect 0.5 mg/l (Sb(+3)) - slight effect 5.0 mg/l (Sb(+3)) - distinct effect 10th to 15th day; Food chain concentration potential: can be concentrated by a factor of 300 by marine life

Cleanup/Disposal: Guide No. 157: Eliminate all ignition sources (no smoking, flares, sparks or flames in immediate area). All equipment used when handling the product must be grounded. Do not touch damaged containers or spilled material unless wearing appropriate protective clothing. Stop leak if you can do it without risk. A vapor suppressing foam may be used to reduce vapors. Do not get water inside containers. Use water spray to reduce vapors or divert vapor cloud drift. Prevent entry into waterways, sewers, basements or confined areas. Small Spills: Cover with dry earth, dry sand, or other non-combustible material followed with plastic sheet to minimize spreading or contact with rain. Use clean non-sparking tools to collect material and place it into loosely covered plastic containers for later disposal.

Environmental Physical Data

BCF: certain forms of marine

BOD: 0.5 mg/l sb+ 0

Regulations

RCRA 40CFR: Not listed

CERCLA: 40CFR 302.4: Listed per CWA Section 311(b)(4) RQ: 1000 lb (453.5 kg)

SARA 40CFR 372.65: Listed as Compound

SARA EHS 40CFR 355: Not listed

TSCA: Listed

ANT6500 CAS #: 10025-91-9

ANTIMONY TRICHLORIDE

RTECS: CC4900000

DOT: UN1733; IMO8.0

EINECS Number: 233-047-2

Molecular Formula: Cl_3Sb

Structured MF: $SbCl_3$

Formula Weight: 228.13

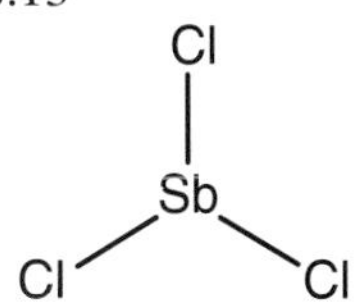

Chemical Structure

Synonyms: ANTIMOINE (TRICHLORURE D'); ANTIMONIO (TRICLORURO DI); ANTIMONIUS CHLORIDE; ANTIMONOUS CHLORIDE; ANTIMONTRICHLORID; ANTIMONY BUTTER; ANTIMONY CHLORIDE; ANTIMONY TRICHLORIDE,LIQUID OR SOLID; ANTIMONY(III) CHLORIDE; ANTIMOONTRICHLRIDE; BUTTER OF ANTIMONY; C.I. 77056; CAUSTIC ANTIMONY; CHLORID ANTIMONITY; CHLORURE ANTIMONIEUX; STIBINE,TRICHLORO-; TRICHLORO STIBINE; TRICHLOROSTIBINE; TRICHLORURE D'ANTIMOINE

Description: colorless rhombic crystals or yellow semi-solid mass; sharp, unpleasant odor

Use: rgnt for aromatic hydrocarbons & vitamin a; for mol wt determinations; in chem microscopy for identification of drugs; in org syntheses, catalyst; bronzing iron, especially gun barrels; mordant for patent leather & in dyeing; coloring zinc black; mfr lakes, particularly from dye woods; furniture polishes; in matches; chlorinating agent; intermed for other antimony cmpd; medication (vet): escharotic, dehorning agent; antimony trichloride in dehorning prepn.

Physical Properties

Boiling Point: 223.5 °C (434 °F)

Freezing Point: 73.4 °C (164.12 °F)

Specific Gravity: 3.14 at 25 °C

Vapor Pressure: 1 mm Hg at 49.2 °C

Water Solubility: Deliquescent

Other Solubilities: Soluble in Carbon Disulfide, Dioxane; Soluble in Carbon Tetrachloride, Ether; Insoluble in Pyridine, Quinoline, organic bases.

pH: Acid

Flash Point: Nonflammable

RTECS Toxicity Data

Acute Oral: Rat LD_{50} Dose: 525 mg/kg. Guinea Pig LD_{50} Dose: 574 mg/kg; Toxic Effects: Peripheral nerve and sensation - Spastic parapysis with or without sensory change; Gastrointestinal - Peritonitis; Gastrointestinal - Necrotic changes.

Acute Inhalation: Man TC_{Lo} Dose: 73 mg/m^3; Toxic Effects: Behavioral - Anorexia (human); Lungs, Thorax, or Respiration - Other changes; Gastrointestinal - Other changes.

Reproductive/Teratogenic: Rat Route: Oral; Dose: 44 mg/kg; Duration: female 1-22D of pregnancy; Effects on Newborn - Growth statistics. Rat Route: Oral; Dose: 4400 ug/kg; Duration: female 1-22D of pregnancy; Effects on Newborn - Delayed effects.

Mutagenic: Hamster Sister Chromatid Exchenge; Cell Type: lung; Dose: 2500 ug/L. Bacteria - B Subtilis DNA Repair; Dose: 10 mmol/L.

Hazard Overviews

Corrosive

Flammable

Health: Corrosive to eyes/skin/respiratory tract. Harmful. Other Acute Effects: harmful if swallowed, inhaled, or absorbed through skin; possible risk of irreversible effects. Chronic Effects: laboratory experiments have shown mutagenic effects.

Fire: Noncombustible. Hazards: water hydrolyzes material liberating acidic gas which in contact with metal surfaces can generate flammable and/or explosive hydrogen gas; emits toxic fumes. Extinguishing agents: noncombustible; use extinguishing media appropriate to surrounding fire conditions; do not use water. Precautions: combustible liquid.

Reactivity: Incompatible with: heat, strong bases, reacts violently with water. Hazardous decomposition products: hydrogen chloride gas, antimony/antimony oxides.

Carcinogenicity: IARC - Not listed; NIOSH - Not listed; NTP - Not listed; ACGIH - Not listed; OSHA - Not listed; EPA - Not listed; MAK - Not listed

Primary Target Organs:

Eyes

Skin

Respiratory System

Exposure Limits

OSHA PEL: TWA: 0.5 mg/m^3; as Sb.

ACGIH TLV: TWA: 0.5 mg/m^3; as Sb.

NIOSH REL: TWA: 0.5 mg/m^3; as Sb.

Environmental

Ecotoxicity: Aquatic toxicity: 17 ppm*/96 hr/fathead minnow/TLm/ fresh (hard) water 9 ppm*/96 hr/fathead minnow/TLm/ fresh (soft) water *as antimony; Food chain concentration potential: High

Cleanup/Disposal: Guide No. 157: Eliminate all ignition sources (no smoking, flares, sparks or flames in immediate area). All equipment used when handling the product must be grounded. Do not touch damaged containers or spilled material unless wearing appropriate protective clothing. Stop leak if you can do it without risk. A vapor suppressing foam may be used to reduce vapors. Do not get water inside containers. Use water spray to reduce vapors or divert vapor cloud drift. Prevent entry into waterways, sewers, basements or confined areas. Small Spills: Cover with dry earth, dry sand, or other non-combustible material followed with plastic sheet to minimize spreading or contact with rain. Use clean non-sparking tools to collect material and place it into loosely covered plastic containers for later disposal.

Environmental Physical Data

BCF: high

Regulations

RCRA 40CFR: Not listed

CERCLA: 40CFR 302.4: Listed per CWA Section 311(b)(4) RQ: 1000 lb (453.5 kg)

SARA 40CFR 372.65: Listed as Compound

SARA EHS 40CFR 355: Not listed

TSCA: Listed

ANT7000 CAS #: 7783-56-4

ANTIMONY TRIFLUORIDE

RTECS: CC5150000

DOT: NA1549; IMO8.0

EINECS Number: 232-009-2

Molecular Formula: F_3Sb

Structured MF: SbF_3

Formula Weight: 178.76

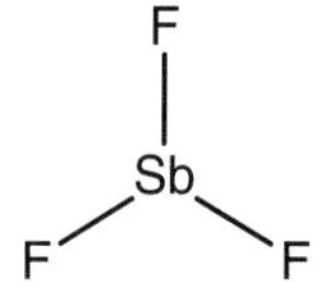

Chemical Structure

Synonyms: ANTIMOINE FLUORURE; ANTIMONIUS FLUORIDE; ANTIMONOUS FLUORIDE; ANTIMONY (111) FLUORIDE(1:3); ANTIMONY FLUORIDE; ANTIMONY TRIFLUORIDE,SOLID OR SOLUTION; ANTIMONY(III) FLUORIDE (1:3); STIBINE,TRIFLUORO-; STIBINE,TRIFLUORO-(9CI); TRIFLUOROANTIMONY; TRIFLUOROSTIBINE

Description: white, grey crystals; odorless

Use: to catalyze fluorinations by hydrogen fluoride; manufacture of chlorofluorides; in dyeing; manufacture of pottery & porcelains; in electroplating

Physical Properties

Boiling Point: 376 °C (709 °F)

Freezing Point: 292 °C (557.6 °F)

Specific Gravity: 4.379 at 20.9 °C

Vapor Pressure: 3 mm Hg at 17 °C

Water Solubility: Dissolves in Water

Other Solubilities: Insoluble in Ammonia.

Ionization Potential (eV): 12.1

Flash Point: Nonflammable

RTECS Toxicity Data

Acute Oral: Mouse LD_{50} Dose: 804 mg/kg.
Acute Dermal: Mouse LD_{Lo} Route: Subcutaneous Dose: 22900 ug/kg; Toxic Effects: Behavioral - Convulsions or effect on seizure threshold; Lungs, Thorax, or Respiration - Dyspnea; Lungs, Thorax, or Respiration - Other changes. Frog LD_{Lo} Route: Subcutaneous Dose: 224 mg/kg.

Hazard Overviews

Fire Diamond

Health: Irritating to eyes/skin/respiratory tract. Also Causes: fluoride poisoning. Chronic Effects: bone changes, shortness of breath, weight/hair loss, skin eruptions, jaundice, heart, lungs, liver, and kidney damage, increased RBC count, decreased WBC count.
Fire: Use water sprays, dry chemical, carbon dioxide, or foams.
Reactivity: Stable, in closed, dry, moisture-proof containers. Hazardous polymerization cannot occur. Incompatible with: potassium; sodium; hot solutions of perchloric acid. Hazardous decomposition products: fumes of antimony; oxides of fluorine.
Carcinogenicity: IARC - Not listed; NIOSH - Not listed; NTP - Not listed; ACGIH - Class A4, Not classifiable as a human carcinogen; OSHA - Not listed; EPA - Not listed; MAK - Not listed
Primary Target Organs:

Eyes

Skin

Respiratory System
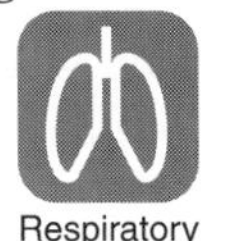
Mucous Membranes

Gastro-intestinal

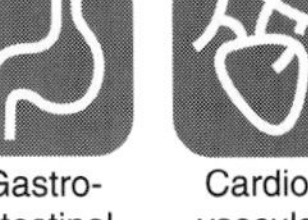
Cardio-vascular

Exposure Limits
OSHA PEL: TWA: 0.5 mg/m^3; as Sb.
ACGIH TLV: TWA: 0.5 mg/m^3; as Sb.
NIOSH REL: TWA: 0.5 mg/m^3; as Sb.

Environmental

Ecotoxicity: Aquatic toxicity: 200 ppm/<24 hr/tinca vulgaris (fish)/ killed/fresh water
Cleanup/Disposal: Guide No. 157: Eliminate all ignition sources (no smoking, flares, sparks or flames in immediate area). All equipment used when handling the product must be grounded. Do not touch damaged containers or spilled material unless wearing appropriate protective clothing. Stop leak if you can do it without risk. A vapor suppressing foam may be used to reduce vapors. Do not get water inside containers. Use water spray to reduce vapors or divert vapor cloud drift. Prevent entry into waterways, sewers, basements or confined areas. Small Spills: Cover with dry earth, dry sand, or other non-combustible material followed with plastic sheet to minimize spreading or contact with rain. Use clean non-sparking tools to collect material and place it into loosely covered plastic containers for later disposal.

Regulations

RCRA 40CFR: Not listed
CERCLA: 40CFR 302.4: Listed per CWA Section 311(b)(4) RQ: 1000 lb (453.5 kg)
SARA 40CFR 372.65: Listed as Compound
SARA EHS 40CFR 355: Not listed
TSCA: Listed

ANT7500 CAS #: 7790-44-5

ANTIMONY TRIIODIDE

DOT: UN1549; IMO6.1
EINECS Number: 232-205-8
Molecular Formula: I_3Sb
Structured MF: I_3Sb
Formula Weight: 502.52

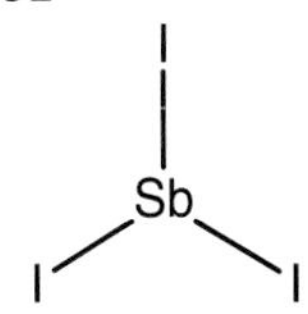

Chemical Structure

Synonyms: ANTIMONY IODIDE; ANTIMONY(III) TRIIODIDE; STIBINE,TRIIODO-
Description: ruby-red, trigonal crystals or yellowish-green crystals

Physical Properties

Boiling Point: 420 °C (788 °F)
Freezing Point: 168 °C (334.4 °F)
Specific Gravity: 4.921 at 17 °C/4 °C
Vapor Pressure: 1 mm Hg at 163.6 °C
Water Solubility: Decomposed by Water
Other Solubilities: Insoluble in Alcohol; Soluble in Benzene, Hydrogen Iodide.
Refraction Index: 2.78
Critical Temperature: 1101 °C
Critical Pressure: 55 atm

Hazard Overviews

Corrosive

Health: Corrosive to eyes/skin/respiratory tract. Other Acute Effects: harmful if swallowed, inhaled, or absorbed through skin; inhalation may result in spasm; inflammation and edema of the larynx and bronchi; chemical pneumonitis; pulmonary edema; exposure may cause burning sensation; coughing; wheezing; laryngitis; shortness of breath; headache; nausea; vomiting; allergic skin reaction; dermatitis; dizziness; weakness; gastrointestinal disturbances; target organs: blood, liver, kidneys, central nervous system. The toxicological properties have not been thoroughly investigated.

Fire: Hazards: emits toxic fumes. Extinguishing agents: carbon dioxide, dry chemical powder or appropriate foam; do not use water. Precautions: combustible liquid.

Reactivity: Incompatible with: acids, strong bases, sensitive to air, reacts violently with water/sodium/potassium. Hazardous decomposition products: iodine, hydrogen iodide, antimony/antimony oxides.

Carcinogenicity: IARC - Not listed; NIOSH - Not listed; NTP - Not listed; ACGIH - Not listed; OSHA - Not listed; EPA - Not listed; MAK - Not listed

Primary Target Organs:

Eyes

Skin

Respiratory System

Liver

Kidneys

Blood

Exposure Limits

OSHA PEL: TWA: 0.5 mg/m^3; as Sb.

ACGIH TLV: TWA: 0.5 mg/m^3; as Sb.

NIOSH REL: TWA: 0.5 mg/m^3; as Sb.

Environmental

Cleanup/Disposal: Guide No. 157: Eliminate all ignition sources (no smoking, flares, sparks or flames in immediate area). All equipment used when handling the product must be grounded. Do not touch damaged containers or spilled material unless wearing appropriate protective clothing. Stop leak if you can do it without risk. A vapor suppressing foam may be used to reduce vapors. Do not get water inside containers. Use water spray to reduce vapors or divert vapor cloud drift. Prevent entry into waterways, sewers, basements or confined areas. Small Spills: Cover with dry earth, dry sand, or other non-combustible material followed with plastic sheet to minimize spreading or contact with rain. Use clean non-sparking tools to collect material and place it into loosely covered plastic containers for later disposal.

Regulations

RCRA 40CFR: Not listed

CERCLA: 40CFR 302.4: Listed as Compound per CWA Section 307(a) per CAA Section 112

SARA 40CFR 372.65: Listed as Compound

SARA EHS 40CFR 355: Not listed

TSCA: Listed

ANT8000 CAS #: 1309-64-4

ANTIMONY TRIOXIDE

RTECS: CC5650000

DOT: NA9201

EINECS Number: 215-175-0

Molecular Formula: O_3Sb_2

Structured MF: Sb_2O_3

Formula Weight: 291.52

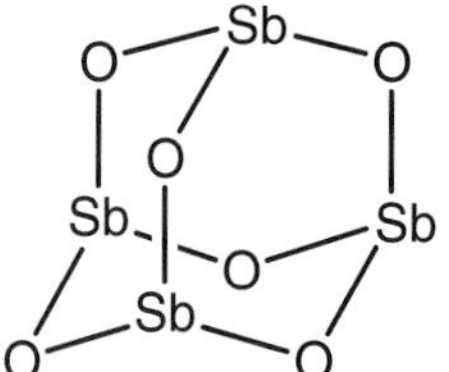

Chemical Structure

Synonyms: A 1530; A 1582; A 1588LP; A1530; A1582; A1588 LP; AMSPEC-KR; ANTIMONIOUS OXIDE; ANTIMONY OXIDE; ANTIMONY(3+) OXIDE; ANTIMONY PEROXIDE; ANTIMONY SESQUIOXIDE; ANTIMONY WHITE; ANTOX; AP 50; AT 3; BLUE STAR; C.I. 77052; C.I. PIGMENT WHITE 11; CHEMETRON FIRE SHIELD; CP 99295; DECHLORANE A-O; DIANTIMONY TRIOXIDE; EXITELITE; EXTREMA; FLOWERS OF ANTIMONY; NA9201; NYACOL A 1530; NYACOL A 1510LP; PATOX C; PATOX H; PATOX L; PATOX M; PATOX S; SENARMONTITE; STIBIOX MS; THERMOGUARD B; THERMOGUARD S; TIMONOX; TWINKLING STAR; VALENTINITE; WEISSPIESSGLANZ; WHITE STAR

Description: white crystalline powder; odorless

Use: manufacture of tartar emetic; as paint pigment; in enamels and glasses; as mordant; in flame-proofing canvas; catalyst; intermediate; staining iron and copper; phosphors

Physical Properties

Boiling Point: 1425 °C (2597 °F)

Freezing Point: 655 °C (1211 °F)

Specific Gravity: Senarmonite 5.2

Density: 5.2 g/mL at 25 °C

Vapor Pressure: 1 mm Hg at 574 °C

Water Solubility: Very Slightly Soluble in Cold Water

Other Solubilities: 95% Ethanol: <1 mg/ml at 20 °C; Acetone: <1 mg/ml at 20 °C; Acetic Acid: Soluble; DMSO: <1 mg/ml at 20 °C; Dilute H2SO4: Slightly Soluble; Dilute HNO3: Slightly Soluble; Hydrochloric Acid: Soluble; Potassium hydroxide: Soluble.

Refraction Index: 2.087

Flash Point: Nonflammable

RTECS Toxicity Data

Acute Oral: Rat LD_{50} Dose: >34600 mg/kg; Toxic Effects: Behavioral - Somnolence (general depressed activity); Skin and appendages - Hair.

Acute Dermal: Rabbit LD_{Lo} Route: Skin; Dose: 2 gm/kg. Rabbit LD_{Lo} Route: Subcutaneous Dose: 2500 ug/kg.

Chronic (Multiple Dose) Inhalation: Rat Dose: 72 ug/m^3/24H/17W-C; Toxic Effects: Blood - Pigmented or nucleated red Blood cells; Biochemical - True cholinesterase; Biochemical - Lipids including transport. Rat Dose: 45 mg/m^3/10W-I; Toxic Effects: Lungs, Thorax, or Respiration - Fibrosis, focal (pneumoconiosis); Liver - Fatty Liver degeneration; DEATH.

Irritation Eye: Rabbit Standard Draize Test Dose: 100 mg; Reaction: mild.

Reproductive/Teratogenic: Rat Route: Inhalation; Dose: 270 ug/m^3; Duration: female 1-21D of pregnancy; Effects on Fertility - Post-implantation mortality; Effects on Embryo or Fetus - Fetal death. Rat Route: Inhalation; Dose: 82 ug/m^3;

Duration: female 1-21D of pregnancy; Effects on Fertility - Pre-implantation mortality; Effects on Embryo or Fetus - Fetotoxicity. Rat Route: Inhalation; Dose: 270 ug/m^3/24 Duration: female 1-21D of pregnancy; Effects on Fertility - Pre-implantation mortality; Post-implantation mortality; Effects on Embryo or Fetus - Fetal death.

Mutagenic: Hamster Sister Chromatid Exchenge; Cell Type: lung; Dose: 90 ug/L. Bacteria - B Subtilis DNA Repair; Dose: 50 mmol/L.

Tumorigenic: Rat Route: Inhalation; Dose: 4200 ug/m^3/52W-I; Toxic Effects: Tumorigenic - Carcinogenic by RTECS criteria; Lungs, Thorax, or Respiration - Tumors; Liver - Tumors. Rat Route: Inhalation; Dose: 4 mg/m^3/1Y-I; Toxic Effects: Tumorigenic - Equivocal tumorigenic agent by RTECS criteria; Lungs, Thorax, or Respiration - Tumors; Liver - Tumors. Rat Route: Inhalation; Dose: 1600 ug/m^3/52W-I; Toxic Effects: Tumorigenic - Neoplastic by RTECS criteria; Liver - Tumors; Skin and appendages - Tumors. Rat Route: Inhalation; Dose: 50 mg/m^3/7H/52W-I; Toxic Effects: Tumorigenic - Carcinogenic by RTECS criteria; Lungs, Thorax, or Respiration - Tumors.

Hazard Overviews

Fire Diamond

Health: Irritating to eyes/skin/respiratory tract. Also Causes: headache, chest pain, shortness of breath, diarrhea, weight loss. Chronic Effects: pneumoconiosis, heart disorders, possible lung cancer.

Fire: Will burn. Use dry chemical, carbon dioxide, water spray, or regular foam.

Reactivity: Stable. Hazardous polymerization cannot occur. Avoid: dust generation; heat; ignition sources. Incompatible with: interhalogens; chlorine trifluoride; bromine pentafluoride; phosphorus pentachloride. Hazardous decomposition products: toxic antimony fumes.

Carcinogenicity: IARC - Group 2B, Possibly carcinogenic to humans; NIOSH - Not listed; NTP - Not listed; ACGIH - Class A2, Suspected human carcinogen; OSHA - Not listed; EPA - Not listed; MAK - Class A2, Unmistakably carcinogenic in animal experimentation only

Primary Target Organs:

Eyes

Skin
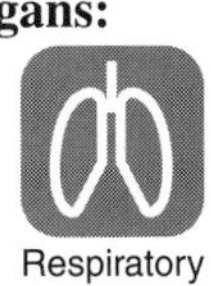
Respiratory System
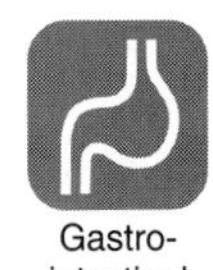
Gastro-intestinal

Cardio-vascular

Exposure Limits

OSHA PEL: TWA: 0.5 mg/m^3; as Sb.

ACGIH TLV: TWA: 0.5 mg/m^3.

NIOSH REL: TWA: 0.5 mg/m^3; as Sb.

Respirator Recommendation

Exposure Range: >0.5 to 5 mg/m^3 Air Purifying, Negative Pressure, Half Mask

Exposure Range: >5 to <50 mg/m^3 Air Purifying, Negative Pressure, Full Face

Exposure Range: 50 to unlimited mg/m^3 Self-contained Breathing Apparatus, Pressure Demand, Full Face

Cartridge Color: magenta (P100)

Environmental

Ecotoxicity: LD_{50} Lepomis macrochirus (bluegill sunfish) > 530 mg/l/96 hr. /Conditions of bioassay not specified LD_{50} Pimephales promelas (fathead minnow) > 833 mg/l/96 hr. /Conditions of bioassay not specified

Cleanup/Disposal: Guide No. 171: Do not touch or walk through spilled material. Stop leak if you can do it without risk. Prevent dust cloud. Avoid inhalation of asbestos dust. Small Dry Spills: With clean shovel place material into clean, dry container and cover loosely; move containers from spill area. Small Spills: Take up with sand or other noncombustible absorbent material and place into containers for later disposal. Large Spills: Dike far ahead of liquid spill for later disposal. Cover powder spill with plastic sheet or tarp to minimize spreading. Prevent entry into waterways, sewers, basements or confined areas.

Environmental Physical Data

BCF: high

BOD: none

Regulations

RCRA 40CFR: Not listed

CERCLA: 40CFR 302.4: Listed per CWA Section 311(b)(4) RQ: 1000 lb (453.5 kg)

SARA 40CFR 372.65: Listed as Compound

SARA EHS 40CFR 355: Not listed

TSCA: Listed

ANT8500 CAS #: 1345-04-6

ANTIMONY TRISULFIDE

RTECS: CC9450000

DOT: NA1325

EINECS Number: 215-713-4

Molecular Formula: S_3Sb_2

Formula Weight: 339.68

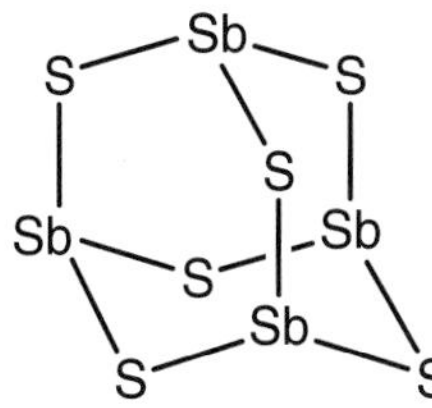

Chemical Structure

Synonyms: ANTIMONOUS SULFIDE; ANTIMONY GLANCE; ANTIMONY ORANGE; ANTIMONY SESQUISULFIDE; ANTIMONY SULFIDE; ANTIMONY(3+) SULFIDE; ANTIMONY TRISULFIDE COLLOID; ANTIMONY VERMILION; BLACK ANTIMONY; C.I. 77060; C.I. PIGMENT RED 107; CRIMSON ANTIMONY; CRIMSON ANTIMONY SULPHIDE; DIANTIMONY TRISULFIDE; LYMPHOSCAN; NEEDLE ANTIMONY; STIBNITE

Description: gray, lustrous, crystalline masses or grayish-black powder or yellowish-reddish, amorphous crystals
Use: manufacture ruby glass, explosives; as pigment & plasticizer in rubber industry; medication (vet): was formerly used as emetic in veterinary medicine; chem intermed in production of metallic antimony & antimony trioxide; vermilion or yellow pigment, antimony salts, pyrotechnics, matches, percussion caps, camouflage paints

Physical Properties

Boiling Point: About 1150 °C (2102 °F)
Freezing Point: 550 °C (1022 °F)
Specific Gravity: 4.12
Water Solubility: 0 g/100 cc at 18 °C
Other Solubilities: Soluble in Alcohol, potassium sulfide, ammonium hydrosulfide; Insoluble in Acetic Acid.

RTECS Toxicity Data

Acute Inhalation: Human TC_{Lo} Dose: 580 ug/m^3/35W; Toxic Effects: Cardiac - EKG changes not diagnostic of above; Vascular - BP elevation not characterized in autonomic section; Gastrointestinal - Ulceration or bleeding from stomach.
Acute Dermal: Mammal LD; Route: Subcutaneous Dose: >139 mg/kg.

Hazard Overviews

Health: Irritating to eyes/skin/respiratory tract. Harmful. Other Acute Effects: harmful if swallowed, inhaled, or absorbed through skin. Chronic Effects: Possible carcinogen.
Fire: Hazards: in powder form capable of creating a dust explosion; emits toxic fumes. Extinguishing agents: carbon dioxide, dry chemical powder or appropriate foam. Precautions: combustible liquid.
Reactivity: Incompatible with: strong oxidizing agents, avoid contact with acid. Hazardous decomposition products: antimony/antimony oxides, sulfur oxides, hydrogen sulfide gas.
Carcinogenicity: IARC - Group 3, Not classifiable as to carcinogenicity to humans; NIOSH - Not listed; NTP - Not listed; ACGIH - Not listed; OSHA - Not listed; EPA - Not listed; MAK - Not listed
Primary Target Organs:

Eyes

Skin

Respiratory System

Exposure Limits
OSHA PEL: TWA: 0.5 mg/m^3; as Sb.
ACGIH TLV: TWA: 0.5 mg/m^3; as Sb.
NIOSH REL: TWA: 0.5 mg/m^3; as Sb.

Environmental

Cleanup/Disposal: Guide No. 133: Eliminate all ignition sources (no smoking, flares, sparks or flames in immediate area). Do not touch or walk through spilled material. Small Dry Spills: With clean shovel place material into clean, dry container and cover loosely; move containers from spill area. Large Spills: Wet down with water and dike for later disposal. Prevent entry into waterways, sewers, basements or confined areas.

Regulations

RCRA 40CFR: Not listed
CERCLA: 40CFR 302.4: Listed as Compound per CWA Section 307(a) per CAA Section 112
SARA 40CFR 372.65: Listed as Compound
SARA EHS 40CFR 355: Not listed
TSCA: Listed

ANT9000 CAS #: 1397-94-0

ANTIMYCIN A

RTECS: CD0350000
Molecular Formula: Unspecified or Variable
Formula Weight: 548.70

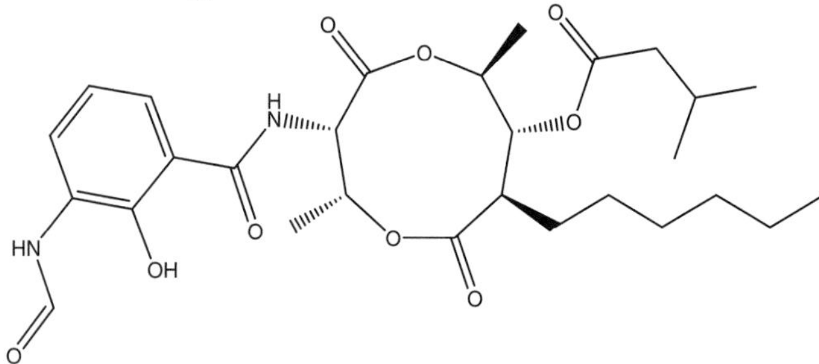

Chemical Structure

Synonyms: ANTIPIRICULLIN; ANTIPRICULLIN; EPA PESTICIDE CHEMICAL CODE 006314; VEROSIN; VIROSIN
Description: crystalline solid

Physical Properties

Freezing Point: 139 °C (282 °F) to 140 °C (284 °F)
Water Solubility: Insoluble in Water
Other Solubilities: Soluble in Acetone Alcohol Chloroform Ether. Slightly soluble in Benzene, Carbon tetrachloride, Petroleum ether

RTECS Toxicity Data

Acute Oral: Rat LD_{50} Dose: 28 mg/kg. Mouse LD_{50} Dose: 55 mg/kg.
Acute Dermal: Rat LD_{50} Route: Subcutaneous Dose: 25 mg/kg. Mouse LD_{50} Route: Subcutaneous Dose: 1600 ug/kg.
Mutagenic: Mouse DNA Inhibition; Cell Type: lymphocyte; Dose: 1 mg/L. Mouse Other Mutation Test Systems; Cell Type: lymphocyte; Dose: 1 mg/L.

Hazard Overviews

Poison

Health: Poison. Other Acute Effects: may be fatal if inhaled, swallowed, or absorbed through skin. Chronic Effects: possible reproductive hazard. The toxicological properties have not been thoroughly investigated.
Fire: Hazards: emits toxic fumes. Extinguishing agents: carbon dioxide, dry chemical powder or appropriate foam. Precautions: combustible liquid.
Reactivity: Stable. Hazardous polymerization will not occur. Hazardous decomposition products: toxic fumes of: carbon monoxide, carbon dioxide, nitrogen oxides.
Carcinogenicity: IARC - Not listed; NIOSH - Not listed; NTP - Not listed; ACGIH - Not listed; OSHA - Not listed; EPA - Not listed; MAK - Not listed

Environmental

Regulations

RCRA 40CFR: Not listed
CERCLA: 40CFR 302.4: Not listed
SARA 40CFR 372.65: Not listed TPQ: 1000/10000 lb
SARA EHS 40CFR 355: Listed TPQ: 1000 lb
TSCA: Not listed

Analytical Methods

Soil: EPA PMD-ANT

APH5000 CAS #: 52-46-0

APHOLATE

RTECS: XX9450000
Molecular Formula: $C_{12}H_{24}N_9P_3$
Formula Weight: 387.36
Synonyms: AFOLAT; APN; AZIRIDINE,1,3,5,2,4,6-TRIAZATRIPHOSPHORINE DERIVATIVE; 1-AZIRIDINYLPHOSPHONITRILE TRIMER; ENT 26,316; HEXA(1-AZIRIDINYL)TRIPHOSPHOTRIAZINE; 2,2,4,4,6,6-HEXAHYDRO-2,2,4,4,6,6-HEXAKIS(1-AZIRIDINYL)-1,3,5,2,4,6-TRIAZATRIPHOSPHORINE; 2,2,4,4,6,6-HEXAKIS(1-AZIRIDINYL)CYCLOTRIPHOSPHAZA-1,3,5-TRIENE; 2,2,4,4,6,6-HEXAKIS(1-AZIRIDINYL)-2,2,4,4,6,6-HEXAHYDRO-1,3,5,2,4,6-TRIAZATRIPHOSPHORINE; HEXAKIS-(1-AZIRIDINYL)PHOSPHONITRILE; HEXAKIS(AZIRIDINYL)PHOSPHOTRIAZINE; MYKO 63; NSC-26812; OLIN MO 2174; OLIN MO. 2174; PHOLATE; PN6; SQ 8388; 1,3,5,2,4,6-TRIAZATRIPHOSPHORINE,2,2,4,4,6,6-HEXAKIS(1-AZIRIDINYL)-2,2,4,4,6,6-HEXAHYDRO-
Description: crystals
Use: exptl insect chemosterilant

Physical Properties

Freezing Point: 147.5 °C (297.5 °F)

RTECS Toxicity Data

Acute Oral: Rat LD_{50} Dose: 98 mg/kg. Mouse LD_{50} Dose: 110 mg/kg.
Acute Dermal: Rat LD_{50} Route: Skin; Dose: 400 mg/kg. Rat LD_{50} Route: Subcutaneous Dose: 21 mg/kg; Toxic Effects: Liver - Hepatitis (hepatocellular necrosis), zonal; Kidney, Ureter, and Bladder - Chgs in tubules (inc acute renal failure, acute tubular necrosis; Blood - Changes in spleen.
Chronic (Multiple Dose) Oral: Rat Dose: 1470 mg/kg/98D-C; Toxic Effects: DEATH. Chicken Dose: 42875 mg/kg/15W-I; Toxic Effects: Blood - Changes in leukocyte (WBC) cell count; Nutritional and gross metabolic - Weight loss or decreased weight gain; DEATH.
Reproductive/Teratogenic: Rat Route: Oral; Dose: 105 mg/kg; Duration: male 21D prior to mating; Paternal Effects - Testes, epididymis, sperm duct. Rat Route: Intraperitoneal; Dose: 10 mg/kg; Duration: female 11D of pregnancy; Effects on Embryo or Fetus - Extra embryonic structures; Fetotoxicity. Rat Route: Intraperitoneal; Dose: 10 mg/kg; Duration: male 1D prior to mating; Effects on Fertility - Male fertility index.
Mutagenic: Human Cytogenetic Analysis; Cell Type: leukocyte; Dose: 100 umol/L. Rat DNA Inhibition; Route: Intraperitoneal; Dose: 5 mg/kg.

Hazard Overviews

Carcinogenicity: IARC - Group 3, Not classifiable as to carcinogenicity to humans; NIOSH - Not listed; NTP - Not listed; ACGIH - Not listed; OSHA - Not listed; EPA - Not listed; MAK - Not listed

Environmental

Regulations

RCRA 40CFR: Not listed
CERCLA: 40CFR 302.4: Not listed
SARA 40CFR 372.65: Not listed
SARA EHS 40CFR 355: Not listed
TSCA: Not listed

APO5000 CAS #: 58-00-4

APOMORPHINE

RTECS: CE0700000
EINECS Number: 200-360-0
Molecular Formula: $C_{17}H_{17}NO_2$
Formula Weight: 267.31
Synonyms: APOMORFIN; (-)-APOMORPHINE; L-APOMORPHINE; R-(-)-APOMORPHINE; APORMORPHINE; 6A BETA-APORPHINE-10,11-DIOL; 4H-DIBENZO(DE,G)QUINOLINE-10,11-DIOL,5,6,6A,7-TETRAHYDRO-6-METHYL-; 4H-DIBENZO(DE,G)QUINOLINE-10,11-DIOL,5,6,6A,7-TETRAHYDRO-6-METHYL-,-; (-)-10,11-DIHYDROXYAPORPHINE
Description: hexagonal plates or rods; odorless
Use: medication: emetic; medication (vet): emetic, expectorant in dogs; medication: expectorant

Physical Properties

Boiling Point: Sublimes
Freezing Point: Decomposes at 195 °C (383 °F)
Water Solubility: Slightly Soluble in Water
Other Solubilities: Soluble in alkali; Slightly Soluble in Hydrochloric Acid.

RTECS Toxicity Data

Acute Oral: Mouse LD_{50} Dose: >100 mg/kg; Toxic Effects: Behavioral - Excitment; Behavioral - Muscle weakness.
Acute Dermal: Rabbit LD_{Lo} Route: Subcutaneous Dose: 5 mg/kg. Mouse LD_{Lo} Route: Subcutaneous Dose: 13 mg/kg.
Reproductive/Teratogenic: Rat Route: Subcutaneous; Dose: 7 mg/kg; Duration: female 13-19D of pregnancy Effects on Newborn - Behavioral.

Hazard Overviews

Carcinogenicity: IARC - Not listed; NIOSH - Not listed; NTP - Not listed; ACGIH - Not listed; OSHA - Not listed; EPA - Not listed; MAK - Not listed

Environmental

Regulations

RCRA 40CFR: Not listed
CERCLA: 40CFR 302.4: Not listed
SARA 40CFR 372.65: Not listed
SARA EHS 40CFR 355: Not listed
TSCA: Not listed

APR5000 CAS #: 77-02-1

APROBARBITAL

RTECS: CP8925000
EINECS Number: 200-997-4
Molecular Formula: $C_{10}H_{14}N_2O_3$
Formula Weight: 210.23

Chemical Structure

Synonyms: ALLIONAL; ALLONAL; 5-ALLYL-5-ISOPROPYLBARBITURATE; 5-ALLYL-5-ISOPROPYLBARBITURIC ACID; ALLYLISOPROPYLBARBITURIC ACID; ALLYLISOPROPYLMALONYLUREA; ALLYLPROPYMAL; ALLYPROPYMAL; ALURATE; ALURATE ELIXIR VERDUM; APROBARBITONE; APROZAL; BARBITURIC ACID,5-ALLYL-5-ISOPROPYL-; ISONAL; 5-ISOPROPYL-5-ALLYLBARBITURIC ACID; ISOPROPYLALLYLBARBITURIC ACID; 5-(1-METHYLETHYL)-5-(2-PROPENYL)-2,4,6(1H,3H,5H)-PYRIMIDINETRIONE; NUMAL; 2,4,6(1H,3H,5H)-PYRIMIDINETRIONE,5-(1-METHYLETHYL)-5-(2-PROPENYL)-
Description: fine, white, crystalline powder
Use: medication: sedative; hypnotic

Physical Properties

Freezing Point: 140 °C (284 °F) to 141.5 °C (286.7 °F)
Water Solubility: Almost Insoluble in Water
Other Solubilities: Lipid solubility is low.
pH: Saturated aqueous solution is acid
Refraction Index: Alpha 1.635

RTECS Toxicity Data

Acute Oral: Rabbit LD_{Lo} Dose: 160 mg/kg.
Acute Dermal: Mouse LD_{50} Route: Subcutaneous Dose: 350 mg/kg.

Hazard Overviews

Health: Harmful. Other Acute Effects: harmful if swallowed; poor judgment; emotional instability; toxic psychosis; nystagmus; dysarthria; ataxia. Chronic Effects: prolonged or repeated exposure can lead to habituation or addiction; target organ: nerves. Possibly causes reproductive effects in humans. The toxicological properties have not been thoroughly investigated.
Fire: Hazards: emits toxic fumes. Extinguishing agents: carbon dioxide, dry chemical powder or appropriate foam; water spray. Precautions: combustible liquid.
Reactivity: Stable. Hazardous polymerization will not occur. Hazardous decomposition products: toxic fumes of: carbon monoxide, carbon dioxide, nitrogen oxides.
Carcinogenicity: IARC - Not listed; NIOSH - Not listed; NTP - Not listed; ACGIH - Not listed; OSHA - Not listed; EPA - Not listed; MAK - Not listed
Primary Target Organs:

Nervous System

Environmental

Regulations

RCRA 40CFR: Not listed
CERCLA: 40CFR 302.4: Not listed
SARA 40CFR 372.65: Not listed
SARA EHS 40CFR 355: Not listed
TSCA: Not listed

AQU5000 CAS #: 8007-56-5

AQUA REGIA

RTECS: MW4100000
Molecular Formula: ClH_2NO_3
Structured MF: $HCl{\bullet}HNO_3$
Formula Weight: 36.46
Synonyms: NITROHYDROCHLORIC ACID
Description: yellow liquid

Physical Properties

Boiling Point: 20.22% HCl 109 °C (228 °F)
Freezing Point: 20.22% HCl -65 °C (-85 °F)
Specific Gravity: 1.5027 at 25 °C
Vapor Density: 1.268
Vapor Pressure: 1 mm Hg at -150.8 °C
Water Solubility: 6 g/100 cc at 60 °C
pH: 1.0 N HCl 0.1

Hazard Overviews

Corrosive

Fire Diamond

Health: Corrosive to eyes/skin/respiratory tract. Also Causes: salivation, headache, fatigue, dizziness, nausea, shock, pulmonary edema; coma, death; tissue destruction of the digestive tract, intense thirst, abdominal pains, vomiting, shock, convulsions, circulatory collapse, laryngeal/glottic spasm, edema. Chronic Effects: Dental discoloration (yellowing) or erosion, chronic bronchitis, pulmonary function changes, dermatitis, conjunctivitis.
Fire: Use dry chemical, carbon dioxide, halon, water spray, fog, or standard foam.
Reactivity: Stable. Hazardous polymerization cannot occur. Incompatible with: acetic anhydride; 2-aminoethanol; ammonium hydroxide; calcium phosphide; chlorosulfonic acid; ethylene diamine; ethylenimine; oleum; perchloric acid; b-propriolactone; propylene oxide; silver perchlorate and carbon tetrachloride; sodium hydroxide; sulfuric acid; uranium phosphide; vinyl acetate; sodium; carbide compounds; bases; metallic powders; carbides; hydrogen sulfide; turpentine. Hazardous decomposition products: fumes of hydrogen chloride; nitric acid; chlorine; nitrogen oxides.
Carcinogenicity: IARC - Not listed; NIOSH - Not listed; NTP - Not listed; ACGIH - Not listed; OSHA - Not listed; EPA - Not listed; MAK - Not listed
Primary Target Organs:

Eyes

Skin

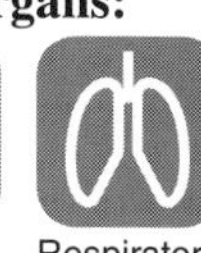
Respiratory System

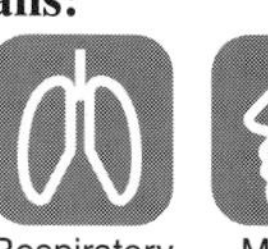
Mucous Membranes

Gastro-intestinal

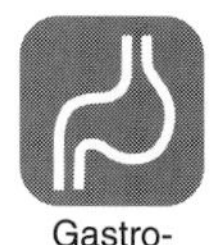
Nervous System

Environmental

Regulations

RCRA 40CFR: Not listed
CERCLA: 40CFR 302.4: Not listed
SARA 40CFR 372.65: Not listed
SARA EHS 40CFR 355: Not listed
TSCA: Not listed

ARA5000 CAS #: 140-57-8

ARAMITE

RTECS: WT2975000
Molecular Formula: $C_{15}H_{23}ClO_4S$
Formula Weight: 334.87
Synonyms: 88-R; 88R; ACARACIDE; ARACIDE; ARAMIT; ARAMITE-15W; ARAMITEARARAMITE-15W; ARATRON; 2-(P-BUTYLPHENOXY)ISOPROPYL 2-CHLOROETHYL SULFITE; 2-(P-TERT-BUTYLPHENOXY)ISOPROPYL 2-CHLOROETHYL SULFITE; BUTYLPHENOXYISOPROPYL CHLOROETHYL SULFITE; 2-(P-T-BUTYLPHENOXY)ISOPROPYL 2'-CHLOROETHYL SULPHITE; 2-(4-T-BUTYLPHENOXY)ISOPROPYL-2-CHLOROETHYL SULFITE; 2-(P-BUTYLPHENOXY)ISOPROPYL-2-CHLOROETHYL SULFITE; 2-(P-T-BUTYLPHENOXY)ISOPROPYL-2'-CHLOROETHYL SULPHITE; 2-(P-T-BUTYLPHENOXY)-1-METHYLETHYL 2-CHLOROETHYL ESTER OFSULPHUROUS ACID; 2-(P-BUTYLPHENOXY)-1-METHYLETHYL 2-CHLOROETHYL SULFITE; 2-(P-T-BUTYLPHENOXY)-1-METHYLETHYL 2'-CHLOROETHYL SULPHITE; 2-(4-TERT-BUTYLPHENOXY)-1-METHYLETHYL 2-CHLOROETHYLSULFITE; 2-(P-TERT-BUTYLPHENOXY)-1-METHYLETHYL 2-CHLOROETHYLSULFITE; 2-(P-T-BUTYLPHENOXY)-1-METHYLETHYL SULPHITE OF 2-CHLOROETHANOL; 2-(P-T-BUTYLPHENOXY)-1-METHYLETHYL SULPHITE OF2-CHLOROETHANOL; 2-(P-BUTYLPHENOXY)-1-METHYLETHYL-2-CHLOROETHYL SULFITE; 2-(P-T-BUTYLPHENOXY)-1-METHYLETHYL-2-CHLOROETHYL SULFITE; CES; 2-CHLOROETHYL 1-METHYL-2-(P-T-BUTYLPHENOXY)ETHYL SULPHATE; 2-CHLOROETHYL SULPHITE OF 1-(P-T-BUTYLPHENOXY)-2-PROPANOL; 2-CHLOROETHYL SULPHITE OF1-(P-TERT-BUTYLPHENOXY)-2-PROPANOL; BETA-CHLOROETHYL-BETA'-(P-T-BUTYLPHENOXY)-ALPHA'-METHYLETHYL SULFITE; BETA-CHLOROETHYL-BETA-(P-T-BUTYLPHENOXY)-ALPHA-METHYLETHYL SULPHITE; BETA-CHLOROETHYL-BETA-(P-TERT-BUTYLPHENOXY)-ALPHA-METHYLETHYL SULPHITE; BETA-CHLOROETHYL-BETA-(P-T-BUTYLPHENOXY)-ALPHA-METHYLETHYLSULPHITE; COMPOUND 88R; ENT 16,519; ESTER OF 2-CHLOROETHANOL WITH2-(P-TERT-BUTYLPHENOXY)-METHYL SULPHITE; ETHANOL,2-CHLORO-,2-(P-T-BUTYLPHENOXY)-1-METHYLETHYLSULFITE; ETHANOL,2-CHLORO-,ESTER WITH 2-(P-TERT-BUTYLPHENOXY)-1-METHYLETHYL SULFITE; ETHANOL,2-CHLORO-,ESTER WITH2-(P-TERT-BUTYLPHENOXY)-1-METHYLETHYL SULFITE; ORTHO-MITE; NIAGARAMITE; 2-PROPANOL,1-(P-T-BUTYLPHENOXY)-,2-CHLOROETHYL SULFITE; SULFUROUS ACID,2-(P-TERT-BUTYLPHENOXY)-1-METHYLETHYL2-CHLOROETHYL ESTER; SULFUROUS ACID,2-(P-T-BUTYLPHENOXY)-1-METHYLETHYL-2-CHLOROETHYL ESTER; SULFUROUS ACID,2-(P-TERT-BUTYLPHENOXY)-1-METHYLETHYL-2-CHLOROETHYL ESTER; SULFUROUS ACID,2-CHLOROETHYL2-(4-(1,1-DIMETHYLETHYL)PHENOXY)-1-METHYLETHYL ESTER; SULFUROUS ACID,2-CHLOROETHYL-,2-(4-(1,1-DIMETHYLETHYL)PHENOXY)-1-METHYLETHYL ESTER; 2-(P-TERC.BUTYLFENOXY)ISOPROPYL-2'-CHLORETHYLESTERKYSELINY SIRICITE
Description: colorless to light-colored, clear oily liquid
Use: antimicrobicide agent (former use); miticide (former use)

Physical Properties

Boiling Point: 175 °C (347 °F) at 0.1 mm Hg
Freezing Point: -31.7 °C (-25.06 °F)
Specific Gravity: 1.145 at 20 °C/20 °C
Water Solubility: Practically Insoluble in Water

Other Solubilities: > 10% in Ethanol; Soluble in Dimethyl Ketone; > 10% in Benzene; > 10% in Acetone; > 10% in Ether.
Refraction Index: 1.51 to 1.5118 at 27 °C/D

RTECS Toxicity Data

Acute Oral: Human LD_{Lo} Dose: 429 mg/kg. Rat LD_{50} Dose: 3900 mg/kg.
Chronic (Multiple Dose) Oral: Rat Dose: 25200 mg/kg/12W-C; Toxic Effects: Nutritional and gross metabolic - Weight loss or decreased weight gain.
Reproductive/Teratogenic: Rat Route: Oral; Dose: 2520 mg/kg; Duration: male 12W prior to mating; Effects on Newborn - Growth statistics.
Tumorigenic: Rat Route: Oral; Dose: 21900 mg/kg/2Y-C; Toxic Effects: Tumorigenic - Carcinogenic by RTECS criteria; Liver - Tumors. Rat Route: Oral; Dose: 8100 mg/kg/2Y-C; Toxic Effects: Tumorigenic - Equivocal tumorigenic agent by RTECS criteria; Liver - Tumors; Liver - Multiple effects. Rat Route: Oral; Dose: 15 gm/kg/2Y-C; Toxic Effects: Tumorigenic - Neoplastic by RTECS criteria; Liver - Tumors.

Hazard Overviews

Carcinogenicity: IARC - Group 2B, Possibly carcinogenic to humans; NIOSH - Not listed; NTP - Not listed; ACGIH - Not listed; OSHA - Not listed; EPA - Class B2, Probable human carcinogen based on animal studies; MAK - Not listed

Environmental

Ecotoxicity: LC_{50} Coturnix japonica (Japanese quail, 14-day old) oral > 5000 ppm (5 day ad libitum diet)
Environmental Fate: If released to soil, it is not expected to leach. If released to water, it may partition significantly from the water column to sediment and suspended material in water. The estimated BCF value of 2265 suggests a potential for significant bioconcentration in aquatic organisms. Insufficient data are available to predict the relative importance or occurrence of chemical or biological degradation processes in soil or water. If released to air, it is expected to exist primarily in the aerosol-phase where it can be physically removed by wet and dry deposition processes.

Environmental Physical Data

Henry's Law Constant: estimated at 4.4×10^{-8}
Sorption Partition Coefficient: $K_{OC} = 1.55 \times 10^{4}$
BCF: estimated at 2265

Regulations

RCRA 40CFR: Not listed
CERCLA: 40CFR 302.4: Not listed
SARA 40CFR 372.65: Not listed
SARA EHS 40CFR 355: Not listed
TSCA: Not listed

Analytical Methods

Soil: SW846 8270B, 8270C
Water / Groundwater: EPA 1625
Plasma: EPA 001, 028, 29; FDA 211.1, 231.1

ARG3000 CAS #: 74-79-3

(L)-ARGININE

RTECS: CF1934200
EINECS Number: 200-811-1
Molecular Formula: $C_6H_{14}N_4O_2$
Formula Weight: 174.20

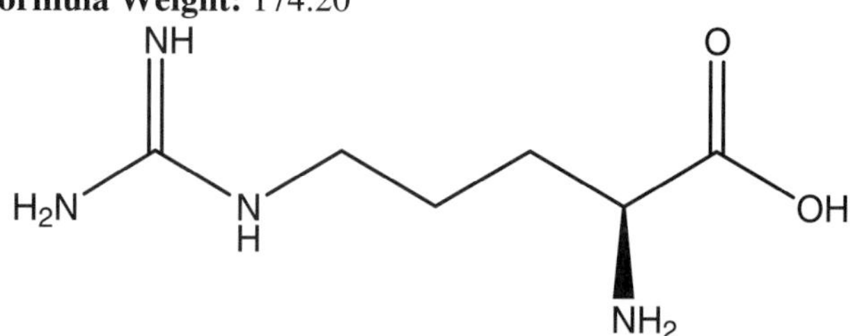

Chemical Structure

Synonyms: ARGININE; ARGININE,L-; L-(+)-ARGININE; PENTANOIC ACID,2-AMINO-5-((AMINOIMINOMETHYL)AMINO-,(S)-
Description: prisms or monoclinic plates
Use: ammonia detoxicant (hepatic failure); biomedical research; pharmaceuticals; dietary supplements

Physical Properties

Freezing Point: Decomposes at 244 °C (471.2 °F)
Water Solubility: Soluble in Water
Other Solubilities: Sparingly Soluble in Alcohol Insoluble in Ether
pH: Strongly alkaline

RTECS Toxicity Data

Mutagenic: Grasshopper Cytogenetic Analysis; Route: Parenteral; Dose: 100 mmol/L.

Hazard Overviews

Health: May be irritating to eyes/skin. Other Acute Effects: may be harmful by inhalation, ingestion, or skin absorption; may cause eye irritation; may cause skin irritation.
Fire: Hazards: emits toxic fumes. Extinguishing agents: water spray; carbon dioxide, dry chemical powder or appropriate foam. Precautions: combustible liquid.
Reactivity: Incompatible with: strong oxidizing agents. Hazardous decomposition products: toxic fumes of: carbon monoxide, carbon dioxide, nitrogen oxides.
Carcinogenicity: IARC - Not listed; NIOSH - Not listed; NTP - Not listed; ACGIH - Not listed; OSHA - Not listed; EPA - Not listed; MAK - Not listed

Environmental

Regulations

RCRA 40CFR: Not listed
CERCLA: 40CFR 302.4: Not listed
SARA 40CFR 372.65: Not listed
SARA EHS 40CFR 355: Not listed

TSCA: Listed

ARG6000 CAS #: 7440-37-1

ARGON

RTECS: CF2300000
DOT: UN1006
EINECS Number: 231-147-0
Molecular Formula: Ar
Structured MF: Ar
Formula Weight: 39.95

Ar

Chemical Structure

Synonyms: ARGON-40
Description: colorless gas, liquid; odorless

Physical Properties

Boiling Point: -185.8 °C (-302 °F)
Freezing Point: -192.2 °C (-313.96 °F)
Specific Gravity: 1.38 Air=1
Vapor Density: 1.38 Air=1
Water Solubility: Slightly Soluble in Water
Other Solubilities: Soluble in organic liquids
Critical Temperature: -122 °C
Critical Pressure: 705 psia
Ionization Potential (eV): 15.75962 +/-0.00001
Flash Point: Will not burn

Hazard Overviews

Compressed Gas

Fire Diamond

Health: Stored as a compressed gas in cylinders. Simple asphyxiant (reduced oxygen available for breathing). Eye and skin contact with the compressed gas can cause frostbite.
Fire: Noncombustible. However, remove cylinder from fire area as high temperatures may cause it to rupture.
Reactivity: Stable. Hazardous polymerization cannot occur. Avoid: cooling with liquid nitrogen.
Carcinogenicity: IARC - Not listed; NIOSH - Not listed; NTP - Not listed; ACGIH - Not listed; OSHA - Not listed; EPA - Not listed; MAK - Not listed
Primary Target Organs:

Eyes

Skin

Nervous System

Environmental

Cleanup/Disposal: Guide No. 121: Do not touch or walk through spilled material. Stop leak if you can do it without risk. Use water spray to reduce vapors or divert vapor cloud drift. Do not direct water at spill or source of leak. If possible, turn leaking containers so that gas escapes rather than liquid. Prevent entry into waterways, sewers, basements or confined areas. Allow substance to evaporate. Ventilate the area.

Regulations

RCRA 40CFR: Not listed
CERCLA: 40CFR 302.4: Not listed
SARA 40CFR 372.65: Not listed
SARA EHS 40CFR 355: Not listed
TSCA: Listed

ARO1000 CAS #: 12674-11-2

AROCLOR 1016

RTECS: TQ1351000
Molecular Formula: Unspecified or Variable

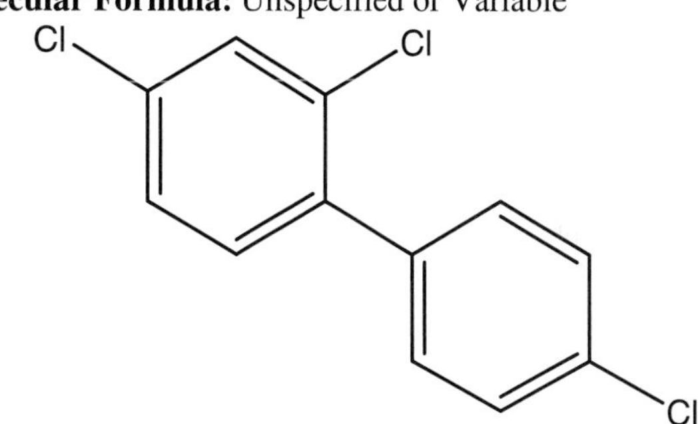

Chemical Structure

Synonyms: AROCHLOR 1016; CHLORODIPHENYL (41% CL); PCB-1016; PCB 1016; POLYCHLORINATED BIPHENYL 1016
Use: production and sale discontinued, but still present in capacitors now in use

Physical Properties

Boiling Point: Distillation 323 °C (613 °F) to 356 °C (673 °F) at 760 mm Hg
Specific Gravity: 1.4
Density: 1.33 g/mL at 25 °C
Vapor Pressure: Estimated 4 $x10^{-4}$ mm Hg at 25 °C
Water Solubility: 906 ppb
Refraction Index: 1.622 to 1.624 at 25 °C
Flash Point: > 141.111 °C

RTECS Toxicity Data

Acute Oral: Rat LD_{50} Dose: 2300 mg/kg.
Reproductive/Teratogenic: Rat Route: Oral; Dose: 31500 ug/kg; Duration: female 1-21D after birth; Specific Developmental Abnormalities - Central nervous system. Rat Route: Oral; Dose: 120 mg/kg; Duration: female 8-21D of pregnancy; Effects on Newborn - Biochemical and metabolic. Monkey Route: Oral; Dose: 18410 ug/kg; Duration: female 30W prior to mating Effects on Newborn - Behavioral.

Hazard Overviews

Reactivity: Stable, but subject to photodechlorination when exposed to sunlight or UV. Hazardous polymerization cannot occur. Avoid: heat and ignition sources. Hazardous decomposition products: highly toxic derivatives (polychlorinated dibenzo-para-dioxins; polychlorinated dibenzofurans); hydrogen chloride; phosgene; other irritants.
Carcinogenicity: IARC - Not listed; NIOSH - Not listed; NTP - Not listed; ACGIH - Not listed; OSHA - Not listed; EPA - Not listed; MAK - Not listed

Environmental

Ecotoxicity: LC_{50} Pteronarcells 610 ug/l/96 hr, NAIAD, at 10 °C (95% confidence limit 424-878 ug/l) /static bioassay LC_{50} Pimephales promelas (Fathead minnow) 28 ug/l/30 day /Conditions of bioassay not given LC_{50} Salmo salar (Atlantic salmon) 134 ug/l/96 hr, wt 5.6 g, at 12 °C (95% confidence limit 113-159 ug/l) /static bioassay LC_{50} Salmo gairdneri (Rainbow trout) 135 ug/l/96 hr, wt 5.6 g, at 12 °C (95% confidence limit 114-159 ug/l) /static bioassay
Environmental Fate: Current evidence suggests that the major source of release to the environment may be an environmental cycling process of material previously introduced into the environment; this cycling process involves volatilization from ground surfaces (water, soil) into the atmosphere with subsequent removal from the atmosphere via wet/dry deposition and then revolatilization. It is a mixture of different congeners of chlorobiphenyl and the relative importance of the environmental fate mechanisms generally depends on the degree of chlorination. In general, the persistence of the PCB congeners increase with an increase in the degree of chlorination. Screening studies have shown that it is biodegraded slowly. Although biodegradation may occur slowly on an environmental basis, no other degradation mechanisms have been shown to be important in natural water and soil systems; therefore, biodegradation may be the ultimate degradation process in water and soil. The PCB composition of the biodegraded Aroclor is different from the original Aroclor. If released to soil, the PCB congeners will become tightly adsorbed to the soil particles. Although the volatilization rate may be low from soil surfaces, the total loss be volatilization over time may be significant because of persistence and stability. Enrichment of the low Cl PCBs occurs in the vapor phase relative to the original Aroclor; the residue will be enriched in the PCBs containing high Cl content. If released to water, adsorption to sediment and suspended matter will be an important fate process. Although adsorption can immobilize it for relatively long periods of time, eventual resolution into the water column has been shown to occur. The PCB composition in water will be enriched in the lower chlorinated PCBs because of their greater water solubility, and the least water soluble PCBs (highest Cl content) will remain adsorbed. In the absence of adsorption, it volatilizes relatively rapidly from water. However, strong PCB adsorption to significantly competes with volatilization which may have a half-life of 2-7 years in typical bodies of water. Although the resulting volatilization rate may be low, the total loss by volatilization over time may be significant because of persistence and stability. It has been shown to bioconcentrate significantly in aquatic organisms. If released to the atmosphere, the PCB congeners will primarily exist in the vapor-phase with enrichment of the most volatile PCBs although a relatively small percentage will partition to the particulate phase. The dominant atmospheric transformation process for these congeners is probably the vapor-phase reaction with hydroxyl radicals which has estimated half-lives ranging from 27.8 days to 3.1 months. Physical removal from the atmosphere, which is important environmentally due to chemical stability, is accomplished by wet and dry deposition.
Cleanup/Disposal: Guide No. 171: Do not touch or walk through spilled material. Stop leak if you can do it without risk. Prevent dust cloud. Avoid inhalation of asbestos dust. Small Dry Spills: With clean shovel place material into clean, dry container and cover loosely; move containers from spill area. Small Spills: Take up with sand or other noncombustible absorbent material and place into containers for later disposal. Large Spills: Dike far ahead of liquid spill for later disposal. Cover powder spill with plastic sheet or tarp to minimize spreading. Prevent entry into waterways, sewers, basements or confined areas.

Environmental Physical Data

Henry's Law Constant: 3.3×10^{-4} to 5.5×10^{-5}
Octanol/Water Partition Coefficient: **log K_{ow}** = 4.38
Sorption Partition Coefficient: **K_{oc}** = 5.21×10^{4} to 1.71×10^{5}
BCF: sheepshead minnow 2.5×10^{4}

Regulations

RCRA 40CFR: Not listed
CERCLA: 40CFR 302.4: Listed per CWA Section 311(b)(4) per CWA Section 307(a) RQ: 1 lb (0.454 kg)
SARA 40CFR 372.65: Listed as Compound
SARA EHS 40CFR 355: Not listed
TSCA: Not listed

Analytical Methods

Soil: CLP LC_PEST, MC_PEST, OHC; SW846 3630B, 8080A, 8081, 8082, 8250A, 8270B, 8270C; EPA 16, 3, PCB-002, PCB-009
Water / Groundwater: EPA PCB-003, PCB-004, 608, 617, 625, 625-S; APHA 6410-B, 6630-C, 6630-D; ASTM D3534
Drinking Water: EPA 505, 508, 525.2; ASTM D5175
Food: FDA 212.1, 232.1; EPA 4
Indoor / Expired Air: NIOSH 5503
Plasma: EPA 001, 29; FDA 211.1, 231.1, 252
Other: EPA P-009-1, 1656

ARO2330 CAS #: 11104-28-2

AROCLOR 1221

RTECS: TQ1352000
Molecular Formula: Unspecified or Variable
Formula Weight: Average 192

Synonyms: AROCHLOR 1221; CHLORODIPHENYL (21% CL); CHLORODIPHENYL (21 PERCENT CL); PCB-1221; PCB 1221; POLYCHLORINATED BIPHENYL 1221

Use: formerly used in gas-transmission turbine hydraulics, rubber plasticizers, adhesives, and electrical capacitors

Physical Properties

Boiling Point: 275 °C (527 °F) to 320 °C (608 °F)
Freezing Point: 1 °C (33.8 °F)
Specific Gravity: 1.182 to 1.192 at 15.5 °C
Vapor Pressure: Average 6.7 x10^{-3} mm Hg at 25 °C
Water Solubility: 3516 ppb
Other Solubilities: miscible in non-polar organic solvents.
Refraction Index: 1.617 to 1.618 at 20 °C
Evaporation Rate: Evaporation loss at 100 °C at 6 hours 1 to 1.5%
Flash Point: > 141.111 °C

RTECS Toxicity Data

Acute Oral: Rat LD_{50} Dose: 3980 mg/kg. Mammal LD_{50} Dose: >750 mg/kg.

Acute Dermal: Rabbit LD_{Lo} Route: Skin; Dose: 3169 mg/kg.

Reproductive/Teratogenic: Rat Route: Subcutaneous; Dose: 1 gm/kg; Duration: female 1D prior to mating; Maternal Effects - Uterus, cervix, vagina. Rat Route: Subcutaneous; Dose: 2 gm/kg; Duration: female 2D prior to mating; Effects on Fertility - Other measures of fertility.

Hazard Overviews

Reactivity: Stable, but subject to photodechlorination when exposed to sunlight or UV. Hazardous polymerization cannot occur. Avoid: heat and ignition sources. Hazardous decomposition products: highly toxic derivatives (polychlorinated dibenzo-para-dioxins; polychlorinated dibenzofurans); hydrogen chloride; phosgene; other irritants.

Carcinogenicity: IARC - Not listed; NIOSH - Not listed; NTP - Not listed; ACGIH - Not listed; OSHA - Not listed; EPA - Not listed; MAK - Not listed

Environmental

Ecotoxicity: LC_{50} ring-necked pheasantS oral greater than 5000 ppm, 10 days old (no mortality at 5000 ppm) LC_{50} Salmo clarki (cutthroat trout) 1,170 ug/l/96 hr, wt 2.7 g, at 9 °C (95% confidence limit 957-1,430 ug/l) /static bioassay LD_{50} Mink oral 0.75-1.0 g/kg

Environmental Fate: It is a mixture of different congeners of chlorobiphenyl and the relative importance of the environmental fate mechanisms generally depends on the degree of chlorination. In general, the persistence of the PCB congeners increase with an increase in the degree of chlorination. In contrast to the more highly chlorinated Aroclors, it appears to be reasonably degradable on an environmental basis. Screening studies have shown that it is readily biodegradable. Biodegradation is probably the ultimate degradation process in both natural water and soil systems since other degradation processes do not appear to be important. The PCB composition of the biodegraded Aroclor is different from the original Aroclor. If released to soil, the PCB congeners will become tightly adsorbed to the soil particles. In the presence of organic solvents PCBs may have a tendency to leach through soil. Significant volatilization may occur from soil surfaces. Enrichment of the low Cl PCBs occurs in the vapor phase relative to the original Aroclor; the residue will be enriched in the PCBs containing high Cl content. If released to water, adsorption to sediment and suspended matter will be an important fate process. Although adsorption may immobilize it for relatively long periods of time eventual resolution into the waste column has been shown to occur. The PCB composition in water will be enriched tin the lower chlorinated PCBs because of their greater water solubility, and the least water soluble PCBs (highest Cl content) will remain adsorbed. In the absence of adsorption, it volatilizes relatively rapidly from water. However, strong PCB adsorption to sediment significantly competes with volatilization which may have a half-life ranging from 2 months to 1 year in typical bodies of water. The PCB congeners have been shown to bioconcentrate significantly in aquatic organisms. If released to the atmosphere, the PCB congeners will exist primarily in the vapor-phase with enrichment of the most volatile PCBs. The dominant atmospheric transformation process for these congeners is probably the vapor-phase reaction with hydroxyl radicals which has estimated half-lives ranging from 12.9 to 27.8 days. Physical removal may come from dry deposition, although wet deposition will be more important then dry deposition.

Cleanup/Disposal: Guide No. 171: Do not touch or walk through spilled material. Stop leak if you can do it without risk. Prevent dust cloud. Avoid inhalation of asbestos dust. Small Dry Spills: With clean shovel place material into clean, dry container and cover loosely; move containers from spill area. Small Spills: Take up with sand or other noncombustible absorbent material and place into containers for later disposal. Large Spills: Dike far ahead of liquid spill for later disposal. Cover powder spill with plastic sheet or tarp to minimize spreading. Prevent entry into waterways, sewers, basements or confined areas.

Environmental Physical Data

Henry's Law Constant: 2.28 x10^{-4}
Octanol/Water Partition Coefficient: log K_{ow} = 4.09
Sorption Partition Coefficient: K_{oc} = 4.04 to 4.88
BCF: calculated at 2.98 to 4.14

Regulations

RCRA 40CFR: Not listed
CERCLA: 40CFR 302.4: Listed per CWA Section 311(b)(4) per CWA Section 307(a) RQ: 1 lb (0.454 kg)
SARA 40CFR 372.65: Listed as Compound
SARA EHS 40CFR 355: Not listed
TSCA: Not listed

Analytical Methods

Soil: CLP LC_PEST, MC_PEST, OHC; SW846 3630B, 8080A, 8081, 8082, 8250A, 8270B, 8270C; EPA 16, 3, PCB-002, PCB-009

Water / Groundwater: EPA PCB-003, PCB-004, 608, 617, 625, 625-S; APHA 6410-B, 6630-C, 6630-D; ASTM D3534

Drinking Water: EPA 505, 508, 525.2; ASTM D5175
Food: FDA 212.1, 232.1; EPA 4
Plasma: EPA 001, 29; FDA 211.1, 231.1, 251.1, 252
Other: EPA P-009-1, 1656

ARO3660 CAS #: 11141-16-5

AROCLOR 1232

RTECS: TQ1354000
Molecular Formula: Unspecified or Variable
Formula Weight: Average 221
Synonyms: AROCHLOR 1232; CHLORODIPHENYL (32% CL); CHLORODIPHENYL (32 PERCENT CL); PCB-1232; PCB 1232; POLYCHLOEINATED BIPHENYL 1232; POLYCHLORINATED BIPHENYL (AROCLOR 1232)
Use: formerly used as hydraulic fluid, rubber plasticizer, adhesive

Physical Properties

Boiling Point: 290 °C (554 °F) to 325 °C (617 °F)
Specific Gravity: 1.27 to 1.28
Vapor Pressure: Average 4.06 x10^{-3} mm Hg at 25 °C
Water Solubility: Solubility in Water is extremely low
Other Solubilities: Soluble in oils and organic solvents
Refraction Index: 1.620 to 1.622 at 20 °C
Evaporation Rate: Evaporation loss at 100 °C at 6 hours 1 to 1.5%
Flash Point: > 141.111 °C

RTECS Toxicity Data

Acute Oral: Rat LD_{50} Dose: 4470 mg/kg.
Acute Dermal: Rabbit LD_{Lo} Route: Skin; Dose: 2 gm/kg.

Hazard Overviews

Reactivity: Stable, but subject to photodechlorination when exposed to sunlight or UV. Hazardous polymerization cannot occur. Avoid: heat and ignition sources. Hazardous decomposition products: highly toxic derivatives (polychlorinated dibenzo-para-dioxins; polychlorinated dibenzofurans); hydrogen chloride; phosgene; other irritants.
Carcinogenicity: IARC - Not listed; NIOSH - Not listed; NTP - Not listed; ACGIH - Not listed; OSHA - Not listed; EPA - Not listed; MAK - Not listed

Environmental

Ecotoxicity: LC_{50} Salmo clarki (Cutthroat trout) 2,500 ug/l/96 hr /Conditions of bioassay not specified LD_{50} Colinus virginianus (Northern bobwhite) oral 3,002 mg/kg diet/5 days on treated diet plus 3 days untreated
Environmental Fate: It is a mixture of different congeners of chlorobiphenyl and the relative importance of the environmental fate mechanisms generally depends on the degree of chlorination. In general, the persistence of the PCB congeners increase with an increase in the degree of chlorination. In contrast to the more highly chlorinated Aroclors, it appears to be reasonably degradable in the environment. One screening study has shown that it is biodegradable. Biodegradation is probably the ultimate degradation process in both natural water and soil systems since other degradation does not appear to be important. The PCB composition of the biodegraded Aroclor is different from the original Aroclor. If released to soil, the PCB congeners will become tightly adsorbed to the soil particles. In the presence of organic solvents, PCBs may have a tendency to leach through soil. Significant volatilization may occur from soil surfaces. Enrichment of the low Cl PCBs occurs in the vapor phase relative to the original Aroclor; the residue will be enriched in the PCBs containing high Cl content. If released to water, adsorption to sediment and suspended matter will be an important fate process. Although adsorption may immobilize it for relatively long periods of time, eventual resolution into the water column has been shown to occur. The PCB composition in water will be enriched in the lower chlorinated PCBs because of their greater water solubility, and the least water soluble PCBs (highest Cl content) will remain adsorbed. In the absence of adsorption, it volatilizes relatively rapidly from water. However, strong PCB adsorption to sediment significantly competes with volatilization which may have a half-life ranging from 2 months to 1 year in typical bodies of water. The PCB congeners have been shown to bioconcentrate significantly in aquatic organisms. If released to the atmosphere, the PCB congeners will primarily exist in the vapor-phase. The dominant atmospheric transformation process for these congeners is probably the vapor-phase reaction with hydroxyl radicals which has estimated half-lives ranging from 12.9 days to 3.1 months. Physical removal from the atmosphere is accomplished by wet and dry deposition with enrichment of the most volatile PCBs, although wet deposition will be more important than dry deposition.
Cleanup/Disposal: Guide No. 171: Do not touch or walk through spilled material. Stop leak if you can do it without risk. Prevent dust cloud. Avoid inhalation of asbestos dust. Small Dry Spills: With clean shovel place material into clean, dry container and cover loosely; move containers from spill area. Small Spills: Take up with sand or other noncombustible absorbent material and place into containers for later disposal. Large Spills: Dike far ahead of liquid spill for later disposal. Cover powder spill with plastic sheet or tarp to minimize spreading. Prevent entry into waterways, sewers, basements or confined areas.

Environmental Physical Data

Henry's Law Constant: 2.28 x10^{-4} to 3.43 x10^{-4}
Octanol/Water Partition Coefficient: log K_{OW} = > 4.54
Sorption Partition Coefficient: K_{OC} = 4.04 to 5.25
BCF: white sucker 5500

Regulations

RCRA 40CFR: Not listed
CERCLA: 40CFR 302.4: Listed per CWA Section 311(b)(4) per CWA Section 307(a) RQ: 1 lb (0.454 kg)
SARA 40CFR 372.65: Listed as Compound
SARA EHS 40CFR 355: Not listed
TSCA: Not listed

Analytical Methods

Soil: CLP LC_PEST, MC_PEST, OHC; SW846 3630B, 8080A, 8081, 8082, 8250A, 8270B, 8270C; EPA 16, 3, PCB-009

Water / Groundwater: EPA PCB-003, 608, 617, 625, 625-S; APHA 6410-B, 6630-C, 6630-D; ASTM D3534

Drinking Water: EPA 505, 508, 525.2; ASTM D5175

Food: EPA 4

Plasma: EPA 29

Other: EPA P-009-1, 1656

ARO4990 CAS #: 53469-21-9

AROCLOR 1242

RTECS: TQ1356000

Molecular Formula: Unspecified or Variable

Structured MF: $C_6H_4ClC_6H_3Cl_2$ (approx)

Formula Weight: 261

Synonyms: AROCHLOR 1242; CHLORIERTE BIPHENYLE,CHLORGEHALT 42%; CHLORODIPHENYL (42% CHLORINE); CHLORODIPHENYL (42% CL); CLORODIFENILI,CLORO 42%; DIPHENYLE CHLORE,42% DE CHLORE; GECHLOREERDEDIFENYL; PCB; PCB-1242; PCB 1242; POLYCHLORINATED BIPHENYL; POLYCHLORINATED BIPHENYL 1242

Use: fomerly in electrical capacitors & transformers, vacuum pumps, gas-transmission turbines; heat transfer fluid, hydraulic fluid, plasticized rubber, carbonless paper, adhesives & wax extenders; prodn & sale was discontinued in 1977; may still be present in transformers & capacitors now in use.

Physical Properties

Boiling Point: 325 °C (617 °F) to 366 °C (691 °F)

Freezing Point: -18.89 °C (-2.002 °F)

Specific Gravity: 1.381 to 1.392 at 25 °C/15.5 °C

Vapor Pressure: Average 50 mm Hg at 25 °C

Water Solubility: 703 ppb

Other Solubilities: Soluble in oils and organic solvents.

Refraction Index: 1.627 to 1.629 at 20 °C

Evaporation Rate: Evaporation loss at 100 °C at 6 hours 0.0 to 0.4%

Flash Point: 176 to 180 °C Cleveland Open Cup

RTECS Toxicity Data

Acute Oral: Rat LD_{50} Dose: 4250 mg/kg; Toxic Effects: Sense organs and special senses - Chromidracryorrhea; Gastrointestinal - Hypermotility, diarrhea; Nutritional and gross metabolic - Weight loss or decreased weight gain. Mammal LD_{50} Dose: >3 gm/kg.

Acute Inhalation: Human TC_{Lo} Dose: 10 mg/m^3; Toxic Effects: Lungs, Thorax, or Respiration - Other changes; Liver - Other changes.

Acute Dermal: Guinea Pig LD_{Lo} Route: Subcutaneous Dose: 345 mg/kg; Toxic Effects: Liver - Other changes; Kidney, Ureter, and Bladder - Other changes; Blood - Changes in spleen.

Reproductive/Teratogenic: Rat Route: Oral; Dose: 945 mg/kg; Duration: female 36W prior to mating Maternal Effects - Other effects on females; Endocrine - Estrogenic. Rat Route: Oral; Dose: 1890 mg/kg; Duration: female 36W prior to mating Effects on Fertility - Female fertility index. Rat Route: Oral; Dose: 1250 mg/kg; Duration: male 5D prior to mating; Effects on Fertility - Post-implantation mortality.

Mutagenic: Mouse Other Mutation Test Systems; Cell Type: other cell types; Dose: 25 ppm/4H.

Hazard Overviews

Reactivity: Stable, subject to photodechlorination when exposed to sunlight or UV. Hazardous polymerization cannot occur. Avoid: heat and ignition sources. Hazardous decomposition products: highly toxic derivatives (polychlorinated dibenzo-para-dioxins; polychlorinated dibenzofurans; hydrogen chloride; phosgene; other irritants).

Carcinogenicity: IARC - Group 2A, Probably carcinogenic to humans; NIOSH - Listed as carcinogen; NTP - Class 2B, Reasonably anticipated to be a carcinogen, sufficient evidence of carcinogenicity from studies in experimental animals; ACGIH - Not listed; OSHA - Not listed; EPA - Class B2, Probable human carcinogen based on animal studies; MAK - Class B, Justifiably suspected of having carcinogenic potential

Exposure Limits

OSHA PEL: TWA: 1 mg/m^3; skin.

ACGIH TLV: TWA: 1 mg/m^3.

NIOSH IDLH: 5 mg/m^3.

DFG MAK: TWA: .1 ppm; 1 mg/m^3.

Respirator Recommendation

Exposure Range: >1 to <5 mg/m^3 Supplied Air, Constant Flow/Pressure Demand, Full Face

Exposure Range: 5 to unlimited mg/m^3 Self-contained Breathing Apparatus, Pressure Demand, Full Face

Note: odor threshold unknown

Environmental

Ecotoxicity: LC_{50} Lepomis macrochirus (Bluegill) 54 ug/l/15 day /Conditions of bioassay not specified LC_{50} Phasianus colchicus (Ring-necked pheasant) oral 2,078 mg/kg diet (5 days on treated diet plus 3 days untreated) LC_{50} Gammarus pseudolimnaeus (Scud) 10 ug/l/96 hr /Conditions of bioassay not specified LC_{50} Ischnura verticalis (Damselfly) 400 ug/l/96 hr /Conditions of bioassay not specified LD_{50} Mustela vison (Mink) ip 1.0 mg/kg LD_{50} Colinus virginianus (Northern bobwhite) oral 2,098 mg/kg diet LC_{50} Oronectes nais (Crayfish) 30 ug/l/7 day /Conditions of bioassay not specified

Environmental Fate: Current evidence suggests that the major source of release to the environment may be an environmental cycling process of material previously introduced into the environment; this cycling process involves volatilization from ground surfaces (water, soil) into the atmosphere with subsequent removal from the atmosphere via wet/dry deposition and then revolatilization. It is a mixture of different congeners of chlorobiphenyl and the relative importance of the environmental fate mechanisms generally depends on the degree of chlorination. In general, the persistence of the PCB congeners increase with an increase in the degree of chlorination. Screening studies have shown that it is biodegraded slowly. Although biodegradation may occur slowly in the environment, no other degradation mechanisms have been shown to be important in natural water and soil systems; therefore, biodegradation may be the ultimate degradation process in water and soil. The PCB composition of the biodegraded Aroclor is different from the original Aroclor. If released to soil, the PCB congeners will become tightly adsorbed to the soil particles. Although the volatilization rate may be low from soil surfaces, the total loss by volatilization over time may be significant because of persistence and stability. Enrichment of the low Cl PCBs occurs in the vapor phase relative to the original The residue will be enriched in the PCBs containing high Cl content. If released to water, adsorption to sediment and suspended matter will be an important fate process. Although adsorption can immobilize it for relatively long periods of time, eventual resolution into the water column has been shown to occur. The PCB composition in water will be enriched in the lower chlorinated PCBs because of their greater water solubility, and the least water soluble PCBs (highest Cl content) will remain adsorbed. In the absence of adsorption, it volatilizes relatively rapidly from water. However, strong PCB adsorption to sediment significantly competes with volatilization which may have a half-life of 2-7 years in typical bodies of water. Although the resulting volatilization rate may be low, the total loss by volatilization over time may be significant because of persistence and stability. It has been shown to bioconcentrate significantly in aquatic organisms. If released to the atmosphere, the PCB congeners will exist primarily in the vapor-phase with enrichment of the most volatile PCBs although a relatively small percentage will partition to the particulate phase. The dominant atmospheric transformation process for these congeners is probably the vapor-phase reaction with hydroxyl radicals which has estimated half-lives ranging from 27.8 days to 4.75 months. Physical removal from the atmosphere, which is important environmentally due to chemical stability, is accomplished by wet and dry deposition.

Cleanup/Disposal: Guide No. 171: Do not touch or walk through spilled material. Stop leak if you can do it without risk. Prevent dust cloud. Avoid inhalation of asbestos dust. Small Dry Spills: With clean shovel place material into clean, dry container and cover loosely; move containers from spill area. Small Spills: Take up with sand or other noncombustible absorbent material and place into containers for later disposal. Large Spills: Dike far ahead of liquid spill for later disposal. Cover powder spill with plastic sheet or tarp to minimize spreading. Prevent entry into waterways, sewers, basements or confined areas.

Environmental Physical Data

Henry's Law Constant: 5 $x10^{-5}$
Octanol/Water Partition Coefficient: log K_{ow} = 4.11
Sorption Partition Coefficient: K_{oc} = 2240 to 1.5 $x10^{5}$
BCF: fathead minnow 2.74 $x10^{5}$

Regulations

RCRA 40CFR: Not listed
CERCLA: 40CFR 302.4: Listed per CWA Section 311(b)(4) per CWA Section 307(a) RQ: 1 lb (0.454 kg)
SARA 40CFR 372.65: Listed as Compound
SARA EHS 40CFR 355: Not listed
TSCA: Not listed

Analytical Methods

Soil: CLP LC_PEST, MC_PEST, OHC; SW846 3630B, 8080A, 8081, 8082, 8250A, 8270B, 8270C; EPA 16, 3, PCB-002, PCB-009
Water / Groundwater: EPA PCB-003, PCB-004, 608, 617, 625, 625-S; APHA 6410-B, 6630-C, 6630-D; ASTM D3534, D4763
Drinking Water: EPA 505, 508, 525.2; ASTM D5175
Food: FDA 212.1, 232.1; EPA 4
Indoor / Expired Air: NIOSH 5503; ASTM D4861
Plasma: EPA 001, 29; FDA 211.1, 231.1, 251.1, 251.2, 252
Other: EPA P-009-1, 1656

ARO6320 **CAS #: 12672-29-6**

AROCLOR 1248

RTECS: TQ1358000
Molecular Formula: Mixture
Formula Weight: Average 288
Synonyms: AROCHLOR 1248; CHLORODIPHENYL (48% CL); CHLORODIPHENYL (48 PERCENT CL); KANECHLOR 400; PCB-1248; PCB 1248; POLYCHLORINATED BIPHENYL 1248
Use: former use in hydraulic fluids, vacuum pumps, rubber plasticizers, synthetic resins, and adhesives

Physical Properties

Boiling Point: 340 °C (644 °F) to 375 °C (707 °F)
Specific Gravity: 1.405 to 1.415 at 15.5 °C
Vapor Pressure: Average 4.94 $x10^{-4}$ mm Hg at 25 °C
Water Solubility: 0.017 ppm at 0 °C
Other Solubilities: Most common organic solvents, oils and fats; Slightly Soluble in Glycerol and glycols
Refraction Index: 1.630 to 1.631 at 20 °C
Evaporation Rate: Evaporation loss at 100 °C at 6 hours 0.0 to 0.3%
Flash Point: 193 to 196 °C Cleveland Open Cup

RTECS Toxicity Data

Acute Oral: Rat LD_{50} Dose: 11 gm/kg.
Acute Dermal: Rabbit LD_{Lo} Route: Skin; Dose: 1269 mg/kg.
Reproductive/Teratogenic: Monkey Route: Oral; Dose: 32 mg/kg; Duration: female 1-23W of pregnancy; Effects on Newborn - Behavioral. Monkey Route: Oral; Dose: 55 mg/kg; Duration: female 26W prior to mating Effects on Newborn - Growth statistics; Physical. Monkey Route: Oral; Dose: 17 mg/kg; Duration: female 26W prior to mating Effects on Fertility - Post-implantation mortality. Monkey Route: Oral; Dose: 35 mg/kg; Duration: female 26W prior to mating Effects on Fertility - Abortion. Monkey Route: Oral; Dose: 24

mg/kg; Duration: female 17W prior to mating Maternal Effects - Menstrual cycle changes or disorders. Monkey Route: Oral; Dose: 83 mg/kg; Duration: female 58W prior to mating Effects on Newborn - Growth statistics; Behavioral; Other postnatal measures or effects.

Hazard Overviews

Reactivity: Stable, but subject to photodechlorination when exposed to sunlight or UV. Hazardous polymerization cannot occur. Avoid: heat and ignition sources. Hazardous decomposition products: highly toxic derivatives (polychlorinated dibenzo-para-dioxins; polychlorinated dibenzofurans); hydrogen chloride; phosgene; other irritants.
Carcinogenicity: IARC - Not listed; NIOSH - Not listed; NTP - Not listed; ACGIH - Not listed; OSHA - Not listed; EPA - Not listed; MAK - Not listed

Environmental

Ecotoxicity: LC_{50} Bobwhite quail, 10 days old, oral 1175 ppm, in 5-day diet, (95% confidence limit 966-1440 ppm) LC_{50} Gammarus pseudolimnaeus (Scud) 52 ug/l/96 hr /Conditions of bioassay not specified LC_{50} Lepomis macrochirus (Bluegill) 10 ug/l/20 day /Conditions of bioassay not specified LC_{50} Gammarus fasciatus (Scud) 52 ug/l/96 hr, mature, at 21 °C /static bioassay LC_{50} Daphnia magna (Cladoceran) 2.6 ug/l/2 wk /Conditions of bioassay not specified
Cleanup/Disposal: Guide No. 171: Do not touch or walk through spilled material. Stop leak if you can do it without risk. Prevent dust cloud. Avoid inhalation of asbestos dust. Small Dry Spills: With clean shovel place material into clean, dry container and cover loosely; move containers from spill area. Small Spills: Take up with sand or other noncombustible absorbent material and place into containers for later disposal. Large Spills: Dike far ahead of liquid spill for later disposal. Cover powder spill with plastic sheet or tarp to minimize spreading. Prevent entry into waterways, sewers, basements or confined areas.

Environmental Physical Data

Octanol/Water Partition Coefficient: log K_{ow} = > 6.11
BCF: catfish 5.64 $x10^4$

Regulations

RCRA 40CFR: Not listed
CERCLA: 40CFR 302.4: Listed per CWA Section 311(b)(4) per CWA Section 307(a) RQ: 1 lb (0.454 kg)
SARA 40CFR 372.65: Listed as Compound
SARA EHS 40CFR 355: Not listed
TSCA: Not listed

Analytical Methods

Soil: CLP LC_PEST, MC_PEST, OHC; SW846 3630B, 8080A, 8081, 8082, 8250A, 8270B, 8270C; EPA 16, 3, PCB-002, PCB-009
Water / Groundwater: EPA PCB-003, PCB-004, 608, 617, 625, 625-S; APHA 6410-B, 6630-C, 6630-D; ASTM D3534
Drinking Water: EPA 505, 508, 525.2; ASTM D5175
Food: FDA 212.1, 232.1; EPA 4
Plasma: EPA 001, 29; FDA 211.1, 231.1, 251.1, 252
Other: EPA P-009-1, 1656

ARO7650 **CAS #: 11097-69-1**

AROCLOR 1254

RTECS: TQ1360000
Molecular Formula: Unspecified or Variable
Structured MF: $C_6H_2Cl_3C_6H_3Cl_2$ (approx)
Formula Weight: Average 327
Synonyms: AROCHLOR 1254; CHLORIERTE BIPHENYLE,CHLORGEHALT 54%; CHLORODIPHENYL (54% CHLORINE); CHLORODIPHENYL (54% CL); CHLORODIPHENYL,54 PERCENT CHLORINE; CLORODIFENILI,CLORO 54%; DIPHENYLE CHLORE,54% DE CHLORE; PCB; PCB-1254; PCB 1254; POLYCHLORINATED BIPHENYL; POLYCHLORINATED BIPHENYL 1254; POLYCHLORINATED BIPHENYL (AROCLOR 1254)
Use: in electrical capacitors, electrical transformers, vacuum pumps, gas-transmission turbines, high-temperature dielectrics for electric wires and electrical equipment, heat-exchange fluids, coatings, inks insecticides, fillers, adhesives, paints and in duplicating papers; as a plasticizer for cellulosics, vinyl resins and chlorinated rubbers; formerly used as hydraulic fluids, fire retardants, wax extenders, dedusting agents, pesticide extenders, lubricants, cutting oils, sealants and caulking compounds

Physical Properties

Boiling Point: Distillation 365 °C (689 °F) to 390 °C (734 °F)
Freezing Point: 10 °C (50 °F)
Specific Gravity: 1.495 to 1.505 at 65 °C/15.5 °C
Vapor Density: 11.2 Air=1
Density: 1.47 to 1.49 g/mL at 90 °C
Vapor Pressure: Average 7.7 $x10^{-5}$ mm Hg at 25 °C
Water Solubility: 70 ppb
Other Solubilities: 95% Ethanol: >=100 mg/ml at 23 °C; Acetone: >=100 mg/ml at 23 °C; DMSO: >=100 mg/ml at 23 °C; Glycerine: Insoluble; Glycols: Insoluble; Oils: Soluble; Organic solvents: Soluble.
Refraction Index: 1.629
Evaporation Rate: Evaporation loss at 100 °C at 6 hours 0.0 to 0.2%
Flash Point: > 141.111 °C

RTECS Toxicity Data

Acute Oral: Rat LD_{50} Dose: 1010 mg/kg. Mammal LD_{50} Dose: 4 gm/kg.
Reproductive/Teratogenic: Rat Route: Oral; Dose: 192 mg/kg; Duration: female 6D after birth; Effects on Newborn - Growth statistics; Physical; Delayed effects. Rat Route: Oral; Dose: 188 mg/kg; Duration: multigenerations; Effects on Newborn - Live birth index. Rat Route: Oral; Dose: 645 mg/kg; Duration: multigenerations; Effects on Newborn - Viability index. Rat Route: Oral; Dose: 90 mg/kg; Duration: female 7-15D of pregnancy; Specific Developmental Abnormalities - Hepatobiliary system. Rat Route: Oral; Dose: 40 mg/kg; Duration: female 5D after birth; Effects on Newborn - Delayed effects. Rat Route: Oral; Dose: 750

mg/kg; Duration: male 5D prior to mating; Effects on Fertility - Pre-implantation mortality.

Mutagenic: Rat DNA Damage; Route: Oral; Dose: 1295 mg/kg. Rat DNA Damage; Route: Intraperitoneal; Dose: 500 mg/kg. Rat DNA Damage; Cell Type: liver; Dose: 300 umol/L.

Tumorigenic: Rat Route: Oral; Dose: 73500 mg/kg/2Y-C; Toxic Effects: Tumorigenic - Carcinogenic by RTECS criteria; Liver - Tumors. Rat Route: Oral; Dose: 1 mg/kg/D-C; Toxic Effects: Tumorigenic - Equivocal tumorigenic agent by RTECS criteria; Gastrointestinal - Tumors. Rat Route: Oral; Dose: 3 mg/kg/D-C; Toxic Effects: Tumorigenic - Equivocal tumorigenic agent by RTECS criteria; Gastrointestinal - Tumors. Rat Route: Oral; Dose: 4 gm/kg/2Y-I; Toxic Effects: Tumorigenic - Equivocal tumorigenic agent by RTECS criteria; Gastrointestinal - Tumors; Liver - Tumors.

Hazard Overviews

Reactivity: Stable, but subject to photodechlorination when exposed to sunlight or UV. Hazardous polymerization cannot occur. Avoid: heat and ignition sources. Hazardous decomposition products: highly toxic derivatives (polychlorinated dibenzo-para-dioxins; polychlorinated dibenzofurans); hydrogen chloride; phosgene; other irritants.

Carcinogenicity: IARC - Group 2A, Probably carcinogenic to humans; NIOSH - Listed as carcinogen; NTP - Class 2B, Reasonably anticipated to be a carcinogen, sufficient evidence of carcinogenicity from studies in experimental animals; ACGIH - Class A3, Animal carcinogen; OSHA - Not listed; EPA - Class B2, Probable human carcinogen based on animal studies; MAK - Class B, Justifiably suspected of having carcinogenic potential

Exposure Limits

OSHA PEL: TWA: 0.5 mg/m^3; skin.

ACGIH TLV: TWA: 0.5 mg/m^3.

NIOSH IDLH: 5 mg/m^3.

DFG MAK: TWA: 0.05 ppm; 0.5 mg/m^3.

Respirator Recommendation

Exposure Range: >0.5 to <5 ppm Supplied Air, Constant Flow/Pressure Demand, Full Face

Exposure Range: 5 to unlimited ppm Self-contained Breathing Apparatus, Pressure Demand, Full Face

Note: odor threshold unknown

Environmental

Ecotoxicity: LC_{50} Macromia (Dragonfly) 800 ug/l/7 days at 21 °C, juvenile /static bioassay LC_{50} Bobwhite quail oral 604 ppm, in 5-day diet, (95% confidence limit 410-840 ppm), age 10 days LC_{50} Gammarus fasciatus (Scud) 2400 ug/l/96 hr at 21 °C, mature /static bioassay LC_{50} Ischnura venticalis (Damselfly) 200 ug/l/96 hr at 15 °C, juvenile /static bioassay LC_{50} Perca flavescens (Yellow perch) greater than 150 ug/l/96 hr at 17 °C, wt 1.0 g /static bioassay LD_{50} Agelaius phoeniceus (Red-winged blackbird) oral 1,500 mg/kg diet/6 day LC_{50} Cyprinodon variegatus (sheepshead minnow), fry 0.1-0.32 ug/l/21 day /Conditions of bioassay not specified LC_{50} Leiostomus xanthurus (spot) 0.5 ug/l/38 day /Conditions of bioassay not specified LD_{50} Mustela vison (mink) oral 4.0 mg/kg LC_{50} Penaeus duorarum (pink shrimp) 1.0 ug/l/12 day /Conditions of bioassay not specified LC_{50} Palaemonetes pugio (grass shrimp) 6.1-7.8 ug/l/96 hr /Conditions of bioassay not specified LC_{50} Hydra oligactis (hydra) 10,000 ug/l/72 hr /Conditions of bioassay not specified LC_{50} Lepomis macrochius (Bluegill) 2740 ug/l/96 hr at 18 °C (95% confidence limit 1294-5810 ug/l), wt 0.8 g /static bioassay LD_{50} Sturnus vulgaris (European starling) oral 1,500 mg/kg diet/96 hr

Environmental Fate: Current evidence suggests that the major source of release to the environment is an environmental cycling process of material previously introduced into the environment; this cycling process involves volatilization from ground surfaces (water, soil) into the atmosphere with subsequent removal from the atmosphere via wet/dry deposition and then revolatilization. It is a mixture of different congeners of chlorobiphenyl and the relative importance of the environmental fate mechanisms generally depends on the degree of chlorination. In general, the persistence of the PCB congeners increase with an increase in the degree of chlorination. Screening studies have shown that it is generally resistant to biodegradation. Although biodegradation may occur slowly in the environmental, no other degradation mechanism have been shown to be important in natural water and soil systems; therefore, biodegradation may be the ultimate degradation process in water and soil. The PCB composition of the biodegraded Aroclor is different from the original Aroclor. If released to soil, the PCB congeners will become tightly adsorbed to the soil particles. In the presence of organic solvents, PCBs may have a tendency to leach through soil. Although the volatilization rate may be low from soil surfaces, the total loss by volatilization over time may be significant because of persistence and stability. Enrichment of the low Cl PCBs occurs in the vapor phase relative to the original Aroclor; the residue will be enriched in the PCBs containing high Cl content. If released to water, adsorption to sediment and suspended matter will be an important fate process. Although adsorption can immobilize it for relatively long periods of time, eventual resolution into the water column has been shown to occur. The PCB composition in water will be enriched in the lower chlorinated PCBs because of their greater water solubility, and the least water soluble PCBs (highest Cl content) will remain adsorbed. In the absence of adsorption, it volatilizes relatively rapidly from water. However, strong PCB adsorption competes with volatilization which may have a half-life in excess of 4 years in typical bodies of water. Although the resulting volatilization rate may be low, the total loss by volatilization over time may be significant because of persistence and stability. It has been shown to bioconcentrate significantly in aquatic organisms. If released to the atmosphere, the PCB congeners will primarily exist in the vapor-phase with enrichment of the most volatile PCBs although a relatively small percentage will partition to the particulate phase. The dominant atmospheric transformation process for these congeners is probably the vapor-phase reaction with hydroxyl radicals which has estimated half-lives ranging for 3.1 months to 1.3 years. Physical removal from the atmosphere, which is very important environmentally due to chemical stability, is accomplished by wet and dry deposition.

Cleanup/Disposal: Guide No. 171: Do not touch or walk through spilled material. Stop leak if you can do it without risk. Prevent dust cloud. Avoid inhalation of asbestos dust. Small Dry Spills: With clean shovel place material into clean, dry container and cover loosely; move containers from spill area. Small Spills: Take up with sand or other noncombustible absorbent material and place into containers for later disposal. Large Spills: Dike far ahead of liquid spill for later disposal. Cover powder spill with plastic sheet or tarp to minimize spreading. Prevent entry into waterways, sewers, basements or confined areas.

Environmental Physical Data

Henry's Law Constant: 5 $x10^{-5}$

Octanol/Water Partition Coefficient: log K_{ow} = estimated at 6.30

Sorption Partition Coefficient: K_{oc} = 4.25 $x10^{4}$

BCF: mullett 1254

Regulations

RCRA 40CFR: Not listed

CERCLA: 40CFR 302.4: Listed per CWA Section 311(b)(4) per CWA Section 307(a) RQ: 1 lb (0.454 kg)

SARA 40CFR 372.65: Listed as Compound

SARA EHS 40CFR 355: Not listed

TSCA: Not listed

Analytical Methods

Soil: CLP LC_PEST, MC_PEST, OHC; SW846 3630B, 8080A, 8081, 8082, 8250A, 8270B, 8270C; EPA 16, 3, 025, PCB-002, PCB-005, PCB-009

Water / Groundwater: EPA PCB-003, PCB-004, 608, 617, 625, 625-S, 022; APHA 6410-B, 6630-C, 6630-D; ASTM D3534, D4763

Drinking Water: EPA 505, 508, 525.2; ASTM D5175

Food: FDA 212.1, 232.1; EPA 4

Indoor / Expired Air: NIOSH 5503; ASTM D4861

Plasma: EPA 001, 29; FDA 211.1, 231.1, 251.1, 251.2, 252; AOAC 990.07

Other: EPA P-009-1, 1656

ARO8980 **CAS #: 11096-82-5**

AROCLOR 1260

RTECS: TQ1362000

Molecular Formula: Unspecified or Variable

Formula Weight: Average 372

Synonyms: AROCHLOR 1260; CHLORODIPHENYL (60% CL); CHLORODIPHENYL (60 PERCENT CL); CLOPHEN A60; KANECHLOR 600; PCB-1260; PCB 1260; PHENOCLOR DP6; POLYCHLORINATED BIPHENYL 1260

Use: formerly in electrical transformers, hydraulic fluids, plasticizer in synthetic resins and dedusting agents

Physical Properties

Boiling Point: 385 °C (725 °F) to 420 °C (788 °F)

Specific Gravity: 1.58 at 25 °C

Vapor Pressure: 4.05 $x10^{-5}$ mm Hg at 25 °C

Water Solubility: 0.08 ml/l at 24 °C

Other Solubilities: Most common organic solvents, oils and fats; Slightly Soluble in Glycerol and glycols

Refraction Index: 1.647 to 1.649 at 20 °C

Evaporation Rate: Evaporation loss at 100 °C at 6 hours 0.0 to 0.1%

Flash Point: > 141.111 °C

RTECS Toxicity Data

Acute Oral: Rat LD_{50} Dose: 1315 mg/kg; Toxic Effects: Behavioral - Somnolence (general depressed activity); Gastrointestinal - Hypermotility, diarrhea.

Acute Dermal: Rabbit LD_{Lo} Route: Skin; Dose: 2 gm/kg.

Reproductive/Teratogenic: Rat Route: Oral; Dose: 1675 mg/kg; Duration: multigenerations; Effects on Newborn - Live birth index. Mouse Route: Oral; Dose: 74 mg/kg; Duration: female 62D prior to mating Maternal Effects - Menstrual cycle changes or disorders; Effects on Fertility - Pre-implantation mortality.

Mutagenic: Rat Cytogenetic Analysis; Route: Oral; Dose: 1080 mg/kg/26W-C.

Tumorigenic: Rat Route: Oral; Dose: 4380 mg/kg/83W-C; Toxic Effects: Tumorigenic - Carcinogenic by RTECS criteria; Liver - Tumors. Rat Route: Oral; Dose: 4992 mg/kg/2Y-C; Toxic Effects: Tumorigenic - Carcinogenic by RTECS criteria; Liver - Tumors. Rat Route: Oral; Dose: 360 mg/kg/17W-C; Toxic Effects: Tumorigenic - Neoplastic by RTECS criteria; Liver - Tumors.

Hazard Overviews

Reactivity: Stable, but subject to photodechlorination when exposed to sunlight or UV. Hazardous polymerization cannot occur. Avoid: heat and ignition sources. Hazardous decomposition products: highly toxic derivatives (polychlorinated dibenzo-para-dioxins; polychlorinated dibenzofurans); hydrogen chloride; phosgene; other irritants.

Carcinogenicity: IARC - Not listed; NIOSH - Not listed; NTP - Listed; ACGIH - Not listed; OSHA - Not listed; EPA - Not listed; MAK - Not listed

Environmental

Ecotoxicity: LC_{50} Perca flavescens (Yellow perch) >200 ug/l/96 hr /Conditions of bioassay not specified LC_{50} Coturnix japonica (Japanese quail) oral 2195 ppm LD_{50} Colinus virginianus (Northern bobwhite) oral 747 mg/kg diet/5 days on treated diet plus 3 days untreated LD_{50} Anas platyrhynchos (Mallard) oral >2 mg/kg LC_{50} Salmo gairdneri (Rainbow trout) 21 ug/l/20 day /Conditions of bioassay not specified

Environmental Fate: Current evidence suggests that the major source of release to the environment is an environmental cycling process of material previously introduced into the environment; this cycling process involved volatilization from ground surfaces (water, soil) into the atmosphere with subsequent removal from the atmosphere via wet/dry deposition and revolatilization. It is a mixture of different congeners of chlorobiphenyl and the relative importance of the environmental fate mechanisms generally depends on the degree of chlorination. In general, the persistence of the PCB congeners increase with an increase in the degree of chlorination. Screening studies have shown that it is resistant to biodegradation. Although biodegradation may occur very slowly in the environment, no other degradation mechanisms have been shown to be important in natural water and soil systems; therefore, biodegradation may be the ultimate degradation process in water and soil. The PCB composition of the biodegraded Aroclor is different from the original Aroclor. If released to soil, the PCB congeners will become tightly adsorbed to the soil particles. In the presence of organic solvents, PCBs may have a tendency to leach through soil. Although the volatilization rate may be low from soil surfaces, the total loss by volatilization over time may be significant because of persistence and stability. Enrichment of the low Cl PCBs occurs in the vapor phase relative to the original Aroclor; the residue will be enriched in the PCBs containing high Cl content. If released to water, adsorption to sediment and suspended matter will be an important fate process. Although adsorption can immobilize it for relatively long periods of time, eventual resolution into the water column has been shown to occur. The PCB composition in water will be enriched in the lower chlorinated PCBs because of their greater water solubility, and the least water soluble PCBs (highest Cl content) will remain adsorbed. In the absence of adsorption, It volatilizes relatively rapidly from water. However, strong PCB adsorption to sediment significantly competes with volatilization which may have a half-life in excess of 60 years in typical bodies of water. It has been shown to bioconcentrate significantly in aquatic organisms. If released to the atmosphere, the PCB congeners will exist primarily in the vapor-phase with enrichment of the most volatile PCBs although a relatively small percentage will partition to the particulate phase. The dominant atmospheric transformation process for these congeners is probably the vapor-phase reaction with hydroxyl radicals which have estimated half-lives ranging from 4.75 months to 1.31 years. Physical removal from the atmosphere, which is very important environmentally due to chemical stability, is accomplished by wet and dry deposition.

Cleanup/Disposal: Guide No. 171: Do not touch or walk through spilled material. Stop leak if you can do it without risk. Prevent dust cloud. Avoid inhalation of asbestos dust. Small Dry Spills: With clean shovel place material into clean, dry container and cover loosely; move containers from spill area. Small Spills: Take up with sand or other noncombustible absorbent material and place into containers for later disposal. Large Spills: Dike far ahead of liquid spill for later disposal. Cover powder spill with plastic sheet or tarp to minimize spreading. Prevent entry into waterways, sewers, basements or confined areas.

Environmental Physical Data

Henry's Law Constant: 5 x10^{-5}

Octanol/Water Partition Coefficient: log K_{ow} = 6.11

Sorption Partition Coefficient: K_{oc} = 6.1 x10^{4} to 7.4 x10^{5}

BCF: fathead minnow 2.7 x10^{5}

Regulations

RCRA 40CFR: Not listed

CERCLA: 40CFR 302.4: Listed per CWA Section 311(b)(4) per CWA Section 307(a) RQ: 1 lb (0.454 kg)

SARA 40CFR 372.65: Not listed

SARA EHS 40CFR 355: Not listed

TSCA: Not listed

Analytical Methods

Soil: CLP LC_PEST, MC_PEST, OHC; SW846 3630B, 8080A, 8081, 8082, 8250A, 8270B, 8270C; EPA 16, 3, PCB-002, PCB-005, PCB-009

Water / Groundwater: EPA PCB-003, PCB-004, 608, 617, 625, 625-S, 022; APHA 6410-B, 6630-C, 6630-D; ASTM D3534

Drinking Water: EPA 505, 508, 525.2; ASTM D5175

Food: FDA 212.1, 232.1; EPA 4

Indoor / Expired Air: ASTM D4861

Plasma: EPA 001, 29; FDA 211.1, 231.1, 251.1, 251.2, 252

Other: EPA P-009-1, 1656

ARS1000 CAS #: 98-50-0

ARSANILIC ACID

RTECS: CF7875000

EINECS Number: 202-674-3

Molecular Formula: $C_6H_8AsNO_3$

Formula Weight: 217.04

Chemical Structure

Synonyms: ACIDE P-ARSANILIQUE; 4-AMINOBENZENEARSONIC ACID; P-AMINOBENZENEARSONIC ACID; AMINOPHENYLARSINE ACID; P-AMINOPHENYLARSINE ACID; (4-AMINOPHENYL)ARSONIC ACID; (P-AMINOPHENYL)ARSONIC ACID; 4-AMINOPHENYLARSONIC ACID; P-AMINOPHENYLARSONIC ACID; P-ANILINEARSONIC ACID; ANTOXYLIC ACID; 4-ARSANILIC ACID; ARSANILIC ACID-100; P-ARSANILIC ACID; ARSONIC ACID,(4-AMINOPHENYL)-; ARSONIC ACID,(4-AMINOPHENYL)-(9CI); AS-101; ATOXYLIC ACID; BENZENEARSONIC ACID,P-AMINO-; KYSELINA ARSANILOVA; PRO-GEN; PRO-GEN 90; PROGEN 90; PRO-GEN 227 PREMIX; R-SONIC

Description: white, monoclinic needles; practically odorless

Use: antiprotozoan additive for animal feeds; vet medicinal for swine; chem int for the amebicide & trichomonacide, carbarsone; chem int for sodium arsanilate; mfr medicinal arsenicals; vet: growth promotant, to improve feed efficiency; vet: coccidiostat; grasshopper bait

Physical Properties

Freezing Point: 232 °C (449.6 °F)
Specific Gravity: 1.9571 at 20 °C
Water Solubility: Slightly Soluble in Cold Water
Other Solubilities: Slightly Soluble in Alcohol, Acetic Acid; Soluble in Amyl Alcohol, solution of alkali carbonates; moderately Soluble in concentrated mineral acids; Insoluble in Acetone, Benzene, Chloroform, Ether, moderately dilute mineral acids
Flash Point: Yields flammable vapors on heating above melting point

RTECS Toxicity Data

Acute Oral: Rat LD_{50} Dose: >1 gm/kg.
Chronic (Multiple Dose) Oral: Rabbit Dose: 243 mg/kg/30D-I; Toxic Effects: Gastrointestinal - Hypermotility, diarrhea; Liver - Hepatitis (hepatocellular necrosis), zonal; Nutritional and gross metabolic - Weight loss or decreased weight gain.

Hazard Overviews

Poison

Flammable

Health: May cause irritation. Poison. Other Acute Effects: toxic by inhalation, in contact with skin and if swallowed; may be fatal if inhaled, swallowed, or absorbed through skin. Chronic Effects: possible risk of irreversible effects. Possible carcinogen.
Fire: Flammable. Hazards: emits toxic fumes. Extinguishing agents: water spray; carbon dioxide, dry chemical powder or appropriate foam. Precautions: combustible liquid.
Carcinogenicity: IARC - Not listed; NIOSH - Not listed; NTP - Not listed; ACGIH - Not listed; OSHA - Not listed; EPA - Not listed; MAK - Not listed
Exposure Limits
OSHA PEL: TWA: 0.5 mg/m^3; as As.
ACGIH TLV: TWA: 0.01 mg/m^3; as As.
NIOSH REL: STEL: 0.002 mg/m^3; Ceiling (15 min) as As.

Environmental

Regulations

RCRA 40CFR: Not listed
CERCLA: 40CFR 302.4: Listed as Compound per CWA Section 307(a) per CAA Section 112
SARA 40CFR 372.65: Listed as Compound
SARA EHS 40CFR 355: Not listed
TSCA: Listed

Analytical Methods

Indoor / Expired Air: NIOSH 5022

ARS1670 CAS #: 1327-52-2

ARSENATE

RTECS: CG0700000
Molecular Formula: AsH_3O_4
Structured MF: H_3AsO_4•$5H_2O$
Formula Weight: 229.8
Synonyms: ACIDE ARSENIQUE LIQUIDE; ARSENIC ACID; CRAB GRASS KILLER; DESICCANT L-10; HI-YIELD DESICCANT H-10; ORTHOARSENIC ACID; ZOTOX; ZOTOX CRAB GRASS KILLER
Description: white to colorless liquid, or crystals; odorless

Physical Properties

Specific Gravity: 2
Water Solubility: Exists only in solution
pH: Acid
Flash Point: Nonflammable

RTECS Toxicity Data

Acute Oral: Rat LD_{50} Dose: 48 mg/kg. Rabbit LD_{Lo} Dose: 5 mg/kg.
Reproductive/Teratogenic: Rat Route: Intraperitoneal; Dose: 30 mg/kg; Duration: female 9D of pregnancy; Effects on Embryo or Fetus - Cytological changes (inc. somatic cell genetic material); Specific Developmental Abnormalities - Central nervous system. Mouse Route: Oral; Dose: 120 mg/kg; Duration: female 7-15D of pregnancy; Effects on Embryo or Fetus - Fetotoxicity; Fetal death.

Hazard Overviews

Fire: Noncombustible.
Carcinogenicity: IARC - Not listed; NIOSH - Not listed; NTP - Not listed; ACGIH - Not listed; OSHA - Not listed; EPA - Not listed; MAK - Not listed
Exposure Limits
ACGIH TLV: TWA: .01 mg/m^3.
NIOSH REL: STEL: .002 mg/m^3; Ceiling, 15 min, Inorganic.

Environmental

Cleanup/Disposal: Guide No. 151: Do not touch damaged containers or spilled material unless wearing appropriate protective clothing. Stop leak if you can do it without risk. Prevent entry into waterways, sewers, basements or confined areas. Cover with plastic sheet to prevent spreading. Absorb or cover with dry earth, sand or other non-combustible material and transfer to containers. Do not get water inside containers.

Environmental Physical Data

BCF: no food chain concentration potential
BOD: none

Regulations

RCRA 40CFR: Not listed
CERCLA: 40CFR 302.4: Not listed
SARA 40CFR 372.65: Not listed
SARA EHS 40CFR 355: Not listed

TSCA: Not listed

ARS2340 CAS #: 7440-38-2

ARSENIC

RTECS: CG0525000
DOT: UN1558; IMO6.1
EINECS Number: 231-148-6
Molecular Formula: As
Structured MF: As_4
Formula Weight: 74.92

As

Chemical Structure

Synonyms: ARSEN; ARSENIA; ARSENIC-75; ARSENIC BLACK; ARSENICALS; COLLOIDAL ARSENIC; GRAY ARSENIC; GREY ARSENIC; METALLIC ARSENIC
Description: silvery, black solid; odorless
Use: alloying constituent; mfr of certain types of glass; in metallurgy for hardening copper, lead alloys; to make gallium arsenide for dipoles & other electronic devices; doping agent in germanium & silicon solid state products; special solders; medicine; component of alloys; component of electrical devices; medication: to mfr arsenical org cmpd for therapeutic use; as radioactive tracer in toxicology; used as a catalyst in the manufacture of ethylene oxide; used in semiconductor devices

Physical Properties

Boiling Point: Sublimes
Freezing Point: 817 °C (1502.6 °F) at 28 atm
Specific Gravity: 5.727 at 14 °C
Vapor Pressure: 1 mm Hg at 372 °C
Water Solubility: Insoluble
Other Solubilities: Insoluble in caustic and nonionizing acids.
Critical Temperature: 1400 °C
Critical Pressure: 22.3 mPa
Ionization Potential (eV): 9.8152
Flash Point: Noncombustible Solid

RTECS Toxicity Data

Acute Oral: Man TD_{Lo} Dose: 7857 mg/kg/55Y; Toxic Effects: Gastrointestinal - Changes in structure or function of esophagus; Blood - Hemorrhage; Skin and appendages - Dermatitis, other. Rat LD_{50} Dose: 763 mg/kg; Toxic Effects: Behavioral - Ataxia; Gastrointestinal - Hypermotility, diarrhea.
Acute Dermal: Rabbit LD_{Lo} Route: Subcutaneous Dose: 300 mg/kg. Guinea Pig LD_{Lo} Route: Subcutaneous Dose: 300 mg/kg.
Reproductive/Teratogenic: Rat Route: Oral; Dose: 605 ug/kg; Duration: female 35W prior to mating Effects on Fertility - Pre-implantation mortality; Post-implantation mortality. Rat Route: Oral; Dose: 580 ug/kg; Duration: female 30W prior to mating Specific Developmental Abnormalities - Musculoskeletal system.
Mutagenic: Human Cytogenetic Analysis; Route: Unreported; Dose: 4286 ug/kg. Mouse Cytogenetic Analysis; Route: Oral; Dose: 280 mg/kg/8W.
Tumorigenic: Man Route: Oral; Dose: 76 mg/kg/12Y-I; Toxic Effects: Tumorigenic - Carcinogenic by RTECS criteria; Liver - Tumors; Blood - Hemorrhage. Rabbit Route: Implant; Dose: 75 mg/kg; Toxic Effects: Tumorigenic - Equivocal tumorigenic agent by RTECS criteria; Lungs, Thorax, or Respiration - Tumors; Liver - Tumors.

Hazard Overviews

Flammable

Fire Diamond

Health: Irritating to eyes/skin/respiratory tract. Also Causes: damage to blood-forming organs, nervous and cardiovascular systems, cancer hazard.
Fire: Flammable as a powder. Use dry chemical, carbon dioxide, water spray, or regular foam.
Reactivity: Stable. Hazardous polymerization cannot occur. Incompatible with: powerful oxidizers (bromates; peroxides; chlorates; iodates; lithium; silver nitrate; potassium nitrate; potassium permanganate; chromium (VI) oxide; halogens; bromine azide; palladium; dirubidium acetylide; zinc; platinum. Hazardous decomposition products: irritating or poisonous gases.
Carcinogenicity: IARC - Group 1, Carcinogenic to humans; NIOSH - Listed as carcinogen; NTP - Class 1, Known to be a carcinogen; ACGIH - Class A1, Confirmed human carcinogen; OSHA - Listed as a carcinogen; EPA - Class A, Human carcinogen; MAK - Class A1, Capable of inducing malignant tumors as shown by experience with humans
Primary Target Organs:

Eyes

Skin

Respiratory System

Mucous Membranes

Nervous System

Blood

Exposure Limits
OSHA PEL: TWA: 0.5 mg/m^3; as As.
ACGIH TLV: TWA: 0.01 mg/m^3; as As.
NIOSH REL: STEL: 0.002 mg/m^3; Ceiling (15 min) as As.
NIOSH IDLH: 5 mg/m^3; as As.
Respirator Recommendation
Exposure Range: >0.01 to 0.1 mg/m^3 Air Purifying, Negative Pressure, Half Mask
Exposure Range: >0.1 to 1 mg/m^3 Air Purifying, Negative Pressure, Full Face
Exposure Range: >1 to <5 mg/m^3 Supplied Air, Constant Flow/Pressure Demand, Full Face
Exposure Range: 5 to unlimited mg/m^3 Self-contained Breathing Apparatus, Pressure Demand, Full Face
Cartridge Color: magenta (P100)

Environmental

Ecotoxicity: Food chain concentration potential: Bioaccumulated by fresh water and marine aquatic organisms

Cleanup/Disposal: Guide No. 152: Do not touch damaged containers or spilled material unless wearing appropriate protective clothing. Stop leak if you can do it without risk. Prevent entry into waterways, sewers, basements or confined areas. Cover with plastic sheet to prevent spreading. Absorb or cover with dry earth, sand or other non-combustible material and transfer to containers. Do not get water inside containers.

Environmental Physical Data

BCF: bioaccumulated by aquatic organisms

BOD: none

Regulations

RCRA 40CFR: Not listed

CERCLA: 40CFR 302.4: Listed per CWA Section 307(a) per CAA Section 112 RQ: 1 lb (0.454 kg)

SARA 40CFR 372.65: Listed

SARA EHS 40CFR 355: Not listed

TSCA: Listed

Analytical Methods

Air: EPA 29, 0060, 108, 1632, 1669, ITM-001; Canada 1AP79-1

Soil: CLP 200.10_M,200.62,200.7_M,202.62,206.2_M,206.3_M,6020_M,ICP-AES; EPA 15,108A,108B,108C,200.7,200.8,PMD-AS; SW846 3005A,3010A,3015,3031,3040A,3050A&B,3051,3052,6010A&B,6020,7000A,7060A,7061A,7062,7063,OSW-A; ASTM D1971; USGS I5062

Water / Groundwater: EPA 200.0, 200.1, 200.12, 200.15, 200.7, 200.9, 206.3, 206.4, 206.5; DOE SO010R; APHA 3113-B, 3114-B, 3120; ASTM D1976, D2972; CEM RD42; USGS E-SPEC, I1060, I1062, I2062, I3062, I4062, I7062; HACH 8013

Drinking Water: AOAC 920.202, 920.205, 993.14; APHA 3500-AS

Food: EPA 14; AOAC 920.13, 920.17, 920.20, 920.22, 920.28, 920.29, 922.03, 924.04, 925.02, 925.03, 963.06

Indoor / Expired Air: NIOSH 7300, 7900

Plasma: EPA 200.11

Other: EPA 1620, 206.2; AOAC 990.08

ARS3010 **CAS #: 7778-39-4**

ARSENIC ACID

RTECS: CG0700000

DOT: UN1553; UN1554; IMO6.1

EINECS Number: 231-901-9

Molecular Formula: AsH_3O_4

Structured MF: As_2O_5

Formula Weight: 141.93

Synonyms: ACIDE ARSENIQUE LIQUIDE; ARSENATE; ARSENIC ACID,LIQUID; CRAB GRASS KILLER; DESICCANT L-10; DESSICANT L-10; EPA PESTICIDE CHEMICAL CODE 006801; HI-YIELD DESICCANT H-10; ORTHOARSENIC ACID; SCORCH; ZOTOX; ZOTOX CRAB GRASS KILLER

Description: white translucent crystals;commercial grade is a very pale yellow, syrupy liquid

Use: mfr arsenates; glass making; wood treating process; defoliant (under special regulations) hemihydrate; soil sterilant hemihydrate; arsenic acid is used to promote the desication of cotton, particularly cotton which is to be stripped; chem int for chromated copper arsenate-wood preservative, fluor chrome arsenate phenol (fcap); agent in glass mfr

Physical Properties

Boiling Point: 160 °C (320 °F)

Freezing Point: 35.5 °C (95.9 °F)

Specific Gravity: 2.2 at 20 °C

Water Solubility: 302 g/100 cc at12.5 °C

Other Solubilities: freely Soluble in Glycerol

pH: Weak acid properties

Flash Point: Nonflammable

RTECS Toxicity Data

Acute Oral: Rat LD_{50} Dose: 48 mg/kg. Rabbit LD_{Lo} Dose: 5 mg/kg.

Reproductive/Teratogenic: Rat Route: Intraperitoneal; Dose: 30 mg/kg; Duration: female 9D of pregnancy; Effects on Embryo or Fetus - Cytological changes (inc. somatic cell genetic material); Specific Developmental Abnormalities - Central nervous system. Mouse Route: Oral; Dose: 120 mg/kg; Duration: female 7-15D of pregnancy; Effects on Embryo or Fetus - Fetotoxicity; Fetal death.

Mutagenic: Human Cytogenetic Analysis; Cell Type: leukocyte; Dose: 7200 nmol/L.

Hazard Overviews

Fire: Noncombustible.

Carcinogenicity: IARC - Group 1, Carcinogenic to humans; NIOSH - Listed as carcinogen; NTP - Class 1, Known to be a carcinogen; ACGIH - Class A1, Confirmed human carcinogen; OSHA - Listed as a carcinogen; EPA - Class A, Human carcinogen; MAK - Class A1, Capable of inducing malignant tumors as shown by experience with humans

Exposure Limits

OSHA PEL: TWA: 0.01 mg/m^3; as As inorganic.

ACGIH TLV: TWA: 0.01 mg/m^3; as As.

NIOSH REL: STEL: 0.002 mg/m^3; ceiling (15 min) as As.

Respirator Recommendation

Exposure Range: >0.01 to <5 mg/m^3 Supplied Air, Constant Flow/Pressure Demand, Full Face

Exposure Range: 5 to unlimited mg/m^3 Self-contained Breathing Apparatus, Pressure Demand, Full Face

Note: as arsenic, inorganic compounds; refer to 29CFR 1910.1018 for more specific respirator recommendations

Environmental

Cleanup/Disposal: Guide No. 154: Eliminate all ignition sources (no smoking, flares, sparks or flames in immediate area). Do not touch damaged containers or spilled material unless wearing appropriate protective clothing. Stop leak if you can do it without risk. Prevent entry into waterways, sewers, basements or confined areas. Absorb or cover with dry earth, sand or other non-combustible material and transfer to containers. Do not get water inside containers.

Environmental Physical Data

BCF: aquatic organisms do not bioaccumulate
BOD: none

Regulations

RCRA 40CFR: Listed Hazardous Waste No. P010 Toxic Waste
CERCLA: 40CFR 302.4: Listed per RCRA Section 3001 RQ: 1 lb (0.454 kg)
SARA 40CFR 372.65: Listed as Compound
SARA EHS 40CFR 355: Not listed
TSCA: Listed

ARS3680 CAS #: 1303-32-8

ARSENIC DISULFIDE

Molecular Formula: As_2S_2
Structured MF: As_2S_2
Formula Weight: 214

As—S
S—As
As—S
S—As

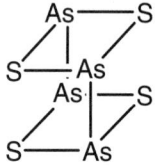

Chemical Structure

Description: red-brown solid; odorless

Physical Properties

Boiling Point: 565 °C (1049 °F)
Freezing Point: 320 °C (608 °F)
Specific Gravity: 3.4
Water Solubility: Practically Insoluble in Water
Flash Point: Nonflammable

Hazard Overviews

Fire: Noncombustible.
Carcinogenicity: IARC - Not listed; NIOSH - Not listed; NTP - Not listed; ACGIH - Not listed; OSHA - Not listed; EPA - Not listed; MAK - Not listed

Environmental

Cleanup/Disposal: Guide No. 152: Do not touch damaged containers or spilled material unless wearing appropriate protective clothing. Stop leak if you can do it without risk. Prevent entry into waterways, sewers, basements or confined areas. Cover with plastic sheet to prevent spreading. Absorb or cover with dry earth, sand or other non-combustible material and transfer to containers. Do not get water inside containers.

Regulations

RCRA 40CFR: Not listed
CERCLA: 40CFR 302.4: Not listed
SARA 40CFR 372.65: Not listed
SARA EHS 40CFR 355: Not listed
TSCA: Not listed

ARS4350 CAS #: 1303-28-2

ARSENIC PENTOXIDE

RTECS: CG2275000
DOT: UN1559; IMO6.1
EINECS Number: 215-116-9
Molecular Formula: As_2O_5
Structured MF: As_2O_5
Formula Weight: 229.84

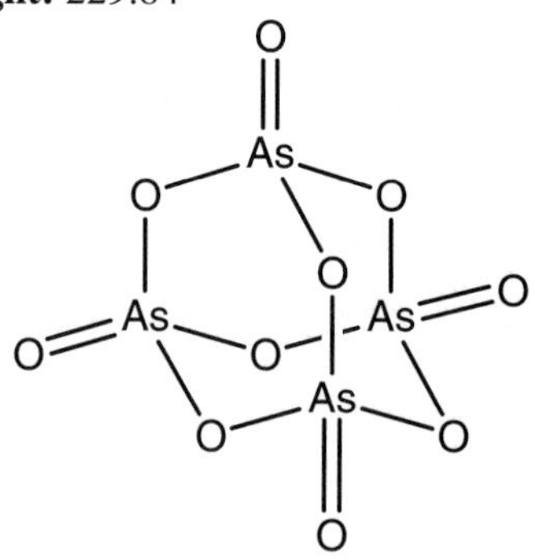

Chemical Structure

Synonyms: ANHYDRIDE ARSENIQUE; ARSENIC ACID ANHYDRIDE; ARSENIC ANHYDRIDE; ARSENIC (V) OXIDE; ARSENIC OXIDE; ARSENIC(V) OXIDE; ARSENIC PENTAOXIDE; DIARSENIC PENTOXIDE; EPA PESTICIDE CHEMICAL CODE 006802
Description: white powder; odorless
Use: chemical intermediate for metal arsenates-eg, lead arsenate; herbicide-esp preharvest cotton defoliant; in formulated wood preservatives; insecticide;soil sterilant; glass ingredient; mfr of colored glass; in adhesives for metals; in weed control; as fungicide; dyeing and printing

Physical Properties

Boiling Point: Decomposes at 1 atm
Freezing Point: 315 °C (599 °F)
Specific Gravity: 4.32
Water Solubility: Combines Very Slowly with Water
Other Solubilities: Soluble in acid and alkali.
Flash Point: Nonflammable

RTECS Toxicity Data

Acute Oral: Rat LD_{50} Dose: 8 mg/kg. Mouse LD_{50} Dose: 55 mg/kg.

Acute Dermal: Rabbit LD_{Lo} Route: Subcutaneous Dose: 12400 ug/kg.
Reproductive/Teratogenic: Rat Route: Intratesticular; Dose: 4597 ug/kg; Duration: male 1D prior to mating; Paternal Effects - Spermatogenesis; Testes, epididymis, sperm duct. Mouse Route: Subcutaneous; Dose: 4597 ug/kg; Duration: male 30D prior to mating; Paternal Effects - Testes, epididymis, sperm duct.
Mutagenic: Human DNA Inhibition; Cell Type: fibroblast; Dose: 100 umol/L. Human Cytogenetic Analysis; Cell Type: leukocyte; Dose: 1200 nmol/L.

Hazard Overviews

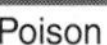
Poison

Fire Diamond

Health: Irritating to eyes/skin/respiratory tract. Poison. Also Causes: Inhalation: laryngitis, bronchitis, rhinitis, nasal ulceration, nausea, vomiting, and diarrhea, pulmonary edema, respiratory failure, constriction of the throat, difficulty in swallowing, excruciating abdominal pain, dehydration, muscular cramps, or liver, cardiovascular, and nervous systems damage. Chronic Effects: hyperkeratosis, pigment changes, liver damage, numbness, muscle weakness, pancytopenia, aplastic anemia, leukopenia, bone marrow depression, damage to the vascular system, necrosis, gastrointestinal disturbances, conjunctivitis, lung, skin, or internal cancer.
Fire: Noncombustible. Use agent suitable for surrounding fire.
Reactivity: Stable. Hazardous polymerization cannot occur. Avoid: heat; ignition sources. Incompatible with: acids; halogens; aluminum; zinc; bromine pentafluoride. Hazardous decomposition products: arsine gas; arsenic oxides.
Carcinogenicity: IARC - Group 1, Carcinogenic to humans; NIOSH - Listed as carcinogen; NTP - Class 1, Known to be a carcinogen; ACGIH - Class A1, Confirmed human carcinogen; OSHA - Listed as a carcinogen; EPA - Class A, Human carcinogen; MAK - Class A1, Capable of inducing malignant tumors as shown by experience with humans
Primary Target Organs:

Eyes

Skin
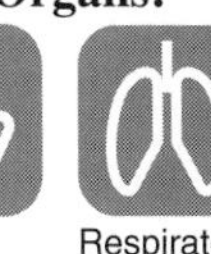

Respiratory System

Nervous System

Liver

Blood

Exposure Limits
OSHA PEL: TWA: 0.01 mg/m^3; as As.
ACGIH TLV: TWA: 0.01 mg/m^3; as As.
NIOSH REL: STEL: 0.002 mg/m^3; ceiling (15 min) as As.
Respirator Recommendation
Exposure Range: >0.01 to <5 mg/m^3 Supplied Air, Constant Flow/Pressure Demand, Full Face
Exposure Range: 5 to unlimited mg/m^3 Self-contained Breathing Apparatus, Pressure Demand, Full Face
Note: as arsenic, inorganic compounds; refer to 29CFR 1910.1018 for more specific respirator recommendations

Environmental

Ecotoxicity: LD_{50} Grasshopper oral 0.9-25.6 ppm/7 days; 1.0-5.5 ppm/14 days
Cleanup/Disposal: Guide No. 151: Do not touch damaged containers or spilled material unless wearing appropriate protective clothing. Stop leak if you can do it without risk. Prevent entry into waterways, sewers, basements or confined areas. Cover with plastic sheet to prevent spreading. Absorb or cover with dry earth, sand or other non-combustible material and transfer to containers. Do not get water inside containers.

Environmental Physical Data

BCF: snail 3

Regulations

RCRA 40CFR: Listed Hazardous Waste No. P011 Toxic Waste
CERCLA: 40CFR 302.4: Listed per CWA Section 311(b)(4) per RCRA Section 3001 RQ: 1 lb (0.454 kg)
SARA 40CFR 372.65: Not listed TPQ: 100/10000 lb
SARA EHS 40CFR 355: Listed TPQ: 1 lb
TSCA: Listed

ARS5020 CAS #: 7784-33-0

ARSENIC TRIBROMIDE

RTECS: CG1375000
DOT: UN1555; IMO6.1
EINECS Number: 232-057-4
Molecular Formula: $AsBr_3$
Structured MF: $AsBr_3$
Formula Weight: 314.63

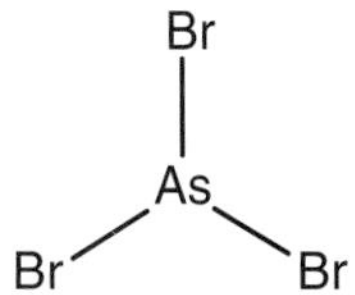

Chemical Structure

Synonyms: ARSENIC BROMIDE; ARSENIC(II) BROMIDE; ARSENIOUS BROMIDE; ARSENOUS BROMIDE; ARSENOUS TRIBROMIDE; TRIBROMOARSINE
Description: colorless to yellow to yellowish-white prisms or crystals
Use: analytical chemistry, medicine

Physical Properties

Boiling Point: 221 °C (430 °F)
Freezing Point: 31.1 °C (87.98 °F)
Specific Gravity: 3.397 at 25 °C/4 °C
Vapor Pressure: 1 mm Hg at 41.8 °C
Water Solubility: Decomposes
Other Solubilities: Soluble in Hydrochloric Acid, hydrocarbons, chlorinated hydrocarbons, Carbon Disulfide.

Hazard Overviews

Poison

Corrosive

Health: Corrosive to eyes/skin/respiratory tract. Poison. Other Acute Effects: may be fatal if inhaled, swallowed, or absorbed through skin; material is extremely destructive to tissue of the mucous membranes and upper respiratory tract; eyes, and skin; inhalation may result in spasm, inflammation and edema of the larynx and bronchi, chemical pneumonitis and pulmonary edema; symptoms of exposure may include burning sensation; coughing; wheezing; laryngitis; shortness of breath; headache; nausea; vomiting. Chronic Effects: may cause heritable genetic damage; target organs: skin, lungs. Carcinogen.

Fire: Hazards: emits toxic fumes. Extinguishing agents: noncombustible; use extinguishing media appropriate to surrounding fire conditions. Precautions: combustible liquid.

Reactivity: Hazardous polymerization will not occur. Incompatible with: acids, bases, oxidizing agents, reacts violently with sodium/potassium/aluminum, may decompose on exposure to moist air or water, sensitive to light. Hazardous decomposition products: arsenic oxides, hydrogen bromide gas.

Carcinogenicity: IARC - Group 1, Carcinogenic to humans; NIOSH - Listed as carcinogen; NTP - Class 1, Known to be a carcinogen; ACGIH - Class A1, Confirmed human carcinogen; OSHA - Listed as a carcinogen; EPA - Class A, Human carcinogen; MAK - Class A1, Capable of inducing malignant tumors as shown by experience with humans

Primary Target Organs:

Eyes

Skin

Respiratory System

Exposure Limits

OSHA PEL: TWA: 0.01 mg/m^3; as As inorganic.

ACGIH TLV: TWA: 0.01 mg/m^3; as As.

NIOSH REL: STEL: 0.002 mg/m^3; ceiling (15 min) as As.

Respirator Recommendation

Exposure Range: >0.01 to <5 mg/m^3 Supplied Air, Constant Flow/Pressure Demand, Full Face

Exposure Range: 5 to unlimited mg/m^3 Self-contained Breathing Apparatus, Pressure Demand, Full Face

Note: as arsenic, inorganic compounds; refer to 29CFR 1910.1018 for more specific respirator recommendations

Environmental

Cleanup/Disposal: Guide No. 151: Do not touch damaged containers or spilled material unless wearing appropriate protective clothing. Stop leak if you can do it without risk. Prevent entry into waterways, sewers, basements or confined areas. Cover with plastic sheet to prevent spreading. Absorb or cover with dry earth, sand or other non-combustible material and transfer to containers. Do not get water inside containers.

Environmental Physical Data

BCF: no biomagnification

Regulations

RCRA 40CFR: Not listed

CERCLA: 40CFR 302.4: Listed as Compound per CWA Section 307(a) per CAA Section 112

SARA 40CFR 372.65: Listed as Compound

SARA EHS 40CFR 355: Not listed

TSCA: Listed

ARS5690 CAS #: 7784-34-1

ARSENIC TRICHLORIDE

RTECS: CG1750000

DOT: UN1560; IMO6.1

EINECS Number: 232-059-5

Molecular Formula: $AsCl_3$

Structured MF: $AsCl_3$

Formula Weight: 181.28

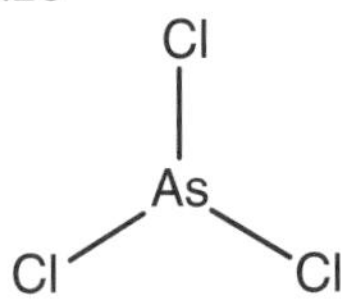

Chemical Structure

Synonyms: ARSENIC BUTTER; ARSENIC CHLORIDE; ARSENIC CHLORIDE,LIQUID; ARSENIC TRICHLORIDE,LIQUID; ARSENIC(III) CHLORIDE; ARSENIC(III) TRICHLORIDE; ARSENIOUS CHLORIDE; ARSENOUS CHLORIDE; ARSENOUS TRICHLORIDE; ARSENOUS TRICHLORIDE (9CI); BUTTER OF ARSENIC; CAUSTIC ARSENIC CHLORIDE; CAUSTIC OIL OF ARSENIC; CHLORURE D'ARSENIC; CHLORURE ARSENIEUX; FUMING LIQUID ARSENIC; TRICHLOROARSINE; TRICHLORURE D'ARSENIC

Description: colorless, pale yellow oily liquid; acrid odor

Use: intermediate for organic arsenicals (pharmaceuticals, insecticides) and in the ceramic industry

Physical Properties

Boiling Point: 130.21 °C (266 °F)

Freezing Point: -16 °C (3.2 °F)

Specific Gravity: 2.1497 at 25 °C/4 °C

Vapor Density: 6.25 Air=1

Saturated Vapor Density: 1.282911071 kg/m^3

Density: 2.15 g/mL at 25 °C

Vapor Pressure: 10 mm Hg at 23.5 °C

Water Solubility: 1 mol can be dissolved in 9 moles of Water

Other Solubilities: 95% Ethanol: >=100 mg/ml at 19 °C; Acetone: >=100 mg/ml at 19 °C; Carbon Tetrachloride, Chloroform: miscible; DMSO: >=100 mg/ml at 19 °C; Ether: miscible; HBr: Soluble; Hydrochloric Acid: Soluble; Iodine: miscible; Oils and fats: miscible.

Surface Tension: Estimated at 20 dynes/cm

pH: Acid

Refraction Index: 1.6006 at 20 °C/D
Flash Point: Nonflammable

RTECS Toxicity Data

Acute Inhalation: Mouse LC_{Lo} Dose: 338 ppm/10M. Cat LC_{Lo} Dose: 200 mg/m^3/20M.
Mutagenic: Human Cytogenetic Analysis; Cell Type: leukocyte; Dose: 600 nmol/L. Hamster Morphological Transformation; Cell Type: embryo; Dose: 3 umol/L.

Hazard Overviews

Corrosive

Fire Diamond

Health: Toxic. Corrosive to eyes/skin/respiratory tract. Also Causes: nasal perforations, cough, chest pain, difficulty breathing, giddiness, headache, weakness, hyperpigmentation, thickening of skin, dermatitis, tissue damage to heart/live/kidney/pancreas/stomach, necrosis, burning lips, throat constriction, difficulty swallowing, abdominal pain, nausea, diarrhea, dehydration, blood, protein and sugar in the urine, elevation of liver enzymes, dizziness, collapse, shock, rapid pulse, cold sweats, coma, peripheral neuropathy. Chronic Effects: garlic odor on breath, sweating, excessive salivation, blood changes, anemia, increased vascularity of bone marrow, lung/skin/liver cancer.
Fire: Noncombustible. Use agent suitable for surrounding fire. Decomposes on contact with water, releasing toxic hydrogen chloride gas.
Reactivity: Unstable, will decompose on exposure to moisture or sunlight. Hazardous polymerization cannot occur. Avoid: exposure to water. Incompatible with: acids; alkalis; water; hexafluoroisopropylideneaminolithium; sodium; potassium; aluminum; common metals (zinc; iron); hydrogen gas. Hazardous decomposition products: arsenic trioxide; hydrogen chloride gas.
Carcinogenicity: IARC - Group 1, Carcinogenic to humans; NIOSH - Listed as carcinogen; NTP - Class 1, Known to be a carcinogen; ACGIH - Class A1, Confirmed human carcinogen; OSHA - Listed as a carcinogen; EPA - Class A, Human carcinogen; MAK - Class A1, Capable of inducing malignant tumors as shown by experience with humans
Primary Target Organs:

Eyes

Skin

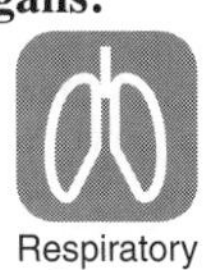
Respiratory System

Nervous System

Liver

Blood

Exposure Limits
OSHA PEL: TWA: 0.01 mg/m^3; as As inorganic.
ACGIH TLV: TWA: 0.01 mg/m^3; as As.
NIOSH REL: STEL: 0.002 mg/m^3; ceiling (15 min) as As.
Respirator Recommendation
Exposure Range: >0.01 to <5 mg/m^3 Supplied Air, Constant Flow/Pressure Demand, Full Face
Exposure Range: 5 to unlimited mg/m^3 Self-contained Breathing Apparatus, Pressure Demand, Full Face
Note: as arsenic, inorganic compounds; refer to 29CFR 1910.1018 for more specific respirator recommendations

Environmental

Cleanup/Disposal: Guide No. 157: Eliminate all ignition sources (no smoking, flares, sparks or flames in immediate area). All equipment used when handling the product must be grounded. Do not touch damaged containers or spilled material unless wearing appropriate protective clothing. Stop leak if you can do it without risk. A vapor suppressing foam may be used to reduce vapors. Do not get water inside containers. Use water spray to reduce vapors or divert vapor cloud drift. Prevent entry into waterways, sewers, basements or confined areas. Small Spills: Cover with dry earth, dry sand, or other non-combustible material followed with plastic sheet to minimize spreading or contact with rain. Use clean non-sparking tools to collect material and place it into loosely covered plastic containers for later disposal.

Environmental Physical Data

BCF: no biomagnification

Regulations

RCRA 40CFR: Not listed
CERCLA: 40CFR 302.4: Listed per CWA Section 311(b)(4) RQ: 1 lb (0.454 kg)
SARA 40CFR 372.65: Not listed TPQ: 500 lb
SARA EHS 40CFR 355: Listed TPQ: 1 lb
TSCA: Listed

ARS6360 **CAS #: 7784-35-2**

ARSENIC TRIFLUORIDE

RTECS: CG5775000
EINECS Number: 232-060-0
Molecular Formula: AsF_3
Structured MF: AsF_3
Formula Weight: 131.91
Synonyms: ARSENIC FLUORIDE; ARSENOUS FLUORIDE; ARSENOUS TRIFLUORIDE; TL 156; TRIFLUOROARSINE
Description: colorless, oily mobile liquid
Use: chem int for fluorine compounds; catalyst for fluorinations; ion implantation; synthesis of arsenic pentafluoride; component in wood treatment products for weather and fungal resistance

Physical Properties

Boiling Point: 57.8 °C (136 °F)
Freezing Point: -5.95 °C (21.29 °F)
Specific Gravity: 2.73 at 15 °C/15 °C
Vapor Pressure: 100 mm Hg at 13.2 °C
Water Solubility: Decomposes in Water
Other Solubilities: Soluble in Ethanol, Ether, Benzene, Ammonium Hydroxide
Ionization Potential (eV): 12.3 +/-0.05

RTECS Toxicity Data

Acute Inhalation: Mouse LC_{Lo} Dose: 2000 mg/m^3/10M.

Hazard Overviews

Corrosive

Carcinogenicity: IARC - Group 1, Carcinogenic to humans; NIOSH - Listed as carcinogen; NTP - Class 1, Known to be a carcinogen; ACGIH - Class A1, Confirmed human carcinogen; OSHA - Listed as a carcinogen; EPA - Class A, Human carcinogen; MAK - Class A1, Capable of inducing malignant tumors as shown by experience with humans

Exposure Limits

OSHA PEL: TWA: 2.5 mg/m^3; as F.

ACGIH TLV: TWA: 0.01 mg/m^3; as As.

NIOSH REL: STEL: 0.002 mg/m^3; Ceiling (15 min) as As.

Environmental

Ecotoxicity: LC_{50} Salvelinus fontinalis (brook trout) 10,440 ug/l/262 hr LC_{50} bay scallop 3,490 ug/l/96 hr LC_{50} Lepomis macrochirus (bluegill, fingerling) 290 ug/l/48 hr LC_{50} Oncorhynchus keta (chum salmon) 8,330 ug/l/48 hr LC_{50} Daphnia magna (cladoceran) 2,850 ug/l/3 wk LC_{50} Nereis diversicolor (polychaete worm) >14,500 ug/l/192 hr LC_{50} Penaeus seliferus (white shrimp, juvenile) 24,700 ug/l/96 hr LC_{100} Oncorhynchus gorbuscha (pink salmon) 12,307 ug/l/96 hr; 7,195 ug/l/day LC_{50} Gastrophryne carolinensis (toad, embryo-larval) 40 ug/l/7 days

Environmental Physical Data

BCF: no biomagnification

Regulations

RCRA 40CFR: Not listed

CERCLA: 40CFR 302.4: Listed as Compound per CWA Section 307(a) per CAA Section 112

SARA 40CFR 372.65: Listed as Compound

SARA EHS 40CFR 355: Not listed

TSCA: Listed

ARS7030 — CAS #: 7784-45-4

ARSENIC TRIIODIDE

RTECS: CG1950000

EINECS Number: 232-068-4

Molecular Formula: AsI_3

Structured MF: AsI_3

Formula Weight: 455.62

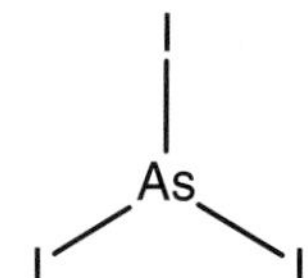

Chemical Structure

Synonyms: ARSENIC TRIIODIDE; ARSENOUS IODIDE; ARSENOUS TRIIODIDE (9CI); TRIIODOARSINE

Physical Properties

Boiling Point: 400 °C (752 °F)

Freezing Point: 141 °C (285.8 °F)

Specific Gravity: 4.39

Water Solubility: Moderately

Hazard Overviews

Poison

Corrosive

Health: Corrosive to eyes/skin/respiratory tract. Poison. Other Acute Effects: may be fatal if inhaled, swallowed, or absorbed through skin; inhalation may result in spasm, inflammation and edema of the larynx and bronchi, chemical pneumonitis and pulmonary edema; symptoms of exposure may include burning sensation; coughing; wheezing; laryngitis; shortness of breath; headache; nausea; vomiting. Chronic Effects: may alter genetic material; prolonged exposure may produce iodism in sensitive individuals; skin rash; running nose; for severe cases, pimples; boils; hives; blisters and black and blue spots; readily diffused across the placenta; neonatal deaths from respiratory distress secondary to goiter have been reported; known to cause drug-induced fevers; target organs: skin, lungs, liver. Carcinogen.

Fire: Hazards: emits toxic fumes. Extinguishing agents: carbon dioxide, dry chemical powder or appropriate foam; do not use water. Precautions: combustible liquid.

Reactivity: Incompatible with: acids, bases, oxidizing agents, aluminum fluorine, heavy metals, metal oxides iron and iron salts, zinc, heat may decompose on exposure to moist air or water, reacts violently with sodium potassium. Hazardous decomposition products: toxic fumes of: iodine, arsenic oxides, hydrogen iodide.

Carcinogenicity: IARC - Group 1, Carcinogenic to humans; NIOSH - Listed as carcinogen; NTP - Class 1, Known to be a carcinogen; ACGIH - Class A1, Confirmed human carcinogen; OSHA - Listed as a carcinogen; EPA - Class A, Human carcinogen; MAK - Class A1, Capable of inducing malignant tumors as shown by experience with humans

Primary Target Organs:

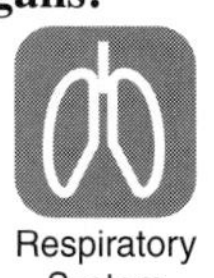

Eyes Skin Respiratory System Liver

Exposure Limits

OSHA PEL: TWA: 0.01 mg/m^3; as As inorganic.

ACGIH TLV: TWA: 0.01 mg/m^3; as As.

NIOSH REL: STEL: 0.002 mg/m^3; ceiling (15 min) as As.

Respirator Recommendation

Exposure Range: >0.01 to <5 mg/m^3 Supplied Air, Constant Flow/Pressure Demand, Full Face

Exposure Range: 5 to unlimited mg/m^3 Self-contained Breathing Apparatus, Pressure Demand, Full Face

Note: as arsenic, inorganic compounds; refer to 29CFR 1910.1018 for more specific respirator recommendations

Environmental

Regulations

RCRA 40CFR: Not listed

CERCLA: 40CFR 302.4: Listed as Compound per CWA Section 307(a) per CAA Section 112

SARA 40CFR 372.65: Listed as Compound

SARA EHS 40CFR 355: Not listed

TSCA: Listed

ARS7700 CAS #: 1327-53-3

ARSENIC TRIOXIDE

RTECS: CG3325000

DOT: UN1561; IMO6.1

EINECS Number: 215-481-4

Molecular Formula: As_2O_3

Structured MF: As_2O_3

Formula Weight: 197.82

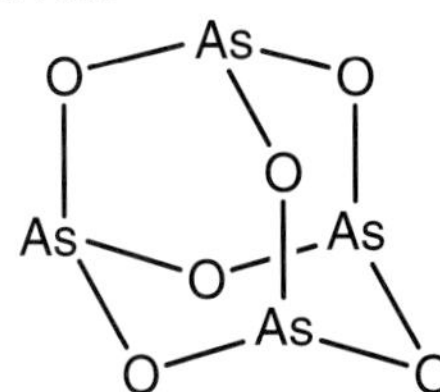

Chemical Structure

Synonyms: ACIDE ARSENIEUX; ANHYDRIDE ARSENIEUX; ARSENIC BLANC; ARSENIC (III) OXIDE; ARSENIC OXIDE; ARSENIC SESQUIOXIDE; ARSENIC TRIOXIDE,SOLID; ARSENIC(III) OXIDE; ARSENICUM ALBUM; ARSENIGEN SAURE; ARSENIOUS ACID; ARSENIOUS ACID ANHYDRIDE; ARSENIOUS OXIDE; ARSENIOUS TRIOXIDE; ARSENITE; ARSENOLITE; ARSENOUS ACID; ARSENOUS ACID ANHYDRIDE; ARSENOUS ANHYDRIDE; ARSENOUS OXIDE; ARSENOUS OXIDE ANHYDRIDE; ARSENTRIOXIDE; ARSODENT; CLAUDELITE; CLAUDETITE; CRUDE ARSENIC; DIARSENIC TRIOXIDE; EPA PESTICIDE CHEMICAL CODE 007001; WHITE ARSENIC

Description: white, transparent powder; odorless

Use: primary for arsenic compounds; manufacturing of glass, paris green, enamels, weed killers, merallic arsenic; for preserving hides; killing rodents, insects; in sheep dips and weed killers; textile mordant

Physical Properties

Boiling Point: 465 °C (869 °F)

Freezing Point: 312.3 °C (594.14 °F)

Specific Gravity: 3.738

Vapor Pressure: 66.1 mm Hg at 312 °C

Water Solubility: Sparingly Soluble in Cold Water

Other Solubilities: Alkali carbon: Soluble; Alkali: Soluble; Hydrochloric Acid: Soluble.

Refraction Index: 1.755

Flash Point: Nonflammable

RTECS Toxicity Data

Acute Oral: Human LD_{Lo} Dose: 1429 ug/kg. Man LD_{Lo} Dose: 29 mg/kg; Toxic Effects: Behavioral - Sleep; Behavioral - Muscle weakness; Gastrointestinal - Hypermotility, diarrhea. Man LD_{Lo} Dose: 286 mg/kg; Toxic Effects: Cardiac - Arrythmias (including changes in conduction); Liver - Liver function tests impaired; Musculoskelital - Other changes. Man LD_{Lo} Dose: 2857 mg/kg; Toxic Effects: Behavioral - Coma; Liver - Fatty Liver degeneration; Kidney, Ureter, and Bladder - Renal function tests depressed. Rat LD_{50} Dose: 14600 ug/kg.

Acute Dermal: Rat LD_{Lo} Route: Subcutaneous Dose: 8 mg/kg; Toxic Effects: Skin and appendages - Corrosive. Mouse LD_{50} Route: Subcutaneous Dose: 9800 ug/kg.

Chronic (Multiple Dose) Oral: Rat Dose: 900 mg/kg/15W-I; Toxic Effects: Behavioral - Change in motor activity (specific assay); Nutritional and gross metabolic - Weight loss or decreased weight gain; DEATH. Rat Dose: 350 ug/kg/30D-I; Toxic Effects: Kidney, Ureter, and Bladder - Other changes in urine composition; Nutritional and gross metabolic - Changes in sodium; Nutritional and gross metabolic - Changes in potassium.

Chronic (Multiple Dose) Inhalation: Rat Dose: 31 ug/m^3/24H/22W-C; Toxic Effects: Brain and coverings - Recordings from specific areas of CNS; Blood - Other changes. Rat Dose: 500 ug/m^3/24H/33D-C; Toxic Effects: Brain and coverings - Recordings from specific areas of CNS; Blood - Changes in serum composition; Biochemical - True cholinesterase.

Reproductive/Teratogenic: Woman Route: Oral; Dose: 600 mg/kg; Duration: female 30W of pregnancy; Effects on Newborn - Apgor score (human only); Other neonatal measures or effects. Mouse Route: Inhalation; Dose: 28500 ug/m^3/ Duration: female 9-12D of pregnancy; Effects on Embryo or Fetus - Cytological changes (inc. somatic cell genetic material); Specific Developmental Abnormalities - Musculoskeletal system. Mouse Route: Inhalation; Dose: 260 ug/m^3/4H Duration: female 9-12D of pregnancy; Effects on Embryo or Fetus - Fetotoxicity.

Mutagenic: Human Unscheduled DNA Synthesis; Cell Type: lung; Dose: 1 umol/L. Human DNA Inhibition; Cell Type: HeLa cell; Dose: 500 umol/L. Human Cytogenetic Analysis;

Cell Type: leukocyte; Dose: 1200 nmol/L. Human Sister Chromatid Exchenge; Cell Type: lymphocyte; Dose: 2 ug/cm^3.

Tumorigenic: Rat Route: Intratracheal; Dose: 16 mg/kg/15W-I; Toxic Effects: Tumorigenic - Equivocal tumorigenic agent by RTECS criteria; Lungs, Thorax, or Respiration - Tumors. Rat Route: Intratracheal; Dose: 75 mg/kg/15W-I; Toxic Effects: Tumorigenic - Equivocal tumorigenic agent by RTECS criteria; Lungs, Thorax, or Respiration - Tumors. Rat Route: Intratracheal; Dose: 167 mg/kg/15W-I; Toxic Effects: Tumorigenic - Equivocal tumorigenic agent by RTECS criteria; Lungs, Thorax, or Respiration - Tumors.

Hazard Overviews

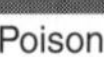
Poison

Fire Diamond

Health: Irritating to eyes/skin/respiratory tract. Poison. Also Causes: headache, coughing, difficulty breathing, chest pains, pulmonary edema, nausea and vomiting, abdominal pain, difficulty swallowing, bloody diarrhea, dehydration, intense thirst, fluid-PPE electrolyte disturbances, hypotension), metabolic acidosis, hemolysis, pancytopenia, anemia, Mee's lines, peripheral neuropathy or brain damage. Chronic Effects: photophobia, hair loss, perforation of the nasal septum, hoarse voice, aplastic anemia, and painful ulceration.

Fire: Noncombustible. Use agent suitable for surrounding fire.

Reactivity: Stable. Hazardous polymerization cannot occur. Avoid: elevated temperatures; dispersion into air. Incompatible with: tannic acid; infusion cinchona; vegetable astringent fusions and decoctions; acids; halogens; aluminum; zinc fillings; sodium nitrate; mercury; metals. Hazardous decomposition products: arsenic trioxide fumes; arsine gas.

Carcinogenicity: IARC - Group 1, Carcinogenic to humans; NIOSH - Listed as carcinogen; NTP - Class 1, Known to be a carcinogen; ACGIH - Class A1, Confirmed human carcinogen; OSHA - Listed as a carcinogen; EPA - Class A, Human carcinogen; MAK - Class A1, Capable of inducing malignant tumors as shown by experience with humans

Primary Target Organs:

Eyes

Skin

Respiratory System

Mucous Membranes

Nervous System

Blood

Exposure Limits

OSHA PEL: TWA: 0.01 mg/m^3; as As inorganic.

ACGIH TLV: TWA: 0.01 mg/m^3; as As.

NIOSH REL: STEL: 0.002 mg/m^3; ceiling (15 min) as As.

Respirator Recommendation

Exposure Range: >0.01 to <5 mg/m^3 Supplied Air, Constant Flow/Pressure Demand, Full Face

Exposure Range: 5 to unlimited mg/m^3 Self-contained Breathing Apparatus, Pressure Demand, Full Face

Note: as arsenic, inorganic compounds; refer to 29CFR 1910.1018 for more specific respirator recommendations

Environmental

Ecotoxicity: LC_{50} Oncorhynchus keta (Chum salmon) 8,330 ug/l/48 hr /Conditions of bioassay not specified LC_{54} Oncorhynchus gorbuscha (pink salmon) 3,787 ug/l/10 days /Conditions of bioassay not specified

Cleanup/Disposal: Guide No. 151: Do not touch damaged containers or spilled material unless wearing appropriate protective clothing. Stop leak if you can do it without risk. Prevent entry into waterways, sewers, basements or confined areas. Cover with plastic sheet to prevent spreading. Absorb or cover with dry earth, sand or other non-combustible material and transfer to containers. Do not get water inside containers.

Environmental Physical Data

Henry's Law Constant: 1.7 x10^{-12}

BCF: 28 days 4

Regulations

RCRA 40CFR: Listed Hazardous Waste No. P012 Toxic Waste

CERCLA: 40CFR 302.4: Listed per CWA Section 311(b)(4) per RCRA Section 3001 RQ: 1 lb (0.454 kg)

SARA 40CFR 372.65: Not listed TPQ: 100/10000 lb

SARA EHS 40CFR 355: Listed TPQ: 1 lb

TSCA: Listed

Analytical Methods

Food: AOAC 920.18, 920.19

Indoor / Expired Air: NIOSH 7901

ARS8370 CAS #: 1303-33-9

ARSENIC TRISULFIDE

RTECS: CG2638000

DOT: NA1557; IMO6.1

EINECS Number: 215-117-4

Molecular Formula: As_2S_3

Structured MF: As_2S_3

Formula Weight: 246.03

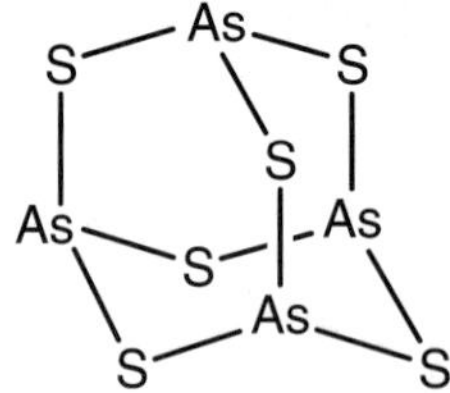

Chemical Structure

Synonyms: ARSENIC RED; ARSENIC SESQUISULFIDE; ARSENIC SESQUISULPHIDE; ARSENIC SULFIDE; ARSENIC SULFIDE YELLOW; ARSENIC SULPHIDE; ARSENIC TERSULFIDE; ARSENIC TERSULPHIDE; ARSENIC TRISULPHIDE; ARSENIC YELLOW; ARSENIOUS SULFIDE; ARSENIOUS SULPHIDE; ARSENOUS SULFIDE; AURIPIGMENT; C.I. 77086; C.I. PIGMENT YELLOW; DIARSENIC TRISULFIDE; DIARSENIC TRISULPHIDE; KING'S GOLD; NATURAL ORPIMENT; ORPIMENT; SHIO; YELLOW ARSENIC SULFIDE

Description: yellow or orange powder, changes to a red form at 170 deg C; odorless
Use: infrared-transmitting glass component; dehairing agent in tanning; agent in pyrotechnics mfr; agent in semiconductors mfr; pigment-eg, in shellac (former use); in electrical photoconductors; manufacture of oil cloth and linoleum; formerly an insecticidal ingredient as dip for goats & sheep; as reducing agent

Physical Properties

Boiling Point: 707 °C (1305 °F)
Freezing Point: 300 °C (572 °F) to 325 °C (617 °F)
Specific Gravity: 3.46
Water Solubility: 0.5 ppm at 18 °C
Other Solubilities: Soluble in Alcohol.
Refraction Index: 2.4
Flash Point: Nonflammable

RTECS Toxicity Data

Acute Oral: Rat LD_{50} Dose: 185 mg/kg; Toxic Effects: Behavioral - Muscle weakness; Lungs, Thorax, or Respiration - Dyspnea; Gastrointestinal - Hypermotility, diarrhea. Mouse LD_{50} Dose: 254 mg/kg; Toxic Effects: Behavioral - Muscle weakness; Lungs, Thorax, or Respiration - Dyspnea; Gastrointestinal - Hypermotility, diarrhea.
Acute Dermal: Rat LD_{50} Route: Skin; Dose: 936 mg/kg.
Tumorigenic: Rat Route: Subcutaneous; Dose: 125 mg/kg; Toxic Effects: Tumorigenic - Equivocal tumorigenic agent by RTECS criteria; Tumorigenic - Tumors at site of application. Mouse Route: Intratracheal; Dose: 73885 ug/kg/15W-C; Toxic Effects: Tumorigenic - Equivocal tumorigenic agent by RTECS criteria; Lungs, Thorax, or Respiration - Tumors.

Hazard Overviews

Poison

Corrosive

Fire Diamond

Health: Irritating to eyes/skin/respiratory tract. Poison. Other Acute Effects: may be fatal if inhaled, swallowed, or absorbed through skin. Chronic Effects: may alter genetic material; target organs: skin, lungs. Carcinogen.
Fire: Noncombustible. Hazards: emits toxic fumes. Extinguishing agents: noncombustible; use extinguishing media appropriate to surrounding fire conditions. Precautions: combustible liquid.
Reactivity: Incompatible with: heat, acids, oxidizing agents, halogens, may decompose on exposure to moist air or water. Hazardous decomposition products: toxic fumes of: arsenic oxides sulfur oxides hydrogen sulfide gas.
Carcinogenicity: IARC - Group 1, Carcinogenic to humans; NIOSH - Listed as carcinogen; NTP - Class 1, Known to be a carcinogen; ACGIH - Class A1, Confirmed human carcinogen; OSHA - Listed as a carcinogen; EPA - Class A, Human carcinogen; MAK - Class A1, Capable of inducing malignant tumors as shown by experience with humans

Primary Target Organs:

Eyes

Skin

Respiratory System

Exposure Limits
OSHA PEL: TWA: 0.01 mg/m^3; as As.
ACGIH TLV: TWA: 0.01 mg/m^3; as As.
NIOSH REL: STEL: 0.002 mg/m^3; ceiling (15 min) as As.
Respirator Recommendation
Exposure Range: >0.01 to <5 mg/m^3 Supplied Air, Constant Flow/Pressure Demand, Full Face
Exposure Range: 5 to unlimited mg/m^3 Self-contained Breathing Apparatus, Pressure Demand, Full Face
Note: as arsenic, inorganic compounds; refer to 29CFR 1910.1018 for more specific respirator recommendations

Environmental

Ecotoxicity: LC_{50} Pimephales promelas (fathead minnow) 82,400 ug/l/96 hr /Conditions of bioassay not specified LC_{50} Penaeus setiferus (white shrimp, juvenile) 24,700 ug/l/96 hr /Conditions of bioassay not specified LC_{50} Oncorhynchus keta (Chum salmon) 8330 ug/l/48 hr /Conditions of bioassay not specified LC_{50} White shrimp 500 mg/l/96 hr /conditions of bioassay not specified
Cleanup/Disposal: Guide No. 152: Do not touch damaged containers or spilled material unless wearing appropriate protective clothing. Stop leak if you can do it without risk. Prevent entry into waterways, sewers, basements or confined areas. Cover with plastic sheet to prevent spreading. Absorb or cover with dry earth, sand or other non-combustible material and transfer to containers. Do not get water inside containers.

Environmental Physical Data

BCF: biomagnification does not occur

Regulations

RCRA 40CFR: Not listed
CERCLA: 40CFR 302.4: Listed per CWA Section 311(b)(4) RQ: 1 lb (0.454 kg)
SARA 40CFR 372.65: Listed as Compound
SARA EHS 40CFR 355: Not listed
TSCA: Listed

ARS9040 CAS #: 7784-42-1

ARSINE

RTECS: CG6475000
DOT: UN2188; IMO2.3
EINECS Number: 232-066-3
Molecular Formula: $As H_3$
Structured MF: AsH_3
Formula Weight: 77.93
Synonyms: AGENT SA; ARSENIC HYDRID; ARSENIC HYDRIDE; ARSENIC TRIHYDRIDE; ARSENIURETTED HYDROGEN; ARSENOUS

HYDRIDE; ARSENOWODOR; ARSENWASSERSTOFF; HYDROGEN ARSENIDE

Description: colorless gas; garlic-like odor

Use: organic synthesis; military poison gas; doping agent for solid state electronic components

Physical Properties

Boiling Point: -55 °C (-67 °F)
Freezing Point: -117 °C (-178.6 °F)
Specific Gravity: Gas 2.695
Vapor Density: 2.7 Air=1
Density: 2.695 g/L
Vapor Pressure: 11000 mm Hg at 20 °C
Water Solubility: 20 ml/100 g Cold Water at 20 °C
Other Solubilities: 95% Ethanol: Slightly Soluble; Alkalies: Slightly Soluble; Benzene: Soluble; Chloroform: Soluble.
pH: Aqueous solution is neutral
Critical Temperature: 100 °C
Ionization Potential (eV): 9.89
Flash Point: Flammable
LEL: 5.1% v/v
UEL: 78% v/v

RTECS Toxicity Data

Acute Inhalation: Rat LC_{50} Dose: 390 mg/m^3/10M. Human LC_{Lo} Dose: 25 ppm/30M; Toxic Effects: Blood - Other hemolysis with or without anemia; Endocrine - Change in GH. Human LC_{Lo} Dose: 300 ppm/5M. Monkey LC_{Lo} Dose: 600 mg/m^3/1hr. Human TC_{Lo} Dose: 3 ppm; Toxic Effects: Blood - Pigmented or nucleated red blood cells; Blood - Other hemolysis with or without anemia. Man TC_{Lo} Dose: 325 ug/m^3; Toxic Effects: Gastrointestinal - Other changes; Kidney, ureter, and Bladder - Hematuria.

Hazard Overviews

Poison

Explosive

Flammable

Compressed Gas

Fire Diamond

Health: Poison. Also Causes: headache, malaise, generalized weakness, muscle cramps, dizziness, pale or bluish face, dyspnea, abdominal pain, nausea, vomiting, red staining of conjunctiva, cold sweat, trembling of arms and legs, convulsions, dark red urine, jaundice, frostbite, bone marrow depression, peripheral neuropathies, pulmonary edema, cardiovascular collapse, kidney failure. Chronic Effects: anemia, shortness of breath, weakness.

Fire: Explosive and flammable. For small fires, let burn unless leak can be stopped. For large fires, use water spray, fog, or regular foam. Apply cooling water to sides of containers until fire is out.

Reactivity: Stable, in closed, cool, dry containers; but can explode upon contact with dry, warm air or when exposed to shock.. Hazardous polymerization cannot occur. Avoid: exposure to heat; shock. Incompatible with: oxidizing materials; acids; halogens; potassium and ammonia; light and moisture. Hazardous decomposition products: arsenic fumes; arsenic oxide fumes.

Carcinogenicity: IARC - Group 1, Carcinogenic to humans; NIOSH - Listed as carcinogen; NTP - Class 1, Known to be a carcinogen; ACGIH - Class A1, Confirmed human carcinogen; OSHA - Listed as a carcinogen; EPA - Class A, Human carcinogen; MAK - Class A1, Capable of inducing malignant tumors as shown by experience with humans

Primary Target Organs:

Nervous System

Liver

Kidneys

Cardio-vascular

Blood

Bone

Exposure Limits
OSHA PEL: TWA: 0.05 ppm; 0.2 mg/m^3.
ACGIH TLV: TWA: 0.5 ppm; 0.16 mg/m^3.
NIOSH REL: STEL: 0.002 mg/m^3; 15-minute.
NIOSH IDLH: 3 ppm.
DFG MAK: TWA: 0.05 ppm; 0.2 mg/m^3.

Respirator Recommendation
Exposure Range: >0.05 to <3 ppm Supplied Air, Constant Flow/Pressure Demand, Full Face
Exposure Range: 3 to unlimited ppm Self-contained Breathing Apparatus, Pressure Demand, Full Face

Environmental

Cleanup/Disposal: Guide No. 119: Eliminate all ignition sources (no smoking, flares, sparks or flames in immediate area). All equipment used when handling the product must be grounded. Fully encapsulating, vapor protective clothing should be worn for spills and leaks with no fire. Do not touch or walk through spilled material. Stop leak if you can do it without risk. Do not direct water at spill or source of leak. Use water spray to reduce vapors or divert vapor cloud drift. For chlorosilanes, use AFFF alcohol-resistant medium expansion foam to reduce vapors. If possible, turn leaking containers so that gas escapes rather than liquid. Prevent entry into waterways, sewers, basements or confined areas. Isolate area until gas has dispersed.

Regulations

RCRA 40CFR: Not listed
CERCLA: 40CFR 302.4: Listed as Compound per CWA Section 307(a) per CAA Section 112
SARA 40CFR 372.65: Not listed TPQ: 100 lb
SARA EHS 40CFR 355: Listed TPQ: 100 lb
TSCA: Listed

Analytical Methods

Air: ASTM D4490
Indoor / Expired Air: NIOSH 6001

ASB1000 CAS #: 1332-21-4

ASBESTOS

RTECS: CI6475000
Molecular Formula: Unspecified or Variable

Structured MF: Hydrated mineral silicates
Formula Weight: Varies
Synonyms: ACTINOLITE; ACTINOLITE ASBESTOS; AMIANTHUS; AMOSITE; AMOSITE (CUMMINGTONITE-GRUNERITE); AMPHIBOLE; ANTHOPHYLLITE; ANTHOPHYLLITE ASBESTOS; ASBEST; ASBESTOS DUST; ASBESTOS FIBER; ASBESTOS FIBRE; ASBESTOSE; ASCARITE; CHRYSOTILE; CROCIDOLITE (RIEBECKITE); FIBROUS GRUNERITE; TREMOLITE; TREMOLITE ASBESTOS
Description: white, greenish, blue, gray-green fibrous solid; odorless
Use: in asbestos cement for pipes, ducts, construction applications; fire resistant textiles, friction materials (ie, brake linings); inert filler; pigment in coatings & sealants; in elastomers; fire & rot resisting material in felts; in paper; thermal & electrical insulation; industrial talcs& greases; in taping cmpd; additive to metals; valve, flange, & pump components; clutch/transmission components; automotive/truck body coatings; electronic motor components; table pads & heat protective mats; molten glass handling equipment; electrical switchboards; laboratory furniture.

Physical Properties

Boiling Point: Decomposes
Freezing Point: Decomposes at 600 °C (1112 °F)
Specific Gravity: 3.0 to 3.3
Vapor Pressure: ~ 0 mm Hg
Water Solubility: Insoluble
Flash Point: Noncombustible

RTECS Toxicity Data

Acute Inhalation: Human TC_{Lo} Dose: 1.2 fb/cc/19Y-C; Toxic Effects: Lungs, Thorax, or Respiration - Pleural effusion; Lungs, Thorax, or Respiration - Dyspnea; Lungs, Thorax, or Respiration - Sputum.
Mutagenic: Bacteria - E Coli Mutations in Microorganisms; Dose: 10 mg/plate (+/-S9).
Tumorigenic: Rat Route: Implant; Dose: 750 mg/kg; Toxic Effects: Tumorigenic - Equivocal tumorigenic agent by RTECS criteria; Tumorigenic - Tumors at site of application. Mouse Route: Intraperitoneal; Dose: 80 mg/kg; Toxic Effects: Tumorigenic - Equivocal tumorigenic agent by RTECS criteria; Tumorigenic - Tumors at site of application.

Hazard Overviews

Fire Diamond

Health: Irritating to eyes/respiratory system at high levels. Chronic Effects: dangerous chronic hazard, can cause lung cancer, asbestosis, and mesothelioma.
Fire: Noncombustible. Use extinguishing agents suitable for surrounding fire.
Reactivity: Stable. Hazardous polymerization cannot occur. Incompatible with: strong acids; glacial acetic acid; hot water.

Carcinogenicity: IARC - Group 1, Carcinogenic to humans; NIOSH - Listed as carcinogen; NTP - Class 1, Known to be a carcinogen; ACGIH - Class A1, Confirmed human carcinogen; OSHA - Listed as a carcinogen; EPA - Class A, Human carcinogen; MAK - Class A1, Capable of inducing malignant tumors as shown by experience with humans
Primary Target Organs:

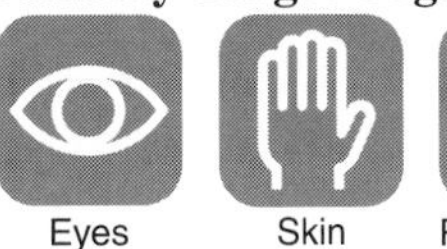
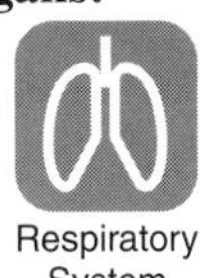
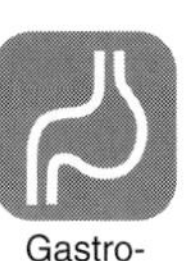

Eyes | Skin | Respiratory System | Gastro-intestinal

Exposure Limits
ACGIH TLV: TWA: 0.5 mg/m^3.
Respirator Recommendation
Exposure Range: >0.1 to 1 f/cc Air Purifying, Negative Pressure, Half Mask
Exposure Range: >1 to 5 f/cc Air Purifying, Negative Pressure, Full Face
Exposure Range: >5 to 10 f/cc Powered Air Purifying Respirator, Half or Full Facepiece or Hood
Exposure Range: >10 to 100 f/cc Supplied Air, Constant Flow/Pressure Demand, Full Face
Exposure Range: >100 to unlimited f/cc Supplied Air, Pressure Demand, with Auxiliary SCBA
Cartridge Color: magenta (P100)

Environmental

Environmental Physical Data

BCF: no evidence

Regulations

RCRA 40CFR: Not listed
CERCLA: 40CFR 302.4: Listed per CWA Section 307(a) per CAA Section 112 RQ: 1 lb (0.454 kg)
SARA 40CFR 372.65: Listed
SARA EHS 40CFR 355: Not listed
TSCA: Listed

Analytical Methods

Soil: NIOSH 9002
Drinking Water: APHA 2570-B
Indoor / Expired Air: NIOSH 7400, 7402; ASTM D4240

ASB5000 CAS #: 12001-29-5

ASBESTOS, CHRYSOTILE

RTECS: CI6478500
DOT: UN2590; IMO9.3
Formula Weight: N/A
Synonyms: 5R04; 5RO4; 7-45 ASBESTOS; ASBESTOS; AVIBEST C; CALIDRIA RG 100; CALIDRIA RG 144; CALIDRIA RG 600; CASSIAR AK; CHRYSOTILE; CHRYSOTILE ASBESTOS; HOOKER NO 1 CHRYSOTILE ASBESTOS; HOOKER NO. 1 CHRYSOTILE ASBESTOS; K 6-30; K6-30; METAXITE; PLASTIBEST 20; RG 600; SERPENTINE; SERPENTINE CHRYSOTILE; SYLODEX; WHITE ASBESTOS; WHITE ASBESTOS,(CHRYSOTILE,ACTINOLITE,ANTHOPHYLLITE,TREMOL ITE)

Use: in cement products, floor tile, paper products, paint & caulking, textiles, plastics; in packings, woven brake linings to clutch facings, & electric insulation; in asbestos cement, roofing products, coatings & cmpds, gaskets, thermal insulation products, & textiles; fireproofing agent; reinforcing agent in cement, paper & plastic products; filtration & reinforcement applications.

Physical Properties

Freezing Point: Decomposes
Density: 2.19 to 2.25
Water Solubility: Dispersible

RTECS Toxicity Data

Acute Inhalation: Human TC_{Lo} Dose: 2.8 fb/cc/5Y; Toxic Effects: Lungs, Thorax, or Respiration - Fibrosis, focal (pneumoconiosis); Lungs, Thorax, or Respiration - Cough; Lungs, Thorax, or Respiration - Dyspnea.
Mutagenic: Human Unscheduled DNA Synthesis; Cell Type: lung; Dose: 10 mg/L. Human Cytogenetic Analysis; Cell Type: lymphocyte; Dose: 10 mg/L. Human Other Mutation Test Systems; Cell Type: fibroblast; Dose: 10 mg/L.
Tumorigenic: Man Route: Inhalation; Dose: 400 mppcf/1Y-C; Toxic Effects: Tumorigenic - Carcinogenic by RTECS criteria; Lungs, Thorax, or Respiration - Fibrosis, focal (pneumoconiosis); Lungs, Thorax, or Respiration - Tumors. Rat Route: Oral; Dose: 7100 mg/kg/39W-C; Toxic Effects: Tumorigenic - Carcinogenic by RTECS criteria; Liver - Tumors; Kidney, Ureter, and Bladder - Kidney tumors. Rat Route: Inhalation; Dose: 11 mg/m^3/26W-I; Toxic Effects: Tumorigenic - Carcinogenic by RTECS criteria; Lungs, Thorax, or Respiration - Tumors. Rat Route: Inhalation; Dose: 12 mg/m^3/13W-I; Toxic Effects: Tumorigenic - Neoplastic by RTECS criteria; Lungs, Thorax, or Respiration - Tumors. Rat Route: Inhalation; Dose: 10 mg/m^3/52W-C; Toxic Effects: Tumorigenic - Equivocal tumorigenic agent by RTECS criteria; Lungs, Thorax, or Respiration - Tumors. Rat Route: Inhalation; Dose: 11 mg/m^3/8H/26W-I; Toxic Effects: Tumorigenic - Carcinogenic by RTECS criteria; Lungs, Thorax, or Respiration - Tumors.

Hazard Overviews

Reactivity: Stable. Hazardous polymerization cannot occur. Incompatible with: strong acids; glacial acetic acid; hot water.
Carcinogenicity: IARC - Group 1, Carcinogenic to humans; NIOSH - Listed as carcinogen; NTP - Class 1, Known to be a carcinogen; ACGIH - Class A1, Confirmed human carcinogen; OSHA - Listed as a carcinogen; EPA - Class A, Human carcinogen; MAK - Class A1, Capable of inducing malignant tumors as shown by experience with humans
Exposure Limits
ACGIH TLV: TWA: 2 mg/m^3; f/cc.
DFG MAK: TWA: 0.25 mg/m^3.
Respirator Recommendation
Exposure Range: >0.1 to 1 f/cc Air Purifying, Negative Pressure, Half Mask
Exposure Range: >1 to 5 f/cc Air Purifying, Negative Pressure, Full Face
Exposure Range: >5 to 10 f/cc Powered Air Purifying Respirator, Half or Full Facepiece or Hood
Exposure Range: >10 to 100 f/cc Supplied Air, Constant Flow/Pressure Demand, Full Face
Exposure Range: >100 to unlimited f/cc Supplied Air, Pressure Demand, with Auxiliary SCBA
Cartridge Color: magenta (P100)

Environmental

Cleanup/Disposal: Guide No. 171: Do not touch or walk through spilled material. Stop leak if you can do it without risk. Prevent dust cloud. Avoid inhalation of asbestos dust. Small Dry Spills: With clean shovel place material into clean, dry container and cover loosely; move containers from spill area. Small Spills: Take up with sand or other noncombustible absorbent material and place into containers for later disposal. Large Spills: Dike far ahead of liquid spill for later disposal. Cover powder spill with plastic sheet or tarp to minimize spreading. Prevent entry into waterways, sewers, basements or confined areas.

Regulations

RCRA 40CFR: Not listed
CERCLA: 40CFR 302.4: Not listed
SARA 40CFR 372.65: Not listed
SARA EHS 40CFR 355: Not listed
TSCA: Not listed

Analytical Methods

Soil: NIOSH 9000

ASB9000 **CAS #: 12001-28-4**

ASBESTOS, CROCIDOLITE

RTECS: CI6479000
Molecular Formula: $Fe_3H_2Na_2O_{45}$
Formula Weight: 765.98
Synonyms: AMORPHOUS CROCIDOLITE ASBESTOS; ASBESTOS; BLUE ASBESTOS; BROWN ASBESTOS; CROCIDOLITE ASBESTOS; FIBROUS CROCIDOLITE ASBESTOS; KROKYDOLITH

Physical Properties

Freezing Point: Decomposes
Water Solubility: Insoluble

RTECS Toxicity Data

Mutagenic: Human DNA Damage; Cell Type: leukocyte; Dose: 50 mg/L. Human Mutations in Mammalian Somatic Cells; Cell Type: other cell types; Dose: 50 mg/L. Human Other Mutation Test Systems; Cell Type: fibroblast; Dose: 10 mg/L.
Tumorigenic: Rat Route: Inhalation; Dose: 11 mg/m^3/1Y-I; Toxic Effects: Tumorigenic - Carcinogenic by RTECS criteria; Lungs, Thorax, or Respiration - Tumors. Rat Route: Inhalation; Dose: 12 mg/m^3/13W-I; Toxic Effects:

Tumorigenic - Neoplastic by RTECS criteria; Lungs, Thorax, or Respiration - Tumors.

Hazard Overviews

Reactivity: Stable. Hazardous polymerization cannot occur. Incompatible with: strong acids; glacial acetic acid; hot water.
Carcinogenicity: IARC - Group 1, Carcinogenic to humans; NIOSH - Listed as carcinogen; NTP - Class 1, Known to be a carcinogen; ACGIH - Class A1, Confirmed human carcinogen; OSHA - Listed as a carcinogen; EPA - Class A, Human carcinogen; MAK - Class A1, Capable of inducing malignant tumors as shown by experience with humans
Exposure Limits
ACGIH TLV: TWA: 0.2 mg/m^3; f/cc.
Respirator Recommendation
Exposure Range: >0.1 to 1 f/cc Air Purifying, Negative Pressure, Half Mask
Exposure Range: >1 to 5 f/cc Air Purifying, Negative Pressure, Full Face
Exposure Range: >5 to 10 f/cc Powered Air Purifying Respirator, Half or Full Facepiece or Hood
Exposure Range: >10 to 100 f/cc Supplied Air, Constant Flow/Pressure Demand, Full Face
Exposure Range: >100 to unlimited f/cc Supplied Air, Pressure Demand, with Auxiliary SCBA
Cartridge Color: magenta (P100)

Environmental

Regulations

RCRA 40CFR: Not listed
CERCLA: 40CFR 302.4: Not listed
SARA 40CFR 372.65: Not listed
SARA EHS 40CFR 355: Not listed
TSCA: Not listed

ASC3000 CAS #: 50-81-7

ASCORBIC ACID

RTECS: CI7650000
EINECS Number: 200-066-2
Molecular Formula: $C_6H_8O_6$
Formula Weight: 176.12

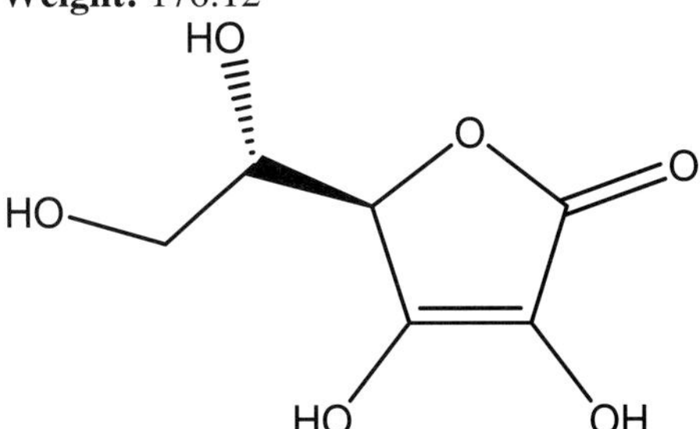

Chemical Structure

Synonyms: AA; ADENEX; ALLERCORB; ANTISCORBIC VITAMIN; ANTISCORBUTIC VITAMIN; ARCO-CEE; ASCOLTIN; ASCOR-B.I.D; ASCORB; ASCORBAJEN; ASCORBATE; L(+)-ASCORBIC ACID; L-(+)-ASCORBIC ACID; L-ASCORBIC ACID; ASCORBICAB; ASCORBICAP; ASCORBIN; ASCORBUTINA; ASCORIN; ASCORTEAL; ASCORVIT; C-LEVEL; C-LONG; C-QUIN; C-SPAN; C-VIMIN; CANTAN; CANTAXIN; CATAVIN C; CE LENT; CEBICURE; CEBID; CEBION; CEBIONE; CECON; CEE-CAPS TD; CEE-VITE; CEGIOLAN; CEGLION; CELASKON; CELIN; CEMAGYL; CE-MI-LIN; CEMILL; CENETONE; CENOLATE; CEREON; CERGONA; CESCORBAT; CETAMID; CETANE; CETANE-CAPS TD; CETEMICAN; CEVALIN; CEVATINE; CEVEX; CEVI-BID; CEVIBID; CEVIMIN; CE-VI-SOL; CEVITAL; CEVITAMIC ACID; CEVITAMIN; CEVITAN; CEVITEX; CEWIN; CIAMIN; CIPCA; CITRISCORB; COLASCOR; CONCEMIN; DAVITAMON C; DORA-C-500; DUOSCORB; HICEE; HYBRIN; IDO-C; 3-KETO-L-GULOFURANOLACTONE; L-3-KETOTHREOHEXURONIC ACID LACTONE; KYSELINA ASKORBOVA; LAROSCORBINE; LEMASCORB; LIQUI-CEE; L-LYXOASCORBIC ACID; MERI-C; NATRASCORB; NATRASCORB INJECTABLE; NSC 33832; 3-OXO-L-GULOFURANOLACTONE; 3-OXO-L-GULOFURANOLACTONE (ENOL FORM); PLANAVIT C; PROSCORBIN; REDOXON; RIBENA; ROSCORBIC; SCORBACID; SCORBU-C; SECORBATE; TESTASCORBIC; L-THREO-HEX-2-ENONIC ACID,GAMMA-LACTONE; VICELAT; VICIN; VICOMIN C; VIFORCIT; VISCORIN; VITACE; VITACEE; VITACIMIN; VITACIN; VITAMIN C; VITAMISIN; VITASCORBOL; XITIX; L-XYLOASCORBIC ACID
Description: white, yellow crystals, needles
Use: in nutrition, color fixing, flavorings and preservatives in meats and other foods; as an oxidant in bread doughs, in abscission of citrus fruit in harvesting, as a reducing agent in analytical chemistry, as an antimicrobial and antioxidant in foodstuffs, in the treatment of vitamin C deficiency, and in veterinary medicine to treat vitamin C deficiency in primates, guinea pigs and fish; a dietary supplement and chemical preservative; cures scurvy and increases resistance to infection It acts as an oxidation-reduction catalyst in the cell

Physical Properties

Boiling Point: Decomposes
Freezing Point: 190 °C (374 °F) to 192 °C (377.6 °F)
Specific Gravity: 1.65
Vapor Pressure: 7.9179 Pa at 465.15 °K
Water Solubility: 80% at 100 °C
Other Solubilities: 0.033 in 95 wt% Ethanol, 0.02 in absolute Ethanol, 0.01 in Glycerol USP, 0.05 in Propylene Glycol.
Surface Tension: 4.039 $x10^{-2}$ N/m
pH: 5 mg/mL 3
Refraction Index: 1.5101 to 1.5204 at 25 °C/D
Critical Temperature: 510 °C
Critical Pressure: 5.29 $x10^6$ Pa
Flash Point: Not available; probably combustible
Autoignition Temperature: 660 °C

RTECS Toxicity Data

Acute Oral: Rat LD_{50} Dose: 11900 mg/kg; Toxic Effects: Sense organs and special senses - Lacrimation; Behavioral - Somnolence (general depressed activity); Gastrointestinal - Hypermotility, diarrhea. Mouse LD_{50} Dose: 3367 mg/kg.
Acute Dermal: Rat LD_{50} Route: Subcutaneous Dose: >10 gm/kg.
Chronic (Multiple Dose) Oral: Mouse Dose: 546 gm/kg/13W-I; Toxic Effects: DEATH.
Reproductive/Teratogenic: Rat Route: Oral; Dose: 2500 mg/kg; Duration: female 1-22D of pregnancy; Effects on Fertility - Post-implantation mortality. Mouse Route:

Intravenous; Dose: 800 mg/kg; Duration: female 8D of pregnancy; Specific Developmental Abnormalities - Central nervous system; Musculoskeletal system.

Mutagenic: Human DNA Damage; Cell Type: fibroblast; Dose: 200 umol/L. Human DNA Damage; Cell Type: other cell types; Dose: 200 umol/L. Human DNA Inhibition; Cell Type: HeLa cell; Dose: 2500 umol/L. Human DNA Inhibition; Cell Type: other cell types; Dose: 200 umol/L. Human DNA Inhibition; Cell Type: other cell types; Dose: 200 mg/L. Human Other Mutation Test Systems; Cell Type: fibroblast; Dose: 200 umol/L. Human Other Mutation Test Systems; Cell Type: other cell types; Dose: 200 umol/L.

Hazard Overviews

Fire Diamond

Health: Mildly irritating to eyes/respiratory tract.

Fire: Will burn. Slight fire hazard when exposed to heat and sparks. Use water fog, dry chemical, alcohol foam, or carbon dioxide to fight fires involving ascorbic acid. Use a water spray to cool fire-exposed tanks or containers. Water or foam may cause frothing.

Reactivity: Stable. Hazardous polymerization cannot occur. Avoid: direct exposure to heat; sparks; open flame; lighted tobacco products; aqueous solutions are rapidly oxidized by air. Incompatible with: strong oxidizers. Hazardous decomposition products: carbon monoxide; carbon dioxide.

Carcinogenicity: IARC - Not listed; NIOSH - Not listed; NTP - Not listed; ACGIH - Not listed; OSHA - Not listed; EPA - Not listed; MAK - Not listed

Primary Target Organs:

Eyes

Respiratory System

Environmental

Environmental Physical Data

Octanol/Water Partition Coefficient: log K_{ow} = -2.15

Regulations

RCRA 40CFR: Not listed
CERCLA: 40CFR 302.4: Not listed
SARA 40CFR 372.65: Not listed
SARA EHS 40CFR 355: Not listed
TSCA: Listed

ASC6000 CAS #: 137-66-6

ASCORBYL PALMITATE

RTECS: CI7671040
EINECS Number: 205-305-4
Molecular Formula: $C_{22}H_{38}O_7$
Formula Weight: 414.54

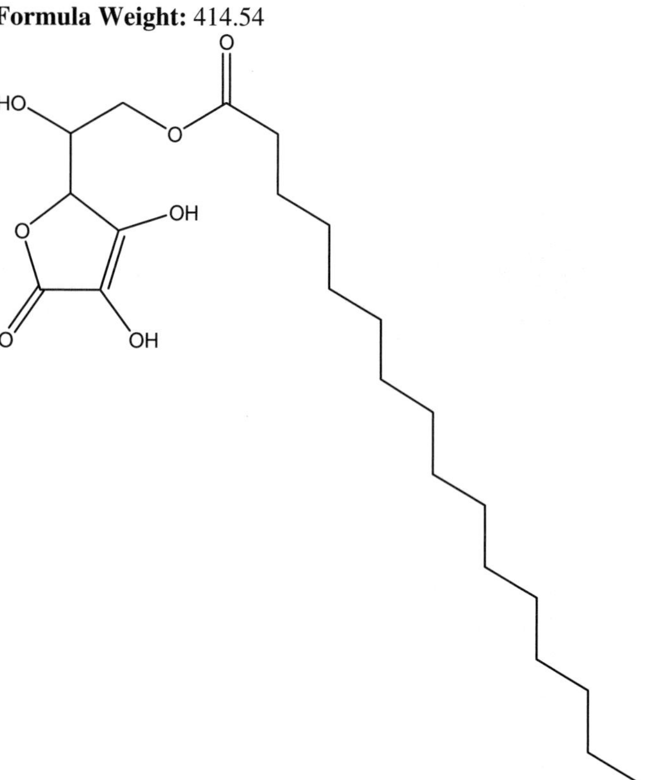

Chemical Structure

Synonyms: ASCORBIC ACID PALMITATE; ASCORBIC ACID PALMITATE (ESTER); L-ASCORBIC ACID,6-HEXADECANOATE; L-ASCORBIC ACID,6-PALMITATE; ASCORBIC PALMITATE; ASCORBYL MONOPALMITATE; L-ASCORBYL MONOPALMITATE; L-ASCORBYL 6-PALMITATE; L-ASCORBYL PALMITATE; ASCORBYLPALMITIC ACID; CETYL ASCORBATE; 6-HEXADECANOYL-L-ASCORBIC ACID; 6-MONOPALMITOYL-L-ASCORBATE; ONDASCORA; PALMITOYL L-ASCORBIC ACID; 6-O-PALMITOYLASCORBIC ACID; 6-PALMITOYLASCORBIC ACID; QUICIFAL

Description: white or yellowish white powder; citrus like odor

Use: chemical preservative food additive; antioxidant used in pharmaceuticals; antioxidant synergist in fats & oils; source of vitamin C; stabilizer; emulsifier

Physical Properties

Freezing Point: 116 °C (240.8 °F)
Water Solubility: Slightly Soluble in Water
Other Solubilities: Soluble in Alcohol, animal oil and vegetable oil

RTECS Toxicity Data

Acute Oral: Mouse LD_{50} Dose: 25 gm/kg.

Hazard Overviews

Health: May cause irritation to eyes/skin/respiratory tract. Other Acute Effects: may be harmful by inhalation, ingestion, or skin absorption.

Fire: Hazards: emits toxic fumes. Extinguishing agents: water spray; carbon dioxide, dry chemical powder or appropriate foam. Precautions: combustible liquid.

Reactivity: Stable. Hazardous polymerization will not occur. Incompatible with: strong oxidizing agents. Hazardous decomposition products: toxic fumes of: carbon monoxide, carbon dioxide.

Carcinogenicity: IARC - Not listed; NIOSH - Not listed; NTP - Not listed; ACGIH - Not listed; OSHA - Not listed; EPA - Not listed; MAK - Not listed

Environmental

Regulations

RCRA 40CFR: Not listed
CERCLA: 40CFR 302.4: Not listed
SARA 40CFR 372.65: Not listed
SARA EHS 40CFR 355: Not listed
TSCA: Listed

ASP1000 CAS #: 22839-47-0

ASPARTAME

RTECS: WM3407000
EINECS Number: 245-261-3
Molecular Formula: $C_{14}H_{18}N_2O_5$
Formula Weight: 294.30

Chemical Structure

Synonyms: 3-AMINO-N-(ALPHA-CARBOXYPHENETHYL)SUCCINAMIC ACID N-METHYLESTER; 3-AMINO-N-(ALPHA-CARBOXYPHENETHYL)SUCCINAMIC ACID N-METHYLESTER,STEREOISOMER; 3-AMINO-N-(ALPHA-METHOXYCARBONYLPHENETHYL) SUCCINAMIC ACID; APM; ASPARTYLPHENYLALANINE METHYL ESTER; L-ASPARTYL-L-PHENYLALANINE METHYL ESTER; N-L-ALPHA-ASPARTYL-L-PHENYLALANINE 1-METHYL ESTER; N-L-ALPHA-ASPARTYL-L-PHENYLALANINE,1-METHYL ESTER; CANDEREL; DIPEPTIDE SWEETENER; EQUAL; METHYL ASPARTYLPHENYLALANATE; METHYL N-L-ALPHA-ASPARTYL-L-PHENYLALANINATE; 1-METHYL N-L-ALPHA-ASPARTYL-L-PHENYLALANINE; NUTRASWEET; L-PHENYLALANINE,N-L-ALPHA-ASPARTYL-,1-METHYL ESTER; L-PHENYLALANINE,N-L-ALPHA-ASPARTYL-,1-METHYL ESTER (9CI); SC 18862; SUCCINAMIC ACID,3-AMINO-N-(ALPHA-CARBOXYPHENETHYL)-,N-METHYL ESTER,STEREOISOMER; SWEET DIPEPTIDE; TRI-SWEET

Description: colorless to off-white crystalline powder or needles; odorless

Use: sweetening agent about 200 times as sweet as sucrose; table sweetener for coffee, tea, breakfast cereals, etc; other food applications identified incl presweetened breakfast cereals, chewing gum, and dry bases for beverages such as instant coffee and tea, gelatins, puddings and fillings, and dairy-product analogue whipped toppings

Physical Properties

Freezing Point: 246 °C (474.8 °F) to 247 °C (476.6 °F)
Water Solubility: 18.2 mg/ml at 25 °C pH=3.72
Other Solubilities: More Soluble in acidic solutions and hot water; slightly Soluble in Alcohol; very slightly Soluble in Chloroform; practically Insoluble in oils.
pH: 0.8% solution in water about 5.3

RTECS Toxicity Data

Acute Oral: Woman TD_{Lo} Dose: 3710 ug/kg; Toxic Effects: Skin and appendages - Dermatitis, allergic.

Reproductive/Teratogenic: Rat Route: Oral; Dose: 275 gm/kg; Duration: male 2W prior to mating; Effects on Newborn - Behavioral. Rat Route: Oral; Dose: 106 gm/kg; Duration: female 12D after birth; Effects on Newborn - Growth statistics. Rat Route: Oral; Dose: 449 gm/kg; Duration: male 2W prior to mating; Effects on Newborn - Weaning or lactation index.

Hazard Overviews

Health: Irritating to eyes/skin. Harmful. Other Acute Effects: may be harmful by inhalation, ingestion, or skin absorption; prolonged or repeated exposure may cause allergic reactions in certain sensitive individuals. Chronic Effects: can release phenylalanine.

Fire: Hazards: emits toxic fumes. Extinguishing agents: water spray; carbon dioxide, dry chemical powder or appropriate foam. Precautions: combustible liquid.

Reactivity: Incompatible with: strong oxidizing agents. Hazardous decomposition products: toxic fumes of: carbon monoxide, carbon dioxide, nitrogen oxides.

Carcinogenicity: IARC - Not listed; NIOSH - Not listed; NTP - Not listed; ACGIH - Not listed; OSHA - Not listed; EPA - Not listed; MAK - Not listed

Primary Target Organs:

Eyes

Skin

Environmental

Environmental Fate: It has been found to be unstable and decomposes to diketopiperazine in water at pH's 7 and above. Therefore, the compound may be unstable in moist soils at these pH's. The structure would suggest that it biodegrades in soil and water, but data on biodegradation are not available. Based on an estimated log K_{oc} value, the adsorption to suspended solids and sediments in water and to soil should be unimportant. Based on an estimated Henry's Law constant, the volatilization from water is not expected to be an important fate process. The estimated bioconcentration factor indicates that bioconcentration in aquatic organisms should not be an important fate process. If present in the atmosphere in the vapor phase, it will react with photochemically produced hydroxyl radicals (estimated half-life 3.6 hours).

Environmental Physical Data

Henry's Law Constant: estimated at 6.36 $x10^{-21}$
Sorption Partition Coefficient: log K_{OC} = estimated at 1.44
BCF: calculated at 3.4

Regulations

RCRA 40CFR: Not listed
CERCLA: 40CFR 302.4: Not listed
SARA 40CFR 372.65: Not listed
SARA EHS 40CFR 355: Not listed
TSCA: Listed

Analytical Methods

Indoor / Expired Air: NIOSH 5031

ASP3000 CAS #: 56-84-8

(L)-ASPARTIC ACID

RTECS: CI9098500
EINECS Number: 200-291-6
Molecular Formula: $C_4H_7NO_4$
Formula Weight: 133.10

HO O OH O NH2

Chemical Structure

Synonyms: (S)-AMINOBUTANEDIOIC ACID; AMINOSUCCINIC ACID; L-AMINOSUCCINIC ACID; ASPARAGIC ACID; L-ASPARAGIC ACID; ASPARAGINIC ACID; L-ASPARAGINIC ACID; ASPARAGINSAURE; (S)-ASPARTIC ACID; ASPARTIC ACID; L-(+)-ASPARTIC ACID; L-ASPARTIC ACID; BUTANEDIOIC ACID,AMINO-,(S)-
Description: colorless crystals, leaflets or rods
Use: biological & clinical studies; prepn of culture media; org intermediate; ingredient of aspartame; detergents; fungicides; germicides; metal complexation aspartic acid; medical use: parenteral nutrition; in flavor reaction products; pharmaceuticals formulations

Physical Properties

Boiling Point: Decomposes at 324 °C (615 °F)
Freezing Point: 270 °C (518 °F) to 271 °C (519.8 °F)
Specific Gravity: 1.661 at 12.5 °C
Water Solubility: 1 g in 2222 ml Water at 20 °C
Other Solubilities: Insoluble in Ethanol, Ethyl Ether, Benzene.

RTECS Toxicity Data

Chronic (Multiple Dose) Oral: Rat Dose: 25079 mg/kg/7D-C; Toxic Effects: Liver - Other changes; Biochemical - Lipids including transport.

Hazard Overviews

Health: May be irritating to eyes/skin. Harmful. Other Acute Effects: may be harmful by inhalation, ingestion, or skin absorption. Chronic Effects: possible risk of irreversible effects; possible mutagen; laboratory experiments have shown mutagenic effects.
Fire: Hazards: emits toxic fumes. Extinguishing agents: water spray; carbon dioxide, dry chemical powder or appropriate foam. Precautions: combustible liquid.
Reactivity: Incompatible with: strong oxidizing agents. Hazardous decomposition products: toxic fumes of: carbon monoxide, carbon dioxide, nitrogen oxides.
Carcinogenicity: IARC - Not listed; NIOSH - Not listed; NTP - Not listed; ACGIH - Not listed; OSHA - Not listed; EPA - Not listed; MAK - Not listed

Environmental

Environmental Physical Data

Octanol/Water Partition Coefficient: log K_{ow} = -3.89

Regulations

RCRA 40CFR: Not listed
CERCLA: 40CFR 302.4: Not listed
SARA 40CFR 372.65: Not listed
SARA EHS 40CFR 355: Not listed
TSCA: Listed

ASP6000 CAS #: 8052-42-4

ASPHALT

RTECS: CI9900000
DOT: UN1999; NA1999; IMO3.0; IMO3.2; IMO3.3
EINECS Number: 232-490-9
Molecular Formula: Unspecified or Variable
Synonyms: ASPHALT CEMENTS; ASPHALT (CUT); ASPHALT FUMES; ASPHALT: ASPHALTUM; ASPHALT,AT OR ABOVE ITS FP; ASPHALTIC BITUMEN; ASPHALT,LIQUID MEDIUM-CURING; ASPHALT,LIQUID RAPID-CURING; ASPHALT,LIQUID SLOW-CURING; ASPHALT,PETROLEUM; ASPHALTUM; BITUMEN; BITUMEN (EUROPEAN TERM); BITUMENS,ASPHALT; BITUMINOUS MATERIALS,ASPHALT; EPA PESTICIDE CHEMICAL CODE 022001; EPA PESTICIDE CHEMICAL CODE 022002; JUDEAN PITCH; MINERAL PITCH; MINERAL RUBBER (VAN); PETROLEUM ASPHALT; PETROLEUM BITUMEN; PETROLEUM PITCH; PETROLEUM REFINING RESIDUES,ASPHALTS; PETROLEUM ROOFING TAR; ROAD ASPHALT; ROAD TAR; ROOFING ASPHALT; TRINIDAD PITCH
Description: brown to black liquid; weak aromatic odor
Use: making roads, roofs, watertight tanks; special paints; adhesive in electrical laminates & hot-melt compositions;

diluent in low-grade rubber products; fluid loss control in oil wells; medium for radioactive waste disposal; pipeline & underground cable coating; rust-preventive hot-dip coatings; base for synthetic turf; water-retaining barrier for sandy soils; supporter of rapid bacterial growth in converting petro components to protein; sound insulation; asphalt blocks.

Physical Properties

Boiling Point: > 371 °C (700 °F)
Freezing Point: -7 °C (19.4 °F) to 43 °C (109.4 °F)
Specific Gravity: 1 to 1.18
Vapor Pressure: Saturated vapor pressure 0.018 lb/sq in at 21 °C
Water Solubility: Insoluble in Water
Other Solubilities: Soluble in Carbon Disulfide.
Surface Tension: 70 dynes/cm
Flash Point: > 204 °C Closed Cup
Autoignition Temperature: 485 °C

RTECS Toxicity Data

Mutagenic: Mouse DNA Adduct; Route: Skin; Dose: 600 mg/kg.
Tumorigenic: Rat Route: Intramuscular; Dose: 5400 mg/kg/24W-I; Toxic Effects: Tumorigenic - Neoplastic by RTECS criteria; Tumorigenic - Facilitates action of known carcinogens. Mouse Route: Skin; Dose: 130 gm/kg/81W-I; Toxic Effects: Tumorigenic - Carcinogenic by RTECS criteria; Lungs, Thorax, or Respiration - Tumors; Skin and appendages - Tumors. Mouse Route: Skin; Dose: 69 gm/kg/43W-I; Toxic Effects: Tumorigenic - Equivocal tumorigenic agent by RTECS criteria; Lungs, Thorax, or Respiration - Tumors; Skin and appendages - Tumors. Mouse Route: Skin; Dose: 905 gm/kg/2Y-I; Toxic Effects: Tumorigenic - Neoplastic by RTECS criteria; Lungs, Thorax, or Respiration - Tumors; Skin and appendages - Tumors.

Hazard Overviews

Fire Diamond

Health: Severe irritation to eyes/skin/respiratory tract. Also Causes: headache, nausea, nervousness, pulmonary edema, coma or death, thermal burn to eyes. Chronic Effects:dermatitis, acne-like lesions, mild keratosis, melanosis, photosensitization, chronic pneumonitis and bronchitis, conjunctiva, eyelid inflammation, weight loss, digestive disturbances, and lack of energy.
Fire: Will burn. Use foam, dry chemical, or carbon dioxide to fight fire. Do not extinguish unless flow can be completely stopped. Water or foam may cause frothing since asphalt is viscous.
Reactivity: Stable. Hazardous polymerization cannot occur. Avoid: contact with heat, flame, or fluorine. Incompatible with: naphtha; volatile solvents. Hazardous decomposition products: carbon monoxide; various aliphatic hydrocarbons; hydrogen sulfide.

Carcinogenicity: IARC - Group 2B, Possibly carcinogenic to humans; NIOSH - Listed as carcinogen; NTP - Not listed; ACGIH - Class A4, Not classifiable as a human carcinogen; OSHA - Not listed; EPA - Not listed; MAK - Class B, Justifiably suspected of having carcinogenic potential
Primary Target Organs:

Eyes

Skin

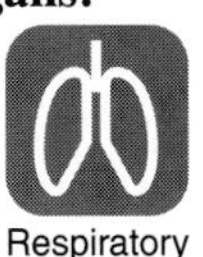
Respiratory System

Mucous Membranes

Exposure Limits
ACGIH TLV: TWA: 5 mg/m^3.
NIOSH REL: STEL: 5 mg/m^3; 15-minute.
Respirator Recommendation
Exposure Range: >5 to 50 mg/m^3 Air Purifying, Negative Pressure, Half Mask
Exposure Range: >50 to 500 mg/m^3 Air Purifying, Negative Pressure, Full Face
Exposure Range: >500 to 5000 mg/m^3 Supplied Air, Constant Flow/Pressure Demand, Full Face
Exposure Range: >5000 to unlimited mg/m^3 Self-contained Breathing Apparatus, Pressure Demand, Full Face
Cartridge Color: black
Note: use dust/mist prefilter

Environmental

Cleanup/Disposal: Guide No. 130: Eliminate all ignition sources (no smoking, flares, sparks or flames in immediate area). All equipment used when handling the product must be grounded. Do not touch or walk through spilled material. Stop leak if you can do it without risk. Prevent entry into waterways, sewers, basements or confined areas. A vapor suppressing foam may be used to reduce vapors. Absorb or cover with dry earth, sand or other non-combustible material and transfer to containers. Use clean non-sparking tools to collect absorbed material. Large Spills: Dike far ahead of liquid spill for later disposal. Water spray may reduce vapor; but may not prevent ignition in closed spaces.

Regulations

RCRA 40CFR: Not listed
CERCLA: 40CFR 302.4: Not listed
SARA 40CFR 372.65: Not listed
SARA EHS 40CFR 355: Not listed
TSCA: Listed

ASP9000 CAS #: 3244-90-4

ASPON

RTECS: XN4550000
EINECS Number: 221-817-0
Molecular Formula: $C_{12}H_{28}O_5P_2S_2$
Formula Weight: 378.46
Synonyms: A 42; ASP 51; BIS-O,O-DI-N-PROPYLPHOSPHOROTHIONIC ANHYDRIDE; E 8573; ENT 16,894; NPD; PROPYL THIOPYROPHOSPHATE; STAUFFER ASP-51; TETRA-

N-PROPYL DITHIONOPYROPHOSPHATE; O,O,O,O,-TETRA-N-PROPYL DITHIOPYROPHOSPHATE; O,O,O,O-TETRAPROPYL DITHIOPYROPHOSPHATE; TETRA-N-PROPYL DITHIOPYROPHOSPHATE; TETRAPROPYL DITHIOPYROPHOSPHATE; TETRAPROPYL ESTER; TETRAPROPYL THIODIPHOSPHATE; TETRAPROPYLDITHIODIFOSFAT; THIOPYROPHOSPHORIC ACID,TETRAPROPYL ESTER

Description: straw to amber-colored liquid; faint aromatic odor

Use: insecticide former use; acaricide former use; a non-systemic insecticide for the control of blissus leucopterus in turf former use

Physical Properties

Boiling Point: 170 °C (338 °F) at 1.0 mm Hg
Freezing Point: < 5 °C (41 °F)
Specific Gravity: 1.119 to 1.123 at 20 °C
Vapor Pressure: 13 mPa at 25 °C
Water Solubility: 30 mg/L at 20 °C
Other Solubilities: miscible with Acetone, Ethanol, Kerosene, 4-methylpentan-2-one, Xylene (Technical product, 93-96% pure).
Refraction Index: 1.471 at 21 °C/D
Flash Point: 149 °C Open Cup

RTECS Toxicity Data

Acute Oral: Rat LD_{50} Dose: 450 mg/kg. Chicken LD_{50} Dose: 436 mg/kg; Toxic Effects: Behavioral - Somnolence (general depressed activity); Behavioral - Convulsions or effect on seizure threshold. Behavioral - Ataxia.

Acute Dermal: Rabbit LD_{50} Route: Skin; Dose: 3830 mg/kg. Rat LD_{50} Route: Skin; Dose: 1800 mg/kg; Toxic Effects: Behavioral - Tremor; Behavioral - Convulsions or effect on seizure threshold; Behavioral - Excitment.

Hazard Overviews

Fire: Will burn.

Carcinogenicity: IARC - Not listed; NIOSH - Not listed; NTP - Not listed; ACGIH - Not listed; OSHA - Not listed; EPA - Not listed; MAK - Not listed

Environmental

Regulations

RCRA 40CFR: Not listed
CERCLA: 40CFR 302.4: Not listed
SARA 40CFR 372.65: Not listed
SARA EHS 40CFR 355: Not listed
TSCA: Not listed

Analytical Methods

Soil: SW846 8141, 8141A
Water / Groundwater: EPA 622.1; SW846 8141, 8141A
Food: FDA 212.1, 232.1, 232.4, 242.1
Plasma: EPA 001; FDA 211.1, 231.1, 252

AST5000 CAS #: 68844-77-9

ASTEMIZOLE

RTECS: DD8968000
EINECS Number: 272-441-9
Molecular Formula: $C_{28}H_{31}FN_4O$
Formula Weight: 458.6

F N NH N N O

Chemical Structure

Synonyms: ASTEMISON; ASTEMIZOL; ASTEMIZOLUM; 1H-BENZIMIDAZOL-2-AMINE,1-((4-FLUOROPHENYL)METHYL)-N-(1-(2-(4-METHOXYPHENYL)ETHL)-4-PIPERIDINYL)-; 1-(P-FLUOROBENZYL)-2-((1-(P-METHOXYPHENETHYL)-4-PIPERIDYL)AMINO)BENZIMIDAZOLE; 1-(P-FLUOROBENZYL)-2-((1-(2-(P-METHOXYPHENYL)ETHYL)PIPERID-4-YL)AMINO)BENZIMIDAZOLE; 1-((4-FLUOROPHENYL)METHYL)-N-(1-(2-(4-METHOXYPHENYL)ETHYL)-4-PIPERIDINYL)-1H-BENZIMIDAZOL-2-AMINE; HESTAZOL; HISAMANAL; HISMANAL; HISTAMEN; HISTAMINOS; KELP; LARIDAL; METODIH; NOVO-MASTIZOL A; PARALERGIN; R 42512; R 45312; RETOLEN; WAREEZOL

Description: crystals

Use: medication: antihistamine

Physical Properties

Freezing Point: 149.1 °C (300.38 °F)

RTECS Toxicity Data

Acute Oral: Woman TD_{Lo} Dose: 5400 ug/kg; Toxic Effects: Behavioral - General anesthetic; Cardiac - EKG changes not diagnostic of above; Cardiac - Pulse rate increased without fall in BP. Woman TD_{Lo} Dose: 4 mg/kg; Toxic Effects: Behavioral - Somnolence (general depressed activity). Woman TD_{Lo} Dose: 4 mg/kg; Toxic Effects: Behavioral - Coma; Cardiac - Arrythmias (including changes in conduction); Gastrointestinal - Nausea or vomiting. Woman TD_{Lo} Dose: 4 mg/kg; Toxic Effects: Cardiac - Change in rate. Rat LD_{50} Dose: >2560 mg/kg.
Acute Dermal: Rat LD_{50} Route: Subcutaneous Dose: 355 mg/kg.

Hazard Overviews

Health: Irritating to eyes/skin/respiratory tract. Harmful. Other Acute Effects: harmful if swallowed; may be harmful if inhaled; may be harmful if absorbed through the skin; exposure can cause an increase in appetite and weight gain.
Fire: Hazards: emits toxic fumes. Extinguishing agents: water spray; carbon dioxide, dry chemical powder or appropriate foam. Precautions: combustible liquid.
Reactivity: Stable. Hazardous polymerization will not occur. Hazardous decomposition products: toxic fumes of: carbon monoxide, carbon dioxide, nitrogen oxides, hydrogen fluoride.
Carcinogenicity: IARC - Not listed; NIOSH - Not listed; NTP - Not listed; ACGIH - Not listed; OSHA - Not listed; EPA - Not listed; MAK - Not listed
Primary Target Organs:

Eyes

Skin

Respiratory System

Environmental

Regulations

RCRA 40CFR: Not listed
CERCLA: 40CFR 302.4: Not listed
SARA 40CFR 372.65: Not listed
SARA EHS 40CFR 355: Not listed
TSCA: Not listed

ATE5000 CAS #: 29122-68-7

ATENOLOL

RTECS: AC3600000
EINECS Number: 249-451-7
Molecular Formula: $C_{14}H_{22}N_2O_3$
Formula Weight: 266.3

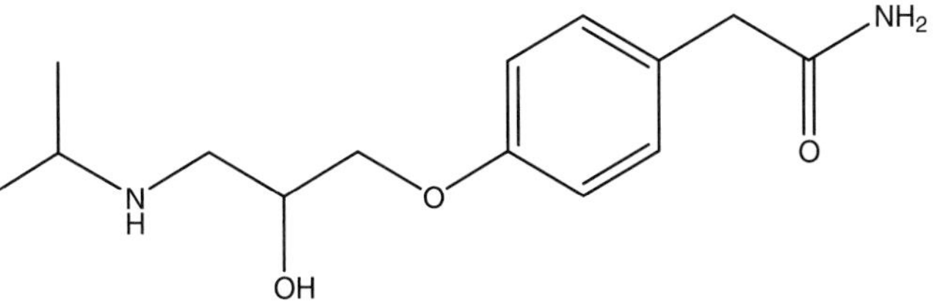

Chemical Structure

Synonyms: ATEHEXAL; ATENOL; BENZENEACETAMIDE,4-(2-HYDROXY-3-((1-METHYLETHYL)AMINO)PROPOXY)-(9CI); 1-P-CARBAMOYLMETHYLPHENOXY-3-ISOPROPYLAMINO-2-PROPANOL; DURAATENOLOL; 2-(P-(2-HYDROXY-3-(ISOPROPYLAMINO)PROPOXY)PHENYL)ACETAMIDE; 2-(P-(2-HYDROXY-3-(ISOPROPYLAMINO)PROPOXY)PHENYLACETAMIDE; 4-(2-HYDROXY-3-((1-METHYLETHYL)AMINO)PROPOXY)BENZENEACETAMIDE; IBINOLO; ICI 66082; MYOCORD; NORMITEN; PRENORMINE; SELES BETA; SELOBLOC; TENOBLOC; TENORMIN; UNIBLOC; UNILOC
Description: colorless to white, crystalline powder or crystals
Use: medication: beta adrenergic blocking agent; medication: antihypertensive, antiangial, antiarrhythmic

Physical Properties

Freezing Point: 146 °C (294.8 °F) to 148 °C (298.4 °F)
Water Solubility: 26.5 mg/ml at 37 °C

RTECS Toxicity Data

Acute Oral: Man TD_{Lo} Dose: 86 mg/kg/60D-I; Toxic Effects: Blood - Changes in serum composition. Man TD_{Lo} Dose: 49 mg/kg/30D-I; Toxic Effects: Behavioral - Excitment. Woman TD_{Lo} Dose: 16 mg/kg; Toxic Effects: Lungs, Thorax, or Respiration - Respiratory obstruction. Woman TD_{Lo} Dose: 1080 mg/kg/78W-I; Toxic Effects: Blood - Leukopenia; Musculoskelital - Other changes; Skin and appendages - Dermatitis, other. Woman TD_{Lo} Dose: 42 mg/kg/3W-I; Toxic Effects: Nutritional and gross metabolic - Changes in calcium.

Hazard Overviews

Health: Irritating. Harmful. Other Acute Effects: harmful if swallowed; may be harmful if inhaled; may be harmful if absorbed through the skin.
Fire: Hazards: emits toxic fumes. Extinguishing agents: water spray; carbon dioxide, dry chemical powder or appropriate foam. Precautions: combustible liquid.
Reactivity: Stable. Hazardous polymerization will not occur. Hazardous decomposition products: toxic fumes of: carbon monoxide, carbon dioxide, nitrogen oxides.
Carcinogenicity: IARC - Not listed; NIOSH - Not listed; NTP - Not listed; ACGIH - Not listed; OSHA - Not listed; EPA - Not listed; MAK - Not listed

Primary Target Organs:

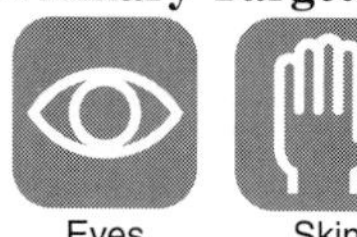
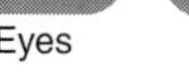

Eyes Skin Respiratory System

Environmental

Regulations

RCRA 40CFR: Not listed
CERCLA: 40CFR 302.4: Not listed
SARA 40CFR 372.65: Not listed
SARA EHS 40CFR 355: Not listed
TSCA: Not listed

ATR3000 CAS #: 1912-24-9

ATRAZINE

RTECS: XY5600000
DOT: UN2763
EINECS Number: 217-617-8
Molecular Formula: $C_8H_{14}ClN_5$
Formula Weight: 215.68

Chemical Structure

Synonyms: A 361; AATRAM 20G; AATREX; AATREX 4L; AATREX 80W; AATREX NINE-0; AATREX NINE-O; ACTINITE PK; 2-AETHYLAMINO-4-CHLOR-6-ISOPROPYLAMINO-1,3,5-TRIAZIN; 2-AETHYLAMINO-4-ISOPROPYLAMINO-6-CHLOR-1,3,5-TRIAZIN; AKTICON; AKTIKON; AKTIKON PK; AKTINIT A; AKTINIT PK; ANELDAZIN; ARGEZIN; ATAZINAX; ATRANEX; ATRASINE; ATRATAF; ATRATOL A; ATRAZIN; ATRED; ATREX; CANDEX; CEKUZINA-T; CET; 2-CHLORO-4-ETHYLAMINEISOPROPYLAMINE-S-TRIAZINE; 1-CHLORO-3-ETHYLAMINO-5-ISOPROPYLAMINO-2,4,6-TRIAZINE; 1-CHLORO-3-ETHYLAMINO-5-ISOPROPYLAMINO-S-TRIAZINE; 2-CHLORO-4-(ETHYLAMINO)-6-(ISOPROPYLAMINO)-S-TRIAZINE; 2-CHLORO-4-(ETHYLAMINO)-6-(ISOPROPYLAMINO)TRIAZINE; 2-CHLORO-4-ETHYLAMINO-6-ISOPROPYLAMINO-1,3,5-TRIAZINE; 2-CHLORO-4-ETHYLAMINO-6-ISOPROPYLAMINO-S-TRIAZINE; 2-CHLORO-4-ETHYLAMINO-6-ISOPROPYLAMINOTRIAZINE; 6-CHLORO-N-ETHYL-N'-(1-METHYLETHYL)-1,3,5-TRIAZINE-2,4-DIAMINE; 2-CHLORO-4-(2-PROPYLAMINO)-6-ETHYLAMINO-S-TRIAZINE; CHROMOZIN; CRISATRINA; CRISAZINE; CYAZIN; EPA PESTICIDE CODE 080803; FARMCO ATRAZINE; FENAMIN; FENAMINE; FENATROL; G 30027; GEIGY 30,027; GESAPRIM; GESAPRIM 50; GESAPRIM 500; GESOPRIM; GIFFEX 4L; GRIFFEX; HERBATOXOL; HUNGAZIN; HUNGAZIN PK; INAKOR; MAIZINA; OLEOGESAPRIM; OLEOGESAPRIM 200; PITEZIN; PRIMATOL; PRIMATOL A; PRIMAZE; RADAZIN; RADIZINE; SHELL ATRAZINE HERBICIDE; STRAZINE; TRIAZINE A 1294; S-TRIAZINE,2-CHLORO-4-(ETHYLAMINO)-6-(ISOPROPYLAMINO)-; S-TRIAZINE,2-CHLORO-4-ETHYLAMINO-6-ISOPROPYLAMINO-; 1,3,5-TRIAZINE-2,4-DIAMINE,6-CHLORO-N-ETHYL-N'-(1-METHYLETHYL)-; 1,3,5-TRIAZINE-2,4-DIAMINE,6-CHLORO-N-ETHYL-N'-(1-METHYLETHYL)-(9CI); VECTAL; VECTAL SC; WEEDEX A; WONUK; ZEAPHOS; ZEAPOS; ZEAZIN; ZEAZINE; ZEOPOS

Description: colorless or white, crystalline powder; odorless
Use: selective herbicide for weed control in agriculture

Physical Properties

Boiling Point: Decomposes
Freezing Point: 171 °C (339.8 °F) to 174 °C (345.2 °F)
Saturated Vapor Density: 1.200000003 kg/m^3
Density: 1.187 g/cu cm at 20 °C
Vapor Pressure: 3 $x10^{-7}$ mm Hg at 20 °C
Water Solubility: 0.003% by weight
Other Solubilities: 12 g/kg Diethyl Ether at 20 °C; 18 g/kg Methanol at 20 °C; 10 g/kg octan-1-ol at 20 °C; 52 g/kg Chloroform at 20 °C; 28 g/kg Ethyl Acetate at 20 °C; 0.36 g/kg n-pentane at 20 °C.
Flash Point: Nonflammable

RTECS Toxicity Data

Acute Oral: Rat LD_{50} Dose: 672 mg/kg. Mouse LD_{50} Dose: 850 mg/kg.
Acute Inhalation: Rat LC_{50} Dose: 5200 mg/m^3/4hr.
Acute Dermal: Rabbit LD_{50} Route: Skin; Dose: 7500 mg/kg. Rat LD_{50} Route: Skin; Dose: >12500 mg/kg.
Irritation Eye: Rabbit Standard Draize Test Dose: 6320 ug; Reaction: severe. Mammal Standard Draize Test Dose: 100 mg; Reaction: severe.
Irritation Skin: Rabbit Open Draize Test Dose: 38 mg open; Reaction: mild. Mammal Standard Draize Test Dose: 500 mg; Reaction: mild.
Reproductive/Teratogenic: Rat Route: Oral; Dose: 7 gm/kg; Duration: female 6-15D of pregnancy; Effects on Embryo or Fetus - Fetotoxicity. Rat Route: Oral; Dose: 700 mg/kg; Duration: female 6-15D of pregnancy; Maternal Effects - Other effects on females; Specific Developmental Abnormalities - Musculoskeletal system. Rat Route: Oral; Dose: 720 mg/kg; Duration: female 10D prior to mating Maternal Effects - Other effects on females; Effects on Newborn - Behavioral.
Mutagenic: Human DNA Damage; Cell Type: lymphocyte; Dose: 100mg/L. Human Unscheduled DNA Synthesis; Cell Type: fibroblast; Dose: 3 mmol/L. Human Cytogenetic Analysis; Cell Type: lymphocyte; Dose: 1 mg/L. Human Cytogenetic Analysis; Cell Type: lymphocyte; Dose: 1 mg/L. Human Cytogenetic Analysis; Cell Type: lymphocyte; Dose: 1 mg/L. Human Cytogenetic Analysis; Cell Type: lymphocyte; Dose: 1 mg/L. Human Cytogenetic Analysis; Cell Type: lymphocyte; Dose: 1 mg/L. Human Cytogenetic Analysis; Cell Type: lymphocyte; Dose: 1 mg/L.
Tumorigenic: Rat Route: Oral; Dose: 33775 mg/kg/2Y-C; Toxic Effects: Tumorigenic - Carcinogenic by RTECS criteria; Blood - Leukemia; Tumorigenic effects - Uterine tumors. Mouse Route: Oral; Dose: 9000 mg/kg/78W-I; Toxic Effects: Tumorigenic - Equivocal tumorigenic agent by

RTECS criteria; Lungs, Thorax, or Respiration - Tumors; Liver - Tumors.

Hazard Overviews

Fire: Noncombustible.

Carcinogenicity: IARC - Group 2B, Possibly carcinogenic to humans; NIOSH - Not listed; NTP - Not listed; ACGIH - Class A4, Not classifiable as a human carcinogen; OSHA - Not listed; EPA - Not listed; MAK - Not listed

Exposure Limits

OSHA PEL Vacated 1989 Limits: TWA: 5 mg/m^3.

ACGIH TLV: TWA: 5 mg/m^3.

NIOSH REL: TWA: 5 mg/m^3.

DFG MAK: TWA: 2 mg/m^3.

Respirator Recommendation

Exposure Range: >5 to 50 mg/m^3 Air Purifying, Negative Pressure, Half Mask

Exposure Range: >50 to 500 mg/m^3 Air Purifying, Negative Pressure, Full Face

Exposure Range: >500 to 5000 mg/m^3 Supplied Air, Constant Flow/Pressure Demand, Full Face

Exposure Range: >5000 to unlimited mg/m^3 Self-contained Breathing Apparatus, Pressure Demand, Full Face

Cartridge Color: black with magenta (P100)

Note: use dust/mist prefilter

Environmental

Ecotoxicity: LD_{50} Pheasant oral > 2000 mg/kg, 3 month old males LC_{50} Lebistes cyanellus (guppies) 4.3 mg/l/96 hr /Conditions of bioassay not specified Microcystis aeruginosa: inhibition of cell multiplication starts at 0.003 mg/l

Environmental Fate: The s- triazine ring is fairly resistant to degradation. 2-Chloro-4- ethyl-amino-6-amino-s-triazine, 2-chloro-4-amino-6-isopropylamino-s-triazine, 2-hydroxy-4-ethylamino-6-isopropyl-amino-s-triazine and 2-hydroxy-4-ethylamino-6-amino-s-triazine have been identified as microbial transformation products of atrazine. Chemical degradation may be more important environmentally than biodegradation. It may hydrolyze fairly rapidly in either acidic or basic environments, yet is fairly resistant to hydrolysis at neutral pHs. Furthermore the rate of hydrolysis was found to drastically increase upon small additions of humic materials, indicating hydrolysis could be catalyzed. For example, the half-life at 25 °C and pH of 4 was 244 days without an additive and 1.73 days with the presence of 2% humic acid. At 25 °C, a 5 mg/l solution of fulvic acid (naturally occurs in soils and most surface waters) resulted in half-lives of 34.8, 174, 398 and 742 days at pHs of 2.9, 4.5, 6.0 and 7.0, respectively. Hydrolysis followed first order kinetics producing hydroxyatrazine as the transformation product. Photolysis did not occur in alcohols and water at wavelengths > 300 nm. However, at wavelengths greater than or equal to 290 nm, the photolysis half-life at a concentration of 10 mg/l in aqueous solution at 15 °C was 25 hr as compared to a half-life of 4.9 hr for identical conditions with an acetone sensitizer added at a concentration of 1 ml/100 ml. It is expected to maintain a very high to medium mobility class in soils and should not strongly absorb to sediments. It is not expected to bioconcentrate or volatilize. However if released to the atmosphere, vapor phase reactions with photochemically produced hydroxyl radicals in the atmosphere may be important (estimated half-life of about 2.6 hr).

Cleanup/Disposal: Guide No. 151: Do not touch damaged containers or spilled material unless wearing appropriate protective clothing. Stop leak if you can do it without risk. Prevent entry into waterways, sewers, basements or confined areas. Cover with plastic sheet to prevent spreading. Absorb or cover with dry earth, sand or other non-combustible material and transfer to containers. Do not get water inside containers.

Environmental Physical Data

Henry's Law Constant: calculated at 2.63 $x10^{-9}$

Octanol/Water Partition Coefficient: log K_{ow} = 2.75

Sorption Partition Coefficient: K_{oc} = estimated at 640

BCF: calculated at 0.3 to 2.0

Regulations

RCRA 40CFR: Not listed

CERCLA: 40CFR 302.4: Not listed

SARA 40CFR 372.65: Listed

SARA EHS 40CFR 355: Not listed

TSCA: Listed

Analytical Methods

Soil: EPA PMD-FLM, PMD-TLC; SW846 8141, 8141A

Water / Groundwater: EPA 619, 022; SW846 8141, 8141A; ASTM D4763; USGS O3106

Drinking Water: EPA 505, 507, 508.1, 525.1, 525.2; AOAC 991.07, 992.14; ASTM D5475

Food: FDA 212.1, 212.2, 232.1, 232.4, 242.1; AOAC 971.08

Indoor / Expired Air: ASTM D4861

Plasma: EPA 001; FDA 211.1, 231.1, 252

Other: EPA 1656

ATR6000 CAS #: 51-55-8

ATROPINE

RTECS: CK0700000

EINECS Number: 200-104-8

Molecular Formula: $C_{17}H_{23}NO_3$

Formula Weight: 289.38

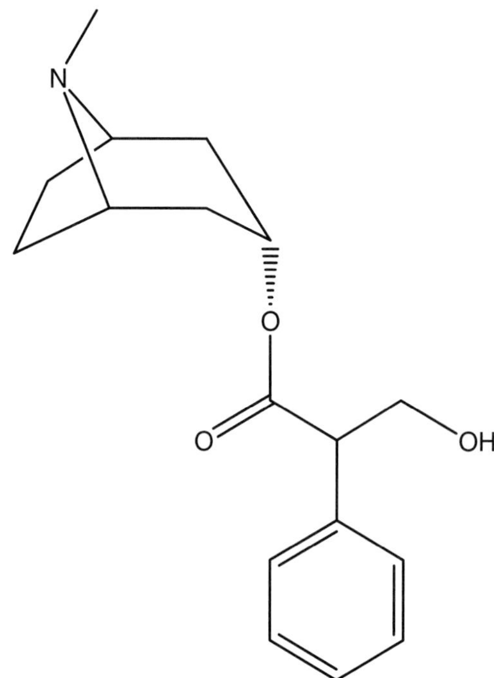

Chemical Structure

Synonyms: ATROPIN; ATROPINA; ATROPIN-FLEXIOLEN; ATROPINOL; ATROPISOL; BENZENEACETIC ACID,ALPHA-(HYDROXYMETHYL)-8-METHYL-8-AZABICYCLO(3,2,1)OCT-3-YLESTER ENDO-(+-)-; DL-HYOSCYAMINE; DL-TROPANYL 2-HYDROXY-1-PHENYLPROPIONATE; DL-TROPYL TROPATE; DL-TROPYLTROPATE; ENDO-(+-)-ALPHA(HYDROXYMETHYL)BENZENEACETIC ACID8-METHYL-8-AZABICYCLO[3,2,1]OCT-3-YL ESTER; EYESULES; ALPHA-(HYDROXYMETHYL)BENZENEACETIC ACID 8-METHYL-8-AZABICYCLO(3.2.1)OCT-3-YL ESTER; ISOPTO-ATROPINE; 2-PHENYLHYDRACRYLIC ACID 3-ALPHA-TROPANYL ESTER; BETA-PHENYL-GAMMA-OXYPROPIONSAEURE-TROPYL-ESTER; BETA-PHENYL-GAMMA-OXYPROPIONSAURE-TROPYL-ESTER; 1ALPHA H,5ALPHA H-TROPAN-3ALPHA-OL (+-)-TROPATE (ESTER); 1-ALPHA-H,5-ALPHA-H-TROPAN-3-ALPHA-OL (+-)-TROPATE (ESTER)(8CI); TROPIC ACID ESTER WITH TROPINE; TROPIC ACID,ESTER WITH TROPINE; TROPIC ACID,3-ALPHA-TROPANYL ESTER; TROPINE TROPATE; TROPINE,TROPATE (ESTER); (+,-)-TROPYL TROPATE; TROYL TROPATE

Description: long, orthorhombic prims or needles; odorless

Use: med: anticholinergic; in treatment for acute myocardial infarction with excessive vagal tone; in ophthalmology, to produce mydriasis & cycloplegia; with opioid for renal colic; prior to administration of general anesthetic, to inhibit salivation & respiratory secretions; bronchodilator; antidote for poisoning due to cholinomimetic alkaloid muscarine in amanita muscari (mushroom).

Physical Properties

Freezing Point: 114 °C (237.2 °F) to 116 °C (240.8 °F)

Water Solubility: 1 g in 455 ml Water

Other Solubilities: 1 g in 5 ml Alcohol, 2.5 ml Glycerin, 2.5 ml boiling Alcohol;.

pH: 0.0015 molar solution 10

RTECS Toxicity Data

Acute Oral: Human TD_{Lo} Dose: 33 ug/kg; Toxic Effects: Sense organs and special senses - Visual field changes; Behavioral - Muscle weakness. Rat LD_{50} Dose: 500 mg/kg.

Acute Dermal: Rabbit LD_{Lo} Route: Subcutaneous Dose: 500 mg/kg. Rat LD_{50} Route: Subcutaneous Dose: 250 mg/kg.

Reproductive/Teratogenic: Woman Route: Intravenous; Dose: 20 ug/kg; Duration: female 26-39W of pregnancy Specific Developmental Abnormalities - Cardiovascular (circulatory) system.

Hazard Overviews

Poison

Health: May cause irritation to eyes/skin/respiratory tract. Poison. Other Acute Effects: may be fatal if inhaled or swallowed; may be harmful if absorbed through the skin; exposure can cause blurred vision, suppressed salivation, vasodilation, hyperpyrexia, excitement, agitation, and delirium; target organs: nerves, eyes, cardiovascular system.

Fire: Hazards: emits toxic fumes. Extinguishing agents: carbon dioxide, dry chemical powder or appropriate foam; water spray. Precautions: combustible liquid.

Reactivity: Stable. Hazardous polymerization will not occur. Incompatible with: alkalis, iodine, mercury salts, tannic acids. Hazardous decomposition products: toxic fumes of: carbon monoxide, carbon dioxide, nitrogen oxides.

Carcinogenicity: IARC - Not listed; NIOSH - Not listed; NTP - Not listed; ACGIH - Not listed; OSHA - Not listed; EPA - Not listed; MAK - Not listed

Primary Target Organs:

Eyes

Nervous System

Cardio-vascular

Environmental

Regulations

RCRA 40CFR: Not listed
CERCLA: 40CFR 302.4: Not listed
SARA 40CFR 372.65: Not listed
SARA EHS 40CFR 355: Not listed
TSCA: Listed

ATT5000 CAS #: 12174-11-7

ATTAPULGITE

RTECS: RT6400000

Molecular Formula: $Al_5H_{12}Mg_5O_{30}Si_8$

Synonyms: 200U/P-RVM; ACTIVATED ATTAPULGITE; ATTACLAY; ATTACLAY X 250; ATTACOTE; ATTAGEL; ATTAGEL 150; ATTAGEL 40; ATTAGEL 50; ATTAPULGUS CLAY; ATTASORB;

DILUEX; MIN-U-GEL 200; MIN-U-GEL 400; MIN-U-GEL FG; PALYGORSCITE; PALYGORSKIT; PALYGORSKITE; PERMAGEL; PHARMASORB-COLLOIDAL; POLYGORSKITE; RVM-FG; X 250; ZEOGEL

Description: gray to yellow solid

Use: med: for diarrhea, activated attapulgite, (vet) in calf scour boluses; in fertilizers, insecticides, fungicides, herbicides, eg, for conditioning & suspending; carrier, thickener; oil-well drilling mud; decolorizing oils & other liqs; floor sweeping cmpd; cosmetics, rubber filler, carrier for catalysts, filtering medium; rgnt for analysis of thiram (tetramethylthiuram disulfide); absorbent in pet litter.

RTECS Toxicity Data

Tumorigenic: Rat Route: Inhalation; Dose: 10 mg/m^3/6H/13W-I; Toxic Effects: Tumorigenic - Equivocal tumorigenic agent by RTECS criteria; Lungs, Thorax, or Respiration - Tumors. Rat Route: Implant; Dose: 200 mg/kg; Toxic Effects: Tumorigenic - Equivocal tumorigenic agent by RTECS criteria; Tumorigenic - Tumors at site of application.

Hazard Overviews

Carcinogenicity: IARC - Group 3, Not classifiable as to carcinogenicity to humans; NIOSH - Not listed; NTP - Not listed; ACGIH - Not listed; OSHA - Not listed; EPA - Not listed; MAK - Class A2, Unmistakably carcinogenic in animal experimentation only

Exposure Limits

OSHA PEL Vacated 1989 Limits: TWA: 2 mg/m^3; as Al soluble.

ACGIH TLV: TWA: 2 mg/m^3; as Al.

NIOSH REL: TWA: 2 mg/m^3; as Al soluble salts, alkyls.

Environmental

Regulations

RCRA 40CFR: Not listed

CERCLA: 40CFR 302.4: Not listed

SARA 40CFR 372.65: Not listed

SARA EHS 40CFR 355: Not listed

TSCA: Not listed

AUR3000 CAS #: 492-80-8

AURAMINE

RTECS: BY3500000

EINECS Number: 207-762-5

Molecular Formula: $C_{17}H_{21}N_3$

Formula Weight: 267.37

Synonyms: ANILINE,4,4'-IMIDOCARBONYLBIS(N,N-DIMETHYL-; ANILINE,4,4'-IMINOCARBONYLBIS(N,N-DIMETHYL-; APYONINE AURAMINE BASE; AURAMINE BASE; AURAMINE (FREE BASE); AURAMINE N BASE; AURAMINE O BASE; AURAMINE OAF; AURAMINE OO; AURAMINE SS; AUREMINE; BENZENAMINE,4,4'-CARBONIMIDOYLBIS(N,N-DIMETHYL-; BENZENAMINE,4,4'-CARBONIMIDOYLBIS(N,N-DIMETHYL-(9CI); BIS(P-DIMETHYLAMINOPHENYL)METHYLENEIMINE; BRILLIANT OIL YELLOW; C.I. 41000B; C.I. BASIC YELLOW 2,FREE BASE; C.I. SOLVENT YELLOW 34; 4,4'-CARBONIMIDOYLBIS(N,N-DIMETHYLBENZENAMINE); 4,4'-DIMETHYLAMINOBENZOPHENONIMIDE; FAT YELLOW A; GLAURAMINE; 4,4'-(IMIDOCARBONYL)BIS(N,N-DIMETHYLANILINE); 4,4-(IMIDOCARBONYL)BIS(N,N-DIMETHYLANILINE); ORIENT OIL YELLOW 101; TETRAMETHYLDIAMINODIPHENYLACETIMINE; TETRAMETHYL-P-DIAMINO-IMIDO-BENZOPHENONE; WAXOLINE YELLOW O; YELLOW PYOCTANINE

Use: to prepare solvent-soluble yellow 34 dye; food dye; smoke dye; dye for paper, cardboard, textiles, leather, oils, waxes, alcoholic solvents, lacquers, pen inks, carbon papers, & typewriter ribbons; fungicide; med: formerly antiseptic for nose & ear surgery, also, in united kingdom, antiseptic for gonorrhea.

Physical Properties

Freezing Point: 136 °C (276.8 °F)

Water Solubility: 11 mg/L

Other Solubilities: Soluble in Alcohol.

pH: Weak base

RTECS Toxicity Data

Acute Oral: Rat LD_{50} Dose: 3 gm/kg; Toxic Effects: Behavioral - Convulsions or effect on seizure threshold; Behavioral - Change in motor activity (specific assay); Behavioral - Muscle weakness.

Chronic (Multiple Dose) Oral: Rat Dose: 12 gm/kg/40D-I; Toxic Effects: Blood - Methemoglobinemia-Carboxyhemoglobin; Blood - Changes in leukocyte (WBC) cell count; Biochemical - True cholinesterase. Rat Dose: 20 mg/kg/17W-I; Toxic Effects: Blood - Changes in serum composition; Biochemical - Catalases.

Mutagenic: Human Unscheduled DNA Synthesis; Cell Type: fibroblast; Dose: 20 mg/L. Human DNA Inhibition; Cell Type: fibroblast; Dose: 1 mmol/L.

Tumorigenic: Rat Route: Oral; Dose: 37 gm/kg/87W-C; Toxic Effects: Tumorigenic - Equivocal tumorigenic agent by RTECS criteria; Tumorigenic - Tumor types after systemic administration not seen spontaneously. Rat Route: Subcutaneous; Dose: 263 mg/kg/21W-C; Toxic Effects: Tumorigenic - Equivocal tumorigenic agent by RTECS criteria; Tumorigenic - Tumors at site of application.

Hazard Overviews

Reactivity: Stable. Hazardous polymerization cannot occur. Avoid: excessive heat. Hazardous decomposition products: carbon; nitrogen oxides.

Carcinogenicity: IARC - Group 2B, Possibly carcinogenic to humans; NIOSH - Not listed; NTP - Not listed; ACGIH - Not listed; OSHA - Not listed; EPA - Not listed; MAK - Class A2, Unmistakably carcinogenic in animal experimentation only

Environmental

Environmental Fate: If released to soil, it will be expected to strongly adsorb to the soil and exhibit only slight mobility based upon an estimated K_{oc}; it will not, therefore, be expected to leach in soil. It may hydrolyze slowly based upon the rate of hydrolysis observed in water. It may slowly biodegrade in soil based upon data from aqueous aerobic activated sludge screening tests. Volatilization should not be an important removal process from near surface soil or other surfaces based upon its estimated Henry's Law constant and estimated vapor pressure. If released to water, it may be subject to direct photolysis based upon its strong absorption of light at wavelengths >290 nm. It may hydrolyze slowly based upon measured hydrolysis rates. It may be subject to photooxidation by various oxidants present in water including hydroxyl and peroxyl radicals and singlet oxygen based upon an estimated and measured rate data for aromatic amines based as a class. It may slowly biodegrade in water based upon data from aerobic activated sludge screening tests. Volatilization should not be an important removal process from water based upon an estimated Henry's Law constant. It will not be expected to bioconcentrate in aquatic organisms based upon an estimated BCF of 228. It will be expected to adsorb to sediment and particulate matter based upon an estimated K_{oc} of 2030. If released to the atmosphere, it will be expected to exist in both the vapor and particulate phases, based upon its estimated vapor pressure. It may be susceptible to direct photolysis in the atmosphere based upon its strong absorption of light at wavelengths >290 nm. It will degrade by reaction with photochemically produced hydroxyl radicals with an estimated half-life of 1.9 hours.

Environmental Physical Data

Henry's Law Constant: calculated at 8.05×10^{-8}
Octanol/Water Partition Coefficient: log K_{ow} = 3.5
Sorption Partition Coefficient: K_{oc} = estimated at 2030
BCF: estimated at 90

Regulations

RCRA 40CFR: Listed Hazardous Waste No. U014 Toxic Waste
CERCLA: 40CFR 302.4: Listed per RCRA Section 3001 RQ: 100 lb (45.35 kg)
SARA 40CFR 372.65: Listed
SARA EHS 40CFR 355: Not listed
TSCA: Listed

AUR6000 **CAS #: 2465-27-2**

AURAMINE HYDROCHLORIDE

RTECS: BY3675000
EINECS Number: 219-567-2
Molecular Formula: $C_{17}H_{22}ClN_3$
Structured MF: $(CH_3)_2NC_6H_4C{=}NHC_6H_4N(CH_3)_2 \cdot HCl$
Formula Weight: 303.66

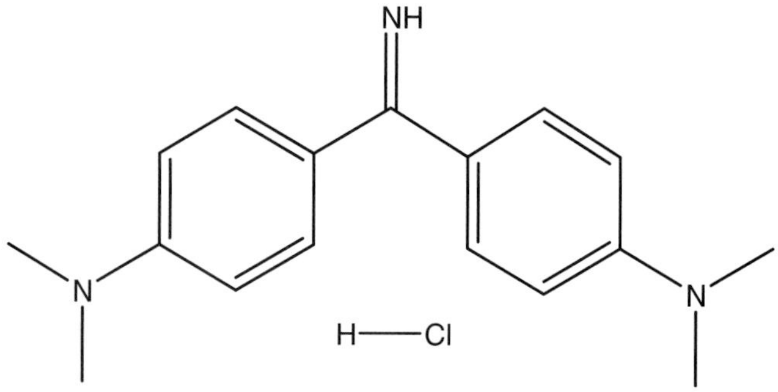

Chemical Structure

Synonyms: ADC AURAMINE O; AIZEN AURAMINE; AIZEN AURAMINE CONC SFA; AIZEN AURAMINE CONC. SFA; AIZEN AURAMINE OH; ANILINE,4,4'-(IMIDOCARBONYL)BIS(N,N-DIMETHYL-,HYDROCHLORIDE; AURAMIN; AURAMINE; AURAMINE 0-100; AURAMINE A1; AURAMINE CHLORIDE; AURAMINE CONC SPECIALLY SOLUBLE IN SPIRIT; AURAMINE CONC. SPECIALLY SOLUBLE IN SPIRIT; AURAMINE EXTRA; AURAMINE EXTRA CONC A; AURAMINE EXTRA CONC. A; AURAMINE FA; AURAMINE FWA; AURAMINE II; AURAMINE LAKE YELLOW O; AURAMINE N; AURAMINE O; AURAMINE O (BIOLOGICAL STAIN); AURAMINE O EXTRA CONC A EXPORT; AURAMINE O EXTRA CONC. A EXPORT; AURAMINE ON; AURAMINE OO; AURAMINE OOO; AURAMINE OS; AURAMINE PURE; AURAMINE SP; AURAMINE YELLOW; BASIC YELLOW 2; BENZENAMINE,4,4'-CARBONIMIDOYLBIS(N,N-DIMETHYL-,MONOHYDROCHLORIDE; BENZENAMINE,4,4'-CARBONIMIDOYLBIS(N,N-DIMETHYL-,MONOHYDROCHLORIDE (9CI); 4,4'-BIS(DIMETHYLAMINO)-BENZHYDRYLIDENIMINE HYDROCHLORIDE; 4,4'-BIS(DIMETHYLAMINO)BENZHYDRYLIDENIMINE HYDROCHLORIDE; 4:4'-BIS(DIMETHYLAMINO)BENZOPHENONE-IMINE HYDROCHLORIDE; 1,1-BIS(P-DIMETHYLAMINOPHENYL)METHYLENIMINE HYDROCHLORIDE; C.I. 41000; C.I. BASIC YELLOW 2; C.I. BASIC YELLOW 2,MONOHYDROCHLORIDE; CALCOZINE YELLOW OX; 4,4'-CARBONIMIDOYLBIS(N,N-DIMETHYLBENZENAMINE) MONOHYDROCHLORIDE; 4,4'-(IMIDOCARBONYL)BIS(N,N-DIMETHYLAMINE) MONOHYDROCHLORIDE; 4,4'-(IMIDOCARBONYL)BIS(N,N-DIMETHYLAMINE),MONOHYDROCHLORIDE; MITSUI AURAMINE O; ZLUT ZASADITA 2
Description: yellow powder
Use: for fluorescent staining of acid-fast bacteria in sputum; tissue stain and coloring of paper, cardboard, some textiles and leather; antiseptic (for use in ear and nose surgery and in the treatment of gonorrhea); and fungicide

Physical Properties

Freezing Point: > 250 °C (482 °F)
Water Solubility: 10 mg/ml in Water
Other Solubilities: 95% Ethanol: <1 mg/ml at 18 °C; Acetone: <1 mg/ml at 18 °C; Chloroform: Soluble; DMSO: <1 mg/ml at 18 °C; Ether: Soluble; Glycerol: Soluble; Methanol: <1 mg/ml at 18 °C; Toluene: <1 mg/ml at 18 °C.
Flash Point: Not available; probably combustible

RTECS Toxicity Data

Mutagenic: Human DNA Damage; Cell Type: fibroblast; Dose: 300 umol/L. Human DNA Inhibition; Cell Type: HeLa cell; Dose: 62 umol/L.

Hazard Overviews

Health: Irritating to eyes/skin. Toxic. Other Acute Effects: harmful if swallowed, inhaled, or absorbed through skin; readily absorbed through skin. Chronic Effects: laboratory experiments have shown mutagenic effects. Possible human carcinogen.
Fire: Will burn. Hazards: emits toxic fumes. Extinguishing agents: water spray; carbon dioxide, dry chemical powder or appropriate foam. Precautions: combustible liquid.
Reactivity: Incompatible with: strong oxidizing agents. Hazardous decomposition products: toxic fumes of: carbon monoxide, carbon dioxide, nitrogen oxides, hydrogen chloride gas.
Carcinogenicity: IARC - Not listed; NIOSH - Not listed; NTP - Not listed; ACGIH - Not listed; OSHA - Not listed; EPA - Not listed; MAK - Class A2, Unmistakably carcinogenic in animal experimentation only
Primary Target Organs:

Eyes

Skin

Environmental

Ecotoxicity: Toxicity to microorganisms: Bacillus subtilis growth inhib. EC_{50}: 55 mg/l
Environmental Fate: In the atmosphere, it should primarily exist in the particulate phase. Vapor phase is expected to degrade rapidly (estimated half-life of 1 hr) by reaction with photochemically produced hydroxyl radicals. Particulate phase may be removed via dry deposition. If released to soil, it may undergo a covalent chemical bonding with humic materials which can result in its chemical alteration to a latent form and prevent leaching. In the absence of covalent bonding, it will be immobile in soil. Photolysis may be important on soil surfaces. Biodegradation does not appear to be an important removal mechanism in soil and water. In water, covalent bonding with humic materials in the water column and sediments may result in partitioning from the water column to sediments. By analogy to the aromatic amine chemical class, it may be susceptible to photolysis and photooxidation via hydroxyl and peroxy radicals in water. Bioconcentration, hydrolysis and volatilization from water should not be important.

Environmental Physical Data

Henry's Law Constant: estimated at 2.8×10^{-16}
Sorption Partition Coefficient: $K_{OC} = 27$
BCF: carp 3.8 to 16

Regulations

RCRA 40CFR: Not listed
CERCLA: 40CFR 302.4: Not listed
SARA 40CFR 372.65: Not listed
SARA EHS 40CFR 355: Not listed
TSCA: Listed

AUT5000 CAS #: 8006-61-9

AUTOMOTIVE GASOLINE, LEAD-FREE

RTECS: LX3300000
EINECS Number: 232-349-1
Molecular Formula: Unspecified or Variable
Structured MF: (Mixture of hydrocarbons)
Formula Weight: N/A
Synonyms: GASOLINE; MOTOR FUEL; MOTOR SPIRITS; NATURAL GASOLINE; PETROL
Description: clear liquid; characteristic odor

Physical Properties

Boiling Point: 38.89 °C (102 °F)
Freezing Point: < 24 °C (75.2 °F)
Specific Gravity: 0.8
Vapor Density: 3 to 4 Air=1
Vapor Pressure: 38 to 300 mm Hg
Water Solubility: Insoluble
Other Solubilities: Soluble in Benzene, Ether, Chloroform
Surface Tension: 19 to 23 dynes/cm at 20 °C
Odor Threshold: 0.005 ppm
Evaporation Rate: > 1 Butyl Acetate=1
Flash Point: -43 °C
Autoignition Temperature: 280 °C
LEL: 1.4% v/v
UEL: 7.6% v/v

RTECS Toxicity Data

Acute Inhalation: Rat LC_{50} Dose: 300 gm/m^3/5M. Man TC_{Lo} Dose: 900 ppm/1hr; Toxic Effects: Sense organs and special senses - Conjunctive irritation; Behavioral - Hallucinations, distorted perceptions; Lungs, Thorax, or Respiration - Cough.
Irritation Eye: Human Standard Draize Test Dose: 140 ppm/8H; Reaction: mild. Man Standard Draize Test Dose: 500 ppm/1H; Reaction: moderate.

Hazard Overviews

Flammable

Fire Diamond

Health: Irritating to eyes/skin/respiratory tract. Also Causes: dizziness, drunkenness, unconsciousness. Chronic Effects: dermatitis, possible cancer hazard.
Fire: Flammable. Can form explosive mixtures in the air. Use dry chemical, carbon dioxide, or foam. Water may be ineffective for extinguishment, but should be used to knock-down vapors and cool containers. Do not use a solid stream of water since it may spread the fuel.

Reactivity: Stable. Hazardous polymerization cannot occur. Avoid: heat; ignition sources. Incompatible with: oxidizing materials (peroxides; nitric acid; perchlorates). Hazardous decomposition products: oxides of carbon.

Carcinogenicity: IARC - Group 2B, Possibly carcinogenic to humans; NIOSH - Listed as carcinogen; NTP - Not listed; ACGIH - Class A3, Animal carcinogen; OSHA - Not listed; EPA - Not listed; MAK - Not listed

Primary Target Organs:

Eyes

Skin

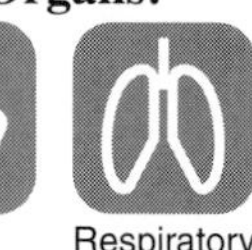
Respiratory System

Nervous System

Liver

Kidneys

Exposure Limits

OSHA PEL Vacated 1989 Limits: TWA: 300 ppm; 900 mg/m^3; STEL: 500 ppm; 1500 mg/m^3.

ACGIH TLV: TWA: 300 ppm; 890 mg/m^3; STEL: 500 ppm; 1480 mg/m^3.

Respirator Recommendation

Exposure Range: >300 to 1000 ppm Air Purifying, Negative Pressure, Half Mask

Exposure Range: >1000 to 15,000 ppm Air Purifying, Negative Pressure, Full Face

Exposure Range: >15,000 to 300,000 ppm Supplied Air, Constant Flow/Pressure Demand, Full Face

Exposure Range: >300,000 to unlimited ppm Self-contained Breathing Apparatus, Pressure Demand, Full Face

Cartridge Color: black

Environmental

Cleanup/Disposal: Guide No. 128: Eliminate all ignition sources (no smoking, flares, sparks or flames in immediate area). All equipment used when handling the product must be grounded. Do not touch or walk through spilled material. Stop leak if you can do it without risk. Prevent entry into waterways, sewers, basements or confined areas. A vapor suppressing foam may be used to reduce vapors. Absorb or cover with dry earth, sand or other non-combustible material and transfer to containers. Use clean non-sparking tools to collect absorbed material. Large Spills: Dike far ahead of liquid spill for later disposal. Water spray may reduce vapor; but may not prevent ignition in closed spaces.

Environmental Physical Data

BOD: 8%, 5 days

Regulations

RCRA 40CFR: Not listed

CERCLA: 40CFR 302.4: Not listed

SARA 40CFR 372.65: Not listed

SARA EHS 40CFR 355: Not listed

TSCA: Listed

Analytical Methods

Soil: SW846 8015B; DOE OS060

AZA1000 CAS #: 115-02-6

AZASERINE

RTECS: VT9625000

EINECS Number: 204-061-6

Molecular Formula: $C_5H_7N_3O_4$

Formula Weight: 173.13

Chemical Structure

Synonyms: ACETIC ACID,DIAZO-,ESTER WITH SERINE; AZASERIN; L-AZASERINE; AZS; C.I. 337; CI-337; CL 337; CN-15,757; CN 15757; DIAZOACETATE (ESTER) L-SERINE; L-DIAZOACETATE (ESTER) SERINE; O-DIAZOACETYL-L-SERINE; NSC-742; P-165; L-SERINE DIAZOACETATE (ESTER); L-SERINE,DIAZOACETATE; L-SERINE,DIAZOACETATE (ESTER); SERINE,DIAZOACETATE (ESTER),L-

Description: pale yellow to green crystals

Use: medication; antifungal; medication (vet): antineoplastic, antibiotic, abortifacient; research chemical for leukemia treatment

Physical Properties

Freezing Point: About 157 °C (314.6 °F)

Water Solubility: 1 x10^5 mg/L at 25 °C

Other Solubilities: Slightly Soluble in Methanol, absolute Ethanol Acetone; Soluble in warm aqueous solution of Methanol, absolute Ethanol Acetone

RTECS Toxicity Data

Acute Oral: Rat LD_{50} Dose: 170 mg/kg; Toxic Effects: Nutritional and gross metabolic - Weight loss or decreased weight gain. Mouse LD_{50} Dose: 150 mg/kg; Toxic Effects: Behavioral - Somnolence (general depressed activity); Nutritional and gross metabolic - Weight loss or decreased weight gain.

Acute Dermal: Mouse LD_{50} Route: Subcutaneous Dose: 50 mg/kg.

Reproductive/Teratogenic: Rat Route: Oral; Dose: 2500 ug/kg; Duration: female 8D of pregnancy; Specific Developmental Abnormalities - Central nervous system; Craniofacial (including nose and tongue); Musculoskeletal system. Rat Route: Oral; Dose: 2500 ug/kg; Duration: female 8D of pregnancy; Effects on Embryo or Fetus - Fetotoxicity; Specific Developmental Abnormalities - Homeostasis.

Mutagenic: Human DNA Damage; Cell Type: fibroblast; Dose: 10 mmol/L. Rat DNA Damage; Route: Subcutaneous; Dose: 10 mg/kg. Rat DNA Damage; Cell Type: liver; Dose: 30 umol/L. Rat DNA Damage; Route: Intraperitoneal; Dose:

3 mg/kg. Rat DNA Damage; Route: Oral; Dose: 100 mg/kg. Rat DNA Damage; Cell Type: other cell types; Dose: 2 mg/L. Rat DNA Damage; Route: Intraperitoneal; Dose: 1 mg/kg.

Tumorigenic: Rat Route: Oral; Dose: 150 mg/kg/5W-C; Toxic Effects: Tumorigenic - Neoplastic by RTECS criteria; Gastrointestinal - Tumors; Kidney, Ureter, and Bladder - Tumors. Rat Route: Intraperitoneal; Dose: 30 mg/kg; Toxic Effects: Tumorigenic - Carcinogenic by RTECS criteria; Gastrointestinal - Tumors; Tumorigenic - Tumors at site of application. Rat Route: Intraperitoneal; Dose: 30 mg/kg; Toxic Effects: Tumorigenic - Carcinogenic by RTECS criteria; Tumorigenic - Tumors at site of application. Rat Route: Intraperitoneal; Dose: 120 mg/kg/24W-I; Toxic Effects: Tumorigenic - Neoplastic by RTECS criteria; Gastrointestinal - Tumors; Kidney, Ureter, and Bladder - Kidney tumors. Rat Route: Intraperitoneal; Dose: 150 mg/kg/5W-I; Toxic Effects: Tumorigenic - Neoplastic by RTECS criteria; Gastrointestinal - Tumors. Rat Route: Intraperitoneal; Dose: 440 mg/kg/13W-I; Toxic Effects: Tumorigenic - Equivocal tumorigenic agent by RTECS criteria; Gastrointestinal - Tumors. Rat Route: Intraperitoneal; Dose: 50 mg/kg; Toxic Effects: Tumorigenic - Equivocal tumorigenic agent by RTECS criteria; Gastrointestinal - Changes in structure or function of exocrine pancreas; Gastrointestinal - Colon tumors. Rat Route: Intraperitoneal; Dose: 60 mg/kg/6W-I; Toxic Effects: Tumorigenic - Equivocal tumorigenic agent by RTECS criteria; Gastrointestinal - Tumors. Rat Route: Intraperitoneal; Dose: 210 mg/kg/6W-I; Toxic Effects: Tumorigenic - Neoplastic by RTECS criteria; Gastrointestinal - Tumors; Kidney, Ureter, and Bladder - Tumors. Rat Route: Intraperitoneal; Dose: 30 mg/kg/6W-I; Toxic Effects: Tumorigenic - Equivocal tumorigenic agent by RTECS criteria; Gastrointestinal - Tumors. Rat Route: Intraperitoneal; Dose: 150 mg/kg/15W-I; Toxic Effects: Tumorigenic - Equivocal tumorigenic agent by RTECS criteria; Gastrointestinal - Tumors. Rat Route: Intraperitoneal; Dose: 260 mg/kg/26W-I; Toxic Effects: Tumorigenic - Carcinogenic by RTECS criteria; Gastrointestinal - Tumors; Kidney, Ureter, and Bladder - Kidney tumors.

Hazard Overviews

Health: Toxic. Other Acute Effects: harmful if swallowed, inhaled, or absorbed through skin; exposure can cause nausea; vomiting; oral lesions; anorexia; leucopenia; lesions of the pancreas, liver and kidney. Chronic Effects: mutagen; target organs: bone marrow, liver. Carcinogen.

Fire: Hazards: emits toxic fumes. Extinguishing agents: water spray; carbon dioxide, dry chemical powder or appropriate foam. Precautions: combustible liquid.

Reactivity: Stable. Hazardous polymerization will not occur. Incompatible with: acids. Hazardous decomposition products: toxic fumes of: carbon monoxide, carbon dioxide, nitrogen oxides.

Carcinogenicity: IARC - Group 2B, Possibly carcinogenic to humans; NIOSH - Not listed; NTP - Not listed; ACGIH - Not listed; OSHA - Not listed; EPA - Not listed; MAK - Not listed

Primary Target Organs:

Liver

Bone

Environmental

Environmental Fate: Information pertaining to the biodegradation in environmental media was not located in the available literature. It should hydrolyze rapidly in acidic soil and water. Hydrolysis half-lives for azaserine in aqueous solution at pHs of 3, 7 and 11 and 25 °C were 2.1 hours, 111 days and 425 days, respectively. It is an amino-acid with a pKa of 8.55 that will dissociate in environmental media in varying proportions that are pH dependent. Ions are not expected to volatilize, nor do they generally adsorb to sediments as strongly as do their neutral counterparts. Volatilization and bioconcentration should not be important environmental fate processes. A low estimated K_{oc} indicates it should not partition from the water column to organic matter contained in sediments and suspended solids. It should be highly mobile in soil and it may leach to ground water. If released to the atmosphere as a gas, reactions with photochemically produced hydroxyl radicals may be important (estimated half-life of 10 hours). However, it is expected to exist almost entirely in the particulate phase in ambient air.

Environmental Physical Data

Henry's Law Constant: 2.56×10^{-16}
Sorption Partition Coefficient: K_{OC} = estimated at 7
BCF: calculated at 0.10

Regulations

RCRA 40CFR: Listed Hazardous Waste No. U015 Toxic Waste
CERCLA: 40CFR 302.4: Listed per RCRA Section 3001 RQ: 1 lb (0.454 kg)
SARA 40CFR 372.65: Not listed
SARA EHS 40CFR 355: Not listed
TSCA: Not listed

AZA3000 CAS #: 96743-08-7

1-AZA-2-SILACYCLOPENTANE

Physical Properties

Specific Gravity: < 1

Hazard Overviews

Carcinogenicity: IARC - Not listed; NIOSH - Not listed; NTP - Not listed; ACGIH - Not listed; OSHA - Not listed; EPA - Not listed; MAK - Not listed

Environmental

Regulations

RCRA 40CFR: Not listed
CERCLA: 40CFR 302.4: Not listed
SARA 40CFR 372.65: Not listed
SARA EHS 40CFR 355: Not listed
TSCA: Not listed

AZA6000 CAS #: 446-86-6

AZATHIOPRINE

RTECS: UO8925000
EINECS Number: 207-175-4
Molecular Formula: $C_9H_7N_7O_2S$
Formula Weight: 277.29

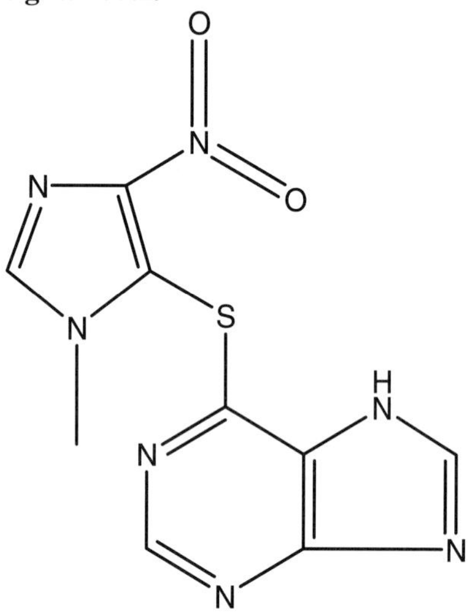

Chemical Structure

Synonyms: AZAMUN; AZANIN; AZATIOPRIN; AZOTHIOPRINE; BW 57-322; CCUCOL; IMURAN; IMUREK; IMUREL; 6-(1'-METHYL-4'-NITRO-5'-IMIDAZOLYL)-MERCAPTOPURINE; METHYLNITROIMIDAZOLYLMERCAPTOPURINE; 6-(1-METHYL-4-NITROIMIDAZOL-5-YLTHIO)PURIN; 6-((1-METHYL-4-NITRO-1H-IMIDAZOL-5-YL)THIO)-1H-PURINE; 6-((1-METHYL-4-NITROIMIDAZOL-5-YL)THIO)PURINE; 6-(1-METHYL-4-NITROIMIDAZOL-5-YLTHIO)PURINE; 6-(1-METHYL-P-NITRO-5-IMIDAZOLYL)-THIOPURINE; 6-(METHYL-P-NITRO-5-IMIDAZOLYL)-THIOPURINE; MURAN; NSC-39084; 1H-PURINE,6-((1-METHYL-4-NITRO-1H-IMIDAZOL-5-YL)THIO)-; 1H-PURINE,6-((1-METHYL-4-NITRO-1H-IMIDAZOL-5-YL)THIO-; RORASUL
Description: pale yellow crystals or yellowish powder
Use: immunosuppressor; treatment of rheumatoid arthritis and administered to inhibit the tendency of the body to reject foreign tissue (kidney and liver transplants)

Physical Properties

Freezing Point: Decomposes at 243 °C (469.4 °F) to 244 °C (471.2 °F)
Water Solubility: < 1 mg/mL at 23 C
Other Solubilities: 95% Ethanol: <1 mg/ml at 23 °C; Acetone: <1 mg/ml at 23 °C; Chloroform: Slightly Soluble; DMSO: 50-100 mg/ml at 23 °C; Dilute alkali solutions: Soluble; Dilute mineral acids: Sparingly Soluble.
Flash Point: Not available; probably combustible

RTECS Toxicity Data

Acute Oral: Man TD_{Lo} Dose: 15 mg/kg/7D-I; Toxic Effects: Lungs, Thorax, or Respiration - Other changes; Kidney, Ureter, and Bladder - Chgs in tubules (inc acute renal failure, acute tubular necrosis; Musculoskelital - Joints. Man TD_{Lo} Dose: 7500 ug/kg/1W-I; Toxic Effects: Vascular - BP lowering not characterized in autonomic section; Kidney, Ureter, and Bladder - Urine volume decreased. Man TD_{Lo} Dose: 50 mg/kg/3W-I; Toxic Effects: Blood - Leukopenia. Woman TD_{Lo} Dose: 42 mg/kg/3W-I; Toxic Effects: Blood - Leukopenia; Blood - Thrombocytopenia; Blood - Changes in cell count (unspecified). Woman TD_{Lo} Dose: 500 ug/kg. Toxic Effects: Gastrointestinal - Hypermotility, diarrhea; Gastrointestinal - Nausea or vomiting; Nutritional and gross metabolic - Body temperature increase. Man LD_{Lo} Dose: 395 mg/kg/56W-I; Toxic Effects: Liver - Other changes. Rat LD_{50} Dose: 535 mg/kg.
Acute Dermal: Rat LD_{50} Route: Subcutaneous Dose: >1 gm/kg. Mouse LD_{50} Route: Subcutaneous Dose: 350 mg/kg; Toxic Effects: Immunological including allergic - Decrease in humoral immune response.
Chronic (Multiple Dose) Dermal: Mouse Route: Subcutaneous; Dose: 780 mg/kg/26W-I; Toxic Effects: Blood - Changes in leukocyte (WBC) cell count.
Reproductive/Teratogenic: Rat Route: Oral; Dose: 15 mg/kg; Duration: female 6-8D of pregnancy; Effects on Fertility - Post-implantation mortality. Rat Route: Oral; Dose: 15 mg/kg; Duration: female 6-8D of pregnancy; Effects on Fertility - Abortion. Rat Route: Oral; Dose: 60 mg/kg; Duration: female 10-15D of pregnancy Effects on Embryo or Fetus - Fetotoxicity; Fetal death; Specific Developmental Abnormalities - Musculoskeletal system. Rat Route: Oral; Dose: 120 mg/kg; Duration: female 10-15D of pregnancy Effects on Newborn - Weaning or lactation index. Rat Route: Oral; Dose: 240 mg/kg; Duration: female 10-15D of pregnancy Effects on Newborn - Growth statistics.
Mutagenic: Human DNA Inhibition; Cell Type: leukocyte; Dose: 500 mg/L. Human Cytogenetic Analysis; Cell Type: leukocyte; Dose: 6 mg/L. Human Cytogenetic Analysis; Cell Type: lymphocyte; Dose: 83 umol/L. Human Cytogenetic Analysis; Route: Unreported; Dose: 11 mg/kg/8D. Woman Cytogenetic Analysis; Route: Unreported; Dose: 15 mg/kg/7W. Woman Cytogenetic Analysis; Route: Unreported; Dose: 59 mg/kg. Human Sister Chromatid Exchenge; Cell Type: lymphocyte; Dose: 100 ng/L. Human Micronucleus Test; Cell Type: lymphocyte; Dose: 50 umol/L. Human Other Mutation Test Systems; Cell Type: bone marrow; Dose: 1 mg/L.
Tumorigenic: Man Route: Oral; Dose: 728 mg/kg/43W-C; Toxic Effects: Tumorigenic - Carcinogenic by RTECS criteria; Blood - Leukemia. Man Route: Oral; Dose: 1565 mg/kg/4Y-C; Toxic Effects: Tumorigenic - Carcinogenic by RTECS criteria; Blood - Leukemia. Man Route: Oral; Dose: 3266 mg/kg/3Y-C; Toxic Effects: Tumorigenic -

Carcinogenic by RTECS criteria; Blood - Leukemia. Woman Route: Oral; Dose: 273 mg/kg/13W-C; Toxic Effects: Tumorigenic - Carcinogenic by RTECS criteria; Blood - Lymphomax including Hodgkin's disease. Woman Route: Oral; Dose: 3 gm/kg/3.5Y-C; Toxic Effects: Tumorigenic - Carcinogenic by RTECS criteria; Kidney, Ureter, and Bladder - Tumors. Woman Route: Oral; Dose: 5460 mg/kg/6Y-I; Toxic Effects: Tumorigenic - Carcinogenic by RTECS criteria; Blood - Leukemia. Woman Route: Oral; Dose: 2 gm/kg/3Y-C; Toxic Effects: Tumorigenic - Carcinogenic by RTECS criteria; Blood - Leukemia. Rat Route: Oral; Dose: 1932 mg/kg/46W-C; Toxic Effects: Tumorigenic - Equivocal tumorigenic agent by RTECS criteria; Gastrointestinal - Tumors. Rat Route: Oral; Dose: 11 gm/kg/W-C; Toxic Effects: Tumorigenic - Carcinogenic by RTECS criteria; Sense organs and special senses - Tumors; Blood - Lymphomax including Hodgkin's disease.

Hazard Overviews

Health: Irritating to eyes/skin/respiratory tract. Toxic. Other Acute Effects: harmful if swallowed; may be harmful if inhaled, or absorbed through the skin; possible sensitizer; prolonged or repeated exposure may cause allergic reactions in certain sensitive individuals; exposure can cause nausea; headache; vomiting; gastrointestinal disturbances; dermatitis; fever; alopecia; shock. Chronic Effects: carcinogenic based on its IARC, OSHA, ACGIH, NTP or EPA classification; may alter genetic material; possible risk of congenital malformation in the fetus; target organs: bone marrow, blood, bladder, liver, cardiovascular system, immune system.

Fire: Will burn. Hazards: emits toxic fumes. Extinguishing agents: water spray; carbon dioxide, dry chemical powder or appropriate foam. Precautions: combustible liquid.

Reactivity: Stable. Hazardous polymerization will not occur. Incompatible with: strong oxidizing agents. Hazardous decomposition products: toxic fumes of: carbon monoxide, carbon dioxide, nitrogen oxides, sulfur oxides.

Carcinogenicity: IARC - Group 1, Carcinogenic to humans; NIOSH - Not listed; NTP - Listed; ACGIH - Not listed; OSHA - Not listed; EPA - Not listed; MAK - Not listed

Primary Target Organs:

Eyes

Skin

Respiratory System

Liver

Cardio-vascular

Blood

Environmental

Regulations

RCRA 40CFR: Not listed
CERCLA: 40CFR 302.4: Not listed
SARA 40CFR 372.65: Not listed
SARA EHS 40CFR 355: Not listed
TSCA: Not listed

AZA9000 CAS #: 54-25-1

6-AZAURIDINE

RTECS: XY8575000
EINECS Number: 200-199-6
Molecular Formula: $C_8H_{11}N_3O_6$
Formula Weight: 245.19

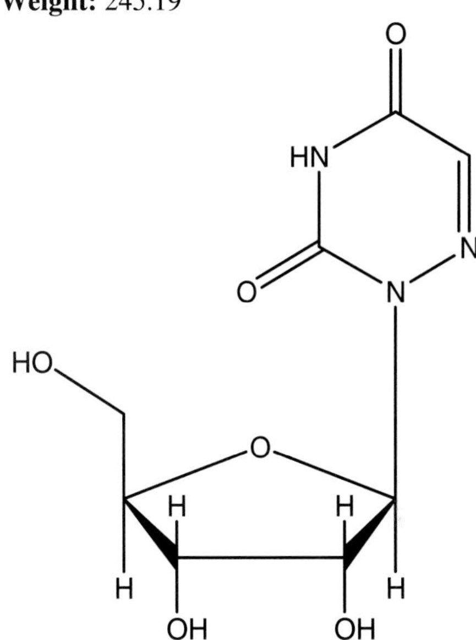

Chemical Structure

Synonyms: 6-AZAURACIL 1-RIBOSIDE; 6-AZAURACIL RIBOSIDE; 6-AZAURACILRIBOSID; 6-AZAURACIL-BETA-D-RIBOSIDE; 6-AZAURACILRIBOSIDE; AZAURIDINE; 6-AZUR; AZUR; 6-AZURIDINE; 3,5-DIOXO-2,3,4,5-TETRAHYDRO-1,2,4-TRIAZINE RIBOSIDE; NSC 32074; RIBO-AZAURACIL; RIBOAZAURACIL; RIBO-AZURACIL; 2-BETA-D-RIBOFURANOSYL-1,2,4-TRIAZIN-3,5(2H,4H)-DION; 2-BETA-D-RIBOFURANOSYL-1,2,4-TRIAZINE-3,5(2H,4H)-DIONE; 2-BETA-D-RIBOFURANOSYL-AS-TRIAZINE-3,5(2H,4H)-DIONE; 1,2,4-TRIAZINE-3,5(2H,4H)-DIONE,2-BETA-D-RIBOFURANOSYL-; AS-TRIAZINE-3,5(2H,4H)-DIONE,2-BETA-D-RIBOFURANOSYL-

Description: crystals

Use: medication: in treatment of mycosis fungoides & polycythemia vera; as antipsoriatic triacetyl deriv; medication: antineoplastic

Physical Properties

Freezing Point: 160 °C (320 °F) to 161 °C (321.8 °F)
Water Solubility: Soluble

RTECS Toxicity Data

Acute Oral: Quail LD_{50} Dose: >316 mg/kg.

Reproductive/Teratogenic: Woman Route: Intravenous; Dose: 150 mg/kg; Duration: female 47D of pregnancy; Effects on Embryo or Fetus - Other effects to embryo or fetus. Rat Route: Intravenous; Dose: 900 mg/kg; Duration: female 7-9D of pregnancy; Effects on Fertility - Post-implantation mortality; Abortion. Rat Route: Intravenous; Dose: 900 mg/kg; Duration: female 12-14D of pregnancy Effects on Embryo or Fetus - Fetal death.

Mutagenic: Rat Sperm Morphology; Route: Intraperitoneal; Dose: 2 gm/kg. Mouse DNA Inhibition; Route: Intraperitoneal; Dose: 3 gm/kg. Mouse DNA Inhibition; Cell Type: leukocyte; Dose: 229 umol/L.

Hazard Overviews

Health: May cause irritation. Harmful. Other Acute Effects: harmful if swallowed, inhaled, or absorbed through skin; may cause nervous system disturbances. Chronic Effects: possible risk of irreversible effects; possible mutagen; laboratory experiments have shown mutagenic effects. The toxicological properties have not been thoroughly investigated.

Fire: Hazards: emits toxic fumes. Extinguishing agents: water spray; carbon dioxide, dry chemical powder or appropriate foam. Precautions: combustible liquid.

Reactivity: Stable. Hazardous polymerization will not occur. Hazardous decomposition products: thermal decomposition may produce carbon monoxide, carbon dioxide, and nitrogen oxides.

Carcinogenicity: IARC - Not listed; NIOSH - Not listed; NTP - Not listed; ACGIH - Not listed; OSHA - Not listed; EPA - Not listed; MAK - Not listed

Environmental

Regulations

RCRA 40CFR: Not listed
CERCLA: 40CFR 302.4: Not listed
SARA 40CFR 372.65: Not listed
SARA EHS 40CFR 355: Not listed
TSCA: Not listed

AZE5000 CAS #: 2133-34-8

(L)-AZETIDINE-2-CARBOXYLIC ACID

RTECS: CM4310500
EINECS Number: 218-362-5
Molecular Formula: $C_4H_7NO_2$
Formula Weight: 101.12

O
OH
NH

Chemical Structure

Synonyms: ACIDE L-AZETIDINE-2-CARBOXYLIC; (S)-AZETIDINE-2-CARBOXYLIC ACID; 2-AZETIDINECARBOXYLIC ACID,(S)-; 2-AZETIDINECARBOXYLIC ACID,L-; AZETIDINE-2-CARBOXYLIC ACID,L-; L-2-AZETIDINECARBOXYLIC ACID; L-AZETIDINE-2-CARBOXYLIC ACID; 2-AZETIDINECARBOXYLIC ACID,(S)-(9CI)

Description: crystals

Physical Properties

Boiling Point: Discolors at 200 °C (392 °F)
Water Solubility: Soluble in Water
Other Solubilities: Practically Insoluble in absolute Ethanol

RTECS Toxicity Data

Acute Dermal: Mouse LD_{50} Route: Subcutaneous Dose: 1 gm/kg.

Reproductive/Teratogenic: Rat Route: Intraperitoneal; Dose: 300 mg/kg; Duration: female 8-10D of pregnancy; Specific Developmental Abnormalities - Musculoskeletal system. Rat Route: Intraperitoneal; Dose: 300 mg/kg; Duration: female 8D of pregnancy; Effects on Embryo or Fetus - Fetotoxicity; Fetal death. Rat Route: Intraperitoneal; Dose: 300 mg/kg; Duration: female 8D of pregnancy; Effects on Fertility - Litter size; Effects on Embryo or Fetus - Fetotoxicity; Fetal death.

Hazard Overviews

Health: Irritating to eyes/skin. Other Acute Effects: may be harmful by inhalation, ingestion, or skin absorption.

Fire: Hazards: emits toxic fumes. Extinguishing agents: water spray; carbon dioxide, dry chemical powder or appropriate foam. Precautions: combustible liquid.

Reactivity: Incompatible with: strong oxidizing agents. Hazardous decomposition products: toxic fumes of: carbon monoxide, carbon dioxide, nitrogen oxides.

Carcinogenicity: IARC - Not listed; NIOSH - Not listed; NTP - Not listed; ACGIH - Not listed; OSHA - Not listed; EPA - Not listed; MAK - Not listed

Primary Target Organs:

Eyes

Skin

Environmental

Regulations

RCRA 40CFR: Not listed
CERCLA: 40CFR 302.4: Not listed
SARA 40CFR 372.65: Not listed
SARA EHS 40CFR 355: Not listed
TSCA: Not listed

AZI1000 CAS #: 86-50-0

AZINOPHOS METHYL (GUTHION)

RTECS: TE1925000
DOT: NA2783
EINECS Number: 201-676-1
Molecular Formula: $C_{10}H_{12}N_3O_3PS_2$
Formula Weight: 317.34

Synonyms: AZINFOS-METHYL; AZINOPHOS-METHYL; AZINPHOS METHYL; AZINPHOS-METHYL; AZINPHOS-METILE; BAY 17147; BAY 9027; BAYER 17147; BAYER 9027; BENZOTRIAZINE DERIVATIVE OF A METHYL DITHIOPHOSPHATE;

BENZOTRIAZINEDITHIOPHOSPHORIC ACID DIMETHOXY ESTER; 1,2,3-BENZOTRIAZIN-4(3H)-ONE; 1,2,3-BENZOTRIAZIN-4(3H)-ONE,3-(MERCAPTOMETHYL)-,O,O-DIMETHYL PHOSPHORODITHIOATE; CARFENE; COTNEON; COTNION; COTNION METHYL; CRYSTHION 2L; CRYSTHYON; DBD; S-(3,4-DIHYDRO-4-OXO-BENZO(ALPHA)(1,2,3)TRIAZIN-3-YLMETHYL) O,O-DIMETHYL; S-(3,4-DIHYDRO-4-OXO-1,2,3-BENZOTRIAZIN-3-YLMETHYL) O,O-DIMETHYL PHOSPHORO-; S-(3,4-DIHYDRO-4-OXO-1,2,3-BENZOTRIAZIN-3-YLMETHYL) O,O-DIMETHYL PHOSPHORODITHIOATE; S-(3,4-DIHYDRO-4-OXO-BENZO(ALPHA)(1,2,3)TRIAZIN-3-YLMETHYL) O,O-DIMETHYL PHOSPHORODITHIOATE; S-(3,4-DIHYDRO-4-OXO-1,2,3-BENZOTRIAZIN-3-YLMETHYL)O,O-DIMETHYL PHOSPHORODITHIOATE; O,O-DIMETHYL S-(3,4-DIHYDRO-4-KETO-1,2,3-BENZOTRIAZINYL-3-METHYL) DITHIO-; O,O-DIMETHYL S-(3,4-DIHYDRO-4-KETO-1,2,3-BENZOTRIAZINYL-3-METHYL) DITHIOPHOSPHATE; O,O-DIMETHYL S-(4-OXO-3H-1,2,3-BENZOTRIAZINE-3-METHYL)PHOSPHORODITHIOATE; O,O-DIMETHYL S-(4-OXO-1,2,3-BENZOTRIAZINO(3)-METHYL) THIOTHIONOPHOSPHATE; O,O-DIMETHYL S-(4-OXOBENZOTRIAZINO-3-METHYL)PHOSPHORODITHIOATE; O,O-DIMETHYL S-(4-OXO-1,2,3-BENZOTRIAZINO(3)-METHYL)THIOTHIONOPHOSPHATE; O,O-DIMETHYL S-4-OXO-1,2,3-BENZOTRIAZIN-3(4H)-YLMETHYL PHOSPHORODITHIOATE; O,O-DIMETHYL S-4-OXO-1,2,3-BENZOTRIAZIN-3(4H)-YLMETHYLPHOSPHORODITHIOATE; O,O-DIMETHYL-S-(BENZAZIMINOMETHYL) DITHIOPHOSPHATE; O,O-DIMETHYL-S-(1,2,3-BENZOTRIAZINYL-4-KETO)METHYL PHOSPHORODITHIOATE; O,O-DIMETHYL-S-(1,2,3-BENZOTRIAZINYL-4-KETO)METHYLPHOSPHORODITHIOATE; DIMETHYLDITHIOPHOSPHORIC ACID N-METHYLBENZAZIMIDE ESTER; O,O-DIMETHYL-S-(4-OXOBENZOTRIAZIN-3-METHYL)-DITHIOPHOSPHAT; O,O-DIMETHYL-S-4-OXO-1,2,3-BENZOTRIAZIN-3(4H)-YL METHYL PHOSPHORODITHIOATE; 0,0-DIMETHYL-S-4-OXO-1,2,3-BENZOTRIAZIN-3(4H)-YLMETHYL PHOSPHORODITHIOATE; O,O-DIMETHYL-S-((4-OXO-3H-1,2,3-BENZOTRIAZIN-3-YL)-METHYL)-DITHIOFOSFAAT; O,O-DIMETHYL-S-((4-OXO-3H-1,2,3-BENZOTRIAZIN-3-YL)-METHYL)-DITHIOPHOSPHAT; O,O-DIMETHYLS-(3,4-DIHYDRO-4-KETO-1,2,3-BENZOTRIAZINYL-3-METHYL)DITHIOPHOSPHATE; O,O-DIMETHYLS-(4-OXO-3H-1,2,3-BENZOTRIAZINE-3-METHYL)PHOSPHORODITHIOATE; O,O-DIMETHYLS-(4-OXOBENZOTRIAZINO-3-METHYL)PHOSPHORODITHIOATE; O,O-DIMETHYLS-(4-OXO-1,2,3-BENZOTRIAZINO(3)-METHYL)THIOTHIONOPHOSPHATE; O,O-DIMETIL-S-((4-OXO-3H-1,2,3-BENZOTRIAZIN-3-IL)-METIL)-DITIOFOSFATO; DITHIOATE; DITHIOATE S-ESTER; ENT 23,233; EPA SHAUGHNESSY #058001; GOTHNION; GUSATHION; GUSATHION-20; GUSATHION 25; GUSATHION K; GUSATHION M; GUSATHION METHYL; GUTHION; 3-(MERCAPTOMETHYL)-1,2,3-BENZOTRIAZIN-4(3H)-ONE O,O-DIMETHYL PHOSPHORO-; 3-(MERCAPTOMETHYL)-1,2,3-BENZOTRIAZIN-4(3H)-ONE O,O-DIMETHYL PHOSPHORODITHIOATE S-ESTER; 3-(MERCAPTOMETHYL)-1,2,3-BENZOTRIAZIN-4(3H)-ONEO,O-DIMETHYL PHOSPHORODITHIOATE S-ESTER; METHYL AZINPHOS; METHYL) ESTER; METHYL GUTHION; METHYLAZINPHOS; N-METHYLBENZAZIMIDE,DIMETHYLDITHIOPHOSPHORIC ACID ESTER; METHYLGUSATHION; METILTRIAZOTION; PHOSPHATE; PHOSPHORODITHIOATE; PHOSPHORODITHIOC ACID O,O-DIMETHYL S-((4-OXO-1,2,3-BENZOTRIAZIN-3(4H)-YL); PHOSPHORODITHIOIC ACID,O,O-DIMETHYL ESTER,S-ESTER WITH 3-(MERCAPTOMETHYL)-; PHOSPHORODITHIOIC ACID,O,O-DIMETHYL ESTER,S-ESTER WITH3-(MERCAPTOMETHYL)-1,2,3-BENZOTRIAZIN-4(3H)-ONE; PHOSPHORODITHIOIC ACID,O,O-DIMETHYLS-((4-OXO-1,2,3-BENZOTRIAZIN-3(4H)-YL)METHYL) ESTER; R 1582

Description: brown, colorless waxy solid, crystals

Use: a nonsystemic, highly persistent insecticide and acaricide chiefly effective against biting and sucking insect pests; on citrus, cotton, grapes, maize, some ornamentals, top fruit and vegetables

Physical Properties

Boiling Point: Decomposes
Freezing Point: 73 °C (163.4 °F) to 74 °C (165.2 °F)
Specific Gravity: 1.44 at 20 °C/4 °C
Density: 1.44 g/mL at 20 °C
Vapor Pressure: < 1 mPa at 20 °C
Water Solubility: 0.003% by weight
Other Solubilities: > 1 kg/l dichloromethane; > 1 kg/l Toluene.
Odor Threshold: Water 0.0002 mg/kg
Refraction Index: 1.6115 at 76 °C/D
Flash Point: Nonflammable

RTECS Toxicity Data

Acute Oral: Rat LD_{50} Dose: 7 mg/kg; Toxic Effects: Biochemical - True cholinesterase. Mouse LD_{50} Dose: 8600 ug/kg.

Acute Inhalation: Rat LC_{50} Dose: 69 mg/m^3/1hr.

Acute Dermal: Rat LD_{50} Route: Skin; Dose: 88 mg/kg. Mouse LD_{50} Route: Skin; Dose: 65 mg/kg; Toxic Effects: Behavioral - Somnolence (general depressed activity); Behavioral - Muscle weakness; Lungs, Thorax, or Respiration - Dyspnea.

Chronic (Multiple Dose) Oral: Rat Dose: 3650 mg/kg/2Y-C; Toxic Effects: Behavioral - Convulsions or effect on seizure threshold; Biochemical - True cholinesterase. Dog Dose: 3094 mg/kg/2Y-C; Toxic Effects: Behavioral - Somnolence (general depressed activity); Behavioral - Muscle weakness; DEATH.

Chronic (Multiple Dose) Inhalation: Rat Dose: 4720 ug/m^3/6H/12W-I; Toxic Effects: Blood - Other changes; Nutritional and gross metabolic - Weight loss or decreased weight gain; Biochemical - True cholinesterase.

Reproductive/Teratogenic: Rat Route: Oral; Dose: 190 mg/kg; Duration: female 6-22D of pregnancy; Effects on Newborn - Viability index; Weaning or lactation index; Growth statistics. Rat Route: Oral; Dose: 12500 ug/kg; Duration: female 6-15D of pregnancy; Specific Developmental Abnormalities - Central nervous system. Rat Route: Oral; Dose: 85 mg/kg; Duration: female 6-22D of pregnancy; Effects on Newborn - Growth statistics.

Mutagenic: Human Unscheduled DNA Synthesis; Cell Type: fibroblast; Dose: 10 umol/L. Human Cytogenetic Analysis; Cell Type: lung; Dose: 120 mg/L. Human Cytogenetic Analysis; Cell Type: other cell types; Dose: 140 mg/L. Human Other Mutation Test Systems; Cell Type: fibroblast; Dose: 120 mg/L.

Tumorigenic: Rat Route: Oral; Dose: 5110 mg/kg/78W-C; Toxic Effects: Tumorigenic - Equivocal tumorigenic agent by RTECS criteria; Gastrointestinal - Tumors; Endocrine - Thyroid tumors. Rat Route: Oral; Dose: 121 gm/kg/78W-C; Toxic Effects: Tumorigenic - Equivocal tumorigenic agent by RTECS criteria; Gastrointestinal - Tumors.

Hazard Overviews

Poison

Fire Diamond

Health: Poison. Causes: watering eyes, chest tightness, wheezing, CNS depression, blurred vision, headache, salivation, cyanosis, nausea, diarrhea, sweating, twitching, convulsions, low blood pressure, cardiac irregularities, bronchorrhea.

Fire: Noncombustible. Use agent suitable for surrounding fire. Container may explode in heat of fire.

Reactivity: Stable. Hazardous polymerization cannot occur. Avoid: excessive dust generation; exposure to heat. Incompatible with: acids; cold alkalis; oxidizers. Hazardous decomposition products: oxides of sulfur, nitrogen, and phosphorus.

Carcinogenicity: IARC - Not listed; NIOSH - Not listed; NTP - Not listed; ACGIH - Class A4, Not classifiable as a human carcinogen; OSHA - Not listed; EPA - Not listed; MAK - Not listed

Primary Target Organs:

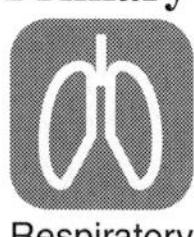

Respiratory System

Nervous System

Cardio-vascular

Exposure Limits

OSHA PEL: TWA: 0.2 mg/m^3; skin.

ACGIH TLV: TWA: 0.2 mg/m^3.

NIOSH REL: TWA: 0.2 mg/m^3.

NIOSH IDLH: 10 mg/m^3.

DFG MAK: TWA: 0.2 mg/m^3.

Respirator Recommendation

Exposure Range: >0.2 to 2 mg/m^3 Air Purifying, Negative Pressure, Half Mask

Exposure Range: >2 to <10 mg/m^3 Air Purifying, Negative Pressure, Full Face

Exposure Range: 10 to unlimited mg/m^3 Self-contained Breathing Apparatus, Pressure Demand, Full Face

Cartridge Color: black with magenta (P100)

Environmental

Ecotoxicity: LC_{50} Ephemerella subvaria 4.5 ug/l/30 day /Conditions of bioassay not specified LC_{50} Gammarus lacustris 0.15 ug/l/96 hr /Conditions of bioassay not specified LC_{50} Gammarus fasciatus 0.10 ug/l/96 hr /Conditions of bioassay not specified LC_{50} Salmo trutta (brown trout) 4 ug/l/96 hr /Conditions of bioassay not specified TLm Crassostrea virginica (American oyster) eggs 620 ppb/48 hr in a static lab bioassay LC_{50} Oncorhynchus kisutch (coho salmon) 17 ug/l/96 hr /Conditions of bioassay not specified TLm Gasterosteus aculeatus (threespine stickleback) 4.8 ppb/96 hr /Conditions of bioassay not specified LC_{50} Salmo gairdnerii (rainbow trout fingerlings) 7.10 mg/l/96 hr /Conditions of bioassay not specified LC_{50} Crangon crangon (European shrimp) 0.33 ug/l/48 hr /Conditions of bioassay not specified LC_{50} Palaemonetes kadiakensis 0.16 ug/l/20 day /Conditions of bioassay not specified LC_{50} Pteronarcys californica 1.5 ug/l/96 hr /Conditions of bioassay not specified

Environmental Fate: Released to soil surfaces it will probably not be persistent. In one field study, 50% applied as an emulsion was lost in 12 days. Biodegradation and volatilization are most likely the primary degradation and transport processes, respectively, for material released to soil surfaces or incorporated in the upper several inches of soil. It will be relatively immobile in soil and thus is not expected to leach extensively. Chemical hydrolysis is probably not important except possibly in alkaline soils. If released to water, it will have a low to medium tendency to sorb to sediments and suspended solids or to bioconcentrate. Volatilization from water is probably not an important transport process since the estimated Henry's Law constant is very low. Biodegradation is probably the most important degradative in natural waters while chemical hydrolysis is probably not significant except in alkaline waters (half life 28 days in pH 8.6 and 25 °C). No information was found on photolysis. An estimated half-life for reaction with photochemically generated hydroxyl radicals in the vapor phase of the atmosphere is 6.1 hours.

Cleanup/Disposal: Guide No. 152: Do not touch damaged containers or spilled material unless wearing appropriate protective clothing. Stop leak if you can do it without risk. Prevent entry into waterways, sewers, basements or confined areas. Cover with plastic sheet to prevent spreading. Absorb or cover with dry earth, sand or other non-combustible material and transfer to containers. Do not get water inside containers.

Environmental Physical Data

Henry's Law Constant: 1.5 x10^{-10}

Octanol/Water Partition Coefficient: log K_{ow} = 2.75

Sorption Partition Coefficient: K_{oc} = 404

BCF: calculated at 72

Regulations

RCRA 40CFR: Not listed

CERCLA: 40CFR 302.4: Listed per CWA Section 311(b)(4) RQ: 1 lb (0.454 kg)

SARA 40CFR 372.65: Not listed TPQ: 10/10000 lb

SARA EHS 40CFR 355: Listed TPQ: 1 lb

TSCA: Not listed

Analytical Methods

Soil: EPA PMD-AZN, PMD-TLC; SW846 8140, 8141, 8141A, 8270B, 8270C

Water / Groundwater: EPA P-005-1, 1657, 614, 622, 022; SW846 8141, 8141A; ASTM D4763

Food: FDA 212.1, 232.1, 232.3, 232.4, 242.1; AOAC 980.09, 989.01

Indoor / Expired Air: NIOSH 5600

Plasma: EPA 001, 027, 028, 29; FDA 211.1, 231.1, 252

AZI5000 CAS #: 2642-71-9

AZINPHOS ETHYL

RTECS: TD8400000
DOT: UN2783; UN2784; UN3017; UN3018; IMO3.2; IMO6.1
EINECS Number: 220-147-6
Molecular Formula: $C_{12}H_{16}N_3O_3PS_2$
Formula Weight: 345.40
Synonyms: ATHYL-GUSATHION; AZINFOS-ETHYL; AZINOPHOS-ETHYL; AZINOS; AZINPHOS-AETHYL; AZINPHOS-ETHYL; AZINPHOSETHYL; AZINPHOS-ETILE; AZINUGEC E; BAY 16255; BAY 16259; BAYER 16259; BENZOTRIAZINE DERIVATIVE OF AN ETHYL DITHIOPHOSPHATE; BIONEX; COTNION-ETHYL; COTNION-ETHYL-METHYL; CRYSTHION; O,O-DIAETHYL-S-(4-OXOBENZOTRIAZIN-3-METHYL)-DITHIOPHOSPHAT; O,O-DIAETHYL-S-((4-OXO-3H-1,2,3-BENZOTRIAZIN-3-YL)-METHYL)-DITHIOPHOSPHAT; O,O-DIETHYL S-(4-OXOBENZOTRIAZINO-3-METHYL)PHOSPHORODITHIOATE; O,O-DIETHYL S-((4-OXO-1,2,3-BENZOTRIAZIN-3(4H)-YL)METHYL)PHOSPHORODITHIOATE; O,O-DIETHYL S(4-OXOBENZYLTRIAZINE-3-METHYL)DITHIOPHOSPHATE; O,O-DIETHYL PHOSPHORODITHIOATE S-ESTER WITH 3-(MERCAPTOMETHYL)-1,2,3-BENZOTRIAZIN-4(3H)-ONE; O,O-DIETHYL PHOSPHORODITHIOATE S-ESTER WITH3-(MERCAPTOMETHYL)-1,2,3-BENZOTRIAZIN-4(3H)-ONE; O,O-DIETHYL-S-(4-OXO-3H-1,2,3-BENZOTRIAZINE-3-YL)-METHYL-DITHIOPHOSPHATE; O,O-DIETHYL-S-((4-OXO-3H-1,2,3-BENZOTRIAZIN-3-YL)-METHYL)-DITHIO FOSFAAT; O,O-DIETHYL-S-((4-OXO-3H-1,2,3-BENZOTRIAZIN-3-YL)-METHYL)-DITHIOFOSFAAT; 3-DIETHYLOXYPHOSPHINOTHIOYLTHIOMETHYL-1,2,3-BENZOTRIAZIN-4(3H)-ONE; O,O-DIETHYLS-(4-OXOBENZOTRIAZINO-3-METHYL)PHOSPHORODITHIOATE; O,O-DIETIL-S-((4-OXO-3H-1,2,3-BENZOTRIAZIN-3-IL)-METIL)-DITIOFOSFATO; S-(3,4-DIHYDRO-4-OXO-1,2,3-BENZOTRIAZIN-3-YLMETHYL) O,O-DIETHYL PHOSPHORODITHIOATE; 3,4-DIHYDRO-4-OXO-3-BENZOTRIAZINYLMETHYL O,O-DIETHYLPHOSPHORODITHIOATE; S-(3,4-DIHYDRO-4-OXO-1,2,3-BENZOTRIAZIN-3-YLMETHYL)O,O-DIETHYL PHOSPHORODITHIOATE; S-3,4-DIHYDRO-4-OXOBENZO(D)(1,2,3)TRIAZIN-3-YLMETHYL)O,O-DIETHYL PHOSPHORODITHIOATE; ENT 22,014; ETHYL AZINPHOS; ETHYL GUSATHION; ETHYL GUTHION; ETHYL HOMOLOG OF GUTHION; GUSATHION; GUSATHION A-M; GUSATHION A; GUSATHION H; GUSATHION H AND K; GUSATHION K; GUSATHION K FORTE; GUSATION A; GUTEX; GUTHION (ETHYL); PHOSPHORODITHIOIC ACID,O,O-DIETHYL ESTER,S-ESTER WITH3-(MERCAPTOMETHYL)-1,2,3-BENZOTRIAZIN-4(3H)-ONE; PHOSPHORODITHIOIC ACID,O,O-DIETHYLS-((4-OXO-1,2,3-BENZOTRIAZIN-3(4H)-YL)METHYL) ESTER; R 1513; SEPIZIN L; TRIAZOTION
Description: colorless crystals or needles
Use: persistent broad spectrum nonsystemic insecticide & acaricide not registered for in usa; for fruits & vegetables, pastures, cotton, cereals, coffee, potatoes, grapes, citrus, tobacco, rice, hops, etc of forest industry; control of chewing & sucking insects, spider mites, beetles, caterpillars & their larvae, aphids, etc.

Physical Properties

Boiling Point: 111 °C (232 °F) at 0.001 mm Hg
Freezing Point: 53 °C (127.4 °F)
Specific Gravity: 1.284 at 20 °C/4 °C
Saturated Vapor Density: 1.200000004 kg/m^3
Vapor Pressure: 2.2 x10^{-7} mm Hg at 20 °C
Water Solubility: Insoluble in Water
Other Solubilities: at 20 °C: greater than 1 kg/l dichloromethane, Toluene; Soluble in 2-propanol. Barely Soluble in n-hexane.
Refraction Index: 1.5928 at 53 °C/D

RTECS Toxicity Data

Acute Oral: Rat LD_{50} Dose: 7 mg/kg. Dog LD_{50} Dose: 12 mg/kg.
Acute Inhalation: Rat LC_{50} Dose: 390 mg/m^3.
Acute Dermal: Rat LD_{50} Route: Skin; Dose: 250 mg/kg.

Hazard Overviews

Carcinogenicity: IARC - Not listed; NIOSH - Not listed; NTP - Not listed; ACGIH - Not listed; OSHA - Not listed; EPA - Not listed; MAK - Not listed

Environmental

Ecotoxicity: LC_{50} Lepomis macrochirus (bluegill) 1.1 ug/l/96 hr at 24 °C (95% confidence limit 0.9-1.2 ug/l), wt 0.8 g. Static bioassay without aeration, pH 7.2-7.5, water hardness 40-50 mg/l as calcium carbonate and alkalinity of 30-35 mg/l EC_{50} Simocephalus, immobilization, 4.2 ug/l/48 hr at 15 °C (95% confidence limit 2.9-6.1 ug/l),first instar. Static bioassay without aeration, pH 7.2-7.5, water hardness 40-50 mg/l as calcium carbonate and alkalinity of 30-35 mg/l EC_{50} Daphnia pulex, immobilization, 3.2 ug/l/48 hr at 15 °C (95% confidence limit 1.8-5.8 ug/l), first instar. Static bioassay without aeration, pH 7.2-7.5, water hardness 40-50 mg/l as calcium carbonate and alkalinity of 30-35 mg/l LC_{50} Pteronarcys 1.5 ug/l/96 hr at 15 °C (95% confidence limit 0.8-2.7 ug/l), second year class. Static bioassay without aeration, pH 7.2-7.5, water hardness 40-50 mg/l as calcium carbonate and alkalinity of 30-35 mg/l
Environmental Fate: When released on soil, it will adsorb strongly to soil, hydrolyze, and probably biodegrade. Its persistence in soil is unknown. If released in water, it will adsorb to sediment and particulate matter in the soil column. It should hydrolyze, especially in alkaline waters, and possibly biodegrade. One study suggests that it persists in rice fields for less than two months. It is not expected to bioconcentrate in fish. In the atmosphere, it will primarily occur as an aerosol and will be removed by gravitational settling. Vapor-phase will have an atmospheric half-life of 2.2 hr due to its reaction with photochemically-produced hydroxyl radicals.
Cleanup/Disposal: Guide No. 131: Fully encapsulating, vapor protective clothing should be worn for spills and leaks with no fire. Eliminate all ignition sources (no smoking, flares, sparks or flames in immediate area). All equipment used when handling the product must be grounded. Do not touch or walk through spilled material. Stop leak if you can do it without risk. Prevent entry into waterways, sewers, basements or confined areas. A vapor suppressing foam may be used to reduce vapors. Small Spills: Absorb with earth, sand or other non-combustible material and transfer to containers for later

disposal. Use clean non-sparking tools to collect absorbed material. Large Spills: Dike far ahead of liquid spill for later disposal. Water spray may reduce vapor; but may not prevent ignition in closed spaces. Guide No. 152: Do not touch damaged containers or spilled material unless wearing appropriate protective clothing. Stop leak if you can do it without risk. Prevent entry into waterways, sewers, basements or confined areas. Cover with plastic sheet to prevent spreading. Absorb or cover with dry earth, sand or other non-combustible material and transfer to containers. Do not get water inside containers.

Environmental Physical Data

Henry's Law Constant: estimated at 9.52 x10^{-9}
Octanol/Water Partition Coefficient: log K_{OW} = 3.40
Sorption Partition Coefficient: K_{OC} = 1200
BCF: estimated at 3.40

Regulations

RCRA 40CFR: Not listed
CERCLA: 40CFR 302.4: Not listed
SARA 40CFR 372.65: Not listed TPQ: 100/10000 lb
SARA EHS 40CFR 355: Listed TPQ: 100 lb
TSCA: Not listed

Analytical Methods

Soil: SW846 8141, 8141A
Water / Groundwater: EPA P-005-1, 1657; SW846 8141, 8141A
Food: FDA 212.1, 212.2, 232.1, 232.3, 232.4, 242.1
Plasma: EPA 001; FDA 211.1, 231.1, 252

AZI9000 CAS #: 1072-52-2

1-AZIRIDINEETHANOL

RTECS: CM7000000
EINECS Number: 214-009-4
Molecular Formula: C_4H_9NO
Formula Weight: 87.14

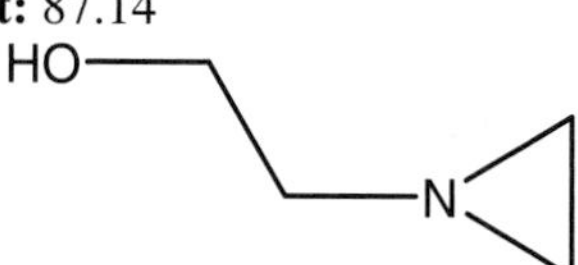

Chemical Structure

Synonyms: 1-AZIRIDINEETHANOL; 2-(1-AZIRIDINYL) ETHANOL; 2-(1-AZIRIDINYL)ETHANOL; 3-HYDROXY-1-ETHYL AZIRIDINE; N-HYDROXYETHYL ETHYLENE IMINE; N-(2-HYDROXYETHYL) ETHYLENEIMINE; 1-(2-HYDROXYETHYL) ETHYLENIMINE; N-(BETA-HYDROXYETHYL)AZIRIDENE; 2-HYDROXY-1-ETHYLAZIRIDINE; BETA-HYDROXY-1-ETHYLAZIRIDINE; N-(2-HYDROXYETHYL)AZIRIDINE; N-(BETA-HYDROXYETHYL)AZIRIDINE; 1-(2-HYDROXYETHYL)ETHYLENIMINE; N-(2-HYDROXYETHYL)ETHYLENIMINE
Description: colorless liquid
Use: as a chemical intermediate

Physical Properties

Boiling Point: 154 °C (309 °F) to 156 °C (313 °F) at 760 mm Hg
Density: 1.088 g/mL
Water Solubility: >= 100 mg/mL at 20 C
Other Solubilities: 95% Ethanol: >=100 mg/ml at 20 °C; Acetone: >=100 mg/ml at 20 °C; DMSO: >=100 mg/ml at 20 °C.
Refraction Index: 1.453 at 25 °C/D
Flash Point: 85 °C Open Cup

RTECS Toxicity Data

Acute Oral: Rat LD_{50} Dose: 74 mg/kg.
Acute Inhalation: Rat LC; Dose: >850 ppm/8hr.
Acute Dermal: Rabbit LD_{50} Route: Skin; Dose: 71 uL/kg.
Chronic (Multiple Dose) Oral: Rat Dose: 50 mg/kg/5D-I; Toxic Effects: Endocrine - Other changes; Blood - Changes in bone marrow not included above; Blood - Changes in spleen.
Irritation Eye: Rabbit Standard Draize Test Dose: 1090 mg; Reaction: severe. Rabbit Standard Draize Test Dose: 750 ug/24H; Reaction: severe.
Irritation Skin: Rabbit Standard Draize Test Dose: 5 mg/24H; Reaction: severe. Rabbit Open Draize Test Dose: 545 mg open; Reaction: moderate.
Mutagenic: Hamster Mutations in Mammalian Somatic Cells; Cell Type: lung; Dose: 100 mg/L. Hamster Sister Chromatid Exchenge; Cell Type: lung; Dose: 100 mg/L.
Tumorigenic: Mouse Route: Subcutaneous; Dose: 900 mg/kg/75W-I; Toxic Effects: Tumorigenic - Neoplastic by RTECS criteria; Tumorigenic - Tumors at site of application.

Hazard Overviews

Poison

Corrosive

Health: Corrosive to eyes/skin/respiratory tract. Poison. Other Acute Effects: may be fatal if inhaled, swallowed, or absorbed through skin; possible risk of irreversible effects; readily absorbed through skin; inhalation may result in spasm, inflammation and edema of the larynx and bronchi, chemical pneumonitis and pulmonary edema; symptoms of exposure may include burning sensation; coughing; wheezing; laryngitis; shortness of breath; headache; nausea; vomiting; may cause nervous system disturbances. Chronic Effects: possible carcinogen.
Fire: Combustible. Hazards: emits toxic fumes. Extinguishing agents: water spray; carbon dioxide, dry chemical powder or appropriate foam. Precautions: combustible liquid.
Reactivity: Incompatible with: strong oxidizing agents, strong acids, acid chlorides, acid anhydrides. Hazardous decomposition products: toxic fumes of: carbon monoxide, carbon dioxide, nitrogen oxides.
Carcinogenicity: IARC - Group 3, Not classifiable as to carcinogenicity to humans; NIOSH - Not listed; NTP - Not listed; ACGIH - Not listed; OSHA - Not listed; EPA - Not listed; MAK - Not listed

Primary Target Organs:

Eyes

Skin

Respiratory System

Environmental

Cleanup/Disposal: Guide No. 153: Eliminate all ignition sources (no smoking, flares, sparks or flames in immediate area). Do not touch damaged containers or spilled material unless wearing appropriate protective clothing. Stop leak if you can do it without risk. Prevent entry into waterways, sewers, basements or confined areas. Absorb or cover with dry earth, sand or other non-combustible material and transfer to containers. Do not get water inside containers.

Regulations

RCRA 40CFR: Not listed
CERCLA: 40CFR 302.4: Not listed
SARA 40CFR 372.65: Not listed
SARA EHS 40CFR 355: Not listed
TSCA: Listed

AZO1000 CAS #: 103-33-3

AZOBENZENE

RTECS: CN1400000
EINECS Number: 203-102-5
Molecular Formula: $C_{12}H_{10}N_2$
Structured MF: $C_6H_5N{=}NC_6H_5$
Formula Weight: 182.22

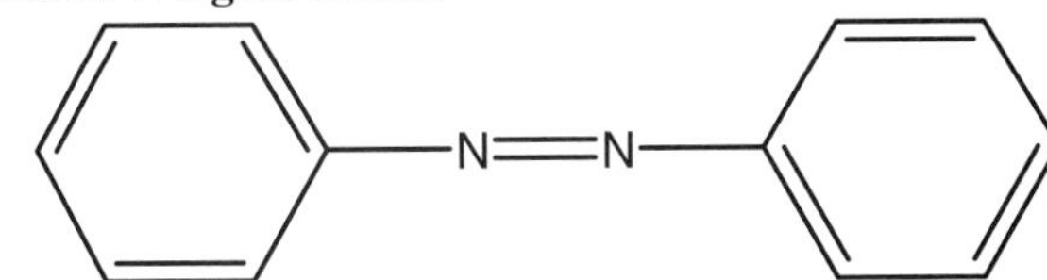

Chemical Structure

Synonyms: AZOBENZEEN; AZOBENZIDE; AZOBENZOL; AZOBISBENZENE; AZODIBENZENE; AZODIBENZENEAZOFUME; AZOFUME; BENZENEAZOBENZENE; BENZENE,AZOBIS-; BENZENE,AZODI; BENZOFUME; DIAZENE,DIPHENYL-; DIAZOBENZENE; 1,2-DIPHENYLDIAZENE; DIPHENYLDIAZENE; DIPHENYLDIIMIDE; ENT 14,611; 1,2-IPHENYLDIAZENE
Description: orange-red leaflets or solid, yellow or orange or orange-red crystals
Use: acaricide; fumigant or smoke in greenhouses for control of mites; intermediate in the production of insecticides and in the manufacture of dyes and rubber accelerators; once used as an intermediate for the production of benzidine and its salts

Physical Properties

Boiling Point: 293 °C (559 °F)
Freezing Point: 68 °C (154.4 °F)
Specific Gravity: 1.203 at 20 °C/4 °C
Saturated Vapor Density: 1.200003003 kg/m^3
Vapor Pressure: 0.00036 mm Hg at 25 °C
Water Solubility: < 0.1 mg/mL at 17.5 C
Other Solubilities: 95% Ethanol: 1-5 mg/ml at 18 °C; Acetone: 10-50 mg/ml at 18 °C; DMSO: 10-50 mg/ml at 18 °C; Ether: Soluble; Glacial Acetic Acid: Soluble; Ligroin: 12 parts/100 parts at 20 °C.
Refraction Index: 1.6266 at 78 °C/D
Flash Point: Not available; probably combustible

RTECS Toxicity Data

Acute Oral: Rat LD_{50} Dose: 1 gm/kg.
Mutagenic: Human DNA Inhibition; Cell Type: lymphocyte; Dose: 40 umol/L. Rat DNA Damage; Cell Type: liver; Dose: 300 umol/L.
Tumorigenic: Rat Route: Oral; Dose: 7350 mg/kg/2Y-C; Toxic Effects: Tumorigenic - Carcinogenic by RTECS criteria; Blood - Tumors. Rat Route: Oral; Dose: 15 gm/kg/2Y-C; Toxic Effects: Tumorigenic - Carcinogenic by RTECS criteria; Blood - Tumors. Rat Route: Oral; Dose: 7280 mg/kg/2Y-C; Toxic Effects: Tumorigenic - Neoplastic by RTECS criteria; Blood - Tumors.

Hazard Overviews

Health: May cause irritation. Toxic. Other Acute Effects: harmful if swallowed, inhaled, or absorbed through skin; absorption into the body leads to the formation of methemoglobin which in sufficient concentration causes cyanosis; onset may be delayed 2 to 4 hours or longer; target organ: liver. Chronic Effects: may alter genetic material. Carcinogen.
Fire: Will burn. Hazards: emits toxic fumes. Extinguishing agents: water spray; carbon dioxide, dry chemical powder or appropriate foam. Precautions: combustible liquid.
Reactivity: Incompatible with: strong oxidizing agents. Hazardous decomposition products: toxic fumes of: carbon monoxide, carbon dioxide, nitrogen oxides.
Carcinogenicity: IARC - Group 3, Not classifiable as to carcinogenicity to humans; NIOSH - Not listed; NTP - Not listed; ACGIH - Not listed; OSHA - Not listed; EPA - Class B2, Probable human carcinogen based on animal studies; MAK - Not listed
Primary Target Organs:

Liver

Environmental

Environmental Fate: If released to the atmosphere, it will degrade in the vapor phase by reaction with photochemically produced hydroxyl radicals (estimated half-life of about 10 days). Photolysis may contribute to its atmospheric degradation. If released to soil, it will adsorb strongly and is not expected to leach. If released to water, it will partition from the water column to sediment. Anaerobic sediment-water studies in 4 different sediments found half-lives ranging from 11.9 to 95.5 hr; test results suggested that degradation was primarily abiotic in nature with biodegradation being a minor process; the exact nature of the abiotic process was not determined, but it was thought to involve adsorption to a reactive site on the sediment where an undetermined reducing process chemically reduced the compound. The test results suggest that it will not be a persistent chemical in soil or water-sediment systems under anaerobic conditions.

Environmental Physical Data

Henry's Law Constant: estimated at 1.35 x10^{-5}
Octanol/Water Partition Coefficient: log K_{ow} = 3.82
Sorption Partition Coefficient: K_{oc} = 1350
BCF: estimated at 220 to 470

Regulations

RCRA 40CFR: Not listed
CERCLA: 40CFR 302.4: Not listed
SARA 40CFR 372.65: Not listed
SARA EHS 40CFR 355: Not listed
TSCA: Listed

AZO4200 CAS #: 123-77-3

1,1'-AZOBIS(FORMAMIDE)

RTECS: LQ1040000
EINECS Number: 204-650-8
Molecular Formula: $C_2H_4N_4O_2$
Structured MF: $H_2NCON{=}NCONH_2$
Formula Weight: 116.08

Chemical Structure

Synonyms: ABFA; AZ; 1,1'-AZOBISCARBAMIDE; AZOBISCARBONAMIDE; AZOBISCARBOXAMIDE; 1,1'-AZOBIS(FORMAMIDE); 1,1'-AZOBISFORMAMIDE; AZODICARBAMIDE; AZODICARBOAMIDE; AZODICARBONAMIDE; AZODICARBOXAMIDE; AZODICARBOXYLIC ACID DIAMIDE; 1,1'-AZODIFORMAMIDE; AZODIFORMAMIDE; AZOFORMAMIDE; DELTA(1,1')-BIUREA; CELOGEN AZ; CELOGEN AZ 130; CELOGEN AZ 199; CELOSEN AZ; CHKHZ 21R; CHKHZ 21; DIAZENEDICARBOXAMIDE; DIAZENEDICARBOXYLIC ACID DIAMIDE; FICEL EP-A; FORMAMIDE,1,1'-AZOBIS-; GENITRON AC; GENITRON AC 2; GENITRON AC 4; GENITRON EPC; KEMPORE; KEMPORE 60/40; KEMPORE 125; KEMPORE R 125; LUCEL ADA; NITROPORE; PINHOLE ACR 3; PINHOLE AK 2; PORAMID K 1; POROFOR 505; POROFOR ADC/R; POROFOR CHKHZ 21R; POROFOR CHKHZ 21; POROFORE 505; POROFOR-LK 1074; UNIFOAM AZ; UNIFOAM AZH 25; UNIFORM AZ; YUNIHOMU AZ
Description: orange-red crystals; yellow powder; odorless
Use: thermoplastic blowing agent for rubbers, foams, plastics, cellular polyolefins, polystyrenes and other unicellular products; aging and bleaching ingredient in cereal flours; and dough conditioner in baking bread

Physical Properties

Boiling Point: Decomposes
Freezing Point: 225 °C (437 °F)
Specific Gravity: 1.65 at 20 °C/20 °C
Density: 1.6 g/cu cm at 20 °C
Vapor Pressure: 7.1 mm Hg at 19.0 °C
Water Solubility: < 0.1 mg/mL at 21 C
Other Solubilities: 95% Ethanol: <1 mg/ml at 16 °C; Acetone: <1 mg/ml at 16 °C; Common organic solvents: Insoluble; DMSO: 10-50 mg/ml at 16 °C; Methanol: <1 mg/ml at 17 °C; Toluene: <1 mg/ml at 17 °C.
pH: Aqueous suspension about 6.5 to 7
Flash Point: 96 °C

RTECS Toxicity Data

Acute Oral: Rat LD_{50} Dose: >6400 mg/kg.
Chronic (Multiple Dose) Oral: Rat Dose: 42 gm/kg/10D-C; Toxic Effects: Endocrine - Evidence of thyroid hypofunction; Endocrine - Changes in thyroid weight. Rat Dose: 233 gm/kg/4W-C; Toxic Effects: Endocrine - Evidence of thyroid hypofunction.
Chronic (Multiple Dose) Inhalation: Rat Dose: 207 mg/m^3/6H/2W-I; Toxic Effects: Liver - Changes in Liver weight; Nutritional and gross metabolic - Weight loss or decreased weight gain. Rat Dose: 204 mg/m^3/6H/13W-I; Toxic Effects: Blood - Changes in serum composition.
Mutagenic: Bacteria - S Typhimurium Mutations in Microorganisms; Dose: 100 ug/plate (+S9), 333 ug/plate (-S9).

Hazard Overviews

Health: May be irritating to eyes/skin/respiratory tract. Harmful. Other Acute Effects: may be harmful by inhalation, ingestion, or skin absorption; may cause allergic respiratory and skin reactions; may cause sensitization by inhalation and skin contact.
Fire: Will burn. Hazards: in powder form capable of creating a dust explosion; emits toxic fumes. Extinguishing agents: carbon dioxide, dry chemical powder or appropriate foam. Precautions: combustible liquid.
Reactivity: Incompatible with: strong oxidizing agents, strong acids, strong bases, heavy metal salts, do not heat above melting point. Hazardous decomposition products: carbon monoxide, carbon dioxide, nitrogen oxides, ammonia.
Carcinogenicity: IARC - Not listed; NIOSH - Not listed; NTP - Not listed; ACGIH - Not listed; OSHA - Not listed; EPA - Not listed; MAK - Not listed

Environmental

Cleanup/Disposal: Guide No. 133: Eliminate all ignition sources (no smoking, flares, sparks or flames in immediate area). Do not touch or walk through spilled material. Small Dry Spills: With clean shovel place material into clean, dry container and cover loosely; move containers from spill area. Large Spills: Wet down with water and dike for later disposal.

Prevent entry into waterways, sewers, basements or confined areas.

Regulations

RCRA 40CFR: Not listed
CERCLA: 40CFR 302.4: Not listed
SARA 40CFR 372.65: Not listed
SARA EHS 40CFR 355: Not listed
TSCA: Listed

AZO5800 CAS #: 78-67-1

2,2'-AZOBIS(ISOBUTYRONITRILE)

RTECS: UG0800000
EINECS Number: 201-132-3
Molecular Formula: $C_8H_{12}N_4$
Formula Weight: 164.2

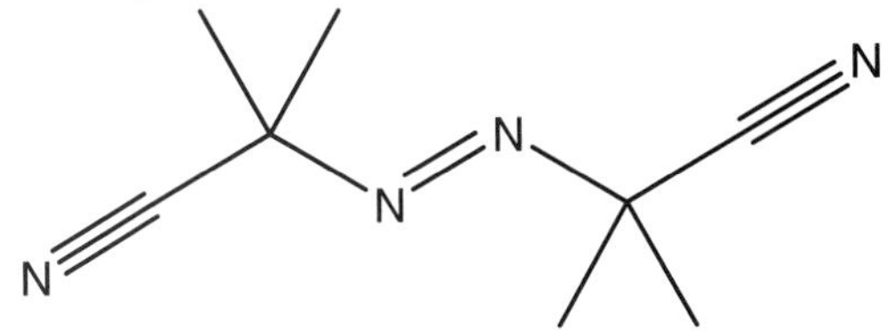

Chemical Structure

Synonyms: ACETO AZIB; AIBN; AIVN; AZDH; ALPHA,ALPHA'-AZOBISISOBUTYLONITRILE; AZOBISISOBUTYLONITRILE; 2,2'-AZOBISISOBUTYRONITRILE; ALPHA,ALPHA'-AZOBIS(ISOBUTYRONITRILE); AZOBISISOBUTYRONITRILE; 2,2'-AZOBIS[2-METHYLPROPANENITRILE]; 2,2'-AZOBIS(2-METHYLPROPIONITRILE); ALPHA,ALPHA'-AZODIISOBUTYRIC ACID DINITRILE; 2,2'-AZODIISOBUTYRONITRILE; ALPHA,ALPHA'-AZODIISOBUTYRONITRILE; AZODIISOBUTYRONITRILE; CHKHZ 57; 2,2'-DICYANO-2,2'-AZOPROPANE; 2,2'-DIMETHYL-2,2'-AZODIPROPIONITRILE; GENITRON; PIANOFOR AN; POLY-ZOLE AZDN; POROFOR-57; POROFOR 57; POROFOR N; POROPHOR N; PROPANENITRILE,2,2'-AZOBIS(2-METHYL-; PROPIONITRILE,2,2'-AZOBIS(2-METHYL-; VAZO; VAZO 64
Description: water; white powder or crystals
Use: blowing agent for elastomers & plastics; initiator for free radical reactions; catalyst for vinyl polymerizations & for curing unsaturated polyester resins; carrier gas for fumigants; molding agent

Physical Properties

Boiling Point: Decomposes at 107 °C (225 °F)
Freezing Point: 105 °C (221 °F)
Water Solubility: Insoluble in Water
Other Solubilities: Soluble in many organic solvents; Soluble in vinyl monomers.
Autoignition Temperature: 64 °C

RTECS Toxicity Data

Acute Oral: Rat LD_{50} Dose: 100 mg/kg; Toxic Effects: Behavioral - General anesthetic; Behavioral - Somnolence (general depressed activity); Behavioral - Ataxia. Mouse LD_{50} Dose: 700 mg/kg.
Acute Inhalation: Rat LC; Dose: >12 gm/m^3/4hr; Toxic Effects: Sense organs and special senses - Conjunctive irritation; Behavioral - Excitment; Nutritional and gross metabolic - Weight loss or decreased weight gain.
Acute Dermal: Rabbit LD_{Lo} Route: Subcutaneous Dose: 50 mg/kg; Toxic Effects: Behavioral - Convulsions or effect on seizure threshold; Lungs, Thorax, or Respiration - Other changes. Rat LD_{Lo} Route: Subcutaneous Dose: 30 mg/kg; Toxic Effects: Behavioral - Convulsions or effect on seizure threshold; Lungs, Thorax, or Respiration - Other changes.
Chronic (Multiple Dose) Oral: Rat Dose: 2200 mg/kg/11D-C; Toxic Effects: Gastrointestinal - Other changes; Nutritional and gross metabolic - Weight loss or decreased weight gain; DEATH.

Hazard Overviews

Health: Irritating to eyes/skin/respiratory tract. Toxic. Other Acute Effects: harmful if swallowed, inhaled, or absorbed through skin; exposure can cause nausea; dizziness; headache; CNS depression.
Fire: Will burn. Hazards: vigorously supports combustion; may be shock-sensitive; may explode when heated; container explosion may occur; in powder form capable of creating a dust explosion; emits toxic fumes. Extinguishing agents: water spray; dry chemical powder; appropriate foam. Precautions: combustible liquid.
Reactivity: Reactive u Incompatible with: strong oxidizing agents, heat, ketones, aldehydes, alcohols, alkali metals, heptane, chemical contamination. Hazardous decomposition products: toxic fumes of: carbon monoxide, carbon dioxide, tetramethylsuccinonitrile.
Carcinogenicity: IARC - Not listed; NIOSH - Not listed; NTP - Not listed; ACGIH - Not listed; OSHA - Not listed; EPA - Not listed; MAK - Not listed
Primary Target Organs:

Eyes

Skin

Respiratory System

Environmental

Regulations

RCRA 40CFR: Not listed
CERCLA: 40CFR 302.4: Not listed
SARA 40CFR 372.65: Not listed
SARA EHS 40CFR 355: Not listed
TSCA: Listed

AZO7400 CAS #: 495-48-7

AZOXYBENZENE

RTECS: CO4025000
EINECS Number: 207-802-1
Molecular Formula: $C_{12}H_{10}N_2O$
Structured MF: $C_6H_5NO=NC_6H_5$

Formula Weight: 198.22

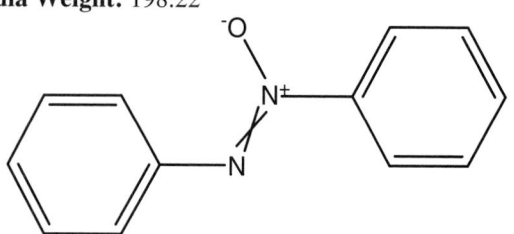

Chemical Structure

Synonyms: AZOBENZENE,OXIDE; AZOSSIBENZENE; AZOXYBENZEEN; AZOXYBENZIDE; AZOXYBENZOL; AZOXYDIBENZENE; BENZENE,AZOXYDI-; DIAZENE,DIPHENYL-,1-OXIDE; DIAZENE,DIPHENYL-,1-OXIDE (9CI); DIPHENYLDIAZENE 1-OXIDE; DIPHENYLDIAZENE OXIDE; FENAZOX; FENTOXAN; ORDINARY AZOXYBENZENE

Description: pale yellow orthorhombic needles

Use: in acaricides and insecticides; as an intermediate in organic synthesis

Physical Properties

Boiling Point: Decomposes
Freezing Point: 36 °C (96.8 °F)
Specific Gravity: 1.159 at 26 °C/4 °C
Water Solubility: Insoluble in Water
Other Solubilities: 95% Ethanol: >=100 mg/ml at 20 °C; Acetone: >=100 mg/ml at 20 °C; Absolute Ethanol: 17.5 parts/100 parts at 16 °C; DMSO: >=100 mg/ml at 20 °C; Ether: Soluble; Ligroin: 43.5 g/100 g at 15 °C.
Refraction Index: 1.652
Flash Point: Not available; probably combustible

RTECS Toxicity Data

Acute Oral: Rat LD_{50} Dose: 620 mg/kg. Mouse LD_{50} Dose: 515 mg/kg; Toxic Effects: Gastrointestinal - Other changes; Liver - Other changes.

Acute Dermal: Rabbit LD_{50} Route: Skin; Dose: 1350 mg/kg. Rabbit LD_{Lo} Route: Subcutaneous Dose: 275 mg/kg; Toxic Effects: Gastrointestinal - Other changes; Kidney, Ureter, and Bladder - Other changes.

Irritation Eye: Rabbit Standard Draize Test Dose: 500 mg/24H; Reaction: mild.

Irritation Skin: Rabbit Standard Draize Test Dose: 500 mg/24H; Reaction: mild. Rabbit Open Draize Test Dose: 10 mg/24H open; Reaction: mild.

Mutagenic: Human Sister Chromatid Exchenge; Cell Type: lymphocyte; Dose: 1 umol/L. Mouse Cytogenetic Analysis; Route: Intraperitoneal; Dose: 792 ug/kg.

Hazard Overviews

Health: Irritating to eyes/skin/respiratory tract. Harmful. Other Acute Effects: harmful if swallowed or absorbed through skin; may be harmful if inhaled; depending on the intensity and duration of exposure, effects may vary from mild irritation to severe destruction of tissue; may cause cyanosis (blue-gray coloring of skin and lips caused by lack of oxygen); exposure can cause damage to the liver, damage to the kidneys, and dermatitis. Chronic Effects: Possible carcinogen. Lab experiments have shown mutagenic effects.

Fire: Will burn. Hazards: emits toxic fumes; container explosion may occur. Extinguishing agents: water spray; carbon dioxide, dry chemical powder or appropriate foam. Precautions: combustible liquid.

Reactivity: Stable. Hazardous polymerization will not occur. Incompatible with: strong oxidizing agents, strong reducing agents, container explosion may occur under fire conditions. Hazardous decomposition products: toxic fumes of: carbon monoxide, carbon dioxide, nitrogen oxides.

Carcinogenicity: IARC - Not listed; NIOSH - Not listed; NTP - Not listed; ACGIH - Not listed; OSHA - Not listed; EPA - Not listed; MAK - Not listed

Primary Target Organs:

Eyes

Skin

Respiratory System

Environmental

Regulations

RCRA 40CFR: Not listed
CERCLA: 40CFR 302.4: Not listed
SARA 40CFR 372.65: Not listed
SARA EHS 40CFR 355: Not listed
TSCA: Listed

AZO9000 **CAS #: 16301-26-1**

AZOXYETHANE

RTECS: HM2970000
Molecular Formula: $C_4H_{10}N_2O$
Formula Weight: 102.16
Synonyms: AZOXYAETHAN; DIAZENE,DIETHYL-,1-OXIDE; ETHANE,AZOXY-

RTECS Toxicity Data

Acute Dermal: Rat LD_{50} Route: Subcutaneous Dose: 240 mg/kg.

Reproductive/Teratogenic: Rat Route: Intravenous; Dose: 25 mg/kg; Duration: female 10D of pregnancy; Specific Developmental Abnormalities - Central nervous system; Eye, ear. Rat Route: Intravenous; Dose: 45 mg/kg; Duration: female 10D of pregnancy; Effects on Embryo or Fetus - Fetal death.

Tumorigenic: Rat Route: Oral; Dose: 500 mg/kg/20W-I; Toxic Effects: Tumorigenic - Equivocal tumorigenic agent by RTECS criteria; Sense organs and special senses - Tumors; Liver - Tumors. Rat Route: Subcutaneous; Dose: 30 mg/kg (11D preg); Toxic Effects: Tumorigenic - Equivocal tumorigenic agent by RTECS criteria; Tumorigenic effects - Transplacental tumorigenesis; Brain and coverings - Tumors. Rat Route: Subcutaneous; Dose: 240 mg/kg/24W-I; Toxic Effects: Tumorigenic - Equivocal tumorigenic agent by

RTECS criteria; Brain and coverings - Tumors. Rat Route: Subcutaneous; Dose: 250 mg/kg; Toxic Effects: Tumorigenic - Equivocal tumorigenic agent by RTECS criteria; Gastrointestinal - Tumors; Tumorigenic - Tumors at site of application. Rat Route: Subcutaneous; Dose: 45 mg/kg/3W-I; Toxic Effects: Tumorigenic - Equivocal tumorigenic agent by RTECS criteria; Gastrointestinal - Tumors.

Hazard Overviews

Carcinogenicity: IARC - Not listed; NIOSH - Not listed; NTP - Not listed; ACGIH - Not listed; OSHA - Not listed; EPA - Not listed; MAK - Not listed

Environmental

Regulations

RCRA 40CFR: Not listed
CERCLA: 40CFR 302.4: Not listed
SARA 40CFR 372.65: Not listed
SARA EHS 40CFR 355: Not listed
TSCA: Not listed

BAC5000	**CAS #: 1405-87-4**
BACITRACIN	

RTECS: CP0175000
EINECS Number: 215-786-2
Molecular Formula: Unspecified or Variable

Chemical Structure

Synonyms: AK-TRACIN; ALTRACIN; AYFIVIN; BACIGUENT; BACI-JEL; BACILIQUIN; BACITEK OINTMENT; BACITRACINA; BACITRACINE; BACITRACIN-NEOMYCIN-POLYMYXIN OINTMENT; BACITRACINUM; BACTINE TRIPLE ANTIBIOTIC; CAMPHO-PHENIQUE TRIPLE PLUS PAIN RELIEVER; EPA PESTICIDE CHEMICAL CODE 006302; FORTRACIN; MYCITRACIN PLUS PAIN RELIEVER; MYCITRACIN TRIPLE ANTIBIOTIC FIRST AID OINTMENT MAXIMUMSTRENGTH; PARENTRACIN; PENITRACIN; SEPTA; SPECTROCIN PLUS; TOPITRACIN; TOPITRASIN; ZUTRACIN
Description: white to pale buff to grayish-white powder; odorless
Use: medication: antibacterial; an additive in feed for various purposes of growth and disease control; medication (vet): antibacterial, usually used locally; growth promotant; enteric infections; food additive permitted in food for human consumption; medication: anti-infective agent

Physical Properties

Freezing Point: 223 °C (433.4 °F)
Water Solubility: Freely Soluble in Water
Other Solubilities: Soluble in Alcohol, methyl Alcohol and glacial Acetic Acid, the solution in the organic solvents usually showing some insoluble residue; practically insoluble in Acetone, Chloroform and Ether
pH: Solution containing 10000 units per ml 5.5 to 7.5

RTECS Toxicity Data

Acute Oral: Mouse LD_{50} Dose: >3750 mg/kg. Guinea Pig LD_{50} Dose: 2 gm/kg.
Acute Dermal: Mouse LD_{50} Route: Subcutaneous Dose: 1300 mg/kg; Toxic Effects: Behavioral - Somnolence (general depressed activity).
Mutagenic: Bacteria - E Coli DNA Adduct; Dose: 50 umol/L.

Hazard Overviews

Health: Irritating. Other Acute Effects: harmful if swallowed or absorbed through skin; prolonged or repeated exposure may cause allergic reactions in certain sensitive individuals; target organ: kidneys.
Fire: Hazards: emits toxic fumes. Extinguishing agents: water spray; carbon dioxide, dry chemical powder or appropriate foam. Precautions: combustible liquid.
Reactivity: Incompatible with: strong oxidizing agents. Hazardous decomposition products: toxic fumes of: carbon monoxide, carbon dioxide, nitrogen oxides, sulfur oxides.
Carcinogenicity: IARC - Not listed; NIOSH - Not listed; NTP - Not listed; ACGIH - Not listed; OSHA - Not listed; EPA - Not listed; MAK - Not listed
Primary Target Organs:

Eyes

Skin

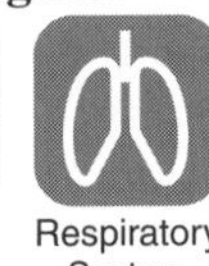
Respiratory System

Kidneys

Environmental

Regulations

RCRA 40CFR: Not listed
CERCLA: 40CFR 302.4: Not listed
SARA 40CFR 372.65: Not listed
SARA EHS 40CFR 355: Not listed
TSCA: Listed

BAN5000	**CAS #: 1689-83-4**
BANTROL	

RTECS: DI4025000
EINECS Number: 216-881-1
Molecular Formula: $C_7H_3I_2NO$
Formula Weight: 370.91

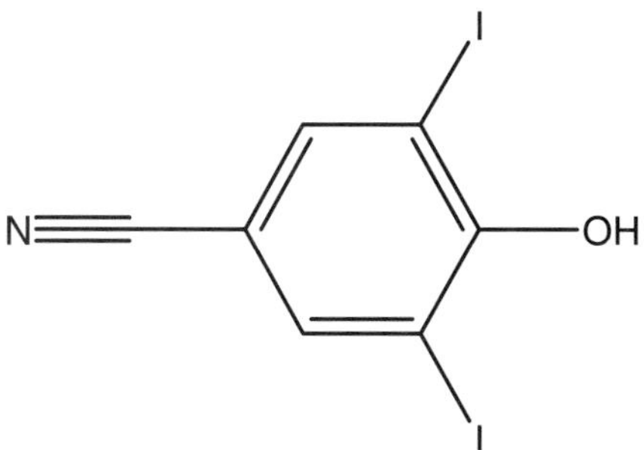

Chemical Structure

Synonyms: ACP 63303; ACTRIL; ACTRILAWN; BENTROL; BENZONITRILE,3,5-DIIODO-4-HYDROXY-; BENZONITRILE,4-HYDROXY-3,5-DIIODO-; CA 69-15; CERTOL; CERTROL; CIPOTRIL; 4-CYANO-2,6-DIIODOPHENOL; 4-CYANO-2,6-DIJODPHENOL; 2,6-DIIODO-4-CYANOPHENOL; 3,5-DIIODO-4-HYDROXYBENZONITRILE; 3,5-DIJOD-4-HYDROXY-BENZONITRIL; 4-HYDROXY-3,5-DIIODOBENZONITRILE; 4-HYDROXY-3,5-DI-IODOPHENYL CYANIDE; IOTOX; IOXYNIL; JOXYNIL; M&B 8873; OXYTRIL; TOTRIL; TREVESPAN

Description: white crystalline solid; faint phenolic odor

Use: postemergence control of fall & spring seedling broadleaf weeds; contact herbicide; molluscicide

Physical Properties

Freezing Point: 212 °C (413.6 °F) to 213.5 °C (416.3 °F)

Vapor Pressure: < $1x10^{-5}$mbar at 20 °C

Water Solubility: 0.005% at 20 °C

Other Solubilities: Slightly Soluble in Chloroform; Soluble in Ether; at 25 °C: in Acetone more than 10 g/100 ml; in Benzene 1.1 g/100 ml; in Carbon Tetrachloride 0.14 g/100 ml; in Methanol 3.3 g/100 ml.

RTECS Toxicity Data

Acute Oral: Human LD_{Lo} Dose: 28 mg/kg; Toxic Effects: Brain and coverings - Other degenerative changes; Lungs, Thorax, or Respiration - Acute pulmonary edema; Blood - Hemorrhage. Rat LD_{50} Dose: 110 mg/kg.

Acute Inhalation: Rat LC_{50} Dose: >3 gm/m^3/6hr.

Acute Dermal: Rat LD_{Lo} Route: Skin; Dose: 210 ug/kg.

Chronic (Multiple Dose) Oral: Rat Dose: 980 mg/kg/14W-C; Toxic Effects: Liver - Fatty liver degeneration; Kidney, Ureter, and Bladder - Changes in kidney weight; Biochemical - Phosphatases. Rat Dose: 21900 mg/kg/2Y-C; Toxic Effects: Liver - Changes in liver weight; Nutritional and gross metabolic - Weight loss or decreased weight gain; Biochemical - Phosphatases. Rat Dose: 980 mg/kg/14W-C; Toxic Effects: Lungs, Thorax, or Respiration - Chronic pulmonary edema; Lungs, Thorax, or Respiration - Changes in lung weight; Liver - Changes in liver weight. Rat Dose: 1365 mg/kg/78W-C; Toxic Effects: Gastrointestinal - Other changes; Liver - Other changes.

Hazard Overviews

Carcinogenicity: IARC - Not listed; NIOSH - Not listed; NTP - Not listed; ACGIH - Not listed; OSHA - Not listed; EPA - Not listed; MAK - Not listed

Environmental

Ecotoxicity: LD_{50} Pheasants oral 1000 mg/kg LC_{50} Harlequin fish 3.3 ppm/48 hr

Regulations

RCRA 40CFR: Not listed

CERCLA: 40CFR 302.4: Not listed

SARA 40CFR 372.65: Not listed

SARA EHS 40CFR 355: Not listed

TSCA: Not listed

Analytical Methods

Soil: EPA P-007-1

Water / Groundwater: EPA P-008-1

Plasma: FDA 221.1

BAR1000 CAS #: 101-27-9

BARBAN

RTECS: FD7700000

EINECS Number: 202-930-4

Molecular Formula: $C_{11}H_9Cl_2NO_2$

Formula Weight: 258.11

Synonyms: A 980; A-980; BARBAMATE; BARBAN; BARBANATE; BARBANE; 2-BUTYN-1-OL,4-CHLORO-,M-CHLOROCARBANILATE; 2-BUTYNYL-4-CHLORO-M-CHLOROCARBANILATE; C-847; CARBAMIC ACID,(3-CHLOROPHENYL)-,4-CHLORO-2-BUTYNYL ESTER; CARBAMIC ACID,(3-CHLOROPHENYL)-,4-CHLORO-2-BUTYNYL ESTER(9CI); CARBANILIC ACID,M-CHLORO-,4-CHLORO-2-BUTYNYL ESTER; CARBIN; CARBYNE; CARBYNE (HERBICIDE); CARYNE; CBN; (4-CHLOOR-BUT-2-YN-YL)-N-(3-CHLOOR-FENYL)-CARBAMAAT; (4-CHLOR-BUT-2-IN-YL)-N-(3-CHLOR-PHENYL)-CARBAMAT; CHLORINAT; N-(3-CHLORO PHENYL) CARBAMATE DE 4-CHLORO 2 BUTYNYLE; N-(3-CHLORO PHENYL) CARBAMATE DE 4-CHLORO 2-BUTYNYLE; CHLORO-2-BUTYNYL M-CHLOROCARBAMATE; 4-CHLORO-2-BUTYNYL 3-CHLOROCARBANILATE; 4-CHLORO-2-BUTYNYL M-CHLOROCARBANILATE; 4-CHLORO-2-BUTYNYL META-CHLOROCARBANILATE; 4-CHLORO-2-BUTYNYL N-(3-CHLOROPHENYL)CARBAMATE; 4-CHLOROBUT-2-YNYL 3-CHLOROPHENYLCARBAMATE; CHLORO-2-BUTYNYL M-CHLOROPHENYLCARBAMATE; 4-CHLORO-2-BUTYNYL-M-CHLOROCARBANILATE; 4-CHLOROBUT-2-YNYL-M-CHLOROCARBANILATE; M-CHLOROCARBANILIC ACID 4-CHLORO-2-BUTYNYL ESTER; M-CHLOROCARBANILIC ACID,4-CHLORO-2-BUTYNYL ESTER; (3-CHLOROPHENYL)CARBAMIC ACID 4-CHLORO-2-BUTYNYL ESTER; (4-CLORO-BUT-2-IN-IL)-N-(3-CLORO-FENIL)-CARBAMMATO; CS-847; FISONS B25; NEOBAN; S-847

Description: colorless, crystalline solid; odorless

Use: selective post-emergence herbicide; former use; herbicide for control of wild oats in field crops; for postemergence control of canary grass in barley, flax, lentils, mustard (grown for oil), peas, safflower, sugar beets, wheat, semi-dwarf wheat, sunflower & soybeans

Physical Properties

Boiling Point: Degrades

Freezing Point: 75 °C (167 °F) to 76 °C (168.8 °F)

Specific Gravity: 1.403 at 25 °C/25 °C

Saturated Vapor Density: 1.200000005 kg/m^3

Vapor Pressure: 3.79 x10^{-7} mm Hg at 25 °C
Water Solubility: 11 oom at 25 °C
Flash Point: 50 °C Closed Cup

RTECS Toxicity Data

Acute Oral: Rat LD_{50} Dose: 527 mg/kg. Mouse LD_{50} Dose: 322 mg/kg.
Acute Inhalation: Rat LC_{50} Dose: 27400 mg/m^3/4hr.
Acute Dermal: Rabbit LD_{50} Route: Skin; Dose: 23000 mg/kg. Rat LD_{50} Route: Skin; Dose: >1600 mg/kg.
Mutagenic: Bacteria - B Subtilis DNA Repair; Dose: 20 ug/disc.

Hazard Overviews

Flammable

Fire: Flammable.
Carcinogenicity: IARC - Not listed; NIOSH - Not listed; NTP - Not listed; ACGIH - Not listed; OSHA - Not listed; EPA - Not listed; MAK - Not listed

Environmental

Ecotoxicity: LC_{50} Rainbow trout 0.6 mg/l (96 hr test) /Conditions of bioassay not specified

Regulations

RCRA 40CFR: Listed Hazardous Waste No. U280 Toxic Waste
CERCLA: 40CFR 302.4: Listed per RCRA Section 3001 RQ: 1 lb (0.454 kg)
SARA 40CFR 372.65: Listed as Compound
SARA EHS 40CFR 355: Not listed
TSCA: Not listed

Analytical Methods

Soil: SW846 8270B, 8270C, 8321A
Water / Groundwater: EPA 632
Drinking Water: AOAC 992.14
Plasma: EPA 028, 29

BAR1500 CAS #: 7440-39-3

BARIUM

RTECS: CQ8370000
DOT: UN1400; IMO4.3
EINECS Number: 231-149-1
Molecular Formula: Ba
Structured MF: Ba
Formula Weight: 137.33

Ba

Chemical Structure

Synonyms: BARIO; BARYUM

Description: silver, white powder
Use: carrier for radium; as getters in electronic tubes

Physical Properties

Boiling Point: 1640 °C (2984 °F)
Freezing Point: 725 °C (1337 °F)
Specific Gravity: 3.51 at 20 °C
Vapor Pressure: 10 mm Hg at 1049 °C
Water Solubility: Decomposes
Other Solubilities: Benzene: Insoluble.
Surface Tension: 224 dynes/cm
Ionization Potential (eV): 5.21170

Hazard Overviews

Explosive

Flammable

Fire Diamond

Health: Severely irritating to eyes/skin/respiratory tract. Also Causes: gastroenteritis, slow pulse, muscle spasm, hypokalemia, bronchial irritation, pneumoconiosis, dermatitis, burns. Chronic Effects: nodular opacities, benign pneumoconiosis (baritosis).
Fire: Explosive and flammable. For small fires, use dry chemical, soda ash, lime, or sand. Do not use water or foam. For large fires, withdraw from area and let fire burn.
Reactivity: Stable. Hazardous polymerization cannot occur. Avoid: heating in hydrogen to about 392 °F/200 °C; exposing free metal to moist air or cold water. Incompatible with: water; carbon tetrachloride; trichloroethylene; fluorotrichloromethane; tetrachloroethylene; acids; trichloroethylene and water; trichlorotrifluoroethane; 1,1,2-trichloro trifluoro ethane; flurotrichloroethane; halogens; ammonia.
Carcinogenicity: IARC - Not listed; NIOSH - Not listed; NTP - Not listed; ACGIH - Class A4, Not classifiable as a human carcinogen; OSHA - Not listed; EPA - Not listed; MAK - Not listed
Primary Target Organs:

Respiratory System

Exposure Limits
OSHA PEL: TWA: 0.5 mg/m^3; as Ba.
ACGIH TLV: TWA: 0.5 mg/m^3; as Ba.
NIOSH REL: TWA: 0.5 mg/m^3; as Ba soluble.
DFG MAK: TWA: 0.5 mg/m^3; as Ba.
Respirator Recommendation
Exposure Range: >0.5 to 5 mg/m^3 Air Purifying, Negative Pressure, Half Mask
Exposure Range: >5 to <50 mg/m^3 Air Purifying, Negative Pressure, Full Face
Exposure Range: 50 to unlimited mg/m^3 Self-contained Breathing Apparatus, Pressure Demand, Full Face
Cartridge Color: dust/mist filter (use P100 or consult supervisor for appropriate dust/mist filter)

Environmental

Cleanup/Disposal: Guide No. 138: Eliminate all ignition sources (no smoking, flares, sparks or flames in immediate area). Do not touch or walk through spilled material. Stop leak if you can do it without risk. Use water spray to reduce vapors or divert vapor cloud drift. Do not get water on spilled substance or inside containers. Small Spills: Cover with dry earth, dry sand, or other non-combustible material followed with plastic sheet to minimize spreading or contact with rain. Dike for later disposal; do not apply water unless directed to do so. Powder Spills: Cover powder spill with plastic sheet or tarp to minimize spreading and keep powder dry. Do not clean-up or dispose of, except under supervision of a specialist.

Regulations

RCRA 40CFR: Not listed
CERCLA: 40CFR 302.4: Not listed
SARA 40CFR 372.65: Listed
SARA EHS 40CFR 355: Not listed
TSCA: Listed

Analytical Methods

Air: EPA 29, 0060, ITM-001
Soil: CLP 200.10_M, 200.62, 200.7_M, 202.62, 208.1_M, 208.2_M, 6020_M, ICP-AES; EPA 200.7, 200.8; SW846 3005A, 3010A, 3015, 3031, 3040A, 3050A, 3050B, 3051, 3052, 6010A, 6010B, 6020, 7000A, 7080A, 7081, OSW-A; ASTM D3974; USGS I5084
Water / Groundwater: EPA 200.0, 200.15, 200.7; APHA 3111-A, 3111-D, 3113-B, 3120; ASTM D3651, D3986, D4382; USGS E-SPEC, I1084, I1472, I3084, I7084; FISON AES-0029; CEM RD42
Drinking Water: AOAC 920.201, 993.14; APHA 3500-BA
Indoor / Expired Air: NIOSH 7056; ASTM D4185
Urine: NIOSH 8310
Other: EPA 1620, 208.1, 208.2; AOAC 990.08

BAR2000 CAS #: 513-77-9

BARIUM CARBONATE

RTECS: CQ8600000
DOT: UN1564; IMO6.1
EINECS Number: 208-167-3
Molecular Formula: $CBaO_3$
Structured MF: $BaCO_3$
Formula Weight: 197.37

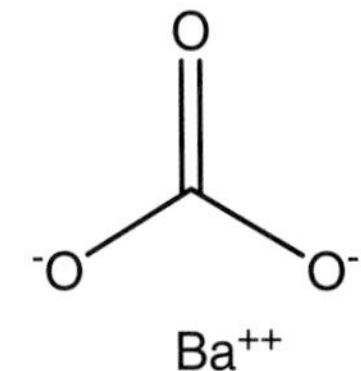

Chemical Structure

Synonyms: BARIUM CARBONATE (1:1); C.I. 77099; C.I. PIGMENT WHITE 10; CARBONIC ACID,BARIUM SALT (1:1); EPA PESTICIDE CHEMICAL CODE 007501
Description: white to greyish-white fine granular powder; odorless
Use: in ceramics, paints, enamels, marble substitutes, rubber; anal rgnt; mfr of paper, barium salts, optical glasses; electrodes; treatment of brines in chlorine-alkali cells to remove sulfates; case-hardening bath; in glass for television tubes; in ferrite mfr; in brick, photographic, glass, & chem mfr industries; oil well drilling mud additive to prevent destabilization by sol materials eg, gypsum; to precipitate sulfates from salt brine; formerly rodenticide.

Physical Properties

Boiling Point: Decomposes at 1300 °C (2372 °F)
Freezing Point: 811 °C (1491.8 °F)
Specific Gravity: 4.43
Vapor Pressure: Negligible
Water Solubility: 1 part /1000 parts at 0 °C
Other Solubilities: Soluble in dilute Hydrochloric Acid, Nitric Acid or Acetic Acid; Soluble in solution of Ammonium Chloride or Ammonium Nitrate; Insoluble in sulfuric acid; Soluble in Ethanol.
Refraction Index: 1.529
Flash Point: Nonflammable

RTECS Toxicity Data

Acute Oral: Human TD_{Lo} Dose: 11 mg/kg; Toxic Effects: Gastrointestinal - Ulceration or bleeding from stomach. Human TD_{Lo} Dose: 29 mg/kg; Toxic Effects: Peripheral nerve and sensation - Flaccid paralysis without anesthesia; Peripheral nerve and sensation - Pareshtesia; Behavioral - Muscle weakness. Woman TD_{Lo} Dose: 800 mg/kg; Toxic Effects: Gastrointestinal - Hypermotility, diarrhea; Gastrointestinal - Nausea or vomiting; Lungs, Thorax, or Respiration - Other changes. Human LD_{Lo} Dose: 17 mg/kg. Man LD_{Lo} Dose: 800 mg/kg; Toxic Effects: Behavioral - Convulsions or effect on seizure threshold; Cardiac - Change in rate; Cardiac - Other changes. Rat LD_{50} Dose: 418 mg/kg.
Chronic (Multiple Dose) Inhalation: Rat Dose: 5200 ug/m^3/17W-I; Toxic Effects: Vascular - BP elevation not characterized in autonomic section; Nutritional and gross metabolic - Changes in phosphorus.
Reproductive/Teratogenic: Rat Route: Inhalation; Dose: 1150 ug/m^3/2 Duration: male 16W prior to mating; Paternal Effects - Spermatogenesis; Testes, epididymis, sperm duct. Rat Route: Inhalation; Dose: 3130 ug/m^3/2 Duration: female 16W prior to mating Maternal Effects - Oogenesis; Ovaries, fallopian tubes.

Hazard Overviews

Fire Diamond

Health: Irritating to the eyes/skin/respiratory tract. Also Causes: ingestion: nausea, vomiting, abdominal pain, skin,

tingling, muscle weakness and paralysis, potassium deficiency, increased blood pressure, slow pulse.
Fire: Noncombustible. Use extinguishing agents suitable for surrounding fire.
Reactivity: Stable. Hazardous polymerization cannot occur. Incompatible with: 2-furanpercarboxylic acid and bromine trifluoride. Hazardous decomposition products: barium oxide; carbon dioxide.
Carcinogenicity: IARC - Not listed; NIOSH - Not listed; NTP - Not listed; ACGIH - Class A4, Not classifiable as a human carcinogen; OSHA - Not listed; EPA - Not listed; MAK - Not listed
Primary Target Organs:

Eyes

Skin
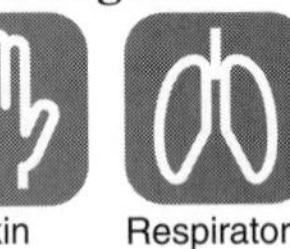
Respiratory System
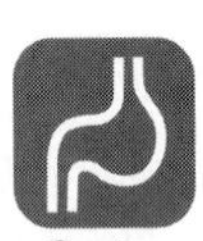
Gastro-intestinal

Nervous System

Cardio-vascular

Exposure Limits
OSHA PEL: TWA: 0.5 mg/m^3; as Ba.
ACGIH TLV: TWA: 0.5 mg/m^3; as Ba.
NIOSH REL: TWA: 0.5 mg/m^3; as Ba soluble.
DFG MAK: TWA: 0.5 mg/m^3; as Ba.

Environmental

Cleanup/Disposal: Guide No. 154: Eliminate all ignition sources (no smoking, flares, sparks or flames in immediate area). Do not touch damaged containers or spilled material unless wearing appropriate protective clothing. Stop leak if you can do it without risk. Prevent entry into waterways, sewers, basements or confined areas. Absorb or cover with dry earth, sand or other non-combustible material and transfer to containers. Do not get water inside containers.

Environmental Physical Data

BCF: marine animals concentrate 7 to 100
BOD: none

Regulations

RCRA 40CFR: Not listed
CERCLA: 40CFR 302.4: Not listed
SARA 40CFR 372.65: Listed as Compound
SARA EHS 40CFR 355: Not listed
TSCA: Listed

BAR3000 CAS #: 10361-37-2

BARIUM CHLORIDE

RTECS: CQ8750000
DOT: UN1564; IMO6.1
EINECS Number: 233-788-1
Molecular Formula: $BaCl_2$
Structured MF: $BaCl_2$
Formula Weight: 208.27

Cl^- Ba^{++} Cl^-

Chemical Structure

Synonyms: BA 0108E; BARIUM DICHLORIDE; SBA 0108E
Description: colorless flat crystals, cubic crystals; odorless
Use: cardiac stimulant; radioactive compound as experimental bone scanning agent; chemical manufacture; ceramic glazing

Physical Properties

Boiling Point: 1560 °C (2840 °F)
Freezing Point: 962.78 °C (1765.004 °F)
Specific Gravity: 3.856 at 24 °C
Density: 3.856 g/mL at 24 °C
Vapor Pressure: Essentially 0 mm Hg
Water Solubility: 31 parts/100 parts Water at 0 °C
Other Solubilities: Hydrochloric Acid: Soluble; Nitric acid: Soluble.
Refraction Index: 1.7303
Flash Point: Probably not flammable

RTECS Toxicity Data

Acute Oral: Rat LD_{50} Dose: 118 mg/kg. Mouse LD_{Lo} Dose: 70 mg/kg.
Acute Dermal: Rabbit LD_{Lo} Route: Subcutaneous Dose: 40 mg/kg. Rat LD_{50} Route: Subcutaneous Dose: 178 mg/kg.
Reproductive/Teratogenic: Rat Route: Intratesticular; Dose: 16659 ug/kg; Duration: male 1D prior to mating; Paternal Effects - Testes, epididymis, sperm duct.
Mutagenic: Yeast - S Cerevisiae Gene Conversion; Dose: 14 mmol/L.

Hazard Overviews

Fire Diamond

Health: Severely irritating to eyes/skin/respiratory tract. Toxic. Also Causes: numbness, muscle tension, slow heart rate, blood vessel constriction, respiratory failure, kidney damage.
Fire: Will burn but doesn't ignite readily. Fight fire with dry chemical, carbon dioxide, water spray, fog, or foam.
Reactivity: Stable. Hazardous polymerization cannot occur. Avoid: exposure to heat; ignition sources. Incompatible with: bromine trifluoride; 2-furan percarboxylic acid. Hazardous decomposition products: chlorine gas.
Carcinogenicity: IARC - Not listed; NIOSH - Not listed; NTP - Not listed; ACGIH - Class A4, Not classifiable as a human carcinogen; OSHA - Not listed; EPA - Not listed; MAK - Not listed
Primary Target Organs:

Eyes

Skin

Respiratory System

Kidneys

Exposure Limits
OSHA PEL: TWA: 0.5 mg/m^3; as Ba.
ACGIH TLV: TWA: 0.5 mg/m^3; as Ba.
NIOSH REL: TWA: 0.5 mg/m^3; as Ba soluble.

NIOSH IDLH: 50 mg/m^3; as Ba.
DFG MAK: TWA: 0.5 mg/m^3; as Ba.
Respirator Recommendation
Exposure Range: >0.5 to 5 mg/m^3 Air Purifying, Negative Pressure, Half Mask
Exposure Range: >5 to <50 mg/m^3 Air Purifying, Negative Pressure, Full Face
Exposure Range: 50 to unlimited mg/m^3 Self-contained Breathing Apparatus, Pressure Demand, Full Face
Cartridge Color: dust/mist filter (use P100 or consult supervisor for appropriate dust/mist filter)

Environmental

Cleanup/Disposal: Guide No. 154: Eliminate all ignition sources (no smoking, flares, sparks or flames in immediate area). Do not touch damaged containers or spilled material unless wearing appropriate protective clothing. Stop leak if you can do it without risk. Prevent entry into waterways, sewers, basements or confined areas. Absorb or cover with dry earth, sand or other non-combustible material and transfer to containers. Do not get water inside containers.

Environmental Physical Data

BCF: marine animals concentrate 7 to 100

Regulations

RCRA 40CFR: Not listed
CERCLA: 40CFR 302.4: Not listed
SARA 40CFR 372.65: Listed as Compound
SARA EHS 40CFR 355: Not listed
TSCA: Listed

BAR3500 CAS #: 10326-27-9

BARIUM CHLORIDE, DIHYDRATE

RTECS: CQ8751000
Molecular Formula: $BaCl_2H_4O_2$
Formula Weight: 244.28

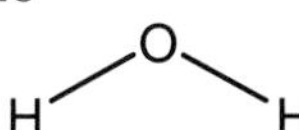

Cl^- Ba^{++} Cl^-

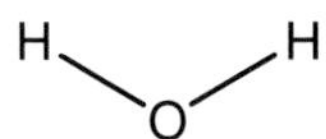

Chemical Structure

Synonyms: BARIUM DICHLORIDE DIHYDRATE
Description: white crystalline solid or crystals
Use: reagent; lube oil additive; boiler compounds; textile dye; pigments; manufacture of white leather

Physical Properties

Boiling Point: 35.7 °C (96 °F)
Freezing Point: 963 °C (1765.4 °F)
Specific Gravity: 3.097 at 24/4 °C
Density: 3.86 g/mL at 19 °C
Water Solubility: >= 100 mg/mL at 20 C
Other Solubilities: 95% Ethanol: <1 mg/ml at 20 °C; Acetone: <1 mg/ml at 20 °C; DMSO: <1 mg/ml at 20 °C; Hydrochloric Acid: Slightly Soluble; Methanol: Soluble; Nitric acid: Slightly Soluble.
Flash Point: Not available; probably combustible

Hazard Overviews

Health: Irritating to eyes/skin/respiratory tract. Toxic. Other Acute Effects: harmful if swallowed, inhaled, or absorbed through skin; exposure can cause stomach pains; vomiting; diarrhea; target organs: heart, nerves, kidneys, g.i. system, bone marrow, spleen, liver.
Fire: Will burn. Hazards: emits toxic fumes. Extinguishing agents: noncombustible; use extinguishing media appropriate to surrounding fire conditions. Precautions: combustible liquid.
Reactivity: Incompatible with: strong oxidizing agents. Hazardous decomposition products: hydrogen chloride gas.
Carcinogenicity: IARC - Not listed; NIOSH - Not listed; NTP - Not listed; ACGIH - Class A4, Not classifiable as a human carcinogen; OSHA - Not listed; EPA - Not listed; MAK - Not listed
Primary Target Organs:

Eyes

Skin
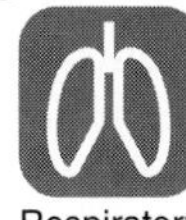
Respiratory System
Gastro-intestinal

Nervous System
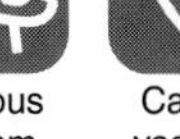

Cardio-vascular

Exposure Limits
OSHA PEL: TWA: 0.5 mg/m^3; as Ba.
ACGIH TLV: TWA: 0.5 mg/m^3; as Ba.
NIOSH REL: TWA: 0.5 mg/m^3; as Ba soluble.
DFG MAK: TWA: 0.5 mg/m^3; as Ba.

Environmental

Regulations

RCRA 40CFR: Not listed
CERCLA: 40CFR 302.4: Not listed
SARA 40CFR 372.65: Listed as Compound
SARA EHS 40CFR 355: Not listed
TSCA: Not listed

BAR4000 CAS #: 10294-40-3

BARIUM CHROMATE

RTECS: CQ8760000
DOT: UN1564; IMO6.1
EINECS Number: 233-660-5
Molecular Formula: $BaCrH_2O_4$

Formula Weight: 253.32

$$O=Cr(=O)(-O^-)-O^- \quad Ba^{++}$$

Chemical Structure

Synonyms: BARIUM CHROMATE (1:1); BARIUM CHROMATE OXIDE; BARIUM CHROMATE (VI); BARYTA YELLOW; C.I. 77103; C.I. PIGMENT YELLOW 31; CHROMIC ACID (H2CRO4),BARIUM SALT (1:1); CHROMIC ACID,BARIUM SALT (1:1); LEMON CHROME; LEMON YELLOW; PERMANENT YELLOW; STEINBUHL YELLOW; ULTRAMARINE YELLOW

Description: yellow heavy monoclinic orthorhombic crystals

Use: as pigment almost entirely in anticorrosion jointing pastes to prevent electro-chem corrosion at junctions of dissimilar metals; artists' colors & in coloring glass, ceramics, porcelain; in metal primers, pyrotechnic compositions; in fuses, safety matches, ignition control devices; in high temperature batteries

Physical Properties

Specific Gravity: 4.498 at 15 °C
Water Solubility: 0 g/100 cc at 16 °C
Other Solubilities: Soluble in strong acids.
Flash Point: Combustible

RTECS Toxicity Data

Mutagenic: Hamster Sister Chromatid Exchenge; Cell Type: ovary; Dose: 100 ug/L.

Hazard Overviews

Health: Irritating to eyes/skin/respiratory tract. Toxic. Other Acute Effects: harmful if swallowed, inhaled, or absorbed through skin; exposure can cause dermatitis; nausea; vomiting; diarrhea; dizziness; convulsions; muscle cramps/spasms; irregular breathing; pulmonary edema; effects may be delayed; exposure has been reported to produce skin and nasal ulcerations with continued exposure leading to perforation of the nasal septa. Chronic Effects: may alter genetic material; may cause heritable genetic damage; target organs: lungs, kidneys, liver. Carcinogen.

Fire: Combustible. Hazards: emits toxic fumes; contact with other material may cause fire. Extinguishing agents: noncombustible; use extinguishing media appropriate to surrounding fire conditions. Precautions: combustible liquid.

Reactivity: Incompatible with: strong reducing agents.

Carcinogenicity: IARC - Group 1, Carcinogenic to humans; NIOSH - Listed as carcinogen; NTP - Class 1, Known to be a carcinogen; ACGIH - Class A1, Confirmed human carcinogen; OSHA - Not listed; EPA - Class A, Human carcinogen; MAK - Class A2, Unmistakably carcinogenic in animal experimentation only

Primary Target Organs:

Eyes

Skin
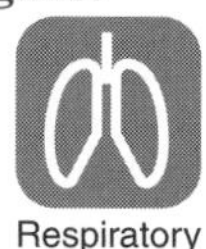
Respiratory System

Liver

Kidneys

Exposure Limits

OSHA PEL: STEL: 0.1 mg/m^3; as CrO_3; Other Values: 0.1 mg/m^3; Clg Cr-VI as CrO_3.

NIOSH REL: TWA: 0.5 mg/m^3; as Cr;Cr-II;Cr-III;Cr(VI)=.001.

Respirator Recommendation

Exposure Range: >0.1 to 1 mg/m^3 Air Purifying, Negative Pressure, Half Mask

Exposure Range: >1 to 10 mg/m^3 Air Purifying, Negative Pressure, Full Face

Exposure Range: >10 to <15 mg/m^3 Supplied Air, Constant Flow/Pressure Demand, Full Face

Exposure Range: 15 to unlimited mg/m^3 Self-contained Breathing Apparatus, Pressure Demand, Full Face

Cartridge Color: magenta (P100)

Note: as chromium VI compounds

Environmental

Cleanup/Disposal: Guide No. 154: Eliminate all ignition sources (no smoking, flares, sparks or flames in immediate area). Do not touch damaged containers or spilled material unless wearing appropriate protective clothing. Stop leak if you can do it without risk. Prevent entry into waterways, sewers, basements or confined areas. Absorb or cover with dry earth, sand or other non-combustible material and transfer to containers. Do not get water inside containers.

Environmental Physical Data

BCF: marine animals concentrate 7 to 100

Regulations

RCRA 40CFR: Not listed
CERCLA: 40CFR 302.4: Listed as Compound per CWA Section 307(a) per CAA Section 112
SARA 40CFR 372.65: Listed as Compound
SARA EHS 40CFR 355: Not listed
TSCA: Listed

BAR4500 **CAS #: 542-62-1**

BARIUM CYANIDE

RTECS: CQ8785000
DOT: UN1565; IMO6.1
EINECS Number: 208-822-3
Molecular Formula: C_2BaN_2
Structured MF: $Ba(CN)_2$
Formula Weight: 189.40

Synonyms: BARIUM DICYANIDE; CIANURO BARICO; CYANURE DE BARYUM

Description: white crystalline powder

Use: electroplating processes; metallurgy; preparation of aluminum, sodium, potassium, lithium, rubidium, cesium, and ammonium cyanides; removing carbonate from potassium baths

Physical Properties

Water Solubility: 80 g/100 cc Water at 14 °C
Other Solubilities: 18 g/100 cc 70% Alcohol at 14 °C.
Flash Point: Nonflammable

Hazard Overviews

Fire: Noncombustible.
Carcinogenicity: IARC - Not listed; NIOSH - Not listed; NTP - Not listed; ACGIH - Class A4, Not classifiable as a human carcinogen; OSHA - Not listed; EPA - Not listed; MAK - Not listed
Exposure Limits
OSHA PEL: TWA: 0.5 mg/m^3; as Ba.
ACGIH TLV: TWA: 0.5 mg/m^3; as Ba.
NIOSH REL: TWA: 0.5 mg/m^3; as Ba soluble.
DFG MAK: TWA: 0.5 mg/m^3; as Ba.

Environmental

Ecotoxicity: TLm Pinfish < 1 mg/l/96 hr /Conditions of the bioassay not specified
Cleanup/Disposal: Guide No. 157: Eliminate all ignition sources (no smoking, flares, sparks or flames in immediate area). All equipment used when handling the product must be grounded. Do not touch damaged containers or spilled material unless wearing appropriate protective clothing. Stop leak if you can do it without risk. A vapor suppressing foam may be used to reduce vapors. Do not get water inside containers. Use water spray to reduce vapors or divert vapor cloud drift. Prevent entry into waterways, sewers, basements or confined areas. Small Spills: Cover with dry earth, dry sand, or other non-combustible material followed with plastic sheet to minimize spreading or contact with rain. Use clean non-sparking tools to collect material and place it into loosely covered plastic containers for later disposal.

Environmental Physical Data

BCF: marine animals concentrate 7 to 100

Regulations

RCRA 40CFR: Listed Hazardous Waste No. P013 Toxic Waste
CERCLA: 40CFR 302.4: Listed per CWA Section 311(b)(4) per RCRA Section 3001 RQ: 10 lb (4.535 kg)
SARA 40CFR 372.65: Listed as Compound
SARA EHS 40CFR 355: Not listed
TSCA: Listed

BAR5000 CAS #: 7787-32-8

BARIUM FLUORIDE

RTECS: CQ9100000
EINECS Number: 232-108-0
Molecular Formula: BaF_2
Structured MF: BaF_2
Formula Weight: 175.34

Ba^{++}

F^- F^-

Chemical Structure

Synonyms: BARYUM FLUORURE
Description: transparent, white cubic crystals, powder; stypic disagreeable odor

Physical Properties

Boiling Point: 2137 °C (3879 °F)
Freezing Point: 1353 °C (2467.4 °F)
Density: 4.89
Water Solubility: 2 g/L at 10 °C
Other Solubilities: Hydrochloric, nitric & hydroflouric acids, ammomium chloride
Refraction Index: 1.475

RTECS Toxicity Data

Acute Oral: Rat LD_{50} Dose: 250 mg/kg; Toxic Effects: Behavioral - Somnolence (general depressed activity); Behavioral - Ataxia; Lungs, Thorax, or Respiration - Respiratory depression.
Acute Dermal: Frog LD_{Lo} Route: Subcutaneous Dose: 1540 mg/kg.
Reproductive/Teratogenic: Rat Route: Inhalation; Dose: 5660 ug/m^3/2 Duration: female 1-21D of pregnancy; Effects on Embryo or Fetus - Fetotoxicity; Fetal death. Mouse Route: Intraperitoneal; Dose: 525 mg/kg; Duration: female 1-21D of pregnancy; Specific Developmental Abnormalities - Other developmental abnormalities. Mouse Route: Intraperitoneal; Dose: 656 mg/kg; Duration: female 1-21D of pregnancy; Effects on Embryo or Fetus - Fetotoxicity.

Hazard Overviews

Fire Diamond

Health: Irritating to eyes/skin/respiratory tract. Also Causes: sore throat, nose bleeds, cough, bronchitis, chest pain, pulmonary edema (fluid in lungs). Chronic Effects: greenish tongue, chronic bronchitis, dermatitis.
Fire: Will burn but doesn't ignite readily. Fight fire with dry chemical, carbon dioxide, water spray, fog, or foam.
Reactivity: Stable. Hazardous polymerization cannot occur. Avoid: excessive dust generation; exposure to heat and ignition sources; contact with strong inorganic acids. Incompatible with: strong inorganic acids. Hazardous decomposition products: hydrogen fluoride gas; barium oxide gas.
Carcinogenicity: IARC - Not listed; NIOSH - Not listed; NTP - Not listed; ACGIH - Class A4, Not classifiable as a human carcinogen; OSHA - Not listed; EPA - Not listed; MAK - Not listed

Primary Target Organs:

Eyes

Skin

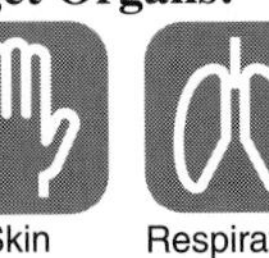
Respiratory System

Bone

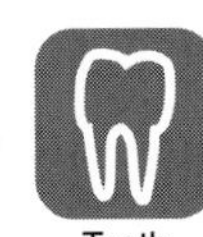
Teeth

Exposure Limits
OSHA PEL: TWA: 0.5 mg/m^3; as Ba.
ACGIH TLV: TWA: 0.5 mg/m^3; as Ba.
NIOSH REL: TWA: 0.5 mg/m^3; as Ba soluble.
DFG MAK: TWA: 0.5 mg/m^3; as Ba.

Environmental

Regulations

RCRA 40CFR: Not listed
CERCLA: 40CFR 302.4: Not listed
SARA 40CFR 372.65: Listed as Compound
SARA EHS 40CFR 355: Not listed
TSCA: Listed

BAR5500 CAS #: 17194-00-2

BARIUM HYDROXIDE

RTECS: CQ9200000
DOT: UN1564; IMO6.1
EINECS Number: 241-234-5
Molecular Formula: BaH_2O_2
Structured MF: $Ba(OH)_2 \cdot 8H_2O$
Formula Weight: 171.38

Ba^{++}

OH^- OH^-

Chemical Structure

Synonyms: BARIUM DIHYDROXIDE; CAUSTIC BARYTA
Description: white powder, crystals; odorless
Use: in mfr of alkali, glass, barium soaps & chem; in rubber vulcanization, in corrosion inhibitors, drilling fluids, pesticides; boiler scale remedy; refining oils; in softening water, & for sulfate removal; fresco painting; refining of beet sugar; agent for water & brine; boiler scale removal; dehairing agent; catalyst in mfr of resins; insecticide & fungicide; puriying agent for caustic soda; steel carbonizing agent; sulfate controlling agent in ceramics; barium salts; anal chemistry; in motor oil detergents; in plastics stabilizers, papermaking additives, sealing compounds, pigment dispersants & self-extinguishing foams.

Physical Properties

Boiling Point: 780 °C (1436 °F)
Freezing Point: 408 °C (766.4 °F)
Specific Gravity: 2.18
Vapor Pressure: 227 mm Hg at 77.9 °C
Water Solubility: Slightly Soluble in Water
Other Solubilities: Insoluble in Acetone.
Refraction Index: 1.471 to 1.502

RTECS Toxicity Data

Acute Oral: Mammal LD_{50} Dose: 308 mg/kg; Toxic Effects: Automatic Nervous System - Other (direct) parasympathomimetic; Behavioral - Muscle weakness; Gastrointestinal - Nausea or vomiting.

Hazard Overviews

Poison

Corrosive

Fire Diamond

Health: Corrosive, causes severe burns to eyes/skin/respiratory tract. Poison. Also Causes: muscle stimulation followed by paralysis; ingestion of large amounts may cause death due to cardiac, respiratory or kidney failure.
Fire: Noncombustible. Use extinguishing agents suitable for surrounding fire.
Reactivity: Unstable, absorbs carbon dioxide from air to form insoluble barium carbonate. Hazardous polymerization cannot occur. Incompatible with: acids; oxidizers; chlorinated rubber; corrosive to metals; oxygen; nitrogen; hydrogen; ammonia; water; halogens; sulfides.
Carcinogenicity: IARC - Not listed; NIOSH - Not listed; NTP - Not listed; ACGIH - Class A4, Not classifiable as a human carcinogen; OSHA - Not listed; EPA - Not listed; MAK - Not listed
Primary Target Organs:

Eyes

Skin

Respiratory System

Nervous System

Kidneys

Cardio-vascular

Exposure Limits
OSHA PEL: TWA: 0.5 mg/m^3; as Ba soluble.
ACGIH TLV: TWA: 0.5 mg/m^3; as Ba.
NIOSH REL: TWA: 0.5 mg/m^3; as Ba soluble.
DFG MAK: TWA: 0.5 mg/m^3; as Ba.

Environmental

Cleanup/Disposal: Guide No. 154: Eliminate all ignition sources (no smoking, flares, sparks or flames in immediate area). Do not touch damaged containers or spilled material unless wearing appropriate protective clothing. Stop leak if you can do it without risk. Prevent entry into waterways, sewers, basements or confined areas. Absorb or cover with dry earth, sand or other non-combustible material and transfer to containers. Do not get water inside containers.

Environmental Physical Data

BCF: marine animals concentrate 7 to 100

Regulations

RCRA 40CFR: Not listed
CERCLA: 40CFR 302.4: Not listed
SARA 40CFR 372.65: Listed as Compound
SARA EHS 40CFR 355: Not listed
TSCA: Listed

BAR6000 CAS #: 22326-55-2

BARIUM HYDROXIDE, MONOHYDRATE

Molecular Formula: BaH_4O_3
Formula Weight: 189.38

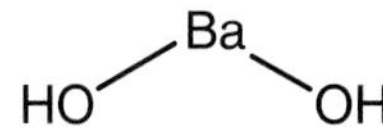

OH_2

Chemical Structure

Physical Properties

Density: 3.743 g/mL at 16 °C
Bulk Density: 65 lbs/cu ft
Vapor Pressure: 227 mm Hg at 77.9 °C
Water Solubility: 6 g/100 cc at 15 °C
Other Solubilities: Dilute Acids
pH: Strongly alkaline
Refraction Index: 1.5
Flash Point: Noncombustible

Hazard Overviews

Reactivity: Unstable, absorbs carbon dioxide from air to form insoluble barium carbonate. Hazardous polymerization cannot occur. Incompatible with: acids; oxidizers; chlorinated rubber; corrosive to metals; oxygen; nitrogen; hydrogen; ammonia; water; halogens; sulfides.
Carcinogenicity: IARC - Not listed; NIOSH - Not listed; NTP - Not listed; ACGIH - Class A4, Not classifiable as a human carcinogen; OSHA - Not listed; EPA - Not listed; MAK - Not listed
Exposure Limits
OSHA PEL: TWA: 0.5 mg/m^3; as Ba soluble.
ACGIH TLV: TWA: 0.5 mg/m^3; as Ba.
NIOSH REL: TWA: 0.5 mg/m^3; as Ba soluble.
DFG MAK: TWA: 0.5 mg/m^3; as Ba.

Environmental

Regulations

RCRA 40CFR: Not listed
CERCLA: 40CFR 302.4: Not listed
SARA 40CFR 372.65: Not listed
SARA EHS 40CFR 355: Not listed
TSCA: Not listed

BAR6500 CAS #: 12230-71-6

BARIUM HYDROXIDE, OCTAHYDRATE

Molecular Formula: $BaH_{18}O_{10}$
Formula Weight: 315.48

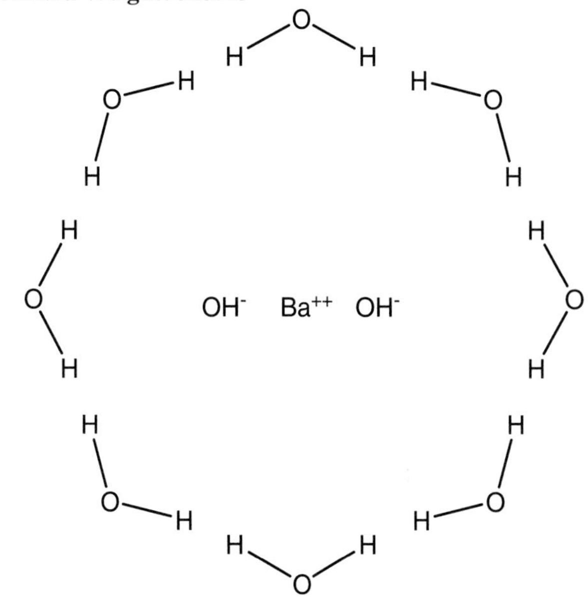

Chemical Structure

Physical Properties

Freezing Point: Octahydrate 78 °C (172.4 °F)
Density: 2.18 g/mL at 16 °C
Bulk Density: 65 lbs/cu ft
Vapor Pressure: 227 mm Hg at 77.9 °C
Water Solubility: 97 g/100 cc at 78 °C
Other Solubilities: Alcohol and Ether; Insoluble in Acetone
pH: Strongly alkaline
Refraction Index: 1.5
Flash Point: Noncombustible

Hazard Overviews

Reactivity: Stable, absorbs carbon dioxide from air to form insoluble barium carbonate. Hazardous polymerization cannot occur. Incompatible with: acids; oxidizers; chlorinated rubber; corrosive to metals; oxygen; nitrogen; hydrogen; ammonia; water; halogens; sulfides.
Carcinogenicity: IARC - Not listed; NIOSH - Not listed; NTP - Not listed; ACGIH - Class A4, Not classifiable as a human carcinogen; OSHA - Not listed; EPA - Not listed; MAK - Not listed
Exposure Limits
OSHA PEL: TWA: 0.5 mg/m^3; as Ba soluble.
ACGIH TLV: TWA: 0.5 mg/m^3; as Ba.
NIOSH REL: TWA: 0.5 mg/m^3; as Ba soluble.
DFG MAK: TWA: 0.5 mg/m^3; as Ba.

Environmental

Regulations

RCRA 40CFR: Not listed
CERCLA: 40CFR 302.4: Not listed
SARA 40CFR 372.65: Not listed
SARA EHS 40CFR 355: Not listed
TSCA: Not listed

BAR7000 CAS #: 1103-38-4

BARIUM LITHOL RED

EINECS Number: 214-160-6
Molecular Formula: $C_{40}H_{28}BaN_4O_8S_2$
Formula Weight: 994.15
Synonyms: BARIUM LITHOL; C.I. PIGMENT RED 49:1; C.I. PIGMENT RED 49,BARIUM SALT (2:1); CALCOTONE RED B; D AND C RED NO 12; DAINICHI LITHOL RED R; ELJON LITHOL RED MS; IRGALITE RED BRL; ISOL RED 3BK; ISOL RED TONER GB; ISOL RED TONER RB; ISOL TOBIAS RED 3BK; ISOL TOBIAS RED GB; ISOL TOBIAS RED RB; LIGHT RED RB; LIGHT RED RCN; LITHOL RED 18959; LITHOL RED 22060; LITHOL RED 27965; LITHOL RED BARIUM TONER; 1-NAPHTHALENESULFONIC ACID,2-((2-HYDROXY-1-NAPHTHALENYL)AZO)-,BARIUM SALT (2:1); PIGMENT RED 49:1; POSTER RED; 1883 RED; RED NO 207; RED TONER YTA; SANYO FAST RED NN; SANYO LACQUER RED RN; SANYO LITHOL RED R; SYMULER RED 2R BA SALT; VULCANOSINE RED RBKX
Use: industrial enamels; toys & dipping enamels; rubber, plastics; in printing inks and paints, having high tinctorial strength; in alkyd resin enamels and lacquers when the solvent fastness is sufficient; linoleum, paper coating, emulsion paints, inks for wrapper, wallpaper and foil printing

Hazard Overviews

Carcinogenicity: IARC - Not listed; NIOSH - Not listed; NTP - Not listed; ACGIH - Class A4, Not classifiable as a human carcinogen; OSHA - Not listed; EPA - Not listed; MAK - Not listed
Exposure Limits
OSHA PEL: TWA: 0.5 mg/m^3; as Ba soluble.
ACGIH TLV: TWA: 0.5 mg/m^3; as Ba.
NIOSH REL: TWA: 0.5 mg/m^3; as Ba soluble.
DFG MAK: TWA: 0.5 mg/m^3; as Ba.

Environmental

Regulations

RCRA 40CFR: Not listed
CERCLA: 40CFR 302.4: Not listed
SARA 40CFR 372.65: Listed as Compound
SARA EHS 40CFR 355: Not listed
TSCA: Listed

BAR7500 CAS #: 10022-31-8

BARIUM NITRATE

RTECS: CQ9625000
DOT: UN1446; IMO5.1
EINECS Number: 233-020-5
Molecular Formula: BaN_2O_6
Structured MF: $Ba(NO_3)_2$
Formula Weight: 261.38

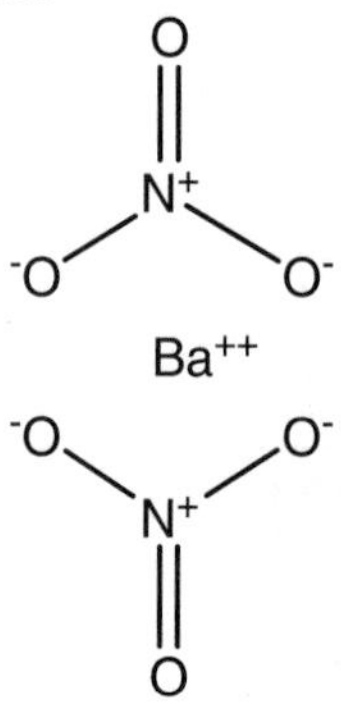

Chemical Structure

Synonyms: BARIUM DINITRATE; BARIUM SALT OF NITRIC ACID; BARIUM(II) NITRATE (1:2); DUSICNAN BARNATY; NITRATE DE BARYUM; NITRATO BARICO; NITRIC ACID,BARIUM SALT; NITROBARITE
Description: white powder, crystal; odorless
Use: manufacture barium oxide; pyrotechnics for green fire; green signal lights; in vacuum tube industry; incendiaries; ceramic glazes; rodenticide (former use); electronics; raw material for manufacture of neon sign lightings; in tracer bullets, primers, and detonators

Physical Properties

Boiling Point: Decomposes
Freezing Point: 575 °C (1067 °F)
Specific Gravity: 3.24 at 23 °C
Vapor Pressure: Low
Water Solubility: 8.7 g/100 cc Water at 20 °C
Other Solubilities: Very Slightly Soluble in Acetone; Insoluble in Alcohol; Slightly Soluble in acid.
Refraction Index: 1.572
Flash Point: Nonflammable

RTECS Toxicity Data

Acute Oral: Rat LD_{50} Dose: 355 mg/kg. Rabbit LD_{Lo} Dose: 150 mg/kg.
Acute Dermal: Mouse LD_{Lo} Route: Subcutaneous Dose: 10 mg/kg.
Irritation Eye: Rabbit Standard Draize Test Dose: 100 mg/24H; Reaction: moderate.
Irritation Skin: Rabbit Standard Draize Test Dose: 500 mg/24H; Reaction: mild.

Hazard Overviews

Fire Diamond

Health: Irritating to eyes/skin/respiratory tract. Also Causes: ulcerations, excessive salivation, nausea, vomiting, colic, diarrhea, tingling in extremities, convulsive tremors, slow pulse, elevated blood pressure, loss of tendon reflexes, general muscular paralysis, and death from respiratory arrest or ventricular fibrillation. Hemorrhage may occur in stomach, intestine, and kidney. Chronic Effects: Repeated skin contact can cause chronic dryness and cracking.

Fire: Noncombustible. However, it is a strong oxidizer capable of igniting combustibles. Use agent suitable for surrounding fire.

Reactivity: Stable. Hazardous polymerization cannot occur. Avoid: contact with combustibles. Incompatible with: magnesium and barium oxide and zinc; aluminum and magnesium alloys; combustibles (paper, oil, wood); acids; oxidizers; mixtures with finely divided aluminum-magnesium alloys. Hazardous decomposition products: nitrogen oxides; barium oxides.

Carcinogenicity: IARC - Not listed; NIOSH - Not listed; NTP - Not listed; ACGIH - Class A4, Not classifiable as a human carcinogen; OSHA - Not listed; EPA - Not listed; MAK - Not listed

Primary Target Organs:

Eyes

Skin

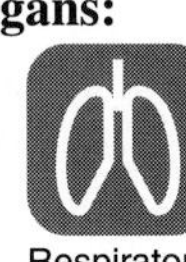
Respiratory System

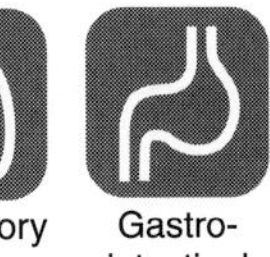
Gastro-intestinal

Nervous System

Cardio-vascular

Exposure Limits

OSHA PEL: TWA: 0.5 mg/m^3; as Ba.

ACGIH TLV: TWA: 0.5 mg/m^3; as Ba.

NIOSH REL: TWA: 0.5 mg/m^3; as Ba soluble.

NIOSH IDLH: 50 mg/m^3; as Ba.

DFG MAK: TWA: 0.5 mg/m^3; as Ba.

Respirator Recommendation

Exposure Range: >0.5 to 5 mg/m^3 Air Purifying, Negative Pressure, Half Mask

Exposure Range: >5 to <50 mg/m^3 Air Purifying, Negative Pressure, Full Face

Exposure Range: 50 to unlimited mg/m^3 Self-contained Breathing Apparatus, Pressure Demand, Full Face

Cartridge Color: dust/mist filter (use P100 or consult supervisor for appropriate dust/mist filter)

Environmental

Ecotoxicity: Aquatic toxicity: 500 ppm/1658 hr/stickle back/average survival/fresh water

Cleanup/Disposal: Guide No. 141: Keep combustibles (wood, paper, oil, etc.) away from spilled material. Do not touch damaged containers or spilled material unless wearing appropriate protective clothing. Stop leak if you can do it without risk. Small Dry Spills: With clean shovel place material into clean, dry container and cover loosely; move containers from spill area. Large Spills: Dike far ahead of spill for later disposal.

Environmental Physical Data

BCF: marine animals concentrate 7 to 100

BOD: none

Regulations

RCRA 40CFR: Not listed

CERCLA: 40CFR 302.4: Not listed

SARA 40CFR 372.65: Listed as Compound

SARA EHS 40CFR 355: Not listed

TSCA: Listed

BAR8000 CAS #: 7787-36-2

BARIUM PERMANGANATE

RTECS: SD6405000

DOT: UN1448; IMO5.1

EINECS Number: 232-110-1

Molecular Formula: BaH_2MnO_4

Structured MF: $Ba(MnO_4)_2$

Formula Weight: 375.22

Synonyms: BARIUM MANGANATE(VII); PERMANGANATE BARICO; PERMANGANATE DE BARYUM; PERMANGANIC ACID (HMNO4),BARIUM SALT; PERMANGANIC ACID,BARIUM SALT

Description: brownish-violet to black crystals; odorless

Use: as dry cell depolarizer; strong disinfectant; manufacturing permanganates

Physical Properties

Boiling Point: Decomposes at 1 atm

Freezing Point: Decomposes at 200 °C (392 °F)

Specific Gravity: 3.77

Water Solubility: 62.5 g/100 ml Water at 11 °C

Other Solubilities: Solution of barium salts yield a white precipitate with 2 N sulfuric acid. This precipitate is Insoluble in Hydrochloric Acid and in nitric acid.

Flash Point: Nonflammable

Hazard Overviews

Fire: Noncombustible.

Carcinogenicity: IARC - Not listed; NIOSH - Not listed; NTP - Not listed; ACGIH - Class A4, Not classifiable as a human carcinogen; OSHA - Not listed; EPA - Not listed; MAK - Not listed

Exposure Limits

OSHA PEL: TWA: 0.5 mg/m^3; as Ba.

OSHA PEL Vacated 1989 Limits: STEL: 5 mg/m^3; Ceiling as Mn.

ACGIH TLV: TWA: 0.5 mg/m^3; as Ba.

NIOSH REL: TWA: 0.5 mg/m^3; as Ba soluble. STEL: 3 mg/m^3; as Mn.

DFG MAK: TWA: 0.5 mg/m^3; as Ba.

Environmental

Ecotoxicity: Aquatic toxicity: 2.2 to 4.1 mg/l of Mn/8 to 18 hr/fish/killed/* *Type of water not specified

Cleanup/Disposal: Guide No. 141: Keep combustibles (wood, paper, oil, etc.) away from spilled material. Do not touch damaged containers or spilled material unless wearing appropriate protective clothing. Stop leak if you can do it without risk. Small Dry Spills: With clean shovel place material into clean, dry container and cover loosely; move containers from spill area. Large Spills: Dike far ahead of spill for later disposal.

Environmental Physical Data

BCF: marine animals concentrate 7 to 100

BOD: none

Regulations

RCRA 40CFR: Not listed

CERCLA: 40CFR 302.4: Listed as Compound per CAA Section 112

SARA 40CFR 372.65: Listed as Compound

SARA EHS 40CFR 355: Not listed

TSCA: Not listed

BAR8500 CAS #: 1304-29-6

BARIUM PEROXIDE

RTECS: CR0175000

DOT: UN1449; IMO5.1

EINECS Number: 215-128-4

Molecular Formula: BaO_2

Structured MF: BaO_2

Formula Weight: 169.36

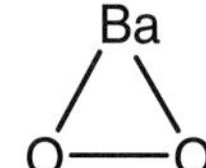

Chemical Structure

Synonyms: BARIO (PEROSSIDO DI); BARIUM BINOXIDE; BARIUM DIOXIDE; BARIUM OXIDE,PER-; BARIUM SUPEROXIDE; BARIUMPEROXID; BARIUMPEROXYDE; DIOXYDE DE BARYUM; PEROXYDE DE BARYUM

Description: grayish-white powder; odorless

Use: bleaching animal substances, vegetable fibers & straw; glass decolorizer; manufacture of hydrogen peroxide & oxygen; in cathodes; dyeing & printing textiles; in powder aluminum in welding; in igniter compositions; oxidizing agent in org synth; used for pyrotechnics and tracer-bullet formulations; as polysulfide curing agent

Physical Properties

Boiling Point: 800 °C (1472 °F)

Freezing Point: 450 °C (842 °F)

Specific Gravity: 4.96

Water Solubility: Very Slightly Soluble in Cold Water

Flash Point: Not flammable

RTECS Toxicity Data

Acute Dermal: Mouse LD_{50} Route: Subcutaneous Dose: 50 mg/kg.

Hazard Overviews

Fire Diamond

Health: Irritating to eyes/skin/respiratory tract. Also Causes: potassium deficiency; may affect function of heart and CNS; excessive salivation, vomiting, severe abdominal pain, violent purging with watery and bloody stools, changes in cardiac rhythm (heartbeat), dilated pupils, confusion and vertigo (dizziness, giddiness), muscle tremors progressing to convulsions, paralysis of the extremities, and ultimately death by respiratory failure or cardiac arrest.

Fire: Noncombustible. However, it is a strong oxidizer capable of igniting combustibles. Use water only! Do not use dry chemical, carbon dioxide, or halon.

Reactivity: Stable. Hazardous polymerization cannot occur. Avoid: moisture; combustible materials. Incompatible with: acetic anhydride; calcium-silicone alloys; powdered aluminum; powdered magnesium; wood; hydrogen sulfide; water; peroxyformic acid; hydroxylamine solution; organic matter; manganese; zinc; barium nitrate; oxygen; nitrogen; hydrogen; ammonia; halogens; sulfides. Hazardous decomposition products: oxygen; barium oxide.

Carcinogenicity: IARC - Not listed; NIOSH - Not listed; NTP - Not listed; ACGIH - Class A4, Not classifiable as a human carcinogen; OSHA - Not listed; EPA - Not listed; MAK - Not listed

Primary Target Organs:

Eyes

Skin

Respiratory System

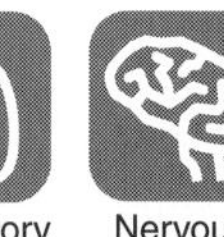
Nervous System

Cardio-vascular

Exposure Limits

OSHA PEL: TWA: 0.5 mg/m^3; as Ba.

ACGIH TLV: TWA: 0.5 mg/m^3; as Ba.

NIOSH REL: TWA: 0.5 mg/m^3; as Ba soluble.

DFG MAK: TWA: 0.5 mg/m^3; as Ba.

Environmental

Cleanup/Disposal: Guide No. 141: Keep combustibles (wood, paper, oil, etc.) away from spilled material. Do not touch damaged containers or spilled material unless wearing appropriate protective clothing. Stop leak if you can do it without risk. Small Dry Spills: With clean shovel place material into clean, dry container and cover loosely; move containers from spill area. Large Spills: Dike far ahead of spill for later disposal.

Environmental Physical Data

BCF: marine animals concentrate 7 to 100

BOD: none

Regulations

RCRA 40CFR: Not listed
CERCLA: 40CFR 302.4: Not listed
SARA 40CFR 372.65: Listed as Compound
SARA EHS 40CFR 355: Not listed
TSCA: Listed

BAR9000 CAS #: 7727-43-7

BARIUM SULFATE

RTECS: CR0600000
DOT: UN1564; IMO6.1
EINECS Number: 231-784-4
Molecular Formula: BaH_2O_4S
Structured MF: $BaSO_4$
Formula Weight: 233.39

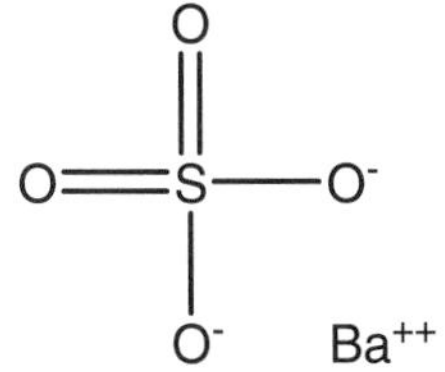

Chemical Structure

Synonyms: ACTYBARYTE; ARTIFICIAL BARITE; ARTIFICIAL HEAVY SPAR; BA (SULFATE); BA147; BAKONTAL; BARAFLAVE; BARICON; BARIDOL; BARII SULPHAS; BARITE; BARITOGEN DELUXE; BARITOP; BARITOP 100; BARITOP G POWDER; BARITOP P; BARIUM 100; BARIUM ANDREU; BARIUM SALT OF SULFURIC ACID; BARIUM SULFATE (1:1); BARIUM SULFURICUM; BARIUM SULPHATE; BAROBAG; BARO-CAT; BAROCAT; BARODENSE; BAROLOID; BAROSPERSE; BAROSPERSE 110; BAROSPERSE II; BAROTRAST; BAR-TEST; BARYTA WHITE; BARYTES; BARYTES 22; BARYTES (NATURAL); BARYTGEN; BARYUM (SULFATE DE); BARYX COLLOIDAL; BARYXINE; BASOFOR; BAYRITES; BF 1 (SALT); BF 10 (SULFATE); BLANC FIXE; C.I. 77120; C.I. PIGMENT WHITE 21; CITOBARYUM; COLONATRAST; DANOBARYT; E-Z PREPARATIONS; E-Z-AC; E-Z-CAT CONCENTRATE; E-Z-HD; E-Z-PAQUE; E-Z-PASTE ESOPHAGEAL CREAM; ENAMEL WHITE; ENECAT; ENEMARK; ENESET; ENTROBAR; EPA PESTICIDE CHEMICAL CODE 007502; EPI-C; EPI-STAT 57; EPI-STAT 61; ESOPHO-CAT; ESOPHOTRAST; ESOPHOTRAST ESOPHAGEAL CREAM; EWEISS; FINEMEAL; GASTROPAQUE-S; GEL-UNIX; HD 85; HD 200 PLUS; HITONE; INTROPAQUE; LACTOBARYT; LIQUIBARINE; LIQUID BAROSPERSE; LIQUID E-Z-PAQUE; LIQUID POLIBAR; LIQUID POLIBAR PLUS; LIQUID SOL-O-PAKE; LIQUIPAKE; MACROPAQUE; MICROBAR; MICROFANOX; MICROPAQUE; MICROPAQUE RD; MICROTRAST; MIKABARIUM B; MIKABARIUM F; MIXOBAR; MIXTURE III; NEOBALGIN; NEOBAR; NOVOPAQUE; OESOBAR; X-OPAC; ORATRAST; PERMANENT WHITE; PIGMENT WHITE 22; POLIBAR; PRECIPITATED BARIUM SULPHATE; PREPCAT; RADIMIX COLON; RADIOBARYT; RADIO-BARYX; RADIOPAQUE; RAYBAR; READI-CAT; READI-CAT 2; RECTO BARIUM; REDI-FLOW; REGLAN; RUGAR; SOLBAR; SOL-O-PAKE; SPARKLE GRANULES; SS 50; STERIPAQUE; SULFURIC ACID,BARIUM SALT (1:1); SUPRAMIKE; SUSPOBAR; TIXOBAR; TOMOCAT 1000 CONCENTRATE; TOMOCAT CONCENTRATE; TONOJUG 2000; TONOPAQUE; TOPCONTRAL; TRAVAD; ULTRA-R; UMBRASOL A; UNIBARYT; UNIT-PAK; VERI-O-PAKE; XYLOCAINE VISCOUS

Description: white, yellowish powder; odorless

Use: in mfr of photographic papers, artificial ivory, cellophane; lithopone; filler for rubber, linoleum, oil cloth, polymeric fibers & resins, paper, lithographic inks; water coloring pigment in wallpaper; in concrete for radioactive shield & hardening agent; paints; filler in cosmetics (eg, lipstick) & delustrant for textiles, plastics, x-ray photography, in battery plate expanders; weighting substance in golf balls.

Physical Properties

Boiling Point: Decomposes at 1600 °C (2912 °F)
Freezing Point: 1580 °C (2876 °F)
Specific Gravity: 4.5 at 15 °C
Vapor Pressure: ~ 0 mm Hg
Water Solubility: 0.00022 g/100 cc Water at 18 °C
Other Solubilities: 0.006 g/100 cc 3% Hydrochloric Acid; Soluble in hot concentrated Sulfuric acid; Practically Insoluble in dilute acids & Alcohol; Practically Insoluble in organic solvents; very slightly Soluble in alkalis and in solution of many salts.
pH: 5% suspension in water is neutral
Refraction Index: 1.637
Flash Point: Noncombustible

RTECS Toxicity Data

Mutagenic: Mouse Micronucleus Test; Route: Intraperitoneal; Dose: 12500 ug/kg.
Tumorigenic: Rat Route: Intrapleural; Dose: 200 mg/kg; Toxic Effects: Tumorigenic - Equivocal tumorigenic agent by RTECS criteria; Lungs, Thorax, or Respiration - Tumors.

Hazard Overviews

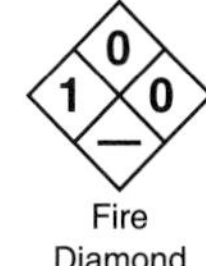

Fire Diamond

Health: Slightly irritating to eyes/skin. Chronic Effects: chronic inhalation causes the benign pneumoconiosis called baritosis.
Fire: Noncombustible. Use agent suitable for surrounding fire.
Reactivity: Stable. Hazardous polymerization cannot occur. Avoid: dusty conditions. Incompatible with: aluminum and heat. Hazardous decomposition products: oxides of sulfur; barium fume.
Carcinogenicity: IARC - Not listed; NIOSH - Not listed; NTP - Not listed; ACGIH - Class A4, Not classifiable as a human carcinogen; OSHA - Not listed; EPA - Not listed; MAK - Not listed
Primary Target Organs:

Eyes

Skin

Exposure Limits
OSHA PEL: TWA: 15 mg/m^3; total dust.
OSHA PEL Vacated 1989 Limits: TWA: 10 mg/m^3. Other Values: respirable mg/m^3; 5.
ACGIH TLV: TWA: 10 mg/m^3.
NIOSH REL: TWA: 10 mg/m^3; as Ba soluble.
DFG MAK: TWA: 0.5 mg/m^3; as Ba.
Respirator Recommendation
Exposure Range: >5 to 50 mg/m^3 Air Purifying, Negative Pressure, Half Mask
Exposure Range: >50 to 500 mg/m^3 Air Purifying, Negative Pressure, Full Face
Exposure Range: >500 to 5000 mg/m^3 Supplied Air, Constant Flow/Pressure Demand, Full Face
Exposure Range: >5000 to unlimited mg/m^3 Self-contained Breathing Apparatus, Pressure Demand, Full Face
Cartridge Color: dust/mist filter (use P100 or consult supervisor for appropriate dust/mist filter)

Environmental

Cleanup/Disposal: Guide No. 154: Eliminate all ignition sources (no smoking, flares, sparks or flames in immediate area). Do not touch damaged containers or spilled material unless wearing appropriate protective clothing. Stop leak if you can do it without risk. Prevent entry into waterways, sewers, basements or confined areas. Absorb or cover with dry earth, sand or other non-combustible material and transfer to containers. Do not get water inside containers.

Environmental Physical Data

BCF: marine animals concentrate 7 to 100

Regulations

RCRA 40CFR: Not listed
CERCLA: 40CFR 302.4: Not listed
SARA 40CFR 372.65: Listed as Compound
SARA EHS 40CFR 355: Not listed
TSCA: Listed

BAS5000 CAS #: 12607-70-4

BASIC NICKEL (II) CARBONATE

RTECS: QR6250000
EINECS Number: 235-715-9
Molecular Formula: $CH_4Ni_3O_7$
Formula Weight: 304.17
Synonyms: BASIC NICKEL(II) CARBONATE; CARBONIC ACID,NICKEL SALT,BASIC; NICKEL CARBONATE HYDROXIDE; NICKEL,(CARBONATO(2-))TETRAHYDROXYTRI-
Description: light green to emerald green crystals or brown powder; odorless
Use: nickel plating; catalyst for hardening of fats; in ceramic colors & glazes; preparation of nickel catalysts

Physical Properties

Freezing Point: Decomposes
Density: Zaratite 2.6
Water Solubility: Insoluble in Water
Other Solubilities: Soluble in acid, Ammonium salts; Soluble in Ammonia, dilute acids with effervescence; Soluble in hot dilute Hydrochloric Acid, Ammonium Hydroxide.
Refraction Index: Zaraite 1.56 to 1.61

Hazard Overviews

Health: Irritating. Toxic. Other Acute Effects: harmful if swallowed, inhaled, or absorbed through skin; exposure can cause gastrointestinal disturbances. Chronic Effects: lung irritation; dermatitis. Carcinogen.
Fire: Hazards: emits toxic fumes. Extinguishing agents: water spray; carbon dioxide, dry chemical powder or appropriate foam. Precautions: combustible liquid.
Reactivity: Incompatible with: strong oxidizing agents, strong acids. Hazardous decomposition products: toxic fumes of: carbon monoxide, carbon dioxide, sulfur oxides.
Carcinogenicity: IARC - Group 1, Carcinogenic to humans; NIOSH - Listed as carcinogen; NTP - Class 2A, Reasonably anticipated to be a carcinogen, limited evidence of carcinogenicity from studies in humans; ACGIH - Not listed; OSHA - Not listed; EPA - Not listed; MAK - Class A1, Capable of inducing malignant tumors as shown by experience with humans
Primary Target Organs:

Eyes

Skin

Respiratory System

Exposure Limits
OSHA PEL: STEL: 1 mg/m^3; as Ni.
OSHA PEL Vacated 1989 Limits: TWA: 1 mg/m^3; insoluble. Other Values: 0.1 mg/m^3; soluble.
ACGIH TLV: TWA: 0.1 mg/m^3; soluble is 1.
NIOSH REL: TWA: 1 mg/m^3; as Ni soluble and insoluble.
Respirator Recommendation
Exposure Range: >1 to <10 mg/m^3 Supplied Air, Constant Flow/Pressure Demand, Half Mask
Exposure Range: 10 to unlimited mg/m^3 Self-contained Breathing Apparatus, Pressure Demand, Full Face
Note: odor threshold unknown; as nickel, soluble

Environmental

Ecotoxicity: LC_{50} Acroneuria lycoria 4 mg/l/96 hr /Nickel ion /Conditions of bioassay not specified LC_{50} Artemia salina 163.0 mg/l/48 hr /Nickel ion/ /Conditions of bioassay not specified LC_{50} Channa punctatus 306.9 mg/l/96-hr /Nickel ion/ /Conditions of bioassay not specified LC_{50} Daphnia magna 0.13 mg/l/3 weeks /Nickel ion/ /Conditions of bioassay not specified

Environmental Physical Data

BCF: not significant

Regulations

RCRA 40CFR: Not listed
CERCLA: 40CFR 302.4: Listed as Compound per CWA Section 307(a) per CAA Section 112

SARA 40CFR 372.65: Listed as Compound
SARA EHS 40CFR 355: Not listed
TSCA: Listed

BEH5000 CAS #: 929-77-1

BEHENIC ACID, METHYL ESTER

EINECS Number: 213-207-8
Molecular Formula: $C_{23}H_{46}O_2$
Formula Weight: 354.62

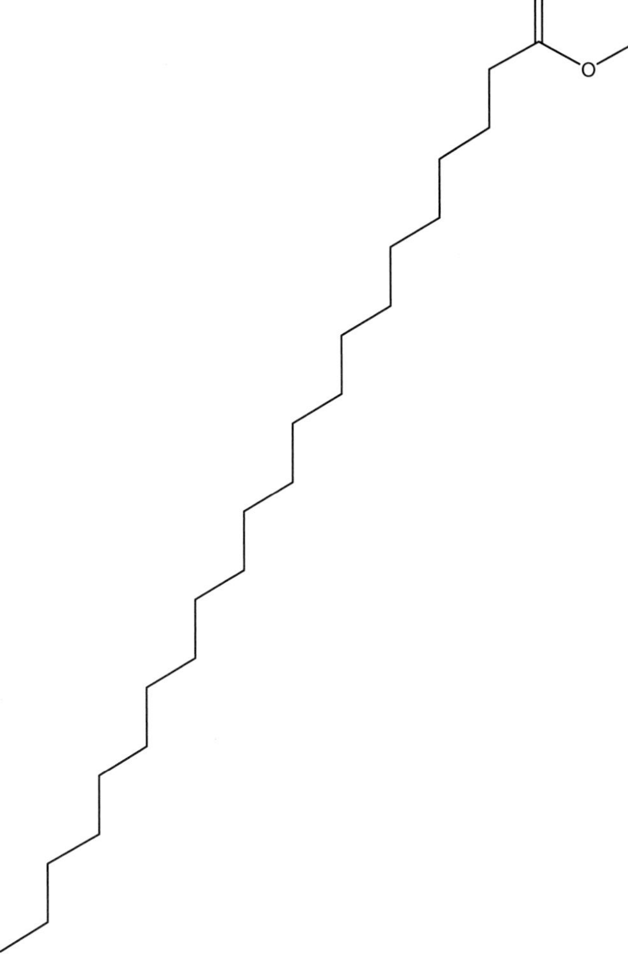

Chemical Structure

Synonyms: DOCOSANOIC ACID,METHYL ESTER; METHYL BEHENATE; METHYL DOCOSANOATE
Description: needles

Physical Properties

Boiling Point: 224 °C (435 °F) to 225 °C (437 °F) at 15 mm Hg
Freezing Point: 54 °C (129.2 °F)
Water Solubility: Insoluble in Water
Other Solubilities: Soluble in Alcohol, Ether
Refraction Index: 1.4339 at 60 °C/D

Hazard Overviews

Health: Irritating to eyes/skin. Other Acute Effects: may be harmful by inhalation, ingestion, or skin.
Fire: Hazards: emits toxic fumes. Extinguishing agents: water spray; carbon dioxide, dry chemical powder or appropriate foam. Precautions: combustible liquid.
Carcinogenicity: IARC - Not listed; NIOSH - Not listed; NTP - Not listed; ACGIH - Not listed; OSHA - Not listed; EPA - Not listed; MAK - Not listed
Primary Target Organs:

Eyes

Skin

Environmental

Regulations

RCRA 40CFR: Not listed
CERCLA: 40CFR 302.4: Not listed
SARA 40CFR 372.65: Not listed
SARA EHS 40CFR 355: Not listed
TSCA: Listed

BEN1000 CAS #: 302-40-9

BENACTYZINE

RTECS: DD2480000
EINECS Number: 206-123-8
Molecular Formula: $C_{20}H_{25}NO_3$
Formula Weight: 327.41
Synonyms: BENACTIZINA; BENACTYZIN; BENZACTYZINE; BENZENEACETIC ACID,ALPHA-HYDROXY-ALPHA-PHENYL-,2-(DIETHYLAMINO)ETHYL ESTER; BENZENEACETIC ACID,ALPHA-HYDROXY-ALPHA-PHENYL-,2-(DIETHYLAMINO)ETHYL ESTER (9CI); BENZILIC ACID BETA-DIETHYLAMINOETHYL ESTER; BENZILIC ACID,2-(DIETHYLAMINO)ETHYL ESTER; DIAZIL; 2-(DIETHYLAMINO)ETHYL BENZILATE; 2-DIETHYLAMINOETHYL BENZILATE; BETA-DIETHYLAMINOETHYL BENZILATE; DIETHYLAMINOETHYL BENZILATE; 2-(DIETHYLAMINO)ETHYL DIPHENYLGLYCOLATE; 2-DIETHYLAMINOETHYL DIPHENYLGLYCOLATE; DIPHENYLGLYCOLIC ACID 2-(DIETHYLAMINO)ETHYL ESTER; ALPHA-HYDROXY-ALPHA-PHENYLBENZENEACETIC ACID 2-(DIETHYLAMINO)ETHYL ESTER
Description: crystals
Use: medication: tranquilizer; anticholinergic

Physical Properties

Freezing Point: 51 °C (123.8 °F)
Water Solubility: 15 g/100 mL Water at 25 °C
Other Solubilities: Soluble in Ethanol; Practically Insoluble in Ether.

RTECS Toxicity Data

Acute Dermal: Mouse LD_{50} Route: Subcutaneous Dose: 159 mg/kg; Toxic Effects: Behavioral - Convulsions or effect on seizure threshold.

Mutagenic: Hamster Cytogenetic Analysis; Route: Intraperitoneal; Dose: 10 mg/kg.

Hazard Overviews

Carcinogenicity: IARC - Not listed; NIOSH - Not listed; NTP - Not listed; ACGIH - Not listed; OSHA - Not listed; EPA - Not listed; MAK - Not listed

Environmental

Regulations

RCRA 40CFR: Not listed
CERCLA: 40CFR 302.4: Not listed
SARA 40CFR 372.65: Not listed
SARA EHS 40CFR 355: Not listed
TSCA: Not listed

BEN1080 CAS #: 22781-23-3

BENDIOCARB

RTECS: FC1140000
DOT: UN2757; UN2758; UN2991; UN2992; IMO3.2; IMO6.1
EINECS Number: 245-216-8
Molecular Formula: $C_{11}H_{13}NO_4$
Formula Weight: 223.23
Synonyms: BENCARBATE; BENDIOCARBE; 1,3-BENZODIOXOLE,2,2-DIMETHYL-4-(N-METHYLAMINOCARBOXYLATO)-; 1,3-BENZODIOXOLE,2,2-DIMETHYL-4-(N-METHYLCARBAMATO)-; 1,3-BENZODIOXOL-4-OL,2,2-DIMETHYL-,METHYLCARBAMATE; CARBAMIC ACID,METHYL-,2,3-(DIMETHYLMETHYLENEDIOXY)PHENYLESTER; CARBAMIC ACID,METHYL-,2,3-(ISOPROPYLIDENEDIOXY)PHENYLESTER; 2,2-DIMETHYL-1,3-BENZDIOXOL-4-YL N-METHYLCARBAMATE; 2,2-DIMETHYL-1,3-BENZODIOXOL-4-OL METHYLCARBAMATE; 2,2-DIMETHYL-1,3-BENZODIOXOL-4-YL METHYLCARBAMATE; 2,2-DIMETHYLBENZO-1,3-DIOXOL-4-YL METHYLCARBAMATE; DYCARB; FICAM; FICAM 80W; FICAM B; FICAM D; FICAM ULV; FICAM ULV' UL; FICAM W; FICAM Z; FISONS NC 6897; GARVOX; 2,3-ISOPROPYLIDENE-DIOXYPHENYL METHYLCARBAMATE; 2,3-ISOPROPYLIDENEDIOXYPHENYL METHYLCARBAMATE; METHYLCARBAMIC ACID 2,3-(ISOPROPYLIDENEDIOXY)PHENYL ESTER; MULTAMAT; MULTIMET; NC 6897; NC6897; NIOMIL; OMS-1394; ROTATE; SEEDOX; SEEDOX 80W; TATOO; TATTOO; TURCAM
Description: white solid
Use: contact & stomach insecticide; to control wide range of nuisance & disease vestor pests, & soil insects, eg, lepidoptera, coleoptera & collembola, flies, wasps, ants, fleas, cockroaches, adult mosquitoes, caterpillars, lacebugs, mealy bugs, etc; in & around the house, on corn & turf.

Physical Properties

Freezing Point: 129 °C (264.2 °F) to 130 °C (266 °F)
Specific Gravity: 1.25000
Saturated Vapor Density: 1.200000053 kg/m^3
Vapor Pressure: 5 x10^{-6} mm Hg at 25 °C
Water Solubility: 40 ppm in Water
Other Solubilities: solubility at 25 °C: less than 1 g/kg odorless Kerosene; 200 g/kg Acetone, Dioxane, Chloroform; 40 g/kg Ethanol, Benzene; in Kerosene 0.03%; in o-Xylene 1.0%; in Dichloromethane or Acetone 20%.

RTECS Toxicity Data

Acute Oral: Rat LD_{50} Dose: 40 mg/kg.

Hazard Overviews

Carcinogenicity: IARC - Not listed; NIOSH - Not listed; NTP - Not listed; ACGIH - Not listed; OSHA - Not listed; EPA - Not listed; MAK - Not listed

Environmental

Environmental Fate: If released to the atmosphere, it will degrade in the vapor phase by reaction with photochemically produced hydroxyl radicals (half-life of about 5 hours). Particulate-phase and aerosols released to air during spray applications will be physically removed from air by dry and wet deposition. It is readily hydrolyzed particularly under neutral or basic conditions with reaction products of 2,3-isopropylidenedioxyphenol, methylamine, and carbon dioxide. Estimated K_{oc} values of 570 and 575 suggest that it will have moderate mobility in soil environments. A Henry's Law constant of 6.6 x10^{-11} atm-cu m/mole at 25 °C indicates that this compound is essentially non-volatile from water. Biodegradation, particularly in cross conditioned soils, will be a major fate process for this compound. In experiments using carbamate pesticide exposed soils, nearly 95-100% removal within 2 to 6 weeks was shown. Using a carbon-depleted sewage inoculum, it was utilized as a carbon source as measured by BOD. A BCF value of 77 was estimated indicating that bioconcentration is not a major fate process within aquatic systems.
Cleanup/Disposal: Guide No. 131: Fully encapsulating, vapor protective clothing should be worn for spills and leaks with no fire. Eliminate all ignition sources (no smoking, flares, sparks or flames in immediate area). All equipment used when handling the product must be grounded. Do not touch or walk through spilled material. Stop leak if you can do it without risk. Prevent entry into waterways, sewers, basements or confined areas. A vapor suppressing foam may be used to reduce vapors. Small Spills: Absorb with earth, sand or other non-combustible material and transfer to containers for later disposal. Use clean non-sparking tools to collect absorbed material. Large Spills: Dike far ahead of liquid spill for later disposal. Water spray may reduce vapor; but may not prevent ignition in closed spaces. Guide No. 151: Do not touch damaged containers or spilled material unless wearing appropriate protective clothing. Stop leak if you can do it without risk. Prevent entry into waterways, sewers, basements or confined areas. Cover with plastic sheet to prevent spreading. Absorb or cover with dry earth, sand or other non-combustible material and transfer to containers. Do not get water inside containers.

Environmental Physical Data

Henry's Law Constant: estimated at 6.6 x10^{-11}

Sorption Partition Coefficient: K_{oc} = estimated at 570
BCF: estimated at 80

Regulations
RCRA 40CFR: Listed Hazardous Waste No. U278 Toxic Waste
CERCLA: 40CFR 302.4: Listed per RCRA Section 3001 RQ: 1 lb (0.454 kg)
SARA 40CFR 372.65: Listed
SARA EHS 40CFR 355: Not listed
TSCA: Not listed

Analytical Methods
Soil: SW846 8321A
Water / Groundwater: EPA 639
Food: AOAC 986.09
Indoor / Expired Air: ASTM D4861
Plasma: FDA 242.2

BEN1160 CAS #: 22961-82-6
BENDIOCARB PHENOL

EINECS Number: 400-900-7
Molecular Formula: $C_9H_{10}O_2$
Formula Weight: 150.18

Hazard Overviews
Carcinogenicity: IARC - Not listed; NIOSH - Not listed; NTP - Not listed; ACGIH - Not listed; OSHA - Not listed; EPA - Not listed; MAK - Not listed

Environmental

Regulations
RCRA 40CFR: Not listed
CERCLA: 40CFR 302.4: Not listed
SARA 40CFR 372.65: Not listed
SARA EHS 40CFR 355: Not listed
TSCA: Not listed

BEN1240 CAS #: 73-48-3
BENDROFLUMETHIAZIDE

RTECS: DK8225000
EINECS Number: 200-800-1
Molecular Formula: $C_{15}H_{14}F_3N_3O_4S_2$
Formula Weight: 421.41

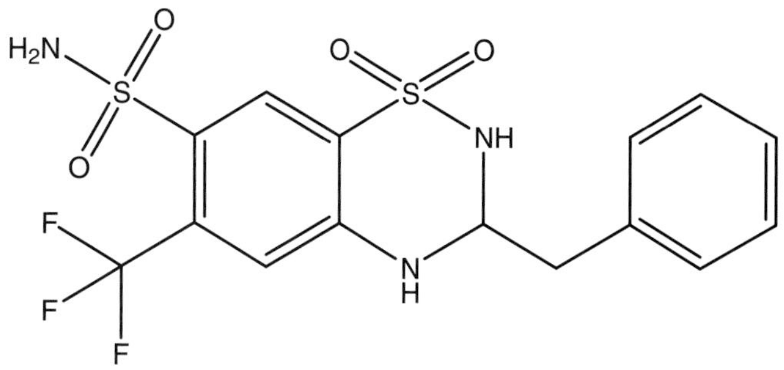

Chemical Structure

Synonyms: APRINOX; BE 724-A; BENDROFLUAZIDE; BENDROFLUMETHAZIDE; BENDROFLUMETHIAZID; BENTRIDE; BENURON; BENZHYDROFLUMETHIAZIDE; 2H-1,2,4-BENZOTHIADIAZINE-7-SULFONAMIDE,3-BENZYL-3,4-DIHYDRO-6-(TRIFLUOROMETHYL)-,1,1-DIOXIDE; 2H-1,2,4-BENZOTHIADIAZINE-7-SULFONAMIDE,3,4-DIHYDRO-3-(PHENYLMETHYL)-6-(TRIFLUOROMETHYL)-,1,1-DIOXIDE; BENZYDROFLUMETHIAZIDE; 3-BENZYL-3,4-DIHYDRO-6-(TRIFLUOROMETHYL)-2H-1,2,4-BENZOTHIADIAZINE-7-SULFONAMIDE 1,1-DIOXIDE; BENZYLHYDROFLUMETHIAZIDE; BENZYLRODIURAN; 3-BENZYL-6-TRIFLUOROMETHYL-7-SULFAMOYL-3,4-DIHYDRO-1,2,4-BENZOTHIADIAZINE,1,1-DIOXIDE; BERKOZIDE; BETA; BHFT; BL H368; BRISTURIC; BRISTURON; CENTYL; FLUMESIL; FT 8; FT 81; INTOLEX; LIVESAN; NATERETIN; NATURETIN; NATURINE; NEO-NACLEX; NEONACLEX; NEO-RONTYL; NIAGARIL; NIKION; ORSILE; PLURYL; PLURYLE; PLUSURIL; POLIURON; RELAN; RELAN BETA; REPICIN; SALURAL; SALURES; SINESALIN; SODIURETIC; THIAZIDICO; 6-TRIFLUOROMETHYL-3-BENZYL-7-SULFAMYL-3,4-DIHYDRO-1,2,4-BENZOTHIADIAZINE,1,1-DIOXIDE; URLEA
Description: white to cream-colored, finely divided, crystalline powder; odorless or has slight, characteristic floral odor
Use: diuretic & antihypertensive agent

Physical Properties
Freezing Point: 221 °C (429.8 °F) to 223 °C (433.4 °F)
Water Solubility: Practically Insoluble in Water
Other Solubilities: freely Soluble in Alcohol & Acetone

RTECS Toxicity Data
Acute Oral: Woman TD_{Lo} Dose: 300 ug/kg/3D-I; Toxic Effects: Nutritional and gross metabolic - Changes in sodium. Mouse LD_{50} Dose: >10 gm/kg.
Chronic (Multiple Dose) Oral: Rat Dose: 22750 mg/kg/26W-I; Toxic Effects: Endocrine - Changes in adrenal weight; Nutritional and gross metabolic - Changes in calcium; Nutritional and gross metabolic - Changes in other metals. Dog Dose: 3640 mg/kg/26W-I; Toxic Effects: Vascular - BP lowering not characterized in autonomic section; Nutritional and gross metabolic - Changes in potassium.
Mutagenic: Hamster Cytogenetic Analysis; Cell Type: lung; Dose: 200 mg/L.

Hazard Overviews
Health: Irritating to eyes/skin/respiratory tract. Other Acute Effects: may be harmful by inhalation, ingestion, or skin absorption; hypersensitivity reactions; hypophosphatemia;

purpura; dermatitis with photosensitivity; depression of formed elements of blood; necrotizing vasculitis; exacerbation of renal and/or heptic insufficiency; disturbance in carbohydrate metabolism; pancreatitis; target organ: kidneys.

Fire: Hazards: emits toxic fumes. Extinguishing agents: water spray; carbon dioxide, dry chemical powder or appropriate foam. Precautions: combustible liquid.

Reactivity: Stable. Hazardous polymerization will not occur. Hazardous decomposition products: carbon monoxide, carbon dioxide, nitrogen oxides.

Carcinogenicity: IARC - Not listed; NIOSH - Not listed; NTP - Not listed; ACGIH - Not listed; OSHA - Not listed; EPA - Not listed; MAK - Not listed

Primary Target Organs:

Eyes

Skin

Respiratory System

Kidneys

Environmental

Regulations

RCRA 40CFR: Not listed
CERCLA: 40CFR 302.4: Not listed
SARA 40CFR 372.65: Not listed
SARA EHS 40CFR 355: Not listed
TSCA: Not listed

BEN1320 CAS #: 1861-40-1

BENEFIN

RTECS: XU4550000
EINECS Number: 217-465-2
Molecular Formula: $C_{13}H_{16}F_3N_3O_4$
Formula Weight: 335.32

Synonyms: BALAN; BALFIN; BANAFINE; BENALAN; BENEFEX; BENEPHIN; BENFLURALIN; BENFLURALINE; BENZENAMINE,N-BUTYL-N-ETHYL-2,6-DINITRO-4-(TRIFLUOROMETHYL)-; BETHRODINE; BINNELL; BLULAN; BONALAN; N-BUTYL-2,6-DINITRO-N-ETHYL-4-(TRIFLUOROMETHYL)ANILINE; N-BUTYL-2,6-DINITRO-N-ETHYL-4-TRIFLUOROMETHYLANILINE; N-BUTYL-N-ETHYL-2,6-DINITRO-4-TRIFLUOROMETHYLANILINE; N-BUTYL-N-ETHYL-2,6-DINITRO-4-(TRIFLUOROMETHYL)BENZENAMINE; N-BUTYL-N-ETHYL-ALPHA,ALPHA,ALPHA-TRI-FLUORO-2,6-DINITRO-P-TOLUIDINE; N-BUTYL-N-ETHYL-ALPHA,ALPHA,ALPHA-TRIFLUORO-2,6-DINITRO-P-TOLUIDINE; N-BUTYLN-ETHYL-2,6-DINITRO-4-(TRIFLUOROMETHYL)BENZENEAMINE; CARPIDOR; EL-110; EMBLEM; FLUBALEX; L 54521; QUILAN; P-TOLUIDINE,N-BUTYL-N-ETHYL-ALPHA,ALPHA,ALPHA-TRIFLUORO-2,6-DINITRO-; ALPHA,ALPHA,ALPHA-TRIFLUORO-2,6-DINITRO-N,N-ETHYLBUTYL-P-TOLUIDINE

Description: yellow-orange crystalline solid; no appreciable odor

Use: selective pre-emergence herbicide; controls annual grasses and broadleaf weeds in seeded alfalfa, birdsfoot trefoil, ladino clover, red clover, direct-seeded lettuce, peanuts, transplant air cured (burley, dark) tobacco, established turf; control of annual grasses and some annual broad-leaved weeds in groundnuts, cucumbers, chicory, endives, field beans, french beans, lentils

Physical Properties

Boiling Point: 148 °C (298 °F) to 149 °C (300 °F) at 7 mm Hg
Freezing Point: 65 °C (149 °F) to 66.5 °C (151.7 °F)
Vapor Pressure: 3.89 $x10^{-4}$ mm Hg at 30 °C
Water Solubility: 70 ppm at 25 °C
Other Solubilities: 2.4 g/100 ml Ethanol at 25 °C.
Flash Point: Not flammable

RTECS Toxicity Data

Acute Oral: Rat LD_{50} Dose: 10 gm/kg. Mouse LD_{50} Dose: 5 gm/kg.

Hazard Overviews

Fire: Noncombustible.

Carcinogenicity: IARC - Not listed; NIOSH - Not listed; NTP - Not listed; ACGIH - Not listed; OSHA - Not listed; EPA - Not listed; MAK - Not listed

Environmental

Ecotoxicity: LC_{50} Bluegill sunfish 0.064 mg/l/96 hr /Conditions of bioassay not specified LD_{50} Anas platyrhynchos (mallard) female oral >2000 mg/kg, 3-4 mo old LC_{50} Gammarus fasciatus (scud) 1.1 mg/l/96 hr at 15 °C (95% confidence limit 0.6-1.9 mg/l), mature. static bioassay without aeration, pH 7.2-7.5, water hardness 40-50 mg/l as calcium carbonate and alkalinity of 30-35 mg/l

Environmental Fate: If released to soil, loss should occur due to biodegradation, photolysis and volatilization. Half-lives in soil range from 0.7-4 months. It is expected to have low mobility in soils. The major pathways for the loss from water may be biodegradation, photolysis and volatilization. Bioconcentration in aquatic organisms should be important. Reaction with photochemically produced hydroxyl radicals may be important for the loss of vapor phase in the atmosphere. The half-life due to this reaction has been estimated to be 3.1 hours. Partial removal will also occur as a result of dry and wet deposition.

Environmental Physical Data

Henry's Law Constant: estimated at 2.91 $x10^{-4}$
Octanol/Water Partition Coefficient: log K_{OW} = 4.69
Sorption Partition Coefficient: K_{OC} = 10
BCF: estimated at 2000 to 2160

Regulations

RCRA 40CFR: Not listed
CERCLA: 40CFR 302.4: Not listed
SARA 40CFR 372.65: Listed
SARA EHS 40CFR 355: Not listed
TSCA: Not listed

Analytical Methods

Soil: SW846 8091
Water / Groundwater: EPA 627
Food: FDA 212.1, 232.1; AOAC 973.13, 973.14

Plasma: EPA 001; FDA 211.1, 231.1, 252
Other: EPA 1656

BEN1400 CAS #: 17804-35-2

BENOMYL

RTECS: DD6475000
DOT: UN2757; UN2758; UN2991; UN2992; IMO3.2; IMO6.1
EINECS Number: 241-775-7
Molecular Formula: $C_{14}H_{18}N_4O_3$
Formula Weight: 290.36

Chemical Structure

Synonyms: AGROCIT; ARBORTRINE; ARILATE; BBC; BC 6597; BENEX; BENLAT; BENLATE; BENLATE 50; BENLATE 50 W; BENOMYL 50W; BENOMYL 50; BENOMYL 50 W; BENOMYL-IMEX; BENOSAN; 2-BENZIMIDAZOLECARBAMIC ACID,1-(BUTYLCARBAMOYL)-,METHYL ESTER; 2-BENZIMIDAZOLECARBAMIC ACID,1-(BUTYLCARBAMOYL)-,METHYLESTER; BNM; 1-(BUTYLCARBAMOYL)-2-BENZIMIDAZOLECARBAMIC ACID,METHYL ESTER; 1-(BUTYLCARBAMOYL)-2-BENZIMIDAZOLECARBAMIC ACID,METHYLESTER; 1-(BUTYLCARBAMOYL)-2-BENZIMIDAZOL-METHYLCARBAMAT; 1-(N-BUTYLCARBAMOYL)-2-(METHOXY-CARBOXAMIDO)-BENZIMIDAZOL; CARBAMIC ACID,(1-((BUTYLAMINO)CARBONYL)-1H-BENZIMIDAZOL-2-YL)-,METHYL ESTER (9CI); CARBAMIC ACID,(1-((BUTYLAMINO)CARBONYL)-1H-BENZIMIDAZOL-2-YL)-,METHYLESTER; CARBAMIC ACID,METHYL-,1-(BUTYLCARBAMOYL)-2-BENZIMIDAZOLE ESTER; CHINOIN-FUNDAZOL; D 1991; DU PONT 1991; F1991; FIBENZOL; FUNDASOL; FUNDAZOL; FUNGICIDE D-1991; FUNGICIDE 1991; FUNGOCHROM; MBC; METHYL 1-((BUTYLAMINO)CARBONYL)-1H-BENZIMIDAZOL-2-YL CARBAMATE; METHYL 1-(BUTYLCARBAMOYL)-2-BENZIMIDAZOLECARBAMATE; METHYL 1-(BUTYLCARBAMOYL)-2-BENZIMIDAZOLYLCARBAMATE; METHYL 1-(BUTYLCARBAMOYL)BENZIMIDAZOL-2-YLCARBAMATE; METHYL1-((BUTYLAMINO)CARBONYL)-1H-BENZIMIDAZOL-2-YLCARBAMATE; NS 02; NS 02 (FUNGICIDE); NS O2; TERSAN; TERSAN 1991; UZGEN
Description: white, crystalline solid; faint, acrid odor
Use: fungicide used in the control of a wide range of diseases of fruit, nuts, vegetables, mushrooms, field crops, ornamentals, turf and trees; as an acaricide to slow the growth of mites; as a veterinary anthelmintic and as an oxidizer in sewage treatment

Physical Properties

Boiling Point: Decomposes
Freezing Point: Decomposes at > 300 °C (572 °F)
Saturated Vapor Density: < 1.200000142 kg/m^3
Vapor Pressure: < 1 x10^{-5} mbar at 20 °C
Water Solubility: 4 mg/kg at 25 °C
Other Solubilities: in some solvents dissociation occurs to form Carbendazim and Butyl Isocyanate.
Flash Point: Nonflammable
Autoignition Temperature: 220 °C
LEL: 0.05 g/l

RTECS Toxicity Data

Acute Oral: Rat LD_{50} Dose: 10 gm/kg.
Acute Inhalation: Rat LC_{50} Dose: >2 gm/m^3/4hr.
Mutagenic: Human Cytogenetic Analysis; Cell Type: liver; Dose: 100 umol/L. Human Cytogenetic Analysis; Cell Type: HeLa cell; Dose: 100 umol/L. Human Cytogenetic Analysis; Cell Type: lymphocyte; Dose: 10 mg/L. Human Sister Chromatid Exchenge; Cell Type: lymphocyte; Dose: 250 ug/L. Human Sex Chromosome Loss; Cell Type: lymphocyte; Dose: 1 mg/L. Human Other Mutation Test Systems; Cell Type: lymphocyte; Dose: 2 mg/L.

Hazard Overviews

Health: Irritating to eyes/skin/respiratory tract. Harmful. Other Acute Effects: harmful if swallowed, inhaled, or absorbed through skin; prolonged or repeated exposure may cause allergic reactions in certain sensitive individuals. Chronic Effects: overexposure may cause reproductive disorder(s) based on tests with laboratory animals; laboratory experiments have shown mutagenic effects; target organ: liver.
Fire: Noncombustible. Hazards: emits toxic fumes. Extinguishing agents: water spray; carbon dioxide, dry chemical powder or appropriate foam. Precautions: combustible liquid.
Reactivity: Stable. Hazardous polymerization will not occur. Incompatible with: strong oxidizing agents. Hazardous decomposition products: toxic fumes of: carbon monoxide, carbon dioxide, nitrogen oxides.
Carcinogenicity: IARC - Not listed; NIOSH - Not listed; NTP - Not listed; ACGIH - Class A4, Not classifiable as a human carcinogen; OSHA - Not listed; EPA - Not listed; MAK - Not listed
Primary Target Organs:

Eyes

Skin

Respiratory System

Liver

Exposure Limits
OSHA PEL: TWA: 15 mg/m^3; total dust.
OSHA PEL Vacated 1989 Limits: TWA: 10 mg/m^3. Other Values: respirable mg/m^3; 5.
ACGIH TLV: TWA: 0.8 ppm; 10 mg/m^3.

Respirator Recommendation

Exposure Range: >5 to 50 mg/m^3 Air Purifying, Negative Pressure, Half Mask

Exposure Range: >50 to 500 mg/m^3 Air Purifying, Negative Pressure, Full Face

Exposure Range: >500 to 5000 mg/m^3 Supplied Air, Constant Flow/Pressure Demand, Full Face

Exposure Range: >5000 to unlimited mg/m^3 Self-contained Breathing Apparatus, Pressure Demand, Full Face

Cartridge Color: black with magenta (P100)

Environmental

Ecotoxicity: LC_{50} Channel catfish 16 ug/l/96 hr at 22 °C (95% confidence limit 11-23 ug/l), wt 0.8 g. Static bioassay without aeration, pH 7.2-7.5, water hardness 40-50 mg/l as calcium carbonate and alkalinity of 30-35 mg/l LC_{50} Coturnix >5000 ppm; no overt signs of toxicity to 5000 ppm

Environmental Fate: Released to soil it will not tend to leach but volatilization from soil may be significant. Hydrolysis in soil is probably the most significant removal process although biodegradation may also be significant, especially for the hydrolysis products. Released to water it will have a low to moderate tendency to sorb to sediments, suspended sediments and biota and will not tend to bioconcentrate to any significant extent. Volatilization from water is probably insignificant. Hydrolysis will probably be the most significant removal process in water (t 1/2 <1 week) although biodegradation and photolysis may also be important. A computer estimated half-life for reaction in the vapor phase with photochemically produced hydroxyl radicals in the atmosphere is 1.6 hours.

Cleanup/Disposal: Guide No. 131: Fully encapsulating, vapor protective clothing should be worn for spills and leaks with no fire. Eliminate all ignition sources (no smoking, flares, sparks or flames in immediate area). All equipment used when handling the product must be grounded. Do not touch or walk through spilled material. Stop leak if you can do it without risk. Prevent entry into waterways, sewers, basements or confined areas. A vapor suppressing foam may be used to reduce vapors. Small Spills: Absorb with earth, sand or other non-combustible material and transfer to containers for later disposal. Use clean non-sparking tools to collect absorbed material. Large Spills: Dike far ahead of liquid spill for later disposal. Water spray may reduce vapor; but may not prevent ignition in closed spaces. Guide No. 151: Do not touch damaged containers or spilled material unless wearing appropriate protective clothing. Stop leak if you can do it without risk. Prevent entry into waterways, sewers, basements or confined areas. Cover with plastic sheet to prevent spreading. Absorb or cover with dry earth, sand or other non-combustible material and transfer to containers. Do not get water inside containers.

Environmental Physical Data

Henry's Law Constant: cannot be calculated

Octanol/Water Partition Coefficient: log K_{ow} = 2.12

Sorption Partition Coefficient: K_{oc} = 2100

BCF: calculated at 3.8

Regulations

RCRA 40CFR: Listed Hazardous Waste No. U271 Toxic Waste

CERCLA: 40CFR 302.4: Listed per RCRA Section 3001 RQ: 1 lb (0.454 kg)

SARA 40CFR 372.65: Listed

SARA EHS 40CFR 355: Not listed

TSCA: Listed

Analytical Methods

Soil: SW846 3640A, 8321A

Water / Groundwater: EPA 631

Food: AOAC 984.09

Plasma: FDA 242.3

BEN1480 CAS #: 741-58-2

BENSULIDE

RTECS: TE0250000

DOT: UN2783; UN2784; UN3017; UN3018; IMO3.2; IMO6.1

EINECS Number: 212-010-4

Molecular Formula: $C_{14}H_{24}NO_4PS_3$

Structured MF: $(CH_3CH_2CH_2O)_2P(S)SCH_2CH_2NHS(O)_2C_6H_5$

Formula Weight: 397.52

Synonyms: BENZENESULFONAMIDE,N-(2-MERCAPTOETHYL)-,S-ESTER WITH O,O-DIISOPROPYLPHOSPHORODITHIOATE; BENZENESULFONAMIDE,N-(2-MERCAPTOETHYL)-,S-ESTER WITHO,O-DIISOPROPYLPHOSPHORODITHIOATE; S-BETA-(BENZENESULFONAMIDO)ETHYL O,O-DIISOPROPYLDITHIOPHOSPHATE; S-2-BENZENESULFONAMIDOETHYL O,O-DI-ISOPROPYLPHOSPHORODITHIOATE; S-2-BENZENESULPHONAMIDOETHYL O,O-DI-ISOPROPYLPHOSPHORODITHIOATE; S-2-BENZENESULPHONAMIDOETHYL O,O-DIISOPROPYLPHOSPHORODITHIOATE; BENZULFIDE; BETAMEC; BETASAN; BETASAN E; BETASAN G; O,O-BIS(1-METHYLETHYL) S-(2-((PHENYLSULFONYL)AMINO)ETHYL)PHEOSPHORODITHIOATE; O,O-BIS(1-METHYLETHYL)-S-[2[(PHENYLSULFONYL)AMINO]ETHYL]PHOSPHORODITHIOATE; O,O-BIS(1-METHYLETHYL)S-(2-((PHENYLSULFONYL)AMINO)ETHYL)PHOSPHORODITHIOATE; O,O-DIISOPROPYL 2-(BENZENESULFONAMIDO)ETHYLDITHIOPHOSPHATE; O,O-DIISOPROPYL S-(2-BENZENESULFONYLAMINOETHYL)PHOSPHORODITHIOATE; O,O-DIISOPROPYL DITHIOPHOSPHATE S-ESTER WITHN-(2-MERCAPTOETHYL)BENZENESULPHONAMIDE; O,O-DI-ISOPROPYL S-2-PHENYLSULPHONYLAMINOETHYLPHOSPHORODITHIOATE; S-(O,O-DIISOPROPYL PHOSPHORODITHIOATE) ESTER OF N-(2-MERCAPTOETHYL)BENZENESULFONAMIDE; S-(O,O-DIISOPROPYL PHOSPHORODITHIOATE) ESTER OFN-(2-MERCAPTOETHYL)BENZENESULFONAMIDE; O,O-DIISOPROPYL PHOSPHORODITHIOATE S-ESTER WITHN-(2-MERCAPTOETHYL)BENZENESULFONAMIDE; N-(2-(O,O-DIISOPROPYLDITHIOPHOSPHORYL)ETHYL)BENZENESULFONAMIDE; N-(BETA-O,O-DIISOPROPYLDITHIOPHOSPHORYLETHYL)BENZENESULFONAMIDE; S-(O,O-DIISOPROPYLPHOSPHORODITHIOATE) OFN-(2-MERCAPTOETHYL)BENZENESULFONAMIDE; DISAN; DISAN (PESTICIDE); (N-(2-ETHYLTHIO)BENZENESULPHONAMIDE-S,O,O-

DIISOPROPYLPHOSPHORODITHIOATE; EXPORSAN; KAYAPHENONE; (N-(2-MERCAPTOETHYL)BENZENESULFONAMIDE); N-(2-MERCAPTOETHYL)BENZENESULFONAMIDE S-(O,O-DIISOPROPYLPHOSPHORODITHIOATE); N-(2-MERCAPTOETHYLBENZENE)SULFONAMIDES-(O,O-DIISOPROPYLPHOSPHORODITHIOATE); PHOSPHORODITHIOIC ACID,O,O-BIS(1-METHYLETHYL) S-(2-((PHENYLSULFONYL)AMINO)ETHYL) ESTER; PHOSPHORODITHIOIC ACID,O,O-BIS(1-METHYLETHYL)S-(2-((PHENYLSULFONYL)AMINO)ETHYL) ESTER; PHOSPHORODITHIOIC ACID,O,O-DIISOPROPYL ESTER,S-ESTERWITH N-(2-MERCAPTOETHYL)BENZENESULFONAMIDE; PREFAR; PREFAR E; PREFER; PRE-SAN; R 4461; R-4461; SAP; SAP (HERBICIDE)

Description: white crystalline solid; amber viscous liquid above 34.4 deg C

Use: selective preemergence herbicide for control of undesirable grasses, crabgrass, annual bluegrass, redroot, pigweed, watergrass, lambsquarters, shepherdspurse, goosegrass, broadleaf weeds & deadnettle in grass & dichondra lawns, cole crops, brassicas, cucurbits, watermelons, cotton, lettuce, tomatoes, carrots, capsicums, rice, & herbage seed crops.

Physical Properties

Boiling Point: Decomposes at 200 °C (392 °F)
Freezing Point: 34.4 °C (93.92 °F)
Specific Gravity: 1.23 at 20 °C/20 °C
Vapor Pressure: < 0.133 mPa at 20 °C
Water Solubility: 25 mg/L at 20 °C
Other Solubilities: miscible with Methyl-isobutyl Ketone.
Refraction Index: 1.5438 at 30 °C/D
Flash Point: 157 °C Open Cup

RTECS Toxicity Data

Acute Oral: Rat LD_{50} Dose: 271 mg/kg.
Acute Dermal: Rat LD_{50} Route: Skin; Dose: 3950 mg/kg.

Hazard Overviews

Fire: Will burn.
Carcinogenicity: IARC - Not listed; NIOSH - Not listed; NTP - Not listed; ACGIH - Not listed; OSHA - Not listed; EPA - Not listed; MAK - Not listed

Environmental

Ecotoxicity: LC_{50} Penaeus azetucus (brown shrimp), loss of equilibrium or death > 1 ppm/96 hr/Conditions of bioassay not specified LC_{50} Common goldfish 1-2 ppm/96 hr /Conditions of bioassay not specified LC_{50} Channel catfish 379 ug/l (96 hr) LC_{50} Gammarus fasciatus 1.4 mg/l/96 hr at 15 °C, mature (95% confidence limit 0.4-5.1 mg/l). static bioassay without aeration, pH 7.2-7.5, water hardness 40-50 mg/l as calcium carbonate and alkalinity of 30-35 mg/l

Cleanup/Disposal: Guide No. 131: Fully encapsulating, vapor protective clothing should be worn for spills and leaks with no fire. Eliminate all ignition sources (no smoking, flares, sparks or flames in immediate area). All equipment used when handling the product must be grounded. Do not touch or walk through spilled material. Stop leak if you can do it without risk. Prevent entry into waterways, sewers, basements or confined areas. A vapor suppressing foam may be used to reduce vapors. Small Spills: Absorb with earth, sand or other non-combustible material and transfer to containers for later disposal. Use clean non-sparking tools to collect absorbed material. Large Spills: Dike far ahead of liquid spill for later disposal. Water spray may reduce vapor; but may not prevent ignition in closed spaces. Guide No. 152: Do not touch damaged containers or spilled material unless wearing appropriate protective clothing. Stop leak if you can do it without risk. Prevent entry into waterways, sewers, basements or confined areas. Cover with plastic sheet to prevent spreading. Absorb or cover with dry earth, sand or other non-combustible material and transfer to containers. Do not get water inside containers.

Regulations

RCRA 40CFR: Not listed
CERCLA: 40CFR 302.4: Not listed
SARA 40CFR 372.65: Not listed
SARA EHS 40CFR 355: Not listed
TSCA: Not listed

Analytical Methods

Soil: EPA PMD-BEL
Water / Groundwater: EPA 636
Food: FDA 212.1, 232.1, 232.3, 232.4, 242.1
Plasma: EPA 001; FDA 211.1, 231.1, 252

BEN1560 **CAS #: 25057-89-0**

BENTAZON

RTECS: DK9900000
EINECS Number: 246-585-8
Molecular Formula: $C_{10}H_{12}N_2O_3S$
Formula Weight: 240.30

Synonyms: BAS 351-07H; BAS 351-H; BAS 3510H; BAS 3512H; BAS 3517H; BAS 351H; BAS 3510; BASAGRAN; BASAGRAN DP; BASAGRAN KV; BASAGRAN M; BASAGRAN-PLUS; BENDIOXIDE; BENTAZONE; 1H-2,1,3-BENZOTHIADIAZIN-4(3H)-ONE-2,2-DIOXIDE,3-ISOPROPYL-; 1H-2,1,3-BENZOTHIADIAZIN-4(3H)-ONE,3-ISOPROPYL-,2,2-DIOXIDE; 1H-2,1,3-BENZOTHIADIAZIN-4(3H)-ONE,3-(1-METHYLETHYL)-,2,2-DIOXIDE; GRAMINON-PLUS; HERBATOX; 3-ISOPROPYL-2,1,3-BENZOTHIADIAZINON-(4)-2,2-DIOXID; 3-ISOPROPYL-(1H)-BENZO-2,1,3-THIADIAZIN-4-ONE-2,2-DIOXIDE; 3-ISOPROPYL-1H-2,1,3-BENZOTHIADIAZIN-4(3H)-ONE-2,2-DIOXIDE; 3-ISOPROPYL-1H-BENZO-2,1,3-THIADIAZIN-4-ONE-2,2-DIOXIDE; 3-ISOPROPYL-1H-2,1,3-BENZOTHIAIAZIN-4(3H)-ONE-2,2-DIOXIDE; LADDOK; LEADER; 3-(1-METHYLETHYL)-(1H)-2,1,3-BENZOTHIADIAZIN-4(3H)-ONE-2,2-DIOXIDE; 3-(1-METHYLETHYL)-1H-2,1,3-BENZOTHIAZAIN-4(3H)-ONE,2,2-DIOXIDE; PENTAZONE; PLEDGE

Description: colorless to white crystalline powder; odorless

Use: selective postemergence contact herbicide; for control of wide range of broadleaved & sedge weeds; in cereals, ornamental turf, potatoes, dry beans, dry peas, snap beans for seed, green (succulent) lima beans, & mint; controls number of primarily by contact actionin graminous & large seeded leguminous crops.

Physical Properties

Boiling Point: Decomposes at 200 °C (392 °F)
Freezing Point: 137 °C (278.6 °F) to 139 °C (282.2 °F)
Saturated Vapor Density: < 2.063124213 kg/m^3
Vapor Pressure: < 0.01 mPa at 20 °C
Water Solubility: 500 mg/kg at 20 °C
Other Solubilities: g/100 g at 20 °C: Ethyl Acetate 65.0, Diethyl Ether 61.6, cyclohexane 0.02.
Flash Point: > 100 °C

RTECS Toxicity Data

Acute Oral: Rat LD_{50} Dose: 1100 mg/kg. Mouse LD_{50} Dose: 1130 mg/kg.
Acute Inhalation: Rat LC_{50} Dose: 5100 mg/m^3/4hr.
Acute Dermal: Rat LD_{50} Route: Skin; Dose: 2500 mg/kg.

Hazard Overviews

Fire: Will burn.
Carcinogenicity: IARC - Not listed; NIOSH - Not listed; NTP - Not listed; ACGIH - Not listed; OSHA - Not listed; EPA - Not listed; MAK - Not listed

Environmental

Ecotoxicity: Fishes: cytotoxicity to goldfish GF-Scale cells NR50 >200 mg/l

Environmental Physical Data

Sorption Partition Coefficient: log K_{oc} = experimental 0
BCF: estimated at 19

Regulations

RCRA 40CFR: Not listed
CERCLA: 40CFR 302.4: Not listed
SARA 40CFR 372.65: Not listed
SARA EHS 40CFR 355: Not listed
TSCA: Listed

Analytical Methods

Soil: SW846 8151
Water / Groundwater: EPA 643; ASTM D5317
Drinking Water: EPA 515.1, 515.2, 555; AOAC 992.32
Food: AOAC 993.02

BEN1640 **CAS #: 1302-78-9**

BENTONITE

RTECS: CT9450000
EINECS Number: 215-108-5
Molecular Formula: Unknown
Formula Weight: N/A

O═Al—O—Al═O   O═Si═O OH_2

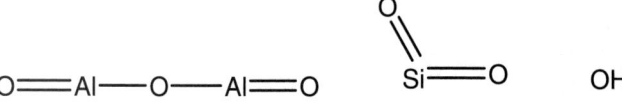

Chemical Structure

Synonyms: ALBAGEL PREMIUM USP 4444; BENTONITE 2073; BENTONITE MAGMA; COLLOIDAL CLAY; HI-JEL; IMVITE I.G.B.A; MAGBOND; MONTMORILLONITE; OTAYLITE; PANTHER CREEK BENTONITE; SOUTHERN BENTONITE; TIXOTON; VOLCALY BENTONITE BC; VOLCLAY; WILKINITE; WILKONITE
Description: cream to pale brown powder, in massive condition varies from yellowish-white to almost black; light yellow or green, cream, pink, gray to black; odorless
Use: emulsifier for oils; base for plasters; pharmaceutic aid; granular carrier; liq clarifier; reagglomerating agent for iron ore; cracking catalyst for petro; agent in drilling mud; in foundry molding sands; lubricant for extrusion of animal feed; absorbent in pet litter, for purifying oils, etc; in sugar purification, brewing, paper industries; cement slurries for oil-wells, canal walls; asphalt modifier; thickener in greases & fireproofing; cosmetics; decolorizing agent; filler in ceramics, etc; polishes & abrasives; food additive.

Physical Properties

Water Solubility: Insoluble in Water or Acids
Other Solubilities: Insoluble in acids.
pH: In presence of water 9

RTECS Toxicity Data

Tumorigenic: Mouse Route: Oral; Dose: 12000 gm/kg/28W-C; Toxic Effects: Tumorigenic - Equivocal tumorigenic agent by RTECS criteria; Liver - Tumors.

Hazard Overviews

Health: Irritating to eyes/skin/respiratory tract. Other Acute Effects: may be harmful by inhalation, ingestion, or skin absorption; Chronic Effects: lung irritation; asthma; target organ: lungs.
Fire: Hazards: emits toxic fumes. Extinguishing agents: noncombustible; use extinguishing media appropriate to surrounding fire conditions. Precautions: combustible liquid.
Reactivity: Incompatible with: strong acids. Hazardous decomposition products: nature of decomposition products not known;.
Carcinogenicity: IARC - Not listed; NIOSH - Not listed; NTP - Not listed; ACGIH - Not listed; OSHA - Not listed; EPA - Not listed; MAK - Not listed
Primary Target Organs:

Eyes

Skin

Respiratory System

Environmental

Regulations

RCRA 40CFR: Not listed
CERCLA: 40CFR 302.4: Not listed
SARA 40CFR 372.65: Not listed
SARA EHS 40CFR 355: Not listed
TSCA: Listed

BEN1720 CAS #: 225-51-4

BENZ(C)ACRIDINE

RTECS: CU2975000
EINECS Number: 205-930-2
Molecular Formula: $C_{17}H_{11}N$
Formula Weight: 229.3
Synonyms: 12-AZABENZ(A)ANTHRACENE; B(C)AC; 3,4-BENZACRIDINE; 7,8-BENZACRIDINE; 3,4-BENZOACRIDINE; ALPHA-CHRYSIDINE; ALPHA-NAPHTHACRIDINE
Description: yellow needles
Use: research chemical

Physical Properties

Freezing Point: 108 °C (226.4 °F)
Water Solubility: Very Slightly Soluble in Water
Other Solubilities: Soluble in Benzene and Acetone

RTECS Toxicity Data

Mutagenic: Hamster Sister Chromatid Exchenge; Cell Type: ovary; Dose: 10 umol/L. Hamster Sister Chromatid Exchenge; Cell Type: lung; Dose: 1 umol/L.
Tumorigenic: Mouse Route: Skin; Dose: 2400 mg/kg/67W-I; Toxic Effects: Tumorigenic - Equivocal tumorigenic agent by RTECS criteria; Skin and appendages - Tumors. Mouse Route: Intraperitoneal; Dose: 9630 mg/kg/3D-I; Toxic Effects: Tumorigenic - Neoplastic by RTECS criteria; Lungs, Thorax, or Respiration - Tumors.

Hazard Overviews

Carcinogenicity: IARC - Group 3, Not classifiable as to carcinogenicity to humans; NIOSH - Not listed; NTP - Not listed; ACGIH - Not listed; OSHA - Not listed; EPA - Not listed; MAK - Not listed

Environmental

Ecotoxicity: LC_{50} Daphnia pulex 0.45 mg/l/24 hr /conditions of bioassay not specified
Environmental Fate: Absorbs sunlight radiation strongly, indicating a potential for direct photolysis in air, in water, or on soil surfaces; however, photolysis rate data are not available for vapor-phase, solution-phase, or adsorbed particulate phase. No information concerning biodegradation in soil or water was located. If released to soil, it is expected to adsorb very strongly and is not expected to leach based on its estimated K_{oc} values of 22500 and 135000. It is not expected to hydrolyze in soil and evaporation from soil surfaces is not expected to be significant. If released to water, it may adsorb strongly to sediments, but is not expected to hydrolyze, volatilize significantly, or react significantly with photochemically produced alkyl peroxy radicals. Based on its estimated log BCF value of 3.27, it may be expected to bioconcentrate; however, bioconcentration may not be appreciable in organisms which have microsomal oxidase, such as fish, as this enzyme enables the organism to metabolize polyaromatic hydrocarbons. If released to the atmosphere, release will be largely associated with particulate matter. Dry deposition may be an important mechanism for its removal from the air. In the vapor-phase in the atmosphere, it will react with photochemically produced hydroxyl radicals with an estimated half-life of 1.25 days, but will not react with ozone.

Environmental Physical Data

Octanol/Water Partition Coefficient: log K_{ow} = 4.607
Sorption Partition Coefficient: K_{oc} = estimated at 2.25 x10^4
BCF: estimated at 3.27

Regulations

RCRA 40CFR: Listed Hazardous Waste No. U016 Toxic Waste
CERCLA: 40CFR 302.4: Listed per RCRA Section 3001 RQ: 100 lb (45.35 kg)
SARA 40CFR 372.65: Not listed
SARA EHS 40CFR 355: Not listed
TSCA: Not listed

BEN1800 CAS #: 100-52-7

BENZALDEHYDE

RTECS: CU4375000
DOT: NA1989; IMO9.0
EINECS Number: 202-860-4
Molecular Formula: C_7H_6O
Structured MF: C_6H_5CHO
Formula Weight: 106.12

O H

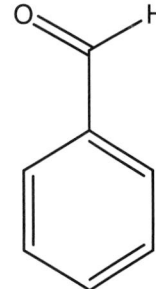

Chemical Structure

Synonyms: ALMOND ARTIFICIAL ESSENTIAL OIL; ARTIFICIAL ALMOND OIL; ARTIFICIAL ESSENTIAL OIL OF ALMOND; BENZALDEHYDE FFC; BENZENE CARBALDEHYDE; BENZENE CARBOXALDEHYDE; BENZENECARBONAL; BENZENECARBOXALDEHYDE; BENZENEMETHYLAL; BENZOIC ALDEHYDE; OIL OF BITTER ALMOND; PHENYLMETHANAL; SYNTHETIC OIL OF BITTER ALMOND
Description: clear to yellow liquid; oil of almond odor
Use: chemical intermediate for dyes, flavoring materials, perfumes and aromatic alcohols; solvent for oils, resins, some cellulose ethers, cellulose acetate and nitrate; manufacture of cinnamic acid, benzoic acid; pharmaceuticals; an photographic chemicals

Physical Properties

Boiling Point: 179 °C (354 °F)

Freezing Point: -26 °C (-14.8 °F)
Specific Gravity: 1.05 at 15 °C/4 °C
Vapor Density: 3.66 Air=1
Vapor Pressure: 1 mm Hg at 26 °C
Water Solubility: 1 parts in 350 parts Water
Other Solubilities: 95% Ethanol: >=100 mg/ml at 20 °C; Acetone: >=100 mg/ml at 20 °C; Benzene: Very Soluble; DMSO: >=100 mg/ml at 20 °C; Ether: miscible; Ligroin: Very Soluble; Oils and fats: miscible.
Surface Tension: 40.0 dynes/cm
Odor Threshold: 0.042 ppm
Refraction Index: 1.5456 at 20 °C/D
Critical Temperature: 178 °C
Critical Pressure: 316 psia
Ionization Potential (eV): 9.4 +/-0.1
Flash Point: 63 °C Closed Cup
Autoignition Temperature: 192 °C
LEL: 1.1% v/v
UEL: 3.7% v/v

RTECS Toxicity Data

Acute Oral: Rat LD_{50} Dose: 1300 mg/kg; Toxic Effects: Behavioral - Somnolence (general depressed activity); Behavioral - Coma. Mouse LD_{50} Dose: 28 mg/kg; Toxic Effects: Behavioral - Somnolence (general depressed activity); Behavioral - Tremor; Lungs, Thorax, or Respiration - Other changes.
Acute Inhalation: Rat LC; Dose: >500 mg/m^3; Toxic Effects: Sense organs and special senses - Conjunctive irritation; Behavioral - Somnolence (general depressed activity); Lungs, Thorax, or Respiration - Other changes. Mouse LC; Dose: >500 mg/m^3; Toxic Effects: Sense organs and special senses - Conjunctive irritation; Behavioral - Somnolence (general depressed activity); Lungs, Thorax, or Respiration - Other changes.
Acute Dermal: Rabbit LD_{50} Route: Subcutaneous Dose: 5 gm/kg. Rat LD_{Lo} Route: Subcutaneous Dose: 5 gm/kg; Toxic Effects: Lungs, Thorax, or Respiration - Respiratory depression; Lungs, Thorax, or Respiration - Other changes.
Chronic (Multiple Dose) Oral: Rat Dose: 9600 mg/kg/16D-I; Toxic Effects: DEATH. Rat Dose: 52 gm/kg/13W-I; Toxic Effects: Brain and coverings - Other degenerative changes; Liver - Fatty liver degeneration; Kidney, Ureter, and Bladder - Chgs in tubules (inc acute renal failure, acute tubular necrosis. Rat Dose: 78 gm/kg/13W-I; Toxic Effects: Kidney, Ureter, and Bladder - Chgs in tubules (inc acute renal failure, acute tubular necrosis; DEATH.
Chronic (Multiple Dose) Inhalation: Rat Dose: 500 ppm/6H/14D-C; Toxic Effects: Liver - Changes in Liver weight; Blood - Changes in other cell count; Nutritional and gross metabolic - Weight loss or decreased weight gain. Rat Dose: 26 mg/m^3/5H/17W-I; Toxic Effects: Blood - Changes in erythrocite (RBC) cell count; Blood - Changes in leukocyte (WBC) cell count; Nutritional and gross metabolic - Weight loss or decreased weight gain.
Irritation Skin: Rabbit Standard Draize Test Dose: 500 mg/24H; Reaction: moderate.
Mutagenic: Human Sister Chromatid Exchenge; Cell Type: lymphocyte; Dose: 1 mmol/L. Mouse Mutations in Mammalian Somatic Cells; Cell Type: lymphocyte; Dose: 400 mg/L.
Tumorigenic: Mouse Route: Oral; Dose: 154 gm/kg/2Y-C; Toxic Effects: Tumorigenic - Neoplastic by RTECS criteria; Gastrointestinal - Tumors.

Hazard Overviews

Fire Diamond

Health: Irritating to eyes/skin/respiratory tract. Also Causes: nausea, abdominal pain, narcotic-like effects, and convulsions, paralysis. Chronic Effects: kidney damage; dermatitis.
Fire: Combustible. Use water spray, dry chemical, foam, or carbon dioxide to extinguish benzaldehyde fires. Use water spray to cool fire-exposed containers.
Reactivity: Stable. Hazardous polymerization cannot occur. Avoid: sources of ignition. Incompatible with: performic acid. Hazardous decomposition products: toxic gases.
Carcinogenicity: IARC - Not listed; NIOSH - Not listed; NTP - Not listed; ACGIH - Not listed; OSHA - Not listed; EPA - Not listed; MAK - Not listed
Primary Target Organs:

Eyes

Skin

Respiratory System
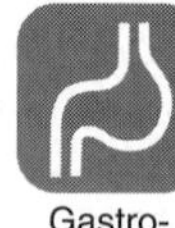
Gastro-intestinal
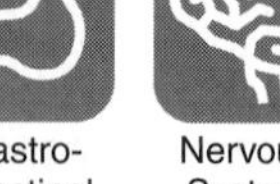
Nervous System
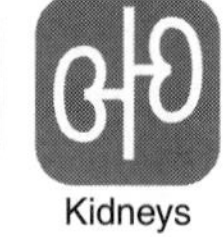
Kidneys

Exposure Limits
AIHA WEEL: TWA: 2 ppm; STEL: 4 ppm 15-min.
Respirator Recommendation
Exposure Range: >2 to 20 mg/m^3 Air Purifying, Negative Pressure, Half Mask
Exposure Range: >20 to 200 mg/m^3 Air Purifying, Negative Pressure, Full Face
Exposure Range: >200 to 2000 mg/m^3 Supplied Air, Constant Flow/Pressure Demand, Full Face
Exposure Range: >2000 to unlimited mg/m^3 Self-contained Breathing Apparatus, Pressure Demand, Full Face
Cartridge Color: black with magenta (P100)

Environmental

Environmental Fate: If released to the atmosphere, it will degrade by reaction with photochemically produced hydroxyl radicals (half-life of 29.8 hr); direct photolysis may contribute to its atmospheric degradation. Physical removal from air by wet deposition can occur. If released to soil or water, the major degradation pathway is expected to be biodegradation. Physical transport from water can occur through volatilization. Estimated K_{oc} values (9-71) suggest that it will leach in soil. Occupational exposure occurs through inhalation of vapor and dermal contact.
Cleanup/Disposal: Guide No. 129: Eliminate all ignition sources (no smoking, flares, sparks or flames in immediate

area). All equipment used when handling the product must be grounded. Do not touch or walk through spilled material. Stop leak if you can do it without risk. Prevent entry into waterways, sewers, basements or confined areas. A vapor suppressing foam may be used to reduce vapors. Absorb or cover with dry earth, sand or other non-combustible material and transfer to containers. Use clean non-sparking tools to collect absorbed material. Large Spills: Dike far ahead of liquid spill for later disposal. Water spray may reduce vapor; but may not prevent ignition in closed spaces.

Environmental Physical Data

Henry's Law Constant: 2.6×10^{-5}
Octanol/Water Partition Coefficient: log K_{ow} = 1.48
Sorption Partition Coefficient: K_{oc} = estimated at 34 to 150
BCF: estimated at 7.8
BOD: 50%, 10 days

Regulations

RCRA 40CFR: Not listed
CERCLA: 40CFR 302.4: Not listed
SARA 40CFR 372.65: Not listed
SARA EHS 40CFR 355: Not listed
TSCA: Listed

Analytical Methods

Air: EPA TO-11, TO-5, 0100
Soil: SW846 8315A
Indoor / Expired Air: EPA IP-1B, IP-6A, IP-6B, IP-6C, 0100

BEN1880 **CAS #: 8001-54-5**

BENZALKONIUM CHLORIDE

RTECS: BO3150000
Molecular Formula: Unknown
Formula Weight: N/A
Synonyms: ALKYL DIMETHYLBENZYL AMMONIUM CHLORIDE; ALKYLDIMETHYLBENZYLAMMONIUM CHLORIDE; ALKYLDIMETHYL(PHENYLMETHYL)QUATERNARY AMMONIUM CHLORIDES; AMMONIUM,ALKYLDIMETHYLBENZYL-,CHLORIDE; AMMONYX; ARQUAD B 100; ARQUAD DMMCB-75; BARQUAT MB-50; BARQUAT MB-80; BAYCLEAN; BENIROL; BENZALKONIUM A; BIONOL; BIO-QUAT 50-24; BIO-QUAT 50-25; BIO-QUAT 50-30; BIO-QUAT 50-40; BIO-QUAT 50-42; BIO-QUAT 50-60; BIO-QUAT 50-65; BIO-QUAT 80-24; BIO-QUAT 80-28; BIO-QUAT 80-40; BIO-QUAT 80-42; BTC; BTC E-8358; BTC 100; BTC 2565; BTC 50; BTC 65; BTC 824; BTC 8248; BTC 8249; BTC 50 USP; BTC 65 USP; CAPITOL; CATAMIN AB; CATAMINE AB; CEQUARTYL; CLEAR EYES; CULVERSAN LC 80; DESITIN DABAWAYS; DESITIN SKIN CARE LOTIONS; DIMANIN A; DISINALL; DODIGEN 226; DRAPOLENE; DRAPOLEX; DREST; E-PILO OPHTHALMIC SOLUTION; ENUCLEN; EPIFRIN OPHTHALMIC SOLUTION; EPINAL; EPPY/N; GARDIQUART SV480; GARDIQUAT 1450; GENAMIN KDS; GERMICIN; GERMINOL; GERMITOL; GERM-I-TOL; HYAMINE 3500; INTEXAN LB-50; INTEXAN LB 50; KATAMIN AB; KATAMIN BAC; KATAMINE AB; KATAMINE BAC; LEDA BENZALKONIUM CHLORIDE; MARINOL; MEFAROL; MURINE FOR THE EYES; MURO'S OPCON A OPHTHALMIC SOLUTION; MURO'S OPCON OPHTHALMIC SOLUTION; MURO TEARS OPHTHALMIC SOLUTION; MUROCOLL-2 OPHTHALMIC SOLUTION; MUROCOLL-19 OPTHALMIC SOLUTION; NEO GERM-I-TOL; ONYX BTC (ONYX OIL & CHEM CO); OSVAN; PARALKAN; PHENEENE GERMICIDAL SOLUTION & TINCTURE; PHENEENE GERMICIDAL SOLUTION AND TINCTURE; QUATERNARY AMMONIUM COMPOUNDS,ALKYLBENZYLDIMETHYL,CHLORIDES; QUATERNIUM-1; QUATERNIUM 1; ROCCAL; RODALON; ROMERGAL CB; TRITON K-60; VIKROL RQ; VISALENS SOAKING/CLEANING SOLUTIONS; VISALENS WETTING SOLUTION; VISINE AC; ZEPHIRAL; ZEPHIRAN; ZEPHIRAN CHLORIDE; ZEPHIROL
Description: white or yellowish-white, amorphous powder or gelatinous pieces or clear, mobile liquid; aromatic odor
Use: cationic surface active agent & germicide against slime mold, algae, fish pathogens & mollusks; in algicides, deodorants, detergent/sanitizers; available as powders, ointments, etc; disinfectants for food plants, etc; cationic surfactants; pharmaceutic aid (preservative); med: antiseptics, bactericides, fungicides, sanitizers & deodorants; astringent; in to preserve sterility of surgical instruments, & ophthalmic soln; antiinfective; (vet) antiseptic; udder wash; spermicide (vaginal).

Physical Properties

Specific Gravity: 0.988
Water Solubility: Very Soluble in Water
Other Solubilities: 1 g of anyhrous for dissolves in about 6 ml Benzene, 100 ml Ether.
pH: Aqueous solution is slightly alkaline
Flash Point: 250 °C Open Cup

RTECS Toxicity Data

Acute Oral: Woman TD_{Lo} Dose: 266 mg/kg; Toxic Effects: Behavioral - Hallucinations, distorted perceptions; Gastrointestinal - Changes in structure or function of esophagus; Gastrointestinal - Hypermotility, diarrhea. Rat LD_{50} Dose: 240 mg/kg; Toxic Effects: Behavioral - Somnolence (general depressed activity); Gastrointestinal - Nausea or vomiting.
Acute Dermal: Rat LD_{50} Route: Subcutaneous Dose: 400 mg/kg; Toxic Effects: Behavioral - Somnolence (general depressed activity). Mouse LD_{50} Route: Subcutaneous Dose: 64 mg/kg; Toxic Effects: Behavioral - Somnolence (general depressed activity).
Irritation Eye: Human Standard Draize Test Dose: 50 ug; Reaction: severe. Monkey Standard Draize Test Dose: 2 mg/24H; Reaction: severe. Rabbit Standard Draize Test Dose: 1 mg/24H; Reaction: severe. Rabbit Standard Draize Test Dose: 10 mg; Reaction: mild.
Irritation Skin: Human Standard Draize Test Dose: 150 ug/3D-I; Reaction: mild. Rabbit Standard Draize Test Dose: 50 mg/24H; Reaction: moderate.
Reproductive/Teratogenic: Rat Route: Intravaginal; Dose: 50 mg/kg; Duration: female 1D of pregnancy; Effects on Fertility - Litter size; Effects on Embryo or Fetus - Fetotoxicity. Rat Route: Intravaginal; Dose: 100 mg/kg; Duration: female 1D of pregnancy; Effects on Fertility - Post-implantation mortality; Effects on Embryo or Fetus - Fetal death. Rat Route: Intravaginal; Dose: 200 mg/kg; Duration: female 1D of pregnancy; Effects on Fertility - Pre-implantation mortality.

Mutagenic: Hamster Sister Chromatid Exchenge; Cell Type: embryo; Dose: 1 mg/L. Bacteria - B Subtilis DNA Repair; Dose: 50 ug/L.

Hazard Overviews

Health: Corrosive to eyes/skin/respiratory tract. Harmful. Other Acute Effects: harmful if swallowed, inhaled, or absorbed through skin; exposure may cause burning sensation; coughing; wheezing; laryngitis; shortness of breath; headache; nausea; vomiting; inhalation may result in spasm; inflammation and edema of the larynx and bronchi; chemical pneumonitis; pulmonary edema.
Fire: Will burn. Hazards: emits toxic fumes. Extinguishing agents: carbon dioxide; dry chemical powder; water spray. Precautions: combustible liquid.
Reactivity: Incompatible with: strong oxidizing agents. Hazardous decomposition products: carbon monoxide, carbon dioxide, nitrogen oxides, hydrogen chloride gas.
Carcinogenicity: IARC - Not listed; NIOSH - Not listed; NTP - Not listed; ACGIH - Not listed; OSHA - Not listed; EPA - Not listed; MAK - Not listed
Primary Target Organs:

Environmental

Regulations

RCRA 40CFR: Not listed
CERCLA: 40CFR 302.4: Not listed
SARA 40CFR 372.65: Not listed
SARA EHS 40CFR 355: Not listed
TSCA: Not listed

Analytical Methods

Soil: EPA PMD-QAC
Food: AOAC 960.14

BEN1960 CAS #: 55-21-0

BENZAMIDE

RTECS: CU8700000
EINECS Number: 200-227-7
Molecular Formula: C_7H_7NO
Formula Weight: 121.14

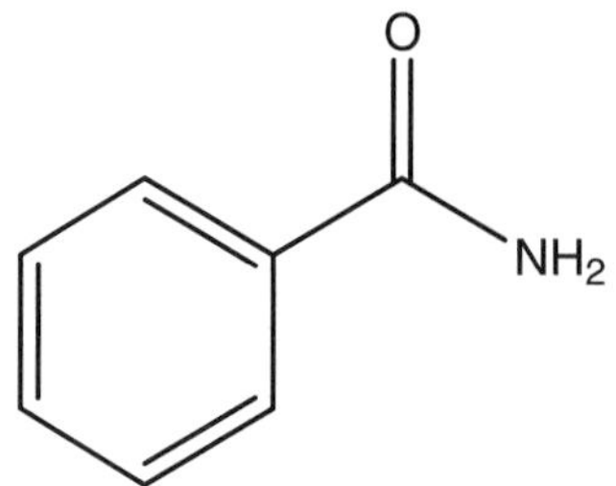

Chemical Structure

Synonyms: AMID KYSELINY BENZOOVE; BENZOIC ACID AMIDE; BENZOYLAMIDE; PHENYL CARBOXYAMIDE; PHENYLCARBOXYAMIDE
Description: colorless crystals, prisms or plates
Use: organic synthesis

Physical Properties

Boiling Point: 288 °C (550 °F)
Freezing Point: 130 °C (266 °F)
Specific Gravity: 1.341
Density: 1.341 g/mL at 4 °C
Vapor Pressure: Estimated 1.65 $x10^{-4}$ mm Hg
Water Solubility: < 1 mg/mL at 22.5 C
Other Solubilities: 95% Ethanol: >=100 mg/ml at 22.5 °C; Acetone: 1-5 mg/ml at 22.5 °C; Ammonia: Soluble; Benzene: Slightly Soluble; Carbon Disulfide: Very Soluble; Carbon Tetrachloride: Very Soluble; DMSO: >=100 mg/ml at 22.5 °C; Ether: Slightly Soluble.
Surface Tension: 47.26 dyne/cm
Ionization Potential (eV): =< 9.45 +/-0.2
Flash Point: Not available; probably combustible

RTECS Toxicity Data

Acute Oral: Mouse LD_{50} Dose: 1160 mg/kg.
Mutagenic: Human Sister Chromatid Exchenge; Cell Type: lymphocyte; Dose: 300 umol/L. Mouse Sister Chromatid Exchenge; Cell Type: leukocyte; Dose: 1 mmol/L.

Hazard Overviews

Health: May be irritating to eyes/skin. Harmful. Other Acute Effects: harmful if swallowed; may be harmful if inhaled; may be harmful if absorbed through the skin.
Fire: Will burn. Hazards: emits toxic fumes. Extinguishing agents: water spray; carbon dioxide, dry chemical powder or appropriate foam. Precautions: combustible liquid.
Reactivity: Incompatible with: strong oxidizing agents, strong bases. Hazardous decomposition products: toxic fumes of: carbon monoxide, carbon dioxide, nitrogen oxides.
Carcinogenicity: IARC - Not listed; NIOSH - Not listed; NTP - Not listed; ACGIH - Not listed; OSHA - Not listed; EPA - Not listed; MAK - Not listed

Environmental

Ecotoxicity: LC_{50} Pimephales promelas (fathead minnow) 661 mg/l/96 hr (confidence limit 590 - 740 mg/l), flow-through bioassay with measured concentrations, 25.1 °C, dissolved oxygen 7.2 mg/l, hardness 47.0 mg/l calcium carbonate, alkalinity 41.3 mg/l calcium carbonate, and pH 7.5

Environmental Fate: If released to soil, it will readily leach into the soil and biodegradation. Biodegradation in the upper layers of soil is rapid, with complete biodegradation reported in 3 to 13 days. Should it leach into aquifers, the biodegradation half-life should be somewhat over a month. If released to surface water, it would not volatilize or adsorb strongly to sediment or particulate matter in the water column. Biodegradation is its likely fate in water, although no information on its half-life in natural water is available. If released in air, it will most likely occur in particulate matter and be subject to gravitational settling. Vapor-phase will react with photochemically produced hydroxyl radicals with an estimated half-life of 18 hr.

Environmental Physical Data

Henry's Law Constant: calculated at 1.95 $x10^{-9}$

Octanol/Water Partition Coefficient: log K_{ow} = 0.640

Sorption Partition Coefficient: K_{oc} = 9 to 57

BCF: calculated at 1.084

Regulations

RCRA 40CFR: Not listed

CERCLA: 40CFR 302.4: Not listed

SARA 40CFR 372.65: Listed

SARA EHS 40CFR 355: Not listed

TSCA: Listed

BEN2040 CAS #: 56-55-3

BENZ(A)ANTHRACENE

RTECS: CV9275000

EINECS Number: 200-280-6

Molecular Formula: $C_{18}H_{12}$

Formula Weight: 228.29

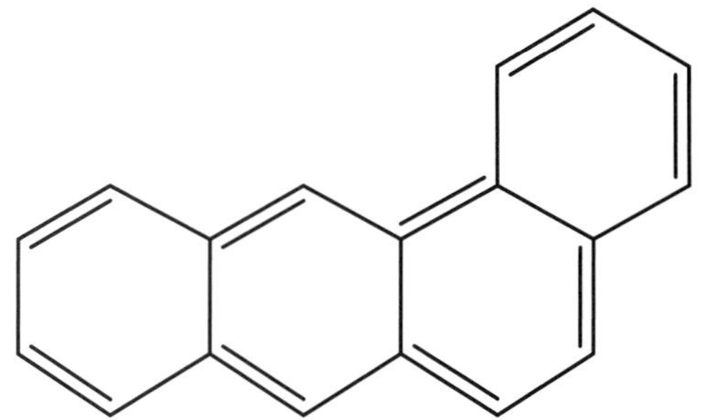

Chemical Structure

Synonyms: B(A)A; BA; BAA; 1,2-BENZ(A)ANTHRACENE; 1,2-BENZANTHRACENE; BENZANTHRACENE; 1,2-BENZANTHRAZEN; 1,2-BENZANTHRENE; BENZANTHRENE; 1,2-BENZOANTHRACENE; BENZO(A)ANTHRACENE; BENZOANTHRACENE; 2,3-BENZOPHENANTHRENE; BENZO(A)PHENANTHRENE; BENZO(B)PHENANTHRENE; 2,3-BENZPHENANTHRENE; NAPHTHANTHRACENE; TETRAPHENE

Description: colorless plates

Use: research chemistry

Physical Properties

Boiling Point: Sublimes at 435 °C (815 °F)

Freezing Point: 162 °C (323.6 °F)

Saturated Vapor Density: 1.2 kg/m^3

Vapor Pressure: 5 $x10^{-9}$ torr at 20 °C

Water Solubility: 0.014 mg/L in Water at 25 °C

Other Solubilities: 95% Ethanol: <1 mg/ml at 20 °C; Acetone: 10-50 mg/ml at 20 °C; Acetic Acid: Slightly Soluble; Benzene: Very Soluble; DMSO: 10-50 mg/ml at 20 °C; Ether: Soluble; Hot Ethanol: Soluble; Organic solvents: Soluble; Toluene: Soluble.

Evaporation Rate: Half life 89 hours

Ionization Potential (eV): 7.44 +/-0.2

Flash Point: Not available; probably combustible

RTECS Toxicity Data

Mutagenic: Human Unscheduled DNA Synthesis; Cell Type: fibroblast; Dose: 100 umol/L. Human Unscheduled DNA Synthesis; Cell Type: HeLa cell; Dose: 100 umol/L. Human DNA Inhibition; Cell Type: other cell types; Dose: 10 umol/L. Human DNA Adduct; Cell Type: lymphocyte; Dose: 30 umol/L. Human Mutations in Mammalian Somatic Cells; Cell Type: lymphocyte; Dose: 9 umol/L.

Tumorigenic: Mouse Route: Skin; Dose: 18 mg/kg; Toxic Effects: Tumorigenic - Neoplastic by RTECS criteria; Skin and appendages - Tumors. Mouse Route: Skin; Dose: 18 mg/kg; Toxic Effects: Tumorigenic - Equivocal tumorigenic agent by RTECS criteria; Skin and appendages - Tumors. Mouse Route: Skin; Dose: 360 mg/kg/56W-I; Toxic Effects: Tumorigenic - Equivocal tumorigenic agent by RTECS criteria; Skin and appendages - Tumors. Mouse Route: Skin; Dose: 240 mg/kg/1W-I; Toxic Effects: Tumorigenic - Neoplastic by RTECS criteria; Skin and appendages - Tumors.

Hazard Overviews

Poison

Health: May cause irritation. Poison. Other Acute Effects: may be fatal if inhaled, swallowed, or absorbed through skin. Chronic Effects: may cause heritable genetic damage; may alter genetic material. Carcinogen.

Fire: Will burn. Extinguishing agents: water spray; carbon dioxide, dry chemical powder or appropriate foam. Precautions: combustible liquid.

Reactivity: Incompatible with: strong oxidizing agents. Hazardous decomposition products: toxic fumes of: carbon monoxide, carbon dioxide.

Carcinogenicity: IARC - Group 2A, Probably carcinogenic to humans; NIOSH - Not listed; NTP - Class 2B, Reasonably anticipated to be a carcinogen, sufficient evidence of carcinogenicity from studies in experimental animals; ACGIH - Class A2, Suspected human carcinogen; OSHA - Not listed; EPA - Class B2, Probable human carcinogen based on animal studies; MAK - Class A2, Unmistakably carcinogenic in animal experimentation only

Environmental

Ecotoxicity: Algae: Anabaena flos-aquae 2w EC_{50} growth +0.014 mg/l NOEC growth +0.003 mg/l

Environmental Fate: When released into water it will rapidly become adsorbed to sediment or particulate matter in the water column, and bioconcentrate into aquatic organisms. In the unadsorbed state, it will degrade by photolysis in a matter of hours to days. Its slow desorption from sediment and particulate matter will maintain a low concentration in the water. Because it is strongly adsorbed to soil it will remain in the upper few centimeters of soil and not leach into groundwater. It will very slowly biodegrade when colonies of microorganisms are acclimated but this is too slow a process (half-life ca 1 year to be significant). In the atmosphere it will be transported long distances and will probably be subject to photolysis and photooxidation although there is little documentation about the rate of these processes in the literature.

Environmental Physical Data

Octanol/Water Partition Coefficient: log K_{ow} = 5.61

Sorption Partition Coefficient: K_{oc} = sediments 55 to 1.87 $x10^6$

BCF: daphnia 4.0

Regulations

RCRA 40CFR: Listed Hazardous Waste No. U018 Toxic Waste

CERCLA: 40CFR 302.4: Listed per RCRA Section 3001 per CWA Section 307(a) RQ: 10 lb (4.535 kg)

SARA 40CFR 372.65: Listed

SARA EHS 40CFR 355: Not listed

TSCA: Listed

Analytical Methods

Air: EPA TO-13; California 429

Soil: CLP LC_SV, MC_SVOA, OHC; EPA 16, 1625, PAH-005, PAH-007, PAH-011, PAH-012, S-004-1; SW846 3630B, 3640A, 3650A, 3650B, 8100, 8250A, 8270B, 8270C, 8275A, 8310, 8410; DOE OS050

Water / Groundwater: EPA PAH-002, PAH-006, S-002-1, 1625, 610, 625, 625-S, 6; APHA 6040-B, 6410-B, 6440-B, 6440-C; ASTM D4657; USGS O3113, O3118

Drinking Water: EPA 525.1, 525.2, 550, 550.1

Indoor / Expired Air: NIOSH 5506, 5515; EPA IP-7-A, IP-7-B

Plasma: EPA 29

Other: EPA PAH-009

BEN2120 CAS #: 82-05-3

BENZANTHRONE

RTECS: CX5075000

EINECS Number: 201-393-3

Molecular Formula: $C_{17}H_{10}O$

Formula Weight: 230.25

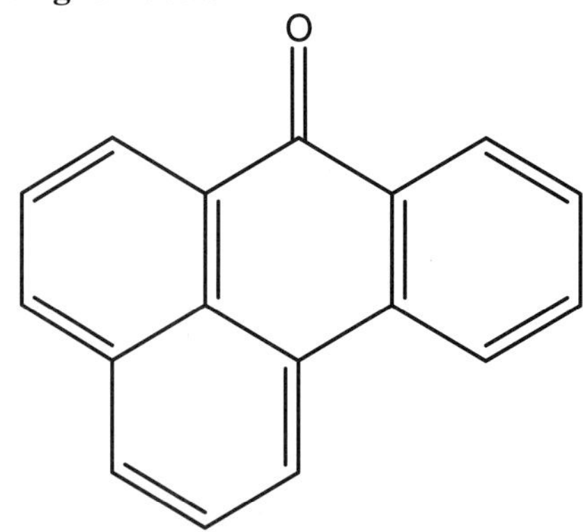

Chemical Structure

Synonyms: BENZANTHRENONE; 1,9-BENZANTHRONE; MS-BENZANTHRONE; BENZANTHRONE DYE; 7H-BENZ(DE)ANTHRACENE-1-ONE; 7H-BENZ(DE)ANTHRACENE-7-ONE; 7H-BENZ(DE)ANTHRACEN-7-ONE; BENZOANTHRONE; 7H-BENZO(DE)ANTHRACEN-7-ONE; DYE,BENZANTHRONE; MESOBENZANTHRONE; NAPHTHANTHRONE; 7-OXOBENZ(DE)ANTHRACENE

Description: pale yellow needles

Use: in dyes

Physical Properties

Freezing Point: 170 °C (338 °F)

Saturated Vapor Density: 1.21095735 kg/m^3

Vapor Pressure: 1 mm Hg at 225.0 °C

Water Solubility: 0.52 g/100g at 20 °C

Other Solubilities: Alcohol: Soluble; Benzene: 1.61 g in 100 g; Chlorobenzene: 2.05 g in 100 g at 20 °C; Glacial Acetic Acid: 0.52 g in 100 g at 20 °C; Other organic solvents: Soluble.

Flash Point: Not available; probably combustible

RTECS Toxicity Data

Chronic (Multiple Dose) Oral: Rat Dose: 15 gm/kg/30D-I; Toxic Effects: Liver - Changes in liver weight; Kidney, Ureter, and Bladder - Changes in kidney weight; Blood - Changes in erythrocite (RBC) cell count.

Chronic (Multiple Dose) Dermal: Mouse Route: Skin; Dose: 117 gm/kg/34W-I; Toxic Effects: Biochemical - Hepatic microsomal mixed oxidase(dealkylation, hyroxylation,etc); Biochemical - Other transferases; Biochemical - Other enzymes.

Irritation Eye: Rabbit Standard Draize Test Dose: 100 mg/24H; Reaction: moderate.

Irritation Skin: Rabbit Standard Draize Test Dose: 500 mg/24H; Reaction: mild.

Mutagenic: Bacteria - S Typhimurium Mutations in Microorganisms; Dose: 200 ug/plate (-S9).

Hazard Overviews

Health: Irritating to eyes/skin/respiratory tract. Other Acute Effects: may be harmful by inhalation, ingestion, or skin absorption.

Fire: Will burn. Extinguishing agents: water spray; carbon dioxide, dry chemical powder or appropriate foam. Precautions: combustible liquid.
Reactivity: Incompatible with: strong oxidizing agents. Hazardous decomposition products: toxic fumes of: carbon monoxide, carbon dioxide.
Carcinogenicity: IARC - Not listed; NIOSH - Not listed; NTP - Not listed; ACGIH - Not listed; OSHA - Not listed; EPA - Not listed; MAK - Not listed
Primary Target Organs:

Eyes

Skin

Respiratory System

Environmental

Environmental Fate: It is not expected to volatilize rapidly from either water or soils; the expected volatilization half-lives from a model river and lake are 840 and 6116 days, respectively. With a K_{oc} of 9.8 $x10^3$, it is not expected to migrate in soils, but is expected to sorb to sediments. Bioaccumulation may also be important because of the BCF value of 181; it has been detected in fish. It will exist in both the vapor and particulate phases in the atmosphere; although the majority will be particle-bound. When present in the atmosphere, it will be subject to oxidation by hydroxyl radicals with an estimated half-life of 21 hours. Particulate-phase may be physically removed from the air by wet and dry deposition.

Environmental Physical Data

Henry's Law Constant: estimated at 6.61 $x10^{-8}$
Octanol/Water Partition Coefficient: log K_{ow} = 4.81
Sorption Partition Coefficient: K_{oc} = estimated at 1.22 $x10^4$
BCF: carp 61 to 181

Regulations

RCRA 40CFR: Not listed
CERCLA: 40CFR 302.4: Not listed
SARA 40CFR 372.65: Not listed
SARA EHS 40CFR 355: Not listed
TSCA: Listed

Analytical Methods

Soil: EPA 1625
Water / Groundwater: EPA 1625

BEN2200 CAS #: 71-43-2

BENZENE

RTECS: CY1400000
DOT: UN1114; IMO3.2
EINECS Number: 200-753-7
Molecular Formula: C_6H_6
Structured MF: C_6H_6
Formula Weight: 78.11

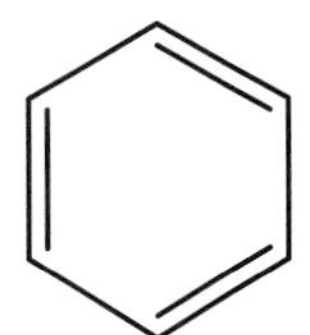

Chemical Structure

Synonyms: (6)ANNULENE; BENZEEN; BENZEN; BENZIN; BENZINE; BENZOL; BENZOL 90; BENZOLE; BENZOLENE; BENZOLO; BICARBURET OF HYDROGEN; CARBON OIL; COAL NAPHTHA; CYCLOHEXATRIENE; EPA PESTICIDE CHEMICAL CODE 008801; FENZEN; MINERAL NAPHTHA; MOTOR BENZOL; NITRATION BENZENE; PHENE; PHENYL HYDRIDE; POLYSTREAM; PYROBENZOL; PYROBENZOLE
Description: colorless liquid; sweet, aromatic odor
Use: as a solvent; in the manufacture of medicines, dyes, artificial leather, linoleum, oil cloth, pesticides, plastics and resins, PCB, aviation fuel, detergents, flavors and perfumes, paints and coatings, airplane dope, varnishes, lacquers, explosives and other organics; in photogravure printing and as a component of high-octane gasoline; to manufacture ethylbenzene, isopropylbenzene, cyclohexane, aniline, maleic anhydride and alkylbenzenes; in veterinary medicine to destroy screwworm larvae in wounds

Physical Properties

Boiling Point: 80.1 °C (176 °F)
Freezing Point: 5.5 °C (41.9 °F)
Specific Gravity: 0.8787 at 15 °C/4 °C
Vapor Density: 2.7 Air=1
Density: 0.905 g/mL at 21 °C
Bulk Density: 7.32 lbs/gal
Vapor Pressure: 100 mm Hg at 26.1 °C
Water Solubility: 0.18 g/100 g of Water at 25 °C
Other Solubilities: miscible with Alcohol, Chloroform, Ether, Carbon Disulfide, Acetone, oils, Carbon Tetrachloride, & Glacial Acetic Acid.
Surface Tension: 28.9 dynes/cm
Odor Threshold: 4.68 ppm
Refraction Index: 1.50108 at 20 °C/D
Evaporation Rate: 2.8 Ether=1
Critical Temperature: 288.9 °C
Critical Pressure: 48.6 atm
Ionization Potential (eV): 9.24
Flash Point: -11 °C Closed Cup
Autoignition Temperature: 562 °C
LEL: 1.3% v/v
UEL: 7.1% v/v

RTECS Toxicity Data

Acute Oral: Man LD_{Lo} Dose: 50 mg/kg. Rat LD_{50} Dose: 930 mg/kg; Toxic Effects: Behavioral - Tremor; Behavioral - Convulsions or effect on seizure threshold.
Acute Inhalation: Rat LC_{50} Dose: 10000 ppm/7hr. Human LC_{Lo} Dose: 2 pph/5M. Human LC_{Lo} Dose: 65 mg/m^3/5Y; Toxic Effects: Blood - Other changes. Human TC_{Lo} Dose: 100 ppm; Toxic Effects: Behavioral - Somnolence (general

depressed activity); Gastrointestinal - Nausea or vomiting; Skin and appendages - Dermatitis, other. Man TC_{Lo} Dose: 150 ppm/1Y-I; Toxic Effects: Blood - Other changes; Nutritional and gross metabolic - Body temperature increase.

Acute Dermal: Rabbit LD_{50} Route: Skin; Dose: >9400 mg/kg. Mouse LD_{50} Route: Skin; Dose: 48 mg/kg.

Chronic (Multiple Dose) Oral: Rat Dose: 6600 mg/kg/27W-I; Toxic Effects: Blood - Leukopenia; Blood - Changes in erythrocite (RBC) cell count. Rat Dose: 17 gm/kg/17W-I; Toxic Effects: Blood - Changes in spleen; Blood - Changes in leukocyte (WBC) cell count.

Chronic (Multiple Dose) Inhalation: Rat Dose: 23 mg/m^3/4H/8D-I; Toxic Effects: Liver - Changes in Liver weight; Endocrine - Changes in pituitary weight. Rat Dose: 300 ppm/6H/13W-I; Toxic Effects: Blood - Leukopenia. Rat Dose: 300 ppm/6H/99W-I; Toxic Effects: Blood - Leukopenia; Blood - Changes in bone marrow not included above; DEATH. Rat Dose: 1000 ppm/7H/28W-I; Toxic Effects: Blood - Changes in other cell count. Rat Dose: 500 ppm/6H/3W-I; Toxic Effects: Blood - Pigmented or nucleated red Blood cells; Blood - Changes in bone marrow not included above; Blood - Changes in leukocyte (WBC) cell count.

Chronic (Multiple Dose) Dermal: Rat Route: Subcutaneous; Dose: 18 mg/kg/21D-I; Toxic Effects: Blood - Changes in erythrocite (RBC) cell count; Blood - Changes in leukocyte (WBC) cell count; Nutritional and gross metabolic - Weight loss or decreased weight gain. Rat Route: Subcutaneous; Dose: 2197 mg/kg/5D-I; Toxic Effects: Blood - Pigmented or nucleated red Blood cells; Blood - Changes in leukocyte (WBC) cell count; Nutritional and gross metabolic - Weight loss or decreased weight gain. Rat Route: Subcutaneous; Dose: 13536 mg/kg/12W-I; Toxic Effects: Liver - Other changes; Kidney, Ureter, and Bladder - Other changes in urine composition; Biochemical - Other hydrolases.

Irritation Eye: Rabbit Standard Draize Test Dose: 2 mg/24H; Reaction: severe. Rabbit Standard Draize Test Dose: 88 mg; Reaction: moderate.

Irritation Skin: Rabbit Standard Draize Test Dose: 20 mg/24H; Reaction: moderate. Rabbit Open Draize Test Dose: 15 mg/24H open; Reaction: mild.

Reproductive/Teratogenic: Rat Route: Inhalation; Dose: 670 mg/m^3/24 Duration: female 15D prior to mating Effects on Fertility - Female fertility index. Rat Route: Inhalation; Dose: 56600 ug/m^3/ Duration: female 1-22D of pregnancy; Effects on Newborn - Biochemical and metabolic. Rat Route: Inhalation; Dose: 50 ppm/24H; Duration: female 7-14D of pregnancy; Effects on Embryo or Fetus - Extra embryonic structures; Fetotoxicity. Rat Route: Inhalation; Dose: 150 ppm/24H; Duration: female 7-14D of pregnancy; Effects on Fertility - Post-implantation mortality; Specific Developmental Abnormalities - Musculoskeletal system.

Mutagenic: Human DNA Inhibition; Cell Type: leukocyte; Dose: 2200 umol/L. Human DNA Inhibition; Cell Type: HeLa cell; Dose: 2200 umol/L. Human Mutations in Mammalian Somatic Cells; Cell Type: lymphocyte; Dose: 1 gm/L. Human Cytogenetic Analysis; Route: Inhalation; Dose: 125 ppm/1Y. Human Cytogenetic Analysis; Cell Type: leukocyte; Dose: 1 mmol/L/72H. Human Cytogenetic Analysis; Cell Type: lymphocyte; Dose: 1 mg/L. Human Cytogenetic Analysis; Route: Unreported; Dose: 10 ppm/4W. Human Sister Chromatid Exchenge; Cell Type: lymphocyte; Dose: 200 umol/L. Human Other Mutation Test Systems; Cell Type: lymphocyte; Dose: 5 umol/L.

Tumorigenic: Human Route: Inhalation; Dose: 10 ppm/8H/10Y-I; Toxic Effects: Tumorigenic - Carcinogenic by RTECS criteria; Blood - Leukemia. Human Route: Inhalation; Dose: 150 ppm/15M/8Y-I; Toxic Effects: Tumorigenic - Carcinogenic by RTECS criteria; Blood - Leukemia. Human Route: Inhalation; Dose: 8 ppb/4W-I; Toxic Effects: Tumorigenic - Carcinogenic by RTECS criteria; Blood - Leukemia. Human Route: Inhalation; Dose: 10 mg/m^3/11Y-I; Toxic Effects: Tumorigenic - Carcinogenic by RTECS criteria; Blood - Leukemia. Man Route: Inhalation; Dose: 200 mg/m^3/78W-I; Toxic Effects: Tumorigenic - Carcinogenic by RTECS criteria; Blood - Leukemia; Blood - Thrombocytopenia. Man Route: Inhalation; Dose: 600 mg/m^3/4Y-I; Toxic Effects: Tumorigenic - Carcinogenic by RTECS criteria; Blood - Leukemia. Man Route: Inhalation; Dose: 150 ppm/11Y-I; Toxic Effects: Tumorigenic - Carcinogenic by RTECS criteria; Blood - Lymphomax including Hodgkin's disease. Rat Route: Oral; Dose: 52 gm/kg/52W-I; Toxic Effects: Tumorigenic - Carcinogenic by RTECS criteria; Endocrine - Tumors; Blood - Leukemia. Rat Route: Inhalation; Dose: 1200 ppm/6H/10W-I; Toxic Effects: Tumorigenic - Equivocal tumorigenic agent by RTECS criteria; Sense organs and special senses - Tumors. Rat Route: Oral; Dose: 52 gm/kg/1Y-I; Toxic Effects: Tumorigenic - Carcinogenic by RTECS criteria; Sense organs and special senses - Tumors; Blood - Leukemia. Rat Route: Oral; Dose: 10 gm/kg/52W-I; Toxic Effects: Tumorigenic - Carcinogenic by RTECS criteria; Endocrine - Tumors; Blood - Leukemia.

Hazard Overviews

Flammable

Fire Diamond

Health: Irritating to eyes/skin/respiratory tract. Toxic. Absorbed through the skin. Also Causes: headache, dizziness, drowsiness. Chronic Effects: dermatitis, leukemia, bone marrow damage, reproductive effects.

Fire: Highly flammable. Can form explosive mixtures in the air. Use water as fog, dry chemical, or carbon dioxide. Water may be ineffective as an extinguishing agent since it can scatter and spread the fire. Benzene will float and may re-ignite on water surface. Use water spray to cool fire-exposed containers, flush spills away from exposures, disperse benzene vapor, and protect personnel attempting to stop an unignited benzene leak.

Reactivity: Stable. Hazardous polymerization cannot occur. Avoid: heat and ignition sources. Incompatible with: diborane; permanganic acid; bromine pentafluoride; peroxodisulfuric acid; peroxomonosulfuric acid; dioxygen difluoride; dioxygenyl tetrafluoroborate; iodine heptafluoride; sodium peroxide and water; iodine pentafluoride; ozone; liquid oxygen; silver perchlorate; nitryl perchlorate; nitric acid; arsenic pentafluoride and potassium methoxide; bromine trifluoride; uranium hexafluoride; hydrogen and Raney nickel [above 410 °F (210 °C)]; oxidizing materials. Hazardous decomposition products: carbon monoxide.

Carcinogenicity: IARC - Group 1, Carcinogenic to humans; NIOSH - Listed as carcinogen; NTP - Class 1, Known to be a carcinogen; ACGIH - Class A2, Suspected human carcinogen; OSHA - Listed as a carcinogen; EPA - Class A, Human carcinogen; MAK - Class A1, Capable of inducing malignant tumors as shown by experience with humans

Primary Target Organs:

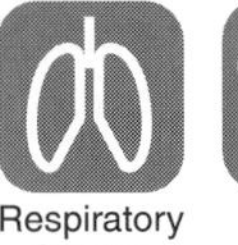

Eyes | Skin | Respiratory System | Nervous System | Blood | Bone

Exposure Limits

OSHA PEL: TWA: 1 ppm; 3 mg/m^3; STEL: 5 ppm; 15 mg/m^3; from Table Z-2.

ACGIH TLV: TWA: 10 ppm; 32 mg/m^3.

NIOSH REL: TWA: 0.1 ppm. STEL: 1 ppm.

NIOSH IDLH: 500 ppm.

Respirator Recommendation

Exposure Range: >1 to 10 ppm Air Purifying, Negative Pressure, Half Mask

Exposure Range: >10 to 100 ppm Air Purifying, Negative Pressure, Full Face

Exposure Range: >100 to 1000 ppm Supplied Air, Constant Flow/Pressure Demand, Full Face

Exposure Range: >1000 to unlimited ppm Self-contained Breathing Apparatus, Pressure Demand, Full Face

Cartridge Color: black

Note: must change cartridge at beginning of each shift

Environmental

Ecotoxicity: LC_{50} Clawed toad (3-4 wk after hatching) 190 mg/l/48 hr /Conditions of bioassay not specified LC_{50} Morone saxatilis (bass) 5.8 to 10.9 ppm/96 hr /Conditions of bioassay not specified LC_{50} Poecilia reticulata (guppy) 63 ppm/14 days /Conditions of bioassay not specified LC_{50} Salmo trutta (brown trout yearlings) 12 mg/l/1 hr (static bioassay) LD_{50} Lepomis macrochirus (bluegill sunfish) 20 mg/l/24 to 48 hr /Conditions of bioassay not specified LC_{100} Tetrahymena pyriformis (ciliate) 12.8 mmole/l/24 hr /Conditions of bioassay not specified LC_{50} Cancer magister (crab larvae) stage 1, 108 ppm/96 hr /Conditions of bioassay not specified LC_{50} Crangon franciscorum (shrimp) 20 ppm/96 hr /Conditions of bioassay not specified

Environmental Fate: If released to soil, it will be subject to rapid volatilization near the surface and that which does not evaporate will be highly to very highly mobile in the soil and may leach to groundwater. It may be subject to biodegradation based on reported biodegradation of 24% and 47% of the initial 20 ppm in a base-rich para-brownish soil in 1 and 10 weeks, respectively. It may be subject to biodegradation in shallow, aerobic groundwaters, but probably not under anaerobic conditions. If released to water, it will be subject to rapid volatilization; the half-life for evaporation in a wind-wave tank with a moderate wind speed of 7.09 m/sec was 5.23 hours; the estimated half-life for volatilization from a model river one meter deep flowing 1 m/sec with a wind velocity of 3 m/sec is estimated to be 2.7 hours at 20 °C. It will not be expected to significantly adsorb to sediment, bioconcentrate in aquatic organisms or hydrolyze. It may be subject to biodegradation based on a reported biodegradation half-life of 16 days in an aerobic river die-away test. In a marine ecosystem biodegradation occurred in 2 days after an acclimation period of 2 days and 2 weeks in the summer and spring, respectively, whereas no degradation occurred in winter. According to one experiment, it has a half-life of 17 days due to photodegradation which could contribute to removal in situations of cold water, poor nutrients, or other conditions less conductive to microbial degradation. If released to the atmosphere, it will exist predominantly in the vapor phase. Gas-phase will not be subject to direct photolysis but it will react with photochemically produced hydroxyl radicals with a half-life of 13.4 days calculated using an experimental rate constant for the reaction. The reaction time in polluted atmospheres which contain nitrogen oxides or sulfur dioxide is accelerated with the half-life being reported as 4-6 hours. Products of photooxidation include phenol, nitrophenols, nitrobenzene, formic acid, and peroxyacetyl nitrate. It is fairly soluble in water and is removed from the atmosphere in rain.

Cleanup/Disposal: Guide No. 130: Eliminate all ignition sources (no smoking, flares, sparks or flames in immediate area). All equipment used when handling the product must be grounded. Do not touch or walk through spilled material. Stop leak if you can do it without risk. Prevent entry into waterways, sewers, basements or confined areas. A vapor suppressing foam may be used to reduce vapors. Absorb or cover with dry earth, sand or other non-combustible material and transfer to containers. Use clean non-sparking tools to collect absorbed material. Large Spills: Dike far ahead of liquid spill for later disposal. Water spray may reduce vapor; but may not prevent ignition in closed spaces.

Environmental Physical Data

Henry's Law Constant: 5.3 x10^{-3}

Octanol/Water Partition Coefficient: log K_{ow} = 2.13

Sorption Partition Coefficient: K_{oc} = woodburn silt loam 31 to 143

BCF: eels 3.5

BOD: 1.2 lb/lb, 10 days

Regulations

RCRA 40CFR: Listed Hazardous Waste No. U019 Toxic Waste Ignitable Waste

CERCLA: 40CFR 302.4: Listed per CWA Section 311(b)(4) per RCRA Section 3001 per CWA Section 307(a) per CAA Section 112 RQ: 10 lb (4.535 kg)

SARA 40CFR 372.65: Listed

SARA EHS 40CFR 355: Not listed

TSCA: Listed

Analytical Methods

Air: EPA 0031, 0040, OA-002-1, VA-001-1, VA-003-1, VA-005-1, VA-006-1, VA-007-1, VA-008-1, VG-006-1, VG-007-1, VG-011-1, TO-1, TO-14, TO-2, TO-3; ASTM D3686, D3687, D4490

Soil: CLP LC_VOA, MC_VOA, OHC; EPA 7, 1624, VG-008-1, VG-010-1, VS-001-1, VS-002-1, VS-005-1, VW-010-1; SW846 1311, 5021, 5032, 5041, 5041A, 8020A, 8021A, 8240B, 8260A, 8260B; DOE OP040R, OS040, OS060

Water / Groundwater: EPA 602, 624, 624-S, VW-001-1, VW-002-1, VW-003-1, VW-004-1, VW-007-1, VW-008-1, VW-014-1; APHA 6210-B, 6210-D, 6220-B, 6230-D; ASTM D3695, D3871, D4763; USGS O3115

Drinking Water: EPA 502.2, 503.1, 524.1, 524.2; APHA 6210-C, 6220-C

Food: EPA 5; AOAC 975.06

Indoor / Expired Air: NIOSH 1500, 1501, 3700; EPA IP-1A, IP-1A-B, IP-1A-C, IP-1B

Plasma: EPA 29

Other: EPA VS-006-1, VW-011-1

BEN2280 CAS #: 608-73-1

BENZENE HEXACHLORIDE

RTECS: GV3150000

DOT: UN2761; UN2762; UN2995; UN2996; IMO3.0; IMO6.1

EINECS Number: 210-168-9

Molecular Formula: $C_6H_6Cl_6$

Formula Weight: 290.80

Synonyms: 666; AGROCIDE; AMBIOCIDE; BENZAHEX; BENZANEX; BENZENEHEXACHLORIDE,MIXED ISOMERS; BENZEX; BHC; COMPOUND-666; CYCLOHEXANE,1,2,3,4,5,6-HEXACHLORO-; CYCLOHEXANE,1,2,3,4,5,6-HEXACHLORO-,(MIXED ISOMERS); DBH; DOL; DOLMIX; ENT 8,601; FBHC; GAMASPRA; GAMMACIDE; GAMMACOID; GAMMEXANE; GAMTOX; GEXANE; GYBEN; HCCH; HCH; HEXABLANC; HEXACHLOR; HEXACHLORAN; HEXACHLORANE; 1,2,3,4,5,6-HEXACHLOROCYCLOHEXANE; HEXACHLOROCYCLOHEXANE; 1,2,3,4,5,6-HEXACHLOROCYCLOHEXANE (MIXTURE OF ISOMERS); HEXAFOR; HEXAMUL; HEXAPOUDRE; HEXYCLAN; HEXYLAN; HILBEECH; ISATOX; KOTOL; LATKA 666; LEXONE; LINTOX; SOPROCIDE; SUBMAR; TBH; TECHNICAL BHC; TECHNICAL HCH; TRI-6; TRIVES-T

Description: white or yellowish powder or flakes or brown-to-white amorphous powder; musty odor reminiscent of the new mown hay

Use: insecticide; control of leafhoppers, stem borers, etc, in lowland rice; seed treatment for reduction of wireworm damage in winter & spring sown cereals; control of pests of cereals, sugar beets & oilseed rape (former use)

Physical Properties

Freezing Point: 65 °C (149 °F)

Specific Gravity: 1.87

Vapor Pressure: ~ 0.5 mm Hg 60 °C

Water Solubility: Insoluble in Water

Other Solubilities: Soluble in Ethyl Alcohol, Chloroform & Ethyl Ether

Flash Point: Not flammable

RTECS Toxicity Data

Acute Oral: Rat LD_{50} Dose: 100 mg/kg. Mouse LD_{50} Dose: 59 mg/kg.

Acute Inhalation: Rat LC_{50} Dose: 690 mg/m^3/4hr. Man TC_{Lo} Dose: 400 ug/kg/3D; Toxic Effects: Behavioral - Headache; Gastrointestinal - Nausea or vomiting; Nutritional and gross metabolic - Body temperature increase.

Acute Dermal: Rabbit LD_{50} Route: Subcutaneous Dose: 75 mg/kg. Rat LD_{50} Route: Skin; Dose: 900 mg/kg.

Chronic (Multiple Dose) Oral: Rat Dose: 2500 mg/kg/90D-C; Toxic Effects: Liver - Other changes; Endocrine - Other changes; Biochemical - Transaminases. Rat Dose: 750 mg/kg/15D-I; Toxic Effects: Liver - Other changes; Kidney, Ureter, and Bladder - Other changes; Biochemical - Other proteins.

Chronic (Multiple Dose) Dermal: Rat Route: Skin; Dose: 3 gm/kg/30D-I; Toxic Effects: Liver - Fatty Liver degeneration; Kidney, Ureter, and Bladder - Other changes; Biochemical - Multiple enzyme effects.

Reproductive/Teratogenic: Rat Route: Oral; Dose: 8100 mg/kg; Duration: male 90D prior to mating; Paternal Effects - Testes, epididymis, sperm duct. Mouse Route: Oral; Dose: 9120 mg/kg; Duration: male 22W prior to mating; Paternal Effects - Spermatogenesis. Mouse Route: Oral; Dose: 5460 mg/kg; Duration: male 91D prior to mating; Paternal Effects - Testes, epididymis, sperm duct.

Mutagenic: Rat Morphological Transformation; Route: Oral; Dose: 875 mg/kg/7W-I. Other Microorganisms Mutations in Microorganisms; Dose: 100 mg/L (-S9).

Tumorigenic: Mouse Route: Oral; Dose: 6720 mg/kg/80W-C; Toxic Effects: Tumorigenic - Carcinogenic by RTECS criteria; Liver - Tumors; Blood - Tumors. Mouse Route: Oral; Dose: 800 mg/kg/80W-I; Toxic Effects: Tumorigenic - Equivocal tumorigenic agent by RTECS criteria; Liver - Tumors; Blood - Tumors. Mouse Route: Oral; Dose: 12600 mg/kg/30W-C; Toxic Effects: Tumorigenic - Carcinogenic by RTECS criteria; Liver - Tumors. Mouse Route: Oral; Dose: 12960 mg/kg/26W-C; Toxic Effects: Tumorigenic - Carcinogenic by RTECS criteria; Liver - Tumors. Mouse Route: Oral; Dose: 21600 mg/kg/52W-C; Toxic Effects: Tumorigenic - Carcinogenic by RTECS criteria; Liver - Tumors. Mouse Route: Oral; Dose: 5400 mg/kg/13W-C; Toxic Effects: Tumorigenic - Neoplastic by RTECS criteria; Liver - Tumors. Mouse Route: Oral; Dose: 7200 mg/kg/17W-C; Toxic Effects: Tumorigenic - Equivocal tumorigenic agent by RTECS criteria; Liver - Tumors. Mouse Route: Oral; Dose: 9 gm/kg/21W-C; Toxic Effects: Tumorigenic - Equivocal tumorigenic agent by RTECS criteria; Liver - Tumors. Mouse Route: Oral; Dose: 10800 mg/kg/26W-C; Toxic Effects: Tumorigenic - Equivocal tumorigenic agent by RTECS criteria; Liver - Tumors.

Hazard Overviews

Fire: Noncombustible.
Carcinogenicity: IARC - Not listed; NIOSH - Not listed; NTP - Listed; ACGIH - Not listed; OSHA - Not listed; EPA - Class B2, Probable human carcinogen based on animal studies; MAK - Not listed

Environmental

Ecotoxicity: EC_{50} Daphnia pulex 680 ug/l/48 hr at 16 °C, 1st instar / static bioassay without aeration, pH 7.2-7.5, water hardness 40-50 mg/l as calcium carbonate and alkalinity of 30-35 mg/l LD_{50} Pheasants oral 118 mg/kg, 3-4 months old female (95% confidence limit 93.6-148 mg/kg) LC_{50} Pteronarcys less than 18 ug/l/96 hr at 15 °C, second year class / static bioassay without aeration, pH 7.2-7.5, water hardness 40-50 mg/l as calcium carbonate and alkalinity of 30-35 mg/l LC_{50} Cutthroat trout 9 ug/l/96 hr at 13 °C, wt 1.0 g (95% confidence limit 8-10 ug/l) / static bioassay without aeration, pH 7.2-7.5, water hardness 40-50 mg/l as calcium carbonate and alkalinity of 30-35 mg/l
Cleanup/Disposal: Guide No. 131: Fully encapsulating, vapor protective clothing should be worn for spills and leaks with no fire. Eliminate all ignition sources (no smoking, flares, sparks or flames in immediate area). All equipment used when handling the product must be grounded. Do not touch or walk through spilled material. Stop leak if you can do it without risk. Prevent entry into waterways, sewers, basements or confined areas. A vapor suppressing foam may be used to reduce vapors. Small Spills: Absorb with earth, sand or other non-combustible material and transfer to containers for later disposal. Use clean non-sparking tools to collect absorbed material. Large Spills: Dike far ahead of liquid spill for later disposal. Water spray may reduce vapor; but may not prevent ignition in closed spaces. Guide No. 151: Do not touch damaged containers or spilled material unless wearing appropriate protective clothing. Stop leak if you can do it without risk. Prevent entry into waterways, sewers, basements or confined areas. Cover with plastic sheet to prevent spreading. Absorb or cover with dry earth, sand or other non-combustible material and transfer to containers. Do not get water inside containers.

Regulations

RCRA 40CFR: Not listed
CERCLA: 40CFR 302.4: Listed as Compound per CWA Section 307(a)
SARA 40CFR 372.65: Not listed
SARA EHS 40CFR 355: Not listed
TSCA: Not listed

Analytical Methods

Soil: EPA 025; SW846 8120A
Water / Groundwater: ASTM D5241
Drinking Water: AOAC 990.06
Food: FDA 212.1, 212.2, 232.1, 232.4, 242.1; AOAC 947.01, 970.52
Plasma: EPA 027; FDA 211.1, 231.1, 251.1, 252

BEN2360 CAS #: 98-05-5

BENZENEARSONIC ACID

RTECS: CY3150000
EINECS Number: 202-631-9
Molecular Formula: $C_6H_7AsO_3$
Formula Weight: 202.03

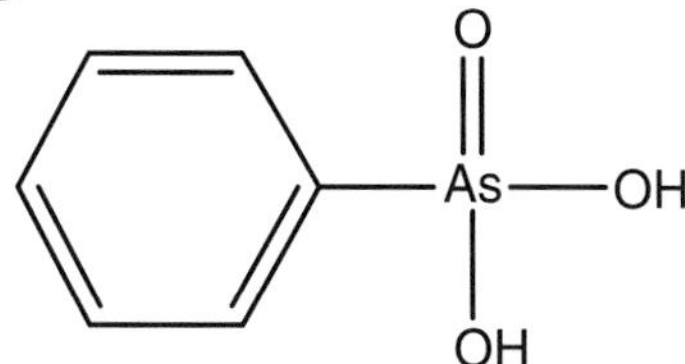

Chemical Structure

Synonyms: ARSONIC ACID,PHENYL-; KYSELINA BENZENARSONOVA; PHENYL ARSENIC ACID; PHENYLARSONIC ACID
Description: crystal powder
Use: reagent for tin; as precipitant in niobium analysis

Physical Properties

Freezing Point: Decomposes at 158 °C (316.4 °F) to 162 °C (323.6 °F)
Specific Gravity: 1.76 at 25 °C
Water Solubility: 1 parts in 40 parts Water
Other Solubilities: Insoluble in Chloroform.

RTECS Toxicity Data

Acute Oral: Rat LD_{Lo} Dose: 50 mg/kg. Mouse LD_{50} Dose: 270 ug/kg.

Hazard Overviews

Poison

Health: Irritating to eyes/skin/respiratory tract. Poison. Other Acute Effects: may be fatal if inhaled, swallowed, or absorbed through skin; exposure can cause: lung irritation, chest pain and edema, which may be fatal; repeated exposure can cause: stomach pains, vomiting, diarrhea. Chronic Effects: possible risk of irreversible effects; causes damage to the liver and the kidneys. Possible carcinogen.
Fire: Hazards: emits toxic fumes. Extinguishing agents: water spray; carbon dioxide, dry chemical powder or appropriate foam. Precautions: combustible liquid.
Carcinogenicity: IARC - Not listed; NIOSH - Not listed; NTP - Not listed; ACGIH - Not listed; OSHA - Not listed; EPA - Not listed; MAK - Not listed

Primary Target Organs:

Eyes Skin Respiratory System

Exposure Limits
OSHA PEL: TWA: 0.5 mg/m^3; as As.
ACGIH TLV: TWA: 0.01 mg/m^3; as As.
NIOSH REL: STEL: 0.002 mg/m^3; Ceiling (15 min) as As.

Environmental

Environmental Physical Data

Octanol/Water Partition Coefficient: log K_{ow} = 0.06

Regulations

RCRA 40CFR: Not listed
CERCLA: 40CFR 302.4: Listed as Compound per CWA Section 307(a) per CAA Section 112
SARA 40CFR 372.65: Not listed TPQ: 10/10000 lb
SARA EHS 40CFR 355: Listed TPQ: 10 lb
TSCA: Listed

BEN2440 CAS #: 108-45-2

1,3-BENZENEDIAMINE

RTECS: SS7700000
DOT: UN1673
EINECS Number: 203-584-7
Molecular Formula: $C_6H_8N_2$
Structured MF: $C_6H_4(NH_2)_2$
Formula Weight: 108.14

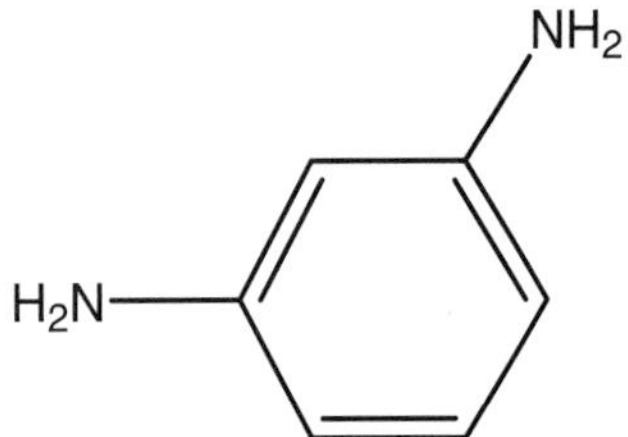

Chemical Structure

Synonyms: M-AMINOALINE; 3-AMINOANILINE; M-AMINOANILINE; META-AMINOANILINE; APCO 2330; M-BENZENEDIAMINE; META-BENZENEDIAMINE; BENZENE,1,3-DIAMINO-; C.I. 76025; C.I. DEVELOPER 11; DEVELOPER 11; DEVELOPER C; DEVELOPER H; DEVELOPER M; 1,3-DIAMINOBENZENE; M-DIAMINOBENZENE; META-DIAMINOBENZENE; DIRECT BROWN BR; DIRECT BROWN GG; M-FENYLENDIAMIN; METAPHENYLENEDIAMINE; MPD; 1,3-PHENYLENEDIAMINE; M-PHENYLENEDIAMINE; META-PHENYLENEDIAMINE; PHENYLENEDIAMINE,META,SOLID
Description: white crystals becoming red on exposure to air or colorless needles;; slight aromatic odor
Use: in dyestuff manufacture and photography, in the detection of nitrite and in textile developing agents, laboratory reagents, vulcanizing agents, ion-exchange resins, block polymers and corrosion inhibitors; in rubber curing agents, decoloring resins, formaldehyde condensates, resinous polyamides, textile fibers, urethanes, petroleum additives, rubber chemicals and reagents for gold and bromine; as a hair dye ingredient and as a curing agent for epoxy resins

Physical Properties

Boiling Point: 284 °C (543 °F) to 287 °C (549 °F)
Freezing Point: 62 °C (143.6 °F) to 63 °C (145.4 °F)
Specific Gravity: 1.139
Vapor Density: 3.7 Air=1
Density: 1.14 g/mL at 15 °C
Vapor Pressure: 1 mm Hg at 99.8 °C
Water Solubility: Soluble in Water
Other Solubilities: DMSO: >=100 mg/ml at 19 °C; Acetone: >=100 mg/ml at 19 °C; Toluene: Very slightly Soluble; Xylene: Very slightly Soluble; Ether: Slightly Soluble; Benzene: Very slightly Soluble; Alcohol: Soluble; Aqueous sodium hydroxide: Soluble.
Refraction Index: 1.6339 at 58 °C/D
Evaporation Rate: < 1 Butyl Acetate=1
Ionization Potential (eV): 7.14
Flash Point: 138 °C
Autoignition Temperature: 560 °C

RTECS Toxicity Data

Acute Oral: Rat LD_{50} Dose: 280 mg/kg. Mouse LD_{50} Dose: 67700 ug/kg.
Acute Dermal: Rabbit LD_{Lo} Route: Skin; Dose: 5 gm/kg. Rabbit LD_{Lo} Route: Subcutaneous Dose: 200 mg/kg; Toxic Effects: Behavioral - Somnolence (general depressed activity); Behavioral - Convulsions or effect on seizure threshold; Lungs, Thorax, or Respiration - Cyanosis.
Reproductive/Teratogenic: Rat Route: Intraperitoneal; Dose: 375 mg/kg; Duration: male 30D prior to mating; Effects on Embryo or Fetus - Fetal death.
Mutagenic: Rat Body Fluid Assay; Indicator Organism: Bacteria - S Typhimurium; Dose: 240 mg/kg. Mouse DNA Inhibition; Route: Oral; Dose: 200 mg/kg.
Tumorigenic: Rat Route: Subcutaneous; Dose: 1485 mg/kg/47W-I; Toxic Effects: Tumorigenic - Equivocal tumorigenic agent by RTECS criteria; Tumorigenic - Tumors at site of application.

Hazard Overviews

Health: Irritating to eyes/skin/respiratory tract. Toxic. Other Acute Effects: harmful if swallowed, inhaled, or absorbed through skin; may cause allergic respiratory and skin reactions; exposure can cause: nausea, dizziness and headache, dermatitis pulmonary edema; effects may be delayed; may cause discoloration of the skin. Chronic Effects: target organs: liver, kidneys, bladder. Causes mutagenic effects in animals.
Fire: Will burn. Hazards: emits toxic fumes; in powder form capable of creating a dust explosion. Extinguishing agents: water spray; carbon dioxide, dry chemical powder or appropriate foam. Precautions: combustible liquid.
Reactivity: Stable. Hazardous polymerization will not occur. Incompatible with: acids, acid chlorides, acid anhydrides,

chloroformates, strong oxidizing agents, may decompose on exposure to light. Hazardous decomposition products: toxic fumes of: carbon monoxide, carbon dioxide, nitrogen oxides.

Carcinogenicity: IARC - Group 3, Not classifiable as to carcinogenicity to humans; NIOSH - Not listed; NTP - Not listed; ACGIH - Class A4, Not classifiable as a human carcinogen; OSHA - Not listed; EPA - Not listed; MAK - Class B, Justifiably suspected of having carcinogenic potential

Primary Target Organs:

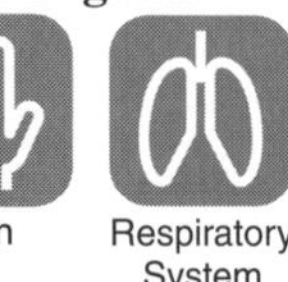
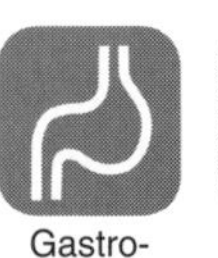

Eyes | Skin | Respiratory System | Gastro-intestinal | Liver | Kidneys

Exposure Limits

ACGIH TLV: TWA: 0.1 mg/m^3.

Respirator Recommendation

Exposure Range: >0.1 to 5 mg/m^3 Supplied Air, Constant Flow/Pressure Demand, Half Mask

Exposure Range: >5 to 100 mg/m^3 Supplied Air, Constant Flow/Pressure Demand, Full Face

Exposure Range: >100 to unlimited mg/m^3 Self-contained Breathing Apparatus, Pressure Demand, Full Face

Note: odor threshold unknown

Environmental

Environmental Fate: If released on land it would probably be lost due to oxidation by air and cations in the soil. It is not adsorbed by soil but it may chemically react with humic material and therefore whether it would leach into groundwater is unknown. If released into water, it may be lost by photolysis in surface layers and oxidation by dissolved oxygen, cations, humic materials and radicals. Biodegradation may also occur. It would not adsorb appreciably to sediment or bioconcentrate in fish but it might chemically react with humic material in the sediment and be found there. Little experimental data on these processes could be found. If released into the atmosphere it will photolyze and react with photochemically produced hydroxyl radicals (estimated half-life 14 hr).

Cleanup/Disposal: Guide No. 153: Eliminate all ignition sources (no smoking, flares, sparks or flames in immediate area). Do not touch damaged containers or spilled material unless wearing appropriate protective clothing. Stop leak if you can do it without risk. Prevent entry into waterways, sewers, basements or confined areas. Absorb or cover with dry earth, sand or other non-combustible material and transfer to containers. Do not get water inside containers.

Environmental Physical Data

Octanol/Water Partition Coefficient: log K_{ow} = calculated at -1.23

BCF: none likely

Regulations

RCRA 40CFR: Not listed

CERCLA: 40CFR 302.4: Not listed

SARA 40CFR 372.65: Listed

SARA EHS 40CFR 355: Not listed

TSCA: Listed

BEN2520 CAS #: 106-50-3

1,4-BENZENEDIAMINE

RTECS: SS8050000

DOT: UN1673; IMO6.1

EINECS Number: 203-404-7

Molecular Formula: $C_6H_8N_2$

Structured MF: $C_6H_4(NH_2)_2$

Formula Weight: 108.14

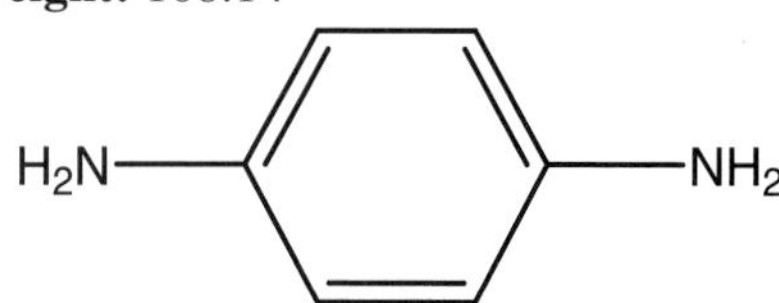

Chemical Structure

Synonyms: 4-AMINOANILINE; P-AMINOANILINE; BASF URSOL D; P-BENZENEDIAMINE; BENZOFUR D; C.I. 76060; C.I. DEVELOPER 13; C.I. OXIDATION BASE 10; DEVELOPER 13; DEVELOPER PF; 1,4-DIAMINOBENZENE; P-DIAMINOBENZENE; DURAFUR BLACK R; P-FENYLENDIAMIN; FENYLENODWUAMINA; FOURAMINE D; FOURRINE 1; FOURRINE D; FOURRINE I; FUR BLACK 41866; FUR BLACK 41867; FUR BLACK R; FUR BROWN 41866; FUR YELLOW; FURRO D; FUTRAMINE D; MAKO H; NAKO H; ORSIN; OXIDATION BASE 10; PARA; PARAPHENYLEN-DIAMINE; PARAPHENYLENEDIAMINE; PELAGOL D; PELAGOL DR; PELAGOL GREY D; PELTOL D; 1,4-PHENYLENE DIAMINE; 1,4-PHENYLENEDIAMINE; P-PHENYLENEDIAMINE; PHENYLENEDIAMINE; PHENYLENEDIAMINE,PARA,SOLID; 6PPD; PPD; RENAL PF; RODOL D; SANTOFLEX IC; SANTOFLEX LC; PARA,SOLID; TERTRAL D; URSOL D; VULKANOX 4020; ZOBA BLACK D

Description: colorless, red upon exposure to air, brown upon exposure to air, black upon exposure to air crystalline solid; mild phenolic odor

Use: azo dye intermediate; photographic developing agent; photochemical measurements; intermediate in manufacture of antioxidants and accelerators for rubber; laboratory reagent; dye hair and fur

Physical Properties

Boiling Point: 267 °C (513 °F)

Freezing Point: 145 °C (293 °F) to 147 °C (296.6 °F)

Specific Gravity: > 1 (Water=1)

Density: 3.72

Vapor Pressure: < 1 mm Hg

Water Solubility: 1 parts in 100 parts Cold Water

Other Solubilities: Alcohol: Soluble; Chloroform: Soluble; Ether: Soluble.

Ionization Potential (eV): 6.89

Flash Point: 156 °C

RTECS Toxicity Data

Acute Oral: Man TD_{Lo} Dose: 71 mg/kg; Toxic Effects: Behavioral - Muscle weakness; Lungs, Thorax, or Respiration

- Acute pulmonary edema; Lungs, Thorax, or Respiration - Dyspnea. Rat LD_{50} Dose: 80 mg/kg.

Acute Inhalation: Rat LC_{50} Dose: 920 mg/m^3/4hr; Toxic Effects: Behavioral - General anesthetic; Gastrointestinal - Nausea or vomiting; Kidney, ureter, and Bladder - Hematuria.

Acute Dermal: Rabbit LD_{Lo} Route: Skin; Dose: 5 gm/kg. Rabbit LD_{Lo} Route: Subcutaneous Dose: 200 mg/kg; Toxic Effects: Cardiac - Pulse rate increased without fall in BP; Lungs, Thorax, or Respiration - Respiratory stimulation; Nutritional and gross metabolic - Body temperature decrease.

Chronic (Multiple Dose) Oral: Rat Dose: 1050 mg/kg/30W-I; Toxic Effects: Brain and coverings - Recordings from specific areas of CNS; Behavioral - Alteration of classical conditioning; Liver - Fatty liver degeneration. Rat Dose: 28 gm/kg/80W-C; Toxic Effects: Endocrine - Changes in spleen weight; Nutritional and gross metabolic - Weight loss or decreased weight gain. Rat Dose: 16800 mg/kg/12W-C; Toxic Effects: Nutritional and gross metabolic - Weight loss or decreased weight gain.

Irritation Skin: Human Standard Draize Test Dose: 250 mg/24H; Reaction: mild. Rabbit Standard Draize Test Dose: 12500 ug/24H; Reaction: mild. Rabbit Standard Draize Test Dose: 250 mg/24H; Reaction: moderate.

Mutagenic: Rat Morphological Transformation; Cell Type: embryo; Dose: 1850 ng/plate. Mouse DNA Inhibition; Route: Oral; Dose: 200 mg/kg.

Tumorigenic: Rat Route: Subcutaneous; Dose: 2625 mg/kg/30W-C; Toxic Effects: Tumorigenic - Equivocal tumorigenic agent by RTECS criteria; Tumorigenic - Tumors at site of application.

Hazard Overviews

Fire Diamond

Health: Irritating to skin/respiratory tract. Toxic. Chronic Effects: asthma.

Fire: Will burn. Powder may form explosive dust-air mixtures. Use dry chemical, foam, carbon dioxide, or water fog. Water or foam may cause frothing. Use water spray to cool fire-exposed tanks/containers.

Reactivity: Stable. Hazardous polymerization cannot occur. Avoid: exposure to heat; ignition sources. Incompatible with: strong oxidizing agents. Hazardous decomposition products: oxides of nitrogen; carbon dioxide; carbon monoxide.

Carcinogenicity: IARC - Group 3, Not classifiable as to carcinogenicity to humans; NIOSH - Not listed; NTP - Not listed; ACGIH - Class A4, Not classifiable as a human carcinogen; OSHA - Not listed; EPA - Not listed; MAK - Class B, Justifiably suspected of having carcinogenic potential

Primary Target Organs:

Skin

Respiratory System

Exposure Limits

OSHA PEL: TWA: 0.1 mg/m^3; skin.

ACGIH TLV: TWA: 0.1 ppm.

NIOSH REL: TWA: 0.1 mg/m^3.

NIOSH IDLH: 25 mg/m^3.

DFG MAK: TWA: 0.1 mg/m^3.

Respirator Recommendation

Exposure Range: >0.1 to 5 mg/m^3 Supplied Air, Constant Flow/Pressure Demand, Half Mask

Exposure Range: >5 to 100 mg/m^3 Supplied Air, Constant Flow/Pressure Demand, Full Face

Exposure Range: >100 to unlimited mg/m^3 Self-contained Breathing Apparatus, Pressure Demand, Full Face

Note: odor threshold unknown

Environmental

Ecotoxicity: Approximate fatal concentration Goldfish 5.74 mg/l/48 hr

Environmental Fate: If released on land it may leach to groundwater; however it may form covalent bonds to humic material which would limit movement through soil. It will not be expected to hydrolyze or evaporate appreciably in soils or on surfaces. If released to water it will not be expected to adsorb to sediments, bioconcentrate in aquatic organisms, hydrolyze, or evaporate appreciably. It will be expected to autooxidize and react with peroxyl radicals and may directly photolyze and biodegrade. If released to the atmosphere it will be subject to direct photolysis and autooxidation. It will react with photochemically produced hydroxyl radicals (estimated half-life of 14.14 hours) and should be subject to significant rainout.

Cleanup/Disposal: Guide No. 153: Eliminate all ignition sources (no smoking, flares, sparks or flames in immediate area). Do not touch damaged containers or spilled material unless wearing appropriate protective clothing. Stop leak if you can do it without risk. Prevent entry into waterways, sewers, basements or confined areas. Absorb or cover with dry earth, sand or other non-combustible material and transfer to containers. Do not get water inside containers.

Environmental Physical Data

Octanol/Water Partition Coefficient: log K_{OW} = -0.25

Sorption Partition Coefficient: log K_{OC} = estimated at 0.58

BCF: estimated at 0.38

Regulations

RCRA 40CFR: Not listed

CERCLA: 40CFR 302.4: Listed per CAA Section 112 RQ: 5000 lb (2268 kg)

SARA 40CFR 372.65: Listed

SARA EHS 40CFR 355: Not listed

TSCA: Listed

Analytical Methods
Soil: SW846 8270C
Plasma: EPA 29

BEN2600 CAS #: 615-28-1

1,2-BENZENEDIAMINE DIHYDROCHLORIDE

RTECS: ST0175000
EINECS Number: 210-418-7
Molecular Formula: $C_6H_{10}Cl_2N_2$
Formula Weight: 181.08

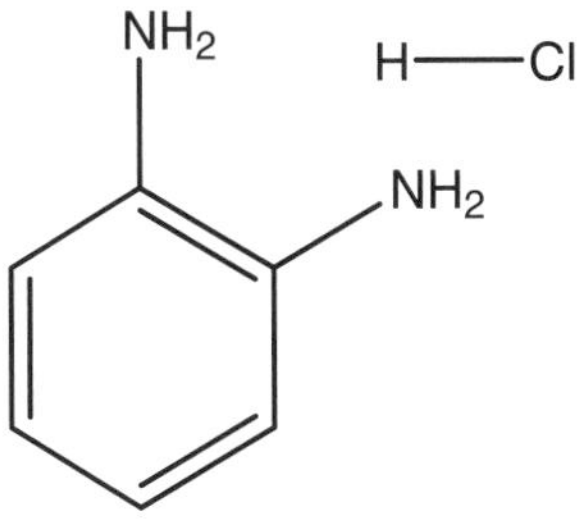

Chemical Structure

Synonyms: O-PHENYLENEDIAMINE,DIHYDROCHLORIDE

Physical Properties
Freezing Point: 250 °C (482 °F)

RTECS Toxicity Data
Acute Dermal: Rabbit LD; Route: Subcutaneous Dose: >500 mg/kg. Rat LD_{Lo} Route: Subcutaneous Dose: 600 mg/kg.
Mutagenic: Bacteria - S Typhimurium Mutations in Microorganisms; Dose: 100 ug/plate (-S9).
Tumorigenic: Rat Route: Oral; Dose: 130 gm/kg/78W-C; Toxic Effects: Tumorigenic - Equivocal tumorigenic agent by RTECS criteria; Liver - Tumors. Mouse Route: Oral; Dose: 260 gm/kg/78W-C; Toxic Effects: Tumorigenic - Carcinogenic by RTECS criteria; Liver - Tumors. Mouse Route: Oral; Dose: 518 gm/kg/78W-C; Toxic Effects: Tumorigenic - Carcinogenic by RTECS criteria; Liver - Tumors.

Hazard Overviews
Health: Irritating to eyes/skin/respiratory tract. Toxic. Other Acute Effects: harmful if swallowed, inhaled, or absorbed through skin; may cause allergic respiratory and skin reactions. Chronic Effects: target organs: bladder, liver. Carcinogen.
Fire: Extinguishing agents: water spray; carbon dioxide, dry chemical powder or appropriate foam. Precautions: combustible liquid.
Reactivity: Stable. Hazardous polymerization will not occur. Incompatible with: strong oxidizing agents. Hazardous decomposition products: toxic fumes of: carbon monoxide, carbon dioxide, nitrogen oxides, hydrogen chloride gas.
Carcinogenicity: IARC - Not listed; NIOSH - Not listed; NTP - Not listed; ACGIH - Not listed; OSHA - Not listed; EPA - Not listed; MAK - Not listed
Primary Target Organs:

Eyes

Skin

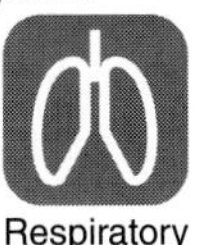
Respiratory System

Liver

Environmental

Regulations
RCRA 40CFR: Not listed
CERCLA: 40CFR 302.4: Listed as Compound per CWA Section 307(a)
SARA 40CFR 372.65: Listed
SARA EHS 40CFR 355: Not listed
TSCA: Listed

BEN2680 CAS #: 541-69-5

1,3-BENZENEDIAMINE DIHYDROCHLORIDE

RTECS: SS9800000
EINECS Number: 208-790-0
Molecular Formula: $C_6H_{10}Cl_2N_2$
Formula Weight: 181.08

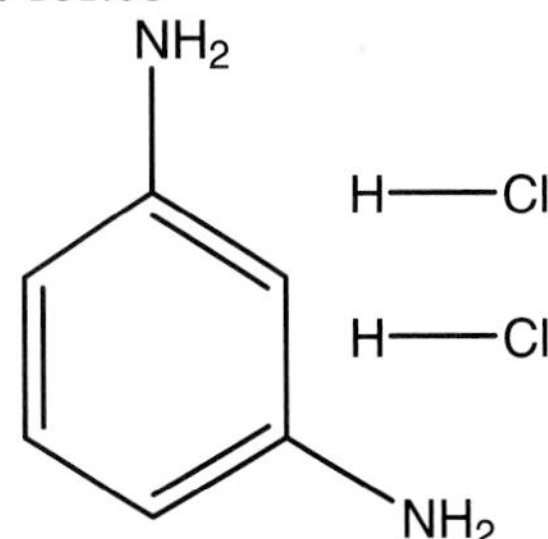

Chemical Structure

Synonyms: 3-AMINOANILINE DIHYDROCHLORIDE; M-AMINOANILINE DIHYDROCHLORIDE; META-AMINOANILINE DIHYDROCHLORIDE; M-BENZENEDIAMINE DIHYDROCHLORIDE; META-BENZENEDIAMINE DIHYDROCHLORIDE; 1,3-BENZENEDIAMINE HYDROCHLORIDE; 1,3-DIAMINOBENZENE DIHYDROCHLORIDE; M-DIAMINOBENZENE DIHYDROCHLORIDE; META-DIAMINOBENZENE DIHYDROCHLORIDE; 1,3-PHENYLENEDIAMINE DIHYDROCHLORIDE; M-PHENYLENEDIAMINE HYDROCHLORIDE; M-PHENYLENEDIAMINE,DIHYDROCHLORIDE
Description: white crystalline powder
Use: analytical reagent for nitrite; used in hair-dye

Physical Properties
Water Solubility: Soluble in Water
Other Solubilities: Soluble in Ethanol

RTECS Toxicity Data

Acute Dermal: Mouse LD_{50} Route: Subcutaneous Dose: 120 mg/kg.
Mutagenic: Bacteria - S Typhimurium Mutations in Microorganisms; Dose: 50 ug/plate (-S9).
Tumorigenic: Rat Route: Subcutaneous; Dose: 1800 mg/kg/21W-I; Toxic Effects: Tumorigenic - Equivocal tumorigenic agent by RTECS criteria; Tumorigenic - Tumors at site of application.

Hazard Overviews

Health: Irritating to eyes/skin/respiratory tract. Toxic. Other Acute Effects: harmful if swallowed, inhaled, or absorbed through skin; may cause allergic skin reaction. Chronic Effects: Laboratory experiments have shown mutagenic effects.
Fire: Hazards: emits toxic fumes. Extinguishing agents: water spray; carbon dioxide, dry chemical powder or appropriate foam. Precautions: combustible liquid.
Reactivity: Incompatible with: strong oxidizing agents. Hazardous decomposition products: toxic fumes of: carbon monoxide, carbon dioxide, nitrogen oxides, hydrogen chloride gas.
Carcinogenicity: IARC - Not listed; NIOSH - Not listed; NTP - Not listed; ACGIH - Not listed; OSHA - Not listed; EPA - Not listed; MAK - Not listed
Primary Target Organs:

Eyes

Skin

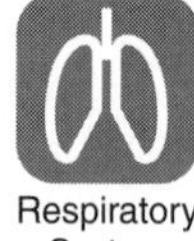
Respiratory System

Environmental

Regulations

RCRA 40CFR: Not listed
CERCLA: 40CFR 302.4: Not listed
SARA 40CFR 372.65: Not listed
SARA EHS 40CFR 355: Not listed
TSCA: Listed

BEN2760 **CAS #: 62654-17-5**

1,4-BENZENEDIAMINE ETHANEDIOATE

Molecular Formula: $C_8H_{10}N_2O_4$
Formula Weight: 198.18

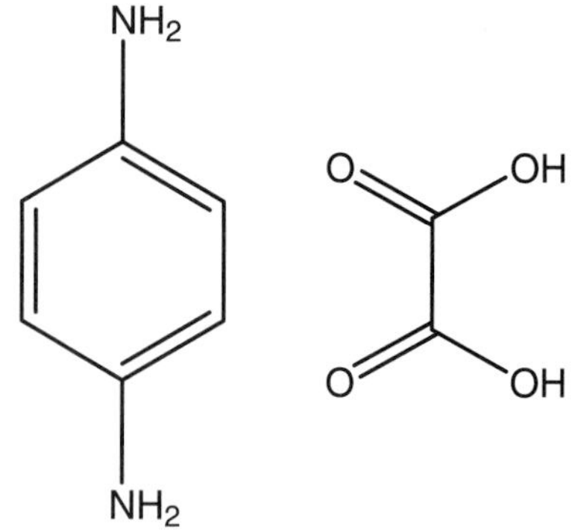

Chemical Structure

Synonyms: 1,4-BENZENEDIAMINE,ETHANEDIOATE (1:1)

Hazard Overviews

Carcinogenicity: IARC - Not listed; NIOSH - Not listed; NTP - Not listed; ACGIH - Not listed; OSHA - Not listed; EPA - Not listed; MAK - Not listed

Environmental

Regulations

RCRA 40CFR: Not listed
CERCLA: 40CFR 302.4: Not listed
SARA 40CFR 372.65: Not listed
SARA EHS 40CFR 355: Not listed
TSCA: Listed

BEN2840 **CAS #: 68966-84-7**

1,3-BENZENEDIAMINE, AR-ETHYL-AR-METHYL

EINECS Number: 273-451-6
Molecular Formula: $C_9H_{14}N_2$
Formula Weight: 150.21
Synonyms: 1,3-BENZENEDIAMINE,AR-ETHYL-AR-METHYL-

Hazard Overviews

Carcinogenicity: IARC - Not listed; NIOSH - Not listed; NTP - Not listed; ACGIH - Not listed; OSHA - Not listed; EPA - Not listed; MAK - Not listed

Environmental

Regulations

RCRA 40CFR: Not listed
CERCLA: 40CFR 302.4: Not listed
SARA 40CFR 372.65: Not listed
SARA EHS 40CFR 355: Not listed
TSCA: Listed

BEN2920 CAS #: 541-70-8

1,3-BENZENEDIAMINE SULFATE

EINECS Number: 208-791-6
Molecular Formula: $C_6H_{10}N_2O_4S$
Formula Weight: 206.21
Synonyms: 1,3-BENZENEDIAMINE,SULFATE (1:1); M-PHENYLENEDIAMINE,SULFATE (1:1)

Hazard Overviews

Carcinogenicity: IARC - Not listed; NIOSH - Not listed; NTP - Not listed; ACGIH - Not listed; OSHA - Not listed; EPA - Not listed; MAK - Not listed

Environmental

Regulations

RCRA 40CFR: Not listed
CERCLA: 40CFR 302.4: Not listed
SARA 40CFR 372.65: Not listed
SARA EHS 40CFR 355: Not listed
TSCA: Listed

BEN3000 CAS #: 16245-77-5

1,4-BENZENEDIAMINE SULFATE

EINECS Number: 240-357-1
Molecular Formula: $C_6H_{10}N_2O_4S$
Formula Weight: 206.21

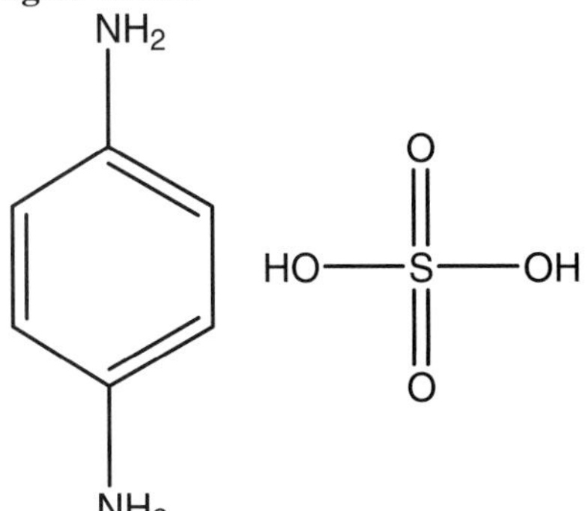

Chemical Structure

Synonyms: 1,4-BENZENEDIAMINE,SULFATE (1:1)

Hazard Overviews

Carcinogenicity: IARC - Not listed; NIOSH - Not listed; NTP - Not listed; ACGIH - Not listed; OSHA - Not listed; EPA - Not listed; MAK - Not listed

Environmental

Regulations

RCRA 40CFR: Not listed
CERCLA: 40CFR 302.4: Not listed
SARA 40CFR 372.65: Not listed
SARA EHS 40CFR 355: Not listed
TSCA: Listed

BEN3080 CAS #: 91-15-6

1,2-BENZENEDICARBONITRILE

RTECS: TI8575000
EINECS Number: 202-044-8
Molecular Formula: $C_8H_4N_2$
Formula Weight: 128.1

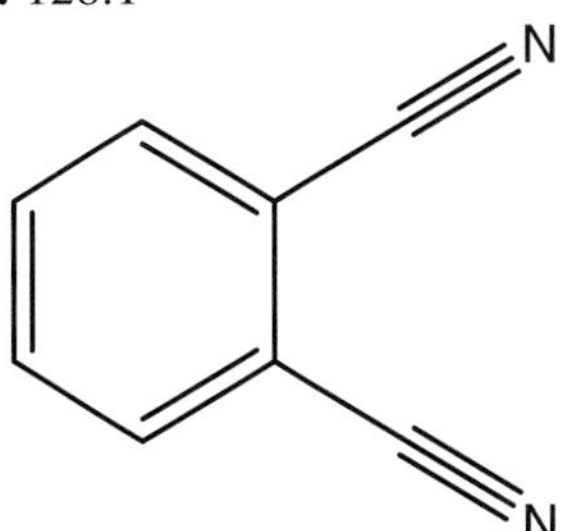

Chemical Structure

Synonyms: 1,2-BENZENDIKARBONITRIL; O-BENZENEDICARBONITRILE; O-BENZENEDINITRILE; 1,2-BENZODINITRILE; O-CYANOBENZONITRILE; 1,2-DICYANOBENZENE; O-DICYANOBENZENE; ORTHO-DICYANOBENZENE; FTALODINITRIL; FTALONITRIL; O-PDN; PHTHALIC ACID DINITRILE; PHTHALIC ACID,DINITRILE; O-PHTHALODINITRILE; PHTHALODINITRILE; PHTHALONITRILE
Description: buff-colored crystals or needles
Use: intermediate in organic synthesis, esp pigments & dyes; base material for high temp lubricants & coatings; insecticide, synergist

Physical Properties

Boiling Point: 150 °C (302 °F)
Freezing Point: 141 °C (285.8 °F)
Specific Gravity: 1.12500
Water Solubility: Slightly Soluble
Other Solubilities: Soluble in Alcohol, Ether, Acetone & Benzene
Ionization Potential (eV): 9.9 +/-0.5
Flash Point: Combustible

RTECS Toxicity Data

Acute Oral: Rat LD_{50} Dose: 125 mg/kg; Toxic Effects: Behavioral - Convulsions or effect on seizure threshold. Mouse LD_{50} Dose: 65 mg/kg; Toxic Effects: Behavioral - Convulsions or effect on seizure threshold.
Acute Dermal: Mouse LD_{50} Route: Subcutaneous Dose: 46 mg/kg; Toxic Effects: Behavioral - Convulsions or effect on seizure threshold.
Tumorigenic: Rat Route: Oral; Dose: 7425 mg/kg/66W-I; Toxic Effects: Tumorigenic - Equivocal tumorigenic agent by RTECS criteria; Liver - Tumors; Blood - Leukemia. Rat Route: Subcutaneous; Dose: 473 mg/kg/21W-I; Toxic

Effects: Tumorigenic - Equivocal tumorigenic agent by RTECS criteria; Liver - Tumors; Blood - Leukemia.

Hazard Overviews

Health: Irritating to eyes/skin/respiratory tract. Toxic. Other Acute Effects: harmful if swallowed, inhaled, or absorbed through skin.
Fire: Combustible. Hazards: emits toxic fumes. Extinguishing agents: water spray; carbon dioxide, dry chemical powder or appropriate foam. Precautions: combustible liquid.
Carcinogenicity: IARC - Not listed; NIOSH - Not listed; NTP - Not listed; ACGIH - Not listed; OSHA - Not listed; EPA - Not listed; MAK - Not listed
Primary Target Organs:

Eyes

Skin

Respiratory System

Environmental

Ecotoxicity: Crustaceans: Daphnia magna Straus 48h EC_0 125 mg/l; Fishes: Oryzias latipes 24h LC_{50} 42.5 mg/l

Environmental Physical Data

Octanol/Water Partition Coefficient: log K_{ow} = calculated at 0.58 to 1.3

Regulations

RCRA 40CFR: Not listed
CERCLA: 40CFR 302.4: Not listed
SARA 40CFR 372.65: Not listed
SARA EHS 40CFR 355: Not listed
TSCA: Listed

BEN3160 CAS #: 99-63-8

1,3-BENZENEDICARBONYL CHLORIDE

RTECS: NT2625000
EINECS Number: 202-774-7
Molecular Formula: $C_8H_4Cl_2O_2$
Structured MF: $C_6H_2(COOH)_2Cl_2$
Formula Weight: 203.03

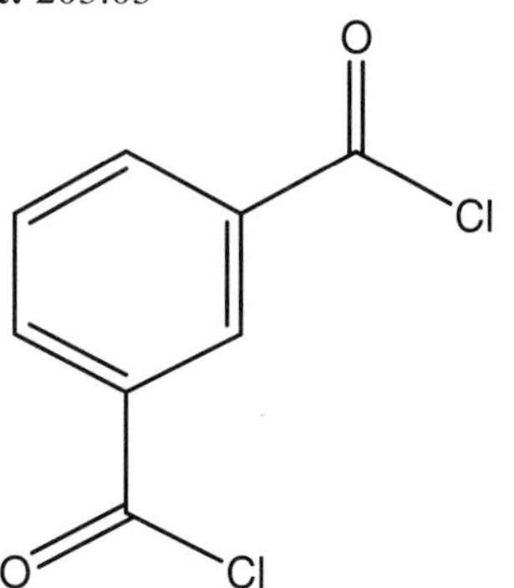

Chemical Structure

Synonyms: M-BENZENEDICARBONYL CHLORIDE; 1,3-BENZENEDICARBONYL DICHLORIDE; 1,3-BENZENEDICARBOXYLIC ACID,DICHLORIDE; DICHLORID KYSELINY ISOFTALOVE; ISOPHTHALIC ACID CHLORIDE; ISOPHTHALIC ACID DICHLORIDE; ISOPHTHALIC CHLORIDE; ISOPHTHALOYL CHLORIDE; ISOPHTHALOYL DICHLORIDE; ISOPHTHALYL CHLORIDE; ISOPHTHALYL DICHLORIDE; ISOTHALOYL CHLORIDE; M-PHTHALIC DICHLORIDE; M-PHTHALOYL CHLORIDE; META-PHTHALYL DICHLORIDE
Description: crystalline solid or prisms
Use: intermediate; dyes; synthetic fibers; resins; films; protective coatings; laboratory reagent; nonthrombogenic glass coating

Physical Properties

Boiling Point: 276 °C (529 °F)
Freezing Point: 43 °C (109.4 °F) to 44 °C (111.2 °F)
Specific Gravity: 1.388 at 17 °C/4 °C
Water Solubility: Reactive with Water and alcohol
Other Solubilities: Slightly Soluble in Alcohol.
Refraction Index: 1.570 at 47 °C/D
Flash Point: 180 °C Open Cup

RTECS Toxicity Data

Acute Oral: Rat LD_{50} Dose: 2200 mg/kg. Mouse LD_{50} Dose: 2221 mg/kg; Toxic Effects: Behavioral - Somnolence (general depressed activity); Behavioral - Food intake (animal); Behavioral - Change in motor activity (specific assay).
Acute Inhalation: Rat LC; Dose: >31 gm/m^3/4hr.
Acute Dermal: Rabbit LD_{50} Route: Skin; Dose: 1410 mg/kg.
Chronic (Multiple Dose) Oral: Rat Dose: 1817 mg/kg/22W-I; Toxic Effects: Blood - Changes in leukocyte (WBC) cell count; Nutritional and gross metabolic - Weight loss or decreased weight gain; DEATH. Rat Dose: 10 gm/kg/2W-I; Toxic Effects: Gastrointestinal - Alteration in gastric secretion; DEATH.
Irritation Eye: Rabbit Standard Draize Test Dose: 40 mg; Reaction: mild.
Irritation Skin: Rabbit Open Draize Test Dose: 200 mg open; Reaction: moderate.

Hazard Overviews

Corrosive

Health: Corrosive to eyes/skin/respiratory tract. Toxic. Other Acute Effects: extremely destructive to tissue of the mucous membranes and upper respiratory tract, eyes and skin, especially in high concentrations; toxic by inhalation, in contact with skin and if swallowed; lachrymator; symptoms of exposure may include burning sensation, coughing, wheezing, laryngitis, shortness of breath, headache, nausea and vomiting; prolonged or repeated exposure may cause allergic reactions in certain sensitive individuals. Chronic Effects: possible sensitizer.

Fire: Will burn. Hazards: water hydrolyzes material liberating acidic gas which in contact with metal surfaces can generate flammable and/or explosive hydrogen gas; emits toxic fumes. Extinguishing agents: carbon dioxide; dry chemical powder; do not use water. Precautions: combustible liquid.
Carcinogenicity: IARC - Not listed; NIOSH - Not listed; NTP - Not listed; ACGIH - Not listed; OSHA - Not listed; EPA - Not listed; MAK - Not listed
Primary Target Organs:

Eyes

Skin

Respiratory System

Environmental

Regulations

RCRA 40CFR: Not listed
CERCLA: 40CFR 302.4: Not listed
SARA 40CFR 372.65: Not listed
SARA EHS 40CFR 355: Not listed
TSCA: Listed

BEN3240 **CAS #: 100-20-9**

1,4-BENZENEDICARBONYL DICHLORIDE

RTECS: WZ1797000
EINECS Number: 202-829-5
Molecular Formula: $C_8H_4Cl_2O_2$
Formula Weight: 203.02

Chemical Structure

Synonyms: 1,4-BENZENEDICARBONYL CHLORIDE; P-PHENYLENEDICARBONYL DICHLORIDE; P-PHTHALOYL CHLORIDE; P-PHTHALOYL DICHLORIDE; TEREPHTHALIC ACID CHLORIDE; TEREPHTHALIC ACID DICHLORIDE; TEREPHTHALIC DICHLORIDE; TEREPHTHALOYL CHLORIDE; TEREPHTHALOYL DICHLORIDE
Description: colorless needles
Use: dye manufacture; synthetic fibers, resins, films; ultraviolet absorption; pharmaceuticals; rubber chemicals; cross-linking agent for polyurethanes and polysulfides

Physical Properties

Boiling Point: 259 °C (498 °F)
Freezing Point: 82 °C (179.6 °F) to 84 °C (183.2 °F)
Water Solubility: Decomposes in Water
Other Solubilities: Soluble in Ether
Flash Point: 180 °C

RTECS Toxicity Data

Acute Oral: Rat LD_{50} Dose: 2500 mg/kg. Mouse LD_{50} Dose: 2140 mg/kg; Toxic Effects: Behavioral - Somnolence (general depressed activity); Lungs, Thorax, or Respiration - Dyspnea.
Acute Inhalation: Rat LC_{50} Dose: 700 mg/m^3/4hr; Toxic Effects: Sense organs and special senses - Lacrimation; Lungs, Thorax, or Respiration - Dyspnea; Lungs, Thorax, or Respiration - Respiratory stimulation.
Acute Dermal: Rabbit LD; Route: Skin; Dose: >200 mg/kg.
Chronic (Multiple Dose) Oral: Rat Dose: 4168 mg/kg/26W-I; Toxic Effects: Liver - Other changes.
Reproductive/Teratogenic: Rat Route: Oral; Dose: 4122 mg/kg; Duration: male 26W prior to mating; Paternal Effects - Spermatogenesis. Rat Route: Oral; Dose: 4122 mg/kg; Duration: female 26W prior to mating Maternal Effects - Menstrual cycle changes or disorders.

Hazard Overviews

Fire: Will burn.
Carcinogenicity: IARC - Not listed; NIOSH - Not listed; NTP - Not listed; ACGIH - Not listed; OSHA - Not listed; EPA - Not listed; MAK - Not listed

Environmental

Regulations

RCRA 40CFR: Not listed
CERCLA: 40CFR 302.4: Not listed
SARA 40CFR 372.65: Not listed
SARA EHS 40CFR 355: Not listed
TSCA: Listed

BEN3320 **CAS #: 84-66-2**

1,2-BENZENEDICARBOXYLIC ACID, DIETHYL ESTER

RTECS: TI1050000
EINECS Number: 201-550-6
Molecular Formula: $C_{12}H_{14}O_4$
Structured MF: $C_6H_4(COOC_2H_5)_2$
Formula Weight: 222.26

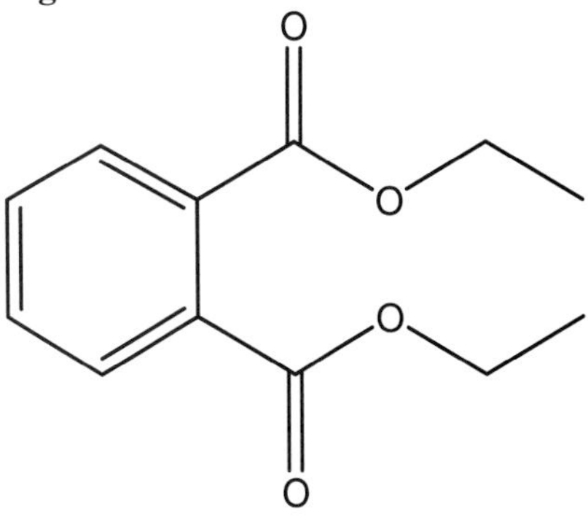

Chemical Structure

Synonyms: ANOZOL; 1,2-BENZENEDICARBOXYLIC ACID DIETHYL ESTER; O-BENZENEDICARBOXYLIC ACID DIETHYL ESTER; DEP; DIETHYL 1,2-BENZENEDICARBOXYLATE; DIETHYL ESTER OF PHTHALIC ACID; DIETHYL O-PHTHALATE; DIETHYL PHTHALATE; DIETHYLESTER KYSELINY FTALOVE; DIETHYL-O-PHTHALATE; DPX-F5384; ESTOL 1550; ETHYL PHTHALATE; NEANTINE; PALATINOL A; PHTHALIC ACID,DIETHYL ESTER; PHTHALOL; PHTHALSAEUREDIAETHYLESTER; PLACIDOL E; SOLVANOL; UNIMOLL DA

Description: clear, colorless liquid; odorless

Use: as a solvent for nitrocellulose and cellulose acetate, wetting agent, camphor substitute, alcohol denaturant (e.g. in surgical spirit), plasticizer in solid rocket propellant, dye application agent and as a diluent in polysulfide dental impression materials; in plasticizer manufacture, insecticidal sprays, plastics manufacture and processing, perfumery as a fixative and solvent, mosquito repellents and manufacture of varnishes and dopes

Physical Properties

Boiling Point: 295 °C (563 °F)
Freezing Point: -40.5 °C (-40.9 °F)
Specific Gravity: 1.232 at 14 °C/4 °C
Vapor Density: 7.66 Air=1
Saturated Vapor Density: 1.200017362 kg/m^3
Density: 1.117 g/mL
Vapor Pressure: 1.65 x10^{-3} mm Hg at 25 °C
Water Solubility: < 1 mg/mL at 19 C
Other Solubilities: miscible with vegetable oils; miscible with Ketones, Esters, aromatic hydrocarbons; partly miscible with aliphatic solvents.
Surface Tension: 37.5 dynes/cm at 20 °C
Refraction Index: 1.5049 at 14 °C/D
Evaporation Rate: < 1 Butyl Acetate=1
Flash Point: 161 °C Open Cup
Autoignition Temperature: 457 °C
LEL: 0.7% v/v

RTECS Toxicity Data

Acute Oral: Rat LD_{50} Dose: 8600 mg/kg; Toxic Effects: Behavioral - Somnolence (general depressed activity); Behavioral - Withdrawal; Nutritional and gross metabolic - Weight loss or decreased weight gain. Mouse LD_{50} Dose: 6172 mg/kg.

Acute Inhalation: Human TC_{Lo} Dose: 1000 mg/m^3; Toxic Effects: Sense organs and special senses - Lacrimation; Lungs, Thorax, or Respiration - Cough; Lungs, Thorax, or Respiration - Other changes.

Acute Dermal: Guinea Pig LD_{50} Route: Subcutaneous Dose: 3 gm/kg.

Chronic (Multiple Dose) Oral: Rat Dose: 44240 mg/kg/14D-C; Toxic Effects: Liver - Changes in liver weight; Nutritional and gross metabolic - Weight loss or decreased weight gain; DEATH - Changes in testicular weight. Rat Dose: 133 gm/kg/6W-C; Toxic Effects: Liver - Changes in liver weight; Nutritional and gross metabolic - Weight loss or decreased weight gain; DEATH - Changes in testicular weight. Rat Dose: 354 gm/kg/16W-C; Toxic Effects: Gastrointestinal - Other changes; Liver - Changes in liver weight; DEATH - Changes in testicular weight. Rat Dose: 25200 mg/kg/3W-C; Toxic Effects: Blood - Changes in serum composition.

Chronic (Multiple Dose) Dermal: Rat Route: Skin; Dose: 24 mL/kg/4W-I; Toxic Effects: Liver - Changes in Liver weight; Kidney, Ureter, and Bladder - Changes in kidney weight. Mouse Route: Skin; Dose: 25 mL/kg/4W-I; Toxic Effects: Liver - Changes in Liver weight; Nutritional and gross metabolic - Weight loss or decreased weight gain.

Reproductive/Teratogenic: Rat Route: Oral; Dose: 25 gm/kg; Duration: female 6-15D of pregnancy; Specific Developmental Abnormalities - Musculoskeletal system. Rat Route: Intraperitoneal; Dose: 506 mg/kg; Duration: female 5-15D of pregnancy; Effects on Fertility - Post-implantation mortality; Effects on Embryo or Fetus - Fetotoxicity; Specific Developmental Abnormalities - Musculoskeletal system.

Mutagenic: Bacteria - S Typhimurium Mutations in Microorganisms; Dose: 200 ug/plate (+S9).

Tumorigenic: Mouse Route: Skin; Dose: 618 mL/kg/2Y-I; Toxic Effects: Tumorigenic - Equivocal tumorigenic agent by RTECS criteria; Liver - Tumors.

Hazard Overviews

Fire Diamond

Health: Irritating to eyes/skin/respiratory tract. Also Causes: narcotic effects. Chronic Effects: bioaccumulative.

Fire: Will burn. Use water spray, dry chemical, carbon dioxide, or foam. Water or foam may cause some frothing. Use water to cool fire-exposed containers.

Reactivity: Stable. Hazardous polymerization cannot occur. Avoid: heat and ignition sources. Incompatible with: strong oxidizers; strong acids; nitric acid; permanganates; water; some forms of plastics. Hazardous decomposition products: carbon monoxide; carbon dioxide; various hydrocarbons.

Carcinogenicity: IARC - Not listed; NIOSH - Not listed; NTP - Not listed; ACGIH - Not listed; OSHA - Not listed; EPA - Class D, Not classifiable as to human carcinogenicity; MAK - Not listed

Primary Target Organs:

Eyes

Skin

Respiratory System

Mucous Membranes

Exposure Limits
OSHA PEL Vacated 1989 Limits: TWA: 5 mg/m^3.
ACGIH TLV: TWA: 5 mg/m^3.
NIOSH REL: TWA: 5 mg/m^3.

Respirator Recommendation
Exposure Range: >5 to 50 mg/m^3 Air Purifying, Negative Pressure, Half Mask
Exposure Range: >50 to 500 mg/m^3 Air Purifying, Negative Pressure, Full Face
Exposure Range: >500 to 5000 mg/m^3 Supplied Air, Constant Flow/Pressure Demand, Full Face

Exposure Range: >5000 to unlimited mg/m^3 Self-contained Breathing Apparatus, Pressure Demand, Full Face
Cartridge Color: dust/mist filter (use P100 or consult supervisor for appropriate dust/mist filter)

Environmental

Ecotoxicity: LC_{50} Sheepshead minnows 30 ppm/96 hr (95% confidence limitS 23-38 ppm) /conditions of bioassay not specified EC_{50} Selenastrum capricornutum (alga) 90,300 ug/l/96 hr, toxic effect: chlorophyll a; 85,600 ug/l/96 hr, toxic effect: cell number LC_{50} Eisenia fetida (earthworm) 850 ug/sq cm of filter paper; 95% confidence interval of 660 to 1,090 ug/sq cm; 48 hr LC_{50} Lepomis macrochirus (bluegill) 110 mg/l/96 hr (undissolved chemical) /conditions of bioassay not specified EC_{50} Skeletonema costatum 65,500 ug/l/96 hr, toxic effect: chlorophyll a
Environmental Fate: If released to soil, it is expected to undergo aerobic biodegradation. Oxidation, chemical hydrolysis and volatilization from wet soil surfaces are not expected to be significant fate processes. May volatilize from dry soil surfaces. If released to water, it is expected to biodegrade (aerobic biodegradation half-life approx. 2 days to >2 weeks). Anaerobic biodegradation would be very slow or not occur at all. Volatilization should not be an important removal process in most bodies of water although it may be important in shallow rivers. Removal by oxidation, chemical hydrolysis, direct photolysis, indirect photolysis or bioaccumulation in aquatic organisms should not be significant. It has accumulated and persisted in the sediments of Chesapeake Bay for over a century. If released to the atmosphere, it is expected to exist in vapor form and as adsorbed matter on airborne particulates. Vapor is expected to react with photochemically generated hydroxyl radical (estimated half-life = 22.2 hours). Physical removal by particulate settling and washout in precipitation will also occur. Degradation by direct photolysis is not expected to be significant.
Cleanup/Disposal: Guide No. 171: Do not touch or walk through spilled material. Stop leak if you can do it without risk. Prevent dust cloud. Avoid inhalation of asbestos dust. Small Dry Spills: With clean shovel place material into clean, dry container and cover loosely; move containers from spill area. Small Spills: Take up with sand or other noncombustible absorbent material and place into containers for later disposal. Large Spills: Dike far ahead of liquid spill for later disposal. Cover powder spill with plastic sheet or tarp to minimize spreading. Prevent entry into waterways, sewers, basements or confined areas.

Environmental Physical Data

Henry's Law Constant: calculated at 4.47 x10^{-7}
Octanol/Water Partition Coefficient: log K_{ow} = 2.47
Sorption Partition Coefficient: K_{OC} = 94 to 526
BCF: bluegill 117

Regulations

RCRA 40CFR: Listed Hazardous Waste No. U088 Toxic Waste
CERCLA: 40CFR 302.4: Listed per RCRA Section 3001 per CWA Section 307(a) RQ: 1000 lb (453.5 kg)
SARA 40CFR 372.65: Listed
SARA EHS 40CFR 355: Not listed
TSCA: Listed

Analytical Methods

Soil: CLP LC_SV, MC_SVOA, OHC; EPA 16, 1625; SW846 3640A, 8060, 8061, 8061A, 8250A, 8270B, 8270C, 8410
Water / Groundwater: EPA S-002-1, 1625, 606, 625, 625-S, 6; APHA 6040-B, 6410-B; ASTM D4763; USGS O3118
Drinking Water: EPA 506, 525.1, 525.2
Food: FDA 212.1, 232.1
Plasma: EPA 001, 29; FDA 211.1, 231.1, 252

BEN3400 **CAS #: 117-84-0**

1,2-BENZENEDICARBOXYLIC ACID, DIOCTYL ESTER

RTECS: TI1925000
EINECS Number: 204-214-7
Molecular Formula: $C_{24}H_{38}O_4$
Structured MF: o-$C_6H_4[COOCH_2CH(C_2H_5)(CH_2)_3CH_3]_2$
Formula Weight: 390.56

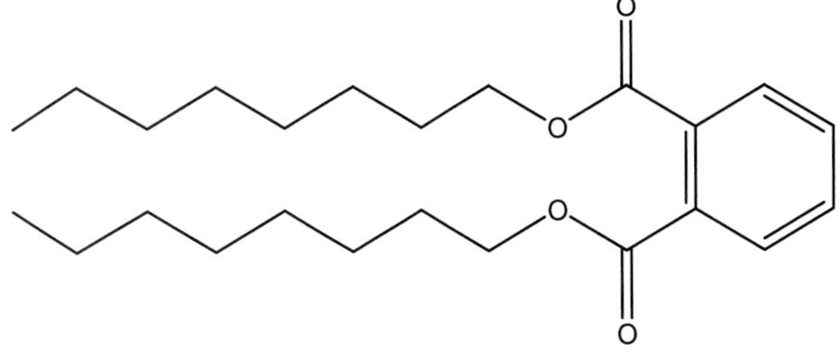

Chemical Structure

Synonyms: 1,2-BENZENEDICARBOXYLIC ACID DIOCTYL ESTER; BENZENEDICARBOXYLIC ACID DI-N-OCTYL ESTER; BENZENEDICARBOXYLIC ACID,DI-N-OCTYL ESTER; O-BENZENEDICARBOXYLIC ACID,DIOCTYL ESTER; CELLUFLEX DOP; DINOPOL NOP; DIOCTYL 1,2-BENZENEDICARBOXYLATE; DIOCTYL O-BENZENEDICARBOXYLATE; DI-N-OCTYL PHTHALATE; DIOCTYL PHTHALATE; N-DIOCTYL PHTHALATE; DIOKTYLESTER KYSELINY FTALOVE; DNOP; N-OCTYL PHTHALATE; OCTYL PHTHALATE; PHTHALIC ACID DIOCTYL ESTER; PHTHALIC ACID,DIOCTYL ESTER; POLYCIZER 162; PX-138; VINICIZER 85
Description: liquid; slight odor
Use: plasticizer in plastics & rubber materials; plasticizer for cellulose ester resins, polystyrene resins, vinyl resins (eg, polyvinyl chloride); dye carrier; for film, wire, cables, and adhesives

Physical Properties

Boiling Point: 220 °C (428 °F) at 4 torr
Freezing Point: -25 °C (-13 °F)
Specific Gravity: 0.978 at 25 °C
Vapor Density: 13.48 Air=1

Density: 0.978 g/mL
Vapor Pressure: < 0.2 mm Hg at 150 °C
Water Solubility: 3 mg/L in Water at 25 °C
Other Solubilities: 95% Ethanol: >=100 mg/ml at 19 °C; Acetone: >=100 mg/ml at 19 °C; DMSO: 50-100 mg/ml at 19 °C.
Surface Tension: Estimated at 15 dynes/cm
Flash Point: 218 °C Open Cup
Autoignition Temperature: 385 °C
LEL: 0.3% v/v

RTECS Toxicity Data

Acute Oral: Rat LD_{50} Dose: 47 gm/kg. Mouse LD_{50} Dose: 6513 mg/kg.
Acute Dermal: Guinea Pig LD_{50} Route: Skin; Dose: >5 gm/kg.
Chronic (Multiple Dose) Oral: Rat Dose: 26850 mg/kg/5D-I; Toxic Effects: Endocrine - Other changes; Immunological including allergic - Decrease in humoral immune response; Immunological including allergic - Decreased immune response. Rat Dose: 25200 mg/kg/21D-C; Toxic Effects: Liver - Fatty liver degeneration; Liver - Changes in liver weight; Biochemical - Phosphatases. Rat Dose: 23 gm/kg/11W-C; Toxic Effects: Liver - Fatty liver degeneration; Liver - Other changes. Rat Dose: 54 gm/kg/90D-C; Toxic Effects: Endocrine - Other changes; Blood - Changes in spleen.
Irritation Eye: Rabbit Standard Draize Test Dose: 20 mg; Reaction: severe. Rabbit Standard Draize Test Dose: 500 mg/24H; Reaction: mild.
Irritation Skin: Rabbit Standard Draize Test Dose: 500 mg/24H; Reaction: mild.
Reproductive/Teratogenic: Rat Route: Intraperitoneal; Dose: 5 gm/kg; Duration: female 5-15D of pregnancy; Effects on Embryo or Fetus - Fetotoxicity; Specific Developmental Abnormalities - Eye, ear; Other developmental abnormalities. Mouse Route: Oral; Dose: 78 gm/kg; Duration: female 7-14D of pregnancy; Effects on Newborn - Live birth index; Growth statistics. Mouse Route: Oral; Dose: 78240 mg/kg; Duration: female 6-13D of pregnancy; Effects on Newborn - Live birth index; Growth statistics.

Hazard Overviews

Fire: Will burn.
Carcinogenicity: IARC - Not listed; NIOSH - Not listed; NTP - Not listed; ACGIH - Not listed; OSHA - Not listed; EPA - Not listed; MAK - Not listed

Environmental

Ecotoxicity: LC_{50} Lepomis microlopus (Redear sunfish) 6.180 ug/l/7-8 days. /Conditions of bioassay not specified LC_{50} Ictalurus punctatus (Channel catfish) 690 ug/l/7 days. /Conditions of bioassay not specified
Environmental Fate: It will adsorb strongly to sediment and particulate matter and slowly biodegrade with acclimation. The half-life for removal from the aqueous phase was reported to be 5 days in an ecosystem study. It bioconcentrates in algae and other aquatic organisms. The data for fish are contradictory but bioconcentration is probably important in species where little or no metabolism occurs. If emitted into the atmosphere as an aerosol it will be subject to gravitational settling and photodegradation by hydroxy radicals (estimated half-life 14 hr.).
Cleanup/Disposal: Guide No. 171: Do not touch or walk through spilled material. Stop leak if you can do it without risk. Prevent dust cloud. Avoid inhalation of asbestos dust. Small Dry Spills: With clean shovel place material into clean, dry container and cover loosely; move containers from spill area. Small Spills: Take up with sand or other noncombustible absorbent material and place into containers for later disposal. Large Spills: Dike far ahead of liquid spill for later disposal. Cover powder spill with plastic sheet or tarp to minimize spreading. Prevent entry into waterways, sewers, basements or confined areas.

Environmental Physical Data

Henry's Law Constant: 2.2 $x10^{-4}$
Octanol/Water Partition Coefficient: log K_{ow} = 5.22
Sorption Partition Coefficient: K_{oc} = estimated at 1.9 $x10^{4}$
BCF: non accumlative in carp

Regulations

RCRA 40CFR: Listed Hazardous Waste No. U107 Toxic Waste
CERCLA: 40CFR 302.4: Listed per RCRA Section 3001 per CWA Section 307(a) RQ: 5000 lb (2268 kg)
SARA 40CFR 372.65: Not listed
SARA EHS 40CFR 355: Not listed
TSCA: Listed

Analytical Methods

Soil: CLP LC_SV, MC_SVOA, OHC; EPA 1625; SW846 3640A, 8060, 8061, 8061A, 8250A, 8270B, 8270C, 8410
Water / Groundwater: EPA 1625, 606, 625, 625-S, S-002-1; APHA 6410-B; USGS O3118
Drinking Water: EPA 506
Food: FDA 212.1, 232.1
Plasma: EPA 001, 29; FDA 211.1, 231.1, 252

BEN3480 **CAS #: 1477-55-0**

1,3-BENZENEDIMETHANAMINE

RTECS: PF8970000
EINECS Number: 216-032-5
Molecular Formula: $C_8H_{12}N_2$
Structured MF: $C_6H_4(CH_2NH_2)_2$
Formula Weight: 136.22

H_2N NH_2

Chemical Structure

Synonyms: 1,3-BIS-AMINOMETHYLBENZEN; 1,3-BIS(AMINOMETHYL)BENZENE; METHYLAMINE,M-PHENYLENEBIS-; MXDA; M-PHENYLENEBIS(METHYLAMINE); M-XYLENE ALPHA,ALPHA'-DIAMINE; M-XYLENE-ALPHA,ALPHA'-DIAMINE; M-XYLYLENDIAMIN; M-XYLYLENEDIAMINE
Description: colorless liquid
Use: curing agent for epoxy resins; a source of m-xylylene diisocyanate

Physical Properties

Boiling Point: 247 °C (477 °F)
Freezing Point: 14.44 °C (57.992 °F)
Specific Gravity: 1.032
Saturated Vapor Density: 1.200175132 kg/m^3
Vapor Pressure: 0.03 mm Hg at 25 °C
Water Solubility: Miscible with Water
Other Solubilities: miscible with Alcohol; partially Soluble in Paraffin hydrocarbon solvents
Flash Point: 133.889 °C Open Cup

RTECS Toxicity Data

Acute Oral: Rat LD_{50} Dose: 930 mg/kg.
Acute Inhalation: Rat LC_{50} Dose: 700 ppm/1hr; Toxic Effects: Sense organs and special senses - Lacrimation; Lungs, Thorax, or Respiration - Respiratory depression.
Acute Dermal: Rabbit LD_{50} Route: Skin; Dose: 2 gm/kg.
Irritation Eye: Rabbit Standard Draize Test Dose: 50 ug/24H; Reaction: severe.
Irritation Skin: Rabbit Standard Draize Test Dose: 750 ug/24H; Reaction: severe.

Hazard Overviews

Corrosive

Health: Corrosive to eyes/skin/respiratory tract. Harmful. Other Acute Effects: harmful if swallowed, inhaled, or absorbed through skin; inhalation may result in spasm, inflammation and edema of the larynx and bronchi, chemical pneumonitis and pulmonary edema; symptoms of exposure may include burning sensation; coughing; wheezing; laryngitis; shortness of breath; headache; nausea; vomiting.
Fire: Will burn. Hazards: emits toxic fumes. Extinguishing agents: water spray; carbon dioxide, dry chemical powder or appropriate foam. Precautions: combustible liquid.
Reactivity: Incompatible with: acids, acid chlorides, acid anhydrides, oxidizing agents, chloroformates. Hazardous decomposition products: thermal decomposition may produce carbon monoxide, carbon dioxide, and nitrogen oxides.
Carcinogenicity: IARC - Not listed; NIOSH - Listed as carcinogen; NTP - Not listed; ACGIH - Not listed; OSHA - Not listed; EPA - Not listed; MAK - Not listed

Primary Target Organs:

Eyes

Skin

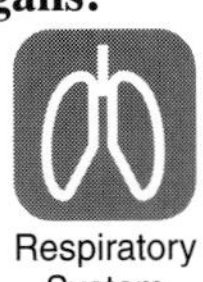
Respiratory System

Exposure Limits
OSHA PEL Vacated 1989 Limits: STEL: 0.1 mg/m^3; Ceiling.
ACGIH TLV: STEL: 0.1 mg/m^3; Ceiling.
NIOSH REL: STEL: 0.1 mg/m^3; skin.
Respirator Recommendation
Exposure Range: >0.1 to 5 mg/m^3 Supplied Air, Constant Flow/Pressure Demand, Half Mask
Exposure Range: >5 to 100 mg/m^3 Supplied Air, Constant Flow/Pressure Demand, Full Face
Exposure Range: >100 to unlimited mg/m^3 Self-contained Breathing Apparatus, Pressure Demand, Full Face
Note: odor threshold unknown

Environmental

Regulations

RCRA 40CFR: Not listed
CERCLA: 40CFR 302.4: Not listed
SARA 40CFR 372.65: Not listed
SARA EHS 40CFR 355: Not listed
TSCA: Listed

BEN3640 CAS #: 98-11-3

BENZENESULFONIC ACID

RTECS: DB4200000
EINECS Number: 202-638-7
Molecular Formula: $C_6H_6O_3S$
Formula Weight: 158.17

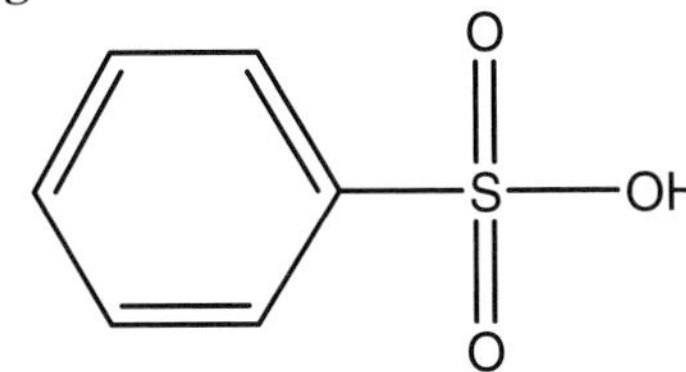

Chemical Structure

Synonyms: BENZENEMONOSULFONIC ACID; BESYLIC ACID; KYSELINA BENZENSULFONOVA; PHENYLSULFONIC ACID
Description: fine, deliquiduescent needles or large plates
Use: manufacture of phenol, resorcinol and other organic syntheses, and as a catalyst

Physical Properties

Boiling Point: Decomposes
Freezing Point: Anhydrous 50 °C (122 °F) to 51 °C (123.8 °F)
Water Solubility: Soluble in Water
Other Solubilities: 95% Ethanol: 5-10 mg/ml at 22 °C; Acetone: 1-5 mg/ml at 22 °C; Acetic Acid: Soluble; Benzene:

Slightly Soluble; Carbon Disulfide: Insoluble; DMSO: 10-50 mg/ml at 22 °C; Ether: Insoluble.
Flash Point: Not available; probably combustible

RTECS Toxicity Data

Acute Oral: Rat LD_{50} Dose: 890 mg/kg.
Acute Dermal: Cat LD_{Lo} Route: Skin; Dose: 10 gm/kg; Toxic Effects: Behavioral - Tremor; Behavioral - Muscle weakness; Gastrointestinal - Changes in structure or function of salivary glands.
Irritation Eye: Rabbit Standard Draize Test Dose: 250 ug/24H; Reaction: severe.
Irritation Skin: Rabbit Standard Draize Test Dose: 2 mg/24H; Reaction: severe.

Hazard Overviews

Corrosive

Health: Corrosive to eyes/skin/respiratory tract. Toxic. Other Acute Effects: toxic by inhalation, in contact with skin and if swallowed; may be fatal if swallowed; extremely destructive to tissue of the mucous membranes and upper respiratory tract, eyes and skin; inhalation may result in spasm, inflammation and edema of the larynx and bronchi, chemical pneumonitis and pulmonary edema; symptoms of exposure may include burning sensation, coughing, wheezing, laryngitis, shortness of breath, headache, nausea and vomiting. Chronic Effects: may cause cancer by inhalation. Human carcinogen.
Fire: Will burn. Hazards: emits toxic fumes; contact with other material may cause fire. Extinguishing agents: water spray; carbon dioxide, dry chemical powder or appropriate foam. Precautions: combustible liquid.
Carcinogenicity: IARC - Not listed; NIOSH - Not listed; NTP - Not listed; ACGIH - Not listed; OSHA - Not listed; EPA - Not listed; MAK - Not listed
Primary Target Organs:

Eyes

Skin

Respiratory System

Environmental

Ecotoxicity: Fishes: Cyprinus carpio: hematological effect after 75 days at 8 mg/l (sodium salt)
Environmental Fate: It is expected to have very high mobility in soil. Volatilization is not expected from either moist or dry soils. In water, it is expected to be essentially non-volatile. Adsorption to sediment, bioconcentration, and hydrolysis are not expected to be important fate processes in aquatic systems. Biodegradation is likely to occur in both aquatic and soil media provided adequate acclimation by microorganisms occurs. It will exist in both the vapor and particulate phases in the ambient atmosphere. If released to the atmosphere, it will degrade by reaction with photochemically produced hydroxyl radicals with an estimated half-life of approximately 29 days. Removal from the atmosphere can occur though wet and dry deposition.
Cleanup/Disposal: Guide No. 171: Do not touch or walk through spilled material. Stop leak if you can do it without risk. Prevent dust cloud. Avoid inhalation of asbestos dust. Small Dry Spills: With clean shovel place material into clean, dry container and cover loosely; move containers from spill area. Small Spills: Take up with sand or other noncombustible absorbent material and place into containers for later disposal. Large Spills: Dike far ahead of liquid spill for later disposal. Cover powder spill with plastic sheet or tarp to minimize spreading. Prevent entry into waterways, sewers, basements or confined areas.

Environmental Physical Data

Henry's Law Constant: estimated at 2.52×10^{-9}
Octanol/Water Partition Coefficient: log K_{ow} = -2.25
Sorption Partition Coefficient: K_{oc} = estimated at 12
BCF: estimated at 1.15

Regulations

RCRA 40CFR: Not listed
CERCLA: 40CFR 302.4: Not listed
SARA 40CFR 372.65: Not listed
SARA EHS 40CFR 355: Not listed
TSCA: Listed

BEN3720 CAS #: 98-09-9

BENZENESULFONYL CHLORIDE

RTECS: DB8750000
DOT: UN2225; IMO8.0
EINECS Number: 202-636-6
Molecular Formula: $C_6H_5ClO_2S$
Structured MF: $C_6H_5SO_2Cl$
Formula Weight: 176.62

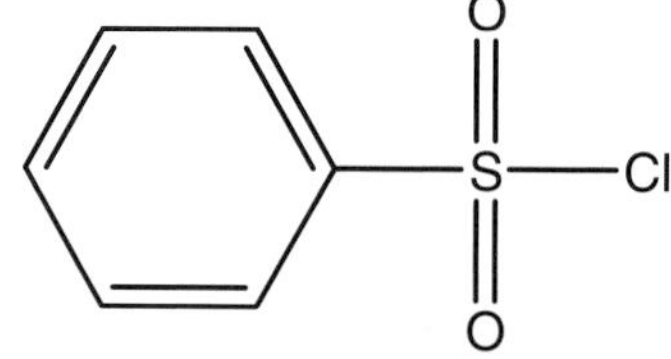

Chemical Structure

Synonyms: BENEZENESULFOCHLORIDE; BENZENE SULFOCHLORIDE; BENZENE SULFONECHLORIDE; BENZENE SULFONYL CHLORIDE; BENZENESULFON CHLORIDE; BENZENESULFONIC (ACID) CHLORIDE; BENZENESULFONIC CHLORIDE; BENZENESULPHONYL CHLORIDE; BENZENOSULFOCHLOREK; BENZENOSULPHOCHLORIDE; BENZOLSULFOCHLORIDE; BSC-REFINE D; PHENYLSULFONYL CHLORIDE
Description: colorless, oily liquid
Use: chem int for benzene sulfonamides, thiophenol, glybuzole (hypoglycemic agent), n-2-chloroethyl amides,

benzonitrile; insecticide; miticide; for fenson acaricide (former use); reagent for friedel-crafts sulfonylation

Physical Properties

Boiling Point: 177 °C (351 °F) at 100 mm Hg
Freezing Point: 14.5 °C (58.1 °F)
Specific Gravity: 1.3842 at 15 °C/15 °C
Vapor Density: Estimate 6.09 Air=1
Saturated Vapor Density: Estimated 1.200546542 kg/m^3
Vapor Pressure: Estimated 0.068 mm Hg at 25 °C
Water Solubility: Insoluble in Water
Other Solubilities: Soluble in Ether, Alcohol
pH: Acid
Flash Point: > 112 °C Closed Cup

RTECS Toxicity Data

Acute Oral: Rat LD_{50} Dose: 1960 mg/kg; Toxic Effects: Behavioral - Somnolence (general depressed activity); Lungs, Thorax, or Respiration - Respiratory depression. Mouse LD_{50} Dose: 828 mg/kg.
Acute Inhalation: Rat LC_{Lo} Dose: 1870 $mg/m^3/1hr$; Toxic Effects: Sense organs and special senses - Chromidracryorrhea; Behavioral - Change in motor activity (specific assay); Lungs, Thorax, or Respiration - Dyspnea.

Hazard Overviews

Corrosive

Health: Corrosive to eyes/skin/respiratory tract. Harmful. Other Acute Effects: harmful by inhalation, in contact with skin and if swallowed; extremely destructive to tissue of the mucous membranes and upper respiratory tract, eyes and skin; inhalation may result in spasm, inflammation and edema of the larynx and bronchi, chemical pneumonitis and pulmonary edema; symptoms of exposure may include burning sensation, coughing, wheezing, laryngitis, shortness of breath, headache, nausea and vomiting.
Fire: Will burn. Hazards: emits toxic fumes; water hydrolyzes material liberating acidic gas which in contact with metal surfaces can generate flammable and/or explosive hydrogen gas. Extinguishing agents: do not use water; carbon dioxide, dry chemical powder or appropriate foam. Precautions: combustible liquid.
Carcinogenicity: IARC - Not listed; NIOSH - Not listed; NTP - Not listed; ACGIH - Not listed; OSHA - Not listed; EPA - Not listed; MAK - Not listed
Primary Target Organs:

Eyes

Skin

Respiratory System

Environmental

Ecotoxicity: LC_{50} Brown trout yearlings 3 mg/l/48 hr /Static bioassay
Environmental Fate: If released to soil, it will be expected to rapidly hydrolyze if the soil is moist, based upon the rapid hydrolysis observed in aqueous solution. Since it rapidly hydrolyzes, adsorption to and volatilization from moist soil are not expected to be significant processes. Based upon an estimated vapor pressure of 0.068 mm Hg at 25 °C, volatilization from dry near-surface soil or other surfaces may be significant processes. If released to water, it will be expected to rapidly hydrolyze with a half-life of 5.1 min at 21 °C. Since it rapidly hydrolyzes, bioconcentration, volatilization, and adsorption to sediment and suspended solids are not expected to be significant processes. No data were located concerning biodegradation, but it probably chemically hydrolyzes significantly faster than it biodegrades. Direct photolysis is not expected to be an important removal process in surface waters. If released to the atmosphere, it will be expected to exist almost entirely in the vapor phase based upon its estimated vapor pressure. It will be susceptible to photooxidation via vapor phase reaction with photochemically hydroxyl radicals. An atmospheric half-life of 7.9 days at an atmospheric concentration of 5 $x10^5$ hydroxyl radicals per cu cm has been estimated for this process based upon an estimated rate constant. Hydrolysis in moist air may be an important removal process based upon its rapid hydrolysis in aqueous solution. Direct photolysis is not expected to be an important removal process in the atmosphere.
Cleanup/Disposal: Guide No. 156: Eliminate all ignition sources (no smoking, flares, sparks or flames in immediate area). All equipment used when handling the product must be grounded. Do not touch damaged containers or spilled material unless wearing appropriate protective clothing. Stop leak if you can do it without risk. A vapor suppressing foam may be used to reduce vapors. For chlorosilanes, use AFFF alcohol-resistant medium expansion foam to reduce vapors. Do not get water on spilled substance or inside containers. Use water spray to reduce vapors or divert vapor cloud drift. Prevent entry into waterways, sewers, basements or confined areas. Small Spills: Cover with dry earth, dry sand, or other non-combustible material followed with plastic sheet to minimize spreading or contact with rain. Use clean non-sparking tools to collect material and place it into loosely covered plastic containers for later disposal.

Environmental Physical Data

BCF: not significant

Regulations

RCRA 40CFR: Listed Hazardous Waste No. U020 Corrosive Waste Reactive Waste
CERCLA: 40CFR 302.4: Listed per RCRA Section 3001 RQ: 100 lb (45.35 kg)
SARA 40CFR 372.65: Not listed
SARA EHS 40CFR 355: Not listed
TSCA: Listed

BEN3800 **CAS #: 121-54-0**

BENZETHONIUM CHLORIDE

RTECS: BO7175000
EINECS Number: 204-479-9
Molecular Formula: $C_{27}H_{42}ClNO_2$
Formula Weight: 448.10

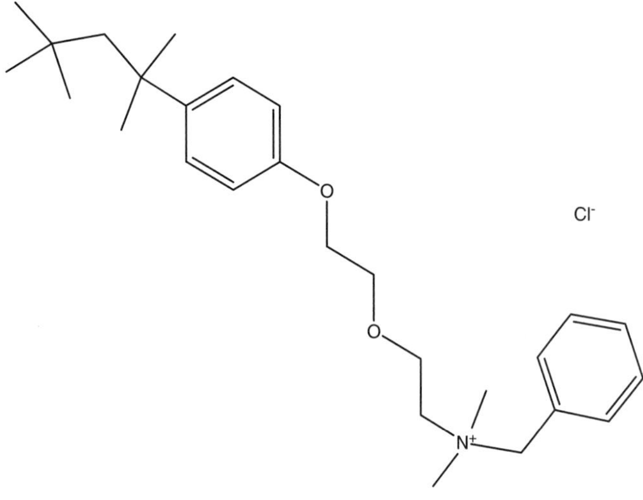

Chemical Structure

Synonyms: AMMONIUM CHLORIDE; AMMONIUM,BENZYLDIMETHYL(2-(2-(P-(1,1,3,3-TETRAMETHYLBUTYL)PHENOXY)ETHOXY)ETHYL)-,CHLORIDE; ANTI-GERM 77; ANTISEPTOL; ANTISEPTOL (QUARTERNARY COMPOUND); BANAGERM; BENZATHONIUM CHLORIDE; BENZENEMETHANAMINIUM CHLORIDE; BENZENEMETHANAMINIUM,N,N-DIMETHYL-N-(2-(2-(4-(1,1,3,3-TETRAMETHYLBUTYL)PHENOXY)ETHOXY)ETHYL)-,CHLORIDE; BENZETHONIICHLORIDUM; BENZETHONIUM; BENZETHONIUM CHLORIDE 1622; BENZETHONIUMCHLORIDE; BENZETONIUM CHLORIDE; BENZYLDIMETHYL(2-(2-(4-(1,1,3,3-TETRAMETHYLBUTYL)PHENOXY)ETHOXY)ETHYL); BENZYLDIMETHYL-P-(1,1,3,3-TETRAMETHYLBUTYL)PHENOXYETHOXYETHYL AMMONIUM; BENZYLDIMETHYL(2-(2-(P-(1,1,3,3-TETRAMETHYLBUTYL)PHENOXY)ETHOXY)ETHYL)AMMONI; BENZYLDIMETHYL(2-(2-(4-(1,1,3,3-TETRAMETHYLBUTYL)PHENOXY)ETHOXY)ETHYL)AMMONIUM CHLORIDE; BENZYLDIMETHYL(2-(2-(P-(1,1,3,3-TETRAMETHYLBUTYL)PHENOXY)ETHOXY)ETHYL)AMMONIUM CHLORIDE; BENZYLDIMETHYL(2-(2-(P-1,1,3,3-TETRAMETHYLBUTYLPHENOXY)ETHOXY)ETHYL)AMMONIUM CHLORIDE; BENZYLDIMETHYL-P-(1,1,3,3-TETRAMETHYLBUTYL)PHENOXYETHOXY-ETHYLAMMONIUM CHLORIDE; BZT; CHLORIDE; DIAPP; P-DIISOBUTYL PHENOXYETHOXYETHYL DIMETHYL BENZYLAMMONIUM CHLORIDE; P-DIISOBUTYL PHENOXYETHOXYETHYL DIMETHYL BENZYLAMMONIUMCHLORIDE; DIISOBUTYLPHENOXYETHOXYETHYL DIMETHYL BENZYL AMMONIUM CHLORIDE; DIISOBUTYLPHENOXYETHOXYETHYL DIMETHYL BENZYL AMMONIUMCHLORIDE; DIISOBUTYLPHENOXYETHOXYETHYLDIMETHYL BENZYL AMMONIUMCHLORIDE; (2-(2-(4-DIISOBUTYLPHENOXY)ETHOXY)ETHYL)DIMETHYLBENZYLAMMONIUMCHLORIDE; (DIISOBUTYLPHENOXYETHOXYETHYL)DIMETHYLBENZYLAMMONIUMCHLORIDE; N,N-DIMETHYL-N-(2-(2-(4-(1,1,3,3-TETRAMETHYLBUTYL)PHENOXY)ETHOXY)ETHYL)-; N,N-DIMETHYL-N-(2-(2-(4-(1,1,3,3-TETRAMETHYLBUTYL)PHENOXY)ETHOXY)ETHYL)-BENZENEMETHANAMINIUM CHLORIDE; DISILYN; FORMULA 144; HYAMINE; HYAMINE 1622; INACTISOL; P-TERT-OCTYLPHENOXYETHOXYETHYLDIMETHYLBENZYLAMMONIUM CHLORIDE; P-TERT-OCTYLPHENOXYETHOXYETHYLDIMETHYLBENZYLAMMONIUM CHLORIDE; PHEMERIDE; PHEMEROL; PHEMEROL CHLORIDE; PHEMERSOL CHLORIDE; PHEMITHYN; POLYMINE D; QAC; QUATRACHLOR; SANIZOL; SOLAMIN; SOLAMINE

Description: colorless crystals; mild odor

Use: topical antiinfective; in veterinary medicine as a topical antiseptic; and as a cationic detergent; in herbicides; and in antiseptics, spermicides, astringents, germicides, disinfectants, and preservatives; to destroy bacteria on skin, surgical instruments, cooking equipment, sickroom supplies, and diapers as well as in hairdressing preparations to control dandruff, control swimming pool algae, and in deodorants

Physical Properties

Freezing Point: 164 °C (327.2 °F) to 166 °C (330.8 °F)

Water Solubility: Very Soluble in Water

Other Solubilities: 95% Ethanol: >=100 mg/ml at 18 °C; Acetone: 1-5 mg/ml at 18 °C; Chloroform: 1:1; DMSO: 10-50 mg/ml at 18 °C; Ether: Slightly Soluble; Light petroleum: Practically Insoluble.

pH: 1% aqueous solution 4.8 to 5.5

Refraction Index: 1.5101 at 25 °C/D

Flash Point: Not available; probably combustible

RTECS Toxicity Data

Acute Oral: Rat LD_{50} Dose: 368 mg/kg. Mouse LD_{50} Dose: 338 mg/kg.

Acute Dermal: Rat LD_{50} Route: Subcutaneous Dose: 119 mg/kg.

Chronic (Multiple Dose) Oral: Rat Dose: 182 gm/kg/2Y-C; Toxic Effects: Gastrointestinal - Other changes; Nutritional and gross metabolic - Weight loss or decreased weight gain; DEATH.

Chronic (Multiple Dose) Dermal: Rat Route: Skin; Dose: 60 mg/kg/16D-I; Toxic Effects: Endocrine - Changes in thymus weight; Skin and appendages - Dermatitis, other; Nutritional and gross metabolic - Weight loss or decreased weight gain. Rat Route: Skin; Dose: 203 mg/kg/13W-I; Toxic Effects: Skin and appendages - Dermatitis, other. Rat Route: Subcutaneous; Dose: 312 mg/kg/1Y-I; Toxic Effects: Nutritional and gross metabolic - Weight loss or decreased weight gain.

Irritation Eye: Rabbit Standard Draize Test Dose: 30 ug; Reaction: severe.

Mutagenic: Hamster Sister Chromatid Exchenge; Cell Type: embryo; Dose: 1 mg/L. Bacteria - E Coli DNA Repair; Dose: 1500 ng/well.

Tumorigenic: Rat Route: Subcutaneous; Dose: 104 mg/kg/1Y-I; Toxic Effects: Tumorigenic - Neoplastic by RTECS criteria; Skin and appendages - Tumors; Tumorigenic - Tumors at site of application.

Hazard Overviews

Corrosive

Health: Severely irritating to eyes; irritating to skin/respiratory tract. Toxic. Other Acute Effects: harmful if swallowed, inhaled, or absorbed through skin.
Fire: Will burn. Hazards: emits toxic fumes. Extinguishing agents: carbon dioxide; dry chemical powder; water spray. Precautions: combustible liquid.
Reactivity: Incompatible with: strong oxidizing agents. Hazardous decomposition products: carbon monoxide, carbon dioxide, nitrogen oxides, hydrogen chloride gas.
Carcinogenicity: IARC - Not listed; NIOSH - Not listed; NTP - Not listed; ACGIH - Not listed; OSHA - Not listed; EPA - Not listed; MAK - Not listed
Primary Target Organs:

Eyes

Skin
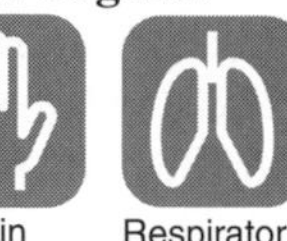
Respiratory System

Environmental

Ecotoxicity: LC_{50} Oncorhynchus kisutch 53000 ug/l/96 hr /Conditions of bioassay not specified LC_{50} Lepomis macrochirus 1400 ug/l/96 hr /Conditions of bioassay not specified
Cleanup/Disposal: Guide No. 171: Do not touch or walk through spilled material. Stop leak if you can do it without risk. Prevent dust cloud. Avoid inhalation of asbestos dust. Small Dry Spills: With clean shovel place material into clean, dry container and cover loosely; move containers from spill area. Small Spills: Take up with sand or other noncombustible absorbent material and place into containers for later disposal. Large Spills: Dike far ahead of liquid spill for later disposal. Cover powder spill with plastic sheet or tarp to minimize spreading. Prevent entry into waterways, sewers, basements or confined areas.

Regulations

RCRA 40CFR: Not listed
CERCLA: 40CFR 302.4: Not listed
SARA 40CFR 372.65: Not listed
SARA EHS 40CFR 355: Not listed
TSCA: Listed

BEN3880 CAS #: 613-94-5

BENZHYDRAZIDE

RTECS: DH1575000
EINECS Number: 210-363-9
Molecular Formula: $C_7H_8N_2O$
Formula Weight: 136.17

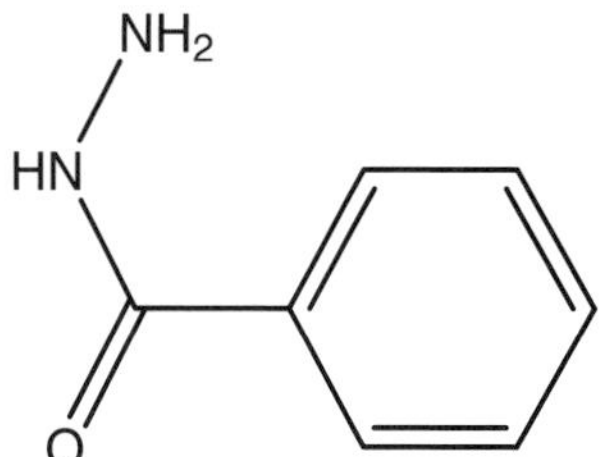

Chemical Structure

Synonyms: BENZOHYDRAZIDE; BENZOHYDRAZINE; BENZOIC ACID,HYDRAZIDE; BENZOIC HYDRAZIDE; BENZOYL HYDRAZIDE; BENZOYLHYDRAZINE; HYDRAZID KYSELINY BENZOOVE; HYDRAZINE,BENZOYL-
Description: plates

Physical Properties

Boiling Point: 267 °C (513 °F)
Freezing Point: 113 °C (235.4 °F) to 117 °C (242.6 °F)
Water Solubility: Soluble in Water
Other Solubilities: Slightly Soluble in Ether, Acetone, Chloroform

RTECS Toxicity Data

Acute Dermal: Rabbit LD_{Lo} Route: Subcutaneous Dose: 102 mg/kg; Toxic Effects: Endocrine - Hypoglycemia. Mouse LD_{50} Route: Subcutaneous Dose: 122 mg/kg; Toxic Effects: Behavioral - Convulsions or effect on seizure threshold.
Tumorigenic: Mouse Route: Oral; Dose: 15 gm/kg/77W-C; Toxic Effects: Tumorigenic - Carcinogenic by RTECS criteria; Lungs, Thorax, or Respiration - Tumors; Blood - Lymphomax including Hodgkin's disease. Mouse Route: Oral; Dose: 13 gm/kg/30W-I; Toxic Effects: Tumorigenic - Neoplastic by RTECS criteria; Lungs, Thorax, or Respiration - Tumors.

Hazard Overviews

Health: Irritating to eyes/skin. Other Acute Effects: may be harmful by inhalation, ingestion, or skin absorption.
Fire: Hazards: emits toxic fumes. Extinguishing agents: water spray; carbon dioxide, dry chemical powder or appropriate foam. Precautions: combustible liquid.
Reactivity: Incompatible with: strong oxidizing agents, strong bases. Hazardous decomposition products: toxic fumes of: carbon monoxide, carbon dioxide, nitrogen oxides.
Carcinogenicity: IARC - Not listed; NIOSH - Not listed; NTP - Not listed; ACGIH - Not listed; OSHA - Not listed; EPA - Not listed; MAK - Not listed
Primary Target Organs:

Eyes

Skin

Environmental

Regulations

RCRA 40CFR: Not listed

CERCLA: 40CFR 302.4: Not listed
SARA 40CFR 372.65: Not listed
SARA EHS 40CFR 355: Not listed
TSCA: Listed

BEN3960 CAS #: 92-87-5

BENZIDINE

RTECS: DC9625000
DOT: UN1885; IMO6.1
EINECS Number: 202-199-1
Molecular Formula: $C_{12}H_{12}N_2$
Structured MF: $NH_2C_6H_4C_6H_4NH_2$
Formula Weight: 184.23

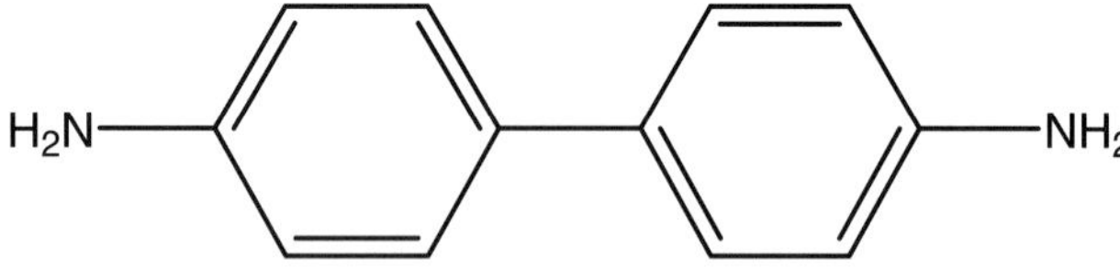

Chemical Structure

Synonyms: BENZIDIN; BENZIDINA; P-BENZIDINE; BENZIDINE BASE; BENZIDINE-BASED DYES; BENZYDYNA; 4,4'-BIANILINE; P,P'-BIANILINE; P,P-BIANILINE; (1,1'-BIPHENYL)-4,4'-DIAMINE; 1,1'-BIPHENYL-4,4'-DIAMINE; 4,4'-BIPHENYLDIAMINE; BIPHENYL-4,4'-DIAMINE; (1,1'-BIPHENYL)-4,4'-DIAMINE (9CI); BIPHENYL,4,4'-DIAMINO-; 4,4'-BIPHENYLENEDIAMINE; C.I. 37225; C.I. AZOIC DIAZO COMPONENT 112; 4,4'-DIAMINOBIPHENY1; 4,4'-DIAMINO-1,1'-BIPHENYL; 4,4'-DIAMINOBIPHENYL; P,P'-DIAMINOBIPHENYL; 4,4'-DIAMINODIPHENYL; P-DIAMINODIPHENYL; P,P'-DIANILINE; 4,4'-DIPHENYLENEDIAMINE; FAST CORINTH BASE B

Description: white, slightly reddish, greyish-yellow solid, crystalline powder; odorless

Use: as a precursor in the synthesis of dyes and pigments used to color textiles, rubber, plastic products, printing inks, paints, lacquers, leathers and paper products, manufacture of a wide variety of organ chemicals, rubber compounding agent, as a reagent for hydrogen peroxide (H2O2 in milk, for detection of blood stains, as a stain in microscopy, in the manufacture of plastic films, in the production of security paper and as a laboratory reagent in determining hydrogen cyanide, sulfate, nicotine and certain sugars

Physical Properties

Boiling Point: About 400 °C (752 °F)
Freezing Point: 115 °C (239 °F) to 120 °C (248 °F)
Specific Gravity: 1.25 at 20 °C/4 °C
Vapor Density: 6.36
Vapor Pressure: Low
Water Solubility: 1 g dissolves in 2500 ml Cold Water
Other Solubilities: 95% Ethanol: 10-50 mg/ml at 20 °C; Acetone: >=100 mg/ml at 20 °C; Absolute Alcohol: 1 g/13 mL; DMSO: >=100 mg/ml at 20 °C; Ether: 1 g/50 mL; Less polar solvents: Readily Soluble.
Critical Temperature: 659.8 °C
Critical Pressure: 479 psia
Flash Point: Combustible

RTECS Toxicity Data

Acute Oral: Rat LD_{50} Dose: 309 mg/kg. Mouse LD_{50} Dose: 214 mg/kg.

Mutagenic: Human DNA Damage; Cell Type: fibroblast; Dose: 3 mmol/L. Human Unscheduled DNA Synthesis; Cell Type: liver; Dose: 10 mg/L. Human Unscheduled DNA Synthesis; Cell Type: HeLa cell; Dose: 100 umol/L. Human Unscheduled DNA Synthesis; Cell Type: fibroblast; Dose: 160 ug/L. Human DNA Inhibition; Cell Type: HeLa cell; Dose: 600 umol/L/30M-C. Human DNA Adduct; Cell Type: lymphocyte; Dose: 30 umol/L. Human Sister Chromatid Exchenge; Cell Type: lymphocyte; Dose: 2 mg/L. Man Sister Chromatid Exchenge; Route: Inhalation; Dose: 7 ug/m^3/27W.

Tumorigenic: Man Route: Inhalation; Dose: 17600 ug/m^3/14Y-C; Toxic Effects: Tumorigenic - Carcinogenic by RTECS criteria; Kidney, Ureter, and Bladder - Hematuria; Kidney, Ureter, and Bladder - Tumors. Rat Route: Oral; Dose: 108 mg/kg/27D-I; Toxic Effects: Tumorigenic - Carcinogenic by RTECS criteria; Skin and appendages - Tumors. Rat Route: Inhalation; Dose: 10 mg/m^3/56W-I; Toxic Effects: Tumorigenic - Equivocal tumorigenic agent by RTECS criteria; Liver - Tumors; Blood - Leukemia. Rat Route: Oral; Dose: 25560 ng/kg/2Y-C; Toxic Effects: Tumorigenic - Equivocal tumorigenic agent by RTECS criteria; Liver - Tumors.

Hazard Overviews

Fire Diamond

Health: Irritating to eyes/skin. Toxic. Chronic Effects: Cancer of the bladder and kidneys. Chronic ingestion may cause swelling of the liver and blood in the urine.

Fire: Not expected to burn readily; if it is involved in a fire, apply an extinguishing agent such as water, carbon dioxide, or dry chemical.

Reactivity: Stable. Hazardous polymerization cannot occur. Avoid: sunlight; air. Hazardous decomposition products: toxic gases; carbon monoxide; benzidine vapor.

Carcinogenicity: IARC - Group 1, Carcinogenic to humans; NIOSH - Listed as carcinogen; NTP - Class 1, Known to be a carcinogen; ACGIH - Class A1, Confirmed human carcinogen; OSHA - Listed as a carcinogen; EPA - Class A, Human carcinogen; MAK - Class A1, Capable of inducing malignant tumors as shown by experience with humans

Primary Target Organs:

Eyes

Skin

Liver

Kidneys

Blood

Respirator Recommendation

Exposure Range: unlimited Self-contained Breathing Apparatus, Pressure Demand, Full Face

Note: TLV not established

Environmental

Ecotoxicity: LC_{50} Notropis lutrensis 2,500 ug/l/96 hr in a static unmeasured bioassay

Environmental Fate: If spilled on soil, it will adsorb to it, especially if the soil is acidic, form complexes with clay particles and be oxidized by metal cations. The rate of degradation in soil in the few studies reported in the literature were 79% degradation in 4 weeks and 10% mineralization in 1 yr. If released in water, it will rapidly adsorb to suspended clay particles, and be oxidized by naturally occurring metal cations such as Fe(III). It will also be lost by reaction with radicals and photolysis. Its half-life in water is approximately 1 day. It will adsorb to sediments and bioconcentrate only moderately in fish. In the atmosphere, it would primarily exist in aerosols, be bound to particulate matter and be subject to gravitational settling and wash-out. It may photolyze and would be readily oxidized by reactive species in the atmosphere such as hydroxyl radicals.

Cleanup/Disposal: Guide No. 153: Eliminate all ignition sources (no smoking, flares, sparks or flames in immediate area). Do not touch damaged containers or spilled material unless wearing appropriate protective clothing. Stop leak if you can do it without risk. Prevent entry into waterways, sewers, basements or confined areas. Absorb or cover with dry earth, sand or other non-combustible material and transfer to containers. Do not get water inside containers.

Environmental Physical Data

Henry's Law Constant: will be very low

Octanol/Water Partition Coefficient: log K_{OW} = 1.34

Sorption Partition Coefficient: K_{OC} = 2.27 $x10^5$ to 8.82 $x10^5$

BCF: fish 1.6

BOD: sewage seed 1.9 to 4.1 lb/lb, 144 hr

Regulations

RCRA 40CFR: Listed Hazardous Waste No. U021 Toxic Waste

CERCLA: 40CFR 302.4: Listed per RCRA Section 3001 per CWA Section 307(a) RQ: 1 lb (0.454 kg)

SARA 40CFR 372.65: Listed

SARA EHS 40CFR 355: Not listed

TSCA: Listed

Analytical Methods

Soil: SW846 3640A, 8250A, 8270B, 8270C; EPA 16

Water / Groundwater: EPA 6, 1625, 553, 605, 625, 625-S; SW846 8325; APHA 6410-B; USGS O3118

Drinking Water: EPA 553

Indoor / Expired Air: NIOSH 5509

Plasma: EPA 29

Urine: NIOSH 8306

BEN4040 CAS #: 134-81-6

BENZIL

RTECS: DD1925000

EINECS Number: 205-157-0

Molecular Formula: $C_{14}H_{10}O_2$

Formula Weight: 210.24

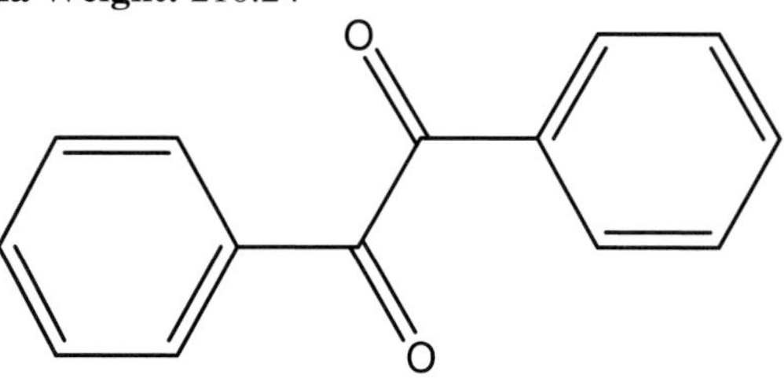

Chemical Structure

Synonyms: BIBENZOYL; DIBENZOYL; DIPHENYL-ALPHA,BETA-DIKETONE; 1,2-DIPHENYLETHANEDIONE; DIPHENYLETHANEDIONE; DIPHENYLGLYOXAL; ETHANEDIONE,DIPHENYL-(9CI); GLYOXAL,DIPHENYL-; WY-20910

Physical Properties

Boiling Point: 346 °C (654.8 °F) to 348 °C (658.4 °F)

Freezing Point: 95 °C (203 °F)

Specific Gravity: 1.08400

Water Solubility: 0.5 g/l at 20 °C

Other Solubilities: Ethanol: Very Soluble; Ether: Very Soluble; Acetone: Soluble; Benzene: Very Soluble; Carbon Tetrachloride: Slightly soluble

Ionization Potential (eV): 8.68 +/-0.7

RTECS Toxicity Data

Acute Oral: Mouse LD_{50} Dose: >3 gm/kg; Toxic Effects: Behavioral - Somnolence (general depressed activity).

Irritation Eye: Rabbit Standard Draize Test Dose: 100 mg/24H; Reaction: moderate.

Hazard Overviews

Health: Irritating to eyes/skin/respiratory tract. Other Acute Effects: may be harmful by inhalation, ingestion, or skin absorption.

Fire: Extinguishing agents: water spray; carbon dioxide, dry chemical powder or appropriate foam. Precautions: combustible liquid.

Reactivity: Incompatible with: strong oxidizing agents. Hazardous decomposition products: toxic fumes of: carbon monoxide, carbon dioxide.

Carcinogenicity: IARC - Not listed; NIOSH - Not listed; NTP - Not listed; ACGIH - Not listed; OSHA - Not listed; EPA - Not listed; MAK - Not listed

Primary Target Organs:

Eyes

Skin

Respiratory System

Environmental

Regulations

RCRA 40CFR: Not listed
CERCLA: 40CFR 302.4: Not listed
SARA 40CFR 372.65: Not listed
SARA EHS 40CFR 355: Not listed
TSCA: Listed

BEN4200 CAS #: 51-17-2

BENZIMIDAZOLE

RTECS: DD5425000
EINECS Number: 200-081-4
Molecular Formula: $C_7H_6N_2$
Formula Weight: 118.14

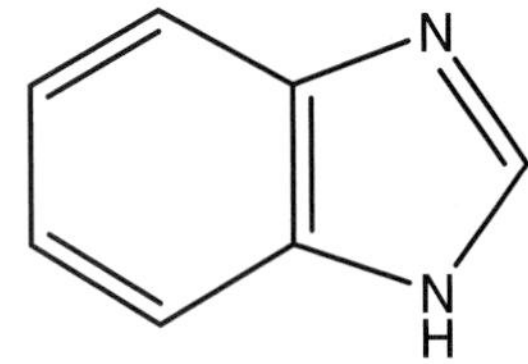

Chemical Structure

Synonyms: 3-AZAINDOLE; AZINDOLE; 1H-BENZIMIDAZOLE; O-BENZIMIDAZOLE; 1H-BENZIMIDAZOLE (9CI); BENZIMINAZOLE; 1,3-BENZODIAZOLE; BENZOGLYOXALINE; BENZOIMIDAZOLE; BZI; 1,3-DIAZAINDENE; N,N'-METHENYL-O-PHENYLENEDIAMINE; NSC 759
Description: tabular crystals or plates
Use: muscle relaxant which causes reversible flaccid paralysis in animals

Physical Properties

Boiling Point: > 360 °C (680 °F) at 760 mm Hg
Freezing Point: 170.5 °C (338.9 °F)
Water Solubility: Miscible with Water
Other Solubilities: 1 g Soluble in 2 g boiling Xylene; Soluble in aqueous solution of acids and strong alkalies.
pH: Weak base
Ionization Potential (eV): 8.0 +/-2.0
Flash Point: Not available; probably combustible

RTECS Toxicity Data

Acute Oral: Rat LD_{Lo} Dose: 500 mg/kg. Mouse LD_{50} Dose: 2910 mg/kg; Toxic Effects: Lungs, Thorax, or Respiration - Other changes.
Mutagenic: Bacteria - E Coli DNA Damage; Dose: 15 mmol/L/48H. Bacteria - E Coli Mutations in Microorganisms; Dose: 1 mg/disc (-S9). Bacteria - S Typhimurium Mutations in Microorganisms; Dose: 250 ug/plate (+S9).

Hazard Overviews

Health: May cause irritation. Harmful. Other Acute Effects: may be harmful by inhalation, ingestion, or skin absorption.
Fire: Will burn. Hazards: emits toxic fumes. Extinguishing agents: water spray; carbon dioxide, dry chemical powder or appropriate foam. Precautions: combustible liquid.
Reactivity: Incompatible with: strong oxidizing agents. Hazardous decomposition products: toxic fumes of: carbon monoxide, carbon dioxide, nitrogen oxides.
Carcinogenicity: IARC - Not listed; NIOSH - Not listed; NTP - Not listed; ACGIH - Not listed; OSHA - Not listed; EPA - Not listed; MAK - Not listed

Environmental

Regulations

RCRA 40CFR: Not listed
CERCLA: 40CFR 302.4: Not listed
SARA 40CFR 372.65: Not listed
SARA EHS 40CFR 355: Not listed
TSCA: Listed

BEN4360 CAS #: 194-69-4

BENZO(C)CHRYSENE

RTECS: DE7350000
Molecular Formula: $C_{22}H_{14}$
Formula Weight: 278.35
Synonyms: 1,2:5,6-DIBENZOPHENANTHRENE; 1,2,5,6-DIBENZPHENANTHRENE
Use: biochemical research

RTECS Toxicity Data

Tumorigenic: Mouse Route: Skin; Dose: 1630 mg/kg/68W-I; Toxic Effects: Tumorigenic - Equivocal tumorigenic agent by RTECS criteria; Skin and appendages - Tumors; Tumorigenic - Tumors at site of application. Mouse Route: Subcutaneous; Dose: 2400 mg/kg/36W-I; Toxic Effects: Tumorigenic - Equivocal tumorigenic agent by RTECS criteria; Tumorigenic - Tumors at site of application.

Hazard Overviews

Carcinogenicity: IARC - Not listed; NIOSH - Not listed; NTP - Not listed; ACGIH - Not listed; OSHA - Not listed; EPA - Not listed; MAK - Not listed

Environmental

Regulations

RCRA 40CFR: Not listed
CERCLA: 40CFR 302.4: Not listed
SARA 40CFR 372.65: Not listed
SARA EHS 40CFR 355: Not listed
TSCA: Not listed

BEN4440 CAS #: 196-78-1

BENZO(G)CHRYSENE

RTECS: DE7525000
Molecular Formula: $C_{22}H_{14}$
Formula Weight: 278.36
Synonyms: 1,2,3,4-DIBENZOPHENANTHRENE; 1,2,3,4-DIBENZPHENANTHRENE
Use: biochemical research

RTECS Toxicity Data

Tumorigenic: Mouse Route: Oral; Dose: 15 gm/kg/74W-I; Toxic Effects: Tumorigenic - Equivocal tumorigenic agent by RTECS criteria; Lungs, Thorax, or Respiration - Tumors; Skin and appendages - Tumors. Mouse Route: Skin; Dose: 720 mg/kg/30W-I; Toxic Effects: Tumorigenic - Equivocal tumorigenic agent by RTECS criteria; Lungs, Thorax, or Respiration - Tumors; Skin and appendages - Tumors.

Hazard Overviews

Carcinogenicity: IARC - Not listed; NIOSH - Not listed; NTP - Not listed; ACGIH - Not listed; OSHA - Not listed; EPA - Not listed; MAK - Not listed

Environmental

Regulations

RCRA 40CFR: Not listed
CERCLA: 40CFR 302.4: Not listed
SARA 40CFR 372.65: Not listed
SARA EHS 40CFR 355: Not listed
TSCA: Not listed

BEN4520 CAS #: 205-99-2

BENZO(B)FLUORANTHENE

RTECS: CU1400000
EINECS Number: 205-911-9
Molecular Formula: $C_{20}H_{12}$
Formula Weight: 252.32

Chemical Structure

Synonyms: B (B) F; B B F; B E F; B(B)F; B(E)F; BBF; BEF; 3,4-BENZ(E)ACEPHENANTHRYLENE; BENZ(E)ACEPHENANTHRYLENE; 2,3-BENZFLUORANTHENE; 3,4-BENZFLUORANTHENE; BENZO(B) FLUORANTHENE; 2,3-BENZOFLUORANTHENE; 3,4-BENZOFLUORANTHENE; BENZO(B)FLUORANTHENE; BENZO(E)FLUORANTHENE; 2,3-BENZOFLUORANTHRENE
Description: colorless needles
Use: research chemical

Physical Properties

Freezing Point: 168 °C (334.4 °F)
Water Solubility: < 1 mg/mL at 19 C
Other Solubilities: 95% Ethanol: <1 mg/ml at 19 °C; Acetone: 10-50 mg/ml at 19 °C; Benzene: Slightly Soluble; DMSO: 10-50 mg/ml at 19 °C.
Flash Point: Not available; probably combustible

RTECS Toxicity Data

Mutagenic: Rat DNA Adduct; Route: Intraperitoneal; Dose: 100 mg/kg. Rat Sister Chromatid Exchenge; Route: Intraperitoneal; Dose: 100 mg/kg.
Tumorigenic: Rat Route: Implant; Dose: 5 mg/kg; Toxic Effects: Tumorigenic - Equivocal tumorigenic agent by RTECS criteria; Lungs, Thorax, or Respiration - Tumors; Tumorigenic - Tumors at site of application. Mouse Route: Skin; Dose: 88 ng/kg/120W-I; Toxic Effects: Tumorigenic - Carcinogenic by RTECS criteria; Skin and appendages - Tumors; Tumorigenic - Tumors at site of application. Mouse Route: Skin; Dose: 72 mg/kg/60W-I; Toxic Effects: Tumorigenic - Equivocal tumorigenic agent by RTECS criteria; Skin and appendages - Tumors; Tumorigenic - Tumors at site of application. Mouse Route: Skin; Dose: 4037 ug/kg/20D-I; Toxic Effects: Tumorigenic - Equivocal tumorigenic agent by RTECS criteria; Skin and appendages - Tumors.

Hazard Overviews

Health: Irritating. Toxic. Other Acute Effects: harmful if swallowed, inhaled, or absorbed through skin. Chronic Effects: may alter genetic material; may cause heritable genetic damage. Carcinogen.
Fire: Will burn. Hazards: emits toxic fumes. Extinguishing agents: water spray; carbon dioxide, dry chemical powder or appropriate foam. Precautions: combustible liquid.
Reactivity: Incompatible with: strong oxidizing agents. Hazardous decomposition products: toxic fumes of: carbon monoxide, carbon dioxide.
Carcinogenicity: IARC - Group 2B, Possibly carcinogenic to humans; NIOSH - Not listed; NTP - Class 2B, Reasonably anticipated to be a carcinogen, sufficient evidence of carcinogenicity from studies in experimental animals; ACGIH - Class A2, Suspected human carcinogen; OSHA - Not listed; EPA - Class B2, Probable human carcinogen based on animal studies; MAK - Class A2, Unmistakably carcinogenic in animal experimentation only

Primary Target Organs:

Eyes

Skin

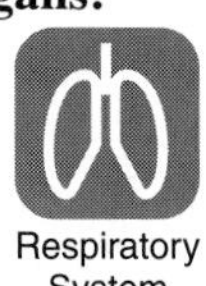
Respiratory System

Respirator Recommendation

Exposure Range: unlimited Self-contained Breathing Apparatus, Pressure Demand, Full Face

Note: odor threshold unknown

Environmental

Environmental Fate: When released to water, adsorption to suspended sediments is expected to remove most from solution. Photolysis and photo-oxidation of the compound which remains in solution is expected to occur but adsorbed compound is expected to resist these processes. Volatilization and biodegradation of dissolved compound may also occur. Bioconcentration in fish may occur; however microsomal oxidase, an enzyme capable of rapidly metabolizing polynuclear aromatic hydrocarbons, is present. Release to the soil may result in some biodegradation. Due to the anticipated strong adsorption to the soil, volatilization, photolysis and leaching to groundwater are not expected to be significant. In the atmosphere it is likely to be adsorbed to particulate matter, and will be subject to wet and dry deposition. In the vapor phase it will react with photochemically generated, atmospheric hydroxyl radicals with an estimated half-life of 1.00 day. Photolysis of vapor phase will be rapid, but the adsorbed compound may not photolyze significantly.

Environmental Physical Data

Henry's Law Constant: estimated at 1.38 $x10^{-4}$

Octanol/Water Partition Coefficient: log K_{ow} = 6.124

Sorption Partition Coefficient: K_{oc} = estimated at 5.88

BCF: expected to bioconcentrate

Regulations

RCRA 40CFR: Not listed

CERCLA: 40CFR 302.4: Listed per CWA Section 307(a) RQ: 1 lb (0.454 kg)

SARA 40CFR 372.65: Listed

SARA EHS 40CFR 355: Not listed

TSCA: Not listed

Analytical Methods

Air: EPA TO-13; California 429

Soil: CLP LC_SV, MC_SVOA, OHC; EPA 16, 1625, PAH-005, PAH-007, PAH-011, PAH-012; SW846 3630B, 3640A, 3650A, 3650B, 8100, 8250A, 8270B, 8270C, 8275A, 8310

Water / Groundwater: EPA PAH-002, PAH-006, 1625, 610, 625, 625-S, 6; APHA 6410-B, 6440-B, 6440-C; ASTM D4657; USGS O3118

Drinking Water: EPA 525.1, 525.2, 550, 550.1

Indoor / Expired Air: NIOSH 5506, 5515; EPA IP-7-A, IP-7-B

Plasma: EPA 29

Other: EPA PAH-009

BEN4600 CAS #: 205-82-3

BENZO(J)FLUORANTHENE

RTECS: DF6300000

EINECS Number: 205-910-3

Molecular Formula: $C_{20}H_{12}$

Formula Weight: 252.32

Synonyms: B(J)F; 10,11-BENZFLUORANTHENE; BENZ(J)FLUORANTHENE; 10,11-BENZOFLUORANTHENE; 7,8-BENZOFLUORANTHENE; BENZO(L)FLUORANTHENE; BENZO-12,13-FLUORANTHENE; DIBENZO(A,JK)FLUORENE

Description: yellow to orange plates or needles

Use: experimental carcinogen; biochemical research

Physical Properties

Freezing Point: 166 °C (330.8 °F)

Water Solubility: Insoluble in Water

Other Solubilities: Soluble in Hydrogen Sulfide on heating.

RTECS Toxicity Data

Mutagenic: Mouse DNA Damage; Route: Skin; Dose: 3760 nmol/kg. Bacteria - S Typhimurium Mutations in Microorganisms; Dose: 10 ug/plate (-S9).

Tumorigenic: Rat Route: Implant; Dose: 25 mg/kg; Toxic Effects: Tumorigenic - Carcinogenic by RTECS criteria; Lungs, Thorax, or Respiration - Tumors; Tumorigenic - Tumors at site of application. Rat Route: Implant; Dose: 5 mg/kg; Toxic Effects: Tumorigenic - Equivocal tumorigenic agent by RTECS criteria; Lungs, Thorax, or Respiration - Tumors; Tumorigenic - Tumors at site of application.

Hazard Overviews

Carcinogenicity: IARC - Group 2B, Possibly carcinogenic to humans; NIOSH - Not listed; NTP - Listed; ACGIH - Not listed; OSHA - Not listed; EPA - Not listed; MAK - Not listed

Environmental

Environmental Fate: It is expected to biodegrade very slowly and is not expected to hydrolyze in the environment. A calculated K_{oc} range of 51,000 to 68,000, indicates it will be highly immobile in soil. In aquatic systems, it partitions from the water column to organic matter contained in sediments and suspended solids. It also has the potential to bioconcentrate in aquatic systems. A Henry's Law constant of 7.39 $x10^{-7}$ atm-cu m/mole at 25 °C, suggests volatilization from environmental waters will be slow. The volatilization half-lives from a model river and model pond, the latter considers the effect of adsorption, have been estimated to be 70 days and over 400 years, respectively. In the atmosphere, the vapor phase reaction with photochemically produced hydroxyl radicals (half-life of 7 hr) may be an important fate process. However, it is expected to exist almost entirely in the particulate phase in ambient air. Nevertheless, it may undergo direct photolysis in the atmosphere. Otherwise, washout by precipitation and gravitational settling may be important atmospheric removal mechanisms.

Environmental Physical Data

Henry's Law Constant: estimated at 7.39 x10^{-7}
Octanol/Water Partition Coefficient: log K_{ow} = 6.12
Sorption Partition Coefficient: K_{oc} = estimated at 5.1 x10^{4}
BCF: estimated at 4.01 to 4.4

Regulations

RCRA 40CFR: Not listed
CERCLA: 40CFR 302.4: Not listed
SARA 40CFR 372.65: Listed
SARA EHS 40CFR 355: Not listed
TSCA: Not listed

Analytical Methods

Soil: SW846 8100

BEN4680 CAS #: 207-08-9

BENZO(K)FLUORANTHENE

RTECS: DF6350000
EINECS Number: 205-916-6
Molecular Formula: $C_{20}H_{12}$
Formula Weight: 252.32

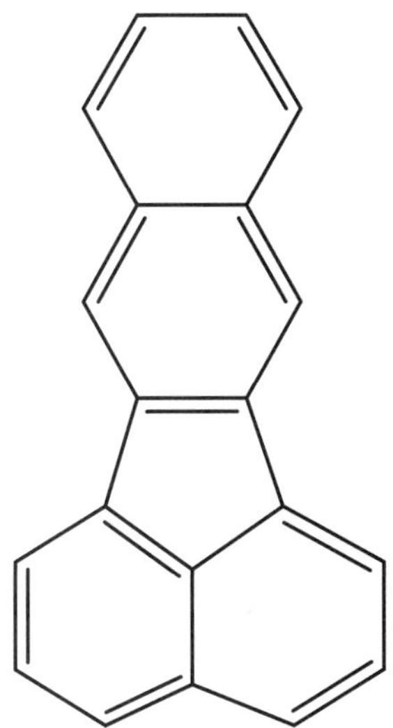

Chemical Structure

Synonyms: B; B (K) F; B K F; 8,9-BENZFLUORANTHENE; BENZO(K) FLUORANTHENE; 11,12-BENZO(K)FLUORANTHENE; 11,12-BENZOFLUORANTHENE; 8,9-BENZOFLUORANTHENE; 11,12-BENZOFLURANTHENE; 2,3,1',8'-BINAPHTHYLENE; 2,3,1',8'-BINAPTHYLENE; BKF; DIBENZO(B,JK)FLUORENE
Description: pale yellow needles
Use: there is no commerical use of this compound

Physical Properties

Boiling Point: 480 °C (896 °F) at 760 mm Hg
Freezing Point: 217 °C (422.6 °F)
Saturated Vapor Density: 1.2 kg/m^3
Vapor Pressure: 0.0000000000959 mm Hg at 25 °C
Water Solubility: Insoluble in Water
Other Solubilities: 95% Ethanol: <1 mg/ml at 20 °C; Acetone: 1-10 mg/ml at 20 °C; Acetic Acid: Soluble; Benzene: Soluble; DMSO: <1 mg/ml at 20 °C; Methanol: <1 mg/ml at 20 °C; Toluene: 5-10 mg/ml at 20 °C.
Flash Point: Not available; probably combustible

RTECS Toxicity Data

Mutagenic: Bacteria - S Typhimurium Mutations in Microorganisms; Dose: 10 ug/plate (-S9).
Tumorigenic: Rat Route: Implant; Dose: 5 mg/kg; Toxic Effects: Tumorigenic - Equivocal tumorigenic agent by RTECS criteria; Lungs, Thorax, or Respiration - Tumors; Tumorigenic - Tumors at site of application. Mouse Route: Skin; Dose: 2820 mg/kg/47W-I; Toxic Effects: Tumorigenic - Equivocal tumorigenic agent by RTECS criteria; Skin and appendages - Tumors; Tumorigenic - Tumors at site of application.

Hazard Overviews

Health: Irritating to eyes/skin/respiratory tract. Toxic. Other Acute Effects: may be harmful by inhalation, ingestion, or skin absorption. Chronic Effects: Probable human carcinogen.
Fire: Will burn. Hazards: emits toxic fumes. Extinguishing agents: water spray; carbon dioxide, dry chemical powder or appropriate foam. Precautions: combustible liquid.
Reactivity: Incompatible with: strong oxidizing agents. Hazardous decomposition products: toxic fumes of: carbon monoxide, carbon dioxide.
Carcinogenicity: IARC - Group 2B, Possibly carcinogenic to humans; NIOSH - Not listed; NTP - Listed; ACGIH - Not listed; OSHA - Not listed; EPA - Class B2, Probable human carcinogen based on animal studies; MAK - Not listed
Primary Target Organs:

Eyes

Skin

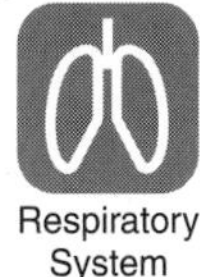
Respiratory System

Environmental

Environmental Fate: Its presence in distant places indicates that it is reasonably stable in the atmosphere and capable of long distant transport. Atmospheric losses are caused by gravitational settling and rainout. On land it is strongly adsorbed to soil and remains in the upper soil layers and should not leach into groundwater. Biodegradation may occur but will be very slow (half-life ca 2 years with acclimated microorganisms). It will get into surface water from dust and precipitation in addition to runoff and effluents. In the water it will sorb to sediment and particulate matter in the water column. It would be expected to bioconcentrate in fish and seafood.

Environmental Physical Data

Henry's Law Constant: estimated at 4.2 x10^{8}
Octanol/Water Partition Coefficient: log K_{ow} = 6.84
Sorption Partition Coefficient: K_{oc} = nearly 1 x10^{6}
BCF: fish 4.97

Regulations

RCRA 40CFR: Not listed
CERCLA: 40CFR 302.4: Listed per CWA Section 307(a) RQ: 5000 lb (2268 kg)
SARA 40CFR 372.65: Listed
SARA EHS 40CFR 355: Not listed
TSCA: Not listed

Analytical Methods

Air: EPA TO-13; California 429
Soil: CLP LC_SV, MC_SVOA, OHC; EPA 16, 1625, PAH-005, PAH-007, PAH-011, PAH-012; SW846 3630B, 3640A, 8100, 8250A, 8270B, 8270C, 8275, 8275A, 8310
Water / Groundwater: EPA PAH-002, PAH-006, 1625, 610, 625, 625-S, 6; APHA 6410-B, 6440-B, 6440-C; ASTM D4657; USGS O3118
Drinking Water: EPA 525.1, 525.2, 550, 550.1
Indoor / Expired Air: NIOSH 5506, 5515; EPA IP-7-A, IP-7-B
Plasma: EPA 29
Other: EPA PAH-009

BEN4760 **CAS #: 206-44-0**

BENZO(J,K)FLUORENE

RTECS: LL4025000
EINECS Number: 205-912-4
Molecular Formula: $C_{16}H_{10}$
Formula Weight: 202.26

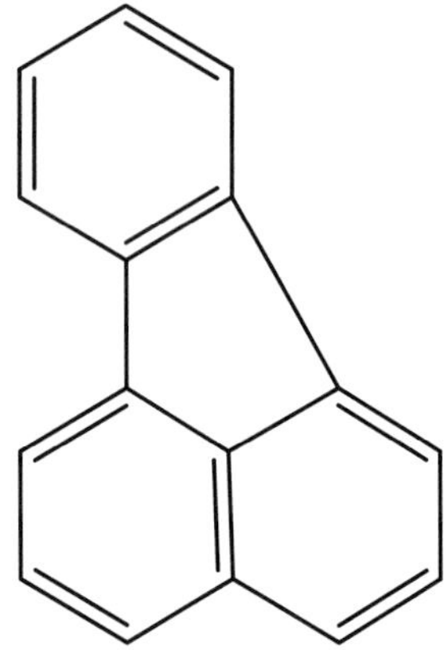

Chemical Structure

Synonyms: 1,2-BENZACENAPHTHENE; BENZENE,1,2-(1,8-NAPHTHALENEDIYL)-; BENZENE,1,2-(1,8-NAPHTHYLENE)-; BENZO (J,K) FLUORENE; BENZO(JK)FLUORENE; IDRYL; 1,2-(1,8-NAPHTHALENE)BENZENE; 1,2-(1,8-NAPHTHALENEDIYL)BENZENE; 1,2-(1,8-NAPHTHYLENE)BENZENE
Description: pale yellow needles or plates

Physical Properties

Boiling Point: About 375 °C (707 °F)
Freezing Point: 111 °C (231.8 °F)
Specific Gravity: 1.252 at 0 °C/4 °C
Saturated Vapor Density: 1.200094334 kg/m^3
Vapor Pressure: 0.01 mm Hg at 20.0 °C
Water Solubility: < 1 mg/mL at 18 C
Other Solubilities: 95% Ethanol: 5-10 mg/ml at 22 °C; Acetone: >=100 mg/ml at 22 °C; Benzene: Soluble; Carbon Disulfide: Soluble; Chloroform: Soluble; DMSO: >=100 mg/ml at 22 °C; Ether: Soluble.
Ionization Potential (eV): 7.95 +/-0.3
Flash Point: Not available; probably combustible

RTECS Toxicity Data

Acute Oral: Rat LD_{50} Dose: 2 gm/kg.
Acute Dermal: Rabbit LD_{50} Route: Skin; Dose: 3180 mg/kg.
Mutagenic: Human Mutations in Mammalian Somatic Cells; Cell Type: lymphocyte; Dose: 2 umol/L. Rat Morphological Transformation; Cell Type: embryo; Dose: 50 mg/L.
Tumorigenic: Mouse Route: Skin; Dose: 280 mg/kg/58W-I; Toxic Effects: Tumorigenic - Equivocal tumorigenic agent by RTECS criteria; Skin and appendages - Tumors; Tumorigenic - Tumors at site of application.

Hazard Overviews

Health: Irritating to eyes/skin/respiratory tract. Harmful. Other Acute Effects: harmful if swallowed, inhaled, or absorbed through skin. Chronic Effects: laboratory experiments have shown mutagenic effects.
Fire: Will burn. Hazards: emits toxic fumes. Extinguishing agents: water spray; carbon dioxide, dry chemical powder or appropriate foam. Precautions: combustible liquid.
Reactivity: Incompatible with: strong oxidizing agents. Hazardous decomposition products: toxic fumes of: carbon monoxide, carbon dioxide.
Carcinogenicity: IARC - Group 3, Not classifiable as to carcinogenicity to humans; NIOSH - Not listed; NTP - Not listed; ACGIH - Not listed; OSHA - Not listed; EPA - Class D, Not classifiable as to human carcinogenicity; MAK - Not listed
Primary Target Organs:

Eyes

Skin

Respiratory System

Environmental

Ecotoxicity: EC_{50} Selenastrum capricornutum (alga) 54,400 ug/l/96 hr, toxic effect: cell numbers; 54,600 ug/l/96 hr, toxic effect: chlorophyll a /Static conditions unmeasured EC_{50} Skeletonema costatum (alga) 45,000 ug/l/96 hr, toxic effect: chlorophyll a; 45,600 ug/l/96 hr, toxic effect: cell numbers /Static conditions unmeasured LC_{50} Cyprinodon variegatus (sheepshead minnow) > 560,000 ug/l/96 hr /Static conditions unmeasured LC_{50} Mysid shrimp 40 ug/l/96 hr. /Static conditions unmeasured LC_{50} Bluegill 3,980 ug/l/96 hr. /Static conditions unmeasured
Environmental Fate: When released into water, it will rapidly become adsorbed to sediment and particulate matter in the water column, and bioconcentrate into aquatic organisms. In fact, concentrations in shellfish such as clams and mussels are

an excellent indicator of pollution in a localized area. In the unadsorbed state it will degrade by photolysis (half-life days to wk). It appears to be stable in sediment for decades or more. Because it is strongly adsorbed to soil, it should remain in the upper few centimeters of the soil. However, its detection in groundwater demonstrates that it can be transported there by some process(es). It should biodegrade in a few years in the presence of acclimated microorganisms. Released in the atmosphere it will photodegrade in the free state (half-life 4-5 days). Aerosols and particulate matter containing sorbed compound is sufficiently stable to be transported long distances while being subject to gravitational settling and rainout. Photochemical smog situations enhance the degradation of both the sorbed molecule and the free vapor.

Environmental Physical Data

Octanol/Water Partition Coefficient: log K_{ow} = 4.90
Sorption Partition Coefficient: K_{oc} = 6.6 x10^4
BCF: rainbow trout 2.58

Regulations

RCRA 40CFR: Listed Hazardous Waste No. U120 Toxic Waste
CERCLA: 40CFR 302.4: Listed per RCRA Section 3001 per CWA Section 307(a) RQ: 100 lb (45.35 kg)
SARA 40CFR 372.65: Not listed
SARA EHS 40CFR 355: Not listed
TSCA: Listed

Analytical Methods

Air: EPA TO-13; California 429
Soil: CLP LC_SV, MC_SVOA, OHC; EPA 16, 1625, PAH-005, PAH-007, PAH-011, PAH-012, S-004-1; SW846 1311, 3630B, 3640A, 8100, 8250A, 8270B, 8270C, 8275A, 8310, 8410; DOE OS050
Water / Groundwater: EPA PAH-002, PAH-006, S-002-1, 1625, 610, 625, 625-S, 6; APHA 6040-B, 6410-B, 6440-B, 6440-C; ASTM D4657, D4763; USGS O3113, O3118
Drinking Water: EPA 550, 550.1
Indoor / Expired Air: NIOSH 5506, 5515; EPA IP-7-A, IP-7-B
Plasma: EPA 29
Other: EPA PAH-009

BEN4840 CAS #: 271-89-6

BENZOFURAN

RTECS: DF6423800
EINECS Number: 205-982-6
Molecular Formula: C_8H_6O
Formula Weight: 118.14

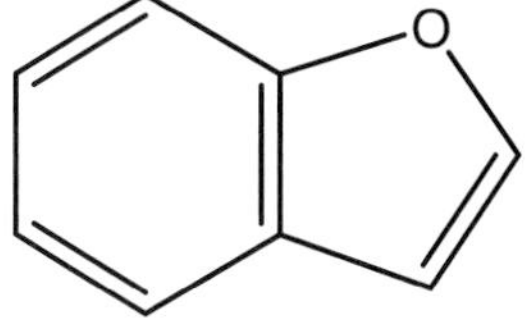

Chemical Structure

Synonyms: 2,3-BENZOFURAN; BENZO(B)FURAN; BENZOFURFURAN; COUMARON; COUMARONE; CUMARONE; 1-OXIDENE; 1-OXINDENE
Description: coloress, oily liquid; aromatic odor
Use: manufacture of coumarone-indene resins

Physical Properties

Boiling Point: 173 °C (343 °F) to 175 °C (347 °F) at 760 mm Hg
Freezing Point: < -18 °C (-0.4 °F)
Specific Gravity: 1.0913 at 22.7 °C/4 °C
Saturated Vapor Density: 1.269403013 kg/m^3
Density: 1.0913 g/mL at 25 °C
Vapor Pressure: 14.3 mm Hg at 25 °C
Water Solubility: Insoluble in Water
Other Solubilities: 95% Ethanol: 10-50 mg/ml at 20 °C; Acetone: >=100 mg/ml at 20 °C; Aqueous alkali solution: Insoluble; Benzene: miscible; DMSO: >=100 mg/ml at 20 °C; Ether: Soluble; Petroleum Ether: miscible.
Refraction Index: 1.56897 at 16.93 °C/D
Ionization Potential (eV): 8.37 +/-2.0
Flash Point: 56 °C

RTECS Toxicity Data

Chronic (Multiple Dose) Oral: Rat Dose: 7 gm/kg/14D-I; Toxic Effects: DEATH. Rat Dose: 16250 mg/kg/13W-I; Toxic Effects: Kidney, Ureter, and Bladder - Chgs in tubules (inc acute renal failure, acute tubular necrosis; Kidney, Ureter, and Bladder - Changes in both tubules and glomeruli.
Mutagenic: Mouse Mutations in Mammalian Somatic Cells; Cell Type: lymphocyte; Dose: 100 mg/L. Mouse Other Mutation Test Systems; Route: Oral; Dose: 100 mg/kg.
Tumorigenic: Rat Route: Oral; Dose: 61800 mg/kg/2Y-C; Toxic Effects: Tumorigenic - Carcinogenic by RTECS criteria; Kidney, Ureter, and Bladder - Kidney tumors. Mouse Route: Oral; Dose: 30900 mg/kg/2Y-C; Toxic Effects: Tumorigenic - Carcinogenic by RTECS criteria; Gastrointestinal - Tumors; Liver - Tumors.

Hazard Overviews

Flammable

Health: Irritating. Toxic. Other Acute Effects: harmful if swallowed, inhaled, or absorbed through skin. Chronic Effects: carcinogen; target organs: liver, kidneys, lungs, pancreas.
Fire: Flammable. Hazards: emits toxic fumes; vapor may travel considerable distance to source of ignition and flash back. Extinguishing agents: carbon dioxide, dry chemical powder or appropriate foam; water spray. Precautions: combustible liquid.

Reactivity: Incompatible with: strong oxidizing agents, may discolor on exposure to light. Hazardous decomposition products: toxic fumes of: carbon monoxide, carbon dioxide.
Carcinogenicity: IARC - Group 2B, Possibly carcinogenic to humans; NIOSH - Not listed; NTP - Not listed; ACGIH - Not listed; OSHA - Not listed; EPA - Not listed; MAK - Not listed
Primary Target Organs:

Eyes

Skin
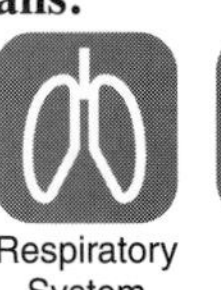
Respiratory System
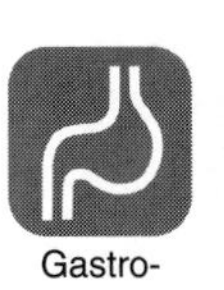
Gastro-intestinal

Liver

Kidneys

Environmental

Ecotoxicity: Fishes: Pimephales promelas: 4d LC_{50} 14 mg/l
Environmental Fate: If released to the atmosphere, it will exist primarily in the vapor phase where it will degrade relatively rapidly by reaction with photochemically formed hydroxyl radicals (estimated half-life of 4.3 hr). Adsorption to airborne dusts has been demonstrated; this mechanism could result in long range airborne transport. If released to water, volatilization may be important. Volatilization half-lives from a model river (1 meter deep) and a model environmental pond have been estimated to be 5 hr and 88 hr, respectively. If released to soil, moderate leaching may be possible based upon an estimated K_{oc} of 288. It has been detected in several ground waters demonstrating that environmental leaching can occur. Insufficient data are available to predict the importance of abiotic or biotic degradation processes in the soil or water environments.
Cleanup/Disposal: Guide No. 128: Eliminate all ignition sources (no smoking, flares, sparks or flames in immediate area). All equipment used when handling the product must be grounded. Do not touch or walk through spilled material. Stop leak if you can do it without risk. Prevent entry into waterways, sewers, basements or confined areas. A vapor suppressing foam may be used to reduce vapors. Absorb or cover with dry earth, sand or other non-combustible material and transfer to containers. Use clean non-sparking tools to collect absorbed material. Large Spills: Dike far ahead of liquid spill for later disposal. Water spray may reduce vapor; but may not prevent ignition in closed spaces.

Environmental Physical Data

Henry's Law Constant: 5.25 x10^{-4}
Octanol/Water Partition Coefficient: log K_{ow} = 2.67
Sorption Partition Coefficient: K_{oc} = estimated at 288
BCF: estimated at 63

Regulations

RCRA 40CFR: Not listed
CERCLA: 40CFR 302.4: Not listed
SARA 40CFR 372.65: Not listed
SARA EHS 40CFR 355: Not listed
TSCA: Listed

Analytical Methods

Drinking Water: APHA 6220-C

BEN4920 CAS #: 65-85-0

BENZOIC ACID

RTECS: DG0875000
EINECS Number: 200-618-2
Molecular Formula: $C_7H_6O_2$
Structured MF: C_6H_5COOH
Formula Weight: 122.13

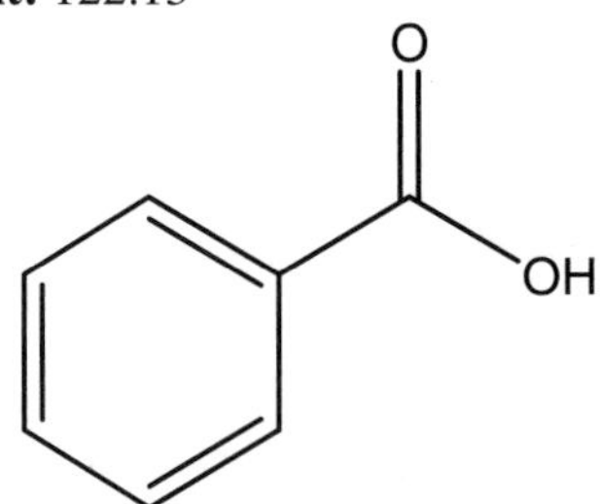

Chemical Structure

Synonyms: ACIDE BENZOIQUE; ACIDO BENZOICO; BENZENE CARBOXYLIC ACID; BENZENE FORMIC ACID; BENZENECARBOXYLIC ACID; BENZENEFORMIC ACID; BENZENEMETHANOIC ACID; BENZENEMETHONIC ACID; BENZOATE; BENZOESAEURE; CARBOXYBENZENE; DIACYCLIC ACID; DRACYLIC ACID; EPA PESTICIDE CHEMICAL CODE 009101; FLOWERS OF BENJAMIN; FLOWERS OF BENZOIN; HA 1; KYSELINA BENZOOVA; PHENYL CARBOXYLIC ACID; PHENYLCARBOXYLIC ACID; PHENYLFORMIC ACID; RETARDED BA; RETARDER BA; RETARDEX; SALVO LIQUID; SALVO POWDER; SOLVO POWDER; TENN-PLAS; TENNPLAS; UNISEPT BZA
Description: colorless to white, tan crystals, powder, platelets, scale; benzoin- or benzaldehyde-like odor
Use: preserving foods, fats, fruit juices, alkaloidal solutions, etc.; manufacture of benzoates, benzoyl compounds, and dyes; as a mordant in calico printing; for curing tobacco; as a standard in volumetric and calorimetric analysis; a pharmaceutic aid (antifungal agent); in plasticizers, perfumes, and dentifrices

Physical Properties

Boiling Point: 249.2 °C (481 °F) at 760 mm Hg
Freezing Point: 122.4 °C (252.32 °F)
Specific Gravity: 1.2659 at 15 °C/4 °C
Vapor Density: 4.21 Air=1
Density: 1.321 g/mL at 19 °C
Vapor Pressure: 1 mm Hg at 96.0 °C
Water Solubility: 1 g/300 ml Water
Other Solubilities: 95% Ethanol: >=100 mg/ml at 20 °C; Acetone: >=100 mg/ml at 20 °C; Benzene: Soluble; Carbon Disulfide: 1 g/30 mL; Carbon Tetrachloride: 1 g/30 mL; Chloroform: 1 g/4.5 mL; DMSO: >=100 mg/ml at 20 °C; Ether: Very Soluble.
Surface Tension: 30 dynes/cm at 130 °C
pH: Saturated solution at 25 °C 2.8
Refraction Index: 1.504 at 132 °C/D
Critical Temperature: 479 °C
Critical Pressure: 45 atm
Ionization Potential (eV): 9.3

Flash Point: 121.111 °C Closed Cup
Autoignition Temperature: 570 °C

RTECS Toxicity Data

Acute Oral: Man LD_{Lo} Dose: 500 mg/kg. Rat LD_{50} Dose: 1700 mg/kg.
Acute Inhalation: Rat LC_{50} Dose: >26 mg/m^3/1hr; Toxic Effects: Sense organs and special senses - Lacrimation; Behavioral - Somnolence (general depressed activity).
Acute Dermal: Human TD_{Lo} Route: Skin; Dose: 6 mg/kg; Toxic Effects: Lungs, Thorax, or Respiration - Dyspnea; Skin and appendages - Dermatitis, allergic. Rabbit LD_{50} Route: Skin; Dose: >10 gm/kg.
Irritation Eye: Rabbit Standard Draize Test Dose: 100 mg; Reaction: severe.
Irritation Skin: Human Standard Draize Test Dose: 22 mg/3D-I; Reaction: moderate. Rabbit Standard Draize Test Dose: 500 mg/24H; Reaction: mild.
Mutagenic: Human DNA Inhibition; Cell Type: lymphocyte; Dose: 5 mmol/L. Bacteria - E Coli Mutations in Microorganisms; Dose: 10 mmol/L (-S9).

Hazard Overviews

Fire Diamond

Health: Severely irritating to eyes/skin/respiratory tract. Also Causes: nausea, GI disorders, allergic response.
Fire: Will burn. Use agent suitable for surrounding fire. High concentrations of benzoic acid dust particles in the air or vapors from hot or molten benzoic acid may form an explosive mixture with air.
Reactivity: Stable. Hazardous polymerization cannot occur. Avoid: exposure to heat and ignition sources. Incompatible with: oxidizers; alkalis; heavy metals. Hazardous decomposition products: carbon oxide(s).
Carcinogenicity: IARC - Not listed; NIOSH - Not listed; NTP - Not listed; ACGIH - Not listed; OSHA - Not listed; EPA - Class D, Not classifiable as to human carcinogenicity; MAK - Not listed
Primary Target Organs:

Eyes

Skin
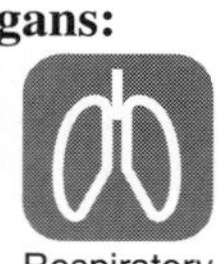
Respiratory System

Mucous Membranes

Environmental

Ecotoxicity: Aquatic toxicity: 200 ppm/7 hr/goldfish/lethal/fresh water 500 ppm/1 hr/sunfish/lethal/fresh water
Environmental Fate: If released on land, it should leach into the ground due to its low soil adsorption and biodegrade (half-life <1 wk). If released in water, it should also readily biodegrade (half-life 0.2-3.6 days). Adsorption to sediment and volatilization should not be significant. While bioconcentration in fish and algae is not important, there is some evidence that bioconcentration in aquatic species like daphnia and snails may be considerable. In the atmosphere, it will be largely associated with aerosols, be subject to gravitational settling, and be scavenged by rain.
Cleanup/Disposal: Guide No. 171: Do not touch or walk through spilled material. Stop leak if you can do it without risk. Prevent dust cloud. Avoid inhalation of asbestos dust. Small Dry Spills: With clean shovel place material into clean, dry container and cover loosely; move containers from spill area. Small Spills: Take up with sand or other noncombustible absorbent material and place into containers for later disposal. Large Spills: Dike far ahead of liquid spill for later disposal. Cover powder spill with plastic sheet or tarp to minimize spreading. Prevent entry into waterways, sewers, basements or confined areas.

Environmental Physical Data

Henry's Law Constant: calculated at 7.0 x10^{-8}
Octanol/Water Partition Coefficient: log K_{ow} = 1.87
BCF: golden ide 10
BOD: 165%, 5 days

Regulations

RCRA 40CFR: Not listed
CERCLA: 40CFR 302.4: Listed per CWA Section 311(b)(4) RQ: 5000 lb (2268 kg)
SARA 40CFR 372.65: Not listed
SARA EHS 40CFR 355: Not listed
TSCA: Listed

Analytical Methods

Soil: EPA 1625; SW846 3640A, 8250A, 8270B, 8270C, 8410
Water / Groundwater: EPA 1625
Plasma: EPA 29

BEN5080 CAS #: 93-97-0

BENZOIC ANHYDRIDE

EINECS Number: 202-291-1
Molecular Formula: $C_{14}H_{10}O_3$
Formula Weight: 226.23

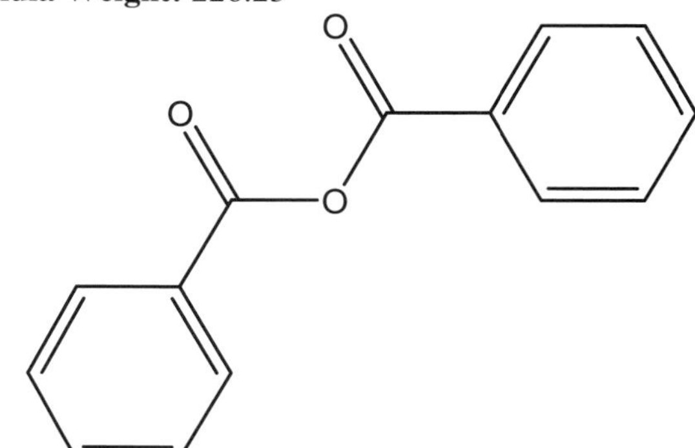

Chemical Structure

Physical Properties

Boiling Point: 360 °C (680 °F)

Freezing Point: 43 °C (109.4 °F)
Specific Gravity: 1.989
Water Solubility: Slightly
Other Solubilities: Ethanol: Soluble; Ether: Soluble; Chloroform: Slightly soluble; ligroin: Insoluble
Refraction Index: 1.5767

Hazard Overviews

Health: Irritating to eyes/skin/respiratory tract. Other Acute Effects: may be harmful by inhalation, ingestion, or skin absorption.
Fire: Extinguishing agents: water spray; carbon dioxide, dry chemical powder or appropriate foam. Precautions: combustible liquid.
Carcinogenicity: IARC - Not listed; NIOSH - Not listed; NTP - Not listed; ACGIH - Not listed; OSHA - Not listed; EPA - Not listed; MAK - Not listed
Primary Target Organs:

Eyes

Skin

Respiratory System

Environmental

Regulations

RCRA 40CFR: Not listed
CERCLA: 40CFR 302.4: Not listed
SARA 40CFR 372.65: Not listed
SARA EHS 40CFR 355: Not listed
TSCA: Not listed

BEN5160 CAS #: 119-53-9

BENZOIN

RTECS: DI1590000
EINECS Number: 204-331-3
Molecular Formula: $C_{14}H_{12}O_2$
Structured MF: $C_6H_5CH(OH)COC_6H_5$
Formula Weight: 212.22

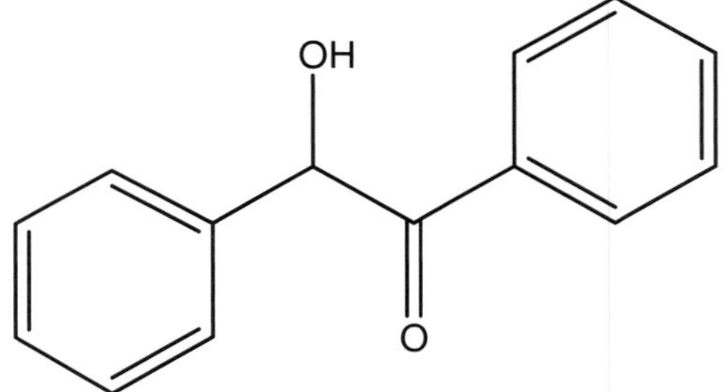

Chemical Structure

Synonyms: ACETOPHENONE,2-HYDROXY-2-PHENYL-; BENZOYL PHENYLCARBINOL; BENZOYLPHENYLCARBINOL; BITTER ALMOND OIL CAMPHOR; BITTER ALMOND-OIL CAMPHOR; BITTER-ALMOND-OIL CAMPHOR; ETHANONE,2-HYDROXY-1,2-DIPHENYL-; FENYL-ALPHA-HYDROXYBENZYLKETON; ALPHA-HYDROXYBENZYL PHENYL KETONE; 2-HYDROXY-1,2-DIPHENYLETHANONE; 2-HYDROXY-2-PHENYLACETOPHENONE; ALPHA-HYDROXY-ALPHA-PHENYLACETOPHENONE; KETONE,ALPHA-HYDROXYBENZYL PHENYL; PHENYLBENZOYL CARBINOL; WY-42956; WY 42956
Description: pale yellow to white crystals; sweet non-descript odor
Use: in organic synthesis; as an intermediate, photopolymerization catalyst and ingredient of inhalants; internally as an expectorant and topically in various preparations as an antiseptic and protective agent; common ingredient in vaporizer fluid marketed for inhalation

Physical Properties

Boiling Point: 344 °C (651 °F)
Freezing Point: 137 °C (278.6 °F)
Specific Gravity: 1.31 at 20/4 °C
Water Solubility: < 0.1 mg/mL at 18 C
Other Solubilities: 95% Ethanol: <1 mg/ml at 20.5 °C; Acetone: 10-50 mg/ml at 20.5 °C; Chloroform: >10%; DMSO: 50-100 mg/ml at 20.5 °C; Ether: Slightly Soluble; Pyridine: Soluble in 5 parts.
Flash Point: Not available; probably combustible

RTECS Toxicity Data

Acute Oral: Rat LD_{50} Dose: 10 gm/kg. Mouse LD_{50} Dose: >3 gm/kg; Toxic Effects: Behavioral - Somnolence (general depressed activity).
Acute Dermal: Rabbit LD_{50} Route: Skin; Dose: 8870 mg/kg.
Mutagenic: Rat DNA Damage; Cell Type: liver; Dose: 63700 ug/L. Rat DNA Damage; Route: Intraperitoneal; Dose: 500 mg/kg.

Hazard Overviews

Health: May be irritating to eyes/skin. Other Acute Effects may be harmful by inhalation, ingestion, or skin absorption.
Fire: Will burn. Hazards: emits toxic fumes. Extinguishing agents: water spray; carbon dioxide, dry chemical powder or appropriate foam. Precautions: combustible liquid.
Reactivity: Incompatible with: strong oxidizing agents. Hazardous decomposition products: toxic fumes of: carbon monoxide, carbon dioxide.
Carcinogenicity: IARC - Not listed; NIOSH - Not listed; NTP - Not listed; ACGIH - Not listed; OSHA - Not listed; EPA - Not listed; MAK - Not listed

Environmental

Regulations

RCRA 40CFR: Not listed
CERCLA: 40CFR 302.4: Not listed
SARA 40CFR 372.65: Not listed
SARA EHS 40CFR 355: Not listed
TSCA: Listed

BEN5320 **CAS #: 100-47-0**

BENZONITRILE

RTECS: DI2450000
DOT: UN2224; IMO6.1
EINECS Number: 202-855-7
Molecular Formula: C_7H_5N
Structured MF: C_6H_5CN
Formula Weight: 103.12

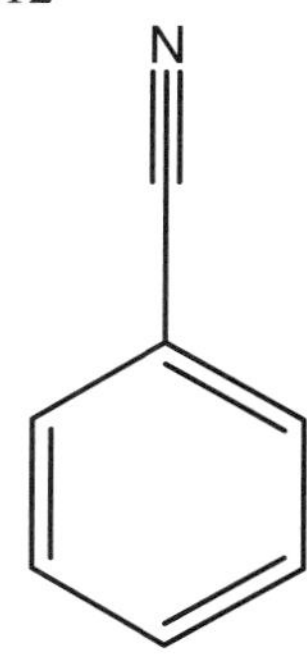

Chemical Structure

Synonyms: BENZENECARBONITRILE; BENZENE,CYANO-; BENZENENITRILE; BENZOIC ACID NITRILE; CYANOBENZENE; FENYLKYANID; PHENYL CYANIDE; PHENYLCYANIDE
Description: colorless liquid; volatile almond oil
Use: solvent and chemical intermediate in the pharmaceutical, dyestuffs and rubber industries; in the manufacture of benzoguanamine, specialty lacquers, many resins and polymers and many anhydrous metallic salts

Physical Properties

Boiling Point: 190.7 °C (375 °F) at 760 mm Hg
Freezing Point: -13 °C (8.6 °F)
Specific Gravity: 1.01 at 15 °C/15 °C
Vapor Density: Vapor Specific Gravity 3.6
Saturated Vapor Density: 1.203099319 kg/m^3
Density: 1.246 g/mL at 20 °C
Vapor Pressure: 0.768 mm Hg at 25 °C
Water Solubility: Slightly Soluble in Cold Water
Other Solubilities: 95% Ethanol: >=100 mg/ml at 23 °C; Acetone: >=100 mg/ml at 23 °C; Alcohol: Soluble; Benzene: Soluble; Chloroform: Soluble; Common organic solvents: miscible; Corn oil: 10 mg/ml; DMSO: >=100 mg/ml at 23 °C; Ether: Soluble.
Surface Tension: 34.7 dynes/cm
Odor Threshold: 2.90 $x10^{-5}$ mg/l
Refraction Index: 1.5289 at 20 °C/D
Critical Temperature: 426.2 °C
Critical Pressure: 41.6 atm
Ionization Potential (eV): 9.69
Flash Point: 75 °C Closed Cup
Autoignition Temperature: 550 °C
LEL: 0.9% v/v
UEL: 11.3% v/v

RTECS Toxicity Data

Acute Oral: Rat LD_{Lo} Dose: 720 mg/kg. Mouse LD_{50} Dose: 971 mg/kg.
Acute Inhalation: Mouse LC_{50} Dose: 1800 mg/m^3. Rat LC_{Lo} Dose: 950 ppm/8hr; Toxic Effects: Lungs, Thorax, or Respiration - Other changes.
Acute Dermal: Rabbit LD_{50} Route: Skin; Dose: 1250 mg/kg. Rabbit LD_{Lo} Route: Subcutaneous Dose: 200 mg/kg; Toxic Effects: Peripheral Nerve and sensation - Spastic parapysis with or without sensory change; Behavioral - Somnolence (general depressed activity); Lungs, Thorax, or Respiration - Dyspnea.
Chronic (Multiple Dose) Inhalation: Rat Dose: 70 mg/m^3/4H/19W-I; Toxic Effects: Liver - Liver function tests impaired; Blood - Normocytic anemia; Biochemical - Cytochrome oxidases (including oxidative phosphorylation). Rabbit Dose: 70 mg/m^3/4H/19W-I; Toxic Effects: Nutritional and gross metabolic - Weight loss or decreased weight gain.
Irritation Skin: Rabbit Standard Draize Test Dose: 500 mg/24H; Reaction: moderate.
Mutagenic: Yeast - S Cerevisiae Sex Chromosome Loss; Dose: 1580 mg/L.

Hazard Overviews

Fire Diamond

Health: Irritating to the skin. Also Causes: headaches, confusion, flushing, cyanide poisoning from decomposition products.
Fire: Combustible. Gives off highly toxic decomposition products. Use foam, dry chemical, or carbon dioxide. Use water spray to cool fire-exposed containers to prevent pressure buildup.
Reactivity: Stable. Hazardous polymerization cannot occur. Incompatible with: oxidizing agents. Hazardous decomposition products: hydrogen cyanide; isocyanates; benzoic acid; carbon monoxide; carbon dioxide; oxides of nitrogen.
Carcinogenicity: IARC - Not listed; NIOSH - Not listed; NTP - Not listed; ACGIH - Not listed; OSHA - Not listed; EPA - Not listed; MAK - Not listed
Primary Target Organs:

Skin

Respiratory System

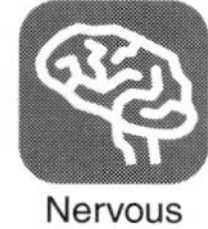
Nervous System

Environmental

Ecotoxicity: Toxicity Threshold (Cell Multiplication Inhibition Test) Pseudomonas putida (bacteria) 11 mg/1 TLm Fathead minnow 78 mg/l/96 hr (hard water) /Conditions of bioassay not specified Toxicity Threshold (Cell Multiplication Inhibition Test) Microcystis aeruginosa (algae) 3.4 mg/l

Environmental Fate: If released to soil or water, biodegradation is expected to be a major fate process. It has been shown to biodegrade readily in various screening studies. Transport from water or soil to the atmosphere can occur through volatilization. Based upon estimated K_{oc} values of 30-70, it may leach readily in soil. If released to the atmosphere, it will degrade by reaction with photochemically produced hydroxyl radicals (estimated half-life of 48 days). Physical removal from air via wet deposition is possible.

Cleanup/Disposal: Guide No. 152: Do not touch damaged containers or spilled material unless wearing appropriate protective clothing. Stop leak if you can do it without risk. Prevent entry into waterways, sewers, basements or confined areas. Cover with plastic sheet to prevent spreading. Absorb or cover with dry earth, sand or other non-combustible material and transfer to containers. Do not get water inside containers.

Environmental Physical Data

Henry's Law Constant: estimated at 5.21×10^{-5}

Octanol/Water Partition Coefficient: log K_{ow} = 1.56

Sorption Partition Coefficient: K_{oc} = estimated at 30 to 70

BCF: estimated at 9

BOD: theoretical 60%, 18 days

Regulations

RCRA 40CFR: Not listed

CERCLA: 40CFR 302.4: Listed per CWA Section 311(b)(4) RQ: 5000 lb (2268 kg)

SARA 40CFR 372.65: Not listed

SARA EHS 40CFR 355: Not listed

TSCA: Listed

Analytical Methods

Water / Groundwater: ASTM D3371, D4763

Indoor / Expired Air: EPA IP-1B

BEN5400 CAS #: 191-24-2

BENZO(G,H,I)PERYLENE

RTECS: DI6200500

EINECS Number: 205-883-8

Molecular Formula: $C_{22}H_{12}$

Formula Weight: 276.34

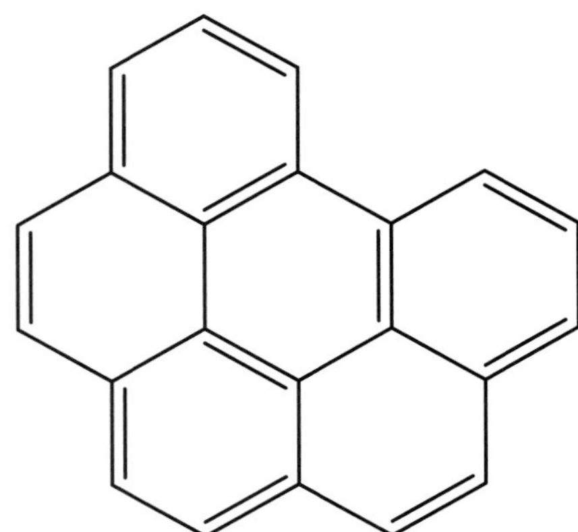

Chemical Structure

Synonyms: BENZO (G,H,I) PERYLENE; BENZO(GHI)PERYLENE; 1,12-BENZOPERYLENE; BENZO(G,H,I)PERYLENE; 1,12-BENZPERYLENE

Description: pale yellow-green large plates

Use: scientific research

Physical Properties

Boiling Point: 550 °C (1022 °F) at 760 mm Hg

Freezing Point: 277 °C (530.6 °F)

Vapor Pressure: 1.0×10^{-10} mm Hg at 25 °C

Water Solubility: 0.00025 to 0.00027 mg/L at 25 °C

Other Solubilities: Soluble in 1,4-dioxane, dichloromethane, Benzene, & Acetone.

Ionization Potential (eV): 7.15

RTECS Toxicity Data

Mutagenic: Mouse DNA Damage; Route: Skin; Dose: 40 umol/kg. Bacteria - S Typhimurium Mutations in Microorganisms; Dose: 2 ug/plate/48H (-S9).

Hazard Overviews

Health: Irritating to eyes/skin. Other Acute Effects: may be harmful by inhalation, ingestion, or skin absorption. The chemical, physical and toxicological properties of this product have not been thoroughly investigated.

Fire: Hazards: emits toxic fumes. Extinguishing agents: water spray; carbon dioxide, dry chemical powder or appropriate foam. Precautions: combustible liquid.

Reactivity: Incompatible with: strong oxidizing agents. Hazardous decomposition products: toxic fumes of: carbon monoxide, carbon dioxide.

Carcinogenicity: IARC - Group 3, Not classifiable as to carcinogenicity to humans; NIOSH - Not listed; NTP - Not listed; ACGIH - Not listed; OSHA - Not listed; EPA - Class D, Not classifiable as to human carcinogenicity; MAK - Not listed

Primary Target Organs:

Eyes Skin

Environmental

Environmental Fate: Biodegrades slowly in the environment. The reported biodegradation half-lives in aerobic soil range from 600 to 650 days. It is not expected to hydrolyze in the environment. A calculated K_{oc} range of 9 $x10^4$ to 4 $x10^5$ indicates it will be highly immobile in soil. In aquatic systems, it partitions from the water column to organic matter contained in sediments and suspended solids. It also has the potential to bioconcentrate in aquatic systems. A Henry's Law constant of 1.6 $x10^{-6}$ atm-cu m/mole at 25 °C suggests volatilization from shallow, fast moving environmental waters may be important. The volatilization half-lives from a model river and a model pond, the latter considers the effect of adsorption, have been estimated to be 38 days and over 1500 years, respectively. In the atmosphere, the vapor phase reaction with photochemically produced hydroxyl radicals (half-life of 2 hr) may be an important fate process. However, it is expected to exist almost entirely in the particulate phase in ambient air. Nevertheless, it may undergo direct photolysis in the atmosphere. Photolytic half-lives adsorbed onto silica gel, alumina, fly ash and carbon black were 7, 22, 29 and greater than 1000 hours, respectively.

Environmental Physical Data

Henry's Law Constant: calculated at 1.6 $x10^{-6}$
Octanol/Water Partition Coefficient: log K_{ow} = 6.58
Sorption Partition Coefficient: K_{oc} = > 1 $x10^6$
BCF: estimated at 4.44 to 4.77

Regulations

RCRA 40CFR: Not listed
CERCLA: 40CFR 302.4: Listed per CWA Section 307(a) RQ: 5000 lb (2268 kg)
SARA 40CFR 372.65: Not listed
SARA EHS 40CFR 355: Not listed
TSCA: Not listed

Analytical Methods

Air: EPA TO-13; California 429
Soil: CLP LC_SV, MC_SVOA, OHC; DOE OS050; EPA 16, 1625, PAH-005, PAH-007, PAH-011, PAH-012; SW846 3630B, 3640A, 8100, 8250A, 8270B, 8270C, 8275A, 8310
Water / Groundwater: EPA PAH-002, PAH-006, 1625, 610, 625, 625-S, 6; APHA 6410-B, 6440-B, 6440-C; ASTM D4657; USGS O3113, O3118
Drinking Water: EPA 525.1, 525.2, 550, 550.1
Indoor / Expired Air: NIOSH 5506, 5515; EPA IP-7-A, IP-7-B
Plasma: EPA 29
Other: EPA PAH-009

BEN5480 CAS #: 119-61-9

BENZOPHENONE

RTECS: DI9950000
EINECS Number: 204-337-6
Molecular Formula: $C_{13}H_{10}O$
Structured MF: $C_6H_5COC_6H_5$
Formula Weight: 182.21

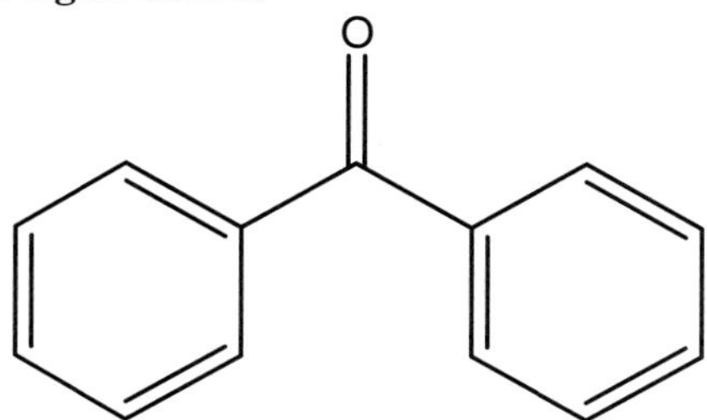

Chemical Structure

Synonyms: BENZENE,BENZOYL-; BENZOYL BENZENE; BENZOYLBENZENE; DIPHENYL KETONE; DIPHENYLKETONE; DIPHENYLMETHANONE; KETONE,DIPHENYL; ALPHA-OXODIPHENYLMETHANE; ALPHA-OXODITANE; PHENYL KETONE
Description: white crystals; rose-like or geranium odor
Use: fixative for heavy perfumes, especially in soaps; in the manufacture of antihistamines, hypnotics; insecticides

Physical Properties

Boiling Point: 305.4 °C (582 °F) at 760 mm Hg
Freezing Point: 48.5 °C (119.3 °F)
Specific Gravity: 1.1108 at 18 °C/4 °C
Water Solubility: Insoluble in Water
Other Solubilities: 95% Ethanol: Soluble (>=10 mg/ml at 25 °C); Acetone: Very Soluble; Acetic Acid: Very Soluble; Benzene: Soluble; CS2: Very Soluble; Chloroform: Very Soluble; DMSO: Soluble (>=10 mg/ml at 25 °C); Ether: Very Soluble; Methanol: Soluble.
Surface Tension: 42 dynes/cm at 50 °C
Refraction Index: 1.5975 at 45.2 °C/D
Ionization Potential (eV): 9.05 +/-0.7
Flash Point: Not available; probably combustible

RTECS Toxicity Data

Acute Oral: Rat LD_{50} Dose: >10 gm/kg. Mouse LD_{50} Dose: 2895 mg/kg; Toxic Effects: Behavioral - Somnolence (general depressed activity); Behavioral - Tremor; Lungs, Thorax, or Respiration - Other changes.
Acute Dermal: Rabbit LD_{50} Route: Skin; Dose: 3535 mg/kg.
Chronic (Multiple Dose) Oral: Rat Dose: 14 gm/kg/28D-C; Toxic Effects: Blood - Pigmented or nucleated red Blood cells; Blood - Changes in serum composition; Blood - Changes in erythrocite (RBC) cell count. Rat Dose: 91 gm/kg/13W-C; Toxic Effects: Nutritional and gross metabolic - Weight loss or decreased weight gain.

Chronic (Multiple Dose) Inhalation: Guinea Pig Dose: 75 mg/kg/15D-I; Toxic Effects: Liver - Other changes; Liver - Hepatitis, fibrous (cirrhosis, post-necrotic scarring).

Hazard Overviews

Fire Diamond

Health: Irritating to eyes/skin. Also Causes: difficulty breathing, respiratory depression, CNS depression, ocular pain, corneal damage, dermatitis, paresthesia.
Fire: Will burn. Extinguish with dry chemicals, alcohol foam, or carbon dioxide. Water may be ineffective on fire.
Reactivity: Stable. Hazardous polymerization cannot occur. Avoid: heat; ignition sources. Incompatible with: oxidizing materials. Hazardous decomposition products: carbon oxides; acrid smoke; irritating fumes.
Carcinogenicity: IARC - Not listed; NIOSH - Not listed; NTP - Not listed; ACGIH - Not listed; OSHA - Not listed; EPA - Not listed; MAK - Not listed
Primary Target Organs:

Eyes

Skin

Nervous System

Exposure Limits
AIHA WEEL: TWA: 5 mg/m^3.
Respirator Recommendation
Exposure Range: >5 to 50 mg/m^3 Air Purifying, Negative Pressure, Half Mask
Exposure Range: >50 to 500 mg/m^3 Air Purifying, Negative Pressure, Full Face
Exposure Range: >500 to 5000 mg/m^3 Supplied Air, Constant Flow/Pressure Demand, Full Face
Exposure Range: >5000 to unlimited mg/m^3 Self-contained Breathing Apparatus, Pressure Demand, Full Face
Cartridge Color: black
Note: use dust/mist prefilter

Environmental

Environmental Fate: If released to soil, it is expected to have low to medium soil mobility (K_{oc}'s of 430 and 517); leaching will be important and adsorption may take place. One aerobic screening study suggests that microbial degradation may be an important fate process in soil and water. Furthermore, one soil column study concludes that anaerobic conditions may inhibit biotic activity. If released to water, photooxidation (half-life of 91 days), photolysis (half-life of greater than 100 days), hydrolysis, and bioconcentration in fish will not be important. Volatilization will be slow; estimated half-life of 26 days from a model river. It may adsorb from the water column to sediment and suspended material and biodegradation in water may occur.
Cleanup/Disposal: Guide No. 128: Eliminate all ignition sources (no smoking, flares, sparks or flames in immediate area). All equipment used when handling the product must be grounded. Do not touch or walk through spilled material. Stop leak if you can do it without risk. Prevent entry into waterways, sewers, basements or confined areas. A vapor suppressing foam may be used to reduce vapors. Absorb or cover with dry earth, sand or other non-combustible material and transfer to containers. Use clean non-sparking tools to collect absorbed material. Large Spills: Dike far ahead of liquid spill for later disposal. Water spray may reduce vapor; but may not prevent ignition in closed spaces.

Environmental Physical Data

Henry's Law Constant: estimated at 1.94 $x10^{-6}$
Octanol/Water Partition Coefficient: log K_{ow} = 3.18
Sorption Partition Coefficient: K_{oc} = 430 to 517
BCF: estimated at 70 to 90
BOD: 12% BODT, 5 days

Regulations

RCRA 40CFR: Not listed
CERCLA: 40CFR 302.4: Not listed
SARA 40CFR 372.65: Not listed
SARA EHS 40CFR 355: Not listed
TSCA: Listed

BEN5560 CAS #: 50-32-8

BENZO(A)PYRENE

RTECS: DJ3675000
EINECS Number: 200-028-5
Molecular Formula: $C_{20}H_{12}$
Formula Weight: 252.30

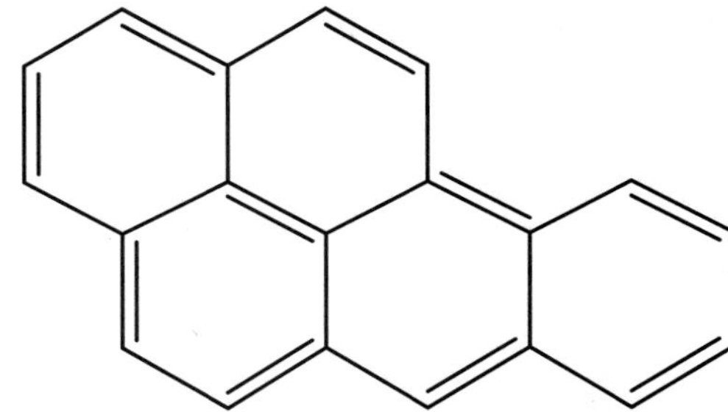

Chemical Structure

Synonyms: B(A)P; BAP; BENZO(D,E,F)CHRYSENE; 3,4-BENZOPIRENE; 1,2-BENZOPYRENE; 3,4-BENZOPYRENE; 6,7-BENZOPYRENE; 3,4-BENZPYREN; 3,4-BENZ(A)PYRENE; 3,4-BENZPYRENE; BENZ(A)PYRENE; 3,4-BENZYLPYRENE; 3,4-BENZYPYRENE; 3,4-BP; BP; COAL TAR PITCH VOLATILES: BENZO(A)PYRENE
Description: pale yellow crystalline solid, powder; aromatic odor
Use: extensively in cancer research

Physical Properties

Boiling Point: > 360 °C (680 °F) at 760 mm Hg
Freezing Point: 179 °C (354.2 °F) to 179.3 °C (354.74 °F)
Specific Gravity: 1.351
Vapor Density: 8.7

Saturated Vapor Density: > 1.212157895 kg/m^3
Density: 1.351 g/mL
Vapor Pressure: > 1 mm Hg at 20 °C
Water Solubility: 0.0038 mg (+/- 000031 mg) in 1 L Water
Other Solubilities: Soluble in Benzene, Toluene, Xylene; Sparingly Soluble in Alcohol, Methanol; Soluble in Ether.
Ionization Potential (eV): 7.12 +/-0.01
Flash Point: Nonflammable

RTECS Toxicity Data

Acute Dermal: Rat LD_{50} Route: Subcutaneous Dose: 50 mg/kg.
Chronic (Multiple Dose) Oral: Rat Dose: 9 gm/kg/90D-I; Toxic Effects: Kidney, Ureter, and Bladder - Chgs in tubules (inc acute renal failure, acute tubular necrosis.
Chronic (Multiple Dose) Dermal: Mouse Route: Subcutaneous; Dose: 400 mg/kg/2W-I; Toxic Effects: Immunological including allergic - Decrease in humoral immune response.
Irritation Skin: Mouse Standard Draize Test Dose: 14 ug; Reaction: mild.
Reproductive/Teratogenic: Rat Route: Oral; Dose: 40 mg/kg; Duration: female 14D of pregnancy; Effects on Embryo or Fetus - Extra embryonic structures; Other effects to embryo or fetus. Rat Route: Oral; Dose: 2 gm/kg; Duration: female 28D prior to mating Effects on Newborn - Stillbirth; Growth statistics. Rat Route: Oral; Dose: 1344 mg/kg; Duration: female 15D prior to mating Effects on Newborn - Live birth index.
Mutagenic: Human DNA Damage; Cell Type: fibroblast; Dose: 10 mg/L. Human DNA Damage; Cell Type: HeLa cell; Dose: 1500 nmol/L. Human DNA Damage; Cell Type: leukocyte; Dose: 500 umol/L. Human Unscheduled DNA Synthesis; Cell Type: other cell types; Dose: 1 umol/L. Human Unscheduled DNA Synthesis; Cell Type: HeLa cell; Dose: 1 mmol/L. Human Unscheduled DNA Synthesis; Cell Type: liver; Dose: 1 umol/L. Human Unscheduled DNA Synthesis; Cell Type: other cell types; Dose: 10 umol/L. Human Unscheduled DNA Synthesis; Cell Type: other cell types; Dose: 3 mg/L. Human Unscheduled DNA Synthesis; Cell Type: other cell types; Dose: 5 umol/L. Human Unscheduled DNA Synthesis; Cell Type: mammary gland; Dose: 10 umol/L. Monkey Unscheduled DNA Synthesis; Cell Type: liver; Dose: 10 mg/L. Human DNA Inhibition; Cell Type: HeLa cell; Dose: 1500 nmol/L. Human DNA Inhibition; Cell Type: fibroblast; Dose: 1 mg/L. Human DNA Adduct; Cell Type: liver; Dose: 100 nmol/L. Human DNA Adduct; Cell Type: lymphocyte; Dose: 1 umol/L/72H. Human DNA Adduct; Cell Type: other cell types; Dose: 1500 nmol/L. Human DNA Adduct; Cell Type: other cell types; Dose: 400 nmol/L. Human DNA Adduct; Cell Type: lung; Dose: 1 umol/L. Human DNA Adduct; Cell Type: other cell types; Dose: 4 umol/L. Monkey DNA Adduct; Cell Type: lung; Dose: 25 umol/L. Monkey DNA Adduct; Cell Type: other cell types; Dose: 1 umol/L. Monkey DNA Adduct; Cell Type: other cell types; Dose: 25 umol/L. Human Mutations in Microorganisms; Cell Type: lymphocyte; Dose: 4 mg/L (+S9). Human Mutations in Mammalian Somatic Cells; Cell Type: other cell types; Dose: 100 nmol/L. Human Mutations in Mammalian Somatic Cells; Cell Type: fibroblast; Dose: 100 nmol/L. Human Mutations in Mammalian Somatic Cells; Cell Type: lymphocyte; Dose: 3 umol/L. Human Mutations in Mammalian Somatic Cells; Cell Type: lymphocyte; Dose: 1200 nmol/L. Human Cytogenetic Analysis; Cell Type: fibroblast; Dose: 40 umol/L/24H. Human Cytogenetic Analysis; Cell Type: leukocyte; Dose: 1 mg/L. Human Cytogenetic Analysis; Cell Type: lymphocyte; Dose: 2400 ug/L. Human Sister Chromatid Exchenge; Cell Type: leukocyte; Dose: 50 umol/L. Human Sister Chromatid Exchenge; Cell Type: lymphocyte; Dose: 1 umol/L/72H. Human Sister Chromatid Exchenge; Cell Type: fibroblast; Dose: 1 mg/L. Human Sister Chromatid Exchenge; Cell Type: liver; Dose: 1 mmol/L. Monkey Sister Chromatid Exchenge; Cell Type: kidney; Dose: 1 umol/L. Human Micronucleus Test; Cell Type: other cell types; Dose: 10 mg/L. Human Micronucleus Test; Cell Type: lymphocyte; Dose: 50 umol/L. Human Morphological Transformation; Cell Type: other cell types; Dose: 10 mg/L. Human Morphological Transformation; Cell Type: fibroblast; Dose: 3200 ug/L.
Tumorigenic: Rat Route: Oral; Dose: 15 mg/kg; Toxic Effects: Tumorigenic - Carcinogenic by RTECS criteria; Gastrointestinal - Tumors; Musculoskelital - Tumors. Monkey Route: Subcutaneous; Dose: 40 mg/kg; Toxic Effects: Tumorigenic - Equivocal tumorigenic agent by RTECS criteria; Lungs, Thorax, or Respiration - Tumors; Tumorigenic - Tumors at site of application.

Hazard Overviews

Fire Diamond

Health: Irritating to eyes/skin/respiratory tract. Also Causes: Gastrointestinal effects, leukoplakia, cancer of the lung, skin, kidneys, bladder, or GI tract.
Fire: Will burn, but does not readily ignite. Use dry chemical, sand, water spray, fog, or foam.
Reactivity: Stable, in closed containers. Hazardous polymerization cannot occur. Avoid: heat and ignition sources. Incompatible with: strong oxidizers (chlorine, bromine, fluorine); oxidizing chemicals (chlorates, perchlorates, permanganates, and nitrates). Hazardous decomposition products: carbon monoxide; carbon dioxide.
Carcinogenicity: IARC - Group 2A, Probably carcinogenic to humans; NIOSH - Listed as carcinogen; NTP - Class 2B, Reasonably anticipated to be a carcinogen, sufficient evidence of carcinogenicity from studies in experimental animals; ACGIH - Class A2, Suspected human carcinogen; OSHA - Not listed; EPA - Class B2, Probable human carcinogen based on animal studies; MAK - Class A2, Unmistakably carcinogenic in animal experimentation only

Primary Target Organs:

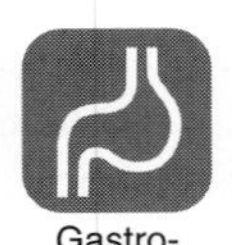

Exposure Limits
OSHA PEL: TWA: 0.2 mg/m^3.
Respirator Recommendation
Exposure Range: unlimited Self-contained Breathing Apparatus, Pressure Demand, Full Face
Note: TLV not established

Environmental

Ecotoxicity: Crustaceans: Daphnia pulex 96h LC_{50} 0.005 mg/l; Fishes: Poeciliopsis lucida 24h LC_{50} 1.2-3.7 mg/l
Environmental Fate: When released to air it may be subject to direct photolysis, although adsorption to particulates apparently can retard this process. It may also be removed by reaction with O3 (half-life 37 min) and NO2 (half-life 7 days), and an estimated half-life for reaction with photochemically produced hydroxyl radicals is 21.49 hr. If released to water, it will adsorb very strongly to sediments and particulate matter, bioconcentrate in aquatic organisms which can not metabolize it, but will not hydrolyze. It may be subject to significant biodegradation, and direct photolysis may be important near the surface of waters; adsorption, however, may significantly retard these two processes. Evaporation may be important with a half-life of 43 days predicted for evaporation from a river 1 m deep, flowing at 1 m/sec with a wind velocity of 3 m/sec; adsorption to sediments and particulates will limit evaporation. If released to soil it will be expected to adsorb very strongly to the soil and will not be expected to appreciably leach to the groundwater, although its presence in some samples of groundwater illustrates that it can be transported there. It will not be expected to hydrolyze or significantly evaporate from soils and surfaces. It may be subject to appreciable biodegradation in soils.

Environmental Physical Data

Octanol/Water Partition Coefficient: log K_{ow} = 6.04
Sorption Partition Coefficient: K_{oc} = 3.95 $x10^6$ to 5.83 $x10^6$
BCF: oysters 3000

Regulations

RCRA 40CFR: Listed Hazardous Waste No. U022 Toxic Waste
CERCLA: 40CFR 302.4: Listed per RCRA Section 3001 per CWA Section 307(a) RQ: 1 lb (0.454 kg)
SARA 40CFR 372.65: Listed
SARA EHS 40CFR 355: Not listed
TSCA: Listed

Analytical Methods

Air: EPA TO-13
Soil: CLP LC_SV, MC_SVOA, OHC; EPA 16, 1625, PAH-005, PAH-007, PAH-011, PAH-012, S-004-1; SW846 3630B, 3640A, 3650A, 3650B, 8100, 8250A, 8270B, 8270C, 8275, 8275A, 8310, 8410; DOE OS050
Water / Groundwater: EPA PAH-002, PAH-006, S-002-1, 1625, 610, 625, 625-S, 6; APHA 6410-B, 6440-B, 6440-C; ASTM D4657, D4763; USGS O3113, O3118
Drinking Water: EPA 525.1, 525.2, 550, 550.1
Indoor / Expired Air: NIOSH 5506, 5515; EPA IP-7-A, IP-7-B
Plasma: EPA 29
Other: EPA PAH-009

BEN5640 CAS #: 192-97-2

BENZO(E)PYRENE

RTECS: DJ4200000
EINECS Number: 205-892-7
Molecular Formula: $C_{20}H_{12}$
Formula Weight: 252.30

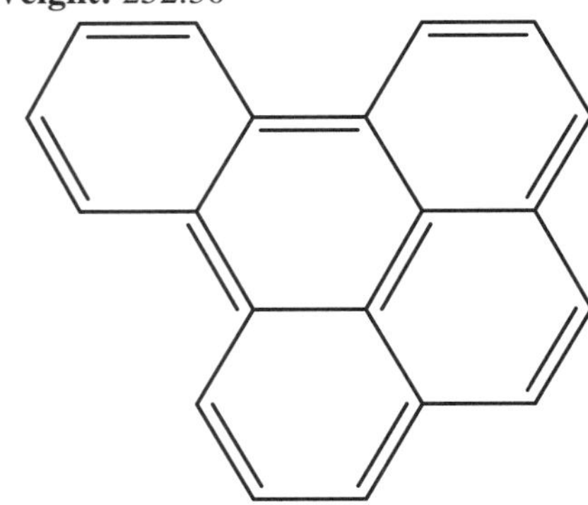

Chemical Structure

Synonyms: B(E)P; 1,2-BENZOPYRENE; 4,5-BENZOPYRENE; 1,2-BENZPYRENE; 4,5-BENZPYRENE
Description: prisms or plates
Use: constituent of coal tars and cigarette smoke and is in the atmosphere as a product of incomplete combustion

Physical Properties

Boiling Point: 3at 10 °C (590 °F) to 312 °C (594 °F) at 10 mm Hg
Freezing Point: 178 °C (352.4 °F) to 179 °C (354.2 °F)
Water Solubility: 2.9 $x10^{-8}$ mol/L dissolve in Water at 25 °C
Other Solubilities: 95% Ethanol: <1 mg/ml at 23 °C; Acetone: <1 mg/ml at 23 °C; Benzene: Soluble; DMSO: <1 mg/ml at 23 °C; Methanol: <1 mg/ml at 22 °C; Toluene: <1 mg/ml at 22 °C; Toluene: Soluble.
Ionization Potential (eV): 7.41
Flash Point: Not available; probably combustible

RTECS Toxicity Data

Reproductive/Teratogenic: Mouse Route: Intraperitoneal; Dose: 300 mg/kg; Duration: female 8D of pregnancy; Effects on Fertility - Post-implantation mortality; Specific Developmental Abnormalities - Musculoskeletal system. Mouse Route: Intraperitoneal; Dose: 150 mg/kg; Duration:

female 8D of pregnancy; Effects on Embryo or Fetus - Fetotoxicity.

Mutagenic: Human Unscheduled DNA Synthesis; Cell Type: HeLa cell; Dose: 1 umol/L. Human Mutations in Microorganisms; Cell Type: lymphocyte; Dose: 2 mg/L (+S9). Human Mutations in Mammalian Somatic Cells; Cell Type: other cell types; Dose: 12 umol/L.

Tumorigenic: Mouse Route: Oral; Dose: 360 mg/kg/43W-I; Toxic Effects: Tumorigenic - Equivocal tumorigenic agent by RTECS criteria; Gastrointestinal - Tumors. Mouse Route: Skin; Dose: 240 mg/kg/30W-I; Toxic Effects: Tumorigenic - Equivocal tumorigenic agent by RTECS criteria; Skin and appendages - Tumors. Mouse Route: Skin; Dose: 516 mg/kg/43W-I; Toxic Effects: Tumorigenic - Equivocal tumorigenic agent by RTECS criteria; Skin and appendages - Tumors.

Hazard Overviews

Health: Irritating to eyes/skin/respiratory tract. Toxic. Other Acute Effects: harmful if swallowed, inhaled, or absorbed through skin. Chronic Effects: laboratory experiments have shown mutagenic effects; teratogen; may cause harm to the unborn child; target organs: skin, lungs. Possible human carcinogen.

Fire: Will burn. Extinguishing agents: water spray; carbon dioxide, dry chemical powder or appropriate foam. Precautions: combustible liquid.

Reactivity: Incompatible with: oxidizing agents. Hazardous decomposition products: carbon monoxide, carbon dioxide.

Carcinogenicity: IARC - Group 3, Not classifiable as to carcinogenicity to humans; NIOSH - Not listed; NTP - Not listed; ACGIH - Not listed; OSHA - Not listed; EPA - Not listed; MAK - Not listed

Primary Target Organs:

Eyes

Skin

Respiratory System

Environmental

Ecotoxicity: Worms: apparent bioconcentration factors in polychaete worms (dry wt/sediment dry wt). Prionospio cirrifera and Spiochaetpoterus costarum BCF: 11.6

Environmental Fate: If released to the atmosphere, it will exist in both the vapor and particulate-phases, although the particulate-phase will usually be dominant. Vapor-phase is rapidly degraded by photochemically produced hydroxyl radicals (estimated half-life of 2.5 hr). Particulate-phase can also be degraded readily by hydroxyl radicals and direct photolysis; however, the rate of degradation depends upon the adsorbing substrate. Substrates containing carbon (>5%) stabilize it and permit long-range atmospheric transport. Physical removal from air by wet and dry deposition is important. If released to soil, it will adsorb strongly and will not leach. If released to water, it will adsorb strongly to sediment and suspended matter and will partition from the water column to sediment. It will bioconcentrate in aquatic organisms that can not metabolize it. Volatilization from water is slow. Limited data suggests that biodegradation are slow.

Environmental Physical Data

Henry's Law Constant: estimated at 1.07×10^{-6}
Sorption Partition Coefficient: K_{oc} = 7.2
BCF: polychaete worms 1.5 to 11.6

Regulations

RCRA 40CFR: Not listed
CERCLA: 40CFR 302.4: Not listed
SARA 40CFR 372.65: Not listed
SARA EHS 40CFR 355: Not listed
TSCA: Not listed

Analytical Methods

Air: EPA TO-13; California 429
Soil: DOE OS050
Indoor / Expired Air: NIOSH 5506, 5515; EPA IP-7-A, IP-7-B

BEN5800 CAS #: 68-76-8

P-BENZOQUINONE, 2,3,5-TRIS(1-AZIRIDINYL)-

RTECS: DK7175000
EINECS Number: 200-692-6
Molecular Formula: $C_{12}H_{13}N_3O_2$
Formula Weight: 231.28

Synonyms: 10257 R.P; AZIRIDINE,1,1',1''-(3,6-DIOXO-1,4-CYCLOHEXADIENE-1,2,4-TRIYL)TRIS-; BAY 3231; BAYER 3231; P-BENZOQUINONE,TRIS(1-AZIRIDINYL)-; 1,1',1''-(3,6-DIOXO-1,4-CYCLOHEXADIENE-1,2,4-TRIYL)TRIS AZIRIDINE; 1,1',1''-(3,6-DIOXO-1,4-CYCLOHEXADIENE-1,2,4-TRIYL)TRISAZIRIDINE; 2,3,5-ETHYLENIMINE-1,4-BENZOQUINONE; NSC-29215; ONCOREDOX; ONCOVEDEX; PRENIMON; RIKER 601; TEIB; TRENIMON; TRENINON; TRIAZICHON; TRIAZIQUINONE; TRIAZIQUINONUM; TRIAZIQUON; TRIAZIQUONE; 2,3,5-TRI-(1-AZIRIDINYL)-P-BENZOQUINONE; 2,3,5-TRIETHYLENEIMINO-1,4-BENZOQUINONE; 2,3,5-TRIETHYLENEIMINO-P-BENZOQUINONE; TRIETHYLENEIMINOBENZOQUINONE; TRIETHYLENIMINOBENZOQUINONE; TRISAETHYLENIMINOBENZOCHINON; 2,3,5-TRIS(1-AZIRIDINE)-P-BENZOQUINONE; 2,3,5-TRIS(AZIRIDINE)-1,4-BENZOQUINONE; 2,3,5-TRIS(1-AZIRIDINO)-P-BENZOQUINONE; 2,3,5-TRIS(AZIRIDINO)-1,4-BENZOQUINONE; 2,3,5-TRIS(1-AZIRIDINYL)-P-BENZOQUINONE; 2,3,5-TRIS(AZIRIDINYL)-1,4-BENZOQUINONE; TRIS(1-AZIRIDINYL)-P-BENZOQUINONE; TRIS(1-AZIRIDINYL)P-BENZOQUINONE; TRIS(AZIRIDINYL)-P-BENZOQUINONE; 2,3,5-TRIS(1-AZIRIDINYL)-2,5-CYCLOHEXADIENE-1,4-DIONE; 2,3,5-TRIS(1-AZIRIDINYL)-2,5-CYLOHEXADIENE-1,4-DIONE; 2,3,5-TRIS(ETHYLENEIMINO)-1,4-BENZOQUINONE; 2,3,5-TRIS(ETHYLENEIMINO)BENZOQUINONE; 2,3,5-TRISETHYLENEIMINOBENZOQUINONE; TRIS(ETHYLENEIMINO)BENZOQUINONE; TRISETHYLENEIMINOQUINONE; 2,3,5-TRIS(ETHYLENIMINO)-P-BENZOQUINONE; TRIS(TRENIMOM)

Description: purple acicular crystals or needles

Use: has been used in treatment of neoplastic diseases and tested as an insect chemosterilant

Physical Properties

Freezing Point: 162.5 °C (324.5 °F) to 163 °C (325.4 °F)
Water Solubility: Sparingly Soluble
Other Solubilities: Acetone: Soluble; Acetic Acid: Soluble in warm; Benzene: Soluble; Chloroform: Soluble; Ethyl Acetate: Soluble; Methanol: Soluble.
Flash Point: Not available; probably combustible

RTECS Toxicity Data

Reproductive/Teratogenic: Mouse Route: Intraperitoneal; Dose: 125 ug/kg; Duration: female 1D prior to mating; Effects on Fertility - Other measures of fertility. Mouse Route: Intraperitoneal; Dose: 1 mg/kg; Duration: female 1D of pregnancy; Maternal Effects - Oogenesis. Mouse Route: Intraperitoneal; Dose: 125 ug/kg; Duration: male 1D prior to mating; Effects on Newborn - Germ cell effects. Mouse Route: Intraperitoneal; Dose: 20 ug/kg; Duration: female 15D of pregnancy; Effects on Embryo or Fetus - Cytological changes (inc. somatic cell genetic material). Mouse Route: Intraperitoneal; Dose: 125 ug/kg; Duration: male 1D prior to mating; Effects on Fertility - Pre-implantation mortality; Effects on Embryo or Fetus - Fetal death. Mouse Route: Intraperitoneal; Dose: 125 ug/kg; Duration: male 1D prior to mating; Effects on Newborn - Weaning or lactation index.
Mutagenic: Human Unscheduled DNA Synthesis; Cell Type: other cell types; Dose: 10 ug/L. Human DNA Inhibition; Cell Type: HeLa cell; Dose: 200 nmol/L/30M-C. Human DNA Inhibition; Cell Type: other cell types; Dose: 100 ug/L. Human DNA Inhibition; Cell Type: other cell types; Dose: 250 ug/L. Human DNA Inhibition; Cell Type: other cell types; Dose: 5 ug/L. Human Cytogenetic Analysis; Cell Type: lymphocyte; Dose: 280 pmol/L. Human Cytogenetic Analysis; Cell Type: fibroblast; Dose: 2500 ug/L. Human Cytogenetic Analysis; Cell Type: leukocyte; Dose: 1 ng/L. Human Sister Chromatid Exchenge; Cell Type: leukocyte; Dose: 10 nmol/L. Human Sister Chromatid Exchenge; Cell Type: lymphocyte; Dose: 10 nmol/L/24H. Human Sister Chromatid Exchenge; Cell Type: fibroblast; Dose: 1 ng/L. Human Micronucleus Test; Cell Type: leukocyte; Dose: 500 nmol/L/8H. Monkey Micronucleus Test; Route: Intravenous; Dose: 62 ug/kg. Monkey Micronucleus Test; Route: Intraperitoneal; Dose: 125 ug/kg. Human Morphological Transformation; Cell Type: HeLa cell; Dose: 1 umol/L. Human Other Mutation Test Systems; Cell Type: other cell types; Dose: 250 ug/L. Human Other Mutation Test Systems; Cell Type: other cell types; Dose: 5 ug/L. Human Other Mutation Test Systems; Cell Type: fibroblast; Dose: 1 ug/L.
Tumorigenic: Rat Route: Intravenous; Dose: 1560 ug/kg/1Y-I; Toxic Effects: Tumorigenic - Carcinogenic by RTECS criteria; Lungs, Thorax, or Respiration - Tumors; Blood - Tumors.

Hazard Overviews

Fire: Will burn.
Carcinogenicity: IARC - Group 3, Not classifiable as to carcinogenicity to humans; NIOSH - Not listed; NTP - Not listed; ACGIH - Not listed; OSHA - Not listed; EPA - Not listed; MAK - Not listed

Environmental

Regulations

RCRA 40CFR: Not listed
CERCLA: 40CFR 302.4: Not listed
SARA 40CFR 372.65: Listed
SARA EHS 40CFR 355: Not listed
TSCA: Not listed

BEN5880 CAS #: 95-16-9

BENZOTHIAZOLE

RTECS: DL0875000
EINECS Number: 202-396-2
Molecular Formula: C_7H_5NS
Formula Weight: 135.18

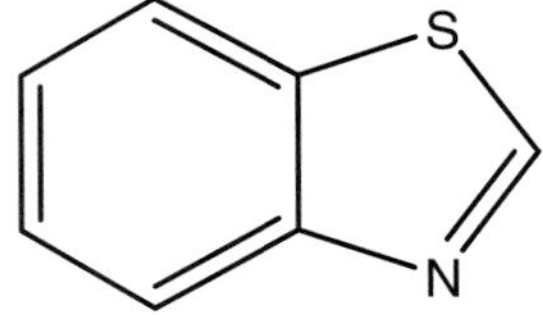

Chemical Structure

Synonyms: BENZOSULFONAZOLE; O-2857; 1-THIA-3-AZAINDENE; VANGARD BT
Description: liquid; odor similar to that of quinoline
Use: flavoring substance or adjuvant; rubber accelerator; antimicrobial agent; component of cyanine dyes

Physical Properties

Boiling Point: 227 °C (441 °F) to 228 °C (442 °F) at 765 mm Hg
Freezing Point: 2 °C (35.6 °F)
Specific Gravity: 1.246 at 20 °C/4 °C
Water Solubility: Slightly Soluble in Water
Other Solubilities: Very Soluble in Ether; Soluble in Acetone.
Refraction Index: 1.6379 at 20 °C/D

RTECS Toxicity Data

Acute Oral: Rat LD_{50} Dose: 466 mg/kg; Toxic Effects: Behavioral - Altered sleep time (including change in righting reflex); Behavioral - Somnolence (general depressed activity); Behavioral - Ataxia. Mouse LD_{50} Dose: 900 mg/kg.
Acute Inhalation: Rat LC; Dose: >1400 mg/m^3/6hr.
Acute Dermal: Rabbit LD_{50} Route: Skin; Dose: >631 mg/kg.

Hazard Overviews

Health: Irritating to eyes/skin/respiratory tract. Toxic. Other Acute Effects: harmful by inhalation, in contact with skin and if swallowed.

Fire: Hazards: emits toxic fumes. Extinguishing agents: water spray; carbon dioxide, dry chemical powder or appropriate foam. Precautions: combustible liquid.
Carcinogenicity: IARC - Not listed; NIOSH - Not listed; NTP - Not listed; ACGIH - Not listed; OSHA - Not listed; EPA - Not listed; MAK - Not listed
Primary Target Organs:

Eyes

Skin
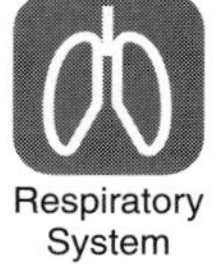
Respiratory System

Environmental

Regulations

RCRA 40CFR: Not listed
CERCLA: 40CFR 302.4: Not listed
SARA 40CFR 372.65: Not listed
SARA EHS 40CFR 355: Not listed
TSCA: Listed

BEN5960 CAS #: 90-16-4

1,2,3-BENZOTRIAZIN-4-(LH)-ONE

RTECS: DM0800000
EINECS Number: 201-971-5
Molecular Formula: $C_7H_5N_3O$
Formula Weight: 147.15

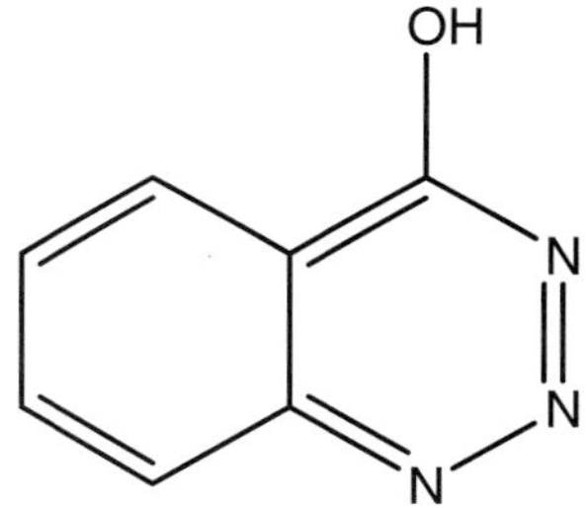

Chemical Structure

Synonyms: BENZAZIMIDE; BENZAZIMIDONE; BENZOKETOTRIAZINE; 1,2,3-BENZOTRIAZIN-4-OL; 1,2,3-BENZOTRIAZIN-4(3H)-ONE; 3H-1,2,3-BENZOTRIAZIN-4-ONE; 4-KETO-(3H)-1,2,3-BENZOTRIAZINE; 4-KETOBENZOTRIAZINE; 4-KETOBENZ-1,2,3-TRIAZINE
Description: tan powder
Use: organic synthesis

Physical Properties

Freezing Point: Decomposes at 210 °C (410 °F)
Saturated Vapor Density: 1.200000001 kg/m^3
Vapor Pressure: 1.23 x10^{-7} mm Hg at 25 °C
Water Solubility: Estimated 7.52 x10^4 mg/l

RTECS Toxicity Data

Acute Oral: Rat LD_{50} Dose: 1530 mg/kg; Toxic Effects: Behavioral - Somnolence (general depressed activity).
Acute Inhalation: Rat LC; Dose: >2 gm/kg/4hr. Mouse LC; Dose: >2 gm/kg/4hr.
Reproductive/Teratogenic: Mouse Route: Oral; Dose: 3480 mg/kg; Duration: male 8D prior to mating; Effects on Fertility - Female fertility index. Mouse Route: Oral; Dose: 2400 mg/kg; Duration: male 15D prior to mating; Effects on Fertility - Male fertility index.

Hazard Overviews

Health: Irritating to eyes/skin/respiratory tract. Other Acute Effects: may be harmful by inhalation, ingestion, or skin absorption.
Fire: Hazards: emits toxic fumes. Extinguishing agents: water spray; carbon dioxide, dry chemical powder or appropriate foam. Precautions: combustible liquid.
Carcinogenicity: IARC - Not listed; NIOSH - Not listed; NTP - Not listed; ACGIH - Not listed; OSHA - Not listed; EPA - Not listed; MAK - Not listed
Primary Target Organs:

Eyes

Skin
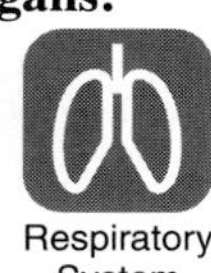
Respiratory System

Environmental

Environmental Fate: If released to soil, it will exhibit a high mobility in soil and, therefore, may leach to groundwater. It may be biodegraded in soil based upon its cometabolism in aqueous screening tests utilizing soil bacteria inoculums and disodium fumarate as an added carbon source. It will not be expected to volatilize from near surface soil. If released to water, it will not be expected to volatilize, adsorb to sediment or suspended particulate matter or to bioconcentrate in aquatic organisms. It may directly photolyze based upon the absorption of UV light at wavelengths >290 nm by the structurally similar pesticide azinphos-methyl and it may be biodegrade in natural waters based upon its moderate cometabolism in aqueous screening tests utilizing soil bacteria inoculums and disodium fumarate as an added carbon source. If released to the atmosphere, it will be expected to exist mostly in the particulate phase. It may directly photolyze in the atmosphere. It will be susceptible to photooxidation via vapor phase reaction with photochemically produced hydroxyl radicals and ozone. An atmospheric half-life of 5.2 hours at an atmospheric concentration of 5 x10^5 hydroxyl radicals per cu cm and 7 x10^{11} ozone molecules per cu cm has been estimated for this process.

Environmental Physical Data

Henry's Law Constant: calculated at 3.17 x10^{-13}
Octanol/Water Partition Coefficient: log K_{ow} = 0.94
Sorption Partition Coefficient: K_{oc} = estimated at 77
BCF: estimated at 3.1

Regulations

RCRA 40CFR: Not listed
CERCLA: 40CFR 302.4: Not listed

SARA 40CFR 372.65: Not listed
SARA EHS 40CFR 355: Not listed
TSCA: Listed

BEN6040 CAS #: 95-14-7

1,2,3-BENZOTRIAZOLE

RTECS: DM1225000
EINECS Number: 202-394-1
Molecular Formula: $C_6H_5N_3$
Formula Weight: 119.12

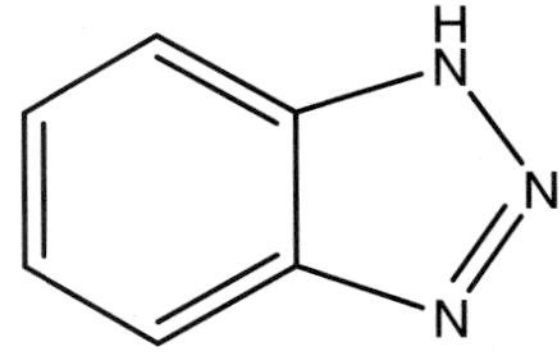

Chemical Structure

Synonyms: 1,2-AMINOAZOPHENYLENE; 1,2,-AMINOZOPHENYLENE; AZIMIDOBENZENE; AZIMINOBENZENE; BENZENE AZIMIDE; BENZISOTRIAZOLE; 1H-1,2,3-BENZOTRIAZOLE; 1H-BENZOTRIAZOLE; BENZOTRIAZOLE; BENZTRIAZOLE; COBRATEC #99; COBRATEC 99; 2,3-DIAZAINDOLE; NSC-3058; 1,2,3-TRIAZA-1H-INDENE; 1,2,3-TRIAZAINDENE; U-6233

Description: white to light tan, crystalline powder or needles; odorless

Use: anticorrosive in metal working; art restoration; tarnish remover and protective coating in construction industry

Physical Properties

Boiling Point: 204 °C (399 °F) at 15 mm Hg
Freezing Point: 98.5 °C (209.3 °F)
Specific Gravity: 1.33
Vapor Density: 4.1 Air=1
Vapor Pressure: 0.04
Water Solubility: Sparingly Soluble in Water
Other Solubilities: 95% Ethanol: >=100 mg/ml at 21 °C; Acetone: >=100 mg/ml at 21 °C; Benzene: Soluble; Chloroform: Soluble; DMF: Soluble; DMSO: >=100 mg/ml at 21 °C; Toluene: Soluble.
Ionization Potential (eV): 9.20 +/-0.05
Flash Point: Not available; probably combustible

RTECS Toxicity Data

Acute Oral: Rat LD_{50} Dose: 600 mg/kg; Toxic Effects: Automatic Nervous System - Other (direct) parasympathomimetic; Behavioral - Somnolence (general depressed activity); Lungs, Thorax, or Respiration - Cyanosis. Mouse LD_{50} Dose: 615 mg/kg.

Chronic (Multiple Dose) Oral: Rat Dose: 109 mg/kg/26W-I; Toxic Effects: Endocrine - Other changes; Blood - Normocytic anemia; Blood - Leukopenia.

Mutagenic: Rat Morphological Transformation; Cell Type: embryo; Dose: 94 ug/plate. Bacteria - E Coli Mutations in Microorganisms; Dose: 33300 ng/plate (+S9), 333 ug/plate (-S9). Bacteria - S Typhimurium Mutations in Microorganisms; Dose: 100 ug/plate (-S9).

Tumorigenic: Rat Route: Oral; Dose: 220 gm/kg/78W-I; Toxic Effects: Tumorigenic - Equivocal tumorigenic agent by RTECS criteria; Brain and coverings - Tumors. Mouse Route: Oral; Dose: 770 gm/kg/78W-I; Toxic Effects: Tumorigenic - Equivocal tumorigenic agent by RTECS criteria; Lungs, Thorax, or Respiration - Tumors; Lungs, Thorax, or Respiration - Bromchiogenic carcinoma.

Hazard Overviews

Health: Irritating to eyes/skin/respiratory tract. Harmful. Other Acute Effects: harmful if swallowed; may be harmful if inhaled or absorbed through the skin.

Fire: Will burn. Hazards: emits toxic fumes; container explosion may occur. Extinguishing agents: water spray; carbon dioxide, dry chemical powder or appropriate foam. Precautions: combustible liquid.

Carcinogenicity: IARC - Not listed; NIOSH - Not listed; NTP - Not listed; ACGIH - Not listed; OSHA - Not listed; EPA - Not listed; MAK - Not listed

Primary Target Organs:

Eyes

Skin

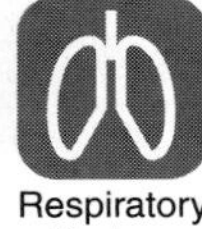
Respiratory System

Environmental

Regulations

RCRA 40CFR: Not listed
CERCLA: 40CFR 302.4: Not listed
SARA 40CFR 372.65: Not listed
SARA EHS 40CFR 355: Not listed
TSCA: Listed

BEN6120 CAS #: 98-07-7

BENZOTRICHLORIDE

RTECS: XT9275000
DOT: UN2226; IMO8.0
EINECS Number: 202-634-5
Molecular Formula: $C_7H_5Cl_3$
Formula Weight: 195.48

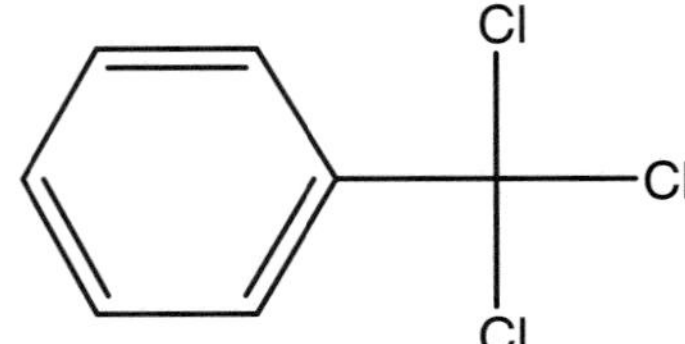

Chemical Structure

Synonyms: BENZENE,(TRICHLOROMETHYL)-; BENZENYL CHLORIDE; BENZENYL TRICHLORIDE; BENZOIC TRICHLORIDE;

BENZOTRICLORURO; BENZYL TRICHLORIDE; BENZYLIDYNE CHLORIDE; CHLORURE DE BENZENYLE; CHLORURE DE BENZYLIDYNE; PHENYL CHLOROFORM; PHENYL TRICHOROMETHANE; PHENYLCHLOROFORM; PHENYLTRICHLOROMETHANE; TOLUENE TRICHLORIDE; TOLUENE,ALPHA,ALPHA,ALPHA-TRICHLORO; TOLUENE,ALPHA,ALPHA,ALPHA-TRICHLORO-; TRICHLOORMETHYLBENZEEN; TRICHLORMETHYLBENZOL; (TRICHLOROMETHYL)BENZENE; 1-(TRICHLOROMETHYL)BENZENE; TRICHLOROMETHYLBENZENE; TRICHLOROPHENYLMETHANE; ALPHA,ALPHA,ALPHA-TRICHLOROTOLUENE; OMEGA,OMEGA,OMEGA-TRICHLOROTOLUENE; TRICLOROMETILBENZENE; TRICLOROTOLUENE

Description: colorless, yellowish liquid; fumes have a penetrating odor

Use: synthetic dyes, organic synthesis

Physical Properties

Boiling Point: 220.8 °C (429 °F) at 760 mm Hg
Freezing Point: -5 °C (23 °F)
Specific Gravity: 1.3756 at 20 °C/4 °C
Vapor Density: 6.77 Air=1
Density: 1.3723 g/mL at 20 °C
Vapor Pressure: 1 mm Hg at 45.8 °C
Water Solubility: Insoluble in Water
Other Solubilities: 95% Ethanol: >=100 mg/ml at 15 °C; Acetone: >=100 mg/ml at 15 °C; Benzene: Soluble; DMSO: >=100 mg/ml at 15 °C; Ether: Soluble.
Surface Tension: 30.03 dynes/cm at 20 °C
Refraction Index: 1.55789 at 20 °C/D
Ionization Potential (eV): =< 9.60
Flash Point: 127 °C Closed Cup
Autoignition Temperature: 211 °C

RTECS Toxicity Data

Acute Oral: Rat LD_{50} Dose: 6 gm/kg. Mouse LD_{50} Dose: 702 mg/kg.

Acute Inhalation: Rat LC_{50} Dose: 19 ppm/2hr; Toxic Effects: Behavioral - Muscle contraction or spasticity; Lungs, Thorax, or Respiration - Respiratory depression; Nutritional and gross metabolic - Weight loss or decreased weight gain. Mouse LC_{50} Dose: 8 ppm/2hr; Toxic Effects: Brain and coverings - Recordings from specific areas of CNS; Behavioral - Change in motor activity (specific assay); Lungs, Thorax, or Respiration - Respiratory depression.

Acute Dermal: Rabbit LD_{50} Route: Skin; Dose: 4 gm/kg; Toxic Effects: Behavioral - Excitment; Liver - Other changes; Skin and appendages - Dermatitis, other. Frog LD_{Lo} Route: Subcutaneous Dose: 2150 mg/kg; Toxic Effects: Automatic Nervous System - Other (direct) parasympathomimetic; Behavioral - Change in motor activity (specific assay).

Chronic (Multiple Dose) Inhalation: Rat Dose: 100 mg/m^3/2H/9W-I; Toxic Effects: Brain and coverings - Other degenerative changes; Vascular - BP lowering not characterized in autonomic section; Blood - Leukopenia.

Irritation Eye: Rabbit Standard Draize Test Dose: 50 ug open; Reaction: severe. Rabbit Standard Draize Test Dose: 50 ug/24H; Reaction: severe.

Irritation Skin: Rabbit Standard Draize Test Dose: 20 mg/24H; Reaction: moderate. Rabbit Open Draize Test Dose: 10 mg/24H open; Reaction: severe.

Mutagenic: Human DNA Damage; Cell Type: other cell types; Dose: 100 ug/L. Rat Cytogenetic Analysis; Route: Inhalation; Dose: 1 ppm/6H/4W-I.

Tumorigenic: Mouse Route: Inhalation; Dose: 1620 ppb/30M/22W-I; Toxic Effects: Tumorigenic - Neoplastic by RTECS criteria; Lungs, Thorax, or Respiration - Tumors; Skin and appendages - Tumors. Mouse Route: Oral; Dose: 1380 mg/kg/25W-I; Toxic Effects: Tumorigenic - Carcinogenic by RTECS criteria; Lungs, Thorax, or Respiration - Tumors; Gastrointestinal - Tumors.

Hazard Overviews

Corrosive

Fire Diamond

Health: Corrosive to eyes/skin/respiratory tract. Also Causes: tearing, pulmonary edema, lung damage, sensitization, CNS depression. Acute Effects: possible cancer hazard.

Fire: Will burn. Use dry chemical, carbon dioxide, fog, or regular foam. Use water only if other extinguishing agents are unavailable since benzotrichloride hydrolyzes into benzoic and hydrochloric acid in presence of moisture.

Reactivity: Stable. Hazardous polymerization cannot occur. Avoid: moisture; acids. Incompatible with: acid; acid vapors. Hazardous decomposition products: carbon dioxide; acidic chloride compounds.

Carcinogenicity: IARC - Group 2B, Possibly carcinogenic to humans; NIOSH - Not listed; NTP - Class 2B, Reasonably anticipated to be a carcinogen, sufficient evidence of carcinogenicity from studies in experimental animals; ACGIH - Not listed; OSHA - Not listed; EPA - Class B2, Probable human carcinogen based on animal studies; MAK - Class A2, Unmistakably carcinogenic in animal experimentation only

Primary Target Organs:

Eyes

Skin

Respiratory System

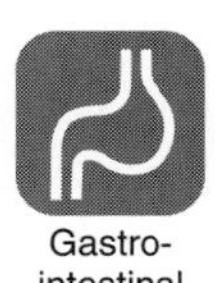
Gastro-intestinal

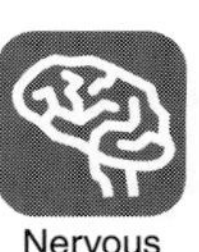
Nervous System

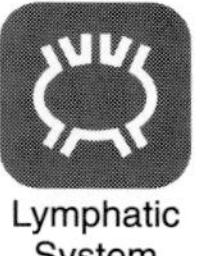
Lymphatic System

Environmental

Ecotoxicity: Toxic threshold concentration Entosiphon sulcatum (protozoan) 56 mg/l/72 hr /25 °C in bi-distilled water; additional conditions of bioassay not specified LC_{50} Daphnia magna (cladoceran) > 100 mg/l /test media: chlorine free tap water, saturated with oxygen, pH 7.6-7.7, temp 20-22 °C, hardness 16 deg (German) LC_{50} Leuciscus idus melanotus (golden orfe) 4,140 mg/l/48 hr /Static bioassay Toxic threshold concentration Microcystis aeruginosa (blue-green alga) 34 mg/l/8 days /cell mutiplication inhibition test, bi-distilled water at pH 7

Environmental Fate: If released to soil, it is expected to hydrolyze in the presence of moisture. Evaporation from dry surfaces is likely to occur. If released to water, it hydrolyzes rapidly to form benzoic acid and hydrochloric acid with a hydrolysis half-life of 11 seconds at 25 °C (pH 7) and 3 minutes at 5.10 °C. Due to the rapid hydrolysis, aquatic volatilization, bioconcentration, and adsorption to sediments are not expected to be important. If released to air, it will react in vapor-phase with photochemically produced hydroxyl radicals with an estimated half-life of 2 days in a typical atmosphere. Atmospheric hydrolysis may occur in the presence of moisture; however, direct photolysis is not expected to occur.

Cleanup/Disposal: Guide No. 156: Eliminate all ignition sources (no smoking, flares, sparks or flames in immediate area). All equipment used when handling the product must be grounded. Do not touch damaged containers or spilled material unless wearing appropriate protective clothing. Stop leak if you can do it without risk. A vapor suppressing foam may be used to reduce vapors. For chlorosilanes, use AFFF alcohol-resistant medium expansion foam to reduce vapors. Do not get water on spilled substance or inside containers. Use water spray to reduce vapors or divert vapor cloud drift. Prevent entry into waterways, sewers, basements or confined areas. Small Spills: Cover with dry earth, dry sand, or other non-combustible material followed with plastic sheet to minimize spreading or contact with rain. Use clean non-sparking tools to collect material and place it into loosely covered plastic containers for later disposal.

Environmental Physical Data

Octanol/Water Partition Coefficient: log K_{ow} = 2.92

BCF: estimated at 98

Regulations

RCRA 40CFR: Listed Hazardous Waste No. U023 Toxic Waste Reactive Waste Corrosive Waste

CERCLA: 40CFR 302.4: Listed per RCRA Section 3001 RQ: 10 lb (4.535 kg)

SARA 40CFR 372.65: Listed TPQ: 100 lb

SARA EHS 40CFR 355: Listed TPQ: 10 lb

TSCA: Listed

Analytical Methods

Soil: SW846 8121

BEN6200 CAS #: 98-08-8

BENZOTRIFLUORIDE

RTECS: XT9450000

EINECS Number: 202-635-0

Molecular Formula: $C_7H_5F_3$

Formula Weight: 146.11

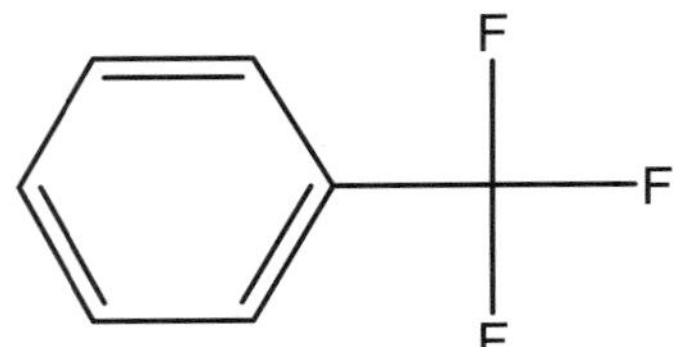

Chemical Structure

Synonyms: BENZENE,(TRIFLUOROMETHYL)-; BENZENYL FLUORIDE; BENZYLIDYNE FLUORIDE; OPHENYLFLUOROFORM; PHENYFLUOROFORM; PHENYLFLUOROFORM; TOLUENE TRIFLUORIDE; TOLUENE,ALPHA,ALPHA,ALPHA-TRIFLUORO-; (TRIFLUOROMETHYL)BENZENE; TRIFLUOROMETHYLBENZENE; ALPHA,ALPHA,ALPHA-TRIFLUOROTOLUENE; OMEGA-TRIFLUOROTOLUENE

Description: water white liquid; aromatic odor

Use: in dye chemistry; in the manufacturing of substituted benzotrifluorides containing an ethylenic group; used in high polymer chemistry; in dielectric fluids, such as transformer oils; vulcanizing agent; insecticide

Physical Properties

Boiling Point: 103.46 °C (218 °F)

Freezing Point: -29.05 °C (-20.29 °F)

Specific Gravity: 1.1886 at 20 °C

Vapor Density: 5.04

Density: 1.199 g/mL

Vapor Pressure: 11 mm Hg at 0 °C

Water Solubility: < 1 mg/mL at 21 C

Other Solubilities: 95% Ethanol: >=100 mg/ml at 21 °C; Acetone: >=100 mg/ml at 21 °C; Benzene: Soluble; Carbon Tetrachloride: miscible; DMSO: >=100 mg/ml at 21 °C; Ether: Soluble; n-Heptane: miscible.

Refraction Index: 1.41486 at 13.3 °C/D

Ionization Potential (eV): 9.69 +/-0.03

Flash Point: 12.222 °C Closed Cup

RTECS Toxicity Data

Acute Oral: Rat LD_{50} Dose: 15 gm/kg; Toxic Effects: Behavioral - General anesthetic; Behavioral - Excitment; Behavioral - Muscle weakness. Mouse LD_{50} Dose: 10000 mg/kg; Toxic Effects: Behavioral - General anesthetic; Behavioral - Excitment; Behavioral - Muscle weakness.

Acute Inhalation: Rat LC_{50} Dose: 70810 mg/m^3/4hr. Mouse LC_{50} Dose: 92240 mg/m^3/2hr.

Acute Dermal: Frog LD_{Lo} Route: Subcutaneous Dose: 870 mg/kg; Toxic Effects: Automatic Nervous System - Other (direct) parasympathomimetic.

Chronic (Multiple Dose) Oral: Rat Dose: 15500 mg/kg/7W-I; Toxic Effects: Kidney, Ureter, and Bladder - Proteinuria; Kidney, Ureter, and Bladder - Other changes in urine composition.

Hazard Overviews

Corrosive

Flammable

Health: Corrosive to eyes/skin/respiratory tract. Other Acute Effects: harmful if swallowed, inhaled, or absorbed through skin; extremely destructive to tissue of the mucous membranes and upper respiratory tract, eyes and skin; inhalation may result in spasm, inflammation and edema of the larynx and bronchi, chemical pneumonitis and pulmonary edema; symptoms of exposure may include burning sensation, coughing, wheezing, laryngitis, shortness of breath, headache, nausea and vomiting.
Fire: Flammable. Hazards: vapor may travel considerable distance to source of ignition and flash back; container explosion may occur; emits toxic fumes. Extinguishing agents: carbon dioxide, dry chemical powder or appropriate foam. Precautions: combustible liquid.
Carcinogenicity: IARC - Not listed; NIOSH - Not listed; NTP - Not listed; ACGIH - Class A4, Not classifiable as a human carcinogen; OSHA - Not listed; EPA - Not listed; MAK - Not listed
Primary Target Organs:

Eyes

Skin

Respiratory System

Exposure Limits
OSHA PEL: TWA: 2.5 mg/m^3; as F.
ACGIH TLV: TWA: 2.5 mg/m^3; as F.
NIOSH REL: TWA: 2.5 mg/m^3; as F.
DFG MAK: TWA: 2.5 mg/m^3; as F.

Environmental

Cleanup/Disposal: Guide No. 131: Fully encapsulating, vapor protective clothing should be worn for spills and leaks with no fire. Eliminate all ignition sources (no smoking, flares, sparks or flames in immediate area). All equipment used when handling the product must be grounded. Do not touch or walk through spilled material. Stop leak if you can do it without risk. Prevent entry into waterways, sewers, basements or confined areas. A vapor suppressing foam may be used to reduce vapors. Small Spills: Absorb with earth, sand or other non-combustible material and transfer to containers for later disposal. Use clean non-sparking tools to collect absorbed material. Large Spills: Dike far ahead of liquid spill for later disposal. Water spray may reduce vapor; but may not prevent ignition in closed spaces.

Regulations

RCRA 40CFR: Not listed
CERCLA: 40CFR 302.4: Not listed
SARA 40CFR 372.65: Not listed
SARA EHS 40CFR 355: Not listed
TSCA: Listed

BEN6280 CAS #: 98-88-4

BENZOYL CHLORIDE

RTECS: DM6600000
DOT: UN1736; IMO8.2
EINECS Number: 202-710-8
Molecular Formula: C_7H_5ClO
Structured MF: C_6H_5COCl
Formula Weight: 140.57

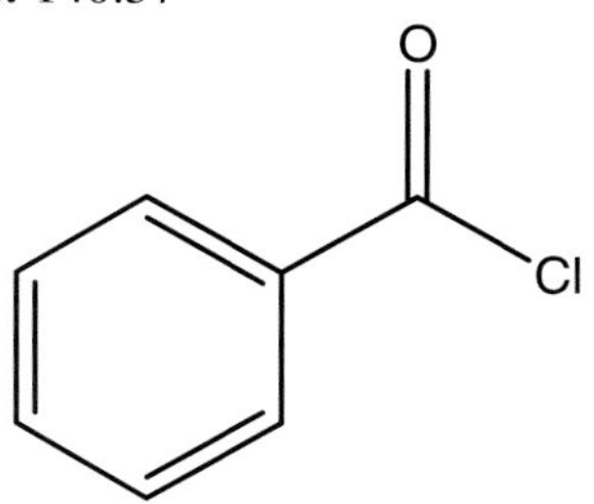

Chemical Structure

Synonyms: BENZALDEHYDE,ALPHA-CHLORO-; BENZENECARBONYL CHLORIDE; BENZOIC ACID CHLORIDE; BENZOIC ACID,CHLORIDE; ALPHA-CHLOROBENZALDEHYDE
Description: clear, colorless liquid; pungent, unpleasant odor
Use: in organic synthesis to produce dye intermediates; as an intermediate to produce a variety of derivatives such as benzoyl peroxide, as a herbicide, as a perfume fixative, as a polymerization catalyst, as a benzolating agent, in the synthesis of aliphatic acid chlorides, in organic analysis for making benzoyl derivatives for identification purposes and for acylation (i.e., the introduction of the benzoyl group into alcohols, phenols and amines (Schotten-Baumann reaction))

Physical Properties

Boiling Point: 197.2 °C (387 °F) at 760 mm Hg
Freezing Point: -1 °C (30.2 °F)
Specific Gravity: 1.207 at 25 °C/4 °C
Vapor Density: 4.9 Air=1
Density: 1.2188 g/mL
Vapor Pressure: 1 mm Hg at 32.1 °C
Water Solubility: Decomposes
Other Solubilities: 95% Ethanol: Decomposes; Alcohol: Decomposes; Benzene: miscible; Carbon Disulfide: miscible; DMSO: Reacts violently; Ether: miscible; Oils: miscible; Toluene: >=100 mg/ml at 22 °C.
Surface Tension: 36.3 dynes/cm
pH: Acid
Refraction Index: 1.55369 at 20 °C/D
Ionization Potential (eV): 9.53 +/-1.0
Flash Point: 72.222 °C Open Cup
Autoignition Temperature: 85 °C
LEL: 1.2% v/v
UEL: 4.9% v/v

RTECS Toxicity Data

Acute Oral: Rat LD_{50} Dose: 1900 mg/kg.

Acute Inhalation: Rat LC_{50} Dose: 1870 mg/m^3/2hr. Human TC_{Lo} Dose: 2 ppm/1M; Toxic Effects: Sense organs and special senses - Other; Lungs, Thorax, or Respiration - Other changes.

Mutagenic: Bacteria - S Typhimurium Mutations in Microorganisms; Dose: 1 umol/plate (+S9). Bacteria - S Typhimurium Mutations in Microorganisms; Dose: 5 mg/plate (-S9).

Tumorigenic: Mouse Route: Skin; Dose: 9200 mg/kg/50W-I; Toxic Effects: Tumorigenic - Equivocal tumorigenic agent by RTECS criteria; Lungs, Thorax, or Respiration - Tumors; Skin and appendages - Tumors. Mouse Route: Skin; Dose: 17600 mg/kg/42W-I; Toxic Effects: Tumorigenic - Equivocal tumorigenic agent by RTECS criteria; Skin and appendages - Tumors. Mouse Route: Skin; Dose: 35200 mg/kg/42W-I; Toxic Effects: Tumorigenic - Equivocal tumorigenic agent by RTECS criteria; Lungs, Thorax, or Respiration - Tumors.

Hazard Overviews

Corrosive

Fire Diamond

Health: Corrosive to eyes/skin/respiratory tract. Also Causes: coughing, difficult breathing, chest tightness, acute pulmonary edema, death.

Fire: Combustible. Use carbon dioxide or dry chemical to put out fire. Never use water or foams since they can react violently with benzoyl chloride.

Reactivity: Stable. Hazardous polymerization cannot occur. Avoid: heat; ignition sources; moisture. Incompatible with: dimethyl sulfoxide; mixture of sodium azide and potassium hydroxide. Hazardous decomposition products: aromatics; oxides of chlorine; toxic carbon monoxide; phosgene.

Carcinogenicity: IARC - Group 3, Not classifiable as to carcinogenicity to humans; NIOSH - Not listed; NTP - Not listed; ACGIH - Not listed; OSHA - Not listed; EPA - Not listed; MAK - Not listed

Primary Target Organs:

Eyes

Skin

Respiratory System

Mucous Membranes

Exposure Limits

ACGIH TLV: STEL: 0.5 ppm; 2.8 mg/m^3; Ceiling.

AIHA WEEL: STEL: 1 ppm 15-min skin.

Respirator Recommendation

Exposure Range: >0.5 to 25 ppm Supplied Air, Constant Flow/Pressure Demand, Half Mask

Exposure Range: >25 to 500 ppm Supplied Air, Constant Flow/Pressure Demand, Full Face

Exposure Range: >500 to unlimited ppm Self-contained Breathing Apparatus, Pressure Demand, Full Face

Environmental

Ecotoxicity: LC_{50} Palaemonetes pugio (grass shrimp) 180 mg/l/96 hr /Conditions of bioassay not specified LC_{50} Pimephales promelas (fathead minnow) 43 mg/l/24 hr; 35 mg/l/48 hr; 35 mg/l/ 96 hr /Conditions of bioassay not specified

Environmental Fate: In the presence of water, the primary fate will be hydrolysis and will occur very rapidly under environmental conditions. Photolysis may occur in the atmosphere, although little is expected to partition there.

Cleanup/Disposal: Guide No. 137: Fully encapsulating, vapor protective clothing should be worn for spills and leaks with no fire. Do not touch damaged containers or spilled material unless wearing appropriate protective clothing. Stop leak if you can do it without risk. Use water spray to reduce vapors; do not put water directly on leak, spill area or inside container. Keep combustibles (wood, paper, oil, etc.) away from spilled material. Small Spills: Cover with dry earth, dry sand, or other non-combustible material followed with plastic sheet to minimize spreading or contact with rain. Use clean non-sparking tools to collect material and place it into loosely covered plastic containers for later disposal. Prevent entry into waterways, sewers, basements or confined areas.

Environmental Physical Data

BCF: not significant

BOD: 165%, 5 days

Regulations

RCRA 40CFR: Not listed

CERCLA: 40CFR 302.4: Listed per CWA Section 311(b)(4) RQ: 1000 lb (453.5 kg)

SARA 40CFR 372.65: Listed

SARA EHS 40CFR 355: Not listed

TSCA: Listed

BEN6360 CAS #: 94-36-0

BENZOYL PEROXIDE

RTECS: DM8575000

DOT: UN2085; UN2086; UN2087; UN2088; UN2089; UN2090; NA9187; IMO5.2

EINECS Number: 202-327-6

Molecular Formula: $C_{14}H_{10}O_4$

Structured MF: $C_6H_5CO\text{-}O\text{-}O\text{-}COC_6H_5$

Formula Weight: 242.24

Chemical Structure

Synonyms: ACETOXYL; ACNEGEL; AZTEC BPO; BENOXYL; BENOXYL (5&10) LOTION; BENZAC; BENZAKNEW; BENZOIC ACID,PEROXIDE; BENZOPEROXIDE; BENZOYL; BENZOYL SUPEROXIDE; BENZOYLPEROXID; BENZOYLPEROXYDE; BZF-60; CADET; CADOX; CADOX BS; CLEARASIL ANTIBACTERIAL ACNE LOTION; CLEARASIL BENZOYL PEROXIDE LOTION; CLEARASIL BP ACNE TREATMENT; CLEARASIL BP ACNE TREATMENT CREAM; CUTICURA ACNE CREAM; DEBROXIDE; DIBENZOYL PEROXIDE; DIBENZOYLPEROXID; DIBENZOYLPEROXYDE; DIPHENYLGLYOXAL PEROXIDE; DRY AND CLEAR; EPI CLEAR ANTISEPTIC LOTION; EPI-CLEAR; FOSTEX; GAROX; INCIDOL; LOROXIDE; LUCIDOL; LUCIDOL 50P; LUCIDOL-70; LUCIDOL B 50; LUCIDOL G 20; LUCIDOL KL 50; LUPERCO; LUPERCO AA; LUPERCO AST; LUPEROX FL; NAYPER B AND BO; NAYPER BO; NOROX BZP-250; NOROX BZP-C-35; NOVADELOX; OXY-10; OXY-5; OXY-10 COVER; OXY WASH; OXYLITE; PANOXYL; PEROSSIDO DI BENZOILE; PEROXIDE,DIBENZOYL; PEROXYDE DE BENZOYLE; PERSADOX; QUINOLOR COMPOUND; STRI-DEX B.P; SULFOXYL; SUPEROX; SUPEROX 744; THERADERM; TOPEX; VANOXIDE; XERAC

Description: white granular, crystalline solid; faint benzaldehyde odor

Use: initiator, curing agent and cross-linking agent in polymerization processes (primarily in the curing of unsaturated polyester resins, production of polystyrene and related resins, styrene polymers and other resins); an oxidizer used to bleach edible oils, flour, bread and other food; a catalyst for radical reactions; as an initiator in dental applications; in medicine in non-prescription drugs for the treatment of acne; as an antiseptic and local anesthetic in the treatment of burns and ulcers; and as a keratolytic; in the embossing of vinyl flooring, as a food additive, in special fast drying printing inks for printing on plastic surfaces, in printing pastes, as a fixing agent in light microscopy, as an initiator in systems used to prepare polymers for roof bolting in mines, as a drying agent for unsaturated oils an as a burn-out agent for cellulose acetate in mixed fabrics with viscose, silk or cotton to produce a lace-like appearance; as an initiator for addition and substitution reactions in organic synthesis; formerly used as a bleaching agent for textiles and paper

Physical Properties

Boiling Point: Decomposes
Freezing Point: 103 °C (217.4 °F) to 106 °C (222.8 °F)
Specific Gravity: 1.334 at 25 °C
Saturated Vapor Density: < 1.201161016 kg/m^3
Density: 1.334 g/mL at 25 °C
Vapor Pressure: < 0.1 torr at 20 °C
Water Solubility: Sparingly Soluble in Water
Other Solubilities: Soluble in Acetone.
Refraction Index: 1.543
Flash Point: 80 °C
Autoignition Temperature: Wet Benzoyl Peroxide 80 °C

RTECS Toxicity Data

Acute Oral: Rat LD_{50} Dose: 7710 mg/kg; Toxic Effects: Lungs, Thorax, or Respiration - Cyanosis; Liver - Other changes; Kidney, Ureter, and Bladder - Other changes in urine composition. Mouse LD_{50} Dose: 5700 mg/kg.
Irritation Eye: Rabbit Standard Draize Test Dose: 500 mg/24H; Reaction: mild.
Mutagenic: Human DNA Damage; Cell Type: other cell types; Dose: 100 umol/L. Human DNA Inhibition; Cell Type: other cell types; Dose: 56 umol/L. Human Mutations in Mammalian Somatic Cells; Cell Type: kidney; Dose: 300 umol/L. Human Other Mutation Test Systems; Cell Type: other cell types; Dose: 56 umol/L.
Tumorigenic: Mouse Route: Skin; Dose: 24 gm/kg/30W-I; Toxic Effects: Tumorigenic - Equivocal tumorigenic agent by RTECS criteria; Skin and appendages - Tumors.

Hazard Overviews

Explosive

Flammable

Fire Diamond

Health: Irritating to eyes/skin/respiratory tract. Also Causes: itchiness of skin/nose/throat/eyes.
Fire: Explosive. Flammable. Fight fire with water from an explosion-proof location. In advanced or large fires, evacuate area. Use water sprays to cool fire-exposed containers.
Reactivity: Stable. Hazardous polymerization cannot occur. Avoid: contaminants; heat; ignition sources; adding hot material to benzoyl peroxide. Incompatible with: strong acids; oxidizing agents; reducing agents; metals; metal oxides; amines; accelerators; methyl methacrylate; organic matter; carbon tetrachloride and ethylene; N,N-dimethylaniline; lithium aluminum hydride. Hazardous decomposition products: carbon dioxide; carbon monoxide.
Carcinogenicity: IARC - Group 3, Not classifiable as to carcinogenicity to humans; NIOSH - Not listed; NTP - Not listed; ACGIH - Class A4, Not classifiable as a human carcinogen; OSHA - Not listed; EPA - Not listed; MAK - Not listed

Primary Target Organs:

Eyes

Skin

Respiratory System

Mucous Membranes

Exposure Limits
OSHA PEL: TWA: 5 mg/m^3.
ACGIH TLV: TWA: 5 mg/m^3.
NIOSH REL: TWA: 5 mg/m^3.
NIOSH IDLH: 1500 mg/m^3.
DFG MAK: TWA: 5 mg/m^3.

Respirator Recommendation
Exposure Range: >5 to 50 mg/m^3 Air Purifying, Negative Pressure, Half Mask
Exposure Range: >50 to 500 mg/m^3 Air Purifying, Negative Pressure, Full Face
Exposure Range: >500 to <1500 mg/m^3 Supplied Air, Constant Flow/Pressure Demand, Full Face
Exposure Range: 1500 to unlimited mg/m^3 Self-contained Breathing Apparatus, Pressure Demand, Full Face
Cartridge Color: black with dust/mist prefilter (use P100 or consult supervisor for appropriate dust/mist prefilter)

Environmental

Cleanup/Disposal: Guide No. 145: Eliminate all ignition sources (no smoking, flares, sparks or flames in immediate area). Keep combustibles (wood, paper, oil, etc.) away from spilled material. Do not touch damaged containers or spilled material unless wearing appropriate protective clothing. Keep substance wet using water spray. Stop leak if you can do it without risk. Small Spills: Take up with inert, damp, noncombustible material using clean non-sparking tools and place into loosely covered plastic containers for later disposal. Large Spills: Wet down with water and dike for later disposal. Prevent entry into waterways, sewers, basements or confined areas. Guide No. 146: Eliminate all ignition sources (no smoking, flares, sparks or flames in immediate area). Keep combustibles (wood, paper, oil, etc.) away from spilled material. Do not touch damaged containers or spilled material unless wearing appropriate protective clothing. Keep substance wet using water spray. Stop leak if you can do it without risk. Small Spills: Take up with inert, damp, noncombustible material using clean non-sparking tools and place into loosely covered plastic containers for later disposal. Large Spills: Wet down with water and dike for later disposal. Prevent entry into waterways, sewers, basements or confined areas. Do not clean-up or dispose of, except under supervision of a specialist.

Environmental Physical Data

BCF: no food chain concentration potential

Regulations

RCRA 40CFR: Not listed
CERCLA: 40CFR 302.4: Not listed
SARA 40CFR 372.65: Listed
SARA EHS 40CFR 355: Not listed
TSCA: Listed

Analytical Methods

Indoor / Expired Air: NIOSH 5009

BEN6440 **CAS #: 85-52-9**

2-BENZOYLBENZOIC ACID

RTECS: DG3600000
EINECS Number: 201-612-2
Molecular Formula: $C_{14}H_{10}O_3$
Formula Weight: 226.2

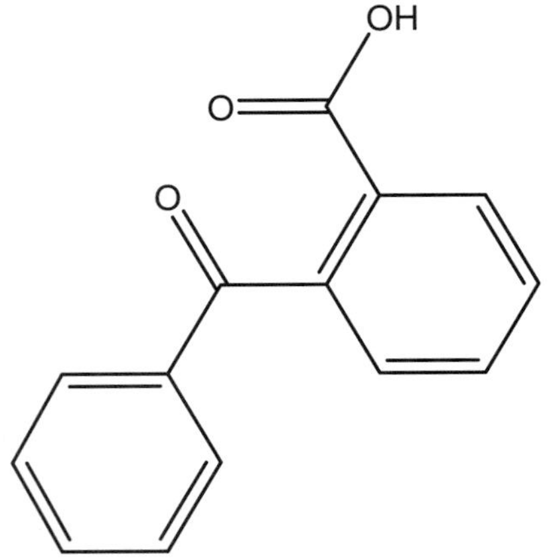

Chemical Structure

Synonyms: BENZOIC ACID,2-BENZOYL-; BENZOIC ACID,O-BENZOYL-; BENZOPHENONE-2-CARBONIC ACID; 2-BENZOPHENONECARBOXYLIC ACID; BENZOPHENONE-2-CARBOXYLIC ACID; 2-BENZOQUINONECARBOXYLIC ACID; O-BENZOYLBENZOIC ACID; 2-CARBOXYBENZOPHENONE
Description: triclinic needles

Physical Properties

Freezing Point: 127 °C (260.6 °F) to 129 °C (264.2 °F)
Water Solubility: Soluble in Hot Water
Other Solubilities: Very Soluble in Alcohol & Ether Soluble in hot Benzene

RTECS Toxicity Data

Acute Oral: Rat LD_{50} Dose: 4600 mg/kg; Toxic Effects: Behavioral - Muscle weakness. Mouse LD_{50} Dose: 880 mg/kg; Toxic Effects: Behavioral - Muscle weakness.

Hazard Overviews

Health: Irritating to eyes/skin/respiratory tract. Other Acute Effects: may be harmful by inhalation, ingestion, or skin absorption.
Fire: Extinguishing agents: water spray; carbon dioxide, dry chemical powder or appropriate foam. Precautions: combustible liquid.
Reactivity: Incompatible with: strong oxidizing agents, strong bases. Hazardous decomposition products: toxic fumes of: carbon monoxide, carbon dioxide.
Carcinogenicity: IARC - Not listed; NIOSH - Not listed; NTP - Not listed; ACGIH - Not listed; OSHA - Not listed; EPA - Not listed; MAK - Not listed
Primary Target Organs:

Eyes

Skin

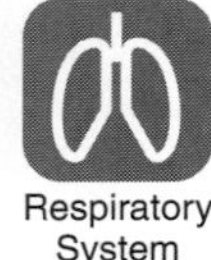
Respiratory System

Environmental

Regulations

RCRA 40CFR: Not listed

CERCLA: 40CFR 302.4: Not listed
SARA 40CFR 372.65: Not listed
SARA EHS 40CFR 355: Not listed
TSCA: Listed

BEN6680 CAS #: 156-08-1

BENZPHETAMINE

RTECS: SG9602000
Molecular Formula: $C_{17}H_{21}N$
Formula Weight: 239.35
Synonyms: BENZENEETHANAMINE,N,ALPHA-DIMETHYL-N-(PHENYLMETHYL)-,(+)-; BENZENEETHANAMINE,N,ALPHA-DIMETHYL-N-(PHENYLMETHYL)-,(+)-(9CI); BENZFETAMINE (INN); (+)-BENZPHETAMINE; (S)-(+)-BENZPHETAMINE; (S)-BENZPHETAMINE; D-BENZPHETAMINE; N-BENZYL-N,ALPHA-DIMETHYLPHENETHYLAMINE; N,ALPHA-DIMETHYL-N-(PHENYLMETHYL)-BENZENEETHANAMINE; D-N-METHYL-N-BENZYL-BETA-PHENYLISOPROPYLAMINE; PHENETHYLAMINE,N-BENZYL-N,ALPHA-DIMETHYL-,(+)-
Description: liquid; odorless
Use: medication as anorexic

Physical Properties

Boiling Point: 127 °C (261 °F) at 0.02 mm Hg
Water Solubility: Practically Insoluble in Water
Other Solubilities: Soluble in 95% Ethanol; 1 g Soluble in 1.5 ml Alcohol, 1.5 ml Chloroform.
Refraction Index: 1.5515 at 19 °C/D

RTECS Toxicity Data

Acute Oral: Man TD_{Lo} Dose: 5357 ug/kg; Toxic Effects: Automatic Nervous System - Sympathomimetic; Vascular - BP elevation not characterized in autonomic section. Rat LD_{50} Dose: 160 mg/kg.

Hazard Overviews

Carcinogenicity: IARC - Not listed; NIOSH - Not listed; NTP - Not listed; ACGIH - Not listed; OSHA - Not listed; EPA - Not listed; MAK - Not listed

Environmental

Regulations

RCRA 40CFR: Not listed
CERCLA: 40CFR 302.4: Not listed
SARA 40CFR 372.65: Not listed
SARA EHS 40CFR 355: Not listed
TSCA: Not listed

BEN6760 CAS #: 91-33-8

BENZTHIAZIDE

RTECS: DK8400000
EINECS Number: 202-061-0
Molecular Formula: $C_{15}H_{14}ClN_3O_4S_3$
Formula Weight: 431.96

Chemical Structure

Synonyms: AQUATAG; 2H-1,2,4-BENZOTHIADIAZINE-7-SULFONAMIDE,3-((BENZYLTHIO)METHYL)-6-CHLORO-,1,1-DIOXIDE; 2H-1,2,4-BENZOTHIADIAZINE-7-SULFONAMIDE,6-CHLORO-3-(((PHENYLMETHYL)THIO)METHYL)-,1,1-DIOXIDE; BENZOTHIAZIDE; 3-BENZYLTHIOMETHYL-6-CHLORO-2H-1,2,4-BENZOTHIADIAZINE-7-SULFONAMIDE 1,1-DIOXIDE; 3-[(BENZYLTHIO)METHYL]-6-CHLORO-2H-1,2,4-BENZOTHIADIAZINE-7-SULFONAMIDE 1,1-DIOXIDE; 3-BENZYLTHIOMETHYL-6-CHLORO-7-SULFAMOYL-1,2,4-BENZOTHIADIAZINE 1,1-DIOXIDE; 3-[(BENZYLTHIO)METHYL]-6-CHLORO-7-SULFAMOYL-2H-BENZO-1,2,4-THIADIAZINE 1,1-DIOXIDE; 6-CHLORO-3-[[(PHENYLMETHYL)THIO]METHYL]-2H-1,2,4-BENZOTHIADIAZINE-7-SULFONAMIDE 1,1-DIOXIDE; 6-CHLORO-7-SULFAMOYL-3-BENZYLTHIOMETHYL-2H-1,2,4-BENZOTHIADIAZINE 1,1-DIOXIDE; DIHYDREX; DIUCENE; EDEMEX; EXNA; EXOSALT; EXVA; FOUANE; FOVANE; FREEURIL; HY-DRINE; LEMAZIDE; P 1393; PFIZER 1393; PROAQUA; REGULON; URESE
Description: fine, white, crystalline powder or crystals; characteristic odor
Use: diuretic & antihypertensive agent

Physical Properties

Freezing Point: 238 °C (460.4 °F) to 239 °C (462.2 °F)
Water Solubility: Practically Insoluble in Water
Other Solubilities: freely Soluble in Dimethylformamide; Slightly Soluble in Acetone; Practically Insoluble in Ether & Chloroform.

RTECS Toxicity Data

Acute Oral: Rat LD_{50} Dose: >10 gm/kg. Mouse LD_{50} Dose: >5 gm/kg.

Hazard Overviews

Health: May be irritating to eyes/skin/respiratory tract. Harmful. Other Acute Effects: may be harmful by inhalation, ingestion, or skin absorption; may cause allergic respiratory and skin reactions; may cause: chloride and potassium loss, lethargy, muscle cramps, acidosis, gastric upset, rash, salivary gland obstruction, psychosis, convulsions, hyperuricemia, pancreatitis, leukopenia, liver disorders, photosensitization, diabetes mellitus, lupus erythematosus, coma. Chronic Effects: target organs: kidneys, blood.
Fire: Extinguishing agents: noncombustible; use extinguishing media appropriate to surrounding fire conditions. Precautions: combustible liquid.
Carcinogenicity: IARC - Not listed; NIOSH - Not listed; NTP - Not listed; ACGIH - Not listed; OSHA - Not listed; EPA - Not listed; MAK - Not listed

Primary Target Organs:

Environmental

Regulations

RCRA 40CFR: Not listed
CERCLA: 40CFR 302.4: Not listed
SARA 40CFR 372.65: Not listed
SARA EHS 40CFR 355: Not listed
TSCA: Not listed

BEN6840 CAS #: 86-13-5

BENZTROPINE

RTECS: YM3100000
Molecular Formula: $C_{21}H_{25}NO$
Formula Weight: 307.47
Synonyms: AKITAN; 8-AZABICYCLO(3.2.1)OCTANE,3-(DIPHENYLMETHOXY)-8-METHYL-,ENDO-; BENZOTROPINE; COBRENTIN; COGENTINE; COGENTINOL; NK-02; 1ALPHA H,5ALPHA H-TROPANE,3ALPHA-(DIPHENYLMETHOXY)-; TROPINE BENZOHYDRYL ETHER
Description: crystals
Use: anticholinergic for treatment of parkinsonism; control of tranquilizer effects on muscles

Physical Properties

Freezing Point: 143 °C (289.4 °F)
Water Solubility: Soluble in Water/ mesylate
pH: About 6

RTECS Toxicity Data

Acute Dermal: Mouse LD_{50} Route: Subcutaneous Dose: 60 mg/kg; Toxic Effects: Sense organs and special senses - Mydriasis (pupilliary dilation); Behavioral - Convulsions or effect on seizure threshold; Lungs, Thorax, or Respiration - Other changes.

Hazard Overviews

Carcinogenicity: IARC - Not listed; NIOSH - Not listed; NTP - Not listed; ACGIH - Not listed; OSHA - Not listed; EPA - Not listed; MAK - Not listed

Environmental

Regulations

RCRA 40CFR: Not listed
CERCLA: 40CFR 302.4: Not listed
SARA 40CFR 372.65: Not listed
SARA EHS 40CFR 355: Not listed
TSCA: Not listed

BEN6920 CAS #: 63-12-7

BENZUINAMIDE

Molecular Formula: $C_{22}H_{32}N_2O_5$
Formula Weight: 404.49
Synonyms: 2-ACETOXY-3-DIETHYLCARBAMYL-9,10-DIMETHOXY-1,2,3,4,6,7-HEXAHYDRO-11B-BENZO(A)QUINOLIZINE; 2-ACETOXY-3-(N,N-DIETHYLCARBOXAMIDO)-9,10-DIMETHOXY-1,2,3,4,6,7-HEXAHYDRO-11BH-BENZOPYRIDOCOLINE; 2-(ACETYLOXY)-N,N-DIETHYL-1,3,4,6,7,11B-HEXAHYDRO-9,10-DIMETHOXY-2H-BENZO(A)QUINOLIZINE-3-CARBOXAMIDE; BENZOCHINAMIDE; BENZOQUINAMIDE; 2H-BENZO(A)QUINOLIZINE-3-CARBOXAMIDE,2-(ACETYLOXY)-N,N-DIETHYL-1,3,4,6,7,11B-HEXAHYDRO-9,10-DIMETHOXY-; 2H-BENZO(A)QUINOLIZINE-3-CARBOXAMIDE,N,N-DIETHYL-1,3,4,6,7,11B-HEXAHYDRO-2-HYDROXY-9,10-DIMETHOXY-,ACETATE (ESTER); BZQ; N,N-DIETHYL-1,3,4,6,7,11B-HEXAHYDRO-2-HYDROXY-9,10-DIMETHOXY-2H-BENZO(A)QUINOLIZINE-3-CARBOXAMIDE ACETATE; EMETE-CON; EMETICON; 2-HYDROXY-3-DIETHYLCARBAMYL-9,10-DIMETHOXY-1,2,3,4,6,7-HEXAHYDRO-11BH-BENZOQUINOLIZINE ACETATE; P-2647; QUANTRIL; QUANTRYL
Description: crystals
Use: medication: tranquilizer, antiemetic; medication: in control of nausea & vomiting

Physical Properties

Freezing Point: 130 °C (266 °F) to 131.5 °C (268.7 °F)
Water Solubility: 1 g/12.5 mL
Other Solubilities: 1 g is Soluble in about 4 ml Alcohol, and 285 ml Acetone
RTECS Toxicity Data

Hazard Overviews

Carcinogenicity: IARC - Not listed; NIOSH - Not listed; NTP - Not listed; ACGIH - Not listed; OSHA - Not listed; EPA - Not listed; MAK - Not listed

Environmental

Regulations

RCRA 40CFR: Not listed
CERCLA: 40CFR 302.4: Not listed
SARA 40CFR 372.65: Not listed
SARA EHS 40CFR 355: Not listed
TSCA: Not listed

BEN7000 CAS #: 140-11-4

BENZYL ACETATE

RTECS: AF5075000
EINECS Number: 205-399-7
Molecular Formula: $C_9H_{10}O_2$
Structured MF: $CH_3COOCH_2C_6H_5$
Formula Weight: 150.17

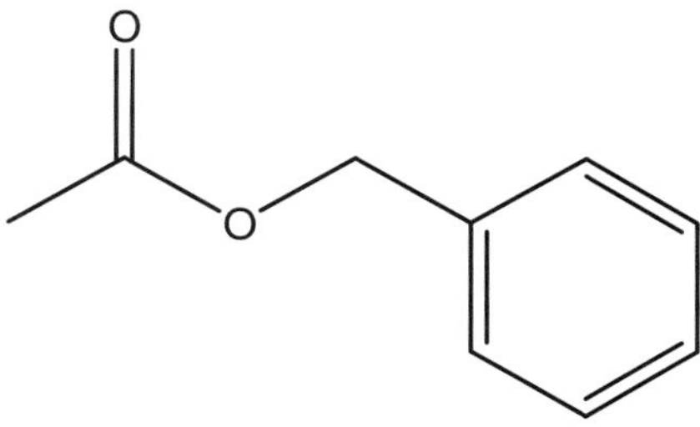

Chemical Structure

Synonyms: ACETIC ACID BENZYL ESTER; ACETIC ACID PHENYLMETHYL ESTER; ACETIC ACID,BENZYL ESTER; ACETIC ACID,PHENYLMETHYL ESTER; (ACETOXYMETHYL)BENZENE; ALPHA-ACETOXYTOLUENE; BENZYL ETHANOATE; BENZYLESTER KYSELINY OCTOVE; PHENYLMETHYL ACETATE

Description: colorless to water-white liquid; pear-like odor

Use: in perfumes, flavor ingredients, solvents for cellulose acetate, cellulose nitrate and natural and synthetic resins; oils, lacquers, polishes, printing inks, varnish removers, food processing and as a food additive in nonalcoholic beverages, ice cream, ices, sweets, baked goods gelatins, puddings and chewing gum

Physical Properties

Boiling Point: 213 °C (415 °F)
Freezing Point: -51 °C (-59.8 °F)
Specific Gravity: 1.05 at 25 °C/4 °C
Vapor Density: 5.1 Air=1
Density: 1.059 to 1.062 g/mL at 15 °C
Vapor Pressure: 1 mm Hg at 45 °C
Water Solubility: Practically Insoluble in Water
Other Solubilities: 95% Ethanol: >=100 mg/ml at 23 °C; Acetone: >=100 mg/ml at 23 °C; Cyclohexane: >=100 mg/ml at 21 °C; DMSO: >=100 mg/ml at 23 °C; Ether: Soluble.
Refraction Index: 1.5232 at 20 °C/D
Flash Point: 90 °C Closed Cup
Autoignition Temperature: 460 °C

RTECS Toxicity Data

Acute Oral: Rat LD_{50} Dose: 2490 mg/kg; Toxic Effects: Behavioral - Somnolence (general depressed activity). Mouse LD_{50} Dose: 830 mg/kg.

Acute Inhalation: Mouse LC_{Lo} Dose: 1300 mg/m^3/22hr; Toxic Effects: Behavioral - General anesthetic. Human TC_{Lo} Dose: 50 ppm; Toxic Effects: Behavioral - Antipsychotic; Lungs, Thorax, or Respiration - Other changes; Kidney, ureter, and Bladder - Other changes.

Acute Dermal: Rabbit LD_{50} Route: Skin; Dose: >5 gm/kg. Rabbit LD_{Lo} Route: Subcutaneous Dose: 4 gm/kg; Toxic Effects: Behavioral - Convulsions or effect on seizure threshold.

Chronic (Multiple Dose) Oral: Rat Dose: 355 gm/kg/13W-C; Toxic Effects: Brain and coverings - Other degenerative changes; Kidney, Ureter, and Bladder - Chgs in tubules (inc acute renal failure, acute tubular necrosis; DEATH. Rat Dose: 65 gm/kg/13W-I; Toxic Effects: DEATH.

Irritation Skin: Rabbit Standard Draize Test Dose: 100 mg/24H; Reaction: moderate.

Mutagenic: Human Mutations in Microorganisms; Cell Type: lymphocyte; Dose: 1500 mg/L (+S9). Mouse Mutations in Microorganisms; Cell Type: lymphocyte; Dose: 500 mg/L (+S9).

Tumorigenic: Rat Route: Oral; Dose: 258 gm/kg/2Y-I; Toxic Effects: Tumorigenic - Neoplastic by RTECS criteria; Gastrointestinal - Tumors. Mouse Route: Oral; Dose: 258 gm/kg/2Y-I; Toxic Effects: Tumorigenic - Neoplastic by RTECS criteria; Liver - Tumors.

Hazard Overviews

Fire Diamond

Health: Irritating to eyes/skin/respiratory tract. Also Causes: skin sensitization, GI tract irritation, nausea/vomiting upon ingestion, urinary system effects

Fire: Combustible. Use dry chemical, carbon dioxide, water spray, or regular foam.

Reactivity: Stable. Hazardous polymerization cannot occur. Avoid: heat; ignition sources; oxidizers. Incompatible with: oxidizers. Hazardous decomposition products: carbon oxide(s).

Carcinogenicity: IARC - Group 3, Not classifiable as to carcinogenicity to humans; NIOSH - Not listed; NTP - Not listed; ACGIH - Class A4, Not classifiable as a human carcinogen; OSHA - Not listed; EPA - Not listed; MAK - Not listed

Primary Target Organs:

Eyes

Skin

Respiratory System

Environmental

Cleanup/Disposal: Guide No. 171: Do not touch or walk through spilled material. Stop leak if you can do it without risk. Prevent dust cloud. Avoid inhalation of asbestos dust. Small Dry Spills: With clean shovel place material into clean, dry container and cover loosely; move containers from spill area. Small Spills: Take up with sand or other noncombustible absorbent material and place into containers for later disposal. Large Spills: Dike far ahead of liquid spill for later disposal. Cover powder spill with plastic sheet or tarp to minimize spreading. Prevent entry into waterways, sewers, basements or confined areas.

Environmental Physical Data

Octanol/Water Partition Coefficient: log K_{ow} = 1.96

Regulations

RCRA 40CFR: Not listed
CERCLA: 40CFR 302.4: Not listed
SARA 40CFR 372.65: Not listed
SARA EHS 40CFR 355: Not listed

TSCA: Listed

BEN7080 **CAS #: 100-51-6**

BENZYL ALCOHOL

RTECS: DN3150000
EINECS Number: 202-859-9
Molecular Formula: C_7H_8O
Structured MF: $C_6H_5CH_2OH$
Formula Weight: 108.13

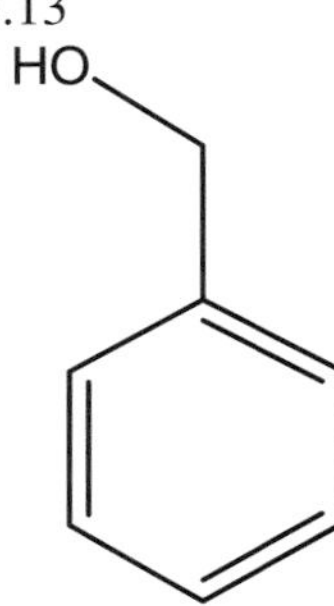

Chemical Structure

Synonyms: ALCOOL BENZYLIQUE; BENZAL ALCOHOL; BENZENECARBINOL; BENZENEMETHANOL; BENZOYL ALCOHOL; BENZYLICUM; EUXYL K 100; (HYDROXYMETHYL)BENZENE; ALPHA-HYDROXYTOLUENE; HYDROXYTOLUENE; METHANOL,PHENYL-; PHENOLCARBINOL; PHENYL CARBINOL; PHENYLCARBINOL; PHENYLMETHANOL; PHENYLMETHYL ALCOHOL; ALPHA-TOLUENOL

Description: colorless to water-white liquid; faint, aromatic odor

Use: as a photographic developer for color movie film and in perfumes, flavor industries, pharmaceuticals as a bacteriostatic, cosmetics, ointments, emulsions, textiles, sheet plastics and inks; a solvent for dyestuffs, cellulose esters, cellulose acetate, casein, gelatin waxes and shellac; as an intermediate for benzyl esters and ether as a surfactant, an insect repellent, local anesthetic, preservative in radio pharmaceuticals and preservative in sterile solutions for intramuscular or intravenous use; in heat-sealing polyethylene films, dyeing nylon filament and in microscopy as an embedding material; in veterinary medicine for relief from pruritus; once was used as an antiseptic

Physical Properties

Boiling Point: 204.7 °C (400 °F)
Freezing Point: -15.19 °C (4.658 °F)
Specific Gravity: 1.04535 at 20 °C/4 °C
Vapor Density: 3.72 Air=1
Saturated Vapor Density: 1.200646252 kg/m^3
Density: 1.043 to 1.046 g/mL at 20 °C
Vapor Pressure: 0.15 mm Hg at 25 °C
Water Solubility: 1 g/25 ml
Other Solubilities: 1 vol dissolved in 1.5 vol of 50% Ethyl Alcohol; miscible with absolute Alcohol & 94% Alcohol, Chloroform; > 10% in Acetone; > 10% in Benzene; > 10% in Ether.
Surface Tension: 39.0 dynes/cm
Odor Threshold: 5.5 ppm
pH: Solution in water is neutral
Refraction Index: 1.5396 at 20 °C/D
Evaporation Rate: 1767 Ether=1
Critical Temperature: 403 °C
Critical Pressure: 663 psia
Ionization Potential (eV): 8.26 +/-0.3
Flash Point: 93 °C Closed Cup
Autoignition Temperature: 436 °C
LEL: 1.3% v/v
UEL: 13% v/v

RTECS Toxicity Data

Acute Oral: Rat LD_{50} Dose: 1230 mg/kg; Toxic Effects: Behavioral - Somnolence (general depressed activity); Behavioral - Excitment; Behavioral - Coma. Mouse LD_{50} Dose: 1360 mg/kg.
Acute Inhalation: Rat LC_{Lo} Dose: 1000 ppm/8hr.
Acute Dermal: Rabbit LD_{50} Route: Skin; Dose: 2 gm/kg. Rat LD_{Lo} Route: Subcutaneous Dose: 1700 mg/kg; Toxic Effects: Sense organs and special senses - Miosis (pupilliary dilation); Behavioral - Coma; Kidney, Ureter, and Bladder - Other changes.
Chronic (Multiple Dose) Oral: Rat Dose: 2100 mg/kg/21D-I; Toxic Effects: Nutritional and gross metabolic - Weight loss or decreased weight gain; DEATH. Rat Dose: 13 gm/kg/13W-I; Toxic Effects: Brain and coverings - Other degenerative changes; DEATH. Rat Dose: 24 mL/kg/12D-I; Toxic Effects: Liver - Other changes; Biochemical - Dehydrogenases.
Irritation Eye: Rabbit Standard Draize Test Dose: 750 ug open; Reaction: severe.
Irritation Skin: Man Standard Draize Test Dose: 16 mg/48H; Reaction: mild. Rabbit Standard Draize Test Dose: 100 mg/24H; Reaction: moderate.
Reproductive/Teratogenic: Mouse Route: Oral; Dose: 6 gm/kg; Duration: female 6-13D of pregnancy; Effects on Newborn - Growth statistics.
Mutagenic: Rat DNA Damage; Cell Type: liver; Dose: 10 mmol/L. Mouse Mutations in Microorganisms; Cell Type: lymphocyte; Dose: 250 mg/L (+S9).

Hazard Overviews

Fire Diamond

Health: Irritating to eyes/skin/respiratory tract. Also Causes: headache, nausea, vomiting, diarrhea; upon severe exposure: respiratory stimulation, muscle paralysis, convulsions, unconsciousness, death.
Fire: Will burn. Fight fires with dry chemical, carbon dioxide, or alcohol-resistant foam. Water may be used but could cause frothing.

Reactivity: Unstable, slowly oxidizes in air. Hazardous polymerization cannot occur. Avoid: heat; ignition sources. Incompatible with: acids, hydrogen bromide ion; oxidizing agents; plastics. Hazardous decomposition products: acrid smoke; fumes.

Carcinogenicity: IARC - Not listed; NIOSH - Not listed; NTP - Not listed; ACGIH - Not listed; OSHA - Not listed; EPA - Not listed; MAK - Not listed

Primary Target Organs:

Eyes

Skin

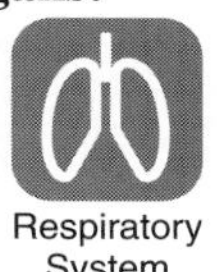
Respiratory System

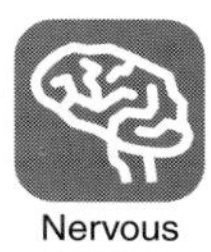
Nervous System

Exposure Limits

AIHA WEEL: TWA: 10 ppm.

Respirator Recommendation

Exposure Range: >10 to 500 ppm Supplied Air, Constant Flow/Pressure Demand, Half Mask

Exposure Range: >500 to 10,000 ppm Supplied Air, Constant Flow/Pressure Demand, Full Face

Exposure Range: >10,000 to unlimited ppm Self-contained Breathing Apparatus, Pressure Demand, Full Face

Note: odor threshold unknown

Environmental

Ecotoxicity: LC_{50} Lepomis macrochirus (bluegill sunfish) 10 ppm/96 hr, static bioassay in fresh water at 23 °C, mild aeration after 24 hr Lepomis macrochirus (bluegill sunfish) static bioassay in fresh water at 23 °C, mild aeration applied after 24 hr: 100% survival after 5 ppm/96 hr, 20% survival after 18 ppm/96 hr, 20% survival after 32 ppm/48 hr LC_{50} Menidia beryllina (tidewater silverside fish) 15 ppm/96 hr, static bioassay in synthetic seawater at 23 °C, mild aeration after 24 hr Menidia beryllina (tidewater silverside fish): static bioassay in synthetic seawater at 23 °C: mild aeration applied after 24 hr: 80% survival after 10 ppm/96 hr, 20% survival after 32 ppm/96 hr

Environmental Fate: If released to soil, it is expected to display high mobility and readily leach through soil. Volatilization from dry soil to the atmosphere may be an important fate process; however, it is not expected to be an important process in moist soils. Microbial degradation in soil may occur, based on limited data. If released to water, it is expected to undergo microbial degradation under aerobic and anaerobic conditions. Neither volatilization to the atmosphere, hydrolysis, direct photolytic degradation, chemical oxidation, bioconcentration in fish and aquatic organisms, nor adsorption to sediment and suspended organic matter are expected to be significant processes in environmental waters. In the atmosphere, it is expected to exist almost entirely in the vapor phase. The estimated half-life for the vapor phase reaction with photochemically produced hydroxyl radicals is 2 days. Its water solubility suggests that it may undergo deposition to the surface by rain washout and other wet deposition processes.

Cleanup/Disposal: Guide No. 171: Do not touch or walk through spilled material. Stop leak if you can do it without risk. Prevent dust cloud. Avoid inhalation of asbestos dust. Small Dry Spills: With clean shovel place material into clean, dry container and cover loosely; move containers from spill area. Small Spills: Take up with sand or other noncombustible absorbent material and place into containers for later disposal. Large Spills: Dike far ahead of liquid spill for later disposal. Cover powder spill with plastic sheet or tarp to minimize spreading. Prevent entry into waterways, sewers, basements or confined areas.

Environmental Physical Data

Henry's Law Constant: calculated at 3.91×10^{-7}

Octanol/Water Partition Coefficient: log K_{ow} = 1.10

Sorption Partition Coefficient: K_{oc} = < 5

BCF: calculated at 4.0

BOD: 155%, 5 days

Regulations

RCRA 40CFR: Not listed

CERCLA: 40CFR 302.4: Not listed

SARA 40CFR 372.65: Not listed

SARA EHS 40CFR 355: Not listed

TSCA: Listed

Analytical Methods

Soil: EPA 1625; SW846 3640A, 8250A, 8270B, 8270C

Water / Groundwater: EPA 1625; ASTM D4763

Plasma: EPA 29

BEN7160 CAS #: 120-51-4

BENZYL BENZOATE

RTECS: DG4200000

EINECS Number: 204-402-9

Molecular Formula: $C_{14}H_{12}O_2$

Structured MF: $C_6H_5COOCH_2C_6H_5$

Formula Weight: 212.24

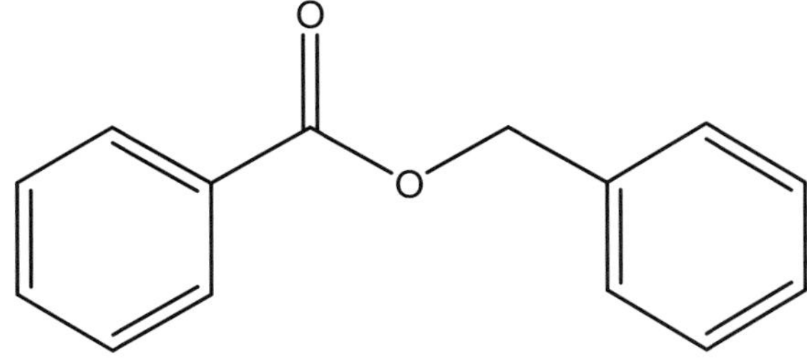

Chemical Structure

Synonyms: ASCABIN; ASCABIOL; BENYLATE; BENZOIC ACID,BENZYL ESTER; BENZOIC ACID,PHENYLMETHYL ESTER; BENZYL ALCOHOL BENZOIC ESTER; BENZYL BENZENECARBOXYLATE; BENZYL PHENYLFORMATE; BENZYLESTER KYSELINY BENZOOVE; BENZYLETS; COLEBENZ; NOVOSCABIN; PERUSCABIN; SCABAGEN; SCABANCA; SCABIDE; SCABIOZON; SCOBENOL; VANZOATE; VENZONATE

Description: leaflets or clear, colorless oily liquid; faint pleasant aromatic odor

Use: camphor substitute in celluloid & plastic pyroxylin compd; pediculicide; acaricide; in synthetic musks, confectionery flavors & chewing gum flavors; fixative; plasticizer for cellulose acetate & nitrocellulose; remedy for scabies; dye carrier; antispasmodic; repellant for chiggers, mosquitoes & ticks on man

Physical Properties

Boiling Point: 323 °C (613 °F) to 324 °C (615 °F)
Freezing Point: 21 °C (69.8 °F)
Specific Gravity: 1.118 at 25 °C/4 °C
Vapor Density: 7.31 Air=1
Vapor Pressure: 15 mm Hg
Water Solubility: Insoluble in Water
Other Solubilities: Soluble in Acetone, Benzene.
Surface Tension: 26.6 dyne/cm at 210.5 °C
pH: Solution in Alcohol is practically neutral to moistened litmus
Refraction Index: 1.5681 at 21 °C/D
Critical Temperature: 547 °C
Critical Pressure: 2.58×10^6 Pa
Flash Point: 148 °C Closed Cup
Autoignition Temperature: 480 °C

RTECS Toxicity Data

Acute Oral: Rat LD_{50} Dose: 1700 uL/kg. Mouse LD_{50} Dose: 1400 uL/kg.
Acute Dermal: Rabbit LD_{50} Route: Skin; Dose: 4 gm/kg. Rat LD_{50} Route: Skin; Dose: 4 mL/kg.
Chronic (Multiple Dose) Dermal: Rabbit Route: Skin; Dose: 180 mL/kg/13W-I; Toxic Effects: Behavioral - Muscle weakness; Nutritional and gross metabolic - Weight loss or decreased weight gain; DEATH.

Hazard Overviews

Health: Irritating to eyes/skin/respiratory tract. Harmful. Other Acute Effects: harmful if swallowed; may be harmful if inhaled or absorbed through the skin.
Fire: Will burn. Hazards: emits toxic fumes. Extinguishing agents: water spray; carbon dioxide, dry chemical powder or appropriate foam. Precautions: combustible liquid.
Reactivity: Stable. Hazardous polymerization will not occur. Incompatible with: strong oxidizing agents, strong bases. Hazardous decomposition products: toxic fumes of: carbon monoxide, carbon dioxide.
Carcinogenicity: IARC - Not listed; NIOSH - Not listed; NTP - Not listed; ACGIH - Not listed; OSHA - Not listed; EPA - Not listed; MAK - Not listed
Primary Target Organs:

Eyes

Skin

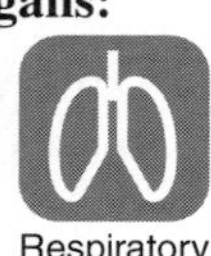
Respiratory System

Environmental

Environmental Physical Data

Octanol/Water Partition Coefficient: log K_{ow} = 3.97

Regulations

RCRA 40CFR: Not listed
CERCLA: 40CFR 302.4: Not listed
SARA 40CFR 372.65: Not listed
SARA EHS 40CFR 355: Not listed
TSCA: Listed

Analytical Methods

Soil: SW846 8061A

BEN7240 CAS #: 100-39-0

BENZYL BROMIDE

RTECS: XS7965000
DOT: UN1737; IMO6.1
EINECS Number: 202-847-3
Molecular Formula: C_7H_7Br
Structured MF: $C_6H_5CH_2Br$
Formula Weight: 171.04

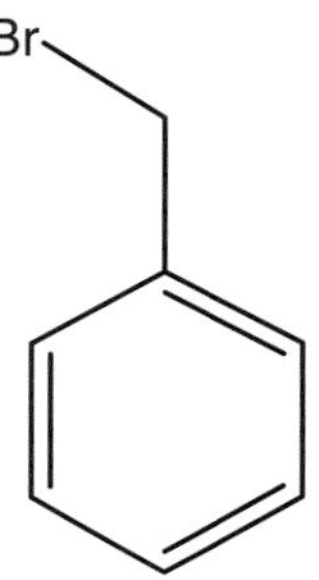

Chemical Structure

Synonyms: BENZENE,(BROMOMETHYL)-; (BROMOMETHYL)BENZENE; P-(BROMOMETHYL)NITROBENZENE; BROMOPHENYLMETHANE; ALPHA-BROMO-TOLUENE; ALPHA-BROMOTOLUENE; BROMOTOLUENE,ALPHA; OMEGA-BROMOTOLUENE; CYCLITE; TOLUENE,ALPHA-BROMO-
Description: clear liquid; pleasant odor
Use: foaming and frothing agents, organic synthesis, benzylating agent

Physical Properties

Boiling Point: 198 °C (388 °F) to 199 °C (390 °F)
Freezing Point: -4 °C (24.8 °F)
Specific Gravity: 1.438 at 22 °C/0 °C
Vapor Density: 5.9 Air=1
Vapor Pressure: 1 mm Hg at 32.2 °C
Water Solubility: Insoluble in Water
Other Solubilities: 95% Ethanol: Reaction; Acetone: >=100 mg/ml at 23 °C; Benzene: Soluble; DMSO: >=100 mg/ml at 23 °C; Ether: Soluble.
Surface Tension: Estimated at 35 dynes/cm
Odor Threshold: 2 ppm as Hydrogen Bromide
pH: Acid
Refraction Index: 1.5752 at 20 °C/D
Ionization Potential (eV): 9.02 +/-0.5
Flash Point: 79 °C Closed Cup

RTECS Toxicity Data

Mutagenic: Hamster Sister Chromatid Exchenge; Cell Type: ovary; Dose: 30 umol/L. Bacteria - E Coli Unscheduled DNA Synthesis; Dose: 1300 umol/L.

Hazard Overviews

Corrosive

Fire Diamond

Health: Corrosive, causes burns to eyes/skin/respiratory tract. Also Causes: CNS depression (high concentrations), pulmonary edema, severe gastrointestinal irritation upon ingestion. Chronic Effects: possible bromism.
Fire: Combustible. Use dry chemical, carbon dioxide, regular foam, or water in flooding quantities as fog. Combustion or contact with water liberates highly toxic hydrogen bromide gas.
Reactivity: Stable. Hazardous polymerization cannot occur. Avoid: storing over a sieve. Incompatible with: water; all common metals. Hazardous decomposition products: toxic hydrogen bromide gas.
Carcinogenicity: IARC - Not listed; NIOSH - Not listed; NTP - Not listed; ACGIH - Not listed; OSHA - Not listed; EPA - Not listed; MAK - Not listed
Primary Target Organs:

Eyes

Skin

Respiratory System

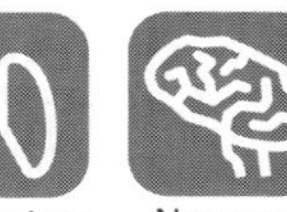
Nervous System

Environmental

Ecotoxicity: Aquatic toxicity: 0.05 mg/l/*/marine fish/no irritant response/salt water 0.1 mg/l/*/marine fish/violent irritant activity/salt water *Time period not specified
Environmental Fate: If released to moist soil, it is expected to hydrolyze as fast, if not faster, than in water. Mobility is expected to be extremely limited. If released to dry soil, this compound is expected to volatilize fairly rapidly. If released to water, the dominant removal mechanism is expected to be chemical hydrolysis (half-life 79 minutes at 25 °C). Due to the reactivity of this compound, volatilization, bioaccumulation in aquatic organisms, and adsorption to suspended solids and sediments in water are not expected to be important fate processes. If released to the atmosphere, the dominant removal mechanism is expected to be reaction with photochemically generated hydroxyl radicals (half-life 6.7 days).
Cleanup/Disposal: Guide No. 156: Eliminate all ignition sources (no smoking, flares, sparks or flames in immediate area). All equipment used when handling the product must be grounded. Do not touch damaged containers or spilled material unless wearing appropriate protective clothing. Stop leak if you can do it without risk. A vapor suppressing foam may be used to reduce vapors. For chlorosilanes, use AFFF alcohol-resistant medium expansion foam to reduce vapors. Do not get water on spilled substance or inside containers. Use water spray to reduce vapors or divert vapor cloud drift. Prevent entry into waterways, sewers, basements or confined areas. Small Spills: Cover with dry earth, dry sand, or other non-combustible material followed with plastic sheet to minimize spreading or contact with rain. Use clean non-sparking tools to collect material and place it into loosely covered plastic containers for later disposal.

Environmental Physical Data

Henry's Law Constant: estimated at 5 x10^{-4}
Octanol/Water Partition Coefficient: log K_{ow} = 2.92
Sorption Partition Coefficient: K_{oc} = estimated at 154 to 923
BCF: estimated at 98

Regulations

RCRA 40CFR: Not listed
CERCLA: 40CFR 302.4: Not listed
SARA 40CFR 372.65: Not listed
SARA EHS 40CFR 355: Not listed
TSCA: Listed

Analytical Methods

Air: ASTM D4490

BEN7320 CAS #: 100-44-7

BENZYL CHLORIDE

RTECS: XS8925000
DOT: UN1738; IMO6.1
EINECS Number: 202-853-6
Molecular Formula: C_7H_7Cl
Structured MF: $C_6H_5CH_2Cl$
Formula Weight: 126.58

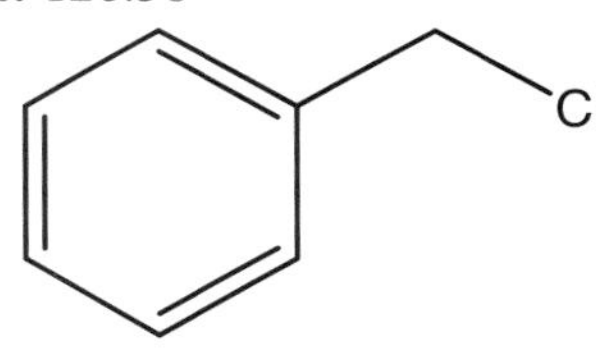

Chemical Structure

Synonyms: BENZENE,(CHLOROMETHYL)-; BENZILE (CLORURO DI); BENZYL CHLORIDE,UNSTABILIZED; BENZYLCHLORID; BENZYLE (CHLORURE DE); (CHLOROMETHYL)BENZENE; CHLOROMETHYLBENZENE; CHLOROPHENYLMETHANE; A-CHLOROTOLUENE; ALPHA-CHLOROTOLUENE; OMEGA-CHLOROTOLUENE; ALPHA-CHLORTOLUOL; CHLORURE DE BENZYLE; TOLUENE,ALPHA-CHLORO-; ALPHA-TOLYL CHLORIDE; TOLYL CHLORIDE
Description: colorless to slightly yellow liquid; sharp, pungent odor
Use: manufacture of benzyl compounds, perfumes, pharmaceutical products, dyes, synthetic tannins, artificial resins, photography, gasoline gum inhibitors and formerly used as irritant gas in chemical warfare

Molecular Formula: $C_8H_7ClO_2$
Structured MF: $C_6H_5CH_2OCOCl$
Formula Weight: 170.60

decomposition products: carbon monoxide, carbon dioxide, hydrogen chloride gas, phosgene gas.
Carcinogenicity: IARC - Not listed; NIOSH - Not listed; NTP - Not listed; ACGIH - Not listed; OSHA - Not listed; EPA - Not listed; MAK - Not listed

Physical Properties

Boiling Point: 179 °C (354 °F)

Health: Corrosive to eyes/skin/respiratory tract. Also Causes: watering eyes, cough, dizziness, difficulty breathing, pulmonary edema (fluid in lungs), permanent eye damage,

Primary Target Organs:

Eyes

Skin

Respiratory System

Environmental

Cleanup/Disposal: Guide No. 137: Fully encapsulating, vapor protective clothing should be worn for spills and leaks with no fire. Do not touch damaged containers or spilled material unless wearing appropriate protective clothing. Stop leak if you can do it without risk. Use water spray to reduce vapors; do not put water directly on leak, spill area or inside container. Keep combustibles (wood, paper, oil, etc.) away from spilled material. Small Spills: Cover with dry earth, dry sand, or other non-combustible material followed with plastic sheet to minimize spreading or contact with rain. Use clean non-sparking tools to collect material and place it into loosely covered plastic containers for later disposal. Prevent entry into waterways, sewers, basements or confined areas.

Environmental Physical Data

BCF: no food chain concentration potential

Regulations

RCRA 40CFR: Not listed
CERCLA: 40CFR 302.4: Not listed
SARA 40CFR 372.65: Not listed
SARA EHS 40CFR 355: Not listed
TSCA: Listed

BEN7480 CAS #: 949-38-2

BENZYL O-CHLOROPHENYL ETHER

Molecular Formula: $C_{13}H_{11}ClO$
Formula Weight: 218.67
Synonyms: BENZENE,1-CHLORO-2-(PHENYLMETHOXY)-; ETHER,BENZYL O-CHLOROPHENYL-

Hazard Overviews

Carcinogenicity: IARC - Not listed; NIOSH - Not listed; NTP - Not listed; ACGIH - Not listed; OSHA - Not listed; EPA - Not listed; MAK - Not listed

Environmental

Regulations

RCRA 40CFR: Not listed
CERCLA: 40CFR 302.4: Listed as Compound per CWA Section 307(a)
SARA 40CFR 372.65: Listed as Compound
SARA EHS 40CFR 355: Not listed
TSCA: Not listed

BEN7560 CAS #: 103-41-3

BENZYL CINNAMATE

RTECS: GD8400000
EINECS Number: 203-109-3
Molecular Formula: $C_{16}H_{14}O_2$
Formula Weight: 238.27

Chemical Structure

Synonyms: BENZYL ALCOHOL,CINNAMATE; BENZYL ALCOHOL,CINNAMIC ESTER; BENZYL BETA-PHENYLACRYLATE; BENZYL GAMMA-PHENYLACRYLATE; BENZYLESTER KYSELINY SKORICOVE; BENZYL-3-PHENYLPROPENOATE; CINNAMEIN; TRANS-CINNAMIC ACID BENZYL ESTER; CINNAMIC ACID,BENZYL ESTER; PHENYLMETHYL 3-PHENYL-2-PROPENOATE; 3-PHENYL-2-PROPENOIC ACID PHENYLMETHYL ESTER; 2-PROPENOIC ACID,3-PHENYL-,PHENYLMETHYL ESTER; 2-PROPENOIC ACID,3-PHENYL-,PHENYLMETHYL ESTER (9CI)
Description: white to pale-yellow, fused, crystalline solid; sweet odor of balsam
Use: artificial flavors, perfumes, mainly as fixative

Physical Properties

Boiling Point: 228 °C (442 °F) to 230 °C (446 °F) at 22 mm Hg
Freezing Point: 39 °C (102.2 °F)
Specific Gravity: 1.109 at 15 °C
Vapor Pressure: 1 mm Hg at 173.8 °C
Water Solubility: Practically Insoluble in Water
Other Solubilities: Soluble in Alcohol, Ether, oils; Practically Insoluble in Propylene Glycol & Glycerin.
Flash Point: > 100 °C

RTECS Toxicity Data

Acute Oral: Rat LD_{50} Dose: 5530 mg/kg; Toxic Effects: Behavioral - Somnolence (general depressed activity); Behavioral - Coma. Guinea Pig LD_{50} Dose: 3760 mg/kg; Toxic Effects: Behavioral - Somnolence (general depressed activity); Gastrointestinal - Gastritis.
Irritation Skin: Rabbit Standard Draize Test Dose: 500 mg/24H; Reaction: mild.

Hazard Overviews

Health: May be irritating to eyes/skin. Other Acute Effects: may be harmful by inhalation, ingestion, or skin absorption.
Fire: Will burn. Extinguishing agents: water spray; carbon dioxide, dry chemical powder or appropriate foam. Precautions: combustible liquid.

Reactivity: Incompatible with: strong oxidizing agents, strong bases. Hazardous decomposition products: toxic fumes of: carbon monoxide, carbon dioxide.
Carcinogenicity: IARC - Not listed; NIOSH - Not listed; NTP - Not listed; ACGIH - Not listed; OSHA - Not listed; EPA - Not listed; MAK - Not listed

Environmental

Regulations

RCRA 40CFR: Not listed
CERCLA: 40CFR 302.4: Not listed
SARA 40CFR 372.65: Not listed
SARA EHS 40CFR 355: Not listed
TSCA: Listed

BEN7640 CAS #: 140-29-4

BENZYL CYANIDE

RTECS: AM1400000
EINECS Number: 205-410-5
Molecular Formula: C_8H_7N
Structured MF: $C_6H_5CH_2CN$
Formula Weight: 117.14

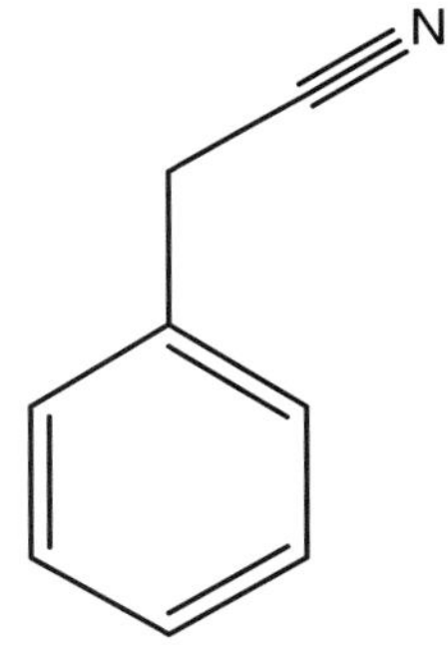

Chemical Structure

Synonyms: ACETIC ACID,PHENYL-NITRILE; ACETONITRILE,PHENYL-; BENZENEACETONITRILE; BENZENEACETONITRILE (9CI); BENZYL NITRILE; BENZYLKYANID; (CYANOMETHYL)BENZENE; ALPHA-CYANOTOLUENE; OMEGA-CYANOTOLUENE; PHENYL ACETYL NITRILE; 2-PHENYLACETONITRILE; PHENYLACETONITRILE; PHENYLACETONITRILE,LIQUID; TOLUENE,ALPHA-CYANO; TOLUENE,ALPHA-CYANO-; ALPHA-TOLUNITRILE
Description: colorless oily liquid; aromatic odor
Use: in organic synthesis, especially penicillin precursors

Physical Properties

Boiling Point: 233.5 °C (452 °F)
Freezing Point: -23.8 °C (-10.84 °F)
Specific Gravity: 1.0214 at 15 °C/15 °C
Density: 1.0157 g/mL
Vapor Pressure: 1 mm Hg at 60 °C
Water Solubility: Insoluble in Water
Other Solubilities: 95% Ethanol: >=100 mg/ml at 17 °C; Acetone: >=100 mg/ml at 17 °C; Alcohol: Soluble; DMSO: >=100 mg/ml at 17 °C; Ether: Soluble.
Refraction Index: 1.52105 at 25 °C/D
Ionization Potential (eV): 9.32 +/-1.0
Flash Point: 101 °C

RTECS Toxicity Data

Acute Oral: Rat LD_{50} Dose: 270 mg/kg; Toxic Effects: Behavioral - Convulsions or effect on seizure threshold; Lungs, Thorax, or Respiration - Dyspnea; Liver - Other changes. Mouse LD_{50} Dose: 45500 ug/kg.
Acute Inhalation: Rat LC_{50} Dose: 430 mg/m^3/2hr; Toxic Effects: Behavioral - Altered sleep time (including change in righting reflex); Behavioral - Muscle contraction or spasticity; Lungs, Thorax, or Respiration - Dyspnea. Mouse LC_{50} Dose: 100 mg/m^3/2hr.
Acute Dermal: Rabbit LD_{50} Route: Skin; Dose: 270 mg/kg. Rabbit LD_{Lo} Route: Subcutaneous Dose: 50 mg/kg; Toxic Effects: Peripheral Nerve and sensation - Spastic parapysis with or without sensory change; Behavioral - Ataxia; Lungs, Thorax, or Respiration - Dyspnea.
Chronic (Multiple Dose) Inhalation: Rat Dose: 60 mg/m^3/2H/4W-I; Toxic Effects: Lungs, Thorax, or Respiration - Structural or functional change in trachea or bronchi; Liver - Other changes; Nutritional and gross metabolic - Weight loss or decreased weight gain.
Irritation Skin: Rabbit Standard Draize Test Dose: 500 mg/24H; Reaction: mild.

Hazard Overviews

Poison

Health: Irritating to eyes/skin/respiratory tract. Poison. Other Acute Effects: may be fatal if inhaled; harmful if swallowed or absorbed through skin; exposure can cause mydriasis; coma; seizures; systemic toxicity is presumed to be a result of the metabolic release of cyanide. Chronic Effects: target organs: blood, cardiovascular system, lungs, central nervous system.
Fire: Will burn. Hazards: emits toxic fumes. Extinguishing agents: water spray; carbon dioxide, dry chemical powder or appropriate foam. Precautions: combustible liquid.
Reactivity: Incompatible with: strong oxidizing agents. Hazardous decomposition products: thermal decomposition may produce carbon monoxide, carbon dioxide, and nitrogen oxides; hydrogen cyanide.
Carcinogenicity: IARC - Not listed; NIOSH - Not listed; NTP - Not listed; ACGIH - Not listed; OSHA - Not listed; EPA - Not listed; MAK - Not listed

Primary Target Organs:

Eyes

Skin

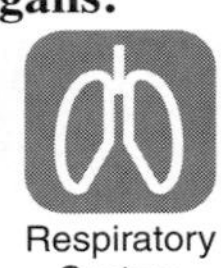
Respiratory System

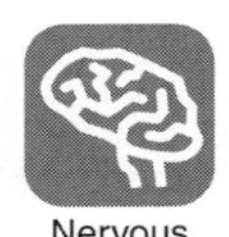
Nervous System

Cardio-vascular

Blood

Environmental

Ecotoxicity: Fishes: Leuciscus idus 48h LC_0 50 mg/l
Cleanup/Disposal: Guide No. 152: Do not touch damaged containers or spilled material unless wearing appropriate protective clothing. Stop leak if you can do it without risk. Prevent entry into waterways, sewers, basements or confined areas. Cover with plastic sheet to prevent spreading. Absorb or cover with dry earth, sand or other non-combustible material and transfer to containers. Do not get water inside containers.

Environmental Physical Data

Octanol/Water Partition Coefficient: log K_{ow} = measured at 1.56

Regulations

RCRA 40CFR: Not listed
CERCLA: 40CFR 302.4: Listed as Compound per CWA Section 307(a) per CAA Section 112
SARA 40CFR 372.65: Not listed TPQ: 500 lb
SARA EHS 40CFR 355: Listed TPQ: 500 lb
TSCA: Listed

BEN7720 CAS #: 98-87-3

BENZYL DICHLORIDE

RTECS: CZ5075000
DOT: UN1886; IMO6.1
EINECS Number: 202-709-2
Molecular Formula: $C_7H_6Cl_2$
Structured MF: $C_6H_5CHCl_2$
Formula Weight: 161.03

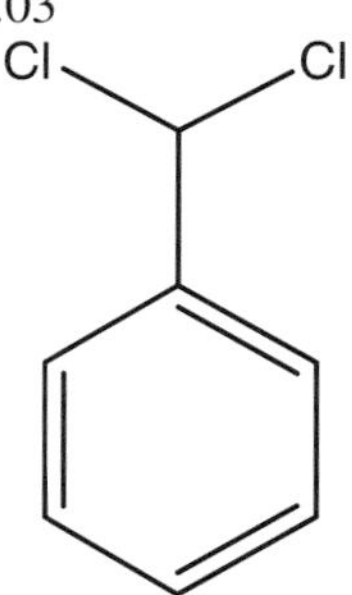

Chemical Structure

Synonyms: AI-28597; BENZAL CHLORIDE; BENZENE,(DICHLOROMETHYL)-; BENZYLENE CHLORIDE; BENZYLIDENE CHLORIDE; CHLOROBENZAL; CHLORURE DE BENZYLIDENE; CLORURO DE BENCILIDENO; (DICHLOROMETHYL)BENZENE; DICHLOROPHENYLMETHANE; ALPHA,ALPHA-DICHLOROTOLUENE; TOLUENE,ALPHA,ALPHA-DICHLORO-

Description: colorless oily liquid; faint aromatic pungent odor
Use: in the manufacture of benzaldehyde, cinnamic acid and benzoyl chloride

Physical Properties

Boiling Point: 205 °C (401 °F)
Freezing Point: -16.4 °C (2.48 °F)
Specific Gravity: 1.26
Vapor Density: Technical Grade 5.6 Air=1
Density: 1.2557 g/mL at 14 °C
Vapor Pressure: 1 mm Hg at 35.4 °C
Water Solubility: Insoluble in Water
Other Solubilities: 95% Ethanol: >=100 mg/ml at 17 °C; Acetone: >=100 mg/ml at 17 °C; DMSO: >=100 mg/ml at 17 °C; Dilute alkali solution: Soluble; Ether: Freely Soluble.
Surface Tension: 20.20 dynes/cm
pH: Expected to be acid
Refraction Index: 1.5502 at 20 °C
Flash Point: 92 °C Closed Cup
Autoignition Temperature: 67 °C

RTECS Toxicity Data

Acute Oral: Rat LD_{50} Dose: 3249 mg/kg. Mouse LD_{50} Dose: 2462 mg/kg.
Acute Inhalation: Rat LC_{50} Dose: 61 ppm/2hr; Toxic Effects: Brain and coverings - Recordings from specific areas of CNS; Behavioral - Excitment; Lungs, Thorax, or Respiration - Respiratory depression. Mouse LC_{50} Dose: 32 ppm/2hr; Toxic Effects: Brain and coverings - Recordings from specific areas of CNS; Behavioral - Excitment; Lungs, Thorax, or Respiration - Respiratory depression.
Mutagenic: Bacteria - B Subtilis DNA Repair; Dose: 31 umol/disc. Bacteria - E Coli Mutations in Microorganisms; Dose: 600 nmol/plate/20M (+S9). Bacteria - S Typhimurium Mutations in Microorganisms; Dose: 600 nmol/plate/20M (-S9).
Tumorigenic: Mouse Route: Skin; Dose: 9200 mg/kg/50W-I; Toxic Effects: Tumorigenic - Carcinogenic by RTECS criteria; Lungs, Thorax, or Respiration - Tumors; Skin and appendages - Tumors. Mouse Route: Skin; Dose: 35200 mg/kg/42W-I; Toxic Effects: Tumorigenic - Neoplastic by RTECS criteria; Lungs, Thorax, or Respiration - Tumors; Skin and appendages - Tumors.

Hazard Overviews

Poison

Corrosive

Health: Corrosive to eyes/skin/respiratory tract. Poison. Other Acute Effects: very toxic by inhalation, in contact with skin and if swallowed; may be fatal if inhaled, swallowed, or absorbed through skin; extremely destructive to tissue of the mucous membranes and upper respiratory tract, eyes and skin; inhalation may result in spasm, inflammation and edema of the larynx and bronchi, chemical pneumonitis and pulmonary edema; lachrymator; symptoms of exposure may include burning sensation, coughing, wheezing, laryngitis, shortness of breath, headache, nausea and vomiting. Chronic Effects: possible risk of irreversible effects. Possible human carcinogen. Possibly causes mutagenic effects in humans.

Fire: Combustible. Hazards: emits toxic fumes. Extinguishing agents: carbon dioxide, dry chemical powder or appropriate foam. Precautions: combustible liquid.

Carcinogenicity: IARC - Group 2B, Possibly carcinogenic to humans; NIOSH - Not listed; NTP - Not listed; ACGIH - Not listed; OSHA - Not listed; EPA - Not listed; MAK - Class A2, Unmistakably carcinogenic in animal experimentation only

Primary Target Organs:

Eyes

Skin

Respiratory System

Environmental

Environmental Fate: If released to water, it will hydrolyze rapidly (half-life of 7.4 minutes at 25 °C and pH 7) and form benzaldehyde as a hydrolysis product. If released to soil hydrolysis is expected to be the dominant fate process in the presence of moisture. Leaching in moist soils is not expected to be significant since it is expected to hydrolyze first. Evaporation from dry surfaces is expected to occur. If released to the atmosphere, it will react in the vapor-phase with photochemically produced hydroxyl radicals at an estimated half-life rate of 1.84 days.

Cleanup/Disposal: Guide No. 156: Eliminate all ignition sources (no smoking, flares, sparks or flames in immediate area). All equipment used when handling the product must be grounded. Do not touch damaged containers or spilled material unless wearing appropriate protective clothing. Stop leak if you can do it without risk. A vapor suppressing foam may be used to reduce vapors. For chlorosilanes, use AFFF alcohol-resistant medium expansion foam to reduce vapors. Do not get water on spilled substance or inside containers. Use water spray to reduce vapors or divert vapor cloud drift. Prevent entry into waterways, sewers, basements or confined areas. Small Spills: Cover with dry earth, dry sand, or other non-combustible material followed with plastic sheet to minimize spreading or contact with rain. Use clean non-sparking tools to collect material and place it into loosely covered plastic containers for later disposal.

Environmental Physical Data

Henry's Law Constant: estimated at 0.0007

Octanol/Water Partition Coefficient: log K_{ow} = 3.217

Sorption Partition Coefficient: K_{oc} = estimated at 510

BCF: estimated at 164

Regulations

RCRA 40CFR: Listed Hazardous Waste No. U017 Toxic Waste

CERCLA: 40CFR 302.4: Listed per RCRA Section 3001 RQ: 5000 lb (2268 kg)

SARA 40CFR 372.65: Listed TPQ: 500 lb

SARA EHS 40CFR 355: Listed TPQ: 5,000 lb

TSCA: Listed

Analytical Methods

Soil: SW846 8121

BEN7800 **CAS #: 100-53-8**

BENZYL MERCAPTAN

RTECS: XT8650000

EINECS Number: 202-862-5

Molecular Formula: C_7H_8S

Structured MF: $C_6H_5CH_2SH$

Formula Weight: 124.20

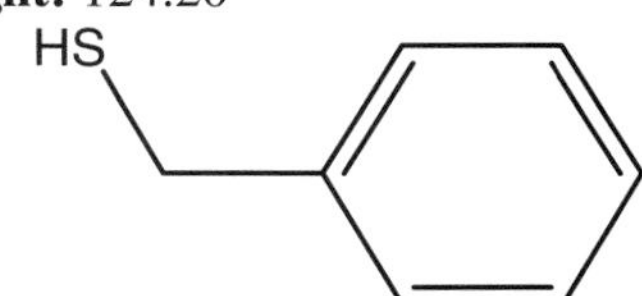

Chemical Structure

Synonyms: BENZENEMETHANETHIOL; BENZYLHYDROSULFIDE; BENZYLTHIOL; (MERCAPTOMETHYL)BENZENE; ALPHA-MERCAPTOTOLUENE; METHANETHIOL,PHENYL-; PHENYLMETHANETHIOL; PHENYLMETHYL MERCAPTAN; THIOBENZYL ALCOHOL; TOLUENE,ALPHA-MERCAPTO-; ALPHA-TOLUENETHIOL; ALPHA-TOLUOLTHIOL; ALPHA-TOLYL MERCAPTAN

Description: colorless to water-white, mobile liquid; repulsive, garlic-like odor or odor of leek

Use: odorant; flavors; in non-alcoholic beverages, ice cream, ices, etc; candy; baked goods; synthetic flavoring substance and adjuvant

Physical Properties

Boiling Point: 194 °C (381 °F) to 195 °C (383 °F)

Freezing Point: -30 °C (-22 °F)

Specific Gravity: 1.05

Vapor Density: 4.28 Air=1

Vapor Pressure: 71.8 mm Hg at 120.766 °C

Water Solubility: Insoluble in Water

Other Solubilities: Very Soluble in Ethanol, Ether. Slightly Soluble in Carbon Tetrachloride.

Odor Threshold: Medium odor threshold 0.00019 ppm

Refraction Index: 1.5751 at 20 °C/D

Ionization Potential (eV): 8.5 +/-0.7

Flash Point: 70 °C Closed Cup

RTECS Toxicity Data

Acute Oral: Rat LD_{50} Dose: 493 mg/kg; Toxic Effects: Behavioral - Somnolence (general depressed activity); Lungs, Thorax, or Respiration - Respiratory depression; Behavioral - Coma.

Acute Inhalation: Mouse LC_{50} Dose: 178 ppm/4hr; Toxic Effects: Behavioral - Muscle weakness; Behavioral - Ataxia; Lungs, Thorax, or Respiration - Cyanosis.

Tumorigenic: Mouse Route: Skin; Dose: 16 gm/kg/26W-I; Toxic Effects: Tumorigenic - Equivocal tumorigenic agent by RTECS criteria; Skin and appendages - Tumors; Tumorigenic - Tumors at site of application.

Hazard Overviews

Health: Irritating to eyes/skin/respiratory tract. Toxic. Other Acute Effects: harmful if swallowed, inhaled, or absorbed through skin; symptoms of exposure may include burning sensation, coughing, wheezing, laryngitis, shortness of breath, headache, nausea and vomiting.

Fire: Combustible. Hazards: emits toxic fumes. Extinguishing agents: water spray; carbon dioxide, dry chemical powder or appropriate foam. Precautions: combustible liquid.

Reactivity: Incompatible with: strong bases, strong oxidizing agents. Hazardous decomposition products: toxic fumes of: carbon monoxide, carbon dioxide, sulfur oxides, hydrogen sulfide gas.

Carcinogenicity: IARC - Not listed; NIOSH - Not listed; NTP - Not listed; ACGIH - Not listed; OSHA - Not listed; EPA - Not listed; MAK - Not listed

Primary Target Organs:

Eyes

Skin

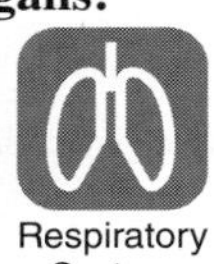
Respiratory System

Environmental

Regulations

RCRA 40CFR: Not listed
CERCLA: 40CFR 302.4: Not listed
SARA 40CFR 372.65: Not listed
SARA EHS 40CFR 355: Not listed
TSCA: Listed

BEN7960 CAS #: 2116-65-6

4-BENZYL PYRIDINE

RTECS: US2625000
EINECS Number: 218-319-0
Molecular Formula: $C_{12}H_{11}N$
Formula Weight: 169.24

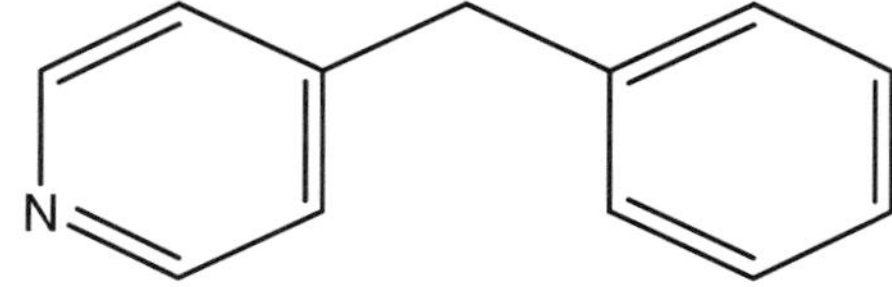

Chemical Structure

Synonyms: BA 33216; 4-BENZYLPYRIDINE; GAMMA-BENZYLPYRIDINE; PYRIDINE,4-(PHENYLMETHYL)-(9CI)

Physical Properties

Boiling Point: 287 °C (548.6 °F)
Freezing Point: 10 °C (50 °F)
Specific Gravity: 1.06120
Water Solubility: Slightly Soluble
Other Solubilities: Ethanol: Soluble; Ether: Very Soluble; Carbon Tetrachloride: Soluble
Refraction Index: 1.5818

RTECS Toxicity Data

Acute Oral: Rat LD_{50} Dose: 560 uL/kg; Toxic Effects: Sense organs and special senses - Chromidracryorrhea; Behavioral - Somnolence (general depressed activity); Gastrointestinal - Hypermotility, diarrhea. Mouse LD_{50} Dose: 630 uL/kg; Toxic Effects: Behavioral - Tremor; Lungs, Thorax, or Respiration - Dyspnea; Gastrointestinal - Ulceration or bleeding from stomach.

Hazard Overviews

Health: Irritating to eyes/skin/respiratory tract. Harmful. Other Acute Effects: harmful if swallowed, inhaled, or absorbed through skin.

Fire: Hazards: emits toxic fumes. Extinguishing agents: water spray; carbon dioxide, dry chemical powder or appropriate foam. Precautions: combustible liquid.

Reactivity: Incompatible with: strong oxidizing agents, strong acids. Hazardous decomposition products: toxic fumes of: carbon monoxide, carbon dioxide, nitrogen oxides.

Carcinogenicity: IARC - Not listed; NIOSH - Not listed; NTP - Not listed; ACGIH - Not listed; OSHA - Not listed; EPA - Not listed; MAK - Not listed

Primary Target Organs:

Eyes

Skin

Respiratory System

Environmental

Regulations

RCRA 40CFR: Not listed
CERCLA: 40CFR 302.4: Not listed
SARA 40CFR 372.65: Not listed
SARA EHS 40CFR 355: Not listed
TSCA: Listed

BEN8040 CAS #: 538-74-9

BENZYL SULFIDE

EINECS Number: 208-703-6
Molecular Formula: $C_{14}H_{14}S$
Structured MF: $(C_6H_5CH_2)_2S$
Formula Weight: 214.32

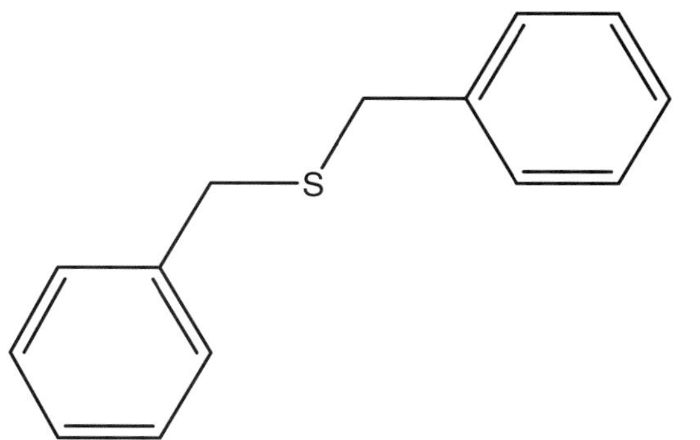

Chemical Structure

Synonyms: BENZENE,1,1'-(THIOBIS(METHYLENE))BIS-; BENZENE,1-1'-[THIOBIS(METHYLENE)]BIS-; BENZYL MONOSULFIDE; BENZYL THIOETHER; DIBENZYL MONOSULFIDE; DIBENZYL SULFIDE; DIBENZYL THIOETHER; SULFIDE,DIBENZYL; 1,1'-(THIOBIS(METHYLENE))BISBENZENE
Description: colorless plates
Use: corrosion inhibitor; organic synthesis

Physical Properties

Boiling Point: Decomposes
Freezing Point: 50 °C (122 °F)
Specific Gravity: 1.0712 at 50 °C/50 °C
Water Solubility: Insoluble in Water
Other Solubilities: 95% Ethanol: 10-50 mg/ml at 21 °C; Acetone: >=100 mg/ml at 21 °C; DMSO: >=100 mg/ml at 21 °C; Ether: Soluble.
Ionization Potential (eV): 8.05 +/-1.0
Flash Point: > 110 °C

Hazard Overviews

Health: May cause irritation. Other Acute Effects: may be harmful by inhalation, ingestion, or skin absorption; exposure can cause nausea, headache and vomiting.
Fire: Will burn. Hazards: emits toxic fumes. Extinguishing agents: water spray; carbon dioxide, dry chemical powder or appropriate foam. Precautions: combustible liquid.
Reactivity: Incompatible with: strong oxidizing agents. Hazardous decomposition products: carbon monoxide, carbon dioxide, sulfur oxides.
Carcinogenicity: IARC - Not listed; NIOSH - Not listed; NTP - Not listed; ACGIH - Not listed; OSHA - Not listed; EPA - Not listed; MAK - Not listed

Environmental

Regulations

RCRA 40CFR: Not listed
CERCLA: 40CFR 302.4: Not listed
SARA 40CFR 372.65: Not listed
SARA EHS 40CFR 355: Not listed
TSCA: Listed

BEN8120 **CAS #: 1694-09-3**

BENZYL VIOLET

RTECS: BQ1140000
EINECS Number: 216-901-9
Molecular Formula: $C_{39}H_{41}N_3NaO_6S_2$
Formula Weight: 734.9
Synonyms: A.F. VIOLET NO 1; A.F. VIOLET NO. 1; ACID FAST VIOLET 5BN; ACID VIOLET; ACID VIOLET 5B; ACID VIOLET 6B; ACID VIOLET 49; ACID VIOLET 5 B; ACID VIOLET 5BN; ACID VIOLET 4BNP; ACID VIOLET 4BNS; ACID VIOLET S; ACILAN VIOLET S4BN; AF VIOLET NO 1; AIZEN ACID VIOLET 5BH; AIZEN FOOD VIOLET NO. 1; AMMONIUM,(4-(P-(DIMETHYLAMINO)-ALPHA-(P-(ETHYL(M-SULFOBENZYL)AMINO)PHENYL)-; ATLANTIC ACID VIOLET 4BNS; BENZENEMETHANAMINIUM,N-(4-((4-(DIMETHYLAMINO)PHENYL)(4-(ETHYL((3-SULFO-; BENZENEMETHANAMINIUM,N-(4-((4-(DIMETHYLAMINO)PHENYL)(4-(ETHYL((3-SULFOPHENYL)METHYL)AMINO)PHENYL)METHYLENE)-2,5-CYCLOHEXADIEN-1-YLIDINE)-N-ETHYL-3-SULFO-,HYDROXIDE, INNER SALT, SODIUM SALT; BENZYL VIOLET 3B; BENZYL VIOLET 4B; BENZYLIDENE)-2,5-CYCLOHEXADIEN-1-YLIDENE)ETHYL(M-SULFOBENZYL)-,HYDROXIDE,; C.I. 42640; C.I. ACID VIOLET 49; C.I. ACID VIOLET 49 (SODIUM SALT); C.I. ACID VIOLET 49,SODIUM SALT; C.I. F FOOD VIOLET 2; C.I. FOOD VIOLET 2; CALCOCID VIOLET 4BNS; COGILOR VIOLET 411.12; COOMASSIE VIOLET; CYCLOHEXADIENIMINE); 2,5-CYCLOHEXADIEN-1-YLIDENE)ETHYL(M-SULFOBENZYL)AMMONIUM HYDROXIDE,INNER; D AND C VIOLET NO 1; D AND C VIOLET NO. 1; (4-(P-(DIMETHYLAMINO)-ALPHA-(P-(ETHYL(M-SULFOBENZYL)AMINO)PHENYL)BENZYLIDENE; DISPERSED VIOLET 12197; ERIOSIN VIOLET 3B; FAST ACID VIOLET 5BN; FD & C VIOLET 1; FD & C VIOLET NO. 1; FD AND C VIOLET 1; FD AND C VIOLET NO 1; FOOD VIOLET 2; FORMYL VIOLET S4BN; HIDACID WOOL VIOLET 5B; INNER SALT,SODIUM SALT; INTRACID VIOLET 4BNS; KITON VIOLET 4BNS; MONOSODIUM SALT OF 4-((N-ETHYL-P-SULFOBENZYLAMINO)PHENYL)-(4(N-ETHYL-P-; ORIENT WATER VIOLET 1; PERGACID VIOLET 2B; PERGACID VIOLET 3B; PHENYL)METHYL)AMINO)PHENYL)METHYLENE)-2,5-CYCLOHEXADIEN-1-YLIDENE)-N-ETHYL; POLAXAL VIOLET 6B; SALT,SODIUM SALT; SERVA VIOLET 49; SOLAR VIOLET 5BN; 3-SULFO-,HYDROXIDE,INNER SALT,SODIUM SALT; SULFONIUMBENZYLAMINO)PHENYL)METHYLENE)-(N,N-DIMETHYL-DELTA(SUP 2.5)-; TERTRACID BRILLIANT VIOLET 6B; TETRACID BRILLIANT VIOLET 6B; 11386 VIOLET; VIOLET 5B; VIOLET 6B; VIOLET 2; VIOLET 5BN; VIOLET KYSELA 49; VIOLET NO 1; VIOLET NO. 1; VIOLET POTRAVINARSKA 2; WOOL VIOLET; WOOL VIOLET 4BN; WOOL VIOLET 5BN
Description: fine powder
Use: dye wool, silk, nylon, leather and anodized aluminum; as a biological stain, as a wood stain, in inks and in coloring paper; as a color additive for food, drugs and cosmetics

Physical Properties

Freezing Point: Decomposes at 245 °C (473 °F) to 250 °C (482 °F)
Water Solubility: Soluble in Water
Other Solubilities: 95% Ethanol: <1 mg/ml at 20 °C; Acetone: <1 mg/ml at 20 °C; DMSO: <1 mg/ml at 20 °C; Methanol: <1 mg/ml at 20 °C; Toluene: <1 mg/ml at 20 °C; Vegetable oils: Insoluble.
Flash Point: Not available; probably combustible

RTECS Toxicity Data

Mutagenic: Bacteria - S Typhimurium Mutations in Microorganisms; Dose: 320 ug/plate (-S9).
Tumorigenic: Rat Route: Oral; Dose: 498 gm/kg/28W-C; Toxic Effects: Tumorigenic - Carcinogenic by RTECS criteria; Sense organs and special senses - Tumors; Skin and appendages - Tumors. Rat Route: Subcutaneous; Dose: 9360 mg/kg/2Y-I; Toxic Effects: Tumorigenic - Equivocal tumorigenic agent by RTECS criteria; Tumorigenic - Tumors at site of application.

Hazard Overviews

Fire: Will burn.
Carcinogenicity: IARC - Group 2B, Possibly carcinogenic to humans; NIOSH - Not listed; NTP - Not listed; ACGIH - Not listed; OSHA - Not listed; EPA - Not listed; MAK - Not listed

Environmental

Regulations

RCRA 40CFR: Not listed
CERCLA: 40CFR 302.4: Not listed
SARA 40CFR 372.65: Not listed
SARA EHS 40CFR 355: Not listed
TSCA: Listed

BEN8200 CAS #: 100-46-9

BENZYLAMINE

RTECS: DP1488500
EINECS Number: 202-854-1
Molecular Formula: C_7H_9N
Structured MF: $C_6H_5CH_2NH_2$
Formula Weight: 107.15

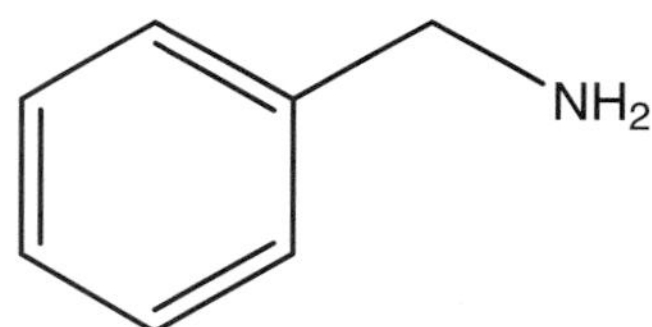

Chemical Structure

Synonyms: ALPHA-AMINO TOLUENE; (AMINOMETHYL)BENZENE; ALPHA-AMINOTOLUENE; AMINOTOLUENE; OMEGA-AMINOTOLUENE; BENZENEMETHANAMINE; BENZENEMETHANAMINE (9CI); MONOBENZYLAMINE; MORINGINE; (PHENYLMETHYL)AMINE; SUMINE 2005; SUMINE 2006
Description: light amber liquid
Use: in organic synth; chemical intermediate for dyes, pharmaceuticals, and polymers; reaction with glyoxal hemiacetal to synthesize 1-substituted isoquinolines

Physical Properties

Boiling Point: 185 °C (365 °F)
Freezing Point: About -46 °C (-50.8 °F)
Specific Gravity: 0.983 at 19 °C/4 °C
Vapor Density: 3.7 Air=1
Saturated Vapor Density: 1.202778509 kg/m^3
Vapor Pressure: 0.653 mm Hg at 25 °C
Water Solubility: Soluble in Water
Other Solubilities: Soluble Alcohol, Ether, Acetone, Benzene
Surface Tension: 39.5 dynes/cm at 20 °C
Refraction Index: 1.5401
Ionization Potential (eV): 8.64 +/-0.7
Flash Point: 63 °C

RTECS Toxicity Data

Acute Oral: Mammal LD_{50} Dose: 700 mg/kg.

Hazard Overviews

Corrosive

Health: Corrosive to eyes/skin/respiratory tract. Harmful Other Acute Effects: harmful if swallowed, inhaled, or absorbed through skin; inhalation may result in spasm, inflammation and edema of the larynx and bronchi, chemical pneumonitis and pulmonary edema; symptoms of exposure may include burning sensation, coughing, wheezing, laryngitis, shortness of breath, headache, nausea and vomiting.
Fire: Combustible. Hazards: emits toxic fumes. Extinguishing agents: carbon dioxide, dry chemical powder or appropriate foam; water spray. Precautions: combustible liquid.
Reactivity: Incompatible with: acids, acid chlorides, acid anhydrides, strong oxidizing agents, absorbs carbon dioxide from air. Hazardous decomposition products: thermal decomposition may produce carbon monoxide, carbon dioxide, and nitrogen oxides.
Carcinogenicity: IARC - Not listed; NIOSH - Not listed; NTP - Not listed; ACGIH - Not listed; OSHA - Not listed; EPA - Not listed; MAK - Not listed
Primary Target Organs:

Eyes

Skin

Respiratory System

Environmental

Ecotoxicity: LC_{50} Pimephales promelas (fathead minnow) 102 mg/l/96 hr (confidence limit 97.9-106 mg/l), flow-through bioassay with measured concentrations, 23.9 °C, dissolved oxygen 6.9 mg/l, hardness 44.7 mg/l calcium carbonate, alkalinity 44.0 mg/l calcium carbonate, and pH 7.9
Environmental Fate: If released to the atmosphere, it will exist primarily in the vapor-phase where it will degrade by reaction with photochemically produced hydroxyl radicals

(estimated half-life of 11.5 hours). Physical removal from the atmosphere may be possible through wet deposition. If released to soil or water, it is expected to biodegrade. A variety of biodegradation screening studies, including die-away tests in sediment and lake water, have demonstrated that it biodegrades readily.

Cleanup/Disposal: Guide No. 132: Fully encapsulating, vapor protective clothing should be worn for spills and leaks with no fire. Eliminate all ignition sources (no smoking, flares, sparks or flames in immediate area). All equipment used when handling the product must be grounded. Do not touch or walk through spilled material. Stop leak if you can do it without risk. Prevent entry into waterways, sewers, basements or confined areas. A vapor suppressing foam may be used to reduce vapors. Absorb with earth, sand or other non-combustible material and transfer to containers (except for Hydrazine). Use clean non-sparking tools to collect absorbed material. Large Spills: Dike far ahead of liquid spill for later disposal. Water spray may reduce vapor; but may not prevent ignition in closed spaces.

Environmental Physical Data

Henry's Law Constant: estimated at 6.12×10^{-7}
Octanol/Water Partition Coefficient: log K_{ow} = 1.09
Sorption Partition Coefficient: K_{oc} = estimated at 7 to 95
BCF: estimated at 4
BOD: 85% BODT, 14 days

Regulations

RCRA 40CFR: Not listed
CERCLA: 40CFR 302.4: Not listed
SARA 40CFR 372.65: Not listed
SARA EHS 40CFR 355: Not listed
TSCA: Listed

Analytical Methods

Water / Groundwater: ASTM D4763

BEN8280 CAS #: 120-32-1

O-BENZYL-P-CHLOROPHENOL

RTECS: GO7175000
EINECS Number: 204-385-8
Molecular Formula: $C_{13}H_{11}ClO$
Structured MF: $C_6H_5CH_2C_6H_3OHCl$
Formula Weight: 218.69

Cl OH

Chemical Structure

Synonyms: 2-BENZYL-4-CHLOROPHENOL; BENZYLCHLOROPHENOL; ORTHO-BENZYL-PARA-CHLOROPHENOL; BIO-CLAVE; 4-CHLORO-2-BENZYLPHENOL; P-CHLORO-O-BENZYLPHENOL; 5-CHLORO-2-HYDROXYDIPHENYLMETHANE; CHLOROPHENE; CHLOROPHENE,USAN; 4-CHLORO-ALPHA-PHENYL-O-CRESOL; 4-CHLORO-ALPHA-PHENYL-ORTHO-CRESOL; 4-CHLORO-2-(PHENYLMETHYL)PHENOL; CLOROFENE; CLOROPHENE; O-CRESOL,4-CHLORO-ALPHA-PHENYL-; KETOLIN-H; KETOLIN H; NEOSABENYL; ORTHOBENZYL-P-CHLOROPHENOL; ORTHOBENZYL-PARA-CHLOROPHENOL; ORTHOBENZYLPARACHLOROPHENOL; PHENOL,4-CHLORO-2-(PHENYLMETHYL)-; PHENOL,4-CHLORO-2-(PHENYLMETHYL)-(9CI); SANTOPHEN; SANTOPHEN 1; SANTOPHEN 1 FLAKE; SANTOPHEN 1 GERMICIDE; SANTOPHEN 1 SOLUTION; SANTOPHEN I; SANTOPHEN I GERMICIDE; SANTOPHEN L GERMICIDE; SEPTIPHENE

Description: white to light tan or pink crystals or flakes; slightly phenolic odor

Use: disinfectant and germicide

Physical Properties

Boiling Point: 160 °C (320 °F) to 162 °C (324 °F) at 3.5 mm Hg
Freezing Point: 48.5 °C (119.3 °F)
Specific Gravity: 1.186 to 1.19 at 55 deg/15.5 deg
Saturated Vapor Density: 1.201032795 kg/m^3
Vapor Pressure: 0.1 mm Hg at 20 °C
Water Solubility: Insoluble in Water
Other Solubilities: 95% Ethanol: >=100 mg/ml at 16 °C; Acetone: >=100 mg/ml at 16 °C; Absolute Ethanol: Soluble; Alkaline solutions: Soluble; DMSO: >=100 mg/ml at 16 °C; Other organic solvents: Soluble.
Flash Point: 187.8 °C

RTECS Toxicity Data

Acute Oral: Rat LD_{50} Dose: 1700 mg/kg. Mouse LD_{50} Dose: 65 mg/kg; Toxic Effects: Behavioral - Somnolence (general depressed activity); Behavioral - Convulsions or effect on seizure threshold; Lungs, Thorax, or Respiration - Dyspnea.

Acute Inhalation: Rat LC; Dose: >13200 mg/m^3/6hr.

Acute Dermal: Rabbit LD_{Lo} Route: Skin; Dose: 5010 mg/kg; Toxic Effects: Behavioral - Muscle weakness; Liver - Change in gall bladder structure or function; Blood - Changes in spleen. Mouse LD_{50} Route: Subcutaneous Dose: 350 mg/kg; Toxic Effects: Behavioral - Somnolence (general depressed activity); Behavioral - Convulsions or effect on seizure threshold; Lungs, Thorax, or Respiration - Dyspnea.

Chronic (Multiple Dose) Oral: Rat Dose: 12 gm/kg/2W-I; Toxic Effects: Kidney, Ureter, and Bladder - Changes in both tubules and glomeruli; Nutritional and gross metabolic - Weight loss or decreased weight gain. Rat Dose: 31200 mg/kg/13W-I; Toxic Effects: Kidney, Ureter, and Bladder - Changes in kidney weight; Endocrine - Changes in thymus weight; Blood - Changes in serum composition.

Chronic (Multiple Dose) Dermal: Rabbit Route: Skin; Dose: 80 mL/kg/2W-I; Toxic Effects: Skin and appendages - Dermatitis, other; DEATH.

Mutagenic: Human Mutations in Mammalian Somatic Cells; Cell Type: lymphocyte; Dose: 40 mg/L. Mouse Mutations in Mammalian Somatic Cells; Cell Type: lymphocyte; Dose: 25 mg/L.

Tumorigenic: Rat Route: Oral; Dose: 87600 mg/kg/2Y-I; Toxic Effects: Tumorigenic - Equivocal tumorigenic agent by RTECS criteria; Kidney, Ureter, and Bladder - Kidney tumors. Mouse Route: Oral; Dose: 175 gm/kg/2Y-I; Toxic Effects: Tumorigenic - Neoplastic by RTECS criteria; Kidney, Ureter, and Bladder - Kidney tumors.

Hazard Overviews

Corrosive

Health: Irritating to eyes/skin/respiratory tract. Other Acute Effects: harmful if swallowed, inhaled, or absorbed through skin.

Fire: Will burn. Hazards: emits toxic fumes. Extinguishing agents: water spray; carbon dioxide, dry chemical powder or appropriate foam. Precautions: combustible liquid.

Reactivity: Stable. Hazardous polymerization will not occur. Incompatible with: strong oxidizing agents. Hazardous decomposition products: toxic fumes of: carbon monoxide, carbon dioxide, hydrogen chloride gas.

Carcinogenicity: IARC - Not listed; NIOSH - Not listed; NTP - Not listed; ACGIH - Not listed; OSHA - Not listed; EPA - Not listed; MAK - Not listed

Primary Target Organs:

Eyes

Skin

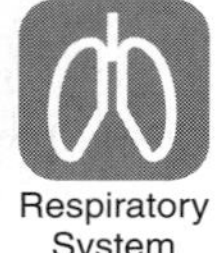
Respiratory System

Environmental

Environmental Fate: Rapidly transformed in aquatic and terrestrial environments through biodegradation and/or photodegradation; in soil, an experimental K_{oc} of 2050 indicates only slight mobility. In the atmosphere, it degrades readily by reaction with photochemically produced hydroxyl radicals (estimated half-life of about 20 hours). Hydrolysis and volatilization are insignificant fate processes for this compound. It has a low potential to bioconcentrate and is rapidly metabolized and excreted in fish.

Cleanup/Disposal: Guide No. 171: Do not touch or walk through spilled material. Stop leak if you can do it without risk. Prevent dust cloud. Avoid inhalation of asbestos dust. Small Dry Spills: With clean shovel place material into clean, dry container and cover loosely; move containers from spill area. Small Spills: Take up with sand or other noncombustible absorbent material and place into containers for later disposal. Large Spills: Dike far ahead of liquid spill for later disposal. Cover powder spill with plastic sheet or tarp to minimize spreading. Prevent entry into waterways, sewers, basements or confined areas.

Environmental Physical Data

Henry's Law Constant: estimated at 9.96×10^{-9}

Octanol/Water Partition Coefficient: log K_{ow} = 3.6

Sorption Partition Coefficient: K_{oc} = 2050

BCF: bluegill 75

BOD: 69% BODT, 28 days

Regulations

RCRA 40CFR: Not listed

CERCLA: 40CFR 302.4: Listed as Compound per CWA Section 307(a)

SARA 40CFR 372.65: Listed as Compound

SARA EHS 40CFR 355: Not listed

TSCA: Listed

Analytical Methods

Soil: EPA PMD-PFH

BEN8360 **CAS #: 122-19-0**

BENZYLDIMETHYLSTEARYLAM-MONIUM CHLORIDE

RTECS: BO7000000

EINECS Number: 204-527-9

Molecular Formula: $C_{27}H_{50}ClN$

Structured MF: $CH_3(CH_2)_{17}N(CH_3)_2CH_2C_6H_5$•Cl

Formula Weight: 424.23

Synonyms: 2B; AMMONIUM,BENZYLDIMETHYLOCTADECYL-,CHLORIDE; AMMONYX 4; AMMONYX 4002; AMMONYX 485; AMMONYX 490; AMMONYX CA SPECIAL; ARQUAD DM18B-90; BARQUAT SB-25; BENZENEMETHANAMINIUM,N,N-DIMETHYL-N-OCTADECYL-,CHLORIDE; BENZENEMETHANAMINIUM,N,N-DIMETHYL-N-OCTADECYL-,CHLORIDE(9CI); BENZYLDIMETHYLOCTADECYLAMMONIUM CHLORIDE; BENZYLOCTADECYLDIMETHYLAMMONIUM CHLORIDE; BENZYLSTEARYLDIMETHYLAMMONIUM CHLORIDE; CARSOQUAT SDQ-25; CARSOQUAT SDQ-85; DEHYQUART STC-25; DIMETHYLBENZYLOCTADECYLAMMONIUM CHLORIDE; N,N-DIMETHYL-N-OCTADECYLBENZENEMETHANAMINIUM CHLORIDE; DIMETHYLOCTADECYLBENZYLAMMONIUM CHLORIDE; INTEXAN SB-85; INTEXSAN SB-85; J SOFT C 4; KATAMINE AB; NISSAN CATION S2-100; N-OCTADECYL-N-BENZYL-N,N-DIMETHYLAMMONIUM CHLORIDE; OCTADECYLDIMETHYLBENZYLAMMONIUM CHLORIDE; ORTHOSAN MB; QUATERNOL 1; STEARALKONIUM CHLORIDE; STEARYLBENZYLDIMETHYL AMMONIUM CHLORIDE; STEARYLDIMETHYLBENZYLAMMONIUM CHLORIDE; STEBAC; STEDBAC; TALLOW BENZYL DIMETHYL AMMONIUM CHLORIDE; TRITON X-40; TRITON X-400; TRITON CG 400; TRITON CG 500; TRITON X 40; TRITON X 400; VARISOFT SDC

Description: white, crystalline powder; white solid or viscous liquid

Use: cationic surfactant used as a germicide and sanitizer;in cosmetics as a surface-active and anti-microbial agent

Physical Properties

Boiling Point: Decomposes at 120 °C (248 °F)

Specific Gravity: Solid > 1.1 at 20 °C

Water Solubility: Soluble in Water

Other Solubilities: 95% Ethanol: >=100 mg/ml at 21 °C; Acetone: 10-50 mg/ml at 21 °C; DMSO: 1-10 mg/ml at 21 °C.
Flash Point: Not available; probably combustible

RTECS Toxicity Data

Acute Oral: Rat LD_{50} Dose: 1250 mg/kg. Mouse LD_{50} Dose: 760 mg/kg.
Chronic (Multiple Dose) Oral: Rat Dose: 364 mg/kg/26W-I; Toxic Effects: Brain and coverings - Recordings from specific areas of CNS; Liver - Liver function tests impaired; Kidney, Ureter, and Bladder - Other changes in urine composition.
Irritation Eye: Rabbit Standard Draize Test Dose: 200 ug; Reaction: severe.
Irritation Skin: Human Standard Draize Test Dose: 3 mg/3D-I; Reaction: mild. Man Standard Draize Test Dose: 125 mg/2D; Reaction: mild.

Hazard Overviews

Fire: Will burn.
Carcinogenicity: IARC - Not listed; NIOSH - Not listed; NTP - Not listed; ACGIH - Not listed; OSHA - Not listed; EPA - Not listed; MAK - Not listed

Environmental

Cleanup/Disposal: Guide No. 171: Do not touch or walk through spilled material. Stop leak if you can do it without risk. Prevent dust cloud. Avoid inhalation of asbestos dust. Small Dry Spills: With clean shovel place material into clean, dry container and cover loosely; move containers from spill area. Small Spills: Take up with sand or other noncombustible absorbent material and place into containers for later disposal. Large Spills: Dike far ahead of liquid spill for later disposal. Cover powder spill with plastic sheet or tarp to minimize spreading. Prevent entry into waterways, sewers, basements or confined areas.

Regulations

RCRA 40CFR: Not listed
CERCLA: 40CFR 302.4: Not listed
SARA 40CFR 372.65: Not listed
SARA EHS 40CFR 355: Not listed
TSCA: Listed

BEN8440 CAS #: 622-08-2

2-(BENZYLOXY) ETHANOL

RTECS: KJ7550000
EINECS Number: 210-719-3
Molecular Formula: $C_9H_{12}O_2$
Formula Weight: 152.20

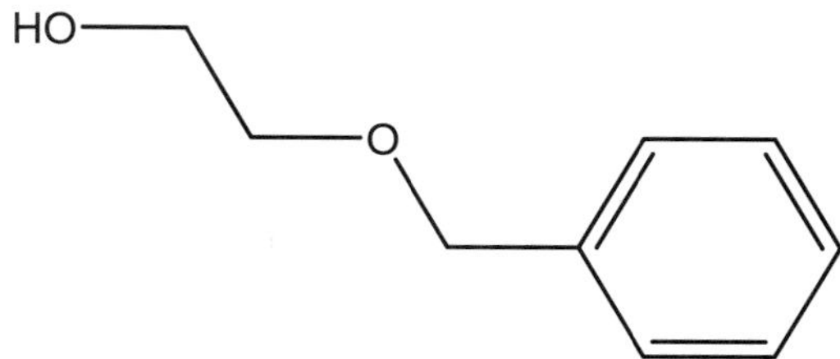

Chemical Structure

Synonyms: BENZYL 'CELLOSOLVE'; BENZYL CELLOSOLVE; BENZYLCELOSOLV; 2-(BENZYLOXY)ETHANOL; 2-BENZYLOXYETHANOL; ETHANOL,2-(BENZYLOXY)-; ETHANOL,2-(PHENYLMETHOXY)-; ETHYLENE GLYCOL MONOBENZYL ETHER; GLYCOL MONOBENZYL ETHER
Description: water white liquid; faint, rose-like odor
Use: laboratory reagent; solvent for cellulose acetate, dyes, inks, resins; perfume fixative; organic synthesis (selective hydroxyethylating agent); coating compositions for leather, paper, & cloth; lacquers

Physical Properties

Boiling Point: 256 °C (493 °F) at 760 mm Hg
Freezing Point: < -75 °C (-103 °F)
Specific Gravity: 1.064 at 20 °C/4 °C
Vapor Density: 5.25 Air=1
Water Solubility: Soluble in Water
Other Solubilities: Soluble in Alcohol, Ether
Refraction Index: 1.5233 at 20 °C/D
Flash Point: 129.444 °C Open Cup
Autoignition Temperature: 352 °C

RTECS Toxicity Data

Acute Oral: Rat LD_{50} Dose: 1190 mg/kg; Toxic Effects: Behavioral - General anesthetic; Gastrointestinal - Other changes; Kidney, Ureter, and Bladder - Other changes.
Irritation Eye: Rabbit Standard Draize Test Dose: 2 mg; Reaction: severe.

Hazard Overviews

Fire Diamond

Health: Irritating to eyes/skin/respiratory tract. Harmful. Other Acute Effects: harmful if swallowed; may be harmful if inhaled; may be harmful if absorbed through the skin.
Fire: Will burn. Hazards: emits toxic fumes. Extinguishing agents: water spray; carbon dioxide, dry chemical powder or appropriate foam. Precautions: combustible liquid.
Reactivity: Incompatible with: strong oxidizing agents. Hazardous decomposition products: toxic fumes of: carbon monoxide, carbon dioxide.
Carcinogenicity: IARC - Not listed; NIOSH - Not listed; NTP - Not listed; ACGIH - Not listed; OSHA - Not listed; EPA - Not listed; MAK - Not listed

Primary Target Organs:

Eyes

Skin

Respiratory System

Environmental

Regulations

RCRA 40CFR: Not listed
CERCLA: 40CFR 302.4: Not listed
SARA 40CFR 372.65: Not listed
SARA EHS 40CFR 355: Not listed
TSCA: Listed

BEN8520 CAS #: 103-16-2

P-(BENZYLOXY)PHENOL

RTECS: SJ7700000
EINECS Number: 203-083-3
Molecular Formula: $C_{13}H_{12}O_2$
Formula Weight: 200.23

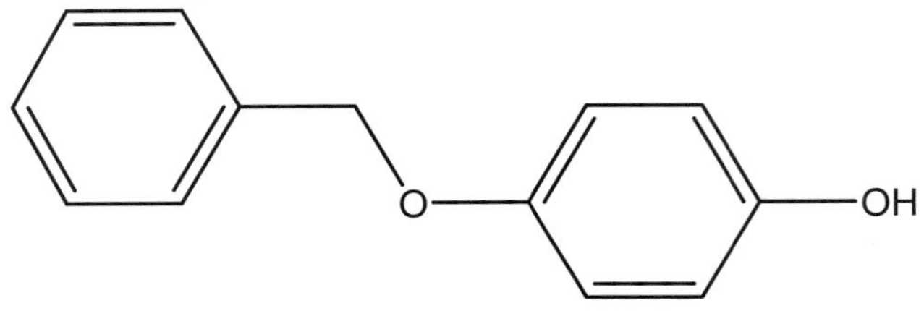

Chemical Structure

Synonyms: AGERITE; AGERITE ALBA; ALBA-DOME; BENOQUIN; BENZOQUIN; BENZYL HYDROQUINONE; BENZYL P-HYDROXYPHENYL ETHER; 4-(BENZYLOXY)PHENOL; 4-BENZYLOXYPHENOL; P-BENZYLOXYPHENOL; CARMIFAL; DEPIGMAN; DERMOCHINONA; HYDROCHINON MONOBENZYLETHER; HYDROQUINONE BENZYL ETHER; HYDROQUINONE MONOBENZYL ETHER; P-HYDROXYPHENYL BENZYL ETHER; LEUCODININE; MONOBENZON; MONOBENZONE; MONOBENZYL ETHER HYDROQUINONE; MONOBENZYL HYDROQUINONE; PHENOL,4-(PHENYLMETHOXY)-; 4-(PHENYLMETHOXY)PHENOL; PIGMEX; SUPERLITE; SUPERLITE (ANTIOXIDANT)
Description: white crystalline powder or lustrous leaflets; odorless
Use: depigmentor; antioxidant for paraffins and polyvinyl chloride, polypropylene and polypropylene oxide; chem int in organic synthesis; rubber antioxidant; stabilizer; polymerization inhibitor

Physical Properties

Freezing Point: 122.5 °C (252.5 °F)
Specific Gravity: 1.26
Water Solubility: 1 g/100ml at 0 °C
Other Solubilities: freely Soluble in Acetone; Practically Insoluble in petroleum hydrocarbons, very Soluble in Benzene and alkalies.
Flash Point: Combustible

RTECS Toxicity Data

Irritation Skin: Guinea Pig Standard Draize Test Dose: 5%/48H; Reaction: mild.
Tumorigenic: Mouse Route: Oral; Dose: 163 gm/kg/78W-I; Toxic Effects: Tumorigenic - Equivocal tumorigenic agent by RTECS criteria; Lungs, Thorax, or Respiration - Tumors; Liver - Tumors. Mouse Route: Subcutaneous; Dose: 1000 mg/kg; Toxic Effects: Tumorigenic - Neoplastic by RTECS criteria; Lungs, Thorax, or Respiration - Tumors; Blood - Tumors.

Hazard Overviews

Health: Irritating to eyes/skin/respiratory tract. Other Acute Effects: may be harmful by inhalation, ingestion, or skin absorption; may cause allergic skin reaction.
Fire: Combustible. Hazards: emits toxic fumes. Extinguishing agents: water spray; carbon dioxide, dry chemical powder or appropriate foam. Precautions: combustible liquid.
Reactivity: Incompatible with: strong oxidizing agents, strong bases. Hazardous decomposition products: toxic fumes of: carbon monoxide, carbon dioxide.
Carcinogenicity: IARC - Not listed; NIOSH - Not listed; NTP - Not listed; ACGIH - Not listed; OSHA - Not listed; EPA - Not listed; MAK - Not listed
Primary Target Organs:

Eyes

Skin

Respiratory System

Environmental

Regulations

RCRA 40CFR: Not listed
CERCLA: 40CFR 302.4: Not listed
SARA 40CFR 372.65: Not listed
SARA EHS 40CFR 355: Not listed
TSCA: Listed

BEN8680 CAS #: 100-87-8

BENZYLSULFONIC ACID

RTECS: DA4635000
EINECS Number: 202-897-6
Molecular Formula: $C_7H_8O_3S$
Formula Weight: 172.21
Synonyms: BENZENEMETHANESULFONIC ACID; BENZYL-ALPHA-SULFONIC ACID; METHANESULFONIC ACID,PHENYL-; PHENYLMETHANESULFONIC ACID; ALPHA-TOLUENESULFONIC ACID

Use: catalyst for furan-type no-bake foundry sand binders; research chemical

RTECS Toxicity Data

Acute Oral: Rat LD_{50} Dose: >1 gm/kg.

Hazard Overviews

Carcinogenicity: IARC - Not listed; NIOSH - Not listed; NTP - Not listed; ACGIH - Not listed; OSHA - Not listed; EPA - Not listed; MAK - Not listed

Environmental

Regulations

RCRA 40CFR: Not listed
CERCLA: 40CFR 302.4: Not listed
SARA 40CFR 372.65: Not listed
SARA EHS 40CFR 355: Not listed
TSCA: Listed

BEP5000 CAS #: 7181-73-9

BEPHENIUM

RTECS: BO7100000
EINECS Number: 230-546-7
Molecular Formula: $C_{17}H_{22}NO$
Formula Weight: 256.36
Synonyms: AMMONIUM,BENZYLDIMETHYL(2-PHENOXYETHYL)-; BENZENEMETHANAMINIUM,N,N-DIMETHYL-N-(2-PHENOXYETHYL)-; N,N-DIMETHYL-N-(2-PHENOXYETHYL)-BENZENEMETHANAMINIUM
Description: odorless
Use: medication: anthelmintic agent; medication (vet): anthelmintic agent

Physical Properties

Freezing Point: 135 °C (275 °F) to 136 °C (276.8 °F)
Water Solubility: Practically Insoluble in Water
Other Solubilities: Soluble in hot Alcohol.

RTECS Toxicity Data

Acute Oral: Mouse LD_{50} Dose: 7880 mg/kg.

Hazard Overviews

Carcinogenicity: IARC - Not listed; NIOSH - Not listed; NTP - Not listed; ACGIH - Not listed; OSHA - Not listed; EPA - Not listed; MAK - Not listed

Environmental

Regulations

RCRA 40CFR: Not listed
CERCLA: 40CFR 302.4: Not listed
SARA 40CFR 372.65: Not listed
SARA EHS 40CFR 355: Not listed
TSCA: Not listed

BER1000 CAS #: 316-41-6

BERBERINE SULFATE

RTECS: DR9867300
EINECS Number: 206-258-2
Molecular Formula: $C_{40}H_{36}N_2O_{12}S$
Formula Weight: 384.40
Synonyms: BENZO(G)-1,3-BENZODIOXOLO(5,6-A)QUINOLIZINIUM,5,6-DIHYDR0-9,10-DIMETHOXY-,SULFATE (2:1); BENZO(G)-1,3-BENZODIOXOLO(5,6-A)QUINOLIZINIUM,5,6-DIHYDRO-9,10-DIMETHOXY-,SULFATE (2:1); BERBERIN SULFATE; BERBERINE SULFATE (2:1); BERBINIUM,7,8,13,13A-TETRADEHYDRO-9,10-DIMETHOXY-2,3-(METHYLENEDIOXY)-,SULFATE (2:1); NEUTRAL BERBERINE SULFATE
Description: yellow crystals
Use: medication: former uses bitter stomachic, antibacterial, antimalarial, antipyretic; mild local anesthetic on mucous membrane

Physical Properties

Water Solubility: Trihydrate 1 30 Parts Water
Other Solubilities: Insoluble in Chloroform

RTECS Toxicity Data

Acute Dermal: Mouse LD_{50} Route: Subcutaneous Dose: 13200 ug/kg.

Hazard Overviews

Carcinogenicity: IARC - Not listed; NIOSH - Not listed; NTP - Not listed; ACGIH - Not listed; OSHA - Not listed; EPA - Not listed; MAK - Not listed

Environmental

Regulations

RCRA 40CFR: Not listed
CERCLA: 40CFR 302.4: Not listed
SARA 40CFR 372.65: Not listed
SARA EHS 40CFR 355: Not listed
TSCA: Not listed

BER1670 CAS #: 12161-82-9

BERTRANDITE

RTECS: DS1225000
EINECS Number: 235-299-9
Molecular Formula: $Be_4H_{12}O_{10}Si_2$
Formula Weight: 264.34
Synonyms: BERYLLIUM SILICATE HYDRATE
Use: raw material for beryllium

Hazard Overviews

Carcinogenicity: IARC - Group 1, Carcinogenic to humans; NIOSH - Listed as carcinogen; NTP - Class 2A, Reasonably anticipated to be a carcinogen, limited evidence of carcinogenicity from studies in humans; ACGIH - Class A2,

Suspected human carcinogen; OSHA - Not listed; EPA - Class B2, Probable human carcinogen based on animal studies; MAK - Class A2, Unmistakably carcinogenic in animal experimentation only

Exposure Limits

OSHA PEL: TWA: 0.002 mg/m^3; as Be. Other Values: 0.025 mg/m^3; as Be 30 min/8 Hr.

ACGIH TLV: TWA: 0.002 mg/m^3; as Be.

NIOSH REL: STEL: 0.0005 mg/m^3; Ceiling as Be.

Environmental

Regulations

RCRA 40CFR: Not listed

CERCLA: 40CFR 302.4: Listed as Compound per CWA Section 307(a) per CAA Section 112

SARA 40CFR 372.65: Listed as Compound

SARA EHS 40CFR 355: Not listed

TSCA: Listed

BER2340 CAS #: 1302-52-9

BERYL

RTECS: DS1400000

EINECS Number: 215-101-7

Molecular Formula: $Al_2Be_3O_{18}Si_6$

Formula Weight: 537.53

Synonyms: BERYL ORE; BERYLLIUM ALUMINIUM SILICATE; BERYLLIUM ALUMINOSILICATE; BERYLLIUM ALUMINUM SILICATE; NATURAL BERYL

Description: colorless, green, blue, yellow or white transparent crystals

Use: chief mineral source of beryllium

Physical Properties

Freezing Point: 1310 °C (2390 °F) to 1510 °C (2750 °F)

Specific Gravity: 2.66

Other Solubilities: Insoluble in acid

Refraction Index: 1.580

RTECS Toxicity Data

Tumorigenic: Rat Route: Inhalation; Dose: 15 mg/m^3/74W-I; Toxic Effects: Tumorigenic - Neoplastic by RTECS criteria; Lungs, Thorax, or Respiration - Tumors. Rat Route: Inhalation; Dose: 15 mg/m^3/6H/73W-I; Toxic Effects: Tumorigenic - Equivocal tumorigenic agent by RTECS criteria; Lungs, Thorax, or Respiration - Tumors.

Hazard Overviews

Carcinogenicity: IARC - Group 1, Carcinogenic to humans; NIOSH - Listed as carcinogen; NTP - Class 2A, Reasonably anticipated to be a carcinogen, limited evidence of carcinogenicity from studies in humans; ACGIH - Class A2, Suspected human carcinogen; OSHA - Not listed; EPA - Class B2, Probable human carcinogen based on animal studies; MAK - Class A2, Unmistakably carcinogenic in animal experimentation only

Exposure Limits

OSHA PEL: TWA: 0.002 mg/m^3; as Be. Other Values: 0.025 mg/m^3; as Be 30 min/8 hr.

OSHA PEL Vacated 1989 Limits: TWA: 2 mg/m^3; as Al soluble.

ACGIH TLV: TWA: 0.002 mg/m^3; as Be.

NIOSH REL: TWA: 2 mg/m^3; as Al soluble salts, alkyls. STEL: 0.0005 mg/m^3; Ceiling as Be.

Environmental

Regulations

RCRA 40CFR: Not listed

CERCLA: 40CFR 302.4: Listed as Compound per CWA Section 307(a) per CAA Section 112

SARA 40CFR 372.65: Listed as Compound

SARA EHS 40CFR 355: Not listed

TSCA: Listed

BER3010 CAS #: 7440-41-7

BERYLLIUM

RTECS: DS1750000

DOT: UN1567; IMO6.1

EINECS Number: 231-150-7

Molecular Formula: Be

Structured MF: Be

Formula Weight: 9.01

Be

Chemical Structure

Synonyms: BERYLLIUM-9; BERYLLIUM DUST; BERYLLIUM METALLIC; BERYLLIUM,METAL POWDER; GLUCINIUM; GLUCINUM

Description: grayish-white powder; odorless

Use: source of neutrons when bombarded with alpha particles; x-ray window; in hardening of copper & prodn of brass; mfr of nonsparking alloy for tools, lightweight alloys & with copper, neutron moderator in nuclear weapons & test reactors; in aerospace & spacecraft mfr (eg, guidance systems, mirrors in optics, heat sink material in aircraft brakes meteorite & heat shielding, solid rocket fuel, navigational systems, aircraft/satellite structures, & missile parts.

Physical Properties

Boiling Point: 2970 °C (5378 °F)

Freezing Point: 1287 °C (2348.6 °F)

Specific Gravity: 1.85 at 20 °C

Vapor Pressure: 10 mm Hg at 1860 °C

Water Solubility: Insoluble in Cold Water

Other Solubilities: Insoluble in Nitric acid.

Flash Point: Combustible solid

RTECS Toxicity Data

Mutagenic: Human DNA Adduct; Cell Type: HeLa cell; Dose: 30 umol/L. Mouse DNA Adduct; Cell Type: Ascites tumor; Dose: 30 umol/L.
Tumorigenic: Rat Route: Intratracheal; Dose: 13 mg/kg; Toxic Effects: Tumorigenic - Neoplastic by RTECS criteria; Lungs, Thorax, or Respiration - Tumors; Lungs, Thorax, or Respiration - Bromchiogenic carcinoma. Rabbit Route: Intravenous; Dose: 20 mg/kg; Toxic Effects: Tumorigenic - Equivocal tumorigenic agent by RTECS criteria; Musculoskelital - Tumors.

Hazard Overviews

Poison

Flammable

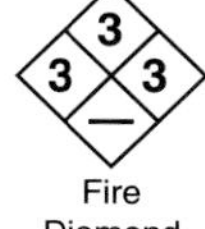

Fire Diamond

Health: Poison. Causes respiratory inflammation, congestion, coughing, pulmonary edema; heavy exposures cause: brain, spleen hemorrhaging, liver inflammation. Chronic Effects: lung, heart, liver, spleen, kidney damage. May cause cancer.
Fire: Flammable. Fight fire with approved dry-powder extinguisher, sand, sodium chloride, or graphite powder. Never use water or carbon dioxide.
Reactivity: Stable. Hazardous polymerization cannot occur. Avoid: heating in air; mixing with carbon dioxide and nitrogen; mixing with carbon tetrachloride or trichloroethylene. Incompatible with: strong bases; phosphorus; fluorine; chlorine; molten lithium metal at 356 °F (180 °C). Hazardous decomposition products: oxide of beryllium fumes.
Carcinogenicity: IARC - Group 1, Carcinogenic to humans; NIOSH - Listed as carcinogen; NTP - Class 2A, Reasonably anticipated to be a carcinogen, limited evidence of carcinogenicity from studies in humans; ACGIH - Class A2, Suspected human carcinogen; OSHA - Not listed; EPA - Class B2, Probable human carcinogen based on animal studies; MAK - Class A2, Unmistakably carcinogenic in animal experimentation only
Primary Target Organs:

Eyes

Skin

Respiratory System

Liver

Kidneys

Cardio-vascular

Exposure Limits
OSHA PEL: TWA: 0.002 mg/m^3; as Be. Other Values: 0.025 mg/m^3; as Be 30 min/8 Hr.
ACGIH TLV: TWA: 0.002 mg/m^3; as Be.
NIOSH REL: STEL: 0.0005 mg/m^3; Ceiling as Be.
NIOSH IDLH: 4 mg/m^3; as Be.
Respirator Recommendation
Exposure Range: >0.002 to 0.02 mg/m^3 Air Purifying, Negative Pressure, Half Mask
Exposure Range: >0.02 to 0.2 mg/m^3 Air Purifying, Negative Pressure, Full Face
Exposure Range: >0.2 to 2 mg/m^3 Supplied Air, Constant Flow/Pressure Demand, Full Face
Exposure Range: >2 to unlimited mg/m^3 Self-contained Breathing Apparatus, Pressure Demand, Full Face
Cartridge Color: magenta (P100)

Environmental

Ecotoxicity: Tlm Pimephales promelas (fathead minnow) 150 ug/l/96 hr (soft water) /Conditions of bioassay not specified
Cleanup/Disposal: Guide No. 134: Fully encapsulating, vapor protective clothing should be worn for spills and leaks with no fire. Eliminate all ignition sources (no smoking, flares, sparks or flames in immediate area). Stop leak if you can do it without risk. Do not touch damaged containers or spilled material unless wearing appropriate protective clothing. Prevent entry into waterways, sewers, basements or confined areas. Use clean non-sparking tools to collect material and place it into loosely covered plastic containers for later disposal.

Regulations

RCRA 40CFR: Listed Hazardous Waste No. P015 Toxic Waste
CERCLA: 40CFR 302.4: Listed per RCRA Section 3001 per CWA Section 307(a) per CAA Section 112 RQ: 10 lb (4.535 kg)
SARA 40CFR 372.65: Listed
SARA EHS 40CFR 355: Not listed
TSCA: Listed

Analytical Methods

Air: EPA 0060, 103, 104, 29, ITM-001
Soil: CLP 200.10_M, 200.62, 200.7_M, 202.62, 210.1_M, 210.2_M, 6020_M, ICP-AES; EPA 13, 200.7, 200.8; SW846 3005A, 3010A, 3015, 3020A, 3031, 3040, 3040A, 3050A, 3050B, 3051, 3052, 6010A, 6010B, 6020, 7000A, 7090, 7091, OSW-A; USGS I5095
Water / Groundwater: EPA 200.0, 200.15, 200.7, 200.9; APHA 3111-A, 3111-D, 3111-E, 3113-B, 3120; ASTM D1976, D3645, D4190; USGS E-SPEC, I1095, I1472, I3095, I7095; FISON AES-0029
Drinking Water: AOAC 993.14; APHA 3500-BE
Food: EPA 14
Indoor / Expired Air: NIOSH 7102, 7300
Plasma: EPA 200.11
Other: EPA 1620, 210.1, 210.2; AOAC 990.08

BER3680 **CAS #: 7787-47-5**

BERYLLIUM CHLORIDE

RTECS: DS2625000
DOT: NA1566; IMO6.1
EINECS Number: 232-116-4
Molecular Formula: $BeCl_2$
Structured MF: $BeCl_2$
Formula Weight: 79.93

Be^{++}

Cl^{-} Cl^{-}

Chemical Structure

Synonyms: BERYLLIUM DICHLORIDE
Description: white, slightly yellow powder; sharp odor
Use: anhydrous form used as acid catalyst in org reactions; chem int for beryllium metal; catalyst for friedel-crafts reaction; chem int for beryllium perchlorate

Physical Properties

Boiling Point: 482.3 °C (900 °F)
Freezing Point: 399.2 °C (750.56 °F)
Specific Gravity: 1.9
Vapor Pressure: 1 mm Hg at 291 °C
Water Solubility: Very Soluble in Water
Other Solubilities: Soluble in Alcohol, Ether, Pyridine, Carbon Disulfide; Insoluble in Benzene, Toluene; Insoluble in Acetone; Ammonia.
pH: Aqueous solution is strongly acid
Flash Point: Nonflammable

RTECS Toxicity Data

Acute Oral: Rat LD_{50} Dose: 86 mg/kg. Mouse LD_{50} Dose: 92 mg/kg.
Reproductive/Teratogenic: Rat Route: Intratracheal; Dose: 1685 ug/kg; Duration: female 3D of pregnancy; Effects on Embryo or Fetus - Fetotoxicity; Specific Developmental Abnormalities - Other developmental abnormalities. Rat Route: Intratracheal; Dose: 1685 ug/kg; Duration: female 5D of pregnancy; Effects on Fertility - Pre-implantation mortality; Specific Developmental Abnormalities - Other developmental abnormalities.
Mutagenic: Hamster Mutations in Mammalian Somatic Cells; Cell Type: lung; Dose: 2 mmol/L. Hamster Sister Chromatid Exchenge; Cell Type: lung; Dose: 31 mg/L.
Tumorigenic: Rat Route: Inhalation; Dose: 20 ug/m^3/1H/17W-I; Toxic Effects: Tumorigenic - Equivocal tumorigenic agent by RTECS criteria; Lungs, Thorax, or Respiration - Tumors.

Hazard Overviews

Fire Diamond

Health: Irritating to eyes/skin/respiratory tract. Toxic. Also Causes: nose bleeds, nasal discharge, headache, fever, fatigue, chest pain, pharyngitis, tracheobronchitis, pneumonitis, decreased pulmonary function, cyanosis, heart failure, photophobia. Chronic Effects: Symptoms of chronic pulmonary granulomatosis or berylliosis include constant hacking cough, shortness of breath, fibrosis, emphysema, decreased lung capacity, chest pain, fatigue, appetite/weight loss. Cyanosis, clubbing of fingers, and heart failure may occur. Chronic inhalation may also produce systemic diseases of the lymph nodes, liver, bones, and kidneys. Contact dermatitis may develop with repeated skin contact.
Fire: Noncombustible. Use agent suitable for surrounding fire. When mixed with large quantities of water, beryllium chloride, in addition to heat, will generate beryllium oxide, hydrogen chloride gas, and hydrochloric acid solution.
Reactivity: Stable, but is deliquescent and readily hydrolyzed. Hazardous polymerization cannot occur. Avoid: contact with moisture. Incompatible with: large quantities of water and heat; metal halides (disulfur dinitride; tetrasulfur tetranitride); metals and water. Hazardous decomposition products: beryllium oxide; chloride fumes.
Carcinogenicity: IARC - Group 1, Carcinogenic to humans; NIOSH - Listed as carcinogen; NTP - Class 2A, Reasonably anticipated to be a carcinogen, limited evidence of carcinogenicity from studies in humans; ACGIH - Class A2, Suspected human carcinogen; OSHA - Not listed; EPA - Class B2, Probable human carcinogen based on animal studies; MAK - Class A2, Unmistakably carcinogenic in animal experimentation only
Primary Target Organs:

Eyes

Skin

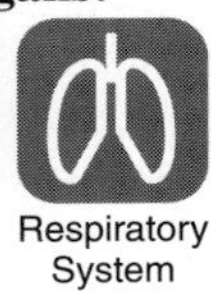
Respiratory System

Liver

Bone

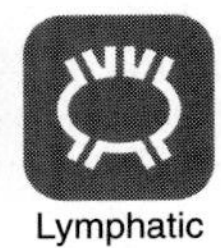
Lymphatic System

Exposure Limits
OSHA PEL: TWA: 0.002 mg/m^3; as Be. Other Values: 0.025 mg/m^3; as Be 30 min/8 Hr.
ACGIH TLV: TWA: 0.002 mg/m^3; as Be.
NIOSH REL: STEL: 0.0005 mg/m^3; Ceiling as Be.

Environmental

Ecotoxicity: LC_{50} Pimephales promelas (fathead minnow) 150 ug/l/96 hr, soft water, 20,000 ug/l/96 hr, hard water. /Beryllium ion/ /Static bioassay EC_{50} Cladoceran (Daphnia magna) 2,500 ug/l/48 hr, effects on reproduction. /Beryllium ion
Cleanup/Disposal: Guide No. 154: Eliminate all ignition sources (no smoking, flares, sparks or flames in immediate area). Do not touch damaged containers or spilled material unless wearing appropriate protective clothing. Stop leak if you can do it without risk. Prevent entry into waterways, sewers, basements or confined areas. Absorb or cover with dry earth, sand or other non-combustible material and transfer to containers. Do not get water inside containers.

Environmental Physical Data

BCF: up to 100-fold under constant exposure
BOD: none

Regulations

RCRA 40CFR: Not listed
CERCLA: 40CFR 302.4: Listed per CWA Section 311(b)(4) RQ: 1 lb (0.454 kg)
SARA 40CFR 372.65: Listed as Compound
SARA EHS 40CFR 355: Not listed
TSCA: Listed

BER4350 **CAS #: 7787-49-7**

BERYLLIUM FLUORIDE

RTECS: DS2800000
DOT: NA1566; IMO6.1
EINECS Number: 232-118-5
Molecular Formula: BeF_2
Structured MF: BeF_2
Formula Weight: 47.01

Be^{++}

F^- F^-

Chemical Structure

Synonyms: BERYLLIUM DIFLUORIDE
Description: colorless glassy mass or white solid; odorless
Use: mfr of beryllium & beryllium alloys; chem int for beryllium metal; component in mfr of nuclear reactors; formerly in glass mfr

Physical Properties

Boiling Point: 1160 °C (2120 °F)
Freezing Point: 555 °C (1031 °F)
Specific Gravity: 1.986 at 25 °C/4 °C
Water Solubility: Soluble in Water
Other Solubilities: Insoluble in anhydrous Hydrogen Fluoride; Soluble in mixture of Alcohol & Ether.
Refraction Index: < 1.33
Ionization Potential (eV): 14.6 +/-0.5
Flash Point: Nonflammable

RTECS Toxicity Data

Acute Oral: Rat LD_{50} Dose: 98 mg/kg. Mouse LD_{50} Dose: 100 mg/kg.
Acute Dermal: Mouse LD_{50} Route: Subcutaneous Dose: 20 mg/kg.
Tumorigenic: Rat Route: Inhalation; Dose: 20 ug/m^3/1H/17W-I; Toxic Effects: Tumorigenic - Equivocal tumorigenic agent by RTECS criteria; Lungs, Thorax, or Respiration - Tumors. Rat Route: Inhalation; Dose: 49 ug/m^3/26W; Toxic Effects: Tumorigenic - Equivocal tumorigenic agent by RTECS criteria; Lungs, Thorax, or Respiration - Tumors.

Hazard Overviews

Poison

Corrosive

Health: Corrosive to eyes/skin/respiratory tract. Poison. Other Acute Effects: may be fatal if inhaled, swallowed, or absorbed through skin; inhalation may result in spasm, inflammation and edema of the larynx and bronchi; chemical pneumonitis; pulmonary edema; exposure may cause burning sensation; coughing; wheezing; laryngitis; shortness of breath; headache; nausea; vomiting. Chronic Effects: Carcinogen.
Fire: Noncombustible. Hazards: emits toxic fumes. Extinguishing agents: noncombustible; use extinguishing media appropriate to surrounding fire conditions. Precautions: combustible liquid.
Reactivity: Incompatible with: strong acids. Hazardous decomposition products: hydrogen fluoride.
Carcinogenicity: IARC - Group 1, Carcinogenic to humans; NIOSH - Listed as carcinogen; NTP - Class 2A, Reasonably anticipated to be a carcinogen, limited evidence of carcinogenicity from studies in humans; ACGIH - Class A2, Suspected human carcinogen; OSHA - Not listed; EPA - Class B2, Probable human carcinogen based on animal studies; MAK - Class A2, Unmistakably carcinogenic in animal experimentation only
Primary Target Organs:

Eyes

Skin

Respiratory System

Exposure Limits
OSHA PEL: TWA: 0.002 mg/m^3; as Be. Other Values: 0.025 mg/m^3; as Be 30 min/8 Hr.
ACGIH TLV: TWA: 0.002 mg/m^3; as Be.
NIOSH REL: STEL: 0.0005 mg/m^3; Ceiling as Be.

Environmental

Ecotoxicity: LC_{50} Pimephales promelas (fathead minnow) 150 ug/l/96 hr, soft water, 20,000 ug/l/96 hr, hard water. /Beryllium ion/ /Static bioassay EC_{50} Cladoceran (Daphnia magna) 2,500 ug/l/48 hr, effects on reproduction. /Beryllium ion
Cleanup/Disposal: Guide No. 154: Eliminate all ignition sources (no smoking, flares, sparks or flames in immediate area). Do not touch damaged containers or spilled material unless wearing appropriate protective clothing. Stop leak if you can do it without risk. Prevent entry into waterways, sewers, basements or confined areas. Absorb or cover with dry earth, sand or other non-combustible material and transfer to containers. Do not get water inside containers.

Environmental Physical Data

BCF: up to 100-fold under constant exposure
BOD: none

Regulations

RCRA 40CFR: Not listed
CERCLA: 40CFR 302.4: Listed per CWA Section 311(b)(4) RQ: 1 lb (0.454 kg)
SARA 40CFR 372.65: Listed as Compound
SARA EHS 40CFR 355: Not listed
TSCA: Listed

BER5020 CAS #: 13327-32-7

BERYLLIUM HYDROXIDE

RTECS: DS3150000
DOT: UN1566; IMO6.1
EINECS Number: 236-368-6
Molecular Formula: BeH_2O_2
Formula Weight: 43.03
Synonyms: BERYLLIUM DIHYDROXIDE; BERYLLIUM HYDRATE
Description: white, amorphous powder or crystals
Use: mfr of beryllium and beryllium oxide

Physical Properties

Boiling Point: Decomposes
Freezing Point: Decomposes
Specific Gravity: 1.92
Water Solubility: Very Slightly Soluble in Water
Other Solubilities: Very Slightly Soluble in dilute alkali; Soluble in hot concentrated Sodium Hydroxide solution, hot concentrated acids

RTECS Toxicity Data

Tumorigenic: Rat Route: Intratracheal; Dose: 1125 ug/kg; Toxic Effects: Tumorigenic - Equivocal tumorigenic agent by RTECS criteria; Lungs, Thorax, or Respiration - Tumors. Rat Route: Intratracheal; Dose: 1785 ug/kg/43W-I; Toxic Effects: Tumorigenic - Equivocal tumorigenic agent by RTECS criteria; Lungs, Thorax, or Respiration - Tumors; Lungs, Thorax, or Respiration - Bromchiogenic carcinoma.

Hazard Overviews

Carcinogenicity: IARC - Group 1, Carcinogenic to humans; NIOSH - Listed as carcinogen; NTP - Class 2A, Reasonably anticipated to be a carcinogen, limited evidence of carcinogenicity from studies in humans; ACGIH - Class A2, Suspected human carcinogen; OSHA - Not listed; EPA - Class B2, Probable human carcinogen based on animal studies; MAK - Class A2, Unmistakably carcinogenic in animal experimentation only
Exposure Limits
OSHA PEL: TWA: 0.002 mg/m^3; as Be. Other Values: 0.025 mg/m^3; as Be 30 min/8 Hr.
ACGIH TLV: TWA: 0.002 mg/m^3; as Be.
NIOSH REL: STEL: 0.0005 mg/m^3; Ceiling as Be.

Environmental

Ecotoxicity: LC_{50} Pimephales promelas (fathead minnow) 150 ug/l/96 hr, soft water, 20,000 ug/l/96 hr, hard water. /Beryllium ion/ /Static bioassay EC_{50} Cladoceran (Daphnia magna) 2,500 ug/l/48 hr, effects on reproduction.
Cleanup/Disposal: Guide No. 154: Eliminate all ignition sources (no smoking, flares, sparks or flames in immediate area). Do not touch damaged containers or spilled material unless wearing appropriate protective clothing. Stop leak if you can do it without risk. Prevent entry into waterways, sewers, basements or confined areas. Absorb or cover with dry earth, sand or other non-combustible material and transfer to containers. Do not get water inside containers.

Regulations

RCRA 40CFR: Not listed
CERCLA: 40CFR 302.4: Listed as Compound per CWA Section 307(a) per CAA Section 112
SARA 40CFR 372.65: Listed as Compound
SARA EHS 40CFR 355: Not listed
TSCA: Listed

BER6360 CAS #: 13597-99-4

BERYLLIUM NITRATE

RTECS: DS3675000
DOT: UN2464; IMO5.1
EINECS Number: 237-062-5
Molecular Formula: $BeH_2N_2O_6$
Structured MF: $Be(NO_3)_2 \cdot 3H_2O$
Formula Weight: 133.03

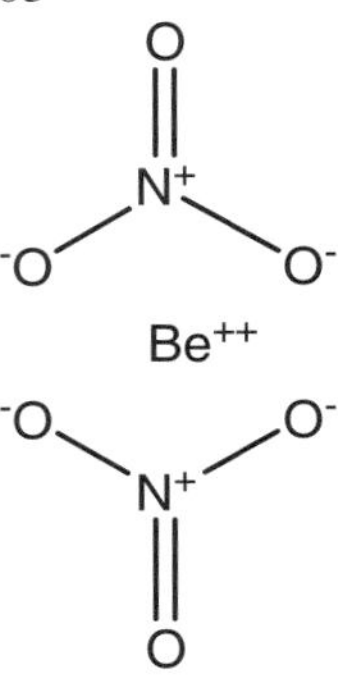

Chemical Structure

Synonyms: BERYLLIUM DINITRATE; BERYLLIUM NITRATE TRIHYDRATE; NITRIC ACID,BERYLLIUM SALT
Description: white solid; odorless
Use: stiffening mantles in gas & acetylene lamps; chemical reagent

Physical Properties

Boiling Point: 142 °C (288 °F)
Freezing Point: 60 °C (140 °F)
Specific Gravity: Solid 1.56 at 20 °C (solid)
Water Solubility: Very Soluble in Water
Other Solubilities: Very Soluble in Alcohol
Flash Point: Not combustible

RTECS Toxicity Data

Acute Dermal: Mouse LD_{Lo} Route: Subcutaneous Dose: 50 mg/kg. Frog LD_{Lo} Route: Subcutaneous Dose: 1041 mg/kg.
Reproductive/Teratogenic: Rat Route: Intratesticular; Dose: 10803 ug/kg; Duration: male 1D prior to mating; Paternal Effects - Spermatogenesis; Testes, epididymis, sperm duct.
Mutagenic: Hamster Sister Chromatid Exchenge; Cell Type: lung; Dose: 31 mg/L. Bacteria - B Subtilis DNA Repair; Dose: 750 ug/disc.

Hazard Overviews

Poison

Corrosive

Health: Corrosive to eyes/skin/respiratory tract. Poison. Other Acute Effects: may be fatal if inhaled; harmful if swallowed or absorbed through skin; inhalation may result in spasm, inflammation and edema of the larynx and bronchi, chemical pneumonitis and pulmonary edema; symptoms of exposure may include burning sensation; coughing; wheezing; laryngitis; shortness of breath; headache; nausea; vomiting; may cause cyanosis (blue-gray coloring of skin, and lips caused by lack of oxygen); allergic skin reaction; anorexia; nose bleeds; fatigue; fever. Chronic Effects: may cause congenital malformation in the fetus; target organs: blood, lungs, spleen, bones, liver, heart, male reproductive system, kidneys, lymph nodes. Carcinogen.

Fire: Noncombustible. Hazards: emits toxic fumes; may accelerate combustion; contact with other material may cause fire. Extinguishing agents: water spray; carbon dioxide, dry chemical powder or appropriate foam. Precautions: combustible liquid.

Reactivity: Stable. Hazardous polymerization will not occur. Incompatible with: oxidizing agents, acids, bases, organic materials, finely powdered metals, chlorinated hydrocarbons. Hazardous decomposition products: toxic fumes of: nitrogen oxides, nitric acid, beryllium oxides.

Carcinogenicity: IARC - Group 1, Carcinogenic to humans; NIOSH - Listed as carcinogen; NTP - Class 2A, Reasonably anticipated to be a carcinogen, limited evidence of carcinogenicity from studies in humans; ACGIH - Class A2, Suspected human carcinogen; OSHA - Not listed; EPA - Class B2, Probable human carcinogen based on animal studies; MAK - Class A2, Unmistakably carcinogenic in animal experimentation only

Primary Target Organs:

Eyes

Skin

Respiratory System

Liver

Cardio-vascular

Blood

Exposure Limits

OSHA PEL: TWA: 0.002 mg/m^3; as Be. Other Values: 0.025 mg/m^3; as Be 30 min/8 Hr.

ACGIH TLV: TWA: 0.002 mg/m^3; as Be.

NIOSH REL: STEL: 0.0005 mg/m^3; Ceiling as Be.

Environmental

Ecotoxicity: LC_{50} Pimephales promelas (fathead minnow) 150 ug/l/96 hr, soft water, 20,000 ug/l/96 hr, hard water. /Beryllium ion/ /Static bioassay EC_{50} Cladoceran (Daphnia magna) 2,500 ug/l/48 hr, effects on reproduction

Cleanup/Disposal: Guide No. 141: Keep combustibles (wood, paper, oil, etc.) away from spilled material. Do not touch damaged containers or spilled material unless wearing appropriate protective clothing. Stop leak if you can do it without risk. Small Dry Spills: With clean shovel place material into clean, dry container and cover loosely; move containers from spill area. Large Spills: Dike far ahead of spill for later disposal.

Environmental Physical Data

BCF: constant exposure 100
BOD: none

Regulations

RCRA 40CFR: Not listed
CERCLA: 40CFR 302.4: Listed per CWA Section 311(b)(4) RQ: 1 lb (0.454 kg)
SARA 40CFR 372.65: Listed as Compound
SARA EHS 40CFR 355: Not listed
TSCA: Listed

BER7030 CAS #: 1304-56-9

BERYLLIUM OXIDE

RTECS: DS4025000
DOT: UN1566; IMO6.1
EINECS Number: 215-133-1
Molecular Formula: BeO
Structured MF: BeO
Formula Weight: 25.01

Be=O

Chemical Structure

Synonyms: BERYLLIA; BERYLLIUM; BERYLLIUM MONOXIDE; BROMELLETE; NATURAL BROMELLITE; THERMALOX; THERMALOX 995

Description: white hexagonal crystals, amorphous powder

Use: agent in glass & ceramics mfr; component of nuclear fuels & moderators; catalyst in organic reactions; refractory material; component of electron & klystron tubes; component of resistor cores & transistor mountings; chem int for other beryllium compounds; electric heat sinks; electrical insulators; microwave oven components; gyroscopes; military vehicle armor; rocket nozzles; crucibles; thermocouple tubing; laser structural components

Physical Properties

Boiling Point: About 3900 °C (7052 °F)
Freezing Point: 2530 °C (4586 °F)
Specific Gravity: 3 at 20 °C
Water Solubility: 2 ug/100 ml in Water at 30 °C
Other Solubilities: slowly Soluble in concentrated acids or fixed alkali hydroxides.
Refraction Index: 1.719
Ionization Potential (eV): 10.1 +/-0.4
Flash Point: Nonflammable

RTECS Toxicity Data

Reproductive/Teratogenic: Rat Route: Intratracheal; Dose: 139 mg/kg; Duration: female 3D of pregnancy; Effects on

Fertility - Pre-implantation mortality; Effects on Embryo or Fetus - Fetotoxicity; Specific Developmental Abnormalities - Other developmental abnormalities.

Tumorigenic: Rat Route: Inhalation; Dose: 28 mg/m^3/17W-C; Toxic Effects: Tumorigenic - Equivocal tumorigenic agent by RTECS criteria; Lungs, Thorax, or Respiration - Tumors. Rat Route: Intratracheal; Dose: 75 mg/kg/15W-I; Toxic Effects: Tumorigenic - Equivocal tumorigenic agent by RTECS criteria; Lungs, Thorax, or Respiration - Tumors. Rat Route: Intratracheal; Dose: 169 mg/kg/2W-I; Toxic Effects: Tumorigenic - Equivocal tumorigenic agent by RTECS criteria; Lungs, Thorax, or Respiration - Chronic pulmonary edema; Biochemical - Multiple enzyme effects. Rat Route: Intratracheal; Dose: 79 mg/kg/15W-I; Toxic Effects: Tumorigenic - Equivocal tumorigenic agent by RTECS criteria; Lungs, Thorax, or Respiration - Tumors.

Hazard Overviews

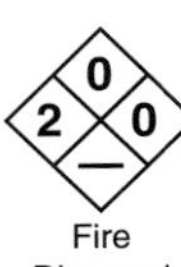

Fire Diamond

Health: Causes:(in high concentrations): acute pneumonitis with chest pain, bronchial spasm, fever, difficulty breathing, cyanosis, cough, blood-tinged sputum, nasal discharge, heart failure, conjunctivitis. Chronic Effects: Symptoms of chronic pulmonary granulomatosis or berylliosis include constant hacking cough, shortness of breath, chest pain, fatigue, appetite/weight loss, fibrosis, emphysema, reduced lung capacity and function. Cyanosis, clubbing of the fingers, heart failure may also occur. May also produce systemic diseases of the lymph nodes, liver, bones, and kidneys. Cancer hazard.

Fire: Noncombustible. Use agent suitable for surrounding fire.

Reactivity: Stable. Hazardous polymerization cannot occur. Avoid: heating with magnesium. Incompatible with: magnesium. Hazardous decomposition products: beryllium oxide.

Carcinogenicity: IARC - Group 1, Carcinogenic to humans; NIOSH - Listed as carcinogen; NTP - Class 2A, Reasonably anticipated to be a carcinogen, limited evidence of carcinogenicity from studies in humans; ACGIH - Class A2, Suspected human carcinogen; OSHA - Not listed; EPA - Class B2, Probable human carcinogen based on animal studies; MAK - Class A2, Unmistakably carcinogenic in animal experimentation only

Primary Target Organs:

Eyes

Respiratory System

Exposure Limits

OSHA PEL: TWA: 0.002 mg/m^3; as Be. Other Values: 0.025 mg/m^3; as Be 30 min/8 Hr.

ACGIH TLV: TWA: 0.002 mg/m^3; as Be.

NIOSH REL: STEL: 0.0005 mg/m^3; Ceiling as Be.

Environmental

Ecotoxicity: LC_{50} Pimephales promelas (fathead minnow) 150 ug/l/96 hr, soft water, 20,000 ug/l/96 hr, hard water /Static bioassay EC_{50} Cladoceran (Daphnia magna) 2,500 ug/l/48 hr, effects on reproduction.

Cleanup/Disposal: Guide No. 154: Eliminate all ignition sources (no smoking, flares, sparks or flames in immediate area). Do not touch damaged containers or spilled material unless wearing appropriate protective clothing. Stop leak if you can do it without risk. Prevent entry into waterways, sewers, basements or confined areas. Absorb or cover with dry earth, sand or other non-combustible material and transfer to containers. Do not get water inside containers.

Environmental Physical Data

BCF: constant exposure 100

BOD: none

Regulations

RCRA 40CFR: Not listed

CERCLA: 40CFR 302.4: Listed as Compound per CWA Section 307(a) per CAA Section 112

SARA 40CFR 372.65: Listed as Compound

SARA EHS 40CFR 355: Not listed

TSCA: Listed

BER7700 CAS #: 13598-15-7

BERYLLIUM PHOSPHATE

RTECS: DS2975000

DOT: UN1566; IMO6.1

Molecular Formula: BeH_3O_4P

Formula Weight: 106.98

Synonyms: BERYLLIUM HYDROGEN PHOSPHATE (1:1); PHOSPHORIC ACID,BERYLLIUM SALT (1:1); PHOSPHOROUS ACID,BERYLLIUM SALT

Physical Properties

Water Solubility: Partly Soluble in Water

RTECS Toxicity Data

Tumorigenic: Rat Route: Inhalation; Dose: 3571 ug/m^3/17W; Toxic Effects: Tumorigenic - Equivocal tumorigenic agent by RTECS criteria; Lungs, Thorax, or Respiration - Tumors. Monkey Route: Inhalation; Dose: 900 ug/kg/17W; Toxic Effects: Tumorigenic - Equivocal tumorigenic agent by RTECS criteria; Lungs, Thorax, or Respiration - Tumors. Monkey Route: Inhalation; Dose: 97 mg/m^3/8D-I; Toxic Effects: Tumorigenic - Equivocal tumorigenic agent by RTECS criteria; Lungs, Thorax, or Respiration - Tumors.

Hazard Overviews

Carcinogenicity: IARC - Group 1, Carcinogenic to humans; NIOSH - Listed as carcinogen; NTP - Class 2A, Reasonably anticipated to be a carcinogen, limited evidence of carcinogenicity from studies in humans; ACGIH - Class A2, Suspected human carcinogen; OSHA - Not listed; EPA -

Class B2, Probable human carcinogen based on animal studies; MAK - Class A2, Unmistakably carcinogenic in animal experimentation only

Exposure Limits

OSHA PEL: TWA: 0.002 mg/m^3; as Be. Other Values: 0.025 mg/m^3; as Be 30 min/8 Hr.

ACGIH TLV: TWA: 0.002 mg/m^3; as Be.

NIOSH REL: STEL: 0.0005 mg/m^3; Ceiling as Be.

Environmental

Ecotoxicity: LC_{50} Pimephales promelas (fathead minnow) 150 ug/l/96 hr, soft water, 20,000 ug/l/96 hr, hard water. /Beryllium ion/ /Static bioassay EC_{50} Cladoceran (daphnia magna) 2,500 ug/l/48 hr, effects on reproduction

Cleanup/Disposal: Guide No. 154: Eliminate all ignition sources (no smoking, flares, sparks or flames in immediate area). Do not touch damaged containers or spilled material unless wearing appropriate protective clothing. Stop leak if you can do it without risk. Prevent entry into waterways, sewers, basements or confined areas. Absorb or cover with dry earth, sand or other non-combustible material and transfer to containers. Do not get water inside containers.

Regulations

RCRA 40CFR: Not listed

CERCLA: 40CFR 302.4: Listed as Compound per CWA Section 307(a) per CAA Section 112

SARA 40CFR 372.65: Listed as Compound

SARA EHS 40CFR 355: Not listed

TSCA: Not listed

BER8370 **CAS #: 13510-49-1**

BERYLLIUM SULFATE

RTECS: DS4800000

DOT: UN1566; IMO6.1

EINECS Number: 236-842-2

Molecular Formula: BeH_2O_4S

Structured MF: $BeSO_4$

Formula Weight: 105.07

Synonyms: BERYLLIUM SULPHATE; SULFURIC ACID,BERYLLIUM SALT (1:1)

Description: colorless crystalline solid; odorless

Use: chemical int for beryllium hydroxide; int for beryllium oxide

Physical Properties

Boiling Point: 540 °C (1004 °F)

Freezing Point: 550 °C (1022 °F)

Specific Gravity: 2.443

Water Solubility: Insoluble in Cold Water

Other Solubilities: Insoluble in alcohol

Flash Point: Nonflammable

RTECS Toxicity Data

Acute Oral: Rat LD_{50} Dose: 82 mg/kg. Mouse LD_{50} Dose: 80 mg/kg.

Acute Inhalation: Rat LC_{Lo} Dose: 10 mg/m^3. Mouse LC_{Lo} Dose: 47 mg/m^3.

Acute Dermal: Rabbit LD_{50} Route: Subcutaneous Dose: 1500 ug/kg. Rat LD_{50} Route: Subcutaneous Dose: 1500 ug/kg.

Mutagenic: Rat Cytogenetic Analysis; Route: Oral; Dose: 910 ug/kg/26W-C. Rat Morphological Transformation; Cell Type: embryo; Dose: 60 ug/L.

Tumorigenic: Rat Route: Inhalation; Dose: 432 ug/m^3/26W; Toxic Effects: Tumorigenic - Equivocal tumorigenic agent by RTECS criteria; Lungs, Thorax, or Respiration - Tumors. Rat Route: Inhalation; Dose: 643 ug/m^3/39W-I; Toxic Effects: Tumorigenic - Equivocal tumorigenic agent by RTECS criteria; Lungs, Thorax, or Respiration - Tumors.

Hazard Overviews

Fire: Noncombustible.

Carcinogenicity: IARC - Group 1, Carcinogenic to humans; NIOSH - Listed as carcinogen; NTP - Class 2A, Reasonably anticipated to be a carcinogen, limited evidence of carcinogenicity from studies in humans; ACGIH - Class A2, Suspected human carcinogen; OSHA - Not listed; EPA - Class B2, Probable human carcinogen based on animal studies; MAK - Class A2, Unmistakably carcinogenic in animal experimentation only

Exposure Limits

OSHA PEL: TWA: 0.002 mg/m^3; as Be. Other Values: 0.025 mg/m^3; as Be 30 min/8 Hr.

ACGIH TLV: TWA: 0.002 mg/m^3; as Be.

NIOSH REL: STEL: 0.0005 mg/m^3; Ceiling as Be.

Environmental

Ecotoxicity: LC_{50} Pimephales promelas (fathead minnow) 150 ug/l/96 hr, soft water, 20,000 ug/l/96 hr, hard water. /Beryllium ion/ /Static bioassay EC_{50} Cladoceran (Daphnia magna) 2,500 ug/l/48 hr, effects on reproduction

Cleanup/Disposal: Guide No. 154: Eliminate all ignition sources (no smoking, flares, sparks or flames in immediate area). Do not touch damaged containers or spilled material unless wearing appropriate protective clothing. Stop leak if you can do it without risk. Prevent entry into waterways, sewers, basements or confined areas. Absorb or cover with dry earth, sand or other non-combustible material and transfer to containers. Do not get water inside containers.

Environmental Physical Data

BCF: up to 100-fold under constant exposure

BOD: none

Regulations

RCRA 40CFR: Not listed

CERCLA: 40CFR 302.4: Listed as Compound per CWA Section 307(a) per CAA Section 112

SARA 40CFR 372.65: Listed as Compound

SARA EHS 40CFR 355: Not listed

TSCA: Listed

BER9040 **CAS #: 7787-56-6**

BERYLLIUM SULFATE TETRAHYDRATE

RTECS: DS5000000
Molecular Formula: BeH_8O_8S
Structured MF: $BeSO_4•4H_2O$
Formula Weight: 177.15

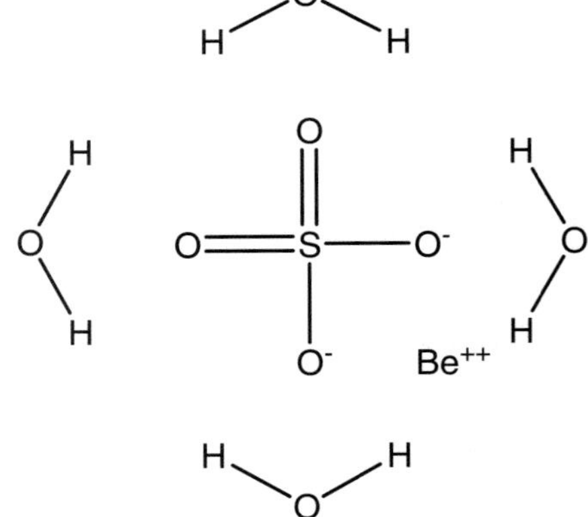

Chemical Structure

Synonyms: BERYLLIUM SULFATE TETRAHYDRATE (1:1:4); BERYLLIUM SULPHATE TETRAHYDRATE; SULFURIC ACID,BERYLLIUM SALT (1:1),TETRAHYDRATE
Description: colorless crystals or white crystalline solid; odorless
Use: chemical intermediate in the processing of beryl and bertrandite ores

Physical Properties

Boiling Point: Decomposes at 580 °C (1076 °F)
Freezing Point: 270 °C (518 °F)
Specific Gravity: 1.713
Density: 1.713 g/mL at 11 °C
Water Solubility: >= 100 mg/mL at 23 C
Other Solubilities: 95% Ethanol: <1 mg/ml at 23 °C; Acetone: <1 mg/ml at 23 °C; Alcohol: Insoluble; Concentrated H_2SO_4: Slightly Soluble; DMSO: <1 mg/ml at 23 °C.
Refraction Index: 1.472
Flash Point: Probably nonflammable

RTECS Toxicity Data

Acute Dermal: Chicken LD_{50} Route: Subcutaneous Dose: 2500 ug/kg; Toxic Effects: Behavioral - Fluid intake; Skin and appendages - Hair.
Mutagenic: Human Cytogenetic Analysis; Cell Type: lymphocyte; Dose: 5 mg/L. Human Sister Chromatid Exchenge; Cell Type: lymphocyte; Dose: 1 mg/L.
Tumorigenic: Rat Route: Inhalation; Dose: 668 ug/m^3/40W-C; Toxic Effects: Tumorigenic - Carcinogenic by RTECS criteria; Lungs, Thorax, or Respiration - Tumors.

Hazard Overviews

Poison

Health: Irritating to eyes/skin/respiratory tract. Poison. Other Acute Effects: may be fatal if inhaled, swallowed, or absorbed through skin. Chronic Effects: Carcinogen.
Fire: Noncombustible. Hazards: emits toxic fumes. Extinguishing agents: noncombustible; use extinguishing media appropriate to surrounding fire conditions. Precautions: combustible liquid.
Reactivity: Incompatible with: strong oxidizing agents. Hazardous decomposition products: sulfur oxides.
Carcinogenicity: IARC - Group 1, Carcinogenic to humans; NIOSH - Listed as carcinogen; NTP - Class 2A, Reasonably anticipated to be a carcinogen, limited evidence of carcinogenicity from studies in humans; ACGIH - Class A2, Suspected human carcinogen; OSHA - Not listed; EPA - Class B2, Probable human carcinogen based on animal studies; MAK - Class A2, Unmistakably carcinogenic in animal experimentation only
Primary Target Organs:

Eyes

Skin

Respiratory System

Exposure Limits
OSHA PEL: TWA: 0.002 mg/m^3; as Be. Other Values: 0.025 mg/m^3; as Be 30 min/8 Hr.
ACGIH TLV: TWA: 0.002 mg/m^3; as Be.
NIOSH REL: STEL: 0.0005 mg/m^3; Ceiling as Be.

Environmental

Cleanup/Disposal: Guide No. 154: Eliminate all ignition sources (no smoking, flares, sparks or flames in immediate area). Do not touch damaged containers or spilled material unless wearing appropriate protective clothing. Stop leak if you can do it without risk. Prevent entry into waterways, sewers, basements or confined areas. Absorb or cover with dry earth, sand or other non-combustible material and transfer to containers. Do not get water inside containers.

Regulations

RCRA 40CFR: Not listed
CERCLA: 40CFR 302.4: Listed as Compound per CWA Section 307(a) per CAA Section 112
SARA 40CFR 372.65: Listed as Compound
SARA EHS 40CFR 355: Not listed
TSCA: Not listed

BET5000 CAS #: 13684-63-4

BETANAL

RTECS: FD9050000
EINECS Number: 237-199-0
Molecular Formula: $C_{16}H_{16}N_2O_4$
Formula Weight: 300.3

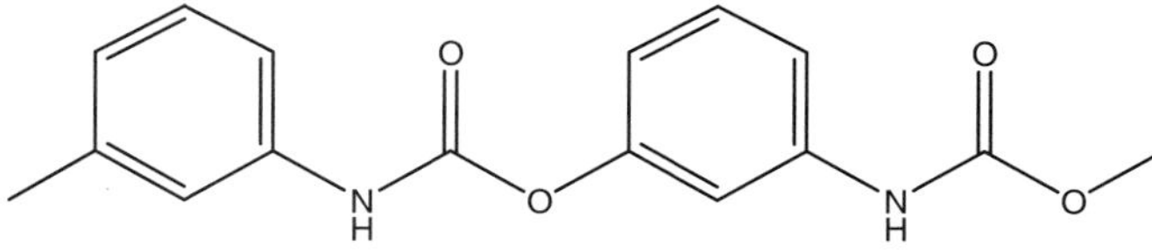

Chemical Structure

Synonyms: CARBAMIC ACID,(3-METHYLPHENYL)-,3-((METHOXYCARBONYL)AMINO)PHENYL ESTER; CARBAMIC ACID,(3-METHYLPHENYL)-,3-((METHOXYCARBONYL)AMINO)PHENYL ESTER (9CI); CARBANILIC ACID,M-HYDROXY,METHYL ESTER,M-METHYLCARBANILATE (ESTER); 3-(CARBOMETHOXYAMINO)PHENYL 3-METHYLCARBANILATE; EP-452; FENMEDIFAM; M-HYDROXYCARBANILIC ACID METHYL ESTER M-METHYLCARBANILATE; M-HYDROXYCARBANILIC ACID,METHYL ESTER,M-METHYLCARBANILATE; KEMIFAM; 3-METHOXYCARBONYLAMINOPHENYL N-3'-METHYLPHENYLCARBAMATE; 3-((METHOXYCARBONYL)AMINO)PHENYLN-(3-METHYLPHENYL)CARBAMATE; 3-METHOXYCARBONYL-N-(3'-METHYLPHENYL)-CARBAMAT; METHYL M-HYDROXYCARBANILATE,M-METHYLCARBANILATE; METHYL 3-(M-TOLYLCARBAMOYLOXY)PHENYLCARBAMATE; METHYL 3-(M-TOYLCARBAMOYLOXY)PHENYLCARBAMATE; METHYL-3-HYDROXYCARBANILATE-3-METHYLCARBANILATE; METHYLN-(3-(N-(3-METHYLPHENYL)CARBAMOYLOXY)PHENYL)CARBAMATE; 3-(METHYLPHENYL) CARBAMIC ACID 3-((METHOXYCARBONYL)AMINO) PHENYL ESTER; 3-(METHYLPHENYL)CARBAMIC ACID 3-((METHOXYCARBONYL)AMINO)PHENYL ESTER; METHYL-3-M-TOLYCARBAMOLOXYPHENYL CARBAMATE; PHENMEDIPHAM; PHENMEDIPHAME; SCHERING-38584; SN-38584; SN 4075; SPIN-AID; SYNBETAN P; VANGARD
Description: colorless crystals; odorless
Use: post emergent herbicide

Physical Properties

Freezing Point: 139 °C (282.2 °F) to 142 °C (287.6 °F)
Density: 0.25 g/cu cm at 20 °C
Vapor Pressure: 1.3 nPa at 25 °C
Water Solubility: 4.7 mg/L Water
Other Solubilities: 95% Ethanol: 5-10 mg/ml at 21 °C; Acetone: >=100 mg/ml at 21 °C; Benzene: 2.5 mg/ml at 25 °C; Chloroform: 20 mg/ml at 25 °C; DMSO: >=100 mg/ml at 21 °C; Hexane: 0.5 mg/ml at 25 °C; Methanol: 50 mg/ml at 25 °C.
Flash Point: 74 °C Closed Cup

RTECS Toxicity Data

Acute Oral: Rat LD_{50} Dose: 4 gm/kg. Mouse LD_{50} Dose: >8 gm/kg.
Acute Dermal: Rabbit LD_{50} Route: Skin; Dose: 10 mL/kg. Rat LD_{50} Route: Skin; Dose: >500 mg/kg.
Mutagenic: Mouse Cytogenetic Analysis; Route: Oral; Dose: 100 mg/kg. Mold - A Nidulans Sex Chromosome Loss; Dose: 40 mg/L.

Hazard Overviews

Fire: Combustible.
Carcinogenicity: IARC - Not listed; NIOSH - Not listed; NTP - Not listed; ACGIH - Not listed; OSHA - Not listed; EPA - Not listed; MAK - Not listed

Environmental

Ecotoxicity: LC_{50} Harlequin fish 16.5 mg/l/96 hr /Conditions of bioassay not specified/emulsifiable preparation
Environmental Fate: If released to the atmosphere, it will degrade rapidly in the vapor phase by reaction with photochemically produced hydroxyl radicals (half-life of about 2.6 hr). Particulate-phase and aerosols released to air during applications of herbicides will be removed from air physically by dry and wet deposition. If released to soil or water, it can degrade through biodegradation and aqueous hydrolysis. The hydrolysis rate increases with increasing pH; at 25 °C, the hydrolysis half-life is about 7.5 days, 18 hr, and 1.8 hr at respective pHs of 6, 7, and 8. It is reported to remain in the top layers of soil (0 to 2 inches) after herbicidal application and has a high K_{oc} value suggesting a low potential to leach. The US Dept of Agriculture's Pesticide Properties Database lists a soil half-life of 30 days; however, the rate could be slower in acidic soil or faster in alkaline soil.

Environmental Physical Data

Henry's Law Constant: estimated at 8.41×10^{-13}
Sorption Partition Coefficient: K_{OC} = estimated at 1860
BCF: estimated at 260
BOD: 52.3% BODT, 5 days

Regulations

RCRA 40CFR: Not listed
CERCLA: 40CFR 302.4: Not listed
SARA 40CFR 372.65: Not listed
SARA EHS 40CFR 355: Not listed
TSCA: Not listed

BIC5000 CAS #: 90-42-6

(1,1'-BICYCLOHEXYL)-2-ONE

RTECS: DT7400000
EINECS Number: 201-991-4
Molecular Formula: $C_{12}H_{20}O$
Formula Weight: 180.3
Synonyms: CYCLOHEXANONE,2-CYCLOHEXYL-; CYCLOHEXANONE,2-CYCLOHEXYL-(6CI); 2-CYCLOHEXYLCYCLOHEXANONE; LAVAMENTHE
Use: fragrance

Physical Properties

Boiling Point: 264 °C (507.2 °F)
Freezing Point: -32 °C (-25.6 °F)
Specific Gravity: 0.96960
Refraction Index: 1.4877

RTECS Toxicity Data

Acute Oral: Rat LD_{50} Dose: 5 gm/kg.
Acute Dermal: Rabbit LD_{50} Route: Skin; Dose: >7800 mg/kg.

Hazard Overviews

Carcinogenicity: IARC - Not listed; NIOSH - Not listed; NTP - Not listed; ACGIH - Not listed; OSHA - Not listed; EPA - Not listed; MAK - Not listed

Environmental

Regulations

RCRA 40CFR: Not listed
CERCLA: 40CFR 302.4: Not listed
SARA 40CFR 372.65: Not listed
SARA EHS 40CFR 355: Not listed
TSCA: Listed

BIF5000 CAS #: 82657-04-3

BIFENTHRIN

RTECS: GZ1227800
Molecular Formula: $C_{23}H_{22}ClF_3O_2$
Formula Weight: 434.89
Synonyms: BIFENTHRINE; BIPHENATE; BIPHENTHRIN; BIPHENTRIN; BRIGADE; CAPTURE; [1ALPHA, 3ALPHA(Z)]-(+-)-3-(2-CHLORO-3,3,3-TRIFLUORO-1-PROPENYL)-2,2-DIMETHYLCYCLOPROPANECARBOXYLIC ACID(2-METHYL[1,1'-BIPHENYL]-3-YL)METHYL ESTER; FMC-54800; FMC 54800; FMC 58000; [1 ALPHA,3 ALPHA(Z)]-(+-)-(2-METHYL[1,1'-BIPHENYL]-3-YL)METHYL 3-(2-CHLORO-3,3,3-TRIFLUORO-1-PROPENYL)-2,2-DIMETHYLCYCLOPROPANECARBOXYLATE; 2-METHYLBIPHENYL-3-YLMETHYL(Z)-(1RS,3RS)-3-(2-CHLORO-3,3,3-TRIFLUOROPROP-1-ENYL)-2,2-DIMETHYLCYCLOPROPANECARBOXYLATE; 2-METHYLBIPHENYL-3-YLMETHYL-(Z)-(1RS)-CIS-3-(2-CHLORO-3,3,3-TRIFLUOROPROP-1-ENYL)-2,2-DIMETHYLCYCLOPROPANECARBOXYLATE; TALSTAR
Description: light brown viscous oil which hardens to solid, light brown mass
Use: insecticide effective against coleoptera, diptera, heteroptera, homoptera, lipidoptera, and othoptera; also controls some species of acarina; control of a wide range of foliar insect pests and mites on cereals, fruit, vines, oilseed rape, potatoes, peas, beans, cotton, ornamentals, and other crops

Physical Properties

Freezing Point: 68 °C (154.4 °F) to 70 °C (158 °F)
Saturated Vapor Density: 1.200000004 kg/m^3
Density: 1.212 g/mL at 25 °C/4 °C
Vapor Pressure: 1.81 x10^{-7} torr at 25 °C
Water Solubility: < 0 mg/L
Other Solubilities: Soluble in methylene chloride, Chloroform, Acetone, Ether, Toluene; slightly Soluble in heptane, Methanol.
Flash Point: 165 °C Open Cup

RTECS Toxicity Data

Acute Oral: Rat LD_{50} Dose: 54500 ug/kg. Duck LD_{50} Dose: >4450 mg/kg.
Acute Dermal: Rabbit LD_{50} Route: Skin; Dose: >2 gm/kg.

Hazard Overviews

Fire: Will burn.
Carcinogenicity: IARC - Not listed; NIOSH - Not listed; NTP - Not listed; ACGIH - Not listed; OSHA - Not listed; EPA - Not listed; MAK - Not listed

Environmental

Ecotoxicity: LC_{50} Mallard duck dietary 1280 mg/kg diet/8 day LC_{50} Rainbow trout 0.00015 mg/l/96 hr /Conditions of bioassay not specified LC_{50} Daphnia 0.0016 mg/l/48 hr /Conditions of bioassay not specified LD_{50} Bobwhite quail oral 1800 mg/kg

Regulations

RCRA 40CFR: Not listed
CERCLA: 40CFR 302.4: Not listed
SARA 40CFR 372.65: Listed
SARA EHS 40CFR 355: Not listed
TSCA: Not listed

Analytical Methods

Food: FDA 212.1, 212.2, 232.1, 232.4, 242.1
Plasma: FDA 211.1, 231.1, 252

BIN5000 CAS #: 485-31-4

BINAPACRYL

RTECS: GQ5600000
DOT: UN2779; UN2780; UN3013; UN3014; IMO3.2; IMO6.1
EINECS Number: 207-612-9
Molecular Formula: $C_{15}H_{18}N_2O_6$
Formula Weight: 322.31
Synonyms: ACRICID; AMBOX; 2-BUTENOIC ACID,3-METHYL-,2-(1-METHYLPROPYL)-4,6-DINITROPHENYL ESTER; 2-BUTENOIC ACID,3-METHYL-,2-(1-METHYLPROPYL)-4,6-DINITROPHENYL ESTER (9CI); 2-SEC-BUTYL-4,6-DINITROPHENYL BETA,BETA-DIMETHYLACRYLATE; 2-SEC-BUTYL-4,6-DINITROPHENYL 3-METHYL-2-BUTENOATE; 2-SEC-BUTYL-4,6-DINITROPHENYL 3-METHYLBUT-2-ENOATE; 2-SEC-BUTYL-4,6-DINITROPHENYL 3-METHYLCROTONATE; 2-SEC-BUTYL-4,6-DINITROPHENYL SENECIOATE; 2-SEC-BUTYL-4,6-DINITROPHENYL-3,3-DIMETHYLACRYLATE; 2-SEC-BUTYL-4,6-DINITROPHENYL-3-METHYL-2-BUTENOATE; CROTONIC ACID,3-METHYL-,2-SEC-BUTYL-4,6-DINITROPHENYLESTER; DAPACRYL; 3,3 DIMETHYL-ACRYLATE DE 2,4-DINITRO-6-(1-METHYLPROPYLE)PHENYLE; 3,3 DIMETHYL-ACRYLATE DE2,4-DINITRO-6-(1-METHYLPROPYLE)PHENYL; 3,3-DIMETHYLACRYLIC ACID 2-SEC-

BUTYL-4,6-DINITROPHENYLESTER; DINAPACRYL; 4,6-DINITRO-2-SEC-BUTYLPHENYL BETA,BETA-DIMETHYLACRYLATE; 2,4-DINITRO-6-SEC-BUTYLPHENYL 2-METHYLCROTONATE; 4,6-DINITROPHENYL-2-SEC-BUTYL-3-METHYL-2-BUTENONATE; DINOSEB METHACRYLATE; DINOSEB,3,3-DIMETHYLACRYL ESTER; ENDOSAN; ENT 25,793; FMC 9044; HOE 2784; HOE 2784 OA; 3-METHYL-2-BUTENOIC ACID 2-SEC-BUTYL-4,6-DINITROPHENYLESTER; 3-METHYL-2-BUTENOIC ACID 2-(1-METHYLPROPYL)-4,6-DINITROPHENYL ESTER; 3-METHYL-2-BUTENOIC ACID2-(1-METHYLPROPYL)-4,6-DINITROPHENYL ESTER; 3-METHYLCROTONIC ACID 2-SEC-BUTYL-4,6-DINITROPHENYL ESTER; (6-(1-METHYL-PROPYL)-2,4-DINITRO-FENYL)-3,3-DIMETHYL-ACRYLAAT; 2-(1-METHYLPROPYL)-4,6-DINITROPHENYL BETA,BETA-DIMETHACRYLATE; 2-(1-METHYLPROPYL)-4,6-DINITROPHENYL 3,3-DIMETHYLACRYLATE; 2-(1-METHYLPROPYL)-4,6-DINITROPHENYL 3-METHYL-2-BUTENOATE; 2-(1-METHYLPROPYL)-4,6-DINITROPHENYLBETA,BETA-DIMETHACRYLATE; (6-(1-METHYL-PROPYL)-2,4-DINITRO-PHENYL)-3,3-DIMETHYL-ACRYLAT; (6-(1-METIL-PROPIL)-2,4-DINITRO-FENIL)-3,3-DIMETIL-ACRILATO; MOROCIDE; MORROCID; NIA 9044; NIAGARA 9044; PHENOL,2-SEC-BUTYL-4,6-DINITRO-,3-METHYLCROTONATE; 2-SEK.BUTYL-4,6-DINITROFENYLESTER KYSELINY 3-METHYLKROTONOVE; SENECIOIC ACID 2-SEC BUTYL-4,6-DINITROPHENYL ESTER; SENECIOIC ACID 2-SEC-BUTYL-4,6-DINITROPHENYL ESTER

Description: colorless to pale yellow to brownish crystals or powder; faint aromatic odor

Use: fungicide and miticide (former use)

Physical Properties

Freezing Point: 66 °C (150.8 °F) to 67 °C (152.6 °F)

Specific Gravity: 1.2307 at 20 °C/4 °C

Vapor Pressure: 1 $x10^{-4}$ mm Hg at 60 °C

Water Solubility: Insoluble in Water

Other Solubilities: Soluble in: 400 mg/l hexane; greater than 500 g/l dichloromethane, Ethyl Acetate, Toluene; 21 g/l Methanol; Soluble in 75% methylene chloride.

RTECS Toxicity Data

Acute Oral: Rat LD_{50} Dose: 58 mg/kg. Mouse LD_{50} Dose: 1600 mg/kg; Toxic Effects: Behavioral - Somnolence (general depressed activity); Behavioral - Convulsions or effect on seizure threshold; Lungs, Thorax, or Respiration - Respiratory stimulation.

Acute Dermal: Rabbit LD_{50} Route: Skin; Dose: 750 mg/kg. Rat LD_{50} Route: Skin; Dose: 720 mg/kg.

Mutagenic: Bacteria - S Typhimurium Mutations in Microorganisms; Dose: 5 mg/plate (+S9).

Hazard Overviews

Carcinogenicity: IARC - Not listed; NIOSH - Not listed; NTP - Not listed; ACGIH - Not listed; OSHA - Not listed; EPA - Not listed; MAK - Not listed

Environmental

Ecotoxicity: LC_{50} Channel catfish 15 ug/l/96 hr at 18 °C, wt 1.4 g. Static bioassay without aeration, pH 7.2-7.5, water hardness 40-50 mg/l as $CaCO_3$ and alkalinity of 30-35 mg/l

Environmental Fate: If released to soil, it may undergo slow hydrolysis in basic soils. It is expected to be essentially immobile in soil. It is known to slowly decompose under the influence of UV light and it may undergo photolysis on sunlit soil surfaces. If released to water, it is expected to undergo slow hydrolysis to dinoseb under alkaline conditions. It may undergo photolysis in aquatic systems. It may significantly bioconcentrate in fish and aquatic organisms and it is not expected to significantly volatilize from aquatic systems to the atmosphere. If released to the atmosphere, it may undergo gas-phase reactions with photochemically produced hydroxyl radicals and with ozone. Estimated half-lives for these processes are 4.63 hours and 3.72 hours, respectively; however, the rate of these processes may be greatly attenuated as binapacryl is expected to exist predominately as a particulate in the atmosphere. It may undergo slow photolytic degradation under the influence of UV light.

Cleanup/Disposal: Guide No. 131: Fully encapsulating, vapor protective clothing should be worn for spills and leaks with no fire. Eliminate all ignition sources (no smoking, flares, sparks or flames in immediate area). All equipment used when handling the product must be grounded. Do not touch or walk through spilled material. Stop leak if you can do it without risk. Prevent entry into waterways, sewers, basements or confined areas. A vapor suppressing foam may be used to reduce vapors. Small Spills: Absorb with earth, sand or other non-combustible material and transfer to containers for later disposal. Use clean non-sparking tools to collect absorbed material. Large Spills: Dike far ahead of liquid spill for later disposal. Water spray may reduce vapor; but may not prevent ignition in closed spaces. Guide No. 153: Eliminate all ignition sources (no smoking, flares, sparks or flames in immediate area). Do not touch damaged containers or spilled material unless wearing appropriate protective clothing. Stop leak if you can do it without risk. Prevent entry into waterways, sewers, basements or confined areas. Absorb or cover with dry earth, sand or other non-combustible material and transfer to containers. Do not get water inside containers.

Environmental Physical Data

Henry's Law Constant: 3.37 $x10^{-7}$

Octanol/Water Partition Coefficient: log K_{ow} = estimated at 4.75

Sorption Partition Coefficient: K_{oc} = calculated at 9100

BCF: calculated at 2400

Regulations

RCRA 40CFR: Not listed

CERCLA: 40CFR 302.4: Not listed

SARA 40CFR 372.65: Not listed

SARA EHS 40CFR 355: Not listed

TSCA: Not listed

Analytical Methods

Soil: EPA PMD-BIN

Food: FDA 212.1, 232.1, 232.4, 242.1

Plasma: EPA 001; FDA 211.1, 231.1, 252

BIO5000 CAS #: 58-85-5

BIOTIN

RTECS: XJ9088200
EINECS Number: 200-399-3
Molecular Formula: $C_{10}H_{16}N_2O_3S$
Formula Weight: 244.31

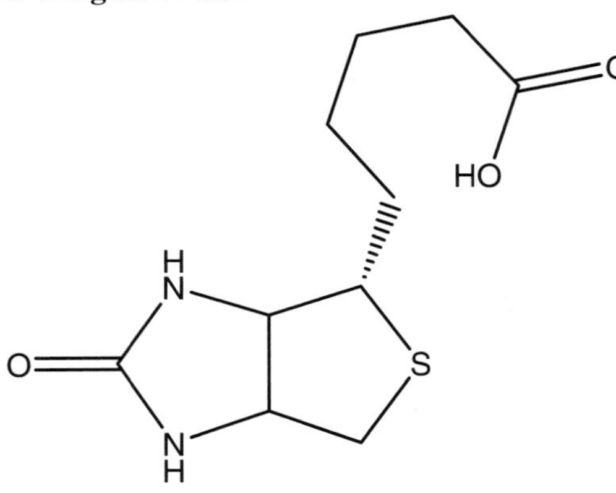

Chemical Structure

Synonyms: BIOEPIDERM; BIOS II; (+)-BIOTIN; D(+)-BIOTIN; D-(+)-BIOTIN; D-BIOTIN; COENZYME R; FACTOR S; FACTOR S (VITAMIN); CIS-HEXAHYDRO-2-OXO-1H-THIENO[3,4]IMIDAZOLE-4-VALERIC ACID; 2'-KETO-3,4-IMIDAZOLIDO-2-TETRAHYDROTHIOPHENE-N-VALERICACID; CIS-TETRAHYDRO-2-OXOTHIENO[3,4-D]-IMIDAZOLINE-4-VALERICACID; 1H-THIENO(3,4-D)IMIDAZOLE-4-PENTANOIC ACID,HEXAHYDRO-2-OXO-,(3AS-(3AALPHA,4BETA,6AALPHA))-; VITAMIN B7; VITAMIN H
Description: colorless, crystalline
Use: nutrient &/or dietary supplement; medicine; component in some multi-vitamin preparations; feed additive for poultry and swine

Physical Properties

Freezing Point: Decomposes at 232 °C (449.6 °F)
Water Solubility: 22 mg/100 mL at 25 °C
Other Solubilities: Slightly Soluble in Chloroform; slightly Soluble in Alcohol (its salts are quite Soluble).
pH: Isolectric point 3.5

RTECS Toxicity Data

Reproductive/Teratogenic: Rat Route: Subcutaneous; Dose: 200 mg/kg; Duration: female 14-15D of pregnancy Maternal Effects - Uterus, cervix, vagina; Effects on Embryo or Fetus - Extra embryonic structures; Fetotoxicity. Rat Route: Subcutaneous; Dose: 100 mg/kg; Duration: male 1D prior to mating; Effects on Fertility - Litter size; Effects on Embryo or Fetus - Extra embryonic structures; Fetotoxicity. Rat Route: Subcutaneous; Dose: 200 mg/kg; Duration: female 1-2D of pregnancy; Effects on Fertility - Post-implantation mortality.

Hazard Overviews

Health: Acute Effects: may be harmful by inhalation, ingestion, or skin absorption; may cause eye and skin irritation.
Fire: Hazards: emits toxic fumes. Extinguishing agents: water spray; carbon dioxide, dry chemical powder or appropriate foam. Precautions: combustible liquid.
Reactivity: Incompatible with: strong oxidizing agents. Hazardous decomposition products: toxic fumes of: carbon monoxide, carbon dioxide.
Carcinogenicity: IARC - Not listed; NIOSH - Not listed; NTP - Not listed; ACGIH - Not listed; OSHA - Not listed; EPA - Not listed; MAK - Not listed

Environmental

Regulations

RCRA 40CFR: Not listed
CERCLA: 40CFR 302.4: Not listed
SARA 40CFR 372.65: Not listed
SARA EHS 40CFR 355: Not listed
TSCA: Listed

BIP1000 CAS #: 92-52-4

BIPHENYL

RTECS: DU8050000
EINECS Number: 202-163-5
Molecular Formula: $C_{12}H_{10}$
Structured MF: $C_6H_5C_6H_5$
Formula Weight: 154.20

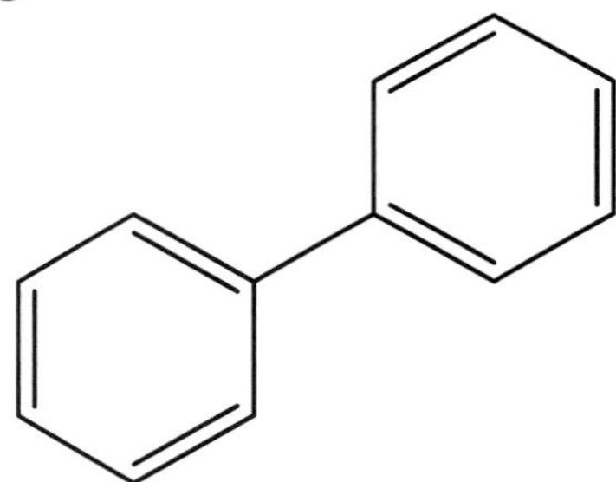

Chemical Structure

Synonyms: BIBENZENE; 1,1'-BIPHENYL; CAROLID AL; 1,1'-DIPHENYL; DIPHENYL; LEMONENE; PHENADOR-X; PHENYL BENZENE; PHENYLBENZENE; PHPH; TETROSIN LY; XENENE
Description: colorless, white to yellow leaf-like crystals; pleasant, peculiar aromatic odor
Use: organic synthesis, heat transfer fluid, preservative in food, fungicide

Physical Properties

Boiling Point: 254 °C (489 °F) to 255 °C (491 °F)
Freezing Point: 69 °C (156.2 °F) to 71 °C (159.8 °F)
Specific Gravity: 1.041 at 20 °C/4 °C
Vapor Density: 5.31 Air=1

Density: 1.041 g/mL
Vapor Pressure: 1 mm Hg at 71 °C
Water Solubility: Insoluble
Other Solubilities: Benzene: Very Soluble; Ether: Soluble.
Odor Threshold: 0.0062 to 0.3 mg/m^3
Refraction Index: 1.588 at 77 °C/D
Evaporation Rate: < 1 Butyl Acetate=1
Critical Temperature: 515.7 °C
Critical Pressure: 37.9 kPa
Ionization Potential (eV): 7.95
Flash Point: 113 °C Closed Cup
Autoignition Temperature: 540 °C
LEL: 0.6% v/v
UEL: 5.8% v/v at 155 °C

RTECS Toxicity Data

Acute Oral: Rat LD_{50} Dose: 2400 mg/kg. Mouse LD_{50} Dose: 1900 mg/kg; Toxic Effects: Behavioral - Somnolence (general depressed activity); Gastrointestinal - Hypermotility, diarrhea.
Acute Inhalation: Rat LC; Dose: >200 mg/m^3. Human TC_{Lo} Dose: 4400 ug/m^3; Toxic Effects: Peripheral Nerve and sensation - Flaccid paralysis without anesthesia; Gastrointestinal - Nausea or vomiting; Gastrointestinal - Other changes.
Acute Dermal: Rabbit LD_{50} Route: Skin; Dose: >5010 mg/kg.
Chronic (Multiple Dose) Oral: Rat Dose: 3300 mg/kg/15D-I; Toxic Effects: Kidney, Ureter, and Bladder - Changes in kidney weight; Blood - Changes in serum composition; Biochemical - Phosphatases. Rat Dose: 41250 mg/kg/24W-C; Toxic Effects: Kidney, Ureter, and Bladder - Chgs in tubules (inc acute renal failure, acute tubular necrosis; Kidney, Ureter, and Bladder - Urine volume increased.
Chronic (Multiple Dose) Inhalation: Rat Dose: 300 mg/m^3/7H/94D-I; Toxic Effects: Sense organs and special senses - Other; Lungs, Thorax, or Respiration - Emphysema; DEATH. Mouse Dose: 5 mg/m^3/7H/92D-I; Toxic Effects: Lungs, Thorax, or Respiration - Emphysema; Lungs, Thorax, or Respiration - Chronic pulmonary edema; DEATH.
Irritation Eye: Rabbit Standard Draize Test Dose: 100 mg; Reaction: mild.
Mutagenic: Rat Unscheduled DNA Synthesis; Route: Oral; Dose: 8400 mg/kg/4W-C. Mouse DNA Damage; Cell Type: lymphocyte; Dose: 50 umol/L.
Tumorigenic: Mouse Route: Oral; Dose: 56 gm/kg; Toxic Effects: Tumorigenic - Equivocal tumorigenic agent by RTECS criteria; Lungs, Thorax, or Respiration - Tumors; Blood - Tumors. Mouse Route: Subcutaneous; Dose: 46 mg/kg; Toxic Effects: Tumorigenic - Neoplastic by RTECS criteria; Lungs, Thorax, or Respiration - Tumors; Liver - Tumors.

Hazard Overviews

Fire Diamond

Health: Irritating to skin. Also Causes: headaches, GI pain, nausea, numbness/aching of limbs, fatigue, malaise. Chronic Effects: dermatitis, liver damage, CNS damage.
Fire: Will burn. Use dry chemical, carbon dioxide, foams, or water fog to put out biphenyl fires. Dust explosions involving biphenyl are significant hazards if a particulate cloud contacts an ignition source.
Reactivity: Stable. Hazardous polymerization cannot occur. Avoid: formation of dust clouds; direct contact with ignition sources. Incompatible with: strong oxidizing agents. Hazardous decomposition products: carbon monoxide aromaticcompounds.
Carcinogenicity: IARC - Not listed; NIOSH - Not listed; NTP - Not listed; ACGIH - Not listed; OSHA - Not listed; EPA - Class D, Not classifiable as to human carcinogenicity; MAK - Not listed
Primary Target Organs:

Eyes

Skin
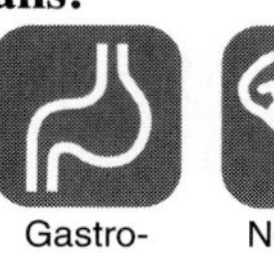
Gastro-intestinal

Nervous System

Liver

Exposure Limits
OSHA PEL: TWA: 0.2 ppm; 1 mg/m^3.
ACGIH TLV: TWA: 0.2 ppm; 1.3 mg/m^3.
NIOSH REL: TWA: 0.2 ppm; 1 mg/m^3.
NIOSH IDLH: 100 mg/m^3.
DFG MAK: TWA: 0.2 ppm; 1 mg/m^3.
Respirator Recommendation
Exposure Range: >0.2 to 2 ppm Air Purifying, Negative Pressure, Half Mask
Exposure Range: >2 to <100 ppm Air Purifying, Negative Pressure, Full Face
Exposure Range: 100 to unlimited ppm Self-contained Breathing Apparatus, Pressure Demand, Full Face
Cartridge Color: black with dust/mist prefilter (use P100 or consult supervisor for appropriate dust/mist prefilter)

Environmental

Ecotoxicity: Tetrahymena pyriformis growth inhib. EC_{50}: 46 mg/l
Environmental Fate: Photolysis and hydrolysis are not expected to be important environmental fate processes. Biodegradation of the compound in both water and soil is expected to be important. Volatilization may also be important in water and soil surfaces. The K_{oc} value indicates that the compound will be low to slightly mobile in soil and its adsorption to sediment and suspended solids in water will be important. It may moderately bioconcentrate in aquatic organisms. The reaction with ozone and nitrate radicals in the atmosphere are not important, but the reaction with hydroxyl radicals has an estimated half-life of 2 days. It may be partially removed from the atmosphere by dry deposition.
Cleanup/Disposal: Guide No. 171: Do not touch or walk through spilled material. Stop leak if you can do it without risk. Prevent dust cloud. Avoid inhalation of asbestos dust. Small Dry Spills: With clean shovel place material into clean, dry container and cover loosely; move containers from spill

area. Small Spills: Take up with sand or other noncombustible absorbent material and place into containers for later disposal. Large Spills: Dike far ahead of liquid spill for later disposal. Cover powder spill with plastic sheet or tarp to minimize spreading. Prevent entry into waterways, sewers, basements or confined areas.

Environmental Physical Data

Henry's Law Constant: estimated at 3.00×10^{-4}
Octanol/Water Partition Coefficient: log K_{ow} = 3.16 to 4.17
Sorption Partition Coefficient: log K_{oc} = 2.94 to 3.52
BCF: golden orfe 280

Regulations

RCRA 40CFR: Not listed
CERCLA: 40CFR 302.4: Listed per CAA Section 112 RQ: 100 lb (45.35 kg)
SARA 40CFR 372.65: Listed
SARA EHS 40CFR 355: Not listed
TSCA: Listed

Analytical Methods

Soil: EPA 1625
Water / Groundwater: EPA 1625, 642
Food: FDA 212.2, 232.4, 242.1; EPA XENO; AOAC 968.25
Indoor / Expired Air: NIOSH 2530

BIP3000 CAS #: 90-41-5

2-BIPHENYLAMINE

RTECS: DU8850000
EINECS Number: 201-990-9
Molecular Formula: $C_{12}H_{11}N$
Structured MF: $C_6H_5C_6H_4NH_2$
Formula Weight: 169.23

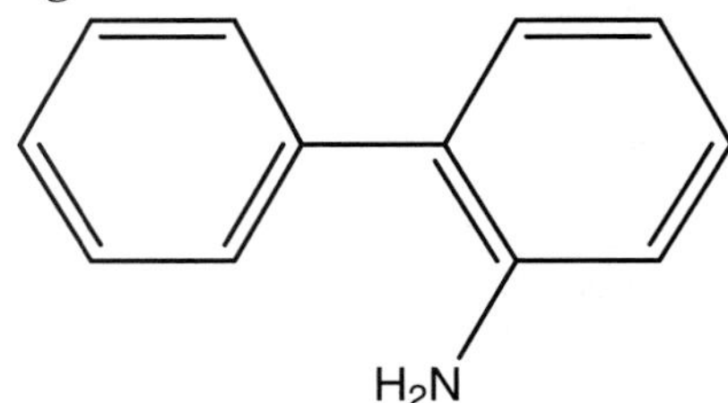

Chemical Structure

Synonyms: 2-AMINOBIFENYL; 2-AMINOBIPHENYL; O-AMINOBIPHENYL; 2-AMINODIPHENYL; O-AMINODIPHENYL; (1,1'-BIPHENYL)-2-AMINE; O-BIPHENYLAMINE; (1,1'-BIPHENYL)-2-AMINE (9CI); 2-PHENYLANILINE; O-PHENYLANILINE
Description: colorless or purplish crystals or leaves
Use: as a dye intermediate and in research and analytical chemistry; a fluorogenic reagent for the detection of carbonyl compounds

Physical Properties

Boiling Point: 299 °C (570 °F) at 760 mm Hg
Freezing Point: 51 °C (123.8 °F) to 53 °C (127.4 °F)
Vapor Density: 5.8 Air=1
Water Solubility: Insoluble in Water
Other Solubilities: 95% Ethanol: >=100 mg/ml at 24 °C; Acetone: >=100 mg/ml at 24 °C; Alcohol: Soluble; Benzene: Soluble; DMSO: >=100 mg/ml at 24 °C; Ether: Soluble.
Flash Point: > 112 °C
Autoignition Temperature: 450 °C

RTECS Toxicity Data

Acute Oral: Rat LD_{50} Dose: 2340 mg/kg; Toxic Effects: Behavioral - Coma; Lungs, Thorax, or Respiration - Dyspnea; Nutritional and gross metabolic - Weight loss or decreased weight gain. Rabbit LD_{50} Dose: 1020 mg/kg; Toxic Effects: Behavioral - Coma; Lungs, Thorax, or Respiration - Dyspnea; Nutritional and gross metabolic - Weight loss or decreased weight gain.
Chronic (Multiple Dose) Oral: Rat Dose: 137 gm/kg/13W-C; Toxic Effects: Kidney, Ureter, and Bladder - Chgs in tubules (inc acute renal failure, acute tubular necrosis; Blood - Changes in spleen; Blood - Changes in leukocyte (WBC) cell count. Rat Dose: 7 gm/kg/14D-C; Toxic Effects: Blood - Changes in spleen. Rat Dose: 328 gm/kg/13W-C; Toxic Effects: Liver - Other changes; Blood - Changes in bone marrow not included above; Blood - Changes in leukocyte (WBC) cell count.
Mutagenic: Rat DNA Damage; Cell Type: liver; Dose: 1 mmol/L. Mouse Morphological Transformation; Cell Type: embryo; Dose: 10 mg/L.

Hazard Overviews

Health: Irritating to eyes/skin/respiratory tract. Harmful. Other Acute Effects: harmful if swallowed, inhaled, or absorbed through skin. Chronic Effects: Possible carcinogen. Lab experiments have shown mutagenic effects.
Fire: Will burn. Hazards: emits toxic fumes. Extinguishing agents: water spray; carbon dioxide, dry chemical powder or appropriate foam. Precautions: combustible liquid.
Carcinogenicity: IARC - Not listed; NIOSH - Not listed; NTP - Not listed; ACGIH - Not listed; OSHA - Not listed; EPA - Not listed; MAK - Not listed
Primary Target Organs:

Eyes

Skin

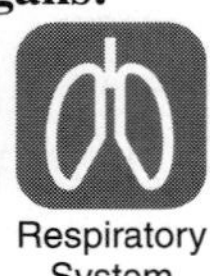
Respiratory System

Environmental

Regulations

RCRA 40CFR: Not listed
CERCLA: 40CFR 302.4: Not listed
SARA 40CFR 372.65: Not listed
SARA EHS 40CFR 355: Not listed
TSCA: Listed

BIP6000 CAS #: 492-17-1

2,4-BIPHENYLDIAMINE

RTECS: DV2100000
Molecular Formula: $C_{12}H_{12}N_2$
Formula Weight: 184.23
Synonyms: O,P'-BIANILINE; ORTHO,PARA'-BIANILINE; (1,1'-BIPHENYL)-2,4'-DIAMINE; 2,4'-DIAMINOBIFENYL; 2,4'-DIAMINO-1,1'-BIPHENYL; 2,4'-DIAMINOBIPHENYL; O,P'-DIAMINOBIPHENYL; ORTHO,PARA'-DIAMINOBIPHENYL; 2,4'-DIAMINODIPHENYL; O,P'-DIANILINE; DIFENYLIN; 2,4'-DIPHENYLDIAMINE; DIPHENYLINE
Description: needles
Use: detection agent for tungsten; chemical intermediate for azo dyes

Physical Properties

Boiling Point: 363 °C (685 °F)
Freezing Point: 45 °C (113 °F)
Water Solubility: Insoluble in Water
Other Solubilities: Very Slightly Soluble in Alcohol or Ether

RTECS Toxicity Data

Acute Oral: Rat LD_{50} Dose: 311 mg/kg.
Mutagenic: Bacteria - S Typhimurium Mutations in Microorganisms; Dose: 100 ug/plate (-S9).
Tumorigenic: Dog Route: Oral; Dose: 7020 mg/kg/5Y-I; Toxic Effects: Tumorigenic - Equivocal tumorigenic agent by RTECS criteria; Lungs, Thorax, or Respiration - Bromchiogenic carcinoma.

Hazard Overviews

Carcinogenicity: IARC - Group 3, Not classifiable as to carcinogenicity to humans; NIOSH - Not listed; NTP - Not listed; ACGIH - Not listed; OSHA - Not listed; EPA - Not listed; MAK - Not listed

Environmental

Regulations

RCRA 40CFR: Not listed
CERCLA: 40CFR 302.4: Not listed
SARA 40CFR 372.65: Not listed
SARA EHS 40CFR 355: Not listed
TSCA: Not listed

BIP9000 CAS #: 366-18-7

2,2-BIPYRIDINE

RTECS: DW1750000
EINECS Number: 206-674-4
Molecular Formula: $C_{10}H_8N_2$
Formula Weight: 156.18

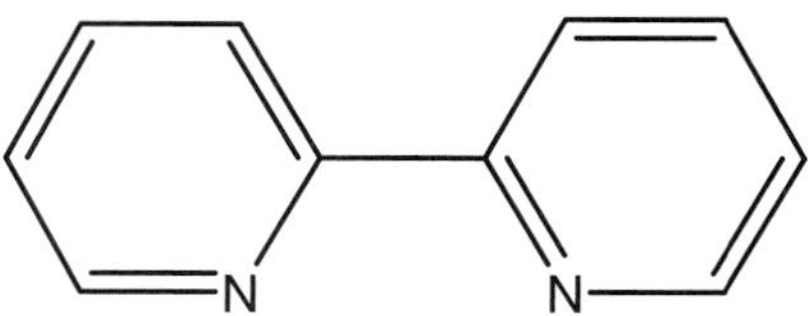

Chemical Structure

Synonyms: 2,2'-BIPYRIDIN; ALPHA,ALPHA'-BIPYRIDINE; BIPYRIDINE; 2,2'-BIPYRIDYL; ALPHA,ALPHA'-BIPYRIDYL; CI-588; 2,2'-DIPYRIDINE; ALPHA,ALPHA'-DIPYRIDINE; 2,2'-DIPYRIDYL; ALPHA,ALPHA'-DIPYRIDYL; ALPHA,ALPHA'-DWUPIRYDYLU; 2-(2-PYRIDYL)PYRIDINE
Description: white crystals or prisms
Use: reagent for the determination of iron

Physical Properties

Boiling Point: 272 °C (522 °F) to 273 °C (523 °F)
Freezing Point: 69.7 °C (157.46 °F)
Water Solubility: 1 parts in about 200 parts Water
Other Solubilities: Very Soluble in Alcohol, Ether, Benzene, Chloroform, & Petroleum Ether
Refraction Index: 1.5841 at 20 °C
Ionization Potential (eV): 8.35 +/-1.2
Flash Point: Combustible

RTECS Toxicity Data

Acute Oral: Rat LD_{50} Dose: 100 mg/kg; Toxic Effects: Behavioral - Muscle weakness; Kidney, Ureter, and Bladder - Other changes in urine composition; Blood - Hemorrhage.
Acute Dermal: Rat LD_{50} Route: Subcutaneous Dose: 131 mg/kg.
Reproductive/Teratogenic: Rat Route: Intraperitoneal; Dose: 60 mg/kg; Duration: female 12D of pregnancy; Specific Developmental Abnormalities - Craniofacial (including nose and tongue); Musculoskeletal system.
Mutagenic: Hamster DNA Damage; Cell Type: lung; Dose: 12500 ug/L. Hamster Mutations in Mammalian Somatic Cells; Cell Type: lung; Dose: 14500 ug/L.
Tumorigenic: Mouse Route: Subcutaneous; Dose: 8000 mg/kg/40W-I; Toxic Effects: Tumorigenic - Equivocal tumorigenic agent by RTECS criteria; Endocrine - Tumors.

Hazard Overviews

Health: Irritating to eyes/skin/respiratory tract. Toxic. Other Acute Effects: harmful if swallowed, inhaled, or absorbed through skin.
Fire: Combustible. Extinguishing agents: water spray; carbon dioxide, dry chemical powder or appropriate foam. Precautions: combustible liquid.
Reactivity: Incompatible with: strong oxidizing agents. Hazardous decomposition products: toxic fumes of: carbon monoxide, carbon dioxide, nitrogen oxides.
Carcinogenicity: IARC - Not listed; NIOSH - Not listed; NTP - Not listed; ACGIH - Not listed; OSHA - Not listed; EPA - Not listed; MAK - Not listed

Primary Target Organs:

Eyes

Skin

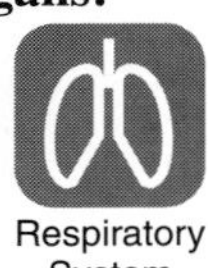
Respiratory System

Environmental

Environmental Fate: Information pertaining to biodegradation in soil and water was not located in the available literature. With a pKa of 4.33, it should exist in environmental media in varying proportions that are pH dependent. Ions generally do not volatilize. A Henry's Law constant of 5.35 x10^{-10} atm-cu m/mole at 25 °C indicates that volatilization from environmental waters and moist soil should not be an important fate process. In aquatic systems, it is not expected to bioconcentrate. It was shown to undergo slow oxidation with photochemically generated hydroxyl radicals in aqueous solution (half-life of about 129 days). A low K_{oc} indicates it should not partition from the water column to organic matter contained in sediments and suspended solids; and it should be highly mobile in soil and it may leach to ground water. However, the bipyridylium cation of diquat has been shown to strongly adsorb humic acid and clays. In the atmosphere, it is expected to exist in both the vapor and particulate phases, and vapor phases reactions with photochemically produced hydroxyl radicals should be important (estimated half-life of 9.4 days). In addition, it has the potential to be physically removed from air by wet deposition.

Environmental Physical Data

Henry's Law Constant: 5.35 x10^{-10}
Sorption Partition Coefficient: K_{OC} = 40
BCF: estimated at 0.70

Regulations

RCRA 40CFR: Not listed
CERCLA: 40CFR 302.4: Not listed
SARA 40CFR 372.65: Not listed
SARA EHS 40CFR 355: Not listed
TSCA: Listed

BIS1000 CAS #: 603-50-9

BISACODYL

RTECS: SM8750000
EINECS Number: 210-044-4
Molecular Formula: $C_{22}H_{19}NO_4$
Formula Weight: 361.38

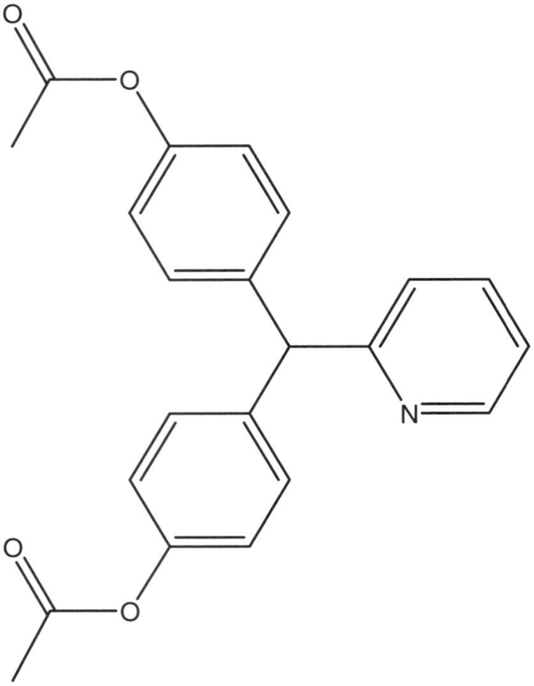

Chemical Structure

Synonyms: BICOL; BIS(P-ACETOXYPHENYL)-2-PYRIDYLMETHANE; BROCALAX; BROXALAX; DAMP; DEFICOL; 2-(4,4'-DIACETOXYDIPHENYLMETHYL)PYRIDINE; (4,4'-DIACETOXYDIPHENYL)(2-PYRIDYL)METHANE; 4,4'-DIACETOXYDIPHENYLPYRID-2-YLMETHANE; DI-(4-ACETOXYPHENYL)-2-PYRIDYLMETHANE; DI-(P-ACETOXYPHENYL)-2-PYRIDYLMETHANE; DULCOLAN; DULCOLAX; DUROLAX; ENDOKOLAT; EULAXAN; FENILAXAN; GODALAX; HILLCOLAX; IVILAX; LA96A; LACO; LAXADIN; LAXANIN N; LAXANS; LAXINE; LAXOREX; NEOLAX; NIGALAX; PERILAX; PHENOL,4,4'-(2-PYRIDINYLMETHYLENE)BIS-,DIACETATE (ESTER); PHENOL,4,4'-(2-PYRIDYLMETHYLENE)DI-,DIACETATE (ESTER); 4,4'-(2-PYRIDYLMETHYLENE)DIPHENOL DIACETATE; PYRILAX; SANVACUAL; SK-BISACODYL; STADALAX; TELEMIN; THERALAX; ULCOLAX; VIDEX; ZETRAX
Description: white to off-white crystalline powder
Use: medication: cathartic

Physical Properties

Freezing Point: 138 °C (280.4 °F)
Water Solubility: Practically Insoluble in Water
Other Solubilities: 1 g Soluble in 210 ml Alcohol, 2.5 ml Chloroform, 275 ml Ether.

RTECS Toxicity Data

Acute Oral: Rat LD_{50} Dose: 4320 mg/kg; Toxic Effects: Behavioral - Altered sleep time (including change in righting reflex); Gastrointestinal - Other changes; Blood - Changes in spleen. Mouse LD_{50} Dose: 17500 mg/kg; Toxic Effects: Behavioral - Ataxia; Lungs, Thorax, or Respiration - Respiratory stimulation; Blood - Hemorrhage.

Hazard Overviews

Health: Irritating to eyes/skin/respiratory tract. Other Acute Effects: harmful if swallowed; overexposure may cause

gastrointestinal disturbances. The toxicological properties have not been thoroughly investigated.
Fire: Hazards: emits toxic fumes. Extinguishing agents: carbon dioxide, dry chemical powder or appropriate foam. Precautions: combustible liquid.
Reactivity: Stable. Hazardous polymerization will not occur. Hazardous decomposition products: toxic fumes of: carbon monoxide, carbon dioxide, nitrogen oxides.
Carcinogenicity: IARC - Not listed; NIOSH - Not listed; NTP - Not listed; ACGIH - Not listed; OSHA - Not listed; EPA - Not listed; MAK - Not listed

Environmental

Regulations
RCRA 40CFR: Not listed
CERCLA: 40CFR 302.4: Not listed
SARA 40CFR 372.65: Not listed
SARA EHS 40CFR 355: Not listed
TSCA: Listed

BIS1110 CAS #: 14220-64-5

BIS(BENZONITRILE)DICHLOROPALLADIUM

EINECS Number: 238-085-3
Molecular Formula: $C_{14}H_{10}N_2Cl_2Pd$
Formula Weight: 383.57

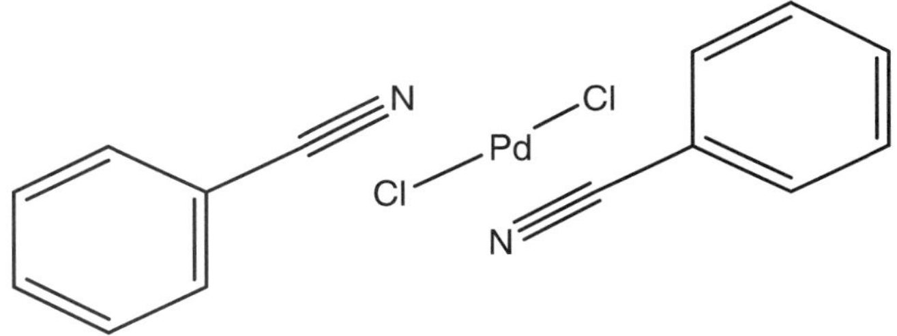

Chemical Structure

Physical Properties
Freezing Point: 131 °C (267.8 °F)
Water Solubility: Insoluble

Hazard Overviews
Health: Irritating to eyes/skin/respiratory tract. Harmful. Other Acute Effects: harmful if swallowed, inhaled, or absorbed through skin.
Fire: Hazards: emits toxic fumes. Extinguishing agents: water spray; carbon dioxide, dry chemical powder or appropriate foam. Precautions: combustible liquid.
Reactivity: Incompatible with: strong oxidizing agents. Hazardous decomposition products: toxic fumes of: carbon monoxide, carbon dioxide, nitrogen oxides, hydrogen chloride gas.
Carcinogenicity: IARC - Not listed; NIOSH - Not listed; NTP - Not listed; ACGIH - Not listed; OSHA - Not listed; EPA - Not listed; MAK - Not listed
Primary Target Organs:

Environmental

Regulations
RCRA 40CFR: Not listed
CERCLA: 40CFR 302.4: Not listed
SARA 40CFR 372.65: Not listed
SARA EHS 40CFR 355: Not listed
TSCA: Not listed

BIS1220 CAS #: 66108-37-0

2,2-BIS(BROMOMETHYL)-3-CHLOROPROPYL BIS(2-CHLORO-1-(CHLORO

EINECS Number: 266-161-6
Molecular Formula: $C_{11}H_{18}Br_2Cl_5O_4P$
Formula Weight: 582.30
Synonyms: PHOSPHORIC ACID,2,2-BIS(BROMOMETHYL)-3-CHLOROPROPYLBIS(2-CHLORO-1-(CHLOROMETHYL)ETHYL) ESTER

Hazard Overviews
Carcinogenicity: IARC - Not listed; NIOSH - Not listed; NTP - Not listed; ACGIH - Not listed; OSHA - Not listed; EPA - Not listed; MAK - Not listed

Environmental

Regulations
RCRA 40CFR: Not listed
CERCLA: 40CFR 302.4: Not listed
SARA 40CFR 372.65: Not listed
SARA EHS 40CFR 355: Not listed
TSCA: Listed

BIS1330 CAS #: 3296-90-0

2,2-BIS(BROMOMETHYL)-1,3-PROPANEDIOL

RTECS: TY3195500
EINECS Number: 221-967-7
Molecular Formula: $C_5H_{10}Br_2O_2$
Structured MF: $HOCH_2C(CH_2Br)_2CH_2OH$
Formula Weight: 261.97

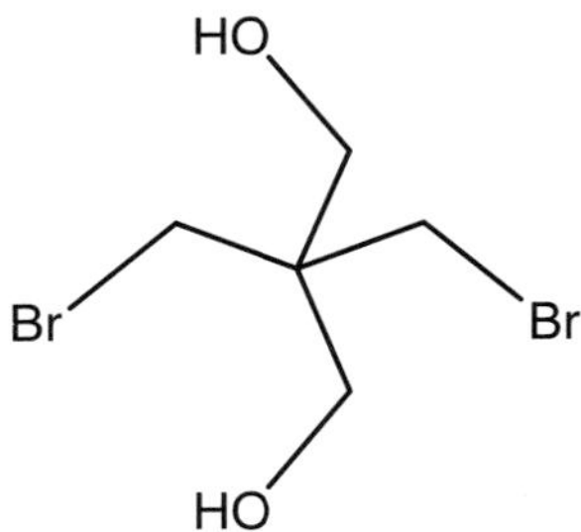

Chemical Structure

Synonyms: 1,3-DIBROMO-2,2-DIMETHYLOLPROPANE; DIBROMOHYDRIN PENTAERYTHRITOL; 2,2-DIBROMOMETHYL-1,3-PROPANEDIOL; DIBROMONEOPENTYL GLYCOL; DIBROMOPENTAERYTHRITOL; FR 1138; PENTAERYTHRITOL DIBROMIDE; PENTAERYTHRITOL DIBROMOHYDRIN; 1,3-PROPANEDIOL,2,2-BIS(BROMOMETHYL)-

Description: off-white powder

Use: constituent of FR-1138, a flame retardant for unsaturated polyester resins and polyurethane foams

Physical Properties

Boiling Point: Decomposes at 235 °C (455 °F)

Freezing Point: 109 °C (228.2 °F) to 110 °C (230 °F)

Specific Gravity: 2.2

Water Solubility: < 1 mg/mL at 19 C

Other Solubilities: DMSO: >=100 mg/ml at 21 °C; 95% Ethanol: >=100 mg/ml at 21 °C; Acetone: >=100 mg/ml at 21 °C; Toluene: 0.6 g/100 g at 25 °C; Corn oil: Insoluble; Carbon Tetrachloride: 0.7 g/100 g at 25 °C; Xylene: 0.5 g/100 g at 25 °C.

Flash Point: Nonflammable

RTECS Toxicity Data

Acute Oral: Rat LD_{50} Dose: 3458 mg/kg; Toxic Effects: Sense organs and special senses - Retinal changes (pigmentary deposition, retinitis, other).

Reproductive/Teratogenic: Mouse Route: Oral; Dose: 124 gm/kg; Duration: male 15W prior to mating; Maternal Effects - Parturition; Effects on Fertility - Litter size; Effects on Newborn - Other postnatal measures or effects. Mouse Route: Oral; Dose: 62 gm/kg; Duration: female 15W prior to mating Effects on Fertility - Female fertility index; Effects on Newborn - Other postnatal measures or effects. Mouse Route: Oral; Dose: 62 gm/kg; Duration: male 15W prior to mating; Paternal Effects - Testes, epididymis, sperm duct; Prostate, seminal vessicle, Cowper's gland, accessory glands.

Mutagenic: Hamster Cytogenetic Analysis; Cell Type: ovary; Dose: 800 mg/L. Bacteria - S Typhimurium Mutations in Microorganisms; Dose: 1 mg/plate (-S9).

Hazard Overviews

Health: Irritating to eyes/skin/respiratory tract. Toxic. Other Acute Effects: harmful if swallowed, inhaled, or absorbed through skin. Chronic Effects: Carcinogen.

Fire: Noncombustible. Hazards: emits toxic fumes. Extinguishing agents: water spray; carbon dioxide, dry chemical powder or appropriate foam. Precautions: combustible liquid.

Reactivity: Incompatible with: strong oxidizing agents. Hazardous decomposition products: toxic fumes of: carbon monoxide, carbon dioxide, hydrogen bromide gas

Carcinogenicity: IARC - Not listed; NIOSH - Not listed; NTP - Not listed; ACGIH - Not listed; OSHA - Not listed; EPA - Not listed; MAK - Not listed

Primary Target Organs:

Eyes

Skin

Respiratory System

Environmental

Regulations

RCRA 40CFR: Not listed

CERCLA: 40CFR 302.4: Not listed

SARA 40CFR 372.65: Not listed

SARA EHS 40CFR 355: Not listed

TSCA: Listed

BIS1440 CAS #: 143-29-3

BIS(2-(2-BUTOXYETHOXY)ETHOXY)-METHANE

RTECS: PA3400000

EINECS Number: 205-598-9

Molecular Formula: $C_{17}H_{36}O_6$

Formula Weight: 336.53

Synonyms: BIS(BUTYLCARBITOL)FORMAL; BUTYLCARBITOL FORMAL; CRYOFLEX; DIBUTYLCARBITOLFORMAL; 5,8,11,13,16,19-HEXAOXATRICOSANE; 5,8,11,13,16,19-HEXAOXATRICOSANE (9CI); METHANE,BIS(2-(2-BUTOXYETHOXY)ETHOXY)-; TP 90B

Use: plasticizer & softener for natural & chloroprene rubbers, nitrile-butadiene rubber, styrene-butadiene rubber; plasticizer & softener for resins-eg, acrylics, urethanes

RTECS Toxicity Data

Acute Oral: Rat LD_{50} Dose: 1746 mg/kg. Mouse LD_{50} Dose: 2700 mg/kg; Toxic Effects: Behavioral - Antipsychotic.

Hazard Overviews

Carcinogenicity: IARC - Not listed; NIOSH - Not listed; NTP - Not listed; ACGIH - Not listed; OSHA - Not listed; EPA - Not listed; MAK - Not listed

Environmental

Regulations

RCRA 40CFR: Not listed

CERCLA: 40CFR 302.4: Not listed

SARA 40CFR 372.65: Not listed

SARA EHS 40CFR 355: Not listed

TSCA: Listed

BIS1550 CAS #: 141-17-3

BIS(2-(2-BUTOXYETHOXY)ETHYL) ADIPATE

RTECS: AU8420000
EINECS Number: 205-465-5
Molecular Formula: $C_{22}H_{42}O_8$
Formula Weight: 434.64
Synonyms: ADIPIC ACID BIS(DIETHYLENE GLYCOL MONOBUTYL ETHER) ESTER; ADIPIC ACID,BIS(2-(2-BUTOXYETHOXY)ETHYL) ESTER; BIS (DIETHYLENE GLYCOL MONOBUTYL ETHER) ADIPATE; DIBUTOXYETHOXYETHYL ADIPATE; HEXANEDIOIC ACID,BIS(2-(2-BUTOXYETHOXY)ETHYL) ESTER; HEXANEDIOIC ACID,BIS(2-(2-BUTOXYETHOXY)ETHYL) ESTER (9CI); TP-95; WAREFLEX
Use: plasticizer for cellulose nitrate & polyvinyl acetate; plasticizer & softener for natural & synthetic rubbers

RTECS Toxicity Data

Acute Oral: Rat LD_{50} Dose: 6 gm/kg.

Hazard Overviews

Health: May cause irritation to eyes/skin. Other Acute Effects: may be harmful by inhalation, ingestion, or skin absorption.
Fire: Extinguishing agents: water spray; carbon dioxide, dry chemical powder or appropriate foam. Precautions: combustible liquid.
Reactivity: Incompatible with: strong oxidizing agents. Hazardous decomposition products: toxic fumes of: carbon monoxide, carbon dioxide.
Carcinogenicity: IARC - Not listed; NIOSH - Not listed; NTP - Not listed; ACGIH - Not listed; OSHA - Not listed; EPA - Not listed; MAK - Not listed

Environmental

Regulations

RCRA 40CFR: Not listed
CERCLA: 40CFR 302.4: Not listed
SARA 40CFR 372.65: Not listed
SARA EHS 40CFR 355: Not listed
TSCA: Listed

BIS1660 CAS #: 117-83-9

BIS(2-BUTOXYETHYL) PHTHALATE

RTECS: TI0175000
EINECS Number: 204-213-1
Molecular Formula: $C_{20}H_{30}O_6$
Formula Weight: 366.50

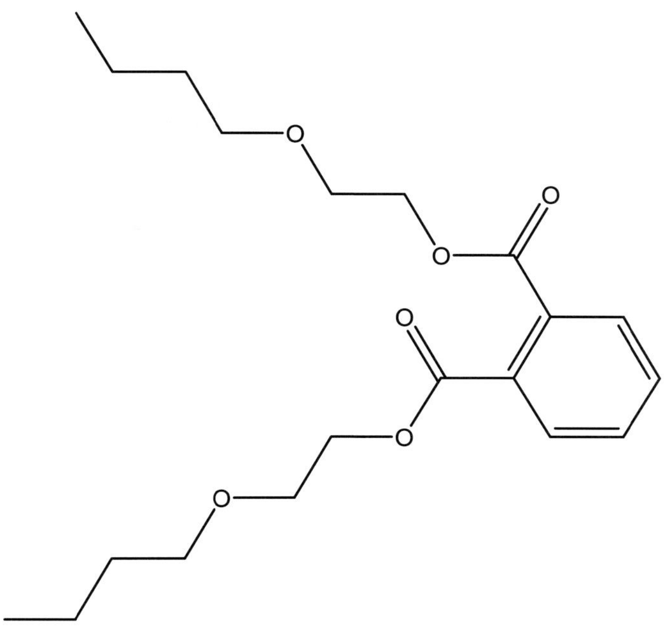

Chemical Structure

Synonyms: 1,2-BENZENEDICARBOXYLIC ACID,BIS(2-BUTOXYETHYL) ESTER; BIS(2-BUTOXYETHYL)PHTHALATE; BETA-BUTOXYETHYL PHTHALATE; BUTYL; BUTYL GLYCOL PHTHALATE; N-BUTYL GLYCOL PHTHALATE; DI-(2-BUTOXYETHYL) ESTER KYSELINY FTALOVE; DIBUTOXYETHYL PHTHALATE; DI-(2-BUTOXYETHYL)ESTER KYSELINY FTALOVE; DI(BUTOXYETHYL)PHTHALATE; DIBUTYL CELLOSOLVE PHTHALATE; DIBUTYLCELLOSOLVE FTALAT; DIBUTYLGLYCOL PHTHALATE; ETHANOL,2-BUTOXY-,PHTHALATE (2:1); KESSCOFLEX; KESSCOFLEX BCP; KRONISOL; PALATINOL K; PHTHALIC ACID,BIS(2-BUTOXYETHYL) ESTER; PLASTHALL DBEP
Description: colorless liquid
Use: plasticizer for polyvinyl chloride, polyvinyl acetate, and other resins; plasticizer for cellulosic, vinyl & acrylic resins; softener & processing aid for chloroprene rubber, nitrile-butadiene rubber, styrene-butadiene rubber

Physical Properties

Boiling Point: 270 °C (518 °F)
Freezing Point: -55 °C (-67 °F)
Specific Gravity: 1.06 at 20 °C
Water Solubility: Relatively low Soluble in Water
Other Solubilities: Soluble to various extents in many common organic solvents and oils.
Refraction Index: 1.40615 at 20 °C
Flash Point: Combustible

RTECS Toxicity Data

Acute Oral: Rat LD_{50} Dose: 8380 mg/kg. Guinea Pig LD_{Lo} Dose: 6 gm/kg.
Irritation Eye: Rabbit Standard Draize Test Dose: 500 mg; Reaction: mild.

Hazard Overviews

Health: May be irritating to eyes/skin/respiratory tract. Harmful. Other Accute Effects: may be harmful by inhalation, ingestion, or skin absorption.
Fire: Combustible. Hazards: emits toxic fumes. Extinguishing agents: water spray; carbon dioxide, dry chemical powder or appropriate foam; foam and water spray are effective but may cause frothing. Precautions: combustible liquid.
Reactivity: Stable. Hazardous polymerization will not occur. Incompatible with: strong oxidizing agents. Hazardous decomposition products: toxic fumes of: carbon monoxide, carbon dioxide.
Carcinogenicity: IARC - Not listed; NIOSH - Not listed; NTP - Not listed; ACGIH - Not listed; OSHA - Not listed; EPA - Not listed; MAK - Not listed

Environmental

Environmental Fate: If released to soil it is expected to display low mobility, and volatilization to the atmosphere is not expected to be an important fate process. No data on biodegradation could be located. If released to water, it is not expected to volatilize to the atmosphere. It may adsorb to sediment and suspended organic matter. Hydrolysis is not expected to be a significant process. No data on the biodegradation of this compound in water could be located. If released to the atmosphere, it is expected to be adsorbed to particulates. Destruction by the vapor phase reaction with photochemically produced hydroxyl radicals may occur.

Environmental Physical Data

Henry's Law Constant: estimated at 2.03×10^{-12}
Octanol/Water Partition Coefficient: log K_{ow} = > 2.12
Sorption Partition Coefficient: K_{oc} = calculated at 670 to 1000
BCF: none likely

Regulations

RCRA 40CFR: Not listed
CERCLA: 40CFR 302.4: Not listed
SARA 40CFR 372.65: Not listed
SARA EHS 40CFR 355: Not listed
TSCA: Listed

Analytical Methods

Plasma: EPA 001

BIS1770 CAS #: 2781-00-2

1,4-BIS(A-(T-BUTYLDIOXY)ISOPROPYL)BENZENE

EINECS Number: 220-479-1
Molecular Formula: $C_{20}H_{34}O_4$
Formula Weight: 338.49
Synonyms: ALPHA,ALPHA'-BIS(TERT-BUTYLDIOXY)-P-DIISOPROPYLBENZENE; 1,4-BIS(ALPHA-(TERT-BUTYLDIOXY)ISOPROPYL)BENZENE; ALPHA,ALPHA'-BIS(TERT-BUTYLPEROXY)-1,4-DIISOPROPYLBENZENE; ALPHA,ALPHA'-BIS(TERT-BUTYLPEROXY)-P-DIISOPROPYLBENZENE; PERKADOX 14/40C; PERKADOX U 14/40; PEROXIDE,(1,4-PHENYLENEBIS(1-METHYLETHYLIDENE))BIS((1,1-DIMETHYLETHYL); PEROXIDE,(P-PHENYLENEDIISOPROPYLIDENE)BIS(TERT-BUTYL; PEROXIMON F 100; PEROXIMON F 40

Hazard Overviews

Carcinogenicity: IARC - Not listed; NIOSH - Not listed; NTP - Not listed; ACGIH - Not listed; OSHA - Not listed; EPA - Not listed; MAK - Not listed

Environmental

Regulations

RCRA 40CFR: Not listed
CERCLA: 40CFR 302.4: Not listed
SARA 40CFR 372.65: Not listed
SARA EHS 40CFR 355: Not listed
TSCA: Listed

BIS1880 CAS #: 65652-41-7

BIS(T-BUTYLPHENYL) PHENYL PHOSPHATE

EINECS Number: 265-859-8
Molecular Formula: $C_{26}H_{31}O_4P$
Formula Weight: 438.5
Synonyms: BUTYL PHENYL PHOSPHATE; DIBUTYL PHENYL PHOSPHATE; DI-TERT-BUTYLPHENYL PHENYL PHOSPHATE; DI(TETR-BUTYLPHENYL)PHENYL PHOSPHATE

Hazard Overviews

Carcinogenicity: IARC - Not listed; NIOSH - Not listed; NTP - Not listed; ACGIH - Not listed; OSHA - Not listed; EPA - Not listed; MAK - Not listed

Environmental

Regulations

RCRA 40CFR: Not listed
CERCLA: 40CFR 302.4: Not listed
SARA 40CFR 372.65: Not listed
SARA EHS 40CFR 355: Not listed
TSCA: Listed

BIS1990 CAS #: 94-17-7

BIS(4-CHLOROBENZOYL) PEROXIDE

RTECS: SD7875000
DOT: UN2113; UN2114; UN2115; IMO5.2
EINECS Number: 202-310-3
Molecular Formula: $C_{14}H_8Cl_2O_4$
Structured MF: $(p\text{-}ClC_6H_4COO)_2$
Formula Weight: 311.12
Synonyms: BIS(P-CHLOROBENZOYL) PEROXIDE; CADOX PS; 4-CHLOROBENZOYL PEROXIDE; P-CHLOROBENZOYL PEROXIDE;

DI-(P-CHLOROBENZOYL) PEROXIDE; P,P'-DICHLOROBENZOYL PEROXIDE; P,P'-DICHLORODIBENZOYL PEROXIDE; LUPERCO BDB; PEROXIDE,BIS(P-CHLOROBENZOYL)-

Description: white, granular; powder; odorless

Use: bleaching agent; polymerization catalyst

Physical Properties

Boiling Point: Decomposes at 1 atm

Freezing Point: Decomposes

Specific Gravity: Solid 1.1 at 20 °C (solid)

Water Solubility: Insoluble in Water

Other Solubilities: Soluble in organic solvents

Flash Point: Flammable

Hazard Overviews

Flammable

Fire: Flammable.

Carcinogenicity: IARC - Not listed; NIOSH - Not listed; NTP - Not listed; ACGIH - Not listed; OSHA - Not listed; EPA - Not listed; MAK - Not listed

Environmental

Cleanup/Disposal: Guide No. 145: Eliminate all ignition sources (no smoking, flares, sparks or flames in immediate area). Keep combustibles (wood, paper, oil, etc.) away from spilled material. Do not touch damaged containers or spilled material unless wearing appropriate protective clothing. Keep substance wet using water spray. Stop leak if you can do it without risk. Small Spills: Take up with inert, damp, noncombustible material using clean non-sparking tools and place into loosely covered plastic containers for later disposal. Large Spills: Wet down with water and dike for later disposal. Prevent entry into waterways, sewers, basements or confined areas. Guide No. 146: Eliminate all ignition sources (no smoking, flares, sparks or flames in immediate area). Keep combustibles (wood, paper, oil, etc.) away from spilled material. Do not touch damaged containers or spilled material unless wearing appropriate protective clothing. Keep substance wet using water spray. Stop leak if you can do it without risk. Small Spills: Take up with inert, damp, noncombustible material using clean non-sparking tools and place into loosely covered plastic containers for later disposal. Large Spills: Wet down with water and dike for later disposal. Prevent entry into waterways, sewers, basements or confined areas. Do not clean-up or dispose of, except under supervision of a specialist.

Regulations

RCRA 40CFR: Not listed

CERCLA: 40CFR 302.4: Not listed

SARA 40CFR 372.65: Not listed

SARA EHS 40CFR 355: Not listed

TSCA: Listed

BIS2100 CAS #: 56428-00-3

BIS(P-CHLOROBENZYL) ETHER

EINECS Number: 260-174-0

Molecular Formula: $C_{14}H_{12}Cl_2O$

Formula Weight: 269.59

Synonyms: BENZENE,1,1'-(OXYBIS(METHYLENE))BIS(4-CHLORO-

Hazard Overviews

Carcinogenicity: IARC - Not listed; NIOSH - Not listed; NTP - Not listed; ACGIH - Not listed; OSHA - Not listed; EPA - Not listed; MAK - Not listed

Environmental

Regulations

RCRA 40CFR: Not listed

CERCLA: 40CFR 302.4: Listed as Compound per CWA Section 307(a)

SARA 40CFR 372.65: Not listed

SARA EHS 40CFR 355: Not listed

TSCA: Not listed

BIS2210 CAS #: 68460-03-7

BIS(2-CHLORO-1-(CHLOROMETHYL)ETHYL) 2,3-DICHLOROPROPYLPHOS

EINECS Number: 270-625-3

Molecular Formula: $C_9H_{15}Cl_6O_4P$

Formula Weight: 430.88

Synonyms: PHOSPHORIC ACID,BIS(2-CHLORO-1-(CHLOROMETHYL)ETHYL)2,3-DICHLOROPROPYL ESTER

Hazard Overviews

Carcinogenicity: IARC - Not listed; NIOSH - Not listed; NTP - Not listed; ACGIH - Not listed; OSHA - Not listed; EPA - Not listed; MAK - Not listed

Environmental

Regulations

RCRA 40CFR: Not listed

CERCLA: 40CFR 302.4: Not listed

SARA 40CFR 372.65: Not listed

SARA EHS 40CFR 355: Not listed

TSCA: Listed

BIS2430 CAS #: 112-26-5

1,2-BIS(2-CHLOROETHOXY)ETHANE

RTECS: KH4900000
EINECS Number: 203-952-7
Molecular Formula: $C_6H_{12}Cl_2O_2$
Formula Weight: 187.07

Chemical Structure

Synonyms: 1,2-BIS(CHLOROETHOXY)ETHANE; BIS(2-CHLOROETHOXY)ETHANE; 2-(2-CHLORETHOXY)ETHYL 2'-CHLORETHYL ETHER; 2-(2-CHLOROETHOXY)ETHYL 2'-CHLOROETHYL ETHER; 1,8-DICHLORO-3,6-DIOXAOCTANE; ETHANE,1,2-BIS(2-CHLOROETHOXY)-; TRIETHYLENE GLYCOL DICHLORIDE; TRIGLYCOL DICHLORIDE; TRIGLYCOL DICHLORIDE ETHER
Description: colorless liquid
Use: solvent for hydrocarbons, oils, etc; extractant; intermediate for resins & insecticides; organic synthesis

Physical Properties

Boiling Point: 241.3 °C (466 °F)
Freezing Point: -31.5 °C (-24.7 °F)
Specific Gravity: 1.1974 at 20 °C/20 °C
Water Solubility: Insoluble in Water
Other Solubilities: Soluble in Carbon Tetrachloride.
Refraction Index: 1.4592 at 25 °C/D
Flash Point: 121 °C Open Cup

RTECS Toxicity Data

Acute Oral: Rat LD_{50} Dose: 250 mg/kg. Guinea Pig LD_{50} Dose: 120 mg/kg; Toxic Effects: Behavioral - General anesthetic; Gastrointestinal - Other changes; Kidney, Ureter, and Bladder - Other changes.
Acute Dermal: Rabbit LD_{50} Route: Skin; Dose: 1410 uL/kg.
Irritation Eye: Rabbit Standard Draize Test Dose: 100 mg; Reaction: severe.
Irritation Skin: Rabbit Open Draize Test Dose: 500 mg open; Reaction: mild.

Hazard Overviews

Health: Severely irritating to eyes, irritating to skin/respiratory tract. Toxic. Other Acute Effects: harmful if swallowed, inhaled, or absorbed through skin; risk of serious damage to eyes.
Fire: Will burn. Hazards: potentially explosive peroxides can form on long time storage in contact with air; light and heat accelerate peroxide formation; emits toxic fumes. Extinguishing agents: water spray; carbon dioxide, dry chemical powder or appropriate foam. Precautions: combustible liquid.
Reactivity: Incompatible with: strong oxidizing agents. Hazardous decomposition products: toxic fumes of: carbon monoxide, carbon dioxide, hydrogen chloride gas.

Carcinogenicity: IARC - Not listed; NIOSH - Not listed; NTP - Not listed; ACGIH - Not listed; OSHA - Not listed; EPA - Not listed; MAK - Not listed
Primary Target Organs:

Environmental

Environmental Fate: If released to the atmosphere, it will degrade in the vapor phase by reaction with photochemically produced hydroxyl radicals (estimated half-life of about 35 hr). If released to soil or water, it will probably biodegrade slowly. No data are available concerning other degradation processes in soil or water. It will leach from municipal and industrial solid wastes of which it is a constituent. An estimated K_{oc} value of 19 indicates that it will be highly mobile in soil.

Environmental Physical Data

Henry's Law Constant: estimated at 7.81×10^{-7}
Sorption Partition Coefficient: K_{oc} **= 19**
BCF: estimated at 2.4

Regulations

RCRA 40CFR: Not listed
CERCLA: 40CFR 302.4: Not listed
SARA 40CFR 372.65: Not listed
SARA EHS 40CFR 355: Not listed
TSCA: Listed

BIS2540 CAS #: 111-91-1

BIS(2-CHLOROETHOXY)METHANE

RTECS: PA3675000
EINECS Number: 203-920-2
Molecular Formula: $C_5H_{10}Cl_2O_2$
Formula Weight: 173.05
Synonyms: BIS(BETACHLOROETHYL) FORMAL; BIS(2-CHLOROETHOXY) METHANE; BIS(2-CHLOROETHOXY)-METHANE; BIS(2-CHLOROETHYL) FORMAL; BIS(2-CHLOROETHYL)FORMAL; BIS(BETA-CHLOROETHYL)FORMAL; BETA,BETA,-DICHLORODIETHYL FORMAL; DICHLORODIETHYL FORMAL; DICHLORODIETHYL METHYLAL; 2,2-DICHLOROETHYL FORMAL; DI-2-CHLOROETHYL FORMAL; DICHLOROETHYL FORMAL; ETHANE,1,1'-(METHYLENEBIS(OXY))BIS(2-CHLORO-; FORMALDEHYDE BIS(2-CHLOROETHYL) ACETAL; FORMALDEHYDE BIS(BETA-CHLOROETHYL) ACETAL; METHANE,BIS(2-CHLOROETHOXY)-; 1,1'-[METHYLENE BIS(OXY)]BIS[2-CHLOROETHANE]
Description: colorless liquid
Use: solvent; intermediate for polysulfide rubber

Physical Properties

Boiling Point: 218.1 °C (425 °F)
Freezing Point: -32.8 °C (-27.04 °F)

Specific Gravity: 1.2339 at 20 °C/20 °C
Saturated Vapor Density: 1.200001098 kg/m^3
Vapor Pressure: 1.4 x10^{-4} mm Hg at 25 °C
Water Solubility: 0.78 g/100 ml Water
Other Solubilities: miscible with most organic solvents.
Flash Point: 110 °C Open Cup

RTECS Toxicity Data

Acute Oral: Rat LD_{50} Dose: 65 mg/kg.
Acute Inhalation: Rat LC_{Lo} Dose: 62 ppm/4hr.
Acute Dermal: Guinea Pig LD_{50} Route: Skin; Dose: 170 mg/kg.
Irritation Eye: Rabbit Standard Draize Test Dose: 500 mg; Reaction: mild.

Hazard Overviews

Fire: Will burn.
Carcinogenicity: IARC - Not listed; NIOSH - Not listed; NTP - Not listed; ACGIH - Not listed; OSHA - Not listed; EPA - Class D, Not classifiable as to human carcinogenicity; MAK - Not listed

Environmental

Environmental Fate: If released to soil, it would be expected to display high to very high mobility. Biodegradation is not expected to be an important fate process. Volatilization from water should be a slow process; the estimated half-life for volatilization from a model pond is 11 years. Hydrolysis can be expected by comparison to other chlorine containing compounds; the half-life for this pH independent process has been estimated at 0.5-2 years. Direct photochemical degradation in the atmosphere or in the upper layers of surface waters should not be an important fate process. The estimated half-life for the atmospheric reaction with photochemically produced hydroxyl radicals is 10 hours.
Cleanup/Disposal: Guide No. 153: Eliminate all ignition sources (no smoking, flares, sparks or flames in immediate area). Do not touch damaged containers or spilled material unless wearing appropriate protective clothing. Stop leak if you can do it without risk. Prevent entry into waterways, sewers, basements or confined areas. Absorb or cover with dry earth, sand or other non-combustible material and transfer to containers. Do not get water inside containers.

Environmental Physical Data

Henry's Law Constant: calculated at 2.6 x10^{-10}
Octanol/Water Partition Coefficient: log K_{ow} = 0.75
Sorption Partition Coefficient: K_{oc} = 7 to 115
BCF: calculated at 0.84 to 2.2

Regulations

RCRA 40CFR: Listed Hazardous Waste No. U024 Toxic Waste
CERCLA: 40CFR 302.4: Listed per RCRA Section 3001 per CWA Section 307(a) RQ: 1000 lb (453.5 kg)
SARA 40CFR 372.65: Listed
SARA EHS 40CFR 355: Not listed
TSCA: Listed

Analytical Methods

Air: EPA 0020
Soil: CLP MC_SVOA, OHC; EPA 16, 1625; SW846 3640A, 8010B, 8110, 8250A, 8270B, 8270C, 8410
Water / Groundwater: EPA 1625, 611, 625, 625-S; APHA 6040-B, 6410-B; USGS O3118
Plasma: EPA 29

BIS2650 CAS #: 111-44-4

BIS(2-CHLOROETHYL) ETHER

RTECS: KN0875000
DOT: UN1916; IMO6.1
EINECS Number: 203-870-1
Molecular Formula: $C_4H_8Cl_2O$
Structured MF: $(ClCH_2CH_2)_2O$
Formula Weight: 143.02

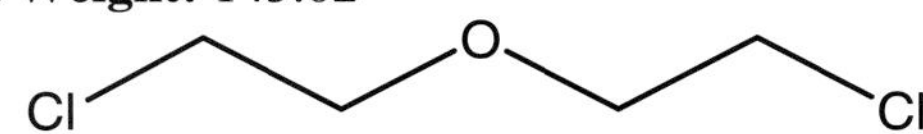

Chemical Structure

Synonyms: BCEE; BIS(BETA-CHLOROETHYL) ETHER; BIS(CHLORO-2-ETHYL) OXIDE; BIS(2-CHLOROETHYL)ETHER; CHLOREX; 1-CHLORO-2-(BETA-CHLOROETHOXY) ETHANE; 1-CHLORO-2-(BETA-CHLOROETHOXY)ETHANE; 2-CHLOROETHYL ETHER; CHLOROETHYL ETHER; 2-CHLOROETHYL ETHER,DICHLOROETHYL OXIDE; CLOREX; DCEE; 2,2'-DICHLOORETHYLETHER; 2,2'-DICHLOR-DIAETHYLAETHER; 2,2'-DICHLORETHYL ETHER; 2,2'-DICHLORODIETHYL ETHER; BETA,BETA-DICHLORODIETHYL ETHER; DICHLOROETHER; 2,2'-DICHLOROETHYL ETHER; BETA,BETA'-DICHLOROETHYL ETHER; DI(2-CHLOROETHYL) ETHER; DI(BETA-CHLOROETHYL) ETHER; DICHLOROETHYL ETHER; SYM-DICHLOROETHYL ETHER; DICHLOROETHYL OXIDE; DI(2-CHLOROETHYL)ETHER; DI(BETA-CHLOROETHYL)ETHER; 1,5-DICHLORO-3-OXAPENTANE; 2,2'-DICLOROETILETERE; DWUCHLORODWUETYLOWY ETER; ENT 4,504; ETHANE,1,1'-OXYBIS(2-CHLORO-; ETHER DICHLORE; ETHER,BIS(2-CHLOROETHYL); 1,1'-OXYBIS(2-CHLORO)ETHANE; OXYDE DE CHLORETHYLE
Description: colorless liquid; pungent, fruity odor
Use: general solvent, selective solvent for production of high-grade lubricating oils; textile scouring and cleansing; wetting and penetrating compounds; organic synthesis; paints, varnishes, lacquers; finish removers; spotting and dry cleaning; soil fumigant; medicinals and pharmaceuticals; manufacture of insecticide and acaricide

Physical Properties

Boiling Point: 178 °C (352 °F)
Freezing Point: -24.5 °C (-12.1 °F)
Specific Gravity: 1.22 at 20 °C/20 °C
Vapor Density: 4.93 Air=1
Vapor Pressure: 0.71 mm Hg at 20.0 °C
Water Solubility: 1% by weight
Other Solubilities: 10% Ethanol: Soluble; 95% Ethanol: >=100 mg/ml at 21 °C; Acetone: >=100 mg/ml at 21 °C; Benzene: Soluble; DMSO: >=100 mg/ml at 21 °C; Ether: Soluble; Most organic solvents: Soluble.

Surface Tension: Estimated at 40 dynes/cm
Odor Threshold: 90 to 2160 mg/m^3
Refraction Index: 1.457 at 20 °C/D
Flash Point: 63 °C Closed Cup
Autoignition Temperature: 369 °C
LEL: 2.7% v/v

RTECS Toxicity Data

Acute Oral: Rat LD_{50} Dose: 75 mg/kg. Mouse LD_{50} Dose: 209 mg/kg; Toxic Effects: Sense organs and special senses - Ptosis; Gastrointestinal - Changes in structure or function of salivary glands; Gastrointestinal - Hypermotility, diarrhea.
Acute Inhalation: Rat LC_{50} Dose: 330 mg/m^3/4hr. Mouse LC_{50} Dose: 650 mg/m^3/2hr.
Acute Dermal: Rabbit LD_{50} Route: Skin; Dose: 90 mg/kg. Guinea Pig LD_{50} Route: Skin; Dose: 300 mg/kg.
Irritation Eye: Rabbit Standard Draize Test Dose: 100 mg; Reaction: severe.
Irritation Skin: Rabbit Open Draize Test Dose: 500 mg open; Reaction: mild.
Mutagenic: Insects - D Melanogaster Specific Locus Test; Route: Parenteral; Dose: 10500 ppm. Bacteria - S Typhimurium Mutations in Microorganisms; Dose: 1457 mg/plate (+S9). Bacteria - S Typhimurium Mutations in Microorganisms; Dose: 1 mg/plate (-S9).
Tumorigenic: Mouse Route: Oral; Dose: 33 gm/kg/79W-C; Toxic Effects: Tumorigenic - Carcinogenic by RTECS criteria; Liver - Tumors; Blood - Lymphomax including Hodgkin's disease. Mouse Route: Subcutaneous; Dose: 2400 mg/kg/60W-I; Toxic Effects: Tumorigenic - Equivocal tumorigenic agent by RTECS criteria; Tumorigenic - Tumors at site of application.

Hazard Overviews

Flammable

Fire Diamond

Health: Severely irritating to eyes/skin/respiratory tract. Toxic. Also Causes: nausea, vomiting, unconsciousness, pulmonary edema, CNS depression, possible liver/kidney injury. Chronic Effects: mild bronchitis, sensitized allergic reaction, possible liver/kidney injury.
Fire: Flammable. Use dry chemical, water spray, fog, or regular foam. Container may explode in heat of fire.
Reactivity: Unstable, forms unstable peroxides when exposed to air. Hazardous polymerization cannot occur. Avoid: heat; ignition sources; sunlight; prolonged exposure to air. Incompatible with: oxidizers; chlorosulfonic acid; fuming sulfuric acid; strong acids; plastics; rubber coatings; moist air. Hazardous decomposition products: carbon oxides; phosgene; hydrogen chloride gases.
Carcinogenicity: IARC - Group 3, Not classifiable as to carcinogenicity to humans; NIOSH - Listed as carcinogen; NTP - Not listed; ACGIH - Class A4, Not classifiable as a human carcinogen; OSHA - Not listed; EPA - Class B2, Probable human carcinogen based on animal studies; MAK - Not listed
Primary Target Organs:

Eyes

Respiratory System

Exposure Limits
OSHA PEL: STEL: 15 ppm; 90 mg/m^3; skin.
OSHA PEL Vacated 1989 Limits: TWA: 5 ppm; 30 mg/m^3; STEL: 10 ppm; 60 mg/m^3.
ACGIH TLV: TWA: 5 ppm; 29 mg/m^3; STEL: 10 ppm; 58 mg/m^3.
NIOSH REL: TWA: 5 ppm; 30 mg/m^3. STEL: 10 ppm; 60 mg/m^3; skin.
NIOSH IDLH: 100 ppm.
DFG MAK: TWA: 10 ppm; 60 mg/m^3.
Respirator Recommendation
Exposure Range: >15 to <100 ppm Supplied Air, Constant Flow/Pressure Demand, Half Mask
Exposure Range: 100 to unlimited ppm Self-contained Breathing Apparatus, Pressure Demand, Full Face
Note: poor warning properties

Environmental

Ecotoxicity: EC_{50} Daphnia magna (cladoceran) 238,000 ug/l/48 hr /in static and unmeasured methodologies
Environmental Fate: Release to water is expected to result in hydrolysis (estimated half-life 40 days) and volatilization. It biodegrades in water following several weeks of acclimation. Aqueous photolysis and photooxidation are not expected to be important processes in aquatic fate. Bioconcentration in aquatic organisms is extremely low. When released to soil, it may hydrolyze and is expected to leach extensively to groundwater. A half-life of 13.44 hr was estimated for the reaction with photochemically produced hydroxyl radicals in the atmosphere. Direct atmospheric photolysis is not expected to be important since it should not absorb light of wavelengths above 290 nm.
Cleanup/Disposal: Guide No. 152: Do not touch damaged containers or spilled material unless wearing appropriate protective clothing. Stop leak if you can do it without risk. Prevent entry into waterways, sewers, basements or confined areas. Cover with plastic sheet to prevent spreading. Absorb or cover with dry earth, sand or other non-combustible material and transfer to containers. Do not get water inside containers.

Environmental Physical Data

Henry's Law Constant: 2.58 x10^{-5}
Octanol/Water Partition Coefficient: log K_{ow} = calculated at 1.58
Sorption Partition Coefficient: log K_{oc} = estimated at 1.38
BCF: bluegill 11

Regulations

RCRA 40CFR: Listed Hazardous Waste No. U025 Toxic Waste

CERCLA: 40CFR 302.4: Listed per RCRA Section 3001 per CWA Section 307(a) RQ: 10 lb (4.535 kg)
SARA 40CFR 372.65: Listed TPQ: 10000 lb
SARA EHS 40CFR 355: Listed TPQ: 10 lb
TSCA: Listed

Analytical Methods

Air: EPA 0020; ASTM D4490
Soil: CLP MC_SVOA, OHC; EPA 16, 1625; SW846 3640A, 8110, 8250A, 8270B, 8270C, 8410, 8430
Water / Groundwater: EPA 1625, 611, 625, 625-S, 6; APHA 6040-B, 6410-B; ASTM D3695; USGS O3118
Drinking Water: AOAC 991.07; ASTM D5475
Indoor / Expired Air: NIOSH 1004
Plasma: EPA 29

BIS2760 CAS #: 154-93-8

BISCHLOROETHYL NITROSOUREA

RTECS: YS2625000
EINECS Number: 205-838-2
Molecular Formula: $C_5H_9Cl_2N_3O_2$
Formula Weight: 214.07
Synonyms: BCNU; BICNU; 1,3-BIS(2-CHLOROETHYL)-1-NITROSOUREA; 1,3-BIS(BETA-CHLOROETHYL)-1-NITROSOUREA; 1,3-BIS-(2-CHLOROETHYL)-1-NITROSOUREA; BIS(2-CHLOROETHYL)NITROSOUREA; BISCHLOROETHYLNITROSOUREA; N,N'-BIS(2-CHLOROETHYL)-N-NITROSOUREA; N,N-BIS(2-CHLOROETHYL)-N-NITROSOUREA; CARMUBRIS; CARMUSTIN; CARMUSTINE; FDA 0345; NITRUMON; NSC-409962; SK 27702; SRI 1720; UREA,N,N'-BIS(2-CHLOROETHYL)-N-NITROSO-(9CI)
Description: orange-yellow solid
Use: antineoplastic drug; used in treating Hodgkin's disease, non-Hodgkin's lymphomas, primary metastatic brain tumors, melanoma and renal cell

Physical Properties

Freezing Point: 30 °C (86 °F) to 32 °C (89.6 °F)
Water Solubility: < 1 mg/mL at 18 C
Other Solubilities: 95% Ethanol: >=100 mg/ml at 18 °C; Acetone: >=100 mg/ml at 18 °C; DMSO: >=100 mg/ml at 18 °C.
Flash Point: Not available; probably combustible

RTECS Toxicity Data

Acute Oral: Rat LD_{50} Dose: 20 mg/kg. Mouse LD_{50} Dose: 19 mg/kg; Toxic Effects: Gastrointestinal - Hypermotility, diarrhea; Liver - Jaundice, other or unclassified; Kidney, Ureter, and Bladder - Urine volume increased.
Acute Dermal: Rat LD_{50} Route: Subcutaneous Dose: 83200 ug/kg; Toxic Effects: Behavioral - Ataxia; Gastrointestinal - Hypermotility, diarrhea; Nutritional and gross metabolic - Weight loss or decreased weight gain. Mouse LD_{50} Route: Subcutaneous Dose: 24 mg/kg; Toxic Effects: Gastrointestinal - Hypermotility, diarrhea; Liver - Jaundice, other or unclassified; Kidney, Ureter, and Bladder - Urine volume increased.
Reproductive/Teratogenic: Rat Route: Intraperitoneal; Dose: 8 mg/kg; Duration: female 6-9D of pregnancy; Specific Developmental Abnormalities - Musculoskeletal system; Cardiovascular (circulatory) system; Urogenital system. Rat Route: Intraperitoneal; Dose: 8 mg/kg; Duration: female 6-9D of pregnancy; Specific Developmental Abnormalities - Central nervous system; Eye, ear; Body wall. Rat Route: Intraperitoneal; Dose: 9 mg/kg; Duration: male 9W prior to mating; Effects on Fertility - Litter size; Pre-implantation mortality. Rat Route: Intraperitoneal; Dose: 4 mg/kg; Duration: female 6-9D of pregnancy; Effects on Embryo or Fetus - Fetotoxicity.
Mutagenic: Human DNA Damage; Cell Type: embryo; Dose: 100 umol/L. Human DNA Damage; Cell Type: lung; Dose: 100 umol/L. Human DNA Damage; Cell Type: other cell types; Dose: 50 umol/L. Human DNA Damage; Cell Type: other cell types; Dose: 60 mg/L. Human DNA Inhibition; Cell Type: leukocyte; Dose: 1 umol/L. Human DNA Inhibition; Cell Type: HeLa cell; Dose: 100 umol/L. Human DNA Inhibition; Cell Type: lymphocyte; Dose: 50 umol/L. Human DNA Adduct; Cell Type: HeLa cell; Dose: 25 umol/L. Human Sister Chromatid Exchenge; Cell Type: lymphocyte; Dose: 5 umol/L. Human Sister Chromatid Exchenge; Cell Type: other cell types; Dose: 4 umol/L. Human Other Mutation Test Systems; Cell Type: HeLa cell; Dose: 10 mg/L.
Tumorigenic: Rat Route: Intravenous; Dose: 16 mg/kg/60W-I; Toxic Effects: Tumorigenic - Equivocal tumorigenic agent by RTECS criteria; Lungs, Thorax, or Respiration - Tumors. Rat Route: Intravenous; Dose: 26 mg/kg/60W-I; Toxic Effects: Tumorigenic - Equivocal tumorigenic agent by RTECS criteria; Gastrointestinal - Tumors. Rat Route: Intravenous; Dose: 45 mg/kg/60W-I; Toxic Effects: Tumorigenic - Equivocal tumorigenic agent by RTECS criteria; Lungs, Thorax, or Respiration - Tumors. Rat Route: Intravenous; Dose: 51 mg/kg/24W-I; Toxic Effects: Tumorigenic - Equivocal tumorigenic agent by RTECS criteria; Lungs, Thorax, or Respiration - Tumors; Gastrointestinal - Tumors.

Hazard Overviews

Poison

Health: Irritating to eyes/skin/respiratory tract. Poison. Other Acute Effects: may be fatal if swallowed; may be harmful if inhaled, or if absorbed through the skin. Chronic Effects: probably carcinogenic based on its IARC, OSHA, ACGIH, NTP or EPA classification; may alter genetic material; may cause congenital malformation in the fetus; cause reproductive disorders; target organs: bone marrow, lungs, liver, kidneys, eyes.
Fire: Will burn. Hazards: emits toxic fumes. Extinguishing agents: water spray; carbon dioxide, dry chemical powder or appropriate foam. Precautions: combustible liquid.
Reactivity: Stable. Hazardous polymerization will not occur. Incompatible with: acids. Hazardous decomposition products:

toxic fumes of: carbon monoxide, carbon dioxide, nitrogen oxides, hydrogen chloride gas.

Carcinogenicity: IARC - Group 2A, Probably carcinogenic to humans; NIOSH - Not listed; NTP - Listed; ACGIH - Not listed; OSHA - Not listed; EPA - Not listed; MAK - Not listed

Primary Target Organs:

Eyes

Skin

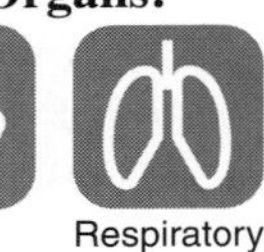
Respiratory System

Liver

Kidneys

Bone

Environmental

Cleanup/Disposal: Guide No. 154: Eliminate all ignition sources (no smoking, flares, sparks or flames in immediate area). Do not touch damaged containers or spilled material unless wearing appropriate protective clothing. Stop leak if you can do it without risk. Prevent entry into waterways, sewers, basements or confined areas. Absorb or cover with dry earth, sand or other non-combustible material and transfer to containers. Do not get water inside containers.

Regulations

RCRA 40CFR: Not listed
CERCLA: 40CFR 302.4: Not listed
SARA 40CFR 372.65: Not listed
SARA EHS 40CFR 355: Not listed
TSCA: Not listed

BIS2870 CAS #: 3944-87-4

1,2-BIS(2-CHLOROETHYLSULFONYL)ETHANE

RTECS: KH5250000
Molecular Formula: $C_6H_{12}Cl_2O_4S_2$
Formula Weight: 283.20
Synonyms: 1,2-BIS((2-CHLOROETHYL)SULFONYL)ETHANE; ETHANE,1,2-BIS((2-CHLOROETHYL)SULFONYL)-

Hazard Overviews

Carcinogenicity: IARC - Not listed; NIOSH - Not listed; NTP - Not listed; ACGIH - Not listed; OSHA - Not listed; EPA - Not listed; MAK - Not listed

Environmental

Regulations

RCRA 40CFR: Not listed
CERCLA: 40CFR 302.4: Listed as Compound per CWA Section 307(a)
SARA 40CFR 372.65: Not listed
SARA EHS 40CFR 355: Not listed
TSCA: Not listed

BIS2980 CAS #: 39638-32-9

BIS(2-CHLOROISOPROPYL) ETHER

DOT: UN2490; IMO6.1
EINECS Number: 254-554-5
Molecular Formula: $C_6H_{12}Cl_2O$
Formula Weight: 171.07
Synonyms: BIS (2-CHLORO-1-METHYLETHYL) ETHER; BIS(2-CHLOROISOPROPYL)ETHER; BIS(2-CHLORO-1-METHYLETHYL)-ETHER; DICHLORODIISOPROPYL ETHER; 2,2'-DICHLOROISOPROPYL ETHER; DICHLOROISOPROPYL ETHER; 2,2'-OXYBIS[2-CHLOROPROPANE]; PROPANE,2,2'-OXYBIS(2-CHLORO-
Use: as solvent for fats, waxes & greases; extractant; in paint & varnish removers; in spotting & cleaning solutions; in textile processing; as chem intermediate; research chemical

Physical Properties

Boiling Point: 187.3 °C (369 °F) at 760 mm Hg
Freezing Point: -101.8 °C (-151.24 °F) to -96.8 °C (-142.24 °F)
Specific Gravity: 1
Vapor Density: 5.9 Air=1
Saturated Vapor Density: 1.205491998 kg/m^3
Vapor Pressure: 0.71 to 0.85 mm Hg at 20 °C
Water Solubility: Water Soluble
Other Solubilities: miscible in organic solvents.
Refraction Index: 1.4451 at 25 °C
Flash Point: 85 °C Open Cup

Hazard Overviews

Fire Diamond

Fire: Combustible.
Carcinogenicity: IARC - Not listed; NIOSH - Not listed; NTP - Not listed; ACGIH - Not listed; OSHA - Not listed; EPA - Not listed; MAK - Not listed

Environmental

Environmental Fate: If released to water or moist soil, it will hydrolyze rapidly based on an estimated hydrolysis half-life of <38.4 sec in water. Therefore, biodegradation, bioconcentration in aquatic organisms and adsorption to soil and sediment are not expected to be significant fate processes. If released to the atmosphere, vapor-phase is degraded by reaction with photochemically produced hydroxyl radicals (estimated half-life of 6.2 days in air).

Cleanup/Disposal: Guide No. 153: Eliminate all ignition sources (no smoking, flares, sparks or flames in immediate area). Do not touch damaged containers or spilled material unless wearing appropriate protective clothing. Stop leak if you can do it without risk. Prevent entry into waterways, sewers, basements or confined areas. Absorb or cover with dry earth, sand or other non-combustible material and transfer to containers. Do not get water inside containers.

Environmental Physical Data

BCF: not significant

Regulations

RCRA 40CFR: Not listed
CERCLA: 40CFR 302.4: Listed as Compound per CWA Section 307(a)
SARA 40CFR 372.65: Not listed
SARA EHS 40CFR 355: Not listed
TSCA: Listed

Analytical Methods

Soil: SW846 8410

BIS3090 **CAS #: 542-88-1**

BIS(CHLOROMETHYL) ETHER

RTECS: KN1575000
DOT: UN2249; IMO6.1
EINECS Number: 208-832-8
Molecular Formula: $C_2H_4Cl_2O$
Structured MF: $(CH_2Cl)_2O$
Formula Weight: 114.97
Synonyms: BCME; BIS-CME; CHLORO(CHLOROMETHOXY)METHANE; CHLOROMETHYL ETHER; DICHLORDIMETHYLAETHER; 1,1'-DICHLORODIMETHYL ETHER; ALPHA,ALPHA'-DICHLORODIMETHYL ETHER; DICHLORODIMETHYL ETHER; SYM-DICHLORO-DIMETHYL ETHER; DICHLORODIMETHYL ETHER,SYMMETRICAL; DICHLOROMETHYL ETHER; SYM-DICHLOROMETHYL ETHER; DIMETHYL-1,1'-DICHLOROETHER; ETHER,BIS(CHLOROMETHYL); METHANE,OXYBIS(CHLORO)-; METHANE,OXYBIS(CHLORO-; OXYBIS(CHLOROMETHANE)
Description: colorless liquid; suffocating odor
Use: lab reagent; monitoring indicator for chloromethyl ether; int in synthesis of anionic exchange strong-base resins of the quaternary ammonium type; research chemical; alkylating agent in the manufacture of polymers

Physical Properties

Boiling Point: 106 °C (223 °F)
Freezing Point: -41.5 °C (-42.7 °F)
Specific Gravity: 1.323 at 15 °C/4 °C
Vapor Density: 4.0 Air=1
Saturated Vapor Density: 1.340422868 kg/m^3
Vapor Pressure: 30 mm Hg at 72 °F
Water Solubility: Reacts
Other Solubilities: miscible with many organic solvents.
Refraction Index: 1.4346 at 20 °C/D
Flash Point: < 19 °C Closed Cup

RTECS Toxicity Data

Acute Oral: Rat LD_{50} Dose: 210 uL/kg.
Acute Inhalation: Rat LC_{50} Dose: 7 ppm/7hr; Toxic Effects: Lungs, Thorax, or Respiration - Chronic pulmonary edema; Blood - Hemorrhage. Man LC_{Lo} Dose: 100 ppm/3M; Toxic Effects: Lungs, Thorax, or Respiration - Other changes. Man TC_{Lo} Dose: 3 ppm; Toxic Effects: Sense organs and special senses - Other; Sense organs and special senses - Conjunctive irritation; Lungs, Thorax, or Respiration - Other changes.
Acute Dermal: Rabbit LD_{50} Route: Skin; Dose: 280 uL/kg.
Mutagenic: Human Unscheduled DNA Synthesis; Cell Type: fibroblast; Dose: 160 ug/L. Mouse Unscheduled DNA Synthesis; Route: Skin; Dose: 360 umol/kg.
Tumorigenic: Rat Route: Inhalation; Dose: 100 ppb/6H/4W-I; Toxic Effects: Tumorigenic - Carcinogenic by RTECS criteria; Sense organs and special senses - Tumors; Lungs, Thorax, or Respiration - Tumors. Rat Route: Inhalation; Dose: 100 ppb/6H/6W-I; Toxic Effects: Tumorigenic - Equivocal tumorigenic agent by RTECS criteria; Lungs, Thorax, or Respiration - Tumors. Rat Route: Inhalation; Dose: 100 ppb/6H/26W-I; Toxic Effects: Tumorigenic - Neoplastic by RTECS criteria; Sense organs and special senses - Tumors; Lungs, Thorax, or Respiration - Tumors. Rat Route: Inhalation; Dose: 75 ppb/6H/2Y-I; Toxic Effects: Tumorigenic - Carcinogenic by RTECS criteria; Sense organs and special senses - Tumors; Lungs, Thorax, or Respiration - Tumors.

Hazard Overviews

Explosive

Flammable

Fire Diamond

Health: Severe irritation to eyes/respiratory tract. Toxic. Also Causes: corneal damage. Chronic Effects: cough, bronchitis, shortness of breath, wheezing, cancer hazard.
Fire: Flammable. Explosive; forms peroxides when heated. For small fires use dry chemical, carbon dioxide, Halon, water spray, or standard foam. For large fires use water spray, fog, or standard foam. Fight fire from maximum distance.
Reactivity: Stable. Hazardous polymerization cannot occur. Avoid: heat; ignition sources; acids. Incompatible with: acids; acid fumes Hazardous decomposition products: highly toxic chloride fumes.
Carcinogenicity: IARC - Group 1, Carcinogenic to humans; NIOSH - Listed as carcinogen; NTP - Class 1, Known to be a carcinogen; ACGIH - Class A2, Suspected human carcinogen; OSHA - Listed as a carcinogen; EPA - Class A, Human carcinogen; MAK - Class A1, Capable of inducing malignant tumors as shown by experience with humans
Primary Target Organs:

Eyes

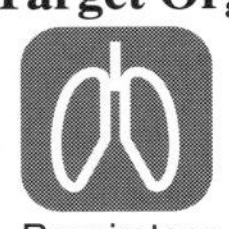
Respiratory System

Exposure Limits
ACGIH TLV: TWA: 0.001 ppm; 0.0047 mg/m^3.
Respirator Recommendation
Exposure Range: >0.001 to 0.05 ppm Self-contained Breathing Apparatus, Pressure Demand, Full Face
Exposure Range: >0.05 to 1 ppm
Exposure Range: >1 to unlimited ppm Self-contained Breathing Apparatus, Pressure Demand, Full Face
Note: odor threshold unknown

Environmental

Environmental Fate: Due to its rapid rate of hydrolysis (half-life < 1 min), it would not be expected to persist in water. It is, however, relatively stable in air (estimated half-life < 1 day for photooxidation and > 18 hr for hydrolysis).

Cleanup/Disposal: Guide No. 153: Eliminate all ignition sources (no smoking, flares, sparks or flames in immediate area). Do not touch damaged containers or spilled material unless wearing appropriate protective clothing. Stop leak if you can do it without risk. Prevent entry into waterways, sewers, basements or confined areas. Absorb or cover with dry earth, sand or other non-combustible material and transfer to containers. Do not get water inside containers.

Environmental Physical Data

Henry's Law Constant: calculated at 2.1 $x10^{-4}$

Octanol/Water Partition Coefficient: log K_{ow} = calculated at -0.38

BCF: bluegill 11

Regulations

RCRA 40CFR: Listed Hazardous Waste No. P016 Toxic Waste

CERCLA: 40CFR 302.4: Listed per RCRA Section 3001 RQ: 10 lb (4.535 kg)

SARA 40CFR 372.65: Listed TPQ: 100 lb

SARA EHS 40CFR 355: Listed TPQ: 10 lb

TSCA: Listed

Analytical Methods

Soil: EPA 7

Water / Groundwater: EPA 6

Food: EPA 5

BIS3200 CAS #: 534-07-6

BISCHLOROMETHYL KETONE

RTECS: UC1430000

DOT: UN2649; IMO6.1

EINECS Number: 208-585-6

Molecular Formula: $C_3H_4Cl_2O$

Formula Weight: 126.97

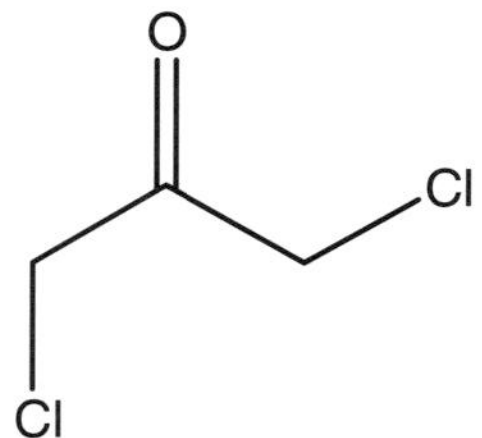

Chemical Structure

Synonyms: ACETONE,1,3-DICHLORO; BIS(CHLOROMETHYL)KETONE; DICHLORO-1,3 ACETONE; 1,3-DICHLOROACETONA; 1,3-DICHLOROACETONE; ALPHA,ALPHA'-DICHLOROACETONE; ALPHA,GAMMA-DICHLOROACETONE; SYM-DICHLOROACETONE; 1,3-DICHLORO-2-PROPANONE; 1,3-DICHLOROPROPANONE; 2-PROPANONE,1,3-DICHLORO-

Description: plates, prisms or needles

Use: vesicant; synthesis of citric acid

Physical Properties

Boiling Point: 173.4 °C (344 °F)

Freezing Point: 45 °C (113 °F)

Specific Gravity: 1.3826 at 46 °C/4 °C

Vapor Density: 4.38 Air=1

Water Solubility: > 10% in Water

Other Solubilities: > 10% in Ethanol; > 10% in Ether.

Refraction Index: 1.4716 at 40 °C

RTECS Toxicity Data

Acute Oral: Rat LD_{50} Dose: 20 mg/kg; Toxic Effects: Behavioral - Somnolence (general depressed activity). Mouse LD_{50} Dose: 18900 ug/kg; Toxic Effects: Sense organs and special senses - Other; Behavioral - Somnolence (general depressed activity).

Acute Inhalation: Rat LC_{50} Dose: 29 mg/m^3/2hr. Mouse LC_{50} Dose: 27 mg/m^3/2hr.

Acute Dermal: Rabbit LD_{50} Route: Skin; Dose: 53 mg/kg; Toxic Effects: Sense organs and special senses - Other; Sense organs and special senses - Iritis; Behavioral - Somnolence (general depressed activity).

Chronic (Multiple Dose) Oral: Rat Dose: 720 mg/kg/90D-C; Toxic Effects: Behavioral - Fluid intake; Blood - Changes in serum composition; Biochemical - Phosphatases.

Irritation Eye: Rabbit Standard Draize Test Dose: 100 mg; Reaction: severe.

Irritation Skin: Rabbit Standard Draize Test Dose: 500 mg; Reaction: severe.

Mutagenic: Hamster Sister Chromatid Exchenge; Cell Type: lung; Dose: 2 umol/L. Non-mammalian species Micronucleus Test; Route: Multiple routes; Dose: 50 ug/L.

Tumorigenic: Mouse Route: Skin; Dose: 37500 ug/kg; Toxic Effects: Tumorigenic - Equivocal tumorigenic agent by RTECS criteria; Skin and appendages - Tumors.

Hazard Overviews

Poison

Corrosive

Health: Corrosive to eyes/skin/respiratory tract. Poison. Other Acute Effects: may be fatal if inhaled, swallowed, or absorbed through skin; material is extremely destructive to tissue of the mucous membranes and upper respiratory tract, eyes and skin; causes blisters on contact with skin; inhalation may result in spasm, inflammation and edema of the larynx and bronchi, chemical pneumonitis and pulmonary edema; symptoms of exposure may include burning sensation, coughing, wheezing, laryngitis, shortness of breath, headache, nausea and vomiting. Chronic Effects: Laboratory experiments have shown mutagenic effects.

Fire: Hazards: emits toxic fumes. Extinguishing agents: water spray; carbon dioxide, dry chemical powder or appropriate foam. Precautions: combustible liquid.

Reactivity: Incompatible with: strong oxidizing agents, strong bases, reducing agents, protect from moisture. Hazardous decomposition products: toxic fumes of: carbon monoxide, carbon dioxide, hydrogen chloride gas.
Carcinogenicity: IARC - Not listed; NIOSH - Not listed; NTP - Not listed; ACGIH - Not listed; OSHA - Not listed; EPA - Not listed; MAK - Not listed
Primary Target Organs:

Eyes

Skin

Respiratory System

Environmental

Cleanup/Disposal: Guide No. 153: Eliminate all ignition sources (no smoking, flares, sparks or flames in immediate area). Do not touch damaged containers or spilled material unless wearing appropriate protective clothing. Stop leak if you can do it without risk. Prevent entry into waterways, sewers, basements or confined areas. Absorb or cover with dry earth, sand or other non-combustible material and transfer to containers. Do not get water inside containers.

Regulations

RCRA 40CFR: Not listed
CERCLA: 40CFR 302.4: Not listed
SARA 40CFR 372.65: Not listed TPQ: 10/10000 lb
SARA EHS 40CFR 355: Listed TPQ: 10 lb
TSCA: Listed

BIS3310 **CAS #: 78-71-7**

3,3-BIS(CHLOROMETHYL)OXETANE

RTECS: RQ6826000
EINECS Number: 201-136-5
Molecular Formula: $C_5H_8Cl_2O$
Formula Weight: 155.03
Synonyms: BCMO; 3,3-DICHLOROMETHYLOXYCYCLOBUTANE; PENTON
Description: finely divided powder for coatings; natural, black, or olive green molding powder
Use: solid & lined valves, pumps, pipe & fittings; monofilament for filter supports & column packing; thermoplastic polymer with high melting temperature of 180 deg c; biological purification of waste waters from mfr of pentaplast; bearings, precision gears, corrosion free coatings, wedges & slot liners for stators in electric motors, films, & rope.

Physical Properties

Boiling Point: 103 °C (217 °F) at 30 mm Hg
Freezing Point: 19 °C (66 °F)
Specific Gravity: 1.4
Water Solubility: very low water absorption.

RTECS Toxicity Data

Acute Oral: Rat LD_{50} Dose: 600 mg/kg; Toxic Effects: Behavioral - Somnolence (general depressed activity); Behavioral - Excitment; Lungs, Thorax, or Respiration - Dyspnea. Mouse LD_{50} Dose: 300 mg/kg; Toxic Effects: Behavioral - Somnolence (general depressed activity); Behavioral - Excitment; Lungs, Thorax, or Respiration - Dyspnea.
Acute Inhalation: Cat LC; Dose: >250 mg/m^3/1hr; Toxic Effects: Sense organs and special senses - Lacrimation; Behavioral - Somnolence (general depressed activity); Behavioral - Food intake (animal). Mouse LC_{50} Dose: 200 mg/m^3/2hr.
Chronic (Multiple Dose) Inhalation: Rat Dose: 5 mg/m^3/4H/22W-I; Toxic Effects: Kidney, Ureter, and Bladder - Chgs in tubules (inc acute renal failure, acute tubular necrosis. Mouse Dose: 5 mg/m^3/4H/22W-I; Toxic Effects: Cardiac - Other changes; Lungs, Thorax, or Respiration - Other changes; Kidney, Ureter, and Bladder - Chgs in tubules (inc acute renal failure, acute tubular necrosis.

Hazard Overviews

Poison

Health: Irritating to eyes/skin/respiratory tract. Poison. Other Acute Effects: may be fatal if inhaled; harmful if swallowed or absorbed through skin; lachrymator; symptoms of exposure may include burning sensation, coughing, wheezing, laryngitis, shortness of breath, headache, nausea and vomiting; target organs: nerves, kidneys, lungs.
Fire: Hazards: emits toxic fumes. Extinguishing agents: water spray; carbon dioxide, dry chemical powder or appropriate foam. Precautions: combustible liquid.
Reactivity: Incompatible with: strong oxidizing agents. Hazardous decomposition products: toxic fumes of: carbon monoxide, carbon dioxide, hydrogen chloride gas.
Carcinogenicity: IARC - Not listed; NIOSH - Not listed; NTP - Not listed; ACGIH - Not listed; OSHA - Not listed; EPA - Not listed; MAK - Not listed
Primary Target Organs:

Eyes

Skin

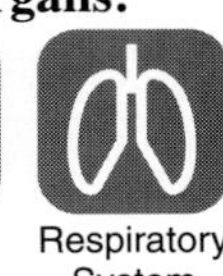
Respiratory System

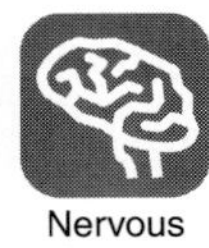
Nervous System

Kidneys

Environmental

Cleanup/Disposal: Guide No. 154: Eliminate all ignition sources (no smoking, flares, sparks or flames in immediate area). Do not touch damaged containers or spilled material unless wearing appropriate protective clothing. Stop leak if you can do it without risk. Prevent entry into waterways, sewers, basements or confined areas. Absorb or cover with dry earth, sand or other non-combustible material and transfer to containers. Do not get water inside containers.

Regulations
RCRA 40CFR: Not listed
CERCLA: 40CFR 302.4: Not listed
SARA 40CFR 372.65: Not listed TPQ: 500 lb
SARA EHS 40CFR 355: Listed TPQ: 500 lb
TSCA: Listed

BIS3420 CAS #: 56960-97-5

1,2-BIS(P-CHLOROPHENYL) ETHANOL

Molecular Formula: $C_{14}H_{12}Cl_2O$
Formula Weight: 267.14
Synonyms: BENZENEETHANOL,4-CHLORO-ALPHA-(4-CHLOROPHENYL)-; ETHANOL,1,2-BIS(P-CHLOROPHENYL)-

Hazard Overviews
Carcinogenicity: IARC - Not listed; NIOSH - Not listed; NTP - Not listed; ACGIH - Not listed; OSHA - Not listed; EPA - Not listed; MAK - Not listed

Environmental

Regulations
RCRA 40CFR: Not listed
CERCLA: 40CFR 302.4: Listed as Compound per CWA Section 307(a)
SARA 40CFR 372.65: Listed as Compound
SARA EHS 40CFR 355: Not listed
TSCA: Not listed

BIS3530 CAS #: 782-74-1

1,2-BIS(2-CHLOROPHENYL) HYDRAZINE

EINECS Number: 212-314-7
Molecular Formula: $C_{12}H_{10}Cl_2N_2$
Formula Weight: 253.12
Synonyms: 2,2'-DICHLOROHYDRAZOBENZENE; HYDRAZINE,1,2-BIS(2-CHLOROPHENYL)-; HYDRAZOBENZENE,2,2'-DICHLORO-

Hazard Overviews
Carcinogenicity: IARC - Not listed; NIOSH - Not listed; NTP - Not listed; ACGIH - Not listed; OSHA - Not listed; EPA - Not listed; MAK - Not listed

Environmental

Regulations
RCRA 40CFR: Not listed
CERCLA: 40CFR 302.4: Listed as Compound per CWA Section 307(a)
SARA 40CFR 372.65: Listed as Compound
SARA EHS 40CFR 355: Not listed
TSCA: Listed

BIS3640 CAS #: 1142-19-4

BIS(P-CHLOROPHENYL)DISULFIDE

RTECS: JO0766000
EINECS Number: 214-531-2
Molecular Formula: $C_{12}H_8Cl_2S_2$
Formula Weight: 287.14

Cl S S Cl

Chemical Structure

Synonyms: BIS(4-CHLOROPHENYL) DISULFIDE; P-CHLOROPHENYL DISULFIDE; DDDS; 4,4'-DICHLORODIPHENYL DISULFIDE; P,P'-DICHLORODIPHENYL DISULFIDE; DI(P-CHLOROPHENYL) DISULFIDE; DISULFIDE,BIS(4-CHLOROPHENYL); DISULFIDE,BIS(P-CHLOROPHENYL)

Physical Properties
Freezing Point: 73 °C (163.4 °F)
Other Solubilities: Chloroform: Soluble

RTECS Toxicity Data
Acute Oral: Mouse LD_{50} Dose: >3 gm/kg.

Hazard Overviews
Carcinogenicity: IARC - Not listed; NIOSH - Not listed; NTP - Not listed; ACGIH - Not listed; OSHA - Not listed; EPA - Not listed; MAK - Not listed

Environmental

Regulations
RCRA 40CFR: Not listed
CERCLA: 40CFR 302.4: Listed as Compound per CWA Section 307(a)
SARA 40CFR 372.65: Listed as Compound
SARA EHS 40CFR 355: Not listed
TSCA: Listed

BIS3860 CAS #: 136-23-2

BIS(DIBUTYLDITHIOCARBAMATO)-ZINC

RTECS: ZH0175000
EINECS Number: 205-232-8
Molecular Formula: $C_{18}H_{36}N_2S_4Zn$
Formula Weight: 474.13

Synonyms: ACCEL BZ; ACETO ZDBD; BIS(N,N-DIBUTYLDITHIOCARBAMATO)ZINC; BUTAZATE; BUTAZATE 50-D; BUTYL ZIMATE; BUTYL ZIRAM; CARBAMIC ACID,DIBUTYLDITHIO-,ZINC COMPLEX; CARBAMODITHIOIC ACID,DIBUTYL-,ZINC SALT; (DIBUTYLDITHIOCARBAMATO)ZINC(II); DIBUTYLDITHIOCARBAMIC ACID ZINC SALT; NOCCELER BZ; SOXINOL BZ; VULCACURE; VULCACURE ZB; VULCAURE ZB; VULKACIT LDB; VULKACIT LDB/C; ZIMATE,BUTYL; ZINC BIBUTYLDITHIOCARBAMATE; ZINC BIS(DIBUTYLDITHIOCARBAMATE); ZINC DIBUTYLDITHIOCARBAMATE; ZINC N,N-DIBUTYLDITHIOCARBAMATE; ZINC,BIS(DIBUTYLCARBAMODITHIOATO-S,S')-,(T-4)-

Description: white powder; pleasant odor

Use: accelerator for rubber vulcanization, latex dispersions, and cements; ultra-accelerator for lubricating oil additive; stabilizer used in food packaging & handling

Physical Properties

Freezing Point: 104 °C (219.2 °F) to 108 °C (226.4 °F)
Specific Gravity: 1.24 at 20 °C/20 °C
Water Solubility: Insoluble in Water
Other Solubilities: Soluble in Carbon Disulfide, Benzene, Chloroform

RTECS Toxicity Data

Tumorigenic: Mouse Route: Oral; Dose: 290 gm/kg/78W-I; Toxic Effects: Tumorigenic - Equivocal tumorigenic agent by RTECS criteria; Lungs, Thorax, or Respiration - Tumors; Liver - Tumors. Mouse Route: Subcutaneous; Dose: 1000 mg/kg; Toxic Effects: Tumorigenic - Equivocal tumorigenic agent by RTECS criteria; Lungs, Thorax, or Respiration - Tumors; Blood - Tumors.

Hazard Overviews

Carcinogenicity: IARC - Not listed; NIOSH - Not listed; NTP - Not listed; ACGIH - Not listed; OSHA - Not listed; EPA - Not listed; MAK - Not listed

Environmental

Regulations

RCRA 40CFR: Not listed
CERCLA: 40CFR 302.4: Listed as Compound per CWA Section 307(a)
SARA 40CFR 372.65: Listed as Compound
SARA EHS 40CFR 355: Not listed
TSCA: Listed

BIS3970 **CAS #: 108-60-1**

BIS(DICHLOROISOPROPYL) ETHER

RTECS: KN1750000
DOT: UN2490; IMO6.1
EINECS Number: 203-598-3
Molecular Formula: $C_6H_{12}Cl_2O$
Structured MF: $[ClCH_2C(CH_3)H]_2O$
Formula Weight: 171.07

Cl Cl O

Chemical Structure

Synonyms: BCIE; BCMEE; BIS(2-CHLOROISOPROPYL) ETHER; BIS(BETA-CHLOROISOPROPYL) ETHER; BIS(BETA-CHLOROISOPROPYL)ETHER; BIS(2-CHLORO-1-METHYLETHYL) ETHER; BIS(1-CHLORO-2-PROPYL) ETHER; (2-CHLORO-1-METHYLETHYL) ETHER; DCIP; DCIP (NEMATOCIDE); 2,2'-DICHLORODIISOPROPYL ETHER; BETA,BETA'-DICHLORODIISOPROPYL ETHER; BETA,BETA-DICHLORODIISOPROPYL ETHER; DICHLORODIISOPROPYL ETHER; 2,2'-DICHLOROISOPROPYL ETHER; 2,2-DICHLOROISOPROPYL ETHER; DICHLOROISOPROPYL ETHER; ETHER,BIS(2-CHLORO-1-METHYLETHYL); ISOPROPYLCHLOREX; NEMAMOL; NEMAMORT; NEMAMORTE; 2,2'-OXYBIS(1-CHLOROPROPANE); PROPANE,2,2'-OXYBIS(1-CHLORO-

Description: colorless liquid

Use: in laboratories and in industrial organic synthesis It is also used in textile treatments, preparation of ion exchange resins, pesticide manufacturing, solvents for polymerization reactions, processing fats, waxes and greases, textile manufacturing, cleaning solution manufacturing, paints and varnishes, spotting agents and control of nematodes

Physical Properties

Boiling Point: 187 °C (369 °F) at 760 mm Hg
Freezing Point: -101.88 °C (-151.384 °F) to -96 °C (-140.8 °F)
Specific Gravity: 1.103 at 20 °C/4 °C
Vapor Density: 6 Air=1
Saturated Vapor Density: 1.200773521 kg/m^3
Vapor Pressure: 0.10 mm Hg at 20 °C
Water Solubility: < 0.1 mg/mL at 22 C
Other Solubilities: 95% Ethanol: >=100 mg/ml at 22 °C; Acetone: >=100 mg/ml at 22 °C; Alcohol: Soluble; Benzene: Soluble; DMSO: >=100 mg/ml at 22 °C; Ether: Soluble; Most oils: miscible; Organic solvents: miscible.
Odor Threshold: Water 3.2 $x10^{-1}$ ppm
Refraction Index: 1.4505 at 20 °C/D
Critical Temperature: 384 °C
Critical Pressure: 413 psia
Flash Point: 85 °C Open Cup

RTECS Toxicity Data

Acute Oral: Rat LD_{50} Dose: 240 mg/kg. Mouse LD_{50} Dose: 296 mg/kg.

Acute Inhalation: Rat LC_{Lo} Dose: 700 ppm/5hr; Toxic Effects: Sense organs and special senses - Iritis; Lungs, Thorax, or Respiration - Dyspnea; Liver - Other changes.

Acute Dermal: Rabbit LD_{50} Route: Skin; Dose: 3 mL/kg. Rat LD_{50} Route: Skin; Dose: >2 gm/kg.

Chronic (Multiple Dose) Oral: Rat Dose: 11250 mg/kg/45D-I; Toxic Effects: Blood - Pigmented or nucleated red Blood cells; Blood - Changes in erythrocite (RBC) cell count; Blood

- Changes in leukocyte (WBC) cell count. Rat Dose: 2340 mg/kg/26W-I; Toxic Effects: Blood - Changes in leukocyte (WBC) cell count; Biochemical - Transaminases; Biochemical - Other enzymes. Rat Dose: 364 gm/kg/2Y-C; Toxic Effects: Behavioral - Food intake (animal); Blood - Changes in spleen; Nutritional and gross metabolic - Weight loss or decreased weight gain.

Irritation Eye: Rabbit Standard Draize Test Dose: 500 mg/24H; Reaction: mild.

Irritation Skin: Rabbit Standard Draize Test Dose: 500 mg/24H; Reaction: mild.

Mutagenic: Mouse Mutations in Mammalian Somatic Cells; Cell Type: lymphocyte; Dose: 250 mg/L. Hamster Cytogenetic Analysis; Cell Type: ovary; Dose: 124 mg/L.

Hazard Overviews

Corrosive

Fire: Combustible.

Carcinogenicity: IARC - Group 3, Not classifiable as to carcinogenicity to humans; NIOSH - Not listed; NTP - Not listed; ACGIH - Not listed; OSHA - Not listed; EPA - Not listed; MAK - Not listed

Respirator Recommendation

Exposure Range: >3 to 150 ppm Supplied Air, Constant Flow/Pressure Demand, Half Mask

Exposure Range: >150 to 3000 ppm Supplied Air, Constant Flow/Pressure Demand, Full Face

Exposure Range: >3000 to unlimited ppm

Note: odor threshold unknown

Environmental

Environmental Fate: If released to the atmosphere, the vapor-phase is expected to degrade by reaction with photochemically produced hydroxyl radicals (estimated half-life of 1.15 days). Based on its relatively high water solubility, long distance transport in water systems may be significant. If released to soil, it is expected to leach significantly (estimated K_{oc} of 73) into groundwater where it may persist for a long period of time. One river die-away study observed no biodegradation; although this study is not specific to soil media, it suggests that biodegradation in soil may be slow. If released to water, volatilization and biodegradation are expected to be the principal removal processes; although these processes may be slow. Volatilization half-lives of 13.9 hr and 6.6 days have been estimated for a model river (one meter deep) and a model environmental pond, respectively. Slow biodegradation may be a significant removal process in water where volatilization is not likely to be important. Hydrolysis, adsorption to sediment and bioconcentration in aquatic organisms are not expected to be environmentally significant removal processes in aquatic systems.

Cleanup/Disposal: Guide No. 153: Eliminate all ignition sources (no smoking, flares, sparks or flames in immediate area). Do not touch damaged containers or spilled material unless wearing appropriate protective clothing. Stop leak if you can do it without risk. Prevent entry into waterways, sewers, basements or confined areas. Absorb or cover with dry earth, sand or other non-combustible material and transfer to containers. Do not get water inside containers.

Environmental Physical Data

Henry's Law Constant: 0.00626

Octanol/Water Partition Coefficient: log K_{ow} = 2.58

Sorption Partition Coefficient: K_{oc} = 73

BCF: calculated at 2.47

Regulations

RCRA 40CFR: Listed Hazardous Waste No. U027 Toxic Waste

CERCLA: 40CFR 302.4: Listed per RCRA Section 3001 per CWA Section 307(a) RQ: 1000 lb (453.5 kg)

SARA 40CFR 372.65: Listed

SARA EHS 40CFR 355: Not listed

TSCA: Listed

Analytical Methods

Air: EPA 0020

Soil: CLP MC_SVOA, OHC; EPA 16, 1625; SW846 3640A, 8110, 8250A, 8270B, 8270C

Water / Groundwater: EPA 1625, 611, 625, 625-S, 6; APHA 6410-B; USGS O3118

Plasma: EPA 29

BIS4080 CAS #: 28076-73-5

BIS(2,4-DICHLOROPHENYL)ETHER

Molecular Formula: $C_{12}H_6Cl_4O$

Formula Weight: 307.79

Synonyms: BENZENE,1,1'-OXYBIS(2,4-DICHLORO-; 2,2',4,4'-TETRACHLORODIPHENYL ETHER; 2,2',4,4'-TETRACHLOROPHENYL OXIDE

Hazard Overviews

Carcinogenicity: IARC - Not listed; NIOSH - Not listed; NTP - Not listed; ACGIH - Not listed; OSHA - Not listed; EPA - Not listed; MAK - Not listed

Environmental

Regulations

RCRA 40CFR: Not listed

CERCLA: 40CFR 302.4: Listed as Compound per CWA Section 307(a)

SARA 40CFR 372.65: Listed as Compound

SARA EHS 40CFR 355: Not listed

TSCA: Not listed

BIS4190 CAS #: 14239-68-0

BIS(DIETHYLDITHIOCARBAMATO)-CADMIUM

RTECS: EU9850000
DOT: UN2570; IMO6.1
EINECS Number: 238-113-4
Molecular Formula: $C_{10}H_{20}CdN_2S_4$
Formula Weight: 408.96
Synonyms: CADMATE; CADMIUM BIS(DIETHYLDITHIOCARBAMATE); CADMIUM BIS(N,N-DIETHYLDITHIOCARBAMATE); CADMIUM DI(DIETHYLDITHIOCARBAMATE); CADMIUM DIETHYL DITHIOCARBAMATE; CADMIUM DIETHYLDITHIOCARBAMATE; CADMIUM SOAP; CADMIUM,BIS(DIETHYLCARBAMODITHIOATO-S,S')-,(T-4)-; CARBAMIC ACID,DIETHYLDITHIO-,CADMIUM SALT; CD DIETHYLDITHIOCARBAMATE; CDEDC; CHROME ORE; CHROMITE HOMOG MINERAL; CHROMITE ORE; ETHYL CADMATE; ETHYL TUADS; HOERNESITE; NATURAL ROESSLERITE; NSC 154470
Description: white to cream colored rods
Use: accelerator for butyl rubber; accelerator for rubber vulcanization; primary accelerator for natural rubbers, and ethylene-propylene diene monomers, and styrene-butadiene rubber

Physical Properties

Freezing Point: 68 °C (154.4 °F) to 76 °C (168.8 °F)
Specific Gravity: 1.39
Water Solubility: Insoluble in Water
Other Solubilities: Soluble in Benzene, Carbon Disulfide, Chloroform; Insoluble in Gasoline

RTECS Toxicity Data

Mutagenic: Bacteria - E Coli DNA Damage; Dose: 1 umol/L. Bacteria - S Typhimurium Mutations in Microorganisms; Dose: 10 ug/plate (+/-S9).
Tumorigenic: Mouse Route: Oral; Dose: 7100 mg/kg/78W-I; Toxic Effects: Tumorigenic - Equivocal tumorigenic agent by RTECS criteria; Lungs, Thorax, or Respiration - Tumors; Liver - Tumors. Mouse Route: Subcutaneous; Dose: 1000 mg/kg; Toxic Effects: Tumorigenic - Equivocal tumorigenic agent by RTECS criteria; Lungs, Thorax, or Respiration - Tumors; Blood - Tumors.

Hazard Overviews

Carcinogenicity: IARC - Group 1, Carcinogenic to humans; NIOSH - Listed as carcinogen; NTP - Class 2A, Reasonably anticipated to be a carcinogen, limited evidence of carcinogenicity from studies in humans; ACGIH - Class A2, Suspected human carcinogen; OSHA - Listed as a carcinogen; EPA - Class B1, Probable human carcinogen based on epidemiologic studies; MAK - Class A2, Unmistakably carcinogenic in animal experimentation only
Exposure Limits
OSHA PEL: TWA: 0.005 mg/m^3; as Cd; see Table Z2.
ACGIH TLV: TWA: 0.002 mg/m^3; as Cd respirable.
NIOSH REL: TWA: 0.1 mg/m^3; as Cd,LOQ; Lowest Feasible Concentration.

Environmental

Cleanup/Disposal: Guide No. 154: Eliminate all ignition sources (no smoking, flares, sparks or flames in immediate area). Do not touch damaged containers or spilled material unless wearing appropriate protective clothing. Stop leak if you can do it without risk. Prevent entry into waterways, sewers, basements or confined areas. Absorb or cover with dry earth, sand or other non-combustible material and transfer to containers. Do not get water inside containers.

Regulations

RCRA 40CFR: Not listed
CERCLA: 40CFR 302.4: Listed as Compound per CWA Section 307(a) per CAA Section 112
SARA 40CFR 372.65: Listed as Compound
SARA EHS 40CFR 355: Not listed
TSCA: Listed

BIS4300 CAS #: 74227-35-3

BIS(4-DIPHENYLSULFONIO)PHENYL

Physical Properties

Water Solubility: Insoluble
Other Solubilities: Soluble in Methanol
Evaporation Rate: < 1 Butyl Acetate=1

Hazard Overviews

Carcinogenicity: IARC - Not listed; NIOSH - Not listed; NTP - Not listed; ACGIH - Not listed; OSHA - Not listed; EPA - Not listed; MAK - Not listed

Environmental

Regulations

RCRA 40CFR: Not listed
CERCLA: 40CFR 302.4: Not listed
SARA 40CFR 372.65: Not listed
SARA EHS 40CFR 355: Not listed
TSCA: Not listed

BIS4410 CAS #: 71786-70-4

BIS(4-DODECYLPHENYL)IODONIUM-HEXA

EINECS Number: 404-420-9

Physical Properties

Boiling Point: > 204 °C (399 °F)
Freezing Point: < -30 °C (-22 °F)
Specific Gravity: 1.16

Vapor Density: > 1 Air=1
Water Solubility: Negligible
Other Solubilities: Soluble in Acetone
Evaporation Rate: Negligible Butyl Acetate=1

Hazard Overviews

Carcinogenicity: IARC - Not listed; NIOSH - Not listed; NTP - Not listed; ACGIH - Not listed; OSHA - Not listed; EPA - Not listed; MAK - Not listed

Environmental

Regulations

RCRA 40CFR: Not listed
CERCLA: 40CFR 302.4: Not listed
SARA 40CFR 372.65: Not listed
SARA EHS 40CFR 355: Not listed
TSCA: Not listed

BIS4520 **CAS #: 2425-79-8**

1,4-BIS(2,3-EPOXYPROPOXY)BUTANE

RTECS: EJ5100000
EINECS Number: 219-371-7
Molecular Formula: $C_{10}H_{18}O_4$
Formula Weight: 202.28

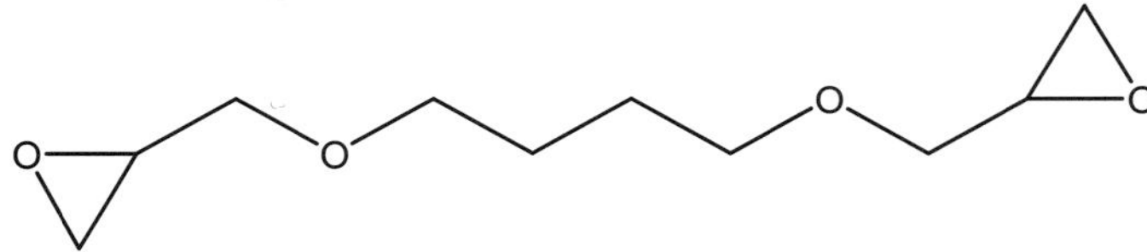

Chemical Structure

Synonyms: ARALDIT DY 026; 1,4-BIS(2,3-EPOXYPROPOXY) BUTANE; 1,4-BIS(2,3-EPOXYPROPYLOXY)BUTANE; 1,4-BIS(GLYCIDYLOXY)BUTANE; 1,4-BUTANE DIGLYCIDYL ETHER; BUTANE,1,4-BIS(2,3-EPOXYPROPOXY)-; 1,4-BUTANEDIOL DIGLYCIDYL ETHER; BUTANE-1:4-DIOL DIGLYCIDYL ETHER; BUTANEDIOL DIGLYCIDYL ETHER; 2,2'-(1,4-BUTANEDIYLBIS(OXYMETHYLENE))BIS OXIRANE; 2,2'-(1,4-BUTANEDIYLBIS(OXYMETHYLENE))BISOXIRANE; CD 15006 A; CHS-RR2; 1,4-DIGLYCIDLOXYBUTANE; 1,4-DIGLYCIDYLOXYBUTANE; GRILONIT RV 1806; OXIRANE,2,2'-(1,4-BUTANEDIYLBIS(OXYMETHYLENE))BIS-; OXIRANE,2,2'-(1,4-BUTANEDIYLBIS(OXYMETHYLENE))BIS-(9CI); TETRAMETHYLENE GLYCOL DIGLYCIDYL ETHER; (TETRAMETHYLENEBIS(OXYMETHYLENE))DIOXIRANE; TK 10352
Description: pale yellow, clear liquid
Use: reactive diluent in epoxy resin systems to reduce viscosity

Physical Properties

Boiling Point: 155 °C (311 °F) to 160 °C (320 °F) at 11 mm Hg
Specific Gravity: 1.09 to 1.1 at 25/4 °C
Density: 1.1 g/cu cm at 25 °C
Water Solubility: 10 to 50 mg/mL at 21 °C
Other Solubilities: 95% Ethanol: >=100 mg/ml at 21 °C; Acetone: >=100 mg/ml at 21 °C; DMSO: >=100 mg/ml at 21 °C.
Refraction Index: 1.4611
Flash Point: > 93.3 °C

RTECS Toxicity Data

Acute Oral: Rat LD_{50} Dose: 1134 mg/kg; Toxic Effects: Sense organs and special senses - Other; Behavioral - Somnolence (general depressed activity); Skin and appendages - Hair. Mouse LD_{50} Dose: 1100 ug/kg.
Acute Inhalation: Rat LC; Dose: >250 ppm/6hr.
Acute Dermal: Rabbit LD_{50} Route: Skin; Dose: 1130 mg/kg.
Irritation Eye: Rabbit Standard Draize Test Dose: 100 mg; Reaction: moderate.
Irritation Skin: Rabbit Standard Draize Test Dose: 10 mg/24H; Reaction: moderate.
Mutagenic: Rat Cytogenetic Analysis; Route: Intraperitoneal; Dose: 100 mg/kg. Hamster Sister Chromatid Exchenge; Cell Type: lung; Dose: 6250 nmol/L.

Hazard Overviews

Corrosive

Health: Corrosive to eyes/skin/respiratory tract. Harmful. Other Acute Effects: harmful if swallowed, inhaled, or absorbed through skin; inhalation may result in spasm, inflammation and edema of the larynx and bronchi, chemical pneumonitis and pulmonary edema; symptoms of exposure may include burning sensation; coughing; wheezing; laryngitis; shortness of breath; headache; nausea; vomiting; may cause allergic reaction.
Fire: Will burn. Hazards: emits toxic fumes. Extinguishing agents: water spray; carbon dioxide, dry chemical powder or appropriate foam. Precautions: combustible liquid.
Reactivity: Incompatible with: acids, bases, oxidizing agents. Hazardous decomposition products: carbon monoxide, carbon dioxide.
Carcinogenicity: IARC - Not listed; NIOSH - Not listed; NTP - Not listed; ACGIH - Not listed; OSHA - Not listed; EPA - Not listed; MAK - Not listed
Primary Target Organs:

Eyes

Skin

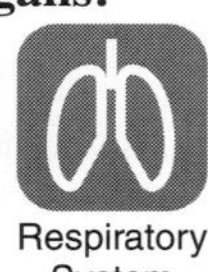
Respiratory System

Environmental

Regulations

RCRA 40CFR: Not listed
CERCLA: 40CFR 302.4: Not listed
SARA 40CFR 372.65: Not listed
SARA EHS 40CFR 355: Not listed

TSCA: Listed

BIS4630 CAS #: 6994-46-3

1,4-BISETHYLAMINOANTHRAQUINONE

EINECS Number: 230-263-9

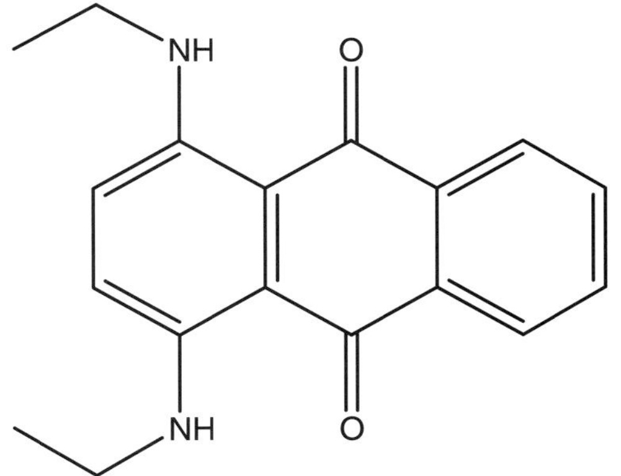

Chemical Structure

Physical Properties

Water Solubility: Negligible

Hazard Overviews

Health: Irritating to eyes/skin/respiratory tract. Other Acute Effects: may be harmful by inhalation, ingestion, or skin absorption.

Fire: Hazards: emits toxic fumes. Extinguishing agents: water spray; carbon dioxide, dry chemical powder or appropriate foam. Precautions: combustible liquid.

Reactivity: Incompatible with: strong oxidizing agents. Hazardous decomposition products: toxic fumes of: carbon monoxide, carbon dioxide, nitrogen oxides.

Carcinogenicity: IARC - Not listed; NIOSH - Not listed; NTP - Not listed; ACGIH - Not listed; OSHA - Not listed; EPA - Not listed; MAK - Not listed

Primary Target Organs:

Eyes

Skin

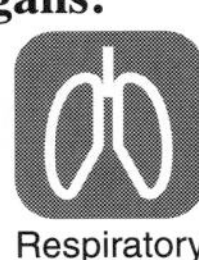
Respiratory System

Environmental

Regulations

RCRA 40CFR: Not listed

CERCLA: 40CFR 302.4: Not listed

SARA 40CFR 372.65: Not listed

SARA EHS 40CFR 355: Not listed

TSCA: Not listed

BIS4740 CAS #: 103-23-1

BIS(2-ETHYLHEXYL) ADIPATE

RTECS: AU9700000

EINECS Number: 203-090-1

Molecular Formula: $C_{22}H_{42}O_4$

Structured MF: $[CH_2CH_2COOCH_2CH(C_2H_5)(CH_2)_3CH_3]_2$

Formula Weight: 370.58

Chemical Structure

Synonyms: ADIPIC ACID,BIS(2-ETHYLHEXYL) ESTER; ADIPOL 2EH; BEHA; BIS(2-ETHYLHEXYL) HEXANEDIOATE; BIS-(2-ETHYLHEXYL)ESTER KYSELINY ADIPOVE; BIS(2-ETHYLHEXYL)HEXANEDIOATE; BISOFLEX DOA; DEHA; DI(2-ETHYLHEXYL) ADIPATE; DI-2-ETHYLHEXYL ADIPATE; DIOCTYL ADIPATE; DOA; EFFEMOLL DOA; EFFOMOLL DOA; ERGOPLAST ADDO; FLEXOL A 26; FLEXOL PLASTICIZER 10-A; FLEXOL PLASTICIZER A-26; HEXANEDIOIC ACID,BIS(2-ETHYLHEXYL) ESTER; HEXANEDIOIC ACID,BIS(2-ETHYLHEXYL) ESTER (9CI); HEXANEDIOIC ACID,DI(2-ETHYLHEXYL) ESTER; HEXANEDIOIC ACID,DIOCTYL ESTER; KEMESTER 5652; KODAFLEX DOA; MOLLAN S; MONOPLEX DOA; MORFLEX 310; NSC 56775; OCTYL ADIPATE; PLASTOMOLL DOA; PX-238; REOMOL DOA; RUCOFLEX PLASTICIZER DOA; SANSOCIZER DOA; SICOL 250; STAFLEX DOA; TRUFLEX DOA; UNIFLEX DOA; VESTINOL OA; WICKENOL 158; WITAMOL 320

Description: colorless or very pale amber oily liquid; slight aromatic odor

Use: a plasticizer; blended with DOP and DIO in processing polyvinyl and other polymers; as a solvent and is found as a component of aircraft lubricants; in plastics manufacture and processing, compounding of vinyl resins, nitrocellulose, synthetic rubber, in plasticizing polyvinyl butyral, cellulose acetate butyrate, polystyrene, dammar wax and in cosmetics; in the compounding of cellulose ester plastics and synthetic elastomers, lubricants, functional (hydraulic) fluids, blends intended for low-temperature uses and polyvinyl chloride resins in film, sheeting, extrusions and plastisols

Physical Properties

Boiling Point: 214 °C (417 °F) at 5 mm Hg

Freezing Point: -67.8 °C (-90.04 °F)

Specific Gravity: 0.922 at 25 °C/4 °C

Vapor Density: 12.8 Air=1

Saturated Vapor Density: 1.200000016 kg/m^3

Density: 0.928 g/mL at 20 °C

Vapor Pressure: 8.5 $x10^{-7}$ mm Hg at 20 °C

Water Solubility: < 0.1 mg/mL at 22 C

Other Solubilities: 95% Ethanol: >=100 mg/ml at 21 °C; Acetone: >=100 mg/ml at 21 °C; Acetic Acid: Soluble; Alcohol: Soluble; DMSO: 10-50 mg/ml at 21 °C; Ether: Soluble.

Surface Tension: Estimated at 15 dynes/cm at 20 °C

Refraction Index: 1.4474 at 20 °C/D
Evaporation Rate: < 1 Butyl Acetate=1
Flash Point: 206 °C Open Cup
Autoignition Temperature: 377 °C
LEL: 0.38% v/v

RTECS Toxicity Data

Acute Oral: Rat LD_{50} Dose: 9100 mg/kg. Mouse LD_{50} Dose: 15 gm/kg.
Chronic (Multiple Dose) Oral: Rat Dose: 25200 mg/kg/3W-C; Toxic Effects: Blood - Changes in serum composition. Mouse Dose: 168 gm/kg/14D-C; Toxic Effects: DEATH.
Irritation Eye: Rabbit Standard Draize Test Dose: 500 mg/24H; Reaction: mild.
Reproductive/Teratogenic: Rat Route: Intraperitoneal; Dose: 15 gm/kg; Duration: female 5-15D of pregnancy; Effects on Embryo or Fetus - Fetotoxicity. Rat Route: Intraperitoneal; Dose: 30 gm/kg; Duration: female 5-15D of pregnancy; Specific Developmental Abnormalities - Other developmental abnormalities.
Mutagenic: Rat Unscheduled DNA Synthesis; Route: Oral; Dose: 378 umol/kg. Mouse Dominant Lethal Test; Route: Intraperitoneal; Dose: 1000 mg/kg.
Tumorigenic: Mouse Route: Oral; Dose: 1038 gm/kg/2Y-C; Toxic Effects: Tumorigenic - Carcinogenic by RTECS criteria; Liver - Tumors. Mouse Route: Oral; Dose: 2163 gm/kg/2Y-C; Toxic Effects: Tumorigenic - Carcinogenic by RTECS criteria; Liver - Tumors. Mouse Route: Oral; Dose: 1048 gm/kg/2Y-C; Toxic Effects: Tumorigenic - Carcinogenic by RTECS criteria; Liver - Tumors.

Hazard Overviews

Health: May be irritating to eyes/skin/respiratory tract. Other Acute Effects: may be harmful by inhalation, ingestion, or skin absorption.
Fire: Will burn. Hazards: static charge buildup can be a potential fire hazard when used in the presence of volatile or flammable mixtures; emits toxic fumes. Extinguishing agents: water spray; carbon dioxide, dry chemical powder or appropriate foam. Precautions: combustible liquid.
Reactivity: Stable. Hazardous polymerization will not occur. Incompatible with: strong oxidizing agents. Hazardous decomposition products: toxic fumes of: carbon monoxide, carbon dioxide.
Carcinogenicity: IARC - Group 3, Not classifiable as to carcinogenicity to humans; NIOSH - Not listed; NTP - Not listed; ACGIH - Not listed; OSHA - Not listed; EPA - Class C, Possible human carcinogen; MAK - Not listed

Environmental

Ecotoxicity: Algae: Scenedesmus subspicatus 72h EC_{20} 400 mg/l; Crustaceans: Daphnia magna Straus 24h EC_{0} 500 mg/l
Environmental Fate: If released to air, it can exist in both vapor and particulate phases. The vapor phase will degrade relatively rapidly by reaction with photochemically produced hydroxyl radicals (estimated half-life of 16 hr). The particulate phase can be physically removed from air by wet and dry deposition. If released to soil or water, it is expected to biodegrade; screening tests have shown that it biodegrades readily. Estimated K_{oc} values of 5004-48,600 suggest that it will be relatively immobile in soil (and not leach) and should partition from the water column to sediment in the aquatic environment.
Cleanup/Disposal: Guide No. 171: Do not touch or walk through spilled material. Stop leak if you can do it without risk. Prevent dust cloud. Avoid inhalation of asbestos dust. Small Dry Spills: With clean shovel place material into clean, dry container and cover loosely; move containers from spill area. Small Spills: Take up with sand or other noncombustible absorbent material and place into containers for later disposal. Large Spills: Dike far ahead of liquid spill for later disposal. Cover powder spill with plastic sheet or tarp to minimize spreading. Prevent entry into waterways, sewers, basements or confined areas.

Environmental Physical Data

Henry's Law Constant: 4.34 $x10^{-7}$
Octanol/Water Partition Coefficient: log K_{ow} = measured at > 6.11
Sorption Partition Coefficient: K_{oc} = 5004
BCF: bluegill 27

Regulations

RCRA 40CFR: Not listed
CERCLA: 40CFR 302.4: Not listed
SARA 40CFR 372.65: Listed
SARA EHS 40CFR 355: Not listed
TSCA: Listed

Analytical Methods

Drinking Water: EPA 506, 525.1, 525.2

BIS4850 CAS #: 3658-48-8

BIS(2-ETHYLHEXYL) HYDROGEN PHOSPHITE

RTECS: SZ6840000
EINECS Number: 222-904-6
Molecular Formula: $C_{16}H_{35}O_3P$
Formula Weight: 306.48

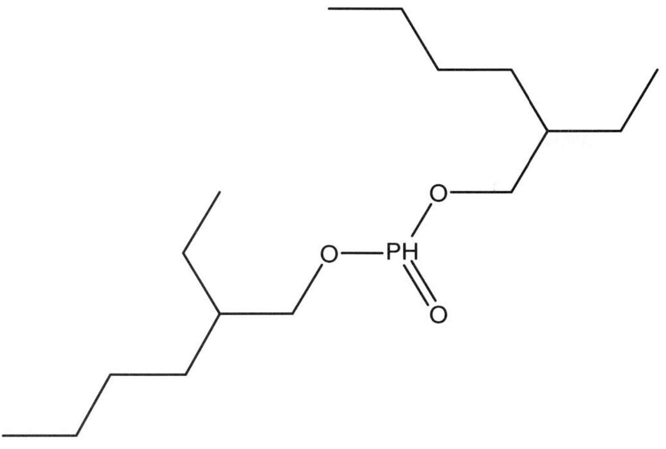

Chemical Structure

Synonyms: BIS(2-ETHYLHEXYL) PHOSPHITE; BIS(2-ETHYLHEXYL) PHOSPHONATE; DI-2-ETHYLHEXYL PHOSPHITE; DIISOOCTYL PHOSPHITE; PHOSPHONIC ACID,BIS(2-ETHYLHEXYL) ESTER

Use: corrosion inhibitor; antioxidant; additive for extreme pressure lubricants & adhesives; textile finishing agent; chem int for organic phosphorus compounds

RTECS Toxicity Data

Acute Oral: Rat LD_{50} Dose: 11900 mg/kg.
Acute Inhalation: Rat LC_{50} Dose: >20 gm/m^3.
Acute Dermal: Rabbit LD_{50} Route: Skin; Dose: 4500 mg/kg.
Irritation Eye: Rabbit Standard Draize Test Dose: 25 mg; Reaction: mild.

Hazard Overviews

Health: Irritating to eyes/skin/respiratory tract. Other Acute Effects: may be harmful by inhalation, ingestion, or skin absorption.
Fire: Hazards: emits toxic fumes.
Reactivity: Hazardous decomposition products: toxic fumes of: carbon monoxide, carbon dioxide; thermal decomposition may produce toxic fumes of phosphorus oxides and/or phosphine.
Carcinogenicity: IARC - Not listed; NIOSH - Not listed; NTP - Not listed; ACGIH - Not listed; OSHA - Not listed; EPA - Not listed; MAK - Not listed
Primary Target Organs:

Eyes

Skin

Respiratory System

Environmental

Environmental Fate: In the environment, it is expected to hydrolyze with half-lives in the order of weeks. Volatilization will probably be an insignificant transport mechanism. Sorption to soils and sediments is expected as is bioconcentration in aquatic organisms. Atmospheric half-life is estimated to be 23 hours. No information was found on photolysis or biodegradation.

Environmental Physical Data

BCF: bioconcentration likely

Regulations

RCRA 40CFR: Not listed
CERCLA: 40CFR 302.4: Not listed
SARA 40CFR 372.65: Not listed
SARA EHS 40CFR 355: Not listed
TSCA: Listed

BIS4960 CAS #: 298-07-7

BIS(2-ETHYLHEXYL) PHOSPHATE

RTECS: TB7875000
DOT: UN1902; NA1902; IMO8.0
EINECS Number: 206-056-4
Molecular Formula: $C_{16}H_{35}O_4P$
Structured MF: $[CH_3CH_2CH_2CH_2CHC_2H_5CH_2O]_2POOH$
Formula Weight: 322.48

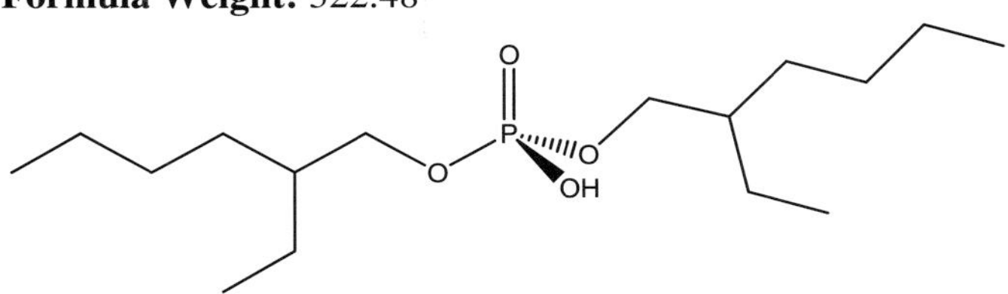

Chemical Structure

Synonyms: BIS(2-ETHYLHEXYL) HYDROGEN PHOSPHATE; BIS(2-ETHYLHEXYL) PHOSPHORIC ACID; BIS(2-ETHYLHEXYL)HYDROGEN PHOSPHATE; BIS(2-ETHYLHEXYL)ORTHOPHOSPHORIC ACID; BIS(2-ETHYLHEXYL)PHOSPHATE; BIS(2-ETHYLHEXYL)PHOSPHORIC ACID; D 2EHPA; DEHPA; DEHPA EXTRACTANT; DI-(2-ETHYLHEXYL) ACID PHOSPHATE; DI-2-ETHYLHEXYL HYDROGEN PHOSPHATE; DI(2-ETHYLHEXYL) ORTHOPHOSPHORIC ACID; DI(2-ETHYLHEXYL) PHOSPHATE; DI(2-ETHYLHEXYL) PHOSPHORIC ACID; DI(2-ETHYLHEXYL)ORTHOPHOSPHORIC ACID; DI(2-ETHYLHEXYL)PHOSPHATE; DI-2(ETHYLHEXYL)PHOSPHORIC ACID; ESCAID 100; 2-ETHYL-1-HEXANOL HYDROGEN PHOSPHATE; HDEHP; HYDROGEN BIS(2-ETHYLHEXYL) PHOSPHATE; KYSELINA DI-(2-ETHYLHEXYL)FOSFORECNA; PHOSPHORIC ACID BIS(ETHYLHEXYL) ESTER; PHOSPHORIC ACID,BIS(2-ETHYLHEXYL) ESTER

Description: amber liquid; odorless

Use: additive to lubrication oils, corrision inhibitor and antioxidant; metal extraction and separation; intermediate for wetting agents and detergents; basic drug extraction; formerly a fire retardant in polymeric materials & other materials

Physical Properties

Boiling Point: Decomposes at 1 atm
Freezing Point: -60 °C (-76 °F)
Specific Gravity: 0.973 at 25 °C/25 °C
Bulk Density: 8.2 lbs/gal
Water Solubility: 0 lb/100 lb at 20 °C

Other Solubilities: Soluble in organic solvents
Surface Tension: Estimated at 30 dynes/cm
pH: Strongly acid
Refraction Index: 1.4420 at 25 °C/D
Flash Point: 196.111 °C Open Cup

RTECS Toxicity Data

Acute Oral: Rat LD_{50} Dose: 4940 uL/kg.
Acute Dermal: Rabbit LD_{50} Route: Skin; Dose: 1250 uL/kg.
Irritation Eye: Rabbit Standard Draize Test Dose: 250 ug/24H; Reaction: severe. Rabbit Standard Draize Test Dose: 5 mg; Reaction: moderate.
Irritation Skin: Rabbit Standard Draize Test Dose: 5 mg/24H; Reaction: severe. Rabbit Open Draize Test Dose: 500 mg open; Reaction: moderate.

Hazard Overviews

Corrosive

Health: Corrosive to eyes/skin/respiratory tract. Toxic. Other Acute Effects: harmful if swallowed, inhaled, or absorbed through skin; inhalation may result in spasm; inflammation and edema of the larynx and bronchi; chemical pneumonitis and pulmonary edema; burning sensation; coughing; wheezing; laryngitis; shortness of breath; headache; nausea; vomiting. Chronic Effects: Carcinogen.
Fire: Will burn. Hazards: emits toxic fumes. Extinguishing agents: water spray; carbon dioxide, dry chemical powder or appropriate foam. Precautions: combustible liquid.
Reactivity: Incompatible with: strong oxidizing agents, strong bases. Hazardous decomposition products: toxic fumes of: carbon monoxide, carbon dioxide; thermal decomposition may produce toxic fumes of phosphorus oxides and/or phosphine.
Carcinogenicity: IARC - Not listed; NIOSH - Not listed; NTP - Not listed; ACGIH - Not listed; OSHA - Not listed; EPA - Not listed; MAK - Not listed
Primary Target Organs:

Eyes

Skin

Respiratory System

Environmental

Environmental Fate: Hydrolysis may be important in the environment at basic pH, but no rate data is available to quantify the relative importance of the process. Data on biodegradation in soil and water were not available. The adsorption to suspended solids and sediments in water and to soil should be moderately strong and the adsorption should become weaker as the pH of the media increases. Based on the estimated Henry's Law constant and vapor pressure, the volatilization of the compound from water and soil should not be important. The estimated bioconcentration factor indicates that bioconcentration in aquatic organisms should be important. In the atmosphere, the reaction with photochemically produced hydroxyl radicals may be the most important process. The half-life of this reaction has been estimated to be 6.2 hr.
Cleanup/Disposal: Guide No. 153: Eliminate all ignition sources (no smoking, flares, sparks or flames in immediate area). Do not touch damaged containers or spilled material unless wearing appropriate protective clothing. Stop leak if you can do it without risk. Prevent entry into waterways, sewers, basements or confined areas. Absorb or cover with dry earth, sand or other non-combustible material and transfer to containers. Do not get water inside containers.

Environmental Physical Data

Henry's Law Constant: estimated at 4.11×10^{-8}
Octanol/Water Partition Coefficient: log K_{ow} = 4.37
Sorption Partition Coefficient: log K_{oc} = estimated at 2.54
BCF: no food chain concentration potential

Regulations

RCRA 40CFR: Not listed
CERCLA: 40CFR 302.4: Not listed
SARA 40CFR 372.65: Not listed
SARA EHS 40CFR 355: Not listed
TSCA: Listed

BIS5070 CAS #: 5810-88-8

BIS(2-ETHYLHEXYL) PHOSPHORODITHIOATE

RTECS: TD5075000
EINECS Number: 227-376-0
Molecular Formula: $C_{16}H_{35}O_2PS_2$
Formula Weight: 354.60
Synonyms: O,O'-BIS(2-ETHYLHEXYL) DITHIOPHOSPHATE; O,O-BIS(2-ETHYLHEXYL) DITHIOPHOSPHATE; BIS(2-ETHYLHEXYL) HYDROGEN DITHIOPHOSPHATE; O,O-BIS(2-ETHYLHEXYL) HYDROGEN PHOSPHOROTHIOATE; O,O-BIS(2-ETHYLHEXYL) PHOSPHORODITHIOATE; O,O'-DI(2-ETHYLHEXYL) DITHIOPHOSPHORIC ACID; DI-2-ETHYLHEXYLPHOSPHORODITHIOIC ACID; PHOSPHORODITHIOIC ACID,O,O-BIS(2-ETHYLHEXYL) ESTER
Use: formerly a lubricating oil additive & rust inhibitor

RTECS Toxicity Data

Acute Oral: Rat LD_{50} Dose: 2140 mg/kg.
Acute Dermal: Rabbit LD_{50} Route: Skin; Dose: 1250 mg/kg; Toxic Effects: Skin and appendages - Primary irritation.
Irritation Eye: Rabbit Standard Draize Test Dose: 750 ug/24H; Reaction: severe.
Irritation Skin: Rabbit Standard Draize Test Dose: 2 mg/24H; Reaction: severe.

Hazard Overviews

Carcinogenicity: IARC - Not listed; NIOSH - Not listed; NTP - Not listed; ACGIH - Not listed; OSHA - Not listed; EPA - Not listed; MAK - Not listed

Environmental

Regulations

RCRA 40CFR: Not listed
CERCLA: 40CFR 302.4: Not listed
SARA 40CFR 372.65: Not listed
SARA EHS 40CFR 355: Not listed
TSCA: Listed

BIS5180 CAS #: 122-62-3

BIS(2-ETHYLHEXYL) SEBACATE

RTECS: VS1000000
EINECS Number: 204-558-8
Molecular Formula: $C_{26}H_{50}O_4$
Formula Weight: 426.66

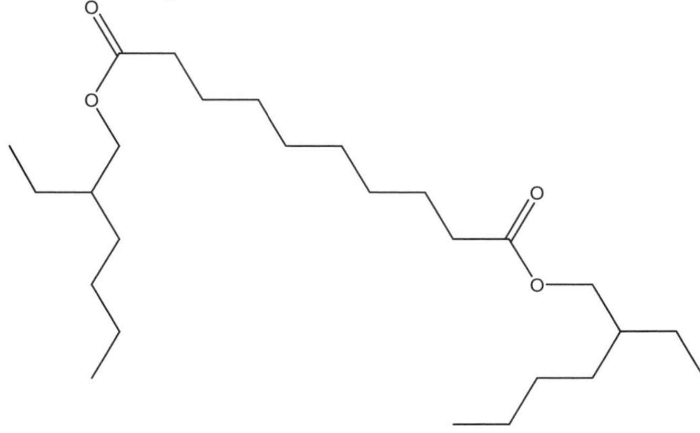

Chemical Structure

Synonyms: BIS(2-ETHYLHEXYL) DECANEDIOATE; BIS(ETHYLHEXYL) SEBACATE; BIS-(2-ETHYLHEXYL)ESTER KYSELINY SEBAKOVE; BIS(2-ETHYLHEXYL)SEBACATE; BISOFLEX; BISOFLEX DOS; DECANEDIOIC ACID,BIS(2-ETHYLHEXYL) ESTER; DEHS; DI(2-ETHYLHEXYL) SEBACATE; DI(2-ETHYLHEXYL)SEBACATE; DI-(2-ETHYLHEXYL)-SEBACATE; DIOCTYL SEBACATE; DOS; EDENOL 888; ERGOPLAST SNO; 2-ETHYLHEXYL SEBACATE; 1-HEXANOL,2-ETHYL-,SEBACATE; MONOPLEX DOS; OCTOIL S; OCTYL SEBACATE; PLEXOL; PLEXOL 201J; PLEXOL 201; PX 438; REOLUBE DOS; SEBACIC ACID DI(2-ETHYLHEXYL) ESTER; SEBACIC ACID,BIS(2-ETHYLHEXYL) ESTER; STAFLEX DOS; UNIFLEX DOS
Description: colorless to pale straw-colored oily liquid; mild odor
Use: plasticizer; in vacuum pumps

Physical Properties

Boiling Point: 256 °C (493 °F) at 5 mm Hg
Freezing Point: -48 °C (-54.4 °F)
Specific Gravity: 0.912 at 25 °C/4 °C
Density: 0.91 g/mL at 25 °C
Water Solubility: Insoluble
Other Solubilities: 95% Ethanol: Soluble; Acetone: Soluble; Benzene: Soluble; DMSO: Soluble.
Refraction Index: 1.451 at 25 °C/D
Flash Point: 210 °C Open Cup

RTECS Toxicity Data

Acute Oral: Mouse LD_{50} Dose: 9500 mg/kg.

Hazard Overviews

Health: May cause irritation to eyes/skin. Other Acute Effects: may be harmful by inhalation, ingestion, or skin absorption.
Fire: Will burn. Extinguishing agents: water spray; carbon dioxide, dry chemical powder or appropriate foam. Precautions: combustible liquid.
Reactivity: Incompatible with: strong oxidizing agents. Hazardous decomposition products: toxic fumes of: carbon monoxide, carbon dioxide.
Carcinogenicity: IARC - Not listed; NIOSH - Not listed; NTP - Not listed; ACGIH - Not listed; OSHA - Not listed; EPA - Not listed; MAK - Not listed

Environmental

Regulations

RCRA 40CFR: Not listed
CERCLA: 40CFR 302.4: Not listed
SARA 40CFR 372.65: Not listed
SARA EHS 40CFR 355: Not listed
TSCA: Listed

BIS5290 CAS #: 577-11-7

BIS(2-ETHYLHEXYL) SODIUM SULFOSUCCINATE

RTECS: WN0525000
EINECS Number: 209-406-4
Molecular Formula: $C_{20}H_{38}NaO_7S$
Structured MF: $C_8H_{17}OOCCH_2CH(SO_3Na)COOC_8H_{17}$
Formula Weight: 445.56

Chemical Structure

Synonyms: AEROSOL GPG; AEROSOL OT; AEROSOL OT-B; AEROSOL OT 75; ALCOPOL O; ALPHASOL OT; BEROL 478; BIS(ETHYLHEXYL) ESTER OF SODIUM SULFOSUCCINIC ACID; 1,4-BIS(2-ETHYLHEXYL) SODIUM SULFOSUCCINATE; BIS(2-ETHYLHEXYL) S-SODIUM SULFOSUCCINATE; BIS(2-ETHYLHEXYL) SULFOSUCCINATE SODIUM SALT; BIS-2-ETHYLHEXYLESTER SULFOJANTARANU SODNEHO; BIS(2-ETHYLHEXYL)SODIUM SULFOSUCCINATE; BUTANEDIOIC ACID,SULFO-,1,4-BIS(2-ETHYLHEXYL) ESTER,SODIUM SALT;

BUTANEDIOIC ACID,SULFO-,1,4-BIS(2-ETHYLHEXYL) ESTER,SODIUM SALT (9CI); CELANOL DOS 75; CLESTOL; COLACE; COMPLEMIX; CONSTONATE; COPROL; D-S-S; DEFILIN; DI-(2-ETHYLHEXYL) SODIUM SULFOSUCCINATE; DIOCTLYN; DIOCTYL ESTER OF SODIUM SULFOSUCCINATE; DIOCTYL ESTER OF SODIUM SULFOSUCCINIC ACID; DIOCTYL SODIUM SULFOSUCCINATE; DIOCTYL SULFOSUCCINATE SODIUM; DIOCTYL SULFOSUCCINATE SODIUM SALT; DIOCTYLAL; DIOCTYL-MEDO FORTE; DIOMEDICONE; DIOSUCCIN; DIOTILAN; DIOVAC; DIOX; DISONATE; DOCUSATE SODIUM; DOXINATE; DOXOL; DSS; DULSIVAC; DUOSOL; 2-ETHYLHEXYL SULFOSUCCINATE SODIUM; HUMIFEN WT 27G; KONLAX; KOSATE; LAXINATE; LAXINATE 100; MANOXAL OT; MANOXOL OT; MERVAMINE; MODANE SOFT; MOLATOC; MOLCER; MOLOFAC; MONAWET MD 70E; MONAWET MO-70; MONAWET MO-84 R2W; MONAWET MO-70 RP; MONOXOL OT; NEKAL WT-27; NEVAX; NIKKOL OTP 70; NORVAL; OBSTON; RAPISOL; REGUTOL; REQUTOL; REVAC; SANMORIN OT 70; SBO; SOBITAL; SODIUM BIS(2-ETHYLHEXYL) SULFOSUCCINATE; SODIUM DI-(2-ETHYLHEXYL) SULFOSUCCINATE; SODIUM DIOCTYL SULFOSUCCINATE; SODIUM DIOCTYL SULPHOSUCCINATE; SODIUM 2-ETHYLHEXYLSULFOSUCCINATE; SODIUM SULFODI-(2-ETHYLHEXYL)-SULFOSUCCINATE; SOFTIL; SOL SODOWEJ SULFOBURSZTYNIANU DWU-2-ETYLOHEKSYLOWEGO; SOLIWAX; SOLUSOL-100%; SOLUSOL-75%; SUCCINIC ACID,SULFO-,1,4-BIS(2-ETHYLHEXYL) ESTER,SODIUMSALT; SULFIMEL DOS; SULFOSUCCINIC ACID BIS(2-ETHYLHEXYL)ESTER SODIUM SALT; SULFOSUCCINIC ACID,BIS(2-ETHYLHEXYL)ESTER SODIUM SALT; SV 102; TEX-WET 1001; TRITON GR-5; TRITON GR 7; VATSOL OT; VELMOL; WAXSOL; WETAID SR

Description: white, wax-like solid; characteristic odor

Use: wetting agent in industrial, pharmaceutical, cosmetic & food applications, & in certain detergents; dispersing & solubilizing agent in foods; adjuvant in tablet formation; food additive (processing aid in sugar industry, stablilizer for hydrophilic colloids); dispersant & emulsifier in various dermatological prepn; in powderless etching systems; found to have potentiating powers, particularly in adding flavor or freshness to canned milks.

Physical Properties

Boiling Point: Decomposes at 1 atm

Freezing Point: Solid Form 155 °C (311 °F)

Specific Gravity: 1.1 at 20 °C

Water Solubility: 15 g/L at 25 °C

Other Solubilities: freely Soluble in Glycerin.

Surface Tension: 0.1% Solution 28.7 dynes/cm

Flash Point: Nonflammable

RTECS Toxicity Data

Acute Oral: Rat LD_{50} Dose: 1900 mg/kg. Mouse LD_{50} Dose: 2643 mg/kg; Toxic Effects: Behavioral - Somnolence (general depressed activity).

Chronic (Multiple Dose) Oral: Rat Dose: 42 gm/kg/12W-C; Toxic Effects: Blood - Changes in serum composition; Biochemical - Lipids including transport. Rat Dose: 12 gm/kg/30D-I; Toxic Effects: DEATH. Rat Dose: 84 mg/kg/24W-I; Toxic Effects: Behavioral - Food intake (animal); Gastrointestinal - Hypermotility, diarrhea; DEATH.

Irritation Eye: Rabbit Standard Draize Test Dose: 1%; Reaction: severe. Rabbit Standard Draize Test Dose: 250 ug; Reaction: mild.

Irritation Skin: Rabbit Standard Draize Test Dose: 10 mg/24H; Reaction: moderate.

Reproductive/Teratogenic: Rat Route: Oral; Dose: 8820 mg/kg; Duration: female multigeneration; Effects on Newborn - Other postnatal measures or effects.

Hazard Overviews

Health: Severely irritating to eyes/skin/respiratory tract. Other Acute Effects: may be harmful by inhalation, ingestion, or skin absorption; eye contact can cause permanent injury; exposure can cause drowsiness; slurred speech; headache; nausea; dizziness; stupor; unconsciousness; target organs: central nervous system, small intestine, large intestine.

Fire: Noncombustible. Hazards: emits toxic fumes; explosive mixtures may form by compounding with strong oxidizing and reducing agents; vapor may travel considerable distance to source of ignition and flash back. Extinguishing agents: water spray; carbon dioxide, dry chemical powder or appropriate foam. Precautions: combustible liquid.

Reactivity: Stable. Hazardous polymerization will not occur. Incompatible with: strong oxidizing agents. Hazardous decomposition products: toxic fumes of: carbon monoxide, carbon dioxide, sulfur oxides.

Carcinogenicity: IARC - Not listed; NIOSH - Not listed; NTP - Not listed; ACGIH - Not listed; OSHA - Not listed; EPA - Not listed; MAK - Not listed

Primary Target Organs:

Eyes

Skin

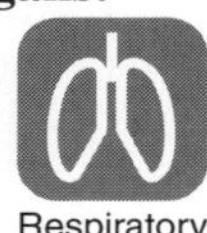
Respiratory System

Gastro-intestinal

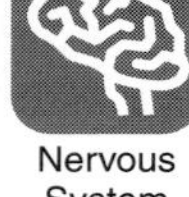
Nervous System

Environmental

Environmental Fate: If released to soil, the expected mobility is low to very high. If released to water, it will be essentially nonvolatile. It has aqueous base-catalyzed hydrolysis half-lives of 243 days at pH of 8 and 6.7 years at pH of 7. Aquatic bioconcentration is not expected to be an important fate process although adsorption to sediment may be possible. Several studies have shown that it biodegrades rapidly. If released to the atmosphere, it will exist primarily in the particulate phase. In the vapor phase, it will degrade in the atmosphere by reaction with photochemically produced hydroxyl radicals with an estimated half-life of 18 hours. Physical removal from air can occur through wet and dry deposition.

Environmental Physical Data

Henry's Law Constant: estimated at 5×10^{-12}

Sorption Partition Coefficient: K_{oc} = 1041

BCF: estimated at 1.13

Regulations

RCRA 40CFR: Not listed

CERCLA: 40CFR 302.4: Not listed

SARA 40CFR 372.65: Not listed

SARA EHS 40CFR 355: Not listed

TSCA: Listed

BIS5400 **CAS #: 6422-86-2**

BIS(2-ETHYLHEXYL) TEREPHTHALATE

RTECS: WZ0883500
EINECS Number: 229-176-9
Molecular Formula: $C_{24}H_{38}O_4$
Formula Weight: 390.54

Chemical Structure

Synonyms: 1,4-BENZENEDICARBOXYLIC ACID,BIS(2-ETHYLHEXYL) ESTER; 1,4-BENZENEDICARBOXYLIC ACID,BIS(2-ETHYLHEXYL)ESTER (9CI); KODAFLEX DOTP; TEREPHTHALIC ACID,BIS(2-ETHYLHEXYL) ESTER
Use: softener for nitrile-butadiene & chloroprene rubber; softener for isobutylene-isoprene rubber; plasticizer compatible with cellulose acetate-butyrate, cellulose nitrate, poly(methyl methacrylate), polystyrene, poly(vinyl butyral), and poly(vinyl chloride) resins

Physical Properties

Boiling Point: 383 °C (721 °F) at 760 mm Hg
Freezing Point: -48 °C (-54.4 °F)
Specific Gravity: 0.9835 at 20/20 °C
Vapor Density: 13.5
Vapor Pressure: 1 mm Hg at 217 °C
Water Solubility: 4 mg/L at 20 °C
Refraction Index: 1.4867 at 25 °C
Flash Point: 238 °C Open Cup

RTECS Toxicity Data

Acute Oral: Mouse LD_{Lo} Dose: 20 gm/kg; Toxic Effects: Behavioral - Somnolence (general depressed activity); Behavioral - Excitment.

Hazard Overviews

Health: Irritating to eyes/skin. Other Acute Effects: may be harmful by inhalation, ingestion, or skin absorption.
Fire: Will burn. Hazards: emits toxic fumes. Extinguishing agents: water spray; carbon dioxide, dry chemical powder or appropriate foam. Precautions: combustible liquid.
Reactivity: Incompatible with: strong oxidizing agents. Hazardous decomposition products: toxic fumes of: carbon monoxide, carbon dioxide.
Carcinogenicity: IARC - Not listed; NIOSH - Not listed; NTP - Not listed; ACGIH - Not listed; OSHA - Not listed; EPA - Not listed; MAK - Not listed
Primary Target Organs:

Eyes

Skin

Environmental

Environmental Fate: If released to the atmosphere, it will exist in both the vapor phase and the particulate phase based on an estimated vapor pressure of 2.14 x10^{-5} mm Hg at 25 °C. In the vapor phase, it should react rapidly with hydroxyl radicals with an estimated half-life of 17.5 hours. Particulate-phase may be removed physically from air by wet and dry deposition. It is expected to be essentially immobile in soil given an estimated K_{oc} of 870,000. Biodegradation in soil is expected to be a major fate process based on results for a similar compound, bis(2-ethylhexyl) phthalate; this compound undergoes aerobic and possibly anaerobic biodegradation. It may volatilize from moist soil surfaces with an estimated Henry's Law constant of 1.02 x10^{-5} atm-cu m/mole. In water, it is expected to bind tightly to particulate matter and sediment in the water column based on its estimated K_{oc} value. It is expected to volatilize slowly from water surfaces based on an estimated Henry's Law constant of 1.02 x10^{-5} atm-cu/mole. The volatilization half-life from a model river was calculated as 7.3 days, from a model lake, 59 days. Biodegradation may also be an important fate process for this compound in the water column. Based on an estimated BCF value of 1,400,000, it should bioconcentrate in aquatic organism. However, a structurally similar compound, bis(2-ethylhexyl) phthalate, had a measured BCF value of 637 with a depuration half-life of 38 days in sheepshead minnows therefore, may be readily metabolized by some organisms.

Environmental Physical Data

Henry's Law Constant: estimated at 1.02 x10^{-5}
Octanol/Water Partition Coefficient: log K_{ow} = 8.39
Sorption Partition Coefficient: K_{oc} = 8.7 x10^{5}
BCF: calculated at 1.4 x10^{6}

Regulations

RCRA 40CFR: Not listed
CERCLA: 40CFR 302.4: Not listed
SARA 40CFR 372.65: Not listed
SARA EHS 40CFR 355: Not listed
TSCA: Listed

BIS5510 **CAS #: 119-47-1**

BIS(2-HYDROXY-3-TERT-BUTYL-5-METHYLPHENYL)METHANE

RTECS: PA3500000
EINECS Number: 204-327-1
Molecular Formula: $C_{23}H_{32}O_2$
Formula Weight: 340.51

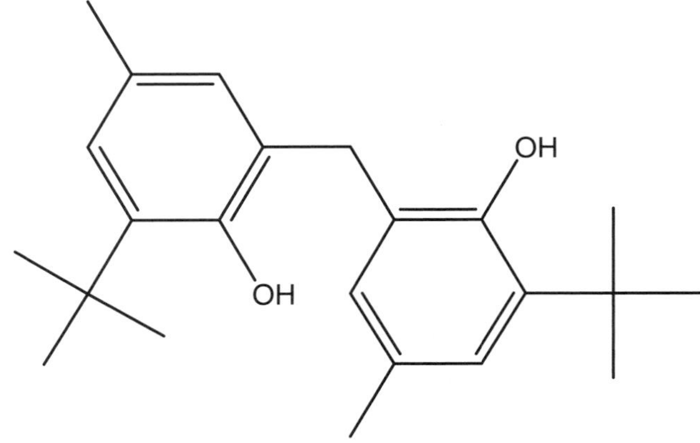

Chemical Structure

Synonyms: 2246; A 22-46; ADVASTAB 405; ANTAGE W 400; ANTI OX; ANTIOXIDANT 1; ANTIOXIDANT 2246; ANTIOXIDANT BKF; ANTIOXIDANT NG-2246; ANTIOXIDANT NG 2246; AO 1; AO 1 (ANTIOXIDANT); AO 2246; AO1; BISAKLOFEN BP; BISALKOFEN BP; BIS(6-HYDROXY-3-METHYL-5-TERT-BUTYLPHENYL)METHANE; 2,2'-BIS-6-TERC.BUTYL-P-KRESYLMETHAN; BKF; CALCO 2246; CAO 14; CAO 5; CATOLIN 14; CHEMANOX 21; P-CRESOL,2,2'-METHYLENEBIS(6-TERT-BUTYL-; CYANOX 2246; LEDERLE 2246; METHANE,2,2'-BIS(6-T-BUTYL-P-CRESYL)-; 2,2'-METHYLENE-BIS(6-TERT-BUTYL-4-METHYLPHENOL); 2,2''-METHYLENEBIS(4-METHYL-6-TERT-BUTYLPHENOL); 2,2'-METHYLENEBIS(4-METHYL-6-TERT-BUTYLPHENOL); NG 2246; NOCRAC NS 6; OXY CHEK 114; PHENOL,2,2'-METHYLENEBIS(6-(1,1-DIMETHYLETHYL)-4-METHYL-; PLASTANOX 2246; SUMILIZER MDP; SYNOX 5LT; VULKANOX BKF
Description: off-white powder
Use: stabilizer used in styrenic and olefin polymers and polyoxymethylene homo and copolymers; antioxidant (ABS, polypropylene, polyacetal, rubber, latex, adhesives)

Physical Properties

Freezing Point: 118 °C (244.4 °F) to 128 °C (262.4 °F)
Density: 1.07 to 1.1 kg/L
Water Solubility: 0 mg/L
Other Solubilities: Soluble in oxygen and aromatic solvents.

RTECS Toxicity Data

Acute Oral: Rat LD_{Lo} Dose: 10 gm/kg. Mouse LD_{50} Dose: 11 gm/kg.
Chronic (Multiple Dose) Oral: Rat Dose: 1623 mg/kg/1W-C; Toxic Effects: Liver - Other changes; Blood - Changes in other cell count; Biochemical - Lipids including transport. Rat Dose: 25200 mg/kg/12W-C; Toxic Effects: Blood - Pigmented or nucleated red Blood cells; Nutritional and gross metabolic - Weight loss or decreased weight gain; Biochemical - Lipids including transport. Rat Dose: 23096 mg/kg/78W-C; Toxic Effects: Liver - Changes in liver weight; Nutritional and gross metabolic - Weight loss or decreased weight gain; DEATH - Changes in testicular weight.
Irritation Eye: Rabbit Standard Draize Test Dose: 100 mg/24H; Reaction: moderate.

Hazard Overviews

Health: Irritating eyes/skin. Other Acute Effects: may be harmful by inhalation, ingestion, or skin absorption.
Fire: Hazards: emits toxic fumes; in powder form capable of creating a dust explosion. Extinguishing agents: water spray; carbon dioxide, dry chemical powder or appropriate foam. Precautions: combustible liquid.
Reactivity: Stable. Hazardous polymerization will not occur. Incompatible with: strong oxidizing agents. Hazardous decomposition products: toxic fumes of: carbon monoxide, carbon dioxide.
Carcinogenicity: IARC - Not listed; NIOSH - Not listed; NTP - Not listed; ACGIH - Not listed; OSHA - Not listed; EPA - Not listed; MAK - Not listed
Primary Target Organs:

Eyes

Skin

Environmental

Environmental Fate: If released to soil, it will be immobile. Volatilization should not be important from moist or dry soil. Based on a MITI biodegradation test, it will not biodegrade in soil. If released to water, it will adsorb to suspended solids and sediment. It may not volatilize from water surfaces based on an estimated Henry's Law constant of 7.9 $x10^{-12}$ atm-cu m/mole. Experimental BCF values of 23-37 and 60-125 suggest that it will bioconcentrate in aquatic organisms. Based on a MITI biodegradation test, it will not biodegrade in water. If released to the atmosphere, it will exist in the particulate phase. Particulate-phase may be physically removed from the air by wet and dry deposition.

Environmental Physical Data

Henry's Law Constant: estimated at 7.9 $x10^{-12}$
Octanol/Water Partition Coefficient: log K_{ow} = 6.25
Sorption Partition Coefficient: K_{oc} = 5.9,8 $x10^4$
BCF: estimated at 3.31 $x10^4$

Regulations

RCRA 40CFR: Not listed
CERCLA: 40CFR 302.4: Not listed
SARA 40CFR 372.65: Not listed
SARA EHS 40CFR 355: Not listed
TSCA: Listed

BIS5620 CAS #: 959-26-2

BIS(2-HYDROXYETHYL) TEREPHTHALATE

EINECS Number: 213-497-6
Molecular Formula: $C_{12}H_{14}O_6$
Formula Weight: 254.24
Synonyms: 1,4-BENZENEDICARBOXYLIC ACID,BIS(2-HYDROXYETHYL) ESTER; BIS(ETHYLENE GLYCOL) TEREPHTHALATE; BIS(BETA-HYDROXYETHYL) TEREPHTHALATE; BIS(HYDROXYETHYL) TEREPHTHALATE; TEREPHTHALIC ACID,BIS(2-HYDROXYETHYL) ESTER
Use: captive chem int in mfr of polyethylene terephthalate

Hazard Overviews

Health: Irritating to eyes/skin/respiratory tract. Other Acute Effects: may be harmful by inhalation, ingestion, or skin absorption.
Fire: Hazards: in powder form capable of creating a dust explosion; emits toxic fumes. Extinguishing agents: water spray; carbon dioxide, dry chemical powder or appropriate foam. Precautions: combustible liquid.
Carcinogenicity: IARC - Not listed; NIOSH - Not listed; NTP - Not listed; ACGIH - Not listed; OSHA - Not listed; EPA - Not listed; MAK - Not listed
Primary Target Organs:

Eyes

Skin

Respiratory System

Environmental

Regulations

RCRA 40CFR: Not listed
CERCLA: 40CFR 302.4: Not listed
SARA 40CFR 372.65: Not listed
SARA EHS 40CFR 355: Not listed
TSCA: Listed

BIS5730 CAS #: 70955-14-5

N,N-BIS(2-HYDROXYETHYL)ALLYLAMINE

Physical Properties

Specific Gravity: 0.91
Water Solubility: Dispersible

Hazard Overviews

Carcinogenicity: IARC - Not listed; NIOSH - Not listed; NTP - Not listed; ACGIH - Not listed; OSHA - Not listed; EPA - Not listed; MAK - Not listed

Environmental

Regulations

RCRA 40CFR: Not listed
CERCLA: 40CFR 302.4: Not listed
SARA 40CFR 372.65: Not listed
SARA EHS 40CFR 355: Not listed
TSCA: Not listed

BIS5840 CAS #: 23746-34-1

BIS(2-HYDROXYETHYL)DITHIOCARBAMIC ACID, POTASSIUM SALT

RTECS: EY9450000
Molecular Formula: $C_5H_{11}KNO_2S_2$
Formula Weight: 220.38
Synonyms: BIS(2-HYDROXYETHYL)CARBAMODITHIOIC ACID,MONOPOTASSIUM SALT; BIS(2-HYDROXYETHYL)DITHIOCARBAMIC ACID,MONOPOTASSIUM SALT; CARBAMIC ACID,BIS(2-HYDROXYETHYL)DITHIO-,MONOPOTASSIUMSALT; CARBAMODITHIOIC ACID,BIS(2-HYDROXYETHYL)-,MONOPOTASSIUMSALT; POTASSIUM BIS(2-HYDROXYETHYL)DITHIOCARBAMATE
Use: analytical reagent for quantitative determination of mercury, palladium, gold, copper, cobalt, & nickel; research chemical

RTECS Toxicity Data

Tumorigenic: Rat Route: Oral; Dose: 82 gm/kg/78W-C; Toxic Effects: Tumorigenic - Equivocal tumorigenic agent by RTECS criteria; Liver - Tumors. Mouse Route: Oral; Dose: 129 gm/kg/79W-C; Toxic Effects: Tumorigenic - Carcinogenic by RTECS criteria; Lungs, Thorax, or Respiration - Tumors; Liver - Tumors.

Hazard Overviews

Carcinogenicity: IARC - Group 3, Not classifiable as to carcinogenicity to humans; NIOSH - Not listed; NTP - Not listed; ACGIH - Not listed; OSHA - Not listed; EPA - Not listed; MAK - Not listed

Environmental

Regulations

RCRA 40CFR: Not listed
CERCLA: 40CFR 302.4: Not listed
SARA 40CFR 372.65: Not listed
SARA EHS 40CFR 355: Not listed
TSCA: Listed

BIS5950 **CAS #: 2948-46-1**

P-BIS(2-HYDROXYISOPROPYL)BENZENE

EINECS Number: 220-964-8
Molecular Formula: $C_{12}H_{18}O_2$
Formula Weight: 194.27

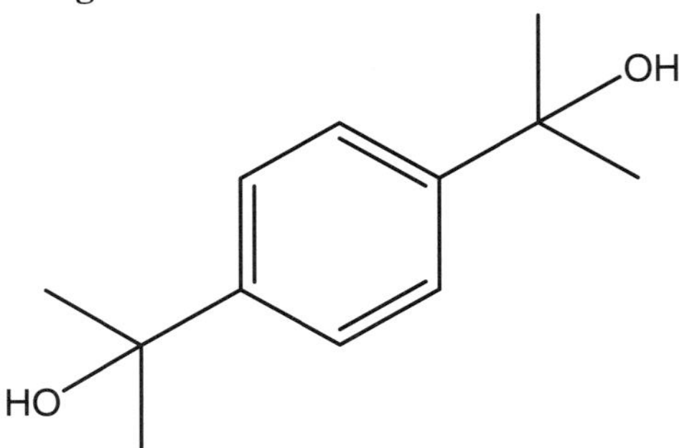

Chemical Structure

Synonyms: 1,4-BENZENEDIMETHANOL,ALPHA,ALPHA,ALPHA',ALPHA'-TETRAMETHYL-; P-BIS(ALPHA-HYDROXYISOPROPYL)BENZENE; 1,4-BIS(2-HYDROXY-2-PROPYL)BENZENE; ALPHA,ALPHA'-DIHYDROXY-P-DIISOPROPYLBENZENE; ALPHA,ALPHA,ALPHA',ALPHA'-TETRAMETHYL-P-BENZENEDIMETHANOL; ALPHA,ALPHA,ALPHA',ALPHA'-TETRAMETHYL-P-XYLYLENEDIOL; P-XYLENE-ALPHA,ALPHA'-DIOL,ALPHA,ALPHA,ALPHA',ALPHA'-TETRAMETHYL-

Hazard Overviews

Health: Irritating to eyes/skin/respiratory tract. Other Acute Effects: may be harmful by inhalation, ingestion, or skin absorption.
Fire: Extinguishing agents: water spray; carbon dioxide, dry chemical powder or appropriate foam. Precautions: combustible liquid.
Reactivity: Incompatible with: acids, acid chlorides, acid anhydrides, oxidizing agents. Hazardous decomposition products: carbon monoxide, carbon dioxide.
Carcinogenicity: IARC - Not listed; NIOSH - Not listed; NTP - Not listed; ACGIH - Not listed; OSHA - Not listed; EPA - Not listed; MAK - Not listed
Primary Target Organs:

Eyes

Skin

Respiratory System

Environmental

Regulations

RCRA 40CFR: Not listed
CERCLA: 40CFR 302.4: Not listed
SARA 40CFR 372.65: Not listed
SARA EHS 40CFR 355: Not listed
TSCA: Listed

BIS6060 **CAS #: 1779-25-5**

BIS(ISOBUTYL)ALUMINUM CHLORIDE

RTECS: BD0560000
EINECS Number: 217-216-8
Molecular Formula: $C_8H_{18}AlCl$
Formula Weight: 176.69

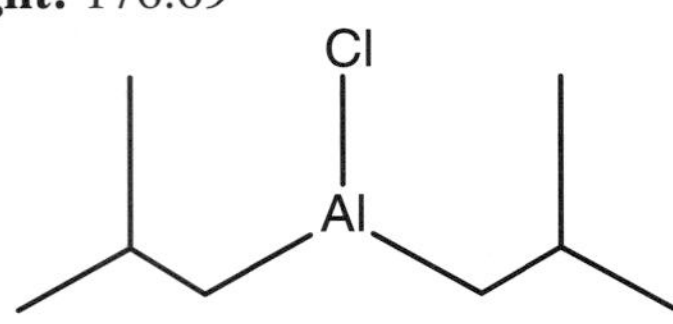

Chemical Structure

Synonyms: ALLUMINIO DIISOBUTIL-MONOCLORURO; ALUMINUM,CHLOROBIS(2-METHYLPROPYL)-; ALUMINUM,CHLOROBIS(2-METHYLPROPYL)-(9CI); ALUMINUM,CHLORODIISOBUTYL-; CHLOROBIS(2-METHYLPROPYL)ALUMINUM; CHLORODIISOBUTYLALUMINUM; DIBAC; DIISOBUTYLALUMINUM CHLORIDE; DIISOBUTYLALUMINUM MONOCHLORIDE; DIISOBUTYLCHLOROALUMINUM
Description: colorless liquid
Use: polyolefin catalyst; polymerization catalyst for E-P-D elastomers

Physical Properties

Boiling Point: 152 °C (305.6 °F)
Freezing Point: -39.5 °C (-39.1 °F)
Density: 0.905 g/mL
Bulk Density: 0.905
Flash Point: Same as solvent

RTECS Toxicity Data

Acute Inhalation: Rat LC_{50} Dose: 67 ppm/1hr. Mouse LC_{Lo} Dose: 680 gm/kg/15M; Toxic Effects: Lungs, Thorax, or Respiration - Acute pulmonary edema; Lungs, Thorax, or Respiration - Dyspnea; Liver - Other changes.

Hazard Overviews

Poison

Corrosive

Health: Corrosive to eyes/skin/respiratory tract. Poison. Other Acute Effects: may be fatal if inhaled, swallowed, or absorbed through skin; inhalation may result in spasm, inflammation and edema of the larynx and bronchi, chemical pneumonitis and pulmonary edema; symptoms of exposure may include burning sensation; coughing; wheezing; laryngitis; shortness of breath; headache; nausea; vomiting.

Fire: Hazards: reacts with water to liberate flammable and/or explosive gas; catches fire if exposed to air. Extinguishing agents: dry chemical powder; do not use water. Precautions: combustible liquid.
Reactivity: Incompatible with: oxygen, oxidizing agents, alcohols, acids, reacts violently with water. Hazardous decomposition products: toxic fumes of: hydrogen chloride gas, aluminum oxide, carbon monoxide, carbon dioxide.
Carcinogenicity: IARC - Not listed; NIOSH - Not listed; NTP - Not listed; ACGIH - Not listed; OSHA - Not listed; EPA - Not listed; MAK - Not listed
Primary Target Organs:

Eyes

Skin

Respiratory System

Exposure Limits
OSHA PEL Vacated 1989 Limits: TWA: 2 mg/m^3; as Al soluble.
ACGIH TLV: TWA: 2 mg/m^3; as Al.
NIOSH REL: TWA: 2 mg/m^3; as Al soluble salts, alkyls.

Environmental

Regulations

RCRA 40CFR: Not listed
CERCLA: 40CFR 302.4: Not listed
SARA 40CFR 372.65: Not listed
SARA EHS 40CFR 355: Not listed
TSCA: Listed

BIS6170 CAS #: 28109-00-4

BIS(ISOPROPYLPHENYL) PHENYL PHOSPHATE

EINECS Number: 248-849-8
Molecular Formula: $C_{24}H_{27}O_4P$
Formula Weight: 410.45
Synonyms: DI(ISOPROPYLPHENYL) PHENYL PHOSPHATE; PHOSPHORIC ACID,BIS((1-METHYLETHYL)PHENYL) PHENYL ESTER(9CI)

Hazard Overviews

Carcinogenicity: IARC - Not listed; NIOSH - Not listed; NTP - Not listed; ACGIH - Not listed; OSHA - Not listed; EPA - Not listed; MAK - Not listed

Environmental

Regulations

RCRA 40CFR: Not listed
CERCLA: 40CFR 302.4: Not listed
SARA 40CFR 372.65: Not listed
SARA EHS 40CFR 355: Not listed
TSCA: Listed

BIS6280 CAS #: 117-82-8

BIS(2-METHOXYETHYL) PHTHALATE

RTECS: TI1400000
EINECS Number: 204-212-6
Molecular Formula: $C_{14}H_{18}O_6$
Formula Weight: 282.32

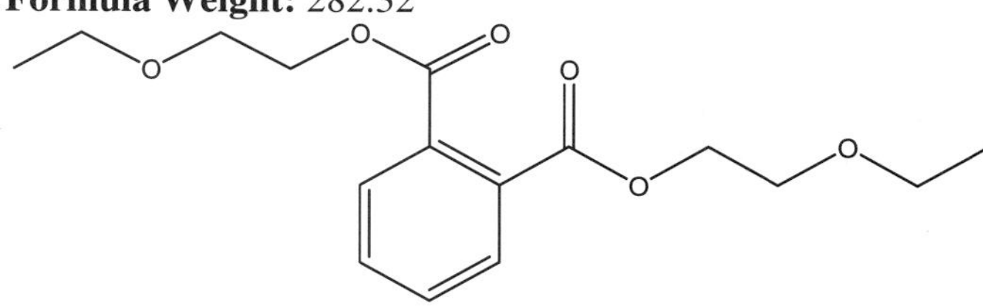

Chemical Structure

Synonyms: 1,2-BENZENEDICARBOXYLIC ACID,BIS(2-METHOXYETHYL) ESTER; 1,2-BENZENEDICARBOXYLIC ACID,BIS(2-METHOXYETHYL) ESTER(9CI); BIS(METHOXYETHYL) PHTHALATE; DIMETHOXY ETHYL PHTHALATE; DI-(2-METHOXYETHYL) ESTER KYSELINY FTALOVE; DI(2-METHOXYETHYL) PHTHALATE; DI-(2-METHOXYETHYL)ESTER KYSELINY FTALOVE; DI(2-METHOXYETHYL)PHTHALATE; DIMETHYL CELLOSOLVE PHTHALATE; DIMETHYL GLYCOL PHTHALATE; DIMETHYLGLYCOL PHTHALATE; DMEP; KESSCOFLEX MCP; KODAFLEX DMEP; METHOX; 2-METHOXYETHYL PHTHALATE; METHYL GLYCOL PHTHALATE; PHTHALIC ACID,BIS(2-METHOXYETHYL) ESTER; PHTHALIC ACID,DI(2-METHOXYETHYL) ESTER; PHTHALIC ACID,DI(METHOXYETHYL) ESTER
Description: practically colorless, oily liquid; very slight odor
Use: plasticizer for cellulose acetate; solvent; molding compositions; adhesives; laminating cements; flash bulb lacquers

Physical Properties

Boiling Point: 352 °C (666 °F) at 760 mm Hg
Freezing Point: -45 °C (-49 °F)
Specific Gravity: 1.1708 at 15 °C
Vapor Pressure: Extremely low
Water Solubility: 0.8% at 20 °C
Other Solubilities: miscible in all proportions in absolute Alcohol; Insoluble in mineral oils.
Refraction Index: 1.5020
Flash Point: 210 °C Open Cup
Autoignition Temperature: 399 °C
LEL: 0.7% v/v

RTECS Toxicity Data

Acute Oral: Rat LD_{Lo} Dose: 2750 mg/kg. Mouse LD_{50} Dose: 3200 mg/kg.
Acute Inhalation: Rat LC_{Lo} Dose: 1595 ppm/6hr; Toxic Effects: Sense organs and special senses - Other.
Acute Dermal: Guinea Pig LD_{50} Route: Skin; Dose: >20 gm/kg.
Irritation Eye: Rabbit Standard Draize Test Dose: 100 mg; Reaction: mild.

Irritation Skin: Guinea Pig Standard Draize Test Dose: 500 mg; Reaction: mild.

Reproductive/Teratogenic: Rat Route: Oral; Dose: 1500 mg/kg; Duration: male 1D prior to mating; Paternal Effects - Spermatogenesis; Testes, epididymis, sperm duct. Rat Route: Oral; Dose: 593 mg/kg; Duration: female 10D of pregnancy; Specific Developmental Abnormalities - Central nervous system; Eye, ear; Cardiovascular (circulatory) system. Rat Route: Oral; Dose: 593 mg/kg; Duration: female 13D of pregnancy; Specific Developmental Abnormalities - Musculoskeletal system; Urogenital system.

Mutagenic: Rat Sperm Morphology; Route: Oral; Dose: 1500 mg/kg. Mouse Dominant Lethal Test; Route: Intraperitoneal; Dose: 1190 mg/kg.

Hazard Overviews

Health: Irritating to eyes/skin/respiratory tract. Toxic. Accute Effects: harmful if swallowed, inhaled, or absorbed through skin. Chronic Effects: may cause congenital malformation in the fetus; may cause reproductive disorders; possible risk of impaired fertility.

Fire: Will burn. Hazards: emits toxic fumes. Extinguishing agents: carbon dioxide, dry chemical powder or appropriate foam. Precautions: combustible liquid.

Reactivity: Incompatible with: strong oxidizing agents. Hazardous decomposition products: toxic fumes of: carbon monoxide, carbon dioxide.

Carcinogenicity: IARC - Not listed; NIOSH - Not listed; NTP - Not listed; ACGIH - Not listed; OSHA - Not listed; EPA - Not listed; MAK - Not listed

Primary Target Organs:

Eyes

Skin

Respiratory System

Environmental

Environmental Fate: If released to soil, it is expected to display high mobility. Volatilization from the soil surface to the atmosphere is not expected to be an important fate process. If released to water, it is not expected to volatilization to the atmosphere, adsorb to sediment and suspended organic matter, bioconcentrate in fish and aquatic organisms, or hydrolyze. No data on biodegradation could be located. If released to the atmosphere, it is expected to be adsorbed to particulates. Destruction by the vapor phase reaction with photochemically produced hydroxyl radicals may occur.

Environmental Physical Data

Henry's Law Constant: estimated at 2.8 $\times 10^{-13}$

Octanol/Water Partition Coefficient: log K_{ow} = 0.989

Sorption Partition Coefficient: K_{oc} = estimated at 82

BCF: none likely

Regulations

RCRA 40CFR: Not listed

CERCLA: 40CFR 302.4: Not listed

SARA 40CFR 372.65: Not listed

SARA EHS 40CFR 355: Not listed

TSCA: Listed

BIS6390 **CAS #: 3001-61-4**

1,3-BIS(METHOXYMETHYL)-4,5-DIHYDROXY CYCLIC ETHYLENEUREA

EINECS Number: 221-082-6

Molecular Formula: $C_7H_{14}N_2O_5$

Formula Weight: 206.23

Synonyms: 2-IMIDAZOLIDINONE,4,5-DIHYDROXY-1,3-BIS(METHOXYMETHYL)-

Hazard Overviews

Carcinogenicity: IARC - Not listed; NIOSH - Not listed; NTP - Not listed; ACGIH - Not listed; OSHA - Not listed; EPA - Not listed; MAK - Not listed

Environmental

Regulations

RCRA 40CFR: Not listed

CERCLA: 40CFR 302.4: Not listed

SARA 40CFR 372.65: Not listed

SARA EHS 40CFR 355: Not listed

TSCA: Listed

BIS6500 **CAS #: 141-07-1**

1,3-BIS(METHOXYMETHYL)UREA

EINECS Number: 205-454-5

Molecular Formula: $C_5H_{12}N_2O_3$

Formula Weight: 148.06

Synonyms: BIS(METHOXYMETHYL)UREA; N,N'-BIS(METHOXYMETHYL)UREA; 1,3-DIMETHOXYMETHYLUREA; DIMETHYLOLUREA DIMETHYL ETHER; N,N'-DIMETHYLOLUREA DIMETHYL ETHER; KAURIT W; UREA,1,3-BIS(METHOXYMETHYL)-; UREA,N,N'-BIS(METHOXYMETHYL)-

Hazard Overviews

Carcinogenicity: IARC - Not listed; NIOSH - Not listed; NTP - Not listed; ACGIH - Not listed; OSHA - Not listed; EPA - Not listed; MAK - Not listed

Environmental

Regulations

RCRA 40CFR: Not listed

CERCLA: 40CFR 302.4: Not listed

SARA 40CFR 372.65: Not listed

SARA EHS 40CFR 355: Not listed

TSCA: Listed

BIS6610 CAS #: 845-52-3

2,4-BIS((3-METHOXYPROPYL)AMINO)-6-(METHYLTHIO)-S-TRIAZINE

RTECS: XY4395000
DOT: UN2763
Molecular Formula: $C_{12}H_{23}N_5O_2S$
Formula Weight: 301.42
Synonyms: 2,4-BIS(3-METHOXYPROPYLAMINO)-6-METHYLTHIO-S-TRIAZINE; N,N'-BIS(3-METHOXYPROPYL)-6-METHYLTHIO-1,3,5-TRIAZINE-2,4-DIAMINE; CP 17029; EPA PESTICIDE CHEMICAL CODE 082701; LAMBAST; 2-METHYLMERCAPTO-4,6-BIS(3-METHOXYPROPYLAMINO)-S-TRIAZINE; MPMT; S-TRIAZINE,2,4-BIS((3-METHOXYPROPYL)AMINO)-6-(METHYLTHIO)-; 1,3,5-TRIAZINE-2,4-DIAMINE,N,N'-BIS(3-METHOXYPROPYL)-6-(METHYLTHIO)-
Description: white crystals or solid
Use: selective herbicide

Physical Properties

Freezing Point: 55 °C (131 °F) to 59 °C (138.2 °F)
Water Solubility: Insoluble in Water
Other Solubilities: Very Soluble in Acetone & Benzene. Slightly Soluble in Ethanol

RTECS Toxicity Data

Acute Oral: Rat LD_{50} Dose: 1400 mg/kg.

Hazard Overviews

Carcinogenicity: IARC - Not listed; NIOSH - Not listed; NTP - Not listed; ACGIH - Not listed; OSHA - Not listed; EPA - Not listed; MAK - Not listed

Environmental

Environmental Fate: If released to soil, it will be expected to exhibit high mobility based upon a K_{oc} estimated from am estimated log Kow and it may, therefore, leach to groundwater. It may be susceptible to hydrolysis in soil and biodegradation based upon data for s-triazine herbicides with similar structures such as prometryn. If released to water, it will not be expected to bioconcentrate in aquatic organisms, adsorb to sediment and suspended particulate matter, or to volatilize based upon an estimated BCF, an estimated K_{oc}, and an estimated Henry's law constant, respectively. Information concerning biodegradation in soil of other s-triazine herbicides such as terbutryne indicates that biodegradation may occur in natural waters. It may be susceptible to slow hydrolysis in water or soil and photooxidation by photochemically produced hydroxyl radicals in water, based upon data for s-triazine herbicides with similar structures such as prometryn. It may be susceptible to direct photolysis based upon the observed degradation of 2-triazines similar in structure such as prometryn in water solutions irradiated with artificial light at wavelengths >290 nm. If released to the atmosphere, it may be subject to vapor-phase reaction with photochemically produced hydroxyl radicals with a half-life of about 2.4 hr estimated for this process.
Cleanup/Disposal: Guide No. 151: Do not touch damaged containers or spilled material unless wearing appropriate protective clothing. Stop leak if you can do it without risk. Prevent entry into waterways, sewers, basements or confined areas. Cover with plastic sheet to prevent spreading. Absorb or cover with dry earth, sand or other non-combustible material and transfer to containers. Do not get water inside containers.

Environmental Physical Data

Henry's Law Constant: calculated at 6.62×10^{-9}
Octanol/Water Partition Coefficient: log K_{ow} = 2.0
Sorption Partition Coefficient: K_{oc} = estimated at 74
BCF: estimated at 19

Regulations

RCRA 40CFR: Not listed
CERCLA: 40CFR 302.4: Not listed
SARA 40CFR 372.65: Not listed
SARA EHS 40CFR 355: Not listed
TSCA: Not listed

BIS6720 CAS #: 10024-74-5

BIS(A-METHYLBENZYL)AMINE

RTECS: HQ7000000
Molecular Formula: $C_{16}H_{19}N$
Formula Weight: 225.34

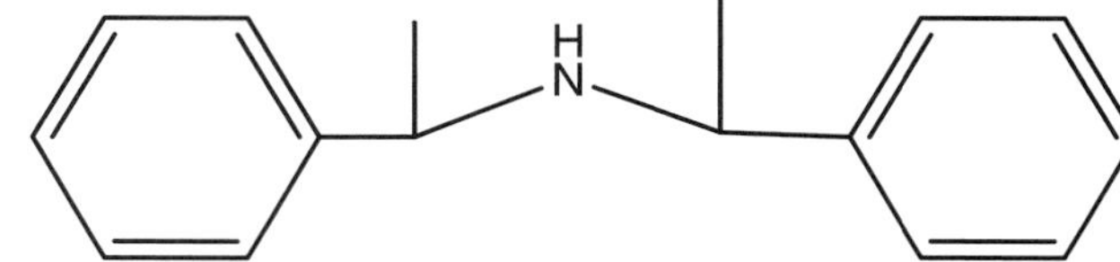

Chemical Structure

Synonyms: AMINE,DIETHYL,1,1'-DIPHENYL; BENZENEMETHANAMINE,ALPHA-METHYL-N-(1-PHENYLETHYL)-; BIS(ALPHA-METHYLBENZYL)AMINE; DIBENZYLAMINE,ALPHA,ALPHA'-DIMETHYL-
Description: yellow crystals

Physical Properties

Boiling Point: 298 °C (568 °F)
Specific Gravity: 1.018 at 13 °C
Water Solubility: Slightly Soluble in Water
Refraction Index: 1.573

RTECS Toxicity Data

Acute Oral: Rat LD_{50} Dose: 2930 mg/kg.
Acute Dermal: Rabbit LD_{50} Route: Skin; Dose: 3970 mg/kg.
Irritation Eye: Rabbit Standard Draize Test Dose: 500 mg/24H; Reaction: mild.
Irritation Skin: Rabbit Standard Draize Test Dose: 5 mg/24H; Reaction: severe.

Hazard Overviews

Health: Severely irritating to skin; irritating to eyes/respiratory tract. Other Acute Effects: may be harmful by inhalation, ingestion, or skin absorption.
Fire: Hazards: emits toxic fumes. Extinguishing agents: carbon dioxide, dry chemical powder or appropriate foam; water spray. Precautions: combustible liquid.
Reactivity: Incompatible with: strong oxidizing agents. Hazardous decomposition products: toxic fumes of: carbon monoxide, carbon dioxide, nitrogen oxides.
Carcinogenicity: IARC - Not listed; NIOSH - Not listed; NTP - Not listed; ACGIH - Not listed; OSHA - Not listed; EPA - Not listed; MAK - Not listed
Primary Target Organs:

Eyes

Skin

Respiratory System

Environmental

Regulations

RCRA 40CFR: Not listed
CERCLA: 40CFR 302.4: Not listed
SARA 40CFR 372.65: Not listed
SARA EHS 40CFR 355: Not listed
TSCA: Not listed

BIS7050 **CAS #: 97-39-2**

N,N'-BIS(2-METHYLPHENYL)GUANIDINE

RTECS: MF1400000
EINECS Number: 202-577-6
Molecular Formula: $C_{15}H_{17}N_3$
Formula Weight: 239.32

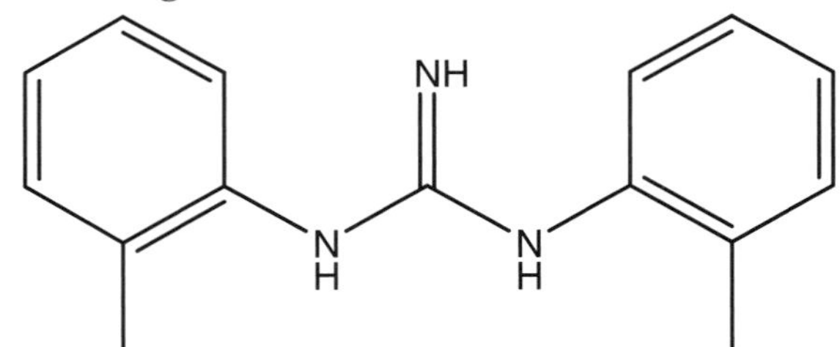

Chemical Structure

Synonyms: 1,3-BIS(O-TOLYL)GUANIDINE; DIORTHOTOLYLGUANIDINE; 1,3-DI-O-TOLYLGUANIDINE; DI-O-TOLYLGUANIDINE; N,N'-DI-O-TOLYLGUANIDINE; DOTG; DOTG ACCELERATOR; EVEITE DOTG; GUANIDINE,N,N'-BIS(2-METHYLPHENYL)-; GUANIDINE,1,3-DI(2-TOLYL)-; GUANIDINE,1,3-DI-O-TOLYL-; VULKACIT DOTG; VULKACIT DOTG/C
Description: white powder or crystals
Use: basic rubber accelerator

Physical Properties

Freezing Point: 179 °C (354.2 °F)
Specific Gravity: 1.1 at 20 °C/4 °C
Water Solubility: Slightly Soluble in Hot Water
Other Solubilities: Slightly Soluble in Alcohol. Very Soluble in Ether. Soluble in Chloroform

RTECS Toxicity Data

Acute Oral: Rat LD_{50} Dose: 500 mg/kg. Rabbit LD_{Lo} Dose: 80 mg/kg.

Hazard Overviews

Health: May cause irritation. Toxic. Other Acute Effects: toxic by inhalation, in contact with skin and if swallowed.
Fire: Extinguishing agents: water spray; carbon dioxide, dry chemical powder or appropriate foam. Precautions: combustible liquid.
Carcinogenicity: IARC - Not listed; NIOSH - Not listed; NTP - Not listed; ACGIH - Not listed; OSHA - Not listed; EPA - Not listed; MAK - Not listed

Environmental

Regulations

RCRA 40CFR: Not listed
CERCLA: 40CFR 302.4: Not listed
SARA 40CFR 372.65: Not listed
SARA EHS 40CFR 355: Not listed
TSCA: Listed

BIS7160 **CAS #: 110-96-3**

N,N-BIS(2-METHYLPROPYL)AMINE

RTECS: TX1750000
EINECS Number: 203-819-3
Molecular Formula: $C_8H_{19}N$
Structured MF: $[(CH_3)_2CHCH_2]_2NH$
Formula Weight: 129.25

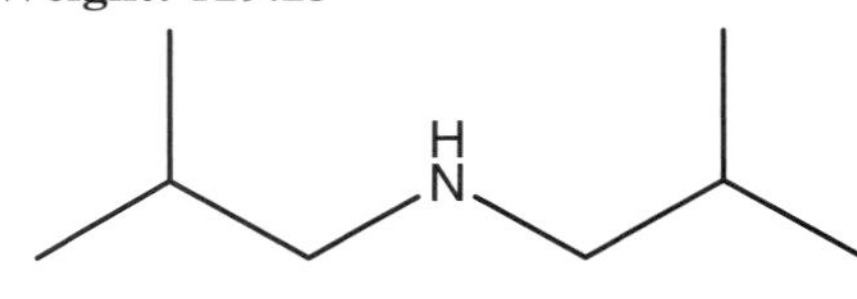

Chemical Structure

Synonyms: AMINE,DIISOBUTYL; BIS(2-METHYLPROPYL)AMINE; BIS(BETA-METHYLPROPYL)AMINE; DIISOBUTYLAMINE; 2-METHYL-N-(2-METHYLPROPYL)-1-PROPANAMINE; 1-PROPANAMINE,2-METHYL-N-(2-METHYLPROPYL)-
Description: water-white liquid; amine odor
Use: organic intermediate for synthesis

Physical Properties

Boiling Point: 140 °C (284 °F) at 760 mm Hg
Freezing Point: -73.5 °C (-100.3 °F)
Specific Gravity: 0.745 at 20 °C

Vapor Density: 4.46 Air=1
Density: 0.75 g/mL
Vapor Pressure: 0.194 kPa at 0 °C
Water Solubility: Slightly Soluble in Water
Other Solubilities: 95% Ethanol: Soluble; Acetone: Soluble; Benzene: Soluble; DMSO: Insoluble; Ether: Soluble.
Surface Tension: 22.58 dynes/cm
Refraction Index: 1.4090 at 20 °C/D
Critical Temperature: 281 °C
Critical Pressure: 370 psia
Ionization Potential (eV): 7.8 +/-2.0
Flash Point: 29 °C
Autoignition Temperature: 290 °C

RTECS Toxicity Data

Acute Oral: Rat LD_{50} Dose: 258 mg/kg. Mouse LD_{50} Dose: 629 mg/kg.

Hazard Overviews

Corrosive

Flammable

Health: Corrosive to eyes/skin/respiratory tract. Other Acute Effects: harmful if swallowed, inhaled, or absorbed through skin; inhalation may result in spasm, inflammation and edema of the larynx and bronchi, chemical pneumonitis and pulmonary edema; exposure can cause coughing; chest pains; difficulty in breathing.
Fire: Flammable. Hazards: vapor may travel considerable distance to source of ignition and flash back. Extinguishing agents: carbon dioxide, dry chemical powder or appropriate foam. Precautions: combustible liquid.
Reactivity: Incompatible with: acids, oxidizing agents. Hazardous decomposition products: thermal decomposition may produce carbon monoxide, carbon dioxide, and nitrogen oxides.
Carcinogenicity: IARC - Not listed; NIOSH - Not listed; NTP - Not listed; ACGIH - Not listed; OSHA - Not listed; EPA - Not listed; MAK - Not listed
Primary Target Organs:

Eyes

Skin

Respiratory System

Environmental

Ecotoxicity: Aquatic toxicity: 20-40 ppm/24 hr critical range/creek chub, minnows
Cleanup/Disposal: Guide No. 132: Fully encapsulating, vapor protective clothing should be worn for spills and leaks with no fire. Eliminate all ignition sources (no smoking, flares, sparks or flames in immediate area). All equipment used when handling the product must be grounded. Do not touch or walk through spilled material. Stop leak if you can do it without risk. Prevent entry into waterways, sewers, basements or confined areas. A vapor suppressing foam may be used to reduce vapors. Absorb with earth, sand or other non-combustible material and transfer to containers (except for Hydrazine). Use clean non-sparking tools to collect absorbed material. Large Spills: Dike far ahead of liquid spill for later disposal. Water spray may reduce vapor; but may not prevent ignition in closed spaces.

Environmental Physical Data

Octanol/Water Partition Coefficient: log K_{ow} = calculated at 2.84 to 3.04

Regulations

RCRA 40CFR: Not listed
CERCLA: 40CFR 302.4: Not listed
SARA 40CFR 372.65: Not listed
SARA EHS 40CFR 355: Not listed
TSCA: Listed

BIS7270 CAS #: 7440-69-9

BISMUTH

RTECS: EB2600000
EINECS Number: 231-177-4
Molecular Formula: Bi
Structured MF: BI
Formula Weight: 208.9804

Bi

Chemical Structure

Synonyms: BISMUTH-209
Description: silver-white, reddish powder; odorless
Use: in fire detection & extinguishing systems; catalyst for making acrylic fibers; in prodn of malleable irons, steels, etc; carrier for radioactive fuel; in type metal for printing; mfr of bismuth salts, fusible boiler plugs, electric fuses, solders; silvering mirrors; in dental technique; in fusible & other alloys; intermed for pharmaceuticals; wire in thermocouplers, galvanometers; hardener for metals; in cosmetics; permanent magnets; semiconductors; coating selenium; thermoelectric materials.

Physical Properties

Boiling Point: 1555 °C (2831 °F) to 1565 °C (2849 °F)
Freezing Point: 271 °C (519.8 °F)
Specific Gravity: 9.78 at 20 °C/4 °C
Vapor Pressure: 1 mm Hg at 1021 °C
Water Solubility: Insoluble in Water
Other Solubilities: Soluble in concentrated Hydrochloric Acid.
Ionization Potential (eV): 7.289
Flash Point: Flammable in powder form

RTECS Toxicity Data

Acute Oral: Rat LD_{50} Dose: 5 gm/kg. Mouse LD_{50} Dose: 10 gm/kg.

Hazard Overviews

Explosive

Flammable

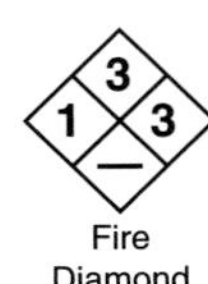

Fire Diamond

Health: May cause mechanical irritation of the eyes/respiratory tract. Chronic Effects: sleeplessness, foul breath, generalized joint pains, exfoliative dermatitis, bismuth line, and anorexia.
Fire: Flammable. Fight fire with dry chemical, carbon dioxide, water spray, or alcohol-resistant foam.
Reactivity: Stable. Hazardous polymerization cannot occur. Avoid: chloric acid; concentrated nitric acid; perchloric acid; fusing ammonium nitrate. Incompatible with: oxidizing agents; nitric, sulfuric, and hydrochloric acids; acids or acid fumes; aluminum; bromine trifluoride; nitrosyl fluoride; ammonium nitrate; chlorine; iodine pentafluoide; bismoth hydroxide and aluminum hydroxide.
Carcinogenicity: IARC - Not listed; NIOSH - Not listed; NTP - Not listed; ACGIH - Not listed; OSHA - Not listed; EPA - Not listed; MAK - Not listed
Primary Target Organs:

Eyes

Skin

Respiratory System

Mucous Membranes

Liver

Kidneys

Environmental

Regulations

RCRA 40CFR: Not listed
CERCLA: 40CFR 302.4: Not listed
SARA 40CFR 372.65: Not listed
SARA EHS 40CFR 355: Not listed
TSCA: Listed

Analytical Methods

Water / Groundwater: APHA 3111-A, 3111-B, 3500-BI; USGS E-SPEC
Indoor / Expired Air: ASTM D4185
Other: EPA 1620

BIS7380 CAS #: 22306-37-2

BISMUTH ACETATE

EINECS Number: 244-904-5
Molecular Formula: $C_6H_9BiO_6$
Formula Weight: 386.12

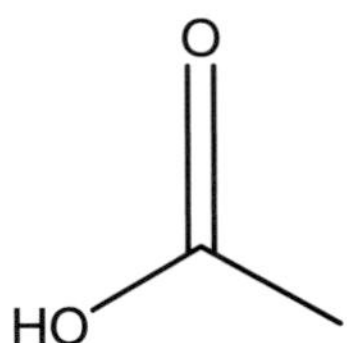

BiH_3

Chemical Structure

Synonyms: ACETIC ACID,BISMUTH(3+) SALT; BISMUTH TRIACETATE
Description: white crystals

Physical Properties

Freezing Point: Decomposes
Water Solubility: Insoluble in Water
Other Solubilities: Soluble in Acetic Acid

Hazard Overviews

Health: Irritating to eyes/skin. Other Acute Effects: may be harmful by inhalation, ingestion, or skin absorption.
Fire: Hazards: emits toxic fumes. Extinguishing agents: water spray; carbon dioxide, dry chemical powder or appropriate foam. Precautions: combustible liquid.
Reactivity: Incompatible with: strong oxidizing agents. Hazardous decomposition products: toxic fumes of: carbon monoxide, carbon dioxide, bismuth oxides.
Carcinogenicity: IARC - Not listed; NIOSH - Not listed; NTP - Not listed; ACGIH - Not listed; OSHA - Not listed; EPA - Not listed; MAK - Not listed
Primary Target Organs:

Eyes

Skin

Environmental

Regulations

RCRA 40CFR: Not listed
CERCLA: 40CFR 302.4: Not listed
SARA 40CFR 372.65: Not listed
SARA EHS 40CFR 355: Not listed
TSCA: Not listed

BIS7490 CAS #: 1304-85-4

BISMUTH SUBNITRATE

RTECS: EB2977000
EINECS Number: 215-136-8
Molecular Formula: $Bi_5H_9N_4O_{22}$
Formula Weight: 1462
Synonyms: BASIC BISMUTH NITRATE; BISMUTH MAGISTERY; BISMUTH NITRATE,BASIC; BISMUTH OXYNITRATE; BISMUTH PAINT; BISMUTH SUBNITRICUM; BISMUTH WHITE; BISMUTHYL NITRATE; BLANC DE FARD; C.I. 77169; C.I. PIGMENT WHITE 17; COMPD WITH NITROGEN OXIDE (N2O5) (6:5),OCTAHYDRATE; COSMETIC WHITE; FLAKE WHITE; MAGISTERY OF BISMUTH; NOVISMUTH; PAINT WHITE; SNOWCAL 5 SW; SNOWCAL 5SW; SPANISH WHITE; VICALIN; VIKALINE
Description: white, heavy, microcrystalline powder; odorless
Use: manufacture bismuth fluxes for enamels; in cosmetics; antacid; (vet) as gastrointestinal protectant; x-ray contrast medium; dusting powder for lesions; in prepn of milk of bismuth; as astringent, adsorbent, & protectant; coating agent in preparations for gastritis; in bleaching creams

Physical Properties

Freezing Point: Decomposes at 260 °C (500 °F)
Specific Gravity: 4.928
Water Solubility: Practically Insoluble in Water
Other Solubilities: Practically Insoluble in organic solvents.
pH: Shows acid to moistened litmus

RTECS Toxicity Data

Acute Oral: Infant TD_{Lo} Dose: 259 mg/kg; Toxic Effects: Blood - Methemoglobinemia-Carboxyhemoglobin. Infant LD_{Lo} Dose: 1 gm/kg.

Hazard Overviews

Health: Irritating to eyes/skin/respiratory tract. Other Acute Effects: may be harmful by inhalation, ingestion, or skin absorption; exposure can cause gastrointestinal disturbances; absorption into the body leads to the formation of methemoglobin which in sufficient concentration causes cyanosis; onset may be delayed 2 to 4 hours or longer; other symptoms of exposure can include anorexia; headache; malaise; skin reactions; discoloration of mucous membranes; mild jaundice; target organ: blood.
Fire: Hazards: emits toxic fumes; organic materials finely powdered metal. Extinguishing agents: carbon dioxide, dry chemical powder or appropriate foam. Precautions: combustible liquid.
Reactivity: Stable. Hazardous polymerization will not occur. Incompatible with: alkaline bicarbonates, soluble iodides, gallic acid, calomel, tannin, salicylic acid, sulfur. Hazardous decomposition products: decomposes to Bi2O3 and nitrogen oxide when heated to red heat;.
Carcinogenicity: IARC - Not listed; NIOSH - Not listed; NTP - Not listed; ACGIH - Not listed; OSHA - Not listed; EPA - Not listed; MAK - Not listed

Primary Target Organs:

Eyes

Skin

Respiratory System

Blood

Environmental

Regulations

RCRA 40CFR: Not listed
CERCLA: 40CFR 302.4: Not listed
SARA 40CFR 372.65: Not listed
SARA EHS 40CFR 355: Not listed
TSCA: Listed

BIS7600 CAS #: 14882-18-9

BISMUTH SUBSALICYLATE

RTECS: EB2985000
EINECS Number: 238-953-1
Molecular Formula: $C_7H_5BiO_4$
Formula Weight: 362.11
Synonyms: BASIC BISMUTH SALICYLATE; BISMOGENOL; BISMUTH OXYSALICYLATE; BISMUTH SALICYLATE,BASIC; BISMUTH,(2-HYDROXYBENZOATO-O(1),O(2))OXO-; BISMUTH,OXO(SALICYLATO)-; (2-HYDROXYBENZOATO-O(1),O(2)OXO-BISMUTH (9CI); 2-HYDROXYBENZOIC ACID BISMUTH (3+) SALT,BASIC; OXO(SALICYLATO)BISMUTH; SALICYLIC ACID BASIC BISMUTH SALT; SALICYLIC ACID,BISMUTH BASIC SALT; STABISOL; TRIS(2-HYDROXYBENZOATO)TRIOXOTRIBISMUTH
Description: white, bulky crystalline powder or prisms; odorless
Use: manufacture of cellulose-base, polystyrene and phenol-formaldehyde resins; in heat-sensitive paper coatings and as stabilizer; to treat Lupus erythematosus

Physical Properties

Boiling Point: Decomposes
Freezing Point: Decomposes
Water Solubility: Almost Insoluble in Water
Other Solubilities: 95% Ethanol: Insoluble (<1 mg/ml at 21.7 °C); DMSO: Insoluble (<1 mg/ml at 21.7 °C).
Flash Point: Not available; probably combustible

Hazard Overviews

Fire: Will burn. Hazards: hazardous polymerization will not occur. Extinguishing agents: water spray; carbon dioxide, dry chemical powder or appropriate foam. Precautions: combustible liquid.
Reactivity: Hazardous polymerization will not occur.
Carcinogenicity: IARC - Not listed; NIOSH - Not listed; NTP - Not listed; ACGIH - Not listed; OSHA - Not listed; EPA - Not listed; MAK - Not listed

Environmental

Regulations

RCRA 40CFR: Not listed
CERCLA: 40CFR 302.4: Not listed
SARA 40CFR 372.65: Not listed
SARA EHS 40CFR 355: Not listed
TSCA: Not listed

BIS7710 CAS #: 1304-82-1

BISMUTH TELLURIDE, UNDOPED

RTECS: EB3110000
EINECS Number: 215-135-2
Molecular Formula: Bi_2Te_3
Structured MF: Bi_2Te_3
Formula Weight: 800.76

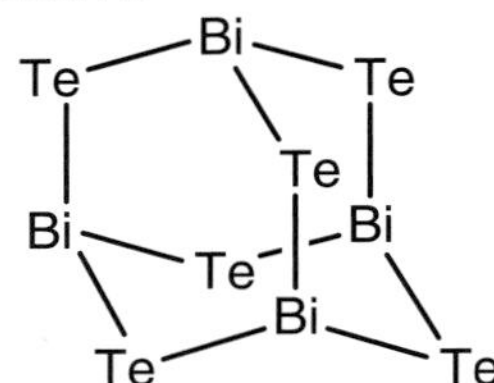

Chemical Structure

Synonyms: BISMUTH SESQUITELLURIDE; BISMUTH TELLURIDE; BISMUTH TRITELLURIDE; DIBISMUTH TRITELLURIDE; TELLUROBISMUTHITE
Description: gray platelets, ingots, crystals

Physical Properties

Freezing Point: 572.78 °C (1063.004 °F)
Specific Gravity: 7.7
Vapor Pressure: ~ 0 mm Hg
Water Solubility: Insoluble
Other Solubilities: Decomposed by nitric acid
Flash Point: Noncombustible Solid

Hazard Overviews

Fire Diamond

Health: Irritating to respiratory tract. Also Causes: upon ingestion: anorexia, nausea, headache, joint pain, sleeplessness, kidney/liver problems. Chronic Effects: bismuth line, peripheral neuritis, bismuth encephalopathy, dysarthria, ataxia, pseudotumor.
Fire: A fire hazard in contact with oxidizers. Use dry chemical, carbon dioxide, water spray, fog, or regular foam.
Reactivity: Stable. Hazardous polymerization cannot occur. Avoid: strong oxidizers; moisture. Incompatible with: strong oxidizers; moisture. Hazardous decomposition products: toxic telluride fumes.

Carcinogenicity: IARC - Not listed; NIOSH - Not listed; NTP - Not listed; ACGIH - Class A4, Not classifiable as a human carcinogen; OSHA - Not listed; EPA - Not listed; MAK - Not listed
Primary Target Organs:

Respiratory System — Mucous Membranes — Nervous System

Exposure Limits
OSHA PEL: TWA: 15 mg/m^3; total dust.
ACGIH TLV: TWA: 10 mg/m^3; as Bi_2Te_3.
NIOSH REL: TWA: 10 mg/m^3; as Te.
DFG MAK: TWA: 0.1 mg/m^3; as Se.
Respirator Recommendation
Exposure Range: >5 to 50 mg/m^3 Air Purifying, Negative Pressure, Half Mask
Exposure Range: >50 to 500 mg/m^3 Air Purifying, Negative Pressure, Full Face
Exposure Range: >500 to 5000 mg/m^3 Supplied Air, Constant Flow/Pressure Demand, Full Face
Exposure Range: >5000 to unlimited mg/m^3 Self-contained Breathing Apparatus, Pressure Demand, Full Face
Cartridge Color: dust/mist filter (use P100 or consult supervisor for appropriate dust/mist filter)

Environmental

Regulations

RCRA 40CFR: Not listed
CERCLA: 40CFR 302.4: Not listed
SARA 40CFR 372.65: Not listed
SARA EHS 40CFR 355: Not listed
TSCA: Listed

BIS7930 CAS #: 80-05-7

BISPHENOL A

RTECS: SL6300000
EINECS Number: 201-245-8
Molecular Formula: $C_{15}H_{16}O_2$
Structured MF: $(CH_3)_2C(C_6H_4OH)_2$
Formula Weight: 228.28

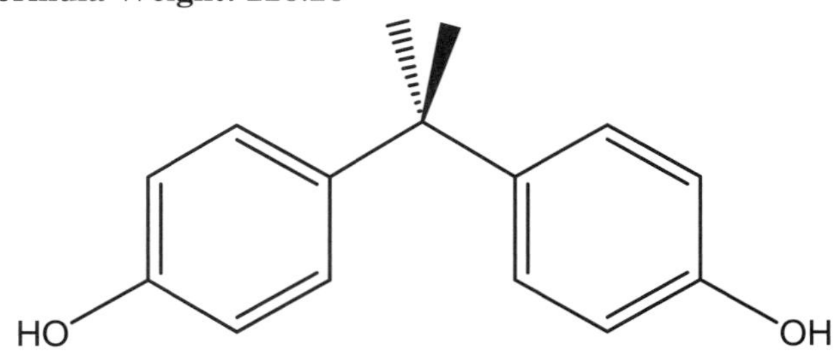

Chemical Structure

Synonyms: BISFEROL A; 2,2-BIS-4'-HYDROXYFENYLPROPAN; BIS(4-HYDROXYPHENYL) DIMETHYLMETHANE; 2,2-BIS(4-

HYDROXYPHENYL)PROPANE; 2,2-BIS(P-HYDROXYPHENYL)PROPANE; BETA,BETA'-BIS(P-HYDROXYPHENYL)PROPANE; BIS(4-HYDROXYPHENYL)PROPANE; BISPHENOL; 4,4'-BISPHENOL A; P,P'-BISPHENOL A; DIAN; DIANO; 4,4'-DIHYDROXDIPHENYLPROPANE; 4,4'-DIHYDROXYDIPHENYLDIMETHYLMETHANE; P,P'-DIHYDROXYDIPHENYLDIMETHYLMETHANE; 2,2-(4,4'-DIHYDROXYDIPHENYL)PROPANE; 2,2-(4,4-DIHYDROXYDIPHENYL)PROPANE; 4,4'-DIHYDROXY-2,2-DIPHENYLPROPANE; 4,4'-DIHYDROXYDIPHENYL-2,2-PROPANE; 4,4'-DIHYDROXYDIPHENYLPROPANE; P,P'-DIHYDROXYDIPHENYLPROPANE; 2,2-DI(4-HYDROXYPHENYL)PROPANE; BETA-DI-P-HYDROXYPHENYLPROPANE; DIMETHYL BIS(P-HYDROXYPHENYL)METHANE; 4,4'-DIMETHYLMETHYLENEDIPHENOL; DIMETHYLMETHYLENE-P,P'-DIPHENOL; 2,2-DI(4-PHENYLOL)PROPANE; DIPHENYLOLPROPANE; IPOGNOX 88; ISOPROPYLIDENEBIS(4-HYDROXYBENZENE); 4,4'-ISOPROPYLIDENEBIS(PHENOL); 4,4'-ISOPROPYLIDENEBISPHENOL; P,P'-ISOPROPYLIDENEBISPHENOL; 4,4'-ISOPROPYLIDENEDIPHENOL; P,P'-ISOPROPYLIDENEDIPHENOL; 4,4'-(1-METHYLETHYLIDENE)BISPHENOL; PARABIS A; PHENOL,4,4'-DIMETHYLMETHYLENEDI-; PHENOL,4,4'-ISOPROPYLIDENEDI-; PHENOL,4,4'-(1-METHYLETHYLIDENE)BIS-; PLURACOL 245; PROPANE,2,2-BIS(P-HYDROXYPHENYL)-; RIKABANOL; UCAR BISPHENOL A; UCAR BISPHENOL HP

Description: white crystals, flakes; mild phenolic odor

Use: an intermediate in the manufacture of polymers, epoxy resins, polycarbonates, fungicides, antioxidants, dyes, phenoxy, polysulfone and certain polyester resins, flame retardants and rubber chemicals

Physical Properties

Boiling Point: 220 °C (428 °F) at 4 mm Hg

Freezing Point: 150 °C (302 °F) to 155 °C (311 °F)

Specific Gravity: 1.195 at 25 °C/25 °C

Saturated Vapor Density: 1.2 kg/m^3

Vapor Pressure: 4 x10^{-8} mm Hg at 25 °C

Water Solubility: < 1 mg/mL at 21.5 C

Other Solubilities: Soluble in aqueous alkaline solution, Alcohol, Acetone; slightly Soluble in Carbon Tetrachloride.

Flash Point: 79.4 °C

Autoignition Temperature: 600 °C

RTECS Toxicity Data

Acute Oral: Rat LD$_{50}$ Dose: 3250 mg/kg. Mouse LD$_{50}$ Dose: 2400 mg/kg; Toxic Effects: Automatic Nervous System - Other (direct) parasympathomimetic; Behavioral - Convulsions or effect on seizure threshold; Behavioral - Ataxia.

Acute Inhalation: Mouse LC; Dose: >1700 mg/m^3/2hr; Toxic Effects: Behavioral - Somnolence (general depressed activity); Behavioral - Ataxia; Lungs, Thorax, or Respiration - Dyspnea.

Acute Dermal: Rabbit LD$_{50}$ Route: Skin; Dose: 3 mL/kg. Mouse LD$_{Lo}$ Route: Subcutaneous Dose: 2500 mg/kg; Toxic Effects: Behavioral - Somnolence (general depressed activity); Liver - Fatty liver degeneration; Nutritional and gross metabolic - Weight loss or decreased weight gain.

Chronic (Multiple Dose) Oral: Rat Dose: 12 mg/kg/12D-I; Toxic Effects: Liver - Fatty liver degeneration; DEATH. Rat Dose: 42 gm/kg/8W-C; Toxic Effects: Nutritional and gross metabolic - Weight loss or decreased weight gain. Rat Dose: 5460 mg/kg/13W-C; Toxic Effects: Gastrointestinal - Other changes; DEATH. Rat Dose: 31500 mg/kg/9W-I; Toxic Effects: Kidney, Ureter, and Bladder - Urine volume increased; Liver - Other changes; Blood - Other hemolysis with or without anemia. Rat Dose: 455 mg/kg/26W-I; Toxic Effects: Behavioral - Alteration of classical conditioning; Liver - Liver function tests impaired; Blood - Other changes.

Chronic (Multiple Dose) Inhalation: Rat Dose: 150 mg/m^3/6H/13W-I; Toxic Effects: Sense organs and special senses - Other; Nutritional and gross metabolic - Weight loss or decreased weight gain. Rat Dose: 47 mg/m^3/2H/19W-I; Toxic Effects: Liver - Changes in Liver weight; Kidney, Ureter, and Bladder - Urine volume decreased; Endocrine - Changes in spleen weight.

Irritation Eye: Rabbit Standard Draize Test Dose: 250 ug/24H; Reaction: severe.

Irritation Skin: Rabbit Standard Draize Test Dose: 500 mg/24H; Reaction: mild. Rabbit Open Draize Test Dose: 250 mg open; Reaction: mild.

Reproductive/Teratogenic: Rat Route: Oral; Dose: 10 gm/kg; Duration: female 6-15D of pregnancy; Effects on Embryo or Fetus - Fetotoxicity. Rat Route: Oral; Dose: 15 gm/kg; Duration: female 6-15D of pregnancy; Effects on Fertility - Post-implantation mortality.

Mutagenic: Rat DNA Adduct; Route: Intraperitoneal; Dose: 200 mg/kg. Rat DNA Adduct; Route: Oral; Dose: 800 mg/kg/4D-C.

Hazard Overviews

Fire Diamond

Health: Irritating to eyes/skin/respiratory tract. Chronic Effects: skin sensitization, rashes, hives, allergic dermatitis, lung injury, and reversible changes in the nasal lining (based on animal studies).

Fire: Slight fire hazard when exposed to heat and an ignition source. May form explosive dust-air mixtures.

Reactivity: Stable. Hazardous polymerization cannot occur. Avoid: sources of ignition or strong oxidizing agents; dust clouds to form during work operations. Incompatible with: strong oxidizing agents. Hazardous decomposition products: carbon monoxide.

Carcinogenicity: IARC - Not listed; NIOSH - Not listed; NTP - Not listed; ACGIH - Not listed; OSHA - Not listed; EPA - Not listed; MAK - Not listed

Primary Target Organs:

Eyes

Skin

Respiratory System

Environmental

Ecotoxicity: LC_{50} Pimephales promelas (fathead minnow) 4.6 (3.6 to 5.4) mg/l/96 hr, flow-through test LC_{50} Menidia menidia (Atlantic silverside) 9.4 (8.3 to 11) mg/l/96 hr /Conditions of bioassay not specified

Environmental Fate: If released to soil it is expected to have moderate to low mobility. This compound may biodegrade under aerobic conditions following acclimation. If released to acclimated water, biodegradation would be the dominant fate process (half-life less than or equal to 4 days). In nonacclimated water, it may biodegrade after a sufficient adaptation period, it may adsorb extensively to suspended solids and sediments or it may photolyze. If released to the atmosphere, it is expected to exist almost entirely in the particulate phase. In particulate form it may be removed from the atmosphere by dry deposition or photolysis. The small fraction which would exist in the vapor phase may react with photochemically generated hydroxyl radicals (half-life 4 hours) or it may photolyze. Photodegradation products are phenol, 4-isopropylphenol, and a semiquinone derivative.

Cleanup/Disposal: Guide No. 156: Eliminate all ignition sources (no smoking, flares, sparks or flames in immediate area). All equipment used when handling the product must be grounded. Do not touch damaged containers or spilled material unless wearing appropriate protective clothing. Stop leak if you can do it without risk. A vapor suppressing foam may be used to reduce vapors. For chlorosilanes, use AFFF alcohol-resistant medium expansion foam to reduce vapors. Do not get water on spilled substance or inside containers. Use water spray to reduce vapors or divert vapor cloud drift. Prevent entry into waterways, sewers, basements or confined areas. Small Spills: Cover with dry earth, dry sand, or other non-combustible material followed with plastic sheet to minimize spreading or contact with rain. Use clean non-sparking tools to collect material and place it into loosely covered plastic containers for later disposal.

Environmental Physical Data

Henry's Law Constant: estimated at 1 $x10^{-10}$

Octanol/Water Partition Coefficient: log K_{ow} = 3.32

Sorption Partition Coefficient: K_{oc} = estimated at 314 to 1524

BCF: carp 100

Regulations

RCRA 40CFR: Not listed
CERCLA: 40CFR 302.4: Not listed
SARA 40CFR 372.65: Listed
SARA EHS 40CFR 355: Not listed
TSCA: Listed

Analytical Methods

Water / Groundwater: ASTM D4763

BIS8040 CAS #: 116-37-0

BISPHENOL A BIS(2-HYDROXYPROPYL) ETHER

RTECS: UB8449895
EINECS Number: 204-137-9
Molecular Formula: $C_{21}H_{28}O_4$
Formula Weight: 344.45
Synonyms: 2,2-BIS(P-(2-HYDROXY-2-METHYLETHOXY)PHENYL)PROPANE; 2,2-BIS(4-(2-HYDROXYPROPOXY)PHENYL)PROPANE; 2,2-BIS(P-(2-HYDROXYPROPOXY)PHENYL)PROPANE; BISPHENOL A BIS(BETA-HYDROXYPROPYL) ETHER; BISPHENOL A-PROPYLENE OXIDE ADDUCT (1:2); DIANOL 33; DOW RESIN 565; HYDROXYPROPYLATED DIPHENYLOLPROPANE; 1,1'-(ISOPROPYLIDENEBIS(P-PHENYLENEOXY))DI-2-PROPANOL; ISOPROPYLIDENEDIPHENOXYPROPANOL; 1,1'-((1-METHYLETHYLIDENE)BIS(4,1-PHENYLENEOXY))BIS-2-PROPANOL; 2-PROPANOL,1,1'-(ISOPROPYLIDENEBIS(P-PHENYLENEOXY))DI-; 2-PROPANOL,1,1'-ISOPROPYLIDENEBIS(P-PHENYLENEOXY)DI-; 2-PROPANOL,1,1'-((1-METHYLETHYLIDENE)BIS(4,1-PHENYLENEOXY))BIS-
Description:
Use: chem int for corrosion-resistant unsaturated polyesters

Hazard Overviews

Carcinogenicity: IARC - Not listed; NIOSH - Not listed; NTP - Not listed; ACGIH - Not listed; OSHA - Not listed; EPA - Not listed; MAK - Not listed

Environmental

Regulations

RCRA 40CFR: Not listed
CERCLA: 40CFR 302.4: Not listed
SARA 40CFR 372.65: Not listed
SARA EHS 40CFR 355: Not listed
TSCA: Listed

BIS8150 CAS #: 1675-54-3

BISPHENOL A DIGLYCIDYL ETHER

RTECS: TX3800000
EINECS Number: 216-823-5
Molecular Formula: $C_{21}H_{24}O_4$
Structured MF: $C_{21}H_{24}O_4$
Formula Weight: 340.45

Chemical Structure

Synonyms: ARALDITE 6005; ARALDITE 6010; 2,2-BIS(4-(2,3-EPOXPROPYLOXY)PHENYL)PROPANE; 4,4'-BIS(2,3-

EPOXYPROPOXY)DIPHENYLDIMETHYLMETHANE; 2,2-BIS(P-(2,3-EPOXYPROPOXY)PHENYL)PROPANE; 2,2-BIS(4-(2,3-EPOXYPROPYLOXY)PHENYL)PROPANE; BIS(4-GLYCIDYLOXYPHENYL)DIMETHYAMETHANE; BIS(4-GLYCIDYLOXYPHENYL)DIMETHYLMETHANE; 2,2-BIS(4'-GLYCIDYLOXYPHENYL)PROPANE; 2,2-BIS(4-GLYCIDYLOXYPHENYL)PROPANE; 2,2-BIS(P-GLYCIDYLOXYPHENYL)PROPANE; BIS(4-HYDROXYPHENYL)DIMETHYLMETHANE DIGLYCIDYL ETHER; BIS(4-HYDROXYPHENYL)DIMETHYLMETHANE,DIGLYCIDYL ETHER; 2,2-BIS(4-HYDROXYPHENYL)PROPANE DIGLYCIDYL ETHER; 2,2-BIS(4-HYDROXYPHENYL)PROPANE,DIGLYCIDYL ETHER; 2,2-BIS(P-HYDROXYPHENYL)PROPANE,DIGLYCIDYL ETHER; BISPHENOL A-EPICHLOROHYDRIN CONDENSATE; BPDGE; D E R 332; D.E.R. 332; DIAN DIGLYCIDYL ETHER; DIAN-BIS-GLYCIDYLETHER; DIGLYCIDYL BISPHENOL A; DIGLYCIDYL BISPHENOL A ETHER; DIGLYCIDYL DIPHENYLOLPROPANE ETHER; DIGLYCIDYL ETHER OF 2,2-BIS(4-HYDROXYPHENYL)PROPANE; DIGLYCIDYL ETHER OF 2,2-BIS(P-HYDROXYPHENYL)PROPANE; DIGLYCIDYL ETHER OF BISPHENOL A; DIGLYCIDYL ETHER OF 4,4'-ISOPROPYLIDENEDIPHENOL; 4,4'-DIHYDROXYDIPHENYLDIMETHYLMETHANE DIGLYCIDYL ETHER; P,P'-DIHYDROXYDIPHENYLDIMETHYLMETHANE DIGLYCIDYL ETHER; 4,4'-DIHYDROXYDIPHENYLDIMETHYLMETHANE,DIGLYCIDYL ETHER; P,P'-DIHYDROXYDIPHENYLDIMETHYLMETHANE,DIGLYCIDYL ETHER; DIOMETHANE DIGLYCIDYL ETHER; EPI-RE2 510; EPI-REZ 508; EPI-REZ 510; EPON 828; EPOTUF 37-140; EPOXIDE A; ERL-2774; GY 6010; 4,4'-ISOPROPYLIDENEBIS(1-(2,3-EPOXYPROPOXY)BENZENE); 4,4'-ISOPROPYLIDENEDIPHENOL DIGLYCIDYL ETHER; 4,4-ISOPROPYLIDENEDIPHENOL EPICHLOROHYDRIN RESIN; 4,4'-ISOPROPYLIDENEDIPHENOL,DIGLYCIDYL ETHER; 2,2'-((1-METHYLETHYLIDENE)BIS(4,1-PHENYLENEOXYMETHYLENE))BISOXIRANE; OLIGOMER 340; OXIRANE,2,2'-((1-METHYLETHYLIDENE)BIS(4,1-PHENYLENEOXYMETHYLENE))BIS-; PROPANE,2,2-BIS (P-(2,3-EPOXYPROPOXY)PHENYL)

Description: yellowish-brown liquid; odorless

Use: making castings and pottings and formulating lightweight foams; as binder in preparation of laminates of paper, polyester cloth, fiberglass cloth and wood sheets

Physical Properties

Boiling Point: Decomposes

Freezing Point: 8 °C (46.4 °F) to 12 °C (53.6 °F)

Density: 1.16 g/mL at 20 °C

Water Solubility: < 1 mg/mL at 19.5 C

Other Solubilities: 95% Ethanol: 50-100 mg/ml at 15 °C; Acetone: >=100 mg/ml at 15 °C; DMSO: 50-100 mg/ml at 15 °C.

Flash Point: 79.444 °C Open Cup

RTECS Toxicity Data

Acute Oral: Rat LD_{50} Dose: 11300 uL/kg. Mouse LD_{50} Dose: 15600 mg/kg; Toxic Effects: Behavioral - Somnolence (general depressed activity); Gastrointestinal - Hypermotility, diarrhea; Nutritional and gross metabolic - Weight loss or decreased weight gain.

Acute Dermal: Rabbit LD_{50} Route: Skin; Dose: 20 gm/kg; Toxic Effects: Behavioral - Somnolence (general depressed activity); Gastrointestinal - Hypermotility, diarrhea; Nutritional and gross metabolic - Weight loss or decreased weight gain.

Irritation Eye: Rabbit Standard Draize Test Dose: 2 mg/24H; Reaction: severe.

Irritation Skin: Rabbit Open Draize Test Dose: 500 mg open; Reaction: mild.

Mutagenic: Bacteria - S Typhimurium Mutations in Microorganisms; Dose: 33 ug/plate (+S9).

Tumorigenic: Mouse Route: Skin; Dose: 166 gm/kg/2Y-I; Toxic Effects: Tumorigenic - Carcinogenic by RTECS criteria; Endocrine - Thyroid tumors. Mouse Route: Skin; Dose: 312 gm/kg/2Y-I; Toxic Effects: Tumorigenic - Carcinogenic by RTECS criteria; Liver - Tumors; Tumorigenic effects - Ovarian tumors. Mouse Route: Skin; Dose: 312 gm/kg/2Y-I; Toxic Effects: Tumorigenic - Carcinogenic by RTECS criteria; Skin and appendages - Tumors.

Hazard Overviews

Health: Irritating to eyes. Other Acute Effects: may be harmful by inhalation, ingestion, or skin absorption; may cause allergic skin reaction.

Fire: Combustible. Hazards: emits toxic fumes. Extinguishing agents: carbon dioxide, dry chemical powder or appropriate foam. Precautions: combustible liquid.

Reactivity: Stable. Hazardous polymerization will not occur, but masses of more than 1 pound of product plus an aliphatic amine will cause irreversibe polymerization with considerable heat buildup. Incompatible with: bases. Hazardous decomposition products: toxic fumes of: carbon monoxide, carbon dioxide.

Carcinogenicity: IARC - Group 3, Not classifiable as to carcinogenicity to humans; NIOSH - Not listed; NTP - Not listed; ACGIH - Not listed; OSHA - Not listed; EPA - Not listed; MAK - Not listed

Primary Target Organs:

Eyes

Environmental

Cleanup/Disposal: Guide No. 171: Do not touch or walk through spilled material. Stop leak if you can do it without risk. Prevent dust cloud. Avoid inhalation of asbestos dust. Small Dry Spills: With clean shovel place material into clean, dry container and cover loosely; move containers from spill area. Small Spills: Take up with sand or other noncombustible absorbent material and place into containers for later disposal. Large Spills: Dike far ahead of liquid spill for later disposal. Cover powder spill with plastic sheet or tarp to minimize spreading. Prevent entry into waterways, sewers, basements or confined areas.

Environmental Physical Data

BCF: no food chain concentration potential

Regulations
RCRA 40CFR: Not listed
CERCLA: 40CFR 302.4: Not listed
SARA 40CFR 372.65: Not listed
SARA EHS 40CFR 355: Not listed
TSCA: Listed

BIS8260 CAS #: 2444-90-8
BISPHENOL A DISODIUM SALT

EINECS Number: 219-488-3
Molecular Formula: $C_{15}H_{16}Na_2O_2$
Formula Weight: 274.28
Synonyms: PHENOL,4,4'-ISOPROPYLIDENEDI-,DISODIUM SALT; PHENOL,4,4'-(1-METHYLETHYLIDENE)BIS-,DISODIUM SALT

Hazard Overviews
Carcinogenicity: IARC - Not listed; NIOSH - Not listed; NTP - Not listed; ACGIH - Not listed; OSHA - Not listed; EPA - Not listed; MAK - Not listed

Environmental

Regulations
RCRA 40CFR: Not listed
CERCLA: 40CFR 302.4: Not listed
SARA 40CFR 372.65: Not listed
SARA EHS 40CFR 355: Not listed
TSCA: Listed

BIS8370 CAS #: 38103-06-9
BISPHENOL A-DIANHYDRIDE

EINECS Number: 253-781-7

Physical Properties
Boiling Point: > 300 °C (572 °F)
Specific Gravity: > 0
Vapor Pressure: < 0.05
Water Solubility: Insoluble

Hazard Overviews
Carcinogenicity: IARC - Not listed; NIOSH - Not listed; NTP - Not listed; ACGIH - Not listed; OSHA - Not listed; EPA - Not listed; MAK - Not listed

Environmental

Regulations
RCRA 40CFR: Not listed
CERCLA: 40CFR 302.4: Not listed
SARA 40CFR 372.65: Not listed
SARA EHS 40CFR 355: Not listed
TSCA: Not listed

BIS8590 CAS #: 56-35-9
BIS(TRIBUTYLIN) OXIDE

RTECS: JN8750000
DOT: UN2786; UN2787; UN3019; UN3020; IMO3.2; IMO6.1
EINECS Number: 200-268-0
Molecular Formula: $C_{24}H_{54}OSn_2$
Formula Weight: 595.62

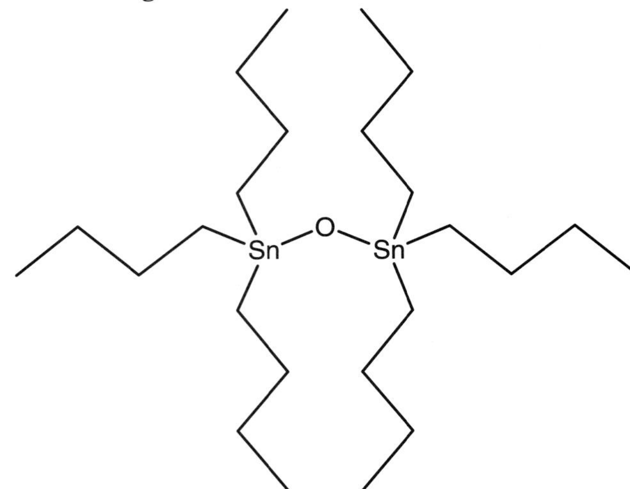

Chemical Structure

Synonyms: BIOMET TBTO; BIS-(TRI-N-BUTYLCIN)OXID; BIS(TRIBUTYLOXIDE) OF TIN; BIS(TRIBUTYLSTANNIUM) OXIDE; BIS(TRIBUTYLSTANNYL) OXIDE; BIS(TRIBUTYLSTANNYL)OXIDE; BIS(TRIBUTYLTIN) OXIDE; BIS(TRI-N-BUTYLTIN)OXIDE; BIS(TRIBUTYLTIN)OXIDE; BIS(TRI-N-BUTYLZINN)-OXYD; BTO; BUTINOX; C-SN-9; ENT 24,979; EPA PESTICIDE CHEMICAL CODE 083001; HEXABUTYLDISTANNIOXAN; HEXABUTYLDISTANNOXANE; HEXABUTYLDITIN; KYSLICNIK TRI-N-BUTYLCINICITY; L.S. 3394; OTBE; 6-OXA-5,7-DISTANNAUNDECANE,5,5,7,7-TETRABUTYL-; OXYBIS(TRIBUTYLTIN); OXYDE DE TRIBUTYLETAIN; STANNANE,TRI-N-BUTYL-,OXIDE; TBOT; TBTO; TIN,BIS(TRIBUTYL)-,OXIDE; TIN,OXYBIS(TRIBUTYL-; TRI-N-BUTYLSTANNANE OXIDE; TRIBUTYLTIN OXIDE; ZK 21995
Description: colorless to yellow liquid; weak odor
Use: fungicide and bactericide, especially in underwater and antifouling paints (q.v.); pesticides

Physical Properties
Boiling Point: 254 °C (489 °F) at 50 mm Hg
Freezing Point: 45 °C (113 °F)
Specific Gravity: 1.17 at 25 °C
Saturated Vapor Density: Much less than 1.230850454 kg/m^3
Vapor Pressure: Much less than 1 mm Hg at 20 °C
Water Solubility: 0.1% in Hot Water
Other Solubilities: 95% Ethanol: 1-10 mg/ml at 21.5 °C; Acetone: 1-10 mg/ml at 18 °C; DMSO: <1 mg/ml at 21.5 °C.
Refraction Index: 1.4864
Flash Point: > 100 °C Closed Cup

RTECS Toxicity Data

Acute Oral: Rat LD_{50} Dose: 87 mg/kg. Mouse LD_{50} Dose: 55 mg/kg.

Acute Inhalation: Guinea Pig LC_{Lo} Dose: 200 mg/m^3; Toxic Effects: Behavioral - Convulsions or effect on seizure threshold; Behavioral - Aggression; Lungs, Thorax, or Respiration - Dyspnea.

Acute Dermal: Rabbit LD_{50} Route: Skin; Dose: 900 mg/kg; Toxic Effects: Behavioral - Somnolence (general depressed activity); Lungs, Thorax, or Respiration - Acute pulmonary edema; Gastrointestinal - Hypermotility, diarrhea. Mouse LD_{50} Route: Skin; Dose: 163 mg/kg; Toxic Effects: Behavioral - Altered sleep time (including change in righting reflex); Behavioral - Change in motor activity (specific assay); Kidney, Ureter, and Bladder - Other changes.

Chronic (Multiple Dose) Oral: Rat Dose: 390 mg/kg/26W-I; Toxic Effects: Endocrine - Other changes; Endocrine - Changes in pituitary weight; Endocrine - Changes in thymus weight. Rat Dose: 462 mg/kg/22W-C; Toxic Effects: Endocrine - Changes in thymus weight; Immunological including allergic - Decreased immune response. Rat Dose: 1400 mg/kg/28D-C; Toxic Effects: Endocrine - Other changes; Endocrine - Changes in spleen weight. Rat Dose: 1700 mg/kg/34D-C; Toxic Effects: Immunological including allergic - Decreased immune response. Monkey Dose: 21120 ug/kg/22W-I; Toxic Effects: Blood - Changes in leukocyte (WBC) cell count.

Chronic (Multiple Dose) Inhalation: Rat Dose: 2800 ug/m^3/4H/32D-I; Toxic Effects: Behavioral - Food intake (animal); Lungs, Thorax, or Respiration - Structural or functional change in trachea or bronchi; DEATH.

Irritation Eye: Rabbit Standard Draize Test Dose: 50 ug/24H; Reaction: severe.

Reproductive/Teratogenic: Rat Route: Oral; Dose: 150 mg/kg; Duration: female 6-20D of pregnancy; Effects on Fertility - Litter size; Effects on Embryo or Fetus - Fetotoxicity; Fetal death. Rat Route: Oral; Dose: 150 mg/kg; Duration: female 6-20D of pregnancy; Specific Developmental Abnormalities - Central nervous system; Effects on Newborn - Viability index; Delayed effects. Rat Route: Oral; Dose: 40 mg/kg; Duration: female 5D after birth; Specific Developmental Abnormalities - Central nervous system; Effects on Newborn - Delayed effects. Rat Route: Oral; Dose: 150 mg/kg; Duration: female 6-20D of pregnancy; Effects on Newborn - Behavioral.

Mutagenic: Mouse Micronucleus Test; Route: Oral; Dose: 60 mg/kg. Hamster Cytogenetic Analysis; Cell Type: ovary; Dose: 5 mg/L.

Tumorigenic: Rat Route: Oral; Dose: 37100 mg/kg/2Y-C; Toxic Effects: Tumorigenic - Carcinogenic by RTECS criteria; Endocrine - Tumors; Endocrine - Thyroid tumors.

Hazard Overviews

Poison

Corrosive

Health: Corrosive to eyes/skin/respiratory tract. Poison. Other Acute Effects: may be fatal if inhaled, swallowed, or absorbed through skin; extremely destructive to tissue of the mucous membranes and upper respiratory tract, eyes and skin; inhalation may result in spasm, inflammation and edema of the larynx and bronchi, chemical pneumonitis and pulmonary edema; symptoms of exposure may include burning sensation, coughing, wheezing, laryngitis, shortness of breath, headache, nausea and vomiting.

Fire: Will burn. Hazards: emits toxic fumes. Extinguishing agents: carbon dioxide, dry chemical powder or appropriate foam. Precautions: combustible liquid.

Reactivity: Incompatible with: strong oxidizing agents. Hazardous decomposition products: toxic fumes of: carbon monoxide, carbon dioxide, tin/tin oxides.

Carcinogenicity: IARC - Not listed; NIOSH - Not listed; NTP - Not listed; ACGIH - Class A4, Not classifiable as a human carcinogen; OSHA - Not listed; EPA - Not listed; MAK - Not listed

Primary Target Organs:

Eyes

Skin

Respiratory System

Exposure Limits

OSHA PEL: TWA: 0.1 mg/m^3; as Sn.

OSHA PEL Vacated 1989 Limits: TWA: 0.1 mg/m^3.

ACGIH TLV: TWA: 2 mg/m^3; see NIOSH.

NIOSH REL: TWA: 2 mg/m^3; inorganic (.1 if organic).

DFG MAK: TWA: 2 mg/m^3; See NIOSH.

Environmental

Ecotoxicity: LC_{50} Pimephales promelas (fathead minnow) 2.7 ug/l/96 hr (confidence limit 2.4 - 3.0 mg/l), flow-through bioassay with measured concentrations, 24.0 °C, dissolved oxygen 7.5 mg/l, hardness 51.5 mg/l calcium carbonate, alkalinity 41.1 mg/l calcium carbonate, and pH 7.5 EC_{50} Pimephales promelas (fathead minnow) 2.7 mg/l/96 hr (confidence limit 2.4 - 3.0 mg/l), flow-through bioassay with measured concentrations, 24.0 °C, dissolved oxygen 7.5 mg/l, hardness 51.5 mg/l calcium carbonate, alkalinity 41.1 mg/l calcium carbonate, and pH 7.5. Effect: behavorial observations were not recorded

Environmental Fate: If released to soil, it will be expected to strongly bind to soil; therefore, it should not leach through soil. It may be susceptible to biodegradation in soil. It should not hydrolyze in soil or volatilize from near-surface soil. If released to water, it will be expected to mainly exist as tributyltin cation. It may react with sulfides present in sediment, which would lead to the formation of bis(tributyltin) sulfide. Tributyltin cation is stable to hydrolysis in water. Abiotic degradation of tributyltin species will be limited to direct photolysis in surface water (half-life >89 days). Tributyltin cation will be expected to strongly bind to soil and sediment and to strongly bioconcentrate in various aquatic organisms. Volatilization from water is not expected to be an important environmental pathway. Tributyltin cation may be susceptible to biodegradation in water with half-lives of between 6 days and 35 weeks reported in water and water-sediment mixtures, many of which had been previously contaminated with tributyltin species. If released to the atmosphere, it can be expected to exist both in the vapor-phase and particulate phases in the ambient atmosphere. It is unknown whether it will directly photolyze in sunlit air, although the direct photolysis of tributyltin species present in water solutions of the compound has been observed. It may be susceptible to photooxidation by photochemically produced hydroxyl radicals in the atmosphere via abstraction of hydrogen from the butyl groups.

Cleanup/Disposal: Guide No. 131: Fully encapsulating, vapor protective clothing should be worn for spills and leaks with no fire. Eliminate all ignition sources (no smoking, flares, sparks or flames in immediate area). All equipment used when handling the product must be grounded. Do not touch or walk through spilled material. Stop leak if you can do it without risk. Prevent entry into waterways, sewers, basements or confined areas. A vapor suppressing foam may be used to reduce vapors. Small Spills: Absorb with earth, sand or other non-combustible material and transfer to containers for later disposal. Use clean non-sparking tools to collect absorbed material. Large Spills: Dike far ahead of liquid spill for later disposal. Water spray may reduce vapor; but may not prevent ignition in closed spaces. Guide No. 153: Eliminate all ignition sources (no smoking, flares, sparks or flames in immediate area). Do not touch damaged containers or spilled material unless wearing appropriate protective clothing. Stop leak if you can do it without risk. Prevent entry into waterways, sewers, basements or confined areas. Absorb or cover with dry earth, sand or other non-combustible material and transfer to containers. Do not get water inside containers.

Environmental Physical Data

Henry's Law Constant: calculated at 1.26×10^{-7}
Sorption Partition Coefficient: $K_{oc} = 9.08 \times 10^{4}$
BCF: carp 589

Regulations

RCRA 40CFR: Not listed
CERCLA: 40CFR 302.4: Not listed
SARA 40CFR 372.65: Listed
SARA EHS 40CFR 355: Not listed
TSCA: Listed

Analytical Methods

Soil: EPA PMD-TQO

BIT5000 CAS #: 4044-65-9

BITOSCANATE

RTECS: NX9150000
EINECS Number: 223-741-3
Molecular Formula: $C_8H_4N_2S_2$
Formula Weight: 192.24

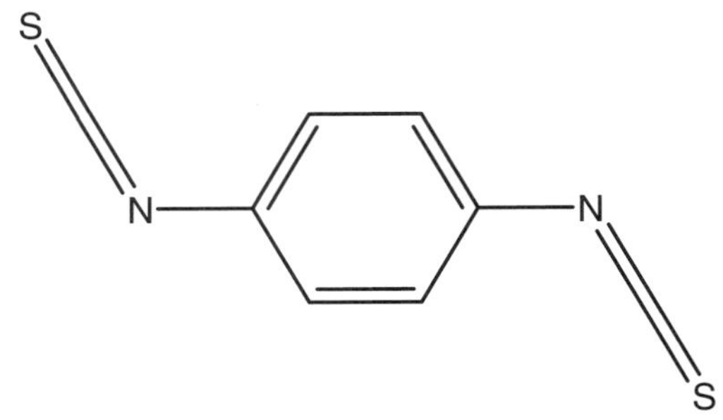

Chemical Structure

Synonyms: 16842; BENZENE,1,4-DIISOTHIOCYANATO-; BISCOMATE; BITOSCANATO; BITOSCANATUM; 1,4-DIISOTHIOCYANATOBENZENE; ISOTHIOCYANIC ACID P-PHENYLENE ESTER; ISOTHIOCYANIC ACID,P-PHENYLENE ESTER; JONIT; PHENYLENE THIOCYANATE; PHENYLENE-1,4-DIISOTHIOCYANATE; WM 842
Description: yellowish-white, crystalline powder; almost odorless
Use: medication: an anthelmintic

Physical Properties

Freezing Point: 132 °C (269.6 °F)
Water Solubility: Practically Insoluble in Water
Other Solubilities: Soluble in Alcohol; Soluble in Chloroform.

RTECS Toxicity Data

Acute Oral: Human TD_{Lo} Dose: 3 mg/kg; Toxic Effects: Behavioral - Hallucinations, distorted perceptions; Gastrointestinal - Nausea or vomiting. Mouse LD_{50} Dose: 230 mg/kg.

Hazard Overviews

Health: Irritating to eyes/skin/respiratory tract. Toxic. Other Acute Effects: harmful if swallowed, inhaled, or absorbed through skin; lachrymator.
Fire: Hazards: emits toxic fumes. Extinguishing agents: carbon dioxide, dry chemical powder or appropriate foam. Precautions: combustible liquid.
Reactivity: Incompatible with: strong oxidizing agents, strong acids, strong bases, may decompose on exposure to moist air or water. Hazardous decomposition products: toxic fumes of: carbon monoxide, carbon dioxide, nitrogen oxides, sulfur oxides.
Carcinogenicity: IARC - Not listed; NIOSH - Not listed; NTP - Not listed; ACGIH - Not listed; OSHA - Not listed; EPA - Not listed; MAK - Not listed

Primary Target Organs:

Eyes

Skin

Respiratory System

Environmental

Regulations

RCRA 40CFR: Not listed
CERCLA: 40CFR 302.4: Not listed
SARA 40CFR 372.65: Not listed TPQ: 500/10000 lb
SARA EHS 40CFR 355: Listed TPQ: 500 lb
TSCA: Listed

BOR1000 CAS #: 11130-12-4

BORAX PENTAHYDRATE

Molecular Formula: $B_4H_{10}NaO_{12}$
Formula Weight: 291.3
Description: white granular solid; odorless

Physical Properties

Freezing Point: 200 °C (392 °F)
Specific Gravity: 1.815
Water Solubility: 4% at 20 °C
pH: 3% solution 9.25

Hazard Overviews

Fire Diamond

Health: Irritating to the eyes, skin, and respiratory tract. Also Causes: cough, nausea, vomiting. Chronic Effects: dermatitis, low blood pressure, restlessness, weakness, seizures.
Fire: Noncombustible. Use agents suitable for surrounding fire.
Reactivity: Stable. Hazardous polymerization cannot occur. Avoid: contact with elemental zirconium. Incompatible with: elemental zirconium.
Carcinogenicity: IARC - Not listed; NIOSH - Not listed; NTP - Not listed; ACGIH - Not listed; OSHA - Not listed; EPA - Not listed; MAK - Not listed
Primary Target Organs:

Eyes

Skin

Respiratory System

Mucous Membranes

Nervous System

Cardio-vascular

Environmental

Regulations

RCRA 40CFR: Not listed
CERCLA: 40CFR 302.4: Not listed
SARA 40CFR 372.65: Not listed
SARA EHS 40CFR 355: Not listed
TSCA: Not listed

BOR1730 CAS #: 10043-35-3

BORIC ACID

RTECS: ED4550000
EINECS Number: 233-139-2
Molecular Formula: BH_3O_3
Formula Weight: 61.84

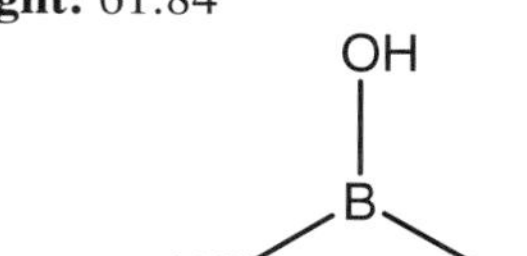

Chemical Structure

Synonyms: BORACIC ACID; BOROFAX; BORON TRIHYDROXIDE; BORSAURE; EPA PESTICIDE CODE 011001; HYDROGEN ORTHOBORATE; ORTHOBORIC ACID; THREE ELEPHANT
Description: white, colorless fine granular powder, crystals; odorless
Use: weatherproofing wood, fireproofing fabrics, as a preservative, manufacture of cements, crockery, porcelain enamels, borosilicate (heat resistant) glass, borates, leather, carpets, hats, soaps and artificial gems; in nickeling bath cosmetics, printing and dyeing, painting, photography, for impregnating wicks, electric condensers, hardening steel, insecticide for cockroaches and black carpet beetles, astringent, antiseptic, glass fibers, metallurgy (welding flux, brazing copper), flame retardant in cellulosic insulation, mattress batting and cotton textile products; fungus control on citrus fruits, to allow talcum powder to flow more freely and as an ingredient of ear drops for the treatment of swimmer's ear, ointment to help heal skin irritations, eye drops to soothe irritated eyes, hemorrhoid ointment and skin cleansers; as mouthwashes, eye lotions, skin lotions, douches for irrigating the bladder and vagina and as hot fomentations for ulcers, whitlows, boils and carbuncles

Physical Properties

Boiling Point: 300 °C (572 °F)
Freezing Point: 168 °C (334.4 °F) to 170 °C (338 °F)
Specific Gravity: 1.435 at 15 °C
Density: 1.435 g/mL at 15 °C
Vapor Pressure: < 0.008 mm Hg
Water Solubility: 20 g/100 cc methanol at 25 °C
Other Solubilities: 1.92 g/100 cc liquid Ammonia at 25 °C; Slightly Soluble in Acetone; 1 g/6 ml Alcohol; 1 g/4 ml Glycerol.
pH: 0.1 molar 5.1
Refraction Index: 1.337
Flash Point: Nonflammable

RTECS Toxicity Data

Acute Oral: Infant TD_{Lo} Dose: 800 mg/kg/4W-I; Toxic Effects: Behavioral - Convulsions or effect on seizure threshold; Gastrointestinal - Hypermotility, diarrhea; Gastrointestinal - Nausea or vomiting. Child TD_{Lo} Dose: 500 mg/kg; Toxic Effects: Gastrointestinal - Nausea or vomiting. Man LD_{Lo} Dose: 429 mg/kg; Toxic Effects: Cardiac - Other changes; Kidney, Ureter, and Bladder - Chgs in tubules (inc acute renal failure, acute tubular necrosis. Woman LD_{Lo} Dose: 200 mg/kg; Toxic Effects: Behavioral - Fluid intake; Gastrointestinal - Hypermotility, diarrhea; Gastrointestinal - Nausea or vomiting. Infant LD_{Lo} Dose: 934 mg/kg. Rat LD_{50} Dose: 2660 mg/kg.

Acute Inhalation: Rat LC_{Lo} Dose: 28 mg/m^3/4hr.

Acute Dermal: Man LD_{Lo} Route: Skin; Dose: 2430 mg/kg; Toxic Effects: Gastrointestinal - Hypermotility, diarrhea; Skin and appendages - Primary irritation; Nutritional and gross metabolic - Body temperature increase. Infant LD_{Lo} Route: Skin; Dose: 1200 mg/kg; Toxic Effects: Behavioral - Convulsions or effect on seizure threshold; Skin and appendages - Dermatitis, other; Nutritional and gross metabolic - Body temperature increase. Infant LD_{Lo} Route: Subcutaneous Dose: 1100 mg/kg; Toxic Effects: Behavioral - Tremor; Gastrointestinal - Hypermotility, diarrhea; Gastrointestinal - Nausea or vomiting. Child LD_{Lo} Route: Skin; Dose: 4 gm/kg/4D. Child LD_{Lo} Route: Skin; Dose: 1500 mg/kg; Toxic Effects: Sense organs and special senses - Conjunctive irritation; Lungs, Thorax, or Respiration - Respiratory depression; Gastrointestinal - Hypermotility, diarrhea.

Irritation Skin: Human Standard Draize Test Dose: 15 mg/3D-I; Reaction: mild.

Reproductive/Teratogenic: Rat Route: Oral; Dose: 6600 mg/kg; Duration: female 1-21D of pregnancy; Effects on Embryo or Fetus - Fetotoxicity; Specific Developmental Abnormalities - Musculoskeletal system; Other developmental abnormalities. Rat Route: Oral; Dose: 45 gm/kg; Duration: male 90D prior to mating; Paternal Effects - Testes, epididymis, sperm duct. Rat Route: Oral; Dose: 5390 mg/kg; Duration: female 6-15D of pregnancy; Effects on Fertility - Post-implantation mortality; Effects on Embryo or Fetus - Fetal death; Specific Developmental Abnormalities - Musculoskeletal system. Rat Route: Oral; Dose: 312 mg/kg; Duration: male 26W prior to mating; Paternal Effects - Testes, epididymis, sperm duct. Rat Route: Oral; Dose: 52 mg/kg; Duration: male 26W prior to mating; Paternal Effects - Spermatogenesis. Rat Route: Inhalation; Dose: 9600 ug/m^3/4 Duration: male 16W prior to mating; Paternal Effects - Spermatogenesis; Testes, epididymis, sperm duct.

Mutagenic: Bacteria - E Coli Mutations in Microorganisms; Dose: 17000 ppm/24H (-S9).

Hazard Overviews

Fire Diamond

Health: Irritating to eyes/skin/respiratory tract. Also Causes: varying degrees of CNS effects, gastrointestinal and kidney toxicity. Chronic Effects: dry skin, skin eruptions, and GI disturbances.

Fire: Noncombustible. Use agents suitable for surrounding fire.

Reactivity: Stable. Hazardous polymerization cannot occur. Incompatible with: alkali carbonates; alkali hydroxides; potassium; acetic anhydride (exploded when heated to 136.4 to 140 °F/58-60 °C). Hazardous decomposition products: carbon oxide(s); boron oxide(s).

Carcinogenicity: IARC - Not listed; NIOSH - Not listed; NTP - Not listed; ACGIH - Not listed; OSHA - Not listed; EPA - Not listed; MAK - Not listed

Primary Target Organs:

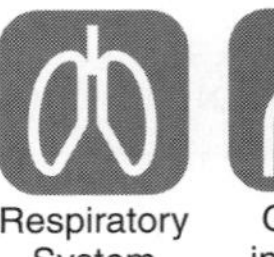
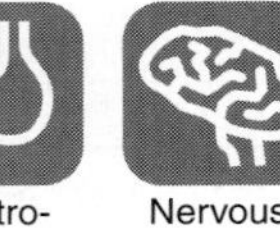

Eyes Skin Respiratory System Gastro-intestinal Nervous System Kidneys

Environmental

Ecotoxicity: LC_{50} Catfish 22 ppm (hard water; exposure was initiated subsequent to fertilization & maintained through 4 days posthatching) /conditions of bioassay not specified LC_{50} Daphnia magna 133 (115-153) mg/l/48 hr /Static bioassay

Cleanup/Disposal: Guide No. 171: Do not touch or walk through spilled material. Stop leak if you can do it without risk. Prevent dust cloud. Avoid inhalation of asbestos dust. Small Dry Spills: With clean shovel place material into clean, dry container and cover loosely; move containers from spill area. Small Spills: Take up with sand or other noncombustible absorbent material and place into containers for later disposal. Large Spills: Dike far ahead of liquid spill for later disposal. Cover powder spill with plastic sheet or tarp to minimize spreading. Prevent entry into waterways, sewers, basements or confined areas.

Environmental Physical Data

BCF: no food chain concentration potential
BOD: none

Regulations

RCRA 40CFR: Not listed
CERCLA: 40CFR 302.4: Not listed
SARA 40CFR 372.65: Not listed
SARA EHS 40CFR 355: Not listed
TSCA: Listed

Analytical Methods

Drinking Water: AOAC 920.202

BOR4650 CAS #: 10043-11-5

BORON NITRIDE POWDER

RTECS: ED7800000
EINECS Number: 233-136-6
Molecular Formula: BN
Structured MF: BN
Formula Weight: 24.82

B≡N

Chemical Structure

Synonyms: BN 40SHP; BORAZON; BORON MONONITRIDE; BZN 550; DENKA BORON NITRIDE GP; DENKA GP; ELBOR; ELBOR LO 10B1-100; ELBOR R; ELBOR RM; ELBORON; GEKSANIT R; HEXANIT R; HEXANITE R; KBN-H10; KUBONIT; KUBONIT KR; SHO BN; SHO BN HPS; SP 1; SP 1 (NITRIDE); SUPER MIGHTY M; UHP-EX; WURZIN
Description: white compressed solid; odorless

Physical Properties

Freezing Point: Sublimes at 3000 °C (5432 °F)
Specific Gravity: 2.34 at 25 °C
Water Solubility: Very Slightly Soluble
Other Solubilities: Hot concentrated alkalies cleave the boron-nitrogen bond, but do not completly solubize the compound
Flash Point: Noncombustible

RTECS Toxicity Data

Acute Oral: Rat LD; Dose: >50 gm/kg.
Acute Dermal: Rabbit LD; Route: Skin; Dose: >20 mL/kg.

Hazard Overviews

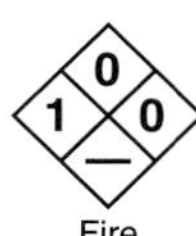

Fire Diamond

Health: Mildly irritating to the eyes,skin, and respiratory tract. Chronic Effects: repeated exposure may cause development of fibrous lung tissue and bronchitis or emphysema.
Fire: Noncombustible. Use extinguishing agents suitable for surrounding fire.
Reactivity: Stable. Hazardous polymerization cannot occur. Avoid: dispersing powder into air. Incompatible with: peroxydisulfuryl difluoride; sodium peroxide; alkali metal hydroxides; carbonates; some acids and organic solvents. Hazardous decomposition products: nitrogen oxide fumes.
Carcinogenicity: IARC - Not listed; NIOSH - Not listed; NTP - Not listed; ACGIH - Not listed; OSHA - Not listed; EPA - Not listed; MAK - Not listed

Primary Target Organs:

Eyes

Skin

Respiratory System

Environmental

Regulations

RCRA 40CFR: Not listed
CERCLA: 40CFR 302.4: Not listed
SARA 40CFR 372.65: Not listed
SARA EHS 40CFR 355: Not listed
TSCA: Listed

BOR5380 CAS #: 1303-86-2

BORON OXIDE

RTECS: ED7900000
EINECS Number: 215-125-8
Molecular Formula: B_2O_3
Structured MF: B_2O_3
Formula Weight: 69.64

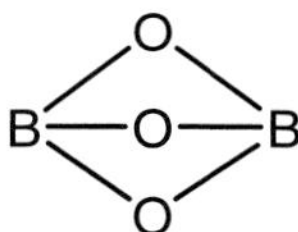

Chemical Structure

Synonyms: BORIC ACID (HBO2),ANHYDRIDE; BORIC ANHYDRIDE; BORIC OXIDE; BORON SESQUIOXIDE; BORON TRIOXIDE; DIBORON TRIOXIDE; FUSED BORIC ACID
Description: colorless, white lumps, crystals; odorless
Use: in herbicides; fire resistant additive for paints, electronics; liq encapsulation techniques; to determine silicon dioxide & alkalies in silicates; in metallurgy; in blowpipe analysis; flux for enamels & glazes; additive for glass fibers; chem intermed for elemental boron, boron master alloys, borides, boron carbide, nitrides & halides; as catalyst; med: mild antiseptics or bacteriostats in eyewashes, mouthwashes, burn dressings, & diaper rash powders.

Physical Properties

Boiling Point: About 1860 °C (3380 °F)
Freezing Point: 448 °C (838.4 °F) to 452 °C (845.6 °F)
Specific Gravity: 2.45 to 2.47
Vapor Pressure: ~ 0 mm Hg
Water Solubility: Slowly Soluble in Water
Other Solubilities: Soluble in Alcohol, Glycerol.
pH: 1% solution 5
Refraction Index: 1.61
Ionization Potential (eV): 13.50
Flash Point: Noncombustible

RTECS Toxicity Data

Acute Oral: Mouse LD_{50} Dose: 3163 mg/kg.
Chronic (Multiple Dose) Inhalation: Rat Dose: 77 mg/m^3/6H/24W-I; Toxic Effects: Kidney, Ureter, and Bladder - Urine volume increased; Kidney, Ureter, and Bladder - Other changes in urine composition.

Hazard Overviews

Fire Diamond

Health: Irritating to eyes/skin/respiratory tract. Chronic Effects: hair loss, swelling around eyes, irritability, anemia, anorexia, convulsions, and kidney damage.
Fire: Noncombustible. Use extinguishing agents suitable for surrounding fire. Avoid using water spray since boron oxide reacts with water to produce heat and form boric acid.
Reactivity: Unstable, reacts slowly with water to release heat and form boric acid; corrosive to metals in the presence of oxygen. Hazardous polymerization cannot occur. Avoid: dust generation; water. Incompatible with: calcium oxide; calcium chloride; bromine pentaflouride.
Carcinogenicity: IARC - Not listed; NIOSH - Listed as carcinogen; NTP - Not listed; ACGIH - Not listed; OSHA - Not listed; EPA - Not listed; MAK - Not listed
Primary Target Organs:

Eyes

Skin
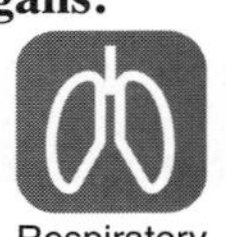
Respiratory System

Nervous System

Kidneys

Blood

Exposure Limits
OSHA PEL: TWA: 15 mg/m^3; total dust.
OSHA PEL Vacated 1989 Limits: TWA: 10 mg/m^3.
ACGIH TLV: TWA: 10 mg/m^3.
NIOSH REL: TWA: 10 mg/m^3.
NIOSH IDLH: 2000 mg/m^3.
DFG MAK: TWA: 15 mg/m^3.
Respirator Recommendation
Exposure Range: >15 to 150 mg/m^3 Air Purifying, Negative Pressure, Half Mask
Exposure Range: >150 to 1500 mg/m^3 Air Purifying, Negative Pressure, Full Face
Exposure Range: >1500 to <2000 mg/m^3 Supplied Air, Constant Flow/Pressure Demand, Full Face
Exposure Range: 2000 to unlimited mg/m^3 Self-contained Breathing Apparatus, Pressure Demand, Full Face
Cartridge Color: dust/mist filter (use P100 or consult supervisor for appropriate dust/mist filter)

Environmental

Regulations

RCRA 40CFR: Not listed
CERCLA: 40CFR 302.4: Not listed
SARA 40CFR 372.65: Not listed
SARA EHS 40CFR 355: Not listed
TSCA: Listed

BOR6110 CAS #: 10294-33-4

BORON TRIBROMIDE

RTECS: ED7400000
DOT: UN2692; IMO8.1
EINECS Number: 233-657-9
Molecular Formula: BBr_3
Structured MF: BBr_3
Formula Weight: 250.57

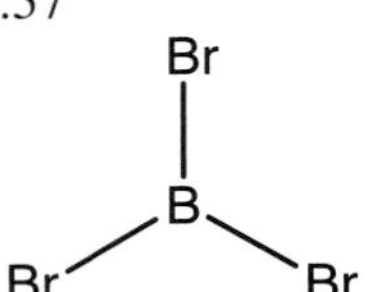

Chemical Structure

Synonyms: BORANE,TRIBROMO-; BORON BROMIDE; TRIBROMOBORANE; TRIBROMOBORON; TRONA
Description: colorless liquid
Use: manufacture of diborane; catalyst in polymerization, alkylation, and acylation; to supply boron to dope semiconductors

Physical Properties

Boiling Point: 90 °C (194 °F)
Freezing Point: -46 °C (-50.8 °F)
Specific Gravity: 2.6431 at 18.4 °C/4 °C
Vapor Density: 8.64 Air=1
Vapor Pressure: 40 mm Hg at 14.0 °C
Water Solubility: Decomposes
Other Solubilities: Soluble in Alcohol; Soluble in Carbon Tetrachloride
Surface Tension: 29.1 dynes/cm at 22 °C
pH: Acid
Refraction Index: 1.5312 at 16.3 °C/D
Critical Temperature: 300 °C
Ionization Potential (eV): 9.70
Flash Point: Nonflammable

Hazard Overviews

Poison

Corrosive

Fire Diamond

Health: Corrosive to eyes/skin/respiratory tract. Poison. Also Causes: tearing, coughing, shortness of breath, swelling of airways, pulmonary edema, respiratory collapse.
Fire: Noncombustible. Do not use water to extinguish fire. Use dry chemical, carbon dioxide, or halon gas.
Reactivity: Stable. Hazardous polymerization cannot occur. Avoid: contact with any other chemical unless chemical compatibility has been determined. Incompatible with: water; steam; alcohol; sodium; potassium. Hazardous decomposition

products: hydrogen bromide gas; oxides of boron; oxides of bromine.

Carcinogenicity: IARC - Not listed; NIOSH - Listed as carcinogen; NTP - Not listed; ACGIH - Not listed; OSHA - Not listed; EPA - Not listed; MAK - Not listed

Primary Target Organs:

Eyes

Skin

Respiratory System

Mucous Membranes

Exposure Limits

OSHA PEL Vacated 1989 Limits: STEL: 1 ppm; 10 mg/m^3; Ceiling.

NIOSH REL: STEL: 1 ppm; 10 mg/m^3.

Respirator Recommendation

Exposure Range: >1 to ppm Self-contained Breathing Apparatus, Pressure Demand, Full Face

Note: poor warning properties

Environmental

Cleanup/Disposal: Guide No. 157: Eliminate all ignition sources (no smoking, flares, sparks or flames in immediate area). All equipment used when handling the product must be grounded. Do not touch damaged containers or spilled material unless wearing appropriate protective clothing. Stop leak if you can do it without risk. A vapor suppressing foam may be used to reduce vapors. Do not get water inside containers. Use water spray to reduce vapors or divert vapor cloud drift. Prevent entry into waterways, sewers, basements or confined areas. Small Spills: Cover with dry earth, dry sand, or other non-combustible material followed with plastic sheet to minimize spreading or contact with rain. Use clean non-sparking tools to collect material and place it into loosely covered plastic containers for later disposal.

Environmental Physical Data

BCF: no food chain concentration potential

BOD: none

Regulations

RCRA 40CFR: Not listed

CERCLA: 40CFR 302.4: Not listed

SARA 40CFR 372.65: Not listed

SARA EHS 40CFR 355: Not listed

TSCA: Listed

BOR6840 CAS #: 10294-34-5

BORON TRICHLORIDE

RTECS: ED1925000

DOT: UN1741; IMO2.2

EINECS Number: 233-658-4

Molecular Formula: BCl_3

Structured MF: BCl_3

Formula Weight: 117.19

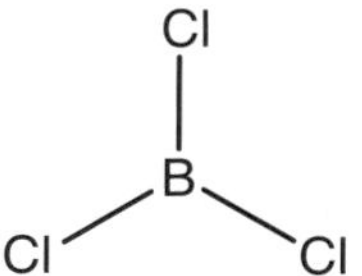

Chemical Structure

Synonyms: BORANE,TRICHLORO-; BORON CHLORIDE; CHLORURE DE BORE; TRICHLOROBORANE; TRICHLOROBORON

Description: colorlees gas; pungent odor

Use: in mfr & purification of boron & metal alloys to remove oxides, nitrides, & carbides; catalyst for org reactions; in semiconductors; in bonding of iron, steels; intermed for advanced composites; agent in depolymerization of paraldehyde; stabilizer for liq sulfur trioxide; for gaseous boriding of metals; soldering flux; making electrical resistors; extinguishing magnesium fires; in refining of aluminum, & in production of optical wave guides; for raising btv value of fuels & rocket propellents.

Physical Properties

Boiling Point: 12.5 °C (55 °F)

Freezing Point: -107 °C (-160.6 °F)

Specific Gravity: 1.35 at 12 °C/4 °C

Vapor Density: 4.03

Vapor Pressure: 1 atm at 12.7 °C

Water Solubility: Decomposes

Other Solubilities: Decomposes in Alcohol; disolves in dioxane with reaction; Soluble in silicone tetrachloride, carbon tetrachloride, titanium tetrachloride, sulfur mono and dichlorides and sulfur dioxide

Surface Tension: 16.7 dynes/cm at 20 °C

pH: pH of solution will be very low

Refraction Index: Alpha 1.4195 at 5.7 °C

Critical Temperature: 178 °C

Critical Pressure: 566 psia

Flash Point: Nonflammable

RTECS Toxicity Data

Acute Inhalation: Rat LC_{50} Dose: 2541 ppm/1hr. Mouse LC_{Lo} Dose: 20 ppm/7hr; Toxic Effects: Lungs, Thorax, or Respiration - Other changes; Skin and appendages - Dermatitis, other; Nutritional and gross metabolic - Weight loss or decreased weight gain.

Hazard Overviews

Corrosive

Fire Diamond

Health: Corrosive, causes severe burns to eyes/skin/respiratory tract. Also Causes: pulmonary edema, digestive tract disturbances. Chronic Effects: possible boron poisoning, possible anorexia, anemia.

Fire: Noncombustible. Use agent suitable for surrounding fire. Reacts with water; do not get water in container.

Reactivity: Stable, in dry, closed containers. Hazardous polymerization cannot occur. Avoid: exposure to heat; water. Incompatible with: water; aniline in the absence of cooling or a diluent; nitrogen; tetraoxide; nitrogen peroxide; nitrogen dioxide; phosphine; fat; grease; alcohol; hexafluorisopropylidene amino lithium; oxygen; organic matter; elastomers; packing materials (Viton; Tygon; Saran; or silastic elastomers); natural or synthetic rubber; hydrogen at 2192 °F (1200 °C); hydrogen sulfide; alkyl mercaptans; ammonia; arsine. Hazardous decomposition products: chlorine gas.

Carcinogenicity: IARC - Not listed; NIOSH - Not listed; NTP - Not listed; ACGIH - Not listed; OSHA - Not listed; EPA - Not listed; MAK - Not listed

Primary Target Organs:

Eyes

Skin

Respiratory System

Mucous Membranes

Environmental

Cleanup/Disposal: Guide No. 125: Fully encapsulating, vapor protective clothing should be worn for spills and leaks with no fire. Do not touch or walk through spilled material. Stop leak if you can do it without risk. If possible, turn leaking containers so that gas escapes rather than liquid. Prevent entry into waterways, sewers, basements or confined areas. Do not direct water at spill or source of leak. Use water spray to reduce vapors or divert vapor cloud drift. Isolate area until gas has dispersed.

Environmental Physical Data

BCF: no food chain concentration potential

BOD: none

Regulations

RCRA 40CFR: Not listed

CERCLA: 40CFR 302.4: Not listed

SARA 40CFR 372.65: Listed TPQ: 500 lb

SARA EHS 40CFR 355: Listed TPQ: 500 lb

TSCA: Listed

BOR7570 CAS #: 7637-07-2

BORON TRIFLUORIDE

RTECS: ED2275000

DOT: UN1008; UN2851; IMO6.1; IMO8.2

EINECS Number: 231-569-5

Molecular Formula: BF_3

Structured MF: BF_3

Formula Weight: 67.82

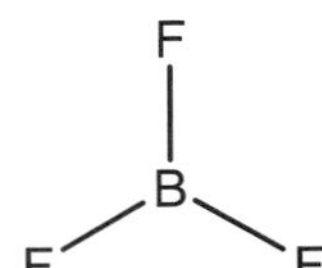

Chemical Structure

Synonyms: ANCA 1040; BORANE,TRIFLUORO-; BORON FLUORIDE; FLUORURE DE BORE; TRIFLUORO BORANE; TRIFLUORO BORON; TRIFLUOROBORANE; TRIFLUOROBORON

Description: colorless gas; pungent suffocating odor

Use: fumigant; in nuclear industry & science (for detection of neutrons, boron-10 enrichment, prodn of neutron absorbing salts); in metallurgy (molten magnesium & alloy antioxidant, flux for soldering magnesium, in gas brazing); catalyst in friedel-crafts type reaction, org synthesis, eg, olefins, alcohols; in operations eg, isomerization, alkylation); prodn of diborane; in casting & heat treating; in electronics ind (treating of separators for electrochem cells; preparing varistors; enhancing surface conductivity of perylene.

Physical Properties

Boiling Point: -99.9 °C (-148 °F)

Freezing Point: -126.8 °C (-196.24 °F)

Vapor Density: 2.38 Air=1

Density: 3.07666 g/L (gas at STP)

Vapor Pressure: 10 mm Hg at -141.3 °C

Water Solubility: 332 g/100 g Water at 0 °C

Other Solubilities: Soluble in concentrated nitric acid, Benzene, dichlorobenzene, Chloroform, Carbon Tetrachloride, Carbon Disulfide; 1.94 g/100 g anhydrous sulfuric acid; Soluble in most saturated & halogenated hydrocarbons & most aromatic compounds.

Odor Threshold: 4.50 to 4.5 mg/m^3

pH: Lewis acid

Critical Temperature: -12.26 °C

Critical Pressure: 49.2 atm

Ionization Potential (eV): 15.50

Flash Point: Nonflammable

RTECS Toxicity Data

Acute Inhalation: Rat LC_{50} Dose: 1180 mg/m^3/4hr. Mouse LC_{50} Dose: 3460 mg/m^3/2hr; Toxic Effects: Behavioral - Somnolence (general depressed activity); Lungs, Thorax, or Respiration - Cyanosis; Blood - Hemorrhage.

Hazard Overviews

Poison

Corrosive

Compressed Gas

Fire Diamond

Health: Corrosive: causes severe irritation of eyes/skin/respiratory tract. Poison. Also Causes: coughing, weak/rapid pulse, chills, chest pain, convulsions, kidney/liver injury. Chronic Effects: mottled teeth, hair loss, anemia.

Fire: Noncombustible. Use extinguishing agents suitable for surrounding fire.

Reactivity: Stable, in closed containers in dry air. Hazardous polymerization cannot occur. Avoid: exposure to moisture. Incompatible with: unsaturated molecules; heat and alkali metals or alkaline earth metals (except magnesium); nitrates; calcium oxide; moist air; lower alcohols; hydrogen sulfide; alkyl mercaptans; alkyl nitrates; ammonia; primary and secondary amines; phosphine; arsine; aluminum chloride; aluminum bromide; slaked lime. Hazardous decomposition products: hydrogen fluoride fumes; boron compounds.
Carcinogenicity: IARC - Not listed; NIOSH - Not listed; NTP - Not listed; ACGIH - Class A4, Not classifiable as a human carcinogen; OSHA - Not listed; EPA - Not listed; MAK - Not listed
Primary Target Organs:

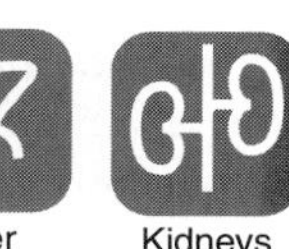

Eyes Skin Respiratory System Nervous System Liver Kidneys

Exposure Limits
OSHA PEL: STEL: 1 ppm; 3 mg/m^3.
ACGIH TLV: TWA: 2.5 mg/m^3; as F.
NIOSH REL: TWA: as F (Flourides). STEL: 1 ppm; 3 mg/m^3.
NIOSH IDLH: 25 ppm.
DFG MAK: TWA: 1 ppm; 3 mg/m^3; as F.
Respirator Recommendation
Exposure Range: >1 to <25 ppm Supplied Air, Constant Flow/Pressure Demand, Full Face
Exposure Range: 25 to unlimited ppm Self-contained Breathing Apparatus, Pressure Demand, Full Face
Note: odor threshold unknown

Environmental

Cleanup/Disposal: Guide No. 125: Fully encapsulating, vapor protective clothing should be worn for spills and leaks with no fire. Do not touch or walk through spilled material. Stop leak if you can do it without risk. If possible, turn leaking containers so that gas escapes rather than liquid. Prevent entry into waterways, sewers, basements or confined areas. Do not direct water at spill or source of leak. Use water spray to reduce vapors or divert vapor cloud drift. Isolate area until gas has dispersed. Guide No. 157: Eliminate all ignition sources (no smoking, flares, sparks or flames in immediate area). All equipment used when handling the product must be grounded. Do not touch damaged containers or spilled material unless wearing appropriate protective clothing. Stop leak if you can do it without risk. A vapor suppressing foam may be used to reduce vapors. Do not get water inside containers. Use water spray to reduce vapors or divert vapor cloud drift. Prevent entry into waterways, sewers, basements or confined areas. Small Spills: Cover with dry earth, dry sand, or other non-combustible material followed with plastic sheet to minimize spreading or contact with rain. Use clean non-sparking tools to collect material and place it into loosely covered plastic containers for later disposal.

Regulations

RCRA 40CFR: Not listed
CERCLA: 40CFR 302.4: Not listed
SARA 40CFR 372.65: Listed TPQ: 500 lb
SARA EHS 40CFR 355: Listed TPQ: 500 lb
TSCA: Listed

BRO1140 CAS #: 314-40-9

BROMACIL

RTECS: YQ9100000
EINECS Number: 206-245-1
Molecular Formula: $C_9H_{13}BrN_2O_2$
Structured MF: $C_9H_{13}BrN_2O_2$
Formula Weight: 261.11

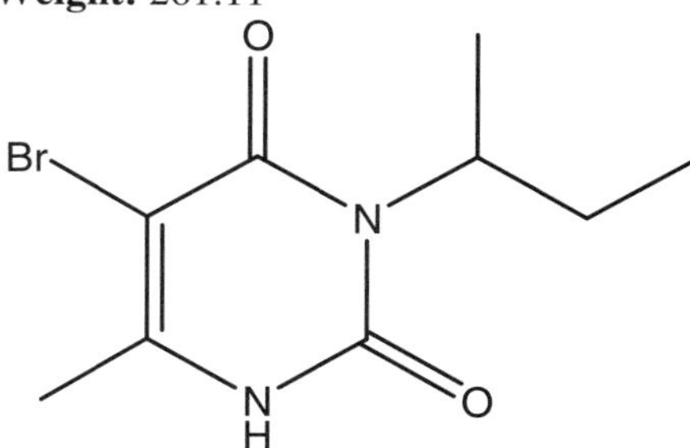

Chemical Structure

Synonyms: BOREA; BOROCIL; BROMAX; BROMAZIL; (5-BROMO-3-SEC-BUTYL-6-METHYLURACIL); 5-BROMO-3-SEC-BUTYL-6-METHYL-URACIL; 5-BROMO-3-SEC-BUTYL-6-METHYLURACIL; 5-BROMO-6-METHYL-3-(1-METHYLPROPYL)-2,4(1H,3H)-PYRIMIDINEDIONE; 5-BROMO-6-METHYL-3-(1-METHYLPROPYL)-2,4(1H,3H)PYRIMIDINEDIONE; 5-BROMO-6-METHYL-3(1-METHYLPROPYL)URACIL; 5-BROMO-6-METHYL-3-(1-METHYLPROPYL)URACIL; CYNOGAN; DU PONT HERBICIDE 976; EEREX GRANULAR WEED KILLER; EEREX WATER SOLUBLE CONCENTRATE WEED KILLER; HERBICIDE 976; HIBOR; HYVAR; HYVAR X-L; HYVAR-X WEED KILLER; HYVAR X-WS; HYVAR X; HYVAR X BROMACIL; HYVAR X WEED KILLER; HYVAREX; HYVAR-XL WEED KILLER; ISTEMUL; KROVAR I WEED KILLER; KROVAR II; KROVAR II WEED KILLER; NALKIL; 2,4(1H,3H)-PYRIMIDINEDIONE,5-BROMO-6-METHYL-3-(1-METHYLPROPYL)-; ROUT G-8; 3-SEK.BUTYL-5-BROM-6-METHYLURACIL; URACIL,5-BROMO-3-SEC-BUTYL-6-METHYL-; URAGAN; URAGON; UREABOR; UROX; UROX B; UROX B WATER SOLUBLE CONCENTRATE WEED KILLER; UROX HA; UROX HX; UROX HX GRANULAR WEED KILLER; UROX-HX
Description: white crystalline solid; odorless
Use: for selective and nonselective control of annual & perennial grasses & broadleaf weeds; for general weed control on non-crop land; for perennial grasses & annual weeds; in pineapple plantations; total weed & brush control on non-crop land; & selective control of annual & perennial weeds & grasses in citrus & pineapple plantations; for selective control of annual & perennial weeds in orange, grapefruit, & lemon orchards.

Physical Properties

Boiling Point: Sublimes
Freezing Point: 157.5 °C (315.5 °F) to 160 °C (320 °F)
Specific Gravity: 1.55 at 25 °C/25 °C
Vapor Pressure: 0.33 mPa at 25 °C

Water Solubility: 815 mg/L at 25 °C
Other Solubilities: Sparingly Soluble in hydrocarbons; Moderately Soluble in strong aq bases.
Flash Point: 44.444 °C Closed Cup

RTECS Toxicity Data

Acute Oral: Rat LD_{50} Dose: 641 mg/kg. Mouse LD_{50} Dose: 3040 mg/kg.
Acute Inhalation: Rat LC_{50} Dose: >4800 mg/m^3.
Acute Dermal: Rabbit LD_{50} Route: Skin; Dose: >5 gm/kg. Rat LD_{50} Route: Skin; Dose: >2500 mg/kg.
Chronic (Multiple Dose) Oral: Rat Dose: 10350 mg/kg/2W-I; Toxic Effects: Behavioral - Ataxia; Gastrointestinal - Changes in structure or function of salivary glands; DEATH. Rat Dose: 45625 mg/kg/2Y-C; Toxic Effects: Nutritional and gross metabolic - Weight loss or decreased weight gain.
Reproductive/Teratogenic: Rat Route: Inhalation; Dose: 38 mg/m^3/2H; Duration: female 7-14D of pregnancy; Effects on Embryo or Fetus - Fetotoxicity; Specific Developmental Abnormalities - Musculoskeletal system.
Mutagenic: Mouse Mutations in Mammalian Somatic Cells; Cell Type: lymphocyte; Dose: 750 mg/L. Insects - D Melanogaster Sex Chromosome Loss; Route: Oral; Dose: 2000 ppm.

Hazard Overviews

Fire Diamond

Health: Mildly irritating to eyes/skin/respiratory tract. Also Causes: vomiting, gastritis, tongue numbness, liver/kidney/thyroid disorder.
Fire: Noncombustible. Use agent suitable for surrounding fire.
Reactivity: Stable. Hazardous polymerization cannot occur. Incompatible with: ammonium sulfamate; No. 2 diesel fuel; xylene; amitrole. Hazardous decomposition products: bromine; carbon oxide(s); nitrogen oxide(s).
Carcinogenicity: IARC - Not listed; NIOSH - Not listed; NTP - Not listed; ACGIH - Class A3, Animal carcinogen; OSHA - Not listed; EPA - Not listed; MAK - Not listed
Primary Target Organs:

Eyes

Skin

Respiratory System

Exposure Limits
OSHA PEL Vacated 1989 Limits: TWA: 1 ppm; 10 mg/m^3.
ACGIH TLV: TWA: 10 mg/m^3.
NIOSH REL: TWA: 1 ppm; 10 mg/m^3.
Respirator Recommendation
Exposure Range: >10 to 100 mg/m^3 Air Purifying, Negative Pressure, Half Mask
Exposure Range: >100 to 1000 mg/m^3 Air Purifying, Negative Pressure, Full Face
Exposure Range: >1000 to 10,000 mg/m^3 Supplied Air, Constant Flow/Pressure Demand, Full Face
Exposure Range: >10,000 to unlimited mg/m^3 Self-contained Breathing Apparatus, Pressure Demand, Full Face
Cartridge Color: black with magenta (P100)

Environmental

Ecotoxicity: LC_{50} Coturnix japonica >5,000 ppm (5 day in diet) /Conditions of bioassay not specified EC_{50} Pimephales promelas (fathead minnow) 167 mg/l/96 hr (confidence limit 162-173 mg/l), flow-through bioassay with measured concentrations, 25.0 °C, dissolved oxygen 8.2 mg/l, hardness 49.4 mg/l calcium carbonate, alkalinity 42.1 mg/l calcium carbonate LC_{50} Salmo gairdneri (Rainbow trout) 56-60 mg active ingredient (as wp)/l/48 hr /Conditions of bioassay not specified
Environmental Fate: If released to the atmosphere, it will degrade rapidly in the vapor phase by reaction with photochemically produced hydroxyl radicals (half-life of about 7.6 hr). Particulate phase will be removed physically from air by wet and dry deposition. If released to soil or water, it can slowly biodegrade. Rapid photo-sensitized photodegradation may occur in natural waters containing sufficient concentrations of sensitizing agents. Numerous K_{oc} measurements, which typically range between 20 and 130, indicate that it can be highly mobile in soil. Detection in ground waters demonstrates that leaching can occur. Results of various persistence studies indicate that it is relatively persistent in soil. The US Dept of Agriculture's Pesticide Properties Database lists a soil half-life of 60 days, but it may typically range from 2 to 8 months.

Environmental Physical Data

Henry's Law Constant: estimated at 5.07 x10^{-11}
Sorption Partition Coefficient: K_{OC} = 23
BCF: fathead minnow 3.2

Regulations

RCRA 40CFR: Not listed
CERCLA: 40CFR 302.4: Not listed
SARA 40CFR 372.65: Listed
SARA EHS 40CFR 355: Not listed
TSCA: Not listed

Analytical Methods

Soil: EPA PMD-TLC; SW846 8321A
Water / Groundwater: EPA 633
Drinking Water: EPA 507, 525.2; AOAC 991.07; ASTM D5475
Food: FDA 212.1, 232.1, 232.4, 242.1
Plasma: EPA 001; FDA 211.1, 231.1, 252
Other: EPA 1656

BRO1280 **CAS #: 53404-19-6**

BROMACIL, LITHIUM SALT

RTECS: UV9021000
EINECS Number: 258-525-8
Molecular Formula: $C_9H_{12}BrLiN_2O_2$
Formula Weight: 267.08

Hazard Overviews

Carcinogenicity: IARC - Not listed; NIOSH - Not listed; NTP - Not listed; ACGIH - Not listed; OSHA - Not listed; EPA - Not listed; MAK - Not listed

Environmental

Regulations

RCRA 40CFR: Not listed
CERCLA: 40CFR 302.4: Not listed
SARA 40CFR 372.65: Listed
SARA EHS 40CFR 355: Not listed
TSCA: Listed

BRO1420 CAS #: 28772-56-7

BROMADIOLONE

RTECS: GN4934700
DOT: UN3024; UN3025; UN3026; UN3027; IMO3.0; IMO6.1
EINECS Number: 249-205-9
Molecular Formula: $C_{30}H_{23}BrO_4$
Formula Weight: 527.42
Synonyms: 2H-1-BENZOPYRAN-2-ONE,3-(3-(4'-BROMO(1,1'-BIPHENYL)-4-YL)-3-HYDROXY-1-PHENYLPROPYL)-4-HYDROXY-; 2H-1-BENZOPYRAN-2-ONE,3-(3-(4-BROMO-(1,1-BIPHENYL)-4-YL)-3-HYDROXY-1-PHENYLPROPYL)-4-HYDROXY-; BOLDO; BOOT HILL; BROMATROL; 3-(3-(4'-BROMO(1,1'-BIPHENYL)-4-YL)3-HYDROXY-1-PHENYLPROPYL)-4-HYDROXY-2H-1-BENZOPYRAN-2-ONE; 3-(3-(4-BROMO(1,1-BIPHENYL)-4-YL)-3-HYDROXY-1-PHENYLPROPYL)-4-HYDROXY-2H-1-BENZOPYRAN-2-ONE; 3-(3-(4-BROMOBIPHENYL-4-YL)-3-HYDROXY-1-PHENYLPROPYL)-4-HYDROXYCOUMARIN; BROMONE; 3-(ALPHA-(P-(P-BROMOPHENYL)-BETA-HYDROXYPHENETHYL)BENZYL)-4-HYDROXYCOUMARIN; 3-(ALPHA-(P-(P-BROMOPHENYL)-BETA-HYDROXYPHENETHYL)BENZYL-4-HYDROXYCOUMARIN; BROMORE; BROPRODIFACOUM; CANADIEN 2000; CONTRAC; CONTRAX; COUMARIN,3-(ALPHA-(P-(P-BROMOPHENYL)-BETA-HYDROXYPHENETHYL)BENZYL)-4-HYDROXY-; COUMARIN,3-(ALPHA-(P-(P-BROMOPHENYL)-BETA-HYDROXYPHENETHYL)BENZYL)-4-HYDROXY-(8CI); EPA PESTICIDE CODE 112001; ERADIC; (HYDROXY-4 COUMARINYL 3)-3 PHENYL-3 (BROMO-4 BIPHENYLYL-4)-1 PROPANOL-1; 3-(4-HYDROXYCOUMARIN-3-YL)-3-PHENYL-1-(4-BROMOBIPHENYL)PROPAN-1-OL; (HYDROXY-4-COUMARINYL-3)-3-PHENYL-3-(BROMO-4-BIPHENYLYL-4)-1-PROPANOL-1; 3-(4-HYDROXY-2-OXOCHROMEN-3-YL)-3-PHENYL-1-(4-BROMOBIPHENYL)PROPAN-1-OL; LM-637; LM 637; MAKI; SUP'OPERATS; RAFIX; RATIMUS; ROTOX; SUPER-CAID; SUPER-CALD; SUPER-ROZOL; TEMUS; TOPIDION
Description: white to off-white powder; odorless
Use: rodenticide (anticoagulant)

Physical Properties

Freezing Point: 200 °C (392 °F) to 210 °C (410 °F)
Saturated Vapor Density: 1.607192021 kg/m^3
Vapor Pressure: 0.002 mPa at 20 °C
Water Solubility: 19 mg/L at 20 °C
Other Solubilities: Slightly Soluble in Chloroform; Practically Insoluble in Diethyl Ether & Hexane; 730 g/l Dimethylformamide at 20 °C; 8.2 g/l Ethanol at 20 °C; 25 g/l Ethyl Acetate at 20 °C.

RTECS Toxicity Data

Acute Oral: Rat LD_{50} Dose: 490 ug/kg. Mouse LD_{50} Dose: 1750 ug/kg; Toxic Effects: Blood - Hemorrhage; Blood - Change in clotting factors.

Hazard Overviews

Carcinogenicity: IARC - Not listed; NIOSH - Not listed; NTP - Not listed; ACGIH - Not listed; OSHA - Not listed; EPA - Not listed; MAK - Not listed

Environmental

Ecotoxicity: LD_{50} Quail oral 1600 mg/kg LC_{50} Rainbow trout 1.4 mg/l/96 hr /Conditions of bioassay not specified
Cleanup/Disposal: Guide No. 131: Fully encapsulating, vapor protective clothing should be worn for spills and leaks with no fire. Eliminate all ignition sources (no smoking, flares, sparks or flames in immediate area). All equipment used when handling the product must be grounded. Do not touch or walk through spilled material. Stop leak if you can do it without risk. Prevent entry into waterways, sewers, basements or confined areas. A vapor suppressing foam may be used to reduce vapors. Small Spills: Absorb with earth, sand or other non-combustible material and transfer to containers for later disposal. Use clean non-sparking tools to collect absorbed material. Large Spills: Dike far ahead of liquid spill for later disposal. Water spray may reduce vapor; but may not prevent ignition in closed spaces. Guide No. 151: Do not touch damaged containers or spilled material unless wearing appropriate protective clothing. Stop leak if you can do it without risk. Prevent entry into waterways, sewers, basements or confined areas. Cover with plastic sheet to prevent spreading. Absorb or cover with dry earth, sand or other non-combustible material and transfer to containers. Do not get water inside containers.

Regulations

RCRA 40CFR: Not listed
CERCLA: 40CFR 302.4: Not listed
SARA 40CFR 372.65: Not listed TPQ: 100/10000 lb
SARA EHS 40CFR 355: Listed TPQ: 100 lb
TSCA: Not listed

Analytical Methods

Soil: EPA PMD-BRA

BRO1840 CAS #: 7726-95-6

BROMINE

RTECS: EF9100000
DOT: UN1744; IMO8.0
EINECS Number: 231-778-1
Molecular Formula: Br_2
Structured MF: Br
Formula Weight: 159.808

Br——Br

Chemical Structure

Synonyms: BROM; BROME; BROMINE SOLUTION; BROMINE SOLUTIONS; BROMO; BROOM; DIBROMINE; EPA PESTICIDE CHEMICAL CODE 008701; MOLECULAR BROMINE

Description: reddish brown liquid; irritating odor

Use: for gold extraction; military gas; bleaching fibers & silk; mfr of anti-knock cmpd (ethylene bromide) for gasoline; fire-retardant for plastics; in photography; shrink-proofing wool; in org synthesis; solvent; intermed for fumigants, ethyl bromide, & other bromine cmpd & salts, in mfr of meds (eg, sedatives, etc), hydraulic fluids, refrigerating, dehumidifying, & disinfecting agents, hair-waving prepn; fire-extinguishing fluid; anal rgnt; formerly antiseptic, deodorant; dyes; oil & gas well drilling & completion fluids.

Physical Properties

Boiling Point: 58.78 °C (138 °F)
Freezing Point: -7.25 °C (18.95 °F)
Specific Gravity: 3.1023 at 25 °C/4 °C
Vapor Density: 5.51 Air=1
Vapor Pressure: 100 mm Hg at 9.3 °C
Water Solubility: 4% by weight
Other Solubilities: Soluble in common organic solvents.
Surface Tension: 41 dynes/cm at 20 °C
Odor Threshold: 0.329 to 24.5 mg/m^3
Refraction Index: 1.661
Critical Temperature: 315 °C
Critical Pressure: 102 atm
Ionization Potential (eV): 10.55
Flash Point: Nonflammable

RTECS Toxicity Data

Acute Oral: Human LD_{Lo} Dose: 14 mg/kg. Rat LD_{50} Dose: 2600 mg/kg.

Acute Inhalation: Rat LC_{50} Dose: 2700 mg/m^3; Toxic Effects: Lungs, Thorax, or Respiration - Other changes. Human LC_{Lo} Dose: 1000 ppm.

Hazard Overviews

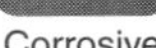
Corrosive

Fire Diamond

Health: Corrosive, causes severe irritation or burns to eyes/skin/respiratory tract. Toxic. Also Causes: dizziness, headache, coughing, and pulmonary edema.

Fire: Noncombustible. However, it is a strong oxidizer capable of igniting combustible materials. Use extinguishing agents suitable for surrounding fire.

Reactivity: Stable. Hazardous polymerization cannot occur. Avoid: combustibles. Incompatible with: acetylene; ammonia; carbonyl compounds (aldehydes, ketones, carboxylic acids); acrylonitrile; reducing materials; metal acetylides; metal carbides; metals; nonmetal hydrides; dimethyl formamide; hydrogen; ethyl phosphine; nickel carbonyl; isobutyrophenone; nitrogen triiodide; oxygen difluoride; ozone; phosphorus; potassium; metal azides (silver azide); methanol and other alcohols; sodium carbide; sodium; lithium; lithium carbide; aluminum; titanium; mercury; germanium; antimony; alkali hydroxides; arsenites; ferrous and mercurous salts; hypophosphites; boron; trialkyl boranes; chlorotrifluoroethylene; cesium acetylene carbide; diethyl ether; phosphine; natural rubber; tetrahydrofuran; mixtures of ethanol and phosphorus; copper hydride; cuprous acetylide; fluorine; magnesium phosphide; diethyl zinc; dimethylformamide; trimethylamine. Hazardous decomposition products: bromine fumes.

Carcinogenicity: IARC - Not listed; NIOSH - Listed as carcinogen; NTP - Not listed; ACGIH - Not listed; OSHA - Not listed; EPA - Not listed; MAK - Not listed

Primary Target Organs:

Eyes

Skin

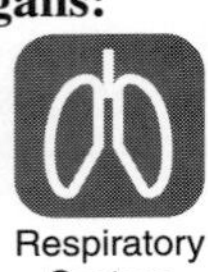
Respiratory System

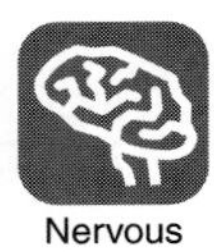
Nervous System

Exposure Limits

OSHA PEL: TWA: 0.1 ppm; 0.7 mg/m^3.

OSHA PEL Vacated 1989 Limits: TWA: 0.1 ppm; 0.7 mg/m^3; STEL: 0.3 ppm; 0.2 mg/m^3.

ACGIH TLV: TWA: 0.1 ppm; 0.66 mg/m^3; STEL: 0.2 ppm; 1.3 mg/m^3.

NIOSH REL: TWA: 0.1 ppm; 0.7 mg/m^3. STEL: 0.3 ppm; 2 mg/m^3.

NIOSH IDLH: 3 ppm.

DFG MAK: TWA: 0.1 ppm; 0.7 mg/m^3.

Respirator Recommendation

Exposure Range: >0.1 to 1 ppm Air Purifying, Negative Pressure, Half Mask

Exposure Range: >1 to <3 ppm Air Purifying, Negative Pressure, Full Face

Exposure Range: 3 to unlimited ppm Self-contained Breathing Apparatus, Pressure Demand, Full Face

Cartridge Color: yellow

Environmental

Ecotoxicity: Aquatic toxicity: 10 ppm/10 hr/cladophora/killed/fresh water 10 ppm/*/fish/irritant/salt water *Time period not specified

Cleanup/Disposal: Guide No. 154: Eliminate all ignition sources (no smoking, flares, sparks or flames in immediate area). Do not touch damaged containers or spilled material unless wearing appropriate protective clothing. Stop leak if you can do it without risk. Prevent entry into waterways, sewers, basements or confined areas. Absorb or cover with dry earth, sand or other non-combustible material and transfer to containers. Do not get water inside containers.

Environmental Physical Data

BCF: no food chain concentration potential
BOD: not pertinent

Regulations

RCRA 40CFR: Not listed
CERCLA: 40CFR 302.4: Not listed

SARA 40CFR 372.65: Listed TPQ: 500 lb
SARA EHS 40CFR 355: Listed TPQ: 500 lb
TSCA: Listed

Analytical Methods

Air: EPA 26, 26A; ASTM D4490
Water / Groundwater: APHA 4500-BR; ASTM D4744
Drinking Water: AOAC 920.202
Food: AOAC 952.24, 974.36
Indoor / Expired Air: NIOSH 6011

BRO1980 CAS #: 13863-41-7

BROMINE CHLORIDE

EINECS Number: 237-601-4
Molecular Formula: BrCl
Structured MF: BrCl
Formula Weight: 115.36
Description: dark red liquid, gas (above 41 F); sharp, penetrating odor

Physical Properties

Boiling Point: 5 °C (41 °F)
Freezing Point: -66 °C (-86.8 °F)
Density: 2.32 at 25 °C
Vapor Pressure: 1750 mm Hg at 20 °C
Water Solubility: 8 wt % at 20 °C
Other Solubilities: Soluble in Ether and Carbon Disulfide
Critical Temperature: 236.5 °C
Flash Point: Noncombustible

Hazard Overviews

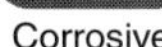
Corrosive

Fire Diamond

Health: Corrosive to eyes/skin/respiratory tract. Also Causes: choking, asphyxia. Chronic Effects: chronic bromine poisoning.
Fire: Noncombustible. However, it is a strong oxidizer capable of igniting combustibles. Use agent suitable for surrounding fire.
Reactivity: Stable. Hazardous polymerization cannot occur. Avoid: exposure to combustible materials; handling in wet atmospheres. Incompatible with: reducing materials; ammonia; corrodes most metals (wet); aluminum; magnesium; mercury; potassium; sodium; titanium; alkalis. Hazardous decomposition products: bromine and chlorine gas.
Carcinogenicity: IARC - Not listed; NIOSH - Not listed; NTP - Not listed; ACGIH - Not listed; OSHA - Not listed; EPA - Not listed; MAK - Not listed

Primary Target Organs:

Eyes

Skin

Respiratory System

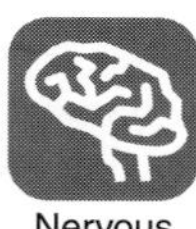
Nervous System

Environmental

Regulations

RCRA 40CFR: Not listed
CERCLA: 40CFR 302.4: Not listed
SARA 40CFR 372.65: Not listed
SARA EHS 40CFR 355: Not listed
TSCA: Listed

BRO2120 CAS #: 7789-30-2

BROMINE PENTAFLUORIDE

RTECS: EF9350000
DOT: UN1745; IMO5.1
EINECS Number: 232-157-8
Molecular Formula: BrF_5
Structured MF: BrF_5
Formula Weight: 174.92
Synonyms: BROMINE FLUORIDE
Description: colorless to pale yellow gas, fuming liquid; pungent odor
Use: as a fluorinating agent to produce fluorocarbons and as an oxidizer in rocket propellant systems

Physical Properties

Boiling Point: 40.76 °C (105 °F)
Freezing Point: -60.5 °C (-76.9 °F)
Specific Gravity: 2.4604 at 25 °C
Vapor Density: 6.05 Air=1 at 20 °C
Saturated Vapor Density: 3.805903448 kg/m^3
Vapor Pressure: 328 torr at 20 °C
Water Solubility: Reacts violently
Other Solubilities: Reacts violently with many acids, alkalis and solvents.
Critical Temperature: 197 °C
Ionization Potential (eV): 13.17 +/-0.01
Flash Point: Will not burn

Hazard Overviews

Poison

Corrosive

Fire Diamond

Health: Corrosive to eyes/skin/respiratory tract. Poison. Also Causes: corrosion and burns of the digestive tract.
Fire: Noncombustible. However, it is a strong oxidizer capable of igniting combustibles. Use extinguishing agents suitable for surrounding fire with one exception; use water only on large fires in flooding quantities as fog.

Reactivity: Stable. Hazardous polymerization cannot occur. Avoid: exposure to water. Incompatible with: acetonitrile; acids; halogens; metal halides; metals; metal oxides; hydrogen-containing materials; perchloryl perchlorate; water/steam; arsenic; carbon monoxide; calcium oxide; grease; paper; wax. Hazardous decomposition products: hydrogen fluoride gas; hydrogen bromide gas.

Carcinogenicity: IARC - Not listed; NIOSH - Not listed; NTP - Not listed; ACGIH - Class A4, Not classifiable as a human carcinogen; OSHA - Not listed; EPA - Not listed; MAK - Not listed

Primary Target Organs:

Eyes

Skin

Respiratory System

Gastro-intestinal

Exposure Limits

OSHA PEL: TWA: 2.5 mg/m^3; as F.

OSHA PEL Vacated 1989 Limits: TWA: 0.1 ppm; 0.7 mg/m^3.

ACGIH TLV: TWA: 0.1 ppm; 0.72 mg/m^3.

NIOSH REL: TWA: 0.1 ppm; 0.7 mg/m^3; as F (Flourides).

DFG MAK: TWA: 2.5 mg/m^3; as F.

Respirator Recommendation

Exposure Range: >0.1 to ppm Self-contained Breathing Apparatus, Pressure Demand, Full Face

Note: odor threshold unknown

Environmental

Cleanup/Disposal: Guide No. 144: Eliminate all ignition sources (no smoking, flares, sparks or flames in immediate area). Do not touch damaged containers or spilled material unless wearing appropriate protective clothing. Stop leak if you can do it without risk. Use water spray to reduce vapors or divert vapor cloud drift. Do not get water on spilled substance or inside containers. Large Spills: Do not clean-up or dispose of, except under supervision of a specialist.

Environmental Physical Data

BCF: no food chain concentration potential

BOD: none

Regulations

RCRA 40CFR: Not listed

CERCLA: 40CFR 302.4: Not listed

SARA 40CFR 372.65: Not listed

SARA EHS 40CFR 355: Not listed

TSCA: Not listed

BRO2260 CAS #: 7787-71-5

BROMINE TRIFLUORIDE

RTECS: EF9360000

DOT: UN1746; IMO8.0

EINECS Number: 232-132-1

Molecular Formula: BrF_3

Structured MF: BrF_3

Formula Weight: 136.92

Synonyms: BROMINE FLUORIDE

Description: colorless to pale yellow liquid, solid; irritating odor

Use: solvent for fluorides; fluorinating agent; electrolytic solvent; in org synth & in forming uranium fluorides for isotopic enrichment & for fuel element reprocessing

Physical Properties

Boiling Point: 125.75 °C (258 °F)

Freezing Point: 8.77 °C (47.786 °F)

Specific Gravity: 2.803 at 25 °C

Vapor Density: 4.7 Air=1

Vapor Pressure: 18 mm Hg at 39 °C

Water Solubility: Insoluble

Surface Tension: 36.3 dynes/cm at 20 °C

pH: Acid

Critical Temperature: Estimated at 164 °C

Ionization Potential (eV): 12.15 +/-0.04

Flash Point: Not flammable

Hazard Overviews

Poison

Corrosive

Fire Diamond

Health: Corrosive, causes severe burns to eyes/skin/respiratory tract. Poison.

Fire: Noncombustible. However, it is a strong oxidizer capable of igniting combustibles. Use dry chemical, dry sand, or carbon dioxide. Do not use water because of the violent reaction that will occur.

Reactivity: Stable. Hazardous polymerization cannot occur. Avoid: exposure to ammonia gas; water. Incompatible with: ammonia gas at 392 °F (200 °C); water. Hazardous decomposition products: fluoride fumes; selenium fumes.

Carcinogenicity: IARC - Not listed; NIOSH - Not listed; NTP - Not listed; ACGIH - Class A4, Not classifiable as a human carcinogen; OSHA - Not listed; EPA - Not listed; MAK - Not listed

Primary Target Organs:

Eyes

Skin
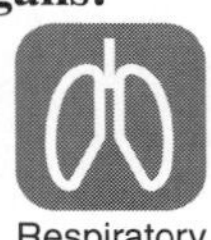
Respiratory System

Mucous Membranes

Exposure Limits

OSHA PEL: TWA: 2.5 mg/m^3; as F.

ACGIH TLV: TWA: 2.5 mg/m^3; as F.

NIOSH REL: TWA: 2.5 mg/m^3; as F (Flourides).

DFG MAK: TWA: 2.5 mg/m^3; as F.

Environmental

Cleanup/Disposal: Guide No. 144: Eliminate all ignition sources (no smoking, flares, sparks or flames in immediate area). Do not touch damaged containers or spilled material unless wearing appropriate protective clothing. Stop leak if you can do it without risk. Use water spray to reduce vapors or divert vapor cloud drift. Do not get water on spilled substance or inside containers. Large Spills: Do not clean-up or dispose of, except under supervision of a specialist.

Environmental Physical Data

BCF: no food chain concentration potential
BOD: none

Regulations

RCRA 40CFR: Not listed
CERCLA: 40CFR 302.4: Not listed
SARA 40CFR 372.65: Not listed
SARA EHS 40CFR 355: Not listed
TSCA: Listed

BRO2540 CAS #: 598-31-2

BROMOACETONE

RTECS: UC0525000
DOT: UN1569; IMO6.1
EINECS Number: 209-928-2
Molecular Formula: C_3H_5BrO
Structured MF: CH_3COCH_2Br
Formula Weight: 136.99
Synonyms: ACETONYL BROMIDE; ACETYL METHYL BROMIDE; ACETYLMETHYL BROMIDE; B-STOFF; BROMOMETHYL METHYL KETONE; 1-BROMO-2-PROPANONE; BROMO-2-PROPANONE; MARTONITE; MONOBROMOACETONE; 2-PROPANONE,1-BROMO-; 2-PROPANONE,BROMO-; 2-PROPANONE,1-BROMO-(9CI)
Description: colorless liquid; rapidly becomes violet; pungent odor
Use: chemical war gas (former use); organic synthesis; tear gas

Physical Properties

Boiling Point: 137 °C (279 °F)
Freezing Point: -36.5 °C (-33.7 °F)
Specific Gravity: 1.634 at 23 °C
Vapor Density: 4.75 Air=1
Saturated Vapor Density: 1.25291706 kg/m^3
Vapor Pressure: 9 mm Hg at 20 °C
Water Solubility: Sparingly Soluble in Water
Other Solubilities: Soluble in Ether, Benzene.
Refraction Index: 1.4697 at 15 °C/D
Ionization Potential (eV): 9.73 +/-2.0
Flash Point: Combustible

RTECS Toxicity Data

Acute Inhalation: Human LC_{Lo} Dose: 572 ppm/10M.

Hazard Overviews

Corrosive

Fire: Combustible.

Carcinogenicity: IARC - Not listed; NIOSH - Not listed; NTP - Not listed; ACGIH - Not listed; OSHA - Not listed; EPA - Not listed; MAK - Not listed

Environmental

Ecotoxicity: Aquatic toxicity: TLm 96: 10-100 ppm
Environmental Fate: If released to the atmosphere, it will degrade via reaction with photochemically produced hydroxyl radicals (estimated half-life of 53 days in an average atmosphere), although degradation via photolysis may be more rapid. Physical removal from air by rainfall is possible. If released to water, photolysis may be an important degradation process. Volatilization from water is not rapid, but may be important in the absence of significant degradation processes. If spilled into a body of water, it will sink to the bottom of the water column, dissolve slowly and turn violet rapidly. If released to soil, significant leaching may be possible. Evaporation from dry surfaces is likely to occur. The potential significance of biodegradation in soil or water is not known.
Cleanup/Disposal: Guide No. 131: Fully encapsulating, vapor protective clothing should be worn for spills and leaks with no fire. Eliminate all ignition sources (no smoking, flares, sparks or flames in immediate area). All equipment used when handling the product must be grounded. Do not touch or walk through spilled material. Stop leak if you can do it without risk. Prevent entry into waterways, sewers, basements or confined areas. A vapor suppressing foam may be used to reduce vapors. Small Spills: Absorb with earth, sand or other non-combustible material and transfer to containers for later disposal. Use clean non-sparking tools to collect absorbed material. Large Spills: Dike far ahead of liquid spill for later disposal. Water spray may reduce vapor; but may not prevent ignition in closed spaces.

Environmental Physical Data

Henry's Law Constant: estimated at 1.2×10^{-5}
Octanol/Water Partition Coefficient: log K_{ow} = 0.485
Sorption Partition Coefficient: K_{oc} = < 50
BCF: not significant

Regulations

RCRA 40CFR: Listed Hazardous Waste No. P017 Toxic Waste
CERCLA: 40CFR 302.4: Listed per RCRA Section 3001 RQ: 1000 lb (453.5 kg)
SARA 40CFR 372.65: Not listed
SARA EHS 40CFR 355: Not listed
TSCA: Not listed

Analytical Methods

Soil: SW846 8240B, 8260A, 8260B
Plasma: EPA 29

BRO2960 **CAS #: 108-86-1**

BROMOBENZENE

RTECS: CY9000000
DOT: UN2514
EINECS Number: 203-623-8
Molecular Formula: C_6H_5Br
Formula Weight: 157.02

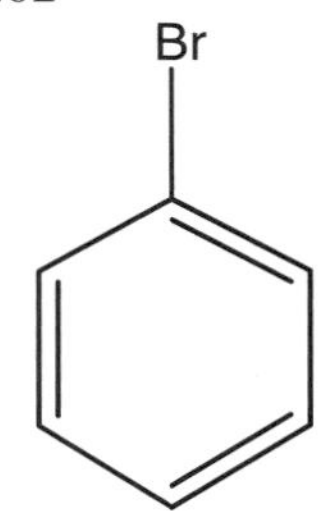

Chemical Structure

Synonyms: BENZENE,BROMO-; MONOBROMOBENZENE; PHENYL BROMIDE; PHENYLBROMIDE
Description: colorless liquid; pungent odor
Use: organic synthesis, especially to make phenyl magnesium bromide; solvent, especially for crystals on a large scale and where a heavy liquid is desirable; and motor oil additive

Physical Properties

Boiling Point: 156.2 °C (313 °F) at 760 mm Hg
Freezing Point: -30.6 °C (-23.08 °F)
Specific Gravity: 1.5083
Vapor Density: 5.41 Air=1
Density: 1.497 g/mL
Vapor Pressure: 5 mm Hg at 27.8 °C
Water Solubility: < 1 mg/mL at 20.5 C
Other Solubilities: 95% Ethanol: >=100 mg/ml at 20.5 °C; Acetone: >=100 mg/ml at 20.5 °C; Benzene: Soluble; Carbon Tetrachloride: Soluble; Chloroform: miscible; DMSO: >=100 mg/ml at 20.5 °C; Ether: Soluble; Petroleum hydrocarbons: miscible.
Surface Tension: 36 dyne/cm at 20 °C
Odor Threshold: Recognition 1.7 to 2.1 mg/m^3
Refraction Index: 1.565 at 15 °C/D
Critical Temperature: 397 °C
Critical Pressure: 33912 mm Hg
Ionization Potential (eV): 9.00 +/-0.02
Flash Point: 51 °C
Autoignition Temperature: 566 °C

RTECS Toxicity Data

Acute Oral: Rat LD_{50} Dose: 2699 mg/kg. Mouse LD_{50} Dose: 2700 mg/kg; Toxic Effects: Behavioral - Somnolence (general depressed activity); Behavioral - Muscle contraction or spasticity.
Acute Inhalation: Rat LC_{50} Dose: 20411 mg/m^3. Mouse LC_{50} Dose: 21 gm/m^3/2hr; Toxic Effects: Behavioral - Somnolence (general depressed activity); Behavioral - Muscle contraction or spasticity.
Acute Dermal: Mouse LD_{50} Route: Subcutaneous Dose: 2 gm/kg; Toxic Effects: Peripheral Nerve and sensation - Flaccid paralysis without anesthesia; Lungs, Thorax, or Respiration - Respiratory stimulation; Lungs, Thorax, or Respiration - Other changes.
Mutagenic: Mouse Micronucleus Test; Route: Intraperitoneal; Dose: 125 mg/kg/24H. Hamster Sister Chromatid Exchenge; Cell Type: ovary; Dose: 500 mg/L.

Hazard Overviews

Flammable

Fire Diamond

Health: Irritating to eyes/skin/respiratory tract. Also Causes: CNS depression, narcosis, changes in blood, liver, and testicular epithelium.
Fire: Flammable. Use dry chemical, carbon dioxide, water spray, fog, or alcohol-resistant foam. Vapors may travel to an ignition source and flash back. Container may explode in heat of fire.
Reactivity: Stable. Hazardous polymerization cannot occur. Avoid: heat; ignition sources. Incompatible with: oxidizers; bromobutane and sodium. Hazardous decomposition products: toxic bromine vapors.
Carcinogenicity: IARC - Not listed; NIOSH - Not listed; NTP - Not listed; ACGIH - Not listed; OSHA - Not listed; EPA - Not listed; MAK - Not listed
Primary Target Organs:

Eyes

Skin

Respiratory System
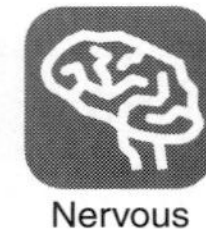
Nervous System
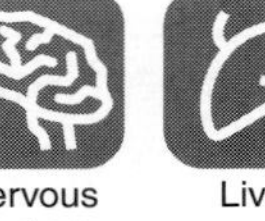

Liver
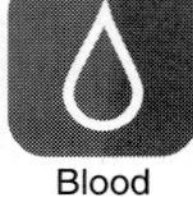
Blood

Environmental

Cleanup/Disposal: Guide No. 129: Eliminate all ignition sources (no smoking, flares, sparks or flames in immediate area). All equipment used when handling the product must be grounded. Do not touch or walk through spilled material. Stop leak if you can do it without risk. Prevent entry into waterways, sewers, basements or confined areas. A vapor suppressing foam may be used to reduce vapors. Absorb or cover with dry earth, sand or other non-combustible material and transfer to containers. Use clean non-sparking tools to collect absorbed material. Large Spills: Dike far ahead of liquid spill for later disposal. Water spray may reduce vapor; but may not prevent ignition in closed spaces.

Environmental Physical Data

Octanol/Water Partition Coefficient: log K_{ow} = 2.99

Regulations

RCRA 40CFR: Not listed
CERCLA: 40CFR 302.4: Not listed
SARA 40CFR 372.65: Not listed
SARA EHS 40CFR 355: Not listed

TSCA: Listed

Analytical Methods

Air: EPA VG-011-1, TO-1; ASTM D4490
Soil: SW846 8010B, 8021A, 8260A
Water / Groundwater: APHA 6040-B, 6210-D, 6230-D
Drinking Water: EPA 502.1, 502.2, 503.1, 524.1, 524.2; APHA 6210-C, 6220-C, 6230-C
Indoor / Expired Air: EPA IP-1B
Plasma: EPA 29

BRO3240 CAS #: 586-76-5

4-BROMOBENZOIC ACID

RTECS: DG4448050
EINECS Number: 209-581-7
Molecular Formula: $C_7H_5BrO_2$
Formula Weight: 201.03

Chemical Structure

Synonyms: BENZOIC ACID,P-BROMO-; BENZOIC ACID,4-BROMO-(9CI); P-BROMOBENZOIC ACID; P-CARBOXYBROMOBENZENE
Description: colorless to red crystals
Use: for the detection of strontium; inorganic synthesis

Physical Properties

Freezing Point: 245.5 °C (473.9 °F)
Specific Gravity: 1.894
Water Solubility: Slightly Soluble
Other Solubilities: Ethanol: Soluble; Ether: Soluble.
Ionization Potential (eV): 9.72 +/-1.0

RTECS Toxicity Data

Acute Oral: Mouse LD_{50} Dose: 1059 mg/kg.

Hazard Overviews

Health: Irritating to eyes/skin/respiratory tract. Harmful. Other Acute Effects: harmful if swallowed; may be harmful if inhaled; may be harmful if absorbed through the skin.
Fire: Hazards: emits toxic fumes. Extinguishing agents: water spray; carbon dioxide, dry chemical powder or appropriate foam. Precautions: combustible liquid.
Reactivity: Incompatible with: strong oxidizing agents. Hazardous decomposition products: toxic fumes of: carbon monoxide, carbon dioxide, hydrogen bromide gas.
Carcinogenicity: IARC - Not listed; NIOSH - Not listed; NTP - Not listed; ACGIH - Not listed; OSHA - Not listed; EPA - Not listed; MAK - Not listed

Primary Target Organs:

Eyes

Skin

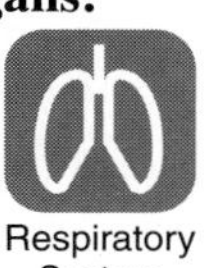
Respiratory System

Environmental

Regulations

RCRA 40CFR: Not listed
CERCLA: 40CFR 302.4: Not listed
SARA 40CFR 372.65: Not listed
SARA EHS 40CFR 355: Not listed
TSCA: Listed

BRO3940 CAS #: 35691-65-7

1-BROMO-1-(BROMOMETHYL)-1,3-PROPANEDICARBONITRILE

RTECS: MA5599000
EINECS Number: 252-681-0
Molecular Formula: $C_6H_6Br_2N_2$
Formula Weight: 265.96
Synonyms: 2-BROMO-2-(BROMOMETHYL)GLUTARONITRILE; 1,2-DIBROMO-2,4-DICYANOBUTANE; METHYLDIBROMOGLUTARONITRILE; PENTANEDINITRILE,2-BROMO-2-(BROMOMETHYL)-

Physical Properties

Freezing Point: 52 °C (125.6 °F)
Water Solubility: Insoluble
Other Solubilities: Acetone: Very Soluble; Benzene: Very Soluble; DMF Very Soluble

RTECS Toxicity Data

Reproductive/Teratogenic: Rat Route: Oral; Dose: 1750 mg/kg; Duration: female 6-15D of pregnancy; Effects on Fertility - Post-implantation mortality.

Hazard Overviews

Carcinogenicity: IARC - Not listed; NIOSH - Not listed; NTP - Not listed; ACGIH - Not listed; OSHA - Not listed; EPA - Not listed; MAK - Not listed

Environmental

Regulations

RCRA 40CFR: Not listed
CERCLA: 40CFR 302.4: Not listed
SARA 40CFR 372.65: Listed
SARA EHS 40CFR 355: Not listed
TSCA: Listed

BRO4220 **CAS #: 353-59-3**

BROMOCHLORODIFLUORO-METHANE

RTECS: PA5270000
DOT: UN1974; IMO2.2
EINECS Number: 206-537-9
Molecular Formula: $CBrClF_2$
Formula Weight: 165.37
Synonyms: BCF; CHLORODIFLUOBROMOMETANO; CHLORODIFLUOROBROMOMETHANE; CHLORODIFLUOROMONOBROMOMETHANE; DAIFLON 12B1; DWUFLUOROCHLOROBROMOMETAN; FLUGEX 12B1; FLUOROCARBON 1211; FREON 12B1; HALON 1211; METHANE,BROMOCHLORODIFLUORO-; R 12B1; R12B1
Description: colorless gas
Use: refrigeration & air conditioning

Physical Properties

Boiling Point: -4 °C (25 °F)
Freezing Point: -161 °C (-257.8 °F)
Specific Gravity: Liquid 1.85 at 15 °C
Vapor Pressure: 41.7 kPa at -25 °C
Ionization Potential (eV): 11.21 +/-0.02

RTECS Toxicity Data

Acute Inhalation: Rat LC_{50} Dose: 20 pph/15M; Toxic Effects: Behavioral - Tremor; Behavioral - Convulsions or effect on seizure threshold; Lungs, Thorax, or Respiration - Respiratory depression. Guinea Pig LC_{Lo} Dose: 30 pph/2hr; Toxic Effects: Behavioral - Convulsions or effect on seizure threshold.
Chronic (Multiple Dose) Inhalation: Rat Dose: 210 ug/m^3/4H/12W-I; Toxic Effects: Blood - Pigmented or nucleated red Blood cells; Blood - Changes in erythrocite (RBC) cell count; Blood - Changes in platelet cell count.
Mutagenic: Bacteria - S Typhimurium Mutations in Microorganisms; Dose: 10 pph (+S9), 5 pph (-S9).

Hazard Overviews

Compressed Gas

Carcinogenicity: IARC - Not listed; NIOSH - Not listed; NTP - Not listed; ACGIH - Not listed; OSHA - Not listed; EPA - Not listed; MAK - Not listed

Environmental

Environmental Fate: If released to the atmosphere, it will diffuse gradually into the stratosphere above the ozone layer where will degrade slowly due to photolysis. The stratospheric lifetime has been estimated to range from 29 to 42 years. The half-life in tropospheric air for the reaction with photochemically produced hydroxyl radicals is estimated to be greater than 44 years. If released to soil, it is expected to have moderate mobility and, therefore, very little leaching into groundwater is expected. Rapid volatilization should occur from terrestrial surfaces. If released to water, volatilization is expected to occur rapidly and be the dominant fate process. Bioconcentration in aquatic organisms may be an important fate process.
Cleanup/Disposal: Guide No. 126: Do not touch or walk through spilled material. Stop leak if you can do it without risk. Do not direct water at spill or source of leak. Use water spray to reduce vapors or divert vapor cloud drift. If possible, turn leaking containers so that gas escapes rather than liquid. Prevent entry into waterways, sewers, basements or confined areas. Allow substance to evaporate. Ventilate the area.

Environmental Physical Data

Henry's Law Constant: 0.094
Octanol/Water Partition Coefficient: log K_{ow} = 2.135
Sorption Partition Coefficient: K_{oc} = estimated at 345.5
BCF: estimated at 24.7

Regulations

RCRA 40CFR: Not listed
CERCLA: 40CFR 302.4: Listed as Compound per CWA Section 307(a)
SARA 40CFR 372.65: Listed
SARA EHS 40CFR 355: Not listed
TSCA: Listed

BRO4640 **CAS #: 74-97-5**

BROMOCHLOROMETHANE

RTECS: PA5250000
DOT: UN1887; IMO6.1
EINECS Number: 200-826-3
Molecular Formula: CH_2BrCl
Structured MF: $BrCH_2Cl$
Formula Weight: 129.38

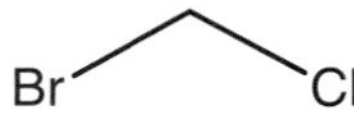

Chemical Structure

Synonyms: CB; CBM; CHLOROBROM; CHLOROBROMOMETHANE; CHLOROMETHYL BROMIDE; FLUOROCARBON 1011; HALON 1011; METHANE,BROMOCHLORO-; METHYL CHLOROBROMIDE; METHYL CHLOROBROMIDE BROMOCHLOROMETHANE; METHYLENE CHLOROBROMIDE; MIL-B-4394-B; MONO-CHLORO-MONO-BROMO-METHANE; MONOCHLOROMONOBROMOMETHANE
Description: colorless, light yellow liquid; chloroform-like odor
Use: in fire extinguishers and in organic synthesis

Physical Properties

Boiling Point: 68.1 °C (155 °F)
Freezing Point: -86.5 °C (-123.7 °F)
Specific Gravity: 1.9344 at 20 °C/4 °C
Vapor Density: 4.46 Air=1

Saturated Vapor Density: 2.047127042 kg/m^3
Density: 1.991 g/mL at 10 °C
Vapor Pressure: 155 to 160 torr at 25 °C
Water Solubility: Insoluble
Other Solubilities: 95% Ethanol: >=100 mg/ml at 20 °C; Acetone: >=100 mg/ml at 20 °C; Alcohol: Soluble; Benzene: Soluble; DMSO: >=100 mg/ml at 20 °C; Ether: Soluble; Organic solvents: Soluble.
Surface Tension: 33.32 dynes/cm at 20 °C
Refraction Index: 1.4838 at 20 °C
Ionization Potential (eV): 10.77
Flash Point: None

RTECS Toxicity Data

Acute Oral: Rat LD_{50} Dose: 5 gm/kg. Mouse LD_{50} Dose: 4300 mg/kg; Toxic Effects: Behavioral - General anesthetic.
Acute Inhalation: Mouse LC_{50} Dose: 12030 mg/m^3/7hr; Toxic Effects: Behavioral - General anesthetic; Behavioral - Muscle weakness; Lungs, Thorax, or Respiration - Dyspnea. Rat LC_{Lo} Dose: 28800 ppm/15M; Toxic Effects: Behavioral - General anesthetic; Behavioral - Altered sleep time (including change in righting reflex); Behavioral - Tremor.
Acute Dermal: Rabbit LD_{50} Route: Skin; Dose: >20 gm/kg.
Chronic (Multiple Dose) Inhalation: Rat Dose: 1000 ppm/7H/16W-I; Toxic Effects: Liver - Fatty Liver degeneration; Kidney, Ureter, and Bladder - Changes in kidney weight; Nutritional and gross metabolic - Changes in other metals. Rat Dose: 500 ppm/6H/26W-I; Toxic Effects: Lungs, Thorax, or Respiration - Fibrosis, focal (pneumoconiosis); Nutritional and gross metabolic - Weight loss or decreased weight gain; Nutritional and gross metabolic - Changes in other metals.
Mutagenic: Hamster Cytogenetic Analysis; Cell Type: lung; Dose: 1 umol/L. Hamster Sister Chromatid Exchenge; Cell Type: lung; Dose: 5 umol/L.

Hazard Overviews

Fire Diamond

Health: Irritating to eyes/skin/respiratory tract. Also Causes: headache, loss of consciousness, gastric upsets, loss in weight, epithelial injury, conjunctival edema, narcosis. Chronic Effects: dermatitis.
Fire: Noncombustible. Use agent suitable for surrounding fire.
Reactivity: Stable. Hazardous polymerization cannot occur. Incompatible with: chemically-active metals (calcium; powdered aluminum; zinc; magnesium); some forms of plastics, rubber and coatings. Hazardous decomposition products: hydrogen chloride; hydrogen bromide; bromine; phosgene; carbon monoxide.
Carcinogenicity: IARC - Not listed; NIOSH - Not listed; NTP - Not listed; ACGIH - Not listed; OSHA - Not listed; EPA - Class D, Not classifiable as to human carcinogenicity; MAK - Not listed

Primary Target Organs:

Eyes

Skin

Respiratory System

Nervous System

Exposure Limits
OSHA PEL: TWA: 200 ppm; 1050 mg/m^3.
ACGIH TLV: TWA: 200 ppm; 1060 mg/m^3.
NIOSH REL: TWA: 200 ppm; 1050 mg/m^3.
NIOSH IDLH: 2000 ppm.
DFG MAK: TWA: 200 ppm; 1050 mg/m^3.
Respirator Recommendation
Exposure Range: >200 to <2000 ppm Supplied Air, Constant Flow/Pressure Demand, Half Mask
Exposure Range: 2000 to unlimited ppm Supplied Air, Constant Flow/Pressure Demand, Full Face
Note: odor threshold unknown

Environmental

Environmental Fate: If released to the soil, it is expected to display high mobility and it has the potential to leach into groundwater. Volatilization from the soil surface to the atmosphere is expected to be a significant process. Limited data suggests that the microbial degradation of this compound may occur in soil under anoxic conditions. If released to water, it is expected to rapidly volatilize to the atmosphere. It is not expected to bioconcentrate in fish and aquatic organisms, nor is it expected to adsorb to sediment and suspended organic matter. Microbial degradation under aerobic conditions may occur, based on limited data. Hydrolysis and direct photochemical degradation are not expected to be significant processes. In the atmosphere, it is expected to exist predominately in the vapor phase. The vapor phase reaction with photochemically produced hydroxyl radicals and direct photochemical degradation are not expected to be significant processes in the atmosphere. The relatively high water solubility of this compound suggests that wet deposition may occur; however, compound deposited by this process would be expected to re-volatilize to the atmosphere.
Cleanup/Disposal: Guide No. 160: Eliminate all ignition sources (no smoking, flares, sparks or flames in immediate area). Stop leak if you can do it without risk. Small Liquid Spills: Take up with sand, earth or other noncombustible absorbent material. Large Spills: Dike far ahead of liquid spill for later disposal. Prevent entry into waterways, sewers, basements or confined areas.

Environmental Physical Data

Henry's Law Constant: calculated at 0.0015
Octanol/Water Partition Coefficient: log K_{ow} = 1.41
Sorption Partition Coefficient: K_{oc} = 21 to 139
BCF: estimated at 3 to 7

Regulations

RCRA 40CFR: Not listed
CERCLA: 40CFR 302.4: Listed as Compound per CWA Section 307(a)
SARA 40CFR 372.65: Not listed

SARA EHS 40CFR 355: Not listed
TSCA: Listed

Analytical Methods

Air: EPA OA-002-1, VA-006-1, VA-008-1; ASTM D3686, D3687, D4490
Soil: SW846 5021, 8021A, 8260A, 8260B
Water / Groundwater: APHA 6210-D, 6230-D
Drinking Water: EPA 502.1, 502.2, 524.1, 524.2; APHA 6210-C, 6230-C
Indoor / Expired Air: NIOSH 1003; EPA IP-1B
Plasma: EPA 29

BRO5480 CAS #: 540-51-2

2-BROMOETHANOL

RTECS: KJ8225000
EINECS Number: 208-748-1
Molecular Formula: C_2H_5BrO
Formula Weight: 124.97

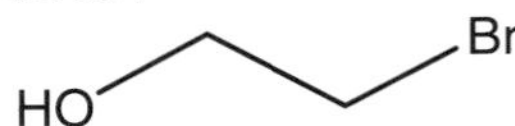

Chemical Structure

Synonyms: BE; BROMOETHANOL; BETA-BROMOETHYL ALCOHOL; ETHYLENEBROMOHYDRIN; GLYCOL BROMOHYDRIN
Description: colorless to dark brown liquid
Use: organic synthesis intermediate

Physical Properties

Boiling Point: Decomposes at 149 °C (300 °F) to 150 °C (302 °F)
Specific Gravity: 1.7629 at 20/4 °C
Density: 1.746 g/cu cm at 26.9 °C
Vapor Pressure: 12 mm Hg at 50 °C
Water Solubility: 10 to 50 mg/mL at 19 °C
Other Solubilities: 95% Ethanol: >=100 mg/ml at 19 °C; Acetone: >=100 mg/ml at 19 °C; Benzene: Soluble; DMSO: >=100 mg/ml at 19 °C; Ether: Very Soluble; Most organic solvents: Soluble; Petroleum Ether: Insoluble.
Refraction Index: 1.492
Ionization Potential (eV): 10.0
Flash Point: 40 °C

RTECS Toxicity Data

Mutagenic: Bacteria - B Subtilis DNA Repair; Dose: 20 uL/disc. Bacteria - E Coli DNA Repair; Dose: 10 umol/plate.
Tumorigenic: Mouse Route: Oral; Dose: 43 gm/kg/80W-C; Toxic Effects: Tumorigenic - Equivocal tumorigenic agent by RTECS criteria; Gastrointestinal - Tumors. Mouse Route: Intraperitoneal; Dose: 150 mg/kg/8W-I; Toxic Effects: Tumorigenic - Neoplastic by RTECS criteria; Lungs, Thorax, or Respiration - Tumors.

Hazard Overviews

Poison

Corrosive

Flammable

Health: Corrosive to eyes/skin/respiratory tract. Poison. Other Acute Effects: may be fatal if inhaled, swallowed, or absorbed through skin; material is extremely destructive to tissue of the mucous membranes and upper respiratory tract, eyes and skin; inhalation may result in spasm, inflammation and edema of the larynx and bronchi, chemical pneumonitis and pulmonary edema; symptoms of exposure may include burning sensation, coughing, wheezing, laryngitis, shortness of breath, headache, nausea and vomiting. Chronic Effects: damage to the liver, kidneys; target organs: liver, kidneys, spleen. Laboratory experiments have shown mutagenic effects.
Fire: Flammable. Hazards: vapor may travel considerable distance to source of ignition and flash back; emits toxic fumes. Extinguishing agents: water spray; carbon dioxide, dry chemical powder or appropriate foam. Precautions: combustible liquid.
Reactivity: Incompatible with: strong acids, strong oxidizing agents, strong reducing agents, acid chlorides, acid anhydrides. Hazardous decomposition products: toxic fumes of: carbon monoxide, carbon dioxide, hydrogen bromide gas.
Carcinogenicity: IARC - Not listed; NIOSH - Not listed; NTP - Not listed; ACGIH - Not listed; OSHA - Not listed; EPA - Not listed; MAK - Not listed
Primary Target Organs:

Eyes

Skin

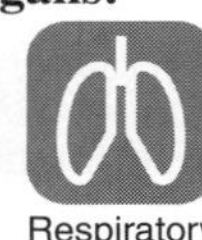
Respiratory System

Liver

Kidneys

Environmental

Cleanup/Disposal: Guide No. 128: Eliminate all ignition sources (no smoking, flares, sparks or flames in immediate area). All equipment used when handling the product must be grounded. Do not touch or walk through spilled material. Stop leak if you can do it without risk. Prevent entry into waterways, sewers, basements or confined areas. A vapor suppressing foam may be used to reduce vapors. Absorb or cover with dry earth, sand or other non-combustible material and transfer to containers. Use clean non-sparking tools to collect absorbed material. Large Spills: Dike far ahead of liquid spill for later disposal. Water spray may reduce vapor; but may not prevent ignition in closed spaces.

Regulations

RCRA 40CFR: Not listed
CERCLA: 40CFR 302.4: Not listed
SARA 40CFR 372.65: Not listed
SARA EHS 40CFR 355: Not listed
TSCA: Not listed

BRO5900 CAS #: 103-63-9

(2-BROMOETHYL)BENZENE

EINECS Number: 203-130-8
Molecular Formula: C_8H_9Br
Formula Weight: 185.07

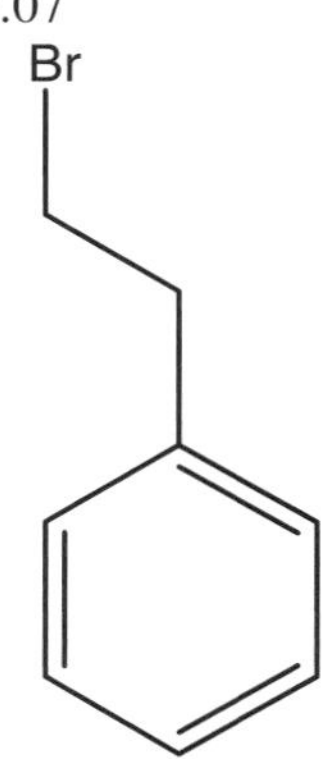

Chemical Structure

Synonyms: BENZENE,(2-BROMOETHYL)-; BETA-BROMOETHYLBENZENE; 1-BROMO-2-PHENYLETHANE; 2-PHENETHYL BROMIDE; BETA-PHENETHYL BROMIDE; PHENETHYL BROMIDE; 2-PHENYL-1-BROMOETHANE; 2-PHENYLETHYL BROMIDE; BETA-PHENYLETHYL BROMIDE; PHENYLETHYL BROMIDE
Use: starting material for the production of various beta-phenethyl derivatives, pharmaceuticals, fragrances, and other fine chemicals

Physical Properties

Boiling Point: 219 °C (426 °F)
Freezing Point: -67.5 °C (-89.5 °F)
Specific Gravity: 1.355
Saturated Vapor Density: 1.202077629 kg/m^3
Vapor Pressure: 0.2445 mm Hg at 25 °C
Water Solubility: 39 mg/L at 25 °C
Other Solubilities: Soluble in Ether, Benzene; slightly Soluble in Carbon Tetrachloride
Refraction Index: 1.5572 at 20 °C

Hazard Overviews

Health: Irritating to eyes/skin/respiratory tract. Other Acute Effects: may be harmful by inhalation, ingestion, or skin absorption.
Fire: Hazards: emits toxic fumes. Extinguishing agents: water spray; carbon dioxide, dry chemical powder or appropriate foam. Precautions: combustible liquid.
Reactivity: Incompatible with: strong oxidizing agents. Hazardous decomposition products: toxic fumes of: carbon monoxide, carbon dioxide, hydrogen bromide gas.
Carcinogenicity: IARC - Not listed; NIOSH - Not listed; NTP - Not listed; ACGIH - Not listed; OSHA - Not listed; EPA - Not listed; MAK - Not listed

Primary Target Organs:

Eyes

Skin

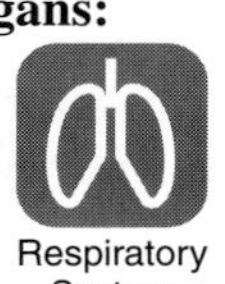
Respiratory System

Environmental

Environmental Fate: Should have low mobility in soil. Volatilization is expected from both moist and dry soils. In water, it is expected to volatilize rapidly with estimated half-lives of 4.8 hours and 5.6 days from a model river and a model lake, respectively. Bioconcentration and adsorption to sediment may be important fate processes in aquatic systems, but hydrolysis is not expected to be important. Insufficient data are available to determine the rate or importance of biodegradation in either soil or aquatic conditions. It will exist in the vapor phase in the ambient atmosphere. If released to the atmosphere, it will degrade by reaction with photochemically produced hydroxyl radicals with an estimated half-life of approximately 2.9 days. Removal from the atmosphere can occur through wet deposition.

Environmental Physical Data

Henry's Law Constant: estimated at 1.52×10^{-3}
Octanol/Water Partition Coefficient: log K_{ow} = 3.09
Sorption Partition Coefficient: K_{oc} = 955
BCF: estimated at 78

Regulations

RCRA 40CFR: Not listed
CERCLA: 40CFR 302.4: Not listed
SARA 40CFR 372.65: Not listed
SARA EHS 40CFR 355: Not listed
TSCA: Listed

BRO6600 CAS #: 52-51-7

2-BROMO-2-NITROPROPANE-1,3-DIOL

RTECS: TY3385000
EINECS Number: 200-143-0
Molecular Formula: $C_3H_6BrNO_4$
Structured MF: $HOCH_2C(Br)(NO_2)CH_2OH$
Formula Weight: 200.01

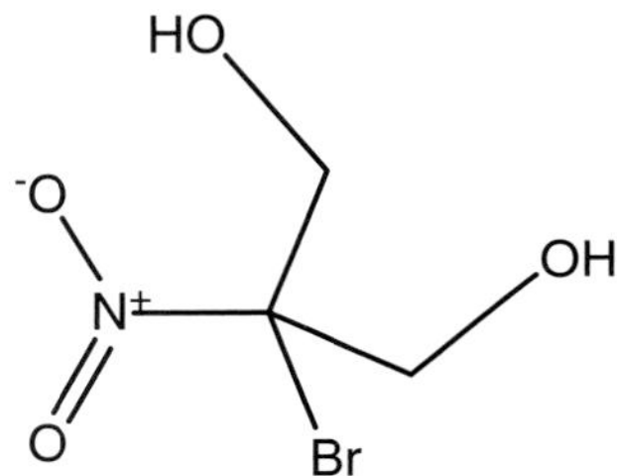

Chemical Structure

Synonyms: 2-BROMO-2-NITROPROPAN-1,3-DIOL; 2-BROMO-2-NITRO-1,3-PROPANEDIOL; BETA-BROMO-BETA-

NITROTRIMETHYLENEGLYCOL; BRONIDIOL; BRONOCOT; BRONOPOL; BRONOPOLU; BRONOSOL; BRONOTAK; MYACIDE AS; 2-NITRO-2-BROMO-1,3-PROPANEDIOL; ONYXIDE 500

Description: colorless to pale brownish-yellow solid or white crystalline solid; odorless or faint odor

Use: as a bactericide, in seed treatment for control of Xanthomonas malvacearum, as a foliar spray for control of a wide range of plant-pathogenic bacteria (especially Erwinia amylovora), in antiseptics and as a preservative in cosmetics and toiletries

Physical Properties

Boiling Point: 151.7 °C (305 °F)
Freezing Point: 130 °C (266 °F) to 133 °C (271.4 °F)
Saturated Vapor Density: 1.200000117 kg/m^3
Vapor Pressure: 0.000012601 mm Hg at 20 °C
Water Solubility: >= 100 mg/mL at 17 C
Other Solubilities: DMSO: >=100 mg/ml at 17 °C; 95% Ethanol: >=100 mg/ml at 17 °C; Acetone: >=100 mg/ml at 17 °C; Benzene: Slightly Soluble; Ether: Slightly Soluble; Oils: Slightly Soluble.
Flash Point: Not available; probably combustible

RTECS Toxicity Data

Acute Oral: Rat LD_{50} Dose: 180 mg/kg. Mouse LD_{50} Dose: 270 mg/kg.
Acute Inhalation: Rat LC_{50} Dose: >5 gm/m^3/6hr.
Acute Dermal: Rat LD_{50} Route: Skin; Dose: 1600 mg/kg; Rat LD_{50} Route: Subcutaneous Dose: 170 mg/kg; Toxic Effects: Behavioral - Somnolence (general depressed activity); Skin and appendages - Hair.
Irritation Skin: Human Standard Draize Test Dose: 10 mg; Reaction: moderate. Rabbit Standard Draize Test Dose: 500 mg/24H; Reaction: mild. Rabbit Standard Draize Test Dose: 80 mg; Reaction: moderate.

Hazard Overviews

Health: Irritating to eyes/skin/respiratory tract. Toxic. Other Acute Effects: harmful if swallowed, inhaled, or absorbed through skin.
Fire: Will burn. Hazards: emits toxic fumes. Extinguishing agents: water spray; carbon dioxide, dry chemical powder or appropriate foam. Precautions: combustible liquid.
Reactivity: Incompatible with: strong oxidizing agents, strong bases, strong reducing agents, acid chlorides, acid anhydrides. Hazardous decomposition products: toxic fumes of: carbon monoxide, carbon dioxide, nitrogen oxides, hydrogen bromide gas.
Carcinogenicity: IARC - Not listed; NIOSH - Not listed; NTP - Not listed; ACGIH - Not listed; OSHA - Not listed; EPA - Not listed; MAK - Not listed
Primary Target Organs:

Eyes

Skin

Respiratory System

Environmental

Cleanup/Disposal: Guide No. 154: Eliminate all ignition sources (no smoking, flares, sparks or flames in immediate area). Do not touch damaged containers or spilled material unless wearing appropriate protective clothing. Stop leak if you can do it without risk. Prevent entry into waterways, sewers, basements or confined areas. Absorb or cover with dry earth, sand or other non-combustible material and transfer to containers. Do not get water inside containers.

Regulations

RCRA 40CFR: Not listed
CERCLA: 40CFR 302.4: Not listed
SARA 40CFR 372.65: Listed
SARA EHS 40CFR 355: Not listed
TSCA: Listed

BRO6740 CAS #: 107-81-3

2-BROMOPENTANE

RTECS: RZ9800000
EINECS Number: 203-521-3
Molecular Formula: $C_5H_{11}Br$
Structured MF: $CH_3CHBr(C_2H_4)CH_3$
Formula Weight: 151.07

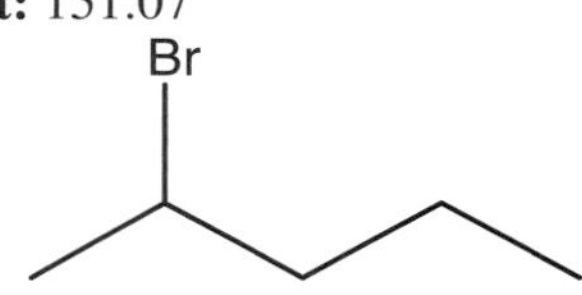

Chemical Structure

Synonyms: 2-PENTYL BROMIDE
Description: colorless liquid

Physical Properties

Boiling Point: 117 °C (243 °F) at 1 atm
Freezing Point: -96 °C (-140 °F)
Specific Gravity: 1.208 at 20 °C
Saturated Vapor Density: 1.354576846 kg/m^3
Vapor Pressure: 3.10 kPa at 25 °C
Other Solubilities: Benzene: Very Soluble; Ether: Very Soluble; Ethanol: Very Soluble; Chloroform: Very Soluble
Refraction Index: 1.4413
Flash Point: 21 °C Closed Cup

RTECS Toxicity Data

Acute Inhalation: Mouse LC_{50} Dose: 33 gm/m^3.
Chronic (Multiple Dose) Inhalation: Rat Dose: 2500 mg/m^3/4H/8W-I; Toxic Effects: Brain and coverings - Other degenerative changes; Liver - Other changes; Kidney, Ureter, and Bladder - Other changes. Rat Dose: 90 mg/m^3/4H/17W-I; Toxic Effects.

Hazard Overviews

Flammable

Health: Irritating to eyes/skin/respiratory tract. Other Acute Effects: may be harmful by inhalation, ingestion, or skin absorption.
Fire: Flammable. Hazards: vapor may travel considerable distance to source of ignition and flash back; container explosion may occur; forms explosive mixtures in air; emits toxic fumes. Extinguishing agents: carbon dioxide, dry chemical powder or appropriate foam; water may be effective for cooling, but may not effect extinguishment. Precautions: combustible liquid.
Reactivity: Incompatible with: strong oxidizing agents, strong bases. Hazardous decomposition products: toxic fumes of: carbon monoxide, carbon dioxide, hydrogen bromide gas.
Carcinogenicity: IARC - Not listed; NIOSH - Not listed; NTP - Not listed; ACGIH - Not listed; OSHA - Not listed; EPA - Not listed; MAK - Not listed
Primary Target Organs:

Eyes

Skin

Respiratory System

Environmental

Cleanup/Disposal: Guide No. 128: Eliminate all ignition sources (no smoking, flares, sparks or flames in immediate area). All equipment used when handling the product must be grounded. Do not touch or walk through spilled material. Stop leak if you can do it without risk. Prevent entry into waterways, sewers, basements or confined areas. A vapor suppressing foam may be used to reduce vapors. Absorb or cover with dry earth, sand or other non-combustible material and transfer to containers. Use clean non-sparking tools to collect absorbed material. Large Spills: Dike far ahead of liquid spill for later disposal. Water spray may reduce vapor; but may not prevent ignition in closed spaces.

Regulations

RCRA 40CFR: Not listed
CERCLA: 40CFR 302.4: Not listed
SARA 40CFR 372.65: Not listed
SARA EHS 40CFR 355: Not listed
TSCA: Listed

BRO7020 CAS #: 101-55-3

4-BROMOPHENYL PHENYL ETHER

EINECS Number: 202-952-4
Molecular Formula: $C_{12}H_9BrO$
Formula Weight: 249.11

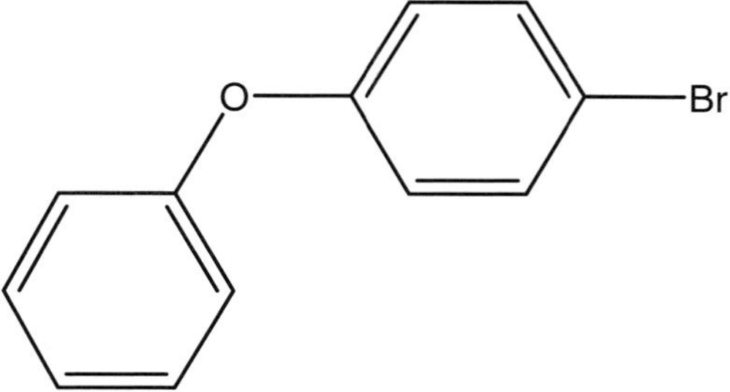

Chemical Structure

Synonyms: BENZENE,1-BROMO-4-PHENOXY-; 4-BROMODIPHENYL ETHER; P-BROMODIPHENYL ETHER; 1-BROMO-4-PHENOXYBENZENE; 4-BROMOPHENOXYBENZENE; P-BROMOPHENOXYBENZENE; P-BROMOPHENYL PHENYL ETHER; 4-BROMOPHENYLPHENYL ETHER; 4-BROMOPHENYLPHENYLETHER; DIPHENYL ETHER,4-BROMO-; ETHER,4-BROMOPHENYL PHENYL; ETHER,P-BROMOPHENYL PHENYL; ETHER,DIPHENYL,4-BROMO-; P-PHENOXYBROMOBENZENE; PHENYL (4-BROMOPHENYL) ETHER
Description: liquid
Use: research chemical; as flame retardant additives in polymers (former use)

Physical Properties

Boiling Point: 310.1 °C (590 °F) at 760 mm Hg
Freezing Point: 18.7 °C (65.66 °F)
Specific Gravity: 1.4208 at 20 °C/4 °C
Saturated Vapor Density: 1.200017976 kg/m^3
Vapor Pressure: 0.0015 torr at 20 °C
Water Solubility: Insoluble in Water
Other Solubilities: Ether: Soluble.
Refraction Index: 1.6084 at 20 °C
Flash Point: > 112 °C

Hazard Overviews

Health: May cause irritation. Other Acute Effects: may be harmful by inhalation, ingestion, or skin absorption.
Fire: Will burn. Hazards: emits toxic fumes. Extinguishing agents: water spray; carbon dioxide, dry chemical powder or appropriate foam. Precautions: combustible liquid.
Reactivity: Incompatible with: strong oxidizing agents. Hazardous decomposition products: toxic fumes of: carbon monoxide, carbon dioxide, hydrogen bromide gas.
Carcinogenicity: IARC - Not listed; NIOSH - Not listed; NTP - Not listed; ACGIH - Not listed; OSHA - Not listed; EPA - Class D, Not classifiable as to human carcinogenicity; MAK - Not listed

Environmental

Ecotoxicity: LC_{50} Lepomis macrochirus (bluegill) 5.9 mg/l/98 hr, age young, wt 0.32-1.2 g, static bioassay, total hardness of water 32-48 mg/l $CaCO_3$, total alkalinity 28-34 mg/l $CaCO_3$, pH 6.7-7.8, dissolved oxygen concentration 7.0-8.8 mg/l, temp 22 + or - °C LC_{50} Daphnia magna (Water flea) 0.36 mg/l/48 hr. /Conditions of bioassay not specified LC_{50} Lepomis macrochiris (bluegill) 4.9 mg/l/96 hr. /Conditions of bioassay not specified

Environmental Fate: If released to water, it may adsorb significantly to sediment and suspended material. If strong adsorption is occurring, volatilization from water may not be important. However, in the absence of strong adsorption, volatilization half lives of 16.5 and 185 hours can be estimated for a model river and environmental pond, respectively. A potential for significant bioconcentration in aquatic organisms may be possible based on an estimated Log Kow of 5.243. If released to soil, significant leaching is not expected to occur due to strong soil adsorption. A single biodegradation study suggests that it is resistant to biodegradation. If released to the atmosphere, it is expected to exist primarily in the gas-phase where it will degrade relatively rapidly by reaction with photochemically formed hydroxyl radicals; the half-life for this reaction can be estimated to be about 1.3 days in average air.

Environmental Physical Data

Henry's Law Constant: estimated at 1.17×10^{-4}
Octanol/Water Partition Coefficient: log K_{ow} = 5.243
Sorption Partition Coefficient: K_{oc} = estimated at 1.7×10^{4}
BCF: estimated at 5690

Regulations

RCRA 40CFR: Listed Hazardous Waste No. U030 Toxic Waste
CERCLA: 40CFR 302.4: Listed per RCRA Section 3001 per CWA Section 307(a) RQ: 100 lb (45.35 kg)
SARA 40CFR 372.65: Not listed
SARA EHS 40CFR 355: Not listed
TSCA: Listed

Analytical Methods

Soil: CLP MC_SVOA, OHC; EPA 16, 1625; SW846 3640A, 8110, 8111, 8250A, 8270B, 8270C, 8275A, 8410
Water / Groundwater: EPA 1625, 611, 625, 625-S, 6; APHA 6040-B, 6410-B; USGS O3118
Plasma: EPA 29

BRO7860 CAS #: 627-18-9

3-BROMO-1-PROPANOL

RTECS: UA7385000
EINECS Number: 210-986-6
Molecular Formula: C_3H_7BrO
Formula Weight: 139.01

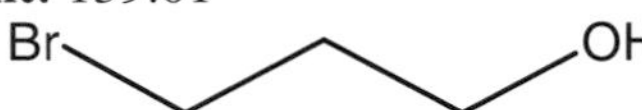

Chemical Structure

Synonyms: 1-BROMO-3-PROPANOL; 3-BROMOPROPANOL; 3-BROMOPROPYL ALCOHOL; 3-HYDROXYPROPYL BROMIDE; TRIMETHYLENE BROMOHYDRIN
Description: colorless, clear liquid

Physical Properties

Boiling Point: 145 °C (293 °F) to 148 °C (298 °F)
Specific Gravity: 1.5374 at 20/4 °C
Water Solubility: 50 to 100 mg/mL at 21 °C
Other Solubilities: 95% Ethanol: >=100 mg/ml at 21 °C; Acetone: >=100 mg/ml at 21 °C; DMSO: >=100 mg/ml at 21 °C; Ether: miscible.
Refraction Index: 1.4834
Flash Point: 93 °C

RTECS Toxicity Data

Mutagenic: Bacteria - S Typhimurium Mutations in Microorganisms; Dose: 177 ug/plate (+S9), 350 ug/plate (-S9).

Hazard Overviews

Health: Irritating to eyes/skin/respiratory tract. Harmful. Other Acute Effects: may be harmful by inhalation, ingestion, or skin absorption.
Fire: Will burn. Hazards: emits toxic fumes. Extinguishing agents: water spray; carbon dioxide, dry chemical powder or appropriate foam. Precautions: combustible liquid.
Reactivity: Incompatible with: strong oxidizing agents. Hazardous decomposition products: toxic fumes of: carbon monoxide, carbon dioxide, hydrogen bromide gas.
Carcinogenicity: IARC - Not listed; NIOSH - Not listed; NTP - Not listed; ACGIH - Not listed; OSHA - Not listed; EPA - Not listed; MAK - Not listed
Primary Target Organs:

Eyes

Skin

Respiratory System

Environmental

Regulations

RCRA 40CFR: Not listed
CERCLA: 40CFR 302.4: Not listed
SARA 40CFR 372.65: Not listed
SARA EHS 40CFR 355: Not listed
TSCA: Listed

BRO9260 CAS #: 76-59-5

BROMTHYMOL BLUE

RTECS: SJ7450000
EINECS Number: 200-971-2
Molecular Formula: $C_{27}H_{28}Br_2O_5S$
Formula Weight: 624.43

Chemical Structure

Synonyms: BROMOTHYMOL BLUE; 3,3'-DIBROMOTHYMOLSULFONPHTHALEIN; DIBROMOTHYMOLSULFOPHTHALEIN; THYMOL,6,6'-(3H-2,1-BENZOXATHIOL-3-YLIDENE)BIS(2-BROMO-,S,S-DIOXIDE (8CI)

Description: off-white, cream-colored crystals, powder

Physical Properties

Freezing Point: Decomposes at 200 °C (392 °F)
Vapor Pressure: Negligible
Water Solubility: Negligible
Other Solubilities: Ether: Very Soluble; Ethanol: Very Soluble

Hazard Overviews

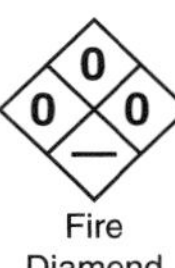

Fire Diamond

Health: This material is relatively nonhazardous in routine industrial situations. Toxic by ingestion, although this possibility is extremely unlikely if recommended personal hygiene procedures are followed. It is not expected to present significant health risks to the workers who use it.
Fire: Will burn. Slight fire hazard when exposed to heat, sparks, and open flame. Use water fog, dry chemical, alcohol foam, or carbon dioxide to fight fires involving bromthymol blue. Use a water spray to cool fire-exposed tanks or containers.
Reactivity: Stable. Hazardous polymerization cannot occur. Avoid: direct exposure to heat; sparks; open flame; lighted tobacco products. Incompatible with: strong oxidizers. Hazardous decomposition products: hydrogen bromide; sulfur oxides; carbon monoxide; carbon dioxide.
Carcinogenicity: IARC - Not listed; NIOSH - Not listed; NTP - Not listed; ACGIH - Not listed; OSHA - Not listed; EPA - Not listed; MAK - Not listed

Environmental

Regulations

RCRA 40CFR: Not listed
CERCLA: 40CFR 302.4: Not listed
SARA 40CFR 372.65: Not listed
SARA EHS 40CFR 355: Not listed
TSCA: Listed

BRU5000 CAS #: 357-57-3

BRUCINE

RTECS: EH8925000
DOT: UN1570; NA2811; IMO6.1
EINECS Number: 206-614-7
Molecular Formula: $C_{23}H_{26}N_2O_4$
Structured MF: $C_{23}H_{26}N_2O_4 \cdot 2H_2O$
Formula Weight: 394.47

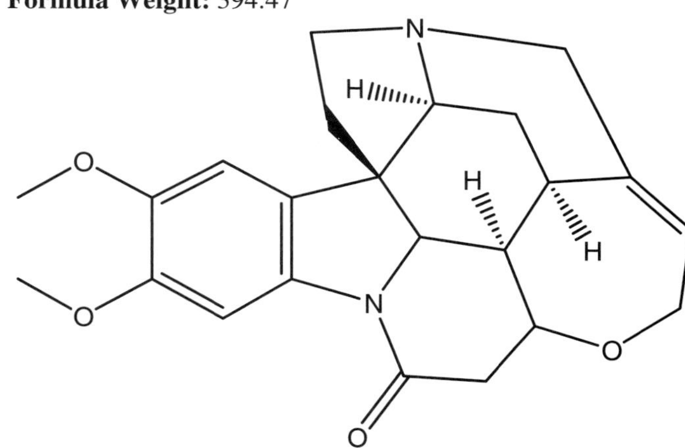

Chemical Structure

Synonyms: BRUCIN; BRUCINA; (-)-BRUCINE; BRUCINE ALKALOID; BRUCINE,SOLID; DIMETHOXY STRYCHNINE; 2,3-DIMETHOXYSTRYCHNIDIN-10-ONE; 10,11-DIMETHOXYSTRYCHNINE; 2,3-DIMETHOXY-STRYCHNINE; 2,3-DIMETHOXYSTRYCHNINE; DIMETHOXYSTRYCHNINE; 10,11-DIMETHYSTRYCHNINE; DOLCO MOUSE CEREAL; PIED PIPER MOUSE SEED; STRYCHNIDIN-10-ONE,2,3-DIMETHOXY-; STRYCHNIDIN-10-ONE,2,3-DIMETHOXY-(9CI); STRYCHNINE,2,3-DIMETHOXY-

Description: white crystalline solid; odorless
Use: denaturing alcohol and oils; in analytical chemistry; for separating racemic mixtures; patented as addition agent for lubricants

Physical Properties

Boiling Point: Decomposes at 230 °C (446 °F)
Freezing Point: 178 °C (352.4 °F)
Specific Gravity: Solid > 1 at 20 °C
Water Solubility: 1 g dissolves in 1320 mL Water
Other Solubilities: Benzene: Slightly Soluble; Chloroform: Very Soluble; Ether: Slightly Soluble; Ethanol: Very Soluble.
pH: Saturated water solution 9.5
Flash Point: Not pertinent (combustible solid)

RTECS Toxicity Data

Acute Oral: Mouse LD_{50} Dose: 150 mg/kg; Toxic Effects: Behavioral - Convulsions or effect on seizure threshold.
Acute Dermal: Mouse LD_{50} Route: Subcutaneous Dose: 60 mg/kg; Toxic Effects: Behavioral - Convulsions or effect on seizure threshold. Pigeon LD_{Lo} Route: Subcutaneous Dose: 58 mg/kg.

Hazard Overviews

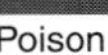
Poison

Fire Diamond

Health: Mildly irritating to eyes/respiratory tract. Poison. Also Causes: CNS effects, nausea, vomiting, ringing in ears, headache, foggy vision, convulsions, paralysis, death.
Fire: Combustible. Use dry chemical, carbon dioxide, water spray, fog, or regular foam. Hazardous combustion products include carbon oxide(s) and nitrogen oxide(s).
Reactivity: Stable. Hazardous polymerization cannot occur. Avoid: elevated temperatures; ignition sources; dispersion of brucine dusts into air. Hazardous decomposition products: carbon; nitrogen oxide(s).
Carcinogenicity: IARC - Not listed; NIOSH - Not listed; NTP - Not listed; ACGIH - Not listed; OSHA - Not listed; EPA - Not listed; MAK - Not listed
Primary Target Organs:

Eyes

Respiratory System

Nervous System

Environmental

Ecotoxicity: LC_{50} Lepomis macrochirus (bluegill sunfish) 36 ppm/96 hr (static bioassay in fresh water at 23 °C, with mild aeration after 24 hr) LC_{50} Menidia beryllina 20 ppm/96 hr (static bioassay in synthetic seawater at 23 °C, with mild aeration after 24 hr)
Environmental Fate: If released to soil, it has the potential to photolyze on soil surfaces. Volatilization and chemical hydrolysis are not expected to be important fate processes in soil. If released to water, it has the potential to photolyze in the upper few meters were there is light penetration. Chemical hydrolysis, volatilization, and bioaccumulation in aquatic organisms are not expected to be important fate processes. If released to the atmosphere, this compound has the potential to be removed by direct photolysis or dry deposition.
Cleanup/Disposal: Guide No. 152: Do not touch damaged containers or spilled material unless wearing appropriate protective clothing. Stop leak if you can do it without risk. Prevent entry into waterways, sewers, basements or confined areas. Cover with plastic sheet to prevent spreading. Absorb or cover with dry earth, sand or other non-combustible material and transfer to containers. Do not get water inside containers. Guide No. 154: Eliminate all ignition sources (no smoking, flares, sparks or flames in immediate area). Do not touch damaged containers or spilled material unless wearing appropriate protective clothing. Stop leak if you can do it without risk. Prevent entry into waterways, sewers, basements or confined areas. Absorb or cover with dry earth, sand or other non-combustible material and transfer to containers. Do not get water inside containers.

Environmental Physical Data

Henry's Law Constant: estimated at 1 x10^{-13}
Octanol/Water Partition Coefficient: log K_{ow} = 0.98
Sorption Partition Coefficient: K_{oc} = estimated at 81
BCF: estimated at 3

Regulations

RCRA 40CFR: Listed Hazardous Waste No. P018 Toxic Waste
CERCLA: 40CFR 302.4: Listed per RCRA Section 3001 RQ: 100 lb (45.35 kg)
SARA 40CFR 372.65: Listed
SARA EHS 40CFR 355: Not listed
TSCA: Listed

Analytical Methods

Water / Groundwater: ASTM D4763

BUT1280 CAS #: 106-99-0

1,3-BUTADIENE

RTECS: EI9275000
DOT: UN1010; IMO2.0
EINECS Number: 203-450-8
Molecular Formula: C_4H_6
Structured MF: CH_2=CHCH=CH_2
Formula Weight: 54.09

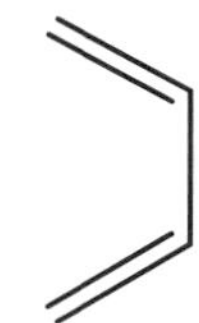

Chemical Structure

Synonyms: BIETHYLENE; BIVINYL; BUTA-1,3-DIEEN; BUTADIEEN; BUTA-1,3-DIEN; BUTADIEN; ALPHA,GAMMA-BUTADIENE; ALPHA-BUTADIENE; ALPHA-GAMMA-BUTADIENE; BUTA-1,3-DIENE; BUTADIENE; BUTADIENE MONOMER; BUTADIENE-1,3-UNINHIBITED; DIVINYL; ERYTHRENE; 1-METHYLALLENE; METHYLALLENE; PYRROLYLENE; VINYLETHYLENE
Description: colorless gas; aromatic odor
Use: polymer component in the manufacture of synthetic rubber, synthetic polymeric elastomers, rocket fuels, plastics, resins, ABS resins, chemical intermediates and latex paints

Physical Properties

Boiling Point: -4.5 °C (24 °F) at 760 mm Hg
Freezing Point: -108.91 °C (-164.038 °F)
Specific Gravity: 0.6211 at 20 °C/4 °C

Vapor Density: 1.87 Air=1
Vapor Pressure: 1840 mm Hg at 21 °C
Water Solubility: Insoluble
Other Solubilities: 95% Ethanol: Soluble; Acetone: Very Soluble; Benzene: Soluble; Ether: Soluble; Organic solvents: Soluble.
Surface Tension: Estimated at 13.4 dynes/cm at 20 °C
Odor Threshold: 0.35 to 2.86 mg/m^3
Refraction Index: 1.4292 at -25 °C/D
Evaporation Rate: > 25
Critical Temperature: 161.8 °C
Critical Pressure: 42.6 atm
Ionization Potential (eV): 9.07
Flash Point: Liquid -76 °C
Autoignition Temperature: 420 °C
LEL: 2.0% v/v
UEL: 11.5% v/v

RTECS Toxicity Data

Acute Oral: Rat LD_{50} Dose: 5480 mg/kg.
Acute Inhalation: Rat LC_{50} Dose: 285 gm/m^3/4hr; Toxic Effects: Behavioral - General anesthetic; Lungs, Thorax, or Respiration - Respiratory depression. Human TC_{Lo} Dose: 2000 ppm/7hr; Toxic Effects: Sense organs and special senses - Other; Behavioral - Hallucinations, distorted perceptions. Human TC_{Lo} Dose: 8000 ppm; Toxic Effects: Sense organs and special senses - Visual field changes; Sense organs and special senses - Conjunctive irritation; Lungs, Thorax, or Respiration - Cough.
Chronic (Multiple Dose) Inhalation: Rat Dose: 1000 ppm/6H/1Y-I; Toxic Effects: Liver - Changes in Liver weight; DEATH. Rat Dose: 2200 mg/m^3/4H/17W-I; Toxic Effects: Blood - Leukopenia; Biochemical - Lipids including transport; Biochemical - Other proteins.
Reproductive/Teratogenic: Rat Route: Inhalation; Dose: 8000 ppm/6H; Duration: female 6-15D of pregnancy; Specific Developmental Abnormalities - Musculoskeletal system. Mouse Route: Inhalation; Dose: 1000 ppm/6H; Duration: female 6-15D of pregnancy; Effects on Fertility - Post-implantation mortality; Effects on Embryo or Fetus - Extra embryonic structures; Fetotoxicity. Mouse Route: Inhalation; Dose: 1000 ppm/6H; Duration: female 6-15D of pregnancy; Maternal Effects - Uterus, cervix, vagina.
Mutagenic: Human Sister Chromatid Exchenge; Cell Type: lymphocyte; Dose: 500 umol/L. Mouse Mutations in Mammalian Somatic Cells; Cell Type: lymphocyte; Dose: 20 pph. Mouse Mutations in Mammalian Somatic Cells; Route: Inhalation; Dose: 625 ppm/6H/2W-I.
Tumorigenic: Rat Route: Inhalation; Dose: 625 ppm/6H/61W; Toxic Effects: Tumorigenic - Carcinogenic by RTECS criteria; Cardiac - Tumors; Lungs, Thorax, or Respiration - Tumors. Rat Route: Inhalation; Dose: 1000 ppm/6H/2Y-I; Toxic Effects: Tumorigenic - Carcinogenic by RTECS criteria; Skin and appendages - Tumors. Rat Route: Inhalation; Dose: 8000 ppm/6H/2Y-I; Toxic Effects: Tumorigenic - Neoplastic by RTECS criteria; Gastrointestinal - Tumors; Endocrine - Thyroid tumors. Rat Route: Inhalation; Dose: 8000 ppm/6H/15W-I; Toxic Effects: Tumorigenic - Carcinogenic by RTECS criteria; Endocrine - Tumors. Rat Route: Inhalation; Dose: 8000 ppm/6H/2Y-I; Toxic Effects: Tumorigenic - Carcinogenic by RTECS criteria; Endocrine - Thyroid tumors; Skin and appendages - Tumors.

Hazard Overviews

Explosive

Flammable

Fire Diamond

Health: Irritating to eyes/respiratory tract. Also Causes: frostbite, drowsiness, possible reproductive/teratogenic effects (based on animal data). Chronic Effects: suspect cancer hazard.
Fire: Explosive and flammable. Try to stop the flow of the gas and use a water spray to cool fire-exposed containers. When mixed with air, butadiene forms potentially explosive peroxides. Fight fire from maximum distance.
Reactivity: Stable. Hazardous polymerization can occur. Avoid: air; heat; sparks; open flame; free radical polymerization agents. Incompatible with: air; phenol; chlorine dioxide; crotonaldehyde; heating; ozone; nitrogen dioxide; copper and its alloys; strong oxidizing agents. Hazardous decomposition products: carbon monoxide; carbon dioxide.
Carcinogenicity: IARC - Group 2A, Probably carcinogenic to humans; NIOSH - Listed as carcinogen; NTP - Class 2B, Reasonably anticipated to be a carcinogen, sufficient evidence of carcinogenicity from studies in experimental animals; ACGIH - Class A2, Suspected human carcinogen; OSHA - Not listed; EPA - Class B2, Probable human carcinogen based on animal studies; MAK - Class A2, Unmistakably carcinogenic in animal experimentation only
Primary Target Organs:

Eyes

Skin

Respiratory System

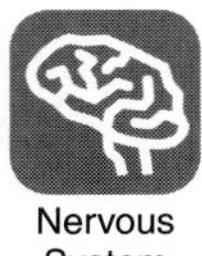
Nervous System

Exposure Limits
OSHA PEL: TWA: 1000 ppm; 2200 mg/m^3.
ACGIH TLV: TWA: 2 ppm; 4.4 mg/m^3.
NIOSH IDLH: 2000 ppm; LEL.
Respirator Recommendation
Exposure Range: >1000 to <2000 ppm Supplied Air, Constant Flow/Pressure Demand, Half Mask
Exposure Range: 2000 to unlimited ppm Self-contained Breathing Apparatus, Pressure Demand, Full Face

Environmental

Ecotoxicity: TLm Pinperch 71.5 mg/l/24 hr /Conditions of bioassay not specified
Environmental Fate: Once in the atmosphere, it will photooxidize by reaction primarily with hydroxyl radicals as well as other species, with an estimated half-life of several hours (less in smog and polluted air). Amounts released into water or land will rapidly decrease due to evaporation and

possibly also due to biodegradation. The compound may leach through soil to groundwater.

Cleanup/Disposal: Guide No. 116: Eliminate all ignition sources (no smoking, flares, sparks or flames in immediate area). All equipment used when handling the product must be grounded. Stop leak if you can do it without risk. Do not touch or walk through spilled material. Do not direct water at spill or source of leak. Use water spray to reduce vapors or divert vapor cloud drift. If possible, turn leaking containers so that gas escapes rather than liquid. Prevent entry into waterways, sewers, basements or confined areas. Isolate area until gas has dispersed.

Environmental Physical Data

Henry's Law Constant: estimated at 6.2×10^{-2}
Octanol/Water Partition Coefficient: log K_{ow} = 1.99
BCF: calculated at 19.1
BOD: not pertinent

Regulations

RCRA 40CFR: Not listed
CERCLA: 40CFR 302.4: Listed per CWA Section 311(b)(4) per RCRA Section 3001 per CAA Section 112 RQ: 10 lb (4.535 kg)
SARA 40CFR 372.65: Listed
SARA EHS 40CFR 355: Not listed
TSCA: Listed

Analytical Methods

Air: EPA 0040; ASTM D2820, D3686, D3687, D4490
Indoor / Expired Air: NIOSH 1024

BUT1350 CAS #: 106-97-8

BUTANE

RTECS: EJ4200000
DOT: UN1011; UN1075; IMO2.1
EINECS Number: 203-448-7
Molecular Formula: C_4H_{10}
Structured MF: $CH_3CH_2CH_2CH_3$
Formula Weight: 58.12

Chemical Structure

Synonyms: A-17; N-BUTANE; BUTANE MIXTURES; BUTANEN; BUTANI; BUTYL HYDRIDE; DIETHYL; HYDROCARBON PROPELLANT A-17; LIQUEFIED PETROLEUM GAS; LIQUIFIED PETROLEUM GAS; LPG; METHYLETHYL METHANE; METHYLETHYLMETHANE; NORMAL-BUTANE; PYROFAX; R 600

Description: colorless gas; faint, disagreeable odor

Use: in the manufacture of synthetic rubber and as an intermediate in the synthesis of 2-methylpropane, butadiene, acetic acid, maleic anhydride and ethylene; as a general purpose food additive a fuel for household and industrial purposes, a solvent, a refrigerant, a standby and enricher gas, a propellant in aerosols, a producer gas and as an extractant; in plastic foam production and for calibrating instruments

Physical Properties

Boiling Point: -0.5 °C (31 °F)
Freezing Point: -138.4 °C (-217.12 °F)
Specific Gravity: 0.6012 at 0 °C/4 °C
Vapor Density: 2.046 Air=1
Density: 0.6011 g/mL at 0 °C
Vapor Pressure: 760 mm Hg at -0.5 °C
Water Solubility: 61 ug/ml Water at 20 °C
Other Solubilities: 95% Ethanol: >10%; Alcohol: 18 to 1 at 17 °C and 770 mm Hg; Chloroform: 30 to 1 at 17 °C; Ether: 25 to 1 at 17 °C; Methylene chloride: Moderately Soluble.
Surface Tension: 0.014 dynes/cm
Odor Threshold: 2.8500 to 14.6300 mg/m^3
Refraction Index: 1.3543 at -13 °C/D
Critical Temperature: 153.2 °C
Critical Pressure: Absolute 525 psia
Ionization Potential (eV): 10.63
Flash Point: -60 °C Closed Cup
Autoignition Temperature: 288 °C
LEL: 1.6% v/v
UEL: 8.4% v/v

RTECS Toxicity Data

Acute Inhalation: Rat LC_{50} Dose: 658 gm/m^3/4hr. Mouse LC_{50} Dose: 680 gm/m^3/2hr.

Hazard Overviews

Flammable

Fire Diamond

Health: May cause burns or frostbite to the eyes and skin. A simple asphyxiant which can displace available oxygen. Also Causes: drowsiness, lightheadedness.
Fire: Flammable. Can form explosive mixtures in the air. Stop flow of gas. Use water to cool fire-exposed tanks. Carbon dioxide can be used to smother flame; however, gas can re-ignite and possibly explode.
Reactivity: Stable. Hazardous polymerization cannot occur. Avoid: heat; flame; oxidizers. Incompatible with: flames; nickel carbonyl and oxygen; oxidizing agents. Hazardous decomposition products: acrid smoke; fumes.
Carcinogenicity: IARC - Not listed; NIOSH - Listed as carcinogen; NTP - Not listed; ACGIH - Not listed; OSHA - Not listed; EPA - Not listed; MAK - Not listed
Primary Target Organs:

Eyes

Skin

Nervous System

Exposure Limits
OSHA PEL Vacated 1989 Limits: TWA: 800 ppm; 1900 mg/m^3.
ACGIH TLV: TWA: 800 ppm; 1900 mg/m^3.

DFG MAK: TWA: 1000 ppm; 2350 mg/m^3.
Respirator Recommendation
Exposure Range: >800 to ppm Self-contained Breathing Apparatus, Pressure Demand, Full Face
Note: poor warning properties

Environmental

Environmental Fate: Photolysis, hydrolysis, and bioconcentration are not expected to be important environmental fate processes. Biodegradation may occur in soil and water; however, volatilization is expected to be the dominant fate process. To a lesser extent adsorption may be important. A K_{oc} range of 450 to 900 indicates a low to medium mobility class in soil. In aquatic systems, it may partition from the water column to organic matter contained in sediments and suspended materials. A Henry's Law constant of 9.47 x10^{-1} atm-cu m/mole at 25 °C suggests extremely rapid volatilization from environmental waters. The volatilization half-lives from a model river and a model pond, the latter considers the effect of adsorption, have been estimated to be 2.2 hr and 2.6 days, respectively. It is expected to exist entirely in the vapor phase in ambient air. Reactions with photochemically produced hydroxyl radicals in the atmosphere have been shown to be important (average half-life of 6 days). Data also suggests the nighttime reactions with radical species and nitrogen oxides may contribute to the atmospheric transformation.
Cleanup/Disposal: Guide No. 115: Eliminate all ignition sources (no smoking, flares, sparks or flames in immediate area). All equipment used when handling the product must be grounded. Do not touch or walk through spilled material. Stop leak if you can do it without risk. If possible, turn leaking containers so that gas escapes rather than liquid. Use water spray to reduce vapors or divert vapor cloud drift. Do not direct water at spill or source of leak. Prevent spreading of vapors through sewers, ventilation systems and confined areas. Isolate area until gas has dispersed.

Environmental Physical Data

Henry's Law Constant: calculated at 9.47 x10^{-1}
Octanol/Water Partition Coefficient: log K_{ow} = 2.89
Sorption Partition Coefficient: K_{oc} = estimated at 450 to 900
BCF: estimated at 1.78
BOD: none

Regulations

RCRA 40CFR: Not listed
CERCLA: 40CFR 302.4: Not listed
SARA 40CFR 372.65: Not listed
SARA EHS 40CFR 355: Not listed
TSCA: Listed

Analytical Methods

Air: ASTM D4490

BUT1560 CAS #: 107-88-0

1,3-BUTANEDIOL

RTECS: EK0440000
EINECS Number: 203-529-7
Molecular Formula: $C_4H_{10}O_2$
Structured MF: $CH_3CH(OH)CH_2CH_2OH$
Formula Weight: 90.12

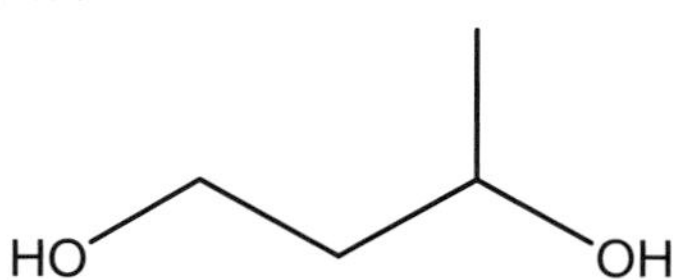

Chemical Structure

Synonyms: 1,3-BUTANDIOL; BUTANE-1,3-DIOL; 1,3-BUTYLENE GLYCOL; BETA-BUTYLENE GLYCOL; 1,3-BUTYLENGLYKOL; 1,3-DIHYDROXYBUTANE; 1-METHYL-1,3-PROPANEDIOL; METHYLTRIMETHYLENE GLYCOL
Description: colorless liquid; little or no odor
Use: intermed in mfr of polyester plasticizers; humectant for cellophane, tobacco; polyurethanes; surface active agents; coupling agent; solvent; food additive & flavoring; in cosmetic & pharmaceutical ind, glycerin substitute; in prodn of plasticizer for cellulosics & poly(vinyl chloride) resins; deicing of aircraft; in special polyester resins; antimicrobial agent, inhibiting gram-neg & gram-pos microorganisms, molds & yeasts.

Physical Properties

Boiling Point: 207.5 °C (406 °F) at 760 mm Hg
Freezing Point: < 50 °C (122 °F)
Specific Gravity: 1.0059 at 20 °C/20 °C
Vapor Density: 3.1 Air=1
Saturated Vapor Density: 1.200066888 kg/m^3
Vapor Pressure: 0.0201 mm Hg at 25 °C
Water Solubility: Practically Insoluble in Water
Other Solubilities: Slightly Soluble in Ether.
Surface Tension: 37.8 dyne/cm at 25 °C
Refraction Index: 1.4401
Critical Temperature: 370 °C
Critical Pressure: 5 x10^6 Pa
Flash Point: 121 °C Cleveland Open Cup
Autoignition Temperature: 394 °C
LEL: 1.9% v/v

RTECS Toxicity Data

Acute Oral: Rat LD_{50} Dose: 18610 mg/kg. Mouse LD_{50} Dose: 12980 mg/kg.
Acute Dermal: Rabbit LD_{50} Route: Skin; Dose: >20 gm/kg. Rat LD_{50} Route: Subcutaneous Dose: 20 gm/kg.
Irritation Eye: Rabbit Standard Draize Test Dose: 500 mg/24H; Reaction: mild. Rabbit Standard Draize Test Dose: 500 mg; Reaction: mild.
Irritation Skin: Rabbit Standard Draize Test Dose: 500 mg/24H; Reaction: mild.

Reproductive/Teratogenic: Rat Route: Oral; Dose: 42360 mg/kg; Duration: female 6-15D of pregnancy; Effects on Newborn - Growth statistics.

Hazard Overviews

Fire Diamond

Health: Mildly irritating to eyes/skin.
Fire: Will burn. Use alcohol foam, dry chemical, carbon dioxide, or water spray or fog. Water spray and fog can cause frothing.
Reactivity: Stable. Hazardous polymerization cannot occur. Avoid: ignition sources; excessive heat. Incompatible with: strong oxidizing agents; epoxides. Hazardous decomposition products: carbon monoxide.
Carcinogenicity: IARC - Not listed; NIOSH - Not listed; NTP - Not listed; ACGIH - Not listed; OSHA - Not listed; EPA - Not listed; MAK - Not listed
Primary Target Organs:

Eyes

Skin

Environmental

Environmental Fate: If released to the atmosphere, it will degrade in the vapor-phase by reaction with photochemically produced hydroxyl radicals (estimated half-life of 1.2 days). If released to soil or water, it will probably biodegrade. Leaching in soil is possible since it is miscible in water.
Cleanup/Disposal: Guide No. 128: Eliminate all ignition sources (no smoking, flares, sparks or flames in immediate area). All equipment used when handling the product must be grounded. Do not touch or walk through spilled material. Stop leak if you can do it without risk. Prevent entry into waterways, sewers, basements or confined areas. A vapor suppressing foam may be used to reduce vapors. Absorb or cover with dry earth, sand or other non-combustible material and transfer to containers. Use clean non-sparking tools to collect absorbed material. Large Spills: Dike far ahead of liquid spill for later disposal. Water spray may reduce vapor; but may not prevent ignition in closed spaces.

Environmental Physical Data

Henry's Law Constant: estimated at 2.30×10^{-7}
BCF: not significant

Regulations

RCRA 40CFR: Not listed
CERCLA: 40CFR 302.4: Not listed
SARA 40CFR 372.65: Not listed
SARA EHS 40CFR 355: Not listed
TSCA: Listed

BUT1630 CAS #: 110-63-4

1,4-BUTANEDIOL

RTECS: EK0525000
EINECS Number: 203-786-5
Molecular Formula: $C_4H_{10}O_2$
Structured MF: $HO(CH_2)_4OH$
Formula Weight: 90.12

Chemical Structure

Synonyms: AGRISYNTH B1D; BUTANE-1,4-DIOL; BUTANEDIOL; 1,4-BUTYLENE GLYCOL; BUTYLENE GLYCOL; 1,4-DIHYDROXYBUTANE; DIOL 14B; SUCOL B; TETRAMETHYLENE 1,4-DIOL; 1,4-TETRAMETHYLENE GLYCOL; TETRAMETHYLENE GLYCOL
Description: colorless viscous liquid; nearly odorless
Use: solvent, humectant, intermediate for plasticizers, pharmaceuticals, crosslinking agent in polyurethane elastomers, manufacture of tetrahydrofuran, and terephthalate plastics

Physical Properties

Boiling Point: 230 °C (446 °F) at 760 mm Hg
Freezing Point: 20.1 °C (68.18 °F)
Specific Gravity: 1.0171 at 20 °C/4 °C
Vapor Density: 3.1 Air=1
Vapor Pressure: < 1 mm Hg at 38 °C
Water Solubility: Soluble in Water
Other Solubilities: 95% Ethanol: >=100 mg/ml at 23 °C; Acetone: >=100 mg/ml at 23 °C; Benzene: 0.3 parts/100 mL solvent; DMSO: >=100 mg/ml at 23 °C; Ether: 3.1 parts/100 mL solvent; Petroleum Ether: 0.9 parts/100 mL solvent.
Surface Tension: 44.6 dyne/cm at 20 °C
Refraction Index: 1.4460 at 20 °C
Critical Temperature: 446 °C
Critical Pressure: 41.2 bar
Flash Point: 121 °C Open Cup
Autoignition Temperature: 402 °C

RTECS Toxicity Data

Acute Oral: Rat LD_{50} Dose: 1525 mg/kg; Toxic Effects: Behavioral - Altered sleep time (including change in righting reflex); Behavioral - Somnolence (general depressed activity); Blood - Other changes. Mouse LD_{50} Dose: 2062 mg/kg; Toxic Effects: Behavioral - Altered sleep time (including change in righting reflex); Behavioral - Somnolence (general depressed activity); Blood - Other changes.
Acute Inhalation: Rat LC_{Lo} Dose: 15 gm/m^3/4hr; Toxic Effects: Sense organs and special senses - Other; Lungs, Thorax, or Respiration - Other changes; Nutritional and gross metabolic - Weight loss or decreased weight gain.

Chronic (Multiple Dose) Oral: Rat Dose: 5460 mg/kg/26W-I; Toxic Effects: Blood - Changes in serum composition; Biochemical - True cholinesterase; Biochemical - Other carbohydrates. Rat Dose: 14 gm/kg/28D-I; Toxic Effects: Blood - Other changes; Biochemical - Dehydrogenases; Biochemical - Other transferases.
Chronic (Multiple Dose) Inhalation: Rat Dose: 5200 mg/m^3/6H/2W-I; Toxic Effects: Blood - Changes in serum composition; Blood - Changes in erythrocite (RBC) cell count; Nutritional and gross metabolic - Weight loss or decreased weight gain.

Hazard Overviews

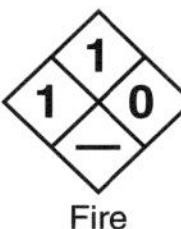

Fire Diamond

Health: Mildly irritating to eyes/skin/respiratory tract. Also Cuases: inebriation, narcosis, incoordination, kidney damage.
Fire: Will burn when exposed to heat or flame. Use dry chemical, carbon dioxide, or alcohol resistant foam. Water or foam may cause frothing of hot liquid. Do not scatter and spread the fire.
Reactivity: Stable. Hazardous polymerization cannot occur. Avoid: heat; ignition sources. Incompatible with: strong inorganic oxidizers; nitric acid; strong hydrogen peroxide; sulfuric acid. Hazardous decomposition products: tetrahydrofuran; carbon oxides.
Carcinogenicity: IARC - Not listed; NIOSH - Not listed; NTP - Not listed; ACGIH - Not listed; OSHA - Not listed; EPA - Not listed; MAK - Not listed
Primary Target Organs:

Eyes

Skin

Environmental

Environmental Fate: If released to the atmosphere, it will degrade by reaction with photochemically produced hydroxyl radicals (estimated half-life of about 1.6 days). If released to soil or water, it is expected to biodegrade. Aquatic oxidation with hydroxyl radicals is very slow with an estimated half-life of 250 days in water at pH 7. Hydrolysis, bioconcentration in aquatic organisms, adsorption to sediment and volatilization from water are not expected to be environmentally important removal mechanisms in aquatic ecosystems.
Cleanup/Disposal: Guide No. 127: Eliminate all ignition sources (no smoking, flares, sparks or flames in immediate area). All equipment used when handling the product must be grounded. Do not touch or walk through spilled material. Stop leak if you can do it without risk. Prevent entry into waterways, sewers, basements or confined areas. A vapor suppressing foam may be used to reduce vapors. Absorb or cover with dry earth, sand or other non-combustible material and transfer to containers. Use clean non-sparking tools to collect absorbed material. Large Spills: Dike far ahead of liquid spill for later disposal. Water spray may reduce vapor; but may not prevent ignition in closed spaces.

Environmental Physical Data

Henry's Law Constant: 1.3×10^{-9}
Octanol/Water Partition Coefficient: log K_{ow} = measured at 0.88
Sorption Partition Coefficient: K_{oc} = estimated at 18
BCF: estimated at 0.4

Regulations

RCRA 40CFR: Not listed
CERCLA: 40CFR 302.4: Not listed
SARA 40CFR 372.65: Not listed
SARA EHS 40CFR 355: Not listed
TSCA: Listed

BUT1910 CAS #: 55-98-1

1,4-BUTANEDIOL DIMETHYLSULFONATE

RTECS: EK1750000
EINECS Number: 200-250-2
Molecular Formula: $C_6H_{14}O_6S_2$
Structured MF: $CH_3SO_2O(CH_2)_4OSO_2CH_3$
Formula Weight: 246.32

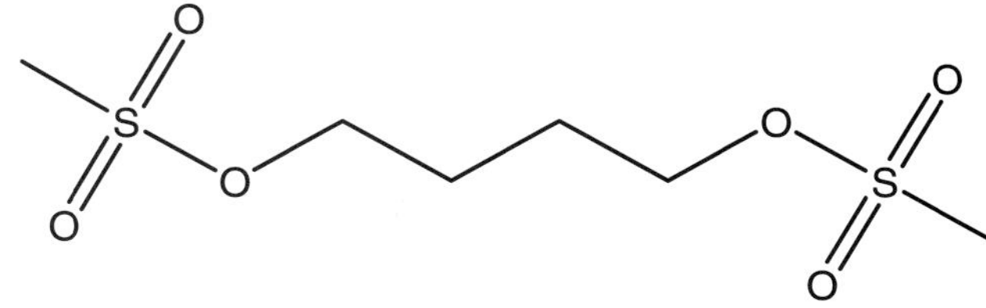

Chemical Structure

Synonyms: 2041 C.B; AN 33501; 1,4-BIS(METHANESULFONOXY)BUTANE; (1,4-BIS(METHANESULFONYLOXY)BUTANE); BUSULFAN; BUSULPHAN; BUSULPHANE; 1,4-BUTANEDIOL DIMETHANESULFONATE; 1,4-BUTANEDIOL DIMETHANESULPHONATE; C.B. 2041; CITOSULFAN; 1,4-DIMESYLOXYBUTANE; 1,4-DIMETHANESULFONOXYBUTANE; 1,4-DIMETHANESULFONOXYLBUTANE; 1,4-DIMETHANESULFONYLOXYBUTANE; 1,4-DIMETHANESULPHONYLOXYBUTANE; 1,4-DIMETHYLSULFONOXYBUTANE; 1,4-DIMETHYLSULFONYLOXYBUTANE; GT 2041; GT 41; LEUCOSULFAN; MABLIN; METHANESULFONIC ACID,TETRAMETHYLENE ESTER; MIELEVCIN; MIELOSAN; MIELUCIN; MILECITAN; MILERAN; MISULBAN; MITOSTAN; MYELEUKON; MYELOLEUKON; MYELOSAN; MYLECYTAN; MYLERAN; NSC-750; NSC-750 SULPHABUTIN; SULFABUTIN; SULPHABUTIN; TETRAMETHYLENE BIS(METHANESULFONATE); TETRAMETHYLENE DIMETHANE SULFONATE; TETRAMETHYLENESTER KYSELINY METHANSULFONOVE; X 149
Description: white crystals or powder
Use: antitumor agent and insect sterilant, treatment of leukemia

Physical Properties

Freezing Point: 114 °C (237.2 °F) to 118 °C (244.4 °F)
Water Solubility: Decomposes
Other Solubilities: 95% Ethanol: <1 mg/ml at 23 °C; Acetone: 1-5 mg/ml at 23 °C; DMSO: 1-5 mg/ml at 23 °C.
Flash Point: Not available; probably combustible

RTECS Toxicity Data

Acute Oral: Man TD_{Lo} Dose: 8 mg/kg/2D-I; Toxic Effects: Behavioral - Convulsions or effect on seizure threshold. Woman TD_{Lo} Dose: 80 mg/kg/8Y; Toxic Effects: Sense organs and special senses - Other; Vascular - Regional or general arteriolar or venous dilation; Gastrointestinal - Changes in structure or function of salivary glands.
Acute Dermal: Rat LD_{50} Route: Subcutaneous Dose: 22 mg/kg; Toxic Effects: Behavioral - Somnolence (general depressed activity); Gastrointestinal - Hypermotility, diarrhea. Mouse LD_{50} Route: Subcutaneous Dose: 63 mg/kg; Toxic Effects: Behavioral - Somnolence (general depressed activity); Gastrointestinal - Hypermotility, diarrhea.
Chronic (Multiple Dose) Oral: Rat Dose: 45 mg/kg/30D-I; Toxic Effects: Endocrine - Changes in thymus weight; Blood - Normocytic anemia; DEATH - Changes in testicular weight. Rabbit Dose: 100 mg/kg/5D-I; Toxic Effects: Blood - Leukopenia; Gastrointestinal - Ulceration or bleeding from duodenum; DEATH.
Reproductive/Teratogenic: Man Route: Oral; Dose: 5400 ug/kg; Duration: male 90D prior to mating; Paternal Effects - Breast Development. Woman Route: Oral; Dose: 17 mg/kg; Duration: female 4-36W of pregnancy; Specific Developmental Abnormalities - Eye, ear; Craniofacial (including nose and tongue); Endocrine system. Woman Route: Oral; Dose: 17 mg/kg; Duration: female 4-36W of pregnancy; Specific Developmental Abnormalities - Urogenital system; Other developmental abnormalities; Effects on Newborn - Growth statistics. Woman Route: Oral; Dose: 8460 ug/kg; Duration: female 22W prior to mating Maternal Effects - Uterus, cervix, vagina. Rat Route: Oral; Dose: 10 mg/kg; Duration: female 13D of pregnancy; Effects on Newborn - Delayed effects. Rat Route: Oral; Dose: 48 mg/kg; Duration: female 7-14D of pregnancy; Effects on Embryo or Fetus - Fetal death. Rat Route: Oral; Dose: 5600 ug/kg; Duration: female 7-14D of pregnancy; Effects on Embryo or Fetus - Fetotoxicity; Specific Developmental Abnormalities - Musculoskeletal system. Rat Route: Oral; Dose: 5 mg/kg; Duration: female 13D of pregnancy; Specific Developmental Abnormalities - Urogenital system. Rat Route: Oral; Dose: 8 mg/kg; Duration: female 11-14D of pregnancy Effects on Embryo or Fetus - Fetotoxicity; Specific Developmental Abnormalities - Craniofacial (including nose and tongue). Rat Route: Oral; Dose: 10 mg/kg; Duration: male 1D prior to mating; Paternal Effects - Spermatogenesis; Testes, epididymis, sperm duct. Rat Route: Oral; Dose: 49 mg/kg; Duration: male 49D prior to mating; Effects on Fertility - Male fertility index. Monkey Route: Intraperitoneal; Dose: 10 mg/kg; Duration: male 1D prior to mating; Paternal Effects - Testes, epididymis, sperm duct; Prostate, seminal vessicle, Cowper's gland, accessory glands.
Mutagenic: Human DNA Inhibition; Cell Type: HeLa cell; Dose: 10 mol/L. Human Cytogenetic Analysis; Cell Type: leukocyte; Dose: 10 mg/L. Human Cytogenetic Analysis; Cell Type: lymphocyte; Dose: 500 ug/L. Human Cytogenetic Analysis; Cell Type: bone marrow; Dose: 1 mg/L. Woman Cytogenetic Analysis; Route: Unreported; Dose: 138 mg/kg. Human Sister Chromatid Exchenge; Route: Oral; Dose: 2 mg/kg/5W. Human Sister Chromatid Exchenge; Cell Type: lymphocyte; Dose: 500 ug/L. Human Sister Chromatid Exchenge; Cell Type: bone marrow; Dose: 1 mg/L. Human Other Mutation Test Systems; Cell Type: leukocyte; Dose: 10 mg/L. Human Other Mutation Test Systems; Cell Type: lymphocyte; Dose: 80 umol/L.
Tumorigenic: Man Route: Oral; Dose: 5684 ug/kg/21W-C; Toxic Effects: Tumorigenic - Carcinogenic by RTECS criteria; Blood - Leukemia. Woman Route: Oral; Dose: 1140 mg/kg/9Y-I; Toxic Effects: Tumorigenic - Carcinogenic by RTECS criteria; Kidney, Ureter, and Bladder - Kidney tumors; Tumorigenic effects - Other reproductive system tumors. Woman Route: Oral; Dose: 16720 ug/kg/2Y-I; Toxic Effects: Tumorigenic - Carcinogenic by RTECS criteria; Tumorigenic effects - Uterine tumors.

Hazard Overviews

Health: Irritating to eyes/skin/respiratory tract. Toxic. Other Acute Effects: harmful if swallowed, inhaled, or absorbed through skin; causes irreversible eye damage which is seen as corneal opacity . Chronic Effects: may alter genetic material; overexposure may cause reproductive disorder(s) based on tests with laboratory animals; possible risk of harm to the unborn child. target organs: bone marrow, eyes. Carcinogen.
Fire: Will burn. Hazards: emits toxic fumes. Extinguishing agents: water spray; carbon dioxide, dry chemical powder or appropriate foam. Precautions: combustible liquid.
Reactivity: Incompatible with: strong oxidizing agents, may decompose on exposure to moist air or water. Hazardous decomposition products: toxic fumes of: carbon monoxide, carbon dioxide, sulfur oxides.
Carcinogenicity: IARC - Group 1, Carcinogenic to humans; NIOSH - Not listed; NTP - Listed; ACGIH - Not listed; OSHA - Not listed; EPA - Not listed; MAK - Not listed
Primary Target Organs:

Eyes

Skin

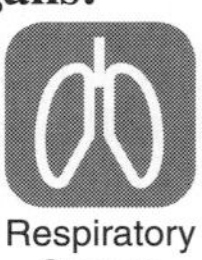
Respiratory System

Bone

Environmental

Cleanup/Disposal: Guide No. 154: Eliminate all ignition sources (no smoking, flares, sparks or flames in immediate area). Do not touch damaged containers or spilled material unless wearing appropriate protective clothing. Stop leak if you can do it without risk. Prevent entry into waterways, sewers, basements or confined areas. Absorb or cover with

dry earth, sand or other non-combustible material and transfer to containers. Do not get water inside containers.

Regulations

RCRA 40CFR: Not listed
CERCLA: 40CFR 302.4: Not listed
SARA 40CFR 372.65: Not listed
SARA EHS 40CFR 355: Not listed
TSCA: Not listed

BUT1980 CAS #: 1703-58-8

1,2,3,4-BUTANETETRACARBOXYLIC ACID

RTECS: EK6100000
EINECS Number: 216-938-0
Molecular Formula: $C_8H_{10}O_8$
Structured MF: $HO_2CCH_2CH(CO_2H)CH(CO_2H)CH_2CO_2H$
Formula Weight: 234.18

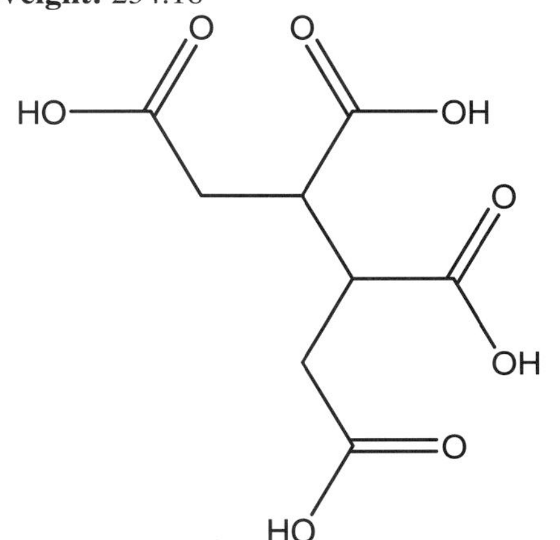

Chemical Structure

Synonyms: BUTANETETRACARBOXYLIC ACID; TCB; 1,2,3,4-TETRACARBOXYBUTANE
Description: white powder or leaflets

Physical Properties

Freezing Point: 196 °C (384.8 °F)
Water Solubility: >= 100 mg/mL at 19 C
Other Solubilities: 95% Ethanol: >=100 mg/ml at 19 °C; Acetone: 10-50 mg/ml at 19 °C; DMSO: >=100 mg/ml at 19 °C.
Flash Point: Not available; probably combustible

RTECS Toxicity Data

Acute Oral: Rat LD_{50} Dose: 1720 mg/kg.
Acute Dermal: Rabbit LD_{Lo} Route: Skin; Dose: 8 gm/kg.

Hazard Overviews

Health: Irritating to eyes/skin/respiratory tract. Harmful. Other Acute Effects: harmful if swallowed; may be harmful if inhaled; may be harmful if absorbed through the skin.
Fire: Will burn. Hazards: emits toxic fumes. Extinguishing agents: water spray; carbon dioxide, dry chemical powder or appropriate foam. Precautions: combustible liquid.
Reactivity: Incompatible with: strong oxidizing agents. Hazardous decomposition products: toxic fumes of: carbon monoxide, carbon dioxide.
Carcinogenicity: IARC - Not listed; NIOSH - Not listed; NTP - Not listed; ACGIH - Not listed; OSHA - Not listed; EPA - Not listed; MAK - Not listed
Primary Target Organs:

Eyes

Skin

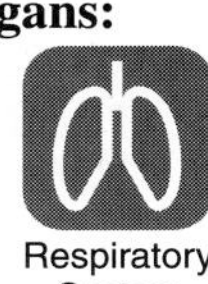
Respiratory System

Environmental

Regulations

RCRA 40CFR: Not listed
CERCLA: 40CFR 302.4: Not listed
SARA 40CFR 372.65: Not listed
SARA EHS 40CFR 355: Not listed
TSCA: Listed

BUT2120 CAS #: 71-36-3

1-BUTANOL

RTECS: EO1400000
DOT: UN1120; NA1120; IMO3.3
EINECS Number: 200-751-6
Molecular Formula: $C_4H_{10}O$
Structured MF: $CH_3CH_2CH_2CH_2OH$
Formula Weight: 74.12

Chemical Structure

Synonyms: ALCOOL BUTYLIQUE; BUTAN-1-OL; BUTANOL; N-BUTAN-1-OL; N-BUTANOL; BUTANOLEN; BUTANOLO; 1-BUTYL ALCOHOL; BUTYL ALCOHOL; N-BUTYL ALCOHOL; BUTYL HYDROXIDE; BUTYLOWY ALKOHOL; BUTYRIC ALCOHOL; BUTYRIC OR NORMAL PRIMARY BUTYL ALCOHOL; CCS 203; HEMOSTYP; 1-HYDROXYBUTANE; METHYLOLPROPANE; NBA; NORMAL PRIMARY BUTYL ALCOHOL; N-PROPYL CARBINOL; PROPYL CARBINOL; PROPYL METHANOL; PROPYLCARBINOL; PROPYLMETHANOL
Description: colorless liquid; fuel oil and banana odor
Use: solvent for vegetable oils, dyes, alkaloids, etc; intermed for pharmaceuticals, ethylene glycol monobutyl ether, butylamines, 2,4-d esters, etc; alkyd resin, fabric coating; mfr of safety glass; food additive (beverages, ice cream, candy, baked goods, etc); in mfr of hydraulic fluids; detergents; dehydrating agent; extractant in mfr of antibiotics, vitamins, & hormones; stabilizing agent; entrainer; in mfr of derivatives of butyl alcohol, incl chems, herbicides, ore flotation agents, etc; med: postoperative pain in otolaryngeal surgery, & anti-hemorrhagic effect in advanced cancer, (vet) bactericide.

Physical Properties

Boiling Point: 117.2 °C (243 °F)
Freezing Point: -89.5 °C (-129.1 °F)
Specific Gravity: 0.8098 at 20 °C/4 °C
Vapor Density: 2.6 Air=1
Saturated Vapor Density: 1.213511434 kg/m^3
Vapor Pressure: 5.5 mm Hg at 20 °C
Water Solubility: 9.1 ml/100 ml Water at 25 °C
Other Solubilities: miscible with many organic solvents; > 10% in Acetone; > 10% in Benzene; > 10% in Ether; > 10% in Ethanol.
Surface Tension: 26.2 dynes/cm
Odor Threshold: 0.3600 to 150.000 mg/m^3
Refraction Index: 1.3993 at 20 °C/D
Evaporation Rate: 0.46 Butyl Acetate=1
Critical Temperature: 289.8 °C
Critical Pressure: 43.6 atm
Ionization Potential (eV): 10.04
Flash Point: 37 °C Closed Cup
Autoignition Temperature: 365 °C
LEL: 1.4% v/v
UEL: 11.2% v/v

RTECS Toxicity Data

Acute Oral: Rat LD_{50} Dose: 790 mg/kg; Toxic Effects: Liver - Fatty Liver degeneration; Kidney, Ureter, and Bladder - Other changes; Blood - Other changes. Mouse LD_{50} Dose: 2680 mg/kg.
Acute Inhalation: Rat LC_{50} Dose: 8000 ppm/4hr. Human TC_{Lo} Dose: 25 ppm; Toxic Effects: Sense organs and special senses - Other; Sense organs and special senses - Conjunctive irritation; Lungs, Thorax, or Respiration - Other changes.
Acute Dermal: Rabbit LD_{50} Route: Skin; Dose: 3400 mg/kg. Mouse LD_{50} Route: Subcutaneous Dose: 3200 mg/kg.
Chronic (Multiple Dose) Dermal: Rat Route: Subcutaneous; Dose: 157 mg/kg/4D-I; Toxic Effects: Behavioral - Ataxia.
Irritation Eye: Rabbit Standard Draize Test Dose: 2 mg/24H; Reaction: severe. Rabbit Standard Draize Test Dose: 2 mg; Reaction: severe.
Irritation Skin: Rabbit Standard Draize Test Dose: 20 mg/24H; Reaction: moderate. Rabbit Standard Draize Test Dose: 405 mg/24H; Reaction: moderate.
Reproductive/Teratogenic: Rat Route: Oral; Dose: 35295 mg/kg; Duration: female 1-15D of pregnancy; Effects on Fertility - Female fertility index; Pre-implantation mortality; Post-implantation mortality. Rat Route: Oral; Dose: 35295 mg/kg; Duration: female 1-15D of pregnancy; Effects on Embryo or Fetus - Fetotoxicity; Effects on Newborn - Biochemical and metabolic. Rat Route: Inhalation; Dose: 6000 ppm/7H; Duration: female 1-19D of pregnancy; Effects on Embryo or Fetus - Fetotoxicity. Rat Route: Inhalation; Dose: 8000 ppm/7H; Duration: female 1-19D of pregnancy; Specific Developmental Abnormalities - Musculoskeletal system.
Mutagenic: Hamster Sex Chromosome Loss; Cell Type: lung; Dose: 100 mmol/L. Mold - A Nidulans Sex Chromosome Loss; Dose: 7000 ppm.

Hazard Overviews

Flammable

Fire Diamond

Health: Irritating to eyes/skin/respiratory tract. Also Causes: headache, dizziness, and drowsiness. Chronic Effects: blurred vision, sensitivity to light; hearing loss and dizziness with concurrent noise exposure
Fire: Highly flammable. Can form explosive mixtures in the air. For small fires, use dry chemical, carbon dioxide, water spray, or alcohol-resistant foam. For large fires, use water spray, fog, or alcohol-resistant foam.
Reactivity: Stable. Hazardous polymerization cannot occur. Avoid: exposure to heat and ignition sources. Incompatible with: aluminum; chromium trioxide; organic peroxides; strong oxidizers; some forms of plastic, rubber, coatings. Hazardous decomposition products: carbon monoxide; acrid smoke.
Carcinogenicity: IARC - Not listed; NIOSH - Not listed; NTP - Not listed; ACGIH - Not listed; OSHA - Not listed; EPA - Class D, Not classifiable as to human carcinogenicity; MAK - Not listed
Primary Target Organs:

Eyes

Skin

Respiratory System

Nervous System

Exposure Limits
OSHA PEL: TWA: 100 ppm; 300 mg/m^3.
OSHA PEL Vacated 1989 Limits: STEL: 50 ppm; 150 mg/m^3; Ceiling.
NIOSH REL: STEL: 50 ppm; 150 mg/m^3; skin.
NIOSH IDLH: 1400 ppm; LEL.
DFG MAK: TWA: 100 ppm; 300 mg/m^3.
Respirator Recommendation
Exposure Range: >100 to 1000 ppm Air Purifying, Negative Pressure, Half Mask
Exposure Range: >1000 to <1400 ppm Supplied Air, Constant Flow/Pressure Demand, Full Face
Exposure Range: 1400 to unlimited ppm Self-contained Breathing Apparatus, Pressure Demand, Full Face
Cartridge Color: black

Environmental

Ecotoxicity: LC_{50} Pimephales promelas (fathead minnow) 1940, 1940, 1940, 1940, & 1940 mg/l at 1, 24, 48, 72, & 96 hr, respectively, at 18 to 22 °C (Static bioassay in reconstituted water) LC_{50} Pimephales promelas (fathead minnow) 1730 mg/l/96 hr (95% confidence limit 1630-1840 mg/l); age 33 days old, water hardness 47.7 mg/l ($CaCO_3$), temp 24.7 °C, pH 7.64, dissolved oxygen 6.3 mg/l, alkalinity 45.5 mg/l ($CaCO_3$) Static bioassay EC_{50} Pimephales promelas (fathead minnow) 1510 mg/l/96 hr; age 33 days old, water hardness 47.7 mg/l ($CaCO_3$), temp 24.7 °C, pH 7.64, dissolved oxygen 6.3 mg/l, alkalinity 45.5 mg/l ($CaCO_3$)

Static bioassay Toxicity Threshold (Cell Multiplication Inhibition Test): Uronema parduczi Chatton-Lwoff (protozoa) 8.0 mg/l EC_{50} Daphnia magna (daphnid) 1983 mg/l/48 hr, toxic effect: lost ability to swim Toxicity Threshold (Cell Multiplication Inhibition Test) Microcystis aeruginosa (algae) 100 mg/l

Environmental Fate: Release to soil may result in volatilization from the soil surface and biodegradation is expected to be significant. It should not bind strongly to soil and so is expected to leach into groundwater. Release to water is expected to result in biodegradation and in volatilization from the water surface. Photooxidation by hydroxyl radicals is expected to be slow. Bioconcentration is not expected to be significant. Vapor phase in the atmosphere is expected to react with photochemically generated hydroxyl radicals with a half-life of 1.2 (experimental)-2.3 (estimated) days.

Cleanup/Disposal: Guide No. 129: Eliminate all ignition sources (no smoking, flares, sparks or flames in immediate area). All equipment used when handling the product must be grounded. Do not touch or walk through spilled material. Stop leak if you can do it without risk. Prevent entry into waterways, sewers, basements or confined areas. A vapor suppressing foam may be used to reduce vapors. Absorb or cover with dry earth, sand or other non-combustible material and transfer to containers. Use clean non-sparking tools to collect absorbed material. Large Spills: Dike far ahead of liquid spill for later disposal. Water spray may reduce vapor; but may not prevent ignition in closed spaces.

Environmental Physical Data

Henry's Law Constant: 5.57 $x10^{-6}$
Octanol/Water Partition Coefficient: log K_{ow} = 0.88
Sorption Partition Coefficient: K_{OC} = estimated at 71.6
BCF: little expected
BOD: 1.1 to 1.92 lb/lb, 5 days

Regulations

RCRA 40CFR: Listed Hazardous Waste No. U031 Ignitable Waste
CERCLA: 40CFR 302.4: Listed per RCRA Section 3001 RQ: 5000 lb (2268 kg)
SARA 40CFR 372.65: Listed
SARA EHS 40CFR 355: Not listed
TSCA: Listed

Analytical Methods

Air: ASTM D3686, D3687, D4490
Soil: EPA 80APP-F2; SW846 1311, 5031, 8015B, 8260A, 8260B
Water / Groundwater: ASTM D3695
Food: AOAC 975.06
Indoor / Expired Air: NIOSH 1401
Plasma: EPA 29
Other: EPA 1666

BUT2190 **CAS #: 78-92-2**

2-BUTANOL

RTECS: EO1750000
DOT: UN1120; NA1120; IMO3.3
EINECS Number: 201-158-5
Molecular Formula: $C_4H_{10}O$
Structured MF: $CH_3CH(OH)CH_2CH_3$
Formula Weight: 74.12

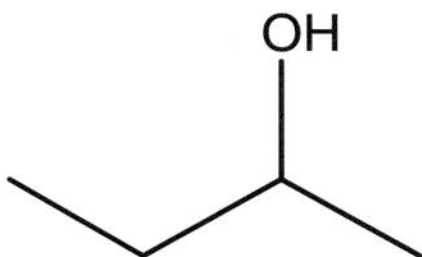

Chemical Structure

Synonyms: ALCOOL BUTYLIQUE SECONDAIRE; BUTAN-2-OL; BUTANOL-2; S-BUTANOL; SEC-BUTANOL; BUTANOL SECONDAIRE; 2-BUTYL ALCOHOL; S-BUTYL ALCOHOL; SEC-BUTYL ALCOHOL; 2-BUTYLALCOHOL; BUTYLENE HYDRATE; CCS 301; DL-2-BUTANOL; ETHYL METHYL CARBINOL; ETHYLMETHYL CARBINOL; 2-HYDROXYBUTANE; METHYL ETHYL CARBINOL; 1-METHYL PROPANOL; METHYLETHYLCARBINOL; 1-METHYL-1-PROPANOL; 1-METHYLPROPYL ALCOHOL; S.B.A; SBA; SECONDARY BUTYL ALCOHOL
Description: colorless liquid; wine odor
Use: in synth of flotation agents; flavors; perfumes; dyestuffs; wetting agents; in industr cleaners, paint removers; solvent for many natural resins, linseed & castor oils; in hydraulic brake fluids, polishes & as chemical intermediate; in dewaxing paraffin; preparation of methyl ethyl ketone; to produce alpha-hydroperoxy alcohols secondary alcohols; used to denature other alcohols

Physical Properties

Boiling Point: 99.44 °C (211 °F)
Freezing Point: -115 °C (-175 °F)
Specific Gravity: 0.8063 at 20 °C/4 °C
Vapor Density: 2.6 Air=1
Vapor Pressure: 23.9 mm Hg at 30 °C
Water Solubility: 15.4 g/100 g Water at 20 °C
Other Solubilities: Soluble in Benzene, Alcohol, Ether, & Acetone.
Surface Tension: 23.5 dynes/cm at 10 °C
Odor Threshold: 131.1500 mg/m^3
Refraction Index: 1.3978
Evaporation Rate: 1.3 Butyl Acetate=1
Critical Temperature: 263 °C
Critical Pressure: 41.4 atm
Ionization Potential (eV): 10.10
Flash Point: 24 °C Closed Cup
Autoignition Temperature: 406 °C
LEL: 1.7% v/v
UEL: 9.85% v/v

RTECS Toxicity Data

Acute Oral: Rat LD_{50} Dose: 6480 mg/kg. Rabbit LD_{50} Dose: 4893 mg/kg; Toxic Effects: Sense organs and special senses - Corneal damage; Cardiac - Pulse rate; Lungs, Thorax, or Respiration - Dyspnea.

Acute Inhalation: Rat LC_{Lo} Dose: 16000 ppm/4hr; Toxic Effects: Sense organs and special senses - Other; Behavioral - Somnolence (general depressed activity); Lungs, Thorax, or Respiration - Other changes.

Irritation Eye: Rabbit Standard Draize Test Dose: 100 mg/24H; Reaction: moderate.

Irritation Skin: Rabbit Standard Draize Test Dose: 500 mg/24H; Reaction: mild.

Reproductive/Teratogenic: Rat Route: Inhalation; Dose: 5000 ppm/7H; Duration: female 1-19D of pregnancy; Effects on Embryo or Fetus - Fetotoxicity. Rat Route: Inhalation; Dose: 7000 ppm/7H; Duration: female 1-19D of pregnancy; Effects on Fertility - Post-implantation mortality; Effects on Embryo or Fetus - Fetal death; Specific Developmental Abnormalities - Musculoskeletal system.

Hazard Overviews

Flammable

Fire Diamond

Health: Mildly irritating to eyes/skin/respiratory tract. Also Causes: CNS depression, headache, nausea, vomiting, diarrhea, muscle weakness, giddiness, ataxia, confusion, delirium, coma. Chronic Effects: drying/cracking skin.

Fire: Flammable. Can form explosive mixtures in the air. Use dry chemical, carbon dioxide, water spray, or alcohol-resistant foam. Fight fire from maximum distance.

Reactivity: Unstable, forms explosive peroxides upon standing via auto-oxidation. Hazardous polymerization cannot occur. Avoid: exposure to heat; ignition sources. Incompatible with: chromium trioxide and other oxidizers; some forms of plastic, rubber, and coatings. Hazardous decomposition products: carbon oxide gases.

Carcinogenicity: IARC - Not listed; NIOSH - Listed as carcinogen; NTP - Not listed; ACGIH - Not listed; OSHA - Not listed; EPA - Not listed; MAK - Not listed

Primary Target Organs:

Eyes

Skin

Nervous System

Exposure Limits

OSHA PEL: TWA: 150 ppm; 450 mg/m^3.

OSHA PEL Vacated 1989 Limits: TWA: 100 ppm; 305 mg/m^3.

ACGIH TLV: TWA: 100 ppm; 300 mg/m^3.

NIOSH REL: TWA: 100 ppm; 305 mg/m^3. STEL: 150 ppm; 455 mg/m^3.

NIOSH IDLH: 2000 ppm.

DFG MAK: TWA: 100 ppm; 300 mg/m^3.

Respirator Recommendation

Exposure Range: >150 to 1000 ppm Air Purifying, Negative Pressure, Half Mask

Exposure Range: >1000 to <2000 ppm Supplied Air, Constant Flow/Pressure Demand, Full Face

Exposure Range: 2000 to unlimited ppm Self-contained Breathing Apparatus, Pressure Demand, Full Face

Cartridge Color: black

Environmental

Ecotoxicity: Toxicity Threshold (Cell Multiplication Inhibition Test) Pseudomonas putida (bacteria) 500 mg/l LD_{50} Carassius auratus (Goldfish) 4300 mg/l/24 hr /Conditions of bioassay not specified LC_{50} Pimephales promelas (fathead minnows) 3670 g/l/96 hr (95% confidence limit 3380-3990 g/l); age 30 days old, water hardness 43.6 mg/l ($CaCO_3$), temp 24.4 °C, pH 7.82, dissolved oxygen 7.5 mg/l, alkalinity 41.4 mg/l ($CaCO_3$) /Type of bioassay not specified Toxicity Threshold (Cell Multiplication Inhibition Test) Microcystis aeruginosa (algae) 312 mg/l

Environmental Fate: If released on soil, it will leach into the ground. It should also volatilize from dry soil. Based on the results of biodegradability screening tests and a few other studies, biodegradation will probably be the key process affecting fate in soil. If released in water, biodegradation will probably also be the primary factor affecting its loss. In one study, half-life was 5 days in river water. Volatilization will only be significant at fairly high temperature (half-life 3.2 days in a model river at 25 °C). Adsorption to sediment and bioconcentration in fish will not be significant transport processes. In the atmosphere, it will be lost by reaction with photochemically produced hydroxyl radicals. Its estimated half-life is 2 days.

Cleanup/Disposal: Guide No. 129: Eliminate all ignition sources (no smoking, flares, sparks or flames in immediate area). All equipment used when handling the product must be grounded. Do not touch or walk through spilled material. Stop leak if you can do it without risk. Prevent entry into waterways, sewers, basements or confined areas. A vapor suppressing foam may be used to reduce vapors. Absorb or cover with dry earth, sand or other non-combustible material and transfer to containers. Use clean non-sparking tools to collect absorbed material. Large Spills: Dike far ahead of liquid spill for later disposal. Water spray may reduce vapor; but may not prevent ignition in closed spaces.

Environmental Physical Data

Henry's Law Constant: estimated at 9.1×10^{-6}

Octanol/Water Partition Coefficient: log K_{OW} = 0.81

Sorption Partition Coefficient: K_{OC} = 5.6

BCF: calculated at 1.7

BOD: 83% BODT, 5 days

Regulations

RCRA 40CFR: Not listed

CERCLA: 40CFR 302.4: Not listed

SARA 40CFR 372.65: Listed

SARA EHS 40CFR 355: Not listed

TSCA: Listed

Analytical Methods

Air: ASTM D3686, D3687
Soil: EPA 80APP-F2
Water / Groundwater: ASTM D3695
Food: AOAC 975.06
Indoor / Expired Air: NIOSH 1401

BUT2260 CAS #: 75-65-0

T-BUTANOL

RTECS: EO1925000
DOT: UN1120; NA1120; IMO3.2
EINECS Number: 200-889-7
Molecular Formula: $C_4H_{10}O$
Structured MF: $(CH_3)_3COH$
Formula Weight: 74.14

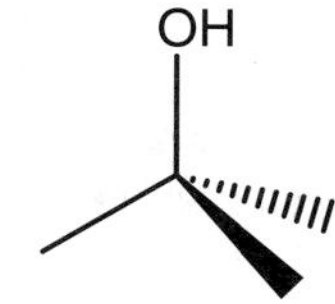

Chemical Structure

Synonyms: ALCOOL BUTYLIQUE TERTIAIRE; TERT-BUTANOL; BUTANOL TERTIAIRE; TERT-BUTYL ALCOHOL; T-BUTYL HYDROXIDE; 1,1-DIMETHYL ETHANOL; 1,1-DIMETHYLETHANOL; METHANOL,TRIMETHYL-; 2-METHYL-2-PROPANOL; 2-METHYLPROPAN-2-OL; 2-METHYLPROPANOL-2; 2-PROPANOL,2-METHYL-; TBA; TRIMETHYL CARBINOL; TRIMETHYL METHANOL; TRIMETHYLCARBINOL; TRIMETHYLMETHANOL
Description: colorless crystals, liquid; camphor-like odor
Use: alcohol denaturant; solvent for pharmaceuticals and perfumes; in paint removers; in manufacture of flotation agents; octane booster in gasoline

Physical Properties

Boiling Point: 82.41 °C (180 °F)
Freezing Point: 25.7 °C (78.26 °F)
Specific Gravity: 0.78
Vapor Density: 2.55
Saturated Vapor Density: 1.303223956 kg/m^3
Density: 0.79 g/mL
Vapor Pressure: 42 mm Hg at 25 °C
Water Solubility: Miscible
Other Solubilities: 95% Ethanol: >=100 mg/ml at 21 °C; Acetone: >=100 mg/ml 21 °C; Chloroform: Soluble; DMSO: >=100 mg/ml at 21 °C; Ether: Very Soluble.
Surface Tension: 20.7 dynes/cm at 25 °C
Odor Threshold: 219 mg/m^3
pH: Essentially neutral
Refraction Index: 1.38468 at 20 °C/D
Critical Temperature: 233 °C
Critical Pressure: 576 psia
Ionization Potential (eV): 9.70
Flash Point: 11 °C
Autoignition Temperature: air 478 °C
LEL: 2.35% v/v
UEL: 8% v/v at 55 °C

RTECS Toxicity Data

Acute Oral: Rat LD_{50} Dose: 3500 mg/kg; Toxic Effects: Behavioral - Ataxia. Rabbit LD_{50} Dose: 3559 mg/kg; Toxic Effects: Sense organs and special senses - Corneal damage; Cardiac - Pulse rate; Lungs, Thorax, or Respiration - Dyspnea.
Chronic (Multiple Dose) Oral: Rat Dose: 150 gm/kg/94D-C; Toxic Effects: Behavioral - Fluid intake; Kidney, Ureter, and Bladder - Urine volume decreased; Nutritional and gross metabolic - Weight loss or decreased weight gain. Mouse Dose: 703 gm/kg/94D-C; Toxic Effects: Behavioral - Fluid intake; Kidney, Ureter, and Bladder - Inflammation, necrosis, or scarring of bladder; Nutritional and gross metabolic - Weight loss or decreased weight gain.
Reproductive/Teratogenic: Rat Route: Inhalation; Dose: 2000 ppm/7H; Duration: female 1-19D of pregnancy; Effects on Embryo or Fetus - Fetotoxicity. Rat Route: Inhalation; Dose: 3500 ppm/7H; Duration: female 1-19D of pregnancy; Specific Developmental Abnormalities - Musculoskeletal system.
Mutagenic: Other Insects Cytogenetic Analysis; Route: Parenteral; Dose: 12500 mg/kg.
Tumorigenic: Rat Route: Oral; Dose: 146 gm/kg/2Y-C; Toxic Effects: Tumorigenic - Neoplastic by RTECS criteria; Kidney, Ureter, and Bladder - Kidney tumors. Mouse Route: Oral; Dose: 1540 gm/kg/2Y-C; Toxic Effects: Tumorigenic - Neoplastic by RTECS criteria; Endocrine - Thyroid tumors.

Hazard Overviews

Flammable

Fire Diamond

Health: Irritating to eyes/respiratory tract. Also Causes: coughing, difficulty breathing, dizziness, fatique, headache, nausea, drowsiness, unconsciousness, dermatitis, vomiting, diarrhea. Chronic Effects: drying dermatitis, damage to the liver, kidney, or bladder.
Fire: Flammable. May form explosive mixtures in the air. For small fires use dry chemical, carbon dioxide, water spray, or alcohol-resistant foam. For large fires use water spray, fog, or alcohol-resistant foam. Fight fire from maximum distance. Fire may be difficult to extinguish and re-ignite.
Reactivity: Stable. Hazardous polymerization cannot occur. Avoid: exposure to heat; ignition sources. Incompatible with: oxidizers; hydrogen peroxide/sulfuric acid (explosive); sodium-potassium alloy; strong mineral acids. Hazardous decomposition products: carbon monoxide; isobutylene gas.
Carcinogenicity: IARC - Not listed; NIOSH - Not listed; NTP - Not listed; ACGIH - Class A4, Not classifiable as a human carcinogen; OSHA - Not listed; EPA - Not listed; MAK - Not listed

Primary Target Organs:

Eyes

Skin

Respiratory System

Nervous System

Exposure Limits

OSHA PEL: TWA: 100 ppm; 300 mg/m^3.

OSHA PEL Vacated 1989 Limits: TWA: 100 ppm; 300 mg/m^3; STEL: 150 ppm; 450 mg/m^3.

ACGIH TLV: TWA: 100 ppm; 303 mg/m^3.

NIOSH REL: TWA: 100 ppm; 300 mg/m^3. STEL: 150 ppm; 450 mg/m^3.

NIOSH IDLH: 1600 ppm.

DFG MAK: TWA: 100 ppm; 300 mg/m^3.

Respirator Recommendation

Exposure Range: >100 to 1000 ppm Air Purifying, Negative Pressure, Half Mask

Exposure Range: >1000 to <1600 ppm Supplied Air, Constant Flow/Pressure Demand, Full Face

Exposure Range: 1600 to unlimited ppm Self-contained Breathing Apparatus, Pressure Demand, Full Face

Cartridge Color: black

Environmental

Ecotoxicity: LD_{100} Semolitus atromaculatus (creek chub) 6000 mg/l/24 hr in Detroit river water /Conditions of bioassay not specified LC_{50} Poecilia reticulata (guppy) 3550 ppm/7 days /Conditions of bioassay not specified

Environmental Fate: Release to the soil is expected to result in volatilization from the soil surface and biodegradation. It is not expected to strongly adsorb to soil; therefore, it is expected to leach to groundwater. When released to water, it is expected to volatilize and it may biodegrade. Aqueous photooxidation will be a slow process. Bioconcentration in fish is not expected to be significant. When released to the atmosphere, it will react with nitrogen oxide with a half-life of above a day. The estimated half-life of the reaction between vapor phase and photochemically generated hydroxyl radicals is 1.09 months.

Cleanup/Disposal: Guide No. 129: Eliminate all ignition sources (no smoking, flares, sparks or flames in immediate area). All equipment used when handling the product must be grounded. Do not touch or walk through spilled material. Stop leak if you can do it without risk. Prevent entry into waterways, sewers, basements or confined areas. A vapor suppressing foam may be used to reduce vapors. Absorb or cover with dry earth, sand or other non-combustible material and transfer to containers. Use clean non-sparking tools to collect absorbed material. Large Spills: Dike far ahead of liquid spill for later disposal. Water spray may reduce vapor; but may not prevent ignition in closed spaces.

Environmental Physical Data

Henry's Law Constant: 1.175 x10^{-5}

Octanol/Water Partition Coefficient: log K_{OW} = 0.35

Sorption Partition Coefficient: K_{OC} = 36.9

BCF: estimated at 0.036

BOD: 0%, 5 days

Regulations

RCRA 40CFR: Not listed

CERCLA: 40CFR 302.4: Not listed

SARA 40CFR 372.65: Listed

SARA EHS 40CFR 355: Not listed

TSCA: Listed

Analytical Methods

Air: ASTM D3686, D3687

Soil: SW846 5031, 8015B, 8260B

Indoor / Expired Air: NIOSH 1400

Other: EPA 1666

BUT2330 CAS #: 78-93-3

2-BUTANONE

RTECS: EL6475000

DOT: UN1193; IMO3.2

EINECS Number: 201-159-0

Molecular Formula: C_4H_8O

Structured MF: $CH_3COCH_2CH_3$

Formula Weight: 72.10

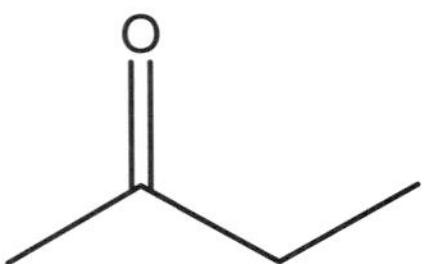

Chemical Structure

Synonyms: ACETONE,METHYL-; AETHYLMETHYLKETON; 3-BUTANONE; BUTANONE; BUTANONE 2; EPA PESTICIDE CHEMICAL CODE 044103; ETHYL METHYL CETONE; ETHYL METHYL KETONE; ETHYLMETHYLCETONE; ETHYLMETHYLKETON; KETONE,ETHYL METHYL; MEETCO; MEK; METHYL ACETONE; METHYL ETHYL KETONE; METILETILCETONA; METILETILCHETONE; METYLOETYLOKETON

Description: colorless liquid; sweet mint or acetone-like odor

Use: as a solvent in nitrocellulose coating and vinyl film manufacture, in smokeless powder manufacture, in cements and adhesives, dewaxing of lubricating oils, "Glyptal" resins, paint removers, organic synthesis, cleaning fluids, acrylic coatings; intermediate in drug manufacture; manufacture of colorless synthetic resins; swelling agent of resins; intermediate in the manufacture of ketones and amines; and printing catalyst and carrier

Physical Properties

Boiling Point: 79.6 °C (175 °F)

Freezing Point: -86.3 °C (-123.34 °F)

Specific Gravity: 0.805 at 20 °C/4 °C

Vapor Density: 2.41 Air=1

Saturated Vapor Density: 1.381864791 kg/m^3

Density: 0.806 g/mL at 20 °C

Bulk Density: 6.71 lbs/gal at 20 °C

Vapor Pressure: 77.5 torr at 20 °C

Water Solubility: 353 g/L Water at 10 °C

Other Solubilities: 95% Ethanol: >=100 mg/ml at 19 °C; Acetone: >=100 mg/ml at 19 °C; Benzene: miscible; DMSO: >=100 mg/ml at 19 °C; Ether: miscible; Industrial organic solvents: Soluble; Oils: miscible.
Surface Tension: 24.6 dynes/cm at 20 °C
Odor Threshold: 0.7375 to 147.5 mg/m^3
Refraction Index: 1.3814 at 15 °C/D
Evaporation Rate: 2.7 Ether=1
Critical Temperature: 262 °C
Critical Pressure: 603 psia
Ionization Potential (eV): 9.54
Flash Point: -9 °C Closed Cup
Autoignition Temperature: 404 °C
LEL: 1.4% v/v
UEL: 11.4% v/v

RTECS Toxicity Data

Acute Oral: Rat LD_{50} Dose: 2737 mg/kg. Mouse LD_{50} Dose: 4050 mg/kg.
Acute Inhalation: Rat LC_{50} Dose: 23500 mg/m^3/8hr. Human TC_{Lo} Dose: 100 ppm/5M; Toxic Effects: Sense organs and special senses - Other; Sense organs and special senses - Conjunctive irritation; Lungs, Thorax, or Respiration - Other changes.
Acute Dermal: Rabbit LD_{50} Route: Skin; Dose: 6480 mg/kg.
Chronic (Multiple Dose) Inhalation: Rat Dose: 5000 ppm/6H/90D-I; Toxic Effects: Liver - Changes in Liver weight; Kidney, Ureter, and Bladder - Urine volume increased; Biochemical - Transaminases. Rat Dose: 750 ppm/7H/7D-I; Toxic Effects: Liver - Liver function tests impaired.
Chronic (Multiple Dose) Dermal: Cat Route: Subcutaneous; Dose: 55500 mg/kg/37W-I; Toxic Effects: DEATH.
Irritation Skin: Rabbit Standard Draize Test Dose: 402 mg/24H; Reaction: mild. Rabbit Standard Draize Test Dose: 500 mg/24H; Reaction: moderate.
Reproductive/Teratogenic: Rat Route: Inhalation; Dose: 3000 ppm/7H; Duration: female 6-15D of pregnancy; Specific Developmental Abnormalities - Craniofacial (including nose and tongue); Urogenital system; Homeostasis. Rat Route: Inhalation; Dose: 1000 ppm/7H; Duration: female 6-15D of pregnancy; Effects on Embryo or Fetus - Fetotoxicity; Specific Developmental Abnormalities - Musculoskeletal system.
Mutagenic: Yeast - S Cerevisiae Sex Chromosome Loss; Dose: 33800 ppm.

Hazard Overviews

Flammable

Fire Diamond

Health: Irritating to eyes/respiratory tract. Also Causes: corneal injury; inhalation may cause dizziness or vomiting. Chronic Effects: dry skin, dermatitis.
Fire: Flammable. Can form explosive mixtures in the air. For small fires, use dry chemical, carbon dioxide, water spray, or alcohol-resistant foam. For large fires, use water spray, fog, or alcohol-resistant foam. Use water to cool fire-exposed containers.
Reactivity: Stable. Hazardous polymerization cannot occur. Avoid: exposure to heat; ignition sources. Incompatible with: chlorosulfonic acid; oleum; potassium-t-butoxide; hydrogen peroxide/nitric acid; 2-propanol; chloroform/alkali; amines; ammonia; inorganic acids; caustics; copper; isocyanates; pyridines; strong oxidizers; some plastics. Hazardous decomposition products: carbon dioxide gas; acrid smoke.
Carcinogenicity: IARC - Not listed; NIOSH - Not listed; NTP - Not listed; ACGIH - Not listed; OSHA - Not listed; EPA - Class D, Not classifiable as to human carcinogenicity; MAK - Not listed
Primary Target Organs:

Eyes

Skin

Respiratory System

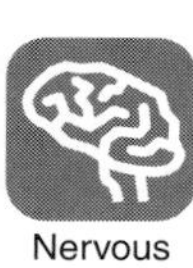
Nervous System

Exposure Limits
OSHA PEL: TWA: 200 ppm; 590 mg/m^3.
OSHA PEL Vacated 1989 Limits: TWA: 200 ppm; 590 mg/m^3; STEL: 300 ppm; 885 mg/m^3.
ACGIH TLV: TWA: 200 ppm; 590 mg/m^3.
NIOSH REL: TWA: 200 ppm; 590 mg/m^3. STEL: 300 ppm; 885 mg/m^3.
NIOSH IDLH: 3000 ppm.
DFG MAK: TWA: 200 ppm; 590 mg/m^3.
Respirator Recommendation
Exposure Range: >200 to 1000 ppm Air Purifying, Negative Pressure, Half Mask
Exposure Range: >1000 to <3000 ppm Air Purifying, Negative Pressure, Full Face
Exposure Range: 3000 to unlimited ppm Supplied Air, Constant Flow/Pressure Demand, Full Face Self-contained Breathing Apparatus, Pressure Demand, Full Face
Cartridge Color: black

Environmental

Ecotoxicity: LC_{50} Pimephales promelas (fathead minnow) 3220 mg/l/96 hr (confidence limit 3130-3320 mg/l) /Conditions of bioassay not specified Toxicity Threshold (Cell Multiplication Inhibition Test) Scenedesmus quadricauda (green algae) 4300 mg/l /Conditions of bioassay not specified Toxicity Threshold (Cell Multiplication Inhibition Test) Entosiphon sulcatum (protoza) 190 mg/l /Conditions of bioassay not specified
Environmental Fate: When discharged into water, it will be lost by evaporation (half-life 3-12 days) or be slowly biodegraded. When released to the atmosphere, it will photodegrade at a moderate rate (half-life 2.3 days or less). It would not be expected to bioconcentrate into aquatic organisms.
Cleanup/Disposal: Guide No. 127: Eliminate all ignition sources (no smoking, flares, sparks or flames in immediate area). All equipment used when handling the product must be grounded. Do not touch or walk through spilled material.

Stop leak if you can do it without risk. Prevent entry into waterways, sewers, basements or confined areas. A vapor suppressing foam may be used to reduce vapors. Absorb or cover with dry earth, sand or other non-combustible material and transfer to containers. Use clean non-sparking tools to collect absorbed material. Large Spills: Dike far ahead of liquid spill for later disposal. Water spray may reduce vapor; but may not prevent ignition in closed spaces.

Environmental Physical Data

Henry's Law Constant: 2.4 x10^{-5}
Octanol/Water Partition Coefficient: log K_{ow} = 0.26 to 0.29
BCF: not significant
BOD: 214%, 5 days

Regulations

RCRA 40CFR: Listed Hazardous Waste No. U159 Toxic Waste Ignitable Waste
CERCLA: 40CFR 302.4: Listed per RCRA Section 3001 RQ: 5000 lb (2268 kg)
SARA 40CFR 372.65: Listed
SARA EHS 40CFR 355: Not listed
TSCA: Listed

Analytical Methods

Air: EPA OA-002-1, VA-005-1, VG-006-1, TO-5; ASTM D3686, D3687, D4490
Soil: CLP MC_VOA, OHC; EPA 1624; SW846 1311, 5031, 5032, 8015A, 8015B, 8240B, 8260A, 8260B; DOE OG015R, OP040R
Water / Groundwater: EPA VW-008-1; ASTM D3695
Drinking Water: EPA 524.2
Indoor / Expired Air: NIOSH 2500
Plasma: EPA 29; NIOSH 8002
Other: EPA VS-006-1

BUT2400 CAS #: 1338-23-4

2-BUTANONE PEROXIDE

RTECS: EL9450000
DOT: UN2127; UN2550; UN2563; UN3068; IMO5.2
EINECS Number: 215-661-2
Molecular Formula: $C_8H_{16}O_4$
Formula Weight: 176.24

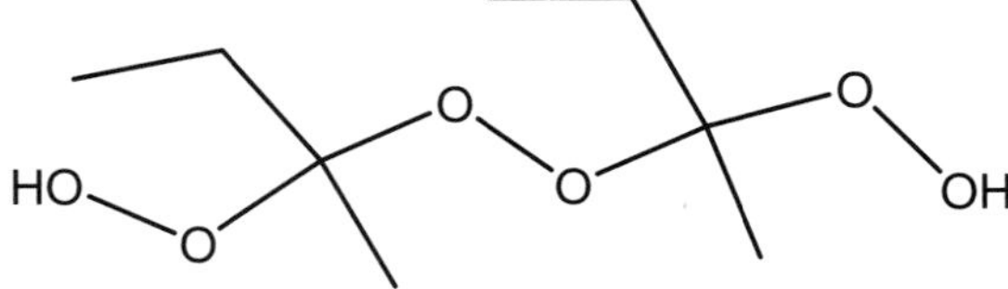

Chemical Structure

Synonyms: 2-BUTANONE,PEROXIDE; BUTANOX LPT; BUTANOX M 105; BUTANOX M 50; CHALOXYD MEKP-HA 1; CHALOXYD MEKP-LA 1; COMPONENT 1: 2-BUTANONE PEROXIDE (45%); COMPONENT 2: DIMETHYL PHTHALATE; ESPERFOAM FR; ETHYL METHYL KETONE PEROXIDE; FR 222; HI-POINT 180; HI-POINT 90; HI-POINT PD-1; KAYAMEK A; KAYAMEK M; KETONOX; LUCIDOL DDM 9; LUCIDOL DELTA X; LUPERSOL DELTA-X; LUPERSOL DDA 30; LUPERSOL DDM; LUPERSOL DEL; LUPERSOL DNF; LUPERSOL DSW; LUPERSOL DELTA X; MEK PEROXIDE; MEKP; MEKPO; MEPOX; METHYL ETHYL KETONE HYDROPEROXIDE; METHYL ETHYL KETONE PEROXIDE; METHYL ETHYL KETONE PEROXIDE,IN SOLUTION WITH >9% BYWEIGHT ACTIVE OXYGEN; METHYLETHYLKETONEHYDROPEROXIDE; METHYLETHYLKETONHYDROPEROXIDE; PERMEK G; PERMEK N; QUICKSET EXTRA; QUICKSET SUPER; SPRAYSET MEKP; SUPEROX 46-710; THERMACURE; TRIGONOX M 50
Description: colorless liquid; ketone odor
Use: hardening agent for glass fiber-reinforced plastics and polyester resins, as a catalyst for processing polystyrene, acrylonitrile-butadiene-styrene polymers and other thermoplastics and as a cross-linking agent and catalyst in the production of other polymers; in the manufacture of acrylic resins, in the manufacture of paints, plastics and rubber, to cure promoted unsaturated polyester resins and vinyl ester resins at ambient temperatures and to cure thermosetting resins; as a solvent and plasticizer for cellulose acetate and cellulose acetate-butyrate compositions, as an insect repellent for personal protection against biting insects, as a plasticizer for nitrocellulose and in the manufacture of resins, rubber, solid rocket propellants, lacquers, plastics, rubber, coating agents, safety glass, molding powders, plasticizers, latex and cellulose-acetate film; as a fluidized bed coating in the manufacture of polyvinylidene fluoride and in perfumes

Physical Properties

Boiling Point: 117.78 °C (244 °F)
Specific Gravity: 1.17
Vapor Density: 6.69 Air=1
Saturated Vapor Density: < 1.200080167 kg/m^3
Vapor Pressure: < 0.01 mm Hg at 20 °C
Water Solubility: Soluble
Other Solubilities: DMSO: >=100 mg/ml at 22 °C; 95% Ethanol: >=100 mg/ml at 22 °C; Acetone: >=100 mg/ml at 22 °C; Alcohol: miscible; Benzene: Soluble; Most organic solvents: Soluble.
pH: Acid
Refraction Index: 1.484
Flash Point: 51.667 to 93.333 °C Micro Open Cup
Autoignition Temperature: 566 °C

RTECS Toxicity Data

Acute Oral: Human TD_{Lo} Dose: 480 mg/kg; Toxic Effects: Gastrointestinal - Changes in structure or function of esophagus; Gastrointestinal - Nausea or vomiting; Gastrointestinal - Other changes. Rat LD_{50} Dose: 484 mg/kg.
Acute Inhalation: Rat LC_{50} Dose: 200 ppm/4hr; Toxic Effects: Lungs, Thorax, or Respiration - Dyspnea. Mouse LC_{50} Dose: 170 ppm/4hr; Toxic Effects: Lungs, Thorax, or Respiration - Dyspnea.
Chronic (Multiple Dose) Oral: Rat Dose: 2033 mg/kg/7W-I; Toxic Effects: Nutritional and gross metabolic - Weight loss or decreased weight gain; DEATH.

Hazard Overviews

Corrosive

Explosive

Flammable

Fire Diamond

Health: Corrosive to eyes/skin/respiratory tract.
Fire: Flammable. Use dry chemical, carbon dioxide, water spray, or alcohol foam. Fires burn vigorously and are difficult to extinguish. Fight fires from a maximum distance; automatic water sprays are recommended. This material can burn slowly at first, but after it is heated it can burn violently in a manner similar to that of burning gasoline.
Reactivity: Unstable. Hazardous polymerization can occur if in contact with incompatible chemicals or are heated. Avoid: acetone; steam pipes; sparks; radiators; flames; heat; ignition sources; contamination. Incompatible with: active mineral acids; amines; strong oxidizing and reducing agents; brass; copper; steel; cobalt naphthenate; cobalt octoate; manganese naphthenate; potassium octoate; diethyl aniline; dimethyl aniline; acetyl acetone. Hazardous decomposition products: carbon monoxide.
Carcinogenicity: IARC - Not listed; NIOSH - Listed as carcinogen; NTP - Not listed; ACGIH - Not listed; OSHA - Not listed; EPA - Not listed; MAK - Not listed
Primary Target Organs:

Eyes

Skin

Respiratory System
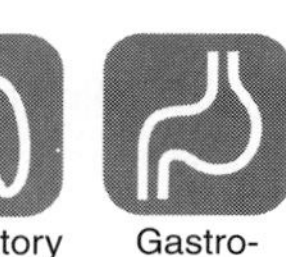
Gastro-intestinal

Exposure Limits
OSHA PEL Vacated 1989 Limits: STEL: 0.07 ppm; 5 mg/m^3; Ceiling.
ACGIH TLV: STEL: 0.2 ppm; 1.5 mg/m^3; Ceiling.
NIOSH REL: STEL: 0.2 ppm; 1.5 mg/m^3.
Respirator Recommendation
Exposure Range: >0.2 to 10 ppm Supplied Air, Constant Flow/Pressure Demand, Half Mask
Exposure Range: >10 to 200 ppm Supplied Air, Constant Flow/Pressure Demand, Full Face
Exposure Range: >200 to unlimited ppm Self-contained Breathing Apparatus, Pressure Demand, Full Face
Note: odor threshold unknown

Environmental

Environmental Fate: If released to the atmosphere, it will degrade by reaction with photochemically produced hydroxyl radicals; the estimated half-life for this reaction is 1.7 days. If released to soil or water, it may react with organic materials since it is a strong oxidizing agent. Leaching in soil may be possible in the absence of oxidation reactions or other degradation processes. Aquatic volatilization, adsorption to sediment, and bioconcentration are not expected to be important fate processes. Insufficient data are available to predict the relative importance of biodegradation.
Cleanup/Disposal: Guide No. 147: Eliminate all ignition sources (no smoking, flares, sparks or flames in immediate area). Keep combustibles (wood, paper, oil, etc.) away from spilled material. Do not touch damaged containers or spilled material unless wearing appropriate protective clothing. Keep substance wet using water spray. Stop leak if you can do it without risk. Small Spills: Take up with inert, damp, noncombustible material using clean non-sparking tools and place into loosely covered plastic containers for later disposal. Large Spills: Wet down with water and dike for later disposal. Prevent entry into waterways, sewers, basements or confined areas. Do not clean-up or dispose of, except under supervision of a specialist.

Environmental Physical Data

Henry's Law Constant: estimated at 1.6 x10^{-8}
Octanol/Water Partition Coefficient: log K_{ow} = 0.914
Sorption Partition Coefficient: K_{oc} = estimated at 74
BCF: estimated at 2.9

Regulations

RCRA 40CFR: Listed Hazardous Waste No. U160 Toxic Waste Reactive Waste
CERCLA: 40CFR 302.4: Listed per RCRA Section 3001 RQ: 10 lb (4.535 kg)
SARA 40CFR 372.65: Not listed
SARA EHS 40CFR 355: Not listed
TSCA: Listed

Analytical Methods

Indoor / Expired Air: NIOSH 3508

BUT2540 **CAS #: 106-98-9**

1-BUTENE

EINECS Number: 203-449-2
Molecular Formula: C_4H_8
Formula Weight: 56.10

Chemical Structure

Synonyms: ALPHA-BUTENE; BUTENE-1; 1-BUTYLENE; ALPHA-BUTYLENE; ETHYLETHYLENE
Description: colorless gas; faint odor
Use: polymer and alkylate gasoline; polybutenes; butadiene; intermediate for C4 and C5 aldehydes, alcohols and other derivatives; production of maleic anhydride by catalytic oxidation

Physical Properties

Boiling Point: -6.47 °C (20 °F)
Freezing Point: -185.35 °C (-301.63 °F)
Specific Gravity: 0.5951 at 20 °C/4 °C
Vapor Density: 1.93 Air=1
Vapor Pressure: 3480 mm Hg at 21 °C
Water Solubility: Insoluble in Water
Other Solubilities: 95% Ethanol: Soluble; Benzene: Soluble; Ether: Soluble; Most organic solvents: Soluble.

Refraction Index: 1.3962 at 20 °C/D
Evaporation Rate: > 1 Butyl Acetate=1
Ionization Potential (eV): 9.55 +/-0.06
Flash Point: -80 °C
Autoignition Temperature: 385 °C
LEL: 1.98% v/v
UEL: 9.65% v/v

Hazard Overviews

Flammable

Compressed Gas

Fire Diamond

Health: Causes: headache, drowsiness, dizziness, lightheadedness, excitation, excess salivation, narcosis, contact with compressed gas can cause burns/frostbite. Exposure to vapor is mildly irritating.
Fire: Flammable. Forms explosive mixtures with air or oxidizing agents. If feasible and without undue risk, shut off flow of gas. Let fire burn if leak can't be stopped. For small fires use dry chemical or carbon dioxide. For large fires use water spray or fog. Vapors may travel to an ignition source and flash back.
Reactivity: Stable. Hazardous polymerization can occur. Avoid: heat; ignition sources. Incompatible with: tetrahydroborate; mixture with oxygen; oxidizers. Hazardous decomposition products: acrid smoke; fumes; carbon oxides; pyrolyzes to C 1 and C 2 alkanes, toluene, and several C 5 and C 6 cyclanes, or cyclenes.
Carcinogenicity: IARC - Not listed; NIOSH - Not listed; NTP - Not listed; ACGIH - Not listed; OSHA - Not listed; EPA - Not listed; MAK - Not listed
Primary Target Organs:

Eyes

Skin

Nervous System

Environmental

Cleanup/Disposal: Guide No. 115: Eliminate all ignition sources (no smoking, flares, sparks or flames in immediate area). All equipment used when handling the product must be grounded. Do not touch or walk through spilled material. Stop leak if you can do it without risk. If possible, turn leaking containers so that gas escapes rather than liquid. Use water spray to reduce vapors or divert vapor cloud drift. Do not direct water at spill or source of leak. Prevent spreading of vapors through sewers, ventilation systems and confined areas. Isolate area until gas has dispersed.

Regulations

RCRA 40CFR: Not listed
CERCLA: 40CFR 302.4: Not listed
SARA 40CFR 372.65: Not listed
SARA EHS 40CFR 355: Not listed
TSCA: Listed

BUT2610 CAS #: 107-01-7

2-BUTENE

RTECS: EM2932000
EINECS Number: 203-452-9
Molecular Formula: C_4H_8
Formula Weight: 56.10

Chemical Structure

Synonyms: BUTENE-2; BETA-BUTYLENE; BUTYLENE-2; DIMETHYLETHYLENE; SYM-DIMETHYLETHYLENE; PSEUDO-BUTYLENE; PSEUDOBUTYLENE
Description: colorless gas; slightly aromatic odor
Use: int for butadiene, heptenes, sec-butyl alc, butylene oxide; prodn of high-density polyethylene & polymer gasolines

Physical Properties

Boiling Point: cis Isomer 3.72 °C (39 °F)
Freezing Point: cis Isomer -138 °C (-216.4 °F)
Specific Gravity: Cis-Isomer 0.6213 at 20 °C
Water Solubility: Insoluble
Other Solubilities: Soluble in Alcohol, Ether, Benzene and most organic solvents

RTECS Toxicity Data

Acute Inhalation: Mouse LC_{50} Dose: 425 ppm.

Hazard Overviews

Flammable

Compressed Gas

Fire Diamond

Health: Mildly irritating to respiratory tract. Also Causes: oxygen displacement, dizziness, headache, fatigue, difficulty breathing, loss of coordination, disorientation, frostbite.
Fire: Flammable. Stop flow of gas. Use dry chemical or carbon dioxide to extinguish flame in order to shut off supply or repair leak. Use water spray or fog to cool surroundings and prevent ignition of other materials.
Reactivity: Stable. Hazardous polymerization cannot occur. Avoid: heat; ignition sources. Incompatible with: oxidizing material. Hazardous decomposition products: carbon oxides; acrid smoke; fumes.
Carcinogenicity: IARC - Not listed; NIOSH - Not listed; NTP - Not listed; ACGIH - Not listed; OSHA - Not listed; EPA - Not listed; MAK - Not listed

Primary Target Organs:

Mucous Membranes

Nervous System

Environmental

Regulations

RCRA 40CFR: Not listed
CERCLA: 40CFR 302.4: Not listed
SARA 40CFR 372.65: Not listed
SARA EHS 40CFR 355: Not listed
TSCA: Listed

BUT2680 CAS #: 590-18-1

CIS-2-BUTENE

EINECS Number: 209-673-7
Molecular Formula: C_4H_8
Formula Weight: 56.10

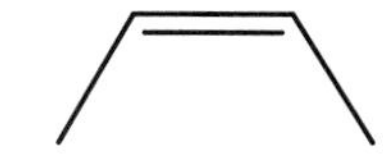

Chemical Structure

Synonyms: (Z)-2-BUTENE; 2-BUTENE,(Z)-; CIS-BUTENE; BETA-CIS-BUTYLENE; CIS-BUTYLENE; CIS-1,2-DIMETHYLETHYLENE; HIGH-BOILING BUTENE-2

Use: solvent; cross-linking agent; polymer gasoline; butadiene synth; chem int for gasoline alkylate & polygas (butylenes mixt); refinery fuel (mixt of butylenes); chem int for butadiene & sec-butanol; cross-linking agent & solvent; component of liquid petroleum gas (heating fuel & feedstock)

Physical Properties

Boiling Point: 3.73 °C (39 °F) at 760 mm Hg
Freezing Point: -139.3 °C (-218.74 °F)
Specific Gravity: 0.6213 at 20 °C/4 °C
Vapor Pressure: 1 mm Hg at 96.4 °C
Water Solubility: Insoluble in Water
Other Solubilities: Very Soluble in Alcohol, Ether; Soluble in Benzene.
Refraction Index: 1.3931 at -25 °C/D
Ionization Potential (eV): 9.11 +/-0.01
Flash Point: -73.333 °C
Autoignition Temperature: 324 °C
LEL: 1.8% v/v
UEL: 9.7% v/v

Hazard Overviews

Reactivity: Stable. Hazardous polymerization cannot occur. Avoid: heat; ignition sources. Incompatible with: oxidizing materials. Hazardous decomposition products: carbon oxides; acrid smoke; fumes.

Carcinogenicity: IARC - Not listed; NIOSH - Not listed; NTP - Not listed; ACGIH - Not listed; OSHA - Not listed; EPA - Not listed; MAK - Not listed

Environmental

Regulations

RCRA 40CFR: Not listed
CERCLA: 40CFR 302.4: Not listed
SARA 40CFR 372.65: Not listed
SARA EHS 40CFR 355: Not listed
TSCA: Listed

Analytical Methods

Air: ASTM D2820

BUT2750 CAS #: 624-64-6

TRANS-2-BUTENE

EINECS Number: 210-855-3
Molecular Formula: C_4H_8
Formula Weight: 56.10

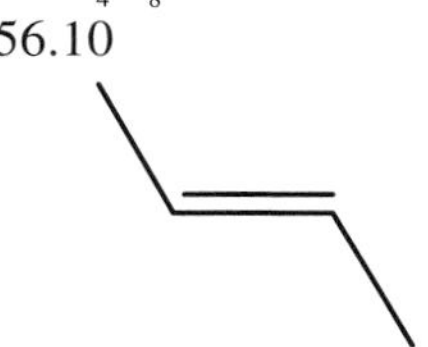

Chemical Structure

Synonyms: (E)-2-BUTENE; 2-BUTENE,(E)-; 2-TRANS-BUTENE; TRANS-BUTENE; BETA-TRANS-BUTYLENE; TRANS-1,2-DIMETHYLETHYLENE; LOW-BOILING BUTENE-2

Use: solvent; cross-linking agent; polymer gasoline; butadiene synth; chem int for gasoline alkylate & polygas; refinery fuel; chem int, eg, for butadiene & sec-butanol; component of liquid petroleum gas (heating fuel & feedstock)

Physical Properties

Boiling Point: 0.88 °C (34 °F) at 760 mm Hg
Freezing Point: -105.55 °C (-157.99 °F)
Specific Gravity: 0.6042 at 20 °C/4 °C
Vapor Pressure: 1 mm Hg at -99.4 °C
Water Solubility: Insoluble in Water
Other Solubilities: Soluble in Benzene.
Odor Threshold: Trans-2 isomer 0.0048 mg/m^3
Refraction Index: 1.3931 at -25 °C/D
Ionization Potential (eV): 9.10 +/-0.01
Flash Point: -73.333 °C
Autoignition Temperature: 324 °C
LEL: 1.8% v/v
UEL: 9.7% v/v

Hazard Overviews

Reactivity: Stable. Hazardous polymerization cannot occur. Avoid: heat and ignition sources. Incompatible with:

oxidizing materials; Hazardous decomposition products: carbon oxides, acrid smoke; fumes.

Carcinogenicity: IARC - Not listed; NIOSH - Not listed; NTP - Not listed; ACGIH - Not listed; OSHA - Not listed; EPA - Not listed; MAK - Not listed

Environmental

Environmental Fate: If released to soil, it is expected to rapidly volatilize to the atmosphere. It is expected to moderately adsorb to soil and it has the potential to biodegrade in soil. If released to water, it is expected to rapidly volatilize to the atmosphere. The half-life for volatilization from a model river is 2.2 hours. It is not expected to significantly bioconcentrate in fish and aquatic organisms and it has the potential to biodegrade in water. If released to the atmosphere, it is expected to be rapidly oxidized by both photochemically produced hydroxyl radicals and ozone. The half-lives for these processes are 6.0 and 1.4 hours, respectively. Night-time degradation by the reaction with nitrate radicals is also expected to be a significant removal process with a calculated half-life of 3.8 hours.

Environmental Physical Data

Henry's Law Constant: 0.154
Octanol/Water Partition Coefficient: log K_{ow} = 2.31
Sorption Partition Coefficient: K_{oc} = 440
BCF: calculated at 33

Regulations

RCRA 40CFR: Not listed
CERCLA: 40CFR 302.4: Not listed
SARA 40CFR 372.65: Not listed
SARA EHS 40CFR 355: Not listed
TSCA: Listed

Analytical Methods

Air: ASTM D2820

BUT2820 CAS #: 126-22-7

BUTONATE

RTECS: ET0175000
DOT: UN2783
EINECS Number: 204-778-4
Molecular Formula: $C_8H_{14}Cl_3O_5P$
Formula Weight: 327.55

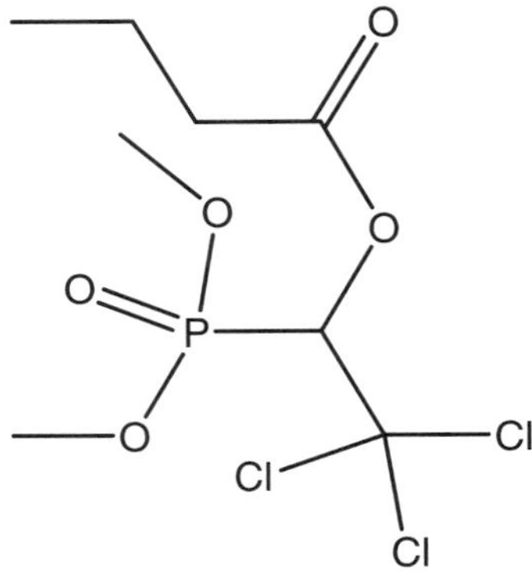

Chemical Structure

Synonyms: BUTANOIC ACID,2,2,2-TRICHLORO-1-(DIMETHOXYPHOSPHINYL)ETHYL ESTE R; BUTANOIC ACID,2,2,2-TRICHLORO-1-(DIMETHOXYPHOSPHINYL)ETHYL ESTER; BUTONAT; BUTYRIC ACID,ESTER WITH DIMETHYL(2,2,2-TRICHLORO-1-HYDROXYETHYL)PHOSPHONATE; BUTYRYLTRICHLORFON; DIMETHOXY-2,2,2-TRICHLORO-L-N-BUTYRYLOXY-ETHYLPHOSPHINEOXIDE; DIMETHOXY-2,2,2-TRICHLORO-1-N-BUTYRYLOXY-ETHYLPHOSPINEOXIDE; DIMETHYL 1-BUTYRYLOXY-2,2,2-TRICHLOROETHYLPHOSPHONATE; 0,0-DIMETHYL 2,2,2-TRICHLORO-1-(N-BUTYRYLOXY)ETHYLPHOSPHONATE; DIMETHYL 2,2,2-TRICHLORO-1-(BUTYRYLOXY)-ETHYLPHOSPHONATE; DIMETHYL 2,2,2-TRICHLORO-1-N-BUTYRYLOXYETHYLPHOSPHONATE; O,O-DIMETHYL 2,2,2-TRICHLORO-1-BUTYRYL-OXYETHYLPHOSPHONATE; O,O-DIMETHYL (2,2,2-TRICHLORO-1-HYDROXYETHYL) PHOSPHONATEESTER OF BUTYRIC ACID; O,O-DIMETHYL2,2,2-TRICHLORO-1-(N-BUTYRYLOXY)ETHYLPHOSPHONATE; O,O-DIMETHYL2,2,2-TRICHLORO-1-N-BUTYRYLOXYETHYLPHOSPHONATE; O,O-DIMETHYL-(1-N-BUTYRYLOXY-2,2,2-TRICHLORAETHYL)-PHOSPHONSAEUREESTER; O,O-DIMETHYL-(1-BUTYRYLOXY-2,2,2-TRICHLOROETHYL)PHOSPHONATE; O,O-DIMETHYL-2,2,2-TRICHLORO-1-PHOSPHONOETHYL BUTYRATE; ENT 20,852; F-139; FEKAMA AT 50; PEDIX-50; PEDIX PE 50; PEDIX-BUTONATE; PHOSPHONIC ACID,(2,2,2-TRICHLORO-1-HYDROXYETHYL)-,DIMETHYL ESTER,BUTYRATE; T-113; TRIBUFON; 2,2,2-TRICHLORO-1-(DIMETHOXYPHOSPHINOYL)ETHYL BUTYRATE; 2,2,2-TRICHLORO-1-(DIMETHOXYPHOSPHINYL)ETHYL BUTANOATE

Description: colorless, somewhat oily liquid; slight ester odor
Use: formerly as an insecticide; medication (vet)

Physical Properties

Boiling Point: 129 °C (264 °F) at 0.5 mm Hg
Density: 11.5 lb/gal
Other Solubilities: deodorized Kerosene dissolves 2-3%.
Refraction Index: 1.4707

RTECS Toxicity Data

Acute Oral: Rat LD_{50} Dose: 1100 mg/kg. Mouse LD_{50} Dose: 760 mg/kg.

Acute Dermal: Rat LD_{50} Route: Skin; Dose: 7 gm/kg. Rat LD_{50} Route: Subcutaneous Dose: 3 gm/kg.

Mutagenic: Mouse DNA Damage; Route: Intraperitoneal; Dose: 200 mg/kg.

Hazard Overviews

Carcinogenicity: IARC - Not listed; NIOSH - Not listed; NTP - Not listed; ACGIH - Not listed; OSHA - Not listed; EPA - Not listed; MAK - Not listed

Environmental

Cleanup/Disposal: Guide No. 152: Do not touch damaged containers or spilled material unless wearing appropriate protective clothing. Stop leak if you can do it without risk. Prevent entry into waterways, sewers, basements or confined areas. Cover with plastic sheet to prevent spreading. Absorb or cover with dry earth, sand or other non-combustible material and transfer to containers. Do not get water inside containers.

Regulations

RCRA 40CFR: Not listed
CERCLA: 40CFR 302.4: Not listed
SARA 40CFR 372.65: Not listed
SARA EHS 40CFR 355: Not listed
TSCA: Not listed

Analytical Methods

Plasma: EPA 028

BUT2890 CAS #: 68310-81-6

T-BUTOXYDIMETHYLISOPROPYL-AMINO

EINECS Number: 269-734-9

Physical Properties

Boiling Point: 158 °C (316 °F)
Specific Gravity: < 1
Water Solubility: Reacts

Hazard Overviews

Carcinogenicity: IARC - Not listed; NIOSH - Not listed; NTP - Not listed; ACGIH - Not listed; OSHA - Not listed; EPA - Not listed; MAK - Not listed

Environmental

Regulations

RCRA 40CFR: Not listed
CERCLA: 40CFR 302.4: Not listed
SARA 40CFR 372.65: Not listed
SARA EHS 40CFR 355: Not listed
TSCA: Not listed

BUT2960 CAS #: 124-16-3

1-(2-BUTOXYETHOXY)-2-PROPANOL

RTECS: UA8050000
EINECS Number: 204-684-3
Molecular Formula: $C_9H_{20}O_3$
Formula Weight: 176.29
Synonyms: 1-(BUTOXYETHOXY)-2-PROPANOL; 1-BUTOXYETHOXY-2-PROPANOL; 2-BUTOXY-1-(2'-HYDROXYPROPOXY)ETHANE; 4,7-DIOXAUNDECAN-2-OL; 2-PROPANOL,1-(2-BUTOXYETHOXY)-
Description: colorless liquid
Use: solvent; in hydraulic fluid components; anti-stall additive for automotive fuels; plasticizer intermediate; solvent for waxes & degreasing solns; chem int

Physical Properties

Boiling Point: 230.3 °C (447 °F)
Freezing Point: -90 °C (-130 °F)
Specific Gravity: 0.931 at 20 °C/20 °C
Water Solubility: Soluble in Water
Flash Point: 121.111 °C

RTECS Toxicity Data

Acute Oral: Rat LD_{50} Dose: 4 mL/kg.
Acute Dermal: Rabbit LD_{50} Route: Skin; Dose: 2830 uL/kg.
Irritation Skin: Rabbit Open Draize Test Dose: 485 mg open; Reaction: mild.

Hazard Overviews

Fire: Will burn.
Carcinogenicity: IARC - Not listed; NIOSH - Not listed; NTP - Not listed; ACGIH - Not listed; OSHA - Not listed; EPA - Not listed; MAK - Not listed

Environmental

Regulations

RCRA 40CFR: Not listed
CERCLA: 40CFR 302.4: Not listed
SARA 40CFR 372.65: Not listed
SARA EHS 40CFR 355: Not listed
TSCA: Listed

BUT3030 CAS #: 1852-16-0

N-(BUTOXYMETHYL)ACRYLAMIDE

RTECS: AS3450000
EINECS Number: 217-442-7
Molecular Formula: $C_8H_{15}NO_2$
Formula Weight: 157.24

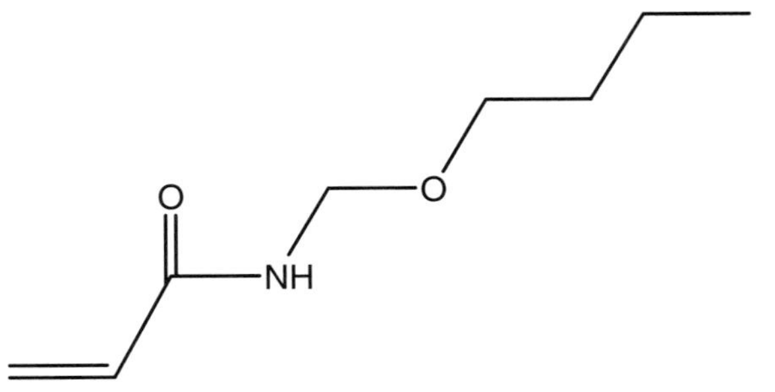

Chemical Structure

Synonyms: ACRYLAMIDE,N-(BUTOXYMETHYL)-; ACRYLAMIDE,N-BUTOXYMETHYL-; N-BUTOXYMETHYLAKRYLAMID; N-(BUTOXYMETHYL)-2-PROPENAMIDE; 2-PROPENAMIDE,N-(BUTOXYMETHYL)-; 2-PROPENAMIDE,N-(BUTOXYMETHYL)-(9CI)

RTECS Toxicity Data

Acute Oral: Rat LD_{50} Dose: 1030 mg/kg.

Hazard Overviews

Health: Irritating to eyes/skin/respiratory tract. Toxic. Other Acute Effects: harmful if swallowed, inhaled, or absorbed through skin; may cause nervous system disturbances; dermatitis. Chronic Effects: may alter genetic material; overexposure may cause reproductive disorder(s) based on tests with laboratory animals; target organs: central nervous system, male reproductive system, liver, lungs, eyes, kidneys. Carcinogen.

Fire: Hazards: emits toxic fumes. Extinguishing agents: water spray; carbon dioxide, dry chemical powder or appropriate foam. Precautions: combustible liquid.

Reactivity: Stable. Hazardous polymerization may occur. Incompatible with: strong oxidizing agents, mineral acids, peroxides, iron and iron salts, copper, aluminum, brass, free radical initiators, acids, bases, polymerizing initiators. Hazardous decomposition products: toxic fumes of: carbon monoxide, carbon dioxide, nitrogen oxides.

Carcinogenicity: IARC - Not listed; NIOSH - Not listed; NTP - Not listed; ACGIH - Not listed; OSHA - Not listed; EPA - Not listed; MAK - Not listed

Primary Target Organs:

Eyes

Skin

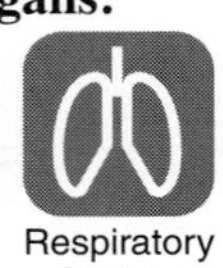
Respiratory System

Nervous System

Liver

Kidneys

Environmental

Regulations

RCRA 40CFR: Not listed
CERCLA: 40CFR 302.4: Not listed
SARA 40CFR 372.65: Not listed
SARA EHS 40CFR 355: Not listed
TSCA: Listed

BUT3100 CAS #: 9003-13-8

BUTOXYPOLYPROPYLENE GLYCOL

RTECS: TR4680000
Molecular Formula: $C_{(3x+4)}H_{(6x+10)}O_{(x+1)}$
Formula Weight: Approximately 800
Synonyms: BUTOXY POLYPROPYLENE GLYCOL; BUTOXYPROPANEDIOL POLYMER; CRAG FLY REPELLENT; ENT 8286; EXP MITICIDE NO 7; EXP. MITICIDE NO. 7; NEWPOL LB3000; OPSB; POLY(OXY(METHYL-1,2-ETHANEDIYL)),ALPHA-BUTYL-OMEGA-HYDROXY-; POLY(OXYPROPYLENE) BUTYL ETHER; POLYOXYPROPYLENE GLYCOL BUTYL MONOETHER; POLYOXYPROPYLENE MONOBUTYL ETHER; POLYPROPYLENE GLYCOL BUTYL ETHER; POLYPROPYLENE GLYCOL MONOBUTYL ETHER; POLYPROPYLENE GLYCOL MONOBUTYLETHER; POLY(PROPYLENE OXIDE),MONOBUTYL ETHER; PPG-14 BUTYL ETHER; PPG-16 BUTYL ETHER; PPG-33 BUTYL ETHER; STABILENE; STABILENE FLY REPELLENT; UCON LB 1800X; UCON LB-250; UCON LB 1145
Description: colorless liquid
Use: repellent component in mfr of oil-base, pressurized, or aerosol sprays; formerly to protect dairy & livestock animals by repelling horse flies, face flies, stable flies, horn flies, etc; not for dairy or meat animals; med: (vet) additive to wound treatment & protective dressing formulations; lubricating agent for metal-to-metal surfaces; emollient in cosmetics (eg, hair conditioners).

Physical Properties

Boiling Point: > 200 °C (392 °F)
Freezing Point: < -17 °C (1.4 °F)
Specific Gravity: 0.99 at 25 °C/25 °C
Vapor Density: > 1 Air=1
Vapor Pressure: 0.001 mm Hg at 30 °C
Water Solubility: Soluble in 100 g Water at 20 °C
Other Solubilities: Soluble in Kerosene & organic solvents
pH: 5 to 8.5
Evaporation Rate: < 1 Butyl Acetate=1
Flash Point: ~ 425 °C Open Cup

RTECS Toxicity Data

Acute Oral: Rat LD_{50} Dose: 9100 mg/kg. Rabbit LD_{50} Dose: 23900 mg/kg.
Acute Dermal: Rabbit LD_{50} Route: Skin; Dose: 21200 uL/kg.
Irritation Skin: Rabbit Open Draize Test Dose: 500 mg open; Reaction: mild.

Hazard Overviews

Health: Irritating to eyes/skin/respiratory tract. Other Acute Effects: may be harmful by inhalation, ingestion, or skin absorption; overexposure to vapor generated at high temperatures may result in eye and respiratory tract irritation; dizziness; nausea; and the inhalation of harmful amounts of material. Chronic Effects: prolonged contact can cause asthma exacerbation.

Fire: Will burn. Hazards: emits toxic fumes. Extinguishing agents: water spray; carbon dioxide, dry chemical powder or

appropriate foam. Precautions: combustible liquid.
Carcinogenicity: IARC - Not listed; NIOSH - Not listed; NTP - Not listed; ACGIH - Not listed; OSHA - Not listed; EPA - Not listed; MAK - Not listed
Primary Target Organs:

Environmental

Regulations

RCRA 40CFR: Not listed
CERCLA: 40CFR 302.4: Not listed
SARA 40CFR 372.65: Not listed
SARA EHS 40CFR 355: Not listed
TSCA: Listed

BUT3170 CAS #: 4223-11-4

1-BUTOXY-2-VINYLOXY ETHANE

RTECS: KH7175000
EINECS Number: 224-172-3
Molecular Formula: $C_8H_{16}O_2$
Formula Weight: 144.24

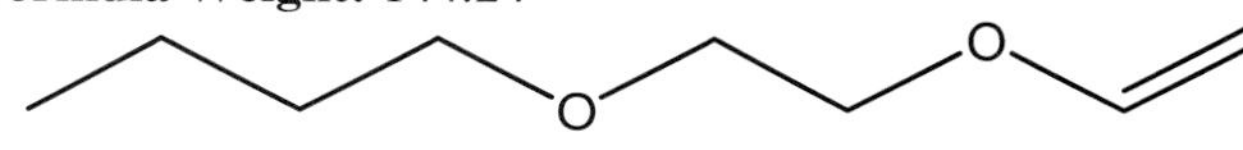

Chemical Structure

Synonyms: BUTANE,1-(2-(ETHENYLOXY)ETHOXY)-; 2-BUTOXYETHYL VINYL ETHER; 2-BUTOXYETHYLVINYL ETHER; 1-BUTOXY-2-(VINYLOXY)ETHANE; BUTYLVINYLETHER ETHYLENGLYKOLU; ETHYLENE GLYCOL BUTYL VINYL ETHER; VINYL 2-(BUTOXYETHYL) ETHER
Description: liquid

RTECS Toxicity Data

Acute Oral: Rat LD_{50} Dose: 3100 mg/kg.
Acute Inhalation: Rat LC_{Lo} Dose: 2000 ppm/8hr.
Acute Dermal: Rabbit LD_{50} Route: Skin; Dose: 3 mL/kg.
Irritation Eye: Rabbit Standard Draize Test Dose: 20 mg open; Reaction: severe. Rabbit Standard Draize Test Dose: 20 mg/24H; Reaction: moderate.
Irritation Skin: Rabbit Standard Draize Test Dose: 500 mg/24H; Reaction: mild. Rabbit Open Draize Test Dose: 10 mg/24H open; Reaction: mild.

Hazard Overviews

Health: Severely irritating to eyes; irritating to skin/respiratory tract. Other Acute Effects: may be harmful by inhalation, ingestion, or skin absorption.
Fire: Hazards: emits toxic fumes. Extinguishing agents: water spray; carbon dioxide, dry chemical powder or appropriate foam. Precautions: combustible liquid.
Reactivity: Incompatible with: strong oxidizing agents. Hazardous decomposition products: toxic fumes of: carbon monoxide, carbon dioxide.
Carcinogenicity: IARC - Not listed; NIOSH - Not listed; NTP - Not listed; ACGIH - Not listed; OSHA - Not listed; EPA - Not listed; MAK - Not listed
Primary Target Organs:

Environmental

Regulations

RCRA 40CFR: Not listed
CERCLA: 40CFR 302.4: Not listed
SARA 40CFR 372.65: Not listed
SARA EHS 40CFR 355: Not listed
TSCA: Not listed

BUT3240 CAS #: 123-86-4

N-BUTYL ACETATE

RTECS: AF7350000
DOT: UN1123; IMO3.2
EINECS Number: 204-658-1
Molecular Formula: $C_6H_{12}O_2$
Structured MF: $CH_3COO(CH_2)_3CH_3$
Formula Weight: 116.16

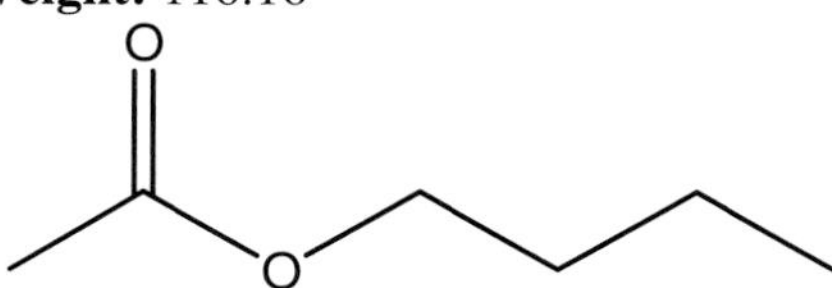

Chemical Structure

Synonyms: ACETATE DE BUTYLE; ACETIC ACID N-BUTYL ESTER; ACETIC ACID,BUTYL ESTER; BUTILE (ACETATI DI); 1-BUTYL ACETATE; BUTYL ACETATE; N-BUTYL ESTER OF ACETIC ACID; BUTYL ETHANOATE; BUTYLACETAT; BUTYLACETATEN; BUTYLE (ACETATE DE); BUTYLESTER KYSELINY OCTOVE; OCTAN N-BUTYLU
Description: nearly colorless liquid; banana-like odor
Use: industrial solvent; component of apple aroma; manufacture of lacquer; artificial perfumes; flavoring extracts; leather, photographic films; plastics, safety glass, mild defatting agent; solvent for natural gums and a dehydrating agent

Physical Properties

Boiling Point: 125 °C (257 °F) to 126 °C (259 °F)
Freezing Point: -77 °C (-106.6 °F)
Specific Gravity: 0.8826 at 20 °C/20 °C
Vapor Density: 4 Air=1
Saturated Vapor Density: 1.247455535 kg/m^3

Vapor Pressure: 10 mm Hg at 20 °C
Water Solubility: 1 parts in about 120 parts Water at 25 °C
Other Solubilities: 95% Ethanol: >=100 mg/ml at 20 °C; Acetone: >=100 mg/ml at 20 °C; Benzene: Soluble; DMSO: >=100 mg/ml at 20 °C; Ether: Soluble; Most hydrocarbons: Soluble.
Surface Tension: Estimated at 57 dynes/cm
Odor Threshold: 33.13 to 94.66 mg/m^3
Refraction Index: 1.3951 at 20 °C/D
Evaporation Rate: 1.0
Critical Temperature: 306 °C
Critical Pressure: 455 psia
Ionization Potential (eV): 10.00
Flash Point: 22 °C Closed Cup
Autoignition Temperature: 425 °C
LEL: 1.7% v/v
UEL: 7.6% v/v

RTECS Toxicity Data

Acute Oral: Rat LD_{50} Dose: 10768 mg/kg; Toxic Effects: Behavioral - Somnolence (general depressed activity); Lungs, Thorax, or Respiration - Other changes; Liver - Other changes. Mouse LD_{50} Dose: 6 gm/kg.
Acute Inhalation: Rat LC_{50} Dose: 2000 ppm/4hr. Human TC_{Lo} Dose: 200 ppm; Toxic Effects: Sense organs and special senses - Other; Sense organs and special senses - Conjunctive irritation; Lungs, Thorax, or Respiration - Other changes.
Acute Dermal: Rabbit LD_{50} Route: Skin; Dose: >17600 mg/kg.
Irritation Eye: Rabbit Standard Draize Test Dose: 100 mg; Reaction: moderate. Rabbit Standard Draize Test Dose: 20 mg open; Reaction: severe.
Irritation Skin: Rabbit Standard Draize Test Dose: 500 mg/24H; Reaction: mild. Rabbit Standard Draize Test Dose: 500 mg/24H; Reaction: moderate.
Reproductive/Teratogenic: Rat Route: Inhalation; Dose: 1500 ppm/7H; Duration: female 7-16D of pregnancy; Effects on Embryo or Fetus - Fetotoxicity; Specific Developmental Abnormalities - Musculoskeletal system.

Hazard Overviews

Flammable

Fire Diamond

Health: Irritating to eyes/skin/respiratory tract. Also Causes: weakness, drowsiness, nausea, and unconsciousness. Chronic Effects: may cause birth defects.
Fire: Highly flammable. Can form explosive mixtures in the air. Use water as fog, dry chemical, carbon dioxide, or alcohol-resistant foam. May float and re-ignite on surface of water. Streams of water scatter and spread fire. However, it can be used to disperse vapors and to cool surroundings and fire-exposed containers.
Reactivity: Stable. Hazardous polymerization cannot occur. Incompatible with: potassium-tert-butoxide; nitrates; strong acids; strong oxidizers; strong alkalies. Hazardous decomposition products: oxidized products; carbon dioxide; carbon monoxide.
Carcinogenicity: IARC - Not listed; NIOSH - Not listed; NTP - Not listed; ACGIH - Class A4, Not classifiable as a human carcinogen; OSHA - Not listed; EPA - Not listed; MAK - Not listed
Primary Target Organs:

Eyes

Skin

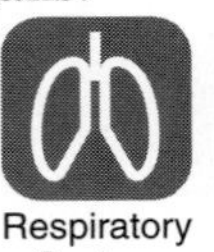
Respiratory System

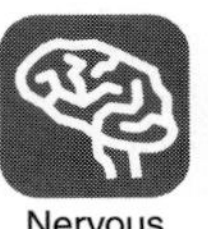
Nervous System

Reproductive

Exposure Limits
OSHA PEL: TWA: 150 ppm; 710 mg/m^3.
OSHA PEL Vacated 1989 Limits: TWA: 150 ppm; 710 mg/m^3; STEL: 200 ppm; 950 mg/m^3.
ACGIH TLV: TWA: 150 ppm; 713 mg/m^3; STEL: 200 ppm; 950 mg/m^3.
NIOSH REL: TWA: 150 ppm; 710 mg/m^3. STEL: 200 ppm; 950 mg/m^3.
NIOSH IDLH: 1700 ppm; LEL.
DFG MAK: TWA: 200 ppm; 950 mg/m^3.
Respirator Recommendation
Exposure Range: >150 to 1500 ppm Air Purifying, Negative Pressure, Half Mask
Exposure Range: >1500 to <1700 ppm Air Purifying, Negative Pressure, Full Face
Exposure Range: 1700 to unlimited ppm Self-contained Breathing Apparatus, Pressure Demand, Full Face
Cartridge Color: black

Environmental

Ecotoxicity: EC_{50} Pimephales promelas (fathead minnow) 18 mg/l/96 hr (confidence limit 17-19 mg/l). Affected fish lost equilibrium prior to death. /Conditions of bioassay not specified TLm Daphnia 44 ppm/48 hr at 23 °C /Conditions of bioassay not specified LC_{50} Menidia beryllina (Island silverside) 185 ppm/96 hr at 23 °C (static bioassay in synthetic seawater, mild aeration applied after 24 hr) TLm Scenedesmus 320 ppm/96 hr at 24 °C /Conditions of bioassay not specified
Environmental Fate: If released to soil, it may be susceptible to significant biodegradation based on its demonstrated biodegradability with a screening test. Chemical hydrolysis in moist alkaline soils (pH approaching 9 or higher) is expected to be important, but not in neutral or acidic soils. It may be subject to moderate-to-high leaching based on estimated K_{oc} values of 34 and 233. Volatilization from dry soil surfaces is likely to be rapid. If released to water, biodegradation and volatilization are expected to be the important removal mechanisms. BOD studies using either a sewage inoculum or a natural river-water inoculum have demonstrated that it is significantly biodegradable. The volatilization half-life from a river one meter deep flowing 1 m/sec with a wind velocity of 3 m/sec has been estimated to be 6.1 hours. The hydrolysis half-lives of n-butyl acetate at pHs 7.0, 8.0, and 9.0 are about 3.1 years, 114 days and 11.4 days, respectively, at 20 °C indicating that hydrolysis will be important only in very

alkaline environmental waters. Aquatic adsorption and bioconcentration are not expected to be significant. If released to air, it will exist almost entirely in the vapor-phase in the ambient atmosphere. The dominant removal mechanism in the atmosphere will be the vapor-phase reaction with photochemically produced hydroxyl radicals which has an estimated half-life of about 6 days in an average atmosphere.

Cleanup/Disposal: Guide No. 129: Eliminate all ignition sources (no smoking, flares, sparks or flames in immediate area). All equipment used when handling the product must be grounded. Do not touch or walk through spilled material. Stop leak if you can do it without risk. Prevent entry into waterways, sewers, basements or confined areas. A vapor suppressing foam may be used to reduce vapors. Absorb or cover with dry earth, sand or other non-combustible material and transfer to containers. Use clean non-sparking tools to collect absorbed material. Large Spills: Dike far ahead of liquid spill for later disposal. Water spray may reduce vapor; but may not prevent ignition in closed spaces.

Environmental Physical Data

Henry's Law Constant: 3.2×10^{-4}
Octanol/Water Partition Coefficient: log K_{ow} = 1.82
Sorption Partition Coefficient: K_{oc} = estimated at 34 to 233
BCF: estimated at 14
BOD: 0.15 to 0.5 lb/lb, 5 days

Regulations

RCRA 40CFR: Not listed
CERCLA: 40CFR 302.4: Listed per CWA Section 311(b)(4) RQ: 5000 lb (2268 kg)
SARA 40CFR 372.65: Not listed
SARA EHS 40CFR 355: Not listed
TSCA: Listed

Analytical Methods

Air: ASTM D3686, D3687, D4490
Water / Groundwater: ASTM D3695
Indoor / Expired Air: NIOSH 1450
Other: EPA 1666

BUT3310 CAS #: 105-46-4

SEC-BUTYL ACETATE

RTECS: AF7380000
DOT: UN1123; IMO3.2
EINECS Number: 203-300-1
Molecular Formula: $C_6H_{12}O_2$
Structured MF: $CH_3COOCH(CH_3)CH_2CH_3$
Formula Weight: 116.16

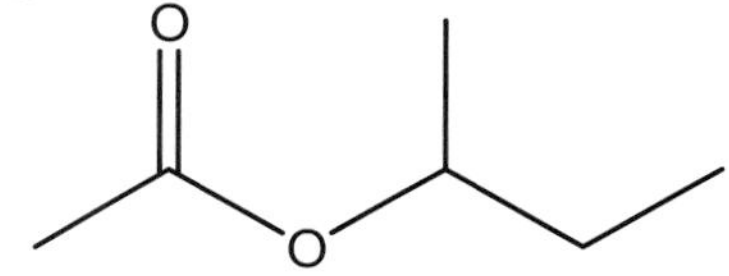

Chemical Structure

Synonyms: ACETATE DE BUTYLE SECONDAIRE; ACETIC ACID SECONDARY BUTYL ESTER; ACETIC ACID,2-BUTOXY ESTER; ACETIC ACID,SEC-BUTYL ESTER; ACETIC ACID,1-METHYLPROPYL ESTER; ACETIC ACID,1-METHYLPROPYL ESTER (9CI); 2-BUTANOL ACETATE; 2-BUTYL ACETATE; SEC-BUTYL ALCOHOL ACETATE; SEC-BUTYL ESTER OF ACETIC ACID; 1-METHYLPROPYL ACETATE
Description: colorless liquid; pleasant, fruity odor
Use: solvent for nitrocellulose lacquers, thinners, nail channels, leather finishes

Physical Properties

Boiling Point: 112 °C (234 °F)
Freezing Point: -73.5 °C (-100.3 °F)
Specific Gravity: 0.862 to 0.866
Vapor Density: 4.0 Air=1
Vapor Pressure: 10 mm Hg
Water Solubility: 0.8% by weight
Other Solubilities: Soluble in Acetone.
Surface Tension: Estimated at 58 dynes/cm
Odor Threshold: ~ 5 ppm
Refraction Index: 1.389 at 20 °C
Evaporation Rate: 2 Butyl Acetate=1
Critical Temperature: 288 °C
Critical Pressure: 32 atm
Ionization Potential (eV): 9.91
Flash Point: 31 °C Open Cup
LEL: 1.7% v/v
UEL: 9.8% v/v

Hazard Overviews

Flammable

Fire Diamond

Health: Irritating to eyes/skin/respiratory tract. Also Causes: excitability, talkativeness, drunken behavior, staggering, lack of coordination, nausea, drowsiness, loss of consciousness, dermatitis, watering eyes, lid inflammation, sensitivity to light, conjunctivitis, gastrointestinal irritation.
Fire: Flammable. Can form explosive mixtures in the air. For small fires use dry chemical, carbon dioxide, water spray, or alcohol-resistant foam. For large fires use water spray, fog, or alcohol-resistant foam.
Reactivity: Stable. Hazardous polymerization cannot occur. Avoid: heat. Incompatible with: nitrates; strong oxidizers; alkalis; acids; plastics; resins. Hazardous decomposition products: carbon dioxide; irritating fumes.
Carcinogenicity: IARC - Not listed; NIOSH - Listed as carcinogen; NTP - Not listed; ACGIH - Not listed; OSHA - Not listed; EPA - Not listed; MAK - Not listed
Primary Target Organs:

Eyes

Skin

Respiratory System
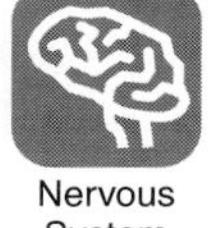
Nervous System

Exposure Limits
OSHA PEL: TWA: 200 ppm; 950 mg/m^3.

ACGIH TLV: TWA: 200 ppm; 950 mg/m^3.
NIOSH REL: TWA: 200 ppm; 950 mg/m^3.
NIOSH IDLH: 1700 ppm; LEL.
DFG MAK: TWA: 200 ppm; 950 mg/m^3.
Respirator Recommendation
Exposure Range: >200 to <1700 ppm Air Purifying, Negative Pressure, Half Mask
Exposure Range: 1700 to unlimited ppm Self-contained Breathing Apparatus, Pressure Demand, Full Face
Cartridge Color: black
Note: use dust/mist prefilter if mist is present

Environmental

Environmental Fate: If released to soil, it may be susceptible to biodegradation based on the demonstrated biodegradability of the similarly structured compounds n-butyl acetate and isobutyl acetate in standard BOD tests. Chemical hydrolysis in moist alkaline soils (pH approaching 9 or higher) may be important, but not in neutral or acidic soils. It may be subject to moderate-to-high leaching based on estimated K_{oc} values of 30 and 158. Volatilization from dry soil surfaces is likely to be rapid. If released to water, volatilization is expected to be an important removal mechanism. The volatilization half-life from a river one meter deep flowing 1 m/sec with a wind velocity of 3 m/sec has been estimated to be 5.4 hours. The hydrolysis half-lives at pHs 7.0, 8.0, and 9.0 are about 12.6 years, 1.26 years and 46 days, respectively, at 25 °C indicating that hydrolysis might only be important in very alkaline environmental waters. Aquatic adsorption and bioconcentration are not expected to be significant. Biodegradation in natural water may be possible based on the demonstrated biodegradability of n-butyl acetate and isobutyl acetate in standard BOD tests. If released to air, it will exist almost entirely in the vapor-phase in the ambient atmosphere. The dominant degradation mechanism in the atmosphere will be the vapor-phase reaction with photochemically produced hydroxyl radicals which has an estimated half-life of about 1.86 days in an average atmosphere.
Cleanup/Disposal: Guide No. 129: Eliminate all ignition sources (no smoking, flares, sparks or flames in immediate area). All equipment used when handling the product must be grounded. Do not touch or walk through spilled material. Stop leak if you can do it without risk. Prevent entry into waterways, sewers, basements or confined areas. A vapor suppressing foam may be used to reduce vapors. Absorb or cover with dry earth, sand or other non-combustible material and transfer to containers. Use clean non-sparking tools to collect absorbed material. Large Spills: Dike far ahead of liquid spill for later disposal. Water spray may reduce vapor; but may not prevent ignition in closed spaces.

Environmental Physical Data

Henry's Law Constant: estimated at 4.2 x10^{-4}
Octanol/Water Partition Coefficient: log K_{OW} = 1.51
Sorption Partition Coefficient: K_{OC} = estimated at 30 to 158
BCF: estimated at 8
BOD: 0.15 to 0.5 lb/lb, 5 days

Regulations

RCRA 40CFR: Not listed
CERCLA: 40CFR 302.4: Listed per CWA Section 311(b)(4) RQ: 5000 lb (2268 kg)
SARA 40CFR 372.65: Not listed
SARA EHS 40CFR 355: Not listed
TSCA: Listed

Analytical Methods

Air: ASTM D3686, D3687
Indoor / Expired Air: NIOSH 1450

BUT3380 CAS #: 540-88-5

TERT-BUTYL ACETATE

RTECS: AF7400000
DOT: UN1123; IMO3.2
EINECS Number: 208-760-7
Molecular Formula: $C_6H_{12}O_2$
Structured MF: $CH_3COOC(CH_3)_3$
Formula Weight: 116.16

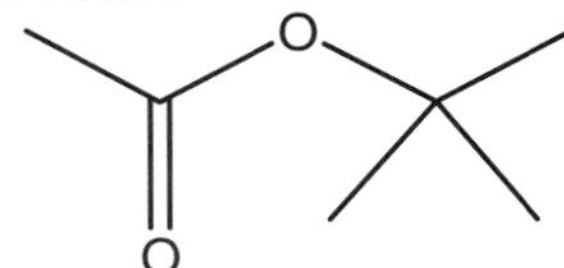

Chemical Structure

Synonyms: ACETIC ACID,TERT-BUTYL ESTER; ACETIC ACID,1,1-DIMETHYLETHYL ESTER; ACETIC ACID,1,1-DIMETHYLETHYL ESTER (9CI); T-BUTYL ACETATE; TERT-BUTYL ESTER OF ACETIC ACID; TEXACO LEAD APPRECIATOR; TLA
Description: colorless liquid; fruity odor
Use: gasoline additive; solvent; antiknock action enhancer of tetraethyl lead, especially at very high octane levels

Physical Properties

Boiling Point: 97 °C (207 °F) to 98 °C (208 °F)
Specific Gravity: 0.8665 at 20 °C/4 °C
Vapor Density: 4 Air=1
Bulk Density: 7.4 lbs/gal
Water Solubility: Practically Insoluble in Water
Other Solubilities: Soluble in Alcohol, Ether & Acetic Acid; miscible in common industrial organic solvents.
Odor Threshold: 0.004 ppm
Refraction Index: 1.3855 at 20 °C
Flash Point: 16.667 to 22.222 °C Closed Cup
LEL: Estimated at 1.5% v/v

Hazard Overviews

Flammable

Fire Diamond

Health: Irritating to eyes/skin/respiratory tract. Also Causes: CNS depression, headache, dizziness, palpitations, gastrointestinal disorders, anemia, liver problems, delayed pulmonary edema. Chronic Effects: neurotoxicity is possible.

Fire: Flammable. Can form explosive mixtures in air. For small fires use dry chemical, carbon dioxide, water spray, or alcohol-resistant foam. For large fires use water spray, fog, or alcohol-resistant foam. Container may explode in heat of fire.

Reactivity: Stable. Hazardous polymerization cannot occur. Avoid: heat; ignition sources. Incompatible with: nitrates; strong alkalis; acids; oxidizers. Hazardous decomposition products: carbon oxide(s).

Carcinogenicity: IARC - Not listed; NIOSH - Listed as carcinogen; NTP - Not listed; ACGIH - Not listed; OSHA - Not listed; EPA - Not listed; MAK - Not listed

Primary Target Organs:

Eyes

Skin

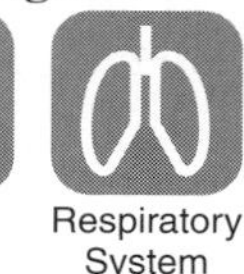
Respiratory System

Nervous System

Liver

Exposure Limits

OSHA PEL: TWA: 200 ppm; 950 mg/m^3.

ACGIH TLV: TWA: 200 ppm; 950 mg/m^3.

NIOSH REL: TWA: 200 ppm; 950 mg/m^3.

NIOSH IDLH: 1500 ppm; LEL.

DFG MAK: TWA: 200 ppm; 950 mg/m^3.

Respirator Recommendation

Exposure Range: >200 to <1500 ppm Air Purifying, Negative Pressure, Half Mask

Exposure Range: 1500 to unlimited ppm Self-contained Breathing Apparatus, Pressure Demand, Full Face

Cartridge Color: black

Environmental

Environmental Fate: If released to soil, chemical hydrolysis in moist alkaline soils (pH approaching 10 or higher) may be important, but hydrolysis in soils of pH 9 or lower is not expected to be important. It may be susceptible to significant leaching in soil based on an estimated K_{oc} value of 134. Volatilization from dry soil surfaces may be rapid. If released to water, volatilization is expected to be an important removal mechanism. The volatilization half-life from a river one meter deep flowing 1 m/sec with a wind velocity of 3 m/sec has been estimated to be 6.0 hours. The hydrolysis half-lives at pHs 7.0, 8.0 and 9.0 are about 135 years, 14.6 years and 1.46 years, respectively, at 25 °C indicating that hydrolysis will be important only in extremely alkaline environmental waters with pHs approaching 10 or higher. Aquatic adsorption and bioconcentration are not expected to be significant. If released to air, it will exist almost entirely in the vapor-phase in the ambient atmosphere. The dominant degradation mechanism in the atmosphere may be the vapor-phase reaction with photochemically produced hydroxyl radicals which has an estimated half-life of about 26 days in an average atmosphere. Physical removal via washout may be possible.

Cleanup/Disposal: Guide No. 129: Eliminate all ignition sources (no smoking, flares, sparks or flames in immediate area). All equipment used when handling the product must be grounded. Do not touch or walk through spilled material. Stop leak if you can do it without risk. Prevent entry into waterways, sewers, basements or confined areas. A vapor suppressing foam may be used to reduce vapors. Absorb or cover with dry earth, sand or other non-combustible material and transfer to containers. Use clean non-sparking tools to collect absorbed material. Large Spills: Dike far ahead of liquid spill for later disposal. Water spray may reduce vapor; but may not prevent ignition in closed spaces.

Environmental Physical Data

Henry's Law Constant: estimated at 3.3 x10^{-4}

Octanol/Water Partition Coefficient: log K_{ow} = 1.38

Sorption Partition Coefficient: K_{oc} = estimated at 134

BCF: estimated at 6.6

BOD: sewage 1.28 lb/lb, 5 days

Regulations

RCRA 40CFR: Not listed

CERCLA: 40CFR 302.4: Listed per CWA Section 311(b)(4) RQ: 5000 lb (2268 kg)

SARA 40CFR 372.65: Not listed

SARA EHS 40CFR 355: Not listed

TSCA: Listed

Analytical Methods

Air: ASTM D3686, D3687

Indoor / Expired Air: NIOSH 1450

BUT3450 CAS #: 141-32-2

N-BUTYL ACRYLATE

RTECS: UD3150000

DOT: UN2348; IMO3.0

EINECS Number: 205-480-7

Molecular Formula: $C_7H_{12}O_2$

Structured MF: $CH_2C{=}CHCO_2(CH_2)_3CH_3$

Formula Weight: 128.17

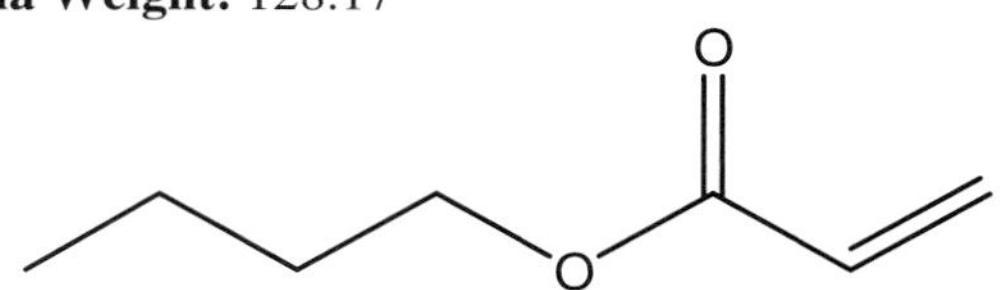

Chemical Structure

Synonyms: ACRYLIC ACID; ACRYLIC ACID N-BUTYL ESTER; ACRYLIC ACID,BUTYL ESTER; BUTYL ACRYLATE; N-BUTYL ESTER; BUTYL ESTER OF ACRYLIC ACID; BUTYL 2-PROPENATE; BUTYL 2-PROPENOATE; BUTYLESTER KYSELINY AKRYLOVE; 2-PROPENOIC ACID,BUTYL ESTER

Description: colorless liquid; sharp, biting odor

Use: manufacture of polymers and resins for textile and leather finishes; paint formulations; intermediate in organic synthesis; adhesives, paint, binders, emulsifier

Physical Properties

Boiling Point: 145 °C (293 °F) at 760 mm Hg

Freezing Point: -64.6 °C (-84.28 °F)

Specific Gravity: 0.8898 at 20 °C/4 °C

Vapor Density: 4.42 Air=1
Density: 0.89 g/mL at 20 °C
Vapor Pressure: 10 mm Hg at 35.5 °C
Water Solubility: 0.1% by weight
Other Solubilities: 95% Ethanol: >=100 mg/ml at 20 °C; Acetone: >=100 mg/ml at 20 °C; DMSO: 10-50 mg/ml at 20 °C; Ether: Soluble.
Surface Tension: Estimated at 20 dynes/cm
Odor Threshold: 0.001 to 0.10 ppm
Refraction Index: 1.4185 at 20 °C/D
Critical Temperature: 327 °C
Critical Pressure: 426 psia
Flash Point: 49 °C Open Cup
Autoignition Temperature: 292 °C
LEL: 1.5% v/v
UEL: 9.9% v/v

RTECS Toxicity Data

Acute Oral: Rat LD_{50} Dose: 900 mg/kg. Mouse LD_{50} Dose: 5880 mg/kg; Toxic Effects: Lungs, Thorax, or Respiration - Other changes.
Acute Inhalation: Rat LC_{50} Dose: 2730 ppm/4hr; Toxic Effects: Sense organs and special senses - Other; Sense organs and special senses - Other; Lungs, Thorax, or Respiration - Dyspnea. Mouse LC_{50} Dose: 7800 mg/m^3/2hr.
Acute Dermal: Rabbit LD_{50} Route: Skin; Dose: 2 mL/kg. Rat LD_{Lo} Route: Skin; Dose: 1700 mg/kg.
Chronic (Multiple Dose) Oral: Rat Dose: 13650 mg/kg/13W-I; Toxic Effects: Liver - Changes in liver weight.
Chronic (Multiple Dose) Inhalation: Rat Dose: 211 ppm/6H/13W-I; Toxic Effects: Liver - Changes in Liver weight; Biochemical - Phosphatases; Nutritional and gross metabolic - Weight loss or decreased weight gain.
Irritation Eye: Rabbit Standard Draize Test Dose: 50 mg; Reaction: mild. Rabbit Standard Draize Test Dose: 500 mg/24H; Reaction: mild.
Irritation Skin: Rabbit Open Draize Test Dose: 10 mg/24H open; Reaction: mild. Rabbit Open Draize Test Dose: 500 mg open; Reaction: mild.
Reproductive/Teratogenic: Rat Route: Inhalation; Dose: 135 ppm/6H; Duration: female 6-15D of pregnancy; Effects on Fertility - Post-implantation mortality.

Hazard Overviews

Flammable

Fire Diamond

Health: Irritating to eyes/skin/respiratory tract. Also Causes: coughing, wheezing, dizziness, headache, nausea, vomiting, narcosis. Chronic Effects: dermatitis, sensitization.
Fire: Flammable. Polymerizes. Explosive in air. Use water spray, dry chemical, alcohol-resistant foam, or carbon dioxide. Use water spray to cool fire-exposed containers. Do not use solid streams of water. Material will float; may re-ignite on water surface.

Reactivity: Stable. Hazardous polymerization can occur by elevated temperature, oxidizers, peroxides, or sunlight.. Avoid: heat; ignition sources. Incompatible with: strong acids, alkalies; oxidizing materials. Hazardous decomposition products: carbon monoxide; carbon dioxide; acrid; irritating fumes.
Carcinogenicity: IARC - Group 3, Not classifiable as to carcinogenicity to humans; NIOSH - Not listed; NTP - Not listed; ACGIH - Class A4, Not classifiable as a human carcinogen; OSHA - Not listed; EPA - Not listed; MAK - Not listed
Primary Target Organs:

Eyes

Skin

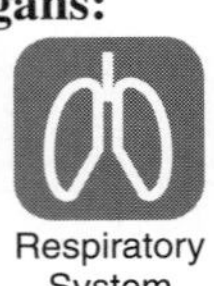
Respiratory System

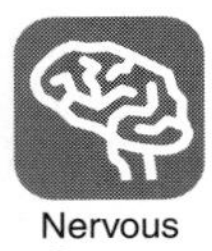
Nervous System

Exposure Limits
OSHA PEL Vacated 1989 Limits: TWA: 10 ppm; 55 mg/m^3.
ACGIH TLV: TWA: 10 ppm; 52 mg/m^3.
NIOSH REL: TWA: 10 ppm; 55 mg/m^3.
DFG MAK: TWA: 10 ppm; 55 mg/m^3.
Respirator Recommendation
Exposure Range: >10 to 100 ppm Air Purifying, Negative Pressure, Half Mask
Exposure Range: >100 to 1000 ppm Air Purifying, Negative Pressure, Full Face
Exposure Range: >1000 to 10,000 ppm Supplied Air, Constant Flow/Pressure Demand, Full Face
Exposure Range: >10,000 to unlimited ppm Self-contained Breathing Apparatus, Pressure Demand, Full Face
Cartridge Color: black

Environmental

Ecotoxicity: Threshold concentration of cell multiplication inhibition of the protozoan Uronema parduczi 21 mg/l
Environmental Fate: If released to soil, it will be expected to exhibit a very high mobility in soil and, therefore, it may leach to groundwater. It may hydrolyze, especially in alkaline soils based upon hydrolysis data for the structurally similar ethyl acrylate. It may biodegrade in soil based upon its biodegradability in aqueous screening tests. It may volatilize from near surface soil and other surfaces. If released to water, it will not be expected to adsorb to sediment or suspended particulate matter or to bioconcentrate in aquatic organisms. Hydrolysis may be a significant process especially in alkaline waters based upon hydrolysis data for the structurally similar ethyl acrylate. It may biodegrade in natural waters based upon its biodegradability in screening tests. It may directly photolyze in sunlight based upon the slight absorption of light at wavelengths > 290 nm by ethyl acrylate and other acrylate esters. It will significantly volatilize from water with an estimated half-life of 5.5 hr for volatilization from a model river. The volatilization half-life from a model pond, which considers the effect of adsorption, has been estimated to be 3.0 days. If released to the atmosphere, it will be expected to exist almost entirely in the vapor phase based upon a reported vapor pressure of 5.45 mm Hg at 25 °C. It will be susceptible

to photooxidation via vapor phase reaction with photochemically produced hydroxyl radicals and ozone. An atmospheric half-life of 12.6 hours at an atmospheric concentration of 5 x10^5 hydroxyl radicals per cu cm and 7X20+11 ozone molecules per cu cm has been estimated for this process.

Cleanup/Disposal: Guide No. 129: Eliminate all ignition sources (no smoking, flares, sparks or flames in immediate area). All equipment used when handling the product must be grounded. Do not touch or walk through spilled material. Stop leak if you can do it without risk. Prevent entry into waterways, sewers, basements or confined areas. A vapor suppressing foam may be used to reduce vapors. Absorb or cover with dry earth, sand or other non-combustible material and transfer to containers. Use clean non-sparking tools to collect absorbed material. Large Spills: Dike far ahead of liquid spill for later disposal. Water spray may reduce vapor; but may not prevent ignition in closed spaces.

Environmental Physical Data

Henry's Law Constant: calculated at 4.60 x10^{-4}
Octanol/Water Partition Coefficient: log K_{ow} = 2.36
Sorption Partition Coefficient: K_{oc} = estimated at 160
BCF: estimated at 37
BOD: 28% BODT, 5 days

Regulations

RCRA 40CFR: Not listed
CERCLA: 40CFR 302.4: Not listed
SARA 40CFR 372.65: Listed
SARA EHS 40CFR 355: Not listed
TSCA: Listed

Analytical Methods

Water / Groundwater: ASTM D3695

BUT3520 CAS #: 94-25-7

BUTYL P-AMINOBENZOATE

RTECS: DG1530000
EINECS Number: 202-317-1
Molecular Formula: $C_{11}H_{15}NO_2$
Formula Weight: 193.24

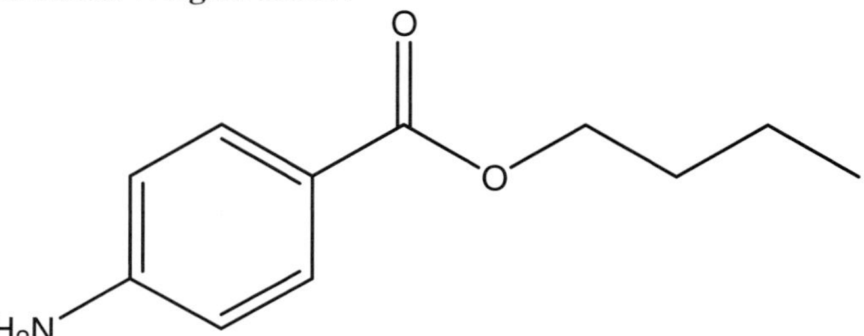

Chemical Structure

Synonyms: 4-AMINOBENZOIC ACID BUTYL ESTER; P-AMINOBENZOIC ACID BUTYL ESTER; BENZOIC ACID,4-AMINO-,BUTYL ESTER; BENZOIC ACID,P-AMINO-,BUTYL ESTER; BUTAMBEN; BUTESIN; BUTESINE; BUTOFORM; 4-(BUTOXYCARBONYL)ANILINE; BUTYL 4-AMINOBENZOATE; BUTYL AMINOBENZOATE; BUTYL PARA-AMINOBENZOATE; N-BUTYL P-AMINOBENZOATE; BUTYL KELOFORM; BUTYLCAINE; BUTYLESTER KYSELINY P-AMINOBENZOOVE; PLANOFORM; SCUROFORM; SCUROFORME

Description: white crystalline powder; odorless
Use: chemical synthesis, ultraviolet absorber in suntan preparations, local anesthetic

Physical Properties

Boiling Point: 174 °C (345 °F) at 8 mm Hg
Freezing Point: 57 °C (134.6 °F)
Water Solubility: 1 g dissolves in about 7 L Water
Other Solubilities: 95% Ethanol: Soluble; Benzene: Soluble; Chloroform: Soluble; Dilute acids: Soluble; Ether: Soluble; Fatty oils: Soluble.
Flash Point: Not available; probably combustible

Hazard Overviews

Health: Irritating to eyes/skin/respiratory tract. Other Acute Effects: may be harmful by inhalation, ingestion, or skin absorption; may cause allergic skin reaction. Chronic Effects: may cause sensitization by skin contact.
Fire: Will burn. Hazards: emits toxic fumes. Extinguishing agents: water spray; carbon dioxide, dry chemical powder or appropriate foam. Precautions: combustible liquid.
Carcinogenicity: IARC - Not listed; NIOSH - Not listed; NTP - Not listed; ACGIH - Not listed; OSHA - Not listed; EPA - Not listed; MAK - Not listed
Primary Target Organs:

Eyes

Skin

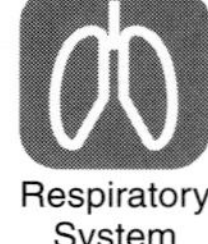
Respiratory System

Environmental

Regulations

RCRA 40CFR: Not listed
CERCLA: 40CFR 302.4: Not listed
SARA 40CFR 372.65: Not listed
SARA EHS 40CFR 355: Not listed
TSCA: Listed

BUT3590 CAS #: 104-51-8

N-BUTYL BENZENE

RTECS: CY9070000
EINECS Number: 203-209-7
Molecular Formula: $C_{10}H_{14}$
Formula Weight: 134.24

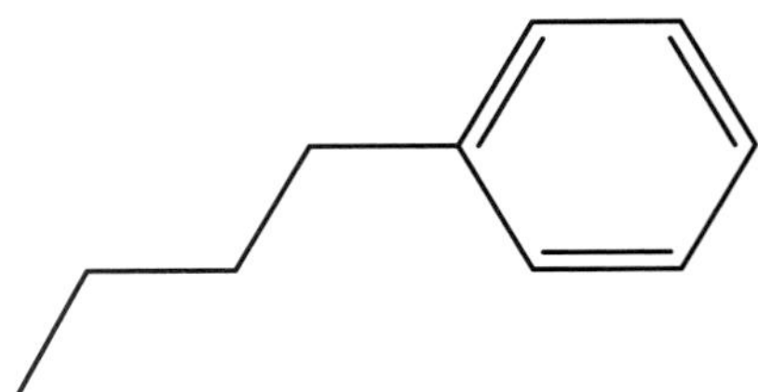

Chemical Structure

Synonyms: 1-BUTYLBENZENE; N-BUTYLBENZENE; 1-PHENYLBUTANE
Description: colorless liquid

Physical Properties

Boiling Point: 183.2 °C (362 °F)
Freezing Point: -81.2 °C (-114.16 °F)
Vapor Density: 4.6
Density: 0.86 at 20/4 °C
Vapor Pressure: 0.150 kPa at 25 °C
Water Solubility: Insoluble
Other Solubilities: Ethanol: Miscible; Ether: Miscible; Acetone: Miscible; Benzene: Miscible; Carbon Tetrachloride: Miscible; Petroleum Ether: Miscible
Refraction Index: 1.4898
Ionization Potential (eV): 8.69 +/-0.01
Flash Point: 71.1 °C Open Cup
Autoignition Temperature: 412 °C
LEL: 0.8% v/v
UEL: 5.8% v/v

RTECS Toxicity Data

Acute Oral: Rat LD_{Lo} Dose: 10 mL/kg; Toxic Effects: Lungs, Thorax, or Respiration - Fibrosis, focal (pneumoconiosis); Lungs, Thorax, or Respiration - Acute pulmonary edema; Blood - Hemorrhage.

Hazard Overviews

Fire Diamond

Health: Also Causes: headache, dizziness, nausea, unconsciousness, suffocation by CNS depression. Chronic Effects: neuropathy, liver damage.
Fire: Combustible. Can form explosive mixtures in the air. For small fires use dry chemical, carbon dioxide, water spray, or regular foam. For large fires use water spray, fog, or regular foam. Vapors may travel to a source of ignition and flash back. Containers may explode in heat of fire.
Reactivity: Stable. Hazardous polymerization cannot occur. Avoid: generation of vapors; exposure to ignition sources. Incompatible with: oxidizing materials. Hazardous decomposition products: carbon monoxide; carbon dioxide; irritating gases; mists; smoke.

Carcinogenicity: IARC - Not listed; NIOSH - Not listed; NTP - Not listed; ACGIH - Not listed; OSHA - Not listed; EPA - Not listed; MAK - Not listed
Primary Target Organs:

Nervous System

Liver

Environmental

Regulations

RCRA 40CFR: Not listed
CERCLA: 40CFR 302.4: Not listed
SARA 40CFR 372.65: Not listed
SARA EHS 40CFR 355: Not listed
TSCA: Listed

Analytical Methods

Air: EPA VG-011-1
Soil: SW846 8021A, 8260A
Water / Groundwater: APHA 6210-D, 6230-D
Drinking Water: EPA 502.2, 503.1, 524.2; APHA 6220-C
Indoor / Expired Air: EPA IP-1B
Plasma: EPA 29

BUT3660 CAS #: 136-60-7

BUTYL BENZOATE

RTECS: DG4925000
EINECS Number: 205-252-7
Molecular Formula: $C_{11}H_{14}O_2$
Formula Weight: 178.22

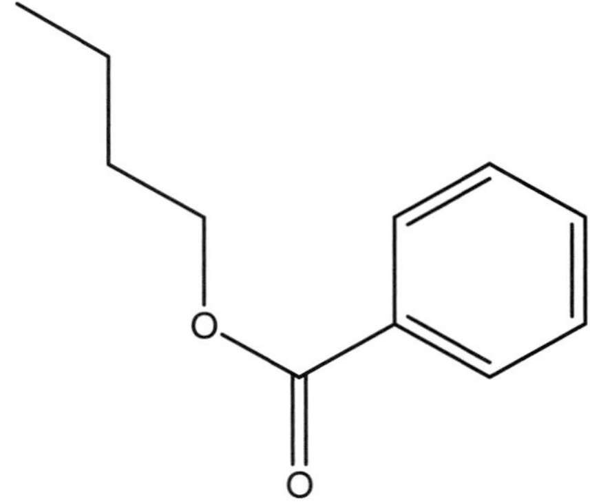

Chemical Structure

Synonyms: ANTHRAPOLE AZ; BENZOIC ACID BUTYL ESTER; BENZOIC ACID N-BUTYL ESTER; BENZOIC ACID,BUTYL ESTER; N-BUTYL BENZOATE; BUTYLESTER KYSELINY BENZOOVE; DAI CARI XBN
Description: colorless, thick, oily liquid; mild odor
Use: plasticizer; dye carrier in the dyeing of polyester fibers; additive in disinfectants; penetrating agent in pesticide formulations; additive in fermentation of soy sauce; solvent

for cellulose ether; perfume ingredient; in food industry; in entomology as selective yellow jacket attractant

Physical Properties

Boiling Point: 250 °C (482 °F)
Freezing Point: -22 °C (-7.6 °F)
Specific Gravity: 1 at 20 °C
Vapor Pressure: 0.153 kPa at 75 °C
Water Solubility: Practically Insoluble in Water
Other Solubilities: Soluble in Alcohol, Ether and Acetone; miscible with oils and hydrocarbons.
Refraction Index: 1.4940 at 25 °C/D
Flash Point: 107 °C Open Cup

RTECS Toxicity Data

Acute Oral: Rat LD_{50} Dose: 735 mg/kg.
Acute Dermal: Rabbit LD_{50} Route: Skin; Dose: 4 gm/kg.
Irritation Eye: Rabbit Standard Draize Test Dose: 500 mg/24H; Reaction: mild.
Irritation Skin: Rabbit Standard Draize Test Dose: 20 mg/24H; Reaction: moderate. Rabbit Open Draize Test Dose: 10 mg/24H open; Reaction: severe.

Hazard Overviews

Health: Irritating to eyes/skin/respiratory tract. Harmful. Other Acute Effects: harmful if swallowed; may be harmful if inhaled or absorbed through the skin.
Fire: Will burn. Hazards: emits toxic fumes. Extinguishing agents: water spray; carbon dioxide, dry chemical powder or appropriate foam. Precautions: combustible liquid.
Reactivity: Incompatible with: strong oxidizing agents, strong bases. Hazardous decomposition products: toxic fumes of: carbon monoxide, carbon dioxide.
Carcinogenicity: IARC - Not listed; NIOSH - Not listed; NTP - Not listed; ACGIH - Not listed; OSHA - Not listed; EPA - Not listed; MAK - Not listed
Primary Target Organs:

Eyes

Skin

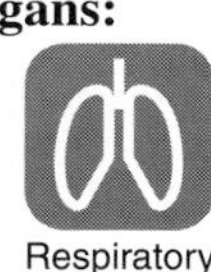
Respiratory System

Environmental

Environmental Fate: If released to the atmosphere, it will degrade by reaction with photochemically produced hydroxyl radicals (estimated half-life of 3.25 days). If released to soil or water, biodegradation is expected to be the major degradation process. Two biodegradation screening studies have found it to be readily biodegradable. Leaching in soil is possible.

Environmental Physical Data

Henry's Law Constant: estimated at 3.97×10^{-5}
Sorption Partition Coefficient: K_{oc} = 125
BCF: estimated at 62

Regulations

RCRA 40CFR: Not listed
CERCLA: 40CFR 302.4: Not listed
SARA 40CFR 372.65: Not listed
SARA EHS 40CFR 355: Not listed
TSCA: Listed

BUT3730 CAS #: 98-73-7

P-TERT-BUTYL BENZOIC ACID

RTECS: DG4708000
EINECS Number: 202-696-3
Molecular Formula: $C_{11}H_{14}O_2$
Formula Weight: 178.2

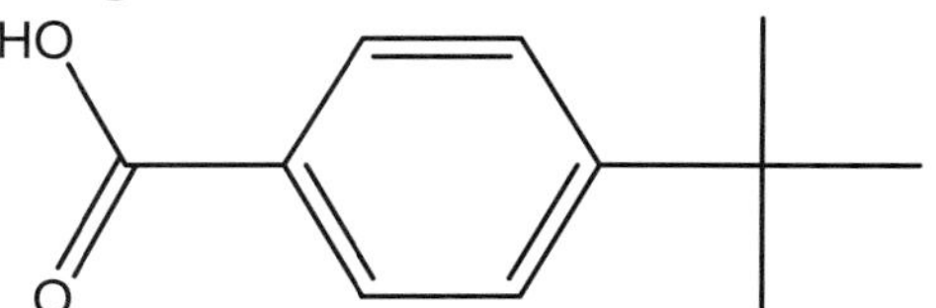

Chemical Structure

Synonyms: BENZOIC ACID,4-TERT-BUTYL-; BENZOIC ACID,P-TERT-BUTYL-; BENZOIC ACID,4-(1,1-DIMETHYLETHYL)-; 4-TERT-BUTYLBENZOIC ACID; P-TERT-BUTYLBENZOIC ACID; KYSELINA P-TERC.BUTYLBENZOOVA; P-TBBA; TBBA
Description: needles

Physical Properties

Freezing Point: 164.5 °C (328.1 °F) to 166.5 °C (331.7 °F)
Water Solubility: Insoluble in Water
Other Solubilities: Very Soluble in Alcohol & Benzene
Ionization Potential (eV): 8.94 +/-0.02

RTECS Toxicity Data

Acute Oral: Rat LD_{50} Dose: 700 mg/kg; Toxic Effects: Behavioral - Change in motor activity (specific assay); Lungs, Thorax, or Respiration - Respiratory depression; Skin and appendages - Hair.
Acute Inhalation: Rat LC_{50} Dose: >1900 mg/m^3/4hr.
Reproductive/Teratogenic: Rat Route: Inhalation; Dose: 106 mg/m^3/6H Duration: male 7D prior to mating; Paternal Effects - Testes, epididymis, sperm duct. Rat Route: Inhalation; Dose: 12500 ug/m^3/ Duration: male 7D prior to mating; Paternal Effects - Spermatogenesis.

Hazard Overviews

Health: May irritate eyes/skin. Toxic. Other Acute Effects: toxic by inhalation, in contact with skin and if swallowed; may cause nervous system disturbance. Chronic Effects: possible risk of irreversible effects; reproductive hazard; neurological hazard; may cause damage to liver and kidneys. Overexposure may cause reproductive disorders based on tests with laboratory animals.
Fire: Extinguishing agents: water spray; carbon dioxide, dry chemical powder or appropriate foam. Precautions: combustible liquid.

Carcinogenicity: IARC - Not listed; NIOSH - Not listed; NTP - Not listed; ACGIH - Not listed; OSHA - Not listed; EPA - Not listed; MAK - Not listed

Environmental

Ecotoxicity: Fishes: goldfish pH 5 24-96h LD_{50} 4 mg/l pH 7 24-96h LD_{50} 33 mg/l

Environmental Physical Data

Octanol/Water Partition Coefficient: log K_{ow} = calculated at 3.86

Regulations

RCRA 40CFR: Not listed
CERCLA: 40CFR 302.4: Not listed
SARA 40CFR 372.65: Not listed
SARA EHS 40CFR 355: Not listed
TSCA: Listed

BUT3800 CAS #: 85-68-7

BUTYL BENZYL PHTHALATE

RTECS: TH9990000
EINECS Number: 201-622-7
Molecular Formula: $C_{19}H_{20}O_4$
Structured MF: $C_6H_5CH_2OOCC_6H_4COOCH_2CH_2CH_2CH_3$
Formula Weight: 312.39

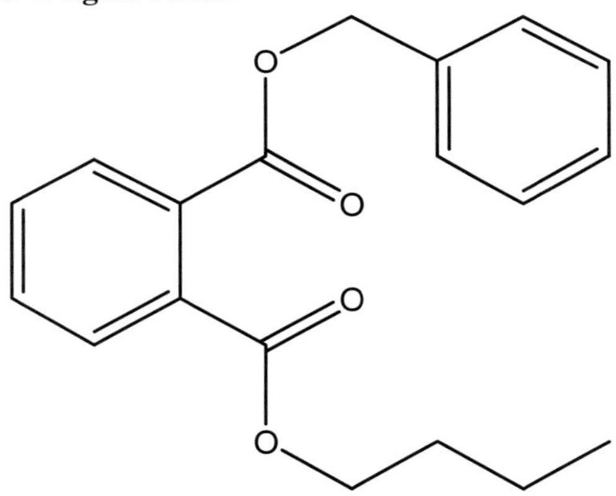

Chemical Structure

Synonyms: BBP; 1,2-BENZENEDICARBOXYLIC ACID,BUTYL PHENYLMETHYL ESTER; BENZYL BUTYL PHTHALATE; BENZYL N-BUTYL PHTHALATE; BENZYL BUTYLPHTHALATE; BENZYL-BUTYLESTER KYSELINY FTALOVE; N-BUTYL BENZYL PHTHALATE; BUTYL BENZYLPHTHALATE; BUTYL PHENYLMETHYL 1,2-BENZENECARBOXYLATE; BUTYL PHENYLMETHYL 1,2-BENZENEDICARBOXYLATE; BUTYLBENZYL PHTHALATE; PALATINOL BB; PHTHALIC ACID,BENZYL BUTYL ESTER; SANTICIZER 160; SICOL 160; UNIMOLL BB
Description: clear, oily liquid; slight odor
Use: as a plasticizer for polyvinyl and cellulosic resins; as an organic intermediate, a solvent and a fixative in perfume

Physical Properties

Boiling Point: 370 °C (698 °F)
Freezing Point: -35 °C (-31 °F)
Specific Gravity: 1.113 to 1.121 at 25 °C/25 °C
Vapor Density: 10.8
Saturated Vapor Density: 1.200000133 kg/m^3
Vapor Pressure: 8.6 $x10^{-6}$ mm Hg at 20 °C
Water Solubility: 2.9 mg/L in Water (temperature not specified)
Other Solubilities: 95% Ethanol: >=100 mg/ml at 23 °C; Acetone: >=100 mg/ml at 23 °C; DMSO: >=100 mg/ml at 23 °C; Most organic solvents: Soluble.
Surface Tension: 39.9 dynes/cm at 25 °C
Refraction Index: 1.535 to 1.540 at 25 °C/D
Flash Point: 198.889 °C Open Cup
Autoignition Temperature: 233 °C
LEL: 1.2% v/v

RTECS Toxicity Data

Acute Oral: Rat LD_{50} Dose: 2330 mg/kg. Mouse LD_{50} Dose: 4170 mg/kg.
Acute Dermal: Rabbit LD_{50} Route: Skin; Dose: >10 gm/kg. Rat LD_{50} Route: Skin; Dose: 6700 mg/kg.
Chronic (Multiple Dose) Oral: Rat Dose: 56 gm/kg/4W-C; Toxic Effects: Behavioral - Food intake (animal); Blood - Hemorrhage; DEATH. Rat Dose: 68250 mg/kg/91D-C; Toxic Effects: Liver - Changes in liver weight. Rat Dose: 38402 mg/kg/13W-C; Toxic Effects: Liver - Changes in liver weight; Kidney, Ureter, and Bladder - Changes in kidney weight. Rat Dose: 21 gm/kg/14D-C; Toxic Effects: Kidney, Ureter, and Bladder - Changes in kidney weight; DEATH - Changes in prostate weight; DEATH - Changes in testicular weight.
Chronic (Multiple Dose) Inhalation: Rat Dose: 2100 mg/m^3/6H/4W-I; Toxic Effects: Nutritional and gross metabolic - Weight loss or decreased weight gain; DEATH. Rat Dose: 789 mg/m^3/6H/13W-I; Toxic Effects: Liver - Changes in Liver weight; Kidney, Ureter, and Bladder - Changes in kidney weight.
Reproductive/Teratogenic: Rat Route: Oral; Dose: 21 gm/kg; Duration: male 14D prior to mating; Paternal Effects - Testes, epididymis, sperm duct; Prostate, seminal vessicle, Cowper's gland, accessory glands. Rat Route: Oral; Dose: 16400 mg/kg; Duration: female 6-15D of pregnancy; Maternal Effects - Uterus, cervix, vagina; Other effects on females; Effects on Embryo or Fetus - Fetal death. Rat Route: Oral; Dose: 16400 mg/kg; Duration: female 6-15D of pregnancy; Effects on Embryo or Fetus - Other effects to embryo or fetus; Specific Developmental Abnormalities - Musculoskeletal system. Rat Route: Oral; Dose: 4900 mg/kg; Duration: female 1-7D of pregnancy; Effects on Fertility - Post-implantation mortality; Specific Developmental Abnormalities - Craniofacial (including nose and tongue); Musculoskeletal system.
Tumorigenic: Rat Route: Oral; Dose: 433 gm/kg/2Y-C; Toxic Effects: Tumorigenic - Carcinogenic by RTECS criteria; Blood - Leukemia; Blood - Lymphomax including Hodgkin's

disease. Rat Route: Oral; Dose: 437 gm/kg/2Y-C; Toxic Effects: Tumorigenic - Carcinogenic by RTECS criteria; Blood - Leukemia.

Hazard Overviews

Health: Irritating to eyes/skin/respiratory tract. Harmful. Other Acute Effects: harmful if swallowed, inhaled, or absorbed through skin. Chronic Effects: target organs: liver, pancreas. Possible carcinogen.

Fire: Will burn. Hazards: emits toxic fumes. Extinguishing agents: water spray; carbon dioxide, dry chemical powder or appropriate foam. Precautions: combustible liquid.

Reactivity: Incompatible with: strong oxidizing agents, strong bases. Hazardous decomposition products: toxic fumes of: carbon monoxide, carbon dioxide.

Carcinogenicity: IARC - Group 3, Not classifiable as to carcinogenicity to humans; NIOSH - Not listed; NTP - Not listed; ACGIH - Not listed; OSHA - Not listed; EPA - Class C, Possible human carcinogen; MAK - Not listed

Primary Target Organs:

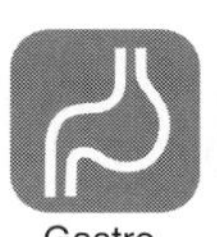

Eyes Skin Respiratory System Gastro-intestinal Liver

Environmental

Ecotoxicity: LC_{50} Lepomis macrochirus (bluegill) 62 mg/l/24 hr and 43 mg/l/96 hr /Conditions of bioassay not specified EC_{50} Selenastrum capricornutum (alga) 110 ug/l/96 hr, toxic effect: chlorophyll a; 130 ug/l/96 hr, toxic effect: cell number EC_{50} Skeletonema costatum (alga) 170 ug/l/96 hr, toxic effect: chlorophyll a; 190 ug/l/96 hr, toxic effect: cell number

Environmental Fate: Released to soil it is expected to adsorb (K_{oc} 65-350) and not to leach extensively although it has been detected in groundwater. Released to aquatic systems it will adsorb to sediments and biota but will not volatilize significantly (Henry's Law constant $<1 \times 10^{-6}$ atm/mol cu m) except under windy conditions or from shallow rivers. Biodegradation appears to be the primary fate mechanism. It is readily biodegraded in activated sludge, semicontinuous activated sludge, salt water, lake water, and under anaerobic conditions. For example, at an initial concentration of 1 mg/l in a lake water microcosm, primary degradation accounted for >95% loss in 7 days; after 28 days, 51-65% had mineralized (ultimate degradation).

Cleanup/Disposal: Guide No. 152: Do not touch damaged containers or spilled material unless wearing appropriate protective clothing. Stop leak if you can do it without risk. Prevent entry into waterways, sewers, basements or confined areas. Cover with plastic sheet to prevent spreading. Absorb or cover with dry earth, sand or other non-combustible material and transfer to containers. Do not get water inside containers.

Environmental Physical Data

Henry's Law Constant: $< 1.0 \times 10^{-6}$

Octanol/Water Partition Coefficient: log K_{ow} = 4.77

Sorption Partition Coefficient: K_{oc} = 68 to 350

BCF: bluegills concentration factor 663

Regulations

RCRA 40CFR: Not listed

CERCLA: 40CFR 302.4: Listed per CWA Section 307(a) RQ: 100 lb (45.35 kg)

SARA 40CFR 372.65: Not listed

SARA EHS 40CFR 355: Not listed

TSCA: Listed

Analytical Methods

Soil: CLP LC_SV, MC_SVOA, OHC; EPA 16, 1625; SW846 3640A, 8060, 8061, 8061A, 8250A, 8270B, 8270C, 8410

Water / Groundwater: EPA S-002-1, 1625, 606, 625, 625-S, 6; APHA 6410-B; USGS O3118

Drinking Water: EPA 506, 525.1, 525.2

Food: FDA 212.1, 232.1

Plasma: EPA 001, 29; FDA 211.1, 231.1, 252

BUT3870 CAS #: 109-65-9

N-BUTYL BROMIDE

RTECS: EJ6225000

DOT: UN1126; IMO3.2

EINECS Number: 203-691-9

Molecular Formula: C_4H_9Br

Structured MF: 1-C_4H_9Br

Formula Weight: 137.03

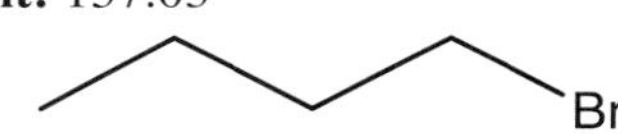

Chemical Structure

Synonyms: 1-BROMOBUTANE; BUTYL BROMIDE

Description: colorless to pale straw-colored liquid

Use: in synthesis

Physical Properties

Boiling Point: 101.3 °C (214 °F) at 760 mm Hg

Freezing Point: -112 °C (-169.6 °F)

Specific Gravity: 1.2686 at 25 °C/4 °C

Vapor Density: 4.72 Air=1

Vapor Pressure: 1.5 psia

Water Solubility: Insoluble in Water

Other Solubilities: Soluble in Acetone, Chloroform.

Surface Tension: 26.5 dynes/cm at 20 °C

Refraction Index: 1.4398 at 20 °C/D

Ionization Potential (eV): 10.12

Flash Point: 18.333 °C Open Cup

Autoignition Temperature: 265 °C

LEL: 2.6% v/v

UEL: 6.6% v/v

RTECS Toxicity Data

Acute Inhalation: Mammal LC_{50} Dose: 25800 mg/m^3.

Hazard Overviews

Flammable

Health: Irritating to eyes/skin/respiratory tract. Other Acute Effects: may be harmful by inhalation, ingestion, or skin absorption; symptoms of exposure may include burning sensation; coughing; wheezing; laryngitis; shortness of breath; headache; nausea; vomiting.
Fire: Flammable. Hazards: vapor may travel considerable distance to source of ignition and flash back; container explosion may occur; forms explosive mixtures in air; emits toxic fumes. Extinguishing agents: carbon dioxide, dry chemical powder or appropriate foam; water may be effective for cooling, but may not effect extinguishment. Precautions: combustible liquid.
Reactivity: Incompatible with: strong oxidizing agents, strong bases, magnesium, sodium, potassium. Hazardous decomposition products: toxic fumes of: carbon monoxide, carbon dioxide, hydrogen bromide gas.
Carcinogenicity: IARC - Not listed; NIOSH - Not listed; NTP - Not listed; ACGIH - Not listed; OSHA - Not listed; EPA - Not listed; MAK - Not listed
Primary Target Organs:

Eyes

Skin

Respiratory System

Environmental

Ecotoxicity: Fishes: Pimephales promelas 4d LC_{50} 36.7 mg/l
Environmental Fate: Expected to have high mobility in soil. Volatilization is expected from both moist and dry soils. In water, it is expected to volatilize rapidly with estimated half-lives of 3.5 hours and 4.7 days from a model river and a model lake, respectively. Neutral hydrolysis may be an important fate process in aquatic systems, but bioconcentration and adsorption to sediment are not expected to be important. Insufficient data are available to determine the rate or importance of biodegradation in soil or aquatic systems. It will exist in the vapor phase in the ambient atmosphere. If released to the atmosphere, it will degrade by reaction with photochemically produced hydroxyl radicals with an estimated half-life of approximately 7.4 days. Removal from the atmosphere can occur through wet deposition.
Cleanup/Disposal: Guide No. 129: Eliminate all ignition sources (no smoking, flares, sparks or flames in immediate area). All equipment used when handling the product must be grounded. Do not touch or walk through spilled material. Stop leak if you can do it without risk. Prevent entry into waterways, sewers, basements or confined areas. A vapor suppressing foam may be used to reduce vapors. Absorb or cover with dry earth, sand or other non-combustible material and transfer to containers. Use clean non-sparking tools to collect absorbed material. Large Spills: Dike far ahead of liquid spill for later disposal. Water spray may reduce vapor; but may not prevent ignition in closed spaces.

Environmental Physical Data

Henry's Law Constant: estimated at 8.71 $x10^{-3}$
Octanol/Water Partition Coefficient: log K_{ow} = 2.75
Sorption Partition Coefficient: K_{oc} = estimated at 81
BCF: estimated at 13.5

Regulations

RCRA 40CFR: Not listed
CERCLA: 40CFR 302.4: Not listed
SARA 40CFR 372.65: Not listed
SARA EHS 40CFR 355: Not listed
TSCA: Listed

BUT3940 CAS #: 78-76-2

SEC-BUTYL BROMIDE

RTECS: EJ6228000
DOT: UN2339; IMO3.2
EINECS Number: 201-140-7
Molecular Formula: C_4H_9Br
Structured MF: 2-C_4H_9Br
Formula Weight: 137.03

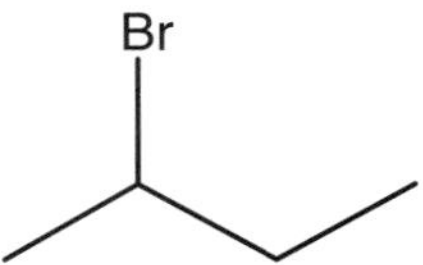

Chemical Structure

Synonyms: 2-BROMOBUTANE; BUTANE,2-BROMO-; 2-BUTYL BROMIDE; METHYLETHYLBROMOMETHANE
Description: clear, colorless liquid; pleasant odor
Use: synthesis; alkylating agent

Physical Properties

Boiling Point: 91.2 °C (196 °F)
Freezing Point: < -50 °C (-58 °F)
Specific Gravity: 1.2425 at 25 °C/25 °C
Saturated Vapor Density: 1.535265517 kg/m^3
Vapor Pressure: 57 mm Hg at 25 °C
Water Solubility: Insoluble in Water
Other Solubilities: Soluble in all proportions in Ether, Acetone; Soluble in Chloroform
Surface Tension: 25.3 dynes/cm at 20 °C
Refraction Index: 1.432 to 1.4344 at 25 °C
Ionization Potential (eV): 10.00
Flash Point: 21.111 °C Open Cup

RTECS Toxicity Data

Tumorigenic: Mouse Route: Intraperitoneal; Dose: 3000 mg/kg/8W-I; Toxic Effects: Tumorigenic - Neoplastic by RTECS criteria; Lungs, Thorax, or Respiration - Tumors.

Hazard Overviews

Health: Irritating to eyes/skin/respiratory tract. Harmful. Other Acute Effects: harmful if swallowed, inhaled, or absorbed through skin; symptoms of exposure may include burning sensation, coughing, wheezing, laryngitis, shortness of breath, headache, nausea and vomiting. Chronic Effects: possible risk of irreversible effects. Possible carcinogen.

Fire: Flammable. Hazards: vapor may travel considerable distance to source of ignition and flash back; container explosion may occur; forms explosive mixtures in air; emits toxic fumes. Extinguishing agents: carbon dioxide, dry chemical powder or appropriate foam; water may be effective for cooling, but may not effect extinguishment. Precautions: combustible liquid.

Reactivity: Incompatible with: strong oxidizing agents, strong bases, magnesium, sodium, potassium. Hazardous decomposition products: toxic fumes of: carbon monoxide, carbon dioxide, hydrogen bromide gas.

Carcinogenicity: IARC - Not listed; NIOSH - Not listed; NTP - Not listed; ACGIH - Not listed; OSHA - Not listed; EPA - Not listed; MAK - Not listed

Primary Target Organs:

Environmental

Environmental Fate: Expected to have high mobility in soil. Volatilization is expected from both moist and dry soils. In water, it is expected to volatilize rapidly with estimated half-lives of 3.49 hours and 4.65 days from a model environmental river and a model environmental lake, respectively. Bioconcentration and adsorption to sediment are not expected to be important fate processes in aquatic systems. It is expected to hydrolyze quickly in aquatic systems and in moist soils based on neutral hydrolysis half-lives of similar compounds such as t-butyl chloride(23 seconds), isopropyl bromide (2.1 days), and n-propyl bromide(26 days) Insufficient data are available to determine the rate or importance of biodegradation in soil or aquatic conditions. It will exist in the vapor phase in the ambient atmosphere. If released to the atmosphere, it will degrade by reaction with photochemically produced hydroxyl radicals with an estimated half-life of approximately 12.09 days. Removal from the atmosphere can occur though wet deposition.

Cleanup/Disposal: Guide No. 130: Eliminate all ignition sources (no smoking, flares, sparks or flames in immediate area). All equipment used when handling the product must be grounded. Do not touch or walk through spilled material. Stop leak if you can do it without risk. Prevent entry into waterways, sewers, basements or confined areas. A vapor suppressing foam may be used to reduce vapors. Absorb or cover with dry earth, sand or other non-combustible material and transfer to containers. Use clean non-sparking tools to collect absorbed material. Large Spills: Dike far ahead of liquid spill for later disposal. Water spray may reduce vapor; but may not prevent ignition in closed spaces.

Environmental Physical Data

Henry's Law Constant: estimated at 0.0158
Sorption Partition Coefficient: K_{oc} = estimated at 68
BCF: estimated at 14

Regulations

RCRA 40CFR: Not listed
CERCLA: 40CFR 302.4: Not listed
SARA 40CFR 372.65: Not listed
SARA EHS 40CFR 355: Not listed
TSCA: Listed

BUT4010 CAS #: 507-19-7

T-BUTYL BROMIDE

RTECS: TX4150000
DOT: UN2342; IMO3.2
EINECS Number: 208-065-9
Molecular Formula: C_4H_9Br
Formula Weight: 137.04

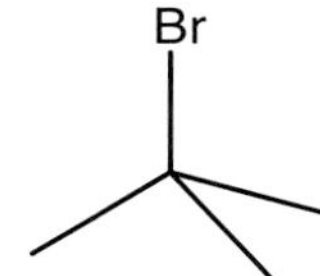

Chemical Structure

Synonyms: 2-BROMOISOBUTANE; 2-BROMO-2-METHYLPROPANE; TERT-BUTYL BROMIDE; PROPANE,2-BROMO-2-METHYL-; TERTIARYBUTYL BROMIDE; TRIMETHYLBROMOMETHANE
Description: colorless liquid
Use: laboratory reagent

Physical Properties

Boiling Point: 73.3 °C (164 °F) at 760 mm Hg
Freezing Point: -16.2 °C (2.84 °F)
Specific Gravity: 1.2125 at 25 °C/4 °C
Vapor Pressure: 5.58 kPa at 0 °C
Water Solubility: Insoluble in Water
Other Solubilities: miscible with organic solvents
Refraction Index: 1.4249 at 25 °C/D
Ionization Potential (eV): 9.87 +/-0.02

RTECS Toxicity Data

Tumorigenic: Mouse Route: Intraperitoneal; Dose: 3000 mg/kg/8W-I; Toxic Effects: Tumorigenic - Neoplastic by RTECS criteria; Lungs, Thorax, or Respiration - Tumors.

Hazard Overviews

Health: May cause irritation to eyes/skin. Harmful. Other Acute Effects: harmful if swallowed, inhaled, or absorbed through skin. Chronic Effects: Possible carcinogen.

Fire: Hazards: vapor may travel considerable distance to source of ignition and flash back; container explosion may occur; forms explosive mixtures in air; emits toxic fumes. Extinguishing agents: carbon dioxide, dry chemical powder or appropriate foam; water may be effective for cooling, but may not effect extinguishment. Precautions: combustible liquid.

Reactivity: Incompatible with: strong oxidizing agents, strong bases. Hazardous decomposition products: toxic fumes of: carbon monoxide, carbon dioxide, hydrogen bromide gas.

Carcinogenicity: IARC - Not listed; NIOSH - Not listed; NTP - Not listed; ACGIH - Not listed; OSHA - Not listed; EPA - Not listed; MAK - Not listed

Environmental

Environmental Fate: If released to the atmosphere, it will exist solely in the vapor-phase based on an experimental vapor pressure of 135.3 mm Hg. In the vapor-phase, it will react slowly with hydroxyl radicals with an estimated half-life of 39 days. Based on an estimated K_{oc} of 566, it has low mobility in soil. In moist soils, hydrolysis should occur based on results from similar compounds; t-butyl chloride, isopropyl bromide, and n-propyl bromide have hydrolysis half-lives of 23 seconds, 2.1 days, and 26 days, respectively. It is expected to volatilize from moist soil surfaces, based on an estimated Henry's Law constant of 4.07×10^{-2} atm-cu m/mole. From dry soil surfaces, it should quickly volatilize due to its high vapor pressure. It is expected to volatilize rapidly from water surfaces based on its Henry's Law constant. The volatilization half-life from a model river was calculated as 3.4 hours, from a model lake, 4.6 days. Hydrolysis is also expected to be a major fate process for this compound in water. It should not bioconcentrate in aquatic organisms based on a BCF of 49.

Cleanup/Disposal: Guide No. 130: Eliminate all ignition sources (no smoking, flares, sparks or flames in immediate area). All equipment used when handling the product must be grounded. Do not touch or walk through spilled material. Stop leak if you can do it without risk. Prevent entry into waterways, sewers, basements or confined areas. A vapor suppressing foam may be used to reduce vapors. Absorb or cover with dry earth, sand or other non-combustible material and transfer to containers. Use clean non-sparking tools to collect absorbed material. Large Spills: Dike far ahead of liquid spill for later disposal. Water spray may reduce vapor; but may not prevent ignition in closed spaces.

Environmental Physical Data

Henry's Law Constant: 4.07×10^{-2}
Octanol/Water Partition Coefficient: log K_{ow} = 2.54
Sorption Partition Coefficient: K_{oc} = 566
BCF: estimated at 49

Regulations

RCRA 40CFR: Not listed
CERCLA: 40CFR 302.4: Not listed
SARA 40CFR 372.65: Not listed
SARA EHS 40CFR 355: Not listed
TSCA: Listed

BUT4080 CAS #: 592-35-8

N-BUTYL CARBAMATE

RTECS: EZ0175000
EINECS Number: 209-751-0
Molecular Formula: $C_5H_{11}NO_2$
Formula Weight: 117.15

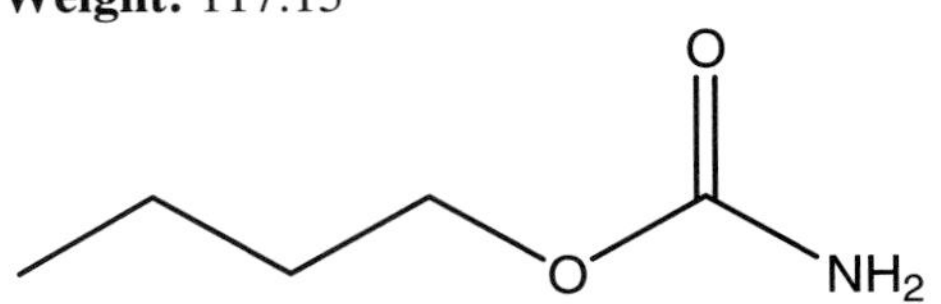

Chemical Structure

Synonyms: BUTYL CARBAMATE; CARBAMIC ACID,BUTYL ESTER
Description: prisms
Use: research chemical

Physical Properties

Boiling Point: Decomposes at 204 °C (399 °F)
Freezing Point: 54 °C (129.2 °F)
Other Solubilities: Very Soluble in Alcohol

RTECS Toxicity Data

Acute Dermal: Mouse LD_{50} Route: Subcutaneous Dose: 540 mg/kg; Toxic Effects: Behavioral - Somnolence (general depressed activity).

Reproductive/Teratogenic: Hamster Route: Intraperitoneal; Dose: 492 mg/kg; Duration: female 8D of pregnancy; Effects on Embryo or Fetus - Fetal death.

Mutagenic: Bacteria - E Coli Mutations in Microorganisms; Dose: 5000 ppm/3H (-S9).

Tumorigenic: Mouse Route: Intraperitoneal; Dose: 1980 mg/kg/6D-C; Toxic Effects: Tumorigenic - Neoplastic by RTECS criteria; Lungs, Thorax, or Respiration - Tumors; Skin and appendages - Tumors.

Hazard Overviews

Health: Irritating. Harmful. Other Acute Effects: may be harmful by inhalation, ingestion, or skin absorption.

Fire: Hazards: emits toxic fumes. Extinguishing agents: water spray; carbon dioxide, dry chemical powder or appropriate foam. Precautions: combustible liquid.

Reactivity: Incompatible with: strong oxidizing agents, strong bases. Hazardous decomposition products: toxic fumes of: carbon monoxide, carbon dioxide, nitrogen oxides.

Carcinogenicity: IARC - Not listed; NIOSH - Not listed; NTP - Not listed; ACGIH - Not listed; OSHA - Not listed; EPA - Not listed; MAK - Not listed

Primary Target Organs:

Eyes Skin Respiratory System

Environmental

Environmental Physical Data

Octanol/Water Partition Coefficient: log K_{ow} = 0.85

Regulations

RCRA 40CFR: Not listed
CERCLA: 40CFR 302.4: Not listed
SARA 40CFR 372.65: Not listed
SARA EHS 40CFR 355: Not listed
TSCA: Listed

BUT4150 CAS #: 109-69-3

N-BUTYL CHLORIDE

RTECS: EJ6300000
DOT: UN1127; IMO3.2
EINECS Number: 203-696-6
Molecular Formula: C_4H_9Cl
Structured MF: $CH_3(CH_2)_3Cl$
Formula Weight: 92.57

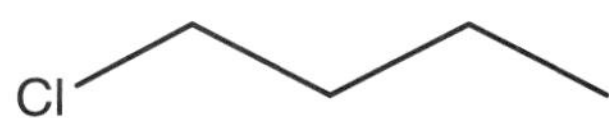

Chemical Structure

Synonyms: BUTANE,1-CHLORO-; BUTYL CHLORIDE; 1-CHLOROBUTANE; CHLORURE DE BUTYLE; N-PROPYLCARBINYL CHLORIDE
Description: colorless liquid; unpleasant odor
Use: as an antihelmintic in veterinary medicine; as an alkylating agent in organic synthesis and as a solvent

Physical Properties

Boiling Point: 78.5 °C (173 °F) at 760 mm Hg
Freezing Point: -123.1 °C (-189.58 °F)
Specific Gravity: 0.88098 at 25 °C/4 °C
Vapor Density: 3.2 Air=1
Density: 0.874 g/mL at 27.8 °C
Vapor Pressure: 40 mm Hg at 5 °C
Water Solubility: 0.066% at 12 °C
Other Solubilities: 95% Ethanol: >=100 mg/ml at 15 °C; Acetone: >=100 mg/ml at 15 °C; DMSO: >=100 mg/ml at 15 °C; Ether: miscible.
Odor Threshold: 3.3352 to 6.3293 mg/m^3
Refraction Index: 1.40223 at 20 °C/D
Critical Temperature: 269 °C
Ionization Potential (eV): 10.46 +/-0.05
Flash Point: -9 °C Closed Cup
Autoignition Temperature: 460 °C
LEL: 1.8% v/v
UEL: 10.1% v/v

RTECS Toxicity Data

Acute Oral: Rat LD_{50} Dose: 2670 mg/kg.
Acute Inhalation: Rat LC_{Lo} Dose: 8000 ppm/4hr.
Acute Dermal: Rabbit LD_{Lo} Route: Skin; Dose: 20 gm/kg.
Chronic (Multiple Dose) Oral: Rat Dose: 364 mg/kg/26W-I; Toxic Effects: Biochemical - True cholinesterase; Biochemical - Phosphatases; Biochemical - Dehydrogenases. Rat Dose: 10500 mg/kg/14D-I; Toxic Effects: Behavioral - Tremor; Behavioral - Convulsions or effect on seizure threshold; DEATH. Rat Dose: 32500 mg/kg/13W-I; Toxic Effects: DEATH.
Irritation Eye: Rabbit Standard Draize Test Dose: 500 mg/24H; Reaction: mild.
Irritation Skin: Rabbit Standard Draize Test Dose: 500 mg/24H; Reaction: mild. Rabbit Open Draize Test Dose: 10 mg/24H open; Reaction: mild.
Reproductive/Teratogenic: Rat Route: Oral; Dose: 13927 mg/kg; Duration: female 1-19D of pregnancy; Effects on Embryo or Fetus - Fetal death.
Mutagenic: Mouse Mutations in Mammalian Somatic Cells; Cell Type: lymphocyte; Dose: 500 mg/L.

Hazard Overviews

Flammable

Health: Irritating to eyes/skin/respiratory tract. Other Acute Effects: may be harmful by inhalation, ingestion, or skin absorption; may cause nervous system disturbances.
Fire: Flammable. Hazards: vapor may travel considerable distance to source of ignition and flash back; container explosion may occur; forms explosive mixtures in air; emits toxic fumes. Extinguishing agents: carbon dioxide, dry chemical powder or appropriate foam; water may be effective for cooling, but may not effect extinguishment. Precautions: combustible liquid.
Reactivity: Stable. Hazardous polymerization will not occur. Incompatible with: strong oxidizing agents, strong bases. Hazardous decomposition products: toxic fumes of: carbon monoxide, carbon dioxide, hydrogen chloride gas.
Carcinogenicity: IARC - Not listed; NIOSH - Not listed; NTP - Not listed; ACGIH - Not listed; OSHA - Not listed; EPA - Class D, Not classifiable as to human carcinogenicity; MAK - Not listed
Primary Target Organs:

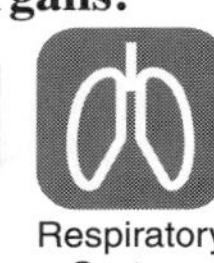

Eyes Skin Respiratory System

Environmental

Ecotoxicity: LC_{50} Poecilia reticulata (guppy) 97 ppm/7 days /Conditions of bioassay not specified

Environmental Fate: Has a high vapor pressure and Henry's Law constant. In addition it has a low adsorptivity to soil. Therefore releases to the land or water will partition, to a large extent, to the atmosphere. Limited data suggests that biodegradation is slow. It's rate of hydrolysis is unknown and estimates of its rate from analogous compounds indicate that hydrolysis could not compete with volatilization as a fate process except in groundwater. If released in surface water, the volatilization half-life in a model river and pond are estimated to be 2.9 hr and 34 hr, respectively. In the atmosphere, it will degrade by reaction with photochemically produced hydroxyl radicals in the atmosphere with a half-life of 7.0 days.

Cleanup/Disposal: Guide No. 130: Eliminate all ignition sources (no smoking, flares, sparks or flames in immediate area). All equipment used when handling the product must be grounded. Do not touch or walk through spilled material. Stop leak if you can do it without risk. Prevent entry into waterways, sewers, basements or confined areas. A vapor suppressing foam may be used to reduce vapors. Absorb or cover with dry earth, sand or other non-combustible material and transfer to containers. Use clean non-sparking tools to collect absorbed material. Large Spills: Dike far ahead of liquid spill for later disposal. Water spray may reduce vapor; but may not prevent ignition in closed spaces.

Environmental Physical Data

Henry's Law Constant: 0.0167
Octanol/Water Partition Coefficient: log K_{ow} = 2.39
Sorption Partition Coefficient: K_{oc} = estimated at 93 to 102
BCF: calculated at 60

Regulations

RCRA 40CFR: Not listed
CERCLA: 40CFR 302.4: Not listed
SARA 40CFR 372.65: Not listed
SARA EHS 40CFR 355: Not listed
TSCA: Listed

Analytical Methods

Water / Groundwater: ASTM D3695

BUT4220 CAS #: 590-02-3

N-BUTYL CHLOROACETATE

EINECS Number: 209-670-0
Molecular Formula: $C_6H_{11}ClO_2$
Formula Weight: 150.61

Chemical Structure

Synonyms: ACETIC ACID,CHLORO-,BUTYL ESTER; BUTYL CHLOROACETATE

Physical Properties

Boiling Point: 183 °C (361 °F)
Specific Gravity: 1.0704 at 20 °C/4 °C
Water Solubility: Insoluble in Water
Other Solubilities: Soluble in Alcohol, Ether
Refraction Index: 1.4297 at 20 °C

Hazard Overviews

Carcinogenicity: IARC - Not listed; NIOSH - Not listed; NTP - Not listed; ACGIH - Not listed; OSHA - Not listed; EPA - Not listed; MAK - Not listed

Environmental

Regulations

RCRA 40CFR: Not listed
CERCLA: 40CFR 302.4: Not listed
SARA 40CFR 372.65: Not listed
SARA EHS 40CFR 355: Not listed
TSCA: Listed

BUT4290 CAS #: 1189-85-1

TERT-BUTYL CHROMATE

RTECS: GB2900000
Molecular Formula: $C_8H_{18}CrO_4$
Structured MF: $((CH_3)_3CO)_2CrO_2$
Formula Weight: 230.26
Synonyms: BIS(TERT-BUTYL)CHROMATE; BIS(1,1-DIMETHYLETHYL) ESTER CHROMIC ACID; T-BUTYL CHROMATE; TERT-BUTYL CHROMATE(VI) (6CI,7CI); CHROMIC ACID (H2CRO4),BIS(1,1-DIMETHYLETHYL) ESTER (9CI); DI-TERT-BUTOXYCHROMYL; DI-TERT-BUTYL CHROMATE; DI-TERT-BUTYL ESTER CHROMIC ACID; DI-TERT-BUTYL ESTER OF CHROMIC ACID
Description: liquid
Use: oxidizing agent; catalyst for alkene polymerisation; corrosion inhibitor

Physical Properties

Freezing Point: -5 °C (23 °F)
Flash Point: Nonflammable

Hazard Overviews

Corrosive

Fire: Noncombustible.
Carcinogenicity: IARC - Group 3, Not classifiable as to carcinogenicity to humans; NIOSH - Listed as carcinogen; NTP - Not listed; ACGIH - Class A4, Not classifiable as a human carcinogen; OSHA - Not listed; EPA - Not listed; MAK - Class A2, Unmistakably carcinogenic in animal experimentation only
Exposure Limits

OSHA PEL: STEL: 0.1 mg/m^3; as CrO_3, skin. Other Values: 0.1 mg/m^3; Clg Cr-VI as CrO_3.
ACGIH TLV: STEL: 0.1 mg/m^3; as $CrC0_3$.
NIOSH REL: TWA: 0.001 mg/m^3; as Cr;Cr-II;Cr-III;Cr(VI)=.001.
NIOSH IDLH: 15 mg/m^3; as Cr(VI).
Respirator Recommendation
Exposure Range: >0.1 to <15 mg/m^3 Air Purifying, Negative Pressure, Half Mask
Exposure Range: 15 to unlimited mg/m^3 Self-contained Breathing Apparatus, Pressure Demand, Full Face Air Purifying, Negative Pressure, Full Face Supplied Air, Constant Flow/Pressure Demand, Full Face Self-contained Breathing Apparatus, Pressure Demand, Full Face
Note: odor threshold unknown

Environmental

Regulations

RCRA 40CFR: Not listed
CERCLA: 40CFR 302.4: Listed as Compound per CWA Section 307(a) per CAA Section 112
SARA 40CFR 372.65: Listed as Compound
SARA EHS 40CFR 355: Not listed
TSCA: Not listed

BUT4430 CAS #: 84-64-0

BUTYL CYCLOHEXYL PHTHALATE

EINECS Number: 201-548-5
Molecular Formula: $C_{18}H_{24}O_4$
Formula Weight: 304.42
Synonyms: 1,2-BENZENEDICARBOXYLIC ACID,BUTYL CYCLOHEXYL ESTER; CYCLOHEXYL BUTYL PHTHALATE; PHTHALIC ACID,BUTYL CYCLOHEXYL ESTER
Description: clear liquid; very mild odor
Use: plasticizer for polymers and elastomers; nitrocellulose lacquers

Physical Properties

Specific Gravity: 1.078
Vapor Pressure: Extremely low
Water Solubility: Insoluble < 1 mg/mL at 20 C
Other Solubilities: 95% Ethanol: Soluble (>=10 mg/ml at 20 °C); DMSO: Soluble (>=10 mg/ml at 20 °C).
Flash Point: > 93.3 °C

Hazard Overviews

Fire: Will burn.
Carcinogenicity: IARC - Not listed; NIOSH - Not listed; NTP - Not listed; ACGIH - Not listed; OSHA - Not listed; EPA - Not listed; MAK - Not listed

Environmental

Environmental Fate: If released to water, volatilization to the atmosphere is not expected to be a significant fate process, nor is hydrolysis expected to occur. Adsorption to sediment and suspended organic matter may be a significant process. No data on the biodegradation of this compound could be located. If released to soil, it is not expected to volatilize to the atmosphere. It is expected to adsorb strongly to soil. In the atmosphere, it is expected to be adsorbed to particulates. Destruction by the vapor phase reaction with photochemically produced hydroxyl radicals may occur.

Environmental Physical Data

Henry's Law Constant: estimated at 1.69 x10^{-7}
BCF: none likely

Regulations

RCRA 40CFR: Not listed
CERCLA: 40CFR 302.4: Not listed
SARA 40CFR 372.65: Not listed
SARA EHS 40CFR 355: Not listed
TSCA: Listed

BUT4500 CAS #: 10108-56-2

N-BUTYL CYCLOHEXYLAMINE

RTECS: GX1050000
EINECS Number: 233-294-6
Molecular Formula: $C_{10}H_{21}N$
Formula Weight: 155.32
Synonyms: (BUTYLAMINO)CYCLOHEXANE; BUTYLCYCLOHEXYLAMINE; N-BUTYLCYCLOHEXYLAMINE; CYCLOHEXANAMINE,N-BUTYL-; CYCLOHEXANE,(BUTYLAMINO)-; CYCLOHEXYLAMINE,N-BUTYL-

Physical Properties

Boiling Point: 207 °C (405 °F)
Freezing Point: 208 °C (406.4 °F)
Specific Gravity: 0.8
Water Solubility: Slightly Soluble in Water
Other Solubilities: Very Soluble in Alcohol, Ether
Flash Point: 93 °C Open Cup

RTECS Toxicity Data

Acute Oral: Rat LD_{50} Dose: 330 mg/kg.
Acute Dermal: Rabbit LD_{50} Route: Skin; Dose: 530 mg/kg; Toxic Effects: Skin and appendages - Primary irritation.

Hazard Overviews

Fire Diamond

Fire: Will burn.
Carcinogenicity: IARC - Not listed; NIOSH - Not listed; NTP - Not listed; ACGIH - Not listed; OSHA - Not listed; EPA - Not listed; MAK - Not listed

Environmental

Regulations

RCRA 40CFR: Not listed
CERCLA: 40CFR 302.4: Not listed
SARA 40CFR 372.65: Not listed
SARA EHS 40CFR 355: Not listed
TSCA: Not listed

BUT4570 CAS #: 106-83-2

BUTYL 9,10-EPOXYSTEARATE

RTECS: RG1575000
EINECS Number: 203-434-0
Molecular Formula: $C_{22}H_{42}O_3$
Formula Weight: 354.64
Synonyms: BUTYL 9,10-EPOXYOCTADECANOATE; BUTYL EPOXYSTEARATE; BUTYL-9,10-EPOXYSTEARATE; 9,10-EPOXYOCTADECANOIC ACID,BUTYL ESTER; 9,10-EPOXYSTEARIC ACID BUTYL ESTER; OCTADECANOIC ACID,9,10-EPOXY-,BUTYL ESTER; OXIRANEOCTANOIC ACID,3-OCTYL-,BUTYL ESTER; TRUFLEX E-74
Description: colorless liquid; mild, slightly fatty, slightly fruity odor
Use: plasticizer for low-temp flexibility improvement of vinyl resins; plasticizer for polyvinyl chloride & cellulosic resins

Physical Properties

Specific Gravity: 0.91 at 20 °C

Hazard Overviews

Carcinogenicity: IARC - Not listed; NIOSH - Not listed; NTP - Not listed; ACGIH - Not listed; OSHA - Not listed; EPA - Not listed; MAK - Not listed

Environmental

Regulations

RCRA 40CFR: Not listed
CERCLA: 40CFR 302.4: Not listed
SARA 40CFR 372.65: Not listed
SARA EHS 40CFR 355: Not listed
TSCA: Listed

BUT4640 CAS #: 142-96-1

N-BUTYL ETHER

RTECS: EK5425000
DOT: UN1149; IMO3.3
EINECS Number: 205-575-3
Molecular Formula: $C_8H_{18}O$
Structured MF: $CH_3(CH_2)_3O(CH_2)_3CH_3$
Formula Weight: 130.26

Chemical Structure

Synonyms: BUTANE,1,1'-OXYBIS-; 1-BUTOXYBUTANE; 1-BUTOXYBUTONE; BUTYL ETHER; BUTYL OXIDE; DI-N-BUTYL ETHER; DIBUTYL ETHER; N-DIBUTYL ETHER; DIBUTYL OXIDE; ETHER BUTYLIQUE; 1,1'-OXYBIS(BUTANE); 1,1'-OXYBISBUTANE
Description: colorless liquid; mild etheral odor
Use: solvent for hydrocarbons and fatty materials; extracting agent used especial for separating metals; solvent purification; and organic synthesis (reaction medium)

Physical Properties

Boiling Point: 142 °C (288 °F) at 760 mm Hg
Freezing Point: -95.3 °C (-139.54 °F)
Specific Gravity: 0.7689 at 20 °C/4 °C
Vapor Density: 4.48 Air=1
Saturated Vapor Density: 1.226463593 kg/m^3
Density: 0.767 g/mL at 20 °C
Vapor Pressure: 4.8 mm Hg at 20 °C
Water Solubility: 0.03 to 0.05% by wt in Water at 20 °C
Other Solubilities: 95% Ethanol: >=100 mg/ml at 22.5 °C; Acetone: >=100 mg/ml at 22.5 °C; DMSO: 10-50 mg/ml at 22.5 °C; Ether: miscible; Most organic solvents: miscible.
Surface Tension: 23 dynes/cm
Odor Threshold: 0.37 to 2.50 mg/m^3
Refraction Index: 1.3992 at 20 °C/D
Flash Point: 37 °C Closed Cup
Autoignition Temperature: 194 °C
LEL: 1.5% v/v
UEL: 7.6% v/v

RTECS Toxicity Data

Acute Oral: Rat LD_{50} Dose: 7400 mg/kg.
Acute Inhalation: Rat LC_{Lo} Dose: 4000 ppm/4hr. Human TC_{Lo} Dose: 200 ppm; Toxic Effects: Sense organs and special senses - Other; Sense organs and special senses - Conjunctive irritation.
Acute Dermal: Rabbit LD_{50} Route: Skin; Dose: 10 mL/kg.
Irritation Eye: Rabbit Standard Draize Test Dose: 500 mg/24H; Reaction: mild.
Irritation Skin: Rabbit Standard Draize Test Dose: 100 mg/24H; Reaction: moderate. Rabbit Open Draize Test Dose: 380 mg open; Reaction: mild.

Hazard Overviews

Flammable

Fire Diamond

Health: Irritating to eyes/skin/respiratory tract. Also Causes: conjunctiva, mouth and stomach irritation.
Fire: Flammable. Can form explosive mixtures in air. Alcohol foam is recommended extinguishing media. For small fires can also use dry chemical, carbon dioxide, water spray, or foam. For large fires can also use water spray, or foam. Water

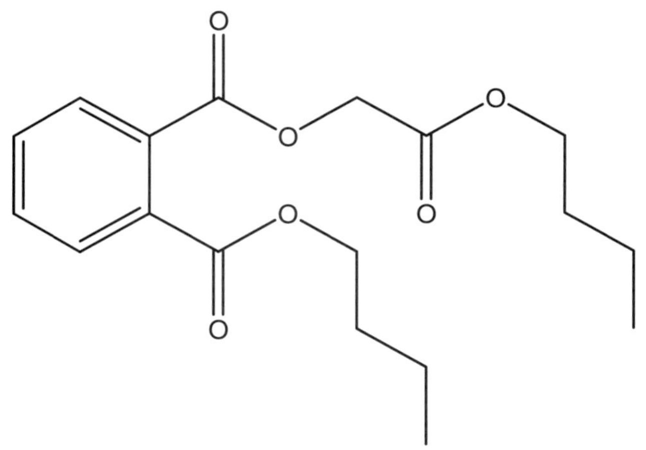

Chemical Structure

Synonyms: 1,2-BENZENEDICARBOXYLIC ACID,2-BUTOXY-2-OXOETHYL BUTYLESTER; BUTOXYCARBONYLMETHYL BUTYL PHTHALATE; BUTYL CARBOBUTOXYMETHYL PHTHALATE; BUTYL PHTHALATE BUTYL GLYCOLATE; BUTYL PHTHALYL BUTYL GLYCOLATE; BUTYLPHTHALYL BUTYL GLYCOLATE; DIBUTYL O-(O-CARBOXYBENZOYL) GLYCOLATE; DIBUTYL O-(O-CARBOXYBENZOYL)GLYCOLATE; DIBUTYL O-CARBOXYBENZOYLOXYACETATE; GLYCOLIC ACID,BUTYL ESTER,BUTYL PHTHALATE; GLYCOLIC ACID,PHTHALATE,DIBUTYL ESTER; PHTHALIC ACID,BUTOXYCARBONYLMETHYL BUTYL ESTER; PHTHALIC ACID,BUTYL ESTER,BUTYL GLYCOLATE; PHTHALIC ACID,BUTYL ESTER,ESTER WITH BUTYL GLYCOLATE; REOMOL 4PG; SANTICIZER B-16; SANTICIZER B 16

Description: colorless liquid; odorless

Use: plasticizer for polyvinyl chloride; in vinyl food wrapping

Physical Properties

Boiling Point: 345 °C (653 °F)
Freezing Point: < -35 °C (-31 °F)
Specific Gravity: 1.097
Vapor Density: 11.6 Air=1
Vapor Pressure: Extremely low
Water Solubility: 120 ppm at 25 °C
Other Solubilities: Soluble to various extents in many common organic solvents and oils.
Flash Point: 199 °C Open Cup

RTECS Toxicity Data

Acute Oral: Rat LD_{50} Dose: 7 gm/kg. Mouse LD_{50} Dose: 12567 mg/kg.

Irritation Eye: Rabbit Standard Draize Test Dose: 500 mg; Reaction: mild.

Reproductive/Teratogenic: Rat Route: Intraperitoneal; Dose: 2296 mg/kg; Duration: female 5-15D of pregnancy; Effects on Fertility - Post-implantation mortality. Rat Route: Intraperitoneal; Dose: 689 mg/kg; Duration: female 5-15D of pregnancy; Effects on Embryo or Fetus - Fetotoxicity; Specific Developmental Abnormalities - Musculoskeletal system.

Mutagenic: Hamster Cytogenetic Analysis; Cell Type: fibroblast; Dose: 125 mg/L/24H.

Hazard Overviews

Health: Irritating to eyes/skin/respiratory tract. Harmful. Other Acute Effects: may be harmful by inhalation, ingestion, or skin absorption; exposure can cause nausea, dizziness and headache. Chronic Effects: target organs: kidneys, central nervous system, eyes, male reproductive system. Possibly causes reproductive disorders.

Fire: Will burn. Hazards: emits toxic fumes. Extinguishing agents: water spray; carbon dioxide, dry chemical powder or appropriate foam. Precautions: combustible liquid.

Reactivity: Incompatible with: oxidizing agents, acids, sensitive to moisture. Hazardous decomposition products: carbon monoxide, carbon dioxide.

Carcinogenicity: IARC - Not listed; NIOSH - Not listed; NTP - Not listed; ACGIH - Not listed; OSHA - Not listed; EPA - Not listed; MAK - Not listed

Primary Target Organs:

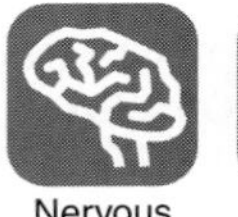

Environmental

Environmental Fate: If released to soil, it is expected to display moderate mobility. Volatilization from the soil surface to the atmosphere is not expected to be an important fate process. If released to water, it is expected to rapidly biodegrade under aerobic conditions. The estimated half-life for volatilization from a model river is 170 days, and therefore, volatilization from water to the atmosphere is not expected to be a significant process. It may adsorb to sediment and suspended organic matter. Bioconcentration of this compound in fish and aquatic organisms is not expected to be a significant process. Hydrolysis in environmental waters is not expected to occur. In the atmosphere, it is expected to be adsorbed to particulates. Destruction by the vapor phase reaction with photochemically produced hydroxyl radicals may occur.

Environmental Physical Data

Henry's Law Constant: 3.88×10^{-7}
Octanol/Water Partition Coefficient: log K_{ow} = > 2.12
Sorption Partition Coefficient: K_{oc} = 314
BCF: calculated at 42

Regulations

RCRA 40CFR: Not listed
CERCLA: 40CFR 302.4: Not listed
SARA 40CFR 372.65: Not listed
SARA EHS 40CFR 355: Not listed
TSCA: Listed

Analytical Methods

Plasma: EPA 001

BUT4850 CAS #: 75-91-2

TERT-BUTYL HYDROPEROXIDE

RTECS: EQ4900000
DOT: UN2094
EINECS Number: 200-915-7
Molecular Formula: $C_4H_{10}O_2$
Structured MF: $(CH_3)_3C{\bullet}O{\bullet}OH$
Formula Weight: 90.12

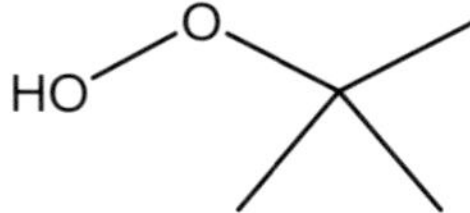

Chemical Structure

Synonyms: TERT-BUTYL HYDROPEROXIDE,>90% WITH WATER; T-BUTYLHYDROPEROXIDE; CADOX TBH; 1,1-DIMETHYLETHYL HYDROPEROXIDE; 1,1-DIMETHYLETHYLHYDROPEROXIDE; ETHYLDIETHYLPEROXIDE; HYDROPEROXIDE,TERT-BUTYL; HYDROPEROXIDE,1,1-DIMETHYLETHYL; HYDROPEROXIDE,1,1-DIMETHYLETHYL-(9CI); HYDROPEROXYDE DE BUTYLE TERTIAIRE; 2-HYDROPEROXY-2-METHYLPROPANE; PERBUTYL H; TBHP-70; TERC BUTYLHYDROPEROXID; TERC. BUTYLHYDROPEROXID; TRIGONOX; TRIGONOX A-75; TRIGONOX A-W70
Description: clear to pale-yellow liquid; odorless
Use: as a catalyst in polymerization, oxidation, and sulfonation reactions; for bleaching and deodorizing; a selective reagent for the introduction of a peroxy group into organic substrates

Physical Properties

Boiling Point: 35 °C (95 °F) at 20 mm Hg
Freezing Point: -8 °C (17.6 °F)
Specific Gravity: 0.896 at 20 °C
Vapor Density: 2.07 Air=1
Density: 0.824 g/mL at 19 °C
Water Solubility: Soluble in Water
Other Solubilities: 95% Ethanol: >=100 mg/ml at 22 °C; Acetone: >=100 mg/ml at 22 °C; Alkali metal hydroxides: very Soluble; Chloroform: Soluble; DMSO: >=100 mg/ml at 22 °C; Ether: Soluble.
Refraction Index: 1.4015 at 20 °C/D
Ionization Potential (eV): =< 10.24 +/-1.0
Flash Point: Commercial 26.67 to 54.44 °C

RTECS Toxicity Data

Acute Oral: Rat LD_{50} Dose: 370 mg/kg; Toxic Effects: Behavioral - Irritability; Gastrointestinal - Alteration in gastric secretion; Blood - Hemorrhage. Mouse LD_{50} Dose: 320 mg/kg; Toxic Effects: Behavioral - Irritability; Gastrointestinal - Alteration in gastric secretion; Blood - Hemorrhage.
Acute Inhalation: Rat LC_{50} Dose: 500 ppm/4hr; Toxic Effects: Lungs, Thorax, or Respiration - Dyspnea. Mouse LC_{50} Dose: 350 ppm/4hr; Toxic Effects: Lungs, Thorax, or Respiration - Dyspnea.
Acute Dermal: Rabbit LD_{50} Route: Skin; Dose: 460 uL/kg; Toxic Effects: Lungs, Thorax, or Respiration - Cyanosis; Liver - Other changes; Kidney, Ureter, and Bladder - Other changes in urine composition. Rat LD_{50} Route: Skin; Dose: 790 mg/kg.
Chronic (Multiple Dose) Inhalation: Rat Dose: 107 mg/m^3/17W-I; Toxic Effects: Sense organs and special senses - Change in sensation of smell; Kidney, Ureter, and Bladder - Other changes in urine composition; Nutritional and gross metabolic - Changes in chlorine. Guinea Pig Dose: 107 mg/m^3/17W-I; Toxic Effects: Blood - Changes in erythrocite (RBC) cell count; Nutritional and gross metabolic - Weight loss or decreased weight gain.
Irritation Eye: Rabbit Standard Draize Test Dose: 100 mg/24H; Reaction: moderate. Rabbit Nonstandard Exposure Dose: 150 mg/1M rinse; Reaction: severe.
Irritation Skin: Rabbit Standard Draize Test Dose: 500 mg/24H; Reaction: severe.
Mutagenic: Rat Cytogenetic Analysis; Route: Inhalation; Dose: 107 mg/m^3/10W. Mouse Cytogenetic Analysis; Route: Inhalation; Dose: 107 mg/m^3/10W.

Hazard Overviews

Corrosive

Explosive

Flammable

Fire Diamond

Health: Corrosive, causes severe burns to eyes/skin/respiratory tract.
Fire: Flammable and explosive. Strong oxidizer capable of igniting combustibles. For small fires use dry chemical, carbon dioxide, water spray, or regular foam. For large fires flood area with water.
Reactivity: Unstable, thermally and shock sensitive. Hazardous polymerization cannot occur. Avoid: distilling to dryness; contact with heat; ignition sources. Incompatible with: hydrogen peroxide/sulfuric acid; acids (even traces); 1,2-dichloroethane; some metals; cobalt, iron, or manganese salts; oxidizable, organic, or flammable materials. Hazardous decomposition products: acrid smoke; methane gas; acetone; t-butyl alcohol.
Carcinogenicity: IARC - Not listed; NIOSH - Not listed; NTP - Not listed; ACGIH - Not listed; OSHA - Not listed; EPA - Not listed; MAK - Not listed
Primary Target Organs:

Eyes

Skin

Respiratory System

Environmental

Cleanup/Disposal: Guide No. 147: Eliminate all ignition sources (no smoking, flares, sparks or flames in immediate area). Keep combustibles (wood, paper, oil, etc.) away from spilled material. Do not touch damaged containers or spilled material unless wearing appropriate protective clothing. Keep substance wet using water spray. Stop leak if you can do it without risk. Small Spills: Take up with inert, damp, noncombustible material using clean non-sparking tools and place into loosely covered plastic containers for later disposal. Large Spills: Wet down with water and dike for later disposal. Prevent entry into waterways, sewers, basements or confined areas. Do not clean-up or dispose of, except under supervision of a specialist.

Environmental Physical Data

BCF: no food chain concentration potential

Regulations

RCRA 40CFR: Not listed
CERCLA: 40CFR 302.4: Not listed
SARA 40CFR 372.65: Not listed
SARA EHS 40CFR 355: Not listed
TSCA: Listed

BUT4920 CAS #: 111-36-4

N-BUTYL ISOCYANATE

RTECS: NQ8250000
EINECS Number: 203-862-8
Molecular Formula: C_5H_9NO
Formula Weight: 99.15

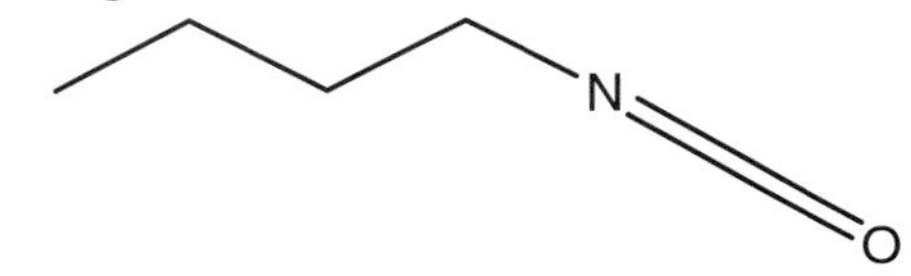

Chemical Structure

Synonyms: BIC; BUTANE,1-ISOCYANATO-; BUTYL ISOCYANATE; 1-ISOCYANATOBUTANE; ISOCYANIC ACID,BUTYL ESTER
Description: colorless liquid
Use: intermediate in the production of carbamate and urea insecticides and fungicides; in the manufacture of sulfonyl urea

Physical Properties

Boiling Point: 115 °C (239 °F)
Freezing Point: < -70 °C (-94 °F)
Specific Gravity: 0.9
Water Solubility: Slightly Soluble in Water
Ionization Potential (eV): =< 10.14 +/-0.05
Flash Point: 19 °C

RTECS Toxicity Data

Acute Oral: Rat LD_{50} Dose: 600 mg/kg; Toxic Effects: Behavioral - Somnolence (general depressed activity); Behavioral - Convulsions or effect on seizure threshold; Lungs, Thorax, or Respiration - Cyanosis. Mouse LD_{50} Dose: 150 mg/kg; Toxic Effects: Behavioral - Somnolence (general depressed activity); Behavioral - Convulsions or effect on seizure threshold; Lungs, Thorax, or Respiration - Cyanosis.
Acute Inhalation: Rat LC_{50} Dose: 3 gm/m^3; Toxic Effects: Behavioral - Somnolence (general depressed activity); Behavioral - Convulsions or effect on seizure threshold; Lungs, Thorax, or Respiration - Cyanosis. Mouse LC_{50} Dose: 680 mg/m^3; Toxic Effects: Behavioral - Somnolence (general depressed activity); Behavioral - Convulsions or effect on seizure threshold; Lungs, Thorax, or Respiration - Cyanosis.
Chronic (Multiple Dose) Inhalation: Rat Dose: 15 mg/m^3/6H/5D-I; Toxic Effects: Lungs, Thorax, or Respiration - Other changes; Lungs, Thorax, or Respiration - Changes in lung weight; Biochemical - Other proteins.

Hazard Overviews

Poison

Flammable

Health: Severely irritating to eyes/skin/respiratory tract. Poison. Other Acute Effects: may be fatal if inhaled, swallowed, or absorbed through skin; lachrymator; possible sensitizer; symptoms of exposure may include burning sensation; coughing; wheezing; laryngitis; shortness of breath; headache; nausea; vomiting; repeated exposure may cause asthma; may cause allergic reaction; prolonged contact can cause dizziness; lung irritation; chest pain and edema which may be fatal.
Fire: Flammable. Hazards: vapor may travel considerable distance to source of ignition and flash back; emits toxic fumes. Extinguishing agents: carbon dioxide; dry chemical powder. Precautions: combustible liquid.
Reactivity: Incompatible with: water, alcohols, strong bases, amines, acids, strong oxidizing agents, heat. Hazardous decomposition products: thermal decomposition may produce carbon monoxide, carbon dioxide, and nitrogen oxides; hydrogen cyanide.
Carcinogenicity: IARC - Not listed; NIOSH - Not listed; NTP - Not listed; ACGIH - Not listed; OSHA - Not listed; EPA - Not listed; MAK - Not listed
Primary Target Organs:

Eyes

Skin

Respiratory System

Environmental

Environmental Fate: If released to the atmosphere, it will degrade by reaction with photochemically produced hydroxyl radicals with an estimated half-life of 100 hours. It is rapidly transformed in aquatic and terrestrial environments via hydrolysis. Some volatilization of this compound from dry soils and other surfaces may also occur because of its moderate vapor pressure. Bioconcentration and biodegradation are not important fate processes because of this compound is hydrolyzed.

Environmental Physical Data

Henry's Law Constant: estimated at 0.002
Octanol/Water Partition Coefficient: log K_{OW} = 2.26
Sorption Partition Coefficient: K_{OC} = estimated at 400

BCF: estimated at 31

Regulations

RCRA 40CFR: Not listed
CERCLA: 40CFR 302.4: Listed as Compound per CWA Section 307(a) per CAA Section 112
SARA 40CFR 372.65: Not listed
SARA EHS 40CFR 355: Not listed
TSCA: Listed

BUT4990 CAS #: 109-19-3

BUTYL ISOVALERATE

RTECS: NY1502000
EINECS Number: 203-654-7
Molecular Formula: $C_9H_{18}O_2$
Formula Weight: 158.27

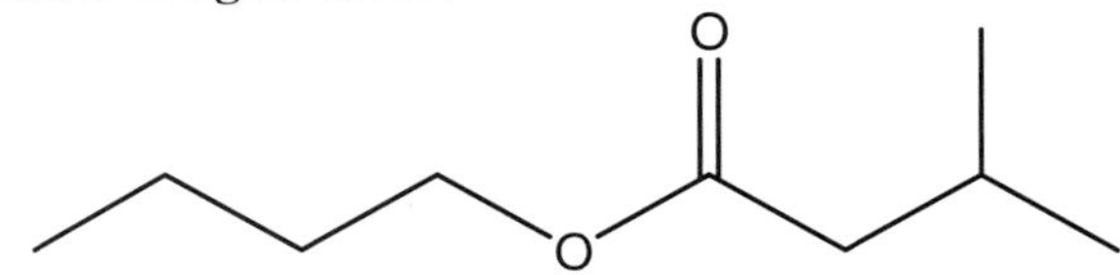

Chemical Structure

Synonyms: BUTANOIC ACID,3-METHYL-,BUTYL ESTER; N-BUTYL ISOPENTANOATE; 1-BUTYL ISOVALERATE; N-BUTYL ISOVALERATE; BUTYL ISOVALERIANATE; BUTYL 3-METHYLBUTYRATE; ISOVALERIC ACID,BUTYL ESTER
Description: liquid

Physical Properties

Boiling Point: 150 °C (302 °F)
Specific Gravity: 0.87
Water Solubility: Insoluble in Water
Other Solubilities: Soluble in most organic solvents.
Refraction Index: 1.4058
Flash Point: 53 °C

RTECS Toxicity Data

Acute Oral: Rat LD_{50} Dose: >5 gm/kg. Rabbit LD_{50} Dose: 8230 mg/kg; Toxic Effects: Sense organs and special senses - Corneal damage; Cardiac - Pulse rate; Lungs, Thorax, or Respiration - Dyspnea.
Acute Dermal: Rabbit LD_{50} Route: Skin; Dose: >5 gm/kg.
Irritation Skin: Rabbit Standard Draize Test Dose: 500 mg/24H; Reaction: mild.

Hazard Overviews

Flammable

Health: Irritating to eyes/skin/respiratory tract. Other Acute Effects: may be harmful by inhalation, ingestion, or skin absorption.
Fire: Flammable. Hazards: emits toxic fumes. Extinguishing agents: carbon dioxide, dry chemical powder or appropriate foam; water spray. Precautions: combustible liquid.
Reactivity: Incompatible with: strong oxidizing agents. Hazardous decomposition products: toxic fumes of: carbon monoxide, carbon dioxide.
Carcinogenicity: IARC - Not listed; NIOSH - Not listed; NTP - Not listed; ACGIH - Not listed; OSHA - Not listed; EPA - Not listed; MAK - Not listed
Primary Target Organs:

Eyes

Skin

Respiratory System

Environmental

Regulations

RCRA 40CFR: Not listed
CERCLA: 40CFR 302.4: Not listed
SARA 40CFR 372.65: Not listed
SARA EHS 40CFR 355: Not listed
TSCA: Listed

BUT5060 CAS #: 138-22-7

N-BUTYL LACTATE

RTECS: OD4025000
EINECS Number: 205-316-4
Molecular Formula: $C_7H_{14}O_3$
Structured MF: $CH_3CH(OH)COOC_4H_9$
Formula Weight: 146.21

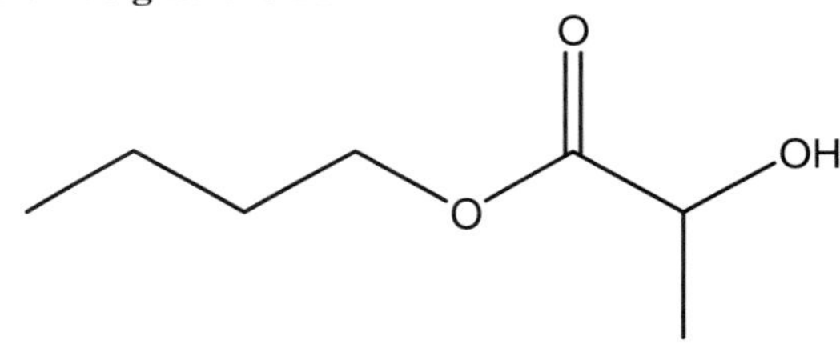

Chemical Structure

Synonyms: BUTYL ESTER OF 2-HYDROXYPROPANOIC ACID; BUTYL ESTER OF LACTIC ACID; BUTYL 2-HYDROXYPROPANOATE; BUTYL ALPHA-HYDROXYPROPIONATE; BUTYL LACTATE; BUTYLESTER KYSELINY MLECNE; 2-PROPANOIC ACID; PROPANOIC ACID,2-HYDROXY-,BUTYL ESTER (9CI)
Description: colorless liquid; mild odor

Physical Properties

Boiling Point: 187.78 °C (370 °F)
Freezing Point: -42.78 °C (-45.004 °F)
Specific Gravity: 0.98
Vapor Density: 5.04 Air=1
Vapor Pressure: 0.4 mm Hg
Water Solubility: Slight

Other Solubilities: Soluble in Alcohol, Ether, laquer solvents, diluents and oils
Odor Threshold: 7 ppm
Refraction Index: 1.4216
Flash Point: 71 °C
Autoignition Temperature: 382 °C
LEL: 1.15% v/v

RTECS Toxicity Data

Acute Oral: Rat LD_{50} Dose: >5 gm/kg.
Acute Dermal: Rabbit LD_{50} Route: Skin; Dose: >5 gm/kg. Rat LD_{50} Route: Subcutaneous Dose: 12 gm/kg.
Irritation Skin: Rabbit Standard Draize Test Dose: 500 mg/24H; Reaction: moderate.

Hazard Overviews

Fire Diamond

Health: Irritating to eyes/skin/respiratory tract. Also Causes: headache, cough, delayed drowsiness, nausea, GI irritation (upon ingestion). May be absorbed through the skin.
Fire: Combustible. For small fires use dry chemical, carbon dioxide, water spray, or regular foam. For large fires use water spray, fog, or regular foam. Vapors may travel to an ignition source and flash back. Containers may explode in heat of fire.
Reactivity: Stable. Hazardous polymerization cannot occur. Avoid: heat; ignition sources. Incompatible with: oxidizers; strong acids; strong bases. Hazardous decomposition products: carbon oxide(s).
Carcinogenicity: IARC - Not listed; NIOSH - Listed as carcinogen; NTP - Not listed; ACGIH - Not listed; OSHA - Not listed; EPA - Not listed; MAK - Not listed
Primary Target Organs:

Eyes

Skin

Respiratory System

Nervous System

Exposure Limits
OSHA PEL Vacated 1989 Limits: TWA: 5 ppm; 25 mg/m^3.
ACGIH TLV: TWA: 5 ppm; 130 mg/m^3.
NIOSH REL: TWA: 5 ppm; 25 mg/m^3.
Respirator Recommendation
Exposure Range: >5 to 250 ppm Supplied Air, Constant Flow/Pressure Demand, Half Mask
Exposure Range: >250 to 5000 ppm Supplied Air, Constant Flow/Pressure Demand, Full Face
Exposure Range: >5000 to unlimited ppm Self-contained Breathing Apparatus, Pressure Demand, Full Face
Note: poor warning properties

Environmental

Cleanup/Disposal: Guide No. 128: Eliminate all ignition sources (no smoking, flares, sparks or flames in immediate area). All equipment used when handling the product must be grounded. Do not touch or walk through spilled material. Stop leak if you can do it without risk. Prevent entry into waterways, sewers, basements or confined areas. A vapor suppressing foam may be used to reduce vapors. Absorb or cover with dry earth, sand or other non-combustible material and transfer to containers. Use clean non-sparking tools to collect absorbed material. Large Spills: Dike far ahead of liquid spill for later disposal. Water spray may reduce vapor; but may not prevent ignition in closed spaces.

Regulations

RCRA 40CFR: Not listed
CERCLA: 40CFR 302.4: Not listed
SARA 40CFR 372.65: Not listed
SARA EHS 40CFR 355: Not listed
TSCA: Listed

BUT5130 **CAS #: 109-79-5**

1-BUTYL MERCAPTAN

RTECS: EK6300000
DOT: UN2347; IMO3.2
EINECS Number: 203-705-3
Molecular Formula: $C_4H_{10}S$
Structured MF: $CH_3(CH_2)_3SH$
Formula Weight: 90.19

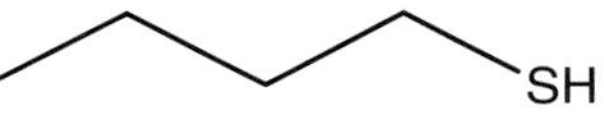

Chemical Structure

Synonyms: 1-BUTANETHIOL; BUTANETHIOL; N-BUTANETHIOL; BUTYL MERCAPTAN; N-BUTYL MERCAPTAN; N-BUTYL THIOALCOHOL; 1-MERCAPTOBUTANE; NORMAL BUTYL THIOALCOHOL; THIOBUTYL ALCOHOL
Description: colorless liquid; strong skunk-like odor
Use: as solvent; int for insecticides, acaricides, herbicides, defoliants

Physical Properties

Boiling Point: 98.4 °C (209 °F) at 760 mm Hg
Freezing Point: -115.7 °C (-176.26 °F)
Specific Gravity: 0.8337 at 20 °C/4 °C
Vapor Density: 6.5 Air=1
Vapor Pressure: 35 mm Hg
Water Solubility: 0.06% by weight
Other Solubilities: Very Soluble in liquid Hydrogen Sulfide.
Surface Tension: Estimated at 30 dynes/cm
Odor Threshold: 0.0001 to 0.001 ppm
Refraction Index: 1.4440
Critical Temperature: 290 °C
Critical Pressure: 572 psia
Ionization Potential (eV): 9.15
Flash Point: 1.667 °C Closed Cup

RTECS Toxicity Data

Acute Oral: Rat LD_{50} Dose: 1500 mg/kg; Toxic Effects: Behavioral - Somnolence (general depressed activity); Lungs, Thorax, or Respiration - Respiratory depression; Behavioral - Coma. Mouse LD_{50} Dose: 3 gm/kg.

Acute Inhalation: Rat LC_{50} Dose: 4020 ppm/4hr; Toxic Effects: Lungs, Thorax, or Respiration - Respiratory stimulation; Behavioral - Somnolence (general depressed activity); Sense organs and special senses - Lacrimation. Mouse LC_{50} Dose: 2500 ppm/4hr; Toxic Effects: Lungs, Thorax, or Respiration - Respiratory stimulation; Behavioral - Somnolence (general depressed activity); Sense organs and special senses - Lacrimation.

Chronic (Multiple Dose) Inhalation: Rat Dose: 1114 ppm/6H/2W-I; Toxic Effects: Kidney, Ureter, and Bladder - Chgs in tubules (inc acute renal failure, acute tubular necrosis; Nutritional and gross metabolic - Weight loss or decreased weight gain; DEATH.

Reproductive/Teratogenic: Mouse Route: Inhalation; Dose: 68 ppm/6H; Duration: female 6-16D of pregnancy; Effects on Fertility - Post-implantation mortality; Specific Developmental Abnormalities - Craniofacial (including nose and tongue); Musculoskeletal system.

Hazard Overviews

Flammable

Fire Diamond

Health: Irritating to eyes/skin/respiratory tract. Also Causes: loss of sense of smell, headache, muscular weakness; convulsions and respiratory paralysis on prolonged exposure, nausea.

Fire: Flammable. For small fires use dry chemical, carbon dioxide, halon, water spray, or standard foam. For large fires use water spray, fog, or standard foam. Fight fire from maximum distance. Use water spray to cool fire-exposed containers.

Reactivity: Stable. Hazardous polymerization cannot occur. Avoid: heat; ignition sources. Incompatible with: acids; acid fumes; oxidizing materials; nitric acid. Hazardous decomposition products: sulfur oxides; hydrogen sulfide.

Carcinogenicity: IARC - Not listed; NIOSH - Listed as carcinogen; NTP - Not listed; ACGIH - Not listed; OSHA - Not listed; EPA - Not listed; MAK - Not listed

Primary Target Organs:

Eyes

Skin

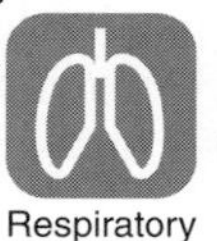
Respiratory System

Nervous System

Exposure Limits

OSHA PEL: TWA: 10 ppm; 35 mg/m^3.

OSHA PEL Vacated 1989 Limits: TWA: 0.5 ppm; 1.5 mg/m^3.

ACGIH TLV: TWA: 0.5 ppm; 1.8 mg/m^3.

NIOSH REL: STEL: 0.5 ppm; 1.8 mg/m^3; 15-minute.

NIOSH IDLH: 500 ppm.

DFG MAK: TWA: 0.5 ppm; 1.5 mg/m^3.

Respirator Recommendation

Exposure Range: >10 to 100 ppm Air Purifying, Negative Pressure, Half Mask

Exposure Range: >100 to <500 ppm Air Purifying, Negative Pressure, Full Face

Exposure Range: 500 to unlimited ppm Self-contained Breathing Apparatus, Pressure Demand, Full Face

Cartridge Color: black

Environmental

Ecotoxicity: Aquatic toxicity: 7.3 mg/l/24 hr/fish/TLm/fresh water

Environmental Fate: If released into water it will primarily be lost by volatilization (half-life 5.0 hr in a typical river). If released on land, it will also evaporate but additionally will leach into the soil. In the atmosphere it will photodegrade with an estimated 1.6 day half-life by reaction with hydroxyl radicals. Little bioconcentration in fish is expected.

Cleanup/Disposal: Guide No. 130: Eliminate all ignition sources (no smoking, flares, sparks or flames in immediate area). All equipment used when handling the product must be grounded. Do not touch or walk through spilled material. Stop leak if you can do it without risk. Prevent entry into waterways, sewers, basements or confined areas. A vapor suppressing foam may be used to reduce vapors. Absorb or cover with dry earth, sand or other non-combustible material and transfer to containers. Use clean non-sparking tools to collect absorbed material. Large Spills: Dike far ahead of liquid spill for later disposal. Water spray may reduce vapor; but may not prevent ignition in closed spaces.

Environmental Physical Data

Henry's Law Constant: calculated at 8.04 x10^{-3}

Octanol/Water Partition Coefficient: log K_{ow} = 2.28

Sorption Partition Coefficient: K_{oc} = estimated at 131

BCF: calculated at 1.5

Regulations

RCRA 40CFR: Not listed

CERCLA: 40CFR 302.4: Not listed

SARA 40CFR 372.65: Not listed

SARA EHS 40CFR 355: Not listed

TSCA: Listed

Analytical Methods

Air: ASTM D4490

Indoor / Expired Air: NIOSH 2542

BUT5200 **CAS #: 513-53-1**

SEC-BUTYL MERCAPTAN

DOT: UN2347; IMO3.2

EINECS Number: 208-165-2

Molecular Formula: $C_4H_{10}S$

Formula Weight: 90.19

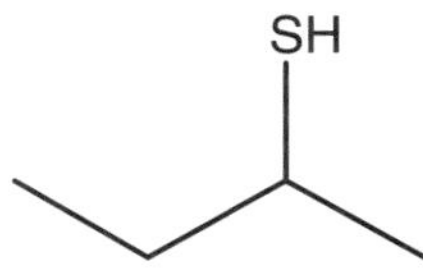

Chemical Structure

Synonyms: 2-BUTANETHIOL; SEC-BUTANETHIOL; 2-BUTYL MERCAPTAN; SEC-BUTYL THIOALCOHOL; SEC-BUTYL THIOL; 2-MERCAPTOBUTANE; 1-METHYL-1-PROPANETHIOL; SECONDARY BUTYLMERCAPTAN
Description: colorless mobile liquid; obnoxious heavy skunk odor
Use: odorant for natural gas; manufacture of cadusafos

Physical Properties

Boiling Point: 84 °C (183 °F) to 85 °C (185 °F)
Freezing Point: -165 °C (-265 °F)
Specific Gravity: 0.8299 at 17 °C
Water Solubility: Slightly Soluble in Water
Other Solubilities: Soluble in Benzene, Petroleum Ether, Alcohol, Ether.
Refraction Index: 1.4363 at 20 °C/D
Ionization Potential (eV): 9.10
Flash Point: -23.333 °C Closed Cup

Hazard Overviews

Flammable

Health: Irritating to eyes/skin/respiratory tract. Other Acute Effects: may be harmful by inhalation, ingestion, or skin absorption; exposure can cause nausea, headache and vomiting.
Fire: Flammable. Hazards: emits toxic fumes; vapor may travel considerable distance to source of ignition and flash back; container explosion may occur; forms explosive mixtures in air. Extinguishing agents: carbon dioxide, dry chemical powder or appropriate foam; water may be effective for cooling, but may not effect extinguishment. Precautions: combustible liquid.
Reactivity: Incompatible with: bases, oxidizing agents, reducing agents, alkali metals. Hazardous decomposition products: toxic fumes of: carbon monoxide, carbon dioxide, sulfur oxides.
Carcinogenicity: IARC - Not listed; NIOSH - Not listed; NTP - Not listed; ACGIH - Not listed; OSHA - Not listed; EPA - Not listed; MAK - Not listed

Primary Target Organs:

Eyes

Skin

Respiratory System

Environmental

Environmental Fate: Should have high mobility in soil. Volatilization is expected from both moist and dry soils. In water, it is expected to volatilize rapidly with estimated half-lives of 2.9 hours and 3.8 days from a model river and a model lake, respectively. Bioconcentration and adsorption to sediment are not expected to be important fate processes in aquatic systems. Insufficient data are available to determine the rate or importance of biodegradation in soil or aquatic conditions. It will exist in the vapor phase in the ambient atmosphere. If released to the atmosphere, it will degrade by reaction with photochemically produced hydroxyl radicals with an experimental half-life of 9-10.8 hours. Removal from the atmosphere can occur though wet deposition.
Cleanup/Disposal: Guide No. 130: Eliminate all ignition sources (no smoking, flares, sparks or flames in immediate area). All equipment used when handling the product must be grounded. Do not touch or walk through spilled material. Stop leak if you can do it without risk. Prevent entry into waterways, sewers, basements or confined areas. A vapor suppressing foam may be used to reduce vapors. Absorb or cover with dry earth, sand or other non-combustible material and transfer to containers. Use clean non-sparking tools to collect absorbed material. Large Spills: Dike far ahead of liquid spill for later disposal. Water spray may reduce vapor; but may not prevent ignition in closed spaces.

Environmental Physical Data

Henry's Law Constant: estimated at 6.11×10^{-3}
Sorption Partition Coefficient: K_{oc} = estimated at 67.7
BCF: estimated at 8.8

Regulations

RCRA 40CFR: Not listed
CERCLA: 40CFR 302.4: Not listed
SARA 40CFR 372.65: Not listed
SARA EHS 40CFR 355: Not listed
TSCA: Listed

BUT5270 CAS #: 75-66-1

T-BUTYL MERCAPTAN

RTECS: TZ7660000
DOT: UN2347; IMO3.2
EINECS Number: 200-890-2
Molecular Formula: $C_4H_{10}S$
Structured MF: $(CH_3)_3CSH$
Formula Weight: 90.19

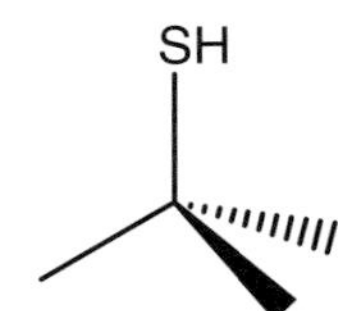

Chemical Structure

Synonyms: TERT-BUTANETHIOL; TERT-BUTYL MERCAPTAN; T-BUTYLMERCAPTAN; TERT-BUTYLMERCAPTAN; TERT-BUTYLTHIOL; 1,1-DIMETHYLETHANETHIOL; 2-ISOBUTANETHIOL; 2-METHYL-2-PROPANETHIOL; 2-PROPANETHIOL,2-METHYL-

Description: colorless liquid; heavy skunk odor

Use: odorant, intermediate; bacterial nutrient; as an odorant for natural gas so that leaks can be readily detected

Physical Properties

Boiling Point: -63.7 °C (-83 °F) at 760 mm Hg
Freezing Point: -0.5 °C (31.1 °F)
Specific Gravity: 0.79426 at 25 deg/4 °C
Vapor Density: 3.1 Air=1
Saturated Vapor Density: 1.803015789 kg/m^3
Vapor Pressure: 181 mm Hg at 25 °C
Water Solubility: Slightly Soluble in Water
Other Solubilities: Soluble in Heptane.
Odor Threshold: 0.001 ppm
Refraction Index: 1.41984 at 25 °C/D
Ionization Potential (eV): 9.03
Flash Point: -28.89 °C

RTECS Toxicity Data

Acute Oral: Rat LD_{50} Dose: 4729 mg/kg; Toxic Effects: Behavioral - Somnolence (general depressed activity); Lungs, Thorax, or Respiration - Respiratory depression; Behavioral - Coma.

Acute Inhalation: Rat LC_{50} Dose: 22200 ppm/4hr; Toxic Effects: Behavioral - Muscle weakness; Behavioral - Ataxia; Lungs, Thorax, or Respiration - Cyanosis. Mouse LC_{50} Dose: 16500 ppm/4hr; Toxic Effects: Behavioral - Muscle weakness; Behavioral - Ataxia; Lungs, Thorax, or Respiration - Cyanosis.

Chronic (Multiple Dose) Inhalation: Rat Dose: 201 ppm/6H/2W-I; Toxic Effects: Liver - Changes in Liver weight; Kidney, Ureter, and Bladder - Changes in kidney weight; Endocrine - Changes in spleen weight.

Hazard Overviews

Flammable

Health: Irritating to eyes/skin/respiratory tract. Other Acute Effects: harmful if inhaled or swallowed; nausea; headache; vomiting.

Fire: Flammable. Hazards: emits toxic fumes; vapor may travel considerable distance to source of ignition and flash back; container explosion may occur; forms explosive mixtures in air. Extinguishing agents: carbon dioxide, dry chemical powder or appropriate foam; water may be effective for cooling, but may not effect extinguishment. Precautions: combustible liquid.

Reactivity: Incompatible with: bases, oxidizing agents, reducing agents, alkali metals. Hazardous decomposition products: toxic fumes of: carbon monoxide, carbon dioxide, sulfur oxides, hydrogen sulfide gas.

Carcinogenicity: IARC - Not listed; NIOSH - Not listed; NTP - Not listed; ACGIH - Not listed; OSHA - Not listed; EPA - Not listed; MAK - Not listed

Primary Target Organs:

Eyes

Skin

Respiratory System

Environmental

Environmental Fate: If released to the atmosphere, it will exist mainly in the vapor phase in the ambient atmosphere based on a vapor pressure of 181 mm Hg at 25 °C. In the vapor-phase, it will react rapidly with hydroxyl radicals with half-lives of 11.0 to 13.3 hours determined from environmental rate constants. An estimated K_{oc} of 350 suggests that this compound will have moderate mobility in soil; natural gas plus the compound passed through a montmorillonite clay gave 85% of the influent odorant in the effluent within the first 100 standard cubic feet of gas. It should volatilize from dry soil surfaces based on its vapor pressure. Given an estimated Henry's Law constant of 6.11 $x10^{-3}$ atm-cu m/mole it should volatilize from moist soil surfaces. In water, based on this Henry's Law constant, it is expected to rapidly volatilize from water surfaces; volatilization half-lives from a model river and model lake were calculated as 2.9 hours and 3.8 days, respectively. This compound is not expected to bioconcentrate in aquatic organisms based on an estimated BCF value of 25.

Cleanup/Disposal: Guide No. 130: Eliminate all ignition sources (no smoking, flares, sparks or flames in immediate area). All equipment used when handling the product must be grounded. Do not touch or walk through spilled material. Stop leak if you can do it without risk. Prevent entry into waterways, sewers, basements or confined areas. A vapor suppressing foam may be used to reduce vapors. Absorb or cover with dry earth, sand or other non-combustible material and transfer to containers. Use clean non-sparking tools to collect absorbed material. Large Spills: Dike far ahead of liquid spill for later disposal. Water spray may reduce vapor; but may not prevent ignition in closed spaces.

Environmental Physical Data

Henry's Law Constant: estimated at 6.11 $x10^{-3}$
Octanol/Water Partition Coefficient: log K_{ow} = 2.14
Sorption Partition Coefficient: K_{oc} = estimated at 350
BCF: estimated at 25

Regulations

RCRA 40CFR: Not listed
CERCLA: 40CFR 302.4: Not listed

SARA 40CFR 372.65: Not listed
SARA EHS 40CFR 355: Not listed
TSCA: Listed

BUT5340 CAS #: 97-88-1

N-BUTYL METHACRYLATE

RTECS: OZ3675000
DOT: UN2227; IMO3.3
EINECS Number: 202-615-1
Molecular Formula: $C_8H_{14}O_2$
Structured MF: $CH_2=C(CH_3)COOCH_2CH_2CH_2CH_3$
Formula Weight: 142.20

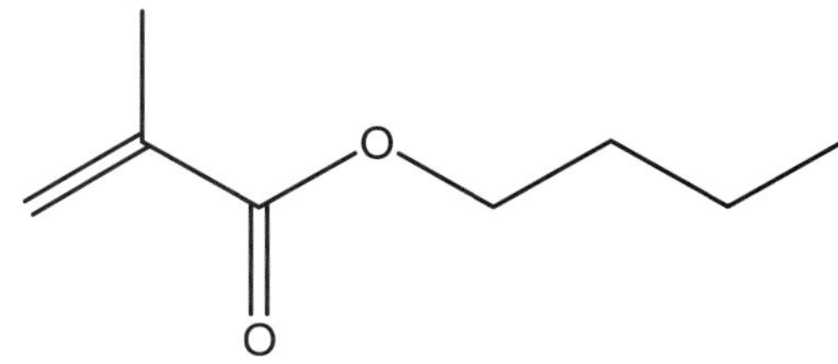

Chemical Structure

Synonyms: BUTIL METACRILATO; BUTYL 2-METHACRYLATE; BUTYL METHACRYLATE; BUTYL 2-METHYL-2-PROPENOATE; BUTYLESTER KYSELINY METHAKRYLOVE; BUTYLMETHACRYLAAT; BUTYL-2-METHYL-2-PROPENOATE; METHACRYLATE DE BUTYLE; METHACRYLIC ACID,BUTYL ESTER; METHACRYLIC ACID,ETHYL ESTER; METHACRYLSAEUREBUTYLESTER; 2-METHYL-BUTYLACRYLAAT; 2-METHYL-BUTYLACRYLAT; 2-METHYL-BUTYLACRYLATE; 2-METHYL-2-PROPENOIC ACID,BUTYL ESTER; 2-PROPENOIC ACID,2-METHYL-,BUTYL ESTER

Description: clear liquid; ester odor

Use: manufacture of methacrylic resins, solvent coatings, adhesives, oil additives emulsions for textiles, leather and paper finishing

Physical Properties

Boiling Point: 160 °C (320 °F) at 760 mm Hg
Freezing Point: < -75 °C (-103 °F)
Specific Gravity: 0.8936 at 20 °C/4 °C
Vapor Density: 4.8 Air=1
Saturated Vapor Density: 1.230200363 kg/m^3
Density: 0.89 g/mL
Vapor Pressure: 4.9 mm Hg at 20 °C
Water Solubility: Insoluble in Water
Other Solubilities: 95% Ethanol: Soluble; DMSO: Insoluble; Ether: Very Soluble.
Surface Tension: 35 dynes/cm
Refraction Index: 1.4240 at 20 °C/D
Flash Point: 18.889 °C Cleveland Open Cup
Autoignition Temperature: 294 °C
LEL: 8% v/v
UEL: Estimated at 8% v/v

RTECS Toxicity Data

Acute Oral: Rat LD_{50} Dose: 16 gm/kg. Mouse LD_{50} Dose: 12900 mg/kg.

Acute Inhalation: Rat LC_{50} Dose: 4910 ppm/4hr; Toxic Effects: Sense organs and special senses - Other; Sense organs and special senses - Other; Lungs, Thorax, or Respiration - Dyspnea.

Acute Dermal: Rabbit LD_{50} Route: Skin; Dose: 11300 uL/kg.

Chronic (Multiple Dose) Oral: Rat Dose: 107 gm/kg/17W-I; Toxic Effects: Lungs, Thorax, or Respiration - Changes in lung weight; Liver - Changes in liver weight; Nutritional and gross metabolic - Weight loss or decreased weight gain.

Reproductive/Teratogenic: Rat Route: Intraperitoneal; Dose: 2304 mg/kg; Duration: female 5-15D of pregnancy; Effects on Fertility - Post-implantation mortality. Rat Route: Intraperitoneal; Dose: 690 mg/kg; Duration: female 5-15D of pregnancy; Effects on Embryo or Fetus - Fetotoxicity; Specific Developmental Abnormalities - Other developmental abnormalities.

Hazard Overviews

Flammable

Fire Diamond

Health: Irritating to eyes/skin/respiratory tract. Also Causes: nausea, pulmonary edema (fluid in lungs), sensitization, watering of eyes or conjunctivitis. Chronic Effects: allergic dermititis, acute pulmonary edema, or asthma.

Fire: Flammable. Can form explosive mixtures in the air. Do not extinguish fire unless flow can be stopped. For small fires use dry chemical, carbon dioxide, water spray, or alcohol-resistant foam. For large fires use water spray, fog, or alcohol-resistant foam. Be aware that solid water streams may spread burning liquid. Remove cylinder from fire.

Reactivity: Stable. Hazardous polymerization can occur in contact with heat, moisture, or oxidizers unless stabilized. Incompatible with: heat; moisture; peroxides; oxidizers. Hazardous decomposition products: carbon dioxide; acrid smoke; irritating fumes.

Carcinogenicity: IARC - Not listed; NIOSH - Not listed; NTP - Not listed; ACGIH - Not listed; OSHA - Not listed; EPA - Not listed; MAK - Not listed

Primary Target Organs:

Eyes

Skin

Respiratory System

Environmental

Environmental Fate: If released to water, it should evaporate into the atmosphere with a volatilization half-life of approximately 13 hours from a model river. In water, some may also partition to sediment. In soil, its K_{oc} of about 878 suggests low soil mobility. No data are available on other potential degradation processes in soil or water. If released to the atmosphere, vapor phase is degraded relatively rapidly by reaction with photochemically produced hydroxyl radicals (estimated half-life of 7.5 hr in air).

Cleanup/Disposal: Guide No. 129: Eliminate all ignition sources (no smoking, flares, sparks or flames in immediate area). All equipment used when handling the product must be grounded. Do not touch or walk through spilled material. Stop leak if you can do it without risk. Prevent entry into waterways, sewers, basements or confined areas. A vapor suppressing foam may be used to reduce vapors. Absorb or cover with dry earth, sand or other non-combustible material and transfer to containers. Use clean non-sparking tools to collect absorbed material. Large Spills: Dike far ahead of liquid spill for later disposal. Water spray may reduce vapor; but may not prevent ignition in closed spaces.

Environmental Physical Data

Henry's Law Constant: estimated at 1.09×10^{-4}
Octanol/Water Partition Coefficient: log K_{ow} = 2.88
Sorption Partition Coefficient: K_{oc} = approximately 878
BCF: estimated at 91

Regulations

RCRA 40CFR: Not listed
CERCLA: 40CFR 302.4: Not listed
SARA 40CFR 372.65: Not listed
SARA EHS 40CFR 355: Not listed
TSCA: Listed

BUT5410 CAS #: 1634-04-4

T-BUTYL METHYL ETHER

RTECS: KN5250000
DOT: UN2398; IMO3.2
EINECS Number: 216-653-1
Molecular Formula: $C_5H_{12}O$
Structured MF: $(CH_3)_3COCH_3(CH_3)_3COCH_3$
Formula Weight: 88.15

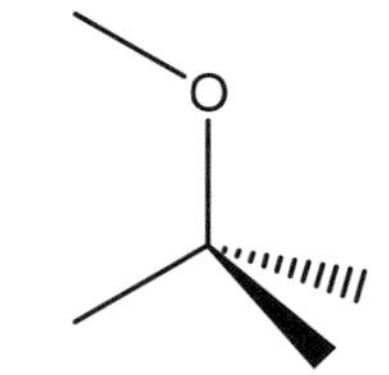

Chemical Structure

Synonyms: TERT-BUTYL METHYL ETHER; ETHER,TERT-BUTYL METHYL; 2-METHOXY-2-METHYL PROPANE; 2-METHOXY-2-METHYLPROPANE; METHYL TERT-BUTYL ETHER; METHYL 1,1-DIMETHYLETHYL ETHER; METHYL-TERT-BUTYL ETHER; 2-METHYL-2-METHOXYPROPANE; MTBE; PROPANE,2-METHOXY-2-METHYL-; PROPANE,2-METHOXY-2-METHYL-(9CI)

Description: clear, colorless liquid; mild mint and hydrocarbon odor

Use: octane booster in gasoline; manufacture of isobutene

Physical Properties

Boiling Point: 55.2 °C (131 °F)
Freezing Point: -109 °C (-164.2 °F)
Specific Gravity: 0.7405 at 20 °C/4 °C
Vapor Density: Calculated 3.0 Air=1
Saturated Vapor Density: 1.989024501 kg/m^3
Bulk Density: 6.18 lbs/gal
Vapor Pressure: 245 mm Hg at 25 °C
Water Solubility: Solubility of Water in methyl-tbutyl ether 2 g/100 g
Refraction Index: 1.3690 at 20 °C/D
Critical Temperature: 224 °C
Critical Pressure: 520 psia
Ionization Potential (eV): 9.24
Flash Point: -26 °C Closed Cup
Autoignition Temperature: 435 °C
LEL: 1.5% v/v

RTECS Toxicity Data

Acute Oral: Rat LD_{50} Dose: 4 gm/kg. Mouse LD_{50} Dose: 5960 uL/kg.

Acute Inhalation: Rat LC_{50} Dose: 23576 ppm/4hr. Mouse LC_{50} Dose: 141 gm/m^3/15M; Toxic Effects: Behavioral - General anesthetic.

Hazard Overviews

Flammable

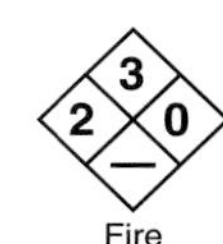

Fire Diamond

Health: Irritating to eyes/skin/respiratory tract. Also Causes: CNS and respiratory depression, aspiration pneumonitis. Chronic Effects: nasal/tracheal inflammation.

Fire: Extremely flammable. Can form explosive mixtures in the air. Use dry chemical, carbon dioxide, halon, water spray, or alcohol foam.

Reactivity: Stable. Hazardous polymerization cannot occur. Avoid: heat; ignition sources. Incompatible with: strong oxidizing agents; strong acids; caustics; amines; aldehydes; ammonia; chlorinated compounds. Hazardous decomposition products: carbon dioxide; water vapor; carbon monoxide; t-butyl formate; acetone, formic acid; methyl radicals.

Carcinogenicity: IARC - Group 2B, Possibly carcinogenic to humans; NIOSH - Not listed; NTP - Not listed; ACGIH - Class A3, Animal carcinogen; OSHA - Not listed; EPA - Not listed; MAK - Not listed

Primary Target Organs:

Eyes

Skin

Respiratory System

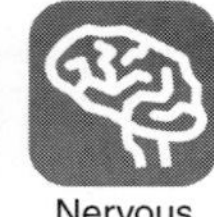
Nervous System

Respirator Recommendation

Exposure Range: >40 to 2000 ppm Supplied Air, Constant Flow/Pressure Demand, Half Mask

Exposure Range: >2000 to 40,000 ppm Supplied Air, Constant Flow/Pressure Demand, Full Face

Exposure Range: >40,000 to unlimited ppm Self-contained Breathing Apparatus, Pressure Demand, Full Face
Note: odor threshold unknown

Environmental

Environmental Fate: If released to soil, it will be subject to volatilization. It will be expected to exhibit very high mobility in soil and, therefore, it may leach to groundwater. It will not be expected to hydrolyze in soil. If released to water, it will not be expected to significantly adsorb to sediment or suspended particulate matter, bioconcentrate in aquatic organisms, hydrolyze, directly photolyze, or photooxidize via reaction with photochemically produced hydroxyl radicals in the water, based upon estimated physical-chemical properties or analogies to other structurally related aliphatic ethers. In surface water it will be subject to rapid volatilization with estimated half-lives of 4.1 hr and 2.0 days for volatilization from a river one meter deep flowing 1 m/sec with a wind velocity of 3 m/sec and a model pond, respectively. It may be resistant to biodegradation in environmental media based upon screening test data from a study using activated sludge inocula. Many ethers are known to be resistant to biodegradation. If released to the atmosphere, it will be expected to exist almost entirely in the vapor phase based on its vapor pressure. It will be susceptible to photooxidation via vapor phase reaction with photochemically produced hydroxyl radicals with an estimated half-life of 5.6 days for this process. Direct photolysis will not be an important removal process since aliphatic ethers do not adsorb light at wavelengths >290 nm.
Cleanup/Disposal: Guide No. 127: Eliminate all ignition sources (no smoking, flares, sparks or flames in immediate area). All equipment used when handling the product must be grounded. Do not touch or walk through spilled material. Stop leak if you can do it without risk. Prevent entry into waterways, sewers, basements or confined areas. A vapor suppressing foam may be used to reduce vapors. Absorb or cover with dry earth, sand or other non-combustible material and transfer to containers. Use clean non-sparking tools to collect absorbed material. Large Spills: Dike far ahead of liquid spill for later disposal. Water spray may reduce vapor; but may not prevent ignition in closed spaces.

Environmental Physical Data

Henry's Law Constant: 5.87 x10^{-4}
Sorption Partition Coefficient: K_{oc} = estimated at 11.2
BCF: carp 1.5

Regulations

RCRA 40CFR: Not listed
CERCLA: 40CFR 302.4: Listed per CAA Section 112 RQ: 1000 lb (453.5 kg)
SARA 40CFR 372.65: Listed
SARA EHS 40CFR 355: Not listed
TSCA: Listed

Analytical Methods

Soil: EPA 80APP-F2
Indoor / Expired Air: NIOSH 1615

BUT5480 **CAS #: 614-45-9**

T-BUTYL PERBENZOATE

RTECS: SD9450000
EINECS Number: 210-382-2
Molecular Formula: $C_{11}H_{14}O_3$
Formula Weight: 194.25

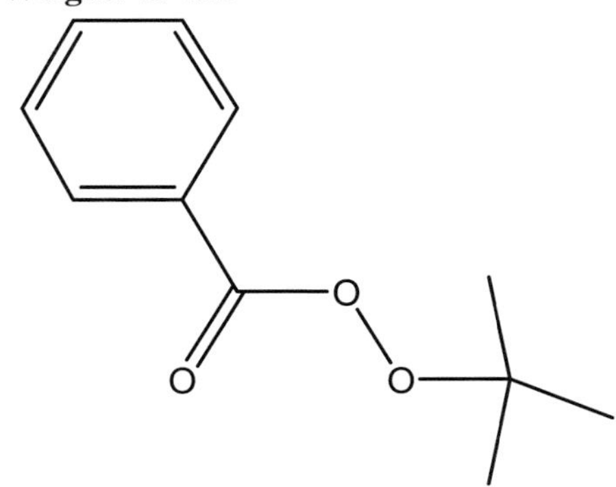

Chemical Structure

Synonyms: BENZENECARBOPEROXOIC ACID,1,1-DIMETHYLETHYL ESTER; BENZOYL TERT-BUTYL PEROXIDE; TERT-BUTYL PERBENZOATE; T-BUTYL PEROXY BENZOATE; TERT-BUTYL PEROXYBENZOATE; CHALOXYD TBPB; ESPEROX 10; NOVOX; PERBENZOATE DE BUTYLE TERTIAIRE; PERBENZOIC ACID,T-BUTYL ESTER; PERBUTYL Z; PEROXYBENZOIC ACID,T-BUTYL ESTER; PEROXYBENZOIC ACID,TERT-BUTYL ESTER; TERC BUTYLPERBENZOAN; TERC.BUTYLESTER KYSELINY PEROXYBENZOOVE; TERC.BUTYLPERBENZOAN; TRIGONOX C
Description: light yellow liquid; mild aromatic odor
Use: polymerization initiator for polyethylene, polystyrene, polyacrylates and polyesters; chemical intermediate

Physical Properties

Boiling Point: > 112 °C (234 °F)
Freezing Point: 8.5 °C (47.3 °F)
Specific Gravity: 1.04 at 25 °C/25 °C
Vapor Density: 6.7 Air=1
Density: 1.02 g/mL
Vapor Pressure: 0.33 mm Hg at 50 °C
Water Solubility: Insoluble in Water
Other Solubilities: 95% Ethanol: >100mg/ml at 20 °C; Acetone: >100mg/ml at 20 °C; DMSO: >100mg/ml at 20 °C; Esters: Soluble; Ketones: Soluble; Organic solvents: Soluble.
Refraction Index: 1.499
Flash Point: 93.333 °C

RTECS Toxicity Data

Acute Oral: Rat LD_{50} Dose: 1012 mg/kg. Mouse LD_{50} Dose: 914 mg/kg; Toxic Effects: Behavioral - Muscle weakness; Lungs, Thorax, or Respiration - Dyspnea; Gastrointestinal - Necrotic changes.
Acute Inhalation: Rat LC; Dose: >57 mg/m^3/4hr. Mouse LC; Dose: >57 mg/m^3/4hr.

Chronic (Multiple Dose) Oral: Rat Dose: 32500 mg/kg/13W-I; Toxic Effects: Gastrointestinal - Other changes; Endocrine - Changes in spleen weight; DEATH - Changes in testicular weight. Rat Dose: 32500 mg/kg/13W-I; Toxic Effects: Gastrointestinal - Other changes; Endocrine - Changes in spleen weight.
Chronic (Multiple Dose) Inhalation: Rat Dose: 6 mg/m^3/4H; Toxic Effects: Kidney, Ureter, and Bladder - Urine volume decreased; Kidney, Ureter, and Bladder - Other changes in urine composition.
Irritation Eye: Rabbit Standard Draize Test Dose: 500 mg/24H; Reaction: mild. Rabbit Nonstandard Exposure Dose: 100 mg/1M rinse; Reaction: mild.
Irritation Skin: Rabbit Standard Draize Test Dose: 500 mg/24H; Reaction: mild.
Mutagenic: Bacteria - S Typhimurium Mutations in Microorganisms; Dose: 67 ug/plate (-S9).

Hazard Overviews

Explosive

Flammable

Fire Diamond

Health: Irritating to eyes/skin/respiratory tract. Also Causes: headache; upon ingestion: GI discomfort, splashes in the eyes may also cause burns.
Fire: Flammable and can be explosive when heated. Strong oxidizer. Use dry chemical, carbon dioxide, water spray, or regular foam.
Reactivity: Stable. Hazardous polymerization cannot occur. Avoid: heat; flames; organic and combustible materials. Incompatible with: organic substances; copper bromide and limonene; cobalt naphthenate; dimethylaniline. Hazardous decomposition products: toxic smoke; flames.
Carcinogenicity: IARC - Not listed; NIOSH - Not listed; NTP - Not listed; ACGIH - Not listed; OSHA - Not listed; EPA - Not listed; MAK - Not listed
Primary Target Organs:

Eyes

Skin

Respiratory System

Mucous Membranes

Environmental

Cleanup/Disposal: Guide No. 146: Eliminate all ignition sources (no smoking, flares, sparks or flames in immediate area). Keep combustibles (wood, paper, oil, etc.) away from spilled material. Do not touch damaged containers or spilled material unless wearing appropriate protective clothing. Keep substance wet using water spray. Stop leak if you can do it without risk. Small Spills: Take up with inert, damp, noncombustible material using clean non-sparking tools and place into loosely covered plastic containers for later disposal. Large Spills: Wet down with water and dike for later disposal. Prevent entry into waterways, sewers, basements or confined areas. Do not clean-up or dispose of, except under supervision of a specialist.

Regulations

RCRA 40CFR: Not listed
CERCLA: 40CFR 302.4: Not listed
SARA 40CFR 372.65: Not listed
SARA EHS 40CFR 355: Not listed
TSCA: Listed

BUT5550 CAS #: 110-05-4

T-BUTYL PEROXIDE

RTECS: ER2450000
DOT: UN2102; UN2255; UN2899; NA9183
EINECS Number: 203-733-6
Molecular Formula: $C_8H_{18}O_2$
Formula Weight: 146.22

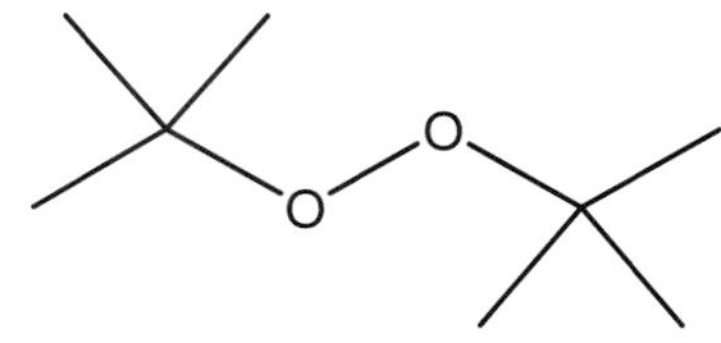

Chemical Structure

Synonyms: BIS(TERT-BUTYL) PEROXIDE; BIS(T-BUTYL)PEROXIDE; BIS(1,1-DIMETHYLETHYL)PEROXIDE; TERT-BUTYL PEROXIDE; CADOX; CADOX TBP; DI-T-BUTYL PEROXIDE; DI-TERT-BUTYL PEROXIDE; DI-TERT-BUTYL PEROXYDE; DI-TERT-BUTYLPEROXID; DTBP; INTEROX DTB; KAYABUTYL D; PERBUTYL D; PEROSSIDO DI BUTILE TERZIARIO; PEROXIDE,BIS(1,1-DIMETHYLETHYL); PEROXIDE,BIS(1,1-DIMETHYLETHYL)-; PEROXYDE DE BUTYLE TERTIAIRE; (TRIBUTYL)PEROXIDE; TRIGONOX B
Description: clear liquid
Use: polymerization catalyst and in organic synthesis

Physical Properties

Boiling Point: 111 °C (232 °F)
Freezing Point: -40 °C (-40 °F)
Specific Gravity: 0.794 at 20 °C/4 °C
Vapor Pressure: 0.751 kPa at 0 °C
Water Solubility: About 0.01%
Other Solubilities: 95% Ethanol: >=100 mg/ml at 21 °C; Acetone: >=100 mg/ml at 21 °C; DMSO: 10-50 mg/ml at 21 °C; Ligroin: Soluble; Styrene: Soluble.
Refraction Index: 1.3890 at 20 °C/D
Ionization Potential (eV): 8.4 +/-0.7
Flash Point: 18.333 °C Open Cup

RTECS Toxicity Data

Acute Oral: Rat LD_{50} Dose: >25 gm/kg. Mouse LD_{50} Dose: 4572 mg/kg.
Acute Inhalation: Rat LC_{50} Dose: >4100 ppm/4hr. Mouse LC_{50} Dose: >4103 ppm/4hr; Toxic Effects: Behavioral - Excitment; Lungs, Thorax, or Respiration - Dyspnea.

Chronic (Multiple Dose) Oral: Rat Dose: 105 gm/kg/7W-I; Toxic Effects: Nutritional and gross metabolic - Weight loss or decreased weight gain; DEATH.
Irritation Eye: Rabbit Standard Draize Test Dose: 500 mg/24H; Reaction: mild. Rabbit Nonstandard Exposure Dose: 200 mg/1M rinse; Reaction: mild.
Reproductive/Teratogenic: Rat Route: Inhalation; Dose: 226 mg/m^3/4H Duration: female 1-19D of pregnancy; Effects on Embryo or Fetus - Fetotoxicity.

Hazard Overviews

Corrosive

Explosive

Flammable

Fire Diamond

Health: Corrosive, causes severe burns to eyes/skin/respiratory tract. Also Causes: headaches, drunkenness, pulmonary edema, angina, grippe, acute respiratory disease, pneumonia.
Fire: Flammable. Explosive. Oxidizer. For small fires use dry chemical, carbon dioxide, water spray, or regular foam. For large fires flood area with water. Fight fire from maximum distance. Foam may be necessary if material is diluted with a low density flammable solvent.
Reactivity: Unstable, shock sensitive, and thermally decomposes above 200 °F (93 °C), process is exothermic and self-ignition may result, especially if oxygen is present. Hazardous polymerization cannot occur. Avoid: heat; ignition sources; shock; friction; reducing materials. Incompatible with: reducing material. Hazardous decomposition products: acetone vapors; ethane gas.
Carcinogenicity: IARC - Not listed; NIOSH - Not listed; NTP - Not listed; ACGIH - Not listed; OSHA - Not listed; EPA - Not listed; MAK - Not listed
Primary Target Organs:

Eyes

Skin
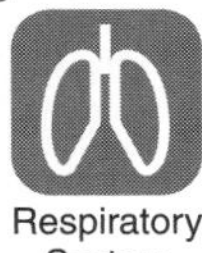
Respiratory System

Environmental

Environmental Fate: If released into the environment it will photodegrade, possibly over the course of several days. Little is known of its behavior in aquatic or terrestrial systems.
Cleanup/Disposal: Guide No. 145: Eliminate all ignition sources (no smoking, flares, sparks or flames in immediate area). Keep combustibles (wood, paper, oil, etc.) away from spilled material. Do not touch damaged containers or spilled material unless wearing appropriate protective clothing. Keep substance wet using water spray. Stop leak if you can do it without risk. Small Spills: Take up with inert, damp, noncombustible material using clean non-sparking tools and place into loosely covered plastic containers for later disposal. Large Spills: Wet down with water and dike for later disposal. Prevent entry into waterways, sewers, basements or confined areas. Guide No. 146: Eliminate all ignition sources (no smoking, flares, sparks or flames in immediate area). Keep combustibles (wood, paper, oil, etc.) away from spilled material. Do not touch damaged containers or spilled material unless wearing appropriate protective clothing. Keep substance wet using water spray. Stop leak if you can do it without risk. Small Spills: Take up with inert, damp, noncombustible material using clean non-sparking tools and place into loosely covered plastic containers for later disposal. Large Spills: Wet down with water and dike for later disposal. Prevent entry into waterways, sewers, basements or confined areas. Do not clean-up or dispose of, except under supervision of a specialist. Guide No. 148: Eliminate all ignition sources (no smoking, flares, sparks or flames in immediate area). Keep combustibles (wood, paper, oil, etc.) away from spilled material. Do not touch or walk through spilled material. Stop leak if you can do it without risk. Small Spills: Take up with inert, damp, noncombustible material using clean non-sparking tools and place into loosely covered plastic containers for later disposal. Large Spills: Dike far ahead of liquid spill for later disposal. Prevent entry into waterways, sewers, basements or confined areas. Do not clean-up or dispose of, except under supervision of a specialist.

Environmental Physical Data

Octanol/Water Partition Coefficient: log K_{ow} = 1 to 4

Regulations

RCRA 40CFR: Not listed
CERCLA: 40CFR 302.4: Not listed
SARA 40CFR 372.65: Not listed
SARA EHS 40CFR 355: Not listed
TSCA: Listed

BUT5620 CAS #: 123-95-5

N-BUTYL STEARATE

RTECS: WI2900000
EINECS Number: 204-666-5
Molecular Formula: $C_{22}H_{44}O_2$
Structured MF: $C_{17}H_{35}COOC_4H_9$
Formula Weight: 340.57

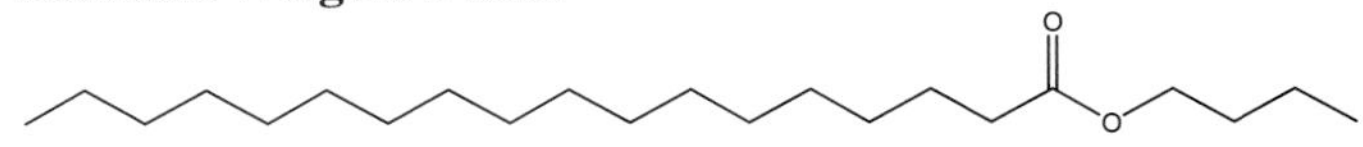

Chemical Structure

Synonyms: APEX 4; BS; BUTYL OCTADECANOATE; N-BUTYL OCTADECANOATE; BUTYL OCTADECYLATE; BUTYL STEARATE; EMEREST 2325; ESTREX 1B 54,1B 55; GROCO 5810; KESSCO BSC; KESSCOFLEX BS; OCTADECANOIC ACID,BUTYL ESTER; OCTADECANOIC ACID,BUTYL ESTER (9CI); POLYCIZER 332; RC PLASTICIZER B-17; STARFOL BS-100; STEARIC ACID,BUTYL ESTER; TEGESTER BUTYL STEARATE; UNIFLEX BYS; WICKENOL 122; WILMAR BUTYL STEARATE; WITCIZER 200; WITCIZER 201
Description: colorless or very pale yellow waxy liquid above 20 deg C; odorless or faintly fatty fruity odor
Use: direct food additive, solvent for flavors--butter, banana; plasticizer for laminated fiber products, rubber hydrochloride,

chlorinated rubber, cable lacquers, cellulose acetate butyrate, polystyrene, etc; lubricant in extrusion & molding of polyvinyl chloride; emollient in creams, lotions, & lipsticks; spreading & softening agent in plastics, textiles, cosmetics, & rubber; in mfr of leather cloth & varnishes; in lubricants for metals; in wax polishes; in carbon paper & inks; water proofing agent, eg, for concrete; in propellants; defoamer.

Physical Properties

Boiling Point: 343 °C (649 °F)
Freezing Point: 27.5 °C (81.5 °F)
Specific Gravity: 0.855 at 20 °C/4 °C
Vapor Pressure: 0.001 kPa at 100 °C
Water Solubility: 0% at 25 °C
Other Solubilities: Soluble in Ether; Soluble in Mineral or vegetable oils.
Refraction Index: 1.4328 at 50 °C/D
Flash Point: 160 °C Closed Cup
Autoignition Temperature: 671 °C

RTECS Toxicity Data

Acute Oral: Rat LD_{50} Dose: 32 gm/kg.
Acute Dermal: Rat LD_{50} Route: Subcutaneous Dose: >32 gm/kg.
Irritation Skin: Rabbit Standard Draize Test Dose: 500 mg; Reaction: moderate.
Reproductive/Teratogenic: Rat Route: Oral; Dose: 418 gm/kg; Duration: male 10W prior to mating; Effects on Newborn - Growth statistics.

Hazard Overviews

Health: Irritating to eyes/skin/respiratory tract. Other Acute Effects: may be harmful by inhalation, ingestion, or skin absorption.The toxicological properties have not been thoroughly investigated.
Fire: Will burn. Hazards: emits toxic fumes. Extinguishing agents: water spray; carbon dioxide, dry chemical powder or appropriate foam. Precautions: combustible liquid.
Reactivity: Stable. Hazardous polymerization will not occur. Hazardous decomposition products: toxic fumes of: carbon monoxide, carbon dioxide.
Carcinogenicity: IARC - Not listed; NIOSH - Not listed; NTP - Not listed; ACGIH - Not listed; OSHA - Not listed; EPA - Not listed; MAK - Not listed
Primary Target Organs:

Eyes

Skin

Respiratory System

Environmental

Environmental Fate: If released to the atmosphere, it will degrade by reaction with photochemically produced hydroxyl radicals (estimated half-life of 15 hours). If released to soil or water, it is expected to biodegrade. Aqueous hydrolysis will be important only in very alkaline environmental media (pH > 9). An estimated K_{oc} of 3.92 x10^5 suggests that it will not leach in soil and will partition to sediment in water. Volatilization from water will be important in the absence of strong adsorption to sediment and bioconcentration is expected to be an important fate process.

Environmental Physical Data

Henry's Law Constant: estimated at 0.03818
Octanol/Water Partition Coefficient: log K_{ow} = 10.2
Sorption Partition Coefficient: K_{oc} = estimated at 3.92 x10^5
BCF: estimated at 3.3 x10^7

Regulations

RCRA 40CFR: Not listed
CERCLA: 40CFR 302.4: Not listed
SARA 40CFR 372.65: Not listed
SARA EHS 40CFR 355: Not listed
TSCA: Listed

BUT5690 CAS #: 7521-80-4

BUTYL TRICHLOROSILANE

RTECS: VV2080000
DOT: UN1747; IMO8.0
EINECS Number: 231-381-3
Molecular Formula: $C_4H_9Cl_3Si$
Structured MF: $CH_3CH_2CH_2CH_2SiCl_3$
Formula Weight: 191.56

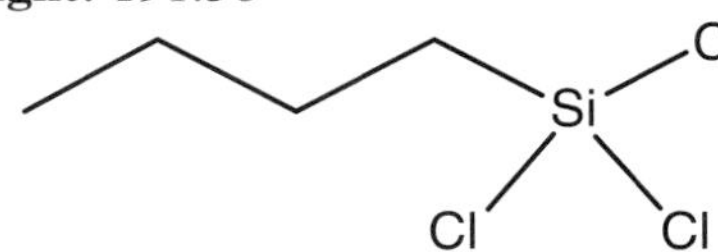

Chemical Structure

Synonyms: N-BUTYL TRICHLOROSILANE; BUTYLSILICON TRICHLORIDE; BUTYLTRICHLOROSILANE; SILANE,BUTYLTRICHLORO-; TRICHLOROALLYLSILANE; TRICHLOROBUTYLSILANE
Description: colorless liquid
Use: chem int for silicones

Physical Properties

Boiling Point: 148.9 °C (300 °F)
Freezing Point: 54 °C (129.2 °F)
Specific Gravity: 1.1606 at 20 °C/4 °C
Vapor Density: 6.4 Air=1
Vapor Pressure: 9.17 kPa at 75 °C
Water Solubility: Decomposes in Water
Other Solubilities: Soluble in Heptane.
Surface Tension: Estimated at 25 dynes/cm
Odor Threshold: as Hydrogen Cloride 0.255 to 10.6 ppm
Refraction Index: 1.4363 at 20 °C/D
Flash Point: 54 °C Open Cup

Hazard Overviews

Health: Corrosive to eyes/skin/respiratory tract. Other Acute Effects: harmful if swallowed, inhaled, or absorbed through skin; inhalation may result in spasm, inflammation and edema of the larynx and bronchi, chemical pneumonitis and pulmonary edema; symptoms of exposure may include burning sensation; coughing; wheezing; laryngitis; shortness of breath; headache; nausea; vomiting.
Fire: Flammable. Hazards: emits toxic fumes. Extinguishing agents: carbon dioxide, dry chemical powder or appropriate foam. Precautions: combustible liquid.
Reactivity: Incompatible with: strong oxidizing agents, reacts violently with water. Hazardous decomposition products: toxic fumes of: carbon monoxide, carbon dioxide, hydrogen chloride gas, silicon oxide.
Carcinogenicity: IARC - Not listed; NIOSH - Not listed; NTP - Not listed; ACGIH - Not listed; OSHA - Not listed; EPA - Not listed; MAK - Not listed
Primary Target Organs:

Environmental

Cleanup/Disposal: Guide No. 155: Eliminate all ignition sources (no smoking, flares, sparks or flames in immediate area). All equipment used when handling the product must be grounded. Do not touch damaged containers or spilled material unless wearing appropriate protective clothing. Stop leak if you can do it without risk. A vapor suppressing foam may be used to reduce vapors. For chlorosilanes, use AFFF alcohol-resistant medium expansion foam to reduce vapors. Do not get water on spilled substance or inside containers. Use water spray to reduce vapors or divert vapor cloud drift. Prevent entry into waterways, sewers, basements or confined areas. Small Spills: Cover with dry earth, dry sand, or other non-combustible material followed with plastic sheet to minimize spreading or contact with rain. Use clean non-sparking tools to collect material and place it into loosely covered plastic containers for later disposal.

Regulations

RCRA 40CFR: Not listed
CERCLA: 40CFR 302.4: Not listed
SARA 40CFR 372.65: Not listed
SARA EHS 40CFR 355: Not listed
TSCA: Listed

BUT5760 CAS #: 111-34-2

BUTYL VINYL ETHER

RTECS: KN5950000
DOT: UN2352; IMO3.2
EINECS Number: 203-860-7
Molecular Formula: $C_6H_{12}O$
Formula Weight: 100.16

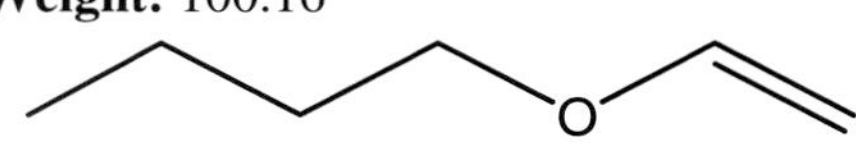

Chemical Structure

Synonyms: AGRISYNTH BVE; BUTANE,1-(ETHENYLOXY)-; BUTIL VINIL ETER; BUTOXYETHENE; N-BUTYL VINYL ETHER; BUTYL VINYL ETHER,INHIBITED; BVE; ETHER BUTYLVINYLIQUE; ETHER,BUTYL VINYL; VINYL BUTYL ETHER; VINYL N-BUTYL ETHER
Description: liquid
Use: in the synthesis of copolymers

Physical Properties

Boiling Point: 93.8 °C (201 °F)
Freezing Point: -92 °C (-133.6 °F)
Specific Gravity: 0.7803 at 20 °C/20 °C
Vapor Density: 1 Air=1
Bulk Density: 7.45 lbs/gal at 20 °C
Vapor Pressure: 1.76 kPa at 0 °C
Water Solubility: Slightly Soluble in Water
Other Solubilities: Soluble in Acetone, Benzene.
Refraction Index: 1.4026 at 20 °C
Flash Point: -9 °C Open Cup
Autoignition Temperature: 255 °C

RTECS Toxicity Data

Acute Oral: Rat LD_{50} Dose: 10 gm/kg.
Acute Inhalation: Mouse LC_{50} Dose: 62 gm/m^3/2hr. Rat LC_{Lo} Dose: 16000 ppm/4hr.
Acute Dermal: Rabbit LD_{50} Route: Skin; Dose: 4240 uL/kg.
Irritation Eye: Rabbit Standard Draize Test Dose: 500 mg/24H; Reaction: mild.
Irritation Skin: Rabbit Open Draize Test Dose: 500 mg open; Reaction: mild.

Hazard Overviews

Health: Irritating to eyes/skin/respiratory tract. Other Acute Effects: may be harmful by inhalation, ingestion, or skin absorption; prolonged exposure can cause narcotic effect.
Fire: Flammable. Hazards: vapor may travel considerable distance to source of ignition and flash back; container explosion may occur; forms explosive mixtures in air. Extinguishing agents: carbon dioxide, dry chemical powder

or appropriate foam; water may be effective for cooling, but may not effect extinguishment. Precautions: combustible liquid.

Reactivity: Incompatible with: oxidizing agents, strong acids. Hazardous decomposition products: toxic fumes of: carbon monoxide, carbon dioxide.

Carcinogenicity: IARC - Not listed; NIOSH - Not listed; NTP - Not listed; ACGIH - Not listed; OSHA - Not listed; EPA - Not listed; MAK - Not listed

Primary Target Organs:

Eyes

Skin

Respiratory System

Environmental

Environmental Fate: If released to soil, it will be subject to volatilization. It will be expected to exhibit high mobility in soil and, therefore, it may leach to groundwater. It may hydrolyze in soil based upon data for hydrolysis in water. If released to water, it will not be expected to significantly adsorb to sediment or suspended particulate matter, bioconcentrate in aquatic organisms, or directly photolyze, based upon estimated physical-chemical properties or analogies to other structurally related aliphatic ethers. It will be susceptible to appreciable hydrolysis in certain environmental waters, especially at acidic pH, with calculated half-lives for hydrolysis of 9.5 hr, 40 days, and 10.9 year at pH 5, 7, and 9, respectively. In surface water, it will be subject to rapid volatilization with estimated half-lives for volatilization of 3.3 hr and 42 hr for volatilization from a river one meter deep flowing 1 m/sec with a wind velocity of 3 m/sec and a model pond, respectively. It is unknown whether it biodegrades in environmental media. If released to the atmosphere, it will be expected to exist almost entirely in the vapor phase based upon the vapor pressure. It will be susceptible to photooxidation via vapor phase reaction with photochemically produced hydroxyl radicals and ozone with an overall atmospheric half-life of 9 hours for these processes based upon estimated rate constants. It will not be expected to photolyze in the atmosphere.

Cleanup/Disposal: Guide No. 127: Eliminate all ignition sources (no smoking, flares, sparks or flames in immediate area). All equipment used when handling the product must be grounded. Do not touch or walk through spilled material. Stop leak if you can do it without risk. Prevent entry into waterways, sewers, basements or confined areas. A vapor suppressing foam may be used to reduce vapors. Absorb or cover with dry earth, sand or other non-combustible material and transfer to containers. Use clean non-sparking tools to collect absorbed material. Large Spills: Dike far ahead of liquid spill for later disposal. Water spray may reduce vapor; but may not prevent ignition in closed spaces.

Environmental Physical Data

Henry's Law Constant: calculated at 2.24 x10^{-3}
Sorption Partition Coefficient: K_{oc} = estimated at 53
BCF: estimated at 6.8

Regulations

RCRA 40CFR: Not listed
CERCLA: 40CFR 302.4: Not listed
SARA 40CFR 372.65: Not listed
SARA EHS 40CFR 355: Not listed
TSCA: Listed

Analytical Methods

Water / Groundwater: ASTM D3695

BUT5830 CAS #: 91-49-6

N-BUTYLACETANILIDE

RTECS: AD9800000
EINECS Number: 202-071-5
Molecular Formula: $C_{12}H_{17}NO$
Formula Weight: 191.28
Synonyms: ACETAMIDE,N-BUTYL-N-PHENYL-; ACETAMIDE,N-BUTYL-N-PHENYL-(9CI); ACETANILIDE,N-BUTYL-; ACETIC ACID,AMIDE,N-BUTYL-N-PHENYL; BAA; BUTYLACETANILIDE; N-BUTYL-N-PHENYLACETAMIDE
Use: a repellant for control of ticks & fleas; a repellant to impregnate clothing

Physical Properties

Boiling Point: 281 °C (538 °F) at 760 mm Hg
Freezing Point: 24.5 °C (76.1 °F)
Specific Gravity: 0.9912 at 20 °C/4 °C
Water Solubility: Insoluble in Water
Other Solubilities: Chloroform: Slightly soluble
Refraction Index: 1.5146 at 20 °C/D
Flash Point: 141.111 °C Closed Cup

RTECS Toxicity Data

Acute Oral: Mouse LD_{50} Dose: 800 mg/kg; Toxic Effects: Behavioral - Somnolence (general depressed activity). Guinea Pig LD_{50} Dose: 300 mg/kg.

Hazard Overviews

Fire: Will burn.
Carcinogenicity: IARC - Not listed; NIOSH - Not listed; NTP - Not listed; ACGIH - Not listed; OSHA - Not listed; EPA - Not listed; MAK - Not listed

Environmental

Regulations

RCRA 40CFR: Not listed
CERCLA: 40CFR 302.4: Not listed
SARA 40CFR 372.65: Not listed
SARA EHS 40CFR 355: Not listed
TSCA: Listed

BUT5900 CAS #: 68969-35-7

BUTYLALLYLMALEATE

EINECS Number: 273-466-8

Physical Properties

Boiling Point: 304.4 °C (580 °F)
Specific Gravity: 1.0345
Vapor Density: < 8.0 Air=1
Vapor Pressure: < 20
Water Solubility: < 0%

Hazard Overviews

Carcinogenicity: IARC - Not listed; NIOSH - Not listed; NTP - Not listed; ACGIH - Not listed; OSHA - Not listed; EPA - Not listed; MAK - Not listed

Environmental

Regulations

RCRA 40CFR: Not listed
CERCLA: 40CFR 302.4: Not listed
SARA 40CFR 372.65: Not listed
SARA EHS 40CFR 355: Not listed
TSCA: Not listed

BUT5970 CAS #: 541-35-5

N-BUTYLAMIDE

EINECS Number: 208-776-4
Molecular Formula: C_4H_9NO
Formula Weight: 87.12

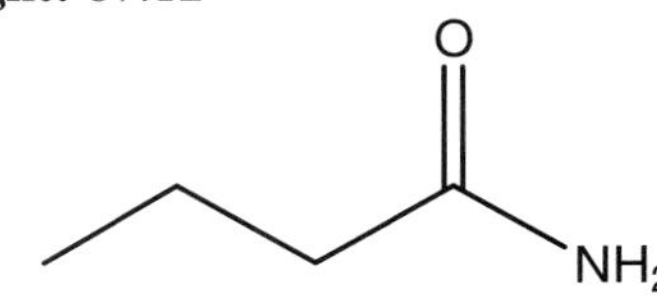

Chemical Structure

Synonyms: BUTANAMIDE; BUTANIMIDIC ACID; BUTANOIC ACID,AMIDE; BUTYRAMIDE; N-BUTYRAMIDE
Description: crystals or leaves

Physical Properties

Boiling Point: 216 °C (421 °F)
Freezing Point: 114.8 °C (238.64 °F)
Specific Gravity: 0.885 at 120 °C
Water Solubility: Soluble in Water
Other Solubilities: Insoluble in Benzene.
Refraction Index: 1.4087 at 130 °C/D

Hazard Overviews

Health: May cause irritation. Other Acute Effects: may be harmful by inhalation, ingestion, or skin absorption.
Fire: Extinguishing agents: water spray; carbon dioxide, dry chemical powder or appropriate foam. Precautions: combustible liquid.
Reactivity: Incompatible with: strong oxidizing agents, acids, bases, strong reducing agents. Hazardous decomposition products: toxic fumes of: carbon monoxide, carbon dioxide, nitrogen oxides.
Carcinogenicity: IARC - Not listed; NIOSH - Not listed; NTP - Not listed; ACGIH - Not listed; OSHA - Not listed; EPA - Not listed; MAK - Not listed

Environmental

Environmental Physical Data

Octanol/Water Partition Coefficient: log K_{ow} = calculated at -0.21

Regulations

RCRA 40CFR: Not listed
CERCLA: 40CFR 302.4: Not listed
SARA 40CFR 372.65: Not listed
SARA EHS 40CFR 355: Not listed
TSCA: Listed

BUT6040 CAS #: 109-73-9

N-BUTYLAMINE

RTECS: EO2975000
DOT: UN1125; IMO3.2
EINECS Number: 203-699-2
Molecular Formula: $C_4H_{11}N$
Structured MF: $CH_3(CH_2)_3NH_2$
Formula Weight: 73.14

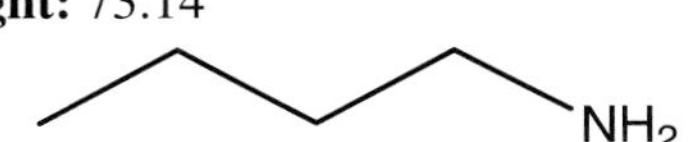

Chemical Structure

Synonyms: 1-AMINO-BUTAAN; 1-AMINOBUTAN; 1-AMINOBUTANE; 1-BUTANAMINE; N-BUTILAMINA; N-BUTYLAMIN; BUTYLAMINE; MONOBUTILAMINA; MONO-N-BUTYLAMINE; MONOBUTYLAMINE; NORRALAMINE; NORVALAMINE; TUTANE
Description: colorless, yellow liquid; ammonia-like odor
Use: intermediate for pharmaceuticals, dyestuffs, rubber chemicals, emulsifying agents, insecticides, synthetic tanning agents

Physical Properties

Boiling Point: 77.8 °C (172 °F) at 760 mm Hg
Freezing Point: -50 °C (-58 °F)
Specific Gravity: 0.7414 at 20 °C/4 °C
Vapor Density: 2.5 Air=1
Density: 0.721 g/cc at 22.5 °C
Vapor Pressure: 82 mm Hg
Water Solubility: Miscible with Water

Other Solubilities: 95% Ethanol: Soluble (>=10 mg/ml); Acetone: Very Soluble; Benzene: Very Soluble; Ether: Very Soluble.
Surface Tension: 19.7 dynes/cm
Odor Threshold: 1.8 ul/l
Refraction Index: 1.4010 at 20 °C/D
Evaporation Rate: 7.3 Butyl Acetate=1
Critical Temperature: 251 °C
Critical Pressure: 603 psia
Ionization Potential (eV): 8.71
Flash Point: -12 °C Closed Cup
Autoignition Temperature: 312 °C
LEL: 1.7% v/v
UEL: 9.8% v/v

RTECS Toxicity Data

Acute Oral: Rat LD_{50} Dose: 366 mg/kg; Toxic Effects: Behavioral - Convulsions or effect on seizure threshold; Behavioral - Ataxia; Lungs, Thorax, or Respiration - Respiratory depression. Mouse LD_{50} Dose: 430 mg/kg.
Acute Inhalation: Mouse LC_{50} Dose: 800 mg/m^3/2hr. Rat LC_{Lo} Dose: 4000 ppm/4hr.
Acute Dermal: Rabbit LD_{50} Route: Skin; Dose: 850 uL/kg. Guinea Pig LD_{50} Route: Skin; Dose: 500 uL/kg.
Irritation Eye: Rabbit Standard Draize Test Dose: 250 ug/24H; Reaction: severe.
Irritation Skin: Rabbit Open Draize Test Dose: 500 mg open; Reaction: severe.
Mutagenic: Rat Cytogenetic Analysis; Route: Oral; Dose: 110 mg/kg.
Tumorigenic: Mouse Route: Intraperitoneal; Dose: 800 mg/kg; Toxic Effects: Tumorigenic - Equivocal tumorigenic agent by RTECS criteria; Lungs, Thorax, or Respiration - Tumors; Skin and appendages - Tumors.

Hazard Overviews

Corrosive

Flammable

Fire Diamond

Health: Corrosive to eyes/skin/respiratory tract. Also Causes: headache, pulmonary edema.
Fire: Flammable. Can form explosive mixtures in air. Use dry chemical, alcohol foam, or carbon dioxide to fight fires. Use a water spray to cool fire-exposed containers, to disperse the vapor of an unignited leak, and protect personnel.
Reactivity: Stable. Hazardous polymerization cannot occur. Avoid: ignition sources. Incompatible with: strong oxidizers; strong acids. Hazardous decomposition products: oxides of nitrogen; carbon monoxide.
Carcinogenicity: IARC - Not listed; NIOSH - Listed as carcinogen; NTP - Not listed; ACGIH - Not listed; OSHA - Not listed; EPA - Not listed; MAK - Not listed

Primary Target Organs:

Eyes

Skin

Respiratory System

Exposure Limits
OSHA PEL: STEL: 5 ppm; 15 mg/m^3; skin.
ACGIH TLV: STEL: 5 ppm; 15 mg/m^3; Ceiling.
NIOSH REL: STEL: 5 ppm; 15 mg/m^3; skin.
NIOSH IDLH: 300 ppm.
DFG MAK: TWA: 5 ppm; 15 mg/m^3.

Environmental

Ecotoxicity: Aquatic toxicity: 30-70 ppm/24 hr/creek chub/critical range/fresh water
Environmental Fate: If released to the soil, it will not adsorb to the soil, and it will be expected to leach rapidly to the groundwater. Hydrolysis will not be a significant removal process. No information on biodegradation in soils or groundwater were found but screening studies suggest that biodegradation may be important. Evaporation from the surface of soils and other surfaces may be a significant removal process. If released to water, it will not be expected to adsorb to the sediment, hydrolyze or bioconcentrate in aquatic organisms. Based on very limited data from laboratory tests of biodegradation with activated sludges and settled sewage seed, it may possibly be subject to biodegradation in natural waters. Evaporation will be expected to be an important removal process with a half-life of 52 hours predicted for evaporation from a river 1 m deep, flowing at 1 m/sec with a wind velocity of 3 m/sec. No information was found on photolysis, but it should not absorb sunlight. Reaction with hydroxyl radicals will be the fastest chemical removal process in the atmosphere (estimated half-life of 5.26 days). Dissolution into rain droplets may be the most important physical removal process in the atmosphere.
Cleanup/Disposal: Guide No. 132: Fully encapsulating, vapor protective clothing should be worn for spills and leaks with no fire. Eliminate all ignition sources (no smoking, flares, sparks or flames in immediate area). All equipment used when handling the product must be grounded. Do not touch or walk through spilled material. Stop leak if you can do it without risk. Prevent entry into waterways, sewers, basements or confined areas. A vapor suppressing foam may be used to reduce vapors. Absorb with earth, sand or other non-combustible material and transfer to containers (except for Hydrazine). Use clean non-sparking tools to collect absorbed material. Large Spills: Dike far ahead of liquid spill for later disposal. Water spray may reduce vapor; but may not prevent ignition in closed spaces.

Environmental Physical Data

Henry's Law Constant: 1.5 x10^{-5}
Octanol/Water Partition Coefficient: log K_{ow} = 0.97
Sorption Partition Coefficient: K_{oc} = estimated at 80
BCF: estimated at 3.2
BOD: theoretical 26.5%, 5 days

Regulations

RCRA 40CFR: Not listed
CERCLA: 40CFR 302.4: Listed per CWA Section 311(b)(4) RQ: 1000 lb (453.5 kg)
SARA 40CFR 372.65: Not listed
SARA EHS 40CFR 355: Not listed
TSCA: Listed

Analytical Methods

Air: ASTM D4490
Indoor / Expired Air: NIOSH 2012

BUT6180 CAS #: 13952-84-6

SEC-BUTYLAMINE

RTECS: EO3325000
EINECS Number: 237-732-7
Molecular Formula: $C_4H_{11}N$
Structured MF: $CH_3CHNH_2C_2H_5$
Formula Weight: 73.14

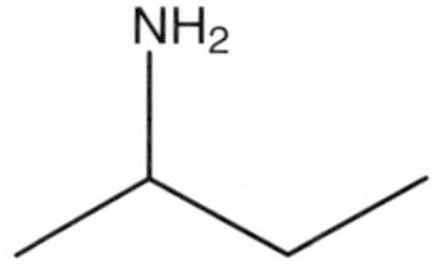

Chemical Structure

Synonyms: 2-AB; (RS)-2-AMINOBUTANE; 2-AMINOBUTANE; 2-AMINOBUTANE BASE; BUTAFUME; 2-BUTANAMINE; (RS)-SEC-BUTYLAMINE; DECCOTANE; FRUCOTE; 1-METHYLPROPYLAMINE; PROPYLAMINE,1-METHYL; TUTANE
Description: colorless liquid; ammoniacal odor
Use: fungicide, intermediate in chemical synthesis, solvent

Physical Properties

Boiling Point: 63 °C (145 °F)
Freezing Point: -104 °C (-155.2 °F)
Specific Gravity: 0.724 at 20 °C/4 °C
Vapor Density: 2.52 Air=1
Saturated Vapor Density: 1.524441016 kg/m^3
Density: 0.72 g/mL
Vapor Pressure: 135 mm Hg at 20 °C
Water Solubility: Miscible with Water
Other Solubilities: 95% Ethanol: Soluble (>=10 mg/ml); Acetone: Very Soluble; Ether: Very Soluble.
Surface Tension: 22.42 dynes/cm at 20 °C
Odor Threshold: 0.24 ppm
Refraction Index: 1.394 at 20 °C/D
Ionization Potential (eV): 8.70
Flash Point: -8.889 °C
Autoignition Temperature: 378 °C
LEL: 1.7% v/v
UEL: 9.8% v/v

RTECS Toxicity Data

Acute Oral: Rat LD_{50} Dose: 152 mg/kg; Toxic Effects: Behavioral - Ataxia; Behavioral - Antipsychotic; Lungs, Thorax, or Respiration - Dyspnea. Dog LD_{50} Dose: 225 mg/kg.
Acute Dermal: Rabbit LD_{50} Route: Skin; Dose: 2500 mg/kg.

Hazard Overviews

Corrosive

Flammable

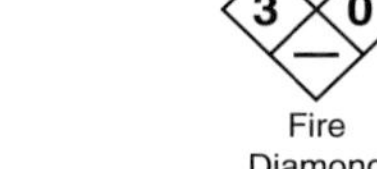

Fire Diamond

Health: Corrosive to eyes/skin/respiratory tract. Toxic. Other Acute Effects: harmful if swallowed, inhaled, or absorbed through skin; inhalation may result in spasm, inflammation and edema of the larynx and bronchi, chemical pneumonitis and pulmonary edema; prolonged contact can cause severe irritation or burns.
Fire: Flammable. Hazards: vapor may travel considerable distance to source of ignition and flash back; container explosion may occur; emits toxic fumes; forms explosive mixtures in air. Extinguishing agents: carbon dioxide, dry chemical powder or appropriate foam; water may be effective for cooling, but may not effect extinguishment. Precautions: combustible liquid.
Reactivity: Incompatible with: acids, acid chlorides, acid anhydrides, strong oxidizing agents, carbon dioxide, sensitive to light. Hazardous decomposition products: thermal decomposition may produce carbon monoxide, carbon dioxide, and nitrogen oxides.
Carcinogenicity: IARC - Not listed; NIOSH - Not listed; NTP - Not listed; ACGIH - Not listed; OSHA - Not listed; EPA - Not listed; MAK - Not listed
Primary Target Organs:

Eyes

Skin

Respiratory System

Exposure Limits
DFG MAK: TWA: 5 ppm; 15 mg/m^3.

Environmental

Ecotoxicity: LC_{50} Bluegill (young) more than 50 mg/l
Cleanup/Disposal: Guide No. 131: Fully encapsulating, vapor protective clothing should be worn for spills and leaks with no fire. Eliminate all ignition sources (no smoking, flares, sparks or flames in immediate area). All equipment used when handling the product must be grounded. Do not touch or walk through spilled material. Stop leak if you can do it without risk. Prevent entry into waterways, sewers, basements or confined areas. A vapor suppressing foam may be used to reduce vapors. Small Spills: Absorb with earth, sand or other non-combustible material and transfer to containers for later disposal. Use clean non-sparking tools to collect absorbed material. Large Spills: Dike far ahead of liquid spill for later

disposal. Water spray may reduce vapor; but may not prevent ignition in closed spaces.

Environmental Physical Data

Octanol/Water Partition Coefficient: log K_{ow} = 0.74

Regulations

RCRA 40CFR: Not listed
CERCLA: 40CFR 302.4: Listed per CWA Section 311(b)(4) RQ: 1000 lb (453.5 kg)
SARA 40CFR 372.65: Not listed
SARA EHS 40CFR 355: Not listed
TSCA: Listed

BUT6250 CAS #: 75-64-9

TERT-BUTYLAMINE

RTECS: EO3330000
EINECS Number: 200-888-1
Molecular Formula: $C_4H_{11}N$
Structured MF: $(CH_3)_3CNH_2$
Formula Weight: 73.14

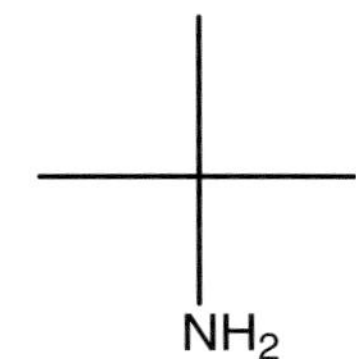

Chemical Structure

Synonyms: 2-AMINOISOBUTANE; 2-AMINO-2-METHYLPROPANE; T-BUTYLAMINE; BUTYLAMINE,TERTIARY; 1,1-DIMETHYLAMINOMETHANE; 1,1-DIMETHYLETHYLAMINE; 2-METHYL-2-AMINOPROPANE; 2-METHYL-2-PROPANAMINE; 2-PROPANAMINE,2-METHYL-; TERTIARY-BUTYLAMINE; TRIMETHYL AMINOMETHANE; TRIMETHYLAMINOMETHANE
Description: colorless liquid; ammonia odor
Use: intermediate for rubber accelerators, insecticides, fungicides, dyestuffs, pharmaceuticals

Physical Properties

Boiling Point: 44 °C (111 °F) to 46 °C (115 °F)
Freezing Point: -72.65 °C (-98.77 °F)
Specific Gravity: 0.6951 at 20 °C/4 °C
Vapor Density: 2.5 Air=1
Saturated Vapor Density: 2.069982577 kg/m^3
Density: 0.7 g/mL
Vapor Pressure: 362 mm Hg at 25 °C
Water Solubility: Very Soluble in Water
Other Solubilities: 95% Ethanol: Very Soluble; DMSO: Soluble; Ether: Very Soluble.
Surface Tension: 19 dynes/cm at 20 °C
Odor Threshold: 0.24 ppm
Refraction Index: 1.37 at 20 °C/D
Ionization Potential (eV): 8.64
Flash Point: 10 °C
Autoignition Temperature: 380 °C
LEL: 1.7% v/v
UEL: 8.9% v/v

RTECS Toxicity Data

Acute Oral: Rat LD_{50} Dose: 78 mg/kg; Toxic Effects: Behavioral - Ataxia; Behavioral - Antipsychotic; Lungs, Thorax, or Respiration - Dyspnea.
Acute Inhalation: Rat LC_{50} Dose: 3800 mg/m^3/4hr; Toxic Effects: Sense organs and special senses - Ptosis; Lungs, Thorax, or Respiration - Dyspnea; Blood - Hemorrhage. Man TC_{Lo} Dose: 40 mg/m^3/8H-I; Toxic Effects: Sense organs and special senses - Visual field changes; Sense organs and special senses - Corneal damage; Sense organs and special senses - Other.
Acute Dermal: Rabbit LD_{Lo} Route: Skin; Dose: 2 gm/kg; Toxic Effects: Behavioral - Somnolence (general depressed activity); Behavioral - Food intake (animal); Gastrointestinal - Other changes.

Hazard Overviews

Corrosive

Flammable

Fire Diamond

Health: Highly irritating to eyes/skin/respiratory tract. Toxic. Also Causes: difficulty breathing, pulmonary edema, possible corneal damage and visual field distortion.
Fire: Flammable. Can form explosive mixtures in air. Use carbon dioxide, water spray as fog, or regular foam. Do not use water spray in direct stream. Do not extinguish fire unless flow can be stopped. Fight fire from maximum distance. Container may explode in heat of fire.
Reactivity: Stable. Hazardous polymerization cannot occur. Avoid: exposure to heat; and ignition sources. Incompatible with: 2,2-dibrom-1,3-dimethylcyclopropanoic acid; some plastics. Hazardous decomposition products: carbon and nitrogen oxides.
Carcinogenicity: IARC - Not listed; NIOSH - Not listed; NTP - Not listed; ACGIH - Not listed; OSHA - Not listed; EPA - Not listed; MAK - Not listed
Primary Target Organs:

Eyes

Skin

Respiratory System

Exposure Limits
DFG MAK: TWA: 5 ppm; 15 mg/m^3.

Environmental

Environmental Fate: If released to soil, it may be susceptible to significant leaching. It has been detected in a leachate from a municipal waste disposal site. It may be expected to evaporate quite rapidly from dry impervious surfaces. Sufficient data are not available to predict the significance of biodegradation in soil or natural water. The results of several screening studies suggest that it may be relatively resistant to environmental biodegradation or may undergo significant biodegradation under properly acclimated conditions. If released to water, it is not expected to significantly adsorb to sediments or to bioconcentrate. The volatilization half-life from a river one meter deep flowing 1 m/sec with a wind velocity of 3 m/sec is estimated to be 1.95 days. The volatilization half-life from a similar river 10 m deep is estimated to be 23.5 days. If released to air, it should exist almost entirely in the vapor-phase due to its high vapor pressure. The half-life for the vapor-phase reaction with photochemically produced hydroxyl radicals has been estimated to be 1.07 months in a typical atmosphere. Due to the complete water solubility, physical removal from the atmosphere by washout or by dissolution into clouds with subsequent rainfall may be possible.

Cleanup/Disposal: Guide No. 131: Fully encapsulating, vapor protective clothing should be worn for spills and leaks with no fire. Eliminate all ignition sources (no smoking, flares, sparks or flames in immediate area). All equipment used when handling the product must be grounded. Do not touch or walk through spilled material. Stop leak if you can do it without risk. Prevent entry into waterways, sewers, basements or confined areas. A vapor suppressing foam may be used to reduce vapors. Small Spills: Absorb with earth, sand or other non-combustible material and transfer to containers for later disposal. Use clean non-sparking tools to collect absorbed material. Large Spills: Dike far ahead of liquid spill for later disposal. Water spray may reduce vapor; but may not prevent ignition in closed spaces.

Environmental Physical Data

Henry's Law Constant: estimated at 1.66×10^{-5}
Octanol/Water Partition Coefficient: log K_{ow} = 0.40
BCF: estimated at 1.2

Regulations

RCRA 40CFR: Not listed
CERCLA: 40CFR 302.4: Listed per RCRA Section 3001 RQ: 1000 lb (453.5 kg)
SARA 40CFR 372.65: Not listed
SARA EHS 40CFR 355: Not listed
TSCA: Listed

BUT6320 CAS #: 3775-90-4

2-(TERT-BUTYLAMINO)ETHYL METHACRYLATE

RTECS: OZ3500000
EINECS Number: 223-228-4
Molecular Formula: $C_{10}H_{19}NO_2$
Formula Weight: 185.25

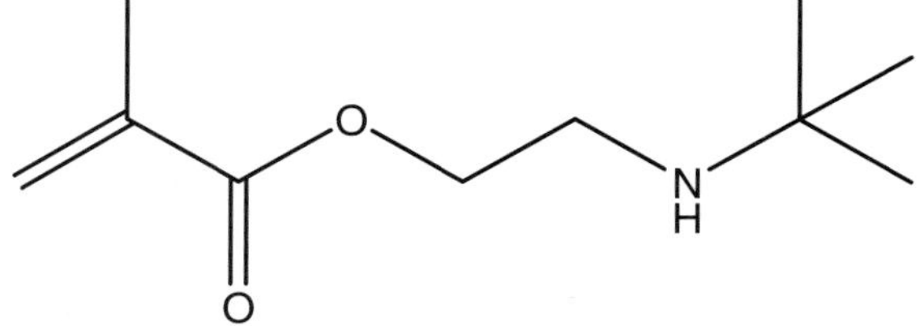

Chemical Structure

Synonyms: AGEFLEX FM-4; N-TERT-BUTYLAMINOETHYL METHACRYLATE; TERT-BUTYLAMINOETHYL METHACRYLATE; ETHANOL,2-(TERT-BUTYLAMINO)-,METHACRYLATE (ESTER); ETHANOL,2-(TERT-BUTYLAMINO)-METHACRYLATE (ESTER); METHACRYLIC ACID,2-(TERT-BUTYLAMINO)ETHYL ESTER; 2-PROPENOIC ACID,2-METHYL-,2-((1,1-DIMETHYLETHYL)AMINO)ETHYL ESTER

Description: liquid

Use: coatings; textile chemicals; dispersing agent for nonaqueous systems; antistatic agent; emulsifying agent; stabilizer for chlorinated polymers; ion exchange resins; cationic precipitating agent; monomer for acrylic resins for floor waxes

Physical Properties

Boiling Point: 100 °C (212 °F) to 105 °C (221 °F) at 12 mm Hg
Freezing Point: < -70 °C (-94 °F)
Specific Gravity: 0.914 at 25 °C
Vapor Density: 5.5 Air=1
Water Solubility: 1.8 g/100 ml Water at 25 °C
Other Solubilities: Chloroform: Soluble
Refraction Index: 1.4440 at 25 °C/D
Flash Point: 11 °C Cleveland Open Cup

Hazard Overviews

Flammable

Health: Irritating to eyes/skin/respiratory tract. Harmful. Other Acute Effects: harmful if swallowed; may be harmful if inhaled; may be harmful if absorbed through the skin.

Fire: Flammable. Hazards: emits toxic fumes; closed containers may rupture and explode during runaway polymerization. Extinguishing agents: carbon dioxide, dry chemical powder or appropriate foam. Precautions: combustible liquid.

Reactivity: Stable. Incompatible with: strong oxidizing agents, catalysts, carbon dioxide, rust, heavy metal salts, water. Hazardous decomposition products: toxic fumes of: carbon monoxide, carbon dioxide, nitrogen oxides.

Carcinogenicity: IARC - Not listed; NIOSH - Not listed; NTP - Not listed; ACGIH - Not listed; OSHA - Not listed; EPA - Not listed; MAK - Not listed

Primary Target Organs:

Eyes

Skin

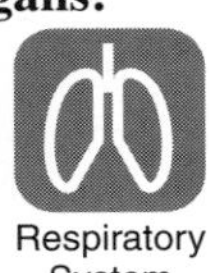
Respiratory System

Environmental

Regulations

RCRA 40CFR: Not listed
CERCLA: 40CFR 302.4: Not listed
SARA 40CFR 372.65: Not listed
SARA EHS 40CFR 355: Not listed
TSCA: Listed

BUT6390 CAS #: 95-31-8

2-(TERT-BUTYLAMINOTHIO)BENZOTHIAZOLE

RTECS: DL6200000
EINECS Number: 202-409-1
Molecular Formula: $C_{11}H_{14}N_2S_2$
Formula Weight: 238.4

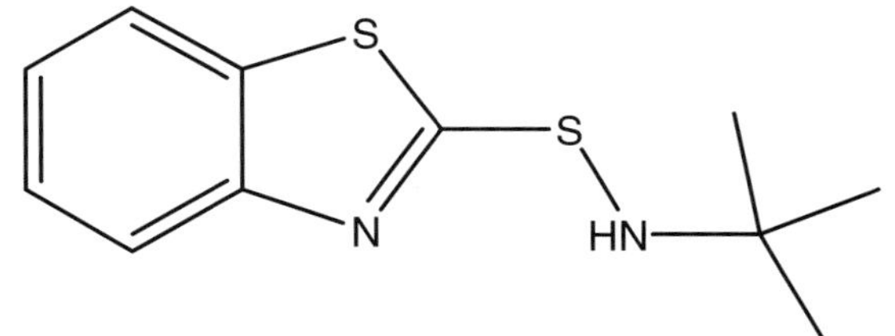

Chemical Structure

Synonyms: ACCEL BNS; 2-BENZOTHIAZOLESULFENAMIDE,N-TERT-BUTYL-; 2-BENZOTHIAZOLESULFENAMIDE,N-(1,1-DIMETHYLETHYL)-; BENZOTHIAZOLESULFENAMIDE,N-(1,1-DIMETHYLETHYL)-(9CI); BENZOTHIAZOLYL-2-TERT-BUTYLSULFENAMIDE; N-TERT-BUTYL-2-BENZOTHIAZOLESULFENAMIDE; N-TERT-BUTYL-2-BENZOTHIAZOLYL SULFENAMIDE; N-TERT-BUTYL-2-BENZOTHIAZYLSULFENAMIDE; N-(1,1-DIMETHYLETHYL)BENZOTHIAZOLESULFENAMIDE; NOCCELER NS; PENNAC TBBS; SANTOCURE NS; SANTOCURE NS VULCANIZATION ACCELERATOR; VANAX NS; VULKACIT NZ

Description: light buff powder or flakes, sometimes colored blue

Use: rubber accelerator; accelerator for natural & synthetic isoprene rubbers; accelerator for butadiene rubbers, eg, styrene-butadiene

Physical Properties

Freezing Point: 104 °C (219.2 °F)
Specific Gravity: 1.29 at 25 °C
Other Solubilities: Soluble in most organic solvents

RTECS Toxicity Data

Acute Oral: Rat LD_{Lo} Dose: 7940 mg/kg; Toxic Effects: Behavioral - Somnolence (general depressed activity); Behavioral - Food intake (animal); Behavioral - Muscle weakness.

Acute Dermal: Rabbit LD_{50} Route: Skin; Dose: >7940 mg/kg.

Mutagenic: Mouse Mutations in Microorganisms; Cell Type: lymphocyte; Dose: 40 mg/L (+S9). Mouse Morphological Transformation; Cell Type: embryo; Dose: 35 mg/L.

Hazard Overviews

Carcinogenicity: IARC - Not listed; NIOSH - Not listed; NTP - Not listed; ACGIH - Not listed; OSHA - Not listed; EPA - Not listed; MAK - Not listed

Environmental

Regulations

RCRA 40CFR: Not listed
CERCLA: 40CFR 302.4: Not listed
SARA 40CFR 372.65: Not listed
SARA EHS 40CFR 355: Not listed
TSCA: Listed

BUT6460 CAS #: 2008-41-5

BUTYLATE

RTECS: EZ7525000
EINECS Number: 217-916-3
Molecular Formula: $C_{11}H_{23}NOS$
Formula Weight: 217.41

Synonyms: ANELDA; ANELDAZINE; BIS(2-METHYLPROPYL)CARBAMOTHIOIC ACID S-ETHYL ESTER; BUTILATE; CARBAMIC ACID,DIISOBUTYLTHIO-,S-ETHYL ESTER; CARBAMOTHIOIC ACID,BIS(2-METHYLPROPYL)-,S-ETHYL ESTER; DIISOBUTYLTHIOCARBAMIC ACID S-ETHYL ESTER; DIISOCARB; S-ETHYL BIS(2-METHYLPROPYL) CARBAMOTHIOATE; S-ETHYL BIS(2-METHYLPROPYL)CARBAMOTHIOATE; S-ETHYL N,N-DIISOBUTYL THIOCARBAMATE; S-ETHYL N,N-DIISOBUTYL THIOLCARBAMATE; ETHYL N,N-DIISOBUTYLTHIOCARBAMATE; S-ETHYL DI-ISOBUTYL(THIOCARBAMATE); S-ETHYL DI-ISOBUTYLTHIOCARBAMATE; S-ETHYL DIISOBUTYLTHIOCARBAMATE; S-ETHYL N,N-DIISOBUTYLTHIOCARBAMATE; ETHYL N,N-DIISOBUTYLTHIOLCARBAMATE; S-ETHYL N,N-DIISOBUTYLTHIOLCARBAMATE; S-ETHYLDIISOBUTYL THIOCARBAMATE; ETHYL-N,N-DIISOBUTYL THIOLCARBAMATE; ETHYL-N,N-DIISOBUTYLTHIOCARBAMATE; GENATE; GENATE PLUS; R 1910; R-1910; R1910; STAUFFER R-1,910; STAUFFER R-1910; SUAZIN; SUTAN; SUTAN 6E; SUTAN +; SUTAR' 85 E; TOMAHAWK

Description: clear to amber liquid; aromatic odor

Use: selective preemergence; against annual grass weed species eg, barnyard grass crabgrass, foxtails & goosegrass; against nutsedges bermudagrass seedlings & johnsongrass seedlings; in maize, by pre-plant soil incorporation; effective control of perennials from seed eg, quackgrass; broadleaf weeds eg, lambsquarters, redroot pigweed, purslane, annual morning glory, florida purslane, & velvetleaf; well tolerated by corn.

Physical Properties

Boiling Point: 138 °C (280 °F) at 21.5 mm Hg
Specific Gravity: 0.9402 at 25 °C/25 °C
Saturated Vapor Density: 1.200133357 kg/m^3
Vapor Pressure: 13 x10^{-3} mm Hg at 25 °C
Water Solubility: 44 ppm in Water at 20 °C
Other Solubilities: miscible with 4-Methylpentan-2-one; miscible with Ethyl Alcohol at 20 °C; miscible with methyl isobutyl ketone.
Refraction Index: 1.4702 at 30 °C/D
Flash Point: 110 °C Tag Open Cup

RTECS Toxicity Data

Acute Oral: Rat LD_{50} Dose: 4 gm/kg.
Acute Inhalation: Mammal LC_{50} Dose: 19 gm/m^3/2hr.
Acute Dermal: Rabbit LD_{50} Route: Skin; Dose: >5 gm/kg.
Mutagenic: Mouse Cytogenetic Analysis; Route: Oral; Dose: 1 gm/kg.

Hazard Overviews

Fire: Will burn.
Carcinogenicity: IARC - Not listed; NIOSH - Not listed; NTP - Not listed; ACGIH - Not listed; OSHA - Not listed; EPA - Not listed; MAK - Not listed

Environmental

Ecotoxicity: LC_{50} Rainbow trout 4.2 mg/l/96 hr /Conditions of bioassay not specified LC_{50} Gammarus fasciatus 11 mg/l/96 hr at 15 °C (95% confidence limit 8-16 mg/l), mature. static bioassay without aeration, pH 7.2-7.5, water hardness 40-50 mg/l as calcium carbonate and alkalinity of 30-35 mg/l LC_{50} Bobwhite quail 40000 mg/kg diet/7 day
Environmental Fate: If released to the atmosphere, it will degrade rapidly in the vapor phase by reaction with photochemically produced hydroxyl radicals (half-life of about 12 hr). Its detection in rainwater samples indicates that physical removal from air via wet deposition can occur. If released to soil, it will dissipate primarily through microbial degradation and volatilization. Immediate soil incorporation can dramatically reduce volatilization. Soils previously exposed to it usually degrade it faster than soils having no prior exposure due to microbial acclimation. It has low to medium mobility in soil. Various persistence studies indicate that it can have a soil half-life ranging from about one to five weeks.

Environmental Physical Data

Henry's Law Constant: estimated at 8.45 x10^{-6}
Sorption Partition Coefficient: K_{OC} = estimated at 185 to 260
BCF: estimated at 73

Regulations

RCRA 40CFR: Listed Hazardous Waste No. U392 Toxic Waste
CERCLA: 40CFR 302.4: Listed per RCRA Section 3001 RQ: 1 lb (0.454 kg)
SARA 40CFR 372.65: Not listed
SARA EHS 40CFR 355: Not listed
TSCA: Listed

Analytical Methods

Soil: EPA PMD-BYA
Water / Groundwater: EPA 634
Drinking Water: EPA 507, 525.2; AOAC 991.07; ASTM D5475
Food: AOAC 974.05

BUT6530 CAS #: 25013-16-5

BUTYLATED HYDROXYANISOLE

RTECS: SL1945000
EINECS Number: 246-563-8
Molecular Formula: $C_{11}H_{16}O_2$
Structured MF: $CH_3OC_6H_3(OH)C(CH_3)_3$
Formula Weight: 180.24

O
HO

Chemical Structure

Synonyms: ANISOLE,BUTYLATED HYDROXY-; ANTIOXYNE B; ANTRANCINE 12; BHA; BOA; BOA (ANTIOXIDANT); 2(3)-TERT-BUTYL-4-HYDROXYANISOLE; BUTYLHYDROXYANISOLE; TERT-BUTYL-4-HYDROXYANISOLE; TERT-BUTYL-P-HYDROXYANISOLE; TERT-BUTYLHYDROXYANISOLE; 2-TERT-BUTYL-4-METHOXYPHENOL; TERT-BUTYL-4-METHOXYPHENOL; TERT-BUTYL-4-METHYLPHENOL; BUTYLOHYDROKSYANIZOL; (1,1-DIMETHYLETHYL)-4-METHOXYPHENOL; EEC NO. E320; EMBANOX; NEPANTIOX 1-F; NIPANTIOX 1-F; PHENOL,TERT-BUTYL-4-METHOXY-; PHENOL,(1,1-DIMETHYLETHYL)-4-METHOXY-; PREMERGE PLUS; PROTEX; SUSTAN 1-F; SUSTANE; SUSTANE 1-F; SUSTANE 1F; TENOX BHA; 2-TERC.BUTYL-4-METHOXYFENOL; VERTAC
Description: white or slightly yellow, waxy solid; faint characteristic odor
Use: antioxidant in fat-containing foods and in edible fats and oils; prevents food from becoming rancid and developing objectionable odors; preservative and antioxidant in cosmetic formulations

Physical Properties

Boiling Point: 264 °C (507 °F) to 270 °C (518 °F) at 733 mm Hg
Freezing Point: 48 °C (118.4 °F) to 55 °C (131 °F)
Water Solubility: Insoluble in Water
Other Solubilities: 1% in Glycerine at 100 °C; 1 g in 4 ml Alcohol, 2 ml Chloroform, 1.2 ml Ether.
Flash Point: 156 °C

RTECS Toxicity Data

Acute Oral: Rat LD_{50} Dose: 2 gm/kg; Toxic Effects: Behavioral - Altered sleep time (including change in righting reflex); Behavioral - Ataxia; Lungs, Thorax, or Respiration - Respiratory stimulation. Mouse LD_{50} Dose: 1100 mg/kg; Toxic Effects: Behavioral - Altered sleep time (including change in righting reflex); Behavioral - Ataxia; Lungs, Thorax, or Respiration - Respiratory stimulation.

Reproductive/Teratogenic: Rat Route: Oral; Dose: 30 gm/kg; Duration: male 2W prior to mating; Effects on Newborn - Growth statistics. Rat Route: Oral; Dose: 36 gm/kg; Duration: male 2W prior to mating; Effects on Newborn - Weaning or lactation index.

Mutagenic: Human DNA Inhibition; Cell Type: HeLa cell; Dose: 400 umol/L. Rat Unscheduled DNA Synthesis; Route: Oral; Dose: 176 gm/kg/21W-C.

Tumorigenic: Rat Route: Oral; Dose: 728 gm/kg/2Y-C; Toxic Effects: Tumorigenic - Carcinogenic by RTECS criteria; Gastrointestinal - Tumors. Rat Route: Oral; Dose: 182 gm/kg/2Y-C; Toxic Effects: Tumorigenic - Equivocal tumorigenic agent by RTECS criteria; Gastrointestinal - Tumors. Rat Route: Oral; Dose: 269 gm/kg/32W-C; Toxic Effects: Tumorigenic - Carcinogenic by RTECS criteria; Kidney, Ureter, and Bladder - Tumors; Tumorigenic - Cells (cultured) transformed. Rat Route: Oral; Dose: 874 gm/kg/2Y-C; Toxic Effects: Tumorigenic - Carcinogenic by RTECS criteria; Gastrointestinal - Tumors; Endocrine - Tumors. Rat Route: Oral; Dose: 4200 mg/kg/10W-C; Toxic Effects: Tumorigenic - Neoplastic by RTECS criteria; Lungs, Thorax, or Respiration - Tumors. Rat Route: Oral; Dose: 874 gm/kg/1Y-C; Toxic Effects: Tumorigenic - Carcinogenic by RTECS criteria; Gastrointestinal - Tumors. Rat Route: Oral; Dose: 202 gm/kg/24W-C; Toxic Effects: Tumorigenic - Neoplastic by RTECS criteria; Gastrointestinal - Tumors.

Hazard Overviews

Health: Irritating to eyes/skin/respiratory tract. Toxic. Other Acute Effects: harmful if swallowed, inhaled, or absorbed through skin; prolonged or repeated exposure may cause allergic reactions in certain sensitive individuals. Chronic Effects: target organ: liver. Carcinogen.

Fire: Will burn. Hazards: emits toxic fumes. Extinguishing agents: water spray; carbon dioxide, dry chemical powder or appropriate foam. Precautions: combustible liquid.

Reactivity: Incompatible with: strong oxidizing agents. Hazardous decomposition products: carbon monoxide, carbon dioxide.

Carcinogenicity: IARC - Group 2B, Possibly carcinogenic to humans; NIOSH - Not listed; NTP - Listed; ACGIH - Not listed; OSHA - Not listed; EPA - Not listed; MAK - Not listed

Primary Target Organs:

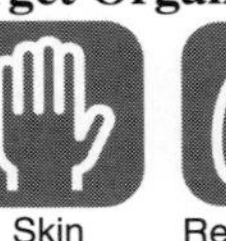

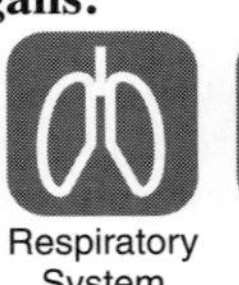

Environmental

Environmental Fate: If released to water, it will volatilize slowly. Adsorption to sediment and bioconcentration may be important fate processes. Insufficient data are available to determine the rate or importance of biodegradation in soil or aquatic conditions. If released to the atmosphere, it will exist primarily in the vapor phase and react with photochemically produced hydroxyl radicals with an estimated half-life of 10.7 hours. Reaction with nitrate radicals may also be important. Removal of atmospheric compound may occur through wet deposition. If released to soil, it is expected to have low soil mobility based on an estimated K_{oc} value of 1390. Volatilization may be slow from dry soils.

Environmental Physical Data

Henry's Law Constant: estimated at 1.17 $x10^{-6}$
Octanol/Water Partition Coefficient: log K_{ow} = 3.5
Sorption Partition Coefficient: K_{oc} = estimated at 1390
BCF: estimated at 269

Regulations

RCRA 40CFR: Not listed
CERCLA: 40CFR 302.4: Not listed
SARA 40CFR 372.65: Not listed
SARA EHS 40CFR 355: Not listed
TSCA: Listed

BUT6600 CAS #: 98-06-6

T-BUTYLBENZENE

RTECS: CY9120000
EINECS Number: 202-632-4
Molecular Formula: $C_{10}H_{14}$
Formula Weight: 134.21

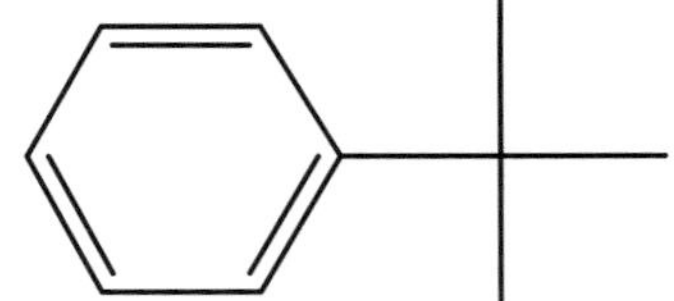

Chemical Structure

Synonyms: BENZENE,TERT-BUTYL-; BENZENE,(1,1-DIMETHYLETHYL)-; BENZENE,(1,1-DIMETHYLETHYL)-(9CI); TERT-BUTYLBENZENE; (1,1-DIMETHYLETHYL)BENZENE; DIMETHYLETHYLBENZENE; 2-METHYL-2-PHENYLPROPANE; PHENYLTRIMETHYLMETHANE; PSEUDOBUTYLBENZENE; TRIMETHYLPHENYLMETHANE

Description: colorless liquid; odorous

Use: in organic synthesis; polymerization solvent; polymer linking agent; solvent in prodn of hydrogen peroxide

Physical Properties

Boiling Point: 168.5 °C (335 °F) at 760 mm Hg
Freezing Point: -58.1 °C (-72.58 °F)
Specific Gravity: 0.8669 at 20 °C/4 °C
Vapor Density: 4.62 Air=1
Vapor Pressure: 5.7 mm Hg at 37.8 °C
Water Solubility: Insoluble in Water
Other Solubilities: Soluble in all proportions in Acetone, Petroleum Ether, Carbon Tetrachloride.
Refraction Index: 1.49235 at 20 °C/D
Ionization Potential (eV): 8.69 +/-0.3
Flash Point: 60 °C Open Cup
Autoignition Temperature: 450 °C
LEL: 0.7% v/v
UEL: 5.7% v/v

RTECS Toxicity Data

Acute Oral: Rat LD_{Lo} Dose: 10 mL/kg; Toxic Effects: Lungs, Thorax, or Respiration - Fibrosis, focal (pneumoconiosis); Lungs, Thorax, or Respiration - Acute pulmonary edema; Blood - Hemorrhage.

Hazard Overviews

Flammable

Health: May cause irritation. Other Acute Effects: may be harmful by inhalation, ingestion, or skin absorption.
Fire: Flammable. Hazards: vapor may travel considerable distance to source of ignition and flash back; container explosion may occur; forms explosive mixtures in air. Extinguishing agents: carbon dioxide, dry chemical powder or appropriate foam; water may be effective for cooling, but may not effect extinguishment. Precautions: combustible liquid.
Carcinogenicity: IARC - Not listed; NIOSH - Not listed; NTP - Not listed; ACGIH - Not listed; OSHA - Not listed; EPA - Not listed; MAK - Not listed

Environmental

Environmental Physical Data

Octanol/Water Partition Coefficient: log K_{ow} = 4.11

Regulations

RCRA 40CFR: Not listed
CERCLA: 40CFR 302.4: Not listed
SARA 40CFR 372.65: Not listed
SARA EHS 40CFR 355: Not listed
TSCA: Listed

Analytical Methods

Air: EPA VG-011-1
Soil: SW846 8021A, 8260A
Water / Groundwater: APHA 6210-D, 6230-D
Drinking Water: EPA 502.2, 503.1, 524.2; APHA 6210-C, 6220-C
Plasma: EPA 29

BUT6670 CAS #: 3663-23-8

4-BUTYL-1,2-BENZENEDIAMINE

EINECS Number: 222-917-7
Molecular Formula: $C_{10}H_{16}N_2$
Formula Weight: 164.25
Synonyms: 1,2-BENZENEDIAMINE,4-BUTYL-; O-PHENYLENEDIAMINE,4-BUTYL-

Hazard Overviews

Carcinogenicity: IARC - Not listed; NIOSH - Not listed; NTP - Not listed; ACGIH - Not listed; OSHA - Not listed; EPA - Not listed; MAK - Not listed

Environmental

Regulations

RCRA 40CFR: Not listed
CERCLA: 40CFR 302.4: Not listed
SARA 40CFR 372.65: Not listed
SARA EHS 40CFR 355: Not listed
TSCA: Listed

BUT6740 CAS #: 98-29-3

4-T-BUTYLCATECHOL

RTECS: UX1400000
EINECS Number: 202-653-9
Molecular Formula: $C_{10}H_{14}O_2$
Formula Weight: 166.24

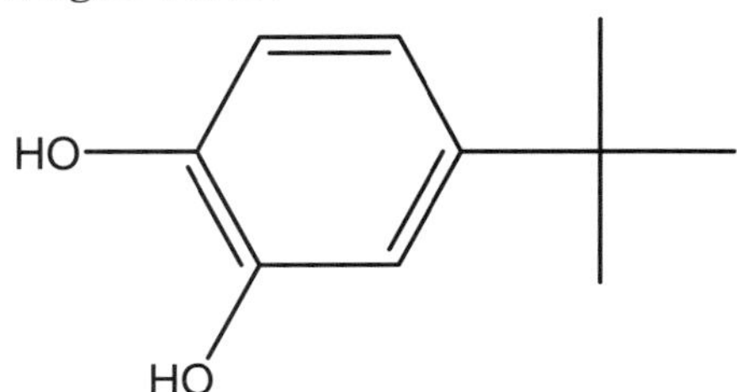

Chemical Structure

Synonyms: 1,2-BENZENEDIOL,4-(1,1-DIMETHYLETHYL)-; 4-TERT-BUTYL-1,2-BENZENEDIOL; 4-T-BUTYLPYROCATECHOL; P-T-BUTYLPYROCATECHOL; 4-TERT-BUTYLPYROKATECHIN; SYNOX TBC
Description: colorless to light-brown crystals

Physical Properties

Boiling Point: 285 °C (545 °F)
Freezing Point: 54 °C (129.2 °F)
Specific Gravity: 1.05
Water Solubility: Negligible
Other Solubilities: Soluble in Acetone, Alcohol and Ether

Flash Point: 130 °C

RTECS Toxicity Data

Acute Oral: Rat LD_{50} Dose: 2820 mg/kg.
Acute Dermal: Rabbit LD_{50} Route: Skin; Dose: 630 uL/kg.
Irritation Eye: Rabbit Standard Draize Test Dose: 50 ug open; Reaction: severe.
Irritation Skin: Rabbit Standard Draize Test Dose: 750 ug/24H; Reaction: severe. Guinea Pig Standard Draize Test Dose: 1%/3W; Reaction: moderate.
Mutagenic: Mouse Mutations in Mammalian Somatic Cells; Cell Type: lymphocyte; Dose: 80 ug/L.

Hazard Overviews

Fire Diamond

Health: Severe irritation to eyes/skin/respiratory tract. Also Causes: burns, sensitization, systemic toxicity (via skin absorption); upon ingestion: corrosion of GI tract. Chronic Effects: irritating dermatitis.
Fire: Will burn. Use dry chemical, carbon dioxide, water mist, fog, or regular foam.
Reactivity: Stable. Hazardous polymerization cannot occur. Avoid: exposure to air; moisture. Incompatible with: oxidizing materials; iron; carbon steel. Hazardous decomposition products: carbon dioxide; acrid fumes; irritating fumes.
Carcinogenicity: IARC - Not listed; NIOSH - Not listed; NTP - Not listed; ACGIH - Not listed; OSHA - Not listed; EPA - Not listed; MAK - Not listed
Primary Target Organs:

Eyes

Skin

Respiratory System

Exposure Limits
AIHA WEEL: TWA: 5 mg/m^3skin.
Respirator Recommendation
Exposure Range: >5 to 50 mg/m^3 Air Purifying, Negative Pressure, Half Mask
Exposure Range: >50 to 500 mg/m^3 Air Purifying, Negative Pressure, Full Face
Exposure Range: >500 to 5000 mg/m^3 Supplied Air, Constant Flow/Pressure Demand, Full Face
Exposure Range: >5000 to unlimited mg/m^3 Self-contained Breathing Apparatus, Pressure Demand, Full Face
Cartridge Color: dust/mist filter (use P100 or consult supervisor for appropriate dust/mist filter)

Environmental

Regulations

RCRA 40CFR: Not listed
CERCLA: 40CFR 302.4: Not listed
SARA 40CFR 372.65: Not listed
SARA EHS 40CFR 355: Not listed
TSCA: Listed

BUT6880 CAS #: 25167-67-3

BUTYLENE

DOT: UN1012; IMO2.1
EINECS Number: 246-689-3
Molecular Formula: C_4H_8
Structured MF: $CH_3CH_2CH{=}{=}CH_2$
Formula Weight: 56.11
Synonyms: 1-BUTENE; BUTENE; N-BUTENE; ALPHA-BUTYLENE; N-BUTYLENE; ETHYLETHYLENE
Description: colorless gas; slight aromatic odor
Use: chem int for butadiene used primarily to mfr elastomers; for sec-butanol used to mfr methyl ethyl ketone; for primary amyl alcohols used to mfr solvents; for butylene oxide used as an acid scavenger; for ethylene-butylene copolymer; production of maleic anhydride by catalytic oxidation of 1-butene

Physical Properties

Boiling Point: Alpha Form -6 °C (21 °F)
Freezing Point: -186 °C (-302.8 °F)
Vapor Density: 1.9 Air=1
Vapor Pressure: Reid 62.5 psia
Water Solubility: Insoluble
Other Solubilities: Soluble in most organic solvents
Surface Tension: 12.5 dynes/cm at 20 °C
Critical Temperature: 146 °C
Critical Pressure: 39.7 atm
Flash Point: 43 °C Closed Cup
Autoignition Temperature: 385 °C
LEL: 1.98% v/v
UEL: 9.65% v/v

Hazard Overviews

Flammable

Compressed Gas

Fire Diamond

Health: A simple asphyxiant which can displace available oxygen. Also Causes: rapid respiration, diminished alertness, impaired coordination, rapid fatigue, emotional instability, nausea, vomiting, prostration, unconsciousness, convulsions, coma, and death.
Fire: Flammable. Extinguish by shutting off the source of the gas. Use water sprays to cool fire-exposed containers and to protect personnel attempting to seal the source of escaping gas. Treat any fire situation involving rapidly escaping and burning butylene gas as an emergency.
Reactivity: Stable. Hazardous polymerization cannot occur. Avoid: exposure to ignition sources (open flame; lighted cigarettes or pipes; uninsulated heating elements; electrical or mechanical sparks); accidental or uncontrollably rapid release

from high-pressure cylinders, tank cars, or pipelines. Incompatible with: strong oxidizing agents. Hazardous decomposition products: carbon dioxide; carbon monoxide.
Carcinogenicity: IARC - Not listed; NIOSH - Not listed; NTP - Not listed; ACGIH - Not listed; OSHA - Not listed; EPA - Not listed; MAK - Not listed

Environmental

Cleanup/Disposal: Guide No. 115: Eliminate all ignition sources (no smoking, flares, sparks or flames in immediate area). All equipment used when handling the product must be grounded. Do not touch or walk through spilled material. Stop leak if you can do it without risk. If possible, turn leaking containers so that gas escapes rather than liquid. Use water spray to reduce vapors or divert vapor cloud drift. Do not direct water at spill or source of leak. Prevent spreading of vapors through sewers, ventilation systems and confined areas. Isolate area until gas has dispersed.

Environmental Physical Data

BOD: none

Regulations

RCRA 40CFR: Not listed
CERCLA: 40CFR 302.4: Not listed
SARA 40CFR 372.65: Not listed
SARA EHS 40CFR 355: Not listed
TSCA: Listed

Analytical Methods

Air: ASTM D2820

BUT6950 CAS #: 1320-66-7

BUTYLENE CHLOROHYDRIN

EINECS Number: 215-305-6
Molecular Formula: C_4H_9ClO
Formula Weight: 108.56
Synonyms: BUTANOL,CHLORO-; BUTENE CHLOROHYDRIN

Hazard Overviews

Carcinogenicity: IARC - Not listed; NIOSH - Not listed; NTP - Not listed; ACGIH - Not listed; OSHA - Not listed; EPA - Not listed; MAK - Not listed

Environmental

Regulations

RCRA 40CFR: Not listed
CERCLA: 40CFR 302.4: Not listed
SARA 40CFR 372.65: Not listed
SARA EHS 40CFR 355: Not listed
TSCA: Listed

BUT7090 CAS #: 584-03-2

1,2-BUTYLENE GLYCOL

RTECS: EK0380000
EINECS Number: 209-527-2
Molecular Formula: $C_4H_{10}O_2$
Formula Weight: 90.12

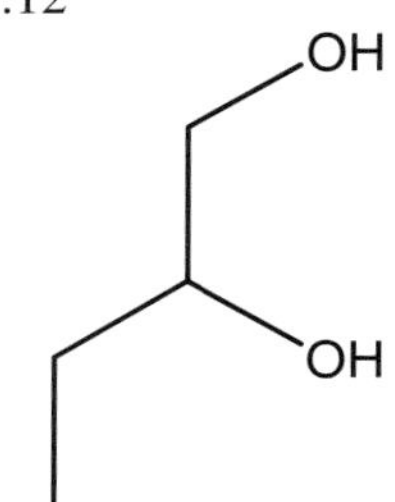

Chemical Structure

Synonyms: ALPHA-BUTYLENEGLYCOL
Description: clear viscous liquid
Use: in indust as intermediate in polyester resins; solvent; intermediate

Physical Properties

Boiling Point: 195 °C (383 °F) to 196.9 °C (386 °F)
Freezing Point: -50 °C (-58 °F)
Specific Gravity: 1.0023 at 20 °C
Vapor Density: 3.1 Air=1
Saturated Vapor Density: Estimated 1.200166721 kg/m^3
Vapor Pressure: Estimated 0.0501 mm Hg at 25 °C
Water Solubility: Miscible with Water
Other Solubilities: Soluble in Alcohol & Acetone.
Refraction Index: 1.4382 at 20 °C
Flash Point: 104 °C Closed Cup

RTECS Toxicity Data

Acute Oral: Rat LD_{50} Dose: 16 gm/kg.

Hazard Overviews

Fire Diamond

Health: Irritating to eyes/skin. Harmful. Acute Effects: harmful if inhaled or swallowed; prolonged exposure can cause gastrointestinal disturbances; nausea; headache; vomiting.
Fire: Will burn. Extinguishing agents: water spray; carbon dioxide, dry chemical powder or appropriate foam. Precautions: combustible liquid.
Reactivity: Incompatible with: acid chlorides, acid anhydrides, oxidizing agents, chloroformates, reducing agents. Hazardous decomposition products: toxic fumes of: carbon monoxide, carbon dioxide.

Carcinogenicity: IARC - Not listed; NIOSH - Not listed; NTP - Not listed; ACGIH - Not listed; OSHA - Not listed; EPA - Not listed; MAK - Not listed

Primary Target Organs:

Eyes

Skin

Environmental

Environmental Fate: If released to the atmosphere, it will degrade by reaction with photochemically produced hydroxyl radicals (estimated half-life of about 25 hours). If released to soil or water, it is expected to biodegrade based on a screening study for the structurally analogous chemical, 1,4-butanediol. Hydrolysis, bioconcentration in aquatic organisms, adsorption to sediment and volatilization from water are not expected to be environmentally important removal mechanisms from aquatic ecosystems.

Environmental Physical Data

Henry's Law Constant: 2.3×10^{-7}
Octanol/Water Partition Coefficient: log K_{ow} = -0.29
Sorption Partition Coefficient: K_{oc} = estimated at 17
BCF: estimated at 0.35

Regulations

RCRA 40CFR: Not listed
CERCLA: 40CFR 302.4: Not listed
SARA 40CFR 372.65: Not listed
SARA EHS 40CFR 355: Not listed
TSCA: Listed

BUT7230 CAS #: 106-88-7

1,2-BUTYLENE OXIDE

RTECS: EK3675000
DOT: UN3022; IMO3.2
EINECS Number: 203-438-2
Molecular Formula: C_4H_8O
Formula Weight: 72.12

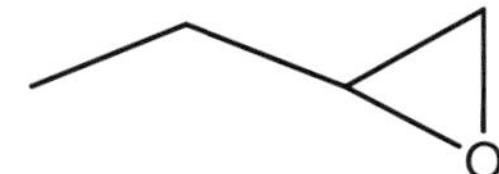

Chemical Structure

Synonyms: BUTANE,1,2-EPOXY-; 1,2-BUTENE OXIDE; 1-BUTENE OXIDE; N-BUTENE-1,2-OXIDE; 1-BUTYLENE OXIDE; ALPHA-BUTYLENE OXIDE; BUTYLENE OXIDE; 1,2-BUTYLENE OXIDE,STABILIZED; 1,2-EPOXYBUTANE; EPOXYBUTANE; ETHYL ETHYLENE OXIDE; ETHYLENE OXIDE,ETHYL-; ETHYLETHYLENE OXIDE; 2-ETHYLOXIRANE; ETHYLOXIRANE; OXIRANE,ETHYL-; PROPYL OXIRANE

Description: colorless liquid; sweet, disagreeable odor

Use: intermediate, especially for various polymers; as a stabilizer for chlorinated solvents

Physical Properties

Boiling Point: 63.3 °C (146 °F)
Freezing Point: -60 °C (-76 °F)
Specific Gravity: 0.837 at 17 °C/4 °C
Vapor Density: 2.49 Air=1
Density: 0.8297 g/mL at 20 °C
Vapor Pressure: 160 mm Hg at 12.8 °C
Water Solubility: >= 100 mg/mL at 17 C
Other Solubilities: 95% Ethanol: >=100 mg/ml at 17 °C; Acetone: >=100 mg/ml at 17 °C; DMSO: >=100 mg/ml at 17 °C; Ether: Soluble; Most organic solvents: miscible.
Refraction Index: 1.3851 at 20 °C
Ionization Potential (eV): =< 10.15
Flash Point: -22 °C Closed Cup
Autoignition Temperature: 515 °C
LEL: 3.1% v/v
UEL: 25.1% v/v

RTECS Toxicity Data

Acute Oral: Rat LD_{50} Dose: 500 mg/kg.
Acute Inhalation: Rat LC_{Lo} Dose: 4000 ppm/4hr. Mouse LC_{Lo} Dose: 398 ppm/4hr.
Acute Dermal: Rabbit LD_{50} Route: Skin; Dose: 2100 mg/kg.
Chronic (Multiple Dose) Inhalation: Rat Dose: 800 ppm/6H/13W-I; Toxic Effects: Liver - Changes in Liver weight. Rat Dose: 1600 ppm/6H/14D-I; Toxic Effects: DEATH. Rat Dose: 600 ppm/6H/13W-I; Toxic Effects: Sense organs and special senses - Other; Kidney, Ureter, and Bladder - Other changes in urine composition; Nutritional and gross metabolic - Weight loss or decreased weight gain.
Irritation Eye: Rabbit Standard Draize Test Dose: 100 mg/24H; Reaction: moderate.
Irritation Skin: Rabbit Standard Draize Test Dose: 500 mg/24H; Reaction: mild.
Reproductive/Teratogenic: Rabbit Route: Inhalation; Dose: 1000 ppm/7H; Duration: female 1-24D of pregnancy; Effects on Fertility - Post-implantation mortality.
Mutagenic: Rat Morphological Transformation; Cell Type: embryo; Dose: 700 mg/L. Mouse Mutations in Microorganisms; Cell Type: lymphocyte; Dose: 400 mg/L (+S9).
Tumorigenic: Rat Route: Inhalation; Dose: 400 ppm/6H/5D/2Y-C; Toxic Effects: Tumorigenic - Carcinogenic by RTECS criteria; Sense organs and special senses - Tumors; Lungs, Thorax, or Respiration - Tumors.

Hazard Overviews

Flammable

Fire Diamond

Health: Irritating to eyes/skin/respiratory tract. Toxic. Also Causes: dizziness, suffocation, allergic sensitization, gastrointestinal tract irritation; upon ingestion: nausea, vomiting.

Fire: Flammable. Can form explosive mixtures in air. Polymerizes if uninhibited. For small fires use dry chemical, carbon dioxide, water spray, or alcohol-resistant foam. For large fires use water spray, fog, or alcohol-resistant foam.

Reactivity: Stablewhen inhibited. Hazardous polymerization can occur if uninhibited and in contact with anhydrous chlorides of iron/tin/aluminum, peroxides, alkali metal hydroxides. Avoid: heat; ignition sources. Incompatible with: anhydrous chlorides of iron, tin, aluminum; peroxides of iron and aluminum; alakli metal hydroxides; acids; alkalis, strong oxidizers; any material having a labile hydrogen atom. Hazardous decomposition products: carbon oxides; acrid smoke.

Carcinogenicity: IARC - Group 3, Not classifiable as to carcinogenicity to humans; NIOSH - Not listed; NTP - Not listed; ACGIH - Not listed; OSHA - Not listed; EPA - Not listed; MAK - Class A2, Unmistakably carcinogenic in animal experimentation only

Primary Target Organs:

Eyes

Skin

Respiratory System

Exposure Limits

AIHA WEEL: TWA: 2 ppm.

Respirator Recommendation

Exposure Range: >2 to ppm Self-contained Breathing Apparatus, Pressure Demand, Full Face

Note: odor threshold unknown

Environmental

Environmental Fate: It is expected to be readily degraded in the environment. If released to the atmosphere, it is degraded by reaction with photochemically produced hydroxyl radicals at an estimated half-life rate of 7.6 days in an average ambient atmosphere. If released to water, it will hydrolyze (estimated half-life of 12.9 days or faster at 25 °C) and volatilize (estimated half-lives of 6.7 hr and 3.2 days from a model river one meter deep and an environmental pond, respectively). If released to soil, it is expected to hydrolyze in the presence of moisture. It is predicted to be highly mobile in soil; however, concurrent hydrolysis is likely to diminish the significance of leaching.

Cleanup/Disposal: Guide No. 127: Eliminate all ignition sources (no smoking, flares, sparks or flames in immediate area). All equipment used when handling the product must be grounded. Do not touch or walk through spilled material. Stop leak if you can do it without risk. Prevent entry into waterways, sewers, basements or confined areas. A vapor suppressing foam may be used to reduce vapors. Absorb or cover with dry earth, sand or other non-combustible material and transfer to containers. Use clean non-sparking tools to collect absorbed material. Large Spills: Dike far ahead of liquid spill for later disposal. Water spray may reduce vapor; but may not prevent ignition in closed spaces.

Environmental Physical Data

Henry's Law Constant: estimated at 1.76×10^{-4}

Sorption Partition Coefficient: **log K_{oc}** = 0.9

BCF: estimated at -0.2

Regulations

RCRA 40CFR: Not listed

CERCLA: 40CFR 302.4: Listed per CAA Section 112 RQ: 100 lb (45.35 kg)

SARA 40CFR 372.65: Listed

SARA EHS 40CFR 355: Not listed

TSCA: Listed

BUT7300 CAS #: 2167-39-7

1,3-BUTYLENE OXIDE

RTECS: EK3750000

Molecular Formula: C_4H_8O

Formula Weight: 72.12

Synonyms: BUTANE,1,3-EPOXY-; 2-METHYLOXETAN; 2-METHYLOXETANE; 1-METHYLTRIMETHYLENE OXIDE; OXETANE,2-METHYL-; OXETANE,2-METHYL-(9CI)

Description: liquid

Use: mfr gasoline additives; as acid scavengers & stabilizers for chlorinated solvents; prodn of corresponding butylene glycols & their derivatives, such as polybutylene glycols, mixed polyglycols & glycol ethers & esters; to make butanolamines, surface-active agents, & gasoline additives; acid scavengers & stabilizers for chlorinated solvents

Physical Properties

Boiling Point: 59 °C (138.2 °F)

Specific Gravity: 0.84100

Refraction Index: 1.3885

Flash Point: Highly flammable

Hazard Overviews

Flammable

Fire Diamond

Fire: Flammable.

Carcinogenicity: IARC - Not listed; NIOSH - Not listed; NTP - Not listed; ACGIH - Not listed; OSHA - Not listed; EPA - Not listed; MAK - Not listed

Environmental

Regulations

RCRA 40CFR: Not listed

CERCLA: 40CFR 302.4: Not listed

SARA 40CFR 372.65: Not listed

SARA EHS 40CFR 355: Not listed

TSCA: Not listed

BUT7370 **CAS #: 3266-23-7**

2,3-BUTYLENE OXIDE

RTECS: EK3855000
EINECS Number: 221-877-8
Molecular Formula: C_4H_8O
Formula Weight: 72.12
Synonyms: 2-BUTENE EXPOXIDE; 2-BUTENE OXIDE; BETA-BUTYLENE OXIDE; 2,3-DIMETHYLETHYLENE OXIDE; 2,3-DIMETHYLOXIRANE; 2,3-EPOXYBUTANE; OXIRANE,2,3-DIMETHYL-(9CI); BETA-OXYBUTENE
Description: colorless, clear liquid
Use: intermediate

Physical Properties

Density: 0.793 g/cu cm at 24 °C
Water Solubility: >= 100 mg/mL at 20 C
Other Solubilities: 95% Ethanol: >=100 mg/ml at 20 °C; Acetone: >=100 mg/ml at 20 °C; Benzene: Soluble; DMSO: >=100 mg/ml at 20 °C; Ether: Very Soluble; Organic solvents: Very Soluble.
Refraction Index: 1.3731
Flash Point: -17 °C

RTECS Toxicity Data

Mutagenic: Bacteria - K Kpneumoniae Mutations in Microorganisms; Dose: 20 mmol/L (-S9). Bacteria - S Typhimurium Mutations in Microorganisms; Dose: 5 umol/plate (+/-S9).

Hazard Overviews

Flammable

Fire: Flammable.
Carcinogenicity: IARC - Not listed; NIOSH - Not listed; NTP - Not listed; ACGIH - Not listed; OSHA - Not listed; EPA - Not listed; MAK - Not listed

Environmental

Regulations

RCRA 40CFR: Not listed
CERCLA: 40CFR 302.4: Not listed
SARA 40CFR 372.65: Not listed
SARA EHS 40CFR 355: Not listed
TSCA: Listed

BUT7440 **CAS #: 26249-20-7**

BUTYLENE OXIDE

RTECS: YP6930000
EINECS Number: 247-545-2
Molecular Formula: C_4H_8O
Formula Weight: 72.10
Synonyms: BUTANE,EPOXY-
Description: water-white liquid; sweetish odor, somewhat like butyric acid
Use: prodn of butanolamines, surface-active agents, gasoline additives, butylene glycols & their derivatives, eg, polybutylene glycols, mixed polyglycols & glycol ethers & esters; acid scavengers & stabilizers for chlorinated solvents; additive in trichloroethylene; corrosion inhibitor in paint & varnish removers; building block for nonionic emulsifiers; chem modification of pinewood; stabilizer for chlorinated solvents.

Physical Properties

Boiling Point: 59 °C (138 °F) to 63 °C (145 °F) at 760 mm Hg
Freezing Point: < -50 °C (-58 °F)
Vapor Density: 183 Air=1
Density: 0.832 at 20 /20 °C
Vapor Pressure: 114 mm Hg at 15 °C
Water Solubility: 9 g/100 g Water at 25 °C
Other Solubilities: miscible with aliphatic and aromatic solvents.
Odor Threshold: 0.0202 to 2.0874 mg/m^3
Flash Point: -15 °C Closed Cup
Autoignition Temperature: 321 °C
LEL: 1.5% v/v
UEL: 18.3% v/v

RTECS Toxicity Data

Acute Oral: Rat LD_{50} Dose: 1410 uL/kg.
Acute Inhalation: Rat LC_{Lo} Dose: 4000 ppm/4hr.
Acute Dermal: Rabbit LD_{50} Route: Skin; Dose: 2100 uL/kg.
Irritation Skin: Rabbit Open Draize Test Dose: 500 mg open; Reaction: mild.

Hazard Overviews

Flammable

Fire Diamond

Fire: Flammable.
Carcinogenicity: IARC - Not listed; NIOSH - Not listed; NTP - Not listed; ACGIH - Not listed; OSHA - Not listed; EPA - Not listed; MAK - Not listed

Environmental

Regulations

RCRA 40CFR: Not listed
CERCLA: 40CFR 302.4: Not listed
SARA 40CFR 372.65: Not listed
SARA EHS 40CFR 355: Not listed
TSCA: Listed

BUT7510 CAS #: 96-70-8

2-T-BUTYL-4-ETHYLPHENOL

EINECS Number: 202-526-8
Molecular Formula: $C_{12}H_{18}O$
Formula Weight: 178.28
Synonyms: BENZENE,2-TERT-BUTYL-4-ETHYL-1-HYDROXY-; 2-TERT-BUTYL-4-ETHYLPHENOL; 4-ETHYL-2-TERT-BUTYLPHENOL; PHENOL,2-TERT-BUTYL-4-ETHYL-; PHENOL,2-(1,1-DIMETHYLETHYL)-4-ETHYL-
Use: chem int for 2,2'-methylenebis(6-t-butyl-4-ethylphenol)

Physical Properties

Boiling Point: 250 °C (482 °F) at 760 mm Hg
Freezing Point: 23 °C (73.4 °F)
Other Solubilities: Soluble in Alcohol Insoluble in alkali

Hazard Overviews

Corrosive

Health: Corrosive to eyes/skin/respiratory tract. Other Acute Effects: harmful if swallowed, inhaled, or absorbed through skin; extremely destructive to tissue of the mucous membranes and upper respiratory tract, eyes and skin; inhalation may result in spasm, inflammation and edema of the larynx and bronchi, chemical pneumonitis and pulmonary edema; symptoms of exposure may include burning sensation, coughing, wheezing, laryngitis, shortness of breath, headache, nausea and vomiting.
Fire: Hazards: emits toxic fumes. Extinguishing agents: carbon dioxide, dry chemical powder or appropriate foam. Precautions: combustible liquid.
Carcinogenicity: IARC - Not listed; NIOSH - Not listed; NTP - Not listed; ACGIH - Not listed; OSHA - Not listed; EPA - Not listed; MAK - Not listed
Primary Target Organs:

Eyes

Skin

Respiratory System

Environmental

Regulations

RCRA 40CFR: Not listed
CERCLA: 40CFR 302.4: Not listed
SARA 40CFR 372.65: Not listed
SARA EHS 40CFR 355: Not listed
TSCA: Listed

BUT7580 CAS #: 2425-74-3

N-T-BUTYLFORMAMIDE

RTECS: LQ1450000
EINECS Number: 219-369-6
Molecular Formula: $C_5H_{11}NO$
Formula Weight: 101.17

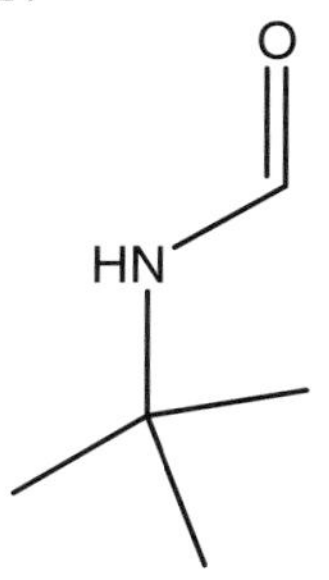

Chemical Structure

Synonyms: N-TERT-BUTYLFORMAMIDE; TERT-BUTYLFORMAMIDE; N-(1,1-DIMETHYL)FORMAMIDE (NLM); FORMAMIDE,N-(TERT-BUTYL)-; FORMAMIDE,N-TERT-BUTYL-; FORMAMIDE,N-(1,1-DIMETHYLETHYL)-
Description: colorless liquid
Use: a solvent & in petroleum additives; intermediate in the manufacture of lubricating oil additives and miscellaneous chemicals

Physical Properties

Boiling Point: High Boiling
Freezing Point: High Boiling
Water Solubility: Soluble in Water
Other Solubilities: Soluble in common hydrocarbon solvents

Hazard Overviews

Health: Irritating to eyes/skin. Other Acute Effects: may be harmful by inhalation, ingestion, or skin absorption.
Fire: Hazards: emits toxic fumes. Extinguishing agents: water spray; carbon dioxide, dry chemical powder or appropriate foam. Precautions: combustible liquid.
Reactivity: Incompatible with: strong oxidizing agents. Hazardous decomposition products: toxic fumes of: carbon monoxide, carbon dioxide, nitrogen oxides.
Carcinogenicity: IARC - Not listed; NIOSH - Not listed; NTP - Not listed; ACGIH - Not listed; OSHA - Not listed; EPA - Not listed; MAK - Not listed
Primary Target Organs:

Eyes

Skin

Environmental

Regulations

RCRA 40CFR: Not listed

CERCLA: 40CFR 302.4: Not listed
SARA 40CFR 372.65: Not listed
SARA EHS 40CFR 355: Not listed
TSCA: Listed

BUT7650 CAS #: 1948-33-0

T-BUTYLHYDROQUINONE

RTECS: MX4375000
EINECS Number: 217-752-2
Molecular Formula: $C_{10}H_{14}O_2$
Structured MF: $(CH_3)_3CC_6H_3(OH)_2$
Formula Weight: 166.24

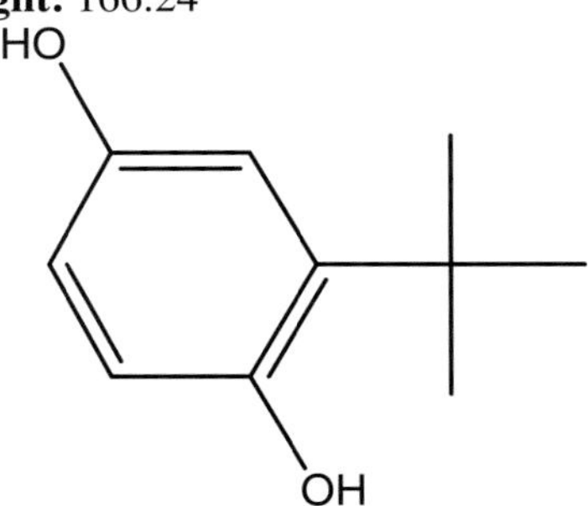

Chemical Structure

Synonyms: BANOX 20BA; 1,4-BENZENEDIOL (1,1-DIMETHYLETHYL)-; 1,4-BENZENEDIOL,2-(1,1-DIMETHYLETHYL)-; 1,4-BENZENEDIOL,2-(1,1-DIMETHYLETHYL)-(9CI); 2-TERT-BUTYL-1,4-BENZENEDIOL; TERT-BUTYL-1,4-BENZENEDIOL; 2-(TERT-BUTYL)-P-HYDROQUINONE; 2-TERT-BUTYLHYDROQUINONE; TERT-BUTYLHYDROQUINONE; 2-(1,1-DIMETHYLETHYL)-1,4-BENZENEDIOL; 2-(1,1-DIMETHYLETHYL)HYDROQUINONE; HYDROQUINONE,T-BUTYL-; HYDROQUINONE,TERT-BUTYL-; MONO-TERT-BUTYLHYDROQUINONE; MONO-TERTIARYBUTYLHYDROQUINONE; MTBHQ; SUSTANE; TBHQ; TENOX TBHQ
Description: white to light tan crystals; very slight odor
Use: intermediate and as a phenolic antioxidant in vegetable fats and oils

Physical Properties

Boiling Point: 295 °C (563 °F)
Freezing Point: 126.5 °C (259.7 °F) to 128.5 °C (263.3 °F)
Specific Gravity: 1.05
Vapor Density: 5.73 Air=1
Saturated Vapor Density: < 1.207472232 kg/m^3
Vapor Pressure: < 1 mm Hg at 20 °C
Water Solubility: < 1 mg/mL at 19 C
Other Solubilities: 95% Ethanol: >=100 mg/ml at 19 °C; Acetone: >=100 mg/ml at 19 °C; Alcohol: Soluble; DMSO: >=100 mg/ml at 19 °C; Ethyl Acetate: Soluble.
Flash Point: 171 °C

RTECS Toxicity Data

Acute Oral: Rat LD_{50} Dose: 700 mg/kg. Mouse LD_{50} Dose: 1 gm/kg.
Acute Inhalation: Rat LC_{Lo} Dose: 2900 mg/m^3/4hr; Toxic Effects: Sense organs and special senses - Other; Behavioral - Ataxia; Lungs, Thorax, or Respiration - Dyspnea.
Mutagenic: Rat Unscheduled DNA Synthesis; Route: Oral; Dose: 33600 mg/kg/4W-C. Mouse Cytogenetic Analysis; Route: Intraperitoneal; Dose: 200 mg/kg.

Hazard Overviews

Health: Irritating to eyes/skin/respiratory tract. Harmful. Other Acute Effects: harmful if swallowed; may be harmful if inhaled; may be harmful if absorbed through the skin.
Fire: Will burn. Hazards: emits toxic fumes. Extinguishing agents: water spray; carbon dioxide, dry chemical powder or appropriate foam. Precautions: combustible liquid.
Reactivity: Incompatible with: strong oxidizing agents, strong bases. Hazardous decomposition products: toxic fumes of: carbon monoxide, carbon dioxide.
Carcinogenicity: IARC - Not listed; NIOSH - Not listed; NTP - Not listed; ACGIH - Not listed; OSHA - Not listed; EPA - Not listed; MAK - Not listed
Primary Target Organs:

Eyes

Skin

Respiratory System

Environmental

Regulations

RCRA 40CFR: Not listed
CERCLA: 40CFR 302.4: Not listed
SARA 40CFR 372.65: Not listed
SARA EHS 40CFR 355: Not listed
TSCA: Listed

BUT7720 CAS #: 121-00-6

3-T-BUTYL-4-HYDROXYANISOLE

RTECS: SK1575000
EINECS Number: 204-442-7
Molecular Formula: $C_{11}H_{16}O_2$
Formula Weight: 180.27

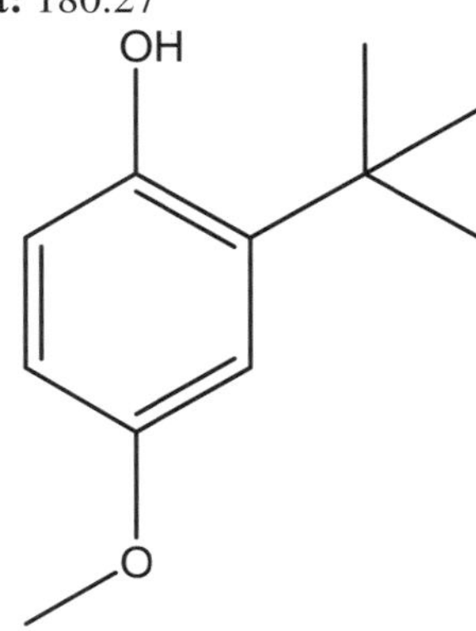

Chemical Structure

Synonyms: 3-BHA; 3-TERT-BUTYL-4-HYDROXYANISOLE; 2-TERT-BUTYL-4-METHOXYPHENOL; 4-METHOXY-2-TERT-BUTYLPHENOL; PHENOL,2-TERT-BUTYL-4-METHOXY-; PHENOL,2-(1,1-DIMETHYLETHYL)-4-METHOXY-
Use: component of BHA (a commonly used food additive); oxidation inhibitor

RTECS Toxicity Data

Acute Oral: Rat LD_{50} Dose: 2910 mg/kg; Toxic Effects: Gastrointestinal - Other changes. Mouse LD_{50} Dose: 1583 mg/kg; Toxic Effects: Gastrointestinal - Other changes.
Mutagenic: Mouse Morphological Transformation; Cell Type: fibroblast; Dose: 30 mg/L. Hamster Cytogenetic Analysis; Cell Type: lung; Dose: 125 mg/L.
Tumorigenic: Hamster Route: Oral; Dose: 134 gm/kg/24W-C; Toxic Effects: Tumorigenic - Neoplastic by RTECS criteria; Gastrointestinal - Tumors. Hamster Route: Oral; Dose: 168 gm/kg/20W-C; Toxic Effects: Tumorigenic - Neoplastic by RTECS criteria; Gastrointestinal - Tumors.

Hazard Overviews

Health: Irritating to eyes/skin/respiratory tract. Toxic. Other Acute Effects: harmful by inhalation, in contact with skin and if swallowed; prolonged or repeated exposure may cause allergic reactions in certain sensitive individuals. Chronic Effects: target organ: liver. Carcinogen. The toxicological properties have not been thoroughly investigated.
Fire: Hazards: emits toxic fumes. Extinguishing agents: water spray; carbon dioxide, dry chemical powder or appropriate foam. Precautions: combustible liquid.
Reactivity: Stable. Hazardous polymerization will not occur. Incompatible with: strong oxidizing agents. Hazardous decomposition products: carbon monoxide, carbon dioxide.
Carcinogenicity: IARC - Not listed; NIOSH - Not listed; NTP - Not listed; ACGIH - Not listed; OSHA - Not listed; EPA - Not listed; MAK - Not listed
Primary Target Organs:

Eyes

Skin

Respiratory System

Liver

Environmental

Regulations

RCRA 40CFR: Not listed
CERCLA: 40CFR 302.4: Not listed
SARA 40CFR 372.65: Not listed
SARA EHS 40CFR 355: Not listed
TSCA: Listed

BUT7860 CAS #: 2409-55-4

2-T-BUTYL-4-METHYLPHENOL

RTECS: GO7000000
EINECS Number: 219-314-6
Molecular Formula: $C_{11}H_{16}O$
Formula Weight: 164.27

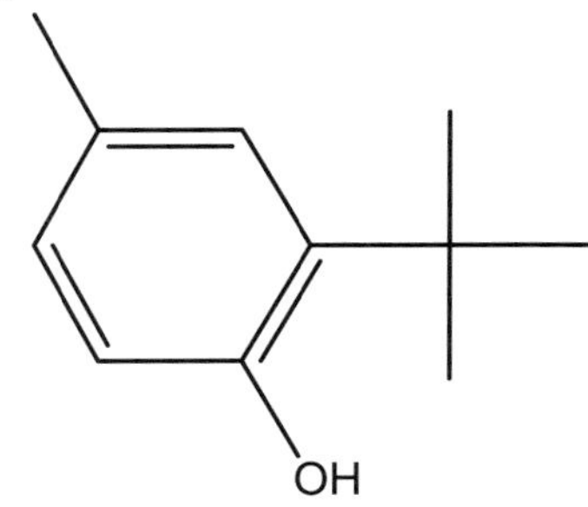

Chemical Structure

Synonyms: 2-T-BUTYL-P-CRESOL; 2-TERT-BUTYL-P-CRESOL; O-TERT-BUTYL-P-CRESOL; 2-TERT-BUTYL-4-METHYLPHENOL; P-CRESOL,2-TERT-BUTYL-; 4-METHYL-2-TERT-BUTYLPHENOL; 4-METHYL-2-(1,1-DIMETHYLETHYL)PHENOL; PHENOL,2-(1,1-DIMETHYLETHYL)-4-METHYL-; 2-TERC BUTYL-P-KRESOL; 2-TERC.BUTYL-P-KRESOL
Use: chem int for aldehyde condensates (rubber antioxidants; for 2,2'-methylenebis(4-methyl-6-t-butylphenol); with sulfur dichloride for rubber antioxidants; for a benzotriazole UV light stabilizer

Physical Properties

Boiling Point: 237 °C (459 °F)
Freezing Point: 55 °C (131 °F)
Saturated Vapor Density: Extrapolated 1.200184124 kg/m^3
Density: 0.9247 g/cc at 75 °C
Vapor Pressure: Extrapolated 0.025 mm Hg at 25 °C
Water Solubility: Insoluble in Water
Other Solubilities: Soluble in oxygenated solvents.
Refraction Index: 1.4969 at 75 °C

RTECS Toxicity Data

Acute Oral: Rat LD_{50} Dose: 2500 mg/kg; Toxic Effects: Sense organs and special senses - Other; Behavioral - Ataxia; Lungs, Thorax, or Respiration - Dyspnea. Mouse LD_{50} Dose: 700 mg/kg; Toxic Effects: Lungs, Thorax, or Respiration - Acute pulmonary edema; Gastrointestinal - Ulceration or bleeding from small intestine.
Acute Dermal: Rabbit LD_{Lo} Route: Skin; Dose: 2200 mg/kg.
Irritation Eye: Rabbit Standard Draize Test Dose: 50 ug/24H; Reaction: severe.
Irritation Skin: Rabbit Standard Draize Test Dose: 2 mg/24H; Reaction: severe.
Mutagenic: Human DNA Inhibition; Cell Type: lymphocyte; Dose: 25 umol/L. Rat Unscheduled DNA Synthesis; Route: Oral; Dose: 336 gm/kg/8W-C.

Tumorigenic: Hamster Route: Oral; Dose: 84 gm/kg/20W-C; Toxic Effects: Tumorigenic - Neoplastic by RTECS criteria; Gastrointestinal - Tumors.

Hazard Overviews

Corrosive

Health: Corrosive to eyes/skin/respiratory tract. Harmful. Other Acute Effects: harmful if swallowed, inhaled, or absorbed through skin; inhalation may result in spasm, inflammation and edema of the larynx and bronchi, chemical pneumonitis and pulmonary edema; symptoms of exposure may include burning sensation; coughing; wheezing; laryngitis; shortness of breath; headache; nausea; vomiting.
Fire: Hazards: under fire conditions, material may decompose to form flammable and/or explosive mixtures in air. Extinguishing agents: water spray; carbon dioxide, dry chemical powder or appropriate foam. Precautions: combustible liquid.
Reactivity: Incompatible with: bases, acid chlorides, acid anhydrides, oxidizing agents, corrodes steel, brass, copper, copper alloys. Hazardous decomposition products: toxic fumes of: carbon monoxide, carbon dioxide.
Carcinogenicity: IARC - Not listed; NIOSH - Not listed; NTP - Not listed; ACGIH - Not listed; OSHA - Not listed; EPA - Not listed; MAK - Not listed
Primary Target Organs:

Eyes

Skin

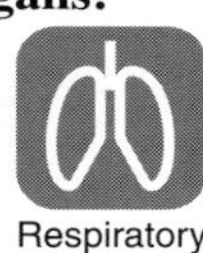

Respiratory System

Environmental

Environmental Fate: If released to soil, it will have no mobility. Volatilization may be important from moist soil surfaces. If released to water, it may adsorb to suspended solids and sediment. It may volatilize from water surfaces with estimated half-lives for a model river and model lake of 31 and 227 days, respectively. An estimated BCF value of 610 suggests that bioconcentration will be high in aquatic organisms. Insufficient data are available to determine the rate or importance of biodegradation in soil or water. If released to the atmosphere, it will exist primarily in the vapor phase in the ambient atmosphere. Vapor-phase is degraded in the atmosphere by reaction with photochemically produced hydroxyl radicals; the half-life for this reaction in air is estimated to be about 7.7 hours. Reaction with nitrate radicals may also be important. Particulate-phase may be physically removed from the air by wet and dry deposition.

Environmental Physical Data

Henry's Law Constant: estimated at 1.5×10^{-6}
Octanol/Water Partition Coefficient: log K_{ow} = 3.97
Sorption Partition Coefficient: K_{oc} = 3200
BCF: estimated at 610

Regulations

RCRA 40CFR: Not listed
CERCLA: 40CFR 302.4: Not listed
SARA 40CFR 372.65: Not listed
SARA EHS 40CFR 355: Not listed
TSCA: Listed

BUT8000 **CAS #: 94-26-8**

BUTYLPARABEN

RTECS: DH1980000
EINECS Number: 202-318-7
Molecular Formula: $C_{11}H_{14}O_3$
Structured MF: $HOC_6H_4COO(CH_2)_3CH_3$
Formula Weight: 194.22

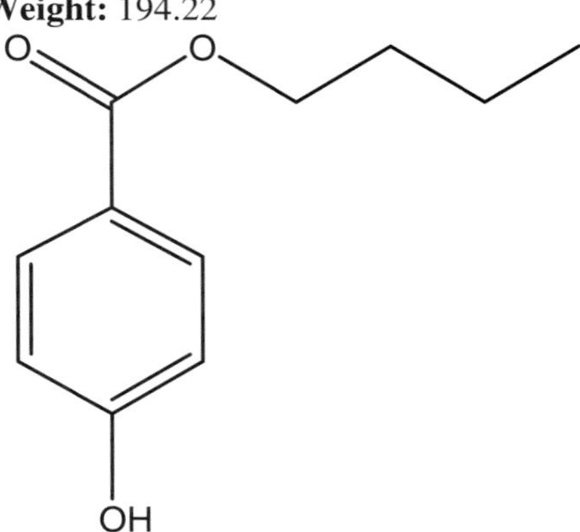

Chemical Structure

Synonyms: ASEPTOFORM BUTYL; BENZOIC ACID,4-HYDROXY-,BUTYL ESTER; BENZOIC ACID,P-HYDROXY-,BUTYL ESTER; BUTOBEN; 4-(BUTOXYCARBONYL)PHENOL; BUTYL BUTEX; BUTYL CHEMOSEPT; BUTYL 4-HYDROXYBENZOATE; BUTYL P-HYDROXYBENZOATE; N-BUTYL HYDROXYBENZOATE; N-BUTYL P-HYDROXYBENZOATE; BUTYL PARABEN; N-BUTYL PARAHYDROXYBENZOATE; BUTYL PARASEPT; BUTYL TEGOSEPT; BUTYL-PARASEPT; 4-HYDROXYBENZOIC ACID BUTYL ESTER; P-HYDROXYBENZOIC ACID BUTYL ESTER; P-HYDROXYBENZOIC ACID N-BUTYL ESTER; NIPABUTYL; PARASEPT; PRESERVAL B; SOLBROL B; SPF; TEGOSEPT B; TEGOSEPT BUTYL
Description: small, colorless crystals or white crystalline powder or finely divided solid; odorless
Use: preservative in many creams, lotions, ointments and other cosmetics, foods, drugs and dentifrices; active against molds, fungi and yeasts, but less active against bacteria

Physical Properties

Freezing Point: 68 °C (154.4 °F) to 69 °C (156.2 °F)
Water Solubility: < 1 mg/mL at 17 C
Other Solubilities: freely Soluble in Acetone, Alcohol, Ether, Chloroform; in Glycerin: 0.3 g/100 g at 25 °C; in Propylene Glycol: 110 g/100 g at 25 °C.
Flash Point: Not available; probably combustible

RTECS Toxicity Data

Acute Oral: Mouse LD_{50} Dose: 13200 mg/kg.
Chronic (Multiple Dose) Oral: Rat Dose: 504 gm/kg/12W-C; Toxic Effects: Nutritional and gross metabolic - Weight loss or decreased weight gain; DEATH.
Irritation Skin: Guinea Pig Standard Draize Test Dose: 5%/48H; Reaction: mild.

Hazard Overviews

Health: Irritating to eyes/skin/respiratory tract. Other Acute Effects: may be harmful by inhalation, ingestion, or skin absorption.
Fire: Will burn. Hazards: emits toxic fumes. Extinguishing agents: water spray; carbon dioxide, dry chemical powder or appropriate foam. Precautions: combustible liquid.
Carcinogenicity: IARC - Not listed; NIOSH - Not listed; NTP - Not listed; ACGIH - Not listed; OSHA - Not listed; EPA - Not listed; MAK - Not listed
Primary Target Organs:

Eyes

Skin

Respiratory System

Environmental

Regulations

RCRA 40CFR: Not listed
CERCLA: 40CFR 302.4: Not listed
SARA 40CFR 372.65: Not listed
SARA EHS 40CFR 355: Not listed
TSCA: Listed

BUT8070 CAS #: 88-18-6

2-TERT-BUTYLPHENOL

RTECS: SJ8921000
DOT: UN2430; UN3145; IMO8.0
EINECS Number: 201-807-2
Molecular Formula: $C_{10}H_{14}O$
Formula Weight: 150.2

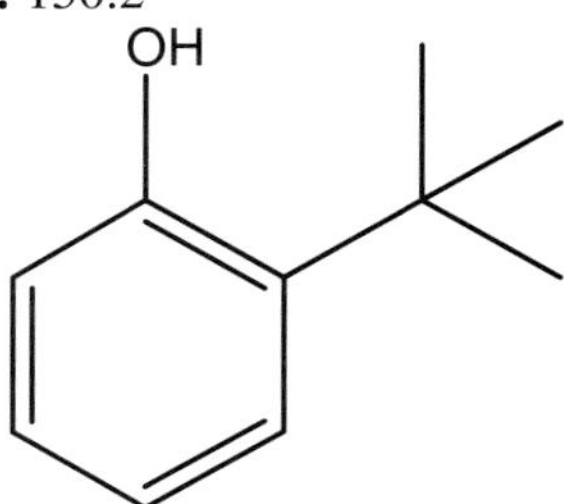

Chemical Structure

Synonyms: BENZENE,1-TERT-BUTYL-2-HYDROXY-; 2-TERT-BUTYL-1-HYDROXYBENZENE; 2-T-BUTYLPHENOL; O-T-BUTYLPHENOL; O-TERT-BUTYLPHENOL; 2-(1,1-DIMETHYLETHYL)PHENOL; PHENOL,O-TERT-BUTYL-; PHENOL,2-(1,1-DIMETHYLETHYL)-
Description: light yellow to amber-colored liquid
Use: intermed for synthetic resins, plasticizers, surface-active agents, perfumes, & other products; permissible antioxidant for aviation gasoline; 2-t-butylphenol starting material for synthesis of antioxidants & agrochems; fragrance compounds made from cis-2-t-butylcyclohexanol which obtained by hydrogenation of 2-t-butylphenol in presence of catalysts; derivatives: 2-t-butylcyclohexyl acetate; dinoterb; ethylene glycol bis(3,3-bis(3-t-butyl-4-hydroxyphenyl)butyrate); 2,2'-methylenebis(4-ethyl-6-t-butyl-phenol).

Physical Properties

Boiling Point: 223 °C (433 °F) at 760 mm Hg
Freezing Point: -6.8 °C (19.76 °F)
Specific Gravity: 0.9783 at 20 °C/4 °C
Saturated Vapor Density: 1.20000099 kg/m^3
Vapor Pressure: 1.5 $x10^{-4}$ mm Hg at 25 °C
Water Solubility: Insoluble in Water
Other Solubilities: Soluble in Isopentane, Toluene & Ethyl Alcohol.
Refraction Index: 1.5160 at 20 °C/D
Ionization Potential (eV): 8.10 +/-0.02
Flash Point: 110 °C Open Cup

Hazard Overviews

Corrosive

Health: Corrosive to eyes/skin/respiratory system. Harmful. Other Acute Effects: harmful if swallowed, inhaled, or absorbed through skin; inhalation may result in spasm, inflammation and edema of the larynx and bronchi, chemical pneumonitis and pulmonary edema; symptoms of exposure may include burning sensation, coughing, wheezing, laryngitis, shortness of breath, headache, nausea and vomiting.
Fire: Will burn. Hazards: under fire conditions, material may decompose to form flammable and/or explosive mixtures in air. Extinguishing agents: water spray; carbon dioxide, dry chemical powder or appropriate foam. Precautions: combustible liquid.
Reactivity: Incompatible with: bases, acid chlorides, acid anhydrides, oxidizing agents, corrodes steel, brass, copper, copper alloys. Hazardous decomposition products: toxic fumes of: carbon monoxide, carbon dioxide.
Carcinogenicity: IARC - Not listed; NIOSH - Not listed; NTP - Not listed; ACGIH - Not listed; OSHA - Not listed; EPA - Not listed; MAK - Not listed

Primary Target Organs:

Eyes

Skin

Respiratory System

Environmental

Environmental Fate: If released to the atmosphere, it will exist in both the vapor and particulate phases based on an experimental vapor pressure of 1.5 $x10^{-4}$ mm Hg at 25 °C. Vapor-phase should be degraded in the atmosphere by reaction with photochemically produced hydroxyl radicals with an estimated half-life of about 9.5 hours. Particulate-phase may be physically removed from the air by wet and dry deposition. It is expected to have low mobility in soil based on an estimated K_{oc} value of 1500. An estimated Henry's Law constant of 1.4 $x10^{-6}$ suggests that volatilization from moist soil surfaces will be slow. Limited data, mainly derived by analogy to 4-t-butylphenol, indicate that it may be resistant to biodegradation in soil and water environments. It (at 30 mg/l) was not biodegraded over a 2 week period using an activated sludge inoculum; in river and sea water 11% of the initial concentration was biodegraded over 3 days. An estimated BCF value of 190 indicates bioconcentration in aquatic organisms may be an important fate process. This compound was rapidly taken up by zebra fish with a steady state concentration reached within 5 hours; the clearance phase required 6 hours. It may volatilize from water surfaces given its estimated Henry's Law constant. Estimated volatilization half-lives for a model river and model lake are 31 and 230 days, respectively.

Cleanup/Disposal: Guide No. 153: Eliminate all ignition sources (no smoking, flares, sparks or flames in immediate area). Do not touch damaged containers or spilled material unless wearing appropriate protective clothing. Stop leak if you can do it without risk. Prevent entry into waterways, sewers, basements or confined areas. Absorb or cover with dry earth, sand or other non-combustible material and transfer to containers. Do not get water inside containers.

Environmental Physical Data

Henry's Law Constant: estimated at 1.4 $x10^{-6}$
Octanol/Water Partition Coefficient: log K_{ow} = 3.31
Sorption Partition Coefficient: K_{oc} = estimated at 1500
BCF: estimated at 190

Regulations

RCRA 40CFR: Not listed
CERCLA: 40CFR 302.4: Not listed
SARA 40CFR 372.65: Not listed
SARA EHS 40CFR 355: Not listed
TSCA: Listed

Analytical Methods

Water / Groundwater: ASTM D4763

BUT8140 **CAS #: 98-54-4**

4-T-BUTYLPHENOL

RTECS: SJ8925000
EINECS Number: 202-679-0
Molecular Formula: $C_{10}H_{14}O$
Structured MF: 1, 4-$(CH_3)_3CC_6H_4OH$
Formula Weight: 150.22

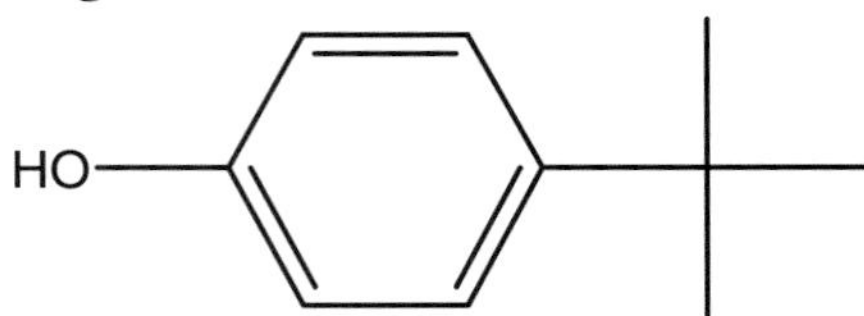

Chemical Structure

Synonyms: BUTYLPHEN; 4-TERT-BUTYLPHENOL; P-TERT-BUTYLPHENOL; PARA-TERT-BUTYLPHENOL; 4-(1,1-DIMETHYLETHYL) PHENOL; 4-(1,1-DIMETHYLETHYL)PHENOL; 1-HYDROXY-4-TERT-BUTYLBENZENE; PHENOL,P-(TERT-BUTYL)-; PHENOL,4-(1,1-DIMETHYLETHYL)-; PTBP; P-TERC.BUTYLFENOL; UCAR BUTYLPHENOL 4-T; UCAR BUTYLPHENOL 4-T FLAKE
Description: white, faintly pink crystals; distinctive phenol odor
Use: intermediate in the manufacture of varnish and lacquer resins; As a soap antioxidant; ingredient in de-emulsifiers; for oil field use; in motor oil additives

Physical Properties

Boiling Point: 239.5 °C (463 °F)
Freezing Point: 101 °C (213.8 °F)
Specific Gravity: 0.908 at 80/4 °C
Vapor Density: 5.1
Vapor Pressure: 1 torr at 70 °C
Water Solubility: Insoluble
Other Solubilities: 95% Ethanol: Soluble; Ether: Soluble.
Refraction Index: 1.4787
Ionization Potential (eV): 7.8
Flash Point: 113 °C Closed Cup

RTECS Toxicity Data

Acute Oral: Rat LD_{50} Dose: 3250 uL/kg. Mammal LD_{50} Dose: 1500 mg/kg.
Acute Inhalation: Rat LC_{Lo} Dose: 5600 mg/m^3/4hr; Toxic Effects: Lungs, Thorax, or Respiration - Respiratory depression; Skin and appendages - Primary irritation; Nutritional and gross metabolic - Weight loss or decreased weight gain.
Acute Dermal: Rabbit LD_{50} Route: Skin; Dose: 2520 uL/kg. Mammal LD_{50} Route: Skin; Dose: 1580 mg/kg.
Irritation Eye: Rabbit Standard Draize Test Dose: 10 mg; Reaction: severe. Rabbit Standard Draize Test Dose: 50 ug/24H; Reaction: severe.
Irritation Skin: Rabbit Standard Draize Test Dose: 500 mg/24H; Reaction: mild. Rabbit Standard Draize Test Dose: 500 mg/4H; Reaction: mild.

Tumorigenic: Hamster Route: Oral; Dose: 252 gm/kg/20W-C; Toxic Effects: Tumorigenic - Neoplastic by RTECS criteria; Gastrointestinal - Tumors.

Hazard Overviews

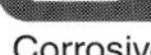
Corrosive

Fire Diamond

Health: Corrosive to eyes/skin/respiratory tract. Also Causes: difficulty breathing, drop in blood pressure, tachycardia, cardiac arrythmias. Chronic Effects: contact dermatitis sensitization.
Fire: Will burn. Use dry chemical, carbon dioxide, foams, or water fog. Dust explosions are significant hazards if a particulate cloud contacts an ignition source.
Reactivity: Stable. Hazardous polymerization cannot occur. Incompatible with: strong oxidizing agents. Hazardous decomposition products: carbon monoxide; aromatic and phenolic compounds.
Carcinogenicity: IARC - Not listed; NIOSH - Not listed; NTP - Not listed; ACGIH - Not listed; OSHA - Not listed; EPA - Not listed; MAK - Not listed
Primary Target Organs:

Eyes

Skin
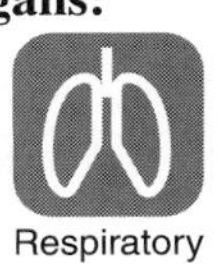
Respiratory System

Mucous Membranes

Exposure Limits
DFG MAK: TWA: 0.08 ppm; 0.5 mg/m^3.

Environmental

Cleanup/Disposal: Guide No. 153: Eliminate all ignition sources (no smoking, flares, sparks or flames in immediate area). Do not touch damaged containers or spilled material unless wearing appropriate protective clothing. Stop leak if you can do it without risk. Prevent entry into waterways, sewers, basements or confined areas. Absorb or cover with dry earth, sand or other non-combustible material and transfer to containers. Do not get water inside containers.

Environmental Physical Data

Octanol/Water Partition Coefficient: log K_{ow} = 3.31
BCF: no food chain concentration potential

Regulations

RCRA 40CFR: Not listed
CERCLA: 40CFR 302.4: Not listed
SARA 40CFR 372.65: Not listed
SARA EHS 40CFR 355: Not listed
TSCA: Listed

Analytical Methods

Soil: EPA PMD-PFH
Water / Groundwater: ASTM D4763

BUT8210 CAS #: 1942-71-8

2-(P-T-BUTYLPHENOXY)CYCLOHEXANOL

EINECS Number: 217-732-3
Molecular Formula: $C_{16}H_{24}O_2$
Formula Weight: 248.37
Synonyms: CYCLOHEXANOL,2-(P-TERT-BUTYLPHENOXY)-

Hazard Overviews

Carcinogenicity: IARC - Not listed; NIOSH - Not listed; NTP - Not listed; ACGIH - Not listed; OSHA - Not listed; EPA - Not listed; MAK - Not listed

Environmental

Regulations

RCRA 40CFR: Not listed
CERCLA: 40CFR 302.4: Not listed
SARA 40CFR 372.65: Not listed
SARA EHS 40CFR 355: Not listed
TSCA: Listed

BUT8280 CAS #: 56803-37-3

T-BUTYLPHENYL DIPHENYL PHOSPHATE

EINECS Number: 260-391-0
Molecular Formula: $C_{22}H_{23}O_4P$
Formula Weight: 382.40
Synonyms: TERT-BUTYLPHENYL DIPHENYL PHOSPHATE; (1,1-DIMETHYLETHYL)PHENYL DIPHENYL ESTER PHOSPHORIC ACID; DIPHENYL MONO(P-TERT-BUTYLPHENYL)PHOSPHATE; PHOSPHORIC ACID,(1,1-DIMETHYLETHYL)PHENYL DIPHENYL ESTER
Description: colorless, clear liquid
Use: flame retardant plasticizer for polyvinyl chloride, for other vinyl polymers, for cellulosics; plasticizer

Physical Properties

Boiling Point: 261 °C (502 °F) at 6 mm Hg
Specific Gravity: 1.18 at 25 °C
Saturated Vapor Density: 1.200000027 kg/m^3
Vapor Pressure: 1.4 x10^{-6} mm Hg at 25 °C
Water Solubility: 3.2 mg/L at 25 °C
Other Solubilities: 95% Ethanol: >=100 mg/ml at 21 °C; Acetone: >=100 mg/ml at 21 °C; DMSO: >=100 mg/ml at 21 °C.
Flash Point: 223.9 °C

Hazard Overviews

Fire: Will burn.

Carcinogenicity: IARC - Not listed; NIOSH - Not listed; NTP - Not listed; ACGIH - Not listed; OSHA - Not listed; EPA - Not listed; MAK - Not listed

Environmental

Environmental Fate: If released to the atmosphere, it will exist in both the vapor phase and the particulate phase. It will degrade in the vapor phase by reaction with hydroxyl radicals with a half-life of 23 hours. In soil, it may be immobile with estimated K_{oc} values of 2300 and 15000. Hydrolysis of the phosphate ester, especially in alkaline environments, is expected to be a major fate process for this compound. Biodegradation by both bacteria and fungi has been shown. The main degradation product produced by fungi was 4-(2-carboxy-2-propyl)triphenyl phosphate with phenolic metabolites as minor components. Half-lives, considering both abiotic and biotic transformations, of 0.44 and 39 days in water and sediment, respectively, were measured for this compound in an artificial pond system. It rapidly binds to particulate matter and sediment in the water column. This compound is also expected to hydrolyze and volatilize from water surfaces. Biodegradation in the water column is probable; in river die-away studies it was completely biodegraded within 11 days. Microcosm studies using water from riverine, estuarine, or lacustrine waters showed from 1.7 to 37.2% mineralized after 8 weeks. Measured BCF values of 528-785 and 1096 for fathead minnows and rainbow trout, respectively, indicate that bioconcentration in aquatic organisms may be possible. However, measured depuration rates of 13.7, 7.8, and 43 hours for rainbow trout, fathead minnows, and chironomid larvae suggest that bioconcentration may not be an important fate process.

Environmental Physical Data

Henry's Law Constant: 8.88×10^{-7}
Octanol/Water Partition Coefficient: log K_{ow} = 5.12
Sorption Partition Coefficient: K_{oc} = 2300 to 1.5×10^{4}
BCF: rainbow trout 1096

Regulations

RCRA 40CFR: Not listed
CERCLA: 40CFR 302.4: Not listed
SARA 40CFR 372.65: Not listed
SARA EHS 40CFR 355: Not listed
TSCA: Listed

BUT8350 **CAS #: 68958-22-5**

P-TERT-BUTYLPHENYLGLYCIDYL ETHER

Physical Properties

Specific Gravity: 1.02
Vapor Density: > 1 Air=1
Vapor Pressure: < 0.1
Water Solubility: Negligible
Evaporation Rate: < 1 Butyl Acetate=1

Hazard Overviews

Carcinogenicity: IARC - Not listed; NIOSH - Not listed; NTP - Not listed; ACGIH - Not listed; OSHA - Not listed; EPA - Not listed; MAK - Not listed

Environmental

Regulations

RCRA 40CFR: Not listed
CERCLA: 40CFR 302.4: Not listed
SARA 40CFR 372.65: Not listed
SARA EHS 40CFR 355: Not listed
TSCA: Not listed

BUT8420 **CAS #: 339-43-5**

1-BUTYL-3-SULFANILYLUREA

RTECS: YS4200000
EINECS Number: 206-424-4
Molecular Formula: $C_{11}H_{17}N_3O_3S$
Formula Weight: 271.37

NH2
O=S=O
H N
NH
O

Chemical Structure

Synonyms: ALENTIN; N'-(4-AMINOBENZENESULFONYL)-N-BUTYLUREA; N-(4-AMINOBENZENESULFONYL)-N'-BUTYLUREA; 4-AMINO-N-((BUTYLAMINO)CARBONYL)BENZENESULFONAMIDE; AMINOPHENUROBUTANE; N'-(4-AMINOPHENYLSULFONYL)-N-BUTYLUREA; BENZENESULFONAMIDE,4-AMINO-N-((BUTYLAMINO)CARBONYL)-; BUCARBAN; BUCROL; BUKARBAN; BURCOL; BUTISULFINA; N'-(BUTYLCARBAMOYL)SULFANILAMIDE; N(SUP 1)-(BUTYLCARBAMOYL)SULFANILAMIDE; N-BUTYL-N'-SULFANILYLUREA; N-BUTYLSULFANILYLUREA; BZ 55; CARBUTAMID; CARBUTAMIDE; CICLORAL; DIABORAL; EMEDAN; GLUCIDORAL; GLUCOFREN; GLYBUTAMIDE; INBUTON; INVENOL; NADISAN; NADIZAN; NORBORAL; ORANIL; ORANYL; ORASULIN; N(SUP 1)-SULFANILYL-N(SUP 2)-BUTYLCARBAMIDE; N(SUP 1)-SULFANILYL-N(SUP 2)-BUTYLUREA; N-SULFANILYL N'BUTYLUREE; N-SULFANILYL-N'-BUTYLUREA; U 6987; UREA,1-BUTYL-3-SULFANILYL-

Description: crystals
Use: medication: hypoglycemic agent; hypoglycemic agent

Physical Properties

Freezing Point: 144 °C (291.2 °F) to 145 °C (293 °F)
Water Solubility: Soluble in Water (pH 6-8)

RTECS Toxicity Data

Acute Oral: Rat LD_{50} Dose: 7800 mg/kg. Mouse LD_{50} Dose: 2800 mg/kg.

Acute Dermal: Mouse LD_{50} Route: Subcutaneous Dose: 2640 mg/kg.

Chronic (Multiple Dose) Oral: Rat Dose: 307 gm/kg/29W-C; Toxic Effects: Nutritional and gross metabolic - Other changes; DEATH. Monkey Dose: 36500 mg/kg/10W-I; Toxic Effects: Lungs, Thorax, or Respiration - Chronic pulmonary edema; Lungs, Thorax, or Respiration - Other changes; DEATH.

Reproductive/Teratogenic: Rat Route: Oral; Dose: 1056 mg/kg; Duration: female 7-14D of pregnancy; Specific Developmental Abnormalities - Musculoskeletal system. Rat Route: Oral; Dose: 1 gm/kg; Duration: female 10D of pregnancy; Effects on Embryo or Fetus - Fetal death; Specific Developmental Abnormalities - Other developmental abnormalities. Rat Route: Oral; Dose: 5 gm/kg; Duration: female 9-10D of pregnancy; Specific Developmental Abnormalities - Eye, ear; Craniofacial (including nose and tongue); Homeostasis. Rat Route: Oral; Dose: 3200 mg/kg; Duration: female 7-14D of pregnancy; Effects on Embryo or Fetus - Fetotoxicity. Rat Route: Oral; Dose: 12 gm/kg; Duration: female 1-12D of pregnancy; Specific Developmental Abnormalities - Eye, ear.

Hazard Overviews

Health: Irritating to eyes/skin/respiratory tract. Other Acute Effects: may be harmful by inhalation, ingestion, or skin absorption; target organs: pancreas.

Fire: Hazards: emits toxic fumes. Extinguishing agents: water spray; carbon dioxide, dry chemical powder or appropriate foam. Precautions: combustible liquid.

Reactivity: Incompatible with: strong oxidizing agents. Hazardous decomposition products: toxic fumes of: carbon monoxide, carbon dioxide, nitrogen oxides, sulfur oxides.

Carcinogenicity: IARC - Not listed; NIOSH - Not listed; NTP - Not listed; ACGIH - Not listed; OSHA - Not listed; EPA - Not listed; MAK - Not listed

Primary Target Organs:

Eyes

Skin

Respiratory System

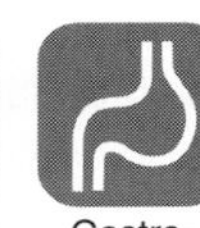
Gastro-intestinal

Environmental

Regulations

RCRA 40CFR: Not listed
CERCLA: 40CFR 302.4: Not listed
SARA 40CFR 372.65: Not listed
SARA EHS 40CFR 355: Not listed
TSCA: Listed

BUT8490 CAS #: 98-51-1

P-TERT-BUTYLTOLUENE

RTECS: XS8400000
DOT: UN2667; IMO6.1
EINECS Number: 202-675-9
Molecular Formula: $C_{11}H_{16}$
Structured MF: $(CH_3)_3CC_6H_4CH_3$
Formula Weight: 148.25

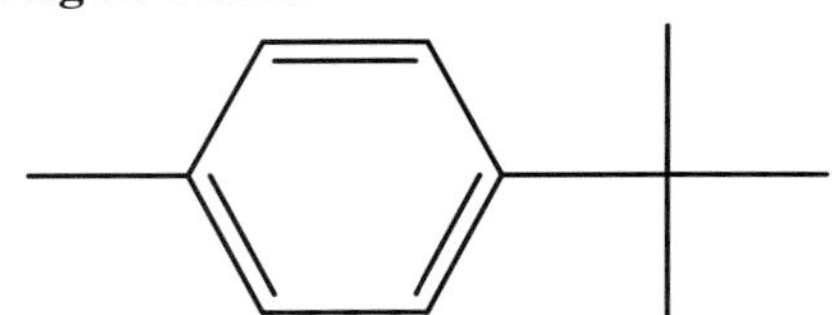

Chemical Structure

Synonyms: BENZENE,1-(1,1-DIMETHYLETHYL)-4-METHYL-; 1-TERT-BUTYL-4-METHYLBENZENE; 4-TERT-BUTYL-1-METHYLBENZENE; 4-TERT-BUTYLTOLUENE; 1-METHYL-4-TERT-BUTYLBENZENE; 4-METHYL-TERT-BUTYLBENZENE; P-METHYL-TERT-BUTYLBENZENE; 8-METHYLPARACYMENE; PTBT; P-TBT; TBT; TOLUENE,P-TERT-BUTYL-

Description: colorless liquid; gasoline-like odor

Use: solvent; intermediate in organic syntheses; solvent for resins

Physical Properties

Boiling Point: 193 °C (379 °F) at 760 mm Hg
Freezing Point: -52 °C (-61.6 °F)
Specific Gravity: 0.8612 at 20 °C/4 °C
Vapor Density: 4.62 Air=1
Saturated Vapor Density: 1.204220281 kg/m^3
Vapor Pressure: 0.65 mm Hg at 25 °C
Water Solubility: Insoluble
Other Solubilities: 95% Ethanol: >=100 mg/ml at 17 °C; Acetone: >=100 mg/ml at 17 °C; Benzene: Soluble; Common industrial solvents: miscible; DMSO: >=100 mg/ml at 17 °C; Ether: Soluble.
Odor Threshold: 30 mg/m^3
Refraction Index: 1.4918 at 20 °C
Ionization Potential (eV): 8.28
Flash Point: 68 °C

RTECS Toxicity Data

Acute Oral: Rat LD_{50} Dose: 1555 mg/kg; Toxic Effects: Behavioral - Somnolence (general depressed activity); Behavioral - Ataxia; Lungs, Thorax, or Respiration - Dyspnea. Mouse LD_{50} Dose: 778 mg/kg; Toxic Effects: Behavioral - Muscle weakness; Behavioral - Ataxia; Lungs, Thorax, or Respiration - Respiratory depression.

Acute Inhalation: Rat LC_{50} Dose: 165 ppm/8hr; Toxic Effects: Sense organs and special senses - Other; Lungs, Thorax, or

Respiration - Dyspnea; Gastrointestinal - Changes in structure or function of salivary glands. Human TC_{Lo} Dose: 10 ppm/3M; Toxic Effects: Gastrointestinal - Nausea or vomiting. Human TC_{Lo} Dose: 20 ppm/5M; Toxic Effects: Sense organs and special senses - Conjunctive irritation; Sense organs and special senses - Change in function; Gastrointestinal - Nausea or vomiting.

Acute Dermal: Rabbit LD_{50} Route: Skin; Dose: 16934 mg/kg; Toxic Effects: Behavioral - Tremor; Behavioral - Convulsions or effect on seizure threshold; Nutritional and gross metabolic - Body temperature decrease.

Chronic (Multiple Dose) Inhalation: Rat Dose: 20 ppm/6H/14D-I; Toxic Effects: Brain and coverings - Recordings from specific areas of CNS. Rat Dose: 850 ppm/1H/5D-I; Toxic Effects: Automatic Nervous System - Other (direct) parasympathomimetic; Lungs, Thorax, or Respiration - Acute pulmonary edema; DEATH. Rat Dose: 50 ppm/2H/26W-I; Toxic Effects: Liver - Changes in Liver weight; Kidney, Ureter, and Bladder - Changes in kidney weight; Blood - Changes in leukocyte (WBC) cell count. Rat Dose: 25 ppm/2H/73D-I; Toxic Effects: Liver - Changes in Liver weight. Rat Dose: 50 ppm/7H/37D-I; Toxic Effects: Liver - Changes in Liver weight.

Irritation Skin: Rabbit Standard Draize Test Dose: 500 mg/24H; Reaction: mild.

Hazard Overviews

Fire Diamond

Health: Irritating to eyes/skin/respiratory tract. Also Causes: nausea, metallic or menthol taste, dizziness, difficult respiration, vomiting, CNS depression. Chronic Effects: nausea, malaise, headache, decreased blood pressure with an increased heart rate, liver damage, red and white blood cell deficiency, prolonged blood clotting time, coronary insufficiency.

Fire: Combustible. Can form explosive mixtures in the air. Use dry chemical, carbon dioxide, water spray, fog, or regular foam.

Reactivity: Stable. Hazardous polymerization cannot occur. Avoid: heat; ignition sources. Incompatible with: oxidizers. Hazardous decomposition products: carbon oxides; acrid smoke.

Carcinogenicity: IARC - Not listed; NIOSH - Listed as carcinogen; NTP - Not listed; ACGIH - Not listed; OSHA - Not listed; EPA - Not listed; MAK - Not listed

Primary Target Organs:

Eyes

Skin

Nervous System

Liver

Cardio-vascular

Blood

Exposure Limits

OSHA PEL: TWA: 10 ppm; 60 mg/m^3.

OSHA PEL Vacated 1989 Limits: TWA: 10 ppm; 60 mg/m^3; STEL: 20 ppm; 120 mg/m^3.

ACGIH TLV: TWA: 1 ppm; 6.1 mg/m^3.

NIOSH REL: TWA: 10 ppm; 60 mg/m^3. STEL: 20 ppm; 120 mg/m^3.

NIOSH IDLH: 100 ppm.

DFG MAK: TWA: 10 ppm; 60 mg/m^3.

Respirator Recommendation

Exposure Range: >10 to <100 ppm Supplied Air, Constant Flow/Pressure Demand, Half Mask

Exposure Range: 100 to unlimited ppm Self-contained Breathing Apparatus, Pressure Demand, Full Face

Note: poor warning properties

Environmental

Ecotoxicity: LD_{50} Goldfish 3 mg/l/24 hr (modified ASTM-D1345) /Conditions of bioassay not specified

Environmental Fate: If released to the atmosphere, it will degrade in the vapor-phase by reaction with photochemically produced hydroxyl radicals (estimated half-life of 2.2 days). If released to water, volatilization will be an important process. Volatilization half-lives ranging from 3.6 hours to 13 days can be estimated for an model environmental river and pond. If released to soil, low mobility is expected since it adsorbs strongly (estimated K_{oc} of 1700-1900). Evaporation from dry surfaces should occur. A single biodegradation screening study suggests that it does not biodegrade easily.

Cleanup/Disposal: Guide No. 131: Fully encapsulating, vapor protective clothing should be worn for spills and leaks with no fire. Eliminate all ignition sources (no smoking, flares, sparks or flames in immediate area). All equipment used when handling the product must be grounded. Do not touch or walk through spilled material. Stop leak if you can do it without risk. Prevent entry into waterways, sewers, basements or confined areas. A vapor suppressing foam may be used to reduce vapors. Small Spills: Absorb with earth, sand or other non-combustible material and transfer to containers for later disposal. Use clean non-sparking tools to collect absorbed material. Large Spills: Dike far ahead of liquid spill for later disposal. Water spray may reduce vapor; but may not prevent ignition in closed spaces.

Environmental Physical Data

Henry's Law Constant: estimated at 0.01535

Sorption Partition Coefficient: K_{OC} = 1700 to 1900

BCF: estimated at 236

Regulations

RCRA 40CFR: Not listed

CERCLA: 40CFR 302.4: Not listed

SARA 40CFR 372.65: Not listed

SARA EHS 40CFR 355: Not listed

TSCA: Listed

Analytical Methods

Air: ASTM D3686, D3687

Indoor / Expired Air: NIOSH 1501

BUT8560 CAS #: 592-31-4

N-BUTYLUREA

RTECS: YS3675000
EINECS Number: 209-748-4
Molecular Formula: $C_5H_{12}N_2O$
Formula Weight: 116.19

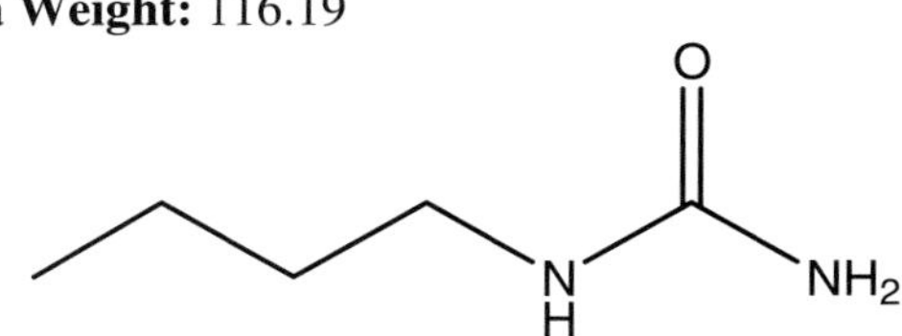

Chemical Structure

Synonyms: 1-BUTYLUREA; BUTYLUREA; N-N-BUTYLUREA; UREA,BUTYL-
Description: white solid, tablets, or needles; odorless
Use: oxidizing agent for copper

Physical Properties

Boiling Point: Decomposes
Freezing Point: 96 °C (204.8 °F)
Water Solubility: Very Soluble in Water
Other Solubilities: 95% Ethanol: >=100 mg/ml at 18 °C; Acetone: 50-100 mg/ml at 18 °C; DMSO: 50-100 mg/ml at 18 °C; Ether: Soluble.
Flash Point: Not available; probably combustible

RTECS Toxicity Data

Acute Oral: Rat LD_{50} Dose: 1255 mg/kg.
Mutagenic: Rat Cytogenetic Analysis; Route: Oral; Dose: 100 mg/kg. Hamster Cytogenetic Analysis; Cell Type: fibroblast; Dose: 4 gm/L/48H.

Hazard Overviews

Health: Irritating to eyes/skin. Harmful. Other Acute Effects: harmful if swallowed, inhaled, or absorbed through skin. Chronic Effects: laboratory experiments have shown mutagenic effects.
Fire: Will burn. Hazards: emits toxic fumes. Extinguishing agents: water spray; carbon dioxide, dry chemical powder or appropriate foam. Precautions: combustible liquid.
Reactivity: Incompatible with: strong oxidizing agents. Hazardous decomposition products: toxic fumes of: carbon monoxide, carbon dioxide, nitrogen oxides.
Carcinogenicity: IARC - Not listed; NIOSH - Not listed; NTP - Not listed; ACGIH - Not listed; OSHA - Not listed; EPA - Not listed; MAK - Not listed

Primary Target Organs:

Eyes

Skin

Environmental

Regulations
RCRA 40CFR: Not listed
CERCLA: 40CFR 302.4: Not listed
SARA 40CFR 372.65: Not listed
SARA EHS 40CFR 355: Not listed
TSCA: Listed

BUT8630 CAS #: 689-11-2

SEC-BUTYLUREA

RTECS: YS3678000
EINECS Number: 211-709-1
Molecular Formula: $C_5H_{12}N_2O$
Formula Weight: 116.16

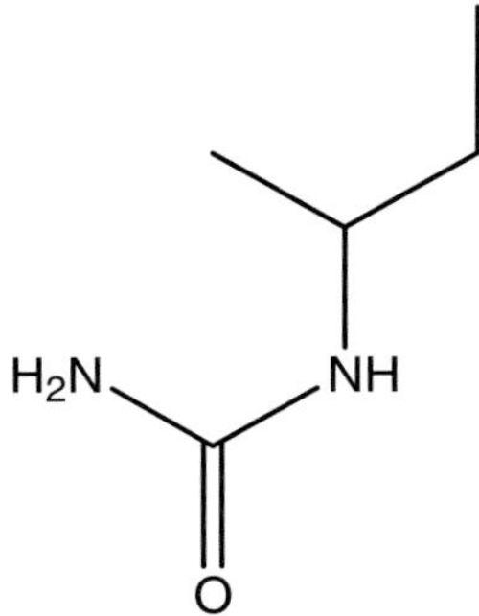

Chemical Structure

Synonyms: N-SEC-BUTYLUREA; UREA,1-SEC-BUTYL-; UREA,SEC-BUTYL-; UREA,(1-METHYLPROPYL)-

Physical Properties

Other Solubilities: Soluble in Alcohol, Ether

RTECS Toxicity Data

Acute Oral: Rat LD_{Lo} Dose: 7500 mg/kg.

Hazard Overviews

Carcinogenicity: IARC - Not listed; NIOSH - Not listed; NTP - Not listed; ACGIH - Not listed; OSHA - Not listed; EPA - Not listed; MAK - Not listed

Environmental

Regulations
RCRA 40CFR: Not listed
CERCLA: 40CFR 302.4: Not listed
SARA 40CFR 372.65: Not listed
SARA EHS 40CFR 355: Not listed

TSCA: Listed

BUT8770 CAS #: 110-65-6

1,4-BUTYNEOIOL

RTECS: ES0525000
DOT: UN2716; IMO6.1
EINECS Number: 203-788-6
Molecular Formula: $C_4H_6O_2$
Structured MF: $HOCH_2C{=}{=}CCH_2OH$
Formula Weight: 86.09

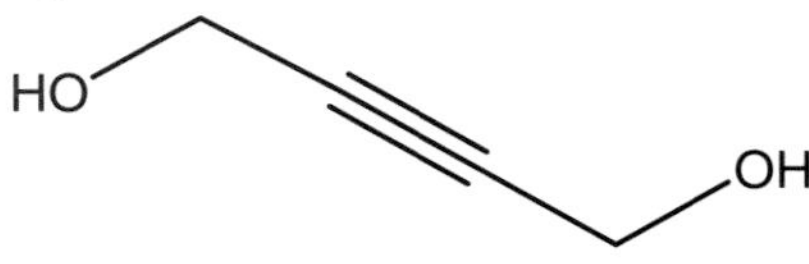

Chemical Structure

Synonyms: AGRISYNTH B3D; BIS(HYDROXYMETHYL)ACETYLENE; 2-BUTIN-1,4-DIOL; 1,4-BUTINODIOL; 1,4-BUTYNEDIOL; 2-BUTYNE-1,4-DIOL; 2-BUTYNEDIOL; BUTYNEDIOL; BUTYNEDIOL-1,4; 1,4-DIHYDROXY-2-BUTYNE
Description: white to light yellow plates
Use: intermediate; corrosion inhibitor; electroplating brightener; defoliant; polymerization accelerator; stabilizer for chlorinated hydrocarbons; cosolvent for paint & varnish removal

Physical Properties

Boiling Point: 238 °C (460 °F) at 760 mm Hg
Freezing Point: 58 °C (136.4 °F)
Specific Gravity: Solid 1.07 at 20 °C
Water Solubility: Soluble in Water
Other Solubilities: Soluble in aqueous acids.
Refraction Index: 1.4804 at 20 °C/D
Flash Point: 128.333 °C Open Cup

RTECS Toxicity Data

Acute Oral: Rat LD_{50} Dose: 105 mg/kg; Toxic Effects: Behavioral - Somnolence (general depressed activity); Lungs, Thorax, or Respiration - Respiratory stimulation; Blood - Other changes. Mouse LD_{50} Dose: 105 mg/kg; Toxic Effects: Behavioral - Somnolence (general depressed activity); Lungs, Thorax, or Respiration - Respiratory stimulation; Blood - Other changes.
Acute Inhalation: Rat LC_{Lo} Dose: 150 mg/m^3/2hr; Toxic Effects: Behavioral - Somnolence (general depressed activity); Lungs, Thorax, or Respiration - Structural or functional change in trachea or bronchi; Blood - Hemorrhage. Mouse LC_{Lo} Dose: 150 mg/m^3/2hr; Toxic Effects: Behavioral - Somnolence (general depressed activity); Lungs, Thorax, or Respiration - Structural or functional change in trachea or bronchi; Blood - Hemorrhage.
Chronic (Multiple Dose) Oral: Rat Dose: 1400 mg/kg/28D-I; Toxic Effects: Liver - Changes in liver weight; Blood - Changes in leukocyte (WBC) cell count; Biochemical - Dehydrogenases. Rat Dose: 364 mg/kg/26W-I; Toxic Effects: Blood - Changes in serum composition; Biochemical - True cholinesterase; Biochemical - Other carbohydrates. Rat Dose: 1400 mg/kg/14D-I; Toxic Effects: Liver - Changes in liver weight; Nutritional and gross metabolic - Weight loss or decreased weight gain.
Chronic (Multiple Dose) Inhalation: Rat Dose: 8 mg/m^3/4H/26W-I; Toxic Effects: Brain and coverings - Recordings from specific areas of CNS; Vascular - BP lowering not characterized in autonomic section; Liver - Liver function tests impaired. Mouse Dose: 90 mg/m^3/2H/30D-I; Toxic Effects: Brain and coverings - Other degenerative changes; Blood - Hemorrhage; DEATH.

Hazard Overviews

Corrosive

Health: Irritating to eyes/skin/respiratory tract. Toxic. Other Acute Effects: harmful if swallowed, inhaled, or absorbed through skin; possible sensitizer; prolonged or repeated exposure may cause allergic reactions in certain sensitive individuals; prolonged exposure can cause narcotic effect; CNS depression.
Fire: Will burn. Hazards: in powder form capable of creating a dust explosion; container explosion may occur. Extinguishing agents: water spray; carbon dioxide, dry chemical powder or appropriate foam. Precautions: combustible liquid.
Reactivity: Incompatible with: strong oxidizing agents, acid chlorides, acid anhydrides, strong bases, strong acids. Hazardous decomposition products: toxic fumes of: carbon monoxide, carbon dioxide.
Carcinogenicity: IARC - Not listed; NIOSH - Not listed; NTP - Not listed; ACGIH - Not listed; OSHA - Not listed; EPA - Not listed; MAK - Not listed
Primary Target Organs:

Eyes

Skin

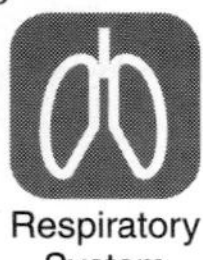
Respiratory System

Environmental

Ecotoxicity: LC_{50} Pimephales promelas (fathead minnow) 53.6 mg/l/96 hr (confidence limit 49.3 - 58.3 mg/l), flow-through bioassay with measured concentrations, 25.1 °C, dissolved oxygen 6.8 mg/l, hardness 46.5 mg/l calcium carbonate, alkalinity 43.5 mg/l calcium carbonate, and pH 7.7
Cleanup/Disposal: Guide No. 153: Eliminate all ignition sources (no smoking, flares, sparks or flames in immediate area). Do not touch damaged containers or spilled material unless wearing appropriate protective clothing. Stop leak if you can do it without risk. Prevent entry into waterways, sewers, basements or confined areas. Absorb or cover with dry earth, sand or other non-combustible material and transfer to containers. Do not get water inside containers.

Environmental Physical Data

Octanol/Water Partition Coefficient: log K_{ow} = measured at 0.73

BCF: no food chain concentration potential

Regulations

RCRA 40CFR: Not listed

CERCLA: 40CFR 302.4: Not listed

SARA 40CFR 372.65: Not listed

SARA EHS 40CFR 355: Not listed

TSCA: Listed

BUT8910 CAS #: 123-72-8

N-BUTYRALDEHYDE

RTECS: ES2275000

DOT: UN1129; IMO3.2

EINECS Number: 204-646-6

Molecular Formula: C_4H_8O

Structured MF: $CH_3CH_2CH_2CHO$

Formula Weight: 72.10

Chemical Structure

Synonyms: ALDEHYDE BUTYRIQUE; ALDEIDE BUTIRRICA; BUTAL; BUTALDEHYDE; BUTALYDE; 1-BUTANAL; BUTANAL; N-BUTANAL; BUTANALDEHYDE; BUTANOL; BUTYL ALDEHYDE; N-BUTYL ALDEHYDE; BUTYLALDEHYDE; BUTYRAL; BUTYRALDEHYD; BUTYRALDEHYDE; BUTYRIC ALDEHYDE; BUTYRYLALDEHYDE

Description: colorless to water-white liquid; suffocating odor

Use: manufacture of plasticizers, rubber accelerators, solvents, high polymers and resins

Physical Properties

Boiling Point: 74.8 °C (167 °F)

Freezing Point: -99 °C (-146.2 °F)

Specific Gravity: 0.8016 at 20 °C/4 °C

Vapor Density: 2.5 Air=1

Saturated Vapor Density: 1.414717786 kg/m^3

Density: 0.803 g/mL at 20 °C

Vapor Pressure: 91.5 mm Hg at 20 °C

Water Solubility: 7.1 (wt %) at 25 °C

Other Solubilities: 95% Ethanol: >=100 mg/ml at 21 °C; Acetone: >=100 mg/ml at 21 °C; Benzene: Soluble; DMSO: >=100 mg/ml at 21 °C; Ether: miscible; Ethyl Acetate: miscible; Many oils: miscible; Many other organic solvents: miscible; Toluene: miscible.

Surface Tension: 24.6 dynes/cm

Odor Threshold: 0.0046 ppm

Refraction Index: 1.379 at 20 °C/D

Evaporation Rate: 7.8 Butyl Acetate=1

Critical Temperature: 251 °C

Critical Pressure: 590 psia

Ionization Potential (eV): 9.84 +/-0.02

Flash Point: -6.67 °C Closed Cup

Autoignition Temperature: 218 °C

LEL: 1.9% v/v

UEL: 12.5% v/v

RTECS Toxicity Data

Acute Oral: Rat LD_{50} Dose: 2490 mg/kg.

Acute Inhalation: Mouse LC_{50} Dose: 44610 mg/m^3/2hr. Rat LC_{Lo} Dose: 8000 ppm/4hr.

Acute Dermal: Rabbit LD_{50} Route: Skin; Dose: 3560 uL/kg. Rat LD_{50} Route: Subcutaneous Dose: 10 gm/kg; Toxic Effects: Behavioral - General anesthetic.

Chronic (Multiple Dose) Oral: Rat Dose: 39 gm/kg/90D-I; Toxic Effects: Lungs, Thorax, or Respiration - Other changes; Biochemical - Transaminases; DEATH. Mouse Dose: 78 gm/kg/90D-I; Toxic Effects: Liver - Changes in liver weight; Kidney, Ureter, and Bladder - Changes in kidney weight; Nutritional and gross metabolic - Weight loss or decreased weight gain.

Irritation Eye: Rabbit Standard Draize Test Dose: 20 mg/24H; Reaction: moderate. Rabbit Standard Draize Test Dose: 75 ug open; Reaction: severe.

Irritation Skin: Rabbit Standard Draize Test Dose: 2 mg/24H; Reaction: severe. Rabbit Open Draize Test Dose: 410 mg open; Reaction: mild.

Mutagenic: Rat Unscheduled DNA Synthesis; Cell Type: liver; Dose: 30 mmol/L. Mouse Sperm Morphology; Route: Intraperitoneal; Dose: 30 mg/kg. Mouse Sperm Morphology; Route: Oral; Dose: 15 gm/kg/50D.

Hazard Overviews

Corrosive

Flammable

Fire Diamond

Health: Corrosive causes severe burns to eyes/skin/respiratory tract. Also Causes: nasal discharge, sneezing, coughing, swelling of eyelids. Narcotic at high concentrations.

Fire: Flammable. Can form explosive mixtures and peroxides in the air. Fight fire from maximum distance. Use dry chemical, foam, or carbon dioxide. Water may be ineffective. Use water spray to keep fire-exposed containers cool.

Reactivity: Unstable, in the absence of inhibitors, it can form explosive peroxides in air; also readily oxidizes in air to butyric acid. Hazardous polymerization can occur in the presence of heat, acids, or alkalis. Avoid: heat; ignition sources. Incompatible with: oxidizing materials; amines; strong alkalis; acids; chlorosulfonic acid; nitric acid; oleum; sulfuric acid.

Carcinogenicity: IARC - Not listed; NIOSH - Not listed; NTP - Not listed; ACGIH - Not listed; OSHA - Not listed; EPA - Not listed; MAK - Not listed

Primary Target Organs:

Eyes

Skin

Respiratory System

Exposure Limits
AIHA WEEL: TWA: 25 ppm.
Respirator Recommendation
Exposure Range: >25 to 250 ppm Air Purifying, Negative Pressure, Half Mask
Exposure Range: >250 to 1000 ppm Air Purifying, Negative Pressure, Full Face
Exposure Range: >1000 to 25,000 ppm Supplied Air, Constant Flow/Pressure Demand, Full Face
Exposure Range: >25,000 to unlimited ppm Self-contained Breathing Apparatus, Pressure Demand, Full Face
Cartridge Color: black

Environmental

Ecotoxicity: Fishes: Pimephales promelas 12h LC_{50} >80 mg/l 24h LC_{50} 20 mg/l
Environmental Fate: If released to the atmosphere, it will degrade readily by reaction with photochemically produced hydroxyl radicals (half-life of 16.4 hr) and direct photolysis. During intense smog-pollution episodes, the natural formation rate can exceed the degradation rate. Physical removal from air by wet deposition can occur. If released to soil or water, the major degradation pathway is expected to be biodegradation. Physical removal from soil surfaces or water can occur through volatilization. Estimated K_{oc} values (9-71) suggest that it will leach in soil.
Cleanup/Disposal: Guide No. 129: Eliminate all ignition sources (no smoking, flares, sparks or flames in immediate area). All equipment used when handling the product must be grounded. Do not touch or walk through spilled material. Stop leak if you can do it without risk. Prevent entry into waterways, sewers, basements or confined areas. A vapor suppressing foam may be used to reduce vapors. Absorb or cover with dry earth, sand or other non-combustible material and transfer to containers. Use clean non-sparking tools to collect absorbed material. Large Spills: Dike far ahead of liquid spill for later disposal. Water spray may reduce vapor; but may not prevent ignition in closed spaces.

Environmental Physical Data

Henry's Law Constant: 1.15 $x10^{-4}$
Octanol/Water Partition Coefficient: log K_{ow} = 0.88
Sorption Partition Coefficient: K_{oc} = estimated at 9 to 71
BCF: estimated at 2.75
BOD: 1.62 lb/lb, 5 days

Regulations

RCRA 40CFR: Not listed
CERCLA: 40CFR 302.4: Not listed
SARA 40CFR 372.65: Listed
SARA EHS 40CFR 355: Not listed
TSCA: Listed

Analytical Methods

Air: EPA TO-11, 0100
Soil: SW846 8315A
Water / Groundwater: ASTM D3695
Drinking Water: EPA 554
Indoor / Expired Air: NIOSH 2539; EPA IP-6A, IP-6B, IP-6C, 0100

BUT8980 **CAS #: 107-92-6**

N-BUTYRIC ACID

RTECS: ES5425000
DOT: UN2820; IMO8.0
EINECS Number: 203-532-3
Molecular Formula: $C_4H_8O_2$
Structured MF: $CH_3CH_2CH_2COOH$
Formula Weight: 88.10

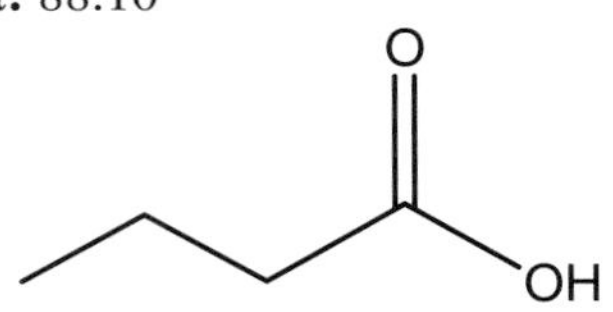

Chemical Structure

Synonyms: BUTANIC ACID; BUTANOIC ACID; N-BUTANOIC ACID; BUTTERSAEURE; BUTYRATE; BUTYRIC ACID; ETHYLACETIC ACID; KYSELINA MASELNA; 1-PROPANECARBOXYLIC ACID; PROPYLFORMIC ACID
Description: colorless liquid; rancid odor
Use: as a synthetic flavoring substance and adjuvant; in the synthesis of butyrate ester perfume and flavor ingredients, in pharmaceuticals, as a deliming agent, in disinfectants, in emulsifying agents, in sweetening gasolines, in artificial flavoring ingredients for certain liqueurs, soda-water, syrups and candies; for varnishes and as a decalcifier of hides

Physical Properties

Boiling Point: 165.5 °C (330 °F) at 760 mm Hg
Freezing Point: -7.9 °C (17.78 °F)
Specific Gravity: 0.9577 at 20 °C/4 °C
Vapor Density: 3.04 Air=1
Saturated Vapor Density: 1.201383648 kg/m^3
Density: 0.964 g/mL
Vapor Pressure: 0.43 mm Hg at 20 °C
Water Solubility: >= 100 mg/mL at 19 C
Other Solubilities: 95% Ethanol: >=100 mg/ml at 19 °C; Acetone: >=100 mg/ml at 19 °C; Alcohol: miscible; DMSO: >=100 mg/ml at 19 °C; Ether: miscible.
Surface Tension: 26.74 dynes/cm
Odor Threshold: 0.001 to 9 mg/m^3
pH: Neutralization value: 636.79
Refraction Index: 1.3980 at 20 °C/D
Critical Temperature: 355 °C
Critical Pressure: 52 atm
Ionization Potential (eV): 10.17 +/-1.0
Flash Point: 72 °C Closed Cup

Autoignition Temperature: 443 °C
LEL: 2.0% v/v
UEL: 10% v/v

RTECS Toxicity Data

Acute Oral: Rat LD_{50} Dose: 2 gm/kg. Mouse LD_{Lo} Dose: 500 mg/kg; Toxic Effects: Gastrointestinal - Necrotic changes; Kidney, Ureter, and Bladder - Other changes; Blood - Changes in spleen.
Acute Dermal: Rabbit LD_{50} Route: Skin; Dose: 530 uL/kg. Mouse LD_{50} Route: Subcutaneous Dose: 3180 mg/kg.
Chronic (Multiple Dose) Oral: Rat Dose: 14 gm/kg/7D-C; Toxic Effects: Gastrointestinal - Other changes. Mouse Dose: 33600 mg/kg/7D-C; Toxic Effects: Gastrointestinal - Other changes.
Irritation Eye: Rabbit Standard Draize Test Dose: 250 ug open; Reaction: severe.
Irritation Skin: Rabbit Standard Draize Test Dose: 20 mg/24H; Reaction: moderate. Rabbit Open Draize Test Dose: 10 mg/24H open; Reaction: severe.
Mutagenic: Human DNA Damage; Cell Type: HeLa cell; Dose: 3 mmol/L. Human DNA Inhibition; Cell Type: lymphocyte; Dose: 4 mmol/L.

Hazard Overviews

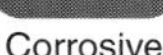
Corrosive

Fire Diamond

Health: Corrosive to eyes/skin/respiratory tract. Also Causes: coughing, difficulty in breathing burns skin/eyes,. vomiting, oral, esophageal, stomach burns.
Fire: Combustible. Can form explosive mixtures in air. Alcohol foam is recommended as an extinguishing agent, but water spray, dry chemical, or carbon dioxide may also be used. Use water spray to cool fire-exposed containers. If the material is on fire or is involved in fire, use water in flooding quantities as a fog because solid streams of water may be ineffective.
Reactivity: Stable. Hazardous polymerization cannot occur. Incompatible with: hydrogen gas; chromium trioxide; oxidizing agents. Hazardous decomposition products: acrid smoke; fumes.
Carcinogenicity: IARC - Not listed; NIOSH - Not listed; NTP - Not listed; ACGIH - Not listed; OSHA - Not listed; EPA - Not listed; MAK - Not listed
Primary Target Organs:

Eyes

Skin

Mucous Membranes

Environmental

Ecotoxicity: TLm Daphnia magna 61 mg/l/48 hr /Conditions of bioassay not specified TLm Limnea macrochirus 200 mg/l/24 hr /Conditions of bioassay not specified
Environmental Fate: If released to soil, it is expected to be relatively mobile, although adsorption may occur by attractive interactions with active sites in the soil. It is not expected to significantly volatilize from either moist or dry soil to the atmosphere. If released to water, it will exist predominately in the dissociated form under environmental conditions. It is expected to biodegrade rapidly under both aerobic and anaerobic conditions. Volatilization from water to the atmosphere is not expected to occur to any significant extent; the half-life for volatilization from a model river is 59 days. It will not significantly adsorb to sediment and suspended organic matter, nor is it expected to significantly bioconcentrate in fish and aquatic organisms. If released to the atmosphere, it is expected to undergo a gas-phase reaction with photochemically produced hydroxyl radicals with a half-life of 8 days. It may also undergo atmospheric removal by wet deposition.
Cleanup/Disposal: Guide No. 153: Eliminate all ignition sources (no smoking, flares, sparks or flames in immediate area). Do not touch damaged containers or spilled material unless wearing appropriate protective clothing. Stop leak if you can do it without risk. Prevent entry into waterways, sewers, basements or confined areas. Absorb or cover with dry earth, sand or other non-combustible material and transfer to containers. Do not get water inside containers.

Environmental Physical Data

Henry's Law Constant: estimated at 5.35 $\times 10^{-7}$
Octanol/Water Partition Coefficient: log K_{ow} = 0.79
Sorption Partition Coefficient: K_{oc} = clastic mud (3.5% organic) 19.1
BCF: calculated at 2.3
BOD: 1.150 lb/lb, 5 days

Regulations

RCRA 40CFR: Not listed
CERCLA: 40CFR 302.4: Listed per CWA Section 311(b)(4) RQ: 5000 lb (2268 kg)
SARA 40CFR 372.65: Not listed
SARA EHS 40CFR 355: Not listed
TSCA: Listed

BUT9050 CAS #: 106-31-0

BUTYRIC ANHYDRIDE

RTECS: ET7090000
EINECS Number: 203-383-4
Molecular Formula: $C_8H_{14}O_3$
Formula Weight: 158.19

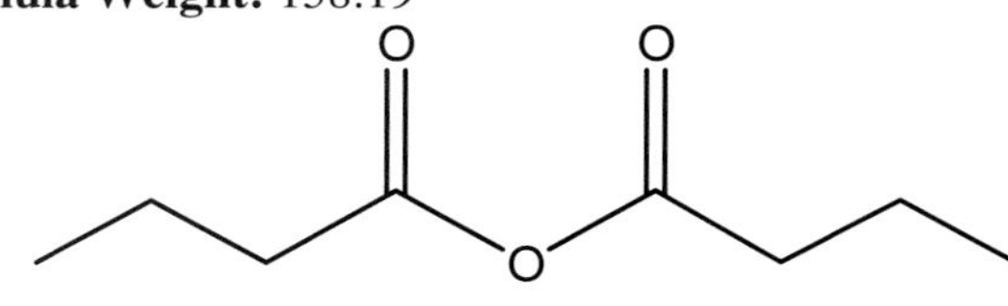

Chemical Structure

Synonyms: ANHYDRID KYSELINY MASELNE; BUTANOIC ACID ANHYDRIDE; BUTANOIC ACID,ANHYDRIDE; BUTANOIC ACID,ANHYDRIDE (9CI); BUTANOIC ANHYDRIDE; BUTYRANHYDRID; BUTYRIC ACID ANHYDRIDE; N-BUTYRIC ACID ANHYDRIDE; N-BUTYRIC ANHYDRIDE; BUTYRIC ANHYDRIDE N; BUTYRYL OXIDE

Description: colorless liquid; pungent odor

Use: prodn of cellulose acetate butyrate, drugs, & tanning agents; chem int for cellulose acetate butyrate resins; chem int for flavoring agents, eg, isopentyl butyrate; chem int for perfumes, eg, geranyl butyrate; chem int for sodium tyropanoate (oral cholecystographic)

Physical Properties

Boiling Point: 199.4 °C (391 °F) to 201.4 °C (395 °F)
Freezing Point: -75 °C (-103 °F)
Specific Gravity: 0.9668 at 20 °C/4 °C
Vapor Pressure: 1.25 kPa at 75 °C
Water Solubility: Soluble in Water
Other Solubilities: Soluble in Alcohol with decomposition; Soluble in Ether
Refraction Index: 1.4070
Flash Point: 54 °C Closed Cup
Autoignition Temperature: 279 °C
LEL: 0.9% v/v
UEL: 5.8% v/v

RTECS Toxicity Data

Acute Oral: Rat LD_{50} Dose: 8790 mg/kg. Mouse LD_{Lo} Dose: 1 gm/kg; Toxic Effects: Behavioral - Somnolence (general depressed activity); Liver - Other changes; Kidney, Ureter, and Bladder - Other changes.

Acute Inhalation: Rat LC; Dose: >50 mg/m^3. Mouse LC; Dose: >50 mg/m^3.

Chronic (Multiple Dose) Inhalation: Mammal Dose: 30 mg/m^3/2H/30D-I; Toxic Effects: Lungs, Thorax, or Respiration - Structural or functional change in trachea or bronchi.

Hazard Overviews

Flammable

Fire: Flammable.

Carcinogenicity: IARC - Not listed; NIOSH - Not listed; NTP - Not listed; ACGIH - Not listed; OSHA - Not listed; EPA - Not listed; MAK - Not listed

Environmental

Regulations

RCRA 40CFR: Not listed
CERCLA: 40CFR 302.4: Not listed
SARA 40CFR 372.65: Not listed
SARA EHS 40CFR 355: Not listed
TSCA: Listed

BUT9120 CAS #: 36536-46-6

B-BUTYROLACTONE

RTECS: RQ8050100
EINECS Number: 253-093-7
Molecular Formula: $C_4H_6O_2$
Formula Weight: 86.10

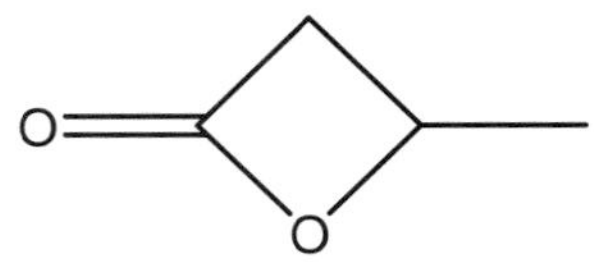

Chemical Structure

Synonyms: (+/-)-BETA-BUTYROLACTONE; (I)-BETA-BUTYROLACTONE; (RS)-BETA-BUTYROLACTONE; DL-3-BUTYROLACTONE; DL-BETA-BUTYROLACTONE; DL-3-HYDROXYBUTANOIC ACID BETA-LACTONE; DL-3-HYDROXYBUTYRIC ACID BETA-LACTONE; DL-3-HYDROXYBUTYRIC ACID LACTONE; DL-HYDROXYBUTYRIC ACID LACTONE; DL-4-METHYL-2-OXETANONE; DL-BETA-METHYL-BETA-PROPIOLACTONE; DL-BETA-METHYLPROPIOLACTONE; (+-)-4-METHYL-2-OXETANONE; (+/-)-4-METHYL-2-OXETANONE; 2-OXETANONE,4-METHYL-,(+/-)

Description: colorless, clear liquid; acetone-like odor

Use: building block for organic synthesis; for the production of beta-oxybutyryl-p-phenetidine

Physical Properties

Boiling Point: 71 °C (160 °F) to 73 °C (163 °F)
Freezing Point: -43.5 °C (-46.3 °F)
Specific Gravity: 1.0555 at 20/20 °C
Density: 1.056 g/mL
Vapor Pressure: 0.5 mm Hg at 42 °C
Water Solubility: >= 100 mg/mL at 23 C
Other Solubilities: 95% Ethanol: >=100 mg/ml at 23 °C; Acetone: >=100 mg/ml at 23 °C; DMSO: >=100 mg/ml at 23 °C; Most organic solvents: miscible.
Refraction Index: 1.4109
Flash Point: 60 °C

RTECS Toxicity Data

Mutagenic: Bacteria - S Typhimurium Mutations in Microorganisms; Dose: 33 ug/plate (+S9), 333 ug/plate (-S9).

Hazard Overviews

Flammable

Health: Irritating to eyes/skin/respiratory tract. Toxic. Other Acute Effects: harmful if swallowed, inhaled, or absorbed through skin. Chronic Effects: Carcinogen.

Fire: Flammable. Hazards: emits toxic fumes. Extinguishing agents: carbon dioxide, dry chemical powder or appropriate foam; water spray. Precautions: combustible liquid.
Reactivity: Incompatible with: strong oxidizing agents, strong bases. Hazardous decomposition products: toxic fumes of: carbon monoxide, carbon dioxide.
Carcinogenicity: IARC - Not listed; NIOSH - Not listed; NTP - Not listed; ACGIH - Not listed; OSHA - Not listed; EPA - Not listed; MAK - Not listed
Primary Target Organs:

Eyes Skin Respiratory System

Environmental

Cleanup/Disposal: Guide No. 128: Eliminate all ignition sources (no smoking, flares, sparks or flames in immediate area). All equipment used when handling the product must be grounded. Do not touch or walk through spilled material. Stop leak if you can do it without risk. Prevent entry into waterways, sewers, basements or confined areas. A vapor suppressing foam may be used to reduce vapors. Absorb or cover with dry earth, sand or other non-combustible material and transfer to containers. Use clean non-sparking tools to collect absorbed material. Large Spills: Dike far ahead of liquid spill for later disposal. Water spray may reduce vapor; but may not prevent ignition in closed spaces.

Regulations

RCRA 40CFR: Not listed
CERCLA: 40CFR 302.4: Not listed
SARA 40CFR 372.65: Not listed
SARA EHS 40CFR 355: Not listed
TSCA: Not listed

BUT9190 CAS #: 96-48-0

BUTYROLACTONE

RTECS: LU3500000
EINECS Number: 202-509-5
Molecular Formula: $C_4H_6O_2$
Formula Weight: 86.09

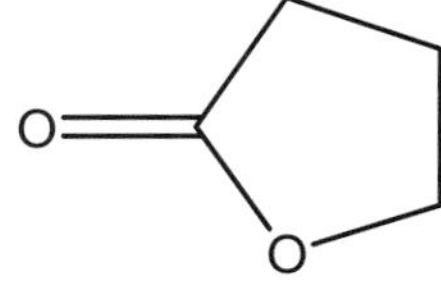

Chemical Structure

Synonyms: 6480; AGRISYNTH BLO; GAMMA BL; GAMMA-BL; BLO; BLON; BUTANOIC ACID,4-HYDROXY-,GAMMA-LACTONE; 1,2-BUTANOLIDE; 1,4-BUTANOLIDE; 4-BUTANOLIDE; BUTYRIC ACID LACTONE; BUTYRIC ACID,4-HYDROXY-,GAMMA-LACTONE; BUTYRL LACTONE; 4-BUTYROLACTONE; ALPHA-BUTYROLACTONE; GAMMA-BUTYROLACTONE; BUTYRYL LACTONE; BUTYRYLACTONE; 4-DEOXYTETRONIC ACID; DIHYDRO-2(3H)-FURANONE; DIHYDRO-2-FURANONE; 2(3H)-FURANONE,DIHYDRO-; GAMMA-6480; 4-HYDROXYBUTANOIC ACID LACTONE; 4-HYDROXYBUTANOIC ACID,GAMMA-LACTONE; GAMMA-HYDROXYBUTYRIC ACID CYCLIC ESTER; 3-HYDROXYBUTYRIC ACID LACTONE; 4-HYDROXYBUTYRIC ACID LACTONE; GAMMA-HYDROXYBUTYRIC ACID LACTONE; 4-HYDROXYBUTYRIC ACID,GAMMA-LACTONE; GAMMA-HYDROXYBUTYROLACTONE; 2-OXOLANONE; TETRAHYDRO-2-FURANONE
Description: colorless, oily liquid; pleasant odor
Use: intermediate for synthesis of butyric acid compounds, polyvinylpyrrolidone, methionine; solvent for acrylate and styrene polymers; ingredient of paint removers and textile assistants

Physical Properties

Boiling Point: 204 °C (399 °F) at 760 mm Hg
Freezing Point: -42 °C (-43.6 °F)
Specific Gravity: 1.1441 at 0 °C/0 °C
Vapor Density: 3 Air=1
Saturated Vapor Density: 1.209946715 kg/m^3
Density: 1.1441 g/mL
Bulk Density: 9.4 lbs/gal
Vapor Pressure: 3.2 mm Hg at 25 °C
Water Solubility: Miscible with Water
Other Solubilities: 95% Ethanol: >=100 mg/ml at 13 °C; Acetone: >=100 mg/ml at 13 °C; Benzene: Soluble; DMSO: >=100 mg/ml at 13 °C; Ether: Soluble; Methanol: Soluble.
Refraction Index: 1.4348 at 25 °C/D
Flash Point: 98.333 °C Open Cup
LEL: 1.4% v/v
UEL: 6.9% v/v

RTECS Toxicity Data

Acute Oral: Rat LD_{50} Dose: 1540 mg/kg; Toxic Effects: Behavioral - Altered sleep time (including change in righting reflex); Behavioral - Somnolence (general depressed activity); Lungs, Thorax, or Respiration - Respiratory depression. Mouse LD_{50} Dose: 1720 mg/kg; Toxic Effects: Behavioral - Altered sleep time (including change in righting reflex); Behavioral - Somnolence (general depressed activity); Lungs, Thorax, or Respiration - Respiratory depression.
Acute Dermal: Guinea Pig LD_{50} Route: Skin; Dose: >5 gm/kg.
Chronic (Multiple Dose) Oral: Rat Dose: 9600 mg/kg/16D-C; Toxic Effects: Nutritional and gross metabolic - Weight loss or decreased weight gain; DEATH. Rat Dose: 40950 mg/kg/13W-C; Toxic Effects: Nutritional and gross metabolic - Weight loss or decreased weight gain.
Reproductive/Teratogenic: Rat Route: Oral; Dose: 25 gm/kg; Duration: male 20D prior to mating; Paternal Effects - Testes, epididymis, sperm duct. Rat Route: Oral; Dose: 500 mg/kg; Duration: female 6-15D of pregnancy; Effects on Embryo or Fetus - Fetotoxicity.
Mutagenic: Hamster Cytogenetic Analysis; Cell Type: ovary; Dose: 2580 mg/L. Hamster Sister Chromatid Exchenge; Cell Type: ovary; Dose: 4940 mg/L.

Tumorigenic: Mouse Route: Oral; Dose: 191 gm/kg/2Y-C; Toxic Effects: Tumorigenic - Equivocal tumorigenic agent by RTECS criteria; Endocrine - Adrenal cortex tumors. Mouse Route: Skin; Dose: 50 gm/kg/42W-I; Toxic Effects: Tumorigenic - Equivocal tumorigenic agent by RTECS criteria; Skin and appendages - Tumors; Tumorigenic - Tumors at site of application.

Hazard Overviews

Health: Severely irritating to eyes; may cause irritation to skin/respiratory tract. Harmful. Other Acute Effects: harmful if swallowed; risk of serious damage to eyes; exposure can cause: nausea, respiratory depression and sedation; narcotic effect. Chronic Effects: possible risk of irreversible effects. Possible human carcinogen.
Fire: Will burn. Hazards: emits toxic fumes. Extinguishing agents: water spray; carbon dioxide, dry chemical powder or appropriate foam. Precautions: combustible liquid.
Carcinogenicity: IARC - Group 3, Not classifiable as to carcinogenicity to humans; NIOSH - Not listed; NTP - Not listed; ACGIH - Not listed; OSHA - Not listed; EPA - Not listed; MAK - Not listed
Primary Target Organs:

Eyes

Environmental

Environmental Fate: If released to soil, it may significantly volatilize from both dry and moist soil to the atmosphere. A calculated soil adsorption coefficient of 53 indicates that it is expected to display moderate to high mobility in soil. In basic soils, the free acid may predominate which may alter both the rate at which it volatilizes from soil and the degree to which it adsorbs. If released to water, it may significantly volatilize from water to the atmosphere. The estimated half-life for volatilization from a model river is 2 days. It is not expected to significantly bioconcentrate in fish and aquatic organisms, nor is it expected to adsorb to sediment and suspended organic matter. In basic waters equilibrium with the free acid may alter the importance of these processes. If released to the atmosphere, it is expected to undergo a gas-phase reaction with photochemically produced hydroxyl radicals; the estimated half life for this process is 6.65 days. It may also undergo atmospheric removal by wet deposition.

Environmental Physical Data

Henry's Law Constant: estimated at 1.81×10^{-5}
Octanol/Water Partition Coefficient: **log K_{ow}** = 0.64
Sorption Partition Coefficient: **K_{oc}** = calculated at 53
BCF: calculated at 1.8

Regulations

RCRA 40CFR: Not listed
CERCLA: 40CFR 302.4: Not listed
SARA 40CFR 372.65: Not listed
SARA EHS 40CFR 355: Not listed

TSCA: Listed

BUT9260 CAS #: 109-74-0

BUTYRONITRILE

RTECS: ET8750000
DOT: UN2411; IMO3.2
EINECS Number: 203-700-6
Molecular Formula: C_4H_7N
Structured MF: $CH_3CH_2CH_2CN$
Formula Weight: 69.10

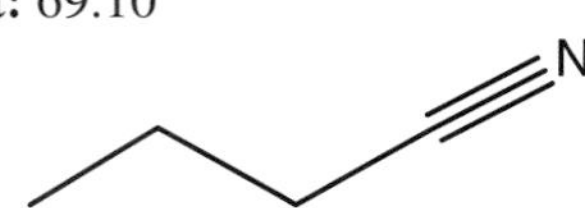

Chemical Structure

Synonyms: BUTANE NITRILE; BUTANENITRILE; N-BUTANENITRILE; N-BUTYL NITRILE; BUTYRIC ACID NITRILE; N-BUTYRONITRILE; BUTYRYLONITRILE; 1-CYANOPROPANE; N-PROPYL CYANIDE; PROPYL CYANIDE; PROPYLKYANID
Description: colorless liquid; sharp, suffocating odor
Use: basic material in indust, chemical & pharmaceutical intermediates & products; poultry medicines; chem int for butyric acid, butyramide & pharmaceuticals, other organics (eg, ketones, esters)

Physical Properties

Boiling Point: 117.5 °C (244 °F) at 760 mm Hg
Freezing Point: -112 °C (-169.6 °F)
Specific Gravity: 0.8091 at 0 °C/4 °C
Vapor Density: 2.4 Air=1
Vapor Pressure: 14 mm Hg
Water Solubility: Sparingly Soluble
Other Solubilities: Soluble in Benzene; 33,000 mg/l at 25 °C (measured).
Surface Tension: 27.33 dyne/cm at 20 °C
Refraction Index: 1.38385 at 20 °C/D
Critical Temperature: 309 °C
Critical Pressure: 549.8 psia
Ionization Potential (eV): 11.67
Flash Point: 26.111 °C Open Cup
Autoignition Temperature: 502 °C
LEL: 1.65% v/v

RTECS Toxicity Data

Acute Oral: Rat LD_{50} Dose: 50 mg/kg; Toxic Effects: Behavioral - Convulsions or effect on seizure threshold; Vascular - Regional or general arteriolar or venous dilation; Lungs, Thorax, or Respiration - Dyspnea. Mouse LD_{50} Dose: 27689 ug/kg.
Acute Inhalation: Mouse LC_{50} Dose: 249 ppm/1hr; Toxic Effects: Liver - Other changes. Rat LC_{Lo} Dose: 1000 ppm/4hr.
Acute Dermal: Rabbit LD_{50} Route: Skin; Dose: 500 uL/kg. Rabbit LD_{Lo} Route: Subcutaneous Dose: 10 mg/kg; Toxic Effects: Peripheral Nerve and sensation - Spastic parapysis

with or without sensory change; Lungs, Thorax, or Respiration - Respiratory depression; Lungs, Thorax, or Respiration - Respiratory stimul

Irritation Eye: Rabbit Standard Draize Test Dose: 500 mg/24H; Reaction: mild.

Irritation Skin: Rabbit Open Draize Test Dose: 395 mg open; Reaction: mild.

Reproductive/Teratogenic: Rat Route: Inhalation; Dose: 200 ppm/6H; Duration: female 6-20D of pregnancy; Effects on Embryo or Fetus - Fetotoxicity.

Hazard Overviews

Poison

Flammable

Fire Diamond

Health: Poison. Irritating to eyes/skin/respiratory tract. Also Causes: dizziness, drowsiness, headache, rapid respiration, decreased blood pressure and pulse, cyanosis.

Fire: Flammable. Use dry chemical, carbon dioxide, water spray, or alcohol-resistant foam. Container may explode in heat of fire.

Reactivity: Stable. Hazardous polymerization cannot occur. Avoid: heat; ignition sources; oxidizers. Incompatible with: oxidizers. Hazardous decomposition products: oxides of nitrogen; cyanide gas.

Carcinogenicity: IARC - Not listed; NIOSH - Listed as carcinogen; NTP - Not listed; ACGIH - Not listed; OSHA - Not listed; EPA - Not listed; MAK - Not listed

Primary Target Organs:

Eyes

Skin

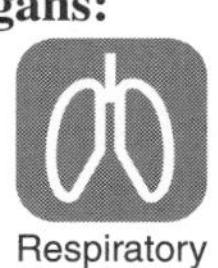
Respiratory System

Nervous System

Exposure Limits

NIOSH REL: TWA: 8 ppm; 22 mg/m^3.

Environmental

Environmental Fate: If released to the atmosphere, it will degrade by reaction with photochemically produced hydroxyl radicals (half-life of 21 days). Physical removal from air via wet deposition is possible since it is relatively soluble in water. If released to soil or water, the major degradation pathway is expected to be biodegradation. By analogy to chemically similar nitriles, biodegradation will occur much more rapidly in the presence of acclimated microbes. High concentrations can be toxic to microbes and prevent biodegradation. An estimated K_{oc} value (14) suggests that it will leach in soil. Physical removal from soil surfaces can occur through volatilization. Volatilization from water may be an important transport mechanism, but it will not occur rapidly.

Cleanup/Disposal: Guide No. 131: Fully encapsulating, vapor protective clothing should be worn for spills and leaks with no fire. Eliminate all ignition sources (no smoking, flares, sparks or flames in immediate area). All equipment used when handling the product must be grounded. Do not touch or walk through spilled material. Stop leak if you can do it without risk. Prevent entry into waterways, sewers, basements or confined areas. A vapor suppressing foam may be used to reduce vapors. Small Spills: Absorb with earth, sand or other non-combustible material and transfer to containers for later disposal. Use clean non-sparking tools to collect absorbed material. Large Spills: Dike far ahead of liquid spill for later disposal. Water spray may reduce vapor; but may not prevent ignition in closed spaces.

Environmental Physical Data

Henry's Law Constant: 5.23 x10^{-5}

Sorption Partition Coefficient: K_{OC} = estimated at 14

BCF: estimated at 2

Regulations

RCRA 40CFR: Not listed

CERCLA: 40CFR 302.4: Not listed

SARA 40CFR 372.65: Not listed

SARA EHS 40CFR 355: Not listed

TSCA: Listed

Analytical Methods

Water / Groundwater: ASTM D3371

BUT9330 **CAS #: 141-75-3**

N-BUTYRYL CHLORIDE

EINECS Number: 205-498-5

Molecular Formula: C_4H_7ClO

Structured MF: C_4H_7ClO

Formula Weight: 106.55

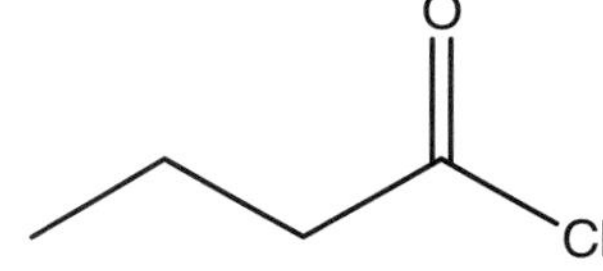

Chemical Structure

Synonyms: BUTANOYL CHLORIDE; BUTYRIC ACID CHLORIDE; BUTYRIC CHLORIDE; BUTYROYL CHLORIDE; BUTYRYL CHLORIDE

Description: colorless liquid; pungent acid chloride odor

Use: in organic synthesis

Physical Properties

Boiling Point: 101 °C (214 °F)

Freezing Point: -89 °C (-128.2 °F)

Specific Gravity: 1.0263 at 20.6 °C/4 °C

Vapor Density: 3.67 Air=1

Density: 1.026 g/mL

Vapor Pressure: 18.8 mm Hg at -2 °C

Water Solubility: Disolves Slowly

Other Solubilities: 95% Ethanol: Reaction; Acetone: >=100 mg/ml at 20 °C; DMSO: >=100 mg/ml at 20 °C; Ether: Very Soluble.

pH: Acid

Refraction Index: 1.412 at 20 °C/D

Flash Point: < 21 °C

Hazard Overviews

Corrosive

Flammable

3 3 2 W
Fire Diamond

Health: Corrosive to eyes/skin/respiratory tract. Other Acute Effects: harmful if swallowed, inhaled, or absorbed through skin; material is extremely destructive to tissue of the mucous membranes and upper respiratory tract, eyes and skin; inhalation may result in spasm, inflammation and edema of the larynx and bronchi; chemical pneumonitis; and pulmonary edema; exposure may cause burning sensation; coughing; wheezing; laryngitis; shortness of breath; headache; nausea; vomiting.

Fire: Flammable. Hazards: vapor may travel considerable distance to source of ignition and flash back; reacts violently with water; water hydrolyzes material liberating acidic gas which in contact with metal surfaces can generate flammable and/or explosive hydrogen gas; emits toxic fumes. Extinguishing agents: carbon dioxide; dry chemical powder; do not use water. Precautions: combustible liquid.

Reactivity: Incompatible with: water, alcohols, oxidizing agents, strong bases. Hazardous decomposition products: carbon monoxide, carbon dioxide, hydrogen chloride gas, phosgene gas.

Carcinogenicity: IARC - Not listed; NIOSH - Not listed; NTP - Not listed; ACGIH - Not listed; OSHA - Not listed; EPA - Not listed; MAK - Not listed

Primary Target Organs:

Eyes

Skin

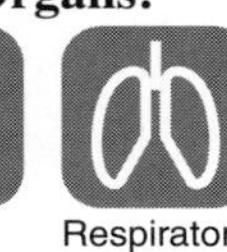
Respiratory System

Environmental

Cleanup/Disposal: Guide No. 132: Fully encapsulating, vapor protective clothing should be worn for spills and leaks with no fire. Eliminate all ignition sources (no smoking, flares, sparks or flames in immediate area). All equipment used when handling the product must be grounded. Do not touch or walk through spilled material. Stop leak if you can do it without risk. Prevent entry into waterways, sewers, basements or confined areas. A vapor suppressing foam may be used to reduce vapors. Absorb with earth, sand or other non-combustible material and transfer to containers (except for Hydrazine). Use clean non-sparking tools to collect absorbed material. Large Spills: Dike far ahead of liquid spill for later disposal. Water spray may reduce vapor; but may not prevent ignition in closed spaces.

Regulations

RCRA 40CFR: Not listed
CERCLA: 40CFR 302.4: Not listed
SARA 40CFR 372.65: Not listed
SARA EHS 40CFR 355: Not listed

TSCA: Listed

CAD1000 CAS #: 523-47-7

B-CADINENE

Molecular Formula: $C_{15}H_{24}$
Formula Weight: 204.39
Synonyms: 3,9-CADINADIENE; CADINA-3,9-DIENE; CADINE-3,9-DIENE; (-)-BETA-CADINENE; BETA-CADINENE,(-)-; CADINENE; NAPHTHALENE,1,2,4A,5,8,8A-HEXAHYDRO-4,7-DIMETHYL-1-(1-METHYLETHYL)-(1S-; NAPHTHALENE,1,2,4A,5,8,8A-HEXAHYDRO-4,7-DIMETHYL-1-(1-METHYLETHYL)-,(1S-(1ALPHA,4ABETA,8AALPHA))-
Description: oil; slight pleasant odor
Use: flavoring agent in candy at a concentration of 4000 ppm, baked good and chewing gum; it is also used as a fragrance in soaps, creams, lotions, perfume and detergents

Physical Properties

Boiling Point: 124 °C (255 °F) at 9 mm Hg
Specific Gravity: 0.9239 at 20 °C/4 °C
Density: 0.92 g/mL
Water Solubility: Insoluble in Water
Other Solubilities: 95% Ethanol: Soluble (>=10 mg/ml at 16 °C); Chloroform: Soluble in hot; DMSO: Slightly Soluble (1-10 mg/ml at 16 °C); Ether: Soluble (Tracor); Ligroin: Soluble.
Refraction Index: 1.5059 at 20 °C/D
Flash Point: > 110 °C

Hazard Overviews

Fire: Will burn.
Carcinogenicity: IARC - Not listed; NIOSH - Not listed; NTP - Not listed; ACGIH - Not listed; OSHA - Not listed; EPA - Not listed; MAK - Not listed

Environmental

Regulations

RCRA 40CFR: Not listed
CERCLA: 40CFR 302.4: Not listed
SARA 40CFR 372.65: Not listed
SARA EHS 40CFR 355: Not listed
TSCA: Not listed

CAD1500 CAS #: 7440-43-9

CADMIUM

RTECS: EU9800000
EINECS Number: 231-152-8
Molecular Formula: Cd
Structured MF: Cd
Formula Weight: 112.41

Cd

Chemical Structure

Synonyms: C I 77180; C.I. 77180; CADMIUM DUST FUME; CADMIUM POWDER; COLLOIDAL CADMIUM; KADMIUM
Description: silver-white, blue-tinged powder; odorless
Use: constituent of easily fusible alloys; soft solder and solder for aluminum; electroplating (major use), deoxidizer in Ni plating; process engraving, electrodes for cadmium vapor lamps, photoelectric cells, photometry of ultraviolet sun-rays, in Ni-Cd storage batteries; as an amalgam (1 Cd : 4 Hg) in dentistry; to charge Jones reductors; fire protection systems; power transmission wire

Physical Properties

Boiling Point: 765 °C (1409 °F)
Freezing Point: 321 °C (609.8 °F)
Specific Gravity: 8.65 at 25 °C
Vapor Pressure: 1 mm Hg at 394 °C
Water Solubility: Insoluble in Water
Other Solubilities: Acid: Soluble; NH_4NO_3: Soluble; Sulfric Acid (hot): Soluble.
Refraction Index: 1.8 at 578 NM & 20 °C/D
Ionization Potential (eV): 8.99367
Flash Point: Flammable
Autoignition Temperature: Layer Cadmium metal dust 250 °C

RTECS Toxicity Data

Acute Oral: Rat LD_{50} Dose: 2330 mg/kg. Mouse LD_{50} Dose: 890 mg/kg.
Acute Inhalation: Rat LC_{50} Dose: 25 mg/m^3/30M; Toxic Effects: Lungs, Thorax, or Respiration - Dyspnea. Human LC_{Lo} Dose: 39 mg/m^3/20M; Toxic Effects: Cardiac - Other changes; Vascular - Thrombosis distant from injection site; Lungs, Thorax, or Respiration - Respiratory depression. Man TC_{Lo} Dose: 88 ug/m^3/8.6Y; Toxic Effects: Kidney, ureter, and Bladder - Proteinuria.
Acute Dermal: Rabbit LD_{Lo} Route: Subcutaneous Dose: 6 mg/kg.
Reproductive/Teratogenic: Rat Route: Oral; Dose: 155 mg/kg; Duration: male 13W prior to mating; Effects on Newborn - Growth statistics; Behavioral. Rat Route: Oral; Dose: 220 mg/kg; Duration: female 1-22D of pregnancy; Effects on Embryo or Fetus - Other effects to embryo or fetus. Rat Route: Oral; Dose: 21500 ug/kg; Duration: multigenerations; Effects on Fertility - Pre-implantation mortality; Effects on Newborn - Germ cell effects. Rat Route: Oral; Dose: 23 mg/kg; Duration: female 1-22D of pregnancy; Specific Developmental Abnormalities - Blood and lymphatic systems (including spleen and marrow).
Mutagenic: Mouse Micronucleus Test; Cell Type: embryo; Dose: 6 umol/L. Hamster Cytogenetic Analysis; Cell Type: ovary; Dose: 1 umol/L.
Tumorigenic: Woman Route: Inhalation; Dose: 129 ug/m^3/20Y-C; Toxic Effects: Tumorigenic - Carcinogenic by RTECS criteria; Lungs, Thorax, or Respiration - Tumors. Rat

Route: Subcutaneous; Dose: 3372 ug/kg; Toxic Effects: Tumorigenic - Carcinogenic by RTECS criteria; Tumorigenic - Tumors at site of application.

Hazard Overviews

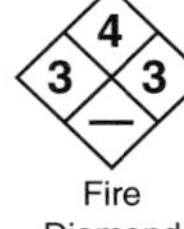

Fire Diamond

Health: Irritating to eyes/skin/respiratory tract. Toxic. Also Causes: nausea, headache, weakness, pulmonary edema. Chronic Effects: emphysema, kidney damage, bone demineralization, possible respiratory cancer.

Fire: Can be flammable as a fine powder. The more finely divided the greater the fire/explosion potential. Use carbon dioxide, dry chemical or sand.

Reactivity: Stable, but powder is pyrophoric. Hazardous polymerization cannot occur. Avoid: creation of dust clouds; exposure to heat and ignition sources. Incompatible with: ammonium nitrate; hydrazoic acid; tellurium; zinc; ammonia; sulfur; selenium; nitryl fluoride; oxidizing agents. Hazardous decomposition products: cadmium oxide fumes.

Carcinogenicity: IARC - Group 1, Carcinogenic to humans; NIOSH - Listed as carcinogen; NTP - Class 2A, Reasonably anticipated to be a carcinogen, limited evidence of carcinogenicity from studies in humans; ACGIH - Class A2, Suspected human carcinogen; OSHA - Listed as a carcinogen; EPA - Class B1, Probable human carcinogen based on epidemiologic studies; MAK - Class A2, Unmistakably carcinogenic in animal experimentation only

Primary Target Organs:

Eyes

Skin

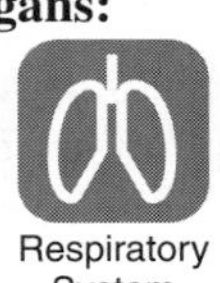
Respiratory System

Liver

Kidneys

Bone

Exposure Limits

OSHA PEL: TWA: 0.005 mg/m^3; as Cd See Table Z2.

ACGIH TLV: TWA: 0.01 mg/m^3; as Cd; Inhalable.

NIOSH REL: TWA: 0.1 mg/m^3; as Cd,LOQ; Lowest Feasible Concentration.

NIOSH IDLH: 9 mg/m^3; as Cd.

Respirator Recommendation

Exposure Range: >0.005 to 0.05 mg/m^3 Air Purifying, Negative Pressure, Half Mask

Exposure Range: >0.05 to 0.5 mg/m^3 Air Purifying, Negative Pressure, Full Face

Exposure Range: >0.5 to 5 mg/m^3 Supplied Air, Constant Flow/Pressure Demand, Full Face

Exposure Range: >5 to unlimited mg/m^3 Self-contained Breathing Apparatus, Pressure Demand, Full Face

Cartridge Color: magenta (P100)

Environmental

Cleanup/Disposal: Guide No. 154: Eliminate all ignition sources (no smoking, flares, sparks or flames in immediate area). Do not touch damaged containers or spilled material unless wearing appropriate protective clothing. Stop leak if you can do it without risk. Prevent entry into waterways, sewers, basements or confined areas. Absorb or cover with dry earth, sand or other non-combustible material and transfer to containers. Do not get water inside containers.

Regulations

RCRA 40CFR: Not listed

CERCLA: 40CFR 302.4: Listed per CWA Section 307(a) RQ: 10 lb (4.535 kg)

SARA 40CFR 372.65: Listed

SARA EHS 40CFR 355: Not listed

TSCA: Listed

Analytical Methods

Air: EPA 29, 0060, 1637, 1638, 1639, 1640, 1669, ITM-001

Soil: CLP 200.10_M, 200.62, 200.7_M, 202.62, 213.1_M, 213.2_M, 6020_M, ICP-AES; EPA 13, 200.7, 200.8, PMD-CD; SW846 1311, 3005A, 3010A, 3015, 3020A, 3031, 3040, 3040A, 3050A&B, 3051, 3052, 6010A&B, 6020, 7000A, 7130, 7131A, OSW-A; ASTM D1971, D3974

Water / Groundwater: EPA 200.0, 200.1, 200.10, 200.12, 200.13, 200.15, 200.7, 200.9; APHA 3111-A, 3111-B, 3111-C, 3113-B, 3120, 3130, 3500-CD; ASTM D1976, D3557, D4190; USGS E-SPEC, I1135, I1136, I1137, I1472, I3135, I3136, I7135, I7136; CEM RD42; FISON AES-0029

Drinking Water: AOAC 974.27, 993.14

Food: EPA 14

Indoor / Expired Air: NIOSH 7048, 7300; ASTM D4185

Plasma: EPA 200.11; NIOSH 8005

Urine: NIOSH 8310; EPA M-01

Other: EPA 1620, 213.1, 213.2; AOAC 990.08

CAD2000 CAS #: 543-90-8

CADMIUM ACETATE

RTECS: EU9810000

DOT: NA2570; IMO6.1

EINECS Number: 208-853-2

Molecular Formula: $C_4H_8CdO_4$

Structured MF: $Cd(C_2H_3O_2)_2 \cdot 2H_2O$

Formula Weight: 230.49

Synonyms: ACETIC ACID,CADMIUM SALT; BIS (ACETOXY) CADMIUM; BIS(ACETOXY)CADMIUM; C.I. 77185; CADMIUM DIACETATE; CADMIUM ETHANOATE; CADMIUM (II) ACETATE

Description: colorless crystals; slight acetic acid odor

Use: to produce iridescent effects on porcelains & pottery; rgnt for determination of sulfur, selenium & tellurium; in cadmium electroplating; in mfr of acetates; assistant in dyeing & printing textiles; in purification of mercaptans from crude oils & gasolines; intermed for cadmium halides-eg, cadmium chloride; showed inhibitory activity in vitro against rice pathogen xanthomonas campestris.

Physical Properties

Boiling Point: Decomposes at 1 atm

Freezing Point: 256 °C (492.8 °F)

Specific Gravity: 2.341

Water Solubility: Very Soluble in Cold Water
Other Solubilities: Soluble in Alcohol
pH: 0.2 molar aqueous solution 7.1
Flash Point: Nonflammable

RTECS Toxicity Data

Acute Oral: Rat LD_{50} Dose: 333 mg/kg.
Chronic (Multiple Dose) Oral: Rat Dose: 328 mg/kg/23W-C; Toxic Effects: Vascular - BP elevation not characterized in autonomic section; Kidney, Ureter, and Bladder - Other changes; Nutritional and gross metabolic - Changes in other metals. Rat Dose: 48889 ug/kg/13W-C; Toxic Effects: Behavioral - Food intake (animal); Blood - Changes in serum composition; Nutritional and gross metabolic - Changes in phosphorus. Rat Dose: 15374 ug/kg/6W-I; Toxic Effects: Liver - Other changes; Blood - Changes in serum composition; Nutritional and gross metabolic - Changes in calcium. Rat Dose: 369 mg/kg/30D-C; Toxic Effects: Brain and coverings - Other degenerative changes; Nutritional and gross metabolic - Changes in other metals; Biochemical - Lipids including transport. Rat Dose: 112 mg/kg/78W-C; Toxic Effects: Cardiac - EKG changes not diagnostic of above; Vascular - BP elevation not characterized in autonomic section; Nutritional and gross metabolic - Changes in other metals.
Reproductive/Teratogenic: Rat Route: Subcutaneous; Dose: 2325 ng/kg; Duration: female 1D prior to mating; Maternal Effects - Ovaries, fallopian tubes. Rat Route: Intraperitoneal; Dose: 2371 ug/kg; Duration: female 12D of pregnancy; Specific Developmental Abnormalities - Central nervous system; Craniofacial (including nose and tongue). Rat Route: Intraperitoneal; Dose: 2371 ug/kg; Duration: female 14D of pregnancy; Effects on Fertility - Post-implantation mortality. Rat Route: Intraperitoneal; Dose: 1 mg/kg; Duration: female 14D of pregnancy; Effects on Embryo or Fetus - Fetotoxicity. Rat Route: Intraperitoneal; Dose: 2 mg/kg; Duration: female 20D of pregnancy; Effects on Newborn - Stillbirth.
Mutagenic: Human Cytogenetic Analysis; Cell Type: lymphocyte; Dose: 10 nmol/L. Hamster Morphological Transformation; Cell Type: embryo; Dose: 1 umol/L.

Hazard Overviews

Health: Irritating. Toxic. Other Acute Effects: harmful if swallowed, inhaled, or absorbed through skin. Chronic Effects: contains cadmium; can cause lung and kidney disease; target organs: lungs, kidneys. Carcinogen.
Fire: Noncombustible. Hazards: emits toxic fumes. Extinguishing agents: water spray; carbon dioxide, dry chemical powder or appropriate foam. Precautions: combustible liquid.
Reactivity: Incompatible with: strong oxidizing agents, strong acids, strong bases. Hazardous decomposition products: toxic fumes of: carbon monoxide, carbon dioxide.
Carcinogenicity: IARC - Group 1, Carcinogenic to humans; NIOSH - Listed as carcinogen; NTP - Class 2A, Reasonably anticipated to be a carcinogen, limited evidence of carcinogenicity from studies in humans; ACGIH - Class A2, Suspected human carcinogen; OSHA - Listed as a carcinogen; EPA - Class B1, Probable human carcinogen based on epidemiologic studies; MAK - Class A2, Unmistakably carcinogenic in animal experimentation only
Primary Target Organs:

Eyes

Skin

Respiratory System

Kidneys

Exposure Limits
OSHA PEL: TWA: 0.005 mg/m^3; as Cd; see Table Z2.
ACGIH TLV: TWA: 0.002 mg/m^3; as Cd respirable.
NIOSH REL: TWA: 0.1 mg/m^3; as Cd,LOQ; Lowest Feasible Concentration.

Environmental

Ecotoxicity: LC_{50} Salmo gairdneri (rainbow trout) 6.2 ug/l/96 hr. /Conditions of bioassay not specified
Cleanup/Disposal: Guide No. 154: Eliminate all ignition sources (no smoking, flares, sparks or flames in immediate area). Do not touch damaged containers or spilled material unless wearing appropriate protective clothing. Stop leak if you can do it without risk. Prevent entry into waterways, sewers, basements or confined areas. Absorb or cover with dry earth, sand or other non-combustible material and transfer to containers. Do not get water inside containers.

Environmental Physical Data

BCF: salmo gairdneri 63
BOD: sewage .43 lb/lb, 5 days

Regulations

RCRA 40CFR: Not listed
CERCLA: 40CFR 302.4: Listed per CWA Section 311(b)(4) RQ: 10 lb (4.535 kg)
SARA 40CFR 372.65: Listed as Compound
SARA EHS 40CFR 355: Not listed
TSCA: Listed

CAD2500 CAS #: 7789-42-6

CADMIUM BROMIDE

RTECS: EU9935000
DOT: UN2570; IMO6.1
EINECS Number: 232-165-1
Molecular Formula: Br_2Cd
Structured MF: $CdBr_2$
Formula Weight: 272.22

$Br^- \quad Cd^{++} \quad Br^-$

Chemical Structure

Synonyms: CADMIUM BROMIDE DIMER; CADMIUM DIBROMIDE
Description: white to yellowish, efflorescent crystalline powder, crystals, or flakes; odorless
Use: photography; process engraving; lithography

Physical Properties

Boiling Point: 863 °C (1585 °F)
Freezing Point: 567 °C (1052.6 °F)
Specific Gravity: 5.192 at 25 °C
Water Solubility: 57 g/100 ml Water at 10 °C
Other Solubilities: 26.6 g/100 ml Alcohol at 15 °C; 0.4 g/100 ml Ether at 15 °C; 1.6 g/100 ml Acetone at 18 °C; Soluble in Hydrochloric Acid.
Flash Point: Nonflammable

RTECS Toxicity Data

Acute Oral: Rat LD_{50} Dose: 322 mg/kg.

Hazard Overviews

Health: May cause irritation to eyes/skin/respiratory tract. Toxic. Other Acute Effects: harmful if swallowed, inhaled, or absorbed through skin; exposure can cause: sweating; weakness. Chronic Effects: can cause lung and kidney disease; target organs: kidneys, lungs, central nervous system, liver, male reproductive system. Carcinogen.
Fire: Noncombustible. Hazards: emits toxic fumes. Extinguishing agents: noncombustible; use extinguishing media appropriate to surrounding fire conditions. Precautions: combustible liquid.
Reactivity: Incompatible with: strong oxidizing agents, protect from moisture, reacts with potassium. Hazardous decomposition products: toxic fumes of: hydrogen bromide gas, cadmium/cadmium oxides.
Carcinogenicity: IARC - Group 1, Carcinogenic to humans; NIOSH - Listed as carcinogen; NTP - Class 2A, Reasonably anticipated to be a carcinogen, limited evidence of carcinogenicity from studies in humans; ACGIH - Class A2, Suspected human carcinogen; OSHA - Listed as a carcinogen; EPA - Class B1, Probable human carcinogen based on epidemiologic studies; MAK - Class A2, Unmistakably carcinogenic in animal experimentation only
Primary Target Organs:

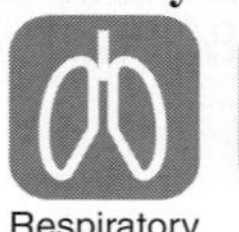
Respiratory System

Nervous System

Liver

Kidneys

Repro-ductive

Exposure Limits
OSHA PEL: TWA: 0.005 mg/m^3; as Cd; see Table Z2.
ACGIH TLV: TWA: 0.002 mg/m^3; as Cd respirable.
NIOSH REL: TWA: 0.1 mg/m^3; as Cd,LOQ; Lowest Feasible Concentration.

Environmental

Ecotoxicity: Food chain concentration potential: Concentrated by shellfish
Cleanup/Disposal: Guide No. 154: Eliminate all ignition sources (no smoking, flares, sparks or flames in immediate area). Do not touch damaged containers or spilled material unless wearing appropriate protective clothing. Stop leak if you can do it without risk. Prevent entry into waterways, sewers, basements or confined areas. Absorb or cover with dry earth, sand or other non-combustible material and transfer to containers. Do not get water inside containers.

Environmental Physical Data

BCF: concentrated by shellfish
BOD: none

Regulations

RCRA 40CFR: Not listed
CERCLA: 40CFR 302.4: Listed per CWA Section 311(b)(4) RQ: 10 lb (4.535 kg)
SARA 40CFR 372.65: Listed as Compound
SARA EHS 40CFR 355: Not listed
TSCA: Listed

CAD3000 CAS #: 513-78-0

CADMIUM CARBONATE

RTECS: FF9320000
DOT: UN2570; IMO6.1
EINECS Number: 208-168-9
Molecular Formula: $CCdO_3$
Formula Weight: 172.42

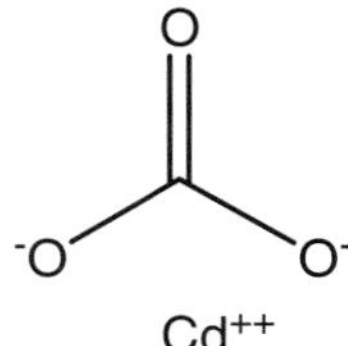

Chemical Structure

Synonyms: CADMIUM MONOCARBONATE; CARBONIC ACID,CADMIUM SALT; CARBONIC ACID,CADMIUM SALT (1:1); CARBONIC ACID,CADMIUM(2+) SALT (1:1); CHEMCARB; EPA PESTICIDE CHEMICAL CODE 012901; KALCIT; MIKROKALCIT; OTAVITE; SUPERMIKROKALCIT
Description: white, trigonal crystals; powder or rhombohedral leaflets
Use: lawn & turf fungicide; intermediate for high purity specialty chem such as phosphors; catalyst in organic reactions; source of cadmium

Physical Properties

Freezing Point: Decomposes at < 500 °C (932 °F)
Specific Gravity: 4.258 at 4 °C
Water Solubility: Insoluble in Water
Other Solubilities: Insoluble in Ammonia; Insoluble in Potassium Cyanide; Soluble in concentrated solution of Ammonium salts.

RTECS Toxicity Data

Acute Oral: Rat LD_{50} Dose: 438 mg/kg. Mouse LD_{50} Dose: 310 mg/kg; Toxic Effects: Behavioral - Tremor; Behavioral - Change in motor activity (specific assay); Nutritional and gross metabolic - Weight loss or decreased weight gain.
Mutagenic: Hamster Sister Chromatid Exchenge; Cell Type: ovary; Dose: 870 nmol/L.

Hazard Overviews

Health: May cause irritation. Poison. Other Acute Effects: harmful if swallowed, inhaled, or absorbed through skin. Chronic Effects: target organs: kidneys, lungs. Carcinogen.
Fire: Hazards: in powder form capable of creating a dust explosion. Extinguishing agents: noncombustible; use extinguishing media appropriate to surrounding fire conditions. Precautions: combustible liquid.
Reactivity: Incompatible with: oxidizing agents. Hazardous decomposition products: toxic fumes;.
Carcinogenicity: IARC - Group 1, Carcinogenic to humans; NIOSH - Listed as carcinogen; NTP - Class 2A, Reasonably anticipated to be a carcinogen, limited evidence of carcinogenicity from studies in humans; ACGIH - Class A2, Suspected human carcinogen; OSHA - Listed as a carcinogen; EPA - Class B1, Probable human carcinogen based on epidemiologic studies; MAK - Class A2, Unmistakably carcinogenic in animal experimentation only
Primary Target Organs:

Exposure Limits
OSHA PEL: TWA: 0.005 mg/m^3; as Cd; see Table Z2.
ACGIH TLV: TWA: 0.002 mg/m^3; as Cd respirable.
NIOSH REL: TWA: 0.1 mg/m^3; as Cd,LOQ; Lowest Feasible Concentration.

Environmental

Cleanup/Disposal: Guide No. 154: Eliminate all ignition sources (no smoking, flares, sparks or flames in immediate area). Do not touch damaged containers or spilled material unless wearing appropriate protective clothing. Stop leak if you can do it without risk. Prevent entry into waterways, sewers, basements or confined areas. Absorb or cover with dry earth, sand or other non-combustible material and transfer to containers. Do not get water inside containers.

Regulations

RCRA 40CFR: Not listed
CERCLA: 40CFR 302.4: Listed as Compound per CWA Section 307(a) per CAA Section 112
SARA 40CFR 372.65: Listed as Compound
SARA EHS 40CFR 355: Not listed
TSCA: Listed

CAD3500 **CAS #: 10108-64-2**

CADMIUM CHLORIDE

RTECS: EV0175000
DOT: NA2570; IMO6.1
EINECS Number: 233-296-7
Molecular Formula: $CdCl_2$
Structured MF: $CdCl_2$
Formula Weight: 183.32

Cl^- Cd^{++} Cl^-

Chemical Structure

Synonyms: CADDY; CADMIUM DICHLORIDE; DICHLOROCADMIUM; KADMIUMCHLORID; VI-CAD
Description: colorless, white crystals; odorless
Use: in photography, in dyeing and calico printing, in the vacuum tube industry, in the manufacture of cadmium yellow, in galvanoplasty, in the manufacture of special mirrors, as an ice-nucleating agent, as a lubricant, in analysis of sulfides to absorb hydrogen sulfide, in testing for pyridine bases, as a fungicide, in the preparation of cadmium sulfide, in analytical chemistry, as an ingredient of electroplating baths, in additions of tinning solutions, in paints and electronic components, as a pesticide, as an insecticide, as a nematocide, as a polymerization catalyst and in pigments, glass and glazes

Physical Properties

Boiling Point: 960 °C (1760 °F)
Freezing Point: 568 °C (1054.4 °F)
Specific Gravity: 4.047 at 25 °C
Density: 4.047 g/mL at 25 °C
Vapor Pressure: 10 mm Hg at 656 °C
Water Solubility: 140 g/100 ml Water at 20 °C
Other Solubilities: Soluble in Acetone; Slightly Soluble in Methanol; Slightly Soluble in Ethanol; Practically Insoluble in Ether.
Refraction Index: 1.65
Flash Point: Nonflammable
Autoignition Temperature: 250 °C

RTECS Toxicity Data

Acute Oral: Woman LD_{Lo} Dose: 3 gm/kg; Toxic Effects: Vascular - BP lowering not characterized in autonomic section; Lungs, Thorax, or Respiration - Acute pulmonary edema; Gastrointestinal - Hypermotility, diarrhea. Rat LD_{50} Dose: 88 mg/kg; Toxic Effects: Gastrointestinal - Changes in structure or function of salivary glands; Gastrointestinal - Hypermotility, diarrhea; Gastrointestinal - Nausea or vomiting.
Acute Inhalation: Dog LC_{90} Dose: 420 mg/m^3/30M; Toxic Effects: Cardiac - Pulse rate; Lungs, Thorax, or Respiration - Acute pulmonary edema; Gastrointestinal - Nausea or vomiting.
Acute Dermal: Rabbit LD_{Lo} Route: Subcutaneous Dose: 18 mg/kg. Rat LD_{50} Route: Subcutaneous Dose: 15166 ug/kg.
Reproductive/Teratogenic: Rat Route: Oral; Dose: 652 mg/kg; Duration: female 7-16D of pregnancy; Maternal Effects - Ovaries, fallopian tubes; Effects on Fertility - Post-implantation mortality; Litter size. Rat Route: Oral; Dose: 326 mg/kg; Duration: female 1-20D of pregnancy; Effects on

Fertility - Pre-implantation mortality; Effects on Embryo or Fetus - Fetotoxicity; Fetal death. Rat Route: Oral; Dose: 48 mg/kg; Duration: female 90D prior to mating Effects on Newborn - Other postnatal measures or effects. Rat Route: Oral; Dose: 14677 ug/kg; Duration: female 6-14D of pregnancy; Effects on Embryo or Fetus - Fetal death; Specific Developmental Abnormalities - Urogenital system. Rat Route: Oral; Dose: 17 mg/kg; Duration: male 6W prior to mating; Specific Developmental Abnormalities - Musculoskeletal system. Rat Route: Oral; Dose: 4890 ng/kg; Duration: male 30D prior to mating; Paternal Effects - Prostate, seminal vessicle, Cowper's gland, accessory glands. Rat Route: Oral; Dose: 280 mg/kg; Duration: female 6-19D of pregnancy; Specific Developmental Abnormalities - Cardiovascular (circulatory) system. Rat Route: Inhalation; Dose: 200 $ug/m^3/24$ Duration: female 1-21D of pregnancy; Effects on Embryo or Fetus - Other effects to embryo or fetus; Maternal Effects - Other effects on females; Nutritional and gross metabolic - Weight loss or decreased weight gain. Rat Route: Inhalation; Dose: 600 $ug/m^3/24$ Duration: female 1-21D of pregnancy; Effects on Embryo or Fetus - Fetotoxicity. Rat Route: Inhalation; Dose: 130 ug/m^3; Duration: female 1-19D of pregnancy; Effects on Embryo or Fetus - Fetotoxicity; Fetal death. Monkey Route: Subcutaneous; Dose: 12 mg/kg; Duration: male 1D prior to mating; Paternal Effects - Testes, epididymis, sperm duct; Prostate, seminal vessicle, Cowper's gland, accessory glands.

Mutagenic: Human DNA Damage; Cell Type: fibroblast; Dose: 1 mmol/L. Human DNA Inhibition; Cell Type: HeLa cell; Dose: 15 umol/L. Human Cytogenetic Analysis; Cell Type: lymphocyte; Dose: 50 umol/L/24H. Human Cytogenetic Analysis; Cell Type: leukocyte; Dose: 1 umol/L. Human Sister Chromatid Exchenge; Cell Type: lymphocyte; Dose: 5 umol/L. Human Sex Chromosome Loss; Cell Type: lymphocyte; Dose: 1 mg/L.

Tumorigenic: Rat Route: Inhalation; Dose: 20 $ug/m^3/23H/78W$-C; Toxic Effects: Tumorigenic - Carcinogenic by RTECS criteria; Lungs, Thorax, or Respiration - Tumors. Rat Route: Inhalation; Dose: 41 $ug/m^3/23H/78W$-C; Toxic Effects: Tumorigenic - Carcinogenic by RTECS criteria; Lungs, Thorax, or Respiration - Tumors. Rat Route: Inhalation; Dose: 82 $ug/m^3/23H/78W$-C; Toxic Effects: Tumorigenic - Carcinogenic by RTECS criteria; Lungs, Thorax, or Respiration - Tumors.

Hazard Overviews

Fire Diamond

Health: Irritating to eyes/skin/respiratory tract. Toxic. Also Causes: chest tightness, cough, dyspnea, headache, vertigo, weakness, chills, respiratory shock, pulmonary edema, wheezing, hemoptysis, burns, nausea, abdominal colic, diarrhea, facial and neck edema. Chronic Effects: rhinits, perforation of nasal septum, anosmia, yellow teeth, anorexia, insomnia, fatigue, pallor, anemia, liver and kidney dysfunction, osteomalacia, toxic kidney stones.

Fire: Noncombustible. Use extinguishing agents suitable for surrounding fire.

Reactivity: Stable. Hazardous polymerization cannot occur. Avoid: generation of dusts. Incompatible with: bromine triflouride; potassium; oxidizers; zinc; selenium; tellurium; hydrogen azide. Hazardous decomposition products: fumes of chloride; cadmium.

Carcinogenicity: IARC - Group 1, Carcinogenic to humans; NIOSH - Listed as carcinogen; NTP - Class 2A, Reasonably anticipated to be a carcinogen, limited evidence of carcinogenicity from studies in humans; ACGIH - Class A2, Suspected human carcinogen; OSHA - Listed as a carcinogen; EPA - Class B1, Probable human carcinogen based on epidemiologic studies; MAK - Class A2, Unmistakably carcinogenic in animal experimentation only

Primary Target Organs:

Skin

Respiratory System

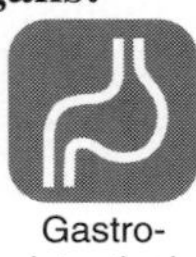
Gastro-intestinal

Liver

Kidneys

Exposure Limits

OSHA PEL: TWA: 0.005 mg/m^3; as Cd; see Table Z2.

ACGIH TLV: TWA: 0.002 mg/m^3; as Cd respirable.

NIOSH REL: TWA: 0.1 mg/m^3; as Cd,LOQ; Lowest Feasible Concentration.

Environmental

Ecotoxicity: LC_{50} Physa integra (snail) 10.4 ug/l/28 days in water hardness of 44-58 mg/l $CaCO_3$. /Conditions of bioassay not specified LC_{50} Mallard duck oral greater than 5000 ppm (no mortality to 1580 ppm, 8% at 5000 ppm) 10 days old LC_{50} Tubifex tubifex (tubificid worm) 320,000 ug/l/48 hr in water hardness of 224 mg/l $CaCO_3$. /Conditions of bioassay not specified LC_{50} Pimephales promelas (fathead minnow) 80.8 ug/l/96 hr in water hardness of 63 mg/l $CaCO_3$; 40.9 ug/l/96 hr in water hardness of 55 mg/l $CaCO_3$ /Conditions of bioassay not specified LC_{50} Ring-necked Pheasant oral 767 ppm/5 days (95% confidence limit, 651-898 ppm) 10 days old LC_{50} Daphnia magna (cladoceran) 14-17 ug/l/72 hr in water hardness of 163 mg/l $CaCO_3$ /Conditions of bioassay not specified LC_{50} Oncorhynchus kisutch (coho salmon, juvenile) 2.0 ug/l/217 hr in water hardness of 22 mg/l $CaCO_3$; (adult) 3.7 ug/l/215 hr in water hardness of 22 mg/l $CaCO_3$ /Conditions of bioassay not specified LC_{50} Salmo gairdneri (rainbow trout), 10-30 ug/l/10 days in water hardness of 125 mg/l $CaCO_3$ in water temperature of 18 °C; 30 ug/l/10 days in water hardness of 125 mg/l $CaCO_3$ in water temperature of 12 °C LC_{50} Palaemonetes pugio (grass shrimp) 50 ug/l/21 days in 20 g/kg salinity /Conditions of bioassay not specified LC_{50} Mysidopsis bahia (mysid) 1 ug/l/17 days in 15-23 g/kg salinity /Conditions of bioassay not specified LC_{50} Crassostrea gigas (pacific oyster) 50 ug/l/23 days /Conditions of bioassay not specified EC_{50} Thalassiosira pseudonana (diatom) 160 ug/l/96 hr, toxic effect: growth rate; Skeletonema costatum (diatom) 175 ug/l/96 hr, toxic effect: growth rate /Conditions of bioassay not specified EC_{50} Gastrophyrne carolinesis (narrow-mouthed toad, embryo larva) 40 ug/l/7 days in water hardness of 195 mg/l $CaCO_3$, toxic effect: death and deformity /Conditions of bioassay not specified LC_{50} Ephemerella sp (may fly) < 3.0 ug/l/28 days in water hardness of 44-48 mg/l $CaCO_3$ /Conditions of bioassay not specified

Cleanup/Disposal: Guide No. 154: Eliminate all ignition sources (no smoking, flares, sparks or flames in immediate area). Do not touch damaged containers or spilled material unless wearing appropriate protective clothing. Stop leak if you can do it without risk. Prevent entry into waterways, sewers, basements or confined areas. Absorb or cover with dry earth, sand or other non-combustible material and transfer to containers. Do not get water inside containers.

Environmental Physical Data

BCF: none likely
BOD: not pertinent

Regulations

RCRA 40CFR: Not listed
CERCLA: 40CFR 302.4: Listed per CWA Section 311(b)(4) RQ: 10 lb (4.535 kg)
SARA 40CFR 372.65: Listed as Compound
SARA EHS 40CFR 355: Not listed
TSCA: Listed

CAD4000 CAS #: 2420-98-6

CADMIUM 2-ETHYLHEXANOATE

EINECS Number: 219-346-0
Molecular Formula: $C_8H_{32}CdO_4$
Formula Weight: 400.83
Synonyms: CADMIUM DI-2-ETHYLHEXYLATE; CADMIUM ETHYLHEXANOATE; CADMIUM 2-ETHYLHEXOATE; HEXANOIC ACID,2-ETHYL-,CADMIUM SALT
Use: to impart transparency to vinyl film (sheet)

Physical Properties

Water Solubility: Low Solubility in Water

Hazard Overviews

Carcinogenicity: IARC - Group 1, Carcinogenic to humans; NIOSH - Listed as carcinogen; NTP - Class 2A, Reasonably anticipated to be a carcinogen, limited evidence of carcinogenicity from studies in humans; ACGIH - Class A2, Suspected human carcinogen; OSHA - Listed as a carcinogen; EPA - Class B1, Probable human carcinogen based on epidemiologic studies; MAK - Class A2, Unmistakably carcinogenic in animal experimentation only

Exposure Limits
OSHA PEL: TWA: 0.005 mg/m^3; as Cd; see Table Z2.
ACGIH TLV: TWA: 0.002 mg/m^3; as Cd respirable.
NIOSH REL: TWA: 0.1 mg/m^3; as Cd,LOQ; Lowest Feasible Concentration.

Environmental

Regulations

RCRA 40CFR: Not listed
CERCLA: 40CFR 302.4: Listed as Compound per CWA Section 307(a) per CAA Section 112
SARA 40CFR 372.65: Listed as Compound
SARA EHS 40CFR 355: Not listed
TSCA: Listed

CAD4500 CAS #: 14486-19-2

CADMIUM FLUOBORATE

RTECS: EV0525000
DOT: UN2570; IMO6.1
EINECS Number: 238-490-5
Molecular Formula: B_2CdF_8
Structured MF: $Cd(BF_4)_2$
Formula Weight: 286.02
Synonyms: BORATE(1-),TETRAFLUORO-,CADMIUM; BORATE(1-),TETRAFLUORO-,CADMIUM (2:1); BORATE(1-),TETRAFLUORO-,CADMIUM (2:1) (9CI); BORATE,(1-)TETRAFLUORO-,CADMIUM SALT (2:1); CADMIUM FLUOBORATE SOLUTION; CADMIUM FLUOROBORATE; CADMIUM TETRAFLUOROBORATE; CADMIUM TETRAFLUOROBORATE (7CI); TL 1026
Description: colorless liquid; odorless
Use: electroplating metals; electrodeposition of cadmium on high strength steels to avoid the problem of hydrogen embrittlement in cyanide plating

Physical Properties

Specific Gravity: 1.6 at 20 °C
Water Solubility: Very Soluble in Water
Other Solubilities: Very Soluble in Alcohol.
Flash Point: Nonflammable

RTECS Toxicity Data

Acute Oral: Rat LD_{Lo} Dose: 250 mg/kg.
Acute Inhalation: Mouse LC_{Lo} Dose: 650 mg/m^3/10M.

Hazard Overviews

Corrosive

Fire: Noncombustible.
Carcinogenicity: IARC - Group 1, Carcinogenic to humans; NIOSH - Listed as carcinogen; NTP - Class 2A, Reasonably anticipated to be a carcinogen, limited evidence of carcinogenicity from studies in humans; ACGIH - Class A2, Suspected human carcinogen; OSHA - Listed as a carcinogen; EPA - Class B1, Probable human carcinogen based on epidemiologic studies; MAK - Class A2, Unmistakably carcinogenic in animal experimentation only

Exposure Limits
OSHA PEL: TWA: 2.5 mg/m^3; as F.

ACGIH TLV: TWA: 0.002 mg/m^3; as Cd respirable.
NIOSH REL: TWA: 0.1 mg/m^3; as Cd,LOQ; Lowest Feasible Concentration.

Environmental

Ecotoxicity: Food chain concentration potential: Concentrated by shellfish
Cleanup/Disposal: Guide No. 154: Eliminate all ignition sources (no smoking, flares, sparks or flames in immediate area). Do not touch damaged containers or spilled material unless wearing appropriate protective clothing. Stop leak if you can do it without risk. Prevent entry into waterways, sewers, basements or confined areas. Absorb or cover with dry earth, sand or other non-combustible material and transfer to containers. Do not get water inside containers.

Environmental Physical Data

BCF: concentrated by shellfish

Regulations

RCRA 40CFR: Not listed
CERCLA: 40CFR 302.4: Listed as Compound per CWA Section 307(a) per CAA Section 112
SARA 40CFR 372.65: Listed as Compound
SARA EHS 40CFR 355: Not listed
TSCA: Listed

CAD5000 CAS #: 1345-09-1

CADMIUM MERCURY SULFIDE

DOT: UN2025; UN2570; IMO6.1
EINECS Number: 215-717-6
Molecular Formula: Unspecified or Variable
Synonyms: C.I. 77201; C.I. PIGMENT RED 113; CADMIUM VERMILION A; MERCURY CADMIUM REDS
Use: pigment in paints, enamels, & printing inks; colorant in plastics, vinyl products, & rubber; colorant in paper & textile printing

Physical Properties

Other Solubilities: When heated with Na_2CO_3yield metallic Hg and are reduced to metal by H2O2 in the presence of alkali hydroxide. Cu, Fe, Zn and many other metals ppt metallic Hg from neutral or slightly acid soln of mercury salts.

Hazard Overviews

Corrosive

Carcinogenicity: IARC - Not listed; NIOSH - Not listed; NTP - Not listed; ACGIH - Not listed; OSHA - Not listed; EPA - Not listed; MAK - Not listed

Environmental

Cleanup/Disposal: Guide No. 151: Do not touch damaged containers or spilled material unless wearing appropriate protective clothing. Stop leak if you can do it without risk. Prevent entry into waterways, sewers, basements or confined areas. Cover with plastic sheet to prevent spreading. Absorb or cover with dry earth, sand or other non-combustible material and transfer to containers. Do not get water inside containers. Guide No. 154: Eliminate all ignition sources (no smoking, flares, sparks or flames in immediate area). Do not touch damaged containers or spilled material unless wearing appropriate protective clothing. Stop leak if you can do it without risk. Prevent entry into waterways, sewers, basements or confined areas. Absorb or cover with dry earth, sand or other non-combustible material and transfer to containers. Do not get water inside containers.

Regulations

RCRA 40CFR: Not listed
CERCLA: 40CFR 302.4: Not listed
SARA 40CFR 372.65: Not listed
SARA EHS 40CFR 355: Not listed
TSCA: Listed

CAD5500 CAS #: 10325-94-7

CADMIUM NITRATE

RTECS: EV1750000
DOT: UN2570; IMO6.1
EINECS Number: 233-710-6
Molecular Formula: $CdH_2N_2O_6$
Structured MF: $Cd(NO_3)_2$
Formula Weight: 236.43
Synonyms: CADMIUM DINITRATE; CADMIUM(II) NITRATE; NITRIC ACID,CADMIUM SALT
Description: white amorphous pieces, hygroscopic needles; odorless
Use: in making other cadmium salts; in photographic emulsions; in reactors to control rate of nuclear fission & reactor poison; coloring glass & porcelain; in mfr of nickel-cadmium batteries, & turf fungicide (tetrahydrate); in fluorescent lamp coatings; impregnation of silica gel plates for improved separation of alkaloids (atropine, scopolamine, strychnine, & others)in pharmaceutical preparations.

Physical Properties

Freezing Point: 350 °C (662 °F)
Specific Gravity: 72
Water Solubility: 109 g/100 ml of Water at 0 °C
Other Solubilities: Soluble in Ethyl Acetate; Soluble in Ammonia; Soluble in Alcohol; Soluble in Ether; Soluble in Acetone.
Flash Point: Nonflammable

RTECS Toxicity Data

Acute Oral: Rat LD_{50} Dose: 300 mg/kg. Mouse LD_{50} Dose: 100 mg/kg.

Reproductive/Teratogenic: Rat Route: Oral; Dose: 40 mg/kg; Duration: female 1-19D of pregnancy; Effects on Fertility - Female fertility index; Pre-implantation mortality; Effects on Embryo or Fetus - Fetotoxicity. Rat Route: Oral; Dose: 40 mg/kg; Duration: female 1-19D of pregnancy; Effects on Embryo or Fetus - Fetal death; Specific Developmental Abnormalities - Cardiovascular (circulatory) system; Effects on Newborn - Weaning or lactation index.

Mutagenic: Hamster Cytogenetic Analysis; Cell Type: ovary; Dose: 1 umol/L. Hamster Sister Chromatid Exchenge; Cell Type: ovary; Dose: 300 nmol/L.

Hazard Overviews

Fire Diamond

Health: Irritating to eyes/skin/respiratory tract. Also Causes: headache, nausea, vomiting, possible kidney/liver damage. Chronic Effects: liver/kidney damage, anorexia, insomnia, fatigue, pallor, anemia, bone demineralization, emphysema, lung fibrosis, cancer of the respiratory tract.

Fire: Noncombustible. Use extinguishing agents suitable for surrounding fire.

Reactivity: Stable. Hazardous polymerization cannot occur. Incompatible with: excessive generation of dusts. Hazardous decomposition products: cadmium fumes; nitrogen oxides.

Carcinogenicity: IARC - Group 1, Carcinogenic to humans; NIOSH - Listed as carcinogen; NTP - Class 2A, Reasonably anticipated to be a carcinogen, limited evidence of carcinogenicity from studies in humans; ACGIH - Class A2, Suspected human carcinogen; OSHA - Listed as a carcinogen; EPA - Class B1, Probable human carcinogen based on epidemiologic studies; MAK - Class A2, Unmistakably carcinogenic in animal experimentation only

Primary Target Organs:

Eyes

Skin

Respiratory System

Liver

Kidneys

Bone

Exposure Limits

OSHA PEL: TWA: 0.005 mg/m^3; as Cd; see Table Z2.

ACGIH TLV: TWA: 0.002 mg/m^3; as Cd respirable.

NIOSH REL: TWA: 0.1 mg/m^3; as Cd,LOQ; Lowest Feasible Concentration.

Environmental

Ecotoxicity: LC_{50} Culex pipiens (mosquito) 765 ug/l/48 hr /Conditions of bioassay not specified LC_{50} Xenopus laevis (african clawed frog) 11,700 ug/l/48 hr in water hardness of 209 mg/l calcium carbonate /Conditions of bioassay not specified LC_{50} Salmo gairdneri (rainbow trout) 55 ug/l/48 hr /Conditions of bioassay not specified EC_{50} Daphnia magna (cladoceran) 160 ug/l/24 hr /Conditions of bioassay not specified LC_{50} Hydra oligactis (hydra) 583 ug/l/48 hr /Conditions of bioassay not specified LC_{50} Lymnaea stagnalis (snail) 583 ug/l/48 hr /Conditions of bioassay not specified LC_{50} Callinectes sapidus (blue crab) juvenile 320 ug/l 4 days in 1 g/kg salinity; 50 ug/l/7 days in 10 g/kg salinity; 150 ug/l/7 days in 30 g/kg salinity. /Conditions of bioassay not specified

Cleanup/Disposal: Guide No. 154: Eliminate all ignition sources (no smoking, flares, sparks or flames in immediate area). Do not touch damaged containers or spilled material unless wearing appropriate protective clothing. Stop leak if you can do it without risk. Prevent entry into waterways, sewers, basements or confined areas. Absorb or cover with dry earth, sand or other non-combustible material and transfer to containers. Do not get water inside containers.

Environmental Physical Data

BCF: fern 960

Regulations

RCRA 40CFR: Not listed

CERCLA: 40CFR 302.4: Listed as Compound per CWA Section 307(a) per CAA Section 112

SARA 40CFR 372.65: Listed as Compound

SARA EHS 40CFR 355: Not listed

TSCA: Listed

CAD6000 CAS #: 10022-68-1

CADMIUM NITRATE TETRAHYDRATE

RTECS: EV1850000

Molecular Formula: $CdH_8N_2O_{10}$

Structured MF: $Cd(NO_3)_2{\bullet}4H_2O$

Formula Weight: 308.5

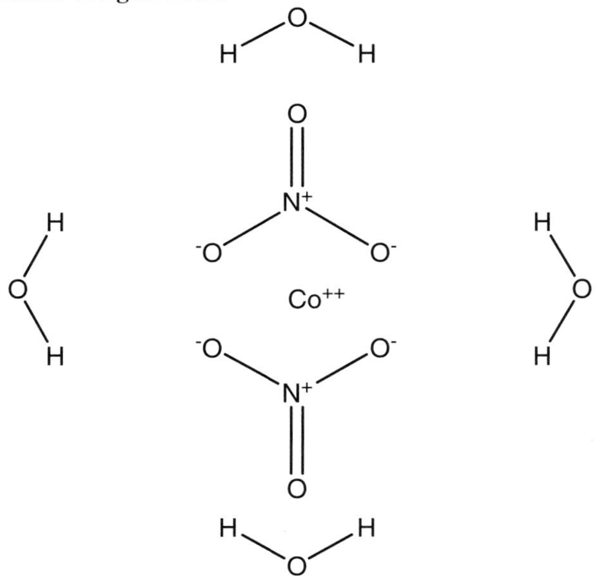

Chemical Structure

Synonyms: C.I. 77192; CADMIUM DINITRATE TETRAHYDRATE; DUSICNAN KADEMNATY; NITRIC ACID,CADMIUM SALT,TETRAHYDRATE

Physical Properties

Boiling Point: 132 °C (270 °F)
Freezing Point: 59.5 °C (139.1 °F)
Specific Gravity: Solid 2.45 at 20 °C
Water Solubility: Anhydrous Form 1 kg/L
Other Solubilities: Soluble in Ammonia Alcohol acids, Acetone, Ether and Ethyl Acetate; Practically Insoluble in nitric acid
Flash Point: Nonflammable

RTECS Toxicity Data

Acute Oral: Rat LD_{50} Dose: 300 mg/kg.
Irritation Eye: Rabbit Standard Draize Test Dose: 20 mg/24H; Reaction: moderate.
Irritation Skin: Rabbit Standard Draize Test Dose: 500 mg/24H; Reaction: severe.
Mutagenic: Rat Unscheduled DNA Synthesis; Cell Type: liver; Dose: 493 nmol/L. Bacteria - E Coli Mutations in Microorganisms; Dose: 6 umol/L (-S9).

Hazard Overviews

Reactivity: Stable. Hazardous polymerization cannot occur. Incompatible with: excessive generation of dusts. Hazardous decomposition products: cadmium fumes; nitrogen oxides.
Carcinogenicity: IARC - Group 1, Carcinogenic to humans; NIOSH - Listed as carcinogen; NTP - Class 2A, Reasonably anticipated to be a carcinogen, limited evidence of carcinogenicity from studies in humans; ACGIH - Class A2, Suspected human carcinogen; OSHA - Listed as a carcinogen; EPA - Class B1, Probable human carcinogen based on epidemiologic studies; MAK - Class A2, Unmistakably carcinogenic in animal experimentation only
Exposure Limits
OSHA PEL: TWA: 0.005 mg/m^3; as Cd; see Table Z2.
ACGIH TLV: TWA: 0.002 mg/m^3; as Cd respirable.
NIOSH REL: TWA: 0.1 mg/m^3; as Cd,LOQ; Lowest Feasible Concentration.

Environmental

Ecotoxicity: Aquatic toxicity: 0.056 ppm*/**/guppy/LD_{50}/fresh water 0.2 ppm/10 days/stickleback/killed/ fresh water *As cadmium **Time period not specified; Food chain concentration potential: Shellfish concentrate 900-1600 times
Cleanup/Disposal: Guide No. 122: Keep combustibles (wood, paper, oil, etc.) away from spilled material. Do not touch or walk through spilled material. Stop leak if you can do it without risk. If possible, turn leaking containers so that gas escapes rather than liquid. Do not direct water at spill or source of leak. Use water spray to reduce vapors or divert vapor cloud drift. Prevent entry into waterways, sewers, basements or confined areas. Allow substance to evaporate. Isolate area until gas has dispersed. Caution: When in contact with refrigerated/cryogenic liquids, many materials become brittle and are likely to break without warning.

Environmental Physical Data

BCF: shellfish 900 to 1600
BOD: none

Regulations

RCRA 40CFR: Not listed
CERCLA: 40CFR 302.4: Not listed
SARA 40CFR 372.65: Not listed
SARA EHS 40CFR 355: Not listed
TSCA: Not listed

CAD6500 CAS #: 49784-44-3

CADMIUM (2+) NTA

DOT: UN2570
EINECS Number: 256-489-8
Synonyms: CADMIUM NITRILOTRIACETATE

Hazard Overviews

Carcinogenicity: IARC - Not listed; NIOSH - Not listed; NTP - Not listed; ACGIH - Not listed; OSHA - Not listed; EPA - Not listed; MAK - Not listed

Environmental

Cleanup/Disposal: Guide No. 154: Eliminate all ignition sources (no smoking, flares, sparks or flames in immediate area). Do not touch damaged containers or spilled material unless wearing appropriate protective clothing. Stop leak if you can do it without risk. Prevent entry into waterways, sewers, basements or confined areas. Absorb or cover with dry earth, sand or other non-combustible material and transfer to containers. Do not get water inside containers.

Regulations

RCRA 40CFR: Not listed
CERCLA: 40CFR 302.4: Not listed
SARA 40CFR 372.65: Not listed
SARA EHS 40CFR 355: Not listed
TSCA: Not listed

CAD7000 CAS #: 1306-19-0

CADMIUM OXIDE

RTECS: EV1925000
DOT: UN2570; IMO6.1
EINECS Number: 215-146-2
Molecular Formula: CdO
Structured MF: CdO
Formula Weight: 128.41

Cd═O

Chemical Structure

Synonyms: ASKA-RID; CADMIUM FUME; CADMIUM MONOXIDE; CADMIUM OXIDE FUME CD: CADMIUM; CADMIUM OXIDE FUME CD: CADMIUM CDO: CADMIUM MONOXIDE; CADMIUM(II) OXIDE; CDO: CADMIUM MONOXIDE; KADMU TLENEK

Description: colorless to white, brownish red powder, crystals; odorless

Use: electroplating, storage battery electrodes, catalyst, semiconductors, manufacture of silver alloys, ceramic glazes, nematocide, anthelminic, phosphors, glass, cadmium electroplating, and an ascaricide in swine

Physical Properties

Boiling Point: 1559 °C (2838 °F)
Freezing Point: 1426.11 °C (2598.998 °F)
Specific Gravity: 8.15
Vapor Pressure: 1 mm Hg at 1000 °C
Water Solubility: Insoluble
Other Solubilities: 95% Ethanol: <1 mg/ml at 20 °C; Acetone: <1 mg/ml at 20 °C; Acids: Soluble; Alkalies: Insoluble; Ammonium salt solutions: Soluble; DMSO: <1 mg/ml at 20 °C.
Refraction Index: 2.49
Flash Point: Nonflammable

RTECS Toxicity Data

Acute Oral: Rat LD_{50} Dose: 72 mg/kg. Mouse LD_{50} Dose: 72 mg/kg.

Acute Inhalation: Monkey LC_{50} Dose: 1500 mg/m^3/10M. Human TC_{Lo} Dose: 8630 ug/m^3/5hr. Man TC_{Lo} Dose: 500 ug/m^3/5Y-I; Toxic Effects: Sense organs and special senses - Change in sensation of smell; Kidney, ureter, and Bladder - Proteinuria. Man TC_{Lo} Dose: 40 ug/m^3; Toxic Effects: Cardiac - Change in rate; Vascular - BP elevation not characterized in autonomic section; Kidney, ureter, and Bladder - Other changes.

Acute Dermal: Mouse LD_{50} Route: Subcutaneous Dose: 94 mg/kg.

Chronic (Multiple Dose) Oral: Rat Dose: 156 ug/kg/26W-I; Toxic Effects: Kidney, Ureter, and Bladder - Proteinuria; Blood - Other changes; Musculoskelital - Other changes.

Chronic (Multiple Dose) Inhalation: Rat Dose: 23 ug/m^3/24H/20W-C; Toxic Effects: Kidney, Ureter, and Bladder - Proteinuria; Nutritional and gross metabolic - Changes in calcium. Rat Dose: 57 ug/m^3/90D-C; Toxic Effects: Lungs, Thorax, or Respiration - Changes in lung weight; Blood - Pigmented or nucleated red Blood cells; Biochemical - Phosphatases. Rat Dose: 300 ug/m^3/6H/2W-I; Toxic Effects: Cardiac - Changes in heart weight; Lungs, Thorax, or Respiration - Other changes; Kidney, Ureter, and Bladder - Changes in kidney weight. Rat Dose: 300 ug/m^3/6H/2W-I; Toxic Effects: Lungs, Thorax, or Respiration - Fibrosis, focal (pneumoconiosis); Lungs, Thorax, or Respiration - Changes in lung weight. Rat Dose: 50 ug/m^3/6H/13W-C; Toxic Effects: Lungs, Thorax, or Respiration - Other changes; Endocrine - Changes in spleen weight; Endocrine - Changes in thymus weight.

Reproductive/Teratogenic: Rat Route: Oral; Dose: 21640 ug/kg; Duration: female 1-19D of pregnancy; Effects on Embryo or Fetus - Maternal-fetal exchange; Fetotoxicity; Fetal death. Rat Route: Oral; Dose: 21640 ug/kg; Duration: female 1-19D of pregnancy; Specific Developmental Abnormalities - Cardiovascular (circulatory) system; Effects on Newborn - Weaning or lactation index; Growth statistics. Rat Route: Inhalation; Dose: 183 ug/m^3/5H Duration: female 21W prior to mating Effects on Newborn - Viability index; Growth statistics. Rat Route: Inhalation; Dose: 23 ug/m^3/5H; Duration: female 21W prior to mating Effects on Newborn - Behavioral. Rat Route: Inhalation; Dose: 91 ug/m^3; Duration: female 1-19D of pregnancy; Effects on Embryo or Fetus - Fetotoxicity. Rat Route: Inhalation; Dose: 2 mg/m^3/6H; Duration: female 4-19D of pregnancy; Effects on Embryo or Fetus - Fetotoxicity; Specific Developmental Abnormalities - Musculoskeletal system. Rat Route: Inhalation; Dose: 16 ug/m^3/24H Duration: female 1-21D of pregnancy; Paternal Effects - Spermatogenesis.

Tumorigenic: Rat Route: Subcutaneous; Dose: 90 mg/kg; Toxic Effects: Tumorigenic - Neoplastic by RTECS criteria; Tumorigenic - Tumors at site of application.

Hazard Overviews

Fire Diamond

Health: Irritating to eyes/respiratory tract. Toxic. Also Causes: cough, headache, weakness, chest tightness, pulmonary edema. Chronic Effects: loss of smell, bronchitis, insomnia, fatigue, kidney damage, bone demineralization.

Fire: Noncombustible. Use extinguishing agents suitable for surrounding fire.

Reactivity: Stable. Hazardous polymerization cannot occur. Avoid: dust generation. Incompatible with: magnesium; metals; hydrogen azide; zinc; selenium; tellurium. Hazardous decomposition products: toxic cadmium fumes.

Carcinogenicity: IARC - Group 1, Carcinogenic to humans; NIOSH - Listed as carcinogen; NTP - Class 2A, Reasonably anticipated to be a carcinogen, limited evidence of carcinogenicity from studies in humans; ACGIH - Class A2, Suspected human carcinogen; OSHA - Listed as a carcinogen; EPA - Class B1, Probable human carcinogen based on epidemiologic studies; MAK - Class A2, Unmistakably carcinogenic in animal experimentation only

Primary Target Organs:

Eyes

Skin

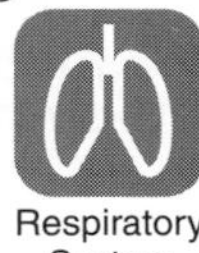
Respiratory System

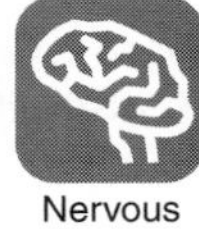
Nervous System

Kidneys

Blood

Exposure Limits

OSHA PEL: TWA: 0.005 mg/m^3; as Cd See Table Z2.
ACGIH TLV: TWA: 0.002 mg/m^3; as Cd respirable.
NIOSH REL: TWA: 0.1 mg/m^3; as Cd,LOQ; Lowest Feasible Concentration.
NIOSH IDLH: 9 mg/m^3; as Cd.

Respirator Recommendation
Exposure Range: >0.005 to 0.05 mg/m^3 Air Purifying, Negative Pressure, Half Mask
Exposure Range: >0.05 to 0.5 mg/m^3 Air Purifying, Negative Pressure, Full Face
Exposure Range: >0.5 to 5 mg/m^3 Supplied Air, Constant Flow/Pressure Demand, Full Face
Exposure Range: >5 to unlimited mg/m^3
Cartridge Color: magenta (P100)

Environmental

Ecotoxicity: Food chain concentration potential: Concentrated by shellfish
Cleanup/Disposal: Guide No. 154: Eliminate all ignition sources (no smoking, flares, sparks or flames in immediate area). Do not touch damaged containers or spilled material unless wearing appropriate protective clothing. Stop leak if you can do it without risk. Prevent entry into waterways, sewers, basements or confined areas. Absorb or cover with dry earth, sand or other non-combustible material and transfer to containers. Do not get water inside containers.

Environmental Physical Data

BCF: concentrated by shellfish
BOD: none

Regulations

RCRA 40CFR: Not listed
CERCLA: 40CFR 302.4: Listed as Compound per CWA Section 307(a) per CAA Section 112
SARA 40CFR 372.65: Not listed TPQ: 100/10000 lb
SARA EHS 40CFR 355: Listed TPQ: 100 lb
TSCA: Listed

Analytical Methods

Indoor / Expired Air: NIOSH 7048

CAD7500 CAS #: 2223-93-0

CADMIUM STEARATE

RTECS: RG1050000
DOT: UN2570; IMO6.1
EINECS Number: 218-743-6
Molecular Formula: $C_{26}H_{72}CdO_4$
Formula Weight: 681.48
Synonyms: ALAIXOL 11; ALAIXOL II; CADMIUM DISTEARATE; CADMIUM OCTADECANOATE; CADMIUM SOAP; CADMIUM SOAPS (STEARATE); CADMIUM(II) STEARATE; KADMIUMSTEARAT; OCTADECANOIC ACID,CADMIUM SALT; SCD; STABILISATOR SCD; STABILIZER SCD; STEARIC ACID,CADMIUM SALT
Description: white powder; slight fatty odor
Use: lubricant and stabilizer in plastics; a commercial adherent; heat stabilizer in PVC plastics

Physical Properties

Freezing Point: 104 °C (219.2 °F)
Specific Gravity: 1.21
Water Solubility: Migrates into distilled Water

RTECS Toxicity Data

Acute Oral: Rat LD_{50} Dose: 1125 mg/kg. Mouse LD_{50} Dose: 590 mg/kg; Toxic Effects: Gastrointestinal - Other changes; Liver - Other changes; Kidney, Ureter, and Bladder - Other changes.
Acute Inhalation: Rat LC_{50} Dose: 130 mg/m^3/2hr; Toxic Effects: Lungs, Thorax, or Respiration - Acute pulmonary edema; Lungs, Thorax, or Respiration - Dyspnea. Human TC_{Lo} Dose: 1800 ug/m^3/2Y; Toxic Effects: Behavioral - Hallucinations, distorted perceptions; Cardiac - Other changes; Nutritional and gross metabolic - Weight loss or decreased weight gain. Woman TC_{Lo} Dose: 147 mg/m^3/35M; Toxic Effects: Behavioral - Hallucinations, distorted perceptions; Gastrointestinal - Nausea or vomiting; Gastrointestinal - Other changes.

Hazard Overviews

Carcinogenicity: IARC - Group 1, Carcinogenic to humans; NIOSH - Listed as carcinogen; NTP - Class 2A, Reasonably anticipated to be a carcinogen, limited evidence of carcinogenicity from studies in humans; ACGIH - Class A2, Suspected human carcinogen; OSHA - Listed as a carcinogen; EPA - Class B1, Probable human carcinogen based on epidemiologic studies; MAK - Class A2, Unmistakably carcinogenic in animal experimentation only
Exposure Limits
OSHA PEL: TWA: 0.005 mg/m^3; as Cd; see Table Z2.
ACGIH TLV: TWA: 0.002 mg/m^3; as Cd respirable.
NIOSH REL: TWA: 0.1 mg/m^3; as Cd,LOQ; Lowest Feasible Concentration.

Environmental

Ecotoxicity: LC_{50} Salmo gairdneri (rainbow trout) 6.0 ug/l/96 hr /Conditions of bioassay not specified
Cleanup/Disposal: Guide No. 154: Eliminate all ignition sources (no smoking, flares, sparks or flames in immediate area). Do not touch damaged containers or spilled material unless wearing appropriate protective clothing. Stop leak if you can do it without risk. Prevent entry into waterways, sewers, basements or confined areas. Absorb or cover with dry earth, sand or other non-combustible material and transfer to containers. Do not get water inside containers.

Regulations

RCRA 40CFR: Not listed
CERCLA: 40CFR 302.4: Listed as Compound per CWA Section 307(a) per CAA Section 112
SARA 40CFR 372.65: Not listed TPQ: 1000/10000 lb
SARA EHS 40CFR 355: Listed TPQ: 1000 lb
TSCA: Listed

CAD8000 CAS #: 10124-36-4

CADMIUM SULFATE

RTECS: EV2700000

DOT: UN2570; IMO6.1
EINECS Number: 233-331-6
Molecular Formula: CdH_2O_4S
Structured MF: $CdSO_4$
Formula Weight: 208.47

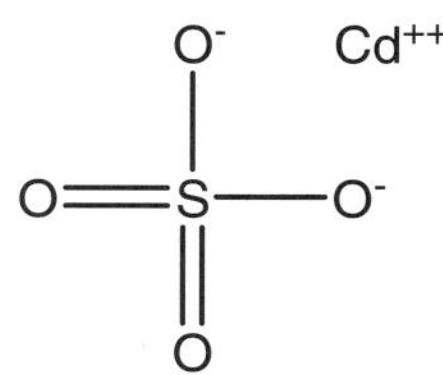

Chemical Structure

Synonyms: CADMIUM MONOSULFATE; CADMIUM SULFATE (1:1); CADMIUM SULPHATE; EPA PESTICIDE CHEMICAL CODE 012905; SULFURIC ACID,CADMIUM SALT (1:1); SULFURIC ACID,CADMIUM(2+) SALT; SULPHURIC ACID,CADMIUM SALT (1:1)
Description: colorless to white crystals, powder; odorless
Use: electrodeposition of cadmium, copper, etc; in phosphors; catalyst in marsh test for arsenic; in determining hydrogen sulfide & fumaric acid; in mfr of cadmium salts of long-chain fatty acids; stabilizer for plastics, esp pvc; in vacuum tubes; fluorescent screens; electrolyte in weston std cell; intermed for cadmium sulfide, etc; fungicide for painting bark of trees; accelerator in cement formation; can absorp hydrochloric acid from waste gases from chem plants.

Physical Properties

Boiling Point: Decomposes at 1 atm
Freezing Point: 1000 °C (1832 °F)
Specific Gravity: 4.691 at 20 °C/4 °C
Vapor Pressure: 1 mm Hg at 394 °C
Water Solubility: 75.5 g/100 ml of Water at 0 °C
Other Solubilities: Insoluble in Alcohol; Insoluble in Acetone; Insoluble in Ammonia.
pH: 5% solution at 25 °C 3.5 to 5
Refraction Index: 1.565
Flash Point: Nonflammable

RTECS Toxicity Data

Acute Oral: Rat LD_{50} Dose: 280 mg/kg. Mouse LD_{50} Dose: 88 mg/kg.
Acute Dermal: Dog LD_{Lo} Route: Subcutaneous Dose: 27 mg/kg. Frog LD_{Lo} Route: Subcutaneous Dose: 105 mg/kg.
Reproductive/Teratogenic: Rat Route: Oral; Dose: 35230 mg/kg; Duration: female 1-19D of pregnancy; Effects on Fertility - Pre-implantation mortality; Post-implantation mortality; Effects on Embryo or Fetus - Maternal-fetal exchange. Rat Route: Oral; Dose: 35230 mg/kg; Duration: female 1-19D of pregnancy; Specific Developmental Abnormalities - Cardiovascular (circulatory) system; Effects on Newborn - Weaning or lactation index; Growth statistics.
Mutagenic: Human DNA Damage; Cell Type: lymphocyte; Dose: 500 umol/L. Rat DNA Damage; Cell Type: liver; Dose: 30 umol/L.

Hazard Overviews

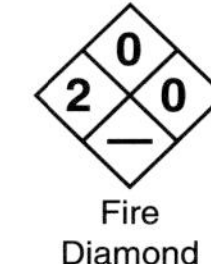

Fire Diamond

Health: Irritating to eyes/skin/respiratory tract. Also Causes: nausea, vomiting, chills, weakness, leg pains, diarrhea, throat dryness, cough, headache, shortness of breath, chest pains, kidney damage, pneumonitis, bronchitis, pulmonary edema, collapse, or shock. Chronic Effects: Chronic bronchitis and rhinitis, gastrointestinal symptoms, impairment or loss of sense of smell, pulmonary fibrosis, emphysema, kidney dysfunction, proteinuria, kidney stones, kidney damage, yellow discoloration of the teeth, and changes in bone.
Fire: Noncombustible. Use agents suitable for surrounding fire.
Reactivity: Stable. Hazardous polymerization cannot occur. Avoid: heat and ignition sources. Incompatible with: oxidizing agents; metals; hydrogen azide; zinc; selenium; tellurium. Hazardous decomposition products: oxides of sulfur; cadmium.
Carcinogenicity: IARC - Group 1, Carcinogenic to humans; NIOSH - Listed as carcinogen; NTP - Class 2A, Reasonably anticipated to be a carcinogen, limited evidence of carcinogenicity from studies in humans; ACGIH - Class A2, Suspected human carcinogen; OSHA - Listed as a carcinogen; EPA - Class B1, Probable human carcinogen based on epidemiologic studies; MAK - Class A2, Unmistakably carcinogenic in animal experimentation only
Primary Target Organs:

Eyes

Skin

Respiratory System

Gastro-intestinal

Kidneys

Bone

Exposure Limits
OSHA PEL: TWA: 0.005 mg/m³; as Cd; see Table Z2.
ACGIH TLV: TWA: 0.002 mg/m³; as Cd respirable.
NIOSH REL: TWA: 0.1 mg/m³; as Cd,LOQ; Lowest Feasible Concentration.

Environmental

Ecotoxicity: LC_{50} Cloeon dipterum (mayfly) 70,600 ug/l/72 hr at 10 °C; 28,600 ug/l/72 hr at 15 °C; 6990 ug/l/72 hr at 25 °C; 930 ug/l/72 hr at 30 °C /Conditions of bioassay not specified LC_{50} Daphnia magna (cladoceran) 3-5 days old 224 ug/l/72 hr at 10 & 15 °C; 12 ug/l/72 hr at 25 °C; 0.1 ug/l/72 hr at 30 °C; Daphnia magna (cladoceran) adult, 479 ug/l/72 hr at 10 °C; 187 ug/l/72 hr at 15 °C; 10.2 ug/l/72 hr at 25 °C; LC_{50} Acanthocyclops viridis (copepod) 0.5 ug/l/72 hr /Conditions of bioassay not specified LC_{50} Idotea baltica (isopod) 10,000 ug/l/5 days (3 g/kg salinity) /Conditions of bioassay not specified
Cleanup/Disposal: Guide No. 154: Eliminate all ignition sources (no smoking, flares, sparks or flames in immediate area). Do not touch damaged containers or spilled material unless wearing appropriate protective clothing. Stop leak if

you can do it without risk. Prevent entry into waterways, sewers, basements or confined areas. Absorb or cover with dry earth, sand or other non-combustible material and transfer to containers. Do not get water inside containers.

Environmental Physical Data

BCF: clam 1752 to 3770
BOD: none

Regulations

RCRA 40CFR: Not listed
CERCLA: 40CFR 302.4: Listed as Compound per CWA Section 307(a) per CAA Section 112
SARA 40CFR 372.65: Listed as Compound
SARA EHS 40CFR 355: Not listed
TSCA: Listed

CAD8500 CAS #: 1306-23-6

CADMIUM SULFIDE

RTECS: EV3150000
DOT: UN2570; IMO6.1
EINECS Number: 215-147-8
Molecular Formula: CdS
Structured MF: CdS
Formula Weight: 144.46

Cd═S

Chemical Structure

Synonyms: AURORA YELLOW; C.I. 77199; C.I. PIGMENT ORANGE 20; C.I. PIGMENT YELLOW 37; CADMIUM GOLDEN 366; CADMIUM LEMON YELLOW 527; CADMIUM MONOSULFIDE; CADMIUM ORANGE; CADMIUM PRIMROSE 819; CADMIUM SULFIDE YELLOW; CADMIUM SULPHIDE; CADMIUM YELLOW; CADMIUM YELLOW 000; CADMIUM YELLOW 892; CADMIUM YELLOW 10G CONC; CADMIUM YELLOW CONC DEEP; CADMIUM YELLOW CONC. DEEP; CADMIUM YELLOW CONC GOLDEN; CADMIUM YELLOW CONC. GOLDEN; CADMIUM YELLOW CONC LEMON; CADMIUM YELLOW CONC. LEMON; CADMIUM YELLOW CONC PRIMROSE; CADMIUM YELLOW CONC. PRIMROSE; CADMIUM YELLOW OZ DARK; CADMIUM YELLOW PRIMROSE 47-4100; CADMOPUR GOLDEN YELLOW N; CADMOPUR YELLOW; CAPSEBON; FERRO LEMON YELLOW; FERRO ORANGE YELLOW; FERRO YELLOW; GREENOCKITE; JAUNE BRILLIANT; ORANGE CADMIUM
Description: light yellow or orange or red-colored cubic or hexagonal crystals; or yellow or brown powder
Use: heat stable, alkali- & hydrogen sulfide-resistant pigments (incl phosphors); for glass yellow, soaps, textiles, paper, rubber; in printing inks, ceramic glazes, fireworks; in x-ray fluorescent screens; body temp detectors; in electronics ind: rectifiers, transistors, photovoltaic cells; photoconductor cells in xerography; stable against oxidn & uv radiation; in lead sealing glass binders to provide smooth durable surface.

Physical Properties

Freezing Point: 1750 °C (3182 °F) at 100 atm
Specific Gravity: 4.82
Water Solubility: 0.00013 g in 100 cc Cold Water
Other Solubilities: Soluble in acid; Very Slightly Soluble in Ammonium Hydroxide; Soluble in concentrated or warm dilute mineral acids with evolution of Hydrogen Sulfide.
Refraction Index: 2.506

RTECS Toxicity Data

Acute Oral: Rat LD_{50} Dose: 7080 mg/kg. Mouse LD_{50} Dose: 1166 mg/kg.
Mutagenic: Human Cytogenetic Analysis; Cell Type: leukocyte; Dose: 62 ug/L. Hamster DNA Damage; Cell Type: ovary; Dose: 10 mg/L.
Tumorigenic: Rat Route: Subcutaneous; Dose: 90 mg/kg; Toxic Effects: Tumorigenic - Carcinogenic by RTECS criteria; Tumorigenic - Tumors at site of application. Rat Route: Subcutaneous; Dose: 135 mg/kg; Toxic Effects: Tumorigenic - Equivocal tumorigenic agent by RTECS criteria; Tumorigenic - Tumors at site of application. Rat Route: Subcutaneous; Dose: 250 mg/kg; Toxic Effects: Tumorigenic - Equivocal tumorigenic agent by RTECS criteria; Blood - Lymphomax including Hodgkin's disease; Tumorigenic - Tumors at site of application.

Hazard Overviews

Health: Irritating to eyes/skin/respiratory tract. Toxic. Other Acute Effects: harmful if inhaled or swallowed; causes irritation; exposure can cause nausea; headache; vomiting. Chronic Effects: may alter genetic material; causes damage to the lungs, liver, kidneys; target organs: kidneys, lungs. Carcinogen.
Fire: Hazards: emits toxic fumes. Extinguishing agents: noncombustible; use extinguishing media appropriate to surrounding fire conditions. Precautions: combustible liquid.
Reactivity: Incompatible with: acids, strong oxidizing agents. Hazardous decomposition products: sulfur oxides, hydrogen sulfide gas.
Carcinogenicity: IARC - Group 1, Carcinogenic to humans; NIOSH - Listed as carcinogen; NTP - Class 2A, Reasonably anticipated to be a carcinogen, limited evidence of carcinogenicity from studies in humans; ACGIH - Class A2, Suspected human carcinogen; OSHA - Listed as a carcinogen; EPA - Class B1, Probable human carcinogen based on epidemiologic studies; MAK - Class A2, Unmistakably carcinogenic in animal experimentation only
Primary Target Organs:

Eyes

Skin

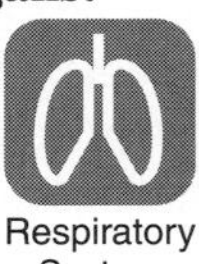
Respiratory System

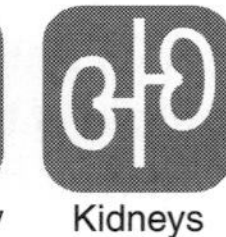
Kidneys

Exposure Limits
OSHA PEL: TWA: 0.005 mg/m^3; as Cd; see Table Z2.
ACGIH TLV: TWA: 0.002 mg/m^3; as Cd respirable.
NIOSH REL: TWA: 0.1 mg/m^3; as Cd,LOQ; Lowest Feasible Concentration.

Environmental

Cleanup/Disposal: Guide No. 154: Eliminate all ignition sources (no smoking, flares, sparks or flames in immediate

area). Do not touch damaged containers or spilled material unless wearing appropriate protective clothing. Stop leak if you can do it without risk. Prevent entry into waterways, sewers, basements or confined areas. Absorb or cover with dry earth, sand or other non-combustible material and transfer to containers. Do not get water inside containers.

Regulations

RCRA 40CFR: Not listed
CERCLA: 40CFR 302.4: Listed as Compound per CWA Section 307(a) per CAA Section 112
SARA 40CFR 372.65: Listed as Compound
SARA EHS 40CFR 355: Not listed
TSCA: Listed

CAF5000 CAS #: 58-08-2

CAFFEINE

RTECS: EV6475000
EINECS Number: 200-362-1
Molecular Formula: $C_8H_{10}N_4O_2$
Formula Weight: 194.19

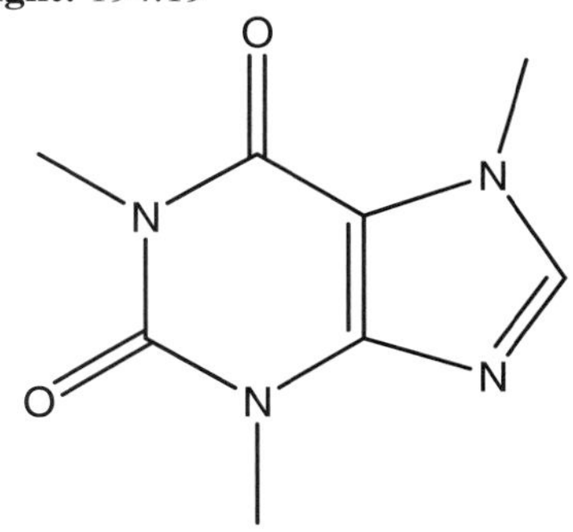

Chemical Structure

Synonyms: ALERT-PEP; ANHYDROUS CAFFEINE; CAFEINA; CAFFEDRINE; CAFFEIN; CAFFEINA; CAFFENIUM; CAFIPEL; COFFEIN; COFFEINE; COFFEINUM; DEXITAC; 3,7-DIHYDRO-1,3,7-TRIMETHYL-1H-PURINE-2,6-DIONE; ELDIATRIC C; GUARANINE; KOFEIN; KOFFEIN; MATEINA; METHYLTHEOBROMIDE; 1-METHYLTHEOBROMINE; METHYLTHEOBROMINE; 7-METHYLTHEOPHYLLINE; NO-DOZ; ORGANEX; 1H-PURINE-2,6-DIONE,3,7-DIHYDRO-1,3,7-TRIMETHYL-; QUICK-PEP; REFRESH'N; STIM; THEIN; THEINE; THEOBROMINE,1-METHYL; THEOBROMINE,1-METHYL-; THEOPHYLLINE,7-METHYL; TIREND; 1,3,7-TRIMETHYL-2,6-DIOXOPURINE; 1,3,7-TRIMETHYL-2,6-DIOXO-1,2,3,6-TETRAHYDROPURINE; 1,3,7-TRIMETHYLXANTHINE; VIVARIN; XANTHINE,1,3,7-TRIMETHYL
Description: white, prism-like crystals; odorless
Use: in beverages and medicines; widely-used stimulant, diuretic and chemosterilant against stored grain pests; in the treatment of shock, asthma and heart disease; in veterinary medicine for cardiac irritation, as a respiratory stimulant and diuretic

Physical Properties

Boiling Point: Sublimes at 178 °C (352 °F)
Freezing Point: Anhydrous 238 °C (460.4 °F)
Specific Gravity: 1.23 at 18 °C/4 °C
Density: 1.23 g/mL at 19 °C
Water Solubility: 10 to 50 mg/mL at 23 °C
Other Solubilities: Soluble in Ethyl Acetate.
pH: 1% solution 6.9
Refraction Index: 1.4936 at 25 °C/D
Flash Point: Not available; probably combustible

RTECS Toxicity Data

Acute Oral: Man TD_{Lo} Dose: 51 mg/kg; Toxic Effects: Cardiac - Change in rate; Kidney, Ureter, and Bladder - Chgs in tubules (inc acute renal failure, acute tubular necrosis; Musculoskelital - Tumors. Man TD_{Lo} Dose: 13 mg/kg; Toxic Effects: Behavioral - Toxic psychosis. Woman TD_{Lo} Dose: 96 mg/kg/1D-I; Toxic Effects: Behavioral - Hallucinations, distorted perceptions; Behavioral - Toxic psychosis; Gastrointestinal - Hypermotility, diarrhea. Infant TD_{Lo} Dose: 14700 ug/kg; Toxic Effects: Behavioral - Muscle contraction or spasticity. Child TD_{Lo} Dose: 140 mg/kg; Toxic Effects: Behavioral - Muscle contraction or spasticity; Cardiac - Pulse rate increased without fall in BP; Blood - Hemorrhage. Human LD_{Lo} Dose: 192 mg/kg. Woman LD_{Lo} Dose: 1 gm/kg; Toxic Effects: Gastrointestinal - Nausea or vomiting. Child LD_{Lo} Dose: 320 mg/kg; Toxic Effects: Behavioral - Convulsions or effect on seizure threshold; Lungs, Thorax, or Respiration - Cyanosis. Rat LD_{50} Dose: 192 mg/kg; Toxic Effects: Brain and coverings - Other degenerative changes; Behavioral - Withdrawal; Kidney, Ureter, and Bladder - Interstitial nephritis.
Acute Dermal: Rabbit LD_{Lo} Route: Subcutaneous Dose: 275 mg/kg. Rat LD_{50} Route: Subcutaneous Dose: 170 mg/kg.
Chronic (Multiple Dose) Oral: Rat Dose: 33210 mg/kg/90D-C; Toxic Effects: Gastrointestinal - Changes in structure or function of salivary glands. Rat Dose: 380 mg/kg/4W-I; Toxic Effects: Endocrine - Changes in pituitary weight; Blood - Changes in other cell count. Rat Dose: 39900 mg/kg/19W-C; Toxic Effects: Nutritional and gross metabolic - Weight loss or decreased weight gain; DEATH - Changes in testicular weight. Rat Dose: 16800 mg/kg/8W-C; Toxic Effects: Behavioral - Food intake (animal); Endocrine - Changes in thymus weight; DEATH - Changes in testicular weight. Rat Dose: 157 gm/kg/75W-C; Toxic Effects: Endocrine - Changes in adrenal weight; Nutritional and gross metabolic - Weight loss or decreased weight gain; DEATH - Changes in testicular weight.
Reproductive/Teratogenic: Woman Route: Oral; Dose: 6750 mg/kg; Duration: female 1-39W of pregnancy; Specific Developmental Abnormalities - Craniofacial (including nose and tongue); Musculoskeletal system. Woman Route: Oral; Dose: 3276 mg/kg; Duration: female 1-39W of pregnancy; Maternal Effects - Parturition; Effects on Newborn - Stillbirth. Woman Route: Oral; Dose: 1092 mg/kg; Duration: female 1-91D of pregnancy; Effects on Fertility - Abortion. Rat Route: Oral; Dose: 627 mg/kg; Duration: female 1-22D of pregnancy; Effects on Newborn - Biochemical and metabolic; Other postnatal measures or effects. Rat Route: Oral; Dose: 85 mg/kg; Duration: female 3-19D of pregnancy; Effects on Newborn - Behavioral; Physical. Rat Route: Oral;

Dose: 200 mg/kg; Duration: female 13-14D of pregnancy Effects on Embryo or Fetus - Extra embryonic structures; Fetotoxicity. Rat Route: Oral; Dose: 1750 mg/kg; Duration: female 15-21D of pregnancy Specific Developmental Abnormalities - Blood and lymphatic systems (including spleen and marrow); Homeostasis. Rat Route: Oral; Dose: 660 mg/kg; Duration: female 1-22D of pregnancy; Effects on Fertility - Female fertility index; Specific Developmental Abnormalities - Urogenital system; Effects on Newborn - Weaning or lactation index. Rat Route: Oral; Dose: 114 mg/kg; Duration: female 1-19D of pregnancy; Specific Developmental Abnormalities - Musculoskeletal system. Rat Route: Oral; Dose: 120 mg/kg; Duration: female 12D of pregnancy; Effects on Embryo or Fetus - Maternal-fetal exchange.

Mutagenic: Human DNA Repair; Cell Type: fibroblast; Dose: 750 umol/L. Human Unscheduled DNA Synthesis; Cell Type: other cell types; Dose: 1 mmol/L. Human DNA Inhibition; Cell Type: HeLa cell; Dose: 1 mmol/L. Human DNA Inhibition; Cell Type: other cell types; Dose: 4 mmol/L. Human Cytogenetic Analysis; Cell Type: leukocyte; Dose: 100 mg/L. Human Cytogenetic Analysis; Cell Type: fibroblast; Dose: 2600 umol/L/24H. Human Cytogenetic Analysis; Cell Type: lymphocyte; Dose: 100 ug/L/24H. Human Cytogenetic Analysis; Cell Type: embryo; Dose: 50 ppm/24H. Human Cytogenetic Analysis; Cell Type: HeLa cell; Dose: 500 mg/L. Human Sister Chromatid Exchenge; Cell Type: lymphocyte; Dose: 1 mmol/L. Human Other Mutation Test Systems; Cell Type: lymphocyte; Dose: 1 mmol/L.

Tumorigenic: Mouse Route: Oral; Dose: 30800 mg/kg/44W-C; Toxic Effects: Tumorigenic - Carcinogenic by RTECS criteria; Skin and appendages - Tumors; Tumorigenic - Increased incidence of tumors in susceptible strains.

Hazard Overviews

Health: Irritating to eyes/skin/respiratory tract. Toxic. Other Acute Effects: harmful if swallowed; may be harmful if inhaled or absorbed through the skin; possible risk of irreversible effects; overexposure by ingestion may result in nervousness; tremors; insomnia; headache; dizziness; drowsiness; incoordination; slowed reaction time; slurred speech; giddiness and unconsciousness; lethargy; convulsions; nausea; vomiting; diarrhea; neurotoxic effects; ataxia; cns stimulation; prolonged or repeated exposure can lead to habituation or addiction; muscle cramps; muscle spasms. Chronic Effects: laboratory experiments have shown mutagenic effects; target organs: central nervous system; heart.

Fire: Will burn. Hazards: emits toxic fumes. Extinguishing agents: carbon dioxide; dry chemical powder; water spray. Precautions: combustible liquid.

Reactivity: Stable. Hazardous polymerization will not occur. Incompatible with: strong oxidizing agents. Hazardous decomposition products: thermal decomposition may produce carbon monoxide, carbon dioxide, and nitrogen oxides.

Carcinogenicity: IARC - Group 3, Not classifiable as to carcinogenicity to humans; NIOSH - Not listed; NTP - Not listed; ACGIH - Not listed; OSHA - Not listed; EPA - Not listed; MAK - Not listed

Primary Target Organs:

Eyes

Skin

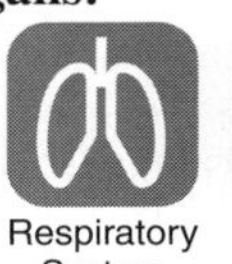
Respiratory System

Nervous System

Cardio-vascular

Environmental

Ecotoxicity: Freshwater tests Streptoxkit F (Streptocephalus proboscideus) test 24 h LC_{50} 447 mg /l

Environmental Fate: If released to soil, it will display very high mobility. It will not volatilize from either moist or dry soil to the atmosphere. Limited data indicate that it has the potential to biodegrade in soil. If release to water, it will not volatilize from water to the atmosphere. It will not bioconcentrate in fish nor will it adsorb to sediment. Limited data indicate that it has the potential to biodegrade in water. If released to the atmosphere, it may undergo a gas-phase reaction with photochemically produced hydroxyl radicals at an estimated half-life of 2.5 hours; however, it will exist predominately adsorbed to particulates in the atmosphere, which may attenuate the rate of this process.

Cleanup/Disposal: Guide No. 151: Do not touch damaged containers or spilled material unless wearing appropriate protective clothing. Stop leak if you can do it without risk. Prevent entry into waterways, sewers, basements or confined areas. Cover with plastic sheet to prevent spreading. Absorb or cover with dry earth, sand or other non-combustible material and transfer to containers. Do not get water inside containers.

Environmental Physical Data

Henry's Law Constant: estimated at 1.9×10^{-19}

Octanol/Water Partition Coefficient: log K_{ow} = -0.07

Sorption Partition Coefficient: K_{oc} = 18 to 22

BCF: calculated at 0.52 to 2.25

Regulations

RCRA 40CFR: Not listed

CERCLA: 40CFR 302.4: Not listed

SARA 40CFR 372.65: Not listed

SARA EHS 40CFR 355: Not listed

TSCA: Listed

Analytical Methods

Soil: SW846 8321A

Water / Groundwater: EPA 553

Drinking Water: EPA 553

CAL1000 CAS #: 20304-47-6

CALACTIN

RTECS: EV7235000

Molecular Formula: $C_{29}H_{40}O_9$

Formula Weight: 532.69

Synonyms: CARD-20(22)-ENOLIDE,3-((4,6-DIDEOXY-BETA-D-ERYTHRO-HEXOPYRANOS-2-ULOS-1-YL)OXY)-2,14-DIHYDROXY-19-OXO-,(2ALPHA,3BETA,5ALPHA)-; PECILOCERIN B; PEKILOCERIN B; POEKILOCERIN B; POKILOCERIN B

Hazard Overviews

Carcinogenicity: IARC - Not listed; NIOSH - Not listed; NTP - Not listed; ACGIH - Not listed; OSHA - Not listed; EPA - Not listed; MAK - Not listed

Environmental

Regulations

RCRA 40CFR: Not listed
CERCLA: 40CFR 302.4: Not listed
SARA 40CFR 372.65: Not listed
SARA EHS 40CFR 355: Not listed
TSCA: Not listed

CAL1320 CAS #: 9007-12-9

CALCITONIN SALMON

RTECS: XP3560000
EINECS Number: 232-693-2
Molecular Formula: Unknown
Formula Weight: 3,600
Synonyms: CALCIMAR (SALMON); CALCITAR; CALCITONIN; TCA; TCT; THYROCALCITONIN
Use: medication: hypocalcemic

RTECS Toxicity Data

Tumorigenic: Mouse Route: Intraperitoneal; Dose: 120 mg/kg/33D-I; Toxic Effects: Tumorigenic - Equivocal tumorigenic agent by RTECS criteria; Peripheral Nerve and sensation - Pareshtesia.

Hazard Overviews

Carcinogenicity: IARC - Not listed; NIOSH - Not listed; NTP - Not listed; ACGIH - Not listed; OSHA - Not listed; EPA - Not listed; MAK - Not listed

Environmental

Regulations

RCRA 40CFR: Not listed
CERCLA: 40CFR 302.4: Not listed
SARA 40CFR 372.65: Not listed
SARA EHS 40CFR 355: Not listed
TSCA: Not listed

CAL1480 CAS #: 7440-70-2

CALCIUM

RTECS: EV8040000
DOT: UN1401; IMO4.3
EINECS Number: 231-179-5
Molecular Formula: Ca
Structured MF: Ca
Formula Weight: 40.08

Ca

Chemical Structure

Synonyms: CALCICAT
Description: silver-white crystalline, lustrous metal; odorless
Use: catalyst for polyester fibers; in metallurgy as deoxidizer for copper, beryllium, steel (together with silicon); reducing agent in prepn of chromium metal powder, thorium, zirconium, uranium; to harden lead for bearings; alloyed with aluminum, copper, & lead, & with cerium to make flints for cigarette & gas lighters; as getter in mfr of electronic vacuum tubes; dehydrating oils; for decarburization & desulfurization of iron & its alloys; separation of nitrogen from argon.

Physical Properties

Boiling Point: 1440 °C (2624 °F)
Freezing Point: 850 °C (1562 °F)
Specific Gravity: 1.54 at 20 °C/4 °C
Vapor Pressure: 10 mm Hg at 983 °C
Water Solubility: Decomposes in Water
Other Solubilities: Insoluble in Benzene, Kerosene.
Flash Point: Flammable solid
Autoignition Temperature: 780 to 810 °C

Hazard Overviews

Corrosive

Flammable

Fire Diamond

Health: Corrosive to eyes/skin/respiratory tract. Also Causes: chemical pneumonitis.
Fire: Flammable solid. Use only graphite powder, soda ash, powdered sodium chloride to extinguish fires. Do not use water, foam, or halogenated hydrocarbons such as Halon or carbon tetrachloride.
Reactivity: Stable, oxidizes in air. Hazardous polymerization cannot occur. Avoid: exposure to moisture; heat; ignition sources. Incompatible with: air; water; mercury (at 734 °F/390 °C); silicon (above 1922 °F/1050 °C); sodium and mixed oxides and heat; alkali metal hydroxides; carbonates; dinitrogen tetraoxide; lead chloride and heat; phosphorus (V) oxide and heat; sulfur and heat; asbestos cement; chlorine fluorides (chlorine trifluoride; chlorine pentafluoride); halogens (fluorine; chlorine); sulfur and vanadium oxide. Hazardous decomposition products: calcium oxides.
Carcinogenicity: IARC - Not listed; NIOSH - Not listed; NTP - Not listed; ACGIH - Not listed; OSHA - Not listed; EPA - Not listed; MAK - Not listed

Primary Target Organs:

Eyes

Skin

Respiratory System

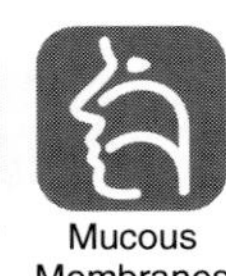
Mucous Membranes

Environmental

Ecotoxicity: Aquatic toxicity: 92 ppm/7 hr/trout/toxic/fresh water; 240 ppm/24 hr/mosquito fish/TLm/fresh water

Cleanup/Disposal: Guide No. 138: Eliminate all ignition sources (no smoking, flares, sparks or flames in immediate area). Do not touch or walk through spilled material. Stop leak if you can do it without risk. Use water spray to reduce vapors or divert vapor cloud drift. Do not get water on spilled substance or inside containers. Small Spills: Cover with dry earth, dry sand, or other non-combustible material followed with plastic sheet to minimize spreading or contact with rain. Dike for later disposal; do not apply water unless directed to do so. Powder Spills: Cover powder spill with plastic sheet or tarp to minimize spreading and keep powder dry. Do not clean-up or dispose of, except under supervision of a specialist.

Environmental Physical Data

BOD: none

Regulations

RCRA 40CFR: Not listed
CERCLA: 40CFR 302.4: Not listed
SARA 40CFR 372.65: Not listed
SARA EHS 40CFR 355: Not listed
TSCA: Listed

Analytical Methods

Air: ASTM D5086

Soil: CLP 200.62, 200.7_M, 202.62, 215.1_M, 6020_M; EPA 200.7; SW846 3005A, 3010A, 3015, 3050A, 3050B, 3051, 3052, 6010A, 6010B, 7000A, 7140; ASTM D3974, D4698; USGS I5152, I5473, I5474; ISWSD 200.6

Water / Groundwater: EPA I-004-1, 200.0, 200.15, 200.7, 215.2; APHA 3111-A, 3111-B, 3120, 3500-CA; ASTM D2332, D511; USGS I1152, I1472, I3152, I3153, I7152; ISWSD 300.7; FISON AES-0029

Drinking Water: AOAC 920.199; APHA 3500-CA

Food: EPA CA-01; AOAC 921.06, 925.55

Indoor / Expired Air: NIOSH 7020, 7300; ASTM D4185

Plasma: EPA CA-02, 200.11

Urine: EPA M-02-MTL

Other: EPA 1620, 215.1; AOAC 990.08

CAL1640 **CAS #: 62-54-4**

CALCIUM ACETATE

RTECS: AF7525000
DOT: UN1401
EINECS Number: 200-540-9
Molecular Formula: $C_4H_6CaO_4$
Formula Weight: 158.17

H O H
O O⁻
Ca⁺⁺
⁻O O

Chemical Structure

Synonyms: ACETIC ACID,CALCIUM SALT; BROWN ACETATE; BROWN ACETATE OF LIME; CALCIUM DIACETATE; GRAY ACETATE; GRAY ACETATE OF LIME; LIME ACETATE; LIME PYROLIGNITE; SORBO-CALCION; TELTOZAN; VINEGAR SALTS

Description: colorless rod-shaped crystals; slight odor of acetic acid

Use: in mfr of acetic acid, acetone; in dying, tanning & curing skins; in lubricants; corrosion inhibitor; antifoam additive in modern formulated antifreezes; mfg of acetate, acetaldehyde; mordant in printing of textiles; stabilizer in resins; additive to calcium soap lubricants; antimold agent in bakery goods, in sausage casings; firming agent for potatoes, stabilizer & thickener for gelatins & puddings; med: skin preparations, antiseptics & blood coagulant tablets.

Physical Properties

Boiling Point: Decomposes > 260 °C (500 °F)
Freezing Point: Decomposes at 160 °C (320 °F)
Specific Gravity: 1.5
Water Solubility: 37.4 g in 100 cc of Water at 0 °C
Other Solubilities: Practically Insoluble in Acetone & Benzene.
pH: 0.2 molar aqueous solution 7.6
Refraction Index: 1.55

RTECS Toxicity Data

Mutagenic: Rat Unscheduled DNA Synthesis; Route: Intraperitoneal; Dose: 1290 umol/kg/5D-I.

Hazard Overviews

Health: May cause irritation. Other Acute Effects: may be harmful by inhalation, ingestion, or skin absorption.

Fire: Hazards: emits toxic fumes. Extinguishing agents: water spray; carbon dioxide, dry chemical powder or appropriate foam. Precautions: combustible liquid.

Reactivity: Incompatible with: strong oxidizing agents, protect from moisture. Hazardous decomposition products: toxic fumes of: carbon monoxide, carbon dioxide.

Carcinogenicity: IARC - Not listed; NIOSH - Not listed; NTP - Not listed; ACGIH - Not listed; OSHA - Not listed; EPA - Not listed; MAK - Not listed

Environmental

Cleanup/Disposal: Guide No. 138: Eliminate all ignition sources (no smoking, flares, sparks or flames in immediate area). Do not touch or walk through spilled material. Stop leak if you can do it without risk. Use water spray to reduce vapors or divert vapor cloud drift. Do not get water on spilled substance or inside containers. Small Spills: Cover with dry earth, dry sand, or other non-combustible material followed with plastic sheet to minimize spreading or contact with rain. Dike for later disposal; do not apply water unless directed to do so. Powder Spills: Cover powder spill with plastic sheet or tarp to minimize spreading and keep powder dry. Do not clean-up or dispose of, except under supervision of a specialist.

Regulations

RCRA 40CFR: Not listed
CERCLA: 40CFR 302.4: Not listed
SARA 40CFR 372.65: Not listed
SARA EHS 40CFR 355: Not listed
TSCA: Listed

CAL1800 CAS #: 9005-35-0

CALCIUM ALGINATE

RTECS: AZ5810000
Molecular Formula: Unknown
Synonyms: ALGINIC ACID,CALCIUM SALT; CA 33; CALGINATE; COMBINACE; KALTOSTAT
Description: white or cream colored powder, filaments, grains, or granules; slight odor
Use: stabilizing & bodying agent in foods; prodn of soluble temporary yarn in hosiery mfr; prodn of swabs for medical applications; prodn of first aid dressings; emulsion stabilizer; suspending agent in soft drinks; dental impression prepn; in drilling muds; in coatings; in flocculation of solids in water treatment; as sizing agent; thickener

Physical Properties

Water Solubility: Insoluble in Water
Other Solubilities: Insoluble in acids, alkaline solutions.

Hazard Overviews

Fire: Hazards: hazardous polymerization will not occur. Extinguishing agents: water spray; carbon dioxide, dry chemical powder or appropriate foam. Precautions: combustible liquid.
Carcinogenicity: IARC - Not listed; NIOSH - Not listed; NTP - Not listed; ACGIH - Not listed; OSHA - Not listed; EPA - Not listed; MAK - Not listed

Environmental

Regulations

RCRA 40CFR: Not listed
CERCLA: 40CFR 302.4: Not listed
SARA 40CFR 372.65: Not listed
SARA EHS 40CFR 355: Not listed
TSCA: Listed

CAL1960 CAS #: 7778-44-1

CALCIUM ARSENATE

RTECS: CG0830000
DOT: UN1573; UN1574; IMO6.1
EINECS Number: 231-904-5
Molecular Formula: $As_2Ca_3H_6O_8$
Structured MF: $Ca_3(AsO_4)_2$
Formula Weight: 398.08

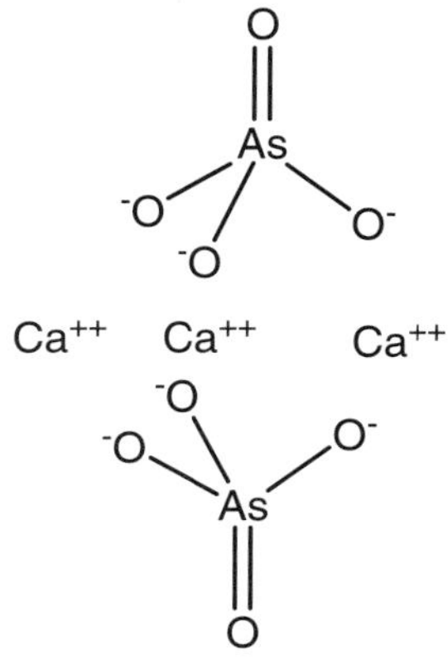

Chemical Structure

Synonyms: ARSENIATE DE CALCIUM; ARSENIC ACID (H3ASO4),CALCIUM SALT (2:3); ARSENIC ACID,CALCIUM SALT(2:3); CALCIUM ORTHOARSENATE; CALCIUM SALT (2:3) OF ARSENIC ACID; CALCIUMARSENAT; CALCIUM-O-ARSENATE; CHIP-CAL; CHIP-CAL GRANULAR; CUCUMBER DUST; EPA PESTICIDE CHEMICAL CODE 013501; FENCAL; FLAC; KALO; KALZIUMARSENIAT; KILMAG; PENCAL; SPRA-CAL; SPRACAL; TRICALCIUM ARSENATE; TRICALCIUM ORTHO-ARSENATE; TRICALCIUMARSENAT; TRICALCIUM-O-ARSENATE; TURF-CAL
Description: colorless to white powder or solid; odorless
Use: herbicide; for pre-emergence treatment of turf & lawns to control crabgrass bluegrass, chickweed, & soil insects incl japanese beetle grubs; insecticide, against japanese beetle grubs; fortifier of bordeaux mixt; formerly for boll weevil control on cotton, leaf-eating insects on tobacco; controls insect grubs; molluscicide.

Physical Properties

Boiling Point: Decomposes
Freezing Point: 1.46 °C (34.628 °F)
Specific Gravity: 3.62
Vapor Pressure: 0 mm Hg
Water Solubility: 0.013 g/100 cc of Water at 25 °C

Other Solubilities: Insoluble in organic solvents.
Flash Point: Nonflammable

RTECS Toxicity Data

Acute Oral: Rat LD_{50} Dose: 20 mg/kg; Toxic Effects: Peripheral nerve and sensation - Spastic parapypsis with or without sensory change; Behavioral - Somnolence (general depressed activity); Nutritional and gross metabolic - Dehydration. Mouse LD_{50} Dose: 794 mg/kg.
Acute Dermal: Rat LD_{50} Route: Skin; Dose: 2400 mg/kg.
Chronic (Multiple Dose) Oral: Rat Dose: 24 gm/kg/2Y-C; Toxic Effects: Kidney, Ureter, and Bladder - Chgs in tubules (inc acute renal failure, acute tubular necrosis; Blood - Changes in spleen; DEATH.
Tumorigenic: Rat Route: Intratracheal; Dose: 1600 ug/kg; Toxic Effects: Tumorigenic - Equivocal tumorigenic agent by RTECS criteria; Lungs, Thorax, or Respiration - Tumors; Tumorigenic - Tumors at site of application. Hamster Route: Intratracheal; Dose: 120 mg/kg/15W-C; Toxic Effects: Tumorigenic - Neoplastic by RTECS criteria; Lungs, Thorax, or Respiration - Tumors. Hamster Route: Intratracheal; Dose: 214 mg/kg/15W-I; Toxic Effects: Tumorigenic - Neoplastic by RTECS criteria; Lungs, Thorax, or Respiration - Tumors.

Hazard Overviews

Fire: Noncombustible.
Carcinogenicity: IARC - Group 1, Carcinogenic to humans; NIOSH - Listed as carcinogen; NTP - Class 1, Known to be a carcinogen; ACGIH - Class A1, Confirmed human carcinogen; OSHA - Listed as a carcinogen; EPA - Class A, Human carcinogen; MAK - Class A1, Capable of inducing malignant tumors as shown by experience with humans
Exposure Limits
OSHA PEL: TWA: 0.01 mg/m^3; as As.
ACGIH TLV: TWA: 0.01 mg/m^3; as As.
NIOSH REL: STEL: 0.002 mg/m^3; Ceiling (15 min) as As.
NIOSH IDLH: 5 mg/m^3; as As.
Respirator Recommendation
Exposure Range: >0.01 to 0.1 mg/m^3 Air Purifying, Negative Pressure, Half Mask
Exposure Range: >0.1 to 1 mg/m^3 Air Purifying, Negative Pressure, Full Face
Exposure Range: >1 to <5 mg/m^3 Supplied Air, Constant Flow/Pressure Demand, Full Face
Exposure Range: 5 to unlimited mg/m^3 Self-contained Breathing Apparatus, Pressure Demand, Full Face
Cartridge Color: magenta (P100)

Environmental

Ecotoxicity: Aquatic toxicity: 1.1 ppm/48 hr/perch/toxic/fresh water 4.3 ppm/264 hr/crabs/toxic/fresh water; Food chain concentration potential: Possible bioaccumulation problem
Cleanup/Disposal: Guide No. 151: Do not touch damaged containers or spilled material unless wearing appropriate protective clothing. Stop leak if you can do it without risk. Prevent entry into waterways, sewers, basements or confined areas. Cover with plastic sheet to prevent spreading. Absorb or cover with dry earth, sand or other non-combustible material and transfer to containers. Do not get water inside containers.

Environmental Physical Data

BCF: biomagnification does not occur
BOD: none

Regulations

RCRA 40CFR: Not listed
CERCLA: 40CFR 302.4: Listed per CWA Section 311(b)(4) RQ: 1 lb (0.454 kg)
SARA 40CFR 372.65: Not listed TPQ: 500/10000 lb
SARA EHS 40CFR 355: Listed TPQ: 1 lb
TSCA: Listed

Analytical Methods

Food: AOAC 921.05

CAL2120 CAS #: 52740-16-6

CALCIUM ARSENITE

RTECS: CH9492050
DOT: UN1574; IMO6.1
EINECS Number: 258-147-3
Molecular Formula: $AsCaH_3O_3$
Structured MF: $CaAsO_3H$
Formula Weight: 164.01
Synonyms: ARSENOUS ACID,CALCIUM SALT; ARSONIC ACID,CALCIUM SALT (1:1); CALCIUM META-ARSENITE; MONO-CALCIUM ARSENITE
Description: white, granular powder; odorless
Use: insecticide, germicide, molluscicide (non usa use); dormant & delayed dormant application to apricots, cherries, & peaches (former usa use); potent in vitro human spermicide; food & pharmaceutical preservatives with less toxicity & side effects (expected for development as vaginal contraceptive agents).

Physical Properties

Water Solubility: Slightly Soluble in Water
Other Solubilities: Soluble in acids
Flash Point: Nonflammable

Hazard Overviews

Corrosive

Fire: Noncombustible.
Carcinogenicity: IARC - Group 1, Carcinogenic to humans; NIOSH - Listed as carcinogen; NTP - Class 1, Known to be a carcinogen; ACGIH - Class A1, Confirmed human carcinogen; OSHA - Listed as a carcinogen; EPA - Class A, Human carcinogen; MAK - Class A1, Capable of inducing malignant tumors as shown by experience with humans
Exposure Limits
OSHA PEL: TWA: 0.01 mg/m^3; as As inorganic.

ACGIH TLV: TWA: 0.01 mg/m^3; as As.
NIOSH REL: STEL: 0.002 mg/m^3; ceiling (15 min) as As.
Respirator Recommendation
Exposure Range: >0.01 to <5 mg/m^3
Exposure Range: 5 to unlimited mg/m^3
Note: as arsenic, inorganic compounds; refer to 29CFR 1910.1018 for more specific respirator recommendations

Environmental

Ecotoxicity: LC_{50} Salvelinus tentinalis (brook trout) 10,440 ug/l/262 hr. /Conditions of bioassay not specified LC_{50} Bay scallop 3,490 ug/l/96 hr. /Conditions of bioassay not specified LC_{50} Lepomis macrochirus (bluegill, fingerling) 290 ug/l/48 hr. /Conditions of bioassay not specified LC_{50} Oncorhynchus keta (chum salmon) 8,330 ug/l/48 hr. /Conditions of bioassay not specified LC_{50} Daphnia magna (cladoceran) 2,850 ug/l/3 wk. /Conditions of bioassay not specified LC_{50} Nereis diversicolor (polychaete worm) >14,500 ug/l/192 hr. /Conditions of bioassay not specified LC_{50} Penaeus seliferus (white shrimp, juvenile) 24,700 ug/l/96 hr. /Conditions of bioassay not specified LC_{100} Oncorhynchus gorbuscha (pink salmon) 12,307 ug/l/96 hr; 7,195 ug/l/day. /Conditions of bioassay not specified LC_{50} Gastrophryne carolinensis (toad, embryo-larval) 40 ug/l/7 days. /Conditions of bioassay not specified
Cleanup/Disposal: Guide No. 151: Do not touch damaged containers or spilled material unless wearing appropriate protective clothing. Stop leak if you can do it without risk. Prevent entry into waterways, sewers, basements or confined areas. Cover with plastic sheet to prevent spreading. Absorb or cover with dry earth, sand or other non-combustible material and transfer to containers. Do not get water inside containers.

Regulations

RCRA 40CFR: Not listed
CERCLA: 40CFR 302.4: Listed per CWA Section 311(b)(4) RQ: 1 lb (0.454 kg)
SARA 40CFR 372.65: Listed as Compound
SARA EHS 40CFR 355: Not listed
TSCA: Not listed

CAL2280 CAS #: 5743-27-1

CALCIUM ASCORBATE

EINECS Number: 227-261-5
Molecular Formula: $C_{12}H_{14}CaO_{12}$
Formula Weight: 390.32
Synonyms: ASCORBIC ACID CALCIUM SALT; L-ASCORBIC ACID,CALCIUM SALT (2:1); CALCI-C
Description: white to yellow crystalline powder or crystals; odorless
Use: antioxidant for foods; vitamin C source

Physical Properties

Water Solubility: Freely Soluble in Water
Other Solubilities: Insoluble in Ether.
pH: Aqueous solution is neutral

Hazard Overviews

Carcinogenicity: IARC - Not listed; NIOSH - Not listed; NTP - Not listed; ACGIH - Not listed; OSHA - Not listed; EPA - Not listed; MAK - Not listed

Environmental

Regulations

RCRA 40CFR: Not listed
CERCLA: 40CFR 302.4: Not listed
SARA 40CFR 372.65: Not listed
SARA EHS 40CFR 355: Not listed
TSCA: Listed

CAL2600 CAS #: 75-20-7

CALCIUM CARBIDE

RTECS: EV9400000
DOT: UN1402; IMO4.3
EINECS Number: 200-848-3
Molecular Formula: C_2Ca
Structured MF: CaC_2
Formula Weight: 64.10

Ca^{++}

$^{-}C \equiv C^{-}$

Chemical Structure

Synonyms: ACETYLENOGEN; CALCIUM ACETYLIDE; CALCIUM DICARBIDE; CARBURE DE CALCIUM; CARBURO CALCICO; ETHYNE,CALCIUM DERIV
Description: greyish-black solid; garlic-like odor
Use: reducing agent; fire signals; welding & cutting metals; mfr of calcium, iron, alloys, lampblack, cyanamide; chemical intermediate for acetylene; in metallurgy for desulfurizing & deoxidizing; generation of acetylene gas, chloroethylenes, vinyl acetate monomer, acetylene chemicals; pineapple flowering agent; desulfurization of flue gas; production of calcium cyanamide, a fertilizer

Physical Properties

Boiling Point: 2300 °C (4172 °F)
Freezing Point: 2300 °C (4172 °F)
Specific Gravity: 2.22
Water Solubility: Decomposes
Refraction Index: 1.75
Flash Point: Nonflammable
LEL: 2.5% v/v
UEL: 82% v/v

Hazard Overviews

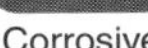
Corrosive

Fire Diamond

Health: Corrosive, causes severe irritation or burns of the eyes/skin/respiratory tract.
Fire: Noncombustible when dry, but forms flammable acetylene gas on exposure to water. Fight surrounding fire with dry chemical, sand, or Met-L-X powder. Do not use water.
Reactivity: Stable. Hazardous polymerization cannot occur. Avoid: contact with water or moisture. Incompatible with: water; selenium; silver nitrate; stannous chloride; sulfur; sodium peroxide; potassium hydroxide and chlorine; chlorine [473 °F (245 °C)]; bromine [662 °F (350 °C)]; iodine [581 °F (305 °C)]; lead fluoride; hydrogen chloride gas and heat; magnesium and heat; silver nitrate solutions; copper and brass; calcium chloride; mixtures with iron (III) chloride, iron (III) oxide, and tin (II) chloride.
Carcinogenicity: IARC - Not listed; NIOSH - Not listed; NTP - Not listed; ACGIH - Not listed; OSHA - Not listed; EPA - Not listed; MAK - Not listed
Primary Target Organs:

Eyes

Skin

Respiratory System
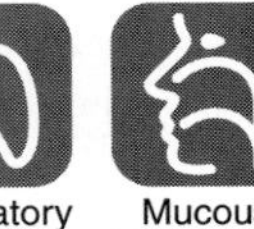
Mucous Membranes

Environmental

Cleanup/Disposal: Guide No. 138: Eliminate all ignition sources (no smoking, flares, sparks or flames in immediate area). Do not touch or walk through spilled material. Stop leak if you can do it without risk. Use water spray to reduce vapors or divert vapor cloud drift. Do not get water on spilled substance or inside containers. Small Spills: Cover with dry earth, dry sand, or other non-combustible material followed with plastic sheet to minimize spreading or contact with rain. Dike for later disposal; do not apply water unless directed to do so. Powder Spills: Cover powder spill with plastic sheet or tarp to minimize spreading and keep powder dry. Do not clean-up or dispose of, except under supervision of a specialist.

Environmental Physical Data

BCF: no food chain concentration potential
BOD: not pertinent

Regulations

RCRA 40CFR: Not listed
CERCLA: 40CFR 302.4: Listed per CWA Section 311(b)(4) RQ: 10 lb (4.535 kg)
SARA 40CFR 372.65: Not listed
SARA EHS 40CFR 355: Not listed
TSCA: Listed

CAL2920 CAS #: 1317-65-3

CALCIUM CARBONATE

RTECS: EV9580000
EINECS Number: 215-279-6
Molecular Formula: Unspecified or Variable
Structured MF: $CaCO_3$
Formula Weight: 100.1
Synonyms: AGRICULTURAL LIMESTONE; AGSTONE; BELL MINE PULVERIZED LIMESTONE; CALCIUM SALT OF CARBONIC ACID; CHALK; DOMOLITE; FRANKLIN; LIMESTONE; LITHOGRAPIC STONE; MARBLE; NATURAL CALCIUM CARBONATE; NATURAL CALCIUM CARBONATE; PORTLAND STONE; SOHNHOFEN STONE
Description: white, colorless powder, crystals; odorless

Physical Properties

Boiling Point: Decomposes
Freezing Point: Decomposes at 825 °C (1517 °F) to 1338.89 °C (2442.002 °F)
Specific Gravity: Aragonite 2.83
Vapor Pressure: ~ 0 mm Hg
Water Solubility: 0.001% by weight
Other Solubilities: Soluble in acids with evolution of carbon dioxide
pH: 8 to 9
Flash Point: Noncombustible

Hazard Overviews

Fire Diamond

Health: Mildly irritating to eyes/skin/respiratory tract. Chronic Effects: hypercalcemia, alkalosis, renal impairment, milk-alkali syndrome.
Fire: Noncombustible. Use agent suitable for surrounding fire.
Reactivity: Stable. Hazardous polymerization cannot occur. Avoid: dust generation; acids. Incompatible with: fluorine; acids; alum; ammonium salts; mercury and hydrogen. Hazardous decomposition products: calcium oxide.
Carcinogenicity: IARC - Not listed; NIOSH - Listed as carcinogen; NTP - Not listed; ACGIH - Not listed; OSHA - Not listed; EPA - Not listed; MAK - Not listed
Primary Target Organs:

Eyes

Skin
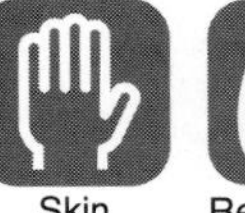
Respiratory System

Nervous System

Kidneys

Exposure Limits
OSHA PEL: TWA: 15 mg/m^3; total dust.
ACGIH TLV: TWA: 10 mg/m^3.
Respirator Recommendation
Exposure Range: >5 to 50 mg/m^3 Air Purifying, Negative Pressure, Half Mask

Exposure Range: >50 to 500 mg/m^3 Air Purifying, Negative Pressure, Full Face
Exposure Range: >500 to 5000 mg/m^3 Supplied Air, Constant Flow/Pressure Demand, Full Face
Exposure Range: >5000 to unlimited mg/m^3 Self-contained Breathing Apparatus, Pressure Demand, Full Face
Cartridge Color: dust/mist filter (use P100 or consult supervisor for appropriate dust/mist filter)

Environmental

Regulations

RCRA 40CFR: Not listed
CERCLA: 40CFR 302.4: Not listed
SARA 40CFR 372.65: Not listed
SARA EHS 40CFR 355: Not listed
TSCA: Listed

CAL3240 CAS #: 10043-52-4

CALCIUM CHLORIDE

RTECS: EV9800000
EINECS Number: 233-140-8
Molecular Formula: $CaCl_2$
Structured MF: $CaCl_2$
Formula Weight: 110.99

Cl——Ca——Cl

Chemical Structure

Synonyms: CALCIUM DICHLORIDE; CALCOSAN; CALPLUS; CALTAC; DOWFLAKE; LIQUIDOW; PELADOW; SNOMELT; SUPERFLAKE ANHYDROUS; URAMINE MC
Description: white crystals, flakes, granules, lumps; odorless
Use: pavement deicer; dehydrating org liq & gases; for antifreeze & refrigerating soln, in fire extinguishers; to preserve wood, stone; in mfr of ice, glues; fireproofing fabrics, coal & ores; coagulant in rubber mfr; for dust control; sizing & finishing fabrics; drainage aid in pulp & paper processing; desiccant in hydrocarbon processing (anhydrous only); coal thawing agent; in oil & gas well fluids, eg, drilling muds; set accelerator in concrete; tire ballast; refrigeration brines; firming agent for fruits; feed additive; in thermal batteries; in blast furnaces; humectant in adhesives; in herbicides, plastics; med: for hypocalcemic tetany; treatment of fluoride poisoning; in pharmaceuticals, eg, blood-replacement prepn.

Physical Properties

Boiling Point: 1670 °C (3038 °F)
Freezing Point: 772 °C (1421.6 °F)
Specific Gravity: 2.152 at 15 °C/4 °C
Water Solubility: Freely Soluble in Water
Other Solubilities: Soluble in Acetone, Acetic Acid.
pH: 8.0 to 9.0
Refraction Index: 1.52
Flash Point: Nonflammable

RTECS Toxicity Data

Acute Oral: Rat LD_{50} Dose: 1 gm/kg. Mouse LD_{50} Dose: 1940 mg/kg.
Acute Dermal: Rabbit LD_{Lo} Route: Subcutaneous Dose: 472 mg/kg. Rat LD_{50} Route: Subcutaneous Dose: 2630 mg/kg.
Mutagenic: Rat Unscheduled DNA Synthesis; Route: Intraperitoneal; Dose: 2500 umol/kg. Rat Cytogenetic Analysis; Cell Type: Ascites tumor; Dose: 3500 mg/kg.
Tumorigenic: Rat Route: Oral; Dose: 112 gm/kg/20W-C; Toxic Effects: Tumorigenic - Equivocal tumorigenic agent by RTECS criteria; Endocrine - Thyroid tumors.

Hazard Overviews

Fire Diamond

Health: Irritating to eyes/skin/respiratory tract. Also Causes: by ingestion: nausea, vomiting, fever, twitching, slow heartbeat, and possible seizures.
Fire: Noncombustible. Use extinguishing agents suitable for surrounding fire. However, be aware that excessive heat will be generated on contact with water.
Reactivity: Stable, but readily absorbs moisture. Hazardous polymerization cannot occur. Avoid: excessive generation of dusts; contact with water. Incompatible with: boric oxide and calcium oxide; bromine trifluoride; zinc; vinyl ether; furan-2-peroxycarbolic acid; water. Hazardous decomposition products: toxic chloride fumes.
Carcinogenicity: IARC - Not listed; NIOSH - Not listed; NTP - Not listed; ACGIH - Not listed; OSHA - Not listed; EPA - Not listed; MAK - Not listed
Primary Target Organs:

Eyes

Skin

Respiratory System

Mucous Membranes

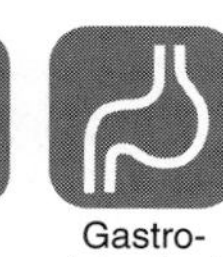
Gastro-intestinal

Environmental

Ecotoxicity: LC_{50} Nitzschia linearia 3,130 mg/l/120 hr in static water
Cleanup/Disposal: Guide No. 171: Do not touch or walk through spilled material. Stop leak if you can do it without risk. Prevent dust cloud. Avoid inhalation of asbestos dust. Small Dry Spills: With clean shovel place material into clean, dry container and cover loosely; move containers from spill area. Small Spills: Take up with sand or other noncombustible absorbent material and place into containers for later disposal. Large Spills: Dike far ahead of liquid spill for later disposal. Cover powder spill with plastic sheet or tarp to minimize spreading. Prevent entry into waterways, sewers, basements or confined areas.

Environmental Physical Data

BCF: none
BOD: none

Regulations

RCRA 40CFR: Not listed
CERCLA: 40CFR 302.4: Not listed
SARA 40CFR 372.65: Not listed
SARA EHS 40CFR 355: Not listed
TSCA: Listed

CAL3400 CAS #: 10035-04-8

CALCIUM CHLORIDE, DIHYDRATE

RTECS: EV9810000
Molecular Formula: $CaCl_2H_4O_2$
Formula Weight: 147.02

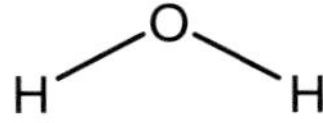

Cl⁻ Ca⁺⁺ Cl⁻

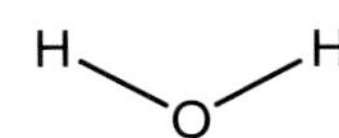

Chemical Structure

Synonyms: CAL PLUS; CALCIUM DICHLORIDE DIHYDRATE; REPLENISHER (CALCIUM)

Physical Properties

Boiling Point: > 1600 °C (2912 °F)
Freezing Point: 176 °C (348.8 °F)
Specific Gravity: 1.85
Water Solubility: Soluble

RTECS Toxicity Data

Mutagenic: Rat Unscheduled DNA Synthesis; Route: Oral; Dose: 500 ug/kg.

Hazard Overviews

Health: Irritating to eyes/skin/respiratory tract. Harmful. Other Acute Effects: harmful if swallowed; may be harmful if inhaled or absorbed through the skin.
Fire: Hazards: emits toxic fumes. Extinguishing agents: noncombustible; use extinguishing media appropriate to surrounding fire conditions. Precautions: combustible liquid.
Reactivity: Stable. Hazardous polymerization will not occur. Incompatible with: strong acids, strong oxidizing agents. Hazardous decomposition products: toxic fumes of: hydrogen chloride gas.
Carcinogenicity: IARC - Not listed; NIOSH - Not listed; NTP - Not listed; ACGIH - Not listed; OSHA - Not listed; EPA - Not listed; MAK - Not listed

Primary Target Organs:

Eyes

Skin

Respiratory System

Environmental

Regulations

RCRA 40CFR: Not listed
CERCLA: 40CFR 302.4: Not listed
SARA 40CFR 372.65: Not listed
SARA EHS 40CFR 355: Not listed
TSCA: Not listed

CAL3560 CAS #: 13765-19-0

CALCIUM CHROMATE

RTECS: GB2750000
EINECS Number: 237-366-8
Molecular Formula: $CaCrH_2O_4$
Structured MF: $CaCrO_4$
Formula Weight: 156.09

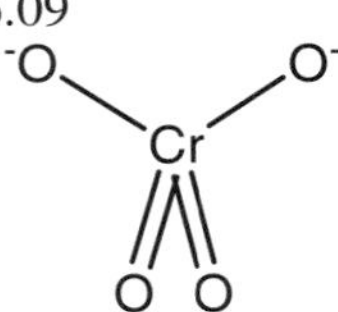

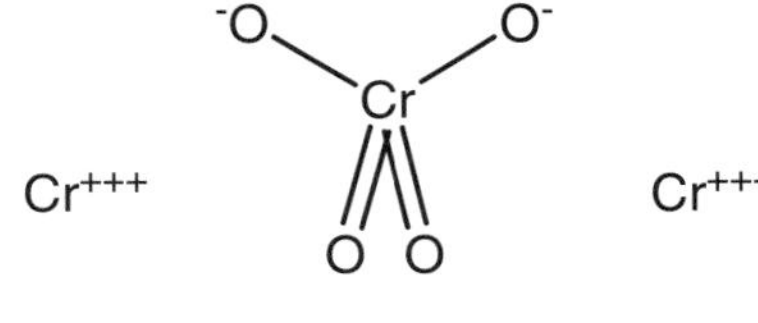

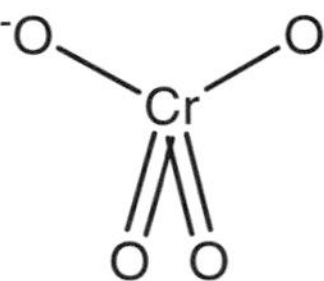

Chemical Structure

Synonyms: C.I. 77223; C.I. PIGMENT YELLOW 33; CALCIUM CHROMATE (VI); CALCIUM CHROMATE(VI); CALCIUM CHROME YELLOW; CALCIUM CHROMIUM OXIDE; CALCIUM MONOCHROMATE; CHROMIC ACID (H2CRO4),CALCIUM SALT (1:1); CHROMIC ACID,CALCIUM SALT (1:1); GELBIN; GELBIN YELLOW ULTRAMARINE; PIGMENT YELLOW 33; STEINBUHL YELLOW; YELLOW ULTRAMARINE
Description: bright yellow powder, crystals
Use: pigment and corrosion inhibitor; in manufacture of chromium; for oxidizing reactions and battery depolarization

Physical Properties

Specific Gravity: 2.89

Density: 2.89 g/mL
Water Solubility: Sparingly Soluble in Water
Other Solubilities: 95% Ethanol: <1 mg/ml at 22 °C; Acetone: <1 mg/ml at 19.5 °C; DMSO: <1 mg/ml at 22 °C; Dilute acids: Soluble.
Flash Point: Nonflammable

RTECS Toxicity Data

Acute Oral: Rat LD_{50} Dose: 327 mg/kg.
Mutagenic: Human Sister Chromatid Exchenge; Cell Type: lymphocyte; Dose: 650 nmol/L. Rat Morphological Transformation; Cell Type: embryo; Dose: 58 ug/L.
Tumorigenic: Rat Route: Implant; Dose: 8 mg/kg; Toxic Effects: Tumorigenic - Carcinogenic by RTECS criteria; Lungs, Thorax, or Respiration - Tumors. Rat Route: Implant; Dose: 50 mg/kg; Toxic Effects: Tumorigenic - Equivocal tumorigenic agent by RTECS criteria; Lungs, Thorax, or Respiration - Tumors. Rat Route: Implant; Dose: 125 mg/kg; Toxic Effects: Tumorigenic - Carcinogenic by RTECS criteria; Tumorigenic - Tumors at site of application. Rat Route: Implant; Dose: 10866 ug/kg; Toxic Effects: Tumorigenic - Carcinogenic by RTECS criteria; Lungs, Thorax, or Respiration - Tumors.

Hazard Overviews

Corrosive

Fire Diamond

Health: Corrosive to eyes/skin/respiratory tract. Also Causes: perforation of nasal septum, pulmonary edema, bronchospasm, leukocytosis, leukopenia, defatting, dermatitis,chrome holes, conjunctivitis, gastroenteritis, circulatory collapse, chronic nephritis, liver damage. Chronic Effects: bronchogenic cancer development.
Fire: Noncombustible. However, it is a strong oxidizer capable of igniting combustibles. Use agent suitable for surrounding fire. Toxic chromium fumes may form in fire.
Reactivity: Stable. Hazardous polymerization cannot occur. Avoid: excessive dust generation. Incompatible with: organic matter or reducing agents (paper; wood; aluminum; plastics; sulfur); acids; ethanol; boron and ignition. Hazardous decomposition products: acrid smoke; chromium fumes.
Carcinogenicity: IARC - Group 1, Carcinogenic to humans; NIOSH - Listed as carcinogen; NTP - Class 1, Known to be a carcinogen; ACGIH - Class A1, Confirmed human carcinogen; OSHA - Not listed; EPA - Class A, Human carcinogen; MAK - Class A2, Unmistakably carcinogenic in animal experimentation only
Primary Target Organs:

Eyes

Skin

Respiratory System

Liver

Kidneys

Blood

Exposure Limits
OSHA PEL: STEL: 0.1 mg/m^3; as CrO_3; Other Values: 0.1 mg/m^3; Clg Cr-VI as CrO_3.
NIOSH REL: TWA: 0.5 mg/m^3; as Cr;Cr-II;Cr-III;Cr(VI)=.001.
Respirator Recommendation
Exposure Range: >0.1 to 1 mg/m^3 Air Purifying, Negative Pressure, Half Mask
Exposure Range: >1 to 10 mg/m^3 Air Purifying, Negative Pressure, Full Face
Exposure Range: >10 to <15 mg/m^3 Supplied Air, Constant Flow/Pressure Demand, Full Face
Exposure Range: 15 to unlimited mg/m^3 Self-contained Breathing Apparatus, Pressure Demand, Full Face
Cartridge Color: magenta (P100)

Environmental

Ecotoxicity: Food chain concentration potential: Bioconcentration up to 2000-fold possible under constant exposure. Not significant under spill conditions
Cleanup/Disposal: Guide No. 140: Keep combustibles (wood, paper, oil, etc.) away from spilled material. Do not touch damaged containers or spilled material unless wearing appropriate protective clothing. Stop leak if you can do it without risk. Do not get water inside containers. Small Dry Spills: With clean shovel place material into clean, dry container and cover loosely; move containers from spill area. Small Liquid Spills: Use a non-combustible material like vermiculite, sand or earth to soak up the product and place into a container for later disposal. Large Spills: Dike far ahead of liquid spill for later disposal. Following product recovery, flush area with water.

Environmental Physical Data

BCF: constant exposure 2000
BOD: none

Regulations

RCRA 40CFR: Listed Hazardous Waste No. U032 Toxic Waste
CERCLA: 40CFR 302.4: Listed per CWA Section 311(b)(4) per RCRA Section 3001 RQ: 10 lb (4.535 kg)
SARA 40CFR 372.65: Listed as Compound
SARA EHS 40CFR 355: Not listed
TSCA: Listed

CAL3880 CAS #: 156-62-7

CALCIUM CYANAMIDE

RTECS: GS6000000
DOT: UN1403; IMO4.3
EINECS Number: 205-861-8
Molecular Formula: CH_2CaN_2
Structured MF: CaNCN
Formula Weight: 80.11

N≡—N=Ca

Chemical Structure

Synonyms: AERO CYANAMID GRANULAR; AERO CYANAMID SPECIAL GRADE; AERO-CYANAMID; ALZODEF; CALCIUM CARBIMIDE; CALCIUM CYANAMID; CALCIUM CYANAMIDE WITH >0.1% OF CALCIUM CARBIDE; CCC; CY-L 500; CYANAMID; CYANAMID GRANULAR; CYANAMID SPECIAL GRADE; CYANAMIDE; CYANAMIDE CALCIQUE; CYANAMIDE GRANULAR; CYANAMIDE SPECIAL GRADE; CYANAMIDE,CALCIUM SALT; CYANAMIDE,CALCIUM SALT (1:1); LIME NITROGEN; LIME-NITROGEN; LIMENITROGEN; NITROGEN LIME; NITROLIM; NITROLIME

Description: gray solid lumps of powder

Use: fertilizer; defoliant; herbicide; pesticide; manufacture and refining of iron; manufacture of calcium cyanide, melamine and dicyandiamide; anthelmintic

Physical Properties

Boiling Point: Sublimes
Freezing Point: 1340 °C (2444 °F)
Specific Gravity: 2.29 at 20 °C/4 °C
Density: 1.083 g/mL
Vapor Pressure: ~ 0 mm Hg
Water Solubility: Essentially Insoluble in Water
Other Solubilities: 95% Ethanol: Decomposes; Acetone: Decomposes; DMSO: Decomposes; No known solvent will bring about solution without decomposition.
Flash Point: Probably noncombustible

RTECS Toxicity Data

Acute Oral: Human LD_{Lo} Dose: 571 mg/kg. Rat LD_{50} Dose: 158 mg/kg.
Acute Inhalation: Rat LC_{50} Dose: >150 mg/m^3/4hr.
Acute Dermal: Rabbit LD_{50} Route: Skin; Dose: 590 mg/kg.
Mutagenic: Bacteria - S Typhimurium Mutations in Microorganisms; Dose: 1 mg/plate (+S9), 100 ug/plate (-S9).
Tumorigenic: Mouse Route: Oral; Dose: 170 gm/kg/2Y-C; Toxic Effects: Tumorigenic - Equivocal tumorigenic agent by RTECS criteria; Vascular - Tumors.

Hazard Overviews

Fire Diamond

Health: Irritating to eyes/skin/respiratory tract. Chronic Effects: Sensitizing dermatitis.
Fire: Noncombustible. Do not use water to extinguish fire. Use dry chemical or carbon dioxide. Flammable acetylene and ammonia gases are produced when material comes in contact with water.
Reactivity: Stable. Hazardous polymerization cannot occur. Avoid: moisture. Incompatible with: moisture. Hazardous decomposition products: acetylene; ammonia gases; carbon monoxide; oxides of nitrogen; oxides of calcium.
Carcinogenicity: IARC - Not listed; NIOSH - Not listed; NTP - Not listed; ACGIH - Class A4, Not classifiable as a human carcinogen; OSHA - Not listed; EPA - Not listed; MAK - Not listed

Primary Target Organs:

Eyes

Skin

Respiratory System

Nervous System

Cardio-vascular

Exposure Limits
OSHA PEL Vacated 1989 Limits: TWA: 0.5 mg/m^3.
ACGIH TLV: TWA: 0.5 mg/m^3.
DFG MAK: TWA: 1 mg/m^3.

Respirator Recommendation
Exposure Range: >0.5 to 5 mg/m^3 Air Purifying, Negative Pressure, Half Mask
Exposure Range: >5 to 50 mg/m^3 Air Purifying, Negative Pressure, Full Face
Exposure Range: >50 to 500 mg/m^3 Supplied Air, Constant Flow/Pressure Demand, Full Face
Exposure Range: >500 to unlimited mg/m^3 Self-contained Breathing Apparatus, Pressure Demand, Full Face
Cartridge Color: dust/mist filter (use P100 or consult supervisor for appropriate dust/mist filter)

Environmental

Cleanup/Disposal: Guide No. 138: Eliminate all ignition sources (no smoking, flares, sparks or flames in immediate area). Do not touch or walk through spilled material. Stop leak if you can do it without risk. Use water spray to reduce vapors or divert vapor cloud drift. Do not get water on spilled substance or inside containers. Small Spills: Cover with dry earth, dry sand, or other non-combustible material followed with plastic sheet to minimize spreading or contact with rain. Dike for later disposal; do not apply water unless directed to do so. Powder Spills: Cover powder spill with plastic sheet or tarp to minimize spreading and keep powder dry. Do not clean-up or dispose of, except under supervision of a specialist.

Regulations

RCRA 40CFR: Not listed
CERCLA: 40CFR 302.4: Listed per CAA Section 112 RQ: 1000 lb (453.5 kg)
SARA 40CFR 372.65: Listed
SARA EHS 40CFR 355: Not listed
TSCA: Listed

CAL4040 CAS #: 592-01-8

CALCIUM CYANIDE

RTECS: EW0700000
DOT: UN1575; IMO6.1
EINECS Number: 209-740-0
Molecular Formula: C_2CaN_2
Structured MF: $Ca(CN)_2$
Formula Weight: 92.12
Synonyms: CALCID; CALCIUM CYANIDE MIXTURE,SOLID; CALCIUM CYANIDE,SOLID; CALCYAN; CALCYANIDE; CYANIDE OF CALCIUM; CYANOGAS; CYANURE DE CALCIUM; CYMAG;

DEGESCH CALCIUM CYANIDE A-DUST; EPA PESTICIDE CHEMICAL CODE 074001

Description: colorless, white white powder or rhombohedric crystals; odor of hydrogen cyanide

Use: fumigant; rodenticide; in stainless-steel manufacturing; in leaching ores of precious metals; stabilizer for cement

Physical Properties

Boiling Point: Decomposes > 250 °C (482 °F)
Freezing Point: Extrapolated 640 °C (1184 °F)
Specific Gravity: Solid 1.853 at 20 °C (solid)
Water Solubility: Decomposes
Other Solubilities: Alcohol: Soluble.
Flash Point: Nonflammable

RTECS Toxicity Data

Acute Oral: Rat LD_{50} Dose: 39 mg/kg.

Hazard Overviews

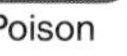
Poison

Fire Diamond

Health: Irritating to eyes/skin/respiratory tract. Poison. Also Causes: chemical asphyxiation, flushing, weakness, confusion, dizziness, difficulty breathing, convulsions, cardiac difficulties, headache, death, nausea/vomiting, salivation, anxiety, confusion, paralysis, coma. Chronic Effects: enlarged thyroid appetite loss, vitamin B-12 and folate abnormalities, insomnia.

Fire: Noncombustible. Use agent suitable for surrounding fire. Do not use water to extinguish fire. Forms flammable hydrogen cyanide gas on contact with water. Noncombustible.

Carcinogenicity: IARC - Not listed; NIOSH - Not listed; NTP - Not listed; ACGIH - Not listed; OSHA - Not listed; EPA - Not listed; MAK - Not listed

Primary Target Organs:

Eyes, Skin, Nervous System, Cardio-vascular, Blood, Glandular System

Exposure Limits
ACGIH TLV: STEL: 5 mg/m^3; Ceiling.
NIOSH REL: STEL: 4.7 ppm; 5 mg/m^3; Ceiling, 10 min.
DFG MAK: TWA: 5 mg/m^3.

Environmental

Ecotoxicity: Aquatic toxicity: 0.12 ppm/96 hr/sunfish/TLm/fresh water >25 ppm/48 hr/cockle/LC_{50}/salt water

Cleanup/Disposal: Guide No. 157: Eliminate all ignition sources (no smoking, flares, sparks or flames in immediate area). All equipment used when handling the product must be grounded. Do not touch damaged containers or spilled material unless wearing appropriate protective clothing. Stop leak if you can do it without risk. A vapor suppressing foam may be used to reduce vapors. Do not get water inside containers. Use water spray to reduce vapors or divert vapor cloud drift. Prevent entry into waterways, sewers, basements or confined areas. Small Spills: Cover with dry earth, dry sand, or other non-combustible material followed with plastic sheet to minimize spreading or contact with rain. Use clean non-sparking tools to collect material and place it into loosely covered plastic containers for later disposal.

Environmental Physical Data

BCF: no food chain concentration potential

Regulations

RCRA 40CFR: Listed Hazardous Waste No. P021 Toxic Waste
CERCLA: 40CFR 302.4: Listed per CWA Section 311(b)(4) per RCRA Section 3001 RQ: 10 lb (4.535 kg)
SARA 40CFR 372.65: Not listed
SARA EHS 40CFR 355: Not listed
TSCA: Listed

CAL4360 **CAS #: 26264-06-2**

CALCIUM DODECYLBENZENESULFONATE

RTECS: DB6750000
EINECS Number: 247-557-8
Molecular Formula: $C_{36}H_{60}CaO_6S_2$
Structured MF: $(CH_3(CH_2)_{11}C_6H_4SO_3)_2Ca$
Formula Weight: 346.54
Synonyms: 1371A; BENZENESULFONIC ACID,DODECYL-,CALCIUM SALT; CALCIUM ALKYLAROMATIC SULFONATE; CALCIUM ALKYLBENZENESULFONATE; CALCIUM BIS(DODECYLBENZENESULFONATE); CALCIUM N-DODECYLBENZENESULFONATE; CALCIUM DODECYLBENZENSULFONATE; CASUL 70HF; DODECYLBENZENESULFONIC ACID CALCIUM SALT; DODECYLBENZENSULFONIC ACID CALCIUM SALT; PRUNE; SINNOZON NCX 70; SOPROFOR S 70

Description: yellowish-brown liquid or light to white granular solid; solvent odor

Use: surfactant (emulsifier & dispersant); foam separation of iron trihydroxide

Physical Properties

Specific Gravity: 1.04 at 25 °C (liquid solution)
Water Solubility: Soluble in Water
Odor Threshold: 0.3 ppm
Flash Point: Noncombustible

RTECS Toxicity Data

Acute Oral: Rat LD_{50} Dose: 4 gm/kg. Mouse LD_{50} Dose: 3680 mg/kg.

Hazard Overviews

Fire: Noncombustible.

Carcinogenicity: IARC - Not listed; NIOSH - Not listed; NTP - Not listed; ACGIH - Not listed; OSHA - Not listed; EPA - Not listed; MAK - Not listed

Environmental

Environmental Physical Data

BOD: activated < 1 lb/lb, 5 days

Regulations

RCRA 40CFR: Not listed
CERCLA: 40CFR 302.4: Listed per CWA Section 311(b)(4) RQ: 1000 lb (453.5 kg)
SARA 40CFR 372.65: Not listed
SARA EHS 40CFR 355: Not listed
TSCA: Listed

CAL4520 CAS #: 7789-75-5

CALCIUM FLUORIDE

RTECS: EW1760000
EINECS Number: 232-188-7
Molecular Formula: CaF_2
Structured MF: CaF_2
Formula Weight: 78.08

Ca^{++}

F^- F^-

Chemical Structure

Synonyms: ACID-SPAR; CALCIUM DIFLUORIDE; FLUORITE; FLUORSPAR; IRTRAN 3; LIPARITE; MET-SPAR; NATURAL FLUORITE
Description: white, colorless powder, crystals
Use: source of fluorine & compd; synthetic fluorspar in optical ind (transmits uv rays); catalyst in dehydration & dehydrogenations; to fluoridate drinking water; in coatings for welding rods; opacifying of glass & enamels; in ferous & nonferrous metal mfr; intermed for hydrofluoric acid; flux for steelmaking, in mfr of iron & steel castings; emory wheels; cements; dentrifices; in phosphor; paint pigment; catalyst in wood preservatives; in spectroscopy & electronics, lasers, & lubricants.

Physical Properties

Boiling Point: 2500 °C (4532 °F)
Freezing Point: 1403 °C (2557.4 °F)
Specific Gravity: 3.18
Water Solubility: 0.0015 g/100 ml Water at 18 °C
Other Solubilities: Soluble in Ammonium salts; Insoluble in Acetone.
Flash Point: Nonflammable

RTECS Toxicity Data

Acute Oral: Rat LD_{50} Dose: 4250 mg/kg; Toxic Effects: Behavioral - Somnolence (general depressed activity); Behavioral - Ataxia; Lungs, Thorax, or Respiration - Respiratory depression. Guinea Pig LD_{Lo} Dose: >5 gm/kg.
Reproductive/Teratogenic: Mouse Route: Intraperitoneal; Dose: 3200 mg/kg; Duration: female 9D of pregnancy; Effects on Fertility - Post-implantation mortality. Mouse Route: Intraperitoneal; Dose: 67200 mg/kg; Duration: female 1-21D of pregnancy; Specific Developmental Abnormalities - Other developmental abnormalities.
Mutagenic: Rat Cytogenetic Analysis; Cell Type: Ascites tumor; Dose: 1 gm/kg.

Hazard Overviews

Fire Diamond

Health: Causes: increase in respiration, coughing, choking, shortness of breath, respiratory arrest, cardiac arrhythmia, hypocalcemia, abdominal pain, difficulty swallowing, nausea, vomiting, increased salivation. Chronic: bronchitis, silicosis, fluorosis, renal damage, decreased blood clotting, characterized by brittle bones, calcified ligaments, musculoskeletal system changes.
Fire: Noncombustible. Use agent suitable for surrounding fire.
Reactivity: Stable. Hazardous polymerization cannot occur. Avoid: excessive dust generation; contact with sulfuric acid. Incompatible with: hot concentrated sulfuric acid. Hazardous decomposition products: fumes of fluoride.
Carcinogenicity: IARC - Not listed; NIOSH - Not listed; NTP - Not listed; ACGIH - Class A4, Not classifiable as a human carcinogen; OSHA - Not listed; EPA - Not listed; MAK - Not listed
Primary Target Organs:

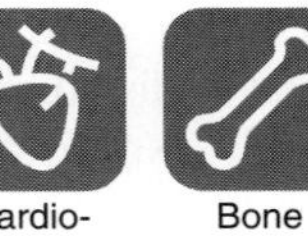
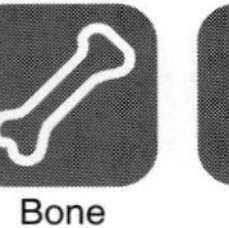

Respiratory System | Kidneys | Cardio-vascular | Bone | Teeth

Exposure Limits
OSHA PEL: TWA: 2.5 mg/m^3; as F.
ACGIH TLV: TWA: 2.5 mg/m^3; as F.
NIOSH REL: TWA: 2.5 mg/m^3; as F (Flourides).
DFG MAK: TWA: 2.5 mg/m^3; as F.

Environmental

Ecotoxicity: Aquatic toxicity: 30000 ppm/time period not specified/tinca vulgaris/lethal/fresh water
Cleanup/Disposal: Guide No. 151: Do not touch damaged containers or spilled material unless wearing appropriate protective clothing. Stop leak if you can do it without risk. Prevent entry into waterways, sewers, basements or confined areas. Cover with plastic sheet to prevent spreading. Absorb or cover with dry earth, sand or other non-combustible material and transfer to containers. Do not get water inside containers.

Regulations

RCRA 40CFR: Not listed

CERCLA: 40CFR 302.4: Not listed
SARA 40CFR 372.65: Not listed
SARA EHS 40CFR 355: Not listed
TSCA: Listed

CAL5160 CAS #: 1305-62-0

CALCIUM HYDROXIDE

RTECS: EW2800000
EINECS Number: 215-137-3
Molecular Formula: CaH_2O_2
Structured MF: $Ca(OH)_2$
Formula Weight: 74.10

HO——Ca——OH

Chemical Structure

Synonyms: BELL MINE; BIOCALC; CALCIUM DIHYDROXIDE; CALCIUM HYDRATE; CALVIT; CALVITAL; CARBOXIDE; CAUSTIC LIME; HYDRATED LIME; KALKHYDRATE; KEMIKAL; LIMBUX; LIME; LIME MILK; LIME WATER; MILK OF LIME; SLAKED LIME
Description: colorless, soft white crystals, granules, powder; odorless
Use: in lubricants, drilling fluid, pesticides, fireproofing coatings, water paint; egg preservative; mfr of paper pulp; in rubber vulcanization; dehairing hides; in binding & paving materials; med: treatment in fluoride poisoning, (vet): antidote to tannin; dehorning paste, depilatory, for wound treatment, fecal waste deodorant, shell-forming agent (poultry), etc; intermed for bleach, eg, calcium hypochlorite; flux in iron & steel mfr; soil treatment; water softening agent; buffering agent; ammonia recovery in gas mfr; disinfectant; food additive; prevents deterioration of apples during storage; deacidification processes involving paper prodn.

Physical Properties

Boiling Point: Decomposes
Freezing Point: 580 °C (1076 °F)
Specific Gravity: 2.24
Vapor Pressure: ~ 0 mm Hg
Water Solubility: 0.185 g/100 cc at 0 °C
Other Solubilities: Soluble in Glycerol, sugar, or Ammonium Chloride solution; Soluble in acids with evolution of much heat.
pH: Aqueous solution saturated at 25 °C 12.4
Refraction Index: Alpha 1.574
Flash Point: Nonflammable

RTECS Toxicity Data

Acute Oral: Rat LD_{50} Dose: 7340 mg/kg. Mouse LD_{50} Dose: 7300 mg/kg.
Irritation Eye: Rabbit Standard Draize Test Dose: 10 mg; Reaction: severe.
Mutagenic: Rat Cytogenetic Analysis; Cell Type: Ascites tumor; Dose: 1200 mg/kg.

Hazard Overviews

Corrosive

Fire Diamond

Health: Corrosive, causes severe burns to eyes/skin/respiratory tract. Chronic Effects: repeated skin contact can cause dermatitis.
Fire: Noncombustible. Use extinguishing agents suitable for surrounding fire.
Reactivity: Stable. Hazardous polymerization cannot occur. Avoid: excessive heat; calcium hydroxide dust. Incompatible with: maleic anhydride; nitroethane; water; nitromethane; nitroparaffins; nitropropane; ammonium salts; phosphorus; polychlorinated phenols; potassium nitrate. Hazardous decomposition products: calcium oxide.
Carcinogenicity: IARC - Not listed; NIOSH - Listed as carcinogen; NTP - Not listed; ACGIH - Not listed; OSHA - Not listed; EPA - Not listed; MAK - Not listed
Primary Target Organs:

Eyes

Skin

Respiratory System

Mucous Membranes

Gastro-intestinal

Exposure Limits
OSHA PEL: TWA: 15 mg/m^3; total dust.
ACGIH TLV: TWA: 5 mg/m^3.
NIOSH REL: TWA: 5 mg/m^3.
Respirator Recommendation
Exposure Range: >5 to 50 mg/m^3 Air Purifying, Negative Pressure, Half Mask
Exposure Range: >50 to 500 mg/m^3 Air Purifying, Negative Pressure, Full Face
Exposure Range: >500 to 5000 mg/m^3 Supplied Air, Constant Flow/Pressure Demand, Full Face
Exposure Range: >5000 to unlimited mg/m^3 Self-contained Breathing Apparatus, Pressure Demand, Full Face
Cartridge Color: dust/mist filter (use P100 or consult supervisor for appropriate dust/mist filter)

Environmental

Ecotoxicity: Aquatic toxicity: 92 ppm/7 hr/trout/toxic/fresh water 240 ppm/24 hr/mosquito fish/TLm/fresh water
Cleanup/Disposal: Guide No. 157: Eliminate all ignition sources (no smoking, flares, sparks or flames in immediate area). All equipment used when handling the product must be grounded. Do not touch damaged containers or spilled material unless wearing appropriate protective clothing. Stop leak if you can do it without risk. A vapor suppressing foam may be used to reduce vapors. Do not get water inside containers. Use water spray to reduce vapors or divert vapor cloud drift. Prevent entry into waterways, sewers, basements or confined areas. Small Spills: Cover with dry earth, dry sand, or other non-combustible material followed with plastic sheet to minimize spreading or contact with rain. Use clean

non-sparking tools to collect material and place it into loosely covered plastic containers for later disposal.

Environmental Physical Data

BCF: none
BOD: none

Regulations

RCRA 40CFR: Not listed
CERCLA: 40CFR 302.4: Not listed
SARA 40CFR 372.65: Not listed
SARA EHS 40CFR 355: Not listed
TSCA: Listed

Analytical Methods

Indoor / Expired Air: NIOSH 7020

CAL5320 CAS #: 7778-54-3

CALCIUM HYPOCHLORITE

RTECS: NH3485000
DOT: UN1748; UN2208; UN2880; IMO5.1
EINECS Number: 231-908-7
Molecular Formula: $CaCl_2H_2O_2$
Structured MF: $Ca(OCl)_2$
Formula Weight: 142.98

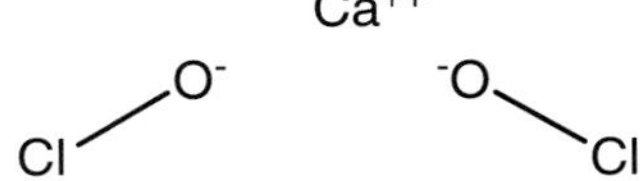

Chemical Structure

Synonyms: B-K POWDER; BK POWDER; BLEACHING POWDER; CALCIUM CHLOROHYDROCHLORITE; CALCIUM CHLOROHYPOCHLORIDE; CALCIUM HYPOCHLORIDE; CALCIUM HYPOCHLORITE MIXTURES DRY WITH >39% AVAILABLECHLORINE; CALCIUM HYPOCHLORITE,DRY; CALCIUM OXYCHLORIDE; CAPORIT; CCH; CHEMICHLON G; CHEMICHLOR G; CHLORIDE OF LIME; CHLORINATED LIME; CHLORINE OF LIME; CHLOROLIME CHEMICAL; EPA PESTICIDE CHEMICAL CODE 014701; HIPOCLORITO CALCICO; HTH; HTH (BLEACHING AGENT); HY-CHLOR; HYPOCHLORITE DE CALCIUM; HYPOCHLOROUS ACID,CALCIUM SALT; HYPOCHLOROUS ACID,CALCIUM SALT,(DRY MIXTURE); LIME CHLORIDE; LO-BAX; LOSANTIN; PERCHLORON; PITTCHLOR; PITTCIDE; PITTCLOR; SENTRY; SOLVOX KS; T-EUSOL

Description: white granules, tablets; strong chlorine odor

Use: algicide; bactericide; deodorant; fungicide; in sugar refining; oxidizing agent; sanitizer in swimming pool, toilet & other applications; bleaching agent in textile, pulp, & paper ind; to cleanup hydrazine spills often with boric acid; chlorinated lime; for disinfection of inanimate objects & drinking water; treatment for arsenides & antimonides; treatment of thiol spills; oxidizer in calico printing, decontaminant for mustard gas, & pesticide for caterpillars.

Physical Properties

Freezing Point: 100 °C (212 °F)
Specific Gravity: 2.35
Water Solubility: Soluble in Cold Water
Other Solubilities: Alcohol: insoluble
Refraction Index: Alpha 1.545
Flash Point: Nonflammable
Autoignition Temperature: 75 °C

RTECS Toxicity Data

Acute Oral: Rat LD_{50} Dose: 850 mg/kg.

Mutagenic: Hamster Cytogenetic Analysis; Cell Type: fibroblast; Dose: 4 gm/L. Bacteria - S Typhimurium Mutations in Microorganisms; Dose: 1 mg/plate (-S9).

Hazard Overviews

Corrosive

Fire Diamond

Health: Corrosive, causes severe irritation or burns to the eyes/skin/respiratory tract. Chronic Effects: dermatitis.

Fire: Noncombustible. However, it is a strong oxidizer capable of igniting combustible materials. Use extinguishing agents suitable for surrounding fire.

Reactivity: Stable, when kept dry and free from contamination. Hazardous polymerization cannot occur. Incompatible with: combustibles; amines; carbon tetrachloride and heat; carbon or charcoal and heat; ethyl alcohol; metal oxides; mercaptons; sulfur; turpentine; strong reducing agents; organic matter; combustible materials; nitromethane; ammonium chloride; N,N-dichloromethylamine and heat; acetic acid and potassium cyanide; ethanol; isobutanethiol; methanol; 1-pro-panethiol; rust; water or steam; sodium carbonate; starch; sodium hydrogen sulfate; nitrogenous bases or acetylene; glycerine algacide; hydroxy compounds (glycerol; diethylene glycol monomethyl ether; phenol); organic sulfur compounds; lubricating oil; acids urea. Hazardous decomposition products: oxygen; chlorine; hydrochloric acid.

Carcinogenicity: IARC - Group 3, Not classifiable as to carcinogenicity to humans; NIOSH - Not listed; NTP - Not listed; ACGIH - Not listed; OSHA - Not listed; EPA - Not listed; MAK - Not listed

Primary Target Organs:

Eyes

Skin

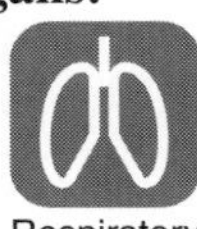
Respiratory System

Mucous Membranes

Gastro-intestinal

Environmental

Ecotoxicity: Aquatic toxicity: 0.5 ppm/*/trout/killed/fresh water *Time period not specified

Cleanup/Disposal: Guide No. 140: Keep combustibles (wood, paper, oil, etc.) away from spilled material. Do not touch damaged containers or spilled material unless wearing appropriate protective clothing. Stop leak if you can do it without risk. Do not get water inside containers. Small Dry Spills: With clean shovel place material into clean, dry container and cover loosely; move containers from spill area.

Small Liquid Spills: Use a non-combustible material like vermiculite, sand or earth to soak up the product and place into a container for later disposal. Large Spills: Dike far ahead of liquid spill for later disposal. Following product recovery, flush area with water.

Environmental Physical Data

BCF: not pertinent
BOD: not pertinent

Regulations

RCRA 40CFR: Not listed
CERCLA: 40CFR 302.4: Listed per CWA Section 311(b)(4) RQ: 10 lb (4.535 kg)
SARA 40CFR 372.65: Not listed
SARA EHS 40CFR 355: Not listed
TSCA: Listed

CAL6280 CAS #: 10124-37-5

CALCIUM NITRATE

RTECS: EW2985000
DOT: UN1454; IMO5.1
EINECS Number: 233-332-1
Molecular Formula: $CaH_2N_2O_6$
Structured MF: $Ca(NO_3)_2$
Formula Weight: 164.10

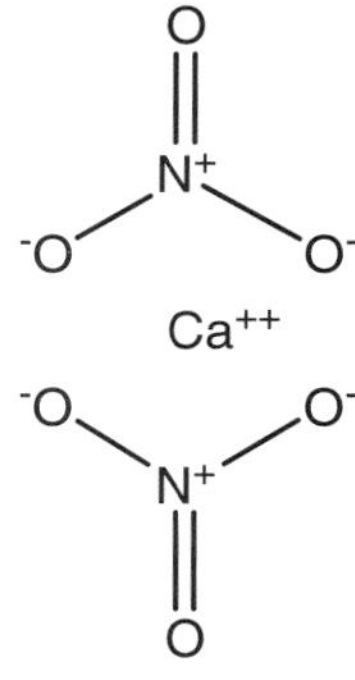

Chemical Structure

Synonyms: CALCIUM DINITRATE; CALCIUM SALTPETER; CALCIUM(II) NITRATE (1:2); LIME NITRATE; LIME SALTPETER; NITRIC ACID,CALCIUM SALT; NITRIC ACID,CALCIUM SALT (8CI,9CI); NITROCALCITE; NORGE SALTPETER; NORWAY SALTPETER; NORWEGIAN SALTPETER; SYNFAT 1006
Description: colorless crystals; odorless
Use: in explosives, fertilizers, matches, pyrotechnics; mfr of incandescent mantles, radio tubes, nitric acid; corrosion inhibitor in diesel fuels; oxidizing agent; source of (14)carbon by nuclear irradiation; stops gelation of animal glue; secondary macronutrient in fertilizer

Physical Properties

Boiling Point: Decomposes
Freezing Point: About 560 °C (1040 °F)
Specific Gravity: 2.504 at 18 °C
Water Solubility: Very Soluble in Water
Other Solubilities: Soluble in anhydrous Ammonia; in Alcohol 14 g/100 cc at 15 °C.
pH: 5% aqueous solution 6
Flash Point: Not flammable

RTECS Toxicity Data

Acute Oral: Rat LD_{50} Dose: 302 mg/kg.

Hazard Overviews

Fire Diamond

Health: Irritating to eyes/skin/respiratory tract. Also Causes: upon ingestion: blue-tinged skin, gastroenteritis, abdominal pains, vomiting, muscular weakness. Chronic Effects: weakness, faintness, headache.
Fire: Noncombustible. However, it is a strong oxidizer capable of igniting combustibles. Use flooding quantities of water. Do not scatter with high pressure water stream. Contact with combustible dusts or vapors may cause an explosion.
Reactivity: Stable. Hazardous polymerization cannot occur. Avoid: exposure to combustibles. Incompatible with: aluminum and ammonium nitrate and formamide and water; ammonium nitrate and hydrocarbon oils; ammonium nitrate and water soluble fuels; organic matter; any combustible material. Hazardous decomposition products: carbon oxide(s); nitrogen oxide(s); calcium oxide.
Carcinogenicity: IARC - Not listed; NIOSH - Not listed; NTP - Not listed; ACGIH - Not listed; OSHA - Not listed; EPA - Not listed; MAK - Not listed
Primary Target Organs:

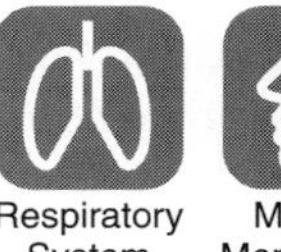

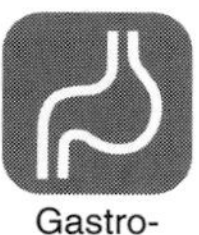

Eyes Skin Respiratory System Mucous Membranes Gastro-intestinal

Environmental

Ecotoxicity: Aquatic toxicity: 10,000 ppm/96 hr/sunfish/TLm/fresh water
Cleanup/Disposal: Guide No. 140: Keep combustibles (wood, paper, oil, etc.) away from spilled material. Do not touch damaged containers or spilled material unless wearing appropriate protective clothing. Stop leak if you can do it without risk. Do not get water inside containers. Small Dry Spills: With clean shovel place material into clean, dry container and cover loosely; move containers from spill area. Small Liquid Spills: Use a non-combustible material like vermiculite, sand or earth to soak up the product and place into a container for later disposal. Large Spills: Dike far ahead of liquid spill for later disposal. Following product recovery, flush area with water.

Environmental Physical Data

BCF: no food chain concentration potential
BOD: none

Regulations

RCRA 40CFR: Not listed
CERCLA: 40CFR 302.4: Not listed
SARA 40CFR 372.65: Not listed
SARA EHS 40CFR 355: Not listed
TSCA: Listed

CAL6440	**CAS #: 1305-78-8**
CALCIUM OXIDE	

RTECS: EW3100000
DOT: UN1910; IMO9.0
EINECS Number: 215-138-9
Molecular Formula: CaO
Structured MF: CaO
Formula Weight: 56.08

Ca=O

Chemical Structure

Synonyms: AIRLOCK; BELL CML(E); BURNED LIME; BURNT LIME; CALCIA; CALCIUM MONOXIDE; CALOXOL CP2; CALOXOL W 3; CALOXOL W3; CALX; CALX USTA; CALXYL; CHAUX VIVE; CML 21; CML 31; DESICAL P; GEBRANNTER KALK; LIME; LIME,BURNED; OXYDE DE CALCIUM; PEBBLE LIME; QUICK LIME; QUICKLIME; RHENOSORB C; RHENOSORB F; UNSLAKED LIME; WAPNIOWY TLENEK

Description: white, grayish-white crystals, granular powder; odorless

Use: in building materials, bricks, plaster, finishing lime etc; in metallurgy, flux in steel prodn; paper pulp prodn; ammonia recovery; in water, industrial waste, & sewage treatment; dietary supplement; food additive; mfr aluminum, magnesium, glass; dehairing hides; flotation of non-ferrous ores; mfr calcium salts, etc; lab: to absorb carbon dioxide; in fungicides, insecticides, drilling fluids, lubricants; clarification of sugar juices; in making chlorinated lime; dehydrating agent; sulfurdioxide removal from stack gases; in poultry feeds.

Physical Properties

Boiling Point: 2850 °C (5162 °F)
Freezing Point: 2572 °C (4661.6 °F)
Specific Gravity: 3.32 to 3.35
Vapor Pressure: ~ 0 mm Hg
Water Solubility: Reacts
Other Solubilities: Soluble in acids, Glycerol,sugar solution; Practically Insoluble in Alcohol
pH: Saturated solution in water is about 12.5
Ionization Potential (eV): 6.5 +/-0.3
Flash Point: Nonflammable

Hazard Overviews

Corrosive

Fire Diamond

Health: Corrosive, causes severe burns to eyes/skin/respiratory tract due to formation of calcium hydroxide in contact with moisture. Chronic Effects: dermatitis, perforation of nasal septum.

Fire: Noncombustible. Use extinguishing agents suitable for surrounding fire. Do not use water unless enough can be used to absorb the significant amounts of heat generated by the calcium oxide and water reaction.

Reactivity: Unstable, will react with water (forming calcium hydroxide) and carbon dioxide [forming calcium carbonate (chalk)] if exposed to air; containers can swell and burst if moisture gets in. Hazardous polymerization cannot occur. Avoid: excessive dust generation; air. Incompatible with: water; carbon dioxide; ethanol; boric oxide and calcium chloride; boron trifluoride; chlorine trifluoride; fluorine; hydrofluoric acid; phosphorus pentoxide; perchlorates; nitrates; permanganates.

Carcinogenicity: IARC - Not listed; NIOSH - Listed as carcinogen; NTP - Not listed; ACGIH - Not listed; OSHA - Not listed; EPA - Not listed; MAK - Not listed

Primary Target Organs:

Eyes

Skin

Respiratory System

Mucous Membranes

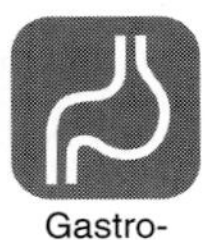
Gastro-intestinal

Exposure Limits
OSHA PEL: TWA: 5 mg/m^3.
ACGIH TLV: TWA: 2 mg/m^3.
NIOSH REL: TWA: 2 mg/m^3.
NIOSH IDLH: 25 mg/m^3.
DFG MAK: TWA: 5 mg/m^3.

Respirator Recommendation
Exposure Range: >5 to <25 mg/m^3 Air Purifying, Negative Pressure, Half Mask
Exposure Range: 25 to unlimited mg/m^3 Self-contained Breathing Apparatus, Pressure Demand, Full Face Self-contained Breathing Apparatus, Pressure Demand, Full Face
Cartridge Color: dust/mist filter (use P100 or consult supervisor for appropriate dust/mist filter)

Environmental

Ecotoxicity: Aquatic toxicity: 92 ppm/7 hr/trout/toxic/fresh water 240 ppm/24 hr/mosquito fish/TLm/fresh water

Cleanup/Disposal: Guide No. 157: Eliminate all ignition sources (no smoking, flares, sparks or flames in immediate area). All equipment used when handling the product must be grounded. Do not touch damaged containers or spilled material unless wearing appropriate protective clothing. Stop leak if you can do it without risk. A vapor suppressing foam may be used to reduce vapors. Do not get water inside containers. Use water spray to reduce vapors or divert vapor cloud drift. Prevent entry into waterways, sewers, basements or confined areas. Small Spills: Cover with dry earth, dry sand, or other non-combustible material followed with plastic sheet to minimize spreading or contact with rain. Use clean non-sparking tools to collect material and place it into loosely covered plastic containers for later disposal.

Environmental Physical Data

BCF: not pertinent
BOD: not pertinent

Regulations

RCRA 40CFR: Not listed
CERCLA: 40CFR 302.4: Not listed
SARA 40CFR 372.65: Not listed
SARA EHS 40CFR 355: Not listed
TSCA: Listed

Analytical Methods

Indoor / Expired Air: NIOSH 7020

CAL7720 CAS #: 1344-95-2

CALCIUM SILICATE

RTECS: VV9150000
EINECS Number: 215-710-8
Molecular Formula: $CaSiO_3$
Structured MF: $CaSiO_3$
Formula Weight: 172.2

O^-, Ca^{++}, Si, ^-O, O^-, O^-, Ca^{++}

Chemical Structure

Synonyms: CALCIUM HYDROSILICATE; CALCIUM METASILICATE; CALCIUM MONOSILICATE; CALCIUM POLYSILICATE; CALCIUM SALT OF SILICIC ACID; CALCIUM SILICATE,SYNTHETIC NONFIBROUS; CALFLO E; CALSIL; CS LAFARGE; FLORITE R; MARIMET 45; MICROCAL 160; MICROCAL ET; MICRO-CEL; MICROCEL T-41; MICRO-CEL A; MICRO-CEL B; MICRO-CEL C; MICRO-CEL E; MICRO-CEL T; MICRO-CEL T 26; MICRO-CEL T 38; MICRO-CEL T26; MICRO-CEL T38; MICRO-CEL T41; PROMAXON P60; SILENE EF; SILICIC ACID,CALCIUM SALT; SILMOS T; SOLEX; STABINEX NW 7PS; STARLEX L; SW 400; TOYOFINE A; WOLLASTONITE (MINERAL)
Description: white or creamy colored free-floating powder
Use: in lime glass, portland cement; filler in elastomers, plastics, paints, paper & paper coatings, & ceramics; absorbent for liqs, gases, vapors; suspension agent, pigment & pigment extender; binder for refractory material; in chromatography; absorbent; anticaking agent in foods (table salt, etc) and pharmaceuticals ;antibiotics, etc); road aggregate; thermal insulator; in cosmetics; conditions powders, eg, ground sulfur to provide free-flowing dust; (vet) antacid.

Physical Properties

Freezing Point: 1540 °C (2804 °F)
Specific Gravity: 2.9
Bulk Density: 15 to 16 lbs/cu ft
Vapor Pressure: ~ 0 mm Hg
Water Solubility: Insoluble in Water
pH: Aqueous slurry 8.0 to 10.0
Refraction Index: 1.616 to 1.631
Flash Point: Noncombustible Solid

Hazard Overviews

Health: Irritating to eyes/respiratory tract. Other Acute Effects: may be harmful by inhalation, ingestion, or skin absorption.
Fire: Noncombustible. Hazards: emits toxic fumes. Extinguishing agents: water spray; carbon dioxide, dry chemical powder or appropriate foam. Precautions: combustible liquid.
Reactivity: Incompatible with: strong oxidizing agents. Hazardous decomposition products: toxic fumes of: silicon oxide.
Carcinogenicity: IARC - Not listed; NIOSH - Not listed; NTP - Not listed; ACGIH - Class A4, Not classifiable as a human carcinogen; OSHA - Not listed; EPA - Not listed; MAK - Not listed
Primary Target Organs:

Eyes

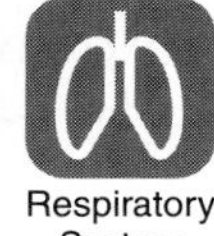
Respiratory System

Exposure Limits
OSHA PEL: TWA: 15 mg/m^3; total dust.
ACGIH TLV: TWA: 10 mg/m^3.
NIOSH REL: TWA: 10 mg/m^3.
Respirator Recommendation
Exposure Range: >5 to 50 mg/m^3 Air Purifying, Negative Pressure, Half Mask
Exposure Range: >50 to 500 mg/m^3 Air Purifying, Negative Pressure, Full Face
Exposure Range: >500 to 5000 mg/m^3 Supplied Air, Constant Flow/Pressure Demand, Full Face
Exposure Range: >5000 to unlimited mg/m^3
Cartridge Color: dust/mist filter (use P100 or consult supervisor for appropriate dust/mist filter)

Environmental

Regulations

RCRA 40CFR: Not listed
CERCLA: 40CFR 302.4: Not listed
SARA 40CFR 372.65: Not listed
SARA EHS 40CFR 355: Not listed
TSCA: Listed

CAL8040 CAS #: 1592-23-0

CALCIUM STEARATE

RTECS: WI3000000
EINECS Number: 216-472-8
Molecular Formula: $C_{36}H_{72}CaO_4$
Formula Weight: 607.00

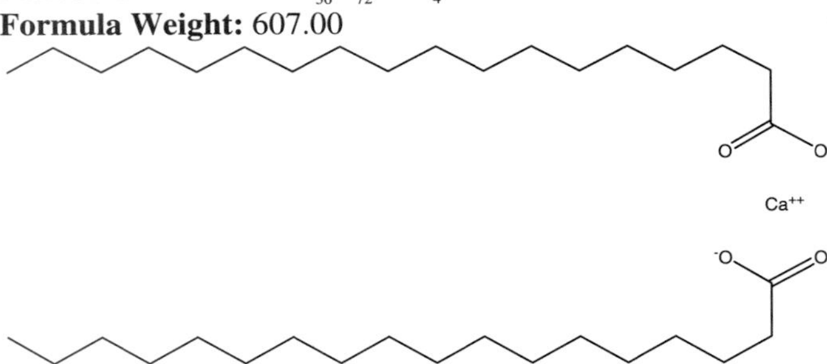

Chemical Structure

Synonyms: AQUACAL; CALCIUM DISTEARATE; CALSTAR; FLEXICHEM; FLEXICHEM CS; G 339S; G 339 S; NOPCOTE C 104; OCTADECANOIC ACID,CALCIUM SALT; STAVINOR 30; STEARIC ACID,CALCIUM SALT; SYNPRO STEARATE; WITCO G 339/S; WITCO G 339S
Description: white to yellowish crystalline powder
Use: plastics additive (stabilizer & internal lubricant esp for pvc); waterproofing fabrics, cement, stucco, explosives; releasing agent for plastic molding powders, pharmaceutical pill & tablet mfr, baked goods; lubricant (eg, for metals); in pencils & wax crayons; conditioning agent in food & pharmaceutical products; water repellent; flatting agent in paints; emulsions; cosmetics; in mfr of paper; moisture barrier coating in foods, drugs, & cosmetics.

Physical Properties

Freezing Point: 180 °C (356 °F)
Specific Gravity: 1.035 Water=1 at 4 °C
Water Solubility: 0.004 g/100 cc Water at 15 °C
Other Solubilities: Slightly Soluble in hot vegetable & mineral oils; quite Soluble in hot Pyridine.

RTECS Toxicity Data

Acute Oral: Rat LD_{50} Dose: >10 gm/kg. Mouse LD_{50} Dose: >10 gm/kg.
Acute Inhalation: Mammal LC; Dose: >1241 mg/m^3/4hr.

Hazard Overviews

Explosive

Fire Diamond

Health: Mildly irritating to respiratory tract. Also Causes: gastrointestinal tract irritation upon ingestion.
Fire: May form explosive dust-air mixtures. Use agents suitable for surrounding fire. Fight fire from maximum distance.
Reactivity: Stable. Hazardous polymerization cannot occur. Avoid: calcium stearate dust in air. Incompatible with: acids; alkalis; strong oxidizers. Hazardous decomposition products: carbon monoxide; calcium oxide.
Carcinogenicity: IARC - Not listed; NIOSH - Not listed; NTP - Not listed; ACGIH - Not listed; OSHA - Not listed; EPA - Not listed; MAK - Not listed
Primary Target Organs:

Respiratory System

Exposure Limits
ACGIH TLV: TWA: 10 mg/m^3.

Environmental

Regulations

RCRA 40CFR: Not listed
CERCLA: 40CFR 302.4: Not listed
SARA 40CFR 372.65: Not listed
SARA EHS 40CFR 355: Not listed
TSCA: Listed

CAL8200 CAS #: 7778-18-9

CALCIUM SULFATE

RTECS: WS6920000
EINECS Number: 231-900-3
Molecular Formula: CaH_2O_4S
Structured MF: $CaSO_4$
Formula Weight: 136.14

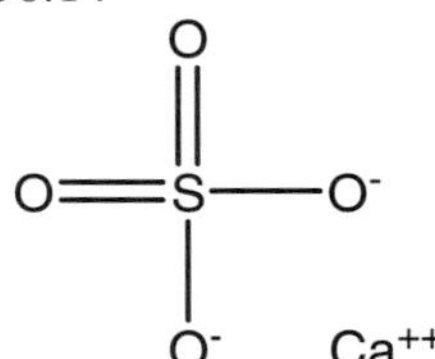

Chemical Structure

Synonyms: ANHYDROUS CALCIUM SULFATE; ANHYDROUS GYPSUM; ANHYDROUS SULFATE OF LIME; CALCIUM SALT OF SULFURIC ACID; CALCIUM SUFATE; CALCIUM SULFATE (1:1); CALCIUM SULPHATE; CRYSALBA; DRIERITE; GIBS; GYPSUM; KARSTENITE; MURIACITE; NATURAL ANHYDRITE; SULFURIC ACID CALCIUM SALT; SULFURIC ACID CALCIUM(2+) SALT (1:1); SULFURIC ACID,CALCIUM SALT (1:1); THIOLITE
Description: white powder, crystals; odorless
Use: soil conditioner; pharmaceutic aid; dietary supplement & food additive; for wall plasters; wall board; tiles, etc; moldings; statuary; in paper ind, hemihydrate; in mfr of portland cement, plaster of paris, yeast, sulfuric acid, calcium carbide, etc; artificial marble; white pigment; filler or glaze in paints, enamels, toothpaste, insecticide dusts; in water treatment, drying agent; in cement; in muds for oilwell drilling; in making tofu.

Physical Properties

Boiling Point: Decomposes
Freezing Point: 1450 °C (2642 °F)
Specific Gravity: 2.96
Vapor Pressure: ~ 0 mm Hg
Water Solubility: Pure Anhydrous Slightly Soluble
Other Solubilities: Soluble in Ammonium salts, Sodium Thiosulfate & Glycerine.
Refraction Index: 1.569
Flash Point: Noncombustible Solid

Hazard Overviews

Fire Diamond

Health: Irritation of the eyes/skin/respiratory tract. Chronic Effects: inflammation of the nasal passages, impaired sense of smell and taste.
Fire: Noncombustible. Use extinguishing agents suitable for surrounding fire.
Reactivity: Stable. Hazardous polymerization cannot occur. Incompatible with: aluminum and heat; phosphorus (at high temperatures); diazomethane. Hazardous decomposition products: fumes of sulfur oxides.
Carcinogenicity: IARC - Not listed; NIOSH - Listed as carcinogen; NTP - Not listed; ACGIH - Not listed; OSHA - Not listed; EPA - Not listed; MAK - Not listed
Primary Target Organs:

Eyes

Skin

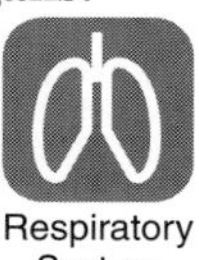
Respiratory System

Exposure Limits
OSHA PEL: TWA: 15 mg/m^3; total dust.
ACGIH TLV: TWA: 10 mg/m^3.
DFG MAK: TWA: 6 mg/m^3.
Respirator Recommendation
Exposure Range: >5 to 50 mg/m^3 Air Purifying, Negative Pressure, Half Mask
Exposure Range: >50 to 500 mg/m^3 Air Purifying, Negative Pressure, Full Face
Exposure Range: >500 to 5000 mg/m^3 Supplied Air, Constant Flow/Pressure Demand, Full Face
Exposure Range: >5000 to unlimited mg/m^3 Self-contained Breathing Apparatus, Pressure Demand, Full Face
Cartridge Color: dust/mist filter (use P100 or consult supervisor for appropriate dust/mist filter)

Environmental

Regulations

RCRA 40CFR: Not listed
CERCLA: 40CFR 302.4: Not listed
SARA 40CFR 372.65: Not listed
SARA EHS 40CFR 355: Not listed
TSCA: Listed

CAM3000 CAS #: 21368-68-3

(+-)-CAMPHOR

RTECS: EX1234000
EINECS Number: 244-350-4
Molecular Formula: $C_{10}H_{16}O$
Formula Weight: 152.26

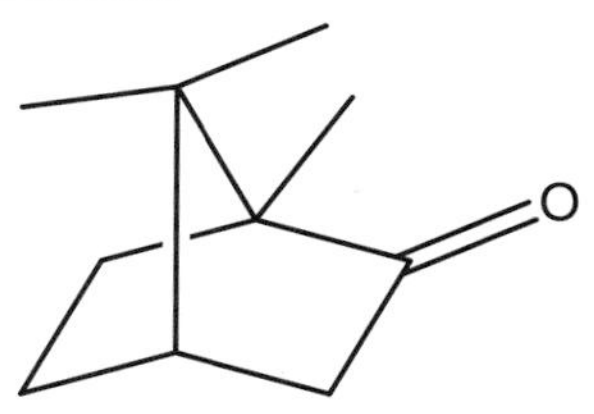

Chemical Structure

Synonyms: BICYCLO(2.2.1)HEPTAN-2-ONE,1,7,7-TRIMETHYL-,(+-)-(9CI); DL-CAMPHOR; (+-)-1,7,7-TRIMETHYLBICYCLO(2.2.1)HEPTAN-2-ONE

Physical Properties

Boiling Point: 204 °C (399 °F)
Freezing Point: Pure 174 °C (345.2 °F)
Specific Gravity: 0.992 at 25/4 °C
Vapor Pressure: 0.18 mm Hg at 20 °C
Water Solubility: 1 g/800 mL at 25 °C
Other Solubilities: Soluble in anline, Carbon Disulfide, Petroleum Ether, liquid Ammonia, concentrated mineral acids in phenol, fixed and volatile oils, higher Alcohols; Soluble in Alcohol, Ether, Benzene, Acetone, Acetic Acid, Chloroform.
Odor Threshold: Variable 0.18 to 16 ppm
Ionization Potential (eV): 8.76
Flash Point: 65.6 °C Closed Cup
Autoignition Temperature: 466 °C
LEL: 0.6% v/v
UEL: 3.5% v/v

RTECS Toxicity Data

Acute Dermal: Rat LD_{50} Route: Subcutaneous Dose: 3040 mg/kg. Mouse LD_{50} Route: Subcutaneous Dose: 3020 mg/kg.

Hazard Overviews

Reactivity: Unstable, sublimes. Hazardous polymerization cannot occur. Avoid: adding salts of any kind to camphor water; exposure to heat and ignition sources. Incompatible with: potassium permanganate; chromic anhydride and naphthalene; oxidizers; steam. Hazardous decomposition products: carbon monoxide.
Carcinogenicity: IARC - Not listed; NIOSH - Not listed; NTP - Not listed; ACGIH - Not listed; OSHA - Not listed; EPA - Not listed; MAK - Not listed

Environmental

Ecotoxicity: Fishes: fathead minnows: static bioassay in Lake Superior water at 18-22 °C: LC_{50},S (1; 24; 48; 72; 96h): 145; 112; 111; 110; 110 mg/l

Regulations

RCRA 40CFR: Not listed
CERCLA: 40CFR 302.4: Not listed
SARA 40CFR 372.65: Not listed
SARA EHS 40CFR 355: Not listed
TSCA: Listed

CAM6000 CAS #: 464-49-3

(1R,4R)-(+)-CAMPHOR

RTECS: EX1260000
EINECS Number: 207-355-2
Molecular Formula: $C_{10}H_{16}O$
Formula Weight: 152.24

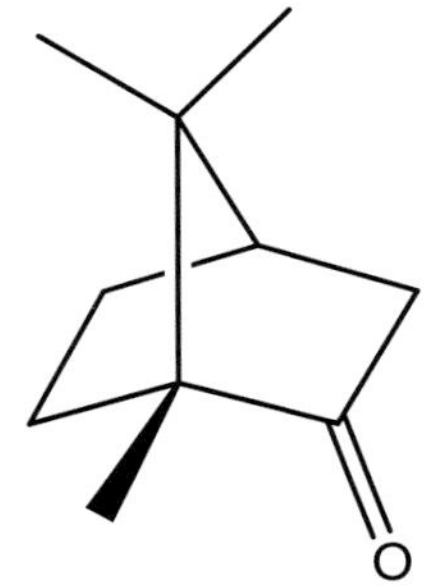

Chemical Structure

Synonyms: ALCANFOR; BICYCLO(2.2.1)HEPTAN-2-ONE,1,7,7-TRIMETHYL-,(1R)-; (+)-2-BORNANONE; D-2-BORNANONE; D-2-CAMPHANONE; (+)-CAMPHOR; CAMPHOR,(+)-; D-(+)-CAMPHOR; D-CAMPHOR; CAMPHOR USP; JAPANESE CAMPHOR; (1R)-1,7,7-TRIMETHYLBICYCLO(2.2.1)HEPTAN-2-ONE

Description:

Use: as an odorant and flavorant in households; as a plasticizer for cellulose esters and ethers, as an insect repellent, as a chemical intermediate, as a plasticizer for cellulose nitrate other explosives and lacquers; and as a respiratory aid and stimulant in camphorated oil; in perfumery, pharmaceutical and industrial products, incense manufacture, varnishes, explosives, embalming fluid, plastics manufacture, medicine (internal and external), insecticides, moth and mildew proofings, tooth powder, flavorings and pyrotechnics; for its counter-irritant and rubefacient properties

Physical Properties

Boiling Point: 204 °C (399 °F)
Freezing Point: 178.8 °C (353.84 °F)
Specific Gravity: 0.992 at 25/4 °C
Vapor Density: 5.24 Air=1
Vapor Pressure: 1 mm Hg at 41.5 °C
Water Solubility: < 1 mg/mL at 15 C
Other Solubilities: DMSO: >=100 mg/ml at 15 °C; Acetone: >=100 mg/ml at 15 °C; Alcohol: Soluble; Benzene: Soluble; Chloroform: Soluble; Ether: Soluble; Fixed oils: Soluble; Glycerol: Practically Insoluble.
Odor Threshold: Variable 0.18 to 16 ppm
Refraction Index: 1.5462
Ionization Potential (eV): 8.76
Flash Point: 64 °C
Autoignition Temperature: 466 °C
LEL: 0.6% v/v
UEL: 3.5% v/v

RTECS Toxicity Data

Acute Oral: Mouse LD_{50} Dose: 1310 mg/kg.
Acute Dermal: Rat LD_{Lo} Route: Subcutaneous Dose: 1700 mg/kg. Mouse LD_{Lo} Route: Subcutaneous Dose: 2200 mg/kg; Toxic Effects: Behavioral - Convulsions or effect on seizure threshold.
Irritation Skin: Rabbit Standard Draize Test Dose: 500 mg/24H; Reaction: moderate.
Reproductive/Teratogenic: Rat Route: Oral; Dose: 4 gm/kg; Duration: female 6-15D of pregnancy; Maternal Effects - Other effects on females.

Hazard Overviews

Reactivity: Unstable, sublimes. Hazardous polymerization cannot occur. Avoid: salts; heat; ignition. Incompatible with: potassium permanganate; chromic anhydride + naphthalene; oxidizers; steam. Hazardous decomposition products: carbon monoxide gas.
Carcinogenicity: IARC - Not listed; NIOSH - Not listed; NTP - Not listed; ACGIH - Not listed; OSHA - Not listed; EPA - Not listed; MAK - Not listed

Environmental

Cleanup/Disposal: Guide No. 133: Eliminate all ignition sources (no smoking, flares, sparks or flames in immediate area). Do not touch or walk through spilled material. Small Dry Spills: With clean shovel place material into clean, dry container and cover loosely; move containers from spill area. Large Spills: Wet down with water and dike for later disposal. Prevent entry into waterways, sewers, basements or confined areas.

Regulations

RCRA 40CFR: Not listed
CERCLA: 40CFR 302.4: Not listed
SARA 40CFR 372.65: Not listed
SARA EHS 40CFR 355: Not listed
TSCA: Listed

CAM9000 CAS #: 76-22-2

CAMPHOR, SYNTHETIC

RTECS: EX1225000
DOT: UN2717
EINECS Number: 200-945-0

Molecular Formula: $C_{10}H_{16}O$
Structured MF: $C_{10}H_{16}O$
Formula Weight: 152.26

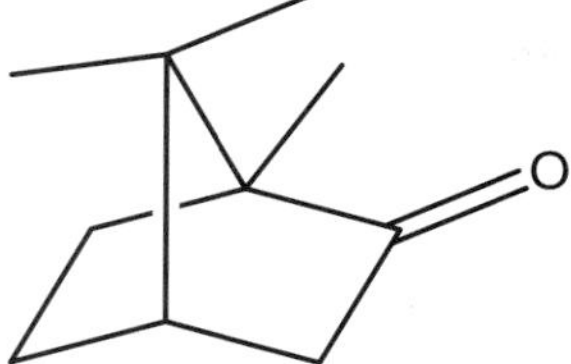

Chemical Structure

Synonyms: BICYCLO(2.2.1)HEPTAN-2-ONE,1,7,7-TRIMETHYL-; BORNANE,2-OXO-; 2-BORNANONE; 2-CAMPHANONE; 2-CAMPHONONE; CAMPHOR,NATURAL; CAMPHOR--NATURAL; FORMOSA CAMPHOR; GUM CAMPHOR; HUILE DE CAMPHRE; JAPAN CAMPHOR; 2-KAMFANON; KAMPFER; 2-KETO-1,7,7-TRIMETHYLNORCAMPHANE; LAUREL CAMPHOR; MATRICARIA CAMPHOR; NORCAMPHOR,1,7,7-TRIMETHYL-; SYNTHETIC CAMPHOR; 1,7,7-TRIMETHYLBICYCLO(2.2.1)-2-HEPTANONE; 1,7,7-TRIMETHYLNORCAMPHOR

Description: translucent, white crystals, granules; penetrating, aromatic odor

Use: in mfr of plastics, cymene, camphorated parachlorophenol, etc; plasticizer; in lacquers, explosives, pyrotechnics, embalming fluid, isolation of cineol, etc; in perfumery; insect repellant; cosmetic (depilatories); dentifrice, deodorantin flavors; japanese white camphor, in beverages, & foods; med: anti-infective & anesthetic; camphorated parachlorophenol in dentistry for infected root canals; counter-irritant; (vet) antipruritic, counterirritant & antiseptic; formerly stimulant & carminative.

Physical Properties

Boiling Point: 204 °C (399 °F)
Freezing Point: 179.75 °C (355.55 °F)
Specific Gravity: 0.992 at 25 °C/4 °C
Vapor Density: 5.24 Air=1
Vapor Pressure: 0.2 mm Hg
Water Solubility: 1 g dissolves (at 25 °C) in: 800 ml Water
Other Solubilities: Liquefies when triturated with chloral hydrate, menthol, resorcinol, beta-naphthol, salol, thymol, phenol, urethane
Odor Threshold: 1.9 ppm
Refraction Index: 1.5462 at 20 °C/D
Ionization Potential (eV): 8.76
Flash Point: 66 °C Closed Cup
Autoignition Temperature: 466 °C
LEL: 0.6% v/v
UEL: 3.5% v/v

RTECS Toxicity Data

Acute Oral: Child TD_{Lo} Dose: 51 mg/kg; Toxic Effects: Behavioral - Somnolence (general depressed activity); Behavioral - Convulsions or effect on seizure threshold. Infant LD_{Lo} Dose: 70 mg/kg; Toxic Effects: Sense organs and special senses - Mydriasis (pupilliary dilation); Behavioral - Convulsions or effect on seizure threshold; Gastrointestinal - Changes in structure or function of salivary glands.

Acute Inhalation: Mouse LC_{Lo} Dose: 400 mg/m^3/3hr; Toxic Effects: Behavioral - Muscle contraction or spasticity.

Acute Dermal: Rat LD_{50} Route: Subcutaneous Dose: 70 mg/kg. Mouse LD_{Lo} Route: Subcutaneous Dose: 200 mg/kg; Toxic Effects: Behavioral - Excitment.

Chronic (Multiple Dose) Inhalation: Mouse Dose: 210 mg/m^3/3H/7W-I; Toxic Effects: Lungs, Thorax, or Respiration - Emphysema. Rabbit Dose: 33 mg/m^3/3H/7W-I; Toxic Effects: Brain and coverings - Other degenerative changes; Cardiac - Other changes; Lungs, Thorax, or Respiration - Emphysema.

Mutagenic: Mouse Sister Chromatid Exchenge; Route: Intraperitoneal; Dose: 80 mg/kg.

Hazard Overviews

Fire Diamond

Health: Irritating to eyes/respiratory tract. Also Causes: nausea, anxiety, headache, dizziness, convulsions, coma, cyanosis, seizures. Chronic Effects: liver damage, granulomatous hepatitis, symptoms similar to Reye's syndrome.

Fire: Combustible. Can form explosive mixtures in the air. Small fires use dry chemical, sand, earth, water spray, or regular foam. Large fires use water spray, fog, or regular foam.

Reactivity: Unstable, sublimes. Hazardous polymerization cannot occur. Avoid:. exposure to heat; ignition sources. Incompatible with: potassium permanganate; chromic anhydride/naphthalene; oxidizers; steam (volatile reaction). Hazardous decomposition products: carbon monoxide.

Carcinogenicity: IARC - Not listed; NIOSH - Not listed; NTP - Not listed; ACGIH - Class A4, Not classifiable as a human carcinogen; OSHA - Not listed; EPA - Not listed; MAK - Not listed

Primary Target Organs:

Eyes

Respiratory System

Nervous System

Liver

Exposure Limits
OSHA PEL: TWA: 2 mg/m^3.
ACGIH TLV: TWA: 2 ppm; 12 mg/m^3; STEL: 3 ppm; 18 mg/m^3.
NIOSH REL: TWA: 2 mg/m^3.
NIOSH IDLH: 200 mg/m^3.
DFG MAK: TWA: 2 ppm; 13 mg/m^3.

Respirator Recommendation

Exposure Range: >2 to 20 ppm Air Purifying, Negative Pressure, Half Mask

Exposure Range: >20 to <200 ppm Air Purifying, Negative Pressure, Full Face

Exposure Range: 200 to unlimited ppm Self-contained Breathing Apparatus, Pressure Demand, Full Face Self-contained Breathing Apparatus, Pressure Demand, Full Face

Cartridge Color: black with magenta (P100)

Environmental

Cleanup/Disposal: Guide No. 133: Eliminate all ignition sources (no smoking, flares, sparks or flames in immediate area). Do not touch or walk through spilled material. Small Dry Spills: With clean shovel place material into clean, dry container and cover loosely; move containers from spill area. Large Spills: Wet down with water and dike for later disposal. Prevent entry into waterways, sewers, basements or confined areas.

Environmental Physical Data

BCF: no food chain concentration potential

Regulations

RCRA 40CFR: Not listed
CERCLA: 40CFR 302.4: Not listed
SARA 40CFR 372.65: Not listed
SARA EHS 40CFR 355: Not listed
TSCA: Listed

Analytical Methods

Air: ASTM D3686, D3687
Indoor / Expired Air: NIOSH 1301

CAN9000 CAS #: 56-25-7

CANTHARIDIN

RTECS: RN8575000
EINECS Number: 200-263-3
Molecular Formula: $C_{10}H_{12}O_4$
Formula Weight: 196.21

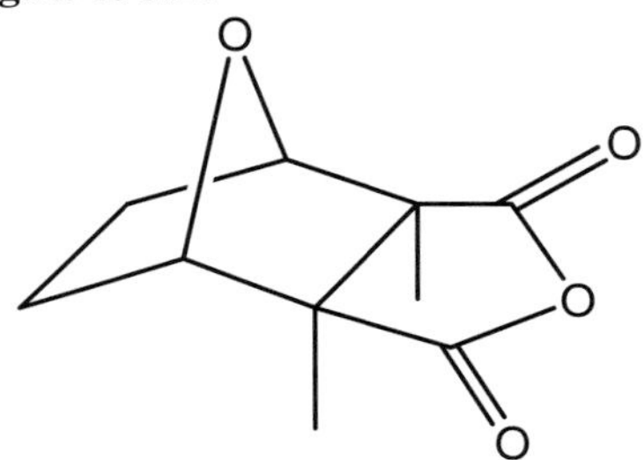

Chemical Structure

Synonyms: CAN; CANTHARIDES CAMPHOR; CANTHARIDINE; CANTHARONE; 1,2-DIMETHYL-3,6-EPOXYPERHYDROPHTHALIC ANHYDRIDE; 2,3-DIMETHYL-7-OXABICYCLO(2.2.1)HEPTANE-2,3-DICARBOXYLICANHYDRIDE; 4,7-EPOXYISOBENZOFURAN-1,3-DIONE,HEXAHYDRO-3A,7A-DIMETHYL-,(3A-ALPHA,4-BETA,7-BETA,7A-ALPHA); 4,7-EPOXYISOBENZOFURAN-1,3-DIONE,HEXAHYDRO-3A,7A-DIMETHYL-,(3AALPHA,4BETA,7BETA,7AALPHA)-; EXO-1,2-CIS-DIMETHYL-3,6-EPOXYHEXAHYDROPHTHALIC ANHYDRIDE; HEXAHYDRO-3AALPHA,7AALPHA-DIMETHYL-4BETA,7BETA-EPOXYISOBENZOFURAN-1,3-DIONE; HEXAHYDRO-3A,7A-DIMETHYL-4,7-EPOXYISOBENZOFURAN-1,3-DIONE; KANTARIDIN; KANTHARIDIN; 7-OXABICYCLO(2.2.1)HEPTANE-2,3-DICARBOXYLIC ANHYDRIDE,2,3-DIMETHYL-

Description: orthorhombic plates, scales; odorless

Use: medication for removal of benign epithelial growth such as warts & molluscum contagiosum; formerly used in veterinary medicine as vesicant and counterirritant

Physical Properties

Freezing Point: 218 °C (424.4 °F)
Water Solubility: Insoluble in Cold Water
Other Solubilities: Soluble in concentrated Sulfuric Acid, Acetic Acid; Slightly Soluble in Benzene, Alcohol.

RTECS Toxicity Data

Acute Oral: Human LD_{Lo} Dose: 428 ug/kg. Rabbit LD_{Lo} Dose: 20 mg/kg; Toxic Effects: Cardiac - Arrythmias (including changes in conduction).

Acute Dermal: Rabbit LD_{Lo} Route: Subcutaneous Dose: 1 mg/kg. Chicken LD_{Lo} Route: Subcutaneous Dose: 75 mg/kg.

Tumorigenic: Mouse Route: Skin; Dose: 25 mg/kg/14W-I; Toxic Effects: Tumorigenic - Neoplastic by RTECS criteria; Skin and appendages - Tumors. Mouse Route: Skin; Dose: 70 mg/kg/52W-I; Toxic Effects: Tumorigenic - Equivocal tumorigenic agent by RTECS criteria; Skin and appendages - Tumors.

Hazard Overviews

Poison

Health: Irritating to skin. Poison. Other Acute Effects: may be fatal if inhaled, swallowed, or absorbed through skin; causes blisters on contact with skin; causes reddening of the skin; possible risk of irreversible effects; vesicant; rubefacient. Chronic Effects: Possible carcinogen.

Fire: Hazards: emits toxic fumes. Extinguishing agents: water spray; carbon dioxide, dry chemical powder or appropriate foam. Precautions: combustible liquid.

Reactivity: Incompatible with: strong oxidizing agents, protect from light. Hazardous decomposition products: toxic fumes of: carbon monoxide, carbon dioxide.

Carcinogenicity: IARC - Group 3, Not classifiable as to carcinogenicity to humans; NIOSH - Not listed; NTP - Not listed; ACGIH - Not listed; OSHA - Not listed; EPA - Not listed; MAK - Not listed

Primary Target Organs:

Skin

Environmental

Cleanup/Disposal: Guide No. 154: Eliminate all ignition sources (no smoking, flares, sparks or flames in immediate area). Do not touch damaged containers or spilled material unless wearing appropriate protective clothing. Stop leak if you can do it without risk. Prevent entry into waterways, sewers, basements or confined areas. Absorb or cover with

dry earth, sand or other non-combustible material and transfer to containers. Do not get water inside containers.

Regulations

RCRA 40CFR: Not listed
CERCLA: 40CFR 302.4: Not listed
SARA 40CFR 372.65: Not listed TPQ: 100/10000 lb
SARA EHS 40CFR 355: Listed TPQ: 100 lb
TSCA: Listed

CAP3000 CAS #: 105-60-2

CAPROLACTAM

RTECS: CM3675000
EINECS Number: 203-313-2
Molecular Formula: $C_6H_{11}NO$
Formula Weight: 113.16

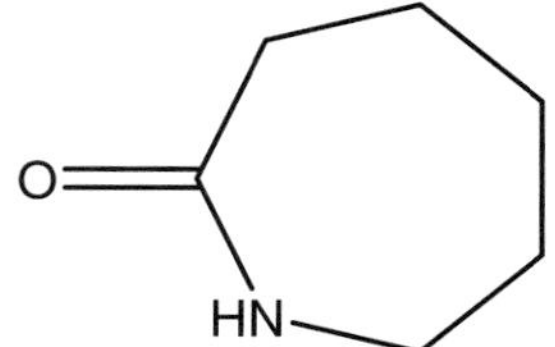

Chemical Structure

Synonyms: A1030; 6-AMINOCAPROIC ACID LACTAM; AMINOCAPROIC LACTAM; 6-AMINOHEXANOIC ACID CYCLIC LACTAM; 1-AZA-2-CYCLOHEPTANONE; 2-AZACYCLOHEPTANONE; 2H-AZEPIN-2-ONE,HEXAHYDRO; 2H-AZEPIN-2-ONE,HEXAHYDRO-; 2H-AZEPIN-7-ONE,HEXAHYDRO; 2H-AZEPIN-7-ONE,HEXAHYDRO-; 6-CAPROLACTAM; GAMMA-CAPROLACTAM; OMEGA-CAPROLACTAM; CAPROLACTAM MONOMER; 6-CAPROLACTAN; E-CAPROLACTUM; OMEGA-CAPROLACTUM; CAPROLATTAME; CAPRON PK4; CYCLOHEXANONE ISO-OXIME; EPSILON-CAPROLACTAM; EPSYLON KAPROLAKTAM; EXTROM 6N; HEXAHYDRO-2-AZEPINONE; HEXAHYDRO-2H-AZEPIN-2-ONE; HEXAHYDRO-2H-AZEPIN-2-ONE (9CI); HEXAMETHYLENIMINE,2-OXO-; 6-HEXANELACTAM; HEXANNIC ACID; HEXANOIC ACID,6-AMINO-,CYCLIC LACTAM; HEXANOIC ACID,6-AMINO-,LACTAM; HEXANOIC ACID-6-AMINO-,LACTAM; HEXANOLACTAM; HEXANONE ISOXIME; HEXANONISOXIM; 1,6-HEXOLACTAM; E-KAPROLAKTAM; KAPROMINE; 2-KETOHEXAMETHYLENEIMINE; 2-KETOHEXAMETHYLENIMINE; 2-OXOHEXAMETHYLENEIMINE; 2-OXOHEXAMETHYLENIMINE; 2-PERHYDROAZEPINONE; STILON
Description: white flakes, fused crystalline solid; unpleasant odor
Use: in the manufacture of synthetic fibers, plastics, bristles, film, coatings, synthetic leather, plasticizers and paint vehicles; cross-linking agent for polyurethanes and it is used in the synthesis of lysine

Physical Properties

Boiling Point: 180 °C (356 °F) at 50 mm Hg
Freezing Point: 70 °C (158 °F)
Specific Gravity: 1.02 at 75/4 °C
Vapor Density: 3.91 Air=1
Saturated Vapor Density: 1.200004582 kg/m^3
Vapor Pressure: 0.001 mm Hg at 20 °C
Water Solubility: Soluble in Water
Other Solubilities: freely Soluble in Methanol, Ethanol,Tetrahydrofurfuryl Alcohol; Ether, Dimethylformamide, Soluble in chlorinated hydrocarbons, Cyclohexene, petroleum fractions.
Surface Tension: Estimated at 20 dynes/cm
Odor Threshold: 0.3 mg/m^3
Refraction Index: 1.4965
Critical Temperature: 507 °C
Critical Pressure: 660 psia
Ionization Potential (eV): 9.07 +/-0.3
Flash Point: 139 °C
Autoignition Temperature: 375 °C
LEL: 1.4% v/v
UEL: 8.0% v/v

RTECS Toxicity Data

Acute Oral: Rat LD_{50} Dose: 1210 mg/kg; Toxic Effects: Sense organs and special senses - Chromidracryorrhea; Behavioral - Convulsions or effect on seizure threshold; Nutritional and gross metabolic - Body temperature decrease. Mouse LD_{50} Dose: 930 mg/kg; Toxic Effects: Behavioral - Muscle contraction or spasticity; Lungs, Thorax, or Respiration - Dyspnea.
Acute Inhalation: Rat LC_{50} Dose: 300 mg/m^3/2hr. Human TC_{Lo} Dose: 100 ppm; Toxic Effects: Lungs, Thorax, or Respiration - Cough.
Acute Dermal: Rabbit LD_{50} Route: Skin; Dose: 1410 uL/kg. Mouse LD_{50} Route: Subcutaneous Dose: 750 mg/kg.
Chronic (Multiple Dose) Oral: Rat Dose: 42 gm/kg/8W-C; Toxic Effects: Nutritional and gross metabolic - Weight loss or decreased weight gain.
Chronic (Multiple Dose) Inhalation: Rat Dose: 5960 mg/kg/82D-C; Toxic Effects: Kidney, Ureter, and Bladder - Other changes in urine composition; Biochemical - True cholinesterase.
Irritation Eye: Rabbit Standard Draize Test Dose: 20 mg/24H; Reaction: moderate.
Irritation Skin: Rabbit Standard Draize Test Dose: 500 mg/24H; Reaction: mild.
Reproductive/Teratogenic: Rat Route: Oral; Dose: 10 gm/kg; Duration: female 6-15D of pregnancy; Effects on Fertility - Post-implantation mortality. Rat Route: Inhalation; Dose: 125 mg/m^3/24 Duration: male 76D prior to mating; Paternal Effects - Spermatogenesis. Rat Route: Inhalation; Dose: 473 mg/m^3/4H Duration: female 1-5D of pregnancy; Effects on Fertility - Pre-implantation mortality; Effects on Embryo or Fetus - Fetal death. Rat Route: Inhalation; Dose: 139 mg/m^3/4H Duration: female 6-12D of pregnancy; Effects on Fertility - Pre-implantation mortality. Rat Route: Inhalation; Dose: 473 mg/m^3/4H Duration: female 60D prior to mating Maternal Effects - Menstrual cycle changes or disorders.
Mutagenic: Human DNA Inhibition; Cell Type: fibroblast; Dose: 1 mmol/L. Human Cytogenetic Analysis; Cell Type: lymphocyte; Dose: 270 mg/L. Human Sex Chromosome Loss; Cell Type: lymphocyte; Dose: 25 mmol/L.

Hazard Overviews

Fire Diamond

Health: Irritating to eyes/skin/respiratory tract. Also Causes: hypotension, leukocytosis, epigastric pain, delayed pulmonary edema, seizures. Chronic Effects: difficulty sleeping, decreased appetite, headache, chest discomfort, mild hypotension, fatigue, lymphocytosis, seizures, nose bleeds, dermatitis, eczema.

Fire: Will burn. Use dry chemical, carbon dioxide, water spray, or foam.

Reactivity: Stable. Hazardous polymerization cannot occur. Avoid: heat; ignition sources. Incompatible with: acetic acid and dinitrogen trioxide; strong mineral acids Hazardous decomposition products: carbon oxides; nitrogen oxides.

Carcinogenicity: IARC - Group 4, Probably not carcinogenic to humans; NIOSH - Not listed; NTP - Not listed; ACGIH - Class A4, Not classifiable as a human carcinogen; OSHA - Not listed; EPA - Not listed; MAK - Not listed

Primary Target Organs:

Eyes

Skin

Respiratory System

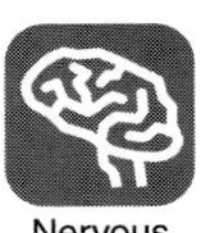
Nervous System

Exposure Limits

OSHA PEL Vacated 1989 Limits: TWA: 5 ppm; 20 mg/m^3; STEL: 10 ppm; 40 mg/m^3.

ACGIH TLV: TWA: 5 ppm; 23 mg/m^3; STEL: 10 ppm; 46 mg/m^3; Vapor, Dust.

NIOSH REL: TWA: 1 mg/m^3. STEL: 3 mg/m^3; dust.

DFG MAK: TWA: 5 mg/m^3.

Respirator Recommendation

Exposure Range: >0.22 to 2.2 ppm Air Purifying, Negative Pressure, Half Mask

Exposure Range: >2.2 to 22 ppm Air Purifying, Negative Pressure, Full Face

Exposure Range: >22 to 220 ppm Supplied Air, Constant Flow/Pressure Demand, Full Face

Exposure Range: >220 to unlimited ppm Self-contained Breathing Apparatus, Pressure Demand, Full Face

Cartridge Color: black with dust/mist prefilter (use P100 or consult supervisor for appropriate dust/mist prefilter)

Environmental

Environmental Fate: If released to soil, it is expected to be rapidly degraded by both microbial and chemical degradation processes. It should leach in soil. If released to water, aerobic biodegradation and chemical degradation would be the dominant removal processes (half-life 5-14 days). Volatilization, bioaccumulation in aquatic organisms, and adsorption to suspended solids and sediments are not expected to be significant fate processes. If released to the atmosphere, it is expected to exist almost entirely in the vapor phase. The dominant removal mechanism is expected to be reaction with photochemically generated hydroxyl radicals (half-life 4.9 hours). This compound may also be removed from the atmosphere in wet deposition.

Cleanup/Disposal: Guide No. 171: Do not touch or walk through spilled material. Stop leak if you can do it without risk. Prevent dust cloud. Avoid inhalation of asbestos dust. Small Dry Spills: With clean shovel place material into clean, dry container and cover loosely; move containers from spill area. Small Spills: Take up with sand or other noncombustible absorbent material and place into containers for later disposal. Large Spills: Dike far ahead of liquid spill for later disposal. Cover powder spill with plastic sheet or tarp to minimize spreading. Prevent entry into waterways, sewers, basements or confined areas.

Environmental Physical Data

Octanol/Water Partition Coefficient: log K_{ow} = -0.19

Sorption Partition Coefficient: log K_{oc} = estimated at 0.8

BCF: estimated at 1

Regulations

RCRA 40CFR: Not listed

CERCLA: 40CFR 302.4: Listed per CAA Section 112 RQ: 5000 lb (2268 kg)

SARA 40CFR 372.65: Not listed

SARA EHS 40CFR 355: Not listed

TSCA: Listed

CAP7000 CAS #: 2425-06-1

CAPTAFOL

RTECS: GW4900000

EINECS Number: 219-363-3

Molecular Formula: $C_{10}H_9Cl_4NO_2S$

Structured MF: $C_{10}H_9Cl_{14}NO_2S$

Formula Weight: 349.06

Synonyms: ALFLOC 7020; ALFLOC 7046; ARBORSEAL; CAPTASPOR; CAPTOFOL; CS 5623; CS5623; 4-CYCLOHEXENE-1,2-DICARBOXIMIDE,N-((1,1,2,2-TETRACHLOROETHYL)THIO)-; DIFOLATAN; DIFOLATAN 4F; DIFOLATAN 80W; DIFOLATAN BOW; DIFOLATAN 4F1; DIFOSAN; FOLCID; FOLTAF; HAIPEN; HAIPEN 50; 1-H-ISOINDOLE-1,3(2H)-DIONE,3A,4,7,7A-TETRAHYDRO-2-((1,1,2,2-TETRACHLOROETHYL)THIO)-; 1H-ISOINDOLE-1,3(2H)-DIONE,3A,4,7,7A-TETRAHYDRO-2-((1,1,2,2-TETRACHLOROETHYL)THIO)-(9CI); KENOFOL; MERPAFOL; MYCODIFOL; NALCO 7046; ORTHO 5865; ORTHO DIFOLATAN 4 FLOWABLE; ORTHO DIFOLATAN 80W; ORTHO-5865; PILLARTAN; PROXEL EF; SANSPOR; SANTAR SM; SULFENIMIDE; SULFONIMIDE; SULPHEIMIDE; TERRAZOL; N-(1,1,2,2-TETRACHLORAETHYLTHIO)-CYCLOHEX-4-EN-1,4-DIACARBOXIMID; N-(1,1,2,2-TETRACHLORAETHYLTHIO)-TETRAHYDROPHTHALAMID; N-1,1,2,2-TETRACHLOROETHYLMERCAPTO-4-CYCLOHEXENE-1,2-CARBOXIMIDE; N-(1,1,2,2-TETRACHLOROETHYLMERCAPTO)-4-CYCLOHEXENE-1,2-DICARBOXIMIDE; N-((1,1,2,2-TETRACHLOROETHYL)SULFENYL)-CIS-4-CYCLOHEXENE-1,2-DICARBOXIMIDE; N-(1,1,2,2-TETRACHLOROETHYLSULFENYL)-CIS-4-CYCLOHEXENE-1,2-DICARBOXIMIDE; N-((1,1,2,2-TETRACHLOROETHYL)THIO)-4-CYCLOHEXENE-1,2-DICARBOXIMIDE; CIS-N-((1,1,2,2-TETRACHLOROETHYL)THIO)-4-CYCLOHEXENE-1,2-DICARBOXIMIDE; N-((1,1,2,2-TETRACHLOROETHYL)THIO)-4-

CYCLOHEXENE-1,2-DICARBOXIMIDE; N-(1,1,2,2-TETRACHLOROETHYLTHIO)-4-CYCLOHEXENE-1,2-DICARBOXIMIDE; N-(1,1,2,2-TETRACHLOROETHYLTHIO)CYCLOHEX-4-ENE-1,2-DICARBOXIMIDE; N-(1,1,2,2-TETRACHLOROETHYLTHIO)-DELTA4-TETRAHYDROPHTHALIMIDE; N-(TETRACHLOROETHYLTHIO)TETRAHYDROPHTHALIMIDE; TETRACHLOROETHYLTHIOTETRAHYDROPHTHALIMIDE; 3A,4,7,7A-TETRAHYDRO-N-(1,1,2,2-TETRACHLOROETHANESULPHENYL)PHTHALIMIDE; 3A,4,7,7A-TETRAHYDRO-2-((1,1,2,2-TETRACHLOROETHYL)THIO)-1H-ISOINDOLE-1,3(2H) -DIONE; 3A,4,7,7A-TETRAHYDRO-2-(1,1,2,2-TETRACHLOROETHYL)THIO-1H-ISOINDOLE-1,3(2H)-DIONE; 1,2,3,6-TETRAHYDRO-N-(1,1,2,2-TETRACHLOROETHYLTHIO)PHTHALIMIDE

Description: colorless or white to pale yellow crystals or crystalline solid; slight, characteristic pungent odor

Use: fungicide on apples, vegetables, potatoes, field crops; protectant-eradicant fungicide on peanuts, citrus, etc; seed protectant on rice, cotton, peanuts; to control foliage & fruit disease of tomatoes, coffee berry disease, potato blight, tapping panel disease of rubber, etc; in lumber industries against wood rot; for control of scab of pome fruit; shot-hole of stone fruit; peach leaf curl; downy mildew & black rot of vines; early & late blights of potatoes; wound protectant for grafting & pruning wounds & cankers on trees.

Physical Properties

Boiling Point: Decomposes
Freezing Point: 160 °C (320 °F) to 161 °C (321.8 °F)
Vapor Pressure: Negligible at room temperature
Water Solubility: 0.0001% by weight
Other Solubilities: Slightly Soluble in most organic solvents; In isopropanol 13, Benzene 25, Toluene 17, Xylene 100, methyl ethyl ketone 44, dimethyl sulphoxide 170 (all in g/kg).
Flash Point: Noncombustible Solid

RTECS Toxicity Data

Acute Oral: Rat LD_{50} Dose: 2500 mg/kg.
Acute Dermal: Rabbit LD_{50} Route: Skin; Dose: 15400 mg/kg.
Reproductive/Teratogenic: Rat Route: Oral; Dose: 11 gm/kg; Duration: multigenerations; Effects on Newborn - Growth statistics. Rat Route: Oral; Dose: 250 mg/kg; Duration: male 5D prior to mating; Effects on Embryo or Fetus - Fetal death.
Mutagenic: Human DNA Inhibition; Cell Type: lymphocyte; Dose: 5 mg/L. Human Cytogenetic Analysis; Cell Type: embryo; Dose: 3400 ug/L. Human Sister Chromatid Exchenge; Cell Type: embryo; Dose: 3400 ug/L. Human Micronucleus Test; Cell Type: embryo; Dose: 3400 ug/L.

Hazard Overviews

Fire: Noncombustible.
Carcinogenicity: IARC - Group 2A, Probably carcinogenic to humans; NIOSH - Listed as carcinogen; NTP - Not listed; ACGIH - Class A4, Not classifiable as a human carcinogen; OSHA - Not listed; EPA - Not listed; MAK - Not listed

Exposure Limits
OSHA PEL Vacated 1989 Limits: TWA: 0.1 mg/m^3.
ACGIH TLV: TWA: 0.1 mg/m^3.
NIOSH REL: TWA: 0.1 mg/m^3.

Respirator Recommendation
Exposure Range: >1 to 10 mg/m^3 Air Purifying, Negative Pressure, Half Mask
Exposure Range: >10 to 100 mg/m^3 Air Purifying, Negative Pressure, Full Face
Exposure Range: >100 to 1000 mg/m^3 Supplied Air, Constant Flow/Pressure Demand, Full Face
Exposure Range: >1000 to unlimited mg/m^3
Cartridge Color: black with magenta (P100)

Environmental

Ecotoxicity: LC_{50} Pheasant >23,070 mg/kg diet/10 day LC_{50} Gammarus lacustris 800 ug/l/96 hr /Conditions of bioassay not specified LC_{50} Bluegill 0.15 mg/l/96 hr /Conditions of bioassay not specified LC_{50} Pteronarcys californica 40 ug/l/96 hr /Conditions of bioassay not specified

Environmental Fate: Strongly adsorbs to soil and should therefore remain in the upper layers of soil. In soil, it is transformed by biodegradation and hydrolysis. The half-life in soil is about 11 days. Both biodegradation and hydrolysis are expected to be the major pathways for the loss in water. The half-life in a river water was estimated to be 0.3 day. Volatilization from water or soil should be negligible. The bioconcentration in aquatic organisms should not be important. Reaction with photochemically produced hydroxyl radicals and ozone will be the important loss processes in the atmosphere. The half-life in air can be estimated to be less than 1.4 hours. Partial removal will also occur as a result of dry and wet deposition.

Environmental Physical Data

Henry's Law Constant: 2.15 $x10^{-9}$
Octanol/Water Partition Coefficient: log K_{ow} = 2.51
Sorption Partition Coefficient: log K_{oc} = 1.724
BCF: estimated at 2.709

Regulations

RCRA 40CFR: Not listed
CERCLA: 40CFR 302.4: Not listed
SARA 40CFR 372.65: Not listed
SARA EHS 40CFR 355: Not listed
TSCA: Not listed

Analytical Methods

Soil: SW846 8081, 8270B, 8270C
Food: FDA 212.1, 212.2, 232.1, 232.4, 242.1
Plasma: EPA 001, 29; FDA 252
Other: EPA 1656

CAP8000 CAS #: 133-06-2

CAPTAN

RTECS: GW5075000
DOT: NA9099
EINECS Number: 205-087-0
Molecular Formula: $C_9H_8Cl_3NO_2S$
Formula Weight: 300.57

Synonyms: AACAPTAN; AGROSOL S; AGROX 2-WAY AND 3-WAY; AMERCIDE; BANGTON; BEAN SEED PROTECTANT; CAPTAB; CAPTAF; CAPTAF 50W; CAPTAF 85W; CAPTAN 50W; CAPTANCAPTENEET 26,538; CAPTANE; CAPTANEX; CAPTAN-STREPTOMYCIN 7.5-0.1 POTATO SEED PIECE PROTECTANT; CAPTEX; 4-CYCLOHEXENE-1,2-DICARBOXIMIDE,N-((TRICHLOROMETHYL)THIO)-; 4-CYCLOHEXENE-1,2-DICARBOXIMIDE,N-(TRICHLOROMETHYL)THIO-; 4-CYCLOHEXENE-1,2-DICARBOXYLIC ACID,IMIDE,N(TRICHLOROMETHYLTHIO)-; ENT 26,538; ESSO FUNGICIDE 406; ESSOFUNGICIDE 406; FLIT 406; FUNGUS BAN TYPE II; GLYODEX 37-22; GLYODEX 3722; GRANOX PFM; GUSTAFSON CAPTAN 30-DD; HEXACAP; 1H-ISOINDOLE-1,3(2H)-DIONE,3A,4,7,7A-TETRAHYDRO-2-((TRICHLOROMETHYL)THIO)-; 1H-ISOINDOLE-1,3-(2H)-DIONE,3A,4,7,7A-TETRAHYDRO-2-((TRICHLOROMETHYL)THIO)-; ISOTOX SEED TREATER; KAPTAN; LE CAPTANE; MALIPUR; MERPAN; MICRO-CHECK 12; NERACID; ORTHOCIDE; ORTHOCIDE-406; ORTHOCIDE 406; ORTHOCIDE 50; ORTHOCIDE 7.5; ORTHOCIDE 75; ORTHOCIDE 83; OSOCIDE; SR-406; SR 406; SR406; STAUFFER CAPTAN; 3A,4,7,7A-TETRAHYDRO-N-(TRICHLOROMETHANESULPHENYL)PHTHALIMIDE; 3A,4,7,7A-TETRAHYDRO-2-((TRICHLOROMETHYL)THIO)-1H-ISOINDOLE-1,3(2H)-DIONE; 1,2,3,6-TETRAHYDRO-N-(TRICHLOROMETHYLTHIO)PHTHALIMIDE; TRICHLORMETHYLTHIOAMID KYSELINY 1,2,3,6-TETRAHYDROFTALOVE; N-(TRICHLOR-METHYLTHIO)-PHTHALIMID; N-(TRICHLOROMETHYLMERCAPTO)-DELTA(SUP 4)-TETRAHYDROPHTHALIMIDE; N-(TRICHLOROMETHYLMERCAPTO)-DELTA(SUP 4)TETRAHYDROPHTHALIMIDE; N-TRICHLOROMETHYLMERCAPTO-4-CYCLOHEXENE-1,2-DICARBOXIMIDE; N-(TRICHLOROMETHYLMERCAPTO)-DELTA(SUP4)-TETRAHYDROPHTHALIMIDE; N-TRICHLOROMETHYLTHIO-CIS-DELTA(SUP 4)-CYCLOHEXENE-1,2-DICARBOXIMIDE; CIS-N-TRICHLOROMETHYLTHIO-4-CYCLOHEXENE-1,2-DICARBOXIMIDE; N-((TRICHLOROMETHYL)THIO)-4-CYCLOHEXENE-1,2-DICARBOXIMIDE; N-(TRICHLOROMETHYL)THIO-4-CYCLOHEXENE-1,2-DICARBOXIMIDE; N-(TRICHLOROMETHYLTHIO)-4-CYCLOHEXENE-1,2-DICARBOXIMIDE; N-(TRICHLOROMETHYLTHIO)CYCLOHEX-4-ENE-1,2-DICARBOXIMIDE; N-TRICHLOROMETHYLTHIOCYCLOHEX-4-ENE-1,2-DICARBOXIMIDE; TRICHLOROMETHYLTHIO-1,2,5,6-TETRAHYDROPHTHALAMIDE; N-((TRICHLOROMETHYL)THIO)TETRAHYDROPHTHALIMIDE; N-TRICHLOROMETHYLTHIO-3A,4,7,7A-TETRAHYDROPHTHALIMIDE; TRIMEGOL; VANCIDE P-75; VANCIDE 89; VANCIDE 89RE; VANGARD K; VANGUARD K; VANICIDE; VONDCAPTAN

Description: white to cream powder; colorless crystals; odorless, but commercial product has a pungent odor

Use: fungicide in paints, plastics, leather and fabrics, in fruit preservation and as a bacteriostat; to control diseases of many fruit, ornamentals and vegetable crops; as a spray, root dip or seed treatment to protect young plants against rot and damping off; controls a wide range of fungal diseases such as pome fruit rot, shot-hole of stone fruit, peach leaf curl, brown rot of cherries, apricots, peaches, plums and citrus fruit, downy mildew and black rot of vines, early and late blights of potatoes and tomatoes, blight and leaf spot in carrots, anthracnose and downy mildew of cucurbits, leaf spot diseases in ornamentals, anthracnose and leaf spot disease of tomatoes and brown patch on turf; in the topical treatment of fungal infections of the skin; in pastes for wallpaper, in paint for greenhouses, in medical facilities, in food packaging, in vinyl coated fabrics and vinyl car roofs, in lacquers, in paper, in rubber stabilizers and in polyethylene for garbage bags and pond liners

Physical Properties

Boiling Point: Decomposes
Freezing Point: 178 °C (352.4 °F)
Specific Gravity: 1.74
Saturated Vapor Density: < 1.200000148 kg/m^3
Density: 1.74 g/mL
Vapor Pressure: < 0.00001 mm Hg at 25 °C
Water Solubility: < 1 mg/mL at 20 C
Other Solubilities: 2.13 g/100 ml Benzene at 26 °C; 0.69 g/100 ml Toluene at 26 °C; 0.29 g/100 ml Ethanol at 26 °C; 0.25 g/100 ml Ether at 26 °C; 20 g/kg Xylene at 25 °C; 21 g/kg Acetone at 25 °C; Insoluble in petroleum oils.
Flash Point: Not available; probably combustible

RTECS Toxicity Data

Acute Oral: Human LD_{Lo} Dose: 1071 mg/kg. Rat LD_{50} Dose: 9 gm/kg.
Acute Inhalation: Rat LC_{50} Dose: >5700 mg/m^3/2hr. Mouse LC_{50} Dose: 4500 mg/m^3/2hr.
Acute Dermal: Rat LD_{50} Route: Skin; Dose: >5 gm/kg.
Chronic (Multiple Dose) Oral: Rat Dose: 64120 mg/kg/14W-I; Toxic Effects: Behavioral - Food intake (animal); Kidney, Ureter, and Bladder - Urine volume increased; DEATH.
Reproductive/Teratogenic: Rat Route: Oral; Dose: 500 mg/kg; Duration: male 5D prior to mating; Effects on Embryo or Fetus - Fetal death. Rat Route: Oral; Dose: 2500 mg/kg; Duration: female 6-10D of pregnancy; Maternal Effects - Uterus, cervix, vagina. Rat Route: Oral; Dose: 25 gm/kg; Duration: female 6-10D of pregnancy; Effects on Newborn - Live birth index.
Mutagenic: Human DNA Damage; Cell Type: fibroblast; Dose: 99 umol/L. Human Unscheduled DNA Synthesis; Cell Type: fibroblast; Dose: 1 umol/L. Human DNA Inhibition; Cell Type: lymphocyte; Dose: 5 mg/L. Human DNA Inhibition; Cell Type: HeLa cell; Dose: 8300 nmol/L. Human Cytogenetic Analysis; Cell Type: lung; Dose: 10 mg/L. Human Sister Chromatid Exchenge; Cell Type: lymphocyte; Dose: 10 umol/L. Human Other Mutation Test Systems; Cell Type: lymphocyte; Dose: 1 mg/L.
Tumorigenic: Mouse Route: Oral; Dose: 1075 gm/kg/80W-C; Toxic Effects: Tumorigenic - Neoplastic by RTECS criteria; Gastrointestinal - Tumors. Mouse Route: Oral; Dose: 540 gm/kg/80W-C; Toxic Effects: Tumorigenic - Equivocal tumorigenic agent by RTECS criteria; Gastrointestinal - Tumors.

Hazard Overviews

Fire: Will burn.
Carcinogenicity: IARC - Group 3, Not classifiable as to carcinogenicity to humans; NIOSH - Not listed; NTP - Not listed; ACGIH - Class A3, Animal carcinogen; OSHA - Not listed; EPA - Not listed; MAK - Not listed
Exposure Limits
OSHA PEL Vacated 1989 Limits: TWA: 0.5 mg/m^3.
ACGIH TLV: TWA: 5 mg/m^3.

NIOSH REL: TWA: 5 mg/m^3.

Respirator Recommendation

Exposure Range: >5 to 250 mg/m^3 Supplied Air, Constant Flow/Pressure Demand, Half Mask

Exposure Range: >250 to 5000 mg/m^3 Supplied Air, Constant Flow/Pressure Demand, Full Face

Exposure Range: >5000 to unlimited mg/m^3 Self-contained Breathing Apparatus, Pressure Demand, Full Face Self-contained Breathing Apparatus, Pressure Demand, Full Face

Note: odor threshold unknown

Environmental

Ecotoxicity: LC_{50} Bluegill sunfish 0.047-0.111 ppm/96 hr. /Conditions of bioassay not specified LC_{50} Daphnia magna 7.06-9.96 ppm/48 hr /Conditions of bioassay not specified LC_{50} Salvelinus namaycush (Lake trout)49.0 ug/l/96 hr at 12 °C (95% confidence limit 40.1-59.9 ug/l), wt 0.4 g, static bioassay LC_{50} Colinus virginianus (Bobwhite quail) oral greater than 2400 ppm, 6-day diet ad libitum, age 14 days LC_{50} Procambarus clarkii (Crayfish) 15631 mg/l/96 hr /conditions of bioassay not specified LC_{50} Oncorhynchus tshawytscha (Chinook salmon) 56.5 ug/l/96 hr at 12 °C (95% confidence limit 52.3-61.0 ug/l) fingerling, static bioassay

Environmental Fate: Released to soil it is not expected to leach extensively, but evaporation from near the surface of soils may be significant. Since captan readily hydrolyzes in water, it will probably also hydrolyze in soil depending upon the pH. Half-lives in moist soil range from 1 to 12 days. Released to water it will have a moderate tendency to sorb to suspended sediments, biota and sediments, and a low to medium tendency to bioconcentrate (BCF=36-900). Volatilization may be significant from shallow rivers and streams but will be slower from lakes and ponds. The primary degradative process in water will be hydrolysis. Hydrolysis half-lives will be on the order of hours. Information about the importance of biodegradation as a competing process to hydrolysis were not found. Direct photolysis is not important in relative clear water; in water with high humic content indirect photooxidation appears to be important. A computer estimated half-life in the vapor phase of the atmosphere based upon reaction with hydroxyl radicals is about one hour. It may also be present in the atmosphere sorbed to particulate matter

Environmental Physical Data

Henry's Law Constant: 7.9 x10^{-6}

Octanol/Water Partition Coefficient: log K_{ow} = 2.35

Sorption Partition Coefficient: K_{oc} = estimated at 196

BCF: > 910

Regulations

RCRA 40CFR: Not listed

CERCLA: 40CFR 302.4: Listed per CWA Section 311(b)(4) RQ: 10 lb (4.535 kg)

SARA 40CFR 372.65: Listed

SARA EHS 40CFR 355: Not listed

TSCA: Listed

Analytical Methods

Air: EPA TO-10

Soil: EPA PMD-TLC; SW846 8081, 8270B, 8270C

Water / Groundwater: EPA 617, 022; APHA 6630-B; ASTM D3086

Food: FDA 212.1, 212.2, 232.1, 232.4, 242.1; AOAC 957.14, 980.06

Indoor / Expired Air: EPA IP-8; ASTM D4861

Plasma: EPA 001, 027, 29; FDA 211.1, 231.1, 252

Other: EPA 1656

CAR1190 CAS #: 51-83-2

CARBACHOL CHLORIDE

RTECS: GA0875000

EINECS Number: 200-127-3

Molecular Formula: $C_6H_{15}ClN_2O_2$

Formula Weight: 182.65

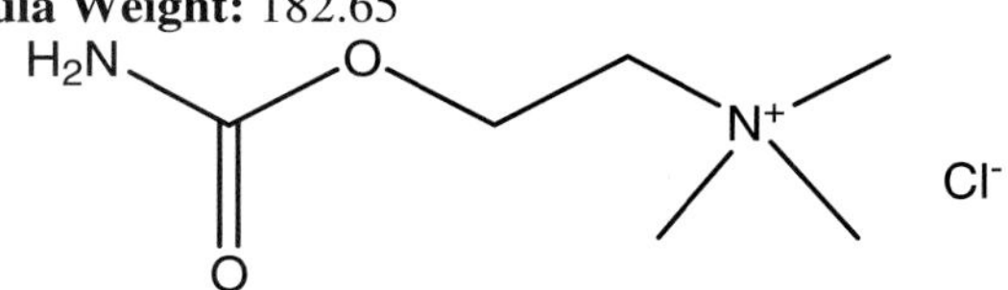

Chemical Structure

Synonyms: 2-((AMINOCARBONYL)OXY)-N,N,N-TRIMETHYLETHANAMINIUM CHLORIDE; 2-((AMINOCARBONYL)OXY)-N,N,N-TRIMETHYLETHANAMINUM CHLORIDE; CARBACH; CARBACHOL; CARBACHOLIN; CARBACHOLINE; CARBACHOLINE CHLORIDE; CARBACHOLUM; CARBACHOLUM CHLORATUM; CARBACOLINA; CARBAMIC ACID,ESTER WITH CHOLINE CHLORIDE; CARBAMINOCHOLINE CHLORIDE; CARBAMINOYLCHOLINE CHLORIDE; CARBAMIOTIN; GAMMA-CARBAMOYL CHOLINE CHLORIDE; CARBAMOYLCHOLINE CHLORIDE; CARBAMOYLCHOLINE-HYDROCHLORIDE; (2-CARBAMOYLOXYETHYL)TRIMETHYLAMMONIUM CHLORIDE; CARBAMYLCHOLINE CHLORIDE; CARBOCHOL; CARBOCHOLIN; CARBOCHOLINE; CARBYL; CARCHOLIN; CHOLINE CARBAMATE CHLORIDE; CHOLINE,CHLORIDE,CARBAMATE; COLETYL; DORYL; DORYL (PHARMACEUTICAL); ETHANAMINIUM,2-((AMINOCARBONYL)OXY)-N,N,N,-TRIMETHYL-,CHLORIDE; ETHANAMINIUM,2-(AMINOCARBONYL)OXY-N,N,N-TRIMETHYL-,CHLORIDE; (2-HYDROXYETHYL)TRIMETHYL AMMONIUM CHLORIDE CARBAMATE; JESTRYL; KARBACHOL; KARBAMOYLCHOLIN CHLORID; LENTIN; LENTINE; MIOSTAT; MORYL; TL 457; VASOPERIF

Description: white or faintly yellow crystals or crystalline powder; odorless or has slight amine-like odor

Use: medication: to produce miosis during ocular surgery, to treat urinary retention of neurogenic origin, for the chronic therapy of noncongestive, wide-angle glaucoma; medication (vet): treatment of rumen atony and impaction in cattle

Physical Properties

Freezing Point: 200 °C (392 °F) to 203 °C (397.4 °F)

Water Solubility: 1 g in 1 ml Water

Other Solubilities: 1 g in 50 ml Alcohol; Very slightly Soluble in dehydrated Alcohol, more readily Soluble on boiling; 1 g in 10 ml Methanol; Practically Insoluble in Chloroform, Ether, Acetone.

pH: Neutral

RTECS Toxicity Data

Acute Oral: Rat LD_{50} Dose: 40 mg/kg. Mouse LD_{50} Dose: 15 mg/kg.
Acute Dermal: Rat LD_{50} Route: Subcutaneous Dose: 4 mg/kg. Mouse LD_{50} Route: Subcutaneous Dose: 3 mg/kg.

Hazard Overviews

Poison

Health: Irritating to eyes/skin/respiratory tract. Poison. Other Acute Effects: may be fatal if swallowed.
Fire: Hazards: emits toxic fumes. Extinguishing agents: carbon dioxide; dry chemical powder; water spray. Precautions: combustible liquid.
Reactivity: Incompatible with: strong oxidizing agents. Hazardous decomposition products: thermal decomposition may produce carbon monoxide, carbon dioxide, and nitrogen oxides; hydrogen chloride gas.
Carcinogenicity: IARC - Not listed; NIOSH - Not listed; NTP - Not listed; ACGIH - Not listed; OSHA - Not listed; EPA - Not listed; MAK - Not listed
Primary Target Organs:

Eyes

Skin

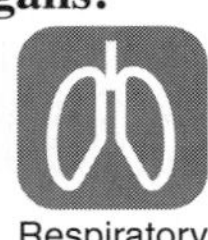
Respiratory System

Environmental

Regulations

RCRA 40CFR: Not listed
CERCLA: 40CFR 302.4: Not listed
SARA 40CFR 372.65: Not listed TPQ: 500/10000 lb
SARA EHS 40CFR 355: Listed TPQ: 500 lb
TSCA: Listed

CAR3090 CAS #: 63-25-2

CARBARYL

RTECS: FC5950000
DOT: UN2757; UN2758; UN2991; UN2992; NA2757; IMO3.2; IMO6.1
EINECS Number: 200-555-0
Molecular Formula: $C_{12}H_{11}NO_2$
Structured MF: $CH_3NHCOOC_{10}H_7$
Formula Weight: 201.22

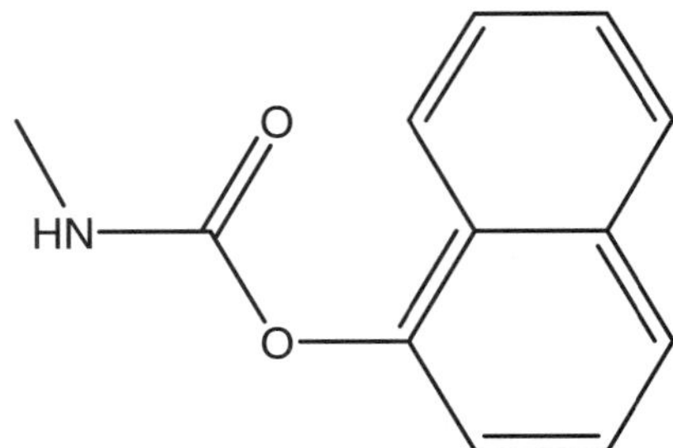

Chemical Structure

Synonyms: ARILAT; ARILATE; ARYLAM; ATOXAN; BERCEMA NMC50; BUG MASTER; CAPROLIN; CARBAMIC ACID,METHYL-,1-NAPHTHYL ESTER; CARBAMINE; CARBARIL; CARBATOX; CARBATOX-60; CARBATOX-75; CARBATOX 75; CARBAVUR; CARBOMATE; CARPOLIN; CARYLDERM; CEKUBARYL; COMPOUND 7744; CRAG SEVIN; CRUNCH; DENAPON; DEVICARB; DICARBAM; DYNA-CARBYL; ENT-23969; ENT 23,969; EXPERIMENTAL INSECTICIDE 7744; GAMONIL; GERMAIN'S; HEXAVIN; KARBARYL; KARBASPRAY; KARBATOX; KARBATOX 75; KARBATOX ZAWIESINOWY; KARBOSEP; LATKA 7744; MENAPHTAM; N-METHYLCARBAMATE DE 1-NAPHTYLE; METHYLCARBAMATE 1-NAPHTHALENOL; METHYLCARBAMATE 1-NAPHTHOL; METHYLCARBAMIC ACID,1-NAPHTHYL ESTER; N-METHYL-1-NAFTYL-CARBAMAAT; N-METHYL-1-NAPHTHYL CARBAMATE; N-METHYL-1-NAPHTHYL-CARBAMAT; N-METHYL-ALPHA-NAPHTHYLCARBAMATE; N-METHYL-ALPHA-NAPHTHYLURETHAN; N-METIL-1-NAFTIL-CARBAMMATO; MONSUR; MUGAN; MURVIN; NAC; NAC (INSECTICIDE); 1-NAFTYLESTER KYSELINY METHYLKARBAMINOVE; ALPHA-NAFTYL-N-METHYLKARBAMAT; 1-NAPHTHALENOL METHYLCARBAMATE; 1-NAPHTHALENOL,METHYLCARBAMATE; 1-NAPHTHALENOL,METHYLCARBAMATE (9CI); 1-NAPHTHALENYL METHYLCARBAMATE; ALPHA-NAPHTHALENYL METHYLCARBAMATE; 1-NAPHTHOL N-METHYLCARBAMATE; 1-NAPHTHYL METHYLCARBAMATE; 1-NAPHTHYL N-METHYL-CARBAMATE; 1-NAPHTHYL N-METHYLCARBAMATE; ALPHA-NAPHTHYL METHYLCARBAMATE; ALPHA-NAPHTHYL N-METHYL-CARBAMATE; ALPHA-NAPHTHYL N-METHYLCARBAMATE; 1-NAPHTHYL-N-METHYL-KARBAMAT; NMC 50; OLITITOX; OLTITOX; OMS-29; PANAM; POMEX; PROSEVOR 85; RAVYON; RYLAM; SAVIT; SEFFEIN; SEPTENE; SEVIMOL; SEVIN; SEVIN 4; SOK; TERCYL; TOXAN; TRICARNAM; UC 7744; UNION CARBIDE 7,744; VETOX; VIOXAN
Description: white, gray solid; odorless
Use: insecticide; acaricide & molluscicide; for agriculture, lawns, & industrial use; to control lepidoptera, coleoptera, & other chewing & sucking insects; in vegetables, tree fruit (incl citrus), soybeans, cotton, ornamentals, forestry, etc; control of earthworms in turf; growth regulator for fruit thinning; in medical facilities, & in sewage treatment plants; med: (vet) animal ectoparasiticide; to control fleas, lice, ticks, & mites on animals, & premises, incl sarcoptic mange on buffaloes; to treat oyster beds against burrowing shrimp.

Physical Properties

Boiling Point: Decomposes
Freezing Point: 145 °C (293 °F)
Specific Gravity: 1.232 at 20 °C/20 °C
Vapor Pressure: $< 4 \times 10^{+5}$ mm Hg at 25 °C
Water Solubility: 120 mg/L Water at 30 °C

Other Solubilities: moderately Soluble in Isophorone; Soluble in Ethanol, Petroleum Ether, Diethyl Ether, freely Soluble in Chloroform; Readily Soluble in polar organic solvents.
Flash Point: Nonflammable

RTECS Toxicity Data

Acute Oral: Man TD_{Lo} Dose: 500 mg/kg; Toxic Effects: Peripheral nerve and sensation - Sensory change involving peripheral nerve; Behavioral - Muscle weakness. Woman LD_{Lo} Dose: 5 gm/kg; Toxic Effects: Automatic Nervous System - Parasympatholytic; Cardiac - Change in rate; Nutritional and gross metabolic - Other changes. Rat LD_{50} Dose: 230 mg/kg; Toxic Effects: Biochemical - True cholinesterase.
Acute Inhalation: Guinea Pig LC; Dose: >390 mg/m^3/4hr.
Acute Dermal: Rabbit LD_{50} Route: Skin; Dose: 2 gm/kg. Rabbit LD; Route: Subcutaneous Dose: >2 gm/kg.
Chronic (Multiple Dose) Oral: Rat Dose: 30 mg/kg/30D-I; Toxic Effects: Liver - Other changes; Biochemical - Hepatic microsomal mixed oxidase(dealkylation, hyroxylation,etc). Rat Dose: 1050 mg/kg/21D-I; Toxic Effects: Liver - Other changes; Blood - Changes in serum composition; Biochemical - Other oxidoreductases. Rat Dose: 10800 mg/kg/96D-C; Toxic Effects: Liver - Changes in liver weight; Kidney, Ureter, and Bladder - Changes in kidney weight; Nutritional and gross metabolic - Weight loss or decreased weight gain. Rat Dose: 25480 mg/kg/1Y-I; Toxic Effects: Endocrine - Evidence of thyroid hypofunction; Nutritional and gross metabolic - Weight loss or decreased weight gain; Biochemical - True cholinesterase. Rat Dose: 1109 mg/kg/22W-I; Toxic Effects: Biochemical - Phosphatases; Biochemical - Phosphokinase; Biochemical - Other enzymes. Rat Dose: 720 mg/kg/10D-I; Toxic Effects: Biochemical - Catecholamine levels in CNS; Biochemical - Dopamine at other sites. Rat Dose: 300 mg/kg/50D-C; Toxic Effects: Brain and coverings - Recordings from specific areas of CNS; Brain and coverings - Other degenerative changes; Biochemical - Other proteins.
Chronic (Multiple Dose) Inhalation: Rat Dose: 214 ug/m^3/24H/13W-I; Toxic Effects: Brain and coverings - Recordings from specific areas of CNS; Biochemical - True cholinesterase; Biochemical - Transaminases. Cat Dose: 63 mg/kg/4W-I; Toxic Effects: Biochemical - True cholinesterase; DEATH.
Irritation Eye: Rabbit Standard Draize Test Dose: 500 mg/24H; Reaction: mild.
Irritation Skin: Rabbit Standard Draize Test Dose: 12 mg/24H; Reaction: severe.
Reproductive/Teratogenic: Rat Route: Oral; Dose: 1370 mg/kg; Duration: female 39W prior to mating Maternal Effects - Menstrual cycle changes or disorders. Rat Route: Oral; Dose: 5500 mg/kg; Duration: multigenerations; Effects on Fertility - Other measures of fertility. Rat Route: Oral; Dose: 27500 ug/kg; Duration: multigenerations; Effects on Newborn - Live birth index; Growth statistics. Rat Route: Oral; Dose: 220 mg/kg; Duration: female 5-15D of pregnancy; Effects on Newborn - Weaning or lactation index. Rat Route: Oral; Dose: 1370 mg/kg; Duration: male 39W prior to mating; Paternal Effects - Spermatogenesis. Rat Route: Oral; Dose: 5475 mg/kg; Duration: female 52W prior to mating Maternal Effects - Ovaries, fallopian tubes; Uterus, cervix, vagina; Effects on Newborn - Live birth index.
Mutagenic: Human Unscheduled DNA Synthesis; Cell Type: fibroblast; Dose: 1 umol/L. Human DNA Inhibition; Cell Type: lung; Dose: 1 mmol/L. Human DNA Inhibition; Cell Type: lymphocyte; Dose: 50 mg/L. Human Cytogenetic Analysis; Cell Type: embryo; Dose: 40 mg/L. Human Other Mutation Test Systems; Cell Type: lung; Dose: 1 mmol/L. Human Other Mutation Test Systems; Cell Type: fibroblast; Dose: 100 mg/kg.
Tumorigenic: Rat Route: Oral; Dose: 5640 mg/kg/94W-I; Toxic Effects: Tumorigenic - Equivocal tumorigenic agent by RTECS criteria; Gastrointestinal - Tumors; Skin and appendages - Tumors. Rat Route: Implant; Dose: 80 mg/kg; Toxic Effects: Tumorigenic - Carcinogenic by RTECS criteria; Skin and appendages - Tumors; Tumorigenic - Tumors at site of application.

Hazard Overviews

Fire Diamond

Health: Severe irritation to eyes/skin/respiratory tract. Toxic. Also Causes: pulmonary edema, miosis, blurred vision, lacrimation, nausea, diarrhea, cyanosis, convulsions. Chronic Effects: possible chronic neurotoxicity.
Fire: Will burn. Use dry chemical, water spray, fog, or regular foam.
Reactivity: Stable. Hazardous polymerization cannot occur. Avoid: heat and ignition sources. Incompatible with: strongly alkaline pesticides (Bordeaux mixture; lime sulfur); strong oxidizers. Hazardous decomposition products: nitrogen oxides; methylamine; carbon monoxide.
Carcinogenicity: IARC - Group 3, Not classifiable as to carcinogenicity to humans; NIOSH - Not listed; NTP - Not listed; ACGIH - Class A4, Not classifiable as a human carcinogen; OSHA - Not listed; EPA - Not listed; MAK - Not listed
Primary Target Organs:

Eyes

Skin

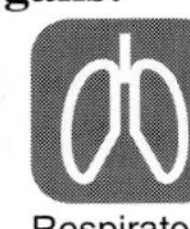
Respiratory System

Nervous System

Cardio-vascular

Exposure Limits
OSHA PEL: TWA: 5 mg/m^3.
ACGIH TLV: TWA: 5 mg/m^3.
NIOSH REL: TWA: 5 mg/m^3.
NIOSH IDLH: 100 mg/m^3.
DFG MAK: TWA: 5 mg/m^3.
Respirator Recommendation
Exposure Range: >5 to 50 mg/m^3 Air Purifying, Negative Pressure, Half Mask
Exposure Range: >50 to 500 mg/m^3 Air Purifying, Negative Pressure, Full Face

Exposure Range: >500 to 5000 mg/m^3 Supplied Air, Constant Flow/Pressure Demand, Full Face
Exposure Range: >500 to unlimited mg/m^3
Cartridge Color: black with magenta (P100)

Environmental

Ecotoxicity: LC_{50} Chinook salmon 2400 ug/l/96 hr (95% confidence limit 1620-3550 ug/l), fingerling / static bioassay LD_{50} Canada goose oral 1790 mg/kg (95% confidence limitS 1480-2180 mg/kg) LD_{50} Pheasant oral 707 mg/kg, 3-4 mo old females LD_{50} Mule deer oral 200-400 mg/kg, 11 mo old females LC_{50} Coho salmon 4340 ug/l/96 hr (95% confidence limit 3310-5690 ug/l), wt 1.0 g / static bioassay LC_{50} Procambarus 1900 ug/l/96 hr (95% confidence limit 1160-3110 ug/l), early instar / static bioassay LC_{50} Lake trout 690 ug/l/96 hr (95% confidence limit 520-910 ug/l), wt 1.7 g / static bioassay EC_{50} Pimephales promelas (fathead minnows) 29 days old 5.29 mg/l/96 hr at 25.8 °C, 6.7 mg/l dissolved oxygen, 45.4 mg/l $CaCO_3$ water hardness, 43.4 mg/l $CaCO_3$ alkalinity, pH 7.7 LD_{50} Bullfrog oral greater than 4000 mg/kg, males LC_{50} Pteronarcys 4.8 ug/l/96 hr (95% confidence limit 3.0-7.7 ug/l), second yr class / static bioassay
Environmental Fate: Release to soil will result in photolysis at the soil surface at a rate dependent upon the soil water content (half-life = 97 hr (dry)-688 hr (wet)). It will hydrolyze relatively rapidly in moist alkaline soil, but only slowly in acidic soil. It may leach to groundwater based on its moderate soil sorption coefficient. Release to water will result in rapid hydrolysis at pH values of 7 and above (half-life = 10.5 days, 1.8 days and 2.5 hours at pH's 7, 8 and 9, respectively, 20 °C). In acidic water, hydrolysis will be slow (half-life = 1500 days at pH 5, 27 °C). Photolysis will be significant (half-life = 52-264 hr). At lower pH values, biodegradation may be significant. Adsorption to high organic content sediments has been demonstrated to be important. Bioconcentration is not expected to be significant. Release to the atmosphere may result in direct photolysis as it absorbs light wavelengths above 290 nm. The estimated half-life for the reaction of vapor phase with photochemically generated, atmospheric hydroxyl radicals is 12.6 hours.
Cleanup/Disposal: Guide No. 131: Fully encapsulating, vapor protective clothing should be worn for spills and leaks with no fire. Eliminate all ignition sources (no smoking, flares, sparks or flames in immediate area). All equipment used when handling the product must be grounded. Do not touch or walk through spilled material. Stop leak if you can do it without risk. Prevent entry into waterways, sewers, basements or confined areas. A vapor suppressing foam may be used to reduce vapors. Small Spills: Absorb with earth, sand or other non-combustible material and transfer to containers for later disposal. Use clean non-sparking tools to collect absorbed material. Large Spills: Dike far ahead of liquid spill for later disposal. Water spray may reduce vapor; but may not prevent ignition in closed spaces. Guide No. 151: Do not touch damaged containers or spilled material unless wearing appropriate protective clothing. Stop leak if you can do it without risk. Prevent entry into waterways, sewers, basements or confined areas. Cover with plastic sheet to prevent spreading. Absorb or cover with dry earth, sand or other non-combustible material and transfer to containers. Do not get water inside containers.

Environmental Physical Data

Henry's Law Constant: 1.28 x10^{-8}
Octanol/Water Partition Coefficient: log K_{ow} = 2.36
Sorption Partition Coefficient: K_{oc} = 370 to 390
BCF: not significant

Regulations

RCRA 40CFR: Listed Hazardous Waste No. U279 Toxic Waste
CERCLA: 40CFR 302.4: Listed per CWA Section 311(b)(4) RQ: 100 lb (45.35 kg)
SARA 40CFR 372.65: Listed
SARA EHS 40CFR 355: Not listed
TSCA: Listed

Analytical Methods

Soil: EPA PMD-CAV; SW846 8270B, 8270C, 8318, 8321A
Water / Groundwater: EPA 553, 632, 022; SW846 8325; ASTM D4763; USGS O3107
Drinking Water: EPA 531.1, 553; AOAC 991.06; ASTM D5315
Food: FDA 232.3, 232.4, 242.1; AOAC 964.18, 968.26, 975.40, 976.04, 985.23
Indoor / Expired Air: NIOSH 5006; ASTM D4861
Plasma: EPA 29; FDA 242.2

CAR3470 CAS #: 10605-21-7

CARBENDAZIM

RTECS: DD6500000
EINECS Number: 234-232-0
Molecular Formula: $C_9H_9N_3O_2$
Formula Weight: 191.21

O
O
N
HN
N
H

Chemical Structure

Synonyms: A 118; A 118 (PESTICIDE); ANTIBAC MF; ARIZIM; BAS-3460; BAS 67054; BAS 3460 F; BATTAL; BAVISTAN; BAVISTIN; BCM; BCM (FUNGICIDE); BENGARD; BENZIMIDAZOLE CARBAMATE DE METHYLE; 2-BENZIMIDAZOLECARBAMIC ACID,METHYL ESTER; BENZIMIDAZOLE-2-CARBAMIC ACID,METHYL ESTER; N-2-(BENZIMIDAZOLYL)CARBAMATE; 1H-BENZIMIDAZOL-2-YLCARBAMIC ACID METHYL ESTER; BERCEMA-BITOSEN; BITOSEN; BMC; BMK; BMK (FUNGICIDE); CARBENDAZIME; CARBENDAZOL; CARBENDAZOLE; CARBENDAZYM; CEKUDAZIM; CTR 6669; CUSTOS; DELSENE; DEROSAL; EK 578; EQUITDAZIN; FALICARBEN; G 665; GARBENDA; HOE 17411; IPO-1250; KEMDAZIN; KOLFUGO; KOLFUGO EXTRA;

LIGNASAN; MBC; MECARZOLE; MEDAMINE; 2-(METHOXY-CARBONYLAMINO)-BENZIMIDAZOL; 2-(METHOXYCARBONYLAMINO)-BENZIMIDAZOLE; METHYL 1H-BENZEMEDAZOL-2-YLCARBAMATE; METHYL 2-BENZIMIDAZOLECARBAMATE; METHYL BENZIMIDAZOLE-2-YL CARBAMATE; METHYL BENZIMIDAZOL-2-YL CARBAMATE; MYCO; PILLARSTIN; PREVENTOL BCM; SPIN; STEIN; STEMPOR; SUPERCARB; THICOPER; TRITICOL; U-32.104

Description: beige to light gray powder

Use: systemic agricultural and horticultural fungicide

Physical Properties

Freezing Point: Decomposes at 302 °C (575.6 °F) to 307 °C (584.6 °F)

Saturated Vapor Density: < 1.200000001 kg/m^3

Density: 1.45 at 20 °C

Vapor Pressure: < 0.000000075 mm Hg at 20 °C

Water Solubility: < 1 mg/mL at 21 C

Other Solubilities: DMSO: 1-10 mg/ml at 21 °C; 95% Ethanol: <1 mg/ml at 21 °C; Acetone: <1 mg/ml at 23 °C; Toluene: <1 mg/ml at 23 °C; Chloroform: 100 mg/L at 20 °C; Benzene: 36 mg/L at 24 °C; Ether: <0.01 g/L at 24 °C; Most solvents: Sparingly Soluble.

Flash Point: Probably nonflammable

RTECS Toxicity Data

Acute Oral: Rat LD_{50} Dose: 6400 mg/kg. Mouse LD_{50} Dose: 7700 mg/kg; Toxic Effects: Automatic Nervous System - Other (direct) parasympathomimetic; Behavioral - Ataxia; Lungs, Thorax, or Respiration - Dyspnea.

Acute Dermal: Rabbit LD_{50} Route: Skin; Dose: 8500 mg/kg. Rat LD_{50} Route: Skin; Dose: 2 gm/kg.

Reproductive/Teratogenic: Rat Route: Oral; Dose: 4 gm/kg; Duration: male 10D prior to mating; Paternal Effects - Testes, epididymis, sperm duct; Effects on Fertility - Male fertility index. Rat Route: Oral; Dose: 153 mg/kg; Duration: female 8-15D of pregnancy; Effects on Embryo or Fetus - Fetotoxicity; Fetal death; Specific Developmental Abnormalities - Central nervous system. Rat Route: Oral; Dose: 150 mg/kg; Duration: male 1D prior to mating; Paternal Effects - Spermatogenesis. Rat Route: Oral; Dose: 153 mg/kg; Duration: female 8-15D of pregnancy; Specific Developmental Abnormalities - Craniofacial (including nose and tongue); Musculoskeletal system. Rat Route: Oral; Dose: 153 mg/kg; Duration: female 8-15D of pregnancy; Effects on Fertility - Post-implantation mortality; Effects on Embryo or Fetus - Fetotoxicity. Rat Route: Inhalation; Dose: 1200 ug/m^3/4 Duration: male 72D prior to mating; Paternal Effects - Spermatogenesis.

Mutagenic: Human Cytogenetic Analysis; Cell Type: liver; Dose: 100 umol/L. Human Cytogenetic Analysis; Cell Type: HeLa cell; Dose: 100 umol/L. Human Sister Chromatid Exchenge; Cell Type: lymphocyte; Dose: 10 mg/L. Human Micronucleus Test; Cell Type: lymphocyte; Dose: 1 umol/L. Human Other Mutation Test Systems; Cell Type: leukocyte; Dose: 1 mg/L. Human Other Mutation Test Systems; Cell Type: lymphocyte; Dose: 20 mg/L.

Hazard Overviews

Fire: Noncombustible.

Carcinogenicity: IARC - Not listed; NIOSH - Not listed; NTP - Not listed; ACGIH - Not listed; OSHA - Not listed; EPA - Not listed; MAK - Not listed

Environmental

Ecotoxicity: Fishes: carp 96h LC_{50} 261 mg/l; rainbow trout 96h LC_{50} 0.36; 2.4 mg/l

Regulations

RCRA 40CFR: Listed Hazardous Waste No. U372 Toxic Waste

CERCLA: 40CFR 302.4: Listed per RCRA Section 3001 RQ: 1 lb (0.454 kg)

SARA 40CFR 372.65: Not listed

SARA EHS 40CFR 355: Not listed

TSCA: Listed

Analytical Methods

Soil: SW846 8321A

Water / Groundwater: EPA 631

Plasma: FDA 242.3

CAR4040 CAS #: 1563-66-2

CARBOFURAN

RTECS: FB9450000

DOT: UN2757; UN2758; UN2991; UN2992; NA2757; IMO3.2; IMO6.1

EINECS Number: 216-353-0

Molecular Formula: $C_{12}H_{15}NO_3$

Structured MF: $C_{12}H_{15}NO_3$

Formula Weight: 221.26

O O NH O

Chemical Structure

Synonyms: BAY 70143; 7-BENZOFURANOL,2,3-DIHYDRO-2,2-DIMETHYL-,METHYLCARBAMATE; 7-BENZOFURANOL,2,3-DIHYDRO-2,2-DIMETHYL-,METHYLCARBAMATE(9CI); BRIFUR; C2292-59A; CARBAMIC ACID,METHYL-,2,3-DIHYDRO-2,2-DIMETHYL-7-BENZOFURANYL ESTER; CARBAMIC ACID,METHYL-,2,2-DIMETHYL-2,3-DIHYDRO-BENZOFURAN-7-YL ESTER; CARBAMIC ACID,METHYL-,2,2-DIMETHYL-2,3-DIHYDROBENZOFURAN-7-YL ESTER; CARBOFURANE; CHINUFUR; CRISFURAN; CURATERR; D 1221; 2,3-DIHYDRO-2,2-DIMETHYL-7-BENZOFURANOL METHYLCARBAMATE; 2,3-DIHYDRO-2,2-DIMETHYL-7-BENZOFURANOL-N-METHYLCARBAMATE; 2,3-

DIHYDRO-2,2-DIMETHYL-7-BENZOFURANYL METHYLCARBAMATE; 2,3-DIHYDRO-2,2-DIMETHYLBENZO-FURAN-7-YL METHYLCARBAMATE; 2,3-DIHYDRO-2,2-DIMETHYLBENZOFURAN-7-YL METHYLCARBAMATE; 2,3-DIHYDRO-2,2-DIMETHYLBENZOFURANYL-7-N-METHYLCARBAMATE; 2,3-DIHYDRO-2,2-DIMTHYLBENZOFURANYL METHYLCARBAMATE; 2,2-DIMETHYL-7-COUMARANYL N-METHYLCARBAMATE; 2,2-DIMETHYL-2,2-DIHYDROBENZOFURANYL-7 N-MEHTYLCARBAMATE; 2,2-DIMETHYL-2,2-DIHYDROBENZOFURANYL-7 N-METHYLCARBAMATE; 2,2-DIMETHYL-2,3-DIHYDRO-7-BENZOFURANYL N-METHYLCARBAMATE; 2,2-DIMETHYL-2,3-DIHYDRO-7-BENZOFURANYL-N-METHYLCARBAMATE; ENT 27,164; ENT 27164; FMC 10242; FURACARB; FURADAN; FURADAN 3G; FURADAN 4F; FURADAN G; FURADAN 75 WP; FURADANE; FURODAN; KARBOFURANU; KENOFURAN; ME F248; METHYL CARBAMIC ACID 2,3-DIHYDRO-2,2-DIMETHYL-7-BENZOFURANYL ESTER; METHYL CARBAMIC ACID2,3-DIHYDRO-2,2-DIMETHYL-7-BENZOFURANYL ESTER; N-METYLOKARBAMINIAN 2,3-DWUWODORO-2,2-DWUMETYLOBENZOFURANYLU-7; NIA 10242; NIA 10242 75 WP; NIAGARA 10242; NIAGARA NIA-10242; OMS-864; OMS 864; PILLARFURAN; YALTOX

Description: white, crystalline solid; odorless or slightly phenolic odor

Use: for systemic and contact insecticide

Physical Properties

Freezing Point: 153 °C (307.4 °F) to 154 °C (309.2 °F)

Specific Gravity: 1.18 at 20 °C/20 °C

Vapor Density: Estimate 7.9 Air=1

Saturated Vapor Density: 1.200000036 kg/m^3

Vapor Pressure: 3.4 x10^{-6} mm Hg at 25 °C

Water Solubility: 700 ppm in Water at 25 °C

Other Solubilities: 1-Methylpyrrolid-z-ong: 30%; Acetone: 15%; Acetonitrile: 14%; Benzene: 4%; Cyclohexane: 9%; Dimethylformamide: 27%; Dimethylsulfoxide: 25%;

Flash Point: Nonflammable

RTECS Toxicity Data

Acute Oral: Rat LD_{50} Dose: 5 mg/kg. Mouse LD_{50} Dose: 2 mg/kg.

Acute Inhalation: Rat LC_{50} Dose: 85 mg/m^3; Toxic Effects: Sense organs and special senses - Other; Gastrointestinal - Changes in structure or function of salivary glands; Gastrointestinal - Nausea or vomiting. Guinea Pig LC_{50} Dose: 43 mg/m^3/4hr; Toxic Effects: Sense organs and special senses - Lacrimation; Behavioral - Somnolence (general depressed activity); Behavioral - Convulsions or effect on seizure threshold.

Acute Dermal: Rabbit LD_{50} Route: Skin; Dose: 885 mg/kg. Rat LD_{50} Route: Skin; Dose: 120 mg/kg.

Mutagenic: Human Cytogenetic Analysis; Cell Type: lymphocyte; Dose: 100 mg/L. Human Sister Chromatid Exchenge; Cell Type: lymphocyte; Dose: 5 mg/L.

Hazard Overviews

Poison

Health: Irritating to eyes/skin/respiratory tract. Poison. Other Acute Effects: may be fatal if inhaled, swallowed, or absorbed through skin; readily absorbed through skin; symptoms of exposure may include burning sensation; coughing; wheezing; laryngitis; shortness of breath; headache; nausea; vomiting; cholinesterase inhibitor; cholinesterase inhibitors can cause heavy salivation and secretion in the lungs; lachrymation; blurred vision; involuntary defecation; diarrhea; tremor; ataxia; sweating; hypothermia; lowered heart rate; and/or a fall in blood pressure; exposure can cause weakness; confusion; gastrointestinal disturbances; dizziness; incoordination; tearing; convulsions; unconsciousness; cyanosis (blue-gray coloring of skin and lips caused by lack of oxygen); target organs: cardiovascular system, central nervous system, blood.

Fire: Noncombustible. Hazards: emits toxic fumes. Extinguishing agents: carbon dioxide, dry chemical powder or appropriate foam. Precautions: combustible liquid.

Reactivity: Stable. Hazardous polymerization will not occur. Incompatible with: strong oxidizing agents. Hazardous decomposition products: toxic fumes of: carbon monoxide, carbon dioxide, nitrogen oxides, cyanides.

Carcinogenicity: IARC - Not listed; NIOSH - Not listed; NTP - Not listed; ACGIH - Class A4, Not classifiable as a human carcinogen; OSHA - Not listed; EPA - Not listed; MAK - Not listed

Primary Target Organs:

Eyes

Skin

Respiratory System

Nervous System

Cardio-vascular

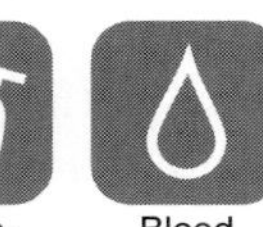

Blood

Exposure Limits

OSHA PEL Vacated 1989 Limits: TWA: 0.1 mg/m^3.

ACGIH TLV: TWA: 0.1 mg/m^3.

NIOSH REL: TWA: 0.1 mg/m^3.

Respirator Recommendation

Exposure Range: >0.1 to 1 mg/m^3 Air Purifying, Negative Pressure, Half Mask

Exposure Range: >1 to 10 mg/m^3 Air Purifying, Negative Pressure, Full Face

Exposure Range: >10 to 100 mg/m^3 Supplied Air, Constant Flow/Pressure Demand, Full Face

Exposure Range: >100 to unlimited mg/m^3 Self-contained Breathing Apparatus, Pressure Demand, Full Face

Cartridge Color: black with magenta (P100)

Environmental

Ecotoxicity: LC_{50} Mallard ducks (Anas platyrhynchos) 10 days old oral (5-day diet ad libitum) 190 ppm (95% confidence limit 156-230 ppm) LC_{50} Yellow perch (Perca flavescens) wt 0.6 g, 147 ug/l/96 hr at 12 °C; static bioassay (95% confidence limit 115-188 ug/l) LC_{50} Sheepshead minnow (Cyprinodon variegatus) 386 ug/l/96 hr; flow-through bioassay LD_{50} Fulvous whistling ducks (Dendrocygna bicolor) females 3 to 6 months old oral (through glass tubing to crop) 0.238 mg/kg (95% confidence limit 0.200-0.283 mg/kg) LC_{50} Coho salmon (Oncorhynchus kisutch) wt 0.6 g,

530 ug/l/96 hr at 12 °C; static bioassay (95% confidence limit 432-650 ug/l)

Environmental Fate: If released to soil, chemical hydrolysis and microbial degradation appear to be the important degradation processes. Chemical hydrolysis is expected to occur more rapidly in alkaline soil as compared to neutral or acidic soils. Results of various degradation studies comparing sterile versus nonsterile soil suggest that soil biodegradation may be important. The rate of degradation in soil is greatly increased by pretreatment. The major metabolites of degradation in soil are 3-hydroxycarbofuran, 3-ketocarbofuran and carbofuran phenol. Experimentally measured K_{oc} values ranging from 14 to 160 indicate that it may leach significantly in many soils; its detection in water table aquifers beneath sandy soils in NY and WI indicate leaching has occurred. Leaching may not occur, however, in very high organic content soils (65% carbon). Volatilization from soil is not expected to be significant, although some evaporation from plants may occur. A review of literature reported the following half-lives disappearance in soil: 2-72 days in laboratory studies, 2-86 days for flooded soils and 26-110 days for field soil. If released to water, it will be subject to significant hydrolysis under alkaline conditions. The hydrolysis half-lives in water at 25 °C are 690, 8.2 and 1.0 weeks at pH 6.0, 7.0 and 8.0, respectively. Direct photolysis and photooxidation (via hydroxyl radicals) may contribute to removal from natural water and may become increasingly important as the acidity of the water increases and the hydrolytic half-life increases. Since it appears to be susceptible to degradation by soil microbes, aquatic microbes may also be able to degrade carbofuran. Aquatic volatilization, adsorption, and bioconcentration are not expected to be important. The half-lives for degradation in river, lake, and seawater from Greece which was irradiated with sunlight were approximately 2, 6, and 12 hours, respectively. The approximated half-lives observed for degradation in sterilized and non-sterilized natural water (pH = 7.8-8.0 and 7.8, respectively) collected from the Holland Marsh, Ontario, were 2.5 and 3 weeks. If released to air, it will react in the vapor-phase with photochemically produced hydroxyl radicals at an estimated half-life of 7.8 hr. Direct photolysis may be important removal process in the atmosphere.

Cleanup/Disposal: Guide No. 131: Fully encapsulating, vapor protective clothing should be worn for spills and leaks with no fire. Eliminate all ignition sources (no smoking, flares, sparks or flames in immediate area). All equipment used when handling the product must be grounded. Do not touch or walk through spilled material. Stop leak if you can do it without risk. Prevent entry into waterways, sewers, basements or confined areas. A vapor suppressing foam may be used to reduce vapors. Small Spills: Absorb with earth, sand or other non-combustible material and transfer to containers for later disposal. Use clean non-sparking tools to collect absorbed material. Large Spills: Dike far ahead of liquid spill for later disposal. Water spray may reduce vapor; but may not prevent ignition in closed spaces. Guide No. 151: Do not touch damaged containers or spilled material unless wearing appropriate protective clothing. Stop leak if you can do it without risk. Prevent entry into waterways, sewers, basements or confined areas. Cover with plastic sheet to prevent spreading. Absorb or cover with dry earth, sand or other non-combustible material and transfer to containers. Do not get water inside containers.

Environmental Physical Data

Henry's Law Constant: calculated at 1.02×10^{-10}
Octanol/Water Partition Coefficient: log K_{ow} = 2.32
Sorption Partition Coefficient: K_{oc} = 60 to 160
BCF: none expected

Regulations

RCRA 40CFR: Listed Hazardous Waste No. P127 Toxic Waste
CERCLA: 40CFR 302.4: Listed per CWA Section 311(b)(4) RQ: 10 lb (4.535 kg)
SARA 40CFR 372.65: Listed TPQ: 10/10000 lb
SARA EHS 40CFR 355: Listed TPQ: 10 lb
TSCA: Listed

Analytical Methods

Soil: SW846 8270B, 8270C, 8318, 8321A
Water / Groundwater: EPA 632, 022; USGS O3107
Drinking Water: EPA 531.1; AOAC 991.06; ASTM D5315
Food: FDA 232.4, 242.1; AOAC 975.40, 985.23, 986.10
Indoor / Expired Air: ASTM D4861
Plasma: EPA 29; FDA 242.2

CAR4230 CAS #: 1563-38-8

CARBOFURAN PHENOL

DOT: NA2757
EINECS Number: 216-350-4
Molecular Formula: $C_{10}H_{12}O_2$
Formula Weight: 164.21

OH
O

Chemical Structure

Synonyms: 7-BENZOFURANOL,2,3-DIHYDRO-2,2-DIMETHYL-; CARBOFURAN 7-PHENOL; NIA 10272

Hazard Overviews

Health: Irritating to eyes/skin/respiratory tract. Other Acute Effects: may be harmful by inhalation, ingestion, or skin absorption.

Fire: Hazards: emits toxic fumes. Extinguishing agents: water spray; carbon dioxide, dry chemical powder or appropriate foam. Precautions: combustible liquid.

Reactivity: Incompatible with: strong oxidizing agents. Hazardous decomposition products: toxic fumes of: carbon monoxide, carbon dioxide.
Carcinogenicity: IARC - Not listed; NIOSH - Not listed; NTP - Not listed; ACGIH - Not listed; OSHA - Not listed; EPA - Not listed; MAK - Not listed
Primary Target Organs:

Eyes

Skin

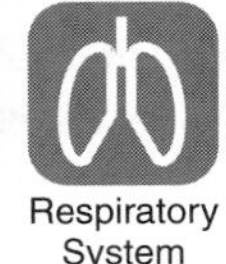
Respiratory System

Environmental

Cleanup/Disposal: Guide No. 151: Do not touch damaged containers or spilled material unless wearing appropriate protective clothing. Stop leak if you can do it without risk. Prevent entry into waterways, sewers, basements or confined areas. Cover with plastic sheet to prevent spreading. Absorb or cover with dry earth, sand or other non-combustible material and transfer to containers. Do not get water inside containers.

Regulations

RCRA 40CFR: Listed Hazardous Waste No. U367 Toxic Waste
CERCLA: 40CFR 302.4: Listed per RCRA Section 3001 RQ: 1 lb (0.454 kg)
SARA 40CFR 372.65: Not listed
SARA EHS 40CFR 355: Not listed
TSCA: Listed

Analytical Methods

Drinking Water: AOAC 992.14

CAR4420 CAS #: 7440-44-0

CARBON

RTECS: FF5250100
DOT: UN1361; UN1362; IMO4.2
EINECS Number: 231-153-3
Molecular Formula: C
Structured MF: C
Formula Weight: 12.01115

C

Chemical Structure

Synonyms: ACETYLENE BLACK; ACTICARBONE; ACTIVATED CARBON; ADSORBIT; AG 3; AG 3 (ADSORBENT); AG 5; AG 5 (ADSORBENT); AK (ADSORBENT); ANTHRASORB; AQUA NUCHAR; AR 3; ART 2; AU 3; BAU; BENZOL BLACK; BG 6080; BLACK 140; BLACK PEARLS; CALCOTONE BLACK; CARBOLAC; CARBON-12; CARBON,ACTIVATED; CARBOPOL EXTRA; CARBOPOL M; CARBOPOL Z 4; CARBOPOL Z EXTRA; CARBOSIEVE; CARBOSORBIT R; CECARBON; CF 8; CF 8 (CARBON); CLF II; CMB 200; CMB 50; COKE POWDER; COLUMBIA LCK; CONDUCTEX; CUZ 3; CWN 2; DARCO; FILTRASORB; FILTRASORB 200; FILTRASORB 400; GROSAFE; HYDRODARCO; IRGALITE 1104; JADO; K 257; MA 100 (CARBON); NORIT; NUCHAR; OU-B; PELIKAN C 11/1431A; SKG; SKT; SKT (ADSORBENT); SU 2000; SU2000; SUCHAR 681; SUPERSORBON IV; SUPERSORBON S 1; SUPERSORBON S1; U 02; WATERCARB; WITCARB 940; XE 340; XF 4175L
Use: reducing agent for purifying metals; in electrodes, electrical devices & steel; jewelry; lubricant (for instrument bearings, etc); dies for hard wires; phonograph needles; in semiconductor research; for lead pencils; refractory crucibles; pigment for printing, etc; cement; matches & explosives; anodes; arc-lamp carbons; electroplating; polishing cmpd, coating for cathode tubes; moderator in nuclear piles; med: activated charcoal antidote, adsorptive, (vet) for diarrhea, wounds; for leather; clarifying, deodorizing, decolorizing & filtering; removal of sulfur dioxide from stack gasses; catalyst for natural gas purification, brewing, chromium electroplating.

Physical Properties

Boiling Point: Sublimes
Freezing Point: 3550 °C (6422 °F)
Specific Gravity: Solid 2 at 20 °C
Vapor Pressure: 1 mm Hg at 3586 °C
Water Solubility: Insoluble
Other Solubilities: Insoluble in acid, alkali /Diamond; Insoluble in acid alkali; Soluble in liquid Iron /Graphite/.
Refraction Index: Graphite 2.1500
Critical Temperature: 6537 °C
Critical Pressure: 2.23 $\times 10^8$ Pa
Ionization Potential (eV): 11.26030
Flash Point: Flammable solid; may ignite spontaneously
Autoignition Temperature: In flowing air 452 to 518 °C

RTECS Toxicity Data

Reproductive/Teratogenic: Rat Route: Subcutaneous; Dose: 167 mg/kg; Duration: female 8D of pregnancy; Effects on Fertility - Post-implantation mortality.

Hazard Overviews

Reactivity: Stable. Hazardous polymerization cannot occur. Avoid: contact with heat; excessive dust generation. Incompatible with: chlorinated paraffins; lead (IV) oxide; manganese (IV) oxide; iron (II) oxide; liquid oxygen; strong oxidizers (chlorates; bromates; nitrates); fatty oils; sodium sulfate; nitric acid. Hazardous decomposition products: carbon dioxide; carbon monoxide; toxic sulphur oxides.
Carcinogenicity: IARC - Not listed; NIOSH - Listed as carcinogen; NTP - Not listed; ACGIH - Not listed; OSHA - Not listed; EPA - Not listed; MAK - Not listed
Exposure Limits
OSHA PEL: TWA: 15 mg/m^3; total dust.
OSHA PEL Vacated 1989 Limits: TWA: 10 mg/m^3. Other Values: respirable mg/m^3; 5.
ACGIH TLV: TWA: 2 mg/m^3.

Environmental

Cleanup/Disposal: Guide No. 133: Eliminate all ignition sources (no smoking, flares, sparks or flames in immediate area). Do not touch or walk through spilled material. Small

Dry Spills: With clean shovel place material into clean, dry container and cover loosely; move containers from spill area. Large Spills: Wet down with water and dike for later disposal. Prevent entry into waterways, sewers, basements or confined areas.

Environmental Physical Data

BCF: no food chain concentration potential
BOD: none

Regulations

RCRA 40CFR: Not listed
CERCLA: 40CFR 302.4: Not listed
SARA 40CFR 372.65: Not listed
SARA EHS 40CFR 355: Not listed
TSCA: Listed

Analytical Methods

Air: EPA 25, 25C
Soil: EPA 440.0
Water / Groundwater: DOE SO010R; ASTM D4129
Food: AOAC 949.12, 972.43
Indoor / Expired Air: NIOSH 5000

CAR4610 CAS #: 1333-86-4

CARBON BLACK

RTECS: FF5800000
DOT: UN1361; NA1361; IMO4.2
EINECS Number: 215-609-9
Molecular Formula: Unspecified or Variable
Structured MF: C
Formula Weight: 12.0

C

Chemical Structure

Synonyms: ACETYLENE BLACK; ARO; AROFLOW; AROGEN; AROTONE; AROVEL; ARROW; ATLANTIC; BLACK KOSMOS 33; BLACK PEARLS; C.I. 77266; C.I. PIGMENT BLACK 6; C.I. PIGMENT BLACK 7; CANCARB; CARBODIS; CARBOLAC; CARBOLAC 1; CARBOMET; CARBON BLACK BV AND V; CARBON BLACK,ACETYLENE; CARBON BLACK,CHANNEL; CARBON BLACK,FURNACE; CARBON BLACK,LAMP; CARBON BLACK,THERMAL; CARBON,AMORPHOUS; CHANNEL BLACK; CK3; COLLOCARB; COLUMBIA CARBON; CONDUCTEX 900; CONTINEX; CORAX A; CORAX P; CROFLEX; CROLAC; DEGUSSA; DELUSSA BLACK FW; DUREX O; EAGLE GERMANTOWN; ELF 78; ELFTEX; ESSEX; EXCELSIOR; EXPLOSION ACETYLENE BLACK; EXPLOSION BLACK; FARBRUSS; FECTO; FLAMRUSS; FURNACE BLACK; FURNAL; FURNEX; FURNEX N 765; GAS BLACK; GAS-FURNACE BLACK; GASTEX; HUBER; HUMENEGRO; IMPINGEMENT BLACK; IMPINGEMENT CARBONS; KETJENBLACK EC; KOSMINK; KOSMOBIL; KOSMOLAK; KOSMOS; KOSMOTHERM; KOSMOVAR; LAMP BLACK; MAGECOL; METANEX; MICRONEX; MIIKE 20; MODULEX; MOGUL; MOGUL L; MOLACCO; MONARCH 1300; MONARCH 700; NEO SPECTRA BEADS AG; NEO-SPECTRA II; NEO-SPECTRA MARK II; NEOTEX; NITERON 55; OIL-FURNACE BLACK; P 33 (CARBON BLACK); P-33; P1250; P68; PEACH BLACK; PELLETEX; PERMABLAK 663; PHILBLACK; PHILBLACK N 550; PHILBLACK N 765; PHILBLACK O; PIGMENT BLACK 7; PRINTEX; PRINTEX 60; RAVEN; RAVEN 30; RAVEN 420; RAVEN 500; RAVEN 8000; REBONEX; REGAL; REGAL 400R; REGAL 300; REGAL 330; REGAL 600; REGAL 99; REGAL SRF; REGENT; ROYAL SPECTRA; SEVACARB; SEVAL; SHAWINIGAN ACETYLENE BLACK; SHELL CARBON; SPECIAL BLACK 1V & V; SPECIAL SCHWARZ; SPHERON; SPHERON 6; STATEX; STATEX N 550; STERLING MT; STERLING N 765; STERLING NS; STERLING SO 1; SUPERBA; SUPER-CARBOVAR; SUPER-SPECTRA; TEXAS; THERMA-ATOMIC BLACK; THERMAL ACETYLENE BLACK; THERMAL BLACK; THERMATOMIC; THERMAX; THERMBLACK; TINOLITE; TM 30; TOKA BLACK 4500; TOKA BLACK 5500; TOKA BLACK 8500; TORCH BRAND; UCET; UKARB; UNITED; VELVETEX; VULCAN; WITCO; WITCOBLAK NO 100; WITCOBLAK NO. 100; WYEX

Description: black powder; odorless

Use: pigment for ink, varnish, rubber tires, eye cosmetics, etc; for leather, stove polish, phonograph records, electrical insulating apparatus; chem reducing agent; carbon black for reinforcing filler in rubber to improve properties (breaking, tear, & abrasion resistances); opacifier; uv light absorber; in carbon paper, typewriter ribbons; cloud seeding; in battery plates; solar energy absorber; plastics stabilizer; for ore reduction & carburizating; carbon brushes & electrodes; thickener in greases.

Physical Properties

Boiling Point: 4200 °C (7592 °F)
Freezing Point: Sublimes
Specific Gravity: 1.8 to 2.1
Vapor Density: 11.6 Air=1
Vapor Pressure: Negligible at 20 °C
Water Solubility: Insoluble
Other Solubilities: Insoluble in any solvents.
Flash Point: < 37.78 °C

RTECS Toxicity Data

Chronic (Multiple Dose) Inhalation: Rat Dose: 50 mg/m^3/6H/90D-I; Toxic Effects: Lungs, Thorax, or Respiration - Other changes.

Mutagenic: Mouse DNA Adduct; Route: Inhalation; Dose: 6200 ug/m^3/16H/12W-I. Bacteria - S Typhimurium Mutations in Microorganisms; Dose: 1 mg/plate (+S9).

Tumorigenic: Rat Route: Inhalation; Dose: 11600 ug/m^3/18H/2Y-I; Toxic Effects: Tumorigenic - Carcinogenic by RTECS criteria; Lungs, Thorax, or Respiration - Tumors.

Hazard Overviews

Explosive

Fire Diamond

Health: Irritating to eyes/skin/respiratory tract. Also Causes: headache, sneezing, coughing, chest pain, blackheads and keratosis. Chronic Effects: decreased lung function due to long-term dust deposition.

Fire: Combustible. May form explosive dust-air mixtures. Use dry chemical, carbon dioxide, water spray, or foam.

Reactivity: Stable. Hazardous polymerization cannot occur. Avoid: heat; dust generation. Incompatible with: chlorinated paraffins; lead (IV) oxide; manganese (IV) oxide; iron (II) oxide; liquid oxygen; strong oxidizers; chlorates; bromates;

nitrates; fatty oils; sodium sulfate; nitric acid. Hazardous decomposition products: carbon dioxide; carbon monoxide; toxic sulfur oxides.

Carcinogenicity: IARC - Group 3, Not classifiable as to carcinogenicity to humans; NIOSH - Listed as carcinogen; NTP - Not listed; ACGIH - Class A4, Not classifiable as a human carcinogen; OSHA - Not listed; EPA - Not listed; MAK - Not listed

Primary Target Organs:

Eyes

Skin

Respiratory System

Exposure Limits

OSHA PEL: TWA: 3.5 mg/m^3.

ACGIH TLV: TWA: 3.5 mg/m^3.

NIOSH REL: TWA: 3.5 mg/m^3.

NIOSH IDLH: 1750 mg/m^3.

Respirator Recommendation

Exposure Range: >3.5 to 35 mg/m^3 Air Purifying, Negative Pressure, Half Mask

Exposure Range: >35 to 350 mg/m^3 Air Purifying, Negative Pressure, Full Face

Exposure Range: >350 to <1750 mg/m^3 Supplied Air, Constant Flow/Pressure Demand, Full Face

Exposure Range: 1750 to unlimited mg/m^3 Self-contained Breathing Apparatus, Pressure Demand, Full Face

Cartridge Color: dust/mist filter (use P100 or consult supervisor for appropriate dust/mist filter)

Environmental

Cleanup/Disposal: Guide No. 133: Eliminate all ignition sources (no smoking, flares, sparks or flames in immediate area). Do not touch or walk through spilled material. Small Dry Spills: With clean shovel place material into clean, dry container and cover loosely; move containers from spill area. Large Spills: Wet down with water and dike for later disposal. Prevent entry into waterways, sewers, basements or confined areas.

Regulations

RCRA 40CFR: Not listed

CERCLA: 40CFR 302.4: Not listed

SARA 40CFR 372.65: Not listed

SARA EHS 40CFR 355: Not listed

TSCA: Listed

CAR4800 CAS #: 124-38-9

CARBON DIOXIDE

RTECS: FF6400000

DOT: UN1013; UN1845; UN2187; IMO2.0; IMO9.0

EINECS Number: 204-696-9

Molecular Formula: CO_2

Structured MF: CO_2

Formula Weight: 44.01

Chemical Structure

Synonyms: AER FIXUS; AFTER-DAMP; ANHYDRIDE CARBONIQUE; CARBON OXIDE; CARBON OXIDE,DI-; CARBONIC ACID ANHYDRIDE; CARBONIC ACID GAS; CARBONIC ANHYDRIDE; CARBONICA; DIOXIDO DE CARBONO; DIOXYDE DE CARBONE; DRY ICE; KHLADON 744; KOHLENDIOXYD; KOHLENSAURE; R 744

Description: colorless, clear, white gas, liquid, solid; odorless

Use: refrigerant; processing, preserving, crusting, cryogenic freezing of foods; prodn of urea, carbonates, methanol, hydrocarbons; provides inert atmosphere for fire extinguishers, refinery & petro products; oil well stimulation; fertilizer; hardening of molds for castings; carbonation of beverages; propellant in aerosols; for 'smoke' on stage; in arc welding; in atmosphere of greenhouses; for inflating life rafts; in mfr of aspirin, golf balls; med: resp stimulant, for hiccups, (vet) wart destruction; in deactivation of soaps; to render livestock insensible prior to slaughter; municipal water treatment; mining, cloud seeding, moderator in nuclear reactors, special lasers, blowing agent, carrier for powdered coal alurry.

Physical Properties

Boiling Point: Sublimes at -78.5 °C (-109 °F)

Freezing Point: -56.6 °C (-69.88 °F) at 5.2 atm

Specific Gravity: 1.527 Gas (Air= 1)

Vapor Density: 1.53 Air=1

Vapor Pressure: 56.5 mm Hg

Water Solubility: 0.145 g/100 ml Water at 25 °C

Other Solubilities: miscible with hydrocarbons and most organic liquids; Quantity dissolved, mL/g (STP) at 20 °C: Acetone: 8.2; Ethanol: 3.6; Benzene: 2.71; Methanol: 4.1; Toluene: 3.0; Xylene: 2.31; heptane: 2.8; mEthyl Acetate: 7.4; Diethyl Ether: 6.3.

Surface Tension: 0.0162 N/m at melting point

pH: Saturated carbon dioxide solutions vary from 3.2 to 3.7

Critical Temperature: 31.3 °C

Critical Pressure: 72.9 atm

Ionization Potential (eV): 13.77

Flash Point: Nonflammable

RTECS Toxicity Data

Acute Inhalation: Human LC_{Lo} Dose: 9 pph/5M. Mammal LC_{Lo} Dose: 90000 ppm/5M.

Chronic (Multiple Dose) Inhalation: Rat Dose: 10000 ppm/24H/30D-C; Toxic Effects: Blood - Other changes. Rabbit Dose: 27000 ppm/24H/30D-C; Toxic Effects: Behavioral - Somnolence (general depressed activity).

Reproductive/Teratogenic: Rat Route: Inhalation; Dose: 6 pph/24H; Duration: female 10D of pregnancy; Specific Developmental Abnormalities - Musculoskeletal system; Cardiovascular (circulatory) system; Respiratory system. Rat Route: Inhalation; Dose: 6 pph/24H; Duration: female 10D of pregnancy; Effects on Newborn - Growth statistics.

Hazard Overviews

Compressed Gas

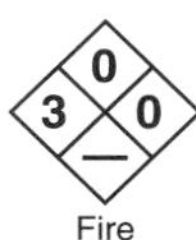

Fire Diamond

Health: Simple asphyxiant (reduces oxygen available for breathing). Eye/skin contact with the compressed gas can cause frostbite.
Fire: Noncombustible. Use extinguishing agents suitable for surrounding fire. Move cylinder from fire area as high temperatures may cause it to rupture.
Reactivity: Stable. Hazardous polymerization cannot occur. Avoid: heat; ignition sources. Incompatible with: magnesium; zirconium; titanium; magnesium-aluminum alloys; aluminum; chromium; manganese; aluminum and sodium oxide; cesium oxide; diethyl magnesium; lithium; magnesium and sodium peroxide; potassium; potassium acetylene carbide; sodium; sodium carbide; sodium potassium; titanium; acrylaldehyde; aziridine; metal acetylides; and sodium peroxide; plastics; rubber; coatings. Hazardous decomposition products: carbon monoxide.
Carcinogenicity: IARC - Not listed; NIOSH - Listed as carcinogen; NTP - Not listed; ACGIH - Not listed; OSHA - Not listed; EPA - Not listed; MAK - Not listed
Primary Target Organs:

Eyes

Skin

Nervous System

Exposure Limits
OSHA PEL: TWA: 5000 ppm; 9000 mg/m^3.
OSHA PEL Vacated 1989 Limits: TWA: 10000 ppm; 18000 mg/m^3; STEL: 30000 ppm; 54000 mg/m^3.
ACGIH TLV: TWA: 5000 ppm; 9000 mg/m^3; STEL: 3000 ppm; 54000 mg/m^3; Ceiling.
NIOSH IDLH: 40000 ppm.
DFG MAK: TWA: 5000 ppm; 9000 mg/m^3.
Respirator Recommendation
Exposure Range: >5000 to <40,000 ppm Supplied Air, Constant Flow/Pressure Demand, Full Face
Exposure Range: 40,000 to unlimited ppm Self-contained Breathing Apparatus, Pressure Demand, Full Face Self-contained Breathing Apparatus, Pressure Demand, Full Face
Note: poor warning properties

Environmental

Ecotoxicity: Rainbow trout 60-240 mg/l/12 hr, toxic effect: lethal Rainbow trout 35 mg/l/96 hr, toxic effect: lethal
Cleanup/Disposal: Guide No. 120: Do not touch or walk through spilled material. Stop leak if you can do it without risk. Use water spray to reduce vapors or divert vapor cloud drift. Do not direct water at spill or source of leak. If possible, turn leaking containers so that gas escapes rather than liquid. Prevent entry into waterways, sewers, basements or confined areas. Allow substance to evaporate. Ventilate the area. Caution: When in contact with refrigerated/cryogenic liquids, many materials become brittle and are likely to break without warning.

Environmental Physical Data

Henry's Law Constant: 1.51 x10^3
BCF: no food chain concentration potential
BOD: none

Regulations

RCRA 40CFR: Not listed
CERCLA: 40CFR 302.4: Not listed
SARA 40CFR 372.65: Not listed
SARA EHS 40CFR 355: Not listed
TSCA: Listed

Analytical Methods

Air: EPA 28A, 3A, 3B, 3C, 6A, 6B, PS-03; ASTM D4490
Soil: APHA 2720-C
Water / Groundwater: EPA 20; APHA 4500-CO2; ASTM D513
Food: AOAC 920.26, 920.28, 920.29, 925.03
Indoor / Expired Air: NIOSH 6603; EPA IP-3A, IP-3B

CAR4990 **CAS #: 75-15-0**

CARBON DISULFIDE

RTECS: FF6650000
DOT: UN1131; IMO6.1
EINECS Number: 200-843-6
Molecular Formula: CS_2
Structured MF: S=C=S
Formula Weight: 76.14

S═════S

Chemical Structure

Synonyms: CARBON BISULFIDE; CARBON BISULPHIDE; CARBON DISULPHIDE; CARBON SULFIDE; CARBON SULPHIDE; CARBONE (SUFURE DE); CARBONE (SULFURE DE); CARBONIO (SOLFURO DI); DITHIOCARBONIC ANHYDRIDE; EPA PESTICIDE CHEMICAL CODE 016401; KOHLENDISULFID (SCHWEFELKOHLENSTOFF); KOOLSTOFDISULFIDE; SCHWEFELKOHLENSTOFF; SOLFURO DI CARBONIO; SULPHOCARBONIC ANHYDRIDE; SULPHURET OF CARBON; WEEVILTOX; WEGLA DWUSIARCZEK
Description: colorless, slightly yellow liquid; chloroform-like odor when pure and a foul, rotten egg odor as the commercial product
Use: as a disinfectant, an insecticide, in the manufacture of rayon and carbon tetrachloride, in electronic vacuum tubes and as a solvent for sulfur, phosphorus, iodine, bromine, rubber intermediates, selenium, fats and resins; in the manufacture of xanthogenates, cellophane, optical glass, matches and paper; as a bactericide, a wood preservative, a nematocide, in the synthesis of dyes, pharmaceuticals and flotation agents, as a solvent in dry spinning PVC and oil wells, in the extraction and processing of oils, fats, resins and waxes and in veterinary medicine as a parasiticide

Physical Properties

Boiling Point: 46.5 °C (116 °F) at 760 mm Hg
Freezing Point: -111.5 °C (-168.7 °F)
Specific Gravity: 1.2632 at 20 °C/4 °C
Vapor Density: 2.67 Air=1
Saturated Vapor Density: 1.962282033 kg/m^3
Density: 1.2632 g/mL
Vapor Pressure: 297 torr at 20 °C
Water Solubility: 0.3% by weight
Other Solubilities: 95% Ethanol: >=100 mg/ml at 20 °C; Acetone: >=100 mg/ml at 20 °C; Alcohol: Soluble; Benzene: miscible; Carbon Tetrachloride: miscible; Chloroform: miscible; DMSO: >=100 mg/ml at 20 °C; Ether: miscible; Fixed and volatile oils: miscible.
Surface Tension: 0.032 dynes/cm
Odor Threshold: 0.0243 to 23.1 mg/m^3
Refraction Index: 1.6319 at 20 °C
Evaporation Rate: 22.6 Butyl Acetate=1
Critical Temperature: 273 °C
Critical Pressure: 1100 psia
Ionization Potential (eV): 10.08
Flash Point: 30 °C Closed Cup
Autoignition Temperature: 90 °C
LEL: 1.3% v/v
UEL: 50% v/v

RTECS Toxicity Data

Acute Oral: Rat LD; Dose: >505 mg/kg. Mouse LD_{50} Dose: 2780 mg/kg.
Acute Inhalation: Rat LC_{50} Dose: 25 gm/m^3/2hr. Human LC_{Lo} Dose: 4000 ppm/30M. Human LC_{Lo} Dose: 2000 ppm/5M.
Chronic (Multiple Dose) Oral: Rat Dose: 5055 mg/kg/4W-I; Toxic Effects: Cardiac - Arrythmias (including changes in conduction); Cardiac - Pulse rate; Nutritional and gross metabolic - Weight loss or decreased weight gain. Rabbit Dose: 25298 mg/kg/26W-I; Toxic Effects: Cardiac - Changes in heart weight; Endocrine - Hyperglycemia.
Chronic (Multiple Dose) Inhalation: Rat Dose: 2000 ppm/4H/2W-I; Toxic Effects: Behavioral - Change in psychophysiological tests; DEATH. Rat Dose: 114 mg/m^3/8H/20W-I; Toxic Effects: Kidney, Ureter, and Bladder - Chgs in tubules (inc acute renal failure, acute tubular necrosis; Blood - Changes in spleen; DEATH. Rat Dose: 800 ppm/6H/12W-I; Toxic Effects: Nutritional and gross metabolic - Weight loss or decreased weight gain; Biochemical - Transaminases; Biochemical - Other enzymes. Rat Dose: 1500 mg/m^3/5H/26W-I; Toxic Effects: Biochemical - Phosphatases; Biochemical - Other enzymes; Biochemical - Lipids including transport. Rat Dose: 800 mg/m^3/5H/1Y-I; Toxic Effects: Biochemical - Phosphatases; Biochemical - Other enzymes. Rat Dose: 100 mg/m^3/17W-I; Toxic Effects: Cardiac - Changes in coronary arteries; Vascular - Structural changes in vessels. Rat Dose: 500 ppm/24H/25W-C; Toxic Effects: Peripheral Nerve and sensation - Recording from afferent nerve; Peripheral Nerve and sensation - Recording from peripheral motor nerve; Nutritional and gross metabolic - Weight loss or decreased weight gain. Rat Dose: 800 ppm/6H/15W-I; Toxic Effects: Brain and coverings - Recordings from specific areas of CNS; Nutritional and gross metabolic - Weight loss or decreased weight gain. Monkey Dose: 1200 mg/m^3/6H/20W-I; Toxic Effects: Endocrine - Hyperglycemia; Blood - Changes in serum composition. Monkey Dose: 160 ppm/6H/36D; Toxic Effects: Sense organs and special senses - Other.
Reproductive/Teratogenic: Man Route: Inhalation; Dose: 40 mg/m^3; Duration: male 91W prior to mating; Paternal Effects - Spermatogenesis. Rat Route: Oral; Dose: 2 gm/kg; Duration: female 6-15D of pregnancy; Effects on Embryo or Fetus - Fetotoxicity. Rat Route: Inhalation; Dose: 200 mg/m^3/24 Duration: female 1-21D of pregnancy; Effects on Fertility - Pre-implantation mortality. Rat Route: Inhalation; Dose: 10 mg/m^3/8H; Duration: female 1-22D of pregnancy; Specific Developmental Abnormalities - Eye, ear; Effects on Newborn - Viability index. Rat Route: Inhalation; Dose: 100 mg/m^3/8H Duration: female 1-22D of pregnancy; Effects on Newborn - Growth statistics. Rat Route: Inhalation; Dose: 100 mg/m^3/8H Duration: female 1-21D of pregnancy; Effects on Embryo or Fetus - Fetal death; Specific Developmental Abnormalities - Craniofacial (including nose and tongue); Homeostasis. Rat Route: Inhalation; Dose: 30 ug/m^3/8H; Duration: female 1-22D of pregnancy; Effects on Newborn - Behavioral. Rat Route: Inhalation; Dose: 600 ppm/6H; Duration: male 50D prior to mating; Paternal Effects - Spermatogenesis; Prostate, seminal vessicle, Cowper's gland, accessory glands.
Mutagenic: Human Sister Chromatid Exchenge; Cell Type: lymphocyte; Dose: 8 mg/L. Human Sister Chromatid Exchenge; Cell Type: lymphocyte; Dose: 10200 ug/L.

Hazard Overviews

Corrosive

Flammable

Fire Diamond

Health: Irritating to skin/eyes/mucous membranes. Also causes: cardiac/CNS damage, headache, dizziness, convulsions. Chronic Effects: pyschosis, liver damage, eye effects, peripheral neuropathies, gastric disturbances, reproductive effects.
Fire: Flammable. Use foam (fluoroprotein, protein, high-expansion, aqueous film-forming), carbon dioxide, or water spray. Do not scatter material with more water than necessary. Container may explode in heat of fire.
Reactivity: Stable. Hazardous polymerization cannot occur. Avoid: exposure to ignition sources. Incompatible with: mercury fulminate; alkali metals; chlorine and other halogens; nitrogen oxide; metal azides; oxidants; aluminum; ethylene diamine; zinc; phenyl copper-triphenylphosphine complexes. Hazardous decomposition products: carbon monoxide; carbon dioxide; toxic sulfur oxides.
Carcinogenicity: IARC - Not listed; NIOSH - Not listed; NTP - Not listed; ACGIH - Not listed; OSHA - Not listed; EPA - Not listed; MAK - Not listed

Primary Target Organs:

Eyes

Skin

Mucous Membranes

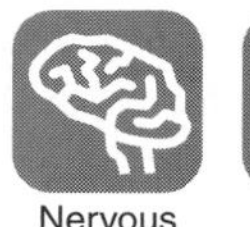
Nervous System

Liver

Cardio-vascular

Exposure Limits

OSHA PEL: TWA: 20 ppm; STEL: 30 ppm; from Table Z-2. Other Values: 100 mg/m^3; 30 min peak 8hr ppm.

OSHA PEL Vacated 1989 Limits: TWA: 4 ppm; 12 mg/m^3; STEL: 12 ppm; 36 mg/m^3.

ACGIH TLV: TWA: 10 ppm; 31 mg/m^3.

NIOSH REL: TWA: 1 ppm; 3 mg/m^3. STEL: 10 ppm; 30 mg/m^3; skin.

NIOSH IDLH: 500 ppm.

DFG MAK: TWA: 10 ppm; 30 mg/m^3.

Respirator Recommendation

Exposure Range: >20 to 200 ppm Air Purifying, Negative Pressure, Half Mask

Exposure Range: >200 to <500 ppm Air Purifying, Negative Pressure, Full Face

Exposure Range: 500 to unlimited ppm Self-contained Breathing Apparatus, Pressure Demand, Full Face

Cartridge Color: black

Environmental

Ecotoxicity: TLm Mosquitofish 162-135 mg/l/24-96 hr /Conditions of bioassay not specified

Environmental Fate: If released on land, it will be primarily lost by volatilization. It may also readily leach into the ground where it may biodegrade. If released into water, it will be primarily lost due to volatilization (half-life 2.6 hr in a model river). Adsorption to sediment and bioconcentration in fish should not be significant. In the atmosphere it degrades by reacting with atomic oxygen and photochemically produced hydroxyl radicals (half-life 6-9 days). The soil may be a natural sink for the chemical by adsorbing and subsequently biodegrading it.

Cleanup/Disposal: Guide No. 131: Fully encapsulating, vapor protective clothing should be worn for spills and leaks with no fire. Eliminate all ignition sources (no smoking, flares, sparks or flames in immediate area). All equipment used when handling the product must be grounded. Do not touch or walk through spilled material. Stop leak if you can do it without risk. Prevent entry into waterways, sewers, basements or confined areas. A vapor suppressing foam may be used to reduce vapors. Small Spills: Absorb with earth, sand or other non-combustible material and transfer to containers for later disposal. Use clean non-sparking tools to collect absorbed material. Large Spills: Dike far ahead of liquid spill for later disposal. Water spray may reduce vapor; but may not prevent ignition in closed spaces.

Environmental Physical Data

Henry's Law Constant: 1.44 x10^{-2}

Octanol/Water Partition Coefficient: log K_{ow} = 0.852

Sorption Partition Coefficient: K_{oc} = estimated at 63

BCF: estimated at 7.9

Regulations

RCRA 40CFR: Listed Hazardous Waste No. P022 Toxic Waste

CERCLA: 40CFR 302.4: Listed per CWA Section 311(b)(4) per RCRA Section 3001 RQ: 100 lb (45.35 kg)

SARA 40CFR 372.65: Listed TPQ: 10000 lb

SARA EHS 40CFR 355: Listed TPQ: 100 lb

TSCA: Listed

Analytical Methods

Air: EPA 15, 0031; ASTM D3686, D3687, D4490

Soil: CLP LC_VOA, MC_VOA, OHC; EPA 1624, VG-008-1; SW846 1311, 5032, 5041, 5041A, 8240B, 8260A, 8260B

Drinking Water: EPA 524.2

Indoor / Expired Air: NIOSH 1600

Plasma: EPA 29

CAR5180 CAS #: 630-08-0

CARBON MONOXIDE

RTECS: FG3500000

DOT: UN1016; IMO2.3

EINECS Number: 211-128-3

Molecular Formula: CO

Structured MF: CO

Formula Weight: 28.01

$^{+}C\equiv O^{-}$

Chemical Structure

Synonyms: CARBON OXIDE; CARBONE (OXYDE DE); CARBONIC OXIDE; CARBONIO (OSSIDO DI); EXHAUST GAS; FLUE GAS; KOHLENMONOXID; KOHLENOXYD; KOOLMONOXYDE; OXYDE DE CARBONE; WEGLA TLENEK

Description: colorless gas; odorless

Use: in metallurgy (reducing agent, in iron ore processing; purification agent for nickel; intermed for metal carbonyls-eg, tungsten carbonyl); fischer-tropsch processes for petro-type products; mfr of metal carbonyls, white pigments; intermed for phosgene, methanol, acetic acid, acrylic acid, etc; in ethylene-carbon monoxide copolymer; in prodn of syngas for synthesis of ammonia, commodity chems & fuels; prodn of acetic acid, formic acid, saturated hydrocarbons & oxygenated compounds, etc.

Physical Properties

Boiling Point: -191.5 °C (-313 °F)

Freezing Point: -205 °C (-337 °F)

Vapor Density: 0.968 Air=1

Density: 1.25 g/L at 0 °C/4 °C

Vapor Pressure: > 1 atm at 20 °C

Water Solubility: 2% by weight

Other Solubilities: Appreciably Soluble in Ethyl Acetate, Chloroform, Acetic Acid; freely absorbed by a concentrated soln of cuprous chloride in Hydrochloric Acid or ammonium hydroxide; solubility in Methanol and Ethanol about 7 times as great as in water

Surface Tension: 9.8 dynes/cm at 93 °C
Refraction Index: Gas 1.0003364 at 0 °C
Critical Temperature: -139 °C
Critical Pressure: 35 atm
Ionization Potential (eV): 14.01
Flash Point: Flammable gas
Autoignition Temperature: 609 °C
LEL: 12.5% v/v
UEL: 74% v/v

RTECS Toxicity Data

Acute Inhalation: Rat LC_{50} Dose: 1807 ppm/4hr. Human LC_{Lo} Dose: 5000 ppm/5M. Man LC_{Lo} Dose: 4000 ppm/30M. Human TC_{Lo} Dose: 600 mg/m^3/10M; Toxic Effects: Behavioral - Headache. Man TC_{Lo} Dose: 650 ppm/45M; Toxic Effects: Blood - Methemoglobinemia-Carboxyhemoglobin; Behavioral - Change in psychophysiological tests.

Chronic (Multiple Dose) Inhalation: Rat Dose: 1800 ppm/1H/14D-I; Toxic Effects: Cardiac - Other changes. Rat Dose: 30 mg/m^3/8H/10W-I; Toxic Effects: Brain and coverings - Other degenerative changes; Behavioral - Muscle contraction or spasticity. Rat Dose: 96 ppm/24H/90D-C; Toxic Effects: Blood - Pigmented or nucleated red Blood cells; Blood - Other changes. Rat Dose: 250 ppm/5H/20D-I; Toxic Effects: Blood - Pigmented or nucleated red Blood cells; Blood - Changes in other cell count; Blood - Changes in erythrocite (RBC) cell count. Rat Dose: 200 mg/m^3/3H/13W-I; Toxic Effects: Brain and coverings - Other degenerative changes; Cardiac - Other changes; Blood - Hemorrhage. Rat Dose: 50 ppm/24H/8W-C; Toxic Effects: Blood - Changes in platelet cell count. Rat Dose: 200 mg/m^3/5H/4W-I; Toxic Effects: Endocrine - Hyperglycemia. Rat Dose: 200 mg/m^3/5H/30W-I; Toxic Effects: Cardiac - Arrythmias (including changes in conduction); Cardiac - EKG changes not diagnostic of above; Cardiac - Pulse rate increased without fall in BP. Rat Dose: 200 ppm/24H/90D-C; Toxic Effects: Blood - Pigmented or nucleated red Blood cells; Blood - Other changes.

Chronic (Multiple Dose) Dermal: Rat Route: Subcutaneous; Dose: 5983 mg/kg/18W-I; Toxic Effects: Blood - Changes in serum composition.

Reproductive/Teratogenic: Rat Route: Inhalation; Dose: 150 ppm/24H; Duration: female 1-22D of pregnancy; Specific Developmental Abnormalities - Cardiovascular (circulatory) system. Rat Route: Inhalation; Dose: 150 ppm/24H; Duration: female 1-22D of pregnancy; Effects on Newborn - Growth statistics; Behavioral. Rat Route: Inhalation; Dose: 1 mg/m^3/24H; Duration: female 72D prior to mating Maternal Effects - Menstrual cycle changes or disorders; Parturition; Effects on Fertility - Female fertility index. Rat Route: Inhalation; Dose: 1 mg/m^3/24H; Duration: female 72D prior to mating Effects on Newborn - Growth statistics; Delayed effects. Rat Route: Inhalation; Dose: 75 ppm/24H; Duration: female 0-20D of pregnancy; Specific Developmental Abnormalities - Immune and reticuloendothelial system.

Mutagenic: Mouse Sister Chromatid Exchenge; Route: Inhalation; Dose: 2500 ppm/10M. Mouse Micronucleus Test; Route: Inhalation; Dose: 1500 ppm/10M.

Hazard Overviews

Poison

Flammable

Compressed Gas

Fire Diamond

Health: Poison. Also Causes: chemical asphyxia, headache, mental confusion, vomiting, giddiness, exhaustion, collapse, unconsciousness, coma, convulsions, death; cerebral edema; angina pectoris; myocardial infarction.

Fire: Very flammable. Let small fire burn unless leak can be stopped immediately. For large fire, use water spray, fog, or regular foam. Fight fire from maximum distance.

Reactivity: Stable. Hazardous polymerization cannot occur. Avoid: oxidizers; halogen compounds; heat; ignition sources. Incompatible with: strong oxidizers; bromine trifluoride; chlorine trifluoride; lithium; iodine heptafluoride; nitrogen trifluoride; silver oxide; cesium oxide; sodium and ammonia; copper perchlorate; liquid dinitrogen oxide. Hazardous decomposition products: carbon; carbon dioxide.

Carcinogenicity: IARC - Not listed; NIOSH - Listed as carcinogen; NTP - Not listed; ACGIH - Not listed; OSHA - Not listed; EPA - Not listed; MAK - Not listed

Primary Target Organs:

Eyes

Skin

Nervous System

Cardio-vascular

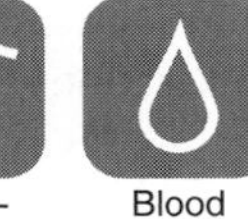
Blood

Exposure Limits
OSHA PEL: TWA: 50 ppm; 55 mg/m^3.
OSHA PEL Vacated 1989 Limits: TWA: 35 ppm; 40 mg/m^3; STEL: 200 ppm; 229 mg/m^3.
ACGIH TLV: TWA: 25 ppm; 29 mg/m^3.
NIOSH REL: TWA: 35 ppm; 40 mg/m^3. STEL: 200 ppm; 229 mg/m^3.
NIOSH IDLH: 1200 ppm.
DFG MAK: TWA: 30 ppm; 33 mg/m^3.
Respirator Recommendation
Exposure Range: >50 to <1200 ppm Supplied Air, Constant Flow/Pressure Demand, Full Face
Exposure Range: 1200 to unlimited ppm Self-contained Breathing Apparatus, Pressure Demand, Full Face
Note: poor odor threshold

Environmental

Ecotoxicity: Aquatic toxicity: 1.5 ppm/1-6 hr/minnows and sunfish/killed/fresh water

Cleanup/Disposal: Guide No. 119: Eliminate all ignition sources (no smoking, flares, sparks or flames in immediate area). All equipment used when handling the product must be grounded. Fully encapsulating, vapor protective clothing should be worn for spills and leaks with no fire. Do not touch or walk through spilled material. Stop leak if you can do it without risk. Do not direct water at spill or source of leak.

Use water spray to reduce vapors or divert vapor cloud drift. For chlorosilanes, use AFFF alcohol-resistant medium expansion foam to reduce vapors. If possible, turn leaking containers so that gas escapes rather than liquid. Prevent entry into waterways, sewers, basements or confined areas. Isolate area until gas has dispersed.

Environmental Physical Data

BCF: no food chain concentration potential
BOD: none

Regulations

RCRA 40CFR: Not listed
CERCLA: 40CFR 302.4: Not listed
SARA 40CFR 372.65: Not listed
SARA EHS 40CFR 355: Not listed
TSCA: Listed

Analytical Methods

Air: EPA 2.6, 10, 10A, 10B, 28A, 3B, 50APP-C, PS-04, PS-04A; ASTM D3162, D3416, D4490; Canada 1RM-15, 1RM-4
Indoor / Expired Air: EPA IP-3A, IP-3B, IP-3C

CAR5370 CAS #: 558-13-4

CARBON TETRABROMIDE

RTECS: FG4725000
DOT: UN2516; IMO6.1
EINECS Number: 209-189-6
Molecular Formula: CBr_4
Structured MF: CBr_4
Formula Weight: 331.63

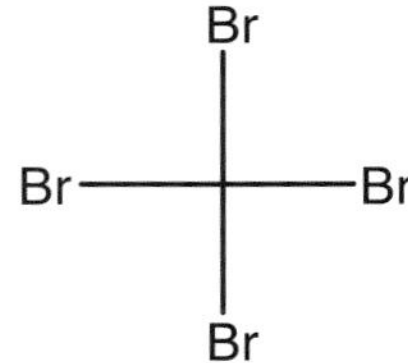

Chemical Structure

Synonyms: BROMID UHLICITY; CARBON BROMIDE; METHANE TETRABROMIDE; METHANE,TETRABROMIDE; METHANE,TETRABROMO-; TETRABROMOMETHANE
Description: colorless when pure, yellow, brown crystals
Use: organic synthesis; chemical intermediate

Physical Properties

Boiling Point: 189.9 °C (374 °F)
Freezing Point: 90 °C (194 °F) to 94 °C (201.2 °F)
Specific Gravity: 2.9609 at 100 °C/4 °C
Vapor Density: 11.4
Vapor Pressure: 40 mm Hg at 96.3 °C
Water Solubility: 0.02% by weight
Other Solubilities: Very Soluble in Carbon Disulfide; Slightly Soluble in liquid Hydrofluoric acid gas.
Refraction Index: Alpha 1.59419 at 100 °C/D (ALPHA)
Ionization Potential (eV): 10.31
Flash Point: Not flammable by standard tests in air

RTECS Toxicity Data

Acute Dermal: Mouse LD_{50} Route: Subcutaneous Dose: 298 mg/kg.

Hazard Overviews

Fire Diamond

Health: Severely irritating to eyes, irritating to the respiratory tract, mildly irritating to skin. Also Causes: narcosis, injury to lungs, lacrimation, hyperemia, edema, mouth irritation, stomach. Chronic Effects: liver/kidney injury.
Fire: Noncombustible. Use agent suitable for surrounding fire.
Reactivity: Stable. Hazardous polymerization cannot occur. Avoid: heat; ignition sources. Incompatible with: carbon tetrabromide; strong oxidizers; hexacyclohexyldilead. Hazardous decomposition products: toxic fumes of bromide.
Carcinogenicity: IARC - Not listed; NIOSH - Listed as carcinogen; NTP - Not listed; ACGIH - Not listed; OSHA - Not listed; EPA - Not listed; MAK - Not listed
Primary Target Organs:

Eyes

Skin

Respiratory System

Liver

Kidneys

Exposure Limits
OSHA PEL Vacated 1989 Limits: TWA: 0.1 ppm; 1.4 mg/m^3; STEL: 0.3 ppm; 4 mg/m^3.
ACGIH TLV: TWA: 0.1 ppm; 1.4 mg/m^3; STEL: 0.3 ppm; 4.1 mg/m^3.
NIOSH REL: TWA: 0.1 ppm; 1.4 mg/m^3. STEL: 0.3 ppm; 4 mg/m^3.
Respirator Recommendation
Exposure Range: >0.1 to 5 ppm Supplied Air, Constant Flow/Pressure Demand, Half Mask
Exposure Range: >5 to 100 ppm Supplied Air, Constant Flow/Pressure Demand, Full Face
Exposure Range: >100 to unlimited ppm Self-contained Breathing Apparatus, Pressure Demand, Full Face
Note: odor threshold unknown

Environmental

Cleanup/Disposal: Guide No. 151: Do not touch damaged containers or spilled material unless wearing appropriate protective clothing. Stop leak if you can do it without risk. Prevent entry into waterways, sewers, basements or confined areas. Cover with plastic sheet to prevent spreading. Absorb or cover with dry earth, sand or other non-combustible material and transfer to containers. Do not get water inside containers.

Regulations

RCRA 40CFR: Not listed

CERCLA: 40CFR 302.4: Not listed
SARA 40CFR 372.65: Not listed
SARA EHS 40CFR 355: Not listed
TSCA: Listed

CAR5560 CAS #: 56-23-5

CARBON TETRACHLORIDE

RTECS: FG4900000
DOT: UN1846; IMO6.1
EINECS Number: 200-262-8
Molecular Formula: CCl_4
Structured MF: CCl_4
Formula Weight: 153.24

Cl
Cl Cl
Cl

Chemical Structure

Synonyms: BENZINOFORM; CARBON CHLORIDE; CARBON TET; CARBONA; CHLORID UHLICITY; CZTEROCHLOREK WEGLA; ENT 27164; ENT 4,705; FASCIOLIN; FLUKOIDS; FREON 10; HALON 104; HALON 1040; METHANE TETRACHLORIDE; METHANE,TETRACHLORO-; NECATORINA; NECATORINE; PERCHLOROMETHANE; R 10; R 10 (REFRIGERANT); TETRACHLOORKOOLSTOF; TETRACHLOORMETAAN; TETRACHLORKOHLENSTOFF,TETRA; TETRACHLORMETHAN; TETRACHLOROCARBON; TETRACHLOROMETHANE; TETRACHLORURE DE CARBONE; TETRACLOROMETANO; TETRACLORURO DI CARBONIO; TETRAFINOL; TETRAFORM; TETRASOL; UNIVERM; VERMOESTRICID
Description: colorless liquid; ethereal odor
Use: in fire extinguishers, in refrigerants, in mixture with potent fumigants to reduce the fire hazard, to render benzene nonflammable and separate xylene isomers as components to reduce flammability; as a metal degreaser, as an agricultural fumigant and as a solvent for lacquers, varnishes, waxes, resins, fats, oils, rubber, organic compounds and rubber cement; as a dry cleaning solvent, in cable manufacture, in the production of semiconductors, in blowing agents, in fluorocarbon propellants and chlorofluoromethanes; in veterinary medicine as an anthelmintic and to treat liver fluke infections in sheep; an azeotropic drying agent for wet spark plugs in automobiles, to extract oil from flowers and seeds, as an extractant and intermediate in many industrial processes, in polymer technology as a reaction medium, catalyst and chain transfer agent, as a solvent for resins and in organic synthesis for chlorination of organic compounds used in soap perfumery and insecticidal industries

Physical Properties

Boiling Point: 76.54 °C (170 °F)
Freezing Point: -23 °C (-9.4 °F)
Specific Gravity: 1.594 at 20 °C/4 °C
Vapor Density: 5.32 Air=1
Saturated Vapor Density: 1.977909256 kg/m^3
Density: 1.597 g/mL at 20 °C
Vapor Pressure: 115 mm Hg at 25 °C
Water Solubility: 0.05% by weight
Other Solubilities: Soluble in Acetone; Soluble in naphtha.
Surface Tension: 27.0 dynes/cm at 20 °C
Odor Threshold: 21.4 ppm
Refraction Index: 1.4607 at 20 °C/D
Evaporation Rate: 12.8 Butyl Acetate=1
Critical Temperature: 283.1 °C
Critical Pressure: 45 atm
Ionization Potential (eV): 11.47
Flash Point: Nonflammable

RTECS Toxicity Data

Acute Oral: Man TD_{Lo} Dose: 1700 mg/kg; Toxic Effects: Behavioral - Tremor; Lungs, Thorax, or Respiration - Other changes; Gastrointestinal - Other changes. Woman TD_{Lo} Dose: 1800 mg/kg; Toxic Effects: Sense organs and special senses - Miosis (pupilliary dilation); Behavioral - Coma; Behavioral - Antipsychotic. Man LD_{Lo} Dose: 429 mg/kg; Toxic Effects: Cardiac - Change in rate; Lungs, Thorax, or Respiration - Cyanosis; Kidney, Ureter, and Bladder - Interstitial nephritis. Rat LD_{50} Dose: 2350 mg/kg.
Acute Inhalation: Rat LC_{50} Dose: 8000 ppm/4hr. Human LC_{Lo} Dose: 1000 ppm. Human LC_{Lo} Dose: 5 pph/5M. Human TC_{Lo} Dose: 20 ppm; Toxic Effects: Gastrointestinal - Nausea or vomiting. Human TC_{Lo} Dose: 45 ppm/3D; Toxic Effects: Behavioral - Somnolence (general depressed activity); Behavioral - Anorexia (human); Gastrointestinal - Nausea or vomiting. Human TC_{Lo} Dose: 317 ppm/30M; Toxic Effects: Gastrointestinal - Nausea or vomiting.
Acute Dermal: Rabbit LD_{50} Route: Skin; Dose: >20 gm/kg. Rabbit LD_{Lo} Route: Subcutaneous Dose: 3 gm/kg.
Chronic (Multiple Dose) Oral: Rat Dose: 1200 mg/kg/12W-I; Toxic Effects: Liver - Changes in liver weight; Biochemical - Hepatic microsomal mixed oxidase(dealkylation, hyroxylation,etc); Biochemical - Transaminases. Rat Dose: 4197 ug/kg/28D-I; Toxic Effects: Liver - Other changes. Rat Dose: 400 mg/kg/10D-I; Toxic Effects: Liver - Changes in liver weight; Nutritional and gross metabolic - Weight loss or decreased weight gain; Biochemical - Transaminases.
Chronic (Multiple Dose) Inhalation: Rat Dose: 41 mg/m^3/4H/8D-I; Toxic Effects: Endocrine - Evidence of thyroid hypofunction. Rat Dose: 61 mg/m^3/90D-C; Toxic Effects: Liver - Fatty Liver degeneration; Liver - Other changes. Rat Dose: 400 ppm/1H/46D-I; Toxic Effects: Liver - Changes in Liver weight. Rat Dose: 200 ppm/7H/27W-I; Toxic Effects: Liver - Hepatitis (hepatocellular necrosis), zonal; Liver - Fatty Liver degeneration; DEATH. Rat Dose: 50 ppm/3H/8W-I; Toxic Effects: Liver - Fatty Liver degeneration. Rat Dose: 400 ppm/8H/46W-I; Toxic Effects: Peripheral Nerve and sensation - Structural change in nerve or sheath; DEATH. Monkey Dose: 515 mg/m^3/8H/6W-I; Toxic Effects: Lungs, Thorax, or Respiration - Fibrosis, interstitial; Nutritional and gross metabolic - Weight loss or decreased weight gain; DEATH. Monkey Dose: 61 mg/m^3/90D-C;

Toxic Effects: Liver - Other changes; Skin and appendages - Hair. Monkey Dose: 200 ppm/8H/46W-I; Toxic Effects: Peripheral Nerve and sensation - Structural change in nerve or sheath.

Chronic (Multiple Dose) Dermal: Rat Route: Subcutaneous; Dose: 31200 uL/kg/12W-I; Toxic Effects: Liver - Hepatitis, fibrous (cirrhosis, post-necrotic scarring).

Irritation Eye: Rabbit Standard Draize Test Dose: 2200 ug/30S; Reaction: mild. Rabbit Standard Draize Test Dose: 500 mg/24H; Reaction: mild.

Irritation Skin: Rabbit Standard Draize Test Dose: 4 mg; Reaction: mild. Rabbit Standard Draize Test Dose: 500 mg/24H; Reaction: mild.

Reproductive/Teratogenic: Rat Route: Oral; Dose: 2 gm/kg; Duration: female 7-8D of pregnancy; Effects on Fertility - Post-implantation mortality. Rat Route: Oral; Dose: 3 gm/kg; Duration: female 14D of pregnancy; Effects on Embryo or Fetus - Extra embryonic structures. Rat Route: Inhalation; Dose: 300 ppm/7H; Duration: female 6-15D of pregnancy; Effects on Embryo or Fetus - Fetotoxicity; Specific Developmental Abnormalities - Musculoskeletal system; Homeostasis. Rat Route: Inhalation; Dose: 250 ppm/8H; Duration: female 10-15D of pregnancy Effects on Newborn - Viability index; Weaning or lactation index.

Mutagenic: Rat DNA Damage; Route: Subcutaneous; Dose: 31 gm/kg/12W-I. Rat DNA Damage; Cell Type: liver; Dose: 3 mmol/L.

Tumorigenic: Rat Route: Subcutaneous; Dose: 15600 mg/kg/12W-I; Toxic Effects: Tumorigenic - Equivocal tumorigenic agent by RTECS criteria; Liver - Tumors. Rat Route: Subcutaneous; Dose: 100 gm/kg/25W-I; Toxic Effects: Tumorigenic - Equivocal tumorigenic agent by RTECS criteria; Liver - Tumors. Rat Route: Subcutaneous; Dose: 31 gm/kg/12W-I; Toxic Effects: Tumorigenic - Equivocal tumorigenic agent by RTECS criteria; Liver - Tumors. Rat Route: Subcutaneous; Dose: 182 gm/kg/70W-I; Toxic Effects: Tumorigenic - Carcinogenic by RTECS criteria; Liver - Tumors; Endocrine - Thyroid tumors.

Hazard Overviews

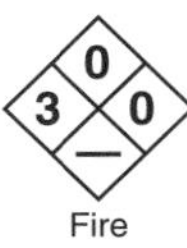

Fire Diamond

Health: Irritating to eyes/skin/respiratory tract. Toxic. Also Causes: incoordination, confusion, liver/kidney damage. Chronic Effects: visual disturbances, aplastic anemia, ulcers, blindness, hearing loss, cancer/reproductive hazard.

Fire: Noncombustible. Use extinguishing agents suitable for surrounding fire.

Reactivity: Stable. Hazardous polymerization cannot occur. Avoid: using as a fire extinguisher on wax fires, uranium fires, or electrical fires. Incompatible with: fluorine gas; alkali metals; aluminum. Hazardous decomposition products: phosgene; hydrogen chloride.

Carcinogenicity: IARC - Group 2B, Possibly carcinogenic to humans; NIOSH - Listed as carcinogen; NTP - Class 2B, Reasonably anticipated to be a carcinogen, sufficient evidence of carcinogenicity from studies in experimental animals; ACGIH - Class A2, Suspected human carcinogen; OSHA - Not listed; EPA - Class B2, Probable human carcinogen based on animal studies; MAK - Class B, Justifiably suspected of having carcinogenic potential

Primary Target Organs:

Eyes

Skin

Respiratory System

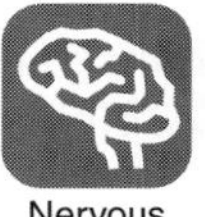
Nervous System

Liver

Kidneys

Exposure Limits

OSHA PEL: TWA: 10 ppm; STEL: 25 ppm; from Table Z-2. Other Values: 200 mg/m^3; 15 min peak 4hr ppm.

OSHA PEL Vacated 1989 Limits: TWA: 2 ppm; 12.6 mg/m^3.

ACGIH TLV: TWA: 5 ppm; 31 mg/m^3; STEL: 10 ppm; 63 mg/m^3.

NIOSH REL: STEL: 2 ppm; 12.6 mg/m^3; 60-minute.

NIOSH IDLH: 200 ppm.

DFG MAK: TWA: 10 ppm; 65 mg/m^3.

Respirator Recommendation

Exposure Range: >10 to 50 ppm Supplied Air, Constant Flow/Pressure Demand, Half Mask

Exposure Range: >50 to <200 ppm Supplied Air, Constant Flow/Pressure Demand, Full Face

Exposure Range: 200 to unlimited ppm Self-contained Breathing Apparatus, Pressure Demand, Full Face

Note: poor warning properties

Environmental

Ecotoxicity: LC_{50} Menidia beryllina 150 ppm/96 hr at 23 °C, static bioassay in fresh water, mild aeration applied after 24 hr Toxicity Threshold (Cell Multiplication Inhibition Test) Scenedesmus quadricauda (green algae) >600 mg/l LC_{50} Poecilia reticulata (Guppy) 67 ppm/14 days /Conditions of bioassay not specified EC_{50} Pimephales promelas (fathead minnow) 20.8 mg/l/96 hr (confidence limit 18.3 - 23.7 mg/l), flow-through bioassay with measured concentrations, 21.7 °C, dissolved oxygen 7.1 mg/l, hardness 49.2 mg/l calcium carbonate Toxicity Threshold (Cell Multiplication Inhibition Test) Uronema parduczi Chatton-Lwoff (protozoa) 616 mg/l

Environmental Fate: In the troposphere, it is extremely stable (residence time of 30-50 years). The primary loss process is by escape to the stratosphere where it photolyzes. As a result of its emission into the atmosphere and slow degradation, the amount in the atmosphere has been increasing. Some released to the atmosphere is expected to partition into the ocean. In water systems, evaporation appears to be the most important removal process, although biodegradation may occur under aerobic and anaerobic conditions (limited data). Releases or spills on soil should result in rapid evaporation due to high vapor pressure and leaching in soil resulting in groundwater contamination due to its low adsorption to soil. Bioconcentration is not significant

Cleanup/Disposal: Guide No. 151: Do not touch damaged containers or spilled material unless wearing appropriate protective clothing. Stop leak if you can do it without risk.

Prevent entry into waterways, sewers, basements or confined areas. Cover with plastic sheet to prevent spreading. Absorb or cover with dry earth, sand or other non-combustible material and transfer to containers. Do not get water inside containers.

Environmental Physical Data

Henry's Law Constant: 3.04 x10^{-2}
Octanol/Water Partition Coefficient: log K_{ow} = 2.83
Sorption Partition Coefficient: K_{oc} = estimated at 71
BCF: low potential
BOD: none

Regulations

RCRA 40CFR: Listed Hazardous Waste No. U211 Toxic Waste
CERCLA: 40CFR 302.4: Listed per CWA Section 311(b)(4) per RCRA Section 3001 per CWA Section 307(a) RQ: 10 lb (4.535 kg)
SARA 40CFR 372.65: Listed
SARA EHS 40CFR 355: Not listed
TSCA: Listed

Analytical Methods

Air: EPA 0031, 0040, OA-002-1, VA-001-1, VA-004-1, VA-005-1, VG-006-1, VG-007-1, VG-011-1, TO-1, TO-14, TO-2, CTM-011; ASTM D3686, D3687, D4490
Soil: CLP LC_VOA, MC_VOA, OHC; EPA 7, 1624, VG-001-1, VG-003-1, VG-004-1, VG-008-1, VS-001-1, VS-002-1, VS-003-1; SW846 1311, 5021, 5032, 5041, 5041A, 8010B, 8021A, 8240B, 8260A, 8260B; DOE OP040R, OS040
Water / Groundwater: EPA 601, 624, 624-S, VW-001-1, VW-002-1, VW-003-1, VW-005-1, VW-008-1, VW-014-1; APHA 6210-B, 6210-D, 6230-B, 6230-D; USGS O3115
Drinking Water: EPA 502.1, 502.2, 524.1, 524.2, 551; APHA 6210-C, 6230-C
Food: EPA 5; AOAC 977.18
Indoor / Expired Air: NIOSH 1003; EPA IP-1A, IP-1A-B, IP-1A-C, IP-1B
Plasma: EPA 29
Other: EPA VS-006-1

CAR5750 CAS #: 353-50-4

CARBONYL FLUORIDE

RTECS: FG6125000
DOT: UN2417; IMO2.0
EINECS Number: 206-534-2
Molecular Formula: CF_2O
Structured MF: COF_2
Formula Weight: 66.01
Synonyms: CARBON DIFLUORIDE OXIDE; CARBON FLUORIDE OXIDE; CARBON OXYFLUORIDE; CARBONIC DIFLUORIDE; CARBONIC DIFLUORIDE (9CI); CARBONYL DIFLUORIDE; DIFLUOROFORMALDEHYDE; DIFLUOROOXOMETHANE; FLUOPHOSGENE; FLUOROFORMYL FLUORIDE; FLUOROPHOSGENE
Description: colorless gas; pungent odor
Use: chem int in org synth, eg, fluorinated alkyl isocyanates; suggested as a military poison gas

Physical Properties

Boiling Point: -83 °C (-117 °F) at 760 mm Hg
Freezing Point: -114 °C (-173.2 °F)
Specific Gravity: 1.139 at -114 °C (liquid)
Vapor Density: 2.28 Air=1
Vapor Pressure: 55.4 mm Hg
Water Solubility: Reacts
Ionization Potential (eV): 13.02
Flash Point: Nonflammable

RTECS Toxicity Data

Acute Inhalation: Rat LC_{50} Dose: 360 ppm/1hr; Toxic Effects: Lungs, Thorax, or Respiration - Consolidation; Lungs, Thorax, or Respiration - Fibrosis, interstitial; Liver - Fatty liver degeneration.

Hazard Overviews

Compressed Gas

Fire Diamond

Health: Severely irritating to eyes/skin/respiratory tract. Compressed gas, may cause frostbite. Chronic Effects: changes in certain fluoride-sensitive enzymes because of poison action of released fluoride anions.
Fire: Noncombustible. Use agent suitable for surrounding fire. Do not use water. Reacts violently with water to produce highly toxic and corrosive hydrogen fluoride and carbon dioxide.
Reactivity: Stable. Hazardous polymerization cannot occur. Avoid: contact with water Incompatible with: moisture (solid ice; steam; moist air; water). Hazardous decomposition products: carbon monoxide; oxides of chlorine; oxides of fluorine.
Carcinogenicity: IARC - Not listed; NIOSH - Not listed; NTP - Not listed; ACGIH - Class A4, Not classifiable as a human carcinogen; OSHA - Not listed; EPA - Not listed; MAK - Not listed
Primary Target Organs:

Eyes

Skin

Respiratory System

Mucous Membranes

Exposure Limits
OSHA PEL: TWA: 2.5 mg/m^3; as F (Flourides).
OSHA PEL Vacated 1989 Limits: TWA: 2 ppm; 5 mg/m^3; STEL: 5 ppm; 15 mg/m^3.
ACGIH TLV: TWA: 2 ppm; 5.4 mg/m^3; STEL: 5 ppm; 13 mg/m^3.
NIOSH REL: TWA: 2 ppm; 5 mg/m^3; as F (Flourides). STEL: 5 ppm; 15 mg/m^3.
DFG MAK: TWA: 2.5 mg/m^3; as F.

Respirator Recommendation
Exposure Range: >2 to 100 ppm Supplied Air, Constant Flow/Pressure Demand, Half Mask
Exposure Range: >100 to 2000 ppm Supplied Air, Constant Flow/Pressure Demand, Full Face
Exposure Range: >2000 to unlimited ppm Self-contained Breathing Apparatus, Pressure Demand, Full Face
Note: odor threshold unknown

Environmental

Cleanup/Disposal: Guide No. 125: Fully encapsulating, vapor protective clothing should be worn for spills and leaks with no fire. Do not touch or walk through spilled material. Stop leak if you can do it without risk. If possible, turn leaking containers so that gas escapes rather than liquid. Prevent entry into waterways, sewers, basements or confined areas. Do not direct water at spill or source of leak. Use water spray to reduce vapors or divert vapor cloud drift. Isolate area until gas has dispersed.

Environmental Physical Data

BCF: no food chain concentration potential
BOD: none

Regulations

RCRA 40CFR: Listed Hazardous Waste No. U033 Toxic Waste Reactive Waste
CERCLA: 40CFR 302.4: Listed per RCRA Section 3001 RQ: 1000 lb (453.5 kg)
SARA 40CFR 372.65: Not listed
SARA EHS 40CFR 355: Not listed
TSCA: Listed

CAR5940 CAS #: 463-58-1

CARBONYL SULFIDE

RTECS: FG6400000
DOT: UN2204; IMO2.3
EINECS Number: 207-340-0
Molecular Formula: COS
Formula Weight: 60.08

S=O

Chemical Structure

Synonyms: CARBON MONOXIDE MONOSULFIDE; CARBON OXIDE SULFIDE; CARBON OXIDE SULFIDE (9CI); CARBON OXYSULFIDE; OXYCARBON SULFIDE; SCO
Description: colorless gas; typical sulfide (rotten eggs) odor
Use: synth of thio organic compounds; chem int, eg, for alkyl carbonates & org sulfur cmpds

Physical Properties

Boiling Point: -50 °C (-58 °F) at 760 mm Hg
Freezing Point: -138 °C (-216.4 °F)
Specific Gravity: 1.028 at 17 °C/4 °C
Vapor Density: 2.1 Air=1
Vapor Pressure: 0.4 kPa at 150 °K
Water Solubility: Soluble in Water
Other Solubilities: Soluble in Toluene; Very Soluble in Carbon Disulfide.
Refraction Index: 1.24
Ionization Potential (eV): 11.185 +/-0.002
LEL: 12% v/v
UEL: 29% v/v

RTECS Toxicity Data

Acute Inhalation: Rat LC_{50} Dose: 1110 ppm/4hr; Toxic Effects: Behavioral - Convulsions or effect on seizure threshold; Behavioral - Ataxia; Lungs, Thorax, or Respiration - Dyspnea. Mouse LC_{50} Dose: 2770 mg/m^3; Toxic Effects: Behavioral - General anesthetic; Behavioral - Convulsions or effect on seizure threshold; Behavioral - Ataxia.

Hazard Overviews

Poison

Flammable

Compressed Gas

Fire Diamond

Health: Irritating to eyes/skin/respiratory tract. Poison. Also Causes: giddiness, headache, rhinitis, pharyngitis, bronchitis, dizziness, amnesia, confusion, unconsciousness, pneumonitis, pulmonary edema, sudden collapse, death, painful conjunctivitis, photophobia, lacrimation, corneal opacity, erythema.
Fire: Flammable. Stop flow of gas. For small fires let fire burn unless leak can be stopped. For large fires use water spray, fog, or regular foam. Vapors may travel to a source of ignition and flash back.
Reactivity: Stable. Hazardous polymerization cannot occur. Avoid: humidity; heat; ignition sources. Incompatible with: oxidizing materials; alkalis. Hazardous decomposition products: carbon monoxide; hydrogen sulfide.
Carcinogenicity: IARC - Not listed; NIOSH - Not listed; NTP - Not listed; ACGIH - Not listed; OSHA - Not listed; EPA - Not listed; MAK - Not listed
Primary Target Organs:

Eyes

Skin

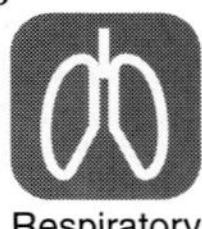
Respiratory System

Nervous System

Cardio-vascular

Environmental

Environmental Fate: If released to soil, it may display high mobility. It is expected to rapidly volatilize from both moist and dry soil to the atmosphere. If released to water, it is expected to rapidly volatilize to the atmosphere. The estimated half-life for volatilization from a model river is 2.3 hr. It is not expected to adsorb to sediment and suspended organic matter nor is it expected to bioconcentrate in fish and aquatic organisms. If released to the atmosphere, it is expected to have a long residence time. Estimates of its atmospheric lifetime range from 200 to 7300 days. Atmospheric removal may occur by a slow gas-phase reaction with photochemically produced hydroxyl radicals or oxygen, direct photolysis, and unknown removal processes invoked to balance the sulfur cycle. Direct photolysis is not expected to occur in the troposphere but it may in the stratosphere.

Cleanup/Disposal: Guide No. 119: Eliminate all ignition sources (no smoking, flares, sparks or flames in immediate area). All equipment used when handling the product must be grounded. Fully encapsulating, vapor protective clothing should be worn for spills and leaks with no fire. Do not touch or walk through spilled material. Stop leak if you can do it without risk. Do not direct water at spill or source of leak. Use water spray to reduce vapors or divert vapor cloud drift. For chlorosilanes, use AFFF alcohol-resistant medium expansion foam to reduce vapors. If possible, turn leaking containers so that gas escapes rather than liquid. Prevent entry into waterways, sewers, basements or confined areas. Isolate area until gas has dispersed.

Environmental Physical Data

Henry's Law Constant: estimated at 4.92 $x10^{-2}$

Octanol/Water Partition Coefficient: log K_{ow} = 0.800

Sorption Partition Coefficient: K_{oc} = 65 to 88

BCF: calculated at 2 to 11

Regulations

RCRA 40CFR: Not listed

CERCLA: 40CFR 302.4: Listed per CAA Section 112 RQ: 100 lb (45.35 kg)

SARA 40CFR 372.65: Listed

SARA EHS 40CFR 355: Not listed

TSCA: Listed

Analytical Methods

Air: EPA 15; ASTM D4490

CAR6130 CAS #: 786-19-6

CARBOPHENOTHION

RTECS: TD5250000

DOT: UN2783; UN2784; UN3017; UN3018; IMO3.2; IMO6.1

EINECS Number: 212-324-1

Molecular Formula: $C_{11}H_{16}ClO_2PS_3$

Formula Weight: 342.85

Synonyms: ACARITHION; AKARITHION; CARBOFENOTHION; CARBOFENTHION; S-(((P-CHLOROPHENYL)THIO)METHYL) O,O-DIETHYLPHOSPHORODITHIOATE; S-((P-CHLOROPHENYLTHIO)METHYL) O,O-DIETHYLPHOSPHORODITHIOATE; S-(4-CHLOROPHENYLTHIOMETHYL)DIETHYL PHOSPHOROTHIOLOTHIONATE; S-(4-CHLOROPHENYLTHIOMETHYL)DIETHYLPHOSPHOROTHIOLOTHIONATE; DAGADIP; O,O-DIAETHYL-S-((4-CHLOR-PHENYL-THIO)-METHYL)DITHIOPHOSPHAT; O,O-DIAETHYL-S-((4-CHLOR-PHENYL-THIO)METHYL)DITHIOPHOSPHAT; O,O-DIETHYL P-CHLOROPHENYLMERCAPTOMETHYL DITHIOPHOSPHATE; O,O-DIETHYL S-(4-CHLOROPHENYLTHIOMETHYL) DITHIOPHOSPHATE; O,O-DIETHYL S-P-CHLOROPHENYLTHIOMETHYL DITHIOPHOSPHATE; O,O-DIETHYL S-(P-CHLOROPHENYLTHIOMETHYL)PHOSPHORODITHIOATE; O,O-DIETHYL DITHIOPHOSPHORIC ACID P-CHLOROPHENYLTHIOMETHYL ESTER; O,O-DIETHYL DITHIOPHOSPHORIC ACID P-CHLOROPHENYLTHIOMETHYLESTER; DIETHYL ESTER; O,O-DIETHYL-S-((4-CHLOOR-FENYL-THIO)-METHYL)-DITHIOFOSFAAT; O,O-DIETHYL-S-(4-CHLOOR-FENYL-THIO)-METHYL)-DITHIOFOSFAAT; O,O-DIETHYL-S-P-CHLORFENYLTHIOMETHYLESTER KYSELINYDITHIOFOSFORECNE; O,O-DIETIL-S-((4-CLORO-FENIL-TIO)-METILE)-DITIOFOSFATO; DITHIOPHOSPHATE DE O,O-DIETHYLE ET DE (4-CHLORO-PHENYL)THIOMETHYLE; ENDYL; ENT 23,708; ETHYL CARBOPHENOTHION; GARRATHION; HEXATHION; KARBOFENOTHION; LETHOX; NEPHOCARP; OLEOAKARITHION; OMS 244; PHOSPHORODITHIOIC ACID,S-(((4-CHLOROPHENYL)THIO)METHYL)O,O-DIETHYL ESTER; PHOSPHORODITHIOIC ACID,S-(((P-CHLOROPHENYL)THIO)METHYL)O,O-DIETHYL ESTER; R 1303; R-1303; STAUFFER R-1,303; TRITHION; TRITHION MITICIDE

Description: colorless liquid; mercaptan-like odor

Use: non-systemic contact & stomach acaricide & insecticide with long residual action; may be used in combination with other insecticides, eg, parathion-methyl, mevinphos; spray to control overwintering aphids, mites & scale insects on dormant deciduous fruit trees; acaricide on citrus trees & cotton; to control mites on grapevines, against mildew oidium species; wheat seed treatment.

Physical Properties

Boiling Point: 82 °C (180 °F) at 0.01 mm Hg

Specific Gravity: 1.271 at 25 °C/4 °C

Saturated Vapor Density: 1.200000005 kg/m^3

Vapor Pressure: 3 $x10^{-7}$ mm Hg at 20 °C

Water Solubility: < 2 ppm at 20 °C

Other Solubilities: Soluble in Kerosene, Xylene, Alcohols, Ketones, & Esters.

pH: 2.43

Refraction Index: 1.597 at 25 °C

RTECS Toxicity Data

Acute Oral: Rat LD_{50} Dose: 6800 ug/kg. Mouse LD_{50} Dose: 218 mg/kg.

Acute Dermal: Rabbit LD_{50} Route: Skin; Dose: 1270 mg/kg. Rat LD_{50} Route: Skin; Dose: 27 mg/kg; Toxic Effects: Behavioral - Tremor; Behavioral - Convulsions or effect on seizure threshold; Behavioral - Excitment.

Chronic (Multiple Dose) Oral: Rat Dose: 7 mg/kg/4W-C; Toxic Effects: Biochemical - True cholinesterase; DEATH. Rat Dose: 630 mg/kg/7D-C; Toxic Effects: Brain and coverings - Other degenerative changes; Biochemical - True cholinesterase; Biochemical - Other esterases.

Mutagenic: Human Sister Chromatid Exchenge; Cell Type: lymphocyte; Dose: 20 ug/L.

Hazard Overviews

Carcinogenicity: IARC - Not listed; NIOSH - Not listed; NTP - Not listed; ACGIH - Not listed; OSHA - Not listed; EPA - Not listed; MAK - Not listed

Environmental

Ecotoxicity: LC_{50} Lepomis macrochirus (Bluegill) 13 ug/l/96 hr at 18 °C, wt 1.1 g (95% confidence limit 10-16 ug/l)

/technical material, 95.3%/. Static bioassay without aeration, pH 7.2-7.5, water hardness 40-50 mg/l as calcium carbonate and alkalinity of 30-35 LC_{50} Gammarus lacustris 5.2 ug/l/96 hr at 21 °C, mature (95% confidence limit 4.1-6.5 ug/l) /technical material, 95.3%/. Static bioassay without aeration, pH 7.2-7.5, water hardness 40-50 mg/l as calcium carbonate and alkalinity of 30-35 mg/l. LD_{50} Mallards oral 121 mg/kg (95% confidence limit 95.9-152 mg/kg), male, 3-4 mo old LD_{50} Sharp-tailed grouse oral 75.6-170 mg/kg, male, adult LC_{50} Palaemonetes 1.2 ug/l/96 hr at 21 °C, mature (95% confidence limit 0.8-1.4 ug/l) /technical material, 95.3%/. Static bioassay without aeration, pH 7.2-7.5, water hardness 40-50 mg/l as calcium carbonate and alkalinity of 30-35 mg/l

Environmental Fate: If released to the atmosphere, it is expected to exist in the vapor and particulate phases. Vapor-phase is expected to rapidly degrade by reaction with photochemically produced hydroxyl radicals (estimated half-life of 1.6 hours). If released to water, hydrolysis may be the primary fate process; however, no rate data under environmental conditions (pH 5-9) were located. It absorbs little light above 290 nm and no light above 310 nm suggesting that photolysis will not be an important fate process. Adsorption to sediment (measured K_{oc}'s of 45,400 and 46,800) and bioconcentration in aquatic organisms will be important (BCF range of about 800 to 6,600). Insufficient data are available to predict the importance of biodegradation in water or soil. In soil, slight mobility may occur and photodegradation on soil surfaces may occur.

Cleanup/Disposal: Guide No. 131: Fully encapsulating, vapor protective clothing should be worn for spills and leaks with no fire. Eliminate all ignition sources (no smoking, flares, sparks or flames in immediate area). All equipment used when handling the product must be grounded. Do not touch or walk through spilled material. Stop leak if you can do it without risk. Prevent entry into waterways, sewers, basements or confined areas. A vapor suppressing foam may be used to reduce vapors. Small Spills: Absorb with earth, sand or other non-combustible material and transfer to containers for later disposal. Use clean non-sparking tools to collect absorbed material. Large Spills: Dike far ahead of liquid spill for later disposal. Water spray may reduce vapor; but may not prevent ignition in closed spaces. Guide No. 152: Do not touch damaged containers or spilled material unless wearing appropriate protective clothing. Stop leak if you can do it without risk. Prevent entry into waterways, sewers, basements or confined areas. Cover with plastic sheet to prevent spreading. Absorb or cover with dry earth, sand or other non-combustible material and transfer to containers. Do not get water inside containers.

Environmental Physical Data

Henry's Law Constant: estimated at 2.15 $x10^{-7}$
Octanol/Water Partition Coefficient: log K_{OW} = 5.33
Sorption Partition Coefficient: K_{OC} = 4.54 $x10^{4}$ to 4.68 $x10^{4}$
BCF: estimated at 6600

Regulations

RCRA 40CFR: Not listed
CERCLA: 40CFR 302.4: Not listed
SARA 40CFR 372.65: Not listed TPQ: 500 lb
SARA EHS 40CFR 355: Listed TPQ: 500 lb
TSCA: Not listed

Analytical Methods

Soil: EPA PMD-TLC, 025; SW846 8141, 8141A, 8270B, 8270C; USGS O5104, O7104
Water / Groundwater: EPA P-005-1, 617, 022; SW846 8141, 8141A; USGS O3104
Food: FDA 212.1, 232.1, 232.2, 232.3, 232.4, 242.1; AOAC 968.24, 974.22
Plasma: EPA 001, 027, 028, 29; FDA 211.1, 231.1, 252
Other: EPA 1656

CAR6320 CAS #: 55285-14-8

CARBOSULFAN

RTECS: EZ3815000
EINECS Number: 259-565-9
Molecular Formula: $C_{20}H_{32}N_2O_3S$
Formula Weight: 380.60
Synonyms: ADVANTAGE; ((DIBUTYLAMINO)THIO)METHYLCARBAMIC ACID,2,2-DIMETHYL-2,3-DIHYDRO-7-BENZOFURANYL ESTER; 2,3-DIHYDRO-2,2-DIMETHYL-7-BENZOFURANYL (DI-N-BUTYLAMINOSULFENYL)METHYLCARBAMATE; 2,3-DIHYDRO-2,2-DIMETHYL-7-BENZOFURANYL((DIBUTYLAMINO)THIO)METHYL CARBAMATE; FMC 35001; MARSHAL; MARSHALL; POSSE

Physical Properties

Boiling Point: 126 °C (258.8 °F)
Specific Gravity: 1.05600
Water Solubility: 0.03 mg/l 25 °C
Other Solubilities: Miscible in organic solvents

RTECS Toxicity Data

Acute Oral: Rat LD_{50} Dose: 51 mg/kg. Mouse LD_{50} Dose: 74 mg/kg.
Acute Inhalation: Rat LC_{50} Dose: 1530 mg/m^3/1hr.
Acute Dermal: Rabbit LD_{50} Route: Skin; Dose: >2 gm/kg. Rat LD_{50} Route: Skin; Dose: >2 gm/kg.
Mutagenic: Human Cytogenetic Analysis; Cell Type: lymphocyte; Dose: 10 ppb. Human Other Mutation Test Systems; Cell Type: lymphocyte; Dose: 500 ppb.

Hazard Overviews

Carcinogenicity: IARC - Not listed; NIOSH - Not listed; NTP - Not listed; ACGIH - Not listed; OSHA - Not listed; EPA - Not listed; MAK - Not listed

Environmental

Regulations

RCRA 40CFR: Listed Hazardous Waste No. P189 Toxic Waste
CERCLA: 40CFR 302.4: Listed per RCRA Section 3001 RQ: 1 lb (0.454 kg)

SARA 40CFR 372.65: Not listed
SARA EHS 40CFR 355: Not listed
TSCA: Not listed

Analytical Methods

Food: FDA 232.4, 242.1

CAS6000 CAS #: 8001-79-4

CASTOR OIL

RTECS: FI4100000
EINECS Number: 232-293-8
Molecular Formula: Unknown
Formula Weight: N/A
Synonyms: AROMATIC CASTOR OIL; CASTOR OIL AROMATIC; COSMETOL; CRYSTAL O; GOLD BOND; NEOLOID; OIL OF PALMA CHRISTI; OLEUM RICINI; OLIO DI RICINO; PHORBYOL; RICINUS OIL; RICIRUS OIL; TANGANTANGAN OIL
Description: pale-yellowish to almost colorless, transparent, liquid; slight, somewhat characteristic odor
Use: additive permitted in food for human consumption, as a purgative, as a plasticizer in lacquers and nitrocellulose, in the production of dibasic acids, lipsticks, in polyurethane coatings, elastomers and adhesives, in fatty acids, in surface-active agents, in hydraulic fluids, in pharmaceuticals, in industrial lubricants and electrical insulating compounds, in the manufacture of Turkey Red oil, as a source of sebacic acid and ricinoleates and as a laxative; as an industrial raw material for the preparation of methane derivatives, in surfactants, in dispersants, in cosmetics, as a basic ingredient in the production of synthetic resins and fibers, as a lubricant in metal drawings, two-cycle engine fuels, as a rubber preservative and mold lubricant, in embalming fluids, in the manufacture of soap, for dyeing and finishing textiles, in alkyds, resinous copolymers, varnishes, oil-based paints, enamel caulks and putties, for plasticizing oilcloth, artificial leather and coated fabrics, to plasticize rosin in the manufacture of sticky fly-paper, in hot melts, duplicating and stencil inks, adhesives and laminates, as a release an anti-sticking agent in the manufacture of hard candy and as a cathartic; in veterinary medicine in the treatment of external wounds, a purgative acting on the small intestine, as a soothing application to the conjunctivae (to allay irritation due to foreign bodies in the eye and for making solutions of alkaloidal bases for ophthalmic purposes)

Physical Properties

Boiling Point: 313 °C (595 °F)
Freezing Point: -12 °C (10.4 °F)
Specific Gravity: 0.961 to 0.963 at 15.5 °C/15.5 °C
Vapor Density: 10 Air=1
Density: 0.953 to 0.965 g/mL at 20 °C
Vapor Pressure: Reid 0.10 psia
Water Solubility: < 1 mg/mL at 20 C
Other Solubilities: Soluble in Chloroform, Glacial Acetic Acid; Soluble in Benzene, Carbon Disulfide.
Surface Tension: 39.0 dynes/cm at 20 °C
Refraction Index: 1.473 to 1.477 at 25 °C/D
Flash Point: 230 °C
Autoignition Temperature: 449 °C

RTECS Toxicity Data

Irritation Eye: Rabbit Standard Draize Test Dose: 500 mg; Reaction: mild.
Irritation Skin: Man Standard Draize Test Dose: 50 mg/48H; Reaction: mild. Rabbit Standard Draize Test Dose: 100 mg/24H; Reaction: severe.

Hazard Overviews

Health: Irritating to eyes/skin/respiratory tract. Other Acute Effects: may be harmful by inhalation, ingestion, or skin absorption; target organ: small intestine.
Fire: Will burn. Extinguishing agents: water spray; carbon dioxide, dry chemical powder or appropriate foam. Precautions: combustible liquid.
Reactivity: Stable. Hazardous polymerization will not occur. Incompatible with: strong oxidizing agents. Hazardous decomposition products: toxic fumes of: carbon monoxide, carbon dioxide.
Carcinogenicity: IARC - Not listed; NIOSH - Not listed; NTP - Not listed; ACGIH - Not listed; OSHA - Not listed; EPA - Not listed; MAK - Not listed
Primary Target Organs:

Eyes

Skin

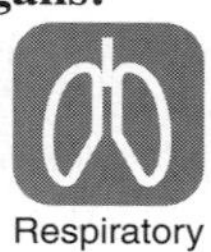
Respiratory System

Gastro-intestinal

Environmental

Cleanup/Disposal: Guide No. 171: Do not touch or walk through spilled material. Stop leak if you can do it without risk. Prevent dust cloud. Avoid inhalation of asbestos dust. Small Dry Spills: With clean shovel place material into clean, dry container and cover loosely; move containers from spill area. Small Spills: Take up with sand or other noncombustible absorbent material and place into containers for later disposal. Large Spills: Dike far ahead of liquid spill for later disposal. Cover powder spill with plastic sheet or tarp to minimize spreading. Prevent entry into waterways, sewers, basements or confined areas.

Regulations

RCRA 40CFR: Not listed
CERCLA: 40CFR 302.4: Not listed
SARA 40CFR 372.65: Not listed
SARA EHS 40CFR 355: Not listed
TSCA: Listed

Analytical Methods

Water / Groundwater: ASTM D4763

CAT3000 CAS #: 120-80-9

CATECHOL

RTECS: UX1050000
EINECS Number: 204-427-5
Molecular Formula: $C_6H_6O_2$
Structured MF: $C_6H_4(OH)_2$
Formula Weight: 110.11

Chemical Structure

Synonyms: 1,2-BENZENE DIOL; BENZENE,O-DIHYDROXY-; 1,2-BENZENEDIOL; O-BENZENEDIOL; ORTHO-BENZENEDIOL; C.I. 76500; C.I. OXIDATION BASE 26; CATECHIN; CATECHIN (PHENOL); CATECHOL (PHENOL); 1,2-DIHYDROXYBENZENE; O-DIHYDROXYBENZENE; ORTHO-DIHYDROXYBENZENE; O-DIOXYBENZENE; ORTHO-DIOXYBENZENE; O-DIPHENOL; DURAFUR DEVELOPER C; FOURAMINE PCH; FOURRINE 68; O-HYDROQUINONE; ORTHO-HYDROQUINONE; 2-HYDROXYPHENOL; O-HYDROXYPHENOL; ORTHO-HYDROXYPHENOL; KATECHOL; NSC 1573; OXYPHENIC ACID; PELAGOL GREY C; O-PHENYLENEDIOL; ORTHO-PHENYLENEDIOL; PHTHALHYDROQUINONE; PYROCATCHUIC ACID; PYROCATECHIN; PYROCATECHINE; PYROCATECHINIC ACID; PYROCATECHOL; PYROCATECHUIC ACID; PYROKATECHIN; PYROKATECHOL
Description: colorless crystalline solid; phenolic odor
Use: antiseptic, photography, dyestuffs, electroplating, specialty inks, antioxidants and light stabilizers, organic synthesis, cosmetics and pharmaceuticals

Physical Properties

Boiling Point: 245.5 °C (474 °F) at 760 mm Hg
Freezing Point: 105 °C (221 °F)
Specific Gravity: 1.344
Vapor Density: 3.79 Air=1
Density: 1.1493 g/mL at 21 °C
Vapor Pressure: 5 mm Hg at 104 °C
Water Solubility: 1 parts in 23 parts Water
Other Solubilities: 95% Ethanol: >=100 mg/ml at 21.5 °C; Acetone: >=100 mg/ml at 21.5 °C; Aqueous alkalies: Soluble; Benzene: Soluble; Carbon Tetrachloride: Soluble; Chloroform: Soluble; DMSO: >=100 mg/ml at 21.5 °C; Ether: Soluble; Pyridine: Soluble.
pH: Weakly acid
Refraction Index: 1.604
Ionization Potential (eV): 8.15 +/-1.0
Flash Point: 127 °C Closed Cup
Autoignition Temperature: 510 °C
LEL: 1.4% v/v

RTECS Toxicity Data

Acute Oral: Rat LD_{50} Dose: 260 mg/kg. Mouse LD_{50} Dose: 260 mg/kg.
Acute Dermal: Rabbit LD_{50} Route: Skin; Dose: 800 mg/kg; Toxic Effects: Skin and appendages - Primary irritation; Nutritional and gross metabolic - Weight loss or decreased weight gain. Rabbit LD_{Lo} Route: Subcutaneous Dose: 225 mg/kg.
Chronic (Multiple Dose) Oral: Chicken Dose: 98 gm/kg/4W-I; Toxic Effects: DEATH.
Reproductive/Teratogenic: Rat Route: Oral; Dose: 1 gm/kg; Duration: female 11D of pregnancy; Effects on Fertility - Litter size. Rat Route: Subcutaneous; Dose: 5 mg/kg; Duration: female 1D prior to mating; Maternal Effects - Ovaries, fallopian tubes.
Mutagenic: Human DNA Damage; Cell Type: lymphocyte; Dose: 100 umol/L. Human DNA Inhibition; Cell Type: HeLa cell; Dose: 200 umol/L. Human Sister Chromatid Exchenge; Cell Type: lymphocyte; Dose: 40 umol/L. Human Other Mutation Test Systems; Cell Type: lymphocyte; Dose: 40 umol/L.
Tumorigenic: Rat Route: Oral; Dose: 437 gm/kg/2Y-C; Toxic Effects: Tumorigenic - Carcinogenic by RTECS criteria; Gastrointestinal - Tumors. Rat Route: Oral; Dose: 60640 mg/kg/24W-C; Toxic Effects: Tumorigenic - Neoplastic by RTECS criteria; Gastrointestinal - Tumors. Rat Route: Oral; Dose: 349 gm/kg/2Y-C; Toxic Effects: Tumorigenic - Carcinogenic by RTECS criteria; Gastrointestinal - Tumors. Rat Route: Oral; Dose: 171 gm/kg/51W-C; Toxic Effects: Tumorigenic - Carcinogenic by RTECS criteria; Gastrointestinal - Tumors.

Hazard Overviews

Corrosive

Fire Diamond

Health: Corrosive, causes severe burns to eyes/skin/respiratory tract. Also Causes: permanent vision damage, eczematous dermatitis, pallor, anorexia, nausea, vomiting, diarrhea, weakness, muscle aches, darkened urine, headache, tinnitus, sweating, convulsions, cyanosis, shock, unconsciousness, respiratory failure, death, blood dyscrasias. Chronic Effects: cough, sputum, occasional sore throat/eye irritation/skin eruptions.
Fire: Will burn. Use water, carbon dioxide, or dry chemical.
Reactivity: Stable. Hazardous polymerization cannot occur. Avoid: heat; ignition sources. Incompatible with: oxidizing sources. Hazardous decomposition products: acrid smoke; irritating fumes; carbon oxides.
Carcinogenicity: IARC - Group 3, Not classifiable as to carcinogenicity to humans; NIOSH - Not listed; NTP - Not listed; ACGIH - Class A3, Animal carcinogen; OSHA - Not listed; EPA - Not listed; MAK - Not listed

Primary Target Organs:

Eyes

Skin

Respiratory System

Nervous System

Blood

Exposure Limits

OSHA PEL Vacated 1989 Limits: TWA: 5 ppm; 20 mg/m^3.
ACGIH TLV: TWA: 5 ppm; 23 mg/m^3.
NIOSH REL: TWA: 5 ppm; 20 mg/m^3.

Respirator Recommendation

Exposure Range: >5 to 50 ppm Air Purifying, Negative Pressure, Half Mask
Exposure Range: >50 to 500 ppm Air Purifying, Negative Pressure, Full Face
Exposure Range: >500 to 5000 ppm Supplied Air, Constant Flow/Pressure Demand, Full Face
Exposure Range: >5000 to unlimited ppm Self-contained Breathing Apparatus, Pressure Demand, Full Face
Cartridge Color: black with dust/mist prefilter (use P100 or consult supervisor for appropriate dust/mist prefilter)

Environmental

Ecotoxicity: LC_{50} Pimephales promelas (fathead minnow) 9.22 mg/l/96 hr (confidence limit 8.62 - 9.87 mg/l), flow through bioassay with measured concentrations, 25.6 °C, dissolved oxygen 6.4 mg/l, hardness 46.0 mg/l calcium carbonate, alkalinity 40.2 mg/l calcium carbonate, and pH 7.7
Environmental Fate: Since the pKa for the initial ionization is 9.23 it will exist in a partially dissociated state in natural waters and moist soil and therefore, its transport and reactivity may be affected by pH. If released to soil, it will be expected to be highly mobile. It may be susceptible to biodegradation in soils and groundwater. It should not volatilize from soil surfaces or hydrolyze under normal environmental conditions. If released to water, it will not be expected to sorb to sediments, hydrolyze, volatilize or bioconcentrate in aquatic organisms. It may be subject to significant biodegradation. Its ionized form but not its nonionized form will be expected to be subject to significant direct photolysis. It may be susceptible to indirect photooxidation by hydroxyl and peroxy radicals in sunlit waters. If released to the atmosphere, it will exist mainly in the vapor phase. Vapor phase is expected to degrade rapidly in air by reaction with photochemically produced hydroxyl radicals with an estimated half-life of 0.654 days for this reaction. Vapor phase reaction with nitrate radicals may be a degradation process during the nighttime.
Cleanup/Disposal: Guide No. 154: Eliminate all ignition sources (no smoking, flares, sparks or flames in immediate area). Do not touch damaged containers or spilled material unless wearing appropriate protective clothing. Stop leak if you can do it without risk. Prevent entry into waterways, sewers, basements or confined areas. Absorb or cover with dry earth, sand or other non-combustible material and transfer to containers. Do not get water inside containers.

Environmental Physical Data

Henry's Law Constant: estimated at 0.810×10^{-10}
Octanol/Water Partition Coefficient: $\log K_{ow} = 0.88$
Sorption Partition Coefficient: K_{oc} = brookston clay loam 118
BCF: estimated at 3
BOD: sewage seed 0.69 lb/lb, 5 days

Regulations

RCRA 40CFR: Not listed
CERCLA: 40CFR 302.4: Listed per RCRA Section 3001 RQ: 100 lb (45.35 kg)
SARA 40CFR 372.65: Listed
SARA EHS 40CFR 355: Not listed
TSCA: Listed

Analytical Methods

Water / Groundwater: ASTM D4763

CEL1000 CAS #: 9004-34-6

CELLULOSE

RTECS: FJ5691460
EINECS Number: 232-674-9
Molecular Formula: $C_{(6x)}H_{(10x)}O_{(5x)}$
Structured MF: $(C_6H_{10}O_5)_n$
Formula Weight: 160,000-560,000
Synonyms: ABICEL; BETA-AMYLOSE; ARBOCEL; ARBOCEL BC 200; ARBOCELL B 600/30; AVICEL; AVICEL 101; AVICEL 102; AVICEL CL 611; AVICEL PH 101; AVICEL PH 105; CELLEX MX; ALPHA-CELLULOSE; CELLULOSE 248; CELLULOSE CRYSTALLINE; CELUFI; CEPO; CEPO CFM; CEPO S 20; CEPO S 40; CHROMEDIA CC 31; CHROMEDIA CF 11; CUPRICELLULOSE; ELCEMA F 150; ELCEMA G 250; ELCEMA P 050; ELCEMA P 100; FRESENIUS D 6; HEWETEN 10; HYDROXYCELLULOSE; KINGCOT; LA 01; MICROCRYSTALLINE CELLULOSE; MN-CELLULOSE; ONOZUKA P 500; PYROCELLULOSE; RAYOPHANE; RAYWEB Q; REXCEL; SIGMACELL; SOLKA-FIL; SOLKA-FLOC; SOLKA-FLOC BW; SOLKA-FLOC BW 100; SOLKA-FLOC BW 20; SOLKA-FLOC BW 200; SOLKA-FLOC BW 2030; SPARTOSE OM-22; SULFITE CELLULOSE; TOMOFAN; TUNICIN; WHATMAN CC-31
Description: white substance; odorless

Physical Properties

Boiling Point: Decomposes
Freezing Point: Decomposes at 260 °C (500 °F)
Vapor Pressure: ~ 0 mm Hg
Water Solubility: Insoluble
Flash Point: Combustible Solid

RTECS Toxicity Data

Acute Oral: Rat LD_{50} Dose: >5 gm/kg.
Acute Inhalation: Rat LC_{50} Dose: >5800 mg/m^3/4hr.
Acute Dermal: Rabbit LD_{50} Route: Skin; Dose: >2 gm/kg.

Hazard Overviews

Health: May cause irritation to eyes/skin. Other Acute Effects: may be harmful if inhaled.
Fire: Combustible. Hazards: in powder form capable of creating a dust explosion; emits toxic fumes; static charge buildup can be a potential fire hazard when used in the

presence of volatile or flammable mixtures. Extinguishing agents: water spray; carbon dioxide, dry chemical powder or appropriate foam. Precautions: combustible liquid.

Carcinogenicity: IARC - Not listed; NIOSH - Listed as carcinogen; NTP - Not listed; ACGIH - Not listed; OSHA - Not listed; EPA - Not listed; MAK - Not listed

Exposure Limits

OSHA PEL: TWA: 15 mg/m^3; total dust.

ACGIH TLV: TWA: 10 mg/m^3.

NIOSH REL: TWA: 10 mg/m^3.

Respirator Recommendation

Exposure Range: >5 to 50 mg/m^3 Air Purifying, Negative Pressure, Half Mask

Exposure Range: >50 to 500 mg/m^3 Air Purifying, Negative Pressure, Full Face

Exposure Range: >500 to 5000 mg/m^3 Supplied Air, Constant Flow/Pressure Demand, Full Face

Exposure Range: >5000 to unlimited mg/m^3 Self-contained Breathing Apparatus, Pressure Demand, Full Face

Cartridge Color: dust/mist filter (use P100 or consult supervisor for appropriate dust/mist filter)

Environmental

Regulations

RCRA 40CFR: Not listed

CERCLA: 40CFR 302.4: Not listed

SARA 40CFR 372.65: Not listed

SARA EHS 40CFR 355: Not listed

TSCA: Not listed

CEL5000 CAS #: 9004-35-7

CELLULOSE ACETATE

Molecular Formula: Unknown

Synonyms: A 432-130B; ACETATE COTTON; ACETATE ESTER OF CELLULOSE; ACETIC ACID,CELLULOSE ESTER; ACETOSE; ACETYL 35; ACETYLCELLULOSE; ALLOGEL; AMPACET C/A; BIODEN; CELLIDOR; CELLIDOR A; CELLIT K 700; CELLIT L 700; CELLULOSE 2,5-ACETATE; CELLULOSE MONOACETATE; CELLULOSE,ACETATE; CELLULOSE,2,5-DIACETATE; CRELLATE; DP 02; DP 06; E 376-40; E 383-40; E 394-30; E 394-40; E 394-45; E 394-60; E 398-10; E 400-25; E-400-25; EASTMAN 298-10; ETROL OEM; MONOACETYLCELLULOSE; NICOLLEMBAL; NIXON C/A; PLASTACELE; PP 612; PP 613; PP 628; STRIPMIX; STRUX; T-CELLIT; TENITE I; VLADIPOR

Description: white flakes, granules; odorless

Use: polymer for acetate fibers, yarn, plastics, staple & tow for cigarette filters, etc; in mfr of rubber & celluloid substitutes, nonflammable photographic & cinema films, airplane dopes, varnishes & lacquers; filaments, phonograph records; waterproofing fabrics & rendering balloons gas-tight; sizing & finishing fabrics; coating skins; insulating electric wires; for transparent sheeting & thermoplastic molding; osmotic cell membrane; in minitype dialyzer; in sewage treatment; in food pacakaging film.

Physical Properties

Freezing Point: About 260 °C (500 °F)

Specific Gravity: 1.27 to 1.34

Water Solubility: Insoluble in Water

Other Solubilities: Soluble in Ethylene Dichloride.

Refraction Index: 1.46 to 1.50

Flash Point: Flammable; moderate fire risk

Hazard Overviews

Explosive

Flammable

Fire Diamond

Health: Mildly irritating to respiratory tract.

Fire: Flammable. Powder forms explosive mixtures in the air. Use dry chemical, carbon dioxide, or water. Apply water as a blanket; high pressure water stream will disperse powder in air and create a severe explosion hazard.

Reactivity: Stable. Hazardous polymerization cannot occur. Avoid: exposure to heat; ignition sources; oxidizers; mixing with nonpolar hydrocarbons (toluene; xylene); adding to any flammable liquid. Incompatible with: oxidizing agents. Hazardous decomposition products: carbon oxides; acetic acid.

Carcinogenicity: IARC - Not listed; NIOSH - Not listed; NTP - Not listed; ACGIH - Not listed; OSHA - Not listed; EPA - Not listed; MAK - Not listed

Primary Target Organs:

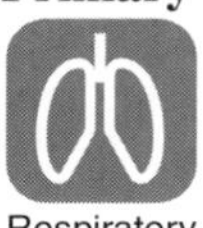
Respiratory System

Environmental

Regulations

RCRA 40CFR: Not listed

CERCLA: 40CFR 302.4: Not listed

SARA 40CFR 372.65: Not listed

SARA EHS 40CFR 355: Not listed

TSCA: Listed

CEL9000 CAS #: 65996-61-4

CELLULOSE FIBER

EINECS Number: 265-995-8

Physical Properties

Specific Gravity: 1.25

Water Solubility: Negligible

Hazard Overviews

Carcinogenicity: IARC - Not listed; NIOSH - Not listed; NTP - Not listed; ACGIH - Not listed; OSHA - Not listed; EPA - Not listed; MAK - Not listed

Environmental

Regulations

RCRA 40CFR: Not listed
CERCLA: 40CFR 302.4: Not listed
SARA 40CFR 372.65: Not listed
SARA EHS 40CFR 355: Not listed
TSCA: Not listed

CER5000 CAS #: 12014-56-1

CERIUM HYDROXIDE

EINECS Number: 234-599-7

HO OH Ce HO OH

Chemical Structure

Physical Properties

Specific Gravity: 5
Other Solubilities: Soluble in concentrated mineral acids

Hazard Overviews

Corrosive

Health: Corrosive to eyes/skin/respiratory tract. Other Acute Effects: harmful if swallowed, inhaled, or absorbed through skin; inhalation may result in spasm, inflammation and edema of the larynx and bronchi; chemical pneumonitis; pulmonary edema; exposure may cause burning sensation; coughing; wheezing; laryngitis; shortness of breath; headache; nausea; vomiting.
Fire: Hazards: emits toxic fumes. Extinguishing agents: dry chemical powder. Precautions: combustible liquid.
Reactivity: Incompatible with: strong oxidizing agents, strong acids. Hazardous decomposition products: nature of decomposition products not known;.
Carcinogenicity: IARC - Not listed; NIOSH - Not listed; NTP - Not listed; ACGIH - Not listed; OSHA - Not listed; EPA - Not listed; MAK - Not listed

Primary Target Organs:

Eyes

Skin

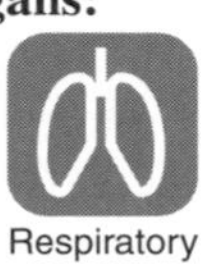
Respiratory System

Environmental

Regulations

RCRA 40CFR: Not listed
CERCLA: 40CFR 302.4: Not listed
SARA 40CFR 372.65: Not listed
SARA EHS 40CFR 355: Not listed
TSCA: Not listed

CES5000 CAS #: 21351-79-1

CESIUM HYDROXIDE

RTECS: FK9800000
DOT: UN1407
EINECS Number: 244-344-1
Molecular Formula: CsHO
Structured MF: CsOH
Formula Weight: 149.92

Cs^+ OH^-

Chemical Structure

Synonyms: CAESIUM HYDROXIDE; CESIUM HYDRATE; CESIUM HYDROXIDE DIMER; CESIUM HYDROXIDE DIMER CESIUM HYDRATE
Description: white, yellowish crystalline mass; odorless

Physical Properties

Freezing Point: 272.22 °C (521.996 °F)
Density: 3.68 g/mL
Vapor Pressure: ~ 0 mm Hg
Water Solubility: 395% at 15 °C
Other Solubilities: Soluble in Alcohol
Flash Point: Noncombustible

RTECS Toxicity Data

Acute Oral: Rat LD_{50} Dose: 570 mg/kg; Toxic Effects: Behavioral - Somnolence (general depressed activity); Behavioral - Muscle contraction or spasticity; Lungs, Thorax, or Respiration - Other changes. Mouse LD_{50} Dose: 800 mg/kg; Toxic Effects: Behavioral - Tetany.
Irritation Eye: Rabbit Nonstandard Exposure Dose: 5 mg/5M rinse; Reaction: severe.
Irritation Skin: Rabbit Standard Draize Test Dose: 5 mg/24H; Reaction: mild.

Hazard Overviews

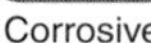
Corrosive

Fire Diamond

Health: Corrosive to eyes/skin/respiratory tract. Also Causes: upon ingestion: burns of the mouth, throat, stomach and may be fatal.

Fire: Noncombustible. Use agent suitable for surrounding fire. Reacts with water to produce excessive heat. It can also react with many metals to produce flammable and explosive hydrogen gas. Avoid splattering or splashing this material.

Reactivity: Stable. Hazardous polymerization cannot occur. Avoid: exposure to water. Incompatible with: glass; carbon dioxide; water; strong acids (hydrochloric; sulfuric; nitric); metals (aluminum; lead; zinc; tin); wet metals.

Carcinogenicity: IARC - Not listed; NIOSH - Listed as carcinogen; NTP - Not listed; ACGIH - Not listed; OSHA - Not listed; EPA - Not listed; MAK - Not listed

Primary Target Organs:

Eyes

Skin

Respiratory System

Mucous Membranes

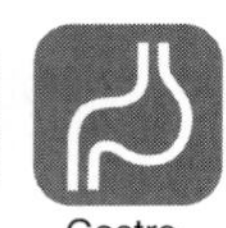
Gastro-intestinal

Exposure Limits

OSHA PEL Vacated 1989 Limits: TWA: 2 mg/m^3.

ACGIH TLV: TWA: 2 mg/m^3.

NIOSH REL: TWA: 2 mg/m^3.

Respirator Recommendation

Exposure Range: >2 to 20 mg/m^3 Air Purifying, Negative Pressure, Half Mask

Exposure Range: >20 to 200 mg/m^3 Air Purifying, Negative Pressure, Full Face

Exposure Range: >200 to 2000 mg/m^3 Supplied Air, Constant Flow/Pressure Demand, Full Face

Exposure Range: >2000 to unlimited mg/m^3 Self-contained Breathing Apparatus, Pressure Demand, Full Face

Cartridge Color: dust/mist filger

Environmental

Cleanup/Disposal: Guide No. 138: Eliminate all ignition sources (no smoking, flares, sparks or flames in immediate area). Do not touch or walk through spilled material. Stop leak if you can do it without risk. Use water spray to reduce vapors or divert vapor cloud drift. Do not get water on spilled substance or inside containers. Small Spills: Cover with dry earth, dry sand, or other non-combustible material followed with plastic sheet to minimize spreading or contact with rain. Dike for later disposal; do not apply water unless directed to do so. Powder Spills: Cover powder spill with plastic sheet or tarp to minimize spreading and keep powder dry. Do not clean-up or dispose of, except under supervision of a specialist.

Regulations

RCRA 40CFR: Not listed

CERCLA: 40CFR 302.4: Not listed

SARA 40CFR 372.65: Not listed

SARA EHS 40CFR 355: Not listed

TSCA: Listed

CHA6000 **CAS #: 64365-11-3**

CHARCOAL, ACTIVATED

EINECS Number: 264-846-4

Molecular Formula: C

Structured MF: C

Formula Weight: 12

Description: black granules, powder; odorless

Physical Properties

Boiling Point: Very High

Freezing Point: > 3500 °C (6332 °F)

Specific Gravity: 2 at 20 °C (solid)

Water Solubility: Insoluble

Flash Point: Flammable solid; may ignite spontaneously in air

Autoignition Temperature: 316 to 399 °C

LEL: Charcoal 0.14 g/l

Hazard Overviews

Explosive

Flammable

Fire Diamond

Health: Irritating to the eyes/respiratory tract. Also Causes: conjunctivitis.

Fire: Flammable. May form explosive dust-air mixtures. For small fires use dry chemical, sand, water spray, or foam. For large fires use water spray, fog, or standard foam. Use water spray to cool fire-exposed containers.

Reactivity: Stable; finely divided carbon particulate may undergo spontaneous combustion when moist, when coated with drying oils, or when freshly calcined. Hazardous polymerization cannot occur. Avoid: heat and ignition sources. Incompatible with: strong oxidizing agents; slowly reacts with oxygen in air at room temperature. Hazardous decomposition products: carbon monoxide (with a deficiency of oxygen).

Carcinogenicity: IARC - Not listed; NIOSH - Not listed; NTP - Not listed; ACGIH - Not listed; OSHA - Not listed; EPA - Not listed; MAK - Not listed

Primary Target Organs:

Eyes

Respiratory System

Environmental

Cleanup/Disposal: Guide No. 133: Eliminate all ignition sources (no smoking, flares, sparks or flames in immediate area). Do not touch or walk through spilled material. Small Dry Spills: With clean shovel place material into clean, dry container and cover loosely; move containers from spill area. Large Spills: Wet down with water and dike for later disposal. Prevent entry into waterways, sewers, basements or confined areas.

Environmental Physical Data

BOD: none

Regulations

RCRA 40CFR: Not listed
CERCLA: 40CFR 302.4: Not listed
SARA 40CFR 372.65: Not listed
SARA EHS 40CFR 355: Not listed
TSCA: Not listed

CHL1030 CAS #: 75-87-6

CHLORAL

RTECS: FM7870000
DOT: UN1760; UN2075; IMO6.1
EINECS Number: 200-911-5
Molecular Formula: C_2HCl_3O
Structured MF: CCl_3CHO
Formula Weight: 147.40

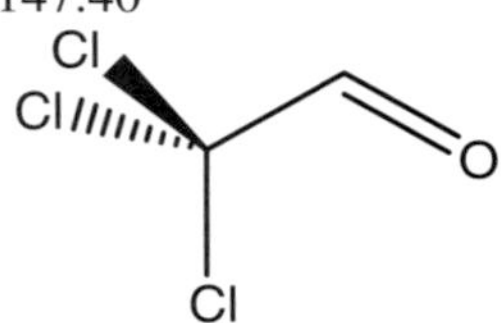

Chemical Structure

Synonyms: ACETALDEHYDE,TRICHLORO-; ACETALDEHYDE,TRICHLORO-(9CI); ANHYDROUS CHLORAL; CHLORAL,ANHYDROUS,INHIBITED; CLORALIO; GRASEX; 2,2,2-TRICHLOROACETALDEHYDE; TRICHLOROACETALDEHYDE; TRICHLOROETHANAL; 2,2,2-TRICHLOROETHANOL
Description: colorless, oily liquid; irritating odor
Use: manufacture of chloral hydrate, DDT; liniment; organic synthesis

Physical Properties

Boiling Point: 97.8 °C (208 °F) at 760 mm Hg
Freezing Point: -57.5 °C (-71.5 °F)
Specific Gravity: 1.5121 at 20 °C/4 °C
Vapor Density: 5.1 Air=1
Saturated Vapor Density: 1.425626134 kg/m^3
Density: 1.5121 g/mL
Vapor Pressure: 35 mm Hg at 20 °C
Water Solubility: Freely Soluble in Water
Other Solubilities: 95% Ethanol: >=10 mg/ml at 22 °C; Acetone: >=10 mg/ml at 22 °C; Chloroform: Soluble; DMSO: >=10 mg/ml at 22 °C; Ether: Soluble.
Surface Tension: 25.34 dynes/cm at 19.4 °C
Odor Threshold: 0.047 ppm
Refraction Index: 1.45572 at 20 °C/D
Ionization Potential (eV): 10.9
Flash Point: 75 °C

RTECS Toxicity Data

Acute Oral: Mammal LD_{50} Dose: 710 mg/kg.
Acute Inhalation: Rat LC_{50} Dose: 440 mg/m^3/4hr; Toxic Effects: Sense organs and special senses - Ptosis; Behavioral - Somnolence (general depressed activity); Lungs, Thorax, or Respiration - Dyspnea. Dog LC_{50} Dose: 5900 mg/m^3/4hr; Toxic Effects: Sense organs and special senses - Lacrimation; Behavioral - Convulsions or effect on seizure threshold; Behavioral - Ataxia.
Chronic (Multiple Dose) Inhalation: Rat Dose: 80 mg/m^3/4H/2W-I; Toxic Effects: Lungs, Thorax, or Respiration - Acute pulmonary edema; Endocrine - Changes in adrenal weight; Nutritional and gross metabolic - Weight loss or decreased weight gain.
Mutagenic: Bacteria - S Typhimurium Mutations in Microorganisms; Dose: 1 mg/plate (+S9), 10 mg/plate (-S9). Yeast - S Cerevisiae Mutations in Microorganisms; Dose: 1 gm/L (+S9).

Hazard Overviews

Corrosive

Health: Irritating to eyes/skin. Toxic. Other Acute Effects: toxic if swallowed.
Fire: Combustible. Hazards: hazardous polymerization will occur. Extinguishing agents: water spray; carbon dioxide, dry chemical powder or appropriate foam. Precautions: combustible liquid.
Reactivity: Hazardous polymerization will occur. Hazardous decomposition products: when heated to decomposition it may emit: toxic fumes; irritating fumes.
Carcinogenicity: IARC - Group 3, Not classifiable as to carcinogenicity to humans; NIOSH - Not listed; NTP - Not listed; ACGIH - Not listed; OSHA - Not listed; EPA - Not listed; MAK - Not listed
Primary Target Organs:

Eyes

Skin

Environmental

Ecotoxicity: Food chain concentration potential: It is estimated that fish in rivers, ponds, lakes, and reservoirs will bioconcentrate chloral 6.7 times the water concentration.
Environmental Fate: When released to soil, it is expected to react with the water therein to yield chloral hydrate. In the event of a large spill, or a spill onto a relatively dry soil, a moderate amount of volatilization would be expected to occur. When released to water, the predominant reaction will be that with water to form chloral hydrate. Since this reaction is dominant it is not possible to realistically assess other fate processes. The half-life for the vapor-phase reaction with photochemically produced hydrol radicals has been estimated

to be 10.77 hr at an atmosphere concentration of 8 x10^5 hydrol radicals per cu cm. It may also react with moisture in the atmosphere to form chloral hydrate which would be subject to rainout

Cleanup/Disposal: Guide No. 153: Eliminate all ignition sources (no smoking, flares, sparks or flames in immediate area). Do not touch damaged containers or spilled material unless wearing appropriate protective clothing. Stop leak if you can do it without risk. Prevent entry into waterways, sewers, basements or confined areas. Absorb or cover with dry earth, sand or other non-combustible material and transfer to containers. Do not get water inside containers. Guide No. 154: Eliminate all ignition sources (no smoking, flares, sparks or flames in immediate area). Do not touch damaged containers or spilled material unless wearing appropriate protective clothing. Stop leak if you can do it without risk. Prevent entry into waterways, sewers, basements or confined areas. Absorb or cover with dry earth, sand or other non-combustible material and transfer to containers. Do not get water inside containers.

Environmental Physical Data

BCF: not significant

Regulations

RCRA 40CFR: Listed Hazardous Waste No. U034 Toxic Waste
CERCLA: 40CFR 302.4: Listed per RCRA Section 3001 RQ: 5000 lb (2268 kg)
SARA 40CFR 372.65: Not listed
SARA EHS 40CFR 355: Not listed
TSCA: Listed

Analytical Methods

Drinking Water: EPA 551

CHL1060 CAS #: 302-17-0

CHLORAL HYDRATE

RTECS: FM8750000
EINECS Number: 206-117-5
Molecular Formula: $C_2H_3Cl_3O_2$
Structured MF: $Cl_3CCH(OH)_2$
Formula Weight: 165.42

Cl Cl Cl H OH OH

Chemical Structure

Synonyms: AQUACHLORAL; BI 3411; CHLORALDURAL; CHLORALDURAT; CHLORALEX; CHLORALI HYDRAS; CHLORAL,MONOHYDRATE; CHLORALVAN; DORMAL; ESCRE; 1,1-ETHANEDIOL,2,2,2-TRICHLORO-; 1,1-ETHANEDIOL,2,2,2-TRICHLORO-(9CI); FELSULES; HS; HYDRAL; HYDRATE DE CHLORAL; KESSODRATE; KLORALHYDRAT; KNOCKOUT DROPS; LORINAL; NOCTEC; NORTEC; NOVOCHLORHYDRATE; NYCOTON; NYCTON; PHALDRONE; RECTULES; SK-CHLORAL HYDRATE; SOMNI SED; SOMNOS; SONTEC; TOSYL; TRAWOTOX; TRICHLORACETALDEHYD-HYDRAT; TRICHLOROACETALDEHYDE HYDRATE; TRICHLOROACETALDEHYDE MONOHYDRATE; TRICHLOROACETALDEHYDE,HYDRATED; 1,1,1-TRICHLORO-2,2-DIHYDROXYETHANE; 1,1,1-TRICHLORO-2,2-ETHANEDIOL; 2,2,2-TRICHLORO-1,1-ETHANEDIOL; 2,2,2-TRICHLOROETHANE-1,1-DIOL; TRICHLOROETHYLIDENE GLYCOL

Description: colorless crystals; acrid odor

Use: as a sedative, a narcotic, an anesthetic and a hypnotic agent; in liniments and in the manufacture of DDT

Physical Properties

Boiling Point: 96.3 °C (205 °F) at 764 mm Hg
Freezing Point: 57 °C (134.6 °F)
Specific Gravity: 1.908 at 20 °C/4 °C
Vapor Pressure: 5 mm Hg at 10 °C
Water Solubility: >= 10 mg/mL at 20.5 C
Other Solubilities: Very Soluble in Pyridine; 1 g/68 g Carbon Disulfide; 1 g/1.3 ml Alcohol; 1 g/1.4 ml olive oil; freely Soluble in Acetone, Methyl Ethyl Ketone; 1 g/2 ml Chloroform; 1 g/1.5 ml Ether; 1 g/0.5 ml Glycerol.
pH: 10% solution in water 3.5 to 4.4
Flash Point: Not available; probably combustible

RTECS Toxicity Data

Acute Oral: Human TD_{Lo} Dose: 300 mg/kg; Toxic Effects: Behavioral - General anesthetic; Cardiac - Arrythmias (including changes in conduction); Vascular - BP lowering not characterized in autonomic section. Man TD_{Lo} Dose: 1 gm/kg; Toxic Effects: Behavioral - Coma; Cardiac - Pulse rate increased without fall in BP; Lungs, Thorax, or Respiration - Cyanosis. Woman TD_{Lo} Dose: 960 mg/kg; Toxic Effects: Behavioral - Somnolence (general depressed activity); Behavioral - Hallucinations, distorted perceptions; Cardiac - Pulse rate increased without fall in BP. Woman TD_{Lo} Dose: 465 mg/kg; Toxic Effects: Sense organs and special senses - Other; Behavioral - Coma; Vascular - BP lowering not characterized in autonomic section. Child TD_{Lo} Dose: 219 mg/kg; Toxic Effects: Cardiac - Arrythmias (including changes in conduction). Human LD_{Lo} Dose: 4 mg/kg; Toxic Effects: Sense organs and special senses - Miosis (pupilliary dilation); Gastrointestinal - Nausea or vomiting. Rat LD_{50} Dose: 479 mg/kg.

Acute Dermal: Rabbit LD_{Lo} Route: Subcutaneous Dose: 1 gm/kg; Toxic Effects: Behavioral - General anesthetic; Lungs, Thorax, or Respiration - Other changes. Rat LD_{50} Route: Skin; Dose: 3030 mg/kg.

Chronic (Multiple Dose) Oral: Rat Dose: 15120 mg/kg/90D-C; Toxic Effects: Liver - Hepatitis (hepatocellular necrosis), zonal; Blood - Changes in serum composition; Biochemical - Transaminases. Rat Dose: 210 mg/kg/30W-I; Toxic Effects: Blood - Other changes; Biochemical - True cholinesterase. Rat Dose: 1413 mg/kg/90D-C; Toxic Effects: Nutritional and gross metabolic - Body temperature decrease.

Reproductive/Teratogenic: Mouse Route: Oral; Dose: 13 gm/kg; Duration: female 3W prior to mating; Effects on Newborn - Behavioral.
Mutagenic: Human Sister Chromatid Exchenge; Cell Type: lymphocyte; Dose: 54 mg/L. Human Sex Chromosome Loss; Cell Type: lymphocyte; Dose: 50 mg/L. Human Other Mutation Test Systems; Cell Type: lymphocyte; Dose: 100 mg/L.
Tumorigenic: Mouse Route: Oral; Dose: 10 mg/kg; Toxic Effects: Tumorigenic - Carcinogenic by RTECS criteria; Liver - Tumors. Mouse Route: Skin; Dose: 960 mg/kg/1W-I; Toxic Effects: Tumorigenic - Equivocal tumorigenic agent by RTECS criteria; Skin and appendages - Tumors.

Hazard Overviews

Fire Diamond

Health: Severely irritating to eyes/skin. Toxic. Also Causes: coma, hypotension, hypothermia, respiratory depression, heart rhythm disturbances, GI irritation, ulceration/perforation of stomach, GI bleeding, liver/kidney dysfunction. Chronic Effects: stomach lining inflammation, skin rashes, kidney damage, addiction.
Fire: Will burn. Gives off toxic fumes of chlorine upon decomposition. Use dry chemical, carbon dioxide, or foam.
Reactivity: Stable. Hazardous polymerization cannot occur. Hazardous decomposition products: toxic fumes of chlorine.
Carcinogenicity: IARC - Group 3, Not classifiable as to carcinogenicity to humans; NIOSH - Not listed; NTP - Not listed; ACGIH - Not listed; OSHA - Not listed; EPA - Not listed; MAK - Not listed
Primary Target Organs:

Eyes

Skin
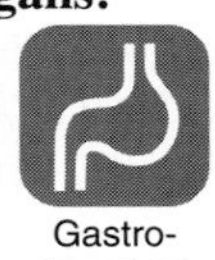
Gastro-intestinal
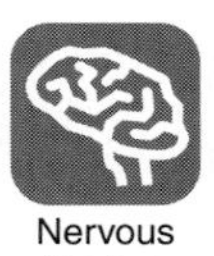
Nervous System

Kidneys

Cardio-vascular

Environmental

Ecotoxicity: algae (Microcystis aeruginosa): 8d EC_0 78 mg/l

Environmental Physical Data

Octanol/Water Partition Coefficient: log K_{ow} = 0.99

Regulations

RCRA 40CFR: Not listed
CERCLA: 40CFR 302.4: Not listed
SARA 40CFR 372.65: Not listed
SARA EHS 40CFR 355: Not listed
TSCA: Listed

Analytical Methods

Soil: SW846 8260A, 8260B
Plasma: EPA 29

CHL1120 CAS #: 133-90-4

CHLORAMBEN

RTECS: DG1925000
EINECS Number: 205-123-5
Molecular Formula: $C_7H_5Cl_2NO_2$
Formula Weight: 206.03

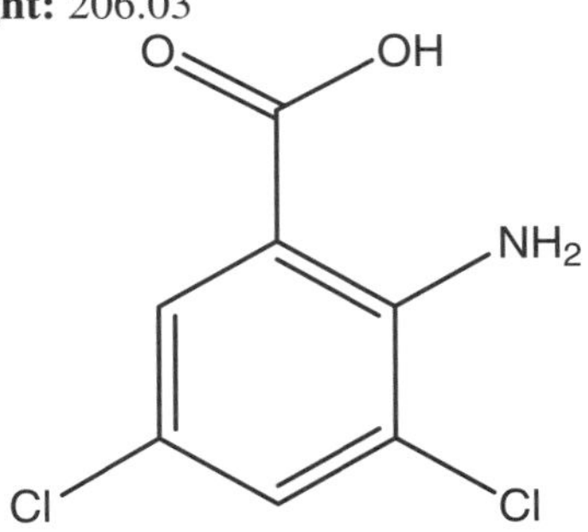

Chemical Structure

Synonyms: ACP-M-728; ACPM-629; AMBIBEN; AMIBEN; AMIBEN DS; AMIBIN; 3-AMINO-2,5-DICHLOROBENZOIC ACID; AMOBEN; BENZOIC ACID,3-AMINO-2,5-DICHLORO-; CHLORAMBED; CHLORAMBENE; 2,5-DICHLORO-3-AMINOBENZOIC ACID; KYSELINA 3-AMINO-2,5-DICHLORBENZOOVA; ORNAMENTAL WEEDER; ORNAMENTAL WEEDER 4G; VEGIBEN
Description: colorless crystalline solid; odorless
Use: weed control herbicide especially in soybean, navy bean, groundnut, maize, sweet potato, asparagus, squash and pumpkin plantings

Physical Properties

Freezing Point: 200 °C (392 °F) to 201 °C (393.8 °F)
Vapor Pressure: 0.007 mm Hg at 100 °C
Water Solubility: 700 mg/L at 25 °C
Other Solubilities: 95% Ethanol: >=100 mg/ml at 21 °C; Acetone: >=100 mg/ml at 21 °C; Benzene: 0.02 g/100 mL; Chloroform: 0.09 gm/100 mL; DMSO: >=100 mg/ml at 21 °C; Methanol: 22.3 gm/100 mL.
Flash Point: Not available; probably combustible

RTECS Toxicity Data

Acute Oral: Rat LD_{50} Dose: 3500 mg/kg. Mouse LD_{50} Dose: 3725 mg/kg.
Acute Dermal: Rabbit LD_{50} Route: Skin; Dose: 3136 mg/kg. Rat LD_{50} Route: Skin; Dose: >2200 mg/kg.
Chronic (Multiple Dose) Oral: Rat Dose: 364 mg/kg/52W-I; Toxic Effects: Biochemical - Transaminases; Biochemical - Other transferases; DEATH.
Mutagenic: Mouse Cytogenetic Analysis; Route: Intraperitoneal; Dose: 58500 ug/kg. Mouse Cytogenetic Analysis; Route: Oral; Dose: 234 mg/kg.
Tumorigenic: Mouse Route: Oral; Dose: 672 gm/kg/80W-C; Toxic Effects: Tumorigenic - Carcinogenic by RTECS criteria; Liver - Tumors. Mouse Route: Oral; Dose: 1344 gm/kg/80W-C; Toxic Effects: Tumorigenic - Carcinogenic by RTECS criteria; Liver - Tumors.

Hazard Overviews

Health: Irritating to eyes/skin/respiratory tract. Toxic. Other Acute Effects: harmful if swallowed, inhaled, or absorbed through skin. Chronic Effects: target organ: liver. Carcinogen.
Fire: Will burn. Hazards: emits toxic fumes. Extinguishing agents: carbon dioxide, dry chemical powder or appropriate foam. Precautions: combustible liquid.
Reactivity: Stable. Hazardous polymerization will not occur. Incompatible with: strong oxidizing agents, acids, bases, sodium hypochlorite. Hazardous decomposition products: toxic fumes of: carbon monoxide, carbon dioxide, nitrogen oxides, hydrogen chloride gas.
Carcinogenicity: IARC - Not listed; NIOSH - Not listed; NTP - Not listed; ACGIH - Not listed; OSHA - Not listed; EPA - Not listed; MAK - Not listed
Primary Target Organs:

Eyes

Skin

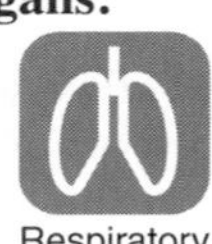
Respiratory System

Liver

Environmental

Ecotoxicity: Algae: Chlorcoccum sp. 10d EC_{50} 50-115 mg/l technical acid 2,225-4,000 mg/l ammonium salt
Environmental Fate: If released to the air, it may exist in the particulate or the vapor phase. Particulate phase may rapidly photodegrade while vapor phase It will degrade relatively rapidly by reaction with photochemically produced hydroxyl radicals (estimated half-life of about 15 hours). The pKa of the -COOH group is approximately 3 and the pKa of the -NH2 group is expected to be <4.6, thereby allowing adsorption of protonated material at low pHs. If released to water, it should not volatilize, bioconcentrate in aquatic organisms, or hydrolyze. In surface water exposed to sunlight, it will photodegrade rapidly (half-life of about 6 hr). In water and soil, it may chemically bind to soil and sediments with higher adsorption occurring either at a low pH with a low organic matter content or at a neutral pH with a high organic matter content. Biodegradation in nonsterile soils may occur via decarboxylation; however, the rate of microbial degradation is expected to depend on the organic content, temperature, and moisture content. On soil surfaces, some may photodegrade.

Environmental Physical Data

Henry's Law Constant: estimated at 2.1×10^{-11}
Sorption Partition Coefficient: K_{oc} = 190 to 21
BCF: estimated at 15.34

Regulations

RCRA 40CFR: Not listed
CERCLA: 40CFR 302.4: Listed per CAA Section 112 RQ: 100 lb (45.35 kg)
SARA 40CFR 372.65: Listed
SARA EHS 40CFR 355: Not listed
TSCA: Listed

Analytical Methods

Soil: SW846 8151
Drinking Water: EPA 515.1, 555
Food: AOAC 971.06
Plasma: FDA 221.1

CHL1150 CAS #: 305-03-3

CHLORAMBUCIL

RTECS: ES7525000
EINECS Number: 206-162-0
Molecular Formula: $C_{14}H_{19}Cl_2NO_2$
Structured MF: $(ClCH_2CH_2)_2NC_6H_4(CH_2)_3COOH$
Formula Weight: 304.23

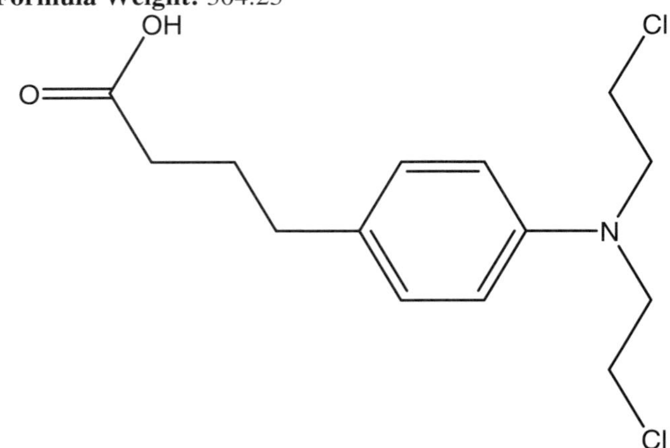

Chemical Structure

Synonyms: AMBOCHLORIN; AMBOCLORIN; BENZENEBUTANOIC ACID,4-(BIS(2-CHLOROETHYL)AMINO)-; 4-(BIS(2-CHLOROETHYL)AMINO)BENZENEBUTANOIC ACID; 4(P-BIS(BETA-CHLOROETHYL)AMINOPHENYL)BUTYRIC ACID; 4-(BIS(2-CHLOROETHYL)AMINO)PHENYLBUTYRIC ACID; 4-(P-(BIS(2-CHLOROETHYL)AMINO)PHENYL)BUTYRIC ACID; 4-(P-BIS(2-CHLOROETHYL)AMINOPHENYL)BUTYRIC ACID; 4-(P-BIS(BETA-CHLOROETHYL)AMINOPHENYL)BUTYRIC ACID; GAMMA-(P-BIS(2-CHLOROETHYL)AMINOPHENYL)BUTYRIC ACID; BUTANOIC ACID,4-(BIS(2-CHLOROETHYL)AMINO)BENZENE-; BUTYRIC ACID,4-(P-(BIS(2-CHLOROETHYL)AMINO)PHENYL); BUTYRIC ACID,4-(P-BIS(2-CHLOROETHYL)AMINOPHENYL)-; CB 1348; CHLORAMINOPHEN; CHLORAMINOPHENE; CHLORBUTIN; CHLOROAMBUCIL; CHLOROBUTIN; CHLOROBUTINE; P-(N,N-DI-2-CHLORETHYLAMINOPHENYL BUTYRIC ACID; N,N-DI-2-CHLOROETHYL-GAMMA-PARA-AMINOPHENYL BUTYRIC ACID; P-(N,N-DI-2-CHLOROETHYL)AMINOPHENYL BUTYRIC ACID; P-N,N-DI-(BETA-CHLOROETHYL)AMINOPHENYL BUTYRIC ACID; PARA-N,N-DI(BETA-CHLOROETHYL)AMINOPHENYL BUTYRIC ACID; GAMMA-(P-DI(2-CHLOROETHYL)AMINOPHENYL)BUTYRIC ACID; N,N-DI-2-CHLOROETHYL-GAMMA-P-AMINOPHENYLBUTYRIC ACID; PARA-(DI(2-CHLOROETHYL)AMINOPHENYL)BUTYRIC ACID; ECLORIL; ELCORIL; ELCORIN; KYSELINA 4-(N,N-BIS-(2-CHLORETHYL)-P-AMINOFENYL)MASELNA; LEUKERAN; LEUKERSAN; LEUKORAN; LINFOLIZIN; LINFOLYSIN; LYMPHOLYSIN; NSC-3088; NSC 3088; PHENYLBUTTERSAEURE-LOST; PHENYLBUTYRIC ACID NITROGEN MUSTARD
Description: off-white, slightly granular powder; slight odor
Use: an antineoplastic agent used in the treatment of malignant diseases such as lymphocytic leukemia, malignant lymphoma (including lymphosarcoma), giant follicular lymphoma,

Hodgkin's disease, primary macroglobulinemia, Kaposi's disease, reticulum-cell sarcoma and sarcoidosis and polycythemia vera; an immunosuppressive agent used in the treatment of systemic lupus erythematosus, acute and chronic glomerular nephritis, nephrotic syndrome, cold hemagglutinic disease, Wegener's granulomatosis and psoriasis; in veterinary medicine in the treatment of leukemias and, to a lesser degree, of solid tissue tumors; in immunosuppressive therapy of chronic hepatitis and of rheumatoid arthritis, and as an insect chemosterilant

Physical Properties

Freezing Point: 64 °C (147.2 °F) to 66 °C (150.8 °F)
Water Solubility: < 0.1 mg/mL at 22 C
Other Solubilities: 2.5 parts in Chloroform at 20 °C; 2 parts in Ethyl Acetate at 20 °C; Soluble in Benzene; readily Soluble in acid or alkali; Soluble in Ether.
pH: pKa= 5.75
Flash Point: Nonflammable

RTECS Toxicity Data

Acute Oral: Man TD_{Lo} Dose: 3571 ug/kg; Toxic Effects: Behavioral - Convulsions or effect on seizure threshold. Woman TD_{Lo} Dose: 82600 ug/kg; Toxic Effects: Lungs, Thorax, or Respiration - Fibrosis, interstitial; Lungs, Thorax, or Respiration - Cough; Lungs, Thorax, or Respiration - Dyspnea. Rat LD_{50} Dose: 76 mg/kg.
Acute Dermal: Rat LD_{Lo} Route: Subcutaneous Dose: 32 mg/kg; Toxic Effects: Tumorigenic - Active as anti-cancer agent. Mouse LD_{50} Route: Subcutaneous Dose: 115 mg/kg.
Reproductive/Teratogenic: Woman Route: Oral; Dose: 13 mg/kg; Duration: female 56D prior to mating Maternal Effects - Menstrual cycle changes or disorders. Woman Route: Oral; Dose: 5160 ug/kg; Duration: female 33-75D of pregnancy Specific Developmental Abnormalities - Urogenital system. Rat Route: Oral; Dose: 8 mg/kg; Duration: female 15D of pregnancy; Specific Developmental Abnormalities - Central nervous system; Eye, ear. Rat Route: Oral; Dose: 6 mg/kg; Duration: female 11D of pregnancy; Specific Developmental Abnormalities - Musculoskeletal system. Rat Route: Oral; Dose: 8 mg/kg; Duration: female 11D of pregnancy; Effects on Fertility - Post-implantation mortality. Rat Route: Oral; Dose: 10 mg/kg; Duration: female 15D of pregnancy; Effects on Embryo or Fetus - Fetal death.
Mutagenic: Human DNA Damage; Cell Type: lymphocyte; Dose: 40 umol/L. Human Unscheduled DNA Synthesis; Cell Type: lymphocyte; Dose: 1 umol/L. Human DNA Inhibition; Cell Type: HeLa cell; Dose: 15 umol/L. Human Cytogenetic Analysis; Route: Unreported; Dose: 3 mg/kg/8W-I. Human Cytogenetic Analysis; Cell Type: lymphocyte; Dose: 250 ug/L. Human Sister Chromatid Exchenge; Route: Unreported; Dose: 8800 ug/kg/44D. Human Sister Chromatid Exchenge; Cell Type: lymphocyte; Dose: 200 nmol/L. Human Sister Chromatid Exchenge; Cell Type: lymphocyte; Dose: 225 ug/L. Human Other Mutation Test Systems; Cell Type: HeLa cell; Dose: 5 mg/L.
Tumorigenic: Human Route: Oral; Dose: 141 mg/kg/5Y-I; Toxic Effects: Tumorigenic - Carcinogenic by RTECS criteria; Blood - Leukemia. Human Route: Oral; Dose: 84 mg/kg/3Y-C; Toxic Effects: Tumorigenic - Carcinogenic by RTECS criteria; Blood - Leukemia. Man Route: Oral; Dose: 84 mg/kg/2.5Y-C; Toxic Effects: Tumorigenic - Carcinogenic by RTECS criteria; Blood - Leukemia. Man Route: Oral; Dose: 59 mg/kg/96W-C; Toxic Effects: Tumorigenic - Carcinogenic by RTECS criteria; Blood - Leukemia. Woman Route: Oral; Dose: 101 mg/kg/82W-C; Toxic Effects: Tumorigenic - Carcinogenic by RTECS criteria; Blood - Leukemia. Woman Route: Oral; Dose: 200 mg/kg/6Y-C; Toxic Effects: Tumorigenic - Carcinogenic by RTECS criteria; Blood - Leukemia. Woman Route: Oral; Dose: 307 mg/kg/7Y-C; Toxic Effects: Tumorigenic - Carcinogenic by RTECS criteria; Blood - Leukemia. Woman Route: Oral; Dose: 180 mg/kg/3Y-I; Toxic Effects: Tumorigenic - Carcinogenic by RTECS criteria; Blood - Leukemia. Woman Route: Oral; Dose: 135 mg/kg/4Y-C; Toxic Effects: Tumorigenic - Carcinogenic by RTECS criteria; Blood - Leukemia. Woman Route: Oral; Dose: 70 mg/kg/94W-C; Toxic Effects: Tumorigenic - Carcinogenic by RTECS criteria; Blood - Leukemia. Child Route: Oral; Dose: 108 mg/kg/77W-C; Toxic Effects: Tumorigenic - Carcinogenic by RTECS criteria; Blood - Leukemia.

Hazard Overviews

Health: Irritating to eyes/skin/respiratory tract. Toxic. Other Acute Effects: harmful if swallowed, inhaled, or absorbed through skin. Chronic Effects: mutagen; possible teratogen; reproductive hazard; may cause heritable genetic damage; possible risk of harm to the unborn child; exposure can cause drug fever; myelosuppression; hepatotoxicity; infertility; seizures; gi toxicity; target organs: bone marrow. Carcinogen.
Fire: Noncombustible. Hazards: emits toxic fumes. Extinguishing agents: carbon dioxide, dry chemical powder or appropriate foam. Precautions: combustible liquid.
Reactivity: Stable. Hazardous polymerization will not occur. Hazardous decomposition products: toxic fumes of: carbon monoxide, carbon dioxide, nitrogen oxides, hydrogen chloride gas.
Carcinogenicity: IARC - Group 1, Carcinogenic to humans; NIOSH - Not listed; NTP - Listed; ACGIH - Not listed; OSHA - Not listed; EPA - Not listed; MAK - Not listed
Primary Target Organs:

Eyes

Skin

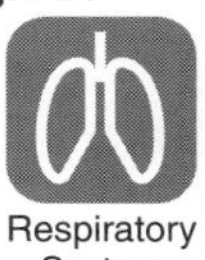
Respiratory System

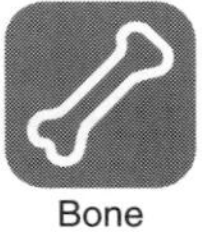
Bone

Environmental

Environmental Fate: If released to the atmosphere, vapor phase is expected to degrade rapidly by reaction with photochemically produced hydroxyl radicals (estimated half-life of 3 hours in air). Particulate phase can be physically removed from air by wet and dry deposition. If released to water, hydrolysis is expected to be the dominant environmental fate process. It has a measured aqueous hydrolysis half-life of 1.7 hours at 25 °C and pH 7. If released to moist soil, it can be expected to hydrolyze. Leaching in soil is not expected to be important due to the relatively rapid rate of hydrolysis. No data are available to predict the biodegradability in soil or water.

Cleanup/Disposal: Guide No. 154: Eliminate all ignition sources (no smoking, flares, sparks or flames in immediate area). Do not touch damaged containers or spilled material unless wearing appropriate protective clothing. Stop leak if you can do it without risk. Prevent entry into waterways, sewers, basements or confined areas. Absorb or cover with dry earth, sand or other non-combustible material and transfer to containers. Do not get water inside containers.

Environmental Physical Data

Henry's Law Constant: estimated at 2.8×10^{-11}
Octanol/Water Partition Coefficient: log K_{ow} = 3.611
Sorption Partition Coefficient: K_{oc} = estimated at 2500
BCF: estimated at 327

Regulations

RCRA 40CFR: Listed Hazardous Waste No. U035 Toxic Waste
CERCLA: 40CFR 302.4: Listed per RCRA Section 3001 RQ: 10 lb (4.535 kg)
SARA 40CFR 372.65: Not listed
SARA EHS 40CFR 355: Not listed
TSCA: Not listed

CHL1240 CAS #: 56-75-7

CHLORAMPHENICOL

RTECS: AB6825000
EINECS Number: 200-287-4
Molecular Formula: $C_{11}H_{12}Cl_2N_2O_5$
Formula Weight: 323.14

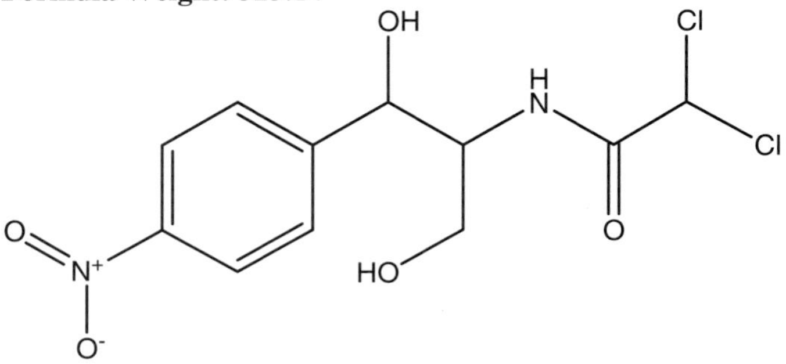

Chemical Structure

Synonyms: ACETAMIDE,2,2-DICHLORO-N-(BETA-HYDROXY-ALPHA-(HYDROXYMETHYL)-P-NITROPHENETHYL)-; ACETAMIDE,2,2-DICHLORO-N-(BETA-HYDROXY-ALPHA-(HYDROXYMETHYL)-P-NITROPHENETHYL)-,D-THREO-(-)-; ACETAMIDE,2,2-DICHLORO-N-(2-HYDROXY-1-(HYDROXYMETHYL)-2-(4-NITROPHENYL)ETHYL)-,(R-(R*,R*))-; ALFICETYN; AMBOFEN; AMPHENICOL; AMPHICOL; AMSECLOR; ANACETIN; AQUAMYCETIN; AUSTRACOL; BIOCETIN; BIOPHENICOL; CAF; CAF (PHARMACEUTICAL); CAM; CAP; CATILAN; CHEMICETIN; CHEMICETINA; CHLOMIN; CHLOMYCOL; CHLORAMEX; CHLORAMFENIKOL; D-CHLORAMPHENICOL; CHLORAMSAAR; CHLORASOL; CHLORA-TABS; CHLORICOL; CHLORMYCETIN R; CHLORNITROMYCIN; CHLORO-25 VETAG; CHLOROAMPHENICOL; CHLOROCAPS; CHLOROCID; CHLOROCID S; CHLOROCIDE; CHLOROCIDIN C; CHLOROCIDIN C TETRAN; CHLOROCOL; CHLOROJECT L; CHLOROMAX; CHLOROMYCETIN; CHLOROMYCETNY; CHLORONITRIN; CHLOROPTIC; CHLOROVULES; CIDOCETINE; CIPLAMYCETIN; CLORAMFICIN; CLORAMICAL; CLORAMICOL; CLORAMIDINA; CLOROAMFENICOLO; CLOROCYN; CLOROMISAN; CLOROSINTEX; COMYCETIN; CPH; CYLPHENICOL; DESPHEN; DETREOMYCIN; DETREOMYCINE; D-(-)-2,2-DICHLORO-N-(BETA-HYDROXY-ALPHA-(HYDROXYMETHYL)-P-NITROPHENYLETHYL)ACETAMIDE; DOCTAMICINA; ECONOCHLOR; EMBACETIN; EMETREN; ENICOL; ENTEROMYCETIN; ERBAPLAST; ERTILEN; FARMICETINA; FENICOL; GLOBENICOL; GLOROUS; HALOMYCETIN; HORTFENICOL; I 337A; INTRAMYCETIN; INTRAMYCTIN; ISICETIN; ISMICETINA; ISOPHENICOL; ISOPTO FENICOL; JUVAMYCETIN; KAMAVER; KEMICETINA; KEMICETINE; KLORITA; KLOROCID S; LEUKOMYAN; LEUKOMYCIN; LEVOMICETINA; LEVOMYCETIN; LOROMISIN; MASTIPHEN; MEDIAMYCETINE; MICLORETIN; MICROCETINA; MYCHEL; MYCINOL; NORMIMYCIN V; NOVOCHLOROCAP; NOVOMYCETIN; NOVOPHENICOL; NSC 3069; OFTALENT; OLEOMYCETIN; OPCLOR; OPHTHOCHLOR; OTACHRON; OTOPHEN; PANTOVERNIL; PARAXIN; PENTAMYCETIN; QUEMICETINA; RIVOMYCIN; ROMPHENIL; SEPTICOL; SIFICETINA; SINTOMICETINA; SINTOMICETINE R; STANOMYCETIN; SYNTHOMYCETIN; SYNTHOMYCETINE; TEVCOCIN; D-(-)-THREO-CHLORAMPHENICOL; D-THREO-CHLORAMPHENICOL; D(-)-THREO-2-DICHLOROACETAMIDO-1-P-NITROPHENYL-1,3-PROPANEDIOL; D-(-)-THREO-2-DICHLOROACETAMIDO-1-P-NITROPHENYL-1,3-PROPANEDIOL; D-THREO-N-DICHLOROACETYL-1-P-NITROPHENYL-2-AMINO-1,3-PROPANEDIOL; D(-)-THREO-2,2-DICHLORO-N-[BETA-HYDROXY-ALPHA-(HYDROXYMETHYL)-P-NITROPHENETHYL]ACETAMIDE; D-(-)-THREO-2,2-DICHLORO-N-(BETA-HYDROXY-ALPHA-(HYDROXYMETHYL))-P-NITROPHENETHYLACETAMIDE; D-THREO-N-(1,1'-DIHYDROXY-1-P-NITROPHENYLISOPROPYL)DICHLOROACETAMIDE; D-(-)-THREO-1-P-NITROPHENYL-2-DICHLORACETAMIDO-1,3-PROPANEDIOL; D-THREO-1-(P-NITROPHENYL)-2-(DICHLOROACETYLAMINO)-1,3-PROPANEDIOL; TIFOMYCIN; TIFOMYCINE; TREOMICETINA; U-6062; UNIMYCETIN; VETICOL

Description: white to greyish-white or yellowish-white fine crystalline powder, fine crystals, needles or elongated plates
Use: medication: antibacterial; antirickettsial; vet: antimicrobial agent; antifungal agent

Physical Properties

Boiling Point: Sublimes at High Vacuum
Freezing Point: 150.5 °C (302.9 °F) to 151.5 °C (304.7 °F)
Vapor Pressure: 1.73×10^{-12} mm Hg
Water Solubility: 1 parts: 25 mg/ml of Water at 25 °C
Other Solubilities: Soluble in Chloroform.
pH: Neutral

RTECS Toxicity Data

Acute Oral: Infant TD_{Lo} Dose: 440 mg/kg; Toxic Effects: Behavioral - Somnolence (general depressed activity); Gastrointestinal - Other changes; Nutritional and gross metabolic - Body temperature decrease. Woman LD_{Lo} Dose: 400 mg/kg; Toxic Effects: Behavioral - Coma; Vascular - Shock; Lungs, Thorax, or Respiration - Cyanosis. Rat LD_{50} Dose: 2500 mg/kg.

Acute Dermal: Rat LD_{50} Route: Subcutaneous Dose: 5 gm/kg; Toxic Effects: Gastrointestinal - Hypermotility, diarrhea. Mouse LD_{50} Route: Subcutaneous Dose: 400 mg/kg.
Chronic (Multiple Dose) Oral: Rat Dose: 60 mg/kg/60D-I; Toxic Effects: Nutritional and gross metabolic - Weight loss or decreased weight gain. Mouse Dose: 17416 mg/kg/4W-I; Toxic Effects: DEATH.
Chronic (Multiple Dose) Dermal: Mouse Route: Subcutaneous; Dose: 2800 mg/kg/2W-I; Toxic Effects: Behavioral - Food intake (animal); Nutritional and gross metabolic - Other changes; DEATH.
Reproductive/Teratogenic: Rat Route: Oral; Dose: 23 gm/kg; Duration: female 1-21D of pregnancy; Effects on Embryo or Fetus - Fetotoxicity; Other effects to embryo or fetus; Specific Developmental Abnormalities - Homeostasis. Rat Route: Oral; Dose: 2500 mg/kg; Duration: female 9D of pregnancy; Specific Developmental Abnormalities - Central nervous system. Rat Route: Oral; Dose: 2500 mg/kg; Duration: female 11D of pregnancy; Effects on Embryo or Fetus - Fetal death. Rat Route: Oral; Dose: 2 gm/kg; Duration: female 8D of pregnancy; Effects on Embryo or Fetus - Fetotoxicity; Fetal death; Specific Developmental Abnormalities - Body wall.
Mutagenic: Human Unscheduled DNA Synthesis; Cell Type: liver; Dose: 1 mmol/L. Human DNA Inhibition; Cell Type: bone marrow; Dose: 1500 umol/L. Human DNA Inhibition; Cell Type: lymphocyte; Dose: 1 mmol/L. Human Cytogenetic Analysis; Cell Type: leukocyte; Dose: 100 mg/L. Human Cytogenetic Analysis; Cell Type: lymphocyte; Dose: 500 mg/L.
Tumorigenic: Man Route: Oral; Dose: 434 mg/kg/W-C; Toxic Effects: Tumorigenic - Carcinogenic by RTECS criteria; Blood - Aplastic anemia; Blood - Leukemia. Woman Route: Oral; Dose: 300 mg/kg/60W-I; Toxic Effects: Tumorigenic - Carcinogenic by RTECS criteria; Blood - Changes in bone marrow not included above; Blood - Leukemia. Woman Route: Oral; Dose: 1680 mg/kg/6W-I; Toxic Effects: Tumorigenic - Carcinogenic by RTECS criteria; Blood - Aplastic anemia; Blood - Leukemia.

Hazard Overviews

Health: May be irritating to eyes/skin/respiratory tract. Toxic. Other Acute Effects: may be harmful by inhalation, ingestion, or skin absorption; may cause allergic reaction; may cause sensitization by inhalation and skin contact; exposure can cause: nausea, headache and vomiting. Chronic Effects: may alter genetic material; possible risk of congenital malformation in the fetus; target organs: blood, bone marrow, nerves, liver. Probable human carcinogen.
Fire: Hazards: emits toxic fumes. Extinguishing agents: water spray; carbon dioxide, dry chemical powder or appropriate foam. Precautions: combustible liquid.
Reactivity: Incompatible with: acids, acid chlorides, acid anhydrides, oxidizing agents. Hazardous decomposition products: carbon monoxide, carbon dioxide, nitrogen oxides, hydrogen chloride gas.
Carcinogenicity: IARC - Group 2A, Probably carcinogenic to humans; NIOSH - Not listed; NTP - Not listed; ACGIH - Not listed; OSHA - Not listed; EPA - Not listed; MAK - Not listed
Primary Target Organs:

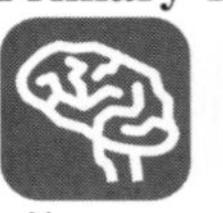

Nervous System Liver Blood Bone

Exposure Limits
AIHA WEEL: TWA: 0.5 mg/m^3.
Respirator Recommendation
Exposure Range: >0.5 to 5 ppm Air Purifying, Negative Pressure, Half Mask
Exposure Range: >5 to 50 ppm Air Purifying, Negative Pressure, Full Face
Exposure Range: >50 to 500 ppm Supplied Air, Constant Flow/Pressure Demand, Full Face
Exposure Range: >500 to unlimited ppm Self-contained Breathing Apparatus, Pressure Demand, Full Face
Cartridge Color: magenta (P100)

Environmental

Ecotoxicity: MicrotoxTM (Photobacterium) test 5 min EC 501,715 mg/l Artoxkit M (Artemia salina) test 24h LC 502,041 mg/l
Environmental Fate: If released to water, it will be essentially non-volatile. Adsorption to sediment and bioconcentration should not be important fate processes. It may biodegrade in soil and water. If released to the atmosphere, it will exist primarily in the particulate phase. Removal of atmospheric material may occur through dry deposition. If released to soil, it is expected to have very high soil mobility based on estimated K_{oc} values of 10 to 47. Volatilization is not expected from either dry or moist soils

Environmental Physical Data

Henry's Law Constant: estimated at 2.29 $x10^{-18}$
Sorption Partition Coefficient: K_{oc} = estimated at 10
BCF: calculated at 6

Regulations

RCRA 40CFR: Not listed
CERCLA: 40CFR 302.4: Not listed
SARA 40CFR 372.65: Not listed
SARA EHS 40CFR 355: Not listed
TSCA: Listed

CHL1330 CAS #: 103-17-3

CHLORBENSIDE

RTECS: WQ2975000
DOT: UN2761
EINECS Number: 203-084-9
Molecular Formula: $C_{13}H_{10}Cl_2S$
Formula Weight: 269.19
Synonyms: BENZENE,1-CHLORO-$-(((4-CHLOROPHENYL)METHYL)THIO)-; BENZENE,1-CHLORO-4-(((4-CHLOROPHENYL)METHYL)THIO)-; CHLOORBENZIDE; (4-CHLOOR-

BENZYL)-(4-CHLOOR-FENYL)-SULFIDE; CHLORACID; CHLORBENSID; (4-CHLOR-BENZYL)-(4-CHLOR-PHENYL)-SULFID; P-CHLOROBENZYL P-CHLOROPHENYL SULFIDE; 4-CHLOROBENZYL 4-CHLOROPHENYL SULPHIDE; P-CHLOROBENZYL P-CHLOROPHENYL SULPHIDE; 1-CHLORO-4-(((4-CHLOROPHENYL)METHYL)THIO)BENZENE; 4-CHLOROPHENYL 4'-CHLOROBENZYL SULFIDE; CHLORPARACIDE; CHLORSULPHACIDE; (4-CLORO-BENZIL)-(4-CLORO-FENIL)-SOLFURO; CP 20; P,P'-DICHLORODIPHENYL SULFIDE; ENT 20,696; HRS 860; MITOX; RD 2195; SULFIDE,P-CHLOROBENZYL P-CHLOROPHENYL; SULFURE DE 4-CHLOROBENZYLE ET DE 4-CHLOROPHENYLE

Description: colorless crystals; almond-like odor

Use: acaricide (former use); control of most varieties of spider mites, particularly in the pre-blossom stage in fruit cultivation & on ornamentals & nursery stock (former use)

Physical Properties

Freezing Point: 75 °C (167 °F) to 76 °C (168.8 °F)

Specific Gravity: 1.421 at 25 °C/4 °C

Saturated Vapor Density: 1.200000158 kg/m^3

Vapor Pressure: 1.21 x10^{-5} mm Hg at 30 °C

Water Solubility: < 1 parts /5000 parts Water

Other Solubilities: in Ethanol about 2.9%. in Kerosene 5-7.5%. Soluble in Acetone, Benzene, Xylene, Toluene, Petroleum Ether

Flash Point: Does not burn or burns with difficulty

RTECS Toxicity Data

Acute Oral: Rat LD_{50} Dose: 2 gm/kg. Mouse LD_{50} Dose: >3 gm/kg.

Hazard Overviews

Fire: Noncombustible.

Carcinogenicity: IARC - Not listed; NIOSH - Not listed; NTP - Not listed; ACGIH - Not listed; OSHA - Not listed; EPA - Not listed; MAK - Not listed

Environmental

Cleanup/Disposal: Guide No. 151: Do not touch damaged containers or spilled material unless wearing appropriate protective clothing. Stop leak if you can do it without risk. Prevent entry into waterways, sewers, basements or confined areas. Cover with plastic sheet to prevent spreading. Absorb or cover with dry earth, sand or other non-combustible material and transfer to containers. Do not get water inside containers.

Regulations

RCRA 40CFR: Not listed

CERCLA: 40CFR 302.4: Not listed

SARA 40CFR 372.65: Not listed

SARA EHS 40CFR 355: Not listed

TSCA: Not listed

Analytical Methods

Water / Groundwater: EPA 022

Food: FDA 212.1, 212.2, 232.1, 232.4, 242.1

Plasma: EPA 001, 027, 028; FDA 211.1, 231.1, 252

CHL1480 CAS #: 57-74-9

CHLORDANE

RTECS: PB9800000

DOT: UN2761; UN2762; UN2995; UN2996; NA2762; IMO3.0; IMO6.1

EINECS Number: 200-349-0

Molecular Formula: $C_{10}H_6Cl_8$

Formula Weight: 409.80

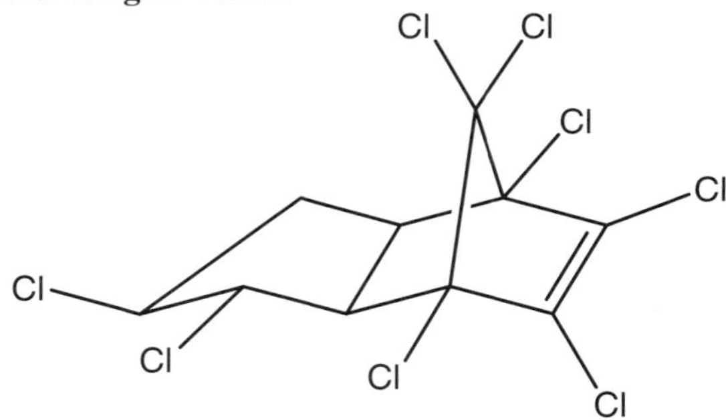

Chemical Structure

Synonyms: 1068; ASPON-CHLORDANE; BELT; CD 68; CHLOORDAAN; CHLOR KIL; CHLORDAN; GAMMA-CHLORDAN; CHLORDANE,LIQUID; CHLORDANO; CHLORINDAN; CHLORODANE; CHLORTOX; CLORDAN; CLORDANO; CORODANE; CORTILAN-NEU; DICHLOROCHLORDENE; DOWCHLOR; ENT 25,552-X; ENT-9932; ENT 9,932; HCS 3260; INTOX 8; KILEX LINDANE; KYPCHLOR; LATKA 1068; M 140; M 410; 4,7-METHANOINDAN,1,2,4,5,6,7,8,8-OCTACHLORO-3A,4,7,7A-TETRAHYDRO-; 4,7-METHANO-1H-INDENE,1,2,4,5,6,7,8,8-OCTACHLORO-2,3,3A,4,7,7A-HEXAHYDRO-; NIRAN; 1,2,4,5,6,7,8,8-OCTACHLOOR-3A,4,7,7A-TETRAHYDRO-4,7-ENDO-METHANO-INDAAN; OCTACHLOR; 1,2,4,5,6,7,8,8-OCTACHLOR-2,3,3A,4,7,7A-HEXAHYDRO-4,7-METHANOINDANE; OCTACHLORO TETRAHYDRO METHANOINDAN; OCTACHLORODIHYDRODICYCLOPENTADIENE; 1,2,4,5,6,7,8,8-OCTACHLORO-2,3,3A,4,7,7A-HEXAHYDRO-4,7-METHANO-1H-INDENE; 1,2,4,5,6,7,8,8-OCTACHLORO-2,3,3A,4,7,7A-HEXAHYDRO-4,7-METHANOINDENE; 1,2,4,5,6,7,8,8-OCTACHLORO-3A,4,7,7A-HEXAHYDRO-4,7-METHYLENE INDANE; 1,2,4,5,6,7,8,8-OCTACHLORO-4,7-METHANE-3A,4,7,7A-TETRAHYDROINDANE; OCTACHLORO-4,7-METHANOHYDROINDANE; 1,2,4,5,6,7,8,8-OCTACHLORO-4,7-METHANO-3A,4,7,7A-TETRAHYDROINDANE; OCTACHLORO-4,7-METHANOTETRAHYDROINDANE; 1,2,4,5,6,7,8,8-OCTACHLORO-3A,4,7,7A-TETRAHYDRO-4,7-METHANOINDAN; 1,2,4,5,6,7,8,8-OCTACHLORO-3A,4,7,7A-TETRAHYDRO-4,7-METHANOINDAN E; 1,2,4,5,6,7,8,8-OCTACHLORO-3A,4,7,7A-TETRAHYDRO-4,7-METHANOINDANE; 1,2,4,5,6,7,10,10-OCTACHLORO-4,7,8,9-TETRAHYDRO-4,7-METHYLENEINDANE; 1,2,4,5,6,7,8,8-OCTACHLOR-3A,4,7,7A-TETRAHYDRO-4,7-ENDO-METHANO-INDAN; 1,2,4,5,6,7,8,8-OCTAHYDRO-2,3,3A,4,7,7A-HEXAHYDRO-4,7-METHANOINDAN; OCTA-KLOR; OKTATERR; OMS 1437; ORTHO-KLOR; 1,2,4,5,6,7,8,8-OTTOCHLORO-3A,4,7,7A-TETRAHYDRO-4,7-ENDO-METHANO--INDAAN; 1,2,4,5,6,7,8,8-OTTOCHLORO-3A,4,7,7A-TETRAIDRO-4,7-ENDO-METANO-INDANO; SD 5532; SHELL SD-5532; STARCHLOR; SYNKLOR; TAT CHLOR 4; TERMI-DED; TOPICHLOR 20; TOPICLOR; TOPICLOR 20; TOXICHLOR; UNEXAN-KOEDER; VELSICOL 1068

Description: solid; chlorine odor

Use: as a non-systemic contact and stomach insecticide with some fumigant action; as an acaricide, a pesticide and wood preservative; in termite control and as a protective treatment for underground cables

Physical Properties

Boiling Point: 175 °C (347 °F) at 2 mm Hg
Freezing Point: 107 °C (224.6 °F) to 108.8 °C (227.84 °F)
Specific Gravity: 1.59 to 1.63 at 25 °C
Vapor Density: 14
Saturated Vapor Density: 1.200000207 kg/m^3
Density: 1.59 to 1.63 g/mL at 25 °C
Vapor Pressure: 1 x10^{-5} mm Hg at 25 °C
Water Solubility: Insoluble in Water
Other Solubilities: DMSO: >=100 mg/ml at 23 °C; 95% Ethanol: 50-100 mg/ml at 23 °C; Acetone: >=100 mg/ml at 23 °C; Petroleum solvents: Soluble; Petroleum hydrocarbons: Soluble; Most organic solvents: Soluble.
Surface Tension: 0.025 dynes/cm
Odor Threshold: 0.0084 to 0.0419 mg/m^3
Refraction Index: 1.56 to 1.57 at 25 °C/D
Flash Point: 107.222 °C Open Cup
Autoignition Temperature: 210 °C
LEL: 0.7% v/v
UEL: 5% v/v

RTECS Toxicity Data

Acute Oral: Man TD_{Lo} Dose: 3071 ug/kg; Toxic Effects: Behavioral - Coma; Lungs, Thorax, or Respiration - Dyspnea; Gastrointestinal - Nausea or vomiting. Human LD_{Lo} Dose: 29 mg/kg; Toxic Effects: Liver - Fatty Liver degeneration. Woman LD_{Lo} Dose: 120 ug/kg; Toxic Effects: Behavioral - Convulsions or effect on seizure threshold; Behavioral - Excitment; Gastrointestinal - Gastritis. Rat LD_{50} Dose: 200 mg/kg.
Acute Inhalation: Cat LC_{50} Dose: 100 mg/m^3/4hr; Toxic Effects: Behavioral - Somnolence (general depressed activity); Behavioral - Convulsions or effect on seizure threshold; Behavioral - Excitement. Mammal LC_{50} Dose: >20 gm/m^3/4hr.
Acute Dermal: Human LD_{Lo} Route: Skin; Dose: 428 mg/kg; Toxic Effects: Behavioral - Tremor; Behavioral - Convulsions or effect on seizure threshold; Behavioral - Ataxia. Rabbit LD_{50} Route: Skin; Dose: 780 mg/kg.
Chronic (Multiple Dose) Oral: Rat Dose: 660 mg/kg/15D-I; Toxic Effects: Behavioral - Convulsions or effect on seizure threshold; DEATH. Rat Dose: 6512 mg/kg/58W-I; Toxic Effects: Liver - Changes in liver weight; Nutritional and gross metabolic - Weight loss or decreased weight gain. Rat Dose: 5475 mg/kg/2Y-C; Toxic Effects: Liver - Hepatitis (hepatocellular necrosis), zonal; Liver - Changes in liver weight; DEATH. Rat Dose: 504 mg/kg/12W-C; Toxic Effects: Blood - Changes in serum composition; Biochemical - True cholinesterase.
Chronic (Multiple Dose) Inhalation: Rat Dose: 28200 ug/m^3/8H/28D-I; Toxic Effects: Liver - Changes in Liver weight; Kidney, Ureter, and Bladder - Changes in kidney weight; Blood - Changes in serum composition. Rat Dose: 10 mg/m^3/8H/90D-I; Toxic Effects: Liver - Changes in Liver weight; Kidney, Ureter, and Bladder - Changes in kidney weight; Nutritional and gross metabolic - Changes in calcium.
Reproductive/Teratogenic: Mouse Route: Oral; Dose: 3360 ug/kg; Duration: female 1-21D of pregnancy; Specific Developmental Abnormalities - Endocrine system; Effects on Newborn - Biochemical and metabolic. Mouse Route: Oral; Dose: 152 mg/kg; Duration: female 1-19D of pregnancy; Specific Developmental Abnormalities - Immune and reticuloendothelial system. Mouse Route: Oral; Dose: 7 mg/kg; Duration: female 15-21D of pregnancy Effects on Newborn - Behavioral. Mouse Route: Oral; Dose: 3040 ug/kg; Duration: female 1-19D of pregnancy; Effects on Newborn - Biochemical and metabolic.
Mutagenic: Human Unscheduled DNA Synthesis; Cell Type: fibroblast; Dose: 1 umol/L. Human Sister Chromatid Exchenge; Cell Type: lymphocyte; Dose: 10 umol/L.
Tumorigenic: Mouse Route: Oral; Dose: 2020 mg/kg/80W-C; Toxic Effects: Tumorigenic - Carcinogenic by RTECS criteria; Liver - Tumors. Mouse Route: Oral; Dose: 3780 mg/kg/80W-C; Toxic Effects: Tumorigenic - Carcinogenic by RTECS criteria; Liver - Tumors.

Hazard Overviews

Fire Diamond

Health: Irritating to eyes/skin/respiratory tract. Also Causes: respiratory failure, confusion, agitation, tremor, incoordination, delirium, convulsions, liver/kidney damage. Chronic Effects: sinusitis, bronchitis, migraine, asthma, neuritis, neuralgia, ovarian and uterine diseases. Possible cancer hazard.
Fire: Noncombustible when pure. However, if it is dissolved in a solvent, it may be combustible or flammable depending on the solvent. Use dry chemical, carbon dioxide, water spray, or alcohol-resistant foam. Container may explode in heat of fire.
Reactivity: Stable. Hazardous polymerization cannot occur. Avoid: heat and ignition sources. Incompatible with: oxidizers; alkalis; some forms of plastics, rubber, and coatings; corrosive to iron and zinc. Hazardous decomposition products: chlorine; phosgene; hydrogen chloride; carbon oxide gases.
Carcinogenicity: IARC - Group 2B, Possibly carcinogenic to humans; NIOSH - Listed as carcinogen; NTP - Not listed; ACGIH - Class A3, Animal carcinogen; OSHA - Not listed; EPA - Class B2, Probable human carcinogen based on animal studies; MAK - Class B, Justifiably suspected of having carcinogenic potential
Primary Target Organs:

Eyes

Skin

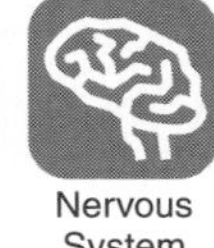
Nervous System

Liver

Kidneys

Blood

Exposure Limits
OSHA PEL: TWA: 0.5 mg/m^3; skin.

ACGIH TLV: TWA: 0.5 mg/m^3.
NIOSH REL: TWA: 0.5 mg/m^3.
NIOSH IDLH: 100 mg/m^3.
DFG MAK: TWA: 0.5 mg/m^3.
Respirator Recommendation
Exposure Range: >0.5 to 25 mg/m^3 Supplied Air, Constant Flow/Pressure Demand, Half Mask
Exposure Range: >25 to <100 mg/m^3 Supplied Air, Constant Flow/Pressure Demand, Full Face
Exposure Range: 100 to unlimited mg/m^3 Self-contained Breathing Apparatus, Pressure Demand, Full Face
Note: odor threshold unknown

Environmental

Ecotoxicity: LC_{50} Bobwhite quail oral 331 ppm in 5 day diet (95% confidence limit 197-497 ppm) LD_{50} Mallard (female) oral 1200 mg/kg (95% confidence limit 954-1510 mg/kg) LC_{50} Planarian 1-5 ppm/10 days /Conditions of bioassay not specified LC_{50} Ictalurus punctatus (channel catfish) 500 ug/l/96 hr /Conditions of bioassay not specified LC_{50} Chironomus plumosus (midge, larva) 10 ug/l/24 hr /Conditions of bioassay not specified Crassostrea virginica (eastern oyster) 10 ug/l/24 hr; Toxic effect: growth affected LC_{50} Mugil curema (white mullet) 5.5 ug/l/48 hr /Conditions of bioassay not specified LC_{50} Palaemonetes kadiakensis (decapod) 4.0 ug/l/96 hr /Conditions of bioassay not specified LC_{50} Pteronarcys californica 15 ug/l/96 hr (95% confidence limit 9-24 ug/l) at 15 °C, second yr class. Static bioassay without aeration, pH 7.2-7.5, water hardness 40-50 mg/l as calcium carbonate and alkalinity of 30-35 mg/l LC_{50} Brown trout 11.1 ug/l/96 hr (95% confidence limit 9.3-13.1 ug/l) at 12 °C, wt 0.6 g. Static bioassay without aeration, pH 7.2-7.5, water hardness 40-50 mg/l as calcium carbonate and alkalinity of 30-35 mg/l LC_{50} Daphnia pulex 24 ug/l/48 hr (95% Confidence limit 20-28 ug/l) at 15 °C, first instar. Static bioassay without aeration, pH 7.2-7.5, water hardness 40-50 mg/l as calcium carbonate and alkalinity of 30-35 mg/l

Environmental Fate: If released to soil, it may persist for long periods of time; under field conditions, the mean degradation rate has been observed to range from 4.05-28.33%/yr with a mean half-life of 3.3 years. It is expected to be generally immobile or only slightly mobile in soil based on field tests, soil column leaching tests and K_{oc} estimation; however, its detection in various groundwaters in NJ and elsewhere indicates that movement to groundwater can occur. Soil volatility tests have found that it can volatilize significantly from soil surfaces on which it has been sprayed, particularly moist soil surfaces; however, shallow incorporation into soil will greatly restrict volatile losses. Although sufficient biodegradation data are not available, it has been suggested that it is very slowly biotransformed in the environment which is consistent with the long persistence periods observed under field conditions. If released to water, it is not expected to undergo significant hydrolysis, oxidation or direct photolysis. The volatilization half-life from a river one meter deep flowing 1 m/sec with a wind velocity of 3 m/sec is estimated to be 7.3-7.9 hours at 23 °C while the volatilization half-lives from a representative environmental pond, river and lake are estimated to be 18-26, 3.6-5.2 and 14.4-20.6 days, respectively. However, adsorption to sediment significantly attenuates the importance of volatilization. Adsorption to sediment is expected to be a major fate process based on soil adsorption data, estimated K_{oc} values (15,500-24,600), and extensive sediment monitoring data. Bioconcentration is expected to be important based on experimental log BCF values which are generally above 3,200. Sensitized photolysis in the water column may be possible. The presence in sediment core samples suggests that it may be very persistent in the adsorbed state in the aquatic environment. If released to the atmosphere it will be expected to exist predominately in the vapor phase. It will react in the vapor-phase with photochemically produced hydroxyl radicals at an estimated half-life rate of 6.2 hr suggesting that this reaction is the dominant chemical removal process. The detection in remote atmospheres (Pacific and Atlantic Oceans; The Arctic) indicates that long range transport occurs. It has been estimated that 96% of the airborne reservoir exists in the sorbed state which may explain why its long range transport is possible without chemical transformation. The detection in rainwater and its observed dry deposition at various rural locations indicates that physical removal via wet and dry deposition occurs in the environment.

Cleanup/Disposal: Guide No. 131: Fully encapsulating, vapor protective clothing should be worn for spills and leaks with no fire. Eliminate all ignition sources (no smoking, flares, sparks or flames in immediate area). All equipment used when handling the product must be grounded. Do not touch or walk through spilled material. Stop leak if you can do it without risk. Prevent entry into waterways, sewers, basements or confined areas. A vapor suppressing foam may be used to reduce vapors. Small Spills: Absorb with earth, sand or other non-combustible material and transfer to containers for later disposal. Use clean non-sparking tools to collect absorbed material. Large Spills: Dike far ahead of liquid spill for later disposal. Water spray may reduce vapor; but may not prevent ignition in closed spaces. Guide No. 151: Do not touch damaged containers or spilled material unless wearing appropriate protective clothing. Stop leak if you can do it without risk. Prevent entry into waterways, sewers, basements or confined areas. Cover with plastic sheet to prevent spreading. Absorb or cover with dry earth, sand or other non-combustible material and transfer to containers. Do not get water inside containers.

Environmental Physical Data

Henry's Law Constant: 8.6 $\times 10^{-4}$
Octanol/Water Partition Coefficient: log K_{ow} = 3.32
Sorption Partition Coefficient: K_{oc} = 1720 to 1.55 $\times 10^4$
BCF: pinfish 6227

Regulations

RCRA 40CFR: Listed Hazardous Waste No. U036 Toxic Waste
CERCLA: 40CFR 302.4: Listed per CWA Section 311(b)(4) per RCRA Section 3001 per CWA Section 307(a) RQ: 1 lb (0.454 kg)

SARA 40CFR 372.65: Listed TPQ: 1000 lb
SARA EHS 40CFR 355: Listed TPQ: 1 lb
TSCA: Not listed

Analytical Methods

Air: EPA TO-10, TO-4, 016
Soil: EPA 16, 3, P-002-1, P-011-1; SW846 3650A, 3650B, 4041, 8080A, 8250A, 8270B, 8270C; DOE OP130R; USGS O5104, O7104
Water / Groundwater: EPA P-003-1, P-004-1, 608, 617, 625, 625-S, 680, 022; APHA 6410-B, 6630-C, 6630-D; ASTM D3086; USGS O3104
Drinking Water: EPA 505, 508; ASTM D5175
Food: FDA 212.1, 212.2, 232.1, 232.4, 242.1; EPA 4; AOAC 962.05, 972.05; USGS O9104
Indoor / Expired Air: NIOSH 5510; EPA IP-8; ASTM D4861, D4947
Plasma: EPA 001, 027, 29; FDA 211.1, 231.1, 251.1, 252

CHL1510 CAS #: 143-50-0

CHLORDECONE

RTECS: PC8575000
DOT: NA2761
EINECS Number: 205-601-3
Molecular Formula: $C_{10}Cl_{10}O$
Formula Weight: 490.68
Synonyms: CIBA 8514; CLORDECONE; COMPOUND 1189; 1,2,3,5,6,7,8,9,10,10-DECACHLORO(5.2.1.0(SUP 2,6).0(SUP 3,9).0(SUP 5,8))-; 1,2,3,5,6,7,8,9,10,10-DECACHLORO(5.2.1.0(SUP 2,6).0(SUP 3,9).0(SUP 5,8))DECANO-4-ONE; 1,1A,3,3A,4,5,5,5A,5B,6-DECACHLOROCTAHYDRO-1,3,4-METHENO-2H-CYCLOBUTA(CD)-; DECACHLOROKETONE; DECACHLORO-1,3,4-METHENO-2H-CYCLOBUTA(CD)PENTALEN-2-ONE; DECACHLOROOCTAHYDRO-KEPONE-2-ONE; DECACHLOROOCTAHYDROKEPONE-2-ONE; 1,1A,3,3A,4,5,5,5A,5B,6-DECACHLORO-OCTAHYDRO-1,3,4-METHENO-2H-CYCLOBUTA[CD] PENTALEN-2-ONE; DECACHLOROOCTAHYDRO-1,3,4-METHENO-2H-CYCLOBUTA(CD)-PENTALEN-2-ON E; 1,1A,3,3A,4,5,5,5A,5B,6-DECACHLOROOCTAHYDRO-1,3,4-METHENO-2H-CYCLOBUTA(CD)PENTALEN-2-ONE; DECACHLORO-OCTAHYDRO-1,3,4-METHENO-2H-CYCLOBUTA(CD)PENTALEN-2-ONE; DECACHLOROOCTAHYDRO-1,3,4-METHENO-2H-CYCLOBUTA(CD)PENTALEN-2-ONE; DECACHLOROPENTACYCLO(5.2.1.0(SUP 2,6).0(SUP 3,9).0(SUP 5,8))DECAN-4-ONE; DECACHLOROPENTACYCLO(5.3.0.0(SUP 2,6).0(SUP 4,10).0(SUP 5,9))DECAN-3-ONE; DECACHLOROPENTACYCLO(5.2.1.0(2,6).0(3,9).0(5),(8))DECAN-4-ONE; DECACHLOROTETRACYCLODECANONE; DECACHLOROTETRAHYDRO-4,7-METHANOINDENEONE; 2,3,3A,4,5,6,7,7A,8,8A-DECACHLORO-3A,4,7,7A-TETRAHYDRO-4,7-METHANOINDEN-1-ONE; DECANO-4-ONE; 1,2,3,4,5,5,6,7,8,9,10,10-DODECACHLOROOCTAHYDRO-1,3,4-METHENO-2-CYCLOBUTA-(C,D) -PENTALONE; ENT-16391; ENT 16,391; GC 1189; GENERAL CHEMICALS 1189; KEPONE; KEPONE-2-ONE,DECACHLOROOCTAHYDRO-; MEREX; 1,3,4-METHENO-2H-CYCLOBUTA(CD)PENTALEN-2-ONE,1,1A,3,3A,4,5,5,5A,5B,6-DECACHLOROOCTAHYDRO-; PENTALEN-2-ONE; PERCHLOROPENTACYCLO[5.3.0.0(2,6).0(3,9).0(4,8) DECAN-5-ONE
Use: as an insecticide, fungicide, pesticide for control of the banana root borer and tobacco wireworm and bait for control of ants an cockroaches

Physical Properties

Boiling Point: Sublimes
Freezing Point: Decomposes at 350 °C (662 °F)
Specific Gravity: 1.61
Vapor Density: 16.94 Air=1
Saturated Vapor Density: < 1.200000008 kg/m^3
Vapor Pressure: < 0.0000003 mm Hg at 25 °C
Water Solubility: < 1 mg/mL at 23 C
Other Solubilities: Soluble in hydrocarbon solvents; Soluble in Alcohols, Ketones, Acetic Acid; Soluble in hexane; Soluble in organic solvents such as Benzene and hexane; Soluble in light petroleum and may be recrystallized from 85-90% aqueous Ethanol.
Flash Point: Not available; probably combustible

RTECS Toxicity Data

Acute Oral: Rat LD_{50} Dose: 91300 ug/kg. Rabbit LD_{50} Dose: 65 mg/kg.
Acute Dermal: Rabbit LD_{50} Route: Skin; Dose: 345 mg/kg. Rat LD_{50} Route: Skin; Dose: >2 gm/kg.
Chronic (Multiple Dose) Oral: Rat Dose: 307 mg/kg/15W-I; Toxic Effects: Behavioral - Alteration of operant conditioning; Nutritional and gross metabolic - Weight loss or decreased weight gain; Nutritional and gross metabolic - Body temperature increase. Rat Dose: 912 mg/kg/2Y-C; Toxic Effects: Liver - Changes in liver weight; Kidney, Ureter, and Bladder - Changes in kidney weight; Nutritional and gross metabolic - Weight loss or decreased weight gain. Rat Dose: 100 mg/kg/10D-I; Toxic Effects: Endocrine - Changes in spleen weight; Blood - Changes in other cell count; Immunological including allergic - Decreased immune response. Rat Dose: 162 mg/kg/90D-C; Toxic Effects: Biochemical - Dopamine in striatum. Rat Dose: 287 mg/kg/35D-C; Toxic Effects: Liver - Other changes; Liver - Changes in liver weight. Rat Dose: 75 mg/kg/3D-I; Toxic Effects: Behavioral - Tremor; Musculoskelital - Other changes; Biochemical - Phosphatases.
Reproductive/Teratogenic: Rat Route: Oral; Dose: 63 mg/kg; Duration: female 2-22D of pregnancy; Effects on Newborn - Biochemical and metabolic. Rat Route: Oral; Dose: 42 mg/kg; Duration: female 2-22D of pregnancy; Effects on Newborn - Other postnatal measures or effects. Rat Route: Oral; Dose: 100 mg/kg; Duration: female 7-16D of pregnancy; Specific Developmental Abnormalities - Central nervous system; Urogenital system; Homeostasis. Rat Route: Oral; Dose: 60 mg/kg; Duration: female 7-16D of pregnancy; Effects on Embryo or Fetus - Fetotoxicity; Specific Developmental Abnormalities - Musculoskeletal system. Rat Route: Oral; Dose: 17 mg/kg; Duration: female 1-22D of pregnancy; Effects on Newborn - Behavioral. Rat Route: Oral; Dose: 18 mg/kg; Duration: male 5D prior to mating; Effects on Fertility - Pre-implantation mortality; Litter size.
Mutagenic: Rat Other Mutation Test Systems; Cell Type: liver; Dose: 100 umol/L. Hamster Sister Chromatid Exchenge; Cell Type: ovary; Dose: 1700 ug/L.
Tumorigenic: Rat Route: Oral; Dose: 200 mg/kg/2Y-C; Toxic Effects: Tumorigenic - Carcinogenic by RTECS criteria;

Liver - Tumors. Rat Route: Oral; Dose: 670 mg/kg/80W-I; Toxic Effects: Tumorigenic - Carcinogenic by RTECS criteria; Liver - Tumors.

Hazard Overviews

Reactivity: Unstable, readily hydrates on exposure to moisture. Hazardous polymerization cannot occur. Avoid: heat; moisture. Hazardous decomposition products: chlorine; hydrogen chloride; carbon oxide(s) gases.

Carcinogenicity: IARC - Group 2B, Possibly carcinogenic to humans; NIOSH - Listed as carcinogen; NTP - Class 2B, Reasonably anticipated to be a carcinogen, sufficient evidence of carcinogenicity from studies in experimental animals; ACGIH - Not listed; OSHA - Not listed; EPA - Not listed; MAK - Class B, Justifiably suspected of having carcinogenic potential

Exposure Limits

NIOSH REL: TWA: 0.001 mg/m^3.

Environmental

Ecotoxicity: EC_{50} Daphnia magna 260 ug/l/48 hr at 17 °C (95% confidence limit 200-345 ug/l), first instar, static bioassay EC_{50} Chironomus 320 ug/l/48 hr at 22 °C (95% confidence limit 220-450 ug/l), fourth instar, static bioassay LC_{50} Trout 0.066 ppm/24 hr; 0.038 ppm/48 hr; 0.02 ppm/96 hr /Conditions of bioassay not specified LC_{50} Cyprinodon variegatus (sheepshead minnow) 69.5 ug/l/96 hr in a flow through bioassay LC_{50} Salmo gardnerii (Rainbow trout) 30 ug/l/96 hr at 12 °C (95% confidence limit 24-38 ug/l), wt 1.1 g, static bioassay EC_{50} Chlorococcum species (algae) 0.35 ug/l/7 days at a temperature of 20 + or - 0.5 °C in a static system experienced growth retardation LC_{50} Brevoortia tyrannus (Atlantic menhaden) 17.4 ug/l/96 hr /Conditions of bioassay not specified LC_{50} Palaemonetes pugio (decapod) 120.9 ug/l/96 hr in a flow-through bioassay LC_{50} Callinectes sapidus (decapod) < 210 ug/l/96 hr in a flow-through bioassay

Environmental Fate: Released to soil it will be expected to adsorb to the soil; However, some leaching to the groundwater may occur, especially in sandy soils and other soils with low organic content. Biodegradation and hydrolysis will not be important fate processes but some evaporation may be observed from the surface of the soil. Released to the water it will be expected to adsorb to the sediment and to bioconcentrate in fish but may not bioconcentrate in crustaceans or other aquatic organisms. It will not be expected to hydrolyze, or biodegrade, and direct photodegradation is not expected to be significant compared to other processes. Evaporation from water also should not be significant, with a half-life of 3.8-46 years predicted for evaporation from a river 1 m deep, flowing at 1 m/sec with a wind velocity of 3 m/sec. Released to the atmosphere it will not be expected to react with photochemically produced hydroxyl radicals or ozone and will not be subject to appreciable direct photodegradation. It should be sorbed to particulate matter in the atmosphere and thus subject to gravitational settling.

Cleanup/Disposal: Guide No. 151: Do not touch damaged containers or spilled material unless wearing appropriate protective clothing. Stop leak if you can do it without risk. Prevent entry into waterways, sewers, basements or confined areas. Cover with plastic sheet to prevent spreading. Absorb or cover with dry earth, sand or other non-combustible material and transfer to containers. Do not get water inside containers.

Environmental Physical Data

Henry's Law Constant: calculated at 1.8 $x10^{-7}$

Sorption Partition Coefficient: K_{oc} = estimated at 2400 to 2600

BCF: fathead minnow 1100 to 2200

Regulations

RCRA 40CFR: Listed Hazardous Waste No. U142 Toxic Waste

CERCLA: 40CFR 302.4: Listed per CWA Section 311(b)(4) per RCRA Section 3001 RQ: 1 lb (0.454 kg)

SARA 40CFR 372.65: Not listed

SARA EHS 40CFR 355: Not listed

TSCA: Not listed

Analytical Methods

Soil: SW846 3640A, 8081, 8081A, 8270B, 8270C

Water / Groundwater: EPA 022

Food: FDA 212.1, 232.1

Indoor / Expired Air: NIOSH 5508

Plasma: EPA 001, 009, 028, 29; FDA 211.1, 231.1, 252

Other: EPA 1656

CHL1630 CAS #: 115-28-6

CHLORENDIC ACID

RTECS: RB9000000

EINECS Number: 204-078-9

Molecular Formula: $C_9H_4Cl_6O_4$

Formula Weight: 388.84

Cl O Cl OH Cl Cl Cl Cl OH

Chemical Structure

Synonyms: 1,4,5,6,7,7-BICYCLO(2.2.1)HEPT-5-ENE-2,3-DICARBOXYLIC ACID; 1,4,5,6,7,7-BICYCLO[2.2.1]-5-HEPTENE-2,3-DICARBOXYLIC ACID; BICYCLO(2.2.1)HEPT-5-ENE-2,3-DICARBOXYLIC ACID,1,4,5,6,7,7-HEXACHLORO-; BICYCLO(2.2.1)HEPT-5-ENE-2,3-DICARBOXYLIC ACID,1,4,5,6,7,7-HEXACHLORO-(9CI); HET ACID; 1,4,5,6,7,7-

HEXACHLOROBICYCLO(2.2.1)-5-HEPTENE-2,3-DICARBOXYLIC ACID; 1,4,5,6,7,7-HEXACHLOROBICYCLO(2.2.1)HEPT-5-ENE-2,3-DICARBOXYLIC ACID; HEXACHLORO-ENDO-METHYLENETETRAHYDROPHTHALIC ACID; HEXACHLOROENDOMETHYLENETETRAHYDROPHTHALIC ACID; 1,4,5,6,7,7-HEXACHLORO-5-NORBORNENE-2,3-DICARBOXYLIC ACID; KYSELINA 3,6-ENDOMETHYLEN-3,4,5,6,7,7-HEXACHLOR-DELTA(SUP4)-TETRAHYDROFTALOVA; KYSELINA 3,6-ENDOMETHYLEN-3,4,5,6,7,7-HEXACHLOR-DELTA(SUP4)-TETRAHYROFTALOVA; KYSELINA HET; KYSELINA 1,2,3,4,7,7-HEXACHLORBICYKLO(2,2,1)HEPT-2-EN-5,6-DIKARBOXYLOVA; 5-NORBORNENE-2,3-DICARBOXYLIC ACID,1,4,5,6,7,7-HEXACHLORO-

Description: white powder

Use: intermediate for synthesis of a wide variety of compounds and it imparts flame resistance to resinous products

Physical Properties

Freezing Point: Decomposes

Density: 0.95 g/mL

Water Solubility: < 1 mg/mL at 21 C

Other Solubilities: DMSO: >=100 mg/ml at 21 °C; 95% Ethanol: >=100 mg/ml at 21 °C; Acetone: >=100 mg/ml at 21 °C; Benzene: 0.81 g/100 g at 25 °C; Carbon Tetrachloride: 0.21 g/100 g at 25 °C; Linseed oil (raw): 9.65 g/100 g at 25 °C; Alcohol: Very Soluble.

RTECS Toxicity Data

Chronic (Multiple Dose) Oral: Rat Dose: 45500 mg/kg/13W-C; Toxic Effects: Liver - Other changes. Mouse Dose: 84 gm/kg/14D-C; Toxic Effects: DEATH.

Irritation Eye: Rabbit Standard Draize Test Dose: 250 ug/24H; Reaction: severe.

Irritation Skin: Rabbit Standard Draize Test Dose: 500 mg/24H; Reaction: mild.

Mutagenic: Mouse Mutations in Mammalian Somatic Cells; Cell Type: lymphocyte; Dose: 1700 mg/L.

Tumorigenic: Rat Route: Oral; Dose: 45063 mg/kg/2Y-C; Toxic Effects: Tumorigenic - Carcinogenic by RTECS criteria; Liver - Tumors. Mouse Route: Oral; Dose: 108 gm/kg/2Y-C; Toxic Effects: Tumorigenic - Carcinogenic by RTECS criteria; Liver - Tumors.

Hazard Overviews

Corrosive

Fire Diamond

Health: Corrosive to eyes; irritating to the skin. Also Causes: sneezing or coughing. Chronic Effects: carcinogenic to liver/pancreas (based on animal data), possible cancer hazard.

Fire: Burns when subjected to direct flame. Use dry chemical, carbon dioxide, water spray, fog, or regular foam. Burning in an open flame may produce hydrogen chloride gas.

Reactivity: Stable. Hazardous polymerization cannot occur. Avoid: heat; excessive dust. Incompatible with: alkaline materials; amines; alkali metals. Hazardous decomposition products: carbon dioxide; toxic hydrogen chloride; chloride fumes.

Carcinogenicity: IARC - Group 2B, Possibly carcinogenic to humans; NIOSH - Not listed; NTP - Listed; ACGIH - Not listed; OSHA - Not listed; EPA - Not listed; MAK - Not listed

Primary Target Organs:

Eyes

Skin

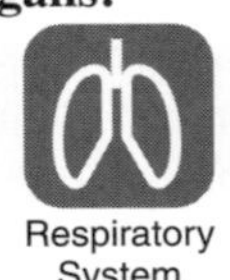

Respiratory System

Mucous Membranes

Environmental

Cleanup/Disposal: Guide No. 171: Do not touch or walk through spilled material. Stop leak if you can do it without risk. Prevent dust cloud. Avoid inhalation of asbestos dust. Small Dry Spills: With clean shovel place material into clean, dry container and cover loosely; move containers from spill area. Small Spills: Take up with sand or other noncombustible absorbent material and place into containers for later disposal. Large Spills: Dike far ahead of liquid spill for later disposal. Cover powder spill with plastic sheet or tarp to minimize spreading. Prevent entry into waterways, sewers, basements or confined areas.

Environmental Physical Data

BCF: none

Regulations

RCRA 40CFR: Not listed

CERCLA: 40CFR 302.4: Not listed

SARA 40CFR 372.65: Listed

SARA EHS 40CFR 355: Not listed

TSCA: Listed

CHL1660 CAS #: 115-27-5

CHLORENDIC ANHYDRIDE

RTECS: RB9080000

EINECS Number: 204-077-3

Molecular Formula: $C_9H_2Cl_6O_3$

Formula Weight: 370.86

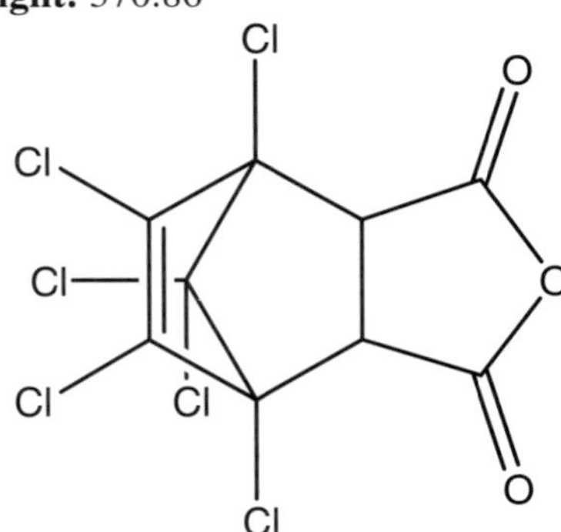

Chemical Structure

Synonyms: CHLORAN 542; HET ANHYDRIDE; 1,4,5,6,7,7-HEXACHLOROBICYCLO-(2,2,1)-5-HEPTENE-2,3-DICARBOXYLIC ANHYDRIDE; 1,4,5,6,7,7-HEXACHLORO-ENDO-

BICYCLO(2.2.1)HEPT-5-ENE-2,3-DICARBOXYLIC ANHYDRIDE; HEXACHLOROENDOMETHYLENE TETRAHYDROPHTHALIC ANHYDRIDE; 1,4,5,6,7,7-HEXACHLORO-ENDO-5-NORBORNENE-2,3-DICARBOXYLICANHYDRIDE; HEXACHLORO-5-NORBORNENE-2,3-DICARBOXYLIC ANHYDRIDE; 4,7-METHANOISOBENZOFURAN-1,3-DIONE,4,5,6,7,8,8-HEXACHLORO-3A,4,7,7A-TETRAHYDRO-; 5-NORBORNENE-2,3-DICARBOXYLIC ANHYDRIDE,1,4,5,6,7,7-HEXACHLORO-

Description: white powder; sharp, pungent chlorine odor

Use: flame retardant in unsaturated polyester resins; chem int for chlorendic acid; hardener for epoxy resins

Physical Properties

Freezing Point: 231 °C (447.8 °F) to 235 °C (455 °F)

Specific Gravity: 1.73

Water Solubility: Slightly Soluble in Water

Other Solubilities: readily Soluble in Acetone, Benzene, Toluene Slightly Soluble in N-Hexane, Carbon Tetrachloride

Flash Point: Nonflammable

RTECS Toxicity Data

Acute Oral: Rat LD_{50} Dose: 2300 mg/kg. Mouse LD_{50} Dose: 2400 mg/kg.

Acute Inhalation: Rat LC_{50} Dose: >1 gm/m^3. Rabbit LC_{50} Dose: >1 gm/m^3.

Hazard Overviews

Fire Diamond

Health: Irritating to eyes/skin/respiratory tract. Also Causes: possible corneal injuries, respiratory distress, allergic sensitization reactions upon occupational skin exposure.

Fire: Does not burn easily. Even if burning, it tends to go out as soon as the ignition source is removed. Use agent suitable for surrounding fire.

Reactivity: Stable. Hazardous polymerization cannot occur. Avoid: dusty working conditions. Incompatible with: chlorendic anhydride; water; acids; strong oxidizing agents; carbonates. Hazardous decomposition products: carbon monoxide; chlorine; hydrogen chloride; maleic acid; toxic hexachlorocyclopentadiene; oxides of chlorine.

Carcinogenicity: IARC - Not listed; NIOSH - Not listed; NTP - Not listed; ACGIH - Not listed; OSHA - Not listed; EPA - Not listed; MAK - Not listed

Primary Target Organs:

Eyes

Skin

Respiratory System

Environmental

Regulations

RCRA 40CFR: Not listed

CERCLA: 40CFR 302.4: Not listed

SARA 40CFR 372.65: Not listed

SARA EHS 40CFR 355: Not listed

TSCA: Listed

CHL1720 CAS #: 470-90-6

CHLORFENVINFOS

RTECS: TB8750000

DOT: UN2783; UN2784; UN3017; UN3018; IMO3.2; IMO6.1

EINECS Number: 207-432-0

Molecular Formula: $C_{12}H_{14}Cl_3O_4P$

Structured MF: $Cl_2C_6H_3C(=CHCl)OP(O)(OC_2H_5)_2$

Formula Weight: 359.56

Synonyms: APACHLOR; BENZYL ALCOHOL,2,4-DICHLORO-ALPHA-(CHLOROMETHYLENE)-,DIETHYL PHOSPHATE; BIRLAN; BIRLANE; BIRLANE 10G; C 8949; C8949; CFV; CFVP; CGA 26351; CHLOFENVINPHOS; O-2-CHLOOR-1-(2,4-DICHLOOR-FENYL)-VINYL-O,O-DIETHYLFOSFAAT; O-2-CHLOR-1-(2,4-DICHLOR-PHENYL)-VINYL-O,O-DIAETHYLPHOSPHAT; CHLORFEHWINFOS; CHLORFENVINPHOS; BETA-2-CHLORO-1-(2',4'-DICHLOROPHENYL) VINYLDIETHYLPHOSPHATE; 2-CHLORO-1-(2,4-DICHLOROPHENYL)-ETHENYL DIETHYL PHOSPHATE; 2-CHLORO-1-(2,4-DICHLOROPHENYL)VINYL DIETHYL PHOSPHATE; 2-CHLORO-1-(2,4-DICHLOROPHENYL)VINYLDIETHYLPHOSPHATE; CHLORPHENVINFOS; CHLORPHENVINPHOS; CLOFENVINFOS; O-2-CLORO-1-(2,4-DICLORO-FENIL)-VINIL-O,O-DIETILFOSFATO; O-2-CLORO-1-(2,4-DICLORO-FENIL)-VINYL-O,O-DIETILFOSFATO; COMPOUND 4072; CVP; CVP (PESTICIDE); DERMATON; O,O-DIAETHYL-O-1-(4,5-DICHLORPHENYL)-2-CHLOR-VINYL-PHOSPHAT; 2,4-DICHLORO-ALPHA-(CHLOROMETHYLENE)BENZYL ALCOHOL DIETHYLPHOSPHATE; 2,4-DICHLORO-ALPHA-(CHLOROMETHYLENE)BENZYLDIETHYLPHOSPHATE; O,O-DIETHYL O-(2-CHLORO-1-(2',4'-DICHLOROPHENYL)VINYL)PHOSPHATE; O,O-DIETHYL O-(2-CHLORO-1-(2,4-DICHLOROPHENYL)VINYL)PHOSPHATE; DIETHYL 1-(2,4-DICHLOROPHENYL)-2-CHLOROVINYL PHOSPHATE; DIETHYL-2-CHLORO-1-(2,4-DICHLOROPHENYL)VINYL PHOSPHATE; O,O-DIETHYL-O-1-(2',4'-DICHLOROPHENYL)-2-CHLOROVINYLPHOSPHATE; O,O-DWUETYLO-O-1-(2,4-DWUCHLOROFENYLO)-2-CHLOROWINYLOFOSFORAN; ENOLOFOS; ENT 24969; GC 4072; HAPTARAX; HAPTASOL; OMS 1328; PHOSPHATE DE O,O-DIETHYLE ET DE O-2-CHLORO-1-(2,4-DICHLOROPHENYL) VINYLE; PHOSPHATE DE O,O-DIETHYLE ET DEO-2-CHLORO-1-(2,4-DICHLOROPHENYL) VINYLE; PHOSPHORIC ACID 2-CHLORO-1-(2,4-DICHLOROPHENYL)ETHENYLDIETHYL ESTER; PHOSPHORIC ACID,2-CHLORO-1-(2,4-DICHLOROPHENYL)ETHENYLDIETHYL ESTER; PHOSPHORIC ACID,2-CHLORO-1-(2,4-DICHLOROPHENYL)VINYLDIETHYL ESTER; SAPECRON; SD-7859; SD 4072; SD 7859; SHELL 4072; STELADONE; SUPONA; SUPONE; TARENE; UNITOX; VINYLPHATE; VINYPHATE

Description: colorless liquid; mild odor

Use: insecticide; acaricide; nematocide, parasiticide; for control of ticks, flies, lice & mites on cattle, sheep; for control of root flies, rootworms & cutworms; foliage insecticide for colorado beetle, leaf hoppers, etc; in maize, sugar cane & rice, breeding places of fly larvae incl dairybarns; soil application or seed treatment for control of fruit flies in maize; against mosquito larvae.

Physical Properties

Boiling Point: 167 °C (333 °F) to 170 °C (338 °F) at 0.5 mm Hg
Freezing Point: -23 °C (-9.4 °F) to -19 °C (-2.2 °F)
Saturated Vapor Density: 1.200000072 kg/m^3
Density: 1.36 g/cu cm at 20 °C
Vapor Pressure: 4.0 x10^{-6} torr at 20 °C
Water Solubility: 145 mg/L Water at 23 °C
Other Solubilities: Chlorfenvinfos is miscible with Acetone Dichloromethane Ethanol Hexane Kerosene Propylene glycol Xylene
Refraction Index: 1.5272 at 25 °C/D
Flash Point: Nonflammable

RTECS Toxicity Data

Acute Oral: Rat LD_{50} Dose: 10 mg/kg. Mouse LD_{50} Dose: 65 mg/kg.
Acute Inhalation: Rat LC_{50} Dose: 50 mg/m^3/4hr.
Acute Dermal: Human TD_{Lo} Route: Skin; Dose: 10 mg/kg; Toxic Effects: Blood - Other changes; Biochemical - True cholinesterase. Rabbit LD_{50} Route: Skin; Dose: 400 mg/kg; Toxic Effects: Sense organs and special senses - Lacrimation; Gastrointestinal - Changes in structure or function of salivary glands; Gastrointestinal - Hypermotility, diarrhea.
Chronic (Multiple Dose) Oral: Rat Dose: 420 mg/kg/12W-C; Toxic Effects: Endocrine - Changes in spleen weight; Biochemical - True cholinesterase. Rat Dose: 49 mg/kg/35W-I; Toxic Effects: Cardiac - Other changes; Liver - Other changes; Kidney, Ureter, and Bladder - Other changes. Rat Dose: 182 mg/kg/26W-I; Toxic Effects: Blood - Changes in leukocyte (WBC) cell count.
Mutagenic: Bacteria - S Typhimurium Mutations in Microorganisms; Dose: 500 ug/plate (+S9).

Hazard Overviews

Fire: Noncombustible.
Carcinogenicity: IARC - Not listed; NIOSH - Not listed; NTP - Not listed; ACGIH - Not listed; OSHA - Not listed; EPA - Not listed; MAK - Not listed

Environmental

Ecotoxicity: LD_{50} (Pheasant) oral 63.5 mg/kg (95% confidence limit 45.8-88.1 mg/kg), 3-4 mo old females /81% beta, 8% alpha isomers LC_{50} Rasbora heteromorpha (Harlequin fish) 0.27 mg/l/48 hr /Conditions of bioassay not specified
Environmental Fate: If released to the atmosphere, it will degrade rapidly in the vapor phase by reaction with photochemically produced hydroxyl radicals (half-life of about 7 hr). Particulate phase and aerosols released to air during spray applications of insecticides will be removed from air physically by dry and wet deposition. If released to soil or water, it will degrade through biodegradation. The importance of microbial degradation has been demonstrated by various persistence studies that compare degradation rates in sterile versus nonsterile soil; in these studies, degradation is much faster in nonsterile soil than in sterile soil. In one 90-day field study, it did not leach in a sandy loam soil. Typical soil half-lives range from roughly 10 to 45 days.
Cleanup/Disposal: Guide No. 131: Fully encapsulating, vapor protective clothing should be worn for spills and leaks with no fire. Eliminate all ignition sources (no smoking, flares, sparks or flames in immediate area). All equipment used when handling the product must be grounded. Do not touch or walk through spilled material. Stop leak if you can do it without risk. Prevent entry into waterways, sewers, basements or confined areas. A vapor suppressing foam may be used to reduce vapors. Small Spills: Absorb with earth, sand or other non-combustible material and transfer to containers for later disposal. Use clean non-sparking tools to collect absorbed material. Large Spills: Dike far ahead of liquid spill for later disposal. Water spray may reduce vapor; but may not prevent ignition in closed spaces. Guide No. 152: Do not touch damaged containers or spilled material unless wearing appropriate protective clothing. Stop leak if you can do it without risk. Prevent entry into waterways, sewers, basements or confined areas. Cover with plastic sheet to prevent spreading. Absorb or cover with dry earth, sand or other non-combustible material and transfer to containers. Do not get water inside containers.

Environmental Physical Data

Henry's Law Constant: estimated at 1.53 x10^{-8}
Octanol/Water Partition Coefficient: log K_{OW} = 3.82
Sorption Partition Coefficient: K_{OC} = 295
BCF: estimated at 470

Regulations

RCRA 40CFR: Not listed
CERCLA: 40CFR 302.4: Not listed
SARA 40CFR 372.65: Not listed TPQ: 500 lb
SARA EHS 40CFR 355: Listed TPQ: 500 lb
TSCA: Not listed

Analytical Methods

Soil: SW846 8141, 8141A, 8270B, 8270C
Water / Groundwater: EPA 1657; SW846 8141, 8141A
Food: FDA 212.1, 232.1, 232.3, 232.4, 242.1
Plasma: EPA 001, 29; FDA 211.1, 231.1, 252

CHL1750 **CAS #: 90982-32-4**

CHLORIMURON ETHYL

RTECS: DG5095000
Molecular Formula: $C_{15}H_{15}ClN_4O_6S$
Formula Weight: 414.82
Synonyms: 2-(((((4-CHLORO-6-METHOXY-2-PYRIMIDINYL)AMINO)CARBONYL)AMINO)SULFONYL)BENZOIC ACIDETHYL ESTER; CLASSIC; DPX-F 6025; DPX-F6025; ETHYL2-(((((4-CHLORO-6-METHOXYPYRIMIDIN-2-YL)-AMINO)CARBONYL)AMINO)SULFONYL) BENZOATE
Description: white crystals or solid
Use: herbicide; annual broadleaf, yellow nutsedge weed control in soybeans

Physical Properties

Freezing Point: 198 °C (388.4 °F) to 201 °C (393.8 °F)
Water Solubility: 1200 ppm
Other Solubilities: Solubility (ppm): Acetone 71,000; acetonitrile 31,000, Benzene 8000, methylene chloride 153,000.

RTECS Toxicity Data

Acute Oral: Rat LD_{50} Dose: 4102 mg/kg.
Acute Inhalation: Rat LC_{50} Dose: >5 gm/m^3/4hr.
Acute Dermal: Rabbit LD_{50} Route: Skin; Dose: >2 gm/kg.

Hazard Overviews

Carcinogenicity: IARC - Not listed; NIOSH - Not listed; NTP - Not listed; ACGIH - Not listed; OSHA - Not listed; EPA - Not listed; MAK - Not listed

Environmental

Regulations

RCRA 40CFR: Not listed
CERCLA: 40CFR 302.4: Not listed
SARA 40CFR 372.65: Listed
SARA EHS 40CFR 355: Not listed
TSCA: Not listed

CHL1900 **CAS #: 7782-50-5**

CHLORINE

RTECS: FO2100000
DOT: UN1017; IMO2.0
EINECS Number: 231-959-5
Molecular Formula: Cl_2
Structured MF: Cl_2
Formula Weight: 70.906

Cl——Cl

Chemical Structure

Synonyms: BERTHOLITE; CHLOOR; CHLOR; CHLORE; CHLORINE MOL; CLORO; EPA PESTICIDE CHEMICAL CODE 020501; MOLECULAR CHLORINE
Description: greenish-yellow gas; suffocating odor
Use: in mfr of pesticides, antifreeze, refrigerants, plastics, resins & other chlorinated hydrocarbons, eg, polyvinyl chloride, chlorobenzene; for bleaching fabric, synthetic rubber & plastics, woodpulp, etc; for purifying water; detinning & dezincing iron; disinfecting; in textile ind (shrink-proofing wool); in flame-retardant; in batteries; food processing; to control biofouling in cooling systems; med: fungicidal foot baths, root canal or tooth extraction, irrigation, wound sterilization; cleaning dairy equipment; disinfectant.

Physical Properties

Boiling Point: -34.6 °C (-30 °F)
Freezing Point: -100.98 °C (-149.764 °F)
Specific Gravity: Liquid 1.5649 at -35 °C
Vapor Density: 2.5 Air=1 at boiling point
Vapor Pressure: 5 atm at 10.3 °C
Water Solubility: 310 cc/100 cc Water at 10 °C
Other Solubilities: Soluble in alkali; Soluble in chlorides and Alcohols
Surface Tension: 26.55 dynes/cm
Odor Threshold: 0.0300 to 15 mg/m^3
Refraction Index: 1.0008
Evaporation Rate: > 1 Butyl Acetate=1
Critical Temperature: 144 °C
Critical Pressure: 76.1 atm
Ionization Potential (eV): 11.48
Flash Point: Nonflammable

RTECS Toxicity Data

Acute Inhalation: Rat LC_{50} Dose: 293 ppm/1hr. Human LC_{Lo} Dose: 2530 mg/m^3/30M; Toxic Effects: Lungs, Thorax, or Respiration - Structural or functional change in trachea or bronchi; Lungs, Thorax, or Respiration - Emphysema. Human LC_{Lo} Dose: 500 ppm/5M.
Reproductive/Teratogenic: Rat Route: Oral; Dose: 565 mg/kg; Duration: male 8W prior to mating; Effects on Newborn - Biochemical and metabolic.
Mutagenic: Human Cytogenetic Analysis; Cell Type: lymphocyte; Dose: 20 ppm. Mouse Sperm Morphology; Route: Oral; Dose: 20 mg/kg/5D-C.
Tumorigenic: Rat Route: Oral; Dose: 5096 mg/kg/2Y-C; Toxic Effects: Tumorigenic - Equivocal tumorigenic agent by RTECS criteria; Blood - Leukemia.

Hazard Overviews

Poison

Corrosive

Compressed Gas

Fire Diamond

Health: Corrosive causes burns to eyes/skin/respiratory tract. Poison. Also Causes: inhalation of high levels can cause difficulty breathing, dizziness, and pulmonary edema which can be fatal. Chronic Effects: tooth erosion, chloracne.
Fire: Noncombustible, but is a strong oxidizer capable of igniting combustible materials. Use extinguishing agents suitable for surrounding fire.
Reactivity: Stable. Hazardous polymerization cannot occur. Avoid: exposure to heat; moisture. Incompatible with: finely divided metals; combustibles; organic materials; moisture; water; steam; acetylene; alcohols; ammonia; arsenic; bismuth; boron; benzene; calcium; activated carbon; carbon disulfide; ether; ethane; ethylene; fluorine; glycerol; hydrazine; hydrocarbons; iodine; methane; oxomonosilane; potassium; polypropylene; silicon; some forms of plastics, rubber, and coatings. Hazardous decomposition products: toxic chlorine products.
Carcinogenicity: IARC - Not listed; NIOSH - Not listed; NTP - Not listed; ACGIH - Class A4, Not classifiable as a human carcinogen; OSHA - Not listed; EPA - Not listed; MAK - Not listed

Primary Target Organs:

Eyes

Skin

Respiratory System

Mucous Membranes

Nervous System

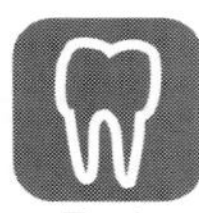
Teeth

Exposure Limits

OSHA PEL: STEL: 1 ppm; 3 mg/m^3.

OSHA PEL Vacated 1989 Limits: TWA: 0.5 ppm; 1.5 mg/m^3; STEL: 1 ppm; 3 mg/m^3.

ACGIH TLV: TWA: 0.5 ppm; 1.5 mg/m^3; STEL: 1 ppm; 2.9 mg/m^3.

NIOSH REL: STEL: 0.5 ppm; 1.45 mg/m^3; 15-minute.

NIOSH IDLH: 10 ppm.

DFG MAK: TWA: 0.5 ppm; 1.5 mg/m^3.

Respirator Recommendation

Exposure Range: >1 to <10 ppm Air Purifying, Negative Pressure, Full Face

Exposure Range: 10 to unlimited ppm Self-contained Breathing Apparatus, Pressure Demand, Full Face

Cartridge Color: white

Environmental

Ecotoxicity: LC_{50} Ictalurus punctatus (channel catfish) (fingerling) 0.07 mg/l/96 hr toxic effect: gill sodium uptake drastically impaired. /Conditions of bioassay not specified LC_{50} Daphnia magna (water flea) 0.017 mg/l/46 hr. /Conditions of bioassay not specified LC_{50} Oncorhynchus kisutch (coho salmon) 208 ug/l/1 hr /Conditions of bioassay not specified TLm Salmo gairdnerii (rainbow trout) 0.08 mg/l/168 hr. /Conditions of bioassay not specified LC_{100} Larval clam 0.5 mg/l/100 hr. /Conditions of bioassay not specified TLm Grass shrimp 0.22 mg/l/96 hr. /Conditions of bioassay not specified

Cleanup/Disposal: Guide No. 124: Fully encapsulating, vapor protective clothing should be worn for spills and leaks with no fire. Do not touch or walk through spilled material. Keep combustibles (wood, paper, oil, etc.) away from spilled material. Stop leak if you can do it without risk. Use water spray to reduce vapors or divert vapor cloud drift. Do not direct water at spill or source of leak. If possible, turn leaking containers so that gas escapes rather than liquid. Prevent entry into waterways, sewers, basements or confined areas. Isolate area until gas has dispersed. Ventilate the area.

Environmental Physical Data

BCF: none

BOD: none

Regulations

RCRA 40CFR: Not listed

CERCLA: 40CFR 302.4: Listed per CWA Section 311(b)(4) RQ: 10 lb (4.535 kg)

SARA 40CFR 372.65: Listed TPQ: 100 lb

SARA EHS 40CFR 355: Listed TPQ: 10 lb

TSCA: Listed

Analytical Methods

Air: EPA 26, 26A, 0050, 0050-0, 0051, 0051-0; ASTM D4490

Soil: SW846 9075, 9076, 9077

Water / Groundwater: EPA 330.1, 330.2, 330.3, 330.4, 330.5; APHA 2350-B, 4500-CL; ASTM D1253, D1291, D4744; HACH 8167, 8168

Food: AOAC 935.08, 935.09, 935.10, 952.24, 974.36

Indoor / Expired Air: NIOSH 6011

CHL1930 CAS #: 10049-04-4

CHLORINE DIOXIDE

RTECS: FO3000000

DOT: NA9191

EINECS Number: 233-162-8

Molecular Formula: ClO_2

Structured MF: ClO_2

Formula Weight: 67.46

Synonyms: ALCIDE; ANTHIUM DIOXCIDE; CHLORINE DIOXIDE (NOT HYDRATE); CHLORINE OXIDE; CHLORINE PEROXIDE; CHLORINE(IV) OXIDE; CHLOROPEROXYL; CHLORYL RADICAL; DOXCIDE 50

Description: red-yellow gas, crystals, liquid; resembles chlorine and nitric acid odor

Use: bleaching cellulose, flour, leather, oils, textiles, beeswax; purification of water; taste & odor control of water; cleaning and detanning leather; mfr of chlorine salts; oxidizing agent; bactericide & antiseptic; aging accelerator in the manufacture of flour; swimming pool water purification; bleaching agent for wood pulp, tallow; wastewater disinfecting agent

Physical Properties

Boiling Point: 11 °C (52 °F)

Freezing Point: -59 °C (-74.2 °F)

Specific Gravity: Liquid 1.642 at 0 °C

Vapor Density: 2.33 Air=1

Vapor Pressure: > 1 atm

Water Solubility: 3.01 g/L at 25 °C and at 345 MM Hg

Other Solubilities: Soluble in alkaline and Sulfuric acid solution.

Ionization Potential (eV): 10.36

Flash Point: Flammable Gas/Combustible Liquid

RTECS Toxicity Data

Acute Oral: Rat LD_{50} Dose: 292 mg/kg; Toxic Effects: Behavioral - Somnolence (general depressed activity); Lungs, Thorax, or Respiration - Respiratory stimulation; Skin and appendages - Hair.

Acute Inhalation: Rat LC_{Lo} Dose: 260 ppm/2hr; Toxic Effects: Sense organs and special senses - Other; Sense organs and special senses - Other.

Irritation Eye: Rabbit Standard Draize Test Dose: 100 mg; Reaction: mild.

Reproductive/Teratogenic: Rat Route: Oral; Dose: 960 mg/kg; Duration: female 76D prior to mating Effects on Fertility - Pre-implantation mortality; Litter size. Rat Route: Oral; Dose: 570 mg/kg; Duration: female 14D prior to mating Effects on Newborn - Behavioral; Physical. Rat Route: Oral;

Dose: 1130 mg/kg; Duration: male 8W prior to mating; Effects on Newborn - Biochemical and metabolic.
Mutagenic: Rat DNA Inhibition; Route: Oral; Dose: 84 mg/kg/12W-C. Mouse Micronucleus Test; Route: Intraperitoneal; Dose: 3200 ug/kg.

Hazard Overviews

Poison

Explosive

Fire Diamond

Health: Severe irritation to eyes/skin/respiratory tract. Poison. Also Causes: pulmonary edema. Chronic Effects: bronchitis.
Fire: Noncombustible. However, it is a strong oxidizer capable of igniting combustibles. Chlorine Dioxide may form explosive mixtures in air. Do not use water to extinguish fire.
Reactivity: Unstable, highly reactive, concentrations stronger than 10% can be detonated easily by sunlight at normal atmospheric pressure. Hazardous polymerization cannot occur. Avoid: exposure to any other substance unless chemical compatibility has been conclusively established, contact with air that contains over 10% by volume chlorine dioxide at a pressure of 0.1 to 1.0 atmospheres; any form of energy (spark, sunlight, heat, etc). Incompatible with: sugar; sulfur; potassium hydroxide; phosphorus; mercury; flourine; difluoroamine; most organic materials. Hazardous decomposition products: oxides of chlorine.
Carcinogenicity: IARC - Not listed; NIOSH - Not listed; NTP - Not listed; ACGIH - Not listed; OSHA - Not listed; EPA - Class D, Not classifiable as to human carcinogenicity; MAK - Not listed
Primary Target Organs:

Eyes

Skin

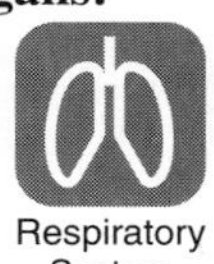
Respiratory System

Mucous Membranes

Exposure Limits
OSHA PEL: TWA: 0.1 ppm; 0.3 mg/m^3.
OSHA PEL Vacated 1989 Limits: TWA: 0.1 ppm; 0.3 mg/m^3; STEL: 0.3 ppm; 0.9 mg/m^3.
ACGIH TLV: TWA: 0.1 ppm; 0.28 mg/m^3; STEL: 0.3 ppm; 0.83 mg/m^3.
NIOSH REL: TWA: 0.1 ppm; 0.3 mg/m^3. STEL: 0.3 ppm; 0.9 mg/m^3.
NIOSH IDLH: 5 ppm.
DFG MAK: TWA: 0.1 ppm; 0.3 mg/m^3.
Respirator Recommendation
Exposure Range: >0.1 to 1 ppm Air Purifying, Negative Pressure, Half Mask
Exposure Range: >1 to <5 ppm Air Purifying, Negative Pressure, Full Face
Exposure Range: 5 to unlimited ppm Self-contained Breathing Apparatus, Pressure Demand, Full Face
Cartridge Color: white

Environmental

Ecotoxicity: LC_{50} Fathead minnow juvenile flow through 0.02 mg/l/96 hr LC_{50} Fathead minnow adult 0.17 mg/l/96 hr /Conditions of bioassay not specified
Cleanup/Disposal: Guide No. 143: Keep combustibles (wood, paper, oil, etc.) away from spilled material. Do not touch damaged containers or spilled material unless wearing appropriate protective clothing. Use water spray to reduce vapors or divert vapor cloud drift. Prevent entry into waterways, sewers, basements or confined areas. Small Spills: Flush area with flooding quantities of water. Large Spills: Do not clean-up or dispose of, except under supervision of a specialist.

Regulations

RCRA 40CFR: Not listed
CERCLA: 40CFR 302.4: Not listed
SARA 40CFR 372.65: Listed
SARA EHS 40CFR 355: Not listed
TSCA: Listed

Analytical Methods

Air: ASTM D4490
Water / Groundwater: APHA 2350-C, 4500-CLO

CHL1990 CAS #: 7790-91-2

CHLORINE TRIFLUORIDE

RTECS: FO2800000
DOT: UN1749; IMO2.3
EINECS Number: 232-230-4
Molecular Formula: ClF_3
Structured MF: ClF_3
Formula Weight: 92.46
Synonyms: CHLORINE FLUORIDE; CHLOROTRIFLUORIDE; TRIFLUORURE DE CHLORE
Description: colorless, yellow-green gas, liquid; sweet and irritating odor
Use: fluorinating agent; in nuclear reactor fuel processing; incendiary; igniter & propellant for rockets; pyrolysis inhibitor for fluorocarbon polymers; oxidizer in propellants; cutting oil well tubes; reprocessing reactor fuels; processing of nuclear fuels to convert uranium to gaseous uranium hexafluoride; etchant for single crystalline silicon; during World War II used by Germany as an incendiary gas

Physical Properties

Boiling Point: 11.75 °C (53 °F)
Freezing Point: -76.34 °C (-105.412 °F)
Vapor Density: > 3 Air=1
Density: 1.825 g/mL at boiling point (liquid)
Vapor Pressure: 10 mm Hg at -71.8 °C
Water Solubility: Decomposes in Water
Surface Tension: 26.6 dynes/cm at 0 °C
Critical Temperature: 154.5 °C
Critical Pressure: 837 psia
Ionization Potential (eV): 13.00

Flash Point: Not flammable

RTECS Toxicity Data

Acute Inhalation: Rat LC_{50} Dose: 299 ppm/1hr. Monkey LC_{50} Dose: 230 ppm/1hr; Toxic Effects: Behavioral - Muscle weakness; Lungs, Thorax, or Respiration - Chronic pulmonary edema; Lungs, Thorax, or Respiration - Cyanosis. Human LC_{Lo} Dose: 50 ppm.

Hazard Overviews

Corrosive

Fire Diamond

Health: Corrosive, causes severe burns to eyes/skin/respiratory tract. Chronic Effects: persistent cough, sneezing, runny nose.

Fire: Noncombustible. However, it is a strong oxidizer capable of igniting combustibles. Use extinguishing agents suitable for surrounding fire except water which produces a violent reaction. If possible, stop gas flow or remove leaking cylinder.

Reactivity: Unstable, stable only in special cylinders. Hazardous polymerization cannot occur. Avoid: oxidizable materials. Incompatible with: water; bis(trifluoromethyl)sulfide or -disulfide; trifluoromethanesulfenyl chloride; polychlorotrifluoroethylene; hydrogen-containing materials; highly chlorinated compounds; nitroaryl compounds; ammonium fluoride; ammonium hydrogen fluoride; iodine; boron-containing materials; fluorinated polymers (with flowing trifluoride); finely divided refractory materials (asbestos; glass wool; sand; tungsten carbide); acids; chromium trioxide; selenium tetrafluoride [above 223 °F (106 °C)]; ruthenium; metals; metal oxides; metal salts; nonmetals; nonmetal salts; organic matter; acetic acid; aluminum; antimony; arsenic; copper; iridium; iron; lead; magnesium; molybdenum; osmium; phosphorus; potassium; rhodium; selenium; silicon; silver; sodium; sulfur; tellurium; tin; tungsten; zinc; oxides; carbon monoxide; graphite; mercury iodide; potassium iodide; rubber; potassium carbonate; silver nitrate; sodium hydroxide; fuels; nitro compounds. Hazardous decomposition products: fumes of fluorine; fumes of chlorine.

Carcinogenicity: IARC - Not listed; NIOSH - Not listed; NTP - Not listed; ACGIH - Class A4, Not classifiable as a human carcinogen; OSHA - Not listed; EPA - Not listed; MAK - Not listed

Primary Target Organs:

Eyes

Skin

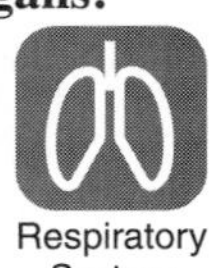
Respiratory System

Exposure Limits

OSHA PEL: STEL: 0.1 ppm; 0.4 mg/m^3.

ACGIH TLV: STEL: 0.1 ppm; 0.38 mg/m^3; Ceiling.

NIOSH REL: TWA: as F (Flourides). STEL: 0.1 ppm; 0.4 mg/m^3.

NIOSH IDLH: 20 ppm.

DFG MAK: TWA: 0.1 ppm; 0.4 mg/m^3; as F.

Respirator Recommendation

Exposure Range: >0.1 to <20 ppm Supplied Air, Constant Flow/Pressure Demand, Full Face

Exposure Range: 20 to unlimited ppm Self-contained Breathing Apparatus, Pressure Demand, Full Face

Note: odor threshold unknown

Environmental

Cleanup/Disposal: Guide No. 124: Fully encapsulating, vapor protective clothing should be worn for spills and leaks with no fire. Do not touch or walk through spilled material. Keep combustibles (wood, paper, oil, etc.) away from spilled material. Stop leak if you can do it without risk. Use water spray to reduce vapors or divert vapor cloud drift. Do not direct water at spill or source of leak. If possible, turn leaking containers so that gas escapes rather than liquid. Prevent entry into waterways, sewers, basements or confined areas. Isolate area until gas has dispersed. Ventilate the area.

Environmental Physical Data

BCF: no food chain concentration potential

BOD: none

Regulations

RCRA 40CFR: Not listed

CERCLA: 40CFR 302.4: Not listed

SARA 40CFR 372.65: Not listed

SARA EHS 40CFR 355: Not listed

TSCA: Listed

CHL2050 CAS #: 24934-91-6

CHLORMEPHOS

RTECS: TD5170000

EINECS Number: 246-538-1

Molecular Formula: $C_5H_{12}ClO_2PS_2$

Formula Weight: 234.7

Synonyms: CHLORMEFOS; CHLORMETHYLFOS; S-(CHLOROMETHY) O,O-DIETHYL PHOSPHORODITHIOATE; S-CHLOROMETHYL O,O-DIETHYL PHOSPHORODITHIOATE; S-(CHLOROMETHYL) O,O-DIETHYLPHOSPHORODITHIOATE; S-CHLOROMETHYL-O,O-DIETHYLPHOSPHOROTHIOLOTHIONATE; DOTAN; EPA PESTICIDE CHEMICAL CODE 295300; MC 2188; PHOSPHORODITHIOIC ACID,S-(CHLOROMETHYL) O,O-DIETHYL ESTER

Description: colorless liquid

Use: contact insecticide, effective when applied to soil for the control of millipedes, white grubs and wireworms in maize and sugar beet

Physical Properties

Boiling Point: 81 °C (178 °F) to 85 °C (185 °F) at 0.1 mm Hg

Specific Gravity: 1.26 at 20 °C

Vapor Pressure: 7.6 Pa at 30 °C

Water Solubility: 60 mg/L Water at 20 °C

Other Solubilities: miscible with most organic solvents.

Refraction Index: 1.5244

RTECS Toxicity Data

Acute Oral: Rat LD_{50} Dose: 7 mg/kg. Chicken LD_{50} Dose: 130 mg/kg.
Acute Inhalation: Rat LC_{50} Dose: 88 mg/m^3/4hr; Toxic Effects: Sense organs and special senses - Other; Behavioral - Tremor; Kidney, ureter, and Bladder - Incontinence.
Acute Dermal: Rabbit LD_{50} Route: Skin; Dose: >1600 mg/kg. Rat LD_{50} Route: Skin; Dose: 27 mg/kg.

Hazard Overviews

Carcinogenicity: IARC - Not listed; NIOSH - Not listed; NTP - Not listed; ACGIH - Not listed; OSHA - Not listed; EPA - Not listed; MAK - Not listed

Environmental

Ecotoxicity: LC_{50} Harlequin fish (Rasbora heteromorpha) 2.5 mg/l/96 hr /Conditions of bioassay not specified LD_{50} Quail oral 260 mg/kg

Regulations

RCRA 40CFR: Not listed
CERCLA: 40CFR 302.4: Not listed
SARA 40CFR 372.65: Not listed TPQ: 500 lb
SARA EHS 40CFR 355: Listed TPQ: 500 lb
TSCA: Not listed

Analytical Methods

Food: FDA 232.4, 242.1

CHL2080 CAS #: 999-81-5

CHLORMEQUAT CHLORIDE

RTECS: BP5250000
EINECS Number: 213-666-4
Molecular Formula: $C_5H_{13}Cl_2N$
Structured MF: $ClCH_2CH_2N(CH_3)_3 \cdot Cl$
Formula Weight: 158.07

Chemical Structure

Synonyms: 5C CYCOCEL; 60-CS-16; AC 38555; AMMONIUM,(2-CHLOROETHYL)TRIMETHYL-,CHLORIDE; ANTYWYLEGACZ; AROTEX EXTRA; BARLEYQUAT; BETTAQUAT; CCC; CCC PLANT GROWTH REGULANT; CE CE CE; CECECE; 2-CHLORAETHYL-TRIMETHYLAMMONIUMCHLORID; CHLORCHOLINCHLORID; CHLORCHOLINCHLORIDE; CHLORCHOLINE CHLORIDE; CHLORMEQUAT; CHLOROCHOLINE CHLORIDE; 2-CHLOROETHYLTRIMETHYL AMMONIUM CHLORIDE; (2-CHLOROETHYL)TRIMETHYLAMMONIUM CHLORIDE; (BETA-CHLOROETHYL)TRIMETHYLAMMONIUM CHLORIDE; 2-CHLOROETHYLTRIMETHYLAMMONIUM CHLORIDE; BETA-CHLOROETHYLTRIMETHYLAMMONIUM CHLORIDE; 2-CHLORO-N,N,N-TRIMETHYLETHANAMINIUM CHLORIDE; CHOLINE DICHLORIDE; CYCLOCEL; CYCOCEL; CYCOCEL 460; CYCOCEL C 5; CYCOCEL-EXTRA; CYCOGAN; CYCOGAN EXTRA; CYOCEL; EI 38,555; ETHANAMINIUM,2-CHLORO-N,N,N-TRIMETHYL-,CHLORIDE; ETHANAMINIUM,2-CHLORO-N,N,N-TRIMETHYL-,CHLORIDE (9CI); FARMACEL; HALLOWEEN; HELSTONE; HICO CCC; HORMOCEL-2CCC; HYQUAT; INCRECEL; LIHOCIN; MIRBEL 5; RETACEL; STABILAN; TITAN; TRIMETHYL-BETA-CHLORETHYLAMMONIUMCHLORID; TRIMETHYL-BETA-CHLOROETHYLAMMONIUM CHLORIDE; TUR; WR62; ZAR
Description: colorless to white crystalline solid; typical amine fish-like odor
Use: a plant growth regulator stated to be effective in shortening the height of wheat and as a ripening agent in sugarcane

Physical Properties

Freezing Point: Decomposes at 245 °C (473 °F)
Saturated Vapor Density: 1.200000001 kg/m^3
Vapor Pressure: 7.5 x10^{-8} mm Hg at 20 °C
Water Solubility: 74% at 20 °C
Other Solubilities: 95% Ethanol: >=100 mg/ml at 23 °C; Acetone: <1 mg/ml at 23 °C; Chloroform: 0.3 g/kg at 20 °C; Cyclohexane: Insoluble; DMSO: <1 mg/ml at 23 °C; Ethanol: 320 g/kg at 20 °C; Ether: Insoluble; Hydrocarbons: Insoluble; Lower Alcohols: Soluble.
Flash Point: Not available; probably combustible

RTECS Toxicity Data

Acute Oral: Human LD_{Lo} Dose: 10 mg/kg; Toxic Effects: Automatic Nervous System - Central sympatholytic; Lungs, Thorax, or Respiration - Respiratory depression. Rat LD_{50} Dose: 600 mg/kg.
Acute Inhalation: Rat LC_{50} Dose: >5200 mg/m^3/4hr.
Acute Dermal: Rabbit LD_{50} Route: Skin; Dose: 232 mg/kg. Rat LD_{50} Route: Skin; Dose: >4 gm/kg.
Chronic (Multiple Dose) Oral: Rat Dose: 7920 mg/kg/72D-I; Toxic Effects: Endocrine - Changes in adrenal weight; Blood - Changes in serum composition; Biochemical - Other oxidoreductases. Rat Dose: 15600 mg/kg/26W-I; Toxic Effects: Behavioral - Alteration of classical conditioning; Blood - Changes in serum composition; Biochemical - Other transferases. Rat Dose: 1920 mg/kg/24D-C; Toxic Effects: Liver - Changes in liver weight; Endocrine - Changes in thymus weight; Immunological including allergic - Decrease in humoral immune response. Rat Dose: 1092 mg/kg/39W-I; Toxic Effects: Endocrine - Changes in adrenal weight; Endocrine - Changes in spleen weight.
Irritation Skin: Rabbit Standard Draize Test Dose: 500 mg/24H; Reaction: mild.
Mutagenic: Mouse DNA Inhibition; Route: Intravaginal; Dose: 5 pph. Bacteria - E Coli Mutations in Microorganisms; Dose: 500 mmol/L;/30M (-S9).
Tumorigenic: Mouse Route: Oral; Dose: 7100 mg/kg/78W-I; Toxic Effects: Tumorigenic - Neoplastic by RTECS criteria; Liver - Tumors.

Hazard Overviews

Poison

Corrosive

Health: Severely irritating to eyes; irritating to skin/respiratory tract. Poison. Other Acute Effects: may be fatal if inhaled, swallowed, or absorbed through skin.
Fire: Will burn. Hazards: emits toxic fumes. Extinguishing agents: carbon dioxide; dry chemical powder; water spray. Precautions: combustible liquid.
Carcinogenicity: IARC - Not listed; NIOSH - Not listed; NTP - Not listed; ACGIH - Not listed; OSHA - Not listed; EPA - Not listed; MAK - Not listed
Primary Target Organs:

Eyes

Skin

Respiratory System

Environmental

Ecotoxicity: LD_{50} Pheasant oral 261 mg/kg LD_{50} Anas platyrhynchos (Mallard) male oral 265 mg/kg (95% confidence limit 211-334 mg/kg) 3 mo old LC_{50} Trout about 4500 mg/l/96 hr /400 g/l formulated product/ /Conditions of bioassay not specified
Environmental Fate: If released to the atmosphere, it will exist in the vapor and particulate phases. Vapor phase will degrade slowly by reaction with photochemically produced hydroxyl radicals (half-life of about 12 days). Particulate phase will be physically removed from air by dry deposition. In soil and water, adsorption and biodegradation will be the primary removal processes. It is a quaternary ammonium compound which exists as a cation in water and moist soils; it is expected to strongly adsorb to various materials, including suspended solids in wastewater treatment facilities, sediments in rivers and lakes, suspended organics and minerals in natural water systems, clays, proteins, and microorganisms. Acclimation enhances biodegradation of quaternary ammonium compounds like this cation and the reduction of biomass or other nutrient materials in natural water may reduce the biodegradation rate of the cation. However, no specific adsorption data were located for the cation and limited biodegradation data are available.
Cleanup/Disposal: Guide No. 154: Eliminate all ignition sources (no smoking, flares, sparks or flames in immediate area). Do not touch damaged containers or spilled material unless wearing appropriate protective clothing. Stop leak if you can do it without risk. Prevent entry into waterways, sewers, basements or confined areas. Absorb or cover with dry earth, sand or other non-combustible material and transfer to containers. Do not get water inside containers.

Environmental Physical Data

Henry's Law Constant: estimated at 1.6 $x10^{-14}$
Octanol/Water Partition Coefficient: log K_{ow} = -3.8
Sorption Partition Coefficient: log K_{oc} = estimated at 0.2 to 2.2
BCF: estimated at 7.6 $x10^{-4}$

Regulations

RCRA 40CFR: Not listed
CERCLA: 40CFR 302.4: Not listed
SARA 40CFR 372.65: Not listed TPQ: 100/10000 lb
SARA EHS 40CFR 355: Listed TPQ: 100 lb
TSCA: Listed

CHL2140 CAS #: 494-03-1

CHLORNAPHAZINE

RTECS: QM2450000
EINECS Number: 207-785-0
Molecular Formula: $C_{14}H_{15}Cl_2N$
Formula Weight: 268.20
Synonyms: ALEUKON; N,N-BIS-(2-CHLORETHYL)-2-NAFTYLAMIN; 2-BIS(2-CHLOROETHYL)AMINONAPHTHALENE; BIS(2-CHLOROETHYL)-BETA-NAPHTHYLAMINE; N,N,-BIS(2-CHLOROETHYL)-2-NAPHTHYLAMINE; N,N-BIS(2-CHLOROETHYL)-2-NAPHTHYLAMINE; CB 1048; CHLORNAFTINA; CHLORNAPHAZIN; CHLORNAPHTHAZINE; CHLORNAPHTHIN; CHLORONAFTINA; CHLORONAPHTHINA; CHLORONAPHTHINE; CLORONAFTINA; 2-N,N-DI(2-CHLOROETHYL)NAPHTHYLAMINE; DI(2-CHLOROETHYL)-BETA-NAPHTHYLAMINE; DICHLOROETHYL-BETA-NAPHTHYLAMINE; N,N-DI(2-CHLOROETHYL)-BETA-NAPHTHYLAMINE; ERYSAN; NAFTICLORINA; 2-NAPHTHALENAMINE,N,N-BIS(2-CHLOROETHYL)-; 2-NAPHTHYLAMINE MUSTARD; NAPHTHYLAMINE MUSTARD; 2-NAPHTHYLAMINE,N,N-BIS(2-CHLOROETHYL)-; 2-NAPHTHYLBIS(2-CHLOROETHYL)AMINE; BETA-NAPHTHYL-BIS-(BETA-CHLOROETHYL)AMINE; BETA-NAPHTHYLBIS(BETA-CHLOROETHYL)AMINE; BETA-NAPHTHYL-DI-(2-CHLOROETHYL)AMINE; BETA-NAPHTHYLDI(2-CHLOROETHYL)AMINE; NSC-62209; NSC-66209; R 48; R48
Description: platelets
Use: antineoplastic

Physical Properties

Boiling Point: 210 °C (410 °F) at 5 mm Hg
Freezing Point: 54 °C (129.2 °F) to 56 °C (132.8 °F)
Water Solubility: Very Sparingly Soluble in Water
Other Solubilities: 95% Ethanol: 5-10 mg/ml at 22 °C; Acetone: <1 mg/ml at 19 °C; Benzene: Very Soluble; DMSO: >=100 mg/ml at 19 °C; Ether: Soluble; Glycerol: Very slightly Soluble; Olive oil: Soluble; Petroleum Ether: Slightly Soluble.
Flash Point: Not available; probably combustible

RTECS Toxicity Data

Mutagenic: Hamster Morphological Transformation; Cell Type: embryo; Dose: 120 ug/L. Insects - D Melanogaster Specific Locus Test; Route: Oral; Dose: 2600 umol/L.
Tumorigenic: Man Route: Oral; Dose: 2468 mg/kg/6Y-I; Toxic Effects: Tumorigenic - Carcinogenic by RTECS criteria; Kidney, Ureter, and Bladder - Tumors. Woman Route: Oral; Dose: 3132 mg/kg/10Y-I; Toxic Effects:

Tumorigenic - Carcinogenic by RTECS criteria; Kidney, Ureter, and Bladder - Tumors. Woman Route: Oral; Dose: 4090 mg/kg/7Y-I; Toxic Effects: Tumorigenic - Carcinogenic by RTECS criteria; Kidney, Ureter, and Bladder - Tumors. Woman Route: Oral; Dose: 3200 mg/kg/11Y-I; Toxic Effects: Tumorigenic - Carcinogenic by RTECS criteria; Kidney, Ureter, and Bladder - Tumors.

Hazard Overviews

Fire: Will burn.

Carcinogenicity: IARC - Group 1, Carcinogenic to humans; NIOSH - Not listed; NTP - Not listed; ACGIH - Not listed; OSHA - Not listed; EPA - Not listed; MAK - Not listed

Environmental

Environmental Fate: If released to the atmosphere, vapor-phase is expected to degrade rapidly by reaction with photochemically produced hydroxyl radicals (estimated half-life of 1.4 hours in typical air). If released to water, hydrolysis is expected to be the major degradation process. It has a measured aqueous hydrolysis half-life of 216 hours at 25 °C and pH 7. If released to moist soil, it can be expected to hydrolyze. An estimated K_{oc} value of 21000 indicates that it should be immobile in soil. No data are available to predict the biodegradability in soil or water.

Environmental Physical Data

Henry's Law Constant: estimated at 1.7×10^{-7}

Octanol/Water Partition Coefficient: log K_{ow} = 4.535

Sorption Partition Coefficient: K_{oc} = estimated at 2.1×10^{4}

BCF: estimated at 1650

Regulations

RCRA 40CFR: Listed Hazardous Waste No. U026 Toxic Waste

CERCLA: 40CFR 302.4: Listed per RCRA Section 3001 RQ: 100 lb (45.35 kg)

SARA 40CFR 372.65: Not listed

SARA EHS 40CFR 355: Not listed

TSCA: Not listed

CHL2170 CAS #: 107-20-0

2-CHLOROACETALDEHYE

RTECS: AB2450000

DOT: UN2232; IMO6.1

EINECS Number: 203-472-8

Molecular Formula: C_2H_3ClO

Structured MF: $ClCH_2CHO$

Formula Weight: 78.50

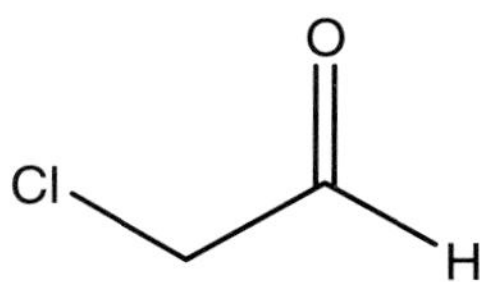

Chemical Structure

Synonyms: ACETALDEHYDE,CHLORO-; 2-CHLOROACETALDEHYDE; ALPHA-CHLOROACETALDEHYDE; CHLOROACETALDEHYDE; CHLOROACETALDEHYDE (40% AQUEOUS SOLUTION); CHLOROACETALDEHYDE MONOMER; CHLOROALDEHYDE; 2-CHLORO-1-ETHANAL; 2-CHLOROETHANAL; CHLOROETHANAL; MONOCHLOROACETALDEHYDE

Description: clear liquid; acrid odor

Use: intermediate for organic synthesis; in the manufacture of 2-aminothiazole and other compounds; as a fungicide and preservative; to facilitate bark removal from tree trunks

Physical Properties

Boiling Point: 85 °C (185 °F) to 86 °C (187 °F) at 760 mm Hg

Freezing Point: -16.3 °C (2.66 °F)

Specific Gravity: 1.19

Vapor Density: 40% aqueous solution 2.7 Air=1

Density: 1.236 g/mL (50% aqueous solution)

Vapor Pressure: 40% Solution 100 mm Hg at 45 °C

Water Solubility: Miscible

Other Solubilities: 95% Ethanol: >=100 mg/ml at 19 °C; Acetone: >=100 mg/ml at 19 °C; DMSO: >=100 mg/ml at 19 °C; Ether: >10%; Methanol: Soluble.

Odor Threshold: 3.0 mg/m^3

Refraction Index: 1.397

Ionization Potential (eV): 10.61

Flash Point: 87.7 °C Closed Cup

Autoignition Temperature: 88 °C

RTECS Toxicity Data

Acute Oral: Rat LD_{50} Dose: 89 mg/kg. Mouse LD_{50} Dose: 82 mg/kg.

Acute Inhalation: Rat LC_{50} Dose: 650 mg/m^3/1hr; Toxic Effects: Vascular - BP elevation not characterized in autonomic section; Lungs, Thorax, or Respiration - Respiratory obstruction; Lungs, Thorax, or Respiration - Other changes.

Acute Dermal: Rabbit LD_{50} Route: Skin; Dose: 267 mg/kg.

Mutagenic: Human DNA Damage; Cell Type: lymphocyte; Dose: 100 umol/L. Rat DNA Damage; Cell Type: liver; Dose: 47 mmol/L.

Hazard Overviews

Corrosive

Fire Diamond

Health: Corrosive to eyes/skin/respiratory tract. Toxic. Also Causes: coughing, choking, bronchial constriction, increased respiratory rate, pulmonary edema, lung damage, CNS and kidney damage, hematologic disturbances, corneal damage, blindness, sore eyelids, burns of the gastrointestinal tract. Chronic Effects: skin allergies, respiratory tract sensitization, asthma attacks with shortness of breath, wheezing, coughing, and/or chest tightness.

Fire: Combustible. Polymerizes on standing. Use dry chemical, water spray, fog, or regular foam.

Reactivity: Stable. Hazardous polymerization can occur on standing, but reverts to the monomer on distillation. Avoid: heat; ignition sources. Incompatible with: oxidizing material; acids. Hazardous decomposition products: carbon monoxide; carbon dioxide; chloride.

Carcinogenicity: IARC - Not listed; NIOSH - Listed as carcinogen; NTP - Not listed; ACGIH - Not listed; OSHA - Not listed; EPA - Not listed; MAK - Not listed

Primary Target Organs:

Eyes

Skin

Respiratory System

Exposure Limits

OSHA PEL: STEL: 1 ppm; 3 mg/m^3.

ACGIH TLV: STEL: 1 ppm; 3.2 mg/m^3; Ceiling.

NIOSH REL: STEL: 1 ppm; 3 mg/m^3.

NIOSH IDLH: 45 ppm.

DFG MAK: TWA: 1 ppm; 3 mg/m^3.

Respirator Recommendation

Exposure Range: >1 to <45 ppm Supplied Air, Constant Flow/Pressure Demand, Full Face

Exposure Range: 45 to unlimited ppm Self-contained Breathing Apparatus, Pressure Demand, Full Face

Note: odor threshold unknown

Environmental

Environmental Fate: If released to moist soil, it is expected to be highly mobile and susceptible to significant leaching. If released to dry soil, this compound is expected to volatilize fairly rapidly from soil surfaces. If released to water, volatilization should be an important, if not the dominant, removal mechanism (half-life from a model river - 42 hours). If released to the atmosphere, it is expected to exist primarily in the vapor phase. Reaction with photochemically generated hydroxyl radicals is expected to be the dominant removal mechanism (half-life 1.7 days). This compound may be susceptible to removal from the atmosphere by wet deposition; however, any removed from the atmosphere by this mechanism has the potential to reenter the atmosphere by volatilization.

Cleanup/Disposal: Guide No. 153: Eliminate all ignition sources (no smoking, flares, sparks or flames in immediate area). Do not touch damaged containers or spilled material unless wearing appropriate protective clothing. Stop leak if you can do it without risk. Prevent entry into waterways, sewers, basements or confined areas. Absorb or cover with dry earth, sand or other non-combustible material and transfer to containers. Do not get water inside containers.

Environmental Physical Data

Henry's Law Constant: estimated at 2 x10^{-5}

Octanol/Water Partition Coefficient: log K_{ow} = 0.39

Sorption Partition Coefficient: K_{oc} = estimated at 39

BCF: estimated at 1

Regulations

RCRA 40CFR: Listed Hazardous Waste No. P023 Toxic Waste

CERCLA: 40CFR 302.4: Listed per RCRA Section 3001 RQ: 1000 lb (453.5 kg)

SARA 40CFR 372.65: Not listed

SARA EHS 40CFR 355: Not listed

TSCA: Listed

Analytical Methods

Soil: SW846 8010B

Indoor / Expired Air: NIOSH 2015

CHL2290 CAS #: 79-11-8

CHLOROACETIC ACID

RTECS: AF8575000

DOT: UN1750; UN1751; IMO8.0

EINECS Number: 201-178-4

Molecular Formula: $C_2H_3ClO_2$

Formula Weight: 94.50

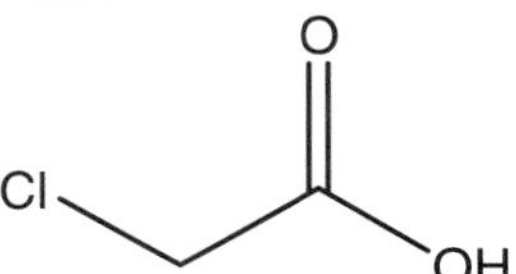

Chemical Structure

Synonyms: ACETIC ACID,CHLORO-; ACIDE CHLORACETIQUE; ACIDE CHLOROACETIQUE; ACIDE MONOCHLORACETIQUE; ACIDOMONOCLOROACETICO; CHLORACETIC ACID; ALPHA-CHLOROACETIC ACID; CHLOROACETIC ACID,SOLID; CHLOROETHANOIC ACID; KYSELINA CHLOROCTOVA; MCA; MONOCHLOORAZIJNZUUR; MONOCHLORACETIC ACID; MONOCHLORESSIGSAEURE; MONOCHLOROACETIC ACID; MONOCHLOROACETIC ACID SOLUTION; MONOCHLOROETHANOIC ACID

Description: colorless, light yellow crystals, liquid; strong vinegar odor

Use: herbicide; preservative; bacteriostat; and intermediate in the production of carboxymethylcellulose, ethyl chloroacetate, glycine, synthetic caffeine, sarcosine, thioglycolic acid, EDTA; 2,4-D; and 2,4,5-T

Physical Properties

Boiling Point: 189 °C (372 °F)

Freezing Point: Alpha 63 °C (145.4 °F)

Specific Gravity: 1.4043 at 40 °C/4 °C

Vapor Density: 3.26 Air=1

Density: 1.58 g/mL at 20 °C

Vapor Pressure: 1 mm Hg at 43.0 °C

Water Solubility: Very Soluble in Water

Other Solubilities: 95% Ethanol: >=100 mg/ml at 20 °C; Acetone: >=100 mg/ml at 20 °C; Benzene: Soluble; Carbon

Disulfide: Soluble; Chloroform: Soluble; DMSO: >=100 mg/ml at 20 °C; Ether: Soluble.
Surface Tension: 33 dynes/cm at 80 °C
Odor Threshold: 0.15 mg/m^3
pH: Acid
Refraction Index: 1.4351 at 55 °C/D
Ionization Potential (eV): 10.7 +/-2.0
Flash Point: 126 °C Closed Cup
Autoignition Temperature: > 500 °C
LEL: 8% v/v

RTECS Toxicity Data

Acute Oral: Rat LD_{50} Dose: 55 mg/kg.
Acute Inhalation: Rat LC_{50} Dose: 180 mg/m^3.
Acute Dermal: Rat LD_{50} Route: Subcutaneous Dose: 5 mg/kg. Mouse LD_{50} Route: Subcutaneous Dose: 250 mg/kg.
Chronic (Multiple Dose) Oral: Rat Dose: 9750 mg/kg/13W-I; Toxic Effects: Cardiac - Other changes; Blood - Changes in erythrocite (RBC) cell count; DEATH. Mouse Dose: 13 gm/kg/13W-I; Toxic Effects: Liver - Other changes; Liver - Changes in liver weight; DEATH. Mouse Dose: 3840 mg/kg/16D-I; Toxic Effects: Sense organs and special senses - Lacrimation; Behavioral - Ataxia; DEATH.
Chronic (Multiple Dose) Inhalation: Rat Dose: 20800 ug/m^3/17W-I; Toxic Effects: Kidney, Ureter, and Bladder - Other changes; Blood - Pigmented or nucleated red Blood cells. Guinea Pig Dose: 20800 ug/m^3/17W-I; Toxic Effects: Kidney, Ureter, and Bladder - Other changes; Blood - Pigmented or nucleated red Blood cells.
Mutagenic: Mouse Mutations in Microorganisms; Cell Type: lymphocyte; Dose: 548 mg/L (+S9). Mouse Mutations in Mammalian Somatic Cells; Cell Type: lymphocyte; Dose: 400 mg/L.
Tumorigenic: Mouse Route: Subcutaneous; Dose: 100 mg/kg; Toxic Effects: Tumorigenic - Equivocal tumorigenic agent by RTECS criteria; Lungs, Thorax, or Respiration - Tumors; Liver - Tumors. Mouse Route: Subcutaneous; Dose: 1300 mg/kg/65W-I; Toxic Effects: Tumorigenic - Equivocal tumorigenic agent by RTECS criteria; Tumorigenic - Tumors at site of application.

Hazard Overviews

Corrosive

Fire Diamond

Health: Corrosive to eyes/skin/respiratory tract. Toxic. Also Causes: convulsions, respiratory depression, anuria, CNS depression, pulmonary edema; ingestion: perforation, peritonitis.
Fire: Will burn. Use water spray, fog, mist, dry chemical, or foam. Choose media appropriate to the surrounding fire. Material may ignite other combustibles such as oil, wood, or paper.
Reactivity: Stable. Hazardous polymerization cannot occur. Incompatible with: metals. Hazardous decomposition products: highly toxic fumes of chlorides and phosgene.

Carcinogenicity: IARC - Not listed; NIOSH - Not listed; NTP - Not listed; ACGIH - Not listed; OSHA - Not listed; EPA - Not listed; MAK - Not listed
Primary Target Organs:

Eyes

Skin

Respiratory System

Mucous Membranes

Nervous System

Exposure Limits
AIHA WEEL: TWA: 0.3 ppm skin; STEL: 1 ppm 15-min skin.
Respirator Recommendation
Exposure Range: >0.3 to 150 ppm Supplied Air, Constant Flow/Pressure Demand, Half Mask
Exposure Range: >150 to 300 ppm Supplied Air, Constant Flow/Pressure Demand, Full Face
Exposure Range: >300 to unlimited ppm Self-contained Breathing Apparatus, Pressure Demand, Full Face
Note: poor warning properties

Environmental

Ecotoxicity: Cyprinus carpio (carp) pertubation concentration, 14 mg/l /Conditions of bioassay not specified Chironomus plumosus (insect) pertubation concentration, 140 mg/l Vorticella campanula (protozoan) pertubation concentration, 9 mg/l
Environmental Fate: If released into surface water, it would biodegrade (73% in 8-10 days). It would not adsorb appreciably to sediment or bioconcentrate in fish. If spilled on land it would biodegrade and leach into the groundwater. Its fate in groundwater is unknown. If released into the air, probably as an aerosol, it will gravitationally settle out and undergo slow photodegradation.
Cleanup/Disposal: Guide No. 153: Eliminate all ignition sources (no smoking, flares, sparks or flames in immediate area). Do not touch damaged containers or spilled material unless wearing appropriate protective clothing. Stop leak if you can do it without risk. Prevent entry into waterways, sewers, basements or confined areas. Absorb or cover with dry earth, sand or other non-combustible material and transfer to containers. Do not get water inside containers.

Environmental Physical Data

Henry's Law Constant: estimated at 1.09×10^{-6}
Octanol/Water Partition Coefficient: log K_{ow} = 0.22
BCF: none expected

Regulations

RCRA 40CFR: Not listed
CERCLA: 40CFR 302.4: Listed per CAA Section 112 RQ: 100 lb (45.35 kg)
SARA 40CFR 372.65: Listed TPQ: 100/10000 lb
SARA EHS 40CFR 355: Listed TPQ: 100 lb
TSCA: Listed

Analytical Methods

Drinking Water: EPA 552, 552.1; APHA 6233-B
Food: AOAC 949.09
Indoor / Expired Air: NIOSH 2008

CHL2350 CAS #: 78-95-5

CHLOROACETONE

RTECS: UC0700000
EINECS Number: 201-161-1
Molecular Formula: C_3H_5ClO
Structured MF: $ClCH_2COCH_3$
Formula Weight: 92.53

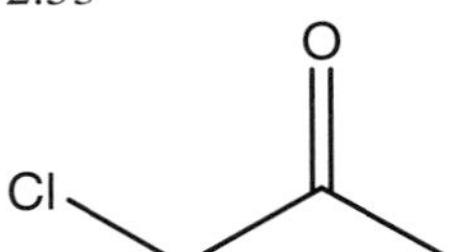

Chemical Structure

Synonyms: A-STOFF; ACETONE,CHLORO-; ACETONYL CHLORIDE; CHLORACETONE; 1-CHLOROACETONE; ALPHA-CHLOROACETONE; 1-CHLORO-2-KETOPROPANE; CHLOROMETHYL METHYL KETONE; 1-CHLORO-2-OXOPROPANE; 1-CHLORO-2-PROPANONE; CHLORO-2-PROPANONE; CHLOROPROPANONE; MONOCHLORACETONE; MONOCHLOROACETONE; 2-PROPANONE,1-CHLORO-; TONITE

Description: colorless liquid; pungent odor

Use: mfr of couplers for color photography; as enzyme inactivator; intermediary in mfr of perfumes, antioxidants, drugs; in insecticide formulations; in photopolymerization of vinyl compounds; reacts with aryl grignard reagents to form stilbenes; catalyst in tetraethyllead production; as selective solvent for separating diolefins; organic synthesis; tear gas

Physical Properties

Boiling Point: 119.7 °C (247 °F)
Freezing Point: -44.5 °C (-48.1 °F)
Specific Gravity: 1.123 at 25 °C/4 °C
Vapor Pressure: 12 mm Hg at 68 °F
Water Solubility: 1 g/10 g
Other Solubilities: > 10% in Ethanol; > 10% in Chloroform.
Surface Tension: 35.27 dynes/cm at 20 °C
Ionization Potential (eV): 9.91 +/-0.03
Flash Point: 40 °C Open Cup

RTECS Toxicity Data

Acute Oral: Rat LD_{50} Dose: 100 mg/kg; Toxic Effects: Behavioral - Somnolence (general depressed activity); Behavioral - Ataxia; Skin and appendages - Hair. Mouse LD_{50} Dose: 127 mg/kg; Toxic Effects: Behavioral - Somnolence (general depressed activity); Behavioral - Ataxia; Skin and appendages - Hair.

Acute Inhalation: Rat LC_{50} Dose: 262 ppm/1hr; Toxic Effects: Behavioral - Somnolence (general depressed activity); Behavioral - Ataxia; Skin and appendages - Hair. Human LC_{Lo} Dose: 605 ppm/10M.

Acute Dermal: Rabbit LD_{50} Route: Skin; Dose: 141 mg/kg; Toxic Effects: Behavioral - Somnolence (general depressed activity); Behavioral - Ataxia; Skin and appendages - Hair. Rat LD_{Lo} Route: Skin; Dose: 100 mg/kg.

Chronic (Multiple Dose) Oral: Rat Dose: 650 mg/kg/17D-I; Toxic Effects: Gastrointestinal - Changes in structure or function of salivary glands; Gastrointestinal - Gastritis; Nutritional and gross metabolic - Weight loss or decreased weight gain.

Mutagenic: Insects - D Melanogaster Sex Chromosome Loss; Route: Inhalation; Dose: 100 pph/6M. Bacteria - S Typhimurium Mutations in Microorganisms; Dose: 6 mg/L (+S9).

Hazard Overviews

Explosive

Flammable

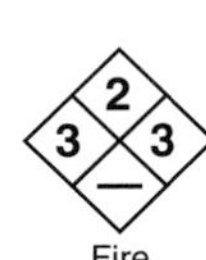

Fire Diamond

Health: Severe irritation to eyes/skin/respiratory tract. Toxic.

Fire: Flammable. Explosive if unstabilized. Gives off toxic phosgene and hydrogen chloride when heated to decomposition. Use dry chemical, carbon dioxide, halon, water spray, or foam. For large fires, waterspray, fog, or standard foam is recommended.

Reactivity: Stable with 0.1% water or 1.0% calcium carbonate. Hazardous polymerization can occur, can polymerize to a rubber-like substance and explode. Avoid: heat and ignition sources. Incompatible with: oxidizing materials; alkalis; sodium ethoxide; any compound with an active hydrogen atom. Hazardous decomposition products: toxic fumes of phosgene; hydrogen chloride.

Carcinogenicity: IARC - Not listed; NIOSH - Not listed; NTP - Not listed; ACGIH - Not listed; OSHA - Not listed; EPA - Not listed; MAK - Not listed

Primary Target Organs:

Eyes

Skin

Mucous Membranes

Exposure Limits

ACGIH TLV: STEL: 1 ppm; 3.8 mg/m^3; Ceiling.

Respirator Recommendation

Exposure Range: >1 to 50 mg/m^3 Air Purifying, Negative Pressure, Half Mask

Exposure Range: >50 to 1000 mg/m^3 Air Purifying, Negative Pressure, Full Face

Exposure Range: >1000 to unlimited mg/m^3 Self-contained Breathing Apparatus, Pressure Demand, Full Face

Note: poor warning properties

Environmental

Ecotoxicity: Fishes: guppy (Poecilia reticulata): log LC_{50}, 14 days: 0.88 mmol/l

Environmental Fate: Little is known about its fate in the environment. From its physical/chemical properties, one can predict that it will hydrolyze, releasing HCl, but will not photolyze, adsorb to soil or bioconcentrate significantly.

Environmental Physical Data

Octanol/Water Partition Coefficient: log K_{ow} = 0.28

BCF: none expected

Regulations

RCRA 40CFR: Not listed
CERCLA: 40CFR 302.4: Not listed
SARA 40CFR 372.65: Not listed
SARA EHS 40CFR 355: Not listed
TSCA: Listed

CHL2410 CAS #: 532-27-4

A-CHLOROACETOPHENONE

RTECS: AM6300000
DOT: UN1697; IMO6.1
EINECS Number: 208-531-1
Molecular Formula: C_8H_7ClO
Structured MF: $C_6H_5COCH_2Cl$
Formula Weight: 154.60

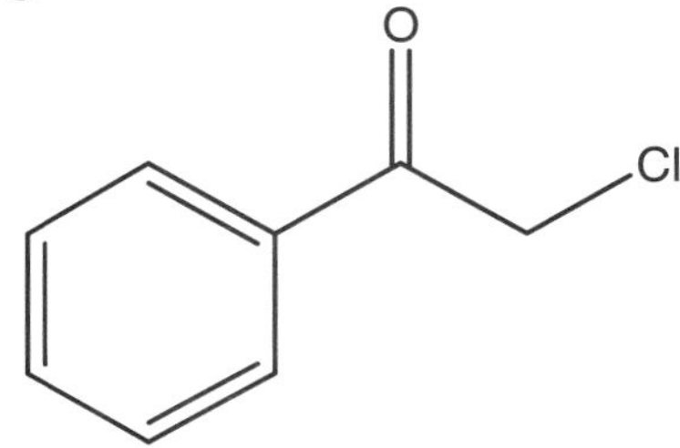

Chemical Structure

Synonyms: ACETOPHENONE,2-CHLORO-; CAF; CAP; CHEMICAL MACE; CHLORACETOPHENONE; 1-CHLOROACETOPHENONE; 2-CHLOROACETOPHENONE; ALPHA-CHLOROACETOPHENONE; CHLOROACETOPHENONE; OMEGA-CHLOROACETOPHENONE; CHLOROACETOPHENONE,LIQUID OR SOLID; CHLOROMETHYL PHENYL KETONE; 2-CHLORO-1-PHENYLETHANONE; CN; CN,LIQUID OR SOLID; ETHANONE,2-CHLORO-1-PHENYL-; MACE; MACE (LACRIMATOR); ALPHA-PHENACYL CHLORIDE; PHENACYL CHLORIDE; PHENACYLCHLORIDE; PHENYL CHLOROMETHYL KETONE; PHENYL CHLOROMETHYLKETONE; PHENYLCHLOROMETHYLKETONE
Description: colorless, white, grey crystals; apple blossom odor
Use: chemical warfare agent with lacrimatory properties; referred to as CN in military circles; principal ingredient of the riot control gas MACE, also called "Chemical Mace"; pharmaceutical intermediate

Physical Properties

Boiling Point: 244 °C (471 °F) to 245 °C (473 °F)
Freezing Point: 56.5 °C (133.7 °F)
Specific Gravity: 1.324 at 15 °C
Vapor Density: 5.2 Air=1
Saturated Vapor Density: 1.200036928 kg/m^3
Vapor Pressure: 0.0054 mm Hg at 20 °C
Water Solubility: Practically Insoluble in Water
Other Solubilities: 95% Ethanol: <1 mg/ml at 19 °C; Acetone: <1 mg/ml at 19 °C; Benzene: Soluble; Carbon Disulfide: Soluble; DMSO: <1 mg/ml at 19 °C; Ether: Soluble; Toluene: <1 mg/ml at 18 °C.
Odor Threshold: 0.1020 to 0.15 mg/m^3
Refraction Index: 1.685
Ionization Potential (eV): 9.44
Flash Point: 118 °C Closed Cup

RTECS Toxicity Data

Acute Oral: Rat LD_{50} Dose: 50 mg/kg. Mouse LD_{50} Dose: 139 mg/kg.
Acute Inhalation: Human LC_{Lo} Dose: 159 mg/m^3/20M. Human TC_{Lo} Dose: 93 mg/m^3/3M; Toxic Effects: Sense organs and special senses - Lacrimation; Sense organs and special senses - Conjunctive irritation; Lungs, Thorax, or Respiration - Dyspnea. Human TC_{Lo} Dose: 20 mg/m^3; Toxic Effects: Sense organs and special senses - Other; Lungs, Thorax, or Respiration - Cough; Lungs, Thorax, or Respiration - Dyspnea.
Chronic (Multiple Dose) Inhalation: Rat Dose: 19 mg/m^3/6H/14D-I; Toxic Effects: DEATH. Mouse Dose: 10 mg/m^3/6H/14D-I; Toxic Effects: DEATH.
Irritation Eye: Rabbit Standard Draize Test Dose: 1 mg; Reaction: mild. Rabbit Standard Draize Test Dose: 3 mg; Reaction: severe.
Irritation Skin: Rabbit Standard Draize Test Dose: 5 mg/24H; Reaction: mild. Rabbit Open Draize Test Dose: 12%/6H open; Reaction: moderate.
Mutagenic: Hamster Cytogenetic Analysis; Cell Type: ovary; Dose: 3 mg/L.
Tumorigenic: Rat Route: Inhalation; Dose: 2 mg/m^3/6H/2Y-I; Toxic Effects: Tumorigenic - Equivocal tumorigenic agent by RTECS criteria; Endocrine - Tumors; Skin and appendages - Tumors. Mouse Route: Skin; Dose: 2400 mg/kg/27W-I; Toxic Effects: Tumorigenic - Neoplastic by RTECS criteria; Skin and appendages - Tumors; Tumorigenic - Tumors at site of application.

Hazard Overviews

Poison

Fire Diamond

Health: Severely irritating to eyes/skin/respiratory tract. Poison. Also Causes: difficulty breathing, watering eyes, pulmonary edema, blurred vision, tingling of nose, rhinorrhea, burning of throat, corneal opacity, GI irritation. Chronic Effects: dermatitis.
Fire: Will burn. Use dry chemical, water spray, fog, or regular foam. Containers may explode in heat of fire.
Reactivity: Stable. Hazardous polymerization cannot occur. Avoid: heat; ignition sources. Incompatible with: strong oxidizers. Hazardous decomposition products: carbon oxide(s); hydrogen chloride; chlorine gas.
Carcinogenicity: IARC - Not listed; NIOSH - Not listed; NTP - Not listed; ACGIH - Class A4, Not classifiable as a human carcinogen; OSHA - Not listed; EPA - Not listed; MAK - Not listed

Primary Target Organs:

Eyes

Skin

Respiratory System

Exposure Limits
OSHA PEL: TWA: 0.05 ppm; 0.3 mg/m^3.
ACGIH TLV: TWA: 0.05 ppm; 0.32 mg/m^3.
NIOSH REL: TWA: 0.05 ppm; 0.3 mg/m^3.
NIOSH IDLH: 15 mg/m^3.
Respirator Recommendation
Exposure Range: >0.3 to 3 ppm Supplied Air, Constant Flow/Pressure Demand, Half Mask
Exposure Range: >3 to <15 ppm Supplied Air, Constant Flow/Pressure Demand, Full Face
Exposure Range: 15 to unlimited ppm Self-contained Breathing Apparatus, Pressure Demand, Full Face
Cartridge Color: black with dust/mist prefilter (use P100 or consult supervisor for appropriate dust/mist prefilter)

Environmental

Environmental Fate: If released to the atmosphere, it will degrade in the vapor-phase by reaction with photochemically produced hydroxyl radicals (estimated half-life of 9.2 days). If released to soil, it will probably leach based on estimated K_{oc} values of 76-325. If released to water, it will volatilize slowly; the estimated volatilization half-lives from a model river (1 m deep) and model environmental pond (2 m deep) are 13.3 and 159 days, respectively. Direct photolysis may contribute to the environmental degradation, but potential rates are unknown. Insufficient data are available to predict the importance of biodegradation in soil or water.
Cleanup/Disposal: Guide No. 153: Eliminate all ignition sources (no smoking, flares, sparks or flames in immediate area). Do not touch damaged containers or spilled material unless wearing appropriate protective clothing. Stop leak if you can do it without risk. Prevent entry into waterways, sewers, basements or confined areas. Absorb or cover with dry earth, sand or other non-combustible material and transfer to containers. Do not get water inside containers.

Environmental Physical Data

Henry's Law Constant: estimated at 3.46 x10^{-6}
Octanol/Water Partition Coefficient: log K_{ow} = 2.09
Sorption Partition Coefficient: K_{oc} = 76 to 325
BCF: estimated at 23

Regulations

RCRA 40CFR: Not listed
CERCLA: 40CFR 302.4: Listed per CAA Section 112 RQ: 100 lb (45.35 kg)
SARA 40CFR 372.65: Listed
SARA EHS 40CFR 355: Not listed
TSCA: Listed

CHL2470 CAS #: 79-04-9

CHLOROACETYL CHLORIDE

RTECS: AO6475000
DOT: UN1752; IMO8.0
EINECS Number: 201-171-6
Molecular Formula: $C_2H_2Cl_2O$
Structured MF: $ClCH_2COCl$
Formula Weight: 112.95

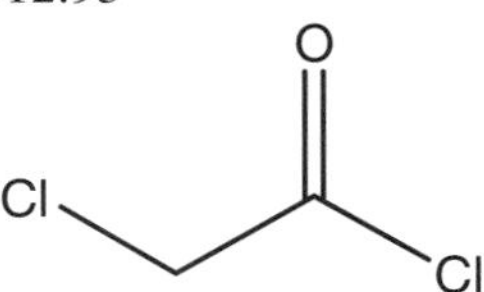

Chemical Structure

Synonyms: ACETYL CHLORIDE,CHLORO-; CHLORACETYL CHLORIDE; CHLORID KYSELINY CHLOROCTOVE; CHLOROACETIC ACID CHLORIDE; CHLOROACETIC CHLORIDE; CHLORURE DE CHLORACETYLE; MONOCHLOROACETYL CHLORIDE
Description: colorless to water-white to slightly yellow liquid; sharp, pungent, irritating strong odor
Use: tear gas; chemical intermediate for insecticides; intermediate in the manufacture of chloroacetophenone and other intermediates; production of chloroacetamide herbicides such as alachlor; in the production of adrenalin, diazepam, chloroacetophenone, chloroacetate esters and chloroacetic anhydride

Physical Properties

Boiling Point: 106 °C (223 °F)
Freezing Point: -21.77 °C (-7.186 °F)
Specific Gravity: 1.4202 at 20 °C/4 °C
Vapor Density: 3.9 Air=1
Saturated Vapor Density: 1.286844828 kg/m^3
Vapor Pressure: 19 mm Hg at 20 °C
Water Solubility: Decomposes
Other Solubilities: Soluble in Ether and Acetone
Surface Tension: Estimated at 25 dynes/cm
Odor Threshold: 0.023 to 0.140 ppm
pH: Acid
Refraction Index: 1.4541 at 20 °C/D
Critical Temperature: 308 °C
Critical Pressure: 5.11 x10^6 Pa
Ionization Potential (eV): 10.30
Flash Point: Nonflammable

RTECS Toxicity Data

Acute Oral: Rat LD_{50} Dose: 208 mg/kg. Mouse LD_{50} Dose: 220 mg/kg.
Acute Inhalation: Rat LC_{50} Dose: 1000 ppm/4hr. Mouse LC_{50} Dose: 1300 ppm/2hr.
Acute Dermal: Rabbit LD_{Lo} Route: Skin; Dose: 316 mg/kg. Rat LD_{50} Route: Skin; Dose: 662 mg/kg; Toxic Effects: Skin and appendages - Corrosive.

Hazard Overviews

Corrosive

Health: Corrosive to eyes/skin/respiratory tract. Toxic. Other Acute Effects: harmful if swallowed, inhaled, or absorbed through skin; extremely destructive to tissue of the mucous membranes and upper respiratory tract, eyes and skin; lachrymator; inhalation may result in spasm, inflammation and edema of the larynx and bronchi, chemical pneumonitis and pulmonary edema; symptoms of exposure may include burning sensation, coughing, wheezing, laryngitis, shortness of breath, headache, nausea and vomiting.
Fire: Noncombustible. Hazards: emits toxic fumes. Extinguishing agents: carbon dioxide, dry chemical powder or appropriate foam. Precautions: combustible liquid.
Reactivity: Incompatible with: strong oxidizing agents, strong bases, alcohols, may decompose on exposure to moist air or water. Hazardous decomposition products: toxic fumes of: carbon monoxide, carbon dioxide, hydrogen chloride gas.
Carcinogenicity: IARC - Not listed; NIOSH - Listed as carcinogen; NTP - Not listed; ACGIH - Not listed; OSHA - Not listed; EPA - Not listed; MAK - Not listed
Primary Target Organs:

Eyes

Skin

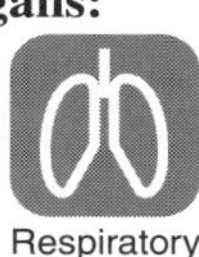
Respiratory System

Exposure Limits
OSHA PEL Vacated 1989 Limits: TWA: 0.05 ppm; 0.2 mg/m^3.
ACGIH TLV: TWA: 0.05 ppm; 0.23 mg/m^3; STEL: 0.15 ppm; 0.69 mg/m^3.
NIOSH REL: TWA: 0.05 ppm; 0.2 mg/m^3.
Respirator Recommendation
Exposure Range: >0.05 to 0.25 mg/m^3 Air Purifying, Negative Pressure, Half Mask
Exposure Range: >0.25 to 50 mg/m^3 Air Purifying, Negative Pressure, Full Face
Exposure Range: >50 to unlimited mg/m^3 Self-contained Breathing Apparatus, Pressure Demand, Full Face Self-contained Breathing Apparatus, Pressure Demand, Full Face
Note: odor threshold unknown

Environmental

Cleanup/Disposal: Guide No. 156: Eliminate all ignition sources (no smoking, flares, sparks or flames in immediate area). All equipment used when handling the product must be grounded. Do not touch damaged containers or spilled material unless wearing appropriate protective clothing. Stop leak if you can do it without risk. A vapor suppressing foam may be used to reduce vapors. For chlorosilanes, use AFFF alcohol-resistant medium expansion foam to reduce vapors. Do not get water on spilled substance or inside containers. Use water spray to reduce vapors or divert vapor cloud drift. Prevent entry into waterways, sewers, basements or confined areas. Small Spills: Cover with dry earth, dry sand, or other non-combustible material followed with plastic sheet to minimize spreading or contact with rain. Use clean non-sparking tools to collect material and place it into loosely covered plastic containers for later disposal.

Environmental Physical Data

BCF: no food chain concentration potential

Regulations

RCRA 40CFR: Not listed
CERCLA: 40CFR 302.4: Not listed
SARA 40CFR 372.65: Not listed
SARA EHS 40CFR 355: Not listed
TSCA: Listed

CHL2530 **CAS #: 4080-31-3**

1-(3-CHLOROALLYL)-3,5,7-TRIAZA-1-AZONIAADAMANTANE CHLORIDE

RTECS: XX8450000
EINECS Number: 223-805-0
Molecular Formula: $C_9H_{16}Cl_2N_4$
Formula Weight: 251.19
Synonyms: CDEC; N-(3-CHLOROALLYL)METHENAMINE; DOWCO 184; DOWICIDE 184; DOWICIDE Q; DOWICIL 100; DOWICIL 75; QUATERNIUM 15
Description: cream colored powder
Use: bactericide used as preservative in latex, paints, floor polishes, joint cements, adhesive, inks, starches, etc.; an antimicrobial agent to serve as a preservative in cosmetic formulations

Physical Properties

Water Solubility: > 100 mg/mL at 22 C
Other Solubilities: 95% Ethanol: 1-10 mg/ml at 21 °C; Acetone: <1 mg/ml at 22 °C; DMSO: 1-10 mg/ml at 21 °C.
Flash Point: Not available; probably combustible

RTECS Toxicity Data

Acute Oral: Rat LD_{50} Dose: 500 mg/kg. Rabbit LD_{50} Dose: 78500 ug/kg.
Acute Dermal: Rabbit LD_{50} Route: Skin; Dose: 565 mg/kg.
Irritation Skin: Rabbit Standard Draize Test Dose: 500 mg/24H; Reaction: mild.
Mutagenic: Bacteria - S Typhimurium Mutations in Microorganisms; Dose: 333 ug/plate (-S9).

Hazard Overviews

Fire: Will burn.
Carcinogenicity: IARC - Not listed; NIOSH - Not listed; NTP - Not listed; ACGIH - Not listed; OSHA - Not listed; EPA - Not listed; MAK - Not listed

Environmental

Regulations

RCRA 40CFR: Not listed
CERCLA: 40CFR 302.4: Not listed
SARA 40CFR 372.65: Listed
SARA EHS 40CFR 355: Not listed
TSCA: Listed

CHL2560 CAS #: 88-51-7

2-CHLORO-4-AMINO TOLUENE-5-SULFONIC ACID

RTECS: XT6330000
EINECS Number: 201-837-6
Molecular Formula: $C_7H_8ClNO_3S$
Formula Weight: 221.7

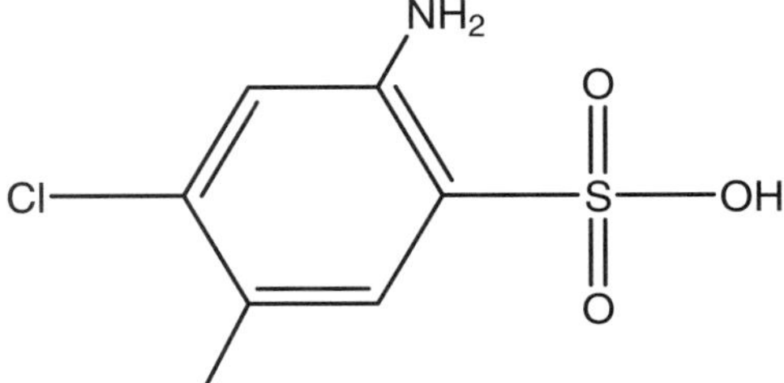

Chemical Structure

Synonyms: 2B ACID; 2-AMINO-4-CHLORO-5-METHYLBENZENESULFONIC ACID; 4-AMINO-2-CHLOROTOLUENE-5-SULFONIC ACID; 6-AMINO-4-CHLORO-M-TOLUENESULFONIC ACID; BENZENESULFONIC ACID,2-AMINO-4-CHLORO-4-CHLORO-5-METHYL-; BENZENESULFONIC ACID,2-AMINO-4-CHLORO-5-METHYL-; BRILLIANT TONING RED AMINE; 2-CHLORO-4-AMINOTOLUENE-5-SULFONIC ACID; 3-CHLORO-4-METHYLANILINE-6-SULFONIC ACID; 2-CHLORO-4-TOLUIDINE-5-SULFONIC ACID; KYSELINA 2-CHLOR-4-TOLUIDIN-5-SULFONOVA; PERMANENT RED 2B AMINE; RED 2B ACID; M-TOLUENESULFONIC ACID,6-AMINO-4-CHLORO-
Description: white to buff powder
Use: intermediate for azo pigments

Physical Properties

Water Solubility: Essentially insoluble as free acid
Other Solubilities: Soluble as ammonium salt.

RTECS Toxicity Data

Acute Oral: Rat LD_{50} Dose: 12300 mg/kg.
Acute Inhalation: Rat LC; Dose: >13 gm/m^3/4hr.
Irritation Eye: Rabbit Standard Draize Test Dose: 500 mg/24H; Reaction: mild.

Hazard Overviews

Health: Irritating to eyes/skin/respiratory tract. Other Acute Effects: may be harmful by inhalation, ingestion, or skin absorption.
Fire: Hazards: emits toxic fumes; in powder form capable of creating a dust explosion. Extinguishing agents: water spray; carbon dioxide, dry chemical powder or appropriate foam. Precautions: combustible liquid.
Reactivity: Stable. Hazardous polymerization will not occur. Incompatible with: strong oxidizing agents. Hazardous decomposition products: toxic fumes of: carbon monoxide, carbon dioxide, nitrogen oxides, sulfur oxides, hydrogen chloride gas.
Carcinogenicity: IARC - Not listed; NIOSH - Not listed; NTP - Not listed; ACGIH - Not listed; OSHA - Not listed; EPA - Not listed; MAK - Not listed
Primary Target Organs:

Eyes

Skin

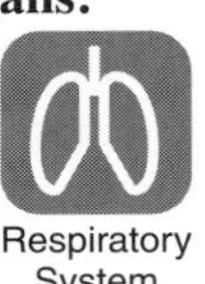
Respiratory System

Environmental

Regulations

RCRA 40CFR: Not listed
CERCLA: 40CFR 302.4: Not listed
SARA 40CFR 372.65: Not listed
SARA EHS 40CFR 355: Not listed
TSCA: Listed

CHL2590 CAS #: 2457-76-3

2-CHLORO-4-AMINOBENZOIC ACID

RTECS: DG1575000
EINECS Number: 219-540-5
Molecular Formula: $C_7H_6ClNO_2$
Formula Weight: 171.59

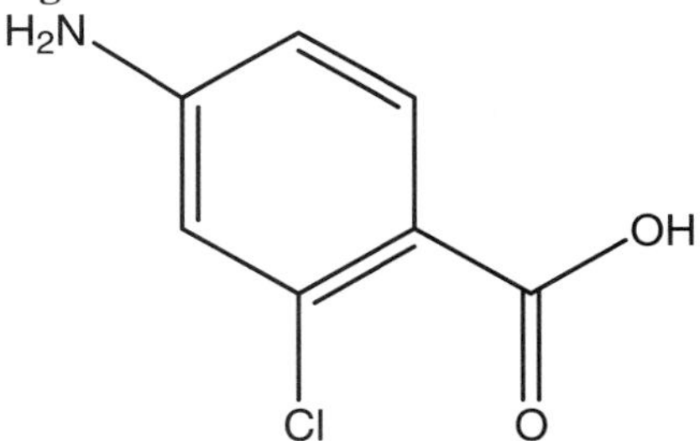

Chemical Structure

Physical Properties

Freezing Point: 211 °C (411.8 °F)
Other Solubilities: Ethanol: Soluble

RTECS Toxicity Data

Mutagenic: Bacteria - E Coli Mutations in Microorganisms; Dose: 500 mg/L (-S9).

Hazard Overviews

Health: Irritating to eyes/skin/respiratory tract. Other Acute Effects: may be harmful by inhalation, ingestion, or skin absorption.
Fire: Hazards: emits toxic fumes. Extinguishing agents: water spray; carbon dioxide, dry chemical powder or appropriate foam. Precautions: combustible liquid.
Reactivity: Incompatible with: strong oxidizing agents. Hazardous decomposition products: toxic fumes of: carbon monoxide, carbon dioxide, nitrogen oxides, hydrogen chloride gas.
Carcinogenicity: IARC - Not listed; NIOSH - Not listed; NTP - Not listed; ACGIH - Not listed; OSHA - Not listed; EPA - Not listed; MAK - Not listed
Primary Target Organs:

Eyes

Skin

Respiratory System

Environmental

Regulations

RCRA 40CFR: Not listed
CERCLA: 40CFR 302.4: Not listed
SARA 40CFR 372.65: Not listed
SARA EHS 40CFR 355: Not listed
TSCA: Listed

CHL2620 CAS #: 108-42-9

M-CHLOROANILINE

RTECS: BX0350000
DOT: UN2018; UN2019; IMO6.1
EINECS Number: 203-581-0
Molecular Formula: C_6H_6ClN
Formula Weight: 127.57

NH_2
Cl

Chemical Structure

Synonyms: 1-AMINO-3-CHLOROBENZENE; M-AMINOCHLOROBENZENE; META-AMINOCHLOROBENZENE; ANILINE,M-CHLORO-; BENZENAMINE,3-CHLORO-; BENZENAMINE,3-CHLORO-(9CI); 3-CHLOORANILINEN; M-CHLORANILINE; 3-CHLOROANILINE; META-CHLOROANILINE; 3-CHLOROBENZENAMINE; 3-CHLOROPHENYLAMINE; M-CHLOROPHENYLAMINE; META-CHLOROPHENYLAMINE; 3-CLOROANILINE; FAST ORANGE GC BASE; ORANGE GC BASE

Description: colorless to light amber liquid; characteristic sweet odor
Use: intermediate for azo dyes and pigments; pharmaceuticals; insecticides; agricultural chemicals

Physical Properties

Boiling Point: 230.5 °C (447 °F)
Freezing Point: -10.4 °C (13.28 °F)
Specific Gravity: 1.215 at 22 °C/4 °C
Vapor Density: 4.41 Air=1
Density: 1.21 g/cu cm at 19.4 °C
Vapor Pressure: < 0.1 mm Hg at 30 °C
Water Solubility: < 1 mg/mL at 18 C
Other Solubilities: 95% Ethanol: >=100 mg/ml at 18 °C; Acetone: >=100 mg/ml at 18 °C; Benzene: Soluble; Carbon Tetrachloride: miscible; DMSO: >=100 mg/ml at 18 °C; Ether: Soluble; Most organic solvents: Soluble.
Refraction Index: 1.5931 at 20 °C/D
Ionization Potential (eV): 8.09 +/-0.1
Flash Point: 123 °C

RTECS Toxicity Data

Acute Oral: Rat LD_{50} Dose: 256 mg/kg. Mouse LD_{50} Dose: 334 mg/kg.
Acute Inhalation: Mouse LC_{50} Dose: 550 mg/m^3/4hr. Rat LD_{50} Dose: 150 ppm/4hr; Toxic Effects: Sense organs and special senses - Corneal damage; Behavioral - Convulsions or effect on seizure threshold; Lungs, Thorax, or Respiration - Cyanosis.
Acute Dermal: Rat LD_{50} Route: Skin; Dose: 250 mg/kg. Guinea Pig LD_{50} Route: Skin; Dose: 100 mg/kg.
Chronic (Multiple Dose) Oral: Rat Dose: 10 gm/kg/13W-I; Toxic Effects: Blood - Methemoglobinemia-Carboxyhemoglobin; Blood - Changes in erythrocite (RBC) cell count; Kidney, Ureter, and Bladder - Other changes in urine composition. Rat Dose: 6125 mg/kg/35W-I; Toxic Effects: Brain and coverings - Recordings from specific areas of CNS.
Mutagenic: Hamster Mutations in Mammalian Somatic Cells; Cell Type: lung; Dose: 300 ug/L. Mold - A Nidulans Mutations in Microorganisms; Dose: 200 mg/L (-S9).

Hazard Overviews

Poison

Health: Irritating to eyes; may be irritating to skin/respiratory tract. Poison. Other Acute Effects: may be fatal if inhaled or absorbed through skin; harmful if swallowed; may cause allergic skin reaction; absorption into the body leads to the formation of methemoglobin which in sufficient concentration causes cyanosis; onset may be delayed 2 to 4 hours or longer; exposure can cause: nausea, headache, vomiting, confusion, weakness, drowsiness, unconsciousness, ataxia conjunctivitis, blurred vision, tearing; target organs: liver, kidneys, blood.

Fire: Will burn. Hazards: emits toxic fumes. Extinguishing agents: water spray; carbon dioxide, dry chemical powder or appropriate foam. Precautions: combustible liquid.
Reactivity: Stable. Hazardous polymerization will not occur. Incompatible with: acids, acid chlorides, acid anhydrides, chloroformates, strong oxidizing agents. Hazardous decomposition products: toxic fumes of: carbon monoxide, carbon dioxide, nitrogen oxides, hydrogen chloride gas.
Carcinogenicity: IARC - Not listed; NIOSH - Not listed; NTP - Not listed; ACGIH - Not listed; OSHA - Not listed; EPA - Not listed; MAK - Not listed
Primary Target Organs:

Eyes

Liver

Kidneys

Blood

Environmental

Ecotoxicity: Algae: Scenedesmus subspicatus inhib. of fluorescence IC_{10} 1.8 mg/l Eisenia andrei: 14d LC_{50}: 568-725 mg/l (estd. conc. in the soil pore water phase); Fishes: Poecilia reticulata 14d LC_{50} 13 mg/l
Environmental Fate: If released to soil, it will undergo covalent chemical bonding with humic materials which can result in its chemical alteration and prevent leaching. Photodegradation will likely occur on soil surfaces exposed to sunlight. It degrades in soil by both chemical and biological processes. In one study, 38 and 18% remained in soil after 2 and 8 weeks, respectively. If released to water, it will bind with humic materials in the water column and sediments. Volatilization is not rapid (half-life of 283 days from a stagnant pond), but may have some importance from shallow, fast-flowing rivers. Photolysis would be expected to occur in clear surface waters. Most biodegradation screening studies indicate that it will readily degrade. It is metabolized in fish and therefore bioconcentration would not be expected. If released to the atmosphere, it will react rapidly with sunlight-produced hydroxyl radicals (estimated half-life of 5.1 hr).
Cleanup/Disposal: Guide No. 152: Do not touch damaged containers or spilled material unless wearing appropriate protective clothing. Stop leak if you can do it without risk. Prevent entry into waterways, sewers, basements or confined areas. Cover with plastic sheet to prevent spreading. Absorb or cover with dry earth, sand or other non-combustible material and transfer to containers. Do not get water inside containers.

Environmental Physical Data

Henry's Law Constant: estimated at 1.68 $x10^{-6}$
Octanol/Water Partition Coefficient: log K_{ow} = 1.88
BCF: carp 0.8 to 2.2

Regulations

RCRA 40CFR: Not listed
CERCLA: 40CFR 302.4: Not listed
SARA 40CFR 372.65: Not listed
SARA EHS 40CFR 355: Not listed
TSCA: Listed

Analytical Methods

Soil: SW846 8131

CHL2650 CAS #: 95-51-2

O-CHLOROANILINE

RTECS: BX0525000
DOT: UN2018; UN2019; IMO6.1
EINECS Number: 202-426-4
Molecular Formula: C_6H_6ClN
Formula Weight: 127.57

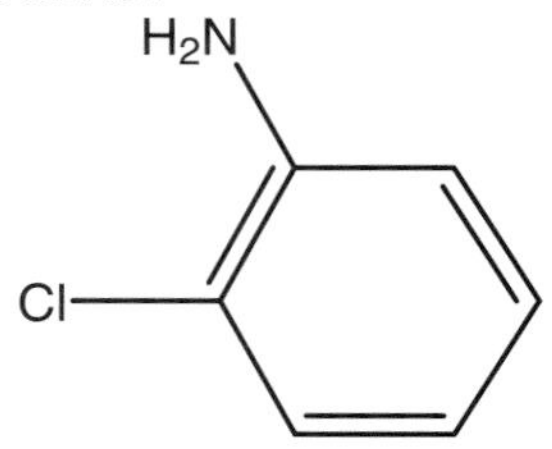

Chemical Structure

Synonyms: 1-AMINO-2-CHLOROBENZENE; ANILINE,O-CHLORO-; BENZENAMINE,2-CHLORO-; BENZENAMINE,2-CHLORO-(9CI); O-CHLORANILINE; 2-CHLOROANILINE; 2-CHLOROBENZENAMINE; 2-CHLOROBENZENEAMINE; 2-CHLOROPHENYLAMINE; FAST YELLOW GC BASE
Description: amber liquid; amine odor
Use: dye intermediate; standards for colorimetric apparatus; manufacture of petroleum solvents and fungicides

Physical Properties

Boiling Point: 208.84 °C (408 °F) at 99.61 mole
Freezing Point: -14 °C (6.8 °F)
Specific Gravity: 1.2114 at 22 °C/4 °C
Vapor Density: 4.4 Air=1
Saturated Vapor Density: 1.200912354 kg/m^3
Density: 1.207 g/cu cm at 25.0 °C
Vapor Pressure: 0.17 mm Hg at 20 °C
Water Solubility: < 1 mg/mL at 20 C
Other Solubilities: 95% Ethanol: >=100 mg/ml at 20 °C; Acetone: >=100 mg/ml at 20 °C; Acids: Soluble; DMSO: >=100 mg/ml at 20 °C; Ether: miscible; Most organic solvents: Soluble.
Surface Tension: 43.66 dyne/cm at 20 °C
Refraction Index: 1.5895
Ionization Potential (eV): 8.50
Flash Point: 97 °C

RTECS Toxicity Data

Acute Oral: Mouse LD_{50} Dose: 256 mg/kg.
Acute Dermal: Cat LD_{Lo} Route: Subcutaneous Dose: 310 mg/kg; Toxic Effects: Behavioral - Convulsions or effect on seizure threshold; Lungs, Thorax, or Respiration - Respiratory depression; Nutritional and gross metabolic - Body temperature decrease. Cat LD_{50} Route: Skin; Dose: 222 mg/kg.

Chronic (Multiple Dose) Oral: Rat Dose: 14560 mg/kg/13W-I; Toxic Effects: Endocrine - Changes in spleen weight; Blood - Pigmented or nucleated red Blood cells; Blood - Methemoglobinemia-Carboxyhemoglobin. Mouse Dose: 14560 mg/kg/13W-I; Toxic Effects: Endocrine - Changes in spleen weight; Blood - Pigmented or nucleated red Blood cells; Blood - Methemoglobinemia-Carboxyhemoglobin.
Mutagenic: Mouse Specific Locus Test; Cell Type: lymphocyte; Dose: 300 mg/L. Hamster Mutations in Mammalian Somatic Cells; Cell Type: lung; Dose: 600 ug/L.

Hazard Overviews

Fire Diamond

Health: Irritating to the eyes. Toxic. Potent central nervous system (CNS) toxin. Also Causes: heart, kidneys, and blood damage; liquid can be absorbed through the skin with systemic toxic effects.
Fire: Combustible. Use dry chemical, water spray, fog, or regular foam.
Reactivity: Stable. Hazardous polymerization cannot occur. Avoid: heat; ignition sources. Incompatible with: oxidizers; nitrous acid. Hazardous decomposition products: chlorine; hydrogen chloride; nitrogen oxide gases.
Carcinogenicity: IARC - Not listed; NIOSH - Not listed; NTP - Not listed; ACGIH - Not listed; OSHA - Not listed; EPA - Not listed; MAK - Not listed
Primary Target Organs:

Eyes

Nervous System

Kidneys

Cardio-vascular

Blood

Environmental

Ecotoxicity: EC_0 Daphnia magna (Daphnids) 0.3 mg/l/48 hr (ability to swim)
Environmental Fate: If released to soil, it will undergo chemical bonding with humic materials which can result in its chemical alteration and prevent leaching. The results of one field study indicate that it is extremely photosensitive on soil surfaces and degrades, in part, by biodegradation within soil. If released to water, it will bind with humic materials in the water column and sediments. Volatilization is not environmentally rapid (half-life of 64 days from a stagnant pond), but may have some importance from shallow rivers. Various screening tests suggest that it is generally resistant to biodegradation or biodegrades slowly. Significant acclimation of microbes may be required for biodegradation to become environmentally important. If released to the atmosphere, it will react rapidly (estimated half-life of 2 days) with sunlight-produced hydroxyl radicals.
Cleanup/Disposal: Guide No. 152: Do not touch damaged containers or spilled material unless wearing appropriate protective clothing. Stop leak if you can do it without risk. Prevent entry into waterways, sewers, basements or confined areas. Cover with plastic sheet to prevent spreading. Absorb or cover with dry earth, sand or other non-combustible material and transfer to containers. Do not get water inside containers.

Environmental Physical Data

Henry's Law Constant: estimated at 4.2×10^{-6}
Octanol/Water Partition Coefficient: log K_{ow} = 1.90
Sorption Partition Coefficient: K_{oc} = soils 96 to 5000
BCF: carp 2.0 to 3.7

Regulations

RCRA 40CFR: Not listed
CERCLA: 40CFR 302.4: Not listed
SARA 40CFR 372.65: Not listed
SARA EHS 40CFR 355: Not listed
TSCA: Listed

Analytical Methods

Soil: SW846 8131

CHL2680 CAS #: 106-47-8

P-CHLOROANILINE

RTECS: BX0700000
DOT: UN2018; UN2019; IMO6.1
EINECS Number: 203-401-0
Molecular Formula: C_6H_6ClN
Structured MF: $ClC_6H_4NH_2$
Formula Weight: 127.57

Cl—(benzene ring)—NH_2

Chemical Structure

Synonyms: 1-AMINO-4-CHLOROBENZENE; 4-AMINO-1-CHLOROBENZENE; 4-AMINOCHLOROBENZENE; P-AMINOCHLOROBENZENE; ANILINE,4-CHLORO-; ANILINE,P-CHLORO-; BENZENAMINE,4-CHLORO-; BENZENEAMINE,4-CHLORO; 4-CHLORANILIN; P-CHLORANILINE; 4-CHLORO-1-AMINOBENZENE; 4-CHLOROANILINE; 4-CHLOROBENZAMINE; 4-CHLOROBENZENAMINE; 4-CHLOROBENZENEAMINE; 4-CHLOROPHENYLAMINE; P-CHLOROPHENYLAMINE; P-CA
Description: colorless, white, pale-yellow crystals
Use: dye intermediate, pharmaceuticals and agricultural chemicals; and intermediate in chemical synthesis

Physical Properties

Boiling Point: 232 °C (450 °F)
Freezing Point: 72.5 °C (162.5 °F)
Specific Gravity: 1.169 at 77 °C/4 °C
Vapor Density: 4.41 Air=1
Saturated Vapor Density: 1.200080502 kg/m^3
Density: 1.17 g/mL
Vapor Pressure: 0.015 mm Hg at 20 °C
Water Solubility: < 1 mg/mL at 23.5 C

Other Solubilities: 95% Ethanol: >=100 mg/ml at 23.5 °C; Acetone: >=100 mg/ml at 23.5 °C; Alcohol: Freely Soluble; Carbon Disulfide: Freely Soluble; DMSO: >=100 mg/ml at 23.5 °C; Ether: Freely Soluble; Organic solvents: Soluble.
Odor Threshold: 1.5 mg/l
pH: Water extract 7
Refraction Index: 1.5546 at 87 °C
Ionization Potential (eV): =< 8.18
Flash Point: > 104.444 °C Open Cup
LEL: 2.2% v/v

RTECS Toxicity Data

Acute Oral: Rat LD_{50} Dose: 300 mg/kg; Toxic Effects: Behavioral - Convulsions or effect on seizure threshold; Behavioral - Muscle weakness; Lungs, Thorax, or Respiration - Respiratory depression. Mouse LD_{50} Dose: 100 mg/kg.
Acute Inhalation: Mouse LC_{12} Dose: 250 mg/m^3/6hr.
Acute Dermal: Rabbit LD_{50} Route: Skin; Dose: 360 mg/kg; Toxic Effects: Skin and appendages - Primary irritation. Rat LD_{50} Route: Skin; Dose: 3200 mg/kg.
Chronic (Multiple Dose) Oral: Rat Dose: 3367 mg/kg/13W-I; Toxic Effects: Blood - Methemoglobinemia-Carboxyhemoglobin; Blood - Changes in erythrocite (RBC) cell count; Kidney, Ureter, and Bladder - Other changes in urine composition.
Chronic (Multiple Dose) Inhalation: Rat Dose: 30 ug/m^3/24H/80D-C; Toxic Effects: Behavioral - Muscle contraction or spasticity.
Chronic (Multiple Dose) Dermal: Rat Route: Subcutaneous; Dose: 1 gm/kg/10D-I; Toxic Effects: Blood - Other changes.
Irritation Eye: Rabbit Standard Draize Test Dose: 250 ug/24H; Reaction: severe.
Irritation Skin: Rabbit Standard Draize Test Dose: 500 mg/24H; Reaction: mild.
Mutagenic: Rat Unscheduled DNA Synthesis; Cell Type: liver; Dose: 5 mg/L. Rat Morphological Transformation; Cell Type: embryo; Dose: 14500 ng/plate.
Tumorigenic: Rat Route: Oral; Dose: 9270 mg/kg/2Y-C; Toxic Effects: Tumorigenic - Carcinogenic by RTECS criteria; Endocrine - Tumors. Rat Route: Oral; Dose: 14 gm/kg/78W-C; Toxic Effects: Tumorigenic - Equivocal tumorigenic agent by RTECS criteria; Vascular - Tumors; Blood - Tumors. Rat Route: Oral; Dose: 18200 mg/kg/2Y-C; Toxic Effects: Tumorigenic - Neoplastic by RTECS criteria; Blood - Tumors.

Hazard Overviews

Fire Diamond

Health: Severely irritating to eyes. Toxic. Also Causes: headache, nausea, vomiting, confusion, incoordination, ringing in the ears, weakness, disorientation, lethargy, drowsiness, methemoglobinemia, heart block, arrhythmias, kidney insufficiency, blood in urine, painful urination, coma, death.
Fire: Will burn. Use dry chemical, carbon dioxde, water spray, or regular foam. Hazardous combustion products may include, carbon oxide(s) and chlorine gas.
Reactivity: Stable. Hazardous polymerization cannot occur. Avoid: heat; ignition sources. Incompatible with: nitrous acid. Hazardous decomposition products: carbon oxides; nitrogen oxides; hydrogen chloride; chlorine gases.
Carcinogenicity: IARC - Group 2B, Possibly carcinogenic to humans; NIOSH - Not listed; NTP - Not listed; ACGIH - Not listed; OSHA - Not listed; EPA - Not listed; MAK - Class A2, Unmistakably carcinogenic in animal experimentation only
Primary Target Organs:

Eyes

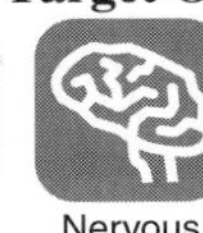
Nervous System

Kidneys

Cardio-vascular

Blood

Environmental

Ecotoxicity: Fishes: Frequency distribution of 24-96h LC_{50} values for fishes (n = 12) rainbow trout: 96h LC_{50},S 14 mg/l bluegill: 96h LC_{50},S 2 mg/l
Environmental Fate: If released on soil, it rapidly combines with soil components forming covalent bonds and polymers. As a consequence of these binding reactions, it is highly resistant to mineralization. It is partially mineralized by chemical and biological action. A few percent will volatilize from the soil. If released into water, it will be lost due to volatilization, photooxidation in surface layers (half-life 0.4 hr), and rapid chemical reactions with humic materials and clay in the water column and sediment. Degradation in air will primarily be due to reaction with hydroxyl radicals (half-life 4.6 hr), although direct photolysis is also possible.
Cleanup/Disposal: Guide No. 152: Do not touch damaged containers or spilled material unless wearing appropriate protective clothing. Stop leak if you can do it without risk. Prevent entry into waterways, sewers, basements or confined areas. Cover with plastic sheet to prevent spreading. Absorb or cover with dry earth, sand or other non-combustible material and transfer to containers. Do not get water inside containers.

Environmental Physical Data

Henry's Law Constant: calculated at 1.16 x10^{-5}
Octanol/Water Partition Coefficient: log K_{ow} = 1.83
Sorption Partition Coefficient: K_{oc} = 5 Belgium soils 230 to 469
BCF: carp 0.18

Regulations

RCRA 40CFR: Listed Hazardous Waste No. P024 Toxic Waste
CERCLA: 40CFR 302.4: Listed per RCRA Section 3001 RQ: 1000 lb (453.5 kg)
SARA 40CFR 372.65: Listed
SARA EHS 40CFR 355: Not listed
TSCA: Listed

Analytical Methods

Air: EPA 0020
Soil: CLP MC_SVOA, OHC; EPA 1625; SW846 3640A, 8131, 8250A, 8270B, 8270C, 8410
Water / Groundwater: EPA 1625; ASTM D4763
Plasma: EPA 29

CHL2800 CAS #: 108-90-7

CHLOROBENZENE

RTECS: CZ0175000
DOT: UN1134; IMO3.3
EINECS Number: 203-628-5
Molecular Formula: C_6H_5Cl
Structured MF: C_6H_5Cl
Formula Weight: 112.56

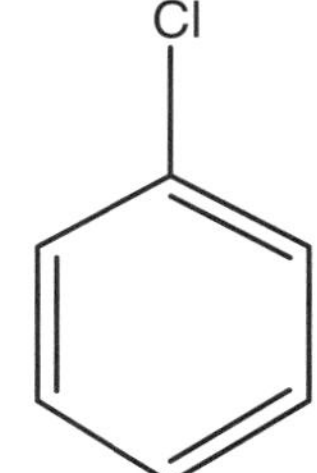

Chemical Structure

Synonyms: BENZENE CHLORIDE; BENZENE,CHLORO-; CHLOORBENZEEN; CHLORBENZENE; CHLORBENZOL; CHLOROBENZEN; CHLOROBENZENU; CHLOROBENZOL; CLOROBENZENE; CP 27; EPA PESTICIDE CHEMICAL CODE 056504; I P CARRIER T 40; MCB; MONOCHLOORBENZEEN; MONOCHLORBENZENE; MONOCHLORBENZOL; MONOCHLOROBENZENE; MONOCLOROBENZENE; PHENYL CHLORIDE; TETROSIN SP
Description: colorless liquid; almond-like odor
Use: manufacture of phenol, chloronitrobenzene, aniline and DDT; a solvent for paints; as a heat transfer medium, pesticide intermediate and a solvent carrier for methylene diisocyanate

Physical Properties

Boiling Point: 132 °C (270 °F)
Freezing Point: -45.6 °C (-50.08 °F)
Specific Gravity: 1.1058 at 20 °C/4 °C
Vapor Density: 3.88 Air=1
Saturated Vapor Density: 1.240036007 kg/m^3
Vapor Pressure: 8.8 mm Hg at 20 °C
Water Solubility: 0.05% by weight
Other Solubilities: 95% Ethanol: >=100 mg/ml at 20 °C; Acetone: >=100 mg/ml at 20 °C; Alcohol: Freely Soluble; Benzene: Freely Soluble; Chloroform: Freely Soluble; DMSO: Reacts; Ether: Freely Soluble.
Surface Tension: 33 dynes/cm at 25 °C
Odor Threshold: 0.98 to 280 mg/m^3
Refraction Index: 1.5241 at 20 °C/D
Evaporation Rate: 1 Butyl Acetate=1
Critical Temperature: 359 °C
Critical Pressure: 44.6 atm
Ionization Potential (eV): 9.07
Flash Point: 29.2 °C Closed Cup
Autoignition Temperature: 638 °C
LEL: 1.8% v/v
UEL: 9.6% v/v

RTECS Toxicity Data

Acute Oral: Rat LD_{50} Dose: 1110 mg/kg; Toxic Effects: Behavioral - Somnolence (general depressed activity); Behavioral - Tremor; Behavioral - Ataxia. Mouse LD_{50} Dose: 2300 mg/kg.
Acute Inhalation: Rat LC_{Lo} Dose: 9000 ppm. Mouse LC_{Lo} Dose: 15 gm/m^3.
Acute Dermal: Rabbit LD; Route: Skin; Dose: >2200 mg/kg. Guinea Pig LD; Route: Skin; Dose: >11 gm/kg.
Chronic (Multiple Dose) Oral: Rat Dose: 14 gm/kg/14D-I; Toxic Effects: Behavioral - Somnolence (general depressed activity); DEATH. Rat Dose: 32500 mg/kg/13W-I; Toxic Effects: Liver - Changes in liver weight; Biochemical - Peptidases; DEATH. Rat Dose: 27300 ug/kg/39W-I; Toxic Effects: Blood - Pigmented or nucleated red Blood cells; Blood - Eosinophilia; Blood - Changes in erythrocite (RBC) cell count.
Chronic (Multiple Dose) Inhalation: Rat Dose: 1 mg/m^3/60D-C; Toxic Effects: Brain and coverings - Recordings from specific areas of CNS; Biochemical - True cholinesterase; Biochemical - Plasma proteins not involving coagulation.
Reproductive/Teratogenic: Rat Route: Inhalation; Dose: 75 ppm/6H; Duration: female 6-15D of pregnancy; Specific Developmental Abnormalities - Musculoskeletal system. Rat Route: Inhalation; Dose: 210 ppm/6H; Duration: female 6-15D of pregnancy; Specific Developmental Abnormalities - Hepatobiliary system.
Mutagenic: Mouse Mutations in Microorganisms; Cell Type: lymphocyte; Dose: 70 mg/L (+S9). Mouse Mutations in Mammalian Somatic Cells; Cell Type: lymphocyte; Dose: 100 mg/L.
Tumorigenic: Rat Route: Oral; Dose: 61800 mg/kg/2Y-I; Toxic Effects: Tumorigenic - Neoplastic by RTECS criteria; Liver - Tumors; Blood - Tumors.

Hazard Overviews

Flammable

Fire Diamond

Health: Irritating to eyes/skin/respiratory tract. Also causes: headache, dizziness, drowsiness, cyanosis, spastic contractions of extremities, and loss of consciousness, depending on the exposure's concentration and duration; pallor, cyanosis, and coma. Chronic Effects: skin burns, headaches, dizziness, somnolence, and dyspeptic disorders, lung, liver, and kidney damage.

Fire: Flammable. Use carbon dioxide, dry chemical, halon, water spray, or standard foam to extinguish fires involving chlorobenzene. Use water in flooding quantities as fog since solid streams of water may spread fire. Fight fire from maximum distance.

Reactivity: Stable. Hazardous polymerization cannot occur. Avoid: heat; ignition sources. Incompatible with: strong oxidizers; dimethyl sulfoxide; silver perchlorate; powdered sodium; phosphorus trichloride and sodium. Hazardous decomposition products: soot; hydrogen chloride; phosgene; carbon monoxide.

Carcinogenicity: IARC - Not listed; NIOSH - Not listed; NTP - Not listed; ACGIH - Class A3, Animal carcinogen; OSHA - Not listed; EPA - Class D, Not classifiable as to human carcinogenicity; MAK - Not listed

Primary Target Organs:

Eyes

Skin

Respiratory System

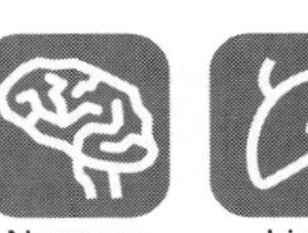
Nervous System

Liver

Kidneys

Exposure Limits

OSHA PEL: TWA: 75 ppm; 350 mg/m^3.

ACGIH TLV: TWA: 10 ppm; 46 mg/m^3.

NIOSH IDLH: 1000 ppm.

DFG MAK: TWA: 10 ppm; 46 mg/m^3.

Respirator Recommendation

Exposure Range: >75 to 750 mg/m^3 Air Purifying, Negative Pressure, Half Mask

Exposure Range: >750 to <1000 mg/m^3 Self-contained Breathing Apparatus, Pressure Demand, Full Face

Exposure Range: 1000 to unlimited mg/m^3 Self-contained Breathing Apparatus, Pressure Demand, Full Face

Cartridge Color: black

Environmental

Ecotoxicity: LC_{50} Poecilia reticulata (guppy) 19 ppm/14 days /Conditions of bioassay not specified LC_{50} Pimephales promelas (fathead minnow) 16.9 mg/l/96 hr (confidence limit 13.8 - 20.6 mg/l), flow-through bioassay with measured concentrations, 25.7 °C, dissolved oxygen 6.2 mg/l, hardness 43.8 mg/l calcium carbonate, alkalinity 43.4 mg/l calcium carbonate LD_{50} Salmo gairdneri (rainbow trout) 1.8 mg/kg/24 hr /Conditions of bioassay not specified

Environmental Fate: Once released it will decrease in concentration due to dilution and photooxidation. Releases into water and onto land will decrease in concentration due to vaporization into the atmosphere and slow biodegradation in the soil or water. It would be expected to percolate into the ground water if soil is sandy and poor in organic matter. Little bioconcentration is expected into fish and food products.

Cleanup/Disposal: Guide No. 130: Eliminate all ignition sources (no smoking, flares, sparks or flames in immediate area). All equipment used when handling the product must be grounded. Do not touch or walk through spilled material. Stop leak if you can do it without risk. Prevent entry into waterways, sewers, basements or confined areas. A vapor suppressing foam may be used to reduce vapors. Absorb or cover with dry earth, sand or other non-combustible material and transfer to containers. Use clean non-sparking tools to collect absorbed material. Large Spills: Dike far ahead of liquid spill for later disposal. Water spray may reduce vapor; but may not prevent ignition in closed spaces.

Environmental Physical Data

Henry's Law Constant: calculated at 3.56×10^{-3}

Octanol/Water Partition Coefficient: log K_{ow} = 2.18 to 2.84

BCF: fish 1 to 2

BOD: 0.3 lb/lb, 5 days

Regulations

RCRA 40CFR: Listed Hazardous Waste No. U037 Toxic Waste

CERCLA: 40CFR 302.4: Listed per CWA Section 311(b)(4) per RCRA Section 3001 per CWA Section 307(a) RQ: 100 lb (45.35 kg)

SARA 40CFR 372.65: Listed

SARA EHS 40CFR 355: Not listed

TSCA: Listed

Analytical Methods

Air: EPA OA-002-1, VA-001-1, VA-003-1, VA-005-1, VG-006-1, VG-007-1, VG-011-1, TO-1, TO-14, TO-3; ASTM D3686, D3687, D4490

Soil: CLP LC_VOA, MC_VOA, OHC; EPA 7, 1624, VG-008-1, VS-001-1, VS-002-1, VW-010-1; SW846 1311, 5021, 5032, 5041, 5041A, 8010B, 8020A, 8021A, 8240B, 8260A, 8260B; DOE OP040R

Water / Groundwater: EPA 601, 602, 624, 624-S, VW-001-1, VW-002-1, VW-003-1, VW-004-1, VW-008-1, VW-014-1; APHA 6040-B, 6210-B, 6210-D, 6220-B, 6230-B, 6230-D; ASTM D3871, D5241; USGS O3115

Drinking Water: EPA 502.1, 502.2, 503.1, 524.1, 524.2; APHA 6210-C, 6220-C, 6230-C

Food: EPA 5

Indoor / Expired Air: NIOSH 1003; EPA IP-1A, IP-1B

Plasma: EPA 29

Other: EPA VS-006-1, VW-011-1

CHL2950 **CAS #: 510-15-6**

CHLOROBENZILATE

RTECS: DD2275000

DOT: UN2761; IMO6.1

EINECS Number: 208-110-2

Molecular Formula: $C_{16}H_{14}Cl_2O_3$

Structured MF: $(ClC_6H_4)_2C(OH)COOCH_2CH_3$

Formula Weight: 325.20

Synonyms: ACAR; ACARABEN; ACARABEN 4E; ACARBEN 4E; AKAR; AKAR 338; AKAR 50; BENZ-O-CHLOR; BENZENEACETIC ACID,4-CHLORO-ALPHA-(4-CHLOROPHENYL)-ALPHA-HYDROXY-,ETHYL ESTER; BENZENEACETIC ACID,4-CHLORO-ALPHA-(4-CHLOROPHENYL)-ALPHA-HYDROXY-,ETHYLESTER; BENZILAN; BENZILIC ACID,4,4'-DICHLORO-,ETHYL ESTER; CHLORBENZILAT; CHLORBENZILATE; CHLORBENZYLATE; CHLOROBENZYLATE; 4-CHLORO-ALPHA-(4-CHLOROPHENYL)-ALPHA-HYDROXYBENZENEACETIC ACID ETHYL ESTER; 4-CHLORO-

ALPHA-(4-CHLOROPHENYL)-ALPHA-HYDROXYBENZENEACETICACID ETHYL ESTER; COMPOUND 338; 4,4'-DICHLORBENZILSAEUREAETHYLESTER; 4,4'-DICHLOROBENZILATE; 4,4'-DICHLOROBENZILIC ACID ETHYL ESTER; ENT 18,596; EPA PESTICIDE CODE 028801; ETHYL 4-CHLORO-ALPHA-(4-CHLOROPHENYL)-ALPHA-HYDROXYBENZENEACETATE; ETHYL 4,4'-DICHLOROBENZILATE; ETHYL P,P'-DICHLOROBENZILATE; ETHYL 4,4'-DICHLORODIPHENYL GLYCOLLATE; ETHYL 4,4'-DICHLOROPHENYL GLYCOLLATE; ETHYL DI(P-CHLOROPHENYL)GLYCOLLATE; ETHYL ESTER OF 4,4'-DICHLOROBENZILIC ACID; ETHYL 2-HYDROXY-2,2-BIS(4-CHLOROPHENYL)ACETATE; ETHYL 2-HYDROXY-2,2-DI(P-CHLOROPHENYL)ACETATE; ETHYL4-CHLORO-ALPHA-(4-CHLOROPHENYL)-ALPHA-HYDROXYBENZENEACETATE; ETHYLDICHLOROBENZILATE; ETHYL-4,4'-DIPHENYLGLYCOLLATE; ETHYLESTER KYSELINY 4,4-DICHLORBENZILOVE; ETHYL-2-HYDROXY-2,2-BIS(4-CHLOROPHENYL)ACETATE; FOLBEX; FOLBEX SMOKE-STRIPS; G 23992; G 338; G23992; GEIGY 338; KOP-MITE

Description: colorless solid (pure)

Use: insecticide, miticide, nonsystemic acaricide and synergist for DDT; to control many species of phytophagous mites on pome fruit, citrus fruit, vines, soybeans, cotton, teat and vegetables; to control bee mites in beehives

Physical Properties

Boiling Point: 146 °C (295 °F) to 148 °C (298 °F) at 0.04 mm Hg

Freezing Point: 36 °C (96.8 °F) to 37.3 °C (99.14 °F)

Saturated Vapor Density: 1.200000035 kg/m^3

Density: 1.2816 g/cu cm at 20 °C

Vapor Pressure: 0.0000022 mm Hg at 20 °C

Water Solubility: 10 mg/L at 20 °C

Other Solubilities: Slightly Soluble in Benzene.

Refraction Index: 1.5727 at 20 °C/D

Flash Point: Not available; probably combustible

RTECS Toxicity Data

Acute Oral: Rat LD_{50} Dose: 700 mg/kg. Mouse LD_{50} Dose: 729 mg/kg.

Acute Dermal: Rabbit LD_{50} Route: Skin; Dose: >1 gm/kg. Rat LD_{50} Route: Skin; Dose: >5 gm/kg.

Chronic (Multiple Dose) Oral: Rat Dose: 4200 mg/kg/17W-I.

Irritation Eye: Rabbit Standard Draize Test Dose: 25 mg; Reaction: moderate.

Irritation Skin: Rabbit Open Draize Test Dose: 125 mg open; Reaction: mild.

Mutagenic: Rat Morphological Transformation; Route: Oral; Dose: 4620 mg/kg/77D-C. Mouse Mutations in Microorganisms; Cell Type: lymphocyte; Dose: 70 mg/L (+S9).

Tumorigenic: Rat Route: Oral; Dose: 5475 mg/kg/2Y-C; Toxic Effects: Tumorigenic - Carcinogenic by RTECS criteria; Blood - Lymphomax including Hodgkin's disease; Skin and appendages - Tumors. Rat Route: Oral; Dose: 17520 mg/kg/2Y-C; Toxic Effects: Tumorigenic - Carcinogenic by RTECS criteria; Blood - Lymphomax including Hodgkin's disease; Skin and appendages - Tumors. Rat Route: Oral; Dose: 72 gm/kg/78W-C; Toxic Effects: Tumorigenic - Equivocal tumorigenic agent by RTECS criteria; Liver - Tumors. Rat Route: Oral; Dose: 1752 mg/kg/2Y-C; Toxic Effects: Tumorigenic - Neoplastic by RTECS criteria; Endocrine - Adrenal cortex tumors; Skin and appendages - Tumors.

Hazard Overviews

Fire: Will burn.

Carcinogenicity: IARC - Group 3, Not classifiable as to carcinogenicity to humans; NIOSH - Not listed; NTP - Not listed; ACGIH - Not listed; OSHA - Not listed; EPA - Not listed; MAK - Not listed

Environmental

Ecotoxicity: LC_{50} Cyprinodon variegatus (sheephead minnow) 1.0 mg/l/48 hr /Conditions of bioassay not specified LC_{50} Colinus virginianus (bobwhite quail) 3375 ppm/7 days LC_{50} Salmo gairdneri (rainbow trout) 0.7 mg/l/96 hr at 13 °C, wt 0.8 g, static bioassay

Environmental Fate: If released to soil, it will be expected to exhibit low mobility in soil and, therefore, will not be expected to leach to groundwater. It will not be expected to volatilize from near surface soil or surfaces. It will be subject to biodegradation. The half life in two fine sandy soils was estimated to be 1.5-5 weeks following application of 0.5-1.0 ppm probably due to biodegradation. If it is released to water, it will be adsorbed by sediment and suspended particulate material. It should not bioconcentrate in aquatic organisms or volatilize. It may be susceptible to biodegradation. In 22 days, 40, 29, and 39% added to sediment free water samples from 3 fresh water lakes was converted to organic products. Degradation in water from a fourth lake occurred only when glucose and inorganic nutrients were added suggesting that it may be metabolized in the lake waters. If released to the atmosphere, it will be susceptible to gas phase reaction with photochemically produced hydroxyl radicals with an estimated half life of about 3.3 hr.

Cleanup/Disposal: Guide No. 151: Do not touch damaged containers or spilled material unless wearing appropriate protective clothing. Stop leak if you can do it without risk. Prevent entry into waterways, sewers, basements or confined areas. Cover with plastic sheet to prevent spreading. Absorb or cover with dry earth, sand or other non-combustible material and transfer to containers. Do not get water inside containers.

Environmental Physical Data

Henry's Law Constant: calculated at 7.24×10^{-8}

Sorption Partition Coefficient: K_{OC} = estimated at 1065

BCF: estimated at 145

Regulations

RCRA 40CFR: Listed Hazardous Waste No. U038 Toxic Waste

CERCLA: 40CFR 302.4: Listed per RCRA Section 3001 RQ: 10 lb (4.535 kg)

SARA 40CFR 372.65: Listed

SARA EHS 40CFR 355: Not listed

TSCA: Listed

Analytical Methods

Soil: EPA 025; SW846 3640A, 8081, 8081A, 8270B, 8270C
Water / Groundwater: EPA 608.1
Drinking Water: EPA 508, 508.1, 525.2; AOAC 990.06
Food: FDA 212.1, 232.1, 232.4, 242.1; AOAC 971.08
Plasma: EPA 001, 027, 028, 29; FDA 211.1, 231.1, 252
Other: EPA 1656

CHL2980 CAS #: 74-11-3

4-CHLOROBENZOIC ACID

RTECS: DG4976010
EINECS Number: 200-805-9
Molecular Formula: $C_7H_5ClO_2$
Formula Weight: 156.57

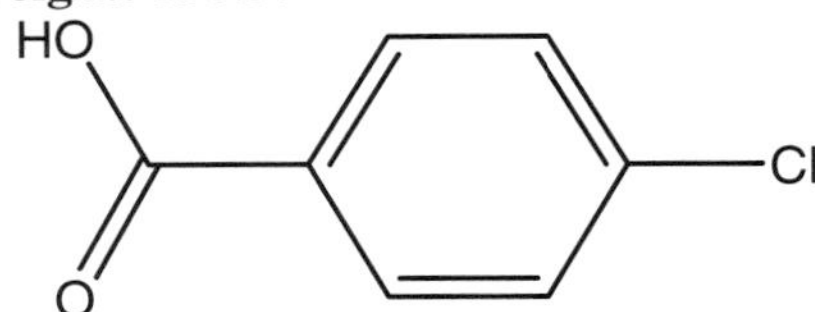

Chemical Structure

Synonyms: ACIDO P-CLOROBENZOICO; BENZOIC ACID,4-CHLORO-; BENZOIC ACID,P-CHLORO-; BENZOIC ACID,4-CHLORO-(9CI); P-CARBOXYCHLOROBENZENE; CHLORADRACYLIC; P-CHLORBENZOIC ACID; P-CHLOROBENZOIC ACID; PARA-CHLOROBENZOIC ACID; CHLORODRACYLIC ACID
Description: nearly white coarse powder; odorless
Use: intermediate for preparation of dyes, fungicides, pharmaceuticals and other organic chemicals; preservative for adhesives and paints

Physical Properties

Boiling Point: Sublimes
Freezing Point: 243 °C (469.4 °F)
Specific Gravity: 1.541 at 24/4 °C
Saturated Vapor Density: 1.20001285 kg/m^3
Vapor Pressure: 1.85 x10^{-3} mm Hg at 25 °C
Water Solubility: 1 parts in 5290 parts Water
Other Solubilities: 95% Ethanol: >=100 mg/ml at 15 °C; Acetone: >=100 mg/ml at 15 °C; Benzene: Insoluble; Carbon Disulfide: Insoluble; Carbon Tetrachloride: Insoluble; DMSO: >=100 mg/ml at 15 °C; Ether: Slightly Soluble; Ligroin: Insoluble; Methanol: Soluble.
pH: < 7
Evaporation Rate: < 1 Butyl Acetate=1
Flash Point: Not available; probably combustible

RTECS Toxicity Data

Acute Oral: Rat LD_{50} Dose: 1170 mg/kg. Mouse LD_{50} Dose: 1170 mg/kg.

Hazard Overviews

Health: Irritating to eyes/skin/respiratory tract. Harmful. Other Acute Effects: harmful if swallowed; may be harmful if inhaled; may be harmful if absorbed through the skin.
Fire: Will burn. Hazards: emits toxic fumes. Extinguishing agents: water spray; carbon dioxide, dry chemical powder or appropriate foam. Precautions: combustible liquid.
Reactivity: Incompatible with: strong oxidizing agents, strong bases. Hazardous decomposition products: toxic fumes of: carbon monoxide, carbon dioxide, hydrogen chloride gas.
Carcinogenicity: IARC - Not listed; NIOSH - Not listed; NTP - Not listed; ACGIH - Not listed; OSHA - Not listed; EPA - Not listed; MAK - Not listed
Primary Target Organs:

Eyes

Skin
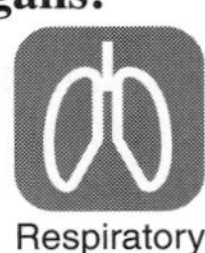
Respiratory System

Environmental

Environmental Fate: It will exist in water and moist soil mainly in the dissociated p-chlorobenzoic acid form under most environmental pH (pH 5-9) based upon a measured pKa of 3.98. If released to soil, it will not be expected to hydrolyze. It may slowly leach through soil. It may be subject to biodegradation in soil based upon data from experiments in natural water. Based upon an estimated Henry's Law constant of 8.03 x10^{-8} at cu m/mole, volatilization will not be an important removal process from surface water or moist soil. If released to water, it will not be expected to hydrolyze. Direct photolysis is probably not a major environmental pathway. Based upon a measured rate constant, reaction with photochemically produced hydroxyl radicals in water will not be a major environmental pathway. It may be subject to biodegradation in natural water with the rate likely to vary greatly depending upon the conditions. Based upon a measured fish BCF of <10, bioconcentration in aquatic organisms is not likely to occur. It should not adsorb to sediment or suspended particulate matter. If released to the atmosphere, it can be expected to exist mainly in the vapor-phase in the ambient atmosphere based upon an estimated vapor pressure of 1.85 x10^{-3} mm Hg at 25 °C. It may be subject to reaction with photochemcially produced hydroxyl radicals in the vapor phase with a rate constant of 0.5625 x10^{-12} cu cm/mole-sec at 25 °C estimated for this process which corresponds to an atmospheric half-life of 28.5 days. Direct photolysis is probably not a major environmental pathway in the atmosphere.

Environmental Physical Data

Henry's Law Constant: estimated at 8.03 x10^{-8}
Octanol/Water Partition Coefficient: log K_{OW} = 2.65
Sorption Partition Coefficient: K_{OC} = estimated at 400
BCF: fish < 10

Regulations

RCRA 40CFR: Not listed
CERCLA: 40CFR 302.4: Not listed

SARA 40CFR 372.65: Not listed
SARA EHS 40CFR 355: Not listed
TSCA: Listed

CHL3010 CAS #: 118-91-2

CHLOROBENZOIC ACID

RTECS: DG4976000
EINECS Number: 204-285-4
Molecular Formula: $C_7H_5ClO_2$
Formula Weight: 156.57

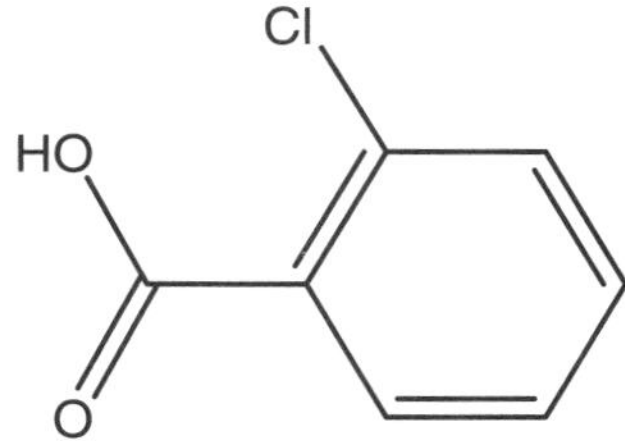

Chemical Structure

Synonyms: BENZOIC ACID,2-CHLORO-; BENZOIC ACID,O-CHLORO-; 2-CBA; 2-CHLOROBENZOIC ACID; O-CHLOROBENZOIC ACID; KYSELINA O-CHLORBENZOOVA
Description: monoclinic prisms
Use: preservative for glues, paints; intermediate in the manufacture of fungicides, dyes, pharmaceuticals and other organic chemicals

Physical Properties

Boiling Point: Sublimes
Freezing Point: 142 °C (287.6 °F)
Specific Gravity: 1.544 at 20 °C/4 °C
Density: 1.544 g/mL at 20 °C
Water Solubility: 1 parts in 900 parts Cold Water
Other Solubilities: 95% Ethanol: >=100 mg/ml at 20 °C; Acetone: 50-100 mg/ml at 20 °C; Benzene: Soluble; DMSO: >=100 mg/ml at 20 °C.
Flash Point: Not available; probably combustible

RTECS Toxicity Data

Acute Oral: Rat LD; Dose: >500 mg/kg.
Irritation Eye: Rabbit Standard Draize Test Dose: 20 mg/24H; Reaction: moderate.
Irritation Skin: Rabbit Standard Draize Test Dose: 500 mg/24H; Reaction: mild.

Hazard Overviews

Health: Irritating to eyes/skin/respiratory tract. Other Acute Effects: may be harmful by inhalation, ingestion, or skin absorption.
Fire: Will burn. Extinguishing agents: water spray; carbon dioxide, dry chemical powder or appropriate foam. Precautions: combustible liquid.
Reactivity: Incompatible with: strong oxidizing agents, strong bases. Hazardous decomposition products: toxic fumes of: carbon monoxide, carbon dioxide, hydrogen chloride gas.
Carcinogenicity: IARC - Not listed; NIOSH - Not listed; NTP - Not listed; ACGIH - Not listed; OSHA - Not listed; EPA - Not listed; MAK - Not listed
Primary Target Organs:

Eyes

Skin

Respiratory System

Environmental

Environmental Fate: If released to soil, it is expected to biodegrade under both aerobic and anaerobic conditions, although there may be a lengthy lag periods. It is not expected to significantly volatilize from either moist or dry soil to the atmosphere. It is expected to display moderate to high mobility in soil, and it will be found predominately in the dissociated form in moist soil under environmental conditions. If released to water, it is expected to biodegrade under both aerobic and anaerobic conditions, although there may be a lengthy lag period. It may undergo direct photolytic degradation in water. It is not expected to significantly volatilize from water to the atmosphere, nor is it expected to significantly adsorb to sediment and suspended organic matter. It will be found predominately in the dissociated form in water under the typical pH's found in the environment. If released to the atmosphere, it may undergo a slow gas-phase reaction with photochemically produced hydroxyl radicals with an estimated half-life of 28.5 days. It may also undergo atmospheric removal by wet deposition processes.

Environmental Physical Data

Henry's Law Constant: estimated at 3.88×10^{-8}
Octanol/Water Partition Coefficient: log K_{ow} = 2.05
Sorption Partition Coefficient: K_{oc} = 65 to 310
BCF: calculated at 8

Regulations

RCRA 40CFR: Not listed
CERCLA: 40CFR 302.4: Not listed
SARA 40CFR 372.65: Not listed
SARA EHS 40CFR 355: Not listed
TSCA: Listed

CHL3220 CAS #: 2698-41-1

O-CHLOROBENZYLIDENE MALONONITRILE

RTECS: OO3675000
EINECS Number: 220-278-9
Molecular Formula: $C_{10}H_5ClN_2$
Structured MF: $ClC_6H_4CH{=}C(CN)_2$
Formula Weight: 188.62

Synonyms: 2-CHLORO BMN; 2-CHLOROBENZALMALONITRILE; (O-CHLOROBENZAL)MALONONITRILE; 2-CHLOROBENZALMALONONITRILE; O-CHLOROBENZYLIDENE MALONITRILE; 2-CHLOROBENZYLIDENE MALONONITRILE; ORTHO-CHLOROBENZYLIDENE MALONONITRILE; 2-CHLOROBMN; ((2-CHLOROPHENYL)METHYLENE)PROPANEDINITRILE; ((2-CHLOROPHENYL)METHYLENE)PROPANENITRILE; CS; CS GAS; CS (LACRIMATOR); BETA,BETA-DICYANO-O-CHLOROSTYRENE; MALONONITRILE,(O-CHLOROBENZYLIDENE)-; OCBM; PROPANEDINITRILE,((2-CHLOROPHENYL)METHYLENE); PROPANEDINITRILE,((2-CHLOROPHENYL)METHYLENE)-; TL 238

Description: white crystalline solid; odor of pepper

Use: riot control agent

Physical Properties

Boiling Point: 310 °C (590 °F) to 315 °C (599 °F)

Freezing Point: 93 °C (199.4 °F) to 95 °C (203 °F)

Vapor Density: 6.52 Air=1

Saturated Vapor Density: 1.200000295 kg/m^3

Vapor Pressure: 3.4 x10^{-5} mm Hg at 20 °C

Water Solubility: Insoluble in Water

Other Solubilities: 95% Ethanol: 1-10 mg/ml at 16 °C; Acetone: 50-100 mg/ml at 16 °C; Benzene: Soluble; DMSO: >=100 mg/ml at 16 °C; Dioxane: Soluble; Ethyl Acetate: Soluble; Methylene chloride: Soluble.

Flash Point: Not available; probably combustible

RTECS Toxicity Data

Acute Oral: Rat LD_{50} Dose: 178 mg/kg; Toxic Effects: Gastrointestinal - Gastritis. Mouse LD_{50} Dose: 282 mg/kg.

Acute Inhalation: Rat LC_{Lo} Dose: 1806 mg/m^3/45M; Toxic Effects: Sense organs and special senses - Other; Lungs, Thorax, or Respiration - Acute pulmonary edema; Blood - Hemorrhage. Human TC_{Lo} Dose: 1500 ug/m^3/90M; Toxic Effects: Sense organs and special senses - Conjunctive irritation; Lungs, Thorax, or Respiration - Cough; Lungs, Thorax, or Respiration - Other changes.

Acute Dermal: Mouse LD; Route: Subcutaneous Dose: >800 mg/kg.

Irritation Eye: Man Standard Draize Test Dose: 5 mg/m3/20S; Reaction: severe. Rabbit Standard Draize Test Dose: 1 mg; Reaction: mild.

Irritation Skin: Human Standard Draize Test Dose: 10 mg/1H; Reaction: mild. Rabbit Open Draize Test Dose: 12%/6H open; Reaction: mild.

Reproductive/Teratogenic: Rat Route: Inhalation; Dose: 6 mg/m^3/5M; Duration: female 6-15D of pregnancy; Effects on Embryo or Fetus - Fetotoxicity.

Mutagenic: Mouse Mutations in Mammalian Somatic Cells; Cell Type: lymphocyte; Dose: 2500 ug/L. Hamster Cytogenetic Analysis; Cell Type: ovary; Dose: 6 mg/L. Hamster Cytogenetic Analysis; Cell Type: lung; Dose: 9400 nmol/L.

Hazard Overviews

Fire: Will burn.

Carcinogenicity: IARC - Not listed; NIOSH - Not listed; NTP - Not listed; ACGIH - Class A4, Not classifiable as a human carcinogen; OSHA - Not listed; EPA - Not listed; MAK - Not listed

Exposure Limits

OSHA PEL: TWA: 0.05 ppm; 0.4 mg/m^3.

OSHA PEL Vacated 1989 Limits: STEL: 0.05 ppm; 0.4 mg/m^3; Ceiling.

ACGIH TLV: STEL: 0.05 ppm; 0.39 mg/m^3; Ceiling.

NIOSH REL: STEL: 0.05 ppm; 0.4 mg/m^3; skin.

NIOSH IDLH: 2 mg/m^3.

Respirator Recommendation

Exposure Range: >0.4 to <2 ppm Supplied Air, Constant Flow/Pressure Demand, Full Face

Exposure Range: 2 to unlimited ppm Self-contained Breathing Apparatus, Pressure Demand, Full Face

Cartridge Color: black with magenta (P100)

Environmental

Ecotoxicity: LC_{50} Rainbow trout 1.28 mg/l/12 hr. /Conditions of bioassay not specified LC_{50} Rainbow trout > 0.1 mg/l < 1 wk

Environmental Fate: If released to the atmosphere as a dust or powder from its use as a riot control agent, it will settle to the ground via dry deposition. If released to water or soil, the major degradation process is expected to be hydrolysis. Aqueous hydrolysis experiments in seawater have determined hydrolysis half-lives of 281.7 min at 0 °C and 14.5 min at 25 °C. However, actual environmental degradation rates may be much slower because the rate at which it dissolves in water can be very slow. Released to water it could float and travel for considerable distances before it dissolves. Insufficient data are available to predict the importance of biodegradation.

Cleanup/Disposal: Guide No. 154: Eliminate all ignition sources (no smoking, flares, sparks or flames in immediate area). Do not touch damaged containers or spilled material unless wearing appropriate protective clothing. Stop leak if you can do it without risk. Prevent entry into waterways, sewers, basements or confined areas. Absorb or cover with dry earth, sand or other non-combustible material and transfer to containers. Do not get water inside containers.

Environmental Physical Data

Henry's Law Constant: estimated at 1.02 x10^{-8}

Octanol/Water Partition Coefficient: log K_{ow} = 1.849

Sorption Partition Coefficient: K_{oc} = 44

BCF: estimated at 41.6

Regulations

RCRA 40CFR: Not listed

CERCLA: 40CFR 302.4: Not listed

SARA 40CFR 372.65: Not listed

SARA EHS 40CFR 355: Not listed

TSCA: Listed

CHL3460 CAS #: 75-68-3

1-CHLORO-1,1-DIFLUOROETHANE

RTECS: KH7650000

DOT: UN2517; IMO2.1
EINECS Number: 200-891-8
Molecular Formula: $C_2H_3ClF_2$
Formula Weight: 100.50

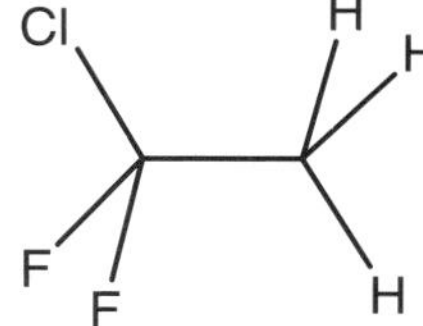

Chemical Structure

Synonyms: CFC 142B; 1,1,1-CHLORODIFLUOROETHANE; CHLORODIFLUOROETHANE; CHLORODIFLUOROETHANES; ALPHA-CHLOROETHYLIDENE FLUORIDE; CHLOROETHYLIDENE FLUORIDE; 1,1-DIFLUORO-1-CHLOROETHANE; DIFLUORO-1-CHLOROETHANE; DIFLUOROCHLOROETHANES; DIFLUOROMONOCHLOROETHANE; FC 142B; FC142B; FLUOROCARBON 142B; FLUOROCARBON FC142B; FREON 142; FREON 142B; GENETRON 142B; GENETRON 101; GENTRON 142B; HYDROCHLOROFLUOROCARBON 142B; R-142B; R142B
Description: colorless gas; nearly odorless
Use: refrigerant; solvent; aerosol propellant for non-food use; chem int for vinylidene fluoride

Physical Properties

Boiling Point: -9.2 °C (15 °F)
Freezing Point: -130.8 °C (-203.44 °F)
Specific Gravity: 1.194 at -9 °C
Vapor Density: 3.7 Air=1
Vapor Pressure: 2528 mm Hg at 25 °C
Water Solubility: 9180 $x10^3$ mg/L at 25 °C
Other Solubilities: Benzene: Soluble
Critical Temperature: 137.1 °C
Critical Pressure: 4.12 mPa
Ionization Potential (eV): 11.98 +/-1.2
Flash Point: Flammable gas
LEL: 6.2% v/v
UEL: 17.9% v/v

RTECS Toxicity Data

Acute Inhalation: Rat LC_{50} Dose: 2050 gm/m^3/4hr. Mouse LC_{50} Dose: 1758 gm/m^3/2hr.
Chronic (Multiple Dose) Inhalation: Rat Dose: 500 gm/m^3/4H/4W-I; Toxic Effects: Endocrine - Evidence of thyroid hypofunction.
Mutagenic: Bacteria - S Typhimurium Mutations in Microorganisms; Dose: 50 pph/24H (-S9).

Hazard Overviews

Flammable

Compressed Gas

Health: Acute Effects: may be harmful.
Fire: Flammable. Hazards: may form explosive mixtures with air; vapor may travel considerable distance to source of ignition and flash back; emits toxic fumes. Extinguishing agents: do not extinguish burning gas if flow cannot be shut off immediately; use water spray or fog nozzle to keep cylinder cool; move cylinder away from fire if there is no risk. Precautions: combustible liquid.
Reactivity: Incompatible with: strong oxidizing agents, magnesium, aluminum and their alloys, brass, steel. Hazardous decomposition products: carbon monoxide, carbon dioxide, hydrogen chloride gas, hydrogen fluoride.
Carcinogenicity: IARC - Not listed; NIOSH - Not listed; NTP - Not listed; ACGIH - Not listed; OSHA - Not listed; EPA - Not listed; MAK - Not listed
Exposure Limits
DFG MAK: TWA: 100 ppm; 4170 mg/m^3.
AIHA WEEL: TWA: 1000 ppm.

Environmental

Environmental Fate: If released to soil, it may rapidly volatilize from soil surfaces or leach through soil possibly into groundwater. If released to water, volatilization (half life of 3 hours from a model river) would be the dominant fate process. If released to the atmosphere, essentially all is expected to exist in the vapor phase. In the troposphere, it would react with photochemically generated hydroxyl radicals (half life of 5-12 years) or diffuse into the stratosphere (half life of 20 years). The overall tropospheric half-life has been estimated to range between 4-7.5 years. In the stratosphere, it would undergo direct photolysis or react with singlet oxygen. Due to its stability, transport long distances from its sources of emissions will take place.
Cleanup/Disposal: Guide No. 115: Eliminate all ignition sources (no smoking, flares, sparks or flames in immediate area). All equipment used when handling the product must be grounded. Do not touch or walk through spilled material. Stop leak if you can do it without risk. If possible, turn leaking containers so that gas escapes rather than liquid. Use water spray to reduce vapors or divert vapor cloud drift. Do not direct water at spill or source of leak. Prevent spreading of vapors through sewers, ventilation systems and confined areas. Isolate area until gas has dispersed.

Environmental Physical Data

Henry's Law Constant: estimated at 0.04
Octanol/Water Partition Coefficient: log K_{ow} = 1.60
Sorption Partition Coefficient: K_{oc} = estimated at 35
BCF: estimated at 42

Regulations

RCRA 40CFR: Not listed
CERCLA: 40CFR 302.4: Listed as Compound per CWA Section 307(a)
SARA 40CFR 372.65: Listed
SARA EHS 40CFR 355: Not listed
TSCA: Listed

Analytical Methods

Air: ASTM D4490

CHL3490 **CAS #: 75-45-6**

CHLORODIFLUOROETHANE

RTECS: PA6390000
DOT: UN1018; IMO2.2
EINECS Number: 200-871-9
Molecular Formula: $CHClF_2$
Structured MF: $CHClF_2$
Formula Weight: 86.47

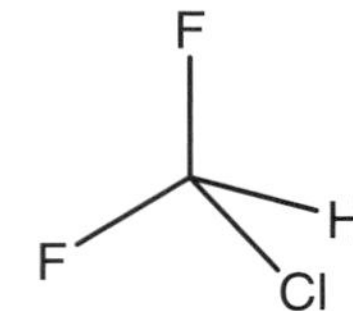

Chemical Structure

Synonyms: ALGEON 22; ALGOFRENE 22; ALGOFRENE TYPE 6; ARCTON 22; ARCTON 4; CFC 22; CHLORODIFLUOROMETHANE; CHLOROFLUOROCARBON 22; DAIFLON 22; DIFLUOROCHLOROMETHANE; DIFLUOROMONOCHLOROMETHANE; DYMEL 22; ELECTRO-CF 22; ESKIMON 22; F 22; FC 22; FLUGENE 22; FLUOROCARBON-22; FORANE 22; FORANE 22 B; FREON 22; FRIGEN; FRIGEN 22; GENETRON 22; HALTRON 22; HYDROCHLOROFLUOROCARBON 22; ISCEON 22; ISOTRON 22; KHALADON 22; KHLADON 22; METHANE,CHLORODIFLUORO-; MONOCHLORODIFLUOROMETHANE; PROPELLANT 22; R 22; R 22 (REFRIGERANT); R-22; R22; REFRIGERANT 22; REFRIGERANT R 22; UCON 22
Description: colorless gas; ether-like odor
Use: refrigerant; low-temperature solvent; fluorocarbon resins, especially tetrafluoroethylene polymers; refrigerant, primarily in food display cases, ice makers, home freezers & heat pumps

Physical Properties

Boiling Point: -40.7 °C (-41 °F) at 760 mm Hg
Freezing Point: -157.4 °C (-251.32 °F)
Specific Gravity: 1.194 at 25 °C
Vapor Density: 2.98 Air=1
Vapor Pressure: 201 kPa at -25 °C
Water Solubility: 0.28 g/L Water at 77 °F & 147 psia
Other Solubilities: > 10% in Acetone; > 10% in Chloroform; > 10% in ethyl Ether; Soluble in Ether, Acetone, and Chloroform.
Surface Tension: Estimated at 15 dynes/cm
Refraction Index: 1.256 at 25 °C
Critical Temperature: 96.0 °C
Critical Pressure: 4.97 mPa
Ionization Potential (eV): 12.45
Flash Point: Nonflammable
Autoignition Temperature: 632 °C

RTECS Toxicity Data

Acute Oral: Rat LD; Dose: >43200 ug/kg.
Acute Inhalation: Rat LC_{50} Dose: 35 pph/15M; Toxic Effects: Behavioral - Altered sleep time (including change in righting reflex); Behavioral - Ataxia; Lungs, Thorax, or Respiration - Respiratory depression. Mouse LC_{50} Dose: 1380 mg/m^3/2hr; Toxic Effects: Behavioral - Somnolence (general depressed activity); Behavioral - Ataxia; Lungs, Thorax, or Respiration - Cyanosis.
Chronic (Multiple Dose) Oral: Rat Dose: 2457 mg/kg/26W-I; Toxic Effects: Brain and coverings - Other degenerative changes; Blood - Changes in other cell count; Nutritional and gross metabolic - Weight loss or decreased weight gain.
Chronic (Multiple Dose) Inhalation: Mouse Dose: 50 gm/m^3/6H/43W-I; Toxic Effects: Brain and coverings - Other degenerative changes; Spinal Cord - Other degenerative changes; Behavioral - Alteration of classical conditioning.
Reproductive/Teratogenic: Rat Route: Inhalation; Dose: 50000 ppm/5H; Duration: male 56D prior to mating; Paternal Effects - Prostate, seminal vessicle, Cowper's gland, accessory glands.
Mutagenic: Bacteria - S Typhimurium Mutations in Microorganisms; Dose: 33 pph/24H-C (+/-S9).

Hazard Overviews

Compressed Gas

Fire Diamond

Health: Severely irritating to respiratory tract. A compressed gas which can cause frostbite. Also Causes: asphyxiation, rapid/irregular breathing, headache, fatigue nausea, vomiting, giddiness, exhaustion, unconsciousness, convulsions, death. Chronic Effects: irregular heartbeat.
Fire: Noncombustible. Use agent suitable for surrounding fire. Cylinder may explode in heat of fire.
Reactivity: Stable. Hazardous polymerization cannot occur. Avoid: exposure to open flames; temperatures above 125 °F (51.5 °C); contact with red-hot metal (forms chlorine; fluorine; phosgene; hydrochloric acid; hydrofluoric acid; carbonyl fluoride). Incompatible with: certain elastomers; alkali or alkaline earth metals; sodium; potassium; powdered aluminum, zinc, beryllium; hydrochloric acid; hydrofluoric acid. Hazardous decomposition products: carbon dioxide; toxic chlorine; fluorine; phosgene; carbonyl fluoride; hydrofluoric acid; hydrochloric acid.
Carcinogenicity: IARC - Group 3, Not classifiable as to carcinogenicity to humans; NIOSH - Not listed; NTP - Not listed; ACGIH - Class A4, Not classifiable as a human carcinogen; OSHA - Not listed; EPA - Not listed; MAK - Not listed
Primary Target Organs:

Eyes

Skin

Respiratory System

Nervous System

Cardio-vascular

Exposure Limits
OSHA PEL Vacated 1989 Limits: TWA: 1000 ppm; 3500 mg/m^3.
ACGIH TLV: TWA: 1000 ppm; 3540 mg/m^3.

NIOSH REL: TWA: 1000 ppm; 3500 mg/m^3. STEL: 1250 ppm; 4375 mg/m^3.
DFG MAK: TWA: 500 ppm; 1800 mg/m^3.
Respirator Recommendation
Exposure Range: >1000 to 50000 mg/m^3 Supplied Air, Constant Flow/Pressure Demand, Half Mask
Exposure Range: >50,000 to <790,000 mg/m^3 Supplied Air, Constant Flow/Pressure Demand, Full Face
Exposure Range: 790,000 to unlimited mg/m^3 Self-contained Breathing Apparatus, Pressure Demand, Full Face
Note: odor threshold unknown

Environmental

Environmental Fate: It is an extremely unreactive gas and losses due to photolysis, photooxidation, hydrolysis, or biodegradation in air and soil will not be significant. If released on land, will volatilize. It is highly mobile in soil and therefore will have a potential for leaching into groundwater. If released in water, it will volatilize. Its half-life in a model river is estimated to be 2.7 hr. In the atmosphere, it is mainly removed by reaction with hydroxyl radicals. It is estimated that the half-life in the troposphere is 11.1 to 17.3 yr. As a result of its long half-life, it will accumulate and disperse all over the globe. It will also be removed from the atmosphere by dry and wet deposition.
Cleanup/Disposal: Guide No. 126: Do not touch or walk through spilled material. Stop leak if you can do it without risk. Do not direct water at spill or source of leak. Use water spray to reduce vapors or divert vapor cloud drift. If possible, turn leaking containers so that gas escapes rather than liquid. Prevent entry into waterways, sewers, basements or confined areas. Allow substance to evaporate. Ventilate the area.

Environmental Physical Data

Henry's Law Constant: 0.0294
Octanol/Water Partition Coefficient: log K_{ow} = 1.08
Sorption Partition Coefficient: K_{oc} = estimated at 58.5
BCF: estimated at 3.9
BOD: none

Regulations

RCRA 40CFR: Not listed
CERCLA: 40CFR 302.4: Listed as Compound per CWA Section 307(a)
SARA 40CFR 372.65: Listed
SARA EHS 40CFR 355: Not listed
TSCA: Listed

Analytical Methods

Air: ASTM D4490
Indoor / Expired Air: NIOSH 1018

CHL3580 CAS #: 698-01-1

O-CHLORO-N,N-DIMETHYLANILINE

Molecular Formula: $C_8H_{10}ClN$
Formula Weight: 155.52
Synonyms: BENZENAMINE,2-CHLORO-N,N-DIMETHYL-; 2-CHLORO-N,N-DIMETHYLANILINE

Physical Properties

Boiling Point: 205 °C (401 °F)
Specific Gravity: 1.10670
Other Solubilities: Benzene: Very Soluble; Ethanol: Very Soluble
Refraction Index: 1.5578

Hazard Overviews

Carcinogenicity: IARC - Not listed; NIOSH - Not listed; NTP - Not listed; ACGIH - Not listed; OSHA - Not listed; EPA - Not listed; MAK - Not listed

Environmental

Regulations

RCRA 40CFR: Not listed
CERCLA: 40CFR 302.4: Not listed
SARA 40CFR 372.65: Not listed
SARA EHS 40CFR 355: Not listed
TSCA: Not listed

CHL3610 CAS #: 17256-39-2

1-CHLORO-N,N-DIMETHYL-2-PROPANAMINE, HYDROCHLORIDE

EINECS Number: 241-289-5
Molecular Formula: $C_5H_{13}Cl_2N$
Formula Weight: 158.06
Synonyms: (BETA-CHLOROISOPROPYL)DIMETHYLAMINE-HYDROCHLORIDE; ETHYLAMINE,2-CHLORO-N,N,1-TRIMETHYL-,HYDROCHLORIDE

Physical Properties

Other Solubilities: Chloroform: Soluble

Hazard Overviews

Carcinogenicity: IARC - Not listed; NIOSH - Not listed; NTP - Not listed; ACGIH - Not listed; OSHA - Not listed; EPA - Not listed; MAK - Not listed

Environmental

Regulations

RCRA 40CFR: Not listed
CERCLA: 40CFR 302.4: Not listed
SARA 40CFR 372.65: Not listed
SARA EHS 40CFR 355: Not listed
TSCA: Listed

CHL3790 CAS #: 1622-32-8

2-CHLOROETHANESULFONYL CHLORIDE

RTECS: KI8060000
EINECS Number: 216-594-1
Molecular Formula: $C_2H_4Cl_2O_2S$
Formula Weight: 163.02

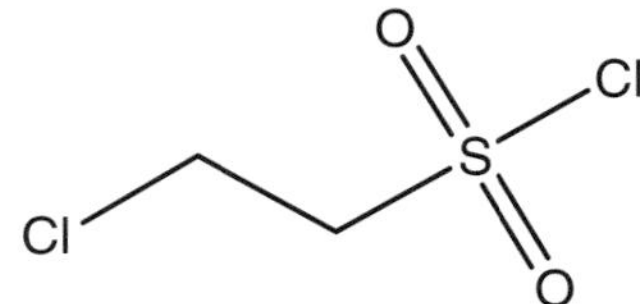

Chemical Structure

Synonyms: 2-CHLOROETHANE SULFOCHLORIDE; BETA-CHLOROETHANESULFONYL CHLORIDE; 2-CHLOROETHYLSULFONYL CHLORIDE; ETHANESULFONYL CHLORIDE,2-CHLORO-
Description: liquid

Physical Properties

Boiling Point: 200 °C (392 °F) to 203 °C (397 °F)
Specific Gravity: 1.555 at 20 °C/4 °C
Water Solubility: Decomposes in Water
Refraction Index: 1.4920

RTECS Toxicity Data

Acute Oral: Rat LD_{50} Dose: 240 mg/kg.
Acute Inhalation: Rat LC_{50} Dose: 420 mg/m^3/4hr. Mouse LC_{50} Dose: 250 mg/m^3/4hr.
Chronic (Multiple Dose) Inhalation: Rat Dose: 10 mg/m^3/4H/17W-I; Toxic Effects: Brain and coverings - Recordings from specific areas of CNS; Vascular - BP lowering not characterized in autonomic section. Rat Dose: 10 mg/m^3/4H/17W-I; Toxic Effects: Vascular - BP lowering not characterized in autonomic section; Liver - Liver function tests impaired; Blood - Changes in serum composition.

Hazard Overviews

Poison

Corrosive

Health: Corrosive to eyes/skin/respiratory tract. Poison. Other Acute Effects: may be fatal if inhaled, swallowed, or absorbed through skin; inhalation may result in spasm, inflammation and edema of the larynx and bronchi, chemical pneumonitis and pulmonary edema; symptoms of exposure may include burning sensation; coughing; wheezing; laryngitis; shortness of breath; headache; nausea; vomiting.
Fire: Hazards: emits toxic fumes. Extinguishing agents: carbon dioxide, dry chemical powder or appropriate foam. Precautions: combustible liquid.
Reactivity: Incompatible with: strong oxidizing agents, strong bases, may decompose on exposure to moist air or water. Hazardous decomposition products: toxic fumes of: carbon monoxide, carbon dioxide, sulfur oxides, hydrogen chloride gas.
Carcinogenicity: IARC - Not listed; NIOSH - Not listed; NTP - Not listed; ACGIH - Not listed; OSHA - Not listed; EPA - Not listed; MAK - Not listed
Primary Target Organs:

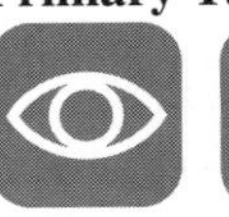
Eyes

Skin

Respiratory System

Environmental

Regulations

RCRA 40CFR: Not listed
CERCLA: 40CFR 302.4: Listed as Compound per CWA Section 307(a)
SARA 40CFR 372.65: Not listed TPQ: 500 lb
SARA EHS 40CFR 355: Listed TPQ: 500 lb
TSCA: Listed

CHL3820 CAS #: 627-11-2

CHLOROETHYL CHLOROFORMATE

RTECS: LQ5950000
EINECS Number: 210-982-4
Molecular Formula: $C_3H_4Cl_2O_2$
Formula Weight: 142.97

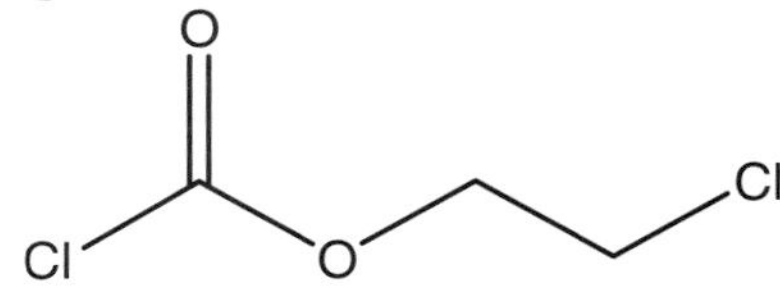

Chemical Structure

Synonyms: CARBONOCHLORIDIC ACID,2-CHLOROETHYL ESTER; 2-CHLORETHYLESTER KYSELINY CHLORMRAVENCI; (2-CHLOROETHOXY) CARBONYL CHLORIDE; BETA-CHLOROETHYL CHLOROCARBONATE; 2-CHLOROETHYL CHLOROFORMATE; CHLOROFORMIC ACID 2-CHLOROETHYL ESTER; CHLOROFORMIC ACID,2-CHLOROETHYL ESTER; CHLOROFORMIC ACID,BETA-CHLOROETHYL ESTER; FORMIC ACID,CHLORO-,2-CHLOROETHYL ESTER; TL 207
Description: colorless liquid
Use: intermediate for synthesis of pesticides, perfumes, drugs, polymers, dyes, and other chemicals

Physical Properties

Boiling Point: 155 °C (311 °F)
Specific Gravity: 1.3847 at 20 °C/4 °C
Vapor Pressure: 13 mm Hg at 48-49 °C
Water Solubility: Insoluble in Cold Water

RTECS Toxicity Data

Acute Oral: Rat LD_{50} Dose: 25 gm/kg. Mouse LD_{50} Dose: >11848 mg/kg; Toxic Effects: Behavioral - Somnolence (general depressed activity).
Acute Inhalation: Rat LC_{50} Dose: 2250 mg/m^3.
Acute Dermal: Rabbit LD_{50} Route: Skin; Dose: 8 gm/kg.

Hazard Overviews

Carcinogenicity: IARC - Not listed; NIOSH - Not listed; NTP - Not listed; ACGIH - Not listed; OSHA - Not listed; EPA - Not listed; MAK - Not listed

Environmental

Regulations

RCRA 40CFR: Not listed
CERCLA: 40CFR 302.4: Not listed
SARA 40CFR 372.65: Not listed
SARA EHS 40CFR 355: Not listed
TSCA: Listed

CHL4270 CAS #: 107-30-2

CHLOROMETHYL METHYL ETHER

RTECS: KN6650000
DOT: UN1239; IMO3.1
EINECS Number: 203-480-1
Molecular Formula: C_2H_5ClO
Structured MF: CH_3OCH_2Cl
Formula Weight: 80.52

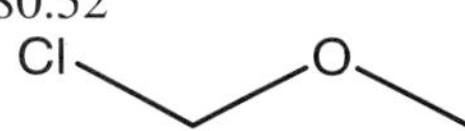

Chemical Structure

Synonyms: CHLORDIMETHYLETHER; CHLORODIMETHYL ETHER; CHLORODIMETHYLETHER; CHLOROMETHOXYMETHANE; CMME; DIMETHYLCHLOROETHER; ETHER METHYLIQUE MONOCHLORE; ETHER,CHLOROMETHYL METHYL; ETHER,DIMETHYL CHLORO; METHANE,CHLOROMETHOXY-; METHOXYCHLOROMETHANE; METHOXYMETHYL CHLORIDE; METHYL CHLOROMETHYL ETHER; METHYLCHLOROMETHYL ETHER; METHYLCHLOROMETHYL ETHER,ANHYDROUS; MONOCHLORODIMETHYL ETHER; MONOCHLOROMETHYL METHYL ETHER
Description: colorless liquid; irritating odor
Use: chem int for dodecylbenzyl chloride; an alkylating agent & solvent used in the manufacture of water repellants, ion-exchange resins, & indust polymers; in synthesis of chloromethylated compounds

Physical Properties

Boiling Point: 59 °C (138 °F) at 760 mm Hg
Freezing Point: -103.5 °C (-154.3 °F)
Specific Gravity: 1.0605 at 20 °C/4 °C
Vapor Density: 2.8 Air=1
Saturated Vapor Density: 1.738575681 kg/m^3
Vapor Pressure: 192 mm Hg at 70 °F
Water Solubility: Reacts
Other Solubilities: Soluble in Acetone, Chloroform, Ether.
Surface Tension: Estimated at 30 dynes/cm
pH: Acid
Refraction Index: 1.3974 at 20 °C/D
Ionization Potential (eV): 10.25
Flash Point: 0 °C Open Cup

RTECS Toxicity Data

Acute Inhalation: Rat LC_{50} Dose: 55 ppm/7hr; Toxic Effects: Lungs, Thorax, or Respiration - Chronic pulmonary edema; Blood - Hemorrhage. Mouse LC_{50} Dose: 1030 mg/m^3/2hr.
Mutagenic: Human DNA Inhibition; Cell Type: lymphocyte; Dose: 5 mL/L. Hamster Morphological Transformation; Cell Type: embryo; Dose: 10 mg/L.
Tumorigenic: Rat Route: Inhalation; Dose: 1 ppm/6H/72W; Toxic Effects: Tumorigenic - Equivocal tumorigenic agent by RTECS criteria; Lungs, Thorax, or Respiration - Tumors; Endocrine - Tumors. Mouse Route: Inhalation; Dose: 2 ppm/82D-I; Toxic Effects: Tumorigenic - Equivocal tumorigenic agent by RTECS criteria; Lungs, Thorax, or Respiration - Tumors.

Hazard Overviews

Corrosive

Explosive

Flammable
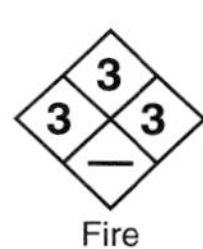

Fire Diamond

Health: Corrosive to eyes/skin/respiratory tract. Also Causes: fever, chills, pneumonia, pulmonary edema, decreases in alertness/consciousness, nausea, vomiting, cardiac irritability. Chronic Effects: cancer hazard.
Fire: Flammable. Explosive; forms peroxides when heated. Use dry chemical, foam, or carbon dioxide. Fight fire from maximum distance. Water may be ineffective on fire, but a water spray may be used to cool fire-exposed containers. When wet forms irritating formaldehyde gas.
Reactivity: Stable. Hazardous polymerization cannot occur. Avoid: heat; ignition sources. Incompatible with: divalent metals. Hazardous decomposition products: toxic fumes of chlorine.
Carcinogenicity: IARC - Group 1, Carcinogenic to humans; NIOSH - Listed as carcinogen; NTP - Class 1, Known to be a carcinogen; ACGIH - Class A2, Suspected human carcinogen; OSHA - Listed as a carcinogen; EPA - Class A, Human carcinogen; MAK - Class A1, Capable of inducing malignant tumors as shown by experience with humans
Primary Target Organs:

Eyes

Skin

Respiratory System

Respirator Recommendation
Exposure Range: unlimited Self-contained Breathing Apparatus, Pressure Demand, Full Face
Note: TLV not established

Environmental

Environmental Fate: It would not be found in waste water due to its very rapid hydrolysis (half-life < 1 sec). Its half-life in humid air has been measured to be 3.5-6 min and 6.5 hr in 2 studies

Cleanup/Disposal: Guide No. 131: Fully encapsulating, vapor protective clothing should be worn for spills and leaks with no fire. Eliminate all ignition sources (no smoking, flares, sparks or flames in immediate area). All equipment used when handling the product must be grounded. Do not touch or walk through spilled material. Stop leak if you can do it without risk. Prevent entry into waterways, sewers, basements or confined areas. A vapor suppressing foam may be used to reduce vapors. Small Spills: Absorb with earth, sand or other non-combustible material and transfer to containers for later disposal. Use clean non-sparking tools to collect absorbed material. Large Spills: Dike far ahead of liquid spill for later disposal. Water spray may reduce vapor; but may not prevent ignition in closed spaces.

Environmental Physical Data

BCF: none likely

Regulations

RCRA 40CFR: Listed Hazardous Waste No. U046 Toxic Waste

CERCLA: 40CFR 302.4: Listed per RCRA Section 3001 RQ: 10 lb (4.535 kg)

SARA 40CFR 372.65: Listed TPQ: 100 lb

SARA EHS 40CFR 355: Listed TPQ: 10 lb

TSCA: Listed

Analytical Methods

Soil: SW846 8010B

CHL4390 CAS #: 100-14-1

1-(CHLOROMETHYL)-4-NITROBENZENE

RTECS: XS9093000

EINECS Number: 202-822-7

Molecular Formula: $C_7H_6ClNO_2$

Formula Weight: 171.58

O, N+, -O, Cl

Chemical Structure

Synonyms: BENZENE,1-(CHLOROMETHYL)-4-NITRO-; BENZENE,1-(CHLOROMETHYL)-4-NITRO-(9CI); BENZYL CHLORIDE,4-NITRO; ALPHA-CHLOR0-P-NITROTOLUENE; 1-CHLOROMETHYL-4-NITROBENZENE; 4-(CHLOROMETHYL)NITROBENZENE; P-(CHLOROMETHYL)NITROBENZENE; ALPHA-CHLORO-4-NITROTOLUENE; ALPHA-CHLORO-NITROTOLUENE; ALPHA-CHLORO-P-NITROTOLUENE; 4-NITROBENZYL CHLORIDE; P-NITROBENZYL CHLORIDE; TOLUENE,ALPHA-CHLORO-P-NITRO-

Description: plates or needles

Use: organic synthesis

Physical Properties

Freezing Point: 71 °C (159.8 °F)

Specific Gravity: 1.5647

Water Solubility: Insoluble

Other Solubilities: 95% Ethanol: Soluble; Acetone: >=10 mg/ml at 19 °C; Benzene: Very Soluble; DMSO: >=10 mg/ml at 19 °C; Ether: Soluble; Ethyl Acetate: Very Soluble; Methanol: Soluble.

Refraction Index: 1.5647 at 62 °C

Flash Point: Low

RTECS Toxicity Data

Acute Oral: Rat LD_{50} Dose: 1809 mg/kg.

Acute Inhalation: Rat LC_{Lo} Dose: 280 mg/m^3/4hr.

Chronic (Multiple Dose) Inhalation: Rat Dose: 20 mg/m^3/6H/2W-I; Toxic Effects: Nutritional and gross metabolic - Weight loss or decreased weight gain; Biochemical - Transaminases.

Mutagenic: Hamster Sister Chromatid Exchenge; Cell Type: ovary; Dose: 100 umol/L. Bacteria - B Subtilis DNA Repair; Dose: 50 ug/disc.

Hazard Overviews

Fire: Will burn.

Carcinogenicity: IARC - Not listed; NIOSH - Not listed; NTP - Not listed; ACGIH - Not listed; OSHA - Not listed; EPA - Not listed; MAK - Not listed

Environmental

Cleanup/Disposal: Guide No. 171: Do not touch or walk through spilled material. Stop leak if you can do it without risk. Prevent dust cloud. Avoid inhalation of asbestos dust. Small Dry Spills: With clean shovel place material into clean, dry container and cover loosely; move containers from spill area. Small Spills: Take up with sand or other noncombustible absorbent material and place into containers for later disposal. Large Spills: Dike far ahead of liquid spill for later disposal. Cover powder spill with plastic sheet or tarp to minimize spreading. Prevent entry into waterways, sewers, basements or confined areas.

Regulations

RCRA 40CFR: Not listed

CERCLA: 40CFR 302.4: Not listed

SARA 40CFR 372.65: Not listed TPQ: 500/10000 lb

SARA EHS 40CFR 355: Listed TPQ: 500 lb

TSCA: Listed

CHL4450 **CAS #: 3653-48-3**

4-CHLORO-2-METHYLPHENOXY ACETATE, SODIUM SALT

RTECS: AG2625000
EINECS Number: 222-895-9
Molecular Formula: $C_9H_8ClNaO_3$
Formula Weight: 222.61

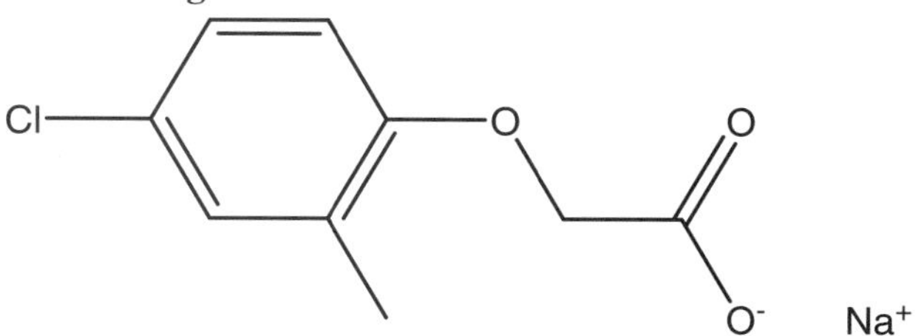

Chemical Structure

Synonyms: 2M-4KH SODIUM SALT; ACETIC ACID,(4-CHLORO-2-METHYLPHENOXY)-,SODIUM SALT; AGROXONE 3; 4-CHLORO-2-METHYLPHENOXYACETIC ACID SODIUM SALT; (P-CHLORO-O-TOLYLOXY)ACETIC ACID SODIUM SALT; CHWASTOKS; CHWASTOX; CHWASTOX 80; DIAMET; DICOTEX 80; DIKOTEKS; DIKOTEX 30; ESTERMINE; HEDONAL M80; MCPA SODIUM SALT; METAXONE; METHOXON; METHOXONE; (2-METHYL-4-CHLOROPHENOXY)ACETIC ACID,SODIUM SALT; PHENOXYLENE; SODIUM (4-CHLORO-2-METHYLPHENOXY)ACETATE; SODIUM MCPA; SODIUM (2-METHYL-4-CHLOROPHENOXY)ACETATE; SYS 67ME; U 46M

RTECS Toxicity Data

Acute Oral: Rat LD_{50} Dose: 800 mg/kg. Mouse LD_{50} Dose: 450 mg/kg.
Acute Inhalation: Rat LC_{Lo} Dose: 520 mg/m^3/4hr; Toxic Effects: Behavioral - Somnolence (general depressed activity); Behavioral - Muscle weakness.
Acute Dermal: Rat LD_{50} Route: Subcutaneous Dose: 500 mg/kg; Toxic Effects: Behavioral - Muscle contraction or spasticity; Gastrointestinal - Hypermotility, diarrhea; Blood - Hemorrhage. Rat LD; Route: Skin; Dose: >2 gm/kg.
Reproductive/Teratogenic: Rat Route: Oral; Dose: 11400 mg/kg; Duration: male 60D prior to mating; Paternal Effects - Spermatogenesis; Testes, epididymis, sperm duct. Rat Route: Oral; Dose: 542 mg/kg; Duration: female 5D of pregnancy; Effects on Fertility - Post-implantation mortality.
Mutagenic: Chicken Sister Chromatid Exchange; Route: Parenteral; Dose: 2800 ug/egg.

Hazard Overviews

Health: Irritating. Toxic. Other Acute Effects: harmful if swallowed, inhaled, or absorbed through skin. Chronic Effects: may alter genetic material. Carcinogen.
Fire: Hazards: emits toxic fumes. Extinguishing agents: water spray; carbon dioxide, dry chemical powder or appropriate foam. Precautions: combustible liquid.
Reactivity: Incompatible with: strong oxidizing agents. Hazardous decomposition products: toxic fumes of: carbon monoxide, carbon dioxide, hydrogen chloride gas.
Carcinogenicity: IARC - Not listed; NIOSH - Not listed; NTP - Not listed; ACGIH - Not listed; OSHA - Not listed; EPA - Not listed; MAK - Not listed
Primary Target Organs:

Eyes

Skin

Respiratory System

Environmental

Regulations

RCRA 40CFR: Not listed
CERCLA: 40CFR 302.4: Not listed
SARA 40CFR 372.65: Listed
SARA EHS 40CFR 355: Not listed
TSCA: Listed

CHL4510 **CAS #: 507-20-0**

2-CHLORO-2-METHYLPROPANE

RTECS: TX5040000
EINECS Number: 208-066-4
Molecular Formula: C_4H_9Cl
Formula Weight: 92.58

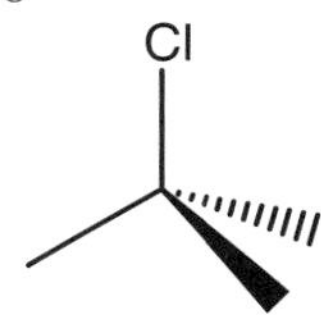

Chemical Structure

Synonyms: TERT-BUTYL CHLORIDE; 2-CHLOROISOBUTANE; TRIMETHYLCHLOROMETHANE

Physical Properties

Boiling Point: 51 °C (123.8 °F)
Freezing Point: -26 °C (-14.8 °F)
Specific Gravity: 0.84200
Saturated Vapor Density: 2.308976138 kg/m^3
Vapor Pressure: 42.7 kPa at 25 °C
Water Solubility: Negligble
Other Solubilities: Ethanol: Miscible; Ether: Miscible; Benzene: Soluble; Carbon Tetrachloride: Soluble; Chloroform: Soluble
Refraction Index: 1.3857
Ionization Potential (eV): 10.61 +/-0.03

RTECS Toxicity Data

Tumorigenic: Mouse Route: Intraperitoneal; Dose: 3000 mg/kg/8W-I; Toxic Effects: Tumorigenic - Neoplastic by RTECS criteria; Lungs, Thorax, or Respiration - Tumors.

Hazard Overviews

Health: May cause irritation to eyes/skin. Other Acute Effects: may be harmful by inhalation, ingestion, or skin absorption.
Fire: Hazards: vapor may travel considerable distance to source of ignition and flash back; container explosion may occur; forms explosive mixtures in air; emits toxic fumes. Extinguishing agents: carbon dioxide, dry chemical powder or appropriate foam; water may be effective for cooling, but may not effect extinguishment. Precautions: combustible liquid.
Reactivity: Incompatible with: strong oxidizing agents, strong bases. Hazardous decomposition products: toxic fumes of: carbon monoxide, carbon dioxide, hydrogen chloride gas.
Carcinogenicity: IARC - Not listed; NIOSH - Not listed; NTP - Not listed; ACGIH - Not listed; OSHA - Not listed; EPA - Class D, Not classifiable as to human carcinogenicity; MAK - Not listed

Environmental

Regulations

RCRA 40CFR: Not listed
CERCLA: 40CFR 302.4: Not listed
SARA 40CFR 372.65: Not listed
SARA EHS 40CFR 355: Not listed
TSCA: Listed

CHL4660 CAS #: 91-58-7

2-CHLORONAPHTHALENE

RTECS: QJ2275000
EINECS Number: 202-079-9
Molecular Formula: $C_{10}H_7Cl$
Formula Weight: 162.61

Cl

Chemical Structure

Synonyms: 2-CHLORNAFTALEN; BETA-CHLORONAPHTHALENE; HALOWAX; NAPHTHALENE,2-CHLORO-
Description: monoclinic plates or leaflets
Use: production of electric condensers; in the insulation of electric cables and wires; as additive to extreme pressure lubricants; supports for storage batteries; coating in foundry use; solvent; and immersion liquid in microscopy

Physical Properties

Boiling Point: 256 °C (493 °F) at 760 mm Hg
Freezing Point: 59.5 °C (139.1 °F)
Specific Gravity: 1.1938 at 90 °C
Saturated Vapor Density: 1.200058051 kg/m^3
Density: 1.2656 g/mL at 16 °C
Vapor Pressure: 7.98 x10^{-3} mm Hg at 25 °C
Water Solubility: < 1 mg/mL at 20.5 C
Other Solubilities: 95% Ethanol: >=100 mg/ml at 20.5 °C; Acetone: >=100 mg/ml at 20.5 °C; Benzene: Soluble; Carbon Disulfide: Soluble; Chloroform: Soluble; DMSO: >=100 mg/ml at 20.5 °C; Ether: Soluble.
Refraction Index: 1.60787 at 70.7 °C/D
Ionization Potential (eV): 8.11 +/-2.0
Flash Point: Not available; probably combustible

RTECS Toxicity Data

Acute Oral: Rat LD_{50} Dose: 2078 mg/kg. Mouse LD_{50} Dose: 886 mg/kg.

Hazard Overviews

Health: Irritating to eyes/skin/respiratory tract. Harmful. Other Acute Effects: harmful if swallowed; may be harmful if inhaled or absorbed through the skin.
Fire: Will burn. Extinguishing agents: water spray; carbon dioxide, dry chemical powder or appropriate foam. Precautions: combustible liquid.
Carcinogenicity: IARC - Not listed; NIOSH - Not listed; NTP - Not listed; ACGIH - Not listed; OSHA - Not listed; EPA - Not listed; MAK - Not listed
Primary Target Organs:

Eyes

Skin

Respiratory System

Environmental

Environmental Fate: Should biodegrade slowly in the environment. Aerobic biodegradation half-lives were 59 and 79 days when contained in a mixture of oil sludge that was added to soil columns along with nitrogen and phosphorus. The hydrolysis is too slow to be environmentally important. It has the potential to undergo direct photolysis in sunlit media. It may also evaporate from dry surfaces. A calculated K_{oc} range of 1130 indicates it will have a low mobility in soil. In aquatic systems, it will bioconcentrate in aquatic organisms and can partition from the water column to organic matter contained in sediments and suspended solids. A Henry's Law constant of 3.15 x10^{-4} atm-cu m/mole at 25 °C suggests volatilization from environmental waters may be important. The volatilization half-lives from a model river and a model pond, the latter considers the effects of adsorption, have been estimated to be 7 hr and 16 days, respectively. It is expected to exist entirely in the vapor phase in ambient air. In the atmosphere, the reaction with photochemically produced hydroxyl radicals (estimated half-life of 23 hr) is likely to be an important fate process. It can also be removed from the atmosphere by wet deposition.
Cleanup/Disposal: Guide No. 171: Do not touch or walk through spilled material. Stop leak if you can do it without risk. Prevent dust cloud. Avoid inhalation of asbestos dust. Small Dry Spills: With clean shovel place material into clean,

dry container and cover loosely; move containers from spill area. Small Spills: Take up with sand or other noncombustible absorbent material and place into containers for later disposal. Large Spills: Dike far ahead of liquid spill for later disposal. Cover powder spill with plastic sheet or tarp to minimize spreading. Prevent entry into waterways, sewers, basements or confined areas.

Environmental Physical Data

Henry's Law Constant: 3.15 x10^{-4}
Sorption Partition Coefficient: K_{OC} = estimated at 1130
BCF: calculated at 3.63

Regulations

RCRA 40CFR: Listed Hazardous Waste No. U047 Toxic Waste
CERCLA: 40CFR 302.4: Listed per RCRA Section 3001 per CWA Section 307(a) RQ: 5000 lb (2268 kg)
SARA 40CFR 372.65: Not listed
SARA EHS 40CFR 355: Not listed
TSCA: Listed

Analytical Methods

Soil: CLP LC_SV, MC_SVOA, OHC; EPA 16, 1625; SW846 3640A, 8120A, 8121, 8250A, 8270B, 8270C, 8410
Water / Groundwater: EPA S-002-1, 1625, 612, 625, 625-S, 6; APHA 6040-B, 6410-B; USGS O3118
Plasma: EPA 29

CHL4690 CAS #: 25586-43-0

CHLORONAPTHALENE

RTECS: QJ2047500
EINECS Number: 247-120-1
Molecular Formula: $C_{10}H_7Cl$
Formula Weight: 162.62
Synonyms: CHLORONAPHTHALENE; HALOWAX 1031; MONOCHLORONAPHTHALENE; NAPHTHALENE,CHLORO-
Description: liquid
Use: raw material for dye production; oil additive to clear sludge & petroleum deposits in engines; electrical insulation and fire-resisting materials; impregnants; sealing compounds; crankcase additive; ingredient in penetrating oils; plasticizer; protective coatings

Physical Properties

Boiling Point: 250 °C (482 °F) at 760 mm Hg
Freezing Point: About -25 °C (-13 °F)
Specific Gravity: 1.2 at 25 °C
Water Solubility: Insoluble in Water
Other Solubilities: good solubility in chlorinated and aromatic solvents, and petroleum Naphthas; limited in Ketones, Ethers, Acetates, mineral oils; Insoluble in Alcohol.
Flash Point: 165 °C Open Cup

RTECS Toxicity Data

Acute Oral: Rat LD_{50} Dose: 2200 mg/kg; Toxic Effects: Behavioral - Somnolence (general depressed activity); Behavioral - Change in motor activity (specific assay). Mouse LD_{50} Dose: 1100 mg/kg; Toxic Effects: Behavioral - Somnolence (general depressed activity); Behavioral - Change in motor activity (specific assay).

Hazard Overviews

Health: Irritating to eyes/skin/respiratory tract. Harmful. Other Acute Effects: harmful if swallowed, inhaled, or absorbed through skin; target organ: liver.
Fire: Will burn. Hazards: emits toxic fumes. Extinguishing agents: water spray; carbon dioxide, dry chemical powder or appropriate foam. Precautions: combustible liquid.
Reactivity: Incompatible with: strong oxidizing agents. Hazardous decomposition products: toxic fumes of: carbon monoxide, carbon dioxide, hydrogen chloride gas.
Carcinogenicity: IARC - Not listed; NIOSH - Not listed; NTP - Not listed; ACGIH - Not listed; OSHA - Not listed; EPA - Not listed; MAK - Not listed
Primary Target Organs:

Eyes

Skin

Respiratory System

Liver

Environmental

Regulations

RCRA 40CFR: Not listed
CERCLA: 40CFR 302.4: Not listed
SARA 40CFR 372.65: Not listed
SARA EHS 40CFR 355: Not listed
TSCA: Listed

Analytical Methods

Plasma: EPA 001

CHL4870 CAS #: 88-73-3

1-CHLORO-2-NITROBENZENE

RTECS: CZ0875000
DOT: UN1578; IMO6.1
EINECS Number: 201-854-9
Molecular Formula: $C_6H_4ClNO_2$
Structured MF: $NO_2C_6H_4Cl$
Formula Weight: 157.56

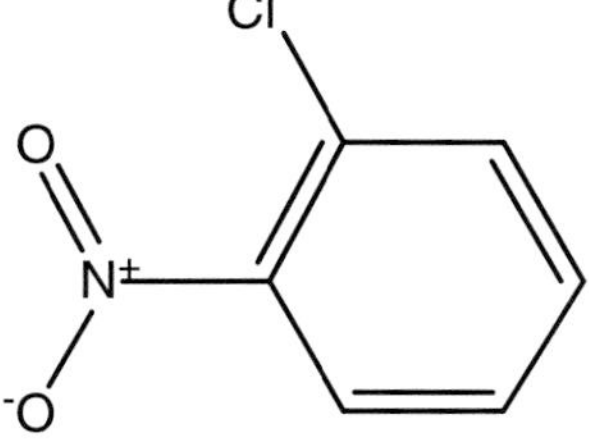

Chemical Structure

Synonyms: BENZENE,1-CHLORO-2-NITRO-; 2-CHLORO-1-NITROBENZENE; 2-CHLORONITROBENZENE; CHLORO-O-NITROBENZENE; O-CHLORONITROBENZENE; CHLORONITROBENZENE,ORTHO,LIQUID; 2-CNB; 1-NITRO-2-CHLOROBENZENE; NITROCHLOROBENZENE,ORTHO-; O-NITROCHLOROBENZENE; ONCB

Description: pale yellow crystals

Use: as an chemical intermediate for carbofuran, o-nitrophenol, 2-chloroaniline and dyes

Physical Properties

Boiling Point: 245.5 °C (474 °F)
Freezing Point: 32 °C (89.6 °F)
Specific Gravity: 1.368 at 22/4 °C
Vapor Density: 5.44 Air=1
Saturated Vapor Density: 1.202799855 kg/m^3
Density: 1.368 g/L at 242 °C
Vapor Pressure: 0.4 mm Hg at 25 °C
Water Solubility: Insoluble in Water
Other Solubilities: 95% Ethanol: >=100 mg/ml at 22 °C; Acetone: >=100 mg/ml at 22 °C; Alcohol: Soluble; Benzene: Soluble; Carbon Tetrachloride: Soluble; DMSO: >=100 mg/ml at 22 °C; Ether: Soluble; Methanol: Soluble; Pyridine: Soluble; Toluene: Soluble.
Surface Tension: 43.63 dynes/cm
Odor Threshold: 0.002 ppm
Critical Temperature: 484 °C
Critical Pressure: 3.98 x10^6 Pa
Flash Point: 127 °C
Autoignition Temperature: 255 °C
LEL: 1.38% v/v
UEL: 8.8% v/v

RTECS Toxicity Data

Acute Oral: Rat LD_{50} Dose: 268 mg/kg. Mouse LD_{50} Dose: 135 mg/kg.

Acute Dermal: Rabbit LD_{50} Route: Skin; Dose: 400 mg/kg.

Chronic (Multiple Dose) Inhalation: Rat Dose: 30 mg/m^3/6H/4W-I; Toxic Effects: Liver - Changes in Liver weight; Endocrine - Changes in spleen weight; Blood - Methemoglobinemia-Carboxyhemoglobin. Rat Dose: 4500 ppb/6H/13W-C; Toxic Effects: Liver - Changes in Liver weight; Blood - Methemoglobinemia-Carboxyhemoglobin; Blood - Changes in erythrocite (RBC) cell count.

Reproductive/Teratogenic: Rat Route: Inhalation; Dose: 18 ppm/6H; Duration: male 13W prior to mating; Paternal Effects - Spermatogenesis. Mouse Route: Inhalation; Dose: 4500 ppb/6H; Duration: male 13W prior to mating; Paternal Effects - Spermatogenesis.

Mutagenic: Bacteria - S Typhimurium Mutations in Microorganisms; Dose: 205 ug/plate (+S9). Bacteria - S Typhimurium Mutations in Microorganisms; Dose: 100 ug/plate (-S9).

Tumorigenic: Rat Route: Oral; Dose: 22 gm/kg/78W-C; Toxic Effects: Tumorigenic - Neoplastic by RTECS criteria; Gastrointestinal - Tumors; Endocrine - Tumors. Mouse Route: Oral; Dose: 140 gm/kg/78W-C; Toxic Effects: Tumorigenic - Carcinogenic by RTECS criteria; Liver - Tumors. Mouse Route: Oral; Dose: 280 gm/kg/78W-C; Toxic Effects: Tumorigenic - Carcinogenic by RTECS criteria; Liver - Tumors.

Hazard Overviews

Corrosive

Fire Diamond

Health: Corrosive to eyes/skin. Poison. Toxic. Also Causes: cyanosis, pulmonary edema, methemoglobinemia, anoxia, headache, fatigue, nausea, dizziness, chest pain, numbness, abdominal pain, aching, palpitations. Chronic Effects: skin sensitization, dermatitis, liver/kidney damage, pancrea disorders.

Fire: Will burn. Use dry chemical, water spray, or regular foam. Fight fire from maximum distance. Hazardous combustion products may include carbon oxide(s), nitrogen oxide(s), hydrogen chloride, phosgene gas.

Reactivity: Stable. Hazardous polymerization cannot occur. Avoid: exposure to heat; ignition sources; dispersing into air. Incompatible with: oxidizers and alkalis. Hazardous decomposition products: nitrogen oxides; phosgene; hydrogen chloride gas.

Carcinogenicity: IARC - Group 3, Not classifiable as to carcinogenicity to humans; NIOSH - Not listed; NTP - Not listed; ACGIH - Not listed; OSHA - Not listed; EPA - Not listed; MAK - Class B, Justifiably suspected of having carcinogenic potential

Primary Target Organs:

Eyes

Skin

Respiratory System

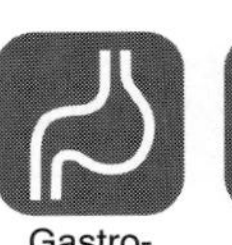
Gastro-intestinal

Liver

Kidneys

Environmental

Environmental Fate: If released to soil, it should be resistant to oxidation and hydrolysis. Leaching may be significant since it is predicted to be moderately mobile in soil. Volatilization from wet soils may be possible; (estimated Henry's Law constant 10-5 atm cu m/mol) however, leaching would lessen the significance of volatilization as a removal mechanism. Volatilization from dry soil surfaces is probably not rapid (estimated VP = 10-2 mm Hg), although it may be a significant removal mechanism under some circumstances. If released to water, it should be resistant to oxidation, hydrolysis and biodegradation. It has the potential to photolyze in water since it absorbs sunlight. Bioconcentration in aquatic organisms and sorption to sediments should not be significant. The volatilization half-life from 1 meter deep in surface water with a current velocity of 1 m/sec and a wind speed of 3 m/sec has been estimated to be 33.5 hours. Volatilization from surface waters with slower current speeds is expected to be less rapid. It has an estimated half-life in river water of 3.2 days based on monitoring data from the Rhine River. In contrast, it was not altered or physically removed during the time period required for transport down 900 miles of the Mississippi River. If released to the atmosphere, in the vapor phase it is predicted to react with photochemically generated hydroxyl radicals with an estimated reaction half-life of 1.97 days at 25 °C. It has the potential to directly photolyze in the atmosphere since it absorbs sunlight. Chloronitrophenols may form as a result of photochemical reaction in air.

Cleanup/Disposal: Guide No. 152: Do not touch damaged containers or spilled material unless wearing appropriate protective clothing. Stop leak if you can do it without risk. Prevent entry into waterways, sewers, basements or confined areas. Cover with plastic sheet to prevent spreading. Absorb or cover with dry earth, sand or other non-combustible material and transfer to containers. Do not get water inside containers.

Environmental Physical Data

Henry's Law Constant: 4.45 $x10^{-5}$
Octanol/Water Partition Coefficient: log K_{ow} = 2.52
Sorption Partition Coefficient: K_{oc} = 155 to 398
BCF: calculated at 100

Regulations

RCRA 40CFR: Not listed
CERCLA: 40CFR 302.4: Listed as Compound per CWA Section 307(a)
SARA 40CFR 372.65: Not listed
SARA EHS 40CFR 355: Not listed
TSCA: Listed

Analytical Methods

Soil: SW846 8091

CHL4930 CAS #: 100-00-5

P-CHLORONITROBENZENE

RTECS: CZ1050000
DOT: UN1578; IMO6.1
EINECS Number: 202-809-6
Molecular Formula: $C_6H_4ClNO_2$
Structured MF: $ClC_6H_4NO_2$
Formula Weight: 157.56

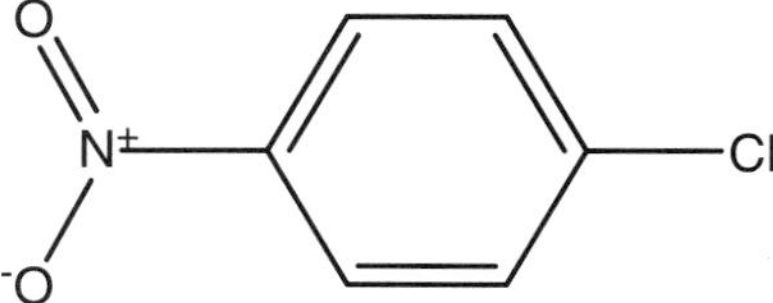

Chemical Structure

Synonyms: BENZENE,1-CHLORO-4-NITRO; BENZENE,1-CHLORO-4-NITRO-; 1-CHLOOR-4-NITROBENZEEN; 1-CHLOR-4-NITROBENZOL; 1-CHLORO-4-NITROBENZENE; 4-CHLORO-1-NITROBENZENE; 4-CHLORONITROBENZENE; P-CHLORONITROBENZENE,SOLID; 1-CLORO-4-NITROBENZENE; P-NITROCHLOORBENZEEN; 1-NITRO-4-CHLOROBENZENE; 4-NITRO-1-CHLOROBENZENE; 4-NITROCHLOROBENZENE; P-NITROCHLOROBENZENE; NITROCHLOROBENZENE,PARA-,SOLID; P-NITROCHLOROBENZOL; P-NITROCLOROBENZENE; P-NITROPHENYL CHLORIDE; PCNB; PNCB

Description: pale yellow crystals; pleasant, aromatic odor
Use: dye chemistry; manufacture of p-nitrophenol, from which parathion is made; agricultural chemicals and rubber chemicals

Physical Properties

Boiling Point: 242 °C (468 °F) at 760 mm Hg
Freezing Point: 82 °C (179.6 °F) to 84 °C (183.2 °F)
Specific Gravity: 1.52
Vapor Density: 5.44 Air=1
Saturated Vapor Density: 1.200657966 kg/m^3
Density: 1.52 g/mL
Vapor Pressure: 0.094 mm Hg at 20 °C
Water Solubility: 225 mg/L at 0 °C
Other Solubilities: 95% Ethanol: 5-10 mg/ml at 22 °C; Acetone: >=100 mg/ml at 22 °C; Acetic Acid: Very Soluble; Carbon Disulfide: Soluble; Chloroform: Very Soluble; DMSO: >=100 mg/ml at 22 °C; Ether: Soluble; Organic solvents: Soluble.
Surface Tension: 3.71 $x10^{-2}$ N/m at 83.65 °C
Refraction Index: 1.5376
Critical Temperature: 478 °C
Critical Pressure: 398 $x10^6$ Pa
Ionization Potential (eV): 9.96
Flash Point: 127.222 °C Closed Cup

RTECS Toxicity Data

Acute Oral: Rat LD_{50} Dose: 420 mg/kg; Toxic Effects: Behavioral - Somnolence (general depressed activity); Liver - Fatty Liver degeneration; Blood - Methemoglobinemia-Carboxyhemoglobin. Mouse LD_{50} Dose: 440 mg/kg.
Acute Inhalation: Rat LC_{Lo} Dose: 16100 mg/m^3/4hr. Cat LC_{Lo} Dose: 25 ppm/7hr.
Acute Dermal: Rabbit LD_{50} Route: Skin; Dose: 3040 mg/kg. Rat LD_{50} Route: Skin; Dose: 16 gm/kg; Toxic Effects: Behavioral - Food intake (animal); Kidney, Ureter, and Bladder - Chgs in tubules (inc acute renal failure, acute tubular necrosis; Blood - Methemoglobinemia-Carboxyhemoglobin.
Chronic (Multiple Dose) Oral: Rat Dose: 1350 mg/kg/2W-I; Toxic Effects: Liver - Hepatitis (hepatocellular necrosis), zonal; Nutritional and gross metabolic - Weight loss or decreased weight gain; DEATH.
Chronic (Multiple Dose) Inhalation: Rat Dose: 46 mg/m^3/6H/4W-I; Toxic Effects: Endocrine - Changes in spleen weight; Blood - Methemoglobinemia-Carboxyhemoglobin; Blood - Changes in erythrocite (RBC) cell count. Rat Dose: 3 ppm/6H/13W-C; Toxic Effects: Liver - Other changes; Blood - Methemoglobinemia-Carboxyhemoglobin; Blood - Changes in platelet cell count.
Chronic (Multiple Dose) Dermal: Rat Route: Subcutaneous; Dose: 1 gm/kg/10D-I; Toxic Effects: Blood - Other changes.
Reproductive/Teratogenic: Rat Route: Inhalation; Dose: 24 ppm/6H; Duration: male 13W prior to mating; Paternal Effects - Spermatogenesis.
Mutagenic: Rat DNA Damage; Cell Type: liver; Dose: 5 umol/L. Mouse DNA Damage; Route: Intraperitoneal; Dose: 60 mg/kg.

Tumorigenic: Mouse Route: Oral; Dose: 194 gm/kg/78W-C; Toxic Effects: Tumorigenic - Carcinogenic by RTECS criteria; Vascular - Tumors; Liver - Tumors. Mouse Route: Oral; Dose: 390 gm/kg/78W-C; Toxic Effects: Tumorigenic - Carcinogenic by RTECS criteria; Vascular - Tumors; Liver - Tumors.

Hazard Overviews

Fire Diamond

Health: Mildly irritating to eyes/skin/respiratory tract. Poison. Toxic. Also Causes: methemoglobin, headache, cyanosis, weakness, dizziness, incoordination, difficulty breathing on mild exertion, rapid heart beat, nausea, vomiting, drowsiness. Chronic Effects: methemoglobinemia, appearance of Heinz bodies, headache, dizziness, eczema.
Fire: Will burn. Use dry chemical, carbon dioxide, or foam. Water may cause frothing.
Reactivity: Stable. Hazardous polymerization cannot occur. Avoid: heat; ignition sources. Incompatible with: alkalis; oxidizers; potassium hydroxide; sodium methoxide and methanol. Hazardous decomposition products: carbon oxide; nitrogen oxide; hydrogen chloride gas.
Carcinogenicity: IARC - Group 3, Not classifiable as to carcinogenicity to humans; NIOSH - Listed as carcinogen; NTP - Not listed; ACGIH - Class A3, Animal carcinogen; OSHA - Not listed; EPA - Not listed; MAK - Class B, Justifiably suspected of having carcinogenic potential
Primary Target Organs:

Eyes

Skin

Blood

Exposure Limits
OSHA PEL: TWA: 1 mg/m^3; skin.
ACGIH TLV: TWA: 0.1 ppm; 0.64 mg/m^3.
NIOSH IDLH: 100 mg/m^3.
Respirator Recommendation
Exposure Range: >1 to 50 mg/m^3 Supplied Air, Constant Flow/Pressure Demand, Half Mask
Exposure Range: >50 to <100 mg/m^3 Supplied Air, Constant Flow/Pressure Demand, Full Face
Exposure Range: 100 to unlimited mg/m^3 Self-contained Breathing Apparatus, Pressure Demand, Full Face
Note: odor threshold unknown

Environmental

Cleanup/Disposal: Guide No. 152: Do not touch damaged containers or spilled material unless wearing appropriate protective clothing. Stop leak if you can do it without risk. Prevent entry into waterways, sewers, basements or confined areas. Cover with plastic sheet to prevent spreading. Absorb or cover with dry earth, sand or other non-combustible material and transfer to containers. Do not get water inside containers.

Environmental Physical Data

Henry's Law Constant: 5.44 $x10^{-5}$
Octanol/Water Partition Coefficient: log K_{ow} = 2.39

Regulations

RCRA 40CFR: Not listed
CERCLA: 40CFR 302.4: Listed as Compound per CWA Section 307(a)
SARA 40CFR 372.65: Not listed
SARA EHS 40CFR 355: Not listed
TSCA: Listed

Analytical Methods

Soil: SW846 8091
Indoor / Expired Air: NIOSH 2005

CHL5020 CAS #: 99-60-5

2-CHLORO-4-NITROBENZOIC ACID

RTECS: DG5425000
EINECS Number: 202-771-0
Molecular Formula: $C_7H_4ClNO_4$
Formula Weight: 201.57

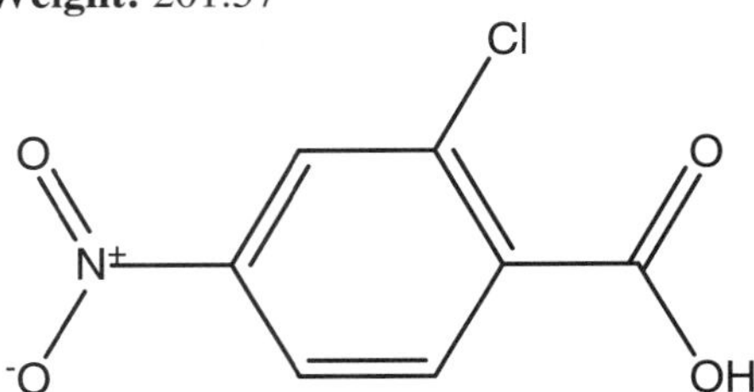

Chemical Structure

Synonyms: KYSELINA 2-CHLORO-4-NITROBENZOOVA

Physical Properties

Freezing Point: 139 °C (282.2 °F)
Water Solubility: 1 g/l
Other Solubilities: Ethanol: Soluble; Ether: Soluble; Benzene: Soluble

RTECS Toxicity Data

Irritation Eye: Rabbit Standard Draize Test Dose: 250 ug/24H; Reaction: severe.

Hazard Overviews

Health: Irritating to eyes/skin/respiratory tract. Other Acute Effects: may be harmful by inhalation, ingestion, or skin absorption.
Fire: Hazards: emits toxic fumes. Extinguishing agents: water spray; carbon dioxide, dry chemical powder or appropriate foam. Precautions: combustible liquid.
Carcinogenicity: IARC - Not listed; NIOSH - Not listed; NTP - Not listed; ACGIH - Not listed; OSHA - Not listed; EPA - Not listed; MAK - Not listed

Primary Target Organs:

Eyes

Skin

Respiratory System

Environmental

Regulations

RCRA 40CFR: Not listed
CERCLA: 40CFR 302.4: Not listed
SARA 40CFR 372.65: Not listed
SARA EHS 40CFR 355: Not listed
TSCA: Listed

CHL5050 **CAS #: 96-99-1**

4-CHLORO-3-NITROBENZOIC ACID

RTECS: DG5425050
EINECS Number: 202-550-9
Molecular Formula: $C_7H_4ClNO_4$
Formula Weight: 201.57

Chemical Structure

Synonyms: KYSELINA 4-CHLORO-3-NITROBENZOOVA

Physical Properties

Freezing Point: 180 °C (356 °F) to 183 °C (361.4 °F)
Specific Gravity: 1.645
Water Solubility: Insoluble
Other Solubilities: Ethanol: Slightly soluble; Acetone: Slightly soluble

RTECS Toxicity Data

Acute Oral: Rat LD_{50} Dose: 3150 mg/kg.
Mutagenic: Bacteria - S Typhimurium Mutations in Microorganisms; Dose: 500 ug/plate (+S9), 1 mg/plate (-S9).

Hazard Overviews

Health: Irritating to eyes/skin/respiratory tract. Other Acute Effects: may be harmful by inhalation, ingestion, or skin absorption; absorption leads to the formation of methemoglobin which, in sufficient concentration, causes cyanosis; onset may be delayed 2 to 4 hours or longer; exposure to and/or consumption of alcohol may increase toxic effects.
Fire: Hazards: emits toxic fumes. Extinguishing agents: water spray; carbon dioxide, dry chemical powder or appropriate foam. Precautions: combustible liquid.
Carcinogenicity: IARC - Not listed; NIOSH - Not listed; NTP - Not listed; ACGIH - Not listed; OSHA - Not listed; EPA - Not listed; MAK - Not listed
Primary Target Organs:

Eyes

Skin

Respiratory System

Environmental

Regulations

RCRA 40CFR: Not listed
CERCLA: 40CFR 302.4: Not listed
SARA 40CFR 372.65: Not listed
SARA EHS 40CFR 355: Not listed
TSCA: Listed

CHL5080 **CAS #: 6307-82-0**

2-CHLORO-5-NITROBENZOIC ACID METHYL ESTER

RTECS: DG5426500
Molecular Formula: $C_8H_6ClNO_4$
Formula Weight: 215.60
Synonyms: METHYLESTER KYSELINY 2-CHLOR-5-NITROBENZOOVE

RTECS Toxicity Data

Acute Oral: Rat LD_{50} Dose: 5360 mg/kg.
Irritation Eye: Rabbit Standard Draize Test Dose: 500 mg/24H; Reaction: mild.
Irritation Skin: Rabbit Standard Draize Test Dose: 500 mg/24H; Reaction: mild.

Hazard Overviews

Carcinogenicity: IARC - Not listed; NIOSH - Not listed; NTP - Not listed; ACGIH - Not listed; OSHA - Not listed; EPA - Not listed; MAK - Not listed

Environmental

Regulations

RCRA 40CFR: Not listed
CERCLA: 40CFR 302.4: Not listed
SARA 40CFR 372.65: Not listed
SARA EHS 40CFR 355: Not listed
TSCA: Not listed

CHL5110 CAS #: 619-08-9

2-CHLORO-4-NITROPHENOL

RTECS: SK5075000
EINECS Number: 210-578-8
Molecular Formula: $C_6H_4ClNO_3$
Formula Weight: 173.56

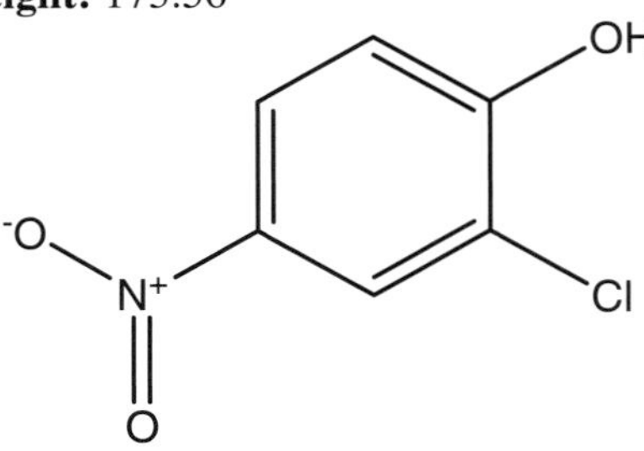

Chemical Structure

Physical Properties

Freezing Point: 106 °C (222.8 °F)
Water Solubility: Soluble
Other Solubilities: Ethanol: Soluble; Ether: Soluble; Benzene: Slightly soluble; Chloroform: Soluble

RTECS Toxicity Data

Acute Oral: Rat LD_{50} Dose: 900 mg/kg; Toxic Effects: Behavioral - Somnolence (general depressed activity); Gastrointestinal - Changes in structure or function of salivary glands; Gastrointestinal - Hypermotility, diarrhea.
Acute Dermal: Rabbit LD_{Lo} Route: Skin; Dose: 2 gm/kg; Toxic Effects: Behavioral - Somnolence (general depressed activity); Behavioral - Food intake (animal); Skin and appendages - Dermatitis, other.

Hazard Overviews

Health: Irritating to eyes/skin/respiratory tract. Harmful. Other Acute Effects: harmful if swallowed, inhaled, or absorbed through skin.
Fire: Hazards: emits toxic fumes. Extinguishing agents: water spray; carbon dioxide, dry chemical powder or appropriate foam. Precautions: combustible liquid.
Reactivity: Incompatible with: strong oxidizing agents, strong bases. Hazardous decomposition products: toxic fumes of: carbon monoxide, carbon dioxide, nitrogen oxides, hydrogen chloride gas.
Carcinogenicity: IARC - Not listed; NIOSH - Not listed; NTP - Not listed; ACGIH - Not listed; OSHA - Not listed; EPA - Not listed; MAK - Not listed

Primary Target Organs:

Eyes

Skin

Respiratory System

Environmental

Ecotoxicity: Toxicity to microorganisms: biodegradation inhibition EC_{50}: 245 mg/l

Regulations

RCRA 40CFR: Not listed
CERCLA: 40CFR 302.4: Listed as Compound per CWA Section 307(a)
SARA 40CFR 372.65: Not listed
SARA EHS 40CFR 355: Not listed
TSCA: Listed

CHL5140 CAS #: 600-25-9

1-CHLORO-1-NITROPROPANE

RTECS: TX5075000
EINECS Number: 209-990-0
Molecular Formula: $C_3H_6ClNO_2$
Structured MF: $CH_3CH_2CH(NO_2)Cl$
Formula Weight: 123.54

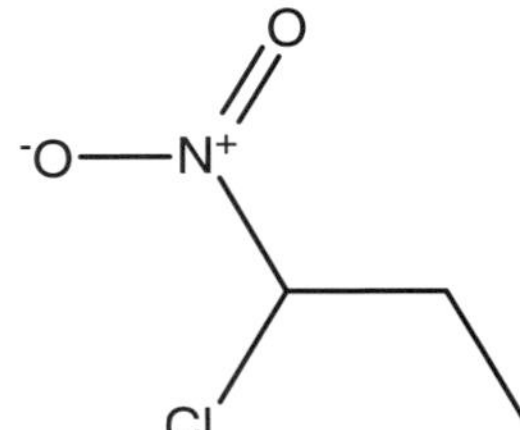

Chemical Structure

Synonyms: CHLORONITROPROPANE; KORAX; LANSTAN; PROPANE,1-CHLORO-1-NITRO-
Description: colorless liquid; unpleasant odor
Use: as a fungicide, in the synthetic rubber industry and as a component in rubber cements

Physical Properties

Boiling Point: 142 °C (288 °F)
Specific Gravity: 1.209 at 20/20 °C
Vapor Density: 4.3 Air=1
Saturated Vapor Density: 1.229854737 kg/m^3
Density: 1.209 g/cu cm at 20 °C
Vapor Pressure: 5.8 mm Hg at 25 °C
Water Solubility: 8000 mg/L
Other Solubilities: 95% Ethanol: >=100 mg/ml at 22 °C; Acetone: >=100 mg/ml at 22 °C; Alcohol: >10%; DMSO: >=100 mg/ml at 22 °C; Ether: >10%; Most organic solvents: miscible.
Refraction Index: 1.4251 at 20 °C/D
Ionization Potential (eV): 9.90

Flash Point: 62 °C Open Cup

RTECS Toxicity Data

Acute Oral: Rat LD_{Lo} Dose: 50 mg/kg. Mouse LD_{50} Dose: 510 mg/kg; Toxic Effects: Behavioral - General anesthetic; Behavioral - Somnolence (general depressed activity).
Acute Inhalation: Mouse LC_{50} Dose: 66 gm/m^3/3hr; Toxic Effects: Behavioral - General anesthetic; Behavioral - Somnolence (general depressed activity). Rabbit LC_{Lo} Dose: 2 gm/m^3/6hr; Toxic Effects: Behavioral - Somnolence (general depressed activity); Lungs, Thorax, or Respiration - Acute pulmonary edema; Lungs, Thorax, or Respiration - Other changes.
Acute Dermal: Mouse LD_{50} Route: Subcutaneous Dose: 165 mg/kg; Toxic Effects: Behavioral - General anesthetic; Behavioral - Somnolence (general depressed activity). Mouse LD; Route: Skin; Dose: >5000 mL/kg.
Mutagenic: Bacteria - S Typhimurium Mutations in Microorganisms; Dose: 333 ug/plate (+/-S9).

Hazard Overviews

Explosive

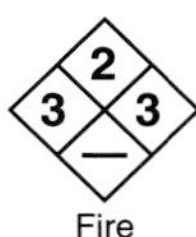

Fire Diamond

Health: Irritating to eyes/skin/respiratory tract. Also Causes: pulmonary edema; liver, kidney and cardiovascular damage (based on animals).
Fire: Combustible and explosive when heated. Use dry chemical, carbon dioxide, water spray, fog, or alcohol-resistant foam. Vapors may travel to an ignition source and flash back.
Reactivity: Unstable, capable of exploding when heated, especially when confined. Hazardous polymerization cannot occur. Avoid: heat; ignition sources. Incompatible with: acids; acid fumes; strong oxidizers. Hazardous decomposition products: hydrogen chloride; phosgene; carbon oxide(s); nitrogen oxide(s).
Carcinogenicity: IARC - Not listed; NIOSH - Listed as carcinogen; NTP - Not listed; ACGIH - Not listed; OSHA - Not listed; EPA - Not listed; MAK - Not listed
Primary Target Organs:

Eyes

Respiratory System

Exposure Limits
OSHA PEL: TWA: 20 ppm; 100 mg/m^3.
ACGIH TLV: TWA: 2 ppm; 10 mg/m^3.
NIOSH REL: TWA: 2 ppm; 10 mg/m^3.
NIOSH IDLH: 100 ppm.
Respirator Recommendation
Exposure Range: >20 to <100 ppm Air Purifying, Negative Pressure, Full Face
Exposure Range: 100 to unlimited ppm Self-contained Breathing Apparatus, Pressure Demand, Full Face
Cartridge Color: black

Environmental

Cleanup/Disposal: Guide No. 171: Do not touch or walk through spilled material. Stop leak if you can do it without risk. Prevent dust cloud. Avoid inhalation of asbestos dust. Small Dry Spills: With clean shovel place material into clean, dry container and cover loosely; move containers from spill area. Small Spills: Take up with sand or other noncombustible absorbent material and place into containers for later disposal. Large Spills: Dike far ahead of liquid spill for later disposal. Cover powder spill with plastic sheet or tarp to minimize spreading. Prevent entry into waterways, sewers, basements or confined areas.

Regulations

RCRA 40CFR: Not listed
CERCLA: 40CFR 302.4: Not listed
SARA 40CFR 372.65: Not listed
SARA EHS 40CFR 355: Not listed
TSCA: Not listed

Analytical Methods

Air: ASTM D4490

CHL5350 CAS #: 76-15-3

CHLOROPENTAFLUOROETHANE

RTECS: KH7877500
DOT: UN1020; IMO2.2
EINECS Number: 200-938-2
Molecular Formula: C_2ClF_5
Structured MF: $CClF_2CF_3$
Formula Weight: 154.47

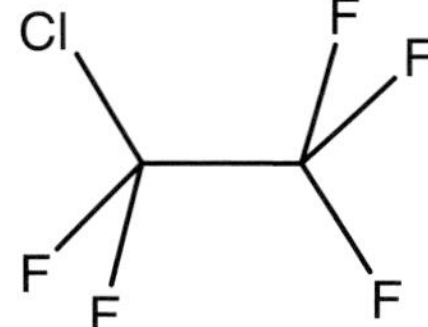

Chemical Structure

Synonyms: CHLOROPENTAFLUORETANO; CHLOROPENTAFLUORETHANE; ETHANE,CHLOROPENTAFLUORO-; F-115; FC 115; FLUOROCARBON-115; FLUOROCARBON 115; FREON 115; GENETRON 115; HALOCARBON 115; MONOCHLOROPENTAFLUOROETHANE; PROPELLANT 115; R 115; R115; REFRIGERANT 115
Description: colorless gas; odorless
Use: refrigerant; former use as propellant in aerosol food preparations; dielectric gas

Physical Properties

Boiling Point: -38 °C (-36 °F) at 760 mm Hg
Freezing Point: -106 °C (-158.8 °F)
Vapor Density: Saturated Vapor at Boiling Point 8.37 g/l
Density: 1.526 kg/L at -20 °C (liquid)
Vapor Pressure: 804.6 kPa
Water Solubility: Insoluble

Other Solubilities: Soluble in Alcohol, Ether
Surface Tension: 5 dyne/cm at 25 °C
Refraction Index: 1.2678
Critical Temperature: 80.0 °C
Critical Pressure: 31.6 bar
Ionization Potential (eV): 12.96
Flash Point: Nonflammable

RTECS Toxicity Data

Acute Inhalation: Rat LC; Dose: >20 pph/2hr.

Hazard Overviews

Compressed Gas

Fire Diamond

Health: Irritating to eyes/skin/respiratory tract. Also Causes: high concentration: confusion, pulmonary irritation, tremors, rarely coma, frost bite. Simple asphyxiant which can displace oxygen.
Fire: Noncombustible. Use agent suitable for surrounding fire.
Reactivity: Stable. Hazardous polymerization cannot occur. Avoid: heat and ignition sources, hot surfaces; contact with water. Incompatible with: powdered aluminum; heated, freshly-exposed aluminum surfaces. Hazardous decomposition products: chlorine; hydrogen fluoride; hydrogen chloride; carbonyl fluoride; phosgene.
Carcinogenicity: IARC - Not listed; NIOSH - Listed as carcinogen; NTP - Not listed; ACGIH - Not listed; OSHA - Not listed; EPA - Not listed; MAK - Not listed
Primary Target Organs:

Eyes

Skin

Respiratory System

Exposure Limits
OSHA PEL Vacated 1989 Limits: TWA: 1000 ppm; 6320 mg/m^3.
ACGIH TLV: TWA: 1000 ppm; 6320 mg/m^3.
NIOSH REL: TWA: 1000 ppm; 6320 mg/m^3.

Environmental

Environmental Fate: If released to the atmosphere, it will diffuse gradually into the stratosphere above the ozone layer where it will degrade slowly due to the photolysis. The stratospheric lifetime has been estimated to range from 230 to 550 years. Decomposition will not occur in tropospheric air. If released to soil, it is expected to have low mobility. It is not expected to degrade in the ambient atmosphere by reaction with photochemically produced hydroxyl radicals as structurally similar compounds have half-lives greater than 100 years for this reaction. Rapid volatilization should occur from terrestrial surfaces. If released to water, volatilization is expected to occur rapidly and be the dominant fate process. Bioconcentration in aquatic organisms is not expected to be important.

Cleanup/Disposal: Guide No. 126: Do not touch or walk through spilled material. Stop leak if you can do it without risk. Do not direct water at spill or source of leak. Use water spray to reduce vapors or divert vapor cloud drift. If possible, turn leaking containers so that gas escapes rather than liquid. Prevent entry into waterways, sewers, basements or confined areas. Allow substance to evaporate. Ventilate the area.

Environmental Physical Data

Henry's Law Constant: estimated at 3.0
Octanol/Water Partition Coefficient: log K_{ow} = 2.4
Sorption Partition Coefficient: K_{oc} = estimated at 708
BCF: not significant

Regulations

RCRA 40CFR: Not listed
CERCLA: 40CFR 302.4: Listed as Compound per CWA Section 307(a)
SARA 40CFR 372.65: Listed
SARA EHS 40CFR 355: Not listed
TSCA: Listed

CHL5380 CAS #: 3691-35-8

CHLOROPHACINONE

RTECS: NK5335000
EINECS Number: 223-003-0
Molecular Formula: $C_{23}H_{15}ClO_3$
Formula Weight: 374.83
Synonyms: ACTOSIN C; AFNOR; BARAAGE; CAID; CHLOORFACINON; 2(2-(4-CHLOOR-FENYL-2-FENYL)-ACETYL)-INDAAN-1,3-DION; CHLORFACINON; 2-(ALPHA-P-CHLOROPHENYLACETYL)INDANE-1,3-DIONE; 2(2-(4-CHLOROPHENYL)-2-PHENYLACETYL)INDAN-1,3-DIONE; 2-((P-CHLOROPHENYL)PHENYLACETYL)-1,3-INDANDIONE; 2-((4-CHLOROPHENYL)PHENYLACETYL)-1H-INDENE-1,3(2H)-DIONE; CHLORPHACINON; ((4-CHLORPHENYL)-1-PHENYL)-ACETYL-1,3-INDANDION; 1-(4-CHLORPHENYL)-1-PHENYL-ACETYL-INDAN-1,3-DION; 2(2-(4-CHLOR-PHENYL-2-PHENYL)ACETYL)INDAN-1,3-DION; 2(2-(4-CLORO-FENIL-2FENIL)-ACETIL)INDAN-1,3-DIONE; DRAT; LEPIT; LIPHADIONE; LM 91; MICROZUL; MURIOL; ORCOMOLEBAIT; 2-(2-PHENYL-2-(4-CHLOROPHENYL)ACETYL)-1,3-INDANDIONE; RATINDAN 3; RATOMET; REDENTIN; ROZOL; SAVIAC; TOPITOX

RTECS Toxicity Data

Acute Oral: Woman TD_{Lo} Dose: 5 mg/kg; Toxic Effects: Vascular - Other changes. Rat LD_{50} Dose: 2100 ug/kg.
Acute Inhalation: Rat LC_{50} Dose: >3 gm/m^3/1hr.
Acute Dermal: Rabbit LD_{50} Route: Skin; Dose: 200 mg/kg.

Hazard Overviews

Carcinogenicity: IARC - Not listed; NIOSH - Not listed; NTP - Not listed; ACGIH - Not listed; OSHA - Not listed; EPA - Not listed; MAK - Not listed

Environmental

Regulations

RCRA 40CFR: Not listed

CERCLA: 40CFR 302.4: Not listed
SARA 40CFR 372.65: Not listed TPQ: 100/10000 lb
SARA EHS 40CFR 355: Listed TPQ: 100 lb
TSCA: Not listed

Analytical Methods

Soil: EPA PMD-CJO

CHL5410 CAS #: 95-57-8

2-CHLOROPHENOL

RTECS: SK2625000
DOT: UN2020; UN2021; IMO6.1
EINECS Number: 202-433-2
Molecular Formula: C_6H_5ClO
Structured MF: ClC_6H_4OH
Formula Weight: 128.56

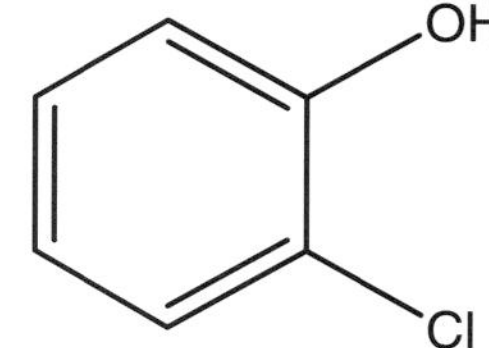

Chemical Structure

Synonyms: P-CHLORFENOL; 1-CHLORO-2-HYDROXYBENZENE; 2-CHLORO-1-HYDROXYBENZENE; O-CHLOROPHENOL; O-CHLORPHENOL; 2-HYDROXYCHLOROBENZENE; PHENOL,2-CHLORO-; PHENOL,O-CHLORO-; PINE-O DISINFECTANT; SEPTI-KLEEN
Description: light amber liquid; unpleasant, medicinal odor
Use: intermediate in the manufacture of higher chlorophenols and phenolic resins and for extracting sulfur and nitrogen compounds from coal; in organic synthesis (dyes)

Physical Properties

Boiling Point: 174.9 °C (347 °F) at 760 mm Hg
Freezing Point: 9.3 °C (48.74 °F)
Specific Gravity: 1.2634 at 20 °C/4 °C
Vapor Density: 4.5 Air=1
Saturated Vapor Density: Calculated 1.211925517 kg/m^3
Density: 1.265 g/mL at 15.5 °C
Vapor Pressure: Calculated 2.2 mm Hg at 20 °C
Water Solubility: Sparingly Soluble in Water
Other Solubilities: 95% Ethanol: >=100 mg/ml at 15 °C; Acetone: >=100 mg/ml at 15 °C; Alcohol: Soluble; Alkalies: Very Soluble; Aqueous sodium hydroxide: Soluble; Benzene: Soluble; Caustic alkali solutions: Soluble; DMSO: >=100 mg/ml at 15 °C; Ether: Soluble.
Surface Tension: 40.3 dynes/cm at 20 °C
Odor Threshold: Chemically pure gas 1.8 $x10^{-4}$ mg/m^3
pH: Weakly acid
Refraction Index: 1.5524 at 20 °C
Flash Point: 64 °C Closed Cup
LEL: 1.7% v/v

RTECS Toxicity Data

Acute Oral: Rat LD_{50} Dose: 670 mg/kg. Mouse LD_{50} Dose: 345 mg/kg; Toxic Effects: Behavioral - Tremor; Behavioral - Convulsions or effect on seizure threshold; Lungs, Thorax, or Respiration - Respiratory stimulation.
Acute Dermal: Rabbit LD_{Lo} Route: Subcutaneous Dose: 950 mg/kg. Rat LD_{50} Route: Subcutaneous Dose: 950 mg/kg.
Reproductive/Teratogenic: Rat Route: Oral; Dose: 4550 mg/kg; Duration: female 70D prior to mating Effects on Fertility - Litter size; Effects on Newborn - Stillbirth.
Mutagenic: Hamster Sex Chromosome Loss; Cell Type: lung; Dose: 800 umol/L.
Tumorigenic: Mouse Route: Skin; Dose: 4800 mg/kg/12W-I; Toxic Effects: Tumorigenic - Equivocal tumorigenic agent by RTECS criteria; Skin and appendages - Tumors.

Hazard Overviews

Corrosive

Fire Diamond

Health: Corrosive. Also Causes: weakness, headache, convulsions, respiratory or cardiac failure, CNS, liver, and kidney toxicity. Chronic Effects: digestive disturbances, nervous disorders, skin eruptions, jaundice, oliguria, uremia.
Fire: Combustible. Use dry chemical, carbon dioxide, water spray, fog, or regular foam. Container may explode in heat of fire.
Reactivity: Stable. Hazardous polymerization cannot occur. Avoid: heat; ignition sources. Incompatible with: strong oxidizers. Hazardous decomposition products: carbon oxide(s); hydrogen chloride; chlorine gas.
Carcinogenicity: IARC - Not listed; NIOSH - Not listed; NTP - Not listed; ACGIH - Not listed; OSHA - Not listed; EPA - Not listed; MAK - Not listed
Primary Target Organs:

Eyes

Skin
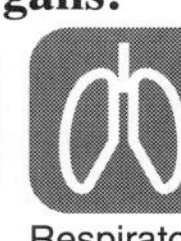
Respiratory System

Nervous System

Liver

Kidneys

Environmental

Ecotoxicity: LD_{50} Redwinged blackbird oral > 113 mg/kg LC_{50} Daphnia magna (Cladoceran) 2580 ug/l/96 hr /Conditions of bioassay not specified LC_{50} Lepomis machrochirus (bluegill) 6,590 ug/l/96 hr /Static, unmeasured boiassay
Environmental Fate: Since the pKa is 8.52 it will exist in water and moist soils in a partially dissociated state which may effect its transport and reactivity in the environment. If it is released to the soil it will be expected to show low to moderate adsorption to the soil based on estimated K_{oc}'s and may leach to the groundwater. Hydrolysis in soil can not be important. Biodegradation in soils may be important with loss of 94% reported for the compound incubated in non-sterile clay loam soil at 4 °C in 6.5 hr vs 1% loss in sterile soil in 12 days. If released to water it may adsorb to sediments. It will

not be expected to bioconcentrate in aquatic organisms and will not chemically hydrolyze. It will be susceptible to photolysis near the surface of waters and biodegradation should be an important fate process with complete removal reported in 13 and 36 days in die-away tests using 2 raw river waters, 15 days in 2 river waters seeded with water from previous die-away tests, and 15 days in acclimated river water. Evaporation from water may be an important transport process with a half-life of 3.3 days estimated for evaporation from a river 1 m deep, flowing at 1 m/sec with a wind velocity of 3 m/sec. If released to the atmosphere it may be susceptible to photolysis and reaction with NO in polluted air. The estimated vapor phase half-life in the atmosphere is 1.96 days mainly as a result of addition of ozone to the aromatic ring.

Cleanup/Disposal: Guide No. 153: Eliminate all ignition sources (no smoking, flares, sparks or flames in immediate area). Do not touch damaged containers or spilled material unless wearing appropriate protective clothing. Stop leak if you can do it without risk. Prevent entry into waterways, sewers, basements or confined areas. Absorb or cover with dry earth, sand or other non-combustible material and transfer to containers. Do not get water inside containers.

Environmental Physical Data

Henry's Law Constant: calculated at 1.3 x10^{-5}
Octanol/Water Partition Coefficient: log K_{ow} = 2.15
Sorption Partition Coefficient: K_{oc} = fine sediments 4890
BCF: bluegill 214

Regulations

RCRA 40CFR: Listed Hazardous Waste No. U048 Toxic Waste
CERCLA: 40CFR 302.4: Listed per RCRA Section 3001 per CWA Section 307(a) RQ: 100 lb (45.35 kg)
SARA 40CFR 372.65: Listed as Compound
SARA EHS 40CFR 355: Not listed
TSCA: Listed

Analytical Methods

Air: EPA 0020
Soil: CLP LC_SV, MC_SVOA, OHC; EPA 16, 1625, S-004-1; SW846 3630B, 3640A, 3650A, 3650B, 8040A, 8041, 8250A, 8270B, 8270C, 8275, 8410
Water / Groundwater: EPA 6, S-002-1, 1625, 604, 625, 625-S; APHA 6410-B, 6420-BA, 6420-BB, 6420-C; ASTM D2580; NCASI CP-86.01; USGS O3117
Drinking Water: EPA 552
Plasma: EPA 29
Other: EPA O-005-1

CHL5470 CAS #: 106-48-9

4-CHLOROPHENOL

RTECS: SK2800000
DOT: UN2020; UN2021; IMO6.1
EINECS Number: 203-402-6
Molecular Formula: C_6H_5ClO
Structured MF: ClC_6H_4OH
Formula Weight: 128.56

OH
Cl

Chemical Structure

Synonyms: APPLIED 3-78; P-CHLORFENOL; 4-CHLORO-1-HYDROXYBENZENE; P-CHLOROPHENOL; 4-HYDROXYCHLOROBENZENE; PARACHLOROPHENOL; PHENOL,4-CHLORO-; PHENOL,P-CHLORO-
Description: white, straw-colored to pink needle-like crystals; phenolic odor
Use: as an intermediate in synthesis of dyes and drugs, as a denaturant for alcohol, as a selective solvent in refining mineral oils, as a topical antiseptic and in the production of 2,4-DCP and the germicide 4-chlorophenol-o-cresol

Physical Properties

Boiling Point: 220 °C (428 °F)
Freezing Point: 43.2 °C (109.76 °F) to 43.7 °C (110.66 °F)
Specific Gravity: 1.2238 at 78 °C/4 °C
Vapor Density: 4.4 Air=1
Saturated Vapor Density: 1.200542069 kg/m^3
Density: 1.26 g/mL at 45 °C
Vapor Pressure: 0.10 mm Hg at 20 °C
Water Solubility: 2.71 parts Soluble in 100 parts Water at 20 °C
Other Solubilities: 95% Ethanol: >=100 mg/ml at 15 °C; Acetone: >=100 mg/ml at 15 °C; Aqueous sodium hydroxide: Soluble; Benzene: Soluble; Chloroform: Soluble; DMSO: >=100 mg/ml at 15 °C; Ether: Soluble; Fixed and volatile oils: Soluble; Glycerol: Soluble.
Odor Threshold: 30 ppm
pH: 1% solution is acid
Refraction Index: 1.5419 at 55 °C/D
Ionization Potential (eV): =< 8.69 +/-2.0
Flash Point: 121 °C Closed Cup

RTECS Toxicity Data

Acute Oral: Rat LD_{50} Dose: 670 mg/kg. Mouse LD_{50} Dose: 367 mg/kg; Toxic Effects: Behavioral - Excitment; Behavioral - Coma.
Acute Inhalation: Rat LC_{50} Dose: 11 mg/m^3. Human TC_{Lo} Dose: 10 gm/m^3/8hr; Toxic Effects: Behavioral - Excitment; Behavioral - Irritability; Nutritional and gross metabolic - Other changes.
Acute Dermal: Rat LD_{50} Route: Skin; Dose: 1500 mg/kg; Toxic Effects: Behavioral - Muscle contraction or spasticity; Skin and appendages - Corrosive. Rat LD_{50} Route: Subcutaneous Dose: 1030 mg/kg.
Chronic (Multiple Dose) Inhalation: Rat Dose: 2 mg/m^3/6H/17W-I; Toxic Effects: Nutritional and gross metabolic - Other changes. Mouse Dose: 24 ug/m^3/24H/17W-

I; Toxic Effects: Blood - Changes in serum composition; Biochemical - True cholinesterase; Biochemical - Transaminases.

Irritation Eye: Rabbit Standard Draize Test Dose: 250 ug/24H; Reaction: severe.

Irritation Skin: Rabbit Standard Draize Test Dose: 2 mg/24H; Reaction: severe.

Reproductive/Teratogenic: Mouse Route: Inhalation; Dose: 760 ug/m^3/24 Duration: male 17W prior to mating; Paternal Effects - Spermatogenesis; Effects on Fertility - Post-implantation mortality; Effects on Embryo or Fetus - Fetal death.

Mutagenic: Rat Cytogenetic Analysis; Route: Oral; Dose: 81 mg/kg. Bacteria - S Typhimurium Mutations in Microorganisms; Dose: 200 ug/plate (+S9). Bacteria - S Typhimurium Mutations in Microorganisms; Dose: 10 ug/plate (-S9).

Hazard Overviews

Poison

Corrosive

Fire Diamond

Health: Corrosive, causes severe irritation or burns to eyes/skin/respiratory tract. Poison. Also Causes: headache, weakness, dizziness, and damage to blood, liver, kidneys, heart. Chronic Effects: may cause cancer.

Fire: Combustible. Use water spray, dry chemical, carbon dioxide or regular foam.

Reactivity: Stable. Hazardous polymerization cannot occur. Avoid: heat; ignition sources. Incompatible with: oxidizers; organic acids; steam. Hazardous decomposition products: chlorine; hydrogen chloride gas.

Carcinogenicity: IARC - Not listed; NIOSH - Not listed; NTP - Not listed; ACGIH - Not listed; OSHA - Not listed; EPA - Not listed; MAK - Not listed

Primary Target Organs:

Eyes

Skin

Respiratory System

Nervous System

Liver

Kidneys

Environmental

Ecotoxicity: LC_{50} Lepomis macrochirus (bluegill) 3830 ug/l/96 hr /Static, unmeasured bioassay LC_{50} Cyprinodon variegatus (sheepshead minnow) 5350 ug/l/96 hr /Static, unmeasured bioassay EC_{50} Selenastrum capricornutum (algae) 4790 ug/l/96 hr, toxic effect: cell production /Conditions of bioassay not specified EC_{50} Skeletonema costatum (alga) 3270 ug/l/96 hr Toxic effect: chlorophyll /Conditions of bioassay not specified LC_{50} Daphnia magna (cladoceran) 4060 ug/l/96 hr /Static, unmeasured bioassay

Environmental Fate: Since the pKa is 9.41, it will exist in water and moist soils in a partially dissociated state which may effect its transport and reactivity in the environment. If released to the soil it will be expected to show low to moderate adsorption to the soil based on estimated K_{oc}'s and it may leach to the groundwater. Hydrolysis in soil will not be important. Biodegradation in soils may be important with loss of 84% reported for material incubated in non-sterile clay loam soil at 4 °C in 6.5 hours vs 0% loss in sterile soil in 12 days. If released to water it may adsorb to sediments based on experimental K_{oc}'s although estimated K_{oc}'s predict this adsorption will be low to moderate. It will not be expected to bioconcentrate in aquatic organisms and will not hydrolyze. It will be susceptible to photolysis near the surface of waters and biodegradation may be an important fate process with complete removal reported in 15 days and 13 days in unacclimated and acclimated river water, respectively. Half-lives of 20 days were reported in water, and 3 days in sediment and seawater, from the estuarine Skidway River, GA at 22 °C. Evaporation from water should not be an important transport process. If released to the atmosphere it may be susceptible to photolysis and reaction with NOX in polluted air. The estimated vapor phase half-life in the atmosphere is 1.96 days as a result of addition of O3 to the aromatic ring. Washout may be an important transport removal process.

Cleanup/Disposal: Guide No. 153: Eliminate all ignition sources (no smoking, flares, sparks or flames in immediate area). Do not touch damaged containers or spilled material unless wearing appropriate protective clothing. Stop leak if you can do it without risk. Prevent entry into waterways, sewers, basements or confined areas. Absorb or cover with dry earth, sand or other non-combustible material and transfer to containers. Do not get water inside containers.

Environmental Physical Data

Henry's Law Constant: calculated at 6.2×10^{-7}

Octanol/Water Partition Coefficient: log K_{ow} = 2.39

Sorption Partition Coefficient: K_{OC} = brookston clay loam soil 70

BCF: goldfish 15

Regulations

RCRA 40CFR: Not listed

CERCLA: 40CFR 302.4: Listed as Compound per CWA Section 307(a)

SARA 40CFR 372.65: Listed as Compound

SARA EHS 40CFR 355: Not listed

TSCA: Listed

Analytical Methods

Soil: SW846 8041, 8410

Water / Groundwater: EPA 1653; ASTM D2580, D4763; NCASI CP-86.01

Indoor / Expired Air: NIOSH 2014

CHL5620 **CAS #: 122-88-3**

P-CHLOROPHENOXYACETIC ACID

RTECS: AG0175000

EINECS Number: 204-581-3

Molecular Formula: $C_8H_7ClO_3$

Formula Weight: 186.60

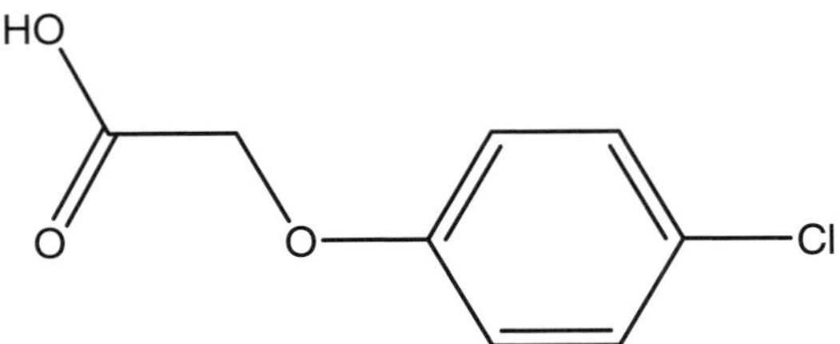

Chemical Structure

Synonyms: ACETIC ACID,(4-CHLOROPHENOXY)-; ACETIC ACID,(P-CHLOROPHENOXY)-; ACETIC ACID,(4-CHLOROPHENOXY)-(9CI); BI 12; (4-CHLOROPHENOXY)ACETIC ACID; 4-CP; 4-CPA; 4CPA; CPA; KYSELINA 4-CHLORFENOXYOCTOVA; MARKS 4-CPA; NSC-8769; NSC 8769; PARACHLOROPHENOXYACETIC ACID; PCPA; SURE-SET; TOMATO FIX; TOMATO FIX CONCENTRATE; TOMATO HOLD; TOMATOTONE

Description: prisms or needles

Use: plant growth regulator; fruit thinner, fruit ripener, bloom improver, fruit setter; a plant growth regulator used to improve tomato set

Physical Properties

Freezing Point: 156 °C (312.8 °F) to 157 °C (314.6 °F)

Water Solubility: Slightly Soluble in Water

Other Solubilities: Chloroform: Slightly soluble

RTECS Toxicity Data

Acute Oral: Rat LD_{50} Dose: 850 mg/kg.

Mutagenic: Human Unscheduled DNA Synthesis; Cell Type: other cell types; Dose: 10 mmol/L.

Hazard Overviews

Health: Irritating to eyes/skin/respiratory tract. Harmful. Other Acute Effects: harmful if swallowed; may be harmful if inhaled or absorbed through the skin.

Fire: Hazards: emits toxic fumes. Extinguishing agents: water spray; carbon dioxide, dry chemical powder or appropriate foam. Precautions: combustible liquid.

Reactivity: Incompatible with: strong oxidizing agents, strong bases. Hazardous decomposition products: toxic fumes of: carbon monoxide, carbon dioxide, hydrogen chloride gas.

Carcinogenicity: IARC - Not listed; NIOSH - Not listed; NTP - Not listed; ACGIH - Not listed; OSHA - Not listed; EPA - Not listed; MAK - Not listed

Primary Target Organs:

Eyes

Skin

Respiratory System

Environmental

Ecotoxicity: LD_{50} Salmo trutta (Sea trout) 147 ppm/24 hr /Conditions of bioassay not specified

Environmental Fate: If released to soil, it is expected to degrade under both aerobic and anaerobic conditions. It will display moderate to high mobility in soil although sorption may increase in highly acidic soils. It is not expected to significantly volatilize from soil. If released to water, it is expected to biodegrade under both aerobic and anaerobic conditions. It will be essentially completely ionized in water under environmentally significant pHs. Neither volatilization to the atmosphere nor bioconcentration in fish and aquatic organisms are expected to be significant process. It may undergo indirect photolytic degradation in the upper layers of natural water systems. If released to the atmosphere, it may undergo vapor-phase oxidation by photochemically produced hydroxyl radicals with an estimated half-life of 2.1 days. It may also undergo atmospheric removal by wet deposition processes.

Environmental Physical Data

Henry's Law Constant: 6.43×10^{-8}

Octanol/Water Partition Coefficient: log K_{ow} = 1.99

Sorption Partition Coefficient: K_{oc} = calculated at 166 to 288

BCF: calculated at 19 to 21

Regulations

RCRA 40CFR: Not listed

CERCLA: 40CFR 302.4: Listed as Compound per CWA Section 307(a)

SARA 40CFR 372.65: Listed as Compound

SARA EHS 40CFR 355: Not listed

TSCA: Listed

Analytical Methods

Plasma: FDA 221.1

CHL5710 CAS #: 80-33-1

P-CHLOROPHENYL P-CHLOROBENZENESULFONATE

RTECS: DB5250000

EINECS Number: 201-270-4

Molecular Formula: $C_{12}H_8Cl_2O_3S$

Formula Weight: 303.16

Chemical Structure

Synonyms: ACARICYDOL E 20; BENZENESULFONIC ACID,4-CHLORO-,4-CHLOROPHENYL ESTER; BENZENESULFONIC ACID,P-CHLORO-,P-CHLOROPHENYL ESTER; BENZOLSULFONAT; C 1,006; CCS; CHLOORFENSON; (4-CHLOOR-FENYL)-4-CHLOOR-BENZEEN-SULFONAAT; CHLORBENZOLSULFONAT; CHLOREFENIZON; CHLORFENSIN; CHLORFENSON; P-CHLORFENYLESTER KYSELINY P-CHLORBENZENSULFONOVE; 4-CHLOROBENZENESULFONATE DE 4-CHLOROPHENYLE; 4-CHLOROBENZENESULFONIC ACID,4-CHLOROPHENYL ESTER; P-CHLOROBENZENESULFONIC ACID,P-CHLOROPHENYL ESTER; CHLOROFENIZON; CHLOROFENSON; CHLOROFENSONE; 4-CHLOROPHENYL 4-CHLOROBENZENESULFONATE; 4-CHLOROPHENYL 4-CHLOROBENZENESULPHONATE; P-CHLOROPHENYL P-CHLOROBENZENESULPHONATE; 4-CHLORPHENYL-4'-CHLORBENZOLSULFONAT; (4-CHLOR-PHENYL)-4-CHLOR-BENZOL-SULFONATE; (4-CLORO-FENIL)-4-CLORO-VENZOL-SOLFONATO; COROTRAN; CPCBS; D 854; DANICUT; DIFENSON; DOW K-6,451; ENT 16,358; EPHIRSULPHONATE; ERYSIT; ESTER SULFONATE; ESTONMITE; ETHERSULFONATE; GENITE 883; K 6451; K-101; LETHALAIRE G-58; MITICIDE K-101; MITRAN; NIAGARATRAN; ONEX; ORTHOTRAN; OTRACID; OVATRAN; OVATRON; OVEX; OVOCHLOR; OVOTOX; OVOTRAN; PARACHLOROPHENYL PARACHLOROBENZENE SULFONATE; PARACHLOROPHENYL-PARACHLOROBENZENE-SULFONATE; PCPCBS; ROZTOCZOL FLUID; SAPPILAN; SAPPIRAN; TRICHLORFENSON

Description: white solid or crystals

Use: former use: miticide; acarid ovicide for cotton, fruit, nut, & ornamental crops; agent for control of powdery mildew

Physical Properties

Freezing Point: 86.5 °C (187.7 °F) to 86.8 °C (188.24 °F)
Vapor Pressure: Negligible at 25 °C
Water Solubility: Practically Insoluble in Water
Other Solubilities: readily Soluble in aromatic solvents; Soluble in 100 g solvent at 25 °C: 1 g in Ethyl Alcohol; 2 g in Kerosene.

RTECS Toxicity Data

Acute Oral: Rat LD_{50} Dose: 2 gm/kg. Mouse LD_{50} Dose: 1475 mg/kg.
Acute Dermal: Rat LD_{50} Route: Skin; Dose: >10 gm/kg.
Chronic (Multiple Dose) Oral: Rat Dose: 1008 mg/kg/12W-C; Toxic Effects: Liver - Changes in liver weight; Endocrine - Changes in thyroid weight; Biochemical - Hepatic microsomal mixed oxidase(dealkylation, hyroxylation,etc).
Tumorigenic: Mouse Route: Oral; Dose: 115 gm/kg/78W-I; Toxic Effects: Tumorigenic - Equivocal tumorigenic agent by RTECS criteria; Liver - Tumors; Blood - Tumors.

Hazard Overviews

Carcinogenicity: IARC - Not listed; NIOSH - Not listed; NTP - Not listed; ACGIH - Not listed; OSHA - Not listed; EPA - Not listed; MAK - Not listed

Environmental

Regulations

RCRA 40CFR: Not listed
CERCLA: 40CFR 302.4: Listed as Compound per CWA Section 307(a)
SARA 40CFR 372.65: Listed as Compound
SARA EHS 40CFR 355: Not listed
TSCA: Not listed

Analytical Methods

Soil: EPA PMD-TLC
Food: FDA 212.1, 212.2, 232.1, 232.4, 242.1
Plasma: EPA 001, 027, 028; FDA 211.1, 231.1, 252

CHL5740 CAS #: 104-12-1

P-CHLOROPHENYL ISOCYANATE

RTECS: NQ8575000
EINECS Number: 203-176-9
Molecular Formula: C_7H_4ClNO
Formula Weight: 153.57

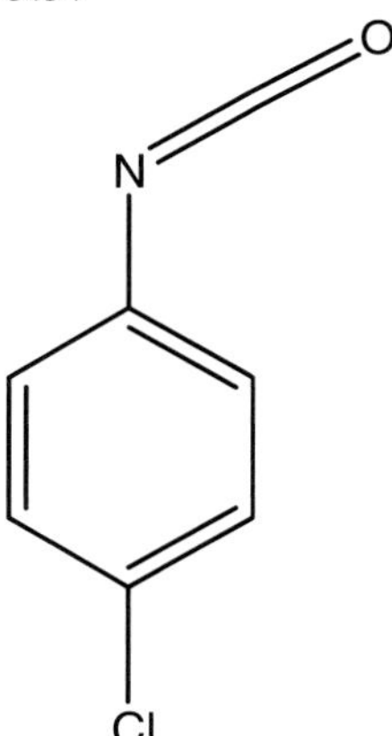

Chemical Structure

Synonyms: BENZENE,1-CHLORO-4-ISOCYANATO; BENZENE,1-CHLORO-4-ISOCYANATO-(9CI); P-CHLORFENYLISOKYANAT; 1-CHLORO-4-ISOCYANATOBENZENE; 4-CHLOROISOCYANATOBENZENE; 4-CHLOROPHENYL ISOCYANATE; PCPI
Description: white solid
Use: chem int for herbicides- difluron, buturon, & monolinuron

Physical Properties

Boiling Point: 115 °C (239 °F) to 117 °C (243 °F) at 45 mm Hg
Freezing Point: 30 °C (86 °F) to 31 °C (87.8 °F)
Saturated Vapor Density: 1.847040544 kg/m^3
Density: 1.25 g/cm at 40 °C
Vapor Pressure: 95.4 mm Hg at 20 °C
Other Solubilities: Soluble in organic solvents
Flash Point: 110 °C

RTECS Toxicity Data

Acute Oral: Mouse LD_{50} Dose: 450 mg/kg.
Acute Inhalation: Mouse LC_{50} Dose: 53 mg/m^3; Toxic Effects: Behavioral - Somnolence (general depressed activity); Behavioral - Excitment; Lungs, Thorax, or Respiration - Dyspnea. Man TC_{Lo} Dose: 800 ug/m^3/1M; Toxic Effects: Sense organs and special senses - Other; Lungs, Thorax, or Respiration - Other changes.
Chronic (Multiple Dose) Inhalation: Rat Dose: 100 ug/m^3/24H/80D-C; Toxic Effects: Liver - Other changes; Kidney, Ureter, and Bladder - Other changes in urine composition; Endocrine - Other changes. Rat Dose: 5 ug/m^3/24H/11W-C; Toxic Effects: Kidney, Ureter, and Bladder - Other changes in urine composition; Blood - Changes in serum composition; Biochemical - Other proteins.
Irritation Eye: Rabbit Standard Draize Test Dose: 250 ug/24H; Reaction: severe.
Irritation Skin: Rabbit Standard Draize Test Dose: 20 mg/24H; Reaction: moderate.

Hazard Overviews

Poison

Corrosive

Health: Corrosive to eyes/skin/respiratory tract. Poison. Other Acute Effects: may be fatal if inhaled, swallowed, or absorbed through skin; inhalation may result in spasm, inflammation and edema of the larynx and bronchi, chemical pneumonitis and pulmonary edema; symptoms of exposure may include burning sensation, coughing, wheezing, laryngitis, shortness of breath, headache, nausea and vomiting; repeated exposure may cause asthma; prolonged or repeated exposure may cause allergic reactions in certain sensitive individuals
Fire: Will burn. Hazards: emits toxic fumes. Extinguishing agents: carbon dioxide; dry chemical powder. Precautions: combustible liquid.
Reactivity: Incompatible with: heat, alcohols, strong bases, amines, acids, strong oxidizing agents, may decompose on exposure to moist air or water. Hazardous decomposition products: thermal decomposition may produce carbon monoxide, carbon dioxide, and nitrogen oxides; hydrogen cyanide, hydrogen chloride gas.
Carcinogenicity: IARC - Not listed; NIOSH - Not listed; NTP - Not listed; ACGIH - Not listed; OSHA - Not listed; EPA - Not listed; MAK - Not listed
Primary Target Organs:

Eyes

Skin

Respiratory System

Environmental

Regulations

RCRA 40CFR: Not listed
CERCLA: 40CFR 302.4: Listed as Compound per CWA Section 307(a) per CAA Section 112
SARA 40CFR 372.65: Listed
SARA EHS 40CFR 355: Not listed
TSCA: Listed

CHL5800 CAS #: 7005-72-3

4-CHLOROPHENYL PHENYL ETHER

EINECS Number: 230-281-7
Molecular Formula: $C_{12}H_9ClO$
Formula Weight: 204.65

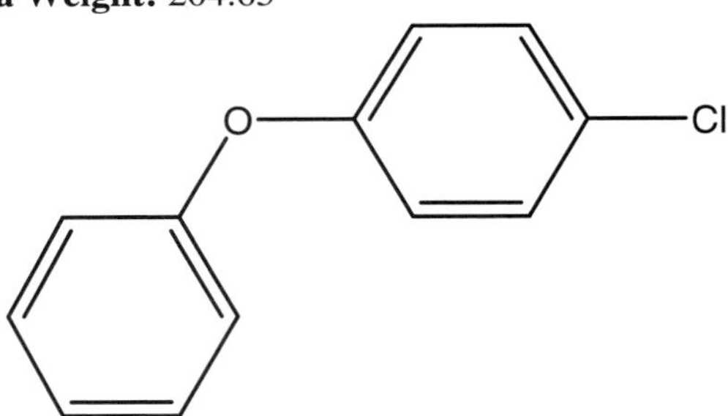

Chemical Structure

Synonyms: BENZENE,1-CHLORO-4-PHENOXY-; 4-CHLORODIPHENYL ETHER; CHLORODIPHENYL ETHER; P-CHLORODIPHENYL ETHER; 1-CHLORO-4-PHENOXYBENZENE; 4-CHLOROPHENOXYBENZENE; P-CHLOROPHENOXYBENZENE; P-CHLOROPHENYL PHENYL ETHER; 4-CHLOROPHENYLPHENYL ETHER; 4-CHLOROPHENYLPHENYLETHER; ETHER,P-CHLOROPHENYL PHENYL; 4-MONOCHLORODIPHENYL OXIDE; MONOCHLORODIPHENYL OXIDE
Description: liquid
Use: dielectric fluid

Physical Properties

Boiling Point: 284 °C (543 °F) to 285 °C (545 °F)
Freezing Point: -8 °C (17.6 °F)

Specific Gravity: 1.2026 at 15 °C
Saturated Vapor Density: 1.200025822 kg/m^3
Vapor Pressure: 0.0027 torr at 25 °C
Water Solubility: 3.3 mg/L at 25 °C
Refraction Index: 1.599
Flash Point: Unknown; combustible

Hazard Overviews

Health: Irritating to eyes/skin. Other Acute Effects: may be harmful by inhalation, ingestion, or skin absorption.
Fire: Combustible. Hazards: emits toxic fumes. Extinguishing agents: water spray; carbon dioxide, dry chemical powder or appropriate foam. Precautions: combustible liquid.
Reactivity: Incompatible with: strong oxidizing agents. Hazardous decomposition products: toxic fumes of: carbon monoxide, carbon dioxide, hydrogen chloride gas.
Carcinogenicity: IARC - Not listed; NIOSH - Not listed; NTP - Not listed; ACGIH - Not listed; OSHA - Not listed; EPA - Not listed; MAK - Not listed
Primary Target Organs:

Eyes

Skin

Environmental

Environmental Fate: If released to the atmosphere, it should react with photochemically produced hydroxyl radicals with an estimated half-life of 1.3 days. Direct photolysis in the atmosphere should be an important fate process, as it has an absorption greater than 290 nm. It should be expected to undergo biodegradation in soil and in water. It should display slight mobility in soil, and volatilization to the atmosphere may be an important process. If released to water, it would be expected to adsorb to sediment and suspended material, can volatilize to the atmosphere, and should bioaccumulate in aquatic organisms. Degradation by direct photolysis in surface water has been estimated to proceed with a half-life of 200-400 days. Volatilization from water to the atmosphere should be an important fate process. The estimated volatilization half-life for a model river is 6 hours, while from a model pond which takes into account adsorption processes, the estimated half-life is 40 days.

Environmental Physical Data

Henry's Law Constant: 8.4 x10^{-4}
Octanol/Water Partition Coefficient: log K_{ow} = 4.08
Sorption Partition Coefficient: K_{oc} = 2260 to 3950
BCF: rainbow trout 736

Regulations

RCRA 40CFR: Not listed
CERCLA: 40CFR 302.4: Listed per CWA Section 307(a) RQ: 5000 lb (2268 kg)
SARA 40CFR 372.65: Listed as Compound
SARA EHS 40CFR 355: Not listed
TSCA: Listed

Analytical Methods

Soil: CLP MC_SVOA, OHC; EPA 16, 1625; SW846 3640A, 8110, 8250A, 8270B, 8270C, 8410
Water / Groundwater: EPA 1625, 611, 625, 625-S, 6; APHA 6040-B, 6410-B; USGS O3118
Indoor / Expired Air: NIOSH 5025
Plasma: EPA 29

CHL6040 CAS #: 95-83-0

4-CHLORO-O-PHENYLENEDIAMINE

RTECS: SS8850000
EINECS Number: 202-456-8
Molecular Formula: $C_6H_7ClN_2$
Structured MF: $ClC_6H_3(NH_2)_2$
Formula Weight: 142.59

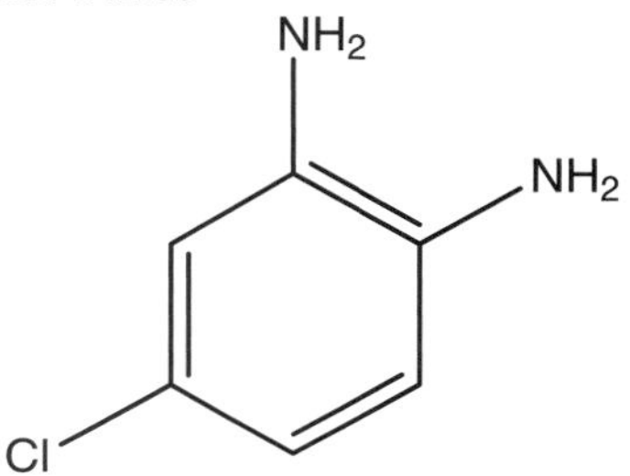

Chemical Structure

Synonyms: 2-AMINO-4-CHLOROANILINE; 1,2-BENZENEDIAMINE,4-CHLORO-(9CI); C.I. 76015; 4-CHLORO-1,2-BENZENEDIAMINE; 1-CHLORO-3,4-DIAMINOBENZENE; 4-CHLORO-1,2-DIAMINOBENZENE; 4-CHLORO-1,2-PHENYLENEDIAMINE; P-CHLORO-1,2-PHENYLENEDIAMINE; P-CHLORO-O-PHENYLENEDIAMINE; 4-CL-O-PD; 1,2-DIAMINO-4-CHLOROBENZENE; 1,2-DIAMINO-4-CHLOROBENZENE; 3,4-DIAMINO-1-CHLOROBENZENE; 3,4-DIAMINOCHLOROBENZENE; URSOL OLIVE 6G
Description: crystalline powder
Use: found in cosmetics and hair dyes; a dye intermediate

Physical Properties

Freezing Point: 76 °C (168.8 °F)
Water Solubility: < 1 mg/mL at 19 C
Other Solubilities: 95% Ethanol: >=100 mg/ml at 19 °C; Acetone: >=100 mg/ml at 19 °C; Benzene: Soluble; DMSO: >=100 mg/ml at 19 °C; Ether: Soluble; Ligroin: Soluble; Petroleum Ether: Soluble; Toluene: 50-100 mg/ml at 23 °C.
Flash Point: Not available; probably combustible

RTECS Toxicity Data

Mutagenic: Human DNA Damage; Cell Type: fibroblast; Dose: 50 umol/L. Rat Unscheduled DNA Synthesis; Cell Type: liver; Dose: 1 umol/L.
Tumorigenic: Rat Route: Oral; Dose: 135 gm/kg/77W-C; Toxic Effects: Tumorigenic - Carcinogenic by RTECS criteria; Liver - Tumors; Kidney, Ureter, and Bladder - Tumors. Rat Route: Oral; Dose: 273 gm/kg/78W-C; Toxic Effects: Tumorigenic - Carcinogenic by RTECS criteria;

Gastrointestinal - Tumors; Kidney, Ureter, and Bladder - Tumors. Rat Route: Oral; Dose: 136 gm/kg/78W-C; Toxic Effects: Tumorigenic - Carcinogenic by RTECS criteria; Gastrointestinal - Tumors; Kidney, Ureter, and Bladder - Tumors. Rat Route: Oral; Dose: 18750 gm/kg/2Y-C; Toxic Effects: Tumorigenic - Carcinogenic by RTECS criteria; Liver - Tumors; Kidney, Ureter, and Bladder - Tumors.

Hazard Overviews

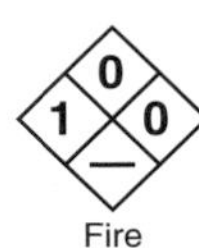

Fire Diamond

Health: Irritating to eyes/respiratory tract. Chronic Effects: possible cancer of the bladder and liver.
Fire: Noncombustible. Use agent suitable for surrounding fire.
Reactivity: Stable. Hazardous polymerization cannot occur. Avoid: heat; ignition sources. Hazardous decomposition products: carbon oxides; nitrogen oxides; chloride.
Carcinogenicity: IARC - Group 2B, Possibly carcinogenic to humans; NIOSH - Not listed; NTP - Class 2B, Reasonably anticipated to be a carcinogen, sufficient evidence of carcinogenicity from studies in experimental animals; ACGIH - Not listed; OSHA - Not listed; EPA - Not listed; MAK - Not listed
Primary Target Organs:

Eyes

Mucous Membranes

Environmental

Environmental Physical Data

Octanol/Water Partition Coefficient: log K_{ow} = 1.28

Regulations

RCRA 40CFR: Not listed
CERCLA: 40CFR 302.4: Listed as Compound per CWA Section 307(a)
SARA 40CFR 372.65: Listed as Compound
SARA EHS 40CFR 355: Not listed
TSCA: Listed

Analytical Methods

Soil: SW846 8270C

CHL6250 CAS #: 5344-82-1

1-(O-CHLOROPHENYL)THIOUREA

RTECS: YS7100000
EINECS Number: 226-291-6
Molecular Formula: $C_7H_7ClN_2S$
Formula Weight: 186.66

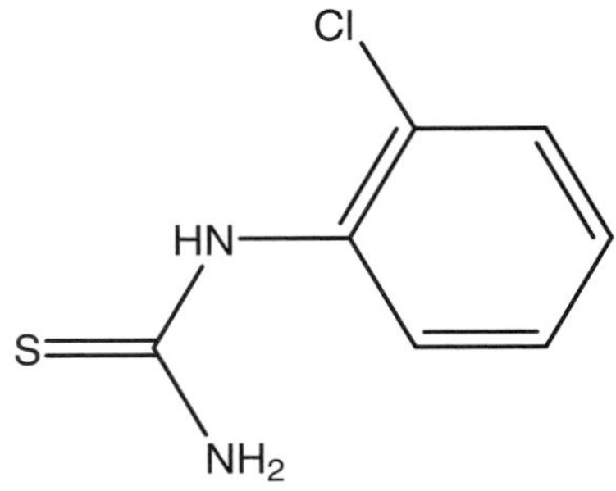

Chemical Structure

Synonyms: 2-CHLOROPHENYL THIOUREA; (O-CHLOROPHENYL)THIOUREA; 1-(2-CHLOROPHENYL)-2-THIOUREA; 1-(2-CHLOROPHENYL)THIOUREA; 2-CHLOROPHENYLTHIOUREA; THIOUREA,(2-CHLOROPHENYL)-; THIOUREA,1-(2-CHLOROPHENYL); UREA,1-(O-CHLOROPHENYL)-2-THIO-
Description: needles or plates
Use: not manufactured or used as end-product industrially in the USA

Physical Properties

Freezing Point: 146 °C (294.8 °F)
Water Solubility: Very Soluble in Water
Other Solubilities: Soluble in Alcohol, Benzene.

RTECS Toxicity Data

Acute Oral: Rat LD_{50} Dose: 4600 ug/kg.

Hazard Overviews

Poison

Health: Irritating. Poison. Other Acute Effects: may be fatal if inhaled, swallowed, or absorbed through skin.
Fire: Hazards: emits toxic fumes. Extinguishing agents: water spray; carbon dioxide, dry chemical powder or appropriate foam. Precautions: combustible liquid.
Reactivity: Incompatible with: strong oxidizing agents, strong acids, strong bases. Hazardous decomposition products: toxic fumes of: carbon monoxide, carbon dioxide, nitrogen oxides, sulfur oxides, hydrogen chloride gas.
Carcinogenicity: IARC - Not listed; NIOSH - Not listed; NTP - Not listed; ACGIH - Not listed; OSHA - Not listed; EPA - Not listed; MAK - Not listed
Primary Target Organs:

Eyes

Skin

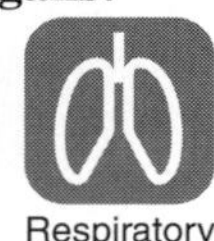
Respiratory System

Environmental

Environmental Fate: No experimental data are available to directly predict environmental fate. It absorbs sunlight relatively weakly which may suggest a potential for direct photolysis. Additional estimations of environmental fate are based on its chemical structure and physical properties. If released to the atmosphere, it should degrade rapidly in the vapor-phase (half-life of 4.3 hr estimated from chemical structure) by reaction with photochemically produced hydroxyl radicals. It may additionally exist in air in the adsorbed-particulate phase. If released to soil or water, covalent bonding to humic materials may be important. The covalent bonding process may represent a mechanism by which it may be converted to a latent form in the biosphere. If bonding does not occur, it should leach through most soils and sorption to sediments in water will not be important. No data are available regarding biodegradation in soil or water.

Environmental Physical Data

Henry's Law Constant: estimated at 1 $x10^{-7}$
Octanol/Water Partition Coefficient: log K_{ow} = 1.20
Sorption Partition Coefficient: K_{oc} = 11 to 107
BCF: estimated at 4.8

Regulations

RCRA 40CFR: Listed Hazardous Waste No. P026 Toxic Waste
CERCLA: 40CFR 302.4: Listed per RCRA Section 3001 RQ: 100 lb (45.35 kg)
SARA 40CFR 372.65: Listed as Compound TPQ: 100/10000 lb
SARA EHS 40CFR 355: Listed TPQ: 100 lb
TSCA: Listed

Analytical Methods

Soil: SW846 3640A
Water / Groundwater: EPA 553; SW846 8325
Drinking Water: EPA 553

CHL6280 CAS #: 2227-13-6

P-CHLOROPHENYL-2,4,5-TRICHLOROPHENYL SULFIDE

RTECS: WQ3850000
EINECS Number: 218-761-4
Molecular Formula: $C_{12}H_6Cl_4S$
Formula Weight: 324.04

Chemical Structure

Synonyms: ANIMERT; ANIMERT V-10; ANIMERT V-101; ANIMERT V-10K; BENZENE,1,2,4-TRICHLORO-5-((4-CHLOROPHENYL)THIO)-; 4-CHLOROPHENYL 2,4,5-TRICHLOROPHENYL SULFIDE; P-CHLOROPHENYL 2,4,5-TRICHLOROPHENYL SULFIDE; ENT 27,115; PHILIPS-DUPHAR V-101; 3,4,6,4'-TETRACHLOR-DIPHENYLSULFID; 2,4,4',5-TETRACHLORODIPHENYL SULFIDE; 2,4,5,4'-TETRACHLORODIPHENYL SULFIDE; TETRASUL; V 101; V-101

RTECS Toxicity Data

Acute Oral: Rat LD_{50} Dose: 3960 mg/kg. Mouse LD_{50} Dose: 5010 mg/kg.
Acute Dermal: Rabbit LD_{50} Route: Skin; Dose: 2 gm/kg. Guinea Pig LD_{50} Route: Skin; Dose: 8200 mg/kg.
Reproductive/Teratogenic: Rat Route: Oral; Dose: 18 gm/kg; Duration: male 17W prior to mating; Paternal Effects - Testes, epididymis, sperm duct; Effects on Fertility - Male fertility index. Rat Route: Oral; Dose: 18 gm/kg; Duration: female 17W prior to mating Maternal Effects - Ovaries, fallopian tubes; Effects on Fertility - Female fertility index.

Hazard Overviews

Carcinogenicity: IARC - Not listed; NIOSH - Not listed; NTP - Not listed; ACGIH - Not listed; OSHA - Not listed; EPA - Not listed; MAK - Not listed

Environmental

Regulations

RCRA 40CFR: Not listed
CERCLA: 40CFR 302.4: Listed as Compound per CWA Section 307(a)
SARA 40CFR 372.65: Listed as Compound
SARA EHS 40CFR 355: Not listed
TSCA: Not listed

Analytical Methods

Food: FDA 212.1, 232.1, 232.4, 242.1; AOAC 976.23
Plasma: FDA 211.1, 231.1, 252

CHL6340 CAS #: 76-06-2

CHLOROPICRIN

RTECS: PB6300000
DOT: UN1580; IMO6.1
EINECS Number: 200-930-9
Molecular Formula: CCl_3NO_2
Structured MF: CCl_3NO_2
Formula Weight: 164.39

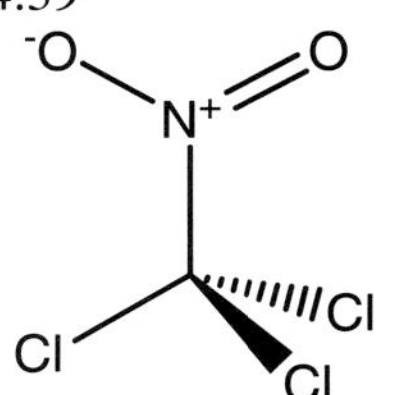

Chemical Structure

Synonyms: ACQUINITE; CHLOORPIKRINE; CHLOROFORM,NITRO-; CHLOROPICRINE; CHLOROPICRIN,LIQUID; CHLOR-O-PIC; CHLORPICRIN; CHLORPIKRIN; CLOROPICRINA; DOJYOPICRIN; DOLOCHLOR; G 25; KLOP; LARVACIDE; LARVACIDE 100; METHANE,TRICHLORONITRO-; MICROLYSIN; NITROCHLOROFORM; NITROTRICHLORO METHANE; NITROTRICHLOROMETHANE; OG 25; PIC-CLOR; PICFUME; PICRIDE; PROFUME A; PS; S 1; TRICHLOORNITROMETHAAN; TRICHLORNITROMETHAN; TRICHLORONITROMETHANE; TRI-CLOR; TRICLORO-NITRO-METANO

Description: colorless liquid; penetrating odor

Use: organic synthesis; dye-stuffs (crystal violet); fumigant, fungicides, insecticides, rat exterminator, tear gas and disinfecting cereals and grains

Physical Properties

Boiling Point: 112 °C (234 °F) at 757 mm Hg

Freezing Point: -64 °C (-83.2 °F)

Specific Gravity: 1.6558 at 20 °C/4 °C

Vapor Density: 5.7 Air=1

Saturated Vapor Density: 1.376974368 kg/m^3

Vapor Pressure: 3.2 kPa at 25 °C

Water Solubility: 0.2272 g/100 ml at 0 °C

Other Solubilities: 95% Ethanol: >=100 mg/ml at 22 °C; Acetone: >=100 mg/ml at 22 °C; Benzene: Soluble; Carbon Disulfide: Soluble; Carbon Tetrachloride: miscible; DMSO: >=100 mg/ml at 22 °C; Ether: Soluble; Methanol: miscible.

Surface Tension: 32.3 dynes/cm at 20 °C

Odor Threshold: Faint Odor at 0.0073 mg/l

Refraction Index: 1.4611 at 20 °C/D

Flash Point: Nonflammable

RTECS Toxicity Data

Acute Oral: Rat LD_{50} Dose: 250 mg/kg.

Acute Inhalation: Rat LC_{50} Dose: 14400 ppb/4hr; Toxic Effects: Lungs, Thorax, or Respiration - Other changes. Human LC_{Lo} Dose: 2000 mg/m^3/10M. Human TC_{Lo} Dose: 2 mg/m^3; Toxic Effects: Sense organs and special senses - Lacrimation; Sense organs and special senses - Conjunctive irritation; Lungs, Thorax, or Respiration - Other changes.

Chronic (Multiple Dose) Oral: Rat Dose: 400 mg/kg/10D-I; Toxic Effects: Endocrine - Changes in thymus weight; Blood - Changes in leukocyte (WBC) cell count; Nutritional and gross metabolic - Weight loss or decreased weight gain. Rat Dose: 2880 mg/kg/90D-I; Toxic Effects: Gastrointestinal - Other changes; Blood - Changes in serum composition; Nutritional and gross metabolic - Weight loss or decreased weight gain.

Mutagenic: Human Sister Chromatid Exchenge; Cell Type: lymphocyte; Dose: 8 mg/L. Bacteria - E Coli Mutations in Microorganisms; Dose: 50 ug/plate (+S9), 100 ug/plate (-S9). Bacteria - S Typhimurium Mutations in Microorganisms; Dose: 50 ug/plate (-S9).

Tumorigenic: Mouse Route: Oral; Dose: 26 gm/kg/78W-I; Toxic Effects: Tumorigenic - Equivocal tumorigenic agent by RTECS criteria; Gastrointestinal - Tumors.

Hazard Overviews

Poison

Explosive

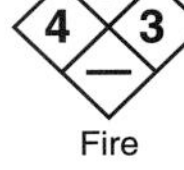

Fire Diamond

Health: Poison. Severely irritating to eyes/skin/respiratory tract. Also Causes: lacrimation, bronchitis, pulmonary edema, nausea, colic, diarrhea, GI irritation.

Fire: Noncombustible, but can be explosive when exposed to shock or heat. Use agent suitable for surrounding fire.

Reactivity: Stable. Hazardous polymerization cannot occur. Incompatible with: mixtures with 3-bromopropyne (shock and heat sensitive explosive); strong oxidizers; alcoholic sodium hydroxide; propargyl bromide; sodium methoxide; aniline/heat. Hazardous decomposition products: very toxic fumes of chlorine and nitrogen oxides.

Carcinogenicity: IARC - Not listed; NIOSH - Not listed; NTP - Not listed; ACGIH - Class A4, Not classifiable as a human carcinogen; OSHA - Not listed; EPA - Not listed; MAK - Not listed

Primary Target Organs:

Eyes

Skin

Respiratory System

Gastro-intestinal

Exposure Limits

OSHA PEL: TWA: 0.1 ppm; 0.7 mg/m^3.

ACGIH TLV: TWA: 0.1 ppm; 0.67 mg/m^3.

NIOSH REL: TWA: 0.1 ppm; 0.7 mg/m^3.

NIOSH IDLH: 2 ppm.

DFG MAK: TWA: 0.1 ppm; 0.7 mg/m^3.

Respirator Recommendation

Exposure Range: >0.1 to 1 ppm Air Purifying, Negative Pressure, Half Mask

Exposure Range: >1 to <2 ppm Air Purifying, Negative Pressure, Full Face

Exposure Range: 2 to unlimited ppm Self-contained Breathing Apparatus, Pressure Demand, Full Face

Cartridge Color: black

Environmental

Environmental Fate: If applied to soil as would be the case in its use as a soil sterilant, it will both rapidly volatilize and leach. It should photolyze on the soils surface. It may degrade in soil by chemical or biological processes. However degradation rates are unknown. It has a high Henry's Law Constant and if released in water would readily volatilize (half-life in a model river and model lake are 4.3 hr and 5.2 days respectively). It will photodegrade in the surface layers of water (half-life about 3 days). Its rate of biodegradation in natural water is unknown, as is its rate of other abiotic dechlorination reactions. It would not be expected to adsorb to sediment or bioconcentrate in fish. If released to the atmosphere, it will photolyze (half-life 20 days), producing phosgene and nitrosyl chloride. Being relatively soluble in water, it may be washed out by rain.

Cleanup/Disposal: Guide No. 154: Eliminate all ignition sources (no smoking, flares, sparks or flames in immediate area). Do not touch damaged containers or spilled material unless wearing appropriate protective clothing. Stop leak if you can do it without risk. Prevent entry into waterways, sewers, basements or confined areas. Absorb or cover with dry earth, sand or other non-combustible material and transfer to containers. Do not get water inside containers.

Environmental Physical Data

Henry's Law Constant: 2.05 $x10^{-3}$
Octanol/Water Partition Coefficient: log K_{ow} = 2.09
Sorption Partition Coefficient: K_{oc} = soils 81
BCF: calculated at 23

Regulations

RCRA 40CFR: Not listed
CERCLA: 40CFR 302.4: Not listed
SARA 40CFR 372.65: Listed
SARA EHS 40CFR 355: Not listed
TSCA: Listed

Analytical Methods

Air: ASTM D4490
Soil: EPA PMD-CKA
Water / Groundwater: EPA 618
Drinking Water: EPA 551

CHL6370 CAS #: 126-99-8

B-CHLOROPRENE

RTECS: EI9625000
DOT: UN1991; IMO3.2
EINECS Number: 204-818-0
Molecular Formula: C_4H_5Cl
Structured MF: $H_2C=CHCCl=CH_2$
Formula Weight: 88.54

Cl

Chemical Structure

Synonyms: 1,3-BUTADIENE,2-CHLORO-; 2-CHLOOR-1,3-BUTADIEEN; 2-CHLOR-1,3-BUTADIEN; 2-CHLORO-1,3-BUTADIENE; 2-CHLOROBUTA-1,3-DIENE; 2-CHLOROBUTADIENE; 2-CHLOROBUTADIENE-1,3; CHLOROBUTADIENE; CHLOROPREEN; CHLOROPREN; BETA-CHLOROPRENE; CHLOROPRENE; CHLOROPRENE,INHIBITED; 2-CLORO-1,3-BUTADIENE; CLOROPRENE; NEOPRENE
Description: colorless liquid; pungent etheral odor
Use: manufacture of artificial rubber (neoprene and duprene) and as a component of adhesives used in food packaging

Physical Properties

Boiling Point: 59.4 °C (139 °F)
Freezing Point: -130 °C (-202 °F)
Specific Gravity: 0.9583 at 20 °C/4 °C
Vapor Density: 3 Air=1
Density: 0.9583 g/cc
Vapor Pressure: 118 mm Hg at 10 °C
Water Solubility: Slight
Other Solubilities: 95% Ethanol: >10%; Acetone: Soluble; Alcohol: Soluble; Benzene: Soluble; Ether: Soluble; Most organic solvents: Soluble.
Odor Threshold: Recognition 0.40 mg/m^3
Refraction Index: 1.4583 at 20 °C/D
Ionization Potential (eV): 8.79
Flash Point: -20 °C Open Cup
Autoignition Temperature: -16 °C
LEL: 4.0% v/v
UEL: 20.0% v/v

RTECS Toxicity Data

Acute Oral: Rat LD_{50} Dose: 450 mg/kg. Mouse LD_{50} Dose: 146 mg/kg.
Acute Inhalation: Rat LC_{50} Dose: 11800 mg/m^3/4hr. Mouse LC_{50} Dose: 2300 mg/m^3.
Acute Dermal: Rat LD_{Lo} Route: Subcutaneous Dose: 500 mg/kg; Toxic Effects: Behavioral - Convulsions or effect on seizure threshold; Lungs, Thorax, or Respiration - Dyspnea; Lungs, Thorax, or Respiration - Cyanosis. Mouse LD_{Lo} Route: Subcutaneous Dose: 1 gm/kg; Toxic Effects: Behavioral - Convulsions or effect on seizure threshold; Lungs, Thorax, or Respiration - Dyspnea; Lungs, Thorax, or Respiration - Cyanosis.
Chronic (Multiple Dose) Oral: Rat Dose: 9100 ug/kg/26W-I; Toxic Effects: Liver - Changes in liver weight; Endocrine - Changes in spleen weight; Biochemical - Dehydrogenases. Rat Dose: 1680 mg/kg/3W-I; Toxic Effects: Liver - Hepatitis (hepatocellular necrosis), diffuse; Liver - Other changes; Biochemical - Hepatic microsomal mixed oxidase(dealkylation, hyroxylation,etc).
Chronic (Multiple Dose) Inhalation: Rat Dose: 161 ppm/6H/4W-I; Toxic Effects: Liver - Changes in Liver weight; Kidney, Ureter, and Bladder - Changes in kidney weight; Nutritional and gross metabolic - Weight loss or decreased weight gain. Rat Dose: 220 ug/m^3/24H/60D-C; Toxic Effects: Brain and coverings - Other degenerative changes; Liver - Other changes; Biochemical - Phosphatases. Rat Dose: 200 mg/m^3/24H/91D-C; Toxic Effects: Brain and coverings - Other degenerative changes; Gastrointestinal - Ulceration or bleeding from large intestine; Blood - Changes in spleen.
Chronic (Multiple Dose) Dermal: Mouse Route: Subcutaneous; Dose: 42 mg/kg/21D-I; Toxic Effects: Immunological including allergic - Decreased immune response; Biochemical - Other proteins.
Reproductive/Teratogenic: Rat Route: Oral; Dose: 9100 ug/kg; Duration: male 26W prior to mating; Paternal Effects - Spermatogenesis. Rat Route: Oral; Dose: 1 mg/kg; Duration: female 11-12D of pregnancy Specific Developmental Abnormalities - Central nervous system. Rat Route: Oral; Dose: 1 mg/kg; Duration: female 11-12D of pregnancy

Effects on Embryo or Fetus - Fetotoxicity. Rat Route: Oral; Dose: 1 mg/kg; Duration: female 9-10D of pregnancy; Specific Developmental Abnormalities - Other developmental abnormalities. Rat Route: Inhalation; Dose: 4 mg/m^3/24H; Duration: female 3-4D of pregnancy; Effects on Embryo or Fetus - Fetal death. Rat Route: Inhalation; Dose: 4 mg/m^3/24H; Duration: female 11-12D of pregnancy Specific Developmental Abnormalities - Central nervous system. Rat Route: Inhalation; Dose: 10 ppm/4H; Duration: female 3-20D of pregnancy; Effects on Fertility - Post-implantation mortality. Rat Route: Inhalation; Dose: 150 ug/m^3/24 Duration: male 19W prior to mating; Paternal Effects - Spermatogenesis. Rat Route: Inhalation; Dose: 500 mg/m^3/5H Duration: female 17W prior to mating Maternal Effects - Menstrual cycle changes or disorders. Rat Route: Inhalation; Dose: 500 mg/m^3/5H Duration: female 30W prior to mating Maternal Effects - Ovaries, fallopian tubes.

Mutagenic: Human Cytogenetic Analysis; Route: Unreported; Dose: 1 mg/m^3. Rat Cytogenetic Analysis; Route: Inhalation; Dose: 1960 ug/m^3/16W.

Hazard Overviews

Flammable

Fire Diamond

Health: Irritating to eyes/skin/respiratory tract. Also Causes:injury to liver, and kidneys, central nervous system depression, drop in blood pressure, skin and mucous membrane irritation, anesthesia, corneal ulcers, corneal necrosis, permanent eye damage. Chronic Effects: headache, irritability, dizziness, insomnia, fatigue, respiratory irritation, chest pain, cardiac palpitations, gastrointestinal disorders, temporary loss of hair (alopecia), dermatitis, conjunctivitis, corneal necrosis, damage the lungs, nervous system, kidneys, liver, spleen, and myocardium (heart wall muscle).

Fire: Flammable. Use alcohol foam. Can form explosive mixtures in the air. Hazardous polymerization can occur. Very dangerous fire hazard when exposed to heat or flame.

Reactivity: Ustable, polymerizes upon standing. Hazardous polymerization can occur upon standing; autoxidation forms an unstable peroxide which catalyzes exothermic polymerization of the monomer. Incompatible with: liquid or gaseous fluorine. Hazardous decomposition products: hydrogen chloride; chlorine.

Carcinogenicity: IARC - Group 3, Not classifiable as to carcinogenicity to humans; NIOSH - Listed as carcinogen; NTP - Not listed; ACGIH - Not listed; OSHA - Not listed; EPA - Not listed; MAK - Not listed

Primary Target Organs:

Eyes

Skin

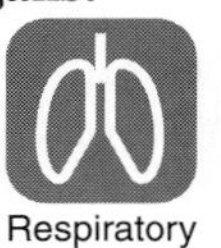
Respiratory System

Nervous System

Liver

Kidneys

Exposure Limits

OSHA PEL: TWA: 25 ppm; 90 mg/m^3.

OSHA PEL Vacated 1989 Limits: TWA: 10 ppm; 35 mg/m^3.

ACGIH TLV: TWA: 10 ppm; 36 mg/m^3.

NIOSH REL: STEL: 1 ppm; 3.6 mg/m^3; 15-minute.

NIOSH IDLH: 300 ppm.

DFG MAK: TWA: 10 ppm; 36 mg/m^3.

Respirator Recommendation

Exposure Range: >25 to <300 ppm Supplied Air, Constant Flow/Pressure Demand, Full Face

Exposure Range: 300 to unlimited ppm Self-contained Breathing Apparatus, Pressure Demand, Full Face

Note: poor warning properties

Environmental

Ecotoxicity: Aquatic toxicity: Finfish/TLm/96 hour = 10 to 100 ppm

Environmental Fate: If released to soil, it should be susceptible to removal by rapid volatilization and transport by leaching into groundwater. Chemical hydrolysis is not expected to occur. If released to water, volatilization is predicted to be the dominant removal mechanism (t1/2 3 hours from a model river 1 m deep with a current speed of 1 m/sec and wind speed of 3 m/sec). In water this compound is not expected to chemically hydrolyze, adsorb significantly to suspended solids or sediments, or bioaccumulate in aquatic organisms. If released to the atmosphere, it is expected to exist almost entirely in the vapor phase. The primary removal mechanism should be reaction with photochemically generated hydroxyl radicals with small amounts being removed by reaction with ozone. The overall reaction half-life has been estimated to be 1.6 hours. Anticipated reaction products include formaldehyde, 1-chloroacrolein, glyoxal, chloroglyoxal, chlorohydroxy acids and aldehydes.

Cleanup/Disposal: Guide No. 131: Fully encapsulating, vapor protective clothing should be worn for spills and leaks with no fire. Eliminate all ignition sources (no smoking, flares, sparks or flames in immediate area). All equipment used when handling the product must be grounded. Do not touch or walk through spilled material. Stop leak if you can do it without risk. Prevent entry into waterways, sewers, basements or confined areas. A vapor suppressing foam may be used to reduce vapors. Small Spills: Absorb with earth, sand or other non-combustible material and transfer to containers for later disposal. Use clean non-sparking tools to collect absorbed material. Large Spills: Dike far ahead of liquid spill for later disposal. Water spray may reduce vapor; but may not prevent ignition in closed spaces.

Environmental Physical Data

Henry's Law Constant: estimated at 3.2 x10^{-2}

Octanol/Water Partition Coefficient: log K_{ow} = estimated at 2.06

Sorption Partition Coefficient: K_{oc} = estimated at 315

BCF: estimated at 22

Regulations

RCRA 40CFR: Not listed

CERCLA: 40CFR 302.4: Listed per CAA Section 112 RQ: 100 lb (45.35 kg)

SARA 40CFR 372.65: Listed

SARA EHS 40CFR 355: Not listed

TSCA: Listed

Analytical Methods

Air: EPA 0031; ASTM D4490
Soil: SW846 8010B, 8240B, 8260A, 8260B
Indoor / Expired Air: NIOSH 1002
Plasma: EPA 29

CHL6610 CAS #: 542-76-7

3-CHLOROPROPIONITRILE

RTECS: UG1400000
EINECS Number: 208-827-0
Molecular Formula: C_3H_4ClN
Formula Weight: 89.53

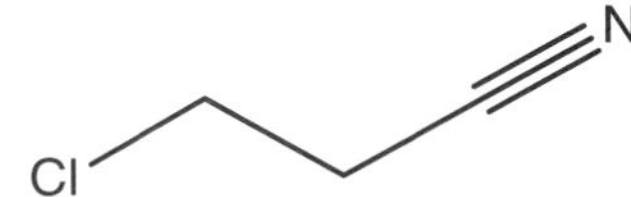

Chemical Structure

Synonyms: 1-CHLORO-2-CYANOETHANE; 3-CHLOROPROPANENITRILE; 3-CHLOROPROPANONITRILE; BETA-CHLOROPROPIONITRILE; 3-CHLORPROPANNITRIL; PROPANENITRILE,3-CHLORO-; PROPIONITRILE,3-CHLORO-
Description: colorless liquid; acrid characteristic odor
Use: in pharmaceutical & polymer synthesis; combines the reactivity of nitrile & an alkyl halide

Physical Properties

Boiling Point: 175 °C (347 °F) to 176 °C (349 °F)
Freezing Point: -51 °C (-59.8 °F)
Specific Gravity: 1.1573 at 20 °C
Water Solubility: 5 g/100 mL at 25 °C
Refraction Index: 1.4341 at 25 °C/D
Flash Point: 75.556 °C Closed Cup

RTECS Toxicity Data

Acute Oral: Rat LD_{50} Dose: 10 mg/kg; Toxic Effects: Behavioral - Somnolence (general depressed activity); Behavioral - Convulsions or effect on seizure threshold; Vascular - Regional or general arteriolar or venous dilation. Mouse LD_{50} Dose: 9 mg/kg; Toxic Effects: Behavioral - General anesthetic.

Hazard Overviews

Poison

Health: Irritating to eyes/skin/respiratory tract. Poison. Other Acute Effects: may be fatal if inhaled, swallowed, or absorbed through skin; symptoms of exposure may include burning sensation, coughing, wheezing, laryngitis, shortness of breath, headache, nausea and vomiting.

Fire: Combustible. Hazards: emits toxic fumes. Extinguishing agents: water spray; carbon dioxide, dry chemical powder or appropriate foam. Precautions: combustible liquid.
Reactivity: Incompatible with: strong acids, strong bases, strong oxidizing agents, strong reducing agents. Hazardous decomposition products: thermal decomposition may produce carbon monoxide, carbon dioxide, and nitrogen oxides; hydrogen chloride gas.
Carcinogenicity: IARC - Not listed; NIOSH - Not listed; NTP - Not listed; ACGIH - Not listed; OSHA - Not listed; EPA - Not listed; MAK - Not listed
Primary Target Organs:

Eyes

Skin

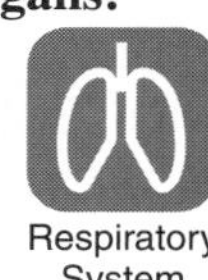
Respiratory System

Environmental

Regulations

RCRA 40CFR: Listed Hazardous Waste No. P027 Toxic Waste
CERCLA: 40CFR 302.4: Listed per RCRA Section 3001 RQ: 1000 lb (453.5 kg)
SARA 40CFR 372.65: Listed TPQ: 1000 lb
SARA EHS 40CFR 355: Listed TPQ: 1,000 lb
TSCA: Listed

Analytical Methods

Soil: SW846 8240B, 8260A, 8260B
Plasma: EPA 29

CHL6640 CAS #: 557-98-2

2-CHLOROPROPYLENE

RTECS: UC7200000
EINECS Number: 209-187-5
Molecular Formula: C_3H_5Cl
Formula Weight: 76.53

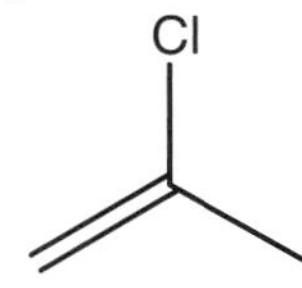

Chemical Structure

Synonyms: 2-CHLORO-1-PROPENE; 2-CHLOROPROPENE; BETA-CHLOROPROPENE; BETA-CHLOROPROPYLENE; ISOPROPENYL CHLORIDE; 1-PROPENE,2-CHLORO-; PROPENE,2-CHLORO-
Description: colorless liquid
Use: chem int in organic synthesis; comonomer (possible use)

Physical Properties

Boiling Point: 22.65 °C (73 °F) at 760 mm Hg
Freezing Point: -137.4 °C (-215.32 °F)
Specific Gravity: 0.9017 at 20 °C/4 °C

Vapor Density: 2.63 Air=1
Vapor Pressure: 12.3 kPa at -25 °C
Water Solubility: Insoluble in Water
Other Solubilities: Soluble in Ether, Acetone, Benzene, Chloroform
Refraction Index: 1.3973 at 20 °C/D
Flash Point: > -20 °C
LEL: 4.5% v/v
UEL: 4.5% v/v

RTECS Toxicity Data

Acute Inhalation: Mouse LC_{50} Dose: 267 gm/m^3; Toxic Effects: Behavioral - General anesthetic; Lungs, Thorax, or Respiration - Respiratory depression; Lungs, Thorax, or Respiration - Other changes.
Chronic (Multiple Dose) Inhalation: Rat Dose: 191 gm/m^3/30M/1W-I; Toxic Effects: Lungs, Thorax, or Respiration - Other changes; Blood - Hemorrhage.
Mutagenic: Bacteria - S Typhimurium Mutations in Microorganisms; Dose: 100 umol/plate (-S9).

Hazard Overviews

Flammable

Fire Diamond

Health: Irritating to eyes/skin/respiratory tract. Also Causes: headache, overexcitement, irregular heartbeat, narcosis, shock, coma, sudden death, cardiac/respiratory failure, skin defatting, vomiting, diarrhea, drowsiness, pulmonary edema.
Fire: For small fires, use dry chemical, carbon dioxide, water spray, or foam. For large fires, use water spray, fog, or foam.
Reactivity: Unstable, becomes a gas at 72.77 °F. Hazardous polymerization data not found. Avoid: heat; ignition sources. Incompatible with: strong oxidizers. Hazardous decomposition products: toxic oxides of carbon; chlorine; chloride fumes; phosgene; hydrogen chloride.
Carcinogenicity: IARC - Not listed; NIOSH - Not listed; NTP - Not listed; ACGIH - Not listed; OSHA - Not listed; EPA - Not listed; MAK - Not listed
Primary Target Organs:

Eyes

Skin

Mucous Membranes

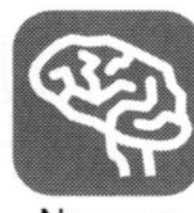
Nervous System

Environmental

Regulations

RCRA 40CFR: Not listed
CERCLA: 40CFR 302.4: Not listed
SARA 40CFR 372.65: Not listed
SARA EHS 40CFR 355: Not listed
TSCA: Listed

CHL6670 CAS #: 109-09-1

2-CHLOROPYRIDINE

RTECS: US5950000
EINECS Number: 203-646-3
Molecular Formula: C_5H_4ClN
Formula Weight: 113.55

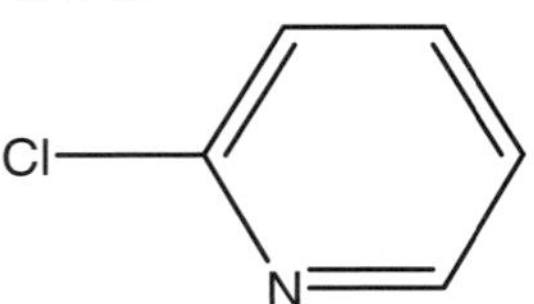

Chemical Structure

Synonyms: ALPHA-CHLOROPYRIDINE; O-CHLOROPYRIDINE; ORTHO-CHLOROPYRIDINE
Description: colorless, clear oily liquid
Use: in the production of antihistamines, germicides, pesticides and agricultural products

Physical Properties

Boiling Point: 170 °C (338 °F)
Freezing Point: -46 °C (-50.8 °F)
Vapor Density: 3.93 Air=1
Density: 1.205 g/mL at 15 °C
Vapor Pressure: 1 mm Hg at 13.3 °C
Water Solubility: 10 to 50 mg/mL at 23 °C
Other Solubilities: 95% Ethanol: >=100 mg/ml at 23 °C; Acetone: >=100 mg/ml at 23 °C; Alcohol: Soluble; DMSO: >=100 mg/ml at 23 °C; Ether: Soluble; Methanol: Not avaialble.
Refraction Index: 1.532
Ionization Potential (eV): 9.0
Flash Point: 65 °C

RTECS Toxicity Data

Acute Oral: Mouse LD_{50} Dose: 110 mg/kg; Toxic Effects: Behavioral - Somnolence (general depressed activity); Behavioral - Antipsychotic; Liver - Other changes.
Acute Inhalation: Rat LC_{Lo} Dose: 100 ppm/4hr; Toxic Effects: Behavioral - Somnolence (general depressed activity); Behavioral - Antipsychotic; Liver - Other changes.
Acute Dermal: Rabbit LD_{50} Route: Skin; Dose: 64 mg/kg; Toxic Effects: Liver - Other changes.
Mutagenic: Mouse Mutations in Microorganisms; Cell Type: lymphocyte; Dose: 400 mg/L (+S9). Mouse Mutations in Mammalian Somatic Cells; Cell Type: lymphocyte; Dose: 1980 mg/L.

Hazard Overviews

Poison

Health: Irritating to eyes/skin/respiratory tract. Poison. Other Acute Effects: may be fatal if inhaled, swallowed, or absorbed through skin.
Fire: Combustible. Hazards: emits toxic fumes; combustible liquid. Extinguishing agents: water spray; carbon dioxide, dry chemical powder or appropriate foam. Precautions: combustible liquid.
Reactivity: Incompatible with: strong oxidizing agents, strong acids. Hazardous decomposition products: toxic fumes of: carbon monoxide, carbon dioxide, nitrogen oxides, hydrogen chloride gas.
Carcinogenicity: IARC - Not listed; NIOSH - Not listed; NTP - Not listed; ACGIH - Not listed; OSHA - Not listed; EPA - Not listed; MAK - Not listed
Primary Target Organs:

Eyes

Skin

Respiratory System

Environmental

Cleanup/Disposal: Guide No. 153: Eliminate all ignition sources (no smoking, flares, sparks or flames in immediate area). Do not touch damaged containers or spilled material unless wearing appropriate protective clothing. Stop leak if you can do it without risk. Prevent entry into waterways, sewers, basements or confined areas. Absorb or cover with dry earth, sand or other non-combustible material and transfer to containers. Do not get water inside containers.

Regulations

RCRA 40CFR: Not listed
CERCLA: 40CFR 302.4: Not listed
SARA 40CFR 372.65: Not listed
SARA EHS 40CFR 355: Not listed
TSCA: Listed

CHL6700 **CAS #: 626-60-8**

3-CHLOROPYRIDINE

RTECS: US6125000
EINECS Number: 210-955-7
Molecular Formula: C_5H_4ClN
Formula Weight: 113.55

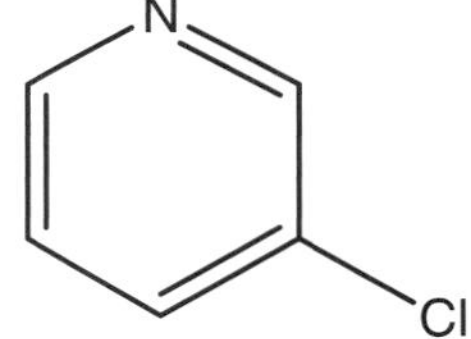

Chemical Structure

Synonyms: M-CHLOROPYRIDINE

Physical Properties

Boiling Point: 148 °C (298.4 °F)
Specific Gravity: 1.194
Water Solubility: 10 g/l at 20 °C
Refraction Index: 1.5304
Ionization Potential (eV): 9.1

RTECS Toxicity Data

Mutagenic: Mouse Mutations in Mammalian Somatic Cells; Cell Type: lymphocyte; Dose: 1672 mg/L. Mouse Cytogenetic Analysis; Cell Type: lymphocyte; Dose: 1433 mg/L.

Hazard Overviews

Health: Irritating to eyes/skin/respiratory tract. Harmful. Other Acute Effects: may be harmful by inhalation, ingestion, or skin absorption.
Fire: Hazards: emits toxic fumes. Extinguishing agents: water spray; carbon dioxide, dry chemical powder or appropriate foam. Precautions: combustible liquid.
Reactivity: Incompatible with: strong oxidizing agents. Hazardous decomposition products: toxic fumes of: carbon monoxide, carbon dioxide, nitrogen oxides, hydrogen chloride gas.
Carcinogenicity: IARC - Not listed; NIOSH - Not listed; NTP - Not listed; ACGIH - Not listed; OSHA - Not listed; EPA - Not listed; MAK - Not listed
Primary Target Organs:

Eyes

Skin

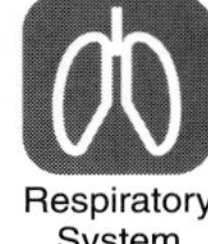
Respiratory System

Environmental

Regulations

RCRA 40CFR: Not listed
CERCLA: 40CFR 302.4: Not listed
SARA 40CFR 372.65: Not listed
SARA EHS 40CFR 355: Not listed
TSCA: Listed

CHL6820 **CAS #: 95-88-5**

4-CHLORORESORCINOL

RTECS: VH0450000
EINECS Number: 202-462-0
Molecular Formula: $C_6H_5ClO_2$
Formula Weight: 144.56

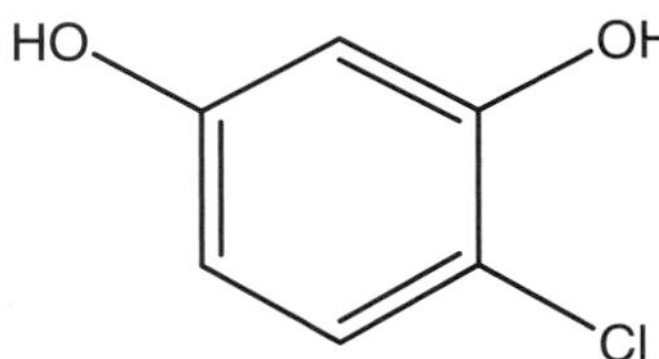

Chemical Structure

Physical Properties

Boiling Point: 147 °C (296.6 °F) at 18 mm Hg
Freezing Point: 106 °C (222.8 °F)
Water Solubility: Very Soluble
Other Solubilities: Ethanol: Very Soluble; Ether: Very Soluble; Acetone: Very Soluble; Benzene: Very Soluble; CS_2: Very Soluble

RTECS Toxicity Data

Acute Oral: Rat LD_{50} Dose: 369 mg/kg; Toxic Effects: Behavioral - Somnolence (general depressed activity); Behavioral - Ataxia; Skin and appendages - Hair.
Irritation Eye: Rabbit Standard Draize Test Dose: 5%; Reaction: mild.
Reproductive/Teratogenic: Rat Route: Oral; Dose: 2 gm/kg; Duration: female 6-15D of pregnancy; Effects on Fertility - Post-implantation mortality.

Hazard Overviews

Health: Irritating to eyes/skin/respiratory tract. Toxic. Other Acute Effects: harmful by inhalation, in contact with skin and if swallowed; depending on the intensity and duration of exposure, effects may vary from mild irritation to severe destruction of tissue; prolonged contact can cause: damage to the eyes, severe irritation or burns.
Fire: Hazards: emits toxic fumes. Extinguishing agents: carbon dioxide, dry chemical powder or appropriate foam. Precautions: combustible liquid.
Carcinogenicity: IARC - Not listed; NIOSH - Not listed; NTP - Not listed; ACGIH - Not listed; OSHA - Not listed; EPA - Not listed; MAK - Not listed
Primary Target Organs:

Eyes

Skin

Respiratory System

Environmental

Regulations

RCRA 40CFR: Not listed
CERCLA: 40CFR 302.4: Not listed
SARA 40CFR 372.65: Not listed
SARA EHS 40CFR 355: Not listed
TSCA: Listed

CHL6880 CAS #: 2039-87-4

O-CHLOROSTYRENE

RTECS: WL4160000
EINECS Number: 218-026-8
Molecular Formula: C_8H_7Cl
Structured MF: $ClC_6H_4CH{=}CH_2$
Formula Weight: 138.60

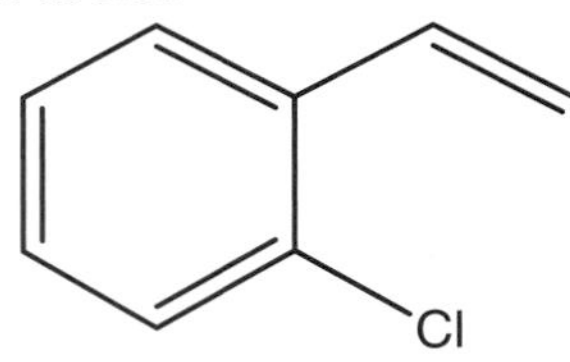

Chemical Structure

Synonyms: 1-CHLORO-2-ETHENYLBENZENE; 2-CHLOROETHENYLBENZENE; O-CHLOROETHENYLBENZENE; 2-CHLOROPHENYLETHYLENE; O-CHLOROPHENYLETHYLENE; 2-CHLOROSTYRENE; ORTHO-CHLOROSTYRENE; 2-CHLOROVINYLBENZENE; O-CHLOROVINYLBENZENE; STRYENE,O-CHLORO-; STYRENE,2-CHLORO-
Description: colorless liquid
Use: organic synthesis

Physical Properties

Boiling Point: 188.7 °C (372 °F) at 760 mm Hg
Freezing Point: -63.1 °C (-81.58 °F)
Specific Gravity: 1.1 at 20 °C
Saturated Vapor Density: 1.205728639 kg/m^3
Vapor Pressure: 9.6 x10^{-1} mm Hg at 25 °C
Water Solubility: Insoluble
Other Solubilities: 95% Ethanol: >=100 mg/ml at 22 °C; Acetone: >=100 mg/ml at 22 °C; Acetic Acid: Soluble; Carbon Tetrachloride: Soluble; DMSO: >=100 mg/ml at 22 °C; Ether: Soluble; Petroleum Ether: miscible.
Refraction Index: 1.5649 at 20 °C/D
Flash Point: 59 °C

Hazard Overviews

Flammable

Health: Irritating to eyes/skin. Harmful. Other Acute Effects: harmful if inhaled or swallowed.
Fire: Flammable. Hazards: container explosion may occur; under fire conditions, material may decompose to form flammable and/or explosive mixtures in air. Extinguishing agents: water spray; carbon dioxide, dry chemical powder or appropriate foam. Precautions: combustible liquid.
Reactivity: Incompatible with: acids, bases, oxidizing agents, halogens may discolor on exposure to light. Hazardous decomposition products: toxic fumes of: carbon monoxide, carbon dioxide, hydrogen chloride gas.

Carcinogenicity: IARC - Not listed; NIOSH - Listed as carcinogen; NTP - Not listed; ACGIH - Not listed; OSHA - Not listed; EPA - Not listed; MAK - Not listed

Primary Target Organs:

Eyes Skin

Exposure Limits

OSHA PEL Vacated 1989 Limits: TWA: 50 ppm; 285 mg/m^3; STEL: 75 ppm; 428 mg/m^3.

ACGIH TLV: TWA: 50 ppm; 283 mg/m^3; STEL: 75 ppm; 425 mg/m^3.

NIOSH REL: TWA: 50 ppm; 285 mg/m^3. STEL: 75 ppm; 428 mg/m^3.

Respirator Recommendation

Exposure Range: >50 to 2500 ppm Supplied Air, Constant Flow/Pressure Demand, Half Mask

Exposure Range: >2500 to 50,000 ppm Supplied Air, Constant Flow/Pressure Demand, Full Face

Exposure Range: >50,000 to unlimited ppm Self-contained Breathing Apparatus, Pressure Demand, Full Face

Note: poor warning properties

Environmental

Environmental Fate: Sufficient data are not available to predict the importance of biodegradation and chemical degradation in the environment. An estimated K_{oc} indicates it will be slightly mobile in soil. In aquatic systems, it should partition from the water column to organic matter contained in sediments and suspended solids. Bioconcentration is not expected to be important in aquatic systems. An estimated Henry's Law constant of 2.39 x10^{-3} atm-cu m/mole at 25 °C suggests volatilization from environmental waters may be rapid. The volatilization half-lives from a model river and a model pond, the latter considers the effect of adsorption, have been estimated to be 4 hr and 15 days, respectively. It is expected to exist entirely in the vapor phase in ambient air. Reactions with photochemically produced hydroxyl radicals (estimated half-life of 14 hr) and ozone (estimated half-life of 13 hr) in the atmosphere are likely to be important fate processes.

Cleanup/Disposal: Guide No. 128: Eliminate all ignition sources (no smoking, flares, sparks or flames in immediate area). All equipment used when handling the product must be grounded. Do not touch or walk through spilled material. Stop leak if you can do it without risk. Prevent entry into waterways, sewers, basements or confined areas. A vapor suppressing foam may be used to reduce vapors. Absorb or cover with dry earth, sand or other non-combustible material and transfer to containers. Use clean non-sparking tools to collect absorbed material. Large Spills: Dike far ahead of liquid spill for later disposal. Water spray may reduce vapor; but may not prevent ignition in closed spaces.

Environmental Physical Data

Henry's Law Constant: estimated at 2.39 x10^{-3}

Octanol/Water Partition Coefficient: log K_{ow} = 3.58

Sorption Partition Coefficient: K_{oc} = 2100

BCF: estimated at 2.49

Regulations

RCRA 40CFR: Not listed

CERCLA: 40CFR 302.4: Not listed

SARA 40CFR 372.65: Not listed

SARA EHS 40CFR 355: Not listed

TSCA: Not listed

CHL7000 CAS #: 7790-94-5

CHLOROSULFONIC ACID

RTECS: FX5730000

DOT: UN1754; IMO8.0

EINECS Number: 232-234-6

Molecular Formula: $ClHO_3S$

Structured MF: $SO_2(OH)Cl$

Formula Weight: 116.53

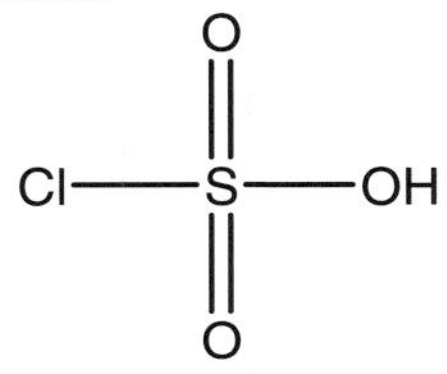

Chemical Structure

Synonyms: CHLOROSULFURIC ACID; MONOCHLOROSULFURIC ACID; SULFONIC ACID,MONOCHLORIDE; SULFURIC CHLOROHYDRIN

Description: colorless to pale-yellow liquid; pungent, acrid odor

Use: chem int for dyes; pesticides; ion-exchange resin; pharmaceuticals incl sulfa drugs & saccharin; alkyl sulfate surfactants; alkylphenol ethoxylate sulfate surfactants; component of military screening smoke; as chlorosulfonating & condensing agent in org syntheses; mfr of synthetic detergents, anhydrous hydrogen chloride & smoke-producing chemicals

Physical Properties

Boiling Point: 151 °C (304 °F) to 152 °C (306 °F) at 755 mm Hg

Freezing Point: -80 °C (-112 °F)

Specific Gravity: 1.76 to 1.77

Vapor Density: 4.02 Air=1

Vapor Pressure: 1 mm Hg at 32 °C

Water Solubility: Explosive Decomposes

Other Solubilities: common solvents include liquid Sulfur Dioxide, Pyridine, and Dichloroethane.

Odor Threshold: 1 to 5 ppm

Refraction Index: 1.437 at 14 °C/D

Flash Point: Nonflammable

Hazard Overviews

Poison

Corrosive

0 4 2 W
Fire Diamond

Health: Corrosive to eyes/skin/respiratory tract. Poison. Also Causes: spasms, pulmonary edema, conjunctivitis, GI tract burns, circulatory collapse. Chronic Effects: dental erosion, respiratory tract damage, dermatitis.

Fire: Noncombustible. However, it is a strong oxidizer capable of igniting combustibles. Do not use water to extinguish fire. Use dry chemical, carbon dioxide, or regular foam to fight surrounding fire.

Reactivity: Stable, but slowly (quickly if moisture is present) evolves flammable hydrogen gas when stored in metal containers. Hazardous polymerization cannot occur. Avoid: contact with water; combustibles. Incompatible with: alcohol; acetic acid; acetic anhydride; acetonitrile; acrolein; acrylic acid; acrylonitrile; allyl alcohol; allyl chloride; 2-amino ethanol; ammonium hydroxide; aniline; n-butyraldehyde; creosote oil; cresol; cumene; dichloroethyl ether; diethylene glycol monomethyl ether; diisobutylene; diisopropyl ether; epichlorohydrin; ethyl acetate; ethyl acrylate; ethylene chlorohydrin; ethylene cyanohydrin; ethylene diamine; ethylene glycol; ethylene glycol monoethyl ether acetate; ethylene imine; glyoxal; hydrofluoric acid; hydrogen peroxide; isoprene; mesityl oxide; metal powders; methyl ethyl ketone; nitric acid; 2-nitropropane; beta-propiolactone; phosphorus; propylene oxide; pyridine; silver nitrate; sodium hydroxide; sulfolane; styrene monomer; vinyl acetate; vinylidene chloride; water; organic material; combustibles. Hazardous decomposition products: hydrogen chloride; sulfur dioxide gas.

Carcinogenicity: IARC - Not listed; NIOSH - Not listed; NTP - Not listed; ACGIH - Not listed; OSHA - Not listed; EPA - Not listed; MAK - Not listed

Primary Target Organs:

Eyes

Skin

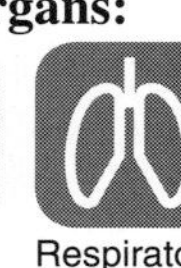
Respiratory System

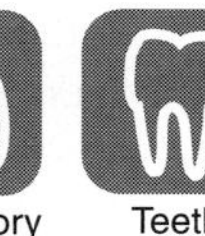
Teeth

Exposure Limits

AIHA WEEL: TWA: 0.3 ppm.

Respirator Recommendation

Exposure Range: >0.3 to 15 ppm Supplied Air, Constant Flow/Pressure Demand, Half Mask

Exposure Range: >15 to 300 ppm Supplied Air, Constant Flow/Pressure Demand, Full Face

Exposure Range: >300 to unlimited ppm Self-contained Breathing Apparatus, Pressure Demand, Full Face

Note: odor threshold unknown

Environmental

Ecotoxicity: Aquatic toxicity: 282 ppm/96 hr/mosquito fish/TLm/fresh water 100-300 ppm/48 hr/shrimp/LC_{50}/salt water

Environmental Fate: If released to soil, it will be expected to rapidly hydrolyze if the soil is moist, based upon the reported violent hydrolysis by water giving hydrochloric and sulfuric acids. Since it rapidly hydrolyzes, biodegradation, adsorption to and volatilization from moist soil are not expected to be significant processes, although no data specifically regarding fate in soil were located. Based upon a measured vapor pressure of 0.75 mm Hg at 20 °C, volatilization from dry near-surface soil or other surfaces may be significant processes. If released to water, it will be hydrolyzed violently by water producing hydrogen chloride and sulfuric acid. Based upon this rapid and violent hydrolysis, bioconcentration, biodegradation, volatilization, and adsorption to sediment and suspended solids are not expected to be significant processes. If released to the atmosphere, it will be expected to exist almost entirely in the vapor phase based upon its vapor pressure. It may be susceptible to hydrolysis in moist air based upon its rapid hydrolysis in aqueous solution and the report that the chemical fumes in air. It will be susceptible to photooxidation via vapor phase reaction with photochemically produced hydroxyl radicals. An atmospheric half-life of 1.2 years at an atmospheric concentration of 5 $x10^5$ hydroxyl radicals per cu cm has been calculated for this process based upon an estimated rate constant.

Cleanup/Disposal: Guide No. 137: Fully encapsulating, vapor protective clothing should be worn for spills and leaks with no fire. Do not touch damaged containers or spilled material unless wearing appropriate protective clothing. Stop leak if you can do it without risk. Use water spray to reduce vapors; do not put water directly on leak, spill area or inside container. Keep combustibles (wood, paper, oil, etc.) away from spilled material. Small Spills: Cover with dry earth, dry sand, or other non-combustible material followed with plastic sheet to minimize spreading or contact with rain. Use clean non-sparking tools to collect material and place it into loosely covered plastic containers for later disposal. Prevent entry into waterways, sewers, basements or confined areas.

Environmental Physical Data

BCF: not significant

BOD: none

Regulations

RCRA 40CFR: Not listed

CERCLA: 40CFR 302.4: Listed per CWA Section 311(b)(4) RQ: 1000 lb (453.5 kg)

SARA 40CFR 372.65: Not listed

SARA EHS 40CFR 355: Not listed

TSCA: Listed

CHL7060 **CAS #: 2837-89-0**

2-CHLORO-1,1,1,2-TETRAFLUOROETHANE

EINECS Number: 220-629-6

Molecular Formula: C_2HClF_4

Formula Weight: 136.48
Synonyms: FREON 124; HFA-124; R 124
Description: colorless gas
Use: as a replacement for chlorofluorocarbons; preparation of hexafluoropropylene

Physical Properties

Boiling Point: -12 °C (10.4 °F)
Vapor Pressure: Estimated 5265 mm Hg at 25 °C
Water Solubility: 253 mg/L at 25 °C (est)

Hazard Overviews

Fire: Noncombustible.
Carcinogenicity: IARC - Not listed; NIOSH - Not listed; NTP - Not listed; ACGIH - Not listed; OSHA - Not listed; EPA - Not listed; MAK - Not listed
Exposure Limits
AIHA WEEL: TWA: 1000 ppm.

Environmental

Environmental Fate: If released to soil, it will rapidly volatilize from both moist and dry soil to the atmosphere. It will display moderate mobility in soil. If released to water, it will rapidly volatilize to the atmosphere. The estimated half-life for volatilization from a model river is 3.4 hours. It will not bioconcentrate in fish and aquatic organisms nor will it adsorb to sediment or suspended organic matter. If released to the atmosphere, it will undergo a slow gas-phase reaction with photochemically produced hydroxyl radicals with an estimated half-life of 1573 days. The atmospheric lifetime has been estimated to range from 5.3-10 years. It may undergo atmospheric removal by wet deposition processes; however, any removed is expected to rapidly re-volatilize to the atmosphere.

Environmental Physical Data

Henry's Law Constant: estimated at 0.540
Octanol/Water Partition Coefficient: log K_{ow} = 1.867
Sorption Partition Coefficient: K_{oc} = 208 to 247
BCF: calculated at 15 to 27

Regulations

RCRA 40CFR: Not listed
CERCLA: 40CFR 302.4: Listed as Compound per CWA Section 307(a)
SARA 40CFR 372.65: Listed
SARA EHS 40CFR 355: Not listed
TSCA: Listed

CHL7090 **CAS #: 63938-10-3**

CHLOROTETRAFLUOROETHANE

DOT: UN1021
EINECS Number: 264-567-8
Molecular Formula: C_2HClF_4
Structured MF: CHF_2CHClF_2
Formula Weight: 136.48
Description: colorless gas
Use: refrigerants, blowing agents, cleaning agents, and fire extinguishants; as blowing agent in polyurethane foams; in aerosols, foams, and refrigerants

Physical Properties

Boiling Point: Boiling Point -12 °C (10 °F)
Freezing Point: -117 °C (-179 °F)
Vapor Density: 4.71 Air=1
Density: Liquid 1.364 g/cu cm at 25 °C
Water Solubility: 1.71% at 24 °C
Critical Temperature: 122.2 °C
Critical Pressure: 3.57 mPa
Flash Point: Nonflammable

Hazard Overviews

Compressed Gas

Fire: Noncombustible.
Carcinogenicity: IARC - Not listed; NIOSH - Not listed; NTP - Not listed; ACGIH - Not listed; OSHA - Not listed; EPA - Not listed; MAK - Not listed

Environmental

Environmental Fate: When released to the atmosphere, it will degrade primarily through reaction with photochemically produced hydroxyl radicals. The atmospheric degradation rate is slow. 1-Chloro- 1,2,2,2-tetrafluoroethane (HCFC-124) has an estimated hydroxyl radical reaction half-life of about 4.4 years and a predicted atmospheric lifetime of 5.3 to 11.2 years. The other isomer, 1-chloro- 1,1,2,2-tetrafluoroethane (HCFC-124a), has a predicted atmospheric lifetime of 57 years. If released to soil or water, volatilization to the atmosphere will be a major fate process; it is a gas under ambient conditions above -10 °C. Volatilization half-lives of 3.4 hr and 40 hours can be estimated for a model river (1 m deep) and a model pond (2 m deep) respectively.
Cleanup/Disposal: Guide No. 126: Do not touch or walk through spilled material. Stop leak if you can do it without risk. Do not direct water at spill or source of leak. Use water spray to reduce vapors or divert vapor cloud drift. If possible, turn leaking containers so that gas escapes rather than liquid. Prevent entry into waterways, sewers, basements or confined areas. Allow substance to evaporate. Ventilate the area.

Environmental Physical Data

Henry's Law Constant: estimated at 0.54
Octanol/Water Partition Coefficient: log K_{ow} = 1.86
Sorption Partition Coefficient: K_{oc} = estimated at 250
BCF: estimated at 16

Regulations

RCRA 40CFR: Not listed
CERCLA: 40CFR 302.4: Listed as Compound per CWA Section 307(a)
SARA 40CFR 372.65: Listed

SARA EHS 40CFR 355: Not listed
TSCA: Not listed

CHL7120 CAS #: 1897-45-6

CHLOROTHALONIL

RTECS: NT2600000
EINECS Number: 217-588-1
Molecular Formula: $C_8Cl_4N_2$
Structured MF: $Cl_4C_6(CN)_2$
Formula Weight: 265.89
Synonyms: 1,3-BENZENEDICARBONITRILE,2,4,5,6-TETRACHLORO-; BRAVO; BRAVO 6F; BRAVO W-75; BRAVO-W-75; BRAVO 500; BRAVO W75; CHLOROALONIL; CHLORTHALONIL; CLORTHALONIL; CLORTOCAF RAMATO; CLORTOSIP; DAC 2787; DACONIL; DACONIL 2787; DACONIL FLOWABLE; DACONIL 2787 FLOWABLE FUNGICIDE; DACONIL 2787 W-75 FUNGICIDE; DACONIL M; DACONIL 2787 W75; DACOSOIL; 1,3-DICYANOTETRACHLOROBENZENE; EXOTHERM; EXOTHERM TERMIL; FABER; FORTURF; ISOPHTHALONITRILE,2,4,5,6-TETRACHLORO-; ISOPHTHALONITRILE,TETRACHLORO-; NOPCOCIDE; NOPCOCIDE N-96; NOPCOCIDE N 96; NOPCOCIDE N40D & N96; REPULSE; SWEEP; TCIN; M-TCPN; META-TCPN; TERMIL; TERRACLACTYL; TETRACHLORISOFTALONITRIL; 2,4,5,6-TETRACHLORO-1,3-BENZENEDICARBONITRILE; 2,4,5,6-TETRACHLORO-3-CYANOBENZONITRILE; 2,4,5,6-TETRACHLOROISOPHTHALONITRILE; TETRACHLOROISOPHTHALONITRILE; TETRACHLOROISOPHTHALONITRITE; META-TETRACHLOROPHTHALODINITRILE; TETRACHLORO-M-PHTHALODINITRILE; TETRACHLORO-META-PHTHALO-DINITRILE; M-TETRACHLOROPHTHALONITRILE; TPN; TPN (PESTICIDE)
Description: white crystalline solid or crystals; odorless
Use: fungicide; preservative in paints and adhesives

Physical Properties

Boiling Point: 350 °C (662 °F) at 760 mm Hg
Freezing Point: 250 °C (482 °F) to 251 °C (483.8 °F)
Specific Gravity: 1.7 at 25 °C/4 °C
Density: 1.8 g/mL (96% pure)
Vapor Pressure: < 0.01 mm Hg at 40 °C
Water Solubility: 0.6 ppm at 0 °C
Other Solubilities: 95% Ethanol: <1 mg/ml at 22 °C; Acetone: 5-10 mg/ml at 22 °C; Benzene: Readily Soluble; Butanone: 20 g/kg; Cyclohexanone: 30 g/kg; DMSO: 5-10 mg/ml at 22 °C; Kerosene: <10 g/kg; Toluene: 10-50 mg/ml at 22 °C; Xylene: 8% at 20 °C.
Flash Point: Nonflammable

RTECS Toxicity Data

Acute Oral: Rat LD_{50} Dose: 10 gm/kg. Mouse LD_{50} Dose: 3700 mg/kg.
Acute Inhalation: Rat LC_{50} Dose: 310 mg/m^3/1hr.
Acute Dermal: Rabbit LD_{50} Route: Skin; Dose: >10 gm/kg. Rat LD_{50} Route: Skin; Dose: >2500 mg/kg.
Mutagenic: Mouse Mutations in Mammalian Somatic Cells; Cell Type: lymphocyte; Dose: 120 ug/L. Hamster Cytogenetic Analysis; Cell Type: ovary; Dose: 500 ug/L.
Tumorigenic: Rat Route: Oral; Dose: 142 gm/kg/80W-C; Toxic Effects: Tumorigenic - Carcinogenic by RTECS criteria; Kidney, Ureter, and Bladder - Kidney tumors.

Hazard Overviews

Fire: Noncombustible.
Carcinogenicity: IARC - Group 3, Not classifiable as to carcinogenicity to humans; NIOSH - Not listed; NTP - Not listed; ACGIH - Not listed; OSHA - Not listed; EPA - Not listed; MAK - Class B, Justifiably suspected of having carcinogenic potential

Environmental

Ecotoxicity: LC_{50} Ictalurus punctatus (Channel catfish) 52 ug/l/96 hr, static bioassy LC_{50} Young mallard duck oral > 21500 mg/kg LC_{50} Mytilus edulis (blue mussels) 5.9 mg/l/96 hr /Bravo 500; conditions of bioassay not specified LC_{50} Bobwhite quail oral 5200 mg/kg
Environmental Fate: If released to the atmosphere, degradation of vapor-phase by reaction with photochemically produced hydroxyl radicals will not be important (estimated half-life of 7 years). Photolysis may be important and particulate-phase will be removed from air via dry deposition. If released to soil, biodegradation may be the primary fate process. Metabolites are isophthalonitrile, mono-, di- and tri-chlorinated isophthalonitriles, 2,5,6-trichloro-4-hydroxyisophthalonitrile and 2,5,6-trichloro-4-methoxyisophthalonitrile. Photolysis may be important on soil surfaces exposed to sunlight and hydrolysis in alkaline moist soils may contribute to the removal from soil. Adsorption will take place (K_{oc} of 1,800) resulting in a low amount of leaching. A degradation half-life of 30 days was estimated in soil. In water, biodegradation (half-lives of 0.18-8.8 days) will be the primary fate process and hydrolysis in alkaline waters (half-life of 38.1 days at pH 9) may be important. Volatilization will not be important.

Environmental Physical Data

Henry's Law Constant: 2×10^{-7}
Octanol/Water Partition Coefficient: log K_{ow} = calculated at 437
Sorption Partition Coefficient: K_{oc} = 1800
BCF: estimated at 820

Regulations

RCRA 40CFR: Not listed
CERCLA: 40CFR 302.4: Not listed
SARA 40CFR 372.65: Listed
SARA EHS 40CFR 355: Not listed
TSCA: Listed

Analytical Methods

Air: EPA TO-10
Soil: EPA PMD-TLC; SW846 8081
Water / Groundwater: EPA 608.2
Drinking Water: EPA 508, 508.1, 525.2; AOAC 990.06
Food: FDA 212.1, 212.2, 232.1, 232.4, 242.1
Indoor / Expired Air: EPA IP-8; ASTM D4861
Plasma: EPA 001; FDA 211.1, 231.1, 252
Other: EPA 1656

CHL7240 CAS #: 108-41-8

M-CHLOROTOLUENE

EINECS Number: 203-580-5
Molecular Formula: C_7H_7Cl
Structured MF: $ClC_6H_4CH_3$
Formula Weight: 126.58

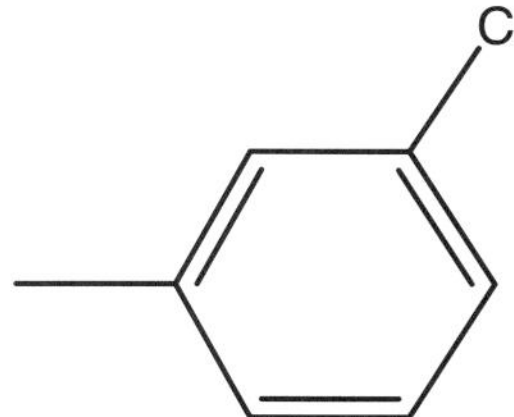

Chemical Structure

Synonyms: 1-CHLORO-3-METHYLBENZENE; 3-CHLOROTOLUENE; META-CHLOROTOLUENE; M-TOLYL CHLORIDE; META-TOLYL CHLORIDE
Description: colorless, clear liquid
Use: solvent; intermediates

Physical Properties

Boiling Point: 162 °C (324 °F)
Freezing Point: -47.8 °C (-54.04 °F)
Specific Gravity: 1.0722
Vapor Density: 4.4 Air=1
Vapor Pressure: 15.1 kPa at 100 °C
Water Solubility: Insoluble
Other Solubilities: Alcohol: Soluble; Benzene: Soluble; Chloroform: Soluble; Ether: Soluble.
Refraction Index: 1.5218
Critical Temperature: 388 °C
Critical Pressure: 567 psia
Ionization Potential (eV): 8.83 +/-0.02
Flash Point: 51 °C Closed Cup
LEL: 1.36% v/v

Hazard Overviews

Flammable

Health: May cause irritation. Harmful. Other Acute Effects: may be harmful by inhalation, ingestion, or skin absorption.
Fire: Flammable. Hazards: emits toxic fumes. Extinguishing agents: water spray; carbon dioxide, dry chemical powder or appropriate foam. Precautions: combustible liquid.
Reactivity: Incompatible with: strong oxidizing agents. Hazardous decomposition products: toxic fumes of: carbon monoxide, carbon dioxide, hydrogen chloride gas.
Carcinogenicity: IARC - Not listed; NIOSH - Not listed; NTP - Not listed; ACGIH - Not listed; OSHA - Not listed; EPA - Not listed; MAK - Not listed

Environmental

Ecotoxicity: Aquatic toxicity: 18 ppm/7d/guppy LD_{50}/fresh water
Cleanup/Disposal: Guide No. 130: Eliminate all ignition sources (no smoking, flares, sparks or flames in immediate area). All equipment used when handling the product must be grounded. Do not touch or walk through spilled material. Stop leak if you can do it without risk. Prevent entry into waterways, sewers, basements or confined areas. A vapor suppressing foam may be used to reduce vapors. Absorb or cover with dry earth, sand or other non-combustible material and transfer to containers. Use clean non-sparking tools to collect absorbed material. Large Spills: Dike far ahead of liquid spill for later disposal. Water spray may reduce vapor; but may not prevent ignition in closed spaces.

Environmental Physical Data

Octanol/Water Partition Coefficient: log K_{ow} = 3.28

Regulations

RCRA 40CFR: Not listed
CERCLA: 40CFR 302.4: Not listed
SARA 40CFR 372.65: Not listed
SARA EHS 40CFR 355: Not listed
TSCA: Not listed

Analytical Methods

Air: EPA OA-002-1
Water / Groundwater: APHA 6040-B; ASTM D5241

CHL7270 CAS #: 95-49-8

O-CHLOROTOLUENE

RTECS: XS9000000
DOT: UN2238; IMO3.3
EINECS Number: 202-424-3
Molecular Formula: C_7H_7Cl
Structured MF: $ClC_6H_4CH_3$
Formula Weight: 126.6

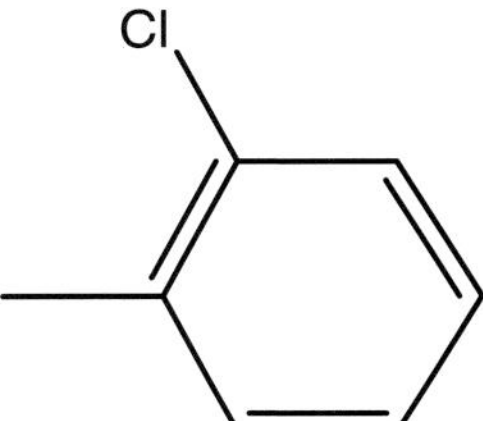

Chemical Structure

Synonyms: BENZENE,1-CHLORO-2-METHYL-(9CI); 1-CHLORO-2-METHYLBENZENE; 2-CHLORO-1-METHYLBENZENE; 2-CHLOROTOLUENE; ORTHO-CHLOROTOLUENE; HALSO 99; 1-METHYL-2-CHLOROBENZENE; 2-METHYLCHLOROBENZENE;

TOLUENE,O-CHLORO-; O-TOLYL CHLORIDE; ORTHO-TOLYL CHLORIDE

Description: colorless liquid; pungent, irritating odor

Use: solvent; dyestuff intermediate; in organic synthesis

Physical Properties

Boiling Point: 158.97 °C (318 °F)
Freezing Point: -35.59 °C (-32.062 °F)
Specific Gravity: 1.0826 at 20 °C/4 °C
Vapor Density: 4.4 Air=1
Saturated Vapor Density: 1.218226933 kg/m^3
Vapor Pressure: 3.43 mm Hg at 25 °C
Water Solubility: 89 mg/L at 25 °C
Other Solubilities: Acetone: Soluble; Alcohol: Soluble; Benzene: Soluble; Chloroform: Soluble; Ether: Soluble.
Surface Tension: 33440 dynes/cm at 20 °C
Odor Threshold: 0.32 ppm
Refraction Index: 1.5258 at 20 °C/D
Critical Temperature: 381.1 °C
Critical Pressure: 567 psia
Ionization Potential (eV): 8.83
Flash Point: 36 °C
LEL: 1.36% v/v

RTECS Toxicity Data

Acute Oral: Rat LD_{50} Dose: 3900 mg/kg; Toxic Effects: Sense organs and special senses - Other; Behavioral - Tremor; Behavioral - Excitement. Mouse LD_{50} Dose: 2500 mg/kg; Toxic Effects: Sense organs and special senses - Other; Behavioral - Tremor; Behavioral - Excitement.

Acute Inhalation: Rat LC_{Lo} Dose: 4400 ppm; Toxic Effects: Behavioral - Convulsions or effect on seizure threshold; Behavioral - Ataxia; Lungs, Thorax, or Respiration - Respiratory stimulation. Mouse LC_{Lo} Dose: 4400 ppm; Toxic Effects: Behavioral - Convulsions or effect on seizure threshold; Behavioral - Ataxia; Lungs, Thorax, or Respiration - Respiratory stimulation.

Hazard Overviews

Flammable

Fire Diamond

Health: Irritating to eyes/skin/respiratory tract.

Fire: Flammable. For small fires use dry chemical, carbon dioxide, water spray, or regular foam. For large fires use water spray, fog, or regular foam. If possible without risk, remove container from fire area.

Reactivity: Stable. Hazardous polymerization cannot occur. Avoid: buildup of vapors; generation of excess heat. Hazardous decomposition products: carbon dioxide; toxic chlorine fumes.

Carcinogenicity: IARC - Not listed; NIOSH - Not listed; NTP - Not listed; ACGIH - Not listed; OSHA - Not listed; EPA - Not listed; MAK - Not listed

Primary Target Organs:

Eyes

Skin

Respiratory System

Exposure Limits

OSHA PEL Vacated 1989 Limits: TWA: 50 ppm; 250 mg/m^3.

ACGIH TLV: TWA: 50 ppm; 259 mg/m^3.

NIOSH REL: TWA: 50 ppm; 250 mg/m^3. STEL: 75 ppm; 375 mg/m^3.

Respirator Recommendation

Exposure Range: >50 to 500 ppm Air Purifying, Negative Pressure, Half Mask

Exposure Range: >500 to 1000 ppm Air Purifying, Negative Pressure, Full Face

Exposure Range: >1000 to 2500 ppm Supplied Air, Constant Flow/Pressure Demand, Half Mask

Exposure Range: >2500 to 50,000 ppm Supplied Air, Constant Flow/Pressure Demand, Full Face

Exposure Range: >50,000 to unlimited ppm Self-contained Breathing Apparatus, Pressure Demand, Full Face

Cartridge Color: black

Environmental

Ecotoxicity: Toxicity threshold (cell multiplication inhibition test): green algae (Scenedesmus quadricauda): 8d EC_0 >100 mg/l

Environmental Fate: If released to the atmosphere, it will exist solely in the vapor phase in the ambient atmosphere based on a measured vapor pressure of 3.43 mm Hg at 25 °C. Vapor-phase is degraded in the atmosphere by reaction with photochemically-produced hydroxyl radicals with a half-life of about 9 days. Measured K_{oc} values from 170-880 indicate that it may have low to moderate mobility in soil. Volatilization from moist and dry soil surfaces may occur based on a measured Henry's Law constant of 3.57 $x10^{-3}$ atm-cu m/mole and its measured vapor pressure, respectively. Based on limited data, it may be resistant to biodegradation. A second order rate constant for the microbial degradation in natural water was experimentally determined to be 2.7 $x10^{-11}$ l/organism-hr. At 100 ppm, it showed only 0-<30% biodegradation in 14 days using an activated sludge inoculum. However, a microbial blend of 10 different bacteria and 2 fungi was able to degrade it at a concentration of 200 mg/l; complete biodegradation occurred in 3 days. It may adsorb to sediment and suspended matter in water based on its measured K_{oc} values. This compound should volatilize from water surfaces given its Henry's Law constant. Estimated half-lives for a model river and model lake are 4 hours and 5 days, respectively. BCF values from 20-112, measured in carp, indicate that it may bioconcentrate in aquatic organisms.

Cleanup/Disposal: Guide No. 130: Eliminate all ignition sources (no smoking, flares, sparks or flames in immediate area). All equipment used when handling the product must be grounded. Do not touch or walk through spilled material.

Stop leak if you can do it without risk. Prevent entry into waterways, sewers, basements or confined areas. A vapor suppressing foam may be used to reduce vapors. Absorb or cover with dry earth, sand or other non-combustible material and transfer to containers. Use clean non-sparking tools to collect absorbed material. Large Spills: Dike far ahead of liquid spill for later disposal. Water spray may reduce vapor; but may not prevent ignition in closed spaces.

Environmental Physical Data

Henry's Law Constant: 3.57×10^{-3}
Octanol/Water Partition Coefficient: log K_{ow} = 3.42
Sorption Partition Coefficient: K_{oc} = 170 to 180
BCF: carp 20 to 112
BOD: < 30% BODT, 14 days

Regulations

RCRA 40CFR: Not listed
CERCLA: 40CFR 302.4: Not listed
SARA 40CFR 372.65: Not listed
SARA EHS 40CFR 355: Not listed
TSCA: Listed

Analytical Methods

Air: EPA OA-002-1, VG-011-1
Soil: SW846 8021A, 8260A
Water / Groundwater: APHA 6210-D, 6230-D; ASTM D5241
Drinking Water: EPA 502.1, 502.2, 503.1, 524.1, 524.2; APHA 6210-C, 6220-C, 6230-C
Plasma: EPA 29

CHL7300 CAS #: 106-43-4

P-CHLOROTOLUENE

RTECS: XS9010000
DOT: UN2238; IMO3.3
EINECS Number: 203-397-0
Molecular Formula: C_7H_7Cl
Structured MF: $CH_3C_6H_4Cl$
Formula Weight: 126.59

Chemical Structure

Synonyms: BENZENE,1-CHLORO-4-METHYL; BENZENE,1-CHLORO-4-METHYL-; 1-CHLORO-4-METHYLBENZENE; 4-CHLORO-1-METHYLBENZENE; 4-CHLOROTOLUENE; PARA-CHLOROTOLUENE; TOLUENE,P-CHLORO; P-TOLYL CHLORIDE; PARA-TOLYL CHLORIDE
Description: colorless liquid; chloro-aromatic odor
Use: solvent and intermediate for organic chemicals and dyes

Physical Properties

Boiling Point: 161.99 °C (324 °F)
Freezing Point: 7.5 °C (45.5 °F)
Specific Gravity: 1.0697 at 20 °C/4 °C
Vapor Density: 4.38 Air=1
Density: 1.07 g/mL
Vapor Pressure: 5 mm Hg at 31 °C
Water Solubility: < 1 mg/mL at 20 C
Other Solubilities: 95% Ethanol: >=100 mg/ml at 20 °C; Acetone: >=100 mg/ml at 20 °C; Benzene: Soluble; Chloroform: Soluble; DMSO: >=100 mg/ml at 20 °C; Ether: Soluble.
Surface Tension: 32.24 dynes/cm at 25 °C
Refraction Index: 1.5184 at 22 °C
Ionization Potential (eV): 8.70 +/-0.02
Flash Point: 60 °C Open Cup

RTECS Toxicity Data

Acute Oral: Rat LD_{50} Dose: 2100 mg/kg; Toxic Effects: Behavioral - Change in motor activity (specific assay); Lungs, Thorax, or Respiration - Cyanosis; Gastrointestinal - Changes in structure or function of salivary glands. Mouse LD_{50} Dose: 1900 mg/kg.
Acute Inhalation: Mouse LC_{50} Dose: 34 gm/m^3/2hr.
Chronic (Multiple Dose) Oral: Rat Dose: 25200 mg/kg/2W-I; Toxic Effects: Behavioral - Food intake (animal); Nutritional and gross metabolic - Weight loss or decreased weight gain; DEATH. Rat Dose: 72 gm/kg/90D-I; Toxic Effects: Liver - Changes in liver weight; Endocrine - Changes in adrenal weight; Blood - Changes in serum composition.

Hazard Overviews

Flammable

Fire: Flammable.
Carcinogenicity: IARC - Not listed; NIOSH - Not listed; NTP - Not listed; ACGIH - Not listed; OSHA - Not listed; EPA - Not listed; MAK - Not listed

Environmental

Ecotoxicity: Aquatic toxicity: 1-10 ppm/96 hour/Finfish/TLm
Environmental Fate: If released to soil, it is expected to have very low mobility and should volatilize fairly rapidly from soil surfaces. Due to a lack of data, the significance of biodegradation in soil or water is not known. However, isolated bacteria have been found to metabolize it to its respective catechol via cis-dihydrodiol. If released to water, volatilization (half-life in a model river 3.5 hours), sensitized photolysis, and adsorption to suspended solids and sediments are predicted to be important fate processes. The relative importance of these fate processes and the rate of compound loss are expected to vary depending upon ambient conditions and characteristics of the water body. Based on monitoring data, the half-life in a river 4-5 m deep during mid-summer was estimated to be 1.2 days. This compound is not expected

to undergo chemical hydrolysis, react with oxidants found in natural waters or bioaccumulate significantly in aquatic organisms. If released to the atmosphere, the dominant removal mechanism is expected to be reaction with photochemically generated hydroxyl radicals (half-life 8.4 days). A slight potential also exists for direct photolysis in the atmosphere.

Cleanup/Disposal: Guide No. 130: Eliminate all ignition sources (no smoking, flares, sparks or flames in immediate area). All equipment used when handling the product must be grounded. Do not touch or walk through spilled material. Stop leak if you can do it without risk. Prevent entry into waterways, sewers, basements or confined areas. A vapor suppressing foam may be used to reduce vapors. Absorb or cover with dry earth, sand or other non-combustible material and transfer to containers. Use clean non-sparking tools to collect absorbed material. Large Spills: Dike far ahead of liquid spill for later disposal. Water spray may reduce vapor; but may not prevent ignition in closed spaces.

Environmental Physical Data

Henry's Law Constant: estimated at 4.1 $x10^{-3}$

Octanol/Water Partition Coefficient: log K_{ow} = 3.33

Sorption Partition Coefficient: K_{oc} = estimated at 446 to 1544

BCF: estimated at 45 to 200

BOD: < 30% BODT, 14 days

Regulations

RCRA 40CFR: Not listed

CERCLA: 40CFR 302.4: Not listed

SARA 40CFR 372.65: Not listed

SARA EHS 40CFR 355: Not listed

TSCA: Listed

Analytical Methods

Air: EPA OA-002-1, VG-011-1

Soil: SW846 8010B, 8021A

Water / Groundwater: APHA 6210-D, 6230-D; ASTM D4763, D5241

Drinking Water: EPA 502.1, 502.2, 503.1, 524.1, 524.2; APHA 6210-C, 6220-C

CHL7420 CAS #: 95-69-2

P-CHLORO-O-TOLUIDINE

RTECS: XU5000000

EINECS Number: 202-441-6

Molecular Formula: C_7H_8ClN

Formula Weight: 141.61

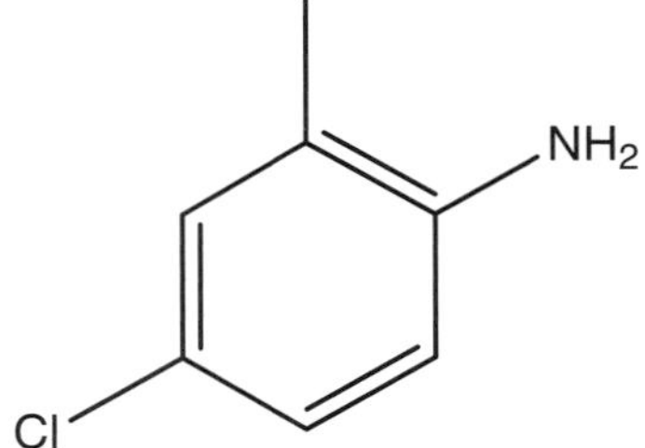

Chemical Structure

Synonyms: AMARTHOL FAST RED TR BASE; 2-AMINO-5-CHLOROTOLUENE; ASYMMETRIC META-CHLORO-ORTHO-TOLUIDINE; AZOENE FAST RED TR BASE; AZOGENE FAST RED TR; AZOIC DIAZO COMPONENT 11,BASE; BENZENAMINE,4-CHLORO-2-METHYL; BENZENAMINE,4-CHLORO-2-METHYL-; BRENTAMINE FAST RED TR BASE; 3-CHLORO-6-AMINOTOLUENE; 5-CHLORO-2-AMINOTOLUENE; 4-CHLORO-2-METHYLANILINE; 4-CHLORO-6-METHYLANILINE; 4-CHLORO-2-METHYLBENZENEAMINE; 4-CHLORO-2-TOLUIDINE; 4-CHLORO-O-TOLUIDINE; PARA-CHLORO-ORTHO-TOLUIDINE; DAITO RED BASE TR; DEAZO FAST RED TRA; DEVAL RED K; DEVAL RED TR; DIAZO FAST RED TRA; FAST RED BASE TR; FAST RED 5CI BASE; FAST RED 5CT BASE; FAST RED TR; FAST RED TR 11; FAST RED TR BASE; FAST RED TR11; FAST RED TRO BASE; KAKO RED TR BASE; KAMBAMINE RED TR; 2-METHYL-4-CHLOROANILINE; MITSUI RED TR BASE; RED BASE CIBA IX; RED BASE IRGA IX; RED BASE NIR; RED BASE NTR; RED TR BASE; RED TRBASE; SANYO FAST RED TR BASE; SONYA FAST RED TR BASE; O-TOLUIDINE,4-CHLORO-; TULA BASE FAST RED TR; TULABASE FAST RED TR

Description: grayish-white crystalline solid; fishy odor

Use: chem int for CI azoic coupling component 8, CI pigment red 7; azoic diazo component for cotton, silk, acetate & nylon; in the manufacture of chlordimeform , an acaricide and insecticide ; as int for prodn of CI pigment yellow 49

Physical Properties

Boiling Point: 241 °C (466 °F) at 760 mm Hg

Freezing Point: 30 °C (86 °F)

Water Solubility: Sparingly Soluble

Other Solubilities: Soluble in Ethanol & dilute acids.

Refraction Index: 1.5848 at 20 °C/D

Flash Point: 99 °C

RTECS Toxicity Data

Acute Oral: Rat LD_{50} Dose: 1058 mg/kg.

Acute Dermal: Cat LD_{Lo} Route: Subcutaneous Dose: 310 mg/kg; Toxic Effects: Behavioral - Ataxia; Gastrointestinal - Nausea or vomiting; Blood - Methemoglobinemia-Carboxyhemoglobin.

Mutagenic: Human Other Mutation Test Systems; Cell Type: HeLa cell; Dose: 1 mmol/L. Rat Unscheduled DNA Synthesis; Cell Type: liver; Dose: 100 umol/L.

Hazard Overviews

Fire Diamond

Health: Irritating to eyes. Also Causes: hematuria, hemorrhagic cystitis, painful/frequent urination, headache, drowsiness, nausea, cyanosis. Chronic Effects: bladder cancer, methemoglobinemia.
Fire: Combustible. Use water or dry chemical.
Reactivity: Stable. Hazardous polymerization cannot occur. Avoid: heat; ignition sources. Incompatible with: copper (II) chloride. Hazardous decomposition products: oxides of nitrogen; chloride fumes.
Carcinogenicity: IARC - Group 2A, Probably carcinogenic to humans; NIOSH - Not listed; NTP - Not listed; ACGIH - Not listed; OSHA - Not listed; EPA - Not listed; MAK - Class A1, Capable of inducing malignant tumors as shown by experience with humans
Primary Target Organs:

Eyes

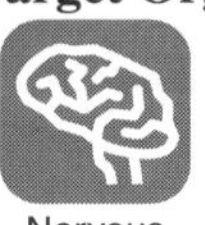
Nervous System

Kidneys

Blood

Environmental

Environmental Fate: If released to the atmosphere, it will degrade in the vapor-phase by reaction with photochemically produced hydroxyl radicals (estimated half-life of 9 hours); it may also undergo direct photolysis. Released to soil it should have moderate mobility with an estimated K_{oc} of 410. The presence of humus or organic matter in the soil may cause strong binding to soil however, due to the high reactivity of the aromatic amino group. Photolysis on soil surfaces may occur. Biodegradation is possible with 20% mineralization of this compound over 6 weeks reported in a soil system. Biologically mediated polymerizing transformations may also occur in soil. Volatilization from water surfaces will be slow based on an estimated Henry's Law constant of 2×10^{-6} atm-cu m/mole. Strong sorption to organic materials in the water column may be expected for some portion released to water. Bioconcentration in aquatic organisms is not a major fate process. Limited biodegradation data suggest that biodegradation in water will be slow. 7% removal of TOC from an industrial wastewater was reported; however GC analysis showed that more extensive degradation of the parent compound occurred.

Environmental Physical Data

Henry's Law Constant: estimated at 2×10^{-6}
Octanol/Water Partition Coefficient: log K_{ow} = 2.27
Sorption Partition Coefficient: K_{oc} = estimated at 410
BCF: calculated at 7.7 to 12

Regulations

RCRA 40CFR: Not listed
CERCLA: 40CFR 302.4: Not listed
SARA 40CFR 372.65: Listed
SARA EHS 40CFR 355: Not listed
TSCA: Listed

Analytical Methods

Water / Groundwater: ASTM D4763

CHL7450 CAS #: 3165-93-3

4-CHLORO-O-TOLUIDINE, HYDROCHLORIDE

RTECS: XU5250000
DOT: UN1579; IMO6.1
EINECS Number: 221-627-8
Molecular Formula: $C_7H_9Cl_2N$
Structured MF: $CH_3C_6H_3(Cl)NH_2$•HCl
Formula Weight: 178.07

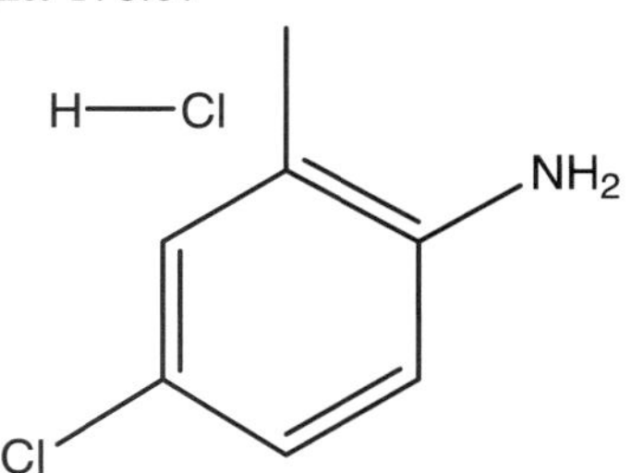

Chemical Structure

Synonyms: AMARTHOL FAST RED TR BASE; AMARTHOL FAST RED TR SALT; 2-AMINO-5-CHLOROTOLUENE HYDROCHLORIDE; ASYMMETRIC META-CHLORO-ORTHO-TOLUIDINE HYDROCHLORIDE; AZANIL RED SALT TRD; AZOENE FAST RED TR SALT; AZOGENE FAST RED TR; AZOIC DIAZO COMPONENT 11 BASE; BENZENAMINE,4-CHLORO-2-METHYL-,HYDROCHLORIDE; BRENTAMINE FAST RED TR SALT; C.I. 37085; C.I. AZOIC DIAZO COMPONENT 11; CHLORHYDRATE DE CHLORO-4 O-TOLUIDINE; CHLORHYDRATE DE 4-CHLOROORTHOTOLUIDINE; 5-CHLORO-2-AMINOTOLUENE HYDROCHLORIDE; 4-CHLORO-2-METHYLANILINE HYDROCHLORIDE; 4-CHLORO-6-METHYLANILINE HYDROCHLORIDE; 4-CHLORO-2-METHYLBENZENAMINE HYDROCHLORIDE; 4-CHLORO-2-TOLUIDINE HYDROCHLORIDE; 4-CHLORO-O-TOLUIDINE HYDROCHLORIDE; P-CHLORO-O-TOLUIDINE HYDROCHLORIDE; CLORHIDRATO DE 4-CLORO-O-TOLUIDINE; DAITO RED SALT TR; DEVOL RED K; DEVOL RED TA SALT; DEVOL RED TR; DIAZO FAST RED TR; DIAZO FAST RED TRA; FAST RED 5CT SALT; FAST RED SALT TR; FAST RED SALT TRA; FAST RED SALT TRN; FAST RED TR SALT; HINDASOL RED TR SALT; KROMON GREEN B; 2-METHYL-4-CHLOROANILINE HYDROCHLORIDE; NATASOL FAST RED TR SALT; NEUTROSEL RED TRVA; OFNA-PERL SALT RRA; RED BASE CIBA IX; RED BASE IRGA IX; RED SALT CIBA IX; RED SALT IRGA IX; RED TRS SALT; SANYO FAST RED SALT TR; O-TOLUIDINE,4-CHLORO-,HYDROCHLORIDE
Description: light orange powder
Use: in the production of azo dyes for cotton, silk, acetate and nylon

Physical Properties

Freezing Point: Decomposes at 92 °C (197.6 °F) to 94 °C (201.2 °F)
Water Solubility: >= 100 mg/mL at 22 C
Other Solubilities: 95% Ethanol: <1 mg/ml at 22 °C; Acetone: <1 mg/ml at 22 °C; DMSO: <1 mg/ml at 22 °C.
Flash Point: Not available; probably combustible

RTECS Toxicity Data

Mutagenic: Mouse Specific Locus Test; Route: Oral; Dose: 300 mg/kg/3D-C. Hamster DNA Damage; Cell Type: lung; Dose: 3 mmol/L.

Tumorigenic: Mouse Route: Oral; Dose: 49 gm/kg/78W-C; Toxic Effects: Tumorigenic - Carcinogenic by RTECS criteria; Vascular - Tumors. Mouse Route: Oral; Dose: 104 gm/kg/99W-C; Toxic Effects: Tumorigenic - Carcinogenic by RTECS criteria; Vascular - Tumors. Mouse Route: Oral; Dose: 108 gm/kg/78W-C; Toxic Effects: Tumorigenic - Carcinogenic by RTECS criteria; Vascular - Tumors. Mouse Route: Oral; Dose: 216 gm/kg/78W-C; Toxic Effects: Tumorigenic - Carcinogenic by RTECS criteria; Liver - Tumors. Mouse Route: Oral; Dose: 97 gm/kg/78W-C; Toxic Effects: Tumorigenic - Carcinogenic by RTECS criteria; Liver - Tumors.

Hazard Overviews

Fire: Will burn.

Carcinogenicity: IARC - Group 2A, Probably carcinogenic to humans; NIOSH - Not listed; NTP - Not listed; ACGIH - Not listed; OSHA - Not listed; EPA - Not listed; MAK - Not listed

Environmental

Cleanup/Disposal: Guide No. 153: Eliminate all ignition sources (no smoking, flares, sparks or flames in immediate area). Do not touch damaged containers or spilled material unless wearing appropriate protective clothing. Stop leak if you can do it without risk. Prevent entry into waterways, sewers, basements or confined areas. Absorb or cover with dry earth, sand or other non-combustible material and transfer to containers. Do not get water inside containers.

Regulations

RCRA 40CFR: Listed Hazardous Waste No. U049 Toxic Waste

CERCLA: 40CFR 302.4: Listed per RCRA Section 3001 RQ: 100 lb (45.35 kg)

SARA 40CFR 372.65: Not listed

SARA EHS 40CFR 355: Not listed

TSCA: Listed

CHL7540 CAS #: 1929-82-4

2-CHLORO-6-TRICHLORO-METHYL PYRIDINE

RTECS: US7525000

EINECS Number: 217-682-2

Molecular Formula: $C_6H_3Cl_4N$

Structured MF: $ClC_5H_3NCCl_3$

Formula Weight: 230.90

Cl Cl Cl N Cl

Chemical Structure

Synonyms: 2-CHLORO-6-TRICHLOROMETHYL PYRIDINE; 2-CHLORO-6-(TRICHLORO-METHYL)PYRIDINE; 2-CHLORO-6-(TRICHLOROMETHYL)PYRIDINE; 2-CHLORO-6-TRICHLOROMETHYLPYRIDINE; DONCO-163; DOWCO-163; N-SERVE; N-SERVE NITROGEN STABILIZER; NITRAPYRIN; 2,2,2,6-TETRACHLORO-2-PICOLINE; ALPHA,ALPHA,ALPHA-6-TETRACHLORO-2-PICOLINE

Description: colorless to off-white crystalline solid or crystals; mild, sweet odor

Use: fertilizer additive to control nitrification and prevent loss of soil nitrogen

Physical Properties

Boiling Point: 136 °C (277 °F) to 137.5 °C (280 °F)

Freezing Point: 62.78 °C (145.004 °F)

Saturated Vapor Density: 1.20003078 kg/m^3

Vapor Pressure: 0.0028 mm Hg at 23 °C

Water Solubility: Insoluble

Other Solubilities: 95% Ethanol: >=100 mg/ml at 21.5 °C; Acetone: >=100 mg/ml at 21.5 °C; Anhydrous Ammonia: Soluble; DMSO: >=100 mg/ml at 21.5 °C; Xylene: Soluble.

Flash Point: Not available; probably combustible

RTECS Toxicity Data

Acute Oral: Rat LD_{50} Dose: 940 mg/kg. Mouse LD_{50} Dose: 710 mg/kg.

Acute Dermal: Rabbit LD_{50} Route: Skin; Dose: 850 mg/kg.

Reproductive/Teratogenic: Rabbit Route: Oral; Dose: 390 mg/kg; Duration: female 6-18D of pregnancy; Specific Developmental Abnormalities - Craniofacial (including nose and tongue).

Mutagenic: Bacteria - S Typhimurium Mutations in Microorganisms; Dose: 100 ug/plate (-S9).

Hazard Overviews

Health: Irritating to eyes/skin/respiratory tract. Toxic. Other Acute Effects: harmful if swallowed, inhaled, or absorbed through skin; readily absorbed through skin. The toxicological properties have not been thoroughly investigated.

Fire: Will burn. Hazards: emits toxic fumes. Extinguishing agents: water spray; carbon dioxide, dry chemical powder or appropriate foam. Precautions: combustible liquid.

Reactivity: Stable. Hazardous polymerization will not occur. Hazardous decomposition products: carbon monoxide, carbon dioxide, nitrogen oxides, hydrogen chloride gas.

Carcinogenicity: IARC - Not listed; NIOSH - Not listed; NTP - Not listed; ACGIH - Class A4, Not classifiable as a human carcinogen; OSHA - Not listed; EPA - Not listed; MAK - Not listed

Primary Target Organs:

Exposure Limits

OSHA PEL: TWA: 15 mg/m^3; total dust.

ACGIH TLV: TWA: 10 mg/m^3; STEL: 20 mg/m^3.

NIOSH REL: TWA: 10 mg/m^3. STEL: 20 mg/m^3; total particulate.

Respirator Recommendation

Exposure Range: >5 to 50 mg/m^3 Air Purifying, Negative Pressure, Half Mask

Exposure Range: >50 to 500 mg/m^3 Air Purifying, Negative Pressure, Full Face

Exposure Range: >500 to 5000 mg/m^3 Supplied Air, Constant Flow/Pressure Demand, Full Face

Exposure Range: >5000 to unlimited mg/m^3 Self-contained Breathing Apparatus, Pressure Demand, Full Face

Cartridge Color: black with magenta (P100)

Environmental

Cleanup/Disposal: Guide No. 154: Eliminate all ignition sources (no smoking, flares, sparks or flames in immediate area). Do not touch damaged containers or spilled material unless wearing appropriate protective clothing. Stop leak if you can do it without risk. Prevent entry into waterways, sewers, basements or confined areas. Absorb or cover with dry earth, sand or other non-combustible material and transfer to containers. Do not get water inside containers.

Regulations

RCRA 40CFR: Not listed

CERCLA: 40CFR 302.4: Not listed

SARA 40CFR 372.65: Listed

SARA EHS 40CFR 355: Not listed

TSCA: Listed

Analytical Methods

Food: FDA 212.1, 232.1

Plasma: FDA 211.1, 231.1, 252

CHL7570 CAS #: 961-11-5

2-CHLORO-1-(2,4,5-TRICHLOROPHENYL)VINYL DIMETHYL PHOSPHATE

RTECS: TB9050000

EINECS Number: 213-506-3

Molecular Formula: $C_{10}H_9Cl_4O_4P$

Formula Weight: 365.96

Chemical Structure

Synonyms: BENZYL ALCOHOL,2,4,5-TRICHLORO-ALPHA-(CHLOROMETHYLENE)-,DIMETHYL PHOSPHATE; 2-CHLORO-1-(2,4,5-TRICHLOROPHENYL)VINYL PHOSPHORIC ACID DIMETHYL ESTER; 2-CHLORO-1-(2,4,5-TRICHLOROPHENYL)VINYL PHOSPHORIC ACIDDIMETHYL ESTER; O,O-DIMETHYL-O-2-CHLOR-1-(2,4,5-TRICHLORPHENYL)-VINYL-PHOSPHAT; GARDONA; IPO 8; PHOSPHORIC ACID,2-CHLORO-1-(2,4,5-TRICHLOROPHENYL)ETHENYLDIMETHYL ESTER; PHOSPHORIC ACID,2-CHLORO-1-(2,4,5-TRICHLOROPHENYL)VINYL DIMETHYL ESTER; RABON; RABOND; 2,4,5-TRICHLORO-ALPHA-(CHLOROMETHYLENE)BENZYL ALCOHOL DIMETHYL PHOSPHATE; 2,4,5-TRICHLORO-ALPHA-(CHLOROMETHYLENE)BENZYL PHOSPHATE

Description: colorless crystals or white powder

Use: insecticide

Physical Properties

Freezing Point: 95 °C (203 °F) to 97 °C (206.6 °F)

Saturated Vapor Density: 1.200000001 kg/m^3

Vapor Pressure: 0.000000041 mm Hg at 20 °C

Water Solubility: < 1 mg/mL at 23 C

Other Solubilities: 95% Ethanol: 10-50 mg/ml at 23 °C; Acetone: 10-50 mg/ml at 23 °C; Absolute Ethanol: Soluble; Chloroform: 400 g/kg at 20 °C; DMSO: 10-50 mg/ml at 23 °C; Dichloromethane: 40 g/100 mL at 20 °C; Xylene: <15 g/100 mL at 20 °C.

Flash Point: Not available; probably combustible

RTECS Toxicity Data

Acute Oral: Rat LD_{50} Dose: 4 gm/kg.

Acute Dermal: Rat LD_{50} Route: Skin; Dose: >10 gm/kg; Toxic Effects: Sense organs and special senses - Lacrimation; Kidney, Ureter, and Bladder - Other changes; Skin and appendages - Hair. Rat LD_{50} Route: Subcutaneous Dose: >15 gm/kg; Toxic Effects: Sense organs and special senses - Lacrimation; Kidney, Ureter, and Bladder - Other changes.

Hazard Overviews

Fire: Will burn.

Carcinogenicity: IARC - Not listed; NIOSH - Not listed; NTP - Not listed; ACGIH - Not listed; OSHA - Not listed; EPA - Not listed; MAK - Not listed

Environmental

Regulations

RCRA 40CFR: Not listed

CERCLA: 40CFR 302.4: Listed as Compound per CWA Section 307(a)
SARA 40CFR 372.65: Listed
SARA EHS 40CFR 355: Not listed
TSCA: Not listed

CHL7600 CAS #: 75-88-7

2-CHLORO-1,1,1-TRIFLUORO-ETHANE

RTECS: KH8008500
EINECS Number: 200-912-0
Molecular Formula: $C_2H_2ClF_3$
Formula Weight: 118.49

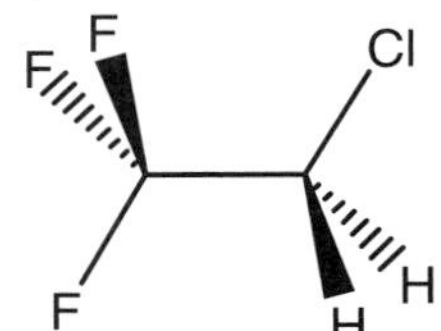

Chemical Structure

Synonyms: CFC 133A; 1-CHLORO-2,2,2-TRIFLUOROETHANE; 2-CHLORO-1,1,1-TRIFLUOROETHANE; FC 133A; FREON 133A; GENETRON 133A; R 133A; 1,1,1-TRIFLUORO-2-CHLOROETHANE; 2,2,2-TRIFLUOROCHLOROETHANE; 1,1,1-TRIFLUOROETHYL CHLORIDE

Physical Properties

Boiling Point: 6.9 °C (44.42 °F)
Freezing Point: -105.5 °C (-157.9 °F)
Specific Gravity: 1.389
Refraction Index: 1.3090

RTECS Toxicity Data

Acute Inhalation: Mouse LC_{50} Dose: 15 pph/1hr; Toxic Effects: Behavioral - General anesthetic; Behavioral - Convulsions or effect on seizure threshold. Guinea Pig LC_{Lo} Dose: 33000 ppm/5hr; Toxic Effects: Lungs, Thorax, or Respiration - Structural or functional change in trachea or bronchi; Lungs, Thorax, or Respiration - Acute pulmonary edema.

Tumorigenic: Rat Route: Oral; Dose: 78 gm/kg/1Y-I; Toxic Effects: Tumorigenic - Carcinogenic by RTECS criteria; Tumorigenic effects - Testicular tumors; Tumorigenic effects - Uterine tumors.

Hazard Overviews

Health: Severely irritating to skin. Harmful. Other Acute Effects: may be harmful if inhaled; can cause rapid suffocation; nausea; dizziness; headache; local frostbite; anesthesia; target organs: nerves, heart. Chronic Effects: possible risk of irreversible effects. Possible carcinogen.

Fire: Hazards: emits toxic fumes; container explosion may occur. Extinguishing agents: use water spray or fog nozzle to keep cylinder cool; move cylinder away from fire if there is no risk. Precautions: combustible liquid.

Reactivity: Incompatible with: store away from heat and direct sunlight. Hazardous decomposition products: toxic fumes of: carbon monoxide, carbon dioxide, hydrogen fluoride, hydrogen chloride gas.

Carcinogenicity: IARC - Group 3, Not classifiable as to carcinogenicity to humans; NIOSH - Not listed; NTP - Not listed; ACGIH - Not listed; OSHA - Not listed; EPA - Not listed; MAK - Not listed

Primary Target Organs:

Skin

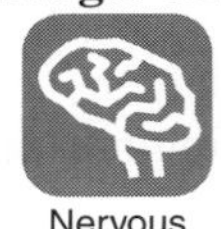
Nervous System

Cardio-vascular

Environmental

Regulations

RCRA 40CFR: Not listed
CERCLA: 40CFR 302.4: Listed as Compound per CWA Section 307(a)
SARA 40CFR 372.65: Listed
SARA EHS 40CFR 355: Not listed
TSCA: Listed

CHL7630 CAS #: 79-38-9

CHLOROTRIFLUOROETHYLENE

RTECS: KV0525000
DOT: UN1082; IMO2.3
EINECS Number: 201-201-8
Molecular Formula: C_2ClF_3
Structured MF: $F_2C{=}CFCl$
Formula Weight: 116.47

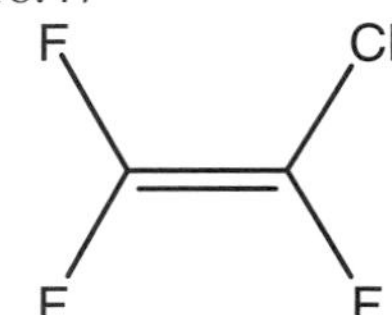

Chemical Structure

Synonyms: CFE; 1-CHLORO-1,2,2-TRIFLUOROETHYLENE; 2-CHLORO-1,1,2-TRIFLUOROETHYLENE; CHLORTRIFLUORAETHYLEN; CTFE; DAIFLON; ETHENE,CHLOROTRIFLUORO-; ETHENE,CHLOROTRIFLUORO-(9CI); ETHYLENE,CHLOROTRIFLUORO-; ETHYLENE,TRIFLUOROCHLORO-; FLUOROPLAST 3; GENETRON 1113; MONOCHLOROTRIFLUOROETHYLENE; R1113; TRIFLUORCHLORETHYLEN; TRIFLUOROCHLORETHYLENE; 1,1,2-TRIFLUORO-2-CHLOROETHYLENE; TRIFLUOROCHLOROETHYLENE; TRIFLUOROCHLOROETHYLENE,INHIBITED; TRIFLUOROMONOCHLOROETHYLENE; TRIFLUOROVINYL CHLORIDE; TRITHENE

Description: colorless gas; faint, ethereal odor

Use: intermed; monomer for chlorotrifluoroethylene resins; to produce kel-f extremely chemly inert material; in mfr of high performance lubricants, plastics & elastomers; to prepare bromotrifluoroethylene; intermed for homopolymers & copolymers, inhalation anesthetic halothane; fluorinating agent to replace hydroxyl groups in steroids & carbohydrates with fluorine; in telomerization with carbon tetrachloride or chloroform for inert fluids, hydraulic fluids, or lubricants.

Physical Properties

Boiling Point: -27.9 °C (-18 °F)
Freezing Point: -158.2 °C (-252.76 °F)
Specific Gravity: Liquid 1.305 at 20 °C
Vapor Density: 4.02 Air=1
Vapor Pressure: 4.592 $x10^{+3}$ mm Hg at 25 °C
Water Solubility: Decomposes in Water
Other Solubilities: Soluble in Benzene, Chloroform
Surface Tension: Estimated at 12 dynes/cm
Refraction Index: 1.38
Critical Temperature: 106.2 °C
Critical Pressure: 592 psia
Ionization Potential (eV): 9.76 +/-2.0
Flash Point: -27.778 °C
LEL: 8.4% v/v
UEL: 38.7% v/v

RTECS Toxicity Data

Acute Oral: Mouse LD_{50} Dose: 268 mg/kg.
Acute Inhalation: Rat LC_{50} Dose: 1000 ppm/4hr; Toxic Effects: Lungs, Thorax, or Respiration - Other changes; Gastrointestinal - Other changes. Mouse LC_{50} Dose: 3000 ppm/7hr.

Hazard Overviews

Flammable

Compressed Gas

Fire Diamond

Health: Causes: asphyxiation, dizziness, nausea, vomiting, frostbite. Chronic Effects: dermatitis.
Fire: Flammable. For small fires use dry chemical or carbon dioxide. For large fires use water spray, fog, or regular foam.
Reactivity: Stable. Hazardous polymerization can occur under certain conditions. Avoid: exposure to heat; flames. Incompatible with: bromine/oxygen (explosive reaction due to the formation of chlorotrifluoroethylene peroxide); chlorine trifluoride/water; chlorine perchlorate; oxygen; ethylene (explosive polymerization); 1,1-dichloroethylene. Hazardous decomposition products: toxic hydrogen chloride; hydrogen fluoride gas.
Carcinogenicity: IARC - Not listed; NIOSH - Not listed; NTP - Not listed; ACGIH - Not listed; OSHA - Not listed; EPA - Not listed; MAK - Not listed

Primary Target Organs:

Eyes

Skin

Respiratory System

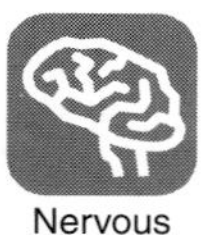
Nervous System

Exposure Limits
AIHA WEEL: TWA: 5 ppm.
Respirator Recommendation
Exposure Range: >5 to 50 ppm Supplied Air, Constant Flow/Pressure Demand, Half Mask
Exposure Range: >50 to 5000 ppm Supplied Air, Constant Flow/Pressure Demand, Full Face
Exposure Range: >5000 to unlimited ppm Self-contained Breathing Apparatus, Pressure Demand, Full Face
Note: odor threshold unknown

Environmental

Environmental Fate: If released to the atmosphere, it is expected to exist solely in the vapor phase based on an experimentally derived vapor pressure of 4600 mm Hg at 25 °C. In the vapor phase, it will degrade via reaction with hydroxyl radicals with experimentally derived half-lives of about 2 days. It is also expected to react with both atomic oxygen and ozone with rate constants of 2.7 $x10^{-11}$ and 1.6 $x10^{-20}$ cu cm/mol sec, respectively. An estimated half-life of 715 years is calculated for the reaction with ozone. If released to soil, it is expected to have moderate mobility based on an estimated K_{oc} of 190. In moist soils, it may decompose. It may volatilize from moist soil surfaces with an estimated Henry's Law constant of 0.31 atm-cu m/mol; it is expected to volatilize rapidly from dry soils due to its high vapor pressure. In water, it is expected to decompose. In a landfill leachate, this compound had a half-life transformation rate of 42 days; in a sulfide-containing buffer, degradation was complete in less than one day. Volatilization from water surfaces may also be a major fate process for this chemical based on its Henry's Law constant. Bioconcentration in aquatic organisms is not an important fate process based on an estimated BCF of 11.
Cleanup/Disposal: Guide No. 119: Eliminate all ignition sources (no smoking, flares, sparks or flames in immediate area). All equipment used when handling the product must be grounded. Fully encapsulating, vapor protective clothing should be worn for spills and leaks with no fire. Do not touch or walk through spilled material. Stop leak if you can do it without risk. Do not direct water at spill or source of leak. Use water spray to reduce vapors or divert vapor cloud drift. For chlorosilanes, use AFFF alcohol-resistant medium expansion foam to reduce vapors. If possible, turn leaking containers so that gas escapes rather than liquid. Prevent entry into waterways, sewers, basements or confined areas. Isolate area until gas has dispersed.

Environmental Physical Data

Henry's Law Constant: estimated at 0.31
Octanol/Water Partition Coefficient: log K_{ow} = 1.65
Sorption Partition Coefficient: K_{oc} = estimated at 190
BCF: estimated at 11

BOD: none

Regulations

RCRA 40CFR: Not listed
CERCLA: 40CFR 302.4: Not listed
SARA 40CFR 372.65: Not listed
SARA EHS 40CFR 355: Not listed
TSCA: Listed

CHL7660 CAS #: 75-72-9

CHLOROTRIFLUOROMETHANE

RTECS: PA6410000
DOT: UN1022; IMO2.2
EINECS Number: 200-894-4
Molecular Formula: $CClF_3$
Structured MF: $CClF_3$
Formula Weight: 104.46

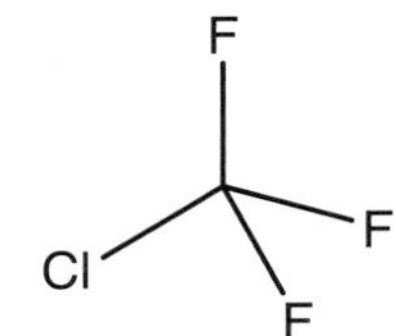

Chemical Structure

Synonyms: ARCTON 3; F 13; FC 13; FLUOROCARBON 13; FREON 13; FRIGEN 13; GENETRON 13; HALOCARBON 13/UCON 13; METHANE,CHLOROTRIFLUORO-; METHANE,MONOCHLOROTRIFLUORO-; MONOCHLOROTRIFLUOROMETHANE; R 13; R13; TRIFLUOROCHLOROMETHANE; TRIFLUOROMETHYL CHLORIDE; TRIFLUOROMONOCHLOROCARBON
Description: colorless gas; ethereal odor
Use: refrigerant in aircraft test chambers; metal & pharmaceutical processing; dielectric & aerospace chemical; coolant; hardening of metals; as etch gas in integrated circuit board manufacture

Physical Properties

Boiling Point: -81.1 °C (-114 °F) at 760 mm Hg
Freezing Point: -181 °C (-293.8 °F)
Specific Gravity: 1.298 at -30 °C
Vapor Density: 3.6 Air=1
Vapor Pressure: 1 mm Hg at -149.5 °C
Surface Tension: 14 dynes/cm at -73.3 °C
Critical Temperature: 28.85 °C
Critical Pressure: 38.2 atm
Ionization Potential (eV): 12.5
Flash Point: Nonflammable

Hazard Overviews

Compressed Gas

Health: Other Acute Effects: may be harmful if inhaled; rapid suffocation; nausea; dizziness; headache; target organ: heart.
Fire: Noncombustible. Hazards: emits toxic fumes. Extinguishing agents: use water spray or fog nozzle to keep cylinder cool; move cylinder away from fire if there is no risk. Precautions: combustible liquid.
Reactivity: Incompatible with: potassium, sodium, aluminum, magnesium, zinc, store away from heat and direct sunlight. Hazardous decomposition products: carbon monoxide, carbon dioxide, hydrogen chloride gas, phosgene gas, hydrogen fluoride.
Carcinogenicity: IARC - Not listed; NIOSH - Not listed; NTP - Not listed; ACGIH - Not listed; OSHA - Not listed; EPA - Not listed; MAK - Not listed
Primary Target Organs:

Cardio-vascular

Exposure Limits
DFG MAK: TWA: 1000 ppm; 4300 mg/m^3.

Environmental

Cleanup/Disposal: Guide No. 126: Do not touch or walk through spilled material. Stop leak if you can do it without risk. Do not direct water at spill or source of leak. Use water spray to reduce vapors or divert vapor cloud drift. If possible, turn leaking containers so that gas escapes rather than liquid. Prevent entry into waterways, sewers, basements or confined areas. Allow substance to evaporate. Ventilate the area.

Regulations

RCRA 40CFR: Not listed
CERCLA: 40CFR 302.4: Listed as Compound per CWA Section 307(a)
SARA 40CFR 372.65: Listed
SARA EHS 40CFR 355: Not listed
TSCA: Listed

CHL7720 CAS #: 460-35-5

3-CHLORO-1,1,1-TRIFLUORO-PROPANE

RTECS: TX6200000
EINECS Number: 207-307-0
Molecular Formula: $C_3H_4ClF_3$
Formula Weight: 132.52
Synonyms: 1-CHLORO-3,3,3-TRIFLUOROPROPANE; 3-CHLORO-1,1,1-TRIFLUOROPROPANE; FREON 253; 1,1,1-TRIFLUORO-3-CHLOROPROPANE

Physical Properties

Boiling Point: 45.1 °C (113.18 °F)
Freezing Point: -107 °C (-160.6 °F)
Specific Gravity: 1.3253
Saturated Vapor Density: 3.162076999 kg/m^3

Vapor Pressure: 46.4 kPa at 25 °C
Refraction Index: 1.3350

RTECS Toxicity Data

Acute Oral: Mouse LD_{50} Dose: 62 mg/kg; Toxic Effects: Behavioral - Sleep; Behavioral - Somnolence (general depressed activity); Behavioral - Ataxia.
Acute Inhalation: Mouse LC_{50} Dose: 800 mg/m^3/2hr. Rat LC_{Lo} Dose: 1800 mg/m^3/2hr.
Chronic (Multiple Dose) Oral: Rat Dose: 1062 mg/kg/30W-I; Toxic Effects: Behavioral - Alteration of classical conditioning. Rabbit Dose: 1062 mg/kg/30W-I; Toxic Effects: Blood - Change in clotting factors; Blood - Other changes.

Hazard Overviews

Carcinogenicity: IARC - Not listed; NIOSH - Not listed; NTP - Not listed; ACGIH - Not listed; OSHA - Not listed; EPA - Not listed; MAK - Not listed

Environmental

Regulations

RCRA 40CFR: Not listed
CERCLA: 40CFR 302.4: Not listed
SARA 40CFR 372.65: Listed
SARA EHS 40CFR 355: Not listed
TSCA: Listed

CHL7810 CAS #: 1982-47-4

CHLOROXURON

RTECS: YS6125000
EINECS Number: 217-843-7
Molecular Formula: $C_{15}H_{15}ClN_2O_2$
Formula Weight: 290.75
Synonyms: C 1983; C-1933; 3-(4-(4-CHLOOR-FENOXY)-FENOXY)-FENYL)-1,1-DIMETHYLUREUM; 3-(4-(4-CHLOOR-FENOXY)-FENYL)-1,1-DIMETHYLUREUM; 3-(4-(4-CHLORO-FENOSSIL)FENIL)-1,1-DIMETIL-UREA; 3-(4-(4-CHLOROPHENOXY)PHENYL-1,1-DIMETHYLUREA; 3-(P-(P-CHLOROPHENOXY)PHENYL)-1,1-DIMETHYLUREA; N'-4-(4-CHLOROPHENOXY)PHENYL-N,N-DIMETHYLUREA; 1-(4-(4-CHLORO-PHENOXY)PHENYL)-3,3-D'METHYLUREE; CHLOROXIFENIDIM; CHLORPHENCARB; 3-(4-(4-CHLOR-PHENOXY)-PHENYL)-1,1-DIMETHYLHARNSTOFF; CIBA 1983; 3-(4-(4-CLORO-FENOSSIL)-1,1-DIMETIL-UREA; GESAMOOS; NOREX; TENORAN; UREA,3-(P-(P-CHLOROPHENOXY)PHENYL)-1,1-DIMETHYL-; UREA,N'-(4-(4-CHLOROPHENOXY)PHENYL)-N,N-DIMETHYL-
Description: colorless to white crystals or powder; odorless
Use: formely, selective pre- & early post-emergent herbicide in peas, beans, carrots, celery, celeriac, onions, leeks, garlic, chives, fennel, parsley, dill, tomatoes, cucurbits, soya beans, strawberries, ornamentals, fruit trees, conifers, & ornamentals; control of annual broad-leaved weeds & grasses; control of mosses in ornamental & sports turf, on paths, non-crop land, & in glasshouses.

Physical Properties

Freezing Point: 151 °C (303.8 °F) to 152 °C (305.6 °F)
Density: 1.34 g/cu cm at 20 °C
Vapor Pressure: 239 nPa at 20 °C
Water Solubility: 4 mg/L at 20 °C
Other Solubilities: Soluble in dimethylformamide and Chloroform. Slightly Soluble in Benzene and Diethyl Ether.
pH: Very low solubility; no significant pH
Flash Point: Nonflammable

RTECS Toxicity Data

Acute Oral: Rat LD_{50} Dose: 3700 mg/kg. Mouse LD_{50} Dose: >1 gm/kg.
Acute Inhalation: Rat LC_{50} Dose: >1350 mg/m^3/6hr.
Acute Dermal: Rabbit LD_{50} Route: Skin; Dose: >10 gm/kg. Rat LD_{50} Route: Skin; Dose: >3 gm/kg.

Hazard Overviews

Fire: Noncombustible.
Carcinogenicity: IARC - Not listed; NIOSH - Not listed; NTP - Not listed; ACGIH - Not listed; OSHA - Not listed; EPA - Not listed; MAK - Not listed

Environmental

Ecotoxicity: LD_{50} Mallard duck oral greater than 2000 mg/kg, 3-4 mo old females LC_{50} Rainbow trout 0.43 mg/l/96 hr at 12 °C (95% confidence limit 0.36-0.51 mg/l), wt 0.7 g. Static bioassay without aeration, pH 7.2-7.5, water hardness 40-50 mg/l as calcium carbonate and alkalinity of 30-35 mg/l

Environmental Physical Data

Octanol/Water Partition Coefficient: log K_{ow} = 4.0

Regulations

RCRA 40CFR: Not listed
CERCLA: 40CFR 302.4: Not listed
SARA 40CFR 372.65: Not listed TPQ: 500/10000 lb
SARA EHS 40CFR 355: Listed TPQ: 500 lb
TSCA: Not listed

Analytical Methods

Soil: EPA PMD-TLC; SW846 8321A
Food: FDA 212.1, 212.2, 232.1, 232.4, 242.1; AOAC 977.06
Plasma: FDA 211.1, 221.1, 231.1, 242.4, 252

CHL8050 CAS #: 2921-88-2

CHLORPYRIFOS

RTECS: TF6300000
DOT: NA2783
EINECS Number: 220-864-4
Molecular Formula: $C_9H_{11}Cl_3NO_3PS$
Formula Weight: 350.59
Synonyms: BRODAN; CHLOROPYRIFOS; CHLOROPYRIPHOS; CHLORPYRIFOS-ETHYL; CHLORPYRIPHOS; CHLORPYRIPHOS-ETHYL; COROBAN; DANUSBAN; DETMOL U.A; DETMOL UA; DHANUSBAN; O,O-DIAETHYL-O-3,5,6-TRICHLOR-2-PYRIDYLMONOTHIOPHOSPHAT; O,O-DIETHYL O-(3,5,6-

TRICHLORO-2-PYRIDINYL)PHOSPHOROTHIOIC ACID ESTER; O,O-DIETHYL O-3,5,6-TRICHLORO-2-PYRIDYL PHOSPHOROTHIOATE; O,O-DIETHYL O-(3,5,6-TRICHLORO-2-PYRIDYL)PHOSPHOROTHIOIC ACID ESTER; O,O-DIETHYLO-(3,5,6-TRICHLORO-2-PYRIDINYL)PHOSPHOROTHIOATE; DOWCO 179; DURMET; DURSBAN; DURSBAN 4E; DURSBAN 10CR; DURSBAN F; DURSBAN R; ENT 27,311; ENT 27311; EPA PESTICIDE CODE 059101; ERADEX; ETHION,DRY; KILLMASTER; LORSBAN; LORSBAN 50SL; OMS-0971; PHOSPHOROTHIOIC ACID,O,O-DIETHYL O-(3,5,6-TRICHLORO-2-PYRIDINYL)ESTER; PHOSPHOROTHIOIC ACID,O,O-DIETHYL O-(3,5,6-TRICHLORO-2-PYRIDYL)ESTER; PHOSPHOROTHIOIC ACID,O,O-DIETHYLO-(3,5,6-TRICHLORO-2-PYRIDINYL) ESTER; PHOSPHOROTHIOIC ACID,O,O-DIETHYLO-(3,5,6-TRICHLORO-2-PYRIDYL) ESTER; PIRIDANE; 2-PYRIDINOL,3,5,6-TRICHLORO-,O-ESTER WITH O,O-DIETHYL PHOSPHOROTHIOATE; 2-PYRIDINOL,3,5,6-TRICHLORO-,O-ESTER WITH O,O-DIETHYLPHOSPHOROTHIOATE; PYRINEX; STIPEND; SUSCON; TERIAL; TERIAL 40L; 3,5,6-TRICHLORO-2-PYRIDINOL O-ESTER WITH O,O-DIETHYL PHOSPHOROTHIOATE; ZIDIL

Description: colorless to white crystalline solid or crystals; mild, mercaptan-like odor

Use: a broad-range contact insecticide and acaricide; in the control of soil insects and some foliar insect pests on a wide range of crops, including pome fruit, stone fruit, citrus fruit, nut crops, strawberries, figs, bananas, vines, vegetables, potatoes, beet, tobacco, soya beans, sunflowers, sweet potatoes, ground nuts, rice, cotton, lucerne, cereal maize, sorghum, asparagus, glasshouse and outdoor ornamentals, mushrooms, and in forestry; in the control of insect pests in stored product household insect pests (including ants and cockroaches), flies and other insects in animal houses, and mosquitoes (adults and larvae); in the control of chinch bugs in Gulf Coast states and in tick control on cattle and sheep (Australia); as an animal ectoparasiticide

Physical Properties

Boiling Point: Decomposes at 160 °C (320 °F)
Freezing Point: 41 °C (105.8 °F) to 42 °C (107.6 °F)
Specific Gravity: 1.398 at 43.5 °C (liquid)
Vapor Density: Calculated 12.09 Air=1
Saturated Vapor Density: 1.200000327 kg/m^3
Vapor Pressure: 1.87 x10^{-5} mm Hg at 25 °C
Water Solubility: 0.7 ppm in Water at 20 °C
Other Solubilities: Solubility at 25 °C: 6.5 kg/kg Acetone; Solubility at 25 °C: 7.9 kg/kg Benzene; Solubility at 25 °C: 6.3 kg/kg Chloroform; Solubility at 25 °C: 450 g/kg Methanol; Soluble in most organic solvents.
Flash Point: 27.778 °C Closed Cup

RTECS Toxicity Data

Acute Oral: Man TD_{Lo} Dose: 300 mg/kg; Toxic Effects: Peripheral nerve and sensation - Pareshtesia; Behavioral - Muscle weakness; Behavioral - Coma.

Acute Dermal: Guinea Pig LD_{Lo} Route: Subcutaneous Dose: 100 mg/kg.

Mutagenic: Human Sister Chromatid Exchenge; Cell Type: lymphocyte; Dose: 2 mg/L. Mouse Cytogenetic Analysis; Cell Type: other cell types; Dose: 500 ug/L.

Hazard Overviews

Flammable

Fire: Flammable.

Carcinogenicity: IARC - Not listed; NIOSH - Not listed; NTP - Not listed; ACGIH - Class A4, Not classifiable as a human carcinogen; OSHA - Not listed; EPA - Not listed; MAK - Not listed

Exposure Limits

OSHA PEL Vacated 1989 Limits: TWA: 0.2 mg/m^3.
ACGIH TLV: TWA: 0.2 mg/m^3.
NIOSH REL: TWA: 0.2 mg/m^3. STEL: 0.6 mg/m^3; skin.

Respirator Recommendation

Exposure Range: >0.2 to 2 mg/m^3 Air Purifying, Negative Pressure, Half Mask
Exposure Range: >2 to 20 mg/m^3 Air Purifying, Negative Pressure, Full Face
Exposure Range: >20 to 200 mg/m^3 Supplied Air, Constant Flow/Pressure Demand, Full Face
Exposure Range: >200 to unlimited mg/m^3 Self-contained Breathing Apparatus, Pressure Demand, Full Face
Cartridge Color: black with magenta (P100)

Environmental

Ecotoxicity: TLm Cymatogaster aggregata (shiner perch) 3.5 ppb/96 hr static lab bioassay TL_{50} Palaemon macrodactylus (korean shrimp) 0.01 (0.002-0.046) ppm/96 hr intermittent flow lab bioassay LC_{90-95} Bactis rhodani (ephimetoptera) 0.01-0.02 ppm/1 hr /Conditions of bioassay not specified LD_{50} Pheasant 3-5 month-old males, oral 8.41 mg/kg (95% confidence limit 2.77-25.5 mg/kg) LD_{50} Quiscalus quiscula (common grackle) oral 13 mg/kg adult LC_{50} Cyprinodon variegatus (sheepshead minnow) juvenile > 1000 ug/l/24 hr at a salinity of 24 g/kg /99% purity; Conditions of bioassay not specified LC_{50} Leiostomus xanthurus (spot) juvenile 7 ug/l/48 hr at a salinity of 26 g/kg /99% purity; Conditions of bioassay not specified LC_{50} Fundulus similis (longnose killifish) 3.2 ug/l/48 hr at a salinity of 24 g/kg /99% purity; Conditions of bioassay not specified LC_{50} Aedes aegypti (mosquito) 2nd instar 0.0011 ug/l/24 hr; 4th instar 0.0014 ug/l/24 hr LC_{50} Coturnix (Japanese quail) oral 293 ppm (95% confidence limit 112-767 ppm) LC_{50} Gammarus lacustris (crustacean) 0.11 ug/l/96 hr /Conditions of bioassay not specified LD_{50} Apis mellifera (honeybee) approx 1.14 ug/bee (as dust) adult worker /Topical application LC_{50} Ephemerella sp (mayfly) 0.33 ug/l/72 hr LD_{50} Rana catesbiana (Bullfrog) males oral more than 400 mg/kg /conditions of bioassay not specified

Environmental Fate: If released to soil, it can degrade by a combination of chemical hydrolysis and microbial degradation. The chemical hydrolysis is clay catalyzed and yields a primary degradation product of 3,5,6-trichloro-2-pyridinol. Volatilization from soil surfaces, is expected to contribute to its loss from soil. It is tightly absorbed by soil and not expected to leach significantly. Although a general soil persistence of 60-120 days has been reported, the persistence can vary greatly depending on soil type, climate, and conditions and has been experimentally measured to range from as little as 2 weeks to over 1 year. If released to water, it partitions significantly from the water column to sediments. The measured hydrolysis half-life at 25 °C at (or near) neutral conditions is 35-78 days. The hydrolysis rate is relatively independent of pH from pH 1 to pH 7, increases significantly under alkaline conditions, decreases 2.5-3 fold with 10 °C temperature decrease, is markedly enhanced by the presence of Cu(+2) ions in sufficient concentration, and is not affected by adsorption to sediments in acidic or neutral water. The hydrolysis products include 3,5,6-trichloro-2-pyridinol and various trichloropyridyl phosphorothioates. The photolysis half-life at the water surface in the US during the mid summer is about 3 to 4 weeks, however, photolysis is not expected to be a very significant removal mechanism in relatively deep waters, in the winter-time, or in any natural waters containing sufficient light attenuating material. Microbial degradation may contribute to removal in some natural waters. The volatilization half-life from a river one meter deep flowing 1m/sec with a wind velocity of 3 m/sec is estimated to be 5.7 days; however, the significance of volatilization may be greatly decreased by aquatic sediment adsorption. Experimental and estimated log BCF values ranging from 2.50 to 3.54 indicate potential significant bioconcentration. The desorption from sediments can contribute to long term residual concentration in the water column (low ppb). If released to air, it will react in the vapor-phase with photochemically produced hydroxyl radical half-life of 13.74 hours, but it is not expected to react with ozone. Photolysis in air may contribute to its transformation.

Cleanup/Disposal: Guide No. 152: Do not touch damaged containers or spilled material unless wearing appropriate protective clothing. Stop leak if you can do it without risk. Prevent entry into waterways, sewers, basements or confined areas. Cover with plastic sheet to prevent spreading. Absorb or cover with dry earth, sand or other non-combustible material and transfer to containers. Do not get water inside containers.

Environmental Physical Data

Henry's Law Constant: estimated at 1.23×10^{-5}
Octanol/Water Partition Coefficient: log K_{ow} = 4.96
Sorption Partition Coefficient: K_{oc} = 4381 to 6129
BCF: mosquito fish 2.67
BOD: degradable

Regulations

RCRA 40CFR: Not listed
CERCLA: 40CFR 302.4: Listed per CWA Section 311(b)(4) RQ: 1 lb (0.454 kg)
SARA 40CFR 372.65: Not listed
SARA EHS 40CFR 355: Not listed
TSCA: Not listed

Analytical Methods

Air: EPA TO-10
Soil: EPA PMD-TLC; SW846 8140, 8141, 8141A
Water / Groundwater: EPA 1657, 622, 022; SW846 8141, 8141A; ASTM D4763
Drinking Water: EPA 508, 525.2
Food: FDA 212.1, 212.2, 232.1, 232.3, 232.4, 242.1; EPA XENO; AOAC 981.03, 985.22
Indoor / Expired Air: NIOSH 5600; EPA IP-8; ASTM D4861
Plasma: EPA 001; FDA 211.1, 231.1, 252

CHL8080 **CAS #: 64902-72-3**

CHLORSULFURON

RTECS: YS6640000
EINECS Number: 265-268-5
Molecular Formula: $C_{12}H_{12}ClN_5O_4S$
Formula Weight: 357.78
Synonyms: BENZENESULFONAMIDE,2-CHLORO-N-(((4-METHOXY-6-METHYL-1,3,5-TRIAZIN-2-YL)AMINO)CARBONYL)-; 2-CHLORO-N-(((4-METHOXY-6-METHYL-1,3,5-TRIAZIN-2-YL)AMINO)CARBONYL) BENZENESULFONAMIDE; 2-CHLORO-N-((4-METHOXY-6-METHYL-1,3,5-TRIAZIN-2-YL)AMINOCARBONYL)-BENZENESULFONAMIDE; 1-((O-CHLOROPHENYL)SULFONYL)-3-(4-METHOXY-6-METHYL-S-TRIAZIN-2-YL)UREA; 1-(2-CHLOROPHENYLSULFONYL)-3-(4-METHOXY-6-METHYL-1,3,5-TRIAZIN-2YL)UREA; 1-(2-CHLOROPHENYLSULPHONYL)-3-(4-METHOXY-6-METHYL-1,3,5-TRIAZIN-2-YL)UREA; CHLORSULFON; DPX 4189; FINESSE; GLEAN; GLEAN C; GLEAN 20DF; TELAR; TRILIXON
Description: colorless to white crystalline solid or crystals; odorless
Use: control of most broadleaf weeds, annual rye grass; some annual grasses in wheat, barley, durum, rye, triticale, oats; herbicide

Physical Properties

Boiling Point: Decomposes at 192 °C (378 °F)
Freezing Point: 174 °C (345.2 °F) to 178 °C (352.4 °F)
Vapor Pressure: 0.61 mPa at 25 °C
Water Solubility: 0.57 g/L at 22 °C
Other Solubilities: Low Soluble in hydrocarbon solvents.
Flash Point: Nonflammable

RTECS Toxicity Data

Acute Oral: Rat LD_{50} Dose: 5545 mg/kg.
Acute Inhalation: Rat LC_{50} Dose: >5900 mg/m^3/4hr.
Acute Dermal: Rabbit LD_{50} Route: Skin; Dose: 3400 mg/kg.

Hazard Overviews

Fire: Noncombustible.
Carcinogenicity: IARC - Not listed; NIOSH - Not listed; NTP - Not listed; ACGIH - Not listed; OSHA - Not listed; EPA - Not listed; MAK - Not listed

Environmental

Ecotoxicity: LD_{50} Mallard duck oral >5000 mg/kg LD_{50} Bobwhite quail oral >5000 mg/kg
Environmental Fate: If released to soil, it will be degraded both due to biodegradation and chemical hydrolysis. The photolysis and volatilization from soil will not be important.

Mobility in soil is expected to be high. Depending on soil and climatological conditions the field half-life in soil has been reported to range from 14 to 168 days. If released to water, it will degrade due to biodegradation and chemical hydrolysis. Photolysis and volatilization of this herbicide do not appear to be important in water. The bioconcentration in aquatic organisms will be negligible. If released to air, vapor phase may react with photochemically produced hydroxyl radicals with an estimated half-life of 5.1 hours. Particulate absorbed compound may be removed from the atmosphere by dry deposition.

Environmental Physical Data

Octanol/Water Partition Coefficient: log K_{ow} = 0.74
Sorption Partition Coefficient: K_{oc} = estimated at 6 to 31
BCF: calculated at 2

Regulations

RCRA 40CFR: Not listed
CERCLA: 40CFR 302.4: Not listed
SARA 40CFR 372.65: Listed
SARA EHS 40CFR 355: Not listed
TSCA: Not listed

Analytical Methods

Soil: EPA PMD-CLV
Food: FDA 212.1, 232.1
Plasma: FDA 211.1, 231.1, 252

CHL8140 CAS #: 21923-23-9

CHLORTHIOPHOS

RTECS: TF1590000
EINECS Number: 244-663-6
Molecular Formula: $C_{11}H_{15}Cl_2O_3PS_2$
Formula Weight: 361.2
Synonyms: CELA S-2957; CELA S 2957; CELAMERCK S-2957; CELAMERCK S 2957; CELATHION; CHLORTHIOPHOS I; CM S 2957; O-(2,5-DICHLORO-4-(METHYLTHIO)PHENYL) O,O-DIETHYLPHOSPHOROTHIOATE; O-(DICHLORO(METHYLTHIO)PHENYL) O,O-DIETHYLPHOSPHOROTHIOATE (3 ISOMERS); O,O-DIETHYL-O-2,4,5-DICHLORO-(METHYLTHIO)PHENYLTHIONOPHOSPHATE; ENT 27635; EPA PESTICIDE CHEMICAL CODE 111811; NSC 195164; OMS 1342; PHOSPHOROTHIOIC ACID,O-(2,5-DICHLORO-4-(METHYLTHIO)PHENYL) O,O-DIETHYL ESTER; S 2957
Description: brown liquid with crystals at low temp
Use: non-systemic stomach and contact insecticide with some acaricidal activity

Physical Properties

Boiling Point: 155 °C (311 °F) at 0.1 mm Hg
Freezing Point: > 25 °C (77 °F)
Specific Gravity: 1.345 at 20 °C Water=1
Saturated Vapor Density: 1.200000072 kg/m^3
Vapor Pressure: 3.98 x10^{-6} mm Hg at 25 °C
Water Solubility: Practically Insoluble in Water
Other Solubilities: Soluble in most organic solvents

RTECS Toxicity Data

Acute Oral: Rat LD_{50} Dose: 13 mg/kg; Toxic Effects: Sense organs and special senses - Other; Behavioral - Convulsions or effect on seizure threshold; Lungs, Thorax, or Respiration - Dyspnea. Mouse LD_{50} Dose: 141 mg/kg; Toxic Effects: Biochemical - Reactivates cholonesterase.
Acute Dermal: Rat LD_{50} Route: Skin; Dose: 58 mg/kg; Toxic Effects: Biochemical - Reactivates cholonesterase.

Hazard Overviews

Carcinogenicity: IARC - Not listed; NIOSH - Not listed; NTP - Not listed; ACGIH - Not listed; OSHA - Not listed; EPA - Not listed; MAK - Not listed

Environmental

Regulations

RCRA 40CFR: Not listed
CERCLA: 40CFR 302.4: Not listed
SARA 40CFR 372.65: Not listed TPQ: 500 lb
SARA EHS 40CFR 355: Listed TPQ: 500 lb
TSCA: Not listed

Analytical Methods

Food: FDA 212.1, 232.1, 232.3, 232.4, 242.1
Plasma: FDA 211.1, 231.1, 252

CHR1000 CAS #: 1066-30-4

CHROMIC ACETATE

RTECS: AG2975000
EINECS Number: 213-909-4
Molecular Formula: $C_6H_{12}O_6Cr$
Structured MF: Cr_3H_2O
Formula Weight: 229.14
Synonyms: ACETIC ACID,CHROMIUM(3+) SALT; CHROMIC ACETATE (III); CHROMIUM ACETATE; CHROMIUM TRIACETATE; CHROMIUM(III) ACETATE
Description: dark green to violet liquid, or solid or powder; acetic acid odor
Use: mordant in dyeing; to improve light stability & dye affinity of textiles & polymers; in tanning; in hardening photographic emulsions; oxidn catalyst; in catalyst for polymerization of olefins; in metal alloys eg, stainless steel; protective coatings on metal; magnetic tapes; & pigments for paints, cement, etc; org chem synthesis, photochem processing, water treatment; med: astringents & antiseptics; in electroplating cleaning agents; cooling waters, fungicides, etc.

Physical Properties

Boiling Point: 100 °C (212 °F) at 1 atm
Specific Gravity: 1.3
Water Solubility: Slightly Soluble in Water
Other Solubilities: 45.4 g/L of Methanol at 15 °C.
Flash Point: Nonflammable

RTECS Toxicity Data

Mutagenic: Human Cytogenetic Analysis; Cell Type: leukocyte; Dose: 16 umol/L. Hamster Mutations in Mammalian Somatic Cells; Cell Type: ovary; Dose: 200 umol/L.

Tumorigenic: Rat Route: Implant; Dose: 1000 mg/kg/56W-I; Toxic Effects: Tumorigenic - Equivocal tumorigenic agent by RTECS criteria; Tumorigenic - Tumors at site of application.

Hazard Overviews

Fire: Noncombustible.

Carcinogenicity: IARC - Group 1, Carcinogenic to humans; NIOSH - Listed as carcinogen; NTP - Class 1, Known to be a carcinogen; ACGIH - Class A1, Confirmed human carcinogen; OSHA - Not listed; EPA - Class A, Human carcinogen; MAK - Class A2, Unmistakably carcinogenic in animal experimentation only

Exposure Limits

OSHA PEL: TWA: 0.5 mg/m^3; as Cr. Other Values: 0.1 mg/m^3; Clg Cr-VI as CrO_3.

NIOSH REL: TWA: 0.5 mg/m^3; as Cr;Cr-II;Cr-III;Cr(VI)=.001.

Respirator Recommendation

Exposure Range: >0.5 to 5 mg/m^3 Air Purifying, Negative Pressure, Half Mask

Exposure Range: >5 to <25 mg/m^3 Air Purifying, Negative Pressure, Full Face

Exposure Range: 25 to unlimited mg/m^3 Self-contained Breathing Apparatus, Pressure Demand, Full Face

Cartridge Color: dust/mist filter (use P100 or consult supervisor for appropriate dust/mist filter)

Note: as chromium III compounds

Environmental

Ecotoxicity: LC_{50} Salmo gairdneri (rainbow trout) 59 mg cr/l/96 hr/conditions of bioassay not specified

Cleanup/Disposal: Guide No. 171: Do not touch or walk through spilled material. Stop leak if you can do it without risk. Prevent dust cloud. Avoid inhalation of asbestos dust. Small Dry Spills: With clean shovel place material into clean, dry container and cover loosely; move containers from spill area. Small Spills: Take up with sand or other noncombustible absorbent material and place into containers for later disposal. Large Spills: Dike far ahead of liquid spill for later disposal. Cover powder spill with plastic sheet or tarp to minimize spreading. Prevent entry into waterways, sewers, basements or confined areas.

Environmental Physical Data

BCF: little tendency

BOD: cr+3 50%, 5 days

Regulations

RCRA 40CFR: Not listed

CERCLA: 40CFR 302.4: Listed per CWA Section 311(b)(4) RQ: 1000 lb (453.5 kg)

SARA 40CFR 372.65: Listed as Compound

SARA EHS 40CFR 355: Not listed

TSCA: Listed

CHR1440 CAS #: 7738-94-5

CHROMIC ACID

RTECS: GB2450000

DOT: UN1463; UN1755; NA1463; IMO5.1; IMO8.0

EINECS Number: 231-801-5

Molecular Formula: CrH_2O_4

Structured MF: CrO_3

Formula Weight: 99.99

Synonyms: ACIDE CHROMIQUE; CHROMIC-ANHYDRIDE; CHROMIC(VI) ACID; CHROMIC(VI)ACID; CHROMIUM TRIOXIDE; CHRONIC ANHYDRIDE

Use: chromium plating intermediate, medicine (caustic), process engraving, anodizing, ceramic glazes, colored glass, metal cleaning, inks, tanning, paints; chrome plating, photography, electric batteries, etching; chromic anhydride (999%) is a strong inorganic acid commercial uses include plating, chemical and dyestuff manufacturing, photography, cement manufacturing, and leather tanning; used as a cauterizing agent, especially to stop nose bleeds

Physical Properties

Boiling Point: Decomposes

Freezing Point: 196 °C (384.8 °F)

Specific Gravity: 1.67 to 2.82

Water Solubility: Soluble in Water

Other Solubilities: Soluble in sulfuric acid and nitric acid.

Flash Point: Noncombustible

RTECS Toxicity Data

Acute Dermal: Dog LD_{Lo} Route: Subcutaneous Dose: 320 mg/kg.

Mutagenic: Bacteria - S Typhimurium Mutations in Microorganisms; Dose: 80 ug/plate (+S9). Yeast - S Cerevisiae DNA Repair; Dose: 1200 nmol/L. Yeast - S Pombe DNA Repair; Dose: 1200 nmol/L. Yeast - S Cerevisiae DNA Repair; Dose: 1200 nmol/L. Yeast - S Pombe DNA Repair; Dose: 1200 nmol/L.

Hazard Overviews

Reactivity: Stable. Hazardous polymerization cannot occur. Avoid: excess heat; contact with combustible or organic materials. Incompatible with: acetic acid; acetic anhydride; acetone; alcohols; alkali metals; ammonia; arsenic; anthracene; benzene; bromine penta fluorine; butyric acid; camphor; chromous sulfide; diethyl ether; glycerol; hydrogen sulfide; methyl alcohol; naphthalene; peroxyformic acid; phosphorus; potassium hexacyanoferrate; pyridine; selenium; sodium; turpentine; ethyl alcohol; hydrocarbons. Hazardous decomposition products: carbon dioxide; smoke; irritating toxic fumes.

Carcinogenicity: IARC - Group 3, Not classifiable as to carcinogenicity to humans; NIOSH - Not listed; NTP - Not listed; ACGIH - Class A4, Not classifiable as a human carcinogen; OSHA - Not listed; EPA - Not listed; MAK - Class A2, Unmistakably carcinogenic in animal experimentation only

Exposure Limits

OSHA PEL: STEL: 0.1 mg/m^3; as CrO_3; Other Values: 0.1 mg/m^3; Clg Cr-VI as CrO_3.

NIOSH REL: TWA: 0.5 mg/m^3; as Cr;Cr-II;Cr-III;Cr(VI)=.001.

Respirator Recommendation

Exposure Range: >0.1 to 1 mg/m^3 Air Purifying, Negative Pressure, Half Mask

Exposure Range: >1 to 10 mg/m^3 Air Purifying, Negative Pressure, Full Face

Exposure Range: >10 to <15 mg/m^3 Supplied Air, Constant Flow/Pressure Demand, Half Mask Supplied Air, Constant Flow/Pressure Demand, Full Face

Exposure Range: 15 to unlimited mg/m^3 Self-contained Breathing Apparatus, Pressure Demand, Full Face

Cartridge Color: magenta (P100)

Note: as chromium VI compounds

Environmental

Cleanup/Disposal: Guide No. 141: Keep combustibles (wood, paper, oil, etc.) away from spilled material. Do not touch damaged containers or spilled material unless wearing appropriate protective clothing. Stop leak if you can do it without risk. Small Dry Spills: With clean shovel place material into clean, dry container and cover loosely; move containers from spill area. Large Spills: Dike far ahead of spill for later disposal. Guide No. 154: Eliminate all ignition sources (no smoking, flares, sparks or flames in immediate area). Do not touch damaged containers or spilled material unless wearing appropriate protective clothing. Stop leak if you can do it without risk. Prevent entry into waterways, sewers, basements or confined areas. Absorb or cover with dry earth, sand or other non-combustible material and transfer to containers. Do not get water inside containers.

Regulations

RCRA 40CFR: Not listed

CERCLA: 40CFR 302.4: Listed per CWA Section 311(b)(4) RQ: 10 lb (4.535 kg)

SARA 40CFR 372.65: Listed as Compound

SARA EHS 40CFR 355: Not listed

TSCA: Listed

Analytical Methods

Air: ASTM D4490

CHR2320 CAS #: 10025-73-7

CHROMIC CHLORIDE

RTECS: GB5425000

EINECS Number: 233-038-3

Molecular Formula: Cl_3Cr

Formula Weight: 158.35

Cl^-

Cr^{+++}

Cl^- Cl^-

Chemical Structure

Synonyms: C.I. 77295; CHROMIUM CHLORIDE; CHROMIUM CHLORIDE,ANHYDROUS; CHROMIUM TRICHLORIDE; PURATRONIC CHROMIUM CHLORIDE; TRICHLOROCHROMIUM

Physical Properties

Freezing Point: 1152 °C (2105.6 °F)

Specific Gravity: 2.87

Water Solubility: Soluble

Other Solubilities: Insoluble in alcohol

RTECS Toxicity Data

Acute Oral: Rat LD_{50} Dose: 1870 mg/kg.

Acute Inhalation: Mouse LC_{50} Dose: 31500 ug/m^3/2hr.

Acute Dermal: Mouse LD_{Lo} Route: Subcutaneous Dose: 800 mg/kg. Guinea Pig LD_{Lo} Route: Skin; Dose: 202 mg/kg.

Reproductive/Teratogenic: Rat Route: Intratesticular; Dose: 12668 ug/kg; Duration: male 1D prior to mating; Paternal Effects - Spermatogenesis; Testes, epididymis, sperm duct. Mouse Route: Subcutaneous; Dose: 450 mg/kg; Duration: female 1-17D of pregnancy; Effects on Embryo or Fetus - Fetotoxicity. Mouse Route: Subcutaneous; Dose: 12668 ug/kg; Duration: male 30D prior to mating; Paternal Effects - Spermatogenesis; Testes, epididymis, sperm duct.

Mutagenic: Human Unscheduled DNA Synthesis; Cell Type: fibroblast; Dose: 100 umol/L. Human Cytogenetic Analysis; Cell Type: lymphocyte; Dose: 2500 ug/L. Human Cytogenetic Analysis; Cell Type: fibroblast; Dose: 100 umol/L. Human Cytogenetic Analysis; Cell Type: other cell types; Dose: 500 mg/L.

Hazard Overviews

Carcinogenicity: IARC - Group 3, Not classifiable as to carcinogenicity to humans; NIOSH - Not listed; NTP - Not listed; ACGIH - Class A4, Not classifiable as a human carcinogen; OSHA - Not listed; EPA - Not listed; MAK - Class A2, Unmistakably carcinogenic in animal experimentation only

Exposure Limits

OSHA PEL: TWA: 0.5 mg/m^3; as Cr. Other Values: 0.1 mg/m^3; Clg Cr-VI as CrO_3.

NIOSH REL: TWA: 0.5 mg/m^3; as Cr;Cr-II;Cr-III;Cr(VI)=.001.

Respirator Recommendation

Exposure Range: >0.5 to 5 mg/m^3 Air Purifying, Negative Pressure, Half Mask

Exposure Range: >5 to <25 mg/m^3 Air Purifying, Negative Pressure, Full Face

Exposure Range: 25 to unlimited mg/m^3 Self-contained Breathing Apparatus, Pressure Demand, Full Face

Cartridge Color: dust/mist filter (use P100 or consult supervisor for appropriate dust/mist filter)
Note: as chromium III compounds

Environmental

Regulations

RCRA 40CFR: Not listed
CERCLA: 40CFR 302.4: Listed as Compound per CWA Section 307(a) per CAA Section 112
SARA 40CFR 372.65: Not listed TPQ: 1/10000 lb
SARA EHS 40CFR 355: Listed TPQ: 1 lb
TSCA: Listed

CHR2760 **CAS #: 10101-53-8**

CHROMIC SULFATE

RTECS: GB7200000
EINECS Number: 233-253-2
Molecular Formula: $Cr_2H_6O_{12}S_3$
Structured MF: $Cr_2(SO_4)_3$
Formula Weight: 392.20

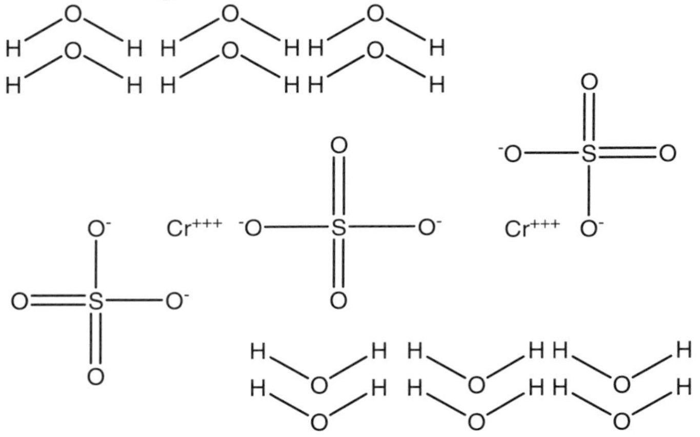

Chemical Structure

Synonyms: BAYCHROM A; BAYCHROM F; C.I. 77305; CHROMIC SULPHATE; CHROMITAN B; CHROMITAN MS; CHROMITAN NA; CHROMIUM (III) SULFATE (2:3); CHROMIUM III SULFATE; CHROMIUM SULFATE; CHROMIUM SULFATE (2:3); CHROMIUM SULPHATE; CHROMIUM SULPHATE (2:3); CROMITAN B; DICHROMIUM SULFATE; DICHROMIUM SULPHATE; DICHROMIUM TRISULFATE; DICHROMIUM TRISULPHATE; KOREON; SULFURIC ACID,CHROMIUM(3+) SALT (3:2)
Description: peach-red to violet powder
Use: in solubilization of gelatin; in catalyst prepn; in tanning of leather; in chrome plating; in mfr of chromium, chromic trioxide & chromium alloys, varnishes, paints, inks, glazes; to improve dispersibility of polymers in water; in agriculture (for improved sugar content & incr yield of grapes; incr cocoon & silk wt of silkworms); in textile ind (dyeing, printing, moth & water proofing); in metal treatment & polishing; in photographic fixing baths; in chem synthesis; corrosion inhibitors; in metal alloys eg, stainless steel; coatings on metal; magnetic tapes; in water treatment; med: astringents & antiseptics; in cooling waters, in fungicides & wood preservatives.

Physical Properties

Boiling Point: Loses H20 at 100 °C (212 °F) at 1 atm
Freezing Point: Decomposes at 100 °C (212 °F)
Specific Gravity: 3.012
Water Solubility: Practically Insoluble in Water
Other Solubilities: Soluble in Alcohol.
pH: Acid
Refraction Index: 1.564
Flash Point: Noncombustible

RTECS Toxicity Data

Mutagenic: Human Cytogenetic Analysis; Cell Type: other cell types; Dose: 500 mg/L. Human Other Mutation Test Systems; Cell Type: other cell types; Dose: 500 mg/L.

Hazard Overviews

Fire Diamond

Health: Mildly irritating to eyes/skin/respiratory tract. Also Causes: possible allergic dermatitis upon subsequent exposure, gastrointestinal irritation upon ingestion.
Fire: Noncombustible. Use agents suitable for surrounding fire.
Reactivity: Stable. Hazardous polymerization cannot occur. Avoid: dispersion of dusts into air. Incompatible with: strong oxidizers; hypophosphites; electrolysis; reducing metals (zinc; magnesium; aluminum in acid solution); basic solutions; hypochlorites; hypobromite; peroxide; oxygen. Hazardous decomposition products: sulfur oxide(s); chromium oxide(s).
Carcinogenicity: IARC - Group 3, Not classifiable as to carcinogenicity to humans; NIOSH - Not listed; NTP - Not listed; ACGIH - Class A4, Not classifiable as a human carcinogen; OSHA - Not listed; EPA - Not listed; MAK - Class A2, Unmistakably carcinogenic in animal experimentation only
Primary Target Organs:

Eyes

Skin

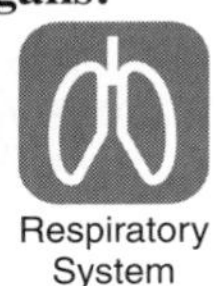
Respiratory System

Exposure Limits
OSHA PEL: TWA: 0.5 mg/m^3; as Cr. Other Values: 0.1 mg/m^3; Clg Cr-VI as CrO_3.
NIOSH REL: TWA: 0.5 mg/m^3; as Cr;Cr-II;Cr-III;Cr(VI)=.001.
Respirator Recommendation
Exposure Range: >0.5 to 5 mg/m^3 Air Purifying, Negative Pressure, Half Mask
Exposure Range: >5 to <25 mg/m^3 Air Purifying, Negative Pressure, Full Face

Exposure Range: 25 to unlimited mg/m^3 Self-contained Breathing Apparatus, Pressure Demand, Full Face
Cartridge Color: dust/mist filter (use P100 or consult supervisor for appropriate dust/mist filter)
Note: as chromium III compounds

Environmental

Ecotoxicity: LC_{50} Coturnix (Japanese quail) > 5000 ppm /Cr2(SO4)3.15H2O; technical grade, 100% AI Selected ranked Species Mean Acute Values for freshwater species are as follows: Hydropsyche betteni (caddisfly) 71,060 ug/l, Daphnia magna (cladoceran) 16,010 ug/l, Lepomis macrochirus (bluegill) 15,020 ug/l, Pimephales promelas (fathead minnow) 10,320 ug/l, Salmo gairdneri (rainbow trout) 9,669 ug/l, Carassius auratus (goldfish) 8,684 ug/l, Poecilla reticulata (guppy) 7,053 ug/l, and Ephemerella subvaria (mayfly) 2,221 ug/l; all at a hardness of 50 mg/l as calcium carbonate
Cleanup/Disposal: Guide No. 154: Eliminate all ignition sources (no smoking, flares, sparks or flames in immediate area). Do not touch damaged containers or spilled material unless wearing appropriate protective clothing. Stop leak if you can do it without risk. Prevent entry into waterways, sewers, basements or confined areas. Absorb or cover with dry earth, sand or other non-combustible material and transfer to containers. Do not get water inside containers.

Environmental Physical Data

BCF: seaweed 1 x10^2
BOD: cr+3 50%, 5 days

Regulations

RCRA 40CFR: Not listed
CERCLA: 40CFR 302.4: Listed per CWA Section 311(b)(4) RQ: 1000 lb (453.5 kg)
SARA 40CFR 372.65: Listed as Compound
SARA EHS 40CFR 355: Not listed
TSCA: Listed

CHR3200 CAS #: 11115-74-5

CHROMIC (VI) ACID

RTECS: GB2450000
Synonyms: ACIDE CHROMIQUE; CHROMIC ACID; CHROMIC(VI) ACID

RTECS Toxicity Data

Acute Dermal: Dog LD_{Lo} Route: Subcutaneous Dose: 320 mg/kg.

Hazard Overviews

Carcinogenicity: IARC - Not listed; NIOSH - Not listed; NTP - Not listed; ACGIH - Not listed; OSHA - Not listed; EPA - Not listed; MAK - Not listed

Exposure Limits

OSHA PEL: STEL: 0.1 mg/m^3; as CrO_3.
ACGIH TLV: TWA: .05 mg/m^3; as Chromium.
NIOSH REL: TWA: .001 mg/m^3. STEL: 10 hour as Chromium VI.

Respirator Recommendation

Exposure Range: >0.1 to 1 mg/m^3 Air Purifying, Negative Pressure, Half Mask
Exposure Range: >1 to 10 mg/m^3 Air Purifying, Negative Pressure, Full Face
Exposure Range: >10 to <15 mg/m^3 Supplied Air, Constant Flow/Pressure Demand, Full Face
Exposure Range: 15 to unlimited mg/m^3 Self-contained Breathing Apparatus, Pressure Demand, Full Face
Cartridge Color: magenta (P100)
Note: as chromium VI compounds

Environmental

Regulations

RCRA 40CFR: Not listed
CERCLA: 40CFR 302.4: Not listed
SARA 40CFR 372.65: Not listed
SARA EHS 40CFR 355: Not listed
TSCA: Not listed

CHR3640 CAS #: 1333-82-0

CHROMIC (VI) OXIDE (1:3)

RTECS: GB6650000
DOT: UN1463; NA1463; IMO5.1
EINECS Number: 215-607-8
Molecular Formula: CrO_3
Structured MF: CrO_3
Formula Weight: 99.99

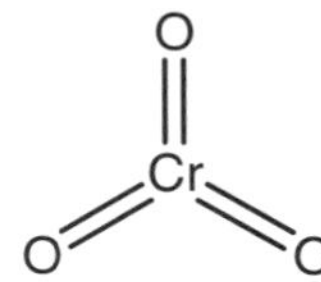

Chemical Structure

Synonyms: ANHYDRIDE CHROMIQUE; ANIDRIDE CROMICA; CHROME (TRIOXYDE DE); CHROMIA; CHROMIC ACID; CHROMIC ACID,SOLID; CHROMIC ACID,SOLUTION; CHROMIC ANHYDRIDE; CHROMIC OXIDE; CHROMIC TRIOXIDE; CHROMIC (VI) ACID; CHROMIC VI ACID; CHROMIC(VI) ACID; CHROMIUM OXIDE; CHROMIUM OXIDE [CRO3]; CHROMIUM (6+) TRIOXIDE; CHROMIUM TRIOXIDE; CHROMIUM(6+) TRIOXIDE; 5CHROMIUM TRIOXIDE,ANHYDROUS; CHROMIUM VI OXIDE; CHROMIUM(VI) OXIDE; CHROMSAEUREANHYDRID; CHROMTRIOXID; CHROOMTRIOXYDE; CHROOMZUURANHYDRIDE; CROMO(TRIOSSIDO DI); MONOCHROMIUM OXIDE; MONOCHROMIUM TRIOXIDE; PURATRONIC; PURATRONIC CHROMIUM TRIOXIDE
Description: dark purplish-red crystals, powder; no detectable odor
Use: in chromium plating: copper stripping; aluminum anodizing; corrosion inhibitor; photography; purifying oil and acetylene; hardening microscopical preparations; oxidant in organic chemistry

Physical Properties

Boiling Point: Decomposes at 250 °C (482 °F)
Freezing Point: 197 °C (386.6 °F)
Specific Gravity: 2.7
Density: 2.7 g/mL
Vapor Pressure: Reid; Very low
Water Solubility: Very Soluble
Other Solubilities: 95% Ethanol: Soluble; Nitric acid: Soluble.
pH: Acid
Ionization Potential (eV): 11.6 +/-0.5
Flash Point: Nonflammable

RTECS Toxicity Data

Acute Oral: Rat LD_{50} Dose: 80 mg/kg; Toxic Effects: Lungs, Thorax, or Respiration - Cyanosis; Gastrointestinal - Hypermotility, diarrhea; Skin and appendages - Hair. Mouse LD_{50} Dose: 127 mg/kg.
Acute Inhalation: Human TC_{Lo} Dose: 110 ug/m^3.
Acute Dermal: Mouse LD_{Lo} Route: Subcutaneous Dose: 20 mg/kg.
Chronic (Multiple Dose) Inhalation: Rat Dose: 29 ug/m^3/20H/90D-I; Toxic Effects: Behavioral - Muscle contraction or spasticity; Lungs, Thorax, or Respiration - Other changes; Biochemical - Carbonic anhydrase.
Reproductive/Teratogenic: Mouse Route: Subcutaneous; Dose: 20 mg/kg; Duration: female 8D of pregnancy; Effects on Embryo or Fetus - Extra embryonic structures; Fetotoxicity. Hamster Route: Intravenous; Dose: 5 mg/kg; Duration: female 8D of pregnancy; Specific Developmental Abnormalities - Homeostasis; Craniofacial (including nose and tongue); Central nervous system. Hamster Route: Intravenous; Dose: 7500 ug/kg; Duration: female 8D of pregnancy; Effects on Fertility - Post-implantation mortality; Effects on Embryo or Fetus - Fetotoxicity; Specific Developmental Abnormalities - Musculoskeletal system. Hamster Route: Intravenous; Dose: 8 mg/kg; Duration: female 8D of pregnancy; Specific Developmental Abnormalities - Body wall.
Mutagenic: Human Cytogenetic Analysis; Cell Type: leukocyte; Dose: 2 mg/L. Human Morphological Transformation; Cell Type: fibroblast; Dose: 100 nmol/L.
Tumorigenic: Human Route: Inhalation; Dose: 110 ug/m^3/3Y-C; Toxic Effects: Tumorigenic - Carcinogenic by RTECS criteria; Sense organs and special senses - Tumors; Lungs, Thorax, or Respiration - Tumors. Rat Route: Implant; Dose: 125 mg/kg; Toxic Effects: Tumorigenic - Carcinogenic by RTECS criteria; Tumorigenic - Tumors at site of application.

Hazard Overviews

Corrosive

Fire Diamond

Health: Corrosive, causes severe burns to eyes/skin/respiratory tract. Toxic Chronic Effects: dermatitis, tooth erosion, and nasal septum perforation.
Fire: Noncombustible. Use extinguishing media suitable for surrounding area.
Reactivity: Stable. Hazardous polymerization cannot occur. Avoid: heat; organic materials; combustibles. Incompatible with: acetic acid; acetic anhydride; acetone; alcohols; alkali metals; ammonia; arsenic; anthracene; benzene; bromine pentafluorine; butyric acid; camphor; chromous sulfide; diethyl ether; glycerol; hydrogen sulfide; methyl alcohol; naphthalene; peroxyformic acid; phosphorus; potassium hexacyanoferrate; pyridine; selenium; sodium; turpentine; ignites ethyl alcohol; hydrocarbons. Hazardous decomposition products: carbon dioxide; smoke; toxic fumes.
Carcinogenicity: IARC - Group 1, Carcinogenic to humans; NIOSH - Listed as carcinogen; NTP - Class 1, Known to be a carcinogen; ACGIH - Class A1, Confirmed human carcinogen; OSHA - Not listed; EPA - Class A, Human carcinogen; MAK - Class A2, Unmistakably carcinogenic in animal experimentation only
Primary Target Organs:

Eyes

Skin

Respiratory System

Liver

Kidneys

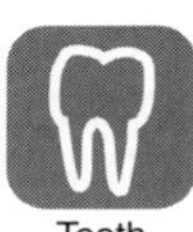
Teeth

Exposure Limits
OSHA PEL: STEL: 0.1 mg/m^3; as CrO_3; Other Values: 0.1 mg/m^3; Clg Cr-VI as CrO_3.
NIOSH REL: TWA: 0.5 mg/m^3; as Cr;Cr-II;Cr-III;Cr(VI)=.001.
Respirator Recommendation
Exposure Range: >0.1 to 1 mg/m^3 Air Purifying, Negative Pressure, Half Mask
Exposure Range: >1 to 10 mg/m^3 Air Purifying, Negative Pressure, Full Face
Exposure Range: >10 to <15 mg/m^3 Supplied Air, Constant Flow/Pressure Demand, Full Face
Exposure Range: 15 to unlimited mg/m^3 Self-contained Breathing Apparatus, Pressure Demand, Full Face
Cartridge Color: magenta (P100)
Note: as chromium VI compounds

Environmental

Ecotoxicity: LC_{50} Ophryotrocha diadema (polychaete worm); 7,500 ug/l as chromium and Ctendrilus seratus (polychaete worm) 4,300 ug/l as chromium; static unmeasured method LC_{50} Capitella captata (polychaete worm) 8,000 ug/l as chromium (larval) and 5,000 ug/l as chromium (adult); static unmeasured method EC_{50} Salmo gairdneri (rainbow trout, embryo larva) 190 ug/l as chromium/28 days, with a water hardness of 101 mg/l as calcium carbonate; Toxic Effect: death and deformity. /Conditions of bioassay not specified Toxic threshold, Daphnia magna 0.016-0.7 ppm
Cleanup/Disposal: Guide No. 141: Keep combustibles (wood, paper, oil, etc.) away from spilled material. Do not touch damaged containers or spilled material unless wearing appropriate protective clothing. Stop leak if you can do it without risk. Small Dry Spills: With clean shovel place material into clean, dry container and cover loosely; move

containers from spill area. Large Spills: Dike far ahead of spill for later disposal.

Environmental Physical Data

BCF: none
BOD: none

Regulations

RCRA 40CFR: Not listed
CERCLA: 40CFR 302.4: Listed as Compound per CWA Section 307(a) per CAA Section 112
SARA 40CFR 372.65: Listed as Compound
SARA EHS 40CFR 355: Not listed
TSCA: Listed

CHR4520 CAS #: 7440-47-3

CHROMIUM

RTECS: GB4200000
EINECS Number: 231-157-5
Molecular Formula: Cr
Structured MF: Cr
Formula Weight: 51.996

Cr

Chemical Structure

Synonyms: CHROM; CHROME; CHROMIUM METAL
Description: steel-gray powder; odorless
Use: for chrome plating of other metals; manufacture of alloys; isotope is used for a blood disease tracer plotting element; in nuclear and high-temperature research

Physical Properties

Boiling Point: 2642 °C (4788 °F)
Freezing Point: 1900 °C (3452 °F)
Specific Gravity: 7.14
Vapor Pressure: 1 mm Hg at 1616 °C
Water Solubility: Insoluble in Water
Other Solubilities: Dilute Hydrochloric Acid: Soluble; Dilute sulfric acid: Soluble.
Surface Tension: 1540 to 1590 nM/m
Flash Point: Noncombustible Solid

RTECS Toxicity Data

Tumorigenic: Rat Route: Intravenous; Dose: 2160 ug/kg/6W-I; Toxic Effects: Tumorigenic - Equivocal tumorigenic agent by RTECS criteria; Gastrointestinal - Tumors; Blood - Lymphomax including Hodgkin's disease. Rat Route: Implant; Dose: 1200 ug/kg/6W-I; Toxic Effects: Tumorigenic - Equivocal tumorigenic agent by RTECS criteria; Blood - Lymphomax including Hodgkin's disease; Tumorigenic - Tumors at site of application.

Hazard Overviews

Explosive

Flammable

Fire Diamond

Health: Irritating to eyes/skin/respiratory tract. Chronic Effects: exposure to chromium fumes can cause fibrosis of the lungs with decreased function.
Fire: Flammable. Use dry chemical or sand.
Reactivity: Stable. Hazardous polymerization cannot occur. Incompatible with: dilute acids (except nitric); strong alkalis; strong oxidizing agents; nitrogen oxide; potassium chlorate; sulfur dioxide; molten lithium at 18 °C; fused ammonium nitrate below 200 °C. Hazardous decomposition products: chromium oxide fumes.
Carcinogenicity: IARC - Group 3, Not classifiable as to carcinogenicity to humans; NIOSH - Not listed; NTP - Listed; ACGIH - Class A4, Not classifiable as a human carcinogen; OSHA - Not listed; EPA - Not listed; MAK - Not listed
Primary Target Organs:

Eyes

Skin

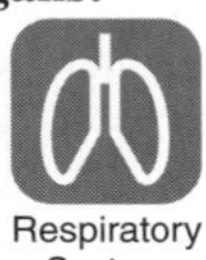
Respiratory System

Exposure Limits
OSHA PEL: TWA: 1 mg/m^3; as Cr. Other Values: 0.1 mg/m^3; Clg Cr-VI as CrO_3.
ACGIH TLV: TWA: 0.5 mg/m^3.
NIOSH REL: TWA: 0.5 mg/m^3; as Cr;Cr-II;Cr-III;Cr(VI)=.001.
NIOSH IDLH: 250 mg/m^3; as Cr.
Respirator Recommendation
Exposure Range: >1 to 10 mg/m^3 Air Purifying, Negative Pressure, Half Mask
Exposure Range: >10 to 100 mg/m^3 Air Purifying, Negative Pressure, Full Face
Exposure Range: >100 to <250 mg/m^3 Supplied Air, Constant Flow/Pressure Demand, Half Mask
Exposure Range: 250 to unlimited mg/m^3 Self-contained Breathing Apparatus, Pressure Demand, Full Face
Cartridge Color: dust/mist filter (use P100 or consult supervisor for appropriate dust/mist filter)

Environmental

Cleanup/Disposal: Guide No. 154: Eliminate all ignition sources (no smoking, flares, sparks or flames in immediate area). Do not touch damaged containers or spilled material unless wearing appropriate protective clothing. Stop leak if you can do it without risk. Prevent entry into waterways, sewers, basements or confined areas. Absorb or cover with dry earth, sand or other non-combustible material and transfer to containers. Do not get water inside containers.

Environmental Physical Data

BCF: snails 1 x10^6
BOD: 62.5 lb/lb, 5 days

Regulations

RCRA 40CFR: Not listed
CERCLA: 40CFR 302.4: Listed per CWA Section 307(a) RQ: 5000 lb (2268 kg)
SARA 40CFR 372.65: Listed
SARA EHS 40CFR 355: Not listed
TSCA: Listed

Analytical Methods

Air: EPA 29, 0060, 306, 306A, CTM-006, ITM-001
Soil: ASTM D1971, D3974; CLP 200.10_M, 200.62, 200.7_M, 202.62, 218.1_M, 218.2_M, 6020_M, ICP-AES; EPA 13, 200.7, 200.8, I-002-1; SW846 1311, 3005A, 3010A, 3015, 3020A, 3031, 3040, 3040A, 3050A&B, 3051, 3052, 6010A&B, 6020, 7000A, 7190, 7191, OSW-A
Water / Groundwater: EPA 200.0, 200.1, 200.12, 200.15, 200.7, 200.9, 218.3; APHA 3111-A, 3111-B, 3111-C, 3113-B, 3120, 3500-CR; ASTM D1687, D1976, D4190; USGS E-SPEC, I1235, I1236, I1238, I3236, I3238, I7236; CEM RD42; FISON AES-0029; HACH 8024
Drinking Water: AOAC 974.27, 993.14; APHA 3500-CR
Food: EPA 14
Indoor / Expired Air: NIOSH 7024, 7300; ASTM D4185
Plasma: EPA 200.11; NIOSH 8005
Urine: NIOSH 8310
Other: EPA 1620, 218.1, 218.2; AOAC 990.08

CHR5840 **CAS #: 7789-02-8**

CHROMIUM (III) NITRATE, NONAHYDRATE

RTECS: GB6300000
Molecular Formula: $CrH_{18}N_3O_{18}$
Formula Weight: 400.21

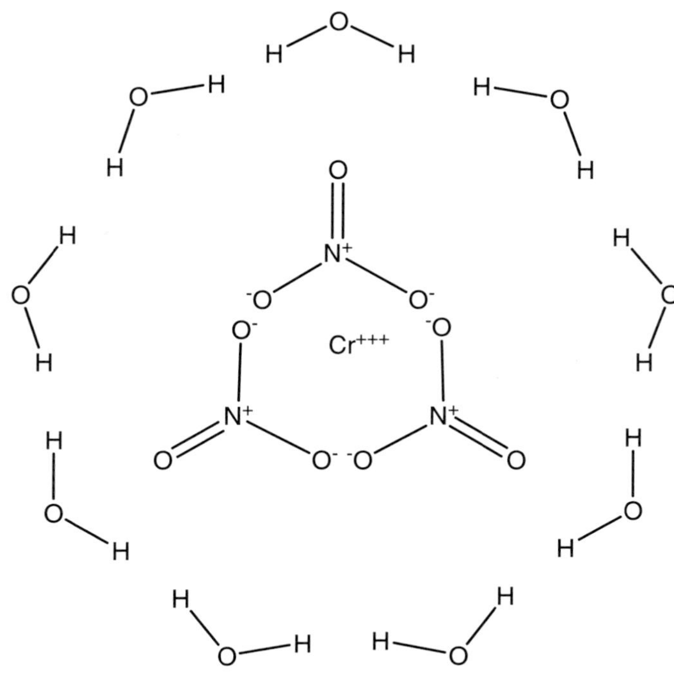

Chemical Structure

Synonyms: CHROMIC NITRATE NONAHYDRATE; CHROMIUM NITRATE NONAHYDRATE; CHROMIUM TRINITRATE NONAHYDRATE; NITRIC ACID,CHROMIUM(3+) SALT,NONAHYDRATE

Physical Properties

Boiling Point: Decomposes at 100 °C (212 °F)
Freezing Point: 60 °C (140 °F)
Density: 1.8 g/cu cm
Water Solubility: Soluble
Other Solubilities: Soluble in Alcohol
Flash Point: Noncombustible

RTECS Toxicity Data

Acute Oral: Rat LD_{50} Dose: 3250 mg/kg.
Mutagenic: Human DNA Inhibition; Cell Type: fibroblast; Dose: 1 mmol/L. Hamster Cytogenetic Analysis; Cell Type: ovary; Dose: 50 mg/L.

Hazard Overviews

Reactivity: Stable. Hazardous polymerization cannot occur. Avoid: exposure to organic and other combustible material. Incompatible with: ether; organic or other easily combustible materials. Hazardous decomposition products: nitrogen oxide(s); chromium oxide(s).
Carcinogenicity: IARC - Group 3, Not classifiable as to carcinogenicity to humans; NIOSH - Not listed; NTP - Not listed; ACGIH - Class A4, Not classifiable as a human carcinogen; OSHA - Not listed; EPA - Not listed; MAK - Class A2, Unmistakably carcinogenic in animal experimentation only

Exposure Limits

OSHA PEL: TWA: 0.5 mg/m^3; as Cr. Other Values: 0.1 mg/m^3; Clg Cr-VI as CrO_3.

NIOSH REL: TWA: 0.5 mg/m^3; as Cr;Cr-II;Cr-III;Cr(VI)=.001.

Respirator Recommendation

Exposure Range: >0.5 to 5 mg/m^3 Air Purifying, Negative Pressure, Half Mask

Exposure Range: >5 to <25 mg/m^3 Air Purifying, Negative Pressure, Full Face

Exposure Range: 25 to unlimited mg/m^3 Self-contained Breathing Apparatus, Pressure Demand, Full Face

Cartridge Color: dust/mist filter (use P100 or consult supervisor for appropriate dust/mist filter)

Note: as chromium III compounds

Environmental

Regulations

RCRA 40CFR: Not listed
CERCLA: 40CFR 302.4: Not listed
SARA 40CFR 372.65: Not listed
SARA EHS 40CFR 355: Not listed
TSCA: Not listed

CHR6280 CAS #: 13548-38-4

CHROMIUM (III) NITRATE, NONAHYDRATE (B)

RTECS: GB6280000
EINECS Number: 236-921-1
Molecular Formula: $CrH_3N_3O_9$
Formula Weight: 238.03

Chemical Structure

Synonyms: CHROMIC NITRATE; CHROMIUM (3+) NITRATE; CHROMIUM NITRATE; CHROMIUM TRINITRATE; NITRIC ACID,CHROMIUM (3+) SALT

Description: pale green, deep violet powder, crystals

Physical Properties

Freezing Point: 60 °C (140 °F)
Density: 1.8 g/cu cm
Water Solubility: Soluble
Other Solubilities: Soluble in Ethyl Acetate, DMSO; Insoluble in Benzene, Carbon Tetrachloride and Chloroform

Flash Point: Noncombustible

RTECS Toxicity Data

Acute Oral: Rat LD_{50} Dose: 3250 mg/kg. Mouse LD_{50} Dose: 2976 mg/kg.

Acute Dermal: Mouse LD_{50} Route: Subcutaneous Dose: 3232 mg/kg.

Mutagenic: Bacteria - B Subtilis DNA Repair; Dose: 160 mmol/L.

Hazard Overviews

Fire Diamond

Health: Mildly irritating to eyes/skin/respiratory tract.

Fire: Noncombustible. Use agents suitable for surrounding fire.

Reactivity: Stable. Hazardous polymerization cannot occur. Avoid: exposure to organic material; combustible material. Incompatible with: ether; easily combustible materials. Hazardous decomposition products: nitrogen oxide(s); chromium oxide(s).

Carcinogenicity: IARC - Group 3, Not classifiable as to carcinogenicity to humans; NIOSH - Not listed; NTP - Not listed; ACGIH - Class A4, Not classifiable as a human carcinogen; OSHA - Not listed; EPA - Not listed; MAK - Class A2, Unmistakably carcinogenic in animal experimentation only

Primary Target Organs:

Eyes

Skin

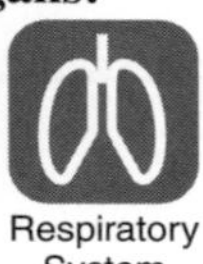
Respiratory System

Exposure Limits

OSHA PEL: TWA: 0.5 mg/m^3; as Cr. Other Values: 0.1 mg/m^3; Clg Cr-VI as CrO_3.

NIOSH REL: TWA: 0.5 mg/m^3; as Cr;Cr-II;Cr-III;Cr(VI)=.001.

Respirator Recommendation

Exposure Range: >0.5 to 5 mg/m^3 Air Purifying, Negative Pressure, Half Mask

Exposure Range: >5 to <25 mg/m^3 Air Purifying, Negative Pressure, Full Face

Exposure Range: 25 to unlimited mg/m^3 Self-contained Breathing Apparatus, Pressure Demand, Full Face

Cartridge Color: dust/mist filter (use P100 or consult supervisor for appropriate dust/mist filter)

Note: as chromium III compounds

Environmental

Regulations

RCRA 40CFR: Not listed
CERCLA: 40CFR 302.4: Listed as Compound per CWA Section 307(a) per CAA Section 112
SARA 40CFR 372.65: Listed as Compound

SARA EHS 40CFR 355: Not listed
TSCA: Listed

CHR6720 CAS #: 1308-38-9

CHROMIUM (III) OXIDE

RTECS: GB6475000
EINECS Number: 215-160-9
Molecular Formula: Cr_2O_3
Structured MF: Cr_2O_3
Formula Weight: 152.02

O^{--} Cr^{+++}

O^{--}

Cr^{+++} O^{--}

Chemical Structure

Synonyms: ANADONIS GREEN; ANHYDRIDE CHROMIQUE; C I 77288; C-GRUN; C.I. 77288; C.I. PIGMENT GREEN 17; CASALIS GREEN; CHROME GREEN; CHROME OCHER; CHROME OCHRE; CHROME OXIDE; CHROME OXIDE GREEN BX; CHROME OXIDE GREEN GN-M; CHROME OXIDE PIGMENT; CHROMIA; CHROMIC ACID GREEN; CHROMIC OXIDE; CHROMIUM OXIDE; CHROMIUM(3+) OXIDE; CHROMIUM OXIDE GREEN; CHROMIUM OXIDE GREENS; CHROMIUM OXIDE PIGMENT; CHROMIUM OXIDE X1134; CHROMIUM SESQUIOXIDE; CHROMIUM(3+) TRIOXIDE; CHROMIUM(III) OXIDE (2:3); COSMETIC HYDROPHOBIC GREEN 9409; COSMETIC MICRO BLEND CHROME OXIDE 9229; DICHROMIUM TRIOXIDE; 11661 GREEN; GREEN CHROME OXIDE; GREEN CHROMIC OXIDE; GREEN CHROMIUM OXIDE; GREEN CINNABAR; GREEN OXIDE OF CHROMIUM; GREEN OXIDE OF CHROMIUM OC-31; GREEN ROUGE; LEAF GREEN; LEVANOX GREEN GA; LEVANOX GREEN GA (HYDRATED CHROMIC OXIDE); OIL GREEN; OXIDE OF CHROMIUM; PURE CHROMIUM OXIDE GREEN 59; ULTRAMARINE GREEN

Description: green crystals, powder

Use: in metallurgy; in mfr of chromium metal, stainless steel & aluminum-chromium alloys, for protective coatings; in abrasives, eg, for polishing jewelry, electric semiconductors; in printing fabrics & banknotes; in dyeing polymers; colorant for latex paints, portland cement; anhyd chromic oxide in ceramics; granules for asphalt roofing, etc; catalyst in mfr of methanol, polyethylene, etc; in refractory brick; hydrated chromic oxide green pigment, esp for automotive finishes; magnetic tapes; org chem synthesis, photochem processing & water treatment; photographic sensitizer; med: astringents & antiseptics.

Physical Properties

Boiling Point: 4000 °C (7232 °F)
Freezing Point: 2435 °C (4415 °F)
Specific Gravity: 5.21
Water Solubility: Practically Insoluble in Water
Other Solubilities: Slightly Soluble in acids, alkalies.
Refraction Index: 2.551

RTECS Toxicity Data

Chronic (Multiple Dose) Inhalation: Rat Dose: 150 mg/m^3/4H/6W-I; Toxic Effects: Lungs, Thorax, or Respiration - Emphysema; Blood - Pigmented or nucleated red Blood cells; Blood - Changes in leukocyte (WBC) cell count.

Mutagenic: Hamster Mutations in Mammalian Somatic Cells; Cell Type: lung; Dose: 50 mg/L. Hamster Sister Chromatid Exchenge; Cell Type: lung; Dose: 34 mg/L.

Tumorigenic: Rat Route: Intrapleural; Dose: 45 mg/kg; Toxic Effects: Tumorigenic - Equivocal tumorigenic agent by RTECS criteria; Lungs, Thorax, or Respiration - Tumors; Tumorigenic - Tumors at site of application. Rat Route: Intratracheal; Dose: 90 mg/kg; Toxic Effects: Tumorigenic - Equivocal tumorigenic agent by RTECS criteria; Lungs, Thorax, or Respiration - Tumors; Tumorigenic - Tumors at site of application.

Hazard Overviews

Fire Diamond

Health: Mildly irritating to eyes/skin/respiratory tract. Chronic Effects: dermatitis, chronic bronchitis, pneumoconiosis.

Fire: Noncombustible. However, it is a strong oxidizer capable of igniting combustibles. Use agent suitable for surrounding fire.

Reactivity: Stable. Hazardous polymerization cannot occur. Avoid: dust generation. Incompatible with: hypophosphites; electrolysis; reducing metals; zinc; magnesium; aluminum in acid solution; lithium and heat; oxygen difluoride; glycerol; chlorine trifluoride; rubidium acetylide.

Carcinogenicity: IARC - Group 3, Not classifiable as to carcinogenicity to humans; NIOSH - Not listed; NTP - Not listed; ACGIH - Class A4, Not classifiable as a human carcinogen; OSHA - Not listed; EPA - Not listed; MAK - Class A2, Unmistakably carcinogenic in animal experimentation only

Primary Target Organs:

Eyes

Skin

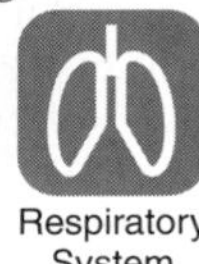
Respiratory System

Exposure Limits

OSHA PEL: TWA: 0.5 mg/m^3; as Cr. Other Values: 0.1 mg/m^3; Clg Cr-VI as CrO_3.

NIOSH REL: TWA: 0.5 mg/m^3; as Cr;Cr-II;Cr-III;Cr(VI)=.001.

Respirator Recommendation

Exposure Range: >0.5 to 5 mg/m^3 Air Purifying, Negative Pressure, Half Mask

Exposure Range: >5 to <25 mg/m^3 Air Purifying, Negative Pressure, Full Face

Exposure Range: 25 to unlimited mg/m^3 Self-contained Breathing Apparatus, Pressure Demand, Full Face

Cartridge Color: dust/mist filter (use P100 or consult supervisor for appropriate dust/mist filter)
Note: as chromium III compounds

Environmental

Environmental Physical Data

BCF: seaweed 1 x10^2

Regulations

RCRA 40CFR: Not listed
CERCLA: 40CFR 302.4: Listed as Compound per CWA Section 307(a) per CAA Section 112
SARA 40CFR 372.65: Listed as Compound
SARA EHS 40CFR 355: Not listed
TSCA: Listed

CHR8480 CAS #: 14977-61-8

CHROMYL CHLORIDE

RTECS: GB5775000
DOT: UN1758; IMO8.0
EINECS Number: 239-056-8
Molecular Formula: Cl_2CrO_2
Structured MF: CrO_2Cl_2
Formula Weight: 154.90

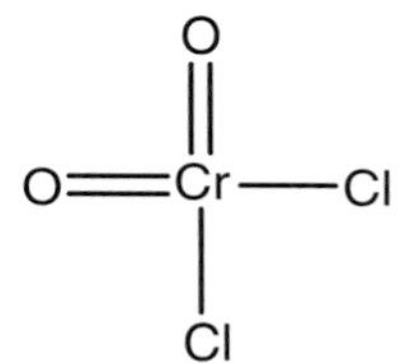

Chemical Structure

Synonyms: CHLOROCHROMIC ANHYDRIDE; CHLORURE DE CHROMYLE; CHROMIC OXYCHLORIDE; CHROMIUM CHLORIDE OXIDE; CHROMIUM DICHLORIDE DIOXIDE; CHROMIUM DIOXIDE DICHLORIDE; CHROMIUM DIOXYCHLORIDE; CHROMIUM OXYCHLORIDE; CHROMIUM (VI) DIOXYCHLORIDE; CHROMIUM,DICHLORODIOXO-; CHROMOXYCHLORID; CHROMYLCHLORID; CHROOMOXYCHLORIDE; CROMILE,CLORURO DI; CROMO,OSSICLORURO DI; DICHLORODIOXOCHROMIUM; DIOXODICHLOROCHROMIUM; OXYCHLORURE CHROMIQUE
Description: dark red liquid; acrid, unpleasant odor
Use: catalyst for polymerization of olefins; for production of aldehydes & ketones; chem rgnt for oxidn of org cmpd, & chlorination of org cmpd; in metal alloys eg, stainless steel; protective coatings on metal; magnetic tapes; & pigments for paints, cement, paper, rubber, composition floor covering & other materials; other incl org chem mfr, photochem processing & indust water treatment; medicine: astringents & antiseptics; in metal finishing, electroplating, cleaning agents, & mordants in textile industry; in cooling waters, & in leather tanning industry, in catalytic mfr, & in fungicides & wood preservatives.

Physical Properties

Boiling Point: 117 °C (243 °F)
Freezing Point: -96.5 °C (-141.7 °F)
Specific Gravity: 1.91 at 25 °C/4 °C
Vapor Density: 0.00698 lb/cu ft at 70 °C
Saturated Vapor Density: 1.337096189 kg/m^3
Vapor Pressure: 20 mm Hg at 20 °C
Water Solubility: Reacts
Other Solubilities: Soluble in Ether, Acetic Acid; dissolves in Chromium Trioxide, yielding powerful oxidant.
Surface Tension: 36.61 dynes/cm
Ionization Potential (eV): 12.60
Flash Point: Not flammable

RTECS Toxicity Data

Mutagenic: Bacteria - S Typhimurium Mutations in Microorganisms; Dose: 50 ug/plate (+S9), 100 ug/plate (-S9).

Hazard Overviews

Corrosive

Fire Diamond

Health: Corrosive to eyes/skin/respiratory tract. Also Causes: cough, wheezing, shortness of breath, pulmonary edema, thrombocytopenia, anemia, methemoglobinemia, dermatitis, irritation, burning, itching, redness, ulceration, lacrimation, sight loss, thirst, dizziness, abdominal pain, vomiting, shock, oliguria, epistaxis. Chronic Effects: chrome holes, nasal congestion, tooth erosion, chest pain, asthma, bronchitis, respiratory tract cancer, conjunctivitis, dermatitis, skin ulceration, liver or kidney damage, acute hepatitis, jaundice.
Fire: Noncombustible. However, it is a strong oxidizer capable of igniting combustibles. Use agent suitable for surrounding fire. Do not use water. It reacts vigorously with water to form chromic acid, chromic chloride, hydrogen chloride, and chlorine gas.
Reactivity: Stable indefinitely in amber glass, aluminum, and stainless steel containers if protected from light and moisture. Hazardous polymerization can occur. Avoid: exposure to water; light. Incompatible with: ammonia; water; moist phosphorus; non metal halides; hydrides; organic solvents (acetone; ethanol; ether; turpentine; urea; flowers of sulfur); sodium azide; reducing agents; fluorine, organic matter; oxidizable materials (paper, wood). Hazardous decomposition products: chromic acid; hydrogen chloride gas; chloride gas.
Carcinogenicity: IARC - Group 3, Not classifiable as to carcinogenicity to humans; NIOSH - Not listed; NTP - Not listed; ACGIH - Class A4, Not classifiable as a human carcinogen; OSHA - Not listed; EPA - Not listed; MAK - Class A2, Unmistakably carcinogenic in animal experimentation only

Primary Target Organs:

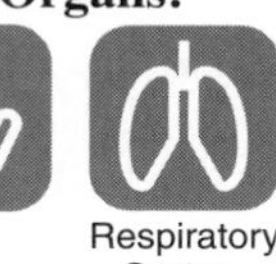

Exposure Limits

OSHA PEL: TWA: 0.5 mg/m^3; as Cr. Other Values: 0.1 mg/m^3; Clg Cr-VI as CrO_3.

ACGIH TLV: TWA: 0.025 ppm; 0.16 mg/m^3.

NIOSH REL: TWA: 0.5 mg/m^3; as Cr;Cr-II;Cr-III;Cr(VI)=.001.

Respirator Recommendation

Exposure Range: >0.001 to 0.05 mg/m^3 Supplied Air, Constant Flow/Pressure Demand, Half Mask

Exposure Range: >0.05 to 1 mg/m^3 Supplied Air, Constant Flow/Pressure Demand, Full Face

Exposure Range: >1 to unlimited mg/m^3 Self-contained Breathing Apparatus, Pressure Demand, Full Face

Note: odor threshold unknown

Environmental

Ecotoxicity: Selected ranked Species Mean Acute Values for saltwater species are as follows: Nassarius obsoletus (mud snail) 105,000 ug/l, Fundulus heteroclitus (mummichog) 74,010 ug/l, Menidia menidia (Atlantic silverside) 15,280 ug/l, Crassostrea gigas (Pacific oyster) 4,538 ug/l, Mytilus edulis (blue mussel) 4,469 ug/l, Mysidopsis bahia (mysid shrimp) 2,030 ug/l, and Nereis virens (polychaete worm) 2,000 ug/l. The saltwater Final Acute Value of 2,158 ug/l for chromium(6+) was calculated from the Genus Mean Acute Values

Cleanup/Disposal: Guide No. 137: Fully encapsulating, vapor protective clothing should be worn for spills and leaks with no fire. Do not touch damaged containers or spilled material unless wearing appropriate protective clothing. Stop leak if you can do it without risk. Use water spray to reduce vapors; do not put water directly on leak, spill area or inside container. Keep combustibles (wood, paper, oil, etc.) away from spilled material. Small Spills: Cover with dry earth, dry sand, or other non-combustible material followed with plastic sheet to minimize spreading or contact with rain. Use clean non-sparking tools to collect material and place it into loosely covered plastic containers for later disposal. Prevent entry into waterways, sewers, basements or confined areas.

Environmental Physical Data

BCF: none

BOD: none

Regulations

RCRA 40CFR: Not listed

CERCLA: 40CFR 302.4: Listed as Compound per CWA Section 307(a) per CAA Section 112

SARA 40CFR 372.65: Listed as Compound

SARA EHS 40CFR 355: Not listed

TSCA: Listed

CHR8920 **CAS #: 218-01-9**

CHRYSENE

RTECS: GC0700000

EINECS Number: 205-923-4

Molecular Formula: $C_{18}H_{12}$

Formula Weight: 228.28

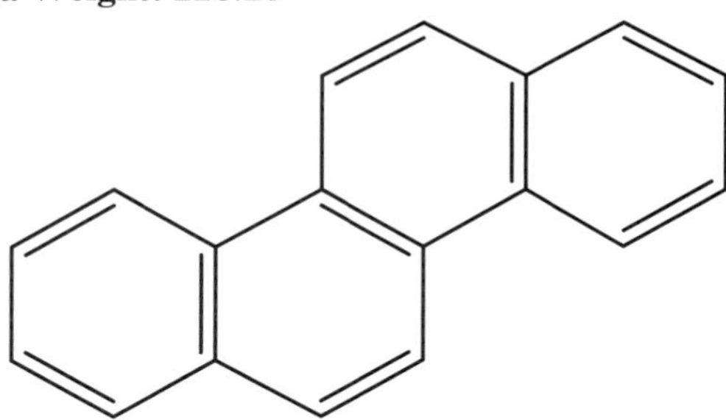

Chemical Structure

Synonyms: BENZO (A) PHENANTHRENE; BENZO[A; 1,2-BENZOPHENANTHRENE; BENZO(A)PHENANTHRENE; 1,2-BENZPHENANTHRENE; BENZ(A)PHENANTHRENE; COAL TAR PITCH VOLATILES: CHRYSENE; 1,2,5,6-DIBENZONAPHTHALENE

Description: red blue fluorescent rhombic plates

Use: organic synthesis; research chemical

Physical Properties

Boiling Point: 448 °C (838 °F)

Freezing Point: 255 °C (491 °F) to 256 °C (492.8 °F)

Specific Gravity: 1.274 at 20 °C/4 °C

Vapor Pressure: 6.3 x10^{-7} mm Hg

Water Solubility: Insoluble 0.0018 mg/kg

Other Solubilities: 95% Ethanol: Slightly Soluble. (1g/1300ml); Acetone: Slightly Soluble; Benzene: Soluble in hot; Carbon Disulfide: Slightly Soluble; Ether: Slightly Soluble; Toluene: Soluble in hot. (1g/480ml).

Ionization Potential (eV): 7.59 +/-0.2

RTECS Toxicity Data

Mutagenic: Human Mutations in Mammalian Somatic Cells; Cell Type: lymphocyte; Dose: 6 umol/L. Human Mutations in Mammalian Somatic Cells; Cell Type: other cell types; Dose: 12 umol/L.

Tumorigenic: Mouse Route: Skin; Dose: 3600 ug/kg; Toxic Effects: Tumorigenic - Neoplastic by RTECS criteria; Skin and appendages - Tumors. Mouse Route: Skin; Dose: 99 mg/kg/31W-I; Toxic Effects: Tumorigenic - Equivocal tumorigenic agent by RTECS criteria; Skin and appendages - Tumors. Mouse Route: Skin; Dose: 40 mg/kg/3W-I; Toxic Effects: Tumorigenic - Equivocal tumorigenic agent by RTECS criteria; Skin and appendages - Tumors. Mouse Route: Skin; Dose: 3600 mg/kg/30W-I; Toxic Effects: Tumorigenic - Equivocal tumorigenic agent by RTECS criteria; Skin and appendages - Tumors. Mouse Route: Skin; Dose: 23 mg/kg; Toxic Effects: Tumorigenic - Neoplastic by RTECS criteria; Skin and appendages - Tumors.

Hazard Overviews

Health: Irritating. Toxic. Other Acute Effects: harmful if swallowed, inhaled, or absorbed through skin. Chronic Effects: may alter genetic material; may cause heritable genetic damage. Carcinogen.

Fire: Hazards: emits toxic fumes. Extinguishing agents: water spray; carbon dioxide, dry chemical powder or appropriate foam. Precautions: combustible liquid.

Reactivity: Incompatible with: strong oxidizing agents. Hazardous decomposition products: toxic fumes of: carbon monoxide, carbon dioxide.

Carcinogenicity: IARC - Group 3, Not classifiable as to carcinogenicity to humans; NIOSH - Listed as carcinogen; NTP - Not listed; ACGIH - Class A3, Animal carcinogen; OSHA - Not listed; EPA - Class B2, Probable human carcinogen based on animal studies; MAK - Class A2, Unmistakably carcinogenic in animal experimentation only

Primary Target Organs:

Eyes

Skin

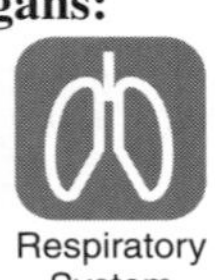
Respiratory System

Exposure Limits

OSHA PEL: TWA: 0.2 mg/m^3.

DFG MAK: TWA: 0.05 mg/m^3.

Respirator Recommendation

Exposure Range: unlimited Self-contained Breathing Apparatus, Pressure Demand, Full Face

Note: TLV not established

Environmental

Environmental Fate: If released to soil it will be expected to adsorb very strongly to the soil and will not be expected to leach appreciably to groundwater. It will not hydrolyze or appreciably evaporate from soils or surfaces, and it may be subject to biodegradation in soils. If released to water, it will adsorb very strongly to sediments and particulate matter, but will not hydrolyze or appreciably evaporate. It will bioconcentrate in species which lack microsomal oxidase. It will be subject to near-surface, direct photolysis with a half-life of 4.4 hours computed for exposure to sunlight at mid-day in midsummer at latitude 40 deg N. The small amount of information available suggests that it may be subject to biodegradation in water systems. Adsorption to various materials may affect the rate of these processes. If released to air, it will be subject to direct photolysis, although adsorption to particulates may affect the rate of this process. The estimated half-life of any gas phase in the atmosphere is 1.25 hours as a result of reaction with photochemically produced hydroxyl radicals.

Environmental Physical Data

Henry's Law Constant: 9.4×10^{-8}

Octanol/Water Partition Coefficient: log K_{ow} = 5.61 to 5.91

Sorption Partition Coefficient: K_{oc} = estimated at 2.51×10^5 to 5.01×10^5

BCF: daphnia manga 2000

Regulations

RCRA 40CFR: Listed Hazardous Waste No. U050 Toxic Waste

CERCLA: 40CFR 302.4: Listed per RCRA Section 3001 per CWA Section 307(a) RQ: 100 lb (45.35 kg)

SARA 40CFR 372.65: Listed

SARA EHS 40CFR 355: Not listed

TSCA: Listed

Analytical Methods

Air: EPA TO-13; California 429

Soil: CLP LC_SV, MC_SVOA, OHC; EPA 16, 1625, PAH-005, PAH-007, PAH-011, PAH-012, S-004-1; SW846 3630B, 3640A, 3650A, 3650B, 8100, 8250A, 8270B, 8270C, 8275A, 8310, 8410; DOE OS050

Water / Groundwater: EPA PAH-002, PAH-006, S-002-1, 1625, 610, 625, 625-S, 6; APHA 6040-B, 6410-B, 6440-B, 6440-C; ASTM D4657, D4763; USGS O3113, O3118

Drinking Water: EPA 525.1, 525.2, 550, 550.1

Indoor / Expired Air: NIOSH 5506, 5515; EPA IP-7-A, IP-7-B

Plasma: EPA 29

Other: EPA PAH-009

CI1130 CAS #: 4680-78-8

C.I. ACID GREEN 3

RTECS: BQ4375000

EINECS Number: 225-132-8

Molecular Formula: $C_{37}H_{36}N_2NaO_6S_2$

Formula Weight: 691.86

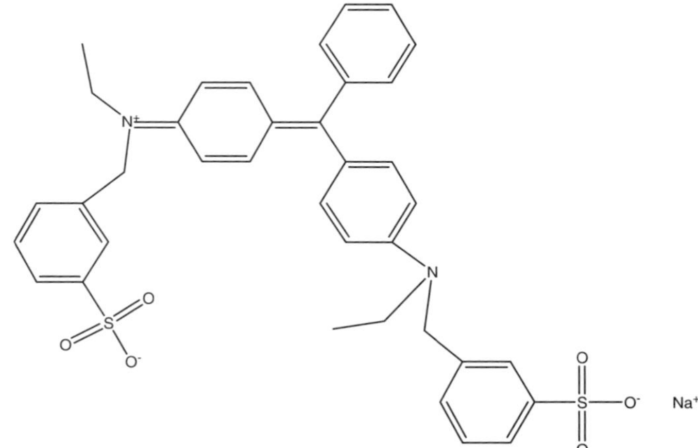

Chemical Structure

Synonyms: A.F. GREEN NO. 1; ACID GREEN; ACID GREEN 2G; ACID GREEN 3; ACID GREEN B; ACID GREEN G; ACID GREEN L; ACID GREEN S; ACID LEATHER GREEN 3G; ACID LEATHER GREEN F; ACIDAL GREEN G; ACILAN GREEN B; AF GREEN NO 1; AMACID GREEN B; BENZENEMETHANAMINIUM,N-ETHYL-N-(4-((4-(ETHYL((3-SULFOPHENYL)METHYL)AMINO)PHENYL)PHENYLMETH-YLENE)-2,5-CYCLOHEXADIEN-1-YLIDENE)-3-SULFO-, HYDROXIDE,INNER SALT, SODIUM SALT; BRILLANT GREEN 3EMBL; BRILLIANT GREEN 3EMBL; BUCACID GUINEA GREEN BA; C.I. 42085; C.I. ACID

GREEN 1; C.I. ACID GREEN 3 (VAN); C.I. ACID GREEN 3,MONOSODIUM SALT; C.I. ACID GREEN 3,MONSODIUM SALT; C.I. ACID GREEN 3,SODIUM SALT; C.I. FOOD GREEN 1; CALCOCID GREEN G; COLOR INDEX NO 42085; FD AND C GREEN NO 1; FD AND C GREEN NO. 1; FDC GREEN 1; FENAZO GREEN L; FOOD GREEN 1; GUINEA GREEN; GUINEA GREEN B; GUINEA GREEN BA; HIDACID EMERALD GREEN; HISPACID GREEN GB; INTRACID GREEN F; JAPAN GREEN 1; KITON GREEN F; KITON GREEN FC; LEATHER GREEN B; LISSAMINE GREEN G; MERANTINE GREEN G; NAPHTHALENE GREEN G; NAPHTHALENE LAKE GREEN G; NAPHTHALENE LEATHER GREEN G; NERAN BRILLIANT GREEN G; NSC 5016; NSC 8684; OXIDE,INNER SALT,SODIUM SALT; PONTACYL GREEN B; PONTACYL GREEN BL; SULFACID BRILLIANT GREEN 1B; SULPHO GREEN 2B; VONDACID GREEN L; ZELEN KYSELA 3; ZELEN POTRAVINARSKA 1

Description: dark green dull powder or bright, crystalline solid

Use: to dye wool, silk, leather, paper; in wood & biological stain & indicator; barium & aluminum salts pigments; formerly to color foods which did not contain fats & oils (eg, gelatine desserts, bakery products), pharmaceuticals (drug capsules), & cosmetics; now considered unsafe, except for cosmetics that do not come into contact with mucous membranes; in japan, in externally applied cosmetics.

Physical Properties

Water Solubility: Soluble in Water
Other Solubilities: slightly Soluble in Ethanol

RTECS Toxicity Data

Acute Oral: Rat LD; Dose: >3 gm/kg.
Mutagenic: Bacteria - S Typhimurium Mutations in Microorganisms; Dose: 320 ug/plate (-S9).
Tumorigenic: Rat Route: Oral; Dose: 660 gm/kg/43W-C; Toxic Effects: Tumorigenic - Equivocal tumorigenic agent by RTECS criteria; Blood - Lymphomax including Hodgkin's disease.

Hazard Overviews

Health: Irritating. Harmful. Other Acute Effects: harmful if swallowed, inhaled, or absorbed through skin. Chronic Effects: Possible carcinogen.
Fire: Hazards: emits toxic fumes. Extinguishing agents: carbon dioxide; dry chemical powder; water spray. Precautions: combustible liquid.
Reactivity: Incompatible with: strong oxidizing agents. Hazardous decomposition products: carbon monoxide, carbon dioxide, nitrogen oxides, sulfur oxides.
Carcinogenicity: IARC - Group 3, Not classifiable as to carcinogenicity to humans; NIOSH - Not listed; NTP - Not listed; ACGIH - Not listed; OSHA - Not listed; EPA - Not listed; MAK - Not listed

Primary Target Organs:

Eyes

Skin

Respiratory System

Environmental

Regulations

RCRA 40CFR: Not listed
CERCLA: 40CFR 302.4: Not listed
SARA 40CFR 372.65: Listed
SARA EHS 40CFR 355: Not listed
TSCA: Listed

CI1390 CAS #: 6459-94-5

C.I. ACID RED 114

RTECS: QJ6475500
EINECS Number: 229-272-0
Molecular Formula: $C_{37}H_{30}N_4Na_2O_{10}S_3$
Formula Weight: 832.86

Chemical Structure

Synonyms: ACID LEATHER RED BG; ACID RED 114; AMACID MILLING RED PRS; BENZYL FAST RED BG; BENZYL RED BR; (1,1'-BIPHENYL)-4-YL)AZO)-7-HYDROXY-1,3-NAPHTHALENEDISULFONIC ACID,; C.I. 23635; C.I. ACID RED 114,DISODIUM SALT; CERVEN KYSELA 114; 8-((3,3'-DIMETHYL-4'-((4-((4-METHYLPHENYL)SULFONYL)OXY)PHENYL)AZO)-; ELCACID MILLING FAST RED RS; ERIONYL RED RS; FENAFOR RED PB; FOLAN RED B; INTRAZONE RED BR; KAYANOL MILLING RED RS; LEATHER FAST RED B; LEVANOL RED GG; MIDLON RED PRS; MILLING FAST RED B; MILLING RED B; MILLING RED BB; MILLING RED SWB; 1,3-NAPHTHALENEDISULFONIC ACID,8-((3,3'-DIMETHYL-4'-((4-(((4-METHYLPHENYL)SULFONYL)OXY)PHENYL)AZO)(1,1'-BIPHENYL)-4-YL)AZO)-7-HYDROXY-, DISODIUM SALT; POLAR RED RS; SANDOLAN RED N-RS; SELLA FAST RED RS; SODIUM SALT; SULPHONOL FAST RED R; SULPHONOL RED R; SUMINOL MILLING RED RS; SUPRANOL FAST RED 3G; SUPRANOL FAST RED GG; SUPRANOL RED PBX-CF; SUPRANOL RED R; TELON FAST RED GG; TERTRACID MILLING RED B; TETRACID MILLING RED B; TETRACID MILLING RED G; VONDAMOL FAST RED RS

Description: dark red powder
Use: dye for textiles and leather

Physical Properties

Water Solubility: < 0.1 mg/mL at 22.5 C

Other Solubilities: 95% Ethanol: <1 mg/ml at 22 °C; Acetone: <1 mg/ml at 22 °C; DMSO: 50-100 mg/ml at 22 °C; Methanol: Soluble.
Flash Point: Not available; probably combustible

RTECS Toxicity Data

Mutagenic: Bacteria - S Typhimurium Mutations in Microorganisms; Dose: 300 nmol/plate (+S9). Bacteria - S Typhimurium Mutations in Microorganisms; Dose: 125 ug/plate (-S9).
Tumorigenic: Rat Route: Oral; Dose: 15 mg/kg/2Y-C; Toxic Effects: Tumorigenic - Carcinogenic by RTECS criteria; Sense organs and special senses - Tumors; Skin and appendages - Tumors.

Hazard Overviews

Health: Irritating. Toxic. Other Acute Effects: harmful if swallowed, inhaled, or absorbed through skin. Chronic Effects: carcinogen; target organs: liver, kidneys, skin, lungs. The toxicological properties have not been thoroughly investigated.
Fire: Will burn. Hazards: emits toxic fumes. Extinguishing agents: water spray; carbon dioxide, dry chemical powder or appropriate foam. Precautions: combustible liquid.
Reactivity: Stable. Hazardous polymerization will not occur. Hazardous decomposition products: carbon monoxide, carbon dioxide, nitrogen oxides, sulfur oxides.
Carcinogenicity: IARC - Group 2B, Possibly carcinogenic to humans; NIOSH - Not listed; NTP - Not listed; ACGIH - Not listed; OSHA - Not listed; EPA - Not listed; MAK - Not listed
Primary Target Organs:

Eyes

Skin
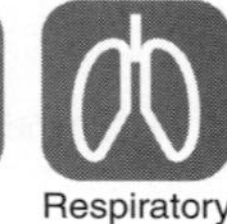
Respiratory System

Liver

Kidneys

Environmental

Regulations

RCRA 40CFR: Not listed
CERCLA: 40CFR 302.4: Not listed
SARA 40CFR 372.65: Listed
SARA EHS 40CFR 355: Not listed
TSCA: Listed

CI2040 CAS #: 989-38-8

C.I. BASIC RED 1

RTECS: DH0175000
EINECS Number: 213-584-9
Molecular Formula: $C_{28}H_{31}ClN_2O_3$
Formula Weight: 479.06

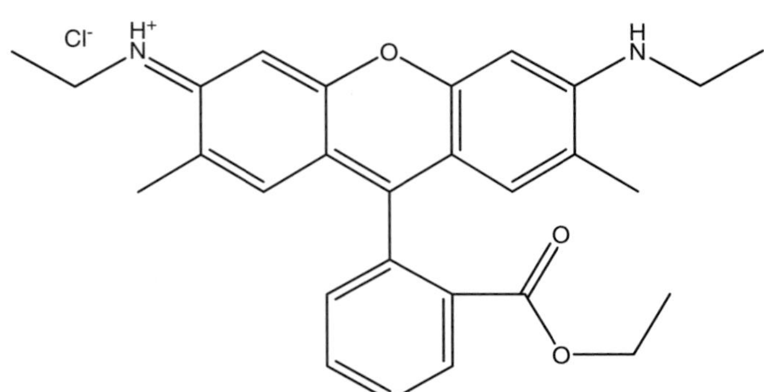

Chemical Structure

Synonyms: AIZEN RHODAMINE 6GCP; BASIC RED 1; BASIC RHODAMINE YELLOW; BASIC RHODAMINIC YELLOW; BENZOIC ACID,2-(6-(ETHYLAMINO)-3-(ETHYLIMINO)-2,7-DIMETHYL-3H-XANTHEN-9-YL)-,ETHYL ESTER,MONOHYDROCHLORIDE; BENZOIC ACID,O-(6-(ETHYLAMINO)-3-(ETHYLIMINO)-2,7-DIMETHYL-3H-XANTHEN-9-YL)-,ETHYL ESTER,MONOHYDROCHLORIDE; C.I. 45160; C.I. BASIC RED 1,MONOHYDROCHLORIDE; CALCOZINE RED 6G; CALCOZINE RHODAMINE 6GX; CERVEN ZASADITA 1; ELCOZINE RHODAMINE 6GDN; ELJON PINK TONER; ETHYL O-(6-(ETHYLAMINO)-3-(ETHYLIMINO)-2,7-DIMETHYL-3H-XANTHEN-9-YL)BENZOATE; FANAL PINK B; FANAL PINK GFK; FANAL RED 25532; FLEXO RED 482; HELIOSTABLE BRILLIANT PINK B EXTRA; MITSUI RHODAMINE; MITSUI RHODAMINE 6GCP; MONOHYDROCHLORIDE; NYCO LIQUID RED GF; RED 169; RH 6G; RHODAMIN 6G; RHODAMINE 6G; RHODAMINE 6 GDN; RHODAMINE 6 GDN EXTRA; RHODAMINE 6G (BIOLOGICAL STAIN); RHODAMINE 590 CHLORIDE; RHODAMINE 6G CHLORIDE; RHODAMINE 69DN EXTRA; RHODAMINE 6G EXTRA; RHODAMINE 6G EXTRA BASE; RHODAMINE F 5GL; RHODAMINE F4G; RHODAMINE F5G; RHODAMINE F5G CHLORIDE; RHODAMINE 6GB; RHODAMINE 6GBN; RHODAMINE 6GCP; RHODAMINE 4GD; RHODAMINE 6GD; RHODAMINE 5GDN; RHODAMINE 6GDN; RHODAMINE GDN; RHODAMINE 6GDN EXTRA; RHODAMINE 6GEX ETHYL ESTER; RHODAMINE 4GH; RHODAMINE 6GH; RHODAMINE 5GL; RHODAMINE 6GO; RHODAMINE 6GX; RHODAMINE J; RHODAMINE 6JH; RHODAMINE 7JH; RHODAMINE 6G LAKE; RHODAMINE LAKE RED 6G; RHODAMINE Y 20-7425; RHODAMINE 6ZH; RHODAMINE ZH; RHODAMINE 6ZH-DN; SILOSUPER PINK B; VALI FAST RED 1308; XANTHYLIUM,9-(2-(ETHOXYCARBONYL)PHENYL)-3,6-BIS(ETHYLAMINO)-2,7-DIMETHYL-,CHLORIDE
Description: bright, bluish pink crystals or reddish purple powder
Use: dye used in tunable lasers; to dye silk, cotton, wool, bast fibers and paper; tracing agent in water pollution studies; in leather dyeing; an adsorption indicator; printing inks

Physical Properties

Vapor Density: 16.5 Air=1
Water Solubility: Soluble in Water
Other Solubilities: 95% Ethanol: >=100 mg/ml at 19.5 °C; Acetone: <1 mg/ml at 19.5 °C; DMSO: <1 mg/ml at 20.5 °C.
Flash Point: Not available; probably combustible

RTECS Toxicity Data

Acute Oral: Rat LD_{Lo} Dose: 125 mg/kg. Mouse LD_{Lo} Dose: 50 mg/kg; Toxic Effects: Cardiac - Other changes; Liver - Other changes; Blood - Other hemolysis with or without anemia.

Mutagenic: Mouse Mutations in Mammalian Somatic Cells; Cell Type: lymphocyte; Dose: 2500 ug/L. Hamster DNA Damage; Cell Type: ovary; Dose: 90 umol/L.

Hazard Overviews

Health: Irritating to eyes/skin. Harmful. Other Acute Effects: harmful if swallowed, inhaled, or absorbed through skin; possible risk of irreversible effects. Chronic Effects: possible carcinogen.

Fire: Will burn. Hazards: emits toxic fumes. Extinguishing agents: water spray; carbon dioxide, dry chemical powder or appropriate foam. Precautions: combustible liquid.

Carcinogenicity: IARC - Group 3, Not classifiable as to carcinogenicity to humans; NIOSH - Not listed; NTP - Not listed; ACGIH - Not listed; OSHA - Not listed; EPA - Not listed; MAK - Not listed

Primary Target Organs:

Eyes

Skin

Environmental

Regulations

RCRA 40CFR: Not listed
CERCLA: 40CFR 302.4: Not listed
SARA 40CFR 372.65: Listed
SARA EHS 40CFR 355: Not listed
TSCA: Listed

CI2300 **CAS #: 569-61-9**

C.I. BASIC RED 9 MONOHYDROCHLORIDE

RTECS: CX9850100
EINECS Number: 209-321-2
Molecular Formula: $C_{19}H_{18}ClN_3$
Formula Weight: 323.82

NH_2^+ Cl^- ... H_2N ... NH_2

Chemical Structure

Synonyms: 4-((4-AMINOPHENYL)(4-IMINO-2,5-CYCLOHEXADIEN-1-YLIDENE)METHYL)BENZENAMINE; BASIC PARAFUCHSINE; BASIC RED 9; BASIC RUBINE; BENZENAMINE,4-((4-AMINOPHENYL)(4-IMINO-2,5-CYCLOHEXADIEN-1-YLIDENE)METHYL),; BENZENAMINE,4-((4-AMINOPHENYL)(4-IMINO-2,5-CYCLOHEXADIEN-1-YLIDENE)METHYL)-,MONOHYDROCHLORIDE; C.I. 42500; C.I. BASIC RED 9; C.I. BASIC RED 9,MONOHYDROCHLORIDE; CALCOZINE MAGENTA N; CERVEN ZASADITA 9; CHLORIDE; P-FUCHSIN; FUCHSIN SP; FUCHSINE DR-001; FUCHSINE SP; FUCHSINE SPC; 4,4'-((4-IMINO-2,5-CYCLOHEXADIEN-1-YLIDENE)METHYLENE)DIANILINE MONOHYDRO-; 4,4'-((4-IMINO-2,5-CYCLOHEXADIEN-1-YLIDENE)METHYLENE)DIANILINE MONOHYDROCHLORIDE; 4,4'-((4-IMINO-2,5-CYCLOHEXADIEN-1-YLIDENE)METHYLENE)DIANILINEMONOHYDROCHLORIDE; PARA MAGENTA; PARA-MAGENTA; MONOHYDROCHLORIDE; ORIENT PARAMAGENTA BASE; PARAFUCHSIN; PARAFUCHSINE; PARAOSANILINE HYDROCHLORIDE; PARAROSANILINE; PARAROSANILINE CHLORIDE; PARAROSANILINE HYDROCHLORIDE; P-ROSANILINE HCL; P-ROSANILINE HYDROCHLORIDE; SCHULTZ-TAB NO 779; SCHULTZ-TAB. NO. 779; 4-TOLUIDINE,ALPHA-(P-AMINOPHENYL)-ALPHA-(4-IMINO-2,5-CYCLOHEXADIEN-1-YLIDENE)-MONOHYDROCHLORIDE; 4,4'4''-TRIAMINOTRIPHENYLMETHAN-HYDROCHLORID

Description: colorless to red crystals or dark-green crystalline powder

Use: dye for silk, acrylics, leather and paper; a pH indicator in the range 1.0 (purple) to 3.1 (red); to stain bacilli (especially influenza and tubercle) in tissue, and for acid-fast bacteria using Methylene Blue as a counterstain

Physical Properties

Freezing Point: Decomposes at 268 °C (514.4 °F) to 270 °C (518 °F)

Water Solubility: Very Slightly Soluble in Water

Other Solubilities: 95% Ethanol: <1 mg/ml at 18 °C; Acetone: <1 mg/ml at 20 °C; DMSO: 10-50 mg/ml at 20 °C; Ether: Very slightly Soluble; Methanol: <1 mg/ml at 19 °C; Sulfuric acid (concentrated): Soluble; Toluene: <1 mg/ml at 19 °C.

Flash Point: Not available; probably combustible

RTECS Toxicity Data

Acute Oral: Mouse LD_{50} Dose: 5 gm/kg.
Chronic (Multiple Dose) Oral: Rat Dose: 18200 mg/kg/13W-C; Toxic Effects: Endocrine - Thyroid weight (goiter); DEATH.
Mutagenic: Human DNA Inhibition; Cell Type: HeLa cell; Dose: 300 umol/L. Rat Unscheduled DNA Synthesis; Cell Type: liver; Dose: 2200 ug/L.
Tumorigenic: Rat Route: Oral; Dose: 728 mg/kg/2Y-C; Toxic Effects: Tumorigenic - Carcinogenic by RTECS criteria; Liver - Tumors; Endocrine - Thyroid tumors. Rat Route: Subcutaneous; Dose: 1714 mg/kg/43W-I; Toxic Effects: Tumorigenic - Equivocal tumorigenic agent by RTECS criteria; Tumorigenic - Tumors at site of application.

Hazard Overviews

Health: Irritating to eyes/skin/respiratory tract. Harmful. Other Acute Effects: may be harmful by inhalation, ingestion, or skin absorption. Chronic Effects: possibly carcinogenic based on its IARC, ACGIH, NTP or EPA classification; laboratory experiments have shown mutagenic effects; thirteen week studies in rats identified the thyroid and pituitary gland as target sites; rats of both sexes showed subcutaneous fibromas, thyroid gland follicular cell adenomas and carcinomas, and zymbal gland carcinomas; male rats showed squamous cell carcinomas, trichoepitheliomas and sebaceous adenomas of the skin; hepatocellular carcinomas appeared in male rats and both sexes of mice; adrenal gland pheochromocytomas or malignant pheochromocytomas and hematopoietic system tumors occurred in female mice; increased incidences of mammary gland tumors occurred in female rats; target organs: thyroid, pituitary.
Fire: Will burn. Hazards: emits toxic fumes. Extinguishing agents: water spray; carbon dioxide, dry chemical powder or appropriate foam. Precautions: combustible liquid.
Reactivity: Incompatible with: strong oxidizing agents. Hazardous decomposition products: toxic fumes of: carbon monoxide, carbon dioxide, hydrogen chloride gas, nitrogen oxides.
Carcinogenicity: IARC - Group 2B, Possibly carcinogenic to humans; NIOSH - Not listed; NTP - Listed; ACGIH - Not listed; OSHA - Not listed; EPA - Not listed; MAK - Not listed
Primary Target Organs:

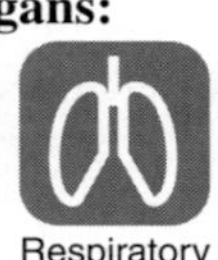

Environmental

Environmental Physical Data

Octanol/Water Partition Coefficient: log K_{ow} = -0.21

Regulations

RCRA 40CFR: Not listed
CERCLA: 40CFR 302.4: Not listed
SARA 40CFR 372.65: Not listed
SARA EHS 40CFR 355: Not listed
TSCA: Listed

CI2560 CAS #: 1937-37-7

C.I. DIRECT BLACK 38

RTECS: QJ6160000
EINECS Number: 217-710-3
Molecular Formula: $C_{34}H_{27}N_9Na_2O_7S_2$
Formula Weight: 783.0

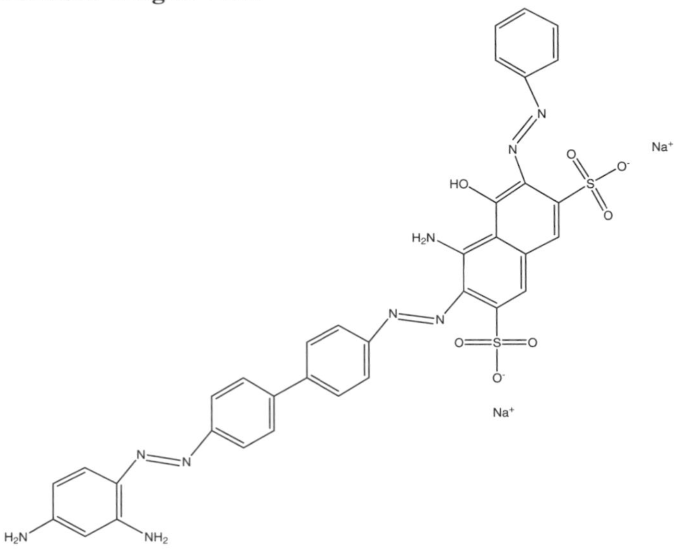

Chemical Structure

Synonyms: AHCO DIRECT BLACK GX; AIREDALE BLACK ED; AIZEN DIRECT DEEP BLACK EH; AIZEN DIRECT DEEP BLACK GH; AIZEN DIRECT DEEP BLACK RH; AMANIL BLACK GL; AMANIL BLACK WD; 4-AMINO-3-((4'-((2,4-DIAMINOPHENYL)AZO)(1,1'-BIPHENYL)-4-YL)AZO)-5-HYDROXY-6; APOMINE BLACK GK; APOMINE BLACK GX; ATLANTIC BLACK BD; ATLANTIC BLACK C; ATLANTIC BLACK E; ATLANTIC BLACK EA; ATLANTIC BLACK GAC; ATLANTIC BLACK GG; ATLANTIC BLACK GXCW; ATLANTIC BLACK GXOO; ATLANTIC BLACK SD; ATUL DIRECT BLACK E; AZINE DEEP BLACK EW; AZOCARD BLACK EW; AZOMINE BLACK EWO; BELAMINE BLACK GX; BENCIDAL BLACK E; BENZAMIL BLACK E; BENZANIL BLACK E; BENZO DEEP BLACK E; BENZO LEATHER BLACK E; BENZOFORM BLACK BCN-CF; BIPHENYL)-4-YL)AZO)-5-HYDROXY-6-(PHENYLAZO)-,DISODIUM SALT; BLACK 2EMBL; BLACK 4EMBL; BRASILAMINA BLACK GN; BRILLIANT CHROME LEATHER BLACK H; C.I. 30235; C.I. DIRECT BLACK 38,DISODIUM SALT; CALCOMINE BLACK; CALCOMINE BLACK EXL; CARBIDE BLACK E; CERN PRIMA 38; CHLORAMINE BLACK C; CHLORAMINE BLACK EC; CHLORAMINE BLACK ERT; CHLORAMINE BLACK EX; CHLORAMINE BLACK EXR; CHLORAMINE BLACK XO; CHLORAMINE CARBON BLACK S; CHLORAMINE CARBON BLACK SJ; CHLORAMINE CARBON BLACK SN; CHLORAZOL BLACK E; CHLORAZOL BLACK E (BIOLOGICAL STAIN); CHLORAZOL BLACK EA; CHLORAZOL BLACK EN; CHLORAZOL BURL BLACK E; CHLORAZOL LEATHER BLACK ENP; CHLORAZOL SILK BLACK G; CHROME LEATHER BLACK E; CHROME LEATHER BLACK EC; CHROME LEATHER BLACK EM; CHROME LEATHER BLACK G; CHROME LEATHER BRILLIANT BLACK ER; COIR DEEP BLACK C; COLUMBIA BLACK EP; COLUMBUS BLACK EP; CORANIL DIRECT BLACK F;

DIACOTTON DEEP BLACK; DIACOTTON DEEP BLACK RX; DIAMINE DEEP BLACK EC; DIAMINE DIRECT BLACK E; DIAPHTAMINE BLACK V; DIAZINE BLACK E; DIAZINE DIRECT BLACK E; DIAZINE DIRECT BLACK G; DIAZOL BLACK 2V; DIPHENYL DEEP BLACK G; DIRECT BLACK META; DIRECT BLACK 3; DIRECT BLACK 38; DIRECT BLACK A; DIRECT BLACK BRN; DIRECT BLACK CX; DIRECT BLACK CXR; DIRECT BLACK E; DIRECT BLACK EW; DIRECT BLACK EX; DIRECT BLACK FR; DIRECT BLACK GAC; DIRECT BLACK GW; DIRECT BLACK GX; DIRECT BLACK GXR; DIRECT BLACK JET; DIRECT BLACK METHYL; DIRECT BLACK N; DIRECT BLACK RX; DIRECT BLACK SD; DIRECT BLACK WS; DIRECT BLACK Z; DIRECT DEEP BLACK E; DIRECT DEEP BLACK E EXTRA; DIRECT DEEP BLACK EAC; DIRECT DEEP BLACK EA-CF; DIRECT DEEP BLACK EW; DIRECT DEEP BLACK EX; DISODIUM4-AMINO-3-((4'-((2,4-DIAMINOPHENYL)AZO)(1,1'-BIPHENYL)-4-AZO)-5-HYDROXY-6-(PHENYLAZO)NAPHTHALENE-2,7-DISULPHONATE; ENIANIL BLACK CN; ERIE BLACK B; ERIE BLACK BF; ERIE BLACK GAC; ERIE BLACK GXOO; ERIE BLACK JET; ERIE BLACK NUG; ERIE BLACK RXOO; ERIE BRILLIANT BLACK S; ERIE FIBRE BLACK VP; FENAMIN BLACK E; FIBRE BLACK VF; FIXANOL BLACK E; FORMALINE BLACK C; FORMIC BLACK BA; FORMIC BLACK C; FORMIC BLACK CW; FORMIC BLACK MTG; FORMIC BLACK TG; HISPAMIN BLACK EF; INTERCHEM DIRECT BLACK Z; KAYAKU DIRECT DEEP BLACK EX; KAYAKU DIRECT DEEP BLACK GX; KAYAKU DIRECT DEEP BLACK S; KAYAKU DIRECT LEATHER BLACK EX; KAYAKU DIRECT SPECIAL BLACK AAX; LURAZOL BLACK BA; META BLACK; MITSUI DIRECT BLACK EX; MITSUI DIRECT BLACK GX; MITSUL DIRECT BLACK EX; MITSUL DIRECT BLACK GX; 2,7-NAPHTHALENEDISULFONIC ACID,4-AMINO-3-((4'-((2,4-DIAMINOPHENYL)AZO)(1,1'; 2,7-NAPHTHALENEDISULFONIC ACID,4-AMINO-3-((4'-((2,4-DIAMINOPHENYL)AZO)(1,1'-BIPHENYL)-4-YL)AZO)--5-HYDROXY-6-(PHENYLAZO)-,DISODIUM SALT; NIPPON DEEP BLACK; NIPPON DEEP BLACK GX; NSC 47756; NSC 8679; PAPER BLACK BA; PAPER BLACK T; PAPER DEEP BLACK C; PARAMINE BLACK B; PARAMINE BLACK E; PEERAMINE BLACK E; PEERAMINE BLACK GXOO; PHENAMINE BLACK BCN-CF; PHENAMINE BLACK CL; PHENAMINE BLACK E; PHENAMINE BLACK E 200; PHENO BLACK EP; PHENO BLACK SGN; (PHENYLAZO)-2,7-NAPHTHALENEDISULFONIC ACID DISODIUM SALT; PONTAMINE BLACK E; PONTAMINE BLACK EBN; SANDOPEL BLACK EX; SERISTAN BLACK B; TELON FAST BLACK E; TERTRODIRECT BLACK E; TETRAZO DEEP BLACK G; TETRODIRECT BLACK E; TETRODIRECT BLACK EFD; UNION BLACK EM; VONDACEL BLACK N

Description: grey-black powder

Use: dye for fabric, leather, cotton, cellulosic materials, paper, wool, silk, bast and hog's hair, plastics, vegetable-ivory buttons, wood flour used as a resin filler, acetate and nylon; to stain wood and biological materials and to produce aqueous inks; reportedly been used in hair dyes; by artists

Physical Properties

Boiling Point: Decomposes

Freezing Point: Decomposes

Water Solubility: Soluble in Water

Other Solubilities: DMSO: 1-5 mg/ml at 20 °C; 95% Ethanol: <1 mg/ml at 20 °C; Acetone: <1 mg/ml at 20 °C; Toluene: <1 mg/ml at 20 °C; Ethanol: Moderately Soluble.

Flash Point: Not available; probably combustible

RTECS Toxicity Data

Acute Oral: Rat LD_{50} Dose: 7600 mg/kg; Toxic Effects: Behavioral - Convulsions or effect on seizure threshold; Behavioral - Ataxia. Rabbit LD_{Lo} Dose: 1262 mg/kg; Toxic Effects: Kidney, Ureter, and Bladder - Other changes in urine composition; Kidney, Ureter, and Bladder - Other changes; Nutritional and gross metabolic - Weight loss or decreased weight gain.

Acute Inhalation: Rat LC_{Lo} Dose: 180 gm/m^3/1hr.

Mutagenic: Rat Unscheduled DNA Synthesis; Route: Oral; Dose: 500 mg/kg/12H-C. Rat Micronucleus Test; Route: Oral; Dose: 500 mg/kg/36H-C.

Hazard Overviews

Health: Irritating to eyes/skin. Toxic. Other Acute Effects: may be harmful by inhalation, ingestion, or skin absorption. Chronic Effects: carcinogen; target organ: liver. The toxicological properties have not been thoroughly investigated.

Fire: Will burn. Extinguishing agents: water spray; carbon dioxide, dry chemical powder or appropriate foam. Precautions: combustible liquid.

Reactivity: Stable. Hazardous polymerization will not occur. Hazardous decomposition products: toxic fumes of: carbon monoxide, carbon dioxide, nitrogen oxides, sulfur oxides.

Carcinogenicity: IARC - Not listed; NIOSH - Not listed; NTP - Listed; ACGIH - Not listed; OSHA - Not listed; EPA - Not listed; MAK - Not listed

Primary Target Organs:

Eyes

Skin

Liver

Environmental

Environmental Fate: If released to soil, the dye will biodegrade with the formation of benzidine and related products. The later products are more persistent in soil than the parent dye. The evaporation of the dye from soil surfaces to air will not be important. It is water soluble which may suggest some soil leaching, but its ionic nature may result in some ion-exchange processes with clay that would retard leaching. When released to water, biodegradation of the dye may be the most important process. There is a paucity of data regarding the abiotic processes (radical oxidation, photolysis and hydrolysis) that may lead to the loss of the dye from water. The loss of the dye from water by evaporation should not be important. The dye particle is likely to be removed from the air by wet and dry deposition. It has been detected in the workplace air of both textile- and paper-dyeing facilities.

Environmental Physical Data

BOD: 8% BODT, 5 days

Regulations

RCRA 40CFR: Not listed

CERCLA: 40CFR 302.4: Not listed

SARA 40CFR 372.65: Listed

SARA EHS 40CFR 355: Not listed

TSCA: Listed

Analytical Methods

Indoor / Expired Air: NIOSH 5013

CI3080 CAS #: 16071-86-6

C.I. DIRECT BROWN 95

RTECS: GL7375000
EINECS Number: 240-221-1
Molecular Formula: $C_{31}H_{18}CuN_6Na_2O_9S$
Formula Weight: 762.15
Synonyms: AIZEN PRIMULA BROWN BRLH; AIZEN PRIMULA BROWN PLH; AMANIL FAST BROWN BRL; AMANIL SUPRA BROWN LBL; ATLANTIC FAST BROWN BRL; ATLANTIC RESIN FAST BROWN BRL; BELAMINE FAST BROWN BRLL; BENZAMIL SUPRA BROWN BRLL; BENZANIL SUPRA BROWN BRLL; BENZANIL SUPRA BROWN BRLN; BIPHENYL)-4-YL)AZO)-2-HYDROXYBENZOATO(2-))COPPER,DISODIUM SALT; (1,1'-BIPHENYL)-4-YL)AZO)-2-HYDROXYBENZOATO(2-))-,DISODIUM SALT; BROWN 4EMBL; C.I. 30145; C.I. DIRECT BROWN; CALCODUR BROWN BRL; CHLORAMINE FAST BROWN BRL; CHLORAMINE FAST BROWN BRLL; CHLORAMINE FAST CUTCH BROWN PL; CHLORANTINE FAST BROWN BRLL; CHROME LEATHER BROWN BRLL; CHROME LEATHER BROWN BRSL; COPPER,(5-((4'-((2,5-DIHYDROXY-4-((2-HYDROXY-5-SULFOPHENYL)AZO)PHENYL)AZO)-; CUPRATE(2-),(5-((4'-((2,6-DIHYDROXY-3-((2-HYDROXY-5-SULFOPHENYL)AZO)PHENYL)AZO)(1,1'-BIPHENYL)-4-YL)AZO)-2-HYDROXYBENZOATO(4-)),DISODIUM; CUPROFIX BROWN GL; DERMA FAST BROWN W-GL; DERMAFIX BROWN PL; DIALUMINOUS BROWN BRS; DIAPHTAMINE LIGHT BROWN BRLL; DIAPHTHAMINE LIGHT BROWN BRLL; DIAZINE FAST BROWN RSL; DIAZOL LIGHT BROWN BRN; DICOREL BROWN LMR; (5-((4'-((2,5-DIHYDROXY-4-((2-HYDROXY-5-SULFOPHENYL)AZO)PHENYL)AZO)(1,1'-; DIPHENYL FAST BROWN BRL; DIRECT BROWN 95; DIRECT BROWN BRL; DIRECT FAST BROWN BRL; DIRECT FAST BROWN LMR; DIRECT LIGHT BROWN BRS; DIRECT SUPRA LIGHT BROWN ML; DURAZOL BROWN BR; DUROFAST BROWN BRL; ELIAMINA LIGHT BROWN BRL; ENIANIL LIGHT BROWN BRL; FASTOLITE BROWN BRL; FASTUSOL BROWN LBRSA; FASTUSOL BROWN LBRSN; FENALUZ BROWN BRL; HELION BROWN BRSL; HISPALUZ BROWN BRL; HNED PRIMA 95; ISMAFAST BROWN BRSL; KAYARUS SUPRA BROWN BRS; KCA LIGHT FAST BROWN; KCA LIGHT FAST BROWN BR; PARANOL FAST BROWN BRL; PEERAMINE FAST BROWN BRL; PONTAMINE FAST BROWN BRL; PONTAMINE FAST BROWN NP; PYRAZOL FAST BROWN BRL; PYRAZOLINE BROWN BRL; SATURN BROWN LBR; SIRIUS SUPRA BROWN BRL; SIRIUS SUPRA BROWN BRS; SOLANTINE BROWN BRL; SOLAR BROWN PL; SOLEX BROWN R; SOLIUS LIGHT BROWN BRLL; SOLIUS LIGHT BROWN BRS; SUMILIGHT SUPRA BROWN BRS; SUPRAZO BROWN BRL; SUPREXCEL BROWN BRL; TERTRODIRECT FAST BROWN BR; TETRAMINE FAST BROWN BRDN EXTRA; TETRAMINE FAST BROWN BRP; TETRAMINE FAST BROWN BRS; TRIANTINE BROWN BRS; TRIANTINE FAST BROWN OG; TRIANTINE FAST BROWN OR; TRIANTINE LIGHT BROWN BRS; TRIANTINE LIGHT BROWN OG
Description: dark brown microcrystals or charcoal black powder
Use: dye cellulose and silk fibers, to stain wool, acetate and nylon fibers, to print cellulose and silk fibers, to dye leather, paper and casein-formaldehyde plastics, and to produce its heavy metal salts which can be used as pigments

Physical Properties

Boiling Point: Decomposes
Freezing Point: Decomposes
Water Solubility: Soluble in Water
Other Solubilities: 95% Ethanol: <1 mg/ml at 19.5 °C; Acetone: <1 mg/ml at 19.5 °C; DMSO: 1-5 mg/ml at 19.5 °C; Ethanol: Slightly Soluble.
Flash Point: Not available; probably combustible

RTECS Toxicity Data

Acute Oral: Rat LD_{50} Dose: 12060 mg/kg. Rabbit LD_{Lo} Dose: 2 gm/kg.
Mutagenic: Rat Unscheduled DNA Synthesis; Route: Oral; Dose: 100 mg/kg. Rat Body Fluid Assay; Indicator Organism: Bacteria - S Typhimurium; Dose: 250 umol/kg.

Hazard Overviews

Fire: Will burn.
Carcinogenicity: IARC - Not listed; NIOSH - Not listed; NTP - Not listed; ACGIH - Not listed; OSHA - Not listed; EPA - Not listed; MAK - Not listed

Environmental

Regulations

RCRA 40CFR: Not listed
CERCLA: 40CFR 302.4: Listed as Compound per CWA Section 307(a)
SARA 40CFR 372.65: Listed
SARA EHS 40CFR 355: Not listed
TSCA: Listed

Analytical Methods

Indoor / Expired Air: NIOSH 5013

CI4380 CAS #: 2832-40-8

C.I. DISPERSE YELLOW 3

RTECS: AC3662000
EINECS Number: 220-600-8
Molecular Formula: $C_{15}H_{15}N_3O_2$
Structured MF: $HOC_6H_3(CH_3)N{=}NC_6H_4NHC(O)CH_3$
Formula Weight: 269.33

HN N=N O HO

Chemical Structure

Synonyms: ACETAMIDE,N-(4((2-HYDROXY-5-METHYLPHENYL)AZO)PHENYL)-; ACETAMIDE,N-(4-((2-HYDROXY-5-METHYLPHENYL)AZO)PHENYL)-; 4-ACETAMIDO-2'-HYDROXY-5'-METHYLAZOBENZENE; ACETAMINE YELLOW CG; ACETATE FAST YELLOW G; ACETOQUINONE LIGHT YELLOW; ACETOQUINONE LIGHT YELLOW 4JLZ; ALTCO SPERSE FAST YELLOW GFN NEW; AMACEL YELLOW G; ARTISIL DIRECT

YELLOW G; ARTISIL YELLOW G; ARTISIL YELLOW 2GN; ATRISIL DIRECT YELLOW G; C.I. 3/11855; C.I. 11855; C.I. SOLVENT YELLOW 77; C.I. SOLVENT YELLOW 92; C.I. SOLVENT YELLOW 99; C.I. SOLVENT YELLOW 14 CYTEMBENA; CALCOSYN YELLOW GC; CALCOSYN YELLOW GCN; CELLITON DISCHARGE YELLOW GL; CELLITON FAST YELLOW G; CELLITON FAST YELLOW GA; CELLITON FAST YELLOW GA-CF; CELLITON YELLOW G; CELLUTATE YELLOW GH; CELUTATE YELLOW GH; CIBACET YELLOW GBA; CIBACET YELLOW 2GC; CIBACETE YELLOW GBA; CILLA FAST YELLOW G; DIACELLITON FAST YELLOW G; DISPERSE FAST YELLOW G; DISPERSE YELLOW 3; DISPERSE YELLOW G; DISPERSE YELLOW Z; DISPERSIVE YELLOW 3T; DISPERSOL FAST YELLOW G; DISPERSOL PRINTING YELLOW G; DISPERSOL YELLOW A-G; DURGACET YELLOW G; DUROSPERSE YELLOW G; EASTONE YELLOW GN; ESTEROQUINONE LIGHT YELLOW 4JL; ESTERQUINONE LIGHT YELLOW 4JL; ESTONE YELLOW GN; FENACET FAST YELLOW G; FENACET YELLOW G; GENACRON YELLOW G; HISPACET FAST YELLOW G; HISPERSE YELLOW G; 4'-((2-HYDROXY-5-METHYLPHENYL)AZO)ACETANILIDE; 4-(2-HYDROXY-5-METHYLPHENYLAZO)ACETANILIDE; N-(4-((2-HYDROXY-5-METHYLPHENYL)AZO)PHENYL)ACETAMIDE; 4'-((6-HYDROXY-M-TOLYL)AZO)ACETANILIDE; INTERCHEM ACETATE YELLOW G; INTERCHEM DISPERSE YELLOW GH; INTERCHEM HISPERSE YELLOW GH; INTRAPERSE YELLOW GBA; INTRASPERSE YELLOW GBA EXTRA; KAYALON FAST YELLOW G; KAYASET YELLOW G; KCA ACETATE FAST YELLOW G; MICROSETILE YELLOW GR; MIKETON FAST YELLOW G; NACELAN FAST YELLOW CG; NOVALON YELLOW 2GN; NYLOQUINONE LIGHT YELLOW 4JL; NYLOQUINONE YELLOW 4J; OSTACET YELLOW P2G; PALACET YELLOW GN; PALANIL YELLOW G; PAMACEL YELLOW G-3; PERLITON YELLOW G; RELITON YELLOW C; RESIREN YELLOW TG; SAFARITONE YELLOW G; SAMARON YELLOW PA3; SERINYL HOSIERY YELLOW GD; SERIPLAS YELLOW GD; SERISOL FAST YELLOW GD; SETACYL YELLOW G; SETACYL YELLOW P-2GL; SETACYL YELLOW 2GN; SILOTRAS YELLOW TSG; SUPRACET FAST YELLOW G; SYNTEN YELLOW 2G; SYNTON YELLOW 2G; TERASIL YELLOW GBA EXTRA; TERASIL YELLOW 2GC; TERTRANESE YELLOW N-2GL; TULADISPERSE FAST YELLOW 2G; VONTERYL YELLOW G; VONTERYL YELLOW R; YELLOW RELITON G; YELLOW Z; ZLUT DISPERZNI 3; ZLUT ROZPOUSTEDLOVA 77

Description: brownish-yellow powder

Use: in dyeing textiles, sheepskins and furs, in coloring polymethyl methacrylate and nylon, and in the surface dyeing of cellulose acetate

Physical Properties

Freezing Point: 268 °C (514.4 °F) to 270 °C (518 °F)

Water Solubility: < 0.1 mg/mL at 18 C

Other Solubilities: 95% Ethanol: 1-5 mg/ml at 22 °C; Acetone: <1 mg/ml at 22 °C; Benzene: Soluble; DMSO: 5-10 mg/ml at 22 °C; Methanol: <1 mg/ml at 17 °C; Toluene: <1 mg/ml at 19 °C.

Flash Point: Not available; probably combustible

RTECS Toxicity Data

Acute Oral: Rat LD; Dose: >14 mg/kg. Mouse LD; Dose: >14 mg/kg.

Mutagenic: Mouse Mutations in Microorganisms; Cell Type: lymphocyte; Dose: 10 mg/L (+S9). Hamster Sister Chromatid Exchenge; Cell Type: ovary; Dose: 5 mg/L.

Tumorigenic: Rat Route: Oral; Dose: 180 gm/kg/2Y-C; Toxic Effects: Tumorigenic - Carcinogenic by RTECS criteria; Liver - Tumors; Blood - Leukemia. Rat Route: Oral; Dose: 216 gm/kg/2Y-C; Toxic Effects: Tumorigenic - Carcinogenic by RTECS criteria; Liver - Tumors; Blood - Leukemia.

Hazard Overviews

Health: Irritating to eyes/skin/respiratory tract. Toxic. Other Acute Effects: harmful if swallowed, inhaled, or absorbed through skin. Chronic Effects: may alter genetic material. Carcinogen.

Fire: Will burn. Hazards: emits toxic fumes. Extinguishing agents: water spray; carbon dioxide, dry chemical powder or appropriate foam. Precautions: combustible liquid.

Reactivity: Incompatible with: strong oxidizing agents. Hazardous decomposition products: toxic fumes of: carbon monoxide, carbon dioxide, nitrogen oxides.

Carcinogenicity: IARC - Group 3, Not classifiable as to carcinogenicity to humans; NIOSH - Not listed; NTP - Not listed; ACGIH - Not listed; OSHA - Not listed; EPA - Not listed; MAK - Not listed

Primary Target Organs:

Eyes

Skin

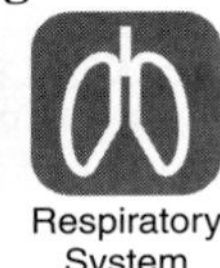
Respiratory System

Environmental

Regulations

RCRA 40CFR: Not listed

CERCLA: 40CFR 302.4: Not listed

SARA 40CFR 372.65: Listed

SARA EHS 40CFR 355: Not listed

TSCA: Listed

CI4770 CAS #: 81-88-9

C.I. FOOD RED 15

RTECS: BP3675000

EINECS Number: 201-383-9

Molecular Formula: $C_{28}H_{31}ClN_2O_3$

Formula Weight: 479.0

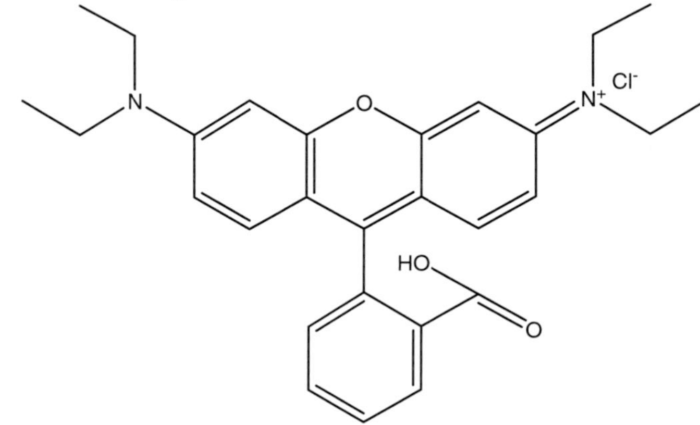

Chemical Structure

Synonyms: ACID BRILLIANT PINK B; ADC RHODAMINE B; AIZEN RHODAMINE BH; AIZEN RHODAMINE BHC; AKIRIKU RHODAMINE B; AMMONIUM,(9-(O-CARBOXYPHENYL)-6-(DIETHYLAMINO)-3H-XANTHEN-3-YLIDENE)DIETHYL-,CHLORIDE; BASIC VIOLET 10; BRILLIANT PINK B; C.I. 45170; C.I. 749; C.I. BASIC VIOLET 10; CALCOZINE RED BX; CALCOZINE RHODAMINE BL; CALCOZINE RHODAMINE BX; CALCOZINE RHODAMINE BXP; 9-O-CARBOXYPHENYL-6-DIETHYLAMINO-3-ETHYLIMINO-3-ISOXANTHENE,3-ETHOCHLORIDE; (9-(O-CARBOXYPHENYL)-6-(DIETHYLAMINO)-3H-XANTHEN-3-YLIDENE) DIETHYLAMMONIUM CHLORIDE; CERISE TONER X1127; CERTIQUAL RHODAMINE; COGILOR RED 321.10; COSMETIC BRILLIANT PINK BLUISH D CONC; D AND C RED NO 19; D+C RED NO 19; D+C RED NO. 19; DIABASIC RHODAMINE B; DIETHYL-M-AMINO-PHENOLPHTHALEIN HYDROCHLORIDE; EDICOL SUPPA ROSE BS; EDICOL SUPRA ROSE B; ELCOZINE RHODAMINE B; ERIOSIN RHODAMINE B; ETHANAMINIUM,N-(9-(2-CARBOXYPHENYL)-6-(DIETHYLAMINO)-3H-XANTHEN-3-YLIDENE)-N-ETHYL-,CHLORIDE; FD AND C RED NO 19; FD&C RED NO. 19; FOOD RED 15; GERANIUM LAKE N; HEXACOL RHODAMINE B EXTRA; IKADA RHODAMINE B; IRAGEN RED L-U; MITSUI RHODAMINE BX; 11411 RED; RED NO 213; RED NO. 213; RH B; RHEONINE B; RHODAMINE; RHODAMINE BS; RHODAMINE B; RHODAMINE B 20-7470; RHODAMINE B 500; RHODAMINE B CHLORIDE; RHODAMINE B EXTRA; RHODAMINE B EXTRA M 310; RHODAMINE B EXTRA S; RHODAMINE B500; RHODAMINE B500 HYDROCHLORIDE; RHODAMINE BA; RHODAMINE BA EXPORT; RHODAMINE BF; RHODAMINE BL; RHODAMINE BN; RHODAMINE BX; RHODAMINE BXL; RHODAMINE BXP; RHODAMINE FB; RHODAMINE FB CI; RHODAMINE LAKE RED B; RHODAMINE O; RHODAMINE S; RHODAMINE,BLUE SHADE; RHODAMINE,TETRAETHYL-; SICILIAN CERISE TONER A-7127; SYMULEX MAGENTA F; SYMULEX PINK F; SYMULEX RHODAMINE B TONER F; TAKAOKA RHODAMINE B; TETRAETHYLDIAMINO-O-CARBOXY-PHENYL-XANTHENYL CHLORIDE; TETRAETHYLRHODAMINE; VIOLET ZASADITA 10

Description: green crystals or reddish-violet powder

Use: as dye, esp for paper; as reagent for antimony, bismuth, cobalt, niobium, gold, manganese, mercury, molybdenum, tantalum, thallium, tungsten; as biological stain; red dye for wool & silk where brilliant fluorescent effects are desired; colorant for plastics

Physical Properties

Boiling Point: 165 °C (329 °F)

Freezing Point: 165 °C (329 °F)

Water Solubility: Very Soluble in Water

Other Solubilities: Slightly Soluble in Hydrochloric Acid & Sodium Hydroxide; Soluble in Benzene.

RTECS Toxicity Data

Acute Oral: Rat LD_{Lo} Dose: 500 mg/kg. Mouse LD_{50} Dose: 887 mg/kg.

Acute Dermal: Mouse LD_{50} Route: Subcutaneous Dose: 180 mg/kg.

Chronic (Multiple Dose) Oral: Rat Dose: 50400 mg/kg/12W-C; Toxic Effects: Liver - Changes in liver weight; Kidney, Ureter, and Bladder - Changes in kidney weight; Nutritional and gross metabolic - Weight loss or decreased weight gain.

Reproductive/Teratogenic: Mouse Route: Intraperitoneal; Dose: 60 mg/kg; Duration: female 7-10D of pregnancy; Effects on Embryo or Fetus - Fetotoxicity.

Mutagenic: Mammal Cytogenetic Analysis; Cell Type: fibroblast; Dose: 2 mg/L. Hamster DNA Damage; Cell Type: ovary; Dose: 900 umol/plate.

Tumorigenic: Rat Route: Subcutaneous; Dose: 3600 mg/kg/68W-I; Toxic Effects: Tumorigenic - Equivocal tumorigenic agent by RTECS criteria; Blood - Lymphomax including Hodgkin's disease; Tumorigenic - Tumors at site of application. Rat Route: Subcutaneous; Dose: 3870 mg/kg/68W-I; Toxic Effects: Tumorigenic - Equivocal tumorigenic agent by RTECS criteria; Tumorigenic - Tumors at site of application.

Hazard Overviews

Health: May be irritating to eyes/skin. Harmful. Other Acute Effects: harmful if swallowed, inhaled, or absorbed through skin. Chronic Effects: ; possible risk of irreversible effects; possible mutagen; laboratory experiments have shown mutagenic effects. Possible carcinogen.

Fire: Hazards: emits toxic fumes. Extinguishing agents: water spray; carbon dioxide, dry chemical powder or appropriate foam. Precautions: combustible liquid.

Reactivity: Incompatible with: strong oxidizing agents. Hazardous decomposition products: thermal decomposition may produce carbon monoxide, carbon dioxide, and nitrogen oxides; hydrogen chloride gas.

Carcinogenicity: IARC - Group 3, Not classifiable as to carcinogenicity to humans; NIOSH - Not listed; NTP - Not listed; ACGIH - Not listed; OSHA - Not listed; EPA - Not listed; MAK - Not listed

Environmental

Regulations

RCRA 40CFR: Not listed

CERCLA: 40CFR 302.4: Not listed

SARA 40CFR 372.65: Listed

SARA EHS 40CFR 355: Not listed

TSCA: Listed

CI4900 CAS #: 3761-53-3

C.I. FOOD RED 5

RTECS: QJ6825000

EINECS Number: 223-178-3

Molecular Formula: $C_{18}H_{16}N_2Na_2O_7S_2$

Formula Weight: 482.4

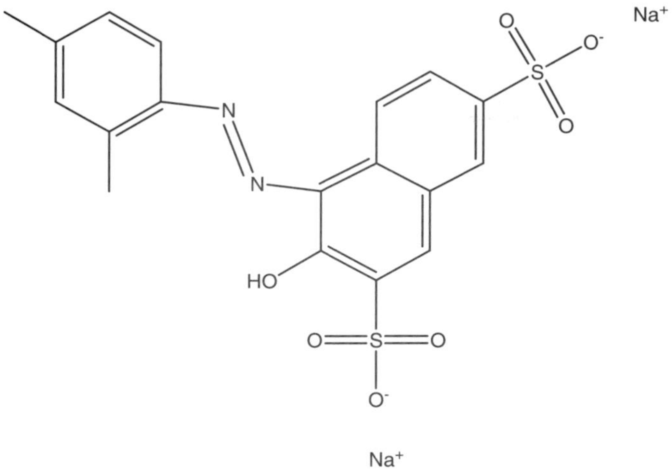

Chemical Structure

Synonyms: ACETACID RED J; ACID LEATHER RED KPR; ACID LEATHER RED P2R; ACID LEATHER SCARLET IRW; ACID PONCEAU R; ACID PONCEAU 2RL; ACID PONCEAU SPECIAL; ACID RED 26; ACID SCARLET; ACID SCARLET 2B; ACID SCARLET 2R; ACID SCARLET 2BN; ACID SCARLET 2R FOR LAKES; ACID SCARLET 2R FOR LAKES BLUISH; ACID SCARLET 2RL; ACID SCARLET 2RN; ACIDAL PONCEAU G; ACILAN PONCEAU RRL; ACILAN PRONCEAU RRL; AHCOCID FAST SCARLET R; AIZEN PONCEAU RH; AMACID LAKE SCARLET 2R; AMACID SCARLET 2R; BRILLIANT PONCEAU G; C.I. 16150; C.I. 79; C.I. ACID RED 26; C.I. ACID RED 26,DISODIUM SALT; C.I. F FOOD RED 5; CALCOCID 2RIL; CALCOCID SCARLET 2R; CALCOCID SCARLET 2RIL; CALCOLAKE SCARLET 2R; CERTICOL PONCEAU MXS; CERVEN KYSELA 26; CERVEN POTRAVINARSKA 5; COLACID PONCEAU SPECIAL; COMACID SCARLET 2R; D AND C RED NO 5; D&C RED NO. 5; 4-((2,4-DIMETHYLPHENYL)AZO)-3-HYDROXY-2,7-NAPHTHALENEDISULFONIC ACID,DISODI; 4-((2,4-DIMETHYLPHENYL)AZO)-3-HYDROXY-2,7-NAPHTHALENEDISULFONIC ACID,DISODIUM SALT; 4-((2,4-DIMETHYLPHENYL)AZO)-3-HYDROXY-2,7-NAPHTHALENEDISULPHONIC ACID,DISODIUM SALT; DISODIUM (2,4-DIMETHYLPHENYLAZO)-2-HYDROXYNAPHTHALENE-3,6-DISULFONATE; DISODIUM (2,4-DIMETHYLPHENYLAZO)-2-HYDROXYNAPHTHALENE-3,6-DISULPHONATE; DISODIUM SALT OF 1-(2,4-XYLYLAZO)-2-NAPHTHOL-3,6-DISULFONIC ACID; DISODIUM SALT OF 1-(2,4-XYLYLAZO)-2-NAPHTHOL-3,6-DISULPHONIC ACID; DISODIUM SALT OF1-(2,4-XYLYLAZO)-2-NAPHTHOL-3,6-DISULFONIC ACID; DISODIUM(2,4-DIMETHYLPHENYLAZO)-2-HYDROXYNAPHTHALENE-3,6-DISULFONATE; DISODIUM(2,4-DIMETHYLPHENYLAZO)-2-HYDROXYNAPHTHALENE-3,6-DISULPHONATE; EDICOL PONCEAU RS; EDICOL SUPRA PONCEAU R; FENAZO SCARLET 2R; FOOD RED NO. 101; HEXACOL PONCEAU 2R; HEXACOL PONCEAU MX; HIDACID SCARLET 2R; HISPACID PONCEAU R; 3-HYDROXY-4-(2,4-XYLYLAZO)-3,7-NAPHTHALENEDISULFONIC ACID,DISODIUM SALT; 3-HYDROXY-4-(2,4-XYLYLAZO)-3,7-NAPHTHALENEDISULPHONICACID,DISODIUM SALT; JAVA PONCEAU 2R; KITON PONCEAU 2R; KITON PONCEAU R; KITON SCARLET 2RC; LAKE PONCEAU; LAKE SCARLET R; LAKE SCARLET 2RBN; NAPHTHALENE LAKE SCARLET R; NAPHTHALENE SCARLET R; 2,7-NAPHTHALENEDISULFONIC ACID,4-((2,4-DIMETHYLPHENYL)AZO)-3-HYDROXY-,DISODIUM SALT; NAPHTHAZINE SCARLET 2R; NEKLACID RED RR; NEW PONCEAU 4R; NSC 10458 (NLM); PAPER RED HRR; PAS KWASOWY 2 RL; PIGMENT PONCEAU R; PONCEAU 2R; PONCEAU RS; PONCEAU 2R (BIOLOGICAL STAIN); PONCEAU BNA; PONCEAU DE XYLIDINE; PONCEAU 2R EXTRA A EXPORT; PONCEAU FR; PONCEAU G; PONCEAU GR; PONCEAU J; PONCEAU MX; PONCEAU NR; PONCEAU PXM; PONCEAU R; PONCEAU R (BIOLOGICAL STAIN); PONCEAU 2RE; PONCEAU RED; PONCEAU RED R; PONCEAU RG; PONCEAU 2RL; PONCEAU RR; PONCEAU RR TYPE 8019; PONCEAU 2RX; PONCEAU XYLIDINE; PONCEAU XYLIDINE (BIOLOGICAL STAIN); PONCEAUX 3R; 1695 RED; RED 101; RED FOR LAKES J; RED NO. 503; RED R; SALT; SCARLET 2R; SCARLET R; SCARLET R (VAN); SCARLET 2RB; SCARLET 2RL; SCARLET 2RL BLUISH; SCARLET RRA; SCHULTZ NO. 95; TERTRACID PONCEAU 2R; TETRACID PONCEAU 2R; XYLIDINE PONCEAU; XYLIDINE PONCEAU 3RS; XYLIDINE RED; 1-XYLYLAZO-2-NAPHTHOL-3,6-DISULFONIC ACID,DISODIUM SALT; 1-(2,4-XYLYLAZO)-2-NAPHTHOL-3,6-DISULPHONIC ACID DISODIUM SALT; 1-XYLYLAZO-2-NAPHTHOL-3,6-DISULPHONIC ACID DISODIUM SALT; 1-(2,4-XYLYLAZO)-2-NAPHTHOL-3,6-DISULPHONIC ACID,DISODIUM SALT; 1-XYLYLAZO-2-NAPHTHOL-3,6-DISULPHONIC ACID,DISODIUM SALT

Description: dark-red crystals

Use: for dyeing wool, silk, inks, paper, pigments (heavy metal salts), wood stains, drugs and cosmetics, leathers, fruit, confectionery and meat products, soaps and face lotions

Physical Properties

Freezing Point: > 300 °C (572 °F)

Water Solubility: Soluble in Water

Other Solubilities: 95% Ethanol: <1 mg/ml at 20 °C; Acetone: <1 mg/ml at 20 °C; DMSO: <1 mg/ml at 20 °C; Ether: Very slightly Soluble; Methanol: <1 mg/ml at 21 °C; Oil: Insoluble; Organic solvents: Insoluble; Toluene: <1 mg/ml at 21 °C.

Flash Point: Not available; probably combustible

RTECS Toxicity Data

Acute Oral: Rat LD_{50} Dose: 23160 mg/kg; Toxic Effects: Behavioral - Muscle weakness; Behavioral - Ataxia. Mouse LD_{50} Dose: >6600 mg/kg.

Mutagenic: Mouse Sister Chromatid Exchenge; Route: Intraperitoneal; Dose: 63 mg/kg. Bacteria - S Typhimurium Mutations in Microorganisms; Dose: 100 ug/plate (-S9).

Hazard Overviews

Health: Irritating. Toxic. Other Acute Effects: harmful if swallowed, inhaled, or absorbed through skin. Chronic Effects: target organ: liver. Carcinogen.

Fire: Will burn. Extinguishing agents: water spray; carbon dioxide, dry chemical powder or appropriate foam. Precautions: combustible liquid.

Reactivity: Incompatible with: strong oxidizing agents. Hazardous decomposition products: toxic fumes of: carbon monoxide, carbon dioxide, nitrogen oxides, sulfur oxides.

Carcinogenicity: IARC - Not listed; NIOSH - Not listed; NTP - Not listed; ACGIH - Not listed; OSHA - Not listed; EPA - Not listed; MAK - Not listed

Primary Target Organs:

Eyes

Skin

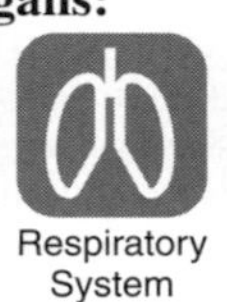
Respiratory System

Liver

Environmental

Regulations

RCRA 40CFR: Not listed
CERCLA: 40CFR 302.4: Not listed
SARA 40CFR 372.65: Listed
SARA EHS 40CFR 355: Not listed
TSCA: Listed

CI6980 CAS #: 37300-23-5

C.I. PIGMENT YELLOW 36

RTECS: GE4600000
Molecular Formula: Unspecified or Variable
Synonyms: PIGMENT YELLOW 36; POTASSIUM ZINC CHROMATE YELLOW; ZINC YELLOW; ZINC YELLOW 1007

Physical Properties

Water Solubility: Practically Insoluble
Flash Point: Will not burn

Hazard Overviews

Reactivity: Stable. Hazardous polymerization cannot occur. Avoid: generating dust or fume. Incompatible with: hydrazine.
Carcinogenicity: IARC - Group 1, Carcinogenic to humans; NIOSH - Listed as carcinogen; NTP - Class 1, Known to be a carcinogen; ACGIH - Class A1, Confirmed human carcinogen; OSHA - Not listed; EPA - Not listed; MAK - Class A1, Capable of inducing malignant tumors as shown by experience with humans
Exposure Limits
ACGIH TLV: TWA: 0.01 mg/m^3; as Cr.

Environmental

Regulations

RCRA 40CFR: Not listed
CERCLA: 40CFR 302.4: Not listed
SARA 40CFR 372.65: Not listed
SARA EHS 40CFR 355: Not listed
TSCA: Listed

CI7630 CAS #: 3118-97-6

C.I. SOLVENT ORANGE 7

RTECS: QL5850000
EINECS Number: 221-490-4
Molecular Formula: $C_{18}H_{16}N_2O$
Formula Weight: 276.32

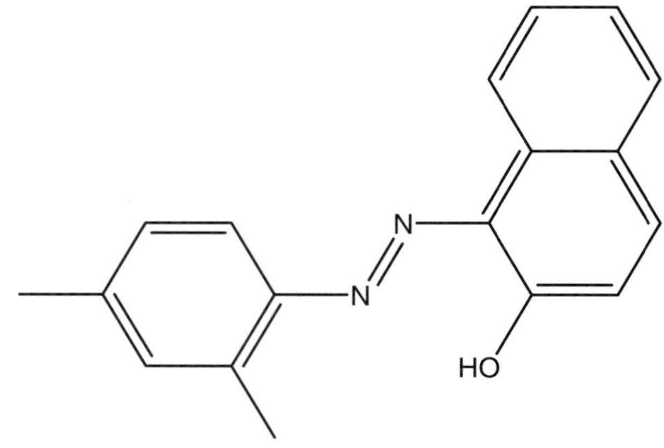

Chemical Structure

Synonyms: A F RED NO 5; A.F. RED NO. 5; AIZEN FOOD RED NO. 5; BRASILAZINA OIL SCARLET 6G; BRILLIANT OIL SCARLET B; C.I. 12140; CALCO OIL SCARLET BL; CALCO OIL SCARLET ZBL; CERES ORANGES RR; CERISOL SCARLET G; CEROTINSCHARLACH G; COLOR INDEX NO: 12140; 1-((2,4-DIMETHYLPHENYL)AZO)-2-NAPHTHALENOL; EXT D & C RED NO 14; EXT D AND C RED NO 14; EXT D AND C RED. NO. 14; EXTRACT D AND C RED NO. 14; FAST OIL ORANGE II; FAT RED (YELLOWISH); FAT SCARLET 2G; FD & C NO 32; FETTORANGE B; GRASAN ORANGE 3R; JAPAN RED 5; JAPAN RED 505; LACQUER ORANGE VR; MOTIROT G; 2-NAPHTHALENOL,1-((2,4-DIMETHYLPHENYL)AZO)-; OIL ORANGE 2R; OIL ORANGE KB; OIL ORANGE N EXTRA; OIL ORANGE R; OIL ORANGE X; OIL ORANGE XO; OIL RED GRO; OIL RED RO; OIL RED XO; OIL SCARLET; OIL SCARLET 6G; OIL SCARLET 371; OIL SCARLET APYO; OIL SCARLET BL; OIL SCARLET L; OIL SCARLET Y; OIL SCARLET YS; ORANGE OIL KB; ORANZ ROZPOUSTEDLOVA 7; PYRONALROT R; RED B; RED NO 5; RED NO. 5; RESIN SCARLET 2R; RESOFORM ORANGE R; ROT B; ROT GG FETTLOESLICH; SOMALIA ORANGE 2R; SOMALIA ORANGE A2R; SUDAN II; SUDAN ORANGE; SUDAN ORANGE RPA; SUDAN ORANGE RRA; SUDAN RED; SUDAN SCARLET 6G; WAXAKOL VERMILION; WAXAKOL VERMILION L; 1-(2,4-XYLYLAZO)-2-NAPHTHOL; 1-(O-XYLYLAZO)-2-NAPHTHOL; 1-XYLYLAZO-2-NAPHTHOL
Description: red to brown-red crystals or needles
Use: reportedly used for coloring oils, waxes; hydrocarbon solvent for polishes, candles and polystyrene resins; to color petroleum products, plastics, shoe polish, cosmetics and drugs applied externally

Physical Properties

Freezing Point: 166 °C (330.8 °F)
Water Solubility: Insoluble in Water
Other Solubilities: Soluble in Ether.

RTECS Toxicity Data

Mutagenic: Bacteria - S Typhimurium Mutations in Microorganisms; Dose: 50 ug/plate (-S9).
Tumorigenic: Mouse Route: Implant; Dose: 80 mg/kg; Toxic Effects: Tumorigenic - Carcinogenic by RTECS criteria; Kidney, Ureter, and Bladder - Tumors.

Hazard Overviews

Health: Irritating to eyes/skin/respiratory tract. Harmful. Other Acute Effects: harmful if swallowed, inhaled, or absorbed through skin; prolonged or repeated exposure may cause allergic reactions in certain sensitive individuals. Chronic Effects: laboratory experiments have shown mutagenic effects; target organ: blood.

Fire: Hazards: emits toxic fumes. Extinguishing agents: water spray; carbon dioxide, dry chemical powder or appropriate foam. Precautions: combustible liquid.

Reactivity: Incompatible with: strong oxidizing agents. Hazardous decomposition products: toxic fumes of: carbon monoxide, carbon dioxide, nitrogen oxides.

Carcinogenicity: IARC - Group 3, Not classifiable as to carcinogenicity to humans; NIOSH - Not listed; NTP - Not listed; ACGIH - Not listed; OSHA - Not listed; EPA - Not listed; MAK - Not listed

Primary Target Organs:

Eyes

Skin

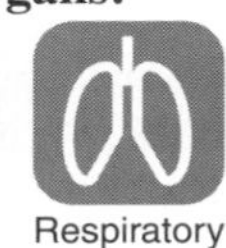
Respiratory System

Blood

Environmental

Regulations

RCRA 40CFR: Not listed
CERCLA: 40CFR 302.4: Not listed
SARA 40CFR 372.65: Listed
SARA EHS 40CFR 355: Not listed
TSCA: Listed

CI7760 CAS #: 1096-48-6

C.I. SOLVENT RED 168

EINECS Number: 214-143-3

Physical Properties

Water Solubility: Insoluble
Other Solubilities: Soluble in Toluene

Hazard Overviews

Carcinogenicity: IARC - Not listed; NIOSH - Not listed; NTP - Not listed; ACGIH - Not listed; OSHA - Not listed; EPA - Not listed; MAK - Not listed

Environmental

Regulations

RCRA 40CFR: Not listed
CERCLA: 40CFR 302.4: Not listed
SARA 40CFR 372.65: Not listed
SARA EHS 40CFR 355: Not listed
TSCA: Not listed

CI7890 CAS #: 842-07-9

C.I. SOLVENT YELLOW 14

RTECS: QL4900000
EINECS Number: 212-668-2
Molecular Formula: $C_{16}H_{12}N_2O$
Formula Weight: 248.3

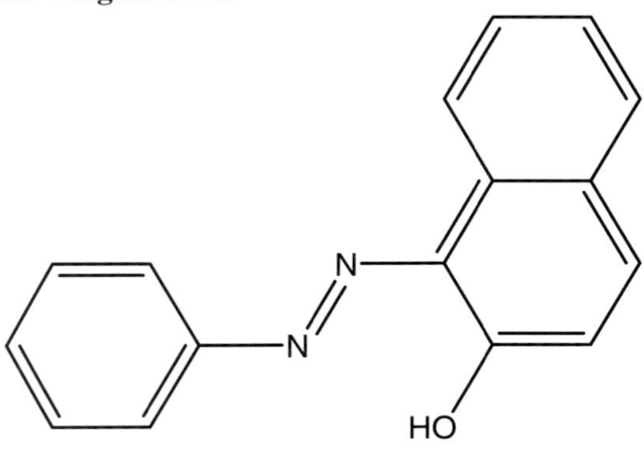

Chemical Structure

Synonyms: ATUL ORANGE R; 1-BENZENEAZO-2-NAPHTHOL; BENZENE-1-AZO-2-NAPHTHOL; BENZENEAZO-BETA-NAPHTHOL; 1-BENZOAZO-2-NAPHTHOL; BRASILAZINA OIL ORANGE; BRILLIANT OIL ORANGE R; C.I. 12055; CALCO OIL ORANGE 7078-Y; CALCO OIL ORANGE Z-7078; CALCO OIL ORANGE 7078; CALCOGAS M; CALCOGAS ORANGE NC; CAMPBELLINE OIL ORANGE; CARMINAPH; CERES ORANGE R; CEROTINORANGE G; DISPERSOL YELLOW PP; DUNKELGELB; ENIAL ORANGE I; FAST OIL ORANGE; FAST OIL ORANGE I; FAST ORANGE; FAT ORANGE 4A; FAT ORANGE RS; FAT ORANGE G; FAT ORANGE I; FAT ORANGE R; FAT SOLUBLE ORANGE; FETTORANGE 4A; FETTORANGE IG; FETTORANGE LG; FETTORANGE R; GRASAL ORANGE; GRASAN ORANGE R; HIDACO OIL ORANGE; 2-HYDROXYNAPHTHYL-1-AZOBENZENE; 2-HYDROXY-1-PHENYLAZONAPHTHALENE; LACQUER ORANGE VG; MORTON ORANGE Y; MOTIORANGE R; 2-NAPHTHALENOL,1-(PHENYLAZO)-; 2-NAPHTHOL,1-(PHENYLAZO)-; 1,2-NAPHTHOQUINONE-1-PHENYLHYDRAZONE; NSC 11227; NSC 51524; OIL ORANGE; OIL ORANGE 2B; OIL ORANGE 7078-V; OIL ORANGE R-14; OIL ORANGE Z-7078; OIL ORANGE 2311; OIL ORANGE 31; OIL ORANGE E; OIL ORANGE EP; OIL ORANGE G; OIL ORANGE PEL; OIL ORANGE PS; OIL ORANGE R; OIL SOLUBLE ORANGE; OLEAL ORANGE R; ORANGE 2 INSOLUBLE; ORANGE A L'HUILE; ORANGE INSOLUBLE OLG; ORANGE PEL; ORANGE R FAT SOLUBLE; ORANGE 3RA SOLUBLE IN GREASE; ORANGE RESENOLE 3; ORANGE RESENOLE NO. 3; ORANGE SOLUBLE A L'HUILE; ORGANOL ORANGE; ORIENT OIL ORANGE PS; PETROL ORANGE Y; 1-(PHENYLAZO)-2-NAPHTHALENOL; 1-(PHENYLAZO)-2-NAPHTHOL; 1-PHENYLAZO-2-NAPHTHOL; 1-PHENYLAZO-BETA-NAPHTHOL; ALPHA-PHENYLAZO-BETA-NAPHTHOL; PLASTORESIN ORANGE F4A; PYRONALORANGE; RESINOL ORANGE R; RESOFORM ORANGE G; SANSEL ORANGE G; SCHARLACH B; SILOTRAS ORANGE TR; SOLVENT YELLOW 14; SOMALIA ORANGE I; SOUDAN I; SPIRIT ORANGE; SPIRIT YELLOW I; STEARIX ORANGE; SUDAN 1; SUDAN I; SUDAN J; SUDAN ORANGE R; SUDAN ORANGE RA; SUDAN ORANGE RA NEW; SUDAN YELLOW; TERTROGRAS ORANGE SV; TOYO OIL ORANGE; WAXAKOL ORANGE GL; WAXOLINE YELLOW I; WAXOLINE YELLOW IM; WAXOLINE YELLOW IP; WAXOLINE YELLOW IS; ZLUT ROZPOUSTEDLOVA 14

Description: brick-red crystals or leaflets; slight odor

Use: dye used for coloring hydrocarbon solvents, oils, fats, waxes, shoe and floor polishes, gasoline, cellulose ester varnishes, styrene resins, soap, colored smokes and plastics

Physical Properties

Boiling Point: Sublimes at > 100 °C (212 °F)
Freezing Point: 134 °C (273.2 °F)
Specific Gravity: 1.69
Water Solubility: Insoluble in Water
Other Solubilities: Soluble in Petroleum Ether, Carbon Disulfide, concentrated Hydrochloric Acid.
pH: Approximately 9
Evaporation Rate: < 1 Butyl Acetate=1
Flash Point: Not available; probably combustible

RTECS Toxicity Data

Chronic (Multiple Dose) Oral: Rat Dose: 4200 mg/kg/14D-C; Toxic Effects: DEATH. Rat Dose: 87360 mg/kg/13W-C; Toxic Effects: DEATH.
Mutagenic: Rat Unscheduled DNA Synthesis; Route: Oral; Dose: 500 mg/kg. Rat Micronucleus Test; Route: Oral; Dose: 250 mg/kg.
Tumorigenic: Rat Route: Oral; Dose: 10815 mg/kg/2Y-C; Toxic Effects: Tumorigenic - Neoplastic by RTECS criteria; Liver - Tumors. Mouse Route: Oral; Dose: 21630 mg/kg/2Y-C; Toxic Effects: Tumorigenic - Neoplastic by RTECS criteria; Liver - Tumors.

Hazard Overviews

Health: Irritating to eyes/skin/respiratory tract. Toxic. Other Acute Effects: harmful if swallowed, inhaled, or absorbed through skin. Chronic Effects: carcinogen.
Fire: Will burn. Extinguishing agents: water spray; carbon dioxide, dry chemical powder or appropriate foam. Precautions: combustible liquid.
Reactivity: Incompatible with: strong oxidizing agents. Hazardous decomposition products: toxic fumes of: carbon monoxide, carbon dioxide, nitrogen oxides.
Carcinogenicity: IARC - Group 3, Not classifiable as to carcinogenicity to humans; NIOSH - Not listed; NTP - Not listed; ACGIH - Not listed; OSHA - Not listed; EPA - Not listed; MAK - Not listed
Primary Target Organs:

Eyes

Skin

Respiratory System

Environmental

Regulations

RCRA 40CFR: Not listed
CERCLA: 40CFR 302.4: Not listed
SARA 40CFR 372.65: Listed
SARA EHS 40CFR 355: Not listed
TSCA: Listed

CI8150 CAS #: 128-66-5

C.I. VAT YELLOW 4

RTECS: HO7030000
EINECS Number: 204-903-2
Molecular Formula: $C_{24}H_{12}O_2$
Formula Weight: 332.36
Synonyms: AHCOVAT PRINTING GOLDEN YELLOW; AHCOVAT PRINTING GOLDEN YELLOW GK; AMANTHRENE GOLDEN YELLOW; AMANTHRENE GOLDEN YELLOW GK; ANTHRAVAT GOLDEN YELLOW; ANTHRAVAT GOLDEN YELLOW GK; ARLANTHRENE GOLDEN YELLOW; ARLANTHRENE GOLDEN YELLOW GK; BENZADONE GOLD YELLOW GK; BENZADONE GOLDEN YELLOW; C.I. 59100; C.I. VAT YELLOW; CALCOLOID GOLDEN YELLOW; CALCOLOID GOLDEN YELLOW GKWP; CALEDON GOLDEN YELLOW; CALEDON GOLDEN YELLOW GK; CALEDON PRINTING YELLOW; CALEDON PRINTING YELLOW GK; CARBANTHRENE GOLDEN YELLOW; CARBANTHRENE GOLDEN YELLOW GK; CIBANONE GOLDEN YELLOW; CIBANONE GOLDEN YELLOW FGK; CIBANONE GOLDEN YELLOW GK; 1,2,6,7-DIBENPYRENE-7,14-QUINONE; DIBENZO(B,DEF)CHRYSENE-7,14-DIONE; 3,4:8,9-DIBENZOPYRENE-5,10-DIONE; DIBENZO(A,B)PYRENE-7,14-DIONE; 2,3,7,8-DIBENZOPYRENE-1,6-QUINONE; 1',2',6',7'-DIBENZPYRENE-7,14-QUINONE; DIBENZPYRENEQUINONE; FEMANTHREN GOLDEN YELLOW; FENANTHREN GOLDEN YELLOW GK; GOLDEN YELLOW; GOLDEN YELLOW ZHKH; HELANTHRENE YELLOW; HELANTHRENE YELLOW GOK; HOSTAVAT GOLDEN YELLOW; HOSTAVAT GOLDEN YELLOW GK; INDANTHREN GOLDEN YELLOW; INDANTHREN GOLDEN YELLOW GK; INDANTHREN PRINTING YELLOW; INDANTHREN PRINTING YELLOW GOK; INDANTHRENE GOLD YELLOW GK; INDANTHRENE GOLDEN YELLOW; INDANTHRENE GOLDEN YELLOW GK; LEUCOSOL GOLDEN YELLOW; LEUCOSOL GOLDEN YELLOW GK; MAYVAT GOLDEN YELLOW; MAYVAT GOLDEN YELLOW GK; MIKETHRENE GOLD YELLOW; MIKETHRENE GOLD YELLOW GK; MIKETHRENE GOLDEN YELLOW; NIHONTHRENE GOLDEN YELLOW; NIHONTHRENE GOLDEN YELLOW GK; NSC 30987; NYANTHRENE GOLDEN YELLOW; NYANTHRENE GOLDEN YELLOW GK; PALANTHRENE GOLDEN YELLOW; PALANTHRENE GOLDEN YELLOW GK; PARADONE GOLDEN YELLOW; PARADONE GOLDEN YELLOW GK; PHARMANTHRENE GOLDEN YELLOW; PHARMANTHRENE GOLDEN YELLOW GK; ROMANTRENE GOLDEN YELLOW; ROMANTRENE GOLDEN YELLOW GOK; SANDOTHRENE GOLDEN YELLOW; SANDOTHRENE GOLDEN YELLOW NGK; SANDOTHRENE PRINTING YELLOW; SANDOTHRENE PRINTING YELLOW NH; SOLANTHRENE BRILLIANT YELLOW; SOLANTHRENE BRILLIANT YELLOW J; TINON GOLDEN YELLOW; TINON GOLDEN YELLOW GK; TYRIAN YELLOW I-GOK; TYRION YELLOW; VAT GOLDEN YELLOW; VAT GOLDEN YELLOW ZHKH; VAT GOLDEN YELLOW ZHKHD; VAT YELLOW 4; VAT YELLOW ZHKH; YELLOW; YELLOW GK BASE
Description: orange, viscous liquid
Use: dye for textiles and paper; by the armed services as a smokescreen and as a signaling agent

Physical Properties

Water Solubility: >= 100 mg/mL at 24 C
Other Solubilities: DMSO: <1 mg/ml at 24 °C; Acetone: <1 mg/ml at 24 °C; Toluene: Slightly Soluble; Alcohol: Slightly Soluble; NitroBenzene: Soluble; Chloroform: Slightly Soluble; Benzene: Slightly Soluble.
Flash Point: > 93.3 °C

RTECS Toxicity Data

Tumorigenic: Mouse Route: Oral; Dose: 7420 gm/kg/2Y-C; Toxic Effects: Tumorigenic - Carcinogenic by RTECS criteria; Blood - Lymphomax including Hodgkin's disease. Mouse Route: Oral; Dose: 2225 gm/kg/106W-C; Toxic Effects: Tumorigenic - Equivocal tumorigenic agent by RTECS criteria; Blood - Lymphomax including Hodgkin's disease.

Hazard Overviews

Fire: Will burn.
Carcinogenicity: IARC - Group 3, Not classifiable as to carcinogenicity to humans; NIOSH - Not listed; NTP - Not listed; ACGIH - Not listed; OSHA - Not listed; EPA - Not listed; MAK - Not listed

Environmental

Regulations

RCRA 40CFR: Not listed
CERCLA: 40CFR 302.4: Not listed
SARA 40CFR 372.65: Listed
SARA EHS 40CFR 355: Not listed
TSCA: Listed

CIO5000 CAS #: 7700-17-6

CIODRIN

RTECS: GQ5075000
EINECS Number: 231-720-5
Molecular Formula: $C_{14}H_{19}O_6P$
Structured MF: $(CH_3O)_2P(O)OC(CH_3)=CHCO_2CH(CH_3)C_6H_5$
Formula Weight: 314.28
Synonyms: 2-BUTENOIC ACID,3-((DIMETHOXYPHOSPHINYL)OXY)-,1-PHENYLETHYL ESTER,(E)-; 2-BUTENOIC ACID,3-((DIMETHOXYPHOSPHINYL)OXY)-,1-PHENYLETHYL ESTER,(E)-(9CI); CIODRIN; CROTONIC ACID,3-HYDROXY-,ALPHA-METHYLBENZYL ESTER,DIMETHYL PHOSPHATE,(E)-; CROTOXYFOS; CROTOXYPHOS; CYODRIN; CYPONA E.C; CYPONA EC; DECROTOX; 3-[(DIMETHOXYPHOSPHINYL)OXY]-2-BUTENOIC ACID 1-PHENYLETHYLESTER; DIMETHYL 2-(ALPHA-METHYLBENZYLOXYCARBONYL)-1-METHYLVINYLPHOSPHATE; O,O-DIMETHYL O-(1-METHYL-2-CARBOXY-ALPHA-PHENYLETHYL)VINYLPHOSPHATE; O,O-DIMETHYL O-[1-METHYL-2-(1-PHENYLCARBETHOXY)VINYL]PHOSPHATE; DIMETHYL PHOSPHATE OF ALPHA-METHYLBENZYL 3-HYDROXY-CIS-CROTONATE; DIMETHYL PHOSPHATE OF ALPHA-METHYLBENZYL3-HYDROXY-CIS-CROTONATE; DIMETHYL-CIS-1-METHYL-2-(1-PHENYLETHOXYCARBONYL)VINYLPHOSPHATE; DUO-KILL; DURAVOS; ENT 24,717; 3-HYDROXYCROTONIC ACID ALPHA-METHYLBENZYL ESTER DIMETHYLPHOSPHATE; KEMDRIN; ALPHA-METHYL BENZYL-3-(DIMETHOXY-PHOSPHINYLOXY)-CIS-CROTONATE; A-METHYL BENZYL-3-HYDROXY CIS CROTONATE DIMETHYL PHOSPHATE; ALPHA-METHYLBENZYL 3-HYDROXYCROTONATE DIMETHYL PHOSPHATE; 1-METHYLBENZYL-3-(DIMETHOXYPHOSPHINYLOXO)ISOCROTONATE; ALPHA-METHYLBENZYL-3-(DIMETHOXY-PHOSPHINYLOXY)-CIS-CROTONATE; ALPHA-METHYLBENZYL(E)-3-HYDROXYCROTONATE ESTER WITHDIMETHYL PHOSPHATE; OMS 239; PANTOZOL 1; CIS-2-(1-PHENYLETHOXY)CARBONYL-1-METHYLVINYLDIMETHYLPHOSPHATE; 1-PHENYLETHYL 3-(DIMETHOXYPHOSPHINOYLOXY)ISOCROTONATE; 1-PHENYLETHYL (E)-3-[(DIMETHOXYPHOSPHINYL)OXY]-2-BUTENOATE; SD 4294; SHELL SD 4294; SIMAX; VOLFAZOL
Description: light straw-colored clear liquid; mild ester odor
Use: insecticide for external use on livestock; control of flies, mites and ticks on cattle and pigs

Physical Properties

Boiling Point: 135 °C (275 °F) at 0.03 mm Hg
Specific Gravity: 1.19 at 25 °C
Saturated Vapor Density: 1.200000217 kg/m^3
Vapor Pressure: 1.4×10^{-5} mm Hg at 20 °C
Water Solubility: 0.001%
Other Solubilities: Soluble in Propan-2-ol, Xylene.
Refraction Index: 1.4988 at 25 °C/D
Flash Point: 79.4 °C Tag Open Cup

RTECS Toxicity Data

Acute Oral: Rat LD_{50} Dose: 38400 ug/kg; Toxic Effects: Biochemical - True cholinesterase.
Acute Dermal: Rat LD_{50} Route: Skin; Dose: 202 mg/kg. Rat LD_{50} Route: Subcutaneous Dose: 47 mg/kg.
Mutagenic: Mouse Mutations in Mammalian Somatic Cells; Cell Type: lymphocyte; Dose: 180 mg/L. Yeast - S Cerevisiae Mutations in Microorganisms; Dose: 5000 ppm (+S9).

Hazard Overviews

Fire: Combustible.
Carcinogenicity: IARC - Not listed; NIOSH - Not listed; NTP - Not listed; ACGIH - Not listed; OSHA - Not listed; EPA - Not listed; MAK - Not listed

Environmental

Ecotoxicity: LC_{50} Salmo clarki (cutthroat trout) 51.0 ug/L/96 hr at 12 °C (95% confidence limit 28.0-91.0 ug/L), wt 1.0 g. Static bioassay without aeration, pH 7.2-7.5, water hardness 40-50 mg/l as calcium carbonate and alkalinity of 30-35 mg/l LD_{50} Coturnix japonica oral 520 ppm LC_{50} Gammarus lacustris 49.0 ug/L/24 hr at 15 °C (95% confidence limit 36.0-67.0 ug/L), mature. Static bioassay without aeration, pH 7.2-7.5, water hardness 40-50 mg/l as calcium carbonate and alkalinity of 30-35 mg/l LD_{50} Mallard oral 790 mg/kg (95% confidence limit 411-1520 mg/kg), 3-4 mo old males LC_{50} Pteronarcys 2.2 ug/L/72 hr at 15 °C, second year class. Static bioassay without aeration, pH 7.2-7.5, water hardness 40-50 mg/l as calcium carbonate and alkalinity of 30-35 mg/l
Environmental Fate: If released on land, it will adsorb moderately to the soil and biodegrade in about a day. Degradation is faster in alkaline than in acid soils. If released in water, it will be lost due to hydrolysis. The hydrolysis half-life is 17 days at pH 6.0 and shorter in more alkaline water. It will also adsorb moderately to sediment and particulate matter in the soil column and biodegrade approximately 2 orders of magnitude faster than the chemical hydrolysis. It would not be expected to volatilize from water or bioconcentrate in fish. If released into the atmosphere during spraying, Aerosols will

be removed by gravitational settling. Vapor-phase will react with photochemically-produced hydroxyl radicals resulting in an estimated atmospheric half-life of 4.2 hr.

Environmental Physical Data

Henry's Law Constant: estimated at 5.8 x10^{-9}
Octanol/Water Partition Coefficient: log K_{ow} = 1.89
Sorption Partition Coefficient: K_{oc} = soils 173
BCF: calculated at 16

Regulations

RCRA 40CFR: Not listed
CERCLA: 40CFR 302.4: Not listed
SARA 40CFR 372.65: Not listed
SARA EHS 40CFR 355: Not listed
TSCA: Not listed

Analytical Methods

Soil: EPA PMD-CRO, PMD-TLC; SW846 8141, 8141A, 8270B, 8270C
Water / Groundwater: EPA 1625, 1657; SW846 8141, 8141A
Food: FDA 212.1, 212.2, 232.1, 232.3, 232.4, 242.1
Plasma: EPA 001, 29; FDA 211.1, 231.1, 252

CIT3000 CAS #: 77-92-9

CITRIC ACID

RTECS: GE7350000
EINECS Number: 201-069-1
Molecular Formula: $C_6H_8O_7$
Structured MF: $HOC(CH_2CO_2H)_2CO_2H$
Formula Weight: 192.12

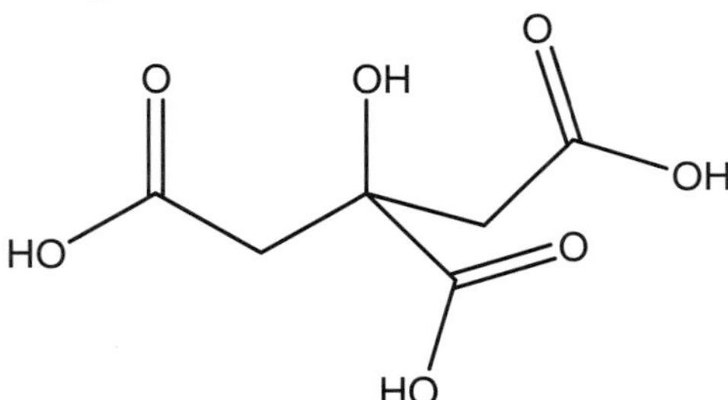

Chemical Structure

Synonyms: ACILETTEN; ANHYDROUS CITRIC ACID; CITRETTEN; CITRIC ACID,ANHYDROUS; CITRO; HYDROCEROL A; 2-HYDROXY-1,2,3-PROPANETRICARBOXYLIC ACID; 2-HYDROXYPROPANETRICARBOXYLIC ACID; 2-HYDROXYTRICARBALLYLIC ACID; BETA-HYDROXYTRICARBALLYLIC ACID; BETA-HYDROXY-TRICARBOXYLIC ACID; KYSELINA CITRONOVA; KYSELINA 2-HYDROXY-1,2,3-PROPANTRIKARBONOVA; NSC 30279; 1,2,3-PROPANETRICARBOXYLIC ACID,2-HYDROXY-
Description: colorless, white crystals, granules, powder; odorless
Use: constituent of fruit drinks, pharmaceutical syrups and other food products; adjusting the pH of foods; an antioxidant in foods; a sequestering agent to remove trace metals; a mordant to brighten colors; electroplating; special inks; reagent for albumin, mucin, glucose and bile pigments; acidification of plasma and urine in vivo and as a component of anticoagulant citrate solutions

Physical Properties

Boiling Point: Decomposes
Freezing Point: 153 °C (307.4 °F)
Specific Gravity: 1.665 at 20 °C/4 °C
Density: 1.542 g/mL at 20 °C
Vapor Pressure: 0.08 mm Hg
Water Solubility: >= 100 mg/mL at 22 C
Other Solubilities: 95% Ethanol: >=100 mg/ml at 22 °C; Acetone: >=100 mg/ml at 22 °C; Amyl acetate: 5.98 g/100 g; Amyl Alcohol: 15.43 g/100 g; DMSO: >=100 mg/ml at 22 °C; Ether: Moderately Soluble; Ethyl Acetate: 5.28 g/100 g; Methanol: 197 g/100 g.
pH: 1 N solution 2.2
Refraction Index: 1.493 at 20 °C
Flash Point: Not pertinent (combustible solid)
Autoignition Temperature: Powder 1010 °C
LEL: 0.28% v/v
UEL: 2.29 kg/m^3 (Dust)

RTECS Toxicity Data

Acute Oral: Rat LD_{50} Dose: 3 gm/kg. Mouse LD_{50} Dose: 5040 mg/kg; Toxic Effects: Lungs, Thorax, or Respiration - Other changes; Musculoskelital - Other changes.
Acute Dermal: Rat LD_{50} Route: Subcutaneous Dose: 5500 mg/kg; Toxic Effects: Lungs, Thorax, or Respiration - Other changes; Musculoskelital - Other changes. Mouse LD_{50} Route: Subcutaneous Dose: 2700 mg/kg; Toxic Effects: Lungs, Thorax, or Respiration - Other changes; Musculoskelital - Other changes.
Irritation Eye: Rabbit Standard Draize Test Dose: 750 ug/24H; Reaction: severe.
Irritation Skin: Rabbit Standard Draize Test Dose: 500 mg/24H; Reaction: mild.

Hazard Overviews

Explosive

Flammable

Fire Diamond

Health: Irritating to skin/respiratory tract; severely irritating to the eyes. Also Causes: GI irritation, hypocalcemia. Chronic Effects: tooth enamel erosion.
Fire: Flammable. Powder can form explosive dust-air mixtures. Fight fire with dry chemical, carbon dioxide, water spray, or regular foam.
Reactivity: Unstable, readily absorbs moisture from air. Hazardous polymerization cannot occur. Avoid: dispersing powder in the air; exposure to heat and ignition sources. Incompatible with: potassium tartrate; alkali and alkaline earth carbonates and bicarbonates; acetates; sulfites; metal nitrates; copper, zinc, aluminum and their alloys. Hazardous decomposition products: acrid; irritating smoke.

Carcinogenicity: IARC - Not listed; NIOSH - Not listed; NTP - Not listed; ACGIH - Not listed; OSHA - Not listed; EPA - Not listed; MAK - Not listed

Primary Target Organs:

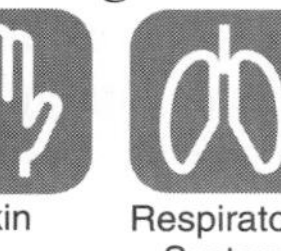
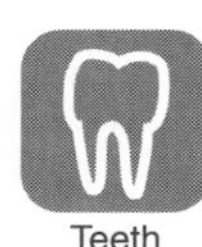

Eyes Skin Respiratory System Teeth

Environmental

Ecotoxicity: Aquatic toxicity: 894 ppm/4 hr/goldfish/killed/fresh water 160 ppm/48 hr/shore crab/TLm/salt water

Cleanup/Disposal: Guide No. 131: Fully encapsulating, vapor protective clothing should be worn for spills and leaks with no fire. Eliminate all ignition sources (no smoking, flares, sparks or flames in immediate area). All equipment used when handling the product must be grounded. Do not touch or walk through spilled material. Stop leak if you can do it without risk. Prevent entry into waterways, sewers, basements or confined areas. A vapor suppressing foam may be used to reduce vapors. Small Spills: Absorb with earth, sand or other non-combustible material and transfer to containers for later disposal. Use clean non-sparking tools to collect absorbed material. Large Spills: Dike far ahead of liquid spill for later disposal. Water spray may reduce vapor; but may not prevent ignition in closed spaces.

Environmental Physical Data

Octanol/Water Partition Coefficient: log K_{ow} = -1.72

BCF: no food chain concentration potential

BOD: 40%, 5 days

Regulations

RCRA 40CFR: Not listed

CERCLA: 40CFR 302.4: Not listed

SARA 40CFR 372.65: Not listed

SARA EHS 40CFR 355: Not listed

TSCA: Listed

CIT9000 CAS #: 6358-53-8

CITRUS RED NO. 2

RTECS: QL3675000

EINECS Number: 228-778-9

Molecular Formula: $C_{18}H_{16}N_2O_3$

Structured MF: $(CH_3O)_2C_6H_3N{=}NC_{10}H_6OH$

Formula Weight: 308.34

Synonyms: C.I. 12156; C.I. SOLVENT RED 80; CERVEN ROZPOUSTEDLOVA 80; CITRUS RED 2; CITRUS RED NO 2; 2,5-DIMETHOXYBENZENEAZO-BETA-NAPHTHOL; 1-((2,5-DIMETHOXYPHENYL)AZO)-2-NAPHTHALENOL; 1-((2,5-DIMETHOXYPHENYL)AZO)-2-NAPHTHOL; 1-(1-(2,5-DIMETHOXYPHENYL)AZO)-2-NAPHTHOL; 1-(2,5-DIMETHOXYPHENYLAZO)-2-NAPHTHOL; 2,5-DIMETHOXY-1-(PHENYLAZO)-2-NAPHTHOL; 2-5-DIMETHOXY-1-(PHENYLAZO)-2-NAPHTHOL; 1-(2,5-DIMETHYLOXYPHENYLAZO)-2-NAPHTHOL; 2-NAPHTHALENOL,1-((2,5-DIMETHOXYPHENYL)AZO)-; 2-NAPHTHOL,1-((2,5-DIMETHOXYPHENYL)AZO)

Description: crystals

Use: coloration of citrus fruits; also used in drugs and cosmetics

Physical Properties

Freezing Point: 155 °C (311 °F) to 157 °C (314.6 °F)

Water Solubility: Slightly Soluble in Water

Other Solubilities: 95% Ethanol: <1 mg/ml at 22 °C; Acetone: <1 mg/ml at 22 °C; DMSO: 1-10 mg/ml at 22 °C; Methanol: <1 mg/ml at 22 °C; Toluene: <1 mg/ml at 22 °C; Vegetable oils: Partially Soluble.

Flash Point: Not available; probably combustible

RTECS Toxicity Data

Mutagenic: Bacteria - S Typhimurium Mutations in Microorganisms; Dose: 67 ug/plate (-S9).

Tumorigenic: Mouse Route: Subcutaneous; Dose: 20 gm/kg/80W-C; Toxic Effects: Tumorigenic - Carcinogenic by RTECS criteria; Lungs, Thorax, or Respiration - Tumors; Liver - Tumors. Mouse Route: Implant; Dose: 80 mg/kg; Toxic Effects: Tumorigenic - Carcinogenic by RTECS criteria; Kidney, Ureter, and Bladder - Tumors.

Hazard Overviews

Fire: Will burn.

Carcinogenicity: IARC - Group 2B, Possibly carcinogenic to humans; NIOSH - Not listed; NTP - Not listed; ACGIH - Not listed; OSHA - Not listed; EPA - Not listed; MAK - Not listed

Environmental

Regulations

RCRA 40CFR: Not listed

CERCLA: 40CFR 302.4: Not listed

SARA 40CFR 372.65: Not listed

SARA EHS 40CFR 355: Not listed

TSCA: Listed

CLO4650 CAS #: 2971-90-6

CLOPIDOL

RTECS: UU7711500

EINECS Number: 221-008-2

Molecular Formula: $C_7H_7Cl_2NO$

Structured MF: $C_7H_7Cl_2NO$

Formula Weight: 192.05

Synonyms: COCCIDIOSTAT C; COYDEN; COYDEN 25; 3,5-DICHLORO-2,6-DIMETHYL-4-PYRIDINOL; FARMCOCCID; LERBEK; METHYLCHLOROPINDOL; METHYLCHLORPINDOL; METICLORPINDOL

Description: white to light-brown, crystalline solid

Physical Properties

Freezing Point: > 315.56 °C (600.008 °F)

Water Solubility: Insoluble
Flash Point: Noncombustible Solid

RTECS Toxicity Data

Acute Oral: Rat LD_{50} Dose: 18 gm/kg. Rabbit LD_{50} Dose: >8 gm/kg.

Hazard Overviews

Fire: Noncombustible.
Carcinogenicity: IARC - Not listed; NIOSH - Not listed; NTP - Not listed; ACGIH - Class A4, Not classifiable as a human carcinogen; OSHA - Not listed; EPA - Not listed; MAK - Not listed
Exposure Limits
OSHA PEL: TWA: 15 mg/m^3; total dust.
ACGIH TLV: TWA: 10 mg/m^3.
NIOSH REL: TWA: 10 mg/m^3. STEL: 20 mg/m^3; total particulate.
Respirator Recommendation
Exposure Range: >5 to 50 mg/m^3 Air Purifying, Negative Pressure, Half Mask
Exposure Range: >50 to 500 mg/m^3 Air Purifying, Negative Pressure, Full Face
Exposure Range: >500 to 5000 mg/m^3 Supplied Air, Constant Flow/Pressure Demand, Full Face
Exposure Range: >5000 to unlimited mg/m^3 Self-contained Breathing Apparatus, Pressure Demand, Full Face
Cartridge Color: dust/mist filter (use P100 or consult supervisor for appropriate dust/mist filter)

Environmental

Regulations

RCRA 40CFR: Not listed
CERCLA: 40CFR 302.4: Not listed
SARA 40CFR 372.65: Not listed
SARA EHS 40CFR 355: Not listed
TSCA: Not listed

COA9000 CAS #: 8001-58-9

COAL TAR CRESOTE

RTECS: GF8615000
DOT: UN1136; UN1137; IMO3.2; IMO3.3
EINECS Number: 232-287-5
Molecular Formula: Unspecified or Variable
Formula Weight: Varies
Synonyms: AWPA #1; BRICK OIL; COAL TAR CREOSOTE; COAL TAR OIL; CREOSOTE; CREOSOTE OIL; CREOSOTE P1; CREOSOTE,FROM COAL TAR; CREOSOTUM; CRESYLIC CREOSOTE; DEAD OIL; EPA PESTICIDE CHEMICAL CODE 025004; HEAVY OIL; LIQUID PITCH OIL; NAPHTHALENE OIL; PRESERV-O-SOTE; SAKRESOTE 100; TAR OIL; WASH OIL
Description: colorless, yellow to black oily liquid; aromatic smoky smell
Use: impregnate wood to protect it from rot and worm lubricant for die molds, agent for water proofing, animal dip and the manufacturing of chemicals; as an insecticide, fungicide, germicide, constituent of fuel oil, pitch for roofing, lampblack and frothing agent in mineral flotation; in veterinary medicine as an intestinal vermicide; in the pharmaceutical industry as an antiseptic, disinfectant, anitpyretic, astringent, styptic and expectorant; as an animal or bird repellent, miticide, herbicide and as a feedstock for the production of carbon blacks

Physical Properties

Boiling Point: 194 °C (381 °F) to 400 °C (752 °F)
Freezing Point: Varies
Specific Gravity: 1.07 to 1.08
Density: 1.06 to 1.1 g/mL
Vapor Pressure: 42 mm Hg at 22 °C
Water Solubility: Slightly Soluble in Water
Other Solubilities: 95% Ethanol: <1 mg/ml at 21.5 °C; Acetone: >=100 mg/ml at 21.5 °C; Alcohol: miscible; Benzene: Soluble; DMSO: >=100 mg/ml at 21.5 °C; Ether: miscible; Fixed alkali hydroxides: Soluble; Glycerin: Soluble; Toluene: Soluble.
Surface Tension: Estimated at 15 dynes/cm
Flash Point: 74 °C Closed Cup
Autoignition Temperature: 336 °C

RTECS Toxicity Data

Acute Oral: Rat LD_{50} Dose: 725 mg/kg; Toxic Effects: Behavioral - Somnolence (general depressed activity). Mouse LD_{50} Dose: 433 mg/kg; Toxic Effects: Behavioral - Convulsions or effect on seizure threshold; Behavioral - Coma; Lungs, Thorax, or Respiration - Respiratory depression.
Reproductive/Teratogenic: Mouse Route: Oral; Dose: 2 gm/kg; Duration: female 5-9D of pregnancy; Maternal Effects - Other effects on females; Effects on Embryo or Fetus - Fetotoxicity.
Mutagenic: Rat Body Fluid Assay; Indicator Organism: Bacteria - S Typhimurium; Dose: 250 mg/kg. Hamster Sister Chromatid Exchenge; Cell Type: ovary; Dose: 10 mg/L.
Tumorigenic: Mouse Route: Skin; Dose: 99 gm/kg/33W-I; Toxic Effects: Tumorigenic - Carcinogenic by RTECS criteria; Skin and appendages - Tumors.

Hazard Overviews

Fire Diamond

Health: Severely irritating to eyes/skin/respiratory tract; may cause corneal burns/scarring. Also Causes: Skin pigment changes, photosensitization, nausea, vomiting, abdominal pain, rapid pulse, respiratory distress, shock, trouble breathing, thready pulse, dizziness, headache, salivation, convulsions, death. Chronic Effects: skin/lung cancer.
Fire: Combustible. For small fires use dry chemical, carbon dioxide, or regular foam. For large fires use fog or regular foam. Use water spray to cool fire-exposed containers. Use

water as an extinguishing agent only when the preferred measures are unavailable.
Reactivity: Stable. Hazardous polymerization cannot occur. Avoid: excessive heat; chlorosulfonic acid. Incompatible with: chlorosulfonic acid. Hazardous decomposition products: oxides of carbon; acrid smoke.
Carcinogenicity: IARC - Group 2A, Probably carcinogenic to humans; NIOSH - Not listed; NTP - Not listed; ACGIH - Not listed; OSHA - Not listed; EPA - Class B1, Probable human carcinogen based on epidemiologic studies; MAK - Not listed
Primary Target Organs:

Eyes

Skin

Respiratory System

Kidneys

Environmental

Ecotoxicity: LD_{50} Colinus virginianus (bob white quail) 1,260 ppm/8 days (60:40 mixture of creosote and coal tar) TL_{50} Carassius auratus (goldfish) 3.51 ppm/24 hr (60:40) mixture of creosote and coal tar) /Conditions of bioassay not specified
Cleanup/Disposal: Guide No. 128: Eliminate all ignition sources (no smoking, flares, sparks or flames in immediate area). All equipment used when handling the product must be grounded. Do not touch or walk through spilled material. Stop leak if you can do it without risk. Prevent entry into waterways, sewers, basements or confined areas. A vapor suppressing foam may be used to reduce vapors. Absorb or cover with dry earth, sand or other non-combustible material and transfer to containers. Use clean non-sparking tools to collect absorbed material. Large Spills: Dike far ahead of liquid spill for later disposal. Water spray may reduce vapor; but may not prevent ignition in closed spaces.

Environmental Physical Data

Octanol/Water Partition Coefficient: log K_{ow} = 1.0
BCF: no food chain concentration potential

Regulations

RCRA 40CFR: Listed Hazardous Waste No. U051 Toxic Waste
CERCLA: 40CFR 302.4: Listed per RCRA Section 3001 RQ: 1 lb (0.454 kg)
SARA 40CFR 372.65: Listed
SARA EHS 40CFR 355: Not listed
TSCA: Listed

Analytical Methods

Soil: SW846 3650B

COB1000 CAS #: 7440-48-4

COBALT

RTECS: GF8750000
EINECS Number: 231-158-0
Molecular Formula: Co
Structured MF: Co
Formula Weight: 58.9332

Co

Chemical Structure

Synonyms: AQUACAT; C.I. 77320; COBALT-59; COBALT METAL DUST; COBALT METAL FUME; KOBALT; SUPER COBALT
Description: gray, black powder; odorless
Use: chemical manufacturing, electroplating, ceramics, lamp filaments, trace element in fertilizers, glass, drier in printing inks, paints and varnishes, colors, principal use in alloys, especially, steels for permanent and soft magnets and high speed tool steels; jet engines, catalyst in sulfur synthesis coordination and complexing agent; treatment of cyanide poisoning; automobile industry, pigment in enamels and glazes; photographic and electrical industries; radioactive material used in medicine for diagnostic aid, biological and medical research, radiation therapy and cancer treatment

Physical Properties

Boiling Point: 2870 °C (5198 °F)
Freezing Point: 1493 °C (2719.4 °F)
Specific Gravity: 8.92 at 20 °C
Density: 8.92 g/mL
Vapor Pressure: 0 mm Hg at 20 °C
Water Solubility: Insoluble in Water
Other Solubilities: 95% Ethanol: <1 mg/ml at 19 °C; Acetone: <1 mg/ml at 19 °C; Acid: Soluble; DMSO: <1 mg/ml at 19 °C; Nitric acid: Readily Soluble; Toluene: <1 mg/ml at 20 °C.
Ionization Potential (eV): 7.8810
Flash Point: Dust is flammable

RTECS Toxicity Data

Acute Oral: Rat LD_{50} Dose: 6171 mg/kg; Toxic Effects: Behavioral - Somnolence (general depressed activity); Behavioral - Ataxia; Gastrointestinal - Hypermotility, diarrhea. Rabbit LD_{Lo} Dose: 750 mg/kg; Toxic Effects: Behavioral - Somnolence (general depressed activity).
Tumorigenic: Rat Route: Intramuscular; Dose: 126 mg/kg; Toxic Effects: Tumorigenic - Neoplastic by RTECS criteria; Gastrointestinal - Tumors; Tumorigenic - Tumors at site of application. Rat Route: Intramuscular; Dose: 126 mg/kg; Toxic Effects: Tumorigenic - Neoplastic by RTECS criteria; Tumorigenic - Tumors at site of application.

Hazard Overviews

Explosive

Flammable

Fire Diamond

Health: Irritating to eyes/skin/respiratory tract. Also Causes: wheezing, hypersensitivity reactions (asthma], dermatitis. Chronic Effects: reduced pulmonary function with lung fibrosis.

Fire: Flammable/pyrophoric as a powder. Use dry chemical or sand. Do not use water.

Reactivity: Stable. Hazardous polymerization cannot occur. Avoid: allowing the powder to accumulate or form a potentially explosive dust cloud. Incompatible with: cold acetylene; fused ammonium nitrate. Hazardous decomposition products: oxides of cobalt.

Carcinogenicity: IARC - Group 2B, Possibly carcinogenic to humans; NIOSH - Not listed; NTP - Not listed; ACGIH - Class A3, Animal carcinogen; OSHA - Not listed; EPA - Not listed; MAK - Class A2, Unmistakably carcinogenic in animal experimentation only

Primary Target Organs:

Eyes

Skin

Respiratory System

Exposure Limits

OSHA PEL: TWA: 0.1 mg/m^3; as Co.

OSHA PEL Vacated 1989 Limits: TWA: 0.05 mg/m^3.

ACGIH TLV: TWA: 0.02 mg/m^3; as Co.

NIOSH REL: TWA: 0.05 mg/m^3.

NIOSH IDLH: 20 mg/m^3; as Co.

Respirator Recommendation

Exposure Range: >0.1 to 1 mg/m^3 Air Purifying, Negative Pressure, Half Mask

Exposure Range: >1 to 10 mg/m^3 Air Purifying, Negative Pressure, Full Face

Exposure Range: >10 to 100 mg/m^3 Supplied Air, Constant Flow/Pressure Demand, Full Face

Exposure Range: >100 to unlimited mg/m^3 Self-contained Breathing Apparatus, Pressure Demand, Full Face

Cartridge Color: magenta (P100)

Environmental

Cleanup/Disposal: Guide No. 133: Eliminate all ignition sources (no smoking, flares, sparks or flames in immediate area). Do not touch or walk through spilled material. Small Dry Spills: With clean shovel place material into clean, dry container and cover loosely; move containers from spill area. Large Spills: Wet down with water and dike for later disposal. Prevent entry into waterways, sewers, basements or confined areas.

Environmental Physical Data

BCF: few plants accumulate cobalt >100 ppm

Regulations

RCRA 40CFR: Not listed

CERCLA: 40CFR 302.4: Listed as Compound per CAA Section 112

SARA 40CFR 372.65: Listed

SARA EHS 40CFR 355: Not listed

TSCA: Listed

Analytical Methods

Air: EPA 29

Soil: CLP 200.10_M, 200.62, 200.7_M, 202.62, 219.1_M, 219.2_M, ICP-AES; EPA 200.7, 200.8; SW846 3005A, 3010A, 3015, 3020A, 3031, 3050A, 3050B, 3051, 3052, 6010A, 6010B, 6020, 7000A, 7200, 7201; ASTM D3974; USGS I5239, I5474

Water / Groundwater: EPA 200.0, 200.10, 200.13, 200.15, 200.7, 200.9; APHA 3111-A, 3111-B, 3111-C, 3113-B, 3120, 3500-CO; ASTM D1976, D3558, D4190; USGS E-SPEC, I1239, I1240, I1241, I1472, I3239, I3240, I7239, I7240; FISON AES-0029

Drinking Water: AOAC 993.14

Indoor / Expired Air: NIOSH 7027, 7300; ASTM D4185

Plasma: NIOSH 8005

Other: EPA 1620, 219.1, 219.2; AOAC 990.08

COB1380 CAS #: 71-48-7

COBALT ACETATE

RTECS: AG3150000

EINECS Number: 200-755-8

Molecular Formula: $C_4H_8CoO_4$

Structured MF: $Co(C_2H_3O_2)_2{\bullet}4H_2O$

Formula Weight: 177.03

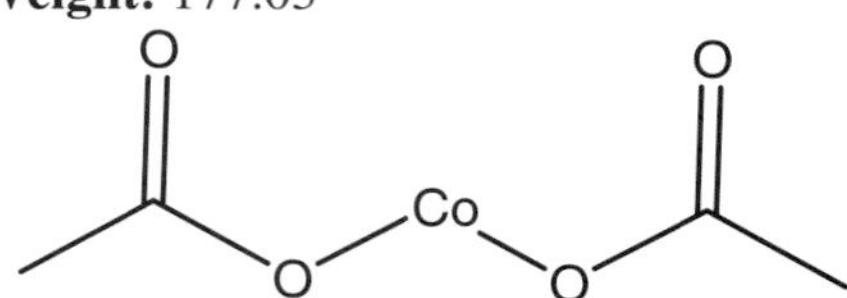

Chemical Structure

Synonyms: ACETIC ACID,COBALT(2+) SALT; BIS(ACETATO)COBALT; COBALT(2+) ACETATE; COBALT DIACETATE; COBALT(II) ACETATE; COBALTOUS ACETATE; COBALTOUS DIACETATE

Description: light-pink, red crystals; vinegar-like odor

Use: formerly foam stabilizers for malt beverages; mineral supplement in cattle feed; bleaching agent & drier for lacquers, varnishes; anodizing agent; esterification catalyst; in inks; oxidn catalyst; eperimentally, cobalt edta, cobalt nitrate & cobalt acetate, simultaneously, effective antidote for cyanide poisoning in albino mice; med: for normochromic, normocytic anemia assoc with severe renal failure.

Physical Properties

Boiling Point: Decomposes at 1 atm

Freezing Point: 140 °C (284 °F)

Specific Gravity: Solid 1.71 at 20 °C

Water Solubility: Readily Soluble in Water

Other Solubilities: 2.1 parts by wt (of the formula wt)/100 parts Methanol by wt at 15 °C.

pH: 0.2 molar aqueous solution 6.8

Refraction Index: Tetrahydrate 1.542

Flash Point: Nonflammable

RTECS Toxicity Data

Acute Oral: Rat LD_{50} Dose: 503 mg/kg; Toxic Effects: Behavioral - Somnolence (general depressed activity); Lungs,

Thorax, or Respiration - Dyspnea; Nutritional and gross metabolic - Weight loss or decreased weight gain.

Chronic (Multiple Dose) Inhalation: Rat Dose: 15 mg/m^3/24H/17W-C; Toxic Effects: Blood - Pigmented or nucleated red Blood cells; Blood - Changes in erythrocite (RBC) cell count; Biochemical - Phosphatases.

Chronic (Multiple Dose) Dermal: Rat Route: Subcutaneous; Dose: 135 mg/kg/9D-I; Toxic Effects: Cardiac - Other changes; Biochemical - Dehydrogenases; Biochemical - Catecholamine levels in sympathetic nerves.

Mutagenic: Hamster Morphological Transformation; Cell Type: embryo; Dose: 200 umol/L.

Hazard Overviews

Fire Diamond

Health: Irritating to the eye/skin/respiratory tract. Also Causes: may cause allergic dermatitis; ingestion can cause facial flushing, ringing in ears, and heart disorders.

Fire: Noncombustible. Use extinguishing agents suitable for surrounding fire.

Reactivity: Stable. Hazardous polymerization cannot occur. Avoid: contact with strong oxidizers. Incompatible with: strong oxidizers. Hazardous decomposition products: irritating cobalt oxide fumes; acrid smoke.

Carcinogenicity: IARC - Group 2B, Possibly carcinogenic to humans; NIOSH - Not listed; NTP - Not listed; ACGIH - Not listed; OSHA - Not listed; EPA - Not listed; MAK - Not listed

Primary Target Organs:

Eyes

Skin

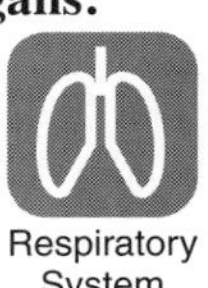
Respiratory System

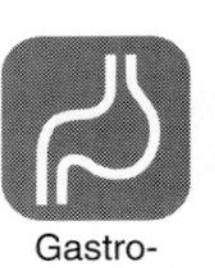
Gastro-intestinal

Nervous System

Cardio-vascular

Exposure Limits

ACGIH TLV: TWA: 0.02 mg/m^3; as Co.

Environmental

Ecotoxicity: Food chain concentration potential: Bioconcentration of 200-1000 fold only under constant exposure. Not significant in spill conditions

Cleanup/Disposal: Guide No. 160: Eliminate all ignition sources (no smoking, flares, sparks or flames in immediate area). Stop leak if you can do it without risk. Small Liquid Spills: Take up with sand, earth or other noncombustible absorbent material. Large Spills: Dike far ahead of liquid spill for later disposal. Prevent entry into waterways, sewers, basements or confined areas.

Environmental Physical Data

BCF: concentration under constant exposure 200 to 1000

Regulations

RCRA 40CFR: Not listed

CERCLA: 40CFR 302.4: Listed as Compound per CAA Section 112

SARA 40CFR 372.65: Listed as Compound

SARA EHS 40CFR 355: Not listed

TSCA: Listed

COB2140 CAS #: 10210-68-1

COBALT CARBONYL

RTECS: GG0300000

EINECS Number: 233-514-0

Molecular Formula: $C_8Co_2O_8$

Structured MF: $C_8Co_2O_8$

Formula Weight: 341.95

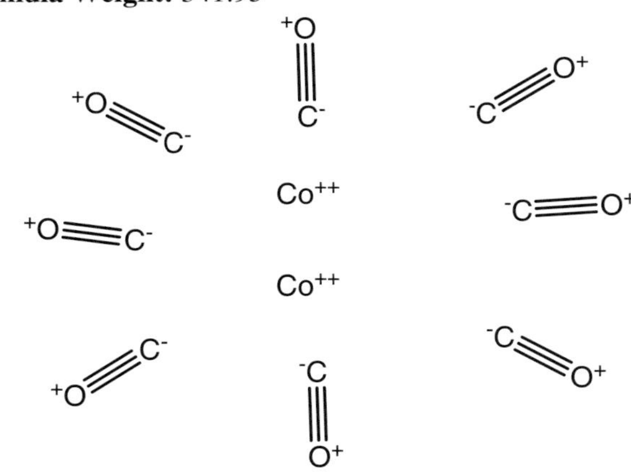

Chemical Structure

Synonyms: COBALT CARBONYL (CO2(CO)8); COBALT OCTACARBONYL; COBALT TETRACARBONYL; COBALT TETRACARBONYL DIMER; COBALT,CARBONYL,TETRA-; COBALT,DI-MU-CARBONYLHEXACARBONYLDI-; DI-MU-CARBONYLHEXACARBONYLDICOBALT; DICOBALT CARBONYL; DICOBALT OCTACARBONYL; OCTACARBONYLDICOBALT

Description: orange, dark brown, white when pure crystalline

Use: catalyst for hydroformylation, hydrogenation, hydrosilation, isomerization, carboxylation, carbonylation, & polymerization reactions; for anti-knock gasoline; converts internal olefins to linear alcohols; for amino acids; for homologation reaction of methanol with carbon monoxide & hydrogen to form ethanol.

Physical Properties

Boiling Point: Decomposes at 52 °C (126 °F)

Freezing Point: 51 °C (123.8 °F)

Specific Gravity: 1.73 at 18 °C

Vapor Pressure: 0.07 mm Hg at 15 °C

Water Solubility: Insoluble in Water

Other Solubilities: Soluble in organic solvents such as Naphtha; Soluble in Ethyl Ether.

Ionization Potential (eV): 8.12 +/-2.0

Flash Point: Noncombustible Solid

RTECS Toxicity Data

Acute Inhalation: Rat LC_{50} Dose: 165 mg/m^3; Toxic Effects: Lungs, Thorax, or Respiration - Acute pulmonary edema. Mouse LC_{50} Dose: 26900 ug/m^3/2hr; Toxic Effects: Sense organs and special senses - Other; Sense organs and special senses - Other; Lungs, Thorax, or Respiration - Dyspnea.

Hazard Overviews

Fire Diamond

Health: Irritating to skin/eyes/respiratory tract. Also Causes: incoordination, drowsiness, fatigue, headache, itchy red skin, breath shortness. Chronic Effects: dermatitis, asthma, pneumoconiosis.
Fire: Use extinguishing media appropriate to the surrounding combustible materials. Under fire conditions, cobalt carbonyl produces highly flammable carbon monoxide.
Reactivity: Unstable, decomposes in air, but stable in an atmosphere of hydrogen and carbon monoxide. Hazardous polymerization cannot occur. Avoid: heating to decomposition. Incompatible with: hydrochloric acid; sulfuric acid; bromine; nitric acid. Hazardous decomposition products: carbon monoxide.
Carcinogenicity: IARC - Group 2B, Possibly carcinogenic to humans; NIOSH - Not listed; NTP - Not listed; ACGIH - Not listed; OSHA - Not listed; EPA - Not listed; MAK - Class A2, Unmistakably carcinogenic in animal experimentation only
Primary Target Organs:

Eyes

Skin

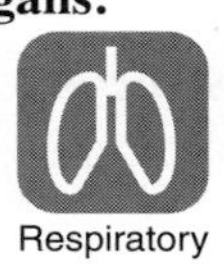
Respiratory System

Mucous Membranes

Kidneys

Exposure Limits
OSHA PEL Vacated 1989 Limits: TWA: 0.1 mg/m^3.
ACGIH TLV: TWA: 0.1 mg/m^3; as Co.
NIOSH REL: TWA: 0.1 mg/m^3.
Respirator Recommendation
Exposure Range: >0.1 to 5 mg/m^3 Supplied Air, Constant Flow/Pressure Demand, Half Mask
Exposure Range: >5 to 100 mg/m^3 Supplied Air, Constant Flow/Pressure Demand, Full Face
Exposure Range: >100 to unlimited mg/m^3 Self-contained Breathing Apparatus, Pressure Demand, Full Face
Note: odor threshold unknown

Environmental

Regulations

RCRA 40CFR: Not listed
CERCLA: 40CFR 302.4: Listed as Compound per CAA Section 112
SARA 40CFR 372.65: Not listed TPQ: 10/10000 lb
SARA EHS 40CFR 355: Listed TPQ: 10 lb
TSCA: Listed

COB2520 CAS #: 16842-03-8

COBALT HYDROCARBONYL

RTECS: GG0900000
Molecular Formula: C_4HCoO_4
Structured MF: $HCo(CO)_4$
Formula Weight: 171.98
Synonyms: COBALT,TETRACARBONYLHYDRO-(8CI,9CI); HYDRIDOTETRACARBONYLCOBALT; HYDROCOBALT TETRACARBONYL; TETRACARBONYLHYDRIDOCOBALT; TETRACARBONYLHYDROCOBALT
Description: gas; offensive odor

Physical Properties

Freezing Point: -26.11 °C (-14.998 °F)
Vapor Density: 5.93 Air=1
Vapor Pressure: > 1 atm
Water Solubility: 0.05% by weight
Ionization Potential (eV): 8.2 +/-0.5
Flash Point: Flammable gas

RTECS Toxicity Data

Acute Inhalation: Rat LC_{50} Dose: 46200 ug/m^3/2hr. Mouse LC_{50} Dose: 17500 ug/m^3/2hr.

Hazard Overviews

Flammable

Fire: Flammable.
Carcinogenicity: IARC - Group 2B, Possibly carcinogenic to humans; NIOSH - Not listed; NTP - Not listed; ACGIH - Not listed; OSHA - Not listed; EPA - Not listed; MAK - Class A2, Unmistakably carcinogenic in animal experimentation only
Exposure Limits
OSHA PEL Vacated 1989 Limits: TWA: 0.1 mg/m^3.
ACGIH TLV: TWA: 0.1 mg/m^3; as Co.
NIOSH REL: TWA: 0.1 mg/m^3.
Respirator Recommendation
Exposure Range: >0.1 to 5 mg/m^3 Supplied Air, Constant Flow/Pressure Demand, Half Mask
Exposure Range: >5 to 100 mg/m^3 Supplied Air, Constant Flow/Pressure Demand, Full Face
Exposure Range: >100 to unlimited mg/m^3 Self-contained Breathing Apparatus, Pressure Demand, Full Face
Note: odor threshold unknown

Environmental

Regulations

RCRA 40CFR: Not listed
CERCLA: 40CFR 302.4: Listed as Compound per CAA Section 112
SARA 40CFR 372.65: Listed as Compound

SARA EHS 40CFR 355: Not listed
TSCA: Not listed

COB2900 CAS #: 62207-76-5

COBALT (II), N,N'-ETHYLENEBIS-(3-FLUOROSALICYCLIDENEIMINAT

RTECS: GG0575000
EINECS Number: 263-458-2
Molecular Formula: $C_{16}H_{12}CoF_2N_2O_2$
Formula Weight: 361.23
Synonyms: COBALT,BIS(3-FLUOROSALICYLALDEHYDE)-ETHYLENEDIIMINE-; N,N'-ETHYLENEBIS(3-FLUOROSALICYLIDENEIMINATO)COBALT (II); FLUOMINE; FLUOMINE DUST
Description: particulate solid

RTECS Toxicity Data

Acute Oral: Rat LD_{50} Dose: 187 mg/kg. Mouse LD_{50} Dose: 123 mg/kg.
Acute Inhalation: Rat LC_{50} Dose: 112 mg/m^3/6hr; Toxic Effects: Sense organs and special senses - Other; Lungs, Thorax, or Respiration - Chronic pulmonary edema; Liver - Other changes. Mouse LC_{50} Dose: 416 mg/m^3/6hr.
Irritation Skin: Rabbit Standard Draize Test Dose: 500 mg/24H; Reaction: moderate.

Hazard Overviews

Carcinogenicity: IARC - Not listed; NIOSH - Not listed; NTP - Not listed; ACGIH - Not listed; OSHA - Not listed; EPA - Not listed; MAK - Not listed
Exposure Limits
ACGIH TLV: TWA: 0.02 mg/m^3; as Co.

Environmental

Regulations

RCRA 40CFR: Not listed
CERCLA: 40CFR 302.4: Listed as Compound per CAA Section 112
SARA 40CFR 372.65: Not listed TPQ: 100/10000 lb
SARA EHS 40CFR 355: Listed TPQ: 100 lb
TSCA: Listed

COB3280 CAS #: 10026-22-9

COBALT (II) NITRATE, HEXAHYDRATE

RTECS: QU7355500
Molecular Formula: $CoH_{12}N_2O_{12}$
Formula Weight: 291.03

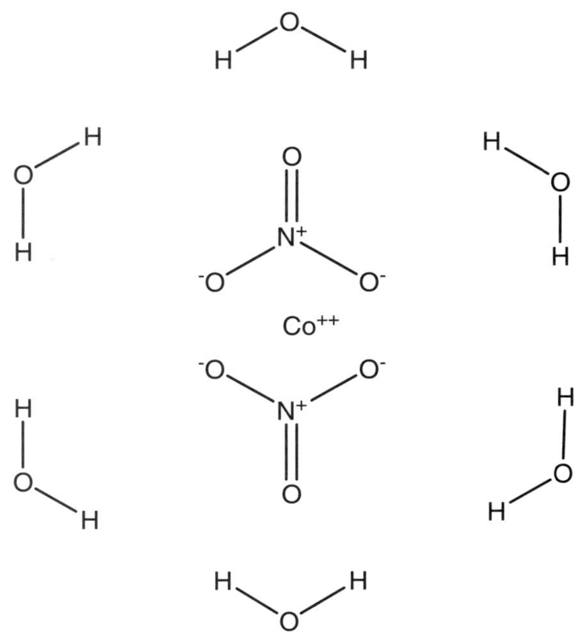

Chemical Structure

Synonyms: COBALT DINITRATE HEXAHYDRATE; COBALT NITRATE HEXAHYDRATE; COBALT(2+) NITRATE HEXAHYDRATE; COBALT(II) NITRATE HEXAHYDRATE; COBALTOUS NITRATE HEXAHYDRATE
Description: red crystals; odorless

Physical Properties

Boiling Point: Decomposes at 74 °C (165 °F)
Freezing Point: 55 °C (131 °F)
Specific Gravity: 1.87 at 25 °C
Water Solubility: 134 g/100 cc at 0 °C
Other Solubilities: Soluble in most organic solvents; Soluble in Acetone; Slightly Soluble in Ammonia
Refraction Index: 1.52
Flash Point: Noncombustible

RTECS Toxicity Data

Acute Oral: Rat LD_{50} Dose: 691 mg/kg; Toxic Effects: Behavioral - Somnolence (general depressed activity); Gastrointestinal - Hypermotility, diarrhea; Nutritional and gross metabolic - Weight loss or decreased weight gain.
Acute Dermal: Mouse LD_{50} Route: Subcutaneous Dose: 171 mg/kg.

Hazard Overviews

Fire Diamond

Health: Irritating to the eyes. Also Causes: allergic dermatitis, excessive intake of cobalt compounds may cause an increase in red blood cells, decreased blood pressure, and heart disease.

Fire: Strong oxidizer capable of igniting combustibles. Toxic and irritating fumes/gases can evolve from this material in a fire situation.

Reactivity: Stable. Hazardous polymerization cannot occur. Avoid: contact with combustibles. Incompatible with: organic substances; ammonium hexacyanoferrate at 428 °F (220 °C); carbon. Hazardous decomposition products: nitrogen oxides.

Carcinogenicity: IARC - Group 2B, Possibly carcinogenic to humans; NIOSH - Not listed; NTP - Not listed; ACGIH - Class A3, Animal carcinogen; OSHA - Not listed; EPA - Not listed; MAK - Class A2, Unmistakably carcinogenic in animal experimentation only

Primary Target Organs:

Eyes

Skin

Cardio-vascular

Blood

Glandular System

Exposure Limits

ACGIH TLV: TWA: 0.02 mg/m^3; as Co.

Environmental

Regulations

RCRA 40CFR: Not listed
CERCLA: 40CFR 302.4: Not listed
SARA 40CFR 372.65: Not listed
SARA EHS 40CFR 355: Not listed
TSCA: Not listed

COB4040 CAS #: 61789-51-3

COBALT NAPHTHENATE SOLUTION

RTECS: QK8925000
EINECS Number: 263-064-0
Molecular Formula: Unspecified or Variable
Formula Weight: N/A
Synonyms: COBALT NAPHTHENATE; COBALT NAPHTHENATES,POWDER; COBALTOUS NAPHTHENATE; NAFTOLITE; NAPHTENATE DE COBALT
Description: blue-violet, purple liquid; mineral spirits odor

Physical Properties

Boiling Point: 157 °C (315 °F) to 193 °C (379 °F)
Specific Gravity: 0.95 at 25 °C
Vapor Density: 3.9
Vapor Pressure: 1 mm Hg at 25 °C
Water Solubility: Insoluble
Other Solubilities: Soluble in Alcohol, Ether and oils
Evaporation Rate: < 1 Butyl Acetate=1
Flash Point: 49 °C
LEL: 0.07% v/v
UEL: 6% v/v

RTECS Toxicity Data

Acute Oral: Rat LD_{50} Dose: 3900 mg/kg.

Irritation Eye: Rabbit Standard Draize Test Dose: 10 mg/24H; Reaction: mild.

Hazard Overviews

Flammable

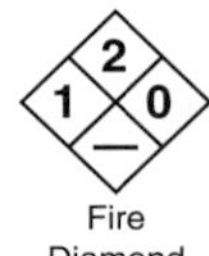

Fire Diamond

Health: Irritating to eyes/skin/respiratory tract. Also Causes: dizziness, headache, nausea, euphoria; upon ingestion: GI irritation, CNS depression, vomiting, diarrhea. Chronic Effects: dermatitis, neurological/lung damage.

Fire: Flammable. For small fires use dry chemical, sand, earth, water spray, or regular foam. For large fires use water spray, fog, or regular foam. Avoid scattering flaming liquid with more water than necessary to put out fire.

Reactivity: Stable. Hazardous polymerization cannot occur. Avoid: exposure to heat; mixing directly with peroxides. Incompatible with: oxidizing agents; organic peroxides; peroxide-containing drying oils. Hazardous decomposition products: acrid smoke; cobalt fumes.

Carcinogenicity: IARC - Group 2B, Possibly carcinogenic to humans; NIOSH - Not listed; NTP - Not listed; ACGIH - Not listed; OSHA - Not listed; EPA - Not listed; MAK - Not listed

Primary Target Organs:

Eyes

Skin

Respiratory System

Nervous System

Environmental

Regulations

RCRA 40CFR: Not listed
CERCLA: 40CFR 302.4: Not listed
SARA 40CFR 372.65: Not listed
SARA EHS 40CFR 355: Not listed
TSCA: Listed

COB5180 CAS #: 7789-43-7

COBALTOUS BROMIDE

RTECS: GF9595000
EINECS Number: 232-166-7
Molecular Formula: Br_2Co
Structured MF: $CoBr_2$
Formula Weight: 218.77

Br^- Co^{++} Br^-

Chemical Structure

Synonyms: COBALT BROMIDE; COBALT DIBROMIDE; COBALT(II) BROMIDE

Description: bright green solid or lustrous green crystal leaflets
Use: in hygrometers; catalyst for organic reactions; normochromic, normocytic anemia assoc with severe renal failure

Physical Properties

Boiling Point: Loses 4 H20 at 1 atm
Freezing Point: 678 °C (1252.4 °F)
Specific Gravity: 4.909 at 25 °C/4 °C
Water Solubility: 112 parts by wt (of formula wt)/100 parts
Other Solubilities: 77.1 g/100 cc of Ethanol at 20 °C; 58.6 g/100 cc of Methanol at 30 °C; Soluble in Ether.
Flash Point: Nonflammable

RTECS Toxicity Data

Acute Oral: Rat LD_{50} Dose: 406 mg/kg; Toxic Effects: Behavioral - Somnolence (general depressed activity); Gastrointestinal - Hypermotility, diarrhea; Nutritional and gross metabolic - Weight loss or decreased weight gain.

Hazard Overviews

Health: Irritating to eyes/skin/respiratory tract. Toxic. Other Acute Effects: harmful if swallowed, inhaled, or absorbed through skin; may cause allergic respiratory and skin reactions. Chronic Effects: target organs: lungs, blood. Carcinogen.
Fire: Noncombustible. Hazards: emits toxic fumes. Extinguishing agents: noncombustible; use extinguishing media appropriate to surrounding fire conditions. Precautions: combustible liquid.
Reactivity: Stable. Hazardous polymerization will not occur. Incompatible with: moisture, oxidizing agents, alkali metals, absorbs ammonia from air. Hazardous decomposition products: toxic fumes of: hydrogen bromide gas.
Carcinogenicity: IARC - Group 2B, Possibly carcinogenic to humans; NIOSH - Not listed; NTP - Not listed; ACGIH - Class A3, Animal carcinogen; OSHA - Not listed; EPA - Not listed; MAK - Class A2, Unmistakably carcinogenic in animal experimentation only
Primary Target Organs:

Eyes

Skin

Respiratory System

Blood

Exposure Limits
ACGIH TLV: TWA: 0.02 mg/m^3; as Co.

Environmental

Ecotoxicity: LC_{50} Fathead minnows (Pimephales promelas) 91.9 mg/l/96 hr @ 22 °C (95% confidence limit 56.3-132.9 mg/l)
Cleanup/Disposal: Guide No. 157: Eliminate all ignition sources (no smoking, flares, sparks or flames in immediate area). All equipment used when handling the product must be grounded. Do not touch damaged containers or spilled material unless wearing appropriate protective clothing. Stop leak if you can do it without risk. A vapor suppressing foam may be used to reduce vapors. Do not get water inside containers. Use water spray to reduce vapors or divert vapor cloud drift. Prevent entry into waterways, sewers, basements or confined areas. Small Spills: Cover with dry earth, dry sand, or other non-combustible material followed with plastic sheet to minimize spreading or contact with rain. Use clean non-sparking tools to collect material and place it into loosely covered plastic containers for later disposal.

Environmental Physical Data

BCF: concentration potential 1000 to 1500

Regulations

RCRA 40CFR: Not listed
CERCLA: 40CFR 302.4: Listed per CWA Section 311(b)(4) RQ: 1000 lb (453.5 kg)
SARA 40CFR 372.65: Listed as Compound
SARA EHS 40CFR 355: Not listed
TSCA: Listed

COB5940 CAS #: 7646-79-9

COBALTOUS CHLORIDE

RTECS: GF9800000
EINECS Number: 231-589-4
Molecular Formula: Cl_2Co
Structured MF: $CoCl_2$
Formula Weight: 129.85

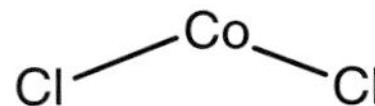

Chemical Structure

Synonyms: COBALT CHLORIDE; COBALT CHLORIDE (COCL2); COBALT DICHLORIDE; COBALT MURIATE; COBALT(II) CHLORIDE; COBALTOUS DICHLORIDE; KOBALT CHLORID
Use: invisible ink; humidity & water indicator; in hygrometers; temp indicator; in electroplating; for painting on glass & porcelain; prepn of catalysts; fertilizer & feed additive; absorbent for military poison gas & ammonia; in mfr of vitamin b12; in gas masks; solid lubricant; in medication: hematinic, treatment of refractory anemias, incl sickle cell, nutritional factor; germicide; for vitamin b12 production; for mossbauer effect (nuclear clock).

Physical Properties

Boiling Point: 1049 °C (1920 °F)
Freezing Point: 735 °C (1355 °F)
Specific Gravity: 3.367 at 25 °C/4 °C
Vapor Pressure: 40 mm Hg at 770 °C
Water Solubility: Soluble in Water
Other Solubilities: 38.5 g/100 cc Methanol; 4.4 g/100 cc Alcohol; 8.6 g/100 cc Acetone.
Refraction Index: 1.625 to 1.67
Flash Point: Nonflammable

RTECS Toxicity Data

Acute Oral: Man TD_{Lo} Dose: 1042 mg/kg/13W-I; Toxic Effects: Sense organs and special senses - Optic nerve neuropathy; Sense organs and special senses - Visual field changes. Child TD_{Lo} Dose: 48 mg/kg; Toxic Effects: Behavioral - Anorexia (human); Endocrine - Thyroid weight (goiter); Nutritional and gross metabolic - Weight loss or decreased weight gain. Child LD_{Lo} Dose: 1500 mg/kg. Rat LD_{50} Dose: 80 mg/kg.

Acute Dermal: Rabbit LD_{Lo} Route: Subcutaneous Dose: 200 mg/kg. Rat LD_{Lo} Route: Skin; Dose: 2 gm/kg; Toxic Effects: Nutritional and gross metabolic - Weight loss or decreased weight gain.

Reproductive/Teratogenic: Rat Route: Oral; Dose: 11 mg/kg; Duration: female 1-22D of pregnancy; Effects on Fertility - Pre-implantation mortality. Rat Route: Intraperitoneal; Dose: 30 gm/kg; Duration: female 15-16D of pregnancy Effects on Fertility - Post-implantation mortality; Effects on Embryo or Fetus - Fetotoxicity.

Mutagenic: Human DNA Inhibition; Cell Type: HeLa cell; Dose: 1 mmol/L. Mammal DNA Damage; Cell Type: lymphocyte; Dose: 750 umol/L.

Tumorigenic: Rat Route: Subcutaneous; Dose: 400 mg/kg/19D-I; Toxic Effects: Tumorigenic - Carcinogenic by RTECS criteria; Tumorigenic - Tumors at site of application.

Hazard Overviews

Reactivity: Stable, unless exposed directly to moisture. Hazardous polymerization cannot occur. Avoid: exposure to moisture. Incompatible with: strong oxidizers; potassium; sodium. Hazardous decomposition products: cobalt oxide; chloride fumes.

Carcinogenicity: IARC - Group 2B, Possibly carcinogenic to humans; NIOSH - Not listed; NTP - Not listed; ACGIH - Class A3, Animal carcinogen; OSHA - Not listed; EPA - Not listed; MAK - Class A2, Unmistakably carcinogenic in animal experimentation only

Exposure Limits

ACGIH TLV: TWA: 0.02 mg/m^3; as Co.

Environmental

Ecotoxicity: Aquatic toxicity: 1000 ppm/30-32 hr/goldfish/killed/fresh (hard) water 10 ppm/168 hr/goldfish/killed/fresh (soft) water 200 ppm/*/mummichogs/no effect/sea water *Time period not specified; Food chain concentration potential: Bioconcentration of 200-1000 fold only under constant exposure. Not significant in spill conditions

Cleanup/Disposal: Guide No. 154: Eliminate all ignition sources (no smoking, flares, sparks or flames in immediate area). Do not touch damaged containers or spilled material unless wearing appropriate protective clothing. Stop leak if you can do it without risk. Prevent entry into waterways, sewers, basements or confined areas. Absorb or cover with dry earth, sand or other non-combustible material and transfer to containers. Do not get water inside containers.

Environmental Physical Data

BCF: constant exposure 200 to 1000

BOD: none

Regulations

RCRA 40CFR: Not listed

CERCLA: 40CFR 302.4: Listed as Compound per CAA Section 112

SARA 40CFR 372.65: Listed as Compound

SARA EHS 40CFR 355: Not listed

TSCA: Listed

COB7080 CAS #: 544-18-3

COBALTOUS FORMATE

DOT: NA9104

EINECS Number: 208-864-2

Molecular Formula: $C_2H_4CoO_4$

Structured MF: $Co(CHOO)_2$

Formula Weight: 148.98

Synonyms: COBALT DIFORMATE; COBALT FORMATE; FORMIC ACID,COBALT(2+) SALT

Description: red crystalline solid

Use: in prepn of cobalt catalysts; manufacture of paint and varnish driers; for normochromic, normocytic anemia assoc with severe renal failure

Physical Properties

Boiling Point: Decomposes at 175 °C (347 °F)

Freezing Point: Decomposes at 175 °C (347 °F)

Specific Gravity: 2.13

Water Solubility: Soluble in Water

Other Solubilities: almost Insoluble in Alcohol

Flash Point: Does not burn or burns with difficulty

Hazard Overviews

Fire: Noncombustible.

Carcinogenicity: IARC - Not listed; NIOSH - Not listed; NTP - Not listed; ACGIH - Not listed; OSHA - Not listed; EPA - Not listed; MAK - Not listed

Exposure Limits

ACGIH TLV: TWA: 0.02 mg/m^3; as Co.

Environmental

Ecotoxicity: Aquatic toxicity: 10 ppm lethal concentration for sticklebacks (as Co). Fish food organisms affected by concentration of 0.5 ppm Co.; Food chain concentration potential: Microorganisms can concentrate cobalt in water up to 1070 to 1500 times.

Cleanup/Disposal: Guide No. 171: Do not touch or walk through spilled material. Stop leak if you can do it without risk. Prevent dust cloud. Avoid inhalation of asbestos dust. Small Dry Spills: With clean shovel place material into clean, dry container and cover loosely; move containers from spill area. Small Spills: Take up with sand or other noncombustible absorbent material and place into containers for later disposal. Large Spills: Dike far ahead of liquid spill for later disposal. Cover powder spill with plastic sheet or tarp to minimize spreading. Prevent entry into waterways, sewers, basements or confined areas.

Environmental Physical Data

BCF: concentration potential 1070 to 1500

Regulations

RCRA 40CFR: Not listed
CERCLA: 40CFR 302.4: Listed per CWA Section 311(b)(4) RQ: 1000 lb (453.5 kg)
SARA 40CFR 372.65: Listed as Compound
SARA EHS 40CFR 355: Not listed
TSCA: Listed

COB8600 CAS #: 14017-41-5

COBALTOUS SULFAMATE

EINECS Number: 237-834-1
Molecular Formula: $CoH_6N_2O_6S_2$
Structured MF: $Co(NH_2SO_3)_2$
Formula Weight: 251.1
Synonyms: SULFAMIC ACID,COBALT(2+) SALT (2:1)
Description: reddish solid
Use: pigment; for electroplating metals; normochromic, normocytic anemia assoc with severe renal failure

Physical Properties

Water Solubility: Soluble in Water

Hazard Overviews

Carcinogenicity: IARC - Group 2B, Possibly carcinogenic to humans; NIOSH - Not listed; NTP - Not listed; ACGIH - Class A3, Animal carcinogen; OSHA - Not listed; EPA - Not listed; MAK - Class A2, Unmistakably carcinogenic in animal experimentation only
Exposure Limits
ACGIH TLV: TWA: 0.02 mg/m^3; as Co.

Environmental

Environmental Physical Data

BCF: few plants accumulate cobalt >100 ppm

Regulations

RCRA 40CFR: Not listed
CERCLA: 40CFR 302.4: Listed per CWA Section 311(b)(4) RQ: 1000 lb (453.5 kg)
SARA 40CFR 372.65: Listed as Compound
SARA EHS 40CFR 355: Not listed
TSCA: Listed

COC4200 CAS #: 61791-10-4

COCO ALKYL BIS(2-HYDROXYETHYL)-

Physical Properties

Boiling Point: > 260 °C (500 °F)
Specific Gravity: 1.071
Vapor Pressure: < 1
Water Solubility: Soluble
Other Solubilities: Soluble in Acetone
pH: 7
Evaporation Rate: < 1 Butyl Acetate=1

Hazard Overviews

Carcinogenicity: IARC - Not listed; NIOSH - Not listed; NTP - Not listed; ACGIH - Not listed; OSHA - Not listed; EPA - Not listed; MAK - Not listed

Environmental

Regulations

RCRA 40CFR: Not listed
CERCLA: 40CFR 302.4: Not listed
SARA 40CFR 372.65: Not listed
SARA EHS 40CFR 355: Not listed
TSCA: Not listed

COL1000 CAS #: 64-86-8

COLCHICINE

RTECS: GH0700000
EINECS Number: 200-598-5
Molecular Formula: $C_{22}H_{25}NO_6$
Formula Weight: 399.43

Chemical Structure

Synonyms: ACETAMIDE; ACETAMIDE, N-(5,6,7,9-TETRAHYDRO-1,2,3,10-TETRAMETHOXY-9-OXOBENZO(ALPHA)HEPTALEN-7-YL)-; ACETAMIDE,N-(5,6,7,9-TETRAHYDRO-1,2,3,10-TETRAMETHOXY-9-OXOBENZO(A)HEPTALEN-7-YL)-,(S)-; 7-ACETAMIDO-6,7-DIHYDRO-1,2,3,10-TETRAMETHOXYBENZO(A)HEPTALEN-9(5H)-ONE; N-ACETYL TRIMETHYLCOLCHICINIC ACID METHYLETHER; N-ACETYLTRIMETHYLCOLCHICINIC ACID METHYL ETHER; BENZO(A)HEPTALEN-9(5H)-ONE; BENZO(A)HEPTALEN-9(5H)-ONE, 7-ACETAMIDO-6,7-DIHYDRO-1,2,3,10-TETRAMETHOXY-; COLCHICEINE METHYL ETHER; COLCHICIN; COLCHICINA; 7-ALPHA-H-COLCHICINE; 7ALPHA-H-COLCHICINE; COLCHINEOS; COLCHISOL; COLCIN; COLSALOID; CONDYLON; NSC 757; N-[(7S)-5,6,7,9-TETRAHYDRO-1,2,3,10-TETRAMETHOXY-9-OXOBENZ(A)HEPTALEN-7-YL)-; N-(5,6,7,9-TETRAHYDRO-1,2,3,10-TETRAMETHOXY-9-OXOBENZO(A)HEPTALEN-7-YL)-; N-(5,6,7,9-TETRAHYDRO-1,2,3,10-TETRAMETHOXY-9-OXOBENZO(ALPHA)HEPTALEN-7-YL)-; N-(5,6,7,9-TETRAHYDRO-1,2,3,10-TETRAMETHOXY-9-OXOBENZO[A]HEPTALEN-7-YL)ACETAMIDE

Description: pale yellow crystals, plates, scales, powder, or needles; odorless or nearly so

Use: in medicine for the treatment of gout and Familial Mediterranean Fever, in research on mitosis and intracellular transport, in research in plant genetics (to induce chromosome doubling), in phytopathology as an experimental tool in the study of normal and abnormal cell division and cell function, and in veterinary medicine as an antineoplastic agent

Physical Properties

Freezing Point: 142 °C (287.6 °F) to 150 °C (302 °F)

Water Solubility: 1 g dissolves in 22 ml Water

Other Solubilities: 95% Ethanol: >=100 mg/ml at 21 °C; Acetone: >=100 mg/ml at 21 °C; Benzene: 1 g/100 mL; Chloroform: Freely Soluble; DMSO: >=100 mg/ml at 21 °C; Ether: 1 g/160 mL at 15.5 °C; Petroleum Ether: Practically Insoluble.

pH: 0.5% solution 5.9

Flash Point: Not available; probably combustible

RTECS Toxicity Data

Acute Oral: Man TD_{Lo} Dose: 12514 ug/kg/2Y-I; Toxic Effects: Behavioral - Muscle weakness. Woman TD_{Lo} Dose: 960 ug/kg; Toxic Effects: Cardiac - Pulse rate increased without fall in BP. Woman TD_{Lo} Dose: 1 mg/kg; Toxic Effects: Behavioral - Hallucinations, distorted perceptions; Cardiac - Pulse rate increased without fall in BP; Blood - Leukopenia. Woman TD_{Lo} Dose: 320 ug/kg; Toxic Effects: Peripheral nerve and sensation - Flaccid paralysis without anesthesia; Behavioral - Muscle weakness; Behavioral - Muscle contraction or spasticity. Human LD_{Lo} Dose: 86 ug/kg; Toxic Effects: Lungs, Thorax, or Respiration - Dyspnea; Gastrointestinal - Other changes; Nutritional and gross metabolic - Body temperature decrease. Man LD_{Lo} Dose: 11 mg/kg; Toxic Effects: Lungs, Thorax, or Respiration - Respiratory stimulation; Kidney, Ureter, and Bladder - Chgs in tubules (inc acute renal failure, acute tubular necrosis.

Acute Dermal: Rabbit LD_{Lo} Route: Subcutaneous Dose: 5 mg/kg; Toxic Effects: Gastrointestinal - Hypermotility, diarrhea. Rat LD_{Lo} Route: Subcutaneous Dose: 4 mg/kg.

Irritation Eye: Rabbit Standard Draize Test Dose: 1%/3D; Reaction: severe.

Reproductive/Teratogenic: Rat Route: Subcutaneous; Dose: 1200 ug/kg; Duration: female 18-20D of pregnancy Specific Developmental Abnormalities - Central nervous system; Effects on Newborn - Growth statistics; Behavioral. Rat Route: Subcutaneous; Dose: 1200 ug/kg; Duration: female 18-20D of pregnancy Effects on Embryo or Fetus - Cytological changes (inc. somatic cell genetic material). Rat Route: Subcutaneous; Dose: 5 mg/kg; Duration: female 8D of pregnancy; Effects on Fertility - Post-implantation mortality.

Mutagenic: Human DNA Damage; Cell Type: fibroblast; Dose: 1 mg/L. Human DNA Inhibition; Cell Type: other cell types; Dose: 100 umol/L. Human DNA Inhibition; Cell Type: lymphocyte; Dose: 1 mg/L. Human DNA Inhibition; Cell Type: HeLa cell; Dose: 230 nmol/L. Human Mutations in Mammalian Somatic Cells; Cell Type: fibroblast; Dose: 10 ug/L. Human Cytogenetic Analysis; Cell Type: HeLa cell; Dose: 50 nmol/L. Human Micronucleus Test; Cell Type: lymphocyte; Dose: 50 nmol/L. Human Other Mutation Test Systems; Cell Type: lymphocyte; Dose: 10 ug/L.

Hazard Overviews

Poison

Health: Irritating to eyes/skin/respiratory tract. Poison. Other Acute Effects: may be fatal if inhaled, swallowed, or absorbed through skin; exposure can cause stomach pains; vomiting; diarrhea. Chronic Effects: carcinogen; may cause congenital malformation in the fetus; possible risk of harm to the unborn child; may alter genetic material; may cause heritable genetic damage; target organs: liver; kidneys; bone marrow; nerves; cardiovascular system.

Fire: Will burn. Hazards: emits toxic fumes. Extinguishing agents: carbon dioxide; dry chemical powder; water spray. Precautions: combustible liquid.

Reactivity: Incompatible with: strong oxidizing agents, sensitive to light. Hazardous decomposition products: thermal decomposition may produce carbon monoxide, carbon dioxide, and nitrogen oxides.

Carcinogenicity: IARC - Not listed; NIOSH - Not listed; NTP - Not listed; ACGIH - Not listed; OSHA - Not listed; EPA - Not listed; MAK - Not listed

Primary Target Organs:

Eyes

Skin

Respiratory System
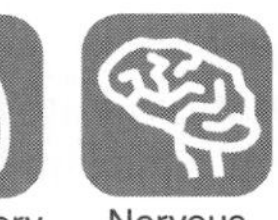
Nervous System

Liver

Kidneys

Environmental

Cleanup/Disposal: Guide No. 151: Do not touch damaged containers or spilled material unless wearing appropriate protective clothing. Stop leak if you can do it without risk. Prevent entry into waterways, sewers, basements or confined areas. Cover with plastic sheet to prevent spreading. Absorb or cover with dry earth, sand or other non-combustible material and transfer to containers. Do not get water inside containers.

Regulations

RCRA 40CFR: Not listed
CERCLA: 40CFR 302.4: Not listed
SARA 40CFR 372.65: Not listed TPQ: 10/10000 lb
SARA EHS 40CFR 355: Listed TPQ: 10 lb
TSCA: Listed

COL9000 CAS #: 9004-70-0

COLLODION

RTECS: QW0970000
DOT: UN2059; IMO3.2; IMO3.3; IMO4.1
Molecular Formula: Unspecified or Variable

Synonyms: BK2-W; BK2-Z; C 2018; CA 80-15; CELEX; CELLOIDIN; CELLULOSE NITRATE; CELLULOSE TETRANITRATE; CELLULOSE,NITRATE; CELLULOSE,NITRATE (9CI); CN 85; COLLODION COTTON; COLLODION WOOL; COLLOXYLIN; COLLOXYLIN VNV; CORIAL EM FINISH F; DAICEL RS 1; E 1440; FLEXIBLE COLLODION; FM-NTS; GUNCOTTON; H 1/2; HX 3/5; KODAK LR 115; LR 115; NITROCEL S; NITROCELLULOSE; R.S.NITROCELLULOSE; RS NITROCELLULOSE; NITROCELLULOSE E950; NITROCOTTON; NITRON; NITRON (NITROCELLULOSE); NIXON N/C; NP 11; NTS 218; NTS 222; NTS 539; NTS 542; NTS 62; PARLODION; PYRALIN; PYROXYLIN; RF 10; RS; RS 1/2; SHADOLAC MT; SOLUBLE GUN COTTON; SYNPOR; TSAPOLAK 964; XYLOIDIN

Description: pale yellow syrupy liquid; ether odor

Use: in photography; in mfr of lacquers, artificial leathers & pearls; in process engraving; in cements; vehicle for application of meds; pyroxylin: in mfr of collodions; celloidin: for embedding sections in microscopy, in electrotechnics, photography; lithography; nitrocellulose: rocket propellant; printing ink base; flashless propellant powder; coating book-binding cloth; high explosives; med: galvanoplasty; corn removers, to seal wounds, in medicated collodions, (vet) skin protectant.

Physical Properties

Boiling Point: 34 °C (93.2 °F)
Freezing Point: Ignites at 170 °C (338 °F)
Specific Gravity: 0.765 to 0.775 at 25 °C/25 °C
Vapor Density: 2.6 Air=1
Vapor Pressure: 450 mm Hg
Water Solubility: Insoluble
Other Solubilities: Soluble in 25 partsof 1 vol Alcohol + 3 vol Ether; Soluble in Methanol, Acetone, Glacial Acetic Acid, Amyl Acetate.
Flash Point: ~ -17.7 °C
Autoignition Temperature: Celluose Nitrate 170 °C
LEL: 1.7% v/v
UEL: 48% v/v

RTECS Toxicity Data

Acute Oral: Rat LD_{50} Dose: >5 gm/kg. Mouse LD_{50} Dose: >5 gm/kg.

Hazard Overviews

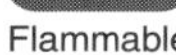
Flammable

Fire Diamond

Health: Irritating to eyes/skin/respiratory tract. Also Causes: ventricular fibrillations, convulsions, respiratory failure upon prolonged exposure or high concentrations. Chronic Effects: weight loss, insomnia, polycythemia, nephritis.

Fire: Flammable. For small fires use dry chemical, earth, sand, water spray, or regular foam. For large fires use water spray, fog, or regular foam.

Reactivity: Unstable, dry nitrocellulose is shock-sensitive, explosive if nitrogen content is greater than 12.6 percent. Hazardous polymerization cannot occur. Avoid: exposure to sunlight; heat; ignition sources; dry states. Incompatible with: acetyl peroxide; bromoazide; chlorine; strong oxidizers; acids; bases; "liquid air". Hazardous decomposition products: carbon oxides; nitrogen oxides; hydrogen cyanide gas.

Carcinogenicity: IARC - Not listed; NIOSH - Not listed; NTP - Not listed; ACGIH - Not listed; OSHA - Not listed; EPA - Not listed; MAK - Not listed

Primary Target Organs:

Eyes

Skin

Respiratory System
Nervous System

Kidneys

Blood

Environmental

Cleanup/Disposal: Guide No. 127: Eliminate all ignition sources (no smoking, flares, sparks or flames in immediate area). All equipment used when handling the product must be grounded. Do not touch or walk through spilled material.

Stop leak if you can do it without risk. Prevent entry into waterways, sewers, basements or confined areas. A vapor suppressing foam may be used to reduce vapors. Absorb or cover with dry earth, sand or other non-combustible material and transfer to containers. Use clean non-sparking tools to collect absorbed material. Large Spills: Dike far ahead of liquid spill for later disposal. Water spray may reduce vapor; but may not prevent ignition in closed spaces.

Regulations

RCRA 40CFR: Not listed
CERCLA: 40CFR 302.4: Not listed
SARA 40CFR 372.65: Not listed
SARA EHS 40CFR 355: Not listed
TSCA: Listed

COP1000 CAS #: 7440-50-8

COPPER

RTECS: GL5325000
EINECS Number: 231-159-6
Molecular Formula: Cu
Structured MF: Cu
Formula Weight: 63.546

Cu

Chemical Structure

Synonyms: ALLBRI NATURAL COPPER; ANAC 110; ARWOOD COPPER; BRONZE POWDER; C.I. 77400; C.I. PIGMENT METAL 2; CDA 101; CDA 102; CDA 110; CDA 122; CE 1110; COPPER BRONZE; COPPER M 1; COPPER METAL DUSTS; COPPER METAL FUMES; COPPER POWDER; COPPER SLAG-AIRBORNE; COPPER SLAG-MILLED; COPPER-AIRBORNE; COPPER,METALLIC POWDER; COPPER-MILLED; CU M3; CUPRUM; E 115 (METAL); EPA PESTICIDE CHEMICAL CODE 022501; 1721 GOLD; GOLD BRONZE; KAFAR COPPER; M 1; M 3; M 4; M1 (COPPER); M2 (COPPER); M3 (COPPER); M3R; M3S; M4 (COPPER); OFHC CU; RANEY COPPER
Description: reddish brown solid, powder; odorless
Use: heating, chem, & pharmaceutical machinery; alloys (monel metal, beryllium-copper); electroplated coatings for nickel, etc; corrosion-resistant piping; catalyst; insulation for liq fuels; in thermal & electrical composites; in works of art; for electrical & electronic products (eg, wire); in transportation ind (eg, automobiles), consumer products (eg, coins, cooking utensils), & in inorganic pigments; intermed for copper chems (eg, cupric sulfate); in intrauterine contraceptive devices; in insecticides, fungicides, & herbicides, anti-fouling paints, nylon, paper products, pollution control catalyst, printing & photo copying, pyrotechnics, & wood preservatives.

Physical Properties

Boiling Point: 2595 °C (4703 °F)
Freezing Point: 1083 °C (1981.4 °F)
Specific Gravity: 8.94
Vapor Pressure: 1 mm Hg at 1628 °C
Water Solubility: Insoluble in Water
Other Solubilities: Soluble in Nitric acid, hot Sulfuric acid; Very Slightly Soluble in Hydrochloric Acid, Ammonium Hydroxide
Flash Point: Noncombustible, except as a powder

RTECS Toxicity Data

Acute Oral: Human TD_{Lo} Dose: 120 ug/kg; Toxic Effects: Gastrointestinal - Nausea or vomiting.
Reproductive/Teratogenic: Rat Route: Oral; Dose: 152 mg/kg; Duration: female 22W prior to mating Effects on Embryo or Fetus - Fetotoxicity; Specific Developmental Abnormalities - Central nervous system. Rat Route: Oral; Dose: 1520 ug/kg; Duration: female 22W prior to mating Specific Developmental Abnormalities - Musculoskeletal system. Rat Route: Oral; Dose: 1210 ug/kg; Duration: female 35W prior to mating Effects on Fertility - Pre-implantation mortality; Post-implantation mortality.
Tumorigenic: Rat Route: Intrapleural; Dose: 100 mg/kg; Toxic Effects: Tumorigenic - Equivocal tumorigenic agent by RTECS criteria; Lungs, Thorax, or Respiration - Fibrosis, focal (pneumoconiosis); Lungs, Thorax, or Respiration - Tumors.

Hazard Overviews

Fire Diamond

Health: Irritating to eyes/skin/respiratory tract. Also Causes: metal fume fever, allergic reaction, high temperature, metallic taste, nausea, coughing, general weakness, muscle aches, exhaustion, skin discoloration, nausea, vomiting, abdominal pain, diarrhea, stomach and intestine ulceration, jaundice, kidney and liver damage. Chronic Effects: mild dermatitis, degeneration of mucous membrane, discolor skin/hair, respiratory disease. Individuals with Wilson's disease (1 in 200,000 individuals) are more susceptible to chronic copper poisoning.
Fire: Finely divided copper burns in air, and in extreme cases ignites spontaneously. Use extinguishing agents suitable for surrounding fire.
Reactivity: Stable. Hazardous polymerization can occur, on long standing a white highly explosive peroxide deposit may form. Avoid: prolonged exposure to air and moisture. Incompatible with: ammonium nitrate; bromates; iodates; chlorates; ethylene oxide; hydrazoic acid; potassium oxide; dimethyl sulfoxide and trichloroacetic acid; hydrogen peroxide; sodium peroxide; sodium azide; sulfuric acid; hydrogen sulfide and air; lead azide; actylenic compounds; chlorine; fluorine [above 250 °F (121 °C)]; chlorine trifluoride; hydrazinium nitrate [above 158 °F (70 °C)];1-bromo-2-propyne; potassium dioxide. Hazardous decomposition products: metallic oxides; copper fumes.
Carcinogenicity: IARC - Not listed; NIOSH - Not listed; NTP - Not listed; ACGIH - Not listed; OSHA - Not listed; EPA - Class D, Not classifiable as to human carcinogenicity; MAK - Not listed

Primary Target Organs:

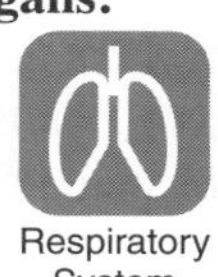

Eyes Skin Respiratory System Mucous Membranes Gastro-intestinal

Exposure Limits

OSHA PEL: TWA: 0.1 mg/m^3; as Cu, fume.
ACGIH TLV: TWA: 0.2 mg/m^3; as Cu; Inhalable.
NIOSH REL: TWA: 1 mg/m^3.
NIOSH IDLH: 100 mg/m^3; as Cu.
DFG MAK: TWA: 0.1 mg/m^3.

Respirator Recommendation

Exposure Range: >0.1 to 1 mg/m^3 Air Purifying, Negative Pressure, Half Mask
Exposure Range: >1 to 10 mg/m^3 Air Purifying, Negative Pressure, Full Face
Exposure Range: >10 to 100 mg/m^3 Supplied Air, Constant Flow/Pressure Demand, Full Face
Exposure Range: >100 to unlimited mg/m^3 Self-contained Breathing Apparatus, Pressure Demand, Full Face
Cartridge Color: magenta (P100)
Note: as a fume; if exposure is as a dust, respirator recommendations are different

Environmental

Regulations

RCRA 40CFR: Not listed
CERCLA: 40CFR 302.4: Listed per CWA Section 307(a) RQ: 5000 lb (2268 kg)
SARA 40CFR 372.65: Listed
SARA EHS 40CFR 355: Not listed
TSCA: Listed

Analytical Methods

Air: EPA 29, 0060, 1638, 1640, 1669, ITM-001
Soil: CLP 200.10_M, 200.62, 200.7_M, 202.62, 220.1_M, 220.2_M, ICP-AES; EPA 13, 200.7, 200.8; SW846 3005A, 3010A, 3015, 3031, 3040, 3040A, 3050A, 3050B, 3051, 3052, 6010A, 6010B, 6020, 7000A, 7210, 7211, OSW-A; ASTM D1971, D3974; USGS I5270, I5474
Water / Groundwater: EPA 200.0, 200.1, 200.10, 200.12, 200.13, 200.15, 200.7, 200.9; APHA 3111-A, 3111-B, 3111-C, 3113-B, 3120, 3500-CU; ASTM D1688, D1976, D2332, D4190; USGS E-SPEC, I1270, I1271, I1272, I1472, I3270, I3271, I7270, I7271; CEM RD42; FISON AES-0029; HACH 8506
Drinking Water: AOAC 974.27, 993.14
Food: EPA 14; AOAC 920.27, 920.28, 920.29, 922.05, 925.03, 964.03, 981.01
Indoor / Expired Air: NIOSH 7029, 7300; ASTM D4185
Plasma: EPA 200.11; NIOSH 8005
Urine: NIOSH 8310
Other: EPA 1620, 220.1, 220.2; AOAC 990.08

COP1640 CAS #: 1344-67-8

COPPER CHLORIDE

RTECS: GL6970000
DOT: UN2802
EINECS Number: 215-704-5
Molecular Formula: Unspecified or Variable
Formula Weight: 134.45
Synonyms: KIRTICOPPER
Description: yellow to brown, light blue-green fine crystals, powder; odorless

Physical Properties

Boiling Point: 993 °C (1819 °F)
Freezing Point: 620 °C (1148 °F)
Specific Gravity: 3
Water Solubility: 71 g/100 mL at 0 °C
Other Solubilities: Soluble in alcohol
pH: 0.2 molar solution 3.6
Flash Point: Noncombustible

Hazard Overviews

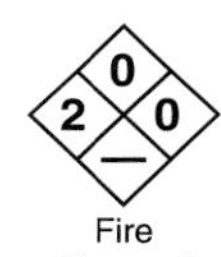

Fire Diamond

Health: Irritating to eyes/skin/respiratory tract. Also Causes: by ingestion: nausea, vomiting, abdominal pain, stomach bleeding, diarrhea. Chronic Effects: mild dermatitis.
Fire: Noncombustible. Use extinguishing agents suitable for surrounding fire.
Reactivity: Stable. Hazardous polymerization cannot occur. Avoid: acids; acid fumes; extreme heat; open flames; aluminum and moisture. Incompatible with: potassium; sodium; cupric chloride; hydrazine; nitromethane; sodium hypobromite. Hazardous decomposition products: toxic chloride fumes.
Carcinogenicity: IARC - Not listed; NIOSH - Not listed; NTP - Not listed; ACGIH - Not listed; OSHA - Not listed; EPA - Not listed; MAK - Not listed
Primary Target Organs:

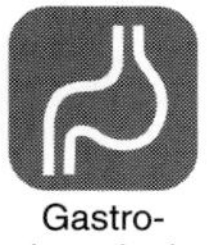

Eyes Skin Respiratory System Mucous Membranes Gastro-intestinal

Environmental

Cleanup/Disposal: Guide No. 154: Eliminate all ignition sources (no smoking, flares, sparks or flames in immediate area). Do not touch damaged containers or spilled material unless wearing appropriate protective clothing. Stop leak if you can do it without risk. Prevent entry into waterways, sewers, basements or confined areas. Absorb or cover with dry earth, sand or other non-combustible material and transfer to containers. Do not get water inside containers.

Regulations

RCRA 40CFR: Not listed
CERCLA: 40CFR 302.4: Not listed
SARA 40CFR 372.65: Not listed
SARA EHS 40CFR 355: Not listed
TSCA: Listed

COP2600 CAS #: 544-92-3

COPPER CYANIDE

RTECS: GL7150000
DOT: UN1587; IMO6.1
EINECS Number: 208-883-6
Molecular Formula: CCuN
Structured MF: CuCN
Formula Weight: 89.56

$N{\equiv}C^{-}$ Cu^{+}

Chemical Structure

Synonyms: COPPER (+1) CYANIDE; COPPER(I) CYANIDE; CUPRICIN; CUPROUS CYANIDE
Description: white to cream, colorless, dark green, dark red powder, orthorhombic crystals, monoclinic crystals; slight, bitter-almond odor
Use: in electroplating copper or iron; as insecticides, fungicide; as antifouling agent in marine paints; as polymerization catalyst

Physical Properties

Boiling Point: Decomposes at 1 atm
Freezing Point: 474 °C (885.2 °F)
Specific Gravity: 2.92 at 20 °C (solid)
Water Solubility: Practically Insoluble in Water
Other Solubilities: 95% Ethanol: Insoluble; HCl: Soluble; KCN: Soluble; NH_3: Slightly Soluble; NH_4OH: Soluble.
Flash Point: Nonflammable

RTECS Toxicity Data

Acute Oral: Rat LD_{50} Dose: 1265 mg/kg.

Hazard Overviews

Corrosive

Fire Diamond

Health: Corrosive, causes severe burns to eyes/skin/respiratory tract. Toxic. Causes chemical asphyxiation (lack of oxygen to tissues).
Fire: Noncombustible. Use extinguishing agents suitable for surrounding fire.
Reactivity: Stable. Hazardous polymerization cannot occur. Avoid: oxidizers; acids. Incompatible with: strong oxidizing agents; acids; magnesium; acetylene gas; 3-methoxy-2-nitrobenzoyl chloride; nitrite salts; metal chlorates; perchlorates; nitrates. Hazardous decomposition products: hydrogen cyanide; nitrogen oxides.
Carcinogenicity: IARC - Not listed; NIOSH - Not listed; NTP - Not listed; ACGIH - Not listed; OSHA - Not listed; EPA - Not listed; MAK - Not listed
Primary Target Organs:

Eyes

Skin

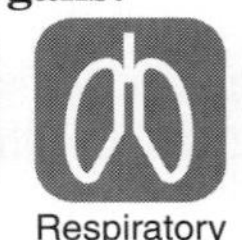
Respiratory System

Nervous System

Environmental

Ecotoxicity: Food chain concentration potential: Copper known to be accumulated by shellfish
Cleanup/Disposal: Guide No. 151: Do not touch damaged containers or spilled material unless wearing appropriate protective clothing. Stop leak if you can do it without risk. Prevent entry into waterways, sewers, basements or confined areas. Cover with plastic sheet to prevent spreading. Absorb or cover with dry earth, sand or other non-combustible material and transfer to containers. Do not get water inside containers.

Environmental Physical Data

BCF: possibly inaquatic organ

Regulations

RCRA 40CFR: Listed Hazardous Waste No. P029 Toxic Waste
CERCLA: 40CFR 302.4: Listed per RCRA Section 3001 RQ: 10 lb (4.535 kg)
SARA 40CFR 372.65: Listed as Compound
SARA EHS 40CFR 355: Not listed
TSCA: Listed

COP2920 CAS #: 137-29-1

COPPER DIMETHYLDITHIOCARBAMATE

RTECS: FA0175000
EINECS Number: 205-287-8
Molecular Formula: $C_6H_{12}CuN_2S_4$
Formula Weight: 303.98
Synonyms: COMPOUND-4018; COPPER(2+) DIMETHYLDITHIOCARBAMATE; COPPER,BIS(DIMETHYLCARBAMODITHIOATO-S,S')-,(SP-4-1) (9CI); COPPER,BIS(DIMETHYLDITHIOCARBAMATO)-; COPPER(II) DIMETHYLDITHIOCARBAMATE; CUMATE; CUPRIC N,N-DIMETHYLDITHIOCARBAMATE; DIMETHYLDITHIOCARBAMATOCOPPER; DIMETHYLDITHIOCARBAMIC ACID COPPER SALT; HERMAT CU; WOLFEN

Physical Properties

Freezing Point: Sublimes at 170 °C (338 °F) to 200 °C (392 °F)

RTECS Toxicity Data

Acute Oral: Rat LD; Dose: >500 mg/kg.
Acute Inhalation: Rat LC_{Lo} Dose: 210 mg/m^3/4hr; Toxic Effects: Lungs, Thorax, or Respiration - Dyspnea.
Mutagenic: Bacteria - S Typhimurium Mutations in Microorganisms; Dose: 2 mg/plate (+S9), 1 mg/plate (-S9).

Hazard Overviews

Carcinogenicity: IARC - Not listed; NIOSH - Not listed; NTP - Not listed; ACGIH - Not listed; OSHA - Not listed; EPA - Not listed; MAK - Not listed

Environmental

Regulations

RCRA 40CFR: Listed Hazardous Waste No. U393 Toxic Waste
CERCLA: 40CFR 302.4: Listed per RCRA Section 3001 RQ: 1 lb (0.454 kg)
SARA 40CFR 372.65: Listed as Compound
SARA EHS 40CFR 355: Not listed
TSCA: Listed

COP4840 CAS #: 10031-43-3

COPPER (II) NITRATE, TRIHYDRATE

RTECS: GL7875000
Molecular Formula: $CuH_6N_2O_9$
Formula Weight: 241.62

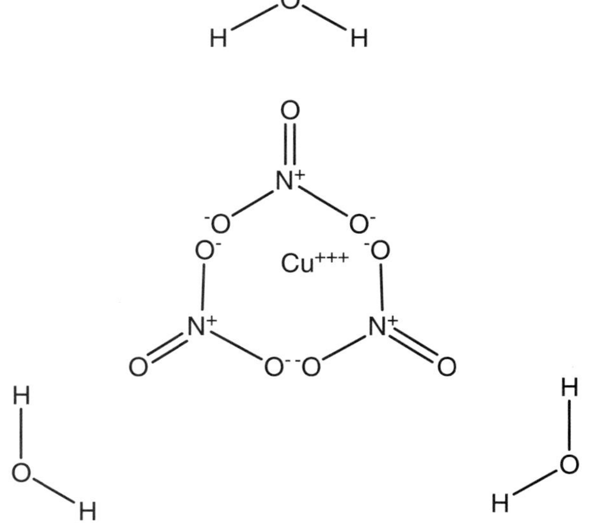

Chemical Structure

Synonyms: COPPER DINITRATE TRIHYDRATE; COPPER NITRATE TRIHYDRATE; COPPER(II) NITRATE TRIHYDRATE; CUPRIC NITRATE TRIHYDRATE; GERHARDITE; NITRIC ACID,COPPER(2+) SALT,TRIHYDRATE

Physical Properties

Boiling Point: Sublimes at 150 °C (302 °F) to 225 °C (437 °F)
Freezing Point: 114.5 °C (238.1 °F)
Density: 2.047 g/L
Water Solubility: Soluble
Other Solubilities: Practically Insoluble in Ethyl Acetate
pH: 0.2 molar solution 4
Flash Point: Noncombustible

RTECS Toxicity Data

Acute Oral: Rat LD_{50} Dose: 940 mg/kg.
Mutagenic: Rat Cytogenetic Analysis; Cell Type: Ascites tumor; Dose: 600 mg/kg.

Hazard Overviews

Reactivity: Stable. Hazardous polymerization cannot occur. Avoid: contact with combustibles. Incompatible with: ammonia and potassium amide; acetic anhydride; potassium ferrocyanide; ammonium or potassium hexacyanoferrate (II) (above 428 °F/220 °C); tin foil. Hazardous decomposition products: copper fumes; nitrogen oxide gas.
Carcinogenicity: IARC - Not listed; NIOSH - Not listed; NTP - Not listed; ACGIH - Not listed; OSHA - Not listed; EPA - Not listed; MAK - Not listed

Environmental

Regulations

RCRA 40CFR: Not listed
CERCLA: 40CFR 302.4: Not listed
SARA 40CFR 372.65: Not listed
SARA EHS 40CFR 355: Not listed
TSCA: Not listed

COP5480 CAS #: 1317-38-0

COPPER (II) OXIDE

RTECS: GL7900000
DOT: UN2775; UN3009; UN3010; IMO6.1
EINECS Number: 215-269-1
Molecular Formula: CuO
Structured MF: CuO
Formula Weight: 79.54

Cu═O

Chemical Structure

Synonyms: BANACOBRU OL; BLACK COPPER OXIDE; BOLIDEN SALT K-33; BOLIDEN-CCA WOOD PRESERVATIVE; C.I. 77403; C.I. PIGMENT BLACK 15; CCA TYPE C WOOD PRESERVATIVE; CHROME BROWN; COPPER BROWN; COPPER MONOOXIDE; COPPER MONOXIDE; COPPER MONOXIDE FUME; COPPER OXIDE; COPPER(2+) OXIDE; COPPER(II) OXIDE; COPPER(II) OXIDE FUME; CUO: BLACK COPPER OXIDE FUME; CUPRIC OXIDE; CUPRIC OXIDE FUME; FARBOIL SUPER TROPICAL ANTI-FOULING 1260; NATURAL TENORITE; OSMOSE K-33 WOOD PRESERVATIVE; OSMOSE K-33-A WOOD PRESERVATIVE; OSMOSE K-33-C WOOD PRESERVATIVE; OSMOSE P-50 WOOD PRESERVATIVE; PARAMELACONITE; WOLMANAC CONCENTRATE

Description: black to brownish-black powder, granules; odorless

Use: pigment in glass, ceramics, artificial gems, etc; in mfr of rayon, in sweetening petro gases; in galvanic electrodes; flux in metallurgy; optical glass polishing; in antifouling paints; exciter in phosphor mixtures; catalyst in ammonia mfr, for org reactions, cloud seeding; in solar energy devices; reduces tar in tobacco smoke; rgnt in anal chemistry; purification of hydrogen; solvent for chromic iron ores; feed supplement; fungicide, insecticide, miticide, molluscicide, & tadpole & shrimp deterrent; for treatment of wood; in fuel additives, food & drugs, nylon, paper products, pigment & dyes, pollution control catalyst, printing & photo copying, pyrotechnics.

Physical Properties

Boiling Point: Decomposes at 1026 °C (1879 °F)
Freezing Point: 1326 °C (2418.8 °F)
Specific Gravity: 6.315 at 14 °C/4 °C
Vapor Pressure: ~ 0 mm Hg
Water Solubility: Practically Insoluble in Water
Other Solubilities: Soluble in Ammonium Chloride, Potassium Cyanide.
Refraction Index: 2.63
Evaporation Rate: < 1 Butyl Acetate=1
Flash Point: Noncombustible Solid

Hazard Overviews

Fire Diamond

Health: Irritating to eyes/skin/respiratory tract. Also Causes: ulceration and necrosis of respiratory passages and skin, necrosis, diarrhea, capillary damage, kidney and liver injury, and central nervous system excitation metal fume fever. Chronic Effects: dermatitis, hemolytic anemia.

Fire: Noncombustible. Use extinguishing media suitable for surrounding area.

Reactivity: Stable. Hazardous polymerization cannot occur. Incompatible with: aluminum; boron; cesium; acetylene carbide; hydrazine; magnesium; phospham; potassium; sodium; titanium; zirconium; rubidium acetylene carbide; barium acetate; yttrium oxide; hydrogen sulfide; anilinium perchlorate; hydrogen, phthalic anhydride; hydroxylamine; dichloromethylsilane; sodium hypobromite solution; acetylene. Hazardous decomposition products: toxic copper fumes.

Carcinogenicity: IARC - Not listed; NIOSH - Listed as carcinogen; NTP - Not listed; ACGIH - Not listed; OSHA - Not listed; EPA - Not listed; MAK - Not listed

Primary Target Organs:

Eyes

Skin

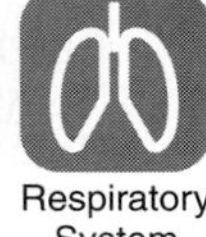
Respiratory System

Exposure Limits
NIOSH IDLH: 100 mg/m^3; as Cu.

Environmental

Cleanup/Disposal: Guide No. 131: Fully encapsulating, vapor protective clothing should be worn for spills and leaks with no fire. Eliminate all ignition sources (no smoking, flares, sparks or flames in immediate area). All equipment used when handling the product must be grounded. Do not touch or walk through spilled material. Stop leak if you can do it without risk. Prevent entry into waterways, sewers, basements or confined areas. A vapor suppressing foam may be used to reduce vapors. Small Spills: Absorb with earth, sand or other non-combustible material and transfer to containers for later disposal. Use clean non-sparking tools to collect absorbed material. Large Spills: Dike far ahead of liquid spill for later disposal. Water spray may reduce vapor; but may not prevent ignition in closed spaces. Guide No. 151: Do not touch damaged containers or spilled material unless wearing appropriate protective clothing. Stop leak if you can do it without risk. Prevent entry into waterways, sewers, basements or confined areas. Cover with plastic sheet to prevent spreading. Absorb or cover with dry earth, sand or other non-combustible material and transfer to containers. Do not get water inside containers.

Regulations

RCRA 40CFR: Not listed
CERCLA: 40CFR 302.4: Listed as Compound per CWA Section 307(a)
SARA 40CFR 372.65: Listed as Compound
SARA EHS 40CFR 355: Not listed
TSCA: Listed

COP6440 CAS #: 1317-40-4

COPPER (II) SULFIDE

RTECS: GL8912000
EINECS Number: 215-271-2
Molecular Formula: CuS
Structured MF: CuS
Formula Weight: 95.61

S═Cu

Chemical Structure

Synonyms: C.I. 77450; C.I. PIGMENT BLUE 34; COPPER BLUE; COPPER MONOSULFIDE; COPPER SULFIDE; COPPER(2+) SULFIDE; COPPER(II) SULFIDE; CUPRIC SULFIDE; HORACE VERNET'S BLUE; MONOCOPPER MONOSULFIDE; NATURAL COVELLITE; OIL BLUE

Description: blue, black crystals, powder

Use: in prepn of mixed catalysts; in textile dye; pigment for varnishes; oxidizing agent, eg, in lithium batteries; in solid state electrodes; biocide in antifouling paints; in agricultural products (insecticides, fungicides, herbicides), corrosion inhibitors, electrolysis & electroplating processes, electronics, fabric & textiles, flameproofing, fuel additives, glass, &

ceramics; in cement, food & drugs, metallurgy, nylon, paper products, pollution control catalyst, printing & photo copying, pyrotechnics, & wood preservatives.

Physical Properties

Freezing Point: Decomposes at 220 °C (428 °F)
Specific Gravity: 4.6
Water Solubility: Practically Insoluble in Water
Other Solubilities: Soluble in hot hydrochloric & sulfuric acids.
Refraction Index: 1.45

RTECS Toxicity Data

Mutagenic: Rat Cytogenetic Analysis; Cell Type: Ascites tumor; Dose: 150 mg/kg. Hamster DNA Damage; Cell Type: ovary; Dose: 10 mg/L.

Hazard Overviews

Fire Diamond

Health: Causes: metal fume fever, an acute flu-like lung reaction.
Fire: Will burn. Use dry chemical, carbon dioxide, water spray, fog, or regular foam. May explode on contact with certain incompatibles such as magnesium chlorate, zinc chlorate, cadmium chlorate.
Reactivity: Stable. Hazardous polymerization can occur when moist air causes copper (II) sulfide to oxidize to copper sulfate. Avoid: moisture. Incompatible with: magnesium chlorate; zinc chlorate; cadmium chlorate; concentrated chloric acid solutions; hydrogen peroxide; ammonium magnesium nitrate and water. Hazardous decomposition products: sulfur oxides.
Carcinogenicity: IARC - Not listed; NIOSH - Not listed; NTP - Not listed; ACGIH - Not listed; OSHA - Not listed; EPA - Not listed; MAK - Not listed
Primary Target Organs:

Respiratory System

Environmental

Regulations

RCRA 40CFR: Not listed
CERCLA: 40CFR 302.4: Listed as Compound per CWA Section 307(a)
SARA 40CFR 372.65: Listed as Compound
SARA EHS 40CFR 355: Not listed
TSCA: Listed

COP8040 **CAS #: 1317-39-1**

COPPER (I) OXIDE

RTECS: GL8050000
DOT: UN2775; UN2776; UN3009; UN3010; IMO3.2; IMO6.1
EINECS Number: 215-270-7
Molecular Formula: Cu_2O
Structured MF: Cu_2O
Formula Weight: 143.08

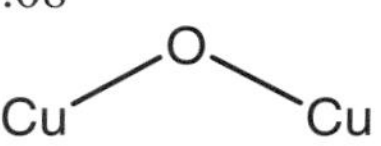

Chemical Structure

Synonyms: BROWN COPPER OXIDE; C.I. 77402; CAOCOBRE; COBRE SANDOZ; COPOX; COPPER HEMIOXIDE; COPPER NORDOX; COPPER (1+) OXIDE; COPPER OXIDE; COPPER(1+) OXIDE; COPPER OXIDE (8CI,9CI); COPPER OXIDE (CU2O); COPPER OXIDE,RED; COPPER PROTOXIDE; COPPER SANDOZ; COPPER SARDEZ; COPPER SUBOXIDE; COPPER-SANDOZ; CP CUPROUS OXIDE PIGMENT GRADE; CUPPER OXIDE; CUPRAMAR; CUPROCIDE; CUPROUS OXIDE; CUPROUS OXIDE TYPE TWO; CUPROUS OXIDE,AA GRADE; DICOPPER MONOXIDE; DICOPPER OXIDE; FUNGIMAR; FUNGI-RHAP CU-75; KUPFEROXYDUL; KUPRITE; NORDOX; OLEO NORDOX; OLEOCUIVRE; OXYDE CUIVREUX; PERECOT; PERENEX; PERENOX; PURPLE COPP 97N; PURPLE COPP 92; PURPLE COPP 97; RED COPP 97N; RED COPP 92; RED COPP 97; RED COPPER OXIDE; YELLOW COMPOUND; YELLOW CUPROCIDE
Description: red to reddish-brown, yellow crystals, powder
Use: antiseptic for fish nets; in marine antifouling paints, photoelectric cells; red pigment for glass, ceramic glazes; in brazing pastes, rectifiers; catalyst; mfr copper salts; electroplating; against fungus diseases on coffee, cocoa, etc; protective fungicide, for seed treatment & for foliage application against blight, downy mildews, & rusts; molluscicide; antioxidant in lubricants; purification of helium; for carbon monoxide absorption.

Physical Properties

Boiling Point: Decomposes at 1800 °C (3272 °F)
Freezing Point: 1232 °C (2249.6 °F)
Specific Gravity: 6 at 25 °C/4 °C
Water Solubility: Practically Insoluble in Water
Other Solubilities: Practically Insoluble in organic solvents; Soluble in dilute Mineral acids, solution of Ammonia and its salts; Soluble in Hydrochloric Acid, Ammonium Chloride, Ammonium Hydroxide; Slightly Soluble in Nitric acid; Insoluble in Alcohol.
Refraction Index: 2.705

RTECS Toxicity Data

Acute Oral: Rat LD_{50} Dose: 470 mg/kg.
Chronic (Multiple Dose) Inhalation: Rat Dose: 1 mg/m^3/24H/14W-C; Toxic Effects: Blood - Changes in serum composition; Blood - Other changes; Blood - Changes in erythrocite (RBC) cell count.

Reproductive/Teratogenic: Rat Route: Inhalation; Dose: 11 ug/m^3/24H Duration: male 14W prior to mating; Effects on Fertility - Male fertility index.

Hazard Overviews

Fire Diamond

Health: Irritating to eyes/skin/respiratory tract. Also Causes: leukocytosis, metallic/sweet taste, hair/skin discoloration, nausea, gastric pain, diarrhea, and possible hemorrhagic gastritis. Can cause metal fume fever. Chronic Effects: hemolytic anemia.

Fire: Noncombustible. Use extinguishing media suitable for surrounding area.

Reactivity: Stable in dry air, but gradually oxidizes in moist air. Hazardous polymerization cannot occur. Incompatible with: aluminum; lithium nitride; peroxyformic acid; hydrazine; acetylene and caustic solution. Hazardous decomposition products: toxic copper fumes.

Carcinogenicity: IARC - Not listed; NIOSH - Not listed; NTP - Not listed; ACGIH - Not listed; OSHA - Not listed; EPA - Not listed; MAK - Not listed

Primary Target Organs:

Eyes

Skin

Respiratory System

Environmental

Cleanup/Disposal: Guide No. 131: Fully encapsulating, vapor protective clothing should be worn for spills and leaks with no fire. Eliminate all ignition sources (no smoking, flares, sparks or flames in immediate area). All equipment used when handling the product must be grounded. Do not touch or walk through spilled material. Stop leak if you can do it without risk. Prevent entry into waterways, sewers, basements or confined areas. A vapor suppressing foam may be used to reduce vapors. Small Spills: Absorb with earth, sand or other non-combustible material and transfer to containers for later disposal. Use clean non-sparking tools to collect absorbed material. Large Spills: Dike far ahead of liquid spill for later disposal. Water spray may reduce vapor; but may not prevent ignition in closed spaces. Guide No. 151: Do not touch damaged containers or spilled material unless wearing appropriate protective clothing. Stop leak if you can do it without risk. Prevent entry into waterways, sewers, basements or confined areas. Cover with plastic sheet to prevent spreading. Absorb or cover with dry earth, sand or other non-combustible material and transfer to containers. Do not get water inside containers.

Regulations

RCRA 40CFR: Not listed

CERCLA: 40CFR 302.4: Listed as Compound per CWA Section 307(a)

SARA 40CFR 372.65: Listed as Compound

SARA EHS 40CFR 355: Not listed

TSCA: Listed

COP8360 CAS #: 147-14-8

COPPER PHTHALOCYANINE BLUE

RTECS: GL8510000

EINECS Number: 205-685-1

Molecular Formula: $C_{32}H_{16}CuN_8$

Formula Weight: 576.084

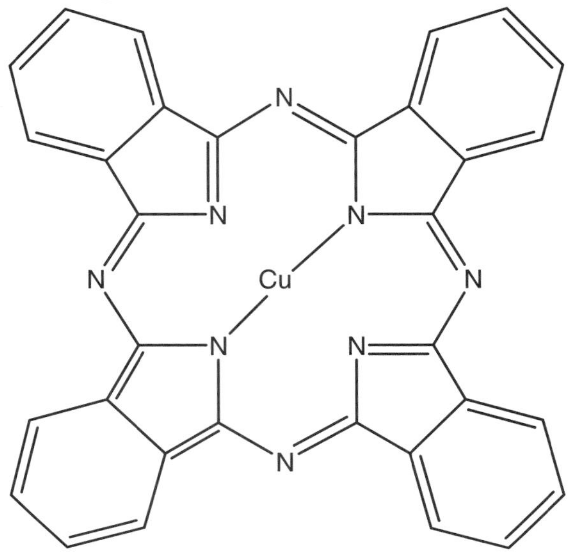

Chemical Structure

Synonyms: ACCOSPERSE CYAN BLUE GT; AQUALINE BLUE; ARLOCYANINE BLUE PS; BAHAMA BLUE BC; BAHAMA BLUE BNC; BAHAMA BLUE LAKE NCNF; BAHAMA BLUE WD; BERMUDA BLUE; BLUE 15B; C.I. 74160; C.I. INGRAIN BLUE 2; C.I. PIGMENT BLUE 15; CALCOTONE BLUE GP; CERES BLUE BHR; CHROMATEX BLUE BN; CONGO BLUE B-4; COPPER PHTHALOCYANIN; COPPER PHTHALOCYANINE; COPPER,(PHTHALOCYANINATO(2-))-; COPPER,(29H,31H-PHTHALOCYANINATO(2-)-N(SUP 29),N(SUP 30),N(SUP 31),N(SUP 32))-,(SP-4-1)-; COPPER,(29H,31H-PHTHALOCYANINATO(2-)-N(29),N(30),N(31),N(32))-,(SP-4-1)-; CROMOPHTAL BLUE 4G; CYAN BLUE BNC 55-3745; CYAN BLUE BNF 55-3753; CYAN BLUE GT 55-3295; CYAN PEACOCK BLUE G; CYANINE BLUE BB; CYANINE BLUE C; DAINICHI CYANINE BLUE B; DAINICHI CYANINE BLUE FPG; DALTOLITE FAST BLUE B; DURATINT BLUE 1001; FASTOGEN BLUE 5110; FASTOGEN BLUE B; FASTOLUX BLUE; FENALAC BLUE B DISP; FRANCONIA BLUE A-4431; GRAPHTOL BLUE BL; GRAPHTOL BLUE 2GLS; HELIO BLUE B; HELIO FAST BLUE BRN; HELIOGEN BLUE B; HELIOGEN BLUE LBG; INDOLEN BLUE 3G; IRGALITE FAST BRILLIANT BLUE BL; ISOL FAST BLUE B; LBX 5; PHTHALO BLUE; (PHTHALOCYANINATO(2-))COPPER; PHTHALOCYANINE BLUE 15

Description: bright blue microcrystals with purple lustre

Use: industrial dye used in paints, lacquers, stains and plastics

Physical Properties

Water Solubility: Practically Insoluble in Water

Other Solubilities: 95% Ethanol: <1 mg/ml at 20 °C; Acetone: <1 mg/ml at 20 °C; DMSO: 10-50 mg/ml at 20 °C.
Flash Point: Not available; probably combustible

RTECS Toxicity Data

Acute Oral: Rat LD; Dose: >15 gm/kg.

Hazard Overviews

Health: Irritating to eyes/skin/respiratory tract. Other Acute Effects: may be harmful by inhalation, ingestion, or skin absorption. Chronic Effects: hepatic cirrhosis; brain damage and demyelination; kidney defects; copper deposition in the cornea as exemplified by humans with Wilson's disease; hemolytic anemia; accelerated arteriosclerosis; baritosis, a benign pneumoconiosis; if ingested, the presence of soluble barium salts as impurities may cause toxic reactions due to bioaccumulation.
Fire: Will burn. Hazards: emits toxic fumes. Extinguishing agents: water spray; carbon dioxide, dry chemical powder or appropriate foam. Precautions: combustible liquid.
Reactivity: Stable. Hazardous polymerization will not occur. Incompatible with: strong oxidizing agents. Hazardous decomposition products: toxic fumes of: carbon monoxide, carbon dioxide, nitrogen oxides.
Carcinogenicity: IARC - Not listed; NIOSH - Not listed; NTP - Not listed; ACGIH - Not listed; OSHA - Not listed; EPA - Not listed; MAK - Not listed
Primary Target Organs:

Eyes

Skin

Respiratory System

Environmental

Regulations

RCRA 40CFR: Not listed
CERCLA: 40CFR 302.4: Listed as Compound per CWA Section 307(a)
SARA 40CFR 372.65: Not listed
SARA EHS 40CFR 355: Not listed
TSCA: Listed

COP8680 **CAS #: 1328-53-6**

COPPER PHTHALOCYANINE GREEN

RTECS: GE4300000
EINECS Number: 215-524-7
Molecular Formula: Unknown
Synonyms: ACCOSPERSE CYAN GREEN G; BRILLIANT GREEN PHTHALOCYANINE; C.I. 74260; C.I. PIGMENT GREEN 42; C.I. PIGMENT GREEN 7; C.I. PIGMENT GREEN 7; CALCOTONE GREEN G; CERES GREEN 3B; CHROMATEX GREEN G; COLANYL GREEN GG; CROMOPHTAL GREEN GF; CROMOPHTHAL GREEN GF; CYAN GREEN 15-3100; CYANINE GREEN GP; CYANINE GREEN NB; CYANINE GREEN T; CYANINE GREEN TONER; DAINICHI CYANINE GREEN FG; DAINICHI CYANINE GREEN FGH; DALTOLITE FAST GREEN GN; DURATINT GREEN 1001; FASTOGEN GREEN 5005; FASTOGEN GREEN B; FASTOLUX GREEN; FENALAC GREEN G; FENALAC GREEN G DISP; GRANADA GREEN LAKE GL; GRAPHTOL GREEN 2GLS; HELIOGEN GREEN 8681K; HELIOGEN GREEN 8682T; HELIOGEN GREEN 8680; HELIOGEN GREEN 8730; HELIOGEN GREEN A; HELIOGEN GREEN G; HELIOGEN GREEN GA; HELIOGEN GREEN GN; HELIOGEN GREEN GNA; HELIOGEN GREEN GTA; HELIOGEN GREEN GV; HELIOGEN GREEN GWS; HOSTAPERM GREEN GG; IRGALITE FAST BRILLIANT GREEN 3GL; IRGALITE FAST BRILLIANT GREEN GL; IRGALITE GREEN GLN; KLONDIKE YELLOW X-2261; LUTETIA FAST EMERALD J; MICROLITH GREEN G-FP; MONARCH GREEN WD; MONASTRAL FAST GREEN BGNA; MONASTRAL FAST GREEN G; MONASTRAL FAST GREEN GD; MONASTRAL FAST GREEN GF; MONASTRAL FAST GREEN GFNP; MONASTRAL FAST GREEN GN; MONASTRAL FAST GREEN GNA; MONASTRAL FAST GREEN GTP; MONASTRAL FAST GREEN GV; MONASTRAL FAST GREEN 2GWD; MONASTRAL FAST GREEN GWD; MONASTRAL FAST GREEN GX; MONASTRAL FAST GREEN GXB; MONASTRAL FAST GREEN GYH; MONASTRAL FAST GREEN LGNA; MONASTRAL GREEN B; MONASTRAL GREEN B PIGMENT; MONASTRAL GREEN G; MONASTRAL GREEN GFN; MONASTRAL GREEN GH; MONASTRAL GREEN GN; MONOLITE FAST GREEN GVSA; NON-FLOCCULATING GREEN G-25; NON-FLOCCULATING GREEN G 25; OPALINE GREEN G-1; OPALINE GREEN G 1; PACIFIC GREEN NO.6491; PERMANENT GREEN TONER GT-376; PHTHALOCYANINE BRILLIANT GREEN; PHTHALOCYANINE GREEN; PHTHALOCYANINE GREEN LX; PHTHALOCYANINE GREEN V; PHTHALOCYANINE GREEN VFT-1080; PHTHALOCYANINE GREEN VFT 1080; PHTHALOCYANINE GREEN WDG 47; PIGMENT FAST GREEN G; PIGMENT FAST GREEN GN; PIGMENT GREEN 7; PIGMENT GREEN PHTHALOCYANINE; PIGMENT GREEN PHTHALOCYANINE V; POLYCHLORO COPPER PHTHALOCYANINE; POLYMO GREEN FBH; POLYMO GREEN FGH; POLYMON GREEN 6G; POLYMON GREEN G; POLYMON GREEN GN; PV-FAST GREEN G; RAMAPO; RESINATED PHTHALOCYANINE GREEN G-5025; SANYO CYANINE GREEN; SANYO PHTHALOCYANINE GREEN F6G; SANYO PHTHALOCYANINE GREEN FB PURE; SEGNALE LIGHT GREEN G; SHERWOOD GREEN A-4436; SHERWOOD GREEN A 4436; SIEGLE FAST GREEN G; SOLFAST GREEN; SOLFAST GREEN 63102; SYNTHALINE GREEN; TERMOSOLIDO GREEN FG SUPRA; THALO GREEN NO.1; THALO GREEN NO 1; THALO GREEN NO. 1; VERSAL GREEN G; VULCAL FAST GREEN F2G; VULCANOSINE FAST GREEN G; VULCOL FAST GREEN F2G; VYNAMON GREEN BE; VYNAMON GREEN BES; VYNAMON GREEN GNA
Description: bright green powder
Use: industrial dye used in paints, lacquers, plastics and a wide range of other products

Physical Properties

Water Solubility: < 1 mg/mL at 21 C
Other Solubilities: 95% Ethanol: <1 mg/ml at 21 °C; Acetone: <1 mg/ml at 21 °C; Concentrated H_2SO_4: Soluble; DMSO: <1 mg/ml at 21 °C.
Flash Point: Not available; probably combustible

RTECS Toxicity Data

Mutagenic: Bacteria - S Typhimurium Mutations in Microorganisms; Dose: 3333 ug/plate (-S9).

Hazard Overviews

Fire: Will burn.
Carcinogenicity: IARC - Not listed; NIOSH - Not listed; NTP - Not listed; ACGIH - Not listed; OSHA - Not listed; EPA - Not listed; MAK - Not listed

Environmental

Regulations

RCRA 40CFR: Not listed
CERCLA: 40CFR 302.4: Listed as Compound per CWA Section 307(a) per CAA Section 112
SARA 40CFR 372.65: Not listed
SARA EHS 40CFR 355: Not listed
TSCA: Listed

COP9000 CAS #: 7758-98-7

COPPER SULFATE

RTECS: GL8800000
DOT: UN2775; UN3009; UN3010; IMO6.1
EINECS Number: 231-847-6
Molecular Formula: CuH_2O_4S
Structured MF: $Cu_2SO_4 \cdot 5H_2O$
Formula Weight: 159.60

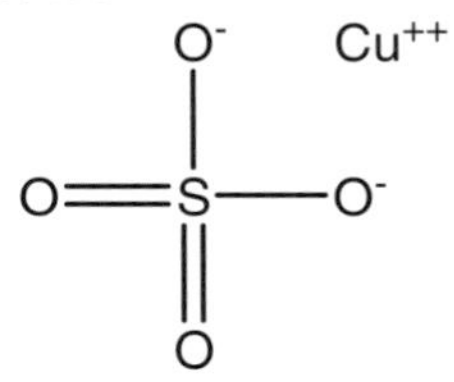

Chemical Structure

Synonyms: ALL CLEAR ROOT DESTROYER; AQUA MAID PERMANENT ALGAECIDE; AQUATRONICS SNAIL-A-CIDE DRI-PAC SNAIL POWDER; BCS COPPER FUNGICIDE; BLUE COPPER; BLUE COPPER AS; BLUE COPPERAS; BLUE STONE; BLUE VICKING; BLUE VITRIOL; BLUESTONE; BONIDE ROOT DESTROYER; CALCANTHITE; COPPER MONOSULFATE; COPPER SULFATE (1:1); COPPER(2+) SULFATE; COPPER(2+) SULFATE (1:1); COPPER SULFATE (CUSO4) PENTAHYDRATE; COPPER SULFATE PENTAHYDRATE; COPPER(2+) SULFATE PENTAHYDRATE; COPPER SULFATE POWDER; COPPER SULPHATE; COPPER(II) SULFATE; COPPER(II) SULFATE PENTAHYDRATE (1:1:5); CP BASIC SULFATE; CSP; CUPRIC SULFATE; CUPRIC SULFATE ANHYDROUS; CUPRIC SULFATE,PENTAHYDRATE; CUPRIC SULPHATE; GRANULAR CRYSTALS COPPER SULFATE; GRIFFIN SUPER CU; INCRACIDE 10A; INCRACIDE E 51; KILCOP 53; KOBASIC; KUPFERSULFAT; KUPFERSULFAT-PENTAHYDRAT; KUPFERVITRIOL; PHELPS TRIANGLE BRAND COPPER SULFATE; PHYTO-BORDEAUX; ROMAN VITRIOL; SA-50 BRAND COPPER SULFATE GRANULAR CRYSTALS; SALZBURG VITRIOL; SNOW CRYSTAL COPPER SULFATE; SULFATE DE CUIVRE; SULFURIC ACID COPPER(2+) SALT (1:1),PENTAHYDRATE; SULFURIC ACID,COPPER(2+) SALT (1:1); SULFURIC ACID,COPPER(2+) SALT,PENTAHYDRATE; TNCS 53; TOBACCO STATES BRAND COPPER SULFATE; TRIANGLE; TRINAGLE
Description: grayish-white, greenish-white crystals, powder
Use: fungicide; bactericide, herbicide, for septic tank lines & sewer lines; against diseases of trees, downy mildew, blights, etc; on berries and nuts; prune/wound treatment; algicide, in ponds, etc; (vet) nutritional feed additive, astringent antiseptic, in foot rot, for ringworm, etc; molluscicide, to control snail hosts of liver flukes; nutrient (gluconate & sulfate) in foods; in fehling's soln; in preserving hides; electroplating soln, & battery electrolyte; in inks, petro refining; flotation agent; pigment; mfr of other salts; in pyrotechnics, adhesives, antirusting compositions; toner in photography; treatment of asphalts; anti-fouling paints, catalysts, textiles, flameproofing, fuel additives, glass, & ceramics; cement, metallurgy, nylon, paper, pollution control catalyst, wood preservatives.

Physical Properties

Boiling Point: Decomposes at 650 °C (1202 °F)
Freezing Point: > 200 °C (392 °F)
Specific Gravity: 2.286 at 15.6 °C/4 °C
Water Solubility: 14.3 g/100 cc Water at 0 °C
Other Solubilities: 1.04 g/100 cc Methanol at 18 °C; 1 g in about 500 ml Alcohol; Practically Insoluble in most organic solvents; 1 g in 3 ml Glycerine; 15.6 g/100 cc Methanol at 18 °C.
pH: 0.2 molar aqueous solution 4
Refraction Index: 1.733
Flash Point: Nonflammable

RTECS Toxicity Data

Acute Oral: Human TD_{Lo} Dose: 11 mg/kg; Toxic Effects: Gastrointestinal - Gastritis; Gastrointestinal - Hypermotility, diarrhea; Gastrointestinal - Nausea or vomiting. Child TD_{Lo} Dose: 150 mg/kg; Toxic Effects: Kidney, Ureter, and Bladder - Chgs in tubules (inc acute renal failure, acute tubular necrosis; Blood - Other hemolysis with or without anemia. Human LD_{Lo} Dose: 50 mg/kg; Toxic Effects: Behavioral - Somnolence (general depressed activity); Kidney, Ureter, and Bladder - Chgs in tubules (inc acute renal failure, acute tubular necrosis; Blood - Hemorrhage. Man LD_{Lo} Dose: 857 mg/kg; Toxic Effects: Gastrointestinal - Nausea or vomiting. Rat LD_{50} Dose: 300 mg/kg.
Acute Dermal: Rat LD_{50} Route: Subcutaneous Dose: 43 mg/kg. Mouse LD_{Lo} Route: Subcutaneous Dose: 500 ug/kg.
Reproductive/Teratogenic: Rat Route: Subcutaneous; Dose: 12768 ug/kg; Duration: male 1D prior to mating; Paternal Effects - Testes, epididymis, sperm duct. Rat Route: Intratesticular; Dose: 3192 ug/kg; Duration: male 1D prior to mating; Paternal Effects - Spermatogenesis; Testes, epididymis, sperm duct.
Mutagenic: Rat DNA Damage; Cell Type: Ascites tumor; Dose: 500 umol/L. Rat DNA Damage; Cell Type: liver; Dose: 1 mmol/L.
Tumorigenic: Chicken Route: Parenteral; Dose: 10 mg/kg; Toxic Effects: Tumorigenic - Equivocal tumorigenic agent by RTECS criteria; Endocrine - Tumors.

Hazard Overviews

Fire Diamond

Health: Highly irritating to the eyes/skin/respiratory tract. Toxic. Also Causes: heavy exposures may cause nausea, vomiting, and serious blood, liver and kidney damage.

Fire: Noncombustible. Use extinguishing agents suitable for surrounding fire.
Reactivity: Stable. Hazardous polymerization cannot occur. Incompatible with: hydroxylamine; magnesium; ammonia (or other caustic) and acetylene; sodium hypobromite; nitromethane; iron; galvanized iron. Hazardous decomposition products: carbon and sulfur oxide(s).
Carcinogenicity: IARC - Not listed; NIOSH - Not listed; NTP - Not listed; ACGIH - Not listed; OSHA - Not listed; EPA - Not listed; MAK - Not listed
Primary Target Organs:

Eyes

Skin
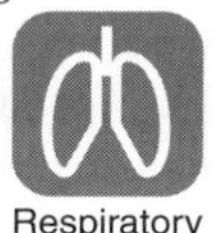
Respiratory System

Liver
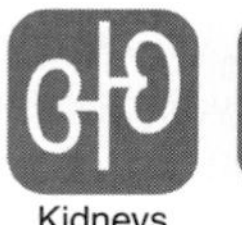
Kidneys

Blood

Environmental

Ecotoxicity: LD_{50} Pheasant oral approx 1,000 ppm LC_{50} Striped bass 1 ppm or lower/96 hr /conditions of bioassay not specified LC_{50} Haliotis cracherodii (abalone) 0.05 mg/l/96 hr /Static system, natural seawater at 14 °C EC_{50} Selenastrum capricornatum (green alga) 85 ug/l/14 days /Cell volume bioassay EC_{50} Thalassiosira pseudonana (alga, saltwater) 5 ug/l/72 hr /Growth rate bioassay LC_{50} Oncorhynchus kisutch (coho salmon) 286 ug/l/96 hr /Conditions of bioassay not specified EC_{10} Salmo gairdneri (rainbow trout; embryo, larvae) 16.5 ug/l/28 days; death and deformity. /Conditions of bioassay not specified LC_{50} Salmo gairdneri (Rainbow trout), wt 1.6 g, 135 ug/l at 13 °C, static bioassay LC_{50} Viviparus bengalensis (Snail) 0.060 ppm copper /96 hr at 32.5 °C; 0.066 ppm copper /96 hr at 24 °C; 0.09 at 27.3 °C; & 0.39 ppm copper/96 hr at 20.3 °C /static bioassay LC_{50} Prawn 0.14 ppm/48 hr, salt water. /Conditions of bioassay not specified
Cleanup/Disposal: Guide No. 131: Fully encapsulating, vapor protective clothing should be worn for spills and leaks with no fire. Eliminate all ignition sources (no smoking, flares, sparks or flames in immediate area). All equipment used when handling the product must be grounded. Do not touch or walk through spilled material. Stop leak if you can do it without risk. Prevent entry into waterways, sewers, basements or confined areas. A vapor suppressing foam may be used to reduce vapors. Small Spills: Absorb with earth, sand or other non-combustible material and transfer to containers for later disposal. Use clean non-sparking tools to collect absorbed material. Large Spills: Dike far ahead of liquid spill for later disposal. Water spray may reduce vapor; but may not prevent ignition in closed spaces. Guide No. 151: Do not touch damaged containers or spilled material unless wearing appropriate protective clothing. Stop leak if you can do it without risk. Prevent entry into waterways, sewers, basements or confined areas. Cover with plastic sheet to prevent spreading. Absorb or cover with dry earth, sand or other non-combustible material and transfer to containers. Do not get water inside containers.

Environmental Physical Data

BCF: not biomagnified
BOD: none

Regulations

RCRA 40CFR: Not listed
CERCLA: 40CFR 302.4: Listed per CWA Section 311(b)(4) RQ: 10 lb (4.535 kg)
SARA 40CFR 372.65: Listed as Compound
SARA EHS 40CFR 355: Not listed
TSCA: Listed

COR5000 CAS #: 8001-30-7

CORN OIL

RTECS: GM4800000
EINECS Number: 232-281-2
Molecular Formula: Unspecified or Variable
Formula Weight: N/A
Synonyms: CORN GERM OIL; LIPOMUL; MAISE OIL; MAYDOL; MAZOLA OIL; OILS,CORN; OILS,GLYCERIDIC,CORN
Description: pale yellow liquid; faint, characteristic odor
Use: as a salad and cooking oil, in preparation of margarine, in the preparation of non-yellowing enamel paint, as a pharmaceutic a (solvent), in foodstuffs, soap, lubricants, leather dressing, factice and hair dressing; in mild cathartics and as a protective for the gastrointestinal tract in case of corrosive poisoning and as an emollient to the skin and mucous membranes

Physical Properties

Freezing Point: -18 °C (-0.4 °F) to -10 °C (14 °F)
Specific Gravity: 0.916 to 0.921 at 25/25 °C
Density: 0.92 g/mL
Vapor Pressure: Negligible
Water Solubility: < 1 mg/mL at 18 C
Other Solubilities: 95% Ethanol: 5-10 mg/ml at 18 °C; Acetone: >=100 mg/ml at 18 °C; Alcohol: Slightly Soluble; Amyl acetate: Soluble; Carbon Disulfide: Soluble; Chloroform: miscible; DMSO: <1 mg/ml at 18 °C; Ether: miscible; Petroleum Ether: miscible.
Refraction Index: 1.47
Evaporation Rate: Negligible
Flash Point: 254 °C
Autoignition Temperature: 393 °C

RTECS Toxicity Data

Acute Oral: Rat LD_{50} Dose: >100 mL/kg.
Irritation Skin: Human Standard Draize Test Dose: 300 mg/3D-I; Reaction: mild.
Reproductive/Teratogenic: Rat Route: Oral; Dose: 12500 mg/kg; Duration: female 15-19D of pregnancy Specific Developmental Abnormalities - Blood and lymphatic systems (including spleen and marrow); Immune and reticuloendothelial system. Rat Route: Oral; Dose: 360 gm/kg; Duration: female 10-22D of pregnancy Effects on Newborn - Biochemical and metabolic. Rat Route: Oral; Dose: 50 mL/kg; Duration: female 6-15D of pregnancy; Effects on Embryo or Fetus - Fetotoxicity; Other effects to embryo or fetus.

Tumorigenic: Rat Route: Oral; Dose: 2600 mL/kg/2Y-I; Toxic Effects: Tumorigenic - Neoplastic by RTECS criteria; Gastrointestinal - Tumors.

Hazard Overviews

Fire Diamond

Health: Corn oil is relatively nonhazardous in routine industrial situations.

Fire: Will burn. Use water fog, dry chemical, alcohol foam, or carbon dioxide to fight fires involving corn oil. Use a water spray to cool tanks or containers that have been exposed to fire.

Reactivity: Stable. Hazardous polymerization cannot occur. Avoid: exposure to heat; sparks; open flame; lighted tobacco products. Incompatible with: strong oxidizers. Hazardous decomposition products: carbon monoxide; carbon dioxide.

Carcinogenicity: IARC - Not listed; NIOSH - Not listed; NTP - Not listed; ACGIH - Not listed; OSHA - Not listed; EPA - Not listed; MAK - Not listed

Environmental

Regulations

RCRA 40CFR: Not listed
CERCLA: 40CFR 302.4: Not listed
SARA 40CFR 372.65: Not listed
SARA EHS 40CFR 355: Not listed
TSCA: Listed

COT5000 CAS #: 8001-29-4

COTTONSEED OIL

RTECS: GN2815000
EINECS Number: 232-280-7
Molecular Formula: Unknown
Formula Weight: N/A
Synonyms: COTTON OIL; DEODORIZED WINTERIZED COTTONSEED OIL; FMC 710; OILS,COTTONSEED; OILS,GLYCERIDIC,COTTONSEED
Description: pale yellow liquid; odorless
Use: manufacture of soaps, oleomargarine, hydrogenated fats, lard substitutes, glycerol, leather dressings, lubricants, cosmetics and salad and cooking oils; in packing fish, as a pharmaceutical aid, in waterproofing compositions and as a dietary supplement

Physical Properties

Boiling Point: Very High
Specific Gravity: 0.915 to 0.921 at 25 deg/25 °C
Vapor Pressure: Reid 0.1 psia
Water Solubility: < 1 mg/mL at 19 C
Other Solubilities: 95% Ethanol: <1 mg/ml at 19 °C; Acetone: >=100 mg/ml at 19 °C; Benzene: Soluble; Carbon Disulfide: miscible; Chloroform: miscible; DMSO: <1 mg/ml at 19 °C; Ether: miscible; Light petroleum: miscible; Petroleum Ether: miscible.
Surface Tension: 35.4 dynes/cm at 20 °C
Refraction Index: 1.4645 to 1.4655 at 40 °C/D
Flash Point: Cottonseed 252 °C Open Cup
Autoignition Temperature: Refined oil 343 °C

RTECS Toxicity Data

Acute Oral: Rat LD_{50} Dose: >90 mL/kg.

Reproductive/Teratogenic: Rat Route: Intraperitoneal; Dose: 2256 mg/kg; Duration: female 5-15D of pregnancy; Effects on Embryo or Fetus - Fetotoxicity; Specific Developmental Abnormalities - Musculoskeletal system. Rat Route: Intraperitoneal; Dose: 10 gm/kg; Duration: female 5-15D of pregnancy; Specific Developmental Abnormalities - Musculoskeletal system.

Tumorigenic: Mouse Route: Oral; Dose: 2940 gm/kg/35W-C; Toxic Effects: Tumorigenic - Equivocal tumorigenic agent by RTECS criteria; Skin and appendages - Tumors.

Hazard Overviews

Fire Diamond

Health: Cottonseed oil is relatively nonhazardous in routine industrial situations.

Fire: Will burn. Use water fog, dry chemical, alcohol foam, or carbon dioxide to fight fires involving cottonseed oil. Use a water spray to cool tanks or containers that have been exposed to fire.

Reactivity: Stable. Hazardous polymerization cannot occur. Avoid: exposure to heat; sparks; open flame; lighted tobacco products. Incompatible with: strong oxidizers. Hazardous decomposition products: carbon monoxide; carbon dioxide.

Carcinogenicity: IARC - Not listed; NIOSH - Not listed; NTP - Not listed; ACGIH - Not listed; OSHA - Not listed; EPA - Not listed; MAK - Not listed

Environmental

Cleanup/Disposal: Guide No. 171: Do not touch or walk through spilled material. Stop leak if you can do it without risk. Prevent dust cloud. Avoid inhalation of asbestos dust. Small Dry Spills: With clean shovel place material into clean, dry container and cover loosely; move containers from spill area. Small Spills: Take up with sand or other noncombustible absorbent material and place into containers for later disposal. Large Spills: Dike far ahead of liquid spill for later disposal. Cover powder spill with plastic sheet or tarp to minimize spreading. Prevent entry into waterways, sewers, basements or confined areas.

Regulations

RCRA 40CFR: Not listed
CERCLA: 40CFR 302.4: Not listed
SARA 40CFR 372.65: Not listed
SARA EHS 40CFR 355: Not listed

TSCA: Listed

Analytical Methods

Water / Groundwater: ASTM D4763

COU1000 CAS #: 56-72-4

COUMAPHOS

RTECS: GN6300000
DOT: UN2783
EINECS Number: 200-285-3
Molecular Formula: $C_{14}H_{16}ClO_5PS$
Formula Weight: 362.78
Synonyms: AGRIDIP; ASANTOL; ASUNTHOL; ASUNTOL; AZUNTHOL; BAY 21/199; BAYER 21/199; BAYMIX; BAYMIX 50; 3-CHLORO-7-DIETHOXYPHOSPHINOTHIOYLOXY-4-METHYLCOUMARIN; 3-CHLORO-7-HYDROXY-4-METHYL-COUMARIN O,O-DIETHYL PHOSPHOROTHIOATE; 3-CHLORO-7-HYDROXY-4-METHYL-COUMARIN O,O-DIETHYLPHOSPHOROTHIOATE; 3-CHLORO-7-HYDROXY-4-METHYL-COUMARIN O-ESTER WITH O,O-DIETHYL PHOSPHOROTHIOATE; 3-CHLORO-7-HYDROXY-4-METHYL-COUMARIN O-ESTER WITHO,O-DIETHYL PHOSPHOROTHIOATE; 3-CHLORO-7-HYDROXY-4-METHYLCOUMARIN O-ESTER WITHO,O-DIETHYL PHOSPHOROTHIOATE; 3-CHLORO-7-HYDROXY-4-METHYL-COUMARIN,O-ESTER WITH O,O-DIETHYLPHOSPHOROTHIOA; 3-CHLORO-4-METHYL-7-COUMARINYL DIETHYL PHOSPHOROTHIOATE; O-3-CHLORO-4-METHYL-7-COUMARINYL O,O-DIETHYLPHOSPHOROTHIOATE; O-3-CHLORO-4-METHYLCOUMARIN-7-YL O,O-DIETHYLPHOSPHOROTHIOATE; 3-CHLORO-4-METHYL-7-HYDROXYCOUMARIN DIETHYL THIOPHOSPHORIC ACID ESTER; 3-CHLORO-4-METHYL-7-HYDROXYCOUMARIN DIETHYL THIOPHOSPHORICACID ESTER; 3-CHLORO-4-METHYL-7-HYDROXYCOUMARIN DIETHYLTHIOPHOSPHORIC ACID ESTER; O-(3-CHLORO-4-METHYL-2-OXO-2H-1-BENZOPYRAN-7-YL)O,O-DIETHYL PHOSPHOROTHIOATE; O-3-CHLORO-4-METHYL-2-OXO-2H-CHROMEN-7-YL-O,O-DIETHYLPHOSPHOROTHIOATE; 3-CHLORO-4-METHYLUMBELLIFERONE O-ESTER WITH O,O-DIETHYL PHOSPHOROTHIOATE; 3-CHLORO-4-METHYLUMBELLIFERONE O-ESTER WITH O,O-DIETHYLPHOSPHOROTHIOATE; CO-RAL; CORAL; COUMAFOS; COUMARIN,3-CHLORO-7-HYDROXY-4-METHYL-,O-ESTER WITHO,O-DIETHYL PHOSPHOROTHIOATE; CUMAFOS; O,O-DIAETHYL-O-(3-CHLOR-4-METHYL-CUMARIN-7-YL)-MONOTHIOPHOSPHAT; O,O-DIETHYL O-(3-CHLORO-4-METHYL-7-COUMARINYL)PHOSPHOROTHIOATE; O,O-DIETHYL O-(3-CHLORO-4-METHYLCOUMARINYL-7)THIOPHOSPHATE; O,O-DIETHYL O-(3-CHLORO-4-METHYL-2-OXO-2H-BENZOPYRAN-7-YL)PHOSPHOROTHIOATE; O,O-DIETHYL 3-CHLORO-4-METHYL-7-UMBELLIFERONETHIOPHOSPHATE; O,O-DIETHYL O-(3-CHLORO-4-METHYLUMBELLIFERONE)THIOPHOSPHATE; DIETHYL 3-CHLORO-4-METHYLUMBELLIFERYL THIONOPHOSPHATE; O,O-DIETHYL O-(3-CHLORO-4-METHYLUMBELLIFERYL)PHOSPHOROTHIOATE; DIETHYL THIOPHOSPHORIC ACID ESTER OF 3-CHLORO-4-METHYL-7-HYDROXYCOUMARIN; DIETHYL THIOPHOSPHORIC ACID ESTER OF3-CHLORO-4-METHYL-7-HYDROXYCOUMARIN; O,O-DIETHYL-O-(3-CHLOOR-4-METHYL-CUMARIN-7-YL)MONOTHIOFOSFAAT; DIETHYL-O(3-CHLORO-4-METHYL-7-COUMARINYL) PHOSPHOROTHIOATE; O,O-DIETHYLO-(3-CHLORO-4-METHYL-7-COUMARINYL)PHOSPHOROTHIOATE; O,O-DIETHYLO-(3-CHLORO-4-METHYLUMBELLIFERYL)PHOSPHOROTHIOATE; O,O-DIETIL-O-(3-CLORO-4-METIL-CUMARIN-7-IL-MONOTIOFOSFATO); DIOLICE; ENT 17,957; EPA PESTICIDE CODE 026501; MELDANE; MELDONE; MUSCATOX; NEGASHUNT; NEGASUNT; OMS 485; PHOSPHOROTHIOIC ACID,O-(3-CHLORO-4-METHYL-2-OXO-2H-1-BENZOPYRAN-7-YL)O,O-DIETHYL ESTER; PHOSPHOROTHIOIC ACID,O,O-DIETHYL ESTER,O-ESTER WITH 3-CHLORO-7-HYDROXY-4-METHYLCOUMARIN; PHOSPHOROTHIOIC ACID,O,O-DIETHYL ESTER,O-ESTER WITH3-CHLORO-7-HYDROXY-4-METHYLCOUMARIN; RESISTOX; RESITOX; SUNTOL; THIOPHOSPHATE DE O,O-DIETHYLE ET DE O-(3-CHLORO-4-METHYL-7-COUMARINYLE); THIOPHOSPHATE DE O,O-DIETHYLE ET DEO-(3-CHLORO-4-METHYL-7-COUMARINYLE); UMBETHION
Description: tan crystalline solid; slight sulfur-like odor
Use: insecticide, acaricide, e.g. for houseflies and cattle flies; nematocide and antihelmintic for animal use

Physical Properties

Boiling Point: 20 °C (68 °F)
Freezing Point: 91 °C (195.8 °F)
Specific Gravity: 1.47 at 20 °C/4 °C
Saturated Vapor Density: 1.200000002 kg/m^3
Vapor Pressure: 1 $x10^{-7}$ mm at 20 °C
Water Solubility: 1.5 ppm at 25 °C
Other Solubilities: 95% Ethanol: <1 mg/ml at 22 °C; Acetone: >=100 mg/ml at 22 °C; Absolute Ethanol: Soluble; Aromatic solvents: Soluble; Chloroform: Soluble; Corn oil: Soluble; DMSO: 50-100 mg/ml at 22 °C.
Odor Threshold: 2.0 $x10^{-2}$ ppm
Flash Point: Not pertinent (combustible solid)

RTECS Toxicity Data

Acute Oral: Rat LD_{50} Dose: 13 mg/kg. Mouse LD_{50} Dose: 28 mg/kg; Toxic Effects: Biochemical - True cholinesterase.
Acute Inhalation: Rat LC_{50} Dose: 303 mg/m^3.
Acute Dermal: Rabbit LD_{50} Route: Skin; Dose: 500 mg/kg. Rat LD_{50} Route: Skin; Dose: 860 mg/kg.
Chronic (Multiple Dose) Oral: Rat Dose: 168 mg/kg/28W-I; Toxic Effects: Biochemical - True cholinesterase.
Mutagenic: Rat Morphological Transformation; Cell Type: embryo; Dose: 1400 ng/plate.

Hazard Overviews

Fire: Combustible.
Carcinogenicity: IARC - Not listed; NIOSH - Not listed; NTP - Not listed; ACGIH - Not listed; OSHA - Not listed; EPA - Not listed; MAK - Not listed

Environmental

Ecotoxicity: LD_{50} Mallards, male age 3-4 mo, oral 29.8 mg/kg (95% confidence limit 21.5-41.3 mg/kg) LC_{50} Lepomis macrochirus (bluegill) 340 ug/l/96 hr at 18 °C, wt 1.3 g, static lab bioassay LC_{50} Gammarus fasciatus (scuds), mature 0.074 ug/l/96 hr at 21 °C (95% confidence limit 0.059-0.092 ug/l), static lab bioassay TLm Crassostrea virginica (eastern oyster) eggs 110 ppb/48 hr static lab bioassay TLm Gasterosteus aculeatus (threespine stickleback) 1470 ppb/96 hr static lab LC_{50} Daphnia magna 1.0 ug/l/48 hr /Conditions of bioassay not specified
Environmental Fate: If spilled on land, it would be expected to adsorb strongly to the soil and very slowly biodegrade (half-life 200 - 300 days). If released as wastewater, it should

adsorb strongly to sediment and particulate matter in the water column. It has been reported to disappear from pond water (pH 5.5) in less than 7 days although the processes contributing to this disappearance are not clear. There is a moderate potential for bioconcentration in aquatic organisms. In the atmosphere, it would most likely be in the form of an aerosol. It will be subject to gravitational settling and may photolyze.

Cleanup/Disposal: Guide No. 152: Do not touch damaged containers or spilled material unless wearing appropriate protective clothing. Stop leak if you can do it without risk. Prevent entry into waterways, sewers, basements or confined areas. Cover with plastic sheet to prevent spreading. Absorb or cover with dry earth, sand or other non-combustible material and transfer to containers. Do not get water inside containers.

Environmental Physical Data

Henry's Law Constant: calculated at 3.2 $x10^{-8}$

Octanol/Water Partition Coefficient: log K_{ow} = 4.13

Sorption Partition Coefficient: K_{oc} = 4230

BCF: aquatic organisms 646

Regulations

RCRA 40CFR: Not listed

CERCLA: 40CFR 302.4: Listed per CWA Section 311(b)(4) RQ: 10 lb (4.535 kg)

SARA 40CFR 372.65: Not listed TPQ: 100/10000 lb

SARA EHS 40CFR 355: Listed TPQ: 10 lb

TSCA: Not listed

Analytical Methods

Soil: EPA PMD-COR, PMD-TLC; SW846 8140, 8141, 8141A, 8270B, 8270C

Water / Groundwater: EPA P-005-1, 1657, 622; SW846 8141, 8141A; ASTM D4763

Food: FDA 212.1, 232.1, 232.3, 232.4, 242.1

Plasma: EPA 001, 027, 028, 29; FDA 211.1, 231.1, 252

COU9000 CAS #: 5836-29-3

COUMATETRALYL

RTECS: GN7630000

DOT: UN3024; UN3025; UN3026; UN3027; IMO3.0; IMO6.1

EINECS Number: 227-424-0

Molecular Formula: $C_{19}H_{16}O_3$

Formula Weight: 292.35

Synonyms: BAY 25634; BAY ENE 11183 B; BAYER 25 634; 2H-1-BENZOPYRAN-2-ONE,4-HYDROXY-3-(1,2,3,4-TETRAHYDRO-1-NAPHTHALENYL)-; 2H-1-BENZOPYRAN-2-ONE,4-HYDROXY-3-(1,2,3,4-TETRAHYDRO-1-NAPHTHALENYL)-(9CI); COUMARIN,4-HYDROXY-3-(1,2,3,4-TETRAHYDRO-1-NAPHTHYL)-; CUMATETRALYL; ENDOX; ENDROCID; ENDROCIDE; ENE 11183 B; 4-HYDROXY-3-(1,2,3,4-TETRAHYDRO-1-NAFTYL)-CUMARINE; 4-HYDROXY-3-(1,2,3,4-TETRAHYDRO-1-NAPHTHALENYL)-2H-1-BENZOPYRAN-2-ONE; 4-HYDROXY-3-(1,2,3,4-TETRAHYDRO-1-NAPHTHYL)COUMARIN; 4-HYDROXY-3-(1,2,3,4-TETRAHYDRO-1-NAPHTHYL)-CUMARIN; 4-IDROSSI-3-(1,2,3,4-TETRAIDRO-1-NAFTIL)-CUMARINA; RACUMIN; RAUCUMIN 57; RODENTIN; 3-(1,2,3,4-TETRAHYDRO-1-NAPHTHYL)-4-HYDROXYCUMARIN; 3-(1,2,3,4-TETRAHYDRO-1-NAPHTYL)-4-HYDROXYCOUMARINE; 3-(ALPHA-TETRAL)-4-OXYCOUMARIN; 3-(ALPHA-TETRALYL)-4-HYDROXYCOUMARIN; 3-(D-TETRALYL)-4-HYDROXYCOUMARIN

Description: colorless powder; odorless

Use: anticoagulant rodenticide; rat killers, effective against brown, and black rats, mouse, field mouse, field vole, and hamster; fumigants for moles

Physical Properties

Freezing Point: 172 °C (341.6 °F) to 176 °C (348.8 °F)

Water Solubility: 4 mg/L at 20 °C

Other Solubilities: Soluble in dilute alkali & most organic solvents; Practically Insoluble in Benzene, moderately Soluble in Alcohols, and readily Soluble Acetone and dioxane.

RTECS Toxicity Data

Acute Oral: Rat LD_{50} Dose: 30 mg/kg. Rabbit LD_{50} Dose: 10 mg/kg.

Acute Dermal: Rat LD_{50} Route: Skin; Dose: 40 mg/kg.

Hazard Overviews

Carcinogenicity: IARC - Not listed; NIOSH - Not listed; NTP - Not listed; ACGIH - Not listed; OSHA - Not listed; EPA - Not listed; MAK - Not listed

Environmental

Ecotoxicity: LC_{50} Fish about 1000 mg/l/96 hr /Conditions of bioassay not specified

Cleanup/Disposal: Guide No. 131: Fully encapsulating, vapor protective clothing should be worn for spills and leaks with no fire. Eliminate all ignition sources (no smoking, flares, sparks or flames in immediate area). All equipment used when handling the product must be grounded. Do not touch or walk through spilled material. Stop leak if you can do it without risk. Prevent entry into waterways, sewers, basements or confined areas. A vapor suppressing foam may be used to reduce vapors. Small Spills: Absorb with earth, sand or other non-combustible material and transfer to containers for later disposal. Use clean non-sparking tools to collect absorbed material. Large Spills: Dike far ahead of liquid spill for later disposal. Water spray may reduce vapor; but may not prevent ignition in closed spaces. Guide No. 151: Do not touch damaged containers or spilled material unless wearing appropriate protective clothing. Stop leak if you can do it without risk. Prevent entry into waterways, sewers, basements or confined areas. Cover with plastic sheet to prevent spreading. Absorb or cover with dry earth, sand or other non-combustible material and transfer to containers. Do not get water inside containers.

Regulations

RCRA 40CFR: Not listed

CERCLA: 40CFR 302.4: Not listed

SARA 40CFR 372.65: Not listed TPQ: 500/10000 lb

SARA EHS 40CFR 355: Listed TPQ: 500 lb

TSCA: Not listed

CRE2140 **CAS #: 120-71-8**

P-CRESIDINE

RTECS: BZ6720000
DOT: UN2431; IMO6.1
EINECS Number: 204-419-1
Molecular Formula: $C_8H_{11}NO$
Structured MF: $CH_3C_6H_3(NH_2)OCH_3$
Formula Weight: 137.2

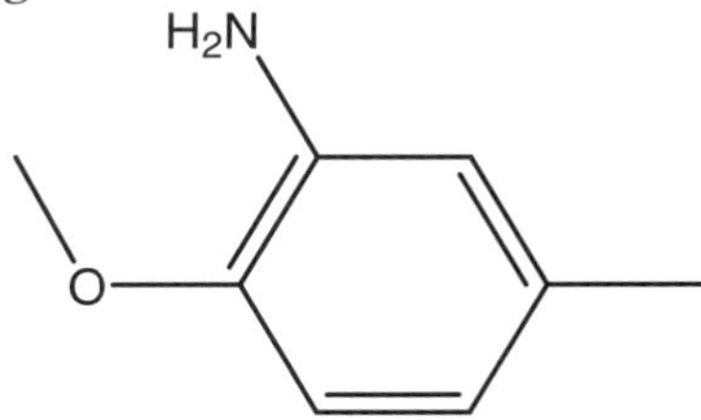

Chemical Structure

Synonyms: 3-AMINO-P-CRESOL METHYL ETHER; M-AMINO-P-CRESOL,METHYL ESTER; 3-AMINO-PARA-CRESOL,METHYL ETHER; META-AMINO-PARA-CRESOL,METHYL ETHER; 1-AMINO-2-METHOXY-5-METHYLBENZENE; 3-AMINO-4-METHOXYTOLUENE; 2-AMINO-4-METHYLANISOLE; O-ANISIDINE,5-METHYL-; ORTHO-ANISIDINE,5-METHYL-; O-ANISIDINE,5-METHYL-(8CI); AZOIC RED 36; BENZENAMINE,2,METHOXY-5-METHYL-O; BENZENAMINE,2-METHOXY-5-METHYL-; BENZENAMINE,2-METHOXY-5-METHYL-(9CI); C.I. AZOIC RED 83; CRESIDINE; P-KRESIDIN; KREZIDIN; KREZIDINE; 2-METHOXY-5-METHYLANILINE; 2-METHOXY-5-METHYL-BENZENAMINE; 2-METHOXY-5-METHYLBENZENAMINE; 4-METHOXY-M-TOLUIDINE; 4-METHOXY-META-TOLUIDINE; 4-METHYL-2-AMINOANISOLE; 5-METHYL-O-ANISIDINE

Description: white crystals; odorless

Use: an intermediate in the production of azo dyes and pigments

Physical Properties

Boiling Point: 235 °C (455 °F)
Freezing Point: 51.5 °C (124.7 °F)
Specific Gravity: 1
Vapor Density: Calculated 4.7 Air=1
Water Solubility: < 1 mg/mL at 22 C
Other Solubilities: 95% Ethanol: >=100 mg/ml at 22 °C; Acetone: >=100 mg/ml at 22 °C; Absolute Ethanol: Soluble; Benzene: >10%; DMSO: >=100 mg/ml at 22 °C; Ether: >10%; Hot Petroleum Ether: Soluble; Organic solvents: Soluble.
Flash Point: > 110 °C

RTECS Toxicity Data

Acute Oral: Rat LD_{50} Dose: 1450 mg/kg.
Irritation Eye: Rabbit Standard Draize Test Dose: 100 mg/24H; Reaction: moderate.
Irritation Skin: Rabbit Standard Draize Test Dose: 500 mg/24H; Reaction: mild.
Mutagenic: Rat Morphological Transformation; Cell Type: embryo; Dose: 31 ug/plate. Bacteria - E Coli Mutations in Microorganisms; Dose: 2 mg/plate (+S9). Bacteria - S Typhimurium Mutations in Microorganisms; Dose: 62500 ng (+S9), 3330 ng/plate (-S9).
Tumorigenic: Rat Route: Oral; Dose: 364 gm/kg/2Y-C; Toxic Effects: Tumorigenic - Carcinogenic by RTECS criteria; Brain and coverings - Tumors; Kidney, Ureter, and Bladder - Tumors. Rat Route: Oral; Dose: 182 gm/kg/2Y-C; Toxic Effects: Tumorigenic - Neoplastic by RTECS criteria; Brain and coverings - Tumors; Kidney, Ureter, and Bladder - Tumors. Rat Route: Oral; Dose: 3640 gm/kg/2Y-C; Toxic Effects: Tumorigenic - Carcinogenic by RTECS criteria; Liver - Tumors; Kidney, Ureter, and Bladder - Tumors. Rat Route: Oral; Dose: 7280 gm/kg/2Y-C; Toxic Effects: Tumorigenic - Carcinogenic by RTECS criteria; Liver - Tumors; Kidney, Ureter, and Bladder - Tumors. Rat Route: Oral; Dose: 364 gm/kg/2Y-C; Toxic Effects: Tumorigenic - Carcinogenic by RTECS criteria; Sense organs and special senses - Tumors. Rat Route: Oral; Dose: 437 gm/kg/2Y-C; Toxic Effects: Tumorigenic - Carcinogenic by RTECS criteria; Sense organs and special senses - Tumors.

Hazard Overviews

Health: Irritating to eyes/skin/respiratory tract. Toxic. Other Acute Effects harmful if swallowed, inhaled, or absorbed through skin; absorption into the body leads to the formation of methemoglobin which in sufficient concentration causes cyanosis; onset may be delayed 2 to 4 hours or longer. Chronic Effects: may alter genetic material; may cause heritable genetic damage. Carcinogen.
Fire: Will burn. Hazards: emits toxic fumes. Extinguishing agents: water spray; carbon dioxide, dry chemical powder or appropriate foam. Precautions: combustible liquid.
Reactivity: Incompatible with: strong oxidizing agents. Hazardous decomposition products: toxic fumes of: carbon monoxide, carbon dioxide, nitrogen oxides.
Carcinogenicity: IARC - Group 2B, Possibly carcinogenic to humans; NIOSH - Not listed; NTP - Class 2B, Reasonably anticipated to be a carcinogen, sufficient evidence of carcinogenicity from studies in experimental animals; ACGIH - Not listed; OSHA - Not listed; EPA - Not listed; MAK - Class A2, Unmistakably carcinogenic in animal experimentation only
Primary Target Organs:

Eyes

Skin

Respiratory System

Environmental

Environmental Fate: If released to soil, it may slowly volatilize from both moist and dry soil to the atmosphere. It may display high mobility in soil; however, the amino group may bind covalently with active sites in soil greatly limiting its mobility. Based on limited data, it is likely to only slowly biodegrade in soil. If released to water, it may slowly volatilize to the atmosphere. The estimated half-lives for volatilization from a model river range from 23-346 days. By analogy to other aromatic amines, it is expected to exist predominately in the ionized form in water which will attenuate the rate of this process. It is not expected to bioconcentrate in fish and aquatic organisms. Based on limited data, it is expected to only slowly biodegrade in water. If released to the atmosphere, it is expected to undergo a rapid gas-phase reaction with photochemically produced hydroxyl radicals with an estimated half-life of 1.8 hours.

Cleanup/Disposal: Guide No. 153: Eliminate all ignition sources (no smoking, flares, sparks or flames in immediate area). Do not touch damaged containers or spilled material unless wearing appropriate protective clothing. Stop leak if you can do it without risk. Prevent entry into waterways, sewers, basements or confined areas. Absorb or cover with dry earth, sand or other non-combustible material and transfer to containers. Do not get water inside containers.

Environmental Physical Data

Henry's Law Constant: estimated at 3.9×10^{-7}
Octanol/Water Partition Coefficient: log K_{ow} = 1.67
Sorption Partition Coefficient: K_{oc} = 42 to 192
BCF: calculated at 100

Regulations

RCRA 40CFR: Not listed
CERCLA: 40CFR 302.4: Not listed
SARA 40CFR 372.65: Listed
SARA EHS 40CFR 355: Not listed
TSCA: Listed

Analytical Methods

Soil: SW846 8270C
Plasma: EPA 29

CRE3280 CAS #: 108-39-4

M-CRESOL

RTECS: GO6125000
DOT: UN2076; IMO6.1
EINECS Number: 203-577-9
Molecular Formula: C_7H_8O
Structured MF: $CH_3C_6H_4OH$
Formula Weight: 108.15

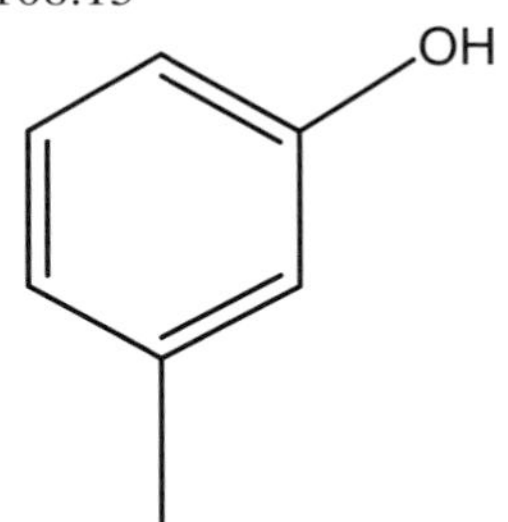

Chemical Structure

Synonyms: BACTICIN; CELCURE DRY MIX (CHEMICALS FOR WOOD PRESERVING); 3-CRESOL; META-CRESOL; M-CRESOLE; M-CRESYLIC ACID; META-CRESYLIC ACID; FRANKLIN CRESOLIS; GALLEX; 1-HYDROXY-3-METHYLBENZENE; 3-HYDROXYTOLUENE; M-HYDROXYTOLUENE; M-KRESOL; 3-METHYL PHENOL; 3-METHYLPHENOL; M-METHYLPHENOL; M-OXYTOLUENE; PHENOL,3-METHYL-; PHENOL,3-METHYL-(9CI); ROVER'S DOG SHAMPOO; M-TOLUOL

Description: colorless or yellowish liquid, solid below 54 deg F; sweet, tarry odor

Use: in disinfectants, in resins, as a raw material for photographic developers, in ore flotation, in fumigation compounds, in explosives, in phenol, as an insecticide, as a wood preservative, in degreasing compounds, in paintbrush cleaners and as an additive to lubricating oils; as an intermediate in the manufacture of chemicals, dyes, plastics and antioxidants; in the manufacture of antiseptics, phosphate esters, herbicides and perfumes, as a solvent, as an engine and metal cleaner and in the textile industry

Physical Properties

Boiling Point: 202 °C (396 °F)
Freezing Point: 11 °C (51.8 °F) to 12 °C (53.6 °F)
Specific Gravity: 1.034 at 20 °C/4 °C
Vapor Density: 3.72 Air=1
Saturated Vapor Density: 1.200172377 kg/m^3
Density: 1.0336 g/mL at 20 °C
Vapor Pressure: 0.04 mm Hg at 20 °C
Water Solubility: 2% by weight
Other Solubilities: Soluble in Carbon Tetrachloride, Acetone, Benzene, other organic solvents; 2.5% in Mineral Oil.
Surface Tension: 41.7 dynes/cm at 20 °C
Odor Threshold: 0.0028 ul/l
pH: 5.5
Refraction Index: 1.5398 at 20 °C/D
Evaporation Rate: 0.015 Butyl Acetate=1
Critical Temperature: 432 °C
Critical Pressure: 45.0 atm
Ionization Potential (eV): 8.98
Flash Point: 86 °C Closed Cup
Autoignition Temperature: 558 °C
LEL: 1.1% v/v

RTECS Toxicity Data

Acute Oral: Rat LD_{50} Dose: 242 mg/kg; Toxic Effects: Behavioral - Somnolence (general depressed activity); Behavioral - Convulsions or effect on seizure threshold; Gastrointestinal - Peritonitis. Mouse LD_{50} Dose: 828 mg/kg.

Acute Inhalation: Rat LC_{50} Dose: >710 mg/m^3/1hr.

Acute Dermal: Rabbit LD_{50} Route: Skin; Dose: 2050 mg/kg; Toxic Effects: Sense organs and special senses - Lacrimation; Behavioral - Convulsions or effect on seizure threshold; Gastrointestinal - Changes in structure or function of salivary glands. Rabbit LD_{Lo} Route: Subcutaneous Dose: 500 mg/kg.

Chronic (Multiple Dose) Oral: Rat Dose: 42 gm/kg/28D-C; Toxic Effects: Brain and coverings - Changes in brain weight; Liver - Changes in liver weight; Kidney, Ureter, and Bladder - Changes in kidney weight. Mouse Dose: 101 gm/kg/28D-C; Toxic Effects: Brain and coverings - Changes in brain weight; Kidney, Ureter, and Bladder - Changes in kidney weight; Nutritional and gross metabolic - Weight loss or decreased weight gain.

Chronic (Multiple Dose) Inhalation: Rabbit Dose: 2 mg/m^3/24H/6W-I; Toxic Effects: Kidney, Ureter, and Bladder - Other changes; Biochemical - Other hydrolases; Biochemical - Other enzymes.
Irritation Eye: Rabbit Standard Draize Test Dose: 103 mg; Reaction: severe.
Irritation Skin: Rabbit Standard Draize Test Dose: 517 mg/24H; Reaction: severe.
Reproductive/Teratogenic: Rabbit Route: Subcutaneous; Dose: 134 gm/kg; Duration: female 6-18D of pregnancy; Effects on Embryo or Fetus - Fetotoxicity.
Mutagenic: Human DNA Inhibition; Cell Type: HeLa cell; Dose: 10 umol/L/4H.
Tumorigenic: Mouse Route: Skin; Dose: 2280 mg/kg/20W-I; Toxic Effects: Tumorigenic - Neoplastic by RTECS criteria; Skin and appendages - Tumors.

Hazard Overviews

Poison

Corrosive

Explosive

Health: Corrosive to eyes/skin/respiratory tract. Poison. Other Acute Effects: may be fatal if inhaled, swallowed, or absorbed through skin; inhalation may result in spasm, inflammation and edema of the larynx and bronchi, chemical pneumonitis and pulmonary edema; symptoms of exposure may include burning sensation, coughing, wheezing, laryngitis, shortness of breath, headache, nausea and vomiting; exposure can cause damage to the kidneys; target organs: central nervous system, lungs ,liver, kidneys.
Fire: Combustible. Extinguishing agents: water spray; carbon dioxide, dry chemical powder or appropriate foam. Precautions: combustible liquid.
Reactivity: Incompatible with: oxidizing agents, bases. Hazardous decomposition products: carbon monoxide, carbon dioxide.
Carcinogenicity: IARC - Not listed; NIOSH - Not listed; NTP - Not listed; ACGIH - Not listed; OSHA - Not listed; EPA - Class C, Possible human carcinogen; MAK - Not listed
Primary Target Organs:

Eyes

Skin

Respiratory System

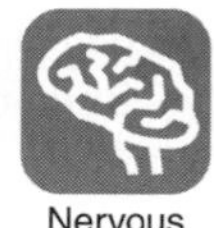
Nervous System

Liver

Kidneys

Exposure Limits
OSHA PEL: TWA: 5 ppm; 22 mg/m^3.
ACGIH TLV: TWA: 5 ppm; 22 mg/m^3.
NIOSH REL: TWA: 2.3 ppm; 10 mg/m^3.
NIOSH IDLH: 250 ppm.

Environmental

Ecotoxicity: LC_{100} Tetrahymena pyriformis (ciliate) 3.5 mmole/l/24 hr /Conditions of bioassay not specified TLm Crucian carp 25 mg/l/24 hr /Conditions of bioassay not specified Toxicity threshold (cell multiplication inhibition test): Scenedesmus quadricauda 15 mg/l TLm Roach 23 mg/l/24 hr TLm Trout embryos 7 mg/l/24 hr /Conditions of bioassay not specified Toxicity threshold (cell multiplication inhibition test): Microcystis aeruginosa 13 mg/l
Environmental Fate: When released to the atmosphere it will react with photochemically produced hydroxyl radicals during the day (half-life 8 hr) and react with nitrate radicals at night (half-life 5 min). It will also be scavenged by rain. Biodegradation will generally be the dominant loss mechanism when it is released into water, and half-lives in most surface waters would range from hours to days. Longer half-lives would occur in oligotrophic lakes and marine waters. Volatilization, bioconcentration in fish, and adsorption to sediment will be unimportant and photolysis is only expected to be important in surface waters of oligotrophic lakes. Its fate in soil has not been well characterized. It is relatively mobile in most soils and will biodegrade (100% in eleven days).
Cleanup/Disposal: Guide No. 153: Eliminate all ignition sources (no smoking, flares, sparks or flames in immediate area). Do not touch damaged containers or spilled material unless wearing appropriate protective clothing. Stop leak if you can do it without risk. Prevent entry into waterways, sewers, basements or confined areas. Absorb or cover with dry earth, sand or other non-combustible material and transfer to containers. Do not get water inside containers.

Environmental Physical Data

Henry's Law Constant: 8.7 x10^{-7}
Octanol/Water Partition Coefficient: log K_{ow} = 1.96
Sorption Partition Coefficient: K_{oc} = brookston clay loam soil 35
BCF: fish 20
BOD: 68 g/g, 5 days

Regulations

RCRA 40CFR: Listed Hazardous Waste No. U052 Toxic Waste
CERCLA: 40CFR 302.4: Listed per CWA Section 311(b)(4) per RCRA Section 3001 RQ: 100 lb (45.35 kg)
SARA 40CFR 372.65: Listed
SARA EHS 40CFR 355: Not listed
TSCA: Listed

Analytical Methods

Air: EPA TO-8
Soil: SW846 3640A, 8041, 8270B, 8270C
Water / Groundwater: ASTM D2580
Plasma: EPA 29

CRE4420 CAS #: 95-48-7

O-CRESOL

RTECS: GO6300000
DOT: UN2076; IMO6.1
EINECS Number: 202-423-8
Molecular Formula: C_7H_8O
Structured MF: $CH_3C_6H_4OH$

Formula Weight: 108.15

OH

Chemical Structure

Synonyms: 2-CRESOL; ORTHO-CRESOL; O-CRESYLIC ACID; 1-HYDROXY-2-METHYLBENZENE; 2-HYDROXYTOLUENE; O-HYDROXYTOLUENE; O-KRESOL; 2-METHYL PHENOL; 2-METHYLPHENOL; O-METHYLPHENOL; O-METHYLPHENYLOL; ORTHOCRESOL; O-OXYTOLUENE; PHENOL,2-METHYL-; PHENOL,2-METHYL-(9CI); O-TOLUOL

Description: colorless to white crystalline solid, liquid above 88 deg F; sweet tarry phenolic odor

Use: as a disinfectant, solvent, resins, metal cleaner, food antioxidant, ore flotation, textile scouring agent, organic intermediate surfactant, cresylic acid constituent, additives to lubricating oil and insecticide; in the manufacturing of perfumes, dyes, plastic herbicides, tricresyl phosphate, salicylaldehyde and coumarin

Physical Properties

Boiling Point: 190.95 °C (376 °F) at 760 mm Hg
Freezing Point: 30.9 °C (87.62 °F)
Specific Gravity: 1.047 at 20 °C/4 °C
Vapor Density: 3.72 Air=1
Density: 1.05 g/mL at 20 °C
Vapor Pressure: 1 mm Hg at 38.2 °C
Water Solubility: 1 parts in about 40 parts Water
Other Solubilities: Soluble in Alcohol, Chloroform,Ether, solution of the fixed alkali hydroxides, Carbon Tetrachloride, Acetone, Benzene, organic solvents, vegetable oils at 30 °C.
Surface Tension: 40.3 dynes/cm
Odor Threshold: 5 ppm
Refraction Index: 1.553 at 20 °C/D
Critical Temperature: 424.4 °C
Critical Pressure: 726.0 psia
Ionization Potential (eV): 8.93
Flash Point: 81 °C Closed Cup
Autoignition Temperature: 599 °C
LEL: 1.35% v/v

RTECS Toxicity Data

Acute Oral: Rat LD$_{50}$ Dose: 121 mg/kg; Toxic Effects: Behavioral - Convulsions or effect on seizure threshold; Lungs, Thorax, or Respiration - Dyspnea; Gastrointestinal - Ulceration or bleeding from stomach. Mouse LD$_{50}$ Dose: 344 mg/kg.

Acute Inhalation: Rat LC$_{50}$ Dose: >1220 mg/m^3/1hr; Toxic Effects: Sense organs and special senses - Lacrimation; Behavioral - Somnolence (general depressed activity). Mouse LC$_{50}$ Dose: 179 mg/m^3/2hr.

Acute Dermal: Rabbit LD$_{50}$ Route: Skin; Dose: 890 mg/kg. Rabbit LD$_{Lo}$ Route: Subcutaneous Dose: 450 mg/kg.

Chronic (Multiple Dose) Oral: Rat Dose: 21 gm/kg/28D-C; Toxic Effects: Liver - Changes in liver weight; Kidney, Ureter, and Bladder - Changes in kidney weight. Rat Dose: 82 gm/kg/13W-C; Toxic Effects: Liver - Changes in liver weight; Kidney, Ureter, and Bladder - Changes in kidney weight; Endocrine - Changes in thymus weight.

Irritation Eye: Rabbit Standard Draize Test Dose: 105 mg; Reaction: severe.

Irritation Skin: Rabbit Standard Draize Test Dose: 524 mg/24H; Reaction: severe.

Mutagenic: Human Sister Chromatid Exchenge; Cell Type: fibroblast; Dose: 8 mmol/L.

Tumorigenic: Mouse Route: Skin; Dose: 4800 mg/kg/12W-I; Toxic Effects: Tumorigenic - Neoplastic by RTECS criteria; Skin and appendages - Tumors.

Hazard Overviews

Poison

Corrosive

Explosive

Health: Corrosive to eyes/skin/respiratory tract. Poison. Other Acute Effects: toxic by inhalation, in contact with skin and if swallowed; material is extremely destructive to tissue of the respiratory tract, eyes and skin; readily absorbed through skin; may be fatal if inhaled, swallowed, or absorbed through skin; inhalation may result in spasm, inflammation and edema of the larynx and bronchi, chemical pneumonitis and pulmonary edema; symptoms of exposure may include burning sensation, coughing, wheezing, laryngitis, shortness of breath, headache, nausea and vomiting; can cause CNS depression; exposure can cause: vomiting, diarrhea, headache, dizziness, gastrointestinal disturbances; target organs: central nervous system, lungs, liver, kidneys, pancreas, spleen, cardiovascular system.

Fire: Combustible. Extinguishing agents: water spray; carbon dioxide, dry chemical powder or appropriate foam. Precautions: combustible liquid.

Reactivity: Incompatible with: oxidizing agents, bases, light sensitive, air sensitive. Hazardous decomposition products: carbon monoxide, carbon dioxide.

Carcinogenicity: IARC - Not listed; NIOSH - Not listed; NTP - Not listed; ACGIH - Not listed; OSHA - Not listed; EPA - Class C, Possible human carcinogen; MAK - Not listed

Primary Target Organs:

Eyes

Skin

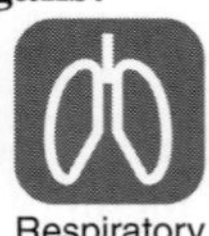
Respiratory System

Nervous System

Liver

Exposure Limits
OSHA PEL: TWA: 5 ppm; 22 mg/m^3.
ACGIH TLV: TWA: 5 ppm; 22 mg/m^3.
NIOSH REL: TWA: 2.3 ppm; 10 mg/m^3.
NIOSH IDLH: 250 ppm.

Environmental

Ecotoxicity: Aquatic toxicity: 49.1-19 ppm/24-96 hr/goldfish/TLm/soft water; 22.2-20.8 ppm/24-96 hr/bluegill/TLm/soft water; 18-13.4 ppm/24-96 hr/fathead minnow/TLm/hard water; 18-50 ppm/24-96 hr/guppy/TLm/hard water ; Waterfowl toxicity: Chronic water fowl toxic limit is 25 ppm

Environmental Fate: When released to the atmosphere it will react with photochemically produced hydroxyl radicals (half-life 9.6 hr) during the day or react with nitrate radicals at night (half-life 2 min). In addition it will be scavenged by rain. When released into water, biodegradation will generally occur within days. However, in surface layers of oligotrophic waters, photolysis may be important. Its fate in soil has not been well characterized; it is mobile and will likely biodegrade, but little evidence is available.

Cleanup/Disposal: Guide No. 153: Eliminate all ignition sources (no smoking, flares, sparks or flames in immediate area). Do not touch damaged containers or spilled material unless wearing appropriate protective clothing. Stop leak if you can do it without risk. Prevent entry into waterways, sewers, basements or confined areas. Absorb or cover with dry earth, sand or other non-combustible material and transfer to containers. Do not get water inside containers.

Environmental Physical Data

Henry's Law Constant: 1.6×10^{-6}
Octanol/Water Partition Coefficient: log K_{ow} = 1.95
Sorption Partition Coefficient: K_{oc} = brookston clay loam soil 22
BCF: calculated at 18
BOD: 1.64 lb/lb, 5 days

Regulations

RCRA 40CFR: Listed Hazardous Waste No. U052 Toxic Waste
CERCLA: 40CFR 302.4: Listed per CWA Section 311(b)(4) per RCRA Section 3001 RQ: 100 lb (45.35 kg)
SARA 40CFR 372.65: Listed TPQ: 1000/10000 lb
SARA EHS 40CFR 355: Listed TPQ: 100 lb
TSCA: Listed

Analytical Methods

Air: EPA TO-8, 0020
Soil: CLP LC_SV, MC_SVOA, OHC; EPA 1625; SW846 1311, 3630B, 3640A, 8041, 8250A, 8270B, 8270C, 8410; DOE OH100R
Water / Groundwater: EPA S-001-1, S-002-1, 1625; ASTM D2580, D4763
Plasma: EPA 29

CRE5560 **CAS #: 106-44-5**

P-CRESOL

RTECS: GO6475000
DOT: UN2076; IMO6.1
EINECS Number: 203-398-6

Molecular Formula: C_7H_8O
Structured MF: $CH_3C_6H_4OH$
Formula Weight: 108.13

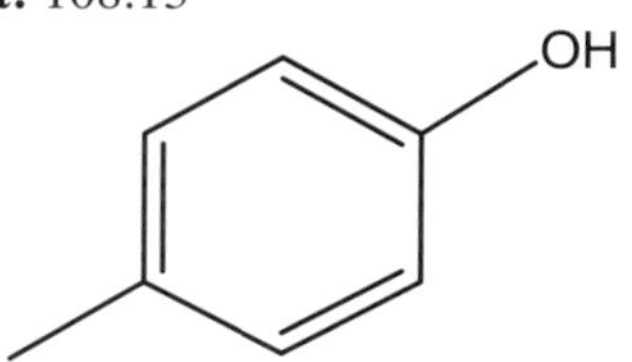

Chemical Structure

Synonyms: 4-CRESOL; PARA-CRESOL; P-CRESYLIC ACID; PARA-CRESYLIC ACID; 1-HYDROXY-4-METHYLBENZENE; 4-HYDROXYTOLUENE; P-HYDROXYTOLUENE; P-KRESOL; 4-METHYL PHENOL; 1-METHYL-4-HYDROXYBENZENE; P-METHYLHYDROXYBENZENE; 4-METHYLPHENOL; P-METHYLPHENOL; P-OXYTOLUENE; PARACRESOL; PARAMETHYL PHENOL; PHENOL,4-METHYL-; PHENOL,4-METHYL-(9CI); P-TOLUOL; P-TOLYL ALCOHOL

Description: colorless; white crystals; prisms or crystalline mass; crystalline solid, liquid above 95 deg F; sweet, tarry odor

Use: in disinfectants, in degreasing compounds, in paintbrush cleaners, as an additive in lubricating oils, in the manufacture of antiseptics, phosphate esters, antioxidants, resins, herbicides, perfumes, explosives and photographic developers, as a solvent, as an engine and metal cleaner and in the textile industry

Physical Properties

Boiling Point: 201.9 °C (395 °F)
Freezing Point: 34.8 °C (94.64 °F)
Specific Gravity: 1.0178 at 20 °C/4 °C
Vapor Density: 3.72 Air=1
Saturated Vapor Density: 1.200172334 kg/m^3
Density: 1.034 g/mL at 20 °C
Vapor Pressure: 0.04 mm Hg at 20 °C
Water Solubility: 2.5 g in 100 ml Water at 50 °C
Other Solubilities: Soluble in aqueous alkali hydroxides; Soluble in organic solvents; Soluble in Alcohol; Soluble in Ether; Soluble in Acetone; Soluble in Benzene; solubility in mineral oil: 0.7%.
Surface Tension: 41.8 dynes/cm at 40 °C
Odor Threshold: 0.2 to 0.46 ppb
pH: 5.5
Refraction Index: 1.5395 at 20 °C/D
Evaporation Rate: 0.011 Butyl Acetate=1
Critical Temperature: 431 °C
Critical Pressure: 746.7 psia
Ionization Potential (eV): 8.97
Flash Point: 86 °C Closed Cup
Autoignition Temperature: 559 °C
LEL: 1.1% v/v

RTECS Toxicity Data

Acute Oral: Rat LD_{50} Dose: 207 mg/kg; Toxic Effects: Sense organs and special senses - Other; Behavioral - Convulsions

or effect on seizure threshold; Gastrointestinal - Ulceration or bleeding from stomach. Mouse LD_{50} Dose: 344 mg/kg.
Acute Inhalation: Rat LC_{50} Dose: >710 mg/m^3/1hr.
Acute Dermal: Rabbit LD_{50} Route: Skin; Dose: 301 mg/kg; Toxic Effects: Behavioral - Tremor; Gastrointestinal - Changes in structure or function of salivary glands; Kidney, Ureter, and Bladder - Other changes. Rabbit LD_{Lo} Route: Subcutaneous Dose: 300 mg/kg.
Chronic (Multiple Dose) Oral: Rat Dose: 21 gm/kg/28D-C; Toxic Effects: Liver - Changes in liver weight; Kidney, Ureter, and Bladder - Changes in kidney weight. Rat Dose: 33600 mg/kg/28D-C; Toxic Effects: Cardiac - Changes in heart weight; Liver - Changes in liver weight; Kidney, Ureter, and Bladder - Changes in kidney weight.
Irritation Eye: Rabbit Standard Draize Test Dose: 103 mg; Reaction: severe.
Irritation Skin: Rabbit Standard Draize Test Dose: 517 mg/24H; Reaction: severe.
Mutagenic: Human DNA Inhibition; Cell Type: lymphocyte; Dose: 25 umol/L. Hamster DNA Inhibition; Cell Type: lung; Dose: 150 umol/L.
Tumorigenic: Mouse Route: Skin; Dose: 2280 mg/kg/20W-I; Toxic Effects: Tumorigenic - Neoplastic by RTECS criteria; Skin and appendages - Tumors.

Hazard Overviews

Corrosive

Fire: Combustible.
Carcinogenicity: IARC - Not listed; NIOSH - Not listed; NTP - Not listed; ACGIH - Not listed; OSHA - Not listed; EPA - Class C, Possible human carcinogen; MAK - Not listed
Exposure Limits
OSHA PEL: TWA: 5 ppm; 22 mg/m^3; skin.
ACGIH TLV: TWA: 5 ppm; 22 mg/m^3.
NIOSH REL: TWA: 2.3 ppm; 10 mg/m^3.
NIOSH IDLH: 250 ppm.

Environmental

Ecotoxicity: LC_{50} Crucian Carp 21 mg/l/24 hr /Conditions of bioassay not specified LC_{50} Roach 17 mg/l/24 hr LC_{100} Tetrahymena pyriformis 3.7 mmole/l/24 hr /Conditions of bioassay not specified LC_{50} Trout embryos 4 mg/l/24 hr /Conditions of bioassay not specified
Environmental Fate: When released to the atmosphere, it will react with photochemically produced hydroxyl radicals during the day (half-life 10 hr) and react with nitrate radicals at night (half-life 4 min). It will also be scavenged by rain. Biodegradation is expected to be the dominant loss mechanism when released into water. Volatilization, bioconcentration in fish, and adsorption to sediment will be unimportant and photolysis is only expected to be significant in surface waters of oligotrophic lakes. Experimental half-lives are only a few hours in eutrophic lakes and ponds but this may be preceded by an acclimation period ranging from hours to days. Half-lives in an oligotrophic lake, marine waters, and in water/sediment ecocores were 6, <4, and <2 days, respectively. Its fate in soil has not been extensively studied; it is mobile and will probably biodegrade.
Cleanup/Disposal: Guide No. 153: Eliminate all ignition sources (no smoking, flares, sparks or flames in immediate area). Do not touch damaged containers or spilled material unless wearing appropriate protective clothing. Stop leak if you can do it without risk. Prevent entry into waterways, sewers, basements or confined areas. Absorb or cover with dry earth, sand or other non-combustible material and transfer to containers. Do not get water inside containers.

Environmental Physical Data

Henry's Law Constant: estimated at 9.6 x10^{-7}
Octanol/Water Partition Coefficient: log K_{ow} = 1.94
Sorption Partition Coefficient: K_{oc} = brookston clay loam soil 49
BCF: calculated at 18
BOD: 1.4 to 1.48 lb/lb, 5 days

Regulations

RCRA 40CFR: Listed Hazardous Waste No. U052 Toxic Waste
CERCLA: 40CFR 302.4: Listed per CWA Section 311(b)(4) per RCRA Section 3001 RQ: 100 lb (45.35 kg)
SARA 40CFR 372.65: Listed
SARA EHS 40CFR 355: Not listed
TSCA: Listed

Analytical Methods

Air: EPA TO-8, 0020
Soil: CLP LC_SV, MC_SVOA, OHC; EPA 1625; SW846 1311, 3630B, 3640A, 8041, 8250A, 8270B, 8270C, 8275, 8410; DOE OH100R
Water / Groundwater: EPA S-001-1, S-002-1, 1625; ASTM D2580, D4763
Plasma: EPA 29
Urine: NIOSH 8305

CRE6700 CAS #: 1319-77-3

CRESOL, ALL ISOMERS

RTECS: GO5950000
DOT: UN2076; IMO6.1
EINECS Number: 215-293-2
Molecular Formula: C_7H_8O
Structured MF: $CH_3C_6H_4OH$
Formula Weight: 108.13

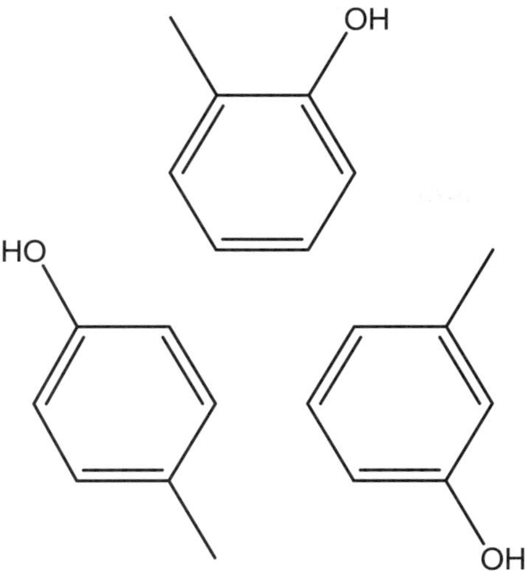

Chemical Structure

Synonyms: ACEDE CRESYLIQUE; ACIDE CRESYLIQUE; BACILLOL; COMPONENT 1 (60%): M-CRESOL; COMPONENT 2 (40%): P-CRESOL; 3-CRESOL; 4-CRESOL; CRESOL; PARA-CRESOL; CRESOLI; CRESOLUM CRUDUM; CRESYLIC ACID; P-CRESYLIC ACID; 1-HYDROXY-4-METHYLBENZENE; HYDROXYMETHYLBENZENE; 4-HYDROXYTOLUENE; P-HYDROXYTOLUENE; HYDROXYTOLUOLE; P-KRESOL; KRESOLE; KRESOLEN; KRESOLUM VENALE; KREZOL; METHYL PHENOL; 1-METHYL-4-HYDROXYBENZENE; P-METHYLHYDROXYBENZENE; 3-METHYLPHENOL; 4-METHYLPHENOL; METHYLPHENOL; P-METHYLPHENOL; P-OXYTOLUENE; PARAMETHYL PHENOL; PHENOL,METHYL-; PHENOL,METHYL-(9CI); TEKRESOL; AR-TOLUENOL; P-TOLUOL; P-TOLYL ALCOHOL; TRICRESOL; TRICRESOLUM; TRIKRESOLUM

Description: colorless, yellow, pinkish liquid; phenolic odor

Use: disinfectants, in solvents, in ore flotation processes, in cleaners, as a motor additive, as an intermediate in the production of phenolic resins and phosphate esters (specifically tricresyl phosphate and cresyl diphenyl phosphate), in fumigants, in photographic developers, in explosives, as a textile scouring agent, as an organic intermediate, in the manufacture of salicylaldehyde, coumarin and herbicides, as a surfactant, in synthetic food flavors, in degreasing compounds, in paint brush cleaners and as an intermediate in the manufacture of chemicals, dyes, plastics and antioxidants

Physical Properties

Boiling Point: 191 °C (376 °F) to 203 °C (397 °F)
Freezing Point: Varies
Specific Gravity: 1.03 to 1.038 at 25 °C/25 °C
Vapor Density: 3.72 Air=1
Density: 1.039 g/cu cm at 17.6 °C
Vapor Pressure: 1 mm Hg at 38-53 °C
Water Solubility: 1 parts in about 50 parts Water
Other Solubilities: miscible with Alcohol, Benzene, Ether, Glycerol, Petroleum Ether; Soluble in solutions of fixed alkali hydroxides; Soluble in vegetable oils; Soluble in Glycol.
Surface Tension: 37 dynes/cm at 20 °C
Odor Threshold: 0.012 to 22.000 mg/m^3
pH: Saturated solutions are neutral or slightly acid
Refraction Index: 1.5353 at 24 °C
Critical Temperature: 432 °C
Critical Pressure: 45 atm
Flash Point: 86 °C
Autoignition Temperature: 593 °C
LEL: 1.1% v/v
UEL: meta or para 1.4% v/v

RTECS Toxicity Data

Acute Oral: Man TD_{Lo} Dose: 177 mg/kg; Toxic Effects: Blood - Other hemolysis with or without anemia; Blood - Other changes. Human LD_{Lo} Dose: 114 mg/kg; Toxic Effects: Vascular - Other changes. Rat LD_{50} Dose: 1454 mg/kg.
Acute Dermal: Rabbit LD_{50} Route: Skin; Dose: 2 gm/kg.

Hazard Overviews

Corrosive

Fire Diamond

Health: Corrosive to eyes/skin/respiratory tract. Also Causes: coma and death, corneal opaqueness; loss of sight; ochronosis, conjunctiva, color change of the tongue (white), thirst, throat swelling, cramps, nausea, vomiting; shock, convulsions, headache, nausea, muscle weakness, pulmonary edema. Chronic Effects: liver, kidney, spleen, pancreatic damage; skin eruptions or dermatitis.

Fire: Combustible. For small fires use dry chemical, carbon dioxide, water spray, or regular foam. For large fires use water spray, fog, or regular foam. Do not scatter material with more water than is necessary to put out fire.

Reactivity: Stable. Hazardous polymerization cannot occur. Avoid: ignition sources. Incompatible with: oxidizers; chlorosulfonic acid; nitric acid; oleum. Hazardous decomposition products: carbon dioxide; toxic cresol fumes.

Carcinogenicity: IARC - Not listed; NIOSH - Not listed; NTP - Not listed; ACGIH - Not listed; OSHA - Not listed; EPA - Not listed; MAK - Not listed

Primary Target Organs:

Eyes

Skin

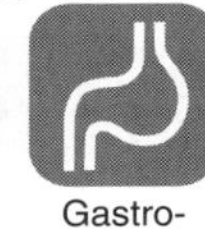
Gastro-intestinal

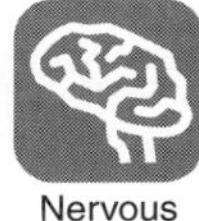
Nervous System

Liver

Kidneys

Exposure Limits
OSHA PEL: TWA: 5 ppm; 22 mg/m^3.
ACGIH TLV: TWA: 5 ppm; 22 mg/m^3.
NIOSH REL: TWA: 2.3 ppm; 10 mg/m^3.
DFG MAK: TWA: 5 ppm; 22 mg/m^3.

Respirator Recommendation
Exposure Range: >5 to 50 ppm Air Purifying, Negative Pressure, Half Mask
Exposure Range: >50 to <250 ppm Air Purifying, Negative Pressure, Full Face
Exposure Range: 250 to unlimited ppm Self-contained Breathing Apparatus, Pressure Demand, Full Face
Cartridge Color: black with dust/mist prefilter (use P100 or consult supervisor for appropriate dust/mist prefilter)

Environmental

Ecotoxicity: LC_{50} Gammarus fasciatus 7.0 mg/l/48 hr (Immature stage) TLm Shrimp 10-100 ppm/48 hr (salt water) /Conditions of bioassay not specified TLm Bluegill 24 mg/l/96 hr (fresh water) /Conditions of bioassay not specified

Environmental Fate: When released to the atmosphere, cresols will degrade by reacting with photochemically produced hydroxyl radicals during the day (half-life 8-10 hr). However, at nighttime reaction with nitrate radicals predominate (half-life 2-5 min). In addition, cresols are soluble compounds and will be scavenged by rain. When released into natural waters, degradation generally occurs within 8 hours after an acclimation period of up to several days. However, in oligotrophic lakes, estuarine, and marine waters the degradation process would be expected to take longer. Volatilization, adsorption, and bioconcentration are not important. Cresols are mobile in soil but biodegradation is probable although data are scant.

Cleanup/Disposal: Guide No. 153: Eliminate all ignition sources (no smoking, flares, sparks or flames in immediate area). Do not touch damaged containers or spilled material unless wearing appropriate protective clothing. Stop leak if you can do it without risk. Prevent entry into waterways, sewers, basements or confined areas. Absorb or cover with dry earth, sand or other non-combustible material and transfer to containers. Do not get water inside containers.

Environmental Physical Data

Henry's Law Constant: calculated at 8.7×10^{-7}

Octanol/Water Partition Coefficient: log K_{ow} = 1.94 to 1.96

Sorption Partition Coefficient: K_{oc} = brookston clay loam soil 22 to 49

BCF: calculated at 18

BOD: meta- 170%, 5 days

Regulations

RCRA 40CFR: Listed Hazardous Waste No. U052 Toxic Waste

CERCLA: 40CFR 302.4: Listed per CWA Section 311(b)(4) per RCRA Section 3001 RQ: 100 lb (45.35 kg)

SARA 40CFR 372.65: Listed

SARA EHS 40CFR 355: Not listed

TSCA: Listed

Analytical Methods

Soil: SW846 3650B, 8040A

Indoor / Expired Air: NIOSH 2546

CRI3000 CAS #: 535-89-7

CRIMIDINE

RTECS: UV8050000

EINECS Number: 208-622-6

Molecular Formula: $C_7H_{10}ClN_3$

Formula Weight: 171.64

Synonyms: CASTRIX; 2-CHLOOR-4-DIMETHYLAMINO-6-METHYL-PYRIMIDINE; 2-CHLOR-4-DIMETHYLAMINO-6-METHYLPYRIMIDIN; 2-CHLORO-4-DIMETHYLAMINO-6-METHYL-PYRIMIDINE; 2-CHLORO-4-METHYL-6-DIMETHYLAMINOPYRIMIDINE; 2-CHLORO-N,N-6-TRIMETHYL-4-PYRIMIDINAMINE; 2-CLORO-4-DIMETILAMINO-6-METIL-PIRIMIDINA; CRIMIDIN; CRIMIDINA; 4-PYRIMIDINAMINE,2-CHLORO-N,N,6-TRIMETHYL-; PYRIMIDINE,2-CHLORO-4-(DIMETHYLAMINO)-6-METHYL-; W 491

Description: colorless crystals

Use: rodenticide

Physical Properties

Boiling Point: 140 °C (284 °F) to 147 °C (297 °F) at 4 mm Hg

Freezing Point: 87 °C (188.6 °F)

Specific Gravity: p-Anisidine 1.07

Saturated Vapor Density: < 1.200000078 kg/m^3

Vapor Pressure: < 1 x10^{-5} mm Hg at 20 °C

Water Solubility: Practically Insoluble in Water

Other Solubilities: Soluble in Alcohol; Soluble in Acetone, Benzene, Chloroform, Diethyl Ether, dilute acids.

RTECS Toxicity Data

Acute Oral: Rat LD_{50} Dose: 1250 ug/kg. Mouse LD_{50} Dose: 1200 ug/kg.

Acute Dermal: Rat LD_{50} Route: Skin; Dose: >1 gm/kg.

Hazard Overviews

Carcinogenicity: IARC - Not listed; NIOSH - Not listed; NTP - Not listed; ACGIH - Not listed; OSHA - Not listed; EPA - Not listed; MAK - Not listed

Environmental

Ecotoxicity: LD_{50} Quail oral about 20 mg/kg LC_{50} Carassius suratus greater than 500 mg/L/(pellets, 0.5%, 96 hr)

Regulations

RCRA 40CFR: Not listed

CERCLA: 40CFR 302.4: Not listed

SARA 40CFR 372.65: Not listed TPQ: 100/10000 lb

SARA EHS 40CFR 355: Listed TPQ: 100 lb

TSCA: Not listed

CRI6000 CAS #: 14464-46-1

CRISTOBALITE

RTECS: VV7325000

EINECS Number: 238-455-4

Molecular Formula: O_2Si

Structured MF: O_2Si

Formula Weight: 60.09

Synonyms: 43-63C; CALCINED DIATOMACEOUS EARTH; CALCINED DIATOMITE; ALPHA-CRYSTOBALITE; CRYSVARL; METACRISTOBALITE; SILICA,CRYSTALLINE-CRISTOBALITE; W 006; WGL 300

Physical Properties

Boiling Point: 2950 °C (5342 °F)

Freezing Point: 1713 °C (3115.4 °F)

Specific Gravity: 2.21

RTECS Toxicity Data

Acute Inhalation: Human TC_{Lo} Dose: 16 mppcf/8H/17.9Y-I; Toxic Effects: Lungs, Thorax, or Respiration - Fibrosis, focal (pneumoconiosis); Lungs, Thorax, or Respiration - Cough; Lungs, Thorax, or Respiration - Dyspnea.

Tumorigenic: Rat Route: Intrapleural; Dose: 90 mg/kg; Toxic Effects: Tumorigenic - Carcinogenic by RTECS criteria; Blood - Lymphomax including Hodgkin's disease. Rat Route: Intrapleural; Dose: 100 mg/kg; Toxic Effects: Tumorigenic - Equivocal tumorigenic agent by RTECS criteria; Blood - Lymphomax including Hodgkin's disease.

Hazard Overviews

Health: Irritating to eyes/skin/respiratory tract. Harmful. Other Acute Effects: harmful if inhaled; may be harmful if swallowed; may be harmful if absorbed through the skin. Chronic Effects: prolonged inhalation may result in silicosis, a disabling pulmonary fibrosis characterized by fibrotic changes and miliary nodules in the lungs; a dry cough; shortness of breath; emphysema; decreased chest expansion; increased susceptibility to tuberculosis; in advanced stages, loss of appetite; pleuritic pain; and total incapacity to work; death due to cardiac failure or destruction of lung tissue; target organ: lungs. Possible carcinogen.

Fire: Extinguishing agents: noncombustible; use extinguishing media appropriate to surrounding fire conditions. Precautions: combustible liquid.

Reactivity: Incompatible with: hydrogen fluoride. Hazardous decomposition products: nature of decomposition products not known;.

Carcinogenicity: IARC - Group 2A, Probably carcinogenic to humans; NIOSH - Listed as carcinogen; NTP - Class 2A, Reasonably anticipated to be a carcinogen, limited evidence of carcinogenicity from studies in humans; ACGIH - Not listed; OSHA - Not listed; EPA - Not listed; MAK - Not listed

Primary Target Organs:

Eyes

Skin

Respiratory System

Exposure Limits

OSHA PEL: TWA: 5 mg/m^3; % respirable SiO_2.

ACGIH TLV: TWA: 0.05 mg/m^3.

DFG MAK: TWA: 0.15 mg/m^3.

Respirator Recommendation

Exposure Range: >0.05 to 0.5 mg/m^3 Air Purifying, Negative Pressure, Half Mask

Exposure Range: >0.5 to 5 mg/m^3 Air Purifying, Negative Pressure, Full Face

Exposure Range: >5 to 50 mg/m^3 Supplied Air, Constant Flow/Pressure Demand, Full Face

Exposure Range: >50 to unlimited mg/m^3 Self-contained Breathing Apparatus, Pressure Demand, Full Face

Cartridge Color: dust/mist filter (use P100 or consult supervisor for appropriate dust/mist filter)

Environmental

Regulations

RCRA 40CFR: Not listed
CERCLA: 40CFR 302.4: Not listed
SARA 40CFR 372.65: Not listed
SARA EHS 40CFR 355: Not listed
TSCA: Listed

Analytical Methods

Indoor / Expired Air: NIOSH 7500, 7601, 7602

CRO3000 CAS #: 123-73-9

(E)-CROTONALDEHYDE

RTECS: GP9625000
DOT: UN1143; IMO3.0
EINECS Number: 204-647-1
Molecular Formula: C_4H_6O
Formula Weight: 70.09

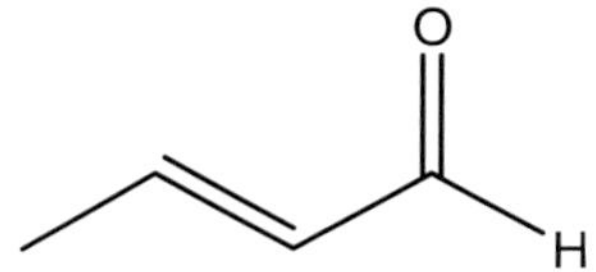

Chemical Structure

Synonyms: ALDEHYDE CROTONIQUE; 2-BUTENAL (TRANS); 2-BUTENAL,(E)-; E-2-BUTENAL; TRANS-2-BUTENAL; CROTENALDEHYDE; CROTONAL; CROTONALDEHYDE; CROTONALDEHYDE,(E)-; TRANS-CROTONALDEHYDE; (E)-CROTONALDEHYDE (IUPAC); 1,2-ETHANEDIOL,DIPROPANOATE (9CI); ETHYLENE DIPROPIONATE; ETHYLENE GLYCOL,DIPROPIONATE (8CI); ETHYLENE PROPIONATE; BETA-METHYL ACROLEIN; TOPANEL; TOPANEL CA

Description: colorless, water-white liquid; pungent suffocating odor

Use: chem int for n-butyl alcohol, quinaldine, crotonic acid, surface-active agents, textile & paper sizes, insecticides & flavoring agents; fuel-gas warning agent; solvent for polyvinyl chloride; alcohol denaturant; leather tanning agent

Physical Properties

Boiling Point: 104 °C (219 °F)
Freezing Point: -74 °C (-101.2 °F)
Specific Gravity: 0.869 at 20 °C/20 °C
Saturated Vapor Density: 1.282580032 kg/m^3
Vapor Pressure: 4.92 kPa at 25 °C
Water Solubility: > 50 g/100ml at 0 °C
Other Solubilities: Soluble in Alcohol, Ether, Acetone, Benzene.
Odor Threshold: 0.1050 to 3.0000 mg/m^3
Refraction Index: 1.4366
Flash Point: 93% Commercial grade 53 °C Open Cup
Autoignition Temperature: 207 °C
LEL: 2.1% v/v
UEL: 15.5% v/v

RTECS Toxicity Data

Acute Oral: Mouse LD_{50} Dose: 240 mg/kg; Toxic Effects: Tumorigenic - Active as anti-cancer agent.
Acute Inhalation: Rat LC_{50} Dose: 4000 mg/m^3/30M; Toxic Effects: Behavioral - Excitment; Behavioral - Convulsions or effect on seizure threshold. Human TC_{Lo} Dose: 12 mg/m^3/10M; Toxic Effects: Sense organs and special senses - Lacrimation; Lungs, Thorax, or Respiration - Other changes.
Acute Dermal: Rabbit LD_{50} Route: Skin; Dose: 380 mg/kg. Rat LD_{50} Route: Subcutaneous Dose: 140 mg/kg; Toxic Effects: Behavioral - Excitment; Behavioral - Convulsions or effect on seizure threshold.
Irritation Skin: Rabbit Open Draize Test Dose: 500 mg open; Reaction: mild.
Mutagenic: Mouse Sperm Morphology; Route: Intraperitoneal; Dose: 30 mg/kg. Mammal DNA Damage; Cell Type: lymphocyte; Dose: 2500 umol/L.

Hazard Overviews

Poison

Corrosive

Explosive

Flammable

Fire Diamond

Health: Corrosive, causes severe burns to eyes/skin/respiratory tract. Also Causes: respiratory tract irritation, watering eyes, delayed pulmonary edema; animal data indicates possible kidney, liver, spleen, blood, thymus damage. Chronic Effects: dermatitis.
Fire: Flammable. May polymerize explosively. Use carbon dioxide, dry chemical, alcohol-resistant foam, and water spray (as fog-solid stream may spread fire). Fight fire from maximum distance.
Reactivity: Unstable, forms crotonic acid and explosive peroxides upon standing. Hazardous polymerization can occur at elevated temperatures or in contact with alkalis. Avoid: heat; ignition sources. Incompatible with: alkalis; ammonia; amines; nitric acid; strong oxidizers; 1,3-butadiene; ethyl acetoacetate; hydrocarbons. Hazardous decomposition products: carbon oxides.
Carcinogenicity: IARC - Not listed; NIOSH - Not listed; NTP - Not listed; ACGIH - Not listed; OSHA - Not listed; EPA - Class C, Possible human carcinogen; MAK - Not listed
Primary Target Organs:

Eyes

Skin

Respiratory System

Exposure Limits
OSHA PEL: TWA: 2 ppm; 6 mg/m^3.
ACGIH TLV: TWA: 2 ppm; 5.7 mg/m^3.

Environmental

Environmental Fate: If released to the atmosphere, it degrades rapidly (typical half-life of 11 hours) via reaction with photochemically produced hydroxyl radicals. If released to water in low concentrations, it can degrade via reaction with photochemically produced oxidants (estimated half-life of 5 days) and volatilize (estimated half-lives of 40 hours from a model river one meter deep and 18.3 days from a pond). If released to soil, it is susceptible to significant leaching. Evaporation from dry surfaces can be expected to occur. If released to the environment in a spill situation, a significant fraction of the spill may polymerize.
Cleanup/Disposal: Guide No. 131: Fully encapsulating, vapor protective clothing should be worn for spills and leaks with no fire. Eliminate all ignition sources (no smoking, flares, sparks or flames in immediate area). All equipment used when handling the product must be grounded. Do not touch or walk through spilled material. Stop leak if you can do it without risk. Prevent entry into waterways, sewers, basements or confined areas. A vapor suppressing foam may be used to reduce vapors. Small Spills: Absorb with earth, sand or other non-combustible material and transfer to containers for later disposal. Use clean non-sparking tools to collect absorbed material. Large Spills: Dike far ahead of liquid spill for later disposal. Water spray may reduce vapor; but may not prevent ignition in closed spaces.

Environmental Physical Data

Henry's Law Constant: 1.94 x10^{-5}
Octanol/Water Partition Coefficient: log K_{ow} = 0.63
Sorption Partition Coefficient: K_{oc} = estimated at 6
BCF: estimated at 0.17
BOD: 37% BODT, 5 days

Regulations

RCRA 40CFR: Listed Hazardous Waste No. U053 Toxic Waste
CERCLA: 40CFR 302.4: Listed per CWA Section 311(b)(4) per RCRA Section 3001 RQ: 100 lb (45.35 kg)
SARA 40CFR 372.65: Not listed TPQ: 1000 lb
SARA EHS 40CFR 355: Listed TPQ: 100 lb
TSCA: Listed

Analytical Methods

Air: EPA 0100
Soil: SW846 5031, 8015B, 8260B, 8315, 8315A
Drinking Water: EPA 554
Indoor / Expired Air: NIOSH 2539; EPA 0100

CRO5000 **CAS #: 4170-30-3**

CROTONALDEHYDE

RTECS: GP9499000
DOT: UN1143; IMO3.0
EINECS Number: 224-030-0
Molecular Formula: C_4H_6O
Structured MF: $CH_3CH=CHCHO$
Formula Weight: 70.09

Chemical Structure

Synonyms: 2-BUTENAL; 2-BUTENAL (9CI); 2-BUTENALDEHYDE; CROTONAL; CROTONALDEHYDE,STABILIZED; CROTONIC ALDEHYDE; CROTYLALDEHYDE; 1,2-ETHANEDIOL,DIPROPANOATE; ETHYLENE DIPROPIONATE; 1-FORMYLPROPENE; KROTONALDEHYD; BETA-METHYL ACROLEIN; METHYL ACROLEIN; BETA-METHYLACROLEIN; PROPYLENE ALDEHYDE; TOPANEL

Use: intermediate in manufacture of n-butanol, crotonic acid and sorbic acids; in resin and rubber antioxidant manufacture; also as a solvent in mineral oil purification; as a warning agent in fuel gas; alcohol denaturant manufacture of butyraldehyde, quinaldine; in locating breaks and leaks in pipes; minor amounts used in manufacture of maleic acid, crotyl alcohol, but chloral hydrate and in rubber accelerators; in organic synthesis; insecticides; and in chemical warfare

Physical Properties

Boiling Point: 102 °C (216 °F)
Freezing Point: -76.5 °C (-105.7 °F)
Specific Gravity: 0.853 at 20 °C/20 °C
Vapor Density: 2.41 Air=1
Saturated Vapor Density: 1.267116152 kg/m^3
Density: 0.8516 at 20 °C
Vapor Pressure: 30 mm Hg at 25 °C
Water Solubility: 18.1 g/100 g of Water at 20 °C
Other Solubilities: 95% Ethanol: >=100 mg/ml at 15 °C; Acetone: >=100 mg/ml at 15 °C; Acid: Soluble; Benzene: Very Soluble; DMSO: >=100 mg/ml at 15 °C; Ether: Very Soluble; Most organic solvents: miscible.
Odor Threshold: 0.105 to 3.0 mg/m^3
Refraction Index: 1.4334 at 17.3 °C/D
Evaporation Rate: 2.7 Butyl Acetate=1
Critical Temperature: 295 °C
Critical Pressure: 43 atm
Ionization Potential (eV): 9.73
Flash Point: 13 °C Closed Cup
Autoignition Temperature: 232 °C
LEL: 2.1% v/v
UEL: 15.5% v/v

RTECS Toxicity Data

Acute Oral: Rat LD_{50} Dose: 206 mg/kg. Mouse LD_{50} Dose: 104 mg/kg.
Acute Inhalation: Rat LC_{50} Dose: 200 mg/m^3/2hr. Mouse LC_{50} Dose: 580 mg/m^3/2hr.
Acute Dermal: Rabbit LD_{50} Route: Skin; Dose: 380 uL/kg. Guinea Pig LD_{50} Route: Skin; Dose: 30 uL/kg.
Mutagenic: Human DNA Adduct; Cell Type: fibroblast; Dose: 100 umol/L. Mammal DNA Adduct; Cell Type: lymphocyte; Dose: 21500 mg/L/16H.
Tumorigenic: Rat Route: Oral; Dose: 2664 mg/kg/2Y-C; Toxic Effects: Tumorigenic - Carcinogenic by RTECS criteria; Liver - Tumors.

Hazard Overviews

Reactivity: Unstable, forms crotonic acid and explosive peroxides upon standing.. Hazardous polymerization can occur at elevated temperatures or in contact with alkalis. Avoid: exposure to heat; ignition sources. Incompatible with: alkalis; ammonia; amines; nitric acid; strong oxidizers; 1,3-butadiene; ethyl acetoacetate; hydrocarbons. Hazardous decomposition products: carbon oxides.
Carcinogenicity: IARC - Group 3, Not classifiable as to carcinogenicity to humans; NIOSH - Not listed; NTP - Not listed; ACGIH - Class A3, Animal carcinogen; OSHA - Not listed; EPA - Class C, Possible human carcinogen; MAK - Class B, Justifiably suspected of having carcinogenic potential

Exposure Limits
OSHA PEL: TWA: 2 ppm; 6 mg/m^3.
ACGIH TLV: TWA: 2 ppm; 5.7 mg/m^3.
NIOSH REL: TWA: 2 ppm; 6 mg/m^3.
NIOSH IDLH: 50 ppm.

Respirator Recommendation
Exposure Range: >2 to 20 ppm Air Purifying, Negative Pressure, Half Mask
Exposure Range: >20 to <50 ppm Air Purifying, Negative Pressure, Full Face
Exposure Range: 50 to unlimited ppm Self-contained Breathing Apparatus, Pressure Demand, Full Face
Cartridge Color: black

Environmental

Ecotoxicity: LC_{50} Bluegill sunfish (Lepomis macrochirus) 3.5 ppm/96 hr static bioassay in fresh water at 23 °C with mild aeration applied after 24 hr (85% aqueous) LC_{50} Menidia beryllina 1.3 ppm/96 hr static bioassay in synthetic seawater at 23 °C with mild aeration applied after 24 hr
Environmental Fate: If released to the atmosphere, it degrades rapidly (typical half-life of 11-12 hr) via reaction with photochemically produced hydroxyl radicals. If released to water, it can degrade via reaction with photochemically produced oxidants (estimated half-life of 120 sunlight hr) and volatilize (estimated half-lives of 40 hours from a model river one meter deep and 15 days from a model lake). If released to soil, it is susceptible to significant leaching. Evaporation from dry surfaces can be expected to occur. Based on limited data, this compound may biodegrade in both soil and water under aerobic and anaerobic conditions. If released to the environment in a spill situation, a significant fraction of the spill may polymerize.
Cleanup/Disposal: Guide No. 131: Fully encapsulating, vapor protective clothing should be worn for spills and leaks with no fire. Eliminate all ignition sources (no smoking, flares, sparks or flames in immediate area). All equipment used when handling the product must be grounded. Do not touch or walk through spilled material. Stop leak if you can do it without risk. Prevent entry into waterways, sewers, basements or confined areas. A vapor suppressing foam may be used to reduce vapors. Small Spills: Absorb with earth, sand or other non-combustible material and transfer to containers for later disposal. Use clean non-sparking tools to collect absorbed material. Large Spills: Dike far ahead of liquid spill for later disposal. Water spray may reduce vapor; but may not prevent ignition in closed spaces.

Environmental Physical Data

Henry's Law Constant: 1.96 x10^{-5}
Octanol/Water Partition Coefficient: log K_{ow} = 0.63
Sorption Partition Coefficient: K_{oc} = estimated at 6
BCF: calculated at 0.17
BOD: 1.3 lb/lb, 10 days

Regulations

RCRA 40CFR: Listed Hazardous Waste No. U053 Toxic Waste
CERCLA: 40CFR 302.4: Listed per CWA Section 311(b)(4) per RCRA Section 3001 RQ: 100 lb (45.35 kg)
SARA 40CFR 372.65: Listed TPQ: 1000 lb
SARA EHS 40CFR 355: Listed TPQ: 100 lb
TSCA: Listed

Analytical Methods

Air: EPA TO-11, TO-5
Water / Groundwater: ASTM D3695
Indoor / Expired Air: NIOSH 3516; EPA IP-6A, IP-6B, IP-6C

CRO7000 CAS #: 3724-65-0

CROTONIC ACID

RTECS: GQ2800000
DOT: UN2823
EINECS Number: 223-077-4
Molecular Formula: $C_4H_6O_2$
Formula Weight: 86.09

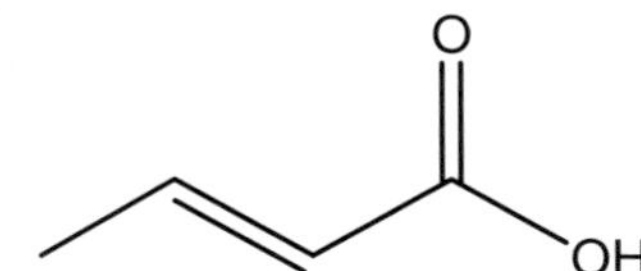

Chemical Structure

Synonyms: ACRYLIC ACID,3-METHYL-; 2-BUTENOIC ACID; ALPHA-BUTENOIC ACID; TRANS-2-BUTENOIC ACID; 2-BUTENOIC ACID (9CI); ALPHA-CROTONIC ACID; CROTONIC ACID,LIQUID OR SOLID; KYSELINA KROTONOVA; BETA-METHACRYLIC ACID; 3-METHYLACRYLIC ACID; BETA-METHYL-ACRYLIC ACID; BETA-METHYLACRYLIC ACID; SOLID CROTONIC ACID
Description: colorless crystalline solid
Use: mfr of copolymers with vinyl acetate used in lacquers & paper sizing; mfr of softening agents for synthetic rubber; in medicinal chemistry; coating, resins, fungicide, corrosion resistance improvement additive; pharmaceuticals, binders, and film

Physical Properties

Boiling Point: 185 °C (365 °F) at 760 mm Hg
Freezing Point: 71.5 °C (160.7 °F) to 71.7 °C (161.06 °F)
Specific Gravity: 1.018 at 15 °C/4 °C
Vapor Density: 2.97 Air=1
Saturated Vapor Density: 1.200590586 kg/m^3
Vapor Pressure: 0.19 mm Hg at 20 °C
Water Solubility: Very Soluble in Water
Other Solubilities: in Ethanol at 25 °C: 52.5% wt/wt & in Toluene at 25 °C: 37.5% wt/wt.
Flash Point: 87.778 °C Open Cup

RTECS Toxicity Data

Acute Oral: Rat LD_{50} Dose: 1 gm/kg. Mouse LD_{50} Dose: 4800 mg/kg; Toxic Effects: Tumorigenic - Active as anti-cancer agent.
Acute Dermal: Rabbit LD_{50} Route: Skin; Dose: 600 mg/kg. Mouse LD_{50} Route: Subcutaneous Dose: 3590 mg/kg.

Hazard Overviews

Corrosive

Fire Diamond

Health: Corrosive to eyes/skin/respiratory tract.
Fire: Combustible. Use alcohol foam, carbon dioxide, or dry chemicals.
Reactivity: Stable. Hazardous polymerization cannot occur. Incompatible with: oxidizing materials. Hazardous decomposition products: acrid smokes; irritating fumes.
Carcinogenicity: IARC - Not listed; NIOSH - Not listed; NTP - Not listed; ACGIH - Not listed; OSHA - Not listed; EPA - Not listed; MAK - Not listed
Primary Target Organs:

Eyes

Skin

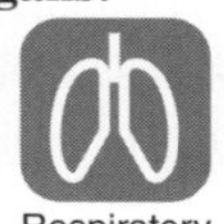
Respiratory System

Mucous Membranes

Environmental

Cleanup/Disposal: Guide No. 153: Eliminate all ignition sources (no smoking, flares, sparks or flames in immediate area). Do not touch damaged containers or spilled material unless wearing appropriate protective clothing. Stop leak if you can do it without risk. Prevent entry into waterways, sewers, basements or confined areas. Absorb or cover with dry earth, sand or other non-combustible material and transfer to containers. Do not get water inside containers.

Regulations

RCRA 40CFR: Not listed
CERCLA: 40CFR 302.4: Not listed
SARA 40CFR 372.65: Not listed
SARA EHS 40CFR 355: Not listed
TSCA: Listed

CRU5000 CAS #: 299-86-5

CRUFOMATE

RTECS: TB3850000
EINECS Number: 206-083-1
Molecular Formula: $C_{12}H_{19}ClNO_3P$

Structured MF: $C_{12}H_{19}ClNO_3P$
Formula Weight: 291.71
Synonyms: AMIDOFOS; AMIDOPHOS; 4-TERT BUTYL 2-CHLOROPHENYL METHYLPHOSPHORAMIDATE DEMETHYLE; 4-TERT. BUTYL 2-CHLOROPHENYL METHYLPHOSPHORAMIDATE DEMETHYLE; O-(4-TERT BUTYL-2-CHLOOR-FENYL)-O-METHYL-FOSFORZUUR-N-METHYL-AMIDE; O-(4-TERT-BUTYL-2-CHLOROPHENYL) O-METHYL N-METHYLAMIDOPHOSPHATE; 4-T-BUTYL-2-CHLOROPHENYL METHYL METHYLPHOSPHORAMIDATE; 4-TERT-BUTYL-2-CHLOROPHENYL METHYL METHYLPHOSPHORAMIDATE; 4-TERT-BUTYL-2-CHLOROPHENYL N-METHYLO-METHYLPHOSPHORAMIDATE; 4-T-BUTYL-2-CHLOROPHENYLMETHYL METHYL PHOSPHORAMIDATE; 4-T-BUTYL-2-CHLOROPHENYLMETHYL METHYLPHOSPHORAMIDATE; O-(4-TERT-BUTYL-2-CHLOR-PHENYL)-O-METHYL-PHOSPHORSAEURE-N-METHYLAMID; 2-CHLORO-4-(1,1-DIMETHYLETHYL)PHENYL METHYLMETHYLPHOSPHORAMIDATE; CRUFOMAT; CRUFOMATE A; DOWCO 132; DOWDO 132; ENT 25,602-X; M-1261; O-METHYL O-2-CHLORO-4-TERT-BUTYLPHENYL N-METHYLAMIDOPHOSPHATE; O-METHYL O-2-CHLORO-4-TERT-BUTYLPHENYLN-METHYLAMIDOPHOSPHATE; O-METHYL-O-(4-TERT-BUTYL-2-CHLOROPHENYL)METHYLPHOSPHORAMIDATE; METHYLPHOSPHORAMIDIC ACID 4-TERT-BUTYL-2-CHLOROPHENYLMETHYL ESTER; METHYLPHOSPHORAMIDIC ACID2-CHLORO-4-(1,1-DIMETHYLETHYL)PHENYL METHYL ESTER; METHYLPHOSPHORAMIDIC ACID,4-T-BUTYL-2-CHLOROPHENYL METHYLESTER; METHYLPHOSPHORAMIDIC ACID,4-T-BUTYL-2-CHLOROPHENYLMETHYL ESTER; MONTREL; PHENOL,4-T-BUTYL-2-CHLORO-,ESTER WITH METHYLMETHYLPHOSPHORAMIDATE; PHOSPHORAMIDIC ACID,4-TERT-BUTYL-2-CHLOROPHENYLPHOSPHORAMIDATE; PHOSPHORAMIDIC ACID,METHYL-,4-TERT-BUTYL-2-CHLOROPHENYLMETHYL ESTER; PHOSPHORAMIDIC ACID,METHYL-,2-CHLORO-4-(1,1-DIMETHYLETHYL)PHENYL METHYL ESTER; RUELENE; RUELENE 25E; RUELENE DRENCH; RULENE; O-(4-TERTBUTYL-2-CHLOOR-FENYL)-O-METHYL-FOSFORZUUR-N-METHYL-AMIDE; O-(4-TERZ.-BUTIL-2-CLORO-FENIL)-O-METIL-FOSFORAMMIDE
Description: colorless to white, crystalline solid in pure form
Use: formerly as insecticide for cattle grubs, horn flies & lice on cattle; medication (vet): anthelmintic

Physical Properties

Boiling Point: Decomposes
Freezing Point: 60 °C (140 °F) to 60.5 °C (140.9 °F)
Specific Gravity: 1.1618 at 70 °C/4 °C
Vapor Pressure: 0.01 mm Hg at 243 °F
Water Solubility: Practically Insoluble in Water
Other Solubilities: readily Soluble in Acetonitrile; Soluble in Ethyl Ether, Methanol, Cyclohexane; Soluble in Alcohol.
Refraction Index: Technical 1.5142 at 20 °C
Flash Point: 60 °C Open Cup

RTECS Toxicity Data

Acute Oral: Rat LD_{50} Dose: 460 mg/kg. Rabbit LD_{50} Dose: 400 mg/kg.
Acute Inhalation: Rat LC_{Lo} Dose: 12 mg/m^3/4hr.
Acute Dermal: Rabbit LD_{50} Route: Skin; Dose: 2 gm/kg. Guinea Pig LD_{Lo} Route: Subcutaneous Dose: 50 mg/kg.
Chronic (Multiple Dose) Oral: Rat Dose: 44 gm/kg/2Y-C; Toxic Effects: Musculoskelital - Other changes; Biochemical - True cholinesterase; DEATH - Changes in testicular weight. Dog Dose: 36 gm/kg/2Y-C; Toxic Effects: Biochemical - True cholinesterase.
Reproductive/Teratogenic: Mouse Route: Skin; Dose: 100 mg/kg; Duration: female 2D prior to mating; Effects on Fertility - Mating Performance; Female fertility index. Mouse Route: Skin; Dose: 100 mg/kg; Duration: female 2D prior to mating; Maternal Effects - Parturition; Effects on Newborn - Weaning or lactation index; Growth statistics. Mouse Route: Skin; Dose: 200 mg/kg; Duration: female 2D prior to mating; Effects on Newborn - Live birth index.

Hazard Overviews

Flammable

Fire: Flammable.
Carcinogenicity: IARC - Not listed; NIOSH - Not listed; NTP - Not listed; ACGIH - Class A4, Not classifiable as a human carcinogen; OSHA - Not listed; EPA - Not listed; MAK - Not listed
Exposure Limits
OSHA PEL Vacated 1989 Limits: TWA: 5 mg/m^3.
ACGIH TLV: TWA: 5 mg/m^3.
NIOSH REL: TWA: 5 mg/m^3. STEL: 20 mg/m^3.
Respirator Recommendation
Exposure Range: >5 to 50 mg/m^3 Air Purifying, Negative Pressure, Half Mask
Exposure Range: >50 to 500 mg/m^3 Air Purifying, Negative Pressure, Full Face
Exposure Range: >500 to 5000 mg/m^3 Supplied Air, Constant Flow/Pressure Demand, Full Face
Exposure Range: >5000 to unlimited mg/m^3 Self-contained Breathing Apparatus, Pressure Demand, Full Face
Cartridge Color: black with magenta (P100)

Environmental

Ecotoxicity: LD_{50} Anas platyrhynchos (Mallard) oral 265 (confidence limits 170-414) mg/kg LC_{50} Gammarus fasciatus 3.7 mg/l/96 hr at 15 °C (95% confidence limit: 3.4-4.1 mg/l), mature static bioassay without aeration, pH 7.2-7.5, water hardness 40-50 mg/l as calcium carbonate and alkalinity of 30-35 mg/l LC_{50} Lepomis macrchirus (bluegill) 1.8 mg/l at 8 °C (95% confidence limit: 1.3-2.4 mg/l), wt 1.0 g static bioassay without aeration, pH 7.2-7.5, water hardness 40-50 mg/l as calcium carbonate and alkalinity of 30-35 mg/l

Regulations

RCRA 40CFR: Not listed
CERCLA: 40CFR 302.4: Not listed
SARA 40CFR 372.65: Not listed
SARA EHS 40CFR 355: Not listed
TSCA: Not listed

Analytical Methods

Soil: EPA PMD-TLC
Water / Groundwater: EPA 022
Food: FDA 212.1, 212.2, 232.1, 232.4, 242.1

Plasma: EPA 001, 028; FDA 211.1, 231.1, 252

CUM1000 CAS #: 98-82-8

CUMENE

RTECS: GR8575000
DOT: UN1918; IMO3.3
EINECS Number: 202-704-5
Molecular Formula: C_9H_{12}
Structured MF: $C_6H_5CH(CH_3)_2$
Formula Weight: 120.19

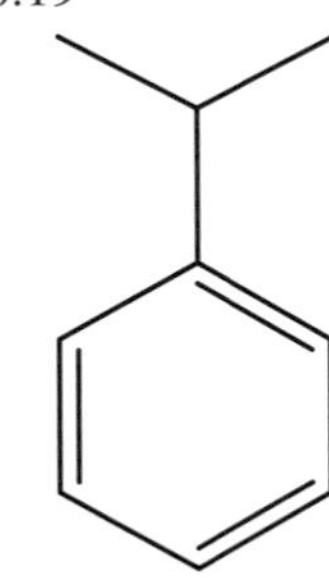

Chemical Structure

Synonyms: BENZENE,ISOPROPYL; BENZENE,(1-METHYLETHYL)-; BENZENE,(1-METHYLETHYL)-(9CI); CUMEEN; CUMOL; 2-FENILPROPANO; 2-FENYL-PROPAAN; ISOPROPILBENZENE; ISOPROPYL BENZENE; ISOPROPYLBENZEEN; ISOPROPYLBENZENE; ISOPROPYL-BENZOL; ISOPROPYLBENZOL; (1-METHYLETHYL)BENZENE; 2-PHENYL PROPANE; 2-PHENYLPROPANE; PROPANE,2-PHENYL
Description: clear, colorless liquid; sharp, aromatic odor
Use: thinner for paints, lacquers & enamels; in petro-based solvents; in high octane aviation fuel; in prodn of styrene, phenol, acetone & alpha-methylstyrene; in mfr of acetophenone, polymerization catalysts (diisopropylbenzene); catalyst for acrylic & polyester type resins; naptha constituent; minor amounts in gasoline blending; raw material for peroxides & oxidn catalysts; intermed for dicumyl peroxide.

Physical Properties

Boiling Point: 152.4 °C (306 °F) at 760 mm Hg
Freezing Point: -96 °C (-140.8 °F)
Specific Gravity: 0.862 at 20 °C/4 °C
Vapor Density: 4.2 Air=1
Vapor Pressure: 10 mm Hg at 38.3 °C
Water Solubility: Insoluble
Other Solubilities: Soluble in all prop with Acetone, Alcohol, Ether, Benzene.
Surface Tension: 54.6 dynes/cm
Odor Threshold: 0.06 mg/m^3
Refraction Index: 1.4915 at 20 °C/D
Critical Temperature: 362.7 °C
Critical Pressure: 31.2 atm
Ionization Potential (eV): 8.75
Flash Point: 38.889 °C Closed Cup
Autoignition Temperature: 424 °C
LEL: 0.9% v/v
UEL: 6.5% v/v

RTECS Toxicity Data

Acute Oral: Rat LD_{50} Dose: 1400 mg/kg; Toxic Effects: Gastrointestinal - Gastritis. Mouse LD_{50} Dose: 12750 mg/kg.
Acute Inhalation: Rat LC_{Lo} Dose: 8000 ppm/4hr. Human TC_{Lo} Dose: 200 ppm; Toxic Effects: Behavioral - Somnolence (general depressed activity); Behavioral - Antipsychotic; Behavioral - Irritability.
Acute Dermal: Rabbit LD_{50} Route: Skin; Dose: 12300 uL/kg.
Irritation Eye: Rabbit Standard Draize Test Dose: 500 mg/24H; Reaction: mild. Rabbit Standard Draize Test Dose: 86 mg; Reaction: mild.
Irritation Skin: Rabbit Standard Draize Test Dose: 100 mg/24H; Reaction: moderate. Rabbit Open Draize Test Dose: 10 mg/24H open; Reaction: mild.

Hazard Overviews

Flammable

Fire Diamond

Health: Irritating to eyes/skin/respiratory tract. Also Causes: headaches, narcosis, and unconsciousness. Chronic Effects: dermatitis.
Fire: Flammable. Can form explosive mixtures in the air. Fight fire from maximum distance. Use carbon dioxide, dry chemical, or foam. Water stream on the burning liquid can scatter flames. Use water spray to cool fire-exposed containers. Its heavier-than-air vapors can flow along surfaces to distant sources and then flash back.
Reactivity: Unstable, forms peroxides upon exposure to air. Hazardous polymerization cannot occur. Avoid: air; heat; ignition sources. Incompatible with: oxidizing agents; nitric acid; fuming sulfuric acid; chlorosulfonic acid. Hazardous decomposition products: carbon oxides.
Carcinogenicity: IARC - Not listed; NIOSH - Not listed; NTP - Not listed; ACGIH - Not listed; OSHA - Not listed; EPA - Not listed; MAK - Not listed
Primary Target Organs:

Eyes

Skin

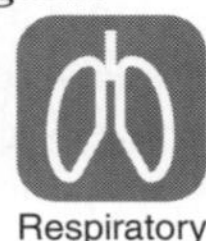
Respiratory System

Mucous Membranes

Nervous System

Exposure Limits
OSHA PEL: TWA: 50 ppm; 245 mg/m^3.
ACGIH TLV: TWA: 50 ppm; 246 mg/m^3.
NIOSH REL: TWA: 50 ppm; 245 mg/m^3.
NIOSH IDLH: 900 ppm; LEL.
DFG MAK: TWA: 50 ppm; 245 mg/m^3.
Respirator Recommendation
Exposure Range: >50 to 500 ppm Air Purifying, Negative Pressure, Half Mask
Exposure Range: >500 to <900 ppm Air Purifying, Negative Pressure, Full Face

Exposure Range: 900 to unlimited ppm Self-contained Breathing Apparatus, Pressure Demand, Full Face
Cartridge Color: black

Environmental

Ecotoxicity: No significant alteration of growth rate was observed in Mytilus edulis (mussel larvae) at concentration of 1 to 50 ppm EC_{50} Pimephales promelas (fathead minnow) 6.32 mg/l/96 hr (confidence limit 6.04 - 6.61 mg/l), flow-through bioassay with measured concentrations, 25.4 °C, dissolved oxygen 6.6 mg/l, hardness 44.3 mg/l calcium carbonate, alkalinity 42.1 mg/l calcium carbonate LD_{50} Agelaius phoeniceus (red-winged blackbird) oral 98 mg/kg LC_{50} Daphnia magna 0.6 ppm/48 hr /Conditions of bioassay not specified

Environmental Fate: When released to soil, it is expected to biodegrade and may volatilize from the soil surface. It is expected to strongly adsorb to soils and is not expected to leach to groundwater. When released to water, it is expected to volatilize with an estimated half-life of 5-14 days and to biodegrade rapidly. Compared to these processes, aqueous photooxidation by hydroxyl radicals (estimated half-life 0.7 years) and peroxy radicals (estimated half-life 2.2 years) are expected to be relatively slow, and so are not expected to be significant fate processes. Adsorption to sediments may occur based on the high soil-sorption coefficient. Bioconcentration is not expected to be significant. When released to the atmosphere, vapor phase will react with photochemically generated hydroxyl radicals with an estimated half-life of 25 h in polluted atmospheres and 49 h in normal atmospheres. The reaction of vapor phase with ozone has an estimated half-life of 3 years and the half-life of direct photolysis was estimated to be 1500 years.

Cleanup/Disposal: Guide No. 131: Fully encapsulating, vapor protective clothing should be worn for spills and leaks with no fire. Eliminate all ignition sources (no smoking, flares, sparks or flames in immediate area). All equipment used when handling the product must be grounded. Do not touch or walk through spilled material. Stop leak if you can do it without risk. Prevent entry into waterways, sewers, basements or confined areas. A vapor suppressing foam may be used to reduce vapors. Small Spills: Absorb with earth, sand or other non-combustible material and transfer to containers for later disposal. Use clean non-sparking tools to collect absorbed material. Large Spills: Dike far ahead of liquid spill for later disposal. Water spray may reduce vapor; but may not prevent ignition in closed spaces.

Environmental Physical Data

Henry's Law Constant: 0.0116
Octanol/Water Partition Coefficient: log K_{ow} = 3.66
Sorption Partition Coefficient: log K_{oc} = estimated at 3.45
BCF: goldfish 35.5
BOD: theoretical 40%, 5 days

Regulations

RCRA 40CFR: Listed Hazardous Waste No. U055 Ignitable Waste
CERCLA: 40CFR 302.4: Listed per RCRA Section 3001 RQ: 5000 lb (2268 kg)
SARA 40CFR 372.65: Listed
SARA EHS 40CFR 355: Not listed
TSCA: Listed

Analytical Methods

Air: EPA VG-011-1, TO-1; ASTM D3686, D3687, D4490
Soil: SW846 8021A, 8260A, 8260B
Water / Groundwater: APHA 6210-D, 6230-D; ASTM D4763
Drinking Water: EPA 502.2, 503.1, 524.2; APHA 6210-C, 6220-C
Indoor / Expired Air: NIOSH 1501; EPA IP-1B
Plasma: EPA 29

CUM5000 CAS #: 80-15-9

CUMENE HYDROPEROXIDE

RTECS: MX2450000
DOT: UN2116; IMO5.2
EINECS Number: 201-254-7
Molecular Formula: $C_9H_{12}O_2$
Structured MF: $C_6H_5C(CH_3)_2OOH$
Formula Weight: 152.21

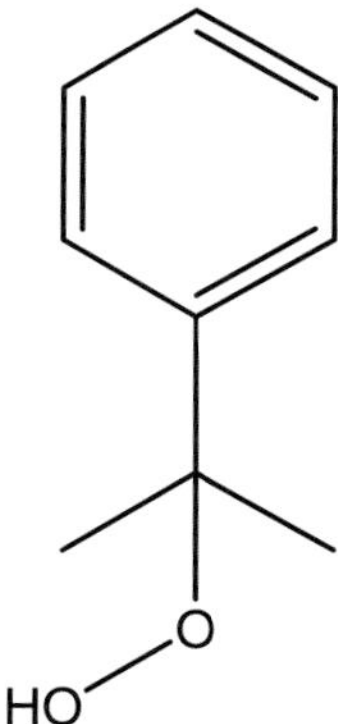

Chemical Structure

Synonyms: CHP; CUMEENHYDROPEROXYDE; ALPHA-CUMENE HYDROPEROXIDE; CUMENT HYDROPEROXIDE; CUMENYL HYDROPEROXIDE; CUMOLHYDROPEROXID; 7-CUMYL HYDROPEROXIDE; ALPHA-CUMYL HYDROPEROXIDE; CUMYL HYDROPEROXIDE; ALPHA,ALPHA-DIMETHYLBENZYL HYDROPEROXIDE; ALPHA,ALPHA-DIMETHYL-BENZYLHYDROPEROXIDE; HYDROPEROXIDE DE CUMENE; HYDROPEROXIDE,ALPHA,ALPHA-DIMETHYLBENZYL-; HYDROPEROXIDE,1-METHYL-1-PHENYLETHYL; HYDROPEROXYDE DE CUMENE; HYDROPEROXYDE DE CUMYLE; 7-HYDROPEROXYKUMEN; HYPERIZ; IDROPEROSSIDO DI CUMENE; IDROPEROSSIDO DI CUMOLO; ISOPROPYLBENZENE HYDROPEROXIDE; KUMENYLHYDROPEROXID; 1-METHYL-1-PHENYLETHYL HYDROPEROXIDE; PERCUMYL H; TRIGONOX K 80; TRIGOROX K 80

Description: colorless to pale yellow liquid; sharp aromatic odor

Use: in the production of acetone and phenol; polymerization catalyst, particularly in redox systems; a convenient probe for studying the mechanism of NADPH dependent lipid peroxidation

Physical Properties

Boiling Point: 153 °C (307 °F)
Freezing Point: < -40 °C (-40 °F)
Specific Gravity: 1.05
Vapor Density: 5 Air=1
Saturated Vapor Density: 1.201610004 kg/m^3
Density: 1.024 g/mL at 20 °C
Vapor Pressure: 0.24 mm Hg at 20 °C
Water Solubility: Slightly Soluble in Water
Other Solubilities: 95% Ethanol: >=100 mg/ml at 18 °C; Acetone: >=100 mg/ml at 18 °C; Alcohol: Soluble; Chlorinated hydrocarbons: Soluble; DMSO: Decomposes; Esters: Soluble; Hydrocarbons: Soluble.
Surface Tension: 25 dynes/cm at 20 °C
pH: Approximately 4
Refraction Index: 1.5210
Evaporation Rate: ~ 0.1 Butyl Acetate=1
Flash Point: 79.444 °C Closed Cup
Autoignition Temperature: ~ 221 °C
LEL: 0.9% v/v
UEL: 6.5% v/v

RTECS Toxicity Data

Acute Oral: Rat LD_{50} Dose: 382 mg/kg; Toxic Effects: Kidney, Ureter, and Bladder - Hematuria. Mouse LD_{Lo} Dose: 5 gm/kg.
Acute Inhalation: Rat LC_{50} Dose: 220 ppm/4hr; Toxic Effects: Lungs, Thorax, or Respiration - Dyspnea. Mouse LC_{50} Dose: 200 ppm/4hr; Toxic Effects: Lungs, Thorax, or Respiration - Dyspnea.
Acute Dermal: Rabbit LD_{Lo} Route: Skin; Dose: 1200 mg/kg; Toxic Effects: Cardiac - Pulse rate increased without fall in BP; Blood - Changes in erythrocite (RBC) cell count; Nutritional and gross metabolic - Body temperature decrease. Rat LD_{50} Route: Skin; Dose: 500 mg/kg; Toxic Effects: Behavioral - Convulsions or effect on seizure threshold; Kidney, Ureter, and Bladder - Hematuria. Rat LD_{50} Route: Subcutaneous Dose: 382 mg/kg.
Chronic (Multiple Dose) Oral: Rat Dose: 1604 mg/kg/7W-I; Toxic Effects: DEATH.
Chronic (Multiple Dose) Inhalation: Rat Dose: 50 ppm/4H/3D-I; Toxic Effects: Lungs, Thorax, or Respiration - Other changes; Kidney, Ureter, and Bladder - Other changes; DEATH.
Irritation Skin: Rabbit Standard Draize Test Dose: 500 mg; Reaction: mild.
Mutagenic: Bacteria - S Typhimurium Mutations in Microorganisms; Dose: 100 ug/plate (+S9). Bacteria - S Typhimurium Mutations in Microorganisms; Dose: 100 ug/plate (-S9).
Tumorigenic: Mouse Route: Subcutaneous; Dose: 8844 mg/kg/67W-I; Toxic Effects: Tumorigenic - Equivocal tumorigenic agent by RTECS criteria; Tumorigenic - Tumors at site of application.

Hazard Overviews

Explosive

Fire Diamond

Health: Severely irritating to eyes/skin/respiratory tract. Toxic. Also Causes: CNS depression, nausea, headache, emphysema, pulmonary edema, possible permanent eye damage. Chronic Effects: dermatitis.
Fire: Combustible. Explosive when heated. Strong oxidizer. Fight fire with dry chemical, carbon dioxide, or regular foam. Use water in flooding quantities as fog; solid water streams may be ineffective and only spread fire.
Reactivity: Unstable, decomposes violently at elevated temperatures (91 and 95 % decomposed at 302 °F/150 °C). Hazardous polymerization cannot occur. Avoid: contact with hot surfaces; ignition sources. Incompatible with: solutions of 1,2-dibromo-1,2-diisocyanatoethane polymers in benzene; cobalt; copper; copper alloys; lead alloys; mineral acids; charcoal; brass; lead; zinc; oxidizing agents such as chlorine and ferric chloride; reducing agents such as sulfur dioxide. Hazardous decomposition products: toxic phenol vapors.
Carcinogenicity: IARC - Not listed; NIOSH - Not listed; NTP - Not listed; ACGIH - Not listed; OSHA - Not listed; EPA - Not listed; MAK - Not listed
Primary Target Organs:

Eyes

Skin

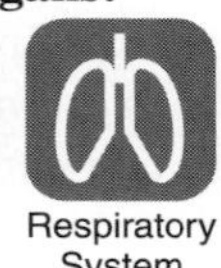
Respiratory System

Nervous System

Exposure Limits
AIHA WEEL: TWA: 1 ppm skin.
Respirator Recommendation
Exposure Range: >1 to 50 ppm Supplied Air, Constant Flow/Pressure Demand, Half Mask
Exposure Range: >50 to 1000 ppm Supplied Air, Constant Flow/Pressure Demand, Full Face
Exposure Range: >1000 to unlimited ppm Self-contained Breathing Apparatus, Pressure Demand, Full Face
Note: odor threshold unknown

Environmental

Cleanup/Disposal: Guide No. 147: Eliminate all ignition sources (no smoking, flares, sparks or flames in immediate area). Keep combustibles (wood, paper, oil, etc.) away from spilled material. Do not touch damaged containers or spilled material unless wearing appropriate protective clothing. Keep substance wet using water spray. Stop leak if you can do it without risk. Small Spills: Take up with inert, damp, noncombustible material using clean non-sparking tools and place into loosely covered plastic containers for later disposal. Large Spills: Wet down with water and dike for later disposal. Prevent entry into waterways, sewers, basements or confined areas. Do not clean-up or dispose of, except under supervision of a specialist.

Environmental Physical Data

BCF: no food chain concentration potential

Regulations

RCRA 40CFR: Listed Hazardous Waste No. U096 Reactive Waste
CERCLA: 40CFR 302.4: Listed per RCRA Section 3001 RQ: 10 lb (4.535 kg)
SARA 40CFR 372.65: Listed
SARA EHS 40CFR 355: Not listed
TSCA: Listed

CUP1000 CAS #: 135-20-6

CUPFERRON

RTECS: NC4725000
EINECS Number: 205-183-2
Molecular Formula: $C_6H_9N_3O_2$
Structured MF: $C_6H_5N(NO)O•NH_4$
Formula Weight: 155.16

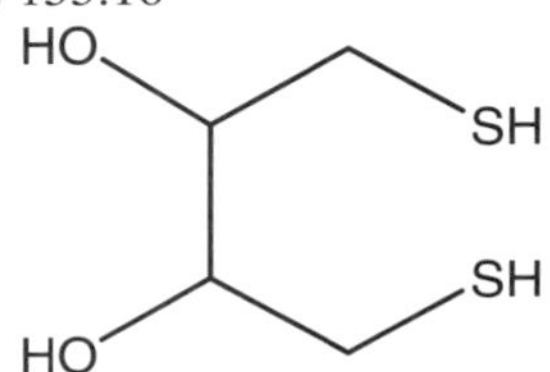

Chemical Structure

Synonyms: AMMONIUM CUPFERRON; AMMONIUM N-NITROSOPHENYLHYDROXYLAMINE; AMMONIUM NITROSO-BETA-PHENYLHYDROXYLAMINE; AMMONIUM NITROSOPHENYLHYDROXYLAMINE; AMMONIUM-N-NITROSOPHENYLHYDROXYLAMINE; BENZENAMINE,N-HYDROXY-N-NITROSO-,AMMONIUM SALT; CUPFERON; CUPFERRON,AMMONIUM SALT; HYDROXYLAMINE,N-NITROSO-N-PHENYL-,AMMONIUM SALT; N-HYDROXY-N-NITROSOBENZENAMINE AMMONIUM SALT; N-HYDROXY-N-NITROSO-BENZENAMINE,AMMONIUM SALT; KUPFERON; KUPFERRON; N-NITROSOFENYLHYDROXYLAMIN AMONNY; N-NITROSO-N-PHENYLHYDROXYAMINE AMMONIUM SALT; N-NITROSOPHENYLHYDROXYLAMIN AMMONIUM SALZ; N-NITROSO-N-PHENYLHYDROXYLAMINE AMMONIUM SALT; N-NITROSOPHENYLHYDROXYLAMINE AMMONIUM SALT
Description: creamy-white crystals
Use: analytical reagent, especially for separation and precipitation of metals, e.g. copper, iron and vanadium; for quantitative analysis and colorimetric estimation of aluminum

Physical Properties

Freezing Point: 163 °C (325.4 °F) to 164 °C (327.2 °F)
Vapor Pressure: Negligible
Water Solubility: Freely Soluble in Water
Other Solubilities: 95% Ethanol: 10-50 mg/ml at 18.5 °C; Acetone: <1 mg/ml at 18.5 °C; Alcohol: Soluble; DMSO: >=100 mg/ml at 18.5 °C.
Evaporation Rate: Negligible n-butyl acetate=1
Flash Point: Not available; probably combustible

RTECS Toxicity Data

Acute Oral: Rat LD_{50} Dose: 199 mg/kg; Toxic Effects: Behavioral - Muscle weakness; Lungs, Thorax, or Respiration - Dyspnea; Skin and appendages - Dermatitis, other.
Irritation Eye: Rabbit Standard Draize Test Dose: 20 mg/24H; Reaction: moderate.
Mutagenic: Human DNA Inhibition; Cell Type: HeLa cell; Dose: 2 mmol/L. Grasshopper Cytogenetic Analysis; Route: Oral; Dose: 1 pph.
Tumorigenic: Rat Route: Oral; Dose: 123 gm/kg/78W-C; Toxic Effects: Tumorigenic - Carcinogenic by RTECS criteria; Vascular - Tumors; Liver - Tumors. Rat Route: Oral; Dose: 9040 mg/kg/65W-C; Toxic Effects: Tumorigenic - Equivocal tumorigenic agent by RTECS criteria; Lungs, Thorax, or Respiration - Tumors.

Hazard Overviews

Health: Irritating to eyes/skin/respiratory tract. Toxic. Other Acute Effects: harmful if swallowed, inhaled, or absorbed through skin; may cause cyanosis. Chronic Effects: dermatitis; target organ: blood. Carcinogen.
Fire: Will burn. Hazards: emits toxic fumes. Extinguishing agents: water spray; carbon dioxide, dry chemical powder or appropriate foam. Precautions: combustible liquid.
Reactivity: Incompatible with: strong oxidizing agents, strong acids, strong bases, protect from moisture. Hazardous decomposition products: toxic fumes of: carbon monoxide, carbon dioxide, nitrogen oxides.
Carcinogenicity: IARC - Not listed; NIOSH - Not listed; NTP - Listed; ACGIH - Not listed; OSHA - Not listed; EPA - Not listed; MAK - Not listed
Primary Target Organs:

Eyes

Skin

Respiratory System

Blood

Environmental

Regulations

RCRA 40CFR: Not listed
CERCLA: 40CFR 302.4: Not listed
SARA 40CFR 372.65: Listed
SARA EHS 40CFR 355: Not listed
TSCA: Listed

CUP1670 CAS #: 142-71-2

CUPRIC ACETATE

RTECS: AG3480000
DOT: UN2775; IMO6.1
EINECS Number: 205-553-3
Molecular Formula: $C_4H_8CuO_4$
Structured MF: $Cu(CH_3COO)_2$
Formula Weight: 181.64

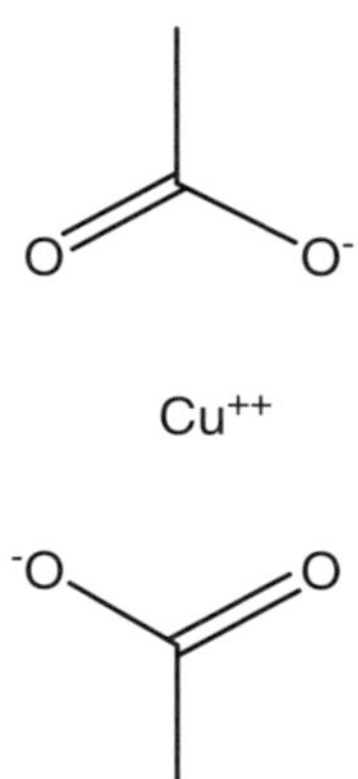

Chemical Structure

Synonyms: ACETATE DE CUIVRE; ACETIC ACID CUPRIC SALT; ACETIC ACID,COPPER(2+) SALT; ACETIC ACID,CUPRIC SALT; COPPER(2+) ACETATE; COPPER ACETATE (CU(C2H3O2)2); COPPER DIACETATE; COPPER(2+) DIACETATE; COPPER (II) ACETATE; COPPER(II) ACETATE; CRYSTALLIZED VERDIGRIS; CRYSTALS OF VENUS; CUPRIC DIACETATE; NEUTRAL VERDIGRIS; OCTAN MEDNATY

Description: bluish-green solid; odorless

Use: intermed for paris green; catalyst for org reactions, eg, rubber aging, polymerization processes involving styrene, acrylonitrile, etc; in textile dyeing; pigment for ceramics; preservative for cellulosic materials; stabilizer for polyurethanes & nylons, copper acetate monohydrate; cupric acetate & cupric nitrate, preservation of plants; in insecticides, fungicides, & herbicides, anti-fouling paints, corrosion inhibitors, electrolysis & electroplating processes, electronics, flameproofing, fuel additives, glass; in cement, food & drugs, metallurgy, paper products, pollution control catalyst, printing & photo copying, pyrotechnics, & wood preservatives; med: (vet) in hematinic mixtures; source of copper; shark repellent.

Physical Properties

Boiling Point: Decomposes at 1 atm
Freezing Point: 115 °C (239 °F)
Specific Gravity: 1.882
Water Solubility: Soluble in Water
Other Solubilities: Soluble in Alcohol; Slightly Soluble in Ether, Glycerol
Refraction Index: 1.545
Flash Point: Nonflammable

RTECS Toxicity Data

Acute Oral: Rat LD_{50} Dose: 501 mg/kg; Toxic Effects: Behavioral - Somnolence (general depressed activity); Behavioral - Convulsions or effect on seizure threshold. Mouse LD_{50} Dose: 196 mg/kg; Toxic Effects: Behavioral - Somnolence (general depressed activity); Behavioral - Convulsions or effect on seizure threshold.

Acute Dermal: Rat LD_{50} Route: Subcutaneous Dose: 350 mg/kg; Toxic Effects: Musculoskelital - Joints; Tumorigenic - Active as anti-cancer agent; Biochemical - Effect on inflammation or mediation of inflammation.

Reproductive/Teratogenic: Rat Route: Subcutaneous; Dose: 40 mg/kg; Duration: female 7-10D of pregnancy; Effects on Fertility - Other measures of fertility.

Hazard Overviews

Health: Irritating to eyes/skin/respiratory tract. Harmful. Other Acute Effects: harmful if swallowed; may be harmful if inhaled or absorbed through the skin. Chronic Effects: hepatic cirrhosis; brain damage and demyelination; kidney defects; and copper deposition in the cornea as exemplified by humans with Wilson's disease; it has also been reported that copper poisoning has lead to hemolytic anemia and accelerates arteriosclerosis.

Fire: Noncombustible. Hazards: emits toxic fumes. Extinguishing agents: water spray; carbon dioxide, dry chemical powder or appropriate foam. Precautions: combustible liquid.

Reactivity: Incompatible with: strong oxidizing agents, strong acids, protect from moisture. Hazardous decomposition products: toxic fumes of: carbon monoxide, carbon dioxide.

Carcinogenicity: IARC - Not listed; NIOSH - Not listed; NTP - Not listed; ACGIH - Not listed; OSHA - Not listed; EPA - Not listed; MAK - Not listed

Primary Target Organs:

Eyes

Skin

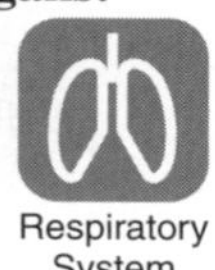
Respiratory System

Environmental

Ecotoxicity: EC_{50} Selenastrum capricornatum (green alga) 85 ug/l/14 days /Cell volume bioassay EC_{50} Thalassiosira pseudonana (alga, saltwater) 5 ug/l/72 hr /Growth rate bioassay EC_{50} Chlorella stigmatophora (alga, saltwater) 70 ug/l/21 days /Cell volume bioassay LC_{50} Corbicula manilensis (Asiatic clam) > 2,600 ug/l/96 hr /Conditions of bioassay not specified LC_{50} Pimephales promelas (fathead minnow) 0.39 mg/l/96 hr (95% confidence limit 0.30-0.51) /Static bioassay LC_{50} Oncorhynchus kisutch (coho salmon) 286 ug/l/96 hr /Conditions of bioassay not specified EC_{10} Salmo gairdneri (rainbow trout; embryo, larvae) 16.5 ug/l/28 days; death and deformity. /Conditions of bioassay not specified LC_{50} Palaemonetes pugio 37.0 mg/l/96 hr (95% confidence limit 12.7-110.9) /Static bioassay

Cleanup/Disposal: Guide No. 151: Do not touch damaged containers or spilled material unless wearing appropriate protective clothing. Stop leak if you can do it without risk. Prevent entry into waterways, sewers, basements or confined areas. Cover with plastic sheet to prevent spreading. Absorb or cover with dry earth, sand or other non-combustible material and transfer to containers. Do not get water inside containers.

Environmental Physical Data

BCF: not biomagnified
BOD: sewage 0.43 lb/lb, 5 days

Regulations

RCRA 40CFR: Not listed
CERCLA: 40CFR 302.4: Listed per CWA Section 311(b)(4) RQ: 100 lb (45.35 kg)
SARA 40CFR 372.65: Listed as Compound
SARA EHS 40CFR 355: Not listed
TSCA: Listed

CUP3010 CAS #: 7447-39-4

CUPRIC CHLORIDE

RTECS: GL7000000
DOT: UN2802; IMO5.1
EINECS Number: 231-210-2
Molecular Formula: Cl_2Cu
Structured MF: $CuCl_{22}H_2O$
Formula Weight: 134.45

Cu^{++}

Cl^- Cl^-

Chemical Structure

Synonyms: COPPER BICHLORIDE; COPPER CHLORIDE; COPPER(2+) CHLORIDE; COPPER DICHLORIDE; COPPER(II) CHLORIDE; CUPRIC DICHLORIDE
Description: yellow to brown, microcrystalline powder; odorless
Use: catalyst for in prodn of chlorine, acetaldehyde, etc; in petro ind, deodorizing, desulfurizing, purifying; mordant for dyeing & printing textiles; in inks; in metallurgy (recovering mercury, refining (copper, etc), metal tinting & electrotyping baths); in photography fixer; in mfr of acrylonitrile, fast black (melanin); in pigments for glass, etc; feed additive, wood preservative, disinfectant; to remove lead from gasoline & oils; in insecticides, fungicides, herbicides, anti-fouling paints, corrosion inhibitors, electronics, cement, food & drugs, nylon, paper products, pollution control catalyst, printing & photo copying, pyrotechnics.

Physical Properties

Boiling Point: Decomposes at 993 °C (1819 °F)
Freezing Point: Extrapolated 630 °C (1166 °F)
Specific Gravity: 3.39 at 25 °C/4 °C
Vapor Pressure: 20 mm Hg at 1970 °C
Water Solubility: Dihydrate 76 parts/100 parts at 25 °C
Other Solubilities: 53 g/100 cc Alcohol at 15 °C; 68 g/100 cc Methanol at 15 °C.
pH: 0.2 molar solution 3.6
Flash Point: Nonflammable

RTECS Toxicity Data

Acute Oral: Rat LD_{50} Dose: 584 mg/kg; Toxic Effects: Behavioral - Somnolence (general depressed activity); Behavioral - Convulsions or effect on seizure threshold. Mouse LD_{50} Dose: 233 mg/kg; Toxic Effects: Behavioral - Somnolence (general depressed activity); Behavioral - Convulsions or effect on seizure threshold.
Reproductive/Teratogenic: Rat Route: Inhalation; Dose: 8 ug/m^3; Duration: female 1-21D of pregnancy; Effects on Fertility - Post-implantation mortality; Effects on Embryo or Fetus - Extra embryonic structures; Fetotoxicity. Mouse Route: Intraperitoneal; Dose: 35 mg/kg; Duration: female 12-18D of pregnancy Effects on Fertility - Post-implantation mortality; Effects on Embryo or Fetus - Fetal death.
Mutagenic: Mammal DNA Damage; Cell Type: lymphocyte; Dose: 2 mmol/L. Yeast - S Cerevisiae Mutations in Microorganisms; Dose: 100 umol/L (-S9).

Hazard Overviews

Health: Irritating to eyes/skin/respiratory tract. Toxic. Other Acute Effects: harmful if inhaled or swallowed; depending on the intensity and duration of exposure, effects may vary from mild irritation to severe destruction of tissue; exposure can cause gastrointestinal disturbances; damage to the eyes, liver, kidneys, lungs. Irritating to eyes/skin/respiratory tract. Toxic. Other Acute Effects: harmful if inhaled or swallowed; depending on the intensity and duration of exposure; effects may vary from mild irritation to severe destruction of tissue; exposure can cause gastrointestinal disturbances; damage to the eyes, liver, kidneys, lungs.
Fire: Noncombustible. Hazards: emits toxic fumes. Extinguishing agents: noncombustible; use extinguishing media appropriate to surrounding fire conditions. Precautions: combustible liquid.
Reactivity: Incompatible with: moisture, alkali metals. Hazardous decomposition products: toxic fumes of: hydrogen chloride gas.
Carcinogenicity: IARC - Not listed; NIOSH - Not listed; NTP - Not listed; ACGIH - Not listed; OSHA - Not listed; EPA - Not listed; MAK - Not listed
Primary Target Organs:

Eyes

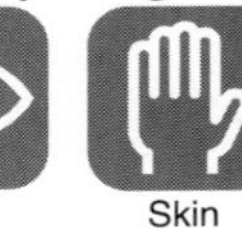
Skin

Respiratory System

Environmental

Ecotoxicity: LC_{50} Corbicula manilensis (asiatic clam) > 2,600 ug/l/96 hr /Conditions of bioassay not specified EC_{50} Selenastrum capricornatum (green alga) 85 ug/l/14 days /Cell volume bioassay EC_{50} Thalassiosira pseudonana (alga, saltwater) 5 ug/l/72 hr /Growth rate bioassay LC_{50} Oncorhynchus kisutch (coho salmon) 286 ug/l/96 hr /Conditions of bioassay not specified EC_{10} Salmo gairdneri (rainbow trout; embryo, larvae) 16.5 ug/l/28 days; death and deformity. /Conditions of bioassay not specified
Cleanup/Disposal: Guide No. 154: Eliminate all ignition sources (no smoking, flares, sparks or flames in immediate area). Do not touch damaged containers or spilled material unless wearing appropriate protective clothing. Stop leak if you can do it without risk. Prevent entry into waterways, sewers, basements or confined areas. Absorb or cover with

dry earth, sand or other non-combustible material and transfer to containers. Do not get water inside containers.

Environmental Physical Data

BCF: not biomagnified
BOD: none

Regulations

RCRA 40CFR: Not listed
CERCLA: 40CFR 302.4: Listed per CWA Section 311(b)(4) RQ: 10 lb (4.535 kg)
SARA 40CFR 372.65: Listed as Compound
SARA EHS 40CFR 355: Not listed
TSCA: Listed

CUP5690 CAS #: 3251-23-8

CUPRIC NITRATE

RTECS: QU7400000
DOT: NA1479
EINECS Number: 221-838-5
Molecular Formula: $CuH_2N_2O_6$
Structured MF: $Cu(NO_3)_2$
Formula Weight: 187.56

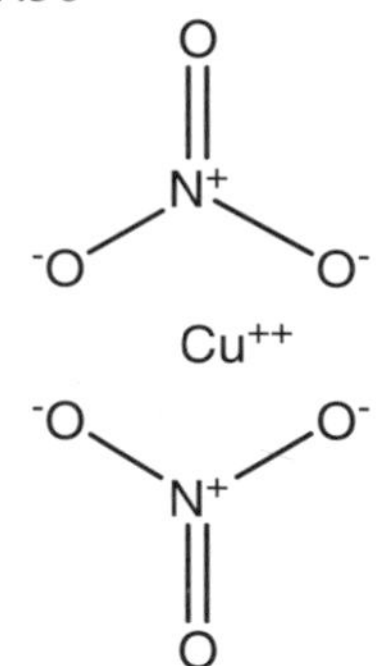

Chemical Structure

Synonyms: COPPER DINITRATE; COPPER (II) NITRATE; COPPER(2+) NITRATE; CUPRIC DINITRATE; NITRIC ACID,COPPER(2+) SALT
Description: bluish-green crystals; odorless
Use: ceramic color; mordant & oxidant in textile dyeing & printing; rgnt for burnishing metals; in nickel plating baths; in aluminum brighteners; in pyrotechnics; catalyst in rocket fuel, for org reactions; nitrating agent; in pharmaceuticals; in ore floatation, drilling mud dispersant; in sarcoma inhibition; corrosion inhibitor; soil additive for pasture or crops; preservation of mosses,etc; in insecticides, fungicides, herbicides, anti-fouling paints, electrolysis & electroplating processes, electronics, flameproofing; in cement, food & drugs, nylon, paper products, pigment & dyes, pollution control catalyst, printing & photo copying, & wood preservatives.

Physical Properties

Boiling Point: 170 °C (338 °F)
Freezing Point: 255 °C (491 °F) to 256 °C (492.8 °F)
Specific Gravity: 2.32
Water Solubility: Soluble in Water
Other Solubilities: practically Insoluble in Ethyl Acetate /Cupric nitrate trihydrate/; 100 g/100 cc Alcohol at 12.5 °C, very slightly Soluble in liq Ammonia /Cupric nitrate trihydrate/; Soluble in Alcohol /Cupric nitrate hexahydrate/.
pH: 0.2 molar aqueous solution 4.0
Flash Point: Nonflammable

RTECS Toxicity Data

Acute Oral: Rat LD_{50} Dose: 794 mg/kg; Toxic Effects: Behavioral - Somnolence (general depressed activity); Behavioral - Convulsions or effect on seizure threshold. Mouse LD_{50} Dose: 430 mg/kg; Toxic Effects: Behavioral - Somnolence (general depressed activity); Behavioral - Convulsions or effect on seizure threshold.
Irritation Eye: Rabbit Standard Draize Test Dose: 100 mg; Reaction: severe. Rabbit Nonstandard Exposure Dose: 100 mg/4S rinse; Reaction: severe.
Irritation Skin: Rabbit Standard Draize Test Dose: 500 mg; Reaction: severe.

Hazard Overviews

Fire Diamond

Health: Severe irritation to eyes/skin/respiratory tract. Also Causes: salivation, nausea, vomiting, diarrhea, hemorrhagic gastritis. Chronic Effects: eczema, cholestasis, biliary cirrhosis, liver damage, kidney, bones, CNS.
Fire: Noncombustible. However, it is a strong oxidizer capable of igniting combustibles. Use extinguishing agents suitable for surrounding fire.
Reactivity: Stable. Hazardous polymerization cannot occur. Avoid: contact with combustibles. Incompatible with: ammonia and potassium amide; acetic anhydride; potassium ferrocyanide; ammonium or potassium hexacyanoferrate (II) (above 428 °F/220 °C); tin foil. Hazardous decomposition products: copper fumes; nitrogen oxide gas.
Carcinogenicity: IARC - Not listed; NIOSH - Not listed; NTP - Not listed; ACGIH - Not listed; OSHA - Not listed; EPA - Not listed; MAK - Not listed
Primary Target Organs:

Eyes

Skin

Respiratory System
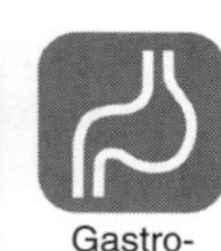
Gastro-intestinal

Environmental

Ecotoxicity: EC_{50} Thalassiosira pseudonana (alga, saltwater) 5 ug/l/72 hr /Growth rate bioassay LC_{50} Corbicula manilensis (Asiatic clam) > 2,600 ug/l/96 hr /Conditions of bioassay not specified TLm Crassostrea gigas (oysters) 1.9 ppm/hr in salt water /Conditions of bioassay not specified EC_{50} Selenastrum capricornatum (green alga) 85 ug/l/14 days /Cell volume bioassay LC_{50} Oncorhynchus kisutch (coho salmon) 286 ug/l/96 hr /Conditions of bioassay not specified EC_{10} Salmo gairdneri (rainbow trout; embryo, larvae) 16.5 ug/l/28 days; death and deformity. /Conditions of bioassay not specified

Cleanup/Disposal: Guide No. 140: Keep combustibles (wood, paper, oil, etc.) away from spilled material. Do not touch damaged containers or spilled material unless wearing appropriate protective clothing. Stop leak if you can do it without risk. Do not get water inside containers. Small Dry Spills: With clean shovel place material into clean, dry container and cover loosely; move containers from spill area. Small Liquid Spills: Use a non-combustible material like vermiculite, sand or earth to soak up the product and place into a container for later disposal. Large Spills: Dike far ahead of liquid spill for later disposal. Following product recovery, flush area with water.

Environmental Physical Data

BCF: not biomagnified
BOD: none

Regulations

RCRA 40CFR: Not listed
CERCLA: 40CFR 302.4: Listed per CWA Section 311(b)(4) RQ: 100 lb (45.35 kg)
SARA 40CFR 372.65: Listed as Compound
SARA EHS 40CFR 355: Not listed
TSCA: Listed

CUP6360 CAS #: 5893-66-3

CUPRIC OXALATE

Molecular Formula: C_2O_4Cu
Structured MF: CuC_2O_4
Formula Weight: 160.6

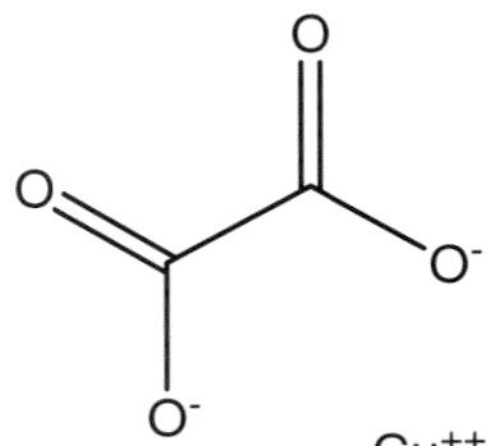

Chemical Structure

Description: bluish white solid; odorless

Physical Properties

Boiling Point: Decomposes at 1 atm
Specific Gravity: Solid > 1 at 20 °C
Water Solubility: Practically Insoluble in Water
Flash Point: Nonflammable

Hazard Overviews

Fire: Noncombustible.

Carcinogenicity: IARC - Not listed; NIOSH - Not listed; NTP - Not listed; ACGIH - Not listed; OSHA - Not listed; EPA - Not listed; MAK - Not listed

Environmental

Ecotoxicity: Food chain concentration potential: Copper known to be accumulated by shellfish. Hazard to humans unknown

Cleanup/Disposal: Guide No. 154: Eliminate all ignition sources (no smoking, flares, sparks or flames in immediate area). Do not touch damaged containers or spilled material unless wearing appropriate protective clothing. Stop leak if you can do it without risk. Prevent entry into waterways, sewers, basements or confined areas. Absorb or cover with dry earth, sand or other non-combustible material and transfer to containers. Do not get water inside containers.

Environmental Physical Data

BCF: known to be accumulated by shellfish
BOD: sewage 0.08 lb/lb, 5 days

Regulations

RCRA 40CFR: Not listed
CERCLA: 40CFR 302.4: Not listed
SARA 40CFR 372.65: Not listed
SARA EHS 40CFR 355: Not listed
TSCA: Not listed

CUP7030 CAS #: 10380-29-7

CUPRIC SULFATE, AMMONIATED

Molecular Formula: $CuH_3NO_{16}S_4$
Structured MF: $Cu(NH_3)_4SO_4$
Formula Weight: 227.73 (anhydrous)

Chemical Structure

Description: dark blue solid; ammonia odor

Physical Properties

Boiling Point: Loses H20 NH at 120 °C (248 °F) at 1 atm
Freezing Point: Decomposes at 110 °C (230 °F)
Specific Gravity: 1.8
Water Solubility: Soluble in Water
Flash Point: Noncombustible

Hazard Overviews

Corrosive

Health: Irritating to eyes/skin/respiratory tract. Other Acute Effects: may be harmful by inhalation, ingestion, or skin absorption.
Fire: Noncombustible. Hazards: emits toxic fumes. Extinguishing agents: carbon dioxide, dry chemical powder or appropriate foam. Precautions: combustible liquid.
Reactivity: Incompatible with: strong oxidizing agents, protect from moisture. Hazardous decomposition products: ammonia, sulfur oxides.
Carcinogenicity: IARC - Not listed; NIOSH - Not listed; NTP - Not listed; ACGIH - Not listed; OSHA - Not listed; EPA - Not listed; MAK - Not listed
Primary Target Organs:

Eyes

Skin

Respiratory System

Environmental

Cleanup/Disposal: Guide No. 171: Do not touch or walk through spilled material. Stop leak if you can do it without risk. Prevent dust cloud. Avoid inhalation of asbestos dust. Small Dry Spills: With clean shovel place material into clean, dry container and cover loosely; move containers from spill area. Small Spills: Take up with sand or other noncombustible absorbent material and place into containers for later disposal. Large Spills: Dike far ahead of liquid spill for later disposal. Cover powder spill with plastic sheet or tarp to minimize spreading. Prevent entry into waterways, sewers, basements or confined areas.

Regulations

RCRA 40CFR: Not listed
CERCLA: 40CFR 302.4: Not listed
SARA 40CFR 372.65: Not listed
SARA EHS 40CFR 355: Not listed
TSCA: Not listed

CUP7700 CAS #: 815-82-7

CUPRIC TARTRATE

EINECS Number: 212-425-0
Molecular Formula: $C_4H_6CuO_6$
Structured MF: $CuC_4H_4O_6$
Formula Weight: 213.64
Synonyms: BUTANEDIOIC ACID,2,3-DIHYDROXY-(R-(R*,R*))-,COPPER(2+)SALT (1:1); 2,3-DIHYDROXYBUTANEDIOIC ACID COPPER SALT; TARTARIC ACID,COPPER(2+) SALT (1:1),(+)-
Description: light blue powder; odorless
Use: in baths for copper electroplating

Physical Properties

Boiling Point: Decomposes at 1 atm
Specific Gravity: > 1
Water Solubility: Slightly Soluble in Water 200 ppm at 15 °C
Other Solubilities: Soluble in acids, alkali solution.
Flash Point: Combustible

Hazard Overviews

Fire: Combustible.
Carcinogenicity: IARC - Not listed; NIOSH - Not listed; NTP - Not listed; ACGIH - Not listed; OSHA - Not listed; EPA - Not listed; MAK - Not listed

Environmental

Ecotoxicity: Aquatic toxicity: The toxicity varies significantly, not only with species, but also with physical and chemical characteristics of the water. Concentrates from 0.015 to 3.0 mg/l cu is toxic to many fish, crustacean, mollusks, insects, and plankton
Cleanup/Disposal: Guide No. 171: Do not touch or walk through spilled material. Stop leak if you can do it without risk. Prevent dust cloud. Avoid inhalation of asbestos dust. Small Dry Spills: With clean shovel place material into clean, dry container and cover loosely; move containers from spill area. Small Spills: Take up with sand or other noncombustible absorbent material and place into containers for later disposal. Large Spills: Dike far ahead of liquid spill for later disposal. Cover powder spill with plastic sheet or tarp to minimize spreading. Prevent entry into waterways, sewers, basements or confined areas.

Environmental Physical Data

BOD: sewage .30 lb/lb, 5 days

Regulations

RCRA 40CFR: Not listed
CERCLA: 40CFR 302.4: Listed per CWA Section 311(b)(4) RQ: 100 lb (45.35 kg)
SARA 40CFR 372.65: Listed as Compound
SARA EHS 40CFR 355: Not listed
TSCA: Listed

CUP9040 CAS #: 7758-89-6

CHLORIDE CUPROUS

RTECS: GL6990000
EINECS Number: 231-842-9
Molecular Formula: ClCu
Structured MF: ClCu
Formula Weight: 98.99

Cu^+ Cl^-

Chemical Structure

Synonyms: CHLORID MEDNY; COPPER CHLORIDE; COPPER(1+) CHLORIDE; COPPER MONOCHLORIDE; CUPROUS DICHLORIDE; DICOPPER DICHLORIDE

Physical Properties

Boiling Point: 1490 °C (2714 °F)
Freezing Point: 430 °C (806 °F)
Specific Gravity: 4.14
Water Solubility: 0.06 g/l at 25 °C
Other Solubilities: Soluble in concentrated HCl, concentrated NH_4OH with complex formation

RTECS Toxicity Data

Acute Oral: Rat LD_{50} Dose: 140 mg/kg. Mouse LD_{50} Dose: 347 mg/kg.
Acute Inhalation: Mouse LC_{50} Dose: 1008 mg/m^3.
Acute Dermal: Guinea Pig LD_{50} Route: Subcutaneous Dose: 100 mg/kg.
Mutagenic: Rat Cytogenetic Analysis; Cell Type: Ascites tumor; Dose: 120 mg/kg.

Hazard Overviews

Health: Irritating to eyes/skin/respiratory tract. Toxic. Other Acute Effects: harmful if inhaled or swallowed; depending on the intensity and duration of exposure, effects may vary from mild irritation to severe destruction of tissue; exposure can cause gastrointestinal disturbances; damage to the eyes, liver, kidneys, lungs.
Fire: Hazards: emits toxic fumes. Extinguishing agents: noncombustible; use extinguishing media appropriate to surrounding fire conditions. Precautions: combustible liquid.
Reactivity: Incompatible with: light sensitive, air sensitive, moisture sensitive, oxidizing agents, alkali metals. Hazardous decomposition products: toxic fumes of: hydrogen chloride gas.
Carcinogenicity: IARC - Not listed; NIOSH - Not listed; NTP - Not listed; ACGIH - Not listed; OSHA - Not listed; EPA - Not listed; MAK - Not listed
Primary Target Organs:

Eyes

Skin

Respiratory System

Environmental

Regulations

RCRA 40CFR: Not listed
CERCLA: 40CFR 302.4: Listed as Compound per CWA Section 307(a)
SARA 40CFR 372.65: Listed as Compound
SARA EHS 40CFR 355: Not listed
TSCA: Listed

CYA1360 CAS #: 21725-46-2

CYANAZINE

RTECS: UG1490000
DOT: UN2763; UN2764; UN2997; UN2998
EINECS Number: 244-544-9
Molecular Formula: $C_9H_{13}ClN_6$
Structured MF: $C_9H_{13}ClN_6$
Formula Weight: 240.7
Synonyms: BLADEX; BLADEX R; BLADEX 80WP; BLANCHOL; 2-CHLORO-4-(1-CYANO-1-METHYLETHYLAMINO)-6-ETHYLAMINE-1,3,5-TRIAZINE; 2-CHLORO-4-(1-CYANO-1-METHYLETHYLAMINO)-6-ETHYLAMINO-1,3,5-TRIAZINE; 2-CHLORO-4-(1-CYANO-1-METHYLETHYLAMINO)-6-ETHYLAMINO-S-TRIAZINE; 2-(4-CHLORO-6-ETHYLAMINO-1,3,5-TRIAZINE-2-YLAMINO)-2-METHYLPROPIONITRILE; 2-(4-CHLORO-6-ETHYLAMINO-S-TRIAZINE-2-YLAMINO)-2-METHYL-PROPIONITRILE; 2-((4-CHLORO-6-(ETHYLAMINO)-1,3,5-TRIAZIN-2-YL)AMINO)-2-METHYLPROPANENITRILE; 2-((4-CHLORO-6-(ETHYLAMINO)-S-TRIAZIN-2-YL)AMINO)-2-METHYLPROPIONITRILE; DW 3418; DW3418; FORTROL; PAYZE; PROPANENITRILE,2-((4-CHLORO-6-(ETHYLAMINO)-1,3,5-TRIAZIN-2-YL)AMINO)-2-METHYL-; SD 15418; SD 45418; S-TRIAZINE,2-CHLORO-4-ETHYLAMINO-6-(1-CYANO-1-METHYL)ETHYLAMINO-; WL 19805
Description: white crystalline solid
Use: a pre and postemergence herbicide for control of annual grasses and broadleaf weeds; early preplant, preemergence or postemergence for field corn weed control on fallow cropland

Physical Properties

Freezing Point: 166.5 °C (331.7 °F) to 167 °C (332.6 °F)
Saturated Vapor Density: 1.2 kg/m^3
Density: Bulk packed 0.45 g/cu cm
Bulk Density: Fluffed 0.35 g/cc
Vapor Pressure: 1.6 x10^{-9} mm Hg at 20 °C
Water Solubility: 171 ppm at 25 °C in Water
Other Solubilities: Solubility at 25 °C (g/l): Benzene 15, Chloroform 210, Ethanol 45, Hexane 15.

RTECS Toxicity Data

Acute Oral: Rat LD_{50} Dose: 149 mg/kg.
Acute Inhalation: Rat LC_{Lo} Dose: >4900 mg/m^3; Toxic Effects: Behavioral - Somnolence (general depressed activity); Lungs, Thorax, or Respiration - Dyspnea.
Mutagenic: Human Cytogenetic Analysis; Cell Type: lymphocyte; Dose: 1 mg/L. Mouse Morphological Transformation; Cell Type: fibroblast; Dose: 50 mg/L.

Hazard Overviews

Carcinogenicity: IARC - Not listed; NIOSH - Not listed; NTP - Not listed; ACGIH - Not listed; OSHA - Not listed; EPA - Not listed; MAK - Not listed

Environmental

Ecotoxicity: LC_{50} Cyprinodon variegatus (Sheepshead minnow) 18 mg/l/48 hr /Conditions of bioassay not specified LC_{50} Gammarus fasciatus 2.0 mg/l/96 hr at 15 °C, mature. Static bioassay without aeration, pH 7.2-7.5, water hardness 40-50 mg/l as calcium carbonate and alkalinity of 30-35 mg/l LC_{50} Salmo gairdneri (Rainbow trout) 9.0 mg/l/96 hr at 13 °C (95% confidence interval 5.6-14.6 mg/l), wt 1.0 g. Static bioassay without aeration, pH 7.2-7.5, water hardness 40-50 mg/l as calcium carbonate and alkalinity of 30-35 mg/l
Environmental Fate: If released to soil, microbial degradation is reported to be the major environmental

degradation process. It has moderate mobility in soil; its detection in various ground waters demonstrates that it can leach. The half-life in soil typically ranges from 12 to 25 days; the USDA lists a soil half-life of 14 days. If released to water, it may degrade through microbial degradation and catalyzed hydrolysis. Uncatalyzed hydrolysis is slow (200 days or more at 25 °C and pH 5.5-9); natural water and soil constituents, such as humic and fulvic acid, may catalyze the chemical hydrolysis. Volatilization from water or soils is not expected to be an important fate process. If released to the atmosphere, it can exist in both the vapor and particulate phases; vapor phase degrades readily by reaction with photochemically produced hydroxyl radicals (estimated half-life of 3 hours). Physical removal from the atmosphere occurs through wet and dry deposition.

Cleanup/Disposal: Guide No. 131: Fully encapsulating, vapor protective clothing should be worn for spills and leaks with no fire. Eliminate all ignition sources (no smoking, flares, sparks or flames in immediate area). All equipment used when handling the product must be grounded. Do not touch or walk through spilled material. Stop leak if you can do it without risk. Prevent entry into waterways, sewers, basements or confined areas. A vapor suppressing foam may be used to reduce vapors. Small Spills: Absorb with earth, sand or other non-combustible material and transfer to containers for later disposal. Use clean non-sparking tools to collect absorbed material. Large Spills: Dike far ahead of liquid spill for later disposal. Water spray may reduce vapor; but may not prevent ignition in closed spaces. Guide No. 151: Do not touch damaged containers or spilled material unless wearing appropriate protective clothing. Stop leak if you can do it without risk. Prevent entry into waterways, sewers, basements or confined areas. Cover with plastic sheet to prevent spreading. Absorb or cover with dry earth, sand or other non-combustible material and transfer to containers. Do not get water inside containers.

Environmental Physical Data

Henry's Law Constant: estimated at 2.57×10^{-10}

Octanol/Water Partition Coefficient: log K_{ow} = 2.22

Sorption Partition Coefficient: K_{oc} = 200

BCF: calculated at 29 to 34

Regulations

RCRA 40CFR: Not listed

CERCLA: 40CFR 302.4: Listed as Compound per CWA Section 307(a) per CAA Section 112

SARA 40CFR 372.65: Listed

SARA EHS 40CFR 355: Not listed

TSCA: Not listed

Analytical Methods

Soil: EPA PMD-CUC

Water / Groundwater: EPA 629; USGS O3106

Drinking Water: EPA 508.1, 525.2; AOAC 992.14

Food: FDA 212.1, 232.1, 232.4, 242.1; AOAC 991.32

Plasma: EPA 001; FDA 252

CYA1720 CAS #: 57-12-5

CYANIDE

RTECS: GS7175000

DOT: UN1588

Molecular Formula: CN

Structured MF: CN

Formula Weight: 26.02

Synonyms: CARBON NITRIDE ION; CYANIDE(1-); CYANIDE ANION; CYANIDE ION; CYANIDE(1-) ION; CYANIDE SOLUTIONS; CYANIDE,DRY; CYANURE; HYDROCYANIC ACID,ION(1-); ISOCYANIDE

Description: varies varies; almond odor

Physical Properties

Boiling Point: Varies

Freezing Point: Varies

Specific Gravity: 2

Vapor Pressure: 0.76 torr at 800 °C

Water Solubility: Soluble in Water

Other Solubilities: Ethyl Alcohol, and Ethyl Ether

pH: Weakly acid, doesn't redden litmus

Flash Point: -17.8 °C Closed Cup

Hazard Overviews

Poison

Fire Diamond

Health: Poison. Irritating to the eyes/respiratory tract. Chronic Effects: skin rash, appetite loss, weakness, dizziness, chest discomfort, nose bleed, hearing changes. Fast acting chemical asphyxiant that prevents tissue utilization of oxygen.

Fire: Will burn. Use water as fog, dry chemical, or foam. Do not use carbon dioxide.

Reactivity: Stable. Hazardous polymerization cannot occur. Avoid: heat. Incompatible with: hypochlorite solutions at pH 10 to 10.3; nitrites at temperatures above 450 °C; chlorates; fluorine; magnesium; nitrates; all inorganic acids. Hazardous decomposition products: carbon dioxide; cyanide vapors.

Carcinogenicity: IARC - Not listed; NIOSH - Not listed; NTP - Not listed; ACGIH - Not listed; OSHA - Not listed; EPA - Class D, Not classifiable as to human carcinogenicity; MAK - Not listed

Primary Target Organs:

Eyes

Skin

Respiratory System

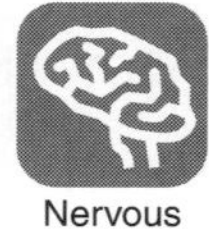
Nervous System

Cardio-vascular

Blood

Exposure Limits

OSHA PEL: TWA: 5 mg/m^3; as CN.

Respirator Recommendation

Exposure Range: >5 to <25 mg/m^3 Supplied Air, Constant Flow/Pressure Demand, Half Mask

Exposure Range: 25 to unlimited mg/m^3 Self-contained Breathing Apparatus, Pressure Demand, Full Face
Note: poor warning properties

Environmental

Cleanup/Disposal: Guide No. 157: Eliminate all ignition sources (no smoking, flares, sparks or flames in immediate area). All equipment used when handling the product must be grounded. Do not touch damaged containers or spilled material unless wearing appropriate protective clothing. Stop leak if you can do it without risk. A vapor suppressing foam may be used to reduce vapors. Do not get water inside containers. Use water spray to reduce vapors or divert vapor cloud drift. Prevent entry into waterways, sewers, basements or confined areas. Small Spills: Cover with dry earth, dry sand, or other non-combustible material followed with plastic sheet to minimize spreading or contact with rain. Use clean non-sparking tools to collect material and place it into loosely covered plastic containers for later disposal.

Regulations

RCRA 40CFR: Listed Hazardous Waste No. P030 Toxic Waste
CERCLA: 40CFR 302.4: Listed per RCRA Section 3001 RQ: 10 lb (4.535 kg)
SARA 40CFR 372.65: Not listed
SARA EHS 40CFR 355: Not listed
TSCA: Listed

Analytical Methods

Air: ASTM D4490
Soil: CLP 335.2, 335.2_M, 335.2_MA, 335.2_MB, 335.2_MC, 335.63; SW846 9010A, 9012, 9012A, 9013, 9213; EPA 8; DOE MU012R; ASTM D5049; USGS I5300, I6302
Water / Groundwater: EPA 335.1, 335.2, 335.3, 335.4; DOE SO010R; APHA 4500-CN; ASTM D2036, D4282, D4374; USGS I1300, I2302, I3300, I4302; HACH 8027
Food: EPA 9

CYA2440 CAS #: 372-09-8

CYANOACETIC ACID

RTECS: AG3675000
EINECS Number: 206-743-9
Molecular Formula: $C_3H_3NO_2$
Structured MF: $CNCH_2COO_4$
Formula Weight: 85.06

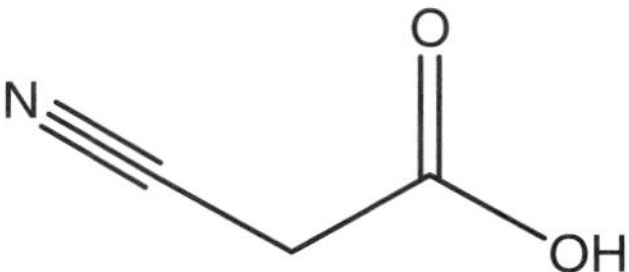

Chemical Structure

Synonyms: ACIDE CYANACETIQUE; CYANESSIGSAEURE; KYSELINA KYANOCTOVA; MALONIC ACID MONONITRILE; MALONIC MONONITRILE; MONOCYANOACETIC ACID
Description: white crystals; unpleasant odor
Use: synthesis of intermediates; mfr barbital; chemical intermediate for malonic acid; diethyl malonate for pharmaceuticals

Physical Properties

Boiling Point: 108 °C (226 °F) at 15 mm Hg
Freezing Point: 66 °C (150.8 °F)
Specific Gravity: Solid > 1.1 at 20 °C
Water Solubility: Soluble in Water
Other Solubilities: Slightly Soluble in Acetic Acid.
Flash Point: 107 °C Closed Cup

RTECS Toxicity Data

Acute Oral: Rat LD_{50} Dose: 1500 mg/kg; Toxic Effects: Behavioral - Change in motor activity (specific assay); Lungs, Thorax, or Respiration - Dyspnea; Skin and appendages - Hair.
Acute Dermal: Rabbit LD_{Lo} Route: Subcutaneous Dose: 1900 mg/kg; Toxic Effects: Peripheral Nerve and sensation - Spastic parapysis with or without sensory change; Lungs, Thorax, or Respiration - Respiratory depression; Nutritional and gross metabolic - Body temperature Frog LD_{Lo} Route: Subcutaneous Dose: 1300 mg/kg; Toxic Effects: Lungs, Thorax, or Respiration - Dyspnea.

Hazard Overviews

Corrosive

Health: Corrosive to eyes/skin/respiratory tract. Harmful. Other Acute Effects: harmful if swallowed, inhaled, or absorbed through skin; material is extremely destructive to tissue of the mucous membranes and upper respiratory tract, eyes and skin; inhalation may result in spasm, inflammation and edema of the larynx and bronchi, chemical pneumonitis and pulmonary edema; symptoms of exposure may include burning sensation, coughing, wheezing, laryngitis, shortness of breath, headache, nausea and vomiting; target organs: cardiovascular system, nerves.
Fire: Will burn. Hazards: emits toxic fumes. Extinguishing agents: water spray; carbon dioxide, dry chemical powder or appropriate foam. Precautions: combustible liquid.
Reactivity: Stable. Hazardous polymerization will not occur. Incompatible with: acids, bases, strong oxidizing agents, strong reducing agents, protect from moisture. Hazardous decomposition products: toxic fumes of: carbon monoxide, carbon dioxide, nitrogen oxides, hydrogen cyanide, acetonitrile.
Carcinogenicity: IARC - Not listed; NIOSH - Not listed; NTP - Not listed; ACGIH - Not listed; OSHA - Not listed; EPA - Not listed; MAK - Not listed

Primary Target Organs:

Environmental

Cleanup/Disposal: Guide No. 157: Eliminate all ignition sources (no smoking, flares, sparks or flames in immediate area). All equipment used when handling the product must be grounded. Do not touch damaged containers or spilled material unless wearing appropriate protective clothing. Stop leak if you can do it without risk. A vapor suppressing foam may be used to reduce vapors. Do not get water inside containers. Use water spray to reduce vapors or divert vapor cloud drift. Prevent entry into waterways, sewers, basements or confined areas. Small Spills: Cover with dry earth, dry sand, or other non-combustible material followed with plastic sheet to minimize spreading or contact with rain. Use clean non-sparking tools to collect material and place it into loosely covered plastic containers for later disposal.

Environmental Physical Data

BCF: no food chain concentration potential

Regulations

RCRA 40CFR: Not listed
CERCLA: 40CFR 302.4: Listed as Compound per CWA Section 307(a) per CAA Section 112
SARA 40CFR 372.65: Not listed
SARA EHS 40CFR 355: Not listed
TSCA: Listed

CYA2800 CAS #: 140-87-4

CYANOACETIC ACID, HYDRAZIDE

RTECS: AG4200000
EINECS Number: 205-437-2
Molecular Formula: $C_3H_5N_3O$
Formula Weight: 99.09

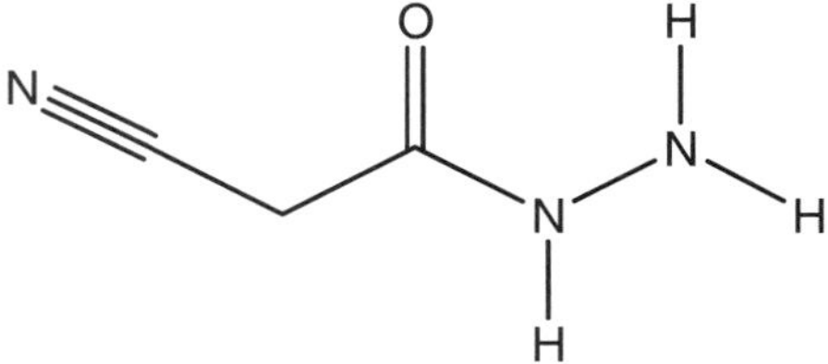

Chemical Structure

Synonyms: AB-42; ARMAZAL; BENECID; CIANAZIL; CYACETACID; CYACETACIDE; CYACETAZID; CYACETAZIDE; CYANACETHYDRAZIDE; CYANACETHYDRAZINE; CYANACETIC ACID HYDRAZIDE; CYANACETOHYDRAZIDE; CYANACETYLHYDRAZIDE; CYANAZIDE; CYANIZIDE; CYANOACETHYDRAZIDE; CYANOACETIC ACID HYDRAZIDE; ALPHA-CYANOACETOHYDRAZIDE; CYANOACETOHYDRAZIDE; CYANOACETYL HYDRAZIDE; CYANOACETYLHYDRAZIDE; CYANOETHYDRAZIDE; CYAZID; CYAZIDE; DICTYCIDE; DICTYZIDE; DIETYCIDE; HELMOX; HIDACIAN; HIDACIANN; KYANACETHYDRAZID; LEANDIN; MACKREAZID; MALONITRILE HYDRAZIDE; MALONONITRILE HYDRAZIDE; NEOHYDRAZID; REACID; REAZID; REAZIDE; TSIAZID
Description: stout prisms
Use: antitubercular; vet: anthelmintic (lungworms)

Physical Properties

Freezing Point: 114.5 °C (238.1 °F) to 115 °C (239 °F)
Water Solubility: Very Soluble in Water
Other Solubilities: Soluble in Alcohol. . Practically Insoluble in Ether
pH: Very slightly acid

RTECS Toxicity Data

Acute Oral: Mouse LD_{50} Dose: 250 mg/kg.
Acute Dermal: Mouse LD_{50} Route: Subcutaneous Dose: 228 mg/kg.

Hazard Overviews

Health: Irritating to eyes/skin/respiratory tract. Toxic. Other Acute Effects: harmful if swallowed, inhaled, or absorbed through skin.
Fire: Hazards: emits toxic fumes. Extinguishing agents: water spray; carbon dioxide, dry chemical powder or appropriate foam. Precautions: combustible liquid.
Reactivity: Incompatible with: strong acids, strong bases, strong oxidizing agents, strong reducing agents. Hazardous decomposition products: thermal decomposition may produce carbon monoxide, carbon dioxide, and nitrogen oxides.
Carcinogenicity: IARC - Not listed; NIOSH - Not listed; NTP - Not listed; ACGIH - Not listed; OSHA - Not listed; EPA - Not listed; MAK - Not listed
Primary Target Organs:

Respiratory System

Environmental

Regulations

RCRA 40CFR: Not listed
CERCLA: 40CFR 302.4: Listed as Compound per CWA Section 307(a) per CAA Section 112
SARA 40CFR 372.65: Not listed
SARA EHS 40CFR 355: Not listed
TSCA: Not listed

CYA4960 CAS #: 2526-62-7

CYANOETHYLTRIMETHOXYSILANE

EINECS Number: 219-764-3

Physical Properties

Boiling Point: 213 °C (415 °F)
Freezing Point: <
Specific Gravity: 1.054
Vapor Density: > 1 Air=1
Vapor Pressure: < 1
Water Solubility: Reacts Slowly
Evaporation Rate: < 1 Butyl Acetate=1

Hazard Overviews

Carcinogenicity: IARC - Not listed; NIOSH - Not listed; NTP - Not listed; ACGIH - Not listed; OSHA - Not listed; EPA - Not listed; MAK - Not listed

Environmental

Regulations

RCRA 40CFR: Not listed
CERCLA: 40CFR 302.4: Listed as Compound per CWA Section 307(a) per CAA Section 112
SARA 40CFR 372.65: Not listed
SARA EHS 40CFR 355: Not listed
TSCA: Not listed

CYA5320 CAS #: 460-19-5

CYANOGEN

RTECS: GT1925000
DOT: UN1026; IMO2.3
EINECS Number: 207-306-5
Molecular Formula: C_2N_2
Structured MF: $(CN)_2$
Formula Weight: 52.04
Synonyms: CARBON NITRIDE; CYANOGENE; CYANOGEN,LIQUEFIED; DICYAN; DICYANOGEN; ETHANEDINITRILE; MONOCYANOGEN; NITRILOACETONITRILE; OXALIC ACID DINITRILE; OXALIC NITRILE; OXALONITRILE; OXALYL CYANIDE; PRUSSITE
Description: colorless gas; pungent, penetrating odor
Use: organic synthesis; fumigant; fuel gas for welding & cutting heat-resistant metals; rocket & missile propellant

Physical Properties

Boiling Point: -21.17 °C (-6 °F)
Freezing Point: -27.9 °C (-18.22 °F)
Specific Gravity: 0.9537 at -21.17 °C/4 °C
Vapor Density: 1.8 Air=1
Vapor Pressure: 400 mm Hg at -33 °C
Water Solubility: 450 cc in 100 ml Water at 20 °C
Other Solubilities: 2300 cc in 100 ml ethyl Alcohol at 20 °C; 500 cc in 100 ml ethyl Ether at 20 °C.
Surface Tension: 22 dynes/cm at 21 °C
Odor Threshold: 500 mg/m^3
Refraction Index: 1.327 at 18 °C (liq)
Critical Temperature: 127 °C
Critical Pressure: 857 psia
Ionization Potential (eV): 13.57
Flash Point: Flammable gas
LEL: 6.6% v/v
UEL: 32% v/v

RTECS Toxicity Data

Acute Inhalation: Rat LC_{50} Dose: 350 ppm/1hr; Toxic Effects: Lungs, Thorax, or Respiration - Respiratory obstruction; Sense organs and special senses - Lacrimation; Behavioral - Somnolence (general depressed activity). Human TC_{Lo} Dose: 16 ppm; Toxic Effects: Sense organs and special senses - Conjunctive irritation; Sense organs and special senses - Change in olfactory nerve.
Acute Dermal: Rabbit LD_{Lo} Route: Subcutaneous Dose: 13 mg/kg; Toxic Effects: Behavioral - Convulsions or effect on seizure threshold. Dog LD_{Lo} Route: Subcutaneous Dose: 15 mg/kg.
Chronic (Multiple Dose) Inhalation: Rat Dose: 25 ppm/6H/26W-I; Toxic Effects: Nutritional and gross metabolic - Weight loss or decreased weight gain.

Hazard Overviews

Poison

Corrosive

Flammable

Compressed Gas

Fire Diamond

Health: Corrosive, causes severe burns to eyes/skin/respiratory tract. Poison. Stored as a compressed gas which can cause frostbite. Also Causes: pulmonary edema
Fire: Dangerously flammable gas with a wide range of explosibility. Try to stop the flow of cyanogen gas and use a water spray to protect personnel attempting to do so. Fight fire from maximum distance.
Reactivity: Stable. Hazardous polymerization cannot occur. Avoid: water; heat; ignition. Incompatible with: acids or their fumes; water; fluorine; chlorine. Hazardous decomposition products: hydrogen cyanide.
Carcinogenicity: IARC - Not listed; NIOSH - Not listed; NTP - Not listed; ACGIH - Not listed; OSHA - Not listed; EPA - Not listed; MAK - Not listed
Primary Target Organs:

Eyes

Skin
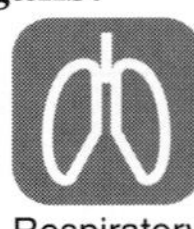
Respiratory System

Mucous Membranes

Exposure Limits
OSHA PEL Vacated 1989 Limits: TWA: 10 ppm; 20 mg/m^3.
ACGIH TLV: TWA: 10 ppm; 21 mg/m^3.
NIOSH REL: TWA: 10 ppm; 20 mg/m^3.
DFG MAK: TWA: 10 ppm; 22 mg/m^3.
Respirator Recommendation
Exposure Range: >10 to 500 ppm Supplied Air, Constant Flow/Pressure Demand, Half Mask
Exposure Range: >500 to 10,000 ppm Supplied Air, Constant Flow/Pressure Demand, Full Face
Exposure Range: >10,000 to unlimited ppm Self-contained Breathing Apparatus, Pressure Demand, Full Face

Note: odor threshold unknown

Environmental

Cleanup/Disposal: Guide No. 119: Eliminate all ignition sources (no smoking, flares, sparks or flames in immediate area). All equipment used when handling the product must be grounded. Fully encapsulating, vapor protective clothing should be worn for spills and leaks with no fire. Do not touch or walk through spilled material. Stop leak if you can do it without risk. Do not direct water at spill or source of leak. Use water spray to reduce vapors or divert vapor cloud drift. For chlorosilanes, use AFFF alcohol-resistant medium expansion foam to reduce vapors. If possible, turn leaking containers so that gas escapes rather than liquid. Prevent entry into waterways, sewers, basements or confined areas. Isolate area until gas has dispersed.

Environmental Physical Data

Henry's Law Constant: 9.91
BCF: no food chain concentration potential

Regulations

RCRA 40CFR: Listed Hazardous Waste No. P031 Toxic Waste
CERCLA: 40CFR 302.4: Listed per RCRA Section 3001 RQ: 100 lb (45.35 kg)
SARA 40CFR 372.65: Not listed
SARA EHS 40CFR 355: Not listed
TSCA: Listed

Analytical Methods

Air: ASTM D4490

CYA5680 CAS #: 506-68-3

CYANOGEN BROMIDE

RTECS: GT2100000
DOT: UN1889; IMO6.1
EINECS Number: 208-051-2
Molecular Formula: CBrN
Structured MF: CNBr
Formula Weight: 105.93

Br———≡N

Chemical Structure

Synonyms: BROMINE CYANIDE; BROMOCYAN; BROMOCYANIDE; BROMOCYANOGEN; BROMURE DE CYANOGEN; CAMPILIT; CYANO BROMIDE; CYANOBROMIDE; CYANOGEN MONOBROMIDE; TL 822
Description: colorless or white needle-like crystals or cubes; penetrating odor
Use: organic synthesis; parasiticide; fumigation compositions; rat exterminants; cyaniding reagent in gold extraction processes

Physical Properties

Boiling Point: 61.4 °C (143 °F) at 760 mm Hg
Freezing Point: 52 °C (125.6 °F)
Specific Gravity: 2.015 at 20 °C/4 °C
Vapor Density: 3.6 Air=1
Vapor Pressure: 20 mm Hg at -1 °C
Water Solubility: > 10% in Water
Other Solubilities: Alcohol: Soluble; Benzene: Soluble; Ether: Soluble.
pH: Acid
Ionization Potential (eV): 11.84 +/-0.01
Flash Point: Nonflammable

RTECS Toxicity Data

Acute Inhalation: Human LC_{Lo} Dose: 92 ppm/10M. Mouse LC_{Lo} Dose: 500 mg/m^3/10M.

Hazard Overviews

Poison

Explosive

Health: Irritating to eyes/skin/respiratory tract. Poison. Other Acute Effects: may be fatal if inhaled, swallowed, or absorbed through skin; symptoms of exposure may include burning sensation; coughing; wheezing; laryngitis; shortness of breath; headache; nausea; vomiting; may cause nervous system disturbances; exposure can cause cyanosis. Chronic Effects: possible carcinogen/mutagen; target organs: blood, liver, pancreas, nerves. Possible human carcinogen.
Fire: Noncombustible. Hazards: may explode when heated; vapor may travel considerable distance to source of ignition and flash back; emits toxic fumes. Extinguishing agents: carbon dioxide, dry chemical powder or appropriate foam. Precautions: combustible liquid.
Reactivity: Incompatible with: strong oxidizing agents, may decompose on exposure to moist air or water. Hazardous decomposition products: toxic fumes of: carbon monoxide, carbon dioxide, nitrogen oxides, hydrogen bromide gas, hydrogen cyanide.
Carcinogenicity: IARC - Not listed; NIOSH - Not listed; NTP - Not listed; ACGIH - Not listed; OSHA - Not listed; EPA - Not listed; MAK - Not listed
Primary Target Organs:

Eyes

Skin

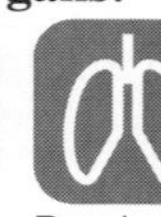
Respiratory System

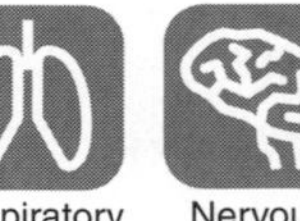
Nervous System

Liver

Blood

Environmental

Cleanup/Disposal: Guide No. 157: Eliminate all ignition sources (no smoking, flares, sparks or flames in immediate area). All equipment used when handling the product must be grounded. Do not touch damaged containers or spilled material unless wearing appropriate protective clothing. Stop leak if you can do it without risk. A vapor suppressing foam may be used to reduce vapors. Do not get water inside containers. Use water spray to reduce vapors or divert vapor cloud drift. Prevent entry into waterways, sewers, basements or confined areas. Small Spills: Cover with dry earth, dry sand, or other non-combustible material followed with plastic sheet to minimize spreading or contact with rain. Use clean non-sparking tools to collect material and place it into loosely covered plastic containers for later disposal.

Environmental Physical Data

BCF: no food chain concentration potential

Regulations

RCRA 40CFR: Listed Hazardous Waste No. U246 Toxic Waste
CERCLA: 40CFR 302.4: Listed per RCRA Section 3001 RQ: 1000 lb (453.5 kg)
SARA 40CFR 372.65: Not listed TPQ: 500/10000 lb
SARA EHS 40CFR 355: Listed TPQ: 1,000 lb
TSCA: Listed

CYA6040 CAS #: 506-77-4

CYANOGEN CHLORIDE

RTECS: GT2275000
DOT: UN1589; IMO2.3
EINECS Number: 208-052-8
Molecular Formula: CClN
Structured MF: CNCl
Formula Weight: 61.48
Synonyms: CHLORCYAN; CHLORINE CYANIDE; CHLOROCYAN; CHLOROCYANIDE; CHLOROCYANOGEN; CHLORURE DE CYANOGENE; CYANOGEN CHLORIDE,INHIBITED; EPA PESTICIDE CODE 025801
Description: colorless gas, liquid; pungent, penetrating odor
Use: chem mfr; military poison gas; warning agent in fumigant gases; in insecticide; in preparation of extremely pure malononitrile; metal cleaner; in ore refining; production of synthetic rubber; derivatives (acrylonitrile, cyanamide, cyanogen chloride, cyanides, & nitroprusside) in industry in fumigating ships & warehouses, in ore-extracting processes.

Physical Properties

Boiling Point: 12.7 °C (55 °F) at 760 mm Hg
Freezing Point: -6 °C (21.2 °F)
Specific Gravity: 1.186 at 20 °C/4 °C
Vapor Density: 2.1 Air=1
Vapor Pressure: 1010 mm Hg
Water Solubility: 7% by weight
Other Solubilities: 10000 cc in 100 ml Ethanol at 20 °C; 5000 cc in 100 ml Ether at 20 °C.
Surface Tension: 24.6 dynes/cm at 10 °C
Odor Threshold: 2 mg/m^3
Ionization Potential (eV): 12.49
Flash Point: Nonflammable

RTECS Toxicity Data

Acute Oral: Cat LD_{50} Dose: 6 mg/kg.
Acute Inhalation: Human TC_{Lo} Dose: 10 mg/m^3; Toxic Effects: Sense organs and special senses - Lacrimation; Sense organs and special senses - Conjunctive irritation; Lungs, Thorax, or Respiration - Chronic pulmonary edema. Man TC_{Lo} Dose: 2 gm/m^3; Toxic Effects: Skin and appendages - Primary irritation.
Acute Dermal: Rabbit LD_{Lo} Route: Subcutaneous Dose: 20 mg/kg. Dog LD_{Lo} Route: Subcutaneous Dose: 5 mg/kg.

Hazard Overviews

Poison

Fire Diamond

Health: Severe irritation to eyes/skin/respiratory tract. Poison. Also Causes: pulmonary edema. Chronic Effects: hoarseness, conjunctivitis, and edema of the eyelid.
Fire: Noncombustible. Use agent suitable for surrounding fire. However, do not use water as it reacts violently with cyanogen chloride.
Reactivity: Stable. Hazardous polymerization cannot occur. Avoid: water; heat. Incompatible with: water; acids; acid fumes; strong oxiding agents. Hazardous decomposition products: hydrogen cyanide.
Carcinogenicity: IARC - Not listed; NIOSH - Not listed; NTP - Not listed; ACGIH - Not listed; OSHA - Not listed; EPA - Not listed; MAK - Not listed
Primary Target Organs:

Eyes

Skin

Respiratory System

Mucous Membranes

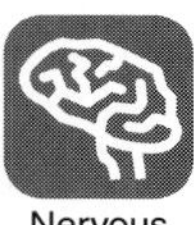
Nervous System

Exposure Limits
OSHA PEL Vacated 1989 Limits: TWA: 0.3 ppm; 0.6 mg/m^3.
ACGIH TLV: STEL: 0.3 ppm; 0.75 mg/m^3; Ceiling.
NIOSH REL: STEL: 0.3 ppm; 0.6 mg/m^3.
Respirator Recommendation
Exposure Range: >0.3 to 15 ppm Supplied Air, Constant Flow/Pressure Demand, Half Mask
Exposure Range: >15 to 300 ppm Supplied Air, Constant Flow/Pressure Demand, Full Face
Exposure Range: >300 to unlimited ppm Self-contained Breathing Apparatus, Pressure Demand, Full Face
Note: odor threshold unknown

Environmental

Ecotoxicity: Aquatic toxicity: 0.08 ppm/*/fish/killed/fresh water *Duration not specified
Cleanup/Disposal: Guide No. 125: Fully encapsulating, vapor protective clothing should be worn for spills and leaks with no fire. Do not touch or walk through spilled material. Stop leak if you can do it without risk. If possible, turn leaking containers so that gas escapes rather than liquid. Prevent entry into waterways, sewers, basements or confined areas. Do not direct water at spill or source of leak. Use water spray to reduce vapors or divert vapor cloud drift. Isolate area until gas has dispersed.

Environmental Physical Data

Octanol/Water Partition Coefficient: log K_{ow} = calculated at 0.64
BCF: no food chain concentration potential

Regulations

RCRA 40CFR: Listed Hazardous Waste No. P033 Toxic Waste
CERCLA: 40CFR 302.4: Listed per CWA Section 311(b)(4) per RCRA Section 3001 RQ: 10 lb (4.535 kg)
SARA 40CFR 372.65: Not listed
SARA EHS 40CFR 355: Not listed
TSCA: Listed

Analytical Methods

Air: ASTM D4490
Water / Groundwater: APHA 4500-CN; ASTM D4165

CYA6400 CAS #: 506-78-5

CYANOGEN IODIDE

RTECS: NN1750000
EINECS Number: 208-053-3
Molecular Formula: CIN
Formula Weight: 152.92

N≡—I

Chemical Structure

Synonyms: IODINE CYANIDE; JODCYAN
Description: white needles; pungent odor
Use: for destroying all lower forms of life; in taxidermy for preserving insects, butterflies, etc

Physical Properties

Boiling Point: Sublimes at 45 °C (113 °F)
Freezing Point: 146 °C (294.8 °F) to 147 °C (296.6 °F)
Specific Gravity: 2.84 at 18 °C
Vapor Pressure: 90.2 kPa at 150 °C
Water Solubility: Soluble in Water
Other Solubilities: Soluble in Alcohol and Ethanol
Ionization Potential (eV): 10.870 +/-0.001

RTECS Toxicity Data

Acute Oral: Cat LD_{Lo} Dose: 18 mg/kg.
Acute Dermal: Mouse LD_{Lo} Route: Subcutaneous Dose: 27 mg/kg. Dog LD_{Lo} Route: Subcutaneous Dose: 19 mg/kg.

Hazard Overviews

Poison

Health: Irritating to eyes/skin/respiratory tract. Poison. Other Acute Effects: may be fatal if inhaled, swallowed, or absorbed through skin.
Fire: Hazards: emits toxic fumes. Extinguishing agents: noncombustible; use extinguishing media appropriate to surrounding fire conditions. Precautions: combustible liquid.
Reactivity: Incompatible with: strong oxidizing agents, strong acids, strong bases, may decompose on exposure to light. Hazardous decomposition products: hydrogen iodide, hydrogen cyanide.
Carcinogenicity: IARC - Not listed; NIOSH - Not listed; NTP - Not listed; ACGIH - Not listed; OSHA - Not listed; EPA - Not listed; MAK - Not listed
Primary Target Organs:

Eyes

Skin

Respiratory System

Environmental

Regulations

RCRA 40CFR: Not listed
CERCLA: 40CFR 302.4: Listed as Compound per CWA Section 307(a) per CAA Section 112
SARA 40CFR 372.65: Not listed TPQ: 1000/10000 lb
SARA EHS 40CFR 355: Listed TPQ: 1000 lb
TSCA: Listed

CYA7480 CAS #: 2636-26-2

CYANOPHOS

RTECS: TF7600000
DOT: UN2783; UN2784; UN3017; UN3018; IMO3.2; IMO6.1
EINECS Number: 220-130-3
Molecular Formula: $C_9H_{10}NO_3PS$
Formula Weight: 243.21
Synonyms: BAY 34727; BAYER 34727; CIAFOS; O-(4-CYANOPHENYL) O,O-DIMETHYL PHOSPHOROTHIOATE; O-P-CYANOPHENYL O,O-DIMETHYL PHOSPHOROTHIOATE; (O-P-CYANOPHENYL-O,O-DIMETHYLPHOSPHOROTHIOATE); CYANOX; CYAP; CYNOCK; 4-(DIMETHOXYPHOSPHINOTHIOYLOXY)BENZONITRILE; O,O,-DIMETHYL O-(4-CYANOPHENYL) PHOSPHOROTHIOATE; O,O-DIMETHYL O-4-CYANOPHENYL PHOSPHOROTHIOATE; O,O-DIMETHYL O-(4-CYANOPHENYL) THIONOPHOSPHATE; O,O-DIMETHYL O-4-CYANOPHENYL THIOPHOSPHATE; O,O-DIMETHYL PHOSPHOROTHIOATE O-ESTER WITHP-HYDROXYBENZONITRILE; O,O-DIMETHYL-O-P-CYANOPHENYL PHOSPHOROTHIOATE; O,O-DIMETHYL-O-(4-CYANO-PHENYL)-MONOTHIOPHOSPHAT; ENT 25,675; EPA PESTICIDE CHEMICAL CODE 268200; MAY & BAKER S-4084; OMS 226; OMS 869; PHOSPHOROTHIOIC ACID O-(4-CYANOPHENYL) O,O-DIMETHYL ESTER; PHOSPHOROTHIOIC ACID O-(4-CYANOPHENYL)O,O-DIMETHYL ESTER; PHOSPHOROTHIOIC ACID O,O-DIMETHYL ESTER,O-ESTER WITHP-HYDROXYBENZONITRILE; PHOSPHOROTHIOIC ACID,O-(4-CYANOPHENYL) O,O-DIMETHYL ESTER; PHOSPHOROTHIOIC ACID,O,O-DIMETHYL ESTER,O-ESTER WITHP-HYDROXYBENZONITRILE; S 4084; SUMITOMO S 4084; SUNITOMO S 4084
Description: yellow to reddish-yellow transparent liquid

Use: insecticide to control lepidopterous pests and sucking insects on fruit, ornamentals and vegetables; locust control insecticide; control of household pests such as cockroaches, houseflies,and mosquitoes; grain protectant

Physical Properties

Boiling Point: Decomposes at 119 °C (246 °F) to 120 °C (248 °F)
Freezing Point: 14 °C (57.2 °F) to 15 °C (59 °F)
Specific Gravity: 1.255 to 1.265 at 25 °C
Vapor Pressure: 105 mPa at 20 °C
Water Solubility: Slightly Soluble in Water
Other Solubilities: In n-hexane 27, Methanol 1000, Xylene 1000 (all in g/kg) at 20 °C; miscible in Benzene, ketones, Toluene, Xylene.
Refraction Index: 1.5404 at 32.5 °C/D

RTECS Toxicity Data

Acute Oral: Rat LD_{50} Dose: 215 mg/kg. Mouse LD_{50} Dose: 720 mg/kg.
Acute Dermal: Rat LD_{50} Route: Skin; Dose: 800 mg/kg. Mouse LD_{50} Route: Skin; Dose: >2 gm/kg.

Hazard Overviews

Carcinogenicity: IARC - Not listed; NIOSH - Not listed; NTP - Not listed; ACGIH - Not listed; OSHA - Not listed; EPA - Not listed; MAK - Not listed

Environmental

Ecotoxicity: LC_{50} Carp 5 mg/l (48 hr) LC_{50} Harlequin fish 36 mg/l (24 hr)
Cleanup/Disposal: Guide No. 131: Fully encapsulating, vapor protective clothing should be worn for spills and leaks with no fire. Eliminate all ignition sources (no smoking, flares, sparks or flames in immediate area). All equipment used when handling the product must be grounded. Do not touch or walk through spilled material. Stop leak if you can do it without risk. Prevent entry into waterways, sewers, basements or confined areas. A vapor suppressing foam may be used to reduce vapors. Small Spills: Absorb with earth, sand or other non-combustible material and transfer to containers for later disposal. Use clean non-sparking tools to collect absorbed material. Large Spills: Dike far ahead of liquid spill for later disposal. Water spray may reduce vapor; but may not prevent ignition in closed spaces. Guide No. 152: Do not touch damaged containers or spilled material unless wearing appropriate protective clothing. Stop leak if you can do it without risk. Prevent entry into waterways, sewers, basements or confined areas. Cover with plastic sheet to prevent spreading. Absorb or cover with dry earth, sand or other non-combustible material and transfer to containers. Do not get water inside containers.

Regulations

RCRA 40CFR: Not listed
CERCLA: 40CFR 302.4: Listed as Compound per CWA Section 307(a) per CAA Section 112
SARA 40CFR 372.65: Not listed TPQ: 1000 lb
SARA EHS 40CFR 355: Listed TPQ: 1000 lb
TSCA: Not listed

Analytical Methods

Food: FDA 232.4, 242.1

CYA8200 CAS #: 108-80-5

CYANURIC ACID

RTECS: XZ1800000
EINECS Number: 203-618-0
Molecular Formula: $C_3H_3N_3O_3$
Formula Weight: 129.08

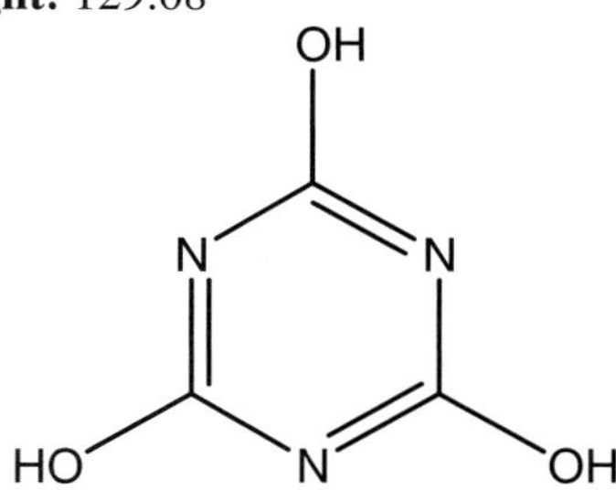

Chemical Structure

Synonyms: ISOCYANURIC ACID; KYSELINA KYANUROVA; NORMAL CYANURIC ACID; PSEUDOCYANURIC ACID; STYM-TRIAZINE-2,4,6-TRIOL; S-2,4,6-TRIAZINETRIOL; S-TRIAZINE-2,4,6-TRIOL; SYM-TRIAZINETRIOL; 1,3,5-TRIAZINE-2,4,6(1H,3H,5H)-TRIONE; S-TRIAZINE-2,4,6(1H,3H,5H)-TRIONE; TRICYANIC ACID; TRIHYDROXYCYANIDINE; 2,4,6-TRIHYDROXY-1,3,5-TRIAZINE; TRIHYDROXYTRIAZINE
Description: crystalline powder; odorless
Use: lab source of cyanic acid gas; selective herbicide; in prepn of melamine, sponge rubber; in chemical syntheses & as intermediate for chlorinated bleaches; whitening agent; in swimming pools to lower the rate of photochemical reduction of chlorine, hypochlorous acid and hypochlorite ion

Physical Properties

Boiling Point: Decomposes
Freezing Point: 360 °C (680 °F)
Specific Gravity: 2.5 at 20 °C/4 °C
Water Solubility: 1 g dissolves in about 200 ml Water
Other Solubilities: 95% Ethanol: Soluble in hot solvent; Acetone: Insoluble; Benzene: Insoluble; Ether: Insoluble.
pH: Saturated aqueous solution at room temperature 4.8

RTECS Toxicity Data

Acute Oral: Rat LD_{50} Dose: 7700 mg/kg. Mouse LD_{50} Dose: 3400 mg/kg.
Chronic (Multiple Dose) Oral: Rat Dose: 5460 mg/kg/26W-I; Toxic Effects: Kidney, Ureter, and Bladder - Other changes; Nutritional and gross metabolic - Weight loss or decreased weight gain.
Irritation Eye: Rabbit Standard Draize Test Dose: 500 mg/24H; Reaction: mild. Rabbit Nonstandard Exposure Dose: 20 mg/24H rinse; Reaction: mild.

Tumorigenic: Rat Route: Oral; Dose: 55 gm/kg/82W-I; Toxic Effects: Tumorigenic - Equivocal tumorigenic agent by RTECS criteria; Liver - Tumors; Skin and appendages - Tumors. Rat Route: Oral; Dose: 60750 mg/kg/81W-I; Toxic Effects: Tumorigenic - Equivocal tumorigenic agent by RTECS criteria; Skin and appendages - Tumors.

Hazard Overviews

Explosive

Health: Severely irritating to eyes/skin/respiratory tract. Harmful. Other Acute Effects: harmful if swallowed, inhaled, or absorbed through skin; symptoms of exposure may include burning sensation; coughing; wheezing; laryngitis; shortness of breath; headache; nausea; vomiting.
Fire: Hazards: emits toxic fumes. Extinguishing agents: water spray; carbon dioxide, dry chemical powder or appropriate foam. Precautions: combustible liquid.
Reactivity: Incompatible with: strong oxidizing agents. Hazardous decomposition products: toxic fumes of: carbon monoxide, carbon dioxide, nitrogen oxides.
Carcinogenicity: IARC - Not listed; NIOSH - Not listed; NTP - Not listed; ACGIH - Not listed; OSHA - Not listed; EPA - Not listed; MAK - Not listed
Primary Target Organs:

Eyes

Skin

Respiratory System

Exposure Limits
AIHA WEEL: TWA: 10 mg/m^3 total; OTHER: 5 mg/m^3 respirable.
Respirator Recommendation
Exposure Range: >5 to 250 mg/m^3 Supplied Air, Constant Flow/Pressure Demand, Half Mask
Exposure Range: >250 to 5000 mg/m^3 Supplied Air, Constant Flow/Pressure Demand, Full Face
Exposure Range: >5000 to unlimited mg/m^3 Self-contained Breathing Apparatus, Pressure Demand, Full Face
Note: odor threshold unknown

Environmental

Environmental Fate: If released to soil, it is expected to be highly mobile. If released to water, it will be essentially nonvolatile. It has a potential to photolyze directly, since it absorbs UV light above 290 nm, but the kinetics of the potential photolysis are not known. It biodegrades readily under a wide variety of natural conditions (1-10 ug/ml cyanuric acid), and particularly well in systems of either low or zero dissolved-oxygen levels, such as anaerobic activated sludge and sewage, soils, mud, and muddy streams and river waters, as well as ordinary aerated activated sludge systems with typically low (1 to 3 ppm) dissolved-oxygen levels. However, at higher concentrations (100 mg) it was determined to be inhibitory to biodegradation. Aquatic bioconcentration and adsorption are not expected to be important fate processes. If released to the atmosphere, it will exist in both the vapor and particulate phases. In the vapor phase, it will degrade in the atmosphere by reaction with photochemically produced hydroxyl radicals with an estimated half-life of approximately 102 days. Physical removal of particulate from air is likely to occur through wet and dry deposition.

Environmental Physical Data

Henry's Law Constant: estimated at 1.36 $x10^{-18}$
Sorption Partition Coefficient: K_{oc} = estimated at 124
BCF: estimated at 8.5

Regulations

RCRA 40CFR: Not listed
CERCLA: 40CFR 302.4: Listed as Compound per CWA Section 307(a) per CAA Section 112
SARA 40CFR 372.65: Not listed
SARA EHS 40CFR 355: Not listed
TSCA: Listed

Analytical Methods

Indoor / Expired Air: NIOSH 5030

CYA8560 CAS #: 108-77-0

CYANURIC CHLORIDE

RTECS: XZ1400000
EINECS Number: 203-614-9
Molecular Formula: $C_3Cl_3N_3$
Formula Weight: 184.41

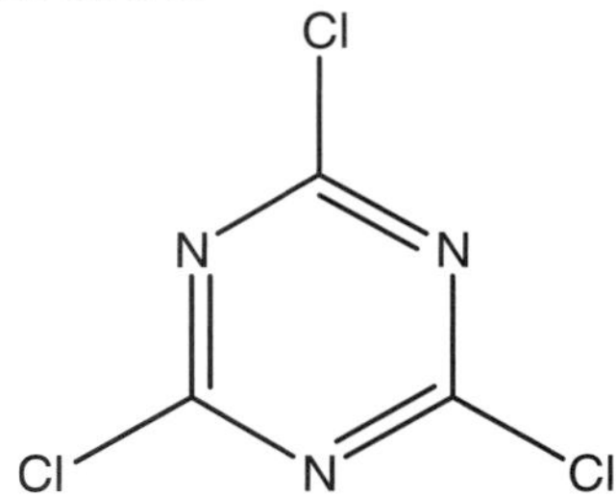

Chemical Structure

Synonyms: CHLOROTRIAZINE; CYANUR CHLORIDE; CYANURCHLORIDE; CYANURIC ACID CHLORIDE; CYANURIC ACID TRICHLORIDE; CYANURIC TRICHLORIDE; CYANURYL CHLORIDE; KYANURCHLORID; SYN-TRICHLOTRIAZIN; S-TRIAZINE TRICHLORIDE; 1,3,5-TRIAZINE,2,4,6-TRICHLORO-; S-TRIAZINE,2,4,6-TRICHLORO-; TRICHLOROCYANIDINE; 1,3,5-TRICHLOROTRIAZINE; 2,4,6-TRICHLORO-1,3,5-TRIAZINE; 2,4,6-TRICHLORO-S-TRIAZINE; 2,4,6-TRICHLOROTRIAZINE; SYM-TRICHLOROTRIAZINE; TRICHLORO-S-TRIAZINE; TRICYANOGEN CHLORIDE
Description: colorless, monoclinic crystals; pungent odor
Use: intermediate in synthesis of chemicals & insecticides; chem int for s-triazine herbicides, triallyl cyanurate, polyesters, optical brighteners, other dyes, plastics, pharmaceuticals, explosives & surfactants

Physical Properties

Boiling Point: 190 °C (374 °F) at 720 mm Hg
Freezing Point: 154 °C (309.2 °F)
Specific Gravity: 1.32
Water Solubility: Insoluble in Water
Other Solubilities: Soluble in Chloroform, Carbon Tetrachloride, hot Ether, Dioxane, Ketones.

RTECS Toxicity Data

Acute Oral: Rat LD_{50} Dose: 485 mg/kg; Toxic Effects: Behavioral - Somnolence (general depressed activity); Nutritional and gross metabolic - Weight loss or decreased weight gain; Nutritional and gross metabolic - Body temperature decrease. Mouse LD_{50} Dose: 350 mg/kg; Toxic Effects: Behavioral - Somnolence (general depressed activity); Behavioral - Food intake (animal); Nutritional and gross metabolic - Weight loss or decreased weight gain.
Acute Inhalation: Mouse LC_{50} Dose: 10 mg/m^3/2hr; Toxic Effects: Automatic Nervous System - Other (direct) parasympathomimetic; Behavioral - Somnolence (general depressed activity); MUSCULOSKELITAL - Other changes.
Chronic (Multiple Dose) Oral: Rat Dose: 2800 mg/kg/28D-C; Toxic Effects: Liver - Changes in liver weight; Blood - Pigmented or nucleated red Blood cells; DEATH.
Chronic (Multiple Dose) Inhalation: Rat Dose: 1880 ug/m^3/4H/11W-I; Toxic Effects: Blood - Changes in erythrocite (RBC) cell count; Nutritional and gross metabolic - Weight loss or decreased weight gain; DEATH.
Irritation Eye: Rabbit Standard Draize Test Dose: 50 ug/24H; Reaction: severe.
Irritation Skin: Rabbit Standard Draize Test Dose: 500 mg/24H; Reaction: moderate.
Tumorigenic: Rat Route: Oral; Dose: 20 gm/kg/73W-I; Toxic Effects: Tumorigenic - Equivocal tumorigenic agent by RTECS criteria; Skin and appendages - Tumors; Tumorigenic effects - Uterine tumors. Rat Route: Multiple routes; Dose: 16 gm/kg/73W-I; Toxic Effects: Tumorigenic - Equivocal tumorigenic agent by RTECS criteria; Skin and appendages - Tumors; Tumorigenic - Tumors at site of application.

Hazard Overviews

Poison

Corrosive

Health: Corrosive to eyes/skin/respiratory tract. Poison. Other Acute Effects: may be fatal if inhaled; harmful if swallowed or absorbed through skin; severe lachrymator; inhalation may result in spasm, inflammation and edema of the larynx and bronchi, chemical pneumonitis and pulmonary edema; symptoms of exposure may include burning sensation; coughing; wheezing; laryngitis; shortness of breath; headache; nausea; vomiting; may cause allergic respiratory and skin reactions; target organs: central nervous system, heart.

Fire: Hazards: emits toxic fumes. Extinguishing agents: carbon dioxide, dry chemical powder or appropriate foam; do not use water. Precautions: combustible liquid.
Reactivity: Stable. Hazardous polymerization will not occur. Incompatible with: strong oxidizing agents, strong acids, alcohols, dimethylformamide, amines, mercaptans, dimethyl sulfoxide (DMSO), reacts violently with water. Hazardous decomposition products: toxic fumes of: carbon monoxide, carbon dioxide, nitrogen oxides, hydrogen chloride gas.
Carcinogenicity: IARC - Not listed; NIOSH - Not listed; NTP - Not listed; ACGIH - Not listed; OSHA - Not listed; EPA - Not listed; MAK - Not listed
Primary Target Organs:

Eyes

Skin

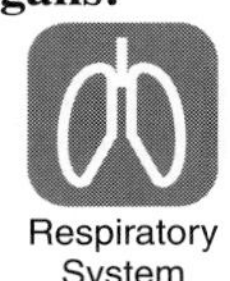
Respiratory System

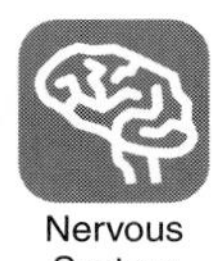
Nervous System

Cardio-vascular

Environmental

Regulations

RCRA 40CFR: Not listed
CERCLA: 40CFR 302.4: Listed as Compound per CWA Section 307(a) per CAA Section 112
SARA 40CFR 372.65: Not listed
SARA EHS 40CFR 355: Not listed
TSCA: Listed

CYA8920 CAS #: 675-14-9

CYANURIC FLUORIDE

RTECS: XZ1750000
EINECS Number: 211-620-8
Molecular Formula: $C_3F_3N_3$
Formula Weight: 135.05

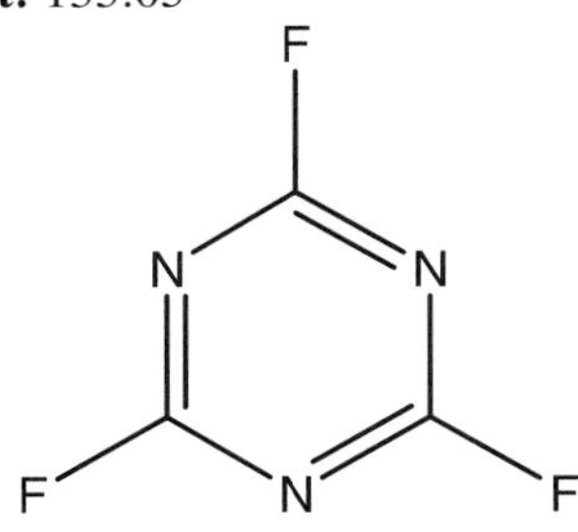

Chemical Structure

Synonyms: 1,3,5-TRIAZINE,2,4,6-TRIFLUORO-; S-TRIAZINE,2,4,6-TRIFLURO-; 2,4,6-TRIFLUORO-1,3,5-TRIAZINE; 2,4,6-TRIFLUORO-S-TRIAZINE
Description: colorless liquid
Use: fiber-reactive dyes

Physical Properties

Boiling Point: 72.4 °C (162 °F) at 101.67 kPa
Freezing Point: -38 °C (-36.4 °F)
Specific Gravity: 1.55 to 1.65 plus or minus 0.05 at 25 °C

Refraction Index: 1.3844 at 24 °C/D
Ionization Potential (eV): 11.3

RTECS Toxicity Data

Acute Inhalation: Rat LC_{50} Dose: 3100 ppb/4hr; Toxic Effects: Behavioral - Excitment; Lungs, Thorax, or Respiration - Chronic pulmonary edema; Lungs, Thorax, or Respiration - Dyspnea.
Acute Dermal: Rabbit LD_{50} Route: Skin; Dose: 160 mg/kg; Toxic Effects: Sense organs and special senses - Lacrimation; Lungs, Thorax, or Respiration - Bronchiolar constriction; Tumorigenic - Carcinogenic by RTECS criteria.

Hazard Overviews

Poison

Fire Diamond

Health: Irritating to eyes/skin/respiratory tract. Poison. May be absorbed through skin in toxic amounts.
Fire: Noncombustible. May react violently with water. Use agent, other than water, suitable for surrounding fire.
Reactivity: Stable in closed containers. Hazardous polymerization cannot occur.Avoid: water. Incompatible with: water. Hazardous decomposition products: toxic fluoride; nirtrogen oxide.
Carcinogenicity: IARC - Not listed; NIOSH - Not listed; NTP - Not listed; ACGIH - Class A4, Not classifiable as a human carcinogen; OSHA - Not listed; EPA - Not listed; MAK - Not listed
Primary Target Organs:

Eyes

Skin

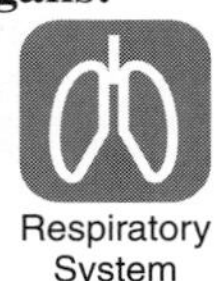
Respiratory System

Exposure Limits
OSHA PEL: TWA: 2.5 mg/m^3; as F.
ACGIH TLV: TWA: 2.5 mg/m^3; as F.
NIOSH REL: TWA: 2.5 mg/m^3; as F.
DFG MAK: TWA: 2.5 mg/m^3; as F.

Environmental

Regulations

RCRA 40CFR: Not listed
CERCLA: 40CFR 302.4: Listed as Compound per CWA Section 307(a) per CAA Section 112
SARA 40CFR 372.65: Not listed TPQ: 100 lb
SARA EHS 40CFR 355: Listed TPQ: 100 lb
TSCA: Listed

CYC1680 CAS #: 1134-23-2

CYCLOATE

RTECS: GU7200000
EINECS Number: 214-482-7
Molecular Formula: $C_{11}H_{21}NOS$
Formula Weight: 215.39
Synonyms: CARBAMIC ACID,CYCLOHEXYLETHYLTHIO-,S-ETHYL ESTER; CARBAMOTHIOIC ACID,CYCLOHEXYLETHYL-,S-ETHYL ESTER; CYCLOHEXANECARBAMIC ACID,N-ETHYLTHIO-,S-ETHYL ESTER; S-ETHYL CYCLOHEXYLETHYL CARBAMOTHIOATE; S-ETHYL CYCLOHEXYLETHYLCARBAMOTHIOATE; S-ETHYL CYCLOHEXYLETHYLTHIOCARBAMATE; S-ETHYL N-CYCLOHEXYL-N-ETHYL(THIOCARBAMATE); S-ETHYL N-ETHYL N-CYCLOHEXYLTHIOLCARBAMATE; S-ETHYL N-ETHYLTHIOCYCLOHEXANECARBAMATE; S-ETHYLETHYLCYCLOHEXYLTHIOCARBAMATE; ETSAN; EUREX; HEXYLTHIOCARBAM; R 2063; R-2063; RO-NEET; RO-NEET 10G; RO-NEET 6-E; RO-NEET E; RONIT; SABET
Description: colorless to amber to yellow; oily; clear liquid; aromatic odor
Use: selective preemergence herbicide for annual grasses, nutgrass, perennial grasses, broadleafweeds, eg, black nightshade, hairy nightshade, henbit, lambsquarters, purslane, redroot pigweed, shepherdspurse, & small stinging nettle; in sugar beets, table beets, spinach, etc.

Physical Properties

Boiling Point: 145 °C (293 °F) at 10 mm Hg
Freezing Point: 11.5 °C (52.7 °F)
Specific Gravity: 1.016 at 30 °C/4 °C
Vapor Pressure: 830 mPa at 25 °C
Water Solubility: 75 mg/L at 20 °C
Other Solubilities: miscible with Methanol, Isopropanol.
Refraction Index: 1.5054 at 30 °C/D
Flash Point: 139 °C Tag Open Cup

RTECS Toxicity Data

Acute Oral: Rat LD_{50} Dose: 1678 mg/kg. Mouse LD_{50} Dose: 1275 mg/kg; Toxic Effects: Behavioral - Somnolence (general depressed activity); Behavioral - Muscle contraction or spasticity; Behavioral - Coma.
Acute Inhalation: Rat LC_{50} Dose: 90 gm/m^3/1hr.
Acute Dermal: Rabbit LD_{50} Route: Skin; Dose: 3 gm/kg. Rat LD_{50} Route: Skin; Dose: 2467 mg/kg.
Chronic (Multiple Dose) Oral: Rat Dose: 27608 mg/kg/17W-I; Toxic Effects: Sense organs and special senses - Hemorrage; Behavioral - Somnolence (general depressed activity); DEATH. Rat Dose: 2192 mg/kg/40D-I; Toxic Effects: Blood - Changes in erythrocite (RBC) cell count; Blood - Changes in leukocyte (WBC) cell count; Biochemical - True cholinesterase. Rat Dose: 1820 mg/kg/26W-I; Toxic Effects: Blood - Pigmented or nucleated red Blood cells; Blood - Changes in erythrocite (RBC) cell count; Blood - Changes in leukocyte (WBC) cell count.
Mutagenic: Mouse Cytogenetic Analysis; Route: Unreported; Dose: 200 mg/kg.

Hazard Overviews

Fire: Will burn.
Carcinogenicity: IARC - Not listed; NIOSH - Not listed; NTP - Not listed; ACGIH - Not listed; OSHA - Not listed; EPA - Not listed; MAK - Not listed

Environmental

Ecotoxicity: LC_{50} Mosquito fish 10 ppm/96 hr /ro-neet 6e/ /conditions of bioassay not specified LC_{50} rainbow trout4.5 mg/l/96 hr /conditions of bioassay not specified LC_{50} Bobwhite quail feed treatment >56000 ppm/7 day /ro-neet 6e

Environmental Fate: If released to soil, the expected mobility is medium to slow and considerable volatilization should occur (65% after 6 hours in one soil). Rapid biodegradation occurs with soil microorganisms. If released to water, it will volatilize slowly. Mineralization does not occur to any great extent in lake water and sewage but it does undergo cometabolism to organic products. Aquatic bioconcentration and adsorption to sediment are not expected to be important fate processes. If released to the atmosphere, it will exist primarily in the vapor phase. In the vapor phase, it will degrade in the atmosphere by reaction with photochemically produced hydroxyl radicals with an estimated half-life of 10 hours.

Environmental Physical Data

Henry's Law Constant: 6.7×10^{-6}
Sorption Partition Coefficient: K_{OC} = estimated at 183
BCF: estimated at 50

Regulations

RCRA 40CFR: Listed Hazardous Waste No. U386 Toxic Waste
CERCLA: 40CFR 302.4: Listed per RCRA Section 3001 RQ: 1 lb (0.454 kg)
SARA 40CFR 372.65: Listed
SARA EHS 40CFR 355: Not listed
TSCA: Not listed

Analytical Methods

Water / Groundwater: EPA 634
Drinking Water: EPA 507, 525.2; AOAC 991.07; ASTM D5475
Food: FDA 212.1, 232.1, 232.4, 242.1; AOAC 974.05
Plasma: FDA 211.1, 231.1, 252

CYC3380 CAS #: 110-82-7

CYCLOHEXANE

RTECS: GU6300000
DOT: UN1145
EINECS Number: 203-806-2
Molecular Formula: C_6H_{12}
Structured MF: C_6H_{12}
Formula Weight: 84.18

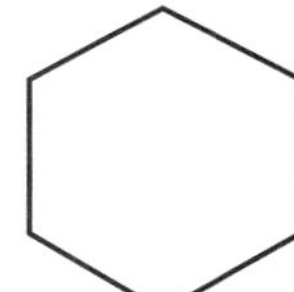

Chemical Structure

Synonyms: ASTM D3055; BENZENE HEXAHYDRIDE; BENZENEHEXAHYDRIDE; BENZENE,HEXAHYDRO-; CICLOESANO; CYCLOHEXAAN; CYCLOHEXAN; CYKLOHEKSAN; GE MATERIAL D5B94; HEXAHYDROBENZENE; HEXAMETHYLENE; HEXANAPHTHENE

Description: colorless liquid; sweet, chloroform odor; pungent when impure

Use: solvent to dissolve cellulose ethers, lacquer resins, fats, waxes, oils, bitumen and crude rubber; in perfume manufacturing, during surface coating operations (lacquers), in synthesis of adipic acid for production of nylon 66 and engineering plastics, during synthesis of caprolactam in nylon 6, paint and varnish remover, in the extraction of essential oils, in analytical chemistry for molecular weight determinations, in the manufacturing of adipic acid, benzene, cyclohexyl chloride, nitrocyclohexane, cyclohexanol and cyclohexanone, in the manufacturing of solid fuel for camp stoves, in fungicidal formulations (possesses fungicidal action) in the industrial recrystallizing of steroids, organic synthesis, recrystallizing medium glass substitutes, solid fuels, in analytic chemistry and in manufacturing of adhesives

Physical Properties

Boiling Point: 80.7 °C (177 °F) at 760 mm Hg
Freezing Point: 6.47 °C (43.646 °F)
Specific Gravity: 0.7781 at 20 °C/4 °C
Vapor Density: 2.9 Air=1
Saturated Vapor Density: 1.485413793 kg/m^3
Density: 0.779 g/mL at 20 °C
Vapor Pressure: 95 mm Hg at 20 °C
Water Solubility: Insoluble in Water
Other Solubilities: Soluble in Alcohol, Ether, Acetone, & Benzene; miscible with Olive Oil; 100 ml of Methanol dissolves 57 g at 20 °C.
Surface Tension: 24.6 dynes/cm
Odor Threshold: 3.56×10^{-5} g/l
Refraction Index: 1.42662 at 20 °C/D
Evaporation Rate: > 1 Butyl Acetate=1
Critical Temperature: 280.4 °C
Critical Pressure: 40 atm
Ionization Potential (eV): 9.88
Flash Point: -18 °C Closed Cup
Autoignition Temperature: 245 °C
LEL: 1.3% v/v
UEL: 8% v/v

RTECS Toxicity Data

Acute Oral: Rat LD_{50} Dose: 12705 mg/kg. Mouse LD_{50} Dose: 813 mg/kg.

Acute Inhalation: Mouse LC_{Lo} Dose: 70 gm/m^3/2hr. Rabbit LC_{Lo} Dose: 89600 mg/m^3/1hr; Toxic Effects: Behavioral - General anesthetic; Behavioral - Tremor; Behavioral - Muscle contraction or spasticity.

Acute Dermal: Rabbit LD; Route: Skin; Dose: >180 gm/kg.

Chronic (Multiple Dose) Inhalation: Rat Dose: 300 ppm/6H/2W-I; Toxic Effects: Brain and coverings - Other degenerative changes; Biochemical - Other oxidoreductases. Rabbit Dose: 7444 ppm/6H/2W-I; Toxic Effects: Automatic

Nervous System - Other (direct) parasympathomimetic; Gastrointestinal - Hypermotility, diarrhea; DEATH. Rabbit Dose: 9220 ppm/6H/5W-I; Toxic Effects: Automatic Nervous System - Other (direct) parasympathomimetic; Behavioral - Tremor; DEATH.

Mutagenic: Bacteria - E Coli DNA Adduct; Dose: 10 umol/L.

Hazard Overviews

Flammable

Fire Diamond

Health: Irritating to eyes/skin/respiratory tract. Also Causes: headache, fatigue, dizziness, staggering, rapid breathing, coughing, nausea, CNS depression, unconsciousness, dryness, gastrointestinal irritation, unconsciousness. Chronic Effects: dermatitis.

Fire: Flammable. Can form explosive mixtures in the air. For small fires, use dry chemical, carbon dioxide, or alcohol-resistant foam. For large fires, use high pressure water fog or alcohol-resistant foam. Material will float and may re-ignite on water surface. Do not scatter with high pressure water stream.

Reactivity: Stable. Hazardous polymerization cannot occur. Avoid: heat; ignition sources. Incompatible with: dinitrogen tetraoxide; oxidizing materials; liquid nitrogen dioxide. Hazardous decomposition products: carbon dioxide; carbon monoxide.

Carcinogenicity: IARC - Not listed; NIOSH - Listed as carcinogen; NTP - Not listed; ACGIH - Not listed; OSHA - Not listed; EPA - Not listed; MAK - Not listed

Primary Target Organs:

Eyes

Skin

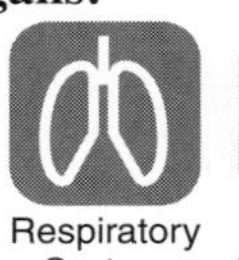

Respiratory System

Mucous Membranes

Gastro-intestinal

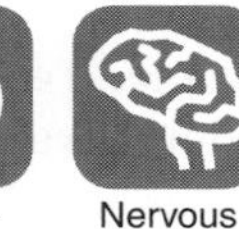

Nervous System

Exposure Limits

OSHA PEL: TWA: 300 ppm; 1050 mg/m^3.

ACGIH TLV: TWA: 300 ppm; 1030 mg/m^3.

NIOSH REL: TWA: 300 ppm; 1050 mg/m^3.

NIOSH IDLH: 1300 ppm; LEL.

DFG MAK: TWA: 300 ppm; 1050 mg/m^3.

Respirator Recommendation

Exposure Range: >300 to <1300 ppm Supplied Air, Constant Flow/Pressure Demand, Half Mask

Exposure Range: 1300 to unlimited ppm Self-contained Breathing Apparatus, Pressure Demand, Full Face

Note: poor warning properties

Environmental

Ecotoxicity: TLm Mosquito fish 15500 ppm/24, 48, 96 hr in lake water LC_{50} Fathead minnow 93 mg/l/72 and 96 hr static bioassay in Lake Superior water at 18-22 °C Coho salmon; No significant mortalities up to 100 ppm after 96 hr in artificial sea water at 8 °C TLm Fathead minnow 30 ppm/96 hr

Environmental Fate: If released on land, it will be lost through volatilization and should leach into groundwater. While resistant to biodegradation, degradation occurs slowly in groundwater in the presence of other petrochemicals. Volatilization from water (estimated half-life 2 hr in a model river) should be the most important fate process occurring in aquatic systems. While bioconcentration in aquatic organisms and adsorption to sediment is estimated to occur to a moderate extent, vaporization should be so rapid that they will not contribute significantly to fate in water. In the atmosphere, it will degrade by reaction with photochemically produced hydroxyl radicals (half-life 52 hr). The half-life is much faster under photochemical smog conditions with half-lives as low as 6 hr being reported.

Cleanup/Disposal: Guide No. 128: Eliminate all ignition sources (no smoking, flares, sparks or flames in immediate area). All equipment used when handling the product must be grounded. Do not touch or walk through spilled material. Stop leak if you can do it without risk. Prevent entry into waterways, sewers, basements or confined areas. A vapor suppressing foam may be used to reduce vapors. Absorb or cover with dry earth, sand or other non-combustible material and transfer to containers. Use clean non-sparking tools to collect absorbed material. Large Spills: Dike far ahead of liquid spill for later disposal. Water spray may reduce vapor; but may not prevent ignition in closed spaces.

Environmental Physical Data

Henry's Law Constant: 7.9

Octanol/Water Partition Coefficient: log K_{ow} = estimated at 3.18

Sorption Partition Coefficient: K_{oc} = estimated at 480

BCF: calculated at 242

Regulations

RCRA 40CFR: Listed Hazardous Waste No. U056 Ignitable Waste

CERCLA: 40CFR 302.4: Listed per CWA Section 311(b)(4) per RCRA Section 3001 RQ: 1000 lb (453.5 kg)

SARA 40CFR 372.65: Listed

SARA EHS 40CFR 355: Not listed

TSCA: Listed

Analytical Methods

Air: ASTM D3686, D3687, D4490

Indoor / Expired Air: NIOSH 1500

Other: EPA 1666

CYC4060 CAS #: 108-93-0

CYCLOHEXANOL

RTECS: GV7875000

EINECS Number: 203-630-6

Molecular Formula: $C_6H_{12}O$

Structured MF: $C_6H_{11}OH$

Formula Weight: 100.16

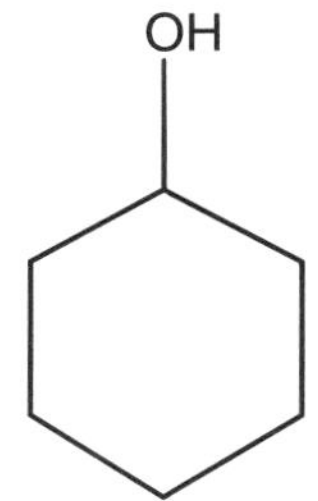

Chemical Structure

Synonyms: ADRONAL; ADRONOL; ANOL; CICLOESANOLO; 1-CYCLOHEXANOL; CYCLOHEXYL ALCOHOL; CYKLOHEKSANOL; HEXAHYDROPHENOL; HEXALIN; HYDRALIN; HYDROPHENOL; HYDROXYCYCLOHEXANE; NAXOL; PHENOL,HEXAHYDRO-

Description: colorless, light yellow viscous liquid; camphor or menthol odor

Use: source of adilpic acid for textile finishing, solvent, blending agent, lacquers, paints and varnishes, plastics and germicides

Physical Properties

Boiling Point: 161.08 °C (322 °F) at 760 mm Hg
Freezing Point: 25.4 °C (77.72 °F)
Specific Gravity: 0.9624 at 20 °C/4 °C
Vapor Density: 3.45 Air=1
Density: 0.96 g/mL
Vapor Pressure: 1 mm Hg
Water Solubility: 3.6% (wt/wt) in Water at 20 °C
Other Solubilities: 95% Ethanol: Very Soluble; Acetone: Soluble; Benzene: Very Soluble; Carbon Disulfide: Very Soluble; Ether: Soluble; Ethyl Acetate: Very Soluble; Linseed Oil: Very Soluble.
Surface Tension: 34.2 dynes/cm at 16.2 °C
Odor Threshold: 0.16 ppm
Refraction Index: 1.4641 at 20 °C/D
Critical Temperature: 352 °C
Critical Pressure: 3.7 mn/sq m
Ionization Potential (eV): 10.00
Flash Point: 62.8 °C Closed Cup
Autoignition Temperature: 300 °C

RTECS Toxicity Data

Acute Oral: Rat LD_{50} Dose: 2060 mg/kg. Rabbit LD_{Lo} Dose: 2200 mg/kg; Toxic Effects: Behavioral - General anesthetic.

Acute Inhalation: Human TC_{Lo} Dose: 75 ppm; Toxic Effects: Sense organs and special senses - Other; Sense organs and special senses - Conjunctive irritation; Lungs, Thorax, or Respiration - Other changes.

Acute Dermal: Rabbit LD_{Lo} Route: Skin; Dose: 12 gm/kg; Toxic Effects: Behavioral - General anesthetic; Behavioral - Convulsions or effect on seizure threshold.

Chronic (Multiple Dose) Oral: Rat Dose: 3185 mg/kg/7D-C; Toxic Effects: Liver - Other changes; Liver - Changes in liver weight.

Chronic (Multiple Dose) Inhalation: Rabbit Dose: 1229 ppm/6H/5W-I; Toxic Effects: Behavioral - General anesthetic; Nutritional and gross metabolic - Weight loss or decreased weight gain; DEATH. Rabbit Dose: 997 ppm/6H/11W-I; Toxic Effects: Behavioral - General anesthetic; Sense organs and special senses - Conjunctive irritation; DEATH.

Irritation Eye: Rabbit Standard Draize Test Dose: 2 mg; Reaction: severe.

Irritation Skin: Rabbit Open Draize Test Dose: 14600 ug/24H open; Reaction: mild.

Reproductive/Teratogenic: Rat Route: Subcutaneous; Dose: 315 mg/kg; Duration: male 21D prior to mating; Paternal Effects - Spermatogenesis; Testes, epididymis, sperm duct; Prostate, seminal vessicle, Cowper's gland, accessory glands. Gerbil Route: Subcutaneous; Dose: 315 mg/kg; Duration: male 21D prior to mating; Paternal Effects - Spermatogenesis; Testes, epididymis, sperm duct; Prostate, seminal vessicle, Cowper's gland, accessory glands.

Mutagenic: Human Cytogenetic Analysis; Cell Type: leukocyte; Dose: 100 umol/L. Mammal DNA Damage; Cell Type: lymphocyte; Dose: 150 mmol/L.

Hazard Overviews

Fire Diamond

Health: Irritating to eyes/skin/respiratory tract. Also Causes: chest heaviness/pain, breathing difficulty, cough, fatigue, bluish face, vomiting/diarrhea, drowsiness, dizziness. Chronic Effects: skin defatting, CNS effects.

Fire: Combustible. Use water spray, dry chemical, alcohol foam, or carbon dioxide. Cool fire-exposed containers with water.

Reactivity: Stable. Hazardous polymerization cannot occur. Avoid: heat; flame; oxidizing material. Incompatible with: oxidizing agents; chromium trioxide. Hazardous decomposition products: carbon monoxide; carbon dioxide; toxic smoke; fumes.

Carcinogenicity: IARC - Not listed; NIOSH - Listed as carcinogen; NTP - Not listed; ACGIH - Not listed; OSHA - Not listed; EPA - Not listed; MAK - Not listed

Primary Target Organs:

Eyes

Skin

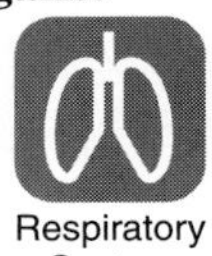
Respiratory System

Nervous System

Exposure Limits
OSHA PEL: TWA: 50 ppm; 200 mg/m^3.
OSHA PEL Vacated 1989 Limits: TWA: 50 ppm; 200 mg/m^3.
ACGIH TLV: TWA: 50 ppm; 206 mg/m^3.
NIOSH REL: TWA: 50 ppm; 200 mg/m^3.
NIOSH IDLH: 400 ppm.
DFG MAK: TWA: 50 ppm; 200 mg/m^3.

Respirator Recommendation
Exposure Range: >50 to <400 ppm Air Purifying, Negative Pressure, Half Mask

Exposure Range: 400 to unlimited ppm Self-contained Breathing Apparatus, Pressure Demand, Full Face
Cartridge Color: black

Environmental

Ecotoxicity: LC_{50} Pimephales promelas (fathead minnow) 732 mg/l/96 hr (confidence limit 696 - 770 mg/l), flow-through bioassay with measured concentrations, 25.4 °C, dissolved oxygen 6.3 mg/l, hardness 46.0 mg/l calcium carbonate, alkalinity 41.0 mg/l calcium carbonate LC_{50} Menidia beryllina 720 ppm/96 hr (static bioassay in synthetic seawater at 23 °C, mild aeration applied after 24 hr)
Environmental Fate: Exhibits high mobility in soil, and if released in soil, would be expected to leach. It would also be expected to volatilize from surface layers of soil. It is readily biodegradable in aerobic biodegradation screening tests and therefore would be expected to biodegrade in soil. If released in water, it would be lost by volatilization. Its estimated half-lives in a model river and model lake are 11.5 hours and 6.6 days, respectively. It may also biodegrade in water based upon its ready biodegradation in aerobic screening tests. It is not expected to adsorb to sediment or particulate matter in the water column or bioconcentrate in aquatic organisms. In the atmosphere, it will degrade by reacting with photochemically-produced hydroxyl radicals (estimated half-life 22 hours).
Cleanup/Disposal: Guide No. 127: Eliminate all ignition sources (no smoking, flares, sparks or flames in immediate area). All equipment used when handling the product must be grounded. Do not touch or walk through spilled material. Stop leak if you can do it without risk. Prevent entry into waterways, sewers, basements or confined areas. A vapor suppressing foam may be used to reduce vapors. Absorb or cover with dry earth, sand or other non-combustible material and transfer to containers. Use clean non-sparking tools to collect absorbed material. Large Spills: Dike far ahead of liquid spill for later disposal. Water spray may reduce vapor; but may not prevent ignition in closed spaces.

Environmental Physical Data

Henry's Law Constant: estimated at 1.2 x10^{-4}
Octanol/Water Partition Coefficient: log K_{ow} = 1.23
Sorption Partition Coefficient: K_{oc} = estimated at 8.1
BCF: estimated at 5.1
BOD: 0.08 lb/lb, 5 days

Regulations

RCRA 40CFR: Not listed
CERCLA: 40CFR 302.4: Not listed
SARA 40CFR 372.65: Listed
SARA EHS 40CFR 355: Not listed
TSCA: Listed

Analytical Methods

Air: ASTM D3686, D3687, D4490
Water / Groundwater: ASTM D3695
Indoor / Expired Air: NIOSH 1402

CYC4230 CAS #: 108-94-1

CYCLOHEXANONE

RTECS: GW1050000
DOT: UN1915; IMO3.3
EINECS Number: 203-631-1
Molecular Formula: $C_6H_{10}O$
Structured MF: $C_6H_{10}O$
Formula Weight: 98.14

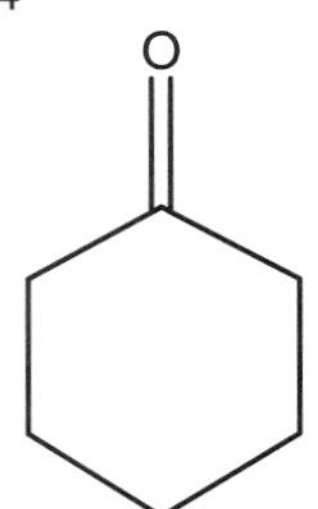

Chemical Structure

Synonyms: ANON; ANONE; CICLOESANONE; CYCLOHEXANON; CYCLOHEXYL KETONE; CYKLOHEKSANON; EPA PESTICIDE CHEMICAL CODE 025902; HEXANON; HYTROL O; KETOCYCLOHEXANE; KETOHEXAMETHYLENE; NADONE; OXOCYCLOHEXANE; PIMELIC KETONE; PIMELIN KETONE; SEXTONE
Description: water white, slightly yellow liquid; acetone or peppermint odor
Use: solvent, swelling agent for PVC, synthetic intermediate, H acceptor in Oppenauer oxidations, degreasing of metals, paint and varnish removers, spot removers, polishes, nylon

Physical Properties

Boiling Point: 155.6 °C (312 °F) at 760 mm Hg
Freezing Point: -31 °C (-23.8 °F)
Specific Gravity: 0.9421 at 25 °C/4 °C
Vapor Density: 3.4 Air=1
Density: 3.4
Vapor Pressure: 5 mm Hg
Water Solubility: 150 g/L in Water at 10 °C
Other Solubilities: 95% Ethanol: >=100 mg/ml at 18 °C; Acetone: >=100 mg/ml at 18 °C; Benzene: Soluble; Chloroform: Soluble; DMSO: >=100 mg/ml at 18 °C; Ether: Soluble; Organic solvents: miscible.
Surface Tension: 34 dynes/cm at 20 °C
Odor Threshold: 0.4800 to 400 mg/m^3
Refraction Index: 1.4507 at 20 °C
Evaporation Rate: 40.6 Ether=1
Critical Temperature: 356 °C
Critical Pressure: 560 psia
Ionization Potential (eV): 9.14
Flash Point: 44 °C Closed Cup
Autoignition Temperature: 420 °C
LEL: 1.1% v/v
UEL: 9.4% v/v

RTECS Toxicity Data

Acute Oral: Rat LD_{50} Dose: 1620 uL/kg. Mouse LD_{50} Dose: 1400 mg/kg.

Acute Inhalation: Rat LC_{50} Dose: 8000 ppm/4hr. Human TC_{Lo} Dose: 75 ppm; Toxic Effects: Sense organs and special senses - Other; Sense organs and special senses - Conjunctive irritation; Lungs, Thorax, or Respiration - Other changes.

Acute Dermal: Rabbit LD_{50} Route: Skin; Dose: 1 mL/kg. Rat LD_{50} Route: Subcutaneous Dose: 2170 mg/kg.

Chronic (Multiple Dose) Inhalation: Rabbit Dose: 3082 ppm/6H/3W-I; Toxic Effects: Behavioral - Ataxia; Nutritional and gross metabolic - Weight loss or decreased weight gain; DEATH.

Irritation Eye: Rabbit Standard Draize Test Dose: 20 mg; Reaction: severe. Rabbit Standard Draize Test Dose: 250 ug/24H; Reaction: severe.

Irritation Skin: Rabbit Open Draize Test Dose: 500 mg open; Reaction: mild.

Reproductive/Teratogenic: Rat Route: Inhalation; Dose: 105 mg/m^3/4H Duration: female 1-20D of pregnancy; Effects on Fertility - Pre-implantation mortality. Mouse Route: Oral; Dose: 11 gm/kg; Duration: female 8-12D of pregnancy; Effects on Newborn - Growth statistics.

Mutagenic: Human Cytogenetic Analysis; Cell Type: leukocyte; Dose: 100 umol/L. Human Cytogenetic Analysis; Cell Type: lymphocyte; Dose: 5 ug/L.

Hazard Overviews

Flammable

Fire Diamond

Health: Irritating to eyes/skin/respiratory tract. Also Causes: inhalation of high concentrations can cause unconsciousness. Chronic Effects: dermatitis.

Fire: Flammable. Can form explosive mixtures in the air. Use water as fog, dry chemical, carbon dioxide, or alcohol foam. Use water spray to cool fire-exposed containers, to flush spills away from sensitive exposures, to disperse the vapor, to dilute spilled cyclohexanone to nonflammable mixtures, and to protect personnel attempting to stop or seal the source of the leaking material.

Reactivity: Stable. Hazardous polymerization cannot occur. Avoid: sources of ignition; plastics; resins, rubbers. Incompatible with: oxidizing agents; nitric acid. Hazardous decomposition products: carbon monoxide.

Carcinogenicity: IARC - Group 3, Not classifiable as to carcinogenicity to humans; NIOSH - Not listed; NTP - Not listed; ACGIH - Class A4, Not classifiable as a human carcinogen; OSHA - Not listed; EPA - Not listed; MAK - Class B, Justifiably suspected of having carcinogenic potential

Primary Target Organs:

Eyes

Skin

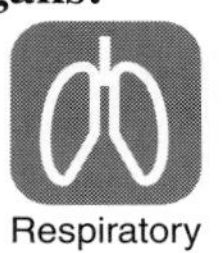
Respiratory System

Nervous System

Exposure Limits

OSHA PEL: TWA: 50 ppm; 200 mg/m^3.

OSHA PEL Vacated 1989 Limits: TWA: 25 ppm; 100 mg/m^3.

ACGIH TLV: TWA: 25 ppm; 100 mg/m^3.

NIOSH REL: TWA: 25 ppm; 100 mg/m^3.

NIOSH IDLH: 700 ppm.

Respirator Recommendation

Exposure Range: >50 to 500 ppm Air Purifying, Negative Pressure, Half Mask

Exposure Range: >500 to <700 ppm Air Purifying, Negative Pressure, Full Face

Exposure Range: 700 to unlimited ppm Self-contained Breathing Apparatus, Pressure Demand, Full Face

Cartridge Color: black

Environmental

Ecotoxicity: LC_{50} Pimephales promelas (fathead minnow) 527 mg/l 96 hr flow-through bioassay, wt 0.12 g, water hardness 45.5 mg/l $CaCO_3$, temp: 25 +/- 1 °C, pH 7.5, dissolved oxygen greater than 60% of saturation

Environmental Fate: Estimated to have high mobility in soil. It would also be expected to volatilize from surface layers of soil. It is readily biodegradable in aerobic biodegradation screening tests and a river die-away test and therefore would be expected to biodegrade in soil. If released in water, it would be slowly lost by volatilization. Its estimated half-lives in a model river and model lake are 4.1 and 33 days, respectively. It also would be expected to biodegrade, but rates in natural water are unavailable. It is not expected to adsorb to sediment or particulate matter in the water column or bioconcentrate in aquatic organisms. In the atmosphere, it will degrade by reacting with photochemically-produced hydroxyl radicals (estimated half-life 1.3 days).

Cleanup/Disposal: Guide No. 127: Eliminate all ignition sources (no smoking, flares, sparks or flames in immediate area). All equipment used when handling the product must be grounded. Do not touch or walk through spilled material. Stop leak if you can do it without risk. Prevent entry into waterways, sewers, basements or confined areas. A vapor suppressing foam may be used to reduce vapors. Absorb or cover with dry earth, sand or other non-combustible material and transfer to containers. Use clean non-sparking tools to collect absorbed material. Large Spills: Dike far ahead of liquid spill for later disposal. Water spray may reduce vapor; but may not prevent ignition in closed spaces.

Environmental Physical Data

Henry's Law Constant: estimated at 9.00 $x10^{-6}$

Octanol/Water Partition Coefficient: log K_{ow} = 0.81

Sorption Partition Coefficient: K_{oc} = estimated at 15

BCF: estimated at 2.4

Regulations

RCRA 40CFR: Listed Hazardous Waste No. U057 Ignitable Waste
CERCLA: 40CFR 302.4: Listed per RCRA Section 3001 RQ: 5000 lb (2268 kg)
SARA 40CFR 372.65: Not listed
SARA EHS 40CFR 355: Not listed
TSCA: Listed

Analytical Methods

Air: ASTM D3686, D3687
Soil: SW846 8315A
Water / Groundwater: ASTM D3695
Drinking Water: EPA 554
Indoor / Expired Air: NIOSH 1300

CYC4570 CAS #: 110-83-8

CYCLOHEXENE

RTECS: GW2500000
DOT: UN2256; IMO3.1
EINECS Number: 203-807-8
Molecular Formula: C_6H_{10}
Structured MF: C_6H_{10}
Formula Weight: 82.14

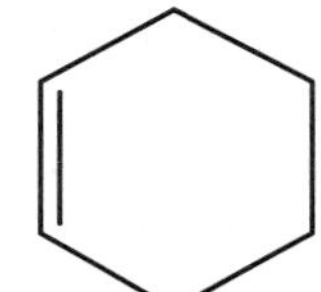

Chemical Structure

Synonyms: BENZENE TETRAHYDRIDE; BENZENETETRAHYDRIDE; BENZENE,TETRAHYDRO-; CYCLOHEX-1-ENE; CYCLOHEXENE RING; CYKLOHEKSEN; HEXANAPHTHYLENE; 1,2,3,4-TETRAHYDROBENZENE; 3,4,5,6-TETRAHYDROBENZENE; TETRAHYDROBENZENE
Description: colorless liquid; sweetish odor
Use: alkylation component; in mfr of adipic acid, maleic acid, hexahydrobenzoic acid & aldehyde; catalyst solvent; oil extraction; chemical intermediate for the miticide omite

Physical Properties

Boiling Point: 83 °C (181 °F) at 760 mm Hg
Freezing Point: -103.5 °C (-154.3 °F)
Specific Gravity: 0.8098 at 20 °C/4 °C
Vapor Pressure: 67 mm Hg
Water Solubility: Insoluble
Other Solubilities: miscible with Ethanol, Ether and Acetone.
Surface Tension: 26.56 mN/m
Odor Threshold: Detection 0.6 mg/m^3
Refraction Index: 1.4465 at 20 °C/D
Critical Temperature: 287.4 °C
Critical Pressure: 4.347 mPa
Ionization Potential (eV): 8.95
Flash Point: -11.667 °C Closed Cup
Autoignition Temperature: 310 °C

RTECS Toxicity Data

Acute Oral: Rat LD_{Lo} Dose: 3200 uL/kg; Toxic Effects: Behavioral - Tremor; Behavioral - Convulsions or effect on seizure threshold; Gastrointestinal - Changes in structure or function of salivary glands.
Acute Inhalation: Rat LC; Dose: >6370 ppm/4hr; Toxic Effects: Behavioral - Tremor; Behavioral - Ataxia; Lungs, Thorax, or Respiration - Other changes.
Chronic (Multiple Dose) Inhalation: Rat Dose: 600 ppm/6H/26W-I; Toxic Effects: Nutritional and gross metabolic - Weight loss or decreased weight gain; Biochemical - Phosphatases.

Hazard Overviews

Flammable

Fire Diamond

Health: Mildly irritating to eyes/respiratory tract. Also Causes: central nervous system (CNS) depression. Chronic Effects: dermatitis.
Fire: Flammable. Can form explosive mixtures in the air. Use foam, carbon dioxide, or dry chemical extinguishing media. Use water spray or mist to cool fire-exposed containers. Use a smothering technique for extinguishing fire. Water may be ineffective.
Reactivity: Stable. Hazardous polymerization cannot occur. Incompatible with: acids; strong oxidizing agents. Hazardous decomposition products: partially oxidized hydrocarbons; carbon dioxide; carbon monoxide.
Carcinogenicity: IARC - Not listed; NIOSH - Listed as carcinogen; NTP - Not listed; ACGIH - Not listed; OSHA - Not listed; EPA - Not listed; MAK - Not listed
Primary Target Organs:

Eyes

Skin

Respiratory System

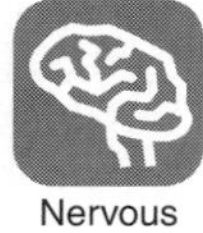

Nervous System

Exposure Limits
OSHA PEL: TWA: 300 ppm; 1015 mg/m^3.
ACGIH TLV: TWA: 300 ppm; 1010 mg/m^3.
NIOSH REL: TWA: 300 ppm; 1015 mg/m^3.
NIOSH IDLH: 2000 ppm.
DFG MAK: TWA: 300 ppm; 1015 mg/m^3.
Respirator Recommendation
Exposure Range: >300 to 1000 ppm Air Purifying, Negative Pressure, Half Mask
Exposure Range: >1000 to <2000 ppm Supplied Air, Constant Flow/Pressure Demand, Half Mask
Exposure Range: 2000 to unlimited ppm Self-contained Breathing Apparatus, Pressure Demand, Full Face
Cartridge Color: black

Environmental

Ecotoxicity: Bacteria: Pseudomonas putida 16h EC_0 17 mg/l; Fishes: young Coho salmon 96h NOLC 100 ppm, in artificial seawater at 8 °C

Environmental Fate: If released to soil, it may biodegrade in aerobic soils. It is expected to rapidly volatilize to the atmosphere and display low to moderate mobility through soil. If released to water, it may biodegrade under aerobic conditions. Bioconcentration and adsorption to sediment and suspended organic matter are not expected to be significant process. Volatilization to the atmosphere is expected to be rapid with an estimated volatilization half-life of 8.3 hours from a model river. If released to the atmosphere, it is expected to undergo a rapid, aerosol-forming reaction with ozone. An experimental rate constants for this reaction of 2.04 $x10^{-16}$ cu-cm/mole-sec at 21 °C translates to an atmospheric half-life of 1.9 hours. Aerosol formation in this reaction, i.e. organic acid formation, was found to increase with increasing humidity. It is also expected to undergo a relatively rapid vapor-phase oxidation with both photochemically produced hydroxyl radicals with an estimated half-life of 8.3 hours and nitrate radicals at night with an estimated half-life of 3.8 hours. It may also undergo atmospheric removal by wet deposition processes.

Cleanup/Disposal: Guide No. 130: Eliminate all ignition sources (no smoking, flares, sparks or flames in immediate area). All equipment used when handling the product must be grounded. Do not touch or walk through spilled material. Stop leak if you can do it without risk. Prevent entry into waterways, sewers, basements or confined areas. A vapor suppressing foam may be used to reduce vapors. Absorb or cover with dry earth, sand or other non-combustible material and transfer to containers. Use clean non-sparking tools to collect absorbed material. Large Spills: Dike far ahead of liquid spill for later disposal. Water spray may reduce vapor; but may not prevent ignition in closed spaces.

Environmental Physical Data

Henry's Law Constant: 4.55 $x10^{-2}$
Octanol/Water Partition Coefficient: log K_{ow} = 2.86
Sorption Partition Coefficient: K_{oc} = calculated at 228 to 856
BCF: calculated at 30 to 87

Regulations

RCRA 40CFR: Not listed
CERCLA: 40CFR 302.4: Not listed
SARA 40CFR 372.65: Not listed
SARA EHS 40CFR 355: Not listed
TSCA: Listed

Analytical Methods

Air: ASTM D3686, D3687
Water / Groundwater: ASTM D3695
Indoor / Expired Air: NIOSH 1500

CYC5080 CAS #: 66-81-9

CYCLOHEXIMIDE

RTECS: MA4375000
EINECS Number: 200-636-0
Molecular Formula: $C_{15}H_{23}NO_4$
Formula Weight: 281.34

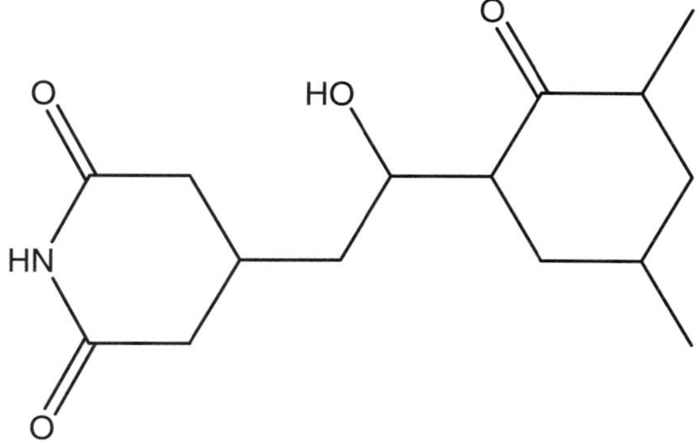

Chemical Structure

Synonyms: ACTI-AID; ACTIDION; ACTIDIONE; ACTIDIONE BR; ACTIDIONE PM; ACTIDIONE TGF; ACTIDONE; ACTISPRAY; AKTIDION; 3-(2-(3,5-DIMETHYL-2-OXOCYCLOHEXYL)-2-HYDROXYETHYL)GLUTARIMIDE; BETA-(2-(3,5-DIMETHYL-2-OXOCYCLOHEXYL)-2-HYDROXYETHYL)GLUTARIMIDE; [1S-[1ALPHA(S*),3ALPHA,5BETA]]-4-[2-(3,5-DIMETHYL-2-OXO-CYCLOHEXYL)]-2-HYDROXYETHYL-2,6-PIPERIDINEDIONE; GLUTARIMIDE,3-(2-(3,5-DIMETHYL-2-OXOCYCLOHEXYL)-2-HYDROXYETHYL)-; HIZAROCIN; KAKEN; NARAMYCIN; NARAMYCIN A; NEOCYCLOHEXIMIDE; NSC-185; 2,6-PIPERIDINEDIONE,4-(2-(3,5-DIMETHYL-2-OXOCYCLOHEXYL)-2-HYDROXYETHYL)-,(1S-(1ALPHA(S*),3ALPHA,5BETA))-; U-4527

Description: colorless crystals or plates

Use: fungicide, antibiotic, abscission of citrus fruit in harvesting, turf disease control, plant growth regulator, inhibiting protein synthesis, and repellent to rodents and other animal pests

Physical Properties

Freezing Point: 119.5 °C (247.1 °F) to 121 °C (249.8 °F)
Water Solubility: 2.1 g/100ml at 2 °C
Other Solubilities: Insoluble in Petroleum Ether; Isopropyl Alcohol 5.5 g/100 ml; Acetone 33.0 g/100 ml; cyclohexanone 19.0 g/100 ml (at 20 °C).
Flash Point: Not available; probably combustible

RTECS Toxicity Data

Acute Oral: Rat LD_{50} Dose: 2 mg/kg. Monkey LD_{50} Dose: 60 mg/kg.

Acute Dermal: Rat LD_{50} Route: Subcutaneous Dose: 2500 ug/kg. Mouse LD_{50} Route: Subcutaneous Dose: 160 mg/kg.

Irritation Skin: Rabbit Standard Draize Test Dose: 1 pph/24H; Reaction: moderate.

Reproductive/Teratogenic: Rat Route: Parenteral; Dose: 750 ug/kg; Duration: female 18D of pregnancy; Effects on Embryo or Fetus - Extra embryonic structures; Fetotoxicity.

Rat Route: Intraperitoneal; Dose: 1 mg/kg; Duration: female 4D of pregnancy; Maternal Effects - Uterus, cervix, vagina. Rat Route: Intraperitoneal; Dose: 300 ug/kg; Duration: female 13D of pregnancy; Effects on Fertility - Post-implantation mortality; Effects on Embryo or Fetus - Fetotoxicity. Rat Route: Intraperitoneal; Dose: 250 ug/kg; Duration: female 10D of pregnancy; Specific Developmental Abnormalities - Central nervous system; Craniofacial (including nose and tongue). Rat Route: Intraperitoneal; Dose: 1 mg/kg; Duration: female 15D of pregnancy; Effects on Embryo or Fetus - Fetotoxicity; Specific Developmental Abnormalities - Hepatobiliary system. Rat Route: Intraperitoneal; Dose: 1 mg/kg; Duration: female 18D of pregnancy; Specific Developmental Abnormalities - Central nervous system; Effects on Newborn - Growth statistics.

Mutagenic: Human DNA Inhibition; Cell Type: other cell types; Dose: 100 mg/L. Human DNA Inhibition; Cell Type: HeLa cell; Dose: 1 umol/L. Human DNA Inhibition; Cell Type: lung; Dose: 10 mg/L. Human DNA Inhibition; Cell Type: other cell types; Dose: 100 mg/L. Human DNA Inhibition; Cell Type: other cell types; Dose: 300 nmol/L. Human Cytogenetic Analysis; Cell Type: other cell types; Dose: 100 mg/L. Human Cytogenetic Analysis; Cell Type: lymphocyte; Dose: 30 mg/L. Human Other Mutation Test Systems; Cell Type: HeLa cell; Dose: 35 umol/L.

Hazard Overviews

Poison

Health: Severely irritating to skin; irritating to eyes/respiratory tract. Poison. Other Acute Effects: may be fatal if inhaled, swallowed, or absorbed through skin; prolonged or repeated exposure may cause allergic reactions in certain sensitive individuals. Chronic Effects: overexposure may cause reproductive disorder(s) based on tests with laboratory animals; possible risk of harm to the unborn child; laboratory experiments have shown mutagenic effects; target organs: liver; kidneys; nerves; g.i. system; pancreas.

Fire: Will burn. Hazards: emits toxic fumes. Extinguishing agents: water spray; carbon dioxide, dry chemical powder or appropriate foam. Precautions: combustible liquid.

Reactivity: Incompatible with: strong oxidizing agents, strong bases, acid chlorides, acid anhydrides. Hazardous decomposition products: toxic fumes of: carbon monoxide, carbon dioxide, nitrogen oxides.

Carcinogenicity: IARC - Not listed; NIOSH - Not listed; NTP - Not listed; ACGIH - Not listed; OSHA - Not listed; EPA - Not listed; MAK - Not listed

Primary Target Organs:

Eyes

Skin

Respiratory System

Nervous System

Liver

Kidneys

Environmental

Ecotoxicity: LD_{50} Mallard duck oral 82.5 mg/kg (95% confidence limit 54.3-126 mg/kg), 3-4-mo old males /sample purity 88.7 LD_{50} Pheasant oral 9.38 mg/kg (95% confidence limit 6.91-12.7 mg/kg), 4-mo old females /sample purity 88.7

Cleanup/Disposal: Guide No. 151: Do not touch damaged containers or spilled material unless wearing appropriate protective clothing. Stop leak if you can do it without risk. Prevent entry into waterways, sewers, basements or confined areas. Cover with plastic sheet to prevent spreading. Absorb or cover with dry earth, sand or other non-combustible material and transfer to containers. Do not get water inside containers.

Environmental Physical Data

Octanol/Water Partition Coefficient: log K_{ow} = 0.55
BCF: calculated at 2

Regulations

RCRA 40CFR: Not listed
CERCLA: 40CFR 302.4: Not listed
SARA 40CFR 372.65: Not listed TPQ: 100/10000 lb
SARA EHS 40CFR 355: Listed TPQ: 100 lb
TSCA: Not listed

CYC5590 CAS #: 108-91-8

CYCLOHEXYLAMINE

RTECS: GX0700000
DOT: UN2357
EINECS Number: 203-629-0
Molecular Formula: $C_6H_{13}N$
Structured MF: $C_6H_{11}NH_2$
Formula Weight: 99.17

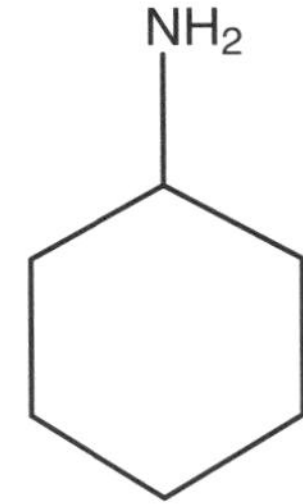

Chemical Structure

Synonyms: AMINOCYCLOHEXANE; AMINOHEXAHYDROBENZENE; ANILINE,HEXAHYDRO-; BENZENAMINE,HEXAHYDRO-; CHA; CYCLOHEXANAMINE; HEXAHYDROANILINE; HEXAHYDROBENZENAMINE

Description: colorless to yellow liquid; unpleasant, fishy odor

Use: in organic synthesis, manufacture of insecticides, plasticizers, corrosion inhibitors, rubber chemicals, dyestuffs, emulsifying agent, dry cleaning soaps, acid gas absorbents; paint, pigment, surfactant, insecticide, oxygen absorber; boiler water treatment, rubber accelerator

Physical Properties

Boiling Point: 134.5 °C (274 °F) at 760 mm Hg
Freezing Point: -17.7 °C (0.14 °F)
Specific Gravity: 0.8647 at 25 °C/25 °C
Vapor Density: 3.42 Air=1
Density: 3.42
Vapor Pressure: 11 mm Hg
Water Solubility: Miscible
Other Solubilities: 95% Ethanol: Very Soluble; Acetone: Very Soluble; Benzene: Very Soluble; DMSO: Very Soluble; Ether: Very Soluble.
Odor Threshold: 2.6 ppm
pH: Strong base
Refraction Index: 1.4565 at 25 °C/D
Critical Temperature: 342 °C
Ionization Potential (eV): 8.37
Flash Point: 31 °C Closed Cup
Autoignition Temperature: 293 °C
LEL: 1.5% v/v
UEL: 9.4% v/v

RTECS Toxicity Data

Acute Oral: Rat LD_{50} Dose: 156 mg/kg; Toxic Effects: Sense organs and special senses - Chromidracryorrhea; Behavioral - Convulsions or effect on seizure threshold; Gastrointestinal - Changes in structure or function of salivary glands. Mouse LD_{50} Dose: 224 mg/kg.
Acute Inhalation: Rat LC_{50} Dose: 7500 mg/m^3; Toxic Effects: Behavioral - Excitment; Behavioral - Muscle contraction or spasticity. Mouse LC_{50} Dose: 1070 mg/m^3; Toxic Effects: Behavioral - Excitment; Behavioral - Muscle contraction or spasticity.
Acute Dermal: Rabbit LD_{50} Route: Skin; Dose: 320 uL/kg. Mouse LD_{50} Route: Subcutaneous Dose: 1150 mg/kg.
Chronic (Multiple Dose) Inhalation: Rat Dose: 700 mg/m^3/2H/9W-I; Toxic Effects: Cardiac - Changes in heart weight; Kidney, Ureter, and Bladder - Changes in kidney weight; DEATH. Rat Dose: 100 mg/m^3/4H/22W-I; Toxic Effects: Kidney, Ureter, and Bladder - Changes in kidney weight; Blood - Changes in cell count (unspecified); Nutritional and gross metabolic - Other changes.
Irritation Eye: Rabbit Standard Draize Test Dose: 50 ug/24H; Reaction: severe.
Irritation Skin: Human Standard Draize Test Dose: 125 mg/48H; Reaction: severe. Rabbit Standard Draize Test Dose: 2 mg/24H; Reaction: severe.
Reproductive/Teratogenic: Rat Route: Oral; Dose: 5600 mg/kg; Duration: male 4W prior to mating; Paternal Effects - Spermatogenesis. Rat Route: Intraperitoneal; Dose: 300 mg/kg; Duration: male 1D prior to mating; Effects on Fertility - Other measures of fertility. Rat Route: Intraperitoneal; Dose: 100 mg/kg; Duration: male 1D prior to mating; Effects on Fertility - Pre-implantation mortality.
Mutagenic: Human DNA Inhibition; Cell Type: HeLa cell; Dose: 100 ug/L. Human Cytogenetic Analysis; Cell Type: leukocyte; Dose: 10 umol/L/5H.

Hazard Overviews

Corrosive

Flammable

Fire Diamond

Health: Corrosive to eyes/skin/respiratory tract. Toxic. Also Causes: drowsiness, dizziness, anxiety, restlessness, nausea, vomiting, slurred speech, dilated pupils.
Fire: Flammable. Use alcohol foam, carbon dioxide, or dry chemical. Do not use a solid stream of water, since stream scatters and spreads fire. Use water spray to cool fire-exposed tanks and containers.
Reactivity: Stable. Hazardous polymerization cannot occur. Avoid: heat; ignition sources; oxidizers. Incompatible with: oxidizing materials; acids; copper alloys; lead. Hazardous decomposition products: toxic fumes of nitrogen oxide.
Carcinogenicity: IARC - Group 3, Not classifiable as to carcinogenicity to humans; NIOSH - Not listed; NTP - Not listed; ACGIH - Class A4, Not classifiable as a human carcinogen; OSHA - Not listed; EPA - Not listed; MAK - Not listed
Primary Target Organs:

Eyes

Skin

Respiratory System

Mucous Membranes

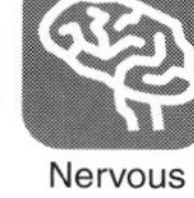
Nervous System

Exposure Limits
OSHA PEL Vacated 1989 Limits: TWA: 10 ppm; 40 mg/m^3.
ACGIH TLV: TWA: 10 ppm; 41 mg/m^3.
NIOSH REL: TWA: 10 ppm; 40 mg/m^3.
DFG MAK: TWA: 10 ppm; 40 mg/m^3.
Respirator Recommendation
Exposure Range: >10 to 100 ppm Air Purifying, Negative Pressure, Half Mask
Exposure Range: >100 to 1000 ppm Air Purifying, Negative Pressure, Full Face
Exposure Range: >1000 to 10,000 ppm Supplied Air, Constant Flow/Pressure Demand, Full Face
Exposure Range: >10,000 to unlimited ppm Self-contained Breathing Apparatus, Pressure Demand, Full Face
Cartridge Color: black

Environmental

Environmental Fate: If released to the atmosphere, it would be expected to photooxidize by reaction with hydroxyl radicals (calculated half-life of 1.82 days). If released on land, it would be subject to evaporation and leaching to groundwater where its fate is unknown. It should not adsorb to soil but may be subject to biodegradation. If released to water, it may be subject to evaporation and hydrolysis but will not adsorb to sediments or bioconcentrate in aquatic organisms. It is biodegraded in river mud sand sewage inocula and therefore may be susceptible to biodegradation in water.

Cleanup/Disposal: Guide No. 132: Fully encapsulating, vapor protective clothing should be worn for spills and leaks with no fire. Eliminate all ignition sources (no smoking, flares, sparks or flames in immediate area). All equipment used when handling the product must be grounded. Do not touch or walk through spilled material. Stop leak if you can do it without risk. Prevent entry into waterways, sewers, basements or confined areas. A vapor suppressing foam may be used to reduce vapors. Absorb with earth, sand or other non-combustible material and transfer to containers (except for Hydrazine). Use clean non-sparking tools to collect absorbed material. Large Spills: Dike far ahead of liquid spill for later disposal. Water spray may reduce vapor; but may not prevent ignition in closed spaces.

Environmental Physical Data

Octanol/Water Partition Coefficient: log K_{OW} = 1.49
Sorption Partition Coefficient: K_{OC} = estimated at 154
BCF: estimated at 7.99

Regulations

RCRA 40CFR: Not listed
CERCLA: 40CFR 302.4: Not listed
SARA 40CFR 372.65: Not listed TPQ: 10000 lb
SARA EHS 40CFR 355: Listed TPQ: 10000 lb
TSCA: Listed

Analytical Methods

Air: ASTM D4490
Water / Groundwater: ASTM D4983

CYC6270 CAS #: 121-82-4

CYCLONITE

RTECS: XY9450000
DOT: UN0072; UN0483; IMO1.1
EINECS Number: 204-500-1
Molecular Formula: $C_3H_6N_6O_6$
Structured MF: $C_3H_6N_6O_6$
Formula Weight: 222.26
Synonyms: CX 84A; CYCLOTRIMETHYLENENITRAMINE; CYCLOTRIMETHYLENETRINITRAMINE; CYKLONIT; ESAIDRO-1,3,5-TRINITRO-1,3,5-TRIAZINA; HEKSOGEN; HEXAHYDRO-1,3,5-TRINITRO-1,3,5-TRIAZIN; HEXAHYDRO-1,3,5-TRINITRO-1,3,5-TRIAZINE; HEXAHYDRO-1,3,5-TRINITRO-S-TRIAZINE; HEXOGEEN; HEXOGEN; HEXOGEN 5W; HEXOGEN (EXPLOSIVE); HEXOLITE; KHP 281; PBX(AF) 108; PBXW 108(E); PE 4; RDX; T4; 1,3,5-TRIAZA-1,3,5-TRINITROCYCLOHEXANE; 1,3,5-TRIAZINE,HEXAHYDRO-1,3,5-TRINITRO-; S-TRIAZINE,HEXAHYDRO-1,3,5-TRINITRO-; 1,3,5-TRIAZINE,HEXAHYDRO-1,3,5-TRINITRO-(9CI); 1,3,5-TRIAZINE,PERHYDRO,1,3,5-TRINITRO-; TRIMETHYLEENTRINITRAMINE; SYM-TRIMETHYLENE TRINITRAMINE; SYM-TRIMETHYLENETRINITRAMINE; TRIMETHYLENETRINITRAMINE; TRINITROCYCLOTRIMETHYLENE TRIAMINE; 1,3,5-TRINITROHEXAHYDRO-1,3,5-TRIAZINE; 1,3,5-TRINITROHEXAHYDRO-S-TRIAZINE; 1,3,5-TRINITROPERHYDRO-1,3,5-TRIAZINE; 1,3,5-TRINITRO-1,3,5-TRIAZACYCLOHEXANE; TRINITROTRIMETHYLENETRIAMINE
Description: white, crystalline powder
Use: high explosive; used as base charge for detonators and as an ingredient of bursting-charges and plastic explosives by the military

Physical Properties

Freezing Point: 205 °C (401 °F) to 206 °C (402.8 °F)
Specific Gravity: 1.82 at 20 °C/4 °C
Vapor Pressure: 0.0004 mm Hg at 230 °F
Water Solubility: Insoluble in Water
Other Solubilities: Slightly Soluble in Ethyl Acetate, Glacial Acetic Acid; Practically Insoluble in Carbon Tetrachloride; Readily Soluble in hot aniline, phenol, warm nitric acid; 1 g dissolves in 25 ml Acetone.
Flash Point: Explodes

RTECS Toxicity Data

Acute Oral: Child TD_{Lo} Dose: 85 mg/kg; Toxic Effects: Behavioral - Convulsions or effect on seizure threshold. Rat LD_{50} Dose: 100 mg/kg; Toxic Effects: Behavioral - Convulsions or effect on seizure threshold.
Chronic (Multiple Dose) Oral: Rat Dose: 3600 mg/kg/90D-C; Toxic Effects: Cardiac - Other changes; Blood - Pigmented or nucleated red Blood cells; Biochemical - Transaminases. Mouse Dose: 28800 mg/kg/90D-C; Toxic Effects: Liver - Changes in liver weight; Nutritional and gross metabolic - Weight loss or decreased weight gain; DEATH.
Reproductive/Teratogenic: Rat Route: Oral; Dose: 10 gm/kg; Duration: male 13W prior to mating; Effects on Newborn - Stillbirth; Live birth index. Rat Route: Oral; Dose: 3 gm/kg; Duration: male 13W prior to mating; Effects on Newborn - Growth statistics. Rat Route: Oral; Dose: 20 mg/kg; Duration: female 6-15D of pregnancy; Effects on Embryo or Fetus - Fetotoxicity.

Hazard Overviews

Flammable

Fire: Flammable.
Carcinogenicity: IARC - Not listed; NIOSH - Not listed; NTP - Not listed; ACGIH - Not listed; OSHA - Not listed; EPA - Class C, Possible human carcinogen; MAK - Not listed
Exposure Limits
OSHA PEL Vacated 1989 Limits: TWA: 1.5 mg/m^3.
ACGIH TLV: TWA: 1.5 mg/m^3.
NIOSH REL: TWA: 1.5 mg/m^3. STEL: 3 mg/m^3; skin.
Respirator Recommendation
Exposure Range: >1.5 to 15 mg/m^3 Air Purifying, Negative Pressure, Half Mask
Exposure Range: >15 to 150 mg/m^3 Air Purifying, Negative Pressure, Full Face

Exposure Range: >150 to 1500 mg/m^3 Supplied Air, Constant Flow/Pressure Demand, Full Face
Exposure Range: >1500 to unlimited mg/m^3 Self-contained Breathing Apparatus, Pressure Demand, Full Face
Cartridge Color: dust/mist filter (use P100 or consult supervisor for appropriate dust/mist filter)

Environmental

Ecotoxicity: LC_{50} Bluegill 3.6 mg/l/96 hr, pH 6.0, static bioassay
Environmental Fate: If released to water, the dominant fate process in translucent waters should be direct photochemical degradation. The half-life for this process is on the order of a few weeks. Aerobic biodegradation in aquatic environments should not occur, although anaerobic degradation under the proper conditions in lakes, ponds, and groundwater may be a significant fate process. Volatilization from water or from soil should not be important. Hydrolysis in terrestrial waters should not occur, although this process is known to proceed slowly in sea water. If released to soil, it should display moderate to high mobility, and can be expected to leach into groundwater. If released to the atmosphere, it can be expected to undergo a rapid reaction with photochemically produced hydroxyl radicals; the half-life for this process can be estimated at 1.5 hours. Direct photochemical degradation in the atmosphere may also be significant.

Environmental Physical Data

Henry's Law Constant: 2.6 x10^{-11}
Octanol/Water Partition Coefficient: log K_{ow} = 0.87
Sorption Partition Coefficient: K_{oc} = estimated at 42 to 167
BCF: bluegill 24.8

Regulations

RCRA 40CFR: Not listed
CERCLA: 40CFR 302.4: Not listed
SARA 40CFR 372.65: Not listed
SARA EHS 40CFR 355: Not listed
TSCA: Listed

Analytical Methods

Soil: SW846 4051, 8330
Water / Groundwater: AOAC 986.22; USGS O3112

CYC6440 CAS #: 111-78-4

CYCLOOCTADIENE

EINECS Number: 203-907-1
Molecular Formula: C_8H_{12}
Formula Weight: 108.19

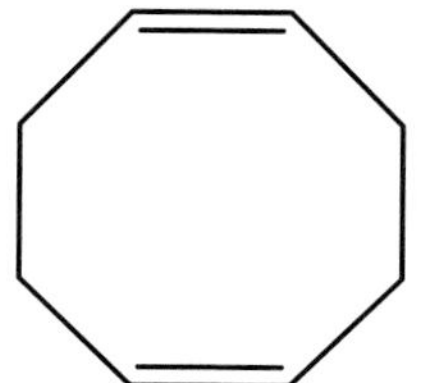

Chemical Structure

Synonyms: CYCLOOCTA-1,5-DIENE
Description: liquid
Use: resin intermediate; third monomer in EPT rubber

Physical Properties

Boiling Point: 150.8 °C (303 °F) at 757 mm Hg
Freezing Point: -69 °C (-92.2 °F) to -70 °C (-94 °F)
Specific Gravity: 0.8818 at 25 °C/4 °C
Vapor Pressure: 8.13 kPa at 75 °C
Water Solubility: Insoluble in Water
Other Solubilities: Soluble in Benzene, Carbon Tetrachloride
Refraction Index: 1.4095 at 25 °C
Ionization Potential (eV): 8.9 +/-0.3
Flash Point: 35 °C

Hazard Overviews

Flammable

Health: Severely irritating to eyes/skin/respiratory tract. Toxic. Other Acute Effects: harmful if swallowed, inhaled, or absorbed through skin; symptoms of exposure may include burning sensation; coughing; wheezing; laryngitis; shortness of breath; headache; nausea; vomiting; may cause allergic respiratory and skin reactions. Chronic Effects: target organ: blood. Carcinogen.
Fire: Flammable. Hazards: vapor may travel considerable distance to source of ignition and flash back. Extinguishing agents: water spray; carbon dioxide, dry chemical powder or appropriate foam. Precautions: combustible liquid.
Reactivity: Incompatible with: strong oxidizing agents. Hazardous decomposition products: toxic fumes of: carbon monoxide, carbon dioxide.
Carcinogenicity: IARC - Not listed; NIOSH - Not listed; NTP - Not listed; ACGIH - Not listed; OSHA - Not listed; EPA - Not listed; MAK - Not listed
Primary Target Organs:

Eyes

Skin

Respiratory System

Blood

Environmental

Ecotoxicity: Fishes: goldfish: 24h LD_{50} 14 mg/l

Regulations

RCRA 40CFR: Not listed
CERCLA: 40CFR 302.4: Not listed
SARA 40CFR 372.65: Not listed
SARA EHS 40CFR 355: Not listed
TSCA: Listed

CYC6780 **CAS #: 542-92-7**

CYCLOPENTADIENE

RTECS: GY1000000
EINECS Number: 208-835-4
Molecular Formula: C_5H_6
Structured MF: C_5H_6
Formula Weight: 66.11
Synonyms: 1,3-CYCLOPENTADIENE; R-PENTINE; PENTOLE; PYROPENTYLENE
Description: colorless liquid; sweetish, turpentine odor
Use: mfr resins; in org synth producing sesquiterpenes, synthetic alkaloids, camphors; starting material for synthetic prostaglandins; chlorinated insecticides; formation of sandwich compounds by chelation

Physical Properties

Boiling Point: 41.5 °C (107 °F) to 42.1 °C (108 °F) at 760 mm Hg
Freezing Point: -85 °C (-121 °F)
Specific Gravity: 0.8021 at 20 °C/4 °C
Vapor Pressure: 400 mm Hg
Water Solubility: Insoluble
Other Solubilities: Soluble in Acetone.
Odor Threshold: 5.0 mg/m^3
Refraction Index: 1.4440 at 20 °C/D
Ionization Potential (eV): 8.56
Flash Point: 25 °C Open Cup
Autoignition Temperature: 640 °C

RTECS Toxicity Data

Acute Inhalation: Rat LC_{50} Dose: 39 gm/m^3; Toxic Effects: Behavioral - Excitment; Behavioral - Muscle contraction or spasticity; Behavioral - Aggression. Mouse LC_{50} Dose: 14 gm/m^3; Toxic Effects: Behavioral - Excitment; Behavioral - Muscle contraction or spasticity; Behavioral - Aggression.
Chronic (Multiple Dose) Inhalation: Rat Dose: 350 mg/m^3/4H/26W-I; Toxic Effects: Blood - Normocytic anemia; Blood - Changes in erythrocite (RBC) cell count.

Hazard Overviews

Explosive

Flammable

Fire Diamond

Health: Irritating to eyes/skin/respiratory tract. Also Causes: liver/kidney/CNS effects (based on animal data), sensitization skin reactions. Chronic Effects: dermatitis.
Fire: Flammable and explosive. Use dry chemical, carbon dioxide, water spray, or regular foam. Fight fire from maximum distance.
Reactivity: Unstable, may explode upon standing. Hazardous polymerization can occur. Avoid: heat; ignition sources. Incompatible with: ammonia; nitrogen oxides; oxygen + ozone; sulfuric acid; nitric acid; potassium hydroxide; dinitrogen tetroxide; other oxidizers. Hazardous decomposition products: carbon oxide(s).
Carcinogenicity: IARC - Not listed; NIOSH - Listed as carcinogen; NTP - Not listed; ACGIH - Not listed; OSHA - Not listed; EPA - Not listed; MAK - Not listed
Primary Target Organs:

Eyes

Skin

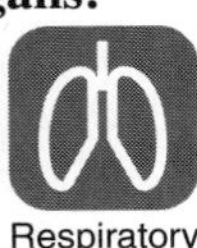
Respiratory System

Exposure Limits
OSHA PEL: TWA: 75 ppm; 200 mg/m^3.
ACGIH TLV: TWA: 75 ppm; 203 mg/m^3.
NIOSH REL: TWA: 75 ppm; 200 mg/m^3.
NIOSH IDLH: 750 ppm.
DFG MAK: TWA: 75 ppm; 200 mg/m^3.
Respirator Recommendation
Exposure Range: >75 to <750 ppm Air Purifying, Negative Pressure, Half Mask
Exposure Range: 750 to unlimited ppm Self-contained Breathing Apparatus, Pressure Demand, Full Face
Cartridge Color: black

Environmental

Environmental Fate: If concentrated solutions are released to soil or water (spilled), this compound is expected to polymerize spontaneously to dicyclopentadiene. If released to soil in dilute amounts, it may undergo extensive leaching or rapid volatilization. If released to water in dilute amounts, it is expected to undergo rapid volatilization (estimated half-life 2.4 hours from a model river). The significance of biodegradation in either soil or water is unknown. Chemical hydrolysis, oxidation, bioaccumulation in aquatic organisms, and adsorption to suspended solids and sediments are not expected to be significant fate processes in water. If released to the atmosphere, it is expected to exist almost entirely in the vapor phase. This compound is expected to rapidly react with ozone molecules, photochemically generated hydroxyl radicals, and possibly nitrate radicals. The atmospheric half-life is estimated to be about 40 minutes.

Environmental Physical Data

Henry's Law Constant: estimated at 2.12 x10^{-2}
Octanol/Water Partition Coefficient: log K_{ow} = 1.89
Sorption Partition Coefficient: K_{oc} = estimated at 71 to 245
BCF: estimated at 9

Regulations

RCRA 40CFR: Not listed
CERCLA: 40CFR 302.4: Not listed
SARA 40CFR 372.65: Not listed
SARA EHS 40CFR 355: Not listed

TSCA: Listed

Analytical Methods

Indoor / Expired Air: NIOSH 2523

CYC6950 CAS #: 287-92-3

CYCLOPENTANE

RTECS: GY2390000
DOT: UN1146; IMO3.1
EINECS Number: 206-016-6
Molecular Formula: C_5H_{10}
Structured MF: C_5H_{10}
Formula Weight: 70.13

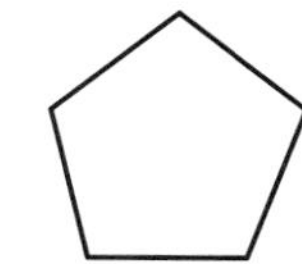

Chemical Structure

Synonyms: PENTAMETHYLENE
Description: colorless liquid; petroleum odor
Use: as solvent; starting material for synth in chemical industry; solvent for cellulose ethers; motor fuel; azeotropic distillation agent; chemical intermediate in prodn of cyclopentadiene (insecticide)

Physical Properties

Boiling Point: 49.2 °C (121 °F) at 760 mm Hg
Freezing Point: -93.9 °C (-137.02 °F)
Specific Gravity: 0.7457 at 20 °C/4 °C
Vapor Density: 2.4 Air=1
Vapor Pressure: 400 mm Hg at 88 °F
Water Solubility: Insoluble in Water
Other Solubilities: Soluble in Alcohol, Benzene, Acetone, Ether.
Surface Tension: 23 dynes/cm at 20 °C
Refraction Index: 1.4065 at 20 °C/D
Critical Temperature: 239 °C
Critical Pressure: 654 psia
Ionization Potential (eV): 10.52
Flash Point: -37 °C
Autoignition Temperature: 361 °C
LEL: 1.1% v/v
UEL: 8.7% v/v

Hazard Overviews

Flammable

Fire Diamond

Health: Irritating to eyes/skin/respiratory tract. Also Causes: CNS depression, excitement, dizziness, confusion, unconsciousness, coma, respiratory failure, redness, GI irritation. Chronic Effects: dermatitis.
Fire: Flammable. Can form explosive mixtures in the air. For small fires use dry chemical, carbon dioxide, water spray, or regular foam. For large fires use water spray, fog, or regular foam. Vapors may travel to an ignition source and flash back. Containers may explode in heat of fire.
Reactivity: Stable. Hazardous polymerization cannot occur. Avoid: heat; ignition sources. Incompatible with: chlorine; bromine; fluorine; sodium hydroxide; potassium hydroxide. Hazardous decomposition products: carbon oxide(s).
Carcinogenicity: IARC - Not listed; NIOSH - Listed as carcinogen; NTP - Not listed; ACGIH - Not listed; OSHA - Not listed; EPA - Not listed; MAK - Not listed
Primary Target Organs:

Eyes

Skin

Respiratory System

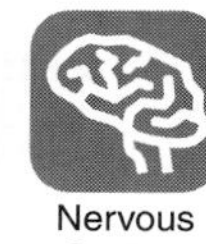

Nervous System

Exposure Limits
OSHA PEL Vacated 1989 Limits: TWA: 600 ppm; 1720 mg/m^3.
ACGIH TLV: TWA: 600 ppm; 1720 mg/m^3.
NIOSH REL: TWA: 600 ppm; 1720 mg/m^3.
Respirator Recommendation
Exposure Range: >600 to 30000 ppm Supplied Air, Constant Flow/Pressure Demand, Half Mask
Exposure Range: >30,000 to 60,000 ppm Supplied Air, Constant Flow/Pressure Demand, Full Face
Exposure Range: >60,000 to unlimited ppm Self-contained Breathing Apparatus, Pressure Demand, Full Face
Note: odor threshold unknown

Environmental

Environmental Fate: Photolysis or hydrolysis are not expected to be important fate processes. Limited data suggests that it is recalcitrant to biodegradation. Varying estimates of K_{oc} indicate a wide range of adsorption characteristics and the mobility class in soil may range from low to medium. In aquatic systems, it may partition from the water column to organic matter in sediments and suspended solids. The potential for bioconcentration in aquatic organisms is low. A Henry's Law constant of 1.88 $x10^{-1}$ atm-cu m/mole at 25 °C suggests rapid volatilization from natural waters and moist soils. The volatilization half-lives from a model river and a model pond, the latter considers the effect of adsorption, have been estimated to be 2.5 hr and 5.2 days, respectively. Based on its vapor pressure, it should rapidly evaporate from dry surfaces, especially when present in high concentration such as in spill situations. It is expected to exist almost entirely in the vapor phase in ambient air. Reactions with photochemically produced hydroxyl radicals in the atmosphere have been shown to be important (average half-life of 3.3 days). Physical removal from air by precipitation and dissolution in clouds may occur; however, the short atmospheric residence time suggests that wet deposition is of limited importance.

Cleanup/Disposal: Guide No. 128: Eliminate all ignition sources (no smoking, flares, sparks or flames in immediate area). All equipment used when handling the product must be grounded. Do not touch or walk through spilled material. Stop leak if you can do it without risk. Prevent entry into waterways, sewers, basements or confined areas. A vapor suppressing foam may be used to reduce vapors. Absorb or cover with dry earth, sand or other non-combustible material and transfer to containers. Use clean non-sparking tools to collect absorbed material. Large Spills: Dike far ahead of liquid spill for later disposal. Water spray may reduce vapor; but may not prevent ignition in closed spaces.

Environmental Physical Data

Henry's Law Constant: calculated at 1.88 x10^{-1}
Octanol/Water Partition Coefficient: log K_{ow} = 3.00
Sorption Partition Coefficient: K_{oc} = 272 to 1020
BCF: estimated at 0.08

Regulations

RCRA 40CFR: Not listed
CERCLA: 40CFR 302.4: Not listed
SARA 40CFR 372.65: Not listed
SARA EHS 40CFR 355: Not listed
TSCA: Listed

CYC7290 CAS #: 120-92-3

CYCLOPENTANONE

RTECS: GY4725000
DOT: UN2245; IMO3.3
EINECS Number: 204-435-9
Molecular Formula: C_5H_8O
Formula Weight: 84.12

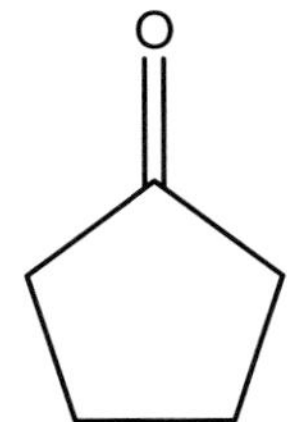

Chemical Structure

Synonyms: ADIPIC KETONE; ADIPINKETON; DUMASIN; KETOCYCLOPENTANE; KETOPENTAMETHYLENE
Description: water-white liquid; ethereal/peppermint odor
Use: intermediate for pharmaceuticals, biologicals, insecticides and rubber chemicals

Physical Properties

Boiling Point: 130.6 °C (267 °F) at 760 mm Hg
Freezing Point: -51.3 °C (-60.34 °F)
Specific Gravity: 0.94869 at 20 °C/4 °C
Saturated Vapor Density: 1.234899139 kg/m^3
Vapor Pressure: 1.55 kPa at 25 °C
Water Solubility: Insoluble
Other Solubilities: Acetone: Soluble; Alcohol: Soluble; Ether: miscible; Hexane: Soluble; Methanol: Soluble.
Odor Threshold: Detection 31 to 1120 mg/m^3
Refraction Index: 1.4366 at 20 °C
Ionization Potential (eV): =< 9.28 +/-0.5
Flash Point: 30.556 °C Closed Cup

RTECS Toxicity Data

Acute Oral: Mammal LD_{50} Dose: 2 gm/kg.
Acute Dermal: Mouse LD_{Lo} Route: Subcutaneous Dose: 2600 mg/kg; Toxic Effects: Automatic Nervous System - Other (direct) parasympathomimetic; Behavioral - Coma; Lungs, Thorax, or Respiration - Other changes. Frog LD_{Lo} Route: Subcutaneous Dose: 3 gm/kg; Toxic Effects: Automatic Nervous System - Other (direct) parasympathomimetic; Behavioral - Change in motor activity (specific assay); Skin and appendages - Sweating.
Irritation Eye: Rabbit Standard Draize Test Dose: 100 mg; Reaction: severe. Rabbit Nonstandard Exposure Dose: 100 mg/4S rinse; Reaction: severe.
Irritation Skin: Rabbit Standard Draize Test Dose: 500 mg; Reaction: mild.

Hazard Overviews

Flammable

Fire Diamond

Health: Severe irritation to eyes/skin/respiratory tract. Moderately Toxic. Also Causes: upon inhalation of high concentrations: CNS depression, narcosis.
Fire: Flammable. Use dry chemical, carbon dioxide, water spray, fog, or alcohol-resistant foam. Vapors may travel to an ignition source and flash back. Containers may explode in the heat of fire.
Reactivity: Stable. Hazardous polymerization can occur in the presence of acid. Avoid: heat; ignition sources. Incompatible with: oxidizing materials; hydrogen peroxide and nitric acid; acids. Hazardous decomposition products: carbon monoxide; carbon dioxide.
Carcinogenicity: IARC - Not listed; NIOSH - Not listed; NTP - Not listed; ACGIH - Not listed; OSHA - Not listed; EPA - Not listed; MAK - Not listed
Primary Target Organs:

Eyes

Skin

Mucous Membranes

Nervous System

Environmental

Ecotoxicity: Toxicity threshold (cell multiplication inhibition test): bacteria (Pseudomonas putida) 16h EC_0 175 mg/l algae (Microcystis aeruginosa) 8d EC_0 63 mg/l
Environmental Fate: It is not expected to undergo hydrolysis or photolysis in the environment. Limited data suggests that it should biodegrade rapidly upon acclimation in soil and water. A low estimated K_{oc} indicates it should have a very high

mobility in soil. In aquatic systems, it should not partition from the water column to organic matter in sediments and suspended solids, nor should it bioconcentrate in aquatic organisms. A Henry's Law constant of 1.0 x10^{-5} atm-cu m/mole at 25 °C suggests that the volatilization from natural waters will be an important fate process. Volatilization half-lives from a model river and a model pond, the latter considers the effect of adsorption, have been estimated to be about 3.5 and 40 days, respectively. Based on its vapor pressure, it should evaporate from dry surfaces, especially when present in high concentration such as in spill situations. It is expected to exist entirely in the vapor phase in ambient air. Vapor phase reactions with photochemically produced hydroxyl radicals in the atmosphere have been shown to be important (half-life of 5.5 days).

Cleanup/Disposal: Guide No. 127: Eliminate all ignition sources (no smoking, flares, sparks or flames in immediate area). All equipment used when handling the product must be grounded. Do not touch or walk through spilled material. Stop leak if you can do it without risk. Prevent entry into waterways, sewers, basements or confined areas. A vapor suppressing foam may be used to reduce vapors. Absorb or cover with dry earth, sand or other non-combustible material and transfer to containers. Use clean non-sparking tools to collect absorbed material. Large Spills: Dike far ahead of liquid spill for later disposal. Water spray may reduce vapor; but may not prevent ignition in closed spaces.

Environmental Physical Data

Henry's Law Constant: 1.0 x10^{-5}
Octanol/Water Partition Coefficient: log K_{ow} = 0.24
Sorption Partition Coefficient: K_{oc} = 30
BCF: calculated at 0.05

Regulations

RCRA 40CFR: Not listed
CERCLA: 40CFR 302.4: Not listed
SARA 40CFR 372.65: Not listed
SARA EHS 40CFR 355: Not listed
TSCA: Listed

CYC7630 CAS #: 50-18-0

CYCLOPHOSPHAMIDE

RTECS: RP5950000
EINECS Number: 200-015-4
Molecular Formula: $C_7H_{15}Cl_2N_2O_2P$
Formula Weight: 261.10
Synonyms: ASTA; ASTA B 518; ASTA B518; B 518; N,N-BIS-(BETA-CHLORAETHYL)-N',O-PROPYLEN-PHOSPHORSAEURE-ESTER-DIAMID; 2-(BIS(2-CHLOROETHYL)AMINO)-2H-1,3,2-OXAZAPHOSPHORINE 2-OXIDE; 1-BIS(2-CHLOROETHYL)AMINO-1-OXO-2-AZA-5-OXAPHOSPHORIDIN; 1-BIS(2-CHLOROETHYL)AMINO-1-OXO-2-AZA-5-OXAPHOSPHORIDINE; 2-(BIS(2-CHLOROETHYL)AMINO)TETRAHYDRO-2H-1,3,2-OXAZAPHOSPHORINE 2-OXIDE; 2-(BIS(2-CHLOROETHYL)AMINO)TETRAHYDRO-2H-1,3,2-OXAZOPHOSPHORINE 2-OXIDE; N,N-BIS(2-CHLOROETHYL)-N'-(3-HYDROXYPROPYL)PHOSPHORODIAMIDIC ACID INTRAMOL; BIS(2-CHLOROETHYL)PHOSPHAMIDE CYCLIC PROPANOLAMIDE ESTER; BIS(2-CHLOROETHYL)PHOSPHORAMIDE CYCLIC PROPANOLAMIDE ESTER; BIS(2-CHLOROETHYL)PHOSPHORAMIDE-CYCLIC PROPANOLAMIDE ESTER; N,N-BIS(2-CHLOROETHYL)-N',O-PROPYLENEPHOSPHORIC ACID ESTER DIAMIDE; N,N-BIS(BETA-CHLOROETHYL)-N',O-PROPYLENEPHOSPHORIC ACID ESTER DIAMIDE; N,N-BIS(2-CHLOROETHYL)-N',O-PROPYLENEPHOSPHORIC ACID ESTERDIAMIDE; N,N-BIS(BETA-CHLOROETHYL)-N',O-PROPYLENEPHOSPHORIC ACIDESTER DIAMIDE; N,N-BIS(2-CHLOROETHYL)TETRAHYDRO-2H-1,3,2-OXAZAPHOSPHORIN-2-AMINE 2-OXIDE; N,N-BIS(BETA-CHLOROETHYL)-N',O-TRIMETHYLENEPHOSPHORIC ACID ESTER DIAMIDE; N,N-BIS(BETA-CHLOROETHYL)-N',O-TRIMETHYLENEPHOSPHORIC ACIDESTER DIAMIDE; N,N-BIS(BETA-CHLOROETHYL)N',O-TRIMETHYLENEPHOSPHORIC ACIDESTER DIAMIDE; CB 4564; CLAFEN; CLAPHENE; CP; CPA; CTX; CY; CYCLOPHOSPHAMID; (-)-CYCLOPHOSPHAMIDE; CYCLOPHOSPHAMIDUM; CYCLOPHOSPHAN; CYCLOPHOSPHANE; CYCLOPHOSPHORAMIDE; CYCLOSTIN; CYKLOFOSFAMID; CYTOPHOSPHAN; CYTOPHOSPHANE; CYTOXAN; 2-(DI(2-CHLOROETHYLAMINO))-1-OXA-3-AZA-2-PHOSPHACYCLOHEXANE 2-OXIDE; N,N-DI(2-CHLOROETHYL)-N,O-PROPYLENE PHOSPHORIC ACID ESTER DIAMIDE; N,N-DI(2-CHLOROETHYL)-N,O-PROPYLENE-PHOSPHORIC ACID ESTERDIAMIDE; ENDOXAN; ENDOXAN R; ENDOXANA; ENDOXANAL; ENDOXAN-ASTA; ENDOXANE; ENDUXAN; ESTER; GENOXAL; HEXADRIN; MITOXAN; NEOSAR; NSC 26271; 2H-1,3,2-OXAZAPHOSPHORIN-2-AMINE,N,N-BIS(2-CHLOROETHYL)TETRAHYDRO-,2-OXIDE; 2H-1,3,2-OXAZAPHOSPHORIN-2-AMINE,N,N-BIS(2-CHLOROETHYL)TETRAHYDRO-,2-OXIDE (9CI); 2-H-1,3,2-OXAZAPHOSPHORINANE; 2H-1,3,2-OXAZAPHOSPHORINE,2-(BIS(2-CHLOROETHYL)AMINO)TETRAHYDRO-,2-OXIDE; PHOSPHORODIAMIDIC ACID,N,N-BIS(2-CHLOROETHYL)-N'-(3-HYDROXYPROPYL)-,INTRAMOL. ESTER; PROCYTOX; SEMDOXAN; SENDOXAN; SENDUXAN; SK 20501; TETRAHYDRO-N,N-BIS(2-CHLOROETHYL)-2H-1,3,2-OXAZAPHOSPHORIN-2-AMINE 2-OXIDE; ZYKLOPHOSPHAMID

Description: fine, white, crystalline powder, liquefies on loss of its water of crystallization; odorless

Use: an antineoplastic agent used to treat malignant lymphoma, multiple myeloma, leukemias and other malignant diseases; treat chronic hepatitis; being tested for use as an insect chemosterilant and for use in the chemical shearing of sheep

Physical Properties

Freezing Point: 49.5 °C (121.1 °F) to 53 °C (127.4 °F)
Water Solubility: 1 part in 25 parts Water
Other Solubilities: 1 in 1 parts Alcohol; Slightly Soluble in Benzene, Carbon Tetrachloride; very slightly Soluble in Ether and Acetone; Soluble in Chloroform, dioxane and glycols and Insoluble in Carbon Tetrachloride and Carbon Disulfide.
Flash Point: Not available; probably combustible

RTECS Toxicity Data

Acute Oral: Human TD_{Lo} Dose: 20 mg/kg; Toxic Effects: Gastrointestinal - Other changes; Kidney, Ureter, and Bladder - Other changes; Skin and appendages - Hair. Man TD_{Lo} Dose: 56 mg/kg/26D-I; Toxic Effects: Blood - Agranulocytosis. Man TD_{Lo} Dose: 56 mg/kg/4W-I; Toxic Effects: Liver - Other changes. Woman TD_{Lo} Dose: 45 mg/kg; Toxic Effects: Kidney, Ureter, and Bladder - Other changes. Child TD_{Lo} Dose: 2500 ug/kg; Toxic Effects: Behavioral -

Convulsions or effect on seizure threshold; Nutritional and gross metabolic - Other changes. Woman LD_{Lo} Dose: 16 mg/kg/4D-I; Toxic Effects: Blood - Other changes Rat LD_{50} Dose: 160 mg/kg.

Acute Dermal: Rat LD_{50} Route: Subcutaneous Dose: 144 mg/kg; Toxic Effects: Blood - Normocytic anemia; Nutritional and gross metabolic - Weight loss or decreased weight gain. Mouse LD_{50} Route: Subcutaneous Dose: 200 mg/kg.

Chronic (Multiple Dose) Oral: Rat Dose: 660 mg/kg/80D-I; Toxic Effects: DEATH. Rat Dose: 87500 ug/kg/5W-C; Toxic Effects: Endocrine - Changes in spleen weight; Blood - Changes in leukocyte (WBC) cell count; Biochemical - Transaminases. Rat Dose: 910 mg/kg/26W-C; Toxic Effects: Endocrine - Changes in spleen weight; Blood - Changes in leukocyte (WBC) cell count; Biochemical - Transaminases. Rat Dose: 15 mg/kg/10D-I; Toxic Effects: Endocrine - Changes in thymus weight; Blood - Changes in leukocyte (WBC) cell count; Immunological including allergic - Decreased immune response.

Reproductive/Teratogenic: Man Route: Oral; Dose: 107 mg/kg; Duration: male 21W prior to mating; Paternal Effects - Spermatogenesis. Woman Route: Oral; Dose: 60 mg/kg; Duration: female 60D prior to mating Maternal Effects - Menstrual cycle changes or disorders. Woman Route: Oral; Dose: 980 mg/kg; Duration: female 2W prior to mating; Specific Developmental Abnormalities - Central nervous system; Musculoskeletal system; Gastrointestinal system. Woman Route: Oral; Dose: 168 ug/kg; Duration: female 20W prior to mating Maternal Effects - Menstrual cycle changes or disorders. Woman Route: Multiple routes; Dose: 382 mg/kg; Duration: female 1-39W of pregnancy; Specific Developmental Abnormalities - Craniofacial (including nose and tongue); Musculoskeletal system. Woman Route: Multiple routes; Dose: 502 mg/kg; Duration: female 1-37W of pregnancy; Specific Developmental Abnormalities - Craniofacial (including nose and tongue); Musculoskeletal system. Woman Route: Multiple routes; Dose: 215 mg/kg; Duration: female 43-70D of pregnancy Specific Developmental Abnormalities - Musculoskeletal system; Cardiovascular (circulatory) system. Rat Route: Oral; Dose: 100 mg/kg; Duration: male 1D prior to mating; Effects on Fertility - Male fertility index. Rat Route: Oral; Dose: 91800 ug/kg; Duration: male 18D prior to mating; Paternal Effects - Spermatogenesis; Effects on Fertility - Post-implantation mortality; Effects on Embryo or Fetus - Fetotoxicity. Rat Route: Oral; Dose: 168 mg/kg; Duration: male 4W prior to mating; Effects on Embryo or Fetus - Cytological changes (inc. somatic cell genetic material). Rat Route: Oral; Dose: 110 mg/kg; Duration: female 7-17D of pregnancy; Effects on Newborn - Stillbirth; Specific Developmental Abnormalities - Central nervous system; Craniofacial (including nose and tongue). Rat Route: Oral; Dose: 110 mg/kg; Duration: female 7-17D of pregnancy; Specific Developmental Abnormalities - Body wall; Musculoskeletal system; Urogenital system. Rat Route: Oral; Dose: 5 mg/kg; Duration: female 11D of pregnancy; Specific Developmental Abnormalities - Gastrointestinal system. Rat Route: Oral; Dose: 122 mg/kg; Duration: male 18D prior to mating; Paternal Effects - Testes, epididymis, sperm duct; Prostate, seminal vessicle, Cowper's gland, accessory glands.

Mutagenic: Human DNA Damage; Route: Intravenous; Dose: 750 mg/kg. Human DNA Damage; Cell Type: leukocyte; Dose: 500 umol/L. Human Unscheduled DNA Synthesis; Cell Type: HeLa cell; Dose: 100 nmol/L. Human Unscheduled DNA Synthesis; Cell Type: lymphocyte; Dose: 1 umol/L. Human Unscheduled DNA Synthesis; Cell Type: other cell types; Dose: 300 mg/L. Human DNA Inhibition; Cell Type: HeLa cell; Dose: 600 umol/L/1H-C. Human DNA Inhibition; Cell Type: other cell types; Dose: 200 mg/L. Human Mutations in Microorganisms; Cell Type: lymphocyte; Dose: 7 mg/L (+S9). Human Mutations in Microorganisms; Cell Type: fibroblast; Dose: 475 nmol/L (+S9). Human Cytogenetic Analysis; Route: Unreported; Dose: 60 mg/kg/8W-I. Human Cytogenetic Analysis; Cell Type: leukocyte; Dose: 1 mmol/L. Human Cytogenetic Analysis; Cell Type: HeLa cell; Dose: 210 mg/L/48H. Human Cytogenetic Analysis; Cell Type: lymphocyte; Dose: 10 ug/L/72H. Human Cytogenetic Analysis; Cell Type: other cell types; Dose: 50 ug/L. Child Cytogenetic Analysis; Route: Unreported; Dose: 4 mg/kg/4W. Monkey Cytogenetic Analysis; Route: Intravenous; Dose: 18400 ug/kg. Human Sister Chromatid Exchenge; Cell Type: other cell types; Dose: 500 mg/L. Human Sister Chromatid Exchenge; Route: Oral; Dose: 31 mg/kg. Human Sister Chromatid Exchenge; Cell Type: lymphocyte; Dose: 10 umol/L. Human Sister Chromatid Exchenge; Cell Type: fibroblast; Dose: 10 umol/L. Human Sister Chromatid Exchenge; Cell Type: leukocyte; Dose: 1 umol/L. Monkey Sister Chromatid Exchenge; Route: Intravenous; Dose: 18400 ug/kg. Monkey Sister Chromatid Exchenge; Cell Type: kidney; Dose: 100 umol/L. Human Micronucleus Test; Cell Type: lymphocyte; Dose: 800 umol/L. Human Body Fluid Assay; Indicator Organism: Yeast - S Cerevisiae; Dose: 6200 ug/kg. Human Body Fluid Assay; Indicator Organism: Bacteria - S Typhimurium; Dose: 14 mg/kg. Human Other Mutation Test Systems; Cell Type: HeLa cell; Dose: 840 mg/L.

Tumorigenic: Human Route: Oral; Dose: 920 mg/kg/3Y-C; Toxic Effects: Tumorigenic - Carcinogenic by RTECS criteria; Kidney, Ureter, and Bladder - Tumors. Man Route: Oral; Dose: 2310 mg/kg/4.5Y-C; Toxic Effects: Tumorigenic - Carcinogenic by RTECS criteria; Gastrointestinal - Tumors. Man Route: Oral; Dose: 1078 mg/kg/3Y-C; Toxic Effects: Tumorigenic - Carcinogenic by RTECS criteria; Kidney, Ureter, and Bladder - Tumors. Man Route: Oral; Dose: 1800 mg/kg/6Y-C; Toxic Effects: Tumorigenic - Carcinogenic by RTECS criteria; Kidney, Ureter, and Bladder - Tumors. Man Route: Oral; Dose: 1190 mg/kg/4Y-I; Toxic Effects: Tumorigenic - Carcinogenic by RTECS criteria; Blood - Leukemia. Woman Route: Oral; Dose: 1890 mg/kg/3Y-I; Toxic Effects: Tumorigenic - Carcinogenic by RTECS criteria; Blood - Leukemia. Woman Route: Oral; Dose: 2700 mg/kg/6Y-C; Toxic Effects: Tumorigenic - Carcinogenic by RTECS criteria; Kidney, Ureter, and Bladder - Tumors. Woman Route: Oral; Dose: 1760 mg/kg/4Y-C; Toxic Effects: Tumorigenic - Carcinogenic by RTECS criteria; Kidney,

Ureter, and Bladder - Tumors. Rat Route: Oral; Dose: 475 mg/kg/100W-I; Toxic Effects: Tumorigenic - Carcinogenic by RTECS criteria; Kidney, Ureter, and Bladder - Tumors; Blood - Lymphomax including Hodgkin's disease. Rat Route: Oral; Dose: 1270 mg/kg/87W-I; Toxic Effects: Tumorigenic - Carcinogenic by RTECS criteria; Kidney, Ureter, and Bladder - Tumors; Blood - Lymphomax including Hodgkin's disease. Rat Route: Oral; Dose: 698 mg/kg/89W-I; Toxic Effects: Tumorigenic - Carcinogenic by RTECS criteria; Kidney, Ureter, and Bladder - Tumors; Blood - Lymphomax including Hodgkin's disease. Rat Route: Oral; Dose: 1075 mg/kg/86W-I; Toxic Effects: Tumorigenic - Carcinogenic by RTECS criteria; Kidney, Ureter, and Bladder - Tumors; Blood - Tumors.

Hazard Overviews

Fire: Will burn.

Carcinogenicity: IARC - Group 1, Carcinogenic to humans; NIOSH - Not listed; NTP - Listed; ACGIH - Not listed; OSHA - Not listed; EPA - Not listed; MAK - Not listed

Environmental

Environmental Fate: If released to the atmosphere, the vapor phase is expected to degrade rapidly by reaction with photochemically produced hydroxyl radicals (estimated half-life of 2.4 hours in typical air). Particulate phase can be physically removed from air by wet and dry deposition. If released to water, hydrolysis may be an important environmental degradation process. Has a measured aqueous hydrolysis half-life of 41 days at 25 °C and pH 7. If released to moist soil, can be expected to hydrolyze. Leaching in soil may be significant. No data are available to predict the biodegradability in soil or water.

Cleanup/Disposal: Guide No. 154: Eliminate all ignition sources (no smoking, flares, sparks or flames in immediate area). Do not touch damaged containers or spilled material unless wearing appropriate protective clothing. Stop leak if you can do it without risk. Prevent entry into waterways, sewers, basements or confined areas. Absorb or cover with dry earth, sand or other non-combustible material and transfer to containers. Do not get water inside containers.

Environmental Physical Data

Sorption Partition Coefficient: K_{OC} = estimated at 13

BCF: estimated at 1.6

Regulations

RCRA 40CFR: Listed Hazardous Waste No. U058 Toxic Waste

CERCLA: 40CFR 302.4: Listed per RCRA Section 3001 RQ: 10 lb (4.535 kg)

SARA 40CFR 372.65: Not listed

SARA EHS 40CFR 355: Not listed

TSCA: Not listed

CYC7970 CAS #: 75-19-4

CYCLOPROPANE

RTECS: GZ0690000

DOT: UN1027; IMO2.0

EINECS Number: 200-847-8

Molecular Formula: C_3H_6

Structured MF: C_3H_6

Formula Weight: 42.08

Chemical Structure

Synonyms: CYCLOPROPANE,LIQUIFIED; TRIMETHYLENE

Description: colorless gas; petroleum ether odor

Use: general anesthetic; medication: anesthetic (inhalation); medication (vet): inhalation anesthetic; organic synthesis

Physical Properties

Boiling Point: -32.7 °C (-27 °F) at 760 mm Hg

Freezing Point: -127.6 °C (-197.68 °F)

Vapor Density: 1.88 Air=1

Density: 1.879 g/L at 0 °C & 1 atm

Vapor Pressure: 4850 mm Hg

Water Solubility: 1 vol per 207 vol Water at 15 °C

Other Solubilities: Soluble in Petroleum Ether, Benzene.

Surface Tension: 22 dynes/cm at 40 °C

Refraction Index: 1.3799 at -42.5 °C/D

Critical Temperature: 125 °C

Critical Pressure: 798 psia

Ionization Potential (eV): 9.86

Flash Point: Flammable gas

Autoignition Temperature: 498 °C

LEL: 2.4% v/v

UEL: 10.3% v/v

RTECS Toxicity Data

Acute Inhalation: Mouse LC_{Lo} Dose: 282 gm/m³/2hr; Toxic Effects: Behavioral - Excitement.

Mutagenic: Chicken Cytogenetic Analysis; Route: Inhalation; Dose: 20 pph/3H.

Hazard Overviews

Flammable

Compressed Gas

Fire Diamond

Health: Mildly irritating to respiratory tract. Also Causes: CNS depression, excitement, stupor, delirium, enlarged pupils, watering and bulging of eyes, cardiac arrhythmias, convulsions, unconsciousness, death, hemorrhagic swelling of the lungs, trachea congestion, and early autolysis (digestion of tissue by the body's own enzymes), frostbite. Chronic

Effects: anorexia, weight loss, insomnia, irritability, kidney inflammation and RBC effects.

Fire: Flammable. Can form explosive mixtures in air. Stop leak. Use dry chemical or carbon dioxide for small fires and water spray or fog for large fires.

Reactivity: Stable. Hazardous polymerization cannot occur. Avoid: oxygen-rich atmospheres (explosive range = 2.5 to 60%); exposure to heat. Incompatible with: oxidizing materials. Hazardous decomposition products: carbon dioxide; acrid smoke.

Carcinogenicity: IARC - Not listed; NIOSH - Not listed; NTP - Not listed; ACGIH - Not listed; OSHA - Not listed; EPA - Not listed; MAK - Not listed

Primary Target Organs:

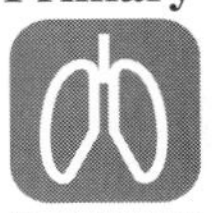

Environmental

Cleanup/Disposal: Guide No. 115: Eliminate all ignition sources (no smoking, flares, sparks or flames in immediate area). All equipment used when handling the product must be grounded. Do not touch or walk through spilled material. Stop leak if you can do it without risk. If possible, turn leaking containers so that gas escapes rather than liquid. Use water spray to reduce vapors or divert vapor cloud drift. Do not direct water at spill or source of leak. Prevent spreading of vapors through sewers, ventilation systems and confined areas. Isolate area until gas has dispersed.

Environmental Physical Data

BCF: no food chain concentration potential
BOD: none

Regulations

RCRA 40CFR: Not listed
CERCLA: 40CFR 302.4: Not listed
SARA 40CFR 372.65: Not listed
SARA EHS 40CFR 355: Not listed
TSCA: Listed

CYC8140 CAS #: 68-41-7

CYCLOSERINE

RTECS: NY2975000
EINECS Number: 200-688-4
Molecular Formula: $C_3H_6N_2O_2$
Formula Weight: 102.09

Chemical Structure

Synonyms: 106-7; D-4-AMINO-3-ISOSSAZOLIDONE; D-4-AMINO-3-ISOXAZOLIDINONE; D-4-AMINO-3-ISOXAZOLIDONE; CICLOSERINA; CLOSINA; CYCLORIN; CYCLOSERIN; (+)-CYCLOSERINE; ALPHA-CYCLOSERINE; CYCLO-D-SERINE; D-CYCLOSERINE; E-733-A; FARMISERINA; FARMISERINE; I-1431; 3-ISOXAZOLIDINONE,4-AMINO-,(+)-; 3-ISOXAZOLIDINONE,4-AMINO-,-; 3-ISOXAZOLIDINONE,4-AMINO-,D-; K-300; MICOSERINA; MIROSERINA; MIROSERYN; NJ-21; NOVOSERIN; ORIENTOMYCIN; D-OXAMICINA; OXAMICINA; D-OXAMYCIN; OXAMYCIN; PA-94; PA 94; RO-1-9213; SEROMYCIN; TISOMYCIN; WASSERINA

Description: white to pale yellow crystals or crystalline powder; odorless or has faint odor

Use: medication: antibacterial (tuberculostatic); agricultural fungicide

Physical Properties

Boiling Point: Decomposes at 155 °C (311 °F) to 156 °C (312.8 °F)
Freezing Point: Decomposes at 155 °C (311 °F) to 156 °C (312.8 °F)
Water Solubility: Soluble in Water
Other Solubilities: Slightly Soluble in Methanol, Propylene Glycol
pH: Aqueous solution is about 6
Refraction Index: Alpha 1.583

RTECS Toxicity Data

Acute Oral: Human TD_{Lo} Dose: 560 mg/kg/4W-I; Toxic Effects: Behavioral - Wakefulness; Behavioral - Tremor. Woman TD_{Lo} Dose: 60 mg/kg;#Toxic Effects: Behavioral - Coma. Rat LD_{50} Dose: >5 gm/kg; Toxic Effects: Behavioral - Somnolence (general depressed activity).

Acute Dermal: Monkey LD_{Lo} Route: Subcutaneous Dose: 4 gm/kg; Toxic Effects: Behavioral - Somnolence (general depressed activity); Behavioral - Ataxia; Gastrointestinal - Nausea or vomiting. Rat LD_{50} Route: Subcutaneous Dose: >3 gm/kg; Toxic Effects: Behavioral - Somnolence (general depressed activity).

Chronic (Multiple Dose) Oral: Monkey Dose: 45 gm/kg/90D-I; Toxic Effects: Gastrointestinal - Hypermotility, diarrhea; Blood - Changes in bone marrow not included above; DEATH.

Chronic (Multiple Dose) Dermal: Dog Route: Subcutaneous; Dose: 15 gm/kg/60D-I; Toxic Effects: Blood - Pigmented or nucleated red Blood cells; Blood - Changes in erythrocite (RBC) cell count; DEATH.

Hazard Overviews

Health: May cause irritation to eyes/skin. Other Acute Effects: may be harmful by inhalation, ingestion, or skin absorption; may cause nervous system disturbances; target organ: nerves.
Fire: Hazards: emits toxic fumes. Extinguishing agents: water spray; carbon dioxide, dry chemical powder or appropriate foam. Precautions: combustible liquid.
Reactivity: Incompatible with: strong oxidizing agents. Hazardous decomposition products: toxic fumes of: carbon monoxide, carbon dioxide, nitrogen oxides.
Carcinogenicity: IARC - Not listed; NIOSH - Not listed; NTP - Not listed; ACGIH - Not listed; OSHA - Not listed; EPA - Not listed; MAK - Not listed
Primary Target Organs:

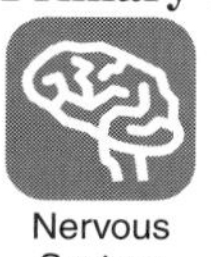

Nervous System

Environmental

Regulations

RCRA 40CFR: Not listed
CERCLA: 40CFR 302.4: Not listed
SARA 40CFR 372.65: Not listed
SARA EHS 40CFR 355: Not listed
TSCA: Not listed

CYC8310 CAS #: 2691-41-0

CYCLOTETRAMETHYLENETETRANITRAMINE

RTECS: XF7450000
DOT: UN0226
EINECS Number: 220-260-0
Molecular Formula: $C_4H_8N_8O_8$
Formula Weight: 296.20

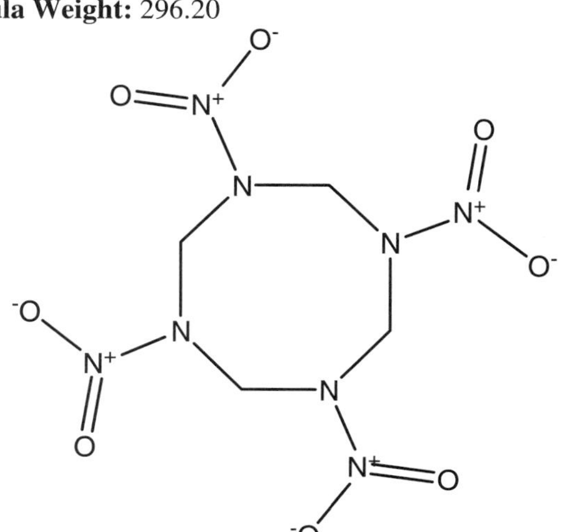

Chemical Structure

Synonyms: HMX; BETA-HMY; HW 4; LX 14-0; OCTOGEN; OKTOGEN; TETRAMETHYLENETETRANITRAMINE; 1,3,5,7-TETRANITRO-1,3,5,7-TETRAAZACYCLOOCTANE; 1,3,5,7-TETRAZOCINE,OCTAHYDRO-1,3,5,7-TETRANITRO-
Use: in the manufacture of explosives

Physical Properties

Freezing Point: 286 °C (546.8 °F)

RTECS Toxicity Data

Acute Oral: Rat LD_{50} Dose: 6490 mg/kg. Mouse LD_{50} Dose: 1500 mg/kg.
Acute Dermal: Rabbit LD_{50} Route: Skin; Dose: 630 mg/kg; Toxic Effects: Behavioral - Convulsions or effect on seizure threshold.
Irritation Skin: Rabbit Standard Draize Test Dose: 500 mg; Reaction: mild.

Hazard Overviews

Carcinogenicity: IARC - Not listed; NIOSH - Not listed; NTP - Not listed; ACGIH - Not listed; OSHA - Not listed; EPA - Not listed; MAK - Not listed

Environmental

Ecotoxicity: LC_{50} Fathead minnows, 7-day old 15 ng/l/96 hr

Regulations

RCRA 40CFR: Not listed
CERCLA: 40CFR 302.4: Not listed
SARA 40CFR 372.65: Not listed
SARA EHS 40CFR 355: Not listed
TSCA: Listed

Analytical Methods

Soil: SW846 8330
Water / Groundwater: AOAC 986.22

CYC8650 CAS #: 2259-96-3

CYCLOTHIAZIDE

RTECS: DK9610000
EINECS Number: 218-859-7
Molecular Formula: $C_{14}H_{16}ClN_3O_4S_2$
Formula Weight: 389.91

Chemical Structure

Synonyms: 35483; ANHYDRON; AQUIREL; 2H-1,2,4-BENZOTHIADIAZINE-7-SULFONAMIDE,3-BICYCLO(2.2.1)HEPT-5-EN-2-YL-6-CHLORO-3,4-DIHYDRO-,1,1-DIOXIDE; 2H-1,2,4-BENZOTHIADIAZINE-7-SULFONAMIDE,6-CHLORO-3,4-DIHYDRO-3-(5-NORBORNEN-2-YL)-,1,1-DIOXIDE; 3-BICYCLO[2.2.1]HEPT-5-EN-2-YL-6-CHLORO-3,4-DIHYDRO-2H-1,2,4-BENZOTHIADIAZINE-7-SULFONAMIDE 1,1-DIOXIDE; 6-CHLORO-3,4-DIHYDRO-3-(2-NORBORNEN-5-YL)-2H-1,2,4-BENZOTHIADIAZINE-7-SULFONAMIDE 1,1-DIOXIDE; 6-CHLORO-3,4-DIHYDRO-3-(2-NORBORNEN-5-YL)-7-SULFAMOYL-1,2,4-BENZOTHIADIAZINE 1,1-DIOXIDE; 6-CHLORO-3-(2-NORBORNEN-5-YL)-7-SULFAMYL-3,4-DIHYDRO-1,2,4-BENZOTHIADIAZINE 1,1-DIOXIDE; DOBURIL; LILLY 35,483; MDI 193; RENAZIDE; VALMIRAN

Description: white to nearly white powder or crystals; practically odorless

Use: medication: diuretic; antihypertensive agent; medication (vet): diuretic agent

Physical Properties

Freezing Point: 234 °C (453.2 °F)

Water Solubility: Practically Insoluble in Water

Other Solubilities: freely Soluble in Acetone, Ethyl Acetate, Methanol; Sparingly Soluble in Alcohol; Practically Insoluble in Chloroform, Ether

RTECS Toxicity Data

Acute Oral: Rat LD_{50} Dose: >5 gm/kg. Mouse LD_{50} Dose: >5 gm/kg.

Hazard Overviews

Health: Irritating to eyes/skin/respiratory tract. Other Acute Effects: may be harmful by inhalation, ingestion, or skin absorption; can cause CNS depression; nausea; headache; vomiting; narcotic effect; damage to the heart; acts as a diuretic; diarrhea; pancreatitis; dizziness; paresthesias. Chronic Effects: hyperglycemia; hyperuricemia; hypersensitivity; anorexia; xanthopsia; leukopenia; agranulocytosis; thrombocytopenia and aplastic anemia; target organs: kidneys, nerves, liver.

Fire: Hazards: emits toxic fumes. Extinguishing agents: water spray; carbon dioxide, dry chemical powder or appropriate foam. Precautions: combustible liquid.

Reactivity: Stable. Hazardous polymerization will not occur. Incompatible with: strong oxidizing agents. Hazardous decomposition products: toxic fumes of: carbon monoxide, carbon dioxide, nitrogen oxides, sulfur oxides, hydrogen chloride gas.

Carcinogenicity: IARC - Not listed; NIOSH - Not listed; NTP - Not listed; ACGIH - Not listed; OSHA - Not listed; EPA - Not listed; MAK - Not listed

Primary Target Organs:

Eyes

Skin

Respiratory System

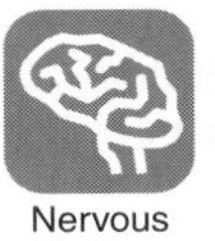
Nervous System

Liver

Kidneys

Environmental

Regulations

RCRA 40CFR: Not listed

CERCLA: 40CFR 302.4: Not listed

SARA 40CFR 372.65: Not listed

SARA EHS 40CFR 355: Not listed

TSCA: Not listed

CYC8820 CAS #: 2163-69-1

CYCLURON

RTECS: YS7875000

EINECS Number: 218-493-8

Molecular Formula: $C_{11}H_{22}N_2O$

Formula Weight: 198.35

Synonyms: 3-CYCLOOCTYL-1,1-DIMETHYLHARNSTOFF; 3-CYCLOOCTYL-1,1-DIMETHYLUREA; N'-CYCLOOCTYL-N,N-DIMETHYLUREA; N-CYCLOOCTYL-N',N'-DIMETHYLUREA; CYCLOURON; HS 61; OMU; UREA,3-CYCLOOCTYL-1,1-DIMETHYL-; UREA,N'-CYCLOOCTYL-N,N-DIMETHYL-

Description: colorless crystals; odorless

Use: pre-emergence herbicide; former use: in forestry, on seed beds before planting out and on beds for transplanting vegetables; former use: pre-emergence weed control in spinach, sugar beet and several vegetable crops

Physical Properties

Freezing Point: 138 °C (280.4 °F)

Saturated Vapor Density: 1.200000001 kg/m^3

Vapor Pressure: 1 x10^{-7} mm Hg at 20 °C

Water Solubility: 1.1 g/L at 20 °C

Other Solubilities: Solubility at 20 °C: 4.4 g/100 g, Ethyl Acetate; 1.6 g/100 g, Diethyl Ether.

RTECS Toxicity Data

Acute Oral: Rat LD_{50} Dose: 1500 mg/kg. Mammal LD_{50} Dose: 2600 mg/kg.

Hazard Overviews

Carcinogenicity: IARC - Not listed; NIOSH - Not listed; NTP - Not listed; ACGIH - Not listed; OSHA - Not listed; EPA - Not listed; MAK - Not listed

Environmental

Environmental Physical Data

Sorption Partition Coefficient: K_{oc} = estimated at 93
BCF: calculated at 12

Regulations

RCRA 40CFR: Not listed
CERCLA: 40CFR 302.4: Not listed
SARA 40CFR 372.65: Not listed
SARA EHS 40CFR 355: Not listed
TSCA: Not listed

CYF5000 CAS #: 68359-37-5

CYFLUTHRIN

RTECS: GZ1253000
EINECS Number: 269-855-7
Molecular Formula: $C_{22}H_{18}Cl_2FNO_3$
Formula Weight: 434.29
Synonyms: BAY FCR 1272; BAY-FCR 1272; BAYTHROID; BAYTHROID H; BAY-VI 1704; (RS)-ALPHA-CYANO-4-FLUORO-3-PHENOXYBENZYL (1RS,3RS: 1RS,3SR)-3-(2,2-DICHLOROVINYL)-2,2-DIMETHYLCYCLOPROPANECARBOXYLATE; (R,S)-ALPHA-CYANO-4-FLUORO-3-PHENOXYBENZYL-(1R,S)-CIS,TRANS-3-(2,2-DICHLOROVINYL)-2,2-DIMETHYLCYCLOPROPANECARBOXYLATE; CYANO(4-FLUORO-3-PHENOXYPHENYL)METHYL3-(2,2-DICHLOROETHENYL)-2,2-DIMETHYL-CYCLOPROPANECARBOXYLATE; CYFLUTHRINE; CYFOXYLATE; 3-(2,2-DICHLOROETHENYL)-2,2-DIETHYLCYCLOPROPANECARBOXYLICACID CYANO(4-FLUORO-3-PHENOXYPHENYL)METHYL ESTER; EULAN SP; FCR 1272; FCR 4545; RESPONSAR; SOLFAC; TEMPO; TEMPO 2
Description: yellowish-brown, partly crystalline oil; aromatic solvent odor
Use: agricultural insecticide; control of chewing and sucking insects on oilseed rape (cabbage stem flea beetleand rape winter stem weevil), cereals (caphids vectors of bydv), ornamentals, maize, cotton, groundnuts, potatoes, rice, lucerne, tobacco, sugar beet, deciduous fruit, and vegetables; control of insect pests, especially houseflies, mosquitos, and cockroaches

Physical Properties

Freezing Point: 60 °C (140 °F)
Vapor Pressure: < 1 mPa at 20 °C
Water Solubility: 2 ug/l at 20 °C
Other Solubilities: >1000 g/l dichloromethane
Refraction Index: 1.5511 at 23 °C/D
Flash Point: Burns with difficulty

RTECS Toxicity Data

Acute Oral: Rat LD_{50} Dose: 900 mg/kg. Mouse LD_{50} Dose: 300 mg/kg.
Acute Inhalation: Rat LC_{50} Dose: 469 gm/m^3/4hr.
Acute Dermal: Rat LD_{50} Route: Skin; Dose: >5 gm/kg.

Hazard Overviews

Fire: Will burn.
Carcinogenicity: IARC - Not listed; NIOSH - Not listed; NTP - Not listed; ACGIH - Not listed; OSHA - Not listed; EPA - Not listed; MAK - Not listed

Environmental

Ecotoxicity: LC_{50} Rainbow trout 0.0006 mg/l/96 hr /Conditions of bioassay not specified

Regulations

RCRA 40CFR: Not listed
CERCLA: 40CFR 302.4: Not listed
SARA 40CFR 372.65: Listed
SARA EHS 40CFR 355: Not listed
TSCA: Not listed

Analytical Methods

Water / Groundwater: EPA 1660
Food: FDA 212.1, 232.1, 232.4, 242.1
Plasma: FDA 252

CYH5000 CAS #: 68085-85-8

CYHALOTHRIN

RTECS: GZ1227770
EINECS Number: 268-450-2
Molecular Formula: $C_{23}H_{19}ClF_3NO_3$
Formula Weight: 449.86
Synonyms: 3-(2-CHLORO-3,3,3-TRIFLUORO-1-PROPENYL)-2,2-DIMETHYLCYCLOPROPANECARBOXYLIC ACIDCYAN(3-PHENOXYPHENYL)METHYL ESTER; (RS)-ALPHA-CYANO-3-PHENOXYBENZYL(Z)-(1RS)-CIS-3-(2-CHLORO-3,3,3-TRIFLUOROPROPENYL)-2,2-DIMETHYLCYCLOPROPANECARBOXYLATE; (RS)-ALPHA-CYANO-3-PHENOXYBENZYL(Z)-(1RS,3RS)-(2-CHLORO-3,3,3-TRIFLUOROPROPENYL)-2,2-DIMETHYLCYCLOPROPANECARBOXYLATE; [1ALPHA,3ALPHA(Z)]-(+-)-CYANO-(3-PHENOXYPHENYL)METHYL3-(2-CHLORO-3,3,3-TRIFLUORO-1-PROPENYL)-2,2-DIMETHYLCYCLOPROPANECARBOXYLATE; CYHALOTHRINE; GRENADE; ICI 146814; ICI 146 814; ICI-PP 563; MATADOR; PP 563; SABER
Description: yellow-brown, viscous liquid; mild odor
Use: insecticide; acaricide; for control of animal ectoparasites, especially boophilus microplus or haematobia irritans on cattle, & lice & ked on sheep; animal dip or spray around animal houses; against wide range of lepidoptera, hemiptera, diptera, & coleoptera species; in public & animal health applications against cockroaches, flies, mosquitos, & ticks, etc; has high activity residual spray on inert surfaces.

Physical Properties

Boiling Point: 187 °C (369 °F) to 190 °C (374 °F) at 0.2 mm Hg
Freezing Point: < 10 °C (50 °F)
Specific Gravity: 1.25 at 25 °C
Saturated Vapor Density: ~ 1.371914082 kg/m^3
Vapor Pressure: ~ 0.001 mPa at 20 °C

Water Solubility: 0.003 mg/L at 20 °C
Other Solubilities: In Acetone, dichloromethane, Methanol, Diethyl Ether, Ethyl Acetate, hexane, Toluene, all >500 g/l at 20 °C.
Refraction Index: 1.534 at 24 °C/D
Flash Point: Burns with difficulty

RTECS Toxicity Data

Acute Oral: Rat LD_{50} Dose: 144 mg/kg. Rabbit LD_{50} Dose: >1 gm/kg.
Acute Inhalation: Rat LC_{50} Dose: 83 mg/m^3/4hr.
Acute Dermal: Rabbit LD_{50} Route: Skin; Dose: >2500 mg/kg.

Hazard Overviews

Fire: Will burn.
Carcinogenicity: IARC - Not listed; NIOSH - Not listed; NTP - Not listed; ACGIH - Not listed; OSHA - Not listed; EPA - Not listed; MAK - Not listed

Environmental

Ecotoxicity: LD_{50} Mallard duck oral >5000 mg/kg LC_{50} Salmo gairdneri (Rainbow trout) (weight= 0.32.1.37 g) at 12 °C 0.54 ug/l/96 hr LC_{50} Daphnia Pulex (stage < 24 hr). Static system at 20 °C, 160 ng/l/ 48 hr
Environmental Fate: If released to soil, it is expected to be immobile and its half-life is 2-4 weeks. If released to water, it will volatilize slowly. It has base catalyzed hydrolysis half-lives of 3.6 years at a pH of 8 and 36 years at a pH of 7. Biodegradation in water may be important based on a half-life of approximately 2-4 weeks in soil. Bioconcentration is expected to be an important fate process along with adsorption to sediment based on high K_{oc} values. If released to the atmosphere, it will exist in the vapor and particulate phases. In the vapor phase, it will degrade in the atmosphere by reaction with photochemically produced hydroxyl radicals with an estimated half-life of 7.6 hours. It will also degrade with reaction with atmospheric ozone with an estimated half-life of 7 days. Physical removal from air can occur through wet and dry deposition.

Environmental Physical Data

Henry's Law Constant: estimated at 1.48 x10^{-6}
Octanol/Water Partition Coefficient: log K_{ow} = 6.9
Sorption Partition Coefficient: K_{oc} = estimated at 4.76 x10^5
BCF: estimated at 1.6364 x10^4

Regulations

RCRA 40CFR: Not listed
CERCLA: 40CFR 302.4: Not listed
SARA 40CFR 372.65: Listed
SARA EHS 40CFR 355: Not listed
TSCA: Not listed

CYM1000 CAS #: 535-77-3

M-CYMENE

RTECS: DA6127310
EINECS Number: 208-617-9
Molecular Formula: $C_{10}H_{14}$
Formula Weight: 134.22

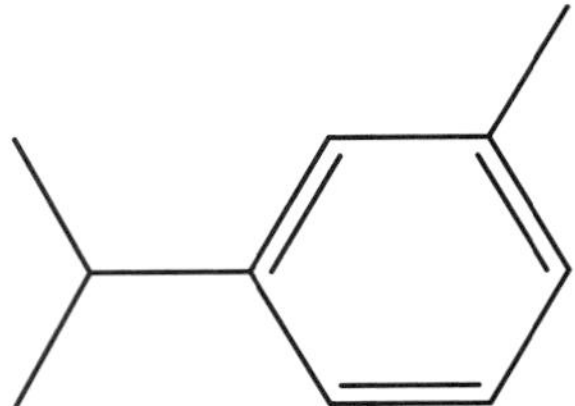

Chemical Structure

Synonyms: BENZENE,1-METHYL-3-(1-METHYLETHYL)-; BETA-CYMENE; M-CYMENE (8CI); M-CYMOL; 1-ISOPROPYL-3-METHYLBENZENE; 3-ISOPROPYLTOLUENE; M-ISOPROPYLTOLUENE; 1-METHYL-3-ISOPROPYLBENZENE; M-METHYLISOPROPYLBENZENE; 1-METHYL-3-(1-METHYLETHYL)BENZENE
Description: transparent liquid

Physical Properties

Boiling Point: 175.14 °C (347 °F)
Freezing Point: -63.75 °C (-82.75 °F)
Specific Gravity: 0.857 at 25 °C/4 °C
Vapor Pressure: 3.53 kPa at 75 °C
Water Solubility: Practically Insoluble in Water
Other Solubilities: Soluble in all proportions in Acetone, Benzene, Petroleum Ether, Carbon Tetrachloride; Soluble in Chloroform.
Refraction Index: 1.4930 at 20 °C/D

RTECS Toxicity Data

Acute Oral: Rat LD_{50} Dose: 2970 mg/kg; Toxic Effects: Behavioral - Convulsions or effect on seizure threshold; Behavioral - Muscle weakness; Nutritional and gross metabolic - Body temperature decrease. Mouse LD_{50} Dose: 3272 mg/kg; Toxic Effects: Behavioral - Convulsions or effect on seizure threshold; Behavioral - Muscle weakness; Nutritional and gross metabolic - Body temperature decrease.
Acute Inhalation: Mouse LC_{50} Dose: 12 gm/m^3; Toxic Effects: Behavioral - Convulsions or effect on seizure threshold; Behavioral - Muscle weakness; Nutritional and gross metabolic - Body temperature decrease.

Hazard Overviews

Health: Irritating to eyes/skin/respiratory tract. Other Acute Effects: may be harmful by inhalation, ingestion, or skin absorption.
Fire: Extinguishing agents: water spray; carbon dioxide, dry chemical powder or appropriate foam. Precautions: combustible liquid.

Reactivity: Incompatible with: strong oxidizing agents. Hazardous decomposition products: toxic fumes of: carbon monoxide, carbon dioxide.
Carcinogenicity: IARC - Not listed; NIOSH - Not listed; NTP - Not listed; ACGIH - Not listed; OSHA - Not listed; EPA - Not listed; MAK - Not listed
Primary Target Organs:

Eyes Skin Respiratory System

Environmental

Regulations

RCRA 40CFR: Not listed
CERCLA: 40CFR 302.4: Not listed
SARA 40CFR 372.65: Not listed
SARA EHS 40CFR 355: Not listed
TSCA: Listed

CYM5000 CAS #: 527-84-4

O-CYMENE

RTECS: DA6127300
EINECS Number: 208-426-0
Molecular Formula: $C_{10}H_{14}$
Formula Weight: 134.22

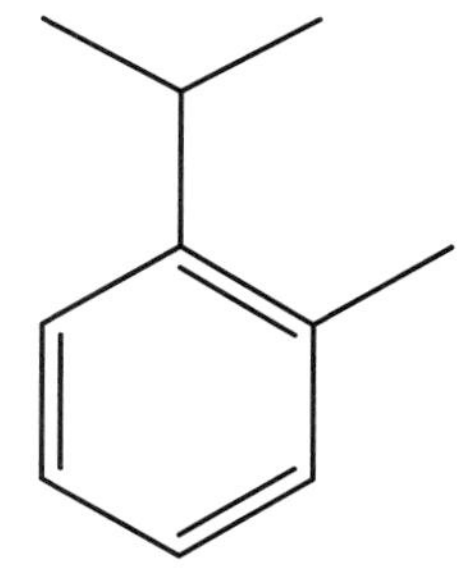

Chemical Structure

Synonyms: BENZENE,1-METHYL-2-(1-METHYLETHYL)-; O-CYMENE (8CI); O-CYMOL; 1-ISOPROPYL-2-METHYLBENZENE; 2-ISOPROPYLTOLUENE; O-ISOPROPYLTOLUENE; 1-METHYL-2-ISOPROPYLBENZENE; 1-METHYL-2-ISOPROPYLBENZOL; 1-METHYL-2-(1-METHYLETHYL)BENZENE
Description: transparent liquid

Physical Properties

Boiling Point: 178.15 °C (353 °F)
Freezing Point: -71.54 °C (-96.772 °F)
Specific Gravity: 0.8726 at 25 °C/4 °C
Vapor Pressure: 3.17 kPa at 75 °C
Water Solubility: Practically Insoluble in Water
Other Solubilities: miscible with organic solvents
Refraction Index: 1.5006 at 20 °C/D

RTECS Toxicity Data

Acute Oral: Rat LD_{50} Dose: 2130 mg/kg; Toxic Effects: Cardiac - Other changes; Lungs, Thorax, or Respiration - Other changes. Mouse LD_{50} Dose: 2024 mg/kg; Toxic Effects: Cardiac - Other changes; Lungs, Thorax, or Respiration - Other changes.
Acute Inhalation: Mouse LC_{50} Dose: 10300 mg/m^3; Toxic Effects: Cardiac - Other changes; Lungs, Thorax, or Respiration - Other changes.

Hazard Overviews

Health: Irritating to eyes/skin/respiratory tract. Other Acute Effects: may be harmful by inhalation, ingestion, or skin absorption.
Fire: Extinguishing agents: water spray; carbon dioxide, dry chemical powder or appropriate foam. Precautions: combustible liquid.
Reactivity: Incompatible with: strong oxidizing agents. Hazardous decomposition products: toxic fumes of: carbon monoxide, carbon dioxide.
Carcinogenicity: IARC - Not listed; NIOSH - Not listed; NTP - Not listed; ACGIH - Not listed; OSHA - Not listed; EPA - Not listed; MAK - Not listed
Primary Target Organs:

Eyes Skin Respiratory System

Environmental

Regulations

RCRA 40CFR: Not listed
CERCLA: 40CFR 302.4: Not listed
SARA 40CFR 372.65: Not listed
SARA EHS 40CFR 355: Not listed
TSCA: Listed

CYM9000 CAS #: 99-87-6

P-CYMENE

RTECS: GZ5950000
DOT: UN2046; IMO3.3
EINECS Number: 202-796-7
Molecular Formula: $C_{10}H_{14}$
Structured MF: p-$CH_3C_6H_4CH(CH_3)_2$
Formula Weight: 134.22

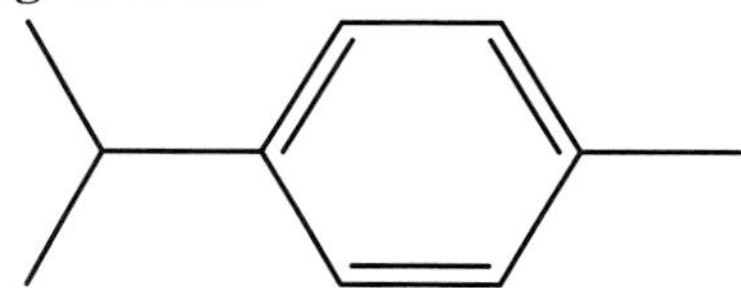

Chemical Structure

Synonyms: BENZENE,1-ISOPROPYL-4-METHYL-; BENZENE,1-METHYL-4-(1-METHYLETHYL)-; CAMPHOGEN; CUMENE,P-METHYL-; CYMENE; CYMOL; P-CYMOL; DOLCYMENE; 1-ISOPROPYL-4-METHYLBENZENE; 4-ISOPROPYL-1-METHYLBENZENE; P-ISOPROPYLMETHYLBENZENE; 4-ISOPROPYLTOLUENE; P-ISOPROPYLTOLUENE; ISOPROPYLTOLUOL; P-METHYLCUMENE; P-METHYLISOPROPYL BENZENE; 1-METHYL-4-ISOPROPYLBENZENE; 4-METHYLISOPROPYLBENZENE; P-METHYLISOPROPYLBENZENE; 1-METHYL-4-(1-METHYLETHYL)BENZENE; METHYLPROPYLBENZENE; PARACYMENE; PARACYMOL; 2-P-TOLYLPROPANE

Description: clear liquid; sweet, aromatic odor

Use: thinner for lacquers & varnishes; mfr para-cresol & carvacrol; chem int for p-cresol & thymol; component of commercial terpene solvent mixtures; heat-exchange medium; solvent; synthetic resin manufacture; metal polishes

Physical Properties

Boiling Point: 177.1 °C (351 °F)
Freezing Point: -67.94 °C (-90.292 °F)
Specific Gravity: 0.8533 at 25 °C/4 °C
Vapor Density: 4.62 Air=1
Vapor Pressure: 1 mm Hg at 17.3 °C
Water Solubility: Practically Insoluble in Water
Other Solubilities: Soluble in all proportions in Acetone, Benzene, Carbon Tetrachloride, Petroleum Ether; Soluble in Chloroform.
Surface Tension: 36.4 dynes/cm
Refraction Index: 1.4885 at 25 °C/D
Ionization Potential (eV): 8.29
Flash Point: 47.222 °C Open Cup
Autoignition Temperature: 436 °C
LEL: 0.7% v/v
UEL: 5.6% v/v

RTECS Toxicity Data

Acute Oral: Rat LD_{50} Dose: 4750 mg/kg; Toxic Effects: Sense organs and special senses - Chromidracryorrhea; Behavioral - Somnolence (general depressed activity); Gastrointestinal - Hypermotility, diarrhea.

Irritation Skin: Rabbit Standard Draize Test Dose: 500 mg/24H; Reaction: moderate.

Hazard Overviews

Flammable

Fire Diamond

Health: Irritating to eyes/skin. Also Causes: headache, lack of coordination, extensive chemical pneumonitis, erythema, dryness, defatting, reddened and smarting skin, nausea, and vomiting.

Fire: Flammable. Can form explosive mixtures in air. For small fires use dry chemical, carbon dioxide, or regular foam. For large fires use fog or regular foam. Floats on water which may make it difficult to extinguish fire. Remove container from fire.

Reactivity: Stable. Hazardous polymerization cannot occur. Avoid: heat; ignition sources. Incompatible with: oxidizers. Hazardous decomposition products: carbon dioxide; acrid smoke; fumes.

Carcinogenicity: IARC - Not listed; NIOSH - Not listed; NTP - Not listed; ACGIH - Not listed; OSHA - Not listed; EPA - Not listed; MAK - Not listed

Primary Target Organs:

Eyes

Skin

Respiratory System

Nervous System

Environmental

Environmental Fate: If released to the atmosphere, it will degrade in the vapor-phase by reaction with photochemically produced hydroxyl radicals (estimated half-life of 25.5 hr). If released to soil or water, it will probably biodegrade. The results of limited biodegradation screening studies suggest that it can biodegrade aerobically in the environment. Biodegradation is the only identifiable degradation process in soil or water. Limited leaching in soil is predicted. Volatilization will be a major removal mechanism from water in the absence of strong adsorption to sediment; however, strong adsorption will greatly reduce the volatilization rate.

Cleanup/Disposal: Guide No. 130: Eliminate all ignition sources (no smoking, flares, sparks or flames in immediate area). All equipment used when handling the product must be grounded. Do not touch or walk through spilled material. Stop leak if you can do it without risk. Prevent entry into waterways, sewers, basements or confined areas. A vapor suppressing foam may be used to reduce vapors. Absorb or cover with dry earth, sand or other non-combustible material and transfer to containers. Use clean non-sparking tools to collect absorbed material. Large Spills: Dike far ahead of liquid spill for later disposal. Water spray may reduce vapor; but may not prevent ignition in closed spaces.

Environmental Physical Data

Henry's Law Constant: estimated at 0.011
Octanol/Water Partition Coefficient: log K_{ow} = 4.10
Sorption Partition Coefficient: K_{oc} = estimated at 770
BCF: estimated at 104

Regulations

RCRA 40CFR: Not listed
CERCLA: 40CFR 302.4: Not listed
SARA 40CFR 372.65: Not listed
SARA EHS 40CFR 355: Not listed
TSCA: Listed

Analytical Methods

Air: EPA VA-006-1, VA-008-1, VG-011-1
Soil: SW846 5021, 8021A, 8260A
Water / Groundwater: EPA 1625; APHA 6210-D, 6230-D; ASTM D4763
Drinking Water: EPA 502.2, 503.1, 524.2; APHA 6220-C
Indoor / Expired Air: EPA IP-1B
Plasma: EPA 29

CYP1000 CAS #: 52315-07-8

CYPERMETHYIN

RTECS: GZ1250000
EINECS Number: 257-842-9
Molecular Formula: $C_{22}H_{19}C_{12}NO_3$
Formula Weight: 416.30
Synonyms: AGROTHRIN; AMBUSH C; AMMO; AMMO PESTICIDE; ANTIBORER 3767; ARDAP; ARRIVO; AVICADE; BARRICADE; CCN 52; CNN52; (+ -)-ALPHA-CYANO-3-PHENOXYBENZYL-(+ -)-CIS,TRANS-3-(2,2-DICHLOROVINYL)-2,2-DIMETHYLCYCLOPROPANECARBOXYLATE; (+-)-ALPHA-CYANO-3-PHENOXYBENZYL 2,2-DIMETHYL-3-(2,2-DICHLOROVINYL)CYCLOPROPANE CARBOXYLATE; (+)ALPHA-CYANO-3-PHENOXYBENZYL-(+)CIS,TRANS-2,2-DICHLOROVINYL-2,2-DIMETHYLCYCLOPROPANECARBOXYLATE; (RS)-ALPHA-CYANO-3-PHENOXYBENZYL(1RS)-CIS,TRANS-3-(2,2-DICHLOROVINYL)-2,2-DIMETHYLCYCLOPROPANECARBOXYLATE; CYANO(3-PHENOXYPHENYL)METHYL3-(2,2-DICHLOROETHENYL)-2,2-DIMETHYLCYCLOPTOPANE-CARBOXYLATE; (CYANO(3-PHENOXYPHENYL)METHYL3-(2,2-DICHLOROVINYL-2,2-DIMETHYLCYCLOPROPANE CARBOXYLATE; CYMBUSH; CYPERCARE; CYPERCOPAL; CYPERKILL; CYPERMETHRIN; CYPERMETHRINE; CYPERMETHRIN-25EC; CYPERMETRYNA; CYPERSECT; CYPOR; CYRUX; DEMON; 3-(2,2-DICHLOROETHENYL)-2,2-DIMETHYLCYCLOPROPANECARBOXYLICACID CYANO(3-PHENOXYPHENYL)-METHYL ESTER; DYSECT; EXP 5598; FASTAC; FENDONA; FENOM; FLECTRON; FMC 30980; FMC 45497; FMC 45806; FOLCORD; HILCYPERIN; IMPERATOR; JF 5705F; KAFIL SUPER; KALIF SUPER; NRDC 149; NRDC 160; NRDC 166; NURELLE; POLYTRIN; PP 383; PP383; RIPCORD; RU 27998; RYCOPEL; SF 06646; SHERPA; SIPERIN; SUPERCYPERMETHRIN; SUPERCYPERMETHRIN FORTE; TOPCLIP PARASOL; TOPPEL; USTAAD; VUCHT 424; WL 43467; WL 8517; WRDC149; YT 305
Description: viscous semi-solid
Use: experimental photostable pyrethroid, insecticide to be used against a wide range of insect pests; insecticide; medication (vet) as an ectoparasiticide

Physical Properties

Freezing Point: 60 °C (140 °F) to 80 °C (176 °F)
Specific Gravity: 1.25000
Vapor Pressure: 0.51 nPa at 70 °C
Water Solubility: 0.01 mg/L at 20 °C
Other Solubilities: Soluble in Methanol and Methylene Dichloride.
Flash Point: Burns with difficulty

RTECS Toxicity Data

Acute Oral: Rat LD_{50} Dose: 57500 ug/kg; Toxic Effects: Behavioral - Somnolence (general depressed activity); Behavioral - Convulsions or effect on seizure threshold; Gastrointestinal - Changes in structure or function of salivary glands. Mouse LD_{50} Dose: 24570 ug/kg; Toxic Effects: Behavioral - Somnolence (general depressed activity); Behavioral - Convulsions or effect on seizure threshold; Gastrointestinal - Changes in structure.
Acute Inhalation: Rat LC_{50} Dose: 7889 mg/m^3/4hr.
Acute Dermal: Rat LD_{50} Route: Skin; Dose: >1600 mg/kg.
Mutagenic: Human DNA Inhibition; Cell Type: lymphocyte; Dose: 67 umol/L. Human Micronucleus Test; Cell Type: lymphocyte; Dose: 200 mg/L.

Hazard Overviews

Fire: Will burn.
Reactivity: Stable in a neutral and weakly acidic media. Hazardous polymerization cannot occur. Avoid: heat and ignition sources. Incompatible with: strong oxidizers; alkaline materials (e.g.., lime and ordinary soaps). Hazardous decomposition products: toxic fumes of cyanide; nitrogen oxides; chloride.
Carcinogenicity: IARC - Not listed; NIOSH - Not listed; NTP - Not listed; ACGIH - Not listed; OSHA - Not listed; EPA - Not listed; MAK - Not listed

Environmental

Ecotoxicity: LC_{50} Rainbow trout formulated product 11 ppb active ingredient/24 hr (static test) LC_{50} Salmo salar (Atlantic salmon) 1.4-12 ug/l/96 hr, juvenile /Conditions of bioassay not specified
Environmental Fate: If released to the atmosphere, it will degrade readily in the vapor phase by reaction with photochemically produced hydroxyl radicals (estimated half-life of 10 hr). Particulate phase will be removed physically from air by wet and dry deposition. If released to soil or water, it can degrade readily through biodegradation, hydrolysis and photodegradation. The photodegradation half-life in sunlight (at the water's surface or as a thin-film on dry surfaces) can be very short (0.5-7 days). Aqueous hydrolysis is expected to become important only in alkaline media (pH > 8). It does not leach in soil. It will partition to sediment from water in aquatic ecosystems; however, field tests have shown that surface applications disperse slowly to subsurface water and sediment. The half-life within soil has been shown to range from several days to over a 100 days (avg. of roughly 30 days).

Environmental Physical Data

Henry's Law Constant: estimated at 1.92 x10^{-7}
Octanol/Water Partition Coefficient: log K_{ow} = 4.47
BCF: golden ide fish 420

Regulations

RCRA 40CFR: Not listed
CERCLA: 40CFR 302.4: Not listed
SARA 40CFR 372.65: Not listed
SARA EHS 40CFR 355: Not listed
TSCA: Not listed

Analytical Methods

Food: FDA 212.1, 212.2, 232.1, 232.4, 242.1; AOAC 985.03, 986.02
Plasma: FDA 211.1, 231.1, 252

CYP3000 **CAS #: 22936-86-3**

CYPRAZINE

RTECS: XY5380000
DOT: UN2763
EINECS Number: 245-338-1
Molecular Formula: $C_9H_{14}ClN_5$
Formula Weight: 227.7
Synonyms: 2-CHLORO-4-CYCLOPROPYLAMINO-6-ISOPROPYLAMINO-1,3,5-TRIAZINE; 2-CHLORO-4-CYCLOPROPYLAMINO-6-ISOPROPYLAMINO-S-TRIAZINE; 6-CHLORO-N-CYCLOPROPYL-N-ISOPROPYL-1,3,5-TRIAZINE-2,4-DIAMINE; CYPROZINE; EPA PESTICIDE CHEMICAL CODE NUMBER 100401; K 6295; OUTFOX; S 6115; S-6115; S-9115; S-TRIAZINE,2-CHLORO-4-(CYCLOPROPYLAMINO)-6-(ISOPROPYLAMINO)-; 1,3,5-TRIAZINE-2,4-DIAMINE,6-CHLORO-N-CYCLOPROPYL-N'-(1-METHYLETHYL)-
Description: white crystals; odorless

Physical Properties

Freezing Point: 167 °C (332.6 °F) to 168 °C (334.4 °F)
Density: 1.24 g/cu cm
Vapor Pressure: 4 $x10^{-5}$ Pa at 20 °C
Water Solubility: 7 ppm at 25 °C
Other Solubilities: 15470.0 ppm in Toluene at 25 °C; 7280.0 ppm in Benzene at 25 °C; 2230.0 ppm in Carbon Tetrachloride at 25 °C; 190710.0 ppm in Acetic Acid at 25 °C; 142430.0 ppm in Acetone at 25 °C; 94290.0 ppm in Ethyl Acetate at 25 °C
Flash Point: 162.778 °C Open Cup

RTECS Toxicity Data

Acute Oral: Rat LD_{50} Dose: 1200 mg/kg.
Acute Dermal: Rabbit LD_{50} Route: Skin; Dose: 7500 mg/kg.

Hazard Overviews

Fire: Will burn.
Carcinogenicity: IARC - Not listed; NIOSH - Not listed; NTP - Not listed; ACGIH - Not listed; OSHA - Not listed; EPA - Not listed; MAK - Not listed

Environmental

Ecotoxicity: TLm Carp 10-20 ppm/48 hr /Conditions of bioassay not specified
Environmental Fate: If released to soil, it will be expected to exhibit medium mobility in soil based upon an estimated K_{oc}, and therefore may leach to groundwater. It may be subject to slow degradation in soil which may be due to both slow chemical hydrolysis and slow biodegradation based upon limited data for it and other structurally related s-triazines. It will not be expected to volatilize from near surface soils or surfaces. If released to water, it will not be expected to bioconcentrate in aquatic organisms, adsorb to sediment and suspended particulate matter, or to volatilize. The s-triazine ring is fairly resistant to microbial attack. It is not known whether biodegradation occurs in natural waters. Chemical hydrolysis may be more important environmentally than biodegradation at low pH or when various catalysts are present. It may be susceptible to direct photolysis and hydrolysis based upon the behavior of other s-triazine herbicides with similar structures. If released to the atmosphere, it may be subject to gas-phase reaction with hydroxyl radicals. The half-life for the vapor-phase reaction with photochemically produced hydroxyl radicals has been estimated to be 2.8 hr.
Cleanup/Disposal: Guide No. 151: Do not touch damaged containers or spilled material unless wearing appropriate protective clothing. Stop leak if you can do it without risk. Prevent entry into waterways, sewers, basements or confined areas. Cover with plastic sheet to prevent spreading. Absorb or cover with dry earth, sand or other non-combustible material and transfer to containers. Do not get water inside containers.

Environmental Physical Data

Henry's Law Constant: calculated at 4.63 $x10^{-10}$
Octanol/Water Partition Coefficient: log K_{ow} = 2.6
Sorption Partition Coefficient: K_{oc} = estimated at 297
BCF: estimated at 60

Regulations

RCRA 40CFR: Not listed
CERCLA: 40CFR 302.4: Not listed
SARA 40CFR 372.65: Not listed
SARA EHS 40CFR 355: Not listed
TSCA: Not listed

Analytical Methods

Food: FDA 232.4, 242.1

CYP5000 **CAS #: 2439-10-3**

CYPREX

RTECS: MF1750000
EINECS Number: 219-459-5
Molecular Formula: $C_{15}H_{33}N_3O_2$
Formula Weight: 287.44
Synonyms: AADODIN; AC 5223; AMERICAN CYANAMID 5,223; AMERICAN CYANAMID 5223; CARPEN; CARPENE; CURITAN; CYPREX 65W; N-DODECYLGUANIDINACETAT; DODECYLGUANIDINE ACETATE; N-DODECYLGUANIDINE ACETATE; DODECYLGUANIDINE MONOACETATE; DODIN; DODINE; DODINE ACETATE; DODINE,MIXTURE WITH GLYODIN; DOGUADINE; DOQUADINE; ENT 16,436; EXPERIMENTAL FUNGICIDE 5223; GUANIDINE,DODECYL-,MONOACETATE; KARPEN; KYSELINA 3-DODECYLGUANIDINOOCTOVA; LAURYLGUANIDINE ACETATE; MELPREX; MELPREX 65W; MELPREX 65; MELPREX LIQUID DODINE; MILPREX; QUESTURAN; RADSPOR; SYLLIT; SYLLIT 65; TSITREX; VENTUROL; VONDODINE
Description: slightly waxy solid or colorless crystals
Use: control of apple scab and cherry leaf spot; algal growth inhibitors, useful in the manufacture of microorganisms-stable oil and water emulsions and in treatment of saline process water; for scab on apples, pears, pecans; leafspot on cherries; foliar diseases of strawberries; bacterial leafspot on peaches; leaf blight of sycamores, black walnuts

Physical Properties

Boiling Point: Dry 210 °C (410 °F)
Freezing Point: 136 °C (276.8 °F)
Saturated Vapor Density: 1.200000002 kg/m^3
Vapor Pressure: 1.5 x10^{-7} mm Hg at 25 °C
Water Solubility: Soluble in Hot Water
Other Solubilities: Insoluble in most organic solvents; in low molecular weight Alcohols ranges from about 7% to 23% at room temperature; Soluble in acids; Readily Soluble in mineral acids; Soluble in Methanol, Ethanol. Practically Insoluble in most organic solvents.

RTECS Toxicity Data

Acute Oral: Rat LD_{50} Dose: 660 mg/kg; Toxic Effects: Behavioral - Somnolence (general depressed activity); Gastrointestinal - Hypermotility, diarrhea; Gastrointestinal - Other changes. Mouse LD_{50} Dose: 266 mg/kg.
Acute Inhalation: Mouse LC_{50} Dose: 129 mg/m^3/2hr.
Acute Dermal: Rabbit LD_{50} Route: Skin; Dose: 2100 mg/kg; Toxic Effects: Gastrointestinal - Ulceration or bleeding from stomach; Blood - Hemorrhage; Skin and appendages - Dermatitis, other. Rat LD_{50} Route: Skin; Dose: >6 gm/kg.
Irritation Eye: Rabbit Standard Draize Test Dose: 100 mg/24H; Reaction: moderate.
Tumorigenic: Mouse Route: Subcutaneous; Dose: 1000 mg/kg; Toxic Effects: Tumorigenic - Equivocal tumorigenic agent by RTECS criteria; Lungs, Thorax, or Respiration - Tumors; Blood - Tumors.

Hazard Overviews

Carcinogenicity: IARC - Not listed; NIOSH - Not listed; NTP - Not listed; ACGIH - Not listed; OSHA - Not listed; EPA - Not listed; MAK - Not listed

Environmental

Ecotoxicity: LC_{50} Gammarus fasciatus 1.1 mg/l/96 hr. Static bioassay without aeration, pH 7.2-7.5, water hardness 40-50 mg/l as calcium carbonate and alkalinity of 30-35 mg/l LD_{50} Mallard duck oral 1142 mg/kg
Environmental Fate: If released to the atmosphere, it will degrade rapidly in the vapor phase by reaction with photochemically produced hydroxyl radicals (half-life of about 1-2 hours). Particulate-phase and aerosols released to air during spray applications will be removed from air physically by dry and wet deposition. It is stable in neutral, acidic, and slightly basic solutions. It is an organic cation and is thus strongly absorbed to soil; an experimental K_{oc} value of 1 x10^5 was measured. A Henry's Law constant of 9.01 x10^{-11} atm-cu m/mole indicates that this compound is essentially non-volatile. Biodegradation may be possible; species of Achromobacter and Flavobacterium, both soil bacteria, have been shown to utilize it for growth when it is available as the sole carbon source. 5% degradation in river mud occurred within 68 days. It was unknown whether this was due to biotic or abiotic conditions. An estimated BCF value of 16 indicates that bioconcentration is not a major fate process.

Environmental Physical Data

Henry's Law Constant: calculated at 9.0 x10^{-11}
Sorption Partition Coefficient: K_{oc} = 1 x10^5
BCF: estimated at 16

Regulations

RCRA 40CFR: Not listed
CERCLA: 40CFR 302.4: Not listed
SARA 40CFR 372.65: Listed
SARA EHS 40CFR 355: Not listed
TSCA: Listed

Analytical Methods

Food: AOAC 964.19, 970.07
Plasma: FDA 221.1

CYP7000 CAS #: 129-03-3

CYPROHEPTADINE

RTECS: TM7000000
EINECS Number: 204-928-9
Molecular Formula: $C_{21}H_{21}N$
Formula Weight: 287.39
Synonyms: 4-(5H-DIBENZO[A,D]CYCLOHEPTEN-5-YLIDENE)-1-METHYLPIPERIDENE; 4-(5H-DIBENZO(A,D)CYCLOHEPTEN-5-YLIDENE)-1-METHYLPIPERIDINE; 4-(5-DIBENZO(A,D)CYCLOHEPTEN-5-YLIDINE)-1-METHYLPIPERIDINE; DRONACTIN; EIPROHEPTADINE; 1-METHYL-4-(5-DIBENZO[A,E]CYCLOHEPTATRIENYLIDENE)PIPERIDINE; 1-METHYL-4-(5H-DIBENZO[A,D]CYCLOHEPTENYLIDENE)PIPERIDINE; 5-(1-METHYLPIPERIDYLIDENE-4)-5H-DIBENZO[A,D]CYCLOPHEPTENE; MK 141; PERIACTIN; PERIACTINE; PERIACTINOL
Description: crystals; odorless or practically ordorless
Use: antihistaminic, antipruritic, appetite stimulant; in various allergic diseases; treatment of pruriticdermatoses; treatment of cold allergy

Physical Properties

Freezing Point: 112.3 °C (234.14 °F) to 113.3 °C (235.94 °F)
Water Solubility: 1 g/275 mL
Other Solubilities: 1 g Soluble in 1.5 ml Methanol, 16 ml Chloroform, 35 ml Alcohol; Practically Insoluble in Ether.

RTECS Toxicity Data

Acute Oral: Rat LD_{50} Dose: 295 mg/kg. Mouse LD_{50} Dose: 106 mg/kg; Toxic Effects: Behavioral - Convulsions or effect on seizure threshold.
Acute Dermal: Mouse LD_{50} Route: Subcutaneous Dose: 107 mg/kg.
Reproductive/Teratogenic: Rat Route: Intraperitoneal; Dose: 10 mg/kg; Duration: female 10D of pregnancy; Effects on Embryo or Fetus - Fetal death; Specific Developmental Abnormalities - Musculoskeletal system; Effects on Newborn - Weaning or lactation index. Rat Route: Intraperitoneal; Dose: 20 mg/kg; Duration: female 13D of pregnancy; Effects on Fertility - Post-implantation mortality. Rat Route:

Intraperitoneal; Dose: 20 mg/kg; Duration: female 10D of pregnancy; Effects on Newborn - Viability index.

Hazard Overviews

Carcinogenicity: IARC - Not listed; NIOSH - Not listed; NTP - Not listed; ACGIH - Not listed; OSHA - Not listed; EPA - Not listed; MAK - Not listed

Environmental

Regulations

RCRA 40CFR: Not listed
CERCLA: 40CFR 302.4: Not listed
SARA 40CFR 372.65: Not listed
SARA EHS 40CFR 355: Not listed
TSCA: Not listed

CYP9000 CAS #: 2759-71-9

CYPROMID

RTECS: GZ1050000
Molecular Formula: $C_{10}H_9Cl_2NO$
Formula Weight: 230.10
Synonyms: CIPROMID; CLOBBER; CYCLOPROPANECARBOXAMIDE,N-(3,4-DICHLOROPHENYL)-; CYCLOPROPANECARBOXANILIDE,3',4'-DICHLORO-; CYPROMIDE; 3,4-DICHLORANILID KYSELINY CYKLOPROPANKARBOXYLOVE; 3',4'-DICHLOROCYCLOPROPANECARBOXANILIDE; N-(3,4-DICHLOROPHENYL)CYCLOPROPANECARBOXAMIDE; N-(3,4-DICHLOROPHENYL)CYCLOPROPANECARBOXYAMIDE; S-6000
Use: herbicide

RTECS Toxicity Data

Acute Oral: Rat LD_{50} Dose: 215 mg/kg. Rabbit LD_{50} Dose: 3028 mg/kg.
Acute Dermal: Rabbit LD_{50} Route: Skin; Dose: 3038 mg/kg.

Hazard Overviews

Carcinogenicity: IARC - Not listed; NIOSH - Not listed; NTP - Not listed; ACGIH - Not listed; OSHA - Not listed; EPA - Not listed; MAK - Not listed

Environmental

Ecotoxicity: TLm: Carp 3.5 ppm/48 hr

Regulations

RCRA 40CFR: Not listed
CERCLA: 40CFR 302.4: Not listed
SARA 40CFR 372.65: Not listed
SARA EHS 40CFR 355: Not listed
TSCA: Not listed

Analytical Methods

Food: FDA 212.1, 232.1
Plasma: EPA 001; FDA 211.1, 231.1, 252

CYR5000 CAS #: 66215-27-8

CYROMAZINE

RTECS: XZ1056500
EINECS Number: 266-257-8
Molecular Formula: $C_6H_{10}N_6$
Formula Weight: 166.18
Synonyms: ARMOR; CGA 72 662; CGA 72662; 2-CYCLOPROPYLAMINO-4,6-DIAMINO-S-TRIAZINE; CYCLOPROPYLMELAMINE; N-CYCLOPROPYL-1,3,5-TRIAZINE-2,4,6-TRIAMINE; CYPROMAZINE; LARVADEX; NEOPREX; NEPOREX; OMS-2014; TRIGARD; VETRAZIN; VETRAZIN (PESTICIDE); VETRAZINE
Description: colorless to white crystalline solid or crystals
Use: insect growth regulator; insecticide; med: (vet) ectoparasiticide; specific activity against dipterous larvae; spray or dip to control lucilia sericata on sheep; addn to chicken diet controls fly larvae in manure; 'vetrazine' water soluble powder for control of blowfly larvae on sheep; trigard wettable powder foliar spray controls against dipterous leafminers in vegetable crops, ornamentals; larvadex, feed premix.

Physical Properties

Freezing Point: 219 °C (426.2 °F) to 222 °C (431.6 °F)
Density: 1.35 g/cu cm at 20 °C
Vapor Pressure: < 0.13 mPa at 20 °C
Water Solubility: 1.1% at 20 °C
Other Solubilities: Methanol 1.7%

RTECS Toxicity Data

Acute Oral: Rat LD_{50} Dose: 3387 mg/kg. Duck LD_{50} Dose: >6 gm/kg.
Acute Inhalation: Rat LC_{50} Dose: >2720 mg/m^3/4hr.
Acute Dermal: Rat LD_{50} Route: Skin; Dose: >3100 mg/kg.

Hazard Overviews

Carcinogenicity: IARC - Not listed; NIOSH - Not listed; NTP - Not listed; ACGIH - Not listed; OSHA - Not listed; EPA - Not listed; MAK - Not listed

Environmental

Ecotoxicity: LC_{50} Bluegill sunfish >90 mg/l/96 hr /Conditions of bioassay not specified LC_{50} Rainbow trout >100 mg/l/96 hr /Conditions of bioassay not specified

Regulations

RCRA 40CFR: Not listed
CERCLA: 40CFR 302.4: Not listed
SARA 40CFR 372.65: Not listed
SARA EHS 40CFR 355: Not listed
TSCA: Not listed

Analytical Methods

Soil: EPA PMD-CYZ

CYS5000 CAS #: 52-90-4

CYSTEINE

RTECS: HA1600000
EINECS Number: 200-158-2
Molecular Formula: $C_3H_7NO_2S$
Formula Weight: 121.16

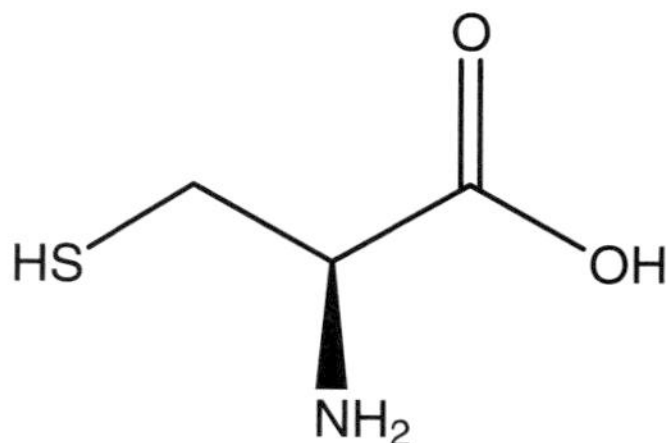

Chemical Structure

Synonyms: L-ALANINE,3-MERCAPTO-; 2-AMINO-3-MERCAPTOPROPIONIC ACID; ALPHA-AMINO-BETA-MERCAPTOPROPIONIC ACID; ALPHA-AMINO-BETA-THIOLPROPIONIC ACID; CYSTEIN; L-(+)-CYSTEINE; L-CYSTEINE; L-CYSTEINE (9CI); HALF CYSTINE; HALF-CYSTEINE; HALF-CYSTINE; BETA-MERCAPTOALANINE; NSC-8746; PROPANOIC ACID,2-AMINO-3-MERCAPTO-,-; THIOSERINE
Description: colorless crystals
Use: vet: detoxicant, radioprotective, antioxidant

Physical Properties

Boiling Point: Decomposes ~ 240 °C (464 °F)
Freezing Point: Decomposes at 240 °C (464 °F)
Water Solubility: Freely Soluble in Water
Other Solubilities: in neutral or slightly alkaline aqueous solution it is oxidized to Cystine by air or by filtration through charcoal.

RTECS Toxicity Data

Acute Oral: Rat LD_{50} Dose: 1890 mg/kg; Toxic Effects: Behavioral - Somnolence (general depressed activity); Lungs, Thorax, or Respiration - Dyspnea; Kidney, Ureter, and Bladder - Other changes in urine composition. Mouse LD_{50} Dose: 660 mg/kg.
Acute Dermal: Rat LD_{50} Route: Subcutaneous Dose: 1550 mg/kg; Toxic Effects: Behavioral - Somnolence (general depressed activity); Behavioral - Ataxia; Lungs, Thorax, or Respiration - Respiratory depression. Mouse LD_{50} Route: Subcutaneous Dose: 1360 mg/kg; Toxic Effects: Behavioral - Somnolence (general depressed activity); Behavioral - Ataxia; Lungs, Thorax, or Respiration - Respiratory depression.
Chronic (Multiple Dose) Oral: Rat Dose: 150 gm/kg/30D-C; Toxic Effects: Behavioral - Convulsions or effect on seizure threshold; Behavioral - Muscle weakness; DEATH. Rat Dose: 308 gm/kg/26W-I; Toxic Effects: Cardiac - Other changes; Lungs, Thorax, or Respiration - Other changes; Kidney, Ureter, and Bladder - Other changes.
Reproductive/Teratogenic: Rat Route: Intraperitoneal; Dose: 100 mg/kg; Duration: female 12D of pregnancy; Effects on Fertility - Litter size. Mouse Route: Oral; Dose: 3600 mg/kg; Duration: female 7-12D of pregnancy; Effects on Newborn - Physical. Mouse Route: Oral; Dose: 27600 mg/kg; Duration: female 23D prior to mating Maternal Effects - Menstrual cycle changes or disorders. Mouse Route: Oral; Dose: 6 gm/kg; Duration: male 30D prior to mating; Paternal Effects - Prostate, seminal vessicle, Cowper's gland, accessory glands.
Mutagenic: Human Unscheduled DNA Synthesis; Cell Type: fibroblast; Dose: 1 mmol/L. Rat DNA Inhibition; Cell Type: liver; Dose: 1 mmol/L.

Hazard Overviews

Health: Irritating to eyes/skin/respiratory tract. Harmful. Other Acute Effects: harmful if swallowed; may be harmful if inhaled; may be harmful if absorbed through the skin.
Fire: Hazards: emits toxic fumes. Extinguishing agents: water spray; carbon dioxide, dry chemical powder or appropriate foam. Precautions: combustible liquid.
Reactivity: Incompatible with: strong oxidizing agents. Hazardous decomposition products: toxic fumes of: carbon monoxide, carbon dioxide, nitrogen oxides, sulfur oxides.
Carcinogenicity: IARC - Not listed; NIOSH - Not listed; NTP - Not listed; ACGIH - Not listed; OSHA - Not listed; EPA - Not listed; MAK - Not listed
Primary Target Organs:

Eyes

Skin

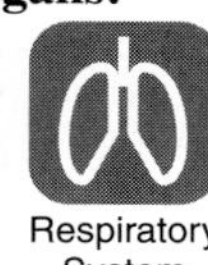
Respiratory System

Environmental

Regulations

RCRA 40CFR: Not listed
CERCLA: 40CFR 302.4: Not listed
SARA 40CFR 372.65: Not listed
SARA EHS 40CFR 355: Not listed
TSCA: Listed

CYT1000 CAS #: 147-94-4

CYTARABINE

RTECS: HA5425000
EINECS Number: 205-705-9
Molecular Formula: $C_9H_{13}N_3O_5$
Formula Weight: 243.22

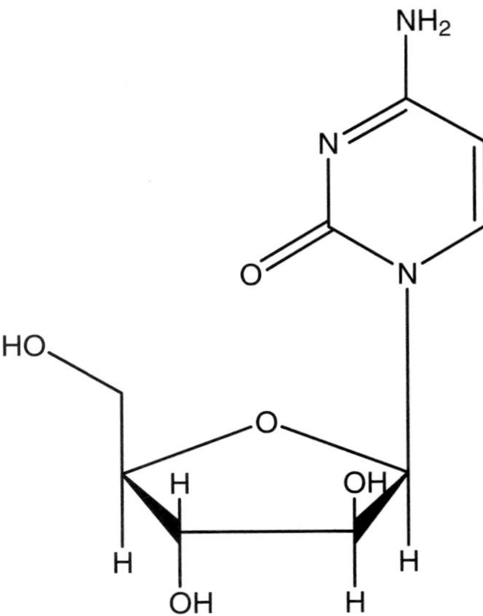

Chemical Structure

Synonyms: AC-1075; ALEXAN; 4-AMINO-1-ARABINOFURANOSYL-2-OXO-1,2-DIHYDROPYRIMIDIN; 4-AMINO-1-ARABINOFURANOSYL-2-OXO-1,2-DIHYDROPYRIMIDINE; 4-AMINO-1-BETA-D-ARABINOFURANOSYL-2(1H)-PYRIMIDINON; 4-AMINO-1-BETA-D-ARABINOFURANOSYL-2(1H)-PYRIMIDINONE; ARA-C; ARABINOCYTIDINE; 1-BETA-D-ARABINOFURANOSYL-4-AMINO-2(1H)PYRIMIDINONE; 1(BETA-D-ARABINOFURANOSYL)CYTOSINE; 1-(BETA-D-ARABINOFURANOSYL)CYTOSINE; 1-ARABINOFURANOSYLCYTOSINE; 1-BETA-ARABINOFURANOSYLCYTOSINE; 1-BETA-D-ARABINOFURANOSYLCYTOSINE; ARABINOFURANOSYLCYTOSINE; 1-BETA-D-ARABINOSYL-4-AMINO-2(1H)PYRIMIDINONE; 1-BETA-D-ARABINOSYLCYTOSINE; ARABINOSYLCYTOSINE; BETA-ARABINOSYLCYTOSINE; BETA-D-ARABINOSYLCYTOSINE; ARABITIN; ARACTIDINE; ARA-CYTIDINE; ARACYTIDINE; ARACYTIN; ARAFCYT; CYTARABIN; CYTARABINA; CYTARABINOSIDE; CYTOSAR; CYTOSAR-U; CYTOSINE ARABINOSIDE; CYTOSINE BETA-D-ARABINOSIDE; CYTOSINE-BETA-D-ARABINOFURANOSIDE; CYTOSINE,1-BETA-D-ARABINOFURANOSYL-; BETA-CYTOSINE,ARABINOSIDE; CYTOSINE-BETA-ARABINOSIDE; CYTOSINEARABINOSIDE; CYTOSINE,1-BETA-D-ARABINOSYL-; NSC 63878; 2(1H)-PYRIMIDINONE,4-AMINO-1-BETA-D-ARABINOFURANOSYL-; 2(1H)-PYRIMIDINONE,4-AMINO-1-BETA-D-ARABINOFURANOSYL-(9CI); SPONGOCYTIDINE; U 19,920A; U-19,920; U-19920 A; UDICIL

Description: white to off-white crystalline powder; odorless

Use: drug useful in combating myelocytic leukemia in adults; antineoplastic; antiviral

Physical Properties

Freezing Point: 212 °C (413.6 °F) to 213 °C (415.4 °F)

Water Solubility: 1 g in about 5 ml Water

Other Solubilities: 1 g in about 500 ml Alcohol; 1 g in about 1000 ml Chloroform & 300 ml Methanol

RTECS Toxicity Data

Acute Oral: Rat LD_{50} Dose: >5 gm/kg. Mouse LD_{50} Dose: 3150 mg/kg.

Acute Dermal: Man TD_{Lo} Route: Subcutaneous Dose: 60 mg/kg/90W-I; Toxic Effects: Sense organs and special senses - Change in acuity; Behavioral - Ataxia; Blood - Changes in spleen. Woman TD_{Lo} Route: Subcutaneous Dose: 6480 ug/kg/12D- Toxic Effects: Blood - Other changes.

Chronic (Multiple Dose) Dermal: Rat Route: Skin; Dose: 875 mg/kg/5W-I; Toxic Effects: Endocrine - Changes in spleen weight; Blood - Changes in erythrocite (RBC) cell count; Blood - Changes in leukocyte (WBC) cell count.

Reproductive/Teratogenic: Rat Route: Subcutaneous; Dose: 75 mg/kg; Duration: female 18-20D of pregnancy Effects on Newborn - Stillbirth; Viability index. Rat Route: Intravenous; Dose: 90 mg/kg; Duration: female 9-14D of pregnancy; Effects on Embryo or Fetus - Fetotoxicity; Specific Developmental Abnormalities - Musculoskeletal system. Rat Route: Intravenous; Dose: 360 mg/kg; Duration: female 9-14D of pregnancy; Effects on Newborn - Live birth index; Weaning or lactation index. Rat Route: Intravenous; Dose: 180 mg/kg; Duration: female 9-14D of pregnancy; Effects on Embryo or Fetus - Fetal death.

Mutagenic: Human DNA Damage; Cell Type: leukocyte; Dose: 10 umol/L. Human DNA Inhibition; Cell Type: other cell types; Dose: 50 ug/L. Human DNA Inhibition; Cell Type: leukocyte; Dose: 20 ug/L. Human DNA Inhibition; Cell Type: other cell types; Dose: 200 nmol/L. Human DNA Inhibition; Cell Type: other cell types; Dose: 10 umol/L. Human DNA Inhibition; Cell Type: other cell types; Dose: 100 umol/L. Human DNA Inhibition; Cell Type: HeLa cell; Dose: 10 nmol/L. Human DNA Inhibition; Cell Type: lymphocyte; Dose: 50 mg/L. Human DNA Inhibition; Cell Type: bone marrow; Dose: 300 ug/L. Human DNA Inhibition; Cell Type: leukocyte; Dose: 100 nmol/L. Human Cytogenetic Analysis; Route: Intravenous; Dose: 239 mg/kg/5D. Human Cytogenetic Analysis; Cell Type: leukocyte; Dose: 50 umol/L/6H. Human Cytogenetic Analysis; Cell Type: lymphocyte; Dose: 3 mg/L/4H. Human Cytogenetic Analysis; Cell Type: lung; Dose: 10 umol/L. Human Sister Chromatid Exchenge; Cell Type: leukocyte; Dose: 6 ug. Human Sister Chromatid Exchenge; Cell Type: lymphocyte; Dose: 400 ug/L. Human Other Mutation Test Systems; Cell Type: other cell types; Dose: 50 mg/L. Human Other Mutation Test Systems; Cell Type: HeLa cell; Dose: 10 mg/L.

Tumorigenic: Rat Route: Intraperitoneal; Dose: 2500 mg/kg/7W-I; Toxic Effects: Tumorigenic - Equivocal tumorigenic agent by RTECS criteria; Blood - Lymphomax including Hodgkin's disease; Skin and appendages - Tumors. Mouse Route: Intraperitoneal; Dose: 4836 mg/kg/26W-I; Toxic Effects: Tumorigenic - Equivocal tumorigenic agent by RTECS criteria; Lungs, Thorax, or Respiration - Tumors; Blood - Lymphomax including Hodgkin's disease.

Hazard Overviews

Health: Irritating to eyes/skin/respiratory tract. Toxic. Other Acute Effects: harmful if swallowed; inhaled; or absorbed through skin; may cause allergic skin reaction; sensitization by skin contact; nervous system disturbances. Chronic Effects: congenital malformation in the fetus; laboratory experiments have shown mutagenic effects; toxic symptoms include gastrointestinal distress; nausea; vomiting; diarrhea;

hepatic dysfunction; fever; dermatitis; myalgia; bone pain; maculopapular rash; conjunctivitis and malaise; target organs: bone marrow, nerves.

Fire: Hazards: emits toxic fumes. Extinguishing agents: water spray; carbon dioxide, dry chemical powder or appropriate foam. Precautions: combustible liquid.

Reactivity: Incompatible with: strong oxidizing agents. Hazardous decomposition products: toxic fumes of: carbon monoxide, carbon dioxide, nitrogen oxides, hydrogen chloride gas.

Carcinogenicity: IARC - Not listed; NIOSH - Not listed; NTP - Not listed; ACGIH - Not listed; OSHA - Not listed; EPA - Not listed; MAK - Not listed

Primary Target Organs:

Eyes

Skin

Respiratory System

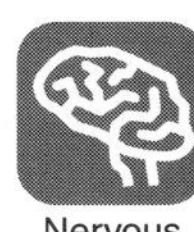
Nervous System

Bone

Environmental

Regulations

RCRA 40CFR: Not listed
CERCLA: 40CFR 302.4: Not listed
SARA 40CFR 372.65: Not listed
SARA EHS 40CFR 355: Not listed
TSCA: Not listed

CYT2600 CAS #: 21739-91-3

CYTEMBENA

RTECS: AS4750000
Molecular Formula: $C_{11}H_9BrNaO_4$
Structured MF: $CH_3OC_6H_4COC(Br)=CHCOO$•Na
Formula Weight: 440.22

Synonyms: ACRYLIC ACID,3-P-ANISOYL-3-BROMO-,SODIUM SALT,(E)-; (E)-3-P-ANISOYL-3-BROMOACRYLIC ACID SODIUM SALT; (E)-3-BROMO-4-(4-METHOXYPHENYL)-4-OXO-2-BUTENOIC ACID SODIUM SALT; 2-BUTENOIC ACID,3-BROMO-4-(4-METHOXYPHENYL)-4-OXO,SODIUM SALT,(E)-; 2-BUTENOIC ACID,3-BROMO-4-(4-METHOXYPHENYL)-4-OXO-,SODIUM SALT,(E)-; 2-BUTENOIC ACID,3-BROMO-4-(4-METHOXYPHENYL)-4-OXO-,SODIUM SALT,(E)-(9CI); CYTEMBENE; MBBA; NSC-104801; NSC 104801; SODNA SUL KYSELINY CIS-BETA-4-METHOXYBENZOYL-BETA-BROMAKRYLOVE

Description: white to off-white powder
Use: folate inhibitor in cancer treatment

Physical Properties

Freezing Point: 260 °C (500 °F) to 263 °C (505.4 °F)
Water Solubility: 50 to 100 mg/mL at 21 °C
Other Solubilities: 95% Ethanol: <1 mg/ml at 21 °C; Acetone: <1 mg/ml at 21 °C; DMSO: >=100 mg/ml at 21 °C.
Flash Point: Not available; probably combustible

RTECS Toxicity Data

Acute Dermal: Rat LD_{50} Route: Subcutaneous Dose: 155 mg/kg. Mouse LD_{50} Route: Subcutaneous Dose: 52 mg/kg.

Mutagenic: Mouse Mutations in Mammalian Somatic Cells; Cell Type: lymphocyte; Dose: 25 mg/L. Hamster Cytogenetic Analysis; Cell Type: ovary; Dose: 25300 ug/L.

Tumorigenic: Rat Route: Intraperitoneal; Dose: 7 mg/kg/2Y-I; Toxic Effects: Tumorigenic - Carcinogenic by RTECS criteria; Skin and appendages - Tumors; Tumorigenic effects - Other reproductive system tumors. Rat Route: Intraperitoneal; Dose: 14 mg/kg/2Y-I; Toxic Effects: Tumorigenic - Carcinogenic by RTECS criteria; Skin and appendages - Tumors; Tumorigenic effects - Other reproductive system tumors.

Hazard Overviews

Fire: Will burn.

Carcinogenicity: IARC - Not listed; NIOSH - Not listed; NTP - Not listed; ACGIH - Not listed; OSHA - Not listed; EPA - Not listed; MAK - Not listed

Environmental

Regulations

RCRA 40CFR: Not listed
CERCLA: 40CFR 302.4: Not listed
SARA 40CFR 372.65: Not listed
SARA EHS 40CFR 355: Not listed
TSCA: Not listed

CYT4200 CAS #: 485-35-8

CYTISINE

RTECS: HA4025000
EINECS Number: 207-616-0
Molecular Formula: $C_{11}H_{14}N_2O$
Formula Weight: 190.24

HN N O

Chemical Structure

Synonyms: BAPTITOXIN; BAPTITOXINE; CITIZIN; CYSTISINE; (-)-CYTISINE; CYTITON; CYTITONE; CYTIZIN; LABURNIN; 1,5-METHANO-8H-PYRIDO(1,2-A)(1,5)DIAZOCIN-8-ONE,1,2,3,4,5,6-HEXAHYDRO-; 1,5-METHANO-8H-PYRIDO(1,2-A)(1,5)DIAZOCIN-8-ONE,1,2,3,4,5,6-HEXAHYDRO-,(1R)-; SOPHORIN; SOPHORINE; TABEX; TSITIZIN; ULEXIN; ULEXINE

Description: orthorhombic prisms
Use: medication: formerly used as an antitussive & antiemetic

Physical Properties

Boiling Point: 218 °C (424 °F) at 2 mm Hg
Freezing Point: 152 °C (305.6 °F) to 153 °C (307.4 °F)
Water Solubility: 1 parts in 13 parts Water
Other Solubilities: Slightly Soluble in Alcohol; Practically Insoluble in Ether.

RTECS Toxicity Data

Acute Oral: Mouse LD_{50} Dose: 101 mg/kg; Toxic Effects: Behavioral - Convulsions or effect on seizure threshold; Behavioral - Rigidity.
Acute Dermal: Rabbit LD_{Lo} Route: Subcutaneous Dose: 30 mg/kg. Rat LD_{Lo} Route: Subcutaneous Dose: 20 mg/kg.

Hazard Overviews

Health: Irritating to eyes/skin/respiratory tract. Toxic. Other Acute Effects: harmful if swallowed, inhaled, or absorbed through skin.
Fire: Hazards: emits toxic fumes. Extinguishing agents: water spray; carbon dioxide, dry chemical powder or appropriate foam. Precautions: combustible liquid.
Reactivity: Incompatible with: strong oxidizing agents. Hazardous decomposition products: toxic fumes of: carbon monoxide, carbon dioxide, nitrogen oxides.
Carcinogenicity: IARC - Not listed; NIOSH - Not listed; NTP - Not listed; ACGIH - Not listed; OSHA - Not listed; EPA - Not listed; MAK - Not listed
Primary Target Organs:

Eyes Skin Respiratory System

Environmental

Regulations

RCRA 40CFR: Not listed
CERCLA: 40CFR 302.4: Not listed
SARA 40CFR 372.65: Not listed
SARA EHS 40CFR 355: Not listed
TSCA: Listed

CYT5800 CAS #: 14930-96-2

CYTOCHALASIN B

RTECS: RO0205000
EINECS Number: 239-000-2
Molecular Formula: $C_{29}H_{37}NO_5$
Formula Weight: 479.67

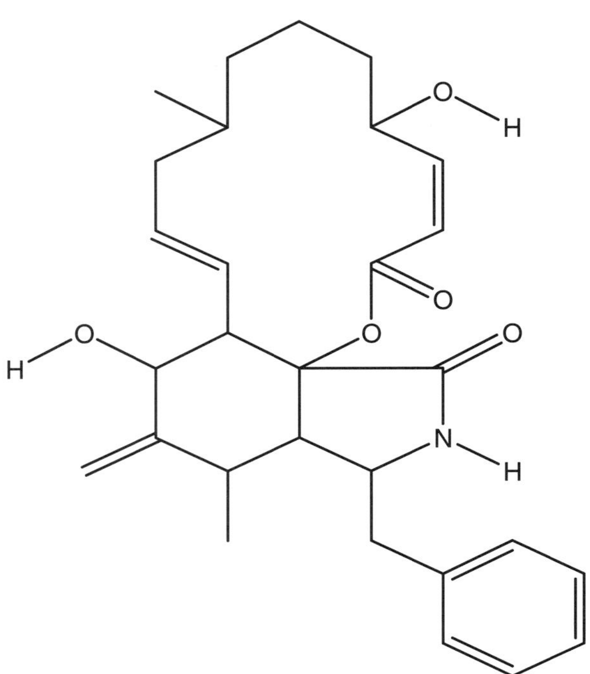

Chemical Structure

Synonyms: (E,E)-(5S,9R,12AS,13S,15S,15AS,16S,18AS)-16-BENZYL-6,7,8,9,10,12A,13,14,15,15A,16,17-DODECAHYDRO-5,13-DIHYDROXY-9,15-DIMETHYL-14-METHYLENE-2H-OXACYCLOTETRADEC[2,3-D]ISOINDOLE-2,18(5H)-DIONE; 7,20-DIHYDROXY-10-PHENYL-5,16-DIMETHYL-24-OXA-[14]CYTOCHALAS-6(12),13,21-TRIEN-23-ONE; 2H-OXACYCLOTETRADEC(2,3-D)ISOINDOLE-2,18(5H)-DIONE,16-BENZYL-6,7,8,9,10,12A,13,14,15,15A,16,17-DODECAHYDRO-5,13-DIHYDROXY- 9,15-DIMETHYL-14-METHYLENE-,(E,E)-(5S,9R,12AS,13S,15S,15AS,16S,18AS)-; 2H-OXACYCLOTETRADEC(2,3-D)ISOINDOLE-2,18(5H)-DIONE,16-BENZYL-6,7,8,9,10,12A,13,14,15,15A,16,17-DODECAHYDRO-5,13-DIHYDROXY-9,15-DIMETHYL-14-METHYLENE-,(E)-(5S,9R,12AS,13S,15S,15AS,16AS,18AS)-; 24-OXA(14)CYTOCHALASA-6(12),13,21-TRIENE-1,23-DIONE,7,20-DIHYDROXY-16-METHYL-10-PHENYL-,(7S,13E,16R,20R,21E)-; PHOMIN
Description: felted needles
Use: in cytological research

Physical Properties

Freezing Point: 218 °C (424.4 °F) to 221 °C (429.8 °F)

RTECS Toxicity Data

Reproductive/Teratogenic: Mouse Route: Oral; Dose: 1500 ug/kg; Duration: female 8D of pregnancy; Specific Developmental Abnormalities - Central nervous system; Musculoskeletal system. Hamster Route: Intraperitoneal; Dose: 5 mg/kg; Duration: female 8D of pregnancy; Effects on Embryo or Fetus - Fetal death; Specific Developmental Abnormalities - Central nervous system.
Mutagenic: Human DNA Inhibition; Cell Type: other cell types; Dose: 1 mg/L. Human DNA Inhibition; Cell Type: HeLa cell; Dose: 1 umol/L. Human Cytogenetic Analysis; Cell Type: lymphocyte; Dose: 6 mg/L. Human Cytogenetic

Analysis; Cell Type: other cell types; Dose: 1 mg/L. Human Micronucleus Test; Cell Type: lymphocyte; Dose: 3 mg/L.

Hazard Overviews

Poison

Health: Irritating. Poison. Other Acute Effects: may be fatal if inhaled, swallowed, or absorbed through skin. Chronic Effects: overexposure may cause reproductive disorder(s) based on tests with laboratory animals; target organ: liver.
Fire: Hazards: emits toxic fumes. Extinguishing agents: water spray; carbon dioxide, dry chemical powder or appropriate foam. Precautions: combustible liquid.
Reactivity: Incompatible with: strong oxidizing agents, strong acids, strong bases. Hazardous decomposition products: toxic fumes of: carbon monoxide, carbon dioxide, nitrogen oxides.
Carcinogenicity: IARC - Not listed; NIOSH - Not listed; NTP - Not listed; ACGIH - Not listed; OSHA - Not listed; EPA - Not listed; MAK - Not listed
Primary Target Organs:

Eyes

Skin

Respiratory System

Liver

Environmental

Regulations

RCRA 40CFR: Not listed
CERCLA: 40CFR 302.4: Not listed
SARA 40CFR 372.65: Not listed
SARA EHS 40CFR 355: Not listed
TSCA: Not listed

CYT9000 CAS #: 36011-19-5

CYTOCHALASINE

RTECS: HA5360000
EINECS Number: 252-835-7
Molecular Formula: $C_{28}H_{33}NO_7$
Formula Weight: 494.61

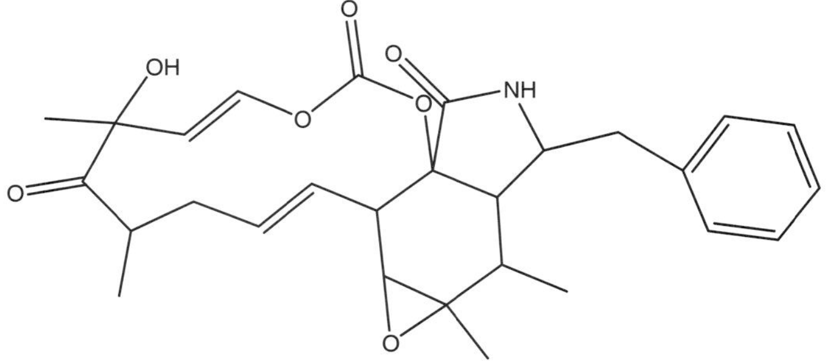

Chemical Structure

Synonyms: 21,23-DIOXA(13)CYTOCHALASA-13,19-DIENE-1,17,22-TRIONE,6,7-EPOXY-18-HYDROXY-16,18-DIMETHYL-10-PHENYL-,(7S,13E,16S,18R,19E)-; 6,7-EPOXY-10-PHENYL-5,6,16,18-TETRAMETHYL-21,23-DIOXA-[13]CYTOCHALAS-13,19-DIENE-17,22-DIONE
Description: crystals
Use: in cytological research

Physical Properties

Freezing Point: Decomposes at 206 °C (402.8 °F) to 208 °C (406.4 °F)
Water Solubility: Sparingly Soluble in Water
Other Solubilities: Soluble in Dichloromethane, Chloroform, Acetone, & Methanol

RTECS Toxicity Data

Acute Oral: Rat LD_{50} Dose: 9100 ug/kg; Toxic Effects: Behavioral - Ataxia; Behavioral - Coma; Lungs, Thorax, or Respiration - Cyanosis.
Reproductive/Teratogenic: Mouse Route: Intraperitoneal; Dose: 8100 ug/kg; Duration: female 7-9D of pregnancy; Effects on Embryo or Fetus - Fetal death; Specific Developmental Abnormalities - Craniofacial (including nose and tongue). Mouse Route: Intraperitoneal; Dose: 3900 ug/kg; Duration: female 7-9D of pregnancy; Effects on Embryo or Fetus - Fetal death. Mouse Route: Intraperitoneal; Dose: 3600 ug/kg; Duration: female 7-9D of pregnancy; Specific Developmental Abnormalities - Central nervous system.

Hazard Overviews

Poison

Health: Irritating. Poison. Other Acute Effects: may be fatal if inhaled, swallowed, or absorbed through skin. Chronic Effects: overexposure may cause reproductive disorder(s) based on tests with laboratory animals; possible terratogen.
Fire: Extinguishing agents: dry chemical powder; carbon dioxide, dry chemical powder or appropriate foam. Precautions: combustible liquid.
Reactivity: Incompatible with: strong oxidizing agents, strong acids, strong bases. Hazardous decomposition products: toxic fumes of: carbon monoxide, carbon dioxide, nitrogen oxides.
Carcinogenicity: IARC - Not listed; NIOSH - Not listed; NTP - Not listed; ACGIH - Not listed; OSHA - Not listed; EPA - Not listed; MAK - Not listed

Primary Target Organs:

Eyes

Skin

Respiratory System

Environmental

Regulations

RCRA 40CFR: Not listed
CERCLA: 40CFR 302.4: Not listed
SARA 40CFR 372.65: Not listed
SARA EHS 40CFR 355: Not listed
TSCA: Not listed

DAC3000 CAS #: 4342-03-4

DACARBAZINE

RTECS: NI3950000
EINECS Number: 224-396-1
Molecular Formula: $C_6H_{10}N_6O$
Formula Weight: 182.18

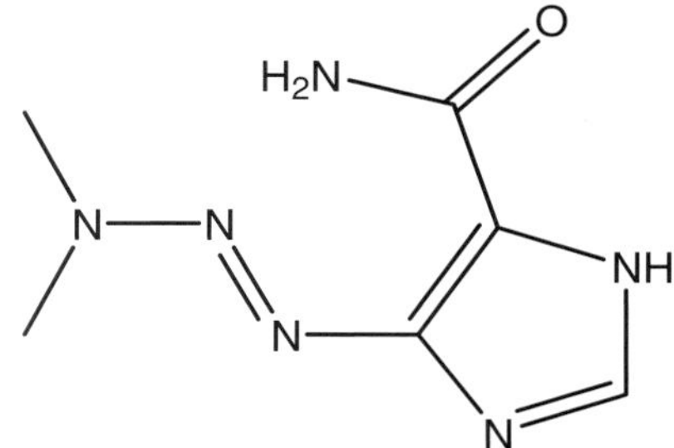

Chemical Structure

Synonyms: BIOCARBAZINE R; DETICENE; DIC; (DIMETHYLTRIAZENO)IMIDAZOLECARBOXAMIDE; 4-(3,3-DIMETHYL-1-TRIAZENO)IMIDAZOLE-5-CARBOXAMIDE; 4-(5)-(3,3-DIMETHYL-1-TRIAZENO)IMIDAZOLE-5(4)-CARBOXAMIDE; 4-(DIMETHYLTRIAZENO)IMIDAZOLE-5-CARBOXAMIDE; 5-(3,3-DIMETHYL-1-TRIAZENO)IMIDAZOLE-4-CARBOXAMIDE; 5-(3,3-DIMETHYLTRIAZENO)IMIDAZOLE-4-CARBOXAMIDE; 5-(DIMETHYLTRIAZENO)IMIDAZOLE-4-CARBOXAMIDE; 5-(3,3-DIMETHYL-1-TRIAZENYL)-1H-IMIDAZOLE-4-CARBOXAMIDE; DTIC; DTIC-AOME; DTIC-DOME; IMIDAZOLE-4-CARBOXAMIDE,5-(3,3-DIMETHYL-1-TRIAZENO1-; 1H-IMIDAZOLE-4-CARBOXAMIDE,5-(3,3-DIMETHYL-1-TRIAZENYL)-; 1H-IMIDAZOLE-4-CARBOXAMIDE,5-(3,3-DIMETHYL-1-TRIAZENYL)-(9CI); NSC-45388; NSC 45388
Description: ivory microcrystaline substance
Use: antineoplastic agent

Physical Properties

Freezing Point: 205 °C (401 °F)
Water Solubility: < 0.1 mg/mL at 15 C
Other Solubilities: 95% Ethanol: <1 mg/ml at 26 °C; Acetone: <1 mg/ml at 26 °C; DMSO: <1 mg/ml at 26 °C; Toluene: <1 mg/ml at 15 °C; Methanol: <1 mg/ml at 15 °C.
Flash Point: Not available; probably combustible

RTECS Toxicity Data

Acute Oral: Rat LD_{50} Dose: 2147 mg/kg; Toxic Effects: Behavioral - Change in motor activity (specific assay); Behavioral - Antipsychotic. Mouse LD_{50} Dose: 2032 mg/kg; Toxic Effects: Behavioral - Change in motor activity (specific assay); Behavioral - Antipsychotic.
Reproductive/Teratogenic: Rat Route: Intraperitoneal; Dose: 50 mg/kg; Duration: female 12D of pregnancy; Effects on Embryo or Fetus - Fetotoxicity. Rat Route: Intraperitoneal; Dose: 200 mg/kg; Duration: female 12D of pregnancy; Specific Developmental Abnormalities - Urogenital system. Rat Route: Intraperitoneal; Dose: 100 mg/kg; Duration: female 12D of pregnancy; Specific Developmental Abnormalities - Musculoskeletal system. Rat Route: Intraperitoneal; Dose: 400 mg/kg; Duration: female 12D of pregnancy; Specific Developmental Abnormalities - Central nervous system; Craniofacial (including nose and tongue).
Mutagenic: Rat DNA Damage; Route: Intraperitoneal; Dose: 6 mg/kg. Rat DNA Inhibition; Route: Intraperitoneal; Dose: 37500 ug/kg. Rat DNA Inhibition; Cell Type: other cell types; Dose: 1 mmol/L.
Tumorigenic: Rat Route: Oral; Dose: 1730 mg/kg/15W-C; Toxic Effects: Tumorigenic - Carcinogenic by RTECS criteria; Blood - Lymphomax including Hodgkin's disease; Skin and appendages - Tumors. Rat Route: Oral; Dose: 3700 mg/kg/14W-C; Toxic Effects: Tumorigenic - Carcinogenic by RTECS criteria; Blood - Lymphomax including Hodgkin's disease; Skin and appendages - Tumors.

Hazard Overviews

Health: Irritating to eyes/skin/respiratory tract. Toxic. Other Acute Effects: harmful if swallowed, inhaled, or absorbed through skin; photosensitizer; exposure can cause anorexia; nausea; vomiting; influenza-like syndrome. Chronic Effects: mutagen; possible teratogen; target organs: blood, liver. Carcinogen.
Fire: Will burn. Extinguishing agents: carbon dioxide, dry chemical powder or appropriate foam. Precautions: combustible liquid.
Reactivity: Stable. Hazardous polymerization will not occur. Hazardous decomposition products: toxic fumes of: carbon monoxide, carbon dioxide, nitrogen oxides.
Carcinogenicity: IARC - Group 2B, Possibly carcinogenic to humans; NIOSH - Not listed; NTP - Not listed; ACGIH - Not listed; OSHA - Not listed; EPA - Not listed; MAK - Not listed
Primary Target Organs:

Eyes

Skin
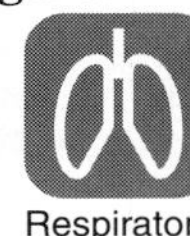
Respiratory System

Liver

Blood

Environmental

Regulations

RCRA 40CFR: Not listed
CERCLA: 40CFR 302.4: Not listed
SARA 40CFR 372.65: Not listed
SARA EHS 40CFR 355: Not listed
TSCA: Not listed

DAC6000 CAS #: 50-76-0

DACTINOMYCIN

RTECS: AU1575000
EINECS Number: 200-063-6

Molecular Formula: $C_{62}H_{86}N_{12}O_{16}$
Formula Weight: 1255.47

Chemical Structure

Synonyms: ACT; ACT D; ACTINOMYCIN 7; ACTINOMYCIN A IV; ACTINOMYCIN C1; ACTINOMYCIN CL; ACTINOMYCIN 11 COSMEGEN; (-)-ACTINOMYCIN D; ACTINOMYCIN D; ACTINOMYCIN I; ACTINOMYCIN I1; ACTINOMYCIN IV; ACTINOMYCIN I(SUB 1); ACTINOMYCIN X 1; ACTINOMYCIN X I; ACTINOMYCINDIOIC D ACID,DILACTONE; ACTINOMYCIN-(THRE-VAL-PRO-SAR-MEVAL); ACTINOMYEIN-THEO-VAL-PRO-SAR-MEVAL; ACTO-D; AD; CHOUNGHWAMYCIN B; COSMEGEN; DACTINOMYCIN D; DACTINOMYEIN D; DILACTONE ACTINOMYCIN D ACID; DILACTONE ACTINOMYCINDIOIC D ACID; HBF 386; LYOVAC COSMEGEN; MERACTINOMYCIN; NSC 3053; ONCOSTATIN K; 3H-PHENOXAZINE-1,9-DICARBOXAMIDE,2-AMINO-N,N'-BIS(HEXADECAHYDRO-2,5,9-TRIMETHYL-6,13-BIS(1-METHYLETHYL)-1,4,7,11,14-PENTAOXO-1H-PYRROLO(2,1-I)(1,4,7,10,13)OXATETRA-AZACYCLOHEXADECIN-10-YL)-4,6 - DIMETHYL-3-OXO-; X 97

Description: bright red crystalline powder or yellow lyophilized powder

Use: antineoplastic agent; antibiotic active against gram-positive bacteria and tumors; used in the treatment of Wilm's tumors, gestational choriocarcinoma, testicular tumors, embryonal rhabdomyosarcomas, lymphomas, Ewing's sarcomas and acute leukemia

Physical Properties

Freezing Point: Decomposes at 245 °C (473 °F) to 248 °C (478.4 °F)

Water Solubility: 1 gram dissolves in 25 ml of Water at 10 °C

Other Solubilities: 95% Ethanol: Very Soluble; Ether: Very slightly Soluble; Propylene Glycol: Soluble; Water + glycol mixtures: Soluble.

Flash Point: Not available; probably combustible

RTECS Toxicity Data

Acute Oral: Rat LD_{50} Dose: 7200 ug/kg; Toxic Effects: Gastrointestinal - Hypermotility, diarrhea; Blood - Other changes. Mouse LD_{50} Dose: 13 mg/kg; Toxic Effects: Gastrointestinal - Hypermotility, diarrhea; Blood - Other changes.

Acute Dermal: Rat LD_{50} Route: Subcutaneous Dose: 800 ug/kg; Toxic Effects: Gastrointestinal - Hypermotility, diarrhea; Blood - Other changes. Mouse LD_{50} Route: Subcutaneous Dose: 500 ug/kg.

Reproductive/Teratogenic: Rat Route: Oral; Dose: 250 ug/kg; Duration: female 13D of pregnancy; Effects on Fertility - Post-implantation mortality. Rat Route: Subcutaneous; Dose: 200 ug/kg; Duration: female 12D of pregnancy; Specific Developmental Abnormalities - Musculoskeletal system; Effects on Embryo or Fetus - Fetal death.

Mutagenic: Human DNA Damage; Cell Type: HeLa cell; Dose: 400 ug/L/15M. Human DNA Damage; Cell Type: other cell types; Dose: 10 ug/L. Monkey DNA Damage; Cell Type: kidney; Dose: 5 mg/L. Human DNA Inhibition; Cell Type: other cell types; Dose: 300 pmol/L. Human DNA Inhibition; Cell Type: other cell types; Dose: 10 ug/L/2H-C. Human DNA Inhibition; Cell Type: HeLa cell; Dose: 700 nmol/L. Human DNA Inhibition; Cell Type: other cell types; Dose: 40 mg/L. Monkey DNA Inhibition; Cell Type: kidney; Dose: 10 ug/L. Human Mutations in Mammalian Somatic Cells; Cell Type: fibroblast; Dose: 1 mg/L. Human Cytogenetic Analysis; Cell Type: lymphocyte; Dose: 200 ug/L/2H. Human Cytogenetic Analysis; Cell Type: leukocyte; Dose: 200 ug/L/24H. Human Cytogenetic Analysis; Cell Type: HeLa cell; Dose: 250 ug/L/1H. Human Sister Chromatid Exchenge; Cell Type: lymphocyte; Dose: 1 nmol/L. Human Other Mutation Test Systems; Cell Type: other cell types; Dose: 40 ug/L. Human Other Mutation Test Systems; Cell Type: HeLa cell; Dose: 250 ug/L.

Tumorigenic: Rat Route: Intraperitoneal; Dose: 2600 ug/kg/17W-I; Toxic Effects: Tumorigenic - Carcinogenic by RTECS criteria; Tumorigenic - Tumors at site of application. Rat Route: Intraperitoneal; Dose: 1700 ug/kg/26W-I; Toxic Effects: Tumorigenic - Neoplastic by RTECS criteria; Tumorigenic - Tumors at site of application.

Hazard Overviews

Poison

Corrosive

Health: Corrosive to eyes/skin/respiratory tract. Poison. Other Acute Effects: may be fatal if inhaled, swallowed, or absorbed through skin; extremely destructive to tissue of the mucous membranes and upper respiratory tract, eyes and skin; inhalation may result in spasm, inflammation and edema of the larynx and bronchi, chemical pneumonitis and pulmonary edema; symptoms of exposure may include burning sensation, coughing, wheezing, laryngitis, shortness of breath, headache, nausea and vomiting; prolonged or

repeated exposure may cause allergic reactions in certain sensitive individuals; exposure can cause anemia; drowsiness; weakness. Chronic Effects: may alter genetic material; may cause congenital malformation in the fetus; target organs: bone marrow, liver, kidneys. Carcinogen.
Fire: Will burn. Hazards: emits toxic fumes. Extinguishing agents: water spray; carbon dioxide, dry chemical powder or appropriate foam. Precautions: combustible liquid.
Reactivity: Incompatible with: strong oxidizing agents, strong acids, strong bases. Hazardous decomposition products: toxic fumes of: carbon monoxide, carbon dioxide, nitrogen oxides.
Carcinogenicity: IARC - Group 3, Not classifiable as to carcinogenicity to humans; NIOSH - Not listed; NTP - Not listed; ACGIH - Not listed; OSHA - Not listed; EPA - Not listed; MAK - Not listed
Primary Target Organs:

Eyes

Skin

Respiratory System

Liver

Kidneys

Bone

Environmental

Cleanup/Disposal: Guide No. 154: Eliminate all ignition sources (no smoking, flares, sparks or flames in immediate area). Do not touch damaged containers or spilled material unless wearing appropriate protective clothing. Stop leak if you can do it without risk. Prevent entry into waterways, sewers, basements or confined areas. Absorb or cover with dry earth, sand or other non-combustible material and transfer to containers. Do not get water inside containers.

Environmental Physical Data

Octanol/Water Partition Coefficient: log K_{ow} = 3.21

Regulations

RCRA 40CFR: Not listed
CERCLA: 40CFR 302.4: Not listed
SARA 40CFR 372.65: Not listed
SARA EHS 40CFR 355: Not listed
TSCA: Not listed

DAN3000 CAS #: 39515-41-8

DANITOL

RTECS: GZ2090000
EINECS Number: 254-485-0
Molecular Formula: $C_{22}H_{23}NO_3$
Formula Weight: 349.46
Synonyms: ALPHA-CYANO-3-PHENOXYBENZYL 2,2,3,3-TETRAMETHYL-1-CYCLOPROPANECARBOXYLATE; DANIMEN; DANITOL; DANITOL FIORI; FENPROPANATE; FENPROPATHRIN; HERALD; MEOTHRIN; OMS-1999; RODY; S 3206; SD 41706; TAME; WL 41706; XE-938

RTECS Toxicity Data

Acute Oral: Rat LD_{50} Dose: 18 mg/kg.
Acute Dermal: Rat LD_{50} Route: Skin; Dose: 870 mg/kg. Rat LD_{50} Route: Subcutaneous Dose: 900 mg/kg.
Mutagenic: Human Micronucleus Test; Cell Type: lymphocyte; Dose: 10 mg/L.

Hazard Overviews

Carcinogenicity: IARC - Not listed; NIOSH - Not listed; NTP - Not listed; ACGIH - Not listed; OSHA - Not listed; EPA - Not listed; MAK - Not listed

Environmental

Regulations

RCRA 40CFR: Not listed
CERCLA: 40CFR 302.4: Listed as Compound per CWA Section 307(a) per CAA Section 112
SARA 40CFR 372.65: Listed
SARA EHS 40CFR 355: Not listed
TSCA: Not listed

Analytical Methods

Food: FDA 212.1, 212.2, 232.1, 232.3
Plasma: FDA 252

DAN6000 CAS #: 7261-97-4

DANTROLENE

RTECS: MU3874000
EINECS Number: 230-684-8
Molecular Formula: $C_{14}H_{10}N_4O_5$
Formula Weight: 314.26
Synonyms: HYDANTOIN,1-((5-(P-NITROPHENYL)FURFURYLIDENE)AMINO-; 2,4-IMIDAZOLIDINEDIONE,1-(((5-(4-NITROPHENYL)-2-FURANYL)METHYLENE)AMINO)-; 1-(((5-(4-NITROPHENYL)-2-FURANYL)METHYLENE)AMINO)-2,4-IMIDAZOLIDINEDIONE; 1-((5-(P-NITROPHENYL)FURFURYLIDENE)AMINO)HYDANTOIN
Description: crystals; odorless
Use: skeletal muscle relaxant

Physical Properties

Boiling Point: Decomposes at 225 °C (437 °F) to 230 °C (446 °F)
Freezing Point: 279 °C (534.2 °F)
Water Solubility: Slightly Soluble in Water
Flash Point: Not available; probably combustible

Hazard Overviews

Fire: Will burn.
Carcinogenicity: IARC - Not listed; NIOSH - Not listed; NTP - Not listed; ACGIH - Not listed; OSHA - Not listed; EPA - Not listed; MAK - Not listed

Environmental

Regulations

RCRA 40CFR: Not listed
CERCLA: 40CFR 302.4: Not listed

SARA 40CFR 372.65: Not listed
SARA EHS 40CFR 355: Not listed
TSCA: Not listed

DAP5000 CAS #: 80-08-0

DAPSONE

RTECS: BY8925000
EINECS Number: 201-248-4
Molecular Formula: $C_{12}H_{12}N_2O_2S$
Structured MF: $(H_2NC_6H_4)_2SO_2$
Formula Weight: 248.30

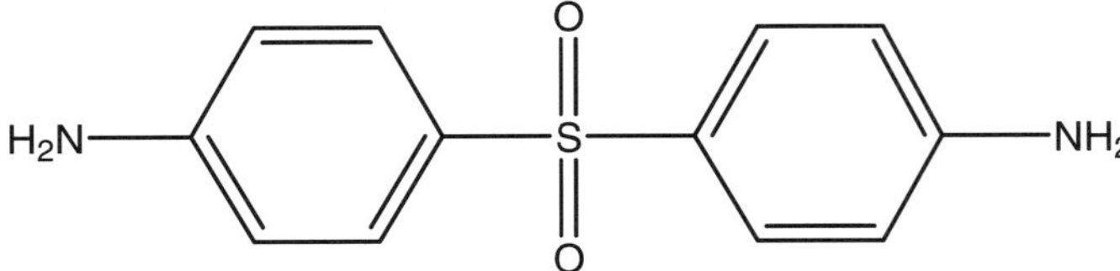

Chemical Structure

Synonyms: 1358F; ANILINE,4,4'-SULFONYLDI-; AVLOSULFON; AVLOSULPHONE; BENZENAMINE,4,4'-SULFONYLBIS-; BENZENAMINE,4,4'-SULFONYLBIS-(9CI); BIS(4-AMINOPHENYL) SULFONE; BIS(P-AMINOPHENYL) SULFONE; BIS(4-AMINOPHENYL)SULPHONE; BIS(P-AMINOPHENYL)SULPHONE; CROYSULFONE; CROYSULPHONE; DADPS; DAPSON; DAPSONUM; DDS; DDS (PHARMACEUTICAL); DDS (VAN); DIAMINODIFENILSULFONA; 4,4-DIAMINODIFENYLSULFON; 4,4'-DIAMINODIPHENYL SULFONE; DIAMINO-4,4'-DIPHENYL SULFONE; P,P'-DIAMINODIPHENYL SULFONE; 4,4'-DIAMINODIPHENYL SULPHONE; DIAMINO-4,4'-DIPHENYL SULPHONE; P,P'-DIAMINODIPHENYL SULPHONE; P,P-DIAMINODIPHENYL SULPHONE; 4,4'-DIAMINOPHENYL SULFONE; DI(4-AMINOPHENYL) SULFONE; DI(P-AMINOPHENYL) SULFONE; DI(4-AMINOPHENYL)SULFONE; DI(4-AMINOPHENYL)SULPHONE; DI(P-AMINOPHENYL)SULPHONE; DIAPHENYLSULFON; DIAPHENYLSULFONE; DIAPHENYLSULPHON; DIAPHENYLSULPHONE; DIPHENASONE; DIPHONE; DISULONE; DSS; DUBRONAX; DUMITONE; EPORAL; F 1358; ICI; MALOPRIM; METABOLITE C; NOVOPHONE; NSC 6091D; NSC-6091; SULFONA; SULFONA-MAE; SULFONE UCB; SULFONE,DIPHENYL,4,4'-DIAMINO-; 1,1'-SULFONYLBIS(4-AMINOBENZENE); 4,4'-SULFONYLBISANILINE; 4,4'-SULFONYLBISBENZAMINE; P,P'-SULFONYLBISBENZAMINE; P,P-SULFONYLBISBENZAMINE; 4,4'-SULFONYLBISBENZENAMINE; P,P-SULFONYLBISBENZENAMINE; 4,4'-SULFONYLDIANILINE; P,P'-SULFONYLDIANILINE; SULPHADIONE; SULPHON-MERE; 1,1'-SULPHONYLBIS(4-AMINOBENZENE); 4,4'-SULPHONYLBISBENZAMINE; P,P'-SULPHONYLBISBENZAMINE; P,P-SULPHONYLBISBENZAMINE; 4,4'-SULPHONYLBISBENZENAMINE; P,P'-SULPHONYLBISBENZENAMINE; P,P-SULPHONYLBISBENZENAMINE; 4,4'-SULPHONYLDIANILINE; P,P'-SULPHONYLDIANILINE; P,P-SULPHONYLDIANILINE; SULPHONYLDIANILINE; SUMICURE S; TARIMYL; UDOLAC; WR 448

Description: white or creamy white crystalline powder, crystals, or leaflets; odorless

Use: an antibacterial drug used in the treatment of leprosy and dermatitis herpetiformis; to treat mycetoma (maduromycosis); has a suppressive action on the malaria parasite; to treat tuberculosis; in polymerization and as a hardening agent in the curing of epoxy resins; in veterinary medicine as an antibacterial and antiprotozoan

Physical Properties

Freezing Point: 175 °C (347 °F) to 176 °C (348.8 °F)
Specific Gravity: 1.33 at 25 °C
Vapor Density: 8.3 Air=1
Water Solubility: < 1 mg/mL at 20 C
Other Solubilities: DMSO: >=100 mg/ml at 19 °C; Acetone: >=100 mg/ml at 19 °C; Benzene: 0.5 mg/ml at room temperature; Fixed oils: Insoluble; Dilute mineral acids: Soluble; Ether: 0.9 mg/ml; Chloroform: 3 mg/ml; Alcohol: 1 in 30 at 20 °C.
Flash Point: > 160 °C

RTECS Toxicity Data

Acute Oral: Human TD_{Lo} Dose: 18 gm/kg/15Y; Toxic Effects: Kidney, Ureter, and Bladder - Other changes; Musculoskelital - Joints. Man TD_{Lo} Dose: 36 mg/kg; Toxic Effects: Blood - Methemoglobinemia-Carboxyhemoglobin. Woman TD_{Lo} Dose: 112 mg/kg/4W-I; Toxic Effects: Behavioral - Muscle weakness; Cardiac - Pulse rate increased without fall in BP; Blood - Oxidant related (GPD deficient) anemia. Woman TD_{Lo} Dose: 300 mg/kg/21W-I; Toxic Effects: Blood - Eosinophilia. Woman TD_{Lo} Dose: 18 mg/kg; Toxic Effects: Lungs, Thorax, or Respiration - Cyanosis; Blood - Methemoglobinemia-Carboxyhemoglobin. Woman TD_{Lo} Dose: 28 mg/kg; Toxic Effects: Blood - Agranulocytosis. Child TD_{Lo} Dose: 5 mg/kg; Toxic Effects: Lungs, Thorax, or Respiration - Cyanosis. Rat LD_{50} Dose: 1 gm/kg.

Acute Dermal: Rabbit LD_{50} Route: Skin; Dose: >4 gm/kg. Mouse LD_{50} Route: Subcutaneous Dose: 250 mg/kg.

Chronic (Multiple Dose) Oral: Rat Dose: 6750 mg/kg/45D-C; Toxic Effects: DEATH. Rat Dose: 910 mg/kg/26W-I; Toxic Effects: Nutritional and gross metabolic - Weight loss or decreased weight gain; Biochemical - Dehydrogenases; Biochemical - Transaminases.

Chronic (Multiple Dose) Inhalation: Rat Dose: 44 mg/m^3/17W-I; Toxic Effects: Brain and coverings - Recordings from specific areas of CNS; Kidney, Ureter, and Bladder - Proteinuria; Nutritional and gross metabolic - Changes in chlorine. Rat Dose: 500 ug/m^3/24H/13W-C; Toxic Effects: Lungs, Thorax, or Respiration - Structural or functional change in trachea or bronchi; Blood - Changes in serum composition; Biochemical - True cholinesterase.

Reproductive/Teratogenic: Rat Route: Oral; Dose: 2100 mg/kg; Duration: male 6W prior to mating; Paternal Effects - Spermatogenesis; Effects on Fertility - Male fertility index.

Mutagenic: Human Cytogenetic Analysis; Cell Type: leukocyte; Dose: 4 mg/L.

Tumorigenic: Rat Route: Oral; Dose: 20 gm/kg/80W-C; Toxic Effects: Tumorigenic - Carcinogenic by RTECS criteria; Gastrointestinal - Tumors; Blood - Tumors. Rat Route: Oral; Dose: 39 gm/kg/78W-C; Toxic Effects: Tumorigenic - Carcinogenic by RTECS criteria; Gastrointestinal - Tumors; Blood - Tumors. Rat Route: Oral; Dose: 52 gm/kg/2Y-I; Toxic Effects: Tumorigenic - Carcinogenic by RTECS criteria; Blood - Changes in spleen; Musculoskelital -

Acute Dermal: Rabbit LD_{50} Route: Skin; Dose: 7 gm/kg. Rat LD_{50} Route: Skin; Dose: 2260 mg/kg; Toxic Effects: Behavioral - Somnolence (general depressed activity); Behavioral - Muscle weakness; Lungs, Thorax, or Respiration - Dyspnea. Rat LD_{50} Route: Subcutaneous Dose: 470 mg/kg; Toxic Effects: Behavioral - Somnolence (general depressed activity); Behavioral - Muscle weakness; Lungs, Thorax, or Respiration - Dyspnea.
Chronic (Multiple Dose) Oral: Rat Dose: 3120 mg/kg/30D-C; Toxic Effects: Liver - Changes in liver weight; Kidney, Ureter, and Bladder - Changes in kidney weight; Nutritional and gross metabolic - Weight loss or decreased weight gain. Rat Dose: 22119 mg/kg/2Y-C; Toxic Effects: Behavioral - Food intake (animal); Liver - Changes in liver weight; Nutritional and gross metabolic - Weight loss or decreased weight gain.
Irritation Eye: Rabbit Standard Draize Test Dose: 500 mg/24H; Reaction: mild.

Hazard Overviews

Fire: Will burn.
Carcinogenicity: IARC - Not listed; NIOSH - Not listed; NTP - Not listed; ACGIH - Not listed; OSHA - Not listed; EPA - Not listed; MAK - Not listed

Environmental

Environmental Fate: If released to soil, it is expected to undergo rapid hydrolysis in moist soils. Hydrolysis is expected to be the primary removal process. It may display high mobility in soil although it is expected to hydrolyze before extensive leaching occurs. If released to water, the dominate fate process is expected to be hydrolysis. Neither bioconcentration, adsorption to sediment, nor volatilization to the atmosphere are expected to be important. If released to the atmosphere, it will exist in both vapor and particulate form. It may undergo atmospheric removal by both wet and dry deposition processes and the vapor-phase will react with photochemically generated hydroxyl radicals with an estimated half-life of 21 mos. It may also undergo hydrolysis during rain events or in clouds.
Cleanup/Disposal: Guide No. 171: Do not touch or walk through spilled material. Stop leak if you can do it without risk. Prevent dust cloud. Avoid inhalation of asbestos dust. Small Dry Spills: With clean shovel place material into clean, dry container and cover loosely; move containers from spill area. Small Spills: Take up with sand or other noncombustible absorbent material and place into containers for later disposal. Large Spills: Dike far ahead of liquid spill for later disposal. Cover powder spill with plastic sheet or tarp to minimize spreading. Prevent entry into waterways, sewers, basements or confined areas.

Environmental Physical Data

Henry's Law Constant: estimated at 4.98×10^{-10}
Sorption Partition Coefficient: K_{OC} = estimated at 90
BCF: calculated at 10

Regulations

RCRA 40CFR: Listed Hazardous Waste No. U366 Toxic Waste
CERCLA: 40CFR 302.4: Listed per RCRA Section 3001 RQ: 1 lb (0.454 kg)
SARA 40CFR 372.65: Listed
SARA EHS 40CFR 355: Not listed
TSCA: Listed

Analytical Methods

Water / Groundwater: EPA 1659

DAZ6000 CAS #: 53404-60-7

DAZOMET, SODIUM SALT

RTECS: XI2840000
Molecular Formula: $C_5H_9N_2NaS_2$
Formula Weight: 184.27

Hazard Overviews

Carcinogenicity: IARC - Not listed; NIOSH - Not listed; NTP - Not listed; ACGIH - Not listed; OSHA - Not listed; EPA - Not listed; MAK - Not listed

Environmental

Regulations

RCRA 40CFR: Not listed
CERCLA: 40CFR 302.4: Not listed
SARA 40CFR 372.65: Listed
SARA EHS 40CFR 355: Not listed
TSCA: Not listed

DB5000 CAS #: 1928-45-6

2,4-D, BUTOXYPROPYL ESTER

RTECS: AG7800000
DOT: UN2765; UN2766; UN2999; UN3000; NA2765; IMO3.2; IMO6.1
Molecular Formula: $C_{15}H_{20}Cl_2O_4$
Formula Weight: 351
Synonyms: ACETIC ACID,(2,4-DICHLOROPHENOXY)-,BUTOXY PROPYLENEDERIV; ACETIC ACID,(2,4-DICHLOROPHENOXY),2-BUTOXYMETHYLETHYLESTER; ACETIC ACID,(2,4-DICHLOROPHENOXY)-,3-BUTOXYPROPYL ESTER; ACETIC ACID,2,4-DICHLOROPHENOXY-,BUTOXYPROPYL ESTER; 2,4-D BUTOXYPROPYL ESTER; 2,4-D PGBE; 2,4-D PROPYLENE GLYCOL BUTYL ETHER ESTER; 2,4-DICHLOROPHENOXYACETIC ACID,PROPYLENE GLYCOL BUTYLETHER ESTER; DOW WEED KILLER X
Description: fuel oil-like odor

Physical Properties

Vapor Pressure: $< 1.5 \times 10^{-4}$ mm Hg at 25 °C
Water Solubility: Generally Immiscible or Insoluble in Water

Other Solubilities: Soluble in non-polar organic solvents such as Hexane, Benzene, Acetone, and Alcohols.
Flash Point: > 79.444 °C Open Cup

RTECS Toxicity Data

Acute Oral: Rat LD_{50} Dose: 500 mg/kg.
Reproductive/Teratogenic: Rat Route: Oral; Dose: 114 mg/kg; Duration: female 6-15D of pregnancy; Effects on Embryo or Fetus - Fetotoxicity; Specific Developmental Abnormalities - Urogenital system; Effects on Newborn - Growth statistics. Rat Route: Oral; Dose: 190 mg/kg; Duration: female 6-15D of pregnancy; Specific Developmental Abnormalities - Musculoskeletal system. Rat Route: Oral; Dose: 758 mg/kg; Duration: female 6-15D of pregnancy; Specific Developmental Abnormalities - Homeostasis.

Hazard Overviews

Fire: Combustible.
Carcinogenicity: IARC - Not listed; NIOSH - Not listed; NTP - Not listed; ACGIH - Not listed; OSHA - Not listed; EPA - Not listed; MAK - Not listed

Environmental

Ecotoxicity: LC_{50} Rainbow trout 1.1 mg/l/48 hr /pH & conditions of bioassay not specified LC_{50} Bluegill 1.0-1.2 mg/l/24 hr /pH & conditions of bioassay not specified
Cleanup/Disposal: Guide No. 131: Fully encapsulating, vapor protective clothing should be worn for spills and leaks with no fire. Eliminate all ignition sources (no smoking, flares, sparks or flames in immediate area). All equipment used when handling the product must be grounded. Do not touch or walk through spilled material. Stop leak if you can do it without risk. Prevent entry into waterways, sewers, basements or confined areas. A vapor suppressing foam may be used to reduce vapors. Small Spills: Absorb with earth, sand or other non-combustible material and transfer to containers for later disposal. Use clean non-sparking tools to collect absorbed material. Large Spills: Dike far ahead of liquid spill for later disposal. Water spray may reduce vapor; but may not prevent ignition in closed spaces. Guide No. 152: Do not touch damaged containers or spilled material unless wearing appropriate protective clothing. Stop leak if you can do it without risk. Prevent entry into waterways, sewers, basements or confined areas. Cover with plastic sheet to prevent spreading. Absorb or cover with dry earth, sand or other non-combustible material and transfer to containers. Do not get water inside containers.

Regulations

RCRA 40CFR: Not listed
CERCLA: 40CFR 302.4: Not listed
SARA 40CFR 372.65: Not listed
SARA EHS 40CFR 355: Not listed
TSCA: Not listed

DDD5000 CAS #: 72-54-8

DDD

RTECS: KI0700000
DOT: UN2761; NA2761; IMO6.1
EINECS Number: 200-783-0
Molecular Formula: $C_{14}H_{10}Cl_4$
Structured MF: $(ClC_6H_4)_2CHCHCl_2$
Formula Weight: 320.05

Cl Cl Cl Cl

Chemical Structure

Synonyms: BENZENE,1,1'-(2,2-DICHLOROETHYLIDENE)BIS(4-CHLORO-; 1,1-BIS(4-CHLOROPHENYL)-2,2-DICHLOROETHANE; 1,1-BIS(P-CHLOROPHENYL)-2,2-DICHLOROETHANE; 2,2-BIS(4-CHLOROPHENYL)-1,1-DICHLOROETHANE; 2,2-BIS(P-CHLOROPHENYL)-1,1-DICHLOROETHANE; 2,2-BIS(P-CHLOROPHENYL-1,1-DICHLOROETHANE; 4,4'-DDD; DDD; P,P'-DDD; 1,1-DICHLOOR-2,2-BIS(4-CHLOOR FENYL)-ETHAAN; 1,1-DICHLOR-2,2-BIS(4-CHLOR-PHENYL)-AETHAN; 1,1-DICHLOR-2,2-BIS(4-CHLORPHENYL)-AETHAN; 1,1-DICHLORO-2,2-BIS(4-CHLOROPHENYL)-ETHANE; 1,1-DICHLORO-2,2-BIS(4-CHLOROPHENYL)ETHANE; 1,1-DICHLORO-2,2-BIS(P-CHLOROPHENYL)ETHANE; 1,1-DICHLORO-2,2-BIS(PARACHLOROPHENYL)ETHANE; 1,1-DICHLORO-2,2-DI(4-CHLOROPHENYL)ETHANE; DICHLORODIPHENYL DICHLOROETHANE; P,P'-DICHLORODIPHENYLDICHLOROETHANE; P,P'-DICHLORODIPHENYL-2,2-DICHLOROETHYLENE; 1,1-DICLORO-2,2-BIS(4-CHLORO-FENIL)-ETANO; 1,1-DICLORO-2,2-BIS(4-CLORO-FENIL)-ETANO; DILENE; ENT 4,225; EPA PESTICIDE CODE 029101; ETHANE,1,1-DICHLORO-2,2-BIS(P-CHLOROPHENYL)-; ME-1700; ME-700; OMS 1078; RHOTHANE; RHOTHANE D-3; ROTHANE; P,P'-TDE; TDE; TETRACHLORODIPHENYLETHANE
Use: as a non-degradable pesticide and a nonsystemic contact and systemic insecticide

Physical Properties

Boiling Point: 193 °C (379 °F) at 1 mm Hg
Freezing Point: 109 °C (228.2 °F)
Specific Gravity: 1.385
Vapor Density: 11
Density: 1.385 g/mL
Vapor Pressure: 10.2 x10^{-7} torr at 30 °C
Water Solubility: 0.005 ppm at 0 °C
Other Solubilities: 95% Ethanol: 5-10 mg/ml at 18.5 °C; Acetone: 10-50 mg/ml at 18.5 °C; DMSO: >=100 mg/ml at 18.5 °C.
Flash Point: Not available; probably combustible

RTECS Toxicity Data

Acute Oral: Rat LD_{50} Dose: 113 mg/kg. Mouse LD_{Lo} Dose: 600 mg/kg; Toxic Effects: Behavioral - Tremor; Behavioral - Convulsions or effect on seizure threshold; Behavioral - Excitement.

Acute Dermal: Rabbit LD_{50} Route: Skin; Dose: 1200 mg/kg; Toxic Effects: Behavioral - Excitment; Behavioral - Convulsions or effect on seizure threshold; Skin and appendages - Primary irritation.

Mutagenic: Rat Cytogenetic Analysis; Cell Type: other cell types; Dose: 10 ug/L. Mouse Host-mediated Assay; Indicator Organism: Bacteria - S Marcescens; Dose: 1500 mg/kg.

Tumorigenic: Rat Route: Oral; Dose: 54 gm/kg/78W-C; Toxic Effects: Tumorigenic - Equivocal tumorigenic agent by RTECS criteria; Endocrine - Thyroid tumors. Mouse Route: Oral; Dose: 39 gm/kg/2Y-C; Toxic Effects: Tumorigenic - Neoplastic by RTECS criteria; Lungs, Thorax, or Respiration - Tumors; Liver - Tumors.

Hazard Overviews

Reactivity: Stable. Hazardous polymerization cannot occur. Avoid: exposure to ignition sources; alkalis. Incompatible with: alkalis. Hazardous decomposition products: chlorine gas.

Carcinogenicity: IARC - Group 2B, Possibly carcinogenic to humans; NIOSH - Not listed; NTP - Not listed; ACGIH - Not listed; OSHA - Not listed; EPA - Class B2, Probable human carcinogen based on animal studies; MAK - Not listed

Environmental

Ecotoxicity: EC_{50} Simocephalus (Daphnid) 4.5 ug/l/48 hr, first instar, at 15 °C (95% confidence limit 3.1-6.6 ug/l), static bioassay LC_{50} Gammarus lacustris 0.64 ug/l/96 hr /Conditions of bioassay not specified LC_{50} Stizostedion vitreum (Walleye)14 ug/l/96 hr, wt 1.0 g, at 18 °C (95% confidence limit 11-19 ug/l), static bioassay LD_{50} Phasianus colchicus (Ring-necked pheasant) female 3-4 months old oral 386 mg/kg (95% confidence limit 270-551 mg/kg) LC_{50} Colinus virginianus (Bobwhite quail) oral 2178 ppm, 5-day diet ad libitum, 23 days old (95% confidence limit 1835-2584 ppm) LC_{50} Gammarus fasciatus (Scud) 0.6 ug/l/96 hr, mature, at 21 °C (95% confidence limit 0.1-1.2 ug/l), static bioassay LC_{50} Monetes kadiakensis (Glass shrimp)2.4 ug/l/96 hr, mature, at 21 °C, static bioassay

Environmental Fate: If released to soil it will adsorb very strongly to the soil and will not be expected to appreciably leach to the groundwater, although its presence in certain groundwater samples illustrates that it can be transported there. It will not hydrolyze in the soil and biodegradation is expected to be slow. Indirect photolysis many be substantial based upon the behavior of the related compound DDT. If released in water it will be expected to strongly adsorb to sediment and to bioconcentrate in aquatic organisms. It will not hydrolyze or directly photodegrade and biodegradation is expected to be slow. It may be subject to evaporation with a half-life of 1.82 days predicted for evaporation from a river 1 m deep, flowing at 1 m/sec with a wind velocity of 3 m/sec; however, the expected adsorption to sediments may retard the evaporation process. If released to the atmosphere, it will not be expected to directly photolyze. The estimated vapor phase half-life in the atmosphere is 1.71 days as a result of reaction with photochemically produced hydroxyl radicals. Fallout and washout will be the major removal mechanisms from the air since it is expected to adsorb to particulate matter.

Cleanup/Disposal: Guide No. 151: Do not touch damaged containers or spilled material unless wearing appropriate protective clothing. Stop leak if you can do it without risk. Prevent entry into waterways, sewers, basements or confined areas. Cover with plastic sheet to prevent spreading. Absorb or cover with dry earth, sand or other non-combustible material and transfer to containers. Do not get water inside containers.

Environmental Physical Data

Henry's Law Constant: calculated at 2.14×10^{-5}
Octanol/Water Partition Coefficient: log K_{ow} = 6.02
Sorption Partition Coefficient: K_{oc} = 8.05×10^{4}
BCF: carp 2710

Regulations

RCRA 40CFR: Listed Hazardous Waste No. U060 Toxic Waste
CERCLA: 40CFR 302.4: Listed per CWA Section 311(b)(4) per RCRA Section 3001 per CWA Section 307(a) RQ: 1 lb (0.454 kg)
SARA 40CFR 372.65: Not listed
SARA EHS 40CFR 355: Not listed
TSCA: Not listed

Analytical Methods

Soil: CLP LC_PEST, MC_PEST, OHC; EPA PMD-TLC, 16, 3, 024, P-002-1, P-011-1; SW846 3630B, 3640A, 4042, 8080A, 8081, 8081A, 8250A, 8270B, 8270C; USGS O5104, O7104
Water / Groundwater: EPA P-003-1, P-004-1, 608, 617, 625, 625-S, 680, 022; APHA 6410-B, 6630-B, 6630-C, 6630-D; ASTM D3086, D4763; USGS O3104
Drinking Water: EPA 508, 508.1, 525.2; AOAC 990.06
Food: FDA 212.1, 212.2, 232.1, 232.4, 242.1; EPA 4; AOAC 970.52; USGS O9104
Plasma: EPA 003, 004, 027, 028, 029, 29; FDA 211.1, 231.1, 251.1, 251.2, 252, 253
Other: EPA P-009-1, 1656

DDE5000 CAS #: 72-55-9

DDE

RTECS: KV9450000
DOT: UN2761; IMO6.1
EINECS Number: 200-784-6
Molecular Formula: $C_{14}H_8Cl_4$
Structured MF: $(ClC_6H_4)_2C{=}CCl_2$
Formula Weight: 318.0

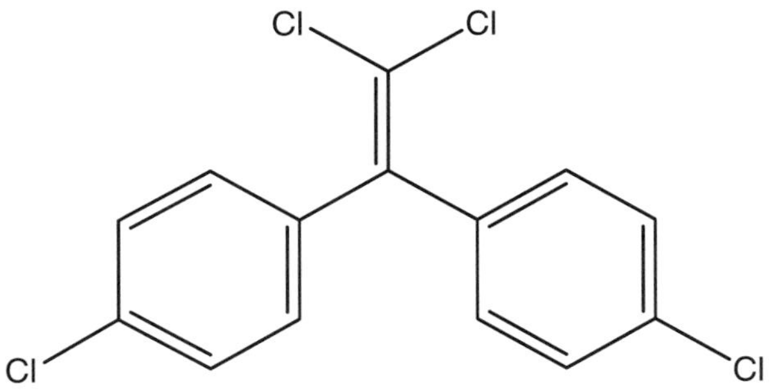

Chemical Structure

Synonyms: BENZENE,1,1'-(DICHLOROETHENYLIDENE)BIS(4-CHLORO-; 2,2-BIS(4-CHLOROPHENYL) 1,1-DICHLOROETHENE; 2,2-BIS(4-CHLOROPHENYL)-1,1-DICHLOROETHENE; 2,2-BIS(P-CHLOROPHENYL)-1,1-DICHLOROETHENE; 1,1-BIS(P-CHLOROPHENYL)-2,2-DICHLOROETHYLENE; 2,2-BIS(4-CHLOROPHENYL)-1,1-DICHLOROETHYLENE; 2,2-BIS(P-CHLOROPHENYL)-1,1-DICHLOROETHYLENE; 4,4'-DDE; P,P'-DDE; DDT DEHYDROCHLORIDE; 1,1-DICHLORO-2,2-BIS(PARA-CHLOROPHENYL) ETHYLENE; 1,1-DICHLORO-2,2-BIS(P-CHLOROPHENYL)ETHYLENE; 1,1-DICHLORO-2,2-DI(P-CHLOROPHENYL)ETHYLENE; P,P'-DICHLORODIPHENYL DICHLOROETHYLENE; DICHLORODIPHENYLDICHLOROETHYLENE; P,P'-DICHLORODIPHENYLDICHLOROETHYLENE; 1,1'-(DICHLOROETHENYLIDENE)BIS(4-CHLOROBENZENE); 1,1'-DICHLOROETHENYLIDENE)BIS(4-CHLOROBENZENE); ETHYLENE,1,1-DICHLORO-2,2-BIS(P-CHLOROPHENYL)-

Description: white, crystalline solid

Use: insecticide and military product

Physical Properties

Boiling Point: 316.5 °C (602 °F)

Freezing Point: 88.4 °C (191.12 °F)

Saturated Vapor Density: 1.200000102 kg/m^3

Vapor Pressure: 6.5 $x10^{-6}$ torr at 20 °C

Water Solubility: 0.01 ppm in Water

Other Solubilities: 95% Ethanol: 10-50 mg/ml at 21 °C; Acetone: 50-100 mg/ml at 21 °C; DMSO: 50-100 mg/ml at 21 °C; Fats: Soluble; Most organic solvents: Soluble.

Flash Point: Not available; probably combustible

RTECS Toxicity Data

Acute Oral: Rat LD_{50} Dose: 880 mg/kg. Mouse LD_{50} Dose: 700 mg/kg.

Chronic (Multiple Dose) Oral: Pigeon Dose: 586 mg/kg/28D-C; Toxic Effects: Behavioral - Food intake (animal); Liver - Changes in liver weight; DEATH. Bird Dose: 76160 ug/kg/8W-C; Toxic Effects: Automatic Nervous System - Sympathomimetic; Liver - Changes in liver weight; Biochemical - Dopamine at other sites.

Reproductive/Teratogenic: Rat Route: Intraperitoneal; Dose: 3500 ug/kg; Duration: female 7D prior to mating; Maternal Effects - Uterus, cervix, vagina.

Mutagenic: Rat DNA Damage; Cell Type: liver; Dose: 300 umol/L. Rat Cytogenetic Analysis; Cell Type: other cell types; Dose: 10 ug/L.

Tumorigenic: Mouse Route: Oral; Dose: 9700 mg/kg/78W-C; Toxic Effects: Tumorigenic - Carcinogenic by RTECS criteria; Liver - Tumors. Mouse Route: Oral; Dose: 28 gm/kg/80W-C; Toxic Effects: Tumorigenic - Neoplastic by RTECS criteria; Liver - Tumors. Mouse Route: Oral; Dose: 17 gm/kg/78W-C; Toxic Effects: Tumorigenic - Carcinogenic by RTECS criteria; Liver - Tumors.

Hazard Overviews

Health: May cause irritation. Toxic. Other Acute Effects: harmful if swallowed, inhaled, or absorbed through skin. Chronic Effects: carcinogen; may alter genetic material; may cause heritable genetic damage.

Fire: Will burn. Extinguishing agents: water spray; carbon dioxide, dry chemical powder or appropriate foam. Precautions: combustible liquid.

Reactivity: Incompatible with: strong oxidizing agents, strong bases. Hazardous decomposition products: toxic fumes of: carbon monoxide, carbon dioxide, hydrogen chloride gas.

Carcinogenicity: IARC - Group 2B, Possibly carcinogenic to humans; NIOSH - Not listed; NTP - Not listed; ACGIH - Not listed; OSHA - Not listed; EPA - Class B2, Probable human carcinogen based on animal studies; MAK - Not listed

Environmental

Ecotoxicity: LC_{50} Phasianus colchicus (ring-necked pheasant) oral (in 5-day diet) 829 ppm (95% confidence limit 746-922 ppm), 10 days old LC_{50} Salmo gairdneri (rainbow trout) 32 ug/l/96 hr at 12 °C, wt 0.8 g, static bioassay (95% confidence limit 26-40 ug/l) LC_{50} Salmo salar (Atlantic salmon) 96 ug/l/96 hr at 12 °C, wt 0.5 g, static bioassay (95% confidence limit 52-177 ug/l)

Environmental Fate: If released to soil it will adsorb very strongly to the soil and will not be expected to leach through soil to groundwater. It will not hydrolyze under normal environmental conditions and will probably not significantly biodegrade. Evaporation from the surface of soils with low organic content (such as sandy soils) may be significant, but adsorption to soils may reduce the rate of evaporation. If released to water it will adsorb very strongly to sediment, bioconcentrate in aquatic organisms, and be subject to photolysis with half-lives of 15-26 hours for photodegradation by environmentally significant wavelengths of light. It will not appreciably hydrolyze or biodegrade in water. Evaporation from water may be important with half-lives of 5.6-6.4 hours predicted for evaporation from a river 1 m deep, flowing at 1 m/sec with a wind velocity of 3 m/sec; however, the expected adsorption to sediments may retard the evaporation process. If released to the atmosphere it may be subject to direct photolysis. The estimated vapor phase half-life in the atmosphere is 4.63 hours as a result of reaction with photochemically produced hydroxyl radicals. The major translocation mechanisms in air is fallout and washout since it should adsorb to particulate matter.

Cleanup/Disposal: Guide No. 151: Do not touch damaged containers or spilled material unless wearing appropriate protective clothing. Stop leak if you can do it without risk. Prevent entry into waterways, sewers, basements or confined areas. Cover with plastic sheet to prevent spreading. Absorb or cover with dry earth, sand or other non-combustible material and transfer to containers. Do not get water inside containers.

Environmental Physical Data

Henry's Law Constant: 0.050 to 0.15
Octanol/Water Partition Coefficient: log K_{ow} = 6.51
Sorption Partition Coefficient: K_{oc} = 5 x10^4
BCF: fish 1.2 x10^4

Regulations

RCRA 40CFR: Not listed
CERCLA: 40CFR 302.4: Listed per CWA Section 307(a) RQ: 1 lb (0.454 kg)
SARA 40CFR 372.65: Not listed
SARA EHS 40CFR 355: Not listed
TSCA: Not listed

Analytical Methods

Air: EPA TO-10, TO-4, 016
Soil: CLP LC_PEST, MC_PEST, OHC; EPA PMD-TLC, 16, 3, 024, 025, P-002-1, P-011-1; SW846 3630B, 3640A, 4042, 8080A, 8081, 8081A, 8270B, 8270C; USGS O5104, O7104
Water / Groundwater: EPA P-003-1, P-004-1, 608, 617, 625, 625-S, 680, 022; APHA 6410-B, 6630-B, 6630-C, 6630-D; ASTM D3086; USGS O3104
Drinking Water: EPA 508, 508.1, 525.2; AOAC 990.06
Food: FDA 212.1, 212.2, 232.1, 232.4, 242.1; EPA 026, 4, XENO; AOAC 970.52; USGS O9104
Indoor / Expired Air: EPA IP-8; ASTM D4861
Plasma: EPA 001, 003, 004, 027, 028, 29; FDA 211.1, 231.1, 251.1, 251.2, 252, 253
Other: EPA P-009-1, 1656

DEC1000 CAS #: 17702-41-9

DECABORANE

RTECS: HD1400000
DOT: UN1868; IMO4.1
EINECS Number: 241-711-8
Molecular Formula: $B_{10}H_{14}$
Structured MF: $B_{10}H_{14}$
Formula Weight: 122.21

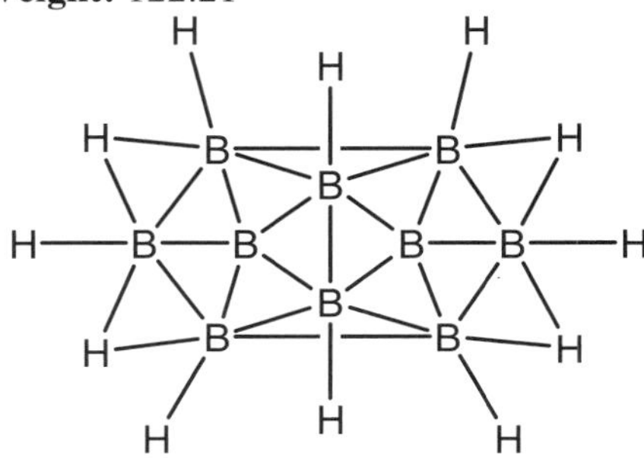

Chemical Structure

Synonyms: BORON HYDRIDE; BORON HYDRIDE (B10H14); DECABORANE(14); DECABORON TETRADECAHYDRIDE; NIDO-DECABORANE

Description: white to light yellow crystalline solid; garlic-like odor
Use: in rocket propellants; catalyst in olefin polymerization; high energy fuels; rubber vulcanizer; to coat metals; manufacture of plastics; oxygen scavenger; mothproofing, dye stripping, reducing & fluxing agent; stabilizer; rayon delustrant; fuel additive; a reducing agent in chemical synthesis

Physical Properties

Boiling Point: 213 °C (415 °F)
Freezing Point: 99.5 °C (211.1 °F)
Specific Gravity: Solid 0.94 at 25 °C; liquid
Vapor Density: 4.2 Air=1 at boiling point
Saturated Vapor Density: 1.200253748 kg/m^3
Vapor Pressure: 0.05 mm Hg at 25 °C
Water Solubility: Slightly Soluble in Cold Water
Other Solubilities: Soluble in Ethyl Acetate, 1-Bromopropane, Ethyl Silicate, Acetic Anhydride, Acetic Acid, Ethyl Borate, Carbon Tetrachloride; Soluble in hexane, Toluene; Soluble in Chloroform.
Odor Threshold: 0.36 mg/m^3
pH: Acid
Ionization Potential (eV): 9.88
Flash Point: 80 °C Closed Cup
Autoignition Temperature: 149 °C

RTECS Toxicity Data

Acute Oral: Rat LD_{50} Dose: 64 mg/kg. Mouse LD_{50} Dose: 41 mg/kg; Toxic Effects: Sense organs and special senses - Corneal damage; Behavioral - Convulsions or effect on seizure threshold; Lungs, Thorax, or Respiration - Cyanosis.
Acute Inhalation: Rat LC_{50} Dose: 46 ppm/4hr; Toxic Effects: Behavioral - Convulsions or effect on seizure threshold; Liver - Other changes; Kidney, ureter, and Bladder - Other changes. Mouse LC_{50} Dose: 12 ppm/4hr; Toxic Effects: Sense organs and special senses - Other; Behavioral - Somnolence (general depressed activity); Behavioral - Ataxia.
Acute Dermal: Rabbit LD_{50} Route: Skin; Dose: 71 mg/kg; Toxic Effects: Behavioral - Altered sleep time (including change in righting reflex); Behavioral - Ataxia; Lungs, Thorax, or Respiration - Dyspnea. Rat LD_{50} Route: Skin; Dose: 740 mg/kg; Toxic Effects: Peripheral Nerve and sensation - Flaccid paralysis without anesthesia; Behavioral - Convulsions or effect on seizure threshold; Lungs, Thorax, or Respiration - Dyspnea.

Hazard Overviews

Explosive

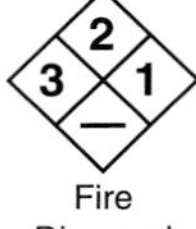

Fire Diamond

Health: Causes: affects nervous system causing narcosis/hyperexcitability; absorbed through skin, possible liver and kidney damage; symptoms may be delayed for 24 to 48 hours.

Fire: Combustible. Explosive potential on contact with heat or flame. Use dry chemical, carbon dioxide, or foam. Do not use water because it slowly forms flammable hydrogen gas. Do not use halogenated fire-extinguishing agents.

Reactivity: Stable. Hazardous polymerization cannot occur. Avoid: heat; ignition sources; halogenated or oxygenated solvents. Incompatible with: oxidizing agents; oxygenated solvents; dimethyl sulfoxide; ethers; halocarbons; carbon tetrachloride; alcohol; ammonia; hot water; moisture. Hazardous decomposition products: toxic fumes of boron oxides; hydrogen.

Carcinogenicity: IARC - Not listed; NIOSH - Listed as carcinogen; NTP - Not listed; ACGIH - Not listed; OSHA - Not listed; EPA - Not listed; MAK - Not listed

Primary Target Organs:

Nervous System

Liver

Kidneys

Exposure Limits

OSHA PEL: TWA: 0.05 ppm; 0.3 mg/m^3.

OSHA PEL Vacated 1989 Limits: TWA: 0.3 ppm; 0.05 mg/m^3; STEL: 0.9 ppm; 0.15 mg/m^3.

ACGIH TLV: TWA: 0.05 ppm; 0.25 mg/m^3; STEL: 0.15 ppm; 0.75 mg/m^3.

NIOSH REL: TWA: 0.05 ppm; 0.3 mg/m^3. STEL: 0.9 mg/m^3; skin.

NIOSH IDLH: 15 mg/m^3.

DFG MAK: TWA: 0.05 ppm; 0.03 mg/m^3.

Respirator Recommendation

Exposure Range: >0.3 to <15 mg/m^3 Supplied Air, Constant Flow/Pressure Demand, Half Mask

Exposure Range: 15 to unlimited mg/m^3 Self-contained Breathing Apparatus, Pressure Demand, Full Face

Note: poor warning properties

Environmental

Cleanup/Disposal: Guide No. 134: Fully encapsulating, vapor protective clothing should be worn for spills and leaks with no fire. Eliminate all ignition sources (no smoking, flares, sparks or flames in immediate area). Stop leak if you can do it without risk. Do not touch damaged containers or spilled material unless wearing appropriate protective clothing. Prevent entry into waterways, sewers, basements or confined areas. Use clean non-sparking tools to collect material and place it into loosely covered plastic containers for later disposal.

Environmental Physical Data

BCF: no food chain concentration potential

Regulations

RCRA 40CFR: Not listed

CERCLA: 40CFR 302.4: Not listed

SARA 40CFR 372.65: Not listed TPQ: 500/10000 lb

SARA EHS 40CFR 355: Listed TPQ: 500 lb

TSCA: Listed

Analytical Methods

Air: ASTM D4490

DEC1440 CAS #: 1163-19-5

DECABROMOBIPHENYL ETHER

RTECS: KN3525000

EINECS Number: 214-604-9

Molecular Formula: $C_{12}Br_{10}O$

Structured MF: $(Br_5C_6)_2O$

Formula Weight: 959.12

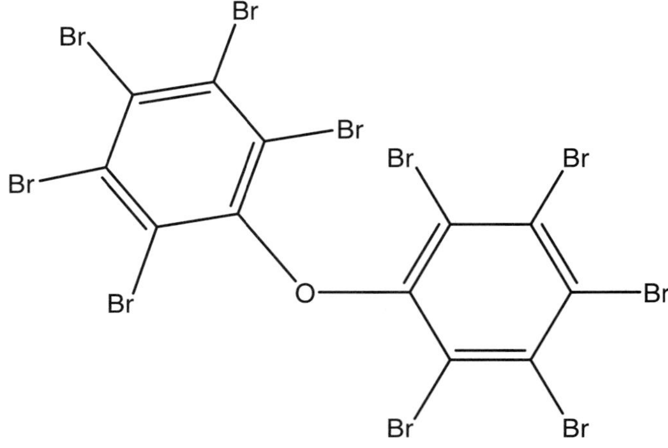

Chemical Structure

Synonyms: BENZENE,1,1'-OXYBIS(2,3,4,5,6-PENTABROMO-; BENZENE,1,1'-OXYBIS(2,3,4,5,6-PENTABROMO-(9CI); BERKFLAM B 10E; BIS(PENTABROMOPHENYL) ETHER; BIS-(PENTABROMOPHENYL ETHER); BIS(PENTABROMOPHENYL)ETHER; BR 55N; BROMKAL 82-0DE; BROMKAL 83-10DE; BROMKAL 82-ODE; DBDPO; DE 83R; DECABROM; DECABROMOBIPHENYL OXIDE; DECABROMODIPHENYL ETHER; DECABROMODIPHENYL OXIDE; DECABROMOPHENYL ETHER; DPBPO; ETHER,BIS(PENTABROMOPHENYL); ETHER,DECABROMODIPHENYL; FR 300; FR 300BA; FRP 53; 1,1'-OXYBIS(2,3,4,5,6-PENTABROMOBENZENE); PENTABROMOPHENYL ETHER; SAYTEX 102E; SAYTEX 102; TARDEX 100

Description: white to off-white powder; chemical odor

Use: flame retardant for plastics, thermoplastic polyesters, textiles and textile blends

Physical Properties

Boiling Point: Decomposes ~ 425 °C (797 °F)

Freezing Point: 295 °C (563 °F) to 305 °C (581 °F)

Specific Gravity: 3

Vapor Pressure: 5 mm Hg at 306 °C

Water Solubility: 20 to 30 ppb

Other Solubilities: 95% Ethanol: <1 mg/ml at 20 °C; Acetone: <1 mg/ml at 20 °C; Corn oil: Insoluble; DMSO: <1 mg/ml at 20 °C; Methanol: <1 mg/ml at 17 °C; Toluene: <1 mg/ml at 17 °C.

Flash Point: Not available; probably combustible

RTECS Toxicity Data

Acute Oral: Rat LD; Dose: >2 gm/kg.
Acute Dermal: Rat LD; Route: Skin; Dose: >3 gm/kg.
Chronic (Multiple Dose) Oral: Rat Dose: 24 gm/kg/30D-C; Toxic Effects: Liver - Other changes; Endocrine - Thyroid weight (goiter).
Reproductive/Teratogenic: Rat Route: Oral; Dose: 100 mg/kg; Duration: female 6-15D of pregnancy; Effects on Fertility - Post-implantation mortality.
Tumorigenic: Rat Route: Oral; Dose: 1092 gm/kg/2Y-C; Toxic Effects: Tumorigenic - Neoplastic by RTECS criteria; Liver - Tumors.

Hazard Overviews

Health: Irritating to eyes/skin/respiratory tract. Harmful. Other Acute Effects: harmful if swallowed, inhaled, or absorbed through skin. Chronic Effects: overexposure may cause reproductive disorder(s) based on tests with laboratory animals; possible risk of irreversible effects; possible teratogen; target organs: liver, kidneys, thyroid, spleen. Possible human carcinogen.
Fire: Will burn. Hazards: emits toxic fumes. Extinguishing agents: water spray; carbon dioxide, dry chemical powder or appropriate foam. Precautions: combustible liquid.
Reactivity: Incompatible with: strong oxidizing agents. Hazardous decomposition products: toxic fumes of: carbon monoxide, carbon dioxide, hydrogen bromide gas.
Carcinogenicity: IARC - Group 3, Not classifiable as to carcinogenicity to humans; NIOSH - Not listed; NTP - Not listed; ACGIH - Not listed; OSHA - Not listed; EPA - Class C, Possible human carcinogen; MAK - Not listed
Primary Target Organs:

Eyes

Skin

Respiratory System

Liver

Kidneys

Glandular System

Exposure Limits
AIHA WEEL: TWA: 5 mg/m^3.
Respirator Recommendation
Exposure Range: >5 to 250 mg/m^3 Supplied Air, Constant Flow/Pressure Demand, Half Mask
Exposure Range: >250 to 1000 mg/m^3 Supplied Air, Constant Flow/Pressure Demand, Full Face
Exposure Range: >1000 to unlimited mg/m^3 Self-contained Breathing Apparatus, Pressure Demand, Full Face
Note: odor threshold unknown

Environmental

Environmental Fate: If released to soil, it would be expected to adsorb strongly. If released to water, it would accumulate in the sediment. Photodegradation on the soil surface and surface layers of water may be fairly rapid. Limited data suggests some biodegradation in water. If released to the atmosphere, dust will be removed by gravitational settling. No bioconcentration in fish has been observed.

Environmental Physical Data

Henry's Law Constant: estimated at 4.45 x10^{-8}
Octanol/Water Partition Coefficient: log K_{ow} = 5.24
Sorption Partition Coefficient: K_{oc} = estimated at 3.3 x10^4
BCF: none in fish

Regulations

RCRA 40CFR: Not listed
CERCLA: 40CFR 302.4: Not listed
SARA 40CFR 372.65: Listed
SARA EHS 40CFR 355: Not listed
TSCA: Listed

DEC1880 **CAS #: 25152-84-5**

TRANS-TRANS-2,4-DECADIENAL

RTECS: HD3000000
EINECS Number: 246-668-9
Molecular Formula: $C_{10}H_{16}O$
Formula Weight: 152.26

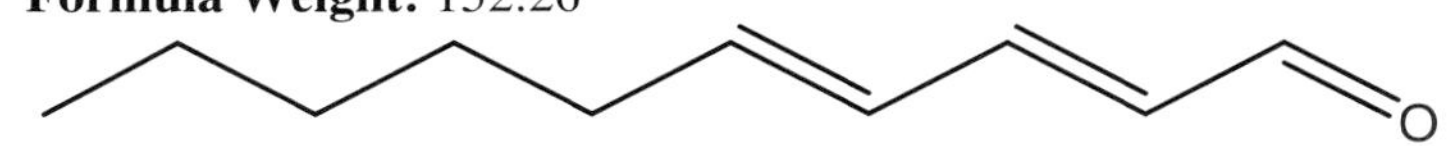

Chemical Structure

Synonyms: (2E,4E)-2,4-DECADIENAL; (2E,4E)-DECADIENAL; (E,E)-2,4-DECADIENAL; TRANS,TRANS-2,4-DECADIENAL

Physical Properties

Boiling Point: 114 °C (237.2 °F) to 116 °C (240.8 °F) at 10 mm Hg
Specific Gravity: 0.857

Hazard Overviews

Health: Irritating to eyes/skin/respiratory tract. Other Acute Effects: may be harmful by inhalation, ingestion, or skin absorption.
Fire: Extinguishing agents: water spray; carbon dioxide, dry chemical powder or appropriate foam. Precautions: combustible liquid.
Reactivity: Incompatible with: strong oxidizing agents, strong bases, reducing agents. Hazardous decomposition products: toxic fumes of: carbon monoxide, carbon dioxide.
Carcinogenicity: IARC - Not listed; NIOSH - Not listed; NTP - Not listed; ACGIH - Not listed; OSHA - Not listed; EPA - Not listed; MAK - Not listed
Primary Target Organs:

Eyes

Skin

Respiratory System

Environmental

Regulations

RCRA 40CFR: Not listed

CERCLA: 40CFR 302.4: Not listed
SARA 40CFR 372.65: Not listed
SARA EHS 40CFR 355: Not listed
TSCA: Listed

DEC2320 CAS #: 91-17-8

DECAHYDRONAPHTHALENE

RTECS: QJ3150000
DOT: UN1147; IMO3.3
EINECS Number: 202-046-9
Molecular Formula: $C_{10}H_{18}$
Structured MF: $C_{10}H_{18}$
Formula Weight: 138.24

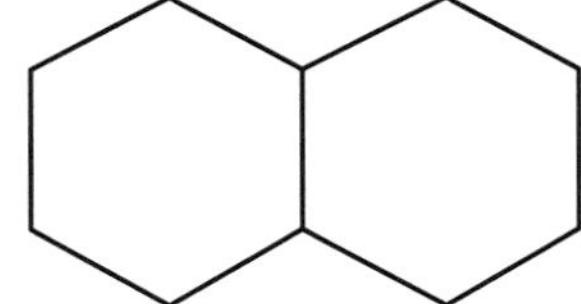

Chemical Structure

Synonyms: BICYCLO(4.4.0)DECANE; BICYCLO(4.4.O)DECANE; DEC; DECALIN; DE-KALIN; DEKALIN; DEKALINA; NAPHTHALANE; NAPHTHALENE,DECAHYDRO-; NAPHTHAN; NAPHTHANE; PERHYDRONAPHTHALENE
Description: clear, colorless liquid; slight methanol odor
Use: solvent for naphthalene, fats, resins, oils, waxes; in lacquers, shoe polishes, floor waxes; in motor fuel & lubricants; patent fuel in stoves; in research as pharmacological, noncarcinogenic vehicle in long-term skin painting & somatic mutation studies; cleaning machinery; stain remover; cleaning fluids

Physical Properties

Boiling Point: Trans 187.3 °C (369 °F)
Freezing Point: Trans -30.7 °C (-23.26 °F)
Vapor Density: 4.8 Air=1
Density: 55.5 lb/cu ft at 70 °F
Vapor Pressure: 0.123 lb/sq inch at 130 °F
Water Solubility: Insoluble
Other Solubilities: Solubility: > 10% in Acetone, Benzene, Ether & Ethanol /Cis- & trans-isomers/; Soluble in Chloroform /cis & trans isomers/; Slightly Soluble in Methanol /trans isomer/.
Surface Tension: 51.5 dynes/cm
Refraction Index: 1.4810
Flash Point: 58 °C Closed Cup
Autoignition Temperature: Trans isomer 250 °C
LEL: 0.7% v/v

RTECS Toxicity Data

Acute Oral: Rat LD_{50} Dose: 4170 mg/kg.
Acute Inhalation: Rat LC_{50} Dose: 710 ppm/4hr. Human TC_{Lo} Dose: 100 ppm; Toxic Effects: Sense organs and special senses - Other; Sense organs and special senses - Conjunctive irritation; Lungs, Thorax, or Respiration - Other changes.
Acute Dermal: Rabbit LD_{50} Route: Skin; Dose: 5900 uL/kg.
Chronic (Multiple Dose) Inhalation: Rat Dose: 5 ppm/24H/90D-C; Toxic Effects: Liver - Changes in Liver weight; Kidney, Ureter, and Bladder - Chgs in tubules (inc acute renal failure, acute tubular necrosis; Nutritional and gross metabolic - Weight loss or decreased weight gain. Rat Dose: 125 ppm/22H/31D-I; Toxic Effects: Liver - Changes in Liver weight; Kidney, Ureter, and Bladder - Chgs in tubules (inc acute renal failure, acute tubular necrosis; Kidney, Ureter, and Bladder - Changes in kidney weight.
Irritation Eye: Rabbit Standard Draize Test Dose: 500 mg/24H; Reaction: mild.
Irritation Skin: Rabbit Standard Draize Test Dose: 100 mg/24H; Reaction: moderate. Rabbit Open Draize Test Dose: 10 mg/24H open; Reaction: mild.
Tumorigenic: Rat Route: Inhalation; Dose: 5 ppm/24H/90D-C; Toxic Effects: Tumorigenic - Neoplastic by RTECS criteria; Endocrine - Tumors. Mouse Route: Inhalation; Dose: 50 ppm/24H/90D-C; Toxic Effects: Tumorigenic - Carcinogenic by RTECS criteria; Endocrine - Tumors.

Hazard Overviews

Flammable

Fire Diamond

Health: Irritating to skin. Also Causes: numbness, headache, CNS depression, vomiting, pulmonary edema, pneumonitis, hemorrhage. Chronic Effects: dermatitis, vesicular eczema.
Fire: Flammable. Use water fog, carbon dioxide, dry chemical, or standard foam. Container may explode in the heat of fire. Vapors may travel to a source of ignition and flash back.
Reactivity: Stable. Hazardous polymerization cannot occur. Avoid: heat; ignition sources. Incompatible with: oxidizing materials; steam. Hazardous decomposition products: acrid smoke; irritating fumes.
Carcinogenicity: IARC - Not listed; NIOSH - Not listed; NTP - Not listed; ACGIH - Not listed; OSHA - Not listed; EPA - Not listed; MAK - Not listed
Primary Target Organs:

Eyes

Skin

Respiratory System

Mucous Membranes

Environmental

Ecotoxicity: Mussel larvae: Mytilus edulis +20% reduction of growth rate at 10 ppm and 50 ppm +5% reduction of growth rate at 100 ppm
Environmental Fate: Should biodegrade in acclimated environments under the proper conditions. It is not expected to undergo hydrolysis or photolysis in the environment. A calculated K_{oc} range of 4700 to 9600 indicates that it will be slightly mobile to immobile in soil. In aquatic systems, it may

partition from the water column to organic matter contained in sediments and suspended solids. It has the potential to bioconcentrate in aquatic systems. A Henry's Law constant of 4.70×10^{-1} atm-cu m/mole at 25 °C suggests volatilization from environmental waters should be rapid. The volatilization half-lives from a model river and model pond, the latter considers the effect of adsorption, have been estimated to be 3.4 hr and 28.1 days, respectively. It is expected to exist entirely in the vapor phase in ambient air. Reaction with photochemically produced hydroxyl radicals (half-life of 20.3 hr) is likely to be an important fate process in the atmosphere.

Cleanup/Disposal: Guide No. 130: Eliminate all ignition sources (no smoking, flares, sparks or flames in immediate area). All equipment used when handling the product must be grounded. Do not touch or walk through spilled material. Stop leak if you can do it without risk. Prevent entry into waterways, sewers, basements or confined areas. A vapor suppressing foam may be used to reduce vapors. Absorb or cover with dry earth, sand or other non-combustible material and transfer to containers. Use clean non-sparking tools to collect absorbed material. Large Spills: Dike far ahead of liquid spill for later disposal. Water spray may reduce vapor; but may not prevent ignition in closed spaces.

Environmental Physical Data

Henry's Law Constant: calculated at 4.70×10^{-1}

Octanol/Water Partition Coefficient: log K_{ow} = 4.79

Sorption Partition Coefficient: K_{oc} = estimated at 4700 to 9600

BCF: estimated at 2.82

Regulations

RCRA 40CFR: Not listed

CERCLA: 40CFR 302.4: Not listed

SARA 40CFR 372.65: Not listed

SARA EHS 40CFR 355: Not listed

TSCA: Listed

DEC2760 **CAS #: 112-31-2**

DECALDEHYDE

RTECS: HD6000000

EINECS Number: 203-957-4

Molecular Formula: $C_{10}H_{20}O$

Structured MF: $CH_3(CH_2)_8CHO$

Formula Weight: 156.27

O

Chemical Structure

Synonyms: ALDEHYDE C-10; ALDEHYDE C10; C-10 ALDEHYDE; CAPRALDEHYDE; CAPRIC ALDEHYDE; CAPRINALDEHYDE; CAPRINIC ALDEHYDE; CAPRYLALDEHYDE; N-DECALDEHYDE; 1-DECANAL; DECANAL; N-DECANAL; DECANALDEHYDE; 1-DECYL ALDEHYDE; DECYL ALDEHYDE; N-DECYL ALDEHYDE; DECYLIC ALDEHYDE

Description: colorless to light yellow liquid; strong orange-rose odor that changes to a fresh citrus odor when diluted

Use: synthetic flavoring substance and adjuvant; fragrance ingredient; for citrus tones; manufacture of synthetic citrus oils; intermediate in the synthesis of pharmaceuticals and in the polymer and pesticide fields; in blossom fragrances (especially to create citrus nuances); production of artificial citrus oils

Physical Properties

Boiling Point: 208.5 °C (407 °F)

Freezing Point: -5 °C (23 °F)

Specific Gravity: 0.83 at 15 °C/4 °C

Vapor Pressure: 0.120 kPa at 50 °C

Water Solubility: Insoluble in Water

Other Solubilities: Soluble in Ethanol, Ether, Acetone; Soluble in mineral oils, fixed oils, volatile oils; Insoluble in Glycerol; Soluble in most organic solvents.

Surface Tension: 28.0 dynes/cm at 24 °C

Odor Threshold: 0.168 ppm

Refraction Index: 1.4287

Flash Point: 84 °C

RTECS Toxicity Data

Acute Oral: Rat LD_{50} Dose: 3730 mg/kg. Mouse LD_{50} Dose: >41750 mg/kg; Toxic Effects: Behavioral - Excitment; Gastrointestinal - Hypermotility, diarrhea; Skin and appendages - Hair.

Acute Dermal: Rabbit LD_{50} Route: Skin; Dose: 5040 mg/kg.

Irritation Skin: Rabbit Standard Draize Test Dose: 500 mg/24H; Reaction: mild. Rabbit Open Draize Test Dose: 14372 ug/24H open; Reaction: severe.

Mutagenic: Bacteria - B Subtilis DNA Repair; Dose: 5 mg/disc.

Hazard Overviews

Health: Severely irritating to eyes/skin/respiratory tract. Other Acute Effects: harmful if swallowed, inhaled, or absorbed through skin; symptoms of exposure may include burning sensation; coughing; wheezing; laryngitis; shortness of breath; headache; nausea; vomiting.

Fire: Combustible. Extinguishing agents: water spray; carbon dioxide, dry chemical powder or appropriate foam. Precautions: combustible liquid.

Reactivity: Incompatible with: strong oxidizing agents, strong bases, strong reducing agents. Hazardous decomposition products: toxic fumes of: carbon monoxide, carbon dioxide.

Carcinogenicity: IARC - Not listed; NIOSH - Not listed; NTP - Not listed; ACGIH - Not listed; OSHA - Not listed; EPA - Not listed; MAK - Not listed

Primary Target Organs:

Eyes

Skin

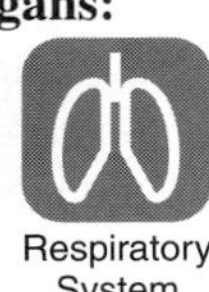
Respiratory System

Environmental

Environmental Fate: If released to the atmosphere, it will mainly exist in the vapor phase based on a experimental vapor pressure of 0.10 mm Hg at 25 °C. Vapor-phase is degraded in the atmosphere by reaction with photochemically produced hydroxyl radicals with an estimated half-life of about 11 hours. An estimated K_{oc} of 200 suggests that it will have moderate mobility in soil. Volatilization from dry and moist soil surfaces may occur. Based on limited data, this compound may biodegrade in both soil and water. After 5 days incubation with a sewage inoculum, 22% of the theoretical BOD was reached. In water, it should not adsorb to suspended matter in the water column based on its K_{oc} value. It will volatilize from water surfaces given a measured Henry's Law constant of 5.87 x10^{-5} atm-cu m/mole. Estimated half-lives for a model river and model lake are 4 hours and 5 days, respectively. Bioconcentration in aquatic organisms may occur based on an estimated BCF value of 420.

Cleanup/Disposal: Guide No. 129: Eliminate all ignition sources (no smoking, flares, sparks or flames in immediate area). All equipment used when handling the product must be grounded. Do not touch or walk through spilled material. Stop leak if you can do it without risk. Prevent entry into waterways, sewers, basements or confined areas. A vapor suppressing foam may be used to reduce vapors. Absorb or cover with dry earth, sand or other non-combustible material and transfer to containers. Use clean non-sparking tools to collect absorbed material. Large Spills: Dike far ahead of liquid spill for later disposal. Water spray may reduce vapor; but may not prevent ignition in closed spaces.

Environmental Physical Data

Henry's Law Constant: 5.87 x10^{-5}
Octanol/Water Partition Coefficient: log K_{ow} = 3.76
Sorption Partition Coefficient: K_{oc} = 200
BCF: estimated at 420
BOD: 22% BODT, 5 days

Regulations

RCRA 40CFR: Not listed
CERCLA: 40CFR 302.4: Not listed
SARA 40CFR 372.65: Not listed
SARA EHS 40CFR 355: Not listed
TSCA: Listed

Analytical Methods

Soil: SW846 8315A
Drinking Water: EPA 554

DEC3200 CAS #: 156-74-1

DECAMETHONIUM

RTECS: BP5948000
Molecular Formula: $C_{16}H_{38}N_2$
Formula Weight: 258.49
Synonyms: AMMONIUM,DECAMETHYLENEBIS(TRIMETHYL-; 1,10-DECANEDIAMINIUM,N,N,N,N',N',N'-HEXAMETHYL-
Description: crystals
Use: medication: skeletal muscle relaxant

Physical Properties

Boiling Point: Decomposes at 268 °C (514 °F) to 270 °C (518 °F)
Freezing Point: Decomposes at 268 °C (514.4 °F) to 270 °C (518 °F)
Water Solubility: Freely Soluble in Water
Other Solubilities: freely Soluble in Alcohol; Slightly Soluble in Chloroform; Insoluble in Ether.

Hazard Overviews

Carcinogenicity: IARC - Not listed; NIOSH - Not listed; NTP - Not listed; ACGIH - Not listed; OSHA - Not listed; EPA - Not listed; MAK - Not listed

Environmental

Regulations

RCRA 40CFR: Not listed
CERCLA: 40CFR 302.4: Not listed
SARA 40CFR 372.65: Not listed
SARA EHS 40CFR 355: Not listed
TSCA: Not listed

DEC3640 CAS #: 541-02-6

DECAMETHYLCYCLOPENTASIL-OXANE

RTECS: GY5945200
EINECS Number: 208-764-9
Molecular Formula: $C_{10}H_{30}O_5Si_5$
Formula Weight: 370.80

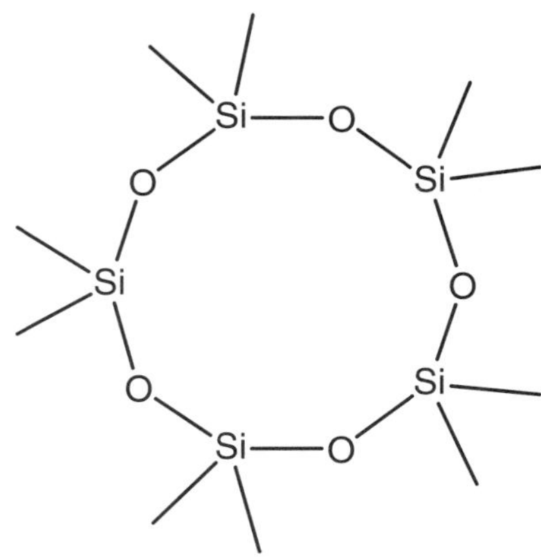

Chemical Structure

Synonyms: CYCLIC DIMETHYLSILOXANE PENTAMER; CYCLOPENTASILOXANE,DECAMETHYL-; DEKAMETHYLCYKLOPENTASILOXAN; DIMETHYLSILOXANE PENTAMER; DOW CORNING 345; DOW CORNING 345 FLUID; KF 995; NUC SILICONE VS 7158; SF 1202; SILICON SF 1202; UNION CARBIDE 7158 SILICONE FLUID; VS 7158

Description: oily liquid

Use: in hair preparations; antiperspirants; intermediates or byproducts in production of silicone fluids, elastomers, and resins; chem int for silicone fluids & elastomers; component of silicone fluid mixtures; high-purity silicone fluid (eg, for cosmetics); a vehicle and end blocking agent in antiperspirants and in aerosol products containing insoluble powders

Physical Properties

Boiling Point: 210 °C (410 °F)
Freezing Point: -38 °C (-36.4 °F)
Specific Gravity: 0.9593 at 20 °C/4 °C
Vapor Pressure: 2.50 kPa at 100 °C
Water Solubility: 0.24 mg/L at 25 °C(est)
Refraction Index: 1.3982 at 20 °C/D

RTECS Toxicity Data

Acute Oral: Rat LD_{50} Dose: >64 mL/kg.
Acute Dermal: Rabbit LD_{50} Route: Skin; Dose: >16 mL/kg.
Chronic (Multiple Dose) Inhalation: Rat Dose: 3060 mg/m^3/6H/4W-I; Toxic Effects: Liver - Changes in Liver weight; Blood - Other changes; Biochemical - Lipids including transport.
Irritation Eye: Rabbit Standard Draize Test Dose: 500 mg/24H; Reaction: mild.
Irritation Skin: Rabbit Standard Draize Test Dose: 500 mg/24H; Reaction: mild.

Hazard Overviews

Health: Irritating to eyes/skin/respiratory tract. Other Acute Effects: may be harmful by inhalation, ingestion, or skin absorption.
Fire: Hazards: emits toxic fumes. Extinguishing agents: carbon dioxide, dry chemical powder or appropriate foam. Precautions: combustible liquid.

Reactivity: Stable. Hazardous polymerization will not occur. Incompatible with: strong oxidizing agents. Hazardous decomposition products: toxic fumes of: carbon monoxide, carbon dioxide, silicon oxide, formaldehyde.
Carcinogenicity: IARC - Not listed; NIOSH - Not listed; NTP - Not listed; ACGIH - Not listed; OSHA - Not listed; EPA - Not listed; MAK - Not listed
Primary Target Organs:

Eyes

Skin
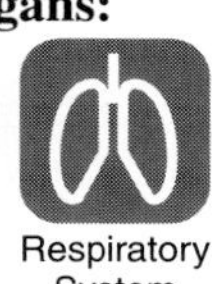
Respiratory System

Environmental

Environmental Fate: If released to soil, it may rapidly volatilize to the atmosphere, although its potential for strong adsorption to soil may attenuate the rate of this process. It is expected to be stable to hydrolysis except in clay soils. If released to water, it is expected to rapidly volatilize to the atmosphere. Its volatilization half-life from a model river has been estimated to be approximately 6 hours. It is resistant to biodegradation and will not hydrolyze in water. An estimated bioconcentration factor of 5000 indicates that it has the potential to bioconcentrate in fish. If released to the atmosphere, it will be oxidized by the gas-phase reaction with photochemically produced hydroxyl radicals with a half-life of 10 days. Oxidation by ozone and nitrate radicals and direct photolysis are not expected to be significant.

Environmental Physical Data

Henry's Law Constant: estimated at 0.4
Octanol/Water Partition Coefficient: log K_{ow} = 5.20
Sorption Partition Coefficient: K_{oc} = estimated at 1.6 $x10^4$
BCF: estimated at 5300

Regulations

RCRA 40CFR: Not listed
CERCLA: 40CFR 302.4: Not listed
SARA 40CFR 372.65: Not listed
SARA EHS 40CFR 355: Not listed
TSCA: Listed

DEC4520 CAS #: 124-18-5

N-DECANE

RTECS: HD6550000
DOT: UN2247; IMO3.3
EINECS Number: 204-686-4
Molecular Formula: $C_{10}H_{22}$
Structured MF: $CH_3(CH_2)_8CH_3$
Formula Weight: 142.28

Chemical Structure

Synonyms: DECANE; DECYL HYDRIDE

Description: colorless liquid; mild aliphatic hydrocarbon odor
Use: organic synthesis; solvent; standardized hydrocarbon; jet fuel research; manufacturing paraffin products; rubber industry; paper processing industry; constituent in polyolefin manufacturing wastes; and gasoline component

Physical Properties

Boiling Point: 174.1 °C (345 °F) at 760 mm Hg
Freezing Point: -29.7 °C (-21.46 °F)
Specific Gravity: 0.73 at 20 °C/4 °C
Vapor Density: 4.9 Air=1
Vapor Pressure: 1 mm Hg at 16.5 °C
Water Solubility: 0.052 mg/L in Water at 25 °C
Other Solubilities: 95% Ethanol: >=100 mg/ml at 21 °C; Acetone: >=100 mg/ml at 21 °C; DMSO: 1-5 mg/ml at 21 °C; Ether: Soluble.
Surface Tension: 25.67 dyne/cm
Odor Threshold: 11.3 mg/m^3
Refraction Index: 1.409 at 25 °C
Critical Temperature: 344.4 °C
Critical Pressure: 20.8 atm
Ionization Potential (eV): 9.65 +/-0.10
Flash Point: 46.111 °C Closed Cup
Autoignition Temperature: 210 °C
LEL: 0.8% v/v
UEL: 5.4% v/v

RTECS Toxicity Data

Acute Inhalation: Mouse LC_{50} Dose: 72300 mg/m^3/2hr; Toxic Effects: Behavioral - Antipsychotic.
Tumorigenic: Mouse Route: Skin; Dose: 25 gm/kg/52W-I; Toxic Effects: Tumorigenic - Equivocal tumorigenic agent by RTECS criteria; Skin and appendages - Tumors; Tumorigenic - Tumors at site of application.

Hazard Overviews

Flammable

Fire Diamond

Health: Irritating to eyes/skin/respiratory tract. Also Causes: drying, cracking, and possible dermatitis
Fire: Flammable. Use carbon dioxide, dry chemical, or foam. Water is ineffective in extinguishing n-decane fires, and a water stream will spread flames; but a water spray should be used to cool fire-exposed containers to prevent pressure rupture. Fight fire from a safe distance. Vapors can flow along surfaces, reach a distant ignition source, and flash back to the original source of the material.
Reactivity: Stable. Hazardous polymerization cannot occur. Avoid: heat; ignition sources; confined areas. Incompatible with: oxidizing materials.
Carcinogenicity: IARC - Not listed; NIOSH - Not listed; NTP - Not listed; ACGIH - Not listed; OSHA - Not listed; EPA - Not listed; MAK - Not listed

Primary Target Organs:

Eyes

Skin

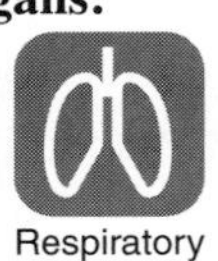
Respiratory System

Nervous System

Environmental

Ecotoxicity: Mussel larvae: Mytilus edulis no significant alteration of growth rate at 10 ppm +80% increase of growth rate at 50 to 100 ppm
Environmental Fate: Photolysis or hydrolysis is not expected to be environmentally important. Biodegradation may occur in soil and water; however, volatilization and adsorption are expected to be far more important environmental fate processes. A high K_{oc} indicates it will be immobile in soil and may partition from the water column to organic matter contained in sediments and suspended solids. Bioconcentration may be important in aquatic systems. A Henry's Law Constant of 5.15 atm-cu m/mole at 25 °C, suggests rapid volatilization from environmental waters. The volatilization half lives from a model river and a model pond, the latter considers the effect of adsorption, have been estimated to be 3.5 hr and 130 days, respectively. It is expected to exist almost entirely in the vapor phase in ambient air. Reactions with photochemically produced hydroxyl radicals in the atmosphere have been shown to be important (estimated half life of 1.4 days).
Cleanup/Disposal: Guide No. 128: Eliminate all ignition sources (no smoking, flares, sparks or flames in immediate area). All equipment used when handling the product must be grounded. Do not touch or walk through spilled material. Stop leak if you can do it without risk. Prevent entry into waterways, sewers, basements or confined areas. A vapor suppressing foam may be used to reduce vapors. Absorb or cover with dry earth, sand or other non-combustible material and transfer to containers. Use clean non-sparking tools to collect absorbed material. Large Spills: Dike far ahead of liquid spill for later disposal. Water spray may reduce vapor; but may not prevent ignition in closed spaces.

Environmental Physical Data

Henry's Law Constant: calculated at 5.15
Octanol/Water Partition Coefficient: log K_{ow} = 5.98
Sorption Partition Coefficient: K_{oc} = estimated at 2.22 x10^4 to 4.27 x10^4
BCF: estimated at 3.52

Regulations

RCRA 40CFR: Not listed
CERCLA: 40CFR 302.4: Not listed
SARA 40CFR 372.65: Not listed
SARA EHS 40CFR 355: Not listed
TSCA: Listed

Analytical Methods

Soil: EPA 1625
Water / Groundwater: EPA 1625

DEC4960 CAS #: 143-10-2

1-DECANETHIOL

EINECS Number: 205-584-2
Molecular Formula: $C_{10}H_{22}S$
Structured MF: $CH_3(CH_2)_9SH$
Formula Weight: 174.3

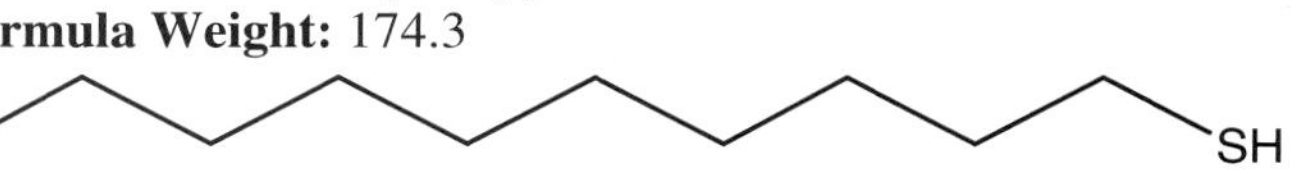

Chemical Structure

Synonyms: DECYLMERCAPTAN; N-DECYLMERCAPTAN; 1-MERCAPTODECANE
Description: colorless liquid; strong odor

Physical Properties

Boiling Point: 240.56 °C (465 °F)
Freezing Point: -26.11 °C (-14.998 °F)
Specific Gravity: 0.84
Vapor Pressure: 2.81 kPa at 125 °C
Water Solubility: Insoluble
Other Solubilities: Ethanol: Soluble; Ether: Soluble
Refraction Index: 1.4509
Flash Point: 98 °C

Hazard Overviews

Health: Irritating to eyes/skin/respiratory tract. Other Acute Effects: nausea; headache; vomiting; may be harmful by inhalation, ingestion, or skin absorption.
Fire: Will burn. Hazards: emits toxic fumes. Extinguishing agents: water spray; carbon dioxide, dry chemical powder or appropriate foam. Precautions: combustible liquid.
Reactivity: Incompatible with: strong oxidizing agents, strong bases. Hazardous decomposition products: toxic fumes of: carbon monoxide, carbon dioxide, sulfur oxides.
Carcinogenicity: IARC - Not listed; NIOSH - Listed as carcinogen; NTP - Not listed; ACGIH - Not listed; OSHA - Not listed; EPA - Not listed; MAK - Not listed
Primary Target Organs:

Eyes

Skin

Respiratory System

Exposure Limits
NIOSH REL: STEL: 0.5 ppm; 3.6 mg/m^3; 15-minute.

Environmental

Cleanup/Disposal: Guide No. 131: Fully encapsulating, vapor protective clothing should be worn for spills and leaks with no fire. Eliminate all ignition sources (no smoking, flares, sparks or flames in immediate area). All equipment used when handling the product must be grounded. Do not touch or walk through spilled material. Stop leak if you can do it without risk. Prevent entry into waterways, sewers, basements or confined areas. A vapor suppressing foam may be used to reduce vapors. Small Spills: Absorb with earth, sand or other non-combustible material and transfer to containers for later disposal. Use clean non-sparking tools to collect absorbed material. Large Spills: Dike far ahead of liquid spill for later disposal. Water spray may reduce vapor; but may not prevent ignition in closed spaces.

Regulations

RCRA 40CFR: Not listed
CERCLA: 40CFR 302.4: Not listed
SARA 40CFR 372.65: Not listed
SARA EHS 40CFR 355: Not listed
TSCA: Not listed

DEC5400 CAS #: 112-30-1

DECANOL

RTECS: HE4375000
EINECS Number: 203-956-9
Molecular Formula: $C_{10}H_{22}O$
Structured MF: $CH_3(CH_2)_8CH_2OH$
Formula Weight: 158.28

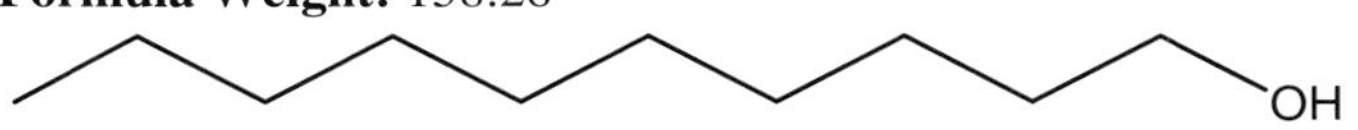

Chemical Structure

Synonyms: AGENT 504; ALCOHOL C-10; ALCOHOL C10; ALFOL 10; ANTAK; C 10 ALCOHOL; CAPRIC ALCOHOL; CAPRINIC ALCOHOL; 1-DECANOL; N-DECAN-1-OL; N-DECANOL; N-DECATYL ALCOHOL; DECYL ALCOHOL; N-DECYL ALCOHOL; DECYLIC ALCOHOL; DYTOL S-91; EPAL 10; LOROL 22; NONYL CARBINOL; N-NONYLCARBINOL; NONYLCARBINOL; PRIMARY DECYL ALCOHOL; ROYALTAC; ROYALTAC M-2; ROYALTAC-85; SIPOL L10; T 148; T-148
Description: colorless to light yellow liquid; pleasant, sweet odor
Use: synthetic flavoring ingredient; antifoaming agent; intermed in mfr of perfumes, detergents, for specialty surfactants in textiles; plant growth regulator; in mfg of plasticizers, synthetic lubricants, petro additives, herbicides, surface active agents, solvents; solvent in electrodes to detect calcium; perfume ind raw material for detergents & defoaming agent; synthetic flavor in non-alcoholic beverages; ice cream, ices, candy, etc.

Physical Properties

Boiling Point: 232.9 °C (451 °F) at 760 mm Hg
Freezing Point: 6.4 °C (43.52 °F)
Specific Gravity: 0.8297 at 20 °C/4 °C
Vapor Density: 5.3 Air=1
Vapor Pressure: 1 mm Hg at 69.5 °C
Water Solubility: Insoluble in Water
Other Solubilities: 1:3 in 60% Alcohol; Soluble in Glacial Acetic Acid.
Surface Tension: 0.029742 N/m at melting point
Odor Threshold: 2.1 x10^{-1} mg/m^3
Refraction Index: 1.4372 at 20 °C/D

Critical Temperature: 427 °C
Critical Pressure: 22 atm
Flash Point: 82.2 °C Open Cup
Autoignition Temperature: 288 °C
LEL: 0.7% v/v
UEL: 5.5% v/v

RTECS Toxicity Data

Acute Oral: Rat LD_{50} Dose: 4720 mg/kg; Toxic Effects: Behavioral - Somnolence (general depressed activity); Lungs, Thorax, or Respiration - Dyspnea. Mouse LD_{50} Dose: 6500 mg/kg.
Acute Inhalation: Mouse LC_{50} Dose: 4 gm/m^3/2hr. Mammal LC_{50} Dose: 3 gm/m^3.
Acute Dermal: Rabbit LD_{50} Route: Skin; Dose: 3560 mg/kg.
Irritation Eye: Rabbit Standard Draize Test Dose: 500 mg/24H; Reaction: mild. Rabbit Standard Draize Test Dose: 83 mg; Reaction: severe.
Irritation Skin: Human Standard Draize Test Dose: 75 mg/3D-I; Reaction: severe. Rabbit Standard Draize Test Dose: 20 mg/24H; Reaction: moderate. Rabbit Standard Draize Test Dose: 2600 mg/kg/24H; Reaction: moderate.
Reproductive/Teratogenic: Rat Route: Oral; Dose: 35295 mg/kg; Duration: female 1-15D of pregnancy; Effects on Fertility - Female fertility index; Pre-implantation mortality; Post-implantation mortality. Rat Route: Oral; Dose: 35295 mg/kg; Duration: female 1-15D of pregnancy; Effects on Embryo or Fetus - Fetotoxicity; Effects on Newborn - Biochemical and metabolic.
Tumorigenic: Mouse Route: Skin; Dose: 12 gm/kg/25W-I; Toxic Effects: Tumorigenic - Equivocal tumorigenic agent by RTECS criteria; Skin and appendages - Tumors; Tumorigenic - Tumors at site of application.

Hazard Overviews

Fire Diamond

Health: Irritating to eyes/skin/respiratory tract. Also Causes: headache, flushed face, nausea, vomiting, CNS depression, mental confusion, lack of coordination. Chronic Effects: dermatitis.
Fire: Combustible. Use foam, carbon dioxide, or dry chemical. Do not use water to extinguish fire because material floats on the water and the water spray may spread the fire. Remove cylinder from fire.
Reactivity: Stable. Hazardous polymerization cannot occur. Avoid: heat; ignition sources; oxidizing materials. Incompatible with: oxidizing materials. Hazardous decomposition products: carbon dioxide; carbon monoxide; irritating smoke; fumes.
Carcinogenicity: IARC - Not listed; NIOSH - Not listed; NTP - Not listed; ACGIH - Not listed; OSHA - Not listed; EPA - Not listed; MAK - Not listed

Primary Target Organs:

Eyes

Skin

Respiratory System

Mucous Membranes

Nervous System

Environmental

Ecotoxicity: LC_{50} Pimephales promelas (fathead minnow) 2.4 mg/l 96 hr flow-through bioassay, wt 0.12 g, water hardness 45.5 mg/l calcium carbonate temp: 25 + or - 1 °C, pH 7.5, dissolved oxygen greater than 60% of saturation.
Environmental Fate: If released to the atmosphere, it will degrade in the vapor phase by reaction with photochemically produced hydroxyl radicals (estimated half-life of about 1 day). If released to soil or water, it is expected to biodegrade. Various biological screening studies have demonstrated that it biodegrades readily. Some transport from water to air may occur via volatilization.
Cleanup/Disposal: Guide No. 127: Eliminate all ignition sources (no smoking, flares, sparks or flames in immediate area). All equipment used when handling the product must be grounded. Do not touch or walk through spilled material. Stop leak if you can do it without risk. Prevent entry into waterways, sewers, basements or confined areas. A vapor suppressing foam may be used to reduce vapors. Absorb or cover with dry earth, sand or other non-combustible material and transfer to containers. Use clean non-sparking tools to collect absorbed material. Large Spills: Dike far ahead of liquid spill for later disposal. Water spray may reduce vapor; but may not prevent ignition in closed spaces.

Environmental Physical Data

Henry's Law Constant: estimated at 4.8 x10^{-5}
Octanol/Water Partition Coefficient: log K_{ow} = 4.57
Sorption Partition Coefficient: K_{oc} = estimated at 600
BCF: estimated at 80
BOD: 29.3%, 5 days

Regulations

RCRA 40CFR: Not listed
CERCLA: 40CFR 302.4: Not listed
SARA 40CFR 372.65: Not listed
SARA EHS 40CFR 355: Not listed
TSCA: Listed

Analytical Methods

Water / Groundwater: ASTM D3695

DEC5840 CAS #: 872-05-9

1-DECENE

EINECS Number: 212-819-2
Molecular Formula: $C_{10}H_{20}$
Structured MF: CH_2=CH$(CH_2)_7CH_3$
Formula Weight: 140.27

Chemical Structure

Synonyms: 1-N-DECENE; ALPHA-DECENE; N-1-DECENE; DECYLENE; N-DECYLENE
Description: colorless liquid; pleasant odor
Use: monomer for polyalpha-olefin synthetic lubricants; in linear alpha-olefin mixt for prodn of primary alcohols; chem int for oxo-process n-undecyl alcohol; organic synthesis of flavors, perfumes, pharmaceuticals, dyes, oils, resins

Physical Properties

Boiling Point: 170.56 °C (339 °F) at 760 mm Hg
Freezing Point: -66.3 °C (-87.34 °F)
Specific Gravity: 0.7408 at 20 °C/4 °C
Vapor Density: 4.84 Air=1
Vapor Pressure: 1 mm Hg at 95.7 °C
Water Solubility: Insoluble in Water
Other Solubilities: Very Soluble in Ether.
Surface Tension: 28 dynes/cm
Odor Threshold: ~ 7 ppm
Refraction Index: 1.4215 at 20 °C/D
Evaporation Rate: < 1 Butyl Acetate=1
Ionization Potential (eV): 9.42 +/-0.01
Flash Point: < 55 °C
Autoignition Temperature: 235 °C

Hazard Overviews

Flammable

Fire Diamond

Health: Irritating to eyes/skin/respiratory tract. Other Acute Effects: may be harmful by inhalation, ingestion, or skin absorption.
Fire: Flammable. Hazards: vapor may travel considerable distance to source of ignition and flash back. Extinguishing agents: carbon dioxide, dry chemical powder or appropriate foam. Precautions: combustible liquid.
Reactivity: Incompatible with: strong oxidizing agents. Hazardous decomposition products: toxic fumes of: carbon monoxide, carbon dioxide.
Carcinogenicity: IARC - Not listed; NIOSH - Not listed; NTP - Not listed; ACGIH - Not listed; OSHA - Not listed; EPA - Not listed; MAK - Not listed
Primary Target Organs:

Eyes

Skin

Respiratory System

Environmental

Cleanup/Disposal: Guide No. 128: Eliminate all ignition sources (no smoking, flares, sparks or flames in immediate area). All equipment used when handling the product must be grounded. Do not touch or walk through spilled material. Stop leak if you can do it without risk. Prevent entry into waterways, sewers, basements or confined areas. A vapor suppressing foam may be used to reduce vapors. Absorb or cover with dry earth, sand or other non-combustible material and transfer to containers. Use clean non-sparking tools to collect absorbed material. Large Spills: Dike far ahead of liquid spill for later disposal. Water spray may reduce vapor; but may not prevent ignition in closed spaces.

Regulations

RCRA 40CFR: Not listed
CERCLA: 40CFR 302.4: Not listed
SARA 40CFR 372.65: Not listed
SARA EHS 40CFR 355: Not listed
TSCA: Listed

DEC6280 CAS #: 25339-53-1

DECENE

EINECS Number: 246-870-7
Molecular Formula: $C_{10}H_{20}$
Formula Weight: 140.27
Use: in the organic synthesis of flavors, perfumes, pharmaceuticals, dyes, oils, and resins

Hazard Overviews

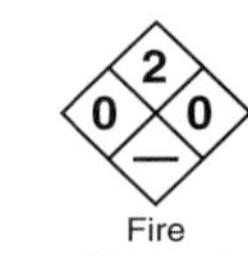

Fire Diamond

Carcinogenicity: IARC - Not listed; NIOSH - Not listed; NTP - Not listed; ACGIH - Not listed; OSHA - Not listed; EPA - Not listed; MAK - Not listed

Environmental

Regulations

RCRA 40CFR: Not listed
CERCLA: 40CFR 302.4: Not listed
SARA 40CFR 372.65: Not listed
SARA EHS 40CFR 355: Not listed
TSCA: Not listed

DEC7160 CAS #: 127-33-3

DECLOMYCIN

RTECS: QI7650000
EINECS Number: 204-834-8
Molecular Formula: $C_{21}H_{21}ClN_2O_8$
Formula Weight: 464.88
Synonyms: BIOTERCICLIN; 7-CHLORO-6-DEMETHYLTETRACYCLINE; CHLORTETRACYCLINE,6-DEMETHYL-; CLORTETRIN; DEGANOL; DEMECLOCYCLINE; DEMECLOR; 6-DEMETHYL-7-CHLOROTETRACYCLINE; 6-

DEMETHYLCHLOROTETRACYCLINE; DEMETHYLCHLOROTETRACYCLINE; DEMETHYLCHLORTETRACYCLIN; 6-DEMETHYL-7-CHLORTETRACYCLINE; 6-DEMETHYLCHLORTETRACYCLINE; DEMETHYLCHLORTETRACYCLINE; DEMETHYLCHLORTETRACYCLINE BASE; 6-DEMETIL-7-CLOROTETRACICLINA; DEMETRACLIN; DIUCICLIN; DMCT; ELKAMICINA; LEDERMYCIN; METHYLCHLORTETRACYCLINE; MEXOCINE; 2-NAPHTHACENECARBOXAMIDE,7-CHLORO-4-(DIMETHYLAMINO)-1,4,4A,5,5A,6,11,12A-OCTAHYDRO-3,6,10,12,12A-PENTAHYDROXY-1,11-DIOXO-; NOVOTRICLINA; PERCICLINA; PERICICLINA; RP 10192; SUMACLINA; TETRACYCLINE,7-CHLORO-6-DEMETHYL-; TRI-DEMETHYLCHLORTETRACYCLINE

Description: yellow crystalline powder; odorless

Use: medication: antibacterial; vet: antimicrobial

Physical Properties

Freezing Point: 174 °C (345.2 °F) to 178 °C (352.4 °F)

Water Solubility: Sparingly Soluble in Water

Other Solubilities: 1 g Soluble in about 980 ml Alcohol; Sparingly Soluble in alkali hydroxides and carbonate solution; Practically Insoluble in Acetone & Chloroform.

pH: About 4.8

RTECS Toxicity Data

Acute Oral: Human TD_{Lo} Dose: 420 mg/kg/6W; Toxic Effects: Endocrine - Diabetis Insipidus (nephrogenic or CNS); Biochemical - Effect on cyclic nucleotides. Human TD_{Lo} Dose: 10 mg/kg; Toxic Effects: Skin and appendages - Dermatitis, other; Skin and appendages - Nails; Immunological including allergic - Othr immed. (hmrl): urticaria, allergic rhinitis, serum sickness. Human TD_{Lo} Dose: 68 mg/kg/8D; Toxic Effects: Kidney, Ureter, and Bladder - Urine volume increased; Kidney, Ureter, and Bladder - Other changes in urine composition. Rat LD_{50} Dose: >6750 mg/kg.

Acute Dermal: Rat LD_{50} Route: Subcutaneous Dose: >2 gm/kg. Mouse LD_{Lo} Route: Subcutaneous Dose: 2500 mg/kg.

Reproductive/Teratogenic: Rat Route: Oral; Dose: 5 gm/kg; Duration: female 1-22D of pregnancy; Specific Developmental Abnormalities - Craniofacial (including nose and tongue); Skin and skin appendages; Effects on Newborn - Weaning or lactation index. Rat Route: Oral; Dose: 5 gm/kg; Duration: female 1-22D of pregnancy; Effects on Newborn - Other postnatal measures or effects.

Mutagenic: Human DNA Inhibition; Cell Type: lymphocyte; Dose: 3750 ug/L.

Hazard Overviews

Carcinogenicity: IARC - Not listed; NIOSH - Not listed; NTP - Not listed; ACGIH - Not listed; OSHA - Not listed; EPA - Not listed; MAK - Not listed

Environmental

Regulations

RCRA 40CFR: Not listed

CERCLA: 40CFR 302.4: Not listed

SARA 40CFR 372.65: Not listed

SARA EHS 40CFR 355: Not listed

TSCA: Not listed

DEC7600 CAS #: 2156-96-9

DECYL ACRYLATE

RTECS: AS7400000

EINECS Number: 218-462-9

Molecular Formula: $C_{13}H_{24}O_2$

Structured MF: $CH_3(CH_2)_9OCOCH{=}CH_2$

Formula Weight: 212.37

Synonyms: ACRYLIC ACID,DECYL ESTER; N-DECYL ACRYLATE; 2-PROPENOIC ACID,DECYL ESTER; 2-PROPENOIC ACID,DECYL ESTER (9CI)

Description: liquid

Use: research chemical

Physical Properties

Boiling Point: 158 °C (316 °F) at 50 mm Hg

Freezing Point: < 0 °C (32 °F)

Specific Gravity: 0.8781 at 20 °C/4 °C

Vapor Density: Estimate 7.3 Air=1

Vapor Pressure: Reid < 0.01 mm Hg

Water Solubility: Very Slightly Soluble in Water

Other Solubilities: React readily with electrophilic, free-radical, and nucleophilic agent

Refraction Index: 1.4400 at 20 °C/D

Flash Point: 227 °C Open Cup

RTECS Toxicity Data

Acute Oral: Rat LD_{50} Dose: 6460 mg/kg.

Acute Dermal: Rabbit LD_{50} Route: Skin; Dose: 6300 mg/kg.

Irritation Skin: Rabbit Open Draize Test Dose: 10 mg/24H open; Reaction: severe.

Hazard Overviews

Fire Diamond

Fire: Will burn.

Carcinogenicity: IARC - Not listed; NIOSH - Not listed; NTP - Not listed; ACGIH - Not listed; OSHA - Not listed; EPA - Not listed; MAK - Not listed

Environmental

Cleanup/Disposal: Guide No. 152: Do not touch damaged containers or spilled material unless wearing appropriate protective clothing. Stop leak if you can do it without risk. Prevent entry into waterways, sewers, basements or confined areas. Cover with plastic sheet to prevent spreading. Absorb or cover with dry earth, sand or other non-combustible material and transfer to containers. Do not get water inside containers.

Regulations

RCRA 40CFR: Not listed

CERCLA: 40CFR 302.4: Not listed

SARA 40CFR 372.65: Not listed
SARA EHS 40CFR 355: Not listed
TSCA: Listed

DEC8040 CAS #: 110-29-2

N-DECYL N-OCTYL ADIPATE

EINECS Number: 203-754-0
Molecular Formula: $C_{24}H_{46}O_4$
Formula Weight: 398.54
Synonyms: ADIPIC ACID,DECYL OCTYL ESTER; ADIPOL ODY; DECYL OCTYL ADIPATE; HERCOFLEX 290; HEXANEDIOIC ACID,DECYL OCTYL ESTER; MONOPLEX NODA; N-OCTYL DECYL ADIPATE; OCTYL DECYL ADIPATE; PX-202; STAFLEX NODA; TRUFLEX 146
Description: liquid; mild odor
Use: low-temperature plasticizer for cellulose acetate butyrate,cellulose acetate & nitrate, ethyl cellulose & polystyrene, polyvinyl butyral & chloride, nitrile-butadiene rubbers, natural rubbers, styrene-butadiene rubbers

Physical Properties

Boiling Point: 220 °C (428 °F) to 254 °C (489 °F) at 4 mm Hg
Freezing Point: -50 °C (-58 °F)
Specific Gravity: 0.92 to 0.98 at 20 °C/20 °C
Refraction Index: 1.447

Hazard Overviews

Carcinogenicity: IARC - Not listed; NIOSH - Not listed; NTP - Not listed; ACGIH - Not listed; OSHA - Not listed; EPA - Not listed; MAK - Not listed

Environmental

Regulations

RCRA 40CFR: Not listed
CERCLA: 40CFR 302.4: Not listed
SARA 40CFR 372.65: Not listed
SARA EHS 40CFR 355: Not listed
TSCA: Listed

DEC8480 CAS #: 104-72-3

DECYLBENZENE

EINECS Number: 203-230-1
Molecular Formula: $C_{16}H_{26}$
Structured MF: $C_6H_5(CH_2)_9CH_3$
Formula Weight: 218.1

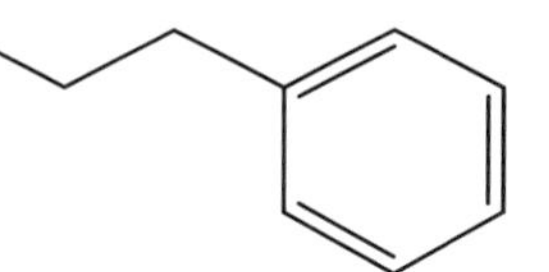

Chemical Structure

Synonyms: BENZENE,DECYL-; DECANE,1-PHENYL-; N-DECYLBENZENE; 1-PHENYLDECANE
Description: colorless liquid

Physical Properties

Boiling Point: 255 °C (491 °F) to 280 °C (536 °F)
Freezing Point: -14 °C (6.8 °F)
Specific Gravity: 0.9
Water Solubility: Insoluble in Water
Other Solubilities: Acetone: Very Soluble; Benzene: Very Soluble; Ether: Very Soluble; Ethanol: Very Soluble
Surface Tension: 29.95 dynes/cm at 20 °C
Refraction Index: 1.4832
Flash Point: 107.222 °C

Hazard Overviews

Health: Irritating to eyes/skin/respiratory tract. Other Acute Effects: may be harmful by inhalation, ingestion, or skin absorption.
Fire: Will burn. Extinguishing agents: water spray; carbon dioxide, dry chemical powder or appropriate foam. Precautions: combustible liquid.
Reactivity: Incompatible with: strong oxidizing agents. Hazardous decomposition products: toxic fumes of: carbon monoxide, carbon dioxide.
Carcinogenicity: IARC - Not listed; NIOSH - Not listed; NTP - Not listed; ACGIH - Not listed; OSHA - Not listed; EPA - Not listed; MAK - Not listed
Primary Target Organs:

Eyes

Skin

Respiratory System

Environmental

Cleanup/Disposal: Guide No. 153: Eliminate all ignition sources (no smoking, flares, sparks or flames in immediate area). Do not touch damaged containers or spilled material unless wearing appropriate protective clothing. Stop leak if you can do it without risk. Prevent entry into waterways, sewers, basements or confined areas. Absorb or cover with dry earth, sand or other non-combustible material and transfer to containers. Do not get water inside containers.

Regulations

RCRA 40CFR: Not listed
CERCLA: 40CFR 302.4: Not listed
SARA 40CFR 372.65: Not listed
SARA EHS 40CFR 355: Not listed
TSCA: Listed

DEC8920 CAS #: 140-60-3

4-DECYLBENZENESULFONIC ACID

EINECS Number: 205-422-0
Molecular Formula: $C_{16}H_{26}O_3S$

Formula Weight: 298.45
Synonyms: BENZENESULFONIC ACID,4-DECYL-; BENZENESULFONIC ACID,P-DECYL-; P-N-DECYLBENZENESULFONATE; P-DECYLBENZENESULFONIC ACID

Hazard Overviews

Carcinogenicity: IARC - Not listed; NIOSH - Not listed; NTP - Not listed; ACGIH - Not listed; OSHA - Not listed; EPA - Not listed; MAK - Not listed

Environmental

Regulations

RCRA 40CFR: Not listed
CERCLA: 40CFR 302.4: Not listed
SARA 40CFR 372.65: Not listed
SARA EHS 40CFR 355: Not listed
TSCA: Listed

DEE5000 CAS #: 134-62-3

DEET

RTECS: XS3675000
EINECS Number: 205-149-7
Molecular Formula: $C_{12}H_{17}NO$
Structured MF: $CH_3C_6H_4CON(C_2H_5)_2$
Formula Weight: 191.26

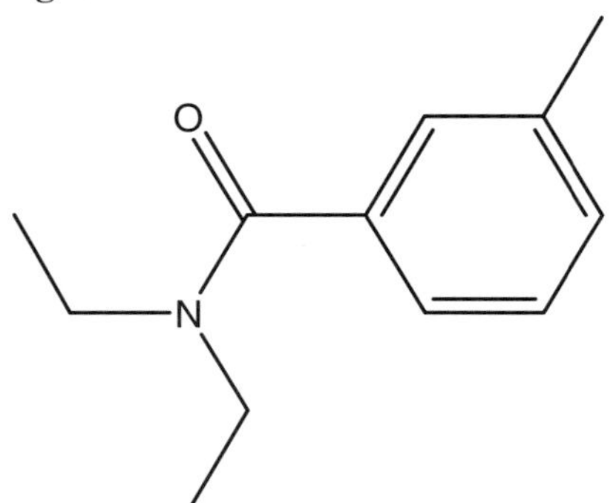

Chemical Structure

Synonyms: AI 3-22542; AUTAN; BAKER'S ANTIFOL; BENZAMIDE,N,N-DIETHYL-3-METHYL-; CHEMFORM; DELPHENE; M-DELPHENE; DET; M-DET; DET (INSECT REPELLANT); DETA; DETA-20; M-DETA; DETAMIDE; DIELTAMID; DIETHYL TOLUAMIDE; N,N-DIETHYL-3-METHYLBENZAMIDE; DIETHYL-M-TOLUAMIDE; DIETHYLTOLUAMIDE; N,N-DIETHYL-M-TOLUAMIDE; ENT 20,218; ENT 20218; ENT 22542; FLYPEL; METADELPHENE; 3-METHYL-N,N-DIETHYLBENZAMIDE; MGK; MGK DIETHYLTOLUAMIDE; NAUGATUCK DET; OFF; REPEL; REPPER-DET; REPUDIN-SPECIAL; M-TOLUAMIDE,N,N-DIETHYL-; M-TOLUIC ACID DIETHYLAMIDE
Description: colorless liquid; mild bland odor
Use: insect repellent, as a resin solvent and in film formers; to impregnate clothing and as a leech repellent

Physical Properties

Boiling Point: 160 °C (320 °F) at 19 mm Hg
Freezing Point: -45 °C (-49 °F)
Specific Gravity: 0.996 at 20 °C/4 °C
Vapor Density: 6.7 Air=1
Density: 0.996 g/cc at 20 °C
Vapor Pressure: 1 mm Hg at 111 °C
Water Solubility: < 1 mg/mL at 20 C
Other Solubilities: miscible with Ethanol, 2-Propanol, Cottonseed Oil, Propylene Glycol.
Refraction Index: 1.5212 at 20 °C/D
Evaporation Rate: < 1 Butyl Acetate=1
Flash Point: 155 °C

RTECS Toxicity Data

Acute Oral: Woman TD_{Lo} Dose: 950 mg/kg; Toxic Effects: Sense organs and special senses - Mydriasis (pupilliary dilation); Behavioral - Coma; Lungs, Thorax, or Respiration - Other changes. Child TD_{Lo} Dose: 4750 mg/kg; Toxic Effects: Behavioral - Convulsions or effect on seizure threshold; Behavioral - Stiffness; Behavioral - Coma. Man LD_{Lo} Dose: 679 mg/kg. Woman LD_{Lo} Dose: 950 mg/kg; Toxic Effects: Behavioral - Coma; Lungs, Thorax, or Respiration - Other changes; Gastrointestinal - Nausea or vomiting. Rat LD_{50} Dose: 1950 mg/kg.
Acute Inhalation: Rat LC_{50} Dose: 5950 mg/m^3; Toxic Effects: Nutritional and gross metabolic - Weight loss or decreased weight gain.
Acute Dermal: Human TD_{Lo} Route: Skin; Dose: 35 mg/kg/5D; Toxic Effects: Skin and appendages - Dermatitis, other. Rabbit LD_{50} Route: Skin; Dose: 3180 uL/kg.
Chronic (Multiple Dose) Oral: Rat Dose: 120 gm/kg/29W-C; Toxic Effects: Liver - Changes in liver weight; Kidney, Ureter, and Bladder - Changes in kidney weight; DEATH - Changes in testicular weight. Rabbit Dose: 7920 mg/kg/15D-I; Toxic Effects: Blood - Changes in serum composition; Nutritional and gross metabolic - Weight loss or decreased weight gain; Nutritional and gross metabolic - Changes in calcium.
Chronic (Multiple Dose) Inhalation: Rat Dose: 1500 mg/m^3/13W-C; Toxic Effects: Liver - Changes in Liver weight; Kidney, Ureter, and Bladder - Changes in kidney weight.
Irritation Eye: Rabbit Standard Draize Test Dose: 10 mg; Reaction: moderate.
Irritation Skin: Rabbit Standard Draize Test Dose: 500 mg; Reaction: moderate.
Reproductive/Teratogenic: Rat Route: Oral; Dose: 7500 mg/kg; Duration: female 6-15D of pregnancy; Maternal Effects - Other effects on females; Effects on Embryo or Fetus - Fetotoxicity. Rat Route: Skin; Dose: 19 gm/kg; Duration: female 1-19D of pregnancy; Effects on Newborn - Viability index.
Mutagenic: Rat Sperm Morphology; Route: Inhalation; Dose: 1500 mg/m^3.

Hazard Overviews

Fire Diamond

Health: Irritating to eyes. Also Causes: CNS disturbances, drop in blood pressure, dermatitis. Toxicity is primarily neurologic and may occur via ingestion or skin exposure.
Fire: Will burn. Use dry chemical, carbon dioxide, water spray, or foam.
Reactivity: Stable. Hazardous polymerization cannot occur. Hazardous decomposition products: toxic fumes of nitrogen oxides.
Carcinogenicity: IARC - Not listed; NIOSH - Not listed; NTP - Not listed; ACGIH - Not listed; OSHA - Not listed; EPA - Not listed; MAK - Not listed
Primary Target Organs:

Eyes

Skin

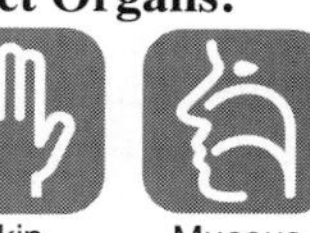
Mucous Membranes

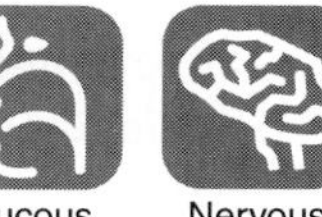
Nervous System

Environmental

Ecotoxicity: LC_{50} Pimephales promelas (fathead minnow) 110 mg/l/96 hr at 25 °C with a water hardness of 45.0 mg/l $CaCO_3$
Environmental Fate: If released to the atmosphere, it will mainly exist in the vapor phase in the ambient atmosphere based on a measured vapor pressure of 5.6 $x10^{-3}$ mm Hg at 20 °C. Vapor-phase is degraded in the atmosphere by reaction with photochemically produced hydroxyl radicals with an estimated half-life of about 15 hours. An estimated K_{oc} of 300 suggests that it will have moderate mobility in soil. Volatilization from dry and moist soil surfaces should not be a major fate process for this compound. Based on limited data, this compound should not readily biodegrade under either aerobic or anaerobic conditions in both soil and water. In water, it may adsorb to suspended matter in the water column based on its K_{oc} value. It is not expected to volatilize from water surfaces given an estimated Henry's Law constant of 7.9 $x10^{-7}$ atm-cu m/mole. Bioconcentration in aquatic organisms should not occur based on BCF values of 0.8-2.4, measured in carp.

Environmental Physical Data

Henry's Law Constant: estimated at 2.1 $x10^{-8}$
Octanol/Water Partition Coefficient: log K_{ow} = 2.02
Sorption Partition Coefficient: K_{oc} = estimated at 300
BCF: carp 0.8 to 2.4

Regulations

RCRA 40CFR: Not listed
CERCLA: 40CFR 302.4: Not listed
SARA 40CFR 372.65: Not listed
SARA EHS 40CFR 355: Not listed
TSCA: Listed

Analytical Methods

Soil: EPA PMD-DEE
Water / Groundwater: EPA 633

DEH1000 CAS #: 1446-61-3

DEHYDROABIETYLAMINE

RTECS: TP8701000
EINECS Number: 215-899-7
Molecular Formula: $C_{20}H_{31}N$
Formula Weight: 315.48

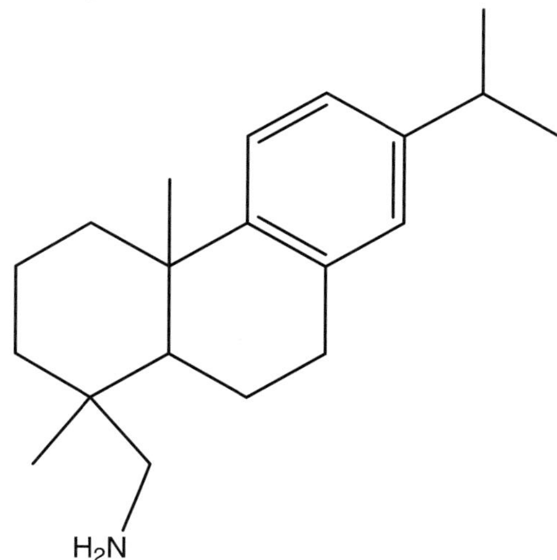

Chemical Structure

Synonyms: AMINE D; 13-ISOPROPYLPODOCARPA-8,11,13-TRIEN-15-AMINE; 1-PHENANTHRENEMETHANAMINE,1,2,3,4,4A,9,10,10A-OCTAHYDRO-1,4A-DIMETHYL-7-(1-METHYLETHYL)-,(1R-(1ALPHA,4ABETA,10AALPHA))-; PODOCARPA-8,11,13-TRIEN-15-AMINE,13-ISOPROPYL-
Description: pale yellow viscous liquid
Use: production of bactericides; fungicides; corrosion inhibitors; asphalt additives; flotation

Physical Properties

Freezing Point: 317.65 °C (603.77 °F)
Specific Gravity: 1.546 at 20 °C/technical
Vapor Pressure: 4.2 $x10^{-6}$ mm Hg
Surface Tension: 3.38 $x10^{-2}$ N/m at 44.65 °C
Critical Temperature: 590 °C
Critical Pressure: 1.7 $x10^{6}$ Pa

Hazard Overviews

Health: Irritating to eyes/skin/respiratory tract. Other Acute Effects: may be harmful by inhalation, ingestion, or skin absorption.
Fire: Hazards: emits toxic fumes. Extinguishing agents: water spray; carbon dioxide, dry chemical powder or appropriate foam. Precautions: combustible liquid.
Reactivity: Incompatible with: strong oxidizing agents. Hazardous decomposition products: toxic fumes of: carbon monoxide, carbon dioxide, nitrogen oxides.

Carcinogenicity: IARC - Not listed; NIOSH - Not listed; NTP - Not listed; ACGIH - Not listed; OSHA - Not listed; EPA - Not listed; MAK - Not listed

Primary Target Organs:

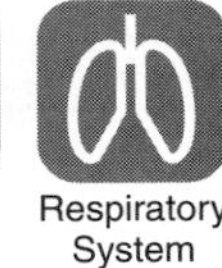

Eyes Skin Respiratory System

Environmental

Regulations

RCRA 40CFR: Not listed
CERCLA: 40CFR 302.4: Not listed
SARA 40CFR 372.65: Not listed
SARA EHS 40CFR 355: Not listed
TSCA: Listed

DEH5000 CAS #: 520-45-6

DEHYDROACETIC ACID

RTECS: UP8050000
EINECS Number: 208-293-9
Molecular Formula: $C_8H_8O_4$
Formula Weight: 168.14

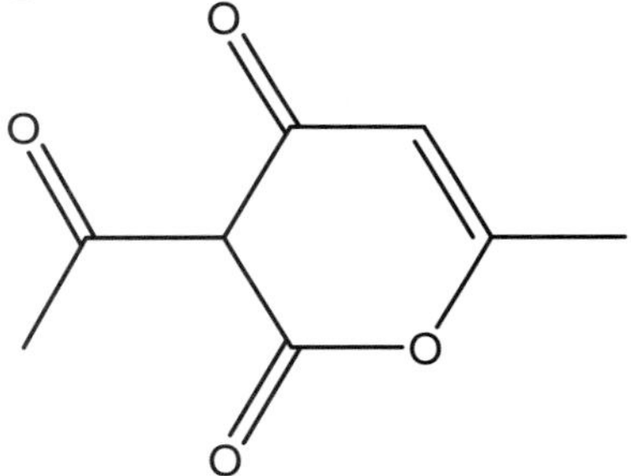

Chemical Structure

Synonyms: ACETIC ACID,DEHYDRO-; 3-ACETYL-4-HYDROXY-6-METHYL-2H-PYRAN-2-ONE; 2-ACETYL-5-HYDROXY-3-OXO-4-HEXENOIC ACID DELTA-LACTONE; 3-ACETYL-6-METHYLDIHYDROPYRANDIONE-2,4; 3-ACETYL-6-METHYL-2,4-PYRANDIONE; 3-ACETYL-6-METHYL-2H-PYRAN-2,4(3H)-DIONE; 3-ACETYL-6-METHYLPYRANDIONE-2,4; 3-ACETYL-6-METHYL-2H-PYRAN-2,4(3H)-DIONE ENOL FORM; 3-ACETYL-6-METHYL-2H-PYRAN-2,4(3H)-DIONE. ENOL FORM; DEHYDRACETIC ACID; DH AA; DHA; DHAA; DHS; 4-HEXENOIC ACID,2-ACETYL-5-HYDROXY-3-OXO,LACTONE; 4-HEXENOIC ACID,2-ACETYL-5-HYDROXY-3-OXO-,DELTA-LACTONE; KYSELINA DEHYDROACETOVA; METHYLACETOPYRONONE; 2H-PYRAN-2,4(3H)-DIONE,3-ACETYL-6-METHYL-; 2H-PYRAN-2-ONE,3-ACETYL-4-HYDROXY-6-METHYL-

Description: colorless to white to cream crystalline powder or prisms; odorless

Use: organic synthesis; as plasticizer; bactericide; fungicide; in antienzyme toothpastes; to reduce pickle bloating; fungicide used in cosmetics & on processed fruits & vegetables; formerly applied as a dip or wrapper impregnant

Physical Properties

Boiling Point: 269.9 °C (518 °F) at 760 mm Hg
Freezing Point: Sublimes at 109 °C (228.2 °F) to 111 °C (231.8 °F)
Water Solubility: 0.1% at 25 °C
Other Solubilities: Soluble in alkalies.
Flash Point: 157 °C Open Cup
Autoignition Temperature: 366 °C

RTECS Toxicity Data

Acute Oral: Rat LD_{50} Dose: 500 mg/kg. Mouse LD_{50} Dose: 1330 mg/kg.

Acute Dermal: Rabbit LD_{Lo} Route: Skin; Dose: 5 gm/kg.

Chronic (Multiple Dose) Oral: Rat Dose: 2100 mg/kg/11D-I; Toxic Effects: Nutritional and gross metabolic - Weight loss or decreased weight gain; DEATH. Rat Dose: 36400 mg/kg/2Y-C; Toxic Effects: Lungs, Thorax, or Respiration - Chronic pulmonary edema; Liver - Changes in liver weight; Nutritional and gross metabolic - Other changes. Monkey Dose: 3300 mg/kg/18D-I; Toxic Effects: Behavioral - Convulsions or effect on seizure threshold; Nutritional and gross metabolic - Weight loss or decreased weight gain; DEATH.

Tumorigenic: Rat Route: Subcutaneous; Dose: 592 mg/kg/37W-I; Toxic Effects: Tumorigenic - Equivocal tumorigenic agent by RTECS criteria; Tumorigenic - Tumors at site of application.

Hazard Overviews

Health: Severely irritating to eyes; irritating to skin/respiratory tract. Toxic. Other Acute Effects: harmful if swallowed, inhaled, or absorbed through skin.

Fire: Will burn. Hazards: in powder form capable of creating a dust explosion. Extinguishing agents: water spray; carbon dioxide, dry chemical powder or appropriate foam. Precautions: combustible liquid.

Reactivity: Incompatible with: strong oxidizing agents, strong bases, strong reducing agents. Hazardous decomposition products: toxic fumes of: carbon monoxide, carbon dioxide.

Carcinogenicity: IARC - Not listed; NIOSH - Not listed; NTP - Not listed; ACGIH - Not listed; OSHA - Not listed; EPA - Not listed; MAK - Not listed

Primary Target Organs:

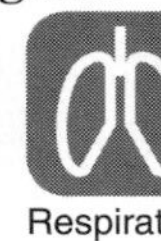

Eyes Skin Respiratory System

Environmental

Regulations

RCRA 40CFR: Not listed
CERCLA: 40CFR 302.4: Not listed
SARA 40CFR 372.65: Not listed
SARA EHS 40CFR 355: Not listed
TSCA: Listed

DEH9000 CAS #: 26400-24-8

DEHYDROHELIOTRIDINE

RTECS: UY8350000
Molecular Formula: $C_8H_{11}NO_2$
Formula Weight: 153.20
Synonyms: 3,8-DEHYDROHELIOTRIDINE; HELIOTRIDINE,3,8-DIDEHYDRO-; 1H-PYRROLIZINE-7-METHANOL,2,3-DIHYDRO-1-HYDROXY-,(S)-

RTECS Toxicity Data

Reproductive/Teratogenic: Rat Route: Intraperitoneal; Dose: 40 mg/kg; Duration: female 14D of pregnancy; Specific Developmental Abnormalities - Central nervous system; Craniofacial (including nose and tongue); Musculoskeletal system. Rat Route: Intraperitoneal; Dose: 60 mg/kg; Duration: female 14D of pregnancy; Effects on Embryo or Fetus - Fetal death. Rat Route: Intraperitoneal; Dose: 30 mg/kg; Duration: female 14D of pregnancy; Effects on Embryo or Fetus - Extra embryonic structures. Rat Route: Intraperitoneal; Dose: 40 mg/kg; Duration: female 14D of pregnancy; Effects on Embryo or Fetus - Fetotoxicity.
Mutagenic: Mouse DNA Inhibition; Cell Type: kidney; Dose: 5 mg/L. Mammal Cytogenetic Analysis; Cell Type: leukocyte; Dose: 500 nmol/L.

Hazard Overviews

Carcinogenicity: IARC - Not listed; NIOSH - Not listed; NTP - Not listed; ACGIH - Not listed; OSHA - Not listed; EPA - Not listed; MAK - Not listed

Environmental

Regulations

RCRA 40CFR: Not listed
CERCLA: 40CFR 302.4: Not listed
SARA 40CFR 372.65: Not listed
SARA EHS 40CFR 355: Not listed
TSCA: Not listed

DEL5000 CAS #: 52918-63-5

DELTAMETHRIN

RTECS: GZ1233000
EINECS Number: 258-256-6
Molecular Formula: $C_{22}H_{19}Br_2NO_3$
Formula Weight: 505.22
Synonyms: BUTOFLIN; BUTOSS; BUTOX; CISLIN; CRACKDOWN; (S)-ALPHA-CYANO-3-PHENOXYBENZYL (1R,3R)-3-(2,2-DIBROMOVINYL)-2,2-DIMETHYL-CYCLOPROPAN-1-CARBOXYLATE; (S)-ALPHA-CYANO-3-PHENOXYBENZYL-(1R)-CIS-3-(2,2-DIBROMOVINYL)-2,2-DIMETHYL-CYCLOPROPANE CARBOXYATE; [1R-[1ALPHA(S*),3ALPHA]]-CYANO(3-PHENOXYPHENYL)METHYL-3-[2,2-DIBROMOETHENYL)-2,2-DIMETHYL-CYCLOPROPANECARBOXYLATE; (1R-(1-ALPHA(S*),3-ALPHA))-CYANO-(3-PHENOXYPHENYL)METHYL-3-(2,2-DIBROMOVINYL)-2,2-DIMETHYLCYCLOPROPANECARBOXYLATE); DECAMETHRIN; DECAMETHRINE; DECIS; DEKAMETRIN; DELTAGRAN; DELTAMETHRINE; DELTAMETRYNA; 3-(2,2-DIBROMOETHENYL)-2,2-DIMETHYLCYCLOPROPANECARBOXYLICACID CYANO(3-PHENOXY-PHENYL)-METHYL ESTER; ESBECYTHRIN; FMC 45498; K-OBIOL; NRDC 161; OMS 1988; K-OTHRIN; K-OTHRINE; RU 22974; RUP 987
Description: colorless to white to slightly beige powder or crystals; odorless
Use: insecticide; to protect stored cereals, beans, etc; in forestry & in public health (e.g., chagas disease & malaria control); control of lepidoptera, homoptera, & coleoptera, flies, mosquitoes (adult & larvae), cockroaches, bedbugs, & other insects in public health; in fruit, vines, vegetables, coffee, etc; control of flying & crawling insects in households, animal houses; wood preservative, & animal ectoparasiticide.

Physical Properties

Boiling Point: Decomposes
Freezing Point: 98 °C (208.4 °F) to 101 °C (213.8 °F)
Saturated Vapor Density: 1.589055383 kg/m^3
Vapor Pressure: 0.002 mPa at 25 °C
Water Solubility: 0.002 mg/L
Other Solubilities: Acetone (500 g/l), Ethanol (15 g/l), cyclohexanone (750 g/l), dioxan (900 g/l), Xylene (250 g/l), Ethyl Acetate; Solubility; in dichloromethane 700, Benzene 450, dimethyl sulphoxide 450, isopropanol 6 (all in g/l at 20 °C).
Refraction Index: 1.34
Flash Point: Burns with difficulty

RTECS Toxicity Data

Acute Oral: Rat LD_{50} Dose: 9360 ug/kg; Toxic Effects: Behavioral - Somnolence (general depressed activity); Behavioral - Convulsions or effect on seizure threshold; Kidney, Ureter, and Bladder - Urine volume increased.
Acute Inhalation: Rat LC_{50} Dose: 785 mg/m^3/2hr.
Mutagenic: Human DNA Inhibition; Cell Type: lymphocyte; Dose: 110 umol/L. Human Sister Chromatid Exchenge; Cell Type: lymphocyte; Dose: 60 mg/L. Human Micronucleus Test; Cell Type: lymphocyte; Dose: 100 mg/L.

Hazard Overviews

Fire: Will burn.
Carcinogenicity: IARC - Group 3, Not classifiable as to carcinogenicity to humans; NIOSH - Not listed; NTP - Not listed; ACGIH - Not listed; OSHA - Not listed; EPA - Not listed; MAK - Not listed

Environmental

Ecotoxicity: LC_{50} Quail dietary >10,000 mg/kg diet/8 day LD_{50} Mallard duck oral >5000 mg/kg LC_{50} Salmo gairdneri (Rainbow trout) (flow system, static conditions) 0.39 ug/l/96 hr LC_{50} Salmo salar 1.97 ug/l/96 hr
Environmental Fate: If released to the atmosphere, it will degrade readily in the vapor phase by reaction with photochemically produced hydroxyl radicals (estimated half-

life of 10 hr). Particulate phase will be removed physically from air by wet and dry deposition. If released to soil or water, it can degrade readily through biodegradation, hydrolysis and photodegradation. The photodegradation half-life in sunlight (at the water's surface or as a thin-film on dry surfaces) can be as fast as several days or less. Aqueous hydrolysis is expected to become important only in alkaline media (pH > 8). It is not expected to leach in soil. It will partition to sediment from water in aquatic ecosystems; however, field tests have shown that surface applications can disperse slowly to subsurface water and sediment. Rapid volatilization has been shown to occur from the microlayer formed on the water's surface by spraying. Volatilization can also be an important fate process for spray applications to soil surfaces and plants. The half-life within soil has been shown to range from several weeks to over a 100 days.

Environmental Physical Data

Henry's Law Constant: estimated at 5×10^{-6}
Octanol/Water Partition Coefficient: log K_{ow} = 5.2
BCF: estimated at 3.7

Regulations

RCRA 40CFR: Not listed
CERCLA: 40CFR 302.4: Not listed
SARA 40CFR 372.65: Not listed
SARA EHS 40CFR 355: Not listed
TSCA: Not listed

Analytical Methods

Food: FDA 212.1, 212.2, 232.1, 232.3, 232.4, 242.1; AOAC 991.03
Plasma: FDA 211.1, 231.1, 252

DEM1000 CAS #: 56-94-0

DEMECARIUM BROMIDE

RTECS: BR3130000
EINECS Number: 200-301-9
Molecular Formula: $C_{32}H_{52}Br_2N_4O_4$
Formula Weight: 716.61
Synonyms: AMMONIUM,(M-HYDROXYPHENYL)TRIMETHYL-,BROMIDE,DECAMETHYLENEBIS(METHYLCARBAMATE) (2:1); BC 48; BENZENAMINIUM,3,3'-(1,10-DECANEDIYLBIS((METHYLIMINO)CARBONYLOXY))BIS(N,N,N-TRIMETHYL-,DIBROMIDE; N,N'-BIS(3-TRIMETHYLAMMONIUMPHENOXYCARBONYL)-N,N'-DIMETHYLDECAMETHYLENEDIAMINE DIBROMIDE; DECAMETHYLENEBIS(M-DIMETHYLAMINOPHENYL-N-METHYLCARBAMATE)DIMETHOBROMIDE; DECAMETHYLENEBIS(N-METHYLCARBAMIC ACIDM-DIMETHYLAMINOPHENYL ESTER) BROMOMETHYLATE; DEMEKARIUM BROMIDE; DEMEKASTIGMINE BROMIDE; FRUMTOSNIL; HUMORSOL; (M-HYDROXYPHENYL)TRIMETHYLAMMONIUM BROMIDEDECAMETHYLENEBIS(METHYLCARBAMATE); TONILEN; TOSMICIL; TOSMILEN; TOSMILENE; VISUMATIC; VISUMIOTIC
Description: white to slightly yellow, crystalline powder
Use: cholinergic (ophthalmic); vet: cholinesterase inhibitor (parasympathomimetic)

Physical Properties

Boiling Point: Decomposes at 162 °C (324 °F) to 167 °C (332.6 °F)
Freezing Point: Decomposes at 162 °C (323.6 °F) to 167 °C (332.6 °F)
Water Solubility: Freely Soluble in Water
Other Solubilities: freely Soluble in Alcohol. Sparingly Soluble in Acetone. Insoluble in Ether
pH: Aqueous solution is neutral

RTECS Toxicity Data

Acute Oral: Mouse LD_{50} Dose: 6490 ug/kg.
Acute Dermal: Mouse LD_{50} Route: Subcutaneous Dose: 24500 ug/kg.

Hazard Overviews

Carcinogenicity: IARC - Not listed; NIOSH - Not listed; NTP - Not listed; ACGIH - Not listed; OSHA - Not listed; EPA - Not listed; MAK - Not listed

Environmental

Regulations

RCRA 40CFR: Not listed
CERCLA: 40CFR 302.4: Not listed
SARA 40CFR 372.65: Not listed
SARA EHS 40CFR 355: Not listed
TSCA: Not listed

DEM5000 CAS #: 8065-48-3

DEMETON

RTECS: TF3150000
DOT: UN2783; UN2784; UN3017; UN3018; IMO3.2; IMO6.1
Molecular Formula: $C_{16}H_{38}O_6P_2S_4$
Structured MF: $(C_2H_5O)_2PSOC_2H_4SC_2H_5$
Formula Weight: 258.34
Synonyms: BAY 10756; BAYER 8169; DEMATON; DEMETON-O + DEMETON-S; DEMOX; DENOX; DIETHOXY THIOPHOSPHORIC ACID ESTER OF 2-ETHYLMERCAPTOETHANOL; DIETHOXY THIOPHOSPHORIC ACID ESTER OF2-ETHYLMERCAPTOETHANOL; DIETHOXYTHIOPHOSPHORIC ACID ESTER OF2-ETHYLMERCAPTOETHANOL; O,O-DIETHYL O(AND S)-2-(ETHYLTHIO)ETHYL PHOSPHOROTHIOATEMIXTURE; O,O-DIETHYL O-(2-(ETHYLMERCAPTO)ETHYL) THIONOPHOSPHATE; O,O-DIETHYL 2-ETHYLMERCAPTOETHYL THIOPHOSPHATE; O-O-DIETHYL-O(AND S)-2-(ETHYLTHIO)ETHYL PHOSPHOROTHIOATE MIXTURE; E 1059; E-1059; ENT 17,295; ETHYL SYSTOX; BETA-ETHYLMERCAPTOETHYL DIETHYL THIONOPHOSPHATE; MERCAPTOFOS; MERCAPTOPHOS; PHOSPHOROTHIOIC ACID,O,O-DIETHYL O-(2-(ETHYLTHIO)ETHYL)ESTER,MIXED WITH O,O-DIETHYL S-(2-(ETHYLTHIO)ETHYL) ESTER(7:3); PHOSPHOROTHIOIC ACID,O,O-DIETHYL O-(2-(ETHYLTHIO)ETHYL)ESTER,MIXT WITH O,O-DIETHYL S-(2-

(ETHYLTHIO)ETHYL)PHOSPHOROTHIOATE; SEPTOX; SYSTEMOX; SYSTOX; ULV

Description: amber, oily liquid; sulfur-like odor

Use: systemic nematocide; as systemic insecticide for ornamentals and nursery stock

Physical Properties

Boiling Point: 134 °C (273 °F) at 2 mm Hg

Freezing Point: < -25 °C (-13 °F)

Specific Gravity: 1.1183 at 20 °C

Vapor Pressure: 3.4 $x10^{-4}$ mbar at 20 °C

Water Solubility: 0.01% by weight

Other Solubilities: strongly catalyzed by polar solvents; Soluble in most organic solvents.

Refraction Index: 1.4875 at 20 °C/D

Flash Point: 45 °C Closed Cup

Autoignition Temperature: xylene solvent 464 °C

LEL: 1.0% v/v

UEL: 5.3% v/v

RTECS Toxicity Data

Acute Oral: Man TD_{Lo} Dose: 144 mg/kg/24D-I; Toxic Effects: Biochemical - True cholinesterase. Human LD_{Lo} Dose: 171 ug/kg.

Acute Inhalation: Rat LC_{Lo} Dose: 15 mg/m^3/4hr.

Mutagenic: Mouse Mutations in Mammalian Somatic Cells; Cell Type: lymphocyte; Dose: 80 mg/L. Hamster Sister Chromatid Exchenge; Cell Type: ovary; Dose: 25 ppm. Hamster Sister Chromatid Exchenge; Cell Type: lung; Dose: 10 mg/L.

Hazard Overviews

Flammable

Fire: Flammable.

Carcinogenicity: IARC - Not listed; NIOSH - Not listed; NTP - Not listed; ACGIH - Not listed; OSHA - Not listed; EPA - Not listed; MAK - Not listed

Exposure Limits

OSHA PEL: TWA: 0.1 mg/m^3; skin.

ACGIH TLV: TWA: 0.01 ppm; 0.11 mg/m^3.

NIOSH REL: TWA: 0.1 mg/m^3.

NIOSH IDLH: 10 mg/m^3.

DFG MAK: TWA: 0.01 ppm; 0.1 mg/m^3.

Respirator Recommendation

Exposure Range: >0.1 to 5 mg/m^3 Supplied Air, Constant Flow/Pressure Demand, Half Mask

Exposure Range: >5 to 100 mg/m^3 Supplied Air, Constant Flow/Pressure Demand, Full Face

Exposure Range: >100 to unlimited mg/m^3 Self-contained Breathing Apparatus, Pressure Demand, Full Face

Note: odor threshold unknown

Environmental

Ecotoxicity: LC_{50} Daphnia pulex 14.0 ug/l/48 hr at 15 °C (95% confidence limit 10.4-18.7 ug/l), 1st instar. Static bioassay without aeration, pH 7.2-7.5, water hardness 40-50 mg/l as calcium carbonate and alkalinity of 30-35 mg/l LC_{50} Coturnix japonica 275 ppm in 5-day diets (95% confidence limit 228-327 ppm), age 12 days LC_{50} Gammarus fasciatus 78 ug/l at 15 °C (95% confidence limit 42-140 ug/l), mature. Static bioassay without aeration, pH 7.2-7.5, water hardness 40-50 mg/l as calcium carbonate and alkalinity of 30-35 mg/l LD_{50} Rana catesbeiana oral 562 mg/kg (95% confidence limit 178-1780 mg/kg), males LC_{50} Micropterus salmoides 148 ug/l at 18 °C in hard water (95% confidence limit 136-162 ug/l), wt 0.7 g. Static bioassay without aeration, pH 7.2-7.5, water hardness 40-50 mg/l as calcium carbonate and alkalinity of 30-35 mg/l LD_{50} Anas platyrhynchos oral 7.19 mg/kg (95% confidence limit 5.19-9.97 mg/kg), 3 mo old males LD_{50} Carpodacus mexicanus oral 2.38 mg/kg, males & females

Cleanup/Disposal: Guide No. 131: Fully encapsulating, vapor protective clothing should be worn for spills and leaks with no fire. Eliminate all ignition sources (no smoking, flares, sparks or flames in immediate area). All equipment used when handling the product must be grounded. Do not touch or walk through spilled material. Stop leak if you can do it without risk. Prevent entry into waterways, sewers, basements or confined areas. A vapor suppressing foam may be used to reduce vapors. Small Spills: Absorb with earth, sand or other non-combustible material and transfer to containers for later disposal. Use clean non-sparking tools to collect absorbed material. Large Spills: Dike far ahead of liquid spill for later disposal. Water spray may reduce vapor; but may not prevent ignition in closed spaces. Guide No. 152: Do not touch damaged containers or spilled material unless wearing appropriate protective clothing. Stop leak if you can do it without risk. Prevent entry into waterways, sewers, basements or confined areas. Cover with plastic sheet to prevent spreading. Absorb or cover with dry earth, sand or other non-combustible material and transfer to containers. Do not get water inside containers.

Environmental Physical Data

BCF: no food chain concentration potential

Regulations

RCRA 40CFR: Not listed

CERCLA: 40CFR 302.4: Not listed

SARA 40CFR 372.65: Not listed TPQ: 500 lb

SARA EHS 40CFR 355: Listed TPQ: 500 lb

TSCA: Not listed

Analytical Methods

Air: ASTM D4490

Soil: SW846 8140, 8141, 8141A

Water / Groundwater: EPA 614, 622; SW846 8141, 8141A

Food: FDA 212.1, 232.1, 232.3, 232.4, 242.1

Indoor / Expired Air: NIOSH 5514

Plasma: EPA 001, 027; FDA 252

DEM9000 **CAS #: 919-86-8**

DEMETON-S-METHYL

RTECS: TG1750000
EINECS Number: 213-052-6
Molecular Formula: $C_6H_{15}O_3PS_2$
Formula Weight: 230.30
Synonyms: BAY 18436; BAYER 25/154; BAYER 18 436; DEMETON-S-METILE; DEMETOX; DEP 836 349; O,O-DIMETHYL S-(2-ETHTHIOETHYL)PHOSPHOROTHIOATE; DIMETHYL S-(2-ETHTHIOETHYL)THIOPHOSPHATE; O,O-DIMETHYL S-ETHYLMERCAPTOETHYL THIOPHOSPHATE; O,O-DIMETHYL 2-ETHYLMERCAPTOETHYL THIOPHOSPHATE,THIOLOISOMER; O,O-DIMETHYL S-(2-(ETHYLTHIO)ETHYL)PHOSPHOROTHIOATE; O,O-DIMETHYL-S-(2-AETHYLTHIO-AETHYL)-MONOTHIOPHOSPHAT; O,O-DIMETHYL-S-(2-ETHYLTHIO-ETHYL)-MONOTHIOFOSFAAT; O,O-DIMETHYL-S-(3-THIA-PENTYL)-MONOTHIOPHOSPHAT; O,O-DIMETIL-S-(2-ETILTIO-ETIL)-MONOTIOFOSFATO; DURATOX; ETHANETHIOL,2-(ETHYLTHIO)-,S-ESTER WITH 0,0-DIMETHYLPHOSPHOROTHIOATE; 2-ETHYLTHIOETHYL DIMETHYL PHOSPHOROTHIOATE; S-(2-(ETHYLTHIO)ETHYL) 0,0-DIMETHYL PHOSPHOROTHIOATE; S-(2-ETHYLTHIOETHYL) O,O-DIMETHYL PHOSPHOROTHIOATE; S-2-ETHYLTHIOETHYL O,O-DIMETHYL PHOSPHOROTHIOATE; S-(2-(ETHYLTHIO)ETHYL) DIMETHYL PHOSPHOROTHIOLATE; S-(2-(ETHYLTHIO)ETHYL) 0,0-DIMETHYL THIOPHOSPHATE; ISOMETASYSTOX; METAISOSEPTOX; METAISOSYSTOX; METASYSTOX (I); METASYSTOX 55; METASYSTOX FORTE; METASYSTOX I; METASYSTOX J; METHYL DEMETON THIOESTER; METHYL ISOSYSTOX; METHYL THIONODEMETON; METHYL-MERCAPTOFOS TEOLERY; METHYL-MERCAPTOFOS TEOLOVY; METHYLMERCAPTOFOSTIOL; MIFATOX; PHOSPHOROTHIOIC ACID,O,O-DIMETHYL S-(2-(ETHYLTHIO)ETHYL)ESTER; PHOSPHOROTHIOIC ACID,S-(2-(ETHYLTHIO)ETHYL) O,O-DIMETHYLESTER; THIOPHOSPHATE DE O,O-DIMETHYLE ET DE S-2-ETHYLTHIO-ETHYLE; USP 2 571 989
Description: colorless to pale yellow oil; unpleasant garlic like odor
Use: a systemic and contact insecticide used on most agricultural and horticultural crops; controls aphid vectors of virus diseases, white-flies, leafhoppers, and sawflies; for the control of aphids, red spider mites, and certain other pests in arable and market garden crops, fruit, and hops

Physical Properties

Boiling Point: 89 °C (192 °F) at 0.15 mm Hg
Specific Gravity: 1.207 at 20 °C/4 °C
Vapor Pressure: 48 mPa at 20 °C
Water Solubility: 3300 ppm in Water at room temp
Other Solubilities: Soluble in organic solvents; 600 g/kg dichloromethane at 20 °C; 600 g/kg propan-2-ol at 20 °C; Readily Soluble in common polar organic solvents (eg, Alcohols, ketones, chlorinated hydrocarbons).
Odor Threshold: Detection in Water 1.0 x10^{-2} mg/l
Refraction Index: 1.5065 at 20 °C/D

RTECS Toxicity Data

Mutagenic: Insects - D Melanogaster Sex Chromosome Loss; Route: Oral; Dose: 80 ppm. Bacteria - E Coli Mutations in Microorganisms; Dose: 5 mg/plate (-S9). Bacteria - S Typhimurium Mutations in Microorganisms; Dose: 5 mg/plate (+S9).

Hazard Overviews

Carcinogenicity: IARC - Not listed; NIOSH - Not listed; NTP - Not listed; ACGIH - Not listed; OSHA - Not listed; EPA - Not listed; MAK - Not listed

Environmental

Regulations

RCRA 40CFR: Not listed
CERCLA: 40CFR 302.4: Not listed
SARA 40CFR 372.65: Not listed TPQ: 500 lb
SARA EHS 40CFR 355: Listed TPQ: 500 lb
TSCA: Listed

Analytical Methods

Air: ASTM D4490
Soil: SW846 8140
Food: FDA 232.4, 242.1

DEO1000 **CAS #: 83-44-3**

DEOXYCHOLIC ACID

RTECS: FZ2100000
EINECS Number: 201-478-5
Molecular Formula: $C_{24}H_{40}O_4$
Formula Weight: 392.56

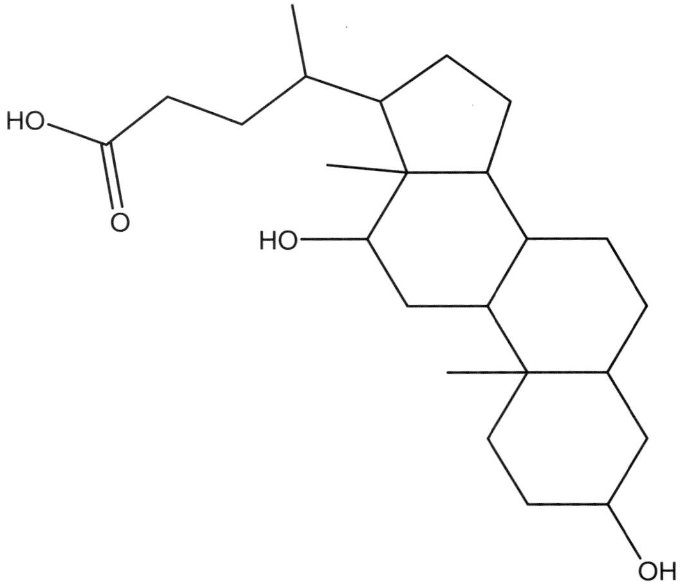

HO O HO OH

Chemical Structure

Synonyms: 5BETA-CHOLAN-24-OIC ACID,3ALPHA,12-ALPHA-DIHYDROXY-; CHOLAN-24-OIC ACID,3,12-DIHYDROXY-,(3-ALPHA,5-BETA,12-ALPHA)-(9CI); CHOLEIC ACID; CHOLEREBIC; CHOLIC ACID,DEOXY-; CHOLOREBIC; DEGALOL; DEOXY CHOLIC ACID; DEOXYCHOLATIC ACID; 5-BETA-DEOXYCHOLIC ACID; 7ALPHA-DEOXYCHOLIC ACID; DESOXYCHOLIC ACID; DESOXYCHOLSAEURE; 3,12-DIHYDROXYCHOLANIC ACID; 3-

ALPHA,12-ALPHA-DIHYDROXYCHOLANIC ACID; 3ALPHA,12ALPHA-DIHYDROXYCHOLANIC ACID; 3-ALPHA,12-ALPHA-DIHYDROXY-5-BETA-CHOLANOIC ACID; 3ALPHA,12ALPHA-DIHYDROXY-5BETA-CHOLANOIC ACID; 3-ALPHA,12-ALPHA-DIHYDROXYCHOLANSAEURE; DROXOLAN; L7-BETA-(1-METHYL-3-CARBOXYPROPYL)-ETIOCHOLANE-3-ALPHA,12-ALPHA-DIOL; 17-BETA-(1-METHYL-3-CARBOXYPROPYL)-ETIOCHOLANE-3-ALPHA,12-ALPHA-DIOL; PYROCHOL; SEPTOCHOL

Description: crystals

Use: emulsifying agent in food; precursor for org synthesis of cortisone; choleretic

Physical Properties

Freezing Point: 176 °C (348.8 °F) to 178 °C (352.4 °F)

Water Solubility: 0.24 g/L at 15 °C

Other Solubilities: at 15 °C: 220.7 g/L in Alcohol, in Ether: 1.16 g/L, in Chloroform: 2.94 g/L, in Benzene: 0.12 g/L, in Acetone: 10.46 g/L, in Glacial Acetic Acid: 9.06 g/L Soluble in solution of alkali hydroxides or carbonates

RTECS Toxicity Data

Acute Oral: Rat LD_{50} Dose: 1 gm/kg; Toxic Effects: Behavioral - Food intake (animal); Nutritional and gross metabolic - Weight loss or decreased weight gain. Mouse LD_{50} Dose: 1 gm/kg.

Reproductive/Teratogenic: Rat Route: Intraperitoneal; Dose: 166 mg/kg; Duration: female 12-19D of pregnancy Effects on Fertility - Post-implantation mortality; Effects on Embryo or Fetus - Fetotoxicity.

Mutagenic: Hamster Morphological Transformation; Cell Type: embryo; Dose: 7250 ug/L. Bacteria - S Typhimurium Mutations in Microorganisms; Dose: 20 mg/L (+S9).

Tumorigenic: Mouse Route: Skin; Dose: 2700 mg/kg/10W-I; Toxic Effects: Tumorigenic - Equivocal tumorigenic agent by RTECS criteria; Skin and appendages - Tumors. Mouse Route: Subcutaneous; Dose: 1120 mg/kg/22W-I; Toxic Effects: Tumorigenic - Equivocal tumorigenic agent by RTECS criteria; Tumorigenic - Tumors at site of application. Mouse Route: Subcutaneous; Dose: 1400 mg/kg/22W-I; Toxic Effects: Tumorigenic - Equivocal tumorigenic agent by RTECS criteria; Tumorigenic - Tumors at site of application.

Hazard Overviews

Health: Irritating to eyes/skin/respiratory tract. Harmful. Other Acute Effects: harmful if swallowed; may be harmful if inhaled or absorbed through the skin.

Fire: Hazards: emits toxic fumes. Extinguishing agents: water spray; carbon dioxide, dry chemical powder or appropriate foam. Precautions: combustible liquid.

Reactivity: Incompatible with: strong oxidizing agents. Hazardous decomposition products: toxic fumes of: carbon monoxide, carbon dioxide.

Carcinogenicity: IARC - Not listed; NIOSH - Not listed; NTP - Not listed; ACGIH - Not listed; OSHA - Not listed; EPA - Not listed; MAK - Not listed

Primary Target Organs:

Eyes

Skin

Respiratory System

Environmental

Regulations

RCRA 40CFR: Not listed

CERCLA: 40CFR 302.4: Not listed

SARA 40CFR 372.65: Not listed

SARA EHS 40CFR 355: Not listed

TSCA: Listed

DEO5000 CAS #: 154-17-6

2-DEOXY-D-GLUCOSE

RTECS: MQ3325000

EINECS Number: 205-823-0

Molecular Formula: $C_6H_{12}O_5$

Formula Weight: 164.18

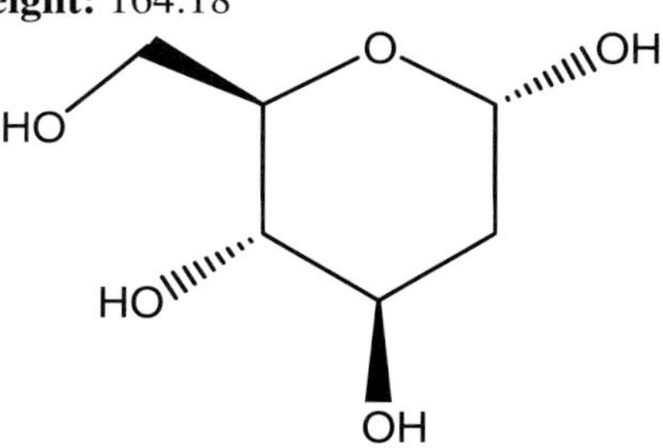

Chemical Structure

Synonyms: 2 DG; D-ARABINO-HEXOSE,2-DEOXY-; BA 2758; 2-DEOXY-D-ARABINO-HEXOSE; 2-DEOXYGLUCOSE; D-2-DEOXYGLUCOSE; 2-DESOXY-D-GLUCOSE; 2-DG; D-GLUCOSE,2-DEOXY-; GLUCOSE,2-DEOXY-; NSC 15193

Use: a glucose antimetabolite, inhibiting glycolysis; research tool to study glucose-dependent or -mediated reactions; research chemical

Physical Properties

Freezing Point: 147 °C (296.6 °F)

RTECS Toxicity Data

Acute Dermal: Rat LD_{50} Route: Subcutaneous Dose: 250 mg/kg.

Reproductive/Teratogenic: Rat Route: Oral; Dose: 2 gm/kg; Duration: female 7-8D of pregnancy; Effects on Fertility - Post-implantation mortality. Rat Route: Oral; Dose: 4 gm/kg; Duration: female 7-14D of pregnancy; Effects on Embryo or Fetus - Fetotoxicity.

Hazard Overviews

Health: Irritating to eyes/skin/respiratory tract. Other Acute Effects: may be harmful by inhalation, ingestion, or skin absorption.

Fire: Hazards: emits toxic fumes. Extinguishing agents: water spray; carbon dioxide, dry chemical powder or appropriate foam. Precautions: combustible liquid.

Reactivity: Stable. Hazardous polymerization will not occur. Hazardous decomposition products: toxic fumes of: carbon monoxide, carbon dioxide.

Carcinogenicity: IARC - Not listed; NIOSH - Not listed; NTP - Not listed; ACGIH - Not listed; OSHA - Not listed; EPA - Not listed; MAK - Not listed

Primary Target Organs:

Eyes

Skin

Respiratory System

Environmental

Regulations

RCRA 40CFR: Not listed
CERCLA: 40CFR 302.4: Not listed
SARA 40CFR 372.65: Not listed
SARA EHS 40CFR 355: Not listed
TSCA: Listed

DEO9000 CAS #: 18883-66-4

2-DEOXY-2-(3-METHYL-3-NITROSOUREIDO)GLUCOPYR-ANOSE

RTECS: LZ5775000
EINECS Number: 242-646-8
Molecular Formula: $C_8H_{15}N_3O_7$
Formula Weight: 265.22

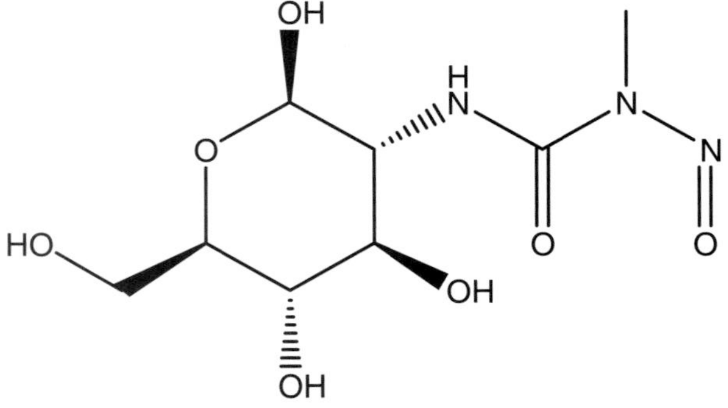

Chemical Structure

Synonyms: 2-DEOXY-2-(((METHYLNITROSOAMINO)CARBONYL)AMINO)-D-GLUCOPYRANOSE; 2-DEOXY-2-(3-METHYL-3-NITROSOUREIDO)-,D-; 2-DEOXY-2-(3-METHYL-3-NITROSOUREIDO)-ALPHA(AND BETA)-D-GLUCOPYRANOSE; 2-DEOXY-2-(3-METHYL-3-NITROSOUREIDO)-ALPHA(ANDBETA)-D-GLUCOPYRANOSE; 2-DEOXY-2-(3-METHYL-3-NITROSOUREIDO)-D-GLUCOPYRANOSE; ESTREPTOZOCINA; D-GLUCOPYRANOSE,2-DEOXY-2-(((METHYLNITROSOAMINO)CARBONYL)AMINO)-; GLUCOPYRANOSE,2-DEOXY-2-(3-METHYL-3-NITROSOUREIDO)-,D-; D-GLUCOSE,2-DEOXY-2-(((METHYLNITROSOAMINO)CARBONYL)AMINO)-; D-GLUCOSE,2-DEOXY-2-(((METHYLNITROSOAMINO)CARBONYL)AMINO)-(9CI); D-GLUCOSE,2-DEOXY-2-(3-METHYL-3-NITROSOUREIDO)-; N-D-GLUCOSYL(2)-N'-NITROSOMETHYLHARNSTOFF; N-D-GLUCOSYL-(2)-N'-NITROSOMETHYLUREA; NSC-85598; NSC-85998; NSC 85998; STR; STREPTOZOCIN; STREPTOZOCINE; STREPTOZOCINIUM; STREPTOZOTICIN; STREPTOZOTOCIN; STRZ; STZ; U 9889; U-9889; ZANOSAR

Description: pale-yellow to ivory-colored crystalline powder, crystales, or platelets

Use: antineoplastic agent; antibiotic

Physical Properties

Freezing Point: 115 °C (239 °F)
Water Solubility: Soluble in Water
Other Solubilities: 95% Ethanol: Soluble; Acetone: Insoluble; DMSO: Slightly Soluble.

RTECS Toxicity Data

Acute Dermal: Mouse LD_{50} Route: Subcutaneous Dose: 335 mg/kg.

Reproductive/Teratogenic: Rat Route: Parenteral; Dose: 40 mg/kg; Duration: female 1D prior to mating; Effects on Fertility - Post-implantation mortality; Effects on Embryo or Fetus - Fetotoxicity; Specific Developmental Abnormalities - Other developmental abnormalities. Rat Route: Intravenous; Dose: 70 mg/kg; Duration: female 1D prior to mating; Maternal Effects - Ovaries, fallopian tubes. Rat Route: Intravenous; Dose: 35 mg/kg; Duration: female 8D of pregnancy; Effects on Embryo or Fetus - Extra embryonic structures; Fetotoxicity. Rat Route: Intravenous; Dose: 30 mg/kg; Duration: female 5D of pregnancy; Effects on Embryo or Fetus - Fetotoxicity; Effects on Newborn - Growth statistics. Rat Route: Intravenous; Dose: 40 mg/kg; Duration: female 5D of pregnancy; Specific Developmental Abnormalities - Endocrine system; Effects on Newborn - Biochemical and metabolic. Rat Route: Intravenous; Dose: 40 mg/kg; Duration: female 1D of pregnancy; Effects on Fertility - Post-implantation mortality; Specific Developmental Abnormalities - Central nervous system; Cardiovascular (circulatory) system. Rat Route: Intravenous; Dose: 60 mg/kg; Duration: female 1D prior to mating; Effects on Fertility - Other measures of fertility.

Mutagenic: Human DNA Damage; Cell Type: other cell types; Dose: 50 mg/L. Human DNA Inhibition; Cell Type: HeLa cell; Dose: 100 umol/L. Human DNA Inhibition; Cell Type: lung; Dose: 1 gm/L/3H-C. Human Cytogenetic Analysis; Cell Type: lung; Dose: 500 mg/L. Human Sister Chromatid Exchenge; Cell Type: other cell types; Dose: 100 umol/L.

Tumorigenic: Rat Route: Intravenous; Dose: 25 mg/kg; Toxic Effects: Tumorigenic - Neoplastic by RTECS criteria; Kidney, Ureter, and Bladder - Kidney tumors. Rat Route:

Intravenous; Dose: 65 mg/kg; Toxic Effects: Tumorigenic - Equivocal tumorigenic agent by RTECS criteria; Gastrointestinal - Tumors; Kidney, Ureter, and Bladder - Kidney tumors. Rat Route: Intravenous; Dose: 50 mg/kg; Toxic Effects: Tumorigenic - Neoplastic by RTECS criteria; Kidney, Ureter, and Bladder - Kidney tumors. Rat Route: Intravenous; Dose: 30 mg/kg; Toxic Effects: Tumorigenic - Neoplastic by RTECS criteria; Gastrointestinal - Tumors; Kidney, Ureter, and Bladder - Kidney tumors. Rat Route: Intravenous; Dose: 50 mg/kg; Toxic Effects: Tumorigenic - Equivocal tumorigenic agent by RTECS criteria; Liver - Tumors; Kidney, Ureter, and Bladder - Kidney tumors.

Hazard Overviews

Health: Irritating. Toxic. Other Acute Effects: harmful if swallowed, inhaled, or absorbed through skin. Chronic Effects: may alter genetic material; may cause congenital malformation in the fetus; anemia; target organs: pancreas, liver, kidneys, blood. Carcinogen.

Fire: Hazards: emits toxic fumes. Extinguishing agents: water spray; carbon dioxide, dry chemical powder or appropriate foam. Precautions: combustible liquid.

Reactivity: Incompatible with: strong oxidizing agents, strong acids, strong bases. Hazardous decomposition products: toxic fumes of: carbon monoxide, carbon dioxide, nitrogen oxides.

Carcinogenicity: IARC - Group 2B, Possibly carcinogenic to humans; NIOSH - Not listed; NTP - Listed; ACGIH - Not listed; OSHA - Not listed; EPA - Not listed; MAK - Not listed

Primary Target Organs:

Eyes

Skin

Respiratory System

Liver

Kidneys

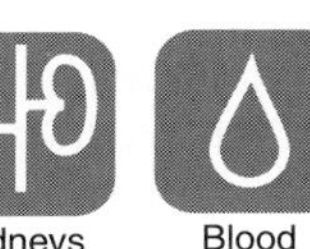
Blood

Environmental

Regulations

RCRA 40CFR: Listed Hazardous Waste No. U206 Toxic Waste

CERCLA: 40CFR 302.4: Listed per RCRA Section 3001 RQ: 1 lb (0.454 kg)

SARA 40CFR 372.65: Not listed

SARA EHS 40CFR 355: Not listed

TSCA: Not listed

DES1000 CAS #: 6190-65-4

DESETHYL ATRAZINE

DOT: UN2763; UN2764; UN2997; UN2998

Molecular Formula: $C_6H_{10}ClN_5$

Formula Weight: 187.63

Synonyms: DEETHYLATRAZINE; G 30033; S-TRIAZINE,2-AMINO-4-CHLORO-6-(ISOPROPYLAMINO)-; 1,3,5-TRIAZINE-2,4-DIAMINE,6-CHLORO-N-(1-METHYLETHYL)-

Hazard Overviews

Carcinogenicity: IARC - Not listed; NIOSH - Not listed; NTP - Not listed; ACGIH - Not listed; OSHA - Not listed; EPA - Not listed; MAK - Not listed

Environmental

Environmental Fate: If released to soil or formed in soil, it will be expected to exhibit high mobility in soil based upon an estimated K_{oc} and, therefore, it may leach to groundwater. It may be subject to slow degradation in soil which may be due to both chemical hydrolysis and biodegradation based upon limited data for other structurally related s-triazines such as atrazine. It will not be expected to volatilize from near surface soils or surfaces. If released to water, it will not be expected to bioconcentrate in aquatic organisms, adsorb to sediment and suspended particulate matter, or to volatilize. The sensitized photolysis, slow hydrolysis and slow biodegradation may occur in natural waters based upon the slow biodegradation observed in soil for other structurally related s-triazines. It may be susceptible to direct photolysis in natural waters based upon the observed degradation of s-triazines similar in structure such as atrazine in water solutions irradiated with artificial light at wavelengths >290 nm. If released to the atmosphere, it may be subject to gas-phase reaction with photochemically produced hydroxyl radicals with a half-life of about 4.6 hr estimated for this process.

Cleanup/Disposal: Guide No. 131: Fully encapsulating, vapor protective clothing should be worn for spills and leaks with no fire. Eliminate all ignition sources (no smoking, flares, sparks or flames in immediate area). All equipment used when handling the product must be grounded. Do not touch or walk through spilled material. Stop leak if you can do it without risk. Prevent entry into waterways, sewers, basements or confined areas. A vapor suppressing foam may be used to reduce vapors. Small Spills: Absorb with earth, sand or other non-combustible material and transfer to containers for later disposal. Use clean non-sparking tools to collect absorbed material. Large Spills: Dike far ahead of liquid spill for later disposal. Water spray may reduce vapor; but may not prevent ignition in closed spaces. Guide No. 151: Do not touch damaged containers or spilled material unless wearing appropriate protective clothing. Stop leak if you can do it without risk. Prevent entry into waterways, sewers, basements or confined areas. Cover with plastic sheet to prevent spreading. Absorb or cover with dry earth, sand or other non-combustible material and transfer to containers. Do not get water inside containers.

Environmental Physical Data

Henry's Law Constant: calculated at 4.63×10^{-10}

Octanol/Water Partition Coefficient: log K_{ow} = 0.725

Sorption Partition Coefficient: K_{oc} = estimated at 4.7

BCF: estimated at 2.1

Regulations

RCRA 40CFR: Not listed

CERCLA: 40CFR 302.4: Not listed

SARA 40CFR 372.65: Not listed

SARA EHS 40CFR 355: Not listed
TSCA: Not listed

DES2600 CAS #: 70-51-9

DESFERRIOXAMINE

RTECS: UG5300000
EINECS Number: 200-738-5
Molecular Formula: $C_{25}H_{48}N_6O_8$
Formula Weight: 560.71
Synonyms: N'-[5-[[4-[[5-(ACETYLHYDROXAMINO)PENTYL]AMINO]-1,4-DIOXOBUTYL]HYDROXYAMINO]PENTYL]-N-(5-AMINOPENTYL)-N-HYDROXYBUTANEDIAMIDE; 1-AMINO-6,17-DIHYDROXY-7,10,18,21-TETRAOXO-27-(N-ACETYLHYDROXYLAMINO)-6,11,17,22-TETRAAZAHEPTAEICOSANE; N-[5-[3-[(5-AMINOPENTYL)HYDROXYCARBAMOYL]PROPIONAMIDO]PENTYL]-3-[[5-(N-HYDROXYACETAMIDO)PENTYL]CARBAMOYL]PROPIONOHYDROXAMIC ACID; 30-AMINO-3,14,25-TRIHYDROXY-3,9,14,20,25-PENTAAZATRIACONTANE -2,10,13,21,24-PENTAONE; 30-AMINO-3,14,25-TRIHYDROXY-3,9,14,20,25-PENTAAZATRIACONTANE-2,10,13,21,24-PENTAONE; BA-29837; BA-33112; BA 33112; N-BENZOYLFERRIOXAMINE B; BUTANEDIAMIDE,N'-(5-((4-((5-(ACETYLHYDROXYAMINO)PENTYL)AMINO)-1,4-DIOXOBUTYL)HYDROXYAMINO)PENTYL)-N-(5-AMINOPENTYL)-N-HYDROXY-; DEFEROXAMIDE B; DEFEROXAMIN; DEFEROXAMINE; DEFEROXAMINE B; DEFEROXAMINUM; DEFERRIOXAMINE; DEFERRIOXAMINE B; DESFERAL; DESFERAN; DESFEREX; DESFERIN; DESFERRAL; DESFERRIFERRIOXAMIN B; DESFERRIN; DESFERRIOXAMINE B; DF B; DFO; DFOA; DFOM; FERRIOXAMINE B,N-BENZOYL-; NSC-527604; 3,9,14,20,25-PENTAAZATRIACONTANE-2,10,13,21,24-PENTONE,30-AMINO-3,14,25-TRIHYDROXY-; PROPIONOHYDROXAMIC ACID,N-(5-(3-((5-AMINOPENTYL)HYDROXYCARBAMOYL)PROPIONAMIDO)PENTYL)-3-((5-(N-HYDROXYACETAMIDO)PENTYL)CARBAMOYL-
Description: crystals
Use: medication: chelating agent for iron

Physical Properties

Freezing Point: 180 °C (356 °F) to 182 °C (359.6 °F)
Water Solubility: Monohydrate 1%

RTECS Toxicity Data

Acute Oral: Mouse LD_{50} Dose: 1340 mg/kg.
Acute Dermal: Human TD_{Lo} Route: Subcutaneous Dose: 37 gm/kg/2Y-I; Toxic Effects: Sense organs and special senses - Optic nerve neuropathy. Child TD_{Lo} Route: Subcutaneous Dose: 12 gm/kg/17W-I; Toxic Effects: Sense organs and special senses - Optic nerve neuropathy; Sense organs and special senses - Change in acuity.
Chronic (Multiple Dose) Oral: Rat Dose: 52 gm/kg/52W-I; Toxic Effects: Lungs, Thorax, or Respiration - Changes in lung weight; Kidney, Ureter, and Bladder - Changes in kidney weight; Endocrine - Changes in adrenal weight.
Mutagenic: Human DNA Inhibition; Cell Type: other cell types; Dose: 50 umol/L. Human DNA Inhibition; Cell Type: liver; Dose: 50 umol/L.

Hazard Overviews

Carcinogenicity: IARC - Not listed; NIOSH - Not listed; NTP - Not listed; ACGIH - Not listed; OSHA - Not listed; EPA - Not listed; MAK - Not listed

Environmental

Regulations

RCRA 40CFR: Not listed
CERCLA: 40CFR 302.4: Not listed
SARA 40CFR 372.65: Not listed
SARA EHS 40CFR 355: Not listed
TSCA: Not listed

DES4200 CAS #: 50-47-5

DESIPRAMINE

RTECS: HO0350000
EINECS Number: 200-040-0
Molecular Formula: $C_{18}H_{22}N_2$
Formula Weight: 266.37
Synonyms: DEMETHYLIMIPRAMINE; DESIMIPRAMINE; DESIMPRAMINE; DESIPRAMIN; DESIPRAMINE (D4); DESMETHYLIMIPRAMINE; 5H-DIBENZ(B,F)AZEPINE,10,11-DIHYDRO-5-(3-(METHYLAMINO)PROPYL)-; 5H-DIBENZ(B,F)AZEPINE-5-PROPANAMINE,10,11-DIHYDRO-N-METHYL-; 10,11-DIHYDRO-5-(3-METHYLAMINOPROPYL)-5H-DIBENZ(B,F)AZEPINE; DIMETHYLIMIPRAMINE; DMI; DMI 50475; IMIPRAMINE,DEMETHYL-; N-(3-METHYLAMINOPROPYL)IMINOBIBENZYL; 5-(GAMMA-METHYLAMINOPROPYL)IMINODIBENZYL; METHYLAMINOPROPYLIMINODIBENZYL; MONODEMETHYLIMIPRAMINE; NORIMIPRAMINE
Description: white crystalline powder; odorless
Use: antidepressant

Physical Properties

Boiling Point: 172 °C (342 °F) to 174 °C (345 °F) at 0.02 mm Hg
Water Solubility: 1 g/20 mL Water
Other Solubilities: 1 g in about 4 ml Chloroform, 20 ml Alcohol; Freely Soluble in Methanol; Insoluble in Ether and HCL.

RTECS Toxicity Data

Acute Oral: Man TD_{Lo} Dose: 1643 mg/kg/1Y-I; Toxic Effects: Behavioral - Convulsions or effect on seizure threshold; Cardiac - Pulse rate increased without fall in BP; Nutritional and gross metabolic - Body temperature increase. Human LD_{Lo} Dose: 30 mg/kg; Toxic Effects: Brain and coverings - Other degenerative changes; Behavioral - Coma; Lungs, Thorax, or Respiration - Cyanosis. Woman LD_{Lo} Dose: 30 mg/kg; Toxic Effects: Brain and coverings - Other degenerative changes; Behavioral - Convulsions or effect on seizure threshold. Child LD_{Lo} Dose: 125 mg/kg; Toxic Effects: Behavioral - Convulsions or effect on seizure threshold; Behavioral - Coma; Lungs, Thorax, or Respiration - Cyanosis. Rat LD_{50} Dose: 375 mg/kg.

Acute Dermal: Rat LD_{50} Route: Subcutaneous Dose: 183 mg/kg. Mouse LD_{50} Route: Subcutaneous Dose: 214 mg/kg.
Reproductive/Teratogenic: Rat Route: Oral; Dose: 500 mg/kg; Duration: female 7-16D of pregnancy; Effects on Embryo or Fetus - Fetotoxicity; Fetal death. Rat Route: Subcutaneous; Dose: 16250 ug/kg; Duration: female 8-20D of pregnancy; Effects on Newborn - Behavioral.
Mutagenic: Other Insects Cytogenetic Analysis; Route: Unreported; Dose: 10 gm/L. Bacteria - E Coli DNA Adduct; Dose: 20 umol/L.

Hazard Overviews

Carcinogenicity: IARC - Not listed; NIOSH - Not listed; NTP - Not listed; ACGIH - Not listed; OSHA - Not listed; EPA - Not listed; MAK - Not listed

Environmental

Regulations

RCRA 40CFR: Not listed
CERCLA: 40CFR 302.4: Not listed
SARA 40CFR 372.65: Not listed
SARA EHS 40CFR 355: Not listed
TSCA: Not listed

DES5800 CAS #: 13684-56-5

DESMEDIPHAM

RTECS: FD0425000
EINECS Number: 237-198-5
Molecular Formula: $C_{16}H_{16}N_2O_4$
Formula Weight: 300.34
Synonyms: 3-(AETHOXYCARBONYLAMINOPHENYL)-N-PHENYL-CARBAMAT; BENTANEX; BETANAL AM; BETANAL AM 11; BETANEX; CARBAMIC ACID,(3-(((PHENYLAMINO)CARBONYL)OXY)PHENYL)-,ETHYL ESTER (9CI); CARBANILIC ACID,M-CARBANILOYLOXY-,ETHYL ESTER; CARBANILIC ACID,M-HYDROXY-,ETHYL ESTER,CARBANILATE(ESTER) (8CI); EP-475; 3-ETHOXYCARBONYLAMINOPHENYL-N-PHENYLCARBAMATE; ETHYL M-HYDROXYCARBANILATE CARBANILATE (ESTER); ETHYL PHENYLCARBAMOYLOXYPHENYLCARBAMATE; SCHERING 38107; SN 38107; SYNBETAN D

Physical Properties

Freezing Point: 120 °C (248 °F)
Water Solubility: 7 mg/l 20 °C
Other Solubilities: 400 g/l acetone, 180 g/l methanol, 400 g/l isophorone, 149 g/l ethyl acetate, 80 g/l chloroform, 0.5 g/l hexane

RTECS Toxicity Data

Acute Oral: Rat LD_{50} Dose: 9600 mg/kg. Mouse LD_{50} Dose: >500 mg/kg.
Acute Dermal: Rabbit LD_{50} Route: Skin; Dose: >2 gm/kg.

Hazard Overviews

Carcinogenicity: IARC - Not listed; NIOSH - Not listed; NTP - Not listed; ACGIH - Not listed; OSHA - Not listed; EPA - Not listed; MAK - Not listed

Environmental

Regulations

RCRA 40CFR: Not listed
CERCLA: 40CFR 302.4: Not listed
SARA 40CFR 372.65: Listed
SARA EHS 40CFR 355: Not listed
TSCA: Not listed

DES7400 CAS #: 1014-69-3

DESMETRYNE

RTECS: XZ0175000
DOT: UN2763
EINECS Number: 213-800-1
Molecular Formula: $C_8H_{15}N_5S$
Formula Weight: 213.3
Synonyms: DESMETRYN; EPA PESTICIDE CHEMICAL CODE 080810; G 34360; GS 34360; 2-ISOPROPILAMINO-4-METILAMINO-6-METILTIO-1,3,5-TRIAZINA; 2-ISOPROPYLAMINO-4-METHYLAMINO-6-METHYLMERCAPTO-S-TRIAZINE; 2-(ISOPROPYLAMINO)-4-(METHYLAMINO)-6-(METHYLTHIO)-S-TRIAZINE; 2-ISOPROPYLAMINO-4-METHYL-AMINO-6-METHYLTHIO-S-TRIAZINE; 2-ISOPROPYLAMINO-4-METHYLAMINO-6-METHYLTHIO-1,3,5-TRIAZINE; 2-METHYLAMINO-4-METHYLTHIO-6-ISOPROPYLAMINO-1,3,5-TRIAZINE; METHYLMERCAPTO-4-ISOPROPYLAMINO-6-METHYLAMINO-S-TRIAZINE; 2-METHYLMERCAPTO-4-METHYLAMINO-6-ISOPROPYLAMINO-S-TRIAZINE; N-METHYL-N'-(1-METHYLETHYL)-6-(METHYLTHIO)-1,3,5-TRIAZINE-2,4-DIAMINE; 2-METHYLTHIO-4-ISOPROPYLAMINO-6-METHYLAMINO-S-TRIAZINE; 2-(METHYLTHIO)-4-(METHYLAMINO)-6-(ISOPROPYLAMINO)-S-TRIAZINE; NORAMETRYNE; SAMURON; SEMERON; SEMERON 25; SEMERON 250; TOPUSYN; 1,3,5-TRIAZINE-2,4-DIAMINE,N-METHYL-N'-(1-METHYLETHYL)-6-(METHYLTHIO)-; S-TRIAZINE,2-(ISOPROPYLAMINO)-4-(METHYLAMINO)-6-(METHYLTHIO)-
Description: colorless to white crystalline powder or solid
Use: selective post-emergence herbicide effective for control of chenopodium album & other broadleaved & grassy weeds in most brassica crops

Physical Properties

Freezing Point: 84 °C (183.2 °F) to 86 °C (186.8 °F)
Density: 1.172 g/cu cm at 20 °C
Vapor Pressure: 0.133 mPa at 20 °C
Water Solubility: 580 mg/L Water at 20 °C
Other Solubilities: 230 g/kg Acetone at 20 °C; 200 g/kg dichloromethane at 20 °C; 2.6 g/kg hexane at 20 °C; 300 g/kg Methanol at 20 °C; 100 g/kg octan-1-ol at 20 °C; 200 mg/kg Toluene at 20 °C.

RTECS Toxicity Data

Acute Oral: Rat LD_{50} Dose: 1390 mg/kg. Mouse LD_{50} Dose: 700 mg/kg.
Acute Inhalation: Rat LC_{50} Dose: 1563 gm/m^3/1hr.
Acute Dermal: Rat LD_{50} Route: Skin; Dose: >1 gm/kg.
Mutagenic: Yeast - S Cerevisiae Gene Conversion; Dose: 50 mg/L.

Hazard Overviews

Carcinogenicity: IARC - Not listed; NIOSH - Not listed; NTP - Not listed; ACGIH - Not listed; OSHA - Not listed; EPA - Not listed; MAK - Not listed

Environmental

Ecotoxicity: Crustaceans: Daphnia magna 48h EC_{50} 26 mg/l
Environmental Fate: If released to soil, it will be expected to exhibit high mobility based upon a K_{oc} estimated from water solubility and, therefore, it may leach to groundwater. It may be susceptible to slow hydrolysis in soil. Volatilization from near-surface soil is not expected to be an important removal mechanism. Greater than 60 weeks time has been reported as the time required for activity to decrease to 50% inhibition of Italian rye grass at an application rate of 4 pounds/acre. It is unknown whether it will biodegrade in soil. If released to water, it will not be expected to bioconcentrate in aquatic organisms, adsorb to sediment and suspended particulate matter, or to volatilize. It may be susceptible to slow hydrolysis in water and photooxidation by photochemically produced hydroxyl radicals in water, based upon data for s-triazine herbicides with similar structures such as prometryn. It is unknown whether it will biodegrade in natural waters. If released to the atmosphere, it may be subject to gas-phase reaction with hydroxyl radicals, a half-life of about 2.7 hr estimated for this process.
Cleanup/Disposal: Guide No. 151: Do not touch damaged containers or spilled material unless wearing appropriate protective clothing. Stop leak if you can do it without risk. Prevent entry into waterways, sewers, basements or confined areas. Cover with plastic sheet to prevent spreading. Absorb or cover with dry earth, sand or other non-combustible material and transfer to containers. Do not get water inside containers.

Environmental Physical Data

Henry's Law Constant: calculated at 4.82 x10^{-10}
Sorption Partition Coefficient: K_{oc} = estimated at 130
BCF: estimated at 17

Regulations

RCRA 40CFR: Not listed
CERCLA: 40CFR 302.4: Not listed
SARA 40CFR 372.65: Not listed
SARA EHS 40CFR 355: Not listed
TSCA: Not listed

DES9000 CAS #: 973-21-7

DESSIN

RTECS: FF9100000
EINECS Number: 213-546-1
Molecular Formula: $C_{14}H_{18}N_2O_7$
Formula Weight: 326.30
Synonyms: ACREX; AKREX; 2-SEC-BUTYL-4,6-DINITROPHENYL ISOPROPYL CARBONATE; CARBONIC ACID,2-SEC-BUTYL-4,6-DINITROPHENYL ISOPROPYLESTER; CARBONIC ACID,1-METHYLETHYL 2-(1-METHYLPROPYL)-4,6-DINITROPHENYL ESTER (9CI); CARBONIC ACID,1-METHYLETHYL2-(1-METHYLPROPYL)-4,6-DINITROPHENYL ESTER; 2,4-DINITRO-6-SEC-BUTYLPHENYL ISOPROPYL CARBONATE; DINITRO-SEC-BUTYLPHENYL ISOPROPYL CARBONATE; 2,4-DINITRO-6-SEK.BUTYL-ISOPROPYLPHENYLCARBONAT; DINOBUTON; DINOFEN; DRAWINOL; DS 18302; ENT 27,244; ISOPHEN; ISOPHEN (PESTICIDE); ISOPROPYL 2,4-DINITRO-6-SEC-BUTYLPHENYL CARBONATE; ISOPROPYL-2-(1-METHYL-N-PROPYL)-4,6-DINITROPHENYLCARBONATE; KASEBON; MC 1053; 1-METHYLETHYL 2-(1-METHYLPROPYL)-4,6-DINITROPHENYLCARBONATE; 2-(1-METHYL-2-PROPYL)-4,6-DINITROPHENYL ISOPROPYLCARBONATE; OMS 1056; PHENOL,2-SEC-BUTYL-4,6-DINITRO-,ISOPROPYLCARBONATE; (2-SEK.BUTYL-4,6-DINITROFENYL)-ISOPROPYLKARBONAT; SYTASOL; TALAN; UC 19786; UNION CARBIDE 19786
Description: pale yellow crystals
Use: non-systemic acaricide & fungicide

Physical Properties

Freezing Point: 56 °C (132.8 °F) to 57 °C (134.6 °F)
Water Solubility: Practically Insoluble in Water
Other Solubilities: Soluble in aliphatic hydrocarbons, Ethanol, fatty oils. highly Soluble in lower aliphatic Ketones, aromatic hydrocarbons

RTECS Toxicity Data

Acute Dermal: Rat LD_{Lo} Route: Skin; Dose: 1500 mg/kg.
Mutagenic: Mouse Cytogenetic Analysis; Route: Oral; Dose: 25 mg/kg.

Hazard Overviews

Carcinogenicity: IARC - Not listed; NIOSH - Not listed; NTP - Not listed; ACGIH - Not listed; OSHA - Not listed; EPA - Not listed; MAK - Not listed

Environmental

Regulations

RCRA 40CFR: Not listed
CERCLA: 40CFR 302.4: Not listed
SARA 40CFR 372.65: Not listed
SARA EHS 40CFR 355: Not listed
TSCA: Not listed

DEU5000 CAS #: 7782-39-0

DEUTERIUM

DOT: UN1957; IMO2.0
EINECS Number: 231-952-7
Molecular Formula: D_2
Structured MF: D_2
Formula Weight: 4.032

D——D

Chemical Structure

Synonyms: D; DEUTERIUM MOLECULE; DIPLOGEN; HEAVY HYDROGEN; HYDROGEN-2; HYDROGEN,ISOTOPE OF MASS 2
Description: colorless gas; odorless
Use: tracer in establishment of rates & kinetics of chemical reactions; explosive in hydrogen bomb; moderator for nuclear reactors

Physical Properties

Boiling Point: -249.49 °C (-417 °F)
Freezing Point: -254.43 °C (-425.974 °F) at 128.5 mm Hg
Specific Gravity: Liquid 0.169 at -250.9 °C
Vapor Pressure: 17.14 kPa
Water Solubility: Slightly Soluble in Cold Water
Critical Temperature: -234.75 °C
Critical Pressure: 16.432 atm
Ionization Potential (eV): 15.4667 +/-0.0001
Autoignition Temperature: 585 °C
LEL: 5% v/v
UEL: 75% v/v

Hazard Overviews

Compressed Gas

Fire Diamond

Health: Acute Effects: may be harmful; exposure can cause nausea; dizziness; headache.
Fire: Will burn. Hazards: emits toxic fumes; do not expose to air and fire; in powder form capable of creating a dust explosion; liberates flammable/explosive hydrogen gas; forms explosive mixtures in air; vapor may travel considerable distance to source of ignition and flash back. Extinguishing agents: do not extinguish burning gas if flow cannot be shut off immediately; use water spray or fog nozzle to keep cylinder cool; move cylinder away from fire if there is no risk. Precautions: combustible liquid.
Reactivity: Incompatible with: air sensitive, moisture sensitive, strong acids, store away from heat and direct sunlight. Hazardous decomposition products: titanium/titanium oxides nickel/nickel oxides liberates flammable/explosive hydrogen gas.
Carcinogenicity: IARC - Not listed; NIOSH - Not listed; NTP - Not listed; ACGIH - Not listed; OSHA - Not listed; EPA - Not listed; MAK - Not listed

Environmental

Cleanup/Disposal: Guide No. 115: Eliminate all ignition sources (no smoking, flares, sparks or flames in immediate area). All equipment used when handling the product must be grounded. Do not touch or walk through spilled material. Stop leak if you can do it without risk. If possible, turn leaking containers so that gas escapes rather than liquid. Use water spray to reduce vapors or divert vapor cloud drift. Do not direct water at spill or source of leak. Prevent spreading of vapors through sewers, ventilation systems and confined areas. Isolate area until gas has dispersed.

Regulations

RCRA 40CFR: Not listed
CERCLA: 40CFR 302.4: Not listed
SARA 40CFR 372.65: Not listed
SARA EHS 40CFR 355: Not listed
TSCA: Listed

DEX1000 CAS #: 50-02-2

DEXAMETHASONE

RTECS: TU3980000
EINECS Number: 200-003-9
Molecular Formula: $C_{22}H_{29}FO_5$
Formula Weight: 392.45

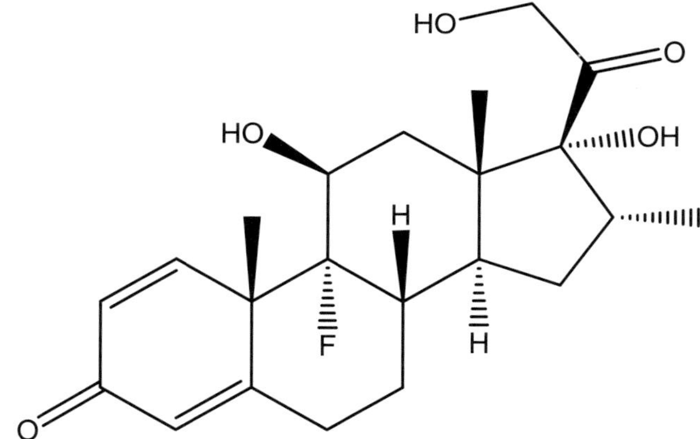

Chemical Structure

Synonyms: AEROSEB-D; AEROSEB-DEX; ANAFLOGISTICO; APHTASOLON; AZIUM; CALONAT; CORSON; CORSONE; CORTISUMMAN; DECACORTIN; DECADERM; DECADRON; DECALIX; DECASONE; DECASPRAY; DECTANCYL; 1-DEHYDRO-16-ALPHA-METHYL-9-ALPHA-FLUOROHYDROCORTISONE; 1-DEHYDRO-16ALPHA-METHYL-0ALPHA-FLUOROHYDROCORTISONE; DEKACORT; DELTAFLUORENE; DERGRAMIN; DERONIL; DESADRENE; DESAMETASONE; DESAMETHASONE; DESAMETON; DESERONIL; DEXA; DEXA MAMALLET; DEXACORT; DEXACORTAL; DEXA-CORTIDELT; DEXACORTIN; DEXA-CORTISYL; DEXADELTONE; DEXAFARMA; DEXALONA; DEXAMETH; DEXAMETHASONE ALCOHOL; DEXAMETHAZONE; DEXAPOLCORT; DEXAPOS; DEXAPROL;

DEXA-SCHEROSON; DEXASCHEROSON; DEXA-SINE; DEXASON; DEXASONE; DEX-IDE; DEXINOLON; DEXINORAL; DEXONE; DEXTELAN; DEZONE; DINORMON; 3,20-DIONE; DIONE; DXMS; FLUORMONE; FLUOROCORT; DELTA(SUP 1)-9-ALPHA-FLUORO-16-ALPHA-METHYLCORTISOL; DELTA(SUP 1)-9-ALPHA-FLUORO-16-ALPHA-METHYLCORTISOL; 9-ALPHA-FLUORO-16-ALPHA-METHYLPREDNISOLONE; 9ALPHA-FLUORO-16ALPHA-METHYLPREDNISOLONE; 9-ALPHA-FLUORO-16-ALPHA-METHYL-1,4-PREGNADIENE-11-BETA,17-ALPHA,21-TRIOL-3,2; 9-ALPHA-FLUORO-16-ALPHA-METHYL-1,4-PREGNADIENE-11-BETA,17-ALPHA,21-TRIOL-3,20-DIONE; 4-ALPHA-FLUORO-16-ALPHA-METHYL-11-BETA,17,21-TRIHYDROXYPREGNA-1,4-DIENE-3,20; 4-ALPHA-FLUORO-16-ALPHA-METHYL-11-BETA,17,21-TRIHYDROXYPREGNA-1,4-DIENE-3,20-DIONE; 9-ALPHA-FLUORO-11-BETA,17-ALPHA,21-TRIHYDROXY-16-ALPHA-METHYL-1,4-PREGNADIEN; 9-ALPHA-FLUORO-11-BETA,17-ALPHA,21-TRIHYDROXY-16-ALPHA-METHYLPREGNA-1,4-DIEN; (11BETA,16ALPHA)-9-FLUORO-11,17,21-TRIHYDROXY-16-METHYLPREGNA-1,4-DIENE-3,20-DIONE; 9-ALPHA-FLUORO-11-BETA,17-ALPHA,21-TRIHYDROXY-16-ALPHA-METHYLPREGNA-1,4-DIENE-3,20-DIONE; 9-FLUORO-11,17,21-TRIHYDROXY-16-METHYLPREGNA-1,4-DIENE-3,20-DIONE; 9-FLUORO-11-BETA,17,21-TRIHYDROXY-16-ALPHA-METHYL-PREGNA-1,4-DIENE-3,20-DIONE; 9-FLUORO-11-BETA,17,21-TRIHYDROXY-16-ALPHA-METHYLPREGNA-1,4-DIENE-3,20-DIONE; FORTECORTIN; GAMMACORTEN; HEXADECADROL; HEXADROL; HL-DEX; ISOPTO-DEX; LOKALISON F; LOVERINE; LUXAZONE; MAXIDEX; 16-ALPHA-METHYL-9-ALPHA-FLUORO-DELTA(SUP 1)-HYDROCORTISONE; 16-ALPHA-METHYL-9-ALPHA-FLUORO-DELTA(SUP 1)HYDROCORTISONE; 16-ALPHA-METHYL-9-ALPHA-FLUORO-1-DEHYDROCORTISOL; 16ALPHA-METHYL-9ALPHA-FLUORO-DELTA1-HYDROCORTISONE; 16-ALPHA-METHYL-9-ALPHA-FLUOROPREDNISOLONE; 16ALPHA-METHYL-9ALPHA-FLUOROPREDNISOLONE; 16-ALPHA-METHYL-9-ALPHA-FLUORO-1,4-PREGNADIENE-11-BETA,17-ALPHA,21-TRIOL-3,2; 16-ALPHA-METHYL-9-ALPHA-FLUORO-1,4-PREGNADIENE-11-BETA,17-ALPHA,21-TRIOL-3,20-DIONE; 16ALPHA-METHYL-9ALPHA-FLUORO-1,4-PREGNADIENE-11BETA,17ALPHA,21-TRIOL-2,20-DIONE; 16-ALPHA-METHYL-9-ALPHA-FLUORO-11-BETA,17-ALPHA,21-TRIHYDROXYPREGNA-1,4-DIEN; 16-ALPHA-METHYL-9-ALPHA-FLUORO-11-BETA,17-ALPHA,21-TRIHYDROXYPREGNA-1,4-DIENE-3,20-DIONE; MEXIDEX; MILLICORTEN; MK 125; OCU-TROL; ORADEXON; PET DERM III; POLICORT; PREDNISOLON F; PREDNISOLONE F; PREGNA-1,4-DIENE-3,20-DIONE,9-FLUORO-11,17,21-TRIHYDROXY-16-METHYL-,(11BETA,16ALPHA)-; PREGNA-1,4-DIENE-3,20-DIONE,9-FLUORO-11BETA,17,21-TRIHYDROXY-16ALPHA-METHYL-; SK-DEXAMETHASONE; SPOLOVEN; SUPERPREDNOL; VISUMETAZONE

Description: white to practically white crystalline powder; odorless

Use: a glucocorticoid and anti-inflammatory agent; in the prevention of the neonatal respiratory distress syndrome, to reduce raised intracranial pressure and to diagnose Cushing's syndrome; in veterinary medicine as an adrenocorticoid steroid

Physical Properties

Freezing Point: 262 °C (503.6 °F) to 264 °C (507.2 °F)

Water Solubility: 10 mg/100 ml at 25 °C

Other Solubilities: Insoluble in Dioxane; slightly Soluble in Alcohol; Insoluble in Ether and Chloroform.

Flash Point: Not available; probably combustible

RTECS Toxicity Data

Acute Oral: Rat LD_{50} Dose: >3 gm/kg.

Acute Dermal: Rabbit LD_{50} Route: Subcutaneous Dose: 7200 ug/kg; Toxic Effects: Sense organs and special senses - Lacrimation; Gastrointestinal - Hypermotility, diarrhea. Rat LD_{50} Route: Subcutaneous Dose: 14 mg/kg; Toxic Effects: Nutritional and gross metabolic - Weight loss or decreased weight gain.

Chronic (Multiple Dose) Oral: Rat Dose: 70 mg/kg/14D-I; Toxic Effects: Endocrine - Changes in spleen weight; Endocrine - Changes in thymus weight. Mouse Dose: 3821 ug/kg/14D-C; Toxic Effects: Endocrine - Changes in spleen weight; Endocrine - Changes in thymus weight; Blood - Other changes.

Chronic (Multiple Dose) Inhalation: Rat Dose: 26700 ug/m^3/30M/21D-I; Toxic Effects: Endocrine - Other changes; Endocrine - Changes in thymus weight; Blood - Changes in erythrocite (RBC) cell count.

Chronic (Multiple Dose) Dermal: Rat Route: Skin; Dose: 7500 ug/kg/30D-I; Toxic Effects: Endocrine - Changes in thymus weight; Nutritional and gross metabolic - Weight loss or decreased weight gain; Biochemical - Transaminases. Rat Route: Subcutaneous; Dose: 250 ug/kg/26W-I; Toxic Effects: Endocrine - Other changes; Blood - Changes in serum composition; DEATH - Changes in testicular weight. Rat Route: Subcutaneous; Dose: 280 ug/kg/28D-I; Toxic Effects: Endocrine - Changes in adrenal weight; Endocrine - Changes in spleen weight; Nutritional and gross metabolic - Weight loss or decreased weight gain.

Reproductive/Teratogenic: Rat Route: Oral; Dose: 6 mg/kg; Duration: female 8-13D of pregnancy; Specific Developmental Abnormalities - Cardiovascular (circulatory) system. Rat Route: Oral; Dose: 2 mg/kg; Duration: female 6-15D of pregnancy; Effects on Embryo or Fetus - Fetal death; Specific Developmental Abnormalities - Craniofacial (including nose and tongue). Rat Route: Oral; Dose: 880 ug/kg; Duration: female 7-17D of pregnancy; Effects on Embryo or Fetus - Extra embryonic structures; Fetotoxicity; Specific Developmental Abnormalities - Eye, ear. Rat Route: Oral; Dose: 8800 ug/kg; Duration: female 7-17D of pregnancy; Effects on Fertility - Litter size; Effects on Embryo or Fetus - Fetal death; Other effects to embryo or fetus. Rat Route: Oral; Dose: 7500 ug/kg; Duration: female 15-17D of pregnancy Specific Developmental Abnormalities - Craniofacial (including nose and tongue); Body wall. Rat Route: Oral; Dose: 8800 ug/kg; Duration: female 7-17D of pregnancy; Maternal Effects - Parturition; Effects on Newborn - Live birth index. Monkey Route: Intramuscular; Dose: 3700 mg/kg; Duration: female 18-23W of pregnancy Maternal Effects - Parturition; Effects on Newborn - Growth statistics; Physical.

Mutagenic: Human Cytogenetic Analysis; Cell Type: lymphocyte; Dose: 10 mg/L. Human Sister Chromatid Exchenge; Cell Type: lymphocyte; Dose: 1 mg/L.

Hazard Overviews

Health: Irritating to eyes/skin/respiratory tract. Harmful. Other Acute Effects may be harmful by inhalation, ingestion, or skin absorption; may cause allergic skin reaction;. Chronic

Effects: possible risk of congenital malformation in the fetus; target organ: pituitary.

Fire: Will burn. Hazards: emits toxic fumes. Extinguishing agents: water spray; carbon dioxide, dry chemical powder or appropriate foam. Precautions: combustible liquid.

Reactivity: Incompatible with: strong oxidizing agents, sensitive to light. Hazardous decomposition products: toxic fumes of: carbon monoxide, carbon dioxide, hydrogen fluoride.

Carcinogenicity: IARC - Not listed; NIOSH - Not listed; NTP - Not listed; ACGIH - Not listed; OSHA - Not listed; EPA - Not listed; MAK - Not listed

Primary Target Organs:

Eyes

Skin

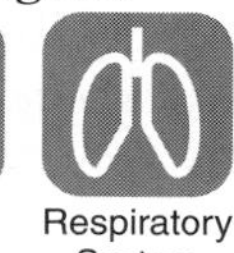
Respiratory System

Glandular System

Environmental

Regulations

RCRA 40CFR: Not listed
CERCLA: 40CFR 302.4: Not listed
SARA 40CFR 372.65: Not listed
SARA EHS 40CFR 355: Not listed
TSCA: Listed

DEX2600 CAS #: 25523-97-1

DEXCHLORPHENIRAMINE

EINECS Number: 247-073-7
Molecular Formula: $C_{16}H_{19}ClN_2$
Formula Weight: 274.80

Synonyms: CHLO-AMINE; D-2-[P-CHLORO-ALPHA-(2-DIMETHYLAMINOETHYL)BENZYL]PYRIDINE; (+)-CHLORPHENIRAMINE; D-CHLORPHENIRAMINE; FORTAMINE; ISOMERINE; PHENDEXTRO; POLARAMINE; PYRIDINE,2-(P-CHLORO-ALPHA-(2-(DIMETHYLAMINO)ETHYL)BENZYL)-; 2-PYRIDINEPROPANAMINE,GAMMA-(4-CHLOROPHENYL)-N,N-DIMETHYL-,(S)-

Description: oily liquid; odorless

Use: antihistamine for allergic & vasomotor rhinitis; antihistamine for mild urticaria and angioedema; antihistamine for insect bite reactions; antihistamine for pruritus of allergic reactions

Physical Properties

Boiling Point: 142 °C (288 °F) at 1.0 mm Hg
Freezing Point: Crystals at 113 °C (235.4 °F) to 115 °C (239 °F)
Water Solubility: 1 g/4 mL Water
Other Solubilities: Soluble in Alcohol, Chloroform; Slightly Soluble in Benzene, Ether
pH: 1% solution 4 to 5
Refraction Index: Alpha 1.509

Hazard Overviews

Carcinogenicity: IARC - Not listed; NIOSH - Not listed; NTP - Not listed; ACGIH - Not listed; OSHA - Not listed; EPA - Not listed; MAK - Not listed

Environmental

Regulations

RCRA 40CFR: Not listed
CERCLA: 40CFR 302.4: Not listed
SARA 40CFR 372.65: Not listed
SARA EHS 40CFR 355: Not listed
TSCA: Not listed

DEX4200 CAS #: 140-56-7

DEXON

RTECS: CZ1750000
EINECS Number: 205-419-4
Molecular Formula: $C_8H_{10}N_3NaO_3S$
Formula Weight: 251.26

Synonyms: BAY 22555; BAY 5072; BAY 72555; BAYER 22,555; BAYER 22555; BAYER 5072; BAYER 22 555; BENZENEDIAZOSULFONIC ACID,P-(DIMETHYLAMINO)-,SODIUM SALT; DAPA; DAPA (PESTICIDE); DAS; DEKSONAL; DEXON 70; DEXOXON; DIAZENESULFONIC ACID,(4-(DIMETHYLAMINO)PHENYL)-,SODIUMSALT; DIAZOBEN; P-DIMETHYLAMINOBENZENE DIAZO SODIUM SULFONATE; P-DIMETHYLAMINOBENZENE DIAZOSULFONIC ACID,SODIUM SALT; P-DIMETHYLAMINOBENZENEDIAZO SODIUM SULFONATE; P-DIMETHYLAMINOBENZENEDIAZOSODIUM SULPHONATE; PARA-DIMETHYLAMINOBENZENEDIAZOSODIUM SULPHONATE; P-(DIMETHYLAMINO)BENZENEDIAZOSULFONATE; PARA-(DIMETHYLAMINO)BENZENEDIAZOSULFONATE; P-(DIMETHYLAMINO)BENZENEDIAZOSULFONIC ACID SODIUM SALT; 4-DIMETHYLAMINOBENZENEDIAZOSULFONIC ACID,SODIUM SALT; P-(DIMETHYLAMINO)BENZENEDIAZOSULFONIC ACID,SODIUM SALT; P-DIMETHYLAMINOBENZENEDIAZOSULFONIC ACID,SODIUM SALT; PARA-(DIMETHYLAMINO)-BENZENE-DIAZOSULFONIC ACID,SODIUM SALT; P-(DIMETHYLAMINO)BENZENEDIAZOSULPHONATE; 4-DIMETHYLAMINOBENZENEDIAZOSULPHONIC ACID,SODIUM SALT; P-(DIMETHYLAMINO)BENZENEDIAZOSULPHONIC ACID,SODIUM SALT; P-DIMETHYLAMINOBENZOLDIAZOSULFONAT (NATRIUMSALZ); (4-(DIMETHYLAMINO)PHENYL)DIAZENE SULFONIC ACID,SODIUM SALT; 4-((DIMETHYLAMINO)PHENYL)DIAZENE SULPHONIC ACID,SODIUMSALT; 4-(DIMETHYLAMINO)PHENYL)DIAZENESULFONATE; (4-(DIMETHYLAMINO)PHENYL)DIAZENESULFONIC ACID,SODIUM SALT; 4-((DIMETHYLAMINO)PHENYL)DIAZENESULFONIC ACID,SODIUM SALT; P-(DIMETHYLAMINO)-PHENYLDIAZO-NATRIUMSULFONAT; N,N-DIMETHYL-P-ANILINEDIAZOSULFONIC ACID SODIUM SALT; ENIAMETHYL ORANGE; FENAMINOSULF; GOLD ORANGE HP; GOLD ORANGE MP; HELIANTHIN; LESAN; METHYL ORANGE; ORANGE III; PEHNAMINOSULF; PHENAMINOSULF; SODIUM 4-(DIMETHYLAMINO)BENZENE DIAZOSULFONATE; SODIUM P-(DIMETHYLAMINO)BENZENE DIAZOSULFONATE; SODIUM 4-(DIMETHYLAMINO)BENZENEDIAZOSULFONATE; SODIUM 4-DIMETHYLAMINOBENZENEDIAZOSULFONATE; SODIUM P-(DIMETHYLAMINO)BENZENEDIAZOSULFONATE; SODIUM 4-

(DIMETHYLAMINO)BENZENEDIAZOSULPHONATE; SODIUM 4-DIMETHYLAMINOBENZENEDIAZOSULPHONATE; SODIUM P-(DIMETHYLAMINO)BENZENEDIAZOSULPHONATE; SODIUM (4-(DIMETHYLAMINO)PHENYL)DIAZENESULFONATE; SODIUM FENAMINOSULF; TROPAEOLIN D

Description: yellowish-brown powder; odorless

Use: fungicide for protection of germinating seed and seedlings

Physical Properties

Freezing Point: Decomposes at 200 °C (392 °F)

Water Solubility: 2 to 3 g/100 ml at 25 °C

Other Solubilities: 95% Ethanol: <1 mg/ml at 22.5 °C; Acetone: <1 mg/ml at 22.5 °C; Benzene: Insoluble; DMSO: 1-10 mg/ml at 22.5 °C; Diethyl Ether: Insoluble; Dimethyl formamide: Soluble; Petroleum oils: Insoluble.

Flash Point: Not available; probably combustible

RTECS Toxicity Data

Acute Oral: Rat LD_{50} Dose: 60 mg/kg. Mouse LD_{Lo} Dose: 140 mg/kg.

Acute Dermal: Rat LD; Route: Skin; Dose: >100 mg/kg.

Reproductive/Teratogenic: Rat Route: Oral; Dose: 25 mg/kg; Duration: female 7-11D of pregnancy; Effects on Embryo or Fetus - Fetal death; Specific Developmental Abnormalities - Musculoskeletal system.

Mutagenic: Human Sister Chromatid Exchenge; Cell Type: lymphocyte; Dose: 10 umol/L. Hamster Cytogenetic Analysis; Cell Type: lung; Dose: 500 umol/L.

Hazard Overviews

Health: Irritating to eyes/skin/respiratory tract. Toxic. Other Acute Effects: harmful if swallowed; may be harmful if inhaled; may be harmful if absorbed through the skin; possible risk of irreversible effects; exposure can cause hyperglycemia; lethargy; possibly death. Chronic Effects: laboratory experiments have shown mutagenic effects; target organs: kidneys, liver.

Fire: Will burn. Hazards: emits toxic fumes. Extinguishing agents: carbon dioxide, dry chemical powder or appropriate foam. Precautions: combustible liquid.

Reactivity: Incompatible with: strong oxidizing agents. Hazardous decomposition products: toxic fumes of: carbon monoxide, carbon dioxide.

Carcinogenicity: IARC - Group 3, Not classifiable as to carcinogenicity to humans; NIOSH - Not listed; NTP - Not listed; ACGIH - Not listed; OSHA - Not listed; EPA - Not listed; MAK - Not listed

Primary Target Organs:

Eyes

Skin

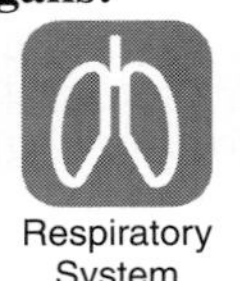
Respiratory System

Liver

Kidneys

Environmental

Ecotoxicity: LC_{50} Rainbow trout more than 60 mg/l/96 hr at 13 °C, wt 1.5 g. Static bioassay without aeration, pH 7.2-7.5, water hardness 40-50 mg/l as calcium carbonate and alkalinity of 30-35 mg/l LC_{50} Gammarus fasciatus 3.7 mg/l/96 hr at 21 °C, (mature stage) (95% confidence limit 2.7-5.0 mg/l). Static bioassay without aeration, pH 7.2-7.5, water hardness 40-50 mg/l as calcium carbonate and alkalinity of 30-35 mg/l LC_{50} Goldfish >10 mg/l/24 hr /Conditions of bioassay not specified LC_{50} Coho salmon more than 100 mg/l/96 hr at 13 °C, wt 1.4 g. Static bioassay without aeration, pH 7.2-7.5, water hardness 40-50 mg/l as calcium carbonate and alkalinity of 30-35 mg/l

Cleanup/Disposal: Guide No. 151: Do not touch damaged containers or spilled material unless wearing appropriate protective clothing. Stop leak if you can do it without risk. Prevent entry into waterways, sewers, basements or confined areas. Cover with plastic sheet to prevent spreading. Absorb or cover with dry earth, sand or other non-combustible material and transfer to containers. Do not get water inside containers.

Regulations

RCRA 40CFR: Not listed

CERCLA: 40CFR 302.4: Not listed

SARA 40CFR 372.65: Not listed

SARA EHS 40CFR 355: Not listed

TSCA: Not listed

DEX5800 CAS #: 81-13-0

DEXPANTHENOL

RTECS: ES4316000

EINECS Number: 201-327-3

Molecular Formula: $C_9H_{19}NO_4$

Formula Weight: 205.25

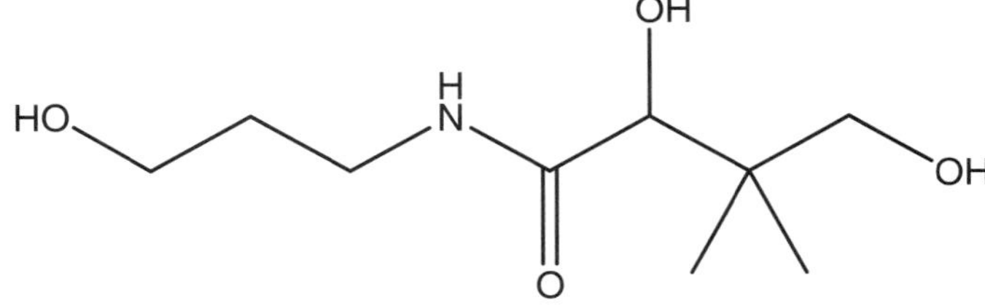

Chemical Structure

Synonyms: ALCOPAN-250; BEPANTHEN; BEPANTHENE; BEPANTOL; BUTANAMIDE,2,4-DIHYDROXY-N-(3-HYDROXYPROPYL)-3,3-DIMETHYL-,-; BUTANAMIDE,2,4-DIHYDROXY-N-(3-HYDROXYPROPYL)-3,3-DIMETHYL-,-(9CI); BUTYRAMIDE,2,4-DIHYDROXY-N-(3-HYDROXYPROPYL)-3,3-DIMETHYL-,D-(+)-; COZYME; D-P-A INJECTION; 2,4-DIHYDROXY-N-(3-HYDROXYPROPYL)3,3-DIMETHYL BUTYRAMIDE; -2,4-DIHYDROXY-N-(3-HYDROXY-PROPYL)-3,3-DIMETHYLBUTANAMIDE; D(+)-ALPHA,GAMMA-DIHYDROXY-N-(3-HYDROXYPROPYL)-BETA,BETA-DIMETHYLBUTYRAMIDE; D-(+)-2,4-DIHYDROXY-N-(3-HYDROXYPROPYL)-3,3-DIMETHYLBUTYRAMIDE; ILOPAN; INTRAPAN; MOTILYN; PANADON; PANTENYL; D(+)-PANTHENOL; D-PANTHENOL; PANTHENOL; PANTHENOL,(+)-; PANTHODERM; PANTOL; D-PANTOTHENOL; PANTOTHENOL; D(+)-PANTOTHENYL ALCOHOL; D-PANTOTHENYL ALCOHOL; PANTOTHENYL ALCOHOL; PANTOTHENYLOL; N-PANTOYL-3-PROPANOLAMINE;

PENTHENOL; PROPANOLAMINE,N-PANTOYL-; PROVITAMIN B; SYNAPAN; THENALTON; ZENTINIC

Description: viscous oily liquid

Use: medication (vet): nutritional factor; dietary source of pantothenic acid; medication: topical treatment of burns, wounds, decubitus ulcers, eczema, contact dermatitis, & pruritus; adjunct therapy in treatment of lupus erythematosus; biochemical research; food additive

Physical Properties

Boiling Point: 118 °C (244 °F) to 120 °C (248 °F) at 0.02 mm Hg

Specific Gravity: 1.2 at 20 °C/20 °C

Water Solubility: Freely Soluble in Water

Other Solubilities: hydrolyzed by alkali & strong acid

pH: About 9.5

Refraction Index: 1.497

RTECS Toxicity Data

Acute Oral: Mouse LD_{50} Dose: 15 gm/kg.

Irritation Eye: Rabbit Standard Draize Test Dose: 500 ug; Reaction: mild.

Irritation Skin: Rabbit Standard Draize Test Dose: 500 mg/4H; Reaction: mild.

Hazard Overviews

Health: May cause irritation. Other Acute Effects: may be harmful by inhalation, ingestion, or skin absorption. The toxicological properties have not been thoroughly investigated.

Fire: Hazards: emits toxic fumes. Extinguishing agents: water spray; carbon dioxide, dry chemical powder or appropriate foam. Precautions: combustible liquid.

Carcinogenicity: IARC - Not listed; NIOSH - Not listed; NTP - Not listed; ACGIH - Not listed; OSHA - Not listed; EPA - Not listed; MAK - Not listed

Environmental

Regulations

RCRA 40CFR: Not listed

CERCLA: 40CFR 302.4: Not listed

SARA 40CFR 372.65: Not listed

SARA EHS 40CFR 355: Not listed

TSCA: Listed

DEX7400 CAS #: 51-64-9

DEXTROAMPHETAMINE

RTECS: SH9100000

EINECS Number: 200-112-1

Molecular Formula: $C_9H_{13}N$

Formula Weight: 135.21

Synonyms: D-AM; D-2-AMINO-1-PHENYLPROPANE; (+)-AMPHETAMINE; (S)-(+)-AMPHETAMINE; (S)-AMPHETAMINE; AMPHETAMINE (D); D-(S)-AMPHETAMINE; D-AMPHETAMINE; AMSUSTAIN; BENZENEETHANAMINE,ALPHA-METHYL-,(S)-; DEPHADREN; DESOXYN; DEXAMPHETAMINE; DEXEDRINE; DEXIDRINE; D-ALPHA-METHYLPHENETHYLAMINE; ALPHA-METHYLPHENETHYLAMINE,D-FORM; (+)-ALPHA-METHYLPHENYLETHYLAMINE; PHENETHYLAMINE,ALPHA-METHYL-,(+)-; PHENETHYLAMINE,ALPHA-METHYL-,D-; D-1-PHENYL-2-AMINOPROPAN; D-1-PHENYL-2-AMINOPROPANE; (S)-ALPHA-PHENYLETHYLAMINE; BETA-PHENYL-ISO-PROPYL AMINE; (S)-(+)-BETA-PHENYLISOPROPYLAMINE; SYMPAMIN

Description: white, crystalline powder, plates, or rods; odorless

Use: drug of abuse; medication: management of children with attention deficit disorder, & as an alternative to methylphenidate in patients with narcolepsy; treatment of obesity, motion sickness, idiopathic edema & depression dextroamphetamine sulfate; medication (vet): sympathomimetic, CNS stimulant

Physical Properties

Boiling Point: 203 °C (397 °F) to 204 °C (399 °F)

Freezing Point: 28 °C (82.4 °F)

Specific Gravity: 0.949 at 15 °C/4 °C

Water Solubility: Slightly Soluble in Water

Other Solubilities: Soluble in Alcohol, Ether.

pH: 5% aqueous solution 5.0 to 6.0

Refraction Index: 1.4704

RTECS Toxicity Data

Acute Oral: Man TD_{Lo} Dose: 42 mg/kg/25W-I; Toxic Effects: Behavioral - Toxic psychosis. Child TD_{Lo} Dose: 3600 ug/kg/10D-I; Toxic Effects: Behavioral - Muscle contraction or spasticity. Monkey LD_{Lo} Dose: 32 mg/kg. Rat LD_{50} Dose: 38 mg/kg.

Acute Dermal: Rat LD_{50} Route: Subcutaneous Dose: 200 mg/kg. Mouse LD_{50} Route: Subcutaneous Dose: 20 mg/kg; Toxic Effects: Behavioral - Excitment; Nutritional and gross metabolic - Body temperature increase.

Reproductive/Teratogenic: Rat Route: Subcutaneous; Dose: 25 mg/kg; Duration: female 5-9D of pregnancy; Effects on Newborn - Behavioral. Rat Route: Subcutaneous; Dose: 8 mg/kg; Duration: female 11-14D of pregnancy Effects on Newborn - Behavioral.

Hazard Overviews

Carcinogenicity: IARC - Not listed; NIOSH - Not listed; NTP - Not listed; ACGIH - Not listed; OSHA - Not listed; EPA - Not listed; MAK - Not listed

Environmental

Regulations

RCRA 40CFR: Not listed

CERCLA: 40CFR 302.4: Not listed

SARA 40CFR 372.65: Not listed

SARA EHS 40CFR 355: Not listed

TSCA: Listed

DEX9000 CAS #: 125-71-3

DEXTROMETHORPHAN

RTECS: QD0194000
EINECS Number: 204-752-2
Molecular Formula: $C_{18}H_{25}NO$
Formula Weight: 271.40
Synonyms: BA 2666; DEXTROMETHORFAN; D-METHORPHAN; DELTA-METHORPHAN; 9ALPHA,13ALPHA,14ALPHA-MORPHINAN,3-METHOXY-17-METHYL-; MORPHINAN,3-METHOXY-17-METHYL-,(9ALPHA,13ALPHA,14ALPHA)-; MORPHINAN,3-METHOXY-17-METHYL-,(9-ALPHA,13-ALPHA,14-ALPHA)-(9CI)
Description: white crystalline powder or crystals; faint odor
Use: antitussive hydrobromide

Physical Properties

Freezing Point: 122 °C (251.6 °F) to 124 °C (255.2 °F)
Water Solubility: About 2% at 25 °C
Other Solubilities: in 95% Ethanol at room temperature (wt/wt); Soluble in Propylene Glycol, Chloroform; Practically Insoluble in Ether; 25% at 85 °C; 10% in Glycerol at room temperature (wt/wt); freely Soluble in Alcohol
pH: 1% aqueous solution 5.2 to 6.5

RTECS Toxicity Data

Acute Oral: Child TD_{Lo} Dose: 16 mg/kg; Toxic Effects: Sense organs and special senses - Other; Behavioral - Change in motor activity (specific assay); Behavioral - Ataxia. Rat LD_{50} Dose: 116 mg/kg; Toxic Effects: Behavioral - Convulsions or effect on seizure threshold; Behavioral - Ataxia; Skin and appendages - Hair.
Acute Dermal: Mouse LD_{50} Route: Subcutaneous Dose: 112 mg/kg.

Hazard Overviews

Carcinogenicity: IARC - Not listed; NIOSH - Not listed; NTP - Not listed; ACGIH - Not listed; OSHA - Not listed; EPA - Not listed; MAK - Not listed

Environmental

Regulations

RCRA 40CFR: Not listed
CERCLA: 40CFR 302.4: Not listed
SARA 40CFR 372.65: Not listed
SARA EHS 40CFR 355: Not listed
TSCA: Not listed

DIA1000 CAS #: 2873-97-4

DIACETONE ACRYLAMIDE

RTECS: AS3475000
EINECS Number: 220-713-2
Molecular Formula: $C_9H_{15}NO_2$
Formula Weight: 169.22

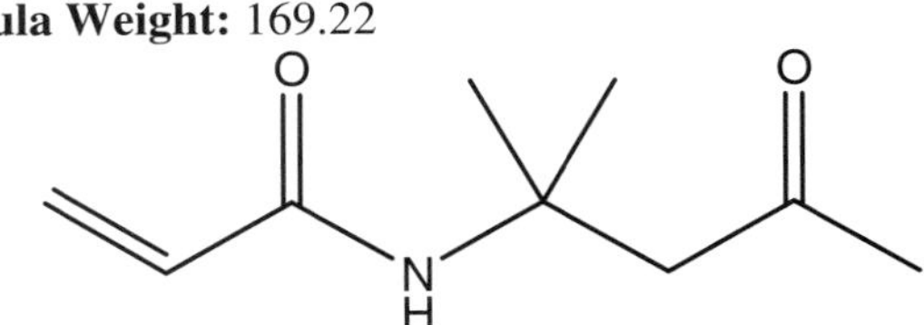

Chemical Structure

Synonyms: ACRYLAMIDE,N,N-DIACETONYL-; ACRYLAMIDE,N-(1,1-DIMETHYL-3-OXOBUTYL)-; DAA; DIALLYL ESTER ACETIC ACID; N-(1,1-DIMETHYL-3-OXOBUTYL)ACRYLAMIDE; N-(1,1-DIMETHYL-3-OXOBUTYL)-2-PROPENAMIDE; N-(2-(2-METHYL-4-OXOPENTYL))ACRYLAMIDE; N-(2-(2-METHYL-4-OXOPENTYL)ACRYLAMIDE; N-[2-(2-METHYL-4-OXOPENTYL)]ACRYLAMIDE; 2-PROPENAMIDE,N,N-BIS(2-OXOPROPYL)-; 2-PROPENAMIDE,N-(1,1-DIMETHYL-3-OXOBUTYL)-
Description: white crystalline solid
Use: in mfr of coatings, laminates, sealers, adhesives, lubricating oils; imparts water tolerance & vapor permeability to copolymer films; cross-linking agent in polyester resins; color photography; orthopedic bandage; in air sprays; hydrophilic monomer, eg, for mfr of soft contact lenses; component of coatings & adhesives; crosslinking agent for polyesters & alkyd molding compds

Physical Properties

Boiling Point: 120 °C (248 °F) at 8 mm Hg
Freezing Point: 57 °C (134.6 °F) to 58 °C (136.4 °F)
Water Solubility: Highly Soluble in Water
Other Solubilities: highly Soluble in most organic solvents

RTECS Toxicity Data

Acute Oral: Rat LD_{50} Dose: 1770 mg/kg; Toxic Effects: Behavioral - Somnolence (general depressed activity); Behavioral - Convulsions or effect on seizure threshold; Lungs, Thorax, or Respiration - Other changes. Mouse LD_{50} Dose: 1303 mg/kg.
Acute Dermal: Rabbit LD_{50} Route: Skin; Dose: >10 gm/kg.

Hazard Overviews

Health: Irritating to eyes/skin. Harmful. Other Acute Effects: harmful if swallowed; may be harmful if inhaled; may be harmful if absorbed through the skin.
Fire: Hazards: emits toxic fumes. Extinguishing agents: water spray; carbon dioxide, dry chemical powder or appropriate foam. Precautions: combustible liquid.
Reactivity: Incompatible with: strong oxidizing agents, protect from moisture, strong bases. Hazardous decomposition products: toxic fumes of: carbon monoxide, carbon dioxide, nitrogen oxides.
Carcinogenicity: IARC - Not listed; NIOSH - Not listed; NTP - Not listed; ACGIH - Not listed; OSHA - Not listed; EPA - Not listed; MAK - Not listed

Primary Target Organs:

Eyes

Skin

Environmental

Regulations

RCRA 40CFR: Not listed
CERCLA: 40CFR 302.4: Not listed
SARA 40CFR 372.65: Not listed
SARA EHS 40CFR 355: Not listed
TSCA: Listed

DIA1150 CAS #: 123-42-2

DIACETONE ALCOHOL

RTECS: SA9100000
DOT: UN1148; IMO3.2
EINECS Number: 204-626-7
Molecular Formula: $C_6H_{12}O_2$
Structured MF: $(CH_3)_2C(OH)CH_2COCH_3$
Formula Weight: 116.16

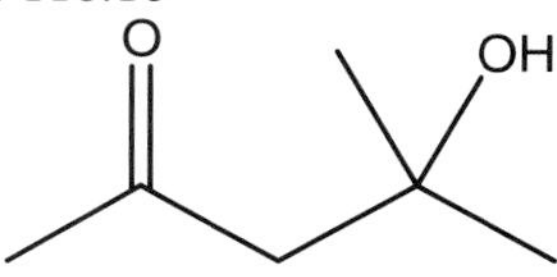

Chemical Structure

Synonyms: ACETONYLDIMETHYLCARBINOL; DIACETONALCOHOL; DIACETONALCOOL; DIACETONALKOHOL; DIACETONE; DIACETONE-ALCOOL; DIACETONYL ALCOHOL; DIKETONE ALCOHOL; DIMETHYL ACETONYL CARBINOL; 4-HYDROXY-2-KETO-4-METHYLPENTANE; 4-HYDROXY-4-METHYL PENTAN-2-ONE; 4-HYDROXY-4-METHYL-PENTAN-2-ON; 4-HYDROXY-4-METHYL-2-PENTANONE; 4-HYDROXY-4-METHYLPENTANONE-2; 4-IDROSSI-4-METIL-PENTAN-2-ONE; 4-METHYL-4-HYDROXY-2-PENTANONE; 2-METHYL-2-PENTANOL-4-ONE; 2-PENTANONE,4-HYDROXY-4-METHYL-; PYRANTON; PYRANTON A; TYRANTON

Description: colorless liquid; mild, pleasant, characteristic odor

Use: intermediate in synth of mesityl oxide, hexalene glycol and other organic chems; solvent for cellulose acetate, nitrocellulose, celluloid, fats, oils, waxes, resins; antifreeze solutions & hydraulic fluids; preservative in pharmaceutical prepn; coating compositions & mercerizations; solvent for certain pigments; manufacture of quick drying inks, photographic film, metal cleaning cmpd; making artificial silk & leather

Physical Properties

Boiling Point: 167.9 °C (334 °F) at 760 mm Hg
Freezing Point: -44 °C (-47.2 °F)
Specific Gravity: 0.9306 at 25 °C/4 °C
Vapor Density: 4.0 Air=1
Vapor Pressure: 1 mm Hg
Water Solubility: Miscible with Water
Other Solubilities: miscible with Alcohol, Ether, other solvents
Odor Threshold: Absolute 0.28 ppm
Refraction Index: 1.4232 at 20 °C/D
Critical Temperature: 334 °C
Critical Pressure: 380 psia
Flash Point: 66 °C Open Cup
Autoignition Temperature: Acetone Free 603 °C
LEL: 1.8% v/v
UEL: 6.9% v/v

RTECS Toxicity Data

Acute Oral: Rat LD_{50} Dose: 4 gm/kg. Mouse LD_{50} Dose: 3950 mg/kg.
Acute Inhalation: Human TC_{Lo} Dose: 100 ppm; Toxic Effects: Sense organs and special senses - Other; Behavioral - Headache; Gastrointestinal - Nausea or vomiting. Human TC_{Lo} Dose: 400 ppm; Toxic Effects: Lungs, Thorax, or Respiration - Other changes.
Acute Dermal: Rabbit LD_{50} Route: Skin; Dose: 13500 mg/kg.
Irritation Eye: Rabbit Standard Draize Test Dose: 20 mg; Reaction: severe.
Irritation Skin: Rabbit Open Draize Test Dose: 500 mg open; Reaction: mild.

Hazard Overviews

Flammable

Fire Diamond

Health: Irritating to eyes/skin/respiratory tract. Also Causes: transient corneal damage, defatting of the skin, dermatitis, narcotic effects (high concentration).
Fire: Flammable. Can form explosive mixtures in the air. Use alcohol foam, water spray, dry chemical, or carbon dioxide. Large fires should be extinguished by alcohol foam or water, using a blanketing effect to smother the fire. Use water spray to reduce the rate of burning and to cool containers.
Reactivity: Stable. Hazardous polymerization cannot occur. Avoid: sources of ignition. Incompatible with: amines; strong oxidizing agents; pyridines; ammonia; isocyanates; inorganic acids; caustics. Hazardous decomposition products: carbon monoxide; mesityl oxide; partial products of oxidation.
Carcinogenicity: IARC - Not listed; NIOSH - Listed as carcinogen; NTP - Not listed; ACGIH - Not listed; OSHA - Not listed; EPA - Not listed; MAK - Not listed
Primary Target Organs:

Eyes

Skin

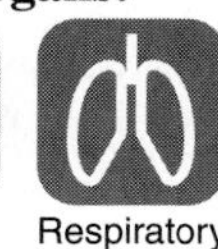
Respiratory System

Mucous Membranes

Exposure Limits
OSHA PEL: TWA: 50 ppm; 240 mg/m^3.
ACGIH TLV: TWA: 50 ppm; 238 mg/m^3.

NIOSH REL: TWA: 50 ppm; 240 mg/m^3.
NIOSH IDLH: 1800 ppm; LEL.
DFG MAK: TWA: 50 ppm; 240 mg/m^3.
Respirator Recommendation
Exposure Range: >50 to 500 ppm Air Purifying, Negative Pressure, Half Mask
Exposure Range: >500 to 1000 ppm Air Purifying, Negative Pressure, Full Face
Exposure Range: >1000 to <1800 ppm Supplied Air, Constant Flow/Pressure Demand, Full Face
Exposure Range: 1800 to unlimited ppm Self-contained Breathing Apparatus, Pressure Demand, Full Face
Cartridge Color: black

Environmental

Ecotoxicity: LD_{50} Goldfish > 5000 mg/l/24 hr. /Conditions of bioassay not specified LC_{50} Lepomis macrochirus 420 ppm/96 hr (static bioassay in fresh water at 23 °C, mild aeration applied after 24 hr) LC_{50} Menidia beryllina 420 ppm/96 hr (static bioassay in synthetic sea water at 23 °C, mild aeration applied after 24 hr)

Environmental Fate: If released to soil, it will be expected to exhibit very high mobility, based upon the reported infinite solubility of the compound in water and an estimated K_{oc} of 21. The compound may, therefore, leach through soil. Although no data were located regarding its biodegradation in environmental media, the compound may be subject to biodegradation in soil based upon results observed in laboratory biodegradation aqueous aerobic screening tests. It should not be subject to volatilization from moist near-surface soil based upon an estimated Henry's Law constant of 4.24 x10^{-9} atm-cu m/mole. However, it may volatilize from dry near-surface soil and other dry surfaces based upon its vapor pressure of 1.71 mm Hg at 25 °C. If released to water, it will not be expected to adsorb to sediment or suspended particulate matter or expected to bioconcentrate in aquatic organisms based upon its estimated K_{oc} and estimated BCF of 0.50, respectively. The compound may be subject to biodegradation in natural waters based upon results observed in laboratory aqueous aerobic biodegradation screening tests using acclimated mixed microbial cultures as inoculum. It should not be subject to volatilization from surface waters based upon the estimated Henry's Law constant. Hydrolysis should not be an important removal process since aliphatic alcohols and ketones (the two functional groups that it contains) generally are resistant to hydrolysis. If released to the atmosphere, it can be expected to exist mainly in the vapor phase in the ambient atmosphere based upon its vapor pressure. The estimated atmospheric half-life for vapor phase reaction with photochemically produced hydroxyl radicals half-life is 12 days at an atmospheric concentration of 5 x10^5 hydroxyl radicals per cu cm. It may be susceptible to direct photolysis in the atmosphere based upon its possible absorption of light at wavelengths > 290 nm. Based upon its high water solubility, the compound may be susceptible to removal from the atmosphere by washout.

Cleanup/Disposal: Guide No. 129: Eliminate all ignition sources (no smoking, flares, sparks or flames in immediate area). All equipment used when handling the product must be grounded. Do not touch or walk through spilled material. Stop leak if you can do it without risk. Prevent entry into waterways, sewers, basements or confined areas. A vapor suppressing foam may be used to reduce vapors. Absorb or cover with dry earth, sand or other non-combustible material and transfer to containers. Use clean non-sparking tools to collect absorbed material. Large Spills: Dike far ahead of liquid spill for later disposal. Water spray may reduce vapor; but may not prevent ignition in closed spaces.

Environmental Physical Data

Henry's Law Constant: estimated at 4.24 x10^{-9}
Octanol/Water Partition Coefficient: log K_{ow} = estimated at -0.098
Sorption Partition Coefficient: K_{oc} = estimated at 21
BCF: estimated at 0.50
BOD: 47% BODT, 5 days

Regulations

RCRA 40CFR: Not listed
CERCLA: 40CFR 302.4: Not listed
SARA 40CFR 372.65: Not listed
SARA EHS 40CFR 355: Not listed
TSCA: Listed

Analytical Methods

Air: ASTM D3686, D3687
Water / Groundwater: ASTM D3695
Indoor / Expired Air: NIOSH 1402

DIA1300 CAS #: 431-03-8

DIACETYL

RTECS: EK2625000
DOT: UN2346; IMO3.2
EINECS Number: 207-069-8
Molecular Formula: $C_4H_6O_2$
Structured MF: $CH_3COCOCH_3$
Formula Weight: 86.09

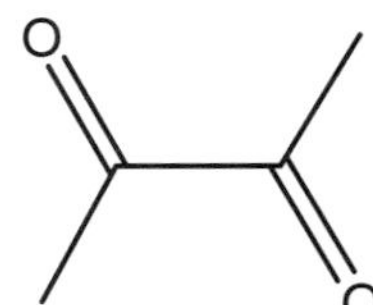

Chemical Structure

Synonyms: BIACETYL; 2,3-BUTADIONE; BUTADIONE; 2,3-BUTANEDIONE; BUTANE-2,3-DIONE; BUTANEDIONE; 2,3-DIKETOBUTANE; DIMETHYL DIKETONE; DIMETHYLGLYOXAL; 2,3-DIOXOBUTANE; GLYOXAL,DIMETHYL-
Description: yellowish green, mobile liquid; quinone odor, vapors have a chlorine-like or rancid butter odor
Use: a synthetic flavoring substance and adjuvant used in the food industry; as a specific reagent for guanidinecontaining compounds (e.g. arginine) and as a carrier of aroma of butter,

vinegar, coffee and other foods; sensitizes photooxidation of alkenes to epoxides

Physical Properties

Boiling Point: 88 °C (190 °F) at 760 mm Hg
Freezing Point: -2.4 °C (27.68 °F)
Specific Gravity: 0.99 at 15 °C/15 °C
Vapor Density: 3 Air=1
Density: 0.981 g/mL
Vapor Pressure: 1.80 kPa at 0 °C
Water Solubility: 1 parts in about 4 parts Water
Other Solubilities: 95% Ethanol: >=100 mg/ml at 22 °C; Acetone: >=100 mg/ml at 22 °C; Alcohol: miscible; Benzene: >10%; DMSO: >=100 mg/ml at 22 °C; Ether: miscible.
Odor Threshold: 8.6 ppb
Refraction Index: 1.3951
Ionization Potential (eV): 9.30
Flash Point: 27 °C Closed Cup
Autoignition Temperature: 365 °C

RTECS Toxicity Data

Acute Oral: Rat LD_{50} Dose: 1580 mg/kg; Toxic Effects: Behavioral - Somnolence (general depressed activity); Behavioral - Convulsions or effect on seizure threshold. Mouse LD_{50} Dose: 250 mg/kg.
Acute Dermal: Rabbit LD_{50} Route: Skin; Dose: >5 gm/kg.
Chronic (Multiple Dose) Oral: Rat Dose: 48600 mg/kg/90D-I; Toxic Effects: Liver - Changes in liver weight; Endocrine - Changes in adrenal weight; Blood - Pigmented or nucleated red Blood cells.
Irritation Skin: Rabbit Standard Draize Test Dose: 500 mg/24H; Reaction: moderate.
Mutagenic: Human Other Mutation Test Systems; Cell Type: embryo; Dose: 20 mg/L. Rat Unscheduled DNA Synthesis; Route: Oral; Dose: 1500 mg/kg.

Hazard Overviews

Flammable

Health: Irritating to eyes/skin/respiratory tract. Toxic. Other Acute Effects: harmful if swallowed, inhaled, or absorbed through skin; exposure can cause nausea, headache and vomiting.
Fire: Flammable. Hazards: vapor may travel considerable distance to source of ignition and flash back; container explosion may occur; forms explosive mixtures in air. Extinguishing agents: carbon dioxide, dry chemical powder or appropriate foam; water may be effective for cooling, but may not effect extinguishment. Precautions: combustible liquid.
Reactivity: Incompatible with: oxidizing agents, strong bases, reducing agents, metals. Hazardous decomposition products: toxic fumes of: carbon monoxide, carbon dioxide.

Carcinogenicity: IARC - Not listed; NIOSH - Not listed; NTP - Not listed; ACGIH - Not listed; OSHA - Not listed; EPA - Not listed; MAK - Not listed
Primary Target Organs:

Eyes

Skin

Respiratory System

Environmental

Environmental Fate: Little is known concerning its fate in the environment other than that it photodegrades fairly rapidly in the atmosphere. The physical/chemical properties would suggest that it will not adsorb strongly to soil or sediment or bioconcentrate in fish. It may biodegrade, but only experimental information is available.
Cleanup/Disposal: Guide No. 127: Eliminate all ignition sources (no smoking, flares, sparks or flames in immediate area). All equipment used when handling the product must be grounded. Do not touch or walk through spilled material. Stop leak if you can do it without risk. Prevent entry into waterways, sewers, basements or confined areas. A vapor suppressing foam may be used to reduce vapors. Absorb or cover with dry earth, sand or other non-combustible material and transfer to containers. Use clean non-sparking tools to collect absorbed material. Large Spills: Dike far ahead of liquid spill for later disposal. Water spray may reduce vapor; but may not prevent ignition in closed spaces.

Environmental Physical Data

Henry's Law Constant: 0.0014
Octanol/Water Partition Coefficient: log K_{ow} = -1.34
BCF: not likely

Regulations

RCRA 40CFR: Not listed
CERCLA: 40CFR 302.4: Not listed
SARA 40CFR 372.65: Not listed
SARA EHS 40CFR 355: Not listed
TSCA: Listed

DIA1450 CAS #: 110-22-5

DIACETYL PEROXIDE

RTECS: AP8500000
EINECS Number: 203-748-8
Molecular Formula: $C_4H_6O_4$
Structured MF: $CH_3CO(O_2)OCCH_3$
Formula Weight: 118.09
Synonyms: ACETYL PEROXIDE; ACETYL PEROXIDE,NOT >25% IN SOLUTION; ACETYL PEROXIDE,SOLID,OR >25% IN SOLUTION; DIACETYL PEROXIDES,SOLID,OR >25% IN SOLUTION; PEROXIDE,DIACETYL
Description: colorless crystals, liquid; strong, pungent odor
Use: initiator & catalyst for resins; catalyst employed to promote polymerization in mfr of certain plastics

Physical Properties

Boiling Point: 63 °C (145 °F) at 21 mm Hg
Freezing Point: 30 °C (86 °F)
Specific Gravity: Liquid 1.2 at 20 °C
Vapor Density: 4.07 Air=1
Water Solubility: Slightly Soluble in Water
Other Solubilities: Very Soluble in Carbon Tetrachloride.
Surface Tension: Estimated at 30 dynes/cm
Flash Point: 45 °C Open Cup
Autoignition Temperature: Explodes

RTECS Toxicity Data

Irritation Eye: Rabbit Nonstandard Exposure Dose: 60 mg/1M rinse; Reaction: severe.

Hazard Overviews

Explosive

Flammable

Fire Diamond

Health: Severely irritating to eyes/skin/respiratory tract
Fire: Explosive. Flammable. Fight fire with carbon dioxide or regular foam. Use water only when other agents unavailable and personnel are prepared for the heat generation that will occur. Strong oxidizer that can ignite combustible material on contact.
Reactivity: Unstable, severe explosion hazard when subjected to shock or exposed to heat. Hazardous polymerization cannot occur. Avoid: hot materials; accelerators; temperatures below 20 °F (-6.6 °C). Incompatible with: organic material; water; steam; reducing material; acid; acid fumes; diethyl ether. Hazardous decomposition products: carbon oxides; irritating smoke.
Carcinogenicity: IARC - Not listed; NIOSH - Not listed; NTP - Not listed; ACGIH - Not listed; OSHA - Not listed; EPA - Not listed; MAK - Not listed
Primary Target Organs:

Eyes

Skin

Respiratory System

Environmental

Cleanup/Disposal: Guide No. 127: Eliminate all ignition sources (no smoking, flares, sparks or flames in immediate area). All equipment used when handling the product must be grounded. Do not touch or walk through spilled material. Stop leak if you can do it without risk. Prevent entry into waterways, sewers, basements or confined areas. A vapor suppressing foam may be used to reduce vapors. Absorb or cover with dry earth, sand or other non-combustible material and transfer to containers. Use clean non-sparking tools to collect absorbed material. Large Spills: Dike far ahead of liquid spill for later disposal. Water spray may reduce vapor; but may not prevent ignition in closed spaces.

Environmental Physical Data

BCF: no food chain concentration potential

Regulations

RCRA 40CFR: Not listed
CERCLA: 40CFR 302.4: Not listed
SARA 40CFR 372.65: Not listed
SARA EHS 40CFR 355: Not listed
TSCA: Listed

DIA1600 CAS #: 613-35-4

4,4'-DIACETYLBENZIDINE

RTECS: DT2800000
EINECS Number: 210-338-2
Molecular Formula: $C_{16}H_{16}N_2O_2$
Formula Weight: 268.32

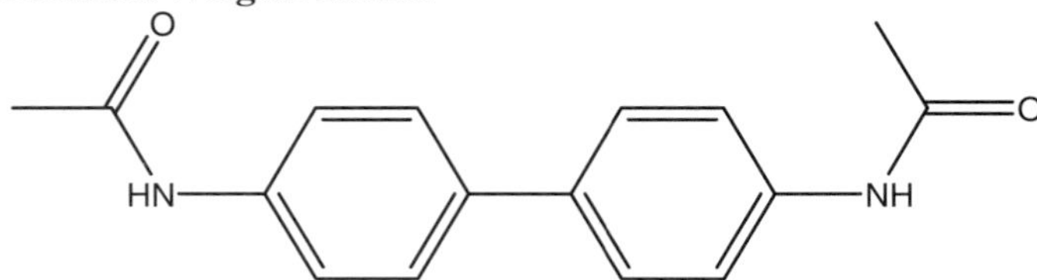

Chemical Structure

Synonyms: ACETAMIDE,N,N'-(1,1'-BIPHENYL)-4,4'-DIYLBIS-; 4',4'''-BIACETANILIDE; BIPHENYL,4,4'-DIACETAMIDO-; N,N'-(1,1'-BIPHENYL)-4,4'-DIYLBIS-ACETAMIDE 4',4'''-BIACETANILIDE; N,N'-(1,1'-BIPHENYL)-4,4'-DIYLBIS-ACETAMIDE4',4'''-BIACETANILIDE; N,N'-4,4'-BIPHENYLYLENEBISACETAMIDE; 4,4'-DIACETAMIDOBIPHENYL; N,N'-DIACETYL BENZIDINE; 4,4'-DIACETYLAMINOBIPHENYL; DIACETYLBENZIDINE; N,N'-DIACETYLBENZIDINE
Description: white needles
Use: to prepare 3,3'-dichlorobenzidine and 3,3'-dinitrobenzidine; research chem (eg, chem int for substituted benzidines)

Physical Properties

Freezing Point: 328.3 °C (622.94 °F)
Other Solubilities: Soluble in Ethanol, Ethyl Acetate.

RTECS Toxicity Data

Mutagenic: Rat DNA Adduct; Cell Type: liver; Dose: 100 mg/L. Bacteria - S Typhimurium Mutations in Microorganisms; Dose: 5 ug/plate (-S9).
Tumorigenic: Rat Route: Oral; Dose: 6300 mg/kg/35W-C; Toxic Effects: Tumorigenic - Equivocal tumorigenic agent by RTECS criteria; Sense organs and special senses - Tumors. Rat Route: Subcutaneous; Dose: 900 mg/kg; Toxic Effects: Tumorigenic - Neoplastic by RTECS criteria; Liver - Hepatitis (hepatocellular necrosis), zonal; Kidney, Ureter, and Bladder - Changes in both tubules and glomeruli. Rat Route: Subcutaneous; Dose: 4350 mg/kg/39W-I; Toxic Effects: Tumorigenic - Equivocal tumorigenic agent by RTECS criteria; Liver - Tumors; Skin and appendages - Tumors.

Hazard Overviews

Poison

Corrosive

Flammable

Health: Corrosive to eyes/skin/respiratory tract. Poison. Other Acute Effects: may be fatal if inhaled, swallowed, or absorbed through skin; readily absorbed through skin; inhalation may result in spasm, inflammation and edema of the larynx and bronchi, chemical pneumonitis and pulmonary edema; symptoms of exposure may include burning sensation; coughing; wheezing; laryngitis; shortness of breath; headache; nausea; vomiting.
Fire: Flammable. Hazards: emits toxic fumes. Extinguishing agents: carbon dioxide, dry chemical powder or appropriate foam; water spray. Precautions: combustible liquid.
Reactivity: Incompatible with: acids, acid chlorides, acid anhydrides, strong oxidizing agents, carbon dioxide. Hazardous decomposition products: thermal decomposition may produce carbon monoxide, carbon dioxide, and nitrogen oxides.
Carcinogenicity: IARC - Not listed; NIOSH - Not listed; NTP - Not listed; ACGIH - Not listed; OSHA - Not listed; EPA - Not listed; MAK - Not listed
Primary Target Organs:

Eyes

Skin

Respiratory System

Environmental

Cleanup/Disposal: Guide No. 131: Fully encapsulating, vapor protective clothing should be worn for spills and leaks with no fire. Eliminate all ignition sources (no smoking, flares, sparks or flames in immediate area). All equipment used when handling the product must be grounded. Do not touch or walk through spilled material. Stop leak if you can do it without risk. Prevent entry into waterways, sewers, basements or confined areas. A vapor suppressing foam may be used to reduce vapors. Small Spills: Absorb with earth, sand or other non-combustible material and transfer to containers for later disposal. Use clean non-sparking tools to collect absorbed material. Large Spills: Dike far ahead of liquid spill for later disposal. Water spray may reduce vapor; but may not prevent ignition in closed spaces.

Regulations

RCRA 40CFR: Not listed
CERCLA: 40CFR 302.4: Not listed
SARA 40CFR 372.65: Not listed
SARA EHS 40CFR 355: Not listed
TSCA: Listed

DIA5500 CAS #: 1904-98-9

2,6-DIAMINOPURINE

EINECS Number: 217-605-2
Molecular Formula: $C_5H_6N_6$
Formula Weight: 150.17

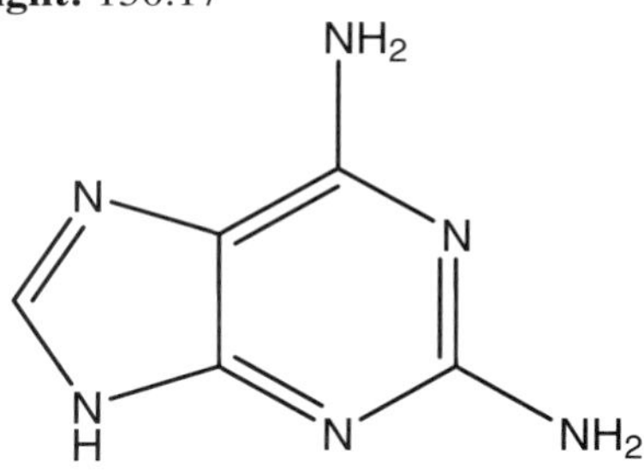

Chemical Structure

Physical Properties

Freezing Point: 302 °C (575.6 °F)

Hazard Overviews

Health: Irritating to eyes/skin/respiratory tract. Harmful. Other Acute Effects: harmful if swallowed, inhaled, or absorbed through skin. Chronic Effects: laboratory experiments have shown mutagenic effects.
Fire: Hazards: emits toxic fumes. Extinguishing agents: water spray; carbon dioxide, dry chemical powder or appropriate foam. Precautions: combustible liquid.
Reactivity: Incompatible with: strong oxidizing agents. Hazardous decomposition products: toxic fumes of: carbon monoxide, carbon dioxide, nitrogen oxides.
Carcinogenicity: IARC - Not listed; NIOSH - Not listed; NTP - Not listed; ACGIH - Not listed; OSHA - Not listed; EPA - Not listed; MAK - Not listed
Primary Target Organs:

Eyes

Skin

Respiratory System

Environmental

Regulations

RCRA 40CFR: Not listed
CERCLA: 40CFR 302.4: Not listed
SARA 40CFR 372.65: Not listed
SARA EHS 40CFR 355: Not listed
TSCA: Listed

DIA5950 CAS #: 54-96-6

3,4-DIAMINOPYRIDINE

RTECS: US7600000

EINECS Number: 200-220-9
Molecular Formula: $C_5H_7N_3$
Formula Weight: 109.15

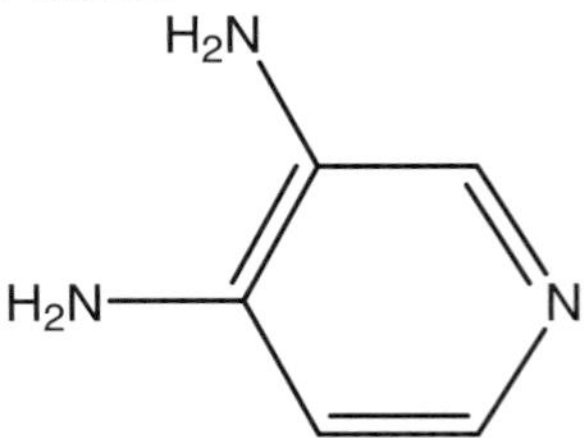

Chemical Structure

Synonyms: DIAMINO-3,4 PYRIDINE; SC10

Physical Properties

Freezing Point: 218 °C (424.4 °F) to 220 °C (428 °F)
Water Solubility: 24 g/l
pH: 11 to 11.5

RTECS Toxicity Data

Acute Dermal: Mouse LD_{50} Route: Subcutaneous Dose: 35 mg/kg; Toxic Effects: Behavioral - Convulsions or effect on seizure threshold; Behavioral - Change in motor activity (specific assay); Skin and appendages - Hair.

Hazard Overviews

Health: Irritating to eyes/skin/respiratory tract. Toxic. Other Acute Effects: harmful if swallowed, inhaled, or absorbed through skin.
Fire: Extinguishing agents: water spray; carbon dioxide, dry chemical powder or appropriate foam. Precautions: combustible liquid.
Reactivity: Incompatible with: strong oxidizing agents. Hazardous decomposition products: toxic fumes of: carbon monoxide, carbon dioxide, nitrogen oxides.
Carcinogenicity: IARC - Not listed; NIOSH - Not listed; NTP - Not listed; ACGIH - Not listed; OSHA - Not listed; EPA - Not listed; MAK - Not listed
Primary Target Organs:

Environmental

Regulations

RCRA 40CFR: Not listed
CERCLA: 40CFR 302.4: Not listed
SARA 40CFR 372.65: Not listed
SARA EHS 40CFR 355: Not listed
TSCA: Not listed

DIA6100 **CAS #: 81-11-8**

4,4-DIAMINO-2,2'-STILBENEDISULFONIC ACID

RTECS: WJ6603000
EINECS Number: 201-325-2
Molecular Formula: $C_{14}H_{14}N_2O_6S_2$
Structured MF: $C_6H_3(NH_2)(SO_3H)CHCHC_6H_3(SO_3H)(NH_2)$
Formula Weight: 370.41

Chemical Structure

Synonyms: AMSONIC ACID; BENZENESULFONIC ACID,2,2'-(1,2-ETHENEDIYL)BIS(5-AMINO-; BENZENESULFONIC ACID,2,2'-(1,2-ETHYLENEDIYL)BIS(5-AMINO-(9CI); DAS; DASD; P,P'-DIAMINODIPHENYLETHYLENE-O,O'-DISULFONIC ACID; 4,4'-DIAMINO-2,2'-STILBENEDISULFONIC ACID; DIAMINOSTILBENEDISULFONIC ACID; P,P'-DIAMINOSTILBENE-O,O'-DISULFONIC ACID; 2,2'-DISULFO-4,4'-STILBENEDIAMINE; 2,2'-(1,2-ETHENEDIYL)BIS[5-AMINO-BENZENESULFONIC ACID]; 2,2'-(1,2-ETHYLENEDIYL)BIS(5-AMINOBENZENESULFONIC ACID); FLAVONIC ACID; 2,2'-STILBENEDISULFONIC ACID,4,4'-DIAMINO-; TINOPAL BHS
Description: yellow needles; odorless
Use: in the manufacture of dyes and in bleaching agents

Physical Properties

Freezing Point: > 325 °C (617 °F)
Water Solubility: Very Slightly Soluble in Water
Other Solubilities: 95% Ethanol: <1 mg/ml at 23 °C; Acetone: <1 mg/ml at 23 °C; Alcohol: Soluble; DMSO: <1 mg/ml at 23 °C; Ether: Soluble; Methanol: <1 mg/ml at 21 °C; Toluene: <1 mg/ml at 21 °C.
pH: Approximately at 30 g/L water (suspension) 4.3
Flash Point: Not available; probably combustible

RTECS Toxicity Data

Acute Oral: Guinea Pig LD_{50} Dose: 47 gm/kg; Toxic Effects: Liver - Liver function tests impaired; Kidney, Ureter, and Bladder - Renal function tests depressed.
Reproductive/Teratogenic: Rat Route: Intraperitoneal; Dose: 300 mg/kg; Duration: female 1D prior to mating; Maternal Effects - Uterus, cervix, vagina.

Hazard Overviews

Corrosive

Health: Corrosive to eyes/skin/respiratory tract. Other Acute Effects: harmful if swallowed, inhaled, or absorbed through skin; extremely destructive to tissue of the mucous membranes and upper respiratory tract, eyes and skin; inhalation may result in spasm, inflammation and edema of the larynx and bronchi, chemical pneumonitis and pulmonary edema; symptoms of exposure may include burning sensation, coughing, wheezing, laryngitis, shortness of breath, headache, nausea and vomiting.
Fire: Will burn. Hazards: emits toxic fumes. Extinguishing agents: carbon dioxide, dry chemical powder or appropriate foam. Precautions: combustible liquid.
Carcinogenicity: IARC - Not listed; NIOSH - Not listed; NTP - Not listed; ACGIH - Not listed; OSHA - Not listed; EPA - Not listed; MAK - Not listed
Primary Target Organs:

Eyes

Skin

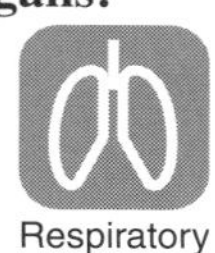
Respiratory System

Environmental

Regulations

RCRA 40CFR: Not listed
CERCLA: 40CFR 302.4: Not listed
SARA 40CFR 372.65: Not listed
SARA EHS 40CFR 355: Not listed
TSCA: Listed

DIA6250 CAS #: 2687-25-4

2,3-DIAMINOTOLUENE

RTECS: XS9550000
EINECS Number: 220-248-5
Molecular Formula: $C_7H_{10}N_2$
Formula Weight: 122.17

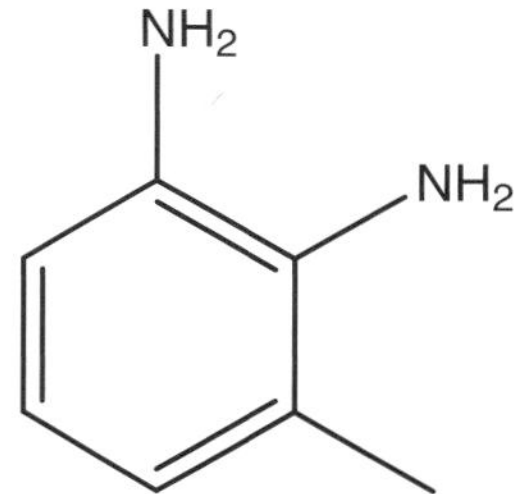

Chemical Structure

Synonyms: 1,2-BENZENEDIAMINE,3-METHYL-; 1,2-BENZENEDIAMINE,3-METHYL-(9CI); 1,2-DIAMINO-3-METHYLBENZENE; 3-METHYL-1,2-BENZENEDIAMINE; 1-METHYL-2,3-PHENYLENEDIAMINE; 3-METHYL-1,2-PHENYLENEDIAMINE; 3-METHYL-O-PHENYLENEDIAMINE; TOLUENE-2,3-DIAMINE; TOLUENE,2,3-DIAMINO-; 2,3-TOLUYLENEDIAMINE; 2,3-TOLYLENEDIAMINE
Description: crystals
Use: chem int for tolyltriazole; corrosion inhibitor; chem int for antioxidants, eg, benzimidazolethiol derivs

Physical Properties

Boiling Point: 255 °C (491 °F)
Freezing Point: 63 °C (145.4 °F) to 64 °C (147.2 °F)
Saturated Vapor Density: Extrapolated 1.20000279 kg/m^3
Vapor Pressure: Extrapolated 0.00055 mm Hg at 25 °C
Water Solubility: Soluble in Water
Other Solubilities: Soluble in Alcohol, Ether

Hazard Overviews

Health: Irritating to eyes/skin/respiratory tract. Harmful. Other Acute Effects: may be harmful by inhalation, ingestion, or skin absorption.
Fire: Hazards: emits toxic fumes. Extinguishing agents: water spray; carbon dioxide, dry chemical powder or appropriate foam. Precautions: combustible liquid.
Reactivity: Incompatible with: strong oxidizing agents, strong acids. Hazardous decomposition products: toxic fumes of: carbon monoxide, carbon dioxide, nitrogen oxides.
Carcinogenicity: IARC - Not listed; NIOSH - Not listed; NTP - Not listed; ACGIH - Not listed; OSHA - Not listed; EPA - Not listed; MAK - Not listed
Primary Target Organs:

Eyes

Skin

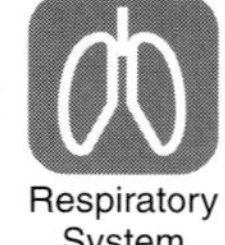
Respiratory System

Environmental

Environmental Fate: If released to the atmosphere, it is expected to degrade rapidly in the vapor-phase (estimated half-life of 1.6 hr) by reaction with photochemically produced hydroxyl radicals. If released to soil, it may undergo a covalent chemical bonding with humic materials which can result in its chemical alteration to a latent form and prevent leaching. In the absence of covalent bonding, leaching may be possible. If released to water, covalent bonding with humic materials in the water column and sediments may result in partitioning from the water column to sediments. By analogy to the aromatic amine chemical class, in the water column it may be susceptible to photooxidation via hydroxyl and peroxy radicals. Insufficient data are available to assess the relative importance of biodegradation in soil or water.

Environmental Physical Data

Henry's Law Constant: estimated at 7.43 x10^{-10}
Octanol/Water Partition Coefficient: log K_{ow} = 0.337
Sorption Partition Coefficient: log K_{oc} = estimated at 1.34
BCF: estimated at 1.06

Regulations

RCRA 40CFR: Not listed
CERCLA: 40CFR 302.4: Not listed
SARA 40CFR 372.65: Not listed
SARA EHS 40CFR 355: Not listed
TSCA: Listed

DIA6400 CAS #: 95-80-7

2,4-DIAMINOTOLUENE

RTECS: XS9625000
DOT: UN1709; NA1709; IMO6.1
EINECS Number: 202-453-1
Molecular Formula: $C_7H_{10}N_2$
Structured MF: $CH_3C_6H_3(NH_2)_2$
Formula Weight: 122.17

NH2

NH2

Chemical Structure

Synonyms: 3-AMINO-P-TOLUIDINE; 5-AMINO-O-TOLUIDINE; AZOGEN DEVELOPER H; 1,3-BENZENEDIAMINE,4-METHYL-; BENZOFUR MT; C.I. 76035; C.I. OXIDATION BASE; C.I. OXIDATION BASE 20; C.I. OXIDATION BASE 200; C.I. OXIDATION BASE 35; DEVELOPER 14; DEVELOPER B; DEVELOPER DB; DEVELOPER DBJ; DEVELOPER H; DEVELOPER MC; DEVELOPER MT; DEVELOPER MT-CF; DEVELOPER MTD; DEVELOPER T; 1,3-DIAMINO-4-METHYLBENZENE; 2,4-DIAMINO-1-METHYLBENZENE; 2,4-DIAMINOTOLUEN; 2,4-DIAMINO-1-TOLUENE; DIAMINOTOLUENE; 2,4-DIAMINOTOLUOL; EUCANINE GB; FOURAMINE; FOURAMINE J; FOURRINE 94; FOURRINE M; 4-METHYL-1,3-BENZENEDIAMINE; 4-METHYL-M-PHENYLENEDIAMINE; MTD; NAKO TMT; PELAGOL GREY J; PELAGOL J; PONTAMINE DEVELOPER TN; RENAL MD; TDA; TERTRAL G; 2,4-TOLAMINE; 2,4-TOLUENEDIAMINE; M-TOLUENEDIAMINE; TOLUENE,2,4-DIAMINE; TOLUENE-2,4-DIAMINE; M-TOLUYENEDIAMINE; M-TOLUYLENDIAMIN; META TOLUYLENE DIAMINE; META-TOLUYLENE DIAMINE; 2,4-TOLUYLENEDIAMINE; M-TOLUYLENEDIAMINE; META-TOLUYLENEDIAMINE; M-TOLYENEDIAMINE; 2,4-TOLYLENEDIAMINE; 4-M-TOLYLENEDIAMINE; M-TOLYLENEDIAMINE; META-TOLYLENEDIAMINE; TOLYLENE-2,4-DIAMINE; ZOBA GKE; ZOGEN DEVELOPER H
Description: colorless to brown needle-shaped crystals, powder
Use: in polymerizations; in dyes used for textiles, leather, furs and hair dye formulations; source of toluene diisocyanate; developer for direct dyes; and chain extender and crosslinker

Physical Properties

Boiling Point: 292 °C (558 °F)
Freezing Point: 99 °C (210.2 °F)
Specific Gravity: 1.045 at 100 °C
Vapor Pressure: 5.52×10^{-5} mm Hg
Water Solubility: 1 to 5 mg/mL at 21 °C
Other Solubilities: 95% Ethanol: 10-50 mg/ml at 21 °C; Acetone: >=100 mg/ml at 21 °C; Aqueous sodium carbonate: Soluble; Benzene: Soluble; DMSO: 50-100 mg/ml at 21 °C; Ether: Soluble.
Flash Point: 149 °C
Autoignition Temperature: > 475 °C

RTECS Toxicity Data

Acute Dermal: Rabbit LD_{Lo} Route: Subcutaneous Dose: 400 mg/kg; Toxic Effects: Brain and coverings - Other degenerative changes; Behavioral - Convulsions or effect on seizure threshold; Lungs, Thorax, or Respiration - Respiratory stimulation. Rat LD_{Lo} Route: Subcutaneous Dose: 350 mg/kg; Toxic Effects: Behavioral - Somnolence (general depressed activity); Cardiac - Pulse rate; Lungs, Thorax, or Respiration - Respiratory stimulation.
Irritation Eye: Rabbit Standard Draize Test Dose: 100 mg/24H; Reaction: moderate.
Irritation Skin: Rabbit Standard Draize Test Dose: 500 mg/24H; Reaction: mild.
Reproductive/Teratogenic: Rat Route: Oral; Dose: 1260 mg/kg; Duration: male 10W prior to mating; Effects on Fertility - Mating Performance. Rat Route: Oral; Dose: 1260 mg/kg; Duration: male 10W prior to mating; Paternal Effects - Spermatogenesis; Prostate, seminal vessicle, Cowper's gland, accessory glands. Rat Route: Oral; Dose: 210 mg/kg; Duration: male 1W prior to mating; Paternal Effects - Spermatogenesis; Testes, epididymis, sperm duct.
Mutagenic: Human DNA Damage; Cell Type: fibroblast; Dose: 100 umol/L. Human Unscheduled DNA Synthesis; Cell Type: liver; Dose: 100 umol/L.
Tumorigenic: Rat Route: Oral; Dose: 2100 mg/kg/90W-C; Toxic Effects: Tumorigenic - Carcinogenic by RTECS criteria; Liver - Tumors; Skin and appendages - Tumors. Rat Route: Oral; Dose: 5030 mg/kg/84W-C; Toxic Effects: Tumorigenic - Carcinogenic by RTECS criteria; Liver - Tumors; Blood - Lymphomax including Hodgkin's disease. Rat Route: Oral; Dose: 9000 mg/kg/36W-C; Toxic Effects: Tumorigenic - Carcinogenic by RTECS criteria; Liver - Tumors.

Hazard Overviews

Fire Diamond

Health: Irritating to eyes/skin/respiratory tract. Also Causes: methemoglobinemia; a blue discoloration of the skin and lips, headache, fatigue, and dizziness, liver damage, nervous system effects, gastritis, rise in blood pressure, tremors, convulsions, coma, keratoconjunctivitis, blepharitis, opacities of the cornea, severe dermatitis, urticaria, deep staining of the skin. Chronic Effects: liver damage, nervous system effects, causing headache, weakness, tremors, fatigue, dizziness, nausea, vomiting, and death.

Fire: Will burn. Use dry chemical, carbon dioxide, water spray, fog, or regular foam.

Reactivity: Stable. Hazardous polymerization cannot occur. Avoid: heat; ignition sources. Incompatible with: acids; acid chlorides; acid anhydrides; strong oxidizers; chloroformates. Hazardous decomposition products: oxides of carbon; oxides of nitrogen.

Carcinogenicity: IARC - Group 2B, Possibly carcinogenic to humans; NIOSH - Listed as carcinogen; NTP - Class 2B, Reasonably anticipated to be a carcinogen, sufficient evidence of carcinogenicity from studies in experimental animals; ACGIH - Not listed; OSHA - Not listed; EPA - Not listed; MAK - Class A2, Unmistakably carcinogenic in animal experimentation only

Primary Target Organs:

Eyes

Skin

Respiratory System

Mucous Membranes

Nervous System

Liver

Respirator Recommendation

Exposure Range: >0.2 to 2 ppm Air Purifying, Negative Pressure, Half Mask

Exposure Range: >2 to 20 ppm Air Purifying, Negative Pressure, Full Face

Exposure Range: >20 to 200 ppm Supplied Air, Constant Flow/Pressure Demand, Full Face

Exposure Range: >200 to unlimited ppm Self-contained Breathing Apparatus, Pressure Demand, Full Face

Cartridge Color: black with magenta (P100)

Environmental

Ecotoxicity: LC_{50} Pimephales promelas (fathead minnow) 1420 mg/l/96 hr (confidence limit not reliable), flow through bioassay with measured concentrations, 24.2 °C, dissolved oxygen 6.5 mg/l, hardness 46.3 mg/l calcium carbonate, alkalinity 123 mg/l calcium carbonate, and pH 7.6

Environmental Fate: Release to the soil will likely result in extensive leaching due to its high estimated water solubility and low estimated soil sorption partition coefficient. Biodegradation in the soil may also occur. Biodegradation and photooxidation are expected to be the major means of degradation in water and soil although photolysis may also occur to some extent. Volatilization from the water and soil are not expected to be significant. Bioconcentration will not be important. In the atmosphere, it may photolyze and will react with hydroxyl radicals with an estimated half-life of 8 hr.

Cleanup/Disposal: Guide No. 151: Do not touch damaged containers or spilled material unless wearing appropriate protective clothing. Stop leak if you can do it without risk. Prevent entry into waterways, sewers, basements or confined areas. Cover with plastic sheet to prevent spreading. Absorb or cover with dry earth, sand or other non-combustible material and transfer to containers. Do not get water inside containers.

Environmental Physical Data

Henry's Law Constant: estimated at 1.2×10^{-9}

Octanol/Water Partition Coefficient: log K_{ow} = estimated at 0.337

Sorption Partition Coefficient: log K_{oc} = 1.56

BCF: fathead minnow 91

Regulations

RCRA 40CFR: Not listed

CERCLA: 40CFR 302.4: Listed per RCRA Section 3001 RQ: 10 lb (4.535 kg)

SARA 40CFR 372.65: Listed

SARA EHS 40CFR 355: Not listed

TSCA: Listed

Analytical Methods

Soil: SW846 3640A, 8270B, 8270C

Water / Groundwater: EPA 1625

Indoor / Expired Air: NIOSH 5516

Plasma: EPA 29

DIA6550 CAS #: 823-40-5

2,6-DIAMINOTOLUENE

RTECS: XS9750000

DOT: NA1709

EINECS Number: 212-513-9

Molecular Formula: $C_7H_{10}N_2$

Formula Weight: 122.17

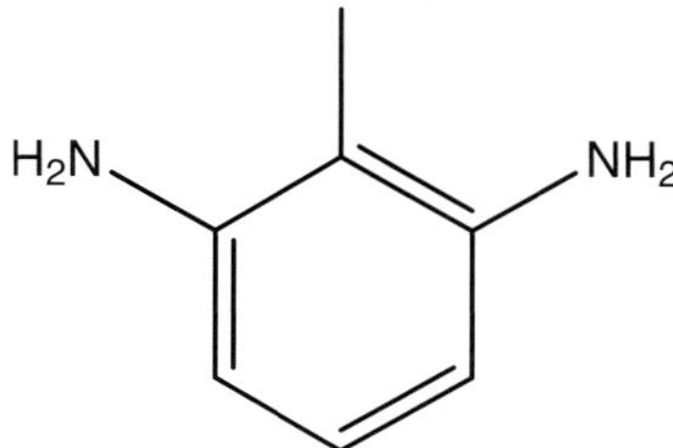

Chemical Structure

Synonyms: 1,3-BENZENEDIAMINE,2-METHYL-; 1,3-DIAMINO-2-METHYLBENZENE; 2,6-DIAMINO-1-METHYLBENZENE; 2-METHYL-1,3-BENZENEDIAMINE; 2-METHYL-1,3-PHENYLENEDIAMINE; 2-METHYL-M-PHENYLENEDIAMINE; TOLUENE-2,6-DIAMINE; TOLUENE,2,6-DIAMINO-; 2,6-TOLUYLENEDIAMINE; 2,6-TOLYLENEDIAMINE

Description: prisms

Use: intermed for toluene diisocyanate, polyether polyols, rubber chems; curing agent for epoxy resins, monomer, chain extender & crosslinker for polymers (mixt with 2,4-isomer); intermed for dyes; intermed for synthesis of toluene diisocyanate; to produce polyurethane foam & elastomers; dyestuff intermed.

Physical Properties

Freezing Point: 106 °C (222.8 °F)
Vapor Pressure: 2.13 kPa at 150 °C
Water Solubility: Soluble in Water
Other Solubilities: 95% Ethanol: Soluble; Acetone: Soluble; Benzene: Soluble; Ether: Soluble.
Flash Point: Not available; probably combustible

RTECS Toxicity Data

Mutagenic: Human Unscheduled DNA Synthesis; Cell Type: liver; Dose: 10 umol/L. Rat Micronucleus Test; Route: Oral; Dose: 300 mg/kg.

Hazard Overviews

Health: Irritating to eyes/skin/respiratory tract. Harmful. Other Acute Effects: may be harmful by inhalation, ingestion, or skin absorption; absorption into the body leads to the formation of methemoglobin which in sufficient concentration causes cyanosis; onset may be delayed 2 to 4 hours or longer. Chronic Effects: overexposure may cause reproductive disorder(s) based on tests with laboratory animals; target organ: blood.
Fire: Will burn. Hazards: emits toxic fumes. Extinguishing agents: water spray; carbon dioxide, dry chemical powder or appropriate foam. Precautions: combustible liquid.
Reactivity: Incompatible with: strong oxidizing agents, strong acids. Hazardous decomposition products: toxic fumes of: carbon monoxide, carbon dioxide, nitrogen oxides.
Carcinogenicity: IARC - Not listed; NIOSH - Not listed; NTP - Not listed; ACGIH - Not listed; OSHA - Not listed; EPA - Not listed; MAK - Not listed
Primary Target Organs:

Eyes

Skin

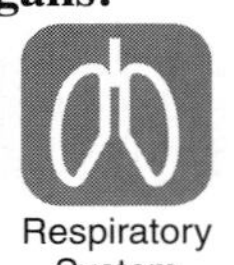
Respiratory System

Blood

Environmental

Cleanup/Disposal: Guide No. 151: Do not touch damaged containers or spilled material unless wearing appropriate protective clothing. Stop leak if you can do it without risk. Prevent entry into waterways, sewers, basements or confined areas. Cover with plastic sheet to prevent spreading. Absorb or cover with dry earth, sand or other non-combustible material and transfer to containers. Do not get water inside containers.

Regulations

RCRA 40CFR: Listed Hazardous Waste No. U221 Toxic Waste
CERCLA: 40CFR 302.4: Listed per RCRA Section 3001 RQ: 10 lb (4.535 kg)
SARA 40CFR 372.65: Not listed
SARA EHS 40CFR 355: Not listed
TSCA: Listed

Analytical Methods

Indoor / Expired Air: NIOSH 5516

DIA6700 CAS #: 108-71-4

3,5-DIAMINOTOLUENE

DOT: NA1709
EINECS Number: 203-609-1
Molecular Formula: $C_7H_{10}N_2$
Formula Weight: 122.17
Synonyms: 1,3-BENZENEDIAMINE,5-METHYL-; 1,3-DIAMINO-5-METHYLBENZENE; 3,5-TOLUENEDIAMINE; TOLUENE-3,5-DIAMINE; TOLUENE,3,5-DIAMINO-TOLUENE
Use: as initiators for rigid urethane production; react with phosgene to give toluene diisocyanate

Physical Properties

Boiling Point: 283 °C (541 °F) to 285 °C (545 °F)
Freezing Point: <
Water Solubility: Very Soluble in Water
Other Solubilities: Soluble in Alcohol, Ether

Hazard Overviews

Carcinogenicity: IARC - Not listed; NIOSH - Not listed; NTP - Not listed; ACGIH - Not listed; OSHA - Not listed; EPA - Not listed; MAK - Not listed

Environmental

Environmental Fate: If released to the atmosphere, it is expected to degrade relatively rapidly (estimated half-life of 1.6 hr) by reaction with photochemically produced hydroxyl radicals. If released to soil, it may undergo a covalent chemical bonding with humic materials which can result in its chemical alteration to a latent form and prevent leaching. In the absence of covalent bonding, leaching may be possible. If released to water, covalent bounding with humic materials in the water column and sediments may result in partitioning from the water column to sediments. By analogy to the aromatic amine chemical class, in the water column it may be susceptible to photooxidation via hydroxyl and peroxy radicals. A single biological screening study has indicated that it is resistant to biodegradation; however, insufficient data are available to assess the relative importance of biodegradation in soil or water.
Cleanup/Disposal: Guide No. 151: Do not touch damaged containers or spilled material unless wearing appropriate protective clothing. Stop leak if you can do it without risk. Prevent entry into waterways, sewers, basements or confined areas. Cover with plastic sheet to prevent spreading. Absorb or cover with dry earth, sand or other non-combustible material and transfer to containers. Do not get water inside containers.

Environmental Physical Data

Henry's Law Constant: estimated at 7.43×10^{-10}
Octanol/Water Partition Coefficient: log K_{ow} = 0.337
Sorption Partition Coefficient: log K_{oc} = estimated at 1.34
BCF: estimated at 1.06

Regulations

RCRA 40CFR: Not listed
CERCLA: 40CFR 302.4: Not listed
SARA 40CFR 372.65: Not listed
SARA EHS 40CFR 355: Not listed
TSCA: Listed

DIA6850 CAS #: 6369-59-1

2,5-DIAMINOTOLUENE SULFATE

RTECS: XT0524000
EINECS Number: 228-871-4
Molecular Formula: $C_7H_{(2x+10)}N_2O_{(4x)}S_x$
Structured MF: $CH_3C_6H_3(NH_2)_2 \cdot H_2SO_4$
Formula Weight: 202.25
Synonyms: 1,4-BENZENEDIAMINE,2-METHYL-,SULFATE; 1,4-BENZENEDIAMINE,2-METHYL-,SULFATE (9CI); C.I. 76043; C.I. 76043; C.I. OXIDATION BASE 4; 2,5-DIAMINOTOLUENE SULPHATE; FOURAMINE STD; 2-METHYL-1,4-BENZENEDIAMINE SULFATE; 2-METHYL-P-PHENYLENEDIAMINE SULPHATE; 2-METHYL-PARA-PHENYLENEDIAMINE SULPHATE; OXIDATION BASE 14; 2,5-TDS; TION O; 2,5-TOLUENDIAMINE SULFATE; 2,4-TOLUENEDIAMINE SULFATE; 2,5-TOLUENEDIAMINE SULFATE; P-TOLUENEDIAMINE SULFATE; P-TOLUENEDIAMINE SULPHATE; PARA-TOLUENEDIAMINE SULPHATE; TOLUENE-2,5-DIAMINE SULPHATE; TOLUENE-2,5-DIAMINE,SULFATE; TOLUYLENE 2,5-DIAMINE SULPHATE; P-TOLUYLENEDIAMINE SULPHATE; PARA-TOLUYLENEDIAMINE SULPHATE; TOLUYLENE-2,5-DIAMINE SULPHATE; P-TOLYLENEDIAMINE SULPHATE; PARA,META-TOLYLENEDIAMINE SULPHATE; PARA-TOLYLENEDIAMINE SULPHATE
Description: off-white powder
Use: intermediate in dye production; dye

Physical Properties

Water Solubility: Soluble in Water
Other Solubilities: 95% Ethanol: <1 mg/ml at 20 °C; Acetone: <1 mg/ml at 20 °C; DMSO: <1 mg/ml at 20 °C.
Flash Point: Not available; probably combustible

RTECS Toxicity Data

Reproductive/Teratogenic: Rat Route: Oral; Dose: 800 mg/kg; Duration: female 6-15D of pregnancy; Effects on Embryo or Fetus - Fetotoxicity. Mouse Route: Oral; Dose: 800 mg/kg; Duration: female 8-12D of pregnancy; Effects on Newborn - Live birth index.
Mutagenic: Mouse DNA Inhibition; Route: Intraperitoneal; Dose: 40 mg/kg. Bacteria - S Typhimurium Mutations in Microorganisms; Dose: 1 mg/plate (+S9). Bacteria - S Typhimurium Mutations in Microorganisms; Dose: 33300 ng/plate (-S9).
Tumorigenic: Mouse Route: Oral; Dose: 66 gm/kg/78W-C; Toxic Effects: Tumorigenic - Equivocal tumorigenic agent by RTECS criteria; Lungs, Thorax, or Respiration - Tumors.

Hazard Overviews

Fire: Will burn.
Carcinogenicity: IARC - Not listed; NIOSH - Not listed; NTP - Not listed; ACGIH - Not listed; OSHA - Not listed; EPA - Not listed; MAK - Not listed

Environmental

Regulations

RCRA 40CFR: Not listed
CERCLA: 40CFR 302.4: Not listed
SARA 40CFR 372.65: Not listed
SARA EHS 40CFR 355: Not listed
TSCA: Listed

DIA7000 CAS #: 14913-33-8

TRANS-DIAMMINEDICHLOROPLATINUM

RTECS: TP2455000
EINECS Number: 238-980-9
Molecular Formula: $Cl_2H_6N_2Pt$
Formula Weight: 300.07

NH_3

Cl——Pt——Cl

NH_3

Chemical Structure

Synonyms: TRANS-DIAMMINEDICHLOROPLATINUM (II); TRANS-DICHLORODIAMMINEPLATINUM (II); NSC 131558; PLATINUM (II),DIAMMINEDICHLORO-,TRANS-; PLATINUM,DIAMMINEDICHLORO-,(SP-4-1); TRANS-PLATINUM(II)DIAMMINEDICHLORIDE; TRANSPLATIN

Physical Properties

Freezing Point: 340 °C (644 °F)

RTECS Toxicity Data

Mutagenic: Human DNA Damage; Cell Type: fibroblast; Dose: 50 umol/L/4H. Human DNA Damage; Cell Type: lung; Dose: 100 umol/L. Human DNA Damage; Cell Type: HeLa cell; Dose: 5 umol/L. Human DNA Damage; Cell Type: other cell types; Dose: 20 mg/L. Human Other Mutation Test Systems; Cell Type: HeLa cell; Dose: 5 umol/L.
Tumorigenic: Mouse Route: Intraperitoneal; Dose: 32408 ug/kg/10W-I; Toxic Effects: Tumorigenic - Equivocal tumorigenic agent by RTECS criteria; Lungs, Thorax, or Respiration - Tumors.

Hazard Overviews

Poison

Health: Irritating to eyes/skin/respiratory tract. Poison. Other Acute Effects: may be fatal if inhaled, swallowed, or absorbed through skin; may cause allergic respiratory reaction; prolonged exposure can cause damage to the liver, kidneys. Chronic Effects: Possible carcinogen.

Fire: Hazards: emits toxic fumes. Extinguishing agents: noncombustible; use extinguishing media appropriate to surrounding fire conditions. Precautions: combustible liquid.

Reactivity: Incompatible with: oxidizing agents. Hazardous decomposition products: nitrogen oxides, hydrogen chloride gas.

Carcinogenicity: IARC - Not listed; NIOSH - Not listed; NTP - Not listed; ACGIH - Not listed; OSHA - Not listed; EPA - Not listed; MAK - Not listed

Primary Target Organs:

Eyes

Skin

Respiratory System

Exposure Limits

OSHA PEL: STEL: 0.002 mg/m^3; as Pt soluble salts.

ACGIH TLV: TWA: 0.002 mg/m^3; as Pt soluble.

NIOSH REL: TWA: 0.002 mg/m^3; as Pt soluble salts.

DFG MAK: TWA: 0.002 mg/m^3; Chloroplat.

Respirator Recommendation

Exposure Range: >0.002 to 0.02 mg/m^3 Air Purifying, Negative Pressure, Half Mask

Exposure Range: >0.02 to 0.2 mg/m^3 Air Purifying, Negative Pressure, Full Face

Exposure Range: >0.2 to 2 mg/m^3 Supplied Air, Constant Flow/Pressure Demand, Full Face

Exposure Range: >2 to unlimited mg/m^3 Self-contained Breathing Apparatus, Pressure Demand, Full Face

Cartridge Color: magenta (P100)

Note: as platinum salts, soluble

Environmental

Regulations

RCRA 40CFR: Not listed

CERCLA: 40CFR 302.4: Not listed

SARA 40CFR 372.65: Not listed

SARA EHS 40CFR 355: Not listed

TSCA: Not listed

DIA7150 CAS #: 7783-28-0

DIAMMONIUM PHOSPHATE

EINECS Number: 231-987-8

Molecular Formula: $H_9N_2O_4P$

Structured MF: $(NH_4)_2HPO_4$

Formula Weight: 132.07

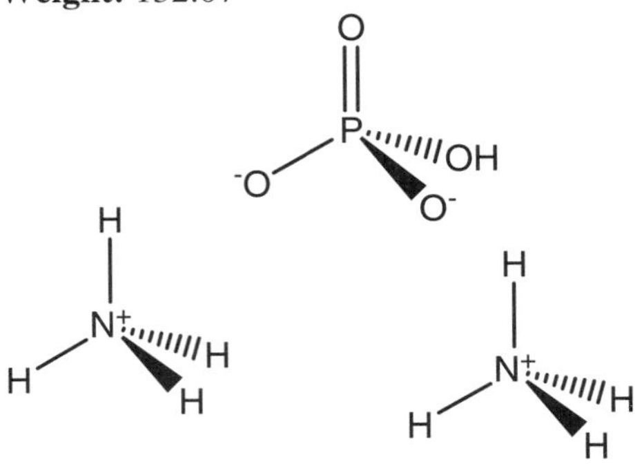

Chemical Structure

Synonyms: AMMONIUM HYDROGEN PHOSPHATE; AMMONIUM MONOHYDROGEN ORTHOPHOSPHATE; AMMONIUM PHOSPHATE; AMMONIUM PHOSPHATE,DIBASIC; DAP; DIAMMONIUM ACID PHOSPHATE; DIAMMONIUM HYDROGEN ORTHOPHOSPHATE; DIAMMONIUM HYDROGEN PHOSPHATE; DIAMMONIUM MONOHYDROGEN PHOSPHATE; DIAMMONIUM ORTHOPHOSPHATE; DIBASIC AMMONIUM PHOSPHATE; FYREX; PELOR; PHOS-CHEK 202; PHOS-CHEK 259; PHOSPHORIC ACID,DIAMMONIUM SALT; SECONDARY AMMONIUM PHOSPHATE

Description: white crystals, powder; weak ammonia odor

Use: fireproofing textiles & vegetable fibers; impregnating lamp wicks; preventing afterglow in matches; flux for soldering tin, copper, brass, & zinc; purifying sugar; in dentifrices, in corrosion inhibitors; plant nutrient soln; halophosphate phosphors; additive in livestock feeds, fertilizer; fire retardant for paper & wood products, & in forest fire control; anticaking ingredient in ammonium nitrate fertilizer; source of nitrogen & phosphorus in yeast foods

Physical Properties

Boiling Point: Decomposes

Freezing Point: Decomposes at 155 °C (311 °F)

Specific Gravity: 1.619

Water Solubility: 1 g dissolves in 17 ml Water

Other Solubilities: Insoluble in Ethanol, Acetone, Ammonia.

pH: About 8

Refraction Index: 1.52

Flash Point: Nonflammable

Hazard Overviews

Fire Diamond

Health: Irritating to eyes/skin/respiratory tract. Also Causes: nausea, diarrhea, mild dermal irritation.
Fire: Noncombustible. Use agent suitable for surrounding fire.
Reactivity: Stable. Hazardous polymerization cannot occur. Incompatible with: sodium hypochlorite. Hazardous decomposition products: fumes of phosphorus oxide; nitrogen oxide; ammonia.
Carcinogenicity: IARC - Not listed; NIOSH - Not listed; NTP - Not listed; ACGIH - Not listed; OSHA - Not listed; EPA - Not listed; MAK - Not listed
Primary Target Organs:

Eyes

Skin

Respiratory System

Environmental

Ecotoxicity: LC_{50} Rainbow trout 160 mg/l/96 hr (95% Confidence interval: 150-171 mg/l) 11 °C, static bioassay Acute toxicity to five freshwater fishes & a scud were determined.in static tests, 96 hr LC_{50} ranged from 40 mg/l for scuds to 1500 mg/l for fingerling fish. in salmon & trout, yolk-sac fry were more susceptible than swim-up fry LC_{50} Grammarus pseudolimnaeus, mature 40 mg/l/96 hr (95% Confidence interval: 32 to 46 mg/l), 18 °C, static bioassay LC_{50} Coho salmon 245 mg/l/96 hr (95% Confidence interval: 216 to 277 mg/l) 11 °C, static bioassay
Cleanup/Disposal: Guide No. 151: Do not touch damaged containers or spilled material unless wearing appropriate protective clothing. Stop leak if you can do it without risk. Prevent entry into waterways, sewers, basements or confined areas. Cover with plastic sheet to prevent spreading. Absorb or cover with dry earth, sand or other non-combustible material and transfer to containers. Do not get water inside containers.

Environmental Physical Data

BCF: no food chain concentration potential

Regulations

RCRA 40CFR: Not listed
CERCLA: 40CFR 302.4: Not listed
SARA 40CFR 372.65: Not listed
SARA EHS 40CFR 355: Not listed
TSCA: Listed

DIA7300 CAS #: 3164-29-2

DIAMMONIUM-L-(+)-TARTRATE

RTECS: WW8050000
EINECS Number: 221-618-9
Molecular Formula: $C_4H_{12}N_2O_6$
Formula Weight: 184.15

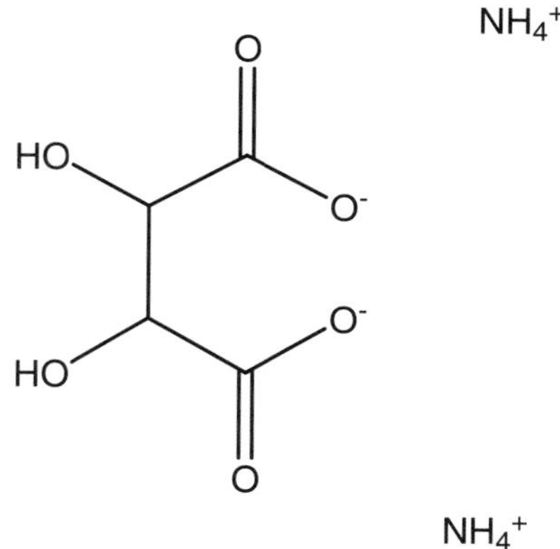

Chemical Structure

Synonyms: AMMONIUM D-TARTRATE; AMMONIUM TARTRATE; BUTANEDIOIC ACID,2,3-DIHYDROXY-(R-(R*,R*))- ,DIAMMONIUMSALT; BUTANEDIOIC ACID,2,3-DIHYDROXY-,(R-(R*,R*))-,DIAMMONIUMSALT; DIAMMONIUM TARTRATE; -2,3-DIHYDROXYBUTANEDIOIC ACID DIAMMONIUM SALT; 2,3-DIHYDROXY-BUTANEDIOIC ACID,DIAMMONIUM SALT; 2,3-DIHYDROXY-BUTANEDIOIC ACID,DIAMMONIUM SALT (9CI); L-TARTARIC ACID AMMONIUM SALT; L-TARTARIC ACID DIAMMONIUM SALT; TARTARIC ACID DIAMMONIUM SALT; L-TARTARIC ACID,AMMONIUM SALT; TARTARIC ACID,DIAMMONIUM SALT
Description: colorless to white granules or crystals
Use: medication: topical ophthalmic agent; textile industry; medicine

Physical Properties

Freezing Point: Decomposes
Specific Gravity: 1.601
Water Solubility: 43.92 g Soluble in 100 g H2O at 0 °C
Other Solubilities: Very Slightly Soluble in Alcohol
pH: 0.2 molar aqueous solution 6.5
Refraction Index: Alpha 1.55

RTECS Toxicity Data

Acute Dermal: Rabbit LD_{50} Route: Subcutaneous Dose: 1130 mg/kg.

Hazard Overviews

Health: Irritating to eyes/skin. Other Acute Effects: may be harmful by inhalation, ingestion, or skin absorption.
Fire: Hazards: emits toxic fumes. Extinguishing agents: water spray; carbon dioxide, dry chemical powder or appropriate foam. Precautions: combustible liquid.
Reactivity: Incompatible with: strong oxidizing agents, strong bases, strong acids. Hazardous decomposition products: toxic fumes of: carbon monoxide, carbon dioxide, nitrogen oxides, ammonia.
Carcinogenicity: IARC - Not listed; NIOSH - Not listed; NTP - Not listed; ACGIH - Not listed; OSHA - Not listed; EPA - Not listed; MAK - Not listed

Primary Target Organs:

Eyes

Skin

Environmental

Regulations

RCRA 40CFR: Not listed
CERCLA: 40CFR 302.4: Listed per CWA Section 311(b)(4) RQ: 5000 lb (2268 kg)
SARA 40CFR 372.65: Not listed
SARA EHS 40CFR 355: Not listed
TSCA: Listed

Analytical Methods

Soil: DOE OM500R

DIA7450 CAS #: 3012-65-5

DIAMONIUM CITRATE

EINECS Number: 221-146-3
Molecular Formula: $C_6H_{14}N_2O_7$
Structured MF: $(NH_4)_2HC_6H_5O_7$
Formula Weight: 226.19

Chemical Structure

Synonyms: AMMONIUM CITRATE; AMMONIUM CITRATE,DIBASIC; AMMONIUM MONOHYDROGEN CITRATE; CITRIC ACID,DIAMMONIUM SALT; DIAMMONIUM CITRATE (SECONDARY); DIAMMONIUM HYDROGEN CITRATE; DIBASIC AMMONIUM CITRATE; 1,2,3-PROPANETRICARBOXYLIC ACID,2-HYDROXY-,DIAMMONIUMSALT
Description: white granules, powder; slight ammonia-like odor
Use: determination of phosphate in fertilizers; pharmaceuticals; rustproofing; cotton printing; plasticizer; source of non-protein nitrogen in chicken food supplements; sequesterant

Physical Properties

Boiling Point: Decomposes at 1 atm
Specific Gravity: 1.48 at 25 °C/4 °C
Water Solubility: Soluble in Water
Other Solubilities: Slightly Soluble in Alcohol
pH: 0.1 molar solution in water 4.3
Flash Point: Not pertinent (combustible solid)

Hazard Overviews

Fire Diamond

Health: Irritating to eyes/skin/respiratory tract. Also Causes: vomiting; high concentrations: temporary blindness, restlessness, chest tightness, cyanosis, rapid/weak pulse, frothy sputum.
Fire: Combustible. Use dry chemical, carbon dioxide, Halon, water spray, fog, or standard foam.
Reactivity: Stable. Hazardous polymerization cannot occur. Avoid: heat. Hazardous decomposition products: ammonia gas.
Carcinogenicity: IARC - Not listed; NIOSH - Not listed; NTP - Not listed; ACGIH - Not listed; OSHA - Not listed; EPA - Not listed; MAK - Not listed
Primary Target Organs:

Eyes

Skin

Respiratory System

Environmental

Cleanup/Disposal: Guide No. 140: Keep combustibles (wood, paper, oil, etc.) away from spilled material. Do not touch damaged containers or spilled material unless wearing appropriate protective clothing. Stop leak if you can do it without risk. Do not get water inside containers. Small Dry Spills: With clean shovel place material into clean, dry container and cover loosely; move containers from spill area. Small Liquid Spills: Use a non-combustible material like vermiculite, sand or earth to soak up the product and place into a container for later disposal. Large Spills: Dike far ahead of liquid spill for later disposal. Following product recovery, flush area with water.

Environmental Physical Data

BCF: no food chain concentration potential
BOD: sewage 0.42 lb/lb, 5 days

Regulations

RCRA 40CFR: Not listed
CERCLA: 40CFR 302.4: Listed per CWA Section 311(b)(4) RQ: 5000 lb (2268 kg)
SARA 40CFR 372.65: Not listed
SARA EHS 40CFR 355: Not listed
TSCA: Listed

Analytical Methods

Soil: DOE OM510R

DIA7600 **CAS #: 131-18-0**

DIAMYL PHTHALATE

RTECS: TI1930000
EINECS Number: 205-017-9
Molecular Formula: $C_{18}H_{26}O_4$
Structured MF: $C_6H_4(COOCH_2CH_2CH_2CH_2CH_3)_2$
Formula Weight: 306.40

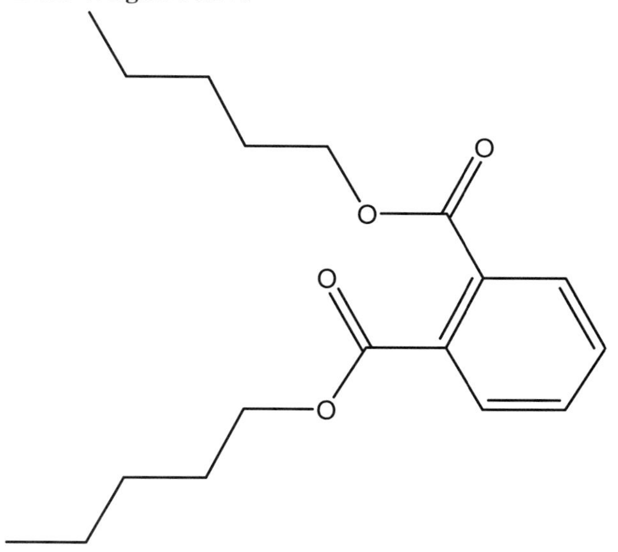

Chemical Structure

Synonyms: AMOIL; AMYL PHTHALATE; 1,2-BENZENEDICARBOXYLIC ACID DIPENTYL ESTER; 1,2-BENZENEDICARBOXYLIC ACID,DIPENTYL ESTER; DIPENTYL 1,2-BENZENEDICARBOXYLATE; DIPENTYL ESTER PHTHALIC ACID; DI-N-PENTYL PHTHALATE; DIPENTYL PHTHALATE; DI-N-PENTYLPHTHALATE; DPP; PHTHALIC ACID DIPENTYL ESTER; PHTHALIC ACID,DIAMYL ESTER; PHTHALIC ACID,DIPENTYL ESTER
Description: oily liquid; nearly odorless
Use: plasticizer for nitrocellulose and resin lacquers; prevention of foam in manufacture of glue; in rubber cements

Physical Properties

Boiling Point: 342 °C (648 °F)
Freezing Point: < -55 °C (-67 °F)
Specific Gravity: 1.022 at 20 °C
Vapor Density: 10.5 Air=1
Density: 1.021 g/cc at 23.3 °C
Vapor Pressure: Extremely low
Water Solubility: About 0.8 mg/L at 25 °C
Other Solubilities: 95% Ethanol: >=100 mg/ml at 20 °C; Acetone: >=100 mg/ml at 20 °C; DMSO: >=100 mg/ml at 20 °C; Organic solvents: Soluble.
Surface Tension: 31.5 dynes/cm at 20 °C
Refraction Index: 1.488 at 25 °C
Flash Point: 118 °C Closed Cup

RTECS Toxicity Data

Reproductive/Teratogenic: Rat Route: Oral; Dose: 6600 mg/kg; Duration: male 3D prior to mating; Paternal Effects - Testes, epididymis, sperm duct. Rat Route: Oral; Dose: 2206 mg/kg; Duration: male 1D prior to mating; Paternal Effects - Other effects on male. Rat Route: Oral; Dose: 2 gm/kg; Duration: male 1D prior to mating; Paternal Effects - Spermatogenesis; Testes, epididymis, sperm duct; Effects on Fertility - Male fertility index.

Hazard Overviews

Fire: Will burn.
Carcinogenicity: IARC - Not listed; NIOSH - Not listed; NTP - Not listed; ACGIH - Not listed; OSHA - Not listed; EPA - Not listed; MAK - Not listed

Environmental

Environmental Fate: If released to soil, it is expected to be essentially immobile. Volatilization from the soil surface to the atmosphere is not expected to be an important process. If released to water, it may bioaccumulate in fish and aquatic organisms, and adsorb to sediment and suspended organic matter. It is not expected to volatilize from water to the atmosphere. Hydrolysis is not expected to be a significant fate process. Limited data indicates that it may biodegrade under aerobic conditions in environmental waters. If released to the atmosphere, it is expected to be adsorbed to particulates. Destruction by the vapor phase reaction with photochemically produced hydroxyl radicals may occur.
Cleanup/Disposal: Guide No. 113: Eliminate all ignition sources (no smoking, flares, sparks or flames in immediate area). All equipment used when handling the product must be grounded. Do not touch or walk through spilled material. Small Spills: Flush area with flooding quantities of water. Large Spills: Wet down with water and dike for later disposal. Keep wetted product wet by slowly adding flooding quantities of water.

Environmental Physical Data

Henry's Law Constant: 8.88×10^{-7}
Octanol/Water Partition Coefficient: log K_{ow} = 4.85
Sorption Partition Coefficient: K_{oc} = 4900
BCF: none likely

Regulations

RCRA 40CFR: Not listed
CERCLA: 40CFR 302.4: Not listed
SARA 40CFR 372.65: Not listed
SARA EHS 40CFR 355: Not listed
TSCA: Listed

Analytical Methods

Plasma: EPA 001

DIA7750 **CAS #: 2050-92-2**

DIAMYLAMINE

RTECS: RZ9100000
EINECS Number: 218-108-3
Molecular Formula: $C_{10}H_{23}N$
Structured MF: $[CH_3(CH_2)_4]_2NH$
Formula Weight: 157.30

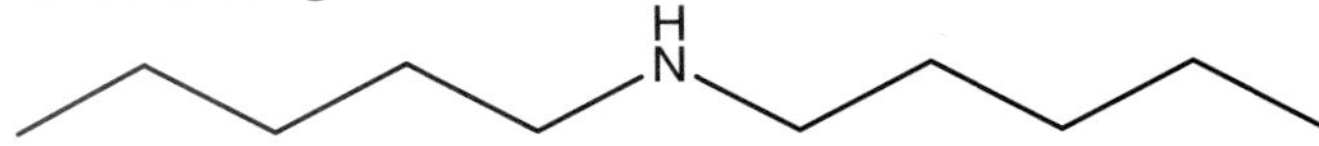

Chemical Structure

Synonyms: AMINE,DIPENTYL; DIAMYL AMINE; DI-N-AMYLAMINE; DI-N-PENTYLAMINE; DIPENTYLAMINE; 1-PENTANAMINE,N-PENTYL-; PENTYLAMINE,PENTYL-; N-PENTYL-1-PENTANAMINE; PENTYLPENTYLAMINE
Description: colorless to pale yellow liquid; fishy, ammonia odor
Use: flotation reagents, corrosion inhibitors, dye-stuffs, rubber accelerators, solvent for oils, resins and some cellulose esters

Physical Properties

Boiling Point: 202 °C (396 °F) at 760 mm Hg
Freezing Point: -7.85 °C (17.87 °F)
Specific Gravity: 0.7771 at 20 °C/4 °C
Vapor Density: 5.4 Air=1
Saturated Vapor Density: 1.201068779 kg/m^3
Density: 0.77 g/mL
Vapor Pressure: 0.153 mm Hg at 25 °C
Water Solubility: Slightly Soluble in Water
Other Solubilities: 95% Ethanol: Soluble (>=10 mg/ml); Acetone: Soluble; Ether: Very Soluble.
Refraction Index: 1.4272 at 20 °C
Flash Point: 51 °C Closed Cup

RTECS Toxicity Data

Acute Oral: Rat LD_{50} Dose: 270 mg/kg.
Acute Inhalation: Rat LC_{Lo} Dose: 63 ppm/4hr.
Acute Dermal: Rabbit LD_{50} Route: Skin; Dose: 350 uL/kg; Toxic Effects: Skin and appendages - Primary irritation.
Irritation Skin: Rabbit Open Draize Test Dose: 500 mg open; Reaction: severe.

Hazard Overviews

Corrosive

Flammable

Fire Diamond

Health: Corrosive to eyes/skin/respiratory tract. Toxic. Also Causes: difficulty breathing, cough, nausea, vomiting, visual disturbances.
Fire: Flammable. For small fires use dry chemical, carbon dioxide, water spray, or regular foam. For large fires use water spray, fog, or regular foam. Be aware that water may cause frothing. Use water only as a spray, never as a direct stream.
Reactivity: Stable. Hazardous polymerization cannot occur. Avoid: heat; ignition sources; oxidizers; acids. Incompatible with: oxidizers; acids. Hazardous decomposition products: carbon; nitrogen oxides.
Carcinogenicity: IARC - Not listed; NIOSH - Not listed; NTP - Not listed; ACGIH - Not listed; OSHA - Not listed; EPA - Not listed; MAK - Not listed
Primary Target Organs:

Eyes

Skin

Respiratory System

Environmental

Environmental Fate: If released to the atmosphere, it will exist mainly in the vapor phase based on a vapor pressure of 0.153 mm Hg at 25 °C, as determined from experimentally derived coefficients. In the vapor-phase, it will react with hydroxyl radicals with an estimated half-life of 4 hours. Based on an estimated K_{oc} of 2600, it should have only low to slight mobility in soil. Volatilization from moist soil surfaces may occur based on an estimated Henry's Law constant of 2.1 $x10^{-4}$ atm-cu m/mole; volatilization from dry soil surfaces may also occur given a vapor pressure of 0.153 mm Hg. In water, it is expected to volatilize from water surfaces based on this compound's Henry's Law constant. It may bind strongly to particulate material and sediment present in the water column. An estimated BCF value of 420 suggests that it may bioconcentrate in aquatic organisms.
Cleanup/Disposal: Guide No. 132: Fully encapsulating, vapor protective clothing should be worn for spills and leaks with no fire. Eliminate all ignition sources (no smoking, flares, sparks or flames in immediate area). All equipment used when handling the product must be grounded. Do not touch or walk through spilled material. Stop leak if you can do it without risk. Prevent entry into waterways, sewers, basements or confined areas. A vapor suppressing foam may be used to reduce vapors. Absorb with earth, sand or other non-combustible material and transfer to containers (except for Hydrazine). Use clean non-sparking tools to collect absorbed material. Large Spills: Dike far ahead of liquid spill for later disposal. Water spray may reduce vapor; but may not prevent ignition in closed spaces.

Environmental Physical Data

Henry's Law Constant: estimated at 2.1 $x10^{-4}$
Octanol/Water Partition Coefficient: log K_{ow} = 3.76
Sorption Partition Coefficient: K_{OC} = 2600
BCF: estimated at 420

Regulations

RCRA 40CFR: Not listed
CERCLA: 40CFR 302.4: Not listed
SARA 40CFR 372.65: Not listed
SARA EHS 40CFR 355: Not listed
TSCA: Listed

DIA7900 CAS #: 68855-54-9

DIATOMACEOUS EARTH

EINECS Number: 272-489-0
Molecular Formula: Unspecified or Variable
Formula Weight: N/A
Synonyms: CELITE; CHROMOSORB; DICALITE; SNOWFLOSS CELITE

Physical Properties

Boiling Point: 2230 °C (4046 °F)
Freezing Point: 1710 °C (3110 °F)
Specific Gravity: 2.35
Water Solubility: Practically Insoluble
pH: 10% slurry 8.5 to 10.5
Flash Point: Noncombustible

Hazard Overviews

Reactivity: Stable in tightly closed containers; exposure to air can cause heat generation. Hazardous polymerization cannot occur. Avoid: high temperatures. Incompatible with: hydrogen fluoride; xenon hexafluoride; heating with alkali carbonates; wet and heated with magnesium; oxygen difluoride; chlorine trifluoride. Hazardous decomposition products: cristobalite; tridymite; crystalline silicas.
Carcinogenicity: IARC - Not listed; NIOSH - Not listed; NTP - Not listed; ACGIH - Not listed; OSHA - Not listed; EPA - Not listed; MAK - Not listed

Environmental

Regulations

RCRA 40CFR: Not listed
CERCLA: 40CFR 302.4: Not listed
SARA 40CFR 372.65: Not listed
SARA EHS 40CFR 355: Not listed
TSCA: Listed

DIA8050 CAS #: 61790-53-2

DIATOMACEOUS EARTH, NATURAL

RTECS: VV7311000
Molecular Formula: Unspecified or Variable
Formula Weight: N/A
Synonyms: AMORPHOUS SILICA; CELITE; KIESELGUHR; SILICA,AMORPHOUS -DIATOMACEOUS EARTH (UNCALCINED); SILICA,AMORPHOUS,DIATOMACEOUS EARTH,CONTAINING >1 %CRYSTALLINE SILICA
Description: light gray powder, brick; odorless

Physical Properties

Boiling Point: 2230 °C (4046 °F)
Freezing Point: 1710 °C (3110 °F)
Specific Gravity: 2.35
Water Solubility: Practically Insoluble
Other Solubilities: Soluble in HF, otherwise insoluble in most acids; soluble in strong alkalies
pH: 10% slurry 8.5 to 10.5
Flash Point: Noncombustible

Hazard Overviews

Fire Diamond

Health: Drying and irritating to eyes/skin/respiratory tract. Also Causes: difficulty breathing, fever, cough, and weight loss. Chronic Effects: decreased chest expansion, extreme difficulty breathing, skin fissures, heart failure.
Fire: Noncombustible. Use extinguishing agents suitable for surrounding fire.
Reactivity: Stable in tightly closed containers; exposure to air can cause heat generation. Hazardous polymerization cannot occur. Avoid: high temperatures. Incompatible with: hydrogen fluoride; xenon hexafluoride; heating with alkali carbonates; wet and heated with magnesium; oxygen difluoride; chlorine trifluoride. Hazardous decomposition products: cristobalite; tridymite; crystalline silicas.
Carcinogenicity: IARC - Group 3, Not classifiable as to carcinogenicity to humans; NIOSH - Not listed; NTP - Not listed; ACGIH - Not listed; OSHA - Not listed; EPA - Not listed; MAK - Not listed
Primary Target Organs:

Eyes

Skin
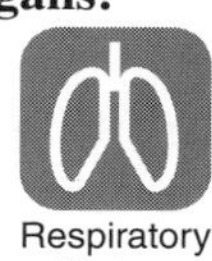
Respiratory System

Mucous Membranes

Exposure Limits
OSHA PEL: TWA: 80 mg/m^3; % SiO_2.
ACGIH TLV: TWA: 10 mg/m^3; Inhalable; 3 mg/m$_3$as respirable.
DFG MAK: TWA: 4 mg/m^3.
Respirator Recommendation
Exposure Range: >6 to 60 mg/m^3 Air Purifying, Negative Pressure, Half Mask
Exposure Range: >60 to 600 mg/m^3 Air Purifying, Negative Pressure, Full Face
Exposure Range: >600 to 6000 mg/m^3 Supplied Air, Constant Flow/Pressure Demand, Full Face
Exposure Range: >6000 to unlimited mg/m^3 Self-contained Breathing Apparatus, Pressure Demand, Full Face
Cartridge Color: dust/mist filter (use P100 or consult supervisor for appropriate dust/mist filter)

Environmental

Regulations

RCRA 40CFR: Not listed
CERCLA: 40CFR 302.4: Not listed
SARA 40CFR 372.65: Not listed
SARA EHS 40CFR 355: Not listed

TSCA: Listed

Analytical Methods

Indoor / Expired Air: NIOSH 7501

DIA8200 CAS #: 117-96-4

DIATRIZOATE

RTECS: DG5950000
EINECS Number: 204-223-6
Molecular Formula: $C_{11}H_9I_3N_2O_4$
Formula Weight: 613.90

Chemical Structure

Synonyms: AMIDOTRIZOATE; AMIDOTRIZOIC ACID; BENZOIC ACID,3,5-BIS(ACETYLAMINO)-2,4,6-TRIIODO-; BENZOIC ACID,3,5-BIS(ACETYLAMINO)-2,4,6-TRIIODO-(9CI); BENZOIC ACID,3,5-DIACETAMIDO-2,4,6-TRIIODO-; 3,5-BIS(ACETYLAMINO)-2,4,6-TRIIODOBENZOIC ACID; 3,5-DIACETAMIDO-2,4,6-TRIIODOBENZOIC ACID; DIAT; DIATRIAZOATE; DIATRIZOESAURE; DIATRIZOIC ACID; ODISTON; 2,4,6-TRIIODO-3,5-DIACETAMIDOBENZOIC ACID; TRIOMBRIN; UROGRAFIN ACID; UROGRANOIC ACID; UROTRAST
Description: white powder or crystals
Use: diagnostic aid (radiopaque medium); vet: x-ray constrast medium; radiopaque medium for diagnostics

Physical Properties

Freezing Point: > 300 °C (572 °F)
Water Solubility: Very Slightly Soluble in Water
Other Solubilities: Very Slightly Soluble in Alcohol; Soluble in Dimethylformamide, alkali hydroxide solutions.
pH: 50% aqueous solution 7.0 to 7.5

Hazard Overviews

Health: Acute Effects: may be harmful by inhalation, ingestion, or skin absorption; exposure can cause: nausea, headache and vomiting other symptoms include sensations of heat, weakness, thirst, coughing, sneezing, itching, pallor, tachycardia and hypotension. The toxicological properties have not been thoroughly investigated.
Fire: Hazards: emits toxic fumes. Extinguishing agents: carbon dioxide, dry chemical powder or appropriate foam. Precautions: combustible liquid.
Reactivity: Stable. Hazardous polymerization will not occur. Hazardous decomposition products: toxic fumes of: carbon monoxide, carbon dioxide, nitrogen oxides, hydrogen iodide.
Carcinogenicity: IARC - Not listed; NIOSH - Not listed; NTP - Not listed; ACGIH - Not listed; OSHA - Not listed; EPA - Not listed; MAK - Not listed

Environmental

Regulations

RCRA 40CFR: Not listed
CERCLA: 40CFR 302.4: Not listed
SARA 40CFR 372.65: Not listed
SARA EHS 40CFR 355: Not listed
TSCA: Not listed

DIA8350 CAS #: 280-57-9

1,4-DIAZABICYCLO(2.2.2)OCTANE

RTECS: HM0354200
EINECS Number: 205-999-9
Molecular Formula: $C_6H_{12}N_2$
Formula Weight: 112.17

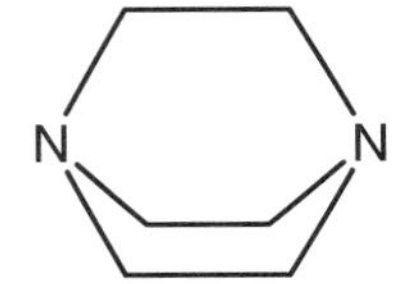

Chemical Structure

Synonyms: BICYCLO(2,2,2)-1,4-DIAZAOCTANE; BICYCLO(2.2.2)-1,4-DIAZAOCTANE; D 33LV; DABCO; DABCO R-8020; DABCO S-25; DABCO CRYSTAL; DABCO EG; DABCO 33LV; 1,4-DIAZABICYCLOOCTANE; N,N'-ENDO-ETHYLENEPIPERAZINE; 1,4-ETHYLENEPIPERAZINE; TED; TEDA; TEXACAT TD 100; THANCAT TD 33; TRIETHYLENEDIAMINE
Description: crystals
Use: catalyst in making urethane foams; singlet oxygen trap used as diagnostic for singlet oxygen; trap for radioiodine

Physical Properties

Boiling Point: 174 °C (345 °F)
Freezing Point: 158 °C (316.4 °F)
Water Solubility: 45 g/100 g Water at 25 °C
Other Solubilities: 13 g/100 g Acetone at 25 °C 51 g/100 g Benzene at 25 °C 77 g/100 g Ethanol at 25 °C 26.1 g/100 g Methyl Ethyl Ketone at 25 °C
Ionization Potential (eV): 7.197 +/-2.0

RTECS Toxicity Data

Acute Oral: Rat LD_{50} Dose: 1700 mg/kg. Rabbit LD_{50} Dose: 1100 mg/kg; Toxic Effects: Behavioral - Somnolence (general depressed activity); Behavioral - Muscle weakness; Behavioral - Muscle contraction or spasticity.

Chronic (Multiple Dose) Oral: Rat Dose: 4680 mg/kg/26W-I; Toxic Effects: Behavioral - Alteration of classical conditioning; Biochemical - True cholinesterase; Biochemical - Transaminases.

Irritation Eye: Rabbit Standard Draize Test Dose: 25 mg; Reaction: moderate.

Irritation Skin: Rabbit Open Draize Test Dose: 2500 ug open; Reaction: mild.

Hazard Overviews

Corrosive

Health: Corrosive to eyes/skin/respiratory tract. Other Acute Effects: harmful if swallowed, inhaled, or absorbed through skin; inhalation may result in spasm; inflammation and edema of the larynx and bronchi; chemical pneumonitis and pulmonary edema; burning sensation; coughing; wheezing; laryngitis; shortness of breath; headache; nausea; vomiting.

Fire: Hazards: emits toxic fumes. Extinguishing agents: carbon dioxide, dry chemical powder or appropriate foam. Precautions: combustible liquid.

Reactivity: Incompatible with: strong oxidizing agents. Hazardous decomposition products: toxic fumes of: carbon monoxide, carbon dioxide, nitrogen oxides.

Carcinogenicity: IARC - Not listed; NIOSH - Not listed; NTP - Not listed; ACGIH - Not listed; OSHA - Not listed; EPA - Not listed; MAK - Not listed

Primary Target Organs:

Eyes

Skin

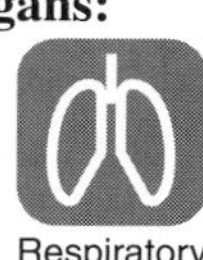
Respiratory System

Environmental

Ecotoxicity: Fishes: Pimephales promelas 4d LC_{50} 1,730 mg/l

Regulations

RCRA 40CFR: Not listed
CERCLA: 40CFR 302.4: Not listed
SARA 40CFR 372.65: Not listed
SARA EHS 40CFR 355: Not listed
TSCA: Listed

DIA8500 CAS #: 439-14-5

DIAZEPAM

RTECS: DF1575000
EINECS Number: 207-122-5

Molecular Formula: $C_{16}H_{13}ClN_2O$
Formula Weight: 284.76

Chemical Structure

Synonyms: ALBORAL; ALISEUM; ALUPRAM; AMIPROL; AN-DING; ANSIOLIN; ANSIOLISINA; APAURIN; APOZEPAM; ASSIVAL; ATENSINE; ATILEN; 2H-1,4-BENZODIAZEPIN-2-ONE,7-CHLORO-1,3-DIHYDRO-1-METHYL-5-PHENYL-; BIALZEPAM; CALMOCITENE; CALMPOSE; CERCINE; CEREGULART; 7-CHLORO-1,3-DIHYDRO-1-METHYL-5-PHENYL-2H-1,4-BENZODIAZEPIN-2-ONE; 7-CHLORO-1-METHYL-5-3H-1,4-BENZIODIAZEPIN-2(1H)-ONE; 7-CHLORO-1-METHYL-5-3H-1,4-BENZODIAZEPIN-2(1H)-ONE; 7-CHLORO-1-METHYL-2-OXO-5-PHENYL-3H-1,4-BENZODIAZEPINE; 7-CHLORO-1-METHYL-5-PHENYL-2H-1,4-BENZODIAZEPIN-2-ONE; 7-CHLORO-1-METHYL-5-PHENYL-3H-1,4-BENZODIAZEPIN-2(1H)-ONE; 7-CHLORO-1-METHYL-5-PHENYL-1,3-DIHYDRO-2H-1,4-BENZODIAZEPIN-2-ONE; CONDITION; DAP; DIACEPAN; DIAPAM; DIAZEMULS; DIAZEPAMU; DIAZEPAN; DIAZETARD; DIENPAX; DIPAM; DIPEZONA; DOMALIUM; DUKSEN; DUXEN; ERIDAN; EVACALM; FAUSTAN; FREUDAL; FRUSTAN; GIHITAN; HORIZON; KABIVITRUM; KIATRIUM; LA-111; LA-III; LEMBROL; LEVIUM; LIBERETAS; METHYL DIAZEPINONE; METHYLDIAZEPINONE (PHARMACEUTICAL); 1-METHYL-5-PHENYL-7-CHLORO-1,3-DIHYDRO-2H-1,4-BENZODIAZEPIN-2-ONE; MOROSAN; NEUROLYTRIL; NOAN; NSC-77518; PACITRAN; E-PAM; PARANTEN; PAXATE; PAXEL; PLIDAN; QUETINIL; QUIATRIL; QUIEVITA; RELAMINAL; RELANIUM; RELAX; RENBORIN; RO 5-2807; S.A. R.L; SAROMET; SEDAPAM; SEDIPAM; SEDUKSEN; SEDUXEN; SERENACK; SERENAMIN; SERENZIN; SETONIL; SIBAZON; SONACON; STESOLID; STESOLIN; TENSOPAM; TRANIMUL; TRANQDYN; TRANQUASE; TRANQUIRIT; TRANQUO-TABLINEN; UMBRIUM; UNISEDIL; USEMPAX AP; VALEO; VALITRAN; VALIUM; VALIUM R; VALRELEASE; VATRAN; VELIUM; VIVAL; VIVOL; WY-3467; ZIPAN

Description: off-white to yellow crystalline powder; practically odorless

Use: in the treatment of anxiety, as a skeletal muscle relaxant in various spastic disorders, as a preanesthetic medication in childbirth and cardioversion, as a hypnotic antiepileptic agent and as a sedative and tranquilizer; in the treatment of psychoneuroti disorders, acute alcohol withdrawal, status epilepticus, schizophrenia, tetan and for preoperative medication; to relieve spasticity and athetosis in patients with cerebral palsy

Physical Properties

Freezing Point: 125 °C (257 °F) to 126 °C (258.8 °F)
Water Solubility: 1 g/2 ml chloroform
Other Solubilities: Sparingly Soluble in Propylene Glycol; 1 g/39 ml Ether; Approx 62.5 mg/ml Alcohol at 25 °C.
Flash Point: Not available; probably combustible

RTECS Toxicity Data

Acute Oral: Man TD_{Lo} Dose: 143 ug/kg; Toxic Effects: Sense organs and special senses - Visual field changes; Sense organs and special senses - Diplopia. Woman TD_{Lo} Dose: 5 mg/kg; Toxic Effects: Behavioral - Somnolence (general depressed activity). Rat LD_{50} Dose: 352 mg/kg; Toxic Effects: Behavioral - Convulsions or effect on seizure threshold; Behavioral - Ataxia; Lungs, Thorax, or Respiration - Dyspnea.
Acute Dermal: Rat LD_{50} Route: Subcutaneous Dose: 6350 ug/kg. Mouse LD_{50} Route: Skin; Dose: 800 mg/kg.
Chronic (Multiple Dose) Oral: Rat Dose: 2250 mg/kg/30D-C; Toxic Effects: Endocrine - Adrenal cortex hyperplasia; Blood - Changes in cell count (unspecified); Nutritional and gross metabolic - Changes in sodium. Rat Dose: 6825 mg/kg/13W-C; Toxic Effects: Kidney, Ureter, and Bladder - Urine volume increased; Blood - Changes in cell count (unspecified); Biochemical - Transaminases. Rat Dose: 7800 mg/kg/26W-I; Toxic Effects: Liver - Other changes; Kidney, Ureter, and Bladder - Other changes; Nutritional and gross metabolic - Other changes. Rat Dose: 17500 mg/kg/35D-C; Toxic Effects: Liver - Changes in liver weight; DEATH; DEATH - Changes in testicular weight. Rat Dose: 930 mg/kg/31W-I; Toxic Effects: Liver - Changes in liver weight; Biochemical - Phosphatases; DEATH - Changes in prostate weight. Rat Dose: 14112 mg/kg/56W-C; Toxic Effects: Gastrointestinal - Changes in pancreatic weight; Liver - Changes in liver weight; Kidney, Ureter, and Bladder - Changes in kidney weight. Monkey Dose: 18 gm/kg/52W-I; Toxic Effects: Liver - Fatty liver degeneration; Kidney, Ureter, and Bladder - Other changes; DEATH.
Reproductive/Teratogenic: Woman Route: Oral; Dose: 22800 ug/kg; Duration: female 25-36W of pregnancy Effects on Newborn - Drug dependence. Woman Route: Oral; Dose: 11600 ug/kg; Duration: female 43D of pregnancy; Specific Developmental Abnormalities - Craniofacial (including nose and tongue). Woman Route: Oral; Dose: 5 mg/kg; Duration: female 36W of pregnancy; Specific Developmental Abnormalities - Cardiovascular (circulatory) system. Woman Route: Intravenous; Dose: 400 ug/kg; Duration: female 39W of pregnancy; Specific Developmental Abnormalities - Cardiovascular (circulatory) system. Rat Route: Oral; Dose: 10500 mg/kg; Duration: female 14D prior to mating Effects on Fertility - Mating Performance; Female fertility index. Rat Route: Oral; Dose: 5 gm/kg; Duration: female 8-17D of pregnancy; Effects on Embryo or Fetus - Fetotoxicity; Effects on Fertility - Post-implantation mortality; Specific Developmental Abnormalities - Musculoskeletal system. Rat Route: Oral; Dose: 675 mg/kg; Duration: female 17-22D of pregnancy Effects on Newborn - Viability index; Growth statistics. Rat Route: Oral; Dose: 40 mg/kg; Duration: female 13-20D of pregnancy Effects on Newborn - Growth statistics; Behavioral. Rat Route: Oral; Dose: 220 mg/kg; Duration: female 1-22D of pregnancy; Maternal Effects - Parturition; Effects on Newborn - Behavioral. Rat Route: Oral; Dose: 14 gm/kg; Duration: male 35D prior to mating; Paternal Effects - Testes, epididymis, sperm duct.
Mutagenic: Human Cytogenetic Analysis; Cell Type: leukocyte; Dose: 10 mg/L. Woman Cytogenetic Analysis; Route: Unreported; Dose: 328 mg/kg/78W. Human Sex Chromosome Loss; Cell Type: lymphocyte; Dose: 25 mg/L.
Tumorigenic: Mouse Route: Oral; Dose: 42 gm/kg/80W-C; Toxic Effects: Tumorigenic - Equivocal tumorigenic agent by RTECS criteria; Liver - Tumors.

Hazard Overviews

Health: Toxic. Other Acute Effects: harmful if swallowed, inhaled, or absorbed through skin; exposure can cause CNS depression, gastrointestinal disturbances; target organ: central nervous system. Chronic Effects: prolonged or repeated exposure can lead to habituation or addiction; possible teratogen;overexposure may cause reproductive disorder(s) based on tests with laboratory animals; possible mutagen.
Fire: Will burn. Hazards: emits toxic fumes. Extinguishing agents: carbon dioxide, dry chemical powder or appropriate foam. Precautions: combustible liquid.
Reactivity: Stable. Hazardous polymerization will not occur. Hazardous decomposition products: toxic fumes of: carbon monoxide, carbon dioxide, nitrogen oxides, hydrogen chloride gas.
Carcinogenicity: IARC - Group 3, Not classifiable as to carcinogenicity to humans; NIOSH - Not listed; NTP - Not listed; ACGIH - Not listed; OSHA - Not listed; EPA - Not listed; MAK - Not listed
Primary Target Organs:

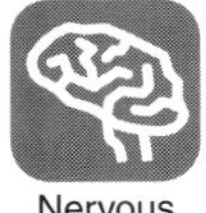

Nervous System

Environmental

Ecotoxicity: Marine tests MicrotoxO (Photobacterium) test 5 min EC_{50} >35,000 Artoxkit M (Artemia salina) test 24h LC_{50} 230

Environmental Physical Data

Octanol/Water Partition Coefficient: log K_{ow} = 2.82

Regulations

RCRA 40CFR: Not listed
CERCLA: 40CFR 302.4: Not listed
SARA 40CFR 372.65: Not listed
SARA EHS 40CFR 355: Not listed
TSCA: Listed

DIA8650 **CAS #: 333-41-5**

DIAZINON

RTECS: TF3325000
DOT: UN2783; NA2783; IMO6.3
EINECS Number: 206-373-8
Molecular Formula: $C_{12}H_{21}N_2O_3PS$
Structured MF: $C_{12}H_{21}N_2O_3PS$
Formula Weight: 304.36
Synonyms: ALFA-TOX; ANTIGAL; BASSADINON; BASUDIN; BASUDIN 10 G; BASUDIN S; BAZUDEN; CIAZINON; COMPASS; DACUTOX; DASSITOX; DAZZEL; DELZINON; O,O-DIAETHYL-O-(2-ISOPROPYL-4-METHYL-PYRIMIDIN-6-YL)-MONOTHIOPHOSPHAT; O,O-DIAETHYL-O-(2-ISOPROPYL-4-METHYL)-6-PYRIMIDYL-THIONOPHOSPHAT; DIANON; DIAZAJET; DIAZIDE; DIAZINON AG 500; DIAZINONE; DIAZITOL; DIAZOL; DICID; DIETHYL 4-(2-ISOPROPYL-6-METHYL PYRIMIDINYL)PHOSPHOROTHIONATE; O,O-DIETHYL O-(2-ISOPROPYL-6-METHYL-4-PYRIMIDINYL)PHOSPHOROTHIOATE; DIETHYL 2-ISOPROPYL-4-METHYL-6-PYRIMIDINYLPHOSPHOROTHIONATE; DIETHYL 4-(2-ISOPROPYL-6-METHYLPYRIMIDINYL)PHOSPHOROTHIONATE; DIETHYL 2-ISOPROPYL-4-METHYL-6-PYRIMIDYL THIONOPHOSPHATE; O,O-DIETHYL O-(2-ISOPROPYL-4-METHYL-6-PYRIMIDYL)THIONOPHOSPHATE; O,O-DIETHYL 2-ISOPROPYL-4-METHYLPYRIMIDYL-6-THIOPHOSPHATE; O,O-DIETHYL O-6-METHYL-2-ISOPROPYL-4-PYRIMIDINYLPHOSPHOROTHIOATE; DIETHYL4-(2-ISOPROPYL-6-METHYLPYRIMIDINYL)PHOSPHOROTHIONATE; O,O-DIETHYL-O-(2-ISOPROPYL-4-METHYL-PYRIMIDIN-6-YL)-MONOTHIOFOSFAAT; O,O-DIETHYL-O-2-ISOPROPYL-4-METHYL-6-PYRIMIDINYL-PHOSPHOROTHIOAT E; O,O-DIETHYL-O-(2-ISOPROPYL-4-METHYL-6-PYRIMIDINYL)-PHOSPHOROTHIOATE; O,O-DIETHYL-O-(2-ISOPROPYL-4-METHYL-6-PYRIMIDYL)PHOSPHOROTHIOATE; O,O-DIETIL-O-(2-ISOPROPIL-4-METIL-PIRIMIDIN-6-IL)-MONOTIOFOSFATO; DIMPYLAT; DIMPYLATE; DIPOFENE; DIZICTOL; DIZINON; DRAWIZON; DYZOL; ENT 19,507; ENT 19507; EXODIN; FLYTROL; G 301; G-24480; GALESAN; GARDEN TOX; GARDENTOX; GEIGY 24480; ISOPROPYLMETHYLPYRIMIDYL DIETHYL THIOPHOSPHATE; O-2-ISOPROPYL-4-METHYLPYRIMIDYL-O,O-DIETHYLPHOSPHOROTHIOATE; KAYAZINON; KAYAZOL; KNOX OUT 2FM; KNOX OUT YELLOW JACKET CONTORL; KNOX-OUT; NEDCIDOL; NEOCIDOL; NEOCIDOL (OIL); NEODINON; NIPSAN; NUCIDOL; OLEODIAZINON; OMS 469; OPTIMIZER; PHOSPHOROTHIOATE,O,O-DIETHYL O-6-(2-ISOPROPYL-4-METHYLPYRIMIDYL); PHOSPHOROTHIOATE,O,O-DIETHYLO-6-(2-ISOPROPYL-4-METHYLPYRIMIDYL); PHOSPHOROTHIOIC ACID,O,O-DIETHYLO-(2-ISOPROPYL-6-METHYL-4-PYRIMIDINYL) ESTER; PHOSPHOROTHIOIC ACID,O,O-DIETHYLO-(ISOPROPYLMETHYLPYRIMIDINYL) ESTER; PHOSPHOROTHIOIC ACID,O,O-DIETHYLO-(6-METHYL-2-(1-METHYLETHYL)-4-PYRIMIDINYL) ESTER; PT 265; 4-PYRIMIDINOL,2-ISOPROPYL-6-METHYL-,O-ESTER WITH O,O-DIETHYL PHOSPHOROTHIOATE; SAROLEX; SPECTRACIDE; SPECTRACIDE 25EC; SROLEX; TERMINATOR; THIOPHOSPHATE DE O,O-DIETHYLE ET DE O-2-ISOPROPYL-4-METHYL-6-PYRIMIDYLE; THIOPHOSPHORIC ACID 2-ISOPROPYL-4-METHYL-6-PYRIMIDYLDIETHYL ESTER
Description: colorless, pale yellow to dark brown liquid; faint ester-like odor
Use: nonsystemic insecticide and acaricide; on a wide variety of agricultural crops, ornamentals, domestic animals, lawns and gardens, and household pests

Physical Properties

Boiling Point: 83 °C (181 °F) to 84 °C (183 °F) at 2X10-3 mm Hg
Specific Gravity: 1.116 to 1.118 at 20 °C
Vapor Density: 1.45 Air=1
Saturated Vapor Density: 1.200002099 kg/m^3
Density: 1.116 to 1.118 g/mL at 20 °C
Vapor Pressure: 1.4 $x10^{-4}$ mm Hg at 20 °C
Water Solubility: 0.004% at 20 °C
Other Solubilities: 95% Ethanol: >=100 mg/ml at 24 °C; Acetone: >=100 mg/ml at 24 °C; Benzene: miscible; Cyclohexane: miscible; Xylene: miscible; DMSO: >=100 mg/ml at 24 °C; Ether: miscible; Petroleum Ether: miscible; Ketones: Soluble.
Surface Tension: Estimated at 35 dynes/cm
Refraction Index: 1.4978 to 1.4981 at 20 °C/D
Flash Point: 82 °C

RTECS Toxicity Data

Acute Oral: Human TD_{Lo} Dose: 214 mg/kg; Toxic Effects: Behavioral - Change in motor activity (specific assay); Behavioral - Muscle weakness; Skin and appendages - Sweating. Rat LD_{50} Dose: 66 mg/kg.
Acute Inhalation: Rat LC_{50} Dose: 3500 mg/m^3/4hr. Mouse LC_{50} Dose: 1600 mg/m^3/4hr.
Acute Dermal: Rabbit LD_{50} Route: Skin; Dose: 3600 mg/kg. Rat LD_{50} Route: Skin; Dose: 180 mg/kg; Toxic Effects: Biochemical - True cholinesterase.
Chronic (Multiple Dose) Oral: Rat Dose: 46 mg/kg/92D-I; Toxic Effects: Blood - Changes in serum composition; Blood - Other changes; Biochemical - True cholinesterase. Rat Dose: 86 mg/kg/30D-C; Toxic Effects: Blood - Changes in serum composition; Blood - Other changes; Biochemical - True cholinesterase. Rat Dose: 91 mg/kg/26W-C; Toxic Effects: Blood - Other changes; Nutritional and gross metabolic - Weight loss or decreased weight gain; Biochemical - True cholinesterase. Rat Dose: 371 mg/kg/14D-C; Toxic Effects: Blood - Change in clotting factors. Rat Dose: 28 mg/kg/28W-I; Toxic Effects: Liver - Fatty liver degeneration; Nutritional and gross metabolic - Weight loss or decreased weight gain. Rat Dose: 1095 mg/kg/2Y-C; Toxic Effects: Endocrine - Changes in thyroid weight; Blood - Other changes; Biochemical - True cholinesterase. Rat Dose: 98 mg/kg/28W-I; Toxic Effects: Nutritional and gross metabolic - Weight loss or decreased weight gain; Biochemical - True cholinesterase; Biochemical - Dopamine at other sites.
Chronic (Multiple Dose) Inhalation: Rat Dose: 201 ug/m^3/24H/13W-C; Toxic Effects: Kidney, Ureter, and Bladder - Other changes in urine composition; Blood - Changes in serum composition; Biochemical - True cholinesterase.
Chronic (Multiple Dose) Dermal: Rabbit Route: Skin; Dose: 210 mg/kg/21D-I; Toxic Effects: Blood - Other changes; Biochemical - True cholinesterase.
Irritation Eye: Rabbit Standard Draize Test Dose: 100 mg; Reaction: severe.

Irritation Skin: Rabbit Open Draize Test Dose: 500 mg open; Reaction: moderate.

Reproductive/Teratogenic: Rat Route: Oral; Dose: 26400 ug/kg; Duration: female 12-15D of pregnancy Specific Developmental Abnormalities - Urogenital system. Rat Route: Oral; Dose: 45 mg/kg; Duration: female 8-12D of pregnancy; Specific Developmental Abnormalities - Musculoskeletal system. Rat Route: Oral; Dose: 63500 ug/kg; Duration: female 10D of pregnancy; Effects on Fertility - Post-implantation mortality.

Mutagenic: Human Cytogenetic Analysis; Cell Type: lymphocyte; Dose: 500 ug/L. Human Other Mutation Test Systems; Cell Type: lymphocyte; Dose: 500 ug/L.

Hazard Overviews

Poison

Fire Diamond

Health: Irritating to eyes/skin/respiratory tract. Poison. Also Causes: weakness, headache, nausea/vomiting, excessive salivation, sweating, dizziness, abdominal cramps, blurred vision, non-reactive pinpoint pupils, incoordination, slurred speech, bronchorrhea which produce moist rales, bronchospasm, pulmonary edema, muscle twitching (especially of the tongue and eyelids), convulsions, death. Gastrointestinal tract irritation may occur followed by symptoms as via inhalation. Chronic Effects: ocular disorders collectively referred to as Saku disease; decreased visual acuity (detail, clarity) in both eyes, narrow field of vision, anomaly of refraction and enhanced tendon reflexes; paresthesia, weakness in the legs, thirst, constipation, headache, constant fatigue, orthostatic anemia (anemia that only occurs when a person is in the upright position and disappears when the person lies down for a short time), rectal disturbances, and impotence.

Fire: Noncombustible when pure. Use dry chemical, carbon dioxide, water spray, fog, or regular foam.

Reactivity: Stable. Hazardous polymerization cannot occur. Avoid: high temperatures. Incompatible with: copper-containing compounds; dilute acids; strong acids; alkalis. Hazardous decomposition products: carbon; nitrogen; sulfur; phosphorus oxides.

Carcinogenicity: IARC - Not listed; NIOSH - Not listed; NTP - Not listed; ACGIH - Class A4, Not classifiable as a human carcinogen; OSHA - Not listed; EPA - Not listed; MAK - Not listed

Primary Target Organs:

Eyes

Skin

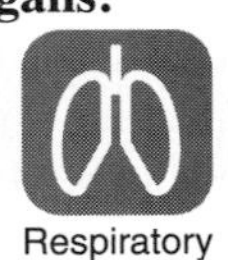

Respiratory System

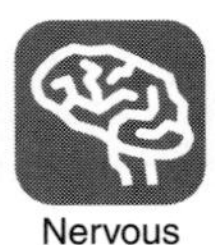

Nervous System

Cardio-vascular

Blood

Exposure Limits

OSHA PEL Vacated 1989 Limits: TWA: 0.1 mg/m^3.

ACGIH TLV: TWA: 0.1 mg/m^3.

NIOSH REL: TWA: 0.1 mg/m^3.

DFG MAK: TWA: 0.1 mg/m^3.

Respirator Recommendation

Exposure Range: >0.1 to 1 mg/m^3 Air Purifying, Negative Pressure, Half Mask

Exposure Range: >1 to 10 mg/m^3 Air Purifying, Negative Pressure, Full Face

Exposure Range: >10 to 1000 mg/m^3 Supplied Air, Constant Flow/Pressure Demand, Full Face

Exposure Range: >1000 to unlimited mg/m^3 Self-contained Breathing Apparatus, Pressure Demand, Full Face

Cartridge Color: dust/mist filter (use P100 or consult supervisor for appropriate dust/mist filter)

Environmental

Ecotoxicity: LC_{50} Coturnix oral 101 ppm LC_{50} Daphnia pulex 0.90 ug/l/96 hr /Conditions of bioassay not specified LD_{50} Mallard, males 3-4 mo oral 3.54 mg/kg (95% confidence limitS 2.37-5.27) LC_{50} Gillia altilis (snail) 40 uM (11 ppm)/96 hr (static renewal bioassay) LC_{50} Pteronarcys (stonefly) 25 ug/l/96 hr /Conditions of bioassay not specified LC_{50} Pimephales promelas (fathead minnow) 31 day old 9.35 mg/l/96 hr. Affected fish lost schooling behavior, were hyperactive and swam in a corkscrew/spiral pattern LD_{50} Bullfrog oral greater than 2000 mg/kg, female LC_{50} Rainbow trout 90 ug/l/96 hr /Conditions of bioassay not specified LC_{50} Bluegill 0.052 ppm/24 hr /Conditions of bioassay not specified

Environmental Fate: If it is released to soil it will not strongly bind to the soil and will be expected to exhibit moderate mobility in the soil. Hydrolysis has been reported to be slow at pH >6, but may be significant in some soils. Biodegradation will be expected to be a major fate process in soils with reported half-lives of <1,2, and 5 weeks in non-sterile soils. Photolysis may be significant on the surface of soils, but evaporation from the surface of soils is not expected to be a significant transport process. If it is released to water it may sorb to sediments moderately but it will not bioconcentrate in aquatic organisms. Hydrolysis may be a significant fate process with reported half-lives of 31 days (pH 5), 185 days (pH 7.4), and 136 days (pH 9.0) at 20 °C and 2-3 weeks in distilled water at pH 6 at room temperature; major products of hydrolysis are 2-isopropyl-4-methyl-6-hydroxypyrimidine and diethyl thiophosphoric acid or diethyl phosphoric acid. Biodegradation and photolysis may be significant fate processes in natural waters. Evaporation may be significant with a half-life of 46 days predicted for evaporation from a river 1 m deep, flowing at 1 m/sec with a wind velocity of 3 m/sec. If released to the atmosphere it may be subject to direct photolysis since it adsorbs light >290 nm. The estimated vapor phase half-life in the atmosphere is 4.83 hours as a result of reaction with photochemically produced hydroxyl radicals.

Cleanup/Disposal: Guide No. 152: Do not touch damaged containers or spilled material unless wearing appropriate protective clothing. Stop leak if you can do it without risk. Prevent entry into waterways, sewers, basements or confined areas. Cover with plastic sheet to prevent spreading. Absorb or cover with dry earth, sand or other non-combustible material and transfer to containers. Do not get water inside containers.

Environmental Physical Data

Henry's Law Constant: calculated at 1.4 x10^{-6}
Octanol/Water Partition Coefficient: log K_{ow} = 1.92 to 3.14
Sorption Partition Coefficient: K_{oc} = sediments 40 to 132
BCF: guppy 17.5

Regulations

RCRA 40CFR: Not listed
CERCLA: 40CFR 302.4: Listed per CWA Section 311(b)(4) RQ: 1 lb (0.454 kg)
SARA 40CFR 372.65: Listed
SARA EHS 40CFR 355: Not listed
TSCA: Listed

Analytical Methods

Air: EPA 016
Soil: EPA PMD-DFN, PMD-TLC, 025; SW846 8140, 8141, 8141A; USGS O5104, O7104
Water / Groundwater: EPA P-005-1, 1657, 614, 622, 022; SW846 8141, 8141A; USGS O3104
Drinking Water: EPA 507, 525.2; AOAC 991.07; ASTM D5475
Food: FDA 212.1, 212.2, 232.1, 232.2, 232.3, 232.4, 242.1; AOAC 968.24, 970.52, 970.53, 971.08, 982.06
Indoor / Expired Air: NIOSH 5600; ASTM D4861
Plasma: EPA 001, 003, 027, 028; FDA 211.1, 231.1, 252, 253

DIA8800 **CAS #: 334-88-3**

DIAZOMETHANE

RTECS: PA7000000
EINECS Number: 206-382-7
Molecular Formula: CH_2N_2
Structured MF: CH_2N_2
Formula Weight: 42.04
Synonyms: ACOMETHYLENE; AZIMETHYLENE; AZOMETHYLENE; DIAZIRINE; DIAZONIUM,METHYLIDE; METHANE,DIAZO-
Description: yellow gas; musty odor
Use: methylating agent for acidic cmpd such as carboxylic acids, phenols, enols

Physical Properties

Boiling Point: -23 °C (-9 °F)
Freezing Point: -145 °C (-229 °F)
Specific Gravity: 1.45
Vapor Density: 1.45 Air=1
Vapor Pressure: > 1 atm
Water Solubility: Reacts
Other Solubilities: freely Soluble in Benzene; Slightly Soluble in Ethyl Alcohol, Ethyl Ether.
Ionization Potential (eV): 9.00
Flash Point: Combustible

RTECS Toxicity Data

Acute Inhalation: Cat LC_{50} Dose: 175 ppm/10M.
Mutagenic: Mold - N Crassa Mutations in Microorganisms; Dose: 250 mmol/L (-S9).
Tumorigenic: Rat Route: Inhalation; Dose: 272 mg/m^3/26W-I; Toxic Effects: Tumorigenic - Equivocal tumorigenic agent by RTECS criteria; Lungs, Thorax, or Respiration - Tumors. Mouse Route: Inhalation; Dose: 272 mg/m^3/26W-I; Toxic Effects: Tumorigenic - Equivocal tumorigenic agent by RTECS criteria; Lungs, Thorax, or Respiration - Tumors.

Hazard Overviews

Corrosive

Explosive

Flammable

Fire Diamond

Health: Corrosive to eyes/skin/respiratory tract. Also Causes: fatigue, fever, cough, shortness of breath, headache, flushed skin, chest pain, wheezing, pneumonitis, pulmonary edema, itching, inflammation, blisters, corneal burns, lesions in eyes, painful swallowing, stomach and abdominal pain, nausea, diarrhea, weak and rapid pulse, cold hands and feet, convulsions. Chronic Effects: asthmatic attacks.
Fire: Explosive and flammable. For small fires use water spray, carbon dioxide, or regular foam. For large fires use water spray, fog, or regular foam. Vapors may travel to an ignition source and flash back. Container may explode in heat of fire. Severe explosion risk if material is introduced to shock.
Reactivity: Unstable, highly reactive; can explode if heated, shocked, or in contact with a rough surface like ground glass. Hazardous polymerization can occur. Avoid: vapor buildup; heat. Incompatible with: calcium sulfate; dimethylaminodimethylarsine; trimethyl tin chloride; alkali metals; acid; acid fumes. Hazardous decomposition products: carbon dioxide; nitrogen oxide fumes.
Carcinogenicity: IARC - Group 3, Not classifiable as to carcinogenicity to humans; NIOSH - Not listed; NTP - Not listed; ACGIH - Class A2, Suspected human carcinogen; OSHA - Not listed; EPA - Not listed; MAK - Class A2, Unmistakably carcinogenic in animal experimentation only
Primary Target Organs:

Eyes

Skin

Respiratory System

Exposure Limits
OSHA PEL: TWA: 0.2 ppm; 0.4 mg/m^3.
ACGIH TLV: TWA: 0.2 ppm; 0.34 mg/m^3.
NIOSH REL: TWA: 0.2 ppm; 0.4 mg/m^3.
NIOSH IDLH: 2 ppm.
Respirator Recommendation
Exposure Range: >0.2 to <2 ppm Supplied Air, Constant Flow/Pressure Demand, Half Mask
Exposure Range: 2 to unlimited ppm Self-contained Breathing Apparatus, Pressure Demand, Full Face

Note: odor threshold unknown

Environmental

Regulations

RCRA 40CFR: Not listed
CERCLA: 40CFR 302.4: Listed per CAA Section 112
RQ: 100 lb (45.35 kg)
SARA 40CFR 372.65: Listed
SARA EHS 40CFR 355: Not listed
TSCA: Not listed

Analytical Methods

Indoor / Expired Air: NIOSH 2515

DIB1000 CAS #: 51-50-3

DIBENAMINE

RTECS: HQ6650000
Molecular Formula: $C_{16}H_{18}ClN$
Formula Weight: 259.80
Synonyms: BENZENEMETHANAMINE,N-(2-CHLOROETHYL)-N-(PHENYLMETHYL)-; 2-CHLOROETHYLDIBENZYLAMINE; N-(2-CHLOROETHYL)DIBENZYLAMINE; DBA; DIBENZYL CHLORETHYLAMINE; DIBENZYLAMINE,N-(2-CHLOROETHYL)-; 2-(DIBENZYLAMINO)ETHYL CHLORIDE; DIBENZYLCHLORETHAMINE; DIBENZYLCHLORETHYLAMINE; N,N-DIBENZYL-2-CHLOROETHYLAMINE; N,N-DIBENZYL-BETA-CHLOROETHYLAMINE; SKF 199; SYMPATHOLYTIN
Description: oily liquid
Use: medication: adrenergic blocker; diagnostic aid (pheochromocytoma)

Physical Properties

Freezing Point: 192 °C (377.6 °F)
Water Solubility: Practically Insoluble in Water
Other Solubilities: Soluble in dilute acid.

RTECS Toxicity Data

Acute Oral: Rat LD_{50} Dose: 2400 mg/kg.
Acute Dermal: Mouse LD_{50} Route: Subcutaneous Dose: 800 mg/kg.
Mutagenic: Mouse DNA Inhibition; Route: Intravaginal; Dose: 2 pph.

Hazard Overviews

Carcinogenicity: IARC - Not listed; NIOSH - Not listed; NTP - Not listed; ACGIH - Not listed; OSHA - Not listed; EPA - Not listed; MAK - Not listed

Environmental

Regulations

RCRA 40CFR: Not listed
CERCLA: 40CFR 302.4: Not listed
SARA 40CFR 372.65: Not listed
SARA EHS 40CFR 355: Not listed
TSCA: Not listed

DIB1110 CAS #: 226-36-8

DIBENZ(A,H)ACRIDINE

RTECS: HN0875000
Molecular Formula: $C_{21}H_{13}N$
Formula Weight: 279.35
Synonyms: 7-AZADIBENZ(A,H)ANTHRACENE; DB(A,H)AC; 1,2,5,6-DIBENZACRIDINE; 1,2:5,6-DIBENZACRIDINE; DIBENZ(A,D)ACRIDINE; DIBENZ(AH)ACRIDINE; 1,2,5,6-DIBENZOACRIDINE; 1,2,5,6-DINAPHTHACRIDINE
Description: yellow crystals
Use: biochemical research

Physical Properties

Boiling Point: 524 °C (975.2 °F)
Freezing Point: 226 °C (438.8 °F)
Other Solubilities: Sparingly Soluble in Ethanol, Soluble in Benzene, Acetone, and Cyclohexane

RTECS Toxicity Data

Mutagenic: Rat DNA Adduct; Route: Intratracheal; Dose: 75 mg/kg. Rat Sister Chromatid Exchenge; Route: Intratracheal; Dose: 150 mg/kg.
Tumorigenic: Rat Route: Implant; Dose: 5 mg/kg; Toxic Effects: Tumorigenic - Carcinogenic by RTECS criteria; Lungs, Thorax, or Respiration - Tumors; Tumorigenic - Tumors at site of application. Rat Route: Implant; Dose: 1500 ug/kg; Toxic Effects: Tumorigenic - Equivocal tumorigenic agent by RTECS criteria; Lungs, Thorax, or Respiration - Tumors; Tumorigenic - Tumors at site of application.

Hazard Overviews

Carcinogenicity: IARC - Group 2B, Possibly carcinogenic to humans; NIOSH - Not listed; NTP - Listed; ACGIH - Not listed; OSHA - Not listed; EPA - Not listed; MAK - Not listed

Environmental

Environmental Fate: Exists solely in the particulate phase in the ambient atmosphere based on an estimated vapor pressure of 7.93 $x10^{-10}$ mm Hg. It may photolyze directly in the atmosphere as it can absorb light above 290 nm. Particulate phase may be removed physically from air by wet and dry deposition. If released to soil, it is expected to be immobile based on measured K_{oc} values of 55,167 for adsorption to an organic rich sludge and 79,400 for adsorption to microbial biomass. An estimated K_{oc} value of 27,000 was calculated from an estimated log Kow. Volatilization of this compound from soil surfaces is unlikely based on an estimated Henry's Law constant of 1.9 $x10^{-9}$ atm-cu m/mol. It may photolyze on soil surfaces. Biodegradation of this compound in soil may occur slowly. In water, it is expected to bind strongly to particulate matter and sediment in the water column based on its measured K_{oc} value. It may biodegrade slowly; using three activated sludges from municipal treatment plants, up to 4.6% of the theoretical oxygen uptake was reached after 144 hours using 2500 mg activated sludge. Using 5000 mg activated sludge, up to 17.3% of the theoretical oxygen uptake was reached after 144 hours. It was highly bioconcentrated in Daphnia pulex with a measured BCF of 3500 but in fathead minnows, a measured BCF of approximately 100 indicates that these fish rapidly metabolize this compound. Volatilization from water surfaces is unlikely based on its estimated Henry's Law constant.

Environmental Physical Data

Henry's Law Constant: estimated at 1.9 x10^{-9}
Octanol/Water Partition Coefficient: log K_{ow} = 5.60
Sorption Partition Coefficient: K_{oc} = 5.5167 x10^{4}
BCF: fathead minnow 100

Regulations

RCRA 40CFR: Not listed
CERCLA: 40CFR 302.4: Not listed
SARA 40CFR 372.65: Listed
SARA EHS 40CFR 355: Not listed
TSCA: Not listed

Analytical Methods

Soil: SW846 8100

DIB1220 CAS #: 224-42-0

DIBENZ(A,J)ACRIDINE

RTECS: HN1050000
Molecular Formula: $C_{21}H_{13}N$
Formula Weight: 279.35
Synonyms: 7-AZADIBENZ(A,J)ANTHRACENE; DB(A,J)AC; 1,2,7,8-DIBENZACRIDINE; 1,2:7,8-DIBENZACRIDINE; 3,4,5,6-DIBENZACRIDINE; DIBENZ(A,F)ACRIDINE; DIBENZO(A,J)ACRIDINE; 3,4,6,7-DINAPHTHACRIDINE
Description: yellow needles or prisms
Use: biochemical research

Physical Properties

Freezing Point: 216 °C (420.8 °F)
Other Solubilities: Soluble in Acetone.

RTECS Toxicity Data

Reproductive/Teratogenic: Mouse Route: Oral; Dose: 2520 mg/kg; Duration: female 21D prior to mating Effects on Fertility - Female fertility index.
Mutagenic: Human Micronucleus Test; Cell Type: lymphocyte; Dose: 5 mg/L. Rat Morphological Transformation; Cell Type: embryo; Dose: 5 mg/L.
Tumorigenic: Mouse Route: Skin; Dose: 99 mg/kg/99W-I; Toxic Effects: Tumorigenic - Carcinogenic by RTECS criteria; Skin and appendages - Tumors. Mouse Route: Skin; Dose: 700 mg/kg/29W-I; Toxic Effects: Tumorigenic - Equivocal tumorigenic agent by RTECS criteria; Skin and appendages - Tumors; Tumorigenic - Tumors at site of application. Mouse Route: Skin; Dose: 590 mg/kg/25W-I; Toxic Effects: Tumorigenic - Equivocal tumorigenic agent by RTECS criteria; Skin and appendages - Tumors; Tumorigenic - Tumors at site of application.

Hazard Overviews

Health: Irritating. Toxic. Other Acute Effects: harmful if swallowed, inhaled, or absorbed through skin. Chronic Effects: may alter genetic material; this material induced skin tumors and increased the incidence of lung tumors in mice; may cause heritable genetic damage; target organ: skin. Carcinogen.
Fire: Hazards: emits toxic fumes. Extinguishing agents: water spray; carbon dioxide, dry chemical powder or appropriate foam. Precautions: combustible liquid.
Reactivity: Incompatible with: strong oxidizing agents. Hazardous decomposition products: toxic fumes of: carbon monoxide, carbon dioxide, nitrogen oxides.
Carcinogenicity: IARC - Group 2B, Possibly carcinogenic to humans; NIOSH - Not listed; NTP - Listed; ACGIH - Not listed; OSHA - Not listed; EPA - Not listed; MAK - Not listed
Primary Target Organs:

Eyes

Skin

Respiratory System

Environmental

Environmental Fate: Exists in both the vapor phase and the particulate phase in the ambient atmosphere based on an estimated vapor pressure of 1.05 x10^{-9} mm Hg. It should degrade in the vapor phase with an estimated half-life of about 14 hours. It may photolyze directly as it absorbs light at environmental wavelengths. Particulate phase may be removed physically from air by wet and dry deposition. If released to soil, it is expected to be immobile based on an estimated K_{oc} of 29,000. Volatilization of this compound from soil surfaces is unlikely based on an estimated Henry's Law constant of 1.9 x10^{-9} atm-cu m/mol. It may photolyze on soil surfaces. Biodegradation of this compound in soil may occur slowly. In water, it is expected to bind strongly to particulate matter and sediment in the water column based on its estimated K_{oc} value. It may biodegrade slowly; using three activated sludges from municipal treatment plants, up to 5% of the theoretical oxygen uptake was reached after 144 hours using 2500 mg/L activated sludge. Using 5000 mg/L activated sludge, up to 10.4% of the theoretical oxygen uptake was reached after 144 hours. A BCF value of 12,000 suggests that it will bioconcentrate significantly in aquatic organisms. Volatilization is unlikely based on its estimated Henry's Law constant.

Environmental Physical Data

Henry's Law Constant: estimated at 1.9 x10^{-9}
Octanol/Water Partition Coefficient: log K_{ow} = 5.67
Sorption Partition Coefficient: K_{oc} = 2.9 x10^{4}
BCF: estimated at 1.2 x10^{4}

Regulations

RCRA 40CFR: Not listed
CERCLA: 40CFR 302.4: Not listed
SARA 40CFR 372.65: Listed
SARA EHS 40CFR 355: Not listed
TSCA: Not listed

Analytical Methods

Soil: SW846 8100, 8250A, 8270B, 8270C
Plasma: EPA 29

DIB1330 CAS #: 224-53-3

DIBENZ(C,H)ACRIDINE

RTECS: HN1225000
Molecular Formula: $C_{21}H_{13}N$
Formula Weight: 279.35
Synonyms: 14-AZADIBENZ(A,J)ANTHRACENE; 1,2,7,8-DIBENZACRIDINE; 3,4:5,6-DIBENZACRIDINE
Use: biochemical research

RTECS Toxicity Data

Mutagenic: Bacteria - S Typhimurium Mutations in Microorganisms; Dose: 4 ug/plate (-S9).
Tumorigenic: Mouse Route: Skin; Dose: 2040 mg/kg/85W-I; Toxic Effects: Tumorigenic - Equivocal tumorigenic agent by RTECS criteria; Skin and appendages - Tumors; Tumorigenic - Tumors at site of application. Mouse Route: Skin; Dose: 300 mg/kg/25W-I; Toxic Effects: Tumorigenic - Equivocal tumorigenic agent by RTECS criteria; Skin and appendages - Tumors; Tumorigenic - Tumors at site of application.

Hazard Overviews

Carcinogenicity: IARC - Not listed; NIOSH - Not listed; NTP - Not listed; ACGIH - Not listed; OSHA - Not listed; EPA - Not listed; MAK - Not listed

Environmental

Regulations

RCRA 40CFR: Not listed
CERCLA: 40CFR 302.4: Not listed
SARA 40CFR 372.65: Not listed
SARA EHS 40CFR 355: Not listed
TSCA: Not listed

DIB1550 CAS #: 53-70-3

DIBENZO(A,H)ANTHRACENE

RTECS: HN2625000
DOT: UN1334; IMO4.1
EINECS Number: 200-181-8
Molecular Formula: $C_{22}H_{14}$
Formula Weight: 278.33

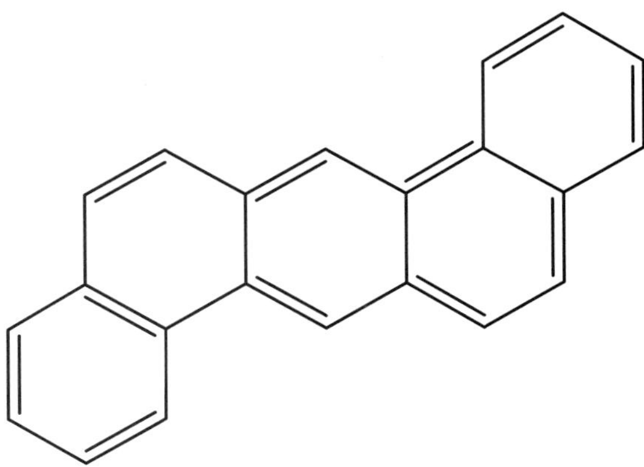

Chemical Structure

Synonyms: 1,2:5,6-BENZANTHRACENE; 1,2,5,6-DBA; DB(A,H)A; DBA; 1,2,5,6-DIBENZANTHRACEEN; 1,2,5,6-DIBENZANTHRACENE; 1,2:5,6-DIBENZ(A)ANTHRACENE; 1,2:5,6-DIBENZANTHRACENE; 1,2:5,6-DIBENZOANTHRACENE; DIBENZO(A,H)ANTHRACENE
Description: colorless crystals, plates or leaflets,
Use: research chemical

Physical Properties

Boiling Point: 524 °C (975 °F)
Freezing Point: 266 °C (510.8 °F)
Specific Gravity: 1.282
Vapor Pressure: 1 $x10^{-10}$ mm Hg
Water Solubility: 0.0005 mg/L in Water at 27 °C
Other Solubilities: Soluble in Petroleum Ether, Benzene, Toluene, Xylene; Soluble in Acetone, or absolute Alcohol.
Ionization Potential (eV): 7.38 +/-0.03
Flash Point: Not available; probably combustible

RTECS Toxicity Data

Mutagenic: Human Unscheduled DNA Synthesis; Cell Type: fibroblast; Dose: 100 umol/L. Human Unscheduled DNA Synthesis; Cell Type: other cell types; Dose: 10 mg/L. Human Unscheduled DNA Synthesis; Cell Type: HeLa cell; Dose: 100 nmol/L. Human DNA Adduct; Cell Type: embryo; Dose: 360 nmol/L.
Tumorigenic: Rat Route: Subcutaneous; Dose: 135 mg/kg/9W-I; Toxic Effects: Tumorigenic - Neoplastic by RTECS criteria; Lungs, Thorax, or Respiration - Tumors; Tumorigenic - Tumors at site of application. Rat Route: Intratracheal; Dose: 100 mg/kg; Toxic Effects: Tumorigenic - Carcinogenic by RTECS criteria; Lungs, Thorax, or Respiration - Tumors.

Hazard Overviews

Health: May cause irritation. Toxic. Other Acute Effects: harmful if swallowed, inhaled, or absorbed through skin. Chronic Effects: may cause heritable genetic damage; mutagen; target organs: liver, lungs. Carcinogen.
Fire: Will burn. Extinguishing agents: water spray; carbon dioxide, dry chemical powder or appropriate foam. Precautions: combustible liquid.

Reactivity: Incompatible with: strong oxidizing agents. Hazardous decomposition products: toxic fumes of: carbon monoxide, carbon dioxide.

Carcinogenicity: IARC - Group 2A, Probably carcinogenic to humans; NIOSH - Not listed; NTP - Listed; ACGIH - Not listed; OSHA - Not listed; EPA - Class B2, Probable human carcinogen based on animal studies; MAK - Not listed

Primary Target Organs:

Respiratory System

Liver

Environmental

Ecotoxicity: TLm Neanthes arenaceodentata > 1 ppm/96 hr at 22 °C in a static bioassay

Environmental Fate: Release to the environment is quite general since it is a ubiquitous product of incomplete combustion. It is largely associated with particulate matter, soils, and sediments. Its presence in places distant from primary sources indicates that it is reasonably stable in the atmosphere and capable of long distance transport. If it is released to soils it will be expected to adsorb very strongly to the soils and will not be expected to leach to the groundwater, hydrolyze or evaporate from soils or surfaces. It will be subject to biodegradation in soils with reported half-lives of 18 and 21 days. If it is released to water it will be expected to adsorb very strongly to sediments and particulate matter and to bioconcentrate in aquatic organisms which lack microsomal oxidase (this enzyme enables the rapid metabolism of certain polycyclic aromatic hydrocarbons). Based on limited data from laboratory screening tests using settled domestic wastewater and activated sludge, it may be subject to biodegradation in natural waters. Since it absorbs solar radiation strongly, it may be subject to direct photolysis in natural waters. However, adsorption may significantly retard photolysis as the photosensitivity of polyaromatic hydrocarbons is strongly dependent upon the nature of the surface upon which the compound is adsorbed. It will not hydrolyze and should not evaporate from water. If released to the atmosphere it will likely be associated with particulate matter and may be subject to moderately long range transport, depending mainly on the particle size distribution and climatic conditions which will determine the rates of wet and dry deposition. Its presence in areas remote from primary sources demonstrates the potential for this long range transport as well as it's considerable stability in the air. It may be subject to direct photolysis in the atmosphere; however, adsorption may significantly retard photolysis as the photosensitivity of polyaromatic hydrocarbons is strongly dependent upon the nature of the surface upon which the compound is adsorbed. The estimated vapor phase half-life in the atmosphere is 1.00 day as a result of reaction with photochemically produced hydroxyl radicals.

Cleanup/Disposal: Guide No. 133: Eliminate all ignition sources (no smoking, flares, sparks or flames in immediate area). Do not touch or walk through spilled material. Small Dry Spills: With clean shovel place material into clean, dry container and cover loosely; move containers from spill area. Large Spills: Wet down with water and dike for later disposal. Prevent entry into waterways, sewers, basements or confined areas.

Environmental Physical Data

Henry's Law Constant: calculated at 7 $x10^{-8}$

Octanol/Water Partition Coefficient: log K_{ow} = 6.50

Sorption Partition Coefficient: K_{oc} = sediments 8.05392 $x10^{5}$ to 3.059425 $x10^{6}$

BCF: daphnia manga 652

Regulations

RCRA 40CFR: Listed Hazardous Waste No. U063 Toxic Waste

CERCLA: 40CFR 302.4: Listed per RCRA Section 3001 per CWA Section 307(a) RQ: 1 lb (0.454 kg)

SARA 40CFR 372.65: Listed

SARA EHS 40CFR 355: Not listed

TSCA: Listed

Analytical Methods

Air: EPA TO-13; California 429

Soil: CLP LC_SV, MC_SVOA, OHC; EPA 16, 1625, PAH-005, PAH-007, PAH-011, PAH-012; SW846 3630B, 3640A, 8100, 8250A, 8270B, 8270C, 8275A, 8310

Water / Groundwater: EPA PAH-006, 1625, 610, 625, 625-S, 6; APHA 6410-B, 6440-B, 6440-C; ASTM D4657, D4763; USGS O3113, O3118

Drinking Water: EPA 525.1, 525.2, 550, 550.1

Indoor / Expired Air: NIOSH 5506, 5515; EPA IP-7-A, IP-7-B

Plasma: EPA 29

Other: EPA PAH-009

DIB1660 **CAS #: 194-59-2**

7H-DIBENZO(C,G)CARBAZOLE

RTECS: HO5600000

EINECS Number: 205-895-3

Molecular Formula: $C_{20}H_{13}N$

Formula Weight: 267.31

Synonyms: 7-AZA-7H-DIBENZO(C,G)FLUORENE; 7H-DB(C,G)C; 3,4,5,6-DIBENZCARBAZOL; 3,4,5,6-DIBENZCARBAZOLE; 3,4,5,6-DIBENZOCARBAZOLE; 3,4,5,6-DINAPHTHACARBAZOLE

Description: needles

Use: there is no known use of this compound

Physical Properties

Freezing Point: 158 °C (316.4 °F)

Saturated Vapor Density: 1.2 kg/m^3

Vapor Pressure: 3.4 $x10^{-9}$ mm Hg at 25 °C

Water Solubility: 63 ug/L

RTECS Toxicity Data

Mutagenic: Human DNA Inhibition; Cell Type: other cell types; Dose: 10 umol/L. Human DNA Adduct; Cell Type:

fibroblast; Dose: 1400 nmol/L. Human Mutations in Mammalian Somatic Cells; Cell Type: fibroblast; Dose: 500 nmol/L. Human Micronucleus Test; Cell Type: lymphocyte; Dose: 50 ug/L.

Tumorigenic: Rat Route: Subcutaneous; Dose: 150 mg/kg/17W-I; Toxic Effects: Tumorigenic - Equivocal tumorigenic agent by RTECS criteria; Tumorigenic - Tumors at site of application. Mouse Route: Oral; Dose: 280 mg/kg/32W-I; Toxic Effects: Tumorigenic - Equivocal tumorigenic agent by RTECS criteria; Gastrointestinal - Tumors; Liver - Tumors.

Hazard Overviews

Carcinogenicity: IARC - Group 2B, Possibly carcinogenic to humans; NIOSH - Not listed; NTP - Listed; ACGIH - Not listed; OSHA - Not listed; EPA - Not listed; MAK - Not listed

Environmental

Environmental Fate: Adsorbs very strongly to soil and if released on soil will remain in the surface layer of soil. It undergoes direct photolysis in water and would be expected to photolyze on the soil surfaces. It is resistant to biodegradation and would be expected to be persistent in the soil. If released in water, it would adsorb strongly to sediment and undergo direct photolysis. A laboratory study suggests that 99% entering natural waters will be adsorbed onto sediment and that the remaining 1% will be photochemically transformed. While the photolysis rate will vary with the time of day and season, the photolysis half-life will be less than a day. It would be expected to bioconcentrate in aquatic organisms. It should exist in the atmosphere as an aerosol and be removed by gravitational settling. It is readily photolyzed in water and may photolyze in atmospheric droplets and while adsorbed to particulate matter.

Environmental Physical Data

Henry's Law Constant: estimated at 2.45×10^{-9}
Octanol/Water Partition Coefficient: log K_{ow} = 5.58
Sorption Partition Coefficient: K_{oc} = 5.5×10^{5} to 4.08×10^{7}
BCF: calculated at 1.02×10^{4}

Regulations

RCRA 40CFR: Not listed
CERCLA: 40CFR 302.4: Not listed
SARA 40CFR 372.65: Listed
SARA EHS 40CFR 355: Not listed
TSCA: Not listed

Analytical Methods

Soil: SW846 8100

DIB1770 CAS #: 217-54-9

DIBENZO(B,K)CHRYSENE

Molecular Formula: $C_{26}H_{16}$
Formula Weight: 328.41
Synonyms: ANTH(2,1-A)ANTHRENE; (ANTHRA-2',1')-1,2-ANTHRACENE; 2',1'-ANTHRA-1,2-ANTHRACENE; ANTHRACENO(2',1':1,2)ANTHRACENE; ANTHRACENO(2,1-A)ANTHRACENE

Physical Properties

Freezing Point: 400 °C (752 °F)
Ionization Potential (eV): 6.98 +/-0.04

Hazard Overviews

Carcinogenicity: IARC - Not listed; NIOSH - Not listed; NTP - Not listed; ACGIH - Not listed; OSHA - Not listed; EPA - Not listed; MAK - Not listed

Environmental

Regulations

RCRA 40CFR: Not listed
CERCLA: 40CFR 302.4: Not listed
SARA 40CFR 372.65: Not listed
SARA EHS 40CFR 355: Not listed
TSCA: Not listed

DIB1880 CAS #: 262-12-4

DIBENZO-P-DIOXIN

RTECS: HP3090000
EINECS Number: 205-974-2
Molecular Formula: $C_{12}H_8O_2$
Formula Weight: 184.20
Synonyms: DIBENZO(1,4)DIOXIN; DIBENZO(B,E)(1,4)DIOXIN; DIBENZO-PARA-DIOXIN; DIBENZODIOXIN; DIPHENYLENE DIOXIDE; OXANTHRENE; PHENODIOXIN
Description: white crystalline solid or crystals
Use: not used commercially in the US

Physical Properties

Freezing Point: 122 °C (251.6 °F) to 123 °C (253.4 °F)
Water Solubility: 1 ppm at 25 °C
Other Solubilities: 95% Ethanol: <1 mg/ml at 18 °C; Acetone: >=10 mg/ml at 18 °C; Benzene: Soluble; Chloroform: Soluble; DMSO: >=10 mg/ml at 18 °C.
Ionization Potential (eV): 7.5
Flash Point: Not available; probably combustible

RTECS Toxicity Data

Acute Oral: Rat LD_{50} Dose: 1220 mg/kg. Mouse LD_{50} Dose: 866 mg/kg.

Tumorigenic: Mouse Route: Skin; Dose: 110 gm/kg/58W-I; Toxic Effects: Tumorigenic - Equivocal tumorigenic agent by RTECS criteria; Skin and appendages - Tumors.

Hazard Overviews

Fire: Will burn.

Carcinogenicity: IARC - Group 3, Not classifiable as to carcinogenicity to humans; NIOSH - Not listed; NTP - Not listed; ACGIH - Not listed; OSHA - Not listed; EPA - Not listed; MAK - Not listed

Environmental

Environmental Fate: If released to the atmosphere, it may exist in both the vapor-phase and particulate phase. Vapor-phase is degraded rapidly in air by reaction with photochemically produced hydroxyl radicals; the half-life for this reaction on average can be estimated to be about 10 hours. The particulate phase material can be physically removed from air by wet and dry deposition. If released to water, photodegradation may be important in surface waters containing small amounts of nitrate ion. Based on estimated K_{oc} values of 4370 to 14400, partitioning from the water column to sediment and suspend aquatic material may be significant. In the presence of significant adsorption, volatilization may not be important. If released to soil, it is expected to be tightly adsorbed in most soils and not susceptible to significant leaching. Photolysis on surfaces exposed to sunlight may occur. Although pure culture studies have shown that it can be metabolized, insufficient data are available to predict the importance of biodegradation in soil and water.

Environmental Physical Data

Henry's Law Constant: estimated at 1 $x10^{-4}$
Octanol/Water Partition Coefficient: log K_{ow} = 4.37
Sorption Partition Coefficient: K_{oc} = 4370 to 1.44 $x10^{4}$
BCF: estimated at 620 to 1230

Regulations

RCRA 40CFR: Not listed
CERCLA: 40CFR 302.4: Not listed
SARA 40CFR 372.65: Not listed
SARA EHS 40CFR 355: Not listed
TSCA: Not listed

DIB1990 CAS #: 5385-75-1

DIBENZO(A,E)FLUORANTHENE

RTECS: HM9799000
Molecular Formula: $C_{24}H_{14}$
Formula Weight: 302.38
Synonyms: 2,3,5,6-DIBENZOFLUORANTHENE; DIBENZO(A,E)FLUORANTHENE

Physical Properties

Freezing Point: 232 °C (449.6 °F)

RTECS Toxicity Data

Mutagenic: Mouse DNA Damage; Cell Type: embryo; Dose: 1 umol/L. Bacteria - S Typhimurium Mutations in Microorganisms; Dose: 500 nmol/L (-S9).
Tumorigenic: Mouse Route: Skin; Dose: 2880 ug/kg/15W-I; Toxic Effects: Tumorigenic - Carcinogenic by RTECS criteria; Skin and appendages - Tumors.

Hazard Overviews

Carcinogenicity: IARC - Group 3, Not classifiable as to carcinogenicity to humans; NIOSH - Not listed; NTP - Not listed; ACGIH - Not listed; OSHA - Not listed; EPA - Not listed; MAK - Not listed

Environmental

Regulations

RCRA 40CFR: Not listed
CERCLA: 40CFR 302.4: Not listed
SARA 40CFR 372.65: Listed
SARA EHS 40CFR 355: Not listed
TSCA: Not listed

DIB2100 CAS #: 132-64-9

DIBENZOFURAN

RTECS: HP4430000
EINECS Number: 205-071-3
Molecular Formula: $C_{12}H_{8}O$
Formula Weight: 168.19

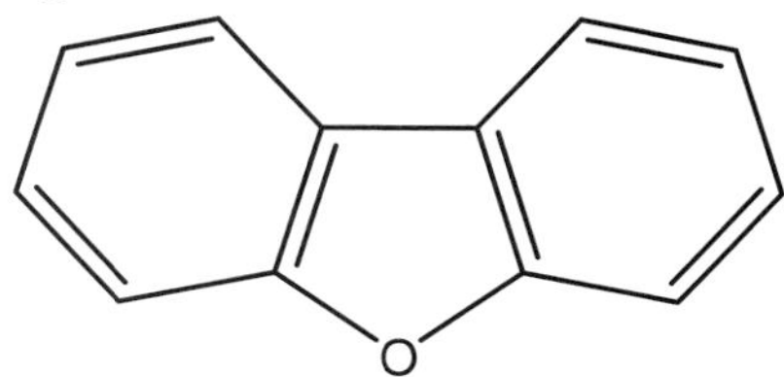

Chemical Structure

Synonyms: (1,1'-BIPHENYL)-2,2'-DIYL OXIDE; 2,2'-BIPHENYLENE OXIDE; 2,2'-BIPHENYLYLENE OXIDE; DIBENZO(B,D)FURAN; DIBENZOL(B,D)FURAN; DIPHENYLENE OXIDE
Description: white, crystalline solid
Use: insecticide

Physical Properties

Boiling Point: 287 °C (549 °F) at 760 mm Hg
Freezing Point: 86 °C (186.8 °F) to 87 °C (188.6 °F)
Specific Gravity: 1.0886 at 99 °C/4 °C
Vapor Density: 5.8 Air=1
Water Solubility: < 1 mg/mL at 20 C
Other Solubilities: 95% Ethanol: 10-50 mg/ml at 20 °C; Acetone: >=100 mg/ml at 20 °C; Acetic Acid: Soluble; Benzene: Slightly Soluble; DMSO: >=100 mg/ml at 20 °C; Ether: Soluble.
Odor Threshold: 0.7752 to 1.6150 mg/m^3
Refraction Index: 1.6079 at 99 °C/D
Ionization Potential (eV): 7.9 +/-1.0
Flash Point: Not available; probably combustible

RTECS Toxicity Data

Mutagenic: Hamster Sister Chromatid Exchenge; Cell Type: ovary; Dose: 10 mg/L.

Hazard Overviews

Health: May cause irritation. Other Acute Effects: may be harmful by inhalation, ingestion, or skin absorption.

Fire: Will burn. Extinguishing agents: water spray; carbon dioxide, dry chemical powder or appropriate foam. Precautions: combustible liquid.

Reactivity: Incompatible with: strong oxidizing agents. Hazardous decomposition products: toxic fumes of: carbon monoxide, carbon dioxide.

Carcinogenicity: IARC - Not listed; NIOSH - Not listed; NTP - Not listed; ACGIH - Not listed; OSHA - Not listed; EPA - Class D, Not classifiable as to human carcinogenicity; MAK - Not listed

Environmental

Environmental Fate: If released to the atmosphere, it will exist primarily in the gas-phase where it will degrade relatively rapidly by reaction with photochemically produced hydroxyl radicals (estimated half-life of 11.3 hr in average air). A small percentage released to air will exist in the particulate phase which may be relatively persistent to atmospheric degradation. Physical removal from air can occur by both wet and dry deposition. If released to water, it may partition significantly from the water column to sediments and suspended material. Volatilization from the water column may be important; however, sorption to sediment may diminish the potential importance of volatilization. If released to soil, it is not expected to leach significantly in most soil types. Biological screening studies have shown that it is biodegraded readily by adapted microbes in the presence of sufficient oxygen. However, in various groundwaters or aquatic sediments where oxygen is limited or lacking, biodegradation may occur very slowly resulting in long periods of persistence.

Environmental Physical Data

Henry's Law Constant: estimated at 9.73 $x10^{-5}$
Octanol/Water Partition Coefficient: log K_{ow} = 4.12
Sorption Partition Coefficient: K_{oc} = 4600 to 6350
BCF: fish 947

Regulations

RCRA 40CFR: Not listed
CERCLA: 40CFR 302.4: Listed per CAA Section 112 RQ: 100 lb (45.35 kg)
SARA 40CFR 372.65: Listed
SARA EHS 40CFR 355: Not listed
TSCA: Listed

Analytical Methods

Soil: CLP MC_SVOA, OHC; EPA 1625; SW846 1311, 3640A, 8250A, 8270B, 8270C, 8275A, 8410; DOE OS050
Water / Groundwater: EPA 1625
Plasma: EPA 29

DIB2210 CAS #: 192-65-4

DIBENZO(A,E)PYRENE

RTECS: QL0175000
EINECS Number: 205-891-1
Molecular Formula: $C_{24}H_{14}$
Formula Weight: 302.4
Synonyms: DB(A,E)P; 1,2,4,5-DIBENZOPYRENE; 1,2:4,5-DIBENZOPYRENE; DIBENZO(A,E)PYRENE; NAPTHO(1,2,3,4-DEF)CHRYSENE
Description: pale yellow needles
Use: experimental carcinogen

Physical Properties

Freezing Point: 233.5 °C (452.3 °F)
Other Solubilities: Slightly Soluble in Alcohol, Acetone, Benzene, Acetic Acid Soluble in concn Sulfuric acid & hot Toluene
Ionization Potential (eV): 7.11

RTECS Toxicity Data

Mutagenic: Human Mutations in Mammalian Somatic Cells; Cell Type: lymphocyte; Dose: 25 nmol/L. Bacteria - S Typhimurium Mutations in Microorganisms; Dose: 5 nmol/plate (-S9).

Tumorigenic: Mouse Route: Skin; Dose: 312 mg/kg/52W-I; Toxic Effects: Tumorigenic - Neoplastic by RTECS criteria; Skin and appendages - Tumors; Tumorigenic - Tumors at site of application. Mouse Route: Subcutaneous; Dose: 72 mg/kg/9W-I; Toxic Effects: Tumorigenic - Equivocal tumorigenic agent by RTECS criteria; Tumorigenic - Tumors at site of application.

Hazard Overviews

Carcinogenicity: IARC - Group 2B, Possibly carcinogenic to humans; NIOSH - Not listed; NTP - Listed; ACGIH - Not listed; OSHA - Not listed; EPA - Not listed; MAK - Not listed

Environmental

Environmental Fate: In the ambient atmosphere, it will exist solely in the particulate phase based on an estimated vapor pressure of 7.03 $x10^{-11}$ mm Hg. There, it may photolyze directly as it absorbs light at environmental wavelengths. Particulate phase may be removed physically from air by wet and dry deposition. If released to soil, it is expected to be immobile based on an estimated K_{oc} of 217,000. Volatilization of this compound from moist soil surfaces is unlikely based on an estimated Henry's Law constant of 1.41 $x10^{-8}$ atm-cu m/mol. It may photolyze on soil surfaces. In water, it is expected to bind strongly to particulate matter and sediment in the water column based on its estimated K_{oc} value. A BCF value of 201,000 suggests that it will bioconcentrate significantly in aquatic organisms. Volatilization from water surfaces is unlikely based on its estimated Henry's Law constant.

Environmental Physical Data

Henry's Law Constant: estimated at 1.41 x10^{-8}
Octanol/Water Partition Coefficient: log K_{ow} = 7.28
Sorption Partition Coefficient: K_{oc} = 2.17 x10^{5}
BCF: estimated at 2.01 x10^{5}

Regulations

RCRA 40CFR: Not listed
CERCLA: 40CFR 302.4: Not listed
SARA 40CFR 372.65: Listed
SARA EHS 40CFR 355: Not listed
TSCA: Not listed

Analytical Methods

Soil: SW846 8100, 8270B, 8270C
Plasma: EPA 29

DIB2320 CAS #: 189-64-0

DIBENZO(A,H)PYRENE

RTECS: HO5775000
EINECS Number: 205-878-0
Molecular Formula: $C_{24}H_{14}$
Formula Weight: 302.38
Synonyms: DB(A,H)P; DIBENZO(B,DEF)CHRYSENE; 1,2,6,7-DIBENZOPYRENE; 3,4,8,9-DIBENZOPYRENE; 3,4:8,9-DIBENZOPYRENE; DIBENZO(A,H)PYRENE; 3,4,8,9-DIBENZPYRENE
Description: golden yellow plates
Use: experimental carcinogen

Physical Properties

Freezing Point: 308 °C (586.4 °F)
Other Solubilities: Soluble in 1,4-dioxane.
Ionization Potential (eV): 6.82

RTECS Toxicity Data

Mutagenic: Human Mutations in Mammalian Somatic Cells; Cell Type: lymphocyte; Dose: 500 nmol/L. Hamster Mutations in Microorganisms; Cell Type: lung; Dose: 30 ug/L (+S9).
Tumorigenic: Rat Route: Parenteral; Dose: 96761 ug/kg; Toxic Effects: Tumorigenic - Equivocal tumorigenic agent by RTECS criteria; Skin and appendages - Tumors. Rat Route: Implant; Dose: 100 mg/kg; Toxic Effects: Tumorigenic - Equivocal tumorigenic agent by RTECS criteria; Tumorigenic - Tumors at site of application.

Hazard Overviews

Carcinogenicity: IARC - Group 2B, Possibly carcinogenic to humans; NIOSH - Not listed; NTP - Listed; ACGIH - Not listed; OSHA - Not listed; EPA - Not listed; MAK - Not listed

Environmental

Environmental Fate: Adsorbs very strongly to soil and if released on soil will remain in the surface layer of soil. It absorbs UV radiation >290 nm and into the visible region and therefore would be subject to direct photolysis. While no photolysis rates are available, photolysis in water may occur based on the behavior of related chemicals. Photolysis may also occur on the soil surface, although experimental verification is lacking for related chemicals. It is expected to be resistant to biodegradation and would therefore be persistent below the soil surface. If released in water, it would adsorb strongly to sediment and may undergo direct photolysis It would be expected to bioconcentrate in aquatic organisms. It should exist in the atmosphere as an aerosol and be removed by gravitational settling. It may photolyze in atmospheric droplets and possibly while adsorbed to particulate matter.

Environmental Physical Data

Henry's Law Constant: estimated at 1.41 x10^{-8}
Octanol/Water Partition Coefficient: log K_{ow} = 7.28
Sorption Partition Coefficient: K_{oc} = estimated at 6.95
BCF: calculated at 2 x10^{5}

Regulations

RCRA 40CFR: Not listed
CERCLA: 40CFR 302.4: Not listed
SARA 40CFR 372.65: Listed
SARA EHS 40CFR 355: Not listed
TSCA: Not listed

Analytical Methods

Soil: SW846 8100

DIB2430 CAS #: 189-55-9

DIBENZO(A,I)PYRENE

RTECS: DI5775000
EINECS Number: 205-877-5
Molecular Formula: $C_{24}H_{14}$
Formula Weight: 302.4

Chemical Structure

Synonyms: BENZO(RST)PENTAPHENE; DB(A,I)P; 1,2,7,8-DIBENZOPYRENE; 3,4:9,10-DIBENZOPYRENE; DIBENZO(A,I)PYRENE; DIBENZO(B,H)PYRENE; 1,2:7,8-DIBENZPYRENE; 3,4:9,10-DIBENZPYRENE; DIBENZ(A,I)PYRENE
Description: greenish-yellow needles, prisms or lamellae

Use: experimental carcinogen; research chemical

Physical Properties

Boiling Point: 275 °C (527 °F) at 0.05 mm Hg
Freezing Point: 280 °C (536 °F)
Ionization Potential (eV): 6.95

RTECS Toxicity Data

Mutagenic: Human Mutations in Mammalian Somatic Cells; Cell Type: lymphocyte; Dose: 3200 nmol/L. Rat DNA Adduct; Route: Intratracheal; Dose: 7500 ug/kg.
Tumorigenic: Rat Route: Parenteral; Dose: 96761 ug/kg; Toxic Effects: Tumorigenic - Carcinogenic by RTECS criteria; Skin and appendages - Tumors. Mouse Route: Skin; Dose: 47 mg/kg/39W-I; Toxic Effects: Tumorigenic - Equivocal tumorigenic agent by RTECS criteria; Skin and appendages - Tumors. Mouse Route: Skin; Dose: 141 mg/kg/47W-I; Toxic Effects: Tumorigenic - Equivocal tumorigenic agent by RTECS criteria; Skin and appendages - Tumors; Tumorigenic - Tumors at site of application.

Hazard Overviews

Health: Toxic. Other Acute Effects: harmful if swallowed, inhaled, or absorbed through skin. Chronic Effects: mutagen; may cause heritable genetic damage; target organs: liver, lungs. Carcinogen.;
Fire: Hazards: emits toxic fumes. Extinguishing agents: water spray; carbon dioxide, dry chemical powder or appropriate foam. Precautions: combustible liquid.
Reactivity: Stable. Hazardous polymerization will not occur. Incompatible with: strong acids, strong oxidizing agents. Hazardous decomposition products: toxic fumes of: carbon monoxide, carbon dioxide.
Carcinogenicity: IARC - Group 2B, Possibly carcinogenic to humans; NIOSH - Not listed; NTP - Listed; ACGIH - Not listed; OSHA - Not listed; EPA - Not listed; MAK - Not listed
Primary Target Organs:

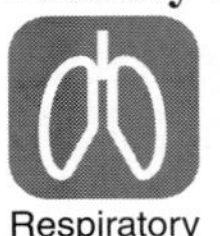
Respiratory System

Liver

Environmental

Environmental Fate: If released to soil, it will be expected to adsorb very strongly to the soil and will not be expected to leach to the groundwater. It will not hydrolyze or appreciably biodegrade or evaporate from soils and surfaces. If released to water, it will strongly adsorb to sediments but will not hydrolyze or appreciably biodegrade or evaporate. It may be subject to direct photolysis in water and may bioconcentrate in species which lack microsomal oxidase. If released to air, it may be subject to direct photolysis. The estimated vapor-phase half-life in the atmosphere is 20.05 hr as a result of reaction with photochemically produced hydroxyl radicals. Adsorption to the particulate matter may retard the rate of its biodegradation, evaporation, bioconcentration, and photolysis.

Environmental Physical Data

Henry's Law Constant: calculated at 4.25 $x10^{-8}$
Octanol/Water Partition Coefficient: log K_{ow} = 7.298
Sorption Partition Coefficient: K_{oc} = estimated at 6.8 $x10^{6}$
BCF: estimated at 2.07 $x10^{5}$

Regulations

RCRA 40CFR: Listed Hazardous Waste No. U064 Toxic Waste
CERCLA: 40CFR 302.4: Listed per RCRA Section 3001 RQ: 10 lb (4.535 kg)
SARA 40CFR 372.65: Listed
SARA EHS 40CFR 355: Not listed
TSCA: Not listed

Analytical Methods

Soil: SW846 8100

DIB2540 CAS #: 191-30-0

DIBENZO(A,L)PYRENE

RTECS: HO6125000
EINECS Number: 205-886-4
Molecular Formula: $C_{24}H_{14}$
Formula Weight: 302.37
Synonyms: BA 51-090462; DB(A,1)P; DB(A,L)P; DIBENZO(DEF,P)CHRYSENE; 1,2,9,10-DIBENZOPYRENE; 1,2:3,4-DIBENZOPYRENE; 2,3:4,5-DIBENZOPYRENE; DIBENZO(A,D)PYRENE; DIBENZO(A,L)PYRENE; 1,2,3,4-DIBENZPYRENE; 4,5,6,7-DIBENZPYRENE
Description: pale yellow plates
Use: experimental carcinogen

Physical Properties

Freezing Point: 162.4 °C (324.32 °F)
Other Solubilities: Soluble in olive oil.

RTECS Toxicity Data

Mutagenic: Human Mutations in Mammalian Somatic Cells; Cell Type: lymphocyte; Dose: 310 pmol/L. Bacteria - S Typhimurium Mutations in Microorganisms; Dose: 5 nmol/plate (-S9).
Tumorigenic: Rat Route: Parenteral; Dose: 96761 ug/kg; Toxic Effects: Tumorigenic - Carcinogenic by RTECS criteria; Skin and appendages - Tumors. Mouse Route: Skin; Dose: 890 mg/kg/37W-I; Toxic Effects: Tumorigenic - Equivocal tumorigenic agent by RTECS criteria; Skin and appendages - Tumors; Tumorigenic - Tumors at site of application.

Hazard Overviews

Carcinogenicity: IARC - Group 2B, Possibly carcinogenic to humans; NIOSH - Not listed; NTP - Listed; ACGIH - Not listed; OSHA - Not listed; EPA - Not listed; MAK - Not listed

Environmental

Environmental Fate: In the ambient atmosphere, it will exist solely in the particulate phase based on an estimated vapor pressure of 7.03 $x10^{-11}$ mm Hg. There, it may photolyze directly as it absorbs light at environmental wavelengths. Particulate phase may be removed physically from air by wet and dry deposition. If released to soil, it is expected to be immobile based on an estimated K_{oc} of 370,000. Biodegradation may occur to a limited extent based on tests using activated sludge as an inoculum. Volatilization of this compound from moist soil surfaces is unlikely given an estimated Henry's Law constant of 1.41 $x10^{-8}$ atm-cu m/mol. It may photolyze on soil surfaces. In water, it is expected to bind strongly to particulate matter and sediment in the water column based on its estimated K_{oc} value. Biodegradation may occur; studies using activated sludge as an inoculum showed that no greater than 6.2% of the theoretical oxygen demand was reached over a 144 hour time period. A BCF value of 426,000 suggests that it will bioconcentrate significantly in aquatic organisms. Volatilization from water surfaces is unlikely based on its estimated Henry's Law constant.

Environmental Physical Data

Henry's Law Constant: estimated at 1.41 $x10^{-8}$
Octanol/Water Partition Coefficient: log K_{ow} = 7.71
Sorption Partition Coefficient: K_{OC} = 3.7 $x10^{5}$
BCF: estimated at 4.26 $x10^{5}$

Regulations

RCRA 40CFR: Not listed
CERCLA: 40CFR 302.4: Not listed
SARA 40CFR 372.65: Listed
SARA EHS 40CFR 355: Not listed
TSCA: Not listed

DIB2650 CAS #: 120-78-5

2,2'-DIBENZOTHIAZYL DISULFIDE

RTECS: DL4550000
EINECS Number: 204-424-9
Molecular Formula: $C_{14}H_8N_2S_4$
Formula Weight: 332.46

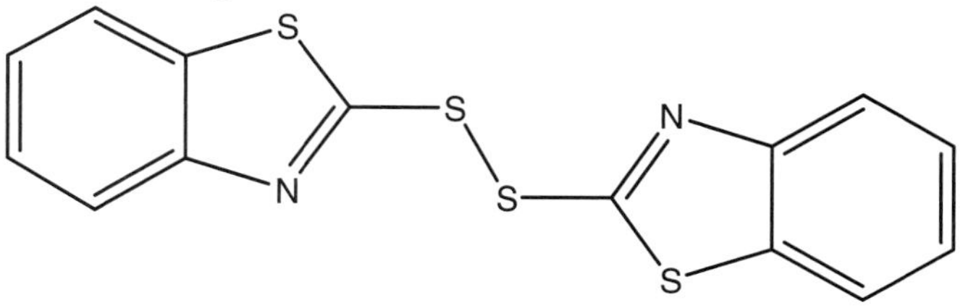

Chemical Structure

Synonyms: ACCEL TM; ALTAX; BENZOTHIAZOLE DISULFIDE; BENZOTHIAZOLE,2,2'-DITHIOBIS-; 2-BENZOTHIAZOLYL DISULFIDE; BENZOTHIAZOLYL DISULFIDE; 2,2'-BENZOTHIAZYL DISULFIDE; 2-BENZOTHIAZYL DISULFIDE; BENZOTHIAZYL DISULFIDE; 2,2'-BIS(BENZOTHIAZOLYL) DISULFIDE; BIS(2-BENZOTHIAZOLYL) DISULFIDE; BIS(BENZOTHIAZOLYL) DISULFIDE; BIS(2-BENZOTHIAZYL) DISULFIDE; DI-2-BENZOTHIAZOLYL DISULFIDE; DIBENZOTHIAZOLYL DISULFIDE; DIBENZOTHIAZOLYL DISULPHIDE; DIBENZOTHIAZYL DISULFIDE; DIBENZOYLTHIAZYL DISULFIDE; DIBENZTHIAZYL DISULFIDE; 2,2'-DITHIOBIS(BENZOTHIAZOLE); DITHIOBIS(BENZOTHIAZOLE); DWUSIARCZEK DWUBENZOTIAZYLU; EKAGOM GS; MBTS; MBTS RUBBER ACCELERATOR; 2-MERCAPTOBENZOTHIAZOLE DISULFIDE; 2-MERCAPTOBENZOTHIAZYL DISULFIDE; MERCAPTOBENZTHIAZYL ETHER; PNEUMAX DM; ROYAL MBTS; THIOFIDE; VULCAFOR MBTS; VULKACIT DM; VULKACIT DM/C; VULKACIT DM/MGC
Description: cream to off-white powder or pellets; odorless
Use: accelerator in the rubber industry; retarder in neoprene

Physical Properties

Freezing Point: 180 °C (356 °F)
Specific Gravity: 1.5
Density: 1.5 g/mL
Water Solubility: Insoluble in Water
Other Solubilities: DMSO: <1 mg/ml at 21 °C; 95% Ethanol: <1 mg/ml at 21 °C; Acetone: <1 mg/ml at 21 °C; Chloroform: Slightly Soluble; Naphtha: <0.5 gm/100 mL at 25 °C; Ether: <0.2 gm/100 mL at 25 °C; Benzene: <0.5 gm/100 mL at 25 °C.
Flash Point: 271 °C

RTECS Toxicity Data

Acute Oral: Rat LD_{50} Dose: >12 gm/kg. Mouse LD_{50} Dose: 7 gm/kg.
Acute Dermal: Rabbit LD_{50} Route: Skin; Dose: >7940 mg/kg.
Reproductive/Teratogenic: Rat Route: Parenteral; Dose: 400 mg/kg; Duration: female 4-11D of pregnancy; Effects on Fertility - Post-implantation mortality; Litter size; Effects on Embryo or Fetus - Fetotoxicity. Rat Route: Parenteral; Dose: 400 mg/kg; Duration: female 4-11D of pregnancy; Effects on Embryo or Fetus - Fetal death. Rat Route: Parenteral; Dose: 800 mg/kg; Duration: male 2D prior to mating; Effects on Fertility - Post-implantation mortality; Litter size; Effects on Embryo or Fetus - Fetotoxicity. Rat Route: Parenteral; Dose: 800 mg/kg; Duration: male 2D prior to mating; Effects on Embryo or Fetus - Fetal death.
Mutagenic: Mouse Mutations in Microorganisms; Cell Type: lymphocyte; Dose: 15 mg/L (+S9). Bacteria - S Typhimurium Mutations in Microorganisms; Dose: 3333 ug/plate (+/-S9).
Tumorigenic: Mouse Route: Oral; Dose: 172 gm/kg/78W-I; Toxic Effects: Tumorigenic - Equivocal tumorigenic agent by RTECS criteria; Lungs, Thorax, or Respiration - Tumors; Blood - Tumors.

Hazard Overviews

Health: Irritating to eyes/skin/respiratory tract. Harmful. Other Acute Effects: may be harmful by inhalation, ingestion, or skin absorption; exposure can cause nausea; headache; vomiting; prolonged or repeated exposure may cause allergic reactions in certain sensitive individuals.
Fire: Will burn. Hazards: emits toxic fumes. Extinguishing agents: water spray; carbon dioxide, dry chemical powder or appropriate foam. Precautions: combustible liquid.

Reactivity: Incompatible with: strong oxidizing agents. Hazardous decomposition products: toxic fumes of: carbon monoxide, carbon dioxide, nitrogen oxides, sulfur oxides.
Carcinogenicity: IARC - Not listed; NIOSH - Not listed; NTP - Not listed; ACGIH - Not listed; OSHA - Not listed; EPA - Not listed; MAK - Not listed
Primary Target Organs:

Eyes

Skin

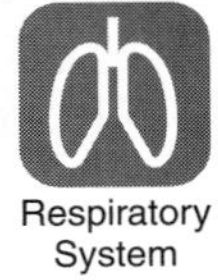
Respiratory System

Environmental

Environmental Fate: If released to soil, it will be immobile and biodegrade slowly (0.8% theoretical BOD in 2 weeks in an aqueous screening test using soil inoculum). If released to water, volatilization is not expected to be an important removal process. However, adsorption to sediment or particulate material in an aquatic system is possible. Insufficient data are available to predict the relative importance or rate of biodegradation in water although one aqueous screening biodegradation test would suggest slow biodegradation. It has the potential to photolyze directly since it absorbs UV light >290 nm, but the kinetics of the potential photolysis are not known. If released to the atmosphere, it will degrade in the ambient atmosphere by reaction with photochemically produced hydroxyl radicals with an estimated half-life of approximately 1.3 hours.

Environmental Physical Data

Henry's Law Constant: estimated at 2.34×10^{-13}
Sorption Partition Coefficient: $K_{oc} = 7.55 \times 10^5$
BCF: carp 1.0 to 7.2

Regulations

RCRA 40CFR: Not listed
CERCLA: 40CFR 302.4: Not listed
SARA 40CFR 372.65: Not listed
SARA EHS 40CFR 355: Not listed
TSCA: Listed

Analytical Methods

Water / Groundwater: EPA 637

DIB2760 CAS #: 103-50-4

DIBENZYL ETHER

RTECS: DQ6125000
EINECS Number: 203-118-2
Molecular Formula: $C_{14}H_{14}O$
Structured MF: $C_6H_5CH_2OCH_2C_6H_5$
Formula Weight: 198.25

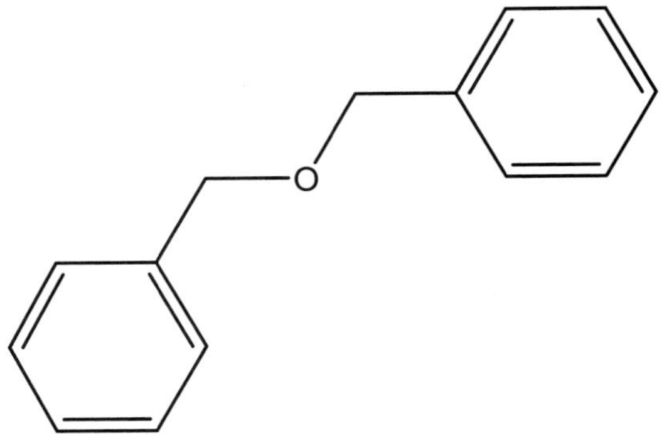

Chemical Structure

Synonyms: BA; BA (PLASTICIZER); BENZENE,1,1'-(OXYBIS(METHYLENE))BIS-; BENZENE,1,1'-[OXYBIS(METHYLENE)]BIS-; BENZYL ETHER; BENZYL OXIDE; DIBENZYLETHER; ETHER,DIBENZYL; 1,1-[OXYBIS(METHYLENE)]BIS[BENZENE]; PLASTIKATOR BA
Description: colorless to pale yellow liquid; faint almond odor
Use: plasticizer for nitrocellulose and as a solve in perfumery

Physical Properties

Boiling Point: Decomposes at 295 °C (563 °F) to 298 °C (568 °F)
Freezing Point: 3.6 °C (38.48 °F)
Specific Gravity: 0.99735 at 25 °C/4 °C
Vapor Density: 6.84 Air=1
Saturated Vapor Density: 1.200009492 kg/m^3
Density: 1.056 g/mL at 16 °C
Vapor Pressure: 1.03×10^{-3} mm Hg at 25 °C
Water Solubility: Practically Insoluble in Water
Other Solubilities: 95% Ethanol: >=100 mg/ml at 21 °C; Acetone: >=100 mg/ml at 21 °C; Alcohol: Soluble; Chloroform: Soluble; DMSO: >=100 mg/ml at 21 °C; Ether: Soluble; Most organic solvents: Soluble.
Refraction Index: 1.54057 at 25 °C
Flash Point: 135 °C Closed Cup

RTECS Toxicity Data

Acute Oral: Rat LD_{50} Dose: 2500 mg/kg.
Irritation Eye: Rabbit Standard Draize Test Dose: 500 mg/24H; Reaction: mild.
Irritation Skin: Rabbit Standard Draize Test Dose: 500 mg/24H; Reaction: mild.

Hazard Overviews

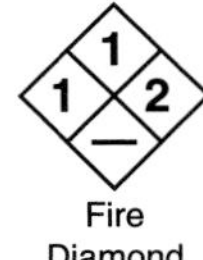

Fire Diamond

Health: Irritating to eyes/skin. Also Causes: CNS depression ranging from headache to unconsciousness, nausea.
Fire: Will burn. Use dry chemical, carbon dioxide, water spray, or regular foam. Caution! Slowly forms shock-sensitive peroxides at ordinary temperatures.

Carcinogenicity: IARC - Not listed; NIOSH - Not listed; NTP - Not listed; ACGIH - Not listed; OSHA - Not listed; EPA - Not listed; MAK - Not listed

Primary Target Organs:

Eyes

Skin
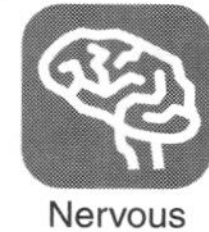
Nervous System

Environmental

Environmental Fate: If released to soil, it will have slight mobility in soil. Volatilization is not expected from either moist or dry soils. According to a MITI biodegradation test, biodegradation is not expected to be an important fate process in soil or water. However, another study with seawater and river water indicated biodegradation may be rapid. If released to water, it would adsorb to suspended solids and sediment. It will be essentially non-volatile from water surfaces. Experimental BCF values of 171-429 and 187-345 suggest that bioconcentration will be high aquatic organisms. If released to the atmosphere it will exist as a vapor. Vapor-phase is degraded in the atmosphere by reaction with photochemically produced hydroxyl radicals with an estimated half-life of about 18 hours.

Cleanup/Disposal: Guide No. 127: Eliminate all ignition sources (no smoking, flares, sparks or flames in immediate area). All equipment used when handling the product must be grounded. Do not touch or walk through spilled material. Stop leak if you can do it without risk. Prevent entry into waterways, sewers, basements or confined areas. A vapor suppressing foam may be used to reduce vapors. Absorb or cover with dry earth, sand or other non-combustible material and transfer to containers. Use clean non-sparking tools to collect absorbed material. Large Spills: Dike far ahead of liquid spill for later disposal. Water spray may reduce vapor; but may not prevent ignition in closed spaces.

Environmental Physical Data

Henry's Law Constant: estimated at 8.3 $x10^{-8}$
Octanol/Water Partition Coefficient: log K_{ow} = 3.31
Sorption Partition Coefficient: K_{oc} = estimated at 1500
BCF: estimated at 190

Regulations

RCRA 40CFR: Not listed
CERCLA: 40CFR 302.4: Not listed
SARA 40CFR 372.65: Not listed
SARA EHS 40CFR 355: Not listed
TSCA: Listed

DIB2870 CAS #: 19287-45-7

DIBORANE

RTECS: HQ9275000
DOT: UN1911; IMO2.1
EINECS Number: 242-940-6
Molecular Formula: B_2H_6
Structured MF: $H_2BH_2BH_2$
Formula Weight: 27.69

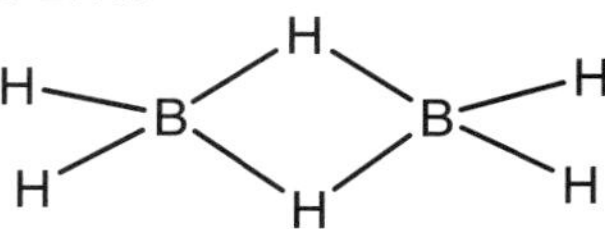

Chemical Structure

Synonyms: BOROETHANE; BORON HYDRIDE; DIBORANE (6); DIBORANE(6); DIBORANE MIXTURES; DIBORON HEXAHYDRIDE
Description: colorless gas; sweet odor
Use: synthesis of organic boron compounds and metal borohydrides; polymerization catalyst for ethylene; fuel for air breathing engines and rockets; reducing agent; doping agent for p-type semiconductors

Physical Properties

Boiling Point: -92.5 °C (-135 °F)
Freezing Point: -165 °C (-265 °F)
Specific Gravity: 0.21 at 15 °C
Vapor Density: 0.96 Air=1
Density: 0.447 g/mL at -183 °C
Vapor Pressure: 224 mm Hg at -112 °C
Water Solubility: Reacts
Other Solubilities: 95% Ethanol: Decomposes; Ammonium hydroxide: Soluble; concentrated Sulfuric acid: Soluble.
Surface Tension: 18.6 dynes/cm at 1.6 °C
Odor Threshold: 2.0000 to 4.0000 mg/m^3
pH: Acid
Critical Temperature: 16.7 °C
Critical Pressure: 39.5 atm
Ionization Potential (eV): 11.38
Flash Point: -90 °C
Autoignition Temperature: 40 to 50 °C
LEL: 0.8% v/v
UEL: 88% v/v

RTECS Toxicity Data

Acute Inhalation: Rat LC_{50} Dose: 40 ppm/4hr. Mouse LC_{50} Dose: 29 ppm/4hr; Toxic Effects: Sense organs and special senses - Other; Lungs, Thorax, or Respiration - Acute pulmonary edema; Lungs, Thorax, or Respiration - Dyspnea.
Mutagenic: Bacteria - S Typhimurium Mutations in Microorganisms; Dose: 2000 ppm (+/-S9). Bacteria - S Typhimurium Mutations in Microorganisms; Dose: 5000 ppm (-S9).

Hazard Overviews

Poison

Flammable
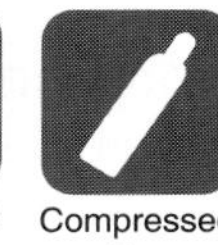
Compressed Gas

Fire Diamond

Health: Poison. Also Causes: respiratory irritation, cough, shortness of breath, chest tightness, headache, dizziness, chills, weakness, muscle twitching. Chronic Effects: asthmatic bronchitis, weight loss, anemia.

Fire: Pyrophoric, readily igniting in air. If possible, stop flow of gas. Allow fire to burn out. Do not try to extinguish with water because it produces a violent reaction. Allow the fire to burn itself out. Use a water spray from an explosion-resistant location or unmanned monitors to cool the fire area and containers.

Reactivity: Unstable, ignites spontaneously in moist air at room temperature. Hazardous polymerization can occur. Avoid: heat; ignition sources; halogenated and oxidizing agents. Incompatible with: water; oxidizing materials; oxygen; dimethyl sulfoxide; tetravinyl lead; benzene vapor; chlorine; aluminum; lithium; fuming nitric acid; borane phosphorus trifluoride (the reaction product of phosphorus trifluoride and diborane); some forms of plastics, rubber, and coatings; ketones; aldehydes; certain organic acid esters. Hazardous decomposition products: boron oxides.

Carcinogenicity: IARC - Not listed; NIOSH - Listed as carcinogen; NTP - Not listed; ACGIH - Not listed; OSHA - Not listed; EPA - Not listed; MAK - Not listed

Primary Target Organs:

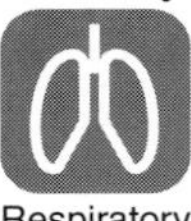
Respiratory System

Nervous System

Blood

Exposure Limits

OSHA PEL: TWA: 0.1 ppm; 0.1 mg/m^3.

ACGIH TLV: TWA: 0.1 ppm; 0.11 mg/m^3.

NIOSH REL: TWA: 0.1 ppm; 0.1 mg/m^3.

NIOSH IDLH: 15 ppm.

DFG MAK: TWA: 0.1 ppm; 0.1 mg/m^3.

Respirator Recommendation

Exposure Range: >0.1 to 5 ppm Supplied Air, Constant Flow/Pressure Demand, Half Mask

Exposure Range: >5 to <15 ppm Supplied Air, Constant Flow/Pressure Demand, Full Face

Exposure Range: 15 to unlimited ppm Self-contained Breathing Apparatus, Pressure Demand, Full Face

Note: poor warning properties

Environmental

Cleanup/Disposal: Guide No. 119: Eliminate all ignition sources (no smoking, flares, sparks or flames in immediate area). All equipment used when handling the product must be grounded. Fully encapsulating, vapor protective clothing should be worn for spills and leaks with no fire. Do not touch or walk through spilled material. Stop leak if you can do it without risk. Do not direct water at spill or source of leak. Use water spray to reduce vapors or divert vapor cloud drift. For chlorosilanes, use AFFF alcohol-resistant medium expansion foam to reduce vapors. If possible, turn leaking containers so that gas escapes rather than liquid. Prevent entry into waterways, sewers, basements or confined areas. Isolate area until gas has dispersed.

Regulations

RCRA 40CFR: Not listed

CERCLA: 40CFR 302.4: Not listed

SARA 40CFR 372.65: Not listed TPQ: 100 lb

SARA EHS 40CFR 355: Listed TPQ: 100 lb

TSCA: Listed

Analytical Methods

Air: ASTM D4490

Indoor / Expired Air: NIOSH 6006

DIB2980 CAS #: 75-81-0

1,2-DIBROMO DICHLOROETHANE

RTECS: KH9325000

EINECS Number: 200-904-7

Molecular Formula: $C_2H_2Br_2Cl_2$

Formula Weight: 256.75

Synonyms: 1,2-DIBROMO-1,1-DICHLOROETHANE; 1,2-DIBROMO-2,2-DICHLOROETHANE; ETHANE,1,2-DIBROMO-1,1-DICHLORO-; NSC 6199

Description: faintly yellow liquid

Physical Properties

Boiling Point: 178.3 °C (353 °F) at 760 mm Hg

Freezing Point: -66.8 °C (-88.24 °F)

Specific Gravity: 2.2623 at 20 °C/4 °C

Saturated Vapor Density: 1.211160163 kg/m^3

Vapor Pressure: 0.9 mm Hg at 25 °C

Water Solubility: 239 mg/L at 25 °C

Other Solubilities: Infinite in common organic solvents.

Refraction Index: 1.5567 at 20 °C

Flash Point: Does not burn or burns with difficulty

RTECS Toxicity Data

Acute Oral: Rat LD_{50} Dose: 205 mg/kg; Toxic Effects: Behavioral - Somnolence (general depressed activity); Behavioral - Convulsions or effect on seizure threshold; Gastrointestinal - Hypermotility, diarrhea.

Acute Inhalation: Rat LC_{Lo} Dose: 83 ppm/6hr; Toxic Effects: Nutritional and gross metabolic - Weight loss or decreased weight gain.

Acute Dermal: Rabbit LD_{50} Route: Skin; Dose: 500 mg/kg; Toxic Effects: Behavioral - Somnolence (general depressed activity); Behavioral - Straub tail; Gastrointestinal - Hypermotility, diarrhea.

Irritation Eye: Rabbit Standard Draize Test Dose: 100 mg; Reaction: moderate.

Irritation Skin: Rabbit Standard Draize Test Dose: 500 mg; Reaction: severe.

Hazard Overviews

Fire: Noncombustible.

Carcinogenicity: IARC - Not listed; NIOSH - Not listed; NTP - Not listed; ACGIH - Not listed; OSHA - Not listed; EPA - Not listed; MAK - Not listed

Environmental

Environmental Fate: If released to soil, it may volatilize fairly rapidly or it may leach readily through soil. This compound may be susceptible to chemical hydrolysis. If released to water, volatilization is probably an important, if not the dominant, removal mechanism (half-life 8 hours from a model river). This compound may also be susceptible to chemical hydrolysis. If released to the atmosphere, it will react with photochemically generated hydroxyl radicals (half-life 697 days) or it will diffuse slowly into the stratosphere, a processes which may take decades, where it will slowly photolyze or react with atomic oxygen. Long distance transport from its emission sources before ultimate removal from the stratosphere is expected.

Environmental Physical Data

Henry's Law Constant: estimated at 4.3 $x10^{-4}$
Octanol/Water Partition Coefficient: log K_{ow} = 3.31
Sorption Partition Coefficient: K_{oc} = estimated at 72

Regulations

RCRA 40CFR: Not listed
CERCLA: 40CFR 302.4: Listed as Compound per CWA Section 307(a)
SARA 40CFR 372.65: Not listed
SARA EHS 40CFR 355: Not listed
TSCA: Listed

DIB3200 CAS #: 3252-43-5

DIBROMOACETONITRILE

RTECS: AL8450000
EINECS Number: 221-843-2
Molecular Formula: C_2HBr_2N
Structured MF: Br_2CHCN
Formula Weight: 198.84

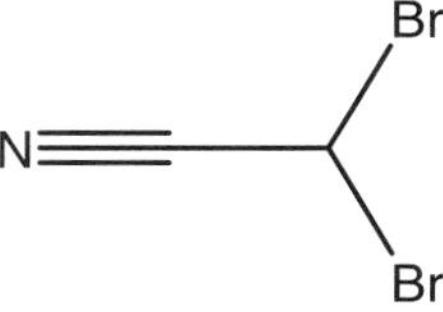

Chemical Structure

Synonyms: ACETONITRILE,DIBROMO-; DIBROMOCYANOMETHANE
Description: amber, clear liquid
Use: with aldehydes or ketones in the preparation of alpha-bromo esters

Physical Properties

Boiling Point: 67 °C (153 °F) to 69 °C (156 °F) at 24 mm Hg
Density: 2.296 g/cu cm at 25 °C
Vapor Pressure: 2 mm Hg at 50 °C
Water Solubility: 5 to 10 mg/mL at 21.5 °C
Other Solubilities: 95% Ethanol: >=100 mg/ml at 21.5 °C; Acetone: >=100 mg/ml at 21.5 °C; DMSO: >=100 mg/ml at 21.5 °C.
Refraction Index: 1.5393
Flash Point: > 93.3 °C

RTECS Toxicity Data

Acute Oral: Rat LD_{50} Dose: 245 mg/kg; Toxic Effects: Behavioral - Somnolence (general depressed activity); Behavioral - Coma; Lungs, Thorax, or Respiration - Respiratory depression. Mouse LD_{50} Dose: 289 mg/kg; Toxic Effects: Behavioral - Somnolence (general depressed activity); Behavioral - Coma; Lungs, Thorax, or Respiration - Respiratory depression.
Reproductive/Teratogenic: Rat Route: Oral; Dose: 750 mg/kg; Duration: female 7-21D of pregnancy; Effects on Newborn - Growth statistics.
Mutagenic: Human DNA Damage; Cell Type: lymphocyte; Dose: 5 umol/L. Rat Unscheduled DNA Synthesis; Route: Oral; Dose: 74100 ug/kg.
Tumorigenic: Mouse Route: Skin; Dose: 2400 mg/kg/2W-I; Toxic Effects: Tumorigenic - Carcinogenic by RTECS criteria; Skin and appendages - Tumors.

Hazard Overviews

Health: Irritating to eyes/skin/respiratory tract. Toxic. Other Acute Effects: harmful if swallowed, inhaled, or absorbed through skin; lachrymator; symptoms of exposure may include burning sensation; coughing; wheezing; laryngitis; shortness of breath; headache; nausea; vomiting.
Fire: Will burn. Hazards: emits toxic fumes. Extinguishing agents: water spray; carbon dioxide, dry chemical powder or appropriate foam. Precautions: combustible liquid.
Reactivity: Incompatible with: strong acids, strong bases, strong oxidizing agents, strong reducing agents. Hazardous decomposition products: thermal decomposition may produce carbon monoxide, carbon dioxide, and nitrogen oxides; hydrogen bromide gas.
Carcinogenicity: IARC - Group 3, Not classifiable as to carcinogenicity to humans; NIOSH - Not listed; NTP - Not listed; ACGIH - Not listed; OSHA - Not listed; EPA - Not listed; MAK - Not listed
Primary Target Organs:

Eyes

Skin

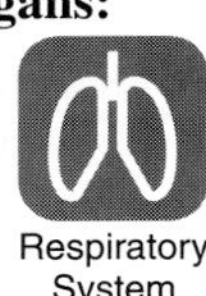
Respiratory System

Environmental

Environmental Fate: If released in water, it will be lost through hydrolysis. Hydrolysis will be faster in alkaline waters and in the presence of chlorine. Roughly 5 and 20% is lost in 10 days at pH 6 and 8, respectively. Volatilization losses are expected to be minimal and bioconcentration in aquatic organisms should not occur. In the atmosphere, it reacts extremely slowly with photochemically-produced hydroxyl radicals (half-life 696 days).
Cleanup/Disposal: Guide No. 153: Eliminate all ignition sources (no smoking, flares, sparks or flames in immediate area). Do not touch damaged containers or spilled material unless wearing appropriate protective clothing. Stop leak if you can do it without risk. Prevent entry into waterways, sewers, basements or confined areas. Absorb or cover with

dry earth, sand or other non-combustible material and transfer to containers. Do not get water inside containers.

Environmental Physical Data

Henry's Law Constant: estimated at 4.06×10^{-7}
Octanol/Water Partition Coefficient: log K_{ow} = 0.47
Sorption Partition Coefficient: K_{oc} = estimated at 12.8
BCF: calculated at 1.34

Regulations

RCRA 40CFR: Not listed
CERCLA: 40CFR 302.4: Not listed
SARA 40CFR 372.65: Not listed
SARA EHS 40CFR 355: Not listed
TSCA: Listed

Analytical Methods

Drinking Water: EPA 551

DIB3310 CAS #: 106-37-6

P-DIBROMOBENZENE

RTECS: CZ1791000
DOT: UN2711; IMO3.3
EINECS Number: 203-390-2
Molecular Formula: $C_6H_4Br_2$
Formula Weight: 235.91

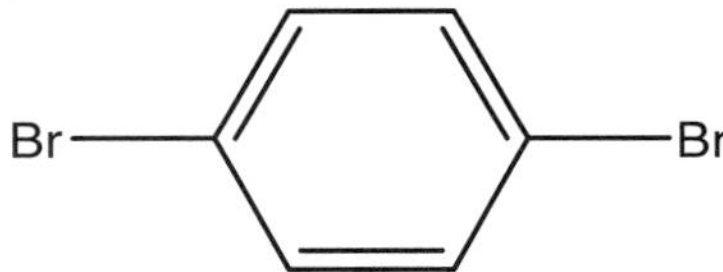

Chemical Structure

Synonyms: BENZENE,1,4-DIBROMO-; BENZENE,P-DIBROMO-; BENZENE,1,4-DIBROMO-(9CI); P-BROMOPHENYL BROMIDE; 1,4-DIBROMOBENZENE
Description: plates; colorless crystals; odor of xylene
Use: organic synthesis of dyestuffs & drugs; manufacture of intermediates

Physical Properties

Boiling Point: 220.4 °C (429 °F)
Freezing Point: 87.3 °C (189.14 °F)
Specific Gravity: 0.9641 at 99.6 °C
Water Solubility: Insoluble in Water
Other Solubilities: Soluble in Chloroform; Soluble in about 70 parts Alcohol; > 10% in Acetone; > 10% in Benzene; > 10% in Ether; > 10% in Ethanol.
Refraction Index: 1.5743 at 99.3 °C/D
Ionization Potential (eV): 8.85 +/-0.02

RTECS Toxicity Data

Acute Oral: Mouse LD_{50} Dose: 3120 mg/kg.

Hazard Overviews

Carcinogenicity: IARC - Not listed; NIOSH - Not listed; NTP - Not listed; ACGIH - Not listed; OSHA - Not listed; EPA - Not listed; MAK - Not listed

Environmental

Cleanup/Disposal: Guide No. 129: Eliminate all ignition sources (no smoking, flares, sparks or flames in immediate area). All equipment used when handling the product must be grounded. Do not touch or walk through spilled material. Stop leak if you can do it without risk. Prevent entry into waterways, sewers, basements or confined areas. A vapor suppressing foam may be used to reduce vapors. Absorb or cover with dry earth, sand or other non-combustible material and transfer to containers. Use clean non-sparking tools to collect absorbed material. Large Spills: Dike far ahead of liquid spill for later disposal. Water spray may reduce vapor; but may not prevent ignition in closed spaces.

Environmental Physical Data

Octanol/Water Partition Coefficient: log K_{ow} = 3.79

Regulations

RCRA 40CFR: Not listed
CERCLA: 40CFR 302.4: Not listed
SARA 40CFR 372.65: Not listed
SARA EHS 40CFR 355: Not listed
TSCA: Listed

DIB3420 CAS #: 110-52-1

1,4-DIBROMOBUTANE

RTECS: EJ7565000
EINECS Number: 203-775-5
Molecular Formula: $C_4H_8Br_2$
Formula Weight: 215.94

Chemical Structure

Synonyms: DBB; 1,4-DIBROMBUTAN

Physical Properties

Boiling Point: 197 °C (386.6 °F)
Freezing Point: -20 °C (-4 °F)
Specific Gravity: 1.789
Vapor Pressure: 3.66 kPa at 100 °C
Water Solubility: Insoluble
Other Solubilities: Carbon Tetrachloride: Slightly soluble; Chloroform: Soluble
Refraction Index: 1.5190
Ionization Potential (eV): 10.15

RTECS Toxicity Data

Mutagenic: Bacteria - S Typhimurium Mutations in Microorganisms; Dose: 10 umol/plate (+S9).

Hazard Overviews

Health: Irritating to eyes/skin/respiratory tract. Harmful. Other Acute Effects: harmful if swallowed, inhaled, or absorbed through skin; lachrymator; symptoms of exposure may include burning sensation; coughing; wheezing; laryngitis; shortness of breath; headache; nausea; vomiting.

Fire: Hazards: emits toxic fumes. Extinguishing agents: water spray; carbon dioxide, dry chemical powder or appropriate foam. Precautions: combustible liquid.

Reactivity: Incompatible with: strong oxidizing agents, strong bases. Hazardous decomposition products: toxic fumes of: carbon monoxide, carbon dioxide, hydrogen bromide gas.

Carcinogenicity: IARC - Not listed; NIOSH - Not listed; NTP - Not listed; ACGIH - Not listed; OSHA - Not listed; EPA - Not listed; MAK - Not listed

Primary Target Organs:

Eyes

Skin

Respiratory System

Environmental

Regulations

RCRA 40CFR: Not listed
CERCLA: 40CFR 302.4: Not listed
SARA 40CFR 372.65: Not listed
SARA EHS 40CFR 355: Not listed
TSCA: Listed

DIB3530 CAS #: 124-48-1

DIBROMOCHLOROMETHANE

RTECS: PA6360000
EINECS Number: 204-704-0
Molecular Formula: $CHBr_2Cl$
Structured MF: $ClCHBr_2$
Formula Weight: 208.28

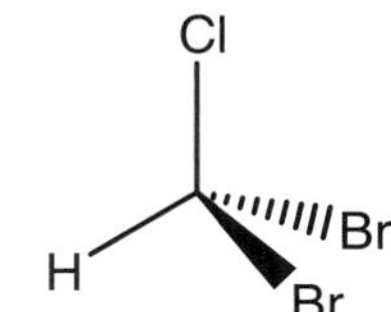

Chemical Structure

Synonyms: CDBM; CHLORODIBROMOMETHANE; METHANE,CHLORODIBROMO-; METHANE,DIBROMOCHLORO-; MONOCHLORODIBROMOMETHANE

Description: clear to yellow liquid

Use: in organic synthesis, in the manufacture of fire extinguishing agents, in refrigerants, aerosol propellants and pesticides

Physical Properties

Boiling Point: 119 °C (246 °F) to 120 °C (248 °F) at 748 mm Hg
Freezing Point: < -20 °C (-4 °F)
Specific Gravity: 2.451 at 20 °C
Vapor Pressure: 45.7 mm Hg at 17 °C
Water Solubility: 4400 ppm at 22 °C
Other Solubilities: 95% Ethanol: >=100 mg/ml at 20 °C; Acetone: >=100 mg/ml at 20 °C; Alcohol: Soluble; Benzene: Soluble; DMSO: >=100 mg/ml at 20 °C; Ether: Soluble; Lipid: Soluble.
Refraction Index: 1.5482 at 20 °C/D
Ionization Potential (eV): 10.5
Flash Point: None

RTECS Toxicity Data

Acute Oral: Rat LD_{50} Dose: 370 mg/kg; Toxic Effects: Peripheral nerve and sensation - Flaccid paralysis without anesthesia; Behavioral - Somnolence (general depressed activity); Behavioral - Tremor. Mouse LD_{50} Dose: 800 mg/kg; Toxic Effects: Brain and coverings - Changes in circulation; Liver - Fatty Liver degeneration; Blood - Hemorrhage.

Chronic (Multiple Dose) Oral: Rat Dose: 14 gm/kg/14D-I; Toxic Effects: Liver - Other changes; Kidney, Ureter, and Bladder - Other changes; DEATH. Rat Dose: 10 gm/kg/4W-C; Toxic Effects: Liver - Changes in liver weight; Kidney, Ureter, and Bladder - Changes in kidney weight; Nutritional and gross metabolic - Weight loss or decreased weight gain. Rat Dose: 8568 mg/kg/9W-I; Toxic Effects: Blood - Changes in erythrocite (RBC) cell count; Nutritional and gross metabolic - Weight loss or decreased weight gain; Biochemical - Transaminases. Rat Dose: 16250 mg/kg/13W-I; Toxic Effects: Liver - Fatty liver degeneration; Kidney, Ureter, and Bladder - Chgs in tubules (inc acute renal failure, acute tubular necrosis; DEATH.

Mutagenic: Human Sister Chromatid Exchenge; Cell Type: lymphocyte; Dose: 400 umol/L. Rat Cytogenetic Analysis; Route: Intraperitoneal; Dose: 20800 ug/kg.

Tumorigenic: Mouse Route: Oral; Dose: 52500 mg/kg/2Y-C; Toxic Effects: Tumorigenic - Carcinogenic by RTECS criteria; Liver - Tumors.

Hazard Overviews

Fire Diamond

Health: Irritating to eyes/skin/respiratory tract. Toxic. Also Causes: dizziness, headache, nausea, vomiting, unconsciousness, methemoglobinemia, depression of rapid eye movement, liver damage.

Fire: Noncombustible. Use agent suitable for surrounding fire. Hazardous combustion products may include carbon oxide(s), hydrogen chloride, hydrogen bromide.
Reactivity: Stable. Hazardous polymerization can occur. Incompatible with: strong bases; oxidizers; magnesium. Hazardous decomposition products: carbon oxide(s); hydrogen chloride; bromine; hydrogen bromide gas(es).
Carcinogenicity: IARC - Group 3, Not classifiable as to carcinogenicity to humans; NIOSH - Not listed; NTP - Not listed; ACGIH - Not listed; OSHA - Not listed; EPA - Class C, Possible human carcinogen; MAK - Not listed
Primary Target Organs:

Eyes

Skin

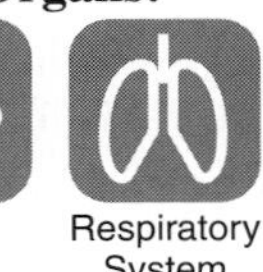
Respiratory System

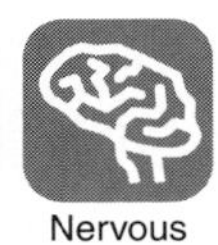
Nervous System

Liver

Environmental

Environmental Fate: If released to surface water, volatilization will be the dominant environmental fate process. The volatilization half-life from rivers and streams has been estimated to range from 43 min to 16.6 days with a typical half-life being 46 hours. In aquatic media where volatilization is not viable (e.g. groundwater), anaerobic biodegradation may be the major removal process. Aquatic hydrolysis, oxidation, direct photolysis, adsorption, and bioconcentration are not environmentally important. If released to soil, volatilization is again likely to be the dominant removal process where exposure to air is possible. It is moderately to highly mobile in soil and can therefore leach into groundwaters. If released to air, the only identifiable transformation process in the troposphere is reaction with hydroxyl radicals which has an estimated half-life of 8.4 months. This relatively persistent half-life indicates that long-range global transport is possible.
Cleanup/Disposal: Guide No. 171: Do not touch or walk through spilled material. Stop leak if you can do it without risk. Prevent dust cloud. Avoid inhalation of asbestos dust. Small Dry Spills: With clean shovel place material into clean, dry container and cover loosely; move containers from spill area. Small Spills: Take up with sand or other noncombustible absorbent material and place into containers for later disposal. Large Spills: Dike far ahead of liquid spill for later disposal. Cover powder spill with plastic sheet or tarp to minimize spreading. Prevent entry into waterways, sewers, basements or confined areas.

Environmental Physical Data

Henry's Law Constant: 8.5 x10^{-4}
Octanol/Water Partition Coefficient: log K_{ow} = 2.24
Sorption Partition Coefficient: K_{oc} = estimated at 95 to 468
BCF: estimated at 0.74 to 1.47

Regulations

RCRA 40CFR: Not listed
CERCLA: 40CFR 302.4: Listed per CWA Section 307(a) RQ: 100 lb (45.35 kg)
SARA 40CFR 372.65: Not listed
SARA EHS 40CFR 355: Not listed
TSCA: Listed

Analytical Methods

Air: EPA VA-001-1, VG-007-1, VG-011-1, 0031
Soil: CLP LC_VOA, MC_VOA, OHC; EPA 7, 1624, VG-008-1, VS-001-1, VS-002-1; SW846 5021, 5032, 5041, 5041A, 8010B, 8021A, 8240B, 8260A, 8260B
Water / Groundwater: EPA 601, 624, 624-S, VW-001-1, VW-002-1, VW-003-1; APHA 6040-B, 6210-B, 6210-D, 6230-B, 6230-D; ASTM D3973; USGS O3115
Drinking Water: EPA 502.1, 502.2, 524.1, 524.2, 551, PART_1, PART_2, PART_3; APHA 6210-C, 6230-C, 6232-B
Food: EPA 5
Plasma: EPA 29

DIB3640 CAS #: 96-12-8

1,2-DIBROMO-3-CHLOROPROPANE

RTECS: TX8750000
DOT: UN2872; IMO6.1
EINECS Number: 202-479-3
Molecular Formula: $C_3H_5Br_2Cl$
Structured MF: $BrCH_2CH(Br)CH_2Cl$
Formula Weight: 236.36

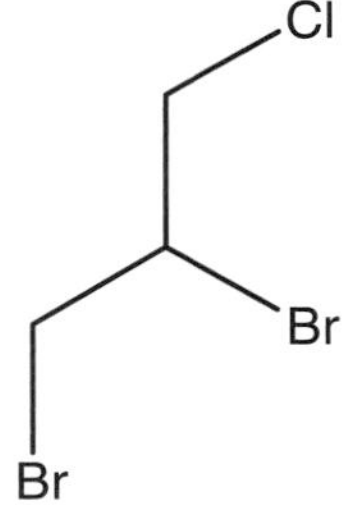

Chemical Structure

Synonyms: BBC 12; BBCP; 1-CHLORO-2,3-DIBROMOPROPANE; 3-CHLORO-1,2-DIBROMOPROPANE; DBCP; 1,2-DIBROM-3-CHLOR-PROPAN; DIBROMCHLORPROPAN; DIBROMOCHLOROPROPANE; 1,2-DIBROMO-3-CHLOROPROPANE (DBCP)-EM; 1,2-DIBROMO-3-CLORO-PROPANO; 1,2-DIBROOM-3-CHLOORPROPAAN; DURHAM NEMATICODE EM 17.1; EPA PESTICIDE CHEMICAL CODE 011301; FUMAGON; FUMAZON 86; FUMAZONE; FUMAZONE 86E; FUMAZONE 86; GRO-TONE NEMATODE GRANULAR; NEMABROM; NEMAFUME; NEMAGON; NEMAGON 20G; NEMAGON 20; NEMAGON 206; NEMAGON 90; NEMAGON SOIL FUMIGANT; NEMAGONE; NEMANAX; NEMANEX; NEMAPAZ; NEMASET; NEMATOCIDE; NEMATOCIDE EM 12.1; NEMATOCIDE EM 15.1; NEMATOCIDE SOLUTION EM 17.1; NEMATOX; NEMAZON; OS 1897; OS1897; OXY DBCP; PROPANE,1-CHLORO-2,3-DIBROMO-; PROPANE,1,2-DIBROMO-3-CHLORO-; SD 1897
Description: amber to brown liquid; pungent odor
Use: as a soil fumigant, nematocide, pesticide and intermediate in organic synthesis

Physical Properties

Boiling Point: 195.5 °C (384 °F) at 760 mm Hg
Freezing Point: 5 °C (41 °F)

Specific Gravity: 2.08 at 20 °C/20 °C
Vapor Density: 2.09
Saturated Vapor Density: 1.209032015 kg/m^3
Density: 2.05 g/mL at 20 °C
Vapor Pressure: 0.8 mm Hg at 21 °C
Water Solubility: 0.1% by weight
Other Solubilities: miscible with oils, Dichloropropane, Isopropyl Alcohol; miscible with Acetone.
Odor Threshold: 0.0965 to 0.2895 mg/m^3
Refraction Index: 1.553 at 14 °C/D
Evaporation Rate: Very much less than 1 Butyl Acetate=1
Flash Point: 76.667 °C Open Cup

RTECS Toxicity Data

Acute Oral: Rat LD_{50} Dose: 170 mg/kg. Mouse LD_{50} Dose: 257 mg/kg.
Acute Inhalation: Rat LC_{50} Dose: 103 ppm/8hr.
Acute Dermal: Rabbit LD_{50} Route: Skin; Dose: 1400 mg/kg. Rat LD_{50} Route: Subcutaneous Dose: 100 mg/kg.
Chronic (Multiple Dose) Oral: Rat Dose: 2025 mg/kg/90D-C; Toxic Effects: Liver - Changes in liver weight; Kidney, Ureter, and Bladder - Changes in kidney weight; Nutritional and gross metabolic - Weight loss or decreased weight gain. Rat Dose: 602 mg/kg/64D-C; Toxic Effects: Behavioral - Fluid intake; Nutritional and gross metabolic - Weight loss or decreased weight gain; DEATH - Changes in testicular weight.
Chronic (Multiple Dose) Inhalation: Rat Dose: 10 ppm/7H/70D-I; Toxic Effects: Liver - Changes in Liver weight; Nutritional and gross metabolic - Weight loss or decreased weight gain; DEATH - Changes in testicular weight. Rat Dose: 25 ppm/6H/13W-I; Toxic Effects: Liver - Hepatitis (hepatocellular necrosis), zonal; Kidney, Ureter, and Bladder - Chgs in tubules (inc acute renal failure, acute tubular necrosis; Blood - Changes in bone marrow not included above. Rat Dose: 10 ppm/24H/14D-C; Toxic Effects: Lungs, Thorax, or Respiration - Other changes; Kidney, Ureter, and Bladder - Chgs in tubules (inc acute renal failure, acute tubular necrosis. Rat Dose: 12 ppm/7H/70D-I; Toxic Effects: DEATH - Changes in testicular weight.
Chronic (Multiple Dose) Dermal: Rat Route: Subcutaneous; Dose: 2400 mg/kg/12W-I; Toxic Effects: Kidney, Ureter, and Bladder - Changes in kidney weight; Blood - Changes in leukocyte (WBC) cell count; DEATH - Changes in testicular weight.
Irritation Eye: Rabbit Standard Draize Test Dose: 1%; Reaction: mild.
Irritation Skin: Rabbit Standard Draize Test Dose: 10 gm; Reaction: severe.
Reproductive/Teratogenic: Rat Route: Oral; Dose: 200 mg/kg; Duration: male 1D prior to mating; Paternal Effects - Other effects on male. Rat Route: Oral; Dose: 500 mg/kg; Duration: female 6-15D of pregnancy; Effects on Embryo or Fetus - Fetotoxicity. Rat Route: Oral; Dose: 250 mg/kg; Duration: male 5D prior to mating; Effects on Fertility - Post-implantation mortality; Litter size. Rat Route: Oral; Dose: 50 mg/kg; Duration: male 5D prior to mating; Effects on Embryo or Fetus - Fetal death. Rat Route: Oral; Dose: 375 mg/kg; Duration: male 75D prior to mating; Paternal Effects - Spermatogenesis; Testes, epididymis, sperm duct; Effects on Fertility - Male fertility index. Rat Route: Inhalation; Dose: 10 ppm/6H; Duration: male 4W prior to mating; Effects on Fertility - Post-implantation mortality. Rat Route: Inhalation; Dose: 1 ppm/6H; Duration: male 4W prior to mating; Paternal Effects - Spermatogenesis.
Mutagenic: Human DNA Inhibition; Cell Type: HeLa cell; Dose: 10 mmol/L. Rat DNA Damage; Cell Type: testes; Dose: 173 mg/L. Rat DNA Damage; Route: Intraperitoneal; Dose: 35 mg/kg. Rat DNA Damage; Route: Intraperitoneal; Dose: 1182 ug/kg. Rat DNA Damage; Cell Type: liver; Dose: 5 umol/L.
Tumorigenic: Rat Route: Oral; Dose: 5475 mg/kg/73W-I; Toxic Effects: Tumorigenic - Carcinogenic by RTECS criteria; Gastrointestinal - Tumors; Skin and appendages - Tumors. Rat Route: Inhalation; Dose: 600 ppb/6H/2Y-I; Toxic Effects: Tumorigenic - Carcinogenic by RTECS criteria; Sense organs and special senses - Tumors. Rat Route: Oral; Dose: 9280 mg/kg/64W-I; Toxic Effects: Tumorigenic - Carcinogenic by RTECS criteria; Gastrointestinal - Tumors; Skin and appendages - Tumors. Rat Route: Inhalation; Dose: 600 ppb/6H/76W-I; Toxic Effects: Tumorigenic - Carcinogenic by RTECS criteria; Sense organs and special senses - Tumors; Endocrine - Adrenal cortex tumors. Rat Route: Inhalation; Dose: 3 ppm/6H/2Y-I; Toxic Effects: Tumorigenic - Carcinogenic by RTECS criteria; Sense organs and special senses - Tumors. Rat Route: Inhalation; Dose: 3 ppm/6H/84W-I; Toxic Effects: Tumorigenic - Carcinogenic by RTECS criteria; Sense organs and special senses - Tumors; Endocrine - Adrenal cortex tumors.

Hazard Overviews

Fire Diamond

Health: Irritating to eyes/skin/respiratory tract. Toxic. Also Causes: pulmonary congestion/edema, conjunctivitis, (CNS) depression with apathy, sluggishness, and ataxia; coma, malaise, headache, ir-ritability; chest, abdominal pain, acute gastrointestinal distress with pulmonary congestion, edema. DBCP can cause sterility even at low exposures. Chronic Effects: necrosis or dermatitis; infertility; severe liver, kidney damage. Possible cancer hazard.
Fire: Combustible. Use carbon dioxide or dry chemical to extinguish fires involving DBCP. Use water spray to cool fire-exposed containers. Do not apply a solid stream of water since it scatters and spreads the flame.
Reactivity: Stable. Hazardous polymerization cannot occur. Avoid: contact with rubber materials and coatings. Incompatible with: magnesium alloys; aluminum alloys; tin alloys. Hazardous decomposition products: hydrogen bromide; hydrogen chloride; carbon monoxide gases and vapor.
Carcinogenicity: IARC - Group 2B, Possibly carcinogenic to humans; NIOSH - Listed as carcinogen; NTP - Class 2B,

Reasonably anticipated to be a carcinogen, sufficient evidence of carcinogenicity from studies in experimental animals; ACGIH - Not listed; OSHA - Listed as a carcinogen; EPA - Not listed; MAK - Class A2, Unmistakably carcinogenic in animal experimentation only

Primary Target Organs:

Eyes

Skin
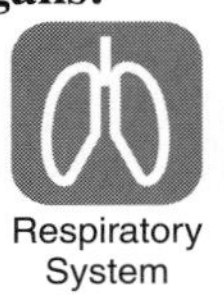
Respiratory System
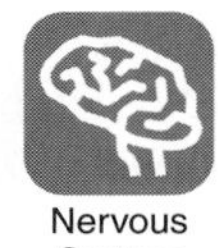
Nervous System

Liver

Kidneys

Respirator Recommendation

Exposure Range: >0.001 to 0.05 ppm Supplied Air, Constant Flow/Pressure Demand, Half Mask

Exposure Range: >0.05 to 1 ppm Supplied Air, Constant Flow/Pressure Demand, Full Face

Exposure Range: >1 to unlimited ppm Self-contained Breathing Apparatus, Pressure Demand, Full Face

Note: poor warning properties

Environmental

Ecotoxicity: LD_{50} Anas platyrhynchos (mallard) 3-5 mo females oral 66.8 mg/kg (95% confidence limit 48.2-92.6 mg/kg) /95% active ingredient LD_{50} Phasianus colchicus (ring-necked pheasant 3-4 mo females 156 mg/kg (95% confidence limit 89.3-271 mg/kg) /95% active ingredient

Environmental Fate: Released to soil it will likely volatilize or leach to groundwater. In alkaline soils, hydrolysis may be significant and biodegradation is possible but is expected to be slow relative to volatilization and leaching to groundwater. In water, it is expected to volatilize rapidly and hydrolyze slowly (half-life = 28 years at 25 °C). In groundwater, it is expected to persist due to its low estimated rate of hydrolysis (half-life = 141 years at 15 °C). Biodegradation may occur, but is expected to be slow relative to the rate of volatilization. Sorption to sediments and bioconcentration are not expected to be significant fate processes. In the atmosphere, vapor phase is expected to react with photochemically produced hydroxyl radicals with an estimated half-life of 12.19 days.

Cleanup/Disposal: Guide No. 159: Do not touch or walk through spilled material. Stop leak if you can do it without risk. Fully encapsulating, vapor protective clothing should be worn for spills and leaks with no fire. Small Spills: Take up with sand or other noncombustible absorbent material and place into containers for later disposal. Large Spills: Dike far ahead of liquid spill for later disposal. Prevent entry into waterways, sewers, basements or confined areas.

Environmental Physical Data

Henry's Law Constant: estimated at 1.47 $x10^{-4}$

Octanol/Water Partition Coefficient: log K_{ow} = calculated at 2.43

Sorption Partition Coefficient: log K_{oc} = 2.11

BCF: estimated at 11

Regulations

RCRA 40CFR: Listed Hazardous Waste No. U066 Toxic Waste

CERCLA: 40CFR 302.4: Listed per RCRA Section 3001 RQ: 1 lb (0.454 kg)

SARA 40CFR 372.65: Listed

SARA EHS 40CFR 355: Not listed

TSCA: Listed

Analytical Methods

Air: EPA VG-011-1

Soil: EPA 1625; SW846 3640A, 5021, 8010B, 8021A, 8081, 8081A, 8240B, 8260A, 8260B, 8270C

Water / Groundwater: EPA 1625, 608.1; APHA 6210-D, 6230-D

Drinking Water: EPA 502.2, 504, 504.1, 524.1, 524.2, 551; SW846 8011; AOAC 993.15; APHA 6210-C, 6230-C, 6231-B; ASTM D5316

Plasma: EPA 028, 29

Other: EPA 1656

DIB3750 CAS #: 3322-93-8

1,2-DIBROMO-4-(1,2-DIBROMOETHYL)CYCLOHEXANE

RTECS: GU8970000

EINECS Number: 222-036-8

Molecular Formula: $C_8H_{12}Br_4$

Formula Weight: 427.82

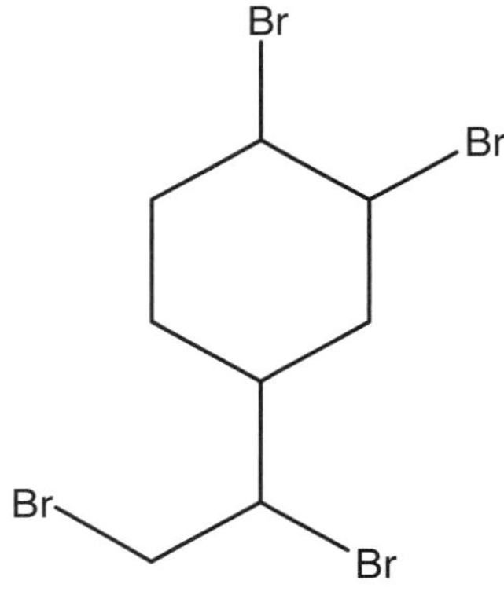

Chemical Structure

Synonyms: CITEX BCL 462; CYCLOHEXANE,1,2-DIBROMO-4-(1,2-DIBROMOETHYL)-; CYCLOHEXANE,1,2-DIBROMO-4-(1,2-DIBROMOMETHYL)-; 1,2-DIBROMO-4-(1,2-DIBROMOETHYL)CYCLOHEXANE; 1-(1,2-DIBROMOETHYL)-3,4-DIBROMOCYCLOHEXANE; 4-(1,2-DIBROMOETHYL)-1,2-DIBROMOCYCLOHEXANE; SAYTEX BCL 462; VINYLCYCLOHEXENE TETRABROMIDE

Description: white crystalline powder; ester-like odor

Use: flame retardant

Physical Properties

Freezing Point: 68 °C (154.4 °F) to 90 °C (194 °F)

Density: 2.28 g/cc

Water Solubility: < 1 mg/mL at 20.5 C

Other Solubilities: 95% Ethanol: <1 mg/ml at 20.5 °C; Acetone: >=100 mg/ml at 20 °C; DMSO: >=100 mg/ml at 20 °C.

Flash Point: Not available; probably combustible

RTECS Toxicity Data

Acute Oral: Rat LD_{50} Dose: 3220 mg/kg; Toxic Effects: Behavioral - Change in motor activity (specific assay); Gastrointestinal - Hypermotility, diarrhea; Skin and appendages - Hair.
Mutagenic: Mouse Specific Locus Test; Cell Type: lymphocyte; Dose: 40 mg/L. Hamster Cytogenetic Analysis; Cell Type: lung; Dose: 125 mg/L.

Hazard Overviews

Fire: Will burn.
Carcinogenicity: IARC - Not listed; NIOSH - Not listed; NTP - Not listed; ACGIH - Not listed; OSHA - Not listed; EPA - Not listed; MAK - Not listed

Environmental

Regulations

RCRA 40CFR: Not listed
CERCLA: 40CFR 302.4: Not listed
SARA 40CFR 372.65: Not listed
SARA EHS 40CFR 355: Not listed
TSCA: Listed

DIB3860 CAS #: 75-61-6

DIBROMODIFLUOROMETHANE

RTECS: PA7525000
DOT: UN1941
EINECS Number: 200-885-5
Molecular Formula: CBr_2F_2
Structured MF: CBr_2F_2
Formula Weight: 209.83

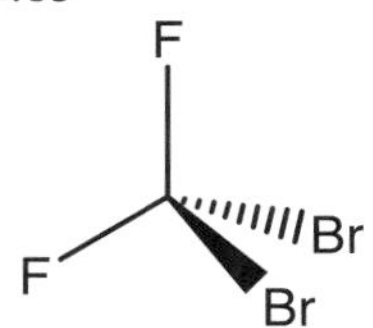

Chemical Structure

Synonyms: DFBM; DIFLUORODIBROMOMETHANE; FREON 12-B2; FREON 12B2; HALON 1202; METHANE,DIBROMODIFLUORO-; R 12B2; R12B2
Description: colorless liquid, gas (above 76 F)
Use: fire-extinguishing agent; synthesis of dyes, pharmaceuticals, quaternary ammonium compd; polymer intermediate; to trace the plume of major power plants

Physical Properties

Boiling Point: 24.5 °C (76 °F) at 760 mm Hg
Freezing Point: -146 °C (-230.8 °F)
Specific Gravity: 2.288 at 15 °C/4 °C
Vapor Density: 7.2 Air=1
Vapor Pressure: 620 mm Hg
Water Solubility: Insoluble
Other Solubilities: Soluble in Methanol.
Refraction Index: 1.399 at 12 °C
Ionization Potential (eV): 11.07
Flash Point: Nonflammable

RTECS Toxicity Data

Acute Inhalation: Mouse LC_{50} Dose: 140 gm/m^3/2hr; Toxic Effects: Behavioral - Rigidity; Liver - Other changes; Kidney, ureter, and Bladder - Other changes.

Hazard Overviews

Fire Diamond

Health: Irritating to eyes/skin/respiratory tract. Also Causes: dermatitis, heart rhythm disturbances, lightheadedness, frostbite. Chronic Effects: heart palpitations, lightheadedness.
Fire: Noncombustible. Use agent suitable for surrounding fire.
Reactivity: Stable. Hazardous polymerization cannot occur. Incompatible with: chemically active metals such as sodium; potassium; calcium; powdered aluminum; zinc; magnesium. Hazardous decomposition products: bromine; fluorine.
Carcinogenicity: IARC - Not listed; NIOSH - Listed as carcinogen; NTP - Not listed; ACGIH - Not listed; OSHA - Not listed; EPA - Not listed; MAK - Not listed
Primary Target Organs:

Eyes

Skin

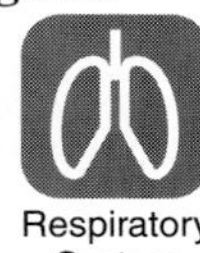
Respiratory System

Exposure Limits
OSHA PEL: TWA: 100 ppm; 860 mg/m^3.
ACGIH TLV: TWA: 100 ppm; 858 mg/m^3.
NIOSH REL: TWA: 100 ppm; 860 mg/m^3.
NIOSH IDLH: 2000 ppm.
DFG MAK: TWA: 100 ppm; 860 mg/m^3.
Respirator Recommendation
Exposure Range: >100 to <2000 ppm Supplied Air, Constant Flow/Pressure Demand, Half Mask
Exposure Range: 2000 to unlimited ppm Self-contained Breathing Apparatus, Pressure Demand, Full Face
Note: odor threshold unknown

Environmental

Cleanup/Disposal: Guide No. 159: Do not touch or walk through spilled material. Stop leak if you can do it without risk. Fully encapsulating, vapor protective clothing should be worn for spills and leaks with no fire. Small Spills: Take up with sand or other noncombustible absorbent material and place into containers for later disposal. Large Spills: Dike far ahead of liquid spill for later disposal. Prevent entry into waterways, sewers, basements or confined areas.

Regulations

RCRA 40CFR: Not listed

CERCLA: 40CFR 302.4: Listed as Compound per CWA Section 307(a)
SARA 40CFR 372.65: Not listed
SARA EHS 40CFR 355: Not listed
TSCA: Listed

Analytical Methods

Indoor / Expired Air: NIOSH 1012

DIB3970 CAS #: 106-93-4

1,2-DIBROMOETHANE

RTECS: KH9275000
DOT: UN1605; IMO6.1
EINECS Number: 203-444-5
Molecular Formula: $C_2H_4Br_2$
Structured MF: $BrCH_2CH_2Br$
Formula Weight: 187.88

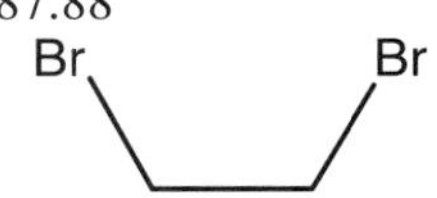

Chemical Structure

Synonyms: AADIBROOM; AETHYLENBROMID; BROMOFUME; BROMURO DI ETILE; CELMIDE; DBE; 1,2-DIBROMAETHAN; 1,2-DIBROMOETANO; ALPHA,BETA-DIBROMOETHANE; DIBROMOETHANE; SYM-DIBROMOETHANE; DIBROMURE D'ETHYLENE; 1,2-DIBROOMETHAAN; DOWFUME W-100; DOWFUME W-8; DOWFUME W-85; DOWFUME W-90; DOWFUME 40; DOWFUME EDB; DOWFUME W 85; DWUBROMOETAN; E-D-BEE; EDB; EDB-85; ENT 15,349; EPA PESTICIDE CHEMICAL CODE 042002; ETHANE,1,2-DIBROMO-; ETHYLENE BROMIDE; 1,2-ETHYLENE DIBROMIDE; ETHYLENE DIBROMIDE; FUMO-GAS; GLYCOL BROMIDE; GLYCOL DIBROMIDE; ISCOBROME D; KOPFUME; NEFIS; NEPHIS; PESTMASTER; PESTMASTER EDB-85; SANHYUUM; SOILBROM; SOILBROM-100; SOILBROM-40; SOILBROM-85; SOILBROM-90; SOILBROME-85; SOILBROM-90EC; SOILBROME-90EC; SOILFUME; UNIFUME
Description: colorless liquid; sweetish chloroform odor
Use: as a scavenger for lead in gasoline, grain, vegetable, fruit and tree crop fumigant, general solvent, waterproofing preparations, organic synthesis, insecticide, medicine, entrainment reagent f conversion of unreactive halides to Grignard reagents, control of nematodes, termites and pine bark beetles, production of fire extinguishing agents and gauge fluids during the manufacture of measuring instruments, organic synthesis in production of dyes, pharmaceuticals and ethylene oxide, and specialty solvent for resins, gums and waxes

Physical Properties

Boiling Point: 131 °C (268 °F) to 132 °C (270 °F)
Freezing Point: 9.8 °C (49.64 °F)
Specific Gravity: 2.172 at 25/25 °C
Vapor Density: 6.48 Air=1
Saturated Vapor Density: 1.295154991 kg/m^3
Density: Liquid 2.172 g/mL
Bulk Density: 18.1 lbs/gal
Vapor Pressure: 11 mm Hg at 20 °C
Water Solubility: 1 parts in about 250 parts Water
Other Solubilities: miscible with most organic solvents; Soluble in Alcohol, Ether, acetate, Benzene.
Surface Tension: 38.75 dynes/cm at 20 °C
Odor Threshold: 76.80 to 62.50 mg/m^3
Refraction Index: 1.5379 at 20 °C/D
Critical Temperature: 309.8 °C
Critical Pressure: 7154 kPa
Ionization Potential (eV): 9.45
Flash Point: Nonflammable

RTECS Toxicity Data

Acute Oral: Woman LD_{Lo} Dose: 90 mg/kg; Toxic Effects: Gastrointestinal - Hypermotility, diarrhea; Gastrointestinal - Nausea or vomiting; Kidney, Ureter, and Bladder - Urine volume decreased. Rat LD_{50} Dose: 108 mg/kg.
Acute Inhalation: Rat LC_{50} Dose: 14300 mg/m^3/30M; Toxic Effects: Peripheral Nerve and sensation - Flaccid paralysis without anesthesia; Behavioral - Somnolence (general depressed activity); Lungs, Thorax, or Respiration - Dyspnea. Guinea Pig LC_{Lo} Dose: 400 ppm/3hr; Toxic Effects: Nutritional and gross metabolic - Conditioned vitamin deficiency; Lungs, Thorax, or Respiration - Chronic pulmonary edema; Liver - Hepatitis (hepatocellular necrosis), zonal.
Acute Dermal: Rabbit LD_{50} Route: Skin; Dose: 300 mg/kg; Toxic Effects: Nutritional and gross metabolic - Body temperature decrease; Skin and appendages - Corrosive. Rat LD_{50} Route: Skin; Dose: 300 mg/kg.
Chronic (Multiple Dose) Oral: Quail Dose: 162 mg/kg/7D-I; Toxic Effects: Liver - Changes in liver weight; Blood - Pigmented or nucleated red Blood cells; Biochemical - Lipids including transport. Quail Dose: 486 mg/kg/21D-I; Toxic Effects: Blood - Changes in serum composition; Biochemical - True cholinesterase; Biochemical - Other transferases.
Chronic (Multiple Dose) Inhalation: Rat Dose: 40 ppm/6H/13W-I; Toxic Effects: Liver - Changes in Liver weight; Blood - Pigmented or nucleated red Blood cells; Nutritional and gross metabolic - Weight loss or decreased weight gain. Mouse Dose: 3 ppm/6H/13W-I; Toxic Effects: DEATH.
Irritation Skin: Human Standard Draize Test Dose: 1538 mg/2H; Reaction: severe. Rabbit Standard Draize Test Dose: 1%/14D; Reaction: severe.
Reproductive/Teratogenic: Man Route: Inhalation; Dose: 88 ppb/8H; Duration: male 5Y prior to mating; Paternal Effects - Spermatogenesis. Rat Route: Oral; Dose: 50 mg/kg; Duration: male 5D prior to mating; Effects on Embryo or Fetus - Fetal death. Rat Route: Inhalation; Dose: 66670 ppb/4H; Duration: female 3-20D of pregnancy; Effects on Newborn - Growth statistics. Rat Route: Inhalation; Dose: 80 ppm/23H; Duration: female 6-15D of pregnancy; Effects on Embryo or Fetus - Fetal death. Rat Route: Inhalation; Dose: 89 ppm/7H; Duration: male 10W prior to mating; Paternal Effects - Testes, epididymis, sperm duct; Prostate, seminal vessicle, Cowper's gland, accessory glands; Effects on Fertility - Male fertility index. Rat Route: Inhalation; Dose: 39 ppm/7H;

Duration: female 3W prior to mating; Effects on Fertility - Mating Performance. Rat Route: Inhalation; Dose: 80 ppm/7H; Duration: female 3W prior to mating; Effects on Fertility - Female fertility index.

Mutagenic: Human DNA Inhibition; Cell Type: lymphocyte; Dose: 5 mL/L. Human Mutations in Mammalian Somatic Cells; Cell Type: lymphocyte; Dose: 5 mg/L. Human Sister Chromatid Exchenge; Cell Type: lymphocyte; Dose: 10 nmol/L. Human Micronucleus Test; Cell Type: lymphocyte; Dose: 1 mmol/L.

Tumorigenic: Rat Route: Oral; Dose: 540 mg/kg/78W-C; Toxic Effects: Tumorigenic - Carcinogenic by RTECS criteria; Vascular - Tumors; Blood - Tumors. Rat Route: Inhalation; Dose: 10 ppm/6H/2Y-I; Toxic Effects: Tumorigenic - Carcinogenic by RTECS criteria; Sense organs and special senses - Tumors; Vascular - Tumors. Rat Route: Oral; Dose: 16 gm/kg/61W-I; Toxic Effects: Tumorigenic - Carcinogenic by RTECS criteria; Gastrointestinal - Tumors; Liver - Tumors. Rat Route: Oral; Dose: 24 gm/kg/57W-I; Toxic Effects: Tumorigenic - Carcinogenic by RTECS criteria; Gastrointestinal - Tumors; Liver - Tumors. Rat Route: Oral; Dose: 6630 mg/kg/34W-I; Toxic Effects: Tumorigenic - Carcinogenic by RTECS criteria; Gastrointestinal - Tumors; Liver - Tumors. Rat Route: Inhalation; Dose: 20 ppm/7H/78W-I; Toxic Effects: Tumorigenic - Carcinogenic by RTECS criteria; Endocrine - Adrenal cortex tumors; Skin and appendages - Tumors. Rat Route: Oral; Dose: 7000 mg/kg/47W-I; Toxic Effects: Tumorigenic - Carcinogenic by RTECS criteria; Gastrointestinal - Tumors; Liver - Tumors. Rat Route: Oral; Dose: 18 gm/kg/44W-I; Toxic Effects: Tumorigenic - Carcinogenic by RTECS criteria; Gastrointestinal - Tumors; Liver - Tumors. Rat Route: Inhalation; Dose: 10 ppm/6H/2Y-I; Toxic Effects: Tumorigenic - Carcinogenic by RTECS criteria; Vascular - Tumors; Lungs, Thorax, or Respiration - Tumors. Rat Route: Inhalation; Dose: 40 ppm/6H/88W-I; Toxic Effects: Tumorigenic - Carcinogenic by RTECS criteria; Sense organs and special senses - Tumors; Vascular - Tumors.

Hazard Overviews

Fire Diamond

Health: Severely irritating to eyes/skin/respiratory tract. Toxic. Also Causes: (high conc): damage to kidneys, liver, heart, lungs, CNS depression. Chronic Effects: reproductive/teratogenic effects, suspect cancer hazard.

Fire: Noncombustible. Use agent suitable for surrounding fire.

Reactivity: Stable. Hazardous polymerization cannot occur. Avoid: light; heat; water; plastics; resins. Incompatible with: aluminum; magnesium; zinc; calcium; sodium; potassium; strong alkalies; strong oxidizing agents; liquid ammonia. Hazardous decomposition products: carbon monoxide; hydrogen bromide.

Carcinogenicity: IARC - Group 2A, Probably carcinogenic to humans; NIOSH - Listed as carcinogen; NTP - Class 2B, Reasonably anticipated to be a carcinogen, sufficient evidence of carcinogenicity from studies in experimental animals; ACGIH - Class A3, Animal carcinogen; OSHA - Not listed; EPA - Class B2, Probable human carcinogen based on animal studies; MAK - Class A2, Unmistakably carcinogenic in animal experimentation only

Primary Target Organs:

Eyes

Skin

Respiratory System

Nervous System

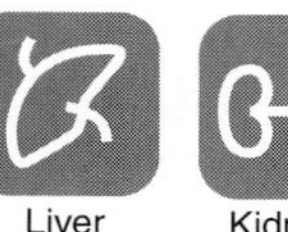
Liver

Kidneys

Exposure Limits

OSHA PEL: TWA: 20 ppm; ceiling.

NIOSH REL: TWA: 0.045 ppm. STEL: 0.13 ppm; 15 mg/m^3; 15-minute.

NIOSH IDLH: 100 ppm.

Respirator Recommendation

Exposure Range: >20 to <100 ppm Supplied Air, Constant Flow/Pressure Demand, Half Mask

Exposure Range: 100 to unlimited ppm Self-contained Breathing Apparatus, Pressure Demand, Full Face

Note: odor threshold unknown

Environmental

Ecotoxicity: LC_{50} Lepomis macrochirus 18 mg/l/48 hr /Conditions of bioassay not specified

Environmental Fate: In the atmosphere, it will degrade by reaction with photochemically produced hydroxyl radicals (half life 32 days). When spilled in water, it, will be removed by evaporation (half life 1-5 days). When spilled on land or applied to land during soil fumigation, it will exhibit low to moderate adsorption and has been found in groundwater. Little bioconcentration into the food chain is expected.

Cleanup/Disposal: Guide No. 154: Eliminate all ignition sources (no smoking, flares, sparks or flames in immediate area). Do not touch damaged containers or spilled material unless wearing appropriate protective clothing. Stop leak if you can do it without risk. Prevent entry into waterways, sewers, basements or confined areas. Absorb or cover with dry earth, sand or other non-combustible material and transfer to containers. Do not get water inside containers.

Environmental Physical Data

Octanol/Water Partition Coefficient: log K_{ow} = calculated at 135

Sorption Partition Coefficient: K_{oc} = 14 to 160

BCF: fish 1

Regulations

RCRA 40CFR: Listed Hazardous Waste No. U067 Toxic Waste

CERCLA: 40CFR 302.4: Listed per CWA Section 311(b)(4) per RCRA Section 3001 RQ: 1 lb (0.454 kg)

SARA 40CFR 372.65: Listed

SARA EHS 40CFR 355: Not listed

TSCA: Listed

Analytical Methods

Air: EPA VG-011-1, TO-1, TO-14; ASTM D3686, D3687, D4490
Soil: CLP LC_VOA; EPA 1624, VG-004-1, VS-003-1; SW846 3640A, 5021, 8010B, 8021A, 8240B, 8260A, 8260B
Water / Groundwater: EPA 618, VW-005-1; APHA 6040-B, 6210-D, 6230-D
Drinking Water: EPA 502.1, 502.2, 504, 504.1, 524.1, 524.2, 551; SW846 8011; AOAC 993.15; APHA 6210-C, 6230-C, 6231-B; ASTM D5316
Food: AOAC 977.18, 986.20
Indoor / Expired Air: NIOSH 1008; EPA IP-1A, IP-1A-B, IP-1A-C, IP-1B
Plasma: EPA 028, 29

DIB4080 CAS #: 24442-57-7

1,2-DIBROMOETHYL ACETATE

EINECS Number: 246-249-0
Molecular Formula: $C_4H_6Br_2O_2$
Formula Weight: 245.91
Synonyms: ACETIC ACID,1,2-DIBROMOETHYL ESTER; 1,2-DIBROMOETHANOL ACETATE; ALPHA,BETA-DIBROMOETHYL ACETATE; ETHANOL,1,2-DIBROMO-,ACETATE

Hazard Overviews

Carcinogenicity: IARC - Not listed; NIOSH - Not listed; NTP - Not listed; ACGIH - Not listed; OSHA - Not listed; EPA - Not listed; MAK - Not listed

Environmental

Ecotoxicity: LC_{50} Bluegill sunfish 1.43 mg/l 24 hr LD_{50} Bluegill sunfish 1.21 mg/l 96 hr

Regulations

RCRA 40CFR: Not listed
CERCLA: 40CFR 302.4: Not listed
SARA 40CFR 372.65: Not listed
SARA EHS 40CFR 355: Not listed
TSCA: Listed

DIB4190 CAS #: 540-49-8

1,2-DIBROMOETHYLENE

RTECS: KV7700000
EINECS Number: 208-747-6
Molecular Formula: $C_2H_2Br_2$
Formula Weight: 185.87

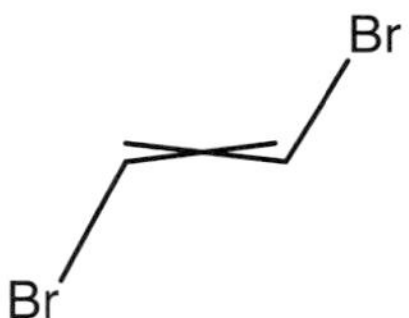

Chemical Structure

Synonyms: ACETYLENE DIBROMIDE; 1,2-DIBROMOETHENE; SYM-DIBROMOETHYLENE; ETHENE,1,2-DIBROMO-; ETHYLENE,1,2-DIBROMO-
Description: colorless liquid
Use: research chemistry

Physical Properties

Specific Gravity: 2.21 at 17 °C/4 °C
Vapor Pressure: cis-isomer 21 mm Hg at 26 °C
Water Solubility: Practically Insoluble in Water
Other Solubilities: Soluble in Alcohol, Ether, Acetone, Benzene & Chloroform.
Refraction Index: 1.5428 at 20 °C/D

RTECS Toxicity Data

Mutagenic: Bacteria - S Typhimurium Mutations in Microorganisms; Dose: 100 ug/plate (+S9), 10 ug/plate (-S9).

Hazard Overviews

Corrosive

Health: Corrosive to eyes/skin/respiratory tract. Toxic. Other Acute Effects: harmful if swallowed, inhaled, or absorbed through skin; material is extremely destructive to tissue of the mucous membranes and upper respiratory tract, eyes and skin; symptoms of exposure may include burning sensation, coughing, wheezing, laryngitis, shortness of breath, headache, nausea and vomiting; inhalation may result in spasm, inflammation and edema of the larynx and bronchi, chemical pneumonitis and pulmonary edema; prolonged exposure can cause narcotic effect.
Fire: Hazards: emits toxic fumes. Extinguishing agents: noncombustible; use extinguishing media appropriate to surrounding fire conditions. Precautions: combustible liquid.
Reactivity: Incompatible with: strong bases, strong oxidizing agents, magnesium. Hazardous decomposition products: toxic fumes of: carbon monoxide, carbon dioxide, hydrogen bromide gas.
Carcinogenicity: IARC - Not listed; NIOSH - Not listed; NTP - Not listed; ACGIH - Not listed; OSHA - Not listed; EPA - Not listed; MAK - Not listed
Primary Target Organs:

Eyes

Skin

Respiratory System

Environmental

Environmental Fate: If released to soil, it would be expected to rapidly volatilize from the soil surface or leach into the soil. Hydrolysis in moist soil is not expected to be important. If released into water, it would rapidly volatilize (half life 4.2 hr in a model river) and probably hydrolyze.

Bioconcentration in aquatic organism or adsorption to sediment would not be significant. In the atmosphere it should degrade by reacting with photochemically produced hydroxyl radicals and ozone (half life 1.9 days).

Environmental Physical Data

Henry's Law Constant: estimated at 0.674×10^{-2}
Octanol/Water Partition Coefficient: log K_{ow} = 1.89
Sorption Partition Coefficient: K_{oc} = 45
BCF: calculated at 16

Regulations

RCRA 40CFR: Not listed
CERCLA: 40CFR 302.4: Not listed
SARA 40CFR 372.65: Not listed
SARA EHS 40CFR 355: Not listed
TSCA: Listed

DIB4300 CAS #: 629-03-8

1,6-DIBROMOHEXANE

RTECS: MO1515000
EINECS Number: 211-067-2
Molecular Formula: $C_6H_{12}Br_2$
Formula Weight: 244.00

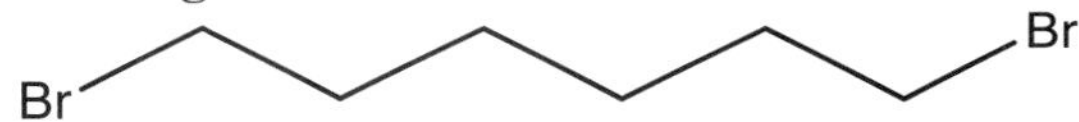

Chemical Structure

Synonyms: 1,6-DIBROMOHEXAN; ALPHA,OMEGA-DIBROMOHEXANE; HEXAMETHYLENE DIBROMIDE

Physical Properties

Boiling Point: 243 °C (469.4 °F)
Freezing Point: -2 °C (28.4 °F)
Specific Gravity: 1.5948
Water Solubility: Insoluble
Other Solubilities: Ether: Soluble; Acetone: Soluble; Carbon Tetrachloride: Slightly soluble; Chloroform: Soluble
Refraction Index: 1.5037

RTECS Toxicity Data

Irritation Eye: Rabbit Standard Draize Test Dose: 100 mg; Reaction: mild. Rabbit Nonstandard Exposure Dose: 100 mg/30S rinse; Reaction: mild.
Irritation Skin: Rabbit Standard Draize Test Dose: 500 mg; Reaction: mild.
Mutagenic: Bacteria - S Typhimurium Mutations in Microorganisms; Dose: 10 umol/plate (+S9).

Hazard Overviews

Health: Irritating to eyes/skin/respiratory tract. Other Acute Effects: may be harmful by inhalation, ingestion, or skin absorption.
Fire: Hazards: emits toxic fumes. Extinguishing agents: water spray; carbon dioxide, dry chemical powder or appropriate foam. Precautions: combustible liquid.
Reactivity: Incompatible with: strong oxidizing agents. Hazardous decomposition products: toxic fumes of: carbon monoxide, carbon dioxide, hydrogen bromide gas.
Carcinogenicity: IARC - Not listed; NIOSH - Not listed; NTP - Not listed; ACGIH - Not listed; OSHA - Not listed; EPA - Not listed; MAK - Not listed
Primary Target Organs:

Eyes

Skin

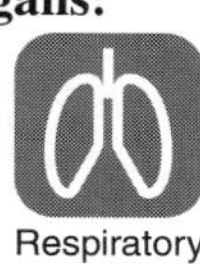
Respiratory System

Environmental

Regulations

RCRA 40CFR: Not listed
CERCLA: 40CFR 302.4: Not listed
SARA 40CFR 372.65: Not listed
SARA EHS 40CFR 355: Not listed
TSCA: Listed

DIB4410 CAS #: 1689-84-5

3,5-DIBROMO-4-HYDROXYBENZONITRILE

RTECS: DI3150000
EINECS Number: 216-882-7
Molecular Formula: $C_7H_3Br_2NO$
Structured MF: $C_7H_3Br_2NO$
Formula Weight: 276.93

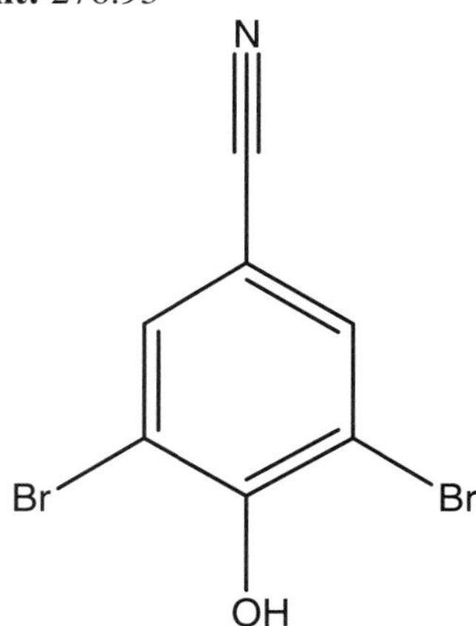

Chemical Structure

Synonyms: BENZONITRILE,3,5-DIBROMO-4-HYDROXY-; BRITTOX; BROMINAL; BROMINAL INDUSTRIAL; BROMINAL M; BROMINAL PLUS; BROMINAL TRIPLE; BROMINEX; BROMINIL; BROMOTRIL; BROMOXYNIL; BRONATE; BROXYNIL; BUCTRIL; BUCTRIL INDUSTRIAL; BUCTRIL M; BUTIL CHLOROFOS; BUTILCHLOROFOS; BUTILCHOROFOS; CERTROL B; CHIPCO BUCTRIL; CHIPCO CRAB-KLEEN; CHIPRO BUCTRIL; CRUSADER S;

4-CYANO-2,6-DIBROMOPHENOL; 2,6-DIBROMO-4-CYANOPHENOL; 3,5-DIBROMO-4-HYDROXYBENZONITRIL; 3,5-DIBROMO-4-HYDROXYPHENYL CYANIDE; 3,5-DIBROMO-4-HYDROXYPHENYLCYANIDE; ENT 20852; ENT 20 852; 4-HYDROXY-3,5-DIBROMOBENZONITRILE; LABUCTRIL; LABUCTRIL 25; LITAROL M; M&B 10 064; M&B 10,064; M&B 10064; M&B 10731; MB-10064; MB 10,064; MB 10064; ME4 BROMINAL; MERIT; NOVACORN; NU-LAWN WEEDER; OXYTRIL M; PARDNER; S 2132; SABRE; TERSET; TOPLAN

Description: colorless to white crystalline solid or light buff to creamy powder; odorless when pure

Use: selective contact herbicide with systemic activity; post-emergence control of annual broad-leaved weeds in cereals, maize, sorghum, flax, onions, garlic, mint, grass-seed crops, turf, & non-crop land; often in combination with other herbicides; for mint, barley, flax forage grasses, oats, onions, rye, sorghum, etc; post-emergent control of common cocklebur, common ragweed; seedling broadleaf weeds eg, blue (purple) mustard, corn gromwell, cowcockle, fiddleneck, field pennycress, etc.

Physical Properties

Boiling Point: Sublimes at 135 °C (275 °F) at 0.2 mbar

Freezing Point: 194 °C (381.2 °F) to 195 °C (383 °F)

Vapor Pressure: < 1 mPa at 20 °C

Water Solubility: 130 ppm (0013%) wt/vol at 20-25 °C

Other Solubilities: in Dimethylformamide 61% wt/vol at 20-25 °C; in Benzene 1% wt/vol at 20-25 °C; in Methanol 9% wt/vol at 20-25 °C; in Acetone 17% wt/vol at 20-25 °C; Solubility in (g/l) at 25 °C: Acetone 170; tetrahydrofuran 410;

RTECS Toxicity Data

Acute Oral: Rat LD_{50} Dose: 190 mg/kg. Mouse LD_{50} Dose: 110 mg/kg.

Acute Dermal: Rabbit LD_{50} Route: Skin; Dose: 3660 mg/kg. Rat LD_{50} Route: Skin; Dose: >2 mg/kg.

Hazard Overviews

Health: Irritating to eyes/skin/respiratory tract. Toxic. Other Acute Effects: harmful if swallowed, inhaled, or absorbed through skin. Chronic Effects: possible reproductive hazard.

Fire: Hazards: emits toxic fumes. Extinguishing agents: water spray; carbon dioxide, dry chemical powder or appropriate foam. Precautions: combustible liquid.

Reactivity: Incompatible with: strong oxidizing agents, strong bases. Hazardous decomposition products: toxic fumes of: carbon monoxide, carbon dioxide, nitrogen oxides, hydrogen bromide gas.

Carcinogenicity: IARC - Not listed; NIOSH - Not listed; NTP - Not listed; ACGIH - Not listed; OSHA - Not listed; EPA - Not listed; MAK - Not listed

Primary Target Organs:

Eyes

Skin

Respiratory System

Environmental

Ecotoxicity: LD_{50} Pheasant oral 50 mg/kg LC_{50} Coturnix japonica >5,000 ppm (5 day in diet) LC_{50} Catfish 0.023 ppm /Conditions of bioassay not specified LC_{50} Golden orfe 0.2 mg/l/96 hr at pH 6.2 /Conditions of bioassay not specified

Environmental Fate: Although it is usually applied as the octanoate ester, the ester hydrolyzes rapidly to bromoxynil. It may also be released as runoff from fields and enter waterways. However, high concentrations appear to occur during rain events within two months of application. It is discharged into the atmosphere as an aerosol during spraying operations and will be removed by gravitational settling and photolysis. If sprayed on fields, it will photolyze on the soil surface and biodegrade. It is expected to adsorb moderately to soil at neutral and alkaline pHs. In field studies, it was completely dissipated from soil in 10 weeks when applied in May and 15 weeks when applied in December. No residues were found in lower layers of soil. If released into water, it will photolyze in surface layers of water and biodegrade. In test ponds, it did not persist in sediment beyond 15 days after treatment. It should not bioconcentrate in aquatic organisms.

Environmental Physical Data

Henry's Law Constant: 9.52×10^{-10}

Sorption Partition Coefficient: K_{oc} = estimated at 300

BCF: estimated at 130

Regulations

RCRA 40CFR: Not listed

CERCLA: 40CFR 302.4: Not listed

SARA 40CFR 372.65: Listed

SARA EHS 40CFR 355: Not listed

TSCA: Not listed

Analytical Methods

Soil: EPA 1625; SW846 8270B, 8270C

Water / Groundwater: EPA P-008-1; EPA 1625, 1661

Indoor / Expired Air: NIOSH 5010

Plasma: EPA 29; FDA 231.1, 252

DIB4520 **CAS #: 10222-01-2**

2,2-DIBROMO-3-NITRILOPROPIONAMIDE

RTECS: AB5956000

EINECS Number: 233-539-7

Molecular Formula: $C_3H_2Br_2N_2O$

Formula Weight: 241.89

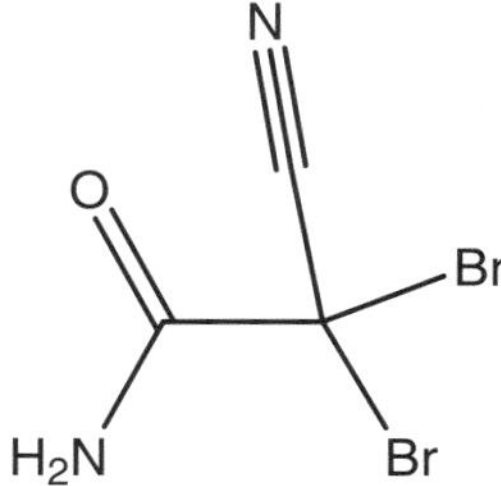

Chemical Structure

Synonyms: ACETAMIDE,2,2-DIBROMO-2-CYANO-(8CI,9CI); DBNPA; 2,2-DIBROMO-2-CYANOACETAMIDE; ALPHA,ALPHA-DIBROMO-ALPHA-CYANOACETAMIDE; DIBROMOCYANOACETAMIDE

RTECS Toxicity Data

Acute Oral: Mammal LD_{50} Dose: 118 mg/kg.
Irritation Eye: Rabbit Standard Draize Test Dose: 100 mg; Reaction: severe.
Irritation Skin: Rabbit Standard Draize Test Dose: 500 mg; Reaction: severe.

Hazard Overviews

Carcinogenicity: IARC - Not listed; NIOSH - Not listed; NTP - Not listed; ACGIH - Not listed; OSHA - Not listed; EPA - Not listed; MAK - Not listed

Environmental

Regulations

RCRA 40CFR: Not listed
CERCLA: 40CFR 302.4: Not listed
SARA 40CFR 372.65: Listed
SARA EHS 40CFR 355: Not listed
TSCA: Listed

DIB4630 CAS #: 827-94-1

2,6-DIBROMO-4-NITROANILINE

RTECS: BX2599000
EINECS Number: 212-577-8
Molecular Formula: $C_6H_4Br_2N_2O_2$
Formula Weight: 295.92

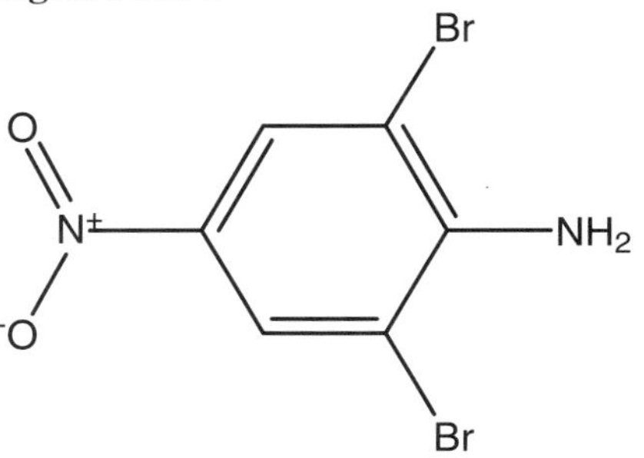

Chemical Structure

Synonyms: ANILINE,2,6-DIBROMO-4-NITRO-; BENZENAMINE,2,6-DIBROMO-4-NITRO-
Description: yellow needles

Physical Properties

Boiling Point: 207 °C (405 °F)
Freezing Point: 207 °C (404.6 °F)
Water Solubility: Slightly Soluble in Water
Other Solubilities: > 10% in Acetic Acid

Hazard Overviews

Health: Irritating to eyes/skin. Harmful. Other Acute Effects: may be harmful by inhalation, ingestion, or skin absorption.
Fire: Hazards: emits toxic fumes. Extinguishing agents: water spray; carbon dioxide, dry chemical powder or appropriate foam. Precautions: combustible liquid.
Reactivity: Incompatible with: acids, acid chlorides, acid anhydrides, chloroformates, strong oxidizing agents. Hazardous decomposition products: toxic fumes of: carbon monoxide, carbon dioxide, nitrogen oxides, hydrogen bromide gas.
Carcinogenicity: IARC - Not listed; NIOSH - Not listed; NTP - Not listed; ACGIH - Not listed; OSHA - Not listed; EPA - Not listed; MAK - Not listed
Primary Target Organs:

Eyes

Skin

Environmental

Regulations

RCRA 40CFR: Not listed
CERCLA: 40CFR 302.4: Not listed
SARA 40CFR 372.65: Not listed
SARA EHS 40CFR 355: Not listed
TSCA: Listed

Analytical Methods

Soil: SW846 8131

DIB4740 CAS #: 111-24-0

1,5-DIBROMOPENTANE

RTECS: SA0320000
EINECS Number: 203-849-7
Molecular Formula: $C_5H_{10}Br_2$
Formula Weight: 229.97

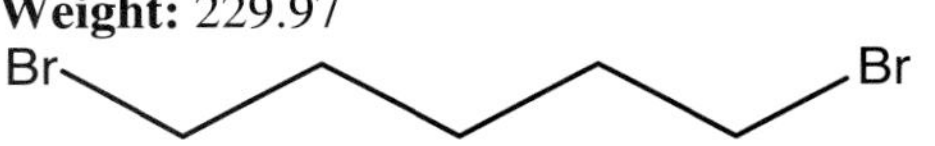

Chemical Structure

Synonyms: PENTAMETHYLENE BROMIDE; PENTAMETHYLENE DIBROMIDE

Physical Properties

Boiling Point: 224 °C (435.2 °F)
Freezing Point: -34 °C (-29.2 °F)
Specific Gravity: 1.7018
Vapor Pressure: 4.45 kPa at 125 °C
Water Solubility: Insoluble
Other Solubilities: Benzene: Soluble; Carbon Tetrachloride: Slightly soluble; Chloroform: Soluble
Refraction Index: 1.5126
Ionization Potential (eV): =< 10.23

RTECS Toxicity Data

Mutagenic: Bacteria - E Coli DNA Repair; Dose: 10 mg/plate. Bacteria - S Typhimurium Mutations in Microorganisms; Dose: 10 umol/plate (+S9).

Hazard Overviews

Health: Irritating to eyes/skin/respiratory tract. Toxic. Other Acute Effects: harmful if swallowed, inhaled, or absorbed through skin. Chronic Effects: may alter genetic material.
Fire: Hazards: emits toxic fumes. Extinguishing agents: water spray; carbon dioxide, dry chemical powder or appropriate foam. Precautions: combustible liquid.
Reactivity: Incompatible with: strong oxidizing agents, strong bases. Hazardous decomposition products: toxic fumes of: carbon monoxide, carbon dioxide, hydrogen bromide gas.
Carcinogenicity: IARC - Not listed; NIOSH - Not listed; NTP - Not listed; ACGIH - Not listed; OSHA - Not listed; EPA - Not listed; MAK - Not listed
Primary Target Organs:

Eyes

Skin

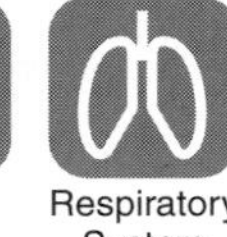
Respiratory System

Environmental

Regulations

RCRA 40CFR: Not listed
CERCLA: 40CFR 302.4: Not listed
SARA 40CFR 372.65: Not listed
SARA EHS 40CFR 355: Not listed
TSCA: Listed

DIB4850 CAS #: 7507-35-9

(3,5-DIBROMOPHENOXY) ACETIC ACID

Molecular Formula: $C_8H_6Br_2O_3$
Formula Weight: 309.94
Synonyms: ACETIC ACID,(3,5-DIBROMOPHENOXY)-

Hazard Overviews

Carcinogenicity: IARC - Not listed; NIOSH - Not listed; NTP - Not listed; ACGIH - Not listed; OSHA - Not listed; EPA - Not listed; MAK - Not listed

Environmental

Regulations

RCRA 40CFR: Not listed
CERCLA: 40CFR 302.4: Not listed
SARA 40CFR 372.65: Not listed
SARA EHS 40CFR 355: Not listed
TSCA: Not listed

DIB4960 CAS #: 96-13-9

2,3-DIBROMOPROPANOL

RTECS: UB0175000
EINECS Number: 202-480-9
Molecular Formula: $C_3H_6Br_2O$
Structured MF: $BrCH_2CHBrCH_2OH$
Formula Weight: 217.91

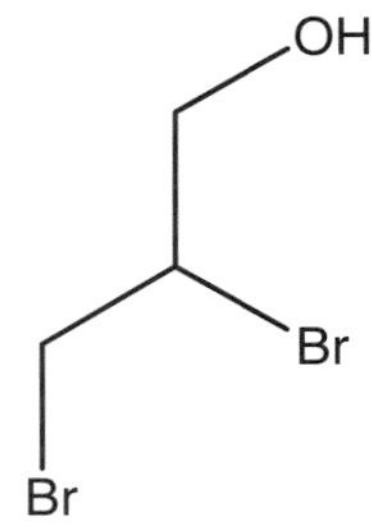

Chemical Structure

Synonyms: ALLYL ALCOHOL DIBROMIDE; BETA-DIBROMOHYDRIN; 1,2-DIBROMOPROPAN-3-OL; 2,3-DIBROMO-1-PROPANOL; DIBROMOPROPANOL; 2,3-DIBROMOPROPYL ALCOHOL; GLYCEROL 1,2-DIBROMOHYDRIN; 1-PROPANOL,2,3-DIBROMO-
Description: colorless liquid
Use: as an intermediate in preparation of flame retardants, insecticides and pharmaceuticals

Physical Properties

Boiling Point: 219 °C (426 °F)
Specific Gravity: 2.12 at 20 °C/4 °C
Density: 2.12 g/mL
Vapor Pressure: 1 mm Hg at 57.0 °C
Water Solubility: 50 to 100 mg/mL at 20 °C
Other Solubilities: 95% Ethanol: >=100 mg/ml at 20 °C; Acetone: >=100 mg/ml at 20 °C; Alcohol: miscible; Benzene: miscible; DMSO: >=100 mg/ml at 20 °C; Ether: miscible.
Refraction Index: 1.5599
Flash Point: > 112 °C

RTECS Toxicity Data

Chronic (Multiple Dose) Dermal: Rat Route: Skin; Dose: 11505 mg/kg/13W-I; Toxic Effects: Kidney, Ureter, and Bladder - Chgs in tubules (inc acute renal failure, acute tubular necrosis.

Mutagenic: Rat DNA Damage; Cell Type: other cell types; Dose: 1 umol/L. Rat Unscheduled DNA Synthesis; Cell Type: liver; Dose: 100 umol/L.

Tumorigenic: Rat Route: Skin; Dose: 47940 mg/kg/51W-I; Toxic Effects: Tumorigenic - Carcinogenic by RTECS criteria; Gastrointestinal - Tumors; Skin and appendages - Tumors. Mouse Route: Skin; Dose: 37170 mg/kg/42W-I; Toxic Effects: Tumorigenic - Carcinogenic by RTECS criteria; Gastrointestinal - Tumors; Liver - Tumors.

Hazard Overviews

Health: Irritating to eyes/skin/respiratory tract. Toxic. Other Acute Effects: harmful by inhalation, in contact with skin and if swallowed. Chronic Effects: possible risk of irreversible effects. Possible human carcinogen. Mutagenic effects shown in laboratory experiments.

Fire: Will burn. Hazards: emits toxic fumes. Extinguishing agents: water spray; carbon dioxide, dry chemical powder or appropriate foam. Precautions: combustible liquid.

Carcinogenicity: IARC - Not listed; NIOSH - Not listed; NTP - Not listed; ACGIH - Not listed; OSHA - Not listed; EPA - Not listed; MAK - Not listed

Primary Target Organs:

Eyes

Skin

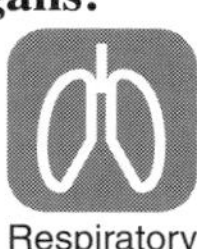
Respiratory System

Environmental

Ecotoxicity: RubisCo test: inhib. of enzym. activity of Ribulose-P2-carboxylase in protoplasts IC_{10}: 654 mg/l. O2 test: inhib. of oxygen production of protoplasts IC_{10}: 327 mg/l.

Regulations

RCRA 40CFR: Not listed
CERCLA: 40CFR 302.4: Not listed
SARA 40CFR 372.65: Not listed
SARA EHS 40CFR 355: Not listed
TSCA: Listed

DIB5180 CAS #: 124-73-2

1,2-DIBROMOTETRAFLUOROETHANE

RTECS: KH9370300
EINECS Number: 204-711-9
Molecular Formula: $C_2Br_2F_4$
Formula Weight: 259.82

Br, F, F, F, F, Br

Chemical Structure

Synonyms: 1,2-DIBROMOPERFLUOROETHANE; 1,2-DIBROMO-1,1,2,2-TETRAFLUOROETHANE; DIBROMOTETRAFLUOROETHANE; SYM-DIBROMOTETRAFLUOROETHANE; ETHANE,1,2-DIBROMO-1,1,2,2-TETRAFLUORO-; ETHANE,1,2-DIBROMOTETRAFLUORO-; ETHANE,1,2-DIBROMO-1,1,2,2-TETRAFLUORO-(9CI); F-114B2; FC 114B2; FLUOBRENE; FREON 114B2; FREON-114B2; HALON 2402; KHLADON 114B2; R 114B2

Description: liquid

Use: refrigerant; fire-extinguishing agent; control fluid

Physical Properties

Boiling Point: 47.3 °C (117 °F)
Freezing Point: -112 °C (-169.6 °F)
Specific Gravity: 2.18 at 21.1 °C
Vapor Pressure: 3.79 kPa at -25 °C
Surface Tension: 18 dyne/cm at 25 °C
Refraction Index: 1.361
Critical Temperature: 214.5 °C
Critical Pressure: 34 atm
Ionization Potential (eV): 11.1 +/-1.0
Flash Point: Nonflammable

RTECS Toxicity Data

Acute Inhalation: Rat LC_{50} Dose: 869 gm/m^3/2hr. Mouse LC_{50} Dose: 300 gm/m^3/2hr; Toxic Effects: Behavioral - Somnolence (general depressed activity); Behavioral - Ataxia; Lungs, Thorax, or Respiration - Cyanosis.

Hazard Overviews

Health: Irritating to eyes/skin/respiratory tract. Other Acute Effects: may be harmful by inhalation, ingestion, or skin absorption; target organ: heart.

Fire: Noncombustible. Hazards: emits toxic fumes. Extinguishing agents: noncombustible; use extinguishing media appropriate to surrounding fire conditions. Precautions: combustible liquid.

Reactivity: Incompatible with: strong oxidizing agents. Hazardous decomposition products: toxic fumes of: carbon monoxide, carbon dioxide, hydrogen bromide gas, hydrogen fluoride.

Carcinogenicity: IARC - Not listed; NIOSH - Not listed; NTP - Not listed; ACGIH - Not listed; OSHA - Not listed; EPA - Not listed; MAK - Not listed

Primary Target Organs:

Environmental

Environmental Fate: It is not expected to react with photochemically produced hydroxyl radicals in the ambient atmosphere as similarly structured compounds have half-lives greater than 44 years for this reaction. If released to soil, it is expected to have low mobility. Rapid volatilization should occur from terrestrial surfaces. If released to water, it is expected to occur rapidly and be the dominant fate process. Bioconcentration in aquatic organisms will not be an important fate process.

Environmental Physical Data

Henry's Law Constant: estimated at 0.1616
Octanol/Water Partition Coefficient: log K_{ow} = 3.13
Sorption Partition Coefficient: K_{oc} = 1202
BCF: estimated at 141.3

Regulations

RCRA 40CFR: Not listed
CERCLA: 40CFR 302.4: Not listed
SARA 40CFR 372.65: Listed
SARA EHS 40CFR 355: Not listed
TSCA: Listed

DIB5290 CAS #: 85-79-0

DIBUCAINE

RTECS: GD3150000
EINECS Number: 201-632-1
Molecular Formula: $C_{20}H_{29}N_3O_2$
Formula Weight: 343.47

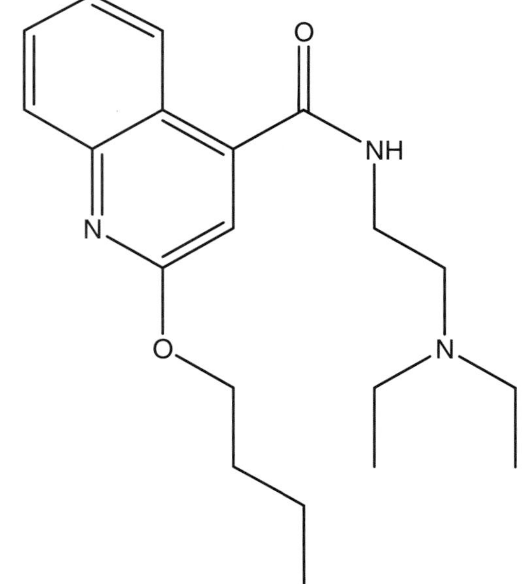

Chemical Structure

Synonyms: 2-BUTOXY-N-(2-(DIETHYLAMINO)ETHYL)CINCHONINAMIDE; 2-BUTOXY-N-(BETA-DIETHYLAMINOETHYL)CINCHONINAMIDE; 2-N-BUTOXY-N-(2-DIETHYLAMINOETHYL)CINCHONINAMIDE; 2-BUTOXYQUINOLINE-4-CARBOXYLIC ACID DIETHYLAMINOETHYLAMIDE; ALPHA-BUTYLOXYCINCHONIC ACID-GAMMA-DIETHYLETHYLENEDIAMINE; ALPHA-BUTYLOXYCINCHONINIC ACID DIETHYLETHYLENEDIAMIDE; CINCHOCAINE; CINCHONINAMIDE,2-BUTOXY-N-(2-(DIETHYLAMINO)ETHYL)-; DERMACAINE; DIBUCAIN; DIBUCAINE BASE; N-(2-(DIETHYLAMINO)ETHYL)-2-BUTOXYCINCHONINAMIDE; NUPERCAINAL; NUPERCAINE; PERCAMINE; 4-QUINOLINECARBOXAMIDE,2-BUTOXY-N-(2-(DIETHYLAMINO)ETHYL-; 4-QUINOLINECARBOXAMIDE,2-BUTOXY-N-(2-(DIETHYLAMINO)ETHYL)-(9CI); SOVCAINE
Description: colorless or almost colorless powder; odorless
Use: medication: local anesthetic; medication (vet): local anesthetic; diagnostic agent

Physical Properties

Boiling Point: Decomposes at 90 °C (194 °F) to 98 °C (208.4 °F)
Freezing Point: 64 °C (147.2 °F)
Water Solubility: Insoluble in Water
Other Solubilities: Soluble in Acetone.
pH: Faintly alkaline (aqueous solution 1 in 20)

RTECS Toxicity Data

Acute Oral: Child LD_{Lo} Dose: 50 mg/kg.
Acute Dermal: Rabbit LD_{50} Route: Subcutaneous Dose: 8500 ug/kg; Toxic Effects: Behavioral - Somnolence (general depressed activity); Behavioral - Convulsions or effect on seizure threshold; Behavioral - Excitment. Mouse LD_{50} Route: Subcutaneous Dose: 28500 ug/kg.

Hazard Overviews

Health: Severely irritating to eyes/skin. Other Acute Effects: may be harmful by inhalation, ingestion, or skin absorption; prolonged or repeated exposure may cause allergic reactions in certain sensitive individuals; extremely potent anesthetic agent; symptoms of overexposure may include headache, nausea, vomiting, hypotension and irregular heart rate. Chronic Effects: target organs: central nervous system, cardiovascular system.
Fire: Hazards: emits toxic fumes. Extinguishing agents: carbon dioxide, dry chemical powder or appropriate foam; water spray. Precautions: combustible liquid.
Reactivity: Stable. Hazardous polymerization will not occur. Hazardous decomposition products: thermal decomposition may produce carbon monoxide, carbon dioxide, and nitrogen oxides.
Carcinogenicity: IARC - Not listed; NIOSH - Not listed; NTP - Not listed; ACGIH - Not listed; OSHA - Not listed; EPA - Not listed; MAK - Not listed

Primary Target Organs:

Eyes

Skin

Nervous System

Cardio-vascular

Environmental

Regulations

RCRA 40CFR: Not listed
CERCLA: 40CFR 302.4: Not listed
SARA 40CFR 372.65: Not listed
SARA EHS 40CFR 355: Not listed
TSCA: Not listed

DIB5400 CAS #: 126-15-8

2,3:4,5-DI(2-BUTENYL)TETRAHYDROFURFURAL

RTECS: HP5075000
EINECS Number: 204-773-7
Molecular Formula: $C_{13}H_{16}O_2$
Formula Weight: 204.29
Synonyms: AC-R-11; 2,3,4,5-BIS(DELTA(SUP 2)-BUTENYLENE)TETRAHYDROFURFURAL; 2,3,4,5-BIS(DELTA(SUP 2)-BUTYLENE)TETRAHYDROFURFURAL; BIS-DELTA(SUP 2)-BUTYLENETETRAHYDROFURFURAL; 2,3:4,5-BIS(2-BUTENE-1,4-DIYL)TETRAHYDROFURFURAL; 2,3,4,5-BIS(2-BUTENYLENE)TETRAHYDROFURFURAL; 2,3,4,5-BIS(DELTA(2)-BUTENYLENE)TETRAHYDROFURFURAL; BISBUTENYLENETETRAHYDROFURFURAL; 2,3,4,5-BIS(2-BUTYLENE TETRAHYDROFURFURAL); 2,3,4,5-BIS(2-BUTYLENE)TETRAHYDRO-2-FURALDEHYDE; 2,3,4,5-BIS(DELTA(2)-BUTYLENE)TETRAHYDROFURFURAL; 2,3:4,5-BIS(2-BUTYLENE)TETRAHYDRO-2-FURFURAL; BUTADIEN-FURFURAL COPOLYMER; 4A(4H)-DIBENZOFURANCARBOXALDEHYDE,1,5A,6,9,9A,9B-HEXAHYDRO-; DIBUTYLENE TETRAFURFURAL; ENT 17,596; 4A-FORMYL-1,4,4A,5A,6,9,9A,9B-OCTAHYDRODIBENZOFURAN; 2-FURALDEHYDE,2,3:4,5-BIS(2-BUTENYLENE)TETRAHYDRO-; 2-FURANCARBOXALDEHYDE,2,3:4,5-BIS(2-BUTENE-1,4-DIYL)TETRAHYDRO-; 1,5A,6,9,9A,9B-HEXAHYDRO-4A(4H)-DIBENZOFURANCARBOXALDEHYDE; MGK 11; MGK REPELLENT 11; MGK REPELLENT II; PHILLIPS R-11; PHILLIPS REPELLENT 11; R-11
Description: pale yellow liquid; fruity odor
Use: insect repellent

Physical Properties

Boiling Point: 307 °C (585 °F)
Freezing Point: -80 °C (-112 °F)
Specific Gravity: 1.1 at 20 °C/4 °C
Water Solubility: Practically Insoluble in Water
Other Solubilities: Practically Insoluble in dilute alkali miscible with petroleum oils, Toluene, Xylene, Ethanol

RTECS Toxicity Data

Acute Oral: Rat LD_{50} Dose: 2500 mg/kg.
Acute Dermal: Rabbit LD_{50} Route: Skin; Dose: >2 gm/kg.

Hazard Overviews

Carcinogenicity: IARC - Not listed; NIOSH - Not listed; NTP - Not listed; ACGIH - Not listed; OSHA - Not listed; EPA - Not listed; MAK - Not listed

Environmental

Regulations

RCRA 40CFR: Not listed
CERCLA: 40CFR 302.4: Not listed
SARA 40CFR 372.65: Not listed
SARA EHS 40CFR 355: Not listed
TSCA: Listed

DIB5510 CAS #: 13170-23-5

DI-T-BUTOXYDIACETOXYSILANE

EINECS Number: 236-112-3

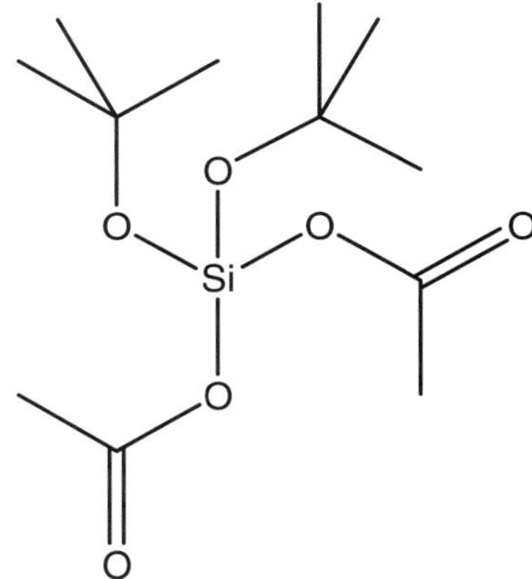

Chemical Structure

Physical Properties

Boiling Point: 102 °C (216 °F)
Freezing Point: 95 °C (203 °F)
Specific Gravity: 1.02
Vapor Density: > 1 Air=1
Water Solubility: Reacts
Odor Threshold: 1 ppm
pH: < 7
Refraction Index: 1.404
Evaporation Rate: < 1 Butyl Acetate=1

Hazard Overviews

Corrosive

Health: Corrosive to eyes/skin/respiratory tract. Other Acute Effects: harmful if swallowed, inhaled, or absorbed through skin; inhalation may result in spasm, inflammation and edema of the larynx and bronchi, chemical pneumonitis and pulmonary edema; symptoms of exposure may include burning sensation; coughing; wheezing; laryngitis; shortness of breath; headache; nausea; vomiting.

Fire: Hazards: emits toxic fumes. Extinguishing agents: carbon dioxide, dry chemical powder or appropriate foam. Precautions: combustible liquid.

Reactivity: Stable. Hazardous polymerization will not occur. Incompatible with: acids, bases, alcohols, amines, strong oxidizing agents, contact with water or moist air generates acetic acid and tert-butyl alcohol. Hazardous decomposition products: toxic fumes of: carbon monoxide, carbon dioxide, silicon oxide.

Carcinogenicity: IARC - Not listed; NIOSH - Not listed; NTP - Not listed; ACGIH - Not listed; OSHA - Not listed; EPA - Not listed; MAK - Not listed

Primary Target Organs:

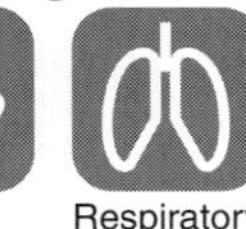

Eyes Skin Respiratory System

Environmental

Regulations

RCRA 40CFR: Not listed
CERCLA: 40CFR 302.4: Not listed
SARA 40CFR 372.65: Not listed
SARA EHS 40CFR 355: Not listed
TSCA: Not listed

DIB5620 CAS #: 112-48-1

1,2-DIBUTOXYETHANE

RTECS: KH9450000
EINECS Number: 203-976-8
Molecular Formula: $C_{10}H_{22}O_2$
Structured MF: $C_4H_9OC_2H_4OC_4H_9$
Formula Weight: 174.32
Synonyms: BUTANE,1,1'-(1,2-ETHANEDIYLBIS(OXY))BIS-; DIBUTYL CELLOSOLVE; DIBUTYL OXITOL; DIBUTYLETHER ETHYLENGLYKOLU; ETHANE,1,2-DIBUTOXY-; ETHYLENE GLYCOL DIBUTYL; ETHYLENE GLYCOL DIBUTYL ETHER
Description: colorless liquid

Physical Properties

Boiling Point: 203.6 °C (398 °F) at 760 mm Hg
Freezing Point: -69 °C (-92.2 °F)
Specific Gravity: 0.837 at 25 °C/25 °C
Vapor Density: 6.01 Air=1
Saturated Vapor Density: 1.200712094 kg/m^3
Bulk Density: 7.0 lbs/gal at 20 °C
Vapor Pressure: 0.09 mm Hg at 20 °C
Water Solubility: Slightly Soluble
Flash Point: 85 °C Open Cup

RTECS Toxicity Data

Acute Oral: Rat LD_{50} Dose: 3250 mg/kg.
Acute Dermal: Rabbit LD_{50} Route: Skin; Dose: 3560 uL/kg.
Irritation Eye: Rabbit Standard Draize Test Dose: 500 mg/24H; Reaction: mild.
Irritation Skin: Rabbit Open Draize Test Dose: 500 mg open; Reaction: mild.

Hazard Overviews

Fire: Combustible.
Carcinogenicity: IARC - Not listed; NIOSH - Not listed; NTP - Not listed; ACGIH - Not listed; OSHA - Not listed; EPA - Not listed; MAK - Not listed

Environmental

Cleanup/Disposal: Guide No. 128: Eliminate all ignition sources (no smoking, flares, sparks or flames in immediate area). All equipment used when handling the product must be grounded. Do not touch or walk through spilled material. Stop leak if you can do it without risk. Prevent entry into waterways, sewers, basements or confined areas. A vapor suppressing foam may be used to reduce vapors. Absorb or cover with dry earth, sand or other non-combustible material and transfer to containers. Use clean non-sparking tools to collect absorbed material. Large Spills: Dike far ahead of liquid spill for later disposal. Water spray may reduce vapor; but may not prevent ignition in closed spaces.

Regulations

RCRA 40CFR: Not listed
CERCLA: 40CFR 302.4: Not listed
SARA 40CFR 372.65: Not listed
SARA EHS 40CFR 355: Not listed
TSCA: Not listed

DIB5730 CAS #: 1929-73-3

2,4-DIBUTOXYETHYL ESTER

RTECS: AG7700000
DOT: UN2765; UN2766; UN2999; UN3000; NA2765; IMO3.2; IMO6.1
EINECS Number: 217-680-1
Molecular Formula: $C_{14}H_{18}Cl_2O_4$
Formula Weight: 311.2
Synonyms: ACETIC ACID,(2,4-DICHLOROPHENOXY)-,2-BUTOXYETHYL ESTER; ACETIC ACID,(2,4-DICHLOROPHENOXY)-,2-BUTOXYETHYL ESTER(8CI,9CI); AQUA-KLEEN; BLADEX-B; BRUSH KILLER 64; BUTOXY-D 3; BUTOXYETHANOL ESTER OF (2,4-DICHLOROPHENOXY) ACETIC ACID; BUTOXYETHYL 2,4-DICHLOROPHENOXYACETATE; 2,4-D BUTOXYETHANOL ESTER; 2,4-D 2-BUTOXYETHYL ESTER; 2,4-D BUTOXYETHYL ESTER; 2,4-D-BEE; 2,4-DBEE; 2,4-DICHLOROPHENOXYACETIC ACID BUTOXYETHANOL ESTER; (2,4-DICHLOROPHENOXY)ACETIC ACID BUTOXYETHYL ESTER; 2,4-DICHLOROPHENOXYACETIC ACID,BUTOXYETHYL ESTER; ESTERON 99 CONCENTRATE; LO-ESTASOL; PLANOTOX; SILVAPROP 1; WEEDONE 638; WEEDONE 100 EMULSIFIABLE; WEEDONE LV-6; WEEDONE LV 4; WEEDONE LV4; WEED-RHAP LV-4D

Description: colorless to amber liquid; odorless when pure

Physical Properties

Boiling Point: 185 °C (365 °F) to 190 °C (374 °F) at 5.5-7 mm Hg
Specific Gravity: 1.225 at 68 °F
Vapor Pressure: 1.7 x10^{-3} at 25 °C
Water Solubility: Generally Immiscible or Insoluble in Water
Other Solubilities: Soluble in oils
Flash Point: > 79.444 °C Open Cup

RTECS Toxicity Data

Acute Oral: Rat LD_{50} Dose: 831 mg/kg.
Acute Dermal: Rabbit LD_{50} Route: Skin; Dose: >2 gm/kg.
Reproductive/Teratogenic: Rat Route: Oral; Dose: 1500 mg/kg; Duration: female 6-15D of pregnancy; Specific Developmental Abnormalities - Musculoskeletal system.

Hazard Overviews

Fire: Combustible.
Carcinogenicity: IARC - Not listed; NIOSH - Not listed; NTP - Not listed; ACGIH - Not listed; OSHA - Not listed; EPA - Not listed; MAK - Not listed

Environmental

Ecotoxicity: LC_{50} Rainbow trout 1.5-1.6 mg/l/24 hr /Static bioassay LC_{50} Phasianus colchicus (ring-necked pheasant) oral greater than 5000 ppm in 5-day diet LC_{50} Pimephales promelas (fathead minnow) 1500 ug/l/48 hr (lethal to eggs) /Conditions of bioassay not specified LC_{50} Palaemonetes kadiakensis (glass shrimp) 1400 ug/l/48 hr /Conditions of bioassay not specified LC_{50} Gammarus lacustris (scud) 440 ug/l/96 hr /Conditions of bioassay not specified
Environmental Fate: If released on land, it should adsorb strongly to the soil and degrade by microbially mediated or chemical hydrolysis to the free acid, 2,4-D. This process may take days or much longer depending on local conditions. If released in water, the ester should rapidly biodegrade without a lag period to the free acid (half-life 1 hr to 1 day). Chemical hydrolysis may be important in alkaline water and photolysis may be important in clear surface waters which have low microbial populations. The ester will adsorb strongly to sediment where it will degrade more slowly. The butoxyethyl ester will be rapidly taken up by plants, algae, and fish but will not bioconcentrate because the ester is readily metabolized and excreted. In the atmosphere, the ester will exist both as an aerosol and as the vapor. The aerosol will be subject to gravitational settling, drift and photolysis (half-life 14 days in summer). The vapor-phase ester will react with photochemically produced hydroxyl radicals (half-life 8.5 hr) and photolyze.
Cleanup/Disposal: Guide No. 131: Fully encapsulating, vapor protective clothing should be worn for spills and leaks with no fire. Eliminate all ignition sources (no smoking, flares, sparks or flames in immediate area). All equipment used when handling the product must be grounded. Do not touch or walk through spilled material. Stop leak if you can do it without risk. Prevent entry into waterways, sewers, basements or confined areas. A vapor suppressing foam may be used to reduce vapors. Small Spills: Absorb with earth, sand or other non-combustible material and transfer to containers for later disposal. Use clean non-sparking tools to collect absorbed material. Large Spills: Dike far ahead of liquid spill for later disposal. Water spray may reduce vapor; but may not prevent ignition in closed spaces. Guide No. 152: Do not touch damaged containers or spilled material unless wearing appropriate protective clothing. Stop leak if you can do it without risk. Prevent entry into waterways, sewers, basements or confined areas. Cover with plastic sheet to prevent spreading. Absorb or cover with dry earth, sand or other non-combustible material and transfer to containers. Do not get water inside containers.

Environmental Physical Data

Sorption Partition Coefficient: K_{oc} = estimated at 1100
BCF: catfish 7 to 55

Regulations

RCRA 40CFR: Not listed
CERCLA: 40CFR 302.4: Listed per CWA Section 311(b)(4) RQ: 100 lb (45.35 kg)
SARA 40CFR 372.65: Listed
SARA EHS 40CFR 355: Not listed
TSCA: Not listed

Analytical Methods

Soil: SW846 8321A
Water / Groundwater: EPA X_89_176, 022
Food: FDA 212.1, 232.1
Plasma: EPA 001; FDA 252

DIB5840 CAS #: 2917-73-9

DIBUTYL AZELATE

EINECS Number: 220-850-8
Molecular Formula: $C_{17}H_{32}O_4$
Formula Weight: 300.44
Synonyms: AZELAIC ACID,DIBUTYL ESTER; ERGOPLAST AZDB; NONANEDIOIC ACID,DIBUTYL ESTER
Description: liquid
Use: constituent of contact lens; plasticizer

Physical Properties

Boiling Point: 336 °C (637 °F)
Water Solubility: Insoluble in Water
Other Solubilities: Chloroform: Slightly soluble

Hazard Overviews

Carcinogenicity: IARC - Not listed; NIOSH - Not listed; NTP - Not listed; ACGIH - Not listed; OSHA - Not listed; EPA - Not listed; MAK - Not listed

Environmental

Regulations

RCRA 40CFR: Not listed

CERCLA: 40CFR 302.4: Not listed
SARA 40CFR 372.65: Not listed
SARA EHS 40CFR 355: Not listed
TSCA: Listed

DIB5950 CAS #: 78-46-6

DIBUTYL BUTYLPHOSPHONATE

RTECS: SZ7000000
EINECS Number: 201-119-2
Molecular Formula: $C_{12}H_{27}O_3P$
Formula Weight: 250.36

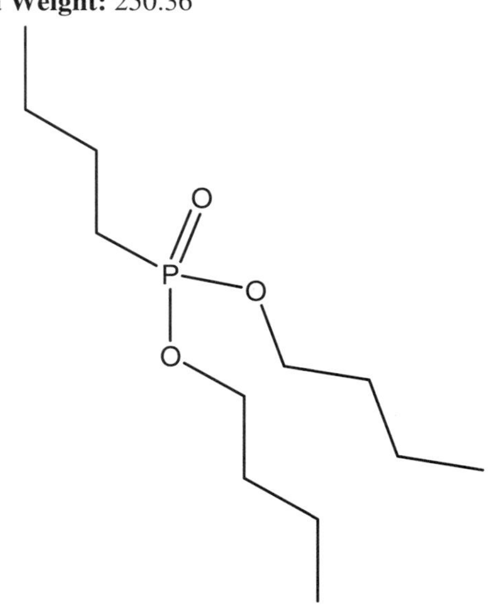

Chemical Structure

Synonyms: DIBUTYL BUTANEPHOSPHONATE; NSC 2666; PHOSPHONIC ACID,BUTYL-,DIBUTYL ESTER
Description: colorless liquid; mild odor
Use: textile conditioner & antistatic agent; solvent extraction agent for uranium & other heavy metals, gasoline additive, antifoam agent, plasticizer, textile conditioner, antistatic agent

Physical Properties

Boiling Point: 127 °C (261 °F) to 128 °C (262 °F) at 2.5 mm Hg
Specific Gravity: 0.948 at 20 °C/4 °C
Vapor Density: 8.62
Vapor Pressure: < .1
Water Solubility: Insoluble in Water
Other Solubilities: miscible with most common organic solvents
Flash Point: 154.444 °C Open Cup

Hazard Overviews

Health: May be irritating to eyes/skin/respiratory tract. Other Acute Effects: may be harmful by inhalation, ingestion, or skin absorption.
Fire: Will burn. Hazards: emits toxic fumes. Extinguishing agents: water spray; carbon dioxide, dry chemical powder or appropriate foam. Precautions: combustible liquid.
Reactivity: Stable. Hazardous polymerization will not occur. Incompatible with: strong oxidizing agents. Hazardous decomposition products: toxic fumes of: carbon monoxide, carbon dioxide, phosphorous oxides.
Carcinogenicity: IARC - Not listed; NIOSH - Not listed; NTP - Not listed; ACGIH - Not listed; OSHA - Not listed; EPA - Not listed; MAK - Not listed

Environmental

Regulations

RCRA 40CFR: Not listed
CERCLA: 40CFR 302.4: Not listed
SARA 40CFR 372.65: Not listed
SARA EHS 40CFR 355: Not listed
TSCA: Listed

DIB6060 CAS #: 94-80-4

2,4-DIBUTYL ESTER

RTECS: AG8050000
DOT: UN2765; UN2766; UN2999; UN3000; NA2765; IMO3.2; IMO6.1
EINECS Number: 202-364-8
Molecular Formula: $C_{12}H_{14}Cl_2O_3$
Formula Weight: 277.1
Synonyms: ABCO W.K-67 2,4-D WEED KILLER; ACETIC ACID,(2,4-DICHLOROPHENOXY)-,BUTYL ESTER; AMOCO 2,4-D WEED KILLER NO 6B; ASSOCIATED SALES 4-POUND 2,4-D ESTER WEED KILLER; BARBER'S 2,4-D ESTER WEED KILLER; BARCO WEED KILLER (ESTER FORMULATION); BUTAPON; 2,4-D-BUTYL; BUTYL 2,4-D; BUTYL 400; BUTYL (2,4-DICHLOROPHENOXY)ACETATE; BUTYL DICHLOROPHENOXYACETATE; 2,4-D BUTYL ESTER; 2,4-D N-BUTYL ESTER; 2,4-D,BUTYL ESTER; BUTYL ESTER 2,4-D; BUTYL ESTER OF 2,4-D; BUTYL(2,4-DICHLOROPHENOXY) ACETATE; N-BUTYLESTER KYSELINY 2,4-DICHLORFENOXYOCTOVE; CHIPMAN 2,4-D BUTYL ESTER 334E; CHIPMAN 2,4-D BUTYL ESTER 4E; CHIPMAN 2,4-D BUTYL ESTER 6E; COMPONENT ORANGE; CROP RIDER 2.67D; CROP RIDER 6D-OS WEED KILLER; CROP RIDER 6D WEED KILLER; 2,4-DBE; DE-PESTER DED-WEED ME-4; DE-PESTER DED-WEED ME-5; DE-PESTER DED-WEED ME-6; DE-PESTER DED-WEED ME-9; DIAMOND SHAMROCK BUTYL 4D; DIAMOND SHAMROCK BUTYL 6D WEED KILLER; 2,4-DICHLOROPHENOXYACETIC ACID BUTYL ESTER; 2,4-DICHLOROPHENOXYACETIC ACID N-BUTYL ESTER; (2,4-DICHLOROPHENOXY)ACETIC ACID,BUTYL ESTER; 2,4-DICHLOROPHENOXYACETIC ACID,BUTYL ESTER; ESSO HERBICIDE 10; ESTERON 99 CONCENTRATE; FELCO BUTYL ESTER 600 2,4-D WEED KILLER; FELCO HV2 WEED KILLER; FELCO HV4 WEED KILLER; FENCE RIDER 4T BRUSH KILLER; FENCE RIDER 6T BRUSH KILLER; FERNESTA; FLORATOX 428 4-POUND 2,4-D ESTER WEED KILLER; FS ESTER 400 WEED KILLER; GENERAL CHEMICAL 2,4-D 3.34 BUTYL ESTER WEED KILLER; GENERAL CHEMICAL 2,4-D 4-BUTYL ESTER WEED KILLER; GENERAL CHEMICAL 2,4-D 6.00 BUTYL ESTER WEED KILLER; GENERAL CHEMICAL 2,4-D BUTYL ESTER WEED KILLER; GREEN CROSS WEED-NO-MORE; HI-ESTER 2,4-D; LINE RIDER 4T; LIRONOX; LIRONOX 326; LO-ESTASOL; MFA 40% BUTYL ESTER WEED KILLER; MFA NO 4 WEED KILLER; MFA NO 6 WEED KILLER; MILLER'S 4# ESTER; MILLER'S 6# ESTER;

MONSANTO 2,4-D BUTYL ESTER; MONSANTO 2,4-D BUTYL ESTER CONCENTRATE; OLIN BUTYL ESTER D267 WEED KILLER; OLIN BUTYL ESTER D4 WEED KILLER; OLIN BUTYL ESTER D6 WEED KILLER; ORANGE II COMPONENT; PARSONS 2,4-D WEED KILLER BUTYL ESTER; PATTERSON'S HI-TEST BUTYL ESTER 2,4-D WEED KILLER; PURPLE COMPONENT; RHODIA 2,4-D BUTYL ESTER 6L; SHELL 40; SURE DEATH 40% BUTYL ESTER TYPE WEEDKILLER; SURE DEATH NO 4 BUTYL ESTER WEED KILLER; SURE DEATH NO 6 BUTYL ESTER WEED KILLER; T-H DED-WEED ME-6; T-H DED-WEED ME-9; TECHNE 40% BUTYL ESTER TYPE WEED KILLER; TECHNE BUTYL ESTER WEED KILLER NO 4; TECHNE BUTYL ESTER WEED KILLER NO 6; THE CROP RIDER; UNICO 2,4-D ESTER WEED KILLER; WEEDONE AERO CONCENTRATE; WEEDONE AERO-CONCENTRATE 96; WEEDONE AERO-CONCENTRATE E; WEEDONE LV-6; WEEDONE LV4; WEED-RHAP B-2.67D; WEED-RHAP B-4D; WEED-RHAP B-6D; WEED-RHAP LV-4D

Description: colorless to light brown, clear liquid; fuel oil-like odor

Use: selective herbicide and defoliant

Physical Properties

Boiling Point: 154 °C (309 °F) to 162 °C (324 °F) at 2 mm Hg
Freezing Point: 9 °C (48.2 °F)
Specific Gravity: 1.235 to 1.245 at 20/20 °C
Saturated Vapor Density: 1.200005268 kg/m^3
Vapor Pressure: 3.9 $x10^{-4}$ mm Hg at 25 °C
Water Solubility: 1 10 mg/L at 25 °C (distilled Water)
Other Solubilities: 95% Ethanol: >=100 mg/ml at 21 °C; Acetone: >=100 mg/ml at 21 °C; DMSO: >=100 mg/ml at 21 °C.
Flash Point: > 79.444 °C Open Cup

RTECS Toxicity Data

Acute Oral: Rat LD_{50} Dose: 600 mg/kg. Mouse LD_{50} Dose: 425 mg/kg.
Acute Dermal: Rabbit LD_{50} Route: Skin; Dose: >2 gm/kg.
Chronic (Multiple Dose) Oral: Rat Dose: 196 mg/kg/28W-I; Toxic Effects: Liver - Fatty liver degeneration; Liver - Other changes.
Chronic (Multiple Dose) Dermal: Rat Route: Subcutaneous; Dose: 600 mg/kg/4D-I; Toxic Effects: Behavioral - Ataxia.
Reproductive/Teratogenic: Rat Route: Oral; Dose: 1500 mg/kg; Duration: female 6-15D of pregnancy; Specific Developmental Abnormalities - Musculoskeletal system. Rat Route: Oral; Dose: 100 ug/kg; Duration: female 10D of pregnancy; Effects on Embryo or Fetus - Fetotoxicity; Fetal death.

Hazard Overviews

Fire: Combustible.
Carcinogenicity: IARC - Not listed; NIOSH - Not listed; NTP - Not listed; ACGIH - Not listed; OSHA - Not listed; EPA - Not listed; MAK - Not listed

Environmental

Ecotoxicity: LC_{50} Salmo clarki (Cutthroat trout) 0.78 mg/l/96 hr (95% confidence limit 0.66-0.92 mg/l) /Conditions of bioassay not specified
Environmental Fate: When released into water, it will disappear by a combination of volatilization, biodegradation, and by hydrolysis and photolysis. The half-life should be < 5 days. The ester would not be expected to bioconcentrate in fish. In the soil the ester will be confined to the surface layers of soil and biodegrade to the acid in a few days at adequate moisture levels.
Cleanup/Disposal: Guide No. 131: Fully encapsulating, vapor protective clothing should be worn for spills and leaks with no fire. Eliminate all ignition sources (no smoking, flares, sparks or flames in immediate area). All equipment used when handling the product must be grounded. Do not touch or walk through spilled material. Stop leak if you can do it without risk. Prevent entry into waterways, sewers, basements or confined areas. A vapor suppressing foam may be used to reduce vapors. Small Spills: Absorb with earth, sand or other non-combustible material and transfer to containers for later disposal. Use clean non-sparking tools to collect absorbed material. Large Spills: Dike far ahead of liquid spill for later disposal. Water spray may reduce vapor; but may not prevent ignition in closed spaces. Guide No. 152: Do not touch damaged containers or spilled material unless wearing appropriate protective clothing. Stop leak if you can do it without risk. Prevent entry into waterways, sewers, basements or confined areas. Cover with plastic sheet to prevent spreading. Absorb or cover with dry earth, sand or other non-combustible material and transfer to containers. Do not get water inside containers.

Environmental Physical Data

BCF: none likely

Regulations

RCRA 40CFR: Not listed
CERCLA: 40CFR 302.4: Listed per CWA Section 311(b)(4) RQ: 100 lb (45.35 kg)
SARA 40CFR 372.65: Listed
SARA EHS 40CFR 355: Not listed
TSCA: Not listed

Analytical Methods

Air: EPA TO-10
Water / Groundwater: EPA 022
Food: FDA 212.1, 232.1
Plasma: EPA 001; FDA 211.1, 231.1, 252

DIB6170 CAS #: 1809-19-4

DIBUTYL HYDROGEN PHOSPHITE

RTECS: HS6475000
EINECS Number: 217-316-1
Molecular Formula: $C_8H_{19}O_3P$
Formula Weight: 194.215

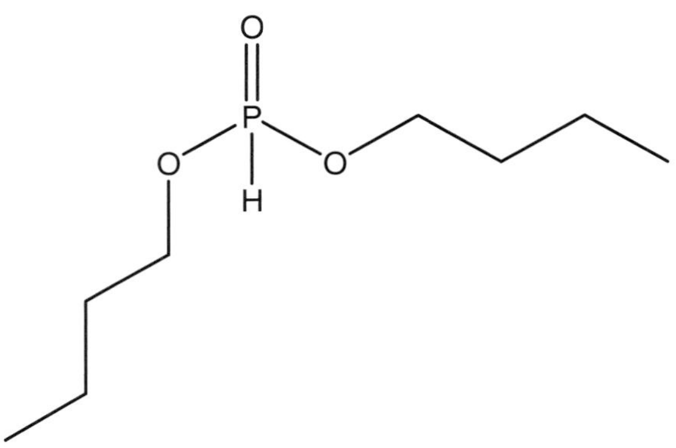

Chemical Structure

Synonyms: BUTYL ALCOHOL,HYDROGEN PHOSPHITE; BUTYL PHOSPHONATE ((BUO)2HPO); DI-N-BUTYL HYDROGEN PHOSPHITE; DIBUTYL HYDROGEN PHOSPHONATE; DIBUTYL PHOSPHITE; DIBUTYL PHOSPHONATE; DIBUTYLFOSFIT; MOBIL DBHP; PHOSPHONIC ACID,DIBUTYL ESTER; PHOSPHOROUS ACID,DIBUTYL ESTER

Description: water-white liquid; penetrating odor

Use: solvent; antioxidant; chemical intermediate

Physical Properties

Boiling Point: 95 °C (203 °F) at 1 mm Hg
Freezing Point: 120 °C (248 °F)
Specific Gravity: 0.986 at 25 °C
Vapor Density: 6.7 Air=1
Vapor Pressure: 1 mm Hg at 95 °C
Water Solubility: 7300 mg/L
Other Solubilities: Soluble in common organic solvents
Refraction Index: 1.4228 at 25 °C/D
Flash Point: 49 °C

RTECS Toxicity Data

Acute Oral: Rat LD_{50} Dose: 3200 mg/kg.
Acute Inhalation: Rat LC_{50} Dose: >20 gm/m^3.
Acute Dermal: Rabbit LD_{50} Route: Skin; Dose: 1990 mg/kg.
Irritation Eye: Rabbit Standard Draize Test Dose: 250 ug open; Reaction: severe.
Irritation Skin: Rabbit Standard Draize Test Dose: 500 mg/24H; Reaction: mild. Rabbit Open Draize Test Dose: 10 mg/24H open; Reaction: mild.

Hazard Overviews

Corrosive

Flammable

Health: Corrosive to eyes/skin/respiratory tract. Other Acute Effects: harmful if swallowed, inhaled, or absorbed through skin; inhalation may result in spasm, inflammation and edema of the larynx and bronchi, chemical pneumonitis and pulmonary edema; symptoms of exposure may include burning sensation; coughing; wheezing; laryngitis; shortness of breath; headache; nausea; vomiting; target organs: nerves, blood.

Fire: Flammable. Hazards: emits toxic fumes. Extinguishing agents: carbon dioxide, dry chemical powder or appropriate foam. Precautions: combustible liquid.

Reactivity: Incompatible with: strong oxidizing agents, strong bases, may decompose on exposure to moist air or water. Hazardous decomposition products: toxic fumes of: carbon monoxide, carbon dioxide; thermal decomposition may produce toxic fumes of phosphorus oxides and/or phosphine.

Carcinogenicity: IARC - Not listed; NIOSH - Not listed; NTP - Not listed; ACGIH - Not listed; OSHA - Not listed; EPA - Not listed; MAK - Not listed

Primary Target Organs:

Eyes

Skin

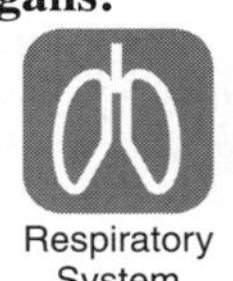
Respiratory System

Nervous System

Blood

Environmental

Environmental Fate: It will hydrolyze in water and moist soil. The hydrolysis rate increases with increasing pH. There is a discrepancy in the hydrolytic half-life at neutral pH derived from the literature. They range from 2.2 days to 60.7 days. It is not known whether it will biodegrade. It would not be expected to adsorb strongly to soil and therefore may leach into the ground. It is not very volatile and therefore would not partition into the atmosphere. It may photodegrade in the atmosphere as well as in surface water, although no environmental rates are available. It would not be expected to bioconcentrate in aquatic organism.

Environmental Physical Data

BCF: none likely

Regulations

RCRA 40CFR: Not listed
CERCLA: 40CFR 302.4: Not listed
SARA 40CFR 372.65: Not listed
SARA EHS 40CFR 355: Not listed
TSCA: Listed

DIB6280 CAS #: 502-56-7

DIBUTYL KETONE

RTECS: RA8230000
EINECS Number: 207-946-5
Molecular Formula: $C_9H_{18}O$
Structured MF: $CH_3(CH_2)_3CO(CH_2)_3CH_3$
Formula Weight: 142.24

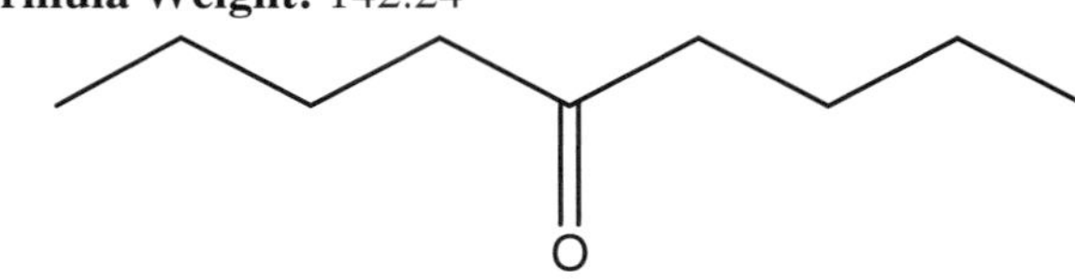

Chemical Structure

Synonyms: BUTYL KETONE; DI-N-BUTYL KETONE; 5-NONANONE; NONAN-5-ONE; 5-OXONONANE
Description: colorless to light yellow liquid

Physical Properties

Boiling Point: 188.4 °C (371 °F)
Freezing Point: -4.8 °C (23.36 °F)
Specific Gravity: 0.8217 at 20 °C/4 °C
Vapor Pressure: 0.049 kPa at 100 °C
Water Solubility: 379 ppm at 25 °C
Other Solubilities: > 10% in Ether; > 10% in Ethanol; > 10% in Chloroform.
Surface Tension: 26.6 dynes/cm at 1.1 °C
Refraction Index: 1.4195 at 20 °C/D
Ionization Potential (eV): 9.07 +/-0.3
Flash Point: 60 °C Closed Cup

RTECS Toxicity Data

Acute Oral: Rat LD_{Lo} Dose: 1 gm/kg.
Chronic (Multiple Dose) Oral: Rat Dose: 76650 mg/kg/14W-I; Toxic Effects: Spinal Cord - Other degenerative changes; Peripheral Nerve and sensation - Spastic parapysis with or without sensory change; Nutritional and gross metabolic - Weight loss or decreased weight gain.

Hazard Overviews

Flammable

Health: Irritating to eyes/skin/respiratory tract. Other Acute Effects: may be harmful by inhalation, ingestion, or skin absorption.
Fire: Flammable. Hazards: emits toxic fumes. Extinguishing agents: carbon dioxide, dry chemical powder or appropriate foam; water spray. Precautions: combustible liquid.
Reactivity: Incompatible with: strong oxidizing agents, strong reducing agents. Hazardous decomposition products: toxic fumes of: carbon monoxide, carbon dioxide.
Carcinogenicity: IARC - Not listed; NIOSH - Not listed; NTP - Not listed; ACGIH - Not listed; OSHA - Not listed; EPA - Not listed; MAK - Not listed
Primary Target Organs:

Eyes

Skin

Respiratory System

Environmental

Ecotoxicity: LC_{50} Pimephales promelas (fathead minnow) 30 days old 31.0 mg/l/96 hr (confidence limit: 29.4-32.6 mg/l) at 25.2 °C (hardness 44.5 mg/l calcium carbonate, pH 7.70) /Purity 98%; conditions of bioassay not specified
Environmental Fate: If released to the atmosphere, it is degraded relatively rapidly by reaction with photochemically produced hydroxyl radicals (estimated half-life of 28 hours in air). If released to water, volatilization is expected to be important. Volatilization half-lives of 7.1 and 80 hr have been estimated for a model river (one meter deep) and a model environmental pond, respectively. If released to soil, it may leach significantly based upon an estimated K_{oc} value of 167. It can be expected to evaporate from dry surfaces. Insufficient experimental data are available to predict the relative importance of biodegradation in soil or water.
Cleanup/Disposal: Guide No. 127: Eliminate all ignition sources (no smoking, flares, sparks or flames in immediate area). All equipment used when handling the product must be grounded. Do not touch or walk through spilled material. Stop leak if you can do it without risk. Prevent entry into waterways, sewers, basements or confined areas. A vapor suppressing foam may be used to reduce vapors. Absorb or cover with dry earth, sand or other non-combustible material and transfer to containers. Use clean non-sparking tools to collect absorbed material. Large Spills: Dike far ahead of liquid spill for later disposal. Water spray may reduce vapor; but may not prevent ignition in closed spaces.

Environmental Physical Data

Henry's Law Constant: estimated at 2.91×10^{-4}
Octanol/Water Partition Coefficient: log K_{ow} = 3.00
Sorption Partition Coefficient: K_{oc} = 167
BCF: estimated at 22

Regulations

RCRA 40CFR: Not listed
CERCLA: 40CFR 302.4: Not listed
SARA 40CFR 372.65: Not listed
SARA EHS 40CFR 355: Not listed
TSCA: Listed

DIB6390 **CAS #: 2528-36-1**

DIBUTYL PHENYL PHOSPHATE

RTECS: TB9626600
EINECS Number: 219-772-7
Molecular Formula: $C_{14}H_{23}O_4P$
Formula Weight: 286.34
Synonyms: DBPP; DIBUTYL PHENYL ESTER PHOSPHORIC ACID; DIBUTYLPHENYL ESTER PHOSPHORIC ACID; PHOSPHORIC ACID,DIBUTYL PHENYL ESTER; PHOSPHORIC ACID,DIBUTYLPHENYL ESTER
Description: clear to slightly yellow liquid; butanolic odor
Use: organic intermediate

Physical Properties

Boiling Point: 131 °C (268 °F) to 132 °C (270 °F)
Specific Gravity: 1.0691 at 25 °C/25 °C
Saturated Vapor Density: 1.200098079 kg/m^3
Vapor Pressure: 0.007 torr at 25 °C
Water Solubility: Very Low Solubility in Water
Other Solubilities: 95% Ethanol: >=100 mg/ml at 22.5 °C; Acetone: >=100 mg/ml at 22.5 °C; DMSO: >=100 mg/ml at 22.5 °C.
Flash Point: 129 °C Closed Cup

RTECS Toxicity Data

Acute Oral: Rat LD_{50} Dose: 2140 mg/kg. Mouse LD_{50} Dose: 1790 mg/kg.
Acute Inhalation: Rat LC; Dose: >7 mg/m^3.
Acute Dermal: Rabbit LD_{50} Route: Skin; Dose: >5 gm/kg.
Reproductive/Teratogenic: Rat Route: Oral; Dose: 48 gm/kg; Duration: male 10W prior to mating; Effects on Newborn - Viability index; Weaning or lactation index; Physical.

Hazard Overviews

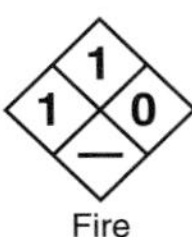

Fire Diamond

Health: Irritating to eyes/skin/respiratory tract. Also Causes: coughing, wheezing. Chronic Effects: dermatitis.
Fire: Will burn. Use dry chemical, carbon dioxide, water spray, fog, or regular foam. Container may explode in heat of fire.
Reactivity: Stable. Hazardous polymerization cannot occur. Avoid: heat; ignition sources; wet metals. Incompatible with: metals when wet. Hazardous decomposition products: carbon oxide(s); chlorine; hydrogen chloride; phosgene gas.
Carcinogenicity: IARC - Not listed; NIOSH - Not listed; NTP - Not listed; ACGIH - Not listed; OSHA - Not listed; EPA - Not listed; MAK - Not listed
Primary Target Organs:

Eyes

Skin

Respiratory System

Exposure Limits
ACGIH TLV: TWA: 0.3 ppm; 0.5 mg/m^3.
Respirator Recommendation
Exposure Range: >0.3 to 3 ppm Air Purifying, Negative Pressure, Half Mask
Exposure Range: >3 to 30 ppm Air Purifying, Negative Pressure, Full Face
Exposure Range: >30 to 300 ppm Supplied Air, Constant Flow/Pressure Demand, Full Face
Exposure Range: >300 to unlimited ppm Self-contained Breathing Apparatus, Pressure Demand, Full Face
Cartridge Color: dust/mist filter (use P100 or consult supervisor for appropriate dust/mist filter)

Environmental

Environmental Fate: If released to water, it will rapidly biodegrade under aerobic conditions (half-life of 3.5 days or less). However, biodegradation in benthic sediments is unclear. Aqueous hydrolysis may become increasingly important with increasing alkalinity; no rate data were located. It will partition from the water column to sediment and it may bioconcentrate in aquatic organisms. If released to soil, biodegradation will be the predominant fate process and aqueous hydrolysis may be important in alkaline soils. It will adsorb strongly to soil and leaching will not be important. If released to the atmosphere, vapor-phase is expected to degrade by reaction with photochemically produced hydroxyl radicals (estimated half-life of 6.5 hours). Particulate will be removed from the atmosphere via wet and dry deposition.

Environmental Physical Data

Henry's Law Constant: estimated at 5.04 x10^{-7}
Octanol/Water Partition Coefficient: log K_{ow} = 4.27
Sorption Partition Coefficient: K_{oc} = estimated at 1360 to 5010
BCF: calculated at 270 to 1100

Regulations

RCRA 40CFR: Not listed
CERCLA: 40CFR 302.4: Not listed
SARA 40CFR 372.65: Not listed
SARA EHS 40CFR 355: Not listed
TSCA: Listed

DIB6500 CAS #: 107-66-4

DIBUTYL PHOSPHATE

RTECS: TB9605000
EINECS Number: 203-509-8
Molecular Formula: $C_8H_{19}O_4P$
Structured MF: $(C_4H_9O)_2(OH)PO$
Formula Weight: 210.21

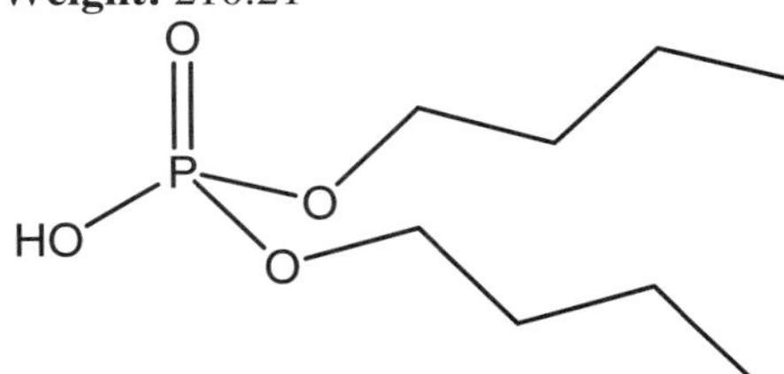

Chemical Structure

Synonyms: DIBUTYL ACID O-PHOSPHATE; DIBUTYL ACID PHOSPHATE; DI-N-BUTYL HYDROGEN PHOSPHATE; DIBUTYL HYDROGEN PHOSPHATE; DI-N-BUTYL PHOSPHATE; DIBUTYL PHOSPHORIC ACID; PHOSPHORIC ACID,DIBUTYL ESTER
Description: colorless to brown liquid; odorless
Use: org catalyst & antifoaming agent; as a catalyst in manufacture of phenolic & urea resins; in metal separation & extraction

Physical Properties

Boiling Point: Decomposes at 100 °C (212 °F)
Freezing Point: Decomposes at > 100 °C (212 °F)
Specific Gravity: 1.06
Vapor Pressure: 1 mm Hg
Water Solubility: Insoluble in Water
pH: Moderately strong monobasic acid
Flash Point: Combustible Liquid

Hazard Overviews

Fire Diamond

Health: Severely irritating to eyes/skin/respiratory tract. Also Causes: headache/irritation/burns of gastrointestinal tract upon ingestion.

Fire: Will burn. Use carbon dioxide, dry chemical, foam, or water fog. Water or foam may cause frothing. Use water spray to cool fire-exposed tanks/containers.

Reactivity: Stable. Hazardous polymerization cannot occur. Incompatible with: strong oxidizers. Hazardous decomposition products: carbon oxides; oxides of phosphorous.

Carcinogenicity: IARC - Not listed; NIOSH - Listed as carcinogen; NTP - Not listed; ACGIH - Not listed; OSHA - Not listed; EPA - Not listed; MAK - Not listed

Primary Target Organs:

Eyes

Skin

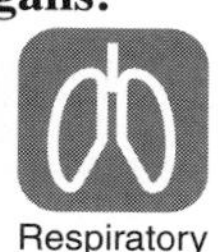
Respiratory System

Exposure Limits

OSHA PEL: TWA: 1 ppm; 5 mg/m^3.

OSHA PEL Vacated 1989 Limits: TWA: 1 ppm; 5 mg/m^3; STEL: 2 ppm; 10 mg/m^3.

ACGIH TLV: TWA: 1 ppm; 8.6 mg/m^3.

NIOSH REL: TWA: 1 ppm; 5 mg/m^3. STEL: 2 ppm; 10 mg/m^3.

NIOSH IDLH: 30 ppm.

Respirator Recommendation

Exposure Range: >1 to <30 ppm Supplied Air, Constant Flow/Pressure Demand, Half Mask

Exposure Range: 30 to unlimited ppm Self-contained Breathing Apparatus, Pressure Demand, Full Face

Note: odor threshold unknown

Environmental

Environmental Fate: If released to the atmosphere, it is expected to exist in both the vapor-phase and particulate phase. It will degrade in the vapor-phase with an estimated half-life of 7.9 hours. Particulate phase may be physically removed from the atmosphere via wet and dry deposition. In soil it is expected to have moderate mobility; some leaching may be possible as it is fairly water soluble. Hydrolysis may occur, particularly in alkaline soils or water. In water, it is not expected to bioconcentrate, volatilize from water surfaces, or adsorb to sediment or particulate matter in the water column.

Environmental Physical Data

Henry's Law Constant: estimated at 4.26×10^{-9}

Octanol/Water Partition Coefficient: log K_{ow} = 2.29

Sorption Partition Coefficient: K_{oc} = 419

BCF: estimated at 32

Regulations

RCRA 40CFR: Not listed

CERCLA: 40CFR 302.4: Not listed

SARA 40CFR 372.65: Not listed

SARA EHS 40CFR 355: Not listed

TSCA: Listed

Analytical Methods

Indoor / Expired Air: NIOSH 5017

DIB6610 CAS #: 84-74-2

DIBUTYL PHTHALATE

RTECS: TI0875000

DOT: NA9095

EINECS Number: 201-557-4

Molecular Formula: $C_{16}H_{22}O_4$

Structured MF: $C_6H_4(COO(CH_2)_3CH_3)_2$

Formula Weight: 278.34

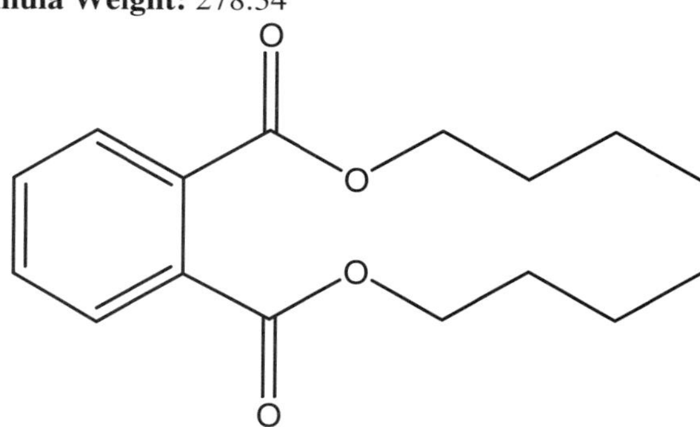

Chemical Structure

Synonyms: AI-3-00283; 1,2-BENZENEDICARBOXYLIC ACID DIBUTYL ESTER; BENZENE-O-DICARBOXYLIC ACID DI-N-BUTYL ESTER; 1,2-BENZENEDICARBOXYLIC ACID,DIBUTYL ESTER; O-BENZENEDICARBOXYLIC ACID,DIBUTYL ESTER; BUTYL PHTHALATE; N-BUTYL PHTHALATE; N-BUTYLPHTHALATE; CELLUFLEX DPB; DBP; DBP (ESTER); DI BUTYL PHTHALATE; DIBUTYL 1,2-BENZENE-DICARBOXYLATE; DIBUTYL 1,2-BENZENEDICARBOXYLATE; DI-N-BUTYL PHTHALATE; DI-N-BUTYLESTER KYSELINY FTALOVE; DIBUTYL-O-PHTHALATE; ELAOL; EPA PESTICIDE CHEMICAL CODE 028001; ERGOPLAST FDB; ERSOPLAST FDA; GENOPLAST B; HATCOL DBP; HEXAPLAS M/B; KODAFLEX DBP; PALATINOL C; PHTHALIC ACID DIBUTYL ESTER; PHTHALIC ACID,DIBUTYL ESTER; POLYCIZER DBP; PX 104; RC PLASTICIZER DBP; STAFLEX DBP; UNIFLEX DBP; UNIMOLL DB; WITCIZER 300

Description: colorless, faint-yellow liquid; ester odor

Use: in plasticizers, cosmetics, safety glass, insecticides, printing inks, paper coatings, adhesives, elastomers and explosives; as a solvent in polysulfide dental impression materials, solvent for perfume oils, perfume fixative, textile lubricating agent and solid rocket propellant

Physical Properties

Boiling Point: 340 °C (644 °F)

Freezing Point: -35 °C (-31 °F)

Specific Gravity: 1.0459 at 20 °C

Vapor Density: 9.58
Density: 1.05 g/mL at 25 °C
Vapor Pressure: 6.4 to 13 x10^{-3} Pa at 20 °C
Water Solubility: < 1 mg/mL at 20 C
Other Solubilities: 95% Ethanol: >=100 mg/ml at 20 °C; Acetone: >=100 mg/ml at 20 °C; Alcohol: Soluble; Benzene: Soluble; Common organic solvents: miscible; DMSO: >=100 mg/ml at 20 °C; Ether: Soluble.
Surface Tension: 34 dynes/cm
Refraction Index: 1.4900 at 20 °C/D
Evaporation Rate: < 1 Butyl Acetate=1
Critical Temperature: 500 °C
Critical Pressure: 250 psia
Flash Point: 157 °C Closed Cup
Autoignition Temperature: 403 °C
LEL: 0.5% v/v

RTECS Toxicity Data

Acute Oral: Human TD_{Lo} Dose: 140 mg/kg; Toxic Effects: Behavioral - Hallucinations, distorted perceptions; Gastrointestinal - Nausea or vomiting; Kidney, Ureter, and Bladder - Other changes. Rat LD_{50} Dose: 8 gm/kg.
Acute Inhalation: Rat LC_{50} Dose: 4250 mg/m^3. Mouse LC_{50} Dose: 25 gm/m^3/2hr.
Acute Dermal: Rabbit LD_{50} Route: Skin; Dose: >20 mL/kg. Rat LD_{Lo} Route: Skin; Dose: 6 gm/kg.
Chronic (Multiple Dose) Oral: Rat Dose: 96750 mg/kg/45D-I; Toxic Effects: Endocrine - Other changes; Nutritional and gross metabolic - Weight loss or decreased weight gain; Biochemical - True cholinesterase. Rat Dose: 28 gm/kg/14D-I; Toxic Effects: DEATH - Changes in testicular weight. Rat Dose: 4 gm/kg/4D-I; Toxic Effects: DEATH - Changes in testicular weight. Rat Dose: 27300 mg/kg/13W-C; Toxic Effects: Liver - Changes in liver weight; Kidney, Ureter, and Bladder - Changes in kidney weight; Blood - Normocytic anemia. Rat Dose: 22750 mg/kg/1Y-C; Toxic Effects: Behavioral - Food intake (animal); DEATH.
Chronic (Multiple Dose) Inhalation: Rat Dose: 900 mg/m^3/6H/35D-I; Toxic Effects: Brain and coverings - Changes in brain weight; Blood - Normocytic anemia; Nutritional and gross metabolic - Weight loss or decreased weight gain.
Reproductive/Teratogenic: Rat Route: Oral; Dose: 2250 mg/kg; Duration: female 7-9D of pregnancy; Effects on Fertility - Post-implantation mortality; Effects on Embryo or Fetus - Fetotoxicity; Specific Developmental Abnormalities - Craniofacial (including nose and tongue). Rat Route: Oral; Dose: 2520 mg/kg; Duration: female 1-21D of pregnancy; Effects on Embryo or Fetus - Extra embryonic structures. Rat Route: Oral; Dose: 8820 mg/kg; Duration: female 7-15D of pregnancy; Effects on Fertility - Post-implantation mortality; Effects on Embryo or Fetus - Fetotoxicity; Fetal death. Rat Route: Oral; Dose: 16800 mg/kg; Duration: male 7D prior to mating; Paternal Effects - Testes, epididymis, sperm duct; Prostate, seminal vessicle, Cowper's gland, accessory glands.
Mutagenic: Hamster Cytogenetic Analysis; Cell Type: fibroblast; Dose: 30 mg/L/24H. Bacteria - S Typhimurium Mutations in Microorganisms; Dose: 100 ug/plate (+S9).

Hazard Overviews

Fire Diamond

Health: Irritating to eyes/skin/respiratory tract. Also Causes: eyelid swelling, photophobia, conjunctivitis, nausea, dizziness, kidney toxicity. Chronic Effects: possible CNS effects (pain, numbness, weakness in extremities).
Fire: Will burn. Use dry chemical, carbon dioxide; water spray or foam may cause frothing. Containers may explode in heat of fire.
Reactivity: Stable. Hazardous polymerization cannot occur. Avoid: exposure to heat; ignition sources. Incompatible with: oxidizers; nitrates; acids; alkalis; liquid chlorine (explosive reaction). Hazardous decomposition products: carbon oxides.
Carcinogenicity: IARC - Not listed; NIOSH - Not listed; NTP - Not listed; ACGIH - Not listed; OSHA - Not listed; EPA - Class D, Not classifiable as to human carcinogenicity; MAK - Not listed
Primary Target Organs:

Eyes

Skin

Respiratory System

Kidneys

Exposure Limits
OSHA PEL: TWA: 5 mg/m^3.
ACGIH TLV: TWA: 5 mg/m^3.
NIOSH REL: TWA: 5 mg/m^3.
NIOSH IDLH: 4000 mg/m^3.
Respirator Recommendation
Exposure Range: >5 to 50 mg/m^3 Air Purifying, Negative Pressure, Half Mask
Exposure Range: >50 to 500 mg/m^3 Air Purifying, Negative Pressure, Full Face
Exposure Range: >500 to <4000 mg/m^3 Supplied Air, Constant Flow/Pressure Demand, Full Face
Exposure Range: 4000 to unlimited mg/m^3 Self-contained Breathing Apparatus, Pressure Demand, Full Face
Cartridge Color: dust/mist filter (use P100 or consult supervisor for appropriate dust/mist filter)

Environmental

Ecotoxicity: EC_{50} Gymnodinium breve (alga) 3.4 ug/l/96 hr, toxic effect: chlorophyll a. /Conditions of bioassay not specified Artemia salina (brine shrimp) 10 mg/l toxic effect: 20% reduction in number of larvae hatched over 24 hr. /Conditions of bioassay not specified Atremia salina (brine shrimp) 50 mg/l toxic effect: 40% reduction in number of larvae hatched over 24 hr. /Conditions of bioassay not specified LC_{50} Gammarus fasciatus (scud) 0.21 mg/l/1500 hr. /Conditions of bioassay not specified Toxic to synchronously developing larvae of the brine shrimp, artemia. the LD_{50} for 24 hr exposure was 30 umol (8 ppm). /conditions of bioassay not specified

Environmental Fate: If released into water it will adsorb moderately to sediment and particulates in the water column. It will disappear in 3-5 days in moderately polluted waters and generally within 3 weeks in cleaner bodies of water. It will not bioconcentrate in fish since it is readily metabolized. If spilled on land it will adsorb moderately to soil and slowly biodegrade (66 and 98% degradation in 26 weeks from two soils). It is found in groundwater under rapid infiltration sites and elsewhere. It has been suggested that its tendency to form complexes with water-soluble fulvic acids, a component of soils, may aid its transport into groundwater. Although it degrades under anaerobic conditions, its fate in groundwater is unknown. If released into air, it is generally associated with the particulate fraction and will be subject to gravitational settling. Vapor phase will degrade by reaction with photochemically produced hydroxyl radicals (estimated half-life 18 hr).

Cleanup/Disposal: Guide No. 171: Do not touch or walk through spilled material. Stop leak if you can do it without risk. Prevent dust cloud. Avoid inhalation of asbestos dust. Small Dry Spills: With clean shovel place material into clean, dry container and cover loosely; move containers from spill area. Small Spills: Take up with sand or other noncombustible absorbent material and place into containers for later disposal. Large Spills: Dike far ahead of liquid spill for later disposal. Cover powder spill with plastic sheet or tarp to minimize spreading. Prevent entry into waterways, sewers, basements or confined areas.

Environmental Physical Data

Henry's Law Constant: 5.3 x10^{-5}
Octanol/Water Partition Coefficient: log K_{ow} = 4.9
Sorption Partition Coefficient: K_{oc} = 160 to 6400
BCF: no food chain concentration potential
BOD: 0.43 lb/lb, 5 days

Regulations

RCRA 40CFR: Listed Hazardous Waste No. U069 Toxic Waste
CERCLA: 40CFR 302.4: Listed per CWA Section 311(b)(4) per RCRA Section 3001 per CWA Section 307(a) RQ: 10 lb (4.535 kg)
SARA 40CFR 372.65: Listed
SARA EHS 40CFR 355: Not listed
TSCA: Listed

Analytical Methods

Soil: CLP LC_SV, MC_SVOA, OHC; EPA 16, 1625; SW846 3640A, 8060, 8061, 8061A, 8250A, 8270B, 8270C, 8410
Water / Groundwater: EPA S-002-1, 1625, 606, 625, 625-S, 6; APHA 6410-B; USGS O3118
Drinking Water: EPA 506, 525.1, 525.2
Food: FDA 212.1, 232.1
Indoor / Expired Air: NIOSH 5020
Plasma: EPA 001, 29; FDA 211.1, 231.1, 252

DIB6720	CAS #: 109-43-3
DIBUTYL SEBACATE	

RTECS: VS1150000
EINECS Number: 203-672-5
Molecular Formula: $C_{18}H_{34}O_4$
Formula Weight: 314.47

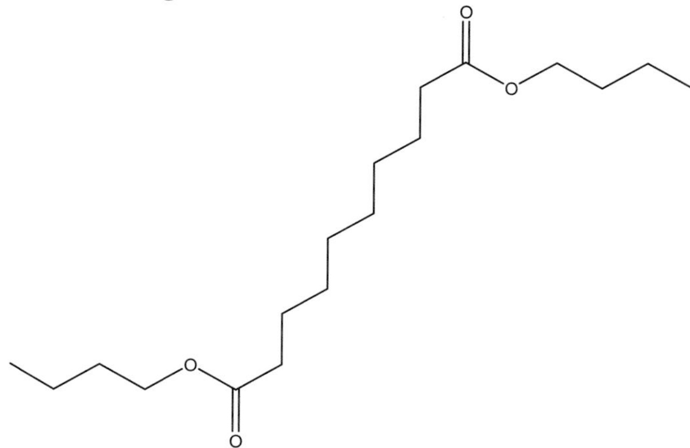

Chemical Structure

Synonyms: BIS(N-BUTYL) SEBACATE; BIS(N-BUTYL)SEBACATE; BUTYL SEBACATE; DECANEDIOIC ACID,DIBUTYL ESTER; DIBUTYL DECANEDIOATE; DIBUTYL 1,8-OCTANEDICARBOXYLATE; DI-N-BUTYL SEBACATE; DIBUTYLESTER KYSELINY SEBAKOVE; DI-N-BUTYLSEBACATE; KODAFLEX DBS; MONOPLEX DBS; POLYCIZER DBS; PX 404; SEBACIC ACID,DIBUTYL ESTER; STAFLEX DBS
Description: clear liquid
Use: synthetic flavoring adjuvant; plasticiser for cellulose acetate butyrate plastics, cellulose acetate propionate plastics, polyvinyl butyral plastics; lubricating ingredient in shaving lotions

Physical Properties

Boiling Point: 344 °C (651 °F) to 345 °C (653 °F)
Freezing Point: -10 °C (14 °F)
Specific Gravity: 0.9405 at 15 °C
Water Solubility: Insoluble in Water
Other Solubilities: Soluble in Ether
Refraction Index: 1.4433 at 15 °C/D
Flash Point: 180 °C

RTECS Toxicity Data

Acute Oral: Rat LD_{50} Dose: 16 gm/kg. Mouse LD_{50} Dose: 19500 mg/kg; Toxic Effects: Lungs, Thorax, or Respiration - Other changes.
Acute Inhalation: Rat LC; Dose: >5400 ug/m^3/4hr. Mouse LC; Dose: >5400 ug/m^3/2hr
Reproductive/Teratogenic: Rat Route: Oral; Dose: 418 gm/kg; Duration: male 10W prior to mating; Effects on Newborn - Growth statistics.

Hazard Overviews

Health: Irritating to eyes/skin/respiratory tract. Other Acute Effects: may be harmful by inhalation, ingestion, or skin absorption.

Fire: Will burn. Hazards: emits toxic fumes. Extinguishing agents: carbon dioxide, dry chemical powder or appropriate foam; use water spray to cool fire-exposed containers; foam and water spray are effective but may cause frothing. Precautions: combustible liquid.

Reactivity: Stable. Hazardous polymerization will not occur. Hazardous decomposition products: carbon monoxide, carbon dioxide.

Carcinogenicity: IARC - Not listed; NIOSH - Not listed; NTP - Not listed; ACGIH - Not listed; OSHA - Not listed; EPA - Not listed; MAK - Not listed

Primary Target Organs:

Eyes

Skin

Respiratory System

Environmental

Regulations

RCRA 40CFR: Not listed
CERCLA: 40CFR 302.4: Not listed
SARA 40CFR 372.65: Not listed
SARA EHS 40CFR 355: Not listed
TSCA: Listed

DIB6830 CAS #: 141-03-7

DIBUTYL SUCCINATE

RTECS: WM6250000
EINECS Number: 205-449-8
Molecular Formula: $C_{12}H_{22}O_4$
Formula Weight: 230.30
Synonyms: B-9; BUTANEDIOIC ACID DIBUTYL ESTER; BUTANEDIOIC ACID,DIBUTYL ESTER; BUTYL BUTANEDIOATE; DIBUTYL BUTANEDIOATE; DI-N-BUTYL SUCCINATE; DI-N-BUTYLESTER KYSELINY JANTAROVE; DI-N-BUTYLSUCCINATE; DNBS; ENT 666; SUCCINIC ACID DI-N-BUTYL ESTER; SUCCINIC ACID DIBUTYL ESTER; SUCCINIC ACID,DIBUTYL ESTER; TABATREX; TABUTREX
Description: colorless liquid
Use: insect repellent especially against biting flies of cattle & household ants & roaches

Physical Properties

Boiling Point: 274.5 °C (526 °F)
Freezing Point: -29 °C (-20.2 °F)
Specific Gravity: 0.9768 at 20 °C/4 °C
Water Solubility: Insoluble in Water
Other Solubilities: Soluble in Alcohol, Ether, & Benzene.
Refraction Index: 1.4299 at 20 °C/D
Flash Point: 135 °C Open Cup

RTECS Toxicity Data

Acute Oral: Rat LD_{50} Dose: 8 gm/kg.

Hazard Overviews

Fire: Will burn.
Carcinogenicity: IARC - Not listed; NIOSH - Not listed; NTP - Not listed; ACGIH - Not listed; OSHA - Not listed; EPA - Not listed; MAK - Not listed

Environmental

Regulations

RCRA 40CFR: Not listed
CERCLA: 40CFR 302.4: Not listed
SARA 40CFR 372.65: Not listed
SARA EHS 40CFR 355: Not listed
TSCA: Listed

Analytical Methods

Soil: EPA PMD-DGL

DIB6940 CAS #: 22673-19-4

DIBUTYL TIN BIS ACETYLACETONATE

EINECS Number: 245-152-0

Physical Properties

Boiling Point: 220 °C (428 °F)
Specific Gravity: 1.23
Water Solubility: Decomposes
Other Solubilities: Soluble in Toluene

Hazard Overviews

Carcinogenicity: IARC - Not listed; NIOSH - Not listed; NTP - Not listed; ACGIH - Not listed; OSHA - Not listed; EPA - Not listed; MAK - Not listed

Environmental

Regulations

RCRA 40CFR: Not listed
CERCLA: 40CFR 302.4: Not listed
SARA 40CFR 372.65: Not listed
SARA EHS 40CFR 355: Not listed
TSCA: Not listed

DIB7270 CAS #: 111-92-2

N-DIBUTYLAMINE

RTECS: HR7780000
DOT: UN2248; IMO8.2
EINECS Number: 203-921-8
Molecular Formula: $C_8H_{19}N$

Structured MF: $(C_4H_9)_2NH$
Formula Weight: 129.24

Chemical Structure

Synonyms: 1-BUTANAMINE,N-BUTYL-; N-BUTYLAMINE; N-BUTYL-1-BUTANAMINE; DIBUTILAMINA; DI-N-BUTYLAMINE; DIBUTYLAMINE

Description: colorless liquid; ammonia-like or fishy odor

Use: corrosion inhibitor; butadiene inhibitor; intermediate for emulsifier rubber accelerators, dyes, insecticides and flotation agents

Physical Properties

Boiling Point: 159 °C (318 °F) to 160 °C (320 °F)
Freezing Point: -59 °C (-74.2 °F) to -60 °C (-76 °F)
Specific Gravity: 0.7601 at 20 °C/4 °C
Vapor Density: 4.5 Air=1
Density: 0.76 g/mL
Vapor Pressure: 0.0 psia
Water Solubility: Soluble in Water
Other Solubilities: 95% Ethanol: Soluble; Acetone: Soluble; Benzene: Soluble; Ether: Very Soluble.
Surface Tension: 24.76 dynes/cm at 20 °C
Odor Threshold: 0.4232 to 2.5392 mg/m^3
pH: > 7
Refraction Index: 1.4177 at 20 °C/D
Evaporation Rate: < 1 Butyl Acetate=1
Critical Temperature: 322.6 °C
Ionization Potential (eV): 7.69 +/-0.3
Flash Point: 51.6 °C Open Cup
LEL: ~ 1.1% v/v

RTECS Toxicity Data

Acute Oral: Rat LD_{50} Dose: 189 mg/kg. Mouse LD_{50} Dose: 290 mg/kg.
Acute Inhalation: Rat LC_{Lo} Dose: 500 ppm/4hr.
Acute Dermal: Rabbit LD_{50} Route: Skin; Dose: 1010 uL/kg. Rat LD_{50} Route: Subcutaneous Dose: 494 mg/kg.
Irritation Eye: Rabbit Standard Draize Test Dose: 250 ug open; Reaction: severe.
Irritation Skin: Rabbit Standard Draize Test Dose: 20 mg/24H; Reaction: moderate. Rabbit Open Draize Test Dose: 10 mg/24H open; Reaction: severe.
Mutagenic: Hamster Cytogenetic Analysis; Cell Type: fibroblast; Dose: 200 mg/L/48H. Bacteria - B Subtilis DNA Repair; Dose: 5 gm/L.

Hazard Overviews

Corrosive Flammable

Fire Diamond

Health: Corrosive, causes severe burns to eyes/skin/respiratory tract. Toxic. Also Causes: headache, coughing, nausea, pulmonary edema, corneal edema, stomach irritation.

Fire: Flammable. Can form explosive mixtures in air. Use water spray, dry chemical, foam, or carbon dioxide. Water may be ineffective because material floats on water.

Reactivity: Stable. Hazardous polymerization cannot occur. Avoid: heat; ignition sources. Incompatible with: acids; oxidizing materials; halogenated compounds; chlorine; hypochlorite; reactive organic compounds; cellulose nitrate. Hazardous decomposition products: carbon monoxide; carbon dioxide; hydrocarbons; nitrogen oxides; amine vapors.

Carcinogenicity: IARC - Not listed; NIOSH - Not listed; NTP - Not listed; ACGIH - Not listed; OSHA - Not listed; EPA - Not listed; MAK - Not listed

Primary Target Organs:

Eyes

Skin

Respiratory System

Mucous Membranes

Exposure Limits
AIHA WEEL: STEL: 5 ppm ceiling skin.

Respirator Recommendation
Exposure Range: >5 to 50 ppm Air Purifying, Negative Pressure, Half Mask
Exposure Range: >50 to 500 ppm Air Purifying, Negative Pressure, Full Face
Exposure Range: >500 to 5000 ppm Supplied Air, Constant Flow/Pressure Demand, Full Face
Exposure Range: >5000 to unlimited ppm Self-contained Breathing Apparatus, Pressure Demand, Full Face
Cartridge Color: black
Note: use dust/mist prefilter if mist is present

Environmental

Environmental Fate: If released on land, it will adsorb strongly to soil. The chemical would be expected to readily biodegrade; however, no estimate of degradative rates in soil are available. If released into water, it will adsorb to sediment and particulate matter in the water column and probably readily biodegrade. Bioconcentration in aquatic organisms will not be appreciable. In the atmosphere, it will react with photochemically produced hydroxyl radicals (estimated half-life 4.4 hr). It should also be scavenged by rain. Dialkylamines are of particular environmental concern because they are precursors of nitroamines. The latter are formed in the atmosphere in the presence of nitrogen oxides but are destroyed by sunlight. In aqueous, acidic solutions, di-n-butylnitrosamine may be formed when nitrite ions are present. Humic acids catalyze this reaction.

Cleanup/Disposal: Guide No. 132: Fully encapsulating, vapor protective clothing should be worn for spills and leaks with no fire. Eliminate all ignition sources (no smoking, flares, sparks or flames in immediate area). All equipment used when handling the product must be grounded. Do not touch or walk through spilled material. Stop leak if you can do it without risk. Prevent entry into waterways, sewers, basements or confined areas. A vapor suppressing foam may be used to reduce vapors. Absorb with earth, sand or other non-combustible material and transfer to containers (except for Hydrazine). Use clean non-sparking tools to collect absorbed material. Large Spills: Dike far ahead of liquid spill for later disposal. Water spray may reduce vapor; but may not prevent ignition in closed spaces.

Environmental Physical Data

Henry's Law Constant: estimated at 9.1 x10^{-5}
Octanol/Water Partition Coefficient: log K_{ow} = 2.83
BCF: estimated at 83

Regulations

RCRA 40CFR: Not listed
CERCLA: 40CFR 302.4: Not listed
SARA 40CFR 372.65: Not listed
SARA EHS 40CFR 355: Not listed
TSCA: Listed

DIB7380 CAS #: 102-81-8

2-N-DIBUTYLAMINOETHANOL

RTECS: KK3850000
EINECS Number: 203-057-1
Molecular Formula: $C_{10}H_{23}NO$
Structured MF: $(C_4H_9)_2NCH_2CH_2OH$
Formula Weight: 173.34

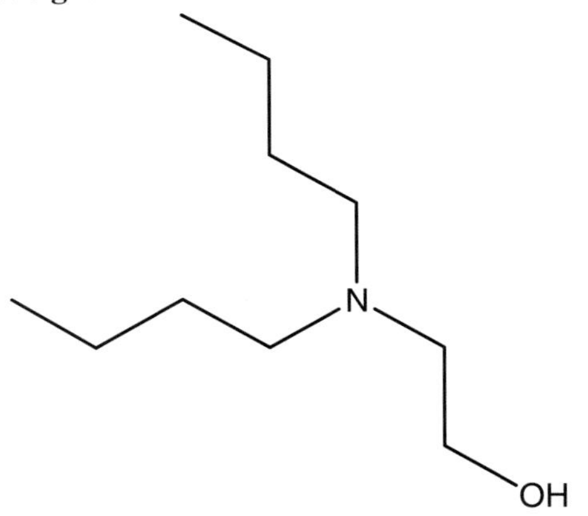

Chemical Structure

Synonyms: BU2AE; DBAE; 2-(DI-N-BUTYLAMINO)ETHANOL; 2-DI-N-BUTYLAMINOETHANOL; 2-DIBUTYLAMINOETHANOL; DIBUTYLAMINOETHANOL; N,N-DI-N-BUTYLAMINOETHANOL; 2-DI-N-BUTYLAMINOETHYL ALCOHOL; BETA-N-DIBUTYLAMINOETHYL ALCOHOL; DIBUTYLAMINOTHANOL; N,N-DIBUTYLETHANOLAMINE; N,N-DIBUTYL-N-(2-HYDROXYETHYL)AMINE
Description: colorless liquid; faint, amine-like odor
Use: synthesis

Physical Properties

Boiling Point: 230 °C (446 °F)
Specific Gravity: 0.86
Vapor Density: 6 Air=1
Vapor Pressure: 0.1 mm Hg
Water Solubility: 0.4% by weight
Other Solubilities: 95% Ethanol: >=100 mg/ml at 18 °C; Acetone: >=100 mg/ml at 18 °C; DMSO: >=100 mg/ml at 18 °C.
pH: Alkaline
Flash Point: 91 °C

RTECS Toxicity Data

Acute Oral: Rat LD_{50} Dose: 1070 mg/kg.
Acute Dermal: Rabbit LD_{50} Route: Skin; Dose: 1680 uL/kg.
Chronic (Multiple Dose) Oral: Rat Dose: 7 gm/kg/5W-C; Toxic Effects: Kidney, Ureter, and Bladder - Changes in kidney weight.
Chronic (Multiple Dose) Inhalation: Rat Dose: 70 ppm/6H/5D-I; Toxic Effects: Liver - Changes in Liver weight; Kidney, Ureter, and Bladder - Changes in kidney weight; Nutritional and gross metabolic - Weight loss or decreased weight gain.
Irritation Eye: Rabbit Standard Draize Test Dose: 20 mg/24H open; Reaction: severe.
Irritation Skin: Rabbit Standard Draize Test Dose: 5 mg/24H; Reaction: severe. Rabbit Open Draize Test Dose: 500 mg open; Reaction: severe.

Hazard Overviews

Fire Diamond

Health: Severe irritation to eyes/skin. Moderately toxic by ingestion/skin absorption. Also Causes: blocks transmission of nerve impulses by inhibiting enzyme, acetylchloinesterase.
Fire: Combustible. For small fires use dry chemical, water spray, or regular foam. For large fires use water spray, fog, or regular foam. Do not scatter material with water from high-pressure hoses.
Reactivity: Stable. Hazardous polymerization cannot occur. Avoid: heat. Incompatible with: oxidizing materials; inorganic acids; organic acids. Hazardous decomposition products: carbon dioxide; toxic nitrogen oxide fumes.
Carcinogenicity: IARC - Not listed; NIOSH - Listed as carcinogen; NTP - Not listed; ACGIH - Not listed; OSHA - Not listed; EPA - Not listed; MAK - Not listed
Primary Target Organs:

Eyes

Skin

Exposure Limits
OSHA PEL Vacated 1989 Limits: TWA: 2 ppm; 14 mg/m^3.
ACGIH TLV: TWA: 0.5 ppm; 3.5 mg/m^3.
NIOSH REL: TWA: 2 ppm; 14 mg/m^3.
Respirator Recommendation
Exposure Range: >0.5 to 25 ppm Supplied Air, Constant Flow/Pressure Demand, Half Mask
Exposure Range: >25 to 500 ppm Supplied Air, Constant Flow/Pressure Demand, Full Face

Exposure Range: >500 to unlimited ppm Self-contained Breathing Apparatus, Pressure Demand, Full Face
Note: odor threshold unknown

Environmental

Cleanup/Disposal: Guide No. 153: Eliminate all ignition sources (no smoking, flares, sparks or flames in immediate area). Do not touch damaged containers or spilled material unless wearing appropriate protective clothing. Stop leak if you can do it without risk. Prevent entry into waterways, sewers, basements or confined areas. Absorb or cover with dry earth, sand or other non-combustible material and transfer to containers. Do not get water inside containers.

Regulations

RCRA 40CFR: Not listed
CERCLA: 40CFR 302.4: Not listed
SARA 40CFR 372.65: Not listed
SARA EHS 40CFR 355: Not listed
TSCA: Listed

Analytical Methods

Indoor / Expired Air: NIOSH 2007

DIB7490 CAS #: 2460-77-7

2,5-DI-T-BUTYL-P-BENZOQUINONE

RTECS: GU5140000
EINECS Number: 219-552-0
Molecular Formula: $C_{14}H_{20}O_2$
Formula Weight: 220.32

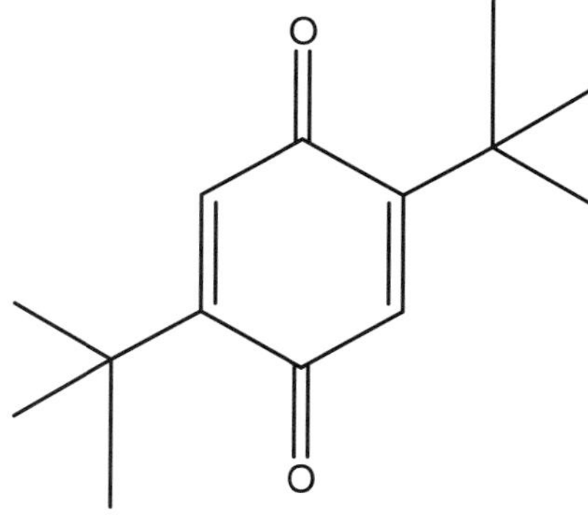

Chemical Structure

Synonyms: P-BENZOQUINONE,2,5-DI-TERT-BUTYL-; 2,5-BIS(1,1-DIMETHYLETHYL)-2,5-CYCLOHEXADIENE-1,4-DIONE; 2,5-CYCLOHEXADIENE-1,4-DIONE,2,5-BIS(1,1-DIMETHYLETHYL)-; 2,5-DI-TERT-BUTYL-1,4-BENZOQUINONE
Description: yellow crystals

Physical Properties

Freezing Point: 152.5 °C (306.5 °F)
Water Solubility: Insoluble in Water
Other Solubilities: Soluble in Ether, Benzene, Acetic Acid, hot Alcohol

RTECS Toxicity Data

Mutagenic: Mouse Morphological Transformation; Cell Type: embryo; Dose: 1500 ug/L.

Hazard Overviews

Health: Irritating to eyes/skin/respiratory tract. Other Acute Effects: may be harmful by inhalation, ingestion, or skin absorption.
Fire: Hazards: emits toxic fumes. Extinguishing agents: water spray; carbon dioxide, dry chemical powder or appropriate foam. Precautions: combustible liquid.
Reactivity: Incompatible with: strong oxidizing agents. Hazardous decomposition products: toxic fumes of: carbon monoxide, carbon dioxide.
Carcinogenicity: IARC - Not listed; NIOSH - Not listed; NTP - Not listed; ACGIH - Not listed; OSHA - Not listed; EPA - Not listed; MAK - Not listed
Primary Target Organs:

Eyes

Skin

Respiratory System

Environmental

Regulations

RCRA 40CFR: Not listed
CERCLA: 40CFR 302.4: Not listed
SARA 40CFR 372.65: Not listed
SARA EHS 40CFR 355: Not listed
TSCA: Listed

DIB7600 CAS #: 719-22-2

2,6-DI-T-BUTYL-P-BENZOQUINONE

RTECS: DK3970000
EINECS Number: 211-946-0
Molecular Formula: $C_{14}H_{20}O_2$
Formula Weight: 220.34

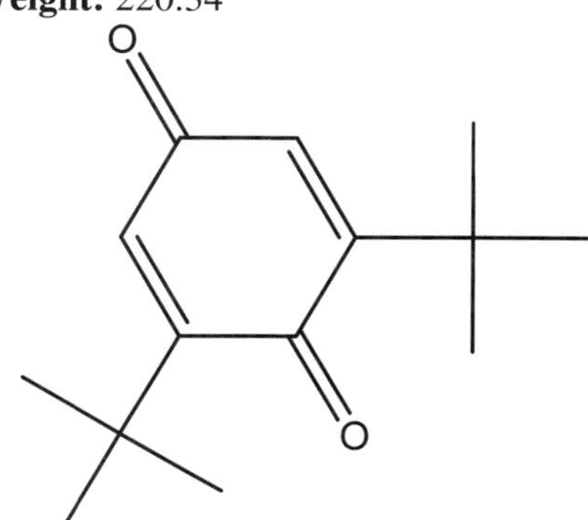

Chemical Structure

Synonyms: P-BENZOQUINONE,2,6-DI-TERT-BUTYL-; 2,5-CYCLOHEXADIENE-1,4-DIONE,2,6-BIS(1,1-DIMETHYLETHYL)-;

DBQ; 2,6-DI-TERT-BUTYL-P-BENZOQUINONE; 2,6-DI-TERT-BUTYLBENZOQUINONE

Description: yellow crystals

Use: oxidant, polymerization catalyst

Physical Properties

Boiling Point: 60 °C (140 °F)

Hazard Overviews

Health: Irritating to eyes/skin/respiratory tract. Other Acute Effects: may be harmful by inhalation, ingestion, or skin absorption.

Fire: Extinguishing agents: water spray; carbon dioxide, dry chemical powder or appropriate foam. Precautions: combustible liquid.

Reactivity: Incompatible with: strong oxidizing agents, strong reducing agents. Hazardous decomposition products: toxic fumes of: carbon monoxide, carbon dioxide.

Carcinogenicity: IARC - Not listed; NIOSH - Not listed; NTP - Not listed; ACGIH - Not listed; OSHA - Not listed; EPA - Not listed; MAK - Not listed

Primary Target Organs:

Eyes

Skin

Respiratory System

Environmental

Environmental Fate: If released to the atmosphere, it should exist mainly in the vapor phase based on an estimated vapor pressure of 4.1 $\times 10^{-4}$ mm Hg. In the vapor phase, it will react rapidly with hydroxyl radicals with an estimated half-life of 17 hours. An estimated K_{oc} of 6000 suggests that it will be immobile in soil. This compound is not expected to volatilize from moist soil surfaces given an estimated Henry's Law constant of 1.6 $\times 10^{-8}$ atm-cu m/mol. Incomplete information is available regarding the biodegradation potential, however, this compound was removed from Rhine River water which underwent bank filtration suggesting that it was either biodegraded or adsorbed to the soil. In water, it is expected to adsorb to organic matter and particulates in the water column based on its K_{oc} value. A half-life of 3.4 days was measured in river water for this compound; a distinction between removal by biodegradation or removal through adsorption to particulate matter was not made. It may bioconcentrate in aquatic organisms given an estimated BCF value of 1400; this compound was detected in Great Lakes fish.

Environmental Physical Data

Henry's Law Constant: estimated at 1.6 $\times 10^{-8}$

Octanol/Water Partition Coefficient: log K_{OW} = 4.42

Sorption Partition Coefficient: K_{OC} = 6100

BCF: estimated at 1400

Regulations

RCRA 40CFR: Not listed

CERCLA: 40CFR 302.4: Not listed

SARA 40CFR 372.65: Not listed

SARA EHS 40CFR 355: Not listed

TSCA: Listed

Analytical Methods

Soil: EPA 1625

Water / Groundwater: EPA 1625

DIB7710 CAS #: 128-37-0

2,6-DI-TERT-BUTYL-P-CRESOL

RTECS: GO7875000

EINECS Number: 204-881-4

Molecular Formula: $C_{15}H_{24}O$

Structured MF: $(C(CH_3)_3)_2CH_3C_6H_2OH$

Formula Weight: 220.34

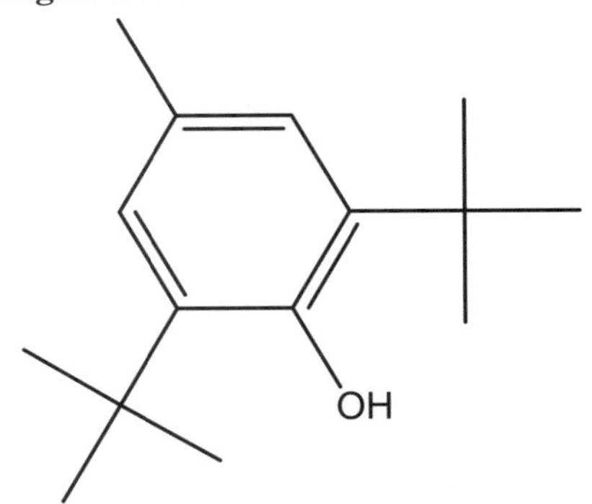

Chemical Structure

Synonyms: ADVASTAB 401; AGIDOL; AGIDOL 1; ALKOFEN BP; ANTIOXIDANT 4K; ANTIOXIDANT 29; ANTIOXIDANT 30; ANTIOXIDANT 4; ANTIOXIDANT DBPC; ANTIOXIDANT KB; ANTRANCINE 8; AO 29; AO 4K; BHT; BHT (FOOD GRADE); 2,6-BIS(1,1-DIMETHYLETHYL)-4-METHYLPHENOL; BUKS; BUTYLATED HYDROXYTOLUENE; BUTYLHYDROXYTOLUENE; 2,6-TERT-BUTYL-4-METHYLPHENOL; BUTYLOHYDROKSYTOLUENU; CAO 1; CAO 3; CATALIN CAO-3; CHEMANOX 11; P-CRESOL,2,6-DI-TERT-BUTYL-; DALPAC; DBMP; DBPC; DBPC (TECHNICAL GRADE); DEENAX; DIBUNOL; DIBUTYLATED HYDROXYTOLUENE; 2,6-DI-T-BUTYL-P-CRESOL; 2,6-DI-TERT-BUTYLCRESOL; DI-TERT-BUTYL-P-CRESOL; DI-TERT-BUTYLCRESOL; O,O'-DI-TERT-BUTYL-P-CRESOL; 2,6-DI-TERT-BUTYL-1-HYDROXY-4-METHYLBENZENE; 3,5-DI-TERT-BUTYL-4-HYDROXYTOLUENE; 2,6-DI-T-BUTYL-4-METHYLPHENOL; 2,6-DI-TERT-BUTYL-4-METHYLPHENOL; 2,6-DI-TERT-BUTYL-P-METHYLPHENOL; 2,6-DI-TERC.BUTYL-P-KRESOL; 4-HYDROXY-3,5-DI-T-BUTYL-TOLUENE; 4-HYDROXY-3,5-DI-TERT-BUTYLTOLUENE; IMPRUVAL; IMPRUVOL; IONOL; IONOL 1; IONOL (ANTIOXIDANT); IONOL CP; IONOLE; KERABIT; 4-METHYL-2,6-TERT BUTYLPHENOL; 4-METHYL-2,6-DI-TERT-BUTYL PHENOL; 4-METHYL-2,6-DI-T-BUTYLPHENOL; 4-METHYL-2,6-DI-TERT-BUTYLPHENOL; METHYLDI-TERT-BUTYLPHENOL; 4-METHYL-2,6-DI-TERC. BUTYLFENOL; NOCRAC 200; NONOX TBC; P 21; PARABAR 441; PARANOX 441; PHENOL,2,6-BIS(1,1-DIMETHYLETHYL)-4-METHYL-; PHENOL,2,6-BIS(1,1-DIMETHYLETHYL)-4-METHYL-(9CI); STAVOX; SUMILIZER BHT; SUSTANE; SUSTANE BHT; SWANOX BHT; TENAMEN 3; TENAMENE 3; TENOX BHT; TOPANOL; TOPANOL O; TOPANOL OC; TOXOLAN P; VANLUBE PC; VANLUBE PCX; VIANOL; VULKANOX KB

Description: white to pale yellow crystals, flakes; slight cresylic odor

Use: antioxidant for food, animal feed, petroleum products, synthetic rubbers, plastics, animal and vegetable oils, soaps; antiskinning agent in paints and inks

Physical Properties

Boiling Point: 265 °C (509 °F)
Freezing Point: 70 °C (158 °F)
Specific Gravity: 1.048 at 20 °C/4 °C
Density: 0.8937 g/mL
Vapor Pressure: 0.01 mm Hg
Water Solubility: Insoluble in Water
Other Solubilities: Insoluble in aqueous alkali; Soluble in naphtha; Insoluble in 10% sodium hydroxide.
Refraction Index: 1.4859 at 75 °C/D
Ionization Potential (eV): =< 7.80
Flash Point: 127 °C

RTECS Toxicity Data

Acute Oral: Woman TD_{Lo} Dose: 80 mg/kg; Toxic Effects: Behavioral - Coma; Gastrointestinal - Gastritis; Gastrointestinal - Nausea or vomiting. Rat LD_{50} Dose: 890 mg/kg.

Chronic (Multiple Dose) Oral: Rat Dose: 109 gm/kg/2Y-C; Toxic Effects: Blood - Changes in serum composition; Blood - Changes in erythrocite (RBC) cell count; Blood - Changes in leukocyte (WBC) cell count. Rat Dose: 27 gm/kg/90D-C; Toxic Effects: Liver - Changes in liver weight; Endocrine - Evidence of thyroid hyperfunction; Endocrine - Changes in thyroid weight. Rat Dose: 37800 mg/kg/6W-C; Toxic Effects: Liver - Changes in liver weight; Endocrine - Changes in adrenal weight; Nutritional and gross metabolic - Weight loss or decreased weight gain. Rat Dose: 5600 mg/kg/16W-C; Toxic Effects: Liver - Changes in liver weight; Kidney, Ureter, and Bladder - Other changes in urine composition. Rat Dose: 275 mg/kg/26W-C; Toxic Effects: Liver - Changes in liver weight; Skin and appendages - Hair; Nutritional and gross metabolic - Weight loss or decreased weight gain. Rat Dose: 112 gm/kg/80W-C; Toxic Effects: Liver - Changes in liver weight; Biochemical - Hepatic microsomal mixed oxidase(dealkylation, hyroxylation,etc); Biochemical - Other proteins. Rat Dose: 5040 mg/kg/7D-C; Toxic Effects: Liver - Changes in liver weight; Blood - Changes in erythrocite (RBC) cell count; Biochemical - Transaminases. Monkey Dose: 14 gm/kg/4W-I; Toxic Effects: Biochemical - Phosphatases.

Chronic (Multiple Dose) Dermal: Mouse Route: Skin; Dose: 5 gm/kg/4W-I; Toxic Effects: Lungs, Thorax, or Respiration - Changes in lung weight; DEATH.

Irritation Eye: Rabbit Standard Draize Test Dose: 100 mg/24H; Reaction: moderate.

Irritation Skin: Human Standard Draize Test Dose: 500 mg/48H; Reaction: mild. Rabbit Standard Draize Test Dose: 500 mg/48H; Reaction: moderate.

Reproductive/Teratogenic: Rat Route: Oral; Dose: 6 gm/kg; Duration: male 13W prior to mating; Effects on Newborn - Growth statistics. Rat Route: Oral; Dose: 18 gm/kg; Duration: male 2W prior to mating; Effects on Newborn - Growth statistics; Behavioral. Rat Route: Oral; Dose: 9 gm/kg; Duration: male 2W prior to mating; Effects on Newborn - Weaning or lactation index. Rat Route: Oral; Dose: 35 gm/kg; Duration: male 10W prior to mating; Effects on Fertility - Pre-implantation mortality.

Mutagenic: Human DNA Inhibition; Cell Type: HeLa cell; Dose: 500 umol/L. Human DNA Inhibition; Cell Type: lymphocyte; Dose: 20 umol/L.

Tumorigenic: Rat Route: Oral; Dose: 134 gm/kg/32W-C; Toxic Effects: Tumorigenic - Carcinogenic by RTECS criteria; Kidney, Ureter, and Bladder - Tumors; Tumorigenic - Cells (cultured) transformed. Rat Route: Oral; Dose: 247 gm/kg/3Y-C; Toxic Effects: Tumorigenic - Carcinogenic by RTECS criteria; Tumorigenic effects - Transplacental tumorigenesis; Liver - Tumors. Rat Route: Oral; Dose: 247 gm/kg/3Y-C; Toxic Effects: Tumorigenic - Neoplastic by RTECS criteria; Tumorigenic effects - Ovarian tumors. Rat Route: Oral; Dose: 247 gm/kg; Toxic Effects: Tumorigenic - Carcinogenic by RTECS criteria; Liver - Tumors. Rat Route: Oral; Dose: 963 gm/kg; Toxic Effects: Tumorigenic - Carcinogenic by RTECS criteria; Gastrointestinal - Tumors.

Hazard Overviews

Fire Diamond

Health: Irritating to eyes/respiratory tract. Also Causes: dizziness, confusion, temporary unconsciousness, gastritis; allergic contact dermatitis. Chronic Effects: muscle weakness.

Fire: Will burn. Use carbon dioxide or dry chemical to fight fires. Cool fire-exposed containers with water spray.

Reactivity: Stable. Hazardous polymerization cannot occur. Avoid: heat; ignition sources; oxidizing agents. Incompatible with: oxidizing agents. Hazardous decomposition products: acrid; toxic fumes.

Carcinogenicity: IARC - Group 3, Not classifiable as to carcinogenicity to humans; NIOSH - Not listed; NTP - Not listed; ACGIH - Class A4, Not classifiable as a human carcinogen; OSHA - Not listed; EPA - Not listed; MAK - Not listed

Primary Target Organs:

Eyes

Skin

Respiratory System

Exposure Limits
OSHA PEL Vacated 1989 Limits: TWA: 10 mg/m^3.
ACGIH TLV: TWA: 10 mg/m^3.
NIOSH REL: TWA: 10 mg/m^3.

Respirator Recommendation
Exposure Range: >10 to 100 mg/m^3 Air Purifying, Negative Pressure, Half Mask
Exposure Range: >100 to 1000 mg/m^3 Air Purifying, Negative Pressure, Full Face
Exposure Range: >1000 to 10,000 mg/m^3 Supplied Air, Constant Flow/Pressure Demand, Full Face

Exposure Range: >10,000 to unlimited mg/m³ Self-contained Breathing Apparatus, Pressure Demand, Full Face
Cartridge Color: dust/mist filter (use P100 or consult supervisor for appropriate dust/mist filter)

Environmental

Ecotoxicity: Algae: Tetrahymena pyriformis 24h EC_{50} 1.7 mg/l; Worms: Tubifex 48h LC_{50} 11 mg/l; Fishes: killifish Oryzias latipes 48h LC_{50} 2.5 mg/l

Regulations

RCRA 40CFR: Not listed
CERCLA: 40CFR 302.4: Not listed
SARA 40CFR 372.65: Not listed
SARA EHS 40CFR 355: Not listed
TSCA: Listed

DIB7820 CAS #: 4835-11-4

N,N'-DIBUTYL-1,6-HEXANEDIAMINE

RTECS: MO1250000
EINECS Number: 225-417-7
Molecular Formula: $C_{14}H_{32}N_2$
Formula Weight: 228.42

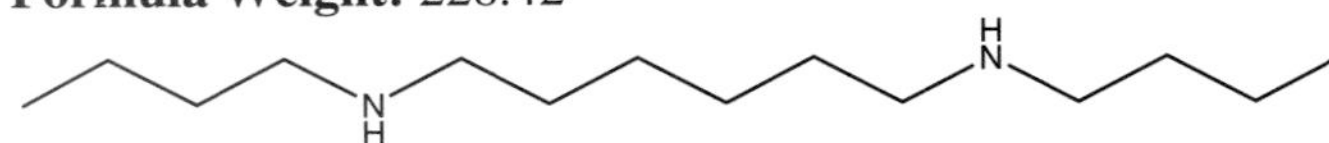

Chemical Structure

Synonyms: DBHMD; DIBUTYLHEXAMETHYLENEDIAMINE; N,N'-DIBUTYLHEXAMETHYLENEDIAMINE; 1,6-N,N'-DIBUTYLHEXANEDIAMINE; 1,6-HEXANEDIAMINE,N,N'-DIBUTYL-

Physical Properties

Boiling Point: 138.5 °C (281.3 °F)
Refraction Index: 1.4470

RTECS Toxicity Data

Acute Inhalation: Rat LC_{50} Dose: 220 mg/m³/4hr; Toxic Effects: Sense organs and special senses - Other; Sense organs and special senses - Conjunctive irritation; Lungs, Thorax, or Respiration - Dyspnea.

Hazard Overviews

Poison

Corrosive

Health: Corrosive to eyes/skin/respiratory tract. Poison. Other Acute Effects: may be fatal if inhaled; harmful if swallowed or absorbed through skin; inhalation may result in spasm, inflammation and edema of the larynx and bronchi, chemical pneumonitis and pulmonary edema; symptoms of exposure may include burning sensation; coughing; wheezing; laryngitis; shortness of breath; headache; nausea; vomiting; target organs: blood, kidneys, liver, thymus, heart.

Fire: Hazards: emits toxic fumes. Extinguishing agents: carbon dioxide, dry chemical powder or appropriate foam; do not use water. Precautions: combustible liquid.
Reactivity: Incompatible with: strong oxidizing agents, reacts violently with water. Hazardous decomposition products: toxic fumes of: carbon monoxide, carbon dioxide, nitrogen oxides.
Carcinogenicity: IARC - Not listed; NIOSH - Not listed; NTP - Not listed; ACGIH - Not listed; OSHA - Not listed; EPA - Not listed; MAK - Not listed
Primary Target Organs:

Eyes

Skin

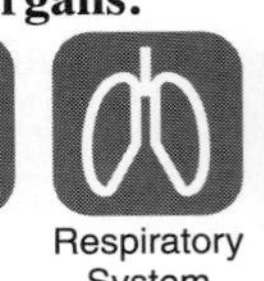
Respiratory System

Liver

Kidneys

Blood

Environmental

Regulations

RCRA 40CFR: Not listed
CERCLA: 40CFR 302.4: Not listed
SARA 40CFR 372.65: Not listed TPQ: 500 lb
SARA EHS 40CFR 355: Listed TPQ: 500 lb
TSCA: Listed

DIB7930 CAS #: 88-58-4

2,5-DI-TERT-BUTYLHYDRO QUINONE

RTECS: MX5160000
EINECS Number: 201-841-8
Molecular Formula: $C_{14}H_{22}O_2$
Formula Weight: 222.36

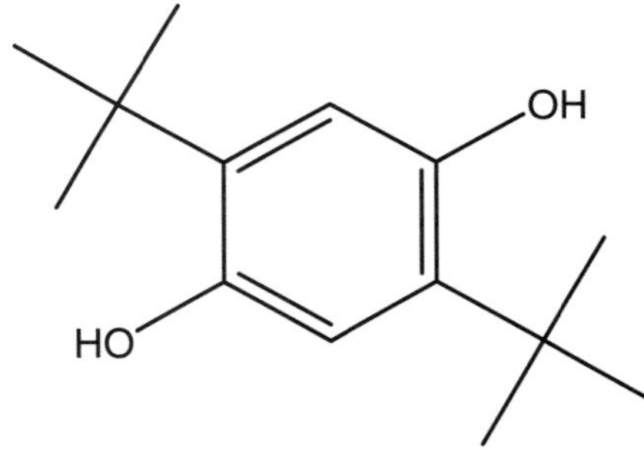

Chemical Structure

Synonyms: 2,5-DI-TERT-BUTYLBENZENE-1,4-DIOL; 2,5-DI-T-BUTYLHYDROQUINONE

Physical Properties

Boiling Point: About 321 °C (610 °F)
Freezing Point: 214 °C (417.2 °F)
Specific Gravity: 1.07
Water Solubility: Negligible

RTECS Toxicity Data

Acute Oral: Rat LD_{Lo} Dose: 800 mg/kg.

Chronic (Multiple Dose) Oral: Rat Dose: 6426 mg/kg/3W-C; Toxic Effects: Blood - Change in clotting factors; DEATH.
Tumorigenic: Hamster Route: Oral; Dose: 134 gm/kg/24W-C; Toxic Effects: Tumorigenic - Neoplastic by RTECS criteria; Gastrointestinal - Tumors.

Hazard Overviews

Health: Irritating to eyes/skin/respiratory tract. Other Acute Effects: may be harmful by inhalation, ingestion, or skin absorption; depending on the intensity and duration of exposure, effects may vary from mild irritation to severe destruction of tissue; prolonged contact can cause damage to the lungs, severe irritation, or burns.
Fire: Extinguishing agents: water spray; carbon dioxide, dry chemical powder or appropriate foam. Precautions: combustible liquid.
Reactivity: Incompatible with: strong oxidizing agents, strong bases. Hazardous decomposition products: toxic fumes of: carbon monoxide, carbon dioxide.
Carcinogenicity: IARC - Not listed; NIOSH - Not listed; NTP - Not listed; ACGIH - Not listed; OSHA - Not listed; EPA - Not listed; MAK - Not listed
Primary Target Organs:

Eyes

Skin

Respiratory System

Environmental

Regulations

RCRA 40CFR: Not listed
CERCLA: 40CFR 302.4: Not listed
SARA 40CFR 372.65: Not listed
SARA EHS 40CFR 355: Not listed
TSCA: Listed

DIB8040 **CAS #: 2082-79-3**

3,5-DI-T-BUTYL-4-HYDROXYHYDROCINNAMIC ACID, OCTADECYL ESTE

EINECS Number: 218-216-0
Molecular Formula: $C_{35}H_{62}O_3$
Formula Weight: 530.89

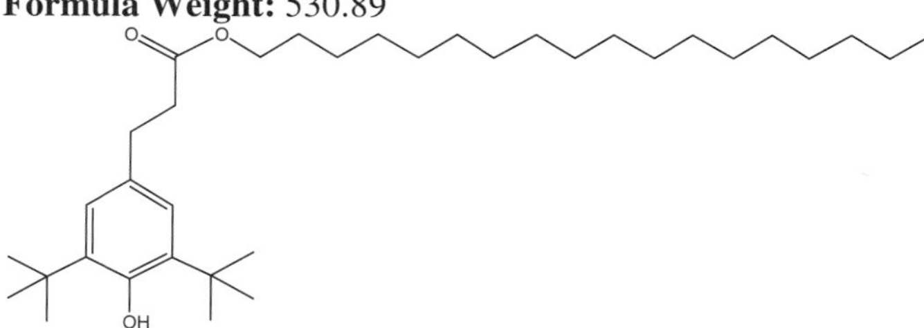

Chemical Structure

Synonyms: ANTIOXIDANT 1076; AO 4; BENZENEPROPANOIC ACID,3,5-BIS(1,1-DIMETHYLETHYL)-4-HYDROXY-,OCTADECYL ESTER; HYDROCINNAMIC ACID,3,5-DI-TERT-BUTYL-4-HYDROXY-,OCTADECYL ESTER; IRGANOX 1076; IRGANOX 1906
Description: white crystalline, free flowing powder; odorless

Hazard Overviews

Health: Irritating to eyes/skin/respiratory tract. Other Acute Effects: may be harmful by inhalation, ingestion, or skin absorption.
Fire: Hazards: emits toxic fumes. Extinguishing agents: water spray; carbon dioxide, dry chemical powder or appropriate foam. Precautions: combustible liquid.
Reactivity: Incompatible with: strong oxidizing agents. Hazardous decomposition products: toxic fumes of: carbon monoxide, carbon dioxide.
Carcinogenicity: IARC - Not listed; NIOSH - Not listed; NTP - Not listed; ACGIH - Not listed; OSHA - Not listed; EPA - Not listed; MAK - Not listed
Primary Target Organs:

Eyes

Skin

Respiratory System

Environmental

Regulations

RCRA 40CFR: Not listed
CERCLA: 40CFR 302.4: Not listed
SARA 40CFR 372.65: Not listed
SARA EHS 40CFR 355: Not listed
TSCA: Listed

DIB8150 **CAS #: 105-76-0**

DIBUTYLMALEATE

RTECS: ON0875000
EINECS Number: 203-328-4
Molecular Formula: $C_{12}H_{20}O_4$
Formula Weight: 228.32

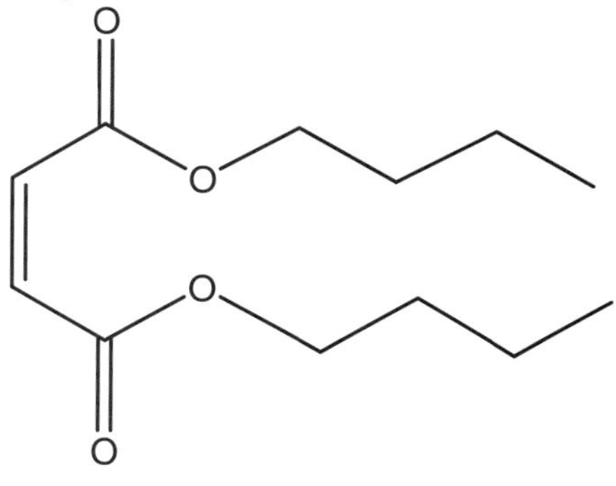

Chemical Structure

Synonyms: 2-BUTENEDIOIC ACID,DIBUTYL ESTER; BUTYL MALEATE; DBM; DIBUTYL MALEATE; DIBUTYLESTER KYSELINY MALEINOVE; PX-504; RC COMONOMER DBM; STAFLEX DBM

Physical Properties

Boiling Point: 281.1 °C (538 °F)
Specific Gravity: 0.991
Saturated Vapor Density: 1.200162838 kg/m^3
Vapor Pressure: 0.002 kPa at 25 °C
Water Solubility: Negligible
Other Solubilities: Soluble in Acetone
Flash Point: 140.5 °C Open Cup

RTECS Toxicity Data

Acute Oral: Rat LD_{50} Dose: 3700 mg/kg.
Acute Dermal: Rabbit LD_{50} Route: Skin; Dose: 10 gm/kg.
Irritation Eye: Rabbit Standard Draize Test Dose: 500 mg/24H; Reaction: mild.
Irritation Skin: Rabbit Open Draize Test Dose: 500 mg open; Reaction: mild.

Hazard Overviews

Fire: Will burn.
Carcinogenicity: IARC - Not listed; NIOSH - Not listed; NTP - Not listed; ACGIH - Not listed; OSHA - Not listed; EPA - Not listed; MAK - Not listed

Environmental

Regulations

RCRA 40CFR: Not listed
CERCLA: 40CFR 302.4: Not listed
SARA 40CFR 372.65: Not listed
SARA EHS 40CFR 355: Not listed
TSCA: Listed

DIB8260 CAS #: 497-39-2

4,6-DI-TERT-BUTYL-3-METHYLPHENOL

RTECS: GO7800000
EINECS Number: 207-847-7
Molecular Formula: $C_{15}H_{24}O$
Formula Weight: 220.34
Synonyms: 2,4-BIS(1,1-DIMETHYLETHYL)-5-METHYLPHENOL; M-CRESOL,4,6-DI-TERT-BUTYL-; DBMC; 4,6-DI-T-BUTYL-M-CRESOL; 4,6-DI-TERT-BUTYL-M-CRESOL; DI-TERT-BUTYL-M-CRESOL; 2,4-DI-T-BUTYL-5-METHYLPHENOL; 3-METHYL-4,6-DI-TERT-BUTYLPHENOL; PHENOL,2,4-BIS(1,1-DIMETHYLETHYL)-5-METHYL-; PHENOL,2,4-BIS(1,1-DIMETHYLETHYL)-5-METHYL-(9CI)
Description: crystals
Use: intermediate in production of rubber chem, modified phenolic resins, synthetic musks of ambrette type; chem int for phenolic resins; chem int for 3-methyl-6-tert-butylphenol; chem int for its phosphite ester; chem int for dealkylation to m-cresol

Physical Properties

Boiling Point: 211 °C (412 °F) at 100 mm Hg
Freezing Point: 62.1 °C (143.78 °F)
Specific Gravity: 0.912 at 80 °C/4 °C
Water Solubility: Practically Insoluble in Water
Other Solubilities: Practically Insoluble in Ethylene Glycol Soluble in Alcohol, Benzene, Carbon Tetrachloride, Ether, Acetone

RTECS Toxicity Data

Acute Oral: Mouse LD_{50} Dose: 1420 mg/kg; Toxic Effects: Lungs, Thorax, or Respiration - Acute pulmonary edema; Gastrointestinal - Ulceration or bleeding from small intestine.

Hazard Overviews

Carcinogenicity: IARC - Not listed; NIOSH - Not listed; NTP - Not listed; ACGIH - Not listed; OSHA - Not listed; EPA - Not listed; MAK - Not listed

Environmental

Regulations

RCRA 40CFR: Not listed
CERCLA: 40CFR 302.4: Not listed
SARA 40CFR 372.65: Not listed
SARA EHS 40CFR 355: Not listed
TSCA: Listed

DIB8370 CAS #: 924-16-3

N,N-DIBUTYLNITROSOAMINE

RTECS: EJ4025000
EINECS Number: 213-101-1
Molecular Formula: $C_8H_{18}N_2O$
Formula Weight: 158.28

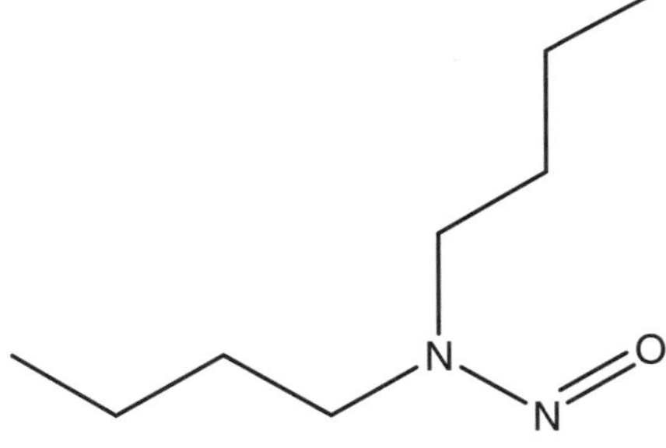

Chemical Structure

Synonyms: 1-BUTANAMINE,N-BUTYL-N-NITROSO-; BUTYLAMINE,N-NITROSODI-; N-BUTYL-N-NITROSO-1-BUTAMINE; DBN; DBNA; DIBUTYLAMINE,N-NITROSO-; DIBUTYLAMINE,N-NITROSO-(6CI); DI-N-BUTYLNITROSAMIN; DI-N-BUTYLNITROSAMINE; DIBUTYLNITROSAMINE; N,N-DI-N-BUTYLNITROSAMINE; NDBA; N-NITROSO-DI-N-BUTYLAMINE; N-NITROSODI-N-BUTYLAMINE; N-NITROSODIBUTYLAMINE; NITROSODIBUTYLAMINE
Description: yellow oil

Use: research chemical, not for commercial purposes; synthesis of di-n-butylhydrazine, although there is no indication that it has been used commercially

Physical Properties

Boiling Point: 116 °C (241 °F) at 14 mm Hg
Specific Gravity: 0.9009 at 20 °C/4 °C
Vapor Pressure: 0.03 mm Hg at 20 °C
Water Solubility: 0.12% at 0 °C
Other Solubilities: organic solvents & vegetable oils
Refraction Index: 1.4475 at 20 °C/D

RTECS Toxicity Data

Acute Oral: Rat LD_{50} Dose: 1200 mg/kg. Hamster LD_{50} Dose: 2150 mg/kg; Toxic Effects: Liver - Other changes; Kidney, Ureter, and Bladder - Chgs in tubules (inc acute renal failure, acute tubular necrosis; Endocrine - Other changes.

Acute Dermal: Rat LD_{50} Route: Subcutaneous Dose: 1200 mg/kg. Hamster LD_{50} Route: Subcutaneous Dose: 561 mg/kg; Toxic Effects: Cardiac - Cardiomyopathy including infarction; Kidney, Ureter, and Bladder - Changes primarily in glomeruli; Endocrine - Other changes.

Reproductive/Teratogenic: Rat Route: Oral; Dose: 1200 mg/kg; Duration: female 12D of pregnancy; Effects on Embryo or Fetus - Fetotoxicity. Rat Route: Oral; Dose: 1200 mg/kg; Duration: female 10D of pregnancy; Effects on Embryo or Fetus - Fetal death.

Mutagenic: Human Unscheduled DNA Synthesis; Cell Type: liver; Dose: 180 umol/L. Human Unscheduled DNA Synthesis; Cell Type: HeLa cell; Dose: 10 umol/L.

Tumorigenic: Rat Route: Oral; Dose: 140 mg/kg/4W-C; Toxic Effects: Tumorigenic - Carcinogenic by RTECS criteria; Gastrointestinal - Tumors. Rat Route: Oral; Dose: 64 gm/kg/20W-C; Toxic Effects: Tumorigenic - Equivocal tumorigenic agent by RTECS criteria; Liver - Tumors. Rat Route: Oral; Dose: 3900 mg/kg/56W-C; Toxic Effects: Tumorigenic - Equivocal tumorigenic agent by RTECS criteria; Gastrointestinal - Tumors. Rat Route: Oral; Dose: 12600 mg/kg/30W-C; Toxic Effects: Tumorigenic - Equivocal tumorigenic agent by RTECS criteria; Gastrointestinal - Tumors; Kidney, Ureter, and Bladder - Tumors. Rat Route: Oral; Dose: 3640 mg/kg/26W-C; Toxic Effects: Tumorigenic - Equivocal tumorigenic agent by RTECS criteria; Liver - Tumors. Rat Route: Oral; Dose: 648 mg/kg/2Y-I; Toxic Effects: Tumorigenic - Carcinogenic by RTECS criteria; Gastrointestinal - Tumors; Liver - Tumors. Rat Route: Oral; Dose: 882 mg/kg/2W-C; Toxic Effects: Tumorigenic - Equivocal tumorigenic agent by RTECS criteria; Kidney, Ureter, and Bladder - Tumors.

Hazard Overviews

Health: Irritating. Toxic. Other Acute Effects: harmful if swallowed, inhaled, or absorbed through skin. Chronic Effects: carcinogen; mutagen; suspected of causing cancers of the lung, nasal sinuses, brain, esophagus, stomach, liver, bladder and kidney.

Fire: Extinguishing agents: carbon dioxide, dry chemical powder or appropriate foam. Precautions: combustible liquid.

Carcinogenicity: IARC - Group 2B, Possibly carcinogenic to humans; NIOSH - Not listed; NTP - Class 2B, Reasonably anticipated to be a carcinogen, sufficient evidence of carcinogenicity from studies in experimental animals; ACGIH - Not listed; OSHA - Not listed; EPA - Class B2, Probable human carcinogen based on animal studies; MAK - Class A2, Unmistakably carcinogenic in animal experimentation only

Primary Target Organs:

Eyes

Skin

Respiratory System

Environmental

Environmental Fate: Volatilization may be an important transport process for material released to the surface of moist soils. In soil it is expected to be relatively mobile and readily transported to groundwater. No information was found on soil biodegradation or hydrolysis. If released to water, it will have very little tendency to sorb to biota, suspended sediments, and sediments. Volatilization from water will probably not be significant. Photolysis may be the most significant removal process in water while hydrolysis is probably not significant. No information was found on biodegradation. A computer estimated half-life in the vapor phase of the atmosphere is 2.8 days.

Environmental Physical Data

Henry's Law Constant: 5.1 $x10^{-6}$
Octanol/Water Partition Coefficient: log K_{ow} = 1.920
Sorption Partition Coefficient: log K_{oc} = 1.920
BCF: calculated at 17

Regulations

RCRA 40CFR: Listed Hazardous Waste No. U172 Toxic Waste
CERCLA: 40CFR 302.4: Listed per RCRA Section 3001 RQ: 10 lb (4.535 kg)
SARA 40CFR 372.65: Listed
SARA EHS 40CFR 355: Not listed
TSCA: Listed

Analytical Methods

Soil: SW846 3640A, 5031, 8015B, 8250A, 8260B, 8270B, 8270C
Water / Groundwater: EPA 1625
Indoor / Expired Air: NIOSH 2522
Plasma: EPA 29

DIB8480 **CAS #: 96-76-4**

2,4-DI-TERT-BUTYLPHENOL

RTECS: SK8260000
EINECS Number: 202-532-0
Molecular Formula: $C_{14}H_{22}O$
Formula Weight: 206.36

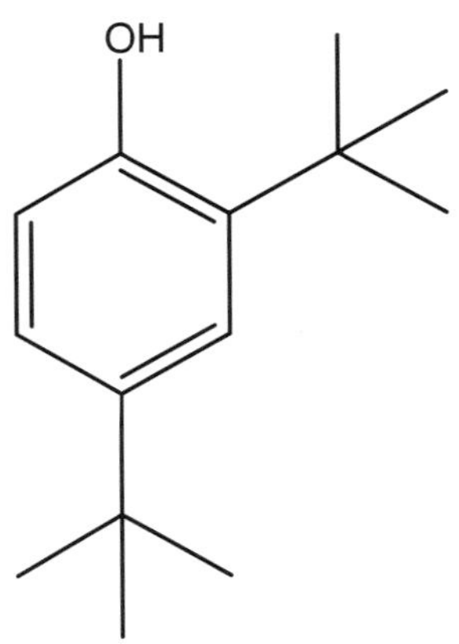

Chemical Structure

Synonyms: ANTIOXIDANT NO. 33; PRODOX 146A-85X; PRODOX 146

Description: white to light strawd crystallized solid; odorless

Physical Properties

Boiling Point: 263 °C (505 °F)
Freezing Point: 57 °C (134.6 °F)
Specific Gravity: 0.92 at 20 °C
Vapor Pressure: Low
Water Solubility: Slightly Soluble
Other Solubilities: Carbon Tetrachloride: Slightly soluble; alkali: Insoluble
Refraction Index: 1.5080
Flash Point: 129 °C

Hazard Overviews

Fire Diamond

Health: Irritating to eyes/skin/respiratory tract. Also Causes: burns on prolonged contact.
Fire: Will burn. Use carbon dioxide, dry chemical, foam, water spray. Use water spray to keep fire-exposed containers cool. Water spray may be used to flush spills away from exposures.
Reactivity: Stable. Hazardous polymerization cannot occur. Avoid: heat; sparks; flames. Incompatible with: strong oxidizing agents; acids; acid chlorides; anhydrides; bases; nitric acid. Hazardous decomposition products: acrid smoke; toxic fumes; carbon monoxide; carbon dioxide.
Carcinogenicity: IARC - Not listed; NIOSH - Not listed; NTP - Not listed; ACGIH - Not listed; OSHA - Not listed; EPA - Not listed; MAK - Not listed

Primary Target Organs:

Eyes

Skin

Respiratory System

Mucous Membranes

Environmental

Regulations

RCRA 40CFR: Not listed
CERCLA: 40CFR 302.4: Not listed
SARA 40CFR 372.65: Not listed
SARA EHS 40CFR 355: Not listed
TSCA: Listed

DIB8590 CAS #: 128-39-2

2,6-DI-TERT-BUTYLPHENOL

RTECS: SK8265000
EINECS Number: 204-884-0
Molecular Formula: $C_{14}H_{22}O$
Formula Weight: 206.33

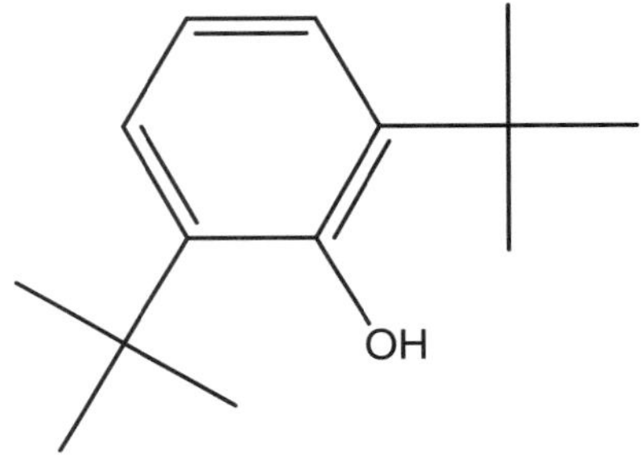

Chemical Structure

Synonyms: 2,6-BIS(TERT-BUTYL)PHENOL; 2,6-BIS(1,1-DIMETHYLETHYL)PHENOL; 2,6-DI-T-BUTYLPHENOL; ETHANOX 701; PHENOL,2,6-BIS(1,1-DIMETHYLETHYL)-; PHENOL,2,6-DI-TERT-BUTYL-

Description: white to light straw crystalline solid; odorless

Use: intermediate, antioxidant; antioxidant in aviation gasoline; antioxidant (oils, petrol, aviation fuels)

Physical Properties

Boiling Point: 253 °C (487 °F)
Freezing Point: 37 °C (98.6 °F)
Specific Gravity: 0.914 at 20 °C
Water Solubility: 2.5 mg/L at 25 °C
Other Solubilities: Soluble in Alcohol and Benzene; Insoluble in water and alkali, Soluble in Acetone, Benzene, carbontetrachloride, ethyl Alcohol, Diethyl Ether, and hydrocarbons.
Refraction Index: 1.5001 at 20 °C
Ionization Potential (eV): 7.70 +/-0.02
Flash Point: 143.33 °C Cleveland Open Cup

Hazard Overviews

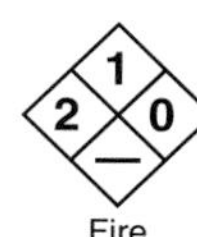

Fire Diamond

Health: Irritating to eyes/skin/respiratory tract. Also Causes: burns on prolonged contact.

Fire: Will burn. Use carbon dioxide, dry chemical, foam, or water spray. Use water spray to keep fire-exposed container cool. Water spray may be used to flush spills away from exposure.

Reactivity: Stable. Hazardous polymerization cannot occur. Avoid: sparks; open flames; excessive heat; elevated temperatures. Incompatible with: strong acids; bases; strong oxidizing; agents; reducing agents. Hazardous decomposition products: toxic fumes; carbon oxides.

Carcinogenicity: IARC - Not listed; NIOSH - Not listed; NTP - Not listed; ACGIH - Not listed; OSHA - Not listed; EPA - Not listed; MAK - Not listed

Primary Target Organs:

Eyes

Skin

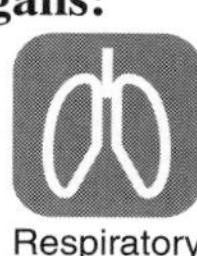
Respiratory System

Mucous Membranes

Environmental

Environmental Fate: If released to the atmosphere, it should mainly exist in the vapor phase based on an estimated vapor pressure of 7.3×10^{-3} mm Hg. Vapor-phase is degraded in the atmosphere by reaction with photochemically produced hydroxyl radicals with an estimated half-life of about 7.8 hours. A measured K_{oc} of 3600 indicates that it will have slight mobility in soil. It may volatilize from moist soils, based on its estimated Henry's Law constant. Photooxidation may be an important fate process on the soil surface. Based on limited data, it seems likely that it will biodegrade in both soil and water. Conversion of 1.1-7.7% of the parent compound to CO2 by an activated sludge inoculum was reported over a 5 day time period; a further 44.9% of this compound was metabolized, but not to completion. In water, it will adsorb to sediment and particulate matter(measured K_{oc} of 3600), volatilize from the surface (estimated half-lives for a model river and lake are 17 and 130 days, respectively), bioaccumulate in aquatic organisms(measured BCF values of 660 and 800), and oxidize in the water column to 2,6-di-tert-butylbenzoquinone. Following 5 days incubation with sludge, 7.7% had volatilized, 46% had biodegraded, and 31.9% was adsorbed to the sludge.

Environmental Physical Data

Henry's Law Constant: estimated at 3.15×10^{-6}

Octanol/Water Partition Coefficient: log K_{OW} = 4.92

Sorption Partition Coefficient: K_{OC} = sediments 3600

BCF: green algae 800

Regulations

RCRA 40CFR: Not listed

CERCLA: 40CFR 302.4: Not listed

SARA 40CFR 372.65: Not listed

SARA EHS 40CFR 355: Not listed

TSCA: Listed

DIB8700 CAS #: 101-96-2

N,N'-DI-SEC-BUTYL-P-PHENYLDIAMINE

RTECS: SS9040000

EINECS Number: 202-992-2

Molecular Formula: $C_{14}H_{24}N_2$

Formula Weight: 220.40

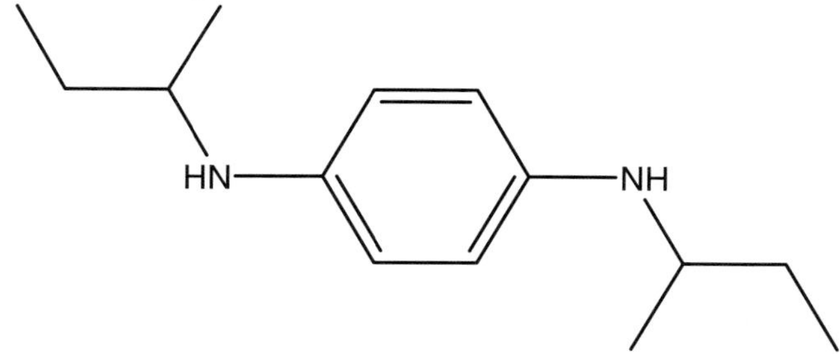

Chemical Structure

Synonyms: ANTIOXIDANT 22; 1,4-BENZENEDIAMINE,N,N'-BIS(1-METHYLPROPYL)-; 1,4-BENZENEDIAMINE,N,N'-BIS(1-METHYLPROPYL)-(9CI); N,N'-BIS(1-METHYLPROPYL)-1,4-BENZENEDIAMINE; CP 40182; N,N'-DI-SEC-BUTYLPARAPHENYLENEDIAMINE; N,N'-DI-S-BUTYL-P-PHENYLENEDIAMINE; N,N'-DI-SEK.BUTYL-P-FENYLENDIAMIN; DU PONT GASOLINE ANTIOXIDANT NO 22; KEROBIT BPD; P-PHENYLENEDIAMINE,N,N'-DI-SEC-BUTYL-; SANTOFLEX 44; TENAMENE 2; TOPANOL M; UOP 5

Description: amber to dark reddish black liquid

Use: oxidation inhibitor and stabilizer in gasoline; antioxidant in aviation gasoline; prevents decomposition of tetraethyl lead in gasoline

Physical Properties

Freezing Point: 17.8 °C (64.04 °F)

Specific Gravity: 0.94 to 0.95 at 24/24 °C

Vapor Pressure: 85.3 mm Hg at 38 °C

Water Solubility: < 1 mg/mL at 20 C

Other Solubilities: 95% Ethanol: >=100 mg/ml at 20 °C; Acetone: >=100 mg/ml at 20 °C; Absolute Alcohol: Soluble; Benzene: Soluble; Caustic solutions: Soluble; DMSO: >=100 mg/ml at 20 °C; Gasoline: Soluble.

Flash Point: 140.556 °C Open Cup

RTECS Toxicity Data

Acute Oral: Rat LD_{50} Dose: 265 mg/kg; Toxic Effects: Sense organs and special senses - Ptosis; Behavioral - Somnolence (general depressed activity); Gastrointestinal - Other changes.

Acute Inhalation: Rat LC_{Lo} Dose: 600 mg/m^3/6hr.

Acute Dermal: Guinea Pig LD_{50} Route: Skin; Dose: 5 gm/kg.

Hazard Overviews

Fire: Will burn.
Carcinogenicity: IARC - Not listed; NIOSH - Not listed; NTP - Not listed; ACGIH - Not listed; OSHA - Not listed; EPA - Not listed; MAK - Not listed

Environmental

Cleanup/Disposal: Guide No. 154: Eliminate all ignition sources (no smoking, flares, sparks or flames in immediate area). Do not touch damaged containers or spilled material unless wearing appropriate protective clothing. Stop leak if you can do it without risk. Prevent entry into waterways, sewers, basements or confined areas. Absorb or cover with dry earth, sand or other non-combustible material and transfer to containers. Do not get water inside containers.

Regulations

RCRA 40CFR: Not listed
CERCLA: 40CFR 302.4: Not listed
SARA 40CFR 372.65: Not listed
SARA EHS 40CFR 355: Not listed
TSCA: Listed

DIB9030 CAS #: 1067-33-0

DIBUTYLTIN DIACETATE

RTECS: WH6880000
DOT: UN2788; IMO6.1
EINECS Number: 213-928-8
Molecular Formula: $C_{12}H_{24}O_4Sn$
Formula Weight: 351.05
Synonyms: BA 2726; BIS(ACETYLOXY)DIBUTYL STANNANE; BIS(ACETYLOXY)DIBUTYLSTANNANE; DIACETOXYBUTYLIN; DIACETOXYBUTYLTIN; DIACETOXYDIBUTLYLTIN; DIACETOXYDIBUTYLSTANNANE; DIACETOXYDIBUTYLTIN; DIBUTLY TIN DIACETATE; DIBUTYL TIN DIACETATE; DIBUTYLDIACETOXYSTANNANE; DIBUTYLSTANNIUM DIACETATE; DI-N-BUTYLTIN DIACETATE; FOMREZ SUL-3; STANNANE,BIS(ACETYLOXY)DIBUTYL-; STANNANE,DIACETOXYDIBUTYL-; T 1; T 1 (CATALYST); TI(CATALYST); TIN,DIBURYL-,DIACETATE; TIN,DIBUTYL-,DIACETATE
Description: colorless to yellow clear liquid; slight acetic acid odor
Use: stabilizer for chlorinated organics; catalyst for condensation reactions

Physical Properties

Boiling Point: 142 °C (288 °F) to 145 °C (293 °F) at 10 atm
Freezing Point: 10 °C (50 °F)
Specific Gravity: 1.31 at 25 °C
Vapor Density: 12.1
Saturated Vapor Density: 1.222794828 kg/m^3
Density: 1.31 g/mL at 25 °C
Vapor Pressure: 1.3 mm Hg at 20 °C
Water Solubility: Insoluble in Water
Other Solubilities: 95% Ethanol: >=100 mg/ml at 23 °C; Acetone: >=100 mg/ml at 23 °C; DMSO: <1 mg/ml at 15 °C ; forms white precipitate.
Odor Threshold: 1.0 ppm
Refraction Index: 1.482 at 20 °C
Evaporation Rate: < 1 Butyl Acetate=1
Flash Point: 143 °C Open Cup

RTECS Toxicity Data

Acute Oral: Rat LD_{50} Dose: 32 mg/kg. Mouse LD_{50} Dose: 46 mg/kg.
Acute Dermal: Rabbit LD_{50} Route: Skin; Dose: 2318 mg/kg; Toxic Effects: Skin and appendages - Dermatitis, other.
Irritation Skin: Rabbit Standard Draize Test Dose: 500 mg/30M; Reaction: severe.
Reproductive/Teratogenic: Rat Route: Oral; Dose: 110 mg/kg; Duration: female 7-17D of pregnancy; Effects on Embryo or Fetus - Fetal death; Specific Developmental Abnormalities - Central nervous system; Craniofacial (including nose and tongue). Rat Route: Oral; Dose: 15200 ug/kg; Duration: female 8D of pregnancy; Specific Developmental Abnormalities - Musculoskeletal system. Rat Route: Oral; Dose: 18700 ug/kg; Duration: female 7-17D of pregnancy; Maternal Effects - Other effects on females. Rat Route: Oral; Dose: 28080 ug/kg; Duration: female 8D of pregnancy; Specific Developmental Abnormalities - Craniofacial (including nose and tongue); Musculoskeletal system.

Hazard Overviews

Poison

Corrosive

Health: Corrosive to eyes/skin/respiratory tract. Poison. Other Acute Effects: may be fatal if inhaled, swallowed, or absorbed through skin; readily absorbed through skin; inhalation may result in spasm, inflammation and edema of the larynx and bronchi, chemical pneumonitis and pulmonary edema; symptoms of exposure may include burning sensation; coughing; wheezing; laryngitis; shortness of breath; headache; nausea; vomiting. target organs: nerves, blood, skin, eyes.
Fire: Will burn. Hazards: emits toxic fumes. Extinguishing agents: water spray; carbon dioxide, dry chemical powder or appropriate foam. Precautions: combustible liquid.
Reactivity: Incompatible with: strong oxidizing agents, strong acids, strong bases, may decompose on exposure to moist air or water, sensitive to light. Hazardous decomposition products: toxic fumes of: carbon monoxide, carbon dioxide, tin/tin oxides.
Carcinogenicity: IARC - Not listed; NIOSH - Not listed; NTP - Not listed; ACGIH - Class A4, Not classifiable as a human carcinogen; OSHA - Not listed; EPA - Not listed; MAK - Not listed

Primary Target Organs:

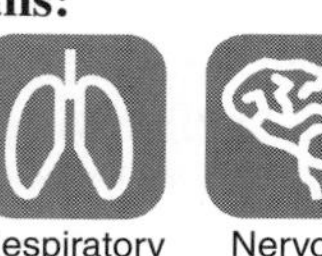

Eyes Skin Respiratory System Nervous System Blood

Exposure Limits
OSHA PEL: TWA: 0.1 mg/m^3; as Sn.
OSHA PEL Vacated 1989 Limits: TWA: 0.1 mg/m^3.
ACGIH TLV: TWA: 2 mg/m^3; see NIOSH.
NIOSH REL: TWA: 2 mg/m^3; inorganic (.1 if organic).
DFG MAK: TWA: 2 mg/m^3; See NIOSH.

Environmental

Cleanup/Disposal: Guide No. 153: Eliminate all ignition sources (no smoking, flares, sparks or flames in immediate area). Do not touch damaged containers or spilled material unless wearing appropriate protective clothing. Stop leak if you can do it without risk. Prevent entry into waterways, sewers, basements or confined areas. Absorb or cover with dry earth, sand or other non-combustible material and transfer to containers. Do not get water inside containers.

Environmental Physical Data

BCF: concentrates to few mg/kg

Regulations

RCRA 40CFR: Not listed
CERCLA: 40CFR 302.4: Not listed
SARA 40CFR 372.65: Not listed
SARA EHS 40CFR 355: Not listed
TSCA: Listed

DIB9140 **CAS #: 683-18-1**

DI-N-BUTYLTIN DICHLORIDE

RTECS: WH7100000
DOT: UN2788; UN3146; IMO6.1
EINECS Number: 211-670-0
Molecular Formula: $C_8H_{18}Cl_2Sn$
Formula Weight: 303.85

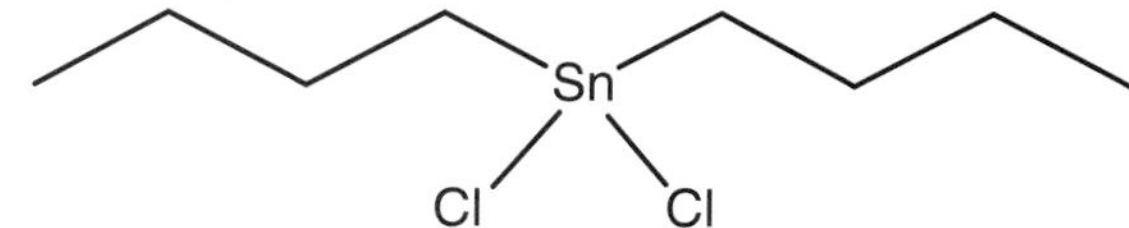

Chemical Structure

Synonyms: CHLORID DI-N-BUTYLCINICITY; D.B.T.C; DBTC; DIBUTYLDICHLOROSTANNANE; DIBUTYLDICHLOROTIN; DIBUTYLSTANNIUM DICHLORIDE; DIBUTYLTIN CHLORIDE; DIBUTYLTIN DICHLORIDE; DI-N-BUTYL-ZINN-DICHLORID; DICHLORODIBUTYLSTANNANE; DICHLORODIBUTYLTIN; STANNANE,DIBUTYLDICHLORO-; TIN,DIBUTYL-,DICHLORIDE
Description: colorless to white or light tan solid, needles, or semisolid
Use: urethane & esterification catalyst for plasticizers, lubricants, & heat-transfer fluids; intermed for organotins, dibutyltin compounds; cocatalyst for mfr of terpene resins; organotin compounds (usually tributyltin oxide or tributyltin fluoride) in antifouling paints; dialkyltin compounds, stabilizers for polyvinyl chloride; catalysts for high resiliency foam in automotive seating.

Physical Properties

Boiling Point: 118 °C (244 °F) to 170 °C (338 °F) at 28 mm Hg
Freezing Point: 142 °C (287.6 °F) at 10 mm Hg
Specific Gravity: 1.36 at 50 °C
Vapor Pressure: 2 mm Hg at 100 °C
Water Solubility: Insoluble
Other Solubilities: Soluble in Ether, Benzene, Alcohol
Refraction Index: 1.499 at 50 °C
Flash Point: 168 °C

RTECS Toxicity Data

Acute Oral: Rat LD_{50} Dose: 100 mg/kg. Mouse LD_{50} Dose: 70 mg/kg.
Acute Dermal: Rabbit LD_{Lo} Route: Skin; Dose: 1360 mg/kg.
Chronic (Multiple Dose) Oral: Rat Dose: 162 mg/kg/8W-C; Toxic Effects: Liver - Other changes; Nutritional and gross metabolic - Weight loss or decreased weight gain. Rat Dose: 378 mg/kg/6W-C; Toxic Effects: Nutritional and gross metabolic - Weight loss or decreased weight gain. Rat Dose: 70 mg/kg/4W-C; Toxic Effects: Immunological including allergic - Decrease in humoral immune response. Rat Dose: 20 mg/kg/10D-I; Toxic Effects: Endocrine - Other changes; Endocrine - Changes in thymus weight; Biochemical - Other proteins.
Chronic (Multiple Dose) Dermal: Rat Route: Skin; Dose: 400 mg/kg/5D-I; Toxic Effects: Liver - Other changes; DEATH.
Irritation Eye: Rabbit Standard Draize Test Dose: 50 ug/24H; Reaction: severe.
Irritation Skin: Rabbit Standard Draize Test Dose: 2 mg/24H; Reaction: severe.
Reproductive/Teratogenic: Rat Route: Oral; Dose: 45 mg/kg; Duration: female 7-15D of pregnancy; Effects on Embryo or Fetus - Fetotoxicity; Specific Developmental Abnormalities - Craniofacial (including nose and tongue); Musculoskeletal system. Rat Route: Oral; Dose: 20 mg/kg; Duration: female 8D of pregnancy; Effects on Fertility - Post-implantation mortality; Effects on Embryo or Fetus - Fetotoxicity; Specific Developmental Abnormalities - Musculoskeletal system. Rat Route: Oral; Dose: 24308 ug/kg; Duration: female 8D of pregnancy; Effects on Embryo or Fetus - Fetotoxicity; Specific Developmental Abnormalities - Craniofacial (including nose and tongue); Musculoskeletal system. Rat Route: Oral; Dose: 20 mg/kg; Duration: female 7-8D of pregnancy; Effects on Fertility - Post-implantation mortality; Litter size; Effects on Embryo or Fetus - Fetotoxicity.
Mutagenic: Rabbit Unscheduled DNA Synthesis; Cell Type: other cell types; Dose: 10 ug/L. Rabbit DNA Inhibition; Cell Type: other cell types; Dose: 100 ug/L.

Hazard Overviews

Poison

Corrosive

Health: Corrosive to eyes/skin/respiratory tract. Poison. Other Acute Effects: harmful if swallowed, inhaled, or absorbed through skin; readily absorbed through skin; inhalation may result in spasm, inflammation and edema of the larynx and bronchi, chemical pneumonitis and pulmonary edema; symptoms of exposure may include burning sensation; coughing; wheezing; laryngitis; shortness of breath; headache; nausea; vomiting.

Fire: Will burn. Hazards: water hydrolyzes material liberating acidic gas which in contact with metal surfaces can generate flammable and/or explosive hydrogen gas; emits toxic fumes. Extinguishing agents: water spray; carbon dioxide, dry chemical powder or appropriate foam. Precautions: combustible liquid.

Reactivity: Incompatible with: heat, may decompose on exposure to moist air or water. Hazardous decomposition products: carbon monoxide, carbon dioxide, tin/tin oxides, hydrogen chloride gas.

Carcinogenicity: IARC - Not listed; NIOSH - Not listed; NTP - Not listed; ACGIH - Class A4, Not classifiable as a human carcinogen; OSHA - Not listed; EPA - Not listed; MAK - Not listed

Primary Target Organs:

Eyes

Skin

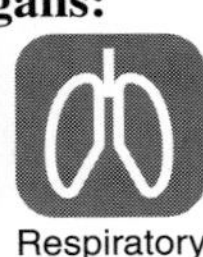
Respiratory System

Exposure Limits

OSHA PEL: TWA: 0.1 mg/m^3; as Sn.

OSHA PEL Vacated 1989 Limits: TWA: 0.1 mg/m^3.

ACGIH TLV: TWA: 2 mg/m^3; see NIOSH.

NIOSH REL: TWA: 2 mg/m^3; inorganic (.1 if organic).

DFG MAK: TWA: 2 mg/m^3; See NIOSH.

Environmental

Environmental Fate: If released to soil, it may adsorb to soil, and biodegrade. Significant adsorption to soil may limit leaching. If released to water, it will behave like a simple protic acid due to the resulting formation of hydronium ions and either dibutyltin dihydroxide or other dibutyltin-hydroxide species. In seawater, the chloride ion may compete effectively with hydroxide ions resulting in the presence of it in the mixture of dibutyltin species. It may be susceptible to direct photolysis, may bioconcentrate in certain tissues of aquatic organisms, and it may biodegrade. It may adsorb to sediments and suspended particulate matter, but, at least in seawater, will probably mainly reside in the solution phase. Dibutyltin compounds may react with sulfides present in sediment, leading to the formation of dibutyltin sulfide. The dissociation and speciation in the environment may affect its fate and transport processes. If released to the atmosphere, it probably will be present both in the vapor and particulate phases. The compound may be susceptible to direct photolysis and may be susceptible to the gas-phase reaction with photochemically produced hydroxyl radicals.

Cleanup/Disposal: Guide No. 153: Eliminate all ignition sources (no smoking, flares, sparks or flames in immediate area). Do not touch damaged containers or spilled material unless wearing appropriate protective clothing. Stop leak if you can do it without risk. Prevent entry into waterways, sewers, basements or confined areas. Absorb or cover with dry earth, sand or other non-combustible material and transfer to containers. Do not get water inside containers.

Environmental Physical Data

BCF: carp 12

Regulations

RCRA 40CFR: Not listed

CERCLA: 40CFR 302.4: Not listed

SARA 40CFR 372.65: Not listed

SARA EHS 40CFR 355: Not listed

TSCA: Listed

DIB9250 CAS #: 77-58-7

DIBUTYLTIN DILAURATE

RTECS: WH7000000

DOT: UN2788; UN3146; IMO6.1

EINECS Number: 201-039-8

Molecular Formula: $C_{32}H_{64}O_4Sn$

Formula Weight: 631.55

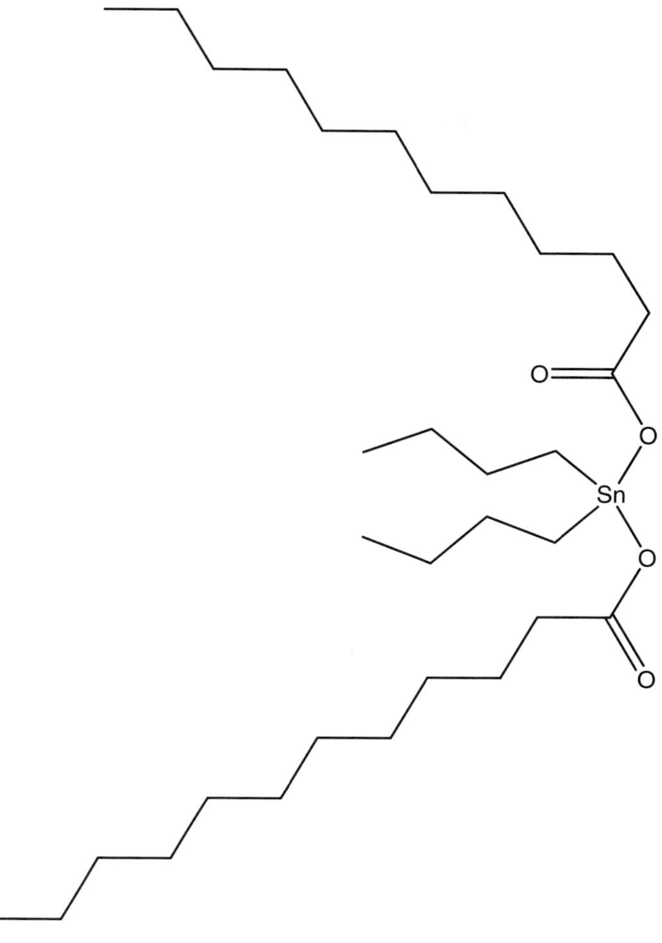

Chemical Structure

Synonyms: ADVASTAB 52; BIS(DODECANOYLOXY)DI-N-BUTYLSTANNANE; BIS(LAUROYLOXY)DI(N-BUTYL)STANNANE; BUTYL NORATE; BUTYNORATE; DAVAINEX; DBTL; DIBUTYLBIS(LAURATO)TIN; DIBUTYLBIS(LAUROXY)STANNANE; DIBUTYLBIS(LAUROYLOXY) TIN; DIBUTYLBIS(LAUROYLOXY)TIN; DIBUTYLSTANNIUM DILAURATE; DIBUTYLSTANNYLENE DILAURATE; DIBUTYLTIN DIDODECANOATE; DIBUTYLTIN N-DODECANOATE; DIBUTYLTIN LAURATE; DIBUTYL-ZINN-DILAURAT; DXR 81; FOMREZ SUL-4; KS 20; LANKROMARK LT 173; LAUDRAN; LAUDRAN DI-N-BUTYLCINICITY; LAURIC ACID,DIBUTYLSTANNYLENE DERIV; LAURIC ACID,DIBUTYLSTANNYLENE DERIVATIVE; LAURIC ACID,DIBUTYLSTANNYLENE SALT; LAURIC ACID,DIBUTYLTIN DERIV; LAURIC ACID,DIBUTYLTIN DERIVATIVE; MARK 1038; SM 2014C; STABILIZER D-22; STANCLERE DBTL; STANNANE,BIS(DODECANOYLOXY) DI-N-BUTYL-; STANNANE,BIS(DODECANOYLOXY)DI-N-BUTYL; STANNANE,BIS(LAUROYLOXY)DIBUTYL-; STANNANE,DIBUTYLBIS(LAUROYLOXY)-; STANNANE,DIBUTYLBIS((1-OXODODECYL)OXY)-; STAVINCOR 1200 SN; STAVINOR 1200 SN; T 12; T 12 (CATALYST); THERM CHEK 820; TIN DIBUTYL DILAURATE; TIN,DIBUTYLBIS(LAUROYLOXY)-; TIN,DI-N-BUTYL-,DI(DODECANOATE); TINOSTAT; TN 12; TN 12 (CATALYST); TVS TIN LAU; TVS-TL 700

Description: soft crystals or yellow, oily liquid

Use: tapeworm drug for chickens; stabilizer for vinyl resins, lacquers, elastomers catalyst for urethane and silicones

Physical Properties

Boiling Point: 205 °C (401 °F) at 1.3 kPa
Freezing Point: 22 °C (71.6 °F) to 24 °C (75.2 °F)
Specific Gravity: 1.066 at 20 °C/20 °C
Vapor Density: 21.8
Density: 21.8
Vapor Pressure: 2.36 x10^{-5} mm Hg
Water Solubility: < 1 mg/mL at 20 C
Other Solubilities: 95% Ethanol: >=100 mg/ml at 20 °C; Acetone: >=100 mg/ml at 20 °C; Benzene: Soluble; Carbon Tetrachloride: Soluble; DMSO: <1 mg/ml at 20 °C; Ether: Soluble; Methanol: Insoluble; Petroleum Ether: Soluble.
Refraction Index: 1.4683 at 20 °C/D
Flash Point: 235 °C Open Cup

RTECS Toxicity Data

Acute Oral: Rat LD_{50} Dose: 175 mg/kg. Mouse LD_{Lo} Dose: 710 mg/kg.

Acute Dermal: Rabbit LD; Route: Skin; Dose: >2 gm/kg; Toxic Effects: Gastrointestinal - Hypermotility, diarrhea; Skin and appendages - Dermatitis, other; Nutritional and gross metabolic - Weight loss or decreased weight gain.

Chronic (Multiple Dose) Oral: Rat Dose: 120 mg/kg/3D-I; Toxic Effects: Brain and coverings - Other degenerative changes; Behavioral - Somnolence (general depressed activity); Biochemical - Lipids including transport. Rat Dose: 262 mg/kg/15D-I; Toxic Effects: Nutritional and gross metabolic - Weight loss or decreased weight gain; Biochemical - Hepatic microsomal mixed oxidase(dealkylation, hyroxylation,etc); DEATH.

Irritation Eye: Rabbit Standard Draize Test Dose: 100 mg/24H; Reaction: moderate.

Irritation Skin: Rabbit Standard Draize Test Dose: 500 mg; Reaction: severe.

Reproductive/Teratogenic: Rat Route: Oral; Dose: 50532 ug/kg; Duration: female 8D of pregnancy; Specific Developmental Abnormalities - Craniofacial (including nose and tongue); Musculoskeletal system.

Hazard Overviews

Corrosive

Health: Corrosive to eyes/skin/respiratory tract. Toxic. Other Acute Effects: harmful if swallowed, inhaled, or absorbed through skin; extremely destructive to tissue of the mucous membranes and upper respiratory tract, eyes and skin; inhalation may result in spasm, inflammation and edema of the larynx and bronchi, chemical pneumonitis and pulmonary edema; symptoms of exposure may include burning sensation, coughing, wheezing, laryngitis, shortness of breath, headache, nausea and vomiting; target organs: G.I. system, liver, nerves.

Fire: Will burn. Hazards: emits toxic fumes. Extinguishing agents: water spray; carbon dioxide, dry chemical powder or appropriate foam. Precautions: combustible liquid.

Reactivity: Incompatible with: strong oxidizing agents, strong bases. Hazardous decomposition products: toxic fumes of: carbon monoxide, carbon dioxide, tin/tin oxides.

Carcinogenicity: IARC - Not listed; NIOSH - Not listed; NTP - Not listed; ACGIH - Class A4, Not classifiable as a human carcinogen; OSHA - Not listed; EPA - Not listed; MAK - Not listed

Primary Target Organs:

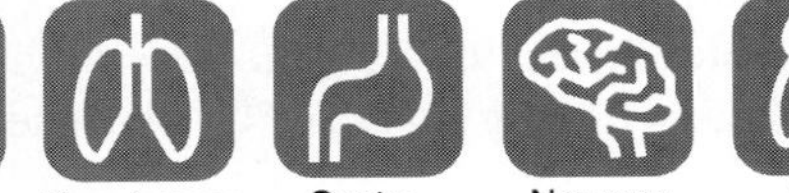
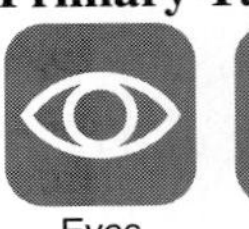

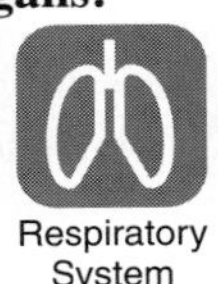
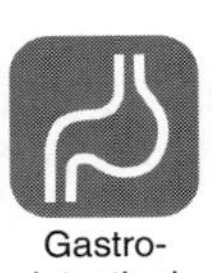

Eyes | Skin | Respiratory System | Gastro-intestinal | Nervous System | Liver

Exposure Limits

OSHA PEL: TWA: 0.1 mg/m^3; as Sn.

OSHA PEL Vacated 1989 Limits: TWA: 0.1 mg/m^3.

ACGIH TLV: TWA: 2 mg/m^3; see NIOSH.

NIOSH REL: TWA: 2 mg/m^3; inorganic (.1 if organic).

DFG MAK: TWA: 2 mg/m^3; See NIOSH.

Environmental

Environmental Fate: If released to soil, it may be subject to adsorption to soil and biodegradation. Significant adsorption to soil may limit leaching. If released to water, it will behave like a simple protic acid due to the resulting formation of hydronium ion and either dibutyltin dihydroxide or other dibutyltin-hydroxide species. In seawater, the chloride ion may compete effectively with hydroxide ions resulting in the presence of dibutyltin dichloride in the mixture of dibutyltin species. It may be susceptible to direct photolysis, may bioconcentrate in certain tissues of aquatic organisms, and it may biodegrade. It may adsorb to sediments and suspended particulate matter, but in seawater it will probably mainly reside in the solution phase. The dissociation and speciation in the environment may affect its fate and transport processes. Dibutyltin compounds may react with sulfides present in sediment, leading to the formation of dibutyltin sulfide. If released to the atmosphere, it will probably be present in both the vapor and particulate phases and may be susceptible to direct photolysis and the gas-phase reaction with photochemically produced hydroxyl radicals.

Cleanup/Disposal: Guide No. 153: Eliminate all ignition sources (no smoking, flares, sparks or flames in immediate area). Do not touch damaged containers or spilled material unless wearing appropriate protective clothing. Stop leak if you can do it without risk. Prevent entry into waterways, sewers, basements or confined areas. Absorb or cover with dry earth, sand or other non-combustible material and transfer to containers. Do not get water inside containers.

Environmental Physical Data

BCF: carp 31

Regulations

RCRA 40CFR: Not listed

CERCLA: 40CFR 302.4: Not listed

SARA 40CFR 372.65: Not listed

SARA EHS 40CFR 355: Not listed

TSCA: Listed

DIC1000 — CAS #: 1918-00-9

DICAMBA

RTECS: DG7525000

EINECS Number: 217-635-6

Molecular Formula: $C_8H_6Cl_2O_3$

Structured MF: $CH_3OC_6H_2Cl_2CO_2H$

Formula Weight: 221.04

Chemical Structure

Synonyms: ACIDO (3,6-DICLORO-2-METOSSI)-BENZOICO; O-ANISIC ACID,3,6-DICHLORO-; BANEX; BANLEN; BANVEL; BANVEL 4S; BANVEL 480; BANVEL CST; BANVEL D; BANVEL D (VELSICOL); BANVEL HERBICIDE; BANVEL II HERBICIDE; BANVEL 4WS; BENZOIC ACID,3,6-DICHLORO-2-METHOXY-; BRUSH BUSTER; COMPOUND B DICAMBA; COMPOUND B (VELSICOL); DIANAT; DIANATE; DICAMBE; 3,6-DICHLOOR-2-METHOXY-BENZOEIZUUR; 3,6-DICHLOR-3-METHOXY-BENZOESAEURE; 3,6-DICHLORO-O-ANISIC ACID; 2,5-DICHLORO-6-METHOXYBENZOIC ACID; 3,6-DICHLORO-2-METHOXYBENZOIC ACID; KYSELINA 3,6-DICHLOR-2-METHOXYBENZOOVA; MDBA; MEDIBEN; 2-METHOXY-3,6-DICHLOROBENZOIC ACID; VELSICOL COMPOUND; VELSICOL 58-CS-11; VELSICOL 58-CS-II

Description: colorless solid; odorless

Use: selective pre- & postemergence herbicide; for spot treatment to control annual & perennial broadleaf weeds in corn, small grains, sugarcane, asparagus, pastures, noncropland areas (eg, fence rows, roadways & wasteland); after harvest of one crop but before planting next crop ; for weed control in cereals (not undersown), usually formulated with one or more phenoxyalkanoic acids; plant growth regulator.

Physical Properties

Freezing Point: 114 °C (237.2 °F) to 116 °C (240.8 °F)

Specific Gravity: 1.57 at 25 °C

Vapor Density: 4.5 mPa at 25 °C

Saturated Vapor Density: 1.200000356 kg/m^3

Vapor Pressure: 3.4 x10^{-5} mm Hg at 25 °C

Water Solubility: Slightly Soluble in Water 4500 ppm at 25 °C

Other Solubilities: Solubility in 100 ml solvent at 25 °C: Ethanol 92.2 g; heavy aromatic naphthanes 6.0 g; Xylene 9.0 g; diAcetone Alcohol 91.0 g; Acetone 81.0 g; methylene chloride 26.0 g; cyclohexanone 91.6 g. Insoluble in pentane.

Odor Threshold: 250.8 ppm

Flash Point: Nonflammable

RTECS Toxicity Data

Acute Oral: Rat LD_{50} Dose: 1039 mg/kg. Mouse LD_{50} Dose: 1190 mg/kg.
Acute Dermal: Rabbit LD_{50} Route: Skin; Dose: >2 gm/kg. Rat LD_{50} Route: Skin; Dose: >1 gm/kg.
Mutagenic: Rat DNA Damage; Route: Intraperitoneal; Dose: 17800 ug/kg. Rat Unscheduled DNA Synthesis; Cell Type: lymphocyte; Dose: 200 mg/L.

Hazard Overviews

Health: Severely irritating to eyes; irritating to skin/respiratory tract. Harmful. Other Acute Effects: harmful if swallowed, inhaled, or absorbed through skin; may cause allergic skin reaction. Chronic Effects: possible mutagen. The toxicological properties have not been thoroughly investigated.
Fire: Noncombustible. Hazards: may form explosive mixtures with air; emits toxic fumes. Extinguishing agents: water spray; carbon dioxide, dry chemical powder or appropriate foam. Precautions: combustible liquid.
Reactivity: Stable. Hazardous polymerization will not occur. Incompatible with: strong oxidizing agents. Hazardous decomposition products: toxic fumes of: carbon monoxide, carbon dioxide, hydrogen chloride gas.
Carcinogenicity: IARC - Not listed; NIOSH - Not listed; NTP - Not listed; ACGIH - Not listed; OSHA - Not listed; EPA - Not listed; MAK - Not listed
Primary Target Organs:

Eyes

Skin

Respiratory System

Environmental

Ecotoxicity: LC_{50} Asellus greater than 100 mg/l/96 hr at 15 °C, mature /static bioassay without aeration, pH 7.2-7.5, water hardness 40-50 mg/l as calcium carbonate and alkalinity of 30-35 mg/l EC_{50} Daphnia magna greater than 100 mg/l/48 hr at 21 °C, first instar / effect: immobilization. static bioassay without aeration, pH 7.2-7.5, water hardness 40-50 mg/l as calcium carbonate and alkalinity of 30-35 mg/l LC_{50} Salmo gairdneri (rainbow trout) 28 mg/l/96 hr at 12 °C, wt 0.8 g /static bioassay without aeration, pH 7.2-7.5, water hardness 40-50 mg/l as calcium carbonate and alkalinity of 30-35 mg/l LC_{50} Paleomonetes greater than 56 mg/l/96 hr at 21 °C, mature /static bioassay without aeration, pH 7.2-7.5, water hardness 40-50 mg/l as calcium carbonate and alkalinity of 30-35 mg/l
Environmental Fate: If released to soil, microbial degradation will probably be the major removal process under most conditions. The principal soil metabolite appears to be 3,6-dichlorosalicylic acid. It is very mobile in most soils and significant leaching is possible. Based on the results of one study, volatilization from soil surfaces may not be an important process, although some volatilization may occur from plant surfaces. The half-life in soil has been observed to vary from 4 to 555 days with the typical half-life being 1 to 4 weeks. Under conditions suitable to rapid metabolism, the half-life is less than two weeks. If released to water, microbial degradation appears to be the important dicamba removal process; photolysis may contribute to its removal from water. Aquatic hydrolysis, volatilization, adsorption to sediment, and bioconcentration are not expected to be significant. If released to the atmosphere, it will probably exist in both the vapor-phase and the adsorbed to particulate phase. The half-life for the vapor-phase reaction with photochemically produced hydroxyl radicals has been estimated to be 6 days. Particulate phase will be subject to wet and dry deposition.
Cleanup/Disposal: Guide No. 151: Do not touch damaged containers or spilled material unless wearing appropriate protective clothing. Stop leak if you can do it without risk. Prevent entry into waterways, sewers, basements or confined areas. Cover with plastic sheet to prevent spreading. Absorb or cover with dry earth, sand or other non-combustible material and transfer to containers. Do not get water inside containers.

Environmental Physical Data

Henry's Law Constant: estimated at 2×10^{-9}
Octanol/Water Partition Coefficient: log K_{ow} = 2.21
Sorption Partition Coefficient: K_{oc} = 4.4
BCF: none observed

Regulations

RCRA 40CFR: Not listed
CERCLA: 40CFR 302.4: Listed per CWA Section 311(b)(4) RQ: 1000 lb (453.5 kg)
SARA 40CFR 372.65: Listed
SARA EHS 40CFR 355: Not listed
TSCA: Not listed

Analytical Methods

Soil: SW846 8150B, 8151, 8321, 8321A
Water / Groundwater: EPA P-008-1, D5317, 1658, 615, X_89_176; ASTM D4763
Drinking Water: EPA 515.1, 515.2, 555; AOAC 992.32
Food: AOAC 969.07, 984.07
Plasma: FDA 221.1

DIC1050 CAS #: 2463-84-5

DICAPTHON

RTECS: TE7875000
Molecular Formula: $C_8H_9ClNO_5PS$
Structured MF: $(CH_3O)_2(P{=}S)O$
Formula Weight: 297.66
Synonyms: AC 4124; ACC 4124; AMERICAN CYANAMID 4,124; AMERICAN CYANAMID 4124; BAY 14981; BAYER 22/190; CAPTEC; O-(4-CHLOOR-3-NITRO-FENYL)-O,O-DIMETHYLMONOTHIOFOSFAAT; O-(4-CHLOR-3-NITRO-PHENYL)-O,O-DIMETHYL-MONOTHIOPHOSPHAT; O-(2-CHLORO-4-NITROPHENYL) O,O-DIMETHYL ESTER; 2-CHLORO-4-NITROPHENYL DIMETHYL PHOSPHOROTHIOATE; O-(2-CHLORO-4-NITROPHENYL) O,O-DIMETHYL PHOSPHOROTHIOATE; O-2-CHLORO-4-NITROPHENYL O,O-DIMETHYL PHOSPHOROTHIOATE; O-(2-CHLORO-4-NITROPHENYL)O,O-DIMETHYL

PHOSPHOROTHIOATE; O-(2-CHLORO-4-NITROPHENYL)O,O-DIMETHYLPHOSPHOROTHIOATE; CHLORTHION; O-(4-CLORO-3-NITRO-FENIL)-O,O-DIMETIL-MONOTIOFOSFATO; DI-CAPTAN; DICAPTAN; DICAPTHION; DICAPTON; O,O-DIMETHYL O-2-CHLORO-4-NITROPHENYL PHOSPHOROTHIOATE; DIMETHYL 2-CHLORO-4-NITROPHENYL THIOPHOSPHATE; O,O-DIMETHYL O-(2-CHLORONITROPHENYL-4-NITROPHENYL)PHOSPHOROTHIOATE; O,O-DIMETHYL-O-(2-CHLORO-4-NITROPHENYL)THIONOPHOSPHATE; ENT 17 035; ENT 17,035; EXPERIMENTAL INSECTICIDE 4124; INSECTICIDE ACC 4124; ISOCHLOORTHION; ISOCHLOROTHION; ISOCHLORTHION; ISOMERIC CHLORTHION; ISOMERIC CLORTHIO; P-NITRO-O-CHLOROPHENYL DIMETHYL THIONOPHOSPHATE; OMS-214; OMS 214; PHENOL,2-CHLORO-4-NITRO-,O-ESTER WITH O,O-DIMETHYLPHOSPHOROTHIOATE; PHOSPHOROTHIOIC ACID O-(2-CHLORO-4-NITROPHENYL)O,O-DIMETHYL ESTER; PHOSPHOROTHIOIC ACID,O-(2-CHLORO-4-NITROPHENYL)O,O-DIMETHYL ESTER; THIOPHOSPHATE DE O,O-DIMETHYLE ET DE O-4-CHLORO-3-NITROPHENYLE; THIOPHOSPHATE DE O,O-DIMETHYLE ET DEO-4-CHLORO-3-NITROPHENYLE

Description: white solid or crystals

Use: insecticide; aphicide; acaracide

Physical Properties

Freezing Point: 53 °C (127.4 °F)
Vapor Pressure: 3.6 x10^{-6} mm Hg
Water Solubility: Practically Insoluble in Water
Other Solubilities: Soluble in Cyclohexane.
Flash Point: Nonflammable

RTECS Toxicity Data

Acute Oral: Rat LD_{50} Dose: 284 mg/kg.
Acute Dermal: Rat LD_{50} Route: Skin; Dose: 790 mg/kg; Toxic Effects: Behavioral - Tremor; Behavioral - Convulsions or effect

Hazard Overviews

Fire: Noncombustible.
Carcinogenicity: IARC - Not listed; NIOSH - Not listed; NTP - Not listed; ACGIH - Not listed; OSHA - Not listed; EPA - Not listed; MAK - Not listed

Environmental

Ecotoxicity: TLm Mercenaria mercenaria (hard clam) 3.34 mg/l/48 hr, eggs /Conditions of bioassay not specified

Environmental Fate: If released to the atmosphere, it will degrade rapidly in the vapor phase by reaction with photochemically produced hydroxyl radicals (half-life of about 6 hr). Particulate phase and aerosols released to air during spray applications of insecticides will be removed from air physically by dry and wet deposition. If released to moist soil or water, it will degrade through aqueous hydrolysis. At 20 °C, aqueous hydrolysis half-lives of 49 days (pH 6.1) and 29 days (pH 7.4) have been determined; hydrolysis increases with alkalinity, so the degradation rate in soil and water should increase with increasing pH. Insufficient data are available to access the relative importance of biodegradation in soil or water. Based upon K_{oc} estimates ranging from 865 to 2100, it is expected to have low mobility in soil.

Environmental Physical Data

Henry's Law Constant: estimated at 9.59 x10^{-8}
Octanol/Water Partition Coefficient: log K_{ow} = 3.58
Sorption Partition Coefficient: K_{oc} = estimated at 995 to 2100
BCF: guppy 891

Regulations

RCRA 40CFR: Not listed
CERCLA: 40CFR 302.4: Not listed
SARA 40CFR 372.65: Not listed
SARA EHS 40CFR 355: Not listed
TSCA: Not listed

Analytical Methods

Soil: EPA PMD-TLC
Food: FDA 212.1, 232.1, 232.3, 232.4, 242.1
Plasma: EPA 001, 028; FDA 211.1, 231.1, 252

DIC1100 CAS #: 2130-56-5

3,3'-DICARBOXYBENZIDINE

RTECS: DV3325000
EINECS Number: 218-348-9
Molecular Formula: $C_{14}H_{12}N_2O_4$
Formula Weight: 272.28
Synonyms: 3,3'-BENZIDINEDICARBOXYLIC ACID; BENZIDINE-M,M'-DICARBOXYLIC ACID; 5,5'-BIANTHRANILIC ACID; (1,1'-BIPHENYL)-3,3'-DICARBOXYLIC ACID,4,4'-DIAMINO-; 3,3'-BIPHENYLDICARBOXYLIC ACID,4,4'-DIAMINO-; 4,4'-DIAMINO-3,3'-BIPHENYLDICARBOXYLIC ACID; 4,4'-DIAMINOBIPHENYL-3,3'-DICARBOXYLIC ACID; 5,5'-DIANTHRANILIC ACID; KWAS BENZYDYNODWUKAROKSYLOWY
Description: white crystals
Use: dyestuff

Physical Properties

Water Solubility: Insoluble in Water
Other Solubilities: Soluble in Alcohol, Ether

RTECS Toxicity Data

Mutagenic: Bacteria - E Coli Phage Inhibition; Dose: 100 mmol/L.
Tumorigenic: Rat Route: Subcutaneous; Dose: 7 gm/kg/77W-I; Toxic Effects: Tumorigenic - Equivocal tumorigenic agent by RTECS criteria; Blood - Tumors; Skin and appendages - Tumors. Mouse Route: Oral; Dose: 22 gm/kg/45W-I; Toxic Effects: Tumorigenic - Equivocal tumorigenic agent by RTECS criteria; Liver - Tumors; Blood - Tumors.

Hazard Overviews

Carcinogenicity: IARC - Not listed; NIOSH - Not listed; NTP - Not listed; ACGIH - Not listed; OSHA - Not listed; EPA - Not listed; MAK - Not listed

Environmental

Regulations

RCRA 40CFR: Not listed
CERCLA: 40CFR 302.4: Not listed
SARA 40CFR 372.65: Not listed
SARA EHS 40CFR 355: Not listed
TSCA: Listed

DIC1150 CAS #: 1345-13-7

DICERIUM TRIOXIDE

RTECS: FK6657000
EINECS Number: 215-718-1
Molecular Formula: Ce_2O_3
Formula Weight: 328.24
Synonyms: C.I. 77280; CERIUM OXIDE; CERIUM SESQUIOXIDE; CERIUM TRIOXIDE; CERIUM(III) OXIDE; CEROUS OXIDE
Description: gray-green, trigonal crystals
Use: constituent of incandescent gas mantles; as hydrocarbon catalyst in self-cleaning ovens; as glass polishing agent

Physical Properties

Freezing Point: 1692 °C (3077.6 °F)
Specific Gravity: 6.86
Water Solubility: Insoluble in Water
Other Solubilities: Soluble in Sulfuric acid.

Hazard Overviews

Carcinogenicity: IARC - Not listed; NIOSH - Not listed; NTP - Not listed; ACGIH - Not listed; OSHA - Not listed; EPA - Not listed; MAK - Not listed

Environmental

Regulations

RCRA 40CFR: Not listed
CERCLA: 40CFR 302.4: Not listed
SARA 40CFR 372.65: Not listed
SARA EHS 40CFR 355: Not listed
TSCA: Listed

DIC1200 CAS #: 1194-65-6

DICHLOBENIL

RTECS: DI3500000
EINECS Number: 214-787-5
Molecular Formula: $C_7H_3Cl_2N$
Structured MF: $C_6H_3Cl_2CN$
Formula Weight: 172.01

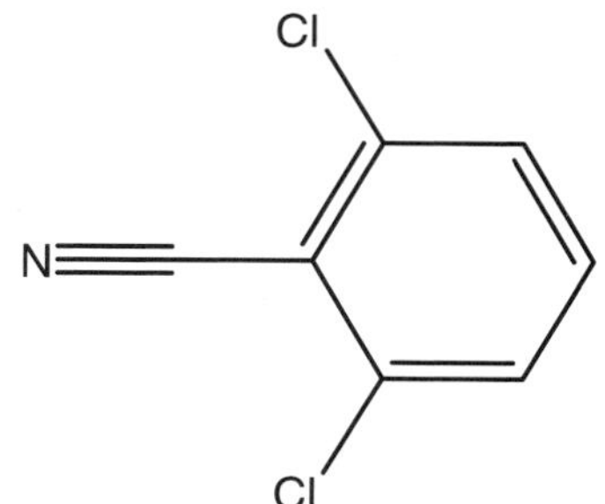

Chemical Structure

Synonyms: BENZONITRILE,2,6-DICHLORO-; CARSORON; CASARON; CASORON; CASORON G-10; CASORON G-4; CASORON W-50; CASORON 133; CASORON G20 SR; CASORON GSR; CODE H 133; CYCLOMEC; 2,6-DBN; DBN (THE HERBICIDE); DCB; DECABANE; 2,6-DICHLORBENZONITRIL; 2,6-DICHLOROBENZONITRILE; 2,6-DICHLOROPHENYL CYANIDE; DU-SPREX; DYCLOMEC; H 1313; H 133; NIA 5996; NIAGARA 5,996; NIAGARA 5006; NOROSAC; PREFIX D
Description: white to off-white crystalline solid; aromatic odor
Use: herbicide; for control of submersed weeds in lakes, ponds & drainage ditches where water not for agricultural or domestic purposes; casoron, dyclomec, for selective weed control in cranberry bogs, ornamentals, nurseries, fruit orchards, vineyards, forest plantations, public green areas; total weed control (industrial sites, railway lines, etc under asphalt); decabane for selective weed control in woody perennial crops.

Physical Properties

Boiling Point: 270 °C (518 °F)
Freezing Point: 144 °C (291.2 °F) to 145 °C (293 °F)
Specific Gravity: > 1
Saturated Vapor Density: 1.200000023 kg/m^3
Vapor Pressure: 3×10^{-6} mm Hg at 20 °C
Water Solubility: 25 ppm at 25 °C
Other Solubilities: Solubility in solvents (w/v) (approx): Acetone 5% at 8 °C; Benzene 5% at 8 °C; cyclohexanone 7% at 15-20 °C; Ethanol 5% at 8 °C; furfural 7% at 8 °C; methylene chloride 10% at 20 °C; methylethylketone 7% at 15-20 °C; tetrahydrofuran.
Flash Point: Nonflammable

RTECS Toxicity Data

Acute Oral: Rat LD_{50} Dose: 2710 mg/kg. Mouse LD_{50} Dose: 2056 mg/kg.
Acute Dermal: Rabbit LD_{50} Route: Skin; Dose: 1350 mg/kg.
Tumorigenic: Mouse Route: Subcutaneous; Dose: 260 ug/kg/39D-I; Toxic Effects: Tumorigenic - Equivocal tumorigenic agent by RTECS criteria; Liver - Tumors; Tumorigenic - Tumors at site of application. Mouse Route: Intraperitoneal; Dose: 260 ug/kg/39D-I; Toxic Effects: Tumorigenic - Equivocal tumorigenic agent by RTECS criteria; Tumorigenic - Tumors at site of application.

Hazard Overviews

Health: Irritating to eyes/skin/respiratory tract. Toxic. Other Acute Effects: harmful if swallowed, inhaled, or absorbed through skin.

Fire: Noncombustible. Hazards: emits toxic fumes. Extinguishing agents: water spray; carbon dioxide, dry chemical powder or appropriate foam. Precautions: combustible liquid.

Reactivity: Incompatible with: strong oxidizing agents. Hazardous decomposition products: toxic fumes of: carbon monoxide, carbon dioxide, nitrogen oxides, hydrogen chloride gas.

Carcinogenicity: IARC - Not listed; NIOSH - Not listed; NTP - Not listed; ACGIH - Not listed; OSHA - Not listed; EPA - Not listed; MAK - Not listed

Primary Target Organs:

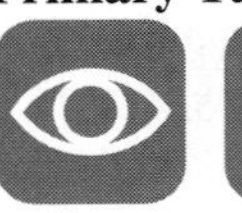
Eyes

Skin

Respiratory System

Environmental

Ecotoxicity: LC_{50} Pheasant 1500 ppm (10 days old LC_{50} Lepomis cyanellus (green sunfish) 5.7 mg/l/96 hr at 18 °C /weight of test organism= 1.1 g /. Static bioassay without aeration, pH 7.2-7.5, water hardness 40-50 mg/l as calcium carbonate and alkalinity of 30-35 mg/l LC_{50} Daphnia pulex (first instar) 3.7 mg/l/96 hr at 15 °C /50% wettable powder/. Static bioassay without aeration, pH 7.2-7.5, water hardness 40-50 mg/l as calcium carbonate and alkalinity of 30-35 mg/l LC_{50} Pteronarcys (second year class) 7.0 mg/l/96 hr at 15 °C /50% wettable powder/. Static bioassay without aeration, pH 7.2-7.5, water hardness 40-50 mg/l as calcium carbonate and alkalinity of 30-35 mg/l

Environmental Fate: Upon release to the soil, it will volatilize, biodegrade and leach. It may persist at low levels, however, for up to seventeen months. Release to water will result in significant biodegradation and volatilization. The compound is not expected to persist in the aqueous environment. It is not expected to bioconcentrate. No experimental information regarding persistence in the atmosphere is available. The compound is not expected to photolyze extensively, however, based on its absorption maximum. The estimated half-life for the reaction with hydroxyl radicals in the atmosphere is 22.27 days.

Cleanup/Disposal: Guide No. 151: Do not touch damaged containers or spilled material unless wearing appropriate protective clothing. Stop leak if you can do it without risk. Prevent entry into waterways, sewers, basements or confined areas. Cover with plastic sheet to prevent spreading. Absorb or cover with dry earth, sand or other non-combustible material and transfer to containers. Do not get water inside containers.

Environmental Physical Data

Henry's Law Constant: estimated at 6.29×10^{-6}

Sorption Partition Coefficient: K_{OC} = estimated at 871

BCF: not significant

Regulations

RCRA 40CFR: Not listed

CERCLA: 40CFR 302.4: Listed per CWA Section 311(b)(4) RQ: 100 lb (45.35 kg)

SARA 40CFR 372.65: Not listed

SARA EHS 40CFR 355: Not listed

TSCA: Not listed

Analytical Methods

Water / Groundwater: ASTM D4763

Food: FDA 212.1, 232.1, 232.4, 242.1; AOAC 979.03

Plasma: EPA 001, 028; FDA 211.1, 231.1, 252

DIC1250 CAS #: 97-17-6

DICHLOFENTHION

RTECS: TF0350000

EINECS Number: 202-564-5

Molecular Formula: $C_{10}H_{13}Cl_2O_3PS$

Formula Weight: 315.17

Synonyms: BROMEX; O,O-DIAETHYL-O-2,4-DICHLOR-PHENYL-MONOTHIOPHOSPHAT; O,O-DIAETHYL-O-2,4-DICHLORPHENYL-THIONOPHOSPHAT; DICHLOFENTION; DICHLORFENTHION; DICHLOROFENTHION; O-(2,4-DICHLOROPHENYL) O,O-DIETHYL PHOSPHOROTHIOATE; O-2,4-DICHLOROPHENYL O,O-DIETHYL PHOSPHOROTHIOATE; 2,4-DICHLORO-PHENYL DIETHYL PHOSPHOROTHIONATE; O,O-DIETHYL O-(2,4-DICHLOROPHENYL) PHOSPHOROTHIOATE; DIETHYL 2,4-DICHLOROPHENYL PHOSPHOROTHIONATE; O,O-DIETHYL O-2,4-DICHLOROPHENYL THIOPHOSPHATE; O,O-DIETHYL-O-(2,4-DICHLOOR-FENYL)-MONOTHIOFOSFAAT; O,O-DIETIL-O-(2,4-DICLORO-FENIL)-MONOTIOFOSFATO; ECP; ENT 17,470; ENT 17470; HEXA-NEMA; HEXANEMA; MOBILAWN; NEMACIDE; NEMACIDE VC-13; PHENOL,2,4-DICHLORO-,O-ESTER WITH O,O-DIETHYLPHOSPHOROTHIOATE; PHOSPHOROTHIOIC ACID,O-(2,4-DICHLOROPHENYL) O,O-DIETHYLESTER; THIOPHOSPHATE DE O-2,4-DICHLOROPHENYLE ET DE O,O-DIETHYLE; TRI-VC 13; V-C 1-13; V-C-13; VC 13; VC-13 NEMACIDE; VC13 NEMACIDE

Description: colorless liquid

Use: nematocide & insecticide

Physical Properties

Boiling Point: 164 °C (327 °F) to 169 °C (336 °F) at 0.1 mm Hg

Specific Gravity: 1.3 at 20 °C

Water Solubility: 0 mg/L at 25 °C

Other Solubilities: Soluble in Ethanol, Ether, Benzene; miscible with Kerosene.

Refraction Index: 1.5291 at 25 °C/D

RTECS Toxicity Data

Acute Oral: Rat LD_{50} Dose: 172 mg/kg. Chicken LD_{50} Dose: 148 mg/kg; Toxic Effects: Peripheral nerve and sensation - Flaccid paralysis without anesthesia; Behavioral - Somnolence (general depressed activity); Behavioral - Convulsions or effect on seizure threshold.

Acute Dermal: Rabbit LD_{50} Route: Skin; Dose: 6 gm/kg. Rat LD_{50} Route: Skin; Dose: 355 mg/kg.

Mutagenic: Human Sister Chromatid Exchenge; Cell Type: lymphocyte; Dose: 2 mg/L.

Hazard Overviews

Carcinogenicity: IARC - Not listed; NIOSH - Not listed; NTP - Not listed; ACGIH - Not listed; OSHA - Not listed; EPA - Not listed; MAK - Not listed

Environmental

Ecotoxicity: LC_{50} Pteronarcys 4.1 ug/L/96 hr at 15 °C, second yr class (95% confidence limit 3.4-5.0 ug/L) LC_{50} Cutthroat trout 640 ug/L/96 hr at 13 °C, wt 1.0 g (95% confidence limit 550-740 ug/L)

Environmental Physical Data

Octanol/Water Partition Coefficient: log K_{ow} = 5.14

Regulations

RCRA 40CFR: Not listed
CERCLA: 40CFR 302.4: Not listed
SARA 40CFR 372.65: Not listed
SARA EHS 40CFR 355: Not listed
TSCA: Not listed

Analytical Methods

Soil: SW846 8141, 8141A
Water / Groundwater: EPA 1657, 622.1, 022; SW846 8141, 8141A
Food: FDA 212.1, 232.1, 232.3
Plasma: EPA 001; FDA 211.1, 231.1, 252

DIC1300 **CAS #: 1085-98-9**

DICHLOFLUANID

RTECS: WO6475000
EINECS Number: 214-118-7
Molecular Formula: $C_9H_{11}Cl_2FN_2O_2S_2$
Formula Weight: 333.24
Synonyms: ANILINE,N-((DICHLOROFLUOROMETHYL)THIO)-N-((DIMETHYLAMINO)SULFONYL)-; BAY 47531; BAYER 47531; DICHLOFLUANIDE; DICHLORFLUANID; N-DICHLORFLUORMETHYLTHIO-N',N'-DIMETHYLAMINOSULFONSAEUREANILID; N-(DICHLOR-FLUOR-METHYL-THIO)-N',N'-DIMETHYL-N-PHENYL-SCHWEFEL-SAEUREDIAMID; 1,1-DICHLORO-N-((DIMETHYLAMINO)SULFONYL)-1-FLUORO-N-PHENYLMETHANE SULFENAMIDE; N-(DICHLOROFLUOROMETHYLTHIO)-N',N'-DIMETHYL-N-PHENYLSULFAMIDE; N-(DICHLOROFLUOROMETHYLTHIO)-N-(DIMETHYLSULFAMOYL)-ANILINE; N,N-DIMETHYL-N'-PHENYL-N'-FLUORODICHLOROMETHYLTHIOSULFAMIDE; DIPAREN; ELVARON; EPAREN; EUPAREN; EUPARENE; KU 13-032-C; KU 13-O32-C; KUE 13032C; METHANESULFENAMIDE,1,1-DICHLORO-N-((DIMETHYLAMINO)SULFONYL)-1-FLUORO-N-PHENYL-; PECUDIN; SULFAMIDE,N-((DICHLOROFLUOROMETHYL)THIO)-N',N'-DIMETHYL-N-PHENYL-
Description: white powder or colorless crystals
Use: acaricide; fungicide; antifouling agent for ships

Physical Properties

Freezing Point: 105.6 °C (222.08 °F)
Saturated Vapor Density: 1.200000017 kg/m^3
Vapor Pressure: 1 x10^{-6} mm Hg at 20 °C
Water Solubility: Insoluble in Water
Other Solubilities: Soluble in Acetone.

RTECS Toxicity Data

Acute Oral: Rat LD_{50} Dose: 500 mg/kg.
Acute Inhalation: Rat LC_{50} Dose: 300 mg/m^3/4hr.
Acute Dermal: Rat LD_{50} Route: Skin; Dose: 1 gm/kg.
Mutagenic: Human DNA Damage; Cell Type: HeLa cell; Dose: 1500 nmol/L. Human DNA Inhibition; Cell Type: HeLa cell; Dose: 1500 nmol/L.

Hazard Overviews

Carcinogenicity: IARC - Not listed; NIOSH - Not listed; NTP - Not listed; ACGIH - Not listed; OSHA - Not listed; EPA - Not listed; MAK - Not listed

Environmental

Ecotoxicity: Fishes: goldfish 48h LC_{50} 0.3 mg/l rainbow trout 96h LC_{50} 0.05 mg/l; Algae: Phaeodactylum tricornutum EC_{50} 300 ppm; Crustaceans: Daphnia magna 24h LC_{50} 280ppm; Fishes: Phoxinus phoxinus 24h LC_{50} 300 ppm

Regulations

RCRA 40CFR: Not listed
CERCLA: 40CFR 302.4: Not listed
SARA 40CFR 372.65: Not listed
SARA EHS 40CFR 355: Not listed
TSCA: Not listed

Analytical Methods

Food: FDA 212.1, 212.2, 232.1, 232.4, 242.1
Plasma: FDA 252

DIC1350 **CAS #: 117-80-6**

DICHLONE

RTECS: QL7525000
DOT: UN2761; NA2761; IMO6.1
EINECS Number: 204-210-5
Molecular Formula: $C_{10}H_4Cl_2O_2$
Structured MF: $C_6H_4(COCCl)_2$
Formula Weight: 227.05

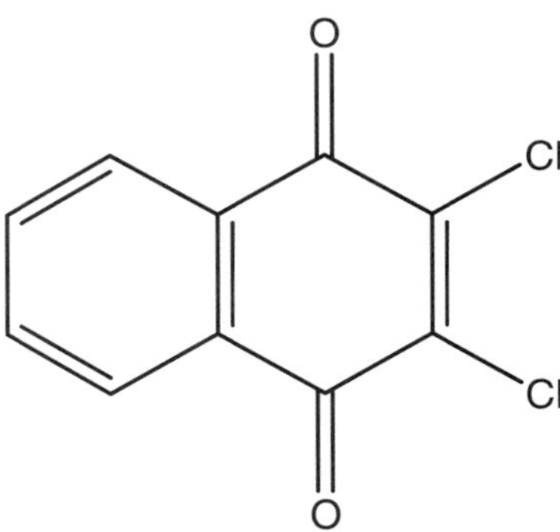

Chemical Structure

Synonyms: ALGISTAT; COMPOUND 604; 2,3-DICHLOR-1,4-NAFTOCHINON; 2,3-DICHLOR-1,4-NAPHTHOCHINON; 2,3-DICHLORO-1,4-NAPHTHALENEDIONE; 2,3-DICHLORO-1,4-NAPHTHAQUINONE; 2,3-DICHLORO-1,4-NAPHTHOQUINONE; 2,3-DICHLORO-ALPHA-NAPHTHOQUINONE; 2,3-DICHLORONAPHTHOQUINONE; 2,3-DICHLORONAPHTHOQUINONE-1,4; DICHLORONAPHTHOQUINONE; DICLONE; ENT 3,776; ENT 3776; LATKA 604; 1,4-NAPHTHALENEDIONE,2,3-DICHLORO-; 1,4-NAPHTHOQUINONE,2,3-DICHLORO-; PHYGON; PHYGON PASTE; PHYGON SEED PROTECTANT; PHYGON XL; QUINTAR; QUINTAR 540F; SANQUINON; U.S. RUBBER 604; UNIROYAL; US RUBBER 604; USR 604

Description: golden yellow needles or leaflets

Use: seed disinfectant; fungicide for foliage and textiles; insecticide; organic catalyst; effective substitute for both sulfur & copper in control of various fungus diseases of fruit trees & on vegetables; to control blue-green algae in ponds, lakes, and swimming pools; former use: chem int-eg, for diathianon; treatment for fruit, vegetable, field crops, ornamentals, resident commercial outdoor areas for brown rot of stone fruit; scab on apples, pears; blossom blights

Physical Properties

Boiling Point: 275 °C (527 °F) at 2 mm Hg
Freezing Point: 195 °C (383 °F)
Specific Gravity: < 1
Vapor Density: 7.84 Air=1
Water Solubility: Practically Insoluble in Water
Other Solubilities: about 4% in Xylene, O-Dichlorobenzene; moderately Soluble in Dioxane; Slightly Soluble in Glacial Acetic Acid; > 10% in Acetone; > 10% in Benzene; > 10% in Ether; > 10% in Chloroform; Slight Soluble in Carbon Tetrachloride.
Ionization Potential (eV): 9.5 +/-2.0
Flash Point: Not flammable

RTECS Toxicity Data

Acute Oral: Rat LD_{50} Dose: 160 mg/kg. Mouse LD_{50} Dose: 440 mg/kg.
Acute Dermal: Rabbit LD_{50} Route: Skin; Dose: 500 mg/kg.
Chronic (Multiple Dose) Oral: Rat Dose: 6 gm/kg/60D-C; Toxic Effects: Liver - Other changes; Biochemical - Glycolytic.
Tumorigenic: Mouse Route: Oral; Dose: 3300 mg/kg/78W-I; Toxic Effects: Tumorigenic - Neoplastic by RTECS criteria; Lungs, Thorax, or Respiration - Tumors; Blood - Tumors. Mouse Route: Subcutaneous; Dose: 22 mg/kg; Toxic Effects: Tumorigenic - Carcinogenic by RTECS criteria; Blood - Tumors.

Hazard Overviews

Health: Irritating to eyes/skin/respiratory tract. Toxic. Other Acute Effects: harmful if swallowed, inhaled, or absorbed through skin.
Fire: Noncombustible. Hazards: emits toxic fumes. Extinguishing agents: water spray; carbon dioxide, dry chemical powder or appropriate foam. Precautions: combustible liquid.
Reactivity: Incompatible with: strong oxidizing agents. Hazardous decomposition products: toxic fumes of: carbon monoxide, carbon dioxide, hydrogen chloride gas.
Carcinogenicity: IARC - Not listed; NIOSH - Not listed; NTP - Not listed; ACGIH - Not listed; OSHA - Not listed; EPA - Not listed; MAK - Not listed
Primary Target Organs:

Eyes

Skin

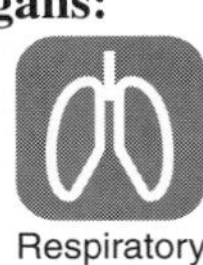
Respiratory System

Environmental

Ecotoxicity: LC_{50} Anas platyrhynchos (Mallard duck), 10 days, oral greater than 5000 ppm LC_{50} Gammarus fasciatus (Scud) (Amphipoda) 1100 ug/l/96 hr at 21 °C in a static bioassay (95% confidence limit 802-1507 ug/l), mature LC_{50} Salmo gairdneri (Rainbow trout) 49 ug/l/96 hr at 12 °C in a static bioassay (95% confidence limit 41-58 ug/l), wt 0.5 g LC_{50} Daphnia magna (crustacean) 25 ug/l/48 hr. /Conditions of bioassay not specified
Environmental Fate: If emitted to the air it will generally be in the form of aerosol and dust and be subject to gravitational settling as well as photooxidation by ozone and hydroxyl radical (estimated half-life in the vapor 1.25 hr). If released into water it will hydrolyze with a half-life of 5 days at neutral pH and faster at alkaline pHs. Photodegradation will also occur in surface waters but the rate of degradation could not be found. Degradation will be fast in moist soil (half-life 1 day) but somewhat slower in dry soil (half-life of months). In field application, it will degrade during the growing season and not leach below about 6 inches.
Cleanup/Disposal: Guide No. 151: Do not touch damaged containers or spilled material unless wearing appropriate protective clothing. Stop leak if you can do it without risk. Prevent entry into waterways, sewers, basements or confined areas. Cover with plastic sheet to prevent spreading. Absorb or cover with dry earth, sand or other non-combustible material and transfer to containers. Do not get water inside containers.

Environmental Physical Data

Octanol/Water Partition Coefficient: log K_{ow} = estimated at 3.16
Sorption Partition Coefficient: K_{oc} = estimated at 1.55×10^4

BCF: calculated at 260

Regulations

RCRA 40CFR: Not listed
CERCLA: 40CFR 302.4: Listed per CWA Section 311(b)(4) RQ: 1 lb (0.454 kg)
SARA 40CFR 372.65: Not listed
SARA EHS 40CFR 355: Not listed
TSCA: Listed

Analytical Methods

Soil: EPA PMD-TLC; SW846 8081, 8270B, 8270C
Water / Groundwater: EPA 022
Food: FDA 212.1, 232.1, 232.3, 232.4, 242.1; AOAC 965.36
Plasma: EPA 001, 027, 028, 29; FDA 211.1, 231.1, 252
Other: EPA 1656

DIC1400 CAS #: 116-52-9

DICHLORAL UREA

RTECS: YS2787000
EINECS Number: 204-144-7
Molecular Formula: $C_5H_6Cl_6N_2O_3$
Formula Weight: 354.83

Chemical Structure

Synonyms: 1,3-BIS(1-HYDROXY-2,2,2-TRICHLOROETHYL)UREA; 1,3-BIS(2,2,2-TRICHLORO-1-HYDROXYETHYL)UREA; CRAG DCU-73W; CRAG EXPERIMENTAL HERBICIDE 2; CRAG HERBICIDE 2; DCM; DCU; DICHLORALUREE; DICLORALUREA; DKHM; EH2; EXPERIMENTAL HERBICIDE 2; RC 9485; SK&F 1995; UREA,1,3-BIS(2,2,2-TRICHLORO-1-HYDROXYETHYL)-; UREA,N,N'-BIS(2,2,2-TRICHLORO-1-HYDROXYETHYL)-
Description: colorless crystals
Use: preemergence herbicide

Physical Properties

Freezing Point: 196 °C (384.8 °F)
Specific Gravity: 1.63 at 25 °C
Water Solubility: 0.0074 ppm Soluble in Water at 25 °C
Other Solubilities: At 25 °C: 18.8 mg/100 g Ethanol, 16.8 mg/100 g Acetone, 2.5 mg/100 g Ether, 5.5 mg/100 g n-hexane, 3.0 mg/100 g Benzene, 1.0 mg/100 g Chloroform.

RTECS Toxicity Data

Acute Oral: Rat LD_{50} Dose: 6400 mg/kg; Toxic Effects: Automatic Nervous System - Other (direct) parasympathomimetic; Behavioral - Somnolence (general depressed activity); Behavioral - Ataxia. Mouse LD_{50} Dose: 6 gm/kg; Toxic Effects: Automatic Nervous System - Other (direct) parasympathomimetic; Behavioral - Somnolence (general depressed activity); Behavioral - Ataxia.
Acute Dermal: Rat LD_{50} Route: Skin; Dose: >5 gm/kg.
Chronic (Multiple Dose) Inhalation: Rat Dose: 59100 ug/m^3/17W-I; Toxic Effects: Blood - Changes in serum composition; Biochemical - Transaminases; Biochemical - Other enzymes. Cat Dose: 59100 ug/m^3/17W-I; Toxic Effects: Blood - Changes in serum composition; Biochemical - Transaminases.
Irritation Skin: Rabbit Open Draize Test Dose: 50 mg open; Reaction: mild.
Mutagenic: Mouse Cytogenetic Analysis; Route: Oral; Dose: 5 gm/kg.

Hazard Overviews

Carcinogenicity: IARC - Not listed; NIOSH - Not listed; NTP - Not listed; ACGIH - Not listed; OSHA - Not listed; EPA - Not listed; MAK - Not listed

Environmental

Regulations

RCRA 40CFR: Not listed
CERCLA: 40CFR 302.4: Not listed
SARA 40CFR 372.65: Not listed
SARA EHS 40CFR 355: Not listed
TSCA: Not listed

DIC1450 CAS #: 51338-27-3

DICHLORFOP-METHYL

RTECS: UF1180000
DOT: UN2765; UN2766; UN2999; UN3000; IMO3.2; IMO6.1
EINECS Number: 257-141-8
Molecular Formula: $C_{16}H_{14}Cl_2O_4$
Formula Weight: 341.19
Synonyms: 2-(4-(2',4'-DICHLOROPHENOXY)-PHENOXY)-METHYL-PROPIONATE; 2-(4-(2,4-DICHLOROPHENOXY)PHENOXY)-METHYL-PROPIONATE; DICLOFOP; DICLOFOP-METHYL; HOE-023408; HOE 23408; HOE 23408 OH; HOE-GRASS; HOEGRASS; HOELON; HOELON 3EC; ILLOXAN; ILOXAN; METHYL 2-(4-(2,4-DICHLOROPHENOXY)PHENOXY)PROPANOATE; METHYL (RS)-2-(4-(2,4-DICHLOROPHENOXY)PHENOXY)PROPIONATE; METHYL 2-(4-(2,4-DICHLOROPHENOXY)PHENOXY)PROPIONATE; (+)-METHYL2-(4-(2,4-DICHLOROPHENOXY)PHENOXY)PROPIONATE; (-)-METHYL2-(4-(2,4-DICHLOROPHENOXY)PHENOXY)PROPIONATE
Description: colorless crystals; odorless
Use: control of broad-leafed plants in row crops; defoliants, general brush control chlorophenoxy compounds; controls annual grassy weeds in wheat, barley, lentils, flax, dry field

peas (including austrian winter peas) and on fallow land for control of annual grassy weeds

Physical Properties

Boiling Point: 175 °C (347 °F) to 177 °C (351 °F) at 0.1 mm Hg
Freezing Point: 39 °C (102.2 °F) to 41 °C (105.8 °F)
Saturated Vapor Density: 5.535824625 kg/m^3
Vapor Pressure: 0.034 mPa at 20 °C
Water Solubility: 3 mg/L at 22 °C
Other Solubilities: Readily Soluble in common organic solvents, eg, Acetone 2490, Diethyl Ether 2280, Xylene 2530, Ethanol 110, light petroleum (BP 60-95 °C) 60 (all in g/l at 20 °C).
Flash Point: 38 °C Closed Cup

RTECS Toxicity Data

Acute Oral: Rat LD_{50} Dose: 512 mg/kg. Mouse LD_{50} Dose: 586 mg/kg.
Acute Inhalation: Rat LC_{50} Dose: 8300 mg/m^3/4hr.
Acute Dermal: Rat LD_{50} Route: Skin; Dose: >2 gm/kg.

Hazard Overviews

Flammable

Fire: Flammable.
Carcinogenicity: IARC - Not listed; NIOSH - Not listed; NTP - Not listed; ACGIH - Not listed; OSHA - Not listed; EPA - Not listed; MAK - Not listed

Environmental

Ecotoxicity: LC_{50} Rainbow trout 0.35 mg/l/96 hr /Conditions of bioassay not specified LC_{50} Coturnix >5000 ppm/5 days
Environmental Fate: If released to soil, it is expected to be slightly mobile to immobile. Several studies have shown that it undergoes rapid microbial hydrolysis in soil to dichlorfop acid, which is fairly stable. If released to water, it will be essentially non-volatile. It has aqueous base-catalyzed half-lives of 6.3 days at pH 9, 63 days at pH of 8, and 1.7 years at pH of 7. It will undergo moderate bioconcentration in aquatic organisms and adsorption to sediment will be possible due to high K_{oc} values (2400 to 16,000). If released to the atmosphere, it will exist in the vapor and particulate phases. In the vapor phase, it will degrade in the atmosphere by reaction with photochemically produced hydroxyl radicals with an estimated half-life of 21 hours. Physical removal from air can occur through wet and dry deposition.
Cleanup/Disposal: Guide No. 131: Fully encapsulating, vapor protective clothing should be worn for spills and leaks with no fire. Eliminate all ignition sources (no smoking, flares, sparks or flames in immediate area). All equipment used when handling the product must be grounded. Do not touch or walk through spilled material. Stop leak if you can do it without risk. Prevent entry into waterways, sewers, basements or confined areas. A vapor suppressing foam may be used to reduce vapors. Small Spills: Absorb with earth, sand or other non-combustible material and transfer to containers for later disposal. Use clean non-sparking tools to collect absorbed material. Large Spills: Dike far ahead of liquid spill for later disposal. Water spray may reduce vapor; but may not prevent ignition in closed spaces. Guide No. 152: Do not touch damaged containers or spilled material unless wearing appropriate protective clothing. Stop leak if you can do it without risk. Prevent entry into waterways, sewers, basements or confined areas. Cover with plastic sheet to prevent spreading. Absorb or cover with dry earth, sand or other non-combustible material and transfer to containers. Do not get water inside containers.

Environmental Physical Data

Henry's Law Constant: estimated at 3.8 $x10^{-8}$
Sorption Partition Coefficient: K_{OC} = estimated at 4000
BCF: estimated at 300

Regulations

RCRA 40CFR: Not listed
CERCLA: 40CFR 302.4: Not listed
SARA 40CFR 372.65: Listed
SARA EHS 40CFR 355: Not listed
TSCA: Not listed

Analytical Methods

Food: FDA 212.1, 232.1, 232.4, 242.1
Plasma: FDA 211.1, 231.1, 252

DIC1500 **CAS #: 1649-08-7**

1,2-DICHLORO DIFLUOROETHANE

RTECS: KI0950000
EINECS Number: 216-714-2
Molecular Formula: $C_2H_2Cl_2F_2$
Formula Weight: 134.94
Synonyms: 1,2-DICHLORO-1,1-DIFLUOROETHANE; HCFC-132B
Description: colorless liquid
Use: potential substitute for ozone depleting chlorofluorocarbons; model for other 1,1,1,2-tetrahaloethanes under consideration as chlorofluorocarbon substitutes; in mechanical vapor compression systems; for refrigeration & air conditioning & account for majority of refrigeration capability in us; refrigerants in home appliances, mobile air conditioning units, retail food refrigeration systems & chillers.

Physical Properties

Boiling Point: 46.8 °C (116 °F) at 760 mm Hg
Freezing Point: -1012 °C (-1789.6 °F)
Specific Gravity: 1.4163 at 20 °C/4 °C
Saturated Vapor Density: Estimated 2.757375681 kg/m^3
Vapor Pressure: Estimated 270 mm Hg at 25 °C
Water Solubility: 999 mg/L at 25 °C (est)
Refraction Index: 1.36193 at 20 °C/D
Ionization Potential (eV): =< 11.8 +/-2.0
Flash Point: Nonflammable

RTECS Toxicity Data

Acute Inhalation: Rat LC_{Lo} Dose: 20000 ppm/4hr; Toxic Effects: Behavioral - General anesthetic; Behavioral - Somnolence (general depressed activity); Liver - Other changes.

Reproductive/Teratogenic: Rat Route: Inhalation; Dose: 2000 ppm/6H; Duration: male 14W prior to mating; Paternal Effects - Spermatogenesis.

Hazard Overviews

Fire: Noncombustible.

Carcinogenicity: IARC - Not listed; NIOSH - Not listed; NTP - Not listed; ACGIH - Not listed; OSHA - Not listed; EPA - Not listed; MAK - Not listed

Environmental

Environmental Fate: If released to soil, it will rapidly volatilize from both moist and dry soil to the atmosphere. It will display moderate to high mobility in soil. If released to water, it will rapidly volatilize to the atmosphere. The estimated half-life for volatilization from a model river is 3.4 hours. It will not bioconcentrate in fish and aquatic organisms nor will it adsorb to sediment or suspended organic matter. If released to the atmosphere, it will undergo a slow gas-phase reaction with photochemically produced hydroxyl radicals with an estimated half-life of 617 days. It may also undergo atmospheric removal by wet deposition processes; however, any removed by this process is expected to rapidly re-volatilize to the atmosphere.

Environmental Physical Data

Henry's Law Constant: estimated at 0.048

Octanol/Water Partition Coefficient: log K_{ow} = 2.154

Sorption Partition Coefficient: K_{oc} = calculated at 98 to 354

BCF: calculated at 12 to 25

Regulations

RCRA 40CFR: Not listed

CERCLA: 40CFR 302.4: Listed as Compound per CWA Section 307(a)

SARA 40CFR 372.65: Listed

SARA EHS 40CFR 355: Not listed

TSCA: Listed

DIC1550 CAS #: 79-02-7

DICHLOROACETALOEHYDE

RTECS: AB2710000

DOT: UN2903; IMO6.1

EINECS Number: 201-169-5

Molecular Formula: $C_2H_2Cl_2O$

Formula Weight: 112.94

Synonyms: ACETALDEHYDE,DICHLORO-; CHLORALDEHYDE; CHLOROALDEHYDE; 2,2-DICHLOROACETALDEHYDE; ALPHA,ALPHA-DICHLOROACETALDEHYDE; NSC 5207

Description: colorless liquid; penetrating pungent odor

Use: manufacture of insecticides

Physical Properties

Boiling Point: 90 °C (194 °F) to 91 °C (196 °F) at 760 mm Hg

Freezing Point: -50 °C (-58 °F)

Specific Gravity: 1.436 at 25 °C

Saturated Vapor Density: 1.428511797 kg/m^3

Density: 1.4 g/mL

Vapor Pressure: 50 mm Hg at 20 °C

Water Solubility: Reacts with Water

Other Solubilities: 95% Ethanol: Soluble.

Ionization Potential (eV): 10.5 +/-5.0

Flash Point: 60 °C Closed Cup

RTECS Toxicity Data

Mutagenic: Human DNA Damage; Cell Type: lymphocyte; Dose: 1 mmol/L. Mouse Dominant Lethal Test; Route: Intraperitoneal; Dose: 176 mg/kg.

Hazard Overviews

Flammable

Fire: Flammable.

Carcinogenicity: IARC - Not listed; NIOSH - Not listed; NTP - Not listed; ACGIH - Not listed; OSHA - Not listed; EPA - Not listed; MAK - Not listed

Environmental

Cleanup/Disposal: Guide No. 131: Fully encapsulating, vapor protective clothing should be worn for spills and leaks with no fire. Eliminate all ignition sources (no smoking, flares, sparks or flames in immediate area). All equipment used when handling the product must be grounded. Do not touch or walk through spilled material. Stop leak if you can do it without risk. Prevent entry into waterways, sewers, basements or confined areas. A vapor suppressing foam may be used to reduce vapors. Small Spills: Absorb with earth, sand or other non-combustible material and transfer to containers for later disposal. Use clean non-sparking tools to collect absorbed material. Large Spills: Dike far ahead of liquid spill for later disposal. Water spray may reduce vapor; but may not prevent ignition in closed spaces.

Regulations

RCRA 40CFR: Not listed

CERCLA: 40CFR 302.4: Not listed

SARA 40CFR 372.65: Not listed

SARA EHS 40CFR 355: Not listed

TSCA: Listed

DIC1600 CAS #: 3018-12-0

DICHLOROACETONITRILE

RTECS: AL8465000

EINECS Number: 221-159-4

Molecular Formula: C_2HCl_2N
Structured MF: Cl_2CHCN
Formula Weight: 109.94

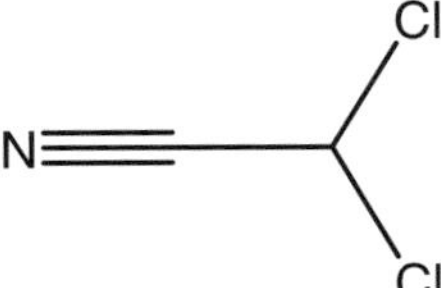

Chemical Structure

Synonyms: ACETONITRILE,DICHLORO-; DICHLOROMETHYL CYANIDE
Description: clear liquid

Physical Properties

Boiling Point: 112.5 °C (235 °F)
Density: 1.369 g/cu cm at 20 °C
Water Solubility: 10 to 50 mg/mL at 21.5 °C
Other Solubilities: 95% Ethanol: >=100 mg/ml at 21.5 °C; Acetone: >=100 mg/ml at 21.5 °C; Alcohol: Soluble; DMSO: >=100 mg/ml at 21.5 °C.
Refraction Index: 1.4391
Flash Point: > 100 °C

RTECS Toxicity Data

Acute Oral: Rat LD_{50} Dose: 330 mg/kg; Toxic Effects: Behavioral - Somnolence (general depressed activity); Behavioral - Coma; Lungs, Thorax, or Respiration - Respiratory depression. Mouse LD_{50} Dose: 270 mg/kg; Toxic Effects: Behavioral - Somnolence (general depressed activity); Behavioral - Coma; Lungs, Thorax, or Respiration - Respiratory depression.
Reproductive/Teratogenic: Rat Route: Oral; Dose: 825 mg/kg; Duration: female 7-21D of pregnancy; Effects on Fertility - Post-implantation mortality; Effects on Newborn - Live birth index. Rat Route: Oral; Dose: 825 mg/kg; Duration: female 7-21D of pregnancy; Effects on Newborn - Weaning or lactation index; Growth statistics. Rat Route: Oral; Dose: 325 mg/kg; Duration: female 6-18D of pregnancy; Specific Developmental Abnormalities - Cardiovascular (circulatory) system; Gastrointestinal system; Urogenital system. Rat Route: Oral; Dose: 325 mg/kg; Duration: female 6-18D of pregnancy; Effects on Fertility - Post-implantation mortality; Effects on Embryo or Fetus - Fetotoxicity.
Mutagenic: Human DNA Damage; Cell Type: lymphocyte; Dose: 10 umol/L. Hamster Sister Chromatid Exchenge; Cell Type: ovary; Dose: 20800 nmol/L.

Hazard Overviews

Corrosive

Health: Corrosive to eyes/skin/respiratory tract. Harmful. Other Acute Effects: harmful if swallowed, inhaled, or absorbed through skin; lachrymator; inhalation may result in spasm, inflammation and edema of the larynx and bronchi, chemical pneumonitis and pulmonary edema; symptoms of exposure may include burning sensation; coughing; wheezing; laryngitis; shortness of breath; headache; nausea; vomiting. Chronic Effects: laboratory experiments have shown mutagenic effects.
Fire: Will burn. Hazards: vapor may travel considerable distance to source of ignition and flash back; emits toxic fumes. Extinguishing agents: carbon dioxide, dry chemical powder or appropriate foam. Precautions: combustible liquid.
Reactivity: Incompatible with: strong oxidizing agents, sensitive to moisture. Hazardous decomposition products: toxic fumes of: carbon monoxide, carbon dioxide, nitrogen oxides, hydrogen chloride gas.
Carcinogenicity: IARC - Group 3, Not classifiable as to carcinogenicity to humans; NIOSH - Not listed; NTP - Not listed; ACGIH - Not listed; OSHA - Not listed; EPA - Not listed; MAK - Not listed
Primary Target Organs:

Eyes

Skin

Respiratory System

Environmental

Environmental Fate: It is very mobile in soil and if released on land it is expected to leach. It may hydrolyze in moist, alkaline soils. If released in water, it will be lost through hydrolysis. Hydrolysis will be faster in alkaline waters and in the presence of chlorine. Roughly 10 and 60% is lost in 10 days at pH 6 and 8, respectively. It will also be slowly lost by volatilization (half-life in a model river is 10 days). Bioconcentration in aquatic organisms should not occur. In the atmosphere, it reacts very slowly with photochemically-produced hydroxyl radicals (half-life 434 days).
Cleanup/Disposal: Guide No. 153: Eliminate all ignition sources (no smoking, flares, sparks or flames in immediate area). Do not touch damaged containers or spilled material unless wearing appropriate protective clothing. Stop leak if you can do it without risk. Prevent entry into waterways, sewers, basements or confined areas. Absorb or cover with dry earth, sand or other non-combustible material and transfer to containers. Do not get water inside containers.

Environmental Physical Data

Henry's Law Constant: estimated at 3.79×10^{-6}
Octanol/Water Partition Coefficient: log K_{ow} = 0.29
Sorption Partition Coefficient: K_{oc} = estimated at 12.8
BCF: calculated at 0.98

Regulations

RCRA 40CFR: Not listed
CERCLA: 40CFR 302.4: Not listed
SARA 40CFR 372.65: Not listed
SARA EHS 40CFR 355: Not listed
TSCA: Not listed

Analytical Methods

Drinking Water: EPA 551; APHA 6230-C

DIC1650 CAS #: 79-36-7

DICHLOROACETYL CHLORIDE

RTECS: AO6650000
DOT: UN1765; IMO8.0
EINECS Number: 201-199-9
Molecular Formula: C_2HCl_3O
Formula Weight: 147.40

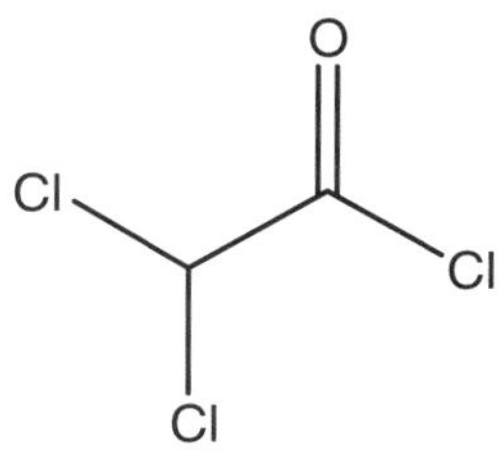

Chemical Structure

Synonyms: ACETYL CHLORIDE,DICHLORO-; CHLORID KYSELINY DICHLOROCTOVE; CHLORURE DE DICHLORACETYLE; DICHLORACETYL CHLORIDE; 2,2-DICHLOROACETYL CHLORIDE; ALPHA,ALPHA-DICHLOROACETYL CHLORIDE; DICHLOROETHANOYL CHLORIDE
Description: liquid; acrid, penetrating odor
Use: intermediate

Physical Properties

Boiling Point: 107 °C (225 °F) to 108 °C (226 °F) at 760 mm Hg
Specific Gravity: 1.5315 at 16 °C/4 °C
Vapor Density: 5.1 Air=1
Saturated Vapor Density: 2.192754991 kg/m^3
Vapor Pressure: 154 mm Hg at 25 °C
Water Solubility: Decomposes
Other Solubilities: 95% Ethanol: Decomposes; Acetone: >=100 mg/ml at 21 °C; DMSO: Decomposes; Ether: miscible.
Refraction Index: 1.4638 at 16 °C/D
Ionization Potential (eV): 11.0 +/-2.0
Flash Point: 66 °C

RTECS Toxicity Data

Acute Oral: Rat LD_{50} Dose: 2460 mg/kg.
Acute Inhalation: Rat LC_{Lo} Dose: 2000 ppm/4hr.
Acute Dermal: Rabbit LD_{50} Route: Skin; Dose: 650 uL/kg.
Irritation Eye: Rabbit Standard Draize Test Dose: 50 ug open; Reaction: severe.
Irritation Skin: Rabbit Standard Draize Test Dose: 2 mg/24H; Reaction: severe. Rabbit Open Draize Test Dose: 10 mg/24H open; Reaction: moderate.
Mutagenic: Bacteria - S Typhimurium Mutations in Microorganisms; Dose: 1 mg/plate (+S9).
Tumorigenic: Mouse Route: Subcutaneous; Dose: 2 mg/kg/80W-I; Toxic Effects: Tumorigenic - Equivocal tumorigenic agent by RTECS criteria; Skin and appendages - Tumors; Tumorigenic - Tumors at site of application.

Hazard Overviews

Corrosive

Health: Corrosive to eyes/skin/respiratory tract. Harmful. Other Acute Effects: harmful if swallowed, inhaled, or absorbed through skin; causes burns; extremely destructive to tissue of the mucous membranes and upper respiratory tract, eyes and skin; lachrymator; inhalation may result in spasm, inflammation and edema of the larynx and bronchi, chemical pneumonitis and pulmonary edema; symptoms of exposure may include burning sensation, coughing, wheezing, laryngitis, shortness of breath, headache, nausea and vomiting.
Fire: Combustible. Hazards: reacts violently with water; water hydrolyzes material liberating acidic gas which in contact with metal surfaces can generate flammable and/or explosive hydrogen gas; emits toxic fumes. Extinguishing agents: carbon dioxide; dry chemical powder; do not use water. Precautions: combustible liquid.
Reactivity: Incompatible with: alcohols, oxidizing agents, strong bases, reacts violently with water. Hazardous decomposition products: carbon monoxide, carbon dioxide, hydrogen chloride gas, phosgene gas.
Carcinogenicity: IARC - Not listed; NIOSH - Not listed; NTP - Not listed; ACGIH - Not listed; OSHA - Not listed; EPA - Not listed; MAK - Not listed
Primary Target Organs:

Eyes Skin Respiratory System

Environmental

Environmental Fate: If released to moist soil, it will hydrolyze upon contact with the water. Evaporation from dry surfaces is expected to occur. If released to water, it will hydrolyze upon contact, forming HCl and dichloroacetic acid. If released to the atmosphere, it will react in the gas-phase with photochemically produced hydroxyl radicals at an estimated half-life rate of 22.8 days. The half-life for the gas-phase tropospheric hydrolysis is in excess of 100 years. Atmospheric wash-out and transformation via rain may be possible.
Cleanup/Disposal: Guide No. 156: Eliminate all ignition sources (no smoking, flares, sparks or flames in immediate area). All equipment used when handling the product must be grounded. Do not touch damaged containers or spilled material unless wearing appropriate protective clothing. Stop leak if you can do it without risk. A vapor suppressing foam may be used to reduce vapors. For chlorosilanes, use AFFF alcohol-resistant medium expansion foam to reduce vapors. Do not get water on spilled substance or inside containers.

Use water spray to reduce vapors or divert vapor cloud drift. Prevent entry into waterways, sewers, basements or confined areas. Small Spills: Cover with dry earth, dry sand, or other non-combustible material followed with plastic sheet to minimize spreading or contact with rain. Use clean non-sparking tools to collect material and place it into loosely covered plastic containers for later disposal.

Environmental Physical Data

BCF: none possible

Regulations

RCRA 40CFR: Not listed
CERCLA: 40CFR 302.4: Not listed
SARA 40CFR 372.65: Not listed
SARA EHS 40CFR 355: Not listed
TSCA: Listed

DIC1700 CAS #: 7572-29-4

DICHLOROACETYLENE

RTECS: AP1080000
DOT: UN1591
Molecular Formula: C_2Cl_2
Structured MF: ClC is equivalent to CCl
Formula Weight: 94.93
Synonyms: DCA; DICHLOROETHYNE; ETHYNE,DICHLORO-(9CI)
Description: liquid

Physical Properties

Boiling Point: Explodes at 32.22 °C (90 °F)
Freezing Point: -66.11 °C (-86.998 °F) to -50 °C (-58 °F)
Specific Gravity: 1.26100
Other Solubilities: Soluble in Acetone Alcohol and Ether
Refraction Index: 1.42790
Ionization Potential (eV): 9.9
Flash Point: Spontaneously combustible

RTECS Toxicity Data

Acute Inhalation: Mouse LC_{50} Dose: 19 ppm/6hr; Toxic Effects: Behavioral - Change in motor activity (specific assay); Lungs, Thorax, or Respiration - Dyspnea; Skin and appendages - Hair. Rabbit LC_{Lo} Dose: 307 ppm/1hr; Toxic Effects: Liver - Fatty liver degeneration; Kidney, ureter, and Bladder - Chgs in tubules (inc acute renal failure, acute tubular necrosis.

Mutagenic: Bacteria - S Typhimurium Mutations in Microorganisms; Dose: 4000 ppm (+/-S9).

Tumorigenic: Rat Route: Inhalation; Dose: 14 ppm/6H/77W-I; Toxic Effects: Tumorigenic - Carcinogenic by RTECS criteria; Vascular - Tumors; Kidney, Ureter, and Bladder - Kidney tumors. Mouse Route: Inhalation; Dose: 2 ppm/24H/77W-I; Toxic Effects: Tumorigenic - Carcinogenic by RTECS criteria; Kidney, Ureter, and Bladder - Kidney tumors; Blood - Tumors.

Hazard Overviews

Poison

Explosive

Flammable

Fire Diamond

Health: Poison. Also Causes: headache, facial herpes, loss of appetite, intense jaw pain, vomiting, cranial nerve palsy, extreme nausea, pulmonary/brain edema. Chronic Effects: nervous system/kidney/liver damage.

Fire: Explosive, pyrophoric. For small fires use carbon dioxide, water spray, or regular foam. For large fires use water spray, regular foam or fog. Do not use dry chemicals. Do not scatter with excess water.

Reactivity: Unstable, self reactive and shock sensitive, gas phase ignites upon exposure to air. Hazardous polymerization cannot occur. Avoid: contact with acids and oxidizing materials; exposure to air; situations where friction or shock could occur. Incompatible with: acid; acid fumes; oxidizing materials. Hazardous decomposition products: carbon dioxide; chloride fumes.

Carcinogenicity: IARC - Group 3, Not classifiable as to carcinogenicity to humans; NIOSH - Listed as carcinogen; NTP - Not listed; ACGIH - Class A3, Animal carcinogen; OSHA - Not listed; EPA - Not listed; MAK - Class A2, Unmistakably carcinogenic in animal experimentation only

Primary Target Organs:

Respiratory System

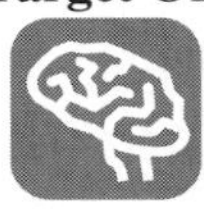
Nervous System

Liver

Kidneys

Exposure Limits

OSHA PEL Vacated 1989 Limits: STEL: 0.1 ppm; 0.4 mg/m^3; Ceiling.

ACGIH TLV: STEL: 0.1 ppm; 0.39 mg/m^3; Ceiling.

Respirator Recommendation

Exposure Range: >0.1 to 5 ppm Supplied Air, Constant Flow/Pressure Demand, Half Mask

Exposure Range: >5 to 100 ppm Supplied Air, Constant Flow/Pressure Demand, Full Face

Exposure Range: >100 to unlimited ppm Self-contained Breathing Apparatus, Pressure Demand, Full Face

Note: odor threshold unknown

Environmental

Cleanup/Disposal: Guide No. 152: Do not touch damaged containers or spilled material unless wearing appropriate protective clothing. Stop leak if you can do it without risk. Prevent entry into waterways, sewers, basements or confined areas. Cover with plastic sheet to prevent spreading. Absorb or cover with dry earth, sand or other non-combustible material and transfer to containers. Do not get water inside containers.

Regulations

RCRA 40CFR: Not listed
CERCLA: 40CFR 302.4: Not listed
SARA 40CFR 372.65: Not listed

SARA EHS 40CFR 355: Not listed
TSCA: Not listed

DIC1750 CAS #: 608-27-5

2,3-DICHLOROANILINE

RTECS: CX9862625
DOT: UN1590; IMO6.1
EINECS Number: 210-157-9
Molecular Formula: $C_6H_5Cl_2N$
Formula Weight: 162.02

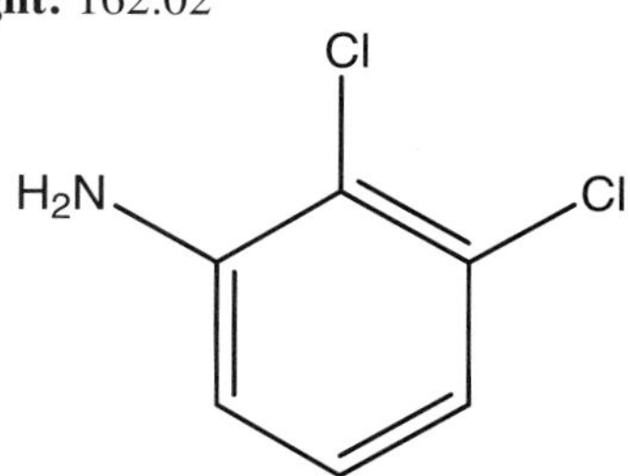

Chemical Structure

Synonyms: ANILINE,2,3-DICHLORO-; ANILINE,2,3-DICHLORO-(7CI,8CI); BENZENAMINE,2,3-DICHLORO-; 2,3-DICHLOROBENZENAMINE
Description: needles
Use: research chemical

Physical Properties

Boiling Point: 252 °C (486 °F)
Freezing Point: 24 °C (75.2 °F)
Saturated Vapor Density: 1.200199168 kg/m^3
Vapor Pressure: 0.0275 mm Hg at 25 °C
Other Solubilities: Soluble in Alcohol, Acetone.
Refraction Index: 1.5969 at 20 °C
Flash Point: > 112 °C Closed Cup

RTECS Toxicity Data

Acute Inhalation: Rat LC_{50} Dose: >8047 mg/m^3/4hr; Toxic Effects: Behavioral - Somnolence (general depressed activity); Behavioral - Tremor; Nutritional and gross metabolic - Weight loss or decreased weight gain.

Hazard Overviews

Poison

Health: Irritating to eyes/skin/respiratory tract. Poison. Other Acute Effects: may be fatal if inhaled, swallowed, or absorbed through skin; may cause allergic skin reaction; absorption into the body leads to the formation of methemoglobin which in sufficient concentration causes cyanosis; onset may be delayed 2 to 4 hours or longer; target organ: blood.

Fire: Will burn. Hazards: container explosion may occur; emits toxic fumes. Extinguishing agents: water spray; carbon dioxide, dry chemical powder or appropriate foam. Precautions: combustible liquid.

Reactivity: Incompatible with: acids, acid chlorides, acid anhydrides, oxidizing agents. Hazardous decomposition products: thermal decomposition may produce carbon monoxide, carbon dioxide, and nitrogen oxides; hydrogen chloride gas.

Carcinogenicity: IARC - Not listed; NIOSH - Not listed; NTP - Not listed; ACGIH - Not listed; OSHA - Not listed; EPA - Not listed; MAK - Not listed

Primary Target Organs:

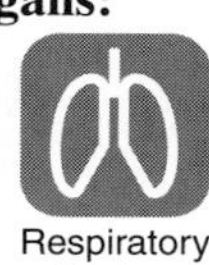
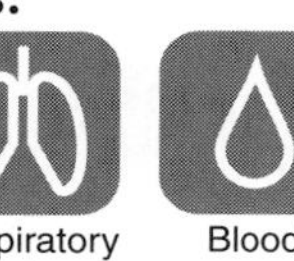

Eyes Skin Respiratory System Blood

Environmental

Environmental Fate: If released to the atmosphere, it is expected to degrade relatively rapidly (estimated half- life of 8.8 hr) by reaction with photochemically produced hydroxyl radicals. If released to soil, it may undergo a covalent chemical bonding with humic materials, which can result in its chemical alteration to a form that will not leach. In the absence of covalent bonding, moderate leaching may be possible. If released to water, covalent bonding with humic materials in the water column and sediments may result in partitioning from the water column to sediments. By analogy to aromatic amine chemical class, in the water column it may be susceptible to photooxidation via hydroxyl and peroxy radicals. Insufficient data are available to assess the importance of biodegradation in soil or water.

Cleanup/Disposal: Guide No. 153: Eliminate all ignition sources (no smoking, flares, sparks or flames in immediate area). Do not touch damaged containers or spilled material unless wearing appropriate protective clothing. Stop leak if you can do it without risk. Prevent entry into waterways, sewers, basements or confined areas. Absorb or cover with dry earth, sand or other non-combustible material and transfer to containers. Do not get water inside containers.

Environmental Physical Data

Henry's Law Constant: estimated at 1.38 $x10^{-6}$
Octanol/Water Partition Coefficient: log K_{ow} = 2.78
Sorption Partition Coefficient: K_{oc} = estimated at 370
BCF: estimated at 1.88

Regulations

RCRA 40CFR: Not listed
CERCLA: 40CFR 302.4: Not listed
SARA 40CFR 372.65: Not listed
SARA EHS 40CFR 355: Not listed
TSCA: Listed

Analytical Methods

Soil: EPA 1625
Water / Groundwater: EPA 1625

DIC1800 CAS #: 554-00-7

2,4-DICHLOROANILINE

RTECS: BX2600000
DOT: UN1590; IMO6.1
EINECS Number: 209-057-8
Molecular Formula: $C_6H_5Cl_2N$
Structured MF: $Cl_2C_6H_3NH_2$
Formula Weight: 162.02

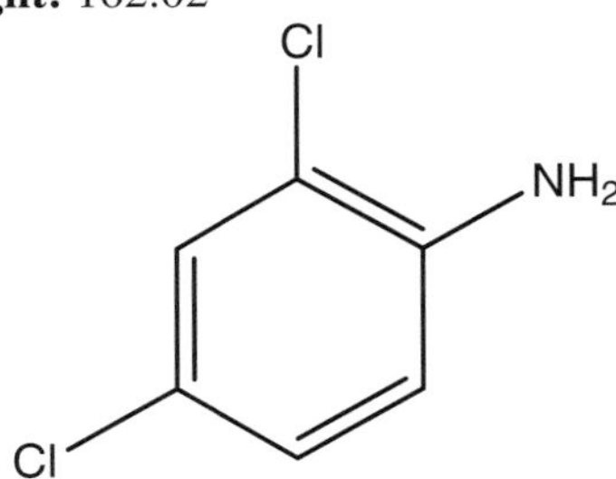

Chemical Structure

Synonyms: ANILINE,2,4-DICHLORO-; BENZENAMINE,2,4-DICHLORO-; BENZENAMINE,2,4-DICHLORO-(9CI); 2,4-DICHLORANILIN; 2,4-DICHLOROBENZENAMINE
Description: prisms or needles
Use: chem intermed for CI pigment yellow 16; laboratory reagent

Physical Properties

Boiling Point: 245 °C (473 °F) at 760 mm Hg
Specific Gravity: 1.567 at 20 °C
Saturated Vapor Density: 1.200027014 kg/m^3
Density: 1.567 g/mL at 20 °C
Vapor Pressure: 0.00373 mm Hg at 25 °C
Water Solubility: Slightly Soluble in Water
Other Solubilities: 95% Ethanol: >=100 mg/ml at 23 °C; Acetone: >=100 mg/ml at 23 °C; DMSO: >=100 mg/ml at 23 °C; Ether: Soluble.
Flash Point: Not available; probably combustible

RTECS Toxicity Data

Acute Oral: Rat LD_{50} Dose: 1600 mg/kg; Toxic Effects: Behavioral - Muscle weakness; Lungs, Thorax, or Respiration - Cyanosis; Lungs, Thorax, or Respiration - Respiratory depression. Mouse LD_{50} Dose: 400 mg/kg; Toxic Effects: Behavioral - Muscle weakness; Lungs, Thorax, or Respiration - Cyanosis; Lungs, Thorax, or Respiration - Respiratory depression.

Hazard Overviews

Poison

Health: Irritating to eyes/skin/respiratory tract. Poison. Other Acute Effects: may be fatal if inhaled, swallowed, or absorbed through skin; may cause allergic skin reaction; absorption into the body leads to the formation of methemoglobin which in sufficient concentration causes cyanosis; onset may be delayed 2 to 4 hours or longer; target organ: blood.
Fire: Will burn. Hazards: container explosion may occur; emits toxic fumes. Extinguishing agents: water spray; carbon dioxide, dry chemical powder or appropriate foam. Precautions: combustible liquid.
Reactivity: Incompatible with: acids, acid chlorides, acid anhydrides, oxidizing agents. Hazardous decomposition products: thermal decomposition may produce carbon monoxide, carbon dioxide, and nitrogen oxides; hydrogen chloride gas.
Carcinogenicity: IARC - Not listed; NIOSH - Not listed; NTP - Not listed; ACGIH - Not listed; OSHA - Not listed; EPA - Not listed; MAK - Not listed
Primary Target Organs:

Eyes

Skin

Respiratory System

Blood

Environmental

Ecotoxicity: LC_{50} Poecilia reticulata (guppy) 11.7 ppm/14 days /Conditions of bioassay not specified
Environmental Fate: If released to the atmosphere, it is expected to degrade relatively rapidly (estimated half- life of 13 hr) by reaction with photochemically produced hydroxyl radicals. If released to soil, it may undergo a covalent chemical bonding with humic materials which can result in its chemical alteration to a form that will not leach and prevent leaching. In the absence of covalent bonding, moderate leaching may be possible. Photolysis on soil surfaces may occur. If released to water, covalent bonding with humic materials in the water column and sediments may result in partitioning from the water column to sediments. A study examining the degradation in an estuarine water exposed to sunlight determined that photolysis was the dominant degradation process in the water's upper surface layer; photolysis half-lives in the natural water ranged from 26 to 82 hr for summer to winter conditions. By analogy to aromatic amine chemical class, in the water column it may be susceptible to photooxidation via hydroxyl and peroxy radicals. Insufficient data are available to assess the relative importance of biodegradation in soil or water
Cleanup/Disposal: Guide No. 153: Eliminate all ignition sources (no smoking, flares, sparks or flames in immediate area). Do not touch damaged containers or spilled material unless wearing appropriate protective clothing. Stop leak if you can do it without risk. Prevent entry into waterways, sewers, basements or confined areas. Absorb or cover with dry earth, sand or other non-combustible material and transfer to containers. Do not get water inside containers.

Environmental Physical Data

Henry's Law Constant: estimated at 1.38×10^{-6}
Octanol/Water Partition Coefficient: log K_{ow} = 2.78

Sorption Partition Coefficient: K_{oc} = 3930
BCF: estimated at 1.88

Regulations

RCRA 40CFR: Not listed
CERCLA: 40CFR 302.4: Not listed
SARA 40CFR 372.65: Not listed
SARA EHS 40CFR 355: Not listed
TSCA: Listed

DIC1850 CAS #: 95-82-9

2,5-DICHLOROANILINE

RTECS: BX2610000
DOT: UN1590; IMO6.1
EINECS Number: 202-455-2
Molecular Formula: $C_6H_5Cl_2N$
Formula Weight: 162.02

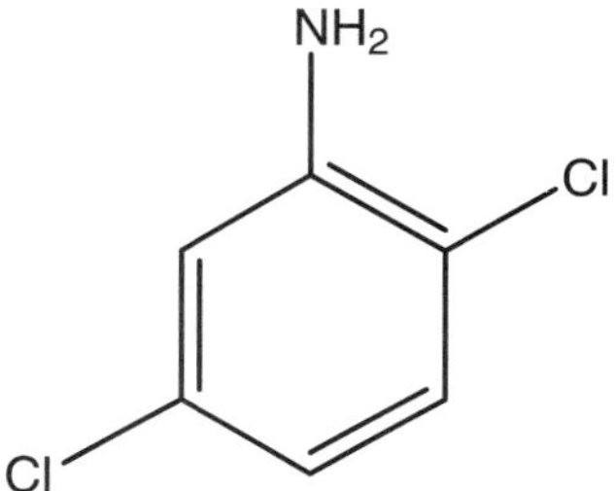

Chemical Structure

Synonyms: AMARTHOL FAST SCARLET GG BASE; AMARTHOL FAST SCARLET GG SALT; AMARTHOL FAST SCARLET GGS BASE; 1-AMINO-2,5-DICHLOROBENZENE; 2-AMINO-1,4-DICHLOROBENZENE; ANILINE,2,5-DICHLORO-; AZOBASE DCA; AZOENE FAST SCARLET 2G BASE; AZOENE FAST SCARLET 2G SALT; AZOGENE FAST SCARLET GG (FREE BASE); AZOGENE FAST SCARLET GGC; AZOIC DIAZO COMPONENT 3; BENZENAMINE,2,5-DICHLORO-; BENZENAMINE,2,5-DICHLORO-(9CI); C.I. 37010; C.I. AZOIC DIAZO COMPONENT 3; DAITO SCARLET BASE GG; DEVOL SCARLET A (FREE BASE); DEVOL SCARLET 2GS BASE; DIAZO FAST SCARLET GG; 2,5-DICHLORANILIN; 2,5-DICHLORO-1-AMINOBENZENE; 2,5-DICHLOROBENZENAMINE; 2,5-DICHLOROBENZENAMINE (9CI); 2,5-DICHLOROBENZENEAMINE; DURGASOL SCARLET GG SALT; FAST RED SGG BASE; FAST SCARLET 2G; FAST SCARLET 2G BASE; FAST SCARLET BASE 2J; FAST SCARLET BASE GGT; FAST SCARLET BASE 2JS; FAST SCARLET DS BASE; FAST SCARLET GG BASE; FAST SCARLET GGS BASE; FAST SCARLET MDC BASE; FAST SCARLET 2G SALT; HILTONIL FAST SCARLET 2G BASE; HILTONIL FAST SCARLET 2GS BASE; HINDAMINE SCARLET GG; KAMBAMINE SCARLET GG BASE; KAYAKU SCARLET GG BASE; LAKE SCARLET GG BASE; MEISEI SCARLET GG SALT; MITSUI SCARLET GG BASE; NAPHTHANIL SCARLET 2G BASE; NAPHTOELAN FAST SCARLET GG BASE; NATASOL SCARLET GG SALT; SANYO FAST SCARLET GG BASE; SCARLET 2G BASE; SCARLET BASE CIBA I; SCARLET BASE GG; SCARLET BASE IRGA I (FREE BASE); SCARLET BASE NGG; SCARLET 2G SALT; SCARLET SALT CIBA I; SPECTROLENE SCARLET 2G; STABAMINE SCARLET GG; SYMULON SCARLET 2G BASE; SYMULON SCARLET 2G SALT
Description: light brown or amber-colored crystalline mass or needles
Use: intermediate for the manufacture of dyes and pesticides

Physical Properties

Boiling Point: 251 °C (484 °F)
Freezing Point: 50 °C (122 °F)
Saturated Vapor Density: 1.200017961 kg/m^3
Vapor Pressure: 0.00248 mm Hg at 25 °C
Water Solubility: Slightly Soluble in Water
Other Solubilities: 95% Ethanol: >=100 mg/ml at 22.5 °C; Acetone: >=100 mg/ml at 22.5 °C; Benzene: Soluble; DMSO: >=100 mg/ml at 22.5 °C; Ether: Soluble.
Flash Point: 112 °C

RTECS Toxicity Data

Acute Oral: Rat LD_{50} Dose: 1600 mg/kg. Mouse LD_{50} Dose: 1600 mg/kg.
Chronic (Multiple Dose) Oral: Rat Dose: 27300 mg/kg/13W-I; Toxic Effects: Endocrine - Changes in spleen weight; Biochemical - True cholinesterase; Biochemical - Catalases.

Hazard Overviews

Poison

Health: Irritating to eyes/skin/respiratory tract. Poison. Other Acute Effects: toxic by inhalation, in contact with skin and if swallowed; may be fatal if inhaled, swallowed, or absorbed through skin; may cause allergic skin reaction; absorption into the body leads to the formation of methemoglobin which, in sufficient concentration, causes cyanosis; onset may be delayed 2 to 4 hours or longer.
Fire: Will burn. Hazards: container explosion may occur; emits toxic fumes. Extinguishing agents: water spray; carbon dioxide, dry chemical powder or appropriate foam. Precautions: combustible liquid.
Carcinogenicity: IARC - Not listed; NIOSH - Not listed; NTP - Not listed; ACGIH - Not listed; OSHA - Not listed; EPA - Not listed; MAK - Not listed
Primary Target Organs:

Eyes

Skin

Respiratory System

Environmental

Ecotoxicity: Crustaceans: Daphnia magna 24h EC_{50} 4.1 mg/l; Fishes: Leuciscus idus 48h LC_0 2 mg/l
Environmental Fate: If released to the atmosphere, it is expected to degrade relatively rapidly (estimated half-life of 8.8 hr) by reaction with photochemically produced hydroxyl radicals. If released to soil, it may undergo a covalent chemical bonding with humic materials which can result in its chemical alteration to a latent form and prevent leaching. A single laboratory persistence study using a loam soil observed a degradation half-life of about 8 weeks. If released to water, covalent bounding with humic materials in the water column and sediments may result in partitioning from the water

column to sediments. By analogy to aromatic amine chemical class, in the water column it may be susceptible to photooxidation via hydroxyl and peroxy radicals. Various biodegradation studies indicate that it is biodegraded slowly.

Cleanup/Disposal: Guide No. 153: Eliminate all ignition sources (no smoking, flares, sparks or flames in immediate area). Do not touch damaged containers or spilled material unless wearing appropriate protective clothing. Stop leak if you can do it without risk. Prevent entry into waterways, sewers, basements or confined areas. Absorb or cover with dry earth, sand or other non-combustible material and transfer to containers. Do not get water inside containers.

Environmental Physical Data

Henry's Law Constant: estimated at 1.38 $x10^{-6}$
Octanol/Water Partition Coefficient: log K_{ow} = 2.75
Sorption Partition Coefficient: K_{oc} = estimated at 350
BCF: estimated at 1.86
BOD: 30% BODT, 14 days

Regulations

RCRA 40CFR: Not listed
CERCLA: 40CFR 302.4: Not listed
SARA 40CFR 372.65: Not listed
SARA EHS 40CFR 355: Not listed
TSCA: Listed

DIC1900 CAS #: 95-76-1

3,4-DICHLOROANILINE

RTECS: BX2625000
DOT: UN1590; IMO6.1
EINECS Number: 202-448-4
Molecular Formula: $C_6H_5Cl_2N$
Structured MF: $NH_2C_6H_3Cl_2$
Formula Weight: 162.03

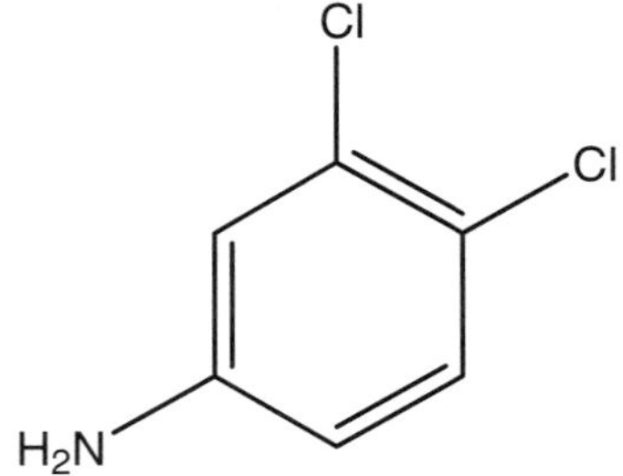

Chemical Structure

Synonyms: 1-AMINO-3,4-DICHLOROBENZENE; ANILINE,3,4-DICHLORO-; BENZENAMINE,3,4-DICHLORO; BENZENAMINE,3,4-DICHLORO-; BENZENAMINE,3,4-DICHLORO-(9CI); 3,4-DCA; DCA; 3,4-DICHLORANILIN; 3,4-DICHLORANILINE; 4,5-DICHLOROANILINE; 3,4-DICHLOROBENZENAMINE
Description: light-tan fine crystals or needles
Use: intermediate for the manufacture of dyes and pesticides; in the manufacture of C.I. Acid Orange 14

Physical Properties

Boiling Point: 272 °C (522 °F)
Freezing Point: 71 °C (159.8 °F) to 72 °C (161.6 °F)
Specific Gravity: 1.33 at 85 °C
Vapor Density: 5.59 Air=1
Saturated Vapor Density: 1.200070619 kg/m^3
Vapor Pressure: 0.00975 mm Hg at 20 °C
Water Solubility: Practically Insoluble in Water
Other Solubilities: 95% Ethanol: >=100 mg/ml at 21.5 °C; Acetone: >=100 mg/ml at 21.5 °C; Alcohol: Soluble; Benzene: Slightly Soluble; DMSO: >=100 mg/ml at 21.5 °C; Ether: Soluble; Most organic solvents: Soluble.
pH: Water extract 9.1
Evaporation Rate: < 1 Butyl Acetate=1
Flash Point: 166 °C Open Cup
Autoignition Temperature: 265 °C
LEL: 2.8% v/v
UEL: 7.2% v/v at 179 °C

RTECS Toxicity Data

Acute Oral: Rat LD_{50} Dose: 545 mg/kg. Mouse LD_{50} Dose: 740 mg/kg.
Acute Inhalation: Rat LC_{Lo} Dose: 65 mg/m^3/4hr.
Acute Dermal: Rabbit LD_{Lo} Route: Skin; Dose: 300 mg/kg. Cat LD_{50} Route: Skin; Dose: 700 mg/kg.
Chronic (Multiple Dose) Oral: Rat Dose: 6370 mg/kg/13W-I; Toxic Effects: Endocrine - Changes in spleen weight; Blood - Changes in erythrocite (RBC) cell count; Biochemical - True cholinesterase. Rat Dose: 3640 mg/kg/26W-I; Toxic Effects: Behavioral - Alteration of classical conditioning; Blood - Methemoglobinemia-Carboxyhemoglobin; Blood - Changes in erythrocite (RBC) cell count.
Chronic (Multiple Dose) Inhalation: Rat Dose: 30 ug/m^3/24H/14W-C; Toxic Effects: Behavioral - Muscle contraction or spasticity.
Irritation Eye: Rabbit Standard Draize Test Dose: 10mg; Reaction: mild. Rabbit Standard Draize Test Dose: 250 ug/24H; Reaction: severe.
Irritation Skin: Rabbit Standard Draize Test Dose: 2 mg/24H; Reaction: severe.
Mutagenic: Human Sister Chromatid Exchenge; Cell Type: lymphocyte; Dose: 125 umol/L. Mold - A Nidulans Mutations in Microorganisms; Dose: 200 mg/L (-S9).

Hazard Overviews

Poison

Corrosive

Health: Corrosive to eyes/skin/respiratory tract. Poison. Other Acute Effects: extremely destructive to tissue of the mucous membranes and upper respiratory tract, eyes and skin; toxic by inhalation, in contact with skin and if swallowed; readily absorbed through skin; may be fatal if inhaled, swallowed, or absorbed through skin; inhalation may result in spasm, inflammation and edema of the larynx and bronchi, chemical pneumonitis and pulmonary edema; symptoms of exposure may include burning sensation, coughing, wheezing, laryngitis, shortness of breath, headache, nausea and vomiting; absorption into the body leads to the formation of methemoglobin which, in sufficient concentration, causes cyanosis; onset may be delayed 2 to 4 hours or longer; may cause allergic respiratory and skin reactions. Chronic Effects: may cause sensitization by inhalation and skin contact; possible risk of irreversible effects; laboratory experiments have shown mutagenic effects; target organ: blood.

Fire: Will burn. Hazards: container explosion may occur; emits toxic fumes. Extinguishing agents: water spray; carbon dioxide, dry chemical powder or appropriate foam. Precautions: combustible liquid.

Carcinogenicity: IARC - Not listed; NIOSH - Not listed; NTP - Not listed; ACGIH - Not listed; OSHA - Not listed; EPA - Not listed; MAK - Not listed

Primary Target Organs:

Eyes

Skin

Respiratory System

Blood

Environmental

Ecotoxicity: LC_{50} Pimephales promelas (fathead minnow) 28 days old 7.0 mg/l/96 hr (Confidence limit: 6.6-7.5 mg/l) at 25.1 °C (98% purity) /Conditions of bioassay not specified LC_{50} Daphnia magna 7.6x10-7 mol/l/96 hr at 20 °C, 10% sea water /Conditions of bioassay not specified LC_{50} Ophryotrocha diadema (larvae) 4 mg/l/96 hr /Conditions of bioassay not specified LC_{50} Palaemonetes varians (larvae) 2.0x10-5 mol/l/96 hr at 15 °C, 100% sea water /Conditions of bioassay not specified

Environmental Fate: If released to the atmosphere, it is expected to degrade relatively rapidly (estimated half-life of 8.8 hr) by reaction with photochemically produced hydroxyl radicals. If released to soil, it may undergo a covalent chemical bonding with humic materials which can result in its chemical alteration to a form and that will not leach. Mineralization in soil occurs slowly and may have a half-life on the order of a year or more. If released to water, direct photolysis is expected to be a major degradation process. A persistence study in outdoor pond waters showed an overall half-life 4.1 to 6.3 days with photolysis being judged as the dominant loss mechanism. Direct photolysis at the water's surface has estimated half-lives ranging from 0.57 hr to 2 days depending upon solar intensity and time of the year. Covalent bonding with humic materials in the water column and sediments may result in partitioning from the water column to sediments. Various biodegradation studies indicate that it is biodegraded slowly.

Cleanup/Disposal: Guide No. 153: Eliminate all ignition sources (no smoking, flares, sparks or flames in immediate area). Do not touch damaged containers or spilled material unless wearing appropriate protective clothing. Stop leak if you can do it without risk. Prevent entry into waterways, sewers, basements or confined areas. Absorb or cover with dry earth, sand or other non-combustible material and transfer to containers. Do not get water inside containers.

Environmental Physical Data

Henry's Law Constant: estimated at 2.26 x10^{-5}
Octanol/Water Partition Coefficient: log K_{ow} = 2.69
Sorption Partition Coefficient: K_{oc} = estimated at 300
BCF: calculated at 1.81
BOD: 30% BODT, 14 days

Regulations

RCRA 40CFR: Not listed
CERCLA: 40CFR 302.4: Not listed
SARA 40CFR 372.65: Not listed
SARA EHS 40CFR 355: Not listed
TSCA: Listed

Analytical Methods

Soil: SW846 8131
Food: FDA 212.1, 232.1, 232.4, 242.1
Plasma: FDA 252

DIC1950 CAS #: 626-43-7

3,5-DICHLOROANILINE

DOT: UN1590; IMO6.1
EINECS Number: 210-948-9
Molecular Formula: $C_6H_5Cl_2N$
Formula Weight: 162.02

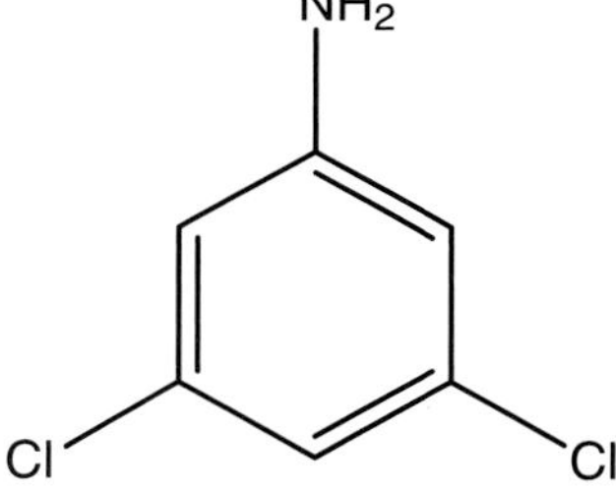

Chemical Structure

Synonyms: ANILINE,3,5-DICHLORO-; BENZENAMINE,3,5-DICHLORO-
Description: needles
Use: chem intermediate for vinclozolin fungicide for grapes & iprodione fungicide for use on turf; chem intermediate for organic synthesis; sumilex, a fungicide very effective against gray mold in many crops, is developed from derivatives of 3,5-dichloroaniline

Physical Properties

Boiling Point: 260 °C (500 °F) at 741 mm Hg
Freezing Point: 51 °C (123.8 °F) to 53 °C (127.4 °F)
Saturated Vapor Density: Estimated 1.20000688 kg/m^3
Vapor Pressure: Estimated 0.00095 mm Hg at 25 °C
Water Solubility: Insoluble in Water
Other Solubilities: Soluble in Alcohol, Ether, Benzene, Chloroform.

Hazard Overviews

Poison

Health: Irritating to eyes/skin/respiratory tract. Poison. Other Acute Effects: may be fatal if inhaled, swallowed, or absorbed through skin; may cause allergic skin reaction; absorption into the body leads to the formation of methemoglobin which in sufficient concentration causes cyanosis; onset may be delayed 2 to 4 hours or longer; target organ: blood. Irritating to eyes/skin/respiratory tract. Poison. Other Acute Effects: may be fatal if inhaled, swallowed, or absorbed through skin; may cause allergic skin reaction; absorption into the body leads to the formation of methemoglobin which in sufficient concentration causes cyanosis; onset may be delayed 2 to 4 hours or longer; target organ: blood.

Fire: Hazards: container explosion may occur; emits toxic fumes. Extinguishing agents: water spray; carbon dioxide, dry chemical powder or appropriate foam. Precautions: combustible liquid.

Reactivity: Incompatible with: acids, acid chlorides, acid anhydrides, oxidizing agents. Hazardous decomposition products: thermal decomposition may produce carbon monoxide, carbon dioxide, and nitrogen oxides; hydrogen chloride gas.

Carcinogenicity: IARC - Not listed; NIOSH - Not listed; NTP - Not listed; ACGIH - Not listed; OSHA - Not listed; EPA - Not listed; MAK - Not listed

Primary Target Organs:

Eyes

Skin

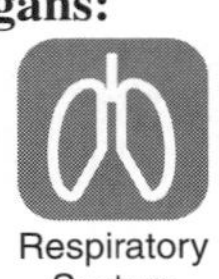

Respiratory System

Blood

Environmental

Ecotoxicity: Fishes: guppy (Poecilia reticulata) 14d LC_{50} 3.9 mg/l

Environmental Fate: If released to the atmosphere, it is expected to degrade relatively rapidly (estimated half-life of 5 hr) by reaction with photochemically produced hydroxyl radicals. If released to soil, it may undergo a covalent chemical bonding with humic materials which can result in its chemical alteration to a form that will not leach and prevent leaching. In the absence of covalent bonding, low to moderate leaching may occur. A single laboratory persistence study using a loam soil observed a degradation half-life of about 5 weeks. If released to water, covalent bonding with humic materials in the water column and sediments may result in partitioning from the water column to sediments. By analogy to aromatic amines as a chemical class, in the water column it may be susceptible to photooxidation via hydroxyl and peroxy radicals. Insufficient data are available to assess the relative importance of biodegradation in soil or water.

Cleanup/Disposal: Guide No. 153: Eliminate all ignition sources (no smoking, flares, sparks or flames in immediate area). Do not touch damaged containers or spilled material unless wearing appropriate protective clothing. Stop leak if you can do it without risk. Prevent entry into waterways, sewers, basements or confined areas. Absorb or cover with dry earth, sand or other non-combustible material and transfer to containers. Do not get water inside containers.

Environmental Physical Data

Henry's Law Constant: estimated at 1.38 $x10^{-6}$

Octanol/Water Partition Coefficient: log K_{ow} = 2.90

Sorption Partition Coefficient: K_{oc} = estimated at 490 to 460

BCF: estimated at 1.97

Regulations

RCRA 40CFR: Not listed

CERCLA: 40CFR 302.4: Not listed

SARA 40CFR 372.65: Not listed

SARA EHS 40CFR 355: Not listed

TSCA: Listed

Analytical Methods

Food: FDA 232.4, 242.1

DIC2000 CAS #: 82-46-2

1,5-DICHLOROANTHRAQUINONE

RTECS: CB6495000

EINECS Number: 201-424-0

Molecular Formula: $C_{14}H_6Cl_2O_2$

Formula Weight: 277.10

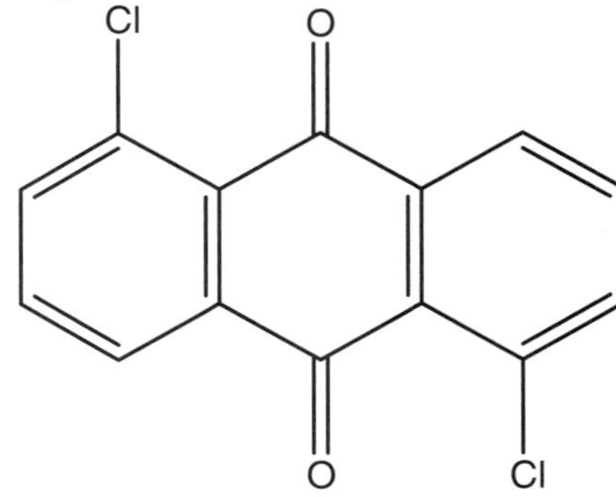

Chemical Structure

Synonyms: 9,10-ANTHRACENEDIONE,1,5-DICHLORO-; 1,5-DICHLORANTHRACHINON; 1,5-DICHLORO-9,10-ANTHRAQUINONE

Physical Properties

Freezing Point: 245 °C (473 °F) to 247 °C (476.6 °F)

Water Solubility: Insoluble

Other Solubilities: Ethanol: Slightly soluble; Acetone: Slightly soluble; Benzene: Slightly soluble $PhNO_2$: Soluble; Concentrated Sulfuric Acid: Soluble; Acetic Acid: Soluble

RTECS Toxicity Data

Irritation Eye: Rabbit Standard Draize Test Dose: 500 mg/24H; Reaction: mild.

Hazard Overviews

Health: Irritating to eyes/skin. Other Acute Effects: may be harmful by inhalation, ingestion, or skin absorption.
Fire: Hazards: emits toxic fumes. Extinguishing agents: water spray; carbon dioxide, dry chemical powder or appropriate foam. Precautions: combustible liquid.
Reactivity: Incompatible with: strong oxidizing agents. Hazardous decomposition products: toxic fumes of: carbon monoxide, carbon dioxide, hydrogen chloride gas.
Carcinogenicity: IARC - Not listed; NIOSH - Not listed; NTP - Not listed; ACGIH - Not listed; OSHA - Not listed; EPA - Not listed; MAK - Not listed
Primary Target Organs:

Eyes

Skin

Environmental

Regulations

RCRA 40CFR: Not listed
CERCLA: 40CFR 302.4: Not listed
SARA 40CFR 372.65: Not listed
SARA EHS 40CFR 355: Not listed
TSCA: Listed

DIC2050 CAS #: 82-43-9

1,8-DICHLOROANTHRAQUINONE

RTECS: CB6496000
EINECS Number: 201-420-9
Molecular Formula: $C_{14}H_6Cl_2O_2$
Formula Weight: 277.10

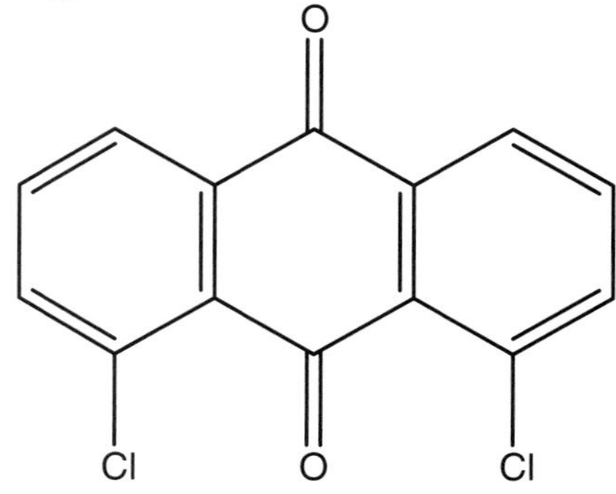

Chemical Structure

Synonyms: 9,10-ANTHRACENEDIONE,1,8-DICHLORO-; 1,8-DICHLORANTHRACHINON; 1,8-DICHLORO-9,10-ANTHRAQUINONE

Physical Properties

Freezing Point: 203 °C (397.4 °F)
Water Solubility: Insoluble
Other Solubilities: Ethanol: Slightly soluble; Benzene: Soluble; Toluene Soluble; $PhNO_2$: Soluble

RTECS Toxicity Data

Irritation Eye: Rabbit Standard Draize Test Dose: 500 mg/24H; Reaction: mild.

Hazard Overviews

Health: Irritating to eyes/skin. Other Acute Effects: may be harmful by inhalation, ingestion, or skin absorption.
Fire: Hazards: emits toxic fumes. Extinguishing agents: water spray; carbon dioxide, dry chemical powder or appropriate foam. Precautions: combustible liquid.
Reactivity: Incompatible with: strong oxidizing agents. Hazardous decomposition products: toxic fumes of: carbon monoxide, carbon dioxide, hydrogen chloride gas.
Carcinogenicity: IARC - Not listed; NIOSH - Not listed; NTP - Not listed; ACGIH - Not listed; OSHA - Not listed; EPA - Not listed; MAK - Not listed
Primary Target Organs:

Eyes

Skin

Environmental

Regulations

RCRA 40CFR: Not listed
CERCLA: 40CFR 302.4: Not listed
SARA 40CFR 372.65: Not listed
SARA EHS 40CFR 355: Not listed
TSCA: Listed

DIC2100 CAS #: 1602-00-2

4,4'-DICHLOROAZOBENZENE

RTECS: CN2330000
Molecular Formula: $C_{12}H_8Cl_2N_2$
Formula Weight: 251.12

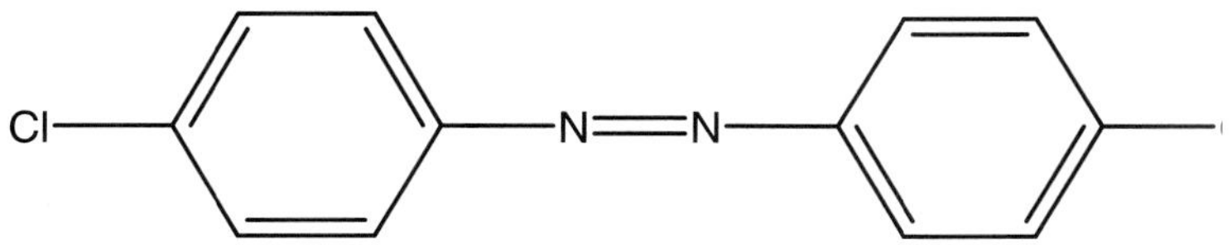

Chemical Structure

Synonyms: AZOBENZENE,4,4'-DICHLORO-; DCAB; DIAZENE,BIS(4-CHLOROPHENYL)-; P,P'-DICHLOROAZOBENZENE

RTECS Toxicity Data

Mutagenic: Bacteria - S Typhimurium Mutations in Microorganisms; Dose: 10 ug/plate (-S9).

Hazard Overviews

Carcinogenicity: IARC - Not listed; NIOSH - Not listed; NTP - Not listed; ACGIH - Not listed; OSHA - Not listed; EPA - Not listed; MAK - Not listed

Environmental

Regulations

RCRA 40CFR: Not listed
CERCLA: 40CFR 302.4: Listed as Compound per CWA Section 307(a)
SARA 40CFR 372.65: Not listed
SARA EHS 40CFR 355: Not listed
TSCA: Not listed

DIC2150 CAS #: 614-26-6

4,4'-DICHLOROAZOXYBENZENE

RTECS: CO4050000
Molecular Formula: $C_{12}H_8Cl_2N_2O$
Formula Weight: 267.12

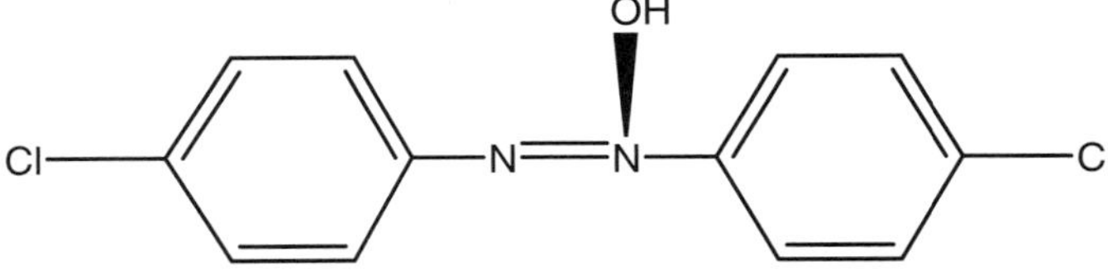

Chemical Structure

Synonyms: AZOXYBENZENE,4,4'-DICHLORO-; BIS(4-CHLOROPHENYL)DIAZENE 1-OXIDE; DCAOB; DIAZENE,BIS(4-CHLOROPHENYL)-,1-OXIDE; DIAZENE,BIS(4-CHLORPHENYL)-,1-OXIDE (9CI); P,P'-DICHLOROAZOXYBENZENE

RTECS Toxicity Data

Mutagenic: Bacteria - S Typhimurium Mutations in Microorganisms; Dose: 50 ug/plate (-S9).

Hazard Overviews

Carcinogenicity: IARC - Not listed; NIOSH - Not listed; NTP - Not listed; ACGIH - Not listed; OSHA - Not listed; EPA - Not listed; MAK - Not listed

Environmental

Regulations

RCRA 40CFR: Not listed
CERCLA: 40CFR 302.4: Listed as Compound per CWA Section 307(a)
SARA 40CFR 372.65: Not listed
SARA EHS 40CFR 355: Not listed
TSCA: Not listed

DIC2250 CAS #: 8023-53-8

DICHLOROBENZALKONIUM CHLORIDE

RTECS: BO3200000
Molecular Formula: Unspecified or Variable
Synonyms: ALKYL(C8H17 TO C18H37) DIMETHYL 3,4-DICHLOROBENZYLAMMONIUM CHLORIDE; AMMONIUM,ALKYL(C8-C18)DIMETHYL 3,4-DICHLOROBENZYL-,CHLORIDE; TETROSAN
Description: crystals
Use: as an antiseptic, germicide, algicide, sanitizer, deodorant

Physical Properties

Water Solubility: Soluble in Water
Other Solubilities: Soluble in Alcohol

RTECS Toxicity Data

Acute Oral: Rat LD_{50} Dose: 730 mg/kg; Toxic Effects: Gastrointestinal - Other changes. Mouse LD_{50} Dose: 2 gm/kg.
Irritation Eye: Rabbit Standard Draize Test Dose: 1%; Reaction: severe.

Hazard Overviews

Carcinogenicity: IARC - Not listed; NIOSH - Not listed; NTP - Not listed; ACGIH - Not listed; OSHA - Not listed; EPA - Not listed; MAK - Not listed

Environmental

Regulations

RCRA 40CFR: Not listed
CERCLA: 40CFR 302.4: Not listed
SARA 40CFR 372.65: Not listed
SARA EHS 40CFR 355: Not listed
TSCA: Not listed

DIC2300 CAS #: 2008-58-4

2,6-DICHLOROBENZAMIDE

EINECS Number: 217-918-4
Molecular Formula: $C_7H_5Cl_2NO$
Formula Weight: 190.02

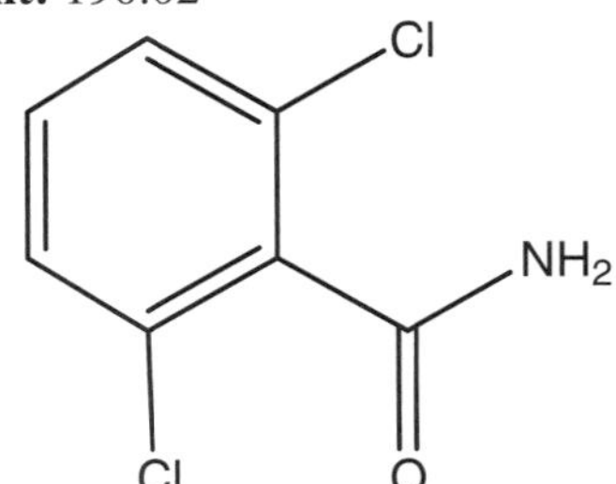

Chemical Structure

Synonyms: BENZAMIDE,2,6-DICHLORO-
Use: formerly in synthesis of dichlobenil

Physical Properties

Freezing Point: 198 °C (388.4 °F) to 200 °C (392 °F)

Hazard Overviews

Health: Irritating to skin. Other Acute Effects: may be harmful by inhalation, ingestion, or skin absorption.
Fire: Hazards: emits toxic fumes. Extinguishing agents: water spray; carbon dioxide, dry chemical powder or appropriate foam. Precautions: combustible liquid.
Reactivity: Incompatible with: strong oxidizing agents. Hazardous decomposition products: toxic fumes of: carbon monoxide, carbon dioxide, nitrogen oxides, hydrogen chloride gas.
Carcinogenicity: IARC - Not listed; NIOSH - Not listed; NTP - Not listed; ACGIH - Not listed; OSHA - Not listed; EPA - Not listed; MAK - Not listed
Primary Target Organs:

Skin

Environmental

Ecotoxicity: Algae: Chlorella fusca BCF (wet wt): 3

Regulations

RCRA 40CFR: Not listed
CERCLA: 40CFR 302.4: Not listed
SARA 40CFR 372.65: Not listed
SARA EHS 40CFR 355: Not listed
TSCA: Not listed

Analytical Methods

Food: FDA 212.1, 212.2, 232.1, 232.4, 242.1
Plasma: FDA 252

DIC2350 CAS #: 95-50-1

1,2-DICHLOROBENZENE

RTECS: CZ4500000
DOT: UN1591; IMO6.1
EINECS Number: 202-425-9
Molecular Formula: $C_6H_4Cl_2$
Structured MF: $C_6H_4Cl_2$
Formula Weight: 147.01

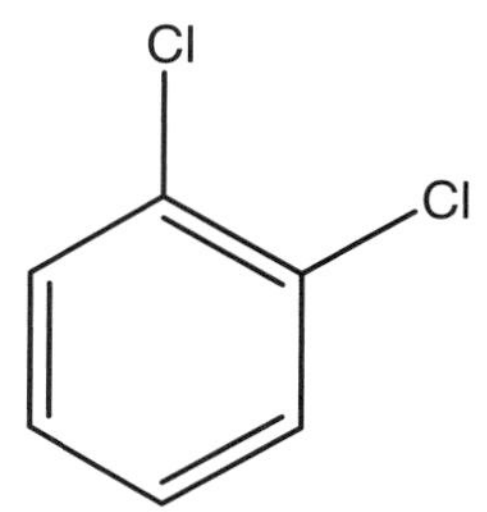

Chemical Structure

Synonyms: BENZENE,1,2-DICHLORO-; BENZENE,O-DICHLORO-; CHLOROBEN; CHLORODEN; CLOROBEN; DCB; O-DCB; O-DICHLOR BENZOL; O-DICHLORBENZENE; O-DICHLOROBENZENE; ORTHO-DICHLOROBENZENE; DICHLOROBENZENE,ORTHO,LIQUID; O-DICHLOROBENZOL; DILANTIN DB; DILATIN DB; DIZENE; DOWTHERM E; EPA PESTICIDE CHEMICAL CODE 059401; ODB; ODCB; ORTHODICHLOROBENZENE; ORTHODICHLOROBENZOL; ORTHOSOL; SPECIAL TERMITE FLUID; TERMITKIL
Description: colorless liquid; disagreeable, aromatic odor
Use: solvent for waxes, gums, resins, tars, rubbers, oils, asphalts and oxides of nonferrous metals; an insecticide for termites and locust borers and a fumigant; in removing sulfur from illuminating gas and as a degreasing agent for metals, leather, paper, dry cleaning, brick, upholstery and wool; an ingredient of metal polishes; as a heat-transfer medium, intermediate in the manufacture of dyes and component of herbicides; in industrial odor control, as a wood and furniture preservative, in the manufacture of 3,4-dichloroaniline and as a solvent carrier in toluene diisocyanate production; in the application or removal of surface coatings, in maintenance of equipment containing heat-transfer agents, in extractive distillation of ethyl benzene from xylene, as a solvent in the applications o motor-oil additive formulations, in paints, in formulations for removing pain in engine cleaning compounds, in firearm cleaners, in rust preventatives, in the dissolution of pitch on paper making felts, as a penetrant, in upper cylinder lubricants and as a magnetic coil coolant

Physical Properties

Boiling Point: 180.5 °C (357 °F) at 760 mm Hg
Freezing Point: -17 °C (1.4 °F)
Specific Gravity: 1.3048 at 20/4 °C
Vapor Density: 5.05
Saturated Vapor Density: 1.209445083 kg/m^3
Density: 1.3048 g/mL at 20 °C/4 °C
Vapor Pressure: 1.47 mm Hg at 25 °C
Water Solubility: 0.01% by weight
Other Solubilities: Soluble in Alcohol, Ether, Acetone & Benzene; 156 ppm at 25 °C.
Surface Tension: 37 dyne/cm at 20 °C
Odor Threshold: 12 to 300 mg/m^3
Refraction Index: 1.5515
Evaporation Rate: < 1 Butyl Acetate=1
Ionization Potential (eV): 9.06
Flash Point: 68.333 °C Open Cup
Autoignition Temperature: 648 °C
LEL: 2% v/v
UEL: 9% v/v

RTECS Toxicity Data

Acute Oral: Rat LD_{50} Dose: 500 mg/kg. Mouse LD_{50} Dose: 4386 mg/kg.

Acute Inhalation: Rat LC_{Lo} Dose: 821 ppm/7hr; Toxic Effects: Behavioral - General anesthetic; Liver - Hepatitis (hepatocellular necrosis), zonal; Sense organs and special senses - Lacrimation. Guinea Pig LC_{Lo} Dose: 800 ppm/24hr; Toxic Effects: Liver - Hepatitis (hepatocellular necrosis), diffuse.

Acute Dermal: Rat LD_{50} Route: Subcutaneous Dose: 5 gm/kg.

Chronic (Multiple Dose) Oral: Rat Dose: 27300 ug/kg/39W-I; Toxic Effects: Kidney, Ureter, and Bladder - Other changes in urine composition; Blood - Agranulocytosis; Blood - Changes in erythrocite (RBC) cell count. Rat Dose: 3 gm/kg/10D-I; Toxic Effects: Liver - Changes in liver weight; Endocrine - Changes in spleen weight; Blood - Changes in leukocyte (WBC) cell count. Rat Dose: 32500 mg/kg/13W-I; Toxic Effects: Liver - Changes in liver weight; Endocrine - Changes in spleen weight; Blood - Changes in erythrocite (RBC) cell count.

Irritation Eye: Rabbit Nonstandard Exposure Dose: 100 mg/30S rinse; Reaction: mild.

Reproductive/Teratogenic: Rat Route: Inhalation; Dose: 200 ppm/6H; Duration: female 6-15D of pregnancy; Specific Developmental Abnormalities - Musculoskeletal system. Rat Route: Intraperitoneal; Dose: 50 mg/kg; Duration: male 1D prior to mating; Paternal Effects - Spermatogenesis.

Mutagenic: Rat Sperm Morphology; Route: Intraperitoneal; Dose: 250 mg/kg. Mouse Mutations in Microorganisms; Cell Type: lymphocyte; Dose: 6500 ug/L (+S9).

Hazard Overviews

Fire Diamond

Health: Irritating to eyes/skin/respiratory tract. Toxic. Also Causes: blisters, burning pain in the stomach, nausea, vomiting, diarrhea. Chronic Effects: headache, anorexia, weight loss, jaundice, cirrhosis.

Fire: Combustible. Can form explosive mixtures in the air. Extinguish fires with water spray, dry chemical, foam, or carbon dioxide.

Reactivity: Stable. Hazardous polymerization cannot occur. Avoid: heat; hot surfaces. Incompatible with: oxidizing agents; aluminum. Hazardous decomposition products: fumes of chlorine.

Carcinogenicity: IARC - Group 3, Not classifiable as to carcinogenicity to humans; NIOSH - Not listed; NTP - Not listed; ACGIH - Class A4, Not classifiable as a human carcinogen; OSHA - Not listed; EPA - Class D, Not classifiable as to human carcinogenicity; MAK - Not listed

Primary Target Organs:

Eyes

Skin

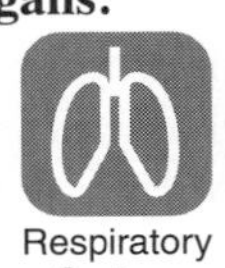
Respiratory System

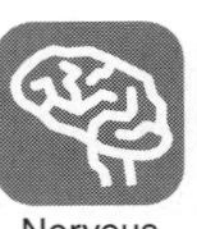
Nervous System

Liver

Kidneys

Exposure Limits

OSHA PEL: STEL: 50 ppm; 300 mg/m^3.

ACGIH TLV: TWA: 25 ppm; 150 mg/m^3; STEL: 50 ppm; 301 mg/m^3.

NIOSH REL: STEL: 50 ppm; 300 mg/m^3.

NIOSH IDLH: 200 ppm.

DFG MAK: TWA: 50 ppm; 300 mg/m^3.

Respirator Recommendation

Exposure Range: >50 to <200 ppm Air Purifying, Negative Pressure, Half Mask

Exposure Range: 200 to unlimited ppm Self-contained Breathing Apparatus, Pressure Demand, Full Face

Cartridge Color: black with dust/mist prefilter (use P100 or consult supervisor for appropriate dust/mist prefilter)

Environmental

Ecotoxicity: Aquatic toxicity: 13 ppm/*/marine plankton/no growth/ salt water *Time period not specified

Environmental Fate: If released to soil, it can be moderately to tightly adsorbed. Leaching from hazardous waste disposal areas has occurred and the detection in various groundwaters indicates that leaching can occur. Volatilization from soil surfaces may be an important transport mechanism. It is possible it will be slowly biodegraded in soil under aerobic conditions. Chemical transformation by hydrolysis, oxidation or direct photolysis are not expected to occur in soil. If released to water, adsorption to sediment will be a major environmental fate process based upon extensive monitoring data in the Great Lakes area and K_{oc} values. Analysis of Lake Ontario sediment cores has indicated the presence and persistence since before 1940. It is volatile from the water column with an estimated half-life of 4.4 hours from a model river one meter deep flowing 1 m/sec with a wind velocity of 3 m/sec at 20 °C; adsorption to sediment will attenuate volatilization. Aerobic biodegradation in water may be possible, however, anaerobic biodegradation is not expected to occur. Experimental BCF values of 66-560 have been reported it has been detected in trout from Lake Ontario. Aquatic hydrolysis, oxidation and direct photolysis are not expected to be important. If released to air, it will exist predominantly in the vapor-phase and will react with photochemically produced hydroxyl radicals at an estimated half-life rate of 24 days in a typical atmosphere. Direct photolysis in the troposphere is not expected to be important. The in rainwater suggests that atmospheric removal via wash-out is possible.

Cleanup/Disposal: Guide No. 152: Do not touch damaged containers or spilled material unless wearing appropriate protective clothing. Stop leak if you can do it without risk. Prevent entry into waterways, sewers, basements or confined areas. Cover with plastic sheet to prevent spreading. Absorb or cover with dry earth, sand or other non-combustible material and transfer to containers. Do not get water inside containers.

Environmental Physical Data

Henry's Law Constant: 0.0024
Octanol/Water Partition Coefficient: log K_{ow} = 3.38
Sorption Partition Coefficient: K_{oc} = 280
BCF: rainbow trout 270 to 560
BOD: theoretical < 0.1 lb/lb, 1/8 days

Regulations

RCRA 40CFR: Listed Hazardous Waste No. U070 Toxic Waste
CERCLA: 40CFR 302.4: Listed per CWA Section 311(b)(4) per RCRA Section 3001 per CWA Section 307(a) RQ: 100 lb (45.35 kg)
SARA 40CFR 372.65: Listed
SARA EHS 40CFR 355: Not listed
TSCA: Listed

Analytical Methods

Air: EPA VG-007-1, VG-011-1, TO-14, 0020; ASTM D3686, D3687, D4490
Soil: CLP LC_VOA, MC_SVOA, OHC; EPA 16, 1625; SW846 3640A, 5021, 8010B, 8020A, 8021A, 8120A, 8121, 8250A, 8260A, 8260B, 8270B, 8270C, 8410
Water / Groundwater: EPA 1625, 601, 602, 612, 624, 625, 625-S, 6, VW-014-1; APHA 6040-B, 6210-B, 6210-D, 6220-B, 6230-B, 6230-D, 6410-B; ASTM D5241; USGS O3118
Drinking Water: EPA 502.1, 502.2, 503.1, 524.1, 524.2; APHA 6210-C, 6220-C, 6230-C
Indoor / Expired Air: NIOSH 1003; EPA IP-1A, IP-1B
Plasma: EPA 29

DIC2400 CAS #: 541-73-1

1,3-DICHLOROBENZENE

RTECS: CZ4499000
EINECS Number: 208-792-1
Molecular Formula: $C_6H_4Cl_2$
Structured MF: $C_6H_4Cl_2$
Formula Weight: 147.00

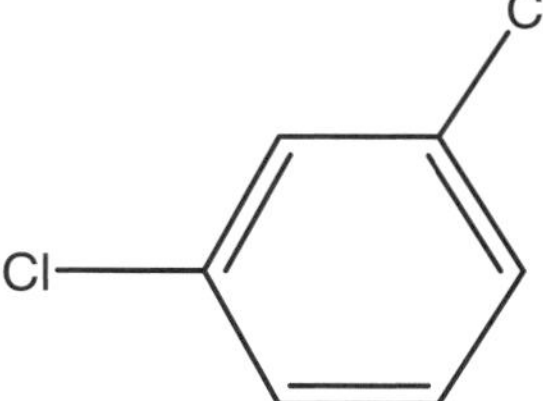

Chemical Structure

Synonyms: BENZENE,1,3-DICHLORO-; BENZENE,M-DICHLORO-; BENZENE,1,3-DICHLORO-(9CI); M-DCB; M-DICHLOROBENZENE; META-DICHLOROBENZENE; M-DICHLOROBENZOL; M-PHENYLENE DICHLORIDE; M-PHENYLENEDICHLORIDE
Description: colorless liquid
Use: fumigant and insecticide

Physical Properties

Boiling Point: 173.53 °C (344 °F)
Freezing Point: -24.7 °C (-12.46 °F)
Specific Gravity: 1.2884 at 20 °C/4 °C
Vapor Density: 5.08 Air=1
Vapor Pressure: 1 mm Hg at 12.1 °C
Water Solubility: < 1 mg/mL at 21 C
Other Solubilities: 95% Ethanol: >=100 mg/ml at 21 °C; Acetone: >=100 mg/ml at 21 °C; Alcohol: Soluble; Benzene: Soluble; Carbon Tetrachloride: Soluble; DMSO: >=100 mg/ml at 21 °C; Ether: Soluble; Ligroin: Soluble.
Surface Tension: 36.01 dynes/cm
Odor Threshold: 0.02 ppm
Refraction Index: 1.5459 at 20 °C/D
Critical Temperature: 415.0 °C
Critical Pressure: 38.3 atm
Ionization Potential (eV): 9.10 +/-0.02
Flash Point: 63 °C
Autoignition Temperature: Estimated at 648 °C
LEL: 2.02% v/v
UEL: 9.2% v/v

RTECS Toxicity Data

Chronic (Multiple Dose) Oral: Rat Dose: 1470 mg/kg/10D-I; Toxic Effects: Liver - Changes in liver weight; Nutritional and gross metabolic - Changes in calcium; Biochemical - Phosphatases. Rat Dose: 3330 mg/kg/90D-I; Toxic Effects: Endocrine - Other changes; Blood - Changes in serum composition; Biochemical - Dehydrogenases.
Mutagenic: Mouse Micronucleus Test; Route: Intraperitoneal; Dose: 175 mg/kg/24H. Yeast - S Cerevisiae Gene Conversion; Dose: 5 ppm.

Hazard Overviews

Fire Diamond

Health: Severely irritating to eyes/skin/respiratory tract; prolonged contact can be corrosive. Also Causes: burning stomach pain, nausea, vomiting, diarrhea, liver and kidney damage.
Fire: Combustible. Use carbon dioxide, dry chemical, water spray, fog or regular foam. Hazardous combustion products may include carbon oxide(s), hydrogen chloride, and chlorine gas.
Reactivity: Stable. Hazardous polymerization cannot occur. Avoid: heat; ignition sources. Incompatible with: oxidizers; magnesium; aluminum; acids; acid fumes. Hazardous decomposition products: carbon oxide(s); hydrogen chloride; chlorine gases.
Carcinogenicity: IARC - Not listed; NIOSH - Not listed; NTP - Not listed; ACGIH - Not listed; OSHA - Not listed; EPA - Class D, Not classifiable as to human carcinogenicity; MAK - Not listed

Primary Target Organs:

Eyes

Skin

Respiratory System

Liver

Kidneys

Environmental

Ecotoxicity: LC_{50} Sheepshead minnow 8.46 mg/l/24 hr /Static bioassay LC_{50} Fathead minnow 12.7 mg/l/96 hr /Static bioassay

Environmental Fate: If released to soil, it can be moderately to tightly adsorbed. Leaching from hazardous waste disposal areas has occurred and the detection in various groundwaters indicates that leaching can occur. Volatilization from soil surfaces may be an important transport mechanism. It is possible that it will be slowly biodegraded in soil under aerobic conditions. Chemical transformation by hydrolysis, oxidation or direct photolysis are not expected to occur in soil. If released to water, adsorption to sediment will be a major environmental fate process based upon extensive monitoring data in the Great Lakes area and K_{oc} values. Analysis of Lake Ontario sediment cores has indicated presence and persistence since before 1940. It is volatile from the water column with an estimated half-life of 4.1 hours from a river one meter deep flowing 1 m/sec with a wind velocity of 3 m/sec at 20 °C; adsorption to sediment will attenuate volatilization. Aerobic biodegradation in water may be possible, however, anaerobic biodegradation is not expected to occur. Experimental BCF values of 89-740 have been reported and it has been detected in trout from Lake Ontario. Hydrolysis, oxidation, and direct photolysis in aquatic environment are not expected to be important. If released to air, it will exist predominantly in the vapor-phase and will react with photochemically produced hydroxyl radicals at an estimated half-life rate of 14 days in a typical atmosphere. Direct photolysis in the troposphere is not expected to be important. The detection in rainwater suggests that atmospheric removal via wash-out is possible.

Environmental Physical Data

Henry's Law Constant: 0.0018
Octanol/Water Partition Coefficient: log K_{ow} = 3.60
Sorption Partition Coefficient: K_{oc} = 293
BCF: rainbow trout 420 to 740

Regulations

RCRA 40CFR: Listed Hazardous Waste No. U071 Toxic Waste
CERCLA: 40CFR 302.4: Listed per RCRA Section 3001 per CWA Section 307(a) RQ: 100 lb (45.35 kg)
SARA 40CFR 372.65: Listed
SARA EHS 40CFR 355: Not listed
TSCA: Listed

Analytical Methods

Air: EPA VG-007-1, VG-011-1, TO-14, 0020
Soil: CLP LC_VOA, MC_SVOA, OHC; EPA 16, 1625; SW846 3640A, 5021, 8010B, 8020A, 8021A, 8120A, 8121, 8250A, 8260A, 8260B, 8270B, 8270C, 8410
Water / Groundwater: EPA 1625, 601, 602, 612, 624, 625, 625-S, 6, VW-014-1; APHA 6040-B, 6210-B, 6210-D, 6220-B, 6230-B, 6230-D, 6410-B; USGS O3118
Drinking Water: EPA 502.1, 502.2, 503.1, 524.1, 524.2; APHA 6210-C, 6220-C, 6230-C
Indoor / Expired Air: EPA IP-1A
Plasma: EPA 001, 29

DIC2450 CAS #: 106-46-7

1,4-DICHLOROBENZENE

RTECS: CZ4550000
DOT: UN1592; IMO6.1
EINECS Number: 203-400-5
Molecular Formula: $C_6H_4Cl_2$
Structured MF: $C_6H_4Cl_2$
Formula Weight: 147.01

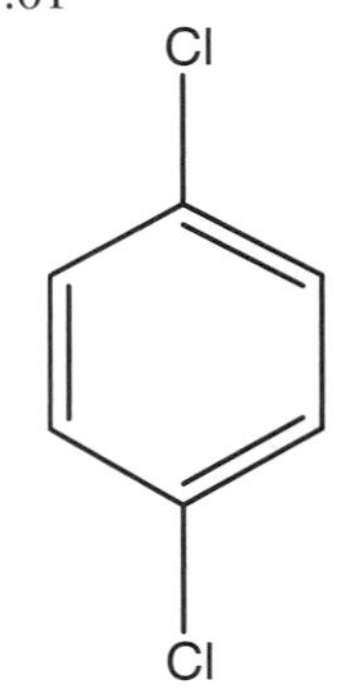

Chemical Structure

Synonyms: BENZENE,1,4-DICHLORO-; BENZENE,P-DICHLORO-; P-CHLOROPHENYL CHLORIDE; PARA CRYSTALS; P-DCB; 1,4-DICHLOORBENZEEN; P-DICHLOORBENZEEN; 1,4-DICHLOR-BENZOL; P-DICHLORBENZOL; DI-CHLORICIDE; P-DICHLOROBENZENE; PARA-DICHLOROBENZENE; DICHLOROBENZENE,PARA,SOLID; P-DICHLOROBENZOL; PARA-DICHLOROBENZOL; DICHLOROCIDE; 1,4-DICLOROBENZENE; P-DICLOROBENZENE; EPA PESTICIDE CHEMICAL CODE 061501; EVOLA; GLOBOL; PARACIDE; PARADI; PARADICHLORBENZOL; PARADICHLOROBENZENE; PARADICHLOROBENZOL; PARADOW; PARAMOTH; PARANUGGETS; PARAZENE; PDB; PDCB; PERSIA-PERAZOL; SANTOCHLOR

Description: white crystals; mothball-like odor

Use: as a moth repellent, general insecticide, pesticide fumigant, germicide, miticide, space odorant, air deodorant, chemical intermediate for dyes and organic chemicals, mildew control agent, disintegrating paste for molding concrete and stoneware, lubricant and disinfectant; in the manufacture of 2,5-dichloroaniline, pharmaceutical manufacture, agriculture (to fumigate soil), manufacture of polyphenylene sulfide resins (used for surface coatings and molding resins) and organic synthesis

Physical Properties

Boiling Point: 174 °C (345 °F) at 760 mm Hg
Freezing Point: 53.1 °C (127.58 °F)
Specific Gravity: 1.2475 at 20/4 °C

Vapor Density: 5.08 Air=1
Saturated Vapor Density: 1.203855136 kg/m^3
Density: 1.2475 g/mL at 20 °C/4 °C
Vapor Pressure: 0.6 mm Hg at 20 °C
Water Solubility: 65.3 mg/L at 25 °C
Other Solubilities: 95% Ethanol: >=100 mg/ml at 23 °C; Acetone: >=100 mg/ml at 23 °C; Alcohol: Soluble; Benzene: Soluble; Carbon Disulfide: Soluble; Chloroform: Soluble; DMSO: >=100 mg/ml at 23 °C; Ether: Soluble; Most organic solvents: Soluble.
Surface Tension: 31.4 dynes/cm at 100 °C
Odor Threshold: 15 to 30 ppm
Refraction Index: 1.5285
Ionization Potential (eV): 8.98
Flash Point: 65.556 °C Closed Cup
Autoignition Temperature: > 482 °C
LEL: 2.5% v/v

RTECS Toxicity Data

Acute Oral: Human TD_{Lo} Dose: 300 mg/kg; Toxic Effects: Sense organs and special senses - Other; Lungs, Thorax, or Respiration - Other changes; Gastrointestinal - Hypermotility, diarrhea. Human LD_{Lo} Dose: 857 mg/kg. Rat LD_{50} Dose: 500 mg/kg.
Acute Inhalation: Mammal LC_{50} Dose: 12 gm/m^3. Cat LC_{Lo} Dose: 37 gm/m^3/30M; Toxic Effects: Behavioral - General anesthetic.
Acute Dermal: Rabbit LD_{50} Route: Skin; Dose: >2 gm/kg. Rat LD_{50} Route: Skin; Dose: >6 gm/kg.
Chronic (Multiple Dose) Oral: Rat Dose: 14 gm/kg/14D-I; Toxic Effects: DEATH. Rat Dose: 58500 mg/kg/13W-I; Toxic Effects: Liver - Changes in liver weight; Kidney, Ureter, and Bladder - Chgs in tubules (inc acute renal failure, acute tubular necrosis; Blood - Changes in erythrocite (RBC) cell count. Rat Dose: 10 gm/kg/4W-I; Toxic Effects: Liver - Hepatitis (hepatocellular necrosis), zonal; Kidney, Ureter, and Bladder - Chgs in tubules (inc acute renal failure, acute tubular necrosis.
Chronic (Multiple Dose) Inhalation: Rat Dose: 158 ppm/7H/31W-I; Toxic Effects: Liver - Changes in Liver weight; Kidney, Ureter, and Bladder - Changes in kidney weight. Rat Dose: 100 gm/m^3/20M/25D-I; Toxic Effects: Behavioral - General anesthetic; Behavioral - Tremor; Blood - Granulocytopenia.
Reproductive/Teratogenic: Rat Route: Oral; Dose: 7500 mg/kg; Duration: female 6-15D of pregnancy; Specific Developmental Abnormalities - Musculoskeletal system. Rat Route: Oral; Dose: 10 gm/kg; Duration: female 6-15D of pregnancy; Effects on Embryo or Fetus - Fetotoxicity.
Mutagenic: Human Sister Chromatid Exchenge; Cell Type: lymphocyte; Dose: 100 ug/L. Rat Sperm Morphology; Route: Intraperitoneal; Dose: 800 mg/kg.
Tumorigenic: Rat Route: Oral; Dose: 155 gm/kg/2Y-I; Toxic Effects: Tumorigenic - Carcinogenic by RTECS criteria; Kidney, Ureter, and Bladder - Kidney tumors. Mouse Route: Oral; Dose: 155 gm/kg/2Y-I; Toxic Effects: Tumorigenic - Carcinogenic by RTECS criteria; Liver - Tumors.

Hazard Overviews

Fire Diamond

Health: Severe irritation to eyes/respiratory tract. Also Causes: hemolytic anemia, jaundice, methemglobinemia, nausea, diarrhea. Chronic Effects: leukemia, lung granulomatosis, liver abnormalities, kidney damage, anemia, cataracts, possible cancer hazard.
Fire: Combustible. May form explosive dust-air mixtures. Use dry chemical, carbon dioxide, alcohol foam, or water spray. Use water spray to cool fire-exposed container, to disperse vapors, or to blanket a pool fire.
Reactivity: Stable. Hazardous polymerization cannot occur. Avoid: heat; ignition sources. Incompatible with: strong oxidizers; oxidizing agents. Hazardous decomposition products: carbon monoxide; chlorides; chlorine.
Carcinogenicity: IARC - Group 2B, Possibly carcinogenic to humans; NIOSH - Listed as carcinogen; NTP - Class 2B, Reasonably anticipated to be a carcinogen, sufficient evidence of carcinogenicity from studies in experimental animals; ACGIH - Class A3, Animal carcinogen; OSHA - Not listed; EPA - Not listed; MAK - Not listed
Primary Target Organs:

Eyes

Skin

Respiratory System

Liver

Kidneys

Blood

Exposure Limits
OSHA PEL: TWA: 75 ppm; 450 mg/m^3.
OSHA PEL Vacated 1989 Limits: TWA: 75 ppm; 450 mg/m^3; STEL: 110 ppm; 675 mg/m^3.
ACGIH TLV: TWA: 10 ppm; 60 mg/m^3.
NIOSH IDLH: 150 ppm.
DFG MAK: TWA: 50 ppm; 300 mg/m^3.
Respirator Recommendation
Exposure Range: >75 to <150 ppm Air Purifying, Negative Pressure, Half Mask
Exposure Range: 150 to unlimited ppm Self-contained Breathing Apparatus, Pressure Demand, Full Face
Cartridge Color: black with dust/mist prefilter (use P100 or consult supervisor for appropriate dust/mist prefilter)

Environmental

Ecotoxicity: LC_{50} Poecilia reticulata (guppy) 4.0 ppm/14 days /Conditions of bioassay not specified LC_{50} Lepomis macrochirus (bluegill sunfish) 4.54 mg/l/24 hr; 4.3 mg/l/48 hr; 4.25 mg/l/96 hr /Static bioassay LC_{50} Sheepshead minnow 7.5-10 mg/l/24 hr; 7.17 mg/l/48 hr; 7.4 mg/l/96 hr /Static bioassay
Environmental Fate: If released to soil, it can be moderately to tightly adsorbed. Leaching from hazardous waste disposal areas has occurred and the detection in various groundwaters indicates that leaching can occur. Volatilization from soil surfaces may be an important transport mechanism. It is

possible it will be slowly biodegraded in soil under aerobic conditions. Chemical transformation by hydrolysis, oxidation or direct photolysis are not expected to occur in soil. If released to water, volatilization may be the dominant removal process. The volatilization half-life from a model river one meter deep flowing one meter/sec with a wind velocity of 3 m/sec is estimated to be 4.3 hours at 20 °C. Adsorption to sediment will be a major environmental fate process based upon extensive monitoring data in the Great Lakes area and K_{oc} values based upon monitoring samples. Analysis of Lake Ontario sediment cores has indicated presence and persistence since before 1940. Adsorption to sediment will attenuate volatilization. Aerobic biodegradation in water may be possible, however, anaerobic biodegradation is not expected to occur. For the most part, experimental BCF values reported in the literature are less than 1000 which suggests that significant bioconcentration will not occur; however, a BCF of 1800 was determined for guppies in one study. Aquatic hydrolysis, oxidation and direct photolysis are not expected to be important. If released to air it will exist predominantly in the vapor-phase and will react with photochemically produced hydroxyl radicals at an estimated half-life rate of 31 days in typical atmosphere. Direct photolysis in the troposphere is not expected to be important. The detection in rain-water suggests that atmospheric removal via wash-out is possible.

Cleanup/Disposal: Guide No. 152: Do not touch damaged containers or spilled material unless wearing appropriate protective clothing. Stop leak if you can do it without risk. Prevent entry into waterways, sewers, basements or confined areas. Cover with plastic sheet to prevent spreading. Absorb or cover with dry earth, sand or other non-combustible material and transfer to containers. Do not get water inside containers.

Environmental Physical Data

Henry's Law Constant: 0.0015
Octanol/Water Partition Coefficient: log K_{ow} = 3.39
Sorption Partition Coefficient: K_{oc} = 273
BCF: increases with log p

Regulations

RCRA 40CFR: Listed Hazardous Waste No. U072 Toxic Waste
CERCLA: 40CFR 302.4: Listed per CWA Section 311(b)(4) per RCRA Section 3001 per CWA Section 307(a) RQ: 100 lb (45.35 kg)
SARA 40CFR 372.65: Listed
SARA EHS 40CFR 355: Not listed
TSCA: Listed

Analytical Methods

Air: EPA VG-007-1, VG-011-1, TO-14, 0020; ASTM D4490
Soil: CLP LC_VOA, MC_SVOA, OHC; EPA 16, 1625, PMD-PAD; SW846 3640A, 5021, 8010B, 8020A, 8021A, 8120A, 8121, 8250A, 8260A, 8260B, 8270B, 8270C, 8410; DOE OP040R, OP130R
Water / Groundwater: EPA 1625, 601, 602, 612, 624, 625, 625-S, 6, VW-014-1; APHA 6040-B, 6210-B, 6210-D, 6220-B, 6230-B, 6230-D, 6410-B; ASTM D5241; USGS O3118
Drinking Water: EPA 502.1, 502.2, 503.1, 524.1, 524.2; APHA 6210-C, 6220-C, 6230-C
Food: FDA 212.1, 232.1
Indoor / Expired Air: NIOSH 1003; EPA IP-1A
Plasma: EPA 001, 29; FDA 211.1, 231.1, 252

DIC2500 CAS #: 25321-22-6

DICHLOROBENZENE

DOT: UN1592
EINECS Number: 246-837-7
Molecular Formula: $C_6H_4Cl_2$
Structured MF: $C_6H_4Cl_2$
Formula Weight: 147.01.
Synonyms: BENZENE,DICHLORO-; DILATIN DBI
Description: colorless liquid; pleasant aromatic odor

Physical Properties

Boiling Point: 179 °C (354 °F)
Freezing Point: -17.6 °C (0.32 °F)
Specific Gravity: 1.3 at 25 °C
Vapor Density: 5.07. Air=1
Vapor Pressure: 1.2 mm Hg at 20 °C
Water Solubility: 0.014% by wt in Water at 20 °C
Other Solubilities: Readily Soluble in natural fats or fat Soluble substances.
Surface Tension: 36.01 dynes/cm at 20 °C
Odor Threshold: 0.005 ppm
Critical Temperature: Estimated at 410.8 °C
Critical Pressure: 562.9 psia
Flash Point: 68.333 °C Closed Cup
Autoignition Temperature: Estimated at 648 °C
LEL: Estimated at 2.02% v/v
UEL: Estimated at 9.2% v/v

Hazard Overviews

Fire Diamond

Fire: Combustible.
Carcinogenicity: IARC - Not listed; NIOSH - Not listed; NTP - Not listed; ACGIH - Not listed; OSHA - Not listed; EPA - Not listed; MAK - Not listed

Environmental

Ecotoxicity: Aquatic toxicity: 10 ppm/48-hour/Zebrafish/LC_{50}
Cleanup/Disposal: Guide No. 152: Do not touch damaged containers or spilled material unless wearing appropriate protective clothing. Stop leak if you can do it without risk. Prevent entry into waterways, sewers, basements or confined areas. Cover with plastic sheet to prevent spreading. Absorb or cover with dry earth, sand or other non-combustible material and transfer to containers. Do not get water inside containers.

Environmental Physical Data

BOD: acclimated < 1 lb/lb, .125 days

Regulations

RCRA 40CFR: Not listed
CERCLA: 40CFR 302.4: Listed per CWA Section 311(b)(4) RQ: 100 lb (45.35 kg)
SARA 40CFR 372.65: Listed
SARA EHS 40CFR 355: Not listed
TSCA: Not listed

Analytical Methods

Air: EPA VA-005-1, VG-006-1
Soil: SW846 3650A, 3650B
Water / Groundwater: EPA S-002-1, VW-008-1
Other: EPA VS-006-1

DIC2550 CAS #: 20103-09-7

2,5-DICHLORO-1,4-BENZENEDIAMINE

RTECS: SS9150000
EINECS Number: 243-512-1
Molecular Formula: $C_6H_6Cl_2N_2$
Formula Weight: 177.04

Chemical Structure

Synonyms: 1,4-BENZENEDIAMINE,2,5-DICHLORO-; BENZENE,2,5-DICHLORO,1,4-DIAMINO; 1,4-DIAMINO-3,6-DICHLOROBENZENE; 2,5-DICHLOR-1,4-FENYLENDIAMIN; 2,5-DICHLORO-P-PHENYLENEDIAMINE; P-PHENYLENEDIAMINE,2,5-DICHLORO-
Description: white prisms

Physical Properties

Freezing Point: 170 °C (338 °F)
Water Solubility: Slightly Soluble in Water
Other Solubilities: Slightly Soluble in Alcohol

RTECS Toxicity Data

Acute Oral: Rat LD_{50} Dose: 1750 mg/kg.

Hazard Overviews

Health: Irritating to eyes/skin/respiratory tract. Other Acute Effects: may be harmful by inhalation, ingestion, or skin absorption; absorption into the body leads to the formation of methemoglobin which in sufficient concentration causes cyanosis; onset may be delayed 2 to 4 hours or longer.
Fire: Hazards: emits toxic fumes. Extinguishing agents: water spray; carbon dioxide, dry chemical powder or appropriate foam. Precautions: combustible liquid.
Reactivity: Incompatible with: acids, acid chlorides, acid anhydrides, chloroformates, strong oxidizing agents. Hazardous decomposition products: toxic fumes of: carbon monoxide, carbon dioxide, nitrogen oxides, hydrogen chloride gas.
Carcinogenicity: IARC - Not listed; NIOSH - Not listed; NTP - Not listed; ACGIH - Not listed; OSHA - Not listed; EPA - Not listed; MAK - Not listed
Primary Target Organs:

Eyes

Skin

Respiratory System

Environmental

Regulations

RCRA 40CFR: Not listed
CERCLA: 40CFR 302.4: Listed as Compound per CWA Section 307(a)
SARA 40CFR 372.65: Not listed
SARA EHS 40CFR 355: Not listed
TSCA: Listed

DIC2600 CAS #: 609-20-1

2,6-DICHLORO-1,4-BENZENEDIAMINE

RTECS: SS9175000
EINECS Number: 210-184-6
Molecular Formula: $C_6H_6Cl_2N_2$
Structured MF: $Cl_2C_6H_2(NH_2)_2$
Formula Weight: 177.04

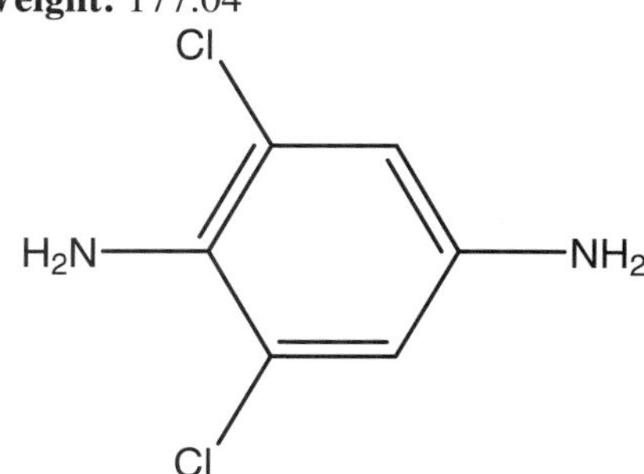

Chemical Structure

Synonyms: 1,4-BENZENEDIAMINE,2,6-DICHLORO-; C.I. 37020; DAITO BROWN SALT RR; 1,4-DIAMINO-2,6-DICHLOROBENZENE; 2,5-DIAMINO-1,3-DICHLOROBENZENE; 2,6-DICHLORO-1,4-PHENYLENEDIAMINE; 2,6-DICHLORO-P-PHENYLENEDIAMINE; 2,6-DICHLORO-PARA-PHENYLENEDIAMINE; 3,5-DICHLORO-1,4-PHENYLENEDIAMINE; FAST BROWN RR SALT; P-PHENYLENEDIAMINE,2,6-DICHLORO-

Description: gray, microcrystalline powder or solid
Use: fungicide, as an intermediate in dye and resin manufacture, in the preparation of certain polyamide fibers and as a curing agent for polyurethane

Physical Properties

Freezing Point: 124 °C (255.2 °F) to 126 °C (258.8 °F)
Water Solubility: < 1 mg/mL at 24 C
Other Solubilities: 95% Ethanol: 10-50 mg/ml at 24 °C; Acetone: 10-50 mg/ml at 24 °C; Acids: >10%; Alcohol: Soluble; Benzene: Soluble; DMSO: >=100 mg/ml at 24 °C; Ethanol: >10%; Ether: Soluble.
Flash Point: Not available; probably combustible

RTECS Toxicity Data

Acute Oral: Rat LD_{50} Dose: 700 mg/kg; Toxic Effects: Liver - Other changes; Blood - Changes in spleen; Nutritional and gross metabolic - Weight loss or decreased weight gain.
Mutagenic: Mouse Mutations in Mammalian Somatic Cells; Cell Type: lymphocyte; Dose: 25 mg/L. Mouse Cytogenetic Analysis; Route: Intraperitoneal; Dose: 100 mg/kg.
Tumorigenic: Mouse Route: Oral; Dose: 87 gm/kg/2Y-C; Toxic Effects: Tumorigenic - Carcinogenic by RTECS criteria; Lungs, Thorax, or Respiration - Tumors; Liver - Tumors. Mouse Route: Oral; Dose: 260 gm/kg/2Y-C; Toxic Effects: Tumorigenic - Carcinogenic by RTECS criteria; Lungs, Thorax, or Respiration - Tumors; Liver - Tumors.

Hazard Overviews

Health: Irritating to eyes/skin/respiratory tract. Toxic. Other Acute Effects: harmful if swallowed, inhaled, or absorbed through skin; absorption into the body leads to the formation of methemoglobin which in sufficient concentration causes cyanosis; onset may be delayed 2 to 4 hours or longer.Chronic Effects: Carcinogen.
Fire: Will burn. Hazards: emits toxic fumes. Extinguishing agents: water spray; carbon dioxide, dry chemical powder or appropriate foam. Precautions: combustible liquid.
Reactivity: Incompatible with: strong oxidizing agents. Hazardous decomposition products: toxic fumes of: carbon monoxide, carbon dioxide, nitrogen oxides, hydrogen chloride gas.
Carcinogenicity: IARC - Group 3, Not classifiable as to carcinogenicity to humans; NIOSH - Not listed; NTP - Not listed; ACGIH - Not listed; OSHA - Not listed; EPA - Not listed; MAK - Not listed
Primary Target Organs:

Eyes

Skin

Respiratory System

Environmental

Regulations

RCRA 40CFR: Not listed
CERCLA: 40CFR 302.4: Listed as Compound per CWA Section 307(a)
SARA 40CFR 372.65: Not listed
SARA EHS 40CFR 355: Not listed
TSCA: Listed

DIC2650 CAS #: 91-94-1

3,3'-DICHLOROBENZIDINE

RTECS: DD0525000
EINECS Number: 202-109-0
Molecular Formula: $C_{12}H_{10}Cl_2N_2$
Structured MF: $NH_2ClC_6H_3C_6H_3ClNH_2$
Formula Weight: 253.13

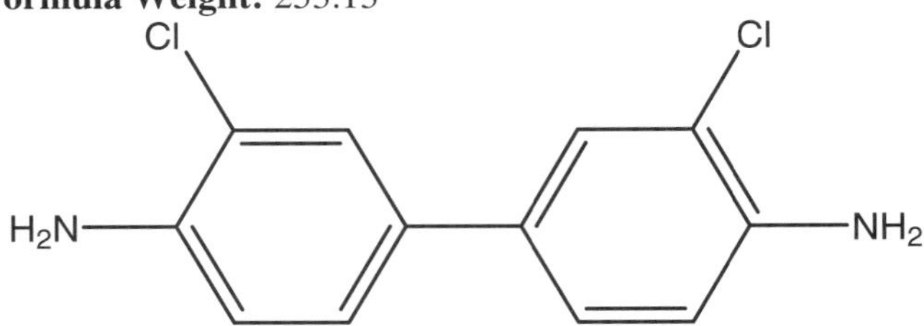

Chemical Structure

Synonyms: BENZIDINE, 3,3'-DICHLORO-; (1,1'-BIPHENYL)-4,4'-DIAMINE, 3,3'-DICHLORO-; C.I. 23060; CURITHANE C126; 4,4'-DIAMINO-3,3'-DICHLOROBIPHENYL; 4,4'-DIAMINO-3,3'-DICHLORODIPHENYL; 3,3'-DICHLORBENZIDIN; 3,3'-DICHLOROBENZIDIN; 3,3'-DICHLOROBENZIDINA; DICHLOROBENZIDINE; O,O'-DICHLOROBENZIDINE; ORTHO,ORTHO'-DICHLOROBENZIDINE; 3,3'-DICHLOROBENZIDINE BASE; DICHLOROBENZIDINE BASE; 3,3'-DICHLORO-4,4'-BIPHENYLDIAMINE; 3,3'-DICHLOROBIPHENYL-4,4'-DIAMINE; 3,3'-DICHLORO-4,4'-DIAMINO(1,1-BIPHENYL); 3,3'-DICHLORO-4,4'-DIAMINOBIPHENYL; 3,3'-DICHLORO-4,4'-DIAMINODIPHENYL
Description: colorless, white needles, crystals; mild odor of aromatic amines
Use: intermediate for dyes and pigments; Curing agent for isocyanate-terminated resins for urethane plastics

Physical Properties

Boiling Point: 402 °C (756 °F)
Freezing Point: 132 °C (269.6 °F) to 133 °C (271.4 °F)
Specific Gravity: 1.16 at 20 °C / 20 °C
Vapor Density: 5.31 Air=1
Saturated Vapor Density: 1.2 kg/m^3
Vapor Pressure: 4.5 $x10^{-9}$ mm Hg at 20 °C
Water Solubility: Insoluble or Slightly Soluble in Water
Other Solubilities: Alcohol: Very Soluble; Benzene: Very Soluble; Ether: Soluble; Glacial Acetic Acid: Very Soluble.
pH: Weak base
Flash Point: Unknown; combustible
Autoignition Temperature: 540 °C
LEL: 0.6% v/v
UEL: 5.8% v/v at 331 °F

RTECS Toxicity Data

Mutagenic: Human Unscheduled DNA Synthesis; Cell Type: HeLa cell; Dose: 100 nmol/L. Rat Body Fluid Assay;

Indicator Organism: Bacteria - S Typhimurium; Dose: 40 mg/kg.

Tumorigenic: Rat Route: Oral; Dose: 17 gm/kg/50W-C; Toxic Effects: Tumorigenic - Carcinogenic by RTECS criteria; Blood - Tumors; Skin and appendages - Tumors. Rat Route: Oral; Dose: 20 gm/kg/52W-I; Toxic Effects: Tumorigenic - Equivocal tumorigenic agent by RTECS criteria; Gastrointestinal - Tumors; Skin and appendages - Tumors. Rat Route: Oral; Dose: 21 gm/kg/50W-C; Toxic Effects: Tumorigenic - Carcinogenic by RTECS criteria; Blood - Tumors; Skin and appendages - Tumors.

Hazard Overviews

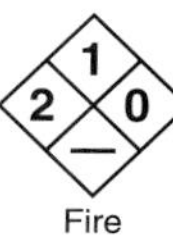

Fire Diamond

Health: Skin contact may cause allergic skin reactions. May be absorbed through skin. Possible cancer hazard. Chronic Effects: bladder cancer.

Fire: Use dry chemical, carbon dioxide, water spray, fog, or foam.

Reactivity: Stable. Hazardous polymerization cannot occur. Avoid: heating to decomposition. Hazardous decomposition products: toxic fumes of hydrochloric acid; chlorine; nitrogen oxides.

Carcinogenicity: IARC - Group 2B, Possibly carcinogenic to humans; NIOSH - Listed as carcinogen; NTP - Class 2B, Reasonably anticipated to be a carcinogen, sufficient evidence of carcinogenicity from studies in experimental animals; ACGIH - Class A3, Animal carcinogen; OSHA - Listed as a carcinogen; EPA - Class B2, Probable human carcinogen based on animal studies; MAK - Class A2, Unmistakably carcinogenic in animal experimentation only

Primary Target Organs:

Skin

Respiratory System

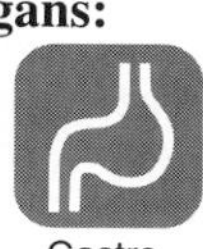
Gastro-intestinal

Liver

Respirator Recommendation

Exposure Range: unlimited Self-contained Breathing Apparatus, Pressure Demand, Full Face

Note: TLV not established

Environmental

Environmental Fate: If released into water, it will rapidly adsorb to sediment and particulate matter where it is very tightly bound, possibly chemically bound, and not be readily dislodged. It will undergo very rapid photooxidation in surface layers of water (half-life 90 sec) forming 3-chlorobenzidine and benzidine. Redox reactions and reactions involving free radicals may also be important. It will bioconcentrate in fish. When released on land, it will tightly bind to the soil and possibly undergo chemical reactions with soil components. Very slow mineralization occurs (2% in 32 weeks). If released to the atmosphere it will most likely be adsorbed to particulate matter and rapidly photodegrade.

Environmental Physical Data

Henry's Law Constant: 4.5 x10^{-8}

Octanol/Water Partition Coefficient: log K_{ow} = 3.02

BCF: bluegill 495 to 507

Regulations

RCRA 40CFR: Listed Hazardous Waste No. U073 Toxic Waste

CERCLA: 40CFR 302.4: Listed per RCRA Section 3001 per CWA Section 307(a) RQ: 1 lb (0.454 kg)

SARA 40CFR 372.65: Listed

SARA EHS 40CFR 355: Not listed

TSCA: Listed

Analytical Methods

Soil: CLP MC_SVOA, OHC; EPA 16, 1625; SW846 3640A, 8250A, 8270B, 8270C

Water / Groundwater: EPA 6, 1625, 553, 605, 625, 625-S; SW846 8325; APHA 6410-B; USGS O3118

Drinking Water: EPA 553

Indoor / Expired Air: NIOSH 5509

Plasma: EPA 29

DIC2700 **CAS #: 612-83-9**

3,3'-DICHLOROBENZIDINE DIHYDROCHLORIDE

RTECS: DD0550000

EINECS Number: 210-323-0

Molecular Formula: $C_{12}H_{12}Cl_4N_2$

Formula Weight: 326.06

Synonyms: BENZIDINE,3,3'-DICHLORO-,DIHYDROCHLORIDE; (1,1'-BIPHENYL)-4,4'-DIAMINE,3,3'-DICHLORO,DIHYDROCHLORIDE; (1,1'-BIPHENYL)-4,4'-DIAMINE,3,3'-DICHLORO-,DIHYDROCHLORIDE; 3,3'-DICHLORO(1,1'-BIPHENYL)-4,4'-DIAMINE DIHYDROCHLORIDE

Description: white to light-gray powder, crystals, or needles; mild odor

Use: manufacture of dyes for printing inks, textiles, plastics and crayons and as a curing agent for solid urethane plastics

Physical Properties

Freezing Point: 132 °C (269.6 °F) to 133 °C (271.4 °F)

Water Solubility: 4 mg/L Water at 22 °C

Other Solubilities: 95% Ethanol: 5-10 mg/ml at 21.5 °C; Acetone: <1 mg/ml at 21.5 °C; DMSO: 10-50 mg/ml at 21.5 °C.

Flash Point: Not available; probably combustible

RTECS Toxicity Data

Acute Oral: Rat LD_{50} Dose: 3820 mg/kg.

Mutagenic: Bacteria - S Typhimurium Mutations in Microorganisms; Dose: 10 ug/plate (+S9), 1 ug/plate (-S9).

Hazard Overviews

Fire: Will burn.

Carcinogenicity: IARC - Not listed; NIOSH - Not listed; NTP - Listed; ACGIH - Not listed; OSHA - Not listed; EPA - Not listed; MAK - Not listed

Environmental

Environmental Physical Data

BCF: fish 495 to 507

Regulations

RCRA 40CFR: Not listed
CERCLA: 40CFR 302.4: Listed as Compound per CWA Section 307(a)
SARA 40CFR 372.65: Listed
SARA EHS 40CFR 355: Not listed
TSCA: Listed

DIC2750 CAS #: 64969-34-2

3,3'-DICHLOROBENZIDINE SULFATE

EINECS Number: 265-293-1
Molecular Formula: $C_{12}H_{14}N_2O_8Cl_2S_2$
Formula Weight: 449.3

Hazard Overviews

Carcinogenicity: IARC - Not listed; NIOSH - Not listed; NTP - Not listed; ACGIH - Not listed; OSHA - Not listed; EPA - Not listed; MAK - Not listed

Environmental

Regulations

RCRA 40CFR: Not listed
CERCLA: 40CFR 302.4: Not listed
SARA 40CFR 372.65: Listed
SARA EHS 40CFR 355: Not listed
TSCA: Not listed

DIC2800 CAS #: 50-84-0

2,4-DICHLOROBENZOIC ACID

RTECS: DG6650000
EINECS Number: 200-067-8
Molecular Formula: $C_7H_4Cl_2O_2$
Structured MF: $Cl_2C_6H_3COOH$
Formula Weight: 191.01

Cl HO Cl O

Chemical Structure

Synonyms: BENZOIC ACID,2,4-DICHLORO-
Description: white to slightly yellowish powder
Use: as an intermediate for antimalarials, dyes, fungicides, pharmaceuticals and other organic chemicals

Physical Properties

Boiling Point: Sublimes
Freezing Point: 157 °C (314.6 °F) to 160 °C (320 °F)
Water Solubility: < 1 mg/mL at 19 C
Other Solubilities: 5% Caustic solution: Soluble; 95% Ethanol: >=100 mg/ml at 19 °C; Acetone: >=100 mg/ml at 19 °C; Alcohol: Soluble; Benzene: >10%; Chloroform: >10%; DMSO: >=100 mg/ml at 19 °C; Ether: >10%; Heptane: Insoluble.
Flash Point: Not available; probably combustible

RTECS Toxicity Data

Acute Oral: Mouse LD_{50} Dose: 830 mg/kg; Toxic Effects: Behavioral - Somnolence (general depressed activity); Behavioral - Ataxia; Lungs, Thorax, or Respiration - Dyspnea.
Acute Dermal: Mouse LD_{50} Route: Subcutaneous Dose: 1200 mg/kg; Toxic Effects: Behavioral - Altered sleep time (including change in righting reflex); Behavioral - Muscle weakness; Lungs, Thorax, or Respiration - Dyspnea.

Hazard Overviews

Health: Irritating to eyes/skin/respiratory tract. Other Acute Effects: may be harmful by inhalation, ingestion, or skin absorption.
Fire: Will burn. Hazards: emits toxic fumes. Extinguishing agents: water spray; carbon dioxide, dry chemical powder or appropriate foam. Precautions: combustible liquid.
Reactivity: Incompatible with: strong oxidizing agents. Hazardous decomposition products: toxic fumes of: carbon monoxide, carbon dioxide, hydrogen chloride gas.
Carcinogenicity: IARC - Not listed; NIOSH - Not listed; NTP - Not listed; ACGIH - Not listed; OSHA - Not listed; EPA - Not listed; MAK - Not listed
Primary Target Organs:

Eyes

Skin

Respiratory System

Environmental

Regulations

RCRA 40CFR: Not listed
CERCLA: 40CFR 302.4: Not listed
SARA 40CFR 372.65: Not listed
SARA EHS 40CFR 355: Not listed
TSCA: Listed

DIC2850 CAS #: 50-30-6

2,6-DICHLOROBENZOIC ACID

RTECS: DG7000000
EINECS Number: 200-025-9
Molecular Formula: $C_7H_4Cl_2O_2$
Structured MF: $Cl_2C_6H_3COOH$
Formula Weight: 191.01

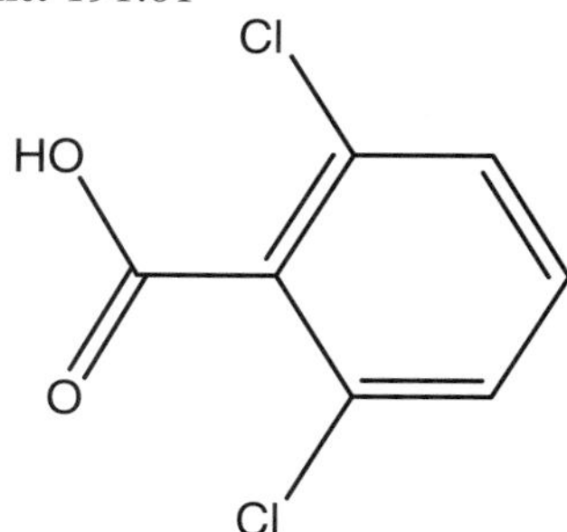

Chemical Structure

Synonyms: BENZOIC ACID,2,6-DICHLORO-
Description: white solid or needles

Physical Properties

Boiling Point: Sublimes
Freezing Point: 144 °C (291.2 °F)
Water Solubility: 1 to 10 mg/mL at 19 °C
Other Solubilities: 95% Ethanol: >=100 mg/ml at 19 °C; Acetone: 1-10 mg/ml at 19 °C; Alcohol: Soluble; Benzene: >10%; DMSO: >=100 mg/ml at 19 °C; Ether: >10%.
Flash Point: Not available; probably combustible

RTECS Toxicity Data

Acute Dermal: Mouse LD_{50} Route: Subcutaneous Dose: 1500 mg/kg; Toxic Effects: Behavioral - Altered sleep time (including change in righting reflex); Behavioral - Muscle weakness; Lungs, Thorax, or Respiration - Dyspnea.

Hazard Overviews

Health: Irritating to eyes/skin/respiratory tract. Other Acute Effects: may be harmful by inhalation, ingestion, or skin absorption.
Fire: Will burn. Hazards: emits toxic fumes. Extinguishing agents: water spray; carbon dioxide, dry chemical powder or appropriate foam. Precautions: combustible liquid.
Reactivity: Incompatible with: strong oxidizing agents. Hazardous decomposition products: toxic fumes of: carbon monoxide, carbon dioxide, hydrogen chloride gas.
Carcinogenicity: IARC - Not listed; NIOSH - Not listed; NTP - Not listed; ACGIH - Not listed; OSHA - Not listed; EPA - Not listed; MAK - Not listed

Primary Target Organs:

Eyes

Skin

Respiratory System

Environmental

Regulations

RCRA 40CFR: Not listed
CERCLA: 40CFR 302.4: Not listed
SARA 40CFR 372.65: Not listed
SARA EHS 40CFR 355: Not listed
TSCA: Not listed

DIC2900 CAS #: 51-44-5

3,4-DICHLOROBENZOIC ACID

RTECS: DG7175000
EINECS Number: 200-099-2
Molecular Formula: $C_7H_4Cl_2O_2$
Formula Weight: 191.01

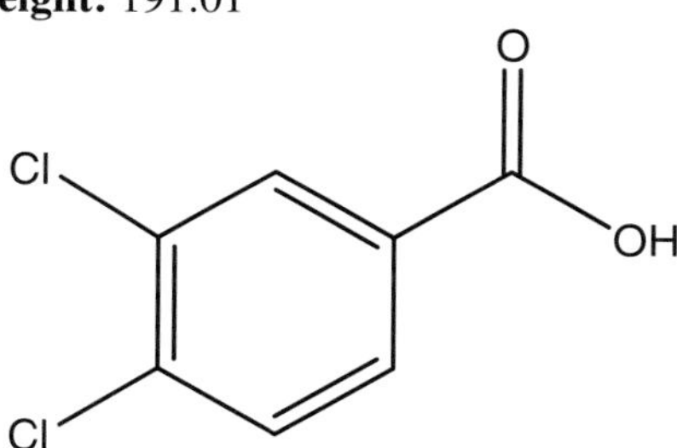

Chemical Structure

Synonyms: SYNSTIGMINE; SYNTOSTIGMIN; SYNTOSTIGMIN (INJECTION); VAGOSTIGMIN

Physical Properties

Freezing Point: 209 °C (408.2 °F)
Water Solubility: Soluble
Other Solubilities: Ethanol: Very Soluble; Ether: Soluble; DMSO: Slightly soluble

RTECS Toxicity Data

Acute Dermal: Mouse LD_{50} Route: Subcutaneous Dose: 400 mg/kg; Toxic Effects: Behavioral - Altered sleep time (including change in righting reflex); Behavioral - Muscle weakness; Lungs, Thorax, or Respiration - Dyspnea.

Hazard Overviews

Health: Irritating to eyes/skin/respiratory tract. Other Acute Effects: may be harmful by inhalation, ingestion, or skin absorption.
Fire: Hazards: emits toxic fumes. Extinguishing agents: water spray; carbon dioxide, dry chemical powder or appropriate foam. Precautions: combustible liquid.

Reactivity: Incompatible with: strong oxidizing agents. Hazardous decomposition products: toxic fumes of: carbon monoxide, carbon dioxide, hydrogen chloride gas.
Carcinogenicity: IARC - Not listed; NIOSH - Not listed; NTP - Not listed; ACGIH - Not listed; OSHA - Not listed; EPA - Not listed; MAK - Not listed
Primary Target Organs:

Eyes

Skin

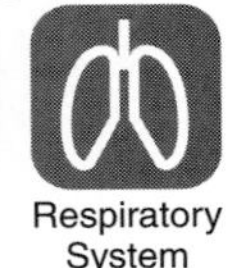
Respiratory System

Environmental

Regulations

RCRA 40CFR: Not listed
CERCLA: 40CFR 302.4: Not listed
SARA 40CFR 372.65: Not listed
SARA EHS 40CFR 355: Not listed
TSCA: Listed

DIC2950 CAS #: 328-84-7

3,4-DICHLOROBENZOTRIFLUORIDE

EINECS Number: 206-337-1
Molecular Formula: $C_7H_3Cl_2F_3$
Formula Weight: 214.99

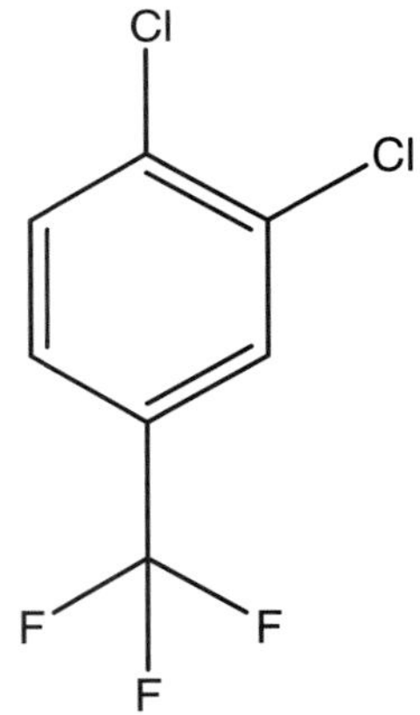

Chemical Structure

Synonyms: BENZENE,1,2-DICHLORO-4-(TRIFLUOROMETHYL)-; 3,4-DICHLOROPHENYLTRIFLUOROMETHANE; 1,2-DICHLORO-4-(TRIFLUOROMETHYL)BENZENE; 3,4-DICHLORO-ALPHA,ALPHA,ALPHA-TRIFLUOROTOLUENE; TOLUENE,3,4-DICHLORO-ALPHA,ALPHA,ALPHA-TRIFLUORO-
Description: colorless, clear liquid; aromatic odor
Use: intermediate for insecticides and herbicides

Physical Properties

Boiling Point: 173 °C (343 °F) to 174 °C (345 °F)
Freezing Point: -13 °C (8.6 °F)to -12 °C (10.4 °F)
Specific Gravity: 1.478
Water Solubility: Negligible
Other Solubilities: Soluble in Petroleum Ether and acetonitrile.
Refraction Index: sodium D line 1.475 at 20 °C
Flash Point: 65 °C

Hazard Overviews

Health: Irritating to eyes/skin/respiratory tract. Other Acute Effects: may be harmful by inhalation, ingestion, or skin absorption.
Fire: Combustible. Hazards: emits toxic fumes. Extinguishing agents: water spray; carbon dioxide, dry chemical powder or appropriate foam. Precautions: combustible liquid.
Reactivity: Incompatible with: strong oxidizing agents. Hazardous decomposition products: toxic fumes of: carbon monoxide, carbon dioxide, hydrogen chloride gas, hydrogen fluoride.
Carcinogenicity: IARC - Not listed; NIOSH - Not listed; NTP - Not listed; ACGIH - Class A4, Not classifiable as a human carcinogen; OSHA - Not listed; EPA - Not listed; MAK - Not listed
Primary Target Organs:

Eyes

Skin

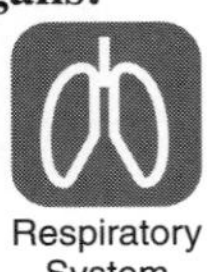
Respiratory System

Exposure Limits
OSHA PEL: TWA: 2.5 mg/m^3; as F.
ACGIH TLV: TWA: 2.5 mg/m^3; as F.
NIOSH REL: TWA: 2.5 mg/m^3; as F.
DFG MAK: TWA: 2.5 mg/m^3; as F.

Environmental

Ecotoxicity: Biouptake by Oligochaete worms Tubifex tubifex and Limnodrilus hoffmeisteri: worm/sediment accumulation factor after 79 days at 8 °C and 0.18 mg/kg sediment: 0.7; t1/2 in worm: <5 days

Environmental Physical Data

Octanol/Water Partition Coefficient: log K_{ow} = estimated at 4.7

Regulations

RCRA 40CFR: Not listed
CERCLA: 40CFR 302.4: Not listed
SARA 40CFR 372.65: Not listed
SARA EHS 40CFR 355: Not listed
TSCA: Listed

DIC3000 CAS #: 2905-60-4

2,3-DICHLOROBENZOYL CHLORIDE

EINECS Number: 220-811-5
Molecular Formula: $C_7H_3Cl_3O$
Formula Weight: 209.46

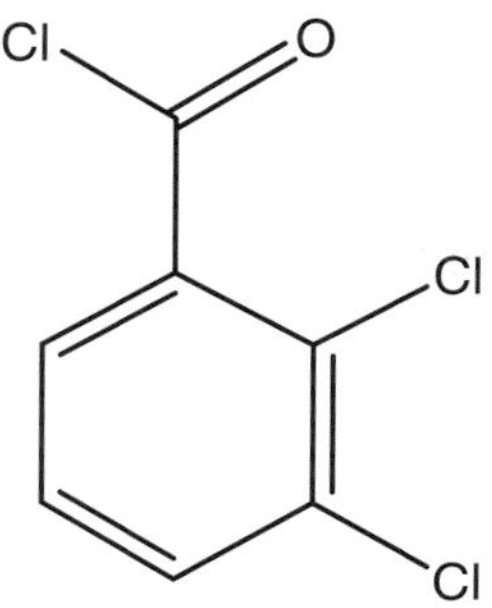

Chemical Structure

Synonyms: BENZOYL CHLORIDE,2,3-DICHLORO-

Hazard Overviews

Carcinogenicity: IARC - Not listed; NIOSH - Not listed; NTP - Not listed; ACGIH - Not listed; OSHA - Not listed; EPA - Not listed; MAK - Not listed

Environmental

Regulations

RCRA 40CFR: Not listed
CERCLA: 40CFR 302.4: Not listed
SARA 40CFR 372.65: Not listed
SARA EHS 40CFR 355: Not listed
TSCA: Listed

DIC3050 CAS #: 2905-61-5

2,5-DICHLOROBENZOYL CHLORIDE

EINECS Number: 220-812-0
Molecular Formula: $C_7H_3Cl_3O$
Formula Weight: 209.46
Synonyms: BENZOYL CHLORIDE,2,5-DICHLORO-

Hazard Overviews

Carcinogenicity: IARC - Not listed; NIOSH - Not listed; NTP - Not listed; ACGIH - Not listed; OSHA - Not listed; EPA - Not listed; MAK - Not listed

Environmental

Regulations

RCRA 40CFR: Not listed
CERCLA: 40CFR 302.4: Not listed
SARA 40CFR 372.65: Not listed
SARA EHS 40CFR 355: Not listed
TSCA: Listed

DIC3200 CAS #: 38721-71-0

DICHLOROBENZYL CHLORIDE

EINECS Number: 254-102-7
Molecular Formula: $C_7H_5Cl_3$
Formula Weight: 195.48
Synonyms: BENZENE,DICHLORO(CHLOROMETHYL)-; TOLUENE,ALPHA,AR,AR-TRICHLORO-
Description: colorless liquid; penetrating odor
Use: intermediate for organic chemicals, pharmaceuticals, & dyes; insecticide; intermediate in the manufacture of benzyl ammonium quarternary algicides used in swimming pool formulations

Physical Properties

Boiling Point: 245 °C (473 °F) to 252 °C (486 °F)
Specific Gravity: 1.415 to 1.42 at 25 °C/15 °C
Water Solubility: Insoluble in Water
Other Solubilities: Soluble in Alcohol, Ether, Acetone

Hazard Overviews

Carcinogenicity: IARC - Not listed; NIOSH - Not listed; NTP - Not listed; ACGIH - Not listed; OSHA - Not listed; EPA - Not listed; MAK - Not listed

Environmental

Environmental Physical Data

Octanol/Water Partition Coefficient: log K_{ow} = estimated at 3.9
BCF: not possibe

Regulations

RCRA 40CFR: Not listed
CERCLA: 40CFR 302.4: Not listed
SARA 40CFR 372.65: Not listed
SARA EHS 40CFR 355: Not listed
TSCA: Listed

DIC3250 CAS #: 75-27-4

DICHLOROBROMOMETHANE

RTECS: PA5310000
EINECS Number: 200-856-7
Molecular Formula: $CHBrCl_2$
Structured MF: $BrCHCl_2$
Formula Weight: 163.83

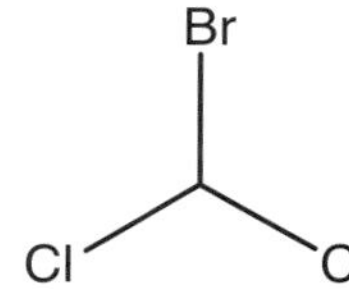

Chemical Structure

Synonyms: BDCM; BROMODICHLOROMETHANE; DICHLOROMETHYLBROMIDE; DICHLOROMONOBROMOMETHANE; METHANE,BROMODICHLORO-; MONOBROMODICHLOROMETHANE
Use: as a fire extinguisher fluid ingredient, a solvent for fats, waxes and resins, a synthesis intermediate, a reagent, a heavy

liquid for mineral and salt separations and in organic synthesis

Physical Properties

Boiling Point: 90.1 °C (194 °F) at 760 mm Hg
Freezing Point: -57.1 °C (-70.78 °F)
Specific Gravity: 1.98 at 20 °C/4 °C
Saturated Vapor Density: 1.567050817 kg/m^3
Density: 1.948 g/cu cm at 21 °C
Vapor Pressure: 50 torr at 20 °C
Water Solubility: 5 to 10 mg/mL at 19 °C
Other Solubilities: 95% Ethanol: >=100 mg/ml at 19 °C; Acetone: >=100 mg/ml at 19 °C; Benzene: Soluble; Chloroform: Soluble; Corn oil: miscible up to 1.50 g/mL; DMSO: >=100 mg/ml at 19 °C; Ether: Soluble; Lipids: Extremely Soluble; Organic solvents: miscible.
Odor Threshold: 1680 mg/m^3
Refraction Index: 1.4964 at 20 °C/D
Ionization Potential (eV): 10.6
Flash Point: Nonflammable

RTECS Toxicity Data

Acute Oral: Rat LD_{50} Dose: 430 mg/kg; Toxic Effects: Behavioral - Somnolence (general depressed activity); Behavioral - Tremor; Liver - Other changes. Mouse LD_{50} Dose: 450 mg/kg; Toxic Effects: Brain and coverings - Changes in circulation; Liver - Fatty Liver degeneration; Blood - Hemorrhage.
Chronic (Multiple Dose) Oral: Rat Dose: 750 mg/kg/5D-I; Toxic Effects: Liver - Changes in liver weight; Kidney, Ureter, and Bladder - Changes in kidney weight; Biochemical - Hepatic microsomal mixed oxidase(dealkylation, hyroxylation,etc). Rat Dose: 5670 mg/kg/4W-C; Toxic Effects: Liver - Changes in liver weight; Blood - Changes in serum composition; Nutritional and gross metabolic - Weight loss or decreased weight gain. Rat Dose: 19500 mg/kg/13W-I; Toxic Effects: DEATH. Rat Dose: 9828 mg/kg/9W-I; Toxic Effects: Blood - Changes in erythrocite (RBC) cell count; Nutritional and gross metabolic - Weight loss or decreased weight gain; Biochemical - Transaminases.
Mutagenic: Human Sister Chromatid Exchenge; Cell Type: lymphocyte; Dose: 400 umol/L. Rat Cytogenetic Analysis; Route: Intraperitoneal; Dose: 164 mg/kg.
Tumorigenic: Rat Route: Oral; Dose: 25500 mg/kg/2Y-C; Toxic Effects: Tumorigenic - Carcinogenic by RTECS criteria; Gastrointestinal - Tumors; Kidney, Ureter, and Bladder - Kidney tumors. Rat Route: Oral; Dose: 51 gm/kg/2Y-C; Toxic Effects: Tumorigenic - Carcinogenic by RTECS criteria; Gastrointestinal - Tumors; Kidney, Ureter, and Bladder - Kidney tumors.

Hazard Overviews

Reactivity: Stable. Hazardous polymerization cannot occur. Avoid: exposure to elevated temperatures. Hazardous decomposition products: chlorine; bromine; phosgene; carbon oxide(s) gases.
Carcinogenicity: IARC - Group 2B, Possibly carcinogenic to humans; NIOSH - Not listed; NTP - Listed; ACGIH - Not listed; OSHA - Not listed; EPA - Class B2, Probable human carcinogen based on animal studies; MAK - Not listed

Environmental

Environmental Fate: If released to surface water, volatilization will be the dominant environmental fate process. The volatilization half-life from rivers and streams has been estimated to range from 33 min to 12 days with a typical half-life being 35 hours. In aquatic regions where volatilization is not viable, anaerobic biodegradation may be the major removal process. Aquatic hydrolysis, oxidation, direct photolysis, adsorption, and bioconcentration are not environmentally important. If released to soil, volatilization is again likely to be the dominant removal process where exposure to air is possible. It is moderately to highly mobile in soil and can therefore leach into groundwaters. If released to air, the only identifiable transformation process in the troposphere is reaction with hydroxyl radicals. The loss through this transformation has an estimated half-life of 6.65 months. This relatively persistent half-life indicates that long-range global transport is possible.
Cleanup/Disposal: Guide No. 171: Do not touch or walk through spilled material. Stop leak if you can do it without risk. Prevent dust cloud. Avoid inhalation of asbestos dust. Small Dry Spills: With clean shovel place material into clean, dry container and cover loosely; move containers from spill area. Small Spills: Take up with sand or other noncombustible absorbent material and place into containers for later disposal. Large Spills: Dike far ahead of liquid spill for later disposal. Cover powder spill with plastic sheet or tarp to minimize spreading. Prevent entry into waterways, sewers, basements or confined areas.

Environmental Physical Data

Henry's Law Constant: 1.6 x10^{-3}
Octanol/Water Partition Coefficient: log K_{ow} = 2.10
Sorption Partition Coefficient: K_{oc} = estimated at 53 to 251
BCF: estimated at 0.72 to 1.37

Regulations

RCRA 40CFR: Not listed
CERCLA: 40CFR 302.4: Listed per CWA Section 307(a) RQ: 5000 lb (2268 kg)
SARA 40CFR 372.65: Listed
SARA EHS 40CFR 355: Not listed
TSCA: Listed

Analytical Methods

Air: EPA 0031, OA-002-1, VA-001-1, VA-006-1, VA-008-1, VG-007-1, VG-011-1
Soil: CLP LC_VOA, MC_VOA, OHC; EPA 7, 1624, VG-008-1, VS-001-1, VS-002-1; SW846 5021, 5032, 5041, 5041A, 8010B, 8021A, 8240B, 8260A, 8260B
Water / Groundwater: EPA 601, 624, 624-S, VW-001-1, VW-002-1, VW-003-1, VW-014-1; APHA 6040-B, 6210-B, 6210-D, 6230-B, 6230-D; ASTM D3973; USGS O3115
Drinking Water: EPA 502.1, 502.2, 524.1, 524.2, 551, PART_1, PART_2, PART_3; APHA 6210-C, 6230-C, 6232-B
Food: EPA 5

Plasma: EPA 29

DIC3300 CAS #: 616-21-7

1,2-DICHLOROBUTANE

EINECS Number: 210-469-5
Molecular Formula: $C_4H_8Cl_2$
Formula Weight: 127.03

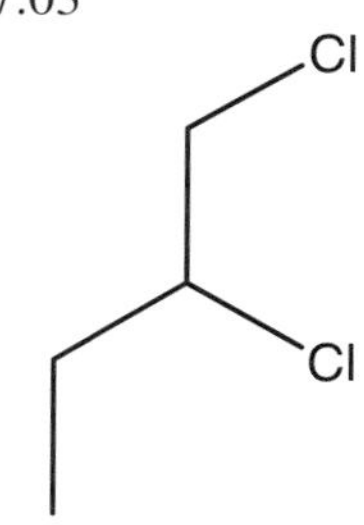

Chemical Structure

Synonyms: BUTANE,1,2-DICHLORO-

Physical Properties

Boiling Point: 124 °C (255 °F) at 760 mm Hg
Specific Gravity: 1.1116 at 25 °C/4 °C
Vapor Density: 4.38 Air=1
Saturated Vapor Density: 1.205337387 kg/m^3
Vapor Pressure: 1 mm Hg at -23.6 °C
Water Solubility: Insoluble in Water
Other Solubilities: Soluble in Ether, Chloroform
Refraction Index: 1.4450 at 20 °C/D
Autoignition Temperature: 275 °C

Hazard Overviews

Health: Irritating to eyes/skin/respiratory tract. Other Acute Effects: may be harmful by inhalation, ingestion, or skin absorption. Chronic Effects: prolonged contact can cause nausea, headache, vomiting; causes dermatitis; target organs: liver, kidneys, central nervous system.
Fire: Will burn. Hazards: vapor may travel considerable distance to source of ignition and flash back; emits toxic fumes. Extinguishing agents: carbon dioxide, dry chemical powder or appropriate foam. Precautions: combustible liquid.
Reactivity: Incompatible with: oxidizing agents, acids, bases, aluminum and its alloys. Hazardous decomposition products: toxic fumes of: hydrogen chloride gas, carbon monoxide, carbon dioxide, phosgene gas.
Carcinogenicity: IARC - Not listed; NIOSH - Not listed; NTP - Not listed; ACGIH - Not listed; OSHA - Not listed; EPA - Not listed; MAK - Not listed

Primary Target Organs:

Eyes

Skin

Respiratory System

Nervous System

Liver

Kidneys

Environmental

Regulations

RCRA 40CFR: Not listed
CERCLA: 40CFR 302.4: Not listed
SARA 40CFR 372.65: Not listed
SARA EHS 40CFR 355: Not listed
TSCA: Listed

DIC3350 CAS #: 926-57-8

1,3-DICHLORO-2-BUTENE

RTECS: EM4760000
DOT: NA2924
EINECS Number: 213-138-3
Molecular Formula: $C_4H_6Cl_2$
Formula Weight: 125.00

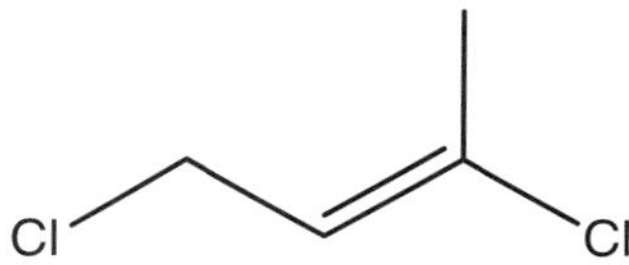

Chemical Structure

Synonyms: 2-BUTENE,1,3-DICHLORO-; 1,3-DICHLOROBUTYLENE
Description: clear to straw-colored liquid
Use: in the pesticide, ddb; intermediate in production of 19-nortestosterone steroids for use as synthetic sex hormones; used as an intermediate in the synthesis of 2,3-dichloro-1,3-butadiene; can be used to produce chloroprene through thermal or catalytic dehalogenation

Physical Properties

Boiling Point: 131 °C (268 °F)
Specific Gravity: 1.161
Vapor Density: 4.31 Air=1
Water Solubility: Insoluble in Water
Other Solubilities: Soluble in Acetone, Benzene, Ether and Ethanol.
Refraction Index: 1.4692 at 20 °C/D
Flash Point: 26.667 °C Closed Cup

RTECS Toxicity Data

Acute Inhalation: Rat LC_{50} Dose: 3930 mg/m^3. Mouse LC_{50} Dose: 4400 mg/m^3.

Hazard Overviews

Flammable

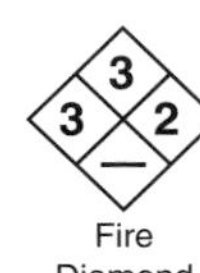

Fire Diamond

Health: Irritating to eyes/skin/respiratory tract. Toxic. Other Acute Effects: harmful if swallowed, inhaled, or absorbed through skin; lachrymator; symptoms of exposure may include burning sensation; coughing; wheezing; laryngitis; shortness of breath; headache; nausea; vomiting.

Fire: Flammable. Hazards: vapor may travel considerable distance to source of ignition and flash back; container explosion may occur; forms explosive mixtures in air; emits toxic fumes. Extinguishing agents: carbon dioxide, dry chemical powder or appropriate foam; water may be effective for cooling, but may not effect extinguishment. Precautions: combustible liquid.

Carcinogenicity: IARC - Not listed; NIOSH - Not listed; NTP - Not listed; ACGIH - Not listed; OSHA - Not listed; EPA - Not listed; MAK - Not listed

Primary Target Organs:

Eyes

Skin

Respiratory System

Environmental

Environmental Fate: If released to the atmosphere, it will degrade in the vapor phase by reaction with photochemically produced hydroxyl radicals (estimated half-life of about 28 hr). If released to moist soil or water, degradation may occur through hydrolysis; similar dichlorobutenes (cis and trans-1,4-dichloro-2-butene) have experimentally-determined hydrolysis half-lives of 3.2 days at 25 °C. An estimated K_{oc} value of 125 suggests that it will have high mobility in soil. Evaporation from dry surfaces will occur.

Cleanup/Disposal: Guide No. 132: Fully encapsulating, vapor protective clothing should be worn for spills and leaks with no fire. Eliminate all ignition sources (no smoking, flares, sparks or flames in immediate area). All equipment used when handling the product must be grounded. Do not touch or walk through spilled material. Stop leak if you can do it without risk. Prevent entry into waterways, sewers, basements or confined areas. A vapor suppressing foam may be used to reduce vapors. Absorb with earth, sand or other non-combustible material and transfer to containers (except for Hydrazine). Use clean non-sparking tools to collect absorbed material. Large Spills: Dike far ahead of liquid spill for later disposal. Water spray may reduce vapor; but may not prevent ignition in closed spaces.

Environmental Physical Data

Henry's Law Constant: estimated at 0.000664
Octanol/Water Partition Coefficient: log K_{ow} = 2.84
Sorption Partition Coefficient: K_{oc} = 125
BCF: estimated at 85

Regulations

RCRA 40CFR: Not listed
CERCLA: 40CFR 302.4: Not listed
SARA 40CFR 372.65: Not listed
SARA EHS 40CFR 355: Not listed
TSCA: Listed

DIC3400 **CAS #: 764-41-0**

1,4-DICHLORO-2-BUTENE

RTECS: EM4900000
DOT: NA2924
EINECS Number: 212-121-8
Molecular Formula: $C_4H_6Cl_2$
Structured MF: $ClCH_2CH=CHCH_2Cl$
Formula Weight: 125.00

Chemical Structure

Synonyms: 2-BUTENE,1,4-DICHLORO-; 2-BUTYLENE DICHLORIDE; 1,4-DCB; 1,4-DICHLOROBUTENE; 1,4-DICHLOROBUTENE-2; 1,4-DICHLORO-2-BUTYLENE

Description: colorless liquid; sweet, pungent odor

Use: for resolving oil-in-water emulsions; intermed for 3,4-dichloro-1-butene (chloroprene intermed), adiponitrile (hexamethylenediamine intermed), 1,4-butanediol (non-us), & tetrahydrofuran; from 1951 to 1983 dupont operated butadiene chlorination process in which 1,4-dichloro-2-butene was converted to 3-hexenedinitrile & then hydrogenated to adiponitrile.

Physical Properties

Boiling Point: 156 °C (313 °F)
Freezing Point: -48 °C (-54.4 °F)
Specific Gravity: 1.1858
Vapor Density: 4 Air=1
Saturated Vapor Density: 85% Technical 1.40907441 kg/m^3
Vapor Pressure: 85% Technical 40 mm Hg at 74-76 °F
Water Solubility: Immiscible with Water
Other Solubilities: Soluble in Ether.
Surface Tension: Estimated at 30 dynes/cm at 20 °C
Refraction Index: 1.4863 at 25 °C
Flash Point: 26.667 °C
LEL: 1.5% v/v
UEL: 4.0% v/v

RTECS Toxicity Data

Acute Oral: Rat LD_{50} Dose: 89 mg/kg. Mouse LD_{50} Dose: 190 mg/kg; Toxic Effects: Behavioral - Muscle weakness; Behavioral - Ataxia; Lungs, Thorax, or Respiration - Respiratory depression.

Acute Inhalation: Mouse LC_{50} Dose: 920 mg/m^3; Toxic Effects: Behavioral - Muscle weakness; Behavioral - Ataxia;

Lungs, Thorax, or Respiration - Respiratory depression. Rat LC_{Lo} Dose: 62 ppm/4hr.

Acute Dermal: Rabbit LD_{50} Route: Skin; Dose: 620 uL/kg.

Chronic (Multiple Dose) Oral: Rat Dose: 18200 ug/kg/26W-I; Toxic Effects: Liver - Liver function tests impaired; Biochemical - Transaminases.

Chronic (Multiple Dose) Inhalation: Rat Dose: 10900 ppb/6H/2W-I; Toxic Effects: Lungs, Thorax, or Respiration - Structural or functional change in trachea or bronchi; Blood - Changes in spleen; Nutritional and gross metabolic - Weight loss or decreased weight gain. Rat Dose: 8700 ug/m^3/17W-I; Toxic Effects: Brain and coverings - Recordings from specific areas of CNS; Liver - Other changes.

Irritation Eye: Rabbit Standard Draize Test Dose: 20 mg open; Reaction: severe. Rabbit Standard Draize Test Dose: 20 mg/24H; Reaction: moderate.

Irritation Skin: Rabbit Standard Draize Test Dose: 20 mg/24H; Reaction: moderate. Rabbit Open Draize Test Dose: 10 mg/24H open; Reaction: severe.

Reproductive/Teratogenic: Rat Route: Oral; Dose: 750 ug/kg; Duration: male 75D prior to mating; Paternal Effects - Spermatogenesis. Rat Route: Inhalation; Dose: 5 ppm/6H; Duration: female 6-15D of pregnancy; Specific Developmental Abnormalities - Musculoskeletal system.

Mutagenic: Rat Cytogenetic Analysis; Route: Inhalation; Dose: 1700 ug/m^3/30D-I. Insects - D Melanogaster Sex Chromosome Loss; Route: Oral; Dose: 2 mmol/L/3D-I.

Tumorigenic: Rat Route: Inhalation; Dose: 1 ppm/6H/82W-I; Toxic Effects: Tumorigenic - Carcinogenic by RTECS criteria; Sense organs and special senses - Tumors. Rat Route: Inhalation; Dose: 100 ppb/6H/82W-I; Toxic Effects: Tumorigenic - Neoplastic by RTECS criteria; Sense organs and special senses - Tumors.

Hazard Overviews

Corrosive

Flammable

Fire Diamond

Health: Corrosive to eyes/skin/respiratory tract. Also Causes: difficulty breathing, possible delayed pulmonary edema, blistering, GI irritation. Chronic Effects: possible nasal tumors.

Fire: Flammable. Can form explosive mixtures in air. For small fires use dry chemical, carbon dioxide, or regular foam. For large fires use fog or regular foam. Water spray may be used, but will form corrosive hydrogen chloride gas. Container may explode in heat of fire.

Reactivity: Stable. Hazardous polymerization cannot occur. Avoid: heat; ignition sources; water; metals. Incompatible with: metal when wet. Hazardous decomposition products: carbon oxide(s); chlorine; hydrogen chloride; phosgene gas.

Carcinogenicity: IARC - Group 3, Not classifiable as to carcinogenicity to humans; NIOSH - Not listed; NTP - Not listed; ACGIH - Class A2, Suspected human carcinogen; OSHA - Not listed; EPA - Not listed; MAK - Class A2, Unmistakably carcinogenic in animal experimentation only

Primary Target Organs:

Eyes

Skin

Respiratory System

Exposure Limits

ACGIH TLV: TWA: 0.005 ppm; 0.025 mg/m^3.

Respirator Recommendation

Exposure Range: >0.005 to 0.25 ppm Supplied Air, Constant Flow/Pressure Demand, Half Mask

Exposure Range: >0.25 to 5 ppm Supplied Air, Constant Flow/Pressure Demand, Full Face

Exposure Range: >5 to unlimited ppm Self-contained Breathing Apparatus, Pressure Demand, Full Face

Note: odor threshold unknown

Environmental

Ecotoxicity: LD_{50} Poecilia reticulata (Guppy) << 40 mg/l/7 day

Environmental Fate: If released to the atmosphere, it will degrade readily in the vapor phase by reaction with photochemically produced hydroxyl radicals (estimated half-life of about 10 hr). If released to moist soil or water, degradation can occur through hydrolysis; It has an experimentally determined hydrolysis half-life of 3.2 days at 25 °C. The results of one soil biological screening study indicate that abiotic degradation is more important than microbial degradation. Soil persistence half-lives of 1.8 to 2.5 days have been observed. A measured K_{oc} value of 215 suggests that it will have moderate mobility in soil and could leach. Evaporation from dry surfaces will occur.

Cleanup/Disposal: Guide No. 132: Fully encapsulating, vapor protective clothing should be worn for spills and leaks with no fire. Eliminate all ignition sources (no smoking, flares, sparks or flames in immediate area). All equipment used when handling the product must be grounded. Do not touch or walk through spilled material. Stop leak if you can do it without risk. Prevent entry into waterways, sewers, basements or confined areas. A vapor suppressing foam may be used to reduce vapors. Absorb with earth, sand or other non-combustible material and transfer to containers (except for Hydrazine). Use clean non-sparking tools to collect absorbed material. Large Spills: Dike far ahead of liquid spill for later disposal. Water spray may reduce vapor; but may not prevent ignition in closed spaces.

Environmental Physical Data

Henry's Law Constant: estimated at 0.000664

Sorption Partition Coefficient: K_{oc} = 215

BCF: estimated at 14

Regulations

RCRA 40CFR: Listed Hazardous Waste No. U074 Toxic Waste Ignitable Waste

CERCLA: 40CFR 302.4: Listed per RCRA Section 3001 RQ: 1 lb (0.454 kg)

SARA 40CFR 372.65: Listed

SARA EHS 40CFR 355: Not listed

TSCA: Listed

Analytical Methods

Soil: SW846 8010B, 8240B

DIC3450 CAS #: 1476-11-5

1,4-DICHLORO-CIS-2-BUTENE

DOT: NA2924
EINECS Number: 216-021-5
Molecular Formula: $C_4H_6Cl_2$
Formula Weight: 125.00

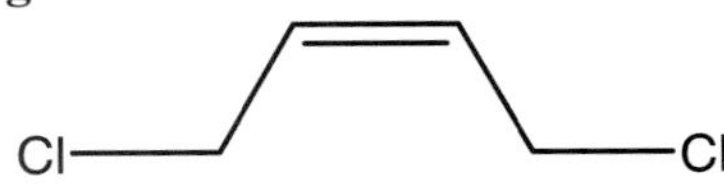

Chemical Structure

Synonyms: 2-BUTENE,1,4-DICHLORO-,(Z)-; 2-BUTENE,1,4-DICHLORO-,CIS-; CIS-1,4-DICHLORO-2-BUTENE
Description: colorless liquid
Use: as nematocides and as intermediates in the manufacture of pesticides dichloroprenes; intermediate in the production of chloropreneis; production of adiponitrile, butane-1,4-diol, and tetrahydrofuran

Physical Properties

Boiling Point: 152.5 °C (307 °F)
Freezing Point: -48 °C (-54.4 °F)
Specific Gravity: 1.188 at 25 °C/4 °C
Saturated Vapor Density: 1.221377858 kg/m^3
Vapor Pressure: 4.09 mm Hg at 25 °C
Water Solubility: Insoluble in Water
Other Solubilities: Soluble in Chloroform, organic solvents.
Surface Tension: 0.045796 N/m at melting point
Refraction Index: 1.4887 at 25 °C/D
Critical Temperature: 367 °C
Critical Pressure: 3.78 x10^6 Pa
Flash Point: Flammable
Autoignition Temperature: 55 °C
LEL: 1.5% v/v
UEL: 4% v/v

Hazard Overviews

Poison

Corrosive

Flammable

Health: Corrosive to eyes/skin/respiratory tract. Poison. Other Acute Effects: may be fatal if inhaled; harmful if swallowed or absorbed through skin; symptoms of exposure may include burning sensation; coughing; wheezing; laryngitis; shortness of breath; headache; nausea; vomiting; inhalation may result in spasm, inflammation and edema of the larynx and bronchi, chemical pneumonitis and pulmonary edema.

Fire: Flammable. Hazards: emits toxic fumes. Extinguishing agents: water spray; carbon dioxide, dry chemical powder or appropriate foam. Precautions: combustible liquid.
Reactivity: Incompatible with: strong oxidizing agents, strong bases, may decompose on exposure to light, may decompose on exposure to moist air or water. Hazardous decomposition products: toxic fumes of: carbon monoxide, carbon dioxide, hydrogen chloride gas.
Carcinogenicity: IARC - Not listed; NIOSH - Not listed; NTP - Not listed; ACGIH - Not listed; OSHA - Not listed; EPA - Not listed; MAK - Not listed
Primary Target Organs:

Eyes

Skin

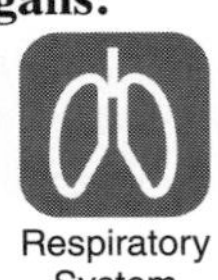
Respiratory System

Environmental

Ecotoxicity: Fishes: guppy (Poecilia reticulata): 7d LC_{50} <<40 mg/l
Environmental Fate: If released to the atmosphere, it will degrade readily in the vapor phase by reaction with photochemically produced hydroxyl radicals (estimated half-life of about 12 hr). If released to moist soil or water, degradation can occur through hydrolysis; it has an experimentally determined hydrolysis half- life of 3.2 days at 25 °C. The results of one soil biological screening study indicate that abiotic degradation is more important than microbial degradation. Soil persistence half-lives of 1.8 to 2.5 days have been observed. A measured K_{oc} value of 215 suggests that it will have moderate mobility in soil and could leach. Evaporation from dry surfaces will occur.
Cleanup/Disposal: Guide No. 132: Fully encapsulating, vapor protective clothing should be worn for spills and leaks with no fire. Eliminate all ignition sources (no smoking, flares, sparks or flames in immediate area). All equipment used when handling the product must be grounded. Do not touch or walk through spilled material. Stop leak if you can do it without risk. Prevent entry into waterways, sewers, basements or confined areas. A vapor suppressing foam may be used to reduce vapors. Absorb with earth, sand or other non-combustible material and transfer to containers (except for Hydrazine). Use clean non-sparking tools to collect absorbed material. Large Spills: Dike far ahead of liquid spill for later disposal. Water spray may reduce vapor; but may not prevent ignition in closed spaces.

Environmental Physical Data

Henry's Law Constant: estimated at 0.00116
Octanol/Water Partition Coefficient: log K_{ow} = calculated at 2.5
Sorption Partition Coefficient: K_{oc} = 215
BCF: estimated at 17

Regulations

RCRA 40CFR: Not listed
CERCLA: 40CFR 302.4: Not listed
SARA 40CFR 372.65: Not listed
SARA EHS 40CFR 355: Not listed

TSCA: Listed

Analytical Methods

Soil: SW846 8260A, 8260B
Indoor / Expired Air: EPA IP-1B
Plasma: EPA 29

DIC3500 CAS #: 110-57-6

1,4-DICHLORO-TRANS-2-BUTENE

RTECS: EM4903000
DOT: NA2924
EINECS Number: 203-779-7
Molecular Formula: $C_4H_6Cl_2$
Formula Weight: 125.00

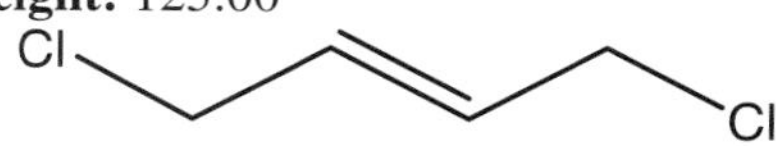

Chemical Structure

Synonyms: 2-BUTENE,1,4-DICHLORO-,(E)-; 2-BUTENE,1,4-DICHLORO-,TRANS-; 2-BUTYLENE DICHLORIDE; 1,4-DICHLOROBUTENE-2 (TRANS); TRANS-1,4-DICHLORO-2-BUTENE
Description: colorless liquid; distinct odor
Use: chemical intermediate for hexamethylenediamine & chloroprene; nematocides and as intermediates in the manufacture of pesticide; intermediate in the production of chloropreneis; production of adiponitrile, butane-1,4-diol, and tetrahydrofuran

Physical Properties

Boiling Point: 155.4 °C (312 °F)
Freezing Point: 2 °C (35.6 °F)
Specific Gravity: 1.183 at 25 °C/4 °C
Vapor Density: 4 Air=1
Water Solubility: Insoluble in Water
Other Solubilities: Soluble in Chloroform, organic solvents.
Surface Tension: 3.8163 x10^{-2} N/m
pH: Acid
Refraction Index: 1.4871 at 25 °C
Critical Temperature: 373 °C
Critical Pressure: 3.78 x10^6 Pa
Flash Point: 53 °C
LEL: 1.5% v/v
UEL: 4% v/v

RTECS Toxicity Data

Acute Inhalation: Rat LC_{50} Dose: 86 ppm/4hr; Toxic Effects: Sense organs and special senses - Lacrimation; Lungs, Thorax, or Respiration - Other changes; Gastrointestinal - Changes in structure or function of salivary glands.
Tumorigenic: Mouse Route: Subcutaneous; Dose: 150 mg/kg/77W-I; Toxic Effects: Tumorigenic - Neoplastic by RTECS criteria; Tumorigenic - Tumors at site of application. Mouse Route: Intraperitoneal; Dose: 150 mg/kg/77W-I; Toxic Effects: Tumorigenic - Equivocal tumorigenic agent by RTECS criteria; Tumorigenic - Tumors at site of application.

Hazard Overviews

Poison

Corrosive

Flammable

Health: Corrosive to eyes/skin/respiratory tract. Poison. Other Acute Effects: may be fatal if inhaled; harmful if swallowed or absorbed through skin; lachrymator; symptoms of exposure may include burning sensation; coughing; wheezing; laryngitis; shortness of breath; headache; nausea; vomiting; inhalation may result in spasm, inflammation and edema of the larynx and bronchi, chemical pneumonitis and pulmonary edema. Chronic Effects: Carcinogen.
Fire: Flammable. Hazards: emits toxic fumes. Extinguishing agents: water spray; carbon dioxide, dry chemical powder or appropriate foam. Precautions: combustible liquid.
Reactivity: Incompatible with: strong oxidizing agents, strong bases, may decompose on exposure to light, may decompose on exposure to moist air or water. Hazardous decomposition products: toxic fumes of: carbon monoxide, carbon dioxide, hydrogen chloride gas.
Carcinogenicity: IARC - Group 3, Not classifiable as to carcinogenicity to humans; NIOSH - Not listed; NTP - Not listed; ACGIH - Not listed; OSHA - Not listed; EPA - Not listed; MAK - Not listed
Primary Target Organs:

Eyes

Skin

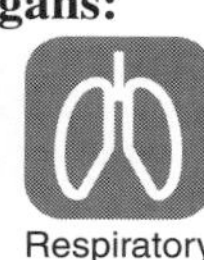
Respiratory System

Environmental

Ecotoxicity: LC_{50} Poecilia reticulata (guppy) << 40 mg/kg/7 days
Environmental Fate: If released to the atmosphere, it will degrade readily in the vapor phase by reaction with photochemically produced hydroxyl radicals (estimated half-life of about 10 hr). If released to moist soil or water, degradation can occur through hydrolysis; it has an experimentally determined hydrolysis half-life of 3.2 days at 25 °C. The results of one soil biological screening study indicate that abiotic degradation is more important than microbial degradation. Soil persistence half-lives of 1.8 to 2.5 days have been observed for the cis-isomer. K_{oc} values of 110-215 suggest that it will have moderate to high mobility in soil and could leach. Evaporation from dry surfaces will occur.
Cleanup/Disposal: Guide No. 132: Fully encapsulating, vapor protective clothing should be worn for spills and leaks with no fire. Eliminate all ignition sources (no smoking, flares, sparks or flames in immediate area). All equipment used when handling the product must be grounded. Do not touch or walk through spilled material. Stop leak if you can do it without risk. Prevent entry into waterways, sewers, basements or confined areas. A vapor suppressing foam may be used to reduce vapors. Absorb with earth, sand or other non-

combustible material and transfer to containers (except for Hydrazine). Use clean non-sparking tools to collect absorbed material. Large Spills: Dike far ahead of liquid spill for later disposal. Water spray may reduce vapor; but may not prevent ignition in closed spaces.

Environmental Physical Data

Henry's Law Constant: estimated at 0.000664
Sorption Partition Coefficient: K_{oc} = 215
BCF: estimated at 14

Regulations

RCRA 40CFR: Not listed
CERCLA: 40CFR 302.4: Not listed
SARA 40CFR 372.65: Listed TPQ: 500 lb
SARA EHS 40CFR 355: Listed TPQ: 500 lb
TSCA: Listed

Analytical Methods

Soil: SW846 3640A, 8260A, 8260B
Drinking Water: EPA 524.2
Plasma: EPA 29

DIC3550 CAS #: 760-23-6

3,4-DICHLORO-1-BUTENE

RTECS: EM4740000
DOT: NA2924
EINECS Number: 212-079-0
Molecular Formula: $C_4H_6Cl_2$
Formula Weight: 125.00

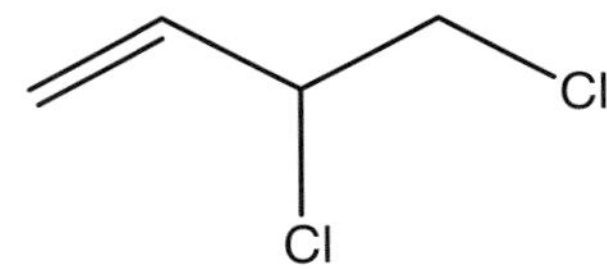

Chemical Structure

Synonyms: 1-BUTENE,3,4-DICHLORO-; 1,2-DICHLORO-3-BUTENE
Description: colorless liquid
Use: chem int for chloroprene; chemical intermediate in the production of 1,4-butanediol

Physical Properties

Boiling Point: 118.6 °C (245 °F)
Freezing Point: -61 °C (-77.8 °F)
Specific Gravity: 1.153 at 25 °C/4 °C
Saturated Vapor Density: 1.314206897 kg/m^3
Vapor Pressure: 21.85 mm Hg at 25 °C
Water Solubility: 420 mg/L of Water
Other Solubilities: Soluble in Ethanol, Ether and Benzene.
Surface Tension: 0.044183 N/m at -61 °C
Refraction Index: 1.4630 at 20 °C/D
Critical Temperature: 316 °C
Critical Pressure: 3.85 x10^6 Pa
Flash Point: 45 °C Closed Cup

RTECS Toxicity Data

Acute Oral: Mouse LD_{50} Dose: 724 mg/kg; Toxic Effects: Behavioral - Somnolence (general depressed activity); Behavioral - Excitment; Lungs, Thorax, or Respiration - Dyspnea.
Acute Inhalation: Rat LC_{50} Dose: 2100 ppm/4hr; Toxic Effects: Behavioral - Somnolence (general depressed activity); Behavioral - Ataxia; Lungs, Thorax, or Respiration - Dyspnea.
Reproductive/Teratogenic: Rat Route: Oral; Dose: 75 ug/kg; Duration: male 75D prior to mating; Paternal Effects - Spermatogenesis.
Mutagenic: Rat Cytogenetic Analysis; Route: Inhalation; Dose: 13700 ug/m^3/30D-I.

Hazard Overviews

Corrosive

Flammable

Fire Diamond

Health: Corrosive to eyes/skin/respiratory tract. Other Acute Effects: harmful if swallowed, inhaled, or absorbed through skin; lachrymator; inhalation may result in spasm, inflammation and edema of the larynx and bronchi, chemical pneumonitis and pulmonary edema; symptoms of exposure may include burning sensation; coughing; wheezing; laryngitis; shortness of breath; headache; nausea; vomiting.
Fire: Flammable. Hazards: vapor may travel considerable distance to source of ignition and flash back; container explosion may occur; forms explosive mixtures in air; emits toxic fumes. Extinguishing agents: carbon dioxide, dry chemical powder or appropriate foam; water may be effective for cooling, but may not effect extinguishment. Precautions: combustible liquid.
Reactivity: Incompatible with: strong oxidizing agents, strong bases. Hazardous decomposition products: toxic fumes of: carbon monoxide, carbon dioxide, hydrogen chloride gas.
Carcinogenicity: IARC - Not listed; NIOSH - Not listed; NTP - Not listed; ACGIH - Not listed; OSHA - Not listed; EPA - Not listed; MAK - Not listed
Primary Target Organs:

Eyes

Skin

Respiratory System

Environmental

Ecotoxicity: Bacteria: Pseudomonas fluorescens 24h EC_0 1,000 mg/l; Fishes: Pimephales promelas 24h LC_{50} 27 mg/l 96h LC_{50},F 7.2; 9.3 mg/l
Environmental Fate: If released to the atmosphere, it will degrade readily in the vapor phase by reaction with photochemically produced hydroxyl radicals (estimated half-life of about 14 hr) and by reaction with ozone (estimated half-life of about 23 hr). If released to moist soil or water, degradation may occur through hydrolysis; other

dichlorobutenes (cis and trans-1,4-dichloro-2-butene) have experimentally determined hydrolysis half-lives of 3.2 days at 25 °C. Estimated K_{oc} values of 130 and 160 suggest that it will have moderate to high mobility in soil and could leach. Evaporation from dry surfaces will occur.

Cleanup/Disposal: Guide No. 132: Fully encapsulating, vapor protective clothing should be worn for spills and leaks with no fire. Eliminate all ignition sources (no smoking, flares, sparks or flames in immediate area). All equipment used when handling the product must be grounded. Do not touch or walk through spilled material. Stop leak if you can do it without risk. Prevent entry into waterways, sewers, basements or confined areas. A vapor suppressing foam may be used to reduce vapors. Absorb with earth, sand or other non-combustible material and transfer to containers (except for Hydrazine). Use clean non-sparking tools to collect absorbed material. Large Spills: Dike far ahead of liquid spill for later disposal. Water spray may reduce vapor; but may not prevent ignition in closed spaces.

Environmental Physical Data

Henry's Law Constant: estimated at 0.00856
Octanol/Water Partition Coefficient: log K_{ow} = measured at 2.0
Sorption Partition Coefficient: K_{oc} = 160
BCF: carp 0.59 to 13.34

Regulations

RCRA 40CFR: Not listed
CERCLA: 40CFR 302.4: Not listed
SARA 40CFR 372.65: Not listed
SARA EHS 40CFR 355: Not listed
TSCA: Listed

Analytical Methods

Indoor / Expired Air: EPA IP-1B

DIC3600 CAS #: 27683-60-9

2,2-DICHLORO-1-(2-CHLOROPHENYL) ETHANOL

EINECS Number: 248-601-9
Molecular Formula: $C_8H_7Cl_3O$
Formula Weight: 225.47
Synonyms: BENZENEMETHANOL,2-CHLORO-ALPHA-(DICHLOROMETHYL)-; BENZYL ALCOHOL,O-CHLORO-ALPHA-(DICHLOROMETHYL)-; O-CHLORO-ALPHA-(DICHLOROMETHYL)BENZYL ALCOHOL

Hazard Overviews

Carcinogenicity: IARC - Not listed; NIOSH - Not listed; NTP - Not listed; ACGIH - Not listed; OSHA - Not listed; EPA - Not listed; MAK - Not listed

Environmental

Regulations

RCRA 40CFR: Not listed
CERCLA: 40CFR 302.4: Listed as Compound per CWA Section 307(a)
SARA 40CFR 372.65: Listed as Compound
SARA EHS 40CFR 355: Not listed
TSCA: Not listed

DIC3650 CAS #: 1201-99-6

2,4-DICHLOROCINNAMIC ACID

RTECS: GD8575000
EINECS Number: 214-860-1
Molecular Formula: $C_9H_6Cl_2O_2$
Formula Weight: 217.05

Chemical Structure

Synonyms: CINNAMIC ACID,2,4-DICHLORO-; 3-(2,4-DICHLOROPHENYL)-2-PROPENOIC ACID; 2-PROPENOIC ACID,3-(2,4-DICHLOROPHENYL)-; 2-PROPENOIC ACID,3-(2,4-DICHLOROPHENYL)-(9CI)

Physical Properties

Freezing Point: 234 °C (453.2 °F)
Other Solubilities: DMSO: Soluble

Hazard Overviews

Health: Irritating to eyes/skin/respiratory tract. Other Acute Effects: may be harmful by inhalation, ingestion, or skin absorption.
Fire: Hazards: emits toxic fumes. Extinguishing agents: water spray; carbon dioxide, dry chemical powder or appropriate foam. Precautions: combustible liquid.
Reactivity: Incompatible with: strong oxidizing agents. Hazardous decomposition products: toxic fumes of: carbon monoxide, carbon dioxide, hydrogen chloride gas.
Carcinogenicity: IARC - Not listed; NIOSH - Not listed; NTP - Not listed; ACGIH - Not listed; OSHA - Not listed; EPA - Not listed; MAK - Not listed
Primary Target Organs:

Eyes

Skin

Respiratory System

Environmental

Regulations

RCRA 40CFR: Not listed
CERCLA: 40CFR 302.4: Not listed

SARA 40CFR 372.65: Not listed
SARA EHS 40CFR 355: Not listed
TSCA: Not listed

DIC3700 CAS #: 2971-38-2

2,4-DICHLOROCROTYL ESTER

RTECS: AG8200000
DOT: UN2765; UN2766; UN2999; UN3000; NA2765; IMO3.2; IMO6.1
Molecular Formula: $C_{12}H_{11}Cl_3O_3$
Formula Weight: 309.58
Synonyms: ACETIC ACID,(2,4-DICHLOROPHENOXY)-,4-CHLORO-2-BUTENYLESTER; CHLOROCROTYL ESTER OF 2,4-D; CROTILIN; CROTILINE; CROTYLIN; 2,4-D,ALPHA-CHLOROCROTYL ESTER; 2,4-DICHLOROPHENOXYACETIC ACID,4-CHLOROCROTONYL ESTER; KROTILIN; KROTILINE
Description: fuel oil-like odor

Physical Properties

Saturated Vapor Density: < 1.200002291 kg/m^3
Vapor Pressure: < 1.5 x10^{-4} mm Hg at 25 °C
Water Solubility: Generally Immiscible or Insoluble in Water
Other Solubilities: Soluble in oils
Flash Point: > 79.444 °C Open Cup

RTECS Toxicity Data

Acute Oral: Rat LD_{50} Dose: 547 mg/kg. Mouse LD_{50} Dose: 489 mg/kg.
Acute Inhalation: Mouse LC_{50} Dose: 2190 mg/m^3/2hr.

Hazard Overviews

Fire: Combustible.
Carcinogenicity: IARC - Not listed; NIOSH - Not listed; NTP - Not listed; ACGIH - Not listed; OSHA - Not listed; EPA - Not listed; MAK - Not listed

Environmental

Cleanup/Disposal: Guide No. 131: Fully encapsulating, vapor protective clothing should be worn for spills and leaks with no fire. Eliminate all ignition sources (no smoking, flares, sparks or flames in immediate area). All equipment used when handling the product must be grounded. Do not touch or walk through spilled material. Stop leak if you can do it without risk. Prevent entry into waterways, sewers, basements or confined areas. A vapor suppressing foam may be used to reduce vapors. Small Spills: Absorb with earth, sand or other non-combustible material and transfer to containers for later disposal. Use clean non-sparking tools to collect absorbed material. Large Spills: Dike far ahead of liquid spill for later disposal. Water spray may reduce vapor; but may not prevent ignition in closed spaces. Guide No. 152: Do not touch damaged containers or spilled material unless wearing appropriate protective clothing. Stop leak if you can do it without risk. Prevent entry into waterways, sewers, basements or confined areas. Cover with plastic sheet to prevent spreading. Absorb or cover with dry earth, sand or other non-combustible material and transfer to containers. Do not get water inside containers.

Regulations

RCRA 40CFR: Not listed
CERCLA: 40CFR 302.4: Listed per CWA Section 311(b)(4) RQ: 100 lb (45.35 kg)
SARA 40CFR 372.65: Listed
SARA EHS 40CFR 355: Not listed
TSCA: Not listed

DIC3750 CAS #: 28434-86-8

3,3'-DICHLORO-4,4'-DIAMINODIPHENYL ETHER

RTECS: KM9625000
Molecular Formula: $C_{12}H_{10}Cl_2N_2O$
Formula Weight: 269.14
Synonyms: ANILINE,4,4'-OXYBIS(2-CHLORO-; BIS(4-AMINO-3-CHLOROPHENYL) ETHER; 3,3'-DICHLOR-4,4'-DIAMINO-DIPHENYLAETHER; 4,4'-OXYBIS(2-CHLOROANILINE); 4,4'-OXYBIS(2-CHLORO-BENZENAMINE); 4,4'-OXYBIS(2-CHLOROBENZENAMINE)

Physical Properties

Freezing Point: 128 °C (262.4 °F) to 129 °C (264.2 °F)
Flash Point: Not available; probably combustible

RTECS Toxicity Data

Acute Dermal: Rat LD_{50} Route: Subcutaneous Dose: >10 gm/kg.
Mutagenic: Rat Unscheduled DNA Synthesis; Cell Type: liver; Dose: 1 umol/L. Bacteria - S Typhimurium Mutations in Microorganisms; Dose: 100 ug/plate (-S9).
Tumorigenic: Rat Route: Subcutaneous; Dose: 11 gm/kg/27W-I; Toxic Effects: Tumorigenic - Carcinogenic by RTECS criteria; Sense organs and special senses - Tumors. Rat Route: Subcutaneous; Dose: 14 gm/kg/96W-I; Toxic Effects: Tumorigenic - Carcinogenic by RTECS criteria; Liver - Tumors.

Hazard Overviews

Fire: Will burn.
Carcinogenicity: IARC - Group 2B, Possibly carcinogenic to humans; NIOSH - Not listed; NTP - Not listed; ACGIH - Not listed; OSHA - Not listed; EPA - Not listed; MAK - Not listed

Environmental

Regulations

RCRA 40CFR: Not listed
CERCLA: 40CFR 302.4: Listed as Compound per CWA Section 307(a)
SARA 40CFR 372.65: Not listed
SARA EHS 40CFR 355: Not listed

TSCA: Not listed

DIC3800 CAS #: 38178-38-0

1,6-DICHLORODIBENZO-P-DIOXIN

RTECS: HP3095800
Molecular Formula: $C_{12}H_4Cl_4O_2$
Formula Weight: 322
Synonyms: DIBENZO(B,E)(1,4)DIOXIN,1,6-DICHLORO-; 1,6-DICHLORODIBENZO-PARA-DIOXIN

Physical Properties

Freezing Point: 305 °C (581 °F) to 306 °C (582.8 °F)
Water Solubility: 19 mg/L
Other Solubilities: Benzene (.57 g/l); Chloroform (.37 g/l); Acetone (.11 g/L) Methanol(.01g/l); lard oil(.04 g/l)

Hazard Overviews

Reactivity: Stable. Hazardous polymerization cannot occur. Avoid: heat and ignition sources. Hazardous decomposition products: toxic fumes of chlorine.
Carcinogenicity: IARC - Group 3, Not classifiable as to carcinogenicity to humans; NIOSH - Not listed; NTP - Not listed; ACGIH - Not listed; OSHA - Not listed; EPA - Not listed; MAK - Not listed

Environmental

Regulations

RCRA 40CFR: Not listed
CERCLA: 40CFR 302.4: Not listed
SARA 40CFR 372.65: Not listed
SARA EHS 40CFR 355: Not listed
TSCA: Not listed

DIC3850 CAS #: 33857-26-0

2,7-DICHLORODIBENZO-P-DIOXIN

RTECS: HP3100000
Molecular Formula: $C_{12}H_6Cl_2O_2$
Formula Weight: 253.08
Synonyms: DCDD; DIBENZO(B,E)(1,4)DIOXIN,2,7-DICHLORO-; DIBENZO-P-DIOXIN,2,7-DICHLORO-; 2,7-DICHLORODIBENZO(B,E)(1,4)DIOXIN; 2,7-DICHLORODIBENZO-PARA-DIOXIN; 2,7-DICHLORODIBENZODIOXIN; 2,7-DICHLORO-P-DIOXIN
Description: colorless crystals
Use: not used commercially in the US

Physical Properties

Freezing Point: 209 °C (408.2 °F) to 210 °C (410 °F)
Saturated Vapor Density: 1.200000014 kg/m^3
Vapor Pressure: 1.125 $x10^{-6}$ mm Hg at 25 °C
Water Solubility: About 0.19 ppm (estimated)
Other Solubilities: 95% Ethanol: <1 mg/ml at 19 °C; DMSO: <1 mg/ml at 19 °C.
Flash Point: Not available; probably combustible

RTECS Toxicity Data

Irritation Eye: Rabbit Standard Draize Test Dose: 2 mg; Reaction: mild.
Reproductive/Teratogenic: Rat Route: Oral; Dose: 5 mg/kg; Duration: female 6-15D of pregnancy; Specific Developmental Abnormalities - Cardiovascular (circulatory) system.
Tumorigenic: Mouse Route: Oral; Dose: 378 gm/kg/90W-C; Toxic Effects: Tumorigenic - Equivocal tumorigenic agent by RTECS criteria; Liver - Tumors; Blood - Leukemia. Mouse Route: Oral; Dose: 756 gm/kg/90W-C; Toxic Effects: Tumorigenic - Equivocal tumorigenic agent by RTECS criteria; Liver - Tumors; Blood - Leukemia.

Hazard Overviews

Fire: Will burn.
Carcinogenicity: IARC - Group 3, Not classifiable as to carcinogenicity to humans; NIOSH - Not listed; NTP - Not listed; ACGIH - Not listed; OSHA - Not listed; EPA - Not listed; MAK - Not listed

Environmental

Environmental Fate: If released to soil, it is expected to be immobile; however, runoff and wind erosion may transport the chemical. A single soil biodegradation study has shown that it is 5% metabolized to CO2 over 10 weeks. If released to water, partitioning from the water column to sediment and suspended material is likely to be significant. In the presence of significant adsorption, volatilization may not be important (estimated half-life of 15.2 days from an environmental pond). Photodegradation may be important in surface waters exposed to sunlight. If released to the atmosphere in particulate phase, physical removal can be expected to occur via wet and dry deposition processes. Vapor-phase should react relatively rapidly in air with photochemically produced hydroxyl radicals; the estimated half-life for this reaction is 1.6 days in avg. air.

Environmental Physical Data

Henry's Law Constant: estimated at 4.88 $x10^{-5}$
Octanol/Water Partition Coefficient: log K_{ow} = 5.75
BCF: calculated at 1580 to 1.38 $x10^4$

Regulations

RCRA 40CFR: Not listed
CERCLA: 40CFR 302.4: Not listed
SARA 40CFR 372.65: Not listed
SARA EHS 40CFR 355: Not listed
TSCA: Not listed

DIC3900 **CAS #: 75-71-8**

DICHLORODIFLUOROMETHANE

RTECS: PA8200000
DOT: UN1028; IMO2.2
EINECS Number: 200-893-9
Molecular Formula: CCl_2F_2
Structured MF: CF_2Cl_2
Formula Weight: 120.91

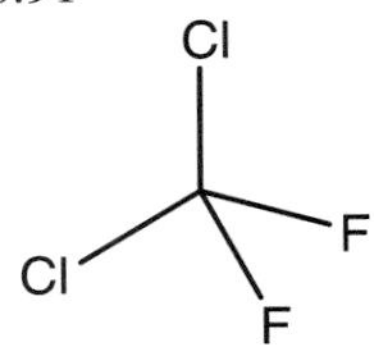

Chemical Structure

Synonyms: ALGOFRENE TYPE 2; ARCTON 12; ARCTON 6; CFC-12; CFC 12; CHLOROFLUOROCARBON 12; DICLORODIFLUOMETANO; DIFLUORODICHLOROMETHANE; DWUCHLORODWUFLUOROMETAN; DYMEL 12; ELECTRO-CF 12; EPA PESTICIDE CHEMICAL CODE 000014; ESKIMON 12; F 12; FC 12; FCC 12; FKW 12; FLUOROCARBON-12; FLUOROCARBON 12; FORANE 12; FREON 12; FREON F-12; FRIGEN 12; GENETRON 12; HALON; HALON 122; ISCEON 122; ISOTRON 12; ISOTRON 2; KAISER CHEMICALS 12; LEDON 12; METHANE,DICHLORODIFLUORO-; PROPELLANT 12; R 12 (REFRIGERANT); R12; REFRIGERANT 12; REFRIGERANT R 12; UCON 12; UCON 12/HALOCARBON 12
Description: colorless gas; nearly odorless
Use: refrigerant, aerosol propellant, rocket propellant, foaming agent, plastics

Physical Properties

Boiling Point: -29.8 °C (-22 °F) at 760 mm Hg
Freezing Point: -158 °C (-252.4 °F)
Specific Gravity: 1.486 at -29.8 °C
Vapor Density: 4.1
Vapor Pressure: 84.8 psia at 70 °F
Water Solubility: 0.28 g/L Water at 25 °C at 1 atm
Other Solubilities: 95% Ethanol: Soluble; Acetic Acid: Soluble; Chloroform: Soluble; Ether: Soluble.
Surface Tension: 9 dynes/cm
Refraction Index: 1.287 at 25 °C/D
Critical Temperature: 112 °C
Critical Pressure: 598 psia
Ionization Potential (eV): 11.75
Flash Point: Nonflammable

RTECS Toxicity Data

Acute Oral: Rat LD; Dose: >5600 ug/kg.
Acute Inhalation: Mouse LC_{50} Dose: 3348 gm/m^3/3hr; Toxic Effects: Behavioral - Sleep; Behavioral - Tremor; Behavioral - Excitement. Human TC_{Lo} Dose: 200000 ppm/30M; Toxic Effects: Sense organs and special senses - Conjunctive irritation; Lungs, Thorax, or Respiration - Fibrosing alveolitis; Liver - Other changes.
Chronic (Multiple Dose) Oral: Rat Dose: 2548 mg/kg/26W-I; Toxic Effects: Behavioral - Alteration of classical conditioning; Blood - Changes in erythrocite (RBC) cell count; Biochemical - True cholinesterase.
Chronic (Multiple Dose) Inhalation: Rat Dose: 4136 mg/m^3/8H/6W-I; Toxic Effects: Lungs, Thorax, or Respiration - Chronic pulmonary edema; DEATH. Rat Dose: 3997 mg/m^3/90D-C; Toxic Effects: Lungs, Thorax, or Respiration - Chronic pulmonary edema; DEATH. Monkey Dose: 3997 mg/m^3/90D-C; Toxic Effects: Lungs, Thorax, or Respiration - Chronic pulmonary edema.

Hazard Overviews

Compressed Gas

Fire Diamond

Health: Mildly irritating to eyes/skin/respiratory tract. Also Causes: asphyxia, nausea,exhaustion, unconsciousness, convulsions, paresthesia, slurred speech, cardiac arrythmia, possible frostbite. Chronic Effects: neurotoxic effects.
Fire: Noncombustible. Use agent suitable for surrounding fire. Cylinder may explode in heat of fire.
Reactivity: Stable. Hazardous polymerization cannot occur. Avoid: exposure to excessive heat. Incompatible with: sodium; potassium; calcium; powdered aluminum; zinc; magnesium. Hazardous decomposition products: carbon dioxide; phosgene; chlorine; hydrogen chloride; hydrogen fluoride; fluorine gases.
Carcinogenicity: IARC - Not listed; NIOSH - Not listed; NTP - Not listed; ACGIH - Class A4, Not classifiable as a human carcinogen; OSHA - Not listed; EPA - Not listed; MAK - Not listed
Primary Target Organs:

Eyes

Skin

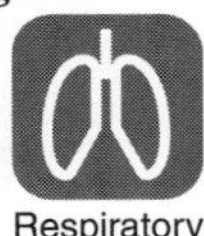
Respiratory System

Nervous System

Cardio-vascular

Exposure Limits
OSHA PEL: TWA: 1000 ppm; 4950 mg/m^3.
ACGIH TLV: TWA: 1000 ppm; 4950 mg/m^3.
NIOSH REL: TWA: 1000 ppm; 4950 mg/m^3.
NIOSH IDLH: 15000 ppm.
DFG MAK: TWA: 1000 ppm; 5000 mg/m^3.
Respirator Recommendation
Exposure Range: >1000 to <15,000 ppm Supplied Air, Constant Flow/Pressure Demand, Half Mask
Exposure Range: 15,000 to unlimited ppm Self-contained Breathing Apparatus, Pressure Demand, Full Face
Note: odor threshold unknown

Environmental

Environmental Fate: If released on land, it will leach into the ground and volatilize from the soil surface. No degradative processes are known to occur in the soil. It is also stable in water and the only removal process will be volatilization. It can enter water bodies from the atmosphere and the in the surface water rapidly reaches equilibrium with the concentration in the air. Ocean currents also carry the chemical long distances and many kilometers below the surface. It is extremely stable in the troposphere and will disperse over the globe and diffuse slowly into the stratosphere where it will be lost by photolysis.

Cleanup/Disposal: Guide No. 126: Do not touch or walk through spilled material. Stop leak if you can do it without risk. Do not direct water at spill or source of leak. Use water spray to reduce vapors or divert vapor cloud drift. If possible, turn leaking containers so that gas escapes rather than liquid. Prevent entry into waterways, sewers, basements or confined areas. Allow substance to evaporate. Ventilate the area.

Environmental Physical Data

Henry's Law Constant: estimated at 0.225
Octanol/Water Partition Coefficient: log K_{ow} = 2.16
Sorption Partition Coefficient: K_{oc} = 200
BCF: possible under constant exposure
BOD: not pertinent

Regulations

RCRA 40CFR: Listed Hazardous Waste No. U075 Toxic Waste
CERCLA: 40CFR 302.4: Listed per RCRA Section 3001 RQ: 5000 lb (2268 kg)
SARA 40CFR 372.65: Listed
SARA EHS 40CFR 355: Not listed
TSCA: Listed

Analytical Methods

Air: EPA TO-14, 0040
Soil: SW846 5032, 8010B, 8021A, 8240B, 8260A, 8260B; EPA 7
Water / Groundwater: EPA 601, 624-S; APHA 6210-D, 6230-B, 6230-D
Drinking Water: EPA 502.1, 502.2, 524.1, 524.2; APHA 6210-C, 6230-C
Food: EPA 5
Indoor / Expired Air: NIOSH 1018; EPA IP-1A
Plasma: EPA 29

DIC3950 CAS #: 56961-05-8

2,6-DICHLORO-N,N-DIMETHYLANILINE

Molecular Formula: $C_8H_9Cl_2N$
Formula Weight: 190.07
Synonyms: BENZENAMINE,2,6-DICHLORO-N,N-DIMETHYL-

Hazard Overviews

Carcinogenicity: IARC - Not listed; NIOSH - Not listed; NTP - Not listed; ACGIH - Not listed; OSHA - Not listed; EPA - Not listed; MAK - Not listed

Environmental

Regulations

RCRA 40CFR: Not listed
CERCLA: 40CFR 302.4: Not listed
SARA 40CFR 372.65: Not listed
SARA EHS 40CFR 355: Not listed
TSCA: Not listed

DIC4000 CAS #: 118-52-5

1,3-DICHLORO-5,5-DIMETHYLHYDANTOIN

RTECS: MU0700000
EINECS Number: 204-258-7
Molecular Formula: $C_5H_6Cl_2N_2O_2$
Formula Weight: 197.03

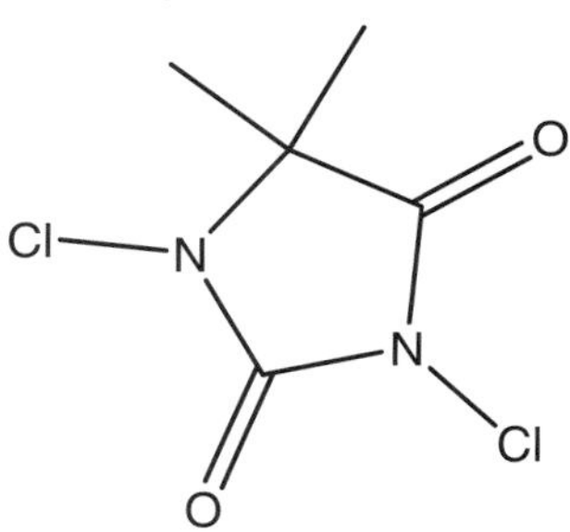

Chemical Structure

Synonyms: DACTIN; DAKTIN; DANTOIN; DCA; DCDMH; DDH; DICHLORANTIN; 1,3-DICHLORO-5,5-DIMETHYL HYDANTOIN; DICHLORO-5,5-DIMETHYLHYDANTOIN; DICHLORODIMETHYLHYDANTOIN; 1,3-DICHLORO-5,5-DIMETHYL-2,4-IMIDAZOLIDINEDIONE; 1,3-DICHLORO-5,5'-METHYLHYDANTOIN; DWUCHLORANTYNY; 1,3-DWUCHLORO-5,5-DWUMETYLOHYDANTOINA; HALANE; HYDAN; HYDAN (ANTISEPTIC); HYDANTOIN,1,3-DICHLORO-5,5-DIMETHYL-; HYDANTOIN,DICHLORODIMETHYL-; 2,4-IMIDAZOLIDINEDIONE,1,3-DICHLORO-5,5-DIMETHYL-; 2,4-IMIDAZOLIDINEDIONE,1,3-DICHLORO-5,5-DIMETHYL-(9CI); OMCHLOR
Description: white powder; mild chlorine odor
Use: chlorinating agent; disinfectant used especially in laundry bleaches; industrial deodorant; intermediate for amino acids, drugs and insecticides; stabilizer for vinyl chloride polymers; polymerization catalyst; and in water treatment

Physical Properties

Boiling Point: Sublimes at 100 °C (212 °F)
Freezing Point: 132 °C (269.6 °F)
Specific Gravity: 1.5 at 20 °C/20 °C
Vapor Density: 6.8 Air=1
Water Solubility: 0.21% at 25 °C, 0.60% at 60 °C
Other Solubilities: DMSO: Decomposes; 95% Ethanol: 5-10 mg/ml at 22 °C; Acetone: Decomposes; Benzene: 9.2% at 25 °C; Highly polar solvents: Freely Soluble; Chloroform: 14%

at 25 °C; Carbon Tetrachloride: 12.5% at 25 °C; Ethylene dichloride: 32% at 25 °C.
pH: Aqueous solution 4.4
Flash Point: 174 °C

RTECS Toxicity Data

Acute Oral: Rat LD_{50} Dose: 542 mg/kg. Rabbit LD_{50} Dose: 1520 mg/kg; Toxic Effects: Behavioral - Somnolence (general depressed activity); Behavioral - Withdrawal; Lungs, Thorax, or Respiration - Dyspnea.
Acute Inhalation: Rat LC_{Lo} Dose: 20 gm/m^3/1hr; Toxic Effects: Behavioral - Tremor; Lungs, Thorax, or Respiration - Respiratory stimulation; Gastrointestinal - Changes in structure or function of salivary glands.
Acute Dermal: Rabbit LD_{50} Route: Skin; Dose: >20 gm/kg.
Chronic (Multiple Dose) Oral: Rat Dose: 8784 mg/kg/28D-I; Toxic Effects: DEATH. Rat Dose: 634 mg/kg/30D-I; Toxic Effects: Brain and coverings - Recordings from specific areas of CNS; Biochemical - True cholinesterase; Biochemical - Transaminases.
Irritation Skin: Rabbit Standard Draize Test Dose: 100 mg/24H; Reaction: severe. Rabbit Standard Draize Test Dose: 500 mg/24H; Reaction: severe.
Mutagenic: Rat Morphological Transformation; Cell Type: embryo; Dose: 6300 ng/plate. Insects - D Melanogaster Sex Chromosome Loss; Route: Parenteral; Dose: 250 ppm.

Hazard Overviews

Fire Diamond

Health: Irritating to eyes/skin/respiratory tract. Also Causes: bronchial irritation, difficulty breathing, stomatitis, nausea, pulmonary edema. Chronic Effects: contact dermatitis.
Fire: Combustible. Contact with water produces toxic fumes of hypochlorous acid. Use dry chemical, foam, or carbon dioxide. Vapors may travel to an ignition source and flash back.
Reactivity: Stable. Hazardous polymerization cannot occur. Avoid: heat; ignition sources; water. Incompatible with: water ; steam; xylene. Hazardous decomposition products: toxic fumes of chlorine; nitrogen oxides.
Carcinogenicity: IARC - Not listed; NIOSH - Listed as carcinogen; NTP - Not listed; ACGIH - Not listed; OSHA - Not listed; EPA - Not listed; MAK - Not listed
Primary Target Organs:

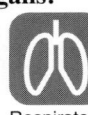

Eyes

Skin

Respiratory System

Exposure Limits
OSHA PEL: TWA: 0.2 mg/m^3.
OSHA PEL Vacated 1989 Limits: TWA: 0.2 mg/m^3; STEL: 0.4 mg/m^3.
ACGIH TLV: TWA: 0.2 mg/m^3.
NIOSH REL: TWA: 0.2 mg/m^3. STEL: 0.4 mg/m^3.
NIOSH IDLH: 5 mg/m^3.
Respirator Recommendation
Exposure Range: >0.2 to 2 mg/m^3 Air Purifying, Negative Pressure, Half Mask
Exposure Range: >2 to <5 mg/m^3 Air Purifying, Negative Pressure, Full Face
Exposure Range: 5 to unlimited mg/m^3 Self-contained Breathing Apparatus, Pressure Demand, Full Face
Cartridge Color: black with dust/mist prefilter (use P100 or consult supervisor for appropriate dust/mist prefilter)

Environmental

Cleanup/Disposal: Guide No. 171: Do not touch or walk through spilled material. Stop leak if you can do it without risk. Prevent dust cloud. Avoid inhalation of asbestos dust. Small Dry Spills: With clean shovel place material into clean, dry container and cover loosely; move containers from spill area. Small Spills: Take up with sand or other noncombustible absorbent material and place into containers for later disposal. Large Spills: Dike far ahead of liquid spill for later disposal. Cover powder spill with plastic sheet or tarp to minimize spreading. Prevent entry into waterways, sewers, basements or confined areas.

Regulations

RCRA 40CFR: Not listed
CERCLA: 40CFR 302.4: Not listed
SARA 40CFR 372.65: Not listed
SARA EHS 40CFR 355: Not listed
TSCA: Listed

DIC4050 CAS #: 95-59-0

2,3-DICHLORO-1,4-DIOXANE

EINECS Number: 202-435-3
Molecular Formula: $C_4H_6Cl_2O_2$
Formula Weight: 157.0
Synonyms: 2,5-DICHLORO-P-DIOXANE; 1,4-DIOXANE,2,3-DICHLORO-; P-DIOXANE,2,3-DICHLORO-
Use: alkylating agent & experimental carcinogen

Physical Properties

Boiling Point: 80 °C (176 °F) to 82 °C (180 °F) at 10 mm Hg
Freezing Point: 30 °C (86 °F)
Specific Gravity: 1.468 at 20 °C/4 °C
Water Solubility: Insoluble in Water
Other Solubilities: Very Soluble in Ether, Acetone, Benzene, Chloroform, Carbon Tetrachloride, Dioxane, Petroleum Ether
Refraction Index: 1.4928 at 20 °C

RTECS Toxicity Data

Hazard Overviews

Carcinogenicity: IARC - Not listed; NIOSH - Not listed; NTP - Not listed; ACGIH - Not listed; OSHA - Not listed; EPA - Not listed; MAK - Not listed

Environmental

Regulations

RCRA 40CFR: Not listed
CERCLA: 40CFR 302.4: Not listed
SARA 40CFR 372.65: Not listed
SARA EHS 40CFR 355: Not listed
TSCA: Listed

DIC4100 CAS #: 80-07-9

4,4'-DICHLORODIPHENYL SULFONE

RTECS: WR3450000
EINECS Number: 201-247-9
Molecular Formula: $C_{12}H_8Cl_2O_2S$
Structured MF: $(ClC_6H_4)_2SO_2$
Formula Weight: 287.16

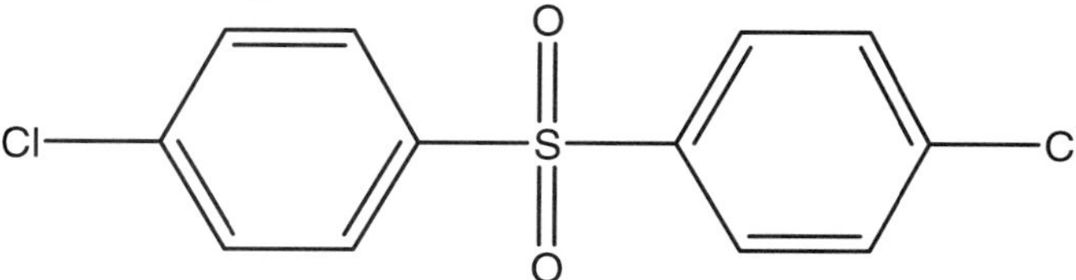

Chemical Structure

Synonyms: BENZENE,1,1'-SULFONYLBIS(4-CHLORO-; BENZENE,1,1'-SULFONYLBIS(4-CHLORO-(9CI); BIS(4-CHLOROPHENYL) SULFONE; BIS(P-CHLOROPHENYL) SULFONE; 4-CHLORO-1-(4-CHLOROPHENYLSULFONYL)BENZENE; 4-CHLOROPHENYL SULFONE; P-CHLOROPHENYL SULFONE; 4,4'-DICHLORODIPHENYL SULFONE; P,P'-DICHLORODIPHENYL SULFONE; 4,4'-DICHLORODIPHENYL SULPHONE; 4,4'-DICHLOROPHENYL SULFONE; DI-4-CHLOROPHENYL SULFONE; DI-P-CHLOROPHENYL SULFONE; SULFONE,BIS(P-CHLOROPHENYL); SULFONE,BIS(P-CHLOROPHENYL)-; 1,1'-SULFONYLBIS(4-CHLOROBENZENE)
Description: off-white powder or crystals

Physical Properties

Boiling Point: Sublimes
Freezing Point: 145 °C (293 °F) to 148 °C (298.4 °F)
Water Solubility: < 1 mg/mL at 20 C
Other Solubilities: 95% Ethanol: <1 mg/ml at 20 °C; Acetone: >=100 mg/ml at 20 °C; DMSO: >=100 mg/ml at 20 °C.
Flash Point: Not available; probably combustible

RTECS Toxicity Data

Acute Oral: Rat LD_{Lo} Dose: 7500 mg/kg; Toxic Effects: Gastrointestinal - Hypermotility, diarrhea. Mouse LD_{50} Dose: 24 gm/kg; Toxic Effects: Behavioral - Tremor.
Chronic (Multiple Dose) Oral: Rat Dose: 364 mg/kg/26W-I; Toxic Effects: Brain and coverings - Recordings from specific areas of CNS; Liver - Liver function tests impaired.

Hazard Overviews

Health: May cause irritation. Toxic. Other Acute Effects: harmful if absorbed through skin; may be harmful if inhaled; may be harmful if swallowed.
Fire: Will burn. Hazards: emits toxic fumes. Extinguishing agents: water spray; carbon dioxide, dry chemical powder or appropriate foam. Precautions: combustible liquid.
Carcinogenicity: IARC - Not listed; NIOSH - Not listed; NTP - Not listed; ACGIH - Not listed; OSHA - Not listed; EPA - Not listed; MAK - Not listed

Environmental

Regulations

RCRA 40CFR: Not listed
CERCLA: 40CFR 302.4: Listed as Compound per CWA Section 307(a)
SARA 40CFR 372.65: Listed as Compound
SARA EHS 40CFR 355: Not listed
TSCA: Listed

DIC4150 CAS #: 80-10-4

DICHLORODIPHENYLSILANE

RTECS: VV3190000
DOT: UN1769; IMO8.0
EINECS Number: 201-251-0
Molecular Formula: $C_{12}H_{10}Cl_2Si$
Structured MF: $(C_6H_5)_2SiCl_2$
Formula Weight: 253.21

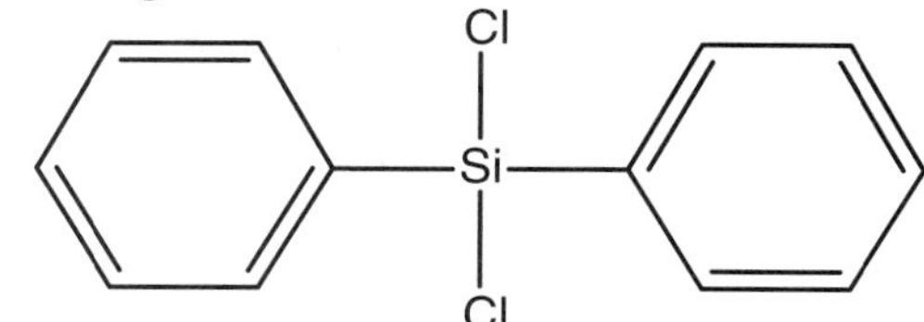

Chemical Structure

Synonyms: DICHLOR-DIFENYLSILAN; DIPHENYL DICHLOROSILANE; DIPHENYLDICHLOROSILANE; DIPHENYLSILICON DICHLORIDE; SILANE,DICHLORODIPHENYL-
Description: colorless liquid; sharp, hydrochloric acid-like, pungent odor
Use: monomer for silicone fluids, rubbers, & resins

Physical Properties

Boiling Point: 302 °C (576 °F) to 305 °C (581 °F) at 757 mm Hg
Freezing Point: -22 °C (-7.6 °F)
Specific Gravity: 1.2216 at 20 °C/4 °C
Vapor Density: 8.45
Water Solubility: Decomposes in Water
Other Solubilities: Ethanol: Soluble; Ether: Soluble; Acetone: Soluble; Benzene: Soluble; Carbon Tetrachloride: Soluble
Surface Tension: Estimated at 26 dynes/cm

pH: Acid
Refraction Index: 1.5819 at 20 °C/D
Flash Point: 142 °C Closed Cup

RTECS Toxicity Data

Irritation Eye: Rabbit Standard Draize Test Dose: 5 mg/24H; Reaction: severe.
Irritation Skin: Rabbit Standard Draize Test Dose: 20 mg/24H; Reaction: moderate.

Hazard Overviews

Corrosive

Health: Corrosive to eyes/skin/respiratory tract. Other Acute Effects: harmful if swallowed, inhaled, or absorbed through skin; extremely destructive to tissue of the mucous membranes and upper respiratory tract, eyes and skin; inhalation may result in spasm, inflammation and edema of the larynx and bronchi, chemical pneumonitis and pulmonary edema; symptoms of exposure may include burning sensation, coughing, wheezing, laryngitis, shortness of breath, headache, nausea and vomiting.
Fire: Will burn. Hazards: emits toxic fumes. Extinguishing agents: appropriate foam; do not use water. Precautions: combustible liquid.
Carcinogenicity: IARC - Not listed; NIOSH - Not listed; NTP - Not listed; ACGIH - Not listed; OSHA - Not listed; EPA - Not listed; MAK - Not listed
Primary Target Organs:

Eyes

Skin

Respiratory System

Environmental

Cleanup/Disposal: Guide No. 156: Eliminate all ignition sources (no smoking, flares, sparks or flames in immediate area). All equipment used when handling the product must be grounded. Do not touch damaged containers or spilled material unless wearing appropriate protective clothing. Stop leak if you can do it without risk. A vapor suppressing foam may be used to reduce vapors. For chlorosilanes, use AFFF alcohol-resistant medium expansion foam to reduce vapors. Do not get water on spilled substance or inside containers. Use water spray to reduce vapors or divert vapor cloud drift. Prevent entry into waterways, sewers, basements or confined areas. Small Spills: Cover with dry earth, dry sand, or other non-combustible material followed with plastic sheet to minimize spreading or contact with rain. Use clean non-sparking tools to collect material and place it into loosely covered plastic containers for later disposal.

Environmental Physical Data

BCF: no food chain concentration potential

Regulations

RCRA 40CFR: Not listed
CERCLA: 40CFR 302.4: Not listed
SARA 40CFR 372.65: Not listed
SARA EHS 40CFR 355: Not listed
TSCA: Listed

Analytical Methods

Water / Groundwater: ASTM D4763

DIC4200 CAS #: 50-29-3

DICHLORODIPHENYLTRICHLORO-ETHANE

RTECS: KJ3325000
DOT: UN2761; NA2761; IMO6.1
EINECS Number: 200-024-3
Molecular Formula: $C_{14}H_9Cl_5$
Structured MF: $(ClC_6H_4)_2CHCCl_3$
Formula Weight: 354.50

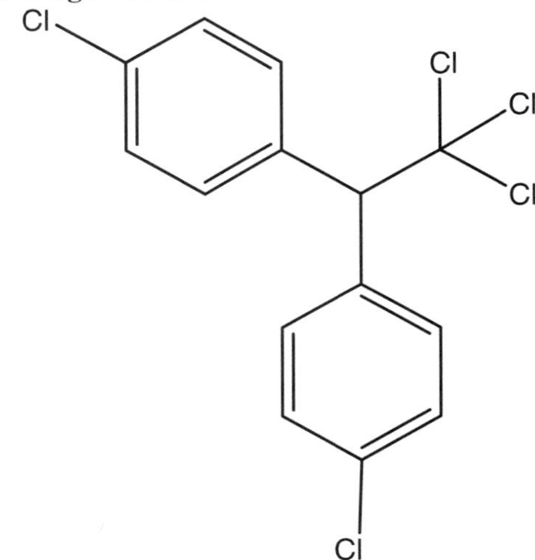

Chemical Structure

Synonyms: AAVERO-EXTRA; AGRITAN; ANOFEX; ARKOTINE; AZOTOX; AZOTOX M-33; BENZENE,1,1'-(2,2,2-TRICHLOROETHYLIDENE)BIS(4-CHLORO-; 2,2-BIS (P-CHLOROPHENYL)-1,1,1-TRICHLOROETHANE; ALPHA,ALPHA-BIS(P-CHLOROPHENYL)-BETA,BETA,BETA-TRICHLORETHANE; 1,1-BIS(P-CHLOROPHENYL)-2,2,2-TRICHLOROETHANE; 1,1-BIS-(P-CHLOROPHENYL)-2,2,2-TRICHLOROETHANE; 2,2-BIS(P-CHLOROPHENYL)-1,1,1-TRICHLOROETHANE; ALPHA,ALPHA-BIS(P-CHLOROPHENYL)-BETA,BETA,BETA-TRICHLOROETHANE; BOSAN SUPRA; BOVIDERMOL; CHLOFENOTAN; CHLOROPHENOTHAN; CHLOROPHENOTHANE; CHLOROPHENOTOXUM; CHLORPHENOTHAN; CHLORPHENOTOXUM; CITOX; CLOFENOTANE; 4,4'-DDT; DDT; P,P'-DDT; P,P-DDT; DEDELO; DEOVAL; DETOX; DETOXAN; DIBOVAN; DIBOVIN; 4,4'-DICHLORODIPHENYLTRICHLOROETHANE; P,P'-DICHLORODIPHENYLTRICHLOROETHANE; DICOPHANE; DIDIGAM; DIDIMAC; DIPHENYLTRICHLOROETHANE; DODAT; DYKOL; ENT-1506; ENT 1,506; EPA PESTICIDE CODE 029201; ESTONATE; ETHANE,1,1,1-TRICHLORO-2,2-BIS(4-CHLOROPHENYL)-; ETHANE,1,1,1-TRICHLORO-2,2-BIS(P-CHLOROPHENYL)-; GENITOX; GESAFID; GESAPON; GESAREX;

GESAROL; GUESAPON; GUESAROL; GYRON; HAVERO-EXTRA; HILDIT; IVORAN; IXODEX; KLORFENOTON; KOPSOL; MICRO DDT 75; MUTOXAN; MUTOXIN; NEOCID; NEOCIDOL; NEOCIDOL (SOLID); OMS 0016; OMS 16; PARACHLOROCIDUM; PEB1; PENTACHLORIN; PENTECH; PENTICIDUM; PPZEIDAN; R50; RUKSEAM; SANTOBANE; TAFIDEX; TECH DDT; 1,1,1-TRICHLOOR-2,2-BIS(4-CHLOOR FENYL)-ETHAAN; 1,1,1-TRICHLOR-2,2-BIS(4-CHLOR-PHENYL)-AETHAN; 1,1,1-TRICHLORO-2,2-BIS(P-CHLOROPHENYL)ETHANE; TRICHLOROBIS(4-CHLOROPHENYL)ETHANE; 1,1,1-TRICHLORO-2,2-DI(4-CHLOROPHENYL)-ETHANE; 1,1,1-TRICHLORO-2,2-DI(4-CHLOROPHENYL)ETHANE; 1,1'-(2,2,2-TRICHLOROETHYLIDENE)BIS(4-CHLOROBENZENE); 1,1,1-TRICLORO-2,2-BIS(4-CLORO-FENIL)-ETANO; 1,1,1-TRICLORO-2,2-BIS(4-CLORO-FENYL)-ETANO; PP'-ZEIDANE; ZEIDANE; ZERDANE

Description: white to gray cystaline solid

Use: as an insecticide, as a pesticide and to control vectors of disease; in veterinary medicine as an insecticide and pediculicide; to kill head lice and disinfect clothing; an ectoparasiticide

Physical Properties

Boiling Point: 260 °C (500 °F)

Freezing Point: 108.5 °C (227.3 °F)

Specific Gravity: 0.98 to 0.99

Saturated Vapor Density: 1.200000003 kg/m^3

Vapor Pressure: 1.5 $x10^{-7}$ mm Hg at 20 °C

Water Solubility: 1 $x10^{-7}$ g/100 ml Water at 27 °C

Other Solubilities: High Solubility in fat (100,000 ppm); g/100 ml solvent: 58 Acetone; 78 Benzene; 42 Benzyl Benzoate; 45 Carbon Tetrachloride; 74 Chloroenzene; 2 95% Alcohol; 28 Ethyl Ether; 10 Gasolene; 3 Isopropanol; 8-10 Kerosene.

Odor Threshold: 5.0725 mg/m^3

Flash Point: 72.222 to 77.222 °C Closed Cup

RTECS Toxicity Data

Acute Oral: Human TD_{Lo} Dose: 16 mg/kg; Toxic Effects: Behavioral - Convulsions or effect on seizure threshold. Human TD_{Lo} Dose: 5 mg/kg; Toxic Effects: Behavioral - General anesthetic; Behavioral - Analgesia. Man TD_{Lo} Dose: 6 mg/kg; Toxic Effects: Behavioral - Headache; Gastrointestinal - Nausea or vomiting; Skin and appendages - Sweating. Human LD_{Lo} Dose: 500 mg/kg; Toxic Effects: Behavioral - Convulsions or effect on seizure threshold; Cardiac - Arrythmias (including changes in conduction); Lungs, Thorax, or Respiration - Other changes. Infant LD_{Lo} Dose: 150 mg/kg; Toxic Effects: Lungs, Thorax, or Respiration - Acute pulmonary edema. Rat LD_{50} Dose: 87 mg/kg. Monkey LD_{50} Dose: 200 mg/kg.

Acute Dermal: Rabbit LD_{50} Route: Skin; Dose: 300 mg/kg; Toxic Effects: Behavioral - Tremor; Behavioral - Muscle weakness; Behavioral - Ataxia. Rabbit LD_{50} Route: Subcutaneous Dose: 250 mg/kg; Toxic Effects: Behavioral - Tremor; Behavioral - Muscle weakness; Behavioral - Ataxia.

Chronic (Multiple Dose) Oral: Rat Dose: 91250 ug/kg/2Y-C; Toxic Effects: Liver - Changes in liver weight. Rat Dose: 754 mg/kg/52D-I; Toxic Effects: Liver - Changes in liver weight; Blood - Changes in serum composition; Biochemical - Transaminases. Rat Dose: 91 mg/kg/26W-I; Toxic Effects: Kidney, Ureter, and Bladder - Other changes in urine composition; Endocrine - Other changes; Endocrine - Changes in adrenal weight. Rat Dose: 4300 mg/kg/14W-I; Toxic Effects: Liver - Changes in liver weight; Endocrine - Changes in adrenal weight; Nutritional and gross metabolic - Weight loss or decreased weight gain. Rat Dose: 462 mg/kg/22W-C; Toxic Effects: Endocrine - Changes in spleen weight; Blood - Changes in serum composition. Rat Dose: 6510 mg/kg/14D-C; Toxic Effects: Behavioral - Tremor; DEATH. Monkey Dose: 11200 mg/kg/32W-C; Toxic Effects: Biochemical - Other hydrolases; Biochemical - Transaminases; DEATH. Monkey Dose: 91 mg/kg/26W-I; Toxic Effects: Biochemical - Hepatic microsomal mixed oxidase(dealkylation, hyroxylation,etc). Monkey Dose: 17500 mg/kg/10W-C; Toxic Effects: Behavioral - Convulsions or effect on seizure threshold; Nutritional and gross metabolic - Other changes; DEATH. Monkey Dose: 1 gm/kg/14W-I; Toxic Effects: Brain and coverings - Other degenerative changes; Biochemical - Lipids including transport.

Chronic (Multiple Dose) Dermal: Rabbit Route: Subcutaneous; Dose: 20 mg/kg/20D-I; Toxic Effects: Lungs, Thorax, or Respiration - Changes in lung weight; Liver - Changes in Liver weight; Endocrine - Changes in spleen weight.

Reproductive/Teratogenic: Rat Route: Oral; Dose: 112 mg/kg; Duration: male 56D prior to mating; Paternal Effects - Spermatogenesis; Testes, epididymis, sperm duct. Rat Route: Oral; Dose: 100 mg/kg; Duration: male 1D prior to mating; Effects on Fertility - Pre-implantation mortality. Rat Route: Oral; Dose: 430 mg/kg; Duration: female 1-22D of pregnancy; Effects on Newborn - Growth statistics. Rat Route: Oral; Dose: 1890 mg/kg; Duration: female 36W prior to mating Effects on Fertility - Female fertility index. Rat Route: Oral; Dose: 250 mg/kg; Duration: female 15-19D of pregnancy Specific Developmental Abnormalities - Urogenital system. Rat Route: Oral; Dose: 50 mg/kg; Duration: male 1D prior to mating; Effects on Fertility - Other measures of fertility.

Mutagenic: Human DNA Inhibition; Cell Type: lymphocyte; Dose: 500 mg/L. Human Cytogenetic Analysis; Cell Type: leukocyte; Dose: 40 mg/L. Human Cytogenetic Analysis; Cell Type: lymphocyte; Dose: 200 ug/L/72H. Human Other Mutation Test Systems; Cell Type: fibroblast; Dose: 100 mg/kg.

Tumorigenic: Rat Route: Oral; Dose: 1225 mg/kg/7W-C; Toxic Effects: Tumorigenic - Carcinogenic by RTECS criteria; Liver - Tumors. Rat Route: Oral; Dose: 12096 mg/kg/3Y-C; Toxic Effects: Tumorigenic - Neoplastic by RTECS criteria; Liver - Tumors. Rat Route: Oral; Dose: 8100 mg/kg/2Y-C; Toxic Effects: Tumorigenic - Equivocal tumorigenic agent by RTECS criteria; Liver - Tumors; Kidney, Ureter, and Bladder - Chgs in tubules (inc acute renal failure, acute tubular necrosis. Rat Route: Oral; Dose: 19 gm/kg/2Y-C; Toxic Effects: Tumorigenic - Neoplastic by RTECS criteria; Liver - Tumors; Blood - Lymphomax including Hodgkin's disease. Rat Route: Oral; Dose: 438 mg/kg/2Y-C; Toxic Effects: Tumorigenic - Neoplastic by RTECS criteria; Liver - Tumors; Blood - Lymphomax including Hodgkin's disease. Rat Route: Oral; Dose: 17976 mg/kg/2Y-C; Toxic Effects: Tumorigenic - Neoplastic by RTECS criteria; Liver - Tumors. Rat Route: Oral; Dose: 24192 mg/kg/3Y-C; Toxic Effects: Tumorigenic - Neoplastic by RTECS criteria; Liver - Tumors.

Hazard Overviews

Fire Diamond

Health: Irritating to eyes/respiratory tract. If ingested, especially in large amounts, central nervous system effects will occur with possible liver damage. Confirmed animal carcinogen and a suspected human carcinogen.

Fire: Noncombustible but is dissolved in a variety of solvents which may be combustible. Use dry chemical, water spray, fog, or regular foam.

Reactivity: Stable. Hazardous polymerization cannot occur. Avoid: exposure to heat and ignition sources. Incompatible with: strong oxidizers; alkaline materials; iron and aluminum salts. Hazardous decomposition products: carbon dioxide.

Carcinogenicity: IARC - Group 2B, Possibly carcinogenic to humans; NIOSH - Listed as carcinogen; NTP - Class 2B, Reasonably anticipated to be a carcinogen, sufficient evidence of carcinogenicity from studies in experimental animals; ACGIH - Class A3, Animal carcinogen; OSHA - Not listed; EPA - Class B2, Probable human carcinogen based on animal studies; MAK - Not listed

Primary Target Organs:

Eyes

Skin

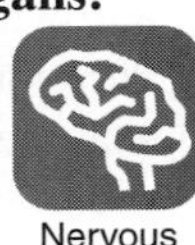
Nervous System

Liver

Exposure Limits

OSHA PEL: TWA: 1 mg/m^3; skin.

ACGIH TLV: TWA: 1 mg/m^3.

NIOSH REL: TWA: 0.5 mg/m^3.

NIOSH IDLH: 500 mg/m^3.

DFG MAK: TWA: 1 mg/m^3.

Respirator Recommendation

Exposure Range: >1 to 50 mg/m^3 Supplied Air, Constant Flow/Pressure Demand, Half Mask

Exposure Range: >50 to 1000 mg/m^3 Supplied Air, Constant Flow/Pressure Demand, Full Face

Exposure Range: >1000 to unlimited mg/m^3 Self-contained Breathing Apparatus, Pressure Demand, Full Face

Note: poor warning properties

Environmental

Ecotoxicity: LC_{50} Micropterus salmoides (Largemouth bass) 1.5 ug/l/96 hr, wt 0.8 g, at 18 °C (95% confidence limit 0.9-2.4 ug/l), static bioassay LD_{50} Callipepla californica (California quail) male 6 months old oral 595 mg/kg (95% confidence limit 430-825 mg/kg) LC_{50} Phasianus colchicus (Pheasant) oral 311 ppm, 5-day diet ad libitum, 2 week old (95% confidence limit 256-374 ppm) LC_{50} Isoperla sp (Stoneflies) 1.2 ug/l/96 hr, juvenile, at 15 °C (95% confidence limit 0.3-4.9 ug/l), static bioassay LD_{50} Grus canadensis (sandhill crane) male and female adult oral >1200 mg/kg /Purity >99% (set point, 105.5 °C) LC_{50} Bufo woodhousei fowleri (toad) tadpole 4-5 wk old 1000 ug/l/96 hr; tadpole 6 wk old 100 ug/l/96 hr; tadpole 7 wk old 30 ug/l/96 hr /From table/ /Conditions of bioassay not specified LC_{50} Pseudacris triseriata (frog) tadpole 400 ug/l/96 hr /From table/ /Conditions of bioassay not specified LC_{50} Leiostomus xanthurus (spot) 1.8 ug/l/2 days /From table/ /Conditions of bioassay not specified LC_{50} Gammarus lacustris (Scuds) 1.0 ug/l/96 hr, mature, at 21 °C (95% confidence limit 0.7-1.5 ug/l), static bioassay LC_{50} Monetes kadiakensis (Glass shrimp)2.3 ug/l/96 hr, mature, at 21 °C (95% confidence limit 1.3-4.9 ug/l), static bioassay LC_{50} Oncorhynchus kisutch (Coho salmon) 4.0 ug/l/96 hr, wt 1.0 g, at 13 °C (95% confidence limit 3.0-6.0 ug/l), static bioassay LC_{50} Orconectes nais (Crayfish) 0.18 ug/l/96 hr, juvenile, at 21 °C in well water (95% confidence limit 0.12-0.30 ug/l), static bioassay LC_{50} Gammarus fasciatus (scud) 3.6 ug/l/48 hr /From table/ /Conditions of bioassay not specified LC_{50} Mugil cephalus (striped mullet) 0.4 ug/l/2 days /From table/ /Conditions of bioassay not specified

Environmental Fate: If released to the terrestrial compartment, it will adsorb very strongly to soil and be subject to evaporation and photodegradation at the surface of soils. It will not leach appreciably to groundwater or hydrolyze but may be subject to biodegradation in flooded soils or under anaerobic conditions. If released to water it will adsorb very strongly to sediments and be subject to evaporation and photooxidation near the surface. It will not hydrolyze and will not significantly biodegrade in most waters. Biodegradation may be significant in sediments. If released to the air it will be subject to direct photodegradation and reaction with photochemically produced hydroxyl radicals. Wet and dry deposition will be major removal mechanisms from the atmospheric compartment.

Cleanup/Disposal: Guide No. 151: Do not touch damaged containers or spilled material unless wearing appropriate protective clothing. Stop leak if you can do it without risk. Prevent entry into waterways, sewers, basements or confined areas. Cover with plastic sheet to prevent spreading. Absorb or cover with dry earth, sand or other non-combustible material and transfer to containers. Do not get water inside containers.

Environmental Physical Data

Henry's Law Constant: 3.8 x10^{-3}

Octanol/Water Partition Coefficient: log K_{ow} = 6.36

Sorption Partition Coefficient: K_{oc} = experimental 2.38 x10^5

BCF: fish 600

BOD: not pertinent

Regulations

RCRA 40CFR: Listed Hazardous Waste No. U061 Toxic Waste

CERCLA: 40CFR 302.4: Listed per CWA Section 311(b)(4) per RCRA Section 3001 per CWA Section 307(a) RQ: 1 lb (0.454 kg)

SARA 40CFR 372.65: Listed as Compound
SARA EHS 40CFR 355: Not listed
TSCA: Listed

Analytical Methods

Air: EPA TO-10, TO-4, 016
Soil: CLP LC_PEST, MC_PEST, OHC; EPA PMD-TLC, 16, 3, 024, 025, P-002-1, P-011-1; SW846 3630B, 3640A, 4042, 8080A, 8081, 8081A, 8250A, 8270B, 8270C; USGS O5104, O7104
Water / Groundwater: EPA P-003-1, P-004-1, 608, 617, 625, 625-S, 680, 022; APHA 6410-B, 6630-B, 6630-C, 6630-D; ASTM D3086, D4763; USGS O3104
Drinking Water: EPA 508, 508.1, 525.2; AOAC 990.06
Food: FDA 212.1, 212.2, 232.1, 232.4, 242.1; EPA 4; AOAC 960.13, 970.52, 972.05, 985.22, 991.04; USGS O9104
Indoor / Expired Air: EPA IP-8; ASTM D4861
Plasma: EPA 001, 003, 004, 027, 029, 29; FDA 211.1, 231.1, 251.1, 251.2, 252, 253
Other: EPA P-009-1, 1656

DIC4250 CAS #: 107-06-2

1,2-DICHLOROETHANE

RTECS: KI0525000
DOT: UN1184; IMO3.2
EINECS Number: 203-458-1
Molecular Formula: $C_2H_4Cl_2$
Structured MF: $ClCH_2CH_2Cl$
Formula Weight: 98.96

Chemical Structure

Synonyms: AETHYLENCHLORID; 1,2-BICHLOROETHANE; BICHLORURE D'ETHYLENE; BORER SOL; BROCIDE; CHLORURE D'ETHYLENE; CLORURO DI ETHENE; 1,2-DCE; DESTRUXOL BORER-SOL; 1,2-DICHLOORETHAAN; 1,2-DICHLOR-AETHAN; DICHLOREMULSION; 1,2-DICHLORETHANE; DI-CHLOR-MULSION; DICHLOR-MULSION; ALPHA,BETA-DICHLOROETHANE; BETA-DICHLOROETHANE; DICHLORO-1,2-ETHANE; SYM-DICHLOROETHANE; DICHLOROETHYLENE; 1,2-DICLOROETANO; DUTCH LIQUID; DUTCH OIL; EDC; ENT 1,656; ETHANE DICHLORIDE; ETHANE,1,2-DICHLORO-; ETHYLEENDICHLORIDE; ETHYLENE CHLORIDE; 1,2-ETHYLENE DICHLORIDE; ETHYLENE DICHLORIDE; FREON 150; GLYCOL DICHLORIDE; NU-G00511; RY DICHLORO-1,2-ETHANE
Description: colorless liquid; sweet, chloroform-like odor
Use: chemical intermediate in the manufacturing of methyl chloroform, perchloroehtylene and ethylene amines, polyvinyl chloride, sulfide compounds, vinyl chloride, acetyl cellulose, trichloroethylene, vinylidene chloride and trichloroethane; as an antiknock additive in leaded fuels, in coatings, pharmaceutical products, color film, pesticides, in extraction of oil from seeds, in the processing of animal fats cleaning agent for textiles, formulation ingredient in grain fumigants, in the extraction of copper from copper ores, solvent for fats, oils, waxes, gums, resins, particularly rubber, tobacco extract, fumigant, paint, varnish and finish removers, metal degreasing, soaps, scouring compounds, wetting and penetrating agents, organic synthesis, ore flotation, fumigant, ingredient in fingernail polish, used in extracting spices, solvent in printing inks, adhesives, asphalt, bitumen, rubber, cellulose, acetate, cellulose ester, degreaser in engineering, textile, petroleum, extracting agent for soybean oil and caffeine, dry cleaning agent, photography, xerography, water softening an cosmetics

Physical Properties

Boiling Point: 83.7 °C (183 °F)
Freezing Point: -35.3 °C (-31.54 °F)
Specific Gravity: 1.2351 at 20 °C
Vapor Density: 3.4 Air=1
Saturated Vapor Density: 1.428544465 kg/m^3
Density: 1.253 g/mL at 20 °C
Vapor Pressure: 60 mm Hg at 20 °C
Water Solubility: 0.869 g/100 ml Water at 20 °C
Other Solubilities: 95% Ethanol: >=100 mg/ml at 19 °C; Acetone: >=100 mg/ml at 19 °C; Alcohol: Soluble; Benzene: Soluble; Chloroform: Soluble; DMSO: >=100 mg/ml at 19 °C; Ether: Soluble.
Surface Tension: Estimated at 30 dynes/cm
Odor Threshold: 24 to 440 mg/m^3
Refraction Index: 1.4448 at 20 °C/D
Critical Temperature: 290 °C
Critical Pressure: 52.90 atm
Ionization Potential (eV): 11.05
Flash Point: 13 °C Closed Cup
Autoignition Temperature: 440 °C
LEL: 6.2% v/v
UEL: 16% v/v

RTECS Toxicity Data

Acute Oral: Human TD_{Lo} Dose: 428 mg/kg; Toxic Effects: Behavioral - Somnolence (general depressed activity); Lungs, Thorax, or Respiration - Cough; Gastrointestinal - Nausea or vomiting. Man TD_{Lo} Dose: 892 mg/kg; Toxic Effects: Gastrointestinal - Hypermotility, diarrhea; Gastrointestinal - Nausea or vomiting; Liver - Jaundice, other or unclassified. Human LD_{Lo} Dose: 286 mg/kg; Toxic Effects: Gastrointestinal - Ulceration or bleeding from stomach; Gastrointestinal - Nausea or vomiting; Liver - Fatty Liver degeneration. Man LD_{Lo} Dose: 714 mg/kg; Toxic Effects: Behavioral - Somnolence (general depressed activity); Cardiac - Change in rate; Lungs, Thorax, or Respiration - Cyanosis. Rat LD_{50} Dose: 670 mg/kg.
Acute Inhalation: Rat LC_{50} Dose: 1000 ppm/7hr; Toxic Effects: Behavioral - Coma; Lungs, Thorax, or Respiration - Cyanosis; Nutritional and gross metabolic - Body temperature decrease. Monkey LC_{50} Dose: 3000 ppm/7hr. Man TC_{Lo} Dose: 4000 ppm/1hr; Toxic Effects: Peripheral Nerve and sensation - Flaccid paralysis without anesthesia; Behavioral - Coma; Gastrointestinal - Nausea or vomiting.
Acute Dermal: Rabbit LD_{50} Route: Skin; Dose: 2800 mg/kg; Toxic Effects: Sense organs and special senses - Lacrimation;

Behavioral - General anesthetic; Behavioral - Ataxia. Rabbit LD_{Lo} Route: Subcutaneous Dose: 1200 mg/kg.

Chronic (Multiple Dose) Oral: Rat Dose: 21840 mg/kg/13W-C; Toxic Effects: Liver - Changes in liver weight; Kidney, Ureter, and Bladder - Changes in kidney weight. Rat Dose: 1 gm/kg/10-I; Toxic Effects: Blood - Changes in serum composition; Liver - Changes in liver weight; DEATH.

Chronic (Multiple Dose) Inhalation: Rat Dose: 200 ppm/7H/6W-I; Toxic Effects: DEATH. Rat Dose: 1500 ppm/7H/5D-I; Toxic Effects: Lungs, Thorax, or Respiration - Respiratory depression; Kidney, Ureter, and Bladder - Chgs in tubules (inc acute renal failure, acute tubular necrosis; DEATH. Monkey Dose: 1000 ppm/7H/9W-I; Toxic Effects: Behavioral - Coma; DEATH.

Irritation Eye: Rabbit Standard Draize Test Dose: 500 mg/24H; Reaction: mild. Rabbit Standard Draize Test Dose: 63 mg; Reaction: severe.

Irritation Skin: Rabbit Standard Draize Test Dose: 500 mg/24H; Reaction: mild. Rabbit Open Draize Test Dose: 625 mg open; Reaction: mild.

Reproductive/Teratogenic: Rat Route: Inhalation; Dose: 300 ppm/7H; Duration: female 6-15D of pregnancy; Effects on Fertility - Post-implantation mortality. Rat Route: Inhalation; Dose: 208 mg/m^3/6H Duration: female 2W prior to mating; Effects on Fertility - Pre-implantation mortality.

Mutagenic: Human DNA Inhibition; Cell Type: lymphocyte; Dose: 5 mL/L. Human Mutations in Mammalian Somatic Cells; Cell Type: lymphocyte; Dose: 100 mg/L.

Tumorigenic: Rat Route: Oral; Dose: 5286 mg/kg/69W-I; Toxic Effects: Tumorigenic - Carcinogenic by RTECS criteria; Gastrointestinal - Tumors; Skin and appendages - Tumors. Rat Route: Inhalation; Dose: 5 ppm/7H/78W-I; Toxic Effects: Tumorigenic - Equivocal tumorigenic agent by RTECS criteria; Blood - Leukemia; Skin and appendages - Tumors. Rat Route: Oral; Dose: 38 gm/kg/78W-I; Toxic Effects: Tumorigenic - Carcinogenic by RTECS criteria; Vascular - Tumors; Gastrointestinal - Tumors. Rat Route: Oral; Dose: 18 gm/kg/78W-I; Toxic Effects: Tumorigenic - Carcinogenic by RTECS criteria; Vascular - Tumors; Gastrointestinal - Tumors.

Hazard Overviews

Flammable

Fire Diamond

Health: Irritating to eyes/skin/respiratory tract. Also Causes: skin defatting, burning; serious injury to eyes. Chronic Effects: injuries to liver, kidneys; weight loss, low blood pressure, jaundice, oliguria, anemia. Suspect cancer hazard.

Fire: Flammable. Use chemical, carbon dioxide, alcohol foam, water spray/fog, or dry sand to fight fires. Use a smothering effect to extinguish fires involving this material. Direct water sprays may be used to cool fire-exposed containers. Fight fire from maximum distance.

Reactivity: Stable. Hazardous polymerization cannot occur. Avoid: heat; ignition sources. Incompatible with: oxidizing agents; liquid ammonia; dimethylaminopropylamine; aluminum; magnesium metal; ethylene dichloride. Hazardous decomposition products: vinyl chloride; chloride fumes; phosgene.

Carcinogenicity: IARC - Group 2B, Possibly carcinogenic to humans; NIOSH - Listed as carcinogen; NTP - Class 2B, Reasonably anticipated to be a carcinogen, sufficient evidence of carcinogenicity from studies in experimental animals; ACGIH - Class A4, Not classifiable as a human carcinogen; OSHA - Not listed; EPA - Class B2, Probable human carcinogen based on animal studies; MAK - Class A2, Unmistakably carcinogenic in animal experimentation only

Primary Target Organs:

Eyes

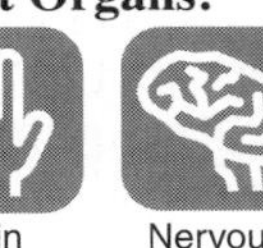
Skin
Nervous System

Liver

Kidneys

Cardio-vascular

Exposure Limits

OSHA PEL: TWA: 50 ppm; STEL: 100 ppm; from Table Z-2. Other Values: 200 mg/m^3; 5 min peak 3hr ppm.

OSHA PEL Vacated 1989 Limits: TWA: 1 ppm; 4 mg/m^3; STEL: 2 ppm; 8 mg/m^3.

NIOSH REL: TWA: 1 ppm; 4 mg/m^3. STEL: 2 ppm; 8 mg/m^3.

NIOSH IDLH: 50 ppm.

Respirator Recommendation

Exposure Range: >10 to <50 ppm Supplied Air, Constant Flow/Pressure Demand, Half Mask

Exposure Range: 50 to unlimited ppm Self-contained Breathing Apparatus, Pressure Demand, Full Face

Note: poor warning properties

Environmental

Ecotoxicity: LC_{50} Mysid shrimp 113,000 ug/l/96 hr in salt water. /Conditions of bioassay not specified LC_{50} Gobius minutus (gobi) 185 mg/l/60 min, 3 hr & up to 96 hr in sea water at 15 °C. /Conditions of bioassay not specified LC_{50} Cyprinodon variegatus (Sheepshead minnows) > 130 ppm but < 230 ppm at 24 hr, 48 hr, 72 hr & 96 hr, static tests, temp 25-31 °C LC_{50} Poecilia reticulata (guppy) 106 ppm/7 days. /Conditions of bioassay not specified LC_{50} Pteronarcys (Stonefly) greater than 100 mg/l/96 hr at 15 °C, second year class, static bioassay LC_{50} Salmo gairdneri (Rainbow trout) 225 mg/l/96 hr at 13 °C, wt 1.8 g, static bioassay Toxicity threshold (cell multiplication inhibition test): bacteria (Pseudomonas putida): 135 mg/l. Algae (Microcystis aeruginosa): 105 mg/l. Green algae (Scenedesmus quadricuda): 719 mg/l. Protozoa (Entosiphon sulcatum): 1127 mg/l

Environmental Fate: Once in the atmosphere, it may be transported long distances and is primarily lost by photooxidation (half-life approx. 1 month). Releases to water will primarily be removed by evaporation (half-life several hours to 10 days). Releases on land will dissipate by volatilization to air and by percolation into groundwater where it is likely to persist for a very long time.

Cleanup/Disposal: Guide No. 129: Eliminate all ignition sources (no smoking, flares, sparks or flames in immediate area). All equipment used when handling the product must be grounded. Do not touch or walk through spilled material. Stop leak if you can do it without risk. Prevent entry into waterways, sewers, basements or confined areas. A vapor suppressing foam may be used to reduce vapors. Absorb or cover with dry earth, sand or other non-combustible material and transfer to containers. Use clean non-sparking tools to collect absorbed material. Large Spills: Dike far ahead of liquid spill for later disposal. Water spray may reduce vapor; but may not prevent ignition in closed spaces.

Environmental Physical Data

Henry's Law Constant: 1.10 $x10^{-3}$
Octanol/Water Partition Coefficient: log K_{ow} = 1.48
Sorption Partition Coefficient: K_{oc} = 33
BCF: bluegill 0.30
BOD: 0.002 lb/lb, 5 days

Regulations

RCRA 40CFR: Listed Hazardous Waste No. U077 Toxic Waste
CERCLA: 40CFR 302.4: Listed per CWA Section 311(b)(4) per RCRA Section 3001 per CWA Section 307(a) RQ: 100 lb (45.35 kg)
SARA 40CFR 372.65: Listed
SARA EHS 40CFR 355: Not listed
TSCA: Listed

Analytical Methods

Air: EPA 0031, OA-002-1, VA-001-1, VA-005-1, VG-006-1, VG-007-1, VG-011-1, TO-1, TO-14, TO-2, TO-3, CTM-011; ASTM D3686, D3687, D4490
Soil: CLP LC_VOA, MC_VOA, OHC; EPA 7, 1624, VG-001-1, VG-008-1, VS-001-1, VS-002-1; SW846 1311, 5021, 5032, 5041, 5041A, 8010B, 8021A, 8240B, 8260A, 8260B; DOE OP040R
Water / Groundwater: EPA 601, 624, 624-S, VW-001-1, VW-002-1, VW-003-1, VW-008-1; APHA 6210-B, 6210-D, 6230-B, 6230-D; ASTM D3695; USGS O3115
Drinking Water: EPA 502.1, 502.2, 524.1, 524.2; APHA 6210-C, 6230-C
Food: EPA 5
Indoor / Expired Air: NIOSH 1003; EPA IP-1A, IP-1A-B, IP-1A-C, IP-1B
Plasma: EPA 29
Other: EPA VS-006-1, VW-012-1, VW-013-1

DIC4300 CAS #: 1300-21-6

DICHLOROETHANE

RTECS: KH9800000
EINECS Number: 215-077-8
Molecular Formula: $C_2H_4Cl_2$
Formula Weight: 98.96
Use: in the production of vinyl chloride monomer 1,2-dichloroethane; an intermediate in the production of 1,1,1-trichloroethane

RTECS Toxicity Data

Acute Oral: Rat LD_{50} Dose: 1120 mg/kg. Mouse LD_{50} Dose: 625 mg/kg.
Acute Inhalation: Mammal LC_{50} Dose: 5 gm/m^3. Mouse LC_{Lo} Dose: 10 gm/m^3.
Acute Dermal: Rabbit LD_{50} Route: Skin; Dose: 3890 mg/kg.
Reproductive/Teratogenic: Rat Route: Inhalation; Dose: 15 mg/m^3/4H; Duration: female 16W prior to mating Effects on Fertility - Pre-implantation mortality; Post-implantation mortality; Effects on Fertility - Female fertility index. Rat Route: Inhalation; Dose: 57 mg/m^3/4H; Duration: female 26W prior to mating Effects on Newborn - Stillbirth; Live birth index; Growth statistics. Rat Route: Inhalation; Dose: 15 mg/m^3/4H; Duration: female 96D prior to mating Maternal Effects - Menstrual cycle changes or disorders. Rat Route: Inhalation; Dose: 15 mg/m^3/4H; Duration: female 96D prior to mating Effects on Embryo or Fetus - Fetal death. Rat Route: Inhalation; Dose: 57 mg/m^3/4H; Duration: female 22W prior to mating Effects on Fertility - Female fertility index; Effects on Embryo or Fetus - Fetal death; Effects on Newborn - Growth statistics.

Hazard Overviews

Carcinogenicity: IARC - Not listed; NIOSH - Not listed; NTP - Not listed; ACGIH - Not listed; OSHA - Not listed; EPA - Not listed; MAK - Not listed

Environmental

Regulations

RCRA 40CFR: Not listed
CERCLA: 40CFR 302.4: Listed as Compound per CWA Section 307(a)
SARA 40CFR 372.65: Not listed
SARA EHS 40CFR 355: Not listed
TSCA: Not listed

DIC4350 CAS #: 10140-87-1

1,2-DICHLOROETHANOL, ACETATE

RTECS: KK4200000
EINECS Number: 233-398-1
Molecular Formula: $C_4H_6Cl_2O_2$
Formula Weight: 157.00
Synonyms: ACETIC ACID 1,2-DICHLOROETHYL ESTER; 1,2-DICHLORETHYLESTER KYSELINY OCTOVE; 1,2-DICHLOROETHYL ACETATE; 1,2-DICHLOROETHYLESTER KYSELINY OCTOVE
Description: water-white liquid
Use: organic synthesis

Physical Properties

Boiling Point: 58 °C (136 °F) to 65 °C (149 °F)
Freezing Point: < -32 °C (-25.6 °F)

Specific Gravity: 1.296 at 20 °C
Water Solubility: Immisicible with Water
Other Solubilities: miscible with Alcohol and ethyl Ether
Refraction Index: 1.444 at 20 °C
Flash Point: About 152 °C

RTECS Toxicity Data

Acute Inhalation: Rat LC_{Lo} Dose: 16 ppm/4hr.

Hazard Overviews

Fire: Will burn.
Carcinogenicity: IARC - Not listed; NIOSH - Not listed; NTP - Not listed; ACGIH - Not listed; OSHA - Not listed; EPA - Not listed; MAK - Not listed

Environmental

Regulations

RCRA 40CFR: Not listed
CERCLA: 40CFR 302.4: Not listed
SARA 40CFR 372.65: Not listed TPQ: 1000 lb
SARA EHS 40CFR 355: Listed TPQ: 1000 lb
TSCA: Not listed

DIC4400 CAS #: 563-43-9

DICHLOROETHYLALUMINUM

RTECS: BD0705000
DOT: UN1924; UN2220; UN2221; UN3052; IMO4.2
EINECS Number: 209-248-6
Molecular Formula: $C_2H_5AlCl_2$
Structured MF: $C_2H_5AlCl_2$
Formula Weight: 126.95

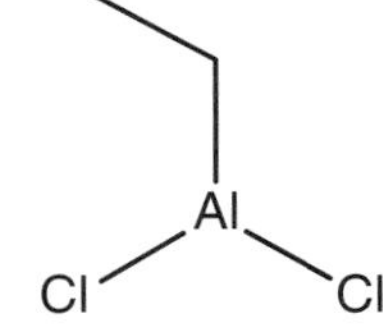

Chemical Structure

Synonyms: ALUMINUM,DICHLOROETHYL-; DICHLOROMONOETHYLALUMINUM; EADC; ETHYL ALUMINUM DICHLORIDE; ETHYLALUMINUM DICHLORIDE; ETHYLDICHLOROALUMINUM
Description: yellow, clear liquid
Use: catalyst for olefin polymerization; aromatic hydrogenation; intermediate

Physical Properties

Boiling Point: 194 °C (381 °F)
Freezing Point: 32 °C (89.6 °F)
Specific Gravity: 1.222
Vapor Density: 4.6 Air=1
Vapor Pressure: 5 mm Hg
Water Solubility: Reacts Violently with Water
Other Solubilities: Soluble in toluene, hexanes
Surface Tension: Estimated at 30 dynes/cm
Flash Point: May ignite spontaneously
Autoignition Temperature: Ignites spontaneously in air

Hazard Overviews

Corrosive

Flammable

Fire Diamond

Health: Corrosive to eyes/skin/respiratory tract. Toxic. Other Acute Effects: harmful if swallowed, inhaled, or absorbed through skin; inhalation may result in spasm, inflammation and edema of the larynx and bronchi, chemical pneumonitis and pulmonary edema; symptoms of exposure may include burning sensation; coughing; wheezing; laryngitis; shortness of breath; headache; nausea; vomiting. Chronic Effects: may cause nervous system disturbances; inhalation studies have demonstrated the development of inflammatory and ulcerous lesions of the penis, prepuce and scrotum in animals; target organs: brain, liver, kidneys, bladder.
Fire: Flammable. Hazards: vapor may travel considerable distance to source of ignition and flash back; reacts with water to liberate flammable and/or explosive gas; emits toxic fumes. Extinguishing agents: carbon dioxide; dry chemical powder; do not use water. Precautions: combustible liquid.
Reactivity: Incompatible with: oxygen, oxidizing agents, alcohols, acids, reacts violently with water, air sensitive. Hazardous decomposition products: toxic fumes of: carbon monoxide, carbon dioxide, hydrogen bromide gas, hydrogen chloride gas; thermal decomposition may produce toxic fumes of phosphorus oxides and/or phosphine.
Carcinogenicity: IARC - Not listed; NIOSH - Not listed; NTP - Not listed; ACGIH - Not listed; OSHA - Not listed; EPA - Not listed; MAK - Not listed
Primary Target Organs:

Eyes

Skin

Respiratory System

Nervous System

Liver

Kidneys

Exposure Limits
OSHA PEL Vacated 1989 Limits: TWA: 2 mg/m^3; as Al soluble.
ACGIH TLV: TWA: 2 mg/m^3; as Al.
NIOSH REL: TWA: 2 mg/m^3; as Al soluble salts, alkyls.

Environmental

Cleanup/Disposal: Guide No. 135: Fully encapsulating, vapor protective clothing should be worn for spills and leak with no fire. Eliminate all ignition sources (no smoking, flares, sparks or flames in immediate area). Do not touch or walk through spilled material. Stop leak if you can do it without risk. Small Spills: Cover with dry earth, dry sand, or other non-combustible material followed with plastic sheet to minimize spreading or contact with rain. Use clean non-sparking tools to collect material and place it into loosely covered plastic containers for later disposal. Prevent entry into waterways, sewers, basements or confined areas.

Environmental Physical Data

BOD: none

Regulations

RCRA 40CFR: Not listed
CERCLA: 40CFR 302.4: Not listed
SARA 40CFR 372.65: Not listed
SARA EHS 40CFR 355: Not listed
TSCA: Listed

DIC4450 CAS #: 75-35-4

1,1-DICHLOROETHYLENE

RTECS: KV9275000
DOT: UN1303; IMO3.1
EINECS Number: 200-864-0
Molecular Formula: $C_2H_2Cl_2$
Structured MF: $H_2C{=}CCl_2$
Formula Weight: 96.94

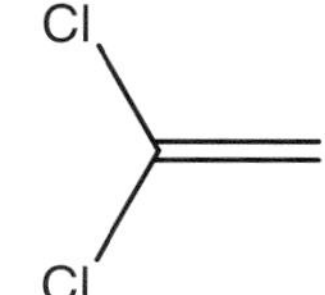

Chemical Structure

Synonyms: ASYM-DICHLOROETHYLENE; CHLORURE DE VINYLIDENE; 1,1-DCE; 1,1-DICHLOROETHENE; 1,1-DICHLOROETHENE (9CI); AS-DICHLOROETHYLENE; ETHENE,1,1-DICHLORO-; ETHYLENE,1,1-DICHLORO-; SCONATEX; VDC; VINYLIDENE CHLORIDE; VINYLIDENE CHLORIDE (II); VINYLIDENE CHLORIDE (INHIBITED); VINYLIDENE CHLORIDE MONOMER; VINYLIDENE CHLORIDE(II); VINYLIDENE CHLORIDE,INHIBITED; VINYLIDENE CHLORIDE,MONOMER; VINYLIDENE DICHLORIDE; VINYLIDINE CHLORIDE
Description: colorless liquid; sweet odor that resembles chloroform
Use: as an intermediate in the production of plastics, as a comonomer with vinyl chloride, acrylonitrile, acrylates, etc., to form various kinds of Saran, in adhesives, in lacquer resins, for concrete and mortar strengthening, in latexes used as barrier coatings and in reinforced polyesters, printing inks and composites for use in furniture and marble

Physical Properties

Boiling Point: 31.7 °C (89 °F) at 760 mm Hg
Freezing Point: -122.5 °C (-188.5 °F)
Specific Gravity: 1.2129 at 20 °C/4 °C
Vapor Density: 3.25 Air=1
Saturated Vapor Density: 3.386163702 kg/m^3
Vapor Pressure: 591 mm Hg at 25 °C
Water Solubility: 0.04% by weight
Other Solubilities: 95% Ethanol: >=100 mg/ml at 21 °C; Acetone: >=100 mg/ml at 21 °C; Benzene: Soluble; Chloroform: Soluble; DMSO: >=100 mg/ml at 21 °C; Ether: Soluble; Organic solvents: Soluble.
Surface Tension: 24 dynes/cm at 15 °C
Odor Threshold: 2000 to 5500 mg/m^3
Refraction Index: 1.4249 at 20 °C/D
Evaporation Rate: Half life of dilute solution (1ppm w/w) 22 minutes
Critical Temperature: 220.8 °C
Critical Pressure: 5.21 mPa
Ionization Potential (eV): 10.00
Flash Point: -16.111 °C Open Cup
Autoignition Temperature: 570 °C
LEL: 7.3% v/v
UEL: 16% v/v

RTECS Toxicity Data

Acute Oral: Rat LD_{50} Dose: 200 mg/kg. Mouse LD_{50} Dose: 194 mg/kg.
Acute Inhalation: Rat LC_{50} Dose: 6350 ppm/4hr; Toxic Effects: Behavioral - Coma. Human TC_{Lo} Dose: 25 ppm; Toxic Effects: Behavioral - General anesthetic; Liver - Other changes; Kidney, ureter, and Bladder - Other changes.
Acute Dermal: Rabbit LD_{Lo} Route: Subcutaneous Dose: 3700 mg/kg. Rabbit LD; Route: Skin; Dose: >2426 mg/kg.
Chronic (Multiple Dose) Oral: Rat Dose: 1300 mg/kg/4W-I; Toxic Effects: Liver - Other changes; Biochemical - Other transferases; Biochemical - Amino acids (including renal excretion).
Chronic (Multiple Dose) Inhalation: Rat Dose: 189 mg/m^3/90D-C; Toxic Effects: Liver - Fatty Liver degeneration; Kidney, Ureter, and Bladder - Chgs in tubules (inc acute renal failure, acute tubular necrosis; Biochemical - Phosphatases. Monkey Dose: 189 mg/m^3/90D-C; Toxic Effects: Liver - Fatty Liver degeneration; Nutritional and gross metabolic - Weight loss or decreased weight gain; DEATH.
Reproductive/Teratogenic: Rat Route: Oral; Dose: 200 mg/kg; Duration: female 6-15D of pregnancy; Effects on Fertility - Other measures of fertility; Effects on Embryo or Fetus - Fetotoxicity. Rat Route: Inhalation; Dose: 80 ppm/7H; Duration: female 6-15D of pregnancy; Specific Developmental Abnormalities - Musculoskeletal system. Rat Route: Inhalation; Dose: 55 ppm/6H; Duration: female 55D prior to mating Effects on Fertility - Female fertility index.
Mutagenic: Rat DNA Damage; Route: Inhalation; Dose: 10 ppm. Mouse DNA Damage; Route: Inhalation; Dose: 50 ppm.
Tumorigenic: Rat Route: Inhalation; Dose: 55 ppm/6H/52W-I; Toxic Effects: Tumorigenic - Equivocal tumorigenic agent by RTECS criteria; Liver - Tumors; Blood - Lymphomax including Hodgkin's disease. Rat Route: Inhalation; Dose: 150 ppm/4H/52W-I; Toxic Effects: Tumorigenic - Equivocal tumorigenic agent by RTECS criteria; Blood - Leukemia; Skin and appendages - Tumors. Rat Route: Inhalation; Dose: 55 ppm/6H/28W-I; Toxic Effects: Tumorigenic - Equivocal tumorigenic agent by RTECS criteria; Liver - Tumors; Skin and appendages - Tumors.

Hazard Overviews

Flammable

Fire Diamond

Health: Irritating to eyes/skin/respiratory tract. Also Causes: narcosis, drunkenness, unconsciousness, conjunctivitis, transient corneal injury, iritis. Chronic Effects: hepatic and renal dysfunction.

Fire: Flammable. Can form explosive mixtures in the air. Combustion by-products include hydrogen chloride and phosgene. Use dry chemical, alcohol foam, or carbon dioxide. Use water to cool fire-exposed containers.

Reactivity: Unstable, self-reactive and rapidly absorbs oxygen to form a violently explosive peroxide. Hazardous polymerization can occur if exposed to sunlight, air, copper, aluminum, or heat.. Incompatible with: chlorosulfonic acid; nitric acid; oleum; oxidizing materials. Hazardous decomposition products: toxic fumes of chlorine and hydrogen chloride.

Carcinogenicity: IARC - Group 3, Not classifiable as to carcinogenicity to humans; NIOSH - Listed as carcinogen; NTP - Not listed; ACGIH - Class A3, Animal carcinogen; OSHA - Not listed; EPA - Class C, Possible human carcinogen; MAK - Class B, Justifiably suspected of having carcinogenic potential

Primary Target Organs:

Eyes

Skin

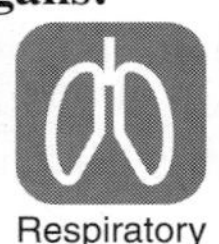
Respiratory System

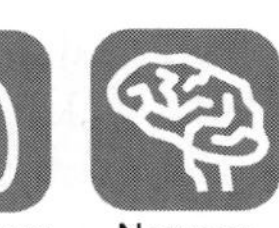
Nervous System

Liver

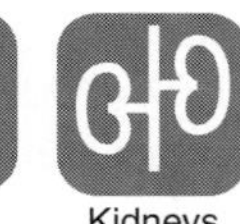
Kidneys

Exposure Limits

OSHA PEL Vacated 1989 Limits: TWA: 1 ppm; 4 mg/m^3.

ACGIH TLV: TWA: 5 ppm; 20 mg/m^3; STEL: 20 ppm; 79 mg/m^3.

DFG MAK: TWA: 2 ppm; 8 mg/m^3.

Respirator Recommendation

Exposure Range: >5 to 250 ppm Supplied Air, Constant Flow/Pressure Demand, Half Mask

Exposure Range: >250 to 5000 ppm Supplied Air, Constant Flow/Pressure Demand, Full Face

Exposure Range: >5000 to unlimited ppm Self-contained Breathing Apparatus, Pressure Demand, Full Face

Note: poor warning properties

Environmental

Ecotoxicity: LC_{50} Cyprinodon variegatus (sheepshead minnow) 249 mg/l/24 hr, 48 hr, 72 hr, 96 hr in a static bioassay using sea water EC_{50} Skeletonema costatum (alga) > 712,000 ug/l/96 hr, Toxic effects: Inhibition chlorophyll synthesis; reduced cell counts. /Conditions of bioassay not specified LC_{50} Lepomis macrochirus (bluegill) 74 mg/l at 24 hr & 96 hr, temp at 21-23 °C, water hardness 32-48 mg/l (calcium carbonate), pH 6.7-7.8, dissolved oxygen concentration 7.0-8.8 mg/l (static bioassay) LC_{50} Mysidopsis bahia (mysid shrimp) > 798 mg/l/24 hr, 48 hr, 72 hr; 224 mg/l/96 hr in a static bioassay using seawater LC_{50} Menidia beryllina (inland silverside) 250 ppm/96 hr in a static bioassay in synthetic seawater at 23 °C with mild aeration

Environmental Fate: Once in the atmosphere it will degrade rapidly by photooxidation with a half-life of 11 hours in relatively clean air or under 2 hours in polluted air. If spilled on land, part will evaporate and part will leach into the groundwater where its fate is unknown, but degradation is expected to be slow based upon microcosm studies. It would not be expected to bioconcentrate into fish.

Cleanup/Disposal: Guide No. 129: Eliminate all ignition sources (no smoking, flares, sparks or flames in immediate area). All equipment used when handling the product must be grounded. Do not touch or walk through spilled material. Stop leak if you can do it without risk. Prevent entry into waterways, sewers, basements or confined areas. A vapor suppressing foam may be used to reduce vapors. Absorb or cover with dry earth, sand or other non-combustible material and transfer to containers. Use clean non-sparking tools to collect absorbed material. Large Spills: Dike far ahead of liquid spill for later disposal. Water spray may reduce vapor; but may not prevent ignition in closed spaces.

Environmental Physical Data

Henry's Law Constant: 2.61 $x10^{-2}$

Octanol/Water Partition Coefficient: log K_{ow} = 1.48

Sorption Partition Coefficient: K_{oc} = estimated at 150

BCF: not significant

Regulations

RCRA 40CFR: Listed Hazardous Waste No. U078 Toxic Waste

CERCLA: 40CFR 302.4: Listed per CWA Section 311(b)(4) per RCRA Section 3001 per CWA Section 307(a) RQ: 100 lb (45.35 kg)

SARA 40CFR 372.65: Listed

SARA EHS 40CFR 355: Not listed

TSCA: Listed

Analytical Methods

Air: EPA VA-001-1, VA-003-1, VA-005-1, VG-006-1, VG-007-1, VG-011-1, TO-14, TO-2, TO-3, 0031, 0040; ASTM D4490

Soil: CLP LC_VOA, MC_VOA, OHC; EPA 7, 1624, VG-008-1, VS-001-1, VS-002-1, VS-003-1; SW846 1311, 5021, 5032, 5041, 5041A, 8010B, 8021A, 8240B, 8260A, 8260B; DOE OP040R

Water / Groundwater: EPA 601, 624, 624-S, VW-001-1, VW-002-1, VW-003-1, VW-004-1, VW-005-1, VW-008-1, VW-014-1; APHA 6210-B, 6210-D, 6230-B, 6230-D; USGS O3115

Drinking Water: EPA 502.1, 502.2, 524.1, 524.2; APHA 6210-C, 6230-C

Food: EPA 5

Indoor / Expired Air: NIOSH 1015; EPA IP-1A, IP-1B

Plasma: EPA 29

Other: EPA VS-006-1, VW-012-1, VW-013-1

DIC4500 **CAS #: 540-59-0**

1,2-DICHLOROETHYLENE

RTECS: KV9360000
DOT: UN1150; IMO3.2
EINECS Number: 208-750-2
Molecular Formula: $C_2H_2Cl_2$
Structured MF: ClCH=CHCl
Formula Weight: 96.95

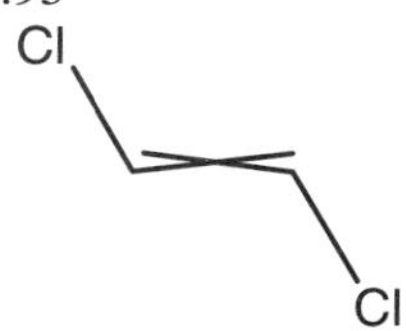

Chemical Structure

Synonyms: ACETYLENE DICHLORIDE; CIS-ACETYLENE DICHLORIDE; TRANS-ACETYLENE DICHLORIDE; 1,2-DCE; 1,2-DICHLOR-AETHEN; 1,2-DICHLOROETHENE; CIS-TRANS-1,2-DICHLOROETHYLENE; DICHLORO-1,2-ETHYLENE; SYM-DICHLOROETHYLENE; TRANS-DICHLOROETHYLENE; DIOFORM; ETHENE,1,2-DICHLORO-; ETHYLENE,1,2-DICHLORO-
Description: colorless liquid; pleasant odor
Use: solvent for fats, phenol, camphor, etc.; used in retarding fermentation, additive to dye and lacquer solutions, perfumes, thermoplastics, organic synthesis, medicine and degreasing agent

Physical Properties

Boiling Point: 47.78 °C (118 °F)
Freezing Point: -50 °C (-58 °F)
Specific Gravity: 1.27 at 25 °C (liquid)
Vapor Density: 3.34 Air=1
Saturated Vapor Density: 1.865934664 kg/m^3
Vapor Pressure: 180 to 265 mm Hg at 20 °C
Water Solubility: 0.4% by weight
Other Solubilities: 95% Ethanol: >=100 mg/ml at 21 °C; Acetone: >=100 mg/ml at 21 °C; Alcohol: Soluble; DMSO: >=100 mg/ml at 21 °C; Ether: Soluble; Most organic solvents: Soluble.
Surface Tension: 24 dynes/cm
Odor Threshold: 0.07 to 18 ppm
Refraction Index: 1.4463
Ionization Potential (eV): 9.65
Flash Point: 2 °C
Autoignition Temperature: 460 °C
LEL: 9.7% v/v
UEL: 12.8% v/v

RTECS Toxicity Data

Acute Oral: Rat LD_{50} Dose: 770 mg/kg.
Acute Inhalation: Frog LC_{Lo} Dose: 117 mg/m^3/1hr; Toxic Effects: Peripheral Nerve and sensation - Flaccid paralysis without anesthesia; Behavioral - Excitment; Lungs, Thorax, or Respiration - Respiratory depression.
Irritation Skin: Rabbit Standard Draize Test Dose: 100 mg/24H; Reaction: moderate.
Mutagenic: Mold - A Nidulans Sex Chromosome Loss; Dose: 750 ppm.

Hazard Overviews

Flammable

Fire Diamond

Health: Irritating to eyes/skin/respiratory tract. Also Causes: narcosis, nausea, tremor, weakness, CNS depression, epigastric cramps.
Fire: Flammable. Use dry chemical, carbon dioxide, halon, water spray, or standard foam. Water may be ineffective unless used to blanket the fire. Vapors may explode when exposed to heat, ignition source, or oxidizers.
Reactivity: Stable. Hazardous polymerization cannot occur. Avoid: addition of hot liquid to cold. Incompatible with: alkalies; nitrogen tetraoxide; difluoromethylene; strong oxidizers; dihy-pofluorite; copper; copper alloys; potassium hydroxide. Hazardous decomposition products: highly toxic fumes of chlorine.
Carcinogenicity: IARC - Not listed; NIOSH - Listed as carcinogen; NTP - Not listed; ACGIH - Not listed; OSHA - Not listed; EPA - Not listed; MAK - Not listed
Primary Target Organs:

Eyes

Skin

Respiratory System

Nervous System

Exposure Limits
OSHA PEL: TWA: 200 ppm; 790 mg/m^3.
ACGIH TLV: TWA: 100 ppm; 793 mg/m^3.
NIOSH REL: TWA: 200 ppm; 790 mg/m^3.
NIOSH IDLH: 1000 ppm.
DFG MAK: TWA: 200 ppm; 790 mg/m^3.
Respirator Recommendation
Exposure Range: >200 to <1000 ppm Air Purifying, Negative Pressure, Half Mask
Exposure Range: 1000 to unlimited ppm Self-contained Breathing Apparatus, Pressure Demand, Full Face
Cartridge Color: black

Environmental

Environmental Fate: If released on soil, it should evaporate readily and leach in soil very slowly. Biodegradation should occur. If released into water, it will be lost mainly through volatilization (half life 3 hr in a model river). Biodegradation, adsorption to sediment, and bioconcentration to aquatic organisms should not be significant. In the atmosphere, cis- and trans- will be lost by reaction with photochemically produced hydroxyl radicals (half lives 8 and 3.6 days, respectively) and scavenged by rain. Because it is relatively long lived in the atmosphere, considerable dispersal from source area should occur.

Cleanup/Disposal: Guide No. 132: Fully encapsulating, vapor protective clothing should be worn for spills and leaks with no fire. Eliminate all ignition sources (no smoking, flares, sparks or flames in immediate area). All equipment used when handling the product must be grounded. Do not touch or walk through spilled material. Stop leak if you can do it without risk. Prevent entry into waterways, sewers, basements or confined areas. A vapor suppressing foam may be used to reduce vapors. Absorb with earth, sand or other non-combustible material and transfer to containers (except for Hydrazine). Use clean non-sparking tools to collect absorbed material. Large Spills: Dike far ahead of liquid spill for later disposal. Water spray may reduce vapor; but may not prevent ignition in closed spaces.

Environmental Physical Data

Henry's Law Constant: 0.00408
Octanol/Water Partition Coefficient: log K_{ow} = calculated at 1.86
Sorption Partition Coefficient: K_{oc} = 36 to 49
BCF: calculated at 15

Regulations

RCRA 40CFR: Not listed
CERCLA: 40CFR 302.4: Not listed
SARA 40CFR 372.65: Listed
SARA EHS 40CFR 355: Not listed
TSCA: Listed

Analytical Methods

Air: EPA VA-002-1, VA-003-1, VA-005-1, VG-006-1; ASTM D3686, D3687, D4490
Soil: EPA VG-002-1, VS-001-1, VS-002-1, VS-005-1; CLP MC_VOA, OHC
Water / Groundwater: EPA 624-S, VW-002-1, VW-004-1, VW-007-1, VW-008-1
Indoor / Expired Air: NIOSH 1003
Other: EPA VS-006-1

DIC4550 **CAS #: 156-59-2**

CIS-1,2-DICHLOROETHYLENE

RTECS: KV9420000
DOT: UN1150; IMO3.2
EINECS Number: 205-859-7
Molecular Formula: $C_2H_2Cl_2$
Structured MF: CHCl=CHCl
Formula Weight: 96.94

Cl
Cl

Chemical Structure

Synonyms: ACETALYNE DICHLORIDE; CIS-1,2-DICHLORETHYLENE; CIS-1,2-DICHLOROETHENE; (Z)-1,2-DICHLOROETHYLENE; 1,2-CIS-DICHLOROETHYLENE; CIS-DICHLOROETHYLENE; ETHENE,1,2-DICHLORO-,(Z)-; ETHYLENE,1,2-DICHLORO-,(Z)-
Description: colorless liquid; sweetish odor
Use: solvent for waxes, resins, fats, phenol, camphor, acetyl cellulose, organic materials and heat-sensitive substances such as caffeine; in rubber manufacture, as a refrigerant, as an additive to dye and lacquer solutions, in retarding fermentation, in organic synthesis, in medicines, in dye extraction, in chlorination reactions and in the manufacture of artificial pearls; a constituent of perfumes and thermoplastics

Physical Properties

Boiling Point: 60.3 °C (141 °F) at 760 mm Hg
Freezing Point: -80.5 °C (-112.9 °F)
Specific Gravity: 1.2837 at 20 °C/4 °C
Vapor Density: 3.34 Air=1
Saturated Vapor Density: 1.939818512 kg/m^3
Density: 1.284 g/cc at 19 °C
Vapor Pressure: 200 mm Hg at 25 °C
Water Solubility: 1 to 5 mg/mL at 16 °C
Other Solubilities: 95% Ethanol: >=100 mg/ml at 17 °C; Acetone: >=100 mg/ml at 17 °C; Alcohol: Soluble; Benzene: Soluble; Chloroform: Soluble; DMSO: >=100 mg/ml at 17 °C; Ether: Soluble; Most organic solvents: Soluble.
Surface Tension: 24 dynes/cm
Odor Threshold: 0.085 ppm
Refraction Index: 1.4490 at 20 °C/D
Critical Temperature: 271 °C
Critical Pressure: 6030 kPa
Ionization Potential (eV): 9.66 +/-0.01
Flash Point: 2.2 to 3.9 °C Closed Cup
Autoignition Temperature: 460 °C
LEL: 9.7% v/v
UEL: 12.8% v/v

RTECS Toxicity Data

Acute Inhalation: Mouse LC_{Lo} Dose: 65 gm/m^3/2hr; Toxic Effects: Behavioral - General anesthetic; Behavioral - Convulsions or effect on seizure threshold; Behavioral - Change in motor activity (specific assay). Cat LC_{Lo} Dose: 20 gm/m^3/6hr; Toxic Effects: Behavioral - General anesthetic; Behavioral - Convulsions or effect on seizure threshold; Nutritional and gross metabolic - Body temperature decrease.
Mutagenic: Rat Unscheduled DNA Synthesis; Cell Type: liver; Dose: 4300 umol/L. Mouse Host-mediated Assay; Indicator Organism: Yeast - S Cerevisiae; Dose: 1300 mg/kg.

Hazard Overviews

Flammable

Health: Irritating to eyes/skin/respiratory tract. Harmful. Other Acute Effects: harmful if swallowed, inhaled, or absorbed through skin; prolonged contact can cause narcotic effect; target organs: central nervous system, liver, kidneys.

Fire: Flammable. Hazards: emits toxic fumes. Extinguishing agents: water spray; carbon dioxide, dry chemical powder or appropriate foam. Precautions: combustible liquid.

Reactivity: Incompatible with: oxidizing agents, bases, may decompose on exposure to air and moisture, may decompose on exposure to light. Hazardous decomposition products: toxic fumes of: hydrogen chloride gas, carbon monoxide, carbon dioxide, phosgene gas.

Carcinogenicity: IARC - Not listed; NIOSH - Not listed; NTP - Not listed; ACGIH - Not listed; OSHA - Not listed; EPA - Class D, Not classifiable as to human carcinogenicity; MAK - Not listed

Primary Target Organs:

Eyes

Skin

Respiratory System

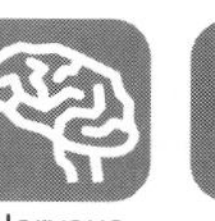
Nervous System

Liver

Kidneys

Exposure Limits

ACGIH TLV: TWA: 200 ppm; 793 mg/m^3.

NIOSH REL: TWA: 200 ppm; 790 mg/m^3.

Environmental

Ecotoxicity: LC_{50} Lepomis machrochirus (bluegill) 135,000 ug/l/96 hr in a static unmeasured bioassay

Environmental Fate: If released on soil, it should evaporate and/or leach into the groundwater where very slow biodegradation should occur. If released into water, it will be lost mainly through volatilization (half life 3 hr in a model river). Biodegradation, adsorption to sediment, and bioconcentration in aquatic organisms should not be significant. In the atmosphere it will be lost by reaction with photochemically produced hydroxyl radicals (half life 8 days) and scavenged by rain. Because it is relatively long lived in the atmosphere, considerable dispersal from source areas should occur.

Cleanup/Disposal: Guide No. 132: Fully encapsulating, vapor protective clothing should be worn for spills and leaks with no fire. Eliminate all ignition sources (no smoking, flares, sparks or flames in immediate area). All equipment used when handling the product must be grounded. Do not touch or walk through spilled material. Stop leak if you can do it without risk. Prevent entry into waterways, sewers, basements or confined areas. A vapor suppressing foam may be used to reduce vapors. Absorb with earth, sand or other non-combustible material and transfer to containers (except for Hydrazine). Use clean non-sparking tools to collect absorbed material. Large Spills: Dike far ahead of liquid spill for later disposal. Water spray may reduce vapor; but may not prevent ignition in closed spaces.

Environmental Physical Data

Henry's Law Constant: estimated at 0.00337

Octanol/Water Partition Coefficient: log K_{ow} = 1.86

Sorption Partition Coefficient: K_{oc} = 49

BCF: calculated at 15

Regulations

RCRA 40CFR: Not listed

CERCLA: 40CFR 302.4: Not listed

SARA 40CFR 372.65: Not listed

SARA EHS 40CFR 355: Not listed

TSCA: Listed

Analytical Methods

Air: EPA VA-006-1, VA-008-1, VG-007-1, VG-011-1, TO-14

Soil: SW846 5021, 8021A, 8260A

Water / Groundwater: APHA 6210-D, 6230-D

Drinking Water: EPA 502.1, 502.2, 524.2; APHA 6210-C, 6230-C

Indoor / Expired Air: EPA IP-1A

Plasma: EPA 29

Other: EPA VW-012-1, VW-013-1

DIC4600 CAS #: 25323-30-2

DICHLOROETHYLENE

RTECS: KV9250000

DOT: UN1150

Molecular Formula: $C_2H_2Cl_2$

Formula Weight: 96.94

Use: for the synthesis of tri- and perchloroethylene

RTECS Toxicity Data

Acute Oral: Mammal LD_{Lo} Dose: 2500 mg/kg; Toxic Effects: Gastrointestinal - Nausea or vomiting.

Acute Inhalation: Mouse LC_{Lo} Dose: 76 gm/m^3/2hr; Toxic Effects: Lungs, Thorax, or Respiration - Dyspnea. Guinea Pig LC_{Lo} Dose: 155 gm/m^3/1hr; Toxic Effects: Behavioral - Convulsions or effect on seizure threshold.

Hazard Overviews

Carcinogenicity: IARC - Not listed; NIOSH - Not listed; NTP - Not listed; ACGIH - Not listed; OSHA - Not listed; EPA - Not listed; MAK - Not listed

Environmental

Cleanup/Disposal: Guide No. 132: Fully encapsulating, vapor protective clothing should be worn for spills and leaks with no fire. Eliminate all ignition sources (no smoking, flares, sparks or flames in immediate area). All equipment used when handling the product must be grounded. Do not touch or walk through spilled material. Stop leak if you can do it without risk. Prevent entry into waterways, sewers, basements or confined areas. A vapor suppressing foam may be used to reduce vapors. Absorb with earth, sand or other non-combustible material and transfer to containers (except for Hydrazine). Use clean non-sparking tools to collect absorbed material. Large Spills: Dike far ahead of liquid spill for later disposal. Water spray may reduce vapor; but may not prevent ignition in closed spaces.

Regulations

RCRA 40CFR: Not listed

CERCLA: 40CFR 302.4: Not listed
SARA 40CFR 372.65: Not listed
SARA EHS 40CFR 355: Not listed
TSCA: Not listed

DIC4650 CAS #: 156-60-5

TRANS-1,2-DICHLOROETHYLENE

RTECS: KV9400000
DOT: UN1150; IMO3.2
EINECS Number: 205-860-2
Molecular Formula: $C_2H_2Cl_2$
Structured MF: ClCH=CHCl
Formula Weight: 96.95

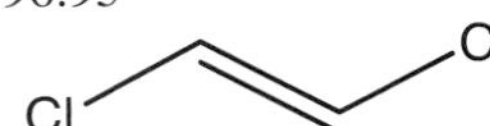

Chemical Structure

Synonyms: TRANS-ACETYLENE DICHLORIDE; 1,2-DICHLORO-,(E)-ETHENE; TRANS-1,2-DICHLOROETHENE; 1,2-DICHLORO-,(E)-ETHYLENE; 1,2-TRANS-DICHLOROETHYLENE; SYM-DICHLOROETHYLENE; TRANS-DICHLOROETHYLENE; ETHYLENE,1,2-DICHLORO-,
Description: colorless, light liquid; sweetish odor
Use: solvent for fats, phenols, camphor, etc.; retards fermentation; rubber manufacturing; refrigerants; additive to dye and lacquer solutions; low temperature solvent for heat sensitive substances; constituent of perfumes, thermoplastics; used in organic synthesis and medicine

Physical Properties

Boiling Point: 48 °C (118 °F) to 48.5 °C (119 °F) at 760 mm Hg
Freezing Point: -50 °C (-58 °F)
Specific Gravity: 1.2565 at 20 °C/4 °C
Vapor Density: 3.34 Air=1
Density: 1.241 g/cc at 22 °C
Vapor Pressure: 200 mm Hg at 14 °C
Water Solubility: < 1 mg/mL at 18 C
Other Solubilities: 95% Ethanol: >=100 mg/ml at 18 °C; Acetone: >=100 mg/ml at 18 °C; Benzene: Soluble; Chloroform: Soluble; DMSO: >=100 mg/ml at 18 °C; Ether: Soluble; Most organic solvents: Soluble.
Surface Tension: 25 $x10^{-3}$ N/m at 20 °C
Odor Threshold: 0.3357 to 1975.00 ppm
Refraction Index: 1.449
Critical Temperature: 244 °C
Critical Pressure: 5510 kPa
Ionization Potential (eV): 9.65 +/-0.02
Flash Point: 6 °C
Autoignition Temperature: 1,2-dichloroethylene 460 °C
LEL: 9.7% v/v
UEL: 12.8% v/v

RTECS Toxicity Data

Acute Oral: Rat LD_{50} Dose: 1235 mg/kg. Mouse LD_{50} Dose: 2122 mg/kg; Toxic Effects: Behavioral - Altered sleep time (including change in righting reflex); Behavioral - Somnolence (general depressed activity); Behavioral - Ataxia.
Acute Inhalation: Mouse LC_{Lo} Dose: 75 gm/m^3/2hr; Toxic Effects: Behavioral - General anesthetic; Behavioral - Convulsions or effect on seizure threshold; Lungs, Thorax, or Respiration - Dyspnea. Human TC_{Lo} Dose: 4800 mg/m^3/10M; Toxic Effects: Behavioral - Sleep; Behavioral - Hallucinations, distorted perceptions.
Acute Dermal: Rabbit LD_{50} Route: Skin; Dose: >5 gm/kg; Toxic Effects: Skin and appendages - Dermatitis, irritative; Nutritional and gross metabolic - Weight loss or decreased weight gain.
Chronic (Multiple Dose) Oral: Rat Dose: 113 gm/kg/90D-C; Toxic Effects: Kidney, Ureter, and Bladder - Changes in kidney weight.
Chronic (Multiple Dose) Inhalation: Rat Dose: 200 ppm/8H/16W-I; Toxic Effects: Lungs, Thorax, or Respiration - Other changes; Liver - Other changes; Endocrine - Changes in endocrine weight (unspecified).
Irritation Eye: Rabbit Standard Draize Test Dose: 10 mg; Reaction: moderate.
Irritation Skin: Rabbit Standard Draize Test Dose: 500 mg/24H; Reaction: moderate.
Reproductive/Teratogenic: Rat Route: Inhalation; Dose: 6000 ppm/6H; Duration: female 7-16D of pregnancy; Effects on Fertility - Post-implantation mortality. Rat Route: Inhalation; Dose: 12000 ppm/6H; Duration: female 7-16D of pregnancy; Effects on Fertility - Post-implantation mortality; Effects on Embryo or Fetus - Fetotoxicity.
Mutagenic: Hamster Sex Chromosome Loss; Cell Type: lung; Dose: 6500 umol/L. Bacteria - S Typhimurium Mutations in Microorganisms; Dose: 50 ug/plate (+S9), 10 ug/plate (-S9).

Hazard Overviews

Flammable

Health: Irritating to eyes/skin/respiratory tract. Harmful. Other Acute Effects: harmful if swallowed, inhaled, or absorbed through skin; prolonged contact can cause narcotic effect; target organs: central nervous system, liver, kidneys.
Fire: Flammable. Hazards: emits toxic fumes. Extinguishing agents: water spray; carbon dioxide, dry chemical powder or appropriate foam. Precautions: combustible liquid.
Reactivity: Incompatible with: oxidizing agents, bases, may decompose on exposure to air and moisture, may decompose on exposure to light. Hazardous decomposition products: toxic fumes of: hydrogen chloride gas, carbon monoxide, carbon dioxide, phosgene gas.
Carcinogenicity: IARC - Not listed; NIOSH - Not listed; NTP - Not listed; ACGIH - Not listed; OSHA - Not listed; EPA - Not listed; MAK - Not listed

Primary Target Organs:

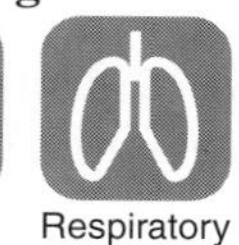

Exposure Limits
ACGIH TLV: TWA: 200 ppm; 793 mg/m^3.
NIOSH REL: TWA: 200 ppm; 790 mg/m^3.

Environmental

Ecotoxicity: LC_{50} Lepomis machrochirus (bluegill) 135,000 ug/l/96 hr in a static unmeasured bioassay
Environmental Fate: If released on soil, it should evaporate and leach into the groundwater where very slow biodegradation should occur. If released into water, it will be lost mainly through volatilization (half-life 3 hr in a model river). Biodegradation, adsorption to sediment, and bioconcentration in aquatic organisms should not be significant. In the atmosphere, it will be lost by reaction with photochemically produced hydroxyl radicals (half-life 3.6 days) and scavenged by rain. Because it is relatively long-lived in the atmosphere, considerable dispersal from source areas should occur.
Cleanup/Disposal: Guide No. 132: Fully encapsulating, vapor protective clothing should be worn for spills and leaks with no fire. Eliminate all ignition sources (no smoking, flares, sparks or flames in immediate area). All equipment used when handling the product must be grounded. Do not touch or walk through spilled material. Stop leak if you can do it without risk. Prevent entry into waterways, sewers, basements or confined areas. A vapor suppressing foam may be used to reduce vapors. Absorb with earth, sand or other non-combustible material and transfer to containers (except for Hydrazine). Use clean non-sparking tools to collect absorbed material. Large Spills: Dike far ahead of liquid spill for later disposal. Water spray may reduce vapor; but may not prevent ignition in closed spaces.

Environmental Physical Data

Henry's Law Constant: estimated at 0.00672
Octanol/Water Partition Coefficient: log K_{ow} = 2.06
Sorption Partition Coefficient: K_{oc} = 36
BCF: calculated at 22

Regulations

RCRA 40CFR: Listed Hazardous Waste No. U079 Toxic Waste
CERCLA: 40CFR 302.4: Listed per RCRA Section 3001 per CWA Section 307(a) RQ: 1000 lb (453.5 kg)
SARA 40CFR 372.65: Not listed
SARA EHS 40CFR 355: Not listed
TSCA: Listed

Analytical Methods

Air: EPA 0031, OA-002-1, VA-001-1, VA-006-1, VA-008-1, VG-007-1, VG-011-1
Soil: CLP LC_VOA; EPA 7, 1624, VG-010-1, VS-003-1, VW-010-1; SW846 5021, 5032, 5041, 5041A, 8010B, 8021A, 8240B, 8260A, 8260B
Water / Groundwater: EPA 601, 624, 624-S, VW-001-1, VW-003-1, VW-005-1, VW-014-1; APHA 6210-B, 6210-D, 6230-B, 6230-D; USGS O3115
Drinking Water: EPA 502.1, 502.2, 524.1, 524.2; APHA 6210-C, 6230-C
Food: EPA 5
Plasma: EPA 29
Other: EPA VW-011-1, VW-012-1, VW-013-1

DIC4750 **CAS #: 1717-00-6**

1,1-DICHLORO-1-FLUOROETHANE

RTECS: KI0997000
Molecular Formula: $C_2H_3Cl_2F$
Formula Weight: 116.95
Synonyms: CFC 141B; CFC-141; FREON-141; HCFC 141B; ISOTRON 141B; R 141B; REFRIGERANT 141B; SOLKANE 141B
Description: colorless liquid; weak ethereal odor
Use: refrigeration & air conditioning

Physical Properties

Boiling Point: 32 °C (90 °F) at 760 mm Hg
Freezing Point: -103.5 °C (-154.3 °F)
Specific Gravity: 1.25000
Saturated Vapor Density: Estimated 3.172889292 kg/m^3
Vapor Pressure: Estimated 412 mm Hg at 25 °C
Water Solubility: 2632 mg/L at 25 °C (est)
Refraction Index: 1.3600 at 10 °C/D
Flash Point: Nonflammable

RTECS Toxicity Data

Acute Oral: Rat LD_{50} Dose: >5 gm/kg.
Acute Inhalation: Rat LC_{50} Dose: 56700 ppm/6hr; Toxic Effects: Behavioral - Tremor; Behavioral - Ataxia. Mouse LC_{50} Dose: 151 gm/m^3/2hr; Toxic Effects: Behavioral - General anesthetic; Behavioral - Ataxia; Lungs, Thorax, or Respiration - Cyanosis.
Acute Dermal: Rabbit LD_{50} Route: Skin; Dose: >2 gm/kg. Rat LD_{50} Route: Skin; Dose: >2 gm/kg; Toxic Effects: Liver - Other changes.
Chronic (Multiple Dose) Inhalation: Rat Dose: 50 gm/m^3/4H/4W-I; Toxic Effects: Endocrine - Evidence of thyroid hypofunction. Rat Dose: 18200 mg/m^3/4H/26W-I; Toxic Effects.
Reproductive/Teratogenic: Rat Route: Inhalation; Dose: 20000 ppm/6H; Duration: female 6-15D of pregnancy; Effects on Fertility - Post-implantation mortality; Litter size; Effects on Embryo or Fetus - Fetotoxicity. Rat Route: Inhalation; Dose: 20000 ppm/6H; Duration: female 6-15D of pregnancy; Effects on Embryo or Fetus - Fetal death.
Mutagenic: Hamster Cytogenetic Analysis; Cell Type: ovary; Dose: 10 pph/4H-C.

Hazard Overviews

Fire: Noncombustible.

Carcinogenicity: IARC - Not listed; NIOSH - Not listed; NTP - Not listed; ACGIH - Not listed; OSHA - Not listed; EPA - Not listed; MAK - Not listed

Exposure Limits

AIHA WEEL: TWA: 500 ppm.

Environmental

Environmental Fate: If released to soil, it will rapidly volatilize from both moist and dry soil to the atmosphere. It will display moderate to high mobility in soil. If released to water, it will rapidly volatilize to the atmosphere. The estimated half-life for volatilization from a model river is 3.2 hours. It will not bioconcentrate in fish and aquatic organisms nor will it adsorb to sediment and suspended organic matter. If released to the atmosphere, it will undergo a slow gas-phase reaction with photochemically produced hydroxyl radicals with an estimated half-life of 1,000 days. It may also undergo atmospheric removal by wet deposition processes; however, any removed is expected to rapidly re-volatilize to the atmosphere.

Environmental Physical Data

Henry's Law Constant: estimated at 0.0241

Octanol/Water Partition Coefficient: log K_{ow} = 2.041

Sorption Partition Coefficient: K_{oc} = calculated at 57 to 307

BCF: calculated at 7 to 21

Regulations

RCRA 40CFR: Not listed

CERCLA: 40CFR 302.4: Listed as Compound per CWA Section 307(a)

SARA 40CFR 372.65: Listed

SARA EHS 40CFR 355: Not listed

TSCA: Listed

DIC4800 CAS #: 75-43-4

DICHLOROFLUOROMETHANE

RTECS: PA8400000

DOT: UN1029; IMO2.2

EINECS Number: 200-869-8

Molecular Formula: $CHCl_2F$

Structured MF: $CHCl_2F$

Formula Weight: 102.92

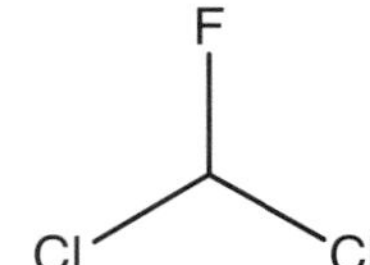

Chemical Structure

Synonyms: ALGOFRENE TYPE 5; ARCTON 7; DICHLOROMONOFLUOROMETHANE; DWUCHLOROFLUOROMETAN; F 21; FC 21; FC-21; FLUORODICHLOROMETHANE; FREON 21; FREON F 21; GENETRON 21; HALON 112; METHANE,DICHLOROFLUORO-; MONOFLUORODICHLOROMETHANE; R 21 (REFRIGERANT); R21; REFRIGERANT 21

Description: colorless gas; slight ethereal odor

Use: blowing agent for cellular polymers; in water, copper, & aluminum purification; in mfr of glass bottles; in regulating devices for leak detection; in thermal expansion valves; & in mfr of materials for electrical applications, insulators & generator windings; refrigerant in aerospace ind; solvent; propellant gas; heat exchange fluid in geothermal energy applications; less usage than other chlorofluorocarbons, eg, trichlorfluoromethane (cfc-11) or dichlorodifluoromethane (cfc-12).

Physical Properties

Boiling Point: 8.9 °C (48 °F) at 760 mm Hg

Freezing Point: -135 °C (-211 °F)

Vapor Density: 3.82 Air=1

Density: 1.405 g/mL at 9 °C

Vapor Pressure: 1360 mm Hg

Water Solubility: 1 g/100 g Water at 20 °C

Other Solubilities: 69 parts/100 parts Acetic Acid; 108 parts/100 parts dioxane; > 10% in ethyl Alcohol; > 10% in ethyl Ether; > 10% in Chloroform.

Surface Tension: 18 dynes/cm at 25 °C

Refraction Index: 1.3724 at 9 °C/D

Critical Temperature: 179 °C

Critical Pressure: 5.18 mPa

Ionization Potential (eV): 12.39

Flash Point: None

Autoignition Temperature: 522 °C

RTECS Toxicity Data

Acute Inhalation: Rat LC_{50} Dose: 49900 ppm/4hr. Mouse LC_{50} Dose: >800 gm/m^3/2hr; Toxic Effects: Behavioral - Tremor; Behavioral - Ataxia; Lungs, Thorax, or Respiration - Other changes.

Chronic (Multiple Dose) Inhalation: Rat Dose: 1 pph/6H/2W-I; Toxic Effects: Liver - Other changes. Rat Dose: 1000 ppm/6H/90D-I; Toxic Effects: Skin and appendages - Hair; DEATH.

Reproductive/Teratogenic: Rat Route: Inhalation; Dose: 1 pph/6H; Duration: female 6-15D of pregnancy; Effects on Fertility - Pre-implantation mortality.

Hazard Overviews

Compressed Gas

Fire Diamond

Health: Irritating to eyes/respiratory tract. Also Causes: frostbite, giddiness, light-headedness, disorientation, nausea, vomiting, narcosis, cardiac dysrhythmias, hypotension, death.

Fire: Use dry chemical or carbon dioxide, water spray, fog, or foam. Vapors may cause dizziness or suffocation and contact with liquid may cause frostbite.

Reactivity: Stable. Hazardous polymerization cannot occur. Avoid: heat and ignition sources. Incompatible with: chemically active metals (sodium; potassium; calcium;

powdered aluminum; zinc; magnesium); acids or acid fumes; chlorofluorocarbons and powdered aluminum or heated aluminum surfaces. Hazardous decomposition products: toxic fumes of chlorine; fluorine; hydrogen chloride; hydrogen fluoride; phosgene.

Carcinogenicity: IARC - Not listed; NIOSH - Listed as carcinogen; NTP - Not listed; ACGIH - Not listed; OSHA - Not listed; EPA - Not listed; MAK - Not listed

Primary Target Organs:

Eyes

Skin

Respiratory System

Cardio-vascular

Exposure Limits

OSHA PEL: TWA: 1000 ppm; 4200 mg/m^3.

OSHA PEL Vacated 1989 Limits: TWA: 10 ppm; 40 mg/m^3.

ACGIH TLV: TWA: 10 ppm; 42 mg/m^3.

NIOSH REL: TWA: 10 ppm; 40 mg/m^3.

NIOSH IDLH: 5000 ppm.

DFG MAK: TWA: 10 ppm; 45 mg/m^3.

Respirator Recommendation

Exposure Range: >1000 to <5000 ppm Supplied Air, Constant Flow/Pressure Demand, Half Mask

Exposure Range: 5000 to unlimited ppm Self-contained Breathing Apparatus, Pressure Demand, Full Face

Note: odor threshold unknown

Environmental

Environmental Fate: It is an extremely unreactive gas and losses due to photolysis, photooxidation, hydrolysis, or biodegradation in air, water, and soil will not be significant. If released on land, most will volatilize into the air. It is highly mobile in soil and therefore will have a potential for leaching into groundwater. If released in water, it will be volatilize into the air. Its half-life in a model river is estimated to be 3.0 hr. In the atmosphere, it is mainly removed by reaction with hydroxyl radicals. The estimated half-life in the troposphere is 2.0 yr. As a result of its long half-life, it will accumulate and disperse all over the globe.

Cleanup/Disposal: Guide No. 126: Do not touch or walk through spilled material. Stop leak if you can do it without risk. Do not direct water at spill or source of leak. Use water spray to reduce vapors or divert vapor cloud drift. If possible, turn leaking containers so that gas escapes rather than liquid. Prevent entry into waterways, sewers, basements or confined areas. Allow substance to evaporate. Ventilate the area.

Environmental Physical Data

Henry's Law Constant: 0.0173

Octanol/Water Partition Coefficient: log K_{ow} = 1.55

Sorption Partition Coefficient: K_{oc} = estimated at 28.3

BCF: estimated at 8.9

BOD: not pertinent

Regulations

RCRA 40CFR: Not listed

CERCLA: 40CFR 302.4: Listed as Compound per CWA Section 307(a)

SARA 40CFR 372.65: Listed

SARA EHS 40CFR 355: Not listed

TSCA: Listed

Analytical Methods

Air: EPA VG-011-1; ASTM D4490

Indoor / Expired Air: NIOSH 2516

DIC4850 CAS #: 96-23-1

DICHLOROHYDRIN

RTECS: UB1400000

EINECS Number: 202-491-9

Molecular Formula: $C_3H_6Cl_2O$

Structured MF: $ClCH_2CH(OH)CH_2Cl$

Formula Weight: 128.99

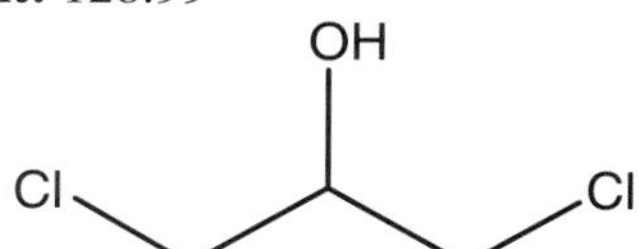

Chemical Structure

Synonyms: 1,3-DICHLOROHYDRIN; ALPHA,GAMMA-DICHLOROHYDRIN; ALPHA-DICHLOROHYDRIN; 1,3-DICHLORO-2-HYDROXYPROPANE; 1,3-DICHLOROISOPROPANOL; 1,3-DICHLOROISOPROPYL ALCOHOL; SYM-DICHLOROISOPROPYL ALCOHOL; 1,3-DICHLORO-2-PROPANOL; 1,3-DICHLOROPROPANOL-2; ENODRIN; GDCH; GLYCEROL 1,3-DICHLOROHYDRIN; GLYCEROL ALPHA,GAMMA-DICHLOROHYDRIN; GLYCEROL DICHLOROHYDRIN; SYM-GLYCEROL DICHLOROHYDRIN; 2-PROPANOL,1,3-DICHLORO-; 2-PROPANOL,1-3-DICHLORO-; ALPHA-PROPENYLDICHLOROHYDRIN; PROPYLENE DICHLOROHYDRIN; U 25,354

Description: colorless, slightly viscous, liquid; ethereal odor

Use: as a general solvent, as an intermediate in organic synthesis and in paints, varnishes, lacquers, water colors, binders and photographic lacquers; as a solvent for hard resins and nitrocellulose, in the manufacture of zapon lacquer and as a cement from celluloid

Physical Properties

Boiling Point: 174.3 °C (346 °F) at 760 mm Hg

Freezing Point: -4 °C (24.8 °F)

Specific Gravity: 1.3506 at 17 °C/4 °C

Vapor Density: 4.4 Air=1

Density: 1.36 to 1.39 g/mL

Vapor Pressure: 7 mm Hg

Water Solubility: 1 parts in 10 parts Water

Other Solubilities: 95% Ethanol: >=100 mg/ml at 23 °C; Acetone: >=100 mg/ml at 23 °C; Alcohol: miscible; DMSO: >=100 mg/ml at 23 °C; Ether: miscible; Most organic solvents: miscible; Vegetable oils: miscible.

Refraction Index: 1.480245 at 17 °C/D

Flash Point: 73.889 °C Open Cup

RTECS Toxicity Data

Acute Oral: Rat LD_{50} Dose: 110 mg/kg. Mouse LD_{50} Dose: 25 mg/kg.
Acute Inhalation: Rat LC_{Lo} Dose: 125 ppm/4hr.
Acute Dermal: Rabbit LD_{50} Route: Skin; Dose: 800 mg/kg; Toxic Effects: Skin and appendages - Primary irritation.
Irritation Skin: Rabbit Open Draize Test Dose: 10 mg/24H open; Reaction: mild.
Mutagenic: Human DNA Inhibition; Cell Type: HeLa cell; Dose: 2500 umol/L. Hamster Sister Chromatid Exchenge; Cell Type: lung; Dose: 250 umol/L.

Hazard Overviews

Health: Irritating to eyes/skin/respiratory tract. Toxic. Other Acute Effects: toxic by inhalation, in contact with skin and if swallowed; risk of serious damage to eyes; readily absorbed through skin;. Chronic Effects: possible risk of irreversible effects; target organs: liver, kidneys and central nervous system. Possibly causes mutagenic effects based on laboratory experiments.
Fire: Combustible. Hazards: emits toxic fumes. Extinguishing agents: water spray; carbon dioxide, dry chemical powder or appropriate foam. Precautions: combustible liquid.
Carcinogenicity: IARC - Not listed; NIOSH - Not listed; NTP - Not listed; ACGIH - Not listed; OSHA - Not listed; EPA - Not listed; MAK - Class A2, Unmistakably carcinogenic in animal experimentation only
Primary Target Organs:

Eyes

Skin

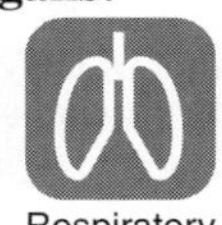
Respiratory System

Nervous System

Liver

Kidneys

Environmental

Ecotoxicity: Toxicity to microorganisms: 30 min Microtox: EC_{50}: 1,800 mg/l
Cleanup/Disposal: Guide No. 153: Eliminate all ignition sources (no smoking, flares, sparks or flames in immediate area). Do not touch damaged containers or spilled material unless wearing appropriate protective clothing. Stop leak if you can do it without risk. Prevent entry into waterways, sewers, basements or confined areas. Absorb or cover with dry earth, sand or other non-combustible material and transfer to containers. Do not get water inside containers.

Regulations

RCRA 40CFR: Not listed
CERCLA: 40CFR 302.4: Not listed
SARA 40CFR 372.65: Not listed
SARA EHS 40CFR 355: Not listed
TSCA: Listed

Analytical Methods

Soil: SW846 3640A, 8010B, 8240B, 8260A, 8260B
Water / Groundwater: EPA 1625
Plasma: EPA 29

DIC4950 CAS #: 2782-57-2

DICHLOROISOCYANURIC ACID

RTECS: XZ1845000
DOT: UN2465; IMO5.1
EINECS Number: 220-487-5
Molecular Formula: $C_3HCl_2N_3O_3$
Formula Weight: 198.98
Synonyms: ACL-59; ACL 70; CDB 60; DICHLOROCYANURIC ACID; DICHLOROISOCYANURATE; DICHLOROISOCYANURIC ACID,DRY OR DICHLOROISOCYANURIC ACIDSALTS; 1,3-DICHLORO-S-TRIAZINE-2,4,6(1H,3H,5H)-TRIONE; 1,3-DICHLORO-S-TRIAZINE-2,4,6-TRIONE; DICHLORO-S-TRIAZINE-2,4,6-TRIONE; FI CLOR 71; HILITE 60; ISOCYANURIC ACID,DICHLORO-; ISOCYANURIC DICHLORIDE; KYSELINA DICHLORISOKYANUROVA; ORCED; 1,3,5-TRIAZINE-2,4,6(1H,3H,5H)-TRIONE,1,3-DICHLORO-; S-TRIAZINE-2,4,6(1H,3H,5H)-TRIONE,1,3-DICHLORO-; TROCLOSENE
Description: white crystalline powder or granules; strong chlorine odor
Use: household dry bleaches; dishwashing compounds; scouring powders; detergent sanitizers; replacement for calcium hypochlorite

Physical Properties

Freezing Point: 225 °C (437 °F)
Density: 34 lb/cu ft (loose bulk, approximately)
Water Solubility: Insoluble < 1 mg/ml at 27 C
Other Solubilities: 95% Ethanol: Insoluble (<1 mg/ml at 27 °C); Acetone: Slightly Soluble (1-10 mg/ml at 21 °C); DMSO: Reacts.
Flash Point: Probably combustible

RTECS Toxicity Data

Acute Oral: Human LD_{Lo} Dose: 3570 mg/kg; Toxic Effects: Gastrointestinal - Ulceration or bleeding from stomach. Rat LD_{50} Dose: 1173 mg/kg.
Irritation Skin: Rabbit Standard Draize Test Dose: 500 mg; Reaction: severe.

Hazard Overviews

Fire: Combustible.
Carcinogenicity: IARC - Not listed; NIOSH - Not listed; NTP - Not listed; ACGIH - Not listed; OSHA - Not listed; EPA - Not listed; MAK - Not listed

Environmental

Cleanup/Disposal: Guide No. 141: Keep combustibles (wood, paper, oil, etc.) away from spilled material. Do not touch damaged containers or spilled material unless wearing appropriate protective clothing. Stop leak if you can do it without risk. Small Dry Spills: With clean shovel place material into clean, dry container and cover loosely; move containers from spill area. Large Spills: Dike far ahead of spill for later disposal.

Regulations

RCRA 40CFR: Not listed

CERCLA: 40CFR 302.4: Listed as Compound per CWA Section 307(a) per CAA Section 112
SARA 40CFR 372.65: Not listed
SARA EHS 40CFR 355: Not listed
TSCA: Listed

DIC5000 CAS #: 149-74-6

DICHLOROMETHYLPHENYLSILANE

RTECS: VV3530000
EINECS Number: 205-746-2
Molecular Formula: $C_7H_8Cl_2Si$
Formula Weight: 191.14

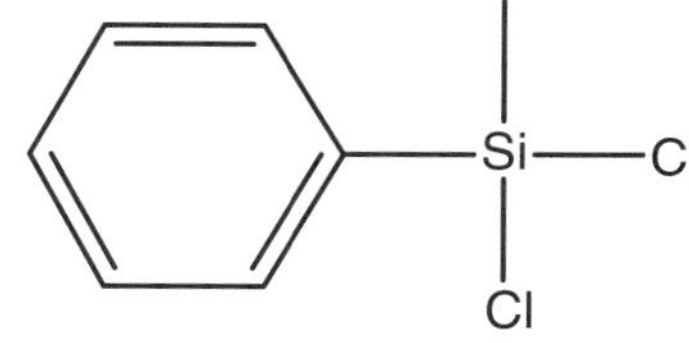

Chemical Structure

Synonyms: DICHLOR-FENYL-METHYLSILANE; METHYLPHENYLDICHLOROSILANE; PHENYLMETHYLDICHLOROSILANE; SILANE,DICHLOROMETHYLPHENYL-
Description: colorless liquid
Use: mfr of silicones; chem int for silicone fluids, resins, & elastomers

Physical Properties

Boiling Point: 82 °C (180 °F) at 13 mm Hg
Freezing Point: 83 °C (181.4 °F)
Specific Gravity: 1.19
Saturated Vapor Density: 1.203973884 kg/m^3
Vapor Pressure: 0.060 kPa at 25 °C
Other Solubilities: Soluble in Benzene, Ether, Methanol
pH: Expected to be acid due to the liberation of hydrochloric acid
Refraction Index: 1.5199 at 25 °C/D
Ionization Potential (eV): =< 9.52
Flash Point: 28.333 °C

RTECS Toxicity Data

Acute Inhalation: Mammal LC_{50} Dose: 150 mg/m^3; Toxic Effects: Lungs, Thorax, or Respiration - Acute pulmonary edema; Liver - Fatty liver degeneration; Blood - Other changes. Mouse LC_{Lo} Dose: 30 mg/m^3/2hr; Toxic Effects: Lungs, Thorax, or Respiration - Other changes; Liver - Fatty liver degeneration; Blood - Changes in spleen.
Acute Dermal: Mouse LD_{Lo} Route: Subcutaneous Dose: 100 mg/kg.
Chronic (Multiple Dose) Inhalation: Rat Dose: 6 mg/m^3/3H/58W-I; Toxic Effects: Brain and coverings - Other degenerative changes; Lungs, Thorax, or Respiration - Acute pulmonary edema; Blood - Changes in spleen.

Hazard Overviews

Poison

Corrosive

Flammable

Health: Corrosive to eyes/skin/respiratory tract. Poison. Other Acute Effects: may be fatal if inhaled; harmful if swallowed or absorbed through skin; inhalation may result in spasm, inflammation and edema of the larynx and bronchi, chemical pneumonitis and pulmonary edema; symptoms of exposure may include burning sensation; coughing; wheezing; laryngitis; shortness of breath; headache; nausea; vomiting. Chronic Effects: may alter genetic material; blood effects; target organs: blood, bone marrow, eyes. Carcinogen.
Fire: Flammable. Hazards: under fire conditions, material may decompose to form flammable and/or explosive mixtures in air; emits toxic fumes. Extinguishing agents: carbon dioxide, dry chemical powder or appropriate foam; do not use water. Precautions: combustible liquid.
Reactivity: Stable. Hazardous polymerization will not occur. Incompatible with: strong acids, strong bases, strong oxidizing agents, alcohols, reacts violently with water, rapidly hydrolyzes to form hydrochloric acid. Hazardous decomposition products: toxic fumes of: carbon monoxide, carbon dioxide, hydrogen gas, hydrogen chloride gas, silicon dioxide, formaldehyde.
Carcinogenicity: IARC - Not listed; NIOSH - Not listed; NTP - Not listed; ACGIH - Not listed; OSHA - Not listed; EPA - Not listed; MAK - Not listed
Primary Target Organs:

Eyes

Skin

Respiratory System

Blood

Bone

Environmental

Regulations

RCRA 40CFR: Not listed
CERCLA: 40CFR 302.4: Not listed
SARA 40CFR 372.65: Not listed TPQ: 1000 lb
SARA EHS 40CFR 355: Listed TPQ: 1000 lb
TSCA: Listed

DIC5050 CAS #: 99-30-9

2,6-DICHLORO-4-NITROANILINE

RTECS: BX2975000
EINECS Number: 202-746-4
Molecular Formula: $C_6H_4Cl_2N_2O_2$
Structured MF: $Cl_2C_6H_2(NO_2)NH_2$
Formula Weight: 207.02

Chemical Structure

Synonyms: AL-50; ALLISAN; ANILINE,2,6-DICHLORO-4-NITRO-; BENZENAMINE,2,6-DICHLORO-4-NITRO-; BENZENAMINE,2,6-DICHLORO-4-NITRO-(9CI); BORTRAN; BOTRAN; BOTRAN 45W; CDNA; CNA; DCNA; DCNA (FUNGICIDE); DCNA(FUNGICIDE); DICHLORAN; DICHLORAN (AMINE FUNGICIDE); DICHLORAN(AMINE FUNGICIDE); 2,6-DICHLOR-4-NITROANILIN; 2,6-DICHLORO-4-NITROBENZENAMINE; DICLORAN; DITRANIL; 4-NITROANILINE,2,6-DICHLORO-; 4-NITRO-2,6-DICHLOROANILINE; RD-6584; RESISAN; U-2069

Description: yellow needles; odorless

Use: as a fungicide and an indicator

Physical Properties

Freezing Point: 191 °C (375.8 °F)

Vapor Pressure: 0.16 mPa at 20 °C

Water Solubility: < 0.1 mg/mL at 25 C

Other Solubilities: 95% Ethanol: <1 mg/ml at 19 °C; Acetone: 10-50 mg/ml at 19 °C; Alcohol: Soluble; Benzene: 4.6 g/L; Chloroform: 12 g/L; Cyclohexane: 6 mg/L; DMSO: >=100 mg/ml at 19 °C; Ethyl Acetate: 19 g/L; Polar organic solvents: Moderately Soluble.

Flash Point: Not available; probably combustible

RTECS Toxicity Data

Acute Oral: Rat LD_{50} Dose: 2400 mg/kg; Toxic Effects: Behavioral - Tremor; Behavioral - Convulsions or effect on seizure threshold; Behavioral - Ataxia. Mouse LD_{50} Dose: 1500 mg/kg.

Acute Inhalation: Rat LC_{50} Dose: >21600 mg/m^3/1hr.

Acute Dermal: Rabbit LD_{50} Route: Skin; Dose: >2 gm/kg. Mouse LD_{50} Route: Skin; Dose: >5 gm/kg.

Chronic (Multiple Dose) Oral: Rat Dose: 12 gm/kg/60D-I; Toxic Effects: Blood - Methemoglobinemia-Carboxyhemoglobin. Rat Dose: 16 gm/kg/26W-I; Toxic Effects: Biochemical - True cholinesterase; Biochemical - Other transferases; Biochemical - Multiple enzyme effects. Rat Dose: 109 gm/kg/2Y-C; Toxic Effects: Liver - Changes in liver weight; Kidney, Ureter, and Bladder - Changes in kidney weight; DEATH.

Mutagenic: Bacteria - S Typhimurium Mutations in Microorganisms; Dose: 2500 ug/plate (+S9). Bacteria - S Typhimurium Mutations in Microorganisms; Dose: 100 ug/plate (-S9).

Tumorigenic: Mouse Route: Oral; Dose: 80 gm/kg/78W-I; Toxic Effects: Tumorigenic - Equivocal tumorigenic agent by RTECS criteria; Liver - Tumors; Blood - Tumors.

Hazard Overviews

Health: Irritating to eyes/skin/respiratory tract. Other Acute Effects: may be harmful by inhalation, ingestion, or skin absorption; prolonged or repeated exposure may cause allergic reactions in certain sensitive individuals; absorption leads to the formation of methemoglobin which, in sufficient concentration, causes cyanosis; onset may be delayed 2 to 4 hours or longer. Chronic Effects: danger of cumulative effects; possible sensitizer; target organs: eyes, liver, blood, central nervous system.

Fire: Will burn. Hazards: emits toxic fumes. Extinguishing agents: water spray; carbon dioxide, dry chemical powder or appropriate foam. Precautions: combustible liquid.

Carcinogenicity: IARC - Not listed; NIOSH - Not listed; NTP - Not listed; ACGIH - Not listed; OSHA - Not listed; EPA - Not listed; MAK - Not listed

Primary Target Organs:

Eyes

Skin

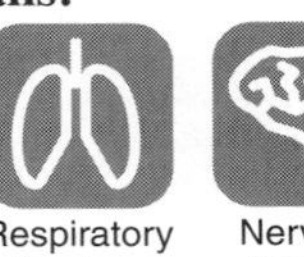
Respiratory System

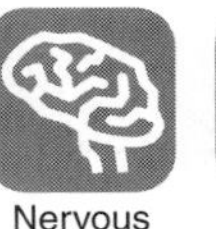
Nervous System

Liver

Blood

Environmental

Ecotoxicity: LD_{50} Anas platyrhynchos (mallard duck) oral 3-4 mo old females >2000 mg/kg /sample purity 97 LD_{50} Phasianus colchicus (ring-necked pheasant) 3-4 mo old females oral 500-1000 mg/kg/sample purity 97

Environmental Fate: Microbial degradation has been shown to be enhanced in soils that have been previously treated with the compound, in flooded soils, and in soils containing amendments such as alfalfa, rice straw, and glucose. Persistence in soil under laboratory conditions can vary, as half-lives of about a day to 30 months have been observed with different conditions. Leaching in soil is not expected to occur. If released to water, biodegradation may be an important degradation process based on analogy to soil studies. It has been shown to be tightly adsorbed to soil; therefore, adsorption to sediment and suspended material in water may be important. If released to the atmosphere, vapor-phase is degraded by reaction with photochemically formed hydroxyl radicals; the half-life for this reaction in air can be estimated to be about 0.8 days. Particulate-phase may be physically removed from air by wet and dry deposition.

Environmental Physical Data

Henry's Law Constant: estimated at 1.3×10^{-8}

Octanol/Water Partition Coefficient: log K_{ow} = 2.76

BCF: estimated at 74

Regulations

RCRA 40CFR: Not listed

CERCLA: 40CFR 302.4: Not listed

SARA 40CFR 372.65: Listed

SARA EHS 40CFR 355: Not listed

TSCA: Listed

Analytical Methods

Soil: EPA PMD-DJA, PMD-TLC; SW846 8131

Water / Groundwater: EPA 1625, 608.2, 617; APHA 6630-B; ASTM D3086
Food: FDA 212.1, 212.2, 232.1, 232.4, 242.1
Indoor / Expired Air: ASTM D4861
Plasma: EPA 001; FDA 231.1, 252

DIC5100 CAS #: 89-61-2

1,4-DICHLORO-2-NITROBENZENE

RTECS: CZ5260000
EINECS Number: 201-923-3
Molecular Formula: $C_6H_3Cl_2NO_2$
Formula Weight: 192.00

Chemical Structure

Synonyms: 2,5-DICHLORNITROBENZEN; 2,5-DICHLORONITROBENZENE; NITRO-P-DICHLOROBENZENE

Physical Properties

Boiling Point: 267 °C (512.6 °F)
Freezing Point: 52.8 °C (127.04 °F)
Specific Gravity: 1.439
Water Solubility: Insoluble
Other Solubilities: Ethanol: Soluble; Ether: Soluble; Benzene: Soluble; Carbon Tetrachloride: Slightly soluble; Chloroform: Soluble; CS_2: Soluble
Refraction Index: 1.4390

RTECS Toxicity Data

Acute Oral: Rat LD_{50} Dose: 1 gm/kg. Mouse LD_{50} Dose: 2850 mg/kg.
Irritation Eye: Rabbit Standard Draize Test Dose: 100 mg/24H; Reaction: moderate.
Irritation Skin: Rabbit Standard Draize Test Dose: 500 mg/24H; Reaction: mild.
Mutagenic: Bacteria - S Typhimurium Mutations in Microorganisms; Dose: 205 ug/plate (+S9). Bacteria - S Typhimurium Mutations in Microorganisms; Dose: 250 ug/plate (-S9).

Hazard Overviews

Health: Irritating to eyes/skin/respiratory tract. Harmful. Other Acute Effects: harmful if swallowed; may be harmful if inhaled or absorbed through the skin; absorption into the body leads to the formation of methemoglobin which in sufficient concentration causes cyanosis; onset may be delayed 2 to 4 hours or longer.
Fire: Extinguishing agents: water spray; carbon dioxide, dry chemical powder or appropriate foam. Precautions: combustible liquid.
Reactivity: Incompatible with: strong oxidizing agents, strong bases. Hazardous decomposition products: toxic fumes of: carbon monoxide, carbon dioxide, nitrogen oxides, hydrogen chloride gas.
Carcinogenicity: IARC - Not listed; NIOSH - Not listed; NTP - Not listed; ACGIH - Not listed; OSHA - Not listed; EPA - Not listed; MAK - Not listed
Primary Target Organs:

Eyes

Skin

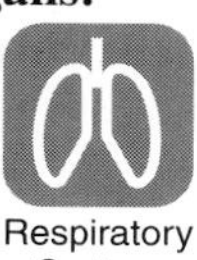
Respiratory System

Environmental

Ecotoxicity: Bacteria: Pseudomonas putida 30 min EC_0 500 mg/l; Fishes: Leuciscus idus 96h LC_{50} 6.3 mg/l 96h LC_{100} 10 mg/l

Environmental Physical Data

Octanol/Water Partition Coefficient: log K_{ow} = calculated at 3.3

Regulations

RCRA 40CFR: Not listed
CERCLA: 40CFR 302.4: Listed as Compound per CWA Section 307(a)
SARA 40CFR 372.65: Not listed
SARA EHS 40CFR 355: Not listed
TSCA: Listed

Analytical Methods

Soil: SW846 8091

DIC5150 CAS #: 3209-22-1

2,3-DICHLORONITROBENZENE

RTECS: CZ5240000
EINECS Number: 221-717-7
Molecular Formula: $C_6H_3Cl_2NO_2$
Formula Weight: 192.00

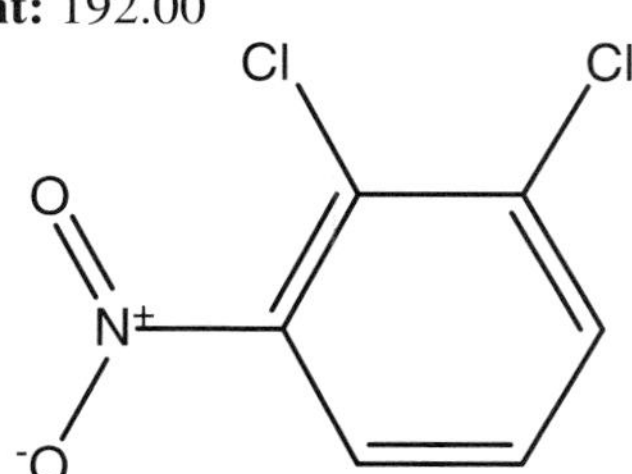

Chemical Structure

Synonyms: BENZENE,1,2-DICHLORO-3-NITRO-; 1,2-DICHLORO-3-NITROBENZENE; 2,3-DICHLORO-1-NITROBENZENE; ORTHO,META-DICHLORO-1-NITROBENZENE
Description: monoclinic needles

Use: research chemical

Physical Properties

Boiling Point: 257 °C (495 °F) to 258 °C (496 °F)
Freezing Point: 61 °C (141.8 °F) to 62 °C (143.6 °F)
Specific Gravity: 1.721 at 14 °C
Saturated Vapor Density: 1.200014643 kg/m^3
Density: 1.721 g/mL at 14/4 °C
Vapor Pressure: 0.00165 mm Hg at 25 °C
Water Solubility: < 0.1 mg/mL at 25 C
Other Solubilities: 95% Ethanol: 50-100 mg/ml at 20 °C; Acetone: >=100 mg/ml at 20 °C; Benzene: Soluble; DMSO: >=100 mg/ml at 20 °C; Ether: Soluble.
Flash Point: 123 °C

RTECS Toxicity Data

Mutagenic: Bacteria - S Typhimurium Mutations in Microorganisms; Dose: 3333 ug/plate (-S9).

Hazard Overviews

Health: Irritating to eyes/skin/respiratory tract. Other Acute Effects: may be harmful by inhalation, ingestion, or skin absorption; absorption into the body leads to the formation of methemoglobin which in sufficient concentration causes cyanosis; onset may be delayed 2 to 4 hours or longer.
Fire: Will burn. Hazards: emits toxic fumes. Extinguishing agents: water spray; carbon dioxide, dry chemical powder or appropriate foam. Precautions: combustible liquid.
Reactivity: Incompatible with: strong oxidizing agents, strong bases. Hazardous decomposition products: toxic fumes of: carbon monoxide, carbon dioxide, nitrogen oxides, hydrogen chloride gas.
Carcinogenicity: IARC - Not listed; NIOSH - Not listed; NTP - Not listed; ACGIH - Not listed; OSHA - Not listed; EPA - Not listed; MAK - Not listed
Primary Target Organs:

Eyes

Skin

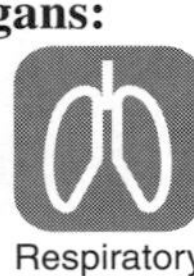
Respiratory System

Environmental

Environmental Fate: If released to water, it is not expected to undergo aqueous hydrolysis. It has been shown to be relatively resistant to biodegradation (half-lives greatly in excess of 4 weeks) in two biodegradation tests using both adapted and non-adapted inoculum. Volatilization half-lives of 45 hours, 56.3 days and 13.6 months can be estimated for a model river (1 meter deep), a model environmental pond, and Lake Zurich (Switzerland), respectively. If released to soil, it is expected to exhibit medium to low soil mobility based upon estimated K_{oc} values of 450 and 690. If released to the atmosphere, it is expected to exist in the gas-phase where it will be degraded slowly (estimated half-life of 315 days) by reaction with photochemically formed hydroxyl radicals. Although it absorbs UV light in the environmental spectra, sufficient data are not available to predict the potential importance of photolysis in air or water.
Cleanup/Disposal: Guide No. 171: Do not touch or walk through spilled material. Stop leak if you can do it without risk. Prevent dust cloud. Avoid inhalation of asbestos dust. Small Dry Spills: With clean shovel place material into clean, dry container and cover loosely; move containers from spill area. Small Spills: Take up with sand or other noncombustible absorbent material and place into containers for later disposal. Large Spills: Dike far ahead of liquid spill for later disposal. Cover powder spill with plastic sheet or tarp to minimize spreading. Prevent entry into waterways, sewers, basements or confined areas.

Environmental Physical Data

Henry's Law Constant: estimated at 3.22×10^{-5}
Octanol/Water Partition Coefficient: log K_{ow} = 3.05
Sorption Partition Coefficient: K_{OC} = 450
BCF: guppy 3.01

Regulations

RCRA 40CFR: Not listed
CERCLA: 40CFR 302.4: Not listed
SARA 40CFR 372.65: Not listed
SARA EHS 40CFR 355: Not listed
TSCA: Listed

Analytical Methods

Soil: SW846 8091
Water / Groundwater: EPA 1625

DIC5200 CAS #: 611-06-3

2,4-DICHLORONITROBENZENE

RTECS: CZ5420000
EINECS Number: 210-248-3
Molecular Formula: $C_6H_3Cl_2NO_2$
Formula Weight: 192.00

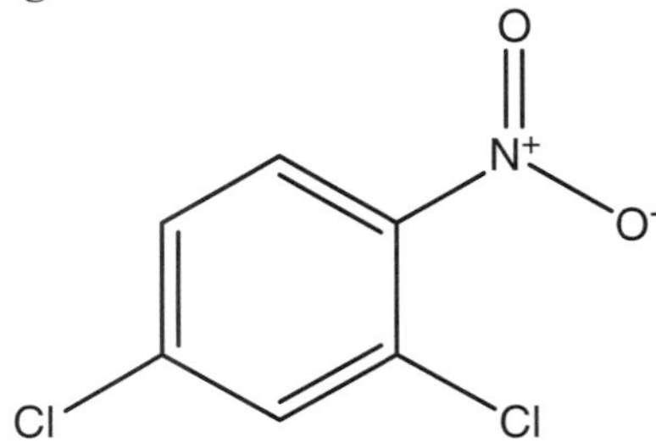

Chemical Structure

Synonyms: BENZENE,2,4-DICHLORO-1-NITRO-; 1,3-DICHLORO-4-NITROBENZENE; 2,4-DICHLORO-1-NITROBENZENE; 1-NITRO-2,4-DICHLOROBENZENE
Description: needles
Use: chemical intermediate for diazoxide (an antihypertensive agent)

Physical Properties

Boiling Point: 258.5 °C (497 °F)

Freezing Point: 34 °C (93.2 °F)
Specific Gravity: 1.439 at 80 °C
Density: 1.479 g/mL
Water Solubility: < 0.1 mg/mL at 16 C
Other Solubilities: 95% Ethanol: >=100 mg/ml at 20 °C; Acetone: >=100 mg/ml at 20 °C; DMSO: >=100 mg/ml at 20 °C; Ether: Soluble.
Refraction Index: Alpha 1.5512 at 78 °C
Flash Point: > 112 °C

RTECS Toxicity Data

Mutagenic: Bacteria - S Typhimurium Mutations in Microorganisms; Dose: 1 mg/plate (+S9). Bacteria - S Typhimurium Mutations in Microorganisms; Dose: 3 ug/plate (-S9).

Hazard Overviews

Health: Irritating to eyes/skin/respiratory tract. Other Acute Effects: may be harmful by inhalation, ingestion, or skin absorption; absorption into the body leads to the formation of methemoglobin which in sufficient concentration causes cyanosis; onset may be delayed 2 to 4 hours or longer.
Fire: Will burn. Hazards: emits toxic fumes. Extinguishing agents: water spray; carbon dioxide, dry chemical powder or appropriate foam. Precautions: combustible liquid.
Reactivity: Incompatible with: strong oxidizing agents, strong bases. Hazardous decomposition products: toxic fumes of: carbon monoxide, carbon dioxide, nitrogen oxides, hydrogen chloride gas.
Carcinogenicity: IARC - Not listed; NIOSH - Not listed; NTP - Not listed; ACGIH - Not listed; OSHA - Not listed; EPA - Not listed; MAK - Not listed
Primary Target Organs:

Eyes

Skin

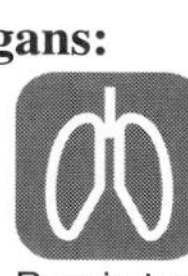
Respiratory System

Environmental

Environmental Fate: If released to water, it is not expected to undergo aqueous hydrolysis. Although it has been shown to be resistant to biodegradation by activated sludge over a 5-day inoculation period, sufficient data are not available to predict the importance of biodegradation in the environment. Volatilization half-lives of 42 hours, 49.3 days and 13.1 months can be estimated for a model river (1 meter deep), a model environmental pond, and Lake Zurich (Switzerland), respectively. If released to soil, it is expected to exhibit low soil mobility based upon an estimated K_{oc} value of 760. If released to the atmosphere, it is expected to exist in the gas-phase where it will be degraded slowly (estimated half-life of 130 days) by reaction with photochemically formed hydroxyl radicals.
Cleanup/Disposal: Guide No. 171: Do not touch or walk through spilled material. Stop leak if you can do it without risk. Prevent dust cloud. Avoid inhalation of asbestos dust. Small Dry Spills: With clean shovel place material into clean, dry container and cover loosely; move containers from spill area. Small Spills: Take up with sand or other noncombustible absorbent material and place into containers for later disposal. Large Spills: Dike far ahead of liquid spill for later disposal. Cover powder spill with plastic sheet or tarp to minimize spreading. Prevent entry into waterways, sewers, basements or confined areas.

Environmental Physical Data

Henry's Law Constant: estimated at 3.22 x10^{-5}
Octanol/Water Partition Coefficient: log K_{ow} = 3.09
Sorption Partition Coefficient: K_{oc} = 760
BCF: guppy 3.02

Regulations

RCRA 40CFR: Not listed
CERCLA: 40CFR 302.4: Listed as Compound per CWA Section 307(a)
SARA 40CFR 372.65: Not listed
SARA EHS 40CFR 355: Not listed
TSCA: Listed

Analytical Methods

Soil: SW846 8091

DIC5250 CAS #: 99-54-7

3,4-DICHLORONITROBENZENE

RTECS: CZ5250000
EINECS Number: 202-764-2
Molecular Formula: $C_6H_3Cl_2NO_2$
Formula Weight: 192.00

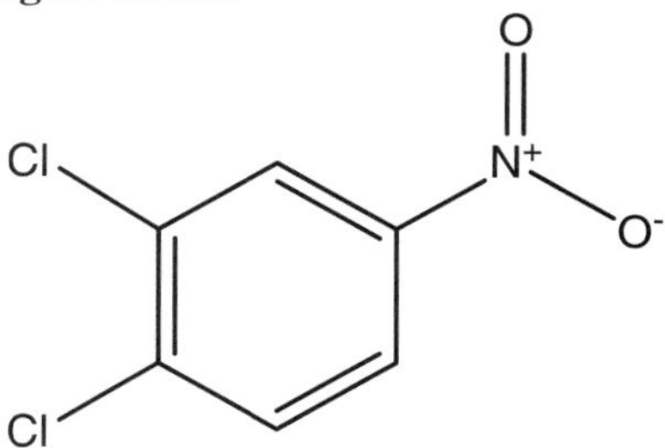

Chemical Structure

Synonyms: BENZENE,1,2-DICHLORO-4-NITRO-; DCNB; 3,4-DICHLORNITROBENZEN; 3,4-DICHLORONITROBENZEN; 1,2-DICHLORO-4-NITROBENZENE; 1-NITRO-3,4-DICHLOROBENZENE
Description: needles; liquid; solid
Use: organic intermediate for synthesis

Physical Properties

Boiling Point: 255 °C (491 °F) to 256 °C (493 °F)
Freezing Point: 43 °C (109.4 °F)
Specific Gravity: 1.4558 at 75 °C/4 °C
Vapor Density: 6.63
Saturated Vapor Density: Extrapolated solid state 1.200088748 kg/m^3
Vapor Pressure: Extrapolated solid state 0.01 mm Hg at 20 °C

Water Solubility: < 0.1 mg/mL at 23.5 C
Other Solubilities: 95% Ethanol: >=100 mg/ml at 20 °C; Acetone: >=100 mg/ml at 20 °C; DMSO: >=100 mg/ml at 20 °C; Ether: Soluble.
Flash Point: 123 °C

RTECS Toxicity Data

Acute Oral: Rat LD_{50} Dose: 953 mg/kg; Toxic Effects: Nutritional and gross metabolic - Weight loss or decreased weight gain. Mouse LD_{50} Dose: 1384 mg/kg.
Acute Inhalation: Rat LC_{50} Dose: 10 gm/m^3/4hr; Toxic Effects: Sense organs and special senses - Other; Behavioral - Somnolence (general depressed activity); Nutritional and gross metabolic - Weight loss or decreased weight gain.
Acute Dermal: Rabbit LD; Route: Skin; Dose: >200 mg/kg; Toxic Effects: Nutritional and gross metabolic - Weight loss or decreased weight gain. Cat LD_{50} Route: Skin; Dose: 790 mg/kg.
Chronic (Multiple Dose) Inhalation: Rat Dose: 10 mg/m^3/4H/17W-I; Toxic Effects: Blood - Thrombocytopenia; Blood - Changes in erythrocite (RBC) cell count; Biochemical - Transaminases. Mouse Dose: 28 mg/m^3/4H/21D-I; Toxic Effects: Blood - Eosinophilia; Blood - Changes in erythrocite (RBC) cell count; Biochemical - Catalases.
Irritation Eye: Rabbit Standard Draize Test Dose: 100 mg/24H; Reaction: moderate.
Irritation Skin: Rabbit Standard Draize Test Dose: 500 mg/24H; Reaction: mild.
Mutagenic: Insects - D Melanogaster Sex Chromosome Loss; Route: Parenteral; Dose: 200 ppm. Bacteria - S Typhimurium Mutations in Microorganisms; Dose: 250 ug/plate (+/-S9).

Hazard Overviews

Health: Irritating to eyes/skin/respiratory tract. Toxic. Other Acute Effects: harmful by inhalation, in contact with skin and if swallowed; readily absorbed through skin; absorption leads to the formation of methemoglobin which, in sufficient concentration, causes cyanosis; onset may be delayed 2 to 4 hours or longer; prolonged or repeated exposure may cause allergic reactions in certain sensitive individuals. Chronic Effects: possible sensitizer.
Fire: Will burn. Hazards: emits toxic fumes. Extinguishing agents: water spray; carbon dioxide, dry chemical powder or appropriate foam. Precautions: combustible liquid.
Carcinogenicity: IARC - Not listed; NIOSH - Not listed; NTP - Not listed; ACGIH - Not listed; OSHA - Not listed; EPA - Not listed; MAK - Not listed
Primary Target Organs:

Eyes

Skin

Respiratory System

Environmental

Ecotoxicity: Bacteria: Photobacterium phosphoreum 30 min EC_{50}10 mg/l; Algae: Haematococcus pluvialis 4h EC_{50} 2 mg/l; Crustaceans: Daphnia magna 24h EC_0 2 mg/l; Fishes: Brachydanio rerio 96h LC_0 5; 9; 10 mg/l
Environmental Fate: If released to water, it is not expected to undergo aqueous hydrolysis. Although biodegradation has been demonstrated under conditions of industrial waste treatment plants, sufficient data are not available to predict the importance of biodegradation in natural water or soil. Volatilization half-lives of 45 hours, 56.3 days and 13.6 months can be estimated for a model river (1 meter deep), a model environmental pond, and Lake Zurich (Switzerland), respectively. If released to soil, leaching may occur based upon an experimentally determined mean K_{oc} of 113 in four silt loam soils. If released to the atmosphere, it is expected to exist in the gas-phase where it will be degraded slowly (estimated half-life of 315 days) by reaction with photochemically formed hydroxyl radicals.
Cleanup/Disposal: Guide No. 171: Do not touch or walk through spilled material. Stop leak if you can do it without risk. Prevent dust cloud. Avoid inhalation of asbestos dust. Small Dry Spills: With clean shovel place material into clean, dry container and cover loosely; move containers from spill area. Small Spills: Take up with sand or other noncombustible absorbent material and place into containers for later disposal. Large Spills: Dike far ahead of liquid spill for later disposal. Cover powder spill with plastic sheet or tarp to minimize spreading. Prevent entry into waterways, sewers, basements or confined areas.

Environmental Physical Data

Henry's Law Constant: estimated at 2.92 x10^{-5}
Octanol/Water Partition Coefficient: log K_{OW} = 3.12
Sorption Partition Coefficient: K_{OC} = 113
BCF: estimated at 138

Regulations

RCRA 40CFR: Not listed
CERCLA: 40CFR 302.4: Listed as Compound per CWA Section 307(a)
SARA 40CFR 372.65: Not listed
SARA EHS 40CFR 355: Not listed
TSCA: Listed

Analytical Methods

Soil: SW846 8091

DIC5300 CAS #: 40188-83-8

3,6-DICHLORO-2-NITROBENZOIC ACID, METHYL ESTER

EINECS Number: 254-832-6
Molecular Formula: $C_8H_5Cl_2NO_4$
Formula Weight: 250.02

Hazard Overviews

Carcinogenicity: IARC - Not listed; NIOSH - Not listed; NTP - Not listed; ACGIH - Not listed; OSHA - Not listed; EPA - Not listed; MAK - Not listed

Environmental

Regulations

RCRA 40CFR: Not listed
CERCLA: 40CFR 302.4: Not listed
SARA 40CFR 372.65: Not listed
SARA EHS 40CFR 355: Not listed
TSCA: Listed

DIC5350 CAS #: 135-12-6

4,4'-DICHLORO-2-NITRODIPHENYL ETHER

RTECS: KN6670000
EINECS Number: 205-176-4
Molecular Formula: $C_{12}H_7Cl_2NO_3$
Formula Weight: 284.10

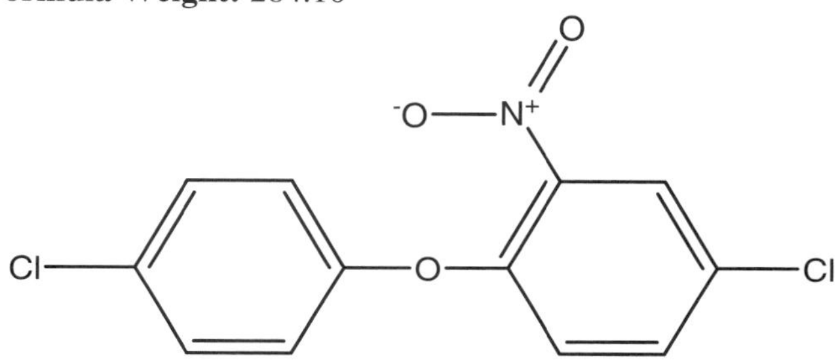

Chemical Structure

Synonyms: BENZENE,4-CHLORO-1-(4-CHLOROPHENOXY)-2-NITRO-; 4-CHLORO-2-NITROPHENYL P-CHLOROPHENYL ETHER; 4-CHLOROPHENYL 4-CHLORO-2-NITROPHENYL ETHER; 4,4'-DICHLOR-2-NITRODIFENYLETHER; ETHER,4-CHLORO-2-NITROPHENYL P-CHLOROPHENYL; ETHER,4-CHLOROPHENYL (4'-CHLORO-2'-NITRO)PHENYL

RTECS Toxicity Data

Irritation Eye: Rabbit Standard Draize Test Dose: 100 mg/24H; Reaction: moderate.
Irritation Skin: Rabbit Standard Draize Test Dose: 500 mg/24H; Reaction: mild.

Hazard Overviews

Carcinogenicity: IARC - Not listed; NIOSH - Not listed; NTP - Not listed; ACGIH - Not listed; OSHA - Not listed; EPA - Not listed; MAK - Not listed

Environmental

Regulations

RCRA 40CFR: Not listed
CERCLA: 40CFR 302.4: Listed as Compound per CWA Section 307(a)
SARA 40CFR 372.65: Not listed
SARA EHS 40CFR 355: Not listed
TSCA: Listed

DIC5400 CAS #: 594-72-9

1,1-DICHLORO-1-NITROETHANE

RTECS: KI1050000
DOT: UN2650; IMO6.1
EINECS Number: 209-854-0
Molecular Formula: $C_2H_3Cl_2NO_2$
Structured MF: $CH_3CCl_2NO_2$
Formula Weight: 143.90
Synonyms: 1,1-DICHLOOR-1-NITROETHAAN; 1,1-DICHLOR-1-NITROAETHAN; DICHLORO-1-NITROETHANE; DICHLORONITROETHANE; 1,1-DICLORO-1-NITROETANO; ETHANE,1,1-DICHLORO-1-NITRO-; ETHIDE
Description: colorless liquid; unpleasant odor
Use: in highly accelerated rubber cements, insecticides solvents and in chemical synthesis; as a fumigant for stored products, produce and grains

Physical Properties

Boiling Point: 124 °C (255 °F)
Freezing Point: 118 °C (244.4 °F) to 120 °C (248 °F)
Specific Gravity: 1.4271 at 20 °C/20 °C
Vapor Density: 5 Air=1
Saturated Vapor Density: 1.300094374 kg/m^3
Vapor Pressure: 16 mm Hg at 25 °C
Water Solubility: 0.25 ml/100 ml Water at 20 °C
Other Solubilities: 95% Ethanol: >=100 mg/ml at 18 °C; Acetone: >=100 mg/ml at 18 °C; DMSO: >=100 mg/ml at 18 °C.
Flash Point: 76 °C Open Cup

RTECS Toxicity Data

Acute Oral: Rat LD_{50} Dose: 410 mg/kg. Rabbit LD_{Lo} Dose: 150 mg/kg; Toxic Effects: Vascular - Other changes; Lungs, Thorax, or Respiration - Acute pulmonary edema; Gastrointestinal - Ulceration or bleeding from stomach.
Acute Inhalation: Rabbit LC_{Lo} Dose: 580 mg/m^3/6hr; Toxic Effects: Sense organs and special senses - Lacrimation; Behavioral - Muscle weakness; Lungs, Thorax, or Respiration - Cough. Guinea Pig LC_{Lo} Dose: 580 mg/m^3/6hr; Toxic Effects: Sense organs and special senses - Lacrimation; Behavioral - Muscle weakness; Lungs, Thorax, or Respiration - Cough.
Mutagenic: Bacteria - S Typhimurium Mutations in Microorganisms; Dose: 333 ug/plate (-S9).

Hazard Overviews

Fire Diamond

Health: Irritating to eyes/skin/respiratory tract. Toxic. Also Causes: pulmonary edema.
Fire: Combustible. Use dry chemical, carbon dioxide, water spray, fog, or regular foam.

Reactivity: Stable. Hazardous polymerization cannot occur. Avoid: heat; ignition sources. Incompatible with: oxidizers. Hazardous decomposition products: carbon oxide(s); nitrogen oxide(s); chlorine; hydrogen chloride gases.
Carcinogenicity: IARC - Not listed; NIOSH - Listed as carcinogen; NTP - Not listed; ACGIH - Not listed; OSHA - Not listed; EPA - Not listed; MAK - Not listed
Primary Target Organs:

Eyes

Skin

Respiratory System

Exposure Limits
OSHA PEL: STEL: 10 ppm; 60 mg/m^3.
OSHA PEL Vacated 1989 Limits: TWA: 2 ppm; 10 mg/m^3.
NIOSH REL: TWA: 2 ppm; 10 mg/m^3.
NIOSH IDLH: 25 ppm.
DFG MAK: TWA: 10 ppm; 60 mg/m^3.
Respirator Recommendation
Exposure Range: >10 to <25 ppm Supplied Air, Constant Flow/Pressure Demand, Half Mask
Exposure Range: 25 to unlimited ppm Self-contained Breathing Apparatus, Pressure Demand, Full Face
Note: odor threshold unknown

Environmental

Environmental Fate: If released to water, volatilization will not be rapid but possibly significant with estimated half-lives of 4 hours and 5 days from a model environmental river and a model lake, respectively. Insufficient data are available to predict the relative importance or rate of biodegradation in soil or water. If released to the atmosphere, it will exist primarily in the vapor phase. Vapor-phase will degrade in the atmosphere by reaction with photochemically produced hydroxyl radicals with an estimated half-life of approximately 107 days. Removal of atmospheric compound may occur through wet deposition. If released to soil, it is expected to have high to very high mobility based on estimated K_{oc} values of 36 to 59. Volatilization is expected from dry soils, and may also be important from moist soils.
Cleanup/Disposal: Guide No. 153: Eliminate all ignition sources (no smoking, flares, sparks or flames in immediate area). Do not touch damaged containers or spilled material unless wearing appropriate protective clothing. Stop leak if you can do it without risk. Prevent entry into waterways, sewers, basements or confined areas. Absorb or cover with dry earth, sand or other non-combustible material and transfer to containers. Do not get water inside containers.

Environmental Physical Data
Henry's Law Constant: estimated at 0.00128
Sorption Partition Coefficient: K_{oc} = 36
BCF: estimated at 7.5

Regulations
RCRA 40CFR: Not listed
CERCLA: 40CFR 302.4: Listed as Compound per CWA Section 307(a)
SARA 40CFR 372.65: Not listed
SARA EHS 40CFR 355: Not listed
TSCA: Not listed

Analytical Methods
Air: ASTM D4490
Indoor / Expired Air: NIOSH 1601

DIC5450 **CAS #: 13474-88-9**

1,1-DICHLORO-1,2,2,3,3-PENTAFLUOROPROPANE

Molecular Formula: $C_3HCl_2F_5$
Formula Weight: 202.94

Hazard Overviews
Carcinogenicity: IARC - Not listed; NIOSH - Not listed; NTP - Not listed; ACGIH - Not listed; OSHA - Not listed; EPA - Not listed; MAK - Not listed

Environmental

Regulations
RCRA 40CFR: Not listed
CERCLA: 40CFR 302.4: Not listed
SARA 40CFR 372.65: Listed
SARA EHS 40CFR 355: Not listed
TSCA: Not listed

DIC5500 **CAS #: 111512-56-2**

1,1-DICHLORO-1,2,3,3,3-PENTAFLUOROPROPANE

Molecular Formula: $C_3HCl_2F_5$
Formula Weight: 202.94

Hazard Overviews
Carcinogenicity: IARC - Not listed; NIOSH - Not listed; NTP - Not listed; ACGIH - Not listed; OSHA - Not listed; EPA - Not listed; MAK - Not listed

Environmental

Regulations
RCRA 40CFR: Not listed
CERCLA: 40CFR 302.4: Not listed
SARA 40CFR 372.65: Listed
SARA EHS 40CFR 355: Not listed
TSCA: Not listed

DIC5550 **CAS #: 422-44-6**

1,2-DICHLORO-1,1,2,3,3-PENTAFLUOROPROPANE

Molecular Formula: $C_3HCl_2F_5$
Formula Weight: 202.94

Hazard Overviews

Carcinogenicity: IARC - Not listed; NIOSH - Not listed; NTP - Not listed; ACGIH - Not listed; OSHA - Not listed; EPA - Not listed; MAK - Not listed

Environmental

Regulations

RCRA 40CFR: Not listed
CERCLA: 40CFR 302.4: Not listed
SARA 40CFR 372.65: Listed
SARA EHS 40CFR 355: Not listed
TSCA: Not listed

DIC5600 **CAS #: 431-86-7**

1,2-DICHLORO-1,1,3,3,3-PENTAFLUOROPROPANE

Molecular Formula: $C_3HCl_2F_5$
Formula Weight: 202.94

Physical Properties

Boiling Point: 50 °C (122 °F)

Hazard Overviews

Carcinogenicity: IARC - Not listed; NIOSH - Not listed; NTP - Not listed; ACGIH - Not listed; OSHA - Not listed; EPA - Not listed; MAK - Not listed

Environmental

Regulations

RCRA 40CFR: Not listed
CERCLA: 40CFR 302.4: Not listed
SARA 40CFR 372.65: Listed
SARA EHS 40CFR 355: Not listed
TSCA: Not listed

DIC5650 **CAS #: 507-55-1**

1,3-DICHLORO-1,1,2,2,3-PENTAFLUOROPROPANE

EINECS Number: 208-076-9
Molecular Formula: $C_4HCl_2F_5$
Formula Weight: 202.94

Hazard Overviews

Carcinogenicity: IARC - Not listed; NIOSH - Not listed; NTP - Not listed; ACGIH - Not listed; OSHA - Not listed; EPA - Not listed; MAK - Not listed

Environmental

Regulations

RCRA 40CFR: Not listed
CERCLA: 40CFR 302.4: Not listed
SARA 40CFR 372.65: Listed
SARA EHS 40CFR 355: Not listed
TSCA: Not listed

DIC5700 **CAS #: 136013-79-1**

1,3-DICHLORO-1,1,2,3,3-PENTAFLUOROPROPANE

Molecular Formula: $C_3HCl_2F_5$
Formula Weight: 202.94

Hazard Overviews

Carcinogenicity: IARC - Not listed; NIOSH - Not listed; NTP - Not listed; ACGIH - Not listed; OSHA - Not listed; EPA - Not listed; MAK - Not listed

Environmental

Regulations

RCRA 40CFR: Not listed
CERCLA: 40CFR 302.4: Not listed
SARA 40CFR 372.65: Listed
SARA EHS 40CFR 355: Not listed
TSCA: Not listed

DIC5750 **CAS #: 128903-21-9**

2,2-DICHLORO-1,1,1,3,3-PENTAFLUOROPROPANE

Molecular Formula: $C_3HCl_2F_5$
Formula Weight: 202.94

Hazard Overviews

Carcinogenicity: IARC - Not listed; NIOSH - Not listed; NTP - Not listed; ACGIH - Not listed; OSHA - Not listed; EPA - Not listed; MAK - Not listed

Environmental

Regulations

RCRA 40CFR: Not listed
CERCLA: 40CFR 302.4: Not listed

SARA 40CFR 372.65: Listed
SARA EHS 40CFR 355: Not listed
TSCA: Not listed

DIC5800 CAS #: 422-48-0

2,3-DICHLORO-1,1,1,2,3-PENTAFLUOROPROPANE

Molecular Formula: $C_3HCl_2F_5$
Formula Weight: 202.94

Hazard Overviews

Carcinogenicity: IARC - Not listed; NIOSH - Not listed; NTP - Not listed; ACGIH - Not listed; OSHA - Not listed; EPA - Not listed; MAK - Not listed

Environmental

Regulations

RCRA 40CFR: Not listed
CERCLA: 40CFR 302.4: Not listed
SARA 40CFR 372.65: Listed
SARA EHS 40CFR 355: Not listed
TSCA: Not listed

DIC5850 CAS #: 422-56-0

3,3-DICHLORO-1,1,1,2,2-PENTAFLUOROPROPANE

EINECS Number: 207-016-9
Molecular Formula: $C_3HCl_2F_5$
Formula Weight: 202.94

Hazard Overviews

Carcinogenicity: IARC - Not listed; NIOSH - Not listed; NTP - Not listed; ACGIH - Not listed; OSHA - Not listed; EPA - Not listed; MAK - Not listed

Environmental

Regulations

RCRA 40CFR: Not listed
CERCLA: 40CFR 302.4: Not listed
SARA 40CFR 372.65: Listed
SARA EHS 40CFR 355: Not listed
TSCA: Not listed

DIC5900 CAS #: 127564-92-5

DICHLOROPENTAFLUOROPROPANE

Molecular Formula: $C_3HCl_2F_5$
Formula Weight: 202.94

Hazard Overviews

Carcinogenicity: IARC - Not listed; NIOSH - Not listed; NTP - Not listed; ACGIH - Not listed; OSHA - Not listed; EPA - Not listed; MAK - Not listed

Environmental

Regulations

RCRA 40CFR: Not listed
CERCLA: 40CFR 302.4: Not listed
SARA 40CFR 372.65: Listed
SARA EHS 40CFR 355: Not listed
TSCA: Not listed

DIC5950 CAS #: 97-23-4

DICHLOROPHENE

RTECS: SM0175000
EINECS Number: 202-567-1
Molecular Formula: $C_{13}H_{10}Cl_2O_2$
Structured MF: $CH_2(C_6H_3(Cl)OH)_2$
Formula Weight: 269.13

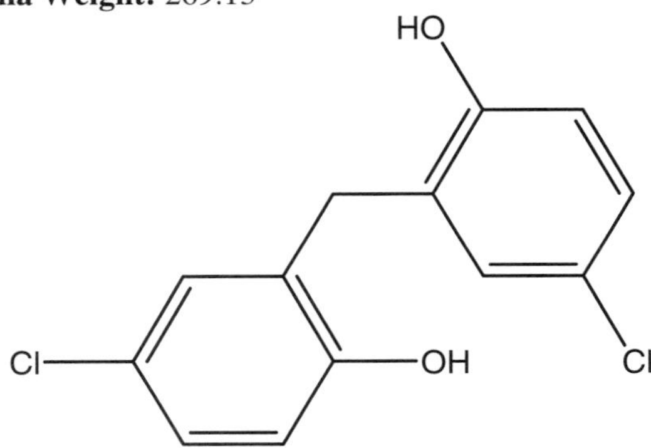

Chemical Structure

Synonyms: ANTHIPHEN; ANTIFEN; ANTIPHEN; BIS(5-CHLOR-2-HYDROXYPHENYL)-METHAN; BIS(5-CHLORO-2-HYDROXYPHENYL)METHANE; BIS(CHLOROHYDROXYPHENYL)METHANE; BIS-2-HYDROXY-5-CHLORFENYLMETHAN; BIS(2-HYDROXY-5-CHLOROPHENYL)METHANE; CORDOCEL; DDDM; DDM; DICESTAL; DICHLOORFEEN; 5,5'-DICHLORO-2,2'-DIHYDROXYDIPHENYLMETHANE; DICHLOROFEN; DI(5-CHLORO-2-HYDROXYPHENYL)METHANE; DI-(5-CHLORO-2-HYDROXYPHENYL)METHANE; 4,4'-DICHLORO-2,2'-METHYLENEDIPHENOL; DICHLOROPHEN; DICHLOROPHEN B; DICHLOROPHENE 10; DICHLORPHEN; DIDROXAN; DIDROXANE; DIFENTAN; 2,2'-DIHYDROXY-5,5'-DICHLORODIPHENYLMETHANE; DIPHENTANE 70; DIPHENTHANE 70; EMBEPHEN; FUNGICIDE FX; FUNGICIDE GM; FUNGICIDE M; G 4; G 4 PURE; G 4 TECHNICAL; GEFIR; GH; GINGIVIT; GIV GARD G 4-40; HALENOL; HYOSAN; KORIUM; O,O-METHYLEEN-BIS(4-CHLOORFENOL); 2,2'-METHYLENEBIS(4-CHLOROPHENOL); O,O-METILEN-BIS(4-CLOROFENOLO); PALACEL; PANACIDE; PARABIS; PHENOL,2,2'-METHYLENEBIS(4-CHLORO-; PLATH-LYSE; PREVENTAL; PREVENTOL; PREVENTOL GD; PREVENTOL GDC; SINDAR G 4; SUPER MOSSTOX; TAENIATOL; TENIATHANE; TENIATOL; TENIOTOL; VERMITHANA; WESPURIL

Description: white crystals; odorless
Use: an anthelmintic used in the treatment of infection by tapeworms (dwarf tapeworm, pork tapeworm); as a fungicide, germicide in soap and shampoo, herbicide, bactericide, antimicrobial (antiprotozoan) and algaecide; to control moss in turf and is the basis of a preparation against athlete's foot; in veterinary medicine to remove ascarids and hookworms and to treat ringworm; some dermatological and cosmetic applications

Physical Properties

Freezing Point: 177 °C (350.6 °F) to 178 °C (352.4 °F)
Saturated Vapor Density: 1.2 kg/m^3
Vapor Pressure: 9.75 x10^{-11} mm Hg at 25 °C
Water Solubility: < 1 mg/mL at 22 C
Other Solubilities: DMSO: >=100 mg/ml at 22 °C; Acetone: >=100 mg/ml at 22 °C; Toluene: Sparingly Soluble; Alcohol: 1 in 1 at 20 °C; Benzene: Slightly Soluble; Alkaline aqueous solutions: Soluble (with decomp).
Flash Point: Not available; probably combustible

RTECS Toxicity Data

Acute Oral: Rat LD_{50} Dose: 1506 mg/kg. Mouse LD_{50} Dose: 1 gm/kg.
Irritation Eye: Rabbit Standard Draize Test Dose: 50 ug/24H; Reaction: severe.
Irritation Skin: Rabbit Standard Draize Test Dose: 500 mg/24H; Reaction: mild.
Mutagenic: Bacteria - S Typhimurium Mutations in Microorganisms; Dose: 50 nmol/plate (+S9). Bacteria - S Typhimurium Mutations in Microorganisms; Dose: 150 ug/plate (-S9).

Hazard Overviews

Health: Irritating to eyes/respiratory tract; may be irritating to skin. Harmful. Other Acute Effects: harmful by inhalation, in contact with skin and if swallowed; risk of serious damage to eyes; causes photosensitivity; exposure to light can result in allergic reactions resulting in dermatologic lesions, which can vary from sunburnlike responses to edematous, vesiculated lesions or bullae; exposure can cause stomach pains, vomiting, and diarrhea; target organ: liver.
Fire: Will burn. Hazards: emits toxic fumes. Extinguishing agents: carbon dioxide, dry chemical powder or appropriate foam; water spray. Precautions: combustible liquid.
Carcinogenicity: IARC - Not listed; NIOSH - Not listed; NTP - Not listed; ACGIH - Not listed; OSHA - Not listed; EPA - Not listed; MAK - Not listed
Primary Target Organs:

Eyes

Skin

Respiratory System

Liver

Environmental

Ecotoxicity: Fishes: harlequin fish (Rasbora heteromorpha): mg/l (sodium salt) 24 hr 48 hr 96 hr3 m (extrap.) LC_{50},F 5.4 4.8 3.6 3.4
Environmental Fate: If released to water, it will be essentially non-volatile. Adsorption to sediment may be an important fate process. Bioconcentration in aquatic organisms should not be an important fate process. It is expected to biodegrade slowly in soil and aquatic conditions. If released to the atmosphere, it will exist primarily in the particulate phase. Removal of atmospheric material may occur through dry deposition. If released to soil, it is expected to have low to immobile soil mobility based on estimated K_{oc} values of 672 to 8 x10^4. Volatilization is not expected from either dry or moist soils.

Environmental Physical Data

Henry's Law Constant: estimated at 1.15 x10^{-12}
Sorption Partition Coefficient: K_{OC} = estimated at 8 x10^4
BCF: calculated at 91

Regulations

RCRA 40CFR: Not listed
CERCLA: 40CFR 302.4: Not listed
SARA 40CFR 372.65: Listed
SARA EHS 40CFR 355: Not listed
TSCA: Listed

Analytical Methods

Soil: EPA PMD-TLC
Water / Groundwater: EPA 604.1

DIC6000 CAS #: 576-24-9

2,3-DICHLOROPHENOL

RTECS: SK8450000
DOT: UN2020; IMO6.1
EINECS Number: 209-399-8
Molecular Formula: $C_6H_4Cl_2O$
Structured MF: $Cl_2C_6H_3OH$
Formula Weight: 163.00

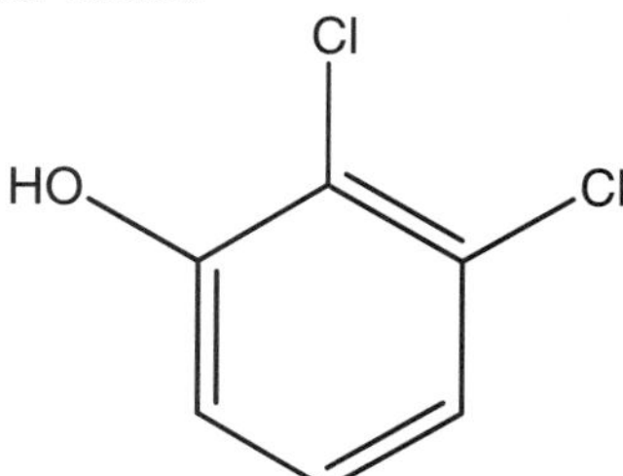

Chemical Structure

Synonyms: PHENOL,2,3-DICHLORO-
Description: crystals
Use: research chemical

Physical Properties

Boiling Point: 206 °C (403 °F)
Freezing Point: 57 °C (134.6 °F) to 59 °C (138.2 °F)
Saturated Vapor Density: Calculated 1.201305953 kg/m^3
Vapor Pressure: Calculated 0.179 mm Hg at 25 °C
Water Solubility: < 1 mg/mL at 20 C
Other Solubilities: 95% Ethanol: >=100 mg/ml at 20 °C; Acetone: >=100 mg/ml at 20 °C; DMSO: >=100 mg/ml at 20 °C; Ether: Soluble.
Odor Threshold: Water 30 ug/l
Flash Point: Not available; probably combustible

RTECS Toxicity Data

Acute Oral: Mouse LD_{50} Dose: 2376 mg/kg; Toxic Effects: Behavioral - Tremor; Behavioral - Convulsions or effect on seizure threshold; Lungs, Thorax, or Respiration - Respiratory stimulation.

Hazard Overviews

Corrosive

Health: Irritating to eyes/skin/respiratory tract. Other Acute Effects: harmful if swallowed, inhaled, or absorbed through skin; depending on the intensity and duration of exposure, effects may vary from mild irritation to severe destruction of tissue; prolonged contact can cause damage to the eyes, severe irritation or burns.
Fire: Will burn. Hazards: emits toxic fumes. Extinguishing agents: carbon dioxide, dry chemical powder or appropriate foam. Precautions: combustible liquid.
Reactivity: Incompatible with: acid chlorides, acid anhydrides, oxidizing agents. Hazardous decomposition products: toxic fumes of: carbon monoxide, carbon dioxide, hydrogen chloride gas.
Carcinogenicity: IARC - Not listed; NIOSH - Not listed; NTP - Not listed; ACGIH - Not listed; OSHA - Not listed; EPA - Not listed; MAK - Not listed
Primary Target Organs:

Eyes

Skin

Respiratory System

Environmental

Ecotoxicity: Crustacean: Daphnia magna 24h EC_{50}4.1; 5.2 mg/l 48h EC_{50}2.6; 3.1 mg/l; Fishes: Pimephales promelas 96h LC_{50}12
Environmental Fate: If released to the atmosphere, degradation can occur by reaction with photochemically formed hydroxyl radicals (estimated avg. half-life of 2.2 days). With a pKa of 7.70, it can exist in both the non-dissociated and ionized forms in environmental soil and water depending upon the pH of the media; in acidic media, the non-dissociated form will predominantly exist; the ionized form will be dominant at pHs above 7.70. If released to soil, the ionized form may adsorb less than the non-dissociated form. An average measured K_{oc} of 426 suggests medium soil mobility; monitoring detection in groundwater wells indicate that leaching can occur. The results of various biodegradation studies suggest that it can degrade under both aerobic and anaerobic conditions with appropriate inocula; however, actual biodegradation in the environment may occur very slowly at sites where the microbes are not adequately acclimated. If released to water, some adsorption to associated sediments may occur (pH may be a determining factor). Photo-oxidation in sunlit natural water may have some importance as a chemical degradation process.
Cleanup/Disposal: Guide No. 153: Eliminate all ignition sources (no smoking, flares, sparks or flames in immediate area). Do not touch damaged containers or spilled material unless wearing appropriate protective clothing. Stop leak if you can do it without risk. Prevent entry into waterways, sewers, basements or confined areas. Absorb or cover with dry earth, sand or other non-combustible material and transfer to containers. Do not get water inside containers.

Environmental Physical Data

Henry's Law Constant: estimated at 4.77 $x10^{-7}$
Octanol/Water Partition Coefficient: log K_{ow} = 2.84 to 3.15
Sorption Partition Coefficient: K_{oc} = 426
BCF: estimated at 85 to 146

Regulations

RCRA 40CFR: Not listed
CERCLA: 40CFR 302.4: Listed as Compound per CWA Section 307(a)
SARA 40CFR 372.65: Listed as Compound
SARA EHS 40CFR 355: Not listed
TSCA: Listed

Analytical Methods

Soil: SW846 8041
Water / Groundwater: ASTM D2580

DIC6050 CAS #: 120-83-2

2,4-DICHLOROPHENOL

RTECS: SK8575000
DOT: UN2020; IMO6.1
EINECS Number: 204-429-6
Molecular Formula: $C_6H_4Cl_2O$
Structured MF: $Cl_2C_6H_3OH$
Formula Weight: 163.00

Chemical Structure

Synonyms: 2,4-DCP; DCP; 2,4-DICHLOROHYDROXYBENZENE; 4,6-DICHLOROPHENOL; PHENOL,2,4-DICHLORO-

Description: colorless to white crystals; strong, medicinal odor

Use: organic synthesis, pesticides, insecticides, manufacture of 2,4-D, wood preservative, antiseptics and seed disinfectants

Physical Properties

Boiling Point: 210 °C (410 °F)

Freezing Point: 45 °C (113 °F)

Specific Gravity: 1.383 at 60 °C/25 °C

Vapor Density: 5.62 Air=1

Vapor Pressure: 1 mm Hg at 53 °C

Water Solubility: 0.45 parts per 100 parts

Other Solubilities: 95% Ethanol: >=100 mg/ml at 18 °C; Acetone: >=100 mg/ml at 18 °C; Alkaline solutions: Highly Soluble; Benzene: Very Soluble; Carbon Tetrachloride: Soluble; Chloroform: Very Soluble; DMSO: >=100 mg/ml at 18 °C; Ether: Very Soluble.

Odor Threshold: 1.4007 mg/m^3

Flash Point: 114 °C Open Cup

RTECS Toxicity Data

Acute Oral: Rat LD_{50} Dose: 580 mg/kg. Mouse LD_{50} Dose: 1276 mg/kg; Toxic Effects: Behavioral - Altered sleep time (including change in righting reflex); Behavioral - Ataxia; Lungs, Thorax, or Respiration - Dyspnea.

Acute Dermal: Rat LD_{50} Route: Subcutaneous Dose: 1730 mg/kg. Mammal LD_{50} Route: Skin; Dose: 790 mg/kg.

Chronic (Multiple Dose) Oral: Rat Dose: 91 gm/kg/13W-C; Toxic Effects: Blood - Changes in bone marrow not included above. Mouse Dose: 31770 mg/kg/90D-C; Toxic Effects: Behavioral - Fluid intake; Blood - Changes in cell count (unspecified); Biochemical - Phosphatases. Mouse Dose: 437 gm/kg/13W-C; Toxic Effects: Kidney, Ureter, and Bladder - Chgs in tubules (inc acute renal failure, acute tubular necrosis; DEATH. Mouse Dose: 67200 mg/kg/14D-C; Toxic Effects: DEATH.

Reproductive/Teratogenic: Rat Route: Oral; Dose: 20 mg/kg; Duration: female 1-20D of pregnancy; Specific Developmental Abnormalities - Other developmental abnormalities. Rat Route: Oral; Dose: 7500 mg/kg; Duration: female 6-15D of pregnancy; Effects on Embryo or Fetus - Fetotoxicity. Rat Route: Oral; Dose: 7500 mg/kg; Duration: female 6-15D of pregnancy; Specific Developmental Abnormalities - Musculoskeletal system.

Mutagenic: Rat DNA Damage; Cell Type: liver; Dose: 200 umol/L. Mouse Mutations in Mammalian Somatic Cells; Cell Type: lymphocyte; Dose: 30 mg/L.

Tumorigenic: Mouse Route: Skin; Dose: 16 gm/kg/39W-I; Toxic Effects: Tumorigenic - Carcinogenic by RTECS criteria; Skin and appendages - Tumors.

Hazard Overviews

Corrosive

Fire Diamond

Health: Corrosive to eyes/skin/respiratory tract. Poison. Also Causes: labored breathing, pulmonary edema, corneal damage, abdominal pain, vomiting, diarrhea, paleness, sweating, weakness, headache, CNS effects (tremor, headache, ringing in ears), unconsciousness; scanty, dark-colored urine, kidney insufficiency, liver damage, methemoglobinemia, hyperbilirubinemia, and hemolytic anemia.

Fire: Will burn. Use dry chemical, carbon dioxide, water spray, or regular foam. Hazardous combustion products may include, carbon oxide(s), hydrogen chloride, chlorine gas.

Reactivity: Stable. Hazardous polymerization cannot occur. Avoid: heat; ignition sources Incompatible with: oxidizers; acids; acid fumes. Hazardous decomposition products: carbon oxide(s); hydrogen chloride; chlorine gases.

Carcinogenicity: IARC - Not listed; NIOSH - Not listed; NTP - Not listed; ACGIH - Not listed; OSHA - Not listed; EPA - Not listed; MAK - Not listed

Primary Target Organs:

Eyes

Skin

Respiratory System

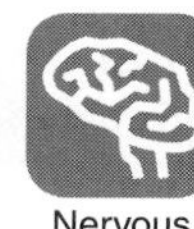
Nervous System

Liver

Blood

Environmental

Ecotoxicity: LC_{50} Poecilia reticulata (guppy) 4.2 ppm/24 hr at pH 7.3 /Conditions of bioassay not specified LC_{50} Lepomis macrochirus (bluegill) 2020 ug/l/96 hr /Static bioassay LC_{50} Pimephales promelas (fathead minnow) 7.75 mg/l/96 hr, (30 days old), confidence limit 7.47-8.05. Test conditions: Water temp= 25.4 °C, dissolved oxygen= 7.9 mg/l, water hardness= 45.2 mg/l calcium carbonate, alkalinity= 42.1 mg/l calcium carbonate LC_{50} Daphnia magna (cladoceran) 2610 ug/l/48 hr /Static bioassay LC_{50} Pimephales promelas (fathead minnow, juvenile) 8230 ug/l/96 hr /Flow-through bioassay Toxicity Threshold (Cell Multiplication Inhibition Test) Entosiphon sulcatum (protozoa) 0.5 mg/l /Conditions of bioassay not specified Toxicity Threshold (Cell Multiplication Inhibition Test) Microcystis aeruginosa (algae) 2 mg/l /Conditions of bioassay not specified

Environmental Fate: If released to the atmosphere, degradation can occur by reaction with photochemically formed hydroxyl radicals (estimated avg. half-life of 5.3 days). Physical removal from air may occur via rainfall. With a pKa of 7.8, it can exist in both the non-dissociated and ionized forms in environmental soil and water depending upon the pH of the media. If released to soil, moderate to slow leaching is possible based on observed K_{oc} values of 200-5000; the ionized form appears more susceptible to leaching than the non-dissociated form. Various biodegradation studies have demonstrated that it is biodegradable under aerobic and anaerobic conditions in both soil and water. If released to water, adsorption to sediments may be important under various conditions determined, in part, by pH. Photodegradation in natural water can occur by direct photolysis or by reaction with sunlight-formed oxidants (singlet oxygen and peroxy radicals).

Cleanup/Disposal: Guide No. 153: Eliminate all ignition sources (no smoking, flares, sparks or flames in immediate area). Do not touch damaged containers or spilled material unless wearing appropriate protective clothing. Stop leak if you can do it without risk. Prevent entry into waterways, sewers, basements or confined areas. Absorb or cover with dry earth, sand or other non-combustible material and transfer to containers. Do not get water inside containers.

Environmental Physical Data

Henry's Law Constant: estimated at 3.57 x10^{-6}
Octanol/Water Partition Coefficient: log K_{ow} = 3.06
Sorption Partition Coefficient: K_{oc} = 200 to 5000
BCF: golden ide fish 100
BOD: 100%, 5 days

Regulations

RCRA 40CFR: Listed Hazardous Waste No. U081 Toxic Waste
CERCLA: 40CFR 302.4: Listed per RCRA Section 3001 per CWA Section 307(a) RQ: 100 lb (45.35 kg)
SARA 40CFR 372.65: Listed
SARA EHS 40CFR 355: Not listed
TSCA: Listed

Analytical Methods

Air: EPA 0020
Soil: CLP LC_SV, MC_SVOA, OHC; EPA 16, 1625, S-004-1; SW846 3630B, 3640A, 8040A, 8041, 8250A, 8270B, 8270C, 8275, 8410
Water / Groundwater: EPA 6, S-002-1, 1625, 1653, 604, 625, 625-S; APHA 6410-B, 6420-BA, 6420-BB, 6420-C; ASTM D2580; NCASI CP-85.01, CP-86.01; USGS O3117
Drinking Water: EPA 552
Plasma: EPA 29
Other: EPA O-005-1

DIC6100 CAS #: 583-78-8

2,5-DICHLOROPHENOL

RTECS: SK8600000
DOT: UN2020; IMO6.1
EINECS Number: 209-520-4
Molecular Formula: $C_6H_4Cl_2O$
Structured MF: $Cl_2C_6H_3OH$
Formula Weight: 163.0

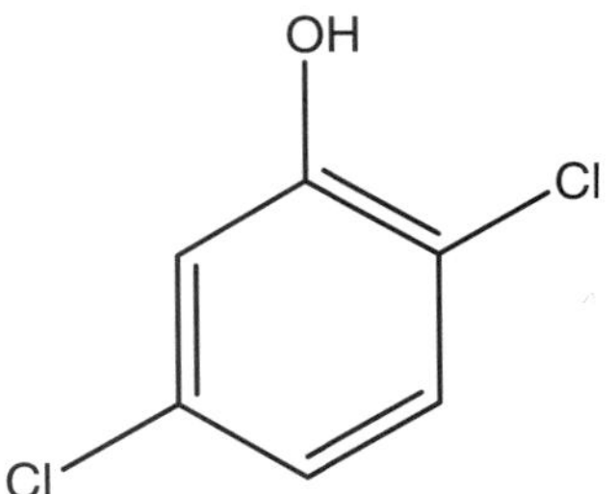

Chemical Structure

Synonyms: 2,5-DCP; 3,6-DICHLOROPHENOL; PHENOL,2,5-DICHLORO-
Description: prisms
Use: chem int for 3,6-dichloro-o-anisic acid, the herbicide dicamba

Physical Properties

Boiling Point: 211 °C (412 °F) at 744 mm Hg
Freezing Point: 59 °C (138.2 °F)
Saturated Vapor Density: Calculated 1.200911978 kg/m^3
Vapor Pressure: Calculated 0.125 mm Hg at 25 °C
Water Solubility: Slightly Soluble in Water
Other Solubilities: 95% Ethanol: >=100 mg/ml at 20 °C; Acetone: >=100 mg/ml at 20 °C; Benzene: Soluble; DMSO: >=100 mg/ml at 20 °C; Ether: Soluble.
Odor Threshold: Water at 30 °C 33 ug/l
Flash Point: Not available; probably combustible

RTECS Toxicity Data

Acute Oral: Rat LD_{50} Dose: 580 mg/kg. Mouse LD_{50} Dose: 946 mg/kg; Toxic Effects: Behavioral - Tremor; Behavioral - Convulsions or effect on seizure threshold; Lungs, Thorax, or Respiration - Respiratory stimulation.
Mutagenic: Mouse Sister Chromatid Exchenge; Route: Intraperitoneal; Dose: 210 mg/kg.

Hazard Overviews

Corrosive

Health: Irritating to eyes/skin/respiratory tract. Harmful. Other Acute Effects: harmful if swallowed, inhaled, or absorbed through skin; depending on the intensity and duration of exposure, effects may vary from mild irritation to severe destruction of tissue; prolonged contact can cause damage to the eyes, severe irritation or burns.
Fire: Will burn. Hazards: emits toxic fumes. Extinguishing agents: carbon dioxide, dry chemical powder or appropriate foam. Precautions: combustible liquid.
Reactivity: Incompatible with: acid chlorides, acid anhydrides, oxidizing agents. Hazardous decomposition products: toxic fumes of: carbon monoxide, carbon dioxide, hydrogen chloride gas.

Carcinogenicity: IARC - Not listed; NIOSH - Not listed; NTP - Not listed; ACGIH - Not listed; OSHA - Not listed; EPA - Not listed; MAK - Not listed

Primary Target Organs:

Eyes

Skin

Respiratory System

Environmental

Environmental Fate: If released to the atmosphere, degradation can occur by reaction with photochemically formed hydroxyl radicals (estimated avg. half-life of 2.2 days). With a pKa of 6.34, it can exist in both the non-dissociated and ionized forms in environmental soil and water depending upon the pH of the media. If released to soil, the ionized form may adsorb less than the non-dissociated form and be more likely to leach. However, sufficient experimental data are not available to accurately estimate leachability in soil. The results of various biodegradation studies suggest that it can degrade under both aerobic and anaerobic conditions with appropriate inocula; however, it is possible that the chemical may resist biodegradation in many environmental media. If released to water, adsorption to sediments may be significant under various conditions since one monitoring study found a much higher concentration in sediments than in the associated water column. Photo-oxidation in sunlit natural water may have some importance as a chemical degradation process.

Cleanup/Disposal: Guide No. 153: Eliminate all ignition sources (no smoking, flares, sparks or flames in immediate area). Do not touch damaged containers or spilled material unless wearing appropriate protective clothing. Stop leak if you can do it without risk. Prevent entry into waterways, sewers, basements or confined areas. Absorb or cover with dry earth, sand or other non-combustible material and transfer to containers. Do not get water inside containers.

Environmental Physical Data

Henry's Law Constant: estimated at 4.77 x10^{-7}

Octanol/Water Partition Coefficient: log K_{ow} = 2.92 to 3.06

Sorption Partition Coefficient: K_{oc} = estimated at 510 to 710

BCF: estimated at 98 to 125

Regulations

RCRA 40CFR: Not listed

CERCLA: 40CFR 302.4: Listed as Compound per CWA Section 307(a)

SARA 40CFR 372.65: Listed as Compound

SARA EHS 40CFR 355: Not listed

TSCA: Listed

Analytical Methods

Soil: SW846 8041

Water / Groundwater: ASTM D2580

DIC6150	CAS #: 87-65-0

2,6-DICHLOROPHENOL

RTECS: SK8750000

DOT: UN2020; IMO6.1

EINECS Number: 201-761-3

Molecular Formula: $C_6H_4Cl_2O$

Structured MF: $Cl_2C_6H_3OH$

Formula Weight: 163.0

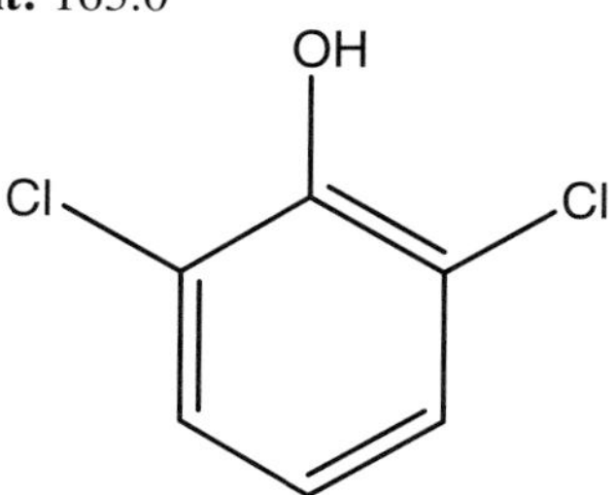

Chemical Structure

Synonyms: 2,6-DCP; 2,6-DICHLORFENOL; PHENOL,2,6-DICHLORO-

Description: needles; strong odor

Use: starting material for the manufacture of trichlorophenols, tetrachlorophenols and pentachlorophenols

Physical Properties

Boiling Point: 219 °C (426 °F) to 220 °C (428 °F) at 740 mm Hg

Freezing Point: 68 °C (154.4 °F) to 69 °C (156.2 °F)

Saturated Vapor Density: 1.20026265 kg/m^3

Vapor Pressure: 0.036 mm Hg at 25 °C

Water Solubility: < 1 mg/mL at 20 C

Other Solubilities: 95% Ethanol: >=100 mg/ml at 20 °C; Acetone: >=100 mg/ml at 20 °C; Benzene: Soluble; DMSO: >=100 mg/ml at 20 °C; Ether: Soluble.

Odor Threshold: 0.003 to 0.2 mg/kg

Ionization Potential (eV): 8.65 +/-0.02

Flash Point: Not available; probably combustible

RTECS Toxicity Data

Acute Oral: Mouse LD_{50} Dose: 2120 mg/kg; Toxic Effects: Behavioral - Tremor; Behavioral - Convulsions or effect on seizure threshold; Lungs, Thorax, or Respiration - Respiratory stimulation.

Irritation Eye: Rabbit Standard Draize Test Dose: 250 ug/24H; Reaction: severe.

Irritation Skin: Rabbit Standard Draize Test Dose: 2 mg/24H; Reaction: severe.

Hazard Overviews

Corrosive

Health: Corrosive to eyes/skin/respiratory system. Harmful. Other Acute Effects: harmful if swallowed, inhaled, or absorbed through skin; inhalation may result in spasm, inflammation and edema of the larynx and bronchi, chemical pneumonitis and pulmonary edema; symptoms of exposure may include burning sensation, coughing, wheezing, laryngitis, shortness of breath, headache, nausea and vomiting; exposure can cause CNS depression and damage to the liver. Chronic Effects: exposure may cause depression of the bone marrow; target organs: central nervous system, liver.

Fire: Will burn. Hazards: emits toxic fumes. Extinguishing agents: carbon dioxide, dry chemical powder or appropriate foam. Precautions: combustible liquid.

Reactivity: Stable. Hazardous polymerization will not occur. Incompatible with: acid chlorides, acid anhydrides, oxidizing agents. Hazardous decomposition products: toxic fumes of: carbon monoxide, carbon dioxide, hydrogen chloride gas.

Carcinogenicity: IARC - Not listed; NIOSH - Not listed; NTP - Not listed; ACGIH - Not listed; OSHA - Not listed; EPA - Not listed; MAK - Not listed

Primary Target Organs:

Eyes

Skin

Respiratory System

Nervous System

Liver

Environmental

Ecotoxicity: LC_{50} Salmo trutta (trout) 4.0 ppm/24 hr at 5 °C, wt 4.5 g (purified material) /Conditions of bioassay not specified LC_{50} Idus idus melanotus (fish) 4 mg/l/48 hr /Conditions of bioassay not specified

Environmental Fate: If released to the atmosphere, degradation can occur by reaction with photochemically formed hydroxyl radicals (estimated avg. half-life of 5.3 days). Physical removal from air may occur via rainfall. With a pKa of 6.8, it can exist in both the non-dissociated and ionized forms in environmental soil and water depending upon the pH of the media. If released to soil, the ionized form may adsorb poorly to soil material, and therefore leach. Sufficient experimental data are not available to accurately estimate the leachability of the non-dissociated form. Various biodegradation studies have demonstrated that it is significantly degradable under aerobic conditions, but only poorly degradable or resistant to degradation under anaerobic conditions. Biodegradation under aerobic conditions may be the most important environmental fate process in soil. If released to water, adsorption to sediments may be significant under various conditions since one monitoring study found a higher concentration in sediments than in the associated water column. Photo-oxidation in sunlit natural water may have some importance as a chemical degradation process.

Cleanup/Disposal: Guide No. 153: Eliminate all ignition sources (no smoking, flares, sparks or flames in immediate area). Do not touch damaged containers or spilled material unless wearing appropriate protective clothing. Stop leak if you can do it without risk. Prevent entry into waterways, sewers, basements or confined areas. Absorb or cover with dry earth, sand or other non-combustible material and transfer to containers. Do not get water inside containers.

Environmental Physical Data

Henry's Law Constant: estimated at 4.77×10^{-7}
Octanol/Water Partition Coefficient: log K_{ow} = 2.64
Sorption Partition Coefficient: K_{oc} = estimated at 270
BCF: calculated at 12

Regulations

RCRA 40CFR: Listed Hazardous Waste No. U082 Toxic Waste
CERCLA: 40CFR 302.4: Listed per RCRA Section 3001 RQ: 100 lb (45.35 kg)
SARA 40CFR 372.65: Listed as Compound
SARA EHS 40CFR 355: Not listed
TSCA: Listed

Analytical Methods

Soil: SW846 3640A, 8040A, 8041, 8250A, 8270B, 8270C
Water / Groundwater: EPA 1625, 1653; ASTM D2580; NCASI CP-85.01, CP-86.01
Plasma: EPA 29

DIC6200 CAS #: 95-77-2

3,4-DICHLOROPHENOL

RTECS: SK8800000
DOT: UN2020; IMO6.1
EINECS Number: 202-450-5
Molecular Formula: $C_6H_4Cl_2O$
Structured MF: $Cl_2C_6H_3OH$
Formula Weight: 163.0

Chemical Structure

Synonyms: 3,4-DCP; 4,5-DICHLOROPHENOL; PHENOL,3,4-DICHLORO-
Description: needles
Use: chem int for 2-chloro-1,4-dihydroxyanthraquinone & 2,3,4-trichlorophenol

Physical Properties

Boiling Point: 253.5 °C (488 °F) at 767 mm Hg
Freezing Point: 68 °C (154.4 °F)
Saturated Vapor Density: Calculated 1.200015832 kg/m^3
Vapor Pressure: Calculated 0.00217 mm Hg at 25 °C
Water Solubility: < 1 mg/mL at 20 C

Other Solubilities: 95% Ethanol: >=100 mg/ml at 20 °C; Acetone: >=100 mg/ml at 20 °C; Benzene: Soluble; DMSO: >=100 mg/ml at 20 °C; Ether: Soluble.
Odor Threshold: Water 100 ug/l
Flash Point: Not available; probably combustible

RTECS Toxicity Data

Acute Oral: Mouse LD_{50} Dose: 1685 mg/kg; Toxic Effects: Behavioral - Tremor; Behavioral - Convulsions or effect on seizure threshold; Lungs, Thorax, or Respiration - Respiratory stimulation.

Hazard Overviews

Corrosive

Health: Severely irritating. Harmful. Other Acute Effects: harmful by inhalation, in contact with skin and if swallowed; high concentrations are extremely destructive to tissues of the mucous membranes and upper respiratory tract, eyes and skin; symptoms of exposure may include: burning sensation, coughing, wheezing, laryngitis, shortness of breath, headache, nausea and vomiting; exposure can cause: tremors, CNS depression; prolonged contact can cause damage to the eyes, severe irritation, or burns.
Fire: Will burn. Hazards: emits toxic fumes. Extinguishing agents: carbon dioxide, dry chemical powder or appropriate foam. Precautions: combustible liquid.
Carcinogenicity: IARC - Not listed; NIOSH - Not listed; NTP - Not listed; ACGIH - Not listed; OSHA - Not listed; EPA - Not listed; MAK - Not listed
Primary Target Organs:

Eyes

Skin

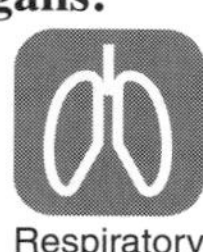

Respiratory System

Environmental

Environmental Fate: If released to the atmosphere, degradation can occur by reaction with photochemically formed hydroxyl radicals (estimated avg. half-life of 2.2 days). With a pKa of 8.59, it can exist in both the non-dissociated and ionized forms in environmental soil and water depending upon the pH of the media; in acidic media, the non-dissociated form will predominantly exist; the ionized form will be dominant at pHs above 8.59. If released to soil, the ionized form may adsorb less than the non-dissociated form and be more likely to leach. Groundwater monitoring at a Finnish sawmill has shown that leaching can occur. The results of various biodegradation studies suggest that it can degrade under both aerobic and anaerobic conditions with appropriate inocula; however, actual biodegradation in the environment may occur very slowly at sites where the microbes are not adequately acclimated. If released to water, adsorption to sediments may be significant for the non-dissociated material. Photo-oxidation in sunlit natural water may have some importance as a chemical degradation process.
Cleanup/Disposal: Guide No. 153: Eliminate all ignition sources (no smoking, flares, sparks or flames in immediate area). Do not touch damaged containers or spilled material unless wearing appropriate protective clothing. Stop leak if you can do it without risk. Prevent entry into waterways, sewers, basements or confined areas. Absorb or cover with dry earth, sand or other non-combustible material and transfer to containers. Do not get water inside containers.

Environmental Physical Data

Henry's Law Constant: estimated at 4.77×10^{-7}
Octanol/Water Partition Coefficient: log K_{ow} = 3.33
Sorption Partition Coefficient: K_{oc} = 1300
BCF: estimated at 200

Regulations

RCRA 40CFR: Not listed
CERCLA: 40CFR 302.4: Listed as Compound per CWA Section 307(a)
SARA 40CFR 372.65: Listed as Compound
SARA EHS 40CFR 355: Not listed
TSCA: Listed

Analytical Methods

Soil: SW846 8041
Water / Groundwater: ASTM D2580; NCASI CP-85.01, CP-86.01

DIC6250	CAS #: 591-35-5

3,5-DICHLOROPHENOL

RTECS: SK8820000
DOT: UN2020; IMO6.1
EINECS Number: 209-714-9
Molecular Formula: $C_6H_4Cl_2O$
Structured MF: $Cl_2C_6H_3OH$
Formula Weight: 163.00

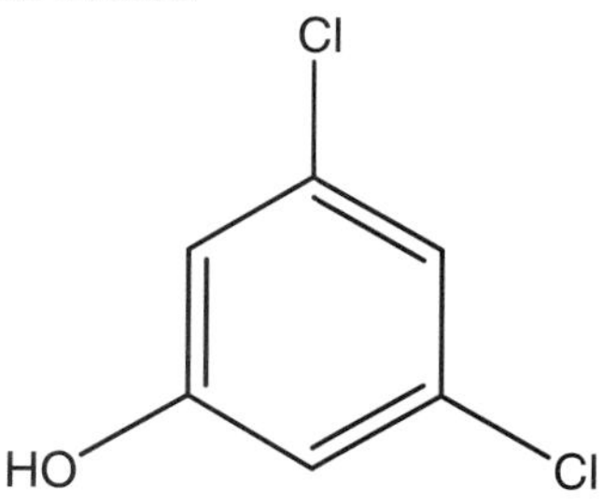

Chemical Structure

Synonyms: 3,5-DCP; PHENOL,3,5-DICHLORO-
Description: prisms
Use: research chemical

Physical Properties

Boiling Point: 233 °C (451 °F) at 757 mm Hg
Freezing Point: 68 °C (154.4 °F)

Saturated Vapor Density: Calculated 1.200060993 kg/m^3
Vapor Pressure: Calculated 0.00836 mm Hg at 25 °C
Water Solubility: Slightly Soluble in Water
Other Solubilities: 95% Ethanol: >=100 mg/ml at 20 °C; Acetone: >=100 mg/ml at 20 °C; DMSO: >=100 mg/ml at 20 °C; Ether: Soluble.
Flash Point: Not available; probably combustible

RTECS Toxicity Data

Acute Oral: Mouse LD_{50} Dose: 2389 mg/kg; Toxic Effects: Behavioral - Tremor; Behavioral - Convulsions or effect on seizure threshold; Lungs, Thorax, or Respiration - Respiratory stimulation.
Reproductive/Teratogenic: Rat Route: Oral; Dose: 2400 mg/kg; Duration: female 6-15D of pregnancy; Maternal Effects - Other effects on females; Effects on Fertility - Post-implantation mortality; Litter size. Rat Route: Oral; Dose: 2400 mg/kg; Duration: female 6-15D of pregnancy; Effects on Embryo or Fetus - Fetotoxicity.

Hazard Overviews

Corrosive

Health: Irritating to eyes/skin/respiratory tract. Other Acute Effects: harmful if swallowed, inhaled, or absorbed through skin; depending on the intensity and duration of exposure, effects may vary from mild irritation to severe destruction of tissue; prolonged contact can cause damage to the eyes severe irritation or burns.
Fire: Will burn. Hazards: emits toxic fumes. Extinguishing agents: carbon dioxide, dry chemical powder or appropriate foam. Precautions: combustible liquid.
Reactivity: Incompatible with: acid chlorides, acid anhydrides, oxidizing agents. Hazardous decomposition products: toxic fumes of: carbon monoxide, carbon dioxide, hydrogen chloride gas.
Carcinogenicity: IARC - Not listed; NIOSH - Not listed; NTP - Not listed; ACGIH - Not listed; OSHA - Not listed; EPA - Not listed; MAK - Not listed
Primary Target Organs:

Eyes

Skin

Respiratory System

Environmental

Environmental Fate: If released to the atmosphere, degradation can occur by reaction with photochemically formed hydroxyl radicals (estimated avg. half-life of 0.9 days). With a pKa of 8.19, it can exist in both the non-dissociated and ionized forms in environmental soil and water depending upon the pH of the media; in acidic media, the non-dissociated form will predominantly exist; the ionized form will be dominant at pHs above 8.19. If released to soil, the ionized form may adsorb less than the non-dissociated form and be more likely to leach. However, sufficient experimental data are not available to accurately estimate leachability in soil. The results of various biodegradation studies suggest that it can degrade under both aerobic and anaerobic condition with appropriate inocula; however, actual biodegradation in the environment may occur very slowly at sites where the microbes are not adequately acclimated. If released to water, adsorption to sediments may be significant for the non-dissociated material. Photo-oxidation in sunlit natural water may have some importance as a chemical degradation process.
Cleanup/Disposal: Guide No. 153: Eliminate all ignition sources (no smoking, flares, sparks or flames in immediate area). Do not touch damaged containers or spilled material unless wearing appropriate protective clothing. Stop leak if you can do it without risk. Prevent entry into waterways, sewers, basements or confined areas. Absorb or cover with dry earth, sand or other non-combustible material and transfer to containers. Do not get water inside containers.

Environmental Physical Data

Henry's Law Constant: estimated at 4.77 x10^{-7}
Octanol/Water Partition Coefficient: log K_{ow} = 3.62 to 3.68
Sorption Partition Coefficient: K_{oc} = 2570 to 2950
BCF: estimated at 330 to 395

Regulations

RCRA 40CFR: Not listed
CERCLA: 40CFR 302.4: Listed as Compound per CWA Section 307(a)
SARA 40CFR 372.65: Listed as Compound
SARA EHS 40CFR 355: Not listed
TSCA: Listed

Analytical Methods

Soil: SW846 8041
Water / Groundwater: NCASI CP-85.01, CP-86.01

DIC6300 **CAS #: 94-75-7**

2,4-DICHLOROPHENOXY ACETIC ACID

RTECS: AG6825000
DOT: UN2765; UN2766; UN2999; UN3000; NA2765; IMO3.2; IMO6.1
EINECS Number: 202-361-1
Molecular Formula: $C_8H_6Cl_2O_3$
Structured MF: $Cl_2C_6H_3OCH_2COOH$
Formula Weight: 221.04

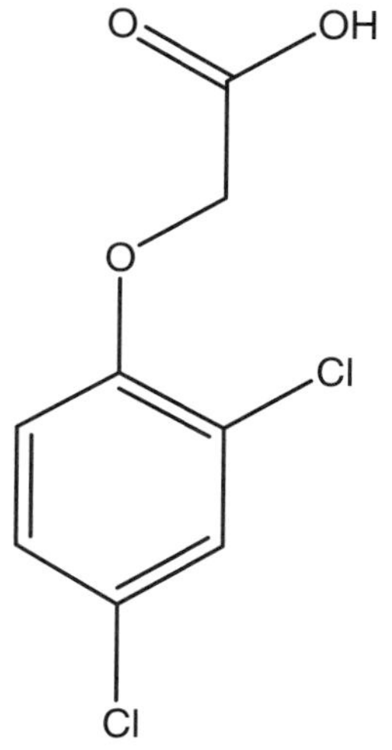

Chemical Structure

Synonyms: ACETIC ACID,(2,4-DICHLOROPHENOXY)-; ACIDE 2,4-DICHLORO PHENOXYACETIQUE; ACIDO(2,4-DICHLORO-FENOSSI)-ACETICO; ACIDO(2,4-DICLORO-FENOSSI)-ACETICO; ACME AMINE 4; ACME BUTYL ESTER 4; ACME LV 4; ACME LV 6; AGRICORN D; AGROTECT; AGROXONE; AMIDOX; AMOXONE; AQUA-KLEEN; BARRAGE; BH 2,4-D; BLADEX-B; BRUSH KILLER 64; BRUSH-RHAP; BUTOXY-D 3: 1 LIQUID EMULSIFIABLE BRUSHKILLER LV96; CHIPCO TURF HERBICIDE; CHLOROXONE; CITRUS FIX; CROP RIDER; CROPRIDER; CROTILIN; 2,4-D; D 50; 2,4-D ACID; D50; DACAMINE; DEBROUSSAILLANT 600; DECAMINE; DED-WEED; DED-WEED LV-69; DEHERBAN; DE-PESTER DED-WEED LV-2; DESORMONE; (2,4-DICHLOOR-FENOXY)-AZIJNZUUR; 2,4-DICHLOOR-FENOXY-AZIJNZUUR; (2,4-DICHLOROPHENOXY)ACETIC ACID; 2,4-DICHLOROPHENOXYACETIC ACID; DICHLOROPHENOXYACETIC ACID; 2,4-DICHLORPHENOXYACETIC ACID; (2,4-DICHLOR-PHENOXY)-ESSIGSAEURE; 2,4-DICHLOR-PHENOXY-ESSIGSAEURE; DICOFUR; DICOPUR; DICOTOX; DINOXOL; DMA-4; DMA 4; DORMON; DORMONE; 2,4-DWUCHLOROFENOKSYOCTOWY KWAS; EMULSAMINE; EMULSAMINE E-3; EMULSAMINE BK; ENT 8,538; ENVERT 171; ENVERT DT; ESTERON; ESTERONE FOUR; ESTONE; FARMCO; FERNESTA; FERNIMINE; FERNOXONE; FERXONE; FOREDEX 75; GREEN CROSS WEED-NO-MORE; HEDONAL; HEDONAL (THE HERBICIDE); HERBIDAL; HIVOL-44; IPANER; KWAS 2,4-DWUCHLOROFENOKSYOCTOWY; KWASU 2,4-DWUCHLOROFENOKSYOCTOWEGO; KYSELINA 2,4-DICHLORFENOXYOCTOVA; LAWN-KEEP; MACONDRAY; MACRONDRAY; MIRACLE; MONOSAN; MOTA MASKROS; MOXON; MOXONE; NETAGRONE; NETAGRONE 600; NSC 423; 2,4-PA; PENNAMINE; PENNAMINE D; PHENOX; PHENOXYACETIC ACID,2,4-DICHLORO; PIELIK; PLANTGARD; R-H WEED RHAP 20; RED DEVIL DRY WEED KILLER; RHODIA; SALVO; SCOTT'S 4-XD WEED CONTROL; B-SELEKTONON; SILVAPROP 1; SPRITZ-HORMIN; SPRITZ-HORMIN/2,4-D; SPRITZ-HORMIT/2,4-D; SUPER D WEEDONE; SUPERORMONE CONCENTRE; TRANSAMINE; TRIBUTON; TRINOXOL; U 46; U 46 D; U 46DP; U-5043; VERGEMASTER; VERTON; VERTON 2D; VERTON 38; VERTON D; VIDON 638; WEED TOX; WEED-AG-BAR; WEEDAR; WEEDAR-64; WEEDATUL; WEEDEZ WONDER BAR; WEED-B-GON; WEEDONE; WEEDONE 100 EMULSIFIABLE; WEEDONE LV4; WEED-RHAP; WEED-RHAP A-4; WEED-RHAP B-266; WEED-RHAP B-4; WEED-RHAP I-3.34; WEED-RHAP LV-4-0; WEEDTROL

Use: selective weed-killer, systemic herbicide and defoliant; to increase the latex output of old rubber trees and in fruit drop control

Physical Properties

Boiling Point: 160 °C (320 °F) at 0.4 mm Hg
Freezing Point: 138 °C (280.4 °F)
Specific Gravity: 1.416 at 25 °C
Vapor Density: 7.63 Air=1
Saturated Vapor Density: 5.366636191 kg/m^3
Density: 1.563 g/mL at 20 °C
Vapor Pressure: 53 Pa at 160 °C
Water Solubility: 540 ppm in Water at 20 °C
Other Solubilities: Insoluble in petroleum oils; 85.0 g/100 g Acetone at 25 °C; 1.07 g/100 g Benzene at 28 °C; 0.5 g/100 g Carbon Disulfide at 29 °C; 0.1 g/100 g Carbon Tetrachloride at 25 °C; 0.10 to 0.35 g/100 g diesel oil and Kerosene at 25 °C
Surface Tension: 66.5 dynes/cm
Odor Threshold: Technical grade 3.13 mg/kg
pH: Weak acid (Ka= 2.3 x10^{-3})
Flash Point: Nonflammable

RTECS Toxicity Data

Acute Oral: Man TD_{Lo} Dose: 2 gm/kg; Toxic Effects: Behavioral - Coma; Lungs, Thorax, or Respiration - Respiratory depression. Man TD_{Lo} Dose: 5714 mg/kg; Toxic Effects: Behavioral - Coma; Cardiac - Change in rate; Lungs, Thorax, or Respiration - Respiratory depression. Human LD_{Lo} Dose: 80 mg/kg; Toxic Effects: Gastrointestinal - Nausea or vomiting; Behavioral - Coma; Behavioral - Somnolence (general depressed activity). Man LD_{Lo} Dose: 93 mg/kg; Toxic Effects: Behavioral - Convulsions or effect on seizure threshold. Rat LD_{50} Dose: 375 mg/kg.
Acute Dermal: Rabbit LD_{50} Route: Skin; Dose: 1400 mg/kg; Toxic Effects: Behavioral - Ataxia; Skin and appendages - Primary irritation. Rat LD_{50} Route: Skin; Dose: 1500 mg/kg.
Chronic (Multiple Dose) Oral: Rat Dose: 13650 mg/kg/13W-C; Toxic Effects: Nutritional and gross metabolic - Weight loss or decreased weight gain. Rat Dose: 200 mg/kg/5W-I; Toxic Effects: Behavioral - Muscle weakness. Rat Dose: 54750 mg/kg/1Y-C; Toxic Effects: Sense organs and special senses - Retinal changes (pigmentary deposition, retinitis, other); Behavioral - Change in motor activity (specific assay).
Irritation Eye: Rabbit Standard Draize Test Dose: 750 ug/24H; Reaction: severe.
Irritation Skin: Rabbit Standard Draize Test Dose: 500 mg/24H; Reaction: mild.
Reproductive/Teratogenic: Rat Route: Oral; Dose: 220 ug/kg; Duration: female 1-22D of pregnancy; Specific Developmental Abnormalities - Blood and lymphatic systems (including spleen and marrow). Rat Route: Oral; Dose: 1 gm/kg; Duration: female 6-15D of pregnancy; Specific Developmental Abnormalities - Musculoskeletal system; Effects on Embryo or Fetus - Fetotoxicity; Fetal death. Rat Route: Oral; Dose: 125 mg/kg; Duration: female 6-15D of pregnancy; Specific Developmental Abnormalities - Musculoskeletal system. Rat Route: Oral; Dose: 500 mg/kg; Duration: female 6-15D of pregnancy; Effects on Embryo or Fetus - Fetotoxicity; Specific Developmental Abnormalities - Central nervous system; Urogenital system. Rat Route: Oral;

Dose: 500 mg/kg; Duration: female 6-15D of pregnancy; Specific Developmental Abnormalities - Homeostasis; Effects on Newborn - Growth statistics.

Mutagenic: Human Unscheduled DNA Synthesis; Cell Type: fibroblast; Dose: 1 umol/L. Human Cytogenetic Analysis; Cell Type: lymphocyte; Dose: 20 ug/L. Human Sister Chromatid Exchenge; Cell Type: lymphocyte; Dose: 10 mg/L.

Hazard Overviews

Reactivity: Stable. Hazardous polymerization cannot occur. Avoid: heat; ignition sources; oxidizers. Incompatible with: strong oxidizers; metals. Hazardous decomposition products: hydrogen chloride; phosgene; carbon monoxide; chlorides.

Carcinogenicity: IARC - Group 2B, Possibly carcinogenic to humans; NIOSH - Not listed; NTP - Not listed; ACGIH - Class A4, Not classifiable as a human carcinogen; OSHA - Not listed; EPA - Not listed; MAK - Not listed

Exposure Limits

OSHA PEL: TWA: 10 mg/m^3.

ACGIH TLV: TWA: 10 mg/m^3.

NIOSH REL: TWA: 10 mg/m^3.

NIOSH IDLH: 100 mg/m^3.

DFG MAK: TWA: 10 mg/m^3.

Respirator Recommendation

Exposure Range: >10 to <100 mg/m^3 Air Purifying, Negative Pressure, Half Mask

Exposure Range: 100 to unlimited mg/m^3 Self-contained Breathing Apparatus, Pressure Demand, Full Face

Cartridge Color: black with magenta (P100)

Environmental

Ecotoxicity: LC_{50} Banded killifish 26.7 mg/l/96 hr static bioassay /SRP TLm Crassostrea virginica (American oyster), egg: 8x10+3 ppb/48 hr; larvae: 740 ppb/14 day; Static lab bioassay LC_{50} Lepomis macrochirus (bluegill fish) 0.9 ppm/48 hr /Conditions of bioassay not specified; SRP LD_{50} Odocoileus hemionus (Mule deer) male & female 8-11 mo old oral 400-800 mg/kg LC_{50} Alburnus alburnus (bleak) embryo 13 mg/l/48 hr /Conditions of bioassay not specified LC_{50} Daphnia magna (water flea) 363-389 mg/l/48 hr /Conditions of bioassay not specified, SRP LD_{50} Columba livia (rock dove) male & female oral 668 mg/kg (95% confidence limit 530-842 mg/kg) LD_{50} Alectoris chukar (chukar) male & female 4 mo old oral 200-400 mg/kg Anabaenopsis raciborskii (blue green algae) at 10 mg/l: Stimulated growth and nitrogen fixation; at 100 mg/l, no significant inhibition of growth; at 1000 mg/l, complete inhibition of growth Toxic dose Tubifex tubifex (worm) 80 mg/l

Environmental Fate: If released on land, it will probably readily biodegrade (typical half-lives <1 day to several weeks). Its adsorption to soils will depend upon organic content and pH of the soil (2,4-D pKa = 2.64). Leaching to groundwater will likely be a significant process in coarse-grained sandy soils with low organic content or with very basic soils. If released to water, it will be lost primarily due to biodegradation (typical half-lives 10 to >50 days). It will be more persistent in oligotrophic waters and where high concentrations are released. Degradation will be rapid in sediments (half-life <1 day). It will not bioconcentrate in aquatic organisms or appreciably adsorb to sediments, especially at basic pH's. If released in air, it will be subject to photooxidation (estimated half-life of 1 day) and rainout..

Cleanup/Disposal: Guide No. 131: Fully encapsulating, vapor protective clothing should be worn for spills and leaks with no fire. Eliminate all ignition sources (no smoking, flares, sparks or flames in immediate area). All equipment used when handling the product must be grounded. Do not touch or walk through spilled material. Stop leak if you can do it without risk. Prevent entry into waterways, sewers, basements or confined areas. A vapor suppressing foam may be used to reduce vapors. Small Spills: Absorb with earth, sand or other non-combustible material and transfer to containers for later disposal. Use clean non-sparking tools to collect absorbed material. Large Spills: Dike far ahead of liquid spill for later disposal. Water spray may reduce vapor; but may not prevent ignition in closed spaces. Guide No. 152: Do not touch damaged containers or spilled material unless wearing appropriate protective clothing. Stop leak if you can do it without risk. Prevent entry into waterways, sewers, basements or confined areas. Cover with plastic sheet to prevent spreading. Absorb or cover with dry earth, sand or other non-combustible material and transfer to containers. Do not get water inside containers.

Environmental Physical Data

Henry's Law Constant: 1.02×10^{-8}

Octanol/Water Partition Coefficient: log K_{ow} = 2.81

Sorption Partition Coefficient: K_{oc} = soils 19.6

BCF: bluegill 1×10^{-5}

Regulations

RCRA 40CFR: Listed Hazardous Waste No. U240 Toxic Waste

CERCLA: 40CFR 302.4: Listed per CWA Section 311(b)(4) per RCRA Section 3001 RQ: 100 lb (45.35 kg)

SARA 40CFR 372.65: Listed

SARA EHS 40CFR 355: Not listed

TSCA: Listed

Analytical Methods

Air: EPA TO-10

Soil: EPA PMD-CPH, PMD-DCA, PMD-TLC; SW846 3640A, 3650A, 3650B, 4015, 8150B, 8151, 8321, 8321A; DOE OH100R; USGS O5105

Water / Groundwater: EPA P-008-1, 1658, 615, X_89_176, 023; APHA 6640-B; ASTM D3478, D4763, D5317; USGS O1105, O3105

Drinking Water: EPA 515.1, 515.2, 555; AOAC 992.32

Food: AOAC 976.03, 978.05, 984.07

Indoor / Expired Air: NIOSH 5001; EPA IP-8; ASTM D4861

Plasma: EPA 028; FDA 221.1

Urine: EPA 008

Other: USGS O7105

DIC6350 CAS #: 94-82-6

4-(2,4-DICHLOROPHENOXY)BUTYRIC ACID

RTECS: ES9100000
DOT: UN2765; UN2766; UN2999; UN3000; IMO3.2; IMO6.1
EINECS Number: 202-366-9
Molecular Formula: $C_{10}H_{10}Cl_2O_3$
Formula Weight: 349.09

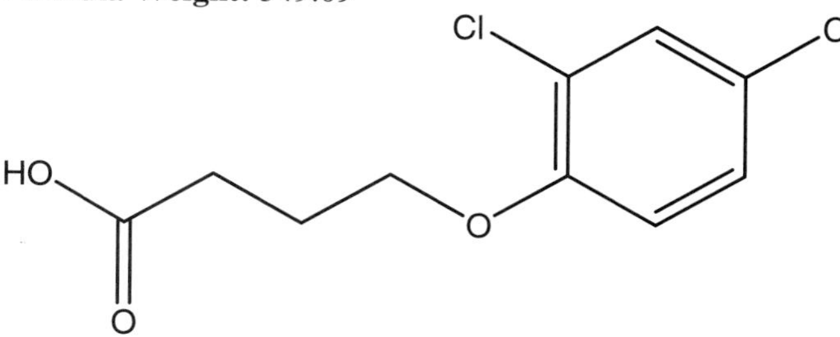

Chemical Structure

Synonyms: BUTIREX; BUTORMONE; BUTOXON; BUTOXONE; BUTOXONE AMINE; BUTOXONE ESTER; BUTYRAC; BUTYRAC 118; BUTYRAC 200; BUTYRAC ESTER; 2,4-D BUTYRIC; 2,4-D BUTYRIC ACID; 2,4-DB; 4(2,4-DB); 4-(2,4-DICHLOROPHENOXY)BUTANOIC ACID; (2,4-DICHLOROPHENOXY)BUTYRIC ACID; GAMMA-(2,4-DICHLOROPHENOXY)BUTYRIC ACID; 2,4-DM; EMBUTONE; EMBUTOX; EMBUTOX E; EMBUTOX KLEAN-UP; KYSELINA 4-(2,4-DICHLORFENOXY)MASELNA; LEGUMEX D
Description: white to light brown crystals; slightly phenolic odor
Use: control of broad-leafed plants in row crops; defoliant; general brush control; post-emergence control of many annual and perennial broad-leafed weeds in lucerne, clovers, undersown cereals, grassland, forage legumes, soya-beans, and groundnuts; selective, translocated hormone-type herbicide

Physical Properties

Freezing Point: 118 °C (244.4 °F) to 120 °C (248 °F)
Saturated Vapor Density: 1.200000061 kg/m^3
Vapor Pressure: 3.5 x10^{-6} mm Hg at 25 °C
Water Solubility: 46 ppm at 25 °C
Other Solubilities: Readily Soluble in Acetone, Ethanol, and Diethyl Ether. Slightly Soluble in Benzene, Toluene, and Kerosene.

RTECS Toxicity Data

Acute Oral: Rat LD_{50} Dose: 700 mg/kg. Mouse LD_{Lo} Dose: 750 mg/kg.
Acute Dermal: Rabbit LD_{50} Route: Skin; Dose: >10 gm/kg. Rat LD_{50} Route: Skin; Dose: 800 mg/kg.
Chronic (Multiple Dose) Oral: Rat Dose: 18928 mg/kg/13W-I; Toxic Effects: Endocrine - Other changes; Blood - Changes in leukocyte (WBC) cell count; DEATH. Rat Dose: 4368 mg/kg/60W-I; Toxic Effects: Liver - Other changes; Blood - Changes in spleen; Nutritional and gross metabolic - Weight loss or decreased weight gain. Rat Dose: 5 gm/kg/10D-I; Toxic Effects: Nutritional and gross metabolic - Weight loss or decreased weight gain; Biochemical - Catalases; DEATH.
Reproductive/Teratogenic: Rat Route: Oral; Dose: 17 mg/kg; Duration: female 1-7D of pregnancy; Specific Developmental Abnormalities - Other developmental abnormalities; Effects on Newborn - Stillbirth. Rat Route: Oral; Dose: 416 mg/kg; Duration: female 5D of pregnancy; Effects on Fertility - Pre-implantation mortality; Effects on Embryo or Fetus - Fetotoxicity. Rat Route: Oral; Dose: 416 mg/kg; Duration: female 9D of pregnancy; Effects on Embryo or Fetus - Fetal death.
Mutagenic: Bacteria - B Subtilis DNA Repair; Dose: 5 mg/disc. Bacteria - E Coli DNA Repair; Dose: 5 mg/disc.

Hazard Overviews

Health: Irritating to eyes/skin/respiratory tract. Toxic. Other Acute Effects: harmful by inhalation, in contact with skin and if swallowed; readily absorbed through skin.
Fire: Hazards: emits toxic fumes. Extinguishing agents: water spray; carbon dioxide, dry chemical powder or appropriate foam. Precautions: combustible liquid.
Carcinogenicity: IARC - Not listed; NIOSH - Not listed; NTP - Not listed; ACGIH - Not listed; OSHA - Not listed; EPA - Not listed; MAK - Not listed
Primary Target Organs:

Eyes

Skin

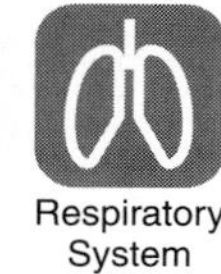
Respiratory System

Environmental

Environmental Fate: If released to soil or water, microbial degradation will be the major environmental degradation process. Microbial degradation occurs through a beta-oxidation mechanism that yields (2,4-dichlorophenoxy)acetic acid (2,4-D). In laboratory persistence studies, it has exhibited a soil half-life of less than 7 days. At normal herbicide application rates, it has a residual activity of about 6 weeks in soil. If released to the atmosphere, it can exist in both the vapor and particulate-phases; vapor-phase degrades readily by reaction with photochemically produced hydroxyl radicals (estimated half-life of 27 hours). Physical removal from the atmosphere may occur through wet and dry deposition.
Cleanup/Disposal: Guide No. 131: Fully encapsulating, vapor protective clothing should be worn for spills and leaks with no fire. Eliminate all ignition sources (no smoking, flares, sparks or flames in immediate area). All equipment used when handling the product must be grounded. Do not touch or walk through spilled material. Stop leak if you can do it without risk. Prevent entry into waterways, sewers, basements or confined areas. A vapor suppressing foam may be used to reduce vapors. Small Spills: Absorb with earth, sand or other non-combustible material and transfer to containers for later disposal. Use clean non-sparking tools to collect absorbed material. Large Spills: Dike far ahead of liquid spill for later

disposal. Water spray may reduce vapor; but may not prevent ignition in closed spaces. Guide No. 152: Do not touch damaged containers or spilled material unless wearing appropriate protective clothing. Stop leak if you can do it without risk. Prevent entry into waterways, sewers, basements or confined areas. Cover with plastic sheet to prevent spreading. Absorb or cover with dry earth, sand or other non-combustible material and transfer to containers. Do not get water inside containers.

Environmental Physical Data

Henry's Law Constant: estimated at 2.29×10^{-9}
Octanol/Water Partition Coefficient: log K_{ow} = 3.53
Sorption Partition Coefficient: K_{oc} = 370
BCF: calculated at 280

Regulations

RCRA 40CFR: Not listed
CERCLA: 40CFR 302.4: Listed as Compound per CWA Section 307(a)
SARA 40CFR 372.65: Listed
SARA EHS 40CFR 355: Not listed
TSCA: Not listed

Analytical Methods

Soil: EPA PMD-CPH, PMD-TLC; SW846 8150B, 8151, 8321, 8321A
Water / Groundwater: EPA P-008-1, 1658, 615, X_89_176; EPA 023; ASTM D5317
Drinking Water: EPA 515.1, 515.2, 555; AOAC 992.32
Plasma: FDA 221.1

DIC6400 CAS #: 97-16-5

2,4-DICHLOROPHENYL BENZENESULFONATE

RTECS: SK9100000
EINECS Number: 202-562-4
Molecular Formula: $C_{12}H_8Cl_2O_3S$
Formula Weight: 303.17
Synonyms: 923; BENZENESULFONIC ACID,2,4-DICHLOROPHENYL ESTER; BENZENESULPHONIC ACID,2,4-DICHLOROPHENYL ESTER; COMPOUND 923; 2,4-DICHLORFENYLESTER KYSELINY BENZENSULFONOVE; 2,4-DICHLOROPHENOL,BENZENESULFONATE; 2,4-DICHLOROPHENYL BENZENESULPHONATE; 2,4-DICHLOROPHENYL ESTER OF BENZENESULFONIC ACID; DPBS; EM 923; GENITE; GENITE 923; GENITE EM-923; GENITE-R99; GENITOL; GENITOL 923; LATKA 923; PHENOL,2,4-DICHLORO-,BENZENESULFONATE
Description: waxy solid
Use: miticide specific for european red mite & clover mite

Physical Properties

Freezing Point: 45 °C (113 °F) to 47 °C (116.6 °F)
Vapor Pressure: 2.7×10^{-4} mm Hg at 30 °C
Water Solubility: Practically Insoluble in Water
Other Solubilities: Soluble in most organic solvents.

RTECS Toxicity Data

Acute Oral: Rat LD_{50} Dose: 1 gm/kg. Rabbit LD_{50} Dose: 700 mg/kg.
Acute Dermal: Rabbit LD_{50} Route: Skin; Dose: >940 mg/kg.
Tumorigenic: Mouse Route: Oral; Dose: 260 gm/kg/78W-I; Toxic Effects: Tumorigenic - Equivocal tumorigenic agent by RTECS criteria; Lungs, Thorax, or Respiration - Tumors; Liver - Tumors. Mouse Route: Subcutaneous; Dose: 1000 mg/kg; Toxic Effects: Tumorigenic - Carcinogenic by RTECS criteria; Blood - Tumors.

Hazard Overviews

Carcinogenicity: IARC - Not listed; NIOSH - Not listed; NTP - Not listed; ACGIH - Not listed; OSHA - Not listed; EPA - Not listed; MAK - Not listed

Environmental

Regulations

RCRA 40CFR: Not listed
CERCLA: 40CFR 302.4: Listed as Compound per CWA Section 307(a)
SARA 40CFR 372.65: Listed as Compound
SARA EHS 40CFR 355: Not listed
TSCA: Not listed

Analytical Methods

Food: FDA 212.1, 232.1, 232.4, 242.1
Plasma: EPA 001, 028; FDA 211.1, 231.1, 252

DIC6450 CAS #: 102-36-3

3,4-DICHLOROPHENYL ISOCYANATE

RTECS: NQ8760000
DOT: UN2550
EINECS Number: 203-026-2
Molecular Formula: $C_7H_3Cl_2NO$
Formula Weight: 188.01

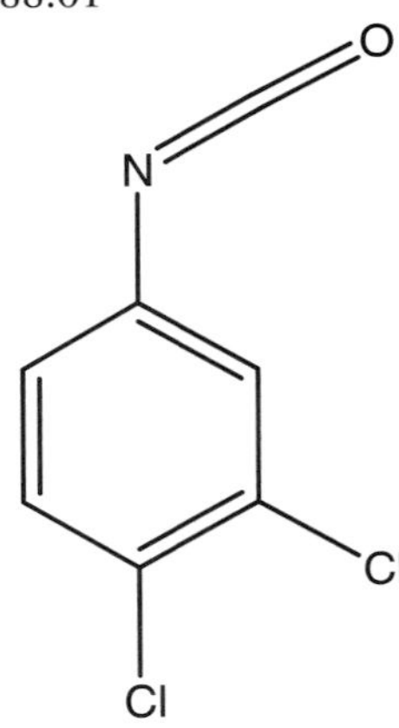

Chemical Structure

Synonyms: BENZENE,1,2-DICHLORO-4-ISOCYANATO-; BENZENE,1,2-DICHLORO-4-ISOCYANATO (9CI); 3,4-

DICHLORFENYLISOKYANAT; 1,2-DICHLORO-4-ISOCYANATOBENZENE; ISOCYANIC ACID,3,4-DICHLOROPHENYL ESTER

Description: white to yellow crystalline solid

Use: chemical intermediate; organic synthesis

Physical Properties

Boiling Point: 113 °C (235 °F) at 9.75 mm Hg

Freezing Point: 43 °C (109.4 °F)

Specific Gravity: o-Anisidine 1.1

Vapor Pressure: 7.43 x10^{+1} Pa at melting point

RTECS Toxicity Data

Acute Oral: Rat LD_{50} Dose: 91 mg/kg.

Acute Inhalation: Rat LC_{50} Dose: 2700 mg/m^3/4hr; Toxic Effects: Sense organs and special senses - Lacrimation; Behavioral - Ataxia; Gastrointestinal - Changes in structure or function of salivary glands. Mouse LC_{50} Dose: 140 mg/m^3/2hr.

Acute Dermal: Rat LD; Route: Skin; Dose: >5 gm/kg.

Hazard Overviews

Poison

Health: Irritating to eyes/skin/respiratory tract. Poison. Other Acute Effects: may be fatal if inhaled, swallowed, or absorbed through skin; symptoms of exposure may include burning sensation, coughing, wheezing, laryngitis, shortness of breath, headache, nausea and vomiting; repeated exposure may cause asthma; prolonged or repeated exposure may cause allergic reactions in certain sensitive individuals.

Fire: Hazards: emits toxic fumes. Extinguishing agents: carbon dioxide; dry chemical powder. Precautions: combustible liquid.

Reactivity: Incompatible with: water, alcohols, strong bases, amines, acids, strong oxidizing agents, heat. Hazardous decomposition products: thermal decomposition may produce carbon monoxide, carbon dioxide, and nitrogen oxides; hydrogen cyanide, hydrogen chloride gas.

Carcinogenicity: IARC - Not listed; NIOSH - Not listed; NTP - Not listed; ACGIH - Not listed; OSHA - Not listed; EPA - Not listed; MAK - Not listed

Primary Target Organs:

Eyes

Skin

Respiratory System

Environmental

Ecotoxicity: Crustaceans: Daphnia magna 24h EC_0 120 mg/l; Fishes: Brachydanio rerio 96h LC_0 35 mg/l 96h LC_{50} 42 mg/l

Environmental Fate: If released to water, it will hydrolyze rapidly. Other fate processes in water will be unimportant by comparison. If released to the atmosphere, it will exist primarily in the vapor phase. Vapor-phase will degrade in the atmosphere by reaction with photochemically produced hydroxyl radicals with an estimated half-life of approximately 36.5 days. Removal of atmospheric compound may occur through wet deposition and hydrolysis. If released to soil, it is expected to have low to slight mobility based on estimated K_{oc} values of 754 to 2557. However, since isocyanates hydrolyze rapidly in water, soil mobility, biodegradation, and other fate processes are not expected to be important. is not expected from moist soils, but may be important from dry soils.

Cleanup/Disposal: Guide No. 147: Eliminate all ignition sources (no smoking, flares, sparks or flames in immediate area). Keep combustibles (wood, paper, oil, etc.) away from spilled material. Do not touch damaged containers or spilled material unless wearing appropriate protective clothing. Keep substance wet using water spray. Stop leak if you can do it without risk. Small Spills: Take up with inert, damp, noncombustible material using clean non-sparking tools and place into loosely covered plastic containers for later disposal. Large Spills: Wet down with water and dike for later disposal. Prevent entry into waterways, sewers, basements or confined areas. Do not clean-up or dispose of, except under supervision of a specialist.

Environmental Physical Data

Henry's Law Constant: estimated at 1.28 x10^{-4}

Sorption Partition Coefficient: K_{OC} = 2557

BCF: estimated at 102

Regulations

RCRA 40CFR: Not listed

CERCLA: 40CFR 302.4: Listed as Compound per CWA Section 307(a) per CAA Section 112

SARA 40CFR 372.65: Listed as Compound TPQ: 500/10000 lb

SARA EHS 40CFR 355: Listed TPQ: 500 lb

TSCA: Listed

DIC6500 CAS #: 1836-75-5

2,4-DICHLOROPHENYL P-NITROPHENYL ETHER

RTECS: KN8400000

EINECS Number: 217-406-0

Molecular Formula: $C_{12}H_7Cl_2NO_3$

Structured MF: $Cl_2C_6H_3OC_6H_4NO_2$

Formula Weight: 284.10

Synonyms: BENZENE,2,4-DICHLORO-1-(4-NITROPHENOXY)-; 2,4-DECHLOROPHENYL P-NITROPHENYL ETHER; 2',4'-DICHLORO-4-NITROBIPHENYL ETHER; 2,4-DICHLORO-4'-NITRODIPHENYL ETHER; 2,4-DICHLORO-1-(4-NITROPHENOXY)BENZENE; 2,4-DICHLORO-4'-NITROPHENYL ETHER; 4-(2,4-DICHLOROPHENOXY)NITROBENZENE; 2,4-DICHLOROPHENYL 4-NITROPHENYL ETHER; 2,4-DICHLORPHENYL-4-NITROPHENYLAETHER; ETHER,2,4-DICHLOROPHENYL P-NITROPHENYL; ETHER,2,4-DICHLOROPHENYL-P-NITROPHENYL; FW 925; MEZOTOX; NICLOFEN; NIP; NITOFEN; NITRAFEN; NITRAPHEN; NITROCHLOR; 4'-NITRO-2,4-DICHLORODIPHENYL ETHER; 4-NITRO-2',4'-DICHLOROPHENYL ETHER; NITROFEN; NITROFENE; NITROPHEN; NITROPHENE; PREPARATION 125; TOK;

TOK E-25; TOK-2; TOK E; TOK E 40; TOK WP-50; TOKKORN; TRIZILIN; TRIZILIN 25

Description: white to yellow crystalline solid

Use: herbicide used on many vegetables, a number of broad-leaved and grass weeds, cereals, rice, sugar beet, some ornamentals, broccoli, cauliflower, cabbage, brussel sprouts, onions, garlic and celery; in nurseries for roses and chrysanthemums

Physical Properties

Boiling Point: 180 °C (356 °F) to 190 °C (374 °F) at 0.25 mm Hg

Freezing Point: 70 °C (158 °F) to 71 °C (159.8 °F)

Specific Gravity: 1.33 at 90 °C

Vapor Pressure: 1.06 mPa at 40 °C

Water Solubility: 0.7 to 1.2 ppm at 22 °C

Other Solubilities: 95% Ethanol: 5-10 mg/ml at 21 °C; Acetone: >=100 mg/ml at 21 °C; Benzene: 2000 gm/kg at 20 °C; DMSO: >=100 mg/ml at 21 °C; Methanol: Soluble; Xylene: Soluble; n-Hexane: 280 gm/kg at 20 °C.

Flash Point: Not available; probably combustible

RTECS Toxicity Data

Acute Oral: Rat LD_{50} Dose: 740 mg/kg; Toxic Effects: Behavioral - Excitment; Lungs, Thorax, or Respiration - Respiratory stimulation; Nutritional and gross metabolic - Body temperature increase. Mouse LD_{50} Dose: 450 mg/kg; Toxic Effects: Behavioral - Excitment; Lungs, Thorax, or Respiration - Respiratory stimulation; Nutritional and gross metabolic - Body temperature increase.

Acute Inhalation: Cat LC_{Lo} Dose: 620 mg/m^3/4hr; Toxic Effects: Behavioral - Excitment; Lungs, Thorax, or Respiration - Dyspnea; Gastrointestinal - Changes in structure or function of salivary glands.

Acute Dermal: Rabbit LD_{Lo} Route: Skin; Dose: 3270 mg/kg. Rat LD_{50} Route: Skin; Dose: 5 gm/kg.

Reproductive/Teratogenic: Rat Route: Oral; Dose: 62500 ug/kg; Duration: female 6-15D of pregnancy; Specific Developmental Abnormalities - Body wall; Respiratory system; Urogenital system. Rat Route: Oral; Dose: 150 mg/kg; Duration: female 7-12D of pregnancy; Effects on Embryo or Fetus - Cytological changes (inc. somatic cell genetic material). Rat Route: Oral; Dose: 37530 ug/kg; Duration: female 6-18D of pregnancy; Effects on Newborn - Physical. Rat Route: Oral; Dose: 62500 ug/kg; Duration: female 6-15D of pregnancy; Effects on Embryo or Fetus - Fetotoxicity. Rat Route: Oral; Dose: 100 mg/kg; Duration: female 11D of pregnancy; Effects on Newborn - Growth statistics. Rat Route: Oral; Dose: 80 mg/kg; Duration: female 10-13D of pregnancy Specific Developmental Abnormalities - Cardiovascular (circulatory) system; Effects on Newborn - Stillbirth; Viability index.

Mutagenic: Rat Morphological Transformation; Cell Type: embryo; Dose: 1500 ng/plate. Bacteria - S Typhimurium Mutations in Microorganisms; Dose: 33300 ng/plate (+S9), 10 ug/plate (-S9).

Tumorigenic: Rat Route: Oral; Dose: 42 gm/kg/94W-C; Toxic Effects: Tumorigenic - Carcinogenic by RTECS criteria; Vascular - Tumors; Liver - Tumors. Mouse Route: Oral; Dose: 24 gm/kg/12W-C; Toxic Effects: Tumorigenic - Carcinogenic by RTECS criteria; Liver - Tumors; Lungs, Thorax, or Respiration - Tumors. Mouse Route: Oral; Dose: 200 gm/kg/78W-C; Toxic Effects: Tumorigenic - Carcinogenic by RTECS criteria; Vascular - Tumors; Liver - Tumors. Mouse Route: Oral; Dose: 308 gm/kg/78W-C; Toxic Effects: Tumorigenic - Carcinogenic by RTECS criteria; Vascular - Tumors; Liver - Tumors. Mouse Route: Oral; Dose: 114 gm/kg/58W-C; Toxic Effects: Tumorigenic - Carcinogenic by RTECS criteria; Vascular - Tumors; Liver - Tumors. Mouse Route: Oral; Dose: 47 gm/kg/12W-C; Toxic Effects: Tumorigenic - Carcinogenic by RTECS criteria; Liver - Tumors; Lungs, Thorax, or Respiration - Tumors.

Hazard Overviews

Fire: Will burn.

Carcinogenicity: IARC - Group 2B, Possibly carcinogenic to humans; NIOSH - Not listed; NTP - Listed; ACGIH - Not listed; OSHA - Not listed; EPA - Not listed; MAK - Not listed

Environmental

Ecotoxicity: Fishes: Rasbora trilineata 96h LC_{50} 0.756 mg/l guppy 96h LC_{50} 1.59 mg/l

Environmental Fate: When applied to soil, it will photolyze on the soil surface and biodegrade. It adsorbs strongly to soil and leaching will be negligible. Biodegradation is fairly rapid in flooded soil (half-life 10 days at 30 °C), but fairly slow in upland soil with substantial residues lasting through two growing seasons in cooler areas. If released in water, it would adsorb strongly to sediment and particulate matter in the water column, photolyze in surface layers (65% degradation in 1 wk) and biodegrade (99% degradation in 50 days). Bioconcentration in fish and aquatic organisms will be appreciable. In the atmosphere, it would exist primarily adsorbed to particulate matter and in aerosols from spraying operations. It will be subject to gravitational settling and rapidly photolyze.

Environmental Physical Data

Henry's Law Constant: 6.74 x10^{-5}

Octanol/Water Partition Coefficient: log K_{ow} = 5.534

Sorption Partition Coefficient: K_{oc} = estimated at 4400

BCF: calculated at 9.46 x10^{4}

Regulations

RCRA 40CFR: Not listed

CERCLA: 40CFR 302.4: Listed as Compound per CWA Section 307(a)

SARA 40CFR 372.65: Listed

SARA EHS 40CFR 355: Not listed

TSCA: Listed

Analytical Methods

Soil: SW846 8081, 8111, 8270B, 8270C

Food: FDA 212.1, 212.2, 232.1, 232.4, 242.1; EPA XENO

Plasma: EPA 001, 003, 29; FDA 211.1, 231.1, 252

Other: EPA 1656

DIC6550 CAS #: 1220-00-4

DI-P-CHLOROPHENYLTHIOUREA

RTECS: YS2660000
Molecular Formula: $C_{13}H_{10}Cl_2N_2S$
Formula Weight: 297.21

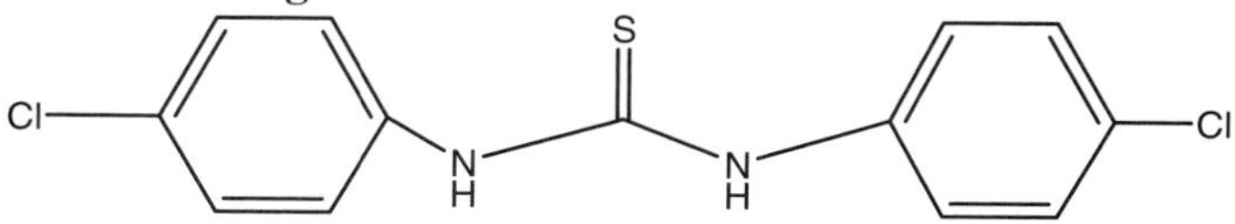

Chemical Structure

Synonyms: 1,3-BIS(P-CHLOROPHENYL)THIOUREA; N,N'-BIS(4-CHLOROPHENYL)THIOUREA; N,N'-BIS(P-CHLOROPHENYL)THIOUREA; CARBANILIDE,4,4'-DICHLOROTHIO-; DI-4-CHLOROPHENYL THIOUREA; THIOUREA,N,N'-BIS(4-CHLOROPHENYL)-; UREA,1,3-BIS(P-CHLOROPHENYL)-2-THIO-

Hazard Overviews

Carcinogenicity: IARC - Not listed; NIOSH - Not listed; NTP - Not listed; ACGIH - Not listed; OSHA - Not listed; EPA - Not listed; MAK - Not listed

Environmental

Regulations

RCRA 40CFR: Not listed
CERCLA: 40CFR 302.4: Listed as Compound per CWA Section 307(a)
SARA 40CFR 372.65: Listed as Compound
SARA EHS 40CFR 355: Not listed
TSCA: Not listed

DIC6600 CAS #: 120-36-5

DICHLOROPROP

RTECS: UF1050000
EINECS Number: 204-390-5
Molecular Formula: $C_9H_8Cl_2O_3$
Formula Weight: 235.07

Chemical Structure

Synonyms: ACIDE 2-(2,4-DICHLORO-PHENOXY) PROPIONIQUE; ACIDE 2-(2,4-DICHLORO-PHENOXY)PROPIONIQUE; ACIDO 2-(2,4-DICLORO-FENOSSI)-PROPIONICO; ACIDO 2-(2,4-DICLORO-FENOSSI)PROPIONICO; BASAGRAN DP; BH 2,4-DP; CANAPUR DP; CELATOX-DP; CORNOX RD; CORNOX RK; CORNOX RK 64; CORNOX RK EXTRA CONCENTRATE; CORNOXYNIL; DESORMONE; 2-(2,4-DICHLOOR-FENOXY)-PROPIONZUUR; 2-(2,4-DICHLOROPHENOXY) PROPIONIC ACID; ALPHA-(2,4-DICHLOROPHENOXY) PROPIONIC ACID; (+ OR -)-2-(2,4-DICHLOROPHENOXY)PROPANOIC ACID; 2-(2,4-DICHLOROPHENOXY)PROPANOIC ACID; (+ OR -)-2-(2,4-DICHLOROPHENOXY)PROPIONIC ACID; 2,4-DICHLOROPHENOXY-ALPHA-PROPIONIC ACID; 2-(2,4-DICHLOROPHENOXY)PROPIONIC ACID; ALPHA-(2,4-DICHLOROPHENOXY)PROPIONIC ACID; 2-(2,4-DICHLOR-PHENOXY)-PROPIONSAEURE; DICHLORPROP; DIKOFAG DP; 2,4-DP; 2-(2,4-DP); ENVERT 171; FERNOXONE; GRAMINON-PLUS; HEDONAL; HEDONAL DP; HERBATOX; HERBITOX; HERBIZID; HERBIZID DP; HORMATOX; KILDIP; KWAS 2,4-DWUCHLOROFENOKSYPROPIONOWY; KYSELINA 2-(2,4-DICHLORFENOXY)PROPIONOVA; MAYCLENE; OXYTRIL P; POLYCLENE; POLYMONE; POLYTOX; PROPANOIC ACID,2-(2,4-DICHLOROPHENOXY)-; PROPIONIC ACID,2-(2,4-DICHLOROPHENOXY)-; RD 406; SERITOX 50; TEXTRONE M; TRI-CORNOX SPRCIAL; U 46 DP FLUID; U 46 DP-FLUID; U46; U46 DP-FLUID; USTILAN NK25; VISKO-RHAP; WEEDONE 170; WEEDONE 2,4-DP; WEEDONE DP

Description: colorless to white to tan, crystalline solid; odorless or faint phenolic odor

Use: selective pre- & post-emergence herbicide for brush control on rangeland, rights-of-way, etc; controls polygonum persicaria, polygonum lapathifolium, aquatic weeds, chickweed, etc; in cereals, pastures, turf; alone or mixed with other hormone type phenoxy herbicides; for control of broad-leaved aquatic weeds; to prevent premature fruit fall in apples; particularly effective for control of bilderdykia convolvulus.

Physical Properties

Boiling Point: 215 °C (419 °F)
Freezing Point: 117.5 °C (243.5 °F) to 118.1 °C (244.58 °F)
Specific Gravity: About 1.42
Saturated Vapor Density: < 1.200000001 kg/m^3
Vapor Pressure: < 1 x10^{-7} mm Hg at 20 °C
Water Solubility: 350 mg/L at 20 °C
Other Solubilities: Chloroform 10.8, Ether 82.1, Ethanol more than 100 (all in g/100 g at 20 °C); solubility at 28 °C: 595 g/L in Acetone, 85 g/L in Benzene, 510 g/L in Isopropanol, 2.1 g/L in Kerosene, 69 g/L in Toluene, 51 g/L in Xylene; Heptane 0.5% wt/vol.

RTECS Toxicity Data

Acute Oral: Rat LD_{50} Dose: 344 mg/kg. Mouse LD_{50} Dose: 309 mg/kg.

Acute Inhalation: Rat LC; Dose: >1600 mg/m^3; Toxic Effects: Behavioral - Somnolence (general depressed activity); Behavioral - Muscle contraction or spasticity.

Acute Dermal: Rat LD_{50} Route: Skin; Dose: 1880 mg/kg; Toxic Effects: Behavioral - Somnolence (general depressed activity); Nutritional and gross metabolic - Body temperature decrease. Mouse LD_{50} Route: Skin; Dose: 1400 mg/kg.

Chronic (Multiple Dose) Oral: Rat Dose: 6188 mg/kg/17W-I; Toxic Effects: DEATH. Rat Dose: 1893 mg/kg/1Y-I; Toxic Effects: Liver - Changes in liver weight; Endocrine - Hypoglycemia; Nutritional and gross metabolic - Changes in sodium.

Chronic (Multiple Dose) Inhalation: Rat Dose: 100 mg/m^3/17W-C; Toxic Effects: Kidney, Ureter, and Bladder - Changes in kidney weight; Biochemical - Other enzymes.
Chronic (Multiple Dose) Dermal: Rabbit Route: Skin; Dose: 4500 mg/kg/30D-I; Toxic Effects: Skin and appendages - Dermatitis, other; DEATH.
Reproductive/Teratogenic: Rat Route: Oral; Dose: 20 mg/kg; Duration: female 4-18D of pregnancy; Effects on Newborn - Behavioral; Physical. Mouse Route: Oral; Dose: 5 gm/kg; Duration: female 6-15D of pregnancy; Effects on Fertility - Post-implantation mortality; Specific Developmental Abnormalities - Craniofacial (including nose and tongue). Mouse Route: Oral; Dose: 3 gm/kg; Duration: female 6-15D of pregnancy; Effects on Embryo or Fetus - Fetotoxicity. Mouse Route: Oral; Dose: 4 gm/kg; Duration: female 6-15D of pregnancy; Specific Developmental Abnormalities - Musculoskeletal system. Mouse Route: Oral; Dose: 20 mg/kg; Duration: female 4-18D of pregnancy; Effects on Fertility - Post-implantation mortality.
Mutagenic: Yeast - S Cerevisiae Mutations in Microorganisms; Dose: 700 mg/L (-S9).

Hazard Overviews

Health: Irritating to eyes/skin/respiratory tract. Toxic. Other Acute Effects harmful if swallowed, inhaled, or absorbed through skin. Chronic Effects: Carcinogen.
Fire: Hazards: emits toxic fumes. Extinguishing agents: water spray; carbon dioxide, dry chemical powder or appropriate foam. Precautions: combustible liquid.
Reactivity: Incompatible with: strong oxidizing agents. Hazardous decomposition products: toxic fumes of: carbon monoxide, carbon dioxide, hydrogen chloride gas.
Carcinogenicity: IARC - Group 2B, Possibly carcinogenic to humans; NIOSH - Not listed; NTP - Not listed; ACGIH - Not listed; OSHA - Not listed; EPA - Not listed; MAK - Not listed
Primary Target Organs:

Environmental

Ecotoxicity: LC_{50} Bluegill sunfish 16 mg/l/48 hr /Conditions of bioassay not specified
Environmental Fate: If released to soil, adsorption will not be an important fate process (experimental K_{oc}'s of 50-62) and, therefore, it is expected to leach. The half-life in soil typically ranges from 8 to greater than 103 days; the US Dept of Agriculture's Pesticide Properties Database lists a soil half-life of 10 days. Insufficient data are available to assess the importance of biodegradation in soil and water. If released to water, volatilization and bioconcentration in fish will not be important. In groundwater, half-lives of 196 to 1,286 days were observed; however, degradation pathways were not specified. If released to the atmosphere, it will exist in the vapor and particulate phases; vapor phase degrades rapidly by reaction with photochemically produced hydroxyl radicals (estimated half-life of 12 hours). Physical removal from air may occur through dry deposition.

Environmental Physical Data

Henry's Law Constant: 1.22 $x10^{-8}$
Sorption Partition Coefficient: K_{oc} = soils 50 to 62
BCF: estimated at 23

Regulations

RCRA 40CFR: Not listed
CERCLA: 40CFR 302.4: Not listed
SARA 40CFR 372.65: Listed
SARA EHS 40CFR 355: Not listed
TSCA: Listed

Analytical Methods

Soil: SW846 8150B, 8151, 8321, 8321A; USGS O5105
Water / Groundwater: EPA 1658, 615, X_89_176, 023; ASTM D5317; USGS O1105, O3105
Drinking Water: EPA 515.1, 515.2, 555; AOAC 992.32
Plasma: FDA 221.1
Other: USGS O7105

DIC6650 CAS #: 78-99-9

1,1-DICHLOROPROPANE

RTECS: TX9450000
EINECS Number: 201-165-3
Molecular Formula: $C_3H_6Cl_2$
Structured MF: C_2H_5•CH•Cl_2
Formula Weight: 112.99

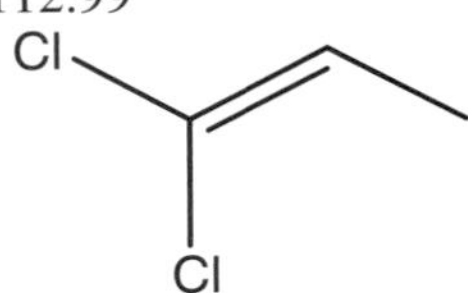

Chemical Structure

Synonyms: ALPHA,ALPHA-DICHLOROPROPANE; PROPANE,1,1-DICHLORO-; ALPHA,ALPHA-PROPYLENE DICHLORIDE; PROPYLIDENE CHLORIDE
Description: liquid; sweet odor
Use: solvent for pesticide formulations

Physical Properties

Boiling Point: 88.1 °C (191 °F) at 760 mm Hg
Specific Gravity: 1.1321 at 20 °C/4 °C
Vapor Density: 3.9 Air=1
Saturated Vapor Density: 1.501357949 kg/m^3
Vapor Pressure: 65.9 torr at 25 °C
Water Solubility: Very Slightly in Water
Other Solubilities: Soluble in many organic solvents.
Surface Tension: Estimated at 26.1 dynes/cm
Refraction Index: 1.4289 at 20 °C/D
Critical Temperature: 266.1 °C
Critical Pressure: 38.3 atm
Flash Point: Estimated at 15.556 °C Open Cup

Autoignition Temperature: Estimated at 557 °C
LEL: 3.4% v/v
UEL: Estimated at 14.5% v/v

RTECS Toxicity Data

Acute Oral: Rat LD_{50} Dose: 6500 mg/kg.
Acute Inhalation: Rat LC_{Lo} Dose: 4000 ppm/4hr.
Acute Dermal: Rabbit LD_{50} Route: Skin; Dose: 14 mL/kg.
Irritation Eye: Rabbit Standard Draize Test Dose: 500 mg/24H; Reaction: mild.
Irritation Skin: Rabbit Standard Draize Test Dose: 500 mg/24H; Reaction: mild.

Hazard Overviews

Flammable

Health: Irritating to eyes/skin/respiratory tract. Harmful. Other Acute Effects: harmful if swallowed, inhaled, or absorbed through skin; can cause CNS depression. Chronic Effects: damage to the liver, kidneys. Possible carcinogen.
Fire: Flammable. Hazards: vapor may travel considerable distance to source of ignition and flash back; emits toxic fumes. Extinguishing agents: water spray; carbon dioxide, dry chemical powder or appropriate foam. Precautions: combustible liquid.
Reactivity: Incompatible with: strong oxidizing agents, strong bases. Hazardous decomposition products: toxic fumes of: carbon monoxide, carbon dioxide, hydrogen chloride gas.
Carcinogenicity: IARC - Not listed; NIOSH - Not listed; NTP - Not listed; ACGIH - Not listed; OSHA - Not listed; EPA - Not listed; MAK - Not listed
Primary Target Organs:

Eyes

Skin
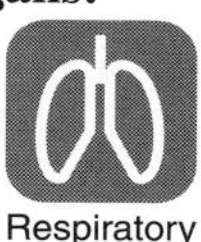
Respiratory System

Environmental

Ecotoxicity: LC_{50} Lepomis macrochirus (bluegill) 97,900 ug/l/96 LC_{50} Daphnia magna (cladorecan) 23000 ug/l/96
Environmental Fate: If released into soil or water during its production and use, it would be lost primarily by volatilization (half-life 3.1 hr in a typical river). Bioconcentration in fish will not be significant. It is poorly adsorbed by soil and may leach. In the air, it will disperse and degrade primarily by reaction with photochemically produced hydroxyl radicals (estimated half-life 8.9 day). It will also be washed out by rain.
Cleanup/Disposal: Guide No. 130: Eliminate all ignition sources (no smoking, flares, sparks or flames in immediate area). All equipment used when handling the product must be grounded. Do not touch or walk through spilled material. Stop leak if you can do it without risk. Prevent entry into waterways, sewers, basements or confined areas. A vapor suppressing foam may be used to reduce vapors. Absorb or cover with dry earth, sand or other non-combustible material and transfer to containers. Use clean non-sparking tools to collect absorbed material. Large Spills: Dike far ahead of liquid spill for later disposal. Water spray may reduce vapor; but may not prevent ignition in closed spaces.

Environmental Physical Data

Henry's Law Constant: calculated at 1.55 $x10^{-2}$
Octanol/Water Partition Coefficient: log K_{ow} = 2.307
Sorption Partition Coefficient: K_{oc} = 125
BCF: estimated at 19.4

Regulations

RCRA 40CFR: Not listed
CERCLA: 40CFR 302.4: Listed per CWA Section 311(b)(4) RQ: 1000 lb (453.5 kg)
SARA 40CFR 372.65: Not listed
SARA EHS 40CFR 355: Not listed
TSCA: Not listed

DIC6700 CAS #: 78-87-5

1,2-DICHLOROPROPANE

RTECS: TX9625000
DOT: UN1279; IMO6.1
EINECS Number: 201-152-2
Molecular Formula: $C_3H_6Cl_2$
Structured MF: $CH_3CH(Cl)CH_2Cl$
Formula Weight: 112.99

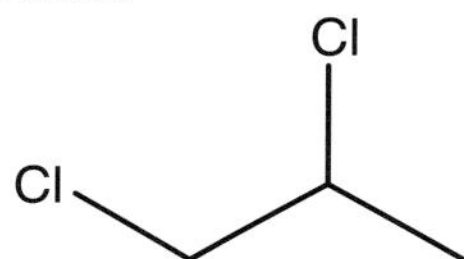

Chemical Structure

Synonyms: BICHLORURE DE PROPYLENE; ALPHA,BETA-DICHLOROPROPANE; DICHLORO-1,2-PROPANE; DWUCHLOROPROPAN; ENT 15,406; PROPANE,1,2-DICHLORO-; PROPYLENE CHLORIDE; ALPHA,BETA-PROPYLENE DICHLORIDE; PROPYLENE DICHLORIDE; PROPYLENEDICHLORIDE
Description: colorless liquid; sweet odor
Use: as an intermediate for perchloroethylene and carbotetrachloride; as a lead scavenger for antiknock fluids, as a solvent for fats, oils, waxes, gums and resins, in solvent mixtures for cellulose esters and ethers, in scouring compounds, in spotting agents, in metal degreasing agents, as a soil fumigant for nematodes, as a grain fumigant and in dry-cleaning fluids

Physical Properties

Boiling Point: 96.4 °C (206 °F)
Freezing Point: -100.4 °C (-148.72 °F)
Specific Gravity: 1.159 at 25 °C/25 °C
Vapor Density: 3.9 Air=1
Saturated Vapor Density: 1.428647913 kg/m^3
Density: 1.1558 g/mL at 20 °C
Bulk Density: 9.6 lbs/gal at 20 °C

Vapor Pressure: 50 mm Hg at 25 °C
Water Solubility: 0.26% by wt at 20 °C
Other Solubilities: 95% Ethanol: >=100 mg/ml at 19 °C; Acetone: >=100 mg/ml at 19 °C; Alcohol: Soluble; Benzene: Soluble; Chloroform: Soluble; DMSO: >=100 mg/ml at 19 °C; Ethanol: Soluble; Ether: Soluble; Most common solvents: miscible; Organic solvents: miscible.
Surface Tension: 29 dynes/cm at 20 °C
Odor Threshold: 50 ppm
Refraction Index: 1.4388 at 20 °C/D
Evaporation Rate: > 1 Butyl Acetate=1
Ionization Potential (eV): 10.87
Flash Point: 21 °C Open Cup
Autoignition Temperature: 557 °C
LEL: 3.4% v/v
UEL: 14.5% v/v

RTECS Toxicity Data

Acute Oral: Rat LD_{50} Dose: 1947 mg/kg. Mouse LD_{50} Dose: 860 mg/kg.
Acute Inhalation: Rat LC_{50} Dose: 14 gm/m^3/8hr. Mouse LC_{Lo} Dose: 1000 ppm/2hr.
Acute Dermal: Rabbit LD_{50} Route: Skin; Dose: 8750 uL/kg.
Chronic (Multiple Dose) Oral: Rat Dose: 28 gm/kg/14D-I; Toxic Effects: Kidney, Ureter, and Bladder - Other changes; DEATH. Rat Dose: 109 mg/kg/26W-I; Toxic Effects: Gastrointestinal - Changes in structure or function of exocrine pancreas; Biochemical - Other hydrolases. Rat Dose: 16250 mg/kg/13W-I; Toxic Effects: Liver - Changes in liver weight; Endocrine - Changes in spleen weight; Biochemical - Transaminases.
Chronic (Multiple Dose) Inhalation: Rat Dose: 9 mg/m^3/24H/94D-C; Toxic Effects: Brain and coverings - Recordings from specific areas of CNS; Lungs, Thorax, or Respiration - Other changes; Biochemical - True cholinesterase. Rat Dose: 10 mg/m^3/24H/60D-I; Toxic Effects: Liver - Other changes; Biochemical - Phosphatases; Biochemical - Dehydrogenases. Rat Dose: 1000 ppm/7H/20W-I; Toxic Effects: Liver - Fatty Liver degeneration; Blood - Changes in spleen; DEATH.
Irritation Eye: Rabbit Standard Draize Test Dose: 500 mg; Reaction: mild.
Reproductive/Teratogenic: Rat Route: Oral; Dose: 1250 mg/kg; Duration: female 6-15D of pregnancy; Maternal Effects - Other effects on females. Rabbit Route: Oral; Dose: 1950 mg/kg; Duration: female 7-19D of pregnancy; Maternal Effects - Other effects on females.
Mutagenic: Mouse Mutations in Microorganisms; Cell Type: lymphocyte; Dose: 11600 ug/L (+S9). Hamster Cytogenetic Analysis; Cell Type: ovary; Dose: 660 mg/L.
Tumorigenic: Mouse Route: Oral; Dose: 130 gm/kg/2Y-I; Toxic Effects: Tumorigenic - Carcinogenic by RTECS criteria; Skin and appendages - Tumors.

Hazard Overviews

Flammable

Fire Diamond

Health: Irritating to eyes/skin/respiratory tract. Also Causes: bloody nose, pulmonary edema, emphysema, bronchopneumonia, tachycardia, corneal epithelium damage, allergic dermatitis, vomiting, abdominal pain, fever, anorexia, black and blue marks, bloody urine, oliguria, night sweats, shock, delirium, cardiovascular collapse, liver/kidney damage, hemolytic anemia. Chronic Effects: dermatitis, liver/kidney problems.
Fire: Flammable. Can form explosive mixtures in the air. Use dry chemical, carbon dioxide, water spray, fog, or regular foam. Fight fire from maximum distance.
Reactivity: Stable. Hazardous polymerization cannot occur. Avoid: exposure to heat; ignition sources. Incompatible with: strong acids; oxidizers; active metals; aluminum. Hazardous decomposition products: hydrogen chloride; chlorine; phosgene gas.
Carcinogenicity: IARC - Group 3, Not classifiable as to carcinogenicity to humans; NIOSH - Listed as carcinogen; NTP - Not listed; ACGIH - Class A4, Not classifiable as a human carcinogen; OSHA - Not listed; EPA - Not listed; MAK - Class B, Justifiably suspected of having carcinogenic potential
Primary Target Organs:

Eyes

Skin

Respiratory System

Nervous System

Liver

Kidneys

Exposure Limits
OSHA PEL: TWA: 75 ppm; 350 mg/m^3.
OSHA PEL Vacated 1989 Limits: TWA: 75 ppm; 350 mg/m^3; STEL: 110 ppm; 510 mg/m^3.
ACGIH TLV: TWA: 75 ppm; 347 mg/m^3; STEL: 110 ppm; 508 mg/m^3.
NIOSH IDLH: 400 ppm.
Respirator Recommendation
Exposure Range: >75 to <400 ppm Air Purifying, Negative Pressure, Half Mask
Exposure Range: 400 to unlimited ppm Self-contained Breathing Apparatus, Pressure Demand, Full Face
Cartridge Color: black

Environmental

Ecotoxicity: LC_{50} Pimephales promelas (fathead minnow) 127 mg/l/96 hr (confidence limit 119 - 135 mg/l), flow-through bioassay with measured concentrations, 24.1 °C, dissolved oxygen 8.9 mg/l, hardness 44.8 mg/l calcium carbonate alkalinity 39.6 mg/l calcium carbonate LC_{50} Poecilia reticulata (guppy) 116 ppm/7 days /Conditions of bioassay not specified LC_{50} Menidia beryllina 240 ppm/96 hr; In synthetic seawater at 23 °C, mild aeration applied after 24 hr LC_{50} Daphnia magna (cladorecan) 52,500 ug/l/96 hr

Environmental Fate: If injected into soil it will be primarily lost by volatilization. It has been detected in groundwater where its fate is unknown. If released into water, it will be lost by volatilization with half-lives ranging from approx. 6 hr for a river to 10 days for a lake. Adsorption to soil and bioconcentration in fish will not be significant. In air it will react with photochemically generated hydroxyl radicals (half-life >23 days) and be washed out by rain.

Cleanup/Disposal: Guide No. 130: Eliminate all ignition sources (no smoking, flares, sparks or flames in immediate area). All equipment used when handling the product must be grounded. Do not touch or walk through spilled material. Stop leak if you can do it without risk. Prevent entry into waterways, sewers, basements or confined areas. A vapor suppressing foam may be used to reduce vapors. Absorb or cover with dry earth, sand or other non-combustible material and transfer to containers. Use clean non-sparking tools to collect absorbed material. Large Spills: Dike far ahead of liquid spill for later disposal. Water spray may reduce vapor; but may not prevent ignition in closed spaces.

Environmental Physical Data

Henry's Law Constant: 2.07 x10^{-3}
Octanol/Water Partition Coefficient: log K_{ow} = 2.28
Sorption Partition Coefficient: K_{oc} = 47
BCF: estimated at 1

Regulations

RCRA 40CFR: Listed Hazardous Waste No. U083 Toxic Waste
CERCLA: 40CFR 302.4: Listed per CWA Section 311(b)(4) per RCRA Section 3001 per CWA Section 307(a) RQ: 1000 lb (453.5 kg)
SARA 40CFR 372.65: Listed
SARA EHS 40CFR 355: Not listed
TSCA: Listed

Analytical Methods

Air: EPA VA-001-1, VA-002-1, VG-007-1, VG-011-1, TO-1, TO-14, 0031, 0040; ASTM D3687
Soil: CLP LC_VOA, MC_VOA, OHC; EPA 7, 1624, VG-002-1, VG-008-1, VS-001-1, VS-002-1; SW846 5021, 5032, 5041, 5041A, 8010B, 8021A, 8240B, 8260A, 8260B
Water / Groundwater: EPA 601, 624, 624-S, VW-001-1, VW-002-1, VW-003-1, VW-014-1; APHA 6210-B, 6210-D, 6230-B, 6230-D; ASTM D3695; USGS O3115
Drinking Water: EPA 502.1, 502.2, 524.1, 524.2; APHA 6210-C, 6230-C
Food: EPA 5
Indoor / Expired Air: NIOSH 1013; EPA IP-1A, IP-1B
Plasma: EPA 29

DIC6750 **CAS #: 142-28-9**

1,3-DICHLOROPROPANE

RTECS: TX9660000
EINECS Number: 205-531-3
Molecular Formula: $C_3H_6Cl_2$
Structured MF: $CH_2ClCH_2CH_2Cl$
Formula Weight: 112.99

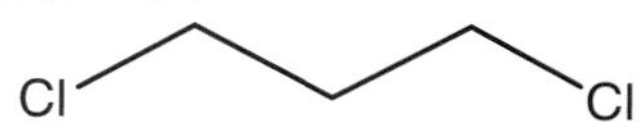

Chemical Structure

Synonyms: PROPANE,1,3-DICHLORO-; TRIMETHYLENE DICHLORIDE
Description: colorless liquid; sweet odor
Use: chem int for cyclopropane (anesthetic)

Physical Properties

Boiling Point: 120.4 °C (249 °F) at 760 mm Hg
Freezing Point: -99.5 °C (-147.1 °F)
Specific Gravity: 1.1876 at 20 °C/4 °C
Vapor Density: 4 Air=1
Vapor Pressure: 40 mm Hg
Water Solubility: Miscible with Water
Surface Tension: 33.93 dynes/cm at 20 °C
Refraction Index: 1.4487 at 20 °C/D
Critical Temperature: 314.3 °C
Critical Pressure: Estimated at 41.75 atm
Ionization Potential (eV): 10.85 +/-0.05
Flash Point: 15.556 °C Open Cup
Autoignition Temperature: ~ 555 °C
LEL: 3.4% v/v
UEL: Estimated at 14.5% v/v

RTECS Toxicity Data

Acute Oral: Dog LD_{Lo} Dose: 3 gm/kg; Toxic Effects: Lungs, Thorax, or Respiration - Chronic pulmonary edema; Liver - Other changes; Blood - Hemorrhage.
Chronic (Multiple Dose) Oral: Rat Dose: 8400 mg/kg/14D-I; Toxic Effects: Liver - Changes in liver weight; Kidney, Ureter, and Bladder - Changes in kidney weight; Biochemical - Other proteins. Rat Dose: 72 gm/kg/90D-I; Toxic Effects: Liver - Changes in liver weight; Nutritional and gross metabolic - Weight loss or decreased weight gain; Biochemical - Phosphatases.
Mutagenic: Hamster Sister Chromatid Exchenge; Cell Type: lung; Dose: 6600 umol/L. Bacteria - S Typhimurium Mutations in Microorganisms; Dose: 10 umol/plate (+/-S9).

Hazard Overviews

Flammable

Health: Irritating to eyes/skin/respiratory tract. Harmful. Other Acute Effects: may be harmful by inhalation, ingestion, or skin absorption; causes dermatitis; prolonged contact can cause nausea; headache; vomiting; CNS depression; repeated exposure can cause damage to the liver, and kidneys.
Fire: Flammable. Hazards: vapor may travel considerable distance to source of ignition and flash back; emits toxic

fumes. Extinguishing agents: carbon dioxide, dry chemical powder or appropriate foam. Precautions: combustible liquid.
Reactivity: Incompatible with: oxidizing agents, acids, bases, aluminum and its alloys. Hazardous decomposition products: toxic fumes of: hydrogen chloride gas, carbon monoxide, carbon dioxide, phosgene gas.
Carcinogenicity: IARC - Not listed; NIOSH - Not listed; NTP - Not listed; ACGIH - Not listed; OSHA - Not listed; EPA - Not listed; MAK - Not listed
Primary Target Organs:

Eyes

Skin

Respiratory System

Environmental

Ecotoxicity: LC_{50} Cyprinodon variegatus (sheepshead minnow) 86,7000 ug/l 96 hr EC_{50} Selenastrum capricornutum 72,200 ug/l/96 hr Toxic Effect: Cell numbers LC_{50} Poecilia reticulata (guppy) 84 ppm/7 days LC_{50} Mysidopsis bahia (mysid shrimp) 10,300 ug/l 96 hr MATC Pimephales promelas 8-16 ug/l (est)
Environmental Fate: If released into the soil or water during its production and use, it would be lost primarily by volatilization (half-life 4 hr in a model river). It is poorly adsorbed by soil and may leach. In the air, it will disperse and degrade primarily by reaction with photochemically produced hydroxyl radicals (half-life 9.5 day). It will also be washed out by rain. Bioconcentration in fish will not be significant.
Cleanup/Disposal: Guide No. 128: Eliminate all ignition sources (no smoking, flares, sparks or flames in immediate area). All equipment used when handling the product must be grounded. Do not touch or walk through spilled material. Stop leak if you can do it without risk. Prevent entry into waterways, sewers, basements or confined areas. A vapor suppressing foam may be used to reduce vapors. Absorb or cover with dry earth, sand or other non-combustible material and transfer to containers. Use clean non-sparking tools to collect absorbed material. Large Spills: Dike far ahead of liquid spill for later disposal. Water spray may reduce vapor; but may not prevent ignition in closed spaces.

Environmental Physical Data

Henry's Law Constant: 9.76×10^{-4}
Octanol/Water Partition Coefficient: log K_{ow} = 2.00
Sorption Partition Coefficient: log K_{oc} = estimated at 1.74
BCF: calculated at 1.06
BOD: 16% BODT, 5 days

Regulations

RCRA 40CFR: Not listed
CERCLA: 40CFR 302.4: Listed per CWA Section 311(b)(4) RQ: 5000 lb (2268 kg)
SARA 40CFR 372.65: Not listed
SARA EHS 40CFR 355: Not listed
TSCA: Listed

Analytical Methods

Air: EPA VG-011-1, TO-1
Soil: SW846 5021, 8021A
Water / Groundwater: APHA 6210-D, 6230-D
Drinking Water: EPA 502.1, 502.2, 524.1, 524.2; APHA 6210-C, 6230-C
Indoor / Expired Air: EPA IP-1B

DIC6800 CAS #: 26638-19-7

DICHLOROPROPANE

RTECS: TX9350000
Molecular Formula: $C_3H_6Cl_2$
Structured MF: $CH_3CHClCH_2Cl$
Formula Weight: 112.99
Synonyms: DICHLORPROPAN; PROPANE DICHLORIDE; PROPANE,DICHLORO-
Description: colorless liquid; chloroform odor
Use: solvent for pesticide formulations

Physical Properties

Boiling Point: 96.2 °C (205 °F)
Freezing Point: -70 °C (-94 °F)
Specific Gravity: 1.2
Vapor Density: 3.9 Air=1
Saturated Vapor Density: 1.428647913 kg/m^3
Vapor Pressure: 50 mm Hg at 25 °C
Water Solubility: Slightly Soluble in Water
Other Solubilities: 540 g/kg in ethanol; soluble greater than 25% in isopropyl alcohol, methyl ethyl ketone, toluene, xylene; soluble in benzene; very soluble in methanol; in acetone 1,700 g/L at 25 °C; sparingly soluble in aromatic solvents
Flash Point: 16 °C
Autoignition Temperature: 556 °C
LEL: 3.4% v/v
UEL: 14.5% v/v

RTECS Toxicity Data

Mutagenic: Rat Cytogenetic Analysis; Route: Subcutaneous; Dose: 21 mmol/L/kg.

Hazard Overviews

Flammable

Fire: Flammable.
Carcinogenicity: IARC - Not listed; NIOSH - Not listed; NTP - Not listed; ACGIH - Not listed; OSHA - Not listed; EPA - Not listed; MAK - Not listed

Environmental

Environmental Fate: If released into soil or water, they will be lost primarily be volatilization (half-life ranging from approx. 6 hr in a typical river to 10 day in a lake). They adsorb poorly to soil and may leach into groundwater where their fate is unknown. Bioconcentration in fish should not be significant. In air, they will react with photochemically produced hydroxyl radicals (half-life 7.1-9.5 day for all isomers except 2,2-dichloropropane which is 1.7 months) and be washed out by rain. Therefore, there will be ample time for dispersion as is evidenced by their presence in air.

Environmental Physical Data

Henry's Law Constant: 2.07 $x10^{-3}$
Octanol/Water Partition Coefficient: log K_{ow} = estimated at 2.0 to 2.3
BCF: fish 11 to 33
BOD: 0%, 20 days

Regulations

RCRA 40CFR: Not listed
CERCLA: 40CFR 302.4: Listed per CWA Section 311(b)(4) RQ: 1000 lb (453.5 kg)
SARA 40CFR 372.65: Not listed
SARA EHS 40CFR 355: Not listed
TSCA: Not listed

Analytical Methods

Air: ASTM D3686, D4490
Water / Groundwater: ASTM D3695

DIC6850 CAS #: 8003-19-8

DICHLOROPROPANE DICHLOROPROPEN

RTECS: TX9800000
DOT: NA2047
Molecular Formula: $C_6H_{10}Cl_4$
Structured MF: $C_3H_4Cl_2$
Formula Weight: 223.96
Synonyms: D-D; DD MIXTURE; DD SOIL FUMIGANT; DICHLOROPROPANE-DICHLOROPROPENE MIXTURE; 1,3-DICHLOROPROPENE AND 1,2-DICHLOROPROPANE MIXTURE; DICHLORPROPAN-DICHLORPROPENGEMISCH; DOWFUME N; ENT 8,420; MIXTURE OF 1,3-DICHLOROPROPANE; 1,3-DICHLOROPROPENE AND RELATED C3 COMPOUNDS; NEMAFENE; PROPANE,DICHLORO-MIXED WITH PROPENE,DICHLORO-; 1-PROPENE,1,3-DICHLORO-,MIXT WITH 1,2-DICHLOROPROPANE; TELONE; VIDDEN D
Description: amber, clear liquid; pungent odor
Use: a pre-plant nematicide effective against nematodes incl root knot, meadow, sting & dagger, spiral & sugar beet nematodes

Physical Properties

Boiling Point: 103 °C (217 °F) to 171 °C (339.98 °F) at 1 atm
Freezing Point: Estimated at -61 °C (-77.8 °F)
Specific Gravity: About 1.4 at 4 °C/20 °C
Vapor Density: 4 Air=1
Vapor Pressure: 4.6 kPa at 20 °C
Water Solubility: About 2 g/kg at 0 °C
Other Solubilities: fully miscible with esters, halogenated solvents, hydrocarbons, ketones
pH: Acid
Flash Point: 19 °C Closed Cup
LEL: 5.3% v/v
UEL: 14.5% v/v

RTECS Toxicity Data

Acute Oral: Rat LD_{50} Dose: 140 mg/kg. Mouse LD_{50} Dose: 3 mg/kg.
Acute Inhalation: Rat LC_{50} Dose: 1000 ppm/4hr; Toxic Effects: Lungs, Thorax, or Respiration - Acute pulmonary edema; Liver - Fatty liver degeneration.
Acute Dermal: Rabbit LD_{50} Route: Skin; Dose: 2100 mg/kg. Rat LD_{50} Route: Skin; Dose: 779 mg/kg.
Irritation Eye: Rabbit Standard Draize Test Dose: 5 mg; Reaction: severe.
Irritation Skin: Rabbit Standard Draize Test Dose: 500 mg/24H; Reaction: severe.
Mutagenic: Bacteria - S Typhimurium Mutations in Microorganisms; Dose: 500 ug/plate (+/-S9).

Hazard Overviews

Flammable

Fire: Flammable.
Carcinogenicity: IARC - Not listed; NIOSH - Not listed; NTP - Not listed; ACGIH - Not listed; OSHA - Not listed; EPA - Not listed; MAK - Not listed

Environmental

Ecotoxicity: LC_{50} Stizostedion vitreum vitreum (walleye) 1.0 mg/l/96 hr at 18 °C, wt 1.3 g /Technical material, 100%/. Static bioassay without aeration, pH 7.2-7.5, water hardness 40-50 mg/l as calcium carbonate and alkalinity of 30-35 mg/l LD_{50} Honeybee above 60 ug/bee
Environmental Fate: If released into soil, it will be lost primarily by volatilization over the course of approximately 4 weeks. It may leach into groundwater. In moist soils, hydrolysis for the isomer may be important. It will also be lost primarily by volatilization within about 10 days. It is less likely to degrade in soil and more likely to leach into groundwater than 1,3-dichloropropene. If released into water, loss of dichloropropane-dichloropropene mixture will also primarily be due to volatilization (half-life 6-8 hr in a typical river). Adsorption to sediment and bioconcentration in fish will not be significant. If released into air it will degrade by reaction with photochemically produced hydroxyl radicals (half-life 29-50 hr for dichloropropene and >23 day for it). It will also be scavenged by rain. Some dispersion from source or target areas would occur, particularly as is evidenced by its presence in ambient air.
Cleanup/Disposal: Guide No. 132: Fully encapsulating, vapor protective clothing should be worn for spills and leaks with no fire. Eliminate all ignition sources (no smoking, flares, sparks or flames in immediate area). All equipment used when handling the product must be grounded. Do not touch

or walk through spilled material. Stop leak if you can do it without risk. Prevent entry into waterways, sewers, basements or confined areas. A vapor suppressing foam may be used to reduce vapors. Absorb with earth, sand or other non-combustible material and transfer to containers (except for Hydrazine). Use clean non-sparking tools to collect absorbed material. Large Spills: Dike far ahead of liquid spill for later disposal. Water spray may reduce vapor; but may not prevent ignition in closed spaces.

Environmental Physical Data

Henry's Law Constant: 0.072 to 0.95
Octanol/Water Partition Coefficient: log K_{ow} = 1.36 to 1.41
Sorption Partition Coefficient: K_{oc} = 14
BCF: estimated at 13

Regulations

RCRA 40CFR: Not listed
CERCLA: 40CFR 302.4: Listed per CWA Section 311(b)(4) RQ: 100 lb (45.35 kg)
SARA 40CFR 372.65: Not listed
SARA EHS 40CFR 355: Not listed
TSCA: Not listed

DIC6900 CAS #: 616-23-9

2,3-DICHLOROPROPANOL

RTECS: UB1225000
EINECS Number: 210-470-0
Molecular Formula: $C_3H_6Cl_2O$
Formula Weight: 128.99

Cl

Cl OH

Chemical Structure

Synonyms: ALPHA,BETA-DICHLOROHYDRIN; BETA-DICHLOROHYDRIN; 1,2-DICHLORO-3-PROPANOL; 1,2-DICHLOROPROPANOL-3; 2,3-DICHLORO-1-PROPANOL; 2,3-DICHLOROPROPYL ALCOHOL; GLYCEROL ALPHA,BETA-DICHLOROHYDRIN; GLYCEROL-2,3-DICHLOROHYDRIN; GLYCEROL-ALPHA,BETA-DICHLOROHYDRIN; 1-PROPANOL,2,3-DICHLORO-
Description: viscous liquid; ethereal odor
Use: solvent for hard resins and nitrocellulose; manufacture of photographic chemicals and lacquer; as a cement for celluloid and a binder for water colors

Physical Properties

Boiling Point: 183 °C (361 °F) to 185 °C (365 °F)
Specific Gravity: 1.3607 at 20 °C/4 °C
Water Solubility: Slightly Soluble in Water
Other Solubilities: 95% Ethanol: >=100 mg/ml at 24.5 °C; Acetone: >=100 mg/ml at 24.5 °C; Benzene: Soluble; DMSO: >=100 mg/ml at 24.5 °C; Ether: Soluble.
Refraction Index: 1.4819 at 20 °C/D
Flash Point: 91.4 °C

RTECS Toxicity Data

Acute Oral: Rat LD_{50} Dose: 90 mg/kg.
Acute Inhalation: Rat LC_{Lo} Dose: 500 ppm/4hr.
Acute Dermal: Rabbit LD_{50} Route: Skin; Dose: 200 mg/kg.
Irritation Eye: Rabbit Standard Draize Test Dose: 20 mg; Reaction: severe.
Mutagenic: Hamster Sister Chromatid Exchenge; Cell Type: lung; Dose: 1 mmol/L. Bacteria - S Typhimurium Mutations in Microorganisms; Dose: 131 ug/plate (+S9). Bacteria - S Typhimurium Mutations in Microorganisms; Dose: 33 ug/plate (-S9).

Hazard Overviews

Fire: Combustible.
Carcinogenicity: IARC - Not listed; NIOSH - Not listed; NTP - Not listed; ACGIH - Not listed; OSHA - Not listed; EPA - Not listed; MAK - Not listed

Environmental

Cleanup/Disposal: Guide No. 153: Eliminate all ignition sources (no smoking, flares, sparks or flames in immediate area). Do not touch damaged containers or spilled material unless wearing appropriate protective clothing. Stop leak if you can do it without risk. Prevent entry into waterways, sewers, basements or confined areas. Absorb or cover with dry earth, sand or other non-combustible material and transfer to containers. Do not get water inside containers.

Regulations

RCRA 40CFR: Not listed
CERCLA: 40CFR 302.4: Not listed
SARA 40CFR 372.65: Not listed
SARA EHS 40CFR 355: Not listed
TSCA: Listed

DIC6950 CAS #: 563-54-2

1,2-DICHLOROPROPENE

RTECS: UC8300000
DOT: UN2047; IMO3.3
EINECS Number: 209-253-3
Molecular Formula: $C_3H_4Cl_2$
Formula Weight: 110.97
Synonyms: DICHLOR; 1,2-DICHLOROPROPYLENE; DICHLORPROPEN-GEMISCH; PDC; 1-PROPENE,1,2-DICHLORO-; PROPENE,1,2-DICHLORO-; PROPYLENE DICHLORIDE
Description: liquid
Use: in dichloropropane-dichloropropene mixture as a soil fumigant for control of nematodes affecting roots of plants

Physical Properties

Boiling Point: 75 °C (167 °F)
Vapor Density: 3.83 Air=1
Density: 1.18 at 20 °C
Vapor Pressure: 90.8 torr

Water Solubility: 4400 mg/L (est)
Other Solubilities: Very Soluble in Alcohol, Methanol, Carbon Tetrachloride /trans-1,2-Dichloropropene/; Soluble in Acetone, Benzene, Chloroform /Cis-1,2-Dichloropropene/.
Refraction Index: 1.4471 at 20 °C

RTECS Toxicity Data

Acute Oral: Rat LD_{50} Dose: 2 gm/kg.
Acute Dermal: Rabbit LD_{50} Route: Skin; Dose: 8750 mg/kg.
Mutagenic: Bacteria - S Typhimurium Mutations in Microorganisms; Dose: 10 uL/plate (+/-S9). Mold - A Nidulans Mutations in Microorganisms; Dose: 10 uL/plate (-S9).

Hazard Overviews

Carcinogenicity: IARC - Not listed; NIOSH - Not listed; NTP - Not listed; ACGIH - Not listed; OSHA - Not listed; EPA - Not listed; MAK - Not listed

Environmental

Environmental Fate: If released in soil, it will primarily be lost by volatilization although there is also a potential for leaching into groundwater. If released into water, it will primarily be lost by volatilization (half-life 3.4 hr in a typical river). Adsorption to soil and bioconcentration in fish will not be significant transport processes. In the atmosphere it will degrade by reaction with photochemically produced hydroxyl radicals and ozone (half-life 10.4 hr).
Cleanup/Disposal: Guide No. 132: Fully encapsulating, vapor protective clothing should be worn for spills and leaks with no fire. Eliminate all ignition sources (no smoking, flares, sparks or flames in immediate area). All equipment used when handling the product must be grounded. Do not touch or walk through spilled material. Stop leak if you can do it without risk. Prevent entry into waterways, sewers, basements or confined areas. A vapor suppressing foam may be used to reduce vapors. Absorb with earth, sand or other non-combustible material and transfer to containers (except for Hydrazine). Use clean non-sparking tools to collect absorbed material. Large Spills: Dike far ahead of liquid spill for later disposal. Water spray may reduce vapor; but may not prevent ignition in closed spaces.

Environmental Physical Data

Henry's Law Constant: estimated at 3.0×10^{-3}
Octanol/Water Partition Coefficient: log K_{ow} = 2.04
Sorption Partition Coefficient: K_{oc} = estimated at 68
BCF: calculated at 1.32

Regulations

RCRA 40CFR: Not listed
CERCLA: 40CFR 302.4: Not listed
SARA 40CFR 372.65: Not listed
SARA EHS 40CFR 355: Not listed
TSCA: Not listed

Analytical Methods

Air: EPA VG-007-1

DIC7000 CAS #: 542-75-6

1,3-DICHLOROPROPENE

RTECS: UC8310000
DOT: UN2047; IMO3.3
EINECS Number: 208-826-5
Molecular Formula: $C_3H_4Cl_2$
Structured MF: $ClCH_2CH{=}CHCl$
Formula Weight: 110.98

Chemical Structure

Synonyms: 3-CHLOROALLYL CHLORIDE; ALPHA-CHLOROALLYL CHLORIDE; GAMMA-CHLOROALLYL CHLORIDE; 3-CHLOROPROPENYL CHLORIDE; 1,3-D; D-D92; DCP; 1,3-DICHLORO-1-PROPENE; 1,3-DICHLORO-2-PROPENE; 1,3-DICHLOROPROPENE-1; CIS,TRANS-1,3-DICHLOROPROPENE; DICHLORO-1,3-PROPENE; DICHLOROPROPENE; 1,3-DICHLORO-1-PROPYLENE; 1,3-DICHLOROPROPYLENE; ALPHA,GAMMA-DICHLOROPROPYLENE; DI-TRAPEX; DI-TRAPEX CP; DORLONE; DORLONE II; EPA PESTICIDE CHEMICAL CODE 029001; NEMATOX; NEMEX; 1-PROPENE,1,3-DICHLORO-; PROPENE,1,3-DICHLORO-; TELONE; TELONE C; TELONE C17; TELONE II; TELONE II SOIL FUMIGANT; VIDDEN D; VORLEX; VORLEX 201
Description: straw, clear liquid; sweet, chloroform odor
Use: soil fumigant, nematocide, pesticide and chemical intermediate; in organic synthesis

Physical Properties

Boiling Point: 108 °C (226 °F)
Freezing Point: < -50 °C (-58 °F)
Specific Gravity: 1.22 at 25 °C
Vapor Density: 3.83 Air=1
Saturated Vapor Density: 1.324532232 kg/m^3
Density: 1.22 g/mL at 20 °C
Vapor Pressure: 27.9 mm Hg at 20 °C
Water Solubility: 0.15% in Water
Other Solubilities: miscible in Acetone, Benzene, Carbon Tetrachloride, N-Heptane, Methanol; miscible with halogenated solvents, esters, and ketones.
Surface Tension: 31.2 dynes/cm at 24 °C
Odor Threshold: 7 of 10 volunteers 1 to 3 ppm
Refraction Index: 1.4735 at 22 °C/D
Evaporation Rate: Evaporation loss of 1ppm solution after 31 hours 50%
Flash Point: cis & trans 35 °C Open Cup
LEL: 2.6% v/v
UEL: 7.8% v/v

RTECS Toxicity Data

Acute Oral: Rat LD_{50} Dose: 470 mg/kg. Mouse LD_{50} Dose: 640 mg/kg.
Acute Inhalation: Mouse LC_{50} Dose: 4650 mg/m^3/2hr. Rat LC_{Lo} Dose: 1000 ppm/2hr; Toxic Effects: Sense organs and special senses - Lacrimation.

Acute Dermal: Rabbit LD_{50} Route: Skin; Dose: 504 mg/kg. Rat LD_{50} Route: Skin; Dose: 775 mg/kg.
Chronic (Multiple Dose) Oral: Rat Dose: 455 mg/kg/26W-I; Toxic Effects: Gastrointestinal - Changes in structure or function of exocrine pancreas; Biochemical - Other hydrolases.
Chronic (Multiple Dose) Inhalation: Rat Dose: 90 ppm/6H/13W-I; Toxic Effects: Lungs, Thorax, or Respiration - Other changes; Kidney, Ureter, and Bladder - Changes in kidney weight; Nutritional and gross metabolic - Weight loss or decreased weight gain. Mouse Dose: 90 ppm/6H/13W-I; Toxic Effects: Lungs, Thorax, or Respiration - Other changes; Liver - Changes in Liver weight; Nutritional and gross metabolic - Weight loss or decreased weight gain.
Mutagenic: Mouse Mutations in Mammalian Somatic Cells; Cell Type: lymphocyte; Dose: 3 mg/L. Hamster Cytogenetic Analysis; Cell Type: ovary; Dose: 49100 ug/L.
Tumorigenic: Rat Route: Oral; Dose: 15600 mg/kg/2Y-I; Toxic Effects: Tumorigenic - Carcinogenic by RTECS criteria; Gastrointestinal - Tumors; Liver - Tumors. Mouse Route: Inhalation; Dose: 60 ppm/6H/5D/2Y; Toxic Effects: Tumorigenic - Neoplastic by RTECS criteria; Lungs, Thorax, or Respiration - Tumors.

Hazard Overviews

Flammable

Fire Diamond

Health: Severe irritation to eyes/skin/respiratory tract. Toxic. Also Causes: headache, dizziness, weakness, nausea, vomiting, blood/liver/kidney dysfunction, pancreas damage, respiratory distress syndrome, irritation to digestive tract. Chronic Effects: dermatitis, possible cancer - Non-Hodgkins lymphoma, acute myelomonocytic leukemia.
Fire: Flammable. Hazardous polymerization can occur if it is unstabilized. Use dry chemical, carbon dioxide, water spray, or regular foam. Fight fire from maximum distance.
Reactivity: Unstable, 1 to 2% epichlorohydrin is added as a stabilizer. Hazardous polymerization can occur if not stabilized. Avoid: heat; ignition sources. Incompatible with: oxidizers; aluminum; aluminum alloys; magnesium; other active metals; some metal salts/halogens; humus-rich soil; milk; potatoes; leather; fur. Hazardous decomposition products: hydrogen chloride; chlorine gas.
Carcinogenicity: IARC - Group 2B, Possibly carcinogenic to humans; NIOSH - Listed as carcinogen; NTP - Class 2B, Reasonably anticipated to be a carcinogen, sufficient evidence of carcinogenicity from studies in experimental animals; ACGIH - Class A3, Animal carcinogen; OSHA - Not listed; EPA - Class B2, Probable human carcinogen based on animal studies; MAK - Class A2, Unmistakably carcinogenic in animal experimentation only

Primary Target Organs:

Eyes

Skin

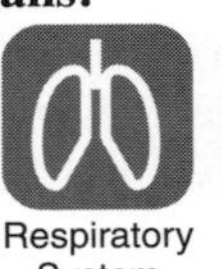
Respiratory System

Nervous System

Exposure Limits
OSHA PEL Vacated 1989 Limits: TWA: 1 ppm; 5 mg/m^3.
NIOSH REL: TWA: 1 ppm; 5 mg/m^3.
Respirator Recommendation
Exposure Range: >1 to 50 ppm Supplied Air, Constant Flow/Pressure Demand, Half Mask
Exposure Range: >50 to 1000 ppm Supplied Air, Constant Flow/Pressure Demand, Full Face
Exposure Range: >1000 to unlimited ppm Self-contained Breathing Apparatus, Pressure Demand, Full Face
Note: odor threshold unknown

Environmental

Ecotoxicity: LC_{50} Anas platyrhynchos (mallard duck) > 10,000 mg/kg in an 8 day diet LC_{50} Cyprinodon variegatus (sheepshead minnow) 1770 mg/l/96 hr /Unmeasured static bioassay EC_{50} Selenastrum capricornutum (freshwater alga) 4950 ug/l/96 hr (effected Chlorophyll A production) /Conditions of bioassay not specified LC_{50} Stizostedion vitreum (Walleye)1080 ug/l/96 hr at 18 °C, wt 1.3 g (95% confidence limit 990-1180 ug/l) /static bioassay without aeration, pH 7.2-7.5, water hardness 40-50 mg/l as calcium carbonate LC_{50} Mysidopsis bahia (mysid shrimp) 790 ug/l/96 hr /Unmeasured static bioassay LC_{50} Daphnia magna (Water flea) 90 ug/l/48 hr at 21 °C, 1st instar (95% confidence limit 63-129 ug/l) / static bioassay without aeration, pH 7.2-7.5, water hardness 40-50 mg/l as calcium carbonate
Environmental Fate: Volatilization occurs from the soil surface and decreases to negligible amounts after approximately 4 weeks. About 5-10% of the fumigant may be lost in the manner following a commercial application to 30 m depth. In spill situations, volatilization will be much more important and rapid. Higher moisture content in the soil will accelerate volatilization since adsorption is much lower to wet soil that dry. It may also leach into groundwater. Although one may expect leaching to be greater in wet soils than dry, because of decreased adsorption, studies show that leaching is lower in wet soils, a situation ascribed to water-blocked passageways in these soils. It hydrolyzes in soil with a half-life of 3 to >69 days. Laboratory studies indicate that even after 10 days in sandy loam soil, <20% of the radioactivity remaining in the soil after an application of radiolabeled compound is as the unchanged parent compound. Significant residues may occur in the soil more than 8 months post treatment, especially in heavy soils. This is the reason why it is recommended waiting 4-6 months after fumigating heavy soil before seeding. Considerable variations in the amounts lost by volatilization and degraded can be expected depending on the method of application, soil type, moisture, and temperature. If released into water, it will be lost primarily through volatilization (half-life from a model river about 4 hours). Adsorption to sediment and bioconcentration in fish

should not be important processes. In the atmosphere, it will degrade by reaction with photochemically produced hydroxyl radicals (half-life 29-50 hr) and may be scavenged by rain. Some dispersion away from source or target areas would be expected. Humans are primarily exposed occupationally and by inhalation near source areas. Lab experiments were conducted concerning the principal factors which influence volatilization from soil. Four independent variables were considered: soil type, soil moisture, soil temperature and dose of the chemical applied to the soil. Soil temperature was the most important variable affecting volatilization loss, followed by soil moisture. Soil type and concentration in the soil had only a small influence on nematicide volatilization. Preliminary experiments on the effect of air flow rate and air humidity, indicated that these variables are additional. factors affecting volatilization loss from the soil.

Cleanup/Disposal: Guide No. 132: Fully encapsulating, vapor protective clothing should be worn for spills and leaks with no fire. Eliminate all ignition sources (no smoking, flares, sparks or flames in immediate area). All equipment used when handling the product must be grounded. Do not touch or walk through spilled material. Stop leak if you can do it without risk. Prevent entry into waterways, sewers, basements or confined areas. A vapor suppressing foam may be used to reduce vapors. Absorb with earth, sand or other non-combustible material and transfer to containers (except for Hydrazine). Use clean non-sparking tools to collect absorbed material. Large Spills: Dike far ahead of liquid spill for later disposal. Water spray may reduce vapor; but may not prevent ignition in closed spaces.

Environmental Physical Data

Henry's Law Constant: calculated at 0.0024
Octanol/Water Partition Coefficient: log K_{ow} = calculated at 1.36
Sorption Partition Coefficient: K_{oc} = 23 to 26
BCF: not significant

Regulations

RCRA 40CFR: Listed Hazardous Waste No. U084 Toxic Waste
CERCLA: 40CFR 302.4: Listed per CWA Section 311(b)(4) per RCRA Section 3001 per CWA Section 307(a) RQ: 100 lb (45.35 kg)
SARA 40CFR 372.65: Listed
SARA EHS 40CFR 355: Not listed
TSCA: Listed

Analytical Methods

Air: EPA 0031
Soil: EPA PMD-CKA
Water / Groundwater: APHA 6040-B

DIC7050 CAS #: 78-88-6

2,3-DICHLOROPROPENE

RTECS: UC8400000
DOT: UN2047; IMO3.3
EINECS Number: 201-153-8
Molecular Formula: $C_3H_4Cl_2$
Structured MF: CH_2CClCH_2Cl
Formula Weight: 110.97

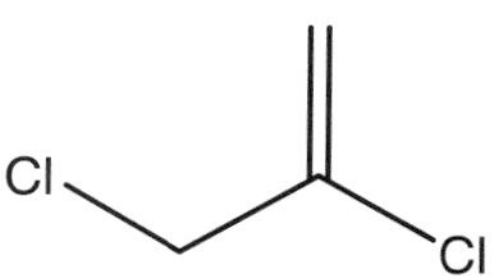

Chemical Structure

Synonyms: 2-CHLOROALLYL CHLORIDE; 1,2-DICHLORO-2-PROPENE; 2,3-DICHLORO-1-PROPENE; 2,3-DICHLOROPROPYLENE; NSC 60520; 1-PROPENE,2,3-DICHLORO; 1-PROPENE,2,3-DICHLORO-; PROPENE,2,3-DICHLORO; PROPENE,2,3-DICHLORO-; PROPYLENE,2,3-DICHLORO
Description: straw-colored liquid; pungent odor
Use: fumigant; chem int for sulfallate herbicide (former use); copolymer and chemical intermediate in agricultural and pharmaceutical products

Physical Properties

Boiling Point: 94 °C (201 °F) at 760 mm Hg
Freezing Point: 10 °C (50 °F)
Specific Gravity: 1.211 at 20 °C/4 °C
Vapor Density: 3.8 Air=1
Saturated Vapor Density: 1.43653775 kg/m^3
Density: 1.222 g/cu cm at 19.9 °C
Vapor Pressure: 53 mm Hg at 25 °C
Water Solubility: 2150 mg/L
Other Solubilities: 95% Ethanol: Decomposes; Acetone: >=100 mg/ml at 22 °C; Benzene: Soluble; Chloroform: Soluble; DMSO: >=100 mg/ml at 22 °C; Ether: Soluble.
Surface Tension: Calculated at 29.9 dynes/cm at 29 °C
Refraction Index: 1.4603 at 20 °C
Evaporation Rate: Evaporation loss of 1ppm solution after 20 hours 50%
Critical Temperature: Estimated at 274.8 °C
Critical Pressure: Estimated at 513 psia
Flash Point: 15 to 16.667 °C Closed Cup
Autoignition Temperature: 557 °C
LEL: 5.3% v/v
UEL: 14.5% v/v

RTECS Toxicity Data

Acute Oral: Rat LD_{50} Dose: 320 mg/kg.
Acute Inhalation: Mouse LC_{50} Dose: 3100 mg/m^3/2hr. Rat LC_{Lo} Dose: 500 ppm/4hr.
Acute Dermal: Rabbit LD_{50} Route: Skin; Dose: 1580 mg/kg; Toxic Effects: Skin and appendages - Primary irritation.
Irritation Skin: Rabbit Open Draize Test Dose: 10 mg/24H open; Reaction: severe.
Mutagenic: Human Unscheduled DNA Synthesis; Cell Type: HeLa cell; Dose: 100 umol/L. Hamster Sister Chromatid Exchenge; Cell Type: lung; Dose: 300 umol/L.

Hazard Overviews

Corrosive Flammable

Health: Corrosive to eyes/skin/respiratory tract. Harmful. Other Acute Effects: harmful if swallowed, inhaled, or absorbed through skin; extremely destructive to tissue of the mucous membranes and upper respiratory tract, eyes and skin; lachrymator; inhalation may result in spasm, inflammation and edema of the larynx and bronchi, chemical pneumonitis and pulmonary edema; symptoms of exposure may include burning sensation, coughing, wheezing, laryngitis, shortness of breath, headache, nausea and vomiting. Chronic Effects: possible mutagen.

Fire: Flammable. Hazards: vapor may travel considerable distance to source of ignition and flash back; container explosion may occur; forms explosive mixtures in air; emits toxic fumes. Extinguishing agents: carbon dioxide, dry chemical powder or appropriate foam; water may be effective for cooling, but may not effect extinguishment. Precautions: combustible liquid.

Reactivity: Incompatible with: strong oxidizing agents, strong bases. Hazardous decomposition products: toxic fumes of: carbon monoxide, carbon dioxide, hydrogen chloride gas.

Carcinogenicity: IARC - Not listed; NIOSH - Not listed; NTP - Not listed; ACGIH - Not listed; OSHA - Not listed; EPA - Not listed; MAK - Not listed

Primary Target Organs:

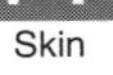

Eyes Skin Respiratory System

Environmental

Ecotoxicity: Aquatic toxicity: 10-100 ppm/96 hr/finfish/TLm 40 ppm/fish toxicity/critical concentration

Environmental Fate: If injected into soil, it will primarily be lost by volatilization over the course of several weeks although for muck soils, release may take many months. There is a potential for leaching. If released into water, it will primarily be lost by volatilization (half-life 3.2 hr in a typical river). Adsorption to soil and bioconcentration in fish will not be important transport processes. In the atmosphere, it will degrade by reaction with photochemically produced hydroxyl radicals and ozone (half-life 8.3 hr).

Cleanup/Disposal: Guide No. 132: Fully encapsulating, vapor protective clothing should be worn for spills and leaks with no fire. Eliminate all ignition sources (no smoking, flares, sparks or flames in immediate area). All equipment used when handling the product must be grounded. Do not touch or walk through spilled material. Stop leak if you can do it without risk. Prevent entry into waterways, sewers, basements or confined areas. A vapor suppressing foam may be used to reduce vapors. Absorb with earth, sand or other non-combustible material and transfer to containers (except for Hydrazine). Use clean non-sparking tools to collect absorbed material. Large Spills: Dike far ahead of liquid spill for later disposal. Water spray may reduce vapor; but may not prevent ignition in closed spaces.

Environmental Physical Data

Henry's Law Constant: calculated at 0.0036
Octanol/Water Partition Coefficient: log K_{ow} = 1.88
Sorption Partition Coefficient: log K_{oc} = estimated at 1.81
BCF: fish 1.20

Regulations

RCRA 40CFR: Not listed
CERCLA: 40CFR 302.4: Listed per CWA Section 311(b)(4) RQ: 100 lb (45.35 kg)
SARA 40CFR 372.65: Listed
SARA EHS 40CFR 355: Not listed
TSCA: Listed

DIC7100 CAS #: 10061-01-5

CIS-1,3-DICHLOROPROPENE

RTECS: UC8325000
DOT: UN2047; IMO3.2
EINECS Number: 233-195-8
Molecular Formula: $C_3H_4Cl_2$
Formula Weight: 110.97

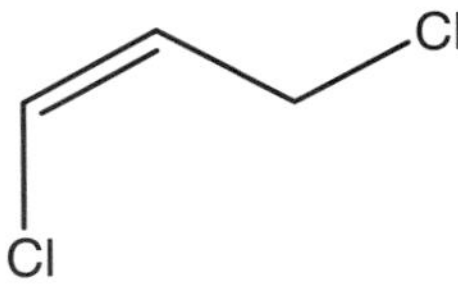

Chemical Structure

Synonyms: CIS-DCP; (Z)-1,3-DICHLORO-1-PROPENE; (Z)-1,3-DICHLOROPROPENE; CIS-1,3-DICHLORO-1-PROPENE; CIS-1,3-DICHLOROPROPYLENE; 1-PROPENE,1,3-DICHLORO-,(Z)-; PROPENE,1,3-DICHLORO-,(Z)-; 1-PROPENE,1,3-DICHLORO-,(Z)-(9 CI); 1-PROPENE,1,3-DICHLORO-,(Z)-(9CI); TELONE II

Description: colorless liquid; sharp, sweet, penetrating, irritating, chloroform-like odor

Use: organic synthesis; soil fumigants; nematicide

Physical Properties

Boiling Point: 104.3 °C (220 °F)
Specific Gravity: 1.217 at 20 °C/4 °C
Vapor Density: 1.4 Air=1 at 37.8 °C
Saturated Vapor Density: 1.391907985 kg/m^3
Density: 1.22 g/mL
Vapor Pressure: 43 mm Hg at 25 °C
Water Solubility: 2700 mg/L in Water at 25 °C
Other Solubilities: 95% Ethanol: >=100 mg/ml at 20.5 °C; Acetone: >=100 mg/ml at 20.5 °C; Benzene: Soluble; Chloroform: Soluble; DMSO: >=100 mg/ml at 20.5 °C; Ether: Soluble; Most organic solvents: Soluble; Octane: Soluble; Toluene: Soluble.
Refraction Index: 1.4730 at 20 °C
Flash Point: cis & trans 35 °C Open Cup

LEL: 2.6% v/v
UEL: 7.8% v/v

RTECS Toxicity Data

Mutagenic: Human Unscheduled DNA Synthesis; Cell Type: HeLa cell; Dose: 100 umol/L. Bacteria - S Typhimurium Mutations in Microorganisms; Dose: 20 ug/plate (+/-S9).
Tumorigenic: Mouse Route: Subcutaneous; Dose: 9240 mg/kg/77W-I; Toxic Effects: Tumorigenic - Neoplastic by RTECS criteria; Tumorigenic - Tumors at site of application.

Hazard Overviews

Corrosive

Flammable

Fire Diamond

Health: Corrosive to eyes/skin/respiratory tract. Toxic. Other Acute Effects: harmful if swallowed, inhaled, or absorbed through skin; sensitizer; inhalation may result in spasm, inflammation and edema of the larynx and bronchi; chemical pneumonitis; pulmonary edema; exposure may cause burning sensation; coughing; wheezing; laryngitis; shortness of breath; headache; nausea; vomiting; allergic respiratory and skin reactions. Chronic Effects: may alter genetic material; may cause heritable genetic damage; target organs: liver, kidneys. Carcinogen.
Fire: Flammable. Hazards: emits toxic fumes; vapor may travel considerable distance to source of ignition and flash back; container explosion may occur. Extinguishing agents: carbon dioxide, dry chemical powder or appropriate foam; water spray; use water spray to cool fire-exposed containers. Precautions: combustible liquid.
Reactivity: Incompatible with: strong oxidizing agents, reacts violently with aluminum/metals/halogens. Hazardous decomposition products: toxic fumes of: carbon monoxide, carbon dioxide, hydrogen chloride gas.
Carcinogenicity: IARC - Not listed; NIOSH - Not listed; NTP - Not listed; ACGIH - Not listed; OSHA - Not listed; EPA - Not listed; MAK - Not listed
Primary Target Organs:

Eyes

Skin

Respiratory System

Liver

Kidneys

Environmental

Environmental Fate: When injected into the soil, volatilization occurs from the soil surface which decreases to insignificant levels after approximately 4 weeks. Low temperatures and heavy rains will lengthen the volatilization period. In spill situations, it will volatilize much faster. It may also leach into groundwater. Most will hydrolyze (half-life 9 to > 69 days) to cis-3-chloroallyl alcohol. In one experiment 12% of the radioactivity from (14)carbon -material remained in soil after 10 days and < 20% of that was unchanged parent. This indicates that most is degraded rather than lost by volatilization. The overall half-life in soil may range from a few days to more than 9 weeks depending upon conditions. Rapid soil disappearance will be associated with moist soil content that promotes hydrolysis and porous soil that promotes volatilization. If released into water, it will be lost primarily through volatilization (half-life from a model river about 4 hours). Adsorption to sediment and bioconcentration in fish are not expected to be important processes. In the atmosphere, this compound will degrade primarily by reaction with photochemically produced hydroxyl radicals (half-life 50 hr) or scavenged by rain. Some dispersion away from source or target areas would be expected.
Cleanup/Disposal: Guide No. 132: Fully encapsulating, vapor protective clothing should be worn for spills and leaks with no fire. Eliminate all ignition sources (no smoking, flares, sparks or flames in immediate area). All equipment used when handling the product must be grounded. Do not touch or walk through spilled material. Stop leak if you can do it without risk. Prevent entry into waterways, sewers, basements or confined areas. A vapor suppressing foam may be used to reduce vapors. Absorb with earth, sand or other non-combustible material and transfer to containers (except for Hydrazine). Use clean non-sparking tools to collect absorbed material. Large Spills: Dike far ahead of liquid spill for later disposal. Water spray may reduce vapor; but may not prevent ignition in closed spaces.

Environmental Physical Data

Henry's Law Constant: calculated at 0.0024
Octanol/Water Partition Coefficient: log K_{ow} = 1.36
Sorption Partition Coefficient: K_{oc} = 23
BCF: estimated at 7

Regulations

RCRA 40CFR: Not listed
CERCLA: 40CFR 302.4: Not listed
SARA 40CFR 372.65: Not listed
SARA EHS 40CFR 355: Not listed
TSCA: Not listed

Analytical Methods

Air: EPA VA-001-1, VG-007-1, VG-011-1, TO-14
Soil: CLP LC_VOA, MC_VOA, OHC; EPA 7, 1624, VG-008-1, VS-001-1, VS-002-1; SW846 5032, 5041, 5041A, 8010B, 8021A, 8240B, 8260A, 8260B
Water / Groundwater: EPA 601, 624, 624-S, VW-001-1, VW-002-1, VW-003-1, VW-014-1; APHA 6210-B, 6230-B; USGS O3115
Drinking Water: EPA 502.1, 502.2, 524.1, 524.2
Food: EPA 5
Indoor / Expired Air: EPA IP-1A
Plasma: EPA 29

DIC7150 CAS #: 26952-23-8

DICHLOROPROPENE

RTECS: UC8280000
DOT: UN2047; IMO3.3
EINECS Number: 248-134-0

Molecular Formula: $C_3H_4Cl_2$
Structured MF: $CHCl{=}CHCH_2Cl$
Formula Weight: 110.98
Synonyms: ALPHA-CHLOROALLYL CHLORIDE; DICHLOROPROPYLENE; 1-PROPENE,DICHLORO-
Description: colorless liquid; chloroform odor
Use: fumigant applied to the soil before crops are planted; preplant for nematode, disease, insect control on a variety of crops; soil treatment before planting of many crops such as cotton, potatoes, sugar beets, tobacco, and vegetable crops; oil and fat solvent

Physical Properties

Boiling Point: 103 °C (217 °F)
Specific Gravity: 1.2
Vapor Density: 3.8 Air=1
Water Solubility: Insoluble in Water
Flash Point: 35 °C

Hazard Overviews

Flammable

Fire: Flammable.
Carcinogenicity: IARC - Not listed; NIOSH - Not listed; NTP - Not listed; ACGIH - Not listed; OSHA - Not listed; EPA - Not listed; MAK - Not listed

Environmental

Environmental Fate: If released into soil, they will be lost primarily by volatilization over the course of approximately 4 wk. They may also leach into groundwater. Hydrolysis should also be important in moist soils. If released into water, loss will also primarily be due to volatilization. If released into air they will degrade by reaction with photochemically produced hydroxyl radicals and ozone, and scavenged by rain. Some dispersion from source or target areas would occur.
Cleanup/Disposal: Guide No. 132: Fully encapsulating, vapor protective clothing should be worn for spills and leaks with no fire. Eliminate all ignition sources (no smoking, flares, sparks or flames in immediate area). All equipment used when handling the product must be grounded. Do not touch or walk through spilled material. Stop leak if you can do it without risk. Prevent entry into waterways, sewers, basements or confined areas. A vapor suppressing foam may be used to reduce vapors. Absorb with earth, sand or other non-combustible material and transfer to containers (except for Hydrazine). Use clean non-sparking tools to collect absorbed material. Large Spills: Dike far ahead of liquid spill for later disposal. Water spray may reduce vapor; but may not prevent ignition in closed spaces.

Environmental Physical Data

Henry's Law Constant: 0.066 to 0.15
BCF: estimated at 13 to 21

Regulations

RCRA 40CFR: Not listed
CERCLA: 40CFR 302.4: Listed per CWA Section 311(b)(4) RQ: 100 lb (45.35 kg)
SARA 40CFR 372.65: Not listed
SARA EHS 40CFR 355: Not listed
TSCA: Listed

DIC7200 CAS #: 10061-02-6

TRANS-1,3-DICHLOROPROPENE

RTECS: UC8320000
DOT: UN2047; IMO3.2
Molecular Formula: $C_3H_4Cl_2$
Formula Weight: 110.97

Cl Cl

Chemical Structure

Synonyms: (E)-1,3-DICHLORO-1-PROPENE; (E)-1,3-DICHLOROPROPENE; TRANS-1,3-DICHLORO-1-PROPENE; TRANS-1,3-DICHLOROPROPYLENE; DI-TRAPEX; 1-PROPENE,1,3-DICHLORO-,(E)-; PROPENE,1,3-DICHLORO-,(E)-; 1-PROPENE,1,3-DICHLORO-,(E)-(9CI); PROPYLENE,1,3-DICHLORO-(TRANS); TELONE II
Description: colorless liquid; sharp, sweet, penetrating chloroform-like odor
Use: together with 1,2-Dichloropropene in the so-called DD mixture as a soil fumigant for control of nematodes

Physical Properties

Boiling Point: 112 °C (234 °F)
Specific Gravity: 1.224 at 20 °C/4 °C
Vapor Density: 1.4 Air=1 at 37.8 °C
Density: 1.22 g/mL
Vapor Pressure: 2.47 kPa
Water Solubility: < 1 mg/mL at 20.5 C
Other Solubilities: 95% Ethanol: >=100 mg/ml at 20.5 °C; Acetone: >=100 mg/ml at 20.5 °C; Benzene: Soluble; DMSO: >=100 mg/ml at 20.5 °C; Ether: Soluble; Most organic solvents: Soluble; Octane: Soluble; Toluene: Soluble.
Refraction Index: 1.4682 at 20 °C
Flash Point: cis & trans 35 °C Open Cup
LEL: 2.6% v/v
UEL: 7.8% v/v

RTECS Toxicity Data

Mutagenic: Human Unscheduled DNA Synthesis; Cell Type: HeLa cell; Dose: 100 umol/L. Bacteria - S Typhimurium Mutations in Microorganisms; Dose: 20 ug/plate (+/-S9).

Hazard Overviews

Corrosive

Flammable

Fire Diamond

Health: Corrosive to eyes/skin/respiratory tract. Toxic. Other Acute Effects: harmful if swallowed, inhaled, or absorbed through skin; sensitizer; inhalation may result in spasm, inflammation and edema of the larynx and bronchi; chemical pneumonitis; pulmonary edema; exposure may cause burning sensation; coughing; wheezing; laryngitis; shortness of breath; headache; nausea; vomiting; allergic respiratory and skin reactions. Chronic Effects: may alter genetic material; may cause heritable genetic damage; target organs: liver, kidneys. Carcinogen.

Fire: Flammable. Hazards: emits toxic fumes; vapor may travel considerable distance to source of ignition and flash back; container explosion may occur. Extinguishing agents: carbon dioxide, dry chemical powder or appropriate foam; water spray; use water spray to cool fire-exposed containers. Precautions: combustible liquid.

Reactivity: Incompatible with: strong oxidizing agents, reacts violently with aluminum/metals/halogens. Hazardous decomposition products: toxic fumes of: carbon monoxide, carbon dioxide, hydrogen chloride gas.

Carcinogenicity: IARC - Not listed; NIOSH - Not listed; NTP - Not listed; ACGIH - Not listed; OSHA - Not listed; EPA - Not listed; MAK - Not listed

Primary Target Organs:

Eyes

Skin
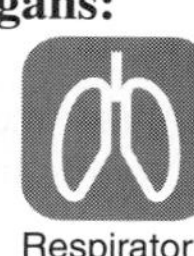
Respiratory System

Liver

Kidneys

Environmental

Environmental Fate: When injected into the soil most will be lost through volatilization over the course of approximately 4 weeks. Higher moisture content in the soil will accelerate volatilization since adsorption is much lower to wet soil than dry. In spill situations, it will volatilize much faster. This compound may also leach. Hydrolysis (half-life ranging behavior 7 to >69 days) to trans-1,3-chloroallyl alcohol will also be important. If released into water, it will be lost primarily through volatilization (half-life from a model river 4 hours). Adsorption to sediment and bioconcentration in fish are not expected to be important processes. In the atmosphere, this compound will degrade primarily by reaction with photochemically produced hydroxyl radicals (half-life 29 hours) or scavenged by rain. Some dispersion away from source or target areas should be expected.

Cleanup/Disposal: Guide No. 132: Fully encapsulating, vapor protective clothing should be worn for spills and leaks with no fire. Eliminate all ignition sources (no smoking, flares, sparks or flames in immediate area). All equipment used when handling the product must be grounded. Do not touch or walk through spilled material. Stop leak if you can do it without risk. Prevent entry into waterways, sewers, basements or confined areas. A vapor suppressing foam may be used to reduce vapors. Absorb with earth, sand or other non-combustible material and transfer to containers (except for Hydrazine). Use clean non-sparking tools to collect absorbed material. Large Spills: Dike far ahead of liquid spill for later disposal. Water spray may reduce vapor; but may not prevent ignition in closed spaces.

Environmental Physical Data

Henry's Law Constant: calculated at 0.0018
Octanol/Water Partition Coefficient: log K_{ow} = 1.41
Sorption Partition Coefficient: K_{oc} = 26
BCF: calculated at 7

Regulations

RCRA 40CFR: Not listed
CERCLA: 40CFR 302.4: Not listed
SARA 40CFR 372.65: Listed
SARA EHS 40CFR 355: Not listed
TSCA: Listed

Analytical Methods

Air: EPA VA-001-1, VG-007-1, VG-011-1, TO-14
Soil: CLP LC_VOA, MC_VOA, OHC; EPA 7, 1624, VG-008-1, VS-001-1, VS-002-1; SW846 5032, 5041, 5041A, 8010B, 8021A, 8240B, 8260A, 8260B
Water / Groundwater: EPA 601, 624, 624-S, VW-001-1, VW-002-1, VW-003-1, VW-014-1; APHA 6210-B, 6230-B; USGS O3115
Drinking Water: EPA 502.1, 502.2, 524.1, 524.2
Food: EPA 5
Indoor / Expired Air: EPA IP-1A
Plasma: EPA 29

DIC7250 CAS #: 75-99-0

2,2-DICHLOROPROPIONIC ACID

RTECS: UF0690000
DOT: NA1760
EINECS Number: 200-923-0
Molecular Formula: $C_3H_4Cl_2O_2$
Structured MF: CH_3CCl_2COOH
Formula Weight: 142.97

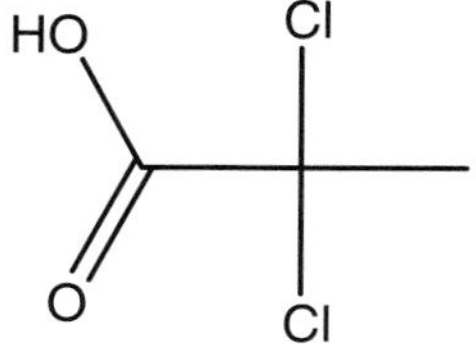

Chemical Structure

Synonyms: ALATEX; BASFAPON; BASFAPON B; BASFAPON/BASFAPON N; BASINEX; BASINEX P; BH DALAPON; CRISAPON; DALAPON; DALAPON 85; DAWPON-RAE; DED-WEED; DEVIPON; 2,2-DICHLOROPROPANOIC ACID; ALPHA,ALPHA-DICHLOROPROPIONIC ACID; ALPHA-DICHLOROPROPIONIC ACID; DOWPON; DOWPON M; 2,2-DPA; DPA; GRAMEVIN; KENAPON;

KYSELINA 2,2-DICHLORPROPIONOVA; LIROPON; PROPANOIC ACID,2,2-DICHLORO-; PROPIONIC ACID,2,2-DICHLORO-; PROPROP; RADAPON; REVENGE; S 1315; S 95 (HERBICIDE); UNIPON

Description: colorless liquid; acrid odor

Use: in non-crop areas such as roadsides, railways, ditches; also in certain established crops such as alfalfa, asparagus, flax, potatoes, rape seed, sugar beets; for quack grass, bermuda grass, johnson grass, other perennial and annual grasses, cattails, rushes often preplant for established perennial grasses in cropland, noncropland areas, irrigation ditch banks translocates to the roots of most species as a growth regulator; selective herbicide; growth regulator; herbicide sodium salt

Physical Properties

Boiling Point: 190 °C (374 °F)
Freezing Point: 20 °C (68 °F)
Specific Gravity: 1.4014 at 20 °C
Vapor Density: 4.9 Air=1
Water Solubility: 50% by weight
Other Solubilities: Very Soluble in alkali solvents.
Odor Threshold: 2500 mg/m^3
pH: Aqueous solution 1.32
Refraction Index: 1.4551 at 20 °C
Flash Point: Nonflammable

RTECS Toxicity Data

Acute Dermal: Rat LD_{50} Route: Skin; Dose: >5 gm/kg.
Mutagenic: Other Microorganisms Mutations in Microorganisms; Dose: 500 ppm (-S9).

Hazard Overviews

Fire Diamond

Health: Irritating to eyes/skin; mildly irritating to respiratory tract. Also Causes: fatigue, vomiting, diarrhea, slow pulse, depression, unbalanced gait, appetite loss. Chronic Effects: liver/kidney damage (in animals).

Fire: Nonflammable, but may form flammable hydrogen gas in enclosed spaces. Use dry chemical, carbon dioxide, water spray, or regular foam.

Reactivity: Stable. Hazardous polymerization cannot occur. Avoid: exposure to heat; ignition sources. Incompatible with: aluminum and copper alloys; water; iron. Hazardous decomposition products: chlorine; hydrogen chloride gas.

Carcinogenicity: IARC - Not listed; NIOSH - Not listed; NTP - Not listed; ACGIH - Not listed; OSHA - Not listed; EPA - Not listed; MAK - Not listed

Primary Target Organs:

Eyes

Skin

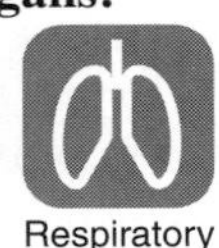
Respiratory System

Nervous System

Exposure Limits

OSHA PEL Vacated 1989 Limits: TWA: 1 ppm; 6 mg/m^3.
NIOSH REL: TWA: 1 ppm; 6 mg/m^3.
DFG MAK: TWA: 1 ppm; 6 mg/m^3.

Respirator Recommendation

Exposure Range: >1 to 50 ppm Supplied Air, Constant Flow/Pressure Demand, Half Mask
Exposure Range: >50 to 1000 ppm Supplied Air, Constant Flow/Pressure Demand, Full Face
Exposure Range: >1000 to unlimited ppm Self-contained Breathing Apparatus, Pressure Demand, Full Face
Note: odor threshold unknown

Environmental

Ecotoxicity: LC_{50} ring-necked pheasant greater than 5000 ppm/5 days (no mortality to 5000 ppm); 10 days old LC_{50} Daphnia pulex 11.0 mg/l/48 hr (95% confidence limit, 8.2-14.7) at 15 °C; first instar. Static bioassay without aeration, pH 7.2-7.5, water hardness 40-50 mg/l as calcium carbonate and alkalinity of 30-35 mg/l LC_{50} Pteronarcys greater than 1,000 mg/l/96 hr at 15 °C; first year class. Static bioassay without aeration, pH 7.2-7.5, water hardness 40-50 mg/l as calcium carbonate and alkalinity of 30-35 mg/l LC_{50} LEpomis macrochiris (bluegill) 105 mg/l/96 hr at 24 °C; 1.0 g. Static bioassay without aeration, pH 7.2-7.5, water hardness 40-50 mg/l as calcium carbonate and alkalinity of 30-35 mg/l

Environmental Fate: If released to soil, microbial degradation and leaching appear to be the important environmental fate processes. It leaches readily in soil; however, under conditions favorable for microbial growth, microbial degradation will probably proceed at a faster rate than leaching. In the absence of microbial action, degradation in soil is slow. The resultant average persistence at recommended rates of application has been reported to be two to four weeks in most agricultural soils during the growing season, although a persistence of six months has been observed in soils of various forests and tree nurseries. If released to water, microbial degradation, hydrolysis, and photolysis are potentially important in the removal of dalapon. The hydrolysis half-life in water is on the order of several months at temperatures less than 25 °C, with the hydrolysis forming pyruvic acid. Under conditions favorable for microbial growth, decomposition via microorganisms will probably be complete within one month which will diminish the importance of chemical hydrolysis. Direct photolysis in water may be possible, although photolytic rates have not been investigated under environmental conditions. Aquatic volatilization, adsorption to sediments, or bioconcentration are not expected to be significant. If released to the atmosphere, it will react in the vapor-phase with photochemically produced hydroxyl radicals at an estimated half-life rate of 72.3 days. Atmospheric removal via washout may be possible since it is extremely water soluble.

Cleanup/Disposal: Guide No. 154: Eliminate all ignition sources (no smoking, flares, sparks or flames in immediate area). Do not touch damaged containers or spilled material unless wearing appropriate protective clothing. Stop leak if you can do it without risk. Prevent entry into waterways, sewers, basements or confined areas. Absorb or cover with

dry earth, sand or other non-combustible material and transfer to containers. Do not get water inside containers.

Environmental Physical Data

Henry's Law Constant: estimated at 6.3 x10^{-8}
Octanol/Water Partition Coefficient: log K_{ow} = 0.76
BCF: fish 3
BOD: unaclimated 0.04 lb/lb, 5 days

Regulations

RCRA 40CFR: Not listed
CERCLA: 40CFR 302.4: Listed per CWA Section 311(b)(4) RQ: 5000 lb (2268 kg)
SARA 40CFR 372.65: Not listed
SARA EHS 40CFR 355: Not listed
TSCA: Listed

Analytical Methods

Soil: EPA PMD-TLC; SW846 8150B, 8151, 8321, 8321A
Water / Groundwater: EPA 1658, 615, X_89_176
Drinking Water: EPA 515.1, 552.1
Food: AOAC 962.06, 984.06
Plasma: FDA 221.1

DIC7300 CAS #: 101-38-2

2,6-DICHLOROQUINONE CHLORIMINE

RTECS: GU5470000
EINECS Number: 202-937-2
Molecular Formula: $C_6H_2Cl_3NO$
Formula Weight: 210.44

Chemical Structure

Synonyms: 4-CHLOROIMINO-2,6-DICHLORO-2,5-CYCLOHEXADIENE-1-ONE; 2,5-CYCLOHEXADIEN-1-ONE,2,6-DICHLORO-4-(CHLOROIMINO)-(8CI,9CI); 2,6-DICHLORO-4-(CHLOROIMINO)-2,5-CYCLOHEXADIEN-1-ONE; 2,6-DICHLOROQUINONE CHLOROIMIDE; 2,6-DICHLOROQUINONECHLOROIMINE; GIBBS REAGENT; N,2,6-TRICHLORO-P-BENZOQUINONE IMINE; N,2,6-TRICHLOROBENZOQUINONE IMINE

Physical Properties

Freezing Point: 65 °C (149 °F) to 67 °C (152.6 °F)
Water Solubility: Moderate Solublity

RTECS Toxicity Data

Acute Oral: Rat LD; Dose: >500 mg/kg.

Hazard Overviews

Health: Irritating to eyes/skin/respiratory tract. Other Acute Effects: may be harmful by inhalation, ingestion, or skin absorption.
Fire: Hazards: may explode when heated; emits toxic fumes. Extinguishing agents: carbon dioxide, dry chemical powder or appropriate foam; water spray. Precautions: combustible liquid.
Reactivity: Incompatible with: heat,-sensitive, strong acids, strong bases. Hazardous decomposition products: thermal decomposition may produce carbon monoxide, carbon dioxide, and nitrogen oxides; hydrogen chloride gas.
Carcinogenicity: IARC - Not listed; NIOSH - Not listed; NTP - Not listed; ACGIH - Not listed; OSHA - Not listed; EPA - Not listed; MAK - Not listed
Primary Target Organs:

Eyes

Skin

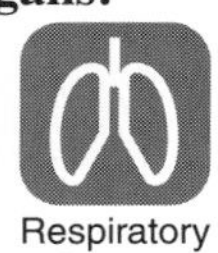
Respiratory System

Environmental

Regulations

RCRA 40CFR: Not listed
CERCLA: 40CFR 302.4: Not listed
SARA 40CFR 372.65: Not listed
SARA EHS 40CFR 355: Not listed
TSCA: Listed

DIC7350 CAS #: 320-72-9

3,5-DICHLOROSALICYLIC ACID

RTECS: VO2450000
EINECS Number: 206-281-8
Molecular Formula: $C_7H_4Cl_2O_3$
Formula Weight: 207.01

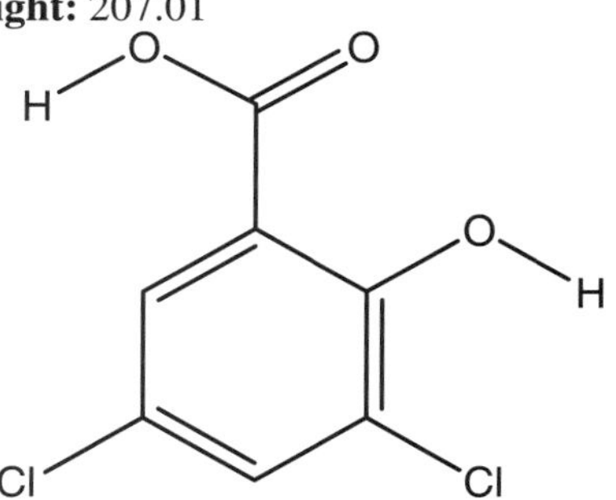

Chemical Structure

Synonyms: BENZOIC ACID,3,5-DICHLORO-2-HYDROXY-; 3,5-DICHLORO-2-HYDROXYBENZOIC ACID; 2-HYDROXY-3,5-DICHLOROBENZOIC ACID; SALICYLIC ACID,3,5-DICHLORO-
Description: needles or prisms

Physical Properties

Boiling Point: Sublimes
Freezing Point: 220 °C (428 °F) to 221 °C (429.8 °F)
Water Solubility: Slightly Soluble in Hot Water
Other Solubilities: Very Soluble in Alcohol. Soluble in Ether

Hazard Overviews

Health: Irritating to eyes/skin/respiratory tract. Other Acute Effects: may be harmful by inhalation, ingestion, or skin absorption.
Fire: Hazards: emits toxic fumes. Extinguishing agents: water spray; carbon dioxide, dry chemical powder or appropriate foam. Precautions: combustible liquid.
Reactivity: Incompatible with: strong oxidizing agents, strong bases. Hazardous decomposition products: toxic fumes of: carbon monoxide, carbon dioxide, hydrogen chloride gas.
Carcinogenicity: IARC - Not listed; NIOSH - Not listed; NTP - Not listed; ACGIH - Not listed; OSHA - Not listed; EPA - Not listed; MAK - Not listed
Primary Target Organs:

Eyes

Skin

Respiratory System

Environmental

Regulations

RCRA 40CFR: Not listed
CERCLA: 40CFR 302.4: Not listed
SARA 40CFR 372.65: Not listed
SARA EHS 40CFR 355: Not listed
TSCA: Listed

DIC7400 CAS #: 4109-96-0

DICHLOROSILANE

RTECS: VV3050000
EINECS Number: 223-888-3
Molecular Formula: Cl_2H_2Si
Formula Weight: 101.01

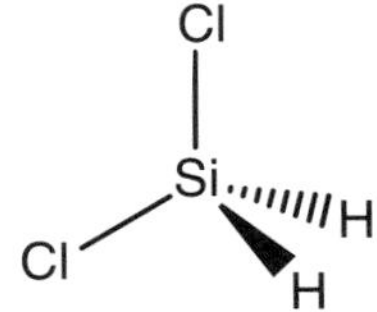

Chemical Structure

Synonyms: CHLOROSILANE; SILICON CHLORIDE HYDRIDE
Description: colorless gas, liquid; acrid odor

Physical Properties

Boiling Point: 8 °C (46 °F)
Freezing Point: -122 °C (-187.6 °F)
Specific Gravity: 1.22
Vapor Density: 3.48
Density: 1.22 at -7 °C
Vapor Pressure: 1230 mm Hg at 20 °C
Water Solubility: Decomposes
Evaporation Rate: Estimated at 82 Butyl Acetate=1
Flash Point: -52 °C
Autoignition Temperature: 41 to 47 °C
LEL: 0.047% v/v
UEL: 0.96% v/v

RTECS Toxicity Data

Acute Inhalation: Rat LC_{50} Dose: 215 ppm; Toxic Effects: Sense organs and special senses - Lacrimation; Lungs, Thorax, or Respiration - Acute pulmonary edema; Lungs, Thorax, or Respiration - Dyspnea. Mouse LC_{50} Dose: 144 ppm/4hr.

Hazard Overviews

Corrosive

Explosive

Flammable

Fire Diamond

Health: Corrosive to eyes/skin/respiratory tract. Also Causes: difficulty breathing, coughing, pulmonary edema, ocular irritation, gastrointestinal burns, nausea, thirst, dizziness, weakness, circulatory collapse, unconsciousness. Chronic Effects: erosion of teeth, bleeding of nose and gums, ulceration of nasal membrane. Forms poisonous gas on contact with water.
Fire: Explosive and flammable. For small fires use dry chemical or carbon dioxide; for large fires use coarse water spray, fog, or foam. If water is used, take precautions against the formation of hydrogen chloride gas. Fight fire from maximum distance. Fire may be difficult to extinguish and may re-ignite.
Reactivity: Unstable, can undergo spontaneous ignition at room temperature. Hazardous polymerization cannot occur. Avoid: exposure to heat; ignition sources; water. Incompatible with: water; acids; alkalis; halocarbons; strong oxidizers; compounds containing active hydrogen atoms (amines; alcohols; ammonia); catalysts (amines; rust; aluminum chloride); some plastics. Hazardous decomposition products: chlorine gas.
Carcinogenicity: IARC - Not listed; NIOSH - Not listed; NTP - Not listed; ACGIH - Not listed; OSHA - Not listed; EPA - Not listed; MAK - Not listed

Primary Target Organs:

Eyes

Skin

Respiratory System

Mucous Membranes

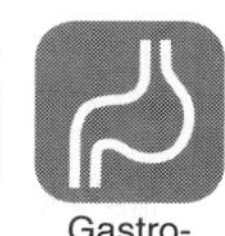
Gastro-intestinal

Environmental

Regulations

RCRA 40CFR: Not listed
CERCLA: 40CFR 302.4: Not listed
SARA 40CFR 372.65: Not listed
SARA EHS 40CFR 355: Not listed
TSCA: Listed

DIC7500 CAS #: 374-07-2

1,1-DICHLORO-1,2,2,2-TETRAFLUOROETHANE

DOT: UN1078; IMO2.2
EINECS Number: 206-774-8
Molecular Formula: $C_2Cl_2F_4$
Formula Weight: 170.92
Synonyms: 1,1-DICHLOROTETRAFLUOROETHANE; DICHLOROTETRAFLUROETHANE; ETHANE,1,1-DICHLORO-1,2,2,2-TETRAFLUORO-; ETHANE,1,1-DICHLOROTETRAFLUORO-; FRIGEN 114A; 1,1,1,2-TETRAFLUORO-2,2-DICHLOROETHANE
Description: colorless gas; odorless
Use: aerosol propellant (possible use); refrigerant & solvent (possible uses)

Physical Properties

Boiling Point: -3.6 °C (26 °F) at 760 mm Hg
Freezing Point: -94 °C (-137.2 °F)
Specific Gravity: 1.455 at 25 °C/4 °C
Vapor Pressure: 30.2 kPa at -25 °C
Water Solubility: 60 mg/L
Refraction Index: 1.3092 at 0 °C/D
Critical Pressure: 3.29 Pa
Flash Point: Nonflammable

Hazard Overviews

Compressed Gas

Fire: Noncombustible.
Carcinogenicity: IARC - Not listed; NIOSH - Not listed; NTP - Not listed; ACGIH - Not listed; OSHA - Not listed; EPA - Not listed; MAK - Not listed

Environmental

Environmental Fate: If released to soil, it would be expected to rapidly volatilize from the soil surface or leach into the soil. If released into water, it would rapidly volatilize (half-life 3.8 hr in a model river). Bioconcentration in aquatic organism or adsorption to sediment would not be significant. It is not expected to degrade in the troposphere and its sole removal process would be slow diffusion to the stratosphere, a process that may take decades, where it will slowly photolyze and degrade by reacting with atomic oxygen.
Cleanup/Disposal: Guide No. 126: Do not touch or walk through spilled material. Stop leak if you can do it without risk. Do not direct water at spill or source of leak. Use water spray to reduce vapors or divert vapor cloud drift. If possible, turn leaking containers so that gas escapes rather than liquid. Prevent entry into waterways, sewers, basements or confined areas. Allow substance to evaporate. Ventilate the area.

Environmental Physical Data

Henry's Law Constant: estimated at 1.69
Octanol/Water Partition Coefficient: log K_{ow} = 2.85
Sorption Partition Coefficient: K_{oc} = 459
BCF: calculated at 86

Regulations

RCRA 40CFR: Not listed
CERCLA: 40CFR 302.4: Listed as Compound per CWA Section 307(a)
SARA 40CFR 372.65: Not listed
SARA EHS 40CFR 355: Not listed
TSCA: Listed

DIC7550 CAS #: 76-14-2

DICHLOROTETRAFLUOROETHANE

RTECS: KI1101000
DOT: UN1958; IMO2.2
EINECS Number: 200-937-7
Molecular Formula: $C_2Cl_2F_4$
Structured MF: $CClF_2CClF_2$
Formula Weight: 170.93

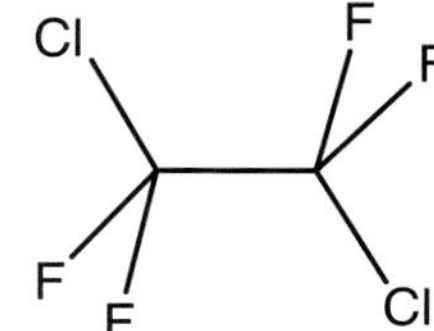

Chemical Structure

Synonyms: ARCTON 114; ARCTON 33; CFC 114; CRIOFLUORANO; CRYOFLUORAN; CRYOFLUORANE; CRYOFLUORANUM; 1,2-DICHLORO-1,1,2,2-TETRAFLUOROETHANE; 1,2-DICHLOROTETRAFLUOROETHANE; SYM-DICHLOROTETRAFLUOROETHANE; EPA PESTICIDE CHEMICAL CODE 326200; ETHANE,1,2-DICHLORO-1,1,2,2-TETRAFLUORO-; ETHANE,1,2-DICHLOROTETRAFLUORO-; F 114; FC 114; FLUORANE 114; FLUOROCARBON 114; FREON 114; FRIGEN 114; FRIGIDERM; GENETRON 114; GENETRON 316; HALOCARBON 114; HALON 242; LEDON 114; PROPELLANT 114; R 114; REFRIGERANT 114; 1,1,2,2-TETRAFLUORO-1,2-DICHLOROETHANE; TETRAFLUORODICHLOROETHANE; UCON 114

Description: colorless liquefied compressed gas; slight ethereal odor

Use: vet: in aerosol skin freezes; for snake & insect bites to retard venom absorption; blowing agent for cellular polymers; solvent & diluent in fluoro-olefins polymerization, cleaning & degreasing printed circuit boards, preparation of uranium tetrafluoride, freons, polymer ints, explosives & volatile substance extraction; fire extinguishing & aerosol foaming agent; to lower aerosol vapor pressure and flammability; refrigerant in indust cooling, air conditioning systems, home appliances, retail food refrigeration systems & chillers; corrosion inhibitor in hydraulic fluids.

Physical Properties

Boiling Point: 4.1 °C (39 °F) at 760 mm Hg
Freezing Point: -94 °C (-137.2 °F)
Specific Gravity: Liquid 1.5312 at 0 °C
Vapor Density: 5.9 Air=1
Vapor Pressure: 2014 mm Hg at 25 °C
Water Solubility: 0.01% by weight
Other Solubilities: Soluble in Alcohol, Ether
Surface Tension: 12 dynes/cm
Refraction Index: 1.3092 at 0 °C/D
Evaporation Rate: > 1 Butyl Acetate=1
Critical Temperature: 145.7 °C
Critical Pressure: 474 psia
Ionization Potential (eV): 12.20
Flash Point: Nonflammable

RTECS Toxicity Data

Acute Inhalation: Rat LC_{50} Dose: 72 pph/30M. Mouse LC_{50} Dose: 70 pph/30M.

Hazard Overviews

Compressed Gas

Fire Diamond

Health: Irritating to eyes/respiratory tract. Also Causes: light-headedness, giddiness, disorientation, cardiac arrhythmias, tremors, coma, frostbite. Chronic Effects: light-headedness, palpitations.

Fire: Noncombustible. However, hazardous decomposition products are given off when it is in contact with heat and ignition sources. Use agent suitable for surrounding fire.

Reactivity: Stable. Hazardous polymerization cannot occur. Avoid: heat and ignition sources Incompatible with: chemically active metals such as sodium, potassium, calcium and powdered zinc; magnesium; aluminum; acids and acid fumes; some forms of plastics, rubber, and coatings. Hazardous decomposition products: hydrogen chloride; phosgene; hydrogen fluoride; carbonyl fluoride.

Carcinogenicity: IARC - Not listed; NIOSH - Not listed; NTP - Not listed; ACGIH - Class A4, Not classifiable as a human carcinogen; OSHA - Not listed; EPA - Not listed; MAK - Not listed

Primary Target Organs:

Eyes

Mucous Membranes

Cardio-vascular

Exposure Limits
OSHA PEL: TWA: 1000 ppm; 7000 mg/m^3.
NIOSH REL: TWA: 1000 ppm; 7000 mg/m^3.
NIOSH IDLH: 15000 ppm.
DFG MAK: TWA: 1000 ppm; 7000 mg/m^3.

Respirator Recommendation
Exposure Range: >1000 to <15,000 ppm Supplied Air, Constant Flow/Pressure Demand, Half Mask
Exposure Range: 15,000 to unlimited ppm Self-contained Breathing Apparatus, Pressure Demand, Full Face
Note: odor threshold unknown

Environmental

Ecotoxicity: Fishes: Leuciscus idus 48h NOEC +2 mg/l

Environmental Fate: If released to soil, it would rapidly volatilize from soil surfaces or leach through soil. If released to water, essentially all is expected to be lost by volatilization (half-life 4 hours from a model river). If released to the atmosphere, it will not degrade in the troposphere, but gradually diffuse into the stratosphere (half-life 20 years). In the stratosphere this compound will slowly photolyze or slowly react with singlet oxygen (stratospheric lifetime 126-310 years). Due to its stability, detection long distances from its sources emissions has occurred. It is not expected to degrade in the ambient atmosphere by reaction with photochemically produced hydroxyl radicals as structurally similar compounds have half-lives greater than 100 years for this reaction.

Cleanup/Disposal: Guide No. 126: Do not touch or walk through spilled material. Stop leak if you can do it without risk. Do not direct water at spill or source of leak. Use water spray to reduce vapors or divert vapor cloud drift. If possible, turn leaking containers so that gas escapes rather than liquid. Prevent entry into waterways, sewers, basements or confined areas. Allow substance to evaporate. Ventilate the area.

Environmental Physical Data

Henry's Law Constant: estimated at 2.8
Octanol/Water Partition Coefficient: log K_{ow} = 2.82
Sorption Partition Coefficient: K_{oc} = estimated at 815 to 300
BCF: estimated at 10 to 82

Regulations

RCRA 40CFR: Not listed
CERCLA: 40CFR 302.4: Listed as Compound per CWA Section 307(a)
SARA 40CFR 372.65: Listed
SARA EHS 40CFR 355: Not listed
TSCA: Listed

Analytical Methods

Air: EPA TO-14, 0040; ASTM D4490
Indoor / Expired Air: NIOSH 1018; EPA IP-1A

DIC7600 CAS #: 4342-61-4

1,2-DICHLOROTETRAMETHYLDISILANE

EINECS Number: 224-400-1

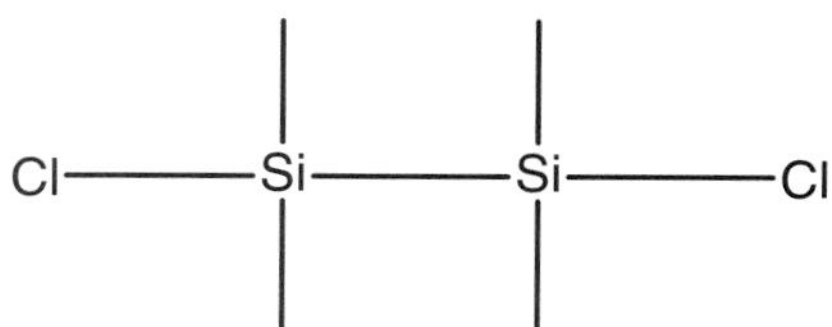

Chemical Structure

Physical Properties

Boiling Point: 151 °C (304 °F)
Specific Gravity: 1.01000
Vapor Pressure: > 2.5
Water Solubility: Reacts
Odor Threshold: 10.0 ppm
Refraction Index: 1.4548
Evaporation Rate: < 1 Butyl Acetate=1

Hazard Overviews

Corrosive

Health: Corrosive to eyes/skin/respiratory tract. Other Acute Effects: harmful if swallowed, inhaled, or absorbed through skin; inhalation may result in spasm, inflammation and edema of the larynx and bronchi, chemical pneumonitis and pulmonary edema; symptoms of exposure may include burning sensation; coughing; wheezing; laryngitis; shortness of breath; headache; nausea; vomiting.
Fire: Hazards: emits toxic fumes. Extinguishing agents: carbon dioxide, dry chemical powder or appropriate foam; do not use water. Precautions: combustible liquid.
Reactivity: Incompatible with: strong oxidizing agents, reacts violently with water. Hazardous decomposition products: toxic fumes of: carbon monoxide, carbon dioxide, hydrogen chloride gas, silicon oxide.
Carcinogenicity: IARC - Not listed; NIOSH - Not listed; NTP - Not listed; ACGIH - Not listed; OSHA - Not listed; EPA - Not listed; MAK - Not listed
Primary Target Organs:

Eyes

Skin

Respiratory System

Environmental

Regulations

RCRA 40CFR: Not listed
CERCLA: 40CFR 302.4: Not listed
SARA 40CFR 372.65: Not listed
SARA EHS 40CFR 355: Not listed
TSCA: Not listed

DIC7650 CAS #: 95-73-8

2,4-DICHLOROTOLUENE

RTECS: XT0730000
DOT: UN2238; IMO3.3
EINECS Number: 202-445-8
Molecular Formula: $C_7H_6Cl_2$
Formula Weight: 161.03

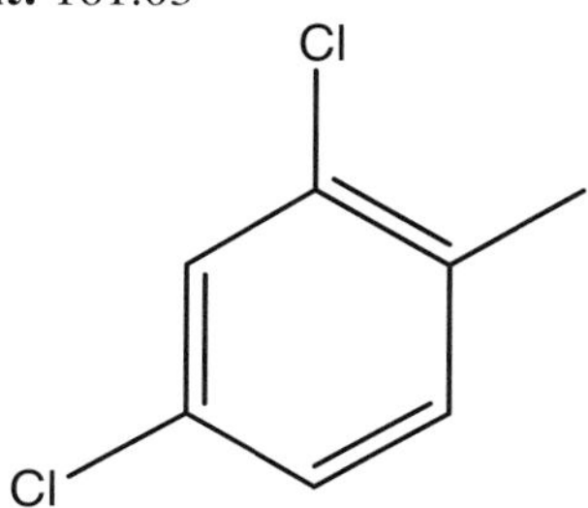

Chemical Structure

Synonyms: BENZENE,2,4-DICHLORO-1-METHYL-; BENZENE,2,4-DICHLORO-1-METHYL-(9CI); 2,4-DICHLORO-1-METHYLBENZENE; 2,4-DICHLOROMETHYLBENZENE; TOLUENE,2,4-DICHLORO-
Description: colorless, clear liquid
Use: high-boiling solvent; intermediate for organic synthesis

Physical Properties

Boiling Point: 196 °C (385 °F) to 197 °C (387 °F) at 760 mm Hg
Freezing Point: -13.5 °C (7.7 °F)
Specific Gravity: 1.2498 at 20 °C/20 °C
Saturated Vapor Density: 1.202990444 kg/m^3
Vapor Pressure: 0.416 mm Hg at 25 °C
Water Solubility: Insoluble in Water
Other Solubilities: Carbon Tetrachloride: Soluble
Surface Tension: 38.29 dyne/cm at 25 °C
Refraction Index: 1.5511 at 20 °C/D
Flash Point: 92.778 °C Open Cup

RTECS Toxicity Data

Acute Oral: Rat LD_{50} Dose: 2400 mg/kg; Toxic Effects: Behavioral - Somnolence (general depressed activity); Behavioral - Tremor; Behavioral - Muscle weakness. Mouse LD_{50} Dose: 2400 mg/kg; Toxic Effects: Behavioral - Somnolence (general depressed activity); Behavioral - Tremor; Behavioral - Muscle weakness.

Chronic (Multiple Dose) Oral: Rat Dose: 6440 mg/kg/2W-I; Toxic Effects: Liver - Other changes; Blood - Other changes; Biochemical - Transaminases.

Hazard Overviews

Health: May cause irritation to eyes/skin. Harmful. Other Acute Effects: harmful by inhalation, in contact with skin and if swallowed.

Fire: Combustible. Hazards: emits toxic fumes. Extinguishing agents: water spray; carbon dioxide, dry chemical powder or appropriate foam. Precautions: combustible liquid.

Carcinogenicity: IARC - Not listed; NIOSH - Not listed; NTP - Not listed; ACGIH - Not listed; OSHA - Not listed; EPA - Not listed; MAK - Not listed

Environmental

Ecotoxicity: Fishes: Poecilia reticulata 14d LC_{50} 4.6 mg/l

Environmental Fate: If released on land, it will adsorb strongly to the soil surface. However, its fate in soil is unknown. If released in water, it will adsorb to sediment and bioconcentrate in aquatic organisms. Loss will occur by volatilization (half- life 4.0 hr in a model river). Its half-life in one river was 1 day based upon monitoring data. In the atmosphere, it should partially exist adsorbed to particulate matter and be subject to gravitational settling. In the vapor phase, it will degrade by reaction with photochemically produced hydroxyl radicals (half-life 11.6 days).

Cleanup/Disposal: Guide No. 130: Eliminate all ignition sources (no smoking, flares, sparks or flames in immediate area). All equipment used when handling the product must be grounded. Do not touch or walk through spilled material. Stop leak if you can do it without risk. Prevent entry into waterways, sewers, basements or confined areas. A vapor suppressing foam may be used to reduce vapors. Absorb or cover with dry earth, sand or other non-combustible material and transfer to containers. Use clean non-sparking tools to collect absorbed material. Large Spills: Dike far ahead of liquid spill for later disposal. Water spray may reduce vapor; but may not prevent ignition in closed spaces.

Environmental Physical Data

Henry's Law Constant: 0.346 $\times 10^{-2}$

Octanol/Water Partition Coefficient: log K_{OW} = 4.24

Sorption Partition Coefficient: K_{OC} = 4800

BCF: calculated at 983

Regulations

RCRA 40CFR: Not listed

CERCLA: 40CFR 302.4: Not listed

SARA 40CFR 372.65: Not listed

SARA EHS 40CFR 355: Not listed

TSCA: Listed

Analytical Methods

Soil: SW846 3640A

DIC7700 CAS #: 118-69-4

2,6-DICHLOROTOLUENE

DOT: UN2238; IMO3.3

EINECS Number: 204-269-7

Molecular Formula: $C_7H_6Cl_2$

Formula Weight: 161.03

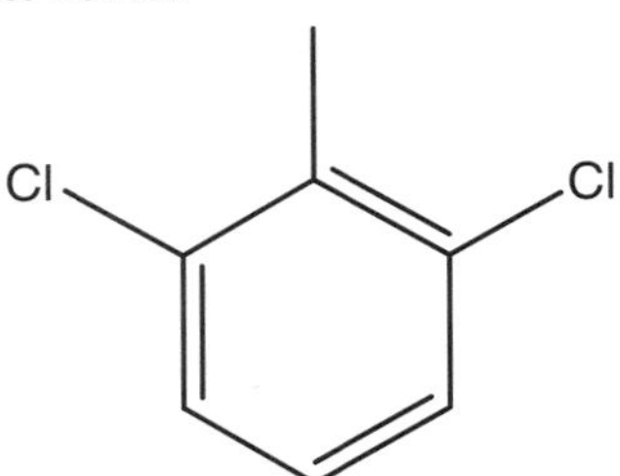

Chemical Structure

Synonyms: BENZENE,1,3-DICHLORO-2-METHYL-; TOLUENE,2,6-DICHLORO-

Use: chem int for dyes; high boiling solvent

Physical Properties

Boiling Point: 198 °C (388 °F) at 760 mm Hg

Freezing Point: 26 °C (78.8 °F)

Specific Gravity: 1268.6 at 20 °C

Water Solubility: Insoluble

Other Solubilities: Soluble in Chloroform

Refraction Index: 1.5507 at 20 °C/D

Flash Point: Combustible

Hazard Overviews

Health: Irritating to eyes/skin/respiratory tract. Harmful. Other Acute Effects: may be harmful by inhalation, ingestion, or skin absorption.

Fire: Combustible. Hazards: emits toxic fumes. Extinguishing agents: water spray; carbon dioxide, dry chemical powder or appropriate foam. Precautions: combustible liquid.

Reactivity: Incompatible with: strong oxidizing agents, strong bases. Hazardous decomposition products: toxic fumes of: carbon monoxide, carbon dioxide, hydrogen chloride gas.

Carcinogenicity: IARC - Not listed; NIOSH - Not listed; NTP - Not listed; ACGIH - Not listed; OSHA - Not listed; EPA - Not listed; MAK - Not listed

Primary Target Organs:

Eyes

Skin

Respiratory System

Environmental

Ecotoxicity: Fishes: Brachydanio rerio 96h LC_0 3.9 mg/l (technical grade) 96h LC_{50} 5.8 mg/l

Environmental Fate: If release on land, it will adsorb strongly to the soil surface. However its fate in soil is unknown. If released in water, it will adsorb to sediment and bioconcentrate in aquatic organisms. Loss by volatilization will be rapid. In the atmosphere, it should partially exist adsorbed to particulate matter and be subject to gravitational settling. It will degrade by reaction with photochemically produced hydroxyl radicals (half-life 11.6 days).

Cleanup/Disposal: Guide No. 130: Eliminate all ignition sources (no smoking, flares, sparks or flames in immediate area). All equipment used when handling the product must be grounded. Do not touch or walk through spilled material. Stop leak if you can do it without risk. Prevent entry into waterways, sewers, basements or confined areas. A vapor suppressing foam may be used to reduce vapors. Absorb or cover with dry earth, sand or other non-combustible material and transfer to containers. Use clean non-sparking tools to collect absorbed material. Large Spills: Dike far ahead of liquid spill for later disposal. Water spray may reduce vapor; but may not prevent ignition in closed spaces.

Environmental Physical Data

Henry's Law Constant: 0.346×10^{-2}
Octanol/Water Partition Coefficient: log K_{ow} = 4.29
Sorption Partition Coefficient: K_{oc} = 5100
BCF: calculated at 1070

Regulations

RCRA 40CFR: Not listed
CERCLA: 40CFR 302.4: Not listed
SARA 40CFR 372.65: Not listed
SARA EHS 40CFR 355: Not listed
TSCA: Listed

DIC7750 CAS #: 16110-89-7

4-(2,4-DICHLORO-1,3,5-TRIAZINYLAMINO)BENZENESULFONIC ACID

EINECS Number: 240-280-3
Molecular Formula: $C_9H_6Cl_2N_4O_3S$
Formula Weight: 321.13
Synonyms: BENZENESULFONIC ACID,4-((4,6-DICHLORO-1,3,5-TRIAZIN-2-YL)AMINO)-; SULFANILIC ACID,N-(4,6-DICHLORO-S-TRIAZIN-2-YL)-

Hazard Overviews

Carcinogenicity: IARC - Not listed; NIOSH - Not listed; NTP - Not listed; ACGIH - Not listed; OSHA - Not listed; EPA - Not listed; MAK - Not listed

Environmental

Regulations

RCRA 40CFR: Not listed
CERCLA: 40CFR 302.4: Listed as Compound per CWA Section 307(a)
SARA 40CFR 372.65: Not listed
SARA EHS 40CFR 355: Not listed
TSCA: Listed

DIC7800 CAS #: 812-04-4

1,1-DICHLORO-1,2,2-TRIFLUOROETHANE

Molecular Formula: $C_2HCl_2F_3$
Formula Weight: 152.93
Use: as alternatives to CFCS in applications such as refrigerants, blowing agents, cleaning agents, and fire extinguishants

Physical Properties

Flash Point: Does not burn or burns with difficulty

Hazard Overviews

Fire: Noncombustible.
Carcinogenicity: IARC - Not listed; NIOSH - Not listed; NTP - Not listed; ACGIH - Not listed; OSHA - Not listed; EPA - Not listed; MAK - Not listed

Environmental

Regulations

RCRA 40CFR: Not listed
CERCLA: 40CFR 302.4: Listed as Compound per CWA Section 307(a)
SARA 40CFR 372.65: Listed
SARA EHS 40CFR 355: Not listed
TSCA: Not listed

DIC7850 CAS #: 354-23-4

1,2-DICHLORO-1,1,2-TRIFLUOROETHANE

RTECS: KI1106000
EINECS Number: 206-549-4
Molecular Formula: $C_2HCl_2F_3$
Formula Weight: 152.93
Synonyms: 1,1,2-TRIFLUORO-1,2-DICHLOROETHANE
Use: as alternatives to CFCS in applications such as refrigerants, blowing agents, cleaning agents, and fire extinguishants

Physical Properties

Boiling Point: 28 °C (82 °F)
Freezing Point: -78 °C (-108.4 °F)
Density: 1.5 g/cu cm at 25 °C
Refraction Index: 1.327 at 20 °C
Ionization Potential (eV): =< 12.00 +/-2.0
Flash Point: Does not burn or burns with difficulty

RTECS Toxicity Data

Acute Inhalation: Mouse LC_{Lo} Dose: 15 pph/2M.

Hazard Overviews

Fire: Noncombustible.
Carcinogenicity: IARC - Not listed; NIOSH - Not listed; NTP - Not listed; ACGIH - Not listed; OSHA - Not listed; EPA - Not listed; MAK - Not listed

Environmental

Environmental Fate: When released to the atmosphere, it will degrade primarily through reaction with photochemically produced hydroxyl radicals. The atmospheric degradation rate is slow. It has an estimated hydroxyl radical reaction half-life of about 3.5 years and a predicted atmospheric lifetime of 8.1 years. If released to soil or water, volatilization to the atmosphere will be a major fate process. Volatilization half-lives of 3.6 hr and 50 hours can be estimated for a model river (1 m deep) and a model pond (2 m deep) respectively.

Environmental Physical Data

Henry's Law Constant: estimated at 0.0955
Octanol/Water Partition Coefficient: log K_{ow} = 2.17
Sorption Partition Coefficient: K_{oc} = estimated at 355
BCF: estimated at 27

Regulations

RCRA 40CFR: Not listed
CERCLA: 40CFR 302.4: Listed as Compound per CWA Section 307(a)
SARA 40CFR 372.65: Listed
SARA EHS 40CFR 355: Not listed
TSCA: Listed

DIC7900 CAS #: 306-83-2

2,2-DICHLORO-1,1,1-TRIFLUOROETHANE

RTECS: KI1108000
EINECS Number: 206-190-3
Molecular Formula: $C_2HCl_2F_3$
Formula Weight: 152.93
Synonyms: FC 123; FREON 123; HCFC 123; R 123; 1,1,1-TRIFLUORO-2,2-DICHLOROETHANE
Use: manufacture of trifluoroacetic acid; for refrigeration & air conditioning

Physical Properties

Boiling Point: 27.82 °C (82.076 °F)
Water Solubility: 1488 mg/L at 25 °C (est)
Ionization Potential (eV): 11.5 +/-2.0

RTECS Toxicity Data

Acute Inhalation: Mouse LC_{50} Dose: 74000 ppm/1hr.

Hazard Overviews

Health: Irritating to eyes/skin/respiratory tract. Other Acute Effects: may be harmful by inhalation, ingestion, or skin absorption. Chronic Effects: heart may develop pressure; target organs: nerves, heart.
Fire: Hazards: emits toxic fumes. Extinguishing agents: water spray; carbon dioxide, dry chemical powder or appropriate foam. Precautions: combustible liquid.
Reactivity: Incompatible with: strong oxidizing agents, sodium, potassium, aluminum, magnesium, zinc. Hazardous decomposition products: toxic fumes of: carbon monoxide, carbon dioxide, hydrogen chloride gas, hydrogen fluoride.
Carcinogenicity: IARC - Not listed; NIOSH - Not listed; NTP - Not listed; ACGIH - Not listed; OSHA - Not listed; EPA - Not listed; MAK - Class B, Justifiably suspected of having carcinogenic potential
Primary Target Organs:

Eyes

Skin

Respiratory System

Nervous System

Cardio-vascular

Environmental

Environmental Fate: If released to soil, it will rapidly volatilize from either moist or dry soil to the atmosphere. It will display moderate mobility in soil. If released to water, it will rapidly volatilize to the atmosphere. The estimated half-life for volatilization from a model river is 3.6 hours. It will not bioconcentrate in fish and aquatic organisms nor will it adsorb to sediment or suspended organic matter. If released to the atmosphere, it will undergo a slow gas-phase reaction with photochemically produced hydroxyl radicals with an estimated half-life of 479 days. The atmospheric life time has been estimated to range from 1.2 to 6.3 years. It may undergo atmospheric removal by wet deposition processes; however, any removed by this process is expected to rapidly re-volatilize to the atmosphere.

Environmental Physical Data

Henry's Law Constant: estimated at 0.0955
Octanol/Water Partition Coefficient: log K_{ow} = 2.307
Sorption Partition Coefficient: K_{oc} = calculated at 430
BCF: calculated at 33

Regulations

RCRA 40CFR: Not listed
CERCLA: 40CFR 302.4: Listed as Compound per CWA Section 307(a)
SARA 40CFR 372.65: Listed
SARA EHS 40CFR 355: Not listed
TSCA: Listed

DIC7950 CAS #: 90454-18-5

DICHLORO-1,1,2-TRIFLUOROETHANE

Molecular Formula: $C_2HCl_2F_3$
Formula Weight: 152.93
Use: as alternatives to CFCS in applications such as refrigerants, blowing agents, cleaning agents, and fire extinguishants

Physical Properties

Boiling Point: 28 °C (82 °F)
Freezing Point: -78 °C (-108.4 °F)
Density: 1.5 g/cu m at 25 °C
Vapor Pressure: 102 kPa at 303.15 °K
Refraction Index: 1.327 at 20 °C
Flash Point: Does not burn or burns with difficulty

Hazard Overviews

Fire: Noncombustible.
Carcinogenicity: IARC - Not listed; NIOSH - Not listed; NTP - Not listed; ACGIH - Not listed; OSHA - Not listed; EPA - Not listed; MAK - Not listed

Environmental

Regulations

RCRA 40CFR: Not listed
CERCLA: 40CFR 302.4: Listed as Compound per CWA Section 307(a)
SARA 40CFR 372.65: Listed
SARA EHS 40CFR 355: Not listed
TSCA: Not listed

DIC8000 CAS #: 34077-87-7

DICHLOROTRIFLUOROETHANE

Molecular Formula: $C_2HCl_2F_3$
Formula Weight: 152.93
Use: refrigerant; blowing agent; cleaning agents; fire extinguishants; substitute for R-11 in foam and refrigerant applications

Physical Properties

Boiling Point: 28.7 °C (84 °F)
Freezing Point: -107 °C (-160.6 °F)
Density: 1.475 g/cu cm at 15 °C
Vapor Pressure: 102 kPa at 303.15 °K
Refraction Index: 1.3332 at 15 °C
Critical Temperature: 185.0 °C
Critical Pressure: 3.79 mPa
Flash Point: Does not burn or burns with difficulty

Hazard Overviews

Fire: Noncombustible.
Carcinogenicity: IARC - Not listed; NIOSH - Not listed; NTP - Not listed; ACGIH - Not listed; OSHA - Not listed; EPA - Not listed; MAK - Not listed

Environmental

Regulations

RCRA 40CFR: Not listed
CERCLA: 40CFR 302.4: Listed as Compound per CWA Section 307(a)
SARA 40CFR 372.65: Listed
SARA EHS 40CFR 355: Not listed
TSCA: Not listed

DIC8050 CAS #: 3615-21-2

4,5-DICHLORO-2-(TRIFLUOROMETHYL)BENZIMID-AZOLE

RTECS: DD7350000
Molecular Formula: $C_8H_3Cl_2F_3N_2$
Formula Weight: 255.03
Synonyms: 1H-BENZIMIDAZOLE,4,5-DICHLORO-2-(TRIFLUOROMETHYL)-; BENZIMIDAZOLE,4,5-DICHLORO-2-(TRIFLUOROMETHYL)-; CHLORFLURAZOLE; CHLOROFLURAZOLE; 4,5-DICHLORO-2-(TRIFLUOROMETHYL)-1H-BENZIMIDAZOLE; 4,5-DICHLORO-2-TRIFLUOROMETHYLBENZIMIDAZOLE; NC 3363
Description: crystalline solid, forming fine white needles when pure
Use: herbicide

Physical Properties

Freezing Point: 213 °C (415.4 °F) to 214 °C (417.2 °F)
Saturated Vapor Density: 1.200000492 kg/m^3
Vapor Pressure: 4 $x10^{-5}$ mm Hg at 22.5 °C
Water Solubility: 69 ppm in Water at 25 °C
Other Solubilities: Sparingly soluble in methylnaphthalene and benzene; Soluble up to 25% in ethanol, ether, chloroform, propylene glycol, glycerol formal, and methylnaphthalene; Very soluble in acetone

RTECS Toxicity Data

Acute Oral: Rat LD_{50} Dose: 13080 ug/kg. Chicken LD_{50} Dose: 34 mg/kg.

Hazard Overviews

Carcinogenicity: IARC - Not listed; NIOSH - Not listed; NTP - Not listed; ACGIH - Not listed; OSHA - Not listed; EPA - Not listed; MAK - Not listed

Environmental

Environmental Physical Data

Octanol/Water Partition Coefficient: log K_{ow} = 3.49

Regulations

RCRA 40CFR: Not listed
CERCLA: 40CFR 302.4: Not listed
SARA 40CFR 372.65: Not listed TPQ: 500/10000 lb
SARA EHS 40CFR 355: Listed TPQ: 500 lb
TSCA: Not listed

DIC8100 CAS #: 133-53-9

2,4-DICHLORO-3,5-XYLENOL

EINECS Number: 205-109-9
Molecular Formula: $C_8H_8Cl_2O$
Formula Weight: 191.06

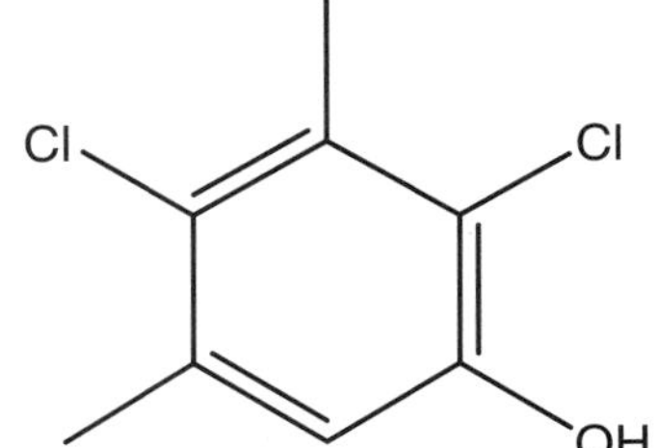

Chemical Structure

Synonyms: BENZENE,2,4-DICHLORO-1,3-DIMETHYL-5-HYDROXY-; DCMX; DECASEPT; 2,4-DICHLORO-3,5-DIMETHYLPHENOL; 2,4-DICHLORO-M,5-XYLENOL; DICHLOROXYLENOL; DIXOL; HEWSOL; PHENOL,2,4-DICHLORO-3,5-DIMETHYL-; PRINSYL; 3,5-XYLENOL,2,4-DICHLORO-
Description: crystals
Use: bacteriostat in soaps; mold inhibitor & preservative; bactericide in disinfectant products (discontinued use)

Physical Properties

Boiling Point: Sublimes
Freezing Point: 95 °C (203 °F) to 96 °C (204.8 °F)
Water Solubility: 1
Other Solubilities: at 15 °C in 100 parts of solvent: 14 parts Benzene; 15 parts Toluene; 73 parts Acetone; 59 parts Diethyl Ketone; 4 parts Petroleum Ether; 10 parts Carbon Tetrachloride; 25 parts Chloroform.

Hazard Overviews

Carcinogenicity: IARC - Not listed; NIOSH - Not listed; NTP - Not listed; ACGIH - Not listed; OSHA - Not listed; EPA - Not listed; MAK - Not listed

Environmental

Regulations

RCRA 40CFR: Not listed
CERCLA: 40CFR 302.4: Not listed
SARA 40CFR 372.65: Not listed
SARA EHS 40CFR 355: Not listed
TSCA: Listed

DIC8150 CAS #: 120-97-8

DICHLORPHENAMIDE

RTECS: CZ9200000
EINECS Number: 204-440-6
Molecular Formula: $C_6H_6Cl_2N_2O_4S_2$
Formula Weight: 305.16

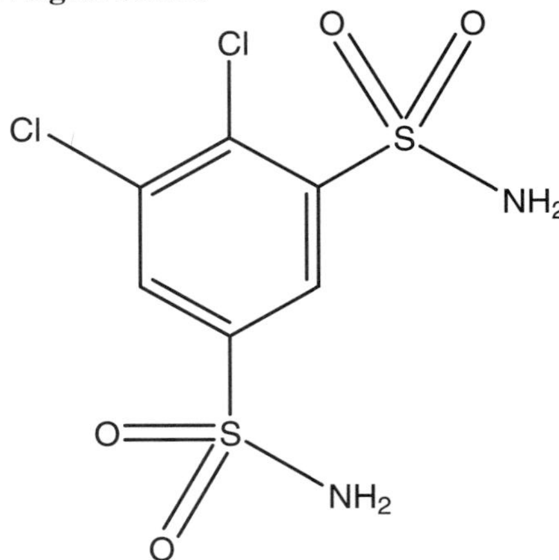

Chemical Structure

Synonyms: ANTIDRASI; 1,3-BENZENEDISULFONAMIDE,4,5-DICHLORO-; M-BENZENEDISULFONAMIDE,4,5-DICHLORO-; CB 8000; CB8000; DARAMIDE; DARANIDE; DASANIDE; DICHLOFENAMID; DICHLOFENAMIDE; 4,5-DICHLORO-M-BENZENEDISULFONAMIDE; 4,5-DICHLORO-1,3-DISULFAMOYLBENZENE; DICHLOROPHENAMIDE; 3,4-DICHLORO-5-SULFAMYLBENZENESULFONAMIDE; DICHLORPHENAMID; DICLOFENAMIDE; 1,3-DISULFAMOYL-4,5-DICHLOROBENZENE; 1,3-DISULFAMYL-4,5-DICHLOROBENZENE; GLAUCOL; ORATROL
Description: white or nearly white, crystalline powder or needles; slight characteristic odor
Use: carbonic anhydrase inhibitor, diuretic; treatment of glaucoma & intraocular tension

Physical Properties

Freezing Point: 239 °C (462.2 °F) to 241 °C (465.8 °F)
Water Solubility: Practically Insoluble in Water
Other Solubilities: freely Soluble in Pyridine & in 1 N NAOH; Soluble in Alcohol; Soluble in 2 N Sodium Carbonate; Slightly Soluble in Ether.

RTECS Toxicity Data

Acute Oral: Rat LD_{50} Dose: 10070 mg/kg. Mouse LD_{50} Dose: 1710 mg/kg.
Reproductive/Teratogenic: Rat Route: Oral; Dose: 4642 mg/kg; Duration: female 1-22D of pregnancy; Specific Developmental Abnormalities - Musculoskeletal system. Mouse Route: Subcutaneous; Dose: 96 mg/kg; Duration:

female 15D of pregnancy; Specific Developmental Abnormalities - Eye, ear.

Hazard Overviews

Health: May cause irritation to eyes/skin/respiratory tract. Harmful. Other Acute Effects: may be harmful by inhalation, ingestion, or skin absorption; target organ: kidneys. Chronic Effects: possible risk of congenital malformation in the fetus; possible risk of harm to the unborn child.

Fire: Extinguishing agents: carbon dioxide, dry chemical powder or appropriate foam; water spray. Precautions: combustible liquid.

Reactivity: Stable. Hazardous polymerization will not occur. Hazardous decomposition products: toxic fumes of: carbon monoxide, carbon dioxide, nitrogen oxides, sulfur oxides, hydrogen chloride gas.

Carcinogenicity: IARC - Not listed; NIOSH - Not listed; NTP - Not listed; ACGIH - Not listed; OSHA - Not listed; EPA - Not listed; MAK - Not listed

Primary Target Organs:

Kidneys

Environmental

Regulations

RCRA 40CFR: Not listed
CERCLA: 40CFR 302.4: Not listed
SARA 40CFR 372.65: Not listed
SARA EHS 40CFR 355: Not listed
TSCA: Not listed

DIC8200 CAS #: 62-73-7

DICHLORVOS

RTECS: TC0350000
DOT: UN2783; UN2784; UN3017; UN3018; NA2783; IMO6.1
EINECS Number: 200-547-7
Molecular Formula: $C_4H_7Cl_2O_4P$
Structured MF: $(CH_3O)_2POOCH{=}CCl_2$
Formula Weight: 220.98

Synonyms: APAVAP; ASTROBOT; ATGARD; ATGARD C; ATGARD V; BAY-19149; BAYER 19149; BENFOS; BIBESOL; BREVINYL; BREVINYL E50; CANOGARD; CEKUSAN; CHLORVINPHOS; CYANOPHOS; CYPONA; DDVF; DDVP; DDVP (INSECTICIDE); DEDEVAP; DENKAVEPON; DERIBAN; DERRIBANTE; DES; DEVIKOL; DICHLOFOS; (2,2-DICHLOOR-VINYL)-DIMETHYL-FOSFAAT; (2,2-DICHLOOR-VINYL)-DIMETHYL-PHOSPHAT; DICHLOORVO; DICHLORFOS; DICHLORMAN; 2,2-DICHLOROETHENOL DIMETHYL PHOSPHATE; 2,2-DICHLOROETHENYL DIMETHYL PHOSPHATE; 2,2-DICHLOROETHENYL PHOSPHORIC ACID DIMETHYL ESTER; DICHLOROPHOS; DICHLOROVAS; (2,2-DICHLORO-VINIL)DIMETIL-FOSFATO; 2,2-DICHLOROVINYL ALCOHOL DIMETHYL PHOSPHATE; 2,2-DICHLOROVINYL DIMETHYL PHOSPHATE; 2,2-DICHLOROVINYL DIMETHYL PHOSPHORIC ACID ESTER; DICHLOROVOS; DICHLOROVOS MIXTURE,DRY; DICHLORPHOS; (2,2-DICHLOR-VINYL)-DIMETHYL-PHOSPHAT; O-(2,2-DICHLORVINYL)-O,O-DIMETHYLPHOSPHAT; DIMETHYL 2,2-DICHLOROETHENYL PHOSPHATE; 0,0-DIMETHYL 0-2,2-DICHLOROVINYL PHOSPHATE; DIMETHYL 2,2-DICHLOROVINYL PHOSPHATE; DIMETHYL DICHLOROVINYL PHOSPHATE; O,O-DIMETHYL 2,2-DICHLOROVINYL PHOSPHATE; O,O-DIMETHYL DICHLOROVINYL PHOSPHATE; O,O-DIMETHYL O-2,2-DICHLOROVINYL PHOSPHATE; O,O-DIMETHYL-O-(2,2-DICHLOR-VINYL)-PHOSPHAT; DIVIPAN; DUO-KILL; DURAVOS; O,O-DWUMETYLO-O-DWUCHLOROWINYLOFOSFORAN; ENT 20738; EQUIGAND; EQUIGARD; EQUIGEL; ESTROSEL; ESTROSOL; ETHENOL,2,2-DICHLORO-,DIMETHYL PHOSPHATE; FECAMA; FEKAMA; FLY FIGHTER; FLY-DIE; HERKAL; HERKOL; INSECTIGAS D; KRECALVIN; LINDAN; MAFU; MAFU STRIP; MARVEX; MOPARI; NEFRAFOS; NERKOL; NOGOS; NOGOS 50; NOGOS 50 EC; NOGOS G; NO-PEST; NO-PEST STRIP; NOVOTOX; NSC-6738; NUVA; NUVAN; NUVAN 7; NUVAN 100EC; OKO; OMS 14; PANAPLATE; PHOSPHATE DE DIMETHYLE ET DE 2,2-DICHLOROVINYLE; PHOSPHORIC ACID,2,2-DICHLOROETHENYL DIMETHYL ESTER; PHOSPHORIC ACID,2,2-DICHLOROVINYL DIMETHYL ESTER; PHOSVIT; RAVAP; SD-1750; SD 1750; SZKLARNIAK; TAP 9VP; TASK; TASK TABS; TENAC; TETRAVOS; UDVF; UNIFOS; UNIFOS 50 EC; UNIFOS (PESTICIDE); UNITOX; VAPONA; VAPONA INSECTICIDE; VAPONITE; VAPORA II; VERDICAN; VERDIPOR; VINYL ALCOHOL; VINYL ALCOHOL,2,2-DICHLORO-,DIMETHYL PHOSPHATE; VINYLOFOS; VINYLOPHOS; WINYLOPHOS

Description: colorless to amber liquid; mild chemical odor

Use: a contact and stomach insecticide used in flea (pest) collars for pets, for control of insects in tobacco and other warehouses, mushroom houses, greenhouses, animal shelters, homes, restaurants and other food-handling establishments, for control of flies, mosquitoes and other disease vectors in aircraft, external parasites on livestock, insects in buildings, outdoor areas and on harvested tomatoes, aphids and some moths and for postharvest treatment of grain; as a penetrant and veterinary anthelmintic, as a household and public health fumigant, for protection of stored products and for crop protection

Physical Properties

Boiling Point: 140 °C (284 °F) at 20 mm Hg
Freezing Point: 84 °C (183.2 °F)
Specific Gravity: 1.415 at 25 °C/4 °C
Saturated Vapor Density: 1.200550854 kg/m^3
Density: 1.415 g/mL at 25 °C
Vapor Pressure: 0.0527 torr at 25 °C
Water Solubility: 0.5% by weight
Other Solubilities: Soluble in Ethanol, Chloroform, Acetone; miscible in aromatic chlorinated hydrocarbon solvents; solubility in Kerosene 2 to 3 g/kg; miscible with most aerosol propellants.
Refraction Index: 1.451 at 25 °C/D
Flash Point: > 79.444 °C Open Cup

RTECS Toxicity Data

Acute Oral: Rat LD_{50} Dose: 17 mg/kg. Mouse LD_{50} Dose: 61 mg/kg.

Acute Inhalation: Rat LC_{50} Dose: 15 mg/m^3/4hr; Toxic Effects: Sense organs and special senses - Lacrimation; Behavioral - Tremor; Gastrointestinal - Changes in structure or function of salivary glands. Mouse LC_{50} Dose: 13 mg/m^3/4hr; Toxic Effects: Sense organs and special senses - Lacrimation; Behavioral - Tremor; Gastrointestinal - Changes in structure or function of salivary glands.

Acute Dermal: Rabbit LD_{50} Route: Skin; Dose: 107 mg/kg. Rat LD_{50} Route: Skin; Dose: 750 ug/kg. Rat LD_{50} Route: Subcutaneous Dose: 10800 ug/kg.

Chronic (Multiple Dose) Oral: Rat Dose: 360 mg/kg/90D-I; Toxic Effects: Liver - Other changes; Blood - Other changes; Biochemical - Lipids including transport. Rat Dose: 56 mg/kg/14D-I; Toxic Effects: Liver - Other changes; Kidney, Ureter, and Bladder - Other changes; Biochemical - Other hydrolases. Rat Dose: 45 mg/kg/6W-I; Toxic Effects: Brain and coverings - Changes in surface EEG; Peripheral Nerve and sensation - Recording from peripheral motor nerve; Biochemical - True cholinesterase.

Chronic (Multiple Dose) Inhalation: Rat Dose: 8200 ug/m^3/45D-I; Toxic Effects: Nutritional and gross metabolic - Weight loss or decreased weight gain; Biochemical - True cholinesterase; DEATH.

Chronic (Multiple Dose) Dermal: Monkey Route: Skin; Dose: 750 mg/kg/10D-I; Toxic Effects: Blood - Other changes; Biochemical - True cholinesterase. Monkey Route: Skin; Dose: 400 mg/kg/4D-I; Toxic Effects: DEATH.

Reproductive/Teratogenic: Rat Route: Oral; Dose: 39200 ug/kg; Duration: female 14-21D of pregnancy Effects on Newborn - Biochemical and metabolic. Rat Route: Intraperitoneal; Dose: 15 mg/kg; Duration: female 11D of pregnancy; Specific Developmental Abnormalities - Body wall.

Mutagenic: Human DNA Damage; Cell Type: lymphocyte; Dose: 62 mg/L. Human DNA Damage; Cell Type: HeLa cell; Dose: 970 umol/L. Human Unscheduled DNA Synthesis; Cell Type: other cell types; Dose: 65 mmol/L. Human DNA Inhibition; Cell Type: lymphocyte; Dose: 62 mg/L. Human Cytogenetic Analysis; Cell Type: lymphocyte; Dose: 1 mg/L.

Tumorigenic: Rat Route: Oral; Dose: 4120 mg/kg/2Y-C; Toxic Effects: Tumorigenic - Neoplastic by RTECS criteria; Lungs, Thorax, or Respiration - Tumors; Gastrointestinal - Tumors. Rat Route: Oral; Dose: 2060 mg/kg/2Y-C; Toxic Effects: Tumorigenic - Carcinogenic by RTECS criteria; Gastrointestinal - Tumors; Blood - Leukemia.

Hazard Overviews

Fire: Combustible.

Carcinogenicity: IARC - Group 2B, Possibly carcinogenic to humans; NIOSH - Not listed; NTP - Not listed; ACGIH - Class A4, Not classifiable as a human carcinogen; OSHA - Not listed; EPA - Class B2, Probable human carcinogen based on animal studies; MAK - Not listed

Exposure Limits

OSHA PEL: TWA: 1 mg/m^3; skin.

ACGIH TLV: TWA: 0.1 ppm; 0.9 mg/m^3.

NIOSH REL: TWA: 1 mg/m^3.

NIOSH IDLH: 100 mg/m^3.

DFG MAK: TWA: 0.1 ppm; 1 mg/m^3.

Respirator Recommendation

Exposure Range: >1 to 50 mg/m^3 Supplied Air, Constant Flow/Pressure Demand, Half Mask

Exposure Range: >50 to <100 mg/m^3 Supplied Air, Constant Flow/Pressure Demand, Full Face

Exposure Range: 100 to unlimited mg/m^3 Self-contained Breathing Apparatus, Pressure Demand, Full Face

Note: odor threshold unknown

Environmental

Ecotoxicity: LC_{50} Cutthroat trout 170 ug/l/96 hr at 12 °C (95% confidence limit 143-203 ug/l), wt 2.5 g in a static bioassay LC_{50} Bluegill 1000 ppm/24 hr /Conditions of bioassay not specified LC_{50} Gammarus faciatus 0.40 ug/l/96 hr /Conditions of bioassay not specified LC_{50} Sphaeroides maculatus (Northern puffer) 2250 ppb/96 hr in a static lab bioassay LC_{50} Crangon septemspinosa (sand shrimp) 4 ppb/96 hr in a static bioassay LC_{50} Palaemonetes vulgaris (grass shrimp) 15 ppb/96 hr in a static bioassay LC_{50} Mugil cephalus (striped mullet) 200 ppb/96 hr in a static lab bioassay LC_{50} Pteronarcys 0.10 ug/l/96 hr at 15 °C (95% confidence limit 0.07-0.15 ug/l), 2ND yr class in a static bioassay

Environmental Fate: If released into water it will hydrolyze with a half-life of approximately 4 days although its half-life varies considerably between pH 4 and 9. It will degrade very slowly at pH 4 and quite rapidly at pH 9. Biodegradation may aid in its disappearance, particularly when acclimated colonies of microorganisms exist or under more acidic conditions when hydrolysis is slower. Bioconcentration in fish will not be significant. The Henry's Law constant indicates that volatilization from environmental waters and moist soil should generally be slow. The volatilization half-lives from a model river and a model pond, the latter considers the effects of adsorption, have been estimated to be 57 and over 400 days, respectively. If released on land, it will leach into the ground water where it will hydrolyze and also degrade through chemical and biological processes with reported half-lives ranging from 1.5-17 days. If released into the atmosphere, it is expected to exist almost entirely in the vapor phase in ambient air. In air, vapor phase will react with photochemically generated hydroxyl radicals and ozone with estimated half-lives of 2 and 320 days, respectively.

Cleanup/Disposal: Guide No. 131: Fully encapsulating, vapor protective clothing should be worn for spills and leaks with no fire. Eliminate all ignition sources (no smoking, flares, sparks or flames in immediate area). All equipment used when handling the product must be grounded. Do not touch or walk through spilled material. Stop leak if you can do it without risk. Prevent entry into waterways, sewers, basements or confined areas. A vapor suppressing foam may be used to reduce vapors. Small Spills: Absorb with earth, sand or other non-combustible material and transfer to containers for later disposal. Use clean non-sparking tools to collect absorbed material. Large Spills: Dike far ahead of liquid spill for later disposal. Water spray may reduce vapor; but may not prevent ignition in closed spaces. Guide No. 152: Do not touch damaged containers or spilled material unless wearing appropriate protective clothing. Stop leak if you can do it without risk. Prevent entry into waterways, sewers, basements or confined areas. Cover with plastic sheet to prevent spreading. Absorb or cover with dry earth, sand or other non-combustible material and transfer to containers. Do not get water inside containers.

Environmental Physical Data

Henry's Law Constant: calculated at 6.3 x10^{-5}
Octanol/Water Partition Coefficient: log K_{ow} = 14.5
Sorption Partition Coefficient: K_{oc} = estimated at 28
BCF: estimated at 0.28
BOD: persists 62 days 20 °C

Regulations

RCRA 40CFR: Not listed
CERCLA: 40CFR 302.4: Listed per CWA Section 311(b)(4) RQ: 10 lb (4.535 kg)
SARA 40CFR 372.65: Listed TPQ: 1000 lb
SARA EHS 40CFR 355: Listed TPQ: 10 lb
TSCA: Listed

Analytical Methods

Air: EPA TO-10; ASTM D4490
Soil: EPA PMD-TLC; SW846 8140, 8141, 8141A, 8270B, 8270C, 8321, 8321A
Water / Groundwater: EPA P-005-1, 1657, 622; SW846 8141, 8141A
Drinking Water: EPA 507, 525.2; AOAC 991.07; ASTM D5475
Food: FDA 212.1, 232.1, 232.3, 232.4, 242.1; AOAC 964.04, 966.07
Indoor / Expired Air: EPA IP-8; ASTM D4861
Plasma: EPA 001, 29; FDA 211.1, 231.1, 252

DIC8250 CAS #: 3116-76-5

DICLOXACILLIN

EINECS Number: 221-488-3
Molecular Formula: $C_{19}H_{17}Cl_2N_3O_5S$
Formula Weight: 470.33
Synonyms: BRL 1702; 6-(3-(2,6-DICHLOROPHENYL)-5-METHYL-4-ISOXAZOLECARBOXAMIDO)-3,3-DIMETHYL-7-OXO-4-THIA-1-AZABICYCLO(3.2.0)HEPTANE-2-CARBOXYL///; 6-(3-(2,6-DICHLOROPHENYL)-5-METHYL-4-ISOXAZOLECARBOXAMIDO)PENICILLANIC ACID; 3-(2,6-DICHLOROPHENYL)-5-METHYL-4-ISOXAZOLYLPENICILLIN; DICLOXACILIN; DICLOXACYCLINE; MACLICINE; METHYLDICHLOROPHENYLISOXAZOLYLPENICILLIN; R-13423; 4-THIA-1-AZABICYCLO(3.2.0)HEPTANE-2-CARBOXYLIC ACID,6-(((3-(2,6-DICHLOROPHENYL)-5-METHYL-4-ISOXAZOLYL)CARBONYL)AMINO)-3,3///
Description: crystals; characteristic odor
Use: antibacterial agent

Physical Properties

Boiling Point: Decomposes at 222 °C (432 °F) to 225 °C (437 °F)
Freezing Point: Decomposes at 222 °C (431.6 °F) to 225 °C (437 °F)
Water Solubility: Crystals; Soluble in Water
Other Solubilities: Soluble in Alcohol; Soluble in Methanol, less Soluble in Butanol; Slightly Soluble in Acetone, organic solvents.

Hazard Overviews

Carcinogenicity: IARC - Not listed; NIOSH - Not listed; NTP - Not listed; ACGIH - Not listed; OSHA - Not listed; EPA - Not listed; MAK - Not listed

Environmental

Regulations

RCRA 40CFR: Not listed
CERCLA: 40CFR 302.4: Not listed
SARA 40CFR 372.65: Not listed
SARA EHS 40CFR 355: Not listed
TSCA: Not listed

DIC8300 CAS #: 115-32-2

DICOFOL

RTECS: DC8400000
DOT: UN2761; NA2761; IMO6.1
EINECS Number: 204-082-0
Molecular Formula: $C_{14}H_9Cl_5O$
Structured MF: $(ClC_6H_4)_2C(OH)CCl_3$
Formula Weight: 370.47
Synonyms: ACARIN; BENZENEMETHANOL,4-CHLORO-ALPHA-(4-CHLOROPHENYL)-ALPHA-(TRICHLOROMETHYL)-; BENZHYDROL,4,4'-DICHLORO-ALPHA-(TRICHLOROMETHYL)-; 1,1-BIS(4-CHLOROPHENYL)-2,2,2-TRICHLOROETHANOL; 1,1-BIS(CHLOROPHENYL)-2,2,2-TRICHLOROETHANOL; 1,1-BIS(P-CHLOROPHENYL)-2,2,2-TRICHLOROETHANOL; CARBAX; CARBOX; CEKUDIFOL; 4-CHLORO-ALPHA-(4-CHLOROPHENYL)-ALPHA-(TRICHLOROMETHYL)BENZENEMETHANOL; 4-CHLORO-ALPHA-(4-CHLOROPHENYL)-ALPHA-(TRICHLOROMETHYL)BENZYL ALCOHOL; CPCA; DECOFOL; DICHLOROKELTHANE; DI-(P-CHLOROPHENYL)TRICHLOROMETHYLCARBINOL; 4,4'-DICHLORO-ALPHA-(TRICHLOROMETHYL)BENZHYDROL; DICOMITE; DTMC; ENT 23,648; EPA PESTICIDE CHEMICAL CODE 010501; ETHANOL,2,2,2-TRICHLORO-1,1-BIS(P-CHLOROPHENYL)-; FW 293; HIFOL; HILFOL; HILFOL 18.5 EC; KELTANE; KELTHANE; PARA,PARA'-KELTHANE; KELTHANE A; KELTHANE DUST BASE; KELTHANETHANOL; MIFOL; MILBOL; MITIGAN; 2,2,2-TRICHLOOR-1,1-BIS(4-CHLOOR FENYL)-ETHANOL; 1,1,1-TRICHLOR-2,2-BIS(4-CHLORPHENYL)-AETHANOL; 2,2,2-TRICHLOR-1,1-BIS(4-CHLOR-PHENYL)-AETHANOL; 2,2,2-TRICHLORO-1,1-BIS(4-CHLOROPHENYL)-ETHANOL; 2,2,2-TRICHLORO-1,1-BIS(4-CHLOROPHENYL)ETHANOL; 2,2,2-TRICHLORO-1,1-BIS(P-CHLOROPHENYL)ETHANOL; 2,2,2-TRICHLORO-1,1-BIS(4-CLORO-FENIL)-ETANOLO; 2,2,2-TRICHLORO-1,1-DI-(4-CHLOROPHENYL)ETHANOL
Description: dark to yellow brown, non-flowable liquid (or waxy solid); slight characteristic odor
Use: a synthetic nonsystemic organochlorine acaricide which has been used primarily for the control of mites on field crops, vegetables, citrus and non-citrus fruits and in greenhouses

Physical Properties

Boiling Point: 180 °C (356 °F) at 0.1 mm Hg
Freezing Point: 77 °C (170.6 °F) to 78 °C (172.4 °F)
Specific Gravity: 1.13 at 20 °C
Density: 1.45 g/mL at 25 °C (technical grade)
Vapor Pressure: Negligible at room temperature
Water Solubility: < 0.1 mg/mL at 22 C
Other Solubilities: 95% Ethanol: >=100 mg/ml at 21 °C; Acetone: >=100 mg/ml at 20 °C; DMSO: >=100 mg/ml at 21 °C; Most aliphatic solvents: Soluble; Most aromatic solvents: Soluble.
Flash Point: 48.889 °C Closed Cup
Autoignition Temperature: 530 °C
LEL: 1.1% v/v
UEL: xylene 7.0% v/v

RTECS Toxicity Data

Acute Oral: Rat LD_{50} Dose: 575 mg/kg. Mouse LD_{50} Dose: 420 mg/kg.
Acute Inhalation: Rat LC_{50} Dose: >5 gm/m^3/4hr.
Acute Dermal: Rabbit LD_{50} Route: Skin; Dose: 1870 mg/kg. Rat LD_{50} Route: Skin; Dose: 100 mg/kg.
Chronic (Multiple Dose) Oral: Rat Dose: 4200 mg/kg/12W-C; Toxic Effects: Liver - Other changes; Liver - Changes in liver weight; Endocrine - Changes in thyroid weight. Rat Dose: 2670 mg/kg/25D-I; Toxic Effects: Brain and coverings - Other degenerative changes; Liver - Other changes; Biochemical - True cholinesterase.
Reproductive/Teratogenic: Rat Route: Oral; Dose: 430 mg/kg; Duration: female 6-15D of pregnancy; Maternal Effects - Other effects on females; Effects on Fertility - Post-implantation mortality; Effects on Embryo or Fetus - Fetotoxicity.
Mutagenic: Human Cytogenetic Analysis; Cell Type: leukocyte; Dose: 20 mg/L. Human Sister Chromatid Exchenge; Cell Type: lymphocyte; Dose: 1 umol/L.
Tumorigenic: Mouse Route: Oral; Dose: 17 gm/kg/78W-C; Toxic Effects: Tumorigenic - Carcinogenic by RTECS criteria; Liver - Tumors. Mouse Route: Oral; Dose: 35 gm/kg/78W-C; Toxic Effects: Tumorigenic - Carcinogenic by RTECS criteria; Liver - Tumors.

Hazard Overviews

Flammable

Fire: Flammable.
Carcinogenicity: IARC - Group 3, Not classifiable as to carcinogenicity to humans; NIOSH - Not listed; NTP - Not listed; ACGIH - Not listed; OSHA - Not listed; EPA - Not listed; MAK - Not listed

Environmental

Ecotoxicity: LD_{50} Pheasant oral 265 mg/kg (95% confidence limit: 211-234 mg/kg), males, age 4 mo LC_{50} Pimephales promelas (fathead minnow) 603 ug/l/96 hr, 32 days old, flow-through bioassay, dissolved oxygen (7.1 mg/l), water hardness (44.2 mg/l $CaCO_3$), alkalinity (42.7 mg/l $CaCO_3$), tank vol (18 l), additions per day (6.8 vol), pH 7.3 LC_{50} Crangon franciscorum (grass shrimp) 0.1 ppm/100 hr /Conditions of bioassay not specified LC_{50} Salmo gairdneri (Rainbow trout) 210 ug/l/96 hr, flow through test in water from lake TLm Salmo gairdneri (rainbow trout) 110 ppm/24 hr, fresh water. /Conditions of bioassay not specified LC_{50} Anas platyhynchos (mallard) oral 1651 mg/kg in 5 day diet (95% confidence limit: 1356-2029 mg/kg), age 10 days old
Environmental Fate: If released to soil, it will be expected to bind to the soil strongly but under some circumstances it may reach groundwater, since it has been detected in groundwater. It is susceptible to hydrolysis in moist soils and evaporation from the surface of moist soils. It may be resistant to biodegradation. If it is released to water it will be expected to bind to the sediments and may bioconcentrate in aquatic organisms. It will be subject to hydrolysis and may directly photodegrade. It may be resistant to biodegradation but may be susceptible to evaporation. If it is released to the atmosphere it may be subject to direct photolysis. The estimated vapor phase half-life in the atmosphere is 2.92 days as a result of reaction with photochemically produced hydroxyl radicals.
Cleanup/Disposal: Guide No. 151: Do not touch damaged containers or spilled material unless wearing appropriate protective clothing. Stop leak if you can do it without risk. Prevent entry into waterways, sewers, basements or confined areas. Cover with plastic sheet to prevent spreading. Absorb or cover with dry earth, sand or other non-combustible material and transfer to containers. Do not get water inside containers.

Environmental Physical Data

Octanol/Water Partition Coefficient: log K_{OW} = 3.54
Sorption Partition Coefficient: K_{OC} = 3950
BCF: fathead minnow 1.5 x10^4

Regulations

RCRA 40CFR: Not listed
CERCLA: 40CFR 302.4: Listed per CWA Section 311(b)(4) RQ: 10 lb (4.535 kg)
SARA 40CFR 372.65: Listed
SARA EHS 40CFR 355: Not listed
TSCA: Not listed

Analytical Methods

Air: EPA TO-10
Soil: EPA PMD-TLC; SW846 8081
Water / Groundwater: EPA 617
Food: FDA 212.1, 212.2, 232.1, 232.4, 242.1; EPA XENO; AOAC 976.02, 986.06
Indoor / Expired Air: EPA IP-8; ASTM D4861
Plasma: EPA 001, 027, 028; FDA 211.1, 231.1, 252
Other: EPA 1656

DIC8350 CAS #: 141-66-2

DICROTOPHOS

RTECS: TC3850000
DOT: UN2783; UN2784; UN3017; UN3018; IMO3.2; IMO6.1
EINECS Number: 205-494-3
Molecular Formula: $C_8H_{16}NO_5P$
Structured MF: $C_8H_{16}NO_5P$
Formula Weight: 237.21
Synonyms: BIDIRL; BIDRIN; C 709; CARBICRON; CARBOMICRON; CIBA 709; CROTONAMIDE,3-HYDROXY-N-N-DIMETHYL-,DIMETHYL PHOSPHATE,(E)-; CROTONAMIDE,3-HYDROXY-N-N-DIMETHYL-,DIMETHYL PHOSPHATE,CIS-; CROTONAMIDE,3-HYDROXY-N,N-DIMETHYL-,CIS-,DIMETHYLPHOSPHATE; DIAPADRIN; DICROTOFOS; 3-(DIMETHOXYPHOSPHINYLOXY)-N,N DIMETHYLISOCROTONAMIDE; 3-(DIMETHOXYPHOSPHINYLOXY)-N,N-DIMETHYL-CIS-CROTONAMIDE; 3-DIMETHOXYPHOSPHINYLOXY-N,N-DIMETHYLISOCROTONAMIDE; DIMETHYL (E)-2-DIMETHYL-CARBAMOYL-1-METHYLVINYL PHOSPHATE; DIMETHYL 2-DIMETHYLCARBAMOYL-1-METHYL-VINYL PHOSPHATE; 2-DIMETHYL CIS-2-DIMETHYL-CARBAMOYL-1-METHYLVINYLPHOSPHATE; 2-DIMETHYL CIS-2-DIMETHYLCARBAMOYL-1-METHYLVINYLPHOSPHATE; O,O-DIMETHYL O-(N,N-DIMETHYLCARBAMOYL-1-METHYLVINYL)PHOSPHATE; DIMETHYL 1-DIMETHYLCARBAMOYL-1-PROPEN-2-YL PHOSPHATE; DIMETHYL ESTER WITH (E)-3-HYDROXY-N,N-DIMETHYLCROTONAMIDEPHOSPHORIC ACID; DIMETHYL PHOSPHATE ESTER WITH 3-HYDROXY-N,N-DIMETHYL-CIS-CROTONAMIDE; DIMETHYL PHOSPHATE ESTER WITH3-HYDROXY-N,N-DIMETHYLCROTONAMIDE; DIMETHYL PHOSPHATE OF 3-HYDROXY-N,N-DIMETHYL-CIS-CROTONAMIDE; DIMETHYL PHOSPHATE OF3-HYDROXY-N,N-DIMETHYL-CIS-CROTONAMIDE; 3-(DIMETHYLAMINO)-1-METHYL-3-OXO-1-PROPENYL DIMETHYLPHOSPHATE; 3-(DIMETHYLAMINO)-1-METHYL-3-OXO-1-PROPENYL DIMETHYLPHOSPHATE (E)-ISOMER; CIS-2-DIMETHYLCARBAMOYL-1-METHYLVINYL DIMETHYLPHOSPHATE; O,O-DIMETHYL-O-(2-DIMETHYL-CARBAMOYL-1-METHYL-VINYL)PHOSPHAT; O,O-DIMETHYL-O-(1,4-DIMETHYL-3-OXO-4-AZA-PENT-1-ENYL)FOSFAAT; O,O-DIMETHYL-O-(1,4-DIMETHYL-3-OXO-4-AZA-PENT-1-ENYL)PHOSPHATE; O,O-DIMETHYL-O-(1-METHYL-2-N,N-DIMETHYL-CARBAMOYL)-VINYL-PHOSPHAT; O,O-DIMETIL-O-(1,4-DIMETIL-3-OXO-4-AZA-PENT-1-ENIL)-FOSFATO; EKTAFOS; ENT 24,482; 3-HYDROXYDIMETHYL CROTONAMIDE DIMETHYL PHOSPHATE; 3-HYDROXY-N,N-DIMETHYL-CIS-CROTONAMIDE DIMETHYL PHOSPHATE; KARBICRON; OLEOBIDRIN; OMS 253; PHOSPHATE DE DIMETHYLE ET DE 2-DIMETHYLCARBAMOYL 1-METHYLVINYLE; PHOSPHORIC ACID 3-(DIMETHYLAMINO)-1-METHYL-3-OXO-1-PROPENYL DIMETHYL ESTER; PHOSPHORIC ACID,DIMETHYL ESTER,ESTER WITH CIS-3-HYDROXY-N,N-DIMETHYLCROTONAMIDE; PHOSPHORIC ACID,DIMETHYL ESTER,ESTER WITH(E)-3-HYDROXY-N,N-DIMETHYLCROTONAMIDE; PHOSPHORIC ACID,DIMETHYL 1-METHYL-N,N-(DIMETHYLAMINO)-3-OXO-1-PROPENYL ESTER,(E)-(9CI); (E)-PHOSPHORIC ACID,3-(DIMETHYLAMINO)-1-METHYL-3-OXO-1-PROPENYL DIMETHYL ESTER; PHOSPHORIC ACID,3-(DIMETHYLAMINO)-1-METHYL-3-OXO-1-PROPENYL DIMETHYLESTER,(E)-; SD 3562; SHELL SD-3562
Description: amber liquid; mild ester odor
Use: systemic insecticide and acaricide for cotton, apples & other crops

Physical Properties

Boiling Point: 400 °C (752 °F) at 760 mm Hg
Specific Gravity: 1.216 at 15 °C/15 °C
Vapor Pressure: 0.0001 mm Hg
Water Solubility: Miscible
Other Solubilities: miscible with Methylene Chloride, Chloroform, Acetonitrile, Xylene, Alcohol; barely Soluble in mineral oils; less than 10 g/kg Diesel Oil, Kerosene; miscible with Isopropyl Oxitol, Phentoxone, Ethyl, Isopropyl & Diacetone Alcohols.
Refraction Index: 1.468 at 23 °C/D
Flash Point: > 93.333 °C Closed Cup

RTECS Toxicity Data

Acute Oral: Rat LD_{50} Dose: 13 mg/kg. Mouse LD_{50} Dose: 11 mg/kg.
Acute Inhalation: Rat LC_{50} Dose: 90 mg/m^3/4hr.
Acute Dermal: Rabbit LD_{50} Route: Skin; Dose: 168 mg/kg. Rat LD_{50} Route: Skin; Dose: 42 mg/kg. Rat LD_{50} Route: Subcutaneous Dose: 8137 ug/kg.
Chronic (Multiple Dose) Oral: Bird Dose: 3500 ug/kg/5D-C; Toxic Effects: Behavioral - Food intake (animal); Nutritional and gross metabolic - Weight loss or decreased weight gain; DEATH.
Mutagenic: Hamster Sister Chromatid Exchenge; Cell Type: ovary; Dose: 300 umol/L. Bacteria - E Coli Mutations in Microorganisms; Dose: 2 mmol/L/4H (-S9). Bacteria - S Typhimurium Mutations in Microorganisms; Dose: 500 ug/plate (-S9).

Hazard Overviews

Poison

Fire Diamond

Health: Severely irritating to eyes/respiratory tract. Poison. Also Causes: nausea, vomiting, abdominal cramps, headache, giddiness, sweating, dizziness, blurred vision, mental confusion, incoordination, slurred speech, muscle twitching, tremors, difficulty breathing, convulsions, coma. Chronic Effects: weakness, memory loss, fatigue, sleep disturbances, appetite loss, disorientation, trembling of hands, sometimes neuritis or partial paralysis.
Fire: Will burn. Use dry chemical, water spray, fog, or regular foam.
Reactivity: Stable, but begins to decompose after 31 days at 75 °C or after 7 days at 90 °C. and is less stable under alkaline conditions. Hazardous polymerization cannot occur. Avoid: exposure to heat; ignition sources; metals; alkalis. Incompatible with: cast iron; mild steel; some stainless steel; brass. Hazardous decomposition products: carbon; nitrogen; phosphorus oxide(s).
Carcinogenicity: IARC - Not listed; NIOSH - Not listed; NTP - Not listed; ACGIH - Class A4, Not classifiable as a human carcinogen; OSHA - Not listed; EPA - Not listed; MAK - Not listed

Primary Target Organs:

Exposure Limits

OSHA PEL Vacated 1989 Limits: TWA: 0.25 mg/m^3.

ACGIH TLV: TWA: 0.25 mg/m^3.

Respirator Recommendation

Exposure Range: >0.25 to 12.5 mg/m^3 Supplied Air, Constant Flow/Pressure Demand, Half Mask

Exposure Range: >12.5 to 250 mg/m^3 Supplied Air, Constant Flow/Pressure Demand, Full Face

Exposure Range: >250 to unlimited mg/m^3 Self-contained Breathing Apparatus, Pressure Demand, Full Face

Note: odor threshold unknown

Environmental

Ecotoxicity: LC_{50} Salmo gairdneri 6.3 mg/l/96 hr at 13 °C, wt 1.0 g. Static bioassay without aeration, pH 7.2-7.5, water hardness 40-50 mg/l as calcium carbonate and alkalinity of 30-35 mg/l LD_{50} Rana catesbeiana oral 2000 mg/kg (95% confidence limit 602-6640 mg/kg), males LD_{50} Columba livia oral 2.00 mg/kg (95% confidence limit 1.53-2.61 mg/kg) LC_{50} Odocoileus hemionus hemionus oral 12.5-25.0 mg/kg, 8-17 mo old males LC_{50} Callipepla californica oral 1.89 mg/kg (95% confidence limit 1.50-2.38 mg/kg), 18 mo old males LC_{50} Pteronarcys californica 0.43 mg/l/96 hr at 15 °C (95% confidence limit 0.34-0.54 mg/l), 2nd yr class. Static bioassay without aeration, pH 7.2-7.5, water hardness 40-50 mg/l as calcium carbonate and alkalinity of 30-35 mg/l

Environmental Fate: If released to the atmosphere, it will degrade rapidly in the vapor-phase by reaction with photochemically produced hydroxyl radicals (half-life of about 4 hr). Particulate-phase and aerosols released to air during applications of insecticides will be removed from air physically by dry and wet deposition. If released to soil or water, it will degrade primarily through biodegradation. Screening studies have demonstrated that it degrades much faster in non-sterile soil as compared to sterilized soil. It has been classified as a non-persistent pesticide with an estimated soil half-life of less than 0.5 months.

Cleanup/Disposal: Guide No. 131: Fully encapsulating, vapor protective clothing should be worn for spills and leaks with no fire. Eliminate all ignition sources (no smoking, flares, sparks or flames in immediate area). All equipment used when handling the product must be grounded. Do not touch or walk through spilled material. Stop leak if you can do it without risk. Prevent entry into waterways, sewers, basements or confined areas. A vapor suppressing foam may be used to reduce vapors. Small Spills: Absorb with earth, sand or other non-combustible material and transfer to containers for later disposal. Use clean non-sparking tools to collect absorbed material. Large Spills: Dike far ahead of liquid spill for later disposal. Water spray may reduce vapor; but may not prevent ignition in closed spaces. Guide No. 152: Do not touch damaged containers or spilled material unless wearing appropriate protective clothing. Stop leak if you can do it without risk. Prevent entry into waterways, sewers, basements or confined areas. Cover with plastic sheet to prevent spreading. Absorb or cover with dry earth, sand or other non-combustible material and transfer to containers. Do not get water inside containers.

Environmental Physical Data

Henry's Law Constant: estimated at 1.2 x10^{-12}

Octanol/Water Partition Coefficient: log K_{ow} = -0.49

Sorption Partition Coefficient: K_{oc} = estimated at 53 to 85

BCF: estimated at 0.1

Regulations

RCRA 40CFR: Not listed

CERCLA: 40CFR 302.4: Not listed

SARA 40CFR 372.65: Not listed TPQ: 100 lb

SARA EHS 40CFR 355: Listed TPQ: 100 lb

TSCA: Not listed

Analytical Methods

Soil: SW846 8141, 8141A, 8270B, 8270C

Water / Groundwater: EPA 1657, 022; SW846 8141, 8141A

Food: FDA 212.1, 232.1, 232.3, 232.4, 242.1

Indoor / Expired Air: NIOSH 5600; ASTM D4861

Plasma: EPA 001, 29; FDA 252

DIC8400 CAS #: 66-76-2

DICUMAROL

RTECS: GN7875000

DOT: UN3024; UN3025; UN3026; UN3027; IMO3.0; IMO6.1

EINECS Number: 200-632-9

Molecular Formula: $C_{19}H_{12}O_6$

Formula Weight: 336.29

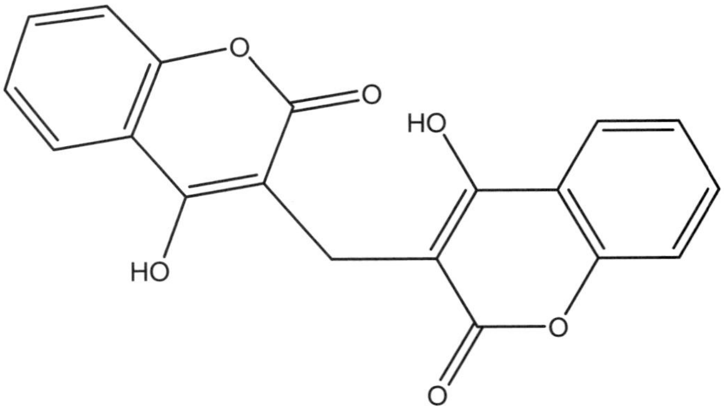

Chemical Structure

Synonyms: ACADYL; ACAVYL; ANTITROMBOSIN; BARACOUMIN; 2H-1-BENZOPYRAN-2-ONE,3,3'-METHYLENEBIS(4-HYDROXY-; 2H-1-BENZOPYRAN-2-ONE,3,3'-METHYLENEBIS(4-HYDROXY-(9CI); BHC; BISHYDROXYCOUMARIN; BIS(4-HYDROXYCOUMARIN-3-YL)METHANE; BIS-3,3'-(4-HYDROXYCOUMARINYL)METHANE; COUMARIN,3,3'-METHYLENEBIS(4-HYDROXY-; CUMA; CUMID; DICOUMAL; DICOUMARIN; DICOUMAROL; DICUMAN; DICUMAOL R; DICUMARINE; DICUMOL; DI-(4-HYDROXY-3-

COUMARINYL)METHANE; 4,4'-DIHYDROXY-3,3'-METHYLENE BIS COUMARIN; DI-4-HYDROXY-3,3'-METHYLENEDICOUMARIN; DIKUMAROL; DUFALONE; DWUKUMAROL; KUMORAN; MELITOXIN; 3,3'-METHYLEEN-BIS(4-HYDROXY-CUMARINE); 3,3'-METHYLEN-BIS(4-HYDROXY-CUMARIN); 3,3'-METHYLENEBIS(4-HYDROXY-2H-1-BENZOPYRAN-2-ONE); 3,3'-METHYLENEBIS(4-HYDROXY-1,2-BENZOPYRONE); 3,3'-METHYLENEBIS(4-HYDROXYCOUMARIN); 3,3'-METHYLENE-BIS(4-HYDROXYCOUMARINE); 3,3'-METILEN-BIS(4-IDROSSI-CUMARINA); TEMPARIN; TROMBOSAN

Description: white or creamy white, crystalline powder, minute crystals, or needles; faint, pleasant odor

Use: medication: anticoagulant; rodenticide

Physical Properties

Freezing Point: 287 °C (548.6 °F) to 293 °C (559.4 °F)

Water Solubility: Practically Insoluble in Water

Other Solubilities: Insoluble in Acetone; Soluble in aqueous alkaline solutions, in Pyridine and similar organic bases; Slightly Soluble in Benzene and Chloroform.

RTECS Toxicity Data

Acute Oral: Rat LD_{50} Dose: 250 mg/kg. Mouse LD_{50} Dose: 233 mg/kg; Toxic Effects: Sense organs and special senses - Ulcerated nasal septum; Gastrointestinal - Ulceration or bleeding from large intestine.

Acute Dermal: Mouse LD_{50} Route: Subcutaneous Dose: 50 mg/kg.

Chronic (Multiple Dose) Oral: Rat Dose: 112 mg/kg/30D-C; Toxic Effects: Lungs, Thorax, or Respiration - Acute pulmonary edema; Blood - Hemorrhage; DEATH. Mouse Dose: 180 mg/kg/30D-C; Toxic Effects: Lungs, Thorax, or Respiration - Acute pulmonary edema; Blood - Hemorrhage; DEATH.

Reproductive/Teratogenic: Woman Route: Oral; Dose: 110 mg/kg; Duration: female 31-40W of pregnancy Effects on Embryo or Fetus - Fetal death; Effects on Newborn - Stillbirth; Other neonatal measures or effects.

Hazard Overviews

Health: May cause irritation. Toxic. Other Acute Effects: harmful if swallowed, inhaled, or absorbed through skin. Chronic Effects: overexposure may cause reproductive disorders based on tests with laboratory animals; possible risk of harm to the unborn child.

Fire: Extinguishing agents: water spray; carbon dioxide, dry chemical powder or appropriate foam. Precautions: combustible liquid.

Reactivity: Incompatible with: strong oxidizing agents. Hazardous decomposition products: toxic fumes of: carbon monoxide, carbon dioxide.

Carcinogenicity: IARC - Not listed; NIOSH - Not listed; NTP - Not listed; ACGIH - Not listed; OSHA - Not listed; EPA - Not listed; MAK - Not listed

Environmental

Ecotoxicity: LC_{50} Pimephales promelas (fathead minnow) 5.11 mg/l/96 hr (95% confidence limit 4.18 - 6.24 mg/l), flow-through bioassay with measured concentrations, 24.5 °C, dissolved oxygen 7.2 mg/l, hardness 44.3 mg/l calcium carbonate, alkalinity 44.0 mg/l calcium carbonate, and pH 7.8

Environmental Fate: If released on soil, would leach. Its degradability in soil and water is unknown. If released in water, it is not expected to volatilize, adsorb to sediment, or bioconcentrate in aquatic organisms. If released in the atmosphere, it would be associated with aerosols and be removed by gravitational settling. Any in the vapor phase would be rapidly oxidized by photochemically-produced hydroxyl radicals and ozone.

Cleanup/Disposal: Guide No. 131: Fully encapsulating, vapor protective clothing should be worn for spills and leaks with no fire. Eliminate all ignition sources (no smoking, flares, sparks or flames in immediate area). All equipment used when handling the product must be grounded. Do not touch or walk through spilled material. Stop leak if you can do it without risk. Prevent entry into waterways, sewers, basements or confined areas. A vapor suppressing foam may be used to reduce vapors. Small Spills: Absorb with earth, sand or other non-combustible material and transfer to containers for later disposal. Use clean non-sparking tools to collect absorbed material. Large Spills: Dike far ahead of liquid spill for later disposal. Water spray may reduce vapor; but may not prevent ignition in closed spaces. Guide No. 151: Do not touch damaged containers or spilled material unless wearing appropriate protective clothing. Stop leak if you can do it without risk. Prevent entry into waterways, sewers, basements or confined areas. Cover with plastic sheet to prevent spreading. Absorb or cover with dry earth, sand or other non-combustible material and transfer to containers. Do not get water inside containers.

Environmental Physical Data

Henry's Law Constant: estimated at 1.36×10^{-13}

Octanol/Water Partition Coefficient: log K_{ow} = 2.07

Sorption Partition Coefficient: K_{oc} = estimated at 24

BCF: calculated at 22

Regulations

RCRA 40CFR: Not listed

CERCLA: 40CFR 302.4: Not listed

SARA 40CFR 372.65: Not listed

SARA EHS 40CFR 355: Not listed

TSCA: Listed

DIC8450 **CAS #: 80-43-3**

DICUMYL PEROXIDE

RTECS: SD8150000

DOT: UN2121; NA2121; IMO5.2

EINECS Number: 201-279-3

Molecular Formula: $C_{18}H_{22}O_2$

Structured MF: $[C_6H_5C(CH_3)_2O]_2$

Formula Weight: 270.40

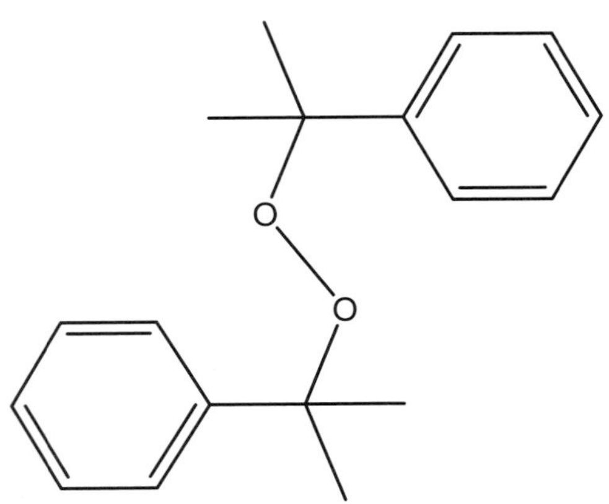

Chemical Structure

Synonyms: ACTIVE DICUMYL PEROXIDE; BIS(ALPHA,ALPHA-DIMETHYLBENZYL) PEROXIDE; BIS(ALPHA,ALPHA-DIMETHYLBENZYL)PEROXIDE; BIS(1-METHYL-1-PHENYLETHYL) PEROXIDE; BIS(2-PHENYL-2-PROPYL) PEROXIDE; CUMENE PEROXIDE; ALPHA-CUMYL PEROXIDE; CUMYL PEROXIDE; DICUMENE HYDROPEROXIDE; DICUMENYL PEROXIDE; ALPHA,ALPHA'-DICUMYL PEROXIDE; DI-ALPHA-CUMYL PEROXIDE; DI-CUP; DI-CUP 40C; DICUP 40; DI-CUP 40HAF; DI-CUP 40 KE; DI-CUP 40KE; DI-CUP R; DI-CUP T; DI-CUPR; DIISOPROPYLBENZENE PEROXIDE; ALPHA,ALPHA-DIMETHYLBENZYL PEROXIDE; ISOPROPYL BENZENE PEROXIDE; ISOPROPYLBENZENE PEROXIDE; KAYACUMYL D; LUPERCO; LUPERCO 500-40C; LUPERCO 500-40KE; LUPEROX; LUPEROX 500R; LUPEROX 500T; LUPEROX 500; LUPERSOL 500; PERCUMYL D; PERCUMYL D 40; PERKADOX 96; PERKADOX B; PERKADOX BC; PERKADOX BC 40; PERKADOX BC 9; PERKADOX BC 95; PERKADOX SB; PEROXIDE,BIS(ALPHA,ALPHA-DIMETHYLBENZYL); PEROXIDE,BIS(1-METHYL-1-PHENYLETHYL); SAMPEROX DCP; VAROX DCP-R; VAROX DCP-T

Description: pale yellow to white granular solid; characteristic odor

Use: polymerization catalyst and vulcanizing agent

Physical Properties

Boiling Point: Decomposes at 130 °C (266 °F)
Freezing Point: 38 °C (100.4 °F)
Specific Gravity: 1.02
Vapor Density: 9.3 Air=1
Water Solubility: < 1 mg/mL at 23 C
Other Solubilities: 95% Ethanol: >=100 mg/ml at 23 °C; Acetone: >=100 mg/ml at 23 °C; DMSO: >=100 mg/ml at 23 °C.
Flash Point: 71.111 °C Closed Cup

RTECS Toxicity Data

Acute Oral: Rat LD_{50} Dose: 4100 mg/kg.

Hazard Overviews

Health: Irritating to eyes/skin/respiratory tract. Other Acute Effects: may be harmful by inhalation, ingestion, or skin absorption; prolonged or repeated inhalation may cause congestion and rhinitis.

Fire: Combustible. Hazards: emits toxic fumes; container explosion may occur. Extinguishing agents: water spray; carbon dioxide, dry chemical powder or appropriate foam. Precautions: combustible liquid.

Carcinogenicity: IARC - Not listed; NIOSH - Not listed; NTP - Not listed; ACGIH - Not listed; OSHA - Not listed; EPA - Not listed; MAK - Not listed

Primary Target Organs:

Eyes Skin Respiratory System

Environmental

Environmental Fate: While its fate in the environment is largely unknown, it should photolyze, adsorb strongly to soil, and probably react with free radicals and hydrolyze. While it is unlikely that it will be very persistent in soil or water, no estimate of its half-life is possible. In the atmosphere it is estimated to have a half-life of the order of a day.

Cleanup/Disposal: Guide No. 145: Eliminate all ignition sources (no smoking, flares, sparks or flames in immediate area). Keep combustibles (wood, paper, oil, etc.) away from spilled material. Do not touch damaged containers or spilled material unless wearing appropriate protective clothing. Keep substance wet using water spray. Stop leak if you can do it without risk. Small Spills: Take up with inert, damp, noncombustible material using clean non-sparking tools and place into loosely covered plastic containers for later disposal. Large Spills: Wet down with water and dike for later disposal. Prevent entry into waterways, sewers, basements or confined areas.

Environmental Physical Data

Octanol/Water Partition Coefficient: log K_{ow} = 3.78
Sorption Partition Coefficient: K_{oc} = 3700
BCF: moderate in fish

Regulations

RCRA 40CFR: Not listed
CERCLA: 40CFR 302.4: Not listed
SARA 40CFR 372.65: Not listed
SARA EHS 40CFR 355: Not listed
TSCA: Listed

DIC8500 CAS #: 504-66-5

DICYANAMIDE

EINECS Number: 207-998-9
Molecular Formula: C_2HN_3
Formula Weight: 67.05
Synonyms: CYANAMIDE,CYANO-; DICYANIMIDE; DITSIANAMID; IMIDODICARBONITRILE
Use: no evidence of commercial use in US

Hazard Overviews

Carcinogenicity: IARC - Not listed; NIOSH - Not listed; NTP - Not listed; ACGIH - Not listed; OSHA - Not listed; EPA - Not listed; MAK - Not listed

Environmental

Regulations

RCRA 40CFR: Not listed
CERCLA: 40CFR 302.4: Listed as Compound per CWA Section 307(a) per CAA Section 112
SARA 40CFR 372.65: Not listed
SARA EHS 40CFR 355: Not listed
TSCA: Listed

DIC8550 CAS #: 626-17-5

1,3-DICYANOBENZENE

RTECS: CZ1900000
EINECS Number: 210-933-7
Molecular Formula: $C_8H_4N_2$
Structured MF: $C_6H_4(CN)_2$
Formula Weight: 128.14

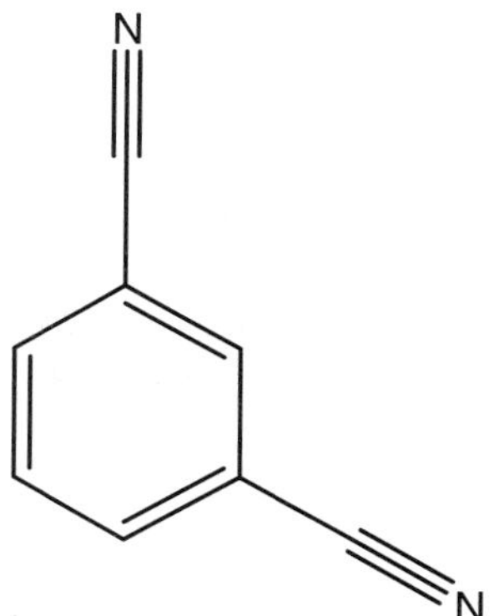

Chemical Structure

Synonyms: 1,3-BENZENDIKARBONITRIL; 1,3-BENZENEDICARBONITRILE; M-BENZENEDINITRILE; 1,3-BENZODINITRILE; 3-CYANOBENZONITRILE; M-DICYANOBENZENE; DINITRILE OF ISOPHTHALIC ACID; IPN; ISOFTALODINITRIL; ISOFTALONITRIL; ISOPHTHALODINITRILE; ISOPHTHALONITRILE; NITRIL KYSELINY ISOFTALOVE; M-PDN; M-PHTHALODINITRILE
Description: colorless to white, crystalline, flaky solid or needles; almond-like odor
Use: chem int for amines-eg, m-xylenediamine

Physical Properties

Boiling Point: Sublimes
Freezing Point: 162 °C (323.6 °F)
Specific Gravity: 4.42
Vapor Pressure: 0.01 mm Hg
Water Solubility: Slightly Soluble in Hot Water
Other Solubilities: Very Soluble in hot Alcohol, Ether, Benzene, Chloroform, Insoluble in Petroleum Ether
Ionization Potential (eV): =< 10.2 +/-0.5
Flash Point: Combustible Solid

RTECS Toxicity Data

Acute Oral: Rat LD_{50} Dose: 860 mg/kg; Toxic Effects: Behavioral - Convulsions or effect on seizure threshold; Lungs, Thorax, or Respiration - Dyspnea; Nutritional and gross metabolic - Body temperature decrease. Mouse LD_{50} Dose: 178 mg/kg.
Acute Inhalation: Rat LC_{50} Dose: >8970 mg/m^3/1hr.
Acute Dermal: Rabbit LD_{50} Route: Skin; Dose: >2 gm/kg. Rat LD_{50} Route: Skin; Dose: >5 gm/kg.
Chronic (Multiple Dose) Oral: Rat Dose: 5250 mg/kg/1W-I; Toxic Effects: Behavioral - Food intake (animal); Nutritional and gross metabolic - Weight loss or decreased weight gain; DEATH. Rat Dose: 4050 mg/kg/30D-I; Toxic Effects: Behavioral - Food intake (animal); Liver - Changes in liver weight; Nutritional and gross metabolic - Weight loss or decreased weight gain. Rat Dose: 4550 mg/kg/26W-I; Toxic Effects: Liver - Liver function tests impaired; Liver - Other changes; Blood - Pigmented or nucleated red Blood cells.
Chronic (Multiple Dose) Inhalation: Rat Dose: 1250 mg/m^3/6H/2W-I; Toxic Effects: Gastrointestinal - Hypermotility, diarrhea; Skin and appendages - Hair; Nutritional and gross metabolic - Weight loss or decreased weight gain.
Irritation Eye: Rabbit Standard Draize Test Dose: 500 mg/24H; Reaction: mild.

Hazard Overviews

Health: Irritating to eyes/skin/respiratory tract. Toxic. Other Acute Effects: harmful if swallowed, inhaled, or absorbed through skin.
Fire: Combustible. Hazards: emits toxic fumes. Extinguishing agents: water spray; carbon dioxide, dry chemical powder or appropriate foam. Precautions: combustible liquid.
Reactivity: Incompatible with: strong acids, strong bases, strong oxidizing agents, strong reducing agents. Hazardous decomposition products: thermal decomposition may produce carbon monoxide, carbon dioxide, and nitrogen oxides.
Carcinogenicity: IARC - Not listed; NIOSH - Listed as carcinogen; NTP - Not listed; ACGIH - Not listed; OSHA - Not listed; EPA - Not listed; MAK - Not listed
Primary Target Organs:

Eyes

Skin

Respiratory System

Exposure Limits
OSHA PEL Vacated 1989 Limits: TWA: 5 mg/m^3. Other Values: total mg/m^3; 10.
ACGIH TLV: TWA: 5 mg/m^3.
NIOSH REL: TWA: 5 mg/m^3.
DFG MAK: TWA: 5 mg/m^3.
Respirator Recommendation
Exposure Range: >5 to 50 mg/m^3 Air Purifying, Negative Pressure, Half Mask
Exposure Range: >50 to 500 mg/m^3 Air Purifying, Negative Pressure, Full Face

Exposure Range: >500 to 5000 mg/m^3 Supplied Air, Constant Flow/Pressure Demand, Full Face
Exposure Range: >5000 to unlimited mg/m^3 Self-contained Breathing Apparatus, Pressure Demand, Full Face
Cartridge Color: dust/mist filter (use P100 or consult supervisor for appropriate dust/mist filter)

Environmental

Regulations

RCRA 40CFR: Not listed
CERCLA: 40CFR 302.4: Listed as Compound per CWA Section 307(a) per CAA Section 112
SARA 40CFR 372.65: Not listed
SARA EHS 40CFR 355: Not listed
TSCA: Listed

DIC8600 CAS #: 2550-40-5

DICYCLOHEXYL DISULFIDE

DOT: NA9188
EINECS Number: 219-851-6
Molecular Formula: $C_{12}H_{22}S_2$
Formula Weight: 230.43

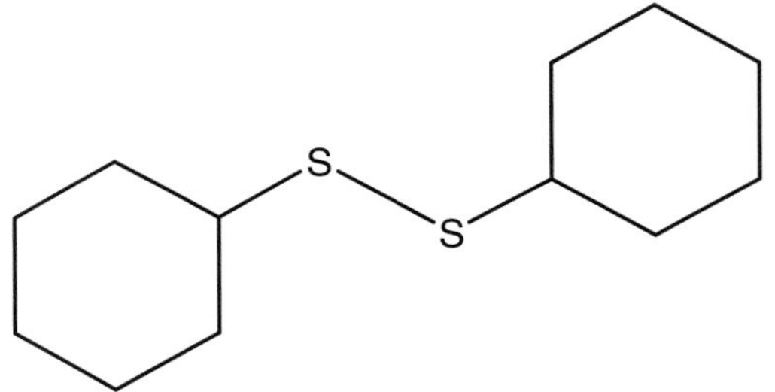

Chemical Structure

Synonyms: BIS(CYCLOHEXYL)DISULFIDE; CYCLOHEXYL DISULFIDE; DISULFIDE,DICYCLOHEXYL
Use: synthetic flavor for food, pharmaceuticals, & chewing gum, particularly for artificial grape flavor; enhances & otherwise agreeably modifies natural and/or artificial aromas & flavors of perfume and tobacco.

Hazard Overviews

Health: Irritating to eyes/skin/respiratory tract. Other Acute Effects: may be harmful by inhalation, ingestion, or skin absorption; exposure can cause nausea; dizziness; headache.
Fire: Hazards: emits toxic fumes. Extinguishing agents: water spray; carbon dioxide, dry chemical powder or appropriate foam. Precautions: combustible liquid.
Reactivity: Incompatible with: strong oxidizing agents, strong bases. Hazardous decomposition products: toxic fumes of: carbon monoxide, carbon dioxide, sulfur oxides.
Carcinogenicity: IARC - Not listed; NIOSH - Not listed; NTP - Not listed; ACGIH - Not listed; OSHA - Not listed; EPA - Not listed; MAK - Not listed

Primary Target Organs:

Eyes

Skin

Respiratory System

Environmental

Cleanup/Disposal: Guide No. 171: Do not touch or walk through spilled material. Stop leak if you can do it without risk. Prevent dust cloud. Avoid inhalation of asbestos dust. Small Dry Spills: With clean shovel place material into clean, dry container and cover loosely; move containers from spill area. Small Spills: Take up with sand or other noncombustible absorbent material and place into containers for later disposal. Large Spills: Dike far ahead of liquid spill for later disposal. Cover powder spill with plastic sheet or tarp to minimize spreading. Prevent entry into waterways, sewers, basements or confined areas.

Regulations

RCRA 40CFR: Not listed
CERCLA: 40CFR 302.4: Not listed
SARA 40CFR 372.65: Not listed
SARA EHS 40CFR 355: Not listed
TSCA: Listed

DIC8650 CAS #: 84-61-7

DICYCLOHEXYL PHTHALATE

RTECS: TI0889000
EINECS Number: 201-545-9
Molecular Formula: $C_{20}H_{26}O_4$
Formula Weight: 330.42

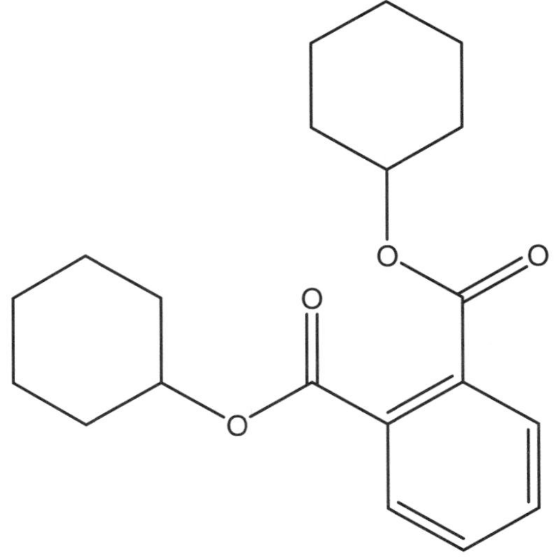

Chemical Structure

Synonyms: 1,2-BENZENEDICARBOXYLIC ACID,DICYCLOHEXYL ESTER; DCHP; DICLOHEXYL 1,2-BENZENEDICARBOXYLATE;

ERGOPLAST FDC; ERGOPLAST.FDC; HF 191; HOWFLEX CP; KP 201; PHTHALIC ACID,DICYCLOHEXYL ESTER; UNIMOLL 66

Description: white, granular solid; mildly aromatic odor

Use: plasticizer for nitrocellulose, ethyl cellulose, chlorinated rubber, polyvinyl acetate; polyvinyl chloride and other polymers

Physical Properties

Boiling Point: 222 °C (432 °F) to 228 °C (442 °F) at 4 mm Hg
Freezing Point: 66 °C (150.8 °F)
Specific Gravity: 1.383 at 20 °C/4 °C
Density: 1.2 g/mL
Vapor Pressure: Extremely low
Water Solubility: Insoluble
Other Solubilities: Ether: Soluble; Most organic solvents: Soluble.
Refraction Index: 1.451 at 20 °C/D
Flash Point: 107 °C

RTECS Toxicity Data

Acute Oral: Rat LD_{50} Dose: 30 mL/kg.
Chronic (Multiple Dose) Oral: Rat Dose: 10500 mg/kg/7D-C; Toxic Effects: Liver - Other changes; Liver - Changes in liver weight.

Hazard Overviews

Health: Irritating to eyes/skin/respiratory tract. Other Acute Effects: may be harmful by inhalation, ingestion, or skin absorption.
Fire: Will burn. Hazards: emits toxic fumes. Extinguishing agents: water spray; carbon dioxide, dry chemical powder or appropriate foam. Precautions: combustible liquid.
Reactivity: Incompatible with: strong oxidizing agents. Hazardous decomposition products: toxic fumes of: carbon monoxide, carbon dioxide.
Carcinogenicity: IARC - Not listed; NIOSH - Not listed; NTP - Not listed; ACGIH - Not listed; OSHA - Not listed; EPA - Not listed; MAK - Not listed
Primary Target Organs:

Eyes

Skin

Respiratory System

Environmental

Environmental Fate: If released to soil, it will display slight mobility in soil. Limited screening data indicates that it may undergo microbial degradation under aerobic conditions in soil. Volatilization from the soil surface to the atmosphere is not expected to be an important process. If released to water, limited data indicates that it may undergo microbial degradation. It is expected to bioaccumulate in fish and aquatic organisms, and adsorb to sediment and suspended organic matter. It is not expected to volatilize from water to the atmosphere, nor is it expected to hydrolyze. If released to the atmosphere, this compound is expected to exist predominately in the particulate form. Destruction by the vapor phase reaction with photochemically produced hydroxyl radicals may occur.

Environmental Physical Data

Henry's Law Constant: 6.58 $x10^{-8}$
Octanol/Water Partition Coefficient: log K_{ow} = > 2.12
Sorption Partition Coefficient: K_{oc} = calculated at 4520
BCF: none likely

Regulations

RCRA 40CFR: Not listed
CERCLA: 40CFR 302.4: Not listed
SARA 40CFR 372.65: Not listed
SARA EHS 40CFR 355: Not listed
TSCA: Listed

Analytical Methods

Plasma: EPA 001

DIC8700 CAS #: 101-83-7

DICYCLOHEXYLAMINE

RTECS: HY4025000
EINECS Number: 202-980-7
Molecular Formula: $C_{12}H_{23}N$
Formula Weight: 181.31

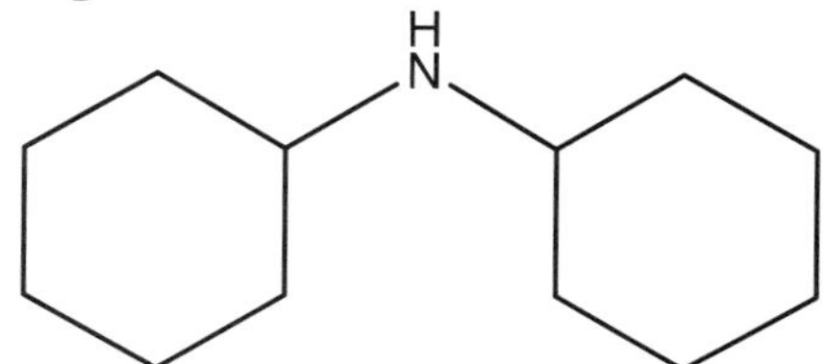

Chemical Structure

Synonyms: N-CYCLOHEXANAMINE; CYCLOHEXANAMINE,N-CYCLOHEXYL-; N-CYCLOHEXYLCYCLOHEXANAMIDE; N-CYCLOHEXYLCYCLOHEXANAMINE; N-CYCLOHEXYL-CYCLOHEXYLAMINE; DCHA; DI-CHA; DICHA; N,N-DICLOHEXYLAMINE; N,N-DICYCLOHEXYLAMINE; DICYKLOHEXYLAMIN; DODECAHYDRODIPHENYLAMINE; PERHYDRODIPHENYLAMINE

Description: colorless liquid; faint fishy odor

Use: organic intermediate, insecticide, plasticizer, corrosion inhibitor, antioxidant, in rubber, lubricating oils, fuels, catalysts for paint, varnishes and inks, detergents and extractants

Physical Properties

Boiling Point: 255.8 °C (492 °F) at 760 mm Hg
Freezing Point: About 20 °C (68 °F)
Specific Gravity: 0.9104 at 25 °C/25 °C
Vapor Density: 6.25 Air=1
Density: 0.91 to 0.92 g/mL
Water Solubility: Sparingly Soluble in Water
Other Solubilities: 95% Ethanol: Soluble; Acetone: Soluble; Benzene: Soluble; Ether: Soluble.
pH: Strong base

Refraction Index: 1.4823 at 25 °C/D
Flash Point: 110 °C

RTECS Toxicity Data

Acute Oral: Rat LD_{50} Dose: 373 mg/kg. Mouse LD_{50} Dose: 500 mg/kg.
Acute Dermal: Rabbit LD_{Lo} Route: Subcutaneous Dose: 500 mg/kg; Toxic Effects: Behavioral - Convulsions or effect on seizure threshold. Mouse LD_{50} Route: Subcutaneous Dose: 135 mg/kg.
Irritation Eye: Rabbit Standard Draize Test Dose: 750 ug/24H; Reaction: severe.
Irritation Skin: Rabbit Standard Draize Test Dose: 2 mg/24H; Reaction: severe.
Mutagenic: Human Cytogenetic Analysis; Cell Type: leukocyte; Dose: 200 ug/L.
Tumorigenic: Rat Route: Oral; Dose: 40 gm/kg/52W-I; Toxic Effects: Tumorigenic - Equivocal tumorigenic agent by RTECS criteria; Gastrointestinal - Tumors; Liver - Tumors. Mouse Route: Subcutaneous; Dose: 2404 mg/kg/48W-I; Toxic Effects: Tumorigenic - Equivocal tumorigenic agent by RTECS criteria; Tumorigenic - Tumors at site of application.

Hazard Overviews

Corrosive

Health: Corrosive to eyes/skin/respiratory tract. Harmful. Other Acute Effects: harmful if swallowed, inhaled, or absorbed through skin; inhalation may result in spasm, inflammation and edema of the larynx and bronchi, chemical pneumonitis and pulmonary edema; symptoms of exposure may include burning sensation, coughing, wheezing, laryngitis, shortness of breath, headache, nausea and vomiting. Chronic Effects: Possible human carcinogen.
Fire: Will burn. Hazards: emits toxic fumes. Extinguishing agents: water spray; carbon dioxide, dry chemical powder or appropriate foam. Precautions: combustible liquid.
Reactivity: Incompatible with: acids, acid chlorides, acid anhydrides, oxidizing agents, chloroformates. Hazardous decomposition products: thermal decomposition may produce carbon monoxide, carbon dioxide, and nitrogen oxides.
Carcinogenicity: IARC - Group 3, Not classifiable as to carcinogenicity to humans; NIOSH - Not listed; NTP - Not listed; ACGIH - Not listed; OSHA - Not listed; EPA - Not listed; MAK - Not listed
Primary Target Organs:

Eyes

Skin

Respiratory System

Environmental

Cleanup/Disposal: Guide No. 153: Eliminate all ignition sources (no smoking, flares, sparks or flames in immediate area). Do not touch damaged containers or spilled material unless wearing appropriate protective clothing. Stop leak if you can do it without risk. Prevent entry into waterways, sewers, basements or confined areas. Absorb or cover with dry earth, sand or other non-combustible material and transfer to containers. Do not get water inside containers.

Regulations

RCRA 40CFR: Not listed
CERCLA: 40CFR 302.4: Not listed
SARA 40CFR 372.65: Not listed
SARA EHS 40CFR 355: Not listed
TSCA: Listed

DIC8750 CAS #: 1212-29-9

N,N'-DICYCLOHEXYLTHIOUREA

RTECS: YS9300000
EINECS Number: 214-920-7
Molecular Formula: $C_{13}H_{24}N_2S$
Formula Weight: 240.45

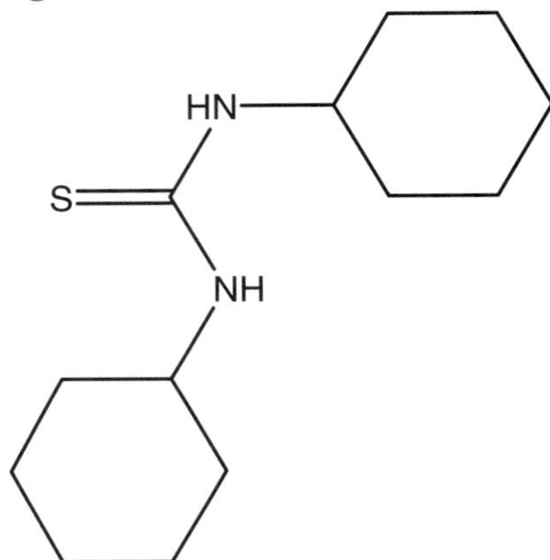

Chemical Structure

Synonyms: 1,3-BIS(CYCLOHEXYL)THIOUREA; DICYCLOHEXYL THIOUREA; N,N'-DICYCLOHEXYLTHIOCARBAMIDE; 1,3-DICYCLOHEXYL-2-THIO-UREA; 1,3-DICYCLOHEXYL-2-THIOUREA; 1,3-DICYCLOHEXYLTHIOUREA; DICYCLOHEXYLTHIOUREA; SYM-DICYCLOHEXYTHIOUREA; THIOUREA,N,N'-DICYCLOHEXYL-; UREA,1,3-DICYCLOHEXYL-2-THIO-
Description: white crystals
Use: chem int for dicyclohexylcarbodiimide (no evidence of current use)

Physical Properties

Water Solubility: < 0.1 mg/mL at 21 C
Other Solubilities: 95% Ethanol: 5-10 mg/ml at 22 °C; Acetone: 5-10 mg/ml at 22 °C; DMSO: 10-50 mg/ml at 22 °C.
Flash Point: Not available; probably combustible

RTECS Toxicity Data

Mutagenic: Rat Morphological Transformation; Cell Type: embryo; Dose: 250 ng/plate. Mouse Mutations in Mammalian Somatic Cells; Cell Type: lymphocyte; Dose: 75 mg/L.

Hazard Overviews

Fire: Will burn.
Carcinogenicity: IARC - Not listed; NIOSH - Not listed; NTP - Not listed; ACGIH - Not listed; OSHA - Not listed; EPA - Not listed; MAK - Not listed

Environmental

Regulations

RCRA 40CFR: Not listed
CERCLA: 40CFR 302.4: Not listed
SARA 40CFR 372.65: Not listed
SARA EHS 40CFR 355: Not listed
TSCA: Listed

DIC8800 CAS #: 77-19-0

DICYCLOMINE

EINECS Number: 201-009-4
Molecular Formula: $C_{19}H_{35}NO_2$
Formula Weight: 308.47
Synonyms: BENTYLOL; (1,1'-BICYCLOHEXYL)-1-CARBOXYLIC ACID,2-(DIETHYLAMINO)ETHYL ESTER; (BICYCLOHEXYL)-1-CARBOXYLIC ACID,2-(DIETHYLAMINO)ETHYLESTER; DICYCLOVERIN; DICYCLOVERINE; DIOCYL; WYOVIN
Description: white, fine crystalline powder or crystals; practically odorless
Use: anticholinergic hydrochloride

Physical Properties

Freezing Point: 164 °C (327.2 °F) to 166 °C (330.8 °F)
Water Solubility: Soluble in Water
Other Solubilities: 1 g in 5 ml Alcohol, 2.5 ml Chloroform, 770 ml Ether; Insoluble in alkaline aqueous medium.

Hazard Overviews

Carcinogenicity: IARC - Not listed; NIOSH - Not listed; NTP - Not listed; ACGIH - Not listed; OSHA - Not listed; EPA - Not listed; MAK - Not listed

Environmental

Regulations

RCRA 40CFR: Not listed
CERCLA: 40CFR 302.4: Not listed
SARA 40CFR 372.65: Not listed
SARA EHS 40CFR 355: Not listed
TSCA: Listed

DIC8850 CAS #: 77-73-6

DICYCLOPENTADIENE

RTECS: PC1050000
DOT: UN2048; IMO3.3
EINECS Number: 201-052-9
Molecular Formula: $C_{10}H_{12}$
Structured MF: $C_{10}H_{12}$
Formula Weight: 132.21

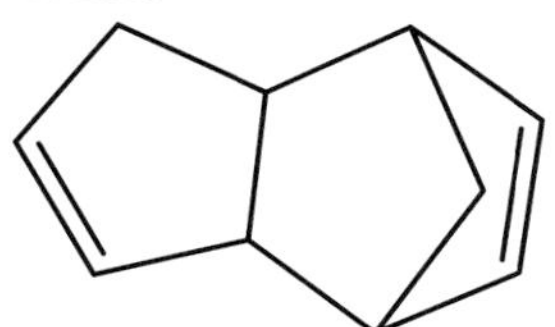

Chemical Structure

Synonyms: BICYCLOPENTADIENE; BISCYCLOPENTADIENE; 1,3-CPD; 1,3-CYCLOPENTADIENE DIMER; CYCLOPENTADIENE DIMER; 1,3-CYCLOPENTADIENE,DIMER; DCPD; 1,3-DICYCLOPENTADIENE DIMER; ALPHA-DICYCLOPENTADIENE (ENDO FORM); DICYKLOPENTADIEN; DIMER CYKLOPENTADIENU; 4,7-METHANO-1H-INDENE,3A,4,7,7A-TETRAHYDRO-; 4,7-METHANOINDENE,3A,4,7,7A-TETRAHYDRO-; 3A,4,7,7A-TETRAHYDRO-4,7-METHANOINDENE; TRICYCLO(5,2,1,0)-3,8-DECADIENE
Description: colorless crystals when nearly pure; camphor-like odor
Use: chemical intermediate for insecticides, EPDM elastomers, metallocenes, paints and varnishes, flame retardants for plastics

Physical Properties

Boiling Point: 64 °C (147 °F) to 65 °C (149 °F) at 14 mm Hg
Freezing Point: 32 °C (89.6 °F)
Specific Gravity: 0.9302 at 35 °C/4 °C
Vapor Density: 4.55 Air=1
Saturated Vapor Density: 1.207867187 kg/m^3
Density: 0.93 g/mL at 35 °C
Bulk Density: 8.2 lbs/gal at 60 °F
Vapor Pressure: 1.4 torr at 20 °C
Water Solubility: 0.02% by weight
Other Solubilities: 95% Ethanol: Soluble; Acetic Acid: Soluble; Carbon Tetrachloride: Soluble; Ether: Very Soluble; Petroleum Ether: Soluble.
Odor Threshold: 0.016 mg/m^3
Refraction Index: 1.505 at 35 °C
Ionization Potential (eV): 8.79 +/-2.0
Flash Point: 32 °C Open Cup
Autoignition Temperature: 503 °C
LEL: 0.8% v/v
UEL: 6.3% v/v

RTECS Toxicity Data

Acute Oral: Rat LD_{50} Dose: 353 mg/kg; Toxic Effects: Behavioral - Convulsions or effect on seizure threshold; Behavioral - Muscle weakness. Mouse LD_{50} Dose: 190 mg/kg.
Acute Inhalation: Mouse LC_{50} Dose: 145 ppm/4hr. Rat LC_{Lo} Dose: 1000 ppm/4hr; Toxic Effects: Behavioral - Tremor; Behavioral - Ataxia; Lungs, Thorax, or Respiration - Dyspnea.
Acute Dermal: Rabbit LD_{50} Route: Skin; Dose: 5080 mg/kg; Toxic Effects: Behavioral - Convulsions or effect on seizure threshold; Behavioral - Muscle weakness.

Chronic (Multiple Dose) Oral: Rat Dose: 182 mg/kg/26W-I; Toxic Effects: Behavioral - Alteration of classical conditioning; Biochemical - Multiple enzyme effects. Rat Dose: 1120 mg/kg/28D-I; Toxic Effects: Kidney, Ureter, and Bladder - Changes in kidney weight; Blood - Changes in serum composition; Biochemical - Transaminases.

Chronic (Multiple Dose) Inhalation: Rat Dose: 250 ppm/6H/2W-I; Toxic Effects: Behavioral - Somnolence (general depressed activity); Nutritional and gross metabolic - Weight loss or decreased weight gain; DEATH. Rat Dose: 35 ppm/7H/18W-I; Toxic Effects: Liver - Changes in Liver weight; Kidney, Ureter, and Bladder - Changes in kidney weight. Rat Dose: 5 ppm/6H/13W-I; Toxic Effects: Kidney, Ureter, and Bladder - Other changes in urine composition; Nutritional and gross metabolic - Changes in sodium; Nutritional and gross metabolic - Changes in phosphorus. Rat Dose: 20 mg/m^3/4H/26W-I; Toxic Effects: Behavioral - Excitment; Vascular - BP elevation not characterized in autonomic section; Kidney, Ureter, and Bladder - Proteinuria.

Irritation Eye: Rabbit Standard Draize Test Dose: 500 mg/24H; Reaction: mild.

Irritation Skin: Rabbit Standard Draize Test Dose: 20 mg/24H; Reaction: moderate. Rabbit Open Draize Test Dose: 10 mg/24H open; Reaction: severe. Rabbit Open Draize Test Dose: 9300 ug/24H open; Reaction: severe.

Hazard Overviews

Flammable

Fire Diamond

Health: Irritating to eyes/skin/respiratory tract. Toxic. Also Causes: nausea, headache, dizziness, olfactory fatigue, CNS depression, dermatitis. Chronic Effects: headaches, kidney damage (based on animal studies).

Fire: Do not extinguish fire unless flow can be stopped. For small fires, use dry chemical, carbon dioxide, water spray, or alcohol-resistant foam. For large fires, use water spray, fog or alcohol-resistant foam. Solid water streams may spread fire, it is better to use flooding quantities as fog. Can form explosive mixtures in air.

Reactivity: Stable. Hazardous polymerization cannot occur, but depolymerization to cyclopentadiene will occur at elevated temperatures. Avoid: exposure to heat and ignition sources. Incompatible with: oxidizers; mineral and organic acids; nonmetal halides; air. Hazardous decomposition products: carbon dioxide; carbon monoxide; cyclopentadiene; acrid smoke and irritating fumes.

Carcinogenicity: IARC - Not listed; NIOSH - Listed as carcinogen; NTP - Not listed; ACGIH - Not listed; OSHA - Not listed; EPA - Not listed; MAK - Not listed

Primary Target Organs:

Eyes

Skin

Respiratory System

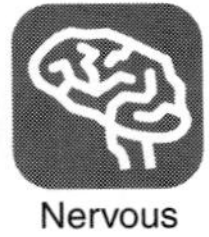
Nervous System

Exposure Limits

OSHA PEL Vacated 1989 Limits: TWA: 5 ppm; 30 mg/m^3.

ACGIH TLV: TWA: 5 ppm; 27 mg/m^3.

DFG MAK: TWA: 0.5 ppm; 3 mg/m^3.

Respirator Recommendation

Exposure Range: >5 to 50 ppm Air Purifying, Negative Pressure, Half Mask

Exposure Range: >50 to 500 ppm Air Purifying, Negative Pressure, Full Face

Exposure Range: >500 to 5000 ppm Supplied Air, Constant Flow/Pressure Demand, Full Face

Exposure Range: >5000 to unlimited ppm Self-contained Breathing Apparatus, Pressure Demand, Full Face

Cartridge Color: black with dust/mist prefilter (use P100 or consult supervisor for appropriate dust/mist prefilter)

Environmental

Ecotoxicity: ECSO Tetrahymena pyriformis (ciliate protozoan) 5.3 mg/l/24 hr (inhibition of cell multiplication) LC_{50} Daphnia species (water flea) 6.9 mg/l/48 hr. /Conditions of bioassay not specified LC_{50} Salmo gairdneri (rainbow trout) 22.86 mg/l/96 hr /Conditions of bioassay not specified

Environmental Fate: If released to the atmosphere, it is expected to undergo rapid destruction by the gas phase reaction with ozone, or with photochemically produced hydroxyl radicals. The half-life for these process can be estimated at 48 min, and 3.1 hr, respectively. Neither direct photochemical degradation, nor wet or dry deposition are expected to be significant. If released to water, the dominant fate process is expected to be volatilization to the atmosphere. The half-life for the volatilization from a model river can be estimated to be 3.4 hr. This process may be attenuated by adsorption of dicyclopentadiene to sediment and suspended organic matter. Bioconcentration in fish and aquatic organisms may occur. Destruction by direct photolysis in water is not expected to occur; however, destruction by photochemically produced oxidants may be significant. Biodegradation in water is not expected to be a significant fate process. If released to soil, volatilization from the soil surface to the atmosphere may be a rapid process. Adsorption to soil, which is expected to occur, may slow this process. Biological degradation in soil is not expected to be a significant fate process based upon limited data.

Cleanup/Disposal: Guide No. 129: Eliminate all ignition sources (no smoking, flares, sparks or flames in immediate area). All equipment used when handling the product must be grounded. Do not touch or walk through spilled material. Stop leak if you can do it without risk. Prevent entry into waterways, sewers, basements or confined areas. A vapor suppressing foam may be used to reduce vapors. Absorb or cover with dry earth, sand or other non-combustible material and transfer to containers. Use clean non-sparking tools to collect absorbed material. Large Spills: Dike far ahead of liquid spill for later disposal. Water spray may reduce vapor; but may not prevent ignition in closed spaces.

Environmental Physical Data

Henry's Law Constant: estimated at 0.0107

Octanol/Water Partition Coefficient: log K_{ow} = 2.89

Sorption Partition Coefficient: K_{oc} = 894
BCF: estimated at 93.2

Regulations

RCRA 40CFR: Not listed
CERCLA: 40CFR 302.4: Not listed
SARA 40CFR 372.65: Listed
SARA EHS 40CFR 355: Not listed
TSCA: Listed

DIC8900 CAS #: 81-21-0

DICYCLOPENTADIENE DIOXIDE

RTECS: PB9625200
EINECS Number: 201-334-1
Molecular Formula: $C_{10}H_{12}O_2$
Formula Weight: 164

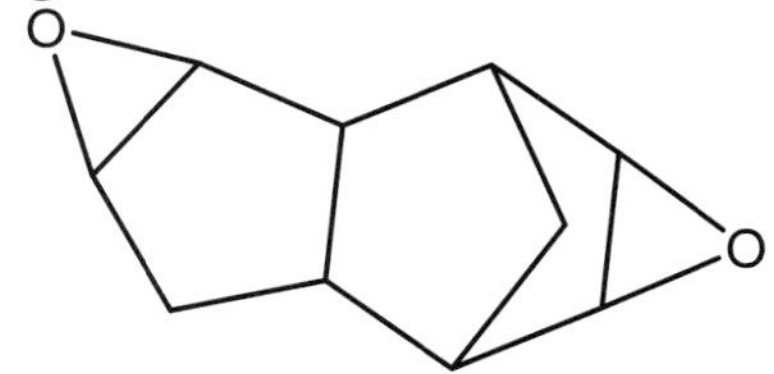

Chemical Structure

Synonyms: BICYCLOPENTADIENE DIOXIDE; DICYCLOPENTADIENE DIEPOXIDE; 1,2:5,6-DIEPOXY-3A,4,5,6,7,7A-HEXAHYDRO-4,7-METHANOINDAN; 1,2:5,6-DIEPOXYHEXAHYDRO-4,7-METHANOINDAN; EP 207; EPOXIDE 207; 2,4-METHANO-2H-BISOXIRENO(A,F)INDENE,OCTAHYDRO-; 4,7-METHANOINDAN,1,2:5,6-DIEPOXY-3A,4,5,6,7,7A-HEXAHYDRO-; 4,7-METHANOINDAN,1,2:5,6-DIEPOXYHEXAHYDRO-; 2,4-METHANO-2H-INDENO(1,2-B:5,6-B')BISOXIRENE,OCTAHYDRO-; UNOX 207X; UNOX 207; UNOX EPOXIDE 207
Description: white crystalline powder; mild terpene like odor
Use: intermediate for epoxy resins, plasticizers, protective coatings; rubber additive for crack prevention & adhesion; for high temp & high performance system, adhesives

Physical Properties

Boiling Point: Sublimes at 120 °C (248 °F) to 135 °C (275 °F) at 10 mm Hg
Freezing Point: 180 °C (356 °F) to 184 °C (363.2 °F)
Specific Gravity: 1.331 at 25 °C
Vapor Density: 4.77 Air=1 at 25 °C
Water Solubility: 0.01%
Other Solubilities: 18.6% in Methanol; 44.7% in Acetone; 18.7% in Ether.

RTECS Toxicity Data

Acute Oral: Rat LD_{50} Dose: 210 mg/kg.
Acute Inhalation: Mammal LC_{50} Dose: 10 gm/m^3/1hr.
Acute Dermal: Rabbit LD_{50} Route: Skin; Dose: 8 gm/kg.
Chronic (Multiple Dose) Inhalation: Rat Dose: 1100 ug/m^3/17W-I; Toxic Effects: Brain and coverings - Recordings from specific areas of CNS; Kidney, Ureter, and Bladder - Proteinuria; Kidney, Ureter, and Bladder - Other changes in urine composition.
Irritation Skin: Rabbit Open Draize Test Dose: 500 mg open; Reaction: mild.

Hazard Overviews

Carcinogenicity: IARC - Not listed; NIOSH - Not listed; NTP - Not listed; ACGIH - Not listed; OSHA - Not listed; EPA - Not listed; MAK - Not listed

Environmental

Regulations

RCRA 40CFR: Not listed
CERCLA: 40CFR 302.4: Not listed
SARA 40CFR 372.65: Not listed
SARA EHS 40CFR 355: Not listed
TSCA: Listed

DIC8950 CAS #: 102-54-5

DICYCLOPENTADIENYL IRON

RTECS: LK0700000
EINECS Number: 203-039-3
Molecular Formula: $C_{10}H_{10}Fe$
Formula Weight: 186.03

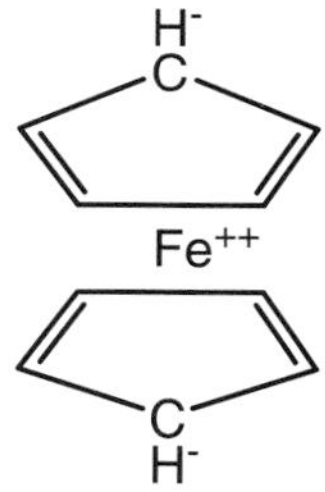

Chemical Structure

Synonyms: BISCYCLOPENTADIENYL IRON; BIS(CYCLOPENTADIENYL)IRON; BISCYCLOPENTADIENYLIRON; DI-PI-CYCLOPENTADIENYL IRON; DI-2,4-CYCLOPENTADIEN-1-YLIRON; DICYCLOPENTADIENYLIRON; FERROCENE; FERROTSEN; IRON BIS (CYCLOPENTADIENIDE); IRON BIS(CYCLOPENTADIENE); IRON BIS(CYCLOPENTADIENIDE); IRON DICYCLOPENTADIENYL; IRON,BIS(ETA(5)-2,4-CYCLOPENTADIEN-1-YL)-
Description: orange crystals; camphor odor
Use: antiknock additive for gasoline; catalyst; fuel oil additive; high temperature lubricant; ultraviolet absorber, as iron fertilizer; electron beam sensitize and coating for missiles and satellites

Physical Properties

Boiling Point: 249 °C (480 °F)
Freezing Point: 173 °C (343.4 °F) to 174 °C (345.2 °F)
Water Solubility: Insoluble
Other Solubilities: dissolves in dilute Nitric & concentrated Sulfuric acids; Soluble in Alcohol, Ether; Practically Insoluble in 10% NaOH and concentrated boiling

Hydrochloric Acid; solubility at 25 °C: 19 g/100 g Benzene; 9 g/100 g Gasoline; 5 g/100 g diesel fuel.
Ionization Potential (eV): 6.88
Flash Point: Not available; probably combustible

RTECS Toxicity Data

Acute Oral: Rat LD_{50} Dose: 1320 mg/kg. Mouse LD_{50} Dose: 832 mg/kg.
Chronic (Multiple Dose) Oral: Dog Dose: 54600 mg/kg/26W-I; Toxic Effects: Liver - Hepatitis, fibrous (cirrhosis, post-necrotic scarring); Endocrine - Other changes; Nutritional and gross metabolic - Changes in iron.
Chronic (Multiple Dose) Inhalation: Rat Dose: 36 mg/m^3/2W-I; Toxic Effects: Liver - Changes in Liver weight; Nutritional and gross metabolic - Weight loss or decreased weight gain. Mouse Dose: 20 mg/m^3/2W-I; Toxic Effects: Liver - Changes in Liver weight; Endocrine - Changes in spleen weight; Nutritional and gross metabolic - Weight loss or decreased weight gain.
Mutagenic: Hamster Sister Chromatid Exchenge; Cell Type: ovary; Dose: 130 ug/L. Insects - D Melanogaster Sex Chromosome Loss; Route: Parenteral; Dose: 100 ppm.
Tumorigenic: Rat Route: Intramuscular; Dose: 5175 mg/kg/2Y-I; Toxic Effects: Tumorigenic - Equivocal tumorigenic agent by RTECS criteria; Tumorigenic - Tumors at site of application.

Hazard Overviews

Flammable

Fire Diamond

Health: Irritating to eyes/skin/respiratory tract. Chronic Effects: dermatitis, sensitization.
Fire: Flammable. Use water spray, carbon dioxide, dry chemical, or alcohol foam.
Reactivity: Stable. Hazardous polymerization cannot occur. Avoid: heat; ignition sources. Incompatible with: ammonium perchlorate; prepared compositions of mercury(II)nitrate; tetranitromethane. Hazardous decomposition products: acrid smoke; irritating fumes.
Carcinogenicity: IARC - Not listed; NIOSH - Listed as carcinogen; NTP - Not listed; ACGIH - Not listed; OSHA - Not listed; EPA - Not listed; MAK - Not listed
Primary Target Organs:

Eyes

Skin

Respiratory System

Exposure Limits
OSHA PEL: TWA: 15 mg/m^3; total dust.
OSHA PEL Vacated 1989 Limits: TWA: 10 mg/m^3. Other Values: respirable mg/m^3; 5.
ACGIH TLV: TWA: 10 mg/m^3.
NIOSH REL: TWA: 10 mg/m^3; as Fe (Salts).

Respirator Recommendation
Exposure Range: >5 to 50 mg/m^3 Air Purifying, Negative Pressure, Half Mask
Exposure Range: >50 to 500 mg/m^3 Air Purifying, Negative Pressure, Full Face
Exposure Range: >500 to 5000 mg/m^3 Supplied Air, Constant Flow/Pressure Demand, Full Face
Exposure Range: >5000 to unlimited mg/m^3 Self-contained Breathing Apparatus, Pressure Demand, Full Face
Cartridge Color: dust/mist filter (use P100 or consult supervisor for appropriate dust/mist filter)

Environmental

Regulations

RCRA 40CFR: Not listed
CERCLA: 40CFR 302.4: Not listed
SARA 40CFR 372.65: Not listed
SARA EHS 40CFR 355: Not listed
TSCA: Listed

DID1000 CAS #: 1254-78-0

DIDECYL PHENYL PHOSPHITE

EINECS Number: 215-012-3
Molecular Formula: $C_{26}H_{47}O_3P$
Formula Weight: 438.63
Synonyms: IRGAPLAST CH 300; PHENYL DIDECYL PHOSPHITE; PHOSPHOROUS ACID,DIDECYL PHENYL ESTER
Description: nearly water-white liquid; odor of alcohol
Use: chemical intermediate; antioxidant; ingredient in stabilizer systems for resins; antioxidant; stabilizer for plastics & elastomers

Physical Properties

Freezing Point: <
Specific Gravity: 0.94 at 25 °C/15.5 °C
Refraction Index: 1.4785 at 25 °C/D
Flash Point: 218.333 °C Open Cup

Hazard Overviews

Fire Diamond

Fire: Will burn.
Carcinogenicity: IARC - Not listed; NIOSH - Not listed; NTP - Not listed; ACGIH - Not listed; OSHA - Not listed; EPA - Not listed; MAK - Not listed

Environmental

Regulations

RCRA 40CFR: Not listed
CERCLA: 40CFR 302.4: Not listed
SARA 40CFR 372.65: Not listed

SARA EHS 40CFR 355: Not listed
TSCA: Listed

DID2600 CAS #: 84-77-5

DIDECYL PHTHALATE

RTECS: TI0900000
EINECS Number: 201-561-6
Molecular Formula: $C_{28}H_{46}O_4$
Formula Weight: 446.74

Chemical Structure

Synonyms: 1,2-BENZENEDICARBOXYLIC ACID,DIDECYL ESTER; DECYL PHTHALATE; DIDECYL 1,2-BENZENEDICARBOXYLATE; DI-N-DECYL PHTHALATE; PHTHALIC ACID,DIDECYL ESTER; VINICIZER 105
Description: colorless to light-colored liquid; practically odorless
Use: plasticizer for vinyl resins; in manufacture of electric cables; for making plastisols

Physical Properties

Boiling Point: 261 °C (502 °F) at 5 mm Hg
Freezing Point: 3 °C (37.4 °F)
Specific Gravity: 0.9675 at 20 °C/20 °C
Vapor Pressure: 0.3 mm Hg at 200 °C
Water Solubility: Insoluble in Water
Other Solubilities: Soluble in hydrocarbons.
Flash Point: 229.444 °C Closed Cup

RTECS Toxicity Data

Acute Dermal: Rabbit LD_{50} Route: Skin; Dose: 16800 mg/kg; Toxic Effects: Skin and appendages - Primary irritation.
Irritation Skin: Rabbit Open Draize Test Dose: 10 mg/24H open; Reaction: mild.

Hazard Overviews

Fire: Will burn.
Carcinogenicity: IARC - Not listed; NIOSH - Not listed; NTP - Not listed; ACGIH - Not listed; OSHA - Not listed; EPA - Not listed; MAK - Not listed

Environmental

Environmental Fate: If released to soil, it is expected to be essentially immobile. Volatilization from the soil surface to the atmosphere is not expected to be an important process. Limited data indicates that if released to water, it may undergo microbial degradation. It is expected to bioaccumulate in fish and aquatic organisms, and adsorb to sediment and suspended organic matter. It is not expected to volatilize from water to the atmosphere. Hydrolysis is not expected to be a significant process. If released to the atmosphere, it is expected to be adsorbed to particulates. It may also undergo destruction by a vapor phase reaction with photochemically produced hydroxyl radicals.

Environmental Physical Data

Henry's Law Constant: 2.81 $x10^{-5}$
Octanol/Water Partition Coefficient: log K_{ow} = > 2.12
Sorption Partition Coefficient: K_{oc} = 8000
BCF: none likely

Regulations

RCRA 40CFR: Not listed
CERCLA: 40CFR 302.4: Not listed
SARA 40CFR 372.65: Not listed
SARA EHS 40CFR 355: Not listed
TSCA: Listed

Analytical Methods

Plasma: EPA 001

DID4200 CAS #: 69655-05-6

DIDEOXYINOSINE

RTECS: NM7460700
Molecular Formula: $C_{10}H_{12}N_4O_3$
Formula Weight: 236.23

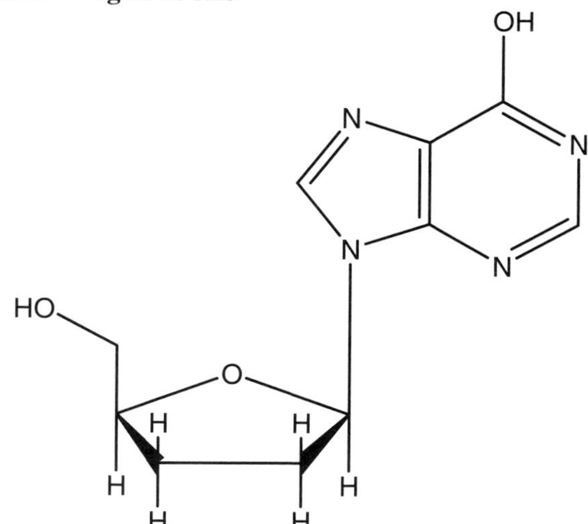

Chemical Structure

Synonyms: BMY 40900; DDI; DDINO; DIDANOSINE; 2',3'-DIDEOXYINOSINE; 2,3-DIDEOXYINOSINE; INOSINE,2',3'-DIDEOXY-; NSC 612049; VIDEX
Description: white crystalline powder; odorless
Use: an antiviral agent; being tested as a possible treatment for acquired immunodeficiency syndrome (AIDS), as it is a potent inhibitor of HIV an antiviral agent; being tested as a possible treatment for acquired immunodeficiency syndrome (AIDS), as it is a potent inhibitor of HIV

Physical Properties

Freezing Point: 160 °C (320 °F) to 163 °C (325.4 °F)
Water Solubility: About 27.3 mg/ml pH=6 at 25 °C
Other Solubilities: 0.02M Potassium phosphate monobasic: Soluble; 0.1% Carboxymethylcellulose: Soluble; 95% Ethanol: <1 mg/ml at 21 °C; Acetone: <1 mg/ml at 21 °C; DMSO: >=100 mg/ml at 21 °C; Methanol: Soluble.
Flash Point: Not available; probably combustible

RTECS Toxicity Data

Acute Oral: Man LD_{Lo} Dose: 2340 mg/kg/34W-I; Toxic Effects: Liver - Fatty Liver degeneration; Liver - Jaundice, cholestatic; Biochemical - Transaminases. Rat LD_{50} Dose: >2 gm/kg.
Mutagenic: Rat DNA Inhibition; Cell Type: liver; Dose: 100 mg/L. Mouse Cytogenetic Analysis; Route: Intraperitoneal; Dose: 6 gm/kg/3D-I.

Hazard Overviews

Health: Irritating. Harmful. Other Acute Effects: harmful if swallowed, inhaled, or absorbed through skin; target organs: pancreas, nerves. Chronic Effects: contains a radioactive isotope which may produce cancer and genetic mutation. The toxicological properties have not been thoroughly investigated.
Fire: Will burn. Hazards: emits toxic fumes. Extinguishing agents: water spray; carbon dioxide, dry chemical powder or appropriate foam. Precautions: combustible liquid.
Reactivity: Stable. Hazardous polymerization will not occur. Incompatible with: strong oxidizing agents. Hazardous decomposition products: thermal decomposition may produce carbon monoxide, carbon dioxide, and nitrogen oxides.
Carcinogenicity: IARC - Not listed; NIOSH - Not listed; NTP - Not listed; ACGIH - Not listed; OSHA - Not listed; EPA - Not listed; MAK - Not listed
Primary Target Organs:

Eyes

Skin

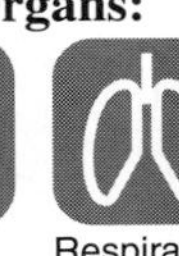
Respiratory System

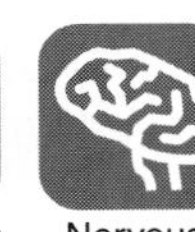
Nervous System

Environmental

Regulations

RCRA 40CFR: Not listed
CERCLA: 40CFR 302.4: Not listed
SARA 40CFR 372.65: Not listed
SARA EHS 40CFR 355: Not listed
TSCA: Not listed

DID7400 **CAS #: 7057-92-3**

DI-N-DODECYL PHOSPHATE

EINECS Number: 230-341-2
Molecular Formula: $C_{24}H_{51}O_4P$
Formula Weight: 434.647

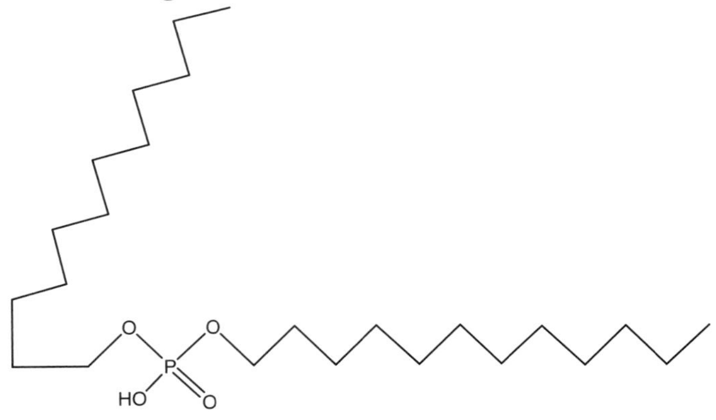

Chemical Structure

Synonyms: DIDODECYL HYDROGEN PHOSPHATE; DIDODECYL PHOSPHATE; DILAURYL ACID PHOSPHATE; DILAURYL PHOSPHATE; ELA; ORTHOLEUM 162; PHOSPHORIC ACID,DIDODECYL ESTER
Use: antistatic agent in textile processing

Hazard Overviews

Carcinogenicity: IARC - Not listed; NIOSH - Not listed; NTP - Not listed; ACGIH - Not listed; OSHA - Not listed; EPA - Not listed; MAK - Not listed

Environmental

Regulations

RCRA 40CFR: Not listed
CERCLA: 40CFR 302.4: Not listed
SARA 40CFR 372.65: Not listed
SARA EHS 40CFR 355: Not listed
TSCA: Listed

DIE1000 **CAS #: 60-57-1**

DIELDRIN

RTECS: IO1750000
DOT: UN2761; UN2995; UN2996; NA2761; IMO6.1
EINECS Number: 200-484-5
Molecular Formula: $C_{12}H_8Cl_6O$
Formula Weight: 380.93

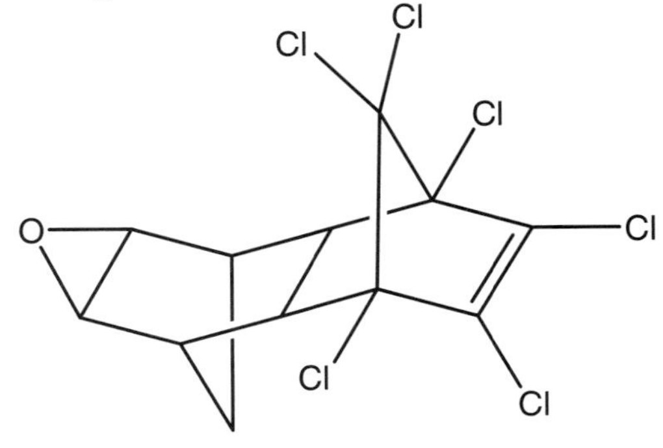

Chemical Structure

Synonyms: ALVIT; ALVIT 55; COMPOUND 497; DIELDREX; DIELDRINE; DIELDRITE; DIELDRIX; DIELMOTH; 1,4:5,8-DIMETHANONAPHTHALENE; 5,8-DIMETHANONAPHTHALENE; 1,4:5,8-DIMETHANONAPHTHALENE,1,2,3,4,10,10-HEXACHLORO-6,7-EPOXY-1,4,4A,; 1,4:5,8-DIMETHANONAPHTHALENE,1,2,3,4,10,10-HEXACHLORO-6,7-EPOXY-1,4,4A,5,6,7,8,8A-OCTAHYDRO-,ENDO,EXO-; DIMETHANONAPHTH(2,3-B)OXIRENE; 2,7:3,6-DIMETHANONAPHTH(2,3-B)OXIRENE,3,4,5,6,9,9-HEXACHLORO-1A,2,2A,3,6,6A,7,7A-OCTAHYDRO-,(1AALPHA,2BETA,2AALPHA,3BETA,6BETA,6AALPHA,7BETA,7AALPHA),; ENDO,EXO-1,2,3,4,10,10-HEXACHLORO-6,7-EPOXY-1,4,4A,5,6,7,8,8A-OCTAHYDRO-; ENDO,EXO-1,2,3,4,10,10-HEXACHLORO-6,7-EPOXY-1,4,4A,5,6,7,8,8A-OCTAHYDRO-1,4:5,8-DIMETHANONAPHTHALENE; ENDO,EXO-3,4,5,6,9,9-HEXACHLORO-1A,2,2A,3,6,6A,7,7A-OCTAHYDRO-2,7:3,6-DIMETHENAPTH(2,3-B) OXIRENE; ENT-16225; ENT 16,225; ENT 16225; EPA PESTICIDE CHEMICAL CODE 045001; EXO-DIELDRIN; HEOD; 1,2,3,4,10,10-HEXACHLORO-6,7-EPOXY-1,4,4A,5,6,7,8,8A-OCTAHYDRO-1 ,4-ENDO,EXO-5,8-DIMETHANONAPHTHALENE; 1,2,3,4,10,10,-HEXACHLORO-6,7-EPOXY-1,4,4A,5,6,7,8,8A-OCTA-HYDRO-ENDO,EXO 1,4:5,8-DIMETHANONAPHTHALENE; 1,2,3,4,10,10-HEXACHLORO-6,7-EPOXY-1,4,4A,5,6,7,8,8A-OCTAHYDRO-1,4-ENDO-EXO-5,8-DIMETHANONAPHTHALENE; 1,2,3,4,10,10-HEXACHLORO-6,7-EPOXY-1,4,4A,5,6,7,8,8A-OCTAHYDRO-ENDO-1,4-EXO-5,8-DIMETHANONAPHTHALENE; HEXACHLOROEPOXYOCTAHYDRO-ENDO,EXO-DIMETHANONAPHTHALENE; 1,2,3,4,10,10-HEXACHLORO-6,7-EPOXY-1,4,4A,5,6,7,8,8A-OCTAHYDRO-EXO-1,4-ENDO-; 1,8,9,10,11,11-HEXACHLORO-4,5-EXOEPOXY-2,3-7,6-ENDO-2,1-7,8-EXO-TETRACYCLO(6.2.1.1 3,6 .02,7) DODEC-9-ENE; 1,2,3,4,10,10-HEXACHLORO-EXO-6,7-EPOXY-1,4,4A,5,6,7,8,8A-OCTAHYDRO-1,4-ENDO,EXO-5,8-DIMETHANONAPHTHALENE; 3,4,5,6,9,9-HEXACHLORO-1A,2,2A,3,6,6A,7,7A-OCTAHYDRO-2,7:3,6-; 3,4,5,6,9,9-HEXACHLORO-1A,2,2A,3,6,6A,7,7A-OCTAHYDRO-2,7:3,6-DIME-THANONAPHTH (2,3-B)OXIRENE; 3,4,5,6,9,9-HEXACHLORO-1A,2,2A,3,6,6A,7,7A-OCTAHYDRO-2,7:3,6-DIMETHANONAPHTH[2,3-B] OXIRENE; (1A ALPHA,2BETA,2A ALPHA,3BETA,6BETA,6A ALPHA,7BETA,7A ALPHA)-3,4,5,6,9,9-HEXACHLORO-1A,2,2A,3,6,6A,7,7A-OCTAHYDRO-2,7:3,6-DIMETHANONAPHTH[2,3-B]OXIRENE; 3,4,5,6,9,9-HEXACHLORO-1A,2,2A,3,6,6A,7,7A-OCTAHYDRO-2,7:3,6-DIMETHANONAPHTH(2,3-B)OXIRENE; (1R,4S,4AS,5R,6R,7S,8S,8AR)-1,2,3,4,10,10-HEXACHLORO-1,4,4A,5,6,7,8,8A-OCTAHYDRO-6,7-EPOXY-1,4:5,8-DIMETHANONAPHTHALENE; 1,2,3,4,10,10-HEXACHLORO-1R,4S,4AS,5R,6R,7S,8S,8AR-OCTAHYDRO-6,7-EPOXY-1,4:5,8-DIMETHANONAPHTHALENE; ILLOXOL; INSECTICIDE NO. 497; LATKA 497; 5,6,7,8,8A-OCTAHYDRO,ENDO,EXO-; OCTALOX; OXRALOX; PANORAM D-31; QUINTOX; RED SHIELD; SD 3417; TERMITOX

Use: as a nonsystemic, persistent insecticide with contact and stomach action

Physical Properties

Boiling Point: Decomposes
Freezing Point: 175 °C (347 °F) to 176 °C (348.8 °F)
Specific Gravity: 1.75
Vapor Density: 13.2
Saturated Vapor Density: 1.200000015 kg/m^3
Density: 1.62 g/cu cm at 20 °C
Vapor Pressure: 7.78 x10^{-7} mm Hg at 25 °C
Water Solubility: 0.02% by weight
Other Solubilities: Slightly Soluble in Petroleum Ether, freely Soluble in Benzene; Insoluble in Methanol and aliphatic hydrocarbons; In Acetone 220, Ethanol 40, dichloromethane 480, Benzene 400, Toluene 410, Carbon Tetrachloride 380, Methanol 10 (all in g/l at 20 °C).
Odor Threshold: 0.041 ppm
Flash Point: Nonflammable

RTECS Toxicity Data

Acute Oral: Man LD_{Lo} Dose: 65 mg/kg. Rat LD_{50} Dose: 38300 ug/kg. Monkey LD_{50} Dose: 3 mg/kg.

Acute Inhalation: Rat LC_{50} Dose: 13 mg/m^3/4hr. Cat LC_{50} Dose: 80 mg/m^3/4hr; Toxic Effects: Behavioral - Somnolence (general depressed activity); Behavioral - Convulsions or effect on seizure threshold; Behavioral - Excitement.

Acute Dermal: Rabbit LD_{50} Route: Skin; Dose: 250 mg/kg. Rabbit LD_{Lo} Route: Subcutaneous Dose: 150 mg/kg.

Chronic (Multiple Dose) Oral: Rat Dose: 75 mg/kg/15D-C; Toxic Effects: Liver - Other changes; Biochemical - Phosphatases; Biochemical - Dehydrogenases. Rat Dose: 88 mg/kg/88D-C; Toxic Effects: Behavioral - Food intake (animal); Nutritional and gross metabolic - Weight loss or decreased weight gain; Biochemical - Phosphatases. Rat Dose: 140 mg/kg/8W-I; Toxic Effects: Liver - Other changes; Blood - Other changes; Biochemical - Other esterases. Rat Dose: 109 mg/kg/2Y-C; Toxic Effects: Liver - Changes in liver weight.

Reproductive/Teratogenic: Rat Route: Oral; Dose: 14 ug/kg; Duration: multigenerations; Effects on Newborn - Behavioral. Mouse Route: Oral; Dose: 30600 ug/kg; Duration: female 6-14D of pregnancy; Specific Developmental Abnormalities - Central nervous system; Eye, ear. Mouse Route: Oral; Dose: 15 mg/kg; Duration: female 9D of pregnancy; Specific Developmental Abnormalities - Craniofacial (including nose and tongue). Mouse Route: Oral; Dose: 2250 ug/kg; Duration: female 6-14D of pregnancy; Effects on Embryo or Fetus - Fetotoxicity. Mouse Route: Oral; Dose: 12500 ug/kg; Duration: male 1D prior to mating; Effects on Fertility - Pre-implantation mortality. Mouse Route: Oral; Dose: 4500 ug/kg; Duration: female 6-14D of pregnancy; Specific Developmental Abnormalities - Musculoskeletal system. Mouse Route: Oral; Dose: 6250 ug/kg; Duration: male 5D prior to mating; Paternal Effects - Other effects on male.

Mutagenic: Human Unscheduled DNA Synthesis; Cell Type: fibroblast; Dose: 1 umol/L. Human DNA Inhibition; Cell Type: lymphocyte; Dose: 100 mg/L. Human DNA Inhibition; Cell Type: HeLa cell; Dose: 400 umol/L.

Tumorigenic: Rat Route: Oral; Dose: 200 mg/kg/2Y-C; Toxic Effects: Tumorigenic - Equivocal tumorigenic agent by RTECS criteria; Lungs, Thorax, or Respiration - Tumors; Skin and appendages - Tumors. Mouse Route: Oral; Dose: 546 mg/kg/65W-C; Toxic Effects: Tumorigenic - Carcinogenic by RTECS criteria; Liver - Tumors. Mouse Route: Oral; Dose: 11 gm/kg/3Y-C; Toxic Effects: Tumorigenic - Neoplastic by RTECS criteria; Lungs, Thorax, or Respiration - Tumors; Liver - Tumors. Mouse Route: Oral; Dose: 610 mg/kg/73W-C; Toxic Effects: Tumorigenic - Neoplastic by RTECS criteria; Lungs, Thorax, or Respiration - Tumors; Liver - Tumors. Mouse Route: Oral; Dose: 714 mg/kg/85W-C; Toxic Effects: Tumorigenic - Carcinogenic by

RTECS criteria; Liver - Tumors. Mouse Route: Oral; Dose: 8 mg/kg/2Y-C; Toxic Effects: Tumorigenic - Equivocal tumorigenic agent by RTECS criteria; Liver - Tumors. Mouse Route: Oral; Dose: 4550 mg/kg/65W-C; Toxic Effects: Tumorigenic - Carcinogenic by RTECS criteria; Lungs, Thorax, or Respiration - Tumors; Liver - Tumors.

Hazard Overviews

Reactivity: Stable. Hazardous polymerization cannot occur. Avoid: exposure to excessive temperatures. Incompatible with: strong oxidizers; active metals (sodium; mineral acids; acid catalysts; phenols). Hazardous decomposition products: hydrogen chloride; chloride gases.

Carcinogenicity: IARC - Group 3, Not classifiable as to carcinogenicity to humans; NIOSH - Listed as carcinogen; NTP - Not listed; ACGIH - Class A4, Not classifiable as a human carcinogen; OSHA - Not listed; EPA - Class B2, Probable human carcinogen based on animal studies; MAK - Not listed

Exposure Limits

OSHA PEL: TWA: 0.25 mg/m^3; skin.

ACGIH TLV: TWA: 0.25 mg/m^3.

NIOSH REL: TWA: 0.25 mg/m^3.

NIOSH IDLH: 50 mg/m^3.

DFG MAK: TWA: 0.25 mg/m^3.

Respirator Recommendation

Exposure Range: >0.25 to 12.5 mg/m^3 Supplied Air, Constant Flow/Pressure Demand, Half Mask

Exposure Range: >12.5 to <50 mg/m^3 Supplied Air, Constant Flow/Pressure Demand, Full Face

Exposure Range: 50 to unlimited mg/m^3 Self-contained Breathing Apparatus, Pressure Demand, Full Face

Note: odor threshold unknown

Environmental

Ecotoxicity: LD_{50} Musca /domestica/ (housefly) female 9.8 ug/fly, 3 days old LC_{50} Pseudacris triseriata (frog, tadpoles) 100 ug/l/96 hr /Conditions of bioassay not specified LD_{50} Passer domesticus (House sparrow) female oral 47.6 mg/kg (95% confidence limit 34.3-66.0 mg/kg) LD_{50} Perdix (Gray partridge) female oral 8.84 mg/kg (95% confidence limit 3.32-23.6 mg/kg) 3-10 mo old LC_{50} Young coturnix (Japanese quail) 60 ppm (95% confidence limit 57-63 ppm) 5 day diet LC_{50} Aedes aegypti (mosquito) late 3rd instar larvae 6 ppb/24 hr /Conditions of bioassay not specified LC_{50} Sphaeroides maculatus (Northern puffer) 34 ppb/96 hr, static lab bioassay (100%) LD_{50} Odocoileus hemionus (Mule deer) male oral 75-150 mg/kg (95% confidence limit) 8-18 mo old LC_{50} Salmo gairdneri (Rainbow trout) 1.2 ug/l/96 hr (95% confidence limit 0.9-1.7 ug/l) wt 1.4 g, water 13 °C, static bioassay without aeration, pH 7.2-7.5, water hardness 40-50 mg/l as calcium carbonate and alkalinity of 30-35 mg/l LC_{50} Pteronarcys 0.5 ug/l/96 hr (95% confidence limit 0.4-0.7 ug/l) 2nd yr class, water temp 15 °C, static bioassay without aeration, pH 7.2-7.5, water hardness 40-50 mg/l as calcium carbonate and alkalinity of 30-35 mg/l

Environmental Fate: It is extremely persistent, but it is known to slowly photo rearrange to photodieldrin (water half-life - 4 months). Released to soil it will persist for long periods (> 7 year), will reach the air either through slow evaporation or adsorption on dust particles, will not leach, and will reach surface water with surface runoff. Once it reaches surface waters it will adsorb strongly to sediments, bioconcentrate in fish and slowly photodegrade. Biodegradation and hydrolysis are unimportant fate processes. Fate in the atmosphere is unknown but monitoring data have demonstrated that it can be carried long distances.

Cleanup/Disposal: Guide No. 131: Fully encapsulating, vapor protective clothing should be worn for spills and leaks with no fire. Eliminate all ignition sources (no smoking, flares, sparks or flames in immediate area). All equipment used when handling the product must be grounded. Do not touch or walk through spilled material. Stop leak if you can do it without risk. Prevent entry into waterways, sewers, basements or confined areas. A vapor suppressing foam may be used to reduce vapors. Small Spills: Absorb with earth, sand or other non-combustible material and transfer to containers for later disposal. Use clean non-sparking tools to collect absorbed material. Large Spills: Dike far ahead of liquid spill for later disposal. Water spray may reduce vapor; but may not prevent ignition in closed spaces. Guide No. 151: Do not touch damaged containers or spilled material unless wearing appropriate protective clothing. Stop leak if you can do it without risk. Prevent entry into waterways, sewers, basements or confined areas. Cover with plastic sheet to prevent spreading. Absorb or cover with dry earth, sand or other non-combustible material and transfer to containers. Do not get water inside containers.

Environmental Physical Data

Octanol/Water Partition Coefficient: log K_{ow} = 6.2

Sorption Partition Coefficient: K_{oc} = 3.87

BCF: fish 3 to 6000

Regulations

RCRA 40CFR: Listed Hazardous Waste No. P037 Toxic Waste

CERCLA: 40CFR 302.4: Listed per CWA Section 311(b)(4) per RCRA Section 3001 per CWA Section 307(a) RQ: 1 lb (0.454 kg)

SARA 40CFR 372.65: Not listed

SARA EHS 40CFR 355: Not listed

TSCA: Not listed

Analytical Methods

Air: EPA TO-10

Soil: CLP LC_PEST, MC_PEST, OHC; EPA PMD-TLC, 16, 3, 024, 025, P-002-1, P-011-1, PCB-005; SW846 3630B, 3640A, 8080A, 8081, 8081A, 8250A, 8270B, 8270C; USGS O5104, O7104

Water / Groundwater: EPA P-003-1, P-004-1, 608, 617, 625, 625-S, 680, 022; APHA 6410-B, 6630-B, 6630-C, 6630-D; ASTM D3086; USGS O3104

Drinking Water: EPA 505, 508, 508.1, 525.2; AOAC 990.06; ASTM D5175

Food: FDA 212.1, 212.2, 232.1, 232.4, 242.1; EPA 026, 4, XENO; AOAC 961.05, 970.52, 972.05, 985.22; USGS O9104

Indoor / Expired Air: EPA IP-8; ASTM D4861
Plasma: EPA 001, 003, 004, 027, 028, 29; FDA 211.1, 231.1, 251.1, 252, 253
Other: EPA P-009-1, 1656

DIE1100 CAS #: 84-17-3

DIENESTROL

RTECS: SL0580000
EINECS Number: 201-519-7
Molecular Formula: $C_{18}H_{18}O_2$
Formula Weight: 266.32

Chemical Structure

Synonyms: AGALDOG; 3,4-BIS(4-HYDROXYPHENYL)-2,4-HEXADIENE; 3,4-BIS(P-HYDROXYPHENYL)-2,4-HEXADIENE; 3,4-BIS(PARA-HYDROXYPHENYL)-2,4-HEXADIENE; CYCLADIENE; DEHYDROSTILBESTROL; DEHYDROSTILBOESTROL; PARA-DIEN; BETA-DIENOESTROL; DIENOESTROL; DIENOL; 4,4'-(1,2-DIETHYLIDENE-1,2-ETHANEDIYL)BISPHENOL; 4,4'-(DIETHYLIDENEETHYLENE)DIPHENOL; P,P'-(DIETHYLIDENEETHYLENE)DIPHENOL; PARA,PARA'-(DIETHYLIDENEETHYLENE)DIPHENOL; 4,4'-DIHYDROXY-GAMMA,DELTA-DIPHENYL-BETA,DELTA-HEXADIENE; DINOVEX; DI(P-OXYPHENYL)-2,4-HEXADIENE; DI(PARA-OXYPHENYL)-2,4-HEXADIENE; DV; ESTRAGARD; ESTRAGUARD; ESTRODIENOL; ESTRORAL; FOLLIDIENE; FOLLORMON; GYNEFOLLIN; 2,4-HEXADIENE,3,4-BIS(4-HYDROXYPHENYL)-; HORMOFEMIN; 4,4'-HYDROXY-GAMMA,DELTA-DIPHENYL-BETA,DELTA-HEXADIENE; ISODIENESTROL; MESOHEXESTROL; OESTRASID; OESTRODIENE; OESTRODIENOL; OESTRORAL; OESTROVIS; PHENOL,4,4'-(1,2-DIETHYLIDENE-1,2-ETHANEDIYL)BIS-(9CI); PHENOL,4,4'-(DIETHYLIDENEETHYLENE)DI-; RESTROL; RETALON; SEXADIEN; SYNESTROL; TESERENE; WILLNESTROL
Description: colorless or white, or practically white needle-like crystalline powder, crystals, or needles; odorless
Use: medication: estrogen; medication (vet): estrogenic hormone therapy

Physical Properties

Freezing Point: 227 °C (440.6 °F) to 228 °C (442.4 °F)
Water Solubility: Practically Insoluble in Water
Other Solubilities: Soluble in fixed oils.

RTECS Toxicity Data

Reproductive/Teratogenic: Rat Route: Oral; Dose: 40 ug/kg; Duration: female 3D of pregnancy; Effects on Fertility - Pre-implantation mortality.
Mutagenic: Human Sister Chromatid Exchenge; Cell Type: fibroblast; Dose: 5 nmol/L. Mouse Unscheduled DNA Synthesis; Route: Subcutaneous; Dose: 400 mg/kg.

Hazard Overviews

Health: May cause irritation. Harmful. Other Acute Effects: may be harmful by inhalation, ingestion, or skin absorption. Chronic Effects: target organ: reproductive system. Possible carcinogen. Possibly causes reproductive effects. The toxicological properties have not been thoroughly investigated.
Fire: Hazards: emits toxic fumes. Extinguishing agents: water spray; carbon dioxide, dry chemical powder or appropriate foam. Precautions: combustible liquid.
Reactivity: Stable. Hazardous polymerization will not occur. Hazardous decomposition products: toxic fumes of: carbon monoxide, carbon dioxide.
Carcinogenicity: IARC - Not listed; NIOSH - Not listed; NTP - Not listed; ACGIH - Not listed; OSHA - Not listed; EPA - Not listed; MAK - Not listed
Primary Target Organs:

Repro-ductive

Environmental

Regulations

RCRA 40CFR: Not listed
CERCLA: 40CFR 302.4: Not listed
SARA 40CFR 372.65: Not listed
SARA EHS 40CFR 355: Not listed
TSCA: Not listed

DIE1200 CAS #: 1464-53-5

DIEPOXYBUTANE

RTECS: EJ8225000
EINECS Number: 215-979-1
Molecular Formula: $C_4H_6O_2$
Formula Weight: 86.09

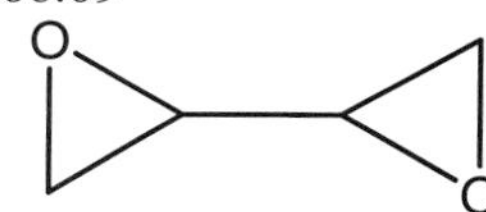

Chemical Structure

Synonyms: BIOXIRAN; 2,2'-BIOXIRANE; BIOXIRANE; 1,1'-BIS(ETHYLENE OXIDE); BUTADIENDIOXYD; 1,3-BUTADIENE DIEPOXIDE; BUTADIENE DIEPOXIDE; BUTADIENE DIOXIDE;

BUTANE DIEPOXIDE; BUTANEDIONE; DEB; 1,2:3,4-DIEPOXYBUTANE; 2,4-DIEPOXYBUTANE; DIOXYBUTADIENE; ENT-26592; ERYTHRITOL ANHYDRIDE

Description: colorless to water-white, low viscosity liquid

Use: research chemical; curing polymers; crosslinking textile fibers

Physical Properties

Boiling Point: 138 °C (280 °F)

Freezing Point: -19 °C (-2.2 °F)

Specific Gravity: 0.962 at 20 °C / 4 °C

Saturated Vapor Density: Calculated 1.221447604 kg/m^3

Vapor Pressure: Calculated 6.9 mm Hg at 25 °C

Water Solubility: Miscible in water

Other Solubilities: Ethanol: Very Soluble

Refraction Index: 1.435

RTECS Toxicity Data

Acute Oral: Rat LD_{50} Dose: 78 mg/kg. Mouse LD_{50} Dose: 72 mg/kg.

Acute Inhalation: Rat LC_{50} Dose: 90 ppm/4hr.

Acute Dermal: Rabbit LD_{50} Route: Skin; Dose: 89 uL/kg.

Irritation Eye: Rabbit Standard Draize Test Dose: 250 ug open; Reaction: severe.

Irritation Skin: Rabbit Standard Draize Test Dose: 5 mg/24H; Reaction: severe. Rabbit Open Draize Test Dose: 10 mg/24H open; Reaction: severe. Rabbit Open Draize Test Dose: 50 mg open; Reaction: severe.

Mutagenic: Human Mutations in Mammalian Somatic Cells; Cell Type: lymphocyte; Dose: 3500 nmol/L. Human Cytogenetic Analysis; Cell Type: lymphocyte; Dose: 100 ug/L. Human Cytogenetic Analysis; Cell Type: bone marrow; Dose: 100 ug/L. Human Sister Chromatid Exchenge; Cell Type: lymphocyte; Dose: 5 mg/L.

Tumorigenic: Rat Route: Intraperitoneal; Dose: 380 mg/kg/13W-I; Toxic Effects: Tumorigenic - Equivocal tumorigenic agent by RTECS criteria; Musculoskelital - Tumors; Tumorigenic - Tumors at site of application. Mouse Route: Skin; Dose: 95 gm/kg/78W-I; Toxic Effects: Tumorigenic - Equivocal tumorigenic agent by RTECS criteria; Skin and appendages - Tumors; Tumorigenic - Tumors at site of application.

Hazard Overviews

Poison

Health: Irritating to eyes/skin/respiratory tract. Poison. Other Acute Effects: may be fatal if inhaled; harmful if swallowed or absorbed through skin; metabolized in the body producing carbon monoxide which increases and sustains carboxyhemoglobin levels in the blood, reducing the oxygen-carrying capacity of the blood; exposure can cause nausea; dizziness; headache; may cause nervous system disturbances. Chronic Effects: may alter genetic material; target organs: liver, pancreas, nerves, cardiovascular system. Carcinogen.

Fire: Hazards: container explosion may occur; emits toxic fumes. Extinguishing agents: water spray; carbon dioxide, dry chemical powder or appropriate foam. Precautions: combustible liquid.

Reactivity: Incompatible with: acids, bases, oxidizing agents. Hazardous decomposition products: carbon monoxide, carbon dioxide.

Carcinogenicity: IARC - Not listed; NIOSH - Not listed; NTP - Listed; ACGIH - Not listed; OSHA - Not listed; EPA - Not listed; MAK - Not listed

Primary Target Organs:

Eyes

Skin

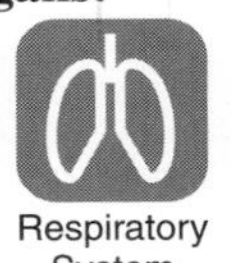

Respiratory System

Nervous System

Liver

Cardio-vascular

Environmental

Environmental Fate: If released to soil, it is predicted to be very mobile in wet soils and should hydrolyze to 1,2,3,4-tetrahydroxybutane. If released to water, it is expected to undergo hydrolysis to 1,2,3,4-tetrahydroxybutane. It is not expected to react with alkyl peroxy radical in water, volatilize from water, bioconcentrate in aquatic organisms or adsorb to sediments. If released to the atmosphere, vapor phase is predicted to be removed primarily by reaction with photochemically generated hydroxyl radicals (half-life 2.4 hours). In addition, removal via wet deposition may also be possible.

Environmental Physical Data

Henry's Law Constant: calculated at 3.63 $x10^{-7}$

Octanol/Water Partition Coefficient: log K_{ow} = -1.84

Sorption Partition Coefficient: K_{oc} = estimated at 2.4

BCF: estimated at 0.024

Regulations

RCRA 40CFR: Listed Hazardous Waste No. U085 Toxic Waste Ignitable Waste

CERCLA: 40CFR 302.4: Listed per RCRA Section 3001 RQ: 10 lb (4.535 kg)

SARA 40CFR 372.65: Listed TPQ: 500 lb

SARA EHS 40CFR 355: Listed TPQ: 10 lb

TSCA: Listed

Analytical Methods

Soil: SW846 8240B, 8260A, 8260B

Water / Groundwater: EPA 1625

Plasma: EPA 29

DIE1300 CAS #: 96-08-2

1,2:8,9-DIEPOXY-P-MENTHANE

RTECS: OS9100000

EINECS Number: 202-475-1

Molecular Formula: $C_{10}H_{16}O_2$

Formula Weight: 168.26

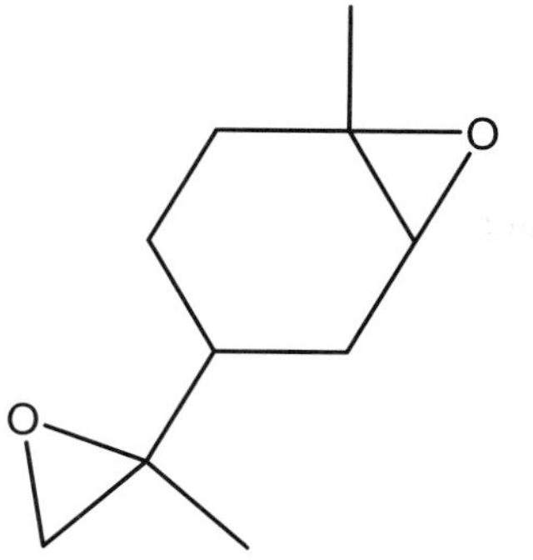

Chemical Structure

Synonyms: 1,2,8,9-DIEPOXYLIMONENE; DIPENTENE DIEPOXIDE; DIPENTENE DIOXIDE; DIPENTENE OXIDE; EPOXIDE 269; 4-(1,2-EPOXY-1-METHYLETHYL)-1-METHYL-7-OXABICYCLO(4.1.0)HEPTANE; ALPHA-LIMONENE DIEPOXIDE; LIMONENE DIEPOXIDE; LIMONENE DIOXIDE; MENTHANE,1,2:8,9-DIEPOXY-; P-MENTHANE,1,2:8,9-DIEPOXY-; 7-OXABICYCLO(4.1.0)HEPTANE,4-(1,2-EPOXY-1-METHYLETHYL)-1-METHYL-; 7-OXABICYCLO(4.1.0)HEPTANE,1-METHYL-4-(2-METHYLOXIRANYL)-; UNOX 269; UNOX EPOXIDE 269; UNOXAT EPOXIDE 269

Description: colorless liquid; mild methanol like ordor

Use: pharmaceticals; suggested as diluent for epoxy resins, as an intermediate in the prepn of modified alkyd resins, a plasticizer, lubricant additive; chemical intermediate

Physical Properties

Boiling Point: 242 °C (468 °F)
Freezing Point: -100 °C (-148 °F)
Specific Gravity: 1.0287 at 20 °C/4 °C
Vapor Density: 7.4
Saturated Vapor Density: 1.200151644 kg/m^3
Vapor Pressure: 0.02 mm Hg at 20 °C
Water Solubility: Slightly Soluble in Water
Other Solubilities: miscible with Methanol and Benzene. miscible with Carbon Tetrachloride and Hexane
Refraction Index: 1.4682 at 25 °C
Flash Point: 118 °C Open Cup

RTECS Toxicity Data

Acute Oral: Rat LD_{50} Dose: 5630 mg/kg.
Acute Inhalation: Rat LC_{50} Dose: 60 gm/m^3/1hr.
Acute Dermal: Rabbit LD_{50} Route: Skin; Dose: 1770 uL/kg; Toxic Effects: Skin and appendages - Primary irritation.
Irritation Skin: Rabbit Open Draize Test Dose: 10 mg/24H open; Reaction: mild.

Hazard Overviews

Fire: Will burn.
Carcinogenicity: IARC - Not listed; NIOSH - Not listed; NTP - Not listed; ACGIH - Not listed; OSHA - Not listed; EPA - Not listed; MAK - Not listed

Environmental

Regulations

RCRA 40CFR: Not listed
CERCLA: 40CFR 302.4: Not listed
SARA 40CFR 372.65: Not listed
SARA EHS 40CFR 355: Not listed
TSCA: Listed

DIE1400 **CAS #: 68334-30-5**

DIESEL FUEL OIL NO. 2-D

RTECS: HZ1800000
EINECS Number: 269-822-7
Molecular Formula: Unspecified or Variable
Formula Weight: N/A
Synonyms: AUTOMOTIVE DIESEL OIL; DIESEL FUEL; DIESEL OIL (PETROLEUM); DIESEL OILS; DIESEL TEST FUEL; FUELS,DIESEL; OLEJ NAPEDOWY III
Description: brown slightly viscous liquid

Physical Properties

Boiling Point: 171 °C (340 °F) to 358 °C (676 °F)
Freezing Point: -34 °C (-29.2 °F)
Specific Gravity: < 0.86
Water Solubility: Insoluble
Surface Tension: 23 to 32 dynes/cm at 20 °C
Odor Threshold: 0.7 ppm
Flash Point: 38 °C Closed Cup
Autoignition Temperature: 177 to 329 °C
LEL: 1.3% v/v
UEL: 7.5% v/v

RTECS Toxicity Data

Acute Oral: Rat LD_{50} Dose: 7500 mg/kg.
Acute Dermal: Rabbit LD; Route: Skin; Dose: >5 mL/kg.
Irritation Skin: Rabbit Standard Draize Test Dose: 500 uL/24H; Reaction: severe.

Hazard Overviews

Flammable

Fire Diamond (Health 0, Flammability 2, Reactivity 0)

Health: Irritating to skin/respiratory tract. Also Causes: increased respiration rate, rapid heart beat, cyanosis; upon ingestion: GI irritation, vomiting, diarrhea, CNS depression (high conc.) Chronic Effects: dermatitis.
Fire: Flammable. Can form explosive mixtures in the air. Use water as fog, dry chemical, or carbon dioxide. Do not use water spray as it may scatter fire. Use a smothering technique for extinguishing fire. Use a water spray to cool fire exposed containers.
Reactivity: Stable. Hazardous polymerization cannot occur. Avoid: heat and ignition sources. Incompatible with: strong

oxidizing agents. Hazardous decomposition products: various hydrocarbons; hydrocarbon derivatives; partial oxidation products (carbon dioxide; carbon monoxide; sulfur dioxide).

Carcinogenicity: IARC - Group 3, Not classifiable as to carcinogenicity to humans; NIOSH - Not listed; NTP - Not listed; ACGIH - Not listed; OSHA - Not listed; EPA - Not listed; MAK - Not listed

Primary Target Organs:

Skin

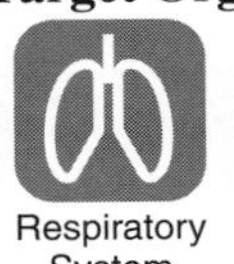
Respiratory System

Cardio-vascular

Environmental

Ecotoxicity: Aquatic toxicity: 204 mg/1/24 hr/juvenile American shad/TLm/salt water; Waterfowl toxicity: 20 mg/kg LD_{50}, mallard

Cleanup/Disposal: Guide No. 128: Eliminate all ignition sources (no smoking, flares, sparks or flames in immediate area). All equipment used when handling the product must be grounded. Do not touch or walk through spilled material. Stop leak if you can do it without risk. Prevent entry into waterways, sewers, basements or confined areas. A vapor suppressing foam may be used to reduce vapors. Absorb or cover with dry earth, sand or other non-combustible material and transfer to containers. Use clean non-sparking tools to collect absorbed material. Large Spills: Dike far ahead of liquid spill for later disposal. Water spray may reduce vapor; but may not prevent ignition in closed spaces.

Regulations

RCRA 40CFR: Not listed
CERCLA: 40CFR 302.4: Not listed
SARA 40CFR 372.65: Not listed
SARA EHS 40CFR 355: Not listed
TSCA: Listed

Analytical Methods

Air: EPA 80APP-G
Soil: EPA O-003-1, O-004-1, O-009-1, 1651, 1662, 1663; SW846 3560, 4030, 8015B, 8440
Water / Groundwater: EPA 418.1; ASTM D3921; USGS O3109
Other: EPA O-006-1

DIE1500 CAS #: 702-54-5

DIETHADION

RTECS: RP6300000
EINECS Number: 211-867-1
Molecular Formula: $C_8H_{13}NO_3$
Formula Weight: 171.22

Synonyms: DIETADIONE; DIETHADIONE; 5,5-DIETHYLDIHYDRO-2H-1,3-OXAZINE-2,4(3H)-DIONE; 5,5-DIETHYL-1,3-OXAZIN-2,4-DIONE; 5,5-DIETHYL-1,3-OXAZINE-2,4-DIONE; 5,5-DIETHYLTETRAHYDRO-2H-1,3-OXAZINE-2,4(3H)-DIONE; 5,5-DIETILDIIDRO-1,3-OSSAZIN-2,4-DIONE; DIETROXIN; DIETROXINE; DIHYDRO-5,5-DIETHYL-2H-1,3-OXAZINE-2,4(3H)-DIONE; DIIDRO-5,5-DIETIL-2H-1,3-OSSAZIN-2,4(3H)-DIONE; DIOXONE; L 1811; LEDOSTEN; LEPTON; 2H-1,3-OXAZINE-2,4(3H)-DIONE,5,5-DIETHYLDIHYDRO-; PERSISTEN; TOCE; TOCEN

Description: crystals
Use: medication: CNS stimulant; analeptic agent

Physical Properties

Freezing Point: 97 °C (206.6 °F) to 98 °C (208.4 °F)

RTECS Toxicity Data

Acute Oral: Rat LD_{50} Dose: 71 mg/kg; Toxic Effects: Behavioral - Convulsions or effect on seizure threshold. Mouse LD_{50} Dose: 81 mg/kg.

Acute Dermal: Rat LD_{50} Route: Subcutaneous Dose: 39 mg/kg; Toxic Effects: Behavioral - Convulsions or effect on seizure threshold. Mouse LD_{50} Route: Subcutaneous Dose: 61 mg/kg.

Hazard Overviews

Carcinogenicity: IARC - Not listed; NIOSH - Not listed; NTP - Not listed; ACGIH - Not listed; OSHA - Not listed; EPA - Not listed; MAK - Not listed

Environmental

Regulations

RCRA 40CFR: Not listed
CERCLA: 40CFR 302.4: Not listed
SARA 40CFR 372.65: Not listed
SARA EHS 40CFR 355: Not listed
TSCA: Listed

DIE1600 CAS #: 111-42-2

DIETHANOLAMINE

RTECS: KL2975000
EINECS Number: 203-868-0
Molecular Formula: $C_4H_{11}NO_2$
Structured MF: $(HOCH_2CH_2)_2NH$
Formula Weight: 105.14

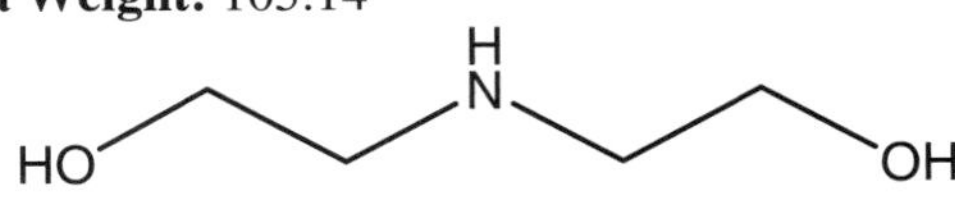

Chemical Structure

Synonyms: AMINE,DIETHYL,2,2-DIHYDROXY-; BIS(2-HYDROXYETHYL)AMINE; BIS(HYDROXYETHYL)AMINE; BIS-(2-HYDROXY)ETHYLAMINE; BIS-2-HYDROXYETHYLAMINE; DEA; DIAETHANOLAMIN; DIETHANOLAMIN; N,N-DIETHANOLAMINE; DIETHYLAMINE,2,2'-DIHYDROXY-; DIETHYLOLAMINE; 2,2'-DIHYDROXYDIETHYAMINE; 2,2'-DIHYDROXYDIETHYLAMINE; DI(2-HYDROXYETHYL)-AMINE; DI(2-HYDROXYETHYL)AMINE; DIOLAMINE; ETHANOL,2,2'-IMINOBIS-; ETHANOL,2,2'-IMINODI-; 2-((2-HYDROXYETHYL)AMINO)ETHANOL; 2-(2-HYDROXYETHYLAMINO)ETHANOL; 2,2'-IMINOBIS(ETHANOL); 2,2'-IMINOBISETHANOL; 2,2'-IMINODI-1-ETHANOL; 2,2'-IMINODIETHANOL; IMINODIETHANOL

Description: colorless solid at room temp., liquid above 82.4 F (28 C); strong ammonia odor

Use: emulsifying and dispersing agent, to solubilizing fusidic acid, for the preparation of salts of iodinated organic acids, to scrub toxic gases from smoke screen, a cation in many water soluble salts of drugs, pesticides, industrial basic solvents, as rubber chemicals, in surface active agents used in textile specialties, herbicides, petroleum demulsifiers, cosmetics, pharmaceuticals, production of lubricants for textile industry, in organic synthesis, cutting oils, shampoos, cleaners and polisher chemical intermediate for resins, plasticizers, absorbent for acid gases, solubilizing 2,4-D and humectant

Physical Properties

Boiling Point: 268.8 °C (516 °F) at 760 mm Hg
Freezing Point: 28 °C (82.4 °F)
Specific Gravity: 1.0966 at 20 °C/4 °C
Vapor Density: 3.65 Air=1
Density: 1.097 g/mL
Vapor Pressure: 5 mm Hg at 138 °C
Water Solubility: 95% by weight
Other Solubilities: miscible with Methanol; Soluble in Benzene: 4.2% at 25 °C; in Ether: 0.8% at 25 °C; in N-Heptane: less than 0.1% at 25 °C; in Carbon Tetrachloride: less than 0.1% at 25 °C.
Odor Threshold: 0.27 ppm
pH: 0.1 N aqueous solution 11
Refraction Index: 1.4753 at 30 °C/D
Critical Temperature: 442 °C
Critical Pressure: 470 psia
Flash Point: 134 °C Open Cup
Autoignition Temperature: 662 °C
LEL: 1.6% v/v
UEL: 9.8% v/v

RTECS Toxicity Data

Acute Oral: Rat LD_{50} Dose: 710 mg/kg. Mouse LD_{50} Dose: 3300 mg/kg; Toxic Effects: Behavioral - Somnolence (general depressed activity); Behavioral - Excitment; Behavioral - Muscle contraction or spasticity.

Acute Dermal: Rabbit LD_{50} Route: Skin; Dose: 12200 mg/kg. Rat LD_{50} Route: Subcutaneous Dose: 2200 mg/kg.

Chronic (Multiple Dose) Oral: Rat Dose: 52 gm/kg/14D-C; Toxic Effects: Kidney, Ureter, and Bladder - Other changes in urine composition; Blood - Normocytic anemia; Biochemical - Dehydrogenases. Rat Dose: 8827 mg/kg/13W-C; Toxic Effects: Kidney, Ureter, and Bladder - Changes in kidney weight; Blood - Normocytic anemia; DEATH - Changes in testicular weight. Rat Dose: 49 gm/kg/7W-C; Toxic Effects: Liver - Other changes; Blood - Normocytic anemia; DEATH.

Chronic (Multiple Dose) Dermal: Rat Route: Skin; Dose: 1008 mg/kg/16D-I; Toxic Effects: Kidney, Ureter, and Bladder - Other changes in urine composition; Blood - Normocytic anemia; Biochemical - Dehydrogenases. Rat Route: Skin; Dose: 8125 mg/kg/13W-I; Toxic Effects: Liver - Changes in Liver weight; Kidney, Ureter, and Bladder - Changes in kidney weight; Blood - Normocytic anemia.

Irritation Eye: Rabbit Standard Draize Test Dose: 5500 mg; Reaction: severe. Rabbit Standard Draize Test Dose: 750 ug/24H; Reaction: severe.

Irritation Skin: Rabbit Standard Draize Test Dose: 500 mg/24H; Reaction: mild. Rabbit Open Draize Test Dose: 50 mg open; Reaction: mild.

Reproductive/Teratogenic: Rat Route: Oral; Dose: 18382 mg/kg; Duration: male 14D prior to mating; Paternal Effects - Spermatogenesis.

Hazard Overviews

Corrosive

Fire Diamond

Health: Corrosive to eyes. Mildly irritating to skin/respiratory tract. Also Causes: sneezing, coughing, tearing, corneal damage; upon ingestion: GI irritation, vomiting, abdominal pain.

Fire: Will burn. Can form explosive mixtures in the air. Water spray may cause frothing, but if applied lightly, can create foaming only on the surface which will then smother fire. Alcohol-resistant foam is also effective.

Reactivity: Stable. Hazardous polymerization cannot occur. Avoid: heat; ignition sources. Incompatible with: oxidizing agents; strong acids; acid anhydrides; halides; copper; zinc; galvanized iron. Hazardous decomposition products: nitrogen oxides.

Carcinogenicity: IARC - Not listed; NIOSH - Listed as carcinogen; NTP - Not listed; ACGIH - Not listed; OSHA - Not listed; EPA - Not listed; MAK - Not listed

Primary Target Organs:

Eyes

Skin

Respiratory System

Gastro-intestinal

Exposure Limits
OSHA PEL Vacated 1989 Limits: TWA: 3 ppm; 15 mg/m^3.
ACGIH TLV: TWA: 0.46 ppm; 2 mg/m^3.
NIOSH REL: TWA: 3 ppm; 15 mg/m^3.

Respirator Recommendation
Exposure Range: >0.46 to 23 ppm Supplied Air, Constant Flow/Pressure Demand, Half Mask
Exposure Range: >23 to 460 ppm Supplied Air, Constant Flow/Pressure Demand, Full Face
Exposure Range: >460 to unlimited ppm Self-contained Breathing Apparatus, Pressure Demand, Full Face
Note: poor warning properties

Environmental

Ecotoxicity: TLm Bluegill sunfish 2100 mg/l/24 hr in tap water /Conditions of bioassay not specified LC_{50} Goldfish 800 mg/l/24 hr at pH 9.6 /Conditions of bioassay not specified TLm Mosquito fish 1800 mg/l/24 hr in turbid Oklahoma water /Conditions of bioassay not specified

Environmental Fate: In soil and water, it is expected to biodegrade fairly rapidly following acclimation (half-life on the order of days to weeks). N-Nitrosodiethanolamine is a metabolite. In soil, it should leach. In the atmosphere, it is expected to exist almost entirely in the vapor phase. Reaction with photochemically generated hydroxyl radicals is expected to be the dominant removal mechanism (half-life 4 hours). This compound may also be removed from the atmosphere in precipitation.

Cleanup/Disposal: Guide No. 171: Do not touch or walk through spilled material. Stop leak if you can do it without risk. Prevent dust cloud. Avoid inhalation of asbestos dust. Small Dry Spills: With clean shovel place material into clean, dry container and cover loosely; move containers from spill area. Small Spills: Take up with sand or other noncombustible absorbent material and place into containers for later disposal. Large Spills: Dike far ahead of liquid spill for later disposal. Cover powder spill with plastic sheet or tarp to minimize spreading. Prevent entry into waterways, sewers, basements or confined areas.

Environmental Physical Data

Henry's Law Constant: estimated at 3.9×10^{-11}
Octanol/Water Partition Coefficient: log K_{ow} = -1.43
Sorption Partition Coefficient: log K_{oc} = estimated at 4
BCF: estimated at 1
BOD: theoretical 10%, 5 days

Regulations

RCRA 40CFR: Not listed
CERCLA: 40CFR 302.4: Listed per CAA Section 112 RQ: 100 lb (45.35 kg)
SARA 40CFR 372.65: Listed
SARA EHS 40CFR 355: Not listed
TSCA: Listed

Analytical Methods

Indoor / Expired Air: NIOSH 3509

DIE1700 CAS #: 143-00-0

DIETHANOLAMINE LAURYL SULFATE

EINECS Number: 205-577-4
Molecular Formula: $C_{16}H_{37}NO_6S$
Formula Weight: 331.2
Synonyms: BIS(2-HYDROXYETHYL)AMMONIUM LAURYL SULFATE; CONDANOL DLS; DEA-LAURYL SULFATE; DODECYL SULFATE DIETHANOLAMINE SALT; LAURYL SULFATE DIETHANOLAMINE SALT; PROPASTE D; SIPON LD; STEPANOL DEA; SULFURIC ACID,MONODODECYL ESTER COMPD WITH2,2'-IMINODIETHANOL (1:1); SULFURIC ACID,MONODODECYL ESTER,COMPD WITH2,2'-IMINOBIS(ETHANOL) (1:1); TEXAPON DLS
Description: pale yellow clear liquid; mild fatty odor
Use: as surfactant in shampoos & cosmetics

Physical Properties

Specific Gravity: Solution 1.01 at 20 °C

Hazard Overviews

Carcinogenicity: IARC - Not listed; NIOSH - Not listed; NTP - Not listed; ACGIH - Not listed; OSHA - Not listed; EPA - Not listed; MAK - Not listed

Environmental

Regulations

RCRA 40CFR: Not listed
CERCLA: 40CFR 302.4: Not listed
SARA 40CFR 372.65: Not listed
SARA EHS 40CFR 355: Not listed
TSCA: Listed

DIE1800 CAS #: 38727-55-8

DIETHATYL ETHYL

RTECS: MB9200000
EINECS Number: 254-105-3
Molecular Formula: $C_{16}H_{22}ClNO_3$
Formula Weight: 311.84
Synonyms: ANTOR; BAY NNT 6867; N-(CHLOROACETYL)-N-(2,6-DIETHYLPHENYL)GLYCINE ETHYL ESTER; N-CHLOROACETYL-N-(2,6-DIETHYLPHENYL)GLYCINE ETHYL ESTER; H 22234; HERCULES 22234

Physical Properties

Freezing Point: 49 °C (120.2 °F) to 50 °C (122 °F)
Other Solubilities: 810 g/kg methanol, 810 g/kg ethanol, 750 g/kg isopropanol, 820 g/kg acetone, 820 g/kg xylene

RTECS Toxicity Data

Acute Oral: Rat LD_{50} Dose: 2300 mg/kg. Mouse LD_{50} Dose: 1650 mg/kg.
Acute Dermal: Rabbit LD_{50} Route: Skin; Dose: 4 gm/kg.

Hazard Overviews

Carcinogenicity: IARC - Not listed; NIOSH - Not listed; NTP - Not listed; ACGIH - Not listed; OSHA - Not listed; EPA - Not listed; MAK - Not listed

Environmental

Regulations

RCRA 40CFR: Not listed
CERCLA: 40CFR 302.4: Not listed
SARA 40CFR 372.65: Listed
SARA EHS 40CFR 355: Not listed
TSCA: Not listed

Analytical Methods

Food: FDA 232.4, 242.1

DIE1900 CAS #: 6175-45-7

DIETHOXYACETOPHENONE

RTECS: MD3284000
EINECS Number: 228-220-4
Molecular Formula: $C_{12}H_{16}O_3$
Formula Weight: 208.28

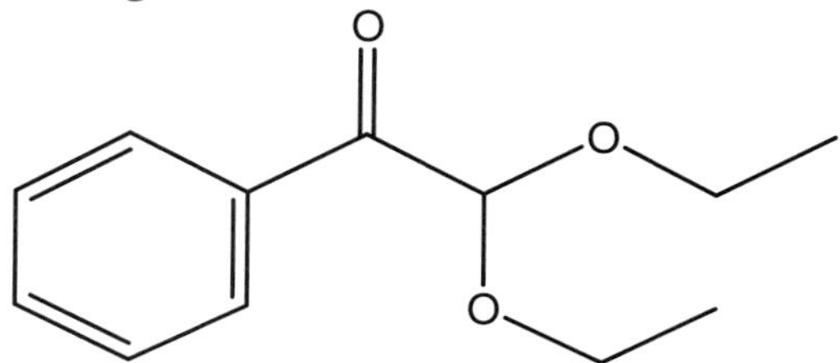

Chemical Structure

Synonyms: 2,2-DIETHOXYACETOPHENONE; ALPHA,ALPHA-DIETHOXYACETOPHENONE; ETHANONE,2,2-DIETHOXY-1-PHENYL-; PHENYLGLYOXAL DIETHYL ACETAL

Physical Properties

Boiling Point: 260 °C (500 °F)
Vapor Pressure: 0.01

RTECS Toxicity Data

Acute Oral: Rat LD_{50} Dose: 5660 uL/kg.
Acute Dermal: Rabbit LD_{50} Route: Skin; Dose: 11300 uL/kg.

Hazard Overviews

Health: Irritating to eyes/skin. Toxic. Other Acute Effects: harmful if swallowed, inhaled, or absorbed through skin. Chronic Effects: Carcinogen.
Fire: Extinguishing agents: carbon dioxide, dry chemical powder or appropriate foam. Precautions: combustible liquid.
Reactivity: Incompatible with: acids, moisture, strong oxidizing agents. Hazardous decomposition products: toxic fumes of: carbon monoxide, carbon dioxide.
Carcinogenicity: IARC - Not listed; NIOSH - Not listed; NTP - Not listed; ACGIH - Not listed; OSHA - Not listed; EPA - Not listed; MAK - Not listed
Primary Target Organs:

Eyes

Skin

Environmental

Regulations

RCRA 40CFR: Not listed
CERCLA: 40CFR 302.4: Not listed
SARA 40CFR 372.65: Not listed
SARA EHS 40CFR 355: Not listed
TSCA: Listed

DIE2200 CAS #: 141-28-6

DIETHYL ADIPATE

RTECS: AV1100000
EINECS Number: 205-477-0
Molecular Formula: $C_{10}H_{18}O_4$
Formula Weight: 202.24

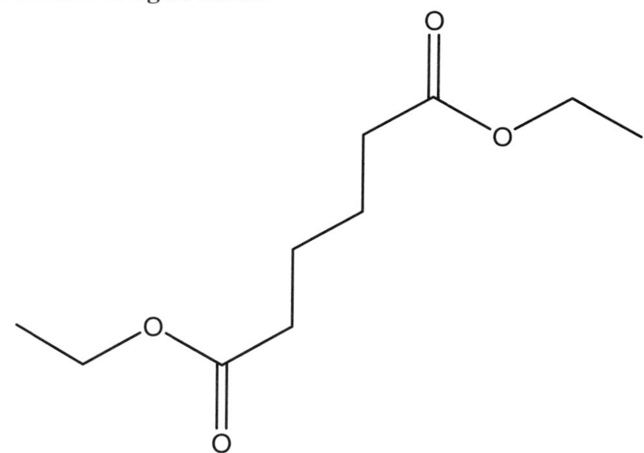

Chemical Structure

Synonyms: ADIPIC ACID,DIETHYL ESTER; DIETHYL HEXANEDIOATE; DIETHYLESTER KYSELINY ADIPOVE; ETHYL ADIPATE; ETHYL DELTA-CARBOETHOXYVALERATE; HEXANEDIOIC ACID,DIETHYL ESTER; HEXANEDIOIC ACID,DIETHYL ESTER (9CI)
Description: colorless liquid
Use: plasticizer; chem int for adipic acid esters by transesterification; plasticizer for cellulose polymers, eg, cellulose acetate; chem int for cyclopentanone; chem int for putrescine (neuroregulator)

Physical Properties

Boiling Point: 245 °C (473 °F)
Freezing Point: -19.8 °C (-3.64 °F)
Specific Gravity: 1.009 at 20 °C
Vapor Pressure: 0.118 kPa at 75 °C
Water Solubility: Insoluble in Water
Other Solubilities: Soluble in Ether.
Refraction Index: 1.426 at 25 °C

RTECS Toxicity Data

Acute Oral: Mouse LD_{50} Dose: 8100 mg/kg.
Reproductive/Teratogenic: Rat Route: Intraperitoneal; Dose: 837 mg/kg; Duration: female 5-15D of pregnancy; Effects on Embryo or Fetus - Fetal death; Specific Developmental Abnormalities - Other developmental abnormalities.
Mutagenic: Mouse Dominant Lethal Test; Route: Intraperitoneal; Dose: 1100 mg/kg.

Hazard Overviews

Health: May cause irritation. Harmful. Other Acute Effects: harmful if swallowed, inhaled, or absorbed through skin;

possible risk of irreversible effects. Chronic Effects: laboratory experiments have shown mutagenic effects.

Fire: Hazards: emits toxic fumes. Extinguishing agents: water spray; carbon dioxide, dry chemical powder or appropriate foam. Precautions: combustible liquid.

Reactivity: Incompatible with: strong oxidizing agents. Hazardous decomposition products: toxic fumes of: carbon monoxide, carbon dioxide.

Carcinogenicity: IARC - Not listed; NIOSH - Not listed; NTP - Not listed; ACGIH - Not listed; OSHA - Not listed; EPA - Not listed; MAK - Not listed

Environmental

Regulations

RCRA 40CFR: Not listed
CERCLA: 40CFR 302.4: Not listed
SARA 40CFR 372.65: Not listed
SARA EHS 40CFR 355: Not listed
TSCA: Listed

DIE2300 CAS #: 105-58-8

DIETHYL CARBONATE

RTECS: FF9800000
EINECS Number: 203-311-1
Molecular Formula: $C_5H_{10}O_3$
Structured MF: $(CH_3CH_2)_2CO_3$
Formula Weight: 118.30

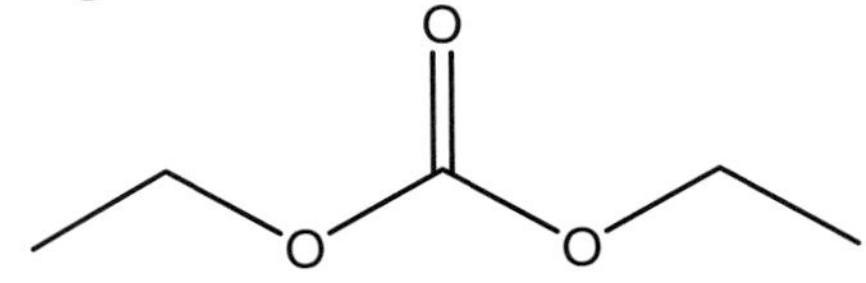

Chemical Structure

Synonyms: CARBONIC ACID,DIETHYL ESTER; CARBONIC ETHER; DEC; DIAETHYLCARBONAT; DIATOL; DIETHYLESTER KYSELINY UHLICITE; DIETHYLKARBONAT; ETHOXYFORMIC ANHYDRIDE; ETHYL CARBONATE; EUFIN

Description: colorless liquid; pleasant ethereal odor

Use: solvent for nitrocellulose, cellulose ethers and many synthetic and natural resins; organic synthesis; adhering rare earths to cathodes

Physical Properties

Boiling Point: 126 °C (259 °F)
Freezing Point: -43 °C (-45.4 °F)
Specific Gravity: 0.9752 at 20 °C/4 °C
Vapor Density: 4.1 Air=1
Saturated Vapor Density: 1.259458352 kg/m^3
Density: 0.98 g/mL
Vapor Pressure: 1.63 kPa at 25 °C
Water Solubility: Insoluble in Water
Other Solubilities: 95% Ethanol: Soluble; Ether: Soluble; hydrocarbons and castor oil.
Surface Tension: 26.3 dynes/cm at 20 °C
pH: Neutral
Refraction Index: 1.3845 at 20 °C/D
Flash Point: 25 °C Closed Cup
Autoignition Temperature: 240 °C

RTECS Toxicity Data

Acute Oral: Rat LD_{Lo} Dose: 15 gm/kg.

Acute Dermal: Rat LD_{50} Route: Subcutaneous Dose: 8500 mg/kg.

Reproductive/Teratogenic: Hamster Route: Intraperitoneal; Dose: 496 mg/kg; Duration: female 8D of pregnancy; Specific Developmental Abnormalities - Central nervous system; Eye, ear; Musculoskeletal system. Hamster Route: Intraperitoneal; Dose: 496 mg/kg; Duration: female 8D of pregnancy; Effects on Embryo or Fetus - Fetotoxicity. Hamster Route: Intraperitoneal; Dose: 1004 mg/kg; Duration: female 8D of pregnancy; Specific Developmental Abnormalities - Craniofacial (including nose and tongue); Body wall.

Tumorigenic: Mouse Route: Oral; Dose: 500 mg/kg; Toxic Effects: Tumorigenic - Equivocal tumorigenic agent by RTECS criteria; Lungs, Thorax, or Respiration - Tumors; Skin and appendages - Tumors. Mouse Route: Intraperitoneal; Dose: 456 mg/kg; Toxic Effects: Tumorigenic - Equivocal tumorigenic agent by RTECS criteria; Lungs, Thorax, or Respiration - Tumors; Skin and appendages - Tumors.

Hazard Overviews

Flammable

Fire Diamond

Health: Irritating to eyes/skin/respiratory tract. Also Causes: headache, dizziness, nausea, weakness, loss of consciousness.

Fire: Flammable. Use foam, carbon dioxide, or dry chemical. Water may be ineffective. Use water spray to cool fire-exposed containers. Containers may explode in the heat of the fire.

Reactivity: Stable. Hazardous polymerization cannot occur. Avoid: heat; ignition sources. Incompatible with: oxidizing materials. Hazardous decomposition products: carbon oxides; acrid smoke; irritating fumes.

Carcinogenicity: IARC - Not listed; NIOSH - Not listed; NTP - Not listed; ACGIH - Not listed; OSHA - Not listed; EPA - Not listed; MAK - Not listed

Primary Target Organs:

Eyes

Skin

Respiratory System

Environmental

Cleanup/Disposal: Guide No. 127: Eliminate all ignition sources (no smoking, flares, sparks or flames in immediate area). All equipment used when handling the product must be grounded. Do not touch or walk through spilled material. Stop leak if you can do it without risk. Prevent entry into waterways, sewers, basements or confined areas. A vapor suppressing foam may be used to reduce vapors. Absorb or cover with dry earth, sand or other non-combustible material and transfer to containers. Use clean non-sparking tools to collect absorbed material. Large Spills: Dike far ahead of liquid spill for later disposal. Water spray may reduce vapor; but may not prevent ignition in closed spaces.

Environmental Physical Data

BCF: no food chain concentration potential

Regulations

RCRA 40CFR: Not listed
CERCLA: 40CFR 302.4: Not listed
SARA 40CFR 372.65: Not listed
SARA EHS 40CFR 355: Not listed
TSCA: Listed

DIE2500 CAS #: 814-49-3

DIETHYL CHLOROPHOSPHATE

RTECS: TD1400000
EINECS Number: 212-396-4
Molecular Formula: $C_4H_{10}ClO_3P$
Formula Weight: 172.55

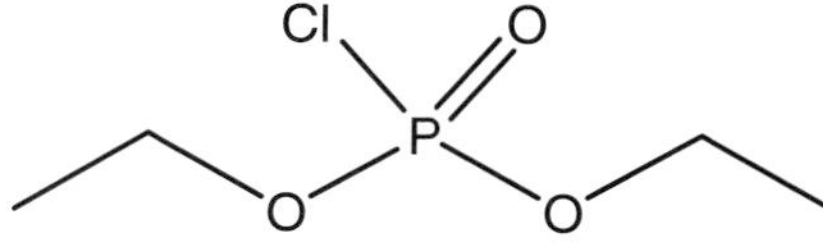

Chemical Structure

Synonyms: CHLOROPHOSPHORIC ACID,DIETHYL ESTER; DIETHOXYPHOSPHORUS OXYCHLORIDE; DIETHYLCHLORFOSFAT; PHOSPHOROCHLORIDIC ACID,DIETHYL ESTER
Description: water-white liquid
Use: intermediate in organic synthesis

Physical Properties

Boiling Point: 60 °C (140 °F) at 2 mm Hg
Specific Gravity: 1.1915 at 25 °C
Vapor Density: 5.94
Other Solubilities: Soluble in Alcohols
Refraction Index: 1.4153 at 25 °C
Flash Point: Combustible

RTECS Toxicity Data

Acute Oral: Rat LD_{50} Dose: 11 mg/kg. Mouse LD_{50} Dose: 100 mg/kg; Toxic Effects: Behavioral - Somnolence (general depressed activity); Behavioral - Convulsions or effect on seizure threshold; Skin and appendages - Hair.
Acute Dermal: Rabbit LD_{50} Route: Skin; Dose: 7900 ug/kg.
Irritation Skin: Rabbit Standard Draize Test Dose: 500 mg/24H; Reaction: mild. Rabbit Open Draize Test Dose: 10 mg/24H open; Reaction: mild.

Hazard Overviews

Poison

Corrosive

Health: Corrosive to eyes/skin/respiratory tract. Poison. Other Acute Effects: may be fatal if inhaled, swallowed, or absorbed through skin; cholinesterase inhibitor; inhalation may result in spasm, inflammation and edema of the larynx and bronchi, chemical pneumonitis and pulmonary edema; symptoms of exposure may include burning sensation; coughing; wheezing; laryngitis; shortness of breath; headache; nausea; vomiting; initial hydrolysis product formed is the highly toxic tetraethylpyro-phosphate.
Fire: Combustible. Hazards: water hydrolyzes material liberating acidic gas which in contact with metal surfaces can generate flammable and/or explosive hydrogen gas; emits toxic fumes. Extinguishing agents: carbon dioxide, dry chemical powder or appropriate foam; do not use water. Precautions: combustible liquid.
Reactivity: Incompatible with: strong bases, strong oxidizing agents, may decompose on exposure to moist air or water. Hazardous decomposition products: toxic fumes of: carbon monoxide, carbon dioxide, hydrogen chloride gas; thermal decomposition may produce toxic fumes of phosphorus oxides and/or phosphine.
Carcinogenicity: IARC - Not listed; NIOSH - Not listed; NTP - Not listed; ACGIH - Not listed; OSHA - Not listed; EPA - Not listed; MAK - Not listed
Primary Target Organs:

Eyes

Skin

Respiratory System

Environmental

Regulations

RCRA 40CFR: Not listed
CERCLA: 40CFR 302.4: Not listed
SARA 40CFR 372.65: Not listed TPQ: 500 lb
SARA EHS 40CFR 355: Listed TPQ: 500 lb
TSCA: Listed

DIE2600 CAS #: 2524-04-1

O,O-DIETHYL CHLOROTHIOPHOSPHONATE

RTECS: TD1780000
EINECS Number: 219-755-4
Molecular Formula: $C_4H_{10}ClO_2PS$
Formula Weight: 188.62

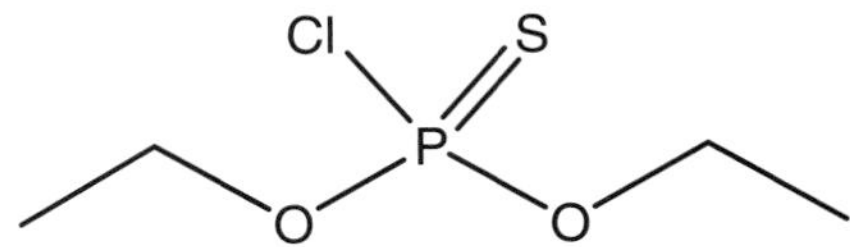

Chemical Structure

Synonyms: DIETHOXYTHIOPHOSPHORYL CHLORIDE; O,O-DIETHYL CHLORIDOPHOSPHOROTHIOATE; O,O-DIETHYL CHLORIDOTHIONOPHOSPHATE; O,O-DIETHYL CHLOROTHIONOPHOSPHATE; DIETHYL CHLOROTHIOPHOSPHATE; O,O-DIETHYL CHLOROTHIOPHOSPHATE; DIETHYL PHOSPHOROCHLORIDOTHIOATE; O,O-DIETHYL PHOSPHOROCHLORIDOTHIOATE; DIETHYL PHOSPHOROCHLORIDOTHIONATE; DIETHYL PHOSPHOROCHLOROTHIOATE; O,O-DIETHYL PHOSPHOROCHLOROTHIOATE; DIETHYL PHOSPHOROTHIOCHLORIDATE; O,O-DIETHYL PHOSPHOROTHIOCHLORIDATE; DIETHYL PHOSPHOROTHIONOCHLORIDATE; O,O-DIETHYL PHOSPHOROTHIONOCHLORIDATE; O,O-DIETHYL THIONOPHOSPHORIC ACID ESTER CHLORIDE; O,O-DIETHYL THIONOPHOSPHORIC CHLORIDE; O,O-DIETHYL THIONOPHOSPHOROCHLORIDATE; O,O-DIETHYL THIONOPHOSPHORYL CHLORIDE; DIETHYL THIOPHOSPHORIC CHLORIDE; DIETHYL THIOPHOSPHORYL CHLORIDE; O,O-DIETHYL THIOPHOSPHORYL CHLORIDE; DIETHYLCHLOROTHIOPHOSPHATE; DIETHYLCHLORTHIOFOSFAT; O,O-DIETHYLPHOSPHOROCHLORIDOTHIOATE; O,O-DIETHYLTHIOPHOSPHOROCHLORIDATE; DIETHYLTHIOPHOSPHORYL CHLORIDE; ETHYL PCT; PHOSPHONOTHIOIC ACID,CHLORO-,O,O-DIETHYL ESTER; PHOSPHOROCHLORIDOTHIOIC ACID,O,O-DIETHYL ESTER

Description: colorless to light amber liquid

Use: intermediate for pesticides; oil & gasoline additives; flame retardants; flotation agents; chem int for parathion

Physical Properties

Boiling Point: 49 °C (120 °F) at <1 mm Hg
Freezing Point: < -75 °C (-103 °F)
Specific Gravity: 1.196 at 25 °C/25 °C
Water Solubility: Insoluble in Water
Other Solubilities: Soluble in most organic solvents
Refraction Index: 1.4705 at 25 °C/D

RTECS Toxicity Data

Acute Oral: Rat LD_{50} Dose: 1340 mg/kg; Toxic Effects: Sense organs and special senses - Lacrimation; Gastrointestinal - Changes in structure or function of salivary glands; Blood - Hemorrhage. Mouse LD_{50} Dose: 800 mg/kg.
Acute Inhalation: Rat LC_{50} Dose: 20 ppm/4hr. Mouse LC_{50} Dose: 725 mg/m^3/2hr.
Acute Dermal: Rabbit LD_{Lo} Route: Skin; Dose: 250 mg/kg.

Hazard Overviews

Poison

Corrosive

Health: Corrosive to eyes/skin/respiratory tract. Poison. Other Acute Effects: may be fatal if inhaled; harmful if swallowed or absorbed through skin; inhibits cholinesterase; inhalation may result in spasm, inflammation and edema of the larynx and bronchi, chemical pneumonitis and pulmonary edema; symptoms of exposure may include burning sensation; coughing; wheezing; laryngitis; shortness of breath; headache; nausea; vomiting; target organ: lungs.
Fire: Hazards: emits toxic fumes. Extinguishing agents: carbon dioxide, dry chemical powder or appropriate foam. Precautions: combustible liquid.
Reactivity: Incompatible with: strong oxidizing agents, may decompose on exposure to moist air or water, strong bases. Hazardous decomposition products: toxic fumes of: carbon monoxide, carbon dioxide, sulfur oxides, hydrogen chloride gas; thermal decomposition may produce toxic fumes of phosphorus oxides and/or phosphine.
Carcinogenicity: IARC - Not listed; NIOSH - Not listed; NTP - Not listed; ACGIH - Not listed; OSHA - Not listed; EPA - Not listed; MAK - Not listed
Primary Target Organs:

Eyes

Skin

Respiratory System

Environmental

Regulations

RCRA 40CFR: Not listed
CERCLA: 40CFR 302.4: Not listed
SARA 40CFR 372.65: Not listed
SARA EHS 40CFR 355: Not listed
TSCA: Listed

DIE2700 **CAS #: 2781-11-5**

DIETHYL ((DIETHANOLAMINO)METHYL)PHOSPHONATE

RTECS: SZ6850000
EINECS Number: 220-482-8
Molecular Formula: $C_9H_{22}NO_5P$
Formula Weight: 255.25

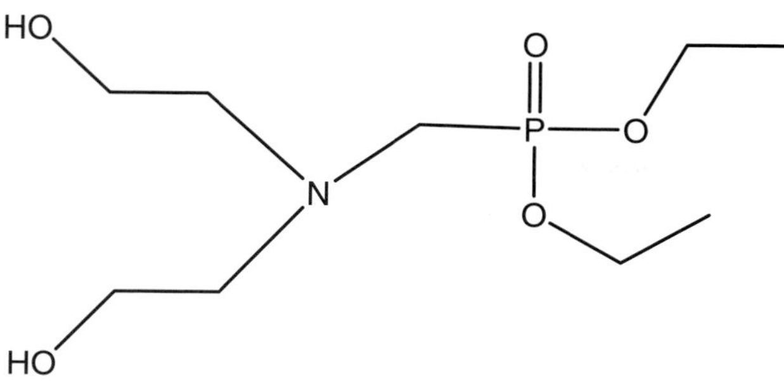

Chemical Structure

Synonyms: ADEKA FC 450; DIETHYL (N,N-BIS(2-HYDROXYETHYL)AMINO)METHANEPHOSPHONATE; DIETHYL ((BIS(2-HYDROXYETHYL)AMINO)METHYL)PHOSPHONATE; DIETHYL ((N,N-BIS(2-HYDROXYETHYL)AMINO)METHYL)PHOSPHONATE; FC 450; FYROL 6; PHOSPHONIC ACID,((BIS(2-HYDROXYETHYL)AMINO)METHYL)-,DIETHYL ESTER
Use: reactive flame retardant for rigid urethane foams

RTECS Toxicity Data

Acute Oral: Rat LD; Dose: >5 gm/kg. Rabbit LD; Dose: >5 gm/kg.
Acute Dermal: Rabbit LD; Route: Skin; Dose: >5 gm/kg. Rat LD; Route: Skin; Dose: >5 gm/kg.

Hazard Overviews

Carcinogenicity: IARC - Not listed; NIOSH - Not listed; NTP - Not listed; ACGIH - Not listed; OSHA - Not listed; EPA - Not listed; MAK - Not listed

Environmental

Regulations

RCRA 40CFR: Not listed
CERCLA: 40CFR 302.4: Not listed
SARA 40CFR 372.65: Not listed
SARA EHS 40CFR 355: Not listed
TSCA: Listed

DIE2800 CAS #: 78-38-6

DIETHYL ETHYLPHOSPHONATE

EINECS Number: 201-111-9
Molecular Formula: $C_6H_{15}O_3P$
Formula Weight: 166.16

Chemical Structure

Synonyms: ETHANEPHOSPHONIC ACID,DIETHYL ESTER; PHOSPHONIC ACID,ETHYL-,DIETHYL ESTER
Description: colorless liquid; sweet odor
Use: heavy metal extraction; gasoline additives; antifoam agent; plasticizer textile conditioner and antistatic agent

Physical Properties

Boiling Point: 198 °C (388 °F) at 760 mm Hg
Freezing Point: 198 °C (388.4 °F)
Specific Gravity: 1.0259 at 20 °C/4 °C
Vapor Density: 5.73 Air=1
Density: 1.03 g/mL
Water Solubility: Slightly Soluble in Water
Other Solubilities: 95% Ethanol: Soluble; Ether: Soluble.
Refraction Index: 1.4163 at 20 °C/D
Flash Point: 105 °C Open Cup

RTECS Toxicity Data

Hazard Overviews
Fire: Will burn.
Carcinogenicity: IARC - Not listed; NIOSH - Not listed; NTP - Not listed; ACGIH - Not listed; OSHA - Not listed; EPA - Not listed; MAK - Not listed

Environmental

Regulations

RCRA 40CFR: Not listed
CERCLA: 40CFR 302.4: Not listed
SARA 40CFR 372.65: Not listed
SARA EHS 40CFR 355: Not listed
TSCA: Listed

DIE2900 CAS #: 623-91-6

DIETHYL FUMARATE

RTECS: EM5950000
EINECS Number: 210-819-7
Molecular Formula: $C_8H_{12}O_4$
Formula Weight: 172.20

Chemical Structure

Synonyms: 2-BUTENEDIOIC ACID (E)-,DIETHYL ESTER; TRANS-2-BUTENEDIOIC ACID DIETHYL ESTER; 2-BUTENEDIOIC ACID,DIETHYL ESTER,(E)-; DIETHYLESTER KYSELINY FUMAROVE; ETHYL FUMARATE; FUMARIC ACID,DIETHYL ESTER
Description: liquid
Use: comonomer for polymers with styrene & vinyl chloride

Physical Properties

Boiling Point: 218.5 °C (425 °F)
Freezing Point: 0.6 °C (33.08 °F)
Specific Gravity: 1.0529
Vapor Density: 5.93 Air=1
Vapor Pressure: 1 mm Hg at 53.2 °C
Water Solubility: Slightly Soluble in Water
Other Solubilities: Acetone: Soluble; Chloroform: Soluble
Refraction Index: 1.4412
Flash Point: 104 °C Closed Cup

RTECS Toxicity Data

Acute Oral: Rat LD_{50} Dose: 1780 mg/kg.

Hazard Overviews

Fire Diamond

Health: Irritating to eyes/skin/respiratory tract. Other Acute Effects: may be harmful by inhalation, ingestion, or skin absorption; may cause allergic skin reaction; strong sensitizer.
Fire: Will burn. Extinguishing agents: carbon dioxide, dry chemical powder or appropriate foam. Precautions: combustible liquid.
Reactivity: Incompatible with: acids, bases, oxidizing agents, reducing agents. Hazardous decomposition products: toxic fumes of: carbon monoxide, carbon dioxide.
Carcinogenicity: IARC - Not listed; NIOSH - Not listed; NTP - Not listed; ACGIH - Not listed; OSHA - Not listed; EPA - Not listed; MAK - Not listed
Primary Target Organs:

Eyes

Skin

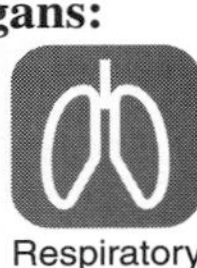
Respiratory System

Environmental

Ecotoxicity: Fishes: eastern mud minnow (Umbra pygmaeo) 96h LC_{50},S8.5 mg/l 96h LC_{50},F 4.2 mg/l

Regulations

RCRA 40CFR: Not listed
CERCLA: 40CFR 302.4: Not listed
SARA 40CFR 372.65: Not listed
SARA EHS 40CFR 355: Not listed
TSCA: Listed

DIE3000 CAS #: 762-04-9

DIETHYL HYDROGEN PHOSPHITE

RTECS: TG7875000
EINECS Number: 212-091-6
Molecular Formula: $C_4H_{11}O_3P$
Formula Weight: 171.049

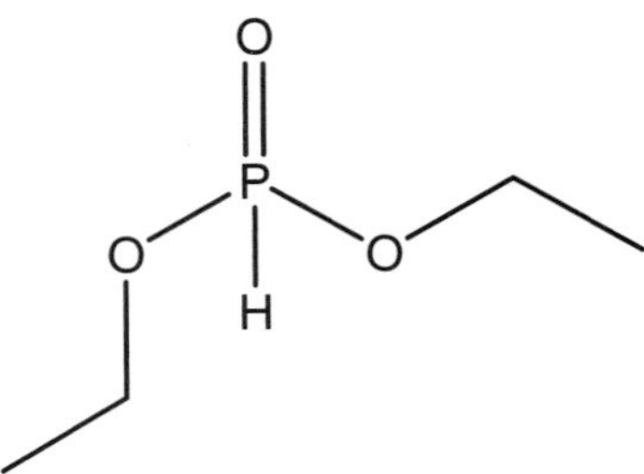

Chemical Structure

Synonyms: DIETHYL ACID PHOSPHITE; DIETHYL PHOSPHITE; DIETHYL PHOSPHONATE; DIETHYLFOSFIT; HYDROGEN DIETHYL PHOSPHITE; PHOSPHONIC ACID,DIETHYL ESTER
Use: textile finishing agent; antioxidant; additive for adhesives & for extreme pressure lubricant; chem int for organic phosphorus compounds

Physical Properties

Boiling Point: 50.5 °C (122.9 °F)
Freezing Point: 90 °C (194 °F)
Specific Gravity: 1.072
Water Solubility: Soluble
Other Solubilities: Carbon Tetrachloride: Soluble
Refraction Index: 1.4061
Ionization Potential (eV): 10.31
Flash Point: 90.5 °C Cleveland Open Cup

RTECS Toxicity Data

Acute Oral: Rat LD_{50} Dose: 3900 mg/kg.
Acute Dermal: Rabbit LD_{50} Route: Skin; Dose: 2020 uL/kg.

Hazard Overviews

Health: Severely irritating to eyes/skin/respiratory tract. Other Acute Effects: harmful if swallowed, inhaled, or absorbed through skin; depending on the intensity and duration of exposure, effects may vary from mild irritation to severe destruction of tissue; exposure can cause nausea; dizziness; headache; target organs: liver, kidneys, bladder, brain.
Fire: Combustible. Hazards: emits toxic fumes. Extinguishing agents: water spray; carbon dioxide, dry chemical powder or appropriate foam. Precautions: combustible liquid.
Reactivity: Incompatible with: strong oxidizing agents, may decompose on exposure to moist air or water, strong bases, strong acids, acid chlorides. Hazardous decomposition products: toxic fumes of: carbon monoxide, carbon dioxide; thermal decomposition may produce toxic fumes of phosphorus oxides and/or phosphine.
Carcinogenicity: IARC - Not listed; NIOSH - Not listed; NTP - Not listed; ACGIH - Not listed; OSHA - Not listed; EPA - Not listed; MAK - Not listed

Primary Target Organs:

Eyes

Skin

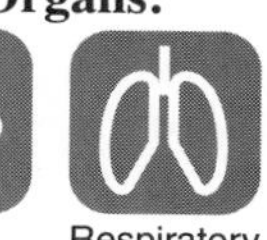
Respiratory System

Nervous System

Liver

Kidneys

Environmental

Ecotoxicity: Algae: Scenedesmus deterioration 250 mg/l; Crustaceans: Daphnia 48h LC_{50} 25 mg/l

Environmental Fate: In soil, it will probably hydrolyze and leach into groundwater (where it will completely hydrolyze). In water, it will hydrolyze but will not evaporate readily nor sorb to sediments or biota. It is not known whether it biodegrades. Photolysis may occur, although no rates are available.

Environmental Physical Data

BCF: will not bioconcentrate
BOD: 0% BODT, 5 days

Regulations

RCRA 40CFR: Not listed
CERCLA: 40CFR 302.4: Not listed
SARA 40CFR 372.65: Not listed
SARA EHS 40CFR 355: Not listed
TSCA: Listed

DIE3100 CAS #: 636-53-3

DIETHYL ISOPHTHALATE

RTECS: NT2500000
EINECS Number: 211-260-1
Molecular Formula: $C_{12}H_{14}O_4$
Formula Weight: 222.24
Synonyms: 1,3-BENZENEDICARBOXYLIC ACID,DIETHYL ESTER; ISOPHTHALIC ACID,DIETHYL ESTER
Use: plasticizer

Physical Properties

Boiling Point: 302 °C (576 °F) at 760 mm Hg
Freezing Point: 11.5 °C (52.7 °F)
Specific Gravity: 1.1239 at 17 °C/4 °C
Water Solubility: Insoluble in Water
Refraction Index: 1.508 at 18 °C/D

Hazard Overviews

Carcinogenicity: IARC - Not listed; NIOSH - Not listed; NTP - Not listed; ACGIH - Not listed; OSHA - Not listed; EPA - Not listed; MAK - Not listed

Environmental

Regulations

RCRA 40CFR: Not listed
CERCLA: 40CFR 302.4: Not listed
SARA 40CFR 372.65: Not listed
SARA EHS 40CFR 355: Not listed
TSCA: Listed

DIE3200 CAS #: 3288-58-2

O,O-DIETHYL S-METHYL DITHIOPHOSPHATE

RTECS: TD9670000
Molecular Formula: $C_5H_{13}O_2PS_2$
Formula Weight: 200.27
Synonyms: O,O-DIETHYL-S-METHYL-DITHIOPHOSPHATE; METHYL ETHYL PHOSPHORODITHIOATE (ETO)-2(MES)PS; PHOSPHORODITHIOIC ACID,O,O-DIETHYL S-METHYL ESTER

RTECS Toxicity Data

Acute Oral: Mouse LD_{50} Dose: 156 mg/kg.

Hazard Overviews

Carcinogenicity: IARC - Not listed; NIOSH - Not listed; NTP - Not listed; ACGIH - Not listed; OSHA - Not listed; EPA - Not listed; MAK - Not listed

Environmental

Regulations

RCRA 40CFR: Listed Hazardous Waste No. U087 Toxic Waste
CERCLA: 40CFR 302.4: Listed per RCRA Section 3001 RQ: 5000 lb (2268 kg)
SARA 40CFR 372.65: Not listed
SARA EHS 40CFR 355: Not listed
TSCA: Not listed

DIE3300 CAS #: 95-92-1

DIETHYL OXALATE

RTECS: RO2800000
EINECS Number: 202-464-1
Molecular Formula: $C_6H_{10}O_4$
Formula Weight: 146.14

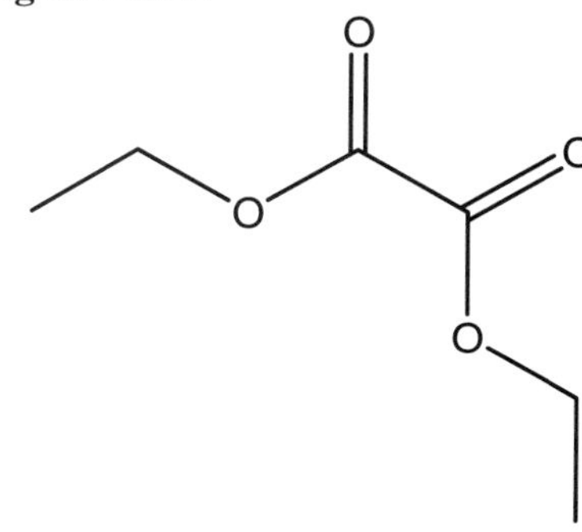

Chemical Structure

Synonyms: DIETHYL ETHANEDIOATE; DIETHYLESTER KYSELINY STAVELOVE; ETHANEDIOIC ACID,DIETHYL ESTER; ETHYL

OXALATE; OXALIC ACID DIETHYL ESTER; OXALIC ACID,DIETHYL ESTER; OXALIC ETHER

Description: colorless, oily liquid; aromatic odor

Use: manufacture of phenobarbital, ethylbenzyl malonate, triethylamine, & similar chemicals; manufacture of plastics, dyestuff intermediates; solvent for cellulose esters; organic syntheses, esp in mfr of pharmaceuticals; solvent for ethers & resins; radio tube cathode fixing lacquers; solvent for perfumes & for natural & synthetic resins (cellulose esters); agent in radio tube cathode fixing lacquers; chem int for pharmaceuticals, plastics & dyes

Physical Properties

Boiling Point: 185.7 °C (366 °F) at 760 mm Hg
Freezing Point: -38.5 °C (-37.3 °F)
Specific Gravity: 1.0785 at 20 °C/4 °C
Vapor Density: 5.04
Bulk Density: ~ 8.96 lbs/gal at 20 °C
Vapor Pressure: 1 mm Hg at 47 °C
Water Solubility: Sparingly Soluble in Water
Other Solubilities: Soluble in all proportions in Alcohol, Ether, Acetone; miscible with Ethyl Acetate.
Odor Threshold: Recognition 0.5 mg/m^3
Refraction Index: 1.4101 at 20 °C/D
Ionization Potential (eV): 9.8 +/-2.0
Flash Point: 75.556 °C Closed Cup

RTECS Toxicity Data

Acute Oral: Mouse LD_{50} Dose: 2 gm/kg; Toxic Effects: Behavioral - Somnolence (general depressed activity); Behavioral - Antipsychotic.
Irritation Skin: Guinea Pig Standard Draize Test Dose: 500 mg/24H; Reaction: mild.

Hazard Overviews

Health: Severely irritating. Harmful. Other Acute Effects: harmful by inhalation, in contact with skin and if swallowed; high concentrations are extremely destructive to tissues of the mucous membranes and upper respiratory tract, eyes and skin; symptoms of exposure may include burning sensation, coughing, wheezing, laryngitis, shortness of breath, headache, nausea and vomiting.
Fire: Combustible. Extinguishing agents: carbon dioxide, dry chemical powder or appropriate foam. Precautions: combustible liquid.
Carcinogenicity: IARC - Not listed; NIOSH - Not listed; NTP - Not listed; ACGIH - Not listed; OSHA - Not listed; EPA - Not listed; MAK - Not listed
Primary Target Organs:

Eyes

Skin

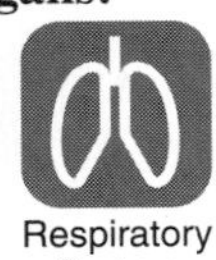
Respiratory System

Environmental

Ecotoxicity: Algae: Microcystis aeruginosa 8d EC_0 9 mg/l

Regulations

RCRA 40CFR: Not listed
CERCLA: 40CFR 302.4: Not listed
SARA 40CFR 372.65: Not listed
SARA EHS 40CFR 355: Not listed
TSCA: Listed

DIE3400 **CAS #: 298-06-6**

O,O-DIETHYL PHOSPHORODITHIOATE

RTECS: TD7350000
EINECS Number: 206-055-9
Molecular Formula: $C_4H_{11}O_2PS_2$
Formula Weight: 186.24

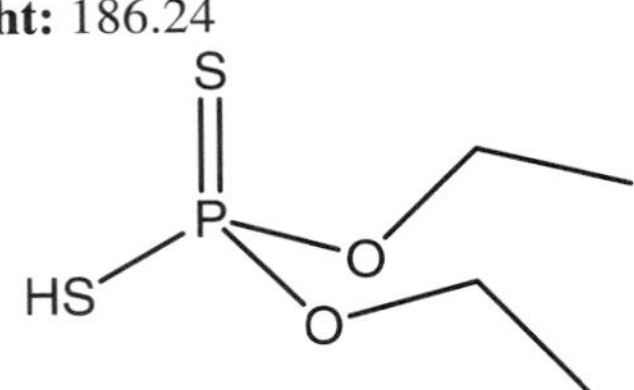

Chemical Structure

Synonyms: O,O'-DIETHYL DITHIOPHOSPHATE; O,O-DIETHYL DITHIOPHOSPHATE; O,O-DIETHYL DITHIOPHOSPHORIC ACID; O,O'-DIETHYL HYDROGEN DITHIOPHOSPHATE; O,O-DIETHYL HYDROGEN PHOSPHORODITHIOATE; O,O-DIETHYL PHOSPHORODITHIOIC ACID; O,O-DIETHYL-S-HYDROGEN PHOSPHORODITHIOATE; KYSELINA O,O-DIETHYLDITHIOFOSFORECNA; PHOSPHORDITHIOIC ACID,O,O-DIETHYL ESTER

Physical Properties

Water Solubility: Soluble

RTECS Toxicity Data

Acute Oral: Rat LD_{50} Dose: 4510 mg/kg.
Irritation Eye: Rabbit Standard Draize Test Dose: 50 ug/24H; Reaction: severe.
Irritation Skin: Rabbit Standard Draize Test Dose: 500 mg/24H; Reaction: mild.

Hazard Overviews

Poison

Corrosive

Health: Corrosive to eyes/skin/respiratory tract. Poison. Other Acute Effects: may be fatal if inhaled, swallowed, or absorbed through skin; inhalation may result in spasm, inflammation and edema of the larynx and bronchi, chemical pneumonitis and pulmonary edema; symptoms of exposure may include burning sensation; coughing; wheezing; laryngitis; shortness of breath; headache; nausea; vomiting; exposure can cause damage to the eyes; nausea; headache;

vomiting. Chronic Effects: will readily react with a range of compounds to form choline esterase inhibitors.

Fire: Hazards: water hydrolyzes material liberating acidic gas which in contact with metal surfaces can generate flammable and/or explosive hydrogen gas; emits toxic fumes. Extinguishing agents: carbon dioxide, dry chemical powder or appropriate foam; do not use water. Precautions: combustible liquid.

Reactivity: Incompatible with: strong bases, strong acids, strong oxidizing agents. Hazardous decomposition products: toxic fumes of: carbon monoxide, carbon dioxide, sulfur oxides; thermal decomposition may produce toxic fumes of phosphorus oxides and/or phosphine; hydrogen sulfide gas.

Carcinogenicity: IARC - Not listed; NIOSH - Not listed; NTP - Not listed; ACGIH - Not listed; OSHA - Not listed; EPA - Not listed; MAK - Not listed

Primary Target Organs:

Eyes

Skin

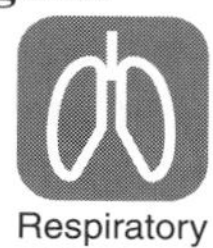
Respiratory System

Environmental

Regulations

RCRA 40CFR: Not listed
CERCLA: 40CFR 302.4: Not listed
SARA 40CFR 372.65: Not listed
SARA EHS 40CFR 355: Not listed
TSCA: Listed

DIE3600 CAS #: 64-67-5

DIETHYL SULFATE

RTECS: WS7875000
DOT: UN1594; IMO6.1
EINECS Number: 200-589-6
Molecular Formula: $C_4H_{10}O_4S$
Structured MF: $(C_2H_5)_2SO_4$
Formula Weight: 154.18

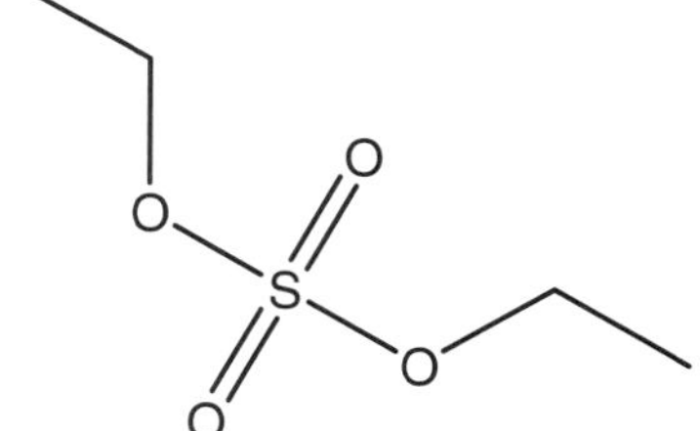

Chemical Structure

Synonyms: DIAETHYLSULFAT; DIETHYL MONOSULFATE; DIETHYL SULPHATE; DIETHYL TETRAOXOSULFATE; DIETHYLESTER KYSELINY SIROVE; DS; ETHYL SULFATE; SULFURIC ACID DIETHYL ESTER

Description: colorless oily liquid; faint ethereal/peppermint odor

Use: ethylating agent; as an accelerator in sulfation of ethylene; in some sulfonations; chem int for ethyl derivatives of phenols, amines, & thiols & as an alkylating agent

Physical Properties

Boiling Point: 96 °C (205 °F) at 15 mm Hg
Freezing Point: -24.5 °C (-12.1 °F)
Specific Gravity: 1.1774 at 20 °C/4 °C
Vapor Density: 5.31 Air=1
Saturated Vapor Density: Extrapolated 1.201976526 kg/m^3
Vapor Pressure: Extrapolated 0.29 mm Hg at 25 °C
Water Solubility: Gradually Decomposed by Water
Other Solubilities: Soluble with Alcohol, Ether
Surface Tension: 33.5 dynes/cm at 20 °C
pH: Acid
Refraction Index: 1.4004 at 20 °C/D
Flash Point: 104.444 °C Closed Cup
Autoignition Temperature: 436 °C
LEL: 4.1% v/v

RTECS Toxicity Data

Acute Oral: Rat LD_{50} Dose: 880 mg/kg. Mouse LD_{50} Dose: 647 mg/kg.

Acute Inhalation: Rat LC_{Lo} Dose: 250 ppm/4hr.

Acute Dermal: Rabbit LD_{50} Route: Skin; Dose: 600 uL/kg. Rat LD_{50} Route: Subcutaneous Dose: 350 mg/kg.

Irritation Eye: Rabbit Standard Draize Test Dose: 2 mg open; Reaction: severe.

Irritation Skin: Rabbit Open Draize Test Dose: 10 mg/24H open; Reaction: severe. Rabbit Open Draize Test Dose: 500 mg open; Reaction: mild.

Reproductive/Teratogenic: Rat Route: Intravenous; Dose: 340 mg/kg; Duration: female 15D of pregnancy; Effects on Embryo or Fetus - Fetal death.

Mutagenic: Human DNA Damage; Cell Type: HeLa cell; Dose: 20 mg. Human DNA Inhibition; Cell Type: HeLa cell; Dose: 600 umol/L. Human Cytogenetic Analysis; Cell Type: lymphocyte; Dose: 300 umol/L. Human Sex Chromosome Loss; Cell Type: lymphocyte; Dose: 100 umol/L.

Tumorigenic: Rat Route: Oral; Dose: 3700 mg/kg/81W-I; Toxic Effects: Tumorigenic - Equivocal tumorigenic agent by RTECS criteria; Gastrointestinal - Tumors. Rat Route: Subcutaneous; Dose: 800 mg/kg/32W-I; Toxic Effects: Tumorigenic - Equivocal tumorigenic agent by RTECS criteria; Tumorigenic - Tumors at site of application.

Hazard Overviews

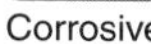
Corrosive

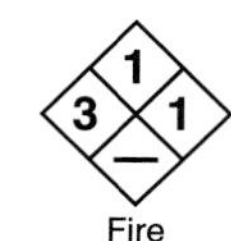

Fire Diamond

Health: Corrosive to eyes/skin/respiratory tract. Also Causes: headache, nausea, vomiting, abdominal pain, collapse. Chronic Effects: Suspect cancer hazard.

Fire: Will burn. Use dry chemical, foam, carbon dioxide, or water spray.

Reactivity: Stable. Hazardous polymerization cannot occur. Avoid: heat and ignition sources. Incompatible with: concentrated nitric acid; strong oxidizing agents (peroxides; peracids); potassium tert-butoxide; 3,8-dinitro-6-phenylphenanthridine and water; iron and water; moisture. Hazardous decomposition products: carbon oxides; ethylene and sulfur oxides; ethyl ether.

Carcinogenicity: IARC - Group 2A, Probably carcinogenic to humans; NIOSH - Not listed; NTP - Class 2A, Reasonably anticipated to be a carcinogen, limited evidence of carcinogenicity from studies in humans; ACGIH - Not listed; OSHA - Not listed; EPA - Not listed; MAK - Class A2, Unmistakably carcinogenic in animal experimentation only

Primary Target Organs:

Eyes

Skin

Respiratory System

Environmental

Ecotoxicity: Aquatic toxicity: TLm 96:100-10 ppm

Environmental Fate: If released to soil, estimated soil adsorption coefficients ranging from 33 to 99 indicate that it may display high to very high mobility in soil; however, it is known to rapidly hydrolyze in water under neutral, basic and acidic conditions and this reaction is expected to dominate its fate in soil. If released to water, it is expected to hydrolyze with an experimental half-life of 1.7 hours at neutral pH. This reaction is catalyzed by both acids and bases. It is not expected to bioconcentrate in fish and aquatic organisms nor is it expected to adsorb to sediment and suspended organic matter. If released to the atmosphere, it is expected to undergo a gas-phase reaction with water vapor with an experimental half-life of <1 day. It may also undergo atmospheric removal by wet deposition during a rain event.

Cleanup/Disposal: Guide No. 152: Do not touch damaged containers or spilled material unless wearing appropriate protective clothing. Stop leak if you can do it without risk. Prevent entry into waterways, sewers, basements or confined areas. Cover with plastic sheet to prevent spreading. Absorb or cover with dry earth, sand or other non-combustible material and transfer to containers. Do not get water inside containers.

Environmental Physical Data

Henry's Law Constant: 8.4×10^{-6}

Octanol/Water Partition Coefficient: log K_{ow} = 1.14

Sorption Partition Coefficient: K_{oc} = 33 to 99

BCF: calculated at 4

Regulations

RCRA 40CFR: Not listed

CERCLA: 40CFR 302.4: Listed per CAA Section 112 RQ: 10 lb (4.535 kg)

SARA 40CFR 372.65: Listed

SARA EHS 40CFR 355: Not listed

TSCA: Listed

Analytical Methods

Soil: SW846 8270C

Plasma: EPA 29

DIE3700 CAS #: 557-20-0

DIETHYL ZINC

RTECS: ZH2070000

DOT: UN1366; UN2845; IMO4.2

EINECS Number: 209-161-3

Molecular Formula: $C_4H_{10}Zn$

Structured MF: $(C_2H_5)_2Zn$

Formula Weight: 123.50

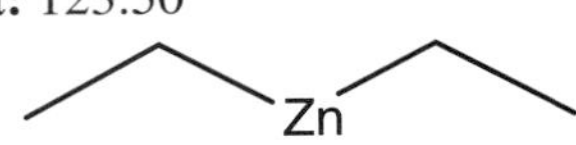

Chemical Structure

Synonyms: DIETHYLZINC; ETHYL ZINC; ZINC ETHIDE; ZINC ETHYL; ZINC,DIETHYL; ZINC,DIETHYL-

Description: colorless liquid; garlic odor

Use: organic synthesis; catalyst for polymerization of olefins; high energy aircraft & missile fuel; production of ethyl mercuric chloride; in paper deacidification

Physical Properties

Boiling Point: 118 °C (244 °F)

Freezing Point: -28 °C (-18.4 °F)

Specific Gravity: 1.2065 at 20 °C/4 °C

Vapor Pressure: Saturated vapor pressure 0.309 lb/sq in at 70 °C

Water Solubility: Decomposes in Cold Water

Other Solubilities: miscible with Ether, Petroleum Ether, Benzene, hydrocarbons

Surface Tension: Estimated at 20 dynes/cm at 20 °C

Refraction Index: Alpha 1.4936 at 20 °C

Ionization Potential (eV): =< 8.6 +/-0.7

Flash Point: Ignites spontaneously

Autoignition Temperature: < -18 °C

Hazard Overviews

Corrosive

Explosive

Flammable

Fire Diamond

Health: Corrosive to eyes/skin/respiratory tract. Also Causes: pulmonary edema, metal fume fever, scarring, nausea, diarrhea, ulceration of digestive tract tissue.

Fire: Flammable. Use dry chemical, soda ash, sand, or lime. Do not use water, water-based foam, carbon tetrachloride, or chlorobromomethane. Decomposes explosively above 248 degrees F. Ignites spontaneously in air and may re-ignite after extinguishment.

Reactivity: Unstable; pyrophoric, ignites when exposed to air or moisture. Hazardous polymerization cannot occur. Avoid:

heat; ignition sources; water. Incompatible with: alkenes + diiodomethane; bromine; chlorine; alcohols; methanol; hydrazine; nitrogen compounds; ozone; sulfur dioxide; water; organic peroxides; arsenic trichloride; phosphorus. Hazardous decomposition products: zinc oxides.

Carcinogenicity: IARC - Not listed; NIOSH - Not listed; NTP - Not listed; ACGIH - Not listed; OSHA - Not listed; EPA - Not listed; MAK - Not listed

Primary Target Organs:

Eyes

Skin

Respiratory System

Mucous Membranes

Gastro-intestinal

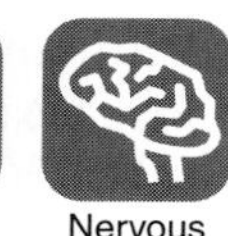
Nervous System

Environmental

Cleanup/Disposal: Guide No. 135: Fully encapsulating, vapor protective clothing should be worn for spills and leak with no fire. Eliminate all ignition sources (no smoking, flares, sparks or flames in immediate area). Do not touch or walk through spilled material. Stop leak if you can do it without risk. Small Spills: Cover with dry earth, dry sand, or other non-combustible material followed with plastic sheet to minimize spreading or contact with rain. Use clean non-sparking tools to collect material and place it into loosely covered plastic containers for later disposal. Prevent entry into waterways, sewers, basements or confined areas.

Environmental Physical Data

BOD: none

Regulations

RCRA 40CFR: Not listed
CERCLA: 40CFR 302.4: Listed as Compound per CWA Section 307(a)
SARA 40CFR 372.65: Listed as Compound
SARA EHS 40CFR 355: Not listed
TSCA: Listed

DIE3800 CAS #: 685-91-6

N,N-DIETHYLACETAMIOE

RTECS: AB7000000
EINECS Number: 211-685-2
Molecular Formula: $C_6H_{13}NO$
Formula Weight: 115.20

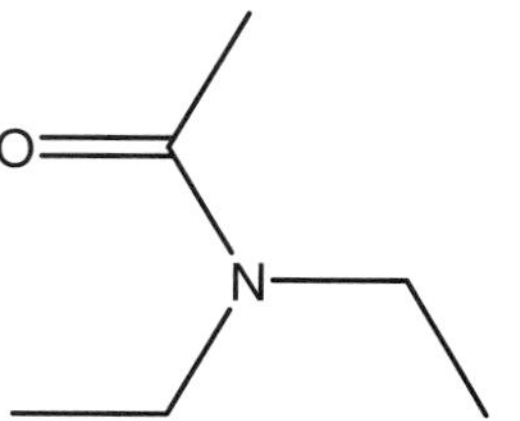

Chemical Structure

Synonyms: ACETAMIDE,N,N-DIETHYL-; ACETIC ACID,AMIDE,N,N-DIETHYL; N,N-DIAETHYLACETAMID; N,N-DIETHYLACETAMIDE
Description: colorless liquid; faint odor

Physical Properties

Boiling Point: 185 °C (365 °F) to 186 °C (367 °F)
Freezing Point: < 20 °C (68 °F)
Specific Gravity: 0.913 at 17.4 °C/4 °C
Vapor Pressure: 2 mm Hg at 35 °C
Water Solubility: Soluble in Water
Other Solubilities: Soluble in Alcohol Soluble in all proportions in Ether, Acetone, Benzene
Refraction Index: 1.4374
Ionization Potential (eV): 8.60 +/-0.02
Flash Point: 76.667 °C

RTECS Toxicity Data

Tumorigenic: Rat Route: Oral; Dose: 910 mg/kg/73W-I; Toxic Effects: Tumorigenic - Equivocal tumorigenic agent by RTECS criteria; Kidney, Ureter, and Bladder - Kidney tumors; Blood - Leukemia.

Hazard Overviews

Health: Severely irritating to eyes/skin/respiratory tract. Harmful. Other Acute Effects: harmful if swallowed, inhaled, or absorbed through skin; symptoms of exposure may include burning sensation; coughing; wheezing; laryngitis; shortness of breath; headache; nausea; vomiting.

Fire: Combustible. Hazards: emits toxic fumes. Extinguishing agents: carbon dioxide, dry chemical powder or appropriate foam; water spray. Precautions: combustible liquid.

Reactivity: Incompatible with: strong oxidizing agents, strong bases, acid chlorides. Hazardous decomposition products: toxic fumes of: carbon monoxide, carbon dioxide, nitrogen oxides.

Carcinogenicity: IARC - Not listed; NIOSH - Not listed; NTP - Not listed; ACGIH - Not listed; OSHA - Not listed; EPA - Not listed; MAK - Not listed

Primary Target Organs:

Eyes

Skin

Respiratory System

Environmental

Regulations

RCRA 40CFR: Not listed
CERCLA: 40CFR 302.4: Not listed
SARA 40CFR 372.65: Not listed
SARA EHS 40CFR 355: Not listed
TSCA: Listed

DIE3900 CAS #: 2235-46-3

N,N-DIETHYLACETOACETAMIDE

RTECS: AK4025000
EINECS Number: 218-792-3
Molecular Formula: $C_8H_{15}NO_2$
Formula Weight: 157.24

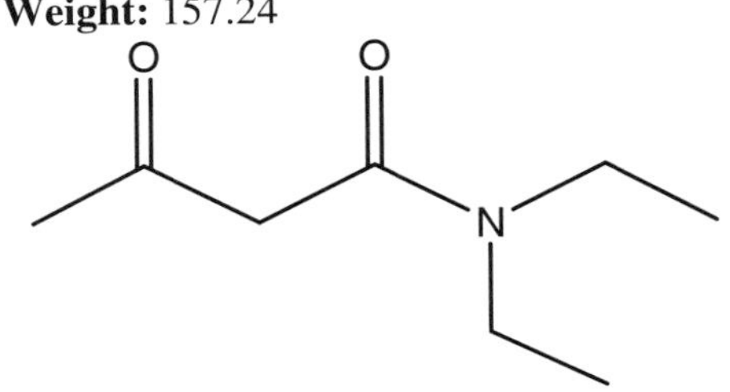

Chemical Structure

Synonyms: BUTANAMIDE,N,N-DIETHYL-3-OXO-; BUTANAMIDE,N,N-DIETHYL-3-OXO-(9CI); DIETHYLACETOACETAMIDE; N,N-DIETHYLACETYLACETAMIDE; DIETHYLAMID KYSELINY ACETOCTOVE; 1-(DIETHYLCARBAMOYL)-2-PROPANONE; N,N-DIETHYL-3-OXOBUTANAMIDE
Description: liquid
Use: precursor in insecticide synthesis (phosphamidon); intermediate for pigments

Physical Properties

Boiling Point: Decomposes
Freezing Point: -70 °C (-94 °F)
Specific Gravity: 0.995 at 20 °C/20 °C
Water Solubility: Soluble in Water
Flash Point: 121.111 °C

RTECS Toxicity Data

Acute Oral: Rat LD_{50} Dose: 4760 mg/kg.
Irritation Skin: Rabbit Standard Draize Test Dose: 500 mg/24H; Reaction: mild.

Hazard Overviews

Health: Irritating to eyes/skin/respiratory tract. Other Acute Effects: may be harmful by inhalation, ingestion, or skin absorption.
Fire: Will burn. Hazards: emits toxic fumes. Extinguishing agents: water spray; carbon dioxide, dry chemical powder or appropriate foam. Precautions: combustible liquid.
Reactivity: Incompatible with: strong oxidizing agents. Hazardous decomposition products: toxic fumes of: carbon monoxide, carbon dioxide, nitrogen oxides.
Carcinogenicity: IARC - Not listed; NIOSH - Not listed; NTP - Not listed; ACGIH - Not listed; OSHA - Not listed; EPA - Not listed; MAK - Not listed

Primary Target Organs:

Eyes

Skin

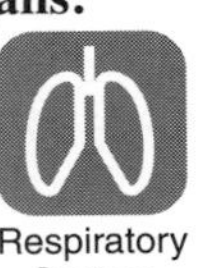
Respiratory System

Environmental

Regulations

RCRA 40CFR: Not listed
CERCLA: 40CFR 302.4: Not listed
SARA 40CFR 372.65: Not listed
SARA EHS 40CFR 355: Not listed
TSCA: Listed

DIE4000 CAS #: 96-10-6

DIETHYLALUMINUM CHLORIDE

RTECS: BD0558000
DOT: UN1101; UN2220; UN2221; UN3051; UN3052; IMO4.2
EINECS Number: 202-477-2
Molecular Formula: $C_4H_{10}AlCl$
Formula Weight: 120.56

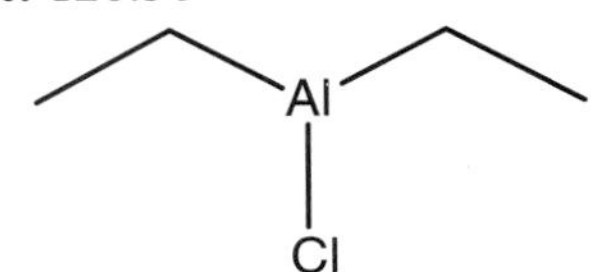

Chemical Structure

Synonyms: ALUMINUM DIETHYL MONOCHLORIDE; ALUMINUM,CHLORODIETHYL-; ALUMINUM,DICHLOROTETRAETHYLDI-; CHLORODIETHYLALUMINUM; DEAC; DIETHYLALUMINIUM CHLORIDE; DIETHYLALUMINUM MONOCHLORIDE; DIETHYLCHLOROALUMINUM
Description: clear, colorless liquid
Use: polyolefin catalyst; intermediate in the production of organometallics

Physical Properties

Boiling Point: 208 °C (406 °F)
Freezing Point: -50 °C (-58 °F)
Specific Gravity: 0.961 at 25 °C
Water Solubility: Reacts and Ignites
Flash Point: Same as solvent

RTECS Toxicity Data

Acute Inhalation: Rat LC_{50} Dose: 11 gm/m^3.

Hazard Overviews

Explosive

Flammable

Fire Diamond: 4 / 3 / 3 / W

Health: Can cause burns to eyes/skin/respiratory tract.

Fire: Explosive/pyrophoric. Flammable. reacts violently and ignites with water. Fight small fires with powdered graphite or sodium chloride, sand, or soda ash. Do not use water, Halon, or foam. If dissolved in solvent; use dry chemical or carbon dioxide. For large fires, withdraw and let burn.

Reactivity: Unstable, ignites spontaneously when exposed to air or water. Hazardous polymerization cannot occur. Avoid: air; water. Incompatible with: oxidizers; water; acids; halogenated hydrocarbons; alcohol; other compounds containing oxygen in their structure; chlorine azide. Hazardous decomposition products: non-toxic aluminum oxide fumes; flammable ethylene; propylene; butylene gases.

Carcinogenicity: IARC - Not listed; NIOSH - Not listed; NTP - Not listed; ACGIH - Not listed; OSHA - Not listed; EPA - Not listed; MAK - Not listed

Primary Target Organs:

Eyes

Skin

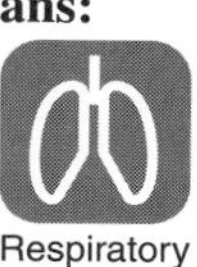
Respiratory System

Exposure Limits

OSHA PEL Vacated 1989 Limits: TWA: 2 mg/m^3; as Al soluble.

ACGIH TLV: TWA: 2 mg/m^3; as Al.

NIOSH REL: TWA: 2 mg/m^3; as Al soluble salts, alkyls.

Environmental

Cleanup/Disposal: Guide No. 135: Fully encapsulating, vapor protective clothing should be worn for spills and leak with no fire. Eliminate all ignition sources (no smoking, flares, sparks or flames in immediate area). Do not touch or walk through spilled material. Stop leak if you can do it without risk. Small Spills: Cover with dry earth, dry sand, or other non-combustible material followed with plastic sheet to minimize spreading or contact with rain. Use clean non-sparking tools to collect material and place it into loosely covered plastic containers for later disposal. Prevent entry into waterways, sewers, basements or confined areas.

Regulations

RCRA 40CFR: Not listed

CERCLA: 40CFR 302.4: Not listed

SARA 40CFR 372.65: Not listed

SARA EHS 40CFR 355: Not listed

TSCA: Listed

DIE4100 CAS #: 109-89-7

DIETHYLAMINE

RTECS: HZ8750000

DOT: UN1154; IMO3.1

EINECS Number: 203-716-3

Molecular Formula: $C_4H_{11}N$

Structured MF: $CH_3CH_2NHCH_2CH_3$

Formula Weight: 73.14

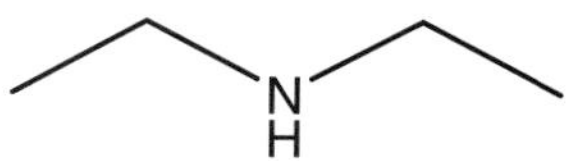

Chemical Structure

Synonyms: DEA; DEN; DIAETHYLAMIN; DIETHAMINE; N,N-DIETHYLAMINE; DIETILAMINA; DWUETYLOAMINA; ETHANAMINE,N-ETHYL; ETHANAMINE,N-ETHYL-; N-ETHYLETHANAMINE

Description: colorless liquid; fishy, ammonia-like odor

Use: in organic synthesis and in the manufacture of diethyldithiocarbamate and thiurams (rubber processing accelerators), diethylaminoethanol (medicinal intermediate), diethylaminopropylamine (epoxy curing agent), N,N-diethyl-m-toluamide and other pesticides, 2-diethylaminoethylmethacrylate and drugs; as a selective solvent; in textile specialties, in rubber and petroleum industries, in flotation agents, in resins, in dyes, in polymerization inhibitors, in electroplating and in corrosion inhibitors

Physical Properties

Boiling Point: 55.5 °C (132 °F)

Freezing Point: -50 °C (-58 °F)

Specific Gravity: 0.7074 at 20 °C/4 °C

Vapor Density: 2.53 Air=1

Saturated Vapor Density: 1.668637024 kg/m^3

Density: 0.7074 g/mL at 20 °C

Vapor Pressure: 195 mm Hg at 20 °C

Water Solubility: Miscible

Other Solubilities: 95% Ethanol: >=100 mg/ml at 17 °C; Acetone: >=100 mg/ml at 17 °C; Alcohol: miscible; DMSO: >=100 mg/ml at 17 °C; Ether: Soluble; Most organic solvents: miscible.

Surface Tension: 20.05 dynes/cm at 20 °C

Odor Threshold: 2.0 x10^{-2} ppm

pH: Strongly alkaline

Refraction Index: 1.3864 at 20 °C/D

Evaporation Rate: 16.9 Butyl Acetate=1

Critical Temperature: 223.3 °C

Critical Pressure: 36.6 atm

Ionization Potential (eV): 8.01

Flash Point: -26 °C Closed Cup

Autoignition Temperature: 312 °C

LEL: 1.8% v/v

UEL: 10.1% v/v

RTECS Toxicity Data

Acute Oral: Rat LD_{50} Dose: 540 mg/kg. Mouse LD_{50} Dose: 500 mg/kg; Toxic Effects: Behavioral - Sleep; Behavioral - Somnolence (general depressed activity).

Acute Inhalation: Rat LC_{50} Dose: 4000 ppm/4hr. Mouse LC_{Lo} Dose: 3 gm/m^3/2hr; Toxic Effects: Brain and coverings - Recordings from specific areas of CNS; Lungs, Thorax, or Respiration - Acute pulmonary edema; Liver - Fatty liver degeneration.

Acute Dermal: Rabbit LD_{50} Route: Skin; Dose: 820 uL/kg.

Chronic (Multiple Dose) Oral: Rat Dose: 4994 mg/kg/11W-I; Toxic Effects: Nutritional and gross metabolic - Weight loss or decreased weight gain.

Chronic (Multiple Dose) Inhalation: Rat Dose: 250 ppm/6H/24W-I; Toxic Effects: Lungs, Thorax, or Respiration - Changes in lung weight; Immunological including allergic - Othr immed. (hmrl): urticaria, allergic rhinitis, serum sickness; Nutritional and gross metabolic - Weight loss or decreased weight gain. Rat Dose: 4 mg/m^3/24H/93D-C; Toxic Effects: Kidney, Ureter, and Bladder - Other changes in urine composition; Blood - Changes in serum composition; Biochemical - True cholinesterase.

Irritation Eye: Rabbit Standard Draize Test Dose: 50 ug open; Reaction: severe.

Irritation Skin: Rabbit Standard Draize Test Dose: 100 mg/24H; Reaction: moderate. Rabbit Open Draize Test Dose: 10 mg/24H open; Reaction: mild. Rabbit Open Draize Test Dose: 500 mg open; Reaction: mild.

Mutagenic: Other Microorganisms Mutations in Microorganisms; Dose: 50 uL/plate (-S9). Other Microorganisms Mutations in Microorganisms; Dose: 400 uL/plate (-S9). Other Microorganisms Mutations in Microorganisms; Dose: 200 ug/plate (-S9).

Hazard Overviews

Corrosive

Flammable

Fire Diamond

Health: Corrosive to eyes/skin/respiratory tract. Also Causes: coughing, nausea, difficulty breathing, pulmonary edema, tissue degeneration of the heart, corneal damage, blindness,swelling of cornea, foggy vision and/or halos around lights, vomiting, diarrhea, collapse, chest pain, heart and liver damage. Chronic Effects: corneal edema, foggy vision, halos around lights, irritation of the lungs, bronchi, throat.

Fire: Flammable. Can form explosive mixtures in the air. Do not try to extinguish unless flow can be stopped. Use water spray, dry chemical, or alcohol resistant foam. Use water spray to cool fire-exposed containers. Aqueous solutions are flammable unless diluted extensively. Fight fire from maximum distance.

Reactivity: Stable. Hazardous polymerization cannot occur. Avoid: heat; ignition sources. Incompatible with: oxidizing materials; acids; halogens; cellulose nitrate; reactive organic compounds; some metals; dicyanofurazan; dicyanofuroxan; plastics; rubber; coatings. Hazardous decomposition products: carbon monoxide; carbon dioxide; hydrocarbons; nitrogen oxides; amine vapors.

Carcinogenicity: IARC - Not listed; NIOSH - Not listed; NTP - Not listed; ACGIH - Class A4, Not classifiable as a human carcinogen; OSHA - Not listed; EPA - Not listed; MAK - Not listed

Primary Target Organs:

Eyes

Skin

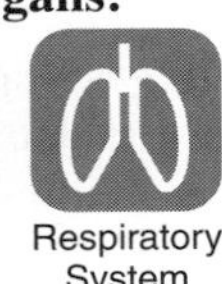
Respiratory System

Exposure Limits

OSHA PEL: TWA: 25 ppm; 75 mg/m^3.

OSHA PEL Vacated 1989 Limits: TWA: 10 ppm; 30 mg/m^3; STEL: 25 ppm; 75 mg/m^3.

ACGIH TLV: TWA: 5 ppm; 15 mg/m^3; STEL: 15 ppm; 45 mg/m^3.

NIOSH REL: TWA: 10 ppm; 30 mg/m^3. STEL: 25 ppm; 75 mg/m^3.

NIOSH IDLH: 200 ppm.

DFG MAK: TWA: 10 ppm; 30 mg/m^3.

Respirator Recommendation

Exposure Range: >25 to <200 ppm Air Purifying, Negative Pressure, Half Mask

Exposure Range: 200 to unlimited ppm Self-contained Breathing Apparatus, Pressure Demand, Full Face

Cartridge Color: black

Environmental

Ecotoxicity: LC_{50} Pimephales promelas (fathead minnow) 30 days old 855 mg/l/96 hr at 24.7 °C /Conditions of bioassay not specified EC_{50} Pimephales promelas (fathead minnow) 30 days old 855 mg/l/96 hr at 24.7 °C /Conditions of bioassay not specified TLm creek chub 85 mg/l/48 hr (fresh water) /Conditions of bioassay not specified

Environmental Fate: If released on land, it should volatilize and leach into the soil. The chemical would be expected to biodegrade, however no estimate of degradative rates in soil are available. If released into water, it will readily biodegrade (half-life 0.9 days at 10 ppm and longer at lower concentrations) and also be removed by volatilization (half-life 31.6 hr in a model river). Adsorption to sediment and bioconcentration in aquatic organisms will not be appreciable. In the atmosphere, it will react with photochemically produced hydroxyl radicals. Degradation half-lives are not available for clean atmospheres, but in polluted atmospheres it completely degrades photochemically within 2 hr. It will also be scavenged by rain. It is a precursor of dimethylnitrosamine. The latter is formed in the atmosphere in the presence of nitrogen oxides but is destroyed by sunlight.

Cleanup/Disposal: Guide No. 132: Fully encapsulating, vapor protective clothing should be worn for spills and leaks with no fire. Eliminate all ignition sources (no smoking, flares, sparks or flames in immediate area). All equipment used when handling the product must be grounded. Do not touch or walk through spilled material. Stop leak if you can do it without risk. Prevent entry into waterways, sewers, basements or confined areas. A vapor suppressing foam may be used to reduce vapors. Absorb with earth, sand or other non-combustible material and transfer to containers (except for Hydrazine). Use clean non-sparking tools to collect absorbed material. Large Spills: Dike far ahead of liquid spill for later disposal. Water spray may reduce vapor; but may not prevent ignition in closed spaces.

Environmental Physical Data

Henry's Law Constant: 1.05×10^{-3}
Octanol/Water Partition Coefficient: log K_{ow} = 0.58
Sorption Partition Coefficient: K_{oc} = 50
BCF: estimated at 1.62
BOD: 59% BODT, 12 days

Regulations

RCRA 40CFR: Not listed
CERCLA: 40CFR 302.4: Listed per CWA Section 311(b)(4) RQ: 100 lb (45.35 kg)
SARA 40CFR 372.65: Not listed
SARA EHS 40CFR 355: Not listed
TSCA: Listed

Analytical Methods

Air: ASTM D4490
Indoor / Expired Air: NIOSH 2010
Other: EPA 1666, 1671

DIE4200 CAS #: 120-21-8

P-(DIETHYLAMINO)BENZALDEHYDE

EINECS Number: 204-377-4
Molecular Formula: $C_{11}H_{15}NO$
Formula Weight: 177.25

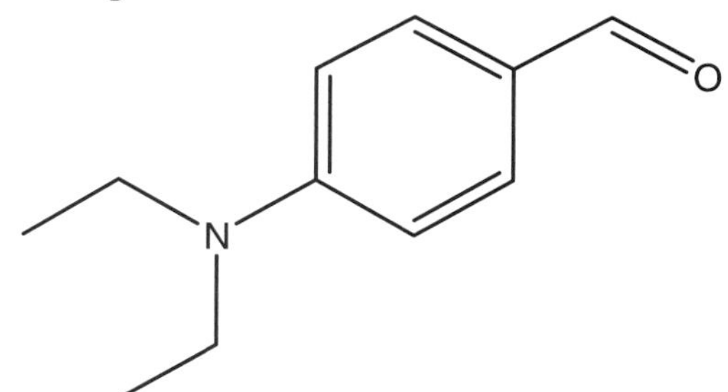

Chemical Structure

Physical Properties

Boiling Point: 174 °C (345 °F)
Freezing Point: 41 °C (105.8 °F)
Water Solubility: Slightly
Other Solubilities: Ethanol: Soluble; Ether: Soluble; Benzene: Soluble; Carbon Tetrachloride: Soluble

Hazard Overviews

Health: Irritating to eyes/skin/respiratory tract. Other Acute Effects: may be harmful by inhalation, ingestion, or skin absorption.
Fire: Hazards: emits toxic fumes. Extinguishing agents: water spray; carbon dioxide, dry chemical powder or appropriate foam. Precautions: combustible liquid.
Reactivity: Incompatible with: strong oxidizing agents, strong bases. Hazardous decomposition products: toxic fumes of: carbon monoxide, carbon dioxide, nitrogen oxides.
Carcinogenicity: IARC - Not listed; NIOSH - Not listed; NTP - Not listed; ACGIH - Not listed; OSHA - Not listed; EPA - Not listed; MAK - Not listed
Primary Target Organs:

Eyes

Skin

Respiratory System

Environmental

Regulations

RCRA 40CFR: Not listed
CERCLA: 40CFR 302.4: Not listed
SARA 40CFR 372.65: Not listed
SARA EHS 40CFR 355: Not listed
TSCA: Not listed

DIE4300 CAS #: 100-37-8

2-DIETHYLAMINOETHANOL

RTECS: KK5075000
DOT: UN2686; IMO3.3
EINECS Number: 202-845-2
Molecular Formula: $C_6H_{15}NO$
Structured MF: $(C_2H_5)_2NCH_2CH_2OH$
Formula Weight: 117.19

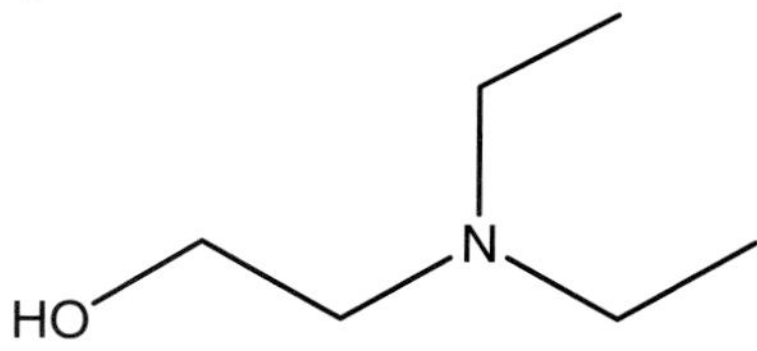

Chemical Structure

Synonyms: DEAE; DIAETHYLAMINOAETHANOL; DIETHYL ETHANOLAMINE; 2-(DIETHYLAMINO) ETHYL ALCOHOL; (DIETHYLAMINO)ETHANOL; 2-(DIETHYLAMINO)ETHANOL; 2-(N,N-DIETHYLAMINO)ETHANOL; 2-N-DIETHYLAMINOETHANOL; BETA-(DIETHYLAMINO)ETHANOL; BETA-DIETHYLAMINOETHANOL; DIETHYLAMINOETHANOL; N,N-DIETHYL-2-AMINOETHANOL; N-DIETHYLAMINOETHANOL; 2-(DIETHYLAMINO)ETHYL ALCOHOL; 2-DIETHYLAMINOETHYL ALCOHOL; BETA-DIETHYLAMINOETHYL ALCOHOL; DIETHYLETHANOLAMINE; N,N-DIETHYLETHANOLAMINE; DIETHYL(2-HYDROXYETHYL)AMINE; DIETHYL-(2-

HYDROXYETHYL)AMINE; N,N-DIETHYL-2-HYDROXYETHYLAMINE; N,N-DIETHYL-N-(BETA-HYDROXYETHYL)AMINE; DIETHYLMONOETHANOLAMINE; N,N-DIETHYLMONOETHANOLAMINE; ETHANOL,2-(DIETHYLAMINO)-; (2-HYDROXYETHYL)DIETHYLAMINE; 2-HYDROXYTRIETHYLAMINE; BETA-HYDROXYTRIETHYLAMINE; PENNAD 150

Description: colorless liquid; nauseating ammonia-like odor

Use: chemical intermediate for production of emulsifiers, detergents, solubilizers cosmetics, textile finishing agents, manufacture of drugs, fatty acid derivative, textile softeners, pharmaceuticals, antirust compositions, curing agent for resins

Physical Properties

Boiling Point: 163 °C (325 °F) at 760 mm Hg

Freezing Point: -70 °C (-94 °F)

Specific Gravity: 0.8921 at 20 °C/4 °C

Vapor Density: 4.03 Air=1

Saturated Vapor Density: 1.300834301 kg/m^3

Vapor Pressure: 21 mm Hg at 20 °C

Water Solubility: Soluble in all Proportions

Other Solubilities: 95% Ethanol: Soluble; Acetone: Soluble; Benzene: Soluble; DMSO: Soluble; Ether: Soluble.

Surface Tension: Estimated at 34.3 dynes/cm

Odor Threshold: Absolute perception limit 0.011 ppm

Refraction Index: 1.4412

Critical Temperature: 377 °C

Critical Pressure: 457.3 psia

Flash Point: 60 °C Open Cup

Autoignition Temperature: 320 °C

LEL: 6.7% v/v

UEL: 11.7% v/v

RTECS Toxicity Data

Acute Oral: Rat LD_{50} Dose: 1300 mg/kg.

Acute Inhalation: Rat LC_{Lo} Dose: 4500 mg/m^3/4hr; Toxic Effects: Brain and coverings - Recordings from specific areas of CNS; Sense organs and special senses - Conjunctive irritation; Behavioral - Convulsions or effect on seizure threshold. Human TC_{Lo} Dose: 200 ppm; Toxic Effects: Gastrointestinal - Nausea or vomiting.

Acute Dermal: Rabbit LD_{50} Route: Skin; Dose: 1260 uL/kg. Mouse LD_{50} Route: Subcutaneous Dose: 650 mg/kg; Toxic Effects: Behavioral - Convulsions or effect on seizure threshold; Behavioral - Ataxia; Lungs, Thorax, or Respiration - Other changes.

Chronic (Multiple Dose) Inhalation: Rat Dose: 500 ppm/6H/5D-I; Toxic Effects: Sense organs and special senses - Other; Lungs, Thorax, or Respiration - Structural or functional change in trachea or bronchi; DEATH. Rat Dose: 200 ppm/6H/26W-I; Toxic Effects: DEATH. Rat Dose: 76 ppm/6H/14W-I; Toxic Effects: Liver - Changes in Liver weight; Kidney, Ureter, and Bladder - Changes in kidney weight; Nutritional and gross metabolic - Weight loss or decreased weight gain. Rat Dose: 301 ppm/6H/2W-I; Toxic Effects: Behavioral - Fluid intake; Nutritional and gross metabolic - Weight loss or decreased weight gain; DEATH.

Irritation Eye: Rabbit Standard Draize Test Dose: 5 mg; Reaction: severe.

Irritation Skin: Rabbit Open Draize Test Dose: 500 mg open; Reaction: mild.

Hazard Overviews

Corrosive

Flammable

Fire Diamond

Health: Corrosive to eyes/skin/respiratory tract. Also Causes: dizziness, nausea, vomiting, diarrhea, chest heaviness/pain, difficulty breathing, coughing, bluish face and lips, wheezing, and pulmonary edema, stomach/abdominal pain; state of shock; possible convulsions.

Fire: Flammable. Can form explosive mixtures in the air. For small fires use dry chemical, carbon dioxide, halon, or alcohol-resistant foam. Do not use water. For large fires use fog or alcohol-resistant foam. Water or foam may cause frothing. Use water to cool fire-exposed container sides, but not to extinguish fire. Be careful not to get water inside containers.

Reactivity: Stable. Hazardous polymerization cannot occur. Avoid: heat. Incompatible with: strong acids; oxidizing materials. Hazardous decomposition products: oxides of carbon; nitrogen oxides.

Carcinogenicity: IARC - Not listed; NIOSH - Listed as carcinogen; NTP - Not listed; ACGIH - Not listed; OSHA - Not listed; EPA - Not listed; MAK - Not listed

Primary Target Organs:

Eyes

Skin

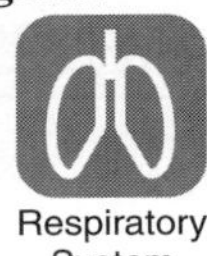
Respiratory System

Mucous Membranes

Exposure Limits

OSHA PEL: TWA: 10 ppm; 50 mg/m^3.

ACGIH TLV: TWA: 2 ppm; 9.6 mg/m^3.

NIOSH REL: TWA: 10 ppm; 50 mg/m^3.

NIOSH IDLH: 100 ppm.

DFG MAK: TWA: 10 ppm; 50 mg/m^3.

Respirator Recommendation

Exposure Range: >10 to <100 ppm Air Purifying, Negative Pressure, Half Mask

Exposure Range: 100 to unlimited ppm Self-contained Breathing Apparatus, Pressure Demand, Full Face

Cartridge Color: black

Environmental

Ecotoxicity: LC_{50} Pimephales promelas (fathead minnow) 1780 mg/l/96 hr (95% confidence limit 1660-1920 mg/l), flow-through bioassay with measured concentrations, 24.9 °C, dissolved oxygen 6.5 mg/l, hardness 40.8 mg/l calcium carbonate, alkalinity 40.9 mg/l calcium carbonate, and pH 7.70

Cleanup/Disposal: Guide No. 132: Fully encapsulating, vapor protective clothing should be worn for spills and leaks with no fire. Eliminate all ignition sources (no smoking, flares,

sparks or flames in immediate area). All equipment used when handling the product must be grounded. Do not touch or walk through spilled material. Stop leak if you can do it without risk. Prevent entry into waterways, sewers, basements or confined areas. A vapor suppressing foam may be used to reduce vapors. Absorb with earth, sand or other non-combustible material and transfer to containers (except for Hydrazine). Use clean non-sparking tools to collect absorbed material. Large Spills: Dike far ahead of liquid spill for later disposal. Water spray may reduce vapor; but may not prevent ignition in closed spaces.

Environmental Physical Data

Octanol/Water Partition Coefficient: log K_{ow} = calculated at 0.31 to 0.46

Regulations

RCRA 40CFR: Not listed
CERCLA: 40CFR 302.4: Not listed
SARA 40CFR 372.65: Not listed
SARA EHS 40CFR 355: Not listed
TSCA: Listed

Analytical Methods

Water / Groundwater: ASTM D4983
Indoor / Expired Air: NIOSH 2007

DIE4400 **CAS #: 2426-54-2**

2-(DIETHYLAMINO)ETHYL ACRYLATE

RTECS: AS8225000
EINECS Number: 219-378-5
Molecular Formula: $C_9H_{17}NO_2$
Formula Weight: 171.24
Synonyms: ACRYLIC ACID,2-(DIETHYLAMINO)ETHYL ESTER; ACRYLIC ACID,N,N-DIETHYLAMINOETHYL ESTER; AGEFLEX FA-2; 2-(DIETHYLAMINO)ETHYL ACRYLATE; BETA-(DIETHYLAMINO)ETHYL ACRYLATE; BETA-DIETHYLAMINOETHYL ACRYLATE; DIETHYLAMINOETHYL ACRYLATE; N,N-DIETHYLAMINOETHYL ACRYLATE; 2-DIETHYLAMINOETHYLESTER KYSELINY AKRYLOVE; 2-PROPENOIC ACID,2-(DIETHYLAMINO)ETHYL ESTER; 2-PROPENOIC ACID,2-(DIETHYLAMINO)ETHYL ESTER (9CI)
Use: chem int for polymeric quaternaries for water treatment; monomer for polymers used in inks & surface coatings

Physical Properties

Boiling Point: 81 °C (178 °F) at 10 mm Hg
Freezing Point: < -60 °C (-76 °F)
Specific Gravity: 0.939 at 20 °C/20 °C
Vapor Density: 5.9 Air=1
Water Solubility: Decomposes in Water
Other Solubilities: Reacts readily with electrophilic, free-radical, and nucleophilic agent
Refraction Index: 1.4376 at 25 °C
Flash Point: 91 °C Open Cup

RTECS Toxicity Data

Acute Oral: Rat LD_{50} Dose: 770 mg/kg.
Acute Dermal: Rabbit LD_{50} Route: Skin; Dose: 200 uL/kg.
Irritation Eye: Rabbit Standard Draize Test Dose: 1 mg; Reaction: mild. Rabbit Standard Draize Test Dose: 250 ug/24H; Reaction: severe.
Irritation Skin: Rabbit Standard Draize Test Dose: 5 mg/24H; Reaction: severe.

Hazard Overviews

Poison

Corrosive

Fire Diamond

Health: Corrosive to eyes/skin/respiratory tract. Poison. Other Acute Effects: may be fatal if inhaled, swallowed, or absorbed through skin; readily absorbed through skin; inhalation may result in spasm, inflammation and edema of the larynx and bronchi, chemical pneumonitis and pulmonary edema; symptoms of exposure may include burning sensation; coughing; wheezing; laryngitis; shortness of breath; headache; nausea; vomiting; may cause allergic skin reaction.
Fire: Combustible. Hazards: emits toxic fumes. Extinguishing agents: water spray; carbon dioxide, dry chemical powder or appropriate foam; use water spray to cool fire-exposed containers. Precautions: combustible liquid.
Reactivity: Stable. Hazardous polymerization may occur. Incompatible with: strong oxidizing agents, catalysts, carbon dioxide, rust, heavy metal salts, water. Hazardous decomposition products: toxic fumes of: carbon monoxide, carbon dioxide, nitrogen oxides.
Carcinogenicity: IARC - Not listed; NIOSH - Not listed; NTP - Not listed; ACGIH - Not listed; OSHA - Not listed; EPA - Not listed; MAK - Not listed
Primary Target Organs:

Eyes

Skin

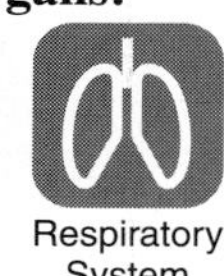
Respiratory System

Environmental

Regulations

RCRA 40CFR: Not listed
CERCLA: 40CFR 302.4: Not listed
SARA 40CFR 372.65: Not listed
SARA EHS 40CFR 355: Not listed
TSCA: Listed

DIE4500 CAS #: 105-16-8

2-(N,N-DIETHYLAMINO)ETHYL METHACRYLATE

RTECS: OZ4150000
EINECS Number: 203-275-7
Molecular Formula: $C_{10}H_{19}NO_2$
Formula Weight: 185.27

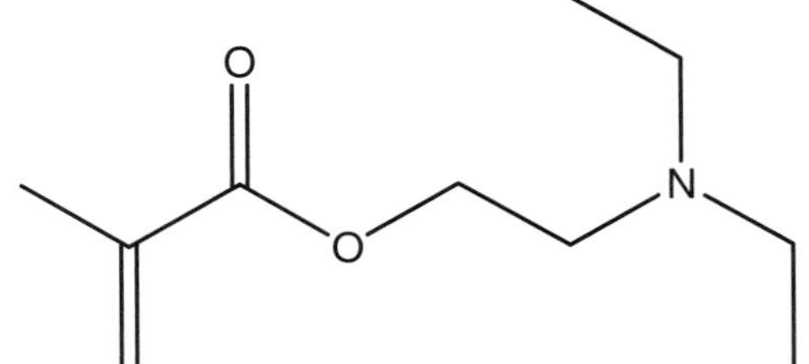

Chemical Structure

Synonyms: DAKTOSE B; 2-(DIETHYLAMINO)ETHYL METHACRYLATE; 2-(N,N-DIETHYLAMINO)ETHYL METHACRYLATE; 2-DIETHYLAMINOETHYL METHACRYLATE; BETA-(DIETHYLAMINO)ETHYL METHACRYLATE; BETA-(N,N-DIETHYLAMINO)ETHYL METHACRYLATE; DIETHYLAMINOETHYL METHACRYLATE; N,N-DIETHYLAMINOETHYL METHACRYLATE; 2-DIETHYLAMINOETHYLESTER KYSELINY METHAKRYLOVE; METHACRYLIC ACID,2-(DIETHYLAMINO)ETHYL ESTER; (2-(METHACRYLOYLOXY)ETHYL)DIETHYLAMINE; 2-PROPENOIC ACID,2-METHYL-,2-(DIETHYLAMINO)ETHYL ESTER; 2-PROPENOIC ACID,2-METHYL-,2-(DIETHYLAMINO)ETHYL ESTER(9CI)

Use: proposed for or used in manufacture of graft polymers, hair sprays, dental materials, treatment of synthetic pelts, preparation of copolymers for waste water purification; monomer for acrylic polymers used in automotive paints; comonomer for polymers used in adhesives; in dental materials, antithrombogenic prosthetics; experimental use in slow-release capsules; cosmetics

Physical Properties

Boiling Point: 49 °C (120 °F) at 0.040 kPa
Refraction Index: 1.442 at 20 °C/D

RTECS Toxicity Data

Acute Oral: Rat LD_{50} Dose: 4696 mg/kg.
Acute Inhalation: Rat LC_{50} Dose: 11 gm/m^3/4hr. Mouse LC_{50} Dose: 12100 mg/m^3/2hr.

Hazard Overviews

Health: Irritating to eyes/skin/respiratory tract. Harmful. Other Acute Effects: harmful if swallowed, inhaled, or absorbed through skin; may cause allergic skin reaction.

Fire: Hazards: emits toxic fumes. Extinguishing agents: water spray; carbon dioxide, dry chemical powder or appropriate foam; use water spray to cool fire-exposed containers. Precautions: combustible liquid.

Reactivity: Stable. Hazardous polymerization may occur. Incompatible with: strong oxidizing agents, catalysts, carbon dioxide, rust, steel, heavy metal salts, copper, water. Hazardous decomposition products: toxic fumes of: carbon monoxide, carbon dioxide, nitrogen oxides.

Carcinogenicity: IARC - Not listed; NIOSH - Not listed; NTP - Not listed; ACGIH - Not listed; OSHA - Not listed; EPA - Not listed; MAK - Not listed

Primary Target Organs:

Eyes

Skin

Respiratory System

Environmental

Regulations

RCRA 40CFR: Not listed
CERCLA: 40CFR 302.4: Not listed
SARA 40CFR 372.65: Not listed
SARA EHS 40CFR 355: Not listed
TSCA: Listed

DIE4600 CAS #: 91-44-1

7-DIETHYLAMINO-4-METHYLCOUMARIN

RTECS: GN6370000
EINECS Number: 202-068-9
Molecular Formula: $C_{14}H_{17}NO_2$
Formula Weight: 231.30

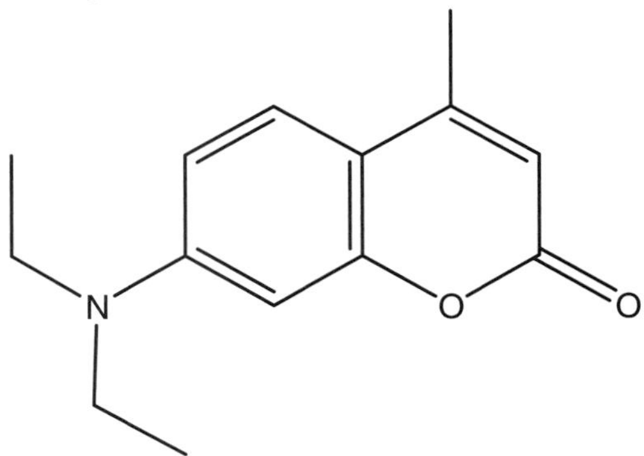

Chemical Structure

Synonyms: 2H-1-BENZOPYRAN-2-ONE,7-(DIETHYLAMINO)-4-METHYL-; 2H-1-BENZOPYRAN-2-ONE,7-(DIETHYLAMINO)-4-METHYL-(9CI); COUMARIN 1; COUMARIN,7-(DIETHYLAMINO)-4-METHYL-; 7-(DIETHYLAMINO)-4-METHYL-2H-1-BENZOPYRAN-2-ONE; 7-(DIETHYLAMINO)-4-METHYLCOUMARIN; 7-DIETHYLAMINO-4-METHYLCUMARIN; MDAC; 4-METHYL-7-(DIETHYLAMINO)COUMARIN

Description: light-tan granular crystals

Use: optical bleach in textile industry; in coatings for paper, labels, book covers, etc.; to lighten plastics, resins, varnishes, lacquers; invisible marking agent

Physical Properties

Freezing Point: 89 °C (192.2 °F)
Water Solubility: Slightly Soluble in Hot Water
Other Solubilities: 95% Ethanol: Soluble; Acetone: Soluble; Ether: Soluble.

RTECS Toxicity Data

Acute Oral: Rat LD_{50} Dose: 5 gm/kg. Mouse LD_{50} Dose: 1780 mg/kg; Toxic Effects: Behavioral - Somnolence (general depressed activity); Behavioral - Ataxia.

Hazard Overviews

Health: Irritating to eyes/skin/respiratory tract. Harmful. Other Acute Effects: harmful if swallowed; may be harmful if inhaled or absorbed through the skin.
Fire: Hazards: emits toxic fumes. Extinguishing agents: water spray; carbon dioxide, dry chemical powder or appropriate foam. Precautions: combustible liquid.
Carcinogenicity: IARC - Not listed; NIOSH - Not listed; NTP - Not listed; ACGIH - Not listed; OSHA - Not listed; EPA - Not listed; MAK - Not listed
Primary Target Organs:

Eyes

Skin

Respiratory System

Environmental

Regulations

RCRA 40CFR: Not listed
CERCLA: 40CFR 302.4: Not listed
SARA 40CFR 372.65: Not listed
SARA EHS 40CFR 355: Not listed
TSCA: Listed

DIE4700 CAS #: 104-78-9

3-DIETHYLAMINOPROPYLAMINE

RTECS: TX7350000
EINECS Number: 203-236-4
Molecular Formula: $C_7H_{18}N_2$
Formula Weight: 130.27

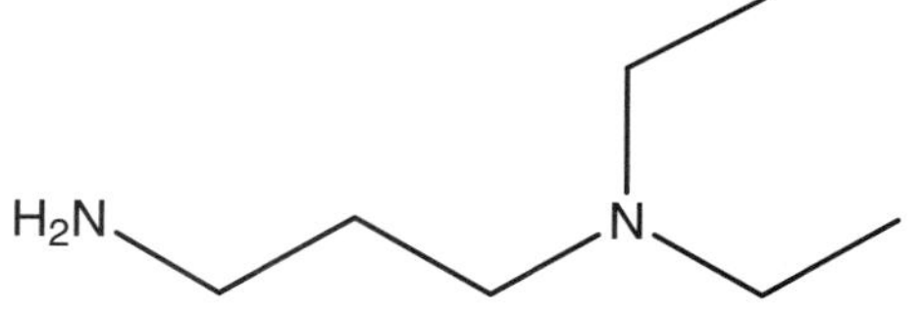

Chemical Structure

Synonyms: 1-AMINO-3-(DIETHYLAMINO)PROPANE; 3-(DIETHYLAMINO)PROPYLAMINE; DIETHYLAMINOPROPYLAMINE; N,N-DIETHYLAMINOPROPYLAMINE; N-(3-DIETHYLAMINOPROPYL)AMINE; DIETHYLAMINOTRIMETHYLENAMINE; N,N-DIETHYL-1,3-DIAMINOPROPANE

Physical Properties

Boiling Point: 159 °C (318.2 °F)
Freezing Point: -60 °C (-76 °F)
Specific Gravity: 0.82200
Vapor Pressure: 0.996 kPa at 50 °C
Water Solubility: Miscible
Refraction Index: 1.443

RTECS Toxicity Data

Acute Oral: Rat LD_{50} Dose: 1410 mg/kg.
Acute Dermal: Rabbit LD_{50} Route: Skin; Dose: 750 uL/kg; Toxic Effects: Skin and appendages - Primary irritation.

Hazard Overviews

Corrosive

Health: Corrosive to eyes/skin/respiratory tract. Harmful. Other Acute Effects: harmful if swallowed, inhaled, or absorbed through skin; inhalation may result in spasm, inflammation and edema of the larynx and bronchi, chemical pneumonitis and pulmonary edema; symptoms of exposure may include burning sensation, coughing, wheezing, laryngitis, shortness of breath, headache, nausea and vomiting.
Fire: Hazards: emits toxic fumes. Extinguishing agents: carbon dioxide, dry chemical powder or appropriate foam; water spray. Precautions: combustible liquid.
Reactivity: Incompatible with: acids, acid chlorides, acid anhydrides, strong oxidizing agents, carbon dioxide. Hazardous decomposition products: thermal decomposition may produce carbon monoxide, carbon dioxide, and nitrogen oxides.
Carcinogenicity: IARC - Not listed; NIOSH - Not listed; NTP - Not listed; ACGIH - Not listed; OSHA - Not listed; EPA - Not listed; MAK - Not listed
Primary Target Organs:

Eyes

Skin

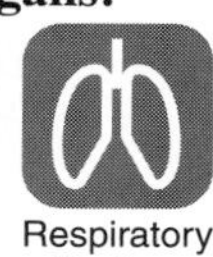
Respiratory System

Environmental

Regulations

RCRA 40CFR: Not listed
CERCLA: 40CFR 302.4: Not listed
SARA 40CFR 372.65: Not listed
SARA EHS 40CFR 355: Not listed
TSCA: Listed

DIE4800 CAS #: 579-66-8

2,6-DIETHYLANILINE

RTECS: BX3500000
EINECS Number: 209-445-7
Molecular Formula: $C_{10}H_{15}N$
Structured MF: $(C_2H_5)_2C_6H_3NH_2$
Formula Weight: 149.26

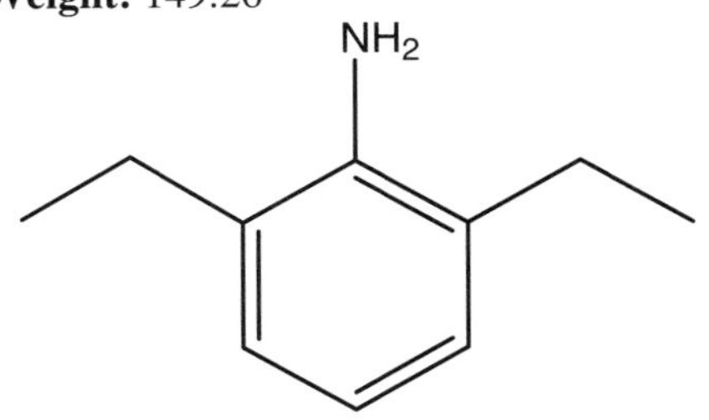

Chemical Structure

Synonyms: ANILINE,2,6-DIETHYL-; BENZENAMINE,2,6-DIETHYL-; BENZENAMINE,2,6-DIETHYL-(9CI); 2,6-DIETHYL ANILINE; 2,6-DIETHYLBENZENAMINE
Description: colorless liquid
Use: synthesis of herbicides

Physical Properties

Boiling Point: 114 °C (237 °F) at 1 atm
Freezing Point: 3 °C (37 °F) to 4 °C (39 °F)
Specific Gravity: 0.906
Vapor Density: 5.15 Air=1
Water Solubility: Sparingly Soluble
Refraction Index: 1.5452
Ionization Potential (eV): =< 7.77
Flash Point: 123 °C Closed Cup

RTECS Toxicity Data

Acute Oral: Rat LD_{50} Dose: 1800 mg/kg; Toxic Effects: Liver - Other changes; Blood - Changes in bone marrow not included above; Blood - Changes in spleen.
Chronic (Multiple Dose) Oral: Rat Dose: 10080 mg/kg/20D-C; Toxic Effects: Nutritional and gross metabolic - Weight loss or decreased weight gain.
Mutagenic: Bacteria - S Typhimurium Mutations in Microorganisms; Dose: 250 ng/plate (-S9).

Hazard Overviews

Health: Irritating to eyes/skin/respiratory tract. Harmful.. Acute Effects: may be harmful by inhalation, ingestion, or skin absorption; absorption into the body leads to the formation of methemoglobin which in sufficient concentration causes cyanosis; onset may be delayed 2 to 4 hours or longer.
Fire: Will burn. Hazards: emits toxic fumes. Extinguishing agents: water spray; carbon dioxide, dry chemical powder or appropriate foam. Precautions: combustible liquid.
Reactivity: Incompatible with: acids, acid chlorides, acid anhydrides, chloroformates, strong oxidizing agents, light sensitive, air sensitive. Hazardous decomposition products: toxic fumes of: carbon monoxide, carbon dioxide, nitrogen oxides.
Carcinogenicity: IARC - Not listed; NIOSH - Not listed; NTP - Not listed; ACGIH - Not listed; OSHA - Not listed; EPA - Not listed; MAK - Not listed
Primary Target Organs:

Eyes

Skin

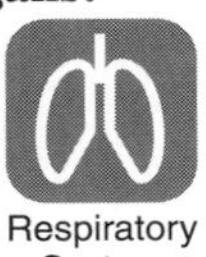
Respiratory System

Environmental

Cleanup/Disposal: Guide No. 132: Fully encapsulating, vapor protective clothing should be worn for spills and leaks with no fire. Eliminate all ignition sources (no smoking, flares, sparks or flames in immediate area). All equipment used when handling the product must be grounded. Do not touch or walk through spilled material. Stop leak if you can do it without risk. Prevent entry into waterways, sewers, basements or confined areas. A vapor suppressing foam may be used to reduce vapors. Absorb with earth, sand or other non-combustible material and transfer to containers (except for Hydrazine). Use clean non-sparking tools to collect absorbed material. Large Spills: Dike far ahead of liquid spill for later disposal. Water spray may reduce vapor; but may not prevent ignition in closed spaces.

Regulations

RCRA 40CFR: Not listed
CERCLA: 40CFR 302.4: Not listed
SARA 40CFR 372.65: Not listed
SARA EHS 40CFR 355: Not listed
TSCA: Listed

DIE4900 CAS #: 91-66-7

N,N-DIETHYLANILINE

RTECS: BX3400000
EINECS Number: 202-088-8
Molecular Formula: $C_{10}H_{15}N$
Formula Weight: 149.23

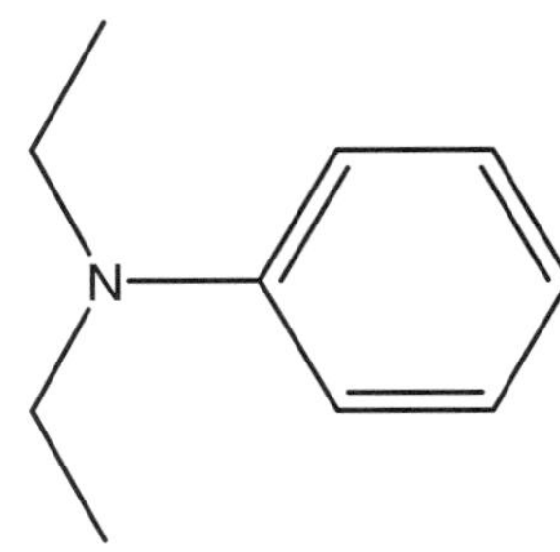

Chemical Structure

Synonyms: ANILINE,N,N-DIETHYL-; BENZENAMINE,N,N-DIETHYL-; BENZENAMINE,N,N-DIETHYL-(9CI); DEA; DIAETHYLANILIN; N,N-DIETHYLAMINOBENZENE; N,N-DIETHYLANILIN; DIETHYLANILINE; N,N-DIETHYLBENZENAMINE; DIETHYLPHENYLAMINE; N-PHENYLDIETHYLAMINE; PHENYLDIETHYLAMINE
Description: clear, light yellow to brown liquid; fish-like or amine odor
Use: dye intermediate

Physical Properties

Boiling Point: 215 °C (419 °F) to 216 °C (421 °F)
Freezing Point: -38 °C (-36.4 °F)
Specific Gravity: 0.9302 at 25 °C/4 °C
Vapor Density: 1 Air=1
Vapor Pressure: 0.145 kPa at 50 °C
Water Solubility: 1 g dissolves in 70 ml of Water at 12 °C
Other Solubilities: 95% Ethanol: >=100 mg/ml at 24.5 °C; Acetone: >=100 mg/ml at 24.5 °C; DMSO: >=100 mg/ml at 24.5 °C; Ether: Soluble.
Refraction Index: 1.5394 at 24 °C/D
Ionization Potential (eV): 6.98 +/-0.02
Flash Point: 85 °C

RTECS Toxicity Data

Acute Inhalation: Rat LC_{50} Dose: 1920 mg/m^3/4hr.

Hazard Overviews

Fire Diamond

Health: Irritating to eyes/skin. Toxic. Also Causes: CNS depression, reduction in the red blood cells' oxygen-carrying capability, blood pressure drop, tachycardia.
Fire: Combustible. Use water fog, dry chemical, carbon dioxide, or foams. Use water sprays to cool fire-exposed containers. Never direct solid streams of water into pools of burning liquid since this tends to scatter and spread the flames.
Reactivity: Stable. Hazardous polymerization cannot occur. Avoid: heat; ignition sources. Incompatible with: strong acids; strong oxidizing agents. Hazardous decomposition products: toxic carbon monoxide; highly toxic aniline vapors.
Carcinogenicity: IARC - Not listed; NIOSH - Not listed; NTP - Not listed; ACGIH - Not listed; OSHA - Not listed; EPA - Not listed; MAK - Not listed
Primary Target Organs:

Eyes

Skin

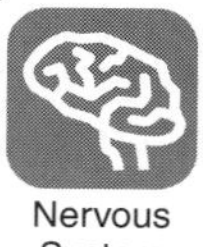
Nervous System

Cardio-vascular

Blood

Environmental

Cleanup/Disposal: Guide No. 153: Eliminate all ignition sources (no smoking, flares, sparks or flames in immediate area). Do not touch damaged containers or spilled material unless wearing appropriate protective clothing. Stop leak if you can do it without risk. Prevent entry into waterways, sewers, basements or confined areas. Absorb or cover with dry earth, sand or other non-combustible material and transfer to containers. Do not get water inside containers.

Regulations

RCRA 40CFR: Not listed
CERCLA: 40CFR 302.4: Listed per CAA Section 112 RQ: 1000 lb (453.5 kg)
SARA 40CFR 372.65: Not listed
SARA EHS 40CFR 355: Not listed
TSCA: Listed

DIE5000 CAS #: 692-42-2

DIETHYLARSINE

Molecular Formula: $C_4H_{11}As$
Formula Weight: 134.05
Synonyms: ARSINE,DIETHYL-

Physical Properties

Boiling Point: 105 °C (221 °F)
Specific Gravity: 1.1338 at 24 °C/4 °C
Saturated Vapor Density: 1.371588022 kg/m^3
Vapor Pressure: 30 mm Hg at 25 °C
Other Solubilities: Very Soluble in Ether, Benzene Soluble in Alcohol, Acetone
Refraction Index: 1.4709 at 20 °C/D
Flash Point: Spontaneously flammable

Hazard Overviews

Flammable

Fire: Flammable.
Carcinogenicity: IARC - Not listed; NIOSH - Not listed; NTP - Not listed; ACGIH - Not listed; OSHA - Not listed; EPA - Not listed; MAK - Not listed
Exposure Limits
OSHA PEL: TWA: 0.5 mg/m^3; as As. Other Values: 0.5 mg/m^3; organic.
ACGIH TLV: TWA: 0.01 mg/m^3; as As.
NIOSH REL: STEL: 0.002 mg/m^3; ceiling (15 min) as As.

Environmental

Environmental Fate: No data concerning fate in soil, water or air were. It may not be very persistent in aerobic soil and water or the atmosphere, based upon a reported comment which indicates that the compound spontaneously flames in air.

Regulations

RCRA 40CFR: Listed Hazardous Waste No. P038 Toxic Waste

CERCLA: 40CFR 302.4: Listed per RCRA Section 3001 RQ: 1 lb (0.454 kg)
SARA 40CFR 372.65: Listed as Compound
SARA EHS 40CFR 355: Not listed
TSCA: Not listed

DIE5100 CAS #: 135-01-3

1,2-DIETHYLBENZENE

RTECS: CZ5640000
DOT: UN2049; IMO3.3
EINECS Number: 205-170-1
Molecular Formula: $C_{10}H_{14}$
Formula Weight: 134.22

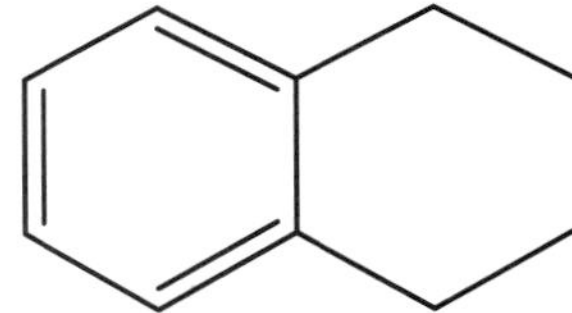

Chemical Structure

Synonyms: BENZENE,1,2-DIETHYL-; BENZENE,O-DIETHYL-; O-DIETHYLBENZENE
Description: colorless; characteristic aromatic, like benzene, odor
Use: intermediate; solvent; manufacture of divinylbenzene; to manufacture naphthalene; in powderless etching

Physical Properties

Boiling Point: 183.4 °C (362 °F) at 760 mm Hg
Freezing Point: -31.2 °C (-24.16 °F)
Saturated Vapor Density: 1.205728857 kg/m^3
Density: 0.88 g/mL at 20 °C
Vapor Pressure: 1 mm Hg at 20.7 °C
Water Solubility: 71 mg/L at 25 °C
Other Solubilities: miscible with Alcohol, Ether, Acetone, Benzene
Refraction Index: 1.5035 at 20 °C
Ionization Potential (eV): =< 8.51
Flash Point: Combustible; moderate fire risk
Autoignition Temperature: Isomeric mixture 430 °C

RTECS Toxicity Data

Acute Oral: Rat LD_{Lo} Dose: 5 gm/kg.
Chronic (Multiple Dose) Oral: Rat Dose: 2400 mg/kg/8W-I; Toxic Effects: Brain and coverings - Recordings from specific areas of CNS; Behavioral - Muscle weakness; Nutritional and gross metabolic - Weight loss or decreased weight gain.

Hazard Overviews

Health: Irritating to eyes/skin/respiratory tract. Other Acute Effects: may be harmful by inhalation, ingestion, or skin absorption; target organs: liver, kidneys.
Fire: Combustible. Extinguishing agents: water spray; carbon dioxide, dry chemical powder or appropriate foam. Precautions: combustible liquid.
Reactivity: Incompatible with: strong oxidizing agents. Hazardous decomposition products: toxic fumes of: carbon monoxide, carbon dioxide.
Carcinogenicity: IARC - Not listed; NIOSH - Not listed; NTP - Not listed; ACGIH - Not listed; OSHA - Not listed; EPA - Not listed; MAK - Not listed
Primary Target Organs:

Eyes

Skin
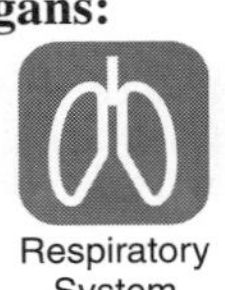
Respiratory System

Liver

Kidneys

Environmental

Ecotoxicity: Toxicity threshold (cell multiplication inhibition test): green algae (Scenedesmus quadricauda) 8d EC_0 >20 mg/l
Environmental Fate: It is not expected to hydrolyze, but has the potential to undergo direct photolysis in sunlit environmental media (it absorbs UV light at wavelengths in the environmentally significant range, >290 nm). Limited aqueous grab sample data for gas oil mixtures containing it suggests it should biodegrade in soil and water. An estimated K_{oc} indicates it should have a medium mobility in soil and it may partition from the water column to organic matter in sediments and suspended solids. The potential for bioconcentration in aquatic organisms is low. A Henry's Law constant of 2.61 x10^{-3} atm-cu m/mole at 25 °C suggests that it should rapidly volatilize from natural waters. The volatilization half-lives from a model river and a model pond, the latter considers the effect of adsorption, have been estimated to be about 3.8 hours and 4.25 days, respectively. Based on its vapor pressure, it should evaporate from dry surfaces, especially when present in high concentration such as in spill situations. It is expected to exist entirely in the vapor phase in ambient air. Vapor phase reactions with photochemically produced hydroxyl radicals in the atmosphere may be important (estimated half-life of 1.9 days). Physical removal from air by rainfall and dissolution in clouds, etc. may occur; however, the short atmospheric residence time suggests that wet deposition is of limited importance.
Cleanup/Disposal: Guide No. 130: Eliminate all ignition sources (no smoking, flares, sparks or flames in immediate area). All equipment used when handling the product must be grounded. Do not touch or walk through spilled material. Stop leak if you can do it without risk. Prevent entry into waterways, sewers, basements or confined areas. A vapor suppressing foam may be used to reduce vapors. Absorb or cover with dry earth, sand or other non-combustible material and transfer to containers. Use clean non-sparking tools to collect absorbed material. Large Spills: Dike far ahead of liquid spill for later disposal. Water spray may reduce vapor; but may not prevent ignition in closed spaces.

Environmental Physical Data

Henry's Law Constant: calculated at 2.61 $x10^{-3}$
Sorption Partition Coefficient: K_{OC} = 418
BCF: calculated at 1.75

Regulations

RCRA 40CFR: Not listed
CERCLA: 40CFR 302.4: Not listed
SARA 40CFR 372.65: Not listed
SARA EHS 40CFR 355: Not listed
TSCA: Not listed

DIE5200 **CAS #: 141-93-5**

1,3-DIETHYLBENZENE

RTECS: CZ5620000
DOT: UN2049; IMO3.3
EINECS Number: 205-511-4
Molecular Formula: $C_{10}H_{14}$
Formula Weight: 134.22
Synonyms: BENZENE,1,3-DIETHYL-; BENZENE,M-DIETHYL-; M-DIETHYLBENZENE
Description: colorless solid
Use: intermediate; solvent; to manufacture divinylbenzene; in powderless etching

Physical Properties

Boiling Point: 181 °C (358 °F) at 760 mm Hg
Freezing Point: -83.89 °C (-119.002 °F)
Specific Gravity: 0.862 at 20 °C/4 °C
Vapor Pressure: 1 mm Hg at 20 °C
Water Solubility: 170 mg/L (est)
Other Solubilities: miscible in Alcohol, Carbon Tetrachloride, Benzene, Ether, Acetone
Refraction Index: 1.4955 at 20 °C
Ionization Potential (eV): 8.49 +/-0.01
Flash Point: Combustible; moderate fire risk
Autoignition Temperature: Isomer mixture 430 °C

RTECS Toxicity Data

Acute Oral: Rat LD_{Lo} Dose: 5 gm/kg.

Hazard Overviews

Health: Irritating to eyes/skin/respiratory tract. Other Acute Effects: may be harmful by inhalation, ingestion, or skin absorption; vapor or mist is irritating; target organs: liver, kidneys.
Fire: Combustible. Extinguishing agents: water spray; carbon dioxide, dry chemical powder or appropriate foam. Precautions: combustible liquid.
Reactivity: Incompatible with: strong oxidizing agents. Hazardous decomposition products: toxic fumes of: carbon monoxide, carbon dioxide.
Carcinogenicity: IARC - Not listed; NIOSH - Not listed; NTP - Not listed; ACGIH - Not listed; OSHA - Not listed; EPA - Not listed; MAK - Not listed

Primary Target Organs:

Eyes

Skin

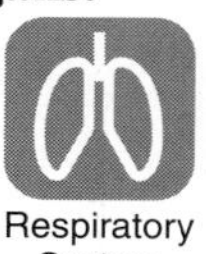
Respiratory System

Liver

Kidneys

Environmental

Ecotoxicity: LC_{50} Pimephales promelas (fathead minnow) 4.15 mg/l/96 hr (confidence limit 4.05 - 4.25 mg/l), flow-through bioassay with measured concentrations, 23.5 °C, dissolved oxygen 7.5 mg/l, hardness 45.8 mg/l calcium carbonate, alkalinity 42.2 mg/l calcium carbonate, and pH 7.28.
Environmental Fate: It is not expected to hydrolyze, but has the potential to undergo direct photolysis in sunlit environmental media (it absorbs UV light at wavelengths in the environmentally significant range, >290 nm). Limited aqueous grab sample data for gas oil mixtures containing it suggests it should biodegrade in soil and water. An estimated K_{oc} indicates it should have a medium mobility in soil and it can partition from the water column to organic matter in sediments and suspended solids. It has the potential to bioconcentrate in aquatic organisms. An estimated Henry's Law constant of 1.17 $x10^{-3}$ atm-cu m/mole at 25 °C suggests that it should rapidly volatilize from natural waters. The volatilization half-lives from a model river and a model pond, the latter considers the effect of adsorption, have been estimated to be about 4.3 hours and 3.7 days, respectively. Based on its vapor pressure, it should evaporate from dry surfaces, especially when present in high concentration such as in spill situations. It is expected to exist entirely in the vapor phase in ambient air. Vapor phase reactions with photochemically produced hydroxyl radicals in the atmosphere may be important (estimated half-life of 1.1 days).
Cleanup/Disposal: Guide No. 130: Eliminate all ignition sources (no smoking, flares, sparks or flames in immediate area). All equipment used when handling the product must be grounded. Do not touch or walk through spilled material. Stop leak if you can do it without risk. Prevent entry into waterways, sewers, basements or confined areas. A vapor suppressing foam may be used to reduce vapors. Absorb or cover with dry earth, sand or other non-combustible material and transfer to containers. Use clean non-sparking tools to collect absorbed material. Large Spills: Dike far ahead of liquid spill for later disposal. Water spray may reduce vapor; but may not prevent ignition in closed spaces.

Environmental Physical Data

Henry's Law Constant: calculated at 1.17 $x10^{-3}$
Octanol/Water Partition Coefficient: log K_{OW} = 4.50
Sorption Partition Coefficient: K_{OC} = 260
BCF: calculated at 3.19

Regulations

RCRA 40CFR: Not listed
CERCLA: 40CFR 302.4: Not listed
SARA 40CFR 372.65: Not listed
SARA EHS 40CFR 355: Not listed
TSCA: Listed

Analytical Methods

Air: ASTM D4490
Water / Groundwater: ASTM D3695

DIE5300 CAS #: 105-05-5

1,4-DIETHYLBENZENE

DOT: UN2049; IMO3.3
EINECS Number: 203-265-2
Molecular Formula: $C_{10}H_{14}$
Formula Weight: 134.22

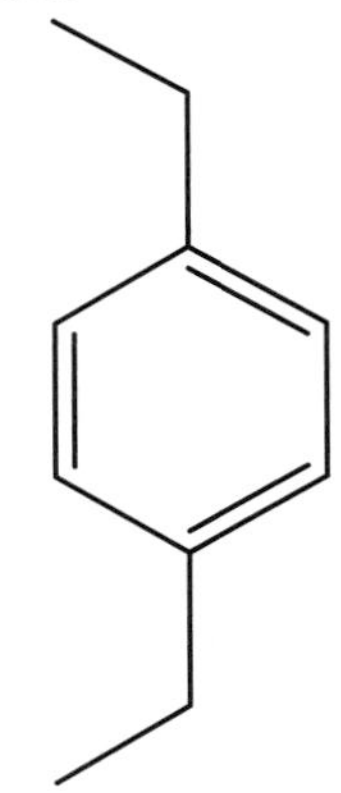

Chemical Structure

Synonyms: BENZENE,1,4-DIETHYL-; BENZENE,P-DIETHYL-; P-DIETHYLBENZENE; P-ETHYLETHYLBENZENE
Description: colorless solid; characteristic odor like benzene or toluene
Use: intermediate; solvent; solvent in the parex process for the production of paraxylene; manufacture of divinylbenzene; powderless etching

Physical Properties

Boiling Point: 183.8 °C (363 °F) at 760 mm Hg
Freezing Point: -42.8 °C (-45.04 °F)
Specific Gravity: 0.862 at 20 °C/4 °C
Vapor Pressure: 1 mm Hg at 20 °C
Water Solubility: 25 mg/L at 25 °C
Other Solubilities: miscible with Alcohol, Carbon Tetrachloride, Benzene, Ether, Acetone
Refraction Index: 1.4967 at 20 °C
Ionization Potential (eV): 8.40
Flash Point: Combustible; moderate fire risk
Autoignition Temperature: Isomeric mixture 430 °C

Hazard Overviews

Health: Irritating to eyes/skin/respiratory tract. Other Acute Effects: may be harmful by inhalation, ingestion, or skin absorption; target organs: liver, kidneys.
Fire: Combustible. Extinguishing agents: water spray; carbon dioxide, dry chemical powder or appropriate foam. Precautions: combustible liquid.
Reactivity: Incompatible with: strong oxidizing agents. Hazardous decomposition products: toxic fumes of: carbon monoxide, carbon dioxide.
Carcinogenicity: IARC - Not listed; NIOSH - Not listed; NTP - Not listed; ACGIH - Not listed; OSHA - Not listed; EPA - Not listed; MAK - Not listed
Primary Target Organs:

Eyes

Skin

Respiratory System

Liver

Kidneys

Environmental

Environmental Fate: It is not expected to hydrolyze, but has the potential to undergo direct photolysis in sunlit environmental media (it absorbs UV light at wavelengths in the environmentally significant range, >290 nm). Limited aqueous grab sample data for gas oil mixtures suggests it should biodegrade in soil and water. A high estimated K_{oc} indicates it should have a low mobility in soil and it should partition from the water column to organic matter in sediments and suspended solids. The potential for bioconcentration in aquatic organisms is low. A Henry's Law constant of 7.54×10^{-3} atm-cu m/mole at 25 °C suggests that it should rapidly volatilize from natural waters. The volatilization half-lives from a model river and a model pond, the latter considers the effect of adsorption, have been estimated to be about 3.5 hours and 6 days, respectively. Based on its vapor pressure, it should evaporate from dry surfaces, especially when present in high concentration such as in spill situations. It is expected to exist entirely in the vapor phase in ambient air. Vapor phase reactions with photochemically produced hydroxyl radicals in the atmosphere may be important (estimated half-life of 1.9 days). The short atmospheric residence time suggests that wet deposition is of limited importance.
Cleanup/Disposal: Guide No. 130: Eliminate all ignition sources (no smoking, flares, sparks or flames in immediate area). All equipment used when handling the product must be grounded. Do not touch or walk through spilled material. Stop leak if you can do it without risk. Prevent entry into waterways, sewers, basements or confined areas. A vapor suppressing foam may be used to reduce vapors. Absorb or cover with dry earth, sand or other non-combustible material and transfer to containers. Use clean non-sparking tools to collect absorbed material. Large Spills: Dike far ahead of liquid spill for later disposal. Water spray may reduce vapor; but may not prevent ignition in closed spaces.

Environmental Physical Data

Henry's Law Constant: calculated at 7.54×10^{-3}
Sorption Partition Coefficient: K_{OC} = 746
BCF: calculated at 2.00

Regulations

RCRA 40CFR: Not listed
CERCLA: 40CFR 302.4: Not listed
SARA 40CFR 372.65: Not listed
SARA EHS 40CFR 355: Not listed

TSCA: Listed

DIE5400 CAS #: 1642-54-2

DIETHYLCARBAMAZINE CITRATE

RTECS: TL1225000
EINECS Number: 216-696-6
Molecular Formula: $C_{16}H_{29}N_3O_8$
Formula Weight: 391.4

Chemical Structure

Synonyms: BANOCIDE; CARICIDE; CARITROL; DICAROCIDE; DIETHYLCARBAMAZANE CITRATE; DIETHYLCARBAMAZINE ACID CITRATE; DIETHYLCARBAMAZINE HYDROGEN CITRATE; DIETHYLCARBAMAZINI CITRAS; 1-DIETHYLCARBAMOYL-4-METHYLPIPERAZINE DIHYDROGEN CITRATE; N,N-DIETHYL-4-METHYL-1-PIPERAZINE CARBOXAMIDE CITRATE; N,N-DIETHYL-4-METHYL-1-PIPERAZINECARBOXAMIDE DIHYDROGENCITRATE; N,N-DIETHYL-4-METHYL-1-PIPERAZINECARBOXAMIDE DIHYDROGENCITRATE (1:1); N,N-DIETHYL-4-METHYLPIPERAZINE-1-CARBOXAMIDE DIHYDROGENCITRATE; DIROCIDE; DITRAZIN; DITRAZIN CITRATE; DITRAZINE; DITRAZINE CITRATE; DITRAZINI CITRAS; DITRAZINUM; ETHODRYL CITRATE; FILARABITS; FILAZINE; FRANOCIDE; FRANOZAN; HETRAZAN; LONGICID; LOXURAN; 1-METHYL-4-DIETHYLCARBAMOYLPIPERAZINE CITRATE; 1-PIPERAZINECARBOXAMIDE,N,N-DIETHYL-4-METHYL-,CITRATE; 1-PIPERAZINECARBOXAMIDE,N,N-DIETHYL-4-METHYL-,2-HYDROXY-1,2,3-PROPANETRICARBOXYLATE; RP 3799
Description: white, crystalline powder; odorless or slight odor
Use: medication; medication (vet)

Physical Properties

Freezing Point: 141 °C (285.8 °F) to 143 °C (289.4 °F)
Water Solubility: > 75% in Water at 20 °C
Other Solubilities: Freely Soluble in hot Alcohol; Sparingly Soluble in cold Alcohol; Practically Insoluble in Acetone, Benzene, Chloroform and Ether.

RTECS Toxicity Data

Acute Oral: Rat LD_{50} Dose: 1400 mg/kg.
Acute Inhalation: Rat LC_{50} Dose: 309 mg/m^3/4hr.

Hazard Overviews

Poison

Health: Irritating to eyes/skin/respiratory tract. Poison. Other Acute Effects: toxic if inhaled; harmful if swallowed; may be harmful if absorbed through the skin; exposure can cause nausea; headache; vomiting; anorexia; dizziness; joint pains; weakness; convulsions.
Fire: Hazards: emits toxic fumes. Extinguishing agents: water spray; carbon dioxide, dry chemical powder or appropriate foam. Precautions: combustible liquid.
Reactivity: Stable. Hazardous polymerization will not occur. Hazardous decomposition products: toxic fumes of: carbon monoxide, carbon dioxide, nitrogen oxides.
Carcinogenicity: IARC - Not listed; NIOSH - Not listed; NTP - Not listed; ACGIH - Not listed; OSHA - Not listed; EPA - Not listed; MAK - Not listed
Primary Target Organs:

Eyes

Skin

Respiratory System

Environmental

Regulations

RCRA 40CFR: Not listed
CERCLA: 40CFR 302.4: Not listed
SARA 40CFR 372.65: Not listed
SARA EHS 40CFR 355: Listed
TSCA: Listed

DIE5500 CAS #: 95-06-7

DIETHYLCARBAMODITHIOIC ACID, 2-CHLORO-2-PROPENYL ESTER

RTECS: EZ5075000
EINECS Number: 202-388-9
Molecular Formula: $C_8H_{14}ClNS_2$
Structured MF: $(CH_3CH_2)_2NC(=S)SCH_2C(Cl)=CH_2$
Formula Weight: 223.79
Synonyms: CARBAMIC ACID,DIETHYLDITHIO-,2-CHLOROALLYL ESTER; CARBAMODITHIOIC ACID,DIETHYL-,2-CHLORO-2-PROPENYL ESTER; CARBAMODITHIOIC ACID,DIETHYL-,2-

CHLORO-2-PROPENYL ESTER(9CI); CDEC; 2-CHLORALLYL DIETHYLDITHIOCARBAMATE; CHLORALLYL DIETHYLDITHIOCARBAMATE; 2-CHLOROALLYL DIETHYLDITHIOCARBAMATE; 2-CHLOROALLYL N,N-DIETHYLDITHIOCARBAMATE; 2-CHLORO-2-PROPENE-1-THIOL DIETHYLDITHIOCARBAMATE; 2-CHLORO-2-PROPENYL DIETHYLCARBAMODITHIOATE; CP 4,742; CP 4572; CP 4742; DIETHYLCARBAMODITHIOIC ACID 2-CHLORO-2-PROPENYL ESTER; DIETHYLDITHIOCARBAMIC ACID 2-CHLOROALLYL ESTER; 2-PROPENE-1-THIOL,2-CHLORO-,DIETHYLDITHIOCARBAMATE; SULFALLATE; THIOALLATE; VEGADEX; VEGADEX SUPER; VEGEDEX

Description: amber oily liquid

Use: pre-emergence herbicide and a pesticide

Physical Properties

Boiling Point: 128 °C (262 °F) to 130 °C (266 °F)

Freezing Point: Decomposes at 150 °C (302 °F)

Specific Gravity: 1.088 at 25 °C

Saturated Vapor Density: 1.200023332 kg/m^3

Density: 1.088 g/mL at 25 °C

Vapor Pressure: 2.2 x10^{-3} mm Hg at 20 °C

Water Solubility: 100 ppm at 25 °C

Other Solubilities: 95% Ethanol: >=100 mg/ml at 22 °C; Acetone: >=100 mg/ml at 22 °C; Alcohol: Soluble; Benzene: Soluble; Chloroform: Soluble; DMSO: >=100 mg/ml at 22 °C; Ether: Soluble; Most organic solvents: Soluble; Toluene: >=100 mg/ml at 22 °C.

Refraction Index: 1.5822 at 25 °C/D

Flash Point: > 93.3 °C

RTECS Toxicity Data

Acute Oral: Rat LD_{50} Dose: 850 mg/kg.

Acute Dermal: Rabbit LD_{50} Route: Skin; Dose: 2200 mg/kg.

Mutagenic: Human DNA Inhibition; Cell Type: lymphocyte; Dose: 500 mg/L. Rat DNA Inhibition; Cell Type: other cell types; Dose: 1 gm/L.

Tumorigenic: Rat Route: Oral; Dose: 6825 mg/kg/78W-C; Toxic Effects: Tumorigenic - Carcinogenic by RTECS criteria; Gastrointestinal - Tumors; Skin and appendages - Tumors. Rat Route: Oral; Dose: 11 gm/kg/78W-I; Toxic Effects: Tumorigenic - Carcinogenic by RTECS criteria; Gastrointestinal - Tumors; Skin and appendages - Tumors.

Hazard Overviews

Fire: Will burn.

Carcinogenicity: IARC - Group 2B, Possibly carcinogenic to humans; NIOSH - Not listed; NTP - Listed; ACGIH - Not listed; OSHA - Not listed; EPA - Not listed; MAK - Not listed

Environmental

Regulations

RCRA 40CFR: Listed Hazardous Waste No. U277 Toxic Waste

CERCLA: 40CFR 302.4: Listed per RCRA Section 3001 RQ: 1 lb (0.454 kg)

SARA 40CFR 372.65: Not listed

SARA EHS 40CFR 355: Not listed

TSCA: Not listed

Analytical Methods

Soil: SW846 8270B, 8270C

Water / Groundwater: EPA 022

Food: FDA 212.1, 232.1, 232.4, 242.1

Plasma: EPA 001, 028, 29; FDA 211.1, 231.1, 252

DIE5600 CAS #: 88-10-8

DIETHYLCARBAMOYL CHLORIDE

RTECS: FD4025000

EINECS Number: 201-798-5

Molecular Formula: $C_5H_{10}ClNO$

Formula Weight: 135.61

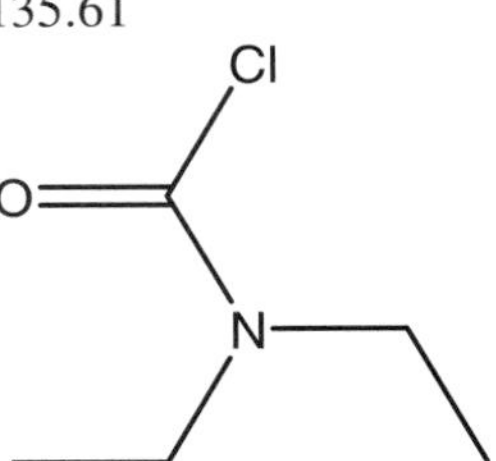

Chemical Structure

Synonyms: CARBAMIC CHLORIDE,DIETHYL-; CARBAMIC CHLORIDE,DIETHYL-(9CI); CARBAMIDOYL CHLORIDE,DIETHYL-; CARBAMOYL CHLORIDE,DIETHYL-; DIETHYL CARBAMYL CHLORIDE; DIETHYLAMID KYSELINY CHLORMRAVENCI; DIETHYLCARBAMIC CHLORIDE; N,N-DIETHYLCARBAMOYL CHLORIDE; DIETHYLCARBAMYL CHLORIDE; DIETHYLCHLOROFORMAMIDE; N,N-DIETHYLCHLOROFORMAMIDE

Description: liquid

Use: intermediate in synthesis of herbicides

Physical Properties

Boiling Point: 187 °C (369 °F) to 190 °C (374 °F)

Freezing Point: -44 °C (-47.2 °F)

Vapor Density: 4.1 Air=1

Water Solubility: Soluble in Water

Flash Point: 162.778 to 172.222 °C Open Cup

RTECS Toxicity Data

Acute Oral: Rat LD_{50} Dose: 2700 mg/kg; Toxic Effects: Behavioral - Coma; Gastrointestinal - Alteration in gastric secretion; Blood - Changes in spleen.

Acute Inhalation: Rat LC_{Lo} Dose: 159 ppm/7hr.

Acute Dermal: Rabbit LD_{Lo} Route: Skin; Dose: 12840 mg/kg; Toxic Effects: Skin and appendages - Dermatitis, other.

Chronic (Multiple Dose) Oral: Rat Dose: 5400 mg/kg/10D-I; Toxic Effects: Gastrointestinal - Hypermotility, diarrhea; DEATH.

Irritation Eye: Rabbit Standard Draize Test Dose: 100 mg; Reaction: moderate.

Tumorigenic: Mouse Route: Skin; Dose: 43200 mg/kg/72W-I; Toxic Effects: Tumorigenic - Carcinogenic by RTECS criteria; Skin and appendages - Tumors; Tumorigenic - Tumors at site of application.

Hazard Overviews

Corrosive

Health: Corrosive to eyes/skin/respiratory system. Toxic. Other Acute Effects: harmful if swallowed, inhaled, or absorbed through skin; inhalation may result in spasm, inflammation and edema of the larynx and bronchi, chemical pneumonitis and pulmonary edema; symptoms of exposure may include burning sensation, coughing, wheezing, laryngitis, shortness of breath, headache, nausea and vomiting. Chronic Effects: may alter genetic material. Carcinogen.
Fire: Will burn. Hazards: emits toxic fumes. Extinguishing agents: carbon dioxide, dry chemical powder or appropriate foam. Precautions: combustible liquid.
Reactivity: Incompatible with: strong oxidizing agents, strong bases, may decompose on exposure to moist air or water. Hazardous decomposition products: toxic fumes of: carbon monoxide, carbon dioxide, nitrogen oxides, hydrogen chloride gas.
Carcinogenicity: IARC - Not listed; NIOSH - Not listed; NTP - Not listed; ACGIH - Not listed; OSHA - Not listed; EPA - Not listed; MAK - Class B, Justifiably suspected of having carcinogenic potential
Primary Target Organs:

Eyes

Skin

Respiratory System

Environmental

Regulations

RCRA 40CFR: Not listed
CERCLA: 40CFR 302.4: Not listed
SARA 40CFR 372.65: Not listed
SARA EHS 40CFR 355: Not listed
TSCA: Listed

DIE5700 CAS #: 1719-53-5

DIETHYLDICHLOROSILANE

RTECS: VV3060000
DOT: UN1767; IMO8.0
EINECS Number: 217-005-0
Molecular Formula: $C_4H_{10}Cl_2Si$
Structured MF: $(CH_3CH_2)_2SiCl_2$
Formula Weight: 157.13

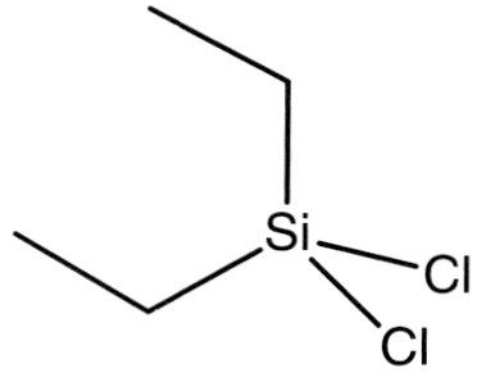

Chemical Structure

Synonyms: DICHLORODIETHYLSILANE; DIETHYLDICHLOROSILICON; SILANE,DICHLORODIETHYL-
Description: colorless liquid
Use: intermediate for silicones

Physical Properties

Boiling Point: 129 °C (264 °F)
Freezing Point: -96 °C (-140.8 °F)
Specific Gravity: 1.0504 at 20 °C/4 °C
Vapor Density: 5.14
Saturated Vapor Density: 1.26976225 kg/m^3
Density: 1.0504 g/mL at 20 °C
Vapor Pressure: 10 mm Hg at 25.4 °C
Water Solubility: Decomposes
Other Solubilities: Acetone: >=100 mg/ml at 19 °C; Alcohol: Decomposes; DMSO: 10-50 mg/ml at 19 °C.
Refraction Index: 1.4809 at 20 °C/D
Flash Point: 77 °C Open Cup

RTECS Toxicity Data

Acute Oral: Rat LD_{Lo} Dose: 1 gm/kg.

Hazard Overviews

Corrosive

Health: Corrosive to eyes/skin/respiratory tract. Other Acute Effects: harmful if swallowed, inhaled, or absorbed through skin; inhalation may result in spasm, inflammation and edema of the larynx and bronchi, chemical pneumonitis and pulmonary edema; symptoms of exposure may include burning sensation; coughing; wheezing; laryngitis; shortness of breath; headache; nausea; vomiting.
Fire: Combustible. Hazards: emits toxic fumes; vapor may travel considerable distance to source of ignition and flash back; water hydrolyzes material liberating acidic gas which in contact with metal surfaces can generate flammable and/or explosive hydrogen gas. Extinguishing agents: carbon dioxide, dry chemical powder or appropriate foam. Precautions: combustible liquid.
Reactivity: Incompatible with: strong oxidizing agents, may decompose on exposure to moist air or water. Hazardous decomposition products: toxic fumes of: carbon monoxide, carbon dioxide, hydrogen chloride gas, silicon oxide
Carcinogenicity: IARC - Not listed; NIOSH - Not listed; NTP - Not listed; ACGIH - Not listed; OSHA - Not listed; EPA - Not listed; MAK - Not listed

Primary Target Organs:

Eyes

Skin

Respiratory System

Environmental

Cleanup/Disposal: Guide No. 155: Eliminate all ignition sources (no smoking, flares, sparks or flames in immediate area). All equipment used when handling the product must be grounded. Do not touch damaged containers or spilled material unless wearing appropriate protective clothing. Stop leak if you can do it without risk. A vapor suppressing foam may be used to reduce vapors. For chlorosilanes, use AFFF alcohol-resistant medium expansion foam to reduce vapors. Do not get water on spilled substance or inside containers. Use water spray to reduce vapors or divert vapor cloud drift. Prevent entry into waterways, sewers, basements or confined areas. Small Spills: Cover with dry earth, dry sand, or other non-combustible material followed with plastic sheet to minimize spreading or contact with rain. Use clean non-sparking tools to collect material and place it into loosely covered plastic containers for later disposal.

Regulations

RCRA 40CFR: Not listed
CERCLA: 40CFR 302.4: Not listed
SARA 40CFR 372.65: Not listed
SARA EHS 40CFR 355: Not listed
TSCA: Listed

DIE5900 CAS #: 1762-27-2

DIETHYLDIMETHYLLEAD

EINECS Number: 217-170-9
Molecular Formula: $C_6H_{16}Pb$
Formula Weight: 295.4
Synonyms: DIETHYLDIMETHYLPLUMBANE; DIMETHYLDIETHYLLEAD; PLUMBANE,DIETHYLDIMETHYL-
Use: gasoline additive

Hazard Overviews

Carcinogenicity: IARC - Not listed; NIOSH - Not listed; NTP - Not listed; ACGIH - Not listed; OSHA - Not listed; EPA - Not listed; MAK - Not listed

Exposure Limits

OSHA PEL: STEL: 0.05 mg/m^3; as Pb inorganic.
ACGIH TLV: TWA: 0.05 mg/m^3; as Pb inorganic.
NIOSH REL: TWA: 0.1 mg/m^3; < inorganic; blood Pb<.06mg/100g.
DFG MAK: TWA: 0.1 mg/m^3; as Pb inorganic.

Environmental

Regulations

RCRA 40CFR: Not listed
CERCLA: 40CFR 302.4: Listed as Compound per CWA Section 307(a) per CAA Section 112
SARA 40CFR 372.65: Listed as Compound
SARA EHS 40CFR 355: Not listed
TSCA: Listed

DIE6000 CAS #: 14324-55-1

DIETHYLDITHIOCARBAMIC ACID ZINC SALT

RTECS: ZH0350000
EINECS Number: 238-270-9
Molecular Formula: $C_{10}H_{20}N_2S_4Zn$
Formula Weight: 363.95

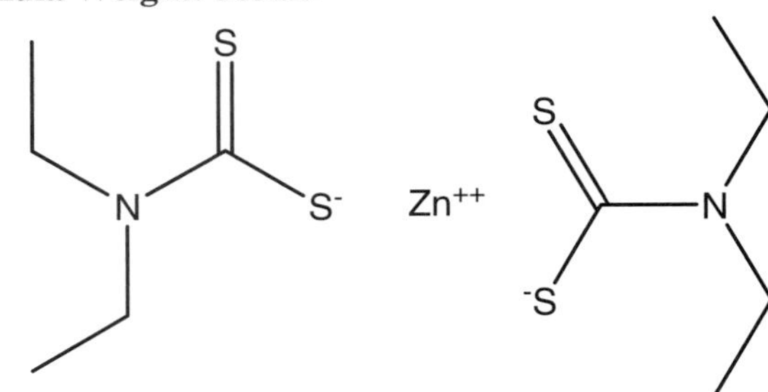

Chemical Structure

Synonyms: BIS(DIETHYLDITHIOCARBAMATO)ZINC; CARBAMODITHIOIC ACID,DIETHYL-,ZINC SALT; ETHAZATE; ETHYL CYMATE; ETHYL ZIMATE; ETHYL ZIRAM; ETHYLZIMATE; HERMAT ZDK; NOCCELER EZ; SOXINOL EZ; VULCACURE ZE; VULKACIT LDA; VULKACIT ZDK; ZIMATE,ETHYL; ZINC BIS(DIETHYLDITHIOCARBAMATE); ZINC DIETHYLCARBAMODITHIOATE; ZINC DIETHYLDITHIOCARBAMATE; ZINC N,N-DIETHYLDITHIOCARBAMATE; ZINC,BIS(DIETHYLCARBAMODITHIOATO-S,S')-,(T-4)-; ZINC,BIS(DIETHYLDITHIOCARBAMATO)
Description: white powder
Use: accelerator for rubber vulcanization; heat stabilizer for polyethylene

Physical Properties

Freezing Point: 172 °C (341.6 °F) to 176 °C (348.8 °F)
Specific Gravity: 1.47 at 20 °C/20 °C
Water Solubility: Insoluble in Water
Other Solubilities: Soluble in Carbon Disulfide, Benzene, Chloroform

RTECS Toxicity Data

Acute Oral: Rat LD_{50} Dose: 700 mg/kg. Mouse LD_{50} Dose: 1400 mg/kg.
Irritation Eye: Rabbit Standard Draize Test Dose: 100 mg/24H; Reaction: moderate.
Mutagenic: Bacteria - S Typhimurium Mutations in Microorganisms; Dose: 25 ug/plate (+/-S9).
Tumorigenic: Mouse Route: Oral; Dose: 28 gm/kg/78W-I; Toxic Effects: Tumorigenic - Equivocal tumorigenic agent by

RTECS criteria; Lungs, Thorax, or Respiration - Tumors; Blood - Tumors. Mouse Route: Subcutaneous; Dose: 464 mg/kg; Toxic Effects: Tumorigenic - Carcinogenic by RTECS criteria; Blood - Tumors.

Hazard Overviews

Health: Severely irritating to eyes; irritating to skin. Toxic. Other Acute Effects: harmful if swallowed, inhaled, or absorbed through skin.
Fire: Hazards: emits toxic fumes. Extinguishing agents: water spray; carbon dioxide, dry chemical powder or appropriate foam. Precautions: combustible liquid.
Reactivity: Incompatible with: strong oxidizing agents. Hazardous decomposition products: toxic fumes of: carbon monoxide, carbon dioxide, sulfur oxides, nitrogen oxides.
Carcinogenicity: IARC - Not listed; NIOSH - Not listed; NTP - Not listed; ACGIH - Not listed; OSHA - Not listed; EPA - Not listed; MAK - Not listed
Primary Target Organs:

Eyes

Skin

Environmental

Regulations

RCRA 40CFR: Listed Hazardous Waste No. U407 Toxic Waste
CERCLA: 40CFR 302.4: Listed per RCRA Section 3001 RQ: 1 lb (0.454 kg)
SARA 40CFR 372.65: Listed as Compound
SARA EHS 40CFR 355: Not listed
TSCA: Listed

DIE6100 CAS #: 111-46-6

DIETHYLENE GLYCOL

RTECS: ID5950000
EINECS Number: 203-872-2
Molecular Formula: $C_4H_{10}O_3$
Structured MF: $(HOCH_2CH_2)_2O$
Formula Weight: 106.14

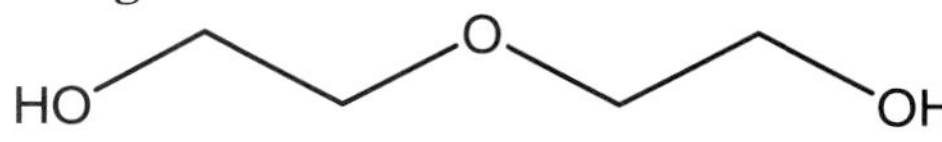

Chemical Structure

Synonyms: BIS(2-HYDROXYETHYL) ETHER; BIS(BETA-HYDROXYETHYL) ETHER; BRECOLANE NDG; CARBITOL; DEACTIVATOR E; DEACTIVATOR H; DEG; DICOL; DIETHYLENGLYKOL; DIGENOS; DIGLYCOL; DIGOL; 2,2'-DIHYDROXYDIETHYL ETHER; BETA,BETA'-DIHYDROXYDIETHYL ETHER; DIHYDROXYDIETHYL ETHER; 2,2'-DIHYDROXYETHYL ETHER; DISSOLVANT APV; ETHANOL,2,2'-OXYBIS-; ETHANOL,2,2'-OXYDI-; ETHYLENE DIGLYCOL; GLYCOL ETHER; GLYCOL ETHYL ETHER; 2-(2-HYDROXYETHOXY)ETHANOL; 2-HYDROXYETHYL ETHER; 3-OXA-1,5-PENTANEDIOL; 3-OXAPENTANE-1,5-DIOL; 2,2'-OXYBIS(ETHANOL); 2,2'-OXYBISETHANOL; 2,2-OXYBISETHANOL; 2,2'-OXYDIETHANOL; 2,2'-OXYETHANOL; TL4N
Description: colorless liquid; odorless
Use: in the production of polyurethane, unsaturated polyester resins and triethylene glycol; as a textile softener, in petroleum solvent extraction, in the dehydration of natural gas, as a plasticizer, in surfactants and as a solvent for nitrocellulose, resins, dyes, oils and many other organic compounds; as a humectant for tobacco, cork, printing ink and glue; in casein, in synthetic sponges and paper products, in book-binding adhesives, as a dyeing assistant, in cosmetics, in antifreeze solutions, in lacquers, in lubricants and in brake fluids

Physical Properties

Boiling Point: 244 °C (471 °F) to 245 °C (473 °F)
Freezing Point: -6.5 °C (20.3 °F)
Specific Gravity: 1.18 at 20 °C/20 °C
Vapor Density: 3.66 Air=1
Saturated Vapor Density: < 1.200042 kg/m^3
Density: 1.118 g/mL
Vapor Pressure: < 0.01 mm Hg at 20 °C
Water Solubility: >= 100 mg/mL at 20 C
Other Solubilities: 95% Ethanol: >=100 mg/ml at 20 °C; Acetone: >=100 mg/ml at 20 °C; Alcohol: miscible; Benzene: Immiscible; Carbon Tetrachloride: Immiscible; DMSO: >=100 mg/ml at 20 °C; Ether: Soluble; Ethylene Glycol: miscible; Toluene: Immiscible.
Surface Tension: 48.5 dynes/cm at 25 °C
Refraction Index: 1.446
Evaporation Rate: < 1 Butyl Acetate=1
Critical Temperature: 408 °C
Critical Pressure: 680 psia
Flash Point: 137.778 to 143.333 °C Open Cup
Autoignition Temperature: 229 °C
LEL: 1.6% v/v
UEL: 10.8% v/v

RTECS Toxicity Data

Acute Oral: Child TD_{Lo} Dose: 2400 mg/kg; Toxic Effects: Behavioral - Somnolence (general depressed activity); Liver - Other changes; Nutritional and gross metabolic - Metabolic acidosis. Human LD_{Lo} Dose: 1 gm/kg. Rat LD_{50} Dose: 12565 mg/kg.
Acute Inhalation: Mouse LC_{Lo} Dose: 130 mg/m^3/2hr; Toxic Effects: Behavioral - General anesthetic; Behavioral - Excitment; Lungs, Thorax, or Respiration - Cyanosis.
Acute Dermal: Rabbit LD_{50} Route: Skin; Dose: 11890 mg/kg. Rat LD_{50} Route: Subcutaneous Dose: 18800 mg/kg.
Chronic (Multiple Dose) Oral: Rat Dose: 297 gm/kg/99D-C; Toxic Effects: Kidney, Ureter, and Bladder - Other changes; DEATH. Dog Dose: 105 gm/kg/18D-I; Toxic Effects: DEATH.
Chronic (Multiple Dose) Inhalation: Rat Dose: 20 mg/m^3/2H/26W-I; Toxic Effects: Vascular - BP lowering not characterized in autonomic section; Lungs, Thorax, or Respiration - Emphysema; DEATH. Mouse Dose: 35

mg/m^3/11W-I; Toxic Effects: Cardiac - Other changes; Liver - Fatty Liver degeneration; DEATH.

Chronic (Multiple Dose) Dermal: Rabbit Route: Skin; Dose: 17300 uL/kg/30D-I; Toxic Effects: Liver - Other changes; Kidney, Ureter, and Bladder - Chgs in tubules (inc acute renal failure, acute tubular necrosis; DEATH.

Irritation Eye: Rabbit Standard Draize Test Dose: 50 mg; Reaction: mild.

Irritation Skin: Human Standard Draize Test Dose: 112 mg/3D-I; Reaction: mild. Rabbit Standard Draize Test Dose: 500 mg; Reaction: mild.

Reproductive/Teratogenic: Rat Route: Oral; Dose: 50 gm/kg; Duration: female 1-20D of pregnancy; Specific Developmental Abnormalities - Musculoskeletal system. Rat Route: Oral; Dose: 76420 mg/kg; Duration: female 6-15D of pregnancy; Effects on Embryo or Fetus - Fetotoxicity. Rat Route: Oral; Dose: 38212 mg/kg; Duration: female 6-15D of pregnancy; Specific Developmental Abnormalities - Musculoskeletal system.

Tumorigenic: Rat Route: Oral; Dose: 890 gm/kg/53W-C; Toxic Effects: Tumorigenic - Carcinogenic by RTECS criteria; Kidney, Ureter, and Bladder - Tumors; Kidney, Ureter, and Bladder - Changes in both tubules and glomeruli. Rat Route: Oral; Dose: 1752 gm/kg/2Y-C; Toxic Effects: Tumorigenic - Equivocal tumorigenic agent by RTECS criteria; Kidney, Ureter, and Bladder - Tumors. Rat Route: Oral; Dose: 584 gm/kg/2Y-C; Toxic Effects: Tumorigenic - Equivocal tumorigenic agent by RTECS criteria; Kidney, Ureter, and Bladder - Tumors. Rat Route: Oral; Dose: 840 mg/kg/81W-I; Toxic Effects: Tumorigenic - Neoplastic by RTECS criteria; Blood - Tumors.

Hazard Overviews

Fire Diamond

Health: Mildly irritating to eyes/skin; if heated or misted, irritating to respiratory tract. Also Causes: pulmonary edema; upon ingestion of large quantities: CNS effects and kidney/liver damage.

Fire: Will burn. Can form explosive mixtures in air. Use alcohol foam, dry chemical, and carbon dioxide. Water may cause frothing, so use with caution.

Reactivity: Stable. Hazardous polymerization cannot occur. Avoid: heat; ignition sources. Incompatible with: oxidizers; sodium hydroxide. Hazardous decomposition products: acrid smoke; carbon dioxide gas.

Carcinogenicity: IARC - Not listed; NIOSH - Not listed; NTP - Not listed; ACGIH - Not listed; OSHA - Not listed; EPA - Not listed; MAK - Not listed

Primary Target Organs:

Eyes

Skin

Respiratory System

Exposure Limits

DFG MAK: TWA: 10 ppm; 44 mg/m^3.

AIHA WEEL: TWA: 50 ppm total; OTHER: 10 mg/m^{3}aerosol only.

Respirator Recommendation

Exposure Range: >55 to 2750 ppm Supplied Air, Constant Flow/Pressure Demand, Half Mask

Exposure Range: >2750 to 55,000 ppm Supplied Air, Constant Flow/Pressure Demand, Full Face

Exposure Range: >55,000 to unlimited ppm Self-contained Breathing Apparatus, Pressure Demand, Full Face

Note: odor threshold unknown

Environmental

Ecotoxicity: Aquatic toxicity: > 32,000 ppm/96 hr/mosquito fish/TLm/ fresh water

Cleanup/Disposal: Guide No. 171: Do not touch or walk through spilled material. Stop leak if you can do it without risk. Prevent dust cloud. Avoid inhalation of asbestos dust. Small Dry Spills: With clean shovel place material into clean, dry container and cover loosely; move containers from spill area. Small Spills: Take up with sand or other noncombustible absorbent material and place into containers for later disposal. Large Spills: Dike far ahead of liquid spill for later disposal. Cover powder spill with plastic sheet or tarp to minimize spreading. Prevent entry into waterways, sewers, basements or confined areas.

Environmental Physical Data

Octanol/Water Partition Coefficient: log K_{ow} = calculated at -1.98

BCF: no food chain concentration potential

BOD: 6%, 5 days

Regulations

RCRA 40CFR: Not listed

CERCLA: 40CFR 302.4: Not listed

SARA 40CFR 372.65: Not listed

SARA EHS 40CFR 355: Not listed

TSCA: Listed

Analytical Methods

Soil: SW846 8430

Water / Groundwater: ASTM D4763

DIE6200 CAS #: 106-75-2

DIETHYLENE GLYCOL, BISCHLOROFORMATE

RTECS: LQ6700000

EINECS Number: 203-430-9

Molecular Formula: $C_6H_8Cl_2O_5$

Formula Weight: 215.04

Synonyms: 1,5-BIS((CHLOROCARBONYL)OXY)-3-OXAPENTANE; CARBONOCHLORIDIC ACID,OXYDI-2,1-ETHANEDIYL ESTER; DIETHYLENE GLYCOL BIS(CHLOROFORMATE); DIGLYCOL CHLORFORMATE; DIGLYCOL CHLOROFORMATE; FORMIC

ACID,CHLORO-,OXYDIETHYLENE ESTER; OXYDIETHYLENE BIS(CHLOROFORMATE); OXYDIETHYLENE CHLOROFORMATE
Description: liquid
Use: prepn of nonvolatile plasticizers or modifying agent

Physical Properties

Boiling Point: 127 °C (261 °F) at 5 mm Hg
Water Solubility: Hydrolyzes Readily in Water
Other Solubilities: Soluble in Acetone, Alcohol, Ether, Chloroform, Benzene
Flash Point: 146 °C Open Cup

RTECS Toxicity Data

Acute Oral: Mouse LD_{50} Dose: 813 mg/kg.
Acute Inhalation: Mouse LC_{50} Dose: 169 ppm/1hr.
Acute Dermal: Mouse LD_{50} Route: Skin; Dose: 3400 mg/kg.

Hazard Overviews

Fire: Will burn.
Carcinogenicity: IARC - Not listed; NIOSH - Not listed; NTP - Not listed; ACGIH - Not listed; OSHA - Not listed; EPA - Not listed; MAK - Not listed

Environmental

Regulations

RCRA 40CFR: Not listed
CERCLA: 40CFR 302.4: Not listed
SARA 40CFR 372.65: Not listed
SARA EHS 40CFR 355: Not listed
TSCA: Listed

DIE6300 CAS #: 2358-84-1

DIETHYLENE GLYCOL BIS(METHACRYLATE)

EINECS Number: 219-099-9
Molecular Formula: $C_{12}H_{18}O_5$
Formula Weight: 242.27
Synonyms: DGM 2; DIETHYLENE GLYCOL DIMETHACRYLATE; METHACRYLIC ACID,OXYDIETHYLENE ESTER; OXYDIETHYLENE METHACRYLATE; 2-PROPENOIC ACID,2-METHYL-,OXYDI-2,1-ETHANEDIYL ESTER; TGM 2
Use: crosslinking monomer for UV curable inks & coatings; crosslinking agent for polymers used in contact lenses and dental materials

Physical Properties

Boiling Point: 120 °C (248 °F) to 125 °C (257 °F) at 0.27 kPa
Specific Gravity: 1.08210
Refraction Index: 1.4550 at 25 °C/D

Hazard Overviews

Health: Irritating to eyes/skin. Other Acute Effects: may be harmful by inhalation, ingestion, or skin absorption; prolonged or repeated exposure may cause allergic reactions in certain sensitive individuals.
Fire: Hazards: emits toxic fumes. Extinguishing agents: carbon dioxide, dry chemical powder or appropriate foam; water spray; use water spray to cool fire-exposed containers. Precautions: combustible liquid.
Reactivity: Stable. Hazardous polymerization may occur. Incompatible with: strong oxidizing agents, reducing agents. Hazardous decomposition products: toxic fumes of: carbon monoxide, carbon dioxide.
Carcinogenicity: IARC - Not listed; NIOSH - Not listed; NTP - Not listed; ACGIH - Not listed; OSHA - Not listed; EPA - Not listed; MAK - Not listed
Primary Target Organs:

Eyes

Skin

Environmental

Regulations

RCRA 40CFR: Not listed
CERCLA: 40CFR 302.4: Not listed
SARA 40CFR 372.65: Not listed
SARA EHS 40CFR 355: Not listed
TSCA: Listed

DIE6400 CAS #: 4074-88-8

DIETHYLENE GLYCOL DIACRYLATE

RTECS: AS9450000
EINECS Number: 223-791-6
Molecular Formula: $C_{10}H_{14}O_5$
Structured MF: $(CH_2{=}CHCO_2CH_2CH_2)_2O$
Formula Weight: 214.24

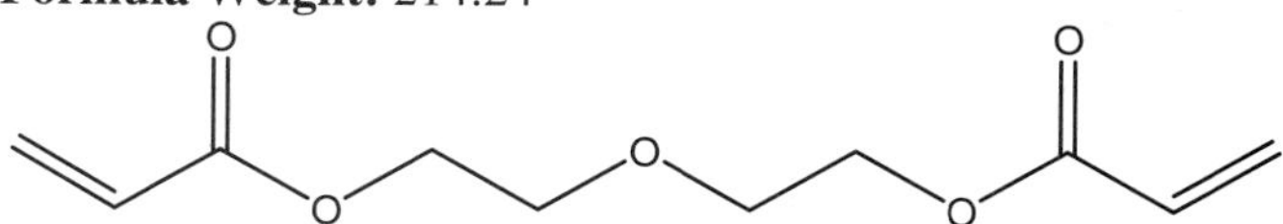

Chemical Structure

Synonyms: ACRYLIC ACID,2-ETHOXYETHANOL DIESTER; ACRYLIC ACID,OXYDIETHYLENE ESTER; ACRYLIC ACID,OXYDIETHYLENE ESTER (8CI); DIACRYLATE DIETHYLENE GLYCOL; NK ESTER A 2G; OXYDIETHYLENE ACRYLATE; OXYDIETHYLENE DIACRYLATE; 2-PROPENOIC ACID,OXYDI-2,1-ETHANEDIYL ESTER; 2-PROPENOIC ACID,OXYDI-2,1-ETHANEDIYL ESTER (9CI); SR 230; TGA 2
Description: colorless, clear liquid; mild, musty odor

Physical Properties

Boiling Point: > 200 °C (392 °F)
Specific Gravity: 1.11
Vapor Density: > 1 Air=1
Density: 1.111 g/mL at 25 °C

Vapor Pressure: 1 mm Hg at 39 °C
Water Solubility: 10 to 50 mg/mL at 18 °C
Other Solubilities: 95% Ethanol: >=100 mg/ml at 18 °C; Acetone: >=100 mg/ml at 18 °C; DMSO: >=100 mg/ml at 18 °C.
Refraction Index: 1.4595
Evaporation Rate: Low
Flash Point: 78 °C

RTECS Toxicity Data

Acute Oral: Rat LD_{50} Dose: 250 mg/kg. Mouse LD_{50} Dose: 550 mg/kg.
Acute Dermal: Rabbit LD_{50} Route: Skin; Dose: 180 uL/kg.
Irritation Eye: Rabbit Standard Draize Test Dose: 100 mg; Reaction: severe.
Irritation Skin: Rabbit Standard Draize Test Dose: 500 mg; Reaction: severe.

Hazard Overviews

Corrosive

Health: Corrosive to eyes/skin/respiratory tract. Toxic. Other Acute Effects: may be fatal if absorbed through skin; harmful if inhaled or swallowed; readily absorbed through skin; inhalation may result in spasm, inflammation and edema of the larynx and bronchi, chemical pneumonitis and pulmonary edema; symptoms of exposure may include burning sensation; coughing; wheezing; laryngitis; shortness of breath; headache; nausea; vomiting; moderate to severe chest tightness; increased mucous discharge; stuffy nose; burning eyes; sneezing; lightheadedness; may cause allergic skin reaction.
Fire: Combustible. Hazards: closed containers may rupture and explode during runaway polymerization; emits toxic fumes. Extinguishing agents: carbon dioxide, dry chemical powder or appropriate foam. Precautions: combustible liquid.
Reactivity: Stable. Incompatible with: strong oxidizing agents, may discolor on exposure to light. Hazardous decomposition products: toxic fumes of: carbon monoxide, carbon dioxide.
Carcinogenicity: IARC - Not listed; NIOSH - Not listed; NTP - Not listed; ACGIH - Not listed; OSHA - Not listed; EPA - Not listed; MAK - Not listed
Primary Target Organs:

Eyes

Skin

Respiratory System

Environmental

Cleanup/Disposal: Guide No. 153: Eliminate all ignition sources (no smoking, flares, sparks or flames in immediate area). Do not touch damaged containers or spilled material unless wearing appropriate protective clothing. Stop leak if you can do it without risk. Prevent entry into waterways, sewers, basements or confined areas. Absorb or cover with dry earth, sand or other non-combustible material and transfer to containers. Do not get water inside containers.

Regulations

RCRA 40CFR: Not listed
CERCLA: 40CFR 302.4: Not listed
SARA 40CFR 372.65: Not listed
SARA EHS 40CFR 355: Not listed
TSCA: Listed

DIE6500 **CAS #: 120-55-8**

DIETHYLENE GLYCOL, DIBENZOATE

RTECS: ID6650000
EINECS Number: 204-407-6
Molecular Formula: $C_{18}H_{18}O_5$
Formula Weight: 314.36

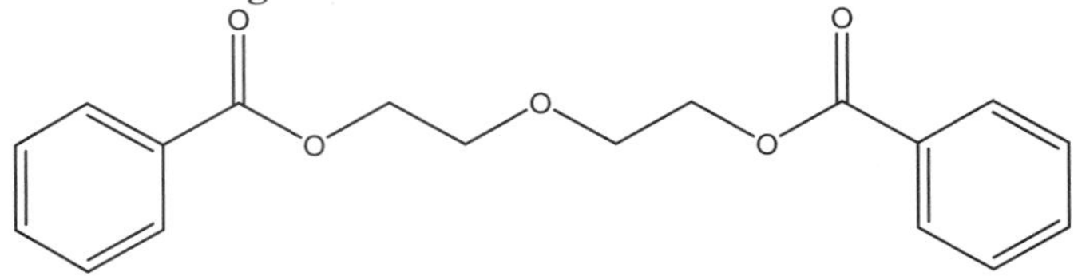

Chemical Structure

Synonyms: BENZO FLEX 2-45; BENZOIC ACID,DIESTER WITH DIETHYLENE GLYCOL; BENZOYLOXYETHOXYETHYL BENZOATE; DIBENZOYLDIETHYLENEGLYCOL ESTER; ETHANOL,2,2'-OXYBIS-,DIBENZOATE
Description: liquid
Use: plasticizer; stabilizer for perfumes; plasticizer for cellulose acetate butyrate resins, cellulose nitrate resins, ethyl cellulose resins, polymethyl methacrylate resins, polyvinyl acetate, butyral & chloride

Physical Properties

Boiling Point: 225 °C (437 °F) to 227 °C (441 °F) at 3 mm Hg
Freezing Point: 34 °C (93.2 °F)
Specific Gravity: 1.2 at 68 °F (20 °C)
Water Solubility: Soluble in Water
Other Solubilities: Ethanol: Very Soluble
Flash Point: 232 °C Closed Cup

RTECS Toxicity Data

Acute Oral: Rat LD_{50} Dose: 2830 mg/kg.
Acute Dermal: Rabbit LD_{50} Route: Skin; Dose: 20 gm/kg.
Irritation Eye: Rabbit Standard Draize Test Dose: 500 mg/24H; Reaction: mild.
Irritation Skin: Rabbit Standard Draize Test Dose: 500 mg/24H; Reaction: mild.

Hazard Overviews

Health: Irritating to eyes/skin. Acute Effects: may be harmful by inhalation, ingestion, or skin absorption.

Fire: Will burn. Hazards: emits toxic fumes. Extinguishing agents: water spray; carbon dioxide, dry chemical powder or appropriate foam. Precautions: combustible liquid.
Reactivity: Incompatible with: strong oxidizing agents. Hazardous decomposition products: toxic fumes of: carbon monoxide, carbon dioxide.
Carcinogenicity: IARC - Not listed; NIOSH - Not listed; NTP - Not listed; ACGIH - Not listed; OSHA - Not listed; EPA - Not listed; MAK - Not listed
Primary Target Organs:

Eyes

Skin

Environmental

Regulations

RCRA 40CFR: Not listed
CERCLA: 40CFR 302.4: Not listed
SARA 40CFR 372.65: Not listed
SARA EHS 40CFR 355: Not listed
TSCA: Listed

DIE6600 CAS #: 5952-26-1

DIETHYLENE GLYCOL, DICARBAMATE

RTECS: KL9890000
Molecular Formula: $C_6H_{12}N_2O_5$
Formula Weight: 192.20
Synonyms: DIETHYLENE GLYCOL,DICARBAMATE (8CI); DIGLYCOLURETHANE; 2,2'-OXYDIETHANOL DICARBAMATE

RTECS Toxicity Data

Acute Oral: Mouse LD_{50} Dose: 8300 mg/kg.

Hazard Overviews

Carcinogenicity: IARC - Not listed; NIOSH - Not listed; NTP - Not listed; ACGIH - Not listed; OSHA - Not listed; EPA - Not listed; MAK - Not listed

Environmental

Regulations

RCRA 40CFR: Listed Hazardous Waste No. U395 Toxic Waste
CERCLA: 40CFR 302.4: Listed per RCRA Section 3001 RQ: 1 lb (0.454 kg)
SARA 40CFR 372.65: Not listed
SARA EHS 40CFR 355: Not listed
TSCA: Listed

DIE6700 CAS #: 112-36-7

DIETHYLENE GLYCOL DIETHYL ETHER

RTECS: KN3160000
EINECS Number: 203-963-7
Molecular Formula: $C_8H_{18}O_3$
Structured MF: $CH_3CH_2(OCH_2CH_2)_2OCH_2CH_3$
Formula Weight: 162.22

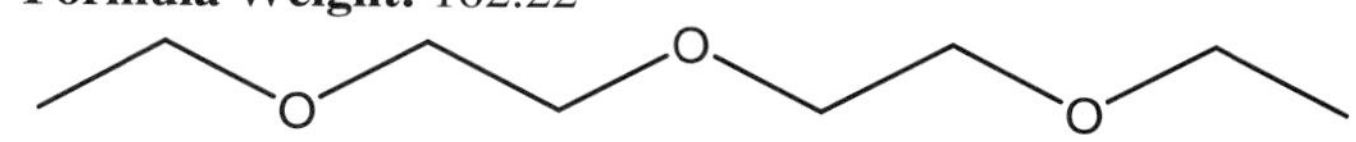

Chemical Structure

Synonyms: BIS(2,2-DIETHOXYETHYL) ETHER; BIS(2-ETHOXYETHYL) ETHER; BIS(2-ETHOXYETHYL)ETHER; DIETHYL CARBITOL; DIETHYLDIETHYLENE GLYCOL; DIETHYLETHER DIETHYLENGLYKOLU; ETHANE,1,1'-OXYBIS(2-ETHOXY-; ETHANOL,2,2'-OXYBIS-,DIETHYL ETHER; ETHER,BIS(2-ETHOXYETHYL); 1-ETHOXY-2-(BETA-ETHOXYETHOXY)ETHANE; 2-(2-ETHOXYETHOXY)-1-ETHOXYETHANE; 2-ETHOXYETHYL ETHER; ETHYL DIGLYME; 2,2'-OXYBISETHANOL DIETHYL ETHER; 1,1'-OXYBIS(2-ETHOXYETHANE); 3,6,9-TRIOXAUNDECANE
Description: colorless liquid
Use: solvent for nitrocellulose, plastic resins and compounds, adhesives, lacquers, sealants and rubber chemicals; in organic synthesis and as a high-boiling reaction medium

Physical Properties

Boiling Point: 189 °C (372 °F) at 760 mm Hg
Freezing Point: -44.3 °C (-47.74 °F)
Specific Gravity: 0.907 at 20 °C/4 °C
Vapor Density: 5.6 Air=1
Saturated Vapor Density: 1.203916813 kg/m^3
Density: 0.909 g/mL
Vapor Pressure: 0.54 mm Hg at 25 °C
Water Solubility: Very Soluble in Water
Other Solubilities: 95% Ethanol: >=100 mg/ml at 19 °C; Acetone: >=100 mg/ml at 19 °C; DMSO: >=100 mg/ml at 19 °C; Ether: Soluble; Hydrocarbons: Soluble; Organic solvents: Soluble.
Refraction Index: 1.4115 at 20 °C/D
Flash Point: 82.222 °C Open Cup
Autoignition Temperature: 205 °C

RTECS Toxicity Data

Acute Oral: Rat LD_{50} Dose: 4970 mg/kg. Guinea Pig LD_{50} Dose: 1850 mg/kg; Toxic Effects: Behavioral - General anesthetic; Gastrointestinal - Other changes; Kidney, Ureter, and Bladder - Other changes.
Acute Dermal: Rabbit LD_{50} Route: Skin; Dose: 6700 uL/kg.
Irritation Eye: Rabbit Standard Draize Test Dose: 50 mg; Reaction: moderate.
Reproductive/Teratogenic: Mouse Route: Oral; Dose: 24 gm/kg; Duration: female 7-14D of pregnancy; Effects on Newborn - Stillbirth; Growth statistics. Mouse Route: Oral;

Dose: 3 gm/kg; Duration: female 6-15D of pregnancy; Effects on Embryo or Fetus - Fetotoxicity.

Hazard Overviews

Fire Diamond

Health: Mildly irritating to eyes/skin/respiratory tract. Also Causes: nausea, vomiting, and diarrhea.

Fire: Combustible. Use dry chemical, carbon dioxide, water spray, fog, or alcohol-resistant foam. Container may explode in heat of fire.

Reactivity: Stable. Hazardous polymerization cannot occur. Avoid: heat; ignition sources; oxidizers. Incompatible with: strong oxidizers. Hazardous decomposition products: carbon oxides; acrid smoke.

Carcinogenicity: IARC - Not listed; NIOSH - Not listed; NTP - Not listed; ACGIH - Not listed; OSHA - Not listed; EPA - Not listed; MAK - Not listed

Primary Target Organs:

Eyes

Skin

Environmental

Cleanup/Disposal: Guide No. 128: Eliminate all ignition sources (no smoking, flares, sparks or flames in immediate area). All equipment used when handling the product must be grounded. Do not touch or walk through spilled material. Stop leak if you can do it without risk. Prevent entry into waterways, sewers, basements or confined areas. A vapor suppressing foam may be used to reduce vapors. Absorb or cover with dry earth, sand or other non-combustible material and transfer to containers. Use clean non-sparking tools to collect absorbed material. Large Spills: Dike far ahead of liquid spill for later disposal. Water spray may reduce vapor; but may not prevent ignition in closed spaces.

Regulations

RCRA 40CFR: Not listed
CERCLA: 40CFR 302.4: Not listed
SARA 40CFR 372.65: Not listed
SARA EHS 40CFR 355: Not listed
TSCA: Listed

DIE6800 **CAS #: 111-96-6**

DIETHYLENE GLYCOL DIMETHYL ETHER

RTECS: KN3339000
EINECS Number: 203-924-4
Molecular Formula: $C_6H_{14}O_3$
Structured MF: $(CH_3OCH_2CH_2)_2O$
Formula Weight: 134.17

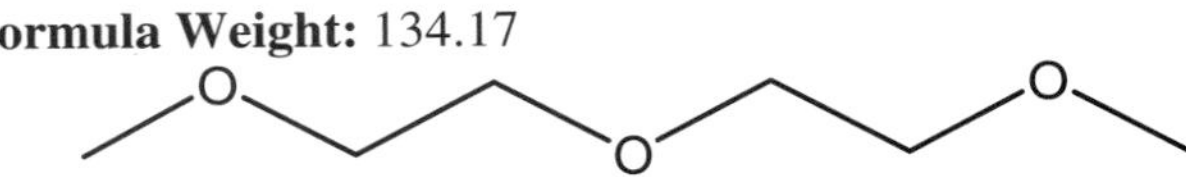

Chemical Structure

Synonyms: BIS(2-METHOXYETHYL) ETHER; BIS(2-METHOXYETHYL)ETHER; DIETHYL GLYCOL DIMETHYL ETHER; DIETHYLENE GLYCOL,DIMETHYL ETHER; DIGLYCOL METHYL ETHER; DIGLYME; DIMETHYL CARBITOL; ETHANE,1,1'-OXYBIS(2-METHOXY-(9CI); ETHANOL,2,2'-OXYBIS-,DIMETHYL ETHER; ETHER,BIS(2-METHOXYETHYL); GLYME-2; GLYME 2; 2-(2-METHOXYETHOXY)-1-METHOXYETHANE; (2-METHOXYETHYL) ETHER; 2-METHOXYETHYL ETHER; METHYLDIGLYME; 2,2'-OXYBISETHANOL DIMETHYL ETHER; 2,2'-OXYBISETHANOL,DIMETHYL ETHER; 1,1'-OXYBIS(2-METHOXY)ETHANE; 1,1'-OXYBIS(2-METHOXYETHANE); POLY SOLV; POLY-SOLV; 2,5,8-TRIOXANONANE

Description: liquid; mild odor

Use: solvent; reaction medium for Grignard and similar syntheses; anhydrous reaction medium for organo-metallic synthesis

Physical Properties

Boiling Point: 162 °C (324 °F) at 760 mm Hg
Freezing Point: -68 °C (-90.4 °F)
Specific Gravity: 0.9451 at 20 °C/20 °C
Vapor Density: 4.62 Air=1
Density: 0.95 g/mL at 25 °C
Vapor Pressure: 0.054 kPa at 0 °C
Water Solubility: Miscible with Water
Other Solubilities: 95% Ethanol: miscible; Ether: miscible.
Refraction Index: 1.4097 at 20 °C/D
Evaporation Rate: > 1 Butyl Acetate=1
Ionization Potential (eV): =< 9.8
Flash Point: 70 °C Open Cup
LEL: 1.2% v/v

RTECS Toxicity Data

Acute Oral: Rat LD_{50} Dose: >1600 mg/kg.

Reproductive/Teratogenic: Rat Route: Oral; Dose: 500 mg/kg; Duration: female 11D of pregnancy; Effects on Embryo or Fetus - Maternal-fetal exchange. Rat Route: Oral; Dose: 13680 mg/kg; Duration: male 20D prior to mating; Paternal Effects - Spermatogenesis; Testes, epididymis, sperm duct; Other effects on male. Rat Route: Inhalation; Dose: 1100 ppm/6H; Duration: male 2W prior to mating; Paternal Effects - Spermatogenesis; Testes, epididymis, sperm duct.

Mutagenic: Rat Dominant Lethal Test; Route: Inhalation; Dose: 1000 ppm/5D-C. Mouse Sperm Morphology; Route: Inhalation; Dose: 1000 ppm/5D-C.

Hazard Overviews

Health: Irritating. Toxic. Other Acute Effects: harmful if swallowed, inhaled, or absorbed through skin. Chronic Effects: may cause congenital malformation in the fetus; may cause reproductive disorders; laboratory studies with this material indicate birth defects, fetotoxicity, embryolethality, anemia, bone marrow damage, hemolysis, immuno-suppression, damage to the male reproductive tissues; target organs: female reproductive system, male reproductive system.

Fire: Combustible. Extinguishing agents: water spray; carbon dioxide, dry chemical powder or appropriate foam. Precautions: combustible liquid.

Reactivity: Incompatible with: strong oxidizing agents, strong acids, strong bases. Hazardous decomposition products: toxic fumes of: carbon monoxide, carbon dioxide.

Carcinogenicity: IARC - Not listed; NIOSH - Not listed; NTP - Not listed; ACGIH - Not listed; OSHA - Not listed; EPA - Not listed; MAK - Not listed

Primary Target Organs:

Eyes

Skin

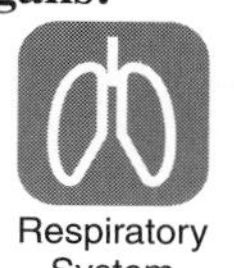
Respiratory System

Repro-ductive

Exposure Limits

DFG MAK: TWA: 5 ppm; 27 mg/m^3.

Environmental

Environmental Fate: It is not expected to undergo hydrolysis or direct photolysis in the environment. The complete miscibility in water suggests that volatilization, adsorption and bioconcentration are not important fate processes. This is supported by an estimated Henry's Law constant of 2.28 x10^{-9} atm-cu m/mole at 25 °C which indicates that volatilization from natural waters and moist soil should be extremely slow. Yet, it may evaporate from dry surfaces, especially when present in high concentration such as in spill situations. A low estimated log BCF suggests it should not bioconcentrate among aquatic organisms. A low K_{oc} indicates it should not partition from the water column to organic matter contained in sediments and suspended solids, and it should be highly mobile in soil. Biodegradation may be an important removal mechanism from aerobic soil and water; however, biodegradation data was not located in the available literature. In the atmosphere, it is expected to exist almost entirely in the vapor phase and reactions with photochemically produced hydroxyl radicals should be important (estimated half-life of 14 hours). Physical removal from air by precipitation and dissolution in clouds may occur; however, its short atmospheric residence time suggests that we deposition is of limited importance.

Cleanup/Disposal: Guide No. 127: Eliminate all ignition sources (no smoking, flares, sparks or flames in immediate area). All equipment used when handling the product must be grounded. Do not touch or walk through spilled material. Stop leak if you can do it without risk. Prevent entry into waterways, sewers, basements or confined areas. A vapor suppressing foam may be used to reduce vapors. Absorb or cover with dry earth, sand or other non-combustible material and transfer to containers. Use clean non-sparking tools to collect absorbed material. Large Spills: Dike far ahead of liquid spill for later disposal. Water spray may reduce vapor; but may not prevent ignition in closed spaces. Guide No. 128: Eliminate all ignition sources (no smoking, flares, sparks or flames in immediate area). All equipment used when handling the product must be grounded. Do not touch or walk through spilled material. Stop leak if you can do it without risk. Prevent entry into waterways, sewers, basements or confined areas. A vapor suppressing foam may be used to reduce vapors. Absorb or cover with dry earth, sand or other non-combustible material and transfer to containers. Use clean non-sparking tools to collect absorbed material. Large Spills: Dike far ahead of liquid spill for later disposal. Water spray may reduce vapor; but may not prevent ignition in closed spaces.

Environmental Physical Data

Henry's Law Constant: calculated at 2.28 x10^{-9}

Octanol/Water Partition Coefficient: log K_{ow} = estimated at -0.06

Sorption Partition Coefficient: K_{oc} = 20

BCF: calculated at 0.28

Regulations

RCRA 40CFR: Not listed

CERCLA: 40CFR 302.4: Not listed

SARA 40CFR 372.65: Not listed

SARA EHS 40CFR 355: Not listed

TSCA: Listed

DIE6900 **CAS #: 112-59-4**

DIETHYLENE GLYCOL HEXYL ETHER

RTECS: KL2625000

EINECS Number: 203-988-3

Molecular Formula: $C_{10}H_{22}O_3$

Structured MF: $CH_3(CH_2)_5OC_2H_4OC_2H_4OH$

Formula Weight: 190.32

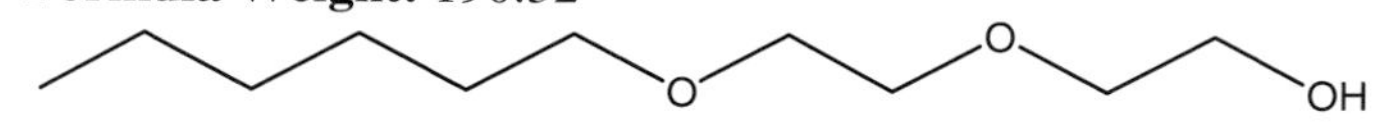

Chemical Structure

Synonyms: DIETHYLENE GLYCOL N-HEXYL ETHER; DIETHYLENE GLYCOL MONO(N-HEXYL) ETHER; DIETHYLENE GLYCOL MONOHEXYL ETHER; 3,6-DIOXA-1-DODECANOL; 3,6-DIOXADODECANOL-1; ETHANOL,2-((2-HEXYLOXY)ETHOXY)-; ETHANOL,2-(2-(HEXYLOXY)ETHOXY)-; ETHANOL,2-[2-(2-HEXYLOXY)ETHOXY]-; N-HEXOXYETHOXYETHANOL; HEXYL CARBITOL; N-HEXYL CARBITOL; N-HEXYL CARBITOL SOLVENT; HEXYLKARBITOL; 2-((2-HEXYLOXY)ETHOXY)ETHANOL

Description: water-white liquid

Use: high boiling solvent; solvent for surface coatings & cleaning solns

Physical Properties

Boiling Point: 259.1 °C (498 °F) at 760 mm Hg

Freezing Point: -33.3 °C (-27.94 °F)

Specific Gravity: 0.935 at 25 °C/4 °C

Saturated Vapor Density: < 1.200087833 kg/m^3

Vapor Pressure: < 0.01 mm Hg at 20 °C

Flash Point: 140.556 °C Open Cup

RTECS Toxicity Data

Acute Oral: Rat LD_{50} Dose: 4920 mg/kg.
Acute Dermal: Rabbit LD_{50} Route: Skin; Dose: 1500 uL/kg.
Irritation Eye: Rabbit Standard Draize Test Dose: 5 mg; Reaction: moderate. Rabbit Standard Draize Test Dose: 750 ug/24H; Reaction: severe.
Irritation Skin: Rabbit Standard Draize Test Dose: 500 mg/24H; Reaction: severe. Rabbit Open Draize Test Dose: 10 mg/24H open; Reaction: mild. Rabbit Open Draize Test Dose: 500 mg open; Reaction: mild.

Hazard Overviews

Health: Severely irritating to eyes/skin/respiratory tract. Toxic. Other Acute Effects: harmful if swallowed, inhaled, or absorbed through skin; symptoms of exposure may include burning sensation; coughing; wheezing; laryngitis; shortness of breath; headache; nausea; vomiting; exposure can cause CNS depression. Chronic Effects: may cause reproductive disorders; laboratory studies with this material indicate birth defects, fetotoxicity, embryolethality, anemia, bone marrow damage, hemolysis, immuno-suppression, damage to the male reproductive system; target organs: kidneys, central nervous system.
Fire: Will burn. Hazards: emits toxic fumes. Extinguishing agents: noncombustible; use extinguishing media appropriate to surrounding fire conditions. Precautions: combustible liquid.
Reactivity: Incompatible with: strong oxidizing agents. Hazardous decomposition products: toxic fumes of: carbon monoxide, carbon dioxide.
Carcinogenicity: IARC - Not listed; NIOSH - Not listed; NTP - Not listed; ACGIH - Not listed; OSHA - Not listed; EPA - Not listed; MAK - Not listed
Primary Target Organs:

Eyes

Skin

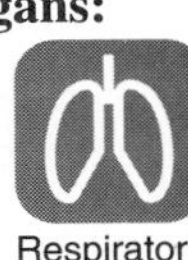
Respiratory System

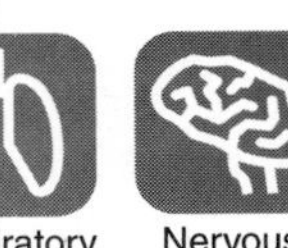
Nervous System

Kidneys

Environmental

Cleanup/Disposal: Guide No. 171: Do not touch or walk through spilled material. Stop leak if you can do it without risk. Prevent dust cloud. Avoid inhalation of asbestos dust. Small Dry Spills: With clean shovel place material into clean, dry container and cover loosely; move containers from spill area. Small Spills: Take up with sand or other noncombustible absorbent material and place into containers for later disposal. Large Spills: Dike far ahead of liquid spill for later disposal. Cover powder spill with plastic sheet or tarp to minimize spreading. Prevent entry into waterways, sewers, basements or confined areas.

Regulations

RCRA 40CFR: Not listed
CERCLA: 40CFR 302.4: Not listed
SARA 40CFR 372.65: Not listed
SARA EHS 40CFR 355: Not listed
TSCA: Listed

DIE7000 CAS #: 112-34-5

DIETHYLENE GLYCOL MONOBUTYL ETHER

RTECS: KJ9100000
EINECS Number: 203-961-6
Molecular Formula: $C_8H_{18}O_3$
Structured MF: $CH_3(CH_2)_3OCH_2CH_2OCH_2CH_2OH$
Formula Weight: 162.23

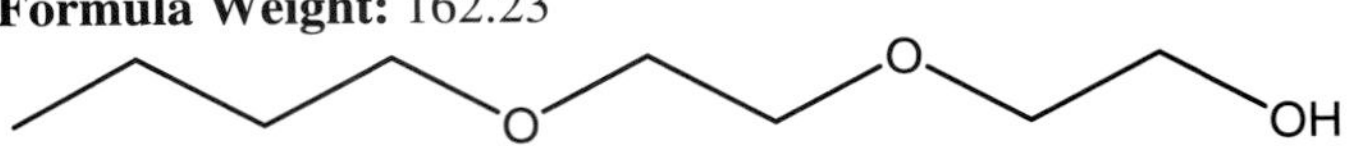

Chemical Structure

Synonyms: BUCB; BUTOXYDIETHYLENE GLYCOL; BUTOXYDIGLYCOL; 2-(2-BUTOXYETHOXY)ETHANOL; BUTOXYETHOXYETHANOL; BUTYL CARBITOL; O-BUTYL DIETHYLENE GLYCOL; BUTYL DIGLYCOL; BUTYL DIGOL; BUTYL DIOXITOL; BUTYL ETHYL CELLOSOLVE; DIEHYLENE DB; DIETHYLENE GLYCOL BUTYL ETHER; DIETHYLENE GLYCOL N-BUTYL ETHER; DIGLYCOL MONOBUTYL ETHER; DOWANOL DB; EKTASOLVE DB; ETHANOL,2-(2-BUTOXYETHOXY)-; ETHANOL,2,2'-OXYBIS-,MONOBUTYL ETHER; GLYCOL ETHER DB; GLYCOL MONOBUTYL ETHER; JEFFERSOL DB; POLY-SOLV DB
Description: colorless liquid; faint, pleasant odor
Use: mosquito repellent and as a solvent for nitrocellulose, oils, gums, dyes, soaps, and polymers; a plasticizer intermediate

Physical Properties

Boiling Point: 230.4 °C (447 °F)
Freezing Point: -68.1 °C (-90.58 °F)
Specific Gravity: 0.9536 at 20 °C/20 °C
Vapor Density: 5.58 Air=1
Saturated Vapor Density: 1.200072539 kg/m^3
Vapor Pressure: 0.01 mm Hg at 20 °C
Water Solubility: Miscible with Water
Other Solubilities: 95% Ethanol: >=100 mg/ml at 20 °C; Acetone: >=100 mg/ml at 20 °C; Benzene: Soluble; DMSO: >=100 mg/ml at 20 °C; Ether: Soluble; Oils: Soluble.
Surface Tension: 34 dynes/cm at 25 °C
Refraction Index: 1.4258 at 27 °C/D
Flash Point: 110 °C Open Cup
Autoignition Temperature: 228 °C
LEL: 1.2% v/v
UEL: 8.5% v/v

RTECS Toxicity Data

Acute Oral: Rat LD_{50} Dose: 5660 mg/kg. Mouse LD_{50} Dose: 2400 mg/kg.
Acute Dermal: Rabbit LD_{50} Route: Skin; Dose: 2700 mg/kg.
Chronic (Multiple Dose) Oral: Rat Dose: 83 gm/kg/13W-I; Toxic Effects: Nutritional and gross metabolic - Weight loss or decreased weight gain; DEATH. Rat Dose: 109 gm/kg/6W-I; Toxic Effects: Behavioral - Food intake

(animal); Blood - Pigmented or nucleated red Blood cells; Nutritional and gross metabolic - Weight loss or decreased weight gain.

Chronic (Multiple Dose) Inhalation: Rat Dose: 5 mg/m^3/24H/17W-C; Toxic Effects: Brain and coverings - Recordings from specific areas of CNS; Blood - Other changes.

Irritation Eye: Rabbit Standard Draize Test Dose: 20 mg/24H; Reaction: moderate. Rabbit Standard Draize Test Dose: 20 mg; Reaction: severe.

Hazard Overviews

Fire Diamond

Health: Irritating to eyes/skin/respiratory tract. Also Causes: temporary corneal damage. Chronic Effects: repeated small exposures can cause cyanosis, acidosis, tachypnea, kidney dysfunction.

Fire: Combustible. Can form explosive mixtures in the air. Fight fire with water, carbon dioxide, dry chemical, or 'alcohol resistant' foam. Be aware that water may cause frothing.

Reactivity: Stable. Hazardous polymerization cannot occur. Avoid: heat; flame; oxidizers. Incompatible with: strong oxidizing agents. Hazardous decomposition products: carbon dioxide; acrid, irritating smoke.

Carcinogenicity: IARC - Not listed; NIOSH - Not listed; NTP - Not listed; ACGIH - Not listed; OSHA - Not listed; EPA - Not listed; MAK - Not listed

Primary Target Organs:

Eyes

Skin

Nervous System

Kidneys

Exposure Limits

DFG MAK: TWA: 100 mg/m^3.

Environmental

Ecotoxicity: LC_{50} Lepomis macrochirus 1300 ppm/96 hr (Static bioassay in fresh water at 23 °C, mild aeration applied after 24 hr) LC_{50} Menidia beryllina 2000 ppm/96 hr (Static bioassay in synthetic seawater at 23 °C, mild aeration applied after 24 hr LC_{50} Goldfish 2700 mg/l/24 hr (modified ASTM-D-1345). /Conditions of bioassay not specified

Environmental Fate: It is not expected to undergo hydrolysis or direct photolysis in the environment. The complete miscibility in water suggests that volatilization, adsorption and bioconcentration are not important fate processes. This is supported by an estimated Henry's Law constant of 1.52 $x10^{-9}$ atm-cu m/mole at 25 °C which indicates that volatilization from natural waters and moist soil should be extremely slow. A low estimated log BCF suggests it should not bioconcentrate among aquatic organisms. A low K_{oc} indicates it should not partition from the water column to organic matter contained in sediments and suspended solids, and it should be highly mobile in soil. Aqueous screening test data indicate that biodegradation may be an important removal mechanism from aerobic soil and water. In the atmosphere, it is expected to exist almost entirely in the vapor phase and reactions with photochemically produced hydroxyl radicals should be important (estimated half-life of 11 hours). Physical removal from air by precipitation and dissolution is clouds may occur; however, its short atmospheric residence time suggests that wet deposition is of limited importance.

Cleanup/Disposal: Guide No. 171: Do not touch or walk through spilled material. Stop leak if you can do it without risk. Prevent dust cloud. Avoid inhalation of asbestos dust. Small Dry Spills: With clean shovel place material into clean, dry container and cover loosely; move containers from spill area. Small Spills: Take up with sand or other noncombustible absorbent material and place into containers for later disposal. Large Spills: Dike far ahead of liquid spill for later disposal. Cover powder spill with plastic sheet or tarp to minimize spreading. Prevent entry into waterways, sewers, basements or confined areas.

Environmental Physical Data

Henry's Law Constant: calculated at 1.52 $x10^{-9}$

Octanol/Water Partition Coefficient: log K_{ow} = 0.91

Sorption Partition Coefficient: K_{oc} = 75

BCF: calculated at 0.46

BOD: theoretical 34%, 5 days

Regulations

RCRA 40CFR: Not listed

CERCLA: 40CFR 302.4: Not listed

SARA 40CFR 372.65: Not listed

SARA EHS 40CFR 355: Not listed

TSCA: Listed

DIE7100 **CAS #: 124-17-4**

DIETHYLENE GLYCOL MONOBUTYL ETHER ACETATE

RTECS: KJ9275000

EINECS Number: 204-685-9

Molecular Formula: $C_{10}H_{20}O_4$

Structured MF: $C_4H_9OCH_2CH_2OCH_2CH_2OCOCH_3$

Formula Weight: 204.30

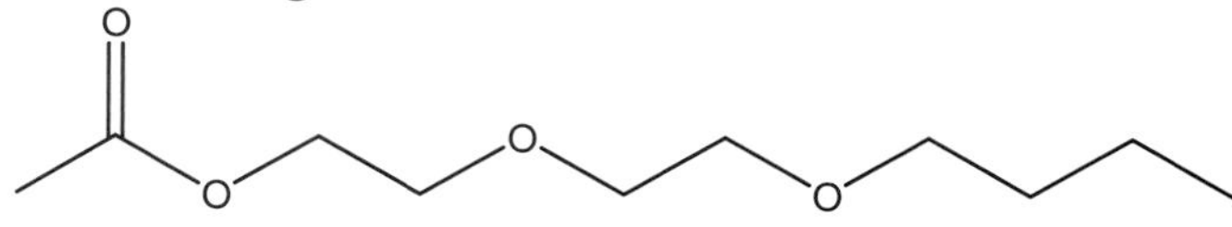

Chemical Structure

Synonyms: ACETIC ACID 2-(2-BUTOXYETHOXY)ETHYL ESTER; 2-(2-BUTOXYETHOXY)ETHANOL ACETATE; 2-(2-BUTOXYETHOXY)ETHYL ACETATE; BUTOXYETHOXYETHYL ACETATE; 2-(2-BUTOXYETHOXY)ETHYLESTER KYSELINY OCTOVE; BUTYL CARBITOL ACETATE; BUTYL DIETHYLENE GLYCOL ACETATE; BUTYLKARBITOLACETAT; DIETHYLENE GLYCOL BUTYL ETHER ACETATE; DIETHYLENE

GLYCOL,MONOBUTYL ETHER,ACETATE; DIETHYLENEGLYCOL MONOBUTYL ETHER ACETATE; DIGLYCOL MONOBUTYL ETHER ACETATE; EKTASOLVE DB ACETATE; GLYCOL ETHER DB ACEATATE

Description: colorless liquid; mildly unpleasant odor

Use: solvent for lacquers and other coatings, antibiotic extractions; coalescing aid for emulsion paints; solvent for cellulose acetate, ester gum, polyvinyl acetate; in paint & lacquer industry

Physical Properties

Boiling Point: 245 °C (473 °F) at 760 mm Hg
Freezing Point: -32 °C (-25.6 °F)
Specific Gravity: 0.985 at 20 °C/4 °C
Vapor Pressure: 1.75 kPa at 125 °C
Water Solubility: Soluble in Water
Other Solubilities: Soluble in all proportions in Alcohol, Ether, Acetone, organic solvents
Surface Tension: Estimated at 22 dynes/cm
Refraction Index: 1.4262 at 20 °C/D
Flash Point: 115.556 °C
Autoignition Temperature: 295 °C
LEL: 0.8% v/v
UEL: 5.0% v/v

RTECS Toxicity Data

Acute Oral: Rat LD_{50} Dose: 6500 mg/kg. Mouse LD_{50} Dose: 6600 uL/kg.

Acute Dermal: Rabbit LD_{50} Route: Skin; Dose: 14500 mg/kg.

Chronic (Multiple Dose) Dermal: Rabbit Route: Skin; Dose: 180 mL/kg/13W-I; Toxic Effects: Behavioral - Muscle weakness; Nutritional and gross metabolic - Weight loss or decreased weight gain; DEATH.

Irritation Eye: Rabbit Standard Draize Test Dose: 500 mg; Reaction: moderate.

Irritation Skin: Rabbit Open Draize Test Dose: 500 mg open; Reaction: mild.

Hazard Overviews

Fire Diamond

Health: Irritating to eyes/skin/respiratory tract. Also Causes: nausea, vomiting, kidney nephrosis (skin absorption of large amounts). Chronic Effects: erythema, exfoliation.

Fire: Will burn. Use carbon dioxide or dry chemical for small fires and 'alcohol-resistant' foam for large fires. Water spray may be ineffective.

Reactivity: Stable. Hazardous polymerization cannot occur. Avoid: heat; ignition sources. Incompatible with: strong acids; alkalies; oxidizers. Hazardous decomposition products: oxides of carbon.

Carcinogenicity: IARC - Not listed; NIOSH - Not listed; NTP - Not listed; ACGIH - Not listed; OSHA - Not listed; EPA - Not listed; MAK - Not listed

Primary Target Organs:

Eyes

Skin

Environmental

Cleanup/Disposal: Guide No. 120: Do not touch or walk through spilled material. Stop leak if you can do it without risk. Use water spray to reduce vapors or divert vapor cloud drift. Do not direct water at spill or source of leak. If possible, turn leaking containers so that gas escapes rather than liquid. Prevent entry into waterways, sewers, basements or confined areas. Allow substance to evaporate. Ventilate the area. Caution: When in contact with refrigerated/cryogenic liquids, many materials become brittle and are likely to break without warning.

Environmental Physical Data

Octanol/Water Partition Coefficient: log K_{ow} = calculated at 1.77
BCF: no food chain concentration potential

Regulations

RCRA 40CFR: Not listed
CERCLA: 40CFR 302.4: Not listed
SARA 40CFR 372.65: Not listed
SARA EHS 40CFR 355: Not listed
TSCA: Listed

Analytical Methods

Water / Groundwater: ASTM D3695

DIE7200 **CAS #: 111-90-0**

DIETHYLENE GLYCOL MONOETHYL ETHER

RTECS: KK8750000
EINECS Number: 203-919-7
Molecular Formula: $C_6H_{14}O_3$
Structured MF: $HOCH_2CH_2OCH_2CH_2OCH_2CH_3$
Formula Weight: 134.17

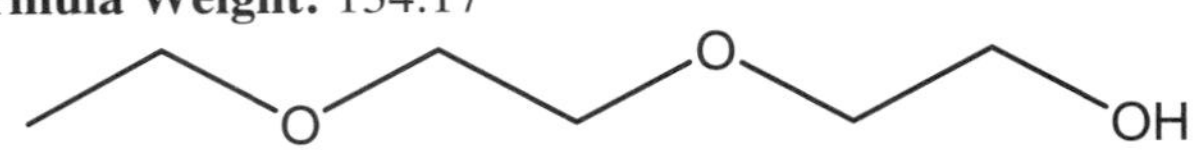

Chemical Structure

Synonyms: AETHYLDIAETHYLENGLYCOL; APV; CARBITOL; CARBITOL CELLOSOLVE; CARBITOL SOLVENT; DIETHYLENE GLYCOL ETHYL ETHER; DIETHYLENE GLYCOL MONOETHYL ESTER; DIGLYCOL; DIGLYCOL MONOETHYL ETHER; 3,6-DIOXA-1-OCTANOL; 3,6-DIOXAOCTAN-1-OL; 3,6-DIOXA-1-OKTANOL; DIOXITOL; DOWANOL; DOWANOL 17; DOWANOL DE; EKTASOLE DE; EKTASOLVE DE; ETHANOL,2-(2-ETHOXYETHOXY)-; ETHANOL,2,2'-OXYBIS-,MONOETHYL ETHER; ETHOXY DIGLYCOL; 2-(2-ETHOXYETHOXY) ETHANOL; 2-(2-ETHOXYETHOXY)ETHANOL; 2-(BETA-ETHOXYETHOXY)ETHANOL; 2-(ETHOXYETHOXY)ETHANOL;

ETHYL CARBITOL; ETHYL CELLOSOLVE; ETHYL DIETHYLENE GLYCOL; ETHYL DIGOL; ETHYLDIGOL; O-ETHYLDIGOL; ETHYLENE DIGLYCOL MONOETHYL ETHER; 1-HYDROXY-3,6-DIOXAOCTANE; KARBITOL; LOSUNGSMITTEL APV; MONOETHYL ETHER OF DIETHYLENE GLYCOL; PM 1799; POLY-SOLV; POLY-SOLV DE; SOLVOLSOL; TRANSCUTOL

Description: clear, colorless liquid; pleasant odor

Use: solvent for dyes, nitrocellulose and resins; in nonaqueous stains for wood, for setting the twist and conditioning yarns and cloth, in textile printing, textile soaps, lacquers, cosmetics and quick-drying varnishes and enamels; in brake fluid diluent and in organic synthesis; to determine saponification values of oils and as neutral solvent for mineral oil-soap and mineral oil-sulphated oil mixtures (giving fine dispersions in water)

Physical Properties

Boiling Point: 196 °C (385 °F)
Freezing Point: -10 °C (14 °F)
Specific Gravity: 0.9855 at 25 °C/4 °C
Vapor Density: 4.62 Air=1
Saturated Vapor Density: 1.200744397 kg/m^3
Density: 0.9981 g/cu cm
Vapor Pressure: 0.13 mm Hg at 25 °C
Water Solubility: >= 100 mg/mL at 20 C
Other Solubilities: 95% Ethanol: >=100 mg/ml at 20 °C; Acetone: >=100 mg/ml at 20 °C; Alcohol: Soluble; Benzene: miscible; Chloroform: miscible; DMSO: >=100 mg/ml at 20 °C; Ether: miscible; Most organic solvents: miscible; Pyridine: miscible.
Surface Tension: 31.8 dynes/cm at 25 °C
Refraction Index: 1.4273 at 20 °C/D
Flash Point: 96 °C Open Cup
Autoignition Temperature: 204 °C
LEL: 1.2% v/v
UEL: 23.5% v/v at 182 °C

RTECS Toxicity Data

Acute Oral: Rat LD_{50} Dose: 5500 mg/kg; Toxic Effects: Behavioral - Altered sleep time (including change in righting reflex); Behavioral - Coma. Mouse LD_{50} Dose: 6600 mg/kg; Toxic Effects: Behavioral - Altered sleep time (including change in righting reflex); Behavioral - Coma.

Acute Dermal: Rabbit LD_{50} Route: Skin; Dose: 8500 uL/kg; Toxic Effects: Behavioral - Somnolence (general depressed activity); Behavioral - Change in motor activity (specific assay); Behavioral - Ataxia. Rat LD_{50} Route: Skin; Dose: 6 gm/kg; Toxic Effects: Behavioral - Somnolence (general depressed activity); Behavioral - Ataxia; Behavioral - Coma. Rat LD_{50} Route: Subcutaneous Dose: 6 gm/kg.

Chronic (Multiple Dose) Oral: Rat Dose: 225 gm/kg/90D-C; Toxic Effects: Kidney, Ureter, and Bladder - Chgs in tubules (inc acute renal failure, acute tubular necrosis; Kidney, Ureter, and Bladder - Changes in kidney weight; Nutritional and gross metabolic - Weight loss or decreased weight gain. Rat Dose: 158 gm/kg/6W-I; Toxic Effects: Behavioral - Somnolence (general depressed activity); Kidney, Ureter, and Bladder - Hematuria; Blood - Changes in erythrocite (RBC) cell count.

Chronic (Multiple Dose) Inhalation: Rat Dose: 5 mg/m^3/24H/17W-C; Toxic Effects: Brain and coverings - Recordings from specific areas of CNS; Blood - Other changes.

Chronic (Multiple Dose) Dermal: Rabbit Route: Skin; Dose: 26700 uL/kg/30D-I; Toxic Effects: Kidney, Ureter, and Bladder - Chgs in tubules (inc acute renal failure, acute tubular necrosis; Kidney, Ureter, and Bladder - Other changes; DEATH.

Irritation Eye: Rabbit Standard Draize Test Dose: 125 mg; Reaction: mild. Rabbit Standard Draize Test Dose: 500 mg; Reaction: moderate.

Irritation Skin: Rabbit Standard Draize Test Dose: 500 mg/24H; Reaction: mild.

Reproductive/Teratogenic: Mouse Route: Oral; Dose: 44 gm/kg; Duration: female 7-14D of pregnancy; Effects on Newborn - Growth statistics. Mouse Route: Oral; Dose: 32 gm/kg; Duration: female 6-13D of pregnancy; Effects on Fertility - Litter size; Effects on Newborn - Live birth index; Growth statistics. Mouse Route: Oral; Dose: 246 gm/kg; Duration: female multigeneration; Specific Developmental Abnormalities - Urogenital system; Effects on Newborn - Other postnatal measures or effects.

Mutagenic: Bacteria - S Typhimurium Mutations in Microorganisms; Dose: 986 mg/plate (+/-S9).

Hazard Overviews

Fire Diamond

Health: Mildly irritating to eyes. Also Causes: kidney damage, nervous system and respiratory depression.

Fire: Will burn. Use dry chemical, foams, water fog, or carbon dioxide to put out fire. Never direct solid streams of water into burning pools of liquid since this can scatter and spread flames.

Reactivity: Stable. Hazardous polymerization cannot occur. Avoid: heat; ignition sources; strong oxidizing agents. Incompatible with: strong oxidizing agents. Hazardous decomposition products: carbon dioxide; toxic carbon monoxide.

Carcinogenicity: IARC - Not listed; NIOSH - Not listed; NTP - Not listed; ACGIH - Not listed; OSHA - Not listed; EPA - Not listed; MAK - Not listed

Primary Target Organs:

Eyes

Exposure Limits

AIHA WEEL: TWA: 25 ppm.

Respirator Recommendation

Exposure Range: >25 to 1250 ppm Supplied Air, Constant Flow/Pressure Demand, Half Mask

Exposure Range: >1250 to 25,000 ppm Supplied Air, Constant Flow/Pressure Demand, Full Face

Exposure Range: >25,000 to unlimited ppm Self-contained Breathing Apparatus, Pressure Demand, Full Face
Note: odor threshold unknown

Environmental

Ecotoxicity: LC_{50} Pimephales promelas (fathead minnow) 26.5 g/l/96 hr (confidence limit 24.2 to 29.0 g/l). Nominal concentrations were not reported. Individual lengths and weights of the test fish were not recorded; however, the measured mean weight was 0.151 g. /Conditions of bioassay not specified LC_{50} Lepomis macrochirus 10,000 ppm/96 hr (static bioassay in fresh water at 23 °C, mild aeration applied after 24 hr)

Environmental Fate: It is not expected to undergo hydrolysis or direct photolysis in the environment. The complete miscibility in water suggests that volatilization, adsorption and bioconcentration are not important fate processes. This is supported by an estimated Henry's Law constant of 8.63 $x10^{-10}$ atm-cu m/mole at 25 °C which indicates that volatilization from natural waters and moist soil should be extremely slow. A low estimated log BCF suggests it should not bioconcentrate among aquatic organisms. A low K_{oc} indicates it should not partition from the water column to organic matter contained in sediments and suspended solids, and it should be highly mobile in soil. Aqueous screening test data indicate that biodegradation is likely to be the most important removal mechanism from aerobic soil and water. In the atmosphere, it is expected to exist almost entirely in the vapor phase and reactions with photochemically produced hydroxyl radicals should be important (estimated half-life of 13 hours). Physical removal from air by precipitation and dissolution in clouds may occur; however, its short atmospheric residence time suggests that wet deposition is of limited importance.

Cleanup/Disposal: Guide No. 171: Do not touch or walk through spilled material. Stop leak if you can do it without risk. Prevent dust cloud. Avoid inhalation of asbestos dust. Small Dry Spills: With clean shovel place material into clean, dry container and cover loosely; move containers from spill area. Small Spills: Take up with sand or other noncombustible absorbent material and place into containers for later disposal. Large Spills: Dike far ahead of liquid spill for later disposal. Cover powder spill with plastic sheet or tarp to minimize spreading. Prevent entry into waterways, sewers, basements or confined areas.

Environmental Physical Data

Henry's Law Constant: calculated at 8.63 $x10^{-10}$
Octanol/Water Partition Coefficient: log K_{ow} = -0.15
Sorption Partition Coefficient: K_{OC} = 20
BCF: calculated at 0.34
BOD: theoretical 34%, 5 days

Regulations

RCRA 40CFR: Not listed
CERCLA: 40CFR 302.4: Not listed
SARA 40CFR 372.65: Not listed
SARA EHS 40CFR 355: Not listed
TSCA: Listed

DIE7300 **CAS #: 112-15-2**

DIETHYLENE GLYCOL MONOETHYL ETHER ACETATE

RTECS: KK8925000
EINECS Number: 203-940-1
Molecular Formula: $C_8H_{16}O_4$
Structured MF: $CH_3COOCH_2OCH_2CH_2OC_2H_5$
Formula Weight: 176.24

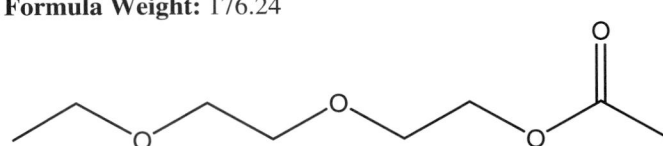

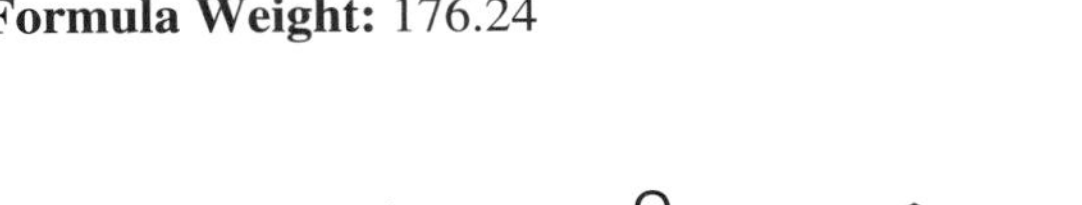

Chemical Structure

Synonyms: ACETIC ACID 2-(2-ETHOXYETHOXY)ETHYL ESTER; CARBITOL ACETATE; DIETHYLENE GLYCOL ETHYL ETHER ACETATE; DIGLYCOL MONOETHYL ETHER ACETATE; EKTASOLVE DE ACETATE; ETHANOL, 2-(2-ETHOXYETHOXY)-,ACETATE; ETHOXYDIGLYCOL ACETATE; 2-(2-ETHOXYETHOXY)ETHANOL ACETATE; 2-(2-ETHOXYETHOXY)ETHYL ACETATE; 2-(2-ETHOXYETHOXY)ETHYLESTER KYSELINY OCTOVE; GLYCOL ETHER DE ACETATE; KARBITOLACETAT
Description: colorless liquid; slight odor
Use: solvent for coatings & lacquers; printing inks; solvent & plasticizer for cellulose esters, gums, resins; solvent for latex paints, lacquers & printing inks

Physical Properties

Boiling Point: 218 °C (424 °F) at 760 mm Hg
Freezing Point: -25 °C (-13 °F)
Specific Gravity: 1.0114 at 20 °C/20 °C
Saturated Vapor Density: 1.201744205 kg/m^3
Vapor Pressure: 0.029 kPa at 25 °C
Water Solubility: Miscible with Water
Other Solubilities: Very Soluble in Acetone, Soluble in ordinary organic solvents.
Refraction Index: 1.4213 at 20 °C/D
Flash Point: 107 °C Open Cup
Autoignition Temperature: 360 °C
LEL: 1.0% v/v
UEL: 19.4% v/v at 375 °F

RTECS Toxicity Data

Acute Oral: Rat LD_{50} Dose: 11 gm/kg. Rabbit LD_{50} Dose: 4400 mg/kg.
Acute Dermal: Rabbit LD_{50} Route: Skin; Dose: 15100 uL/kg.
Irritation Eye: Rabbit Standard Draize Test Dose: 500 mg; Reaction: moderate.
Irritation Skin: Rabbit Open Draize Test Dose: 500 mg open; Reaction: mild.

Hazard Overviews

Health: Irritating to eyes/skin/respiratory tract. Harmful. Other Acute Effects: harmful if swallowed, inhaled, or absorbed through skin. Chronic Effects: may cause reproductive disorders; may impair fertility; laboratory studies with this material indicate birth defects, fetotoxici ty, embryolethality, anemia, bone marrow damage, hemolysis, immuno-suppression, damage to the male reproductive tissues; target organs: liver, kidneys, central nervous system, female reproductive system, male reproductive system.
Fire: Will burn. Hazards: emits toxic fumes. Extinguishing agents: water spray; carbon dioxide, dry chemical powder or appropriate foam. Precautions: combustible liquid.
Reactivity: Incompatible with: strong oxidizing agents. Hazardous decomposition products: toxic fumes of: carbon monoxide, carbon dioxide.
Carcinogenicity: IARC - Not listed; NIOSH - Not listed; NTP - Not listed; ACGIH - Not listed; OSHA - Not listed; EPA - Not listed; MAK - Not listed
Primary Target Organs:

Eyes

Skin

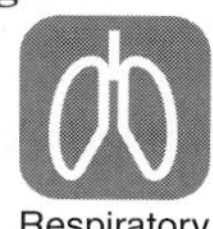
Respiratory System

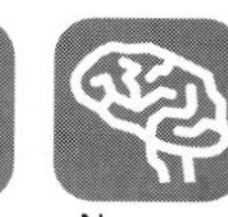
Nervous System

Liver

Kidneys

Environmental

Cleanup/Disposal: Guide No. 171: Do not touch or walk through spilled material. Stop leak if you can do it without risk. Prevent dust cloud. Avoid inhalation of asbestos dust. Small Dry Spills: With clean shovel place material into clean, dry container and cover loosely; move containers from spill area. Small Spills: Take up with sand or other noncombustible absorbent material and place into containers for later disposal. Large Spills: Dike far ahead of liquid spill for later disposal. Cover powder spill with plastic sheet or tarp to minimize spreading. Prevent entry into waterways, sewers, basements or confined areas.

Regulations

RCRA 40CFR: Not listed
CERCLA: 40CFR 302.4: Not listed
SARA 40CFR 372.65: Not listed
SARA EHS 40CFR 355: Not listed
TSCA: Listed

DIE7400 CAS #: 3088-31-1

DIETHYLENE GLYCOL MONOLAURYL ETHER SODIUM SULFATE

EINECS Number: 221-416-0
Molecular Formula: $C_{16}H_{34}NaO_6S$
Formula Weight: 377.47
Synonyms: DIETHYLENE GLYCOL MONODODECYL ETHER SODIUM SULFATE; DIETHYLENE GLYCOL MONODODECYL ETHER SULFATE SODIUM SALT; DIETHYLENE GLYCOL MONOLAURYL ETHER SULFATE SODIUM SALT; 2-(2-DODECYLOXYETHOXY)ETHYL SODIUM SULFATE; ETHANOL,2-(2-(DODECYLOXY)ETHOXY)-,HYDROGEN SULFATE,SODIUM SALT; ETHANOL,2-(2-(DODECYLOXY)ETHOXY)-,HYDROGEN SULFATESODIUM SALT; LAURISTYL DIGLYCOL ETHER SULFATE SODIUM SALT; LAURYL DIETHYLENE GLYCOL ETHER SULFONATE SODIUM; SODIUM DIETHYLENE GLYCOL DODECYL ETHER SULFATE; SODIUM DIOXYETHYLENEDODECYL ETHER SULFATE; SODIUM LAURYL ALCOHOL DIGLYCOL ETHER SULFATE; SODIUM LAURYLOXYETHOXYETHYL SULFATE; SODIUMLAURYLGLYCOLETHER SULFATE; SULFURIC ACID MONO(2-(2-(DODECYLOXY)ETHOXY)ETHYL) ETHERSODIUM SALT; TERGENTOL

Hazard Overviews

Carcinogenicity: IARC - Not listed; NIOSH - Not listed; NTP - Not listed; ACGIH - Not listed; OSHA - Not listed; EPA - Not listed; MAK - Not listed

Environmental

Regulations

RCRA 40CFR: Not listed
CERCLA: 40CFR 302.4: Not listed
SARA 40CFR 372.65: Not listed
SARA EHS 40CFR 355: Not listed
TSCA: Listed

DIE7500 CAS #: 111-77-3

DIETHYLENE GLYCOL MONOMETHYL ETHER

RTECS: KL6125000
EINECS Number: 203-906-6
Molecular Formula: $C_5H_{12}O_3$
Structured MF: $CH_3OCH_2CH_2OCH_2CH_2OH$
Formula Weight: 120.15

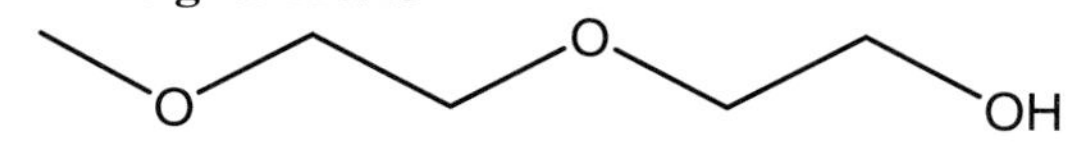

Chemical Structure

Synonyms: DIETHYLENE GLYCOL METHYL ETHER; DIGLYCOL MONOMETHYL ETHER; 3,6-DIOXA-1-HEPTANOL; DOWANOL 16; DOWANOL DM; EKTASOLVE DM; ETHANOL,2-(2-METHOXYETHOXY)-; ETHANOL,2,2'-OXYBIS-,MONOMETHYL ETHER; ETHYLENE DIGLYCOL MONOMETHYL ETHER; GLYCOL ETHER DM; HICOTOL CAR; JEFFERSOL DM; MECB; METHOXYDIGLYCOL; 2-(2-METHOXYETHOXY)ETHANOL; METHOXYETHOXYETHANOL; BETA-METHOXY-BETA'-HYDROXYDIETHYL ETHER; 2-BETA-METHYL CARBITOL; METHYL CARBITOL; METHYL DIGOL; METHYL DIOXITOL; METHYL KARBITOL; 2,2'-OXYBISETHANOL MONOMETHYL ETHER; POLY-SOLV DM
Description: colorless liquid; mild, pleasant odor
Use: solvent (used where a solvent with a higher boiling point is required), brake fluid component and intermediate

Physical Properties

Boiling Point: 193 °C (379 °F)
Freezing Point: < -84 °C (-119.2 °F)
Specific Gravity: 1.035 at 20 °C/4 °C
Vapor Density: 4.14 Air=1
Saturated Vapor Density: 1.200992559 kg/m^3
Density: 1.025 g/mL at 20 °C
Vapor Pressure: 0.2 mm Hg at 20 °C
Water Solubility: Miscible with Water
Other Solubilities: 95% Ethanol: >=100 mg/ml at 19 °C; Acetone: >=100 mg/ml at 19 °C; Benzene: miscible; Carbon Tetrachloride: miscible; DMSO: >=100 mg/ml at 19 °C; Dimethylformamide: miscible; Ether: miscible; Glycerol: miscible; Ketones: miscible.
Surface Tension: 34.8 dynes/cm at 25 °C
Refraction Index: 1.4264 at 27 °C/D
Evaporation Rate: 0.02 n-butyl acetate=1
Flash Point: 93.333 °C Open Cup
Autoignition Temperature: 204 °C
LEL: 1.2% v/v

RTECS Toxicity Data

Acute Oral: Rat LD_{50} Dose: 4 mL/kg. Mouse LD_{50} Dose: 8222 mg/kg; Toxic Effects: Automatic Nervous System - Other (direct) parasympathomimetic; Behavioral - Somnolence (general depressed activity); Lungs, Thorax, or Respiration - Cyanosis.
Acute Inhalation: Rat LC; Dose: >2 gm/m^3/1hr.
Acute Dermal: Rabbit LD_{50} Route: Skin; Dose: 6540 uL/kg.
Chronic (Multiple Dose) Oral: Rat Dose: 107 gm/kg/6W-I; Toxic Effects: Behavioral - Food intake (animal); Nutritional and gross metabolic - Weight loss or decreased weight gain. Rat Dose: 5052 mg/kg/30D-C; Toxic Effects: Blood - Pigmented or nucleated red Blood cells; Biochemical - Phosphatases; Biochemical - Other proteins. Rat Dose: 40 gm/kg/4W-I; Toxic Effects: Brain and coverings - Other degenerative changes; Liver - Other changes; Biochemical - Peptidases. Rat Dose: 40 gm/kg/20D-I; Toxic Effects: Liver - Changes in liver weight; Endocrine - Changes in thymus weight; DEATH - Changes in testicular weight.
Chronic (Multiple Dose) Dermal: Guinea Pig Route: Skin; Dose: 65 gm/kg/13W-I; Toxic Effects: Endocrine - Changes in spleen weight; Kidney, Ureter, and Bladder - Other changes in urine composition; Biochemical - Dehydrogenases.
Irritation Eye: Rabbit Standard Draize Test Dose: 500 mg/24H; Reaction: mild. Rabbit Standard Draize Test Dose: 500 mg; Reaction: moderate.
Reproductive/Teratogenic: Rat Route: Oral; Dose: 19800 mg/kg; Duration: female 7-17D of pregnancy; Maternal Effects - Parturition; Effects on Fertility - Post-implantation mortality; Specific Developmental Abnormalities - Cardiovascular (circulatory) system. Rat Route: Oral; Dose: 21650 mg/kg; Duration: female 7-16D of pregnancy; Effects on Fertility - Litter size; Effects on Embryo or Fetus - Fetotoxicity; Specific Developmental Abnormalities - Cardiovascular (circulatory) system. Rat Route: Oral; Dose: 19800 mg/kg; Duration: female 7-17D of pregnancy; Effects on Fertility - Post-implantation mortality; Effects on Embryo or Fetus - Fetotoxicity; Specific Developmental Abnormalities - Urogenital system. Rat Route: Oral; Dose: 6600 mg/kg; Duration: female 7-17D of pregnancy; Effects on Embryo or Fetus - Fetotoxicity; Specific Developmental Abnormalities - Musculoskeletal system; Effects on Newborn - Physical.

Hazard Overviews

Health: Irritating to eyes/skin/respiratory tract. Other Acute Effects: may be harmful by inhalation, ingestion, or skin absorption; readily absorbed through skin; exposure can cause nausea; dizziness; headache. Chronic Effects: laboratory studies with this material indicate birth defects, fetotoxicity, embryolethality, anemia, bone marrow damage, hemolysis, immuno-suppression, damage to the male reproductive tissues; target organs: female reproductive system, male reproductive system.
Fire: Will burn. Extinguishing agents: carbon dioxide, dry chemical powder or appropriate foam. Precautions: combustible liquid.
Reactivity: Incompatible with: oxidizing agents. Hazardous decomposition products: toxic fumes of: carbon monoxide, carbon dioxide.
Carcinogenicity: IARC - Not listed; NIOSH - Not listed; NTP - Not listed; ACGIH - Not listed; OSHA - Not listed; EPA - Not listed; MAK - Not listed
Primary Target Organs:

Eyes

Skin

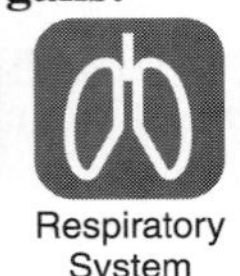
Respiratory System

Repro-ductive

Environmental

Ecotoxicity: LC_{50} Lepomis macrochirus 7500 ppm/96 hr (static bioassy in fresh water at 23 °C, mild aeration applied after 24 hr)
Environmental Fate: It is not expected to undergo hydrolysis or direct photolysis in the environment. The complete miscibility in water suggests that volatilization, adsorption and bioconcentration are not important fate processes. This is supported by an estimated Henry's Law constant of 6.50 x10^{-10} atm-cu m/mole at 25 °C which indicates that volatilization from natural waters and moist soil should be extremely slow. A low estimated log BCF suggests it should not bioconcentrate among aquatic organisms. A low K_{oc} indicates it should not partition from the water column to organic matter contained in sediments and suspended solids, and it should be highly mobile in soil. Limited aqueous screening test data indicate that biodegradation may be an important removal mechanism from aerobic soil and water. In the atmosphere, it is expected to exist almost entirely in the vapor phase and reactions with photochemically produced hydroxyl radicals should be important (estimated half-life of 16 hours). Physical removal from air by precipitation and dissolution in

clouds may occur; however, its short atmospheric residence time suggests that wet deposition is of limited importance.

Cleanup/Disposal: Guide No. 153: Eliminate all ignition sources (no smoking, flares, sparks or flames in immediate area). Do not touch damaged containers or spilled material unless wearing appropriate protective clothing. Stop leak if you can do it without risk. Prevent entry into waterways, sewers, basements or confined areas. Absorb or cover with dry earth, sand or other non-combustible material and transfer to containers. Do not get water inside containers.

Environmental Physical Data

Henry's Law Constant: calculated at 6.50×10^{-10}
Octanol/Water Partition Coefficient: log K_{ow} = -0.68
Sorption Partition Coefficient: K_{oc} = 10
BCF: calculated at 0.75
BOD: theoretical 34%, 5 days

Regulations

RCRA 40CFR: Not listed
CERCLA: 40CFR 302.4: Not listed
SARA 40CFR 372.65: Not listed
SARA EHS 40CFR 355: Not listed
TSCA: Listed

DIE7600 CAS #: 25928-94-3

DIETHYLENE GLYCOL, POLYMER WITH 1-CHLORO-2,3-EPOXYPROPANE

RTECS: ID8970000
Molecular Formula: $C_{(7x)}H_{(15x)}Cl_xO_{(4x)}$
Synonyms: DEG-1; ETHANOL,2,2'-OXYBIS-,POLYMER WITH (CHLOROMETHYL)OXIRANE(9CI)

Physical Properties

Boiling Point: Varies
Specific Gravity: Varies
Vapor Pressure: Low
Water Solubility: Varies
Flash Point: Generally > 93.3 °C

RTECS Toxicity Data

Acute Oral: Mouse LD_{50} Dose: 2200 mg/kg; Toxic Effects: Behavioral - Convulsions or effect on seizure threshold; Lungs, Thorax, or Respiration - Dyspnea.

Hazard Overviews

Reactivity: Stable. Polymerization can occur exothermically when in contact with curing agents (organic amines; acid anhydrides; polyamides). Avoid: exposure to heat and ignition sources. Incompatible with: oxidizing agents; strong acids; mercaptans; bases; bulk contact with curing agents. Hazardous decomposition products: carbon oxides; various hydrocarbons.

Carcinogenicity: IARC - Not listed; NIOSH - Not listed; NTP - Not listed; ACGIH - Not listed; OSHA - Not listed; EPA - Not listed; MAK - Not listed

Environmental

Regulations

RCRA 40CFR: Not listed
CERCLA: 40CFR 302.4: Not listed
SARA 40CFR 372.65: Not listed
SARA EHS 40CFR 355: Not listed
TSCA: Not listed

DIE7700 CAS #: 111-40-0

DIETHYLENETRIAMINE

RTECS: IE1225000
DOT: UN2079; IMO8.0
EINECS Number: 203-865-4
Molecular Formula: $C_4H_{13}N_3$
Structured MF: $(NH_2CH_2CH_2)_2NH$
Formula Weight: 103.17

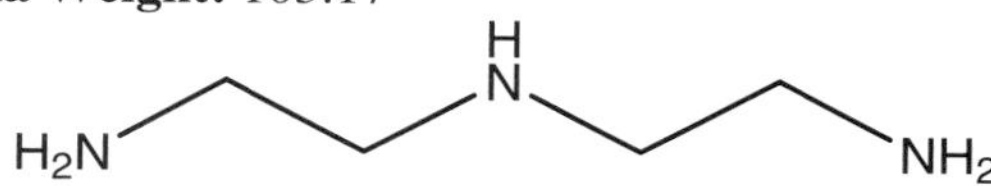

Chemical Structure

Synonyms: AMINOETHYLETHANDIAMINE; AMINOETHYLETHANEDIAMINE; N-(2-AMINOETHYL)-1,2-ETHANEDIAMINE; N-(2-AMINOETHYL)1,2-ETHANEDIAMINE; N-(2-AMINOETHYL)ETHYLENEDIAMINE; 3-AZAPEN-TANE-1,5-DIAMINE; 3-AZAPENTANE-1,5-DIAMINE; BIS(2-AMINOETHYL)AMINE; BIS(BETA-AMINOETHYL)AMINE; N,N-BIS(2-AMINOETHYL)AMINE; CHS-P 1; DETA; 2,2'-DIAMINODIETHYLAMINE; BETA,BETA'-DIAMINODIETHYLAMINE; DIETHYLAMINE,2,2'-DIAMINO-; 1,2-ETHANEDIAMINE,N-(2-AMINOETHYL)-; ETHYLAMINE,2,2'-IMINOBIS-; ETHYLAMINE,2,2'-IMINOBIS-ETHYLENEDIAMINE,N-(2-AMINOETHYL)-; ETHYLENEDIAMINE,N-(2-AMINOETHYL)-; 2,2'-IMINOBIS(ETHANAMINE); 2,2'-IMINOBISETHYLAMINE; 1,4,7-TRIAZAHEPTANE

Description: yellow liquid; ammonia-like odor

Use: solvent for sulfur, acidic gas, resin and dye intermediates for organic synthesis; saponification agent for acidic materials; fuel component

Physical Properties

Boiling Point: 207 °C (405 °F) at 760 mm Hg
Freezing Point: -39 °C (-38.2 °F)
Specific Gravity: 0.9586 at 20 °C/20 °C
Vapor Density: 3.48 Air=1
Saturated Vapor Density: 1.200888425 kg/m^3
Density: 0.95 g/mL at 20 °C
Vapor Pressure: 0.22 mm Hg at 20 °C
Water Solubility: Soluble in all Proportions
Other Solubilities: 95% Ethanol: Very Soluble; Benzene: Very Soluble; Ether: Very Soluble; hydrocarbons: Soluble; Ligroin: Soluble.

Surface Tension: 40.8 dynes/cm
Odor Threshold: 10 ppm
pH: Strongly alkaline
Refraction Index: 1.4810 at 25 °C/D
Flash Point: 101.7 °C Open Cup
Autoignition Temperature: 399 °C
LEL: 1% v/v
UEL: Calculated at 10% v/v

RTECS Toxicity Data

Acute Oral: Rat LD_{50} Dose: 1080 mg/kg; Toxic Effects: Behavioral - Convulsions or effect on seizure threshold.
Acute Dermal: Rabbit LD_{50} Route: Skin; Dose: 1090 mg/kg. Guinea Pig LD_{50} Route: Skin; Dose: 170 uL/kg.
Chronic (Multiple Dose) Oral: Rabbit Dose: 1820 mg/kg/26W-I; Toxic Effects: Biochemical - Transaminases.
Irritation Eye: Rabbit Standard Draize Test Dose: 750 ug open; Reaction: severe.
Irritation Skin: Rabbit Open Draize Test Dose: 10 mg/24H open; Reaction: severe. Rabbit Open Draize Test Dose: 500 mg open; Reaction: moderate.

Hazard Overviews

Corrosive

Fire Diamond

Health: Corrosive, causes severe burns to eyes/skin/respiratory tract. Also Causes: cough, nausea, vomiting, corneal injury. Chronic Effects: asthmatic response/sensitization, skin sensitization.
Fire: Will burn when exposed to heat, sparks, or flames. Use dry chemical, carbon dioxide, water spray, fog, or regular foam. Do not scatter material with water.
Reactivity: Stable. Hazardous polymerization cannot occur. Avoid: ignition sources. Incompatible with: copper and its alloys; cellulose nitrate; nitromethane; acids; chlorine; halogenated compounds; oxidizing materials; reactive organic compounds. Hazardous decomposition products: carbon monoxide; carbon dioxide; hydrocarbons; amine vapors; toxic nitrogen oxide fumes.
Carcinogenicity: IARC - Not listed; NIOSH - Listed as carcinogen; NTP - Not listed; ACGIH - Not listed; OSHA - Not listed; EPA - Not listed; MAK - Not listed
Primary Target Organs:

Eyes

Skin

Respiratory System

Exposure Limits
OSHA PEL Vacated 1989 Limits: TWA: 1 ppm; 4 mg/m^3.
ACGIH TLV: TWA: 1 ppm; 4.2 mg/m^3.
NIOSH REL: TWA: 1 ppm; 4 mg/m^3.
Respirator Recommendation
Exposure Range: >1 to 50 ppm Supplied Air, Constant Flow/Pressure Demand, Half Mask
Exposure Range: >50 to 1000 ppm Supplied Air, Constant Flow/Pressure Demand, Full Face
Exposure Range: >1000 to unlimited ppm Self-contained Breathing Apparatus, Pressure Demand, Full Face
Note: odor threshold unknown

Environmental

Ecotoxicity: Aquatic toxicity: 710 ppm/24 hr/brine shrimp/TLm; Food chain concentration potential: 23% of theoretical
Environmental Fate: If released to the atmosphere, it would be expected to photooxidize by reaction with hydroxyl radicals (estimated half-life 2.7 hr). If released on land, it would be expected to be highly mobile and leach. It is resistant to biodegradation. Its fate in surface waters is largely unknown; however, based upon its high water solubility it would not appreciably adsorb to sediment, volatilize or bioconcentrate in fish.
Cleanup/Disposal: Guide No. 154: Eliminate all ignition sources (no smoking, flares, sparks or flames in immediate area). Do not touch damaged containers or spilled material unless wearing appropriate protective clothing. Stop leak if you can do it without risk. Prevent entry into waterways, sewers, basements or confined areas. Absorb or cover with dry earth, sand or other non-combustible material and transfer to containers. Do not get water inside containers.

Environmental Physical Data

Henry's Law Constant: 1 $x10^{-14}$
Octanol/Water Partition Coefficient: log K_{ow} = calculated at -1.3
Sorption Partition Coefficient: K_{oc} = estimated at 88
BCF: none likely

Regulations

RCRA 40CFR: Not listed
CERCLA: 40CFR 302.4: Not listed
SARA 40CFR 372.65: Not listed
SARA EHS 40CFR 355: Not listed
TSCA: Listed

Analytical Methods

Indoor / Expired Air: NIOSH 2540

DIE7800 CAS #: 103-24-2

DI-2-ETHYLHEXYL AZELATE

RTECS: CM2000000
EINECS Number: 203-091-7
Molecular Formula: $C_{25}H_{48}O_4$
Formula Weight: 412.73
Synonyms: AZELAIC ACID DI(2-ETHYLHEXYL) ESTER; AZELAIC ACID,BIS(2-ETHYLHEXYL) ESTER; AZELAIC ACID,DI(2-ETHYLHEXYL)ESTER; BIS(2-ETHYLHEXYL) AZELATE; BIS(2-ETHYLHEXYL)AZELATE; BIS-(2-ETHYLHEXYL)ESTER KYSELINY AZELAOVE; DI(2-ETHYLHEXYL) AZELATE; DIOCTYL AZELATE; DOZ; NONANEDIOIC ACID,BIS(2-ETHYLHEXYL) ESTER; OCTYL AZELATE; PLASTOLEIN 9058; PLASTOLEIN 9058 DOZ; PLASTOLEIN 9058DOZ; STAFLEX DOX; TRUFLEX DOX

Use: plasticizer for cellulosics, polystyrene & vinyl plastics

Physical Properties

Boiling Point: 237 °C (459 °F) at 5 mm Hg
Freezing Point: -78 °C (-108.4 °F)
Specific Gravity: 0.915 at 25 °C/4 °C
Water Solubility: Insoluble in Water
Other Solubilities: Soluble in Alcohol, Acetone, Benzene
Refraction Index: 1.446 at 25 °C/D
Evaporation Rate: < 1 Butyl Acetate=1

RTECS Toxicity Data

Acute Oral: Rat LD_{50} Dose: 8720 mg/kg.
Acute Dermal: Rabbit LD_{50} Route: Skin; Dose: 20 gm/kg.
Irritation Skin: Rabbit Open Draize Test Dose: 10 mg/24H open; Reaction: mild.

Hazard Overviews

Health: May be irritating to eyes/skin. Other Acute Effects: may be harmful by inhalation, ingestion, or skin absorption.
Fire: Extinguishing agents: water spray; carbon dioxide, dry chemical powder or appropriate foam. Precautions: combustible liquid.
Reactivity: Incompatible with: strong oxidizing agents. Hazardous decomposition products: toxic fumes of: carbon monoxide, carbon dioxide.
Carcinogenicity: IARC - Not listed; NIOSH - Not listed; NTP - Not listed; ACGIH - Not listed; OSHA - Not listed; EPA - Not listed; MAK - Not listed

Environmental

Regulations

RCRA 40CFR: Not listed
CERCLA: 40CFR 302.4: Not listed
SARA 40CFR 372.65: Not listed
SARA EHS 40CFR 355: Not listed
TSCA: Listed

DIE7900 CAS #: 1928-43-4

2,4-DIETHYLHEXYL ESTER

RTECS: AG8525000
EINECS Number: 217-673-3
Molecular Formula: $C_{16}H_{22}Cl_2O_3$
Formula Weight: 333.28
Synonyms: (2,4-DICHLOROPHENOXY)ACETIC ACID 2-ETHYLHEXYL ESTER; 2,4-D 2-ETHYLHEXYL ESTER

RTECS Toxicity Data

Acute Oral: Rat LD_{50} Dose: 300 mg/kg.

Hazard Overviews

Carcinogenicity: IARC - Not listed; NIOSH - Not listed; NTP - Not listed; ACGIH - Not listed; OSHA - Not listed; EPA - Not listed; MAK - Not listed

Environmental

Regulations

RCRA 40CFR: Not listed
CERCLA: 40CFR 302.4: Not listed
SARA 40CFR 372.65: Listed
SARA EHS 40CFR 355: Not listed
TSCA: Not listed

Analytical Methods

Soil: SW846 8321, 8321A
Water / Groundwater: EPA X_89_176
Food: FDA 212.1, 232.1
Plasma: EPA 001; FDA 211.1, 231.1, 252

DIE8000 CAS #: 142-16-5

DI-2-ETHYLHEXYL MALEATE

RTECS: ON0160000
EINECS Number: 205-524-5
Molecular Formula: $C_{20}H_{36}O_4$
Formula Weight: 340.56

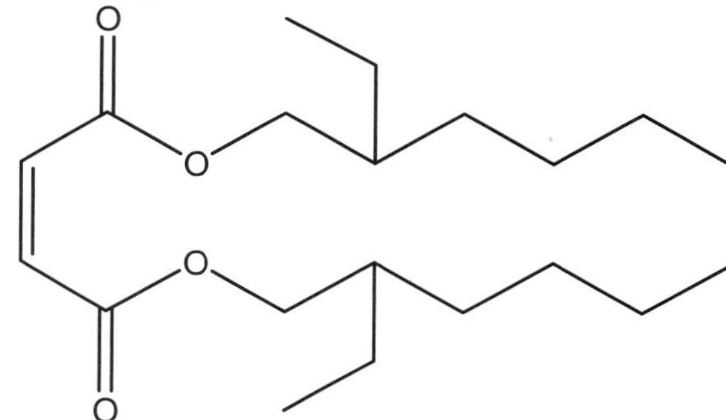

Chemical Structure

Synonyms: BIS(2-ETHYLHEXYL) MALEATE; BIS-(2-ETHYLHEXYL)ESTER KYSELINY MALEINOVE; BIS(2-ETHYLHEXYL)MALEATE; 2-BUTENEDIOIC ACID (Z)-,BIS(2-ETHYLHEXYL) ESTER; DI(2-ETHYLHEXYL)MALEATE; DI-(2-ETHYLHEXYL)MALEATE; DIOCTYL MALEATE; DOM; MALEIC ACID,BIS(2-ETHYLHEXYL) ESTER; MALEIC ACID,BIS(2-ETHYLHEXYL) ESTER (6CI,7CI,8CI); RC COMONOMER DOM
Description: liquid
Use: in copolymers; intermediate in chemical reactions; comonomer & internal plasticizer with vinyl acetate; comonomer with acrylates & methacrylates for polymers; comonomer with n-vinyl-n-methylacetamide for polymer; chem int for sulfosuccinate anionic surfactants

Physical Properties

Boiling Point: 164 °C (327 °F) at 10 mm Hg
Freezing Point: -60 °C (-76 °F)
Specific Gravity: 0.9436
Water Solubility: Insoluble in Water
Flash Point: 185 °C Closed Cup

RTECS Toxicity Data

Acute Oral: Rat LD_{50} Dose: 14 gm/kg. Mouse LD; Dose: >20 gm/kg.
Acute Dermal: Rabbit LD_{50} Route: Skin; Dose: 15 mL/kg.
Irritation Eye: Rabbit Standard Draize Test Dose: 500 mg/24H; Reaction: mild.
Irritation Skin: Rabbit Standard Draize Test Dose: 500 mg/24H; Reaction: mild. Rabbit Open Draize Test Dose: 10 mg/24H open; Reaction: mild.

Hazard Overviews

Health: Irritating to eyes/skin/respiratory tract. Other Acute Effects: may be harmful by inhalation, ingestion, or skin absorption.
Fire: Will burn. Extinguishing agents: water spray; carbon dioxide, dry chemical powder or appropriate foam. Precautions: combustible liquid.
Reactivity: Stable. Hazardous polymerization will not occur. Incompatible with: strong oxidizing agents. Hazardous decomposition products: toxic fumes of: carbon monoxide, carbon dioxide.
Carcinogenicity: IARC - Not listed; NIOSH - Not listed; NTP - Not listed; ACGIH - Not listed; OSHA - Not listed; EPA - Not listed; MAK - Not listed
Primary Target Organs:

Eyes

Skin

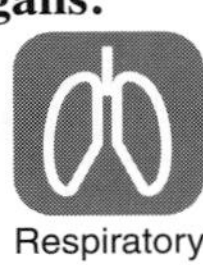
Respiratory System

Environmental

Regulations

RCRA 40CFR: Not listed
CERCLA: 40CFR 302.4: Not listed
SARA 40CFR 372.65: Not listed
SARA EHS 40CFR 355: Not listed
TSCA: Listed

DIE8100 CAS #: 117-81-7

DI(2-ETHYLHEXYL)PHTHALATE

RTECS: TI0350000
EINECS Number: 204-211-0
Molecular Formula: $C_{24}H_{38}O_4$
Structured MF: $C_6H_4[COOCH_2CH(C_2H_5)(CH_2)_3CH_3]_2$
Formula Weight: 390.54

Chemical Structure

Synonyms: BEHP; 1,2-BENZENEDICARBOXYLIC ACID,BIS(2-ETHYLHEXYL) ESTER; 1,2-BENZENEDICARBOXYLIC ACID,BIS(ETHYLHEXYL) ESTER; BIS(2-ETHYLHEXYL) 1,2-BENZENEDICARBOXYLATE; BIS(2-ETHYLHEXYL)-1,2-BENZENEDICARBOXYLATE; BIS-(2-ETHYLHEXYL)-1,2-BENZENEDICARBOXYLATE; BIS-(2-ETHYLHEXYL)ESTER KYSELINY FTALOVE; BIS(2-ETHYLHEXYL)ESTER PHTHALIC ACID; BIS(2-ETHYLHEXYL)PHTHALATE; BIS-(2-ETHYLHEXYL)PHTHALATE; BISOFLEX 81; BISOFLEX DOP; COMPOUND 889; DAF 68; DEHP; DI(2-ETHYLHEXYL) PHTHALATE; DI(ETHYLHEXYL) PHTHALATE; DIETHYLHEXYL PHTHALATE; DI(2-ETHYLHEXYL)ORTHOPHTHALATE; DI-2-ETHYLHEXYLPHTHALATE; DI-SEC-OCTYL PHTHALATE; DIOCTYL PHTHALATE; DOF; DOP; ERGOPLAST FDO; ERGOPLAST FDO-S; 2-ETHYLHEXYL PHTHALATE; ETHYLHEXYL PHTHALATE; EVIPLAST 80; EVIPLAST 81; FLEXIMEL; FLEXOL DOP; FLEXOL PLASTICIZER DOP; GOOD-RITE GP 264; HATCOL DOP; HERCOFLEX 260; JAYFLEX DOP; KODAFLEX DOP; MOLLAN O; NUOPLAZ DOP; OCTOIL; OCTYL PHTHALATE; PALATINOL AH; PHTHALIC ACID DIOCTYL ESTER; PHTHALIC ACID,BIS(2-ETHYLHEXYL) ESTER; PITTSBURGH PX-138; PLATINOL AH; PLATINOL DOP; RC PLASTICIZER DOP; REOMOL D 79P; REOMOL DOP; SICOL 150; STAFLEX DOP; TRUFLEX DOP; VESTINOL AH; VINICIZER 80; WITCIZER 312
Description: light colored liquid; slight odor
Use: in vacuum pumps; as a plasticizer for polyvinyl chloride, especially in the manufacture of medical devices, and as a plasticizer for resins and elastomers; a solvent in erasable ink and dielectric fluid; as an acaricide for use in orchards, an inert ingredient in pesticides, a detector for leaks in respirators, testing of air filtration systems and component in cosmetic products

Physical Properties

Boiling Point: 230 °C (446 °F) at 5 mm Hg
Freezing Point: -50 °C (-58 °F)
Specific Gravity: 0.9861 at 20 °C/20 °C
Vapor Density: 13.45 Air=1
Density: 0.9732 g/mL at 24 °C
Vapor Pressure: 1.32 mm Hg at 200 °C
Water Solubility: < 0.01% at 25 °C
Other Solubilities: 95% Ethanol: >=100 mg/ml at 22 °C; Acetone: >=100 mg/ml at 22 °C; DMSO: 10-50 mg/ml at 22 °C; Hexane: miscible; Mineral oil: miscible.
Surface Tension: Estimated at 15 dynes/cm

Refraction Index: 1.4836 at 20 °C/D
Flash Point: 215 °C Open Cup
Autoignition Temperature: 391 °C
LEL: 0.3% v/v

RTECS Toxicity Data

Acute Oral: Man TD_{Lo} Dose: 143 mg/kg; Toxic Effects: Gastrointestinal - Other changes. Rat LD_{50} Dose: 30 gm/kg.
Acute Dermal: Rabbit LD_{50} Route: Skin; Dose: 25 gm/kg. Rat LD_{Lo} Route: Skin; Dose: 4 gm/kg; Toxic Effects: Skin and appendages - Other glands.
Chronic (Multiple Dose) Oral: Rat Dose: 139 gm/kg/2Y-C; Toxic Effects: Liver - Changes in liver weight; Kidney, Ureter, and Bladder - Changes in kidney weight; Nutritional and gross metabolic - Weight loss or decreased weight gain. Rat Dose: 17500 mg/kg/7D-I; Toxic Effects: Liver - Changes in liver weight; Biochemical - Dehydrogenases; DEATH - Changes in testicular weight. Rat Dose: 59388 mg/kg/6W-C; Toxic Effects: Kidney, Ureter, and Bladder - Changes in kidney weight; Skin and appendages - Hair; Nutritional and gross metabolic - Weight loss or decreased weight gain. Rat Dose: 168 gm/kg/17W-C; Toxic Effects: Blood - Other changes; Skin and appendages - Hair; Nutritional and gross metabolic - Weight loss or decreased weight gain. Rat Dose: 19796 mg/kg/12W-C; Toxic Effects: Liver - Changes in liver weight; Skin and appendages - Hair; DEATH - Changes in testicular weight. Rat Dose: 25200 mg/kg/21D-C; Toxic Effects: Liver - Changes in liver weight; Biochemical - Phosphatases; Biochemical - Catalases. Rat Dose: 14 gm/kg/14D-I; Toxic Effects: Liver - Other changes; Liver - Changes in liver weight.
Chronic (Multiple Dose) Inhalation: Rat Dose: 940 mg/m^3/6H/4W-I; Toxic Effects: Lungs, Thorax, or Respiration - Changes in lung weight; Liver - Changes in Liver weight; Blood - Changes in serum composition.
Irritation Eye: Rabbit Standard Draize Test Dose: 500 mg/24H; Reaction: mild. Rabbit Standard Draize Test Dose: 500 mg; Reaction: mild.
Irritation Skin: Rabbit Standard Draize Test Dose: 500 mg/24H; Reaction: mild.
Reproductive/Teratogenic: Rat Route: Oral; Dose: 7140 mg/kg; Duration: female 1-21D of pregnancy; Effects on Embryo or Fetus - Fetotoxicity. Rat Route: Oral; Dose: 35 mg/kg; Duration: female 14D prior to mating Maternal Effects - Ovaries, fallopian tubes. Rat Route: Oral; Dose: 6 gm/kg; Duration: male 3D prior to mating; Paternal Effects - Testes, epididymis, sperm duct; Prostate, seminal vessicle, Cowper's gland, accessory glands. Rat Route: Oral; Dose: 17200 mg/kg; Duration: multigenerations; Effects on Fertility - Post-implantation mortality. Rat Route: Oral; Dose: 10 gm/kg; Duration: female 6-15D of pregnancy; Effects on Embryo or Fetus - Fetotoxicity; Other effects to embryo or fetus; Specific Developmental Abnormalities - Hepatobiliary system. Rat Route: Oral; Dose: 9766 mg/kg; Duration: female 12D of pregnancy; Specific Developmental Abnormalities - Musculoskeletal system; Cardiovascular (circulatory) system; Urogenital system.
Mutagenic: Human Cytogenetic Analysis; Cell Type: leukocyte; Dose: 6 mg/L. Human Sister Chromatid Exchenge; Cell Type: lymphocyte; Dose: 50 umol/L.
Tumorigenic: Rat Route: Oral; Dose: 216 gm/kg/2Y-C; Toxic Effects: Tumorigenic - Carcinogenic by RTECS criteria; Liver - Tumors; Endocrine - Tumors. Rat Route: Oral; Dose: 433 gm/kg/2Y-C; Toxic Effects: Tumorigenic - Carcinogenic by RTECS criteria; Liver - Tumors; Tumorigenic effects - Testicular tumors. Rat Route: Oral; Dose: 524 gm/kg/2Y-C; Toxic Effects: Tumorigenic - Carcinogenic by RTECS criteria; Liver - Tumors. Rat Route: Oral; Dose: 438 gm/kg/2Y-C; Toxic Effects: Tumorigenic - Carcinogenic by RTECS criteria; Liver - Tumors.

Hazard Overviews

Fire Diamond

Health: Mildly irritating to eyes/skin/respiratory tract. Also Causes: conjunctivitis, keratitis, bronchial irritation, eczema, staggering, abdominal cramps, nausea, diarrhea, CNS depression. Chronic Effects: possible cancer hazard.
Fire: Will burn. Use dry powder, carbon dioxide, or foam to fight a fire involving di(2-ethylhexyl)phthalate. Water or foam may cause frothing.
Reactivity: Stable. Hazardous polymerization cannot occur. Incompatible with: strong oxidizing materials. Hazardous decomposition products: acrid smoke; fumes; carbon dioxide; carbon monoxide.
Carcinogenicity: IARC - Group 2B, Possibly carcinogenic to humans; NIOSH - Listed as carcinogen; NTP - Class 2B, Reasonably anticipated to be a carcinogen, sufficient evidence of carcinogenicity from studies in experimental animals; ACGIH - Class A3, Animal carcinogen; OSHA - Not listed; EPA - Class B2, Probable human carcinogen based on animal studies; MAK - Not listed
Primary Target Organs:

Eyes

Skin

Respiratory System

Gastro-intestinal

Nervous System

Exposure Limits
OSHA PEL: TWA: 5 mg/m^3.
OSHA PEL Vacated 1989 Limits: TWA: 5 mg/m^3; STEL: 10 mg/m^3.
ACGIH TLV: TWA: 5 mg/m^3.
NIOSH REL: TWA: 5 mg/m^3. STEL: 10 mg/m^3.
NIOSH IDLH: 5000 mg/m^3.
DFG MAK: TWA: 10 mg/m^3.
Respirator Recommendation
Exposure Range: >5 to 50 mg/m^3 Air Purifying, Negative Pressure, Half Mask
Exposure Range: >50 to 500 mg/m^3 Air Purifying, Negative Pressure, Full Face
Exposure Range: >500 to <5000 mg/m^3 Supplied Air, Constant Flow/Pressure Demand, Half Mask

Exposure Range: 5000 to unlimited mg/m^3 Self-contained Breathing Apparatus, Pressure Demand, Full Face
Cartridge Color: dust/mist filter (use P100 or consult supervisor for appropriate dust/mist filter)

Environmental

Ecotoxicity: LC_{50} Gammarus pseudolimnaeus more than 32 mg/l/96 hr at 21 °C; juvenile /static bioassay LC_{50} Ictalurus punctatus (channel catfish) more than 100 mg/l/96 hr at 20 °C; wt 1.5 g /static bioassay EC_{50} Gymnodinium breve growth rate 3.1% vol/vol/96 hr /Conditions of bioassay not specified LC_{50} Oncorhynchus kisutch (coho salmon) more than 100 mg/l/96 hr at 16 °C; wt 1.5 g /static bioassay LC_{50} Daphnia magna: 1,000-5,000 ug/l/48 hr /Conditions of bioassay not specified LC_{50} Chironomus plumosus (Midge): > 18,000 ug/l/48 hr /Conditions of bioassay not specified
Environmental Fate: In water it will biodegrade (half-life 2-3 wk), adsorb to sediments and bioconcentrate in aquatic organisms. Atmospheric material will be carried long distances and be removed by rain.
Cleanup/Disposal: Guide No. 171: Do not touch or walk through spilled material. Stop leak if you can do it without risk. Prevent dust cloud. Avoid inhalation of asbestos dust. Small Dry Spills: With clean shovel place material into clean, dry container and cover loosely; move containers from spill area. Small Spills: Take up with sand or other noncombustible absorbent material and place into containers for later disposal. Large Spills: Dike far ahead of liquid spill for later disposal. Cover powder spill with plastic sheet or tarp to minimize spreading. Prevent entry into waterways, sewers, basements or confined areas.

Environmental Physical Data

Henry's Law Constant: 1 x10^{-4}
Octanol/Water Partition Coefficient: log K_{ow} = 4.89
Sorption Partition Coefficient: K_{oc} = 4 to 5
BCF: fish 2
BOD: acclimated < 1 lb/lb, 5 days

Regulations

RCRA 40CFR: Listed Hazardous Waste No. U028 Toxic Waste
CERCLA: 40CFR 302.4: Listed per RCRA Section 3001 per CWA Section 307(a) RQ: 100 lb (45.35 kg)
SARA 40CFR 372.65: Listed
SARA EHS 40CFR 355: Not listed
TSCA: Listed

Analytical Methods

Soil: CLP MC_SVOA, OHC; EPA 16, 1625; SW846 3640A, 8060, 8061, 8061A, 8250A, 8270B, 8270C, 8410
Water / Groundwater: EPA 1625, 609, 625, 625-S, 6; APHA 6410-B; USGS O3118
Drinking Water: EPA 506, 525.1, 525.2
Food: FDA 212.1, 232.1
Indoor / Expired Air: NIOSH 5020
Plasma: EPA 001, 29; FDA 211.1, 231.1, 252

DIE8200 CAS #: 7699-31-2

1,2-DIETHYLHYDRAZINE

RTECS: MV2295000
Molecular Formula: $C_4H_{12}N_2$
Structured MF: $C_2H_5NHNHC_2H_5 \cdot 2HCl$
Formula Weight: 88.15

H—Cl
H
N
N
H
H—Cl

Chemical Structure

Synonyms: 1,2-DIAETHYLHYDRAZIN; N,N'-DIETHYLHYDRAZINE; SYM-DIETHYLHYDRAZINE; 1,2-DIETHYLHYDRAZINE DIHYDROCHLORIDE; HYDRAZINE,1,2-DIETHYL-; HYDRAZINE,1,2-DIETHYL-,DIHYDROCHLORIDE; HYDRAZOETHANE; HYDROAZOETHANE; SDEH; SYMMETRICAL-DIETHYLHYDRAZINE
Description: colorless liquid
Use: in chemical lab for synth of symmetrical di-mannich-bases; research chemical

Physical Properties

Boiling Point: 85 °C (185 °F) to 86 °C (187 °F)
Freezing Point: Decomposes at 169 °C (336.2 °F)
Specific Gravity: 0.797 at 20 °C/4 °C
Saturated Vapor Density: Estimated 1.470522686 kg/m^3
Vapor Pressure: Estimated 84 mm Hg at 25 °C
Water Solubility: >= 100 mg/mL at 18 C
Other Solubilities: 95% Ethanol: <1 mg/ml at 18 °C; Acetone: <1 mg/ml at 18 °C; DMSO: <1 mg/ml at 18 °C.
Refraction Index: 1.4214 at 20 °C
Flash Point: Not available; probably combustible

RTECS Toxicity Data

Tumorigenic: Rat Route: Intravenous; Dose: 50 mg/kg (15D preg); Toxic Effects: Tumorigenic - Equivocal tumorigenic agent by RTECS criteria; Tumorigenic effects - Transplacental tumorigenesis; Brain and coverings - Tumors.

Hazard Overviews

Health: Severely irritating to eyes/skin/respiratory tract. Toxic. Acute Effects: harmful if swallowed, inhaled, or absorbed through skin; symptoms of exposure may include burning sensation; coughing; wheezing; laryngitis; shortness of breath; headache; nausea; vomiting. Chronic Effects: Carcinogen.
Fire: Will burn. Hazards: emits toxic fumes. Extinguishing agents: water spray; carbon dioxide, dry chemical powder or appropriate foam. Precautions: combustible liquid.
Reactivity: Incompatible with: strong oxidizing agents, strong acids, strong bases. Hazardous decomposition products: toxic

fumes of: carbon monoxide, carbon dioxide, nitrogen oxides, hydrogen chloride gas.

Carcinogenicity: IARC - Not listed; NIOSH - Not listed; NTP - Not listed; ACGIH - Not listed; OSHA - Not listed; EPA - Not listed; MAK - Not listed

Primary Target Organs:

Eyes

Skin

Respiratory System

Environmental

Environmental Fate: If released to soil or water, under aerobic conditions and in the presence of small amounts of heavy metal ions, it is expected to undergo rapid oxidation to azoethane. It is assumed that this reaction is easily reversible and does not, in itself, result in destruction, but serves as a mechanism for removal via degradation and transport of azoethane. It is not expected to bioaccumulate significantly in aquatic organisms. If released to air, this compound is expected to exist almost entirely in the vapor phase. Vapor may be subject to photolysis or reaction with photochemically generated hydroxyl radicals (half life 3 hours).

Cleanup/Disposal: Guide No. 171: Do not touch or walk through spilled material. Stop leak if you can do it without risk. Prevent dust cloud. Avoid inhalation of asbestos dust. Small Dry Spills: With clean shovel place material into clean, dry container and cover loosely; move containers from spill area. Small Spills: Take up with sand or other noncombustible absorbent material and place into containers for later disposal. Large Spills: Dike far ahead of liquid spill for later disposal. Cover powder spill with plastic sheet or tarp to minimize spreading. Prevent entry into waterways, sewers, basements or confined areas.

Environmental Physical Data

Henry's Law Constant: estimated at 1.85 $x10^{-5}$
Octanol/Water Partition Coefficient: log K_{ow} = -0.31
Sorption Partition Coefficient: K_{oc} = sediments 7.71
BCF: estimated at 0.3

Regulations

RCRA 40CFR: Not listed
CERCLA: 40CFR 302.4: Not listed
SARA 40CFR 372.65: Not listed
SARA EHS 40CFR 355: Not listed
TSCA: Not listed

DIE8300 CAS #: 1615-80-1

N,N'-DIETHYLHYDRAZINE

RTECS: MV2275000
EINECS Number: 216-567-4
Molecular Formula: $C_4H_{12}N_2$
Formula Weight: 88.18

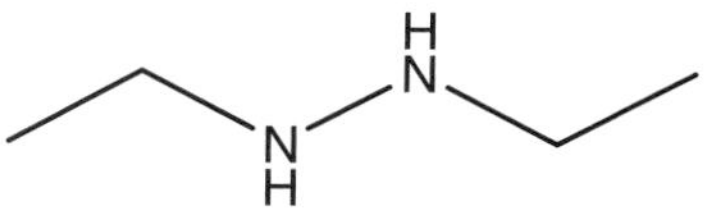

Chemical Structure

Synonyms: 1,2-DIAETHYLHYDRAZIN; 1,2-DIETHYLHYDRAZINE; N-N'-DIETHYLHYDRAZINE; SYM-DIETHYLHYDRAZINE; HYDRAZOETHANE; HYDROAZOETHANE; SDEH

Physical Properties

Boiling Point: 85 °C (185 °F)
Specific Gravity: 0.79700
Water Solubility: Low permeability
Other Solubilities: Benzene: Very Soluble; Ether: Very Soluble; Ethanol: Very Soluble
Refraction Index: 1.4204

RTECS Toxicity Data

Reproductive/Teratogenic: Rat Route: Intravenous; Dose: 10 mg/kg; Duration: female 15D of pregnancy; Specific Developmental Abnormalities - Eye, ear. Rat Route: Intravenous; Dose: 500 mg/kg; Duration: female 15D of pregnancy; Effects on Embryo or Fetus - Fetal death.

Mutagenic: Bacteria - S Typhimurium Mutations in Microorganisms; Dose: 83200 nmol/plate (+/-S9).

Tumorigenic: Rat Route: Subcutaneous; Dose: 700 mg/kg/28W-I; Toxic Effects: Tumorigenic - Equivocal tumorigenic agent by RTECS criteria; Brain and coverings - Tumors; Blood - Leukemia. Rat Route: Intravenous; Dose: 50 mg/kg (15D preg); Toxic Effects: Tumorigenic - Carcinogenic by RTECS criteria; Tumorigenic effects - Transplacental tumorigenesis; Brain and coverings - Tumors. Rat Route: Intravenous; Dose: 150 mg/kg; Toxic Effects: Tumorigenic - Carcinogenic by RTECS criteria; Endocrine - Thyroid tumors; Blood - Leukemia. Rat Route: Intravenous; Dose: 50 mg/kg (15D preg); Toxic Effects: Tumorigenic - Equivocal tumorigenic agent by RTECS criteria; Tumorigenic effects - Transplacental tumorigenesis; Brain and coverings - Tumors. Rat Route: Intravenous; Dose: 1850 mg/kg/37W-I; Toxic Effects: Tumorigenic - Equivocal tumorigenic agent by RTECS criteria; Brain and coverings - Tumors.

Hazard Overviews

Carcinogenicity: IARC - Group 2B, Possibly carcinogenic to humans; NIOSH - Not listed; NTP - Not listed; ACGIH - Not listed; OSHA - Not listed; EPA - Not listed; MAK - Not listed

Environmental

Regulations

RCRA 40CFR: Listed Hazardous Waste No. U086 Toxic Waste
CERCLA: 40CFR 302.4: Listed per RCRA Section 3001 RQ: 10 lb (4.535 kg)
SARA 40CFR 372.65: Not listed

SARA EHS 40CFR 355: Not listed
TSCA: Not listed

DIE8400 CAS #: 3710-84-7

DIETHYLHYDROXYLAMINE

RTECS: NC3500000
EINECS Number: 223-055-4
Molecular Formula: $C_4H_{11}NO$
Formula Weight: 89.16

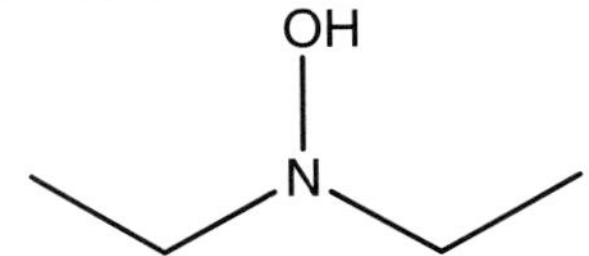

Chemical Structure

Synonyms: N,N-DIETHYLHYDROXYAMINE; N,N-DIETHYLHYDROXYLAMINE; N-HYDROXYDIETHYLAMINE

Physical Properties

Boiling Point: 125 °C (257 °F)
Freezing Point: -6 °C (21.2 °F)
Specific Gravity: 0.86
Vapor Density: 3.1 Air=1
Water Solubility: Soluble
Refraction Index: 1.4195
Flash Point: Combustible

RTECS Toxicity Data

Acute Oral: Rat LD_{Lo} Dose: 1600 mg/kg. Mouse LD_{80} Dose: 2150 mg/kg.
Acute Dermal: Rabbit LD_{Lo} Route: Skin; Dose: 2 gm/kg. Rat LD_{Lo} Route: Skin; Dose: 100 mg/kg.
Reproductive/Teratogenic: Rat Route: Parenteral; Dose: 180 mg/kg; Duration: male 1D prior to mating; Effects on Fertility - Pre-implantation mortality. Rat Route: Parenteral; Dose: 18 mg/kg; Duration: male 1D prior to mating; Effects on Fertility - Post-implantation mortality.
Mutagenic: Human Unscheduled DNA Synthesis; Cell Type: leukocyte; Dose: 4000 ppm. Rat Dominant Lethal Test; Route: Parenteral; Dose: 180 mg/kg.

Hazard Overviews

Health: Irritating to eyes/skin/respiratory tract. Harmful. Other Acute Effects: harmful if absorbed through skin; may be harmful if swallowed; may be harmful if inhaled; prolonged or repeated exposure may cause allergic reactions in certain sensitive individuals. Chronic Effects: laboratory experiments have shown mutagenic effects.
Fire: Combustible. Hazards: emits toxic fumes; vapor may travel considerable distance to source of ignition and flash back. Extinguishing agents: water spray; carbon dioxide, dry chemical powder or appropriate foam. Precautions: combustible liquid.
Reactivity: Stable. Hazardous polymerization will not occur. Incompatible with: strong oxidizing agents, strong acids, sensitive to moisture. Hazardous decomposition products: toxic fumes of: carbon monoxide, carbon dioxide, nitrogen oxides.
Carcinogenicity: IARC - Not listed; NIOSH - Not listed; NTP - Not listed; ACGIH - Not listed; OSHA - Not listed; EPA - Not listed; MAK - Not listed
Primary Target Organs:

Eyes

Skin

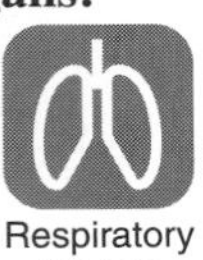
Respiratory System

Environmental

Regulations

RCRA 40CFR: Not listed
CERCLA: 40CFR 302.4: Not listed
SARA 40CFR 372.65: Not listed
SARA EHS 40CFR 355: Not listed
TSCA: Listed

DIE8500 CAS #: 53404-37-8

2,4-DIETHYL-4-METHYLPENTYL ESTER

RTECS: AG8537000
Molecular Formula: $C_{16}H_{22}Cl_2O_3$
Formula Weight: 333.28

Hazard Overviews

Carcinogenicity: IARC - Not listed; NIOSH - Not listed; NTP - Not listed; ACGIH - Not listed; OSHA - Not listed; EPA - Not listed; MAK - Not listed

Environmental

Regulations

RCRA 40CFR: Not listed
CERCLA: 40CFR 302.4: Not listed
SARA 40CFR 372.65: Listed
SARA EHS 40CFR 355: Not listed
TSCA: Not listed

DIE8600 CAS #: 311-45-5

DIETHYL-P-NITROPHENYL PHOSPHATE

RTECS: TC2275000
EINECS Number: 206-221-0
Molecular Formula: $C_{10}H_{14}NO_6P$
Structured MF: $O_2NC_6H_4OP(O)(OC_2H_5)_2$

Formula Weight: 275.22

Chemical Structure

Synonyms: CHINORTA; CHINORTO; O,O'-DIAETHYL-P-NITROPHENYLPHOSPHAT; O,O'DIAETHYL-P-NITROPHENYLPHOSPHAT; DIAETHYL-P-NITROPHENYLPHOSPHORSAEUREESTER; DIETHYL 4-NITROPHENYL PHOSPHATE; DIETHYL P-NITROPHENYL PHOSPHATE; O,O-DIETHYL O-P-NITROPHENYL PHOSPHATE; DIETHYL PARAOXON; O,O-DIETHYL PHOSPHORIC ACID O-P-NITROPHENYL ESTER; DIETHYL-P-NITROFENYL ESTER KYSELINY FOSFORECNE; O,O-DIETYL-O-P-NITROFENYLFOSFAT; E 600; E 600 (PESTICIDE); ENT 16,087; ESTER 25; ETHYL P-NITROPHENYL ETHYLPHOSPHATE; ETHYL PARAOXON; ETICOL; FOSFACOL; FOSFAKOL; HC 2072; MINTACO; MINTACOL; MINTISAL; MIOTISAL; MIOTISAL A; P-NITROPHENYL DIETHYL PHOSPHATE; P-NITROPHENYL DIETHYLPHOSPHATE; P-NITROPHENYLDIETHYL PHOSPHATE; PARA-OXON; OXYPARATHION; PARAOXON; PARAOXONE; PAROXAN; PESTOX 101; PHENOL,P-NITRO-,ESTER WITH DIETHYL PHOSPHATE; PHOSPHACHOLE; PHOSPHACOL; PHOSPHAKOL; PHOSPHONOTHIOIC ACID,DIETHYLPARANITROPHENYL ESTER; PHOSPHORIC ACID DIETHYL 4-NITROPHENYL ESTER; PHOSPHORIC ACID,DIETHYL 4-NITROPHENYL ESTER; PHOSPHORIC ACID,DIETHYL P-NITROPHENYL ESTER; SOLUGLACIT; SOLUGLAUCIT; TS 219

Description: reddish-yellow oily liquid; slight odor

Use: cholinergic, miotic, insecticide

Physical Properties

Boiling Point: 169 °C (336 °F) at 1.0 mm Hg

Freezing Point: Decomposes

Specific Gravity: 1.2683 at 20 °C/4 °C

Water Solubility: 2400 ug/ml in Water at 25 °C

Other Solubilities: Ether: Freely Soluble; Organic solvents: Freely Soluble.

Refraction Index: 1.50959 at 20 °C/D

Flash Point: Not available; probably combustible

RTECS Toxicity Data

Acute Oral: Man TD_{Lo} Dose: 14 mg/kg; Toxic Effects: Sense organs and special senses - Miosis (pupilliary dilation); Behavioral - Convulsions or effect on seizure threshold; Behavioral - Coma. Rat LD_{50} Dose: 1800 ug/kg.

Acute Dermal: Rabbit LD_{50} Route: Skin; Dose: 5 mg/kg; Toxic Effects: Automatic Nervous System - Other (direct) parasympathomimetic. Rabbit LD_{50} Route: Subcutaneous Dose: 230 ug/kg.

Chronic (Multiple Dose) Dermal: Rat Route: Subcutaneous; Dose: 1300 ug/kg/13D-I; Toxic Effects: Kidney, Ureter, and Bladder - Urine volume increased. Rat Route: Subcutaneous; Dose: 6 mg/kg/48D-I; Toxic Effects: Brain and coverings - Other degenerative changes; Biochemical - True cholinesterase; DEATH.

Mutagenic: Hamster Sister Chromatid Exchenge; Cell Type: ovary; Dose: 100 umol/L. Yeast - S Pombe Mutations in Microorganisms; Dose: 5 mmol/L (-S9).

Hazard Overviews

Poison

Health: Poison. Other Acute Effects: may be fatal if inhaled, swallowed, or absorbed through skin; highly toxic cholinesterase inhibitor; may cause nervous system disturbances. Chronic Effects: laboratory experiments have shown mutagenic effects; possible risk of irreversible effects; target organs: nerves, blood.

Fire: Will burn. Hazards: emits toxic fumes. Extinguishing agents: carbon dioxide, dry chemical powder or appropriate foam. Precautions: combustible liquid.

Reactivity: Incompatible with: strong bases, strong oxidizing agents. Hazardous decomposition products: toxic fumes of: carbon monoxide, carbon dioxide, nitrogen oxides, phosphorous oxides.

Carcinogenicity: IARC - Not listed; NIOSH - Not listed; NTP - Not listed; ACGIH - Not listed; OSHA - Not listed; EPA - Class D, Not classifiable as to human carcinogenicity; MAK - Not listed

Primary Target Organs:

Nervous System

Blood

Environmental

Cleanup/Disposal: Guide No. 152: Do not touch damaged containers or spilled material unless wearing appropriate protective clothing. Stop leak if you can do it without risk. Prevent entry into waterways, sewers, basements or confined areas. Cover with plastic sheet to prevent spreading. Absorb or cover with dry earth, sand or other non-combustible material and transfer to containers. Do not get water inside containers.

Environmental Physical Data

Octanol/Water Partition Coefficient: log K_{ow} = 1.59

Regulations

RCRA 40CFR: Listed Hazardous Waste No. P041 Toxic Waste

CERCLA: 40CFR 302.4: Listed per RCRA Section 3001 RQ: 100 lb (45.35 kg)

SARA 40CFR 372.65: Not listed

SARA EHS 40CFR 355: Not listed

TSCA: Not listed

Analytical Methods

Water / Groundwater: EPA 022
Food: FDA 212.1, 232.1, 232.3, 232.4, 242.1; AOAC 974.22
Plasma: EPA 001; FDA 211.1, 231.1, 252

DIE8700 CAS #: 2481-94-9

DIETHYL-4-(PHENYLAZO)BENZENAMINE

RTECS: CX9871000
EINECS Number: 219-616-8
Molecular Formula: $C_{16}H_{19}N_3$
Formula Weight: 253.38

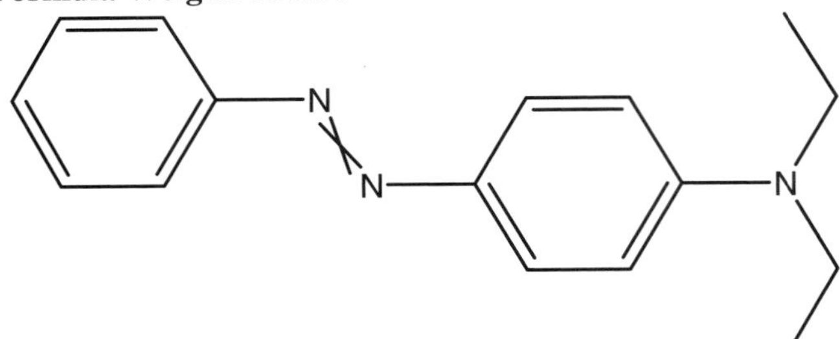

Chemical Structure

Synonyms: ANILINE,N,N-DIETHYL-P-(PHENYLAZO)-; C-299; C.I. 11021; C.I. SOLVENT YELLOW 56; CERES YELLOW GGN; DIETHYL YELLOW; 4-(DIETHYLAMINO)AZOBENZENE; N,N-DIETHYL-4-AMINOAZOBENZENE; P-(DIETHYLAMINO)AZOBENZENE; N,N-DIETHYL-P-(PHENYLAZO)ANILINE; N,N-DIETHYL-4-(PHENYLAZO)BENZENAMINE; FAST OIL YELLOW 64403; FAT YELLOW GGN; OIL YELLOW 2635; OIL YELLOW DE; OIL YELLOW DEA; OIL YELLOW E190; OIL YELLOW ENC; OIL YELLOW GA; OIL YELLOW NB; ORIENT OIL YELLOW GGS; SICO FAT YELLOW P; SUDAN YELLOW GGN; WAXOLINE YELLOW ED

Physical Properties

Water Solubility: Insoluble
pH: Approximately 6.7
Evaporation Rate: < 1 Butyl Acetate=1

RTECS Toxicity Data

Mutagenic: Mouse Mutations in Mammalian Somatic Cells; Cell Type: lymphocyte; Dose: 200 mg/L. Hamster Morphological Transformation; Cell Type: kidney; Dose: 2500 ug/L.

Hazard Overviews

Health: Irritating to eyes/skin/respiratory tract. Harmful. Other Acute Effects: harmful if swallowed, inhaled, or absorbed through skin. Chronic Effects: Possible carcinogen.
Fire: Hazards: emits toxic fumes. Extinguishing agents: water spray; carbon dioxide, dry chemical powder or appropriate foam. Precautions: combustible liquid.
Reactivity: Incompatible with: strong oxidizing agents. Hazardous decomposition products: toxic fumes of: carbon monoxide, carbon dioxide, nitrogen oxides.
Carcinogenicity: IARC - Not listed; NIOSH - Not listed; NTP - Not listed; ACGIH - Not listed; OSHA - Not listed; EPA - Not listed; MAK - Not listed
Primary Target Organs:

Eyes

Skin

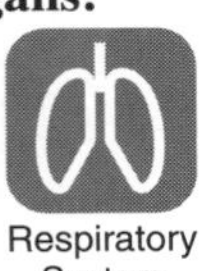
Respiratory System

Environmental

Regulations

RCRA 40CFR: Not listed
CERCLA: 40CFR 302.4: Not listed
SARA 40CFR 372.65: Not listed
SARA EHS 40CFR 355: Not listed
TSCA: Listed

DIE8800 CAS #: 93-05-0

N,N-DIETHYL-P-PHENYLENEDIAMINE

RTECS: SS9275000
EINECS Number: 202-214-1
Molecular Formula: $C_{10}H_{16}N_2$
Formula Weight: 164.3

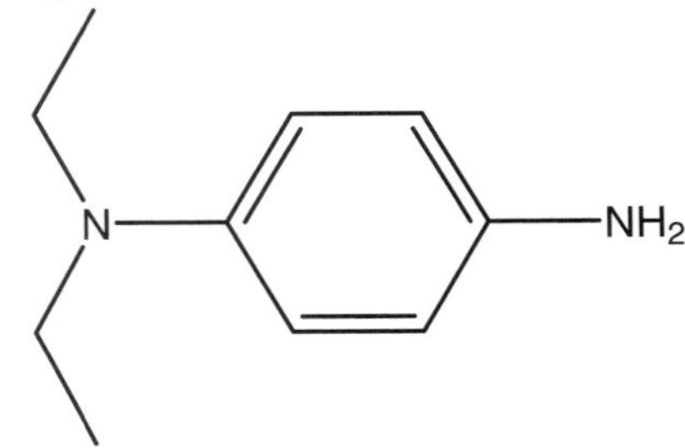

Chemical Structure

Synonyms: 4-AMINO-N,N-DIETHYLANILINE; P-AMINO-N,N-DIETHYLANILINE; P-AMINODIETHYLANILINE; PARA-AMINODIETHYLANILINE; 1,4-BENZENEDIAMINE,N,N-DIETHYL-; 4-(DIETHYLAMINO)ANILINE; DIETHYLAMINOANILINE; N,N-DIETHYL-4-AMINOANILINE; P-(DIETHYLAMINO)ANILINE; N,N'-DIETHYL-P-FENYLENDIAMIN; DIETHYL-PARA-PHENYLENEDIAMINE; N,N'-DIETHYL-P-PHENYLENEDIAMINE; N,N-DIETHYL-1,4-PHENYLENEDIAMINE; N,N-DIETHYL-PARA-PHENYLENEDIAMINE; DPD; PARAAMINO-DIETHYLANILINE; P-PHENYLENEDIAMINE,N,N-DIETHYL-
Description: liquid
Use: dye intermediate; source of diazonium compounds in diazo copying process

Physical Properties

Boiling Point: 260 °C (500 °F) to 262 °C (504 °F)
Freezing Point: 26 °C (78.8 °F) to 28 °C (82.4 °F)

Density: 0.988 g/mL
Water Solubility: Insoluble in Water
Other Solubilities: 95% Ethanol: >=100 mg/ml at 19 °C; Acetone: >=100 mg/ml at 19 °C; DMSO: >=100 mg/ml at 19 °C; Ether: Soluble.
Flash Point: > 110 °C

RTECS Toxicity Data

Acute Oral: Rabbit LD_{Lo} Dose: 450 mg/kg; Toxic Effects: Behavioral - Somnolence (general depressed activity); Behavioral - Convulsions or effect on seizure threshold; Behavioral - Coma. Cat LD_{Lo} Dose: 300 mg/kg; Toxic Effects: Behavioral - Convulsions or effect on seizure threshold; Cardiac - Pulse rate increased without fall in BP; Lungs, Thorax, or Respiration - Respiratory stimulation.
Acute Dermal: Human TD_{Lo} Route: Skin; Dose: 73 ug/kg; Toxic Effects: Blood - Hemorrhage; Skin and appendages - Primary irritation; Skin and appendages - Dermatitis, allergic. Rabbit LD_{Lo} Route: Skin; Dose: 125 mg/kg; Toxic Effects: Behavioral - General anesthetic; Behavioral - Convulsions or effect on seizure threshold; Lungs, Thorax, or Respiration - Cyanosis. Rabbit LD_{Lo} Route: Subcutaneous Dose: 250 mg/kg; Toxic Effects: Behavioral - Somnolence (general depressed activity); Behavioral - Convulsions or effect on seizure threshold; Behavioral - Coma.
Mutagenic: Hamster Cytogenetic Analysis; Cell Type: lung; Dose: 2500 ug/L. Bacteria - S Typhimurium Mutations in Microorganisms; Dose: 666 ug/plate (+S9), 16 ug/plate (-S9).

Hazard Overviews

Poison

Health: Irritating to eyes/skin/respiratory tract. Poison. Other Acute Effects: may be fatal if inhaled, swallowed, or absorbed through skin.
Fire: Will burn. Hazards: emits toxic fumes. Extinguishing agents: water spray; carbon dioxide, dry chemical powder or appropriate foam. Precautions: combustible liquid.
Carcinogenicity: IARC - Not listed; NIOSH - Not listed; NTP - Not listed; ACGIH - Not listed; OSHA - Not listed; EPA - Not listed; MAK - Not listed
Primary Target Organs:

Eyes

Skin

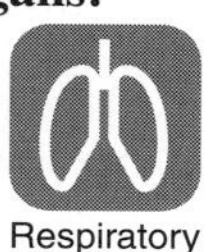
Respiratory System

Environmental

Cleanup/Disposal: Guide No. 171: Do not touch or walk through spilled material. Stop leak if you can do it without risk. Prevent dust cloud. Avoid inhalation of asbestos dust. Small Dry Spills: With clean shovel place material into clean, dry container and cover loosely; move containers from spill area. Small Spills: Take up with sand or other noncombustible absorbent material and place into containers for later disposal. Large Spills: Dike far ahead of liquid spill for later disposal. Cover powder spill with plastic sheet or tarp to minimize spreading. Prevent entry into waterways, sewers, basements or confined areas.

Regulations

RCRA 40CFR: Not listed
CERCLA: 40CFR 302.4: Not listed
SARA 40CFR 372.65: Not listed
SARA EHS 40CFR 355: Not listed
TSCA: Listed

DIE8900 CAS #: 115-76-4

2,2-DIETHYL-1,3-PROPANEDIOL

RTECS: TY5250000
EINECS Number: 204-103-3
Molecular Formula: $C_7H_{16}O_2$
Formula Weight: 132.23

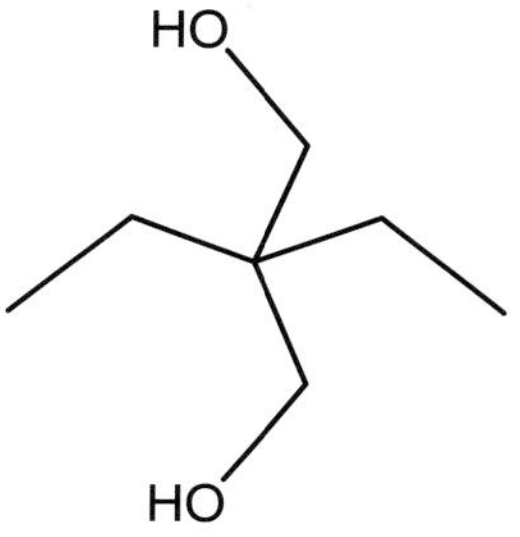

Chemical Structure

Synonyms: 3,3-BIS(HYDROXYMETHYL)PENTANE; DEP; DI-AETHYL-PROPANEDIOL; 2,2-DIETHYLPROPANE-1,3-DIOL; 2,2-DIETHYLPROPANEDIOL-1,3; DIETILPROPANDIOLO; MC 1415; PENDEROL; PRENDEROL; PRENDIOL

Physical Properties

Boiling Point: 160 °C (320 °F) at 50 mm Hg
Specific Gravity: 1.05000
Water Solubility: Soluble
Other Solubilities: Ethanol: Very Soluble; Ether: Very Soluble; Chloroform: Soluble
Refraction Index: 1.4574

RTECS Toxicity Data

Acute Oral: Rat LD_{50} Dose: 850 mg/kg. Mouse LD_{50} Dose: 1550 mg/kg; Toxic Effects: Behavioral - Altered sleep time (including change in righting reflex); Behavioral - Excitment; Nutritional and gross metabolic - Body temperature decrease.
Acute Dermal: Rabbit LD_{50} Route: Skin; Dose: 4240 uL/kg.
Irritation Eye: Rabbit Standard Draize Test Dose: 20 mg open; Reaction: severe. Rabbit Standard Draize Test Dose: 20 mg/24H; Reaction: moderate.
Irritation Skin: Rabbit Open Draize Test Dose: 500 mg open; Reaction: mild.

Hazard Overviews

Health: Irritating to eyes/skin/respiratory tract. Harmful. Other Acute Effects: harmful if swallowed; may be harmful if inhaled or absorbed through the skin.

Fire: Extinguishing agents: water spray; carbon dioxide, dry chemical powder or appropriate foam. Precautions: combustible liquid.

Reactivity: Incompatible with: strong acids, strong oxidizing agents, strong reducing agents, acid chlorides, acid anhydrides. Hazardous decomposition products: toxic fumes of: carbon monoxide, carbon dioxide.

Carcinogenicity: IARC - Not listed; NIOSH - Not listed; NTP - Not listed; ACGIH - Not listed; OSHA - Not listed; EPA - Not listed; MAK - Not listed

Primary Target Organs:

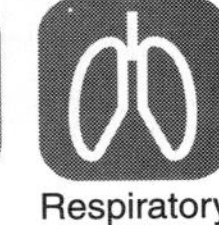

Environmental

Regulations

RCRA 40CFR: Not listed
CERCLA: 40CFR 302.4: Not listed
SARA 40CFR 372.65: Not listed
SARA EHS 40CFR 355: Not listed
TSCA: Not listed

DIE9000 CAS #: 90-84-6

DIETHYLPROPION

RTECS: UG9450000
EINECS Number: 202-019-1
Molecular Formula: $C_{13}H_{19}NO$
Formula Weight: 205.30

Synonyms: ADIPOSON; AMFEPRAMON; AMFEPRAMONE; AMPHEPRAMON; AMPHEPRAMONE; ANOREX; ALPHA-BENZOYLTRIETHYLAMINE; CEGRAMINE; DANYLEN; DERFON; 2-(DIETHYLAMINO)-1-PHENYL-1-PROPANONE; 2-(DIETHYLAMINO)PROPIOPHENONE; 2-DIETHYLAMINOPROPIOPHENONE; ALPHA-DIETHYLAMINOPROPIOPHENONE; DIETHYLPROPIONE; DOBESIN; FENYL-(1-DIETHYLAMINOETHYL)KETON; FREKENTINE; MAGRENE; NEOBES; NOPROPIOPHENONE; OBESITEX; 1-PHENYL-2-DIETHYLAMINO-1-PROPANONE; 1-PROPANONE,2-(DIETHYLAMINO)-1-PHENYL-; PROPIOPHENONE,2-(DIETHYLAMINO)-; SILUTIN; TEPANIL; TYLINAL; UR 1423

Description: white or creamy white, small crystals or crystalline powder; characteristic, mildly aromatic odor

Use: medication as anorexic

Physical Properties

Boiling Point: Decomposes at 168 °C (334 °F)
Freezing Point: Decomposes at 168 °C (334.4 °F)
Water Solubility: About 1 g/0.6 ml water
Other Solubilities: 1 g Soluble in about 0.6 ml Chloroform, 1 ml absolute Methanol, 1 ml Alcohol; Practically Insoluble in Ether

RTECS Toxicity Data

Acute Oral: Man TD_{Lo} Dose: 11 mg/kg; Toxic Effects: Automatic Nervous System - Sympathomimetic; Behavioral - Somnolence (general depressed activity); Vascular - BP elevation not characterized in autonomic section. Rat LD_{50} Dose: >400 mg/kg.

Hazard Overviews

Carcinogenicity: IARC - Not listed; NIOSH - Not listed; NTP - Not listed; ACGIH - Not listed; OSHA - Not listed; EPA - Not listed; MAK - Not listed

Environmental

Regulations

RCRA 40CFR: Not listed
CERCLA: 40CFR 302.4: Not listed
SARA 40CFR 372.65: Not listed
SARA EHS 40CFR 355: Not listed
TSCA: Not listed

DIE9100 CAS #: 56-53-1

DIETHYLSTILBESTROL

RTECS: WJ5600000
EINECS Number: 200-278-5
Molecular Formula: $C_{18}H_{20}O_2$
Structured MF: $HOC_6H_4C(C_2H_5)=C(C_2H_5)C_6H_4OH$
Formula Weight: 268.38

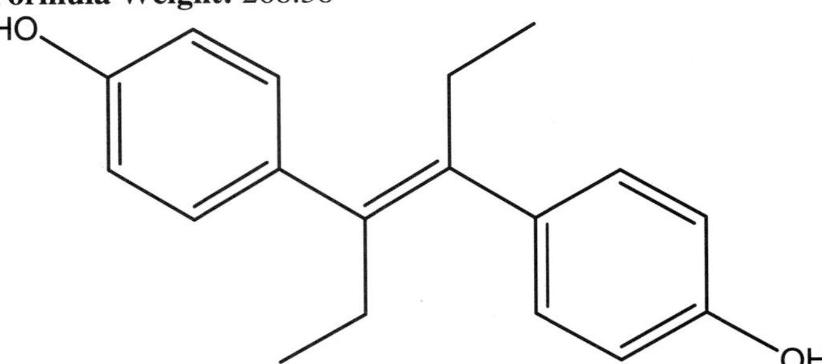

Chemical Structure

Synonyms: ACNESTROL; AGOSTILBEN; ANTIGESTIL; BIO-DES; 3,4-BIS(P-HYDROXYPHENYL)-3-HEXENE; 3,4-BIS(PARAHYDROXYPHENYL)-3-HEXENE; BUFON; CLIMATERINE; COMESTROL; COMESTROL ESTROBENE; CYREN; CYREN A; DAWE'S DESTROL; DEB; DES; DES (SYNTHETIC ESTROGEN); DESMA; DESTROL; DIASTYL; DIBESTROL; DIBESTROL '2' PREMIX; DICORVIN; DI-ESTRYL; (E)-4,4'-(1,2-DIETHYL-1,2-ETHENEDIYL)BISPHENOL; 4,4'-(1,2-DIETHYL-1,2-ETHENEDIYL)BIS-PHENOL; TRANS-4,4'-(1,2-DIETHYL-1,2-ETHENEDIYL)BISPHENOL; (E)-ALPHA,ALPHA'-DIETHYL-4,4'-STILBENEDIOL; 2,2'-DIETHYL-4,4'-STILBENEDIOL; ALPHA,ALPHA'-DIETHYL-(E)-4,4'-STILBENEDIOL; ALPHA,ALPHA'-DIETHYL-4,4'-STILBENEDIOL; ALPHA,ALPHA'-DIETHYLSTILBENEDIOL; TRANS-ALPHA,ALPHA'-

DIETHYL-4,4'-STILBENEDIOL; DIETHYLSTILBESTEROL; TRANS-DIETHYLSTILBESTEROL; TRANS-DIETHYLSTILBESTROL; DIETHYLSTILBOESTEROL; TRANS-DIETHYLSTILBOESTEROL; DIETILESTILBESTROL; 4,4'-DIHYDROXY-ALPHA,BETA-DIETHYLSTILBENE; 4,4'-DIHYDROXYDIETHYLSTILBENE; 3,4'(4,4'-DIHYDROXYPHENYL)HEX-3-ENE; DISTILBENE; DOMESTROL; DYESTROL; ESTILBEN; ESTILBIN; ESTILBIN "MCO"; ESTRIL; ESTROBENE; ESTROGEN; ESTROMENIN; ESTROSYN; FOLLIDIENE; FONATOL; GRAFESTROL; GYNOPHARM; 3-HEXENE,3,4-BIS(P-HYDROXYPHENYL)-; HI-BESTROL; HIBESTROL; IDROESTRIL; ISCOVESCO; MAKAROL; MENOSTILBEEN; MICREST; MICROEST; MILESTROL; NEO-OESTRANOL 1; NEO-OESTRANOL I; NSC-3070; OEKOLP; OESTROGENINE; OESTROL VETAG; OESTROLMENSIL; OESTROMENIN; OESTROMENSIL; OESTROMENSYL; OESTROMIENIN; OESTROMON; PABESTROL; PALESTROL; PERCUTATRINE OESTROGENIQUE ISCOVESCO; PHENOL 4,4'-(1,2-DIETHYL-1,2-ETHENEDIYL)BIS-,(E)-; PHENOL,4,4'-(1,2-DIETHYL-1,2-ETHENEDIYL)BIS-,(E)-; PROTECTONA; RUMESTROL 1; RUMESTROL 2; SEDESTRAN; SERRAL; SEXOCRETIN; SIBOL; SINTESTROL; STIBILIUM; STIL; 4,4'-STILBENEDIOL,2,2'-DIETHYL-; 4,4'-STILBENEDIOL,ALPHA,ALPHA'-DIETHYL-; 4,4'-STILBENEDIOL,ALPHA,ALPHA'-DIETHYL-,(E)-; STILBESTROL; STILBESTROL,DIETHYL-; STILBESTRONE; STILBETIN; STILBOEFRAL; STILBOESTROFORM; STILBOESTROL; STILBOFOLIN; STILBOFOLLIN; STILBOL; STILKAP; STIL-ROL; SYNESTRIN; SYNTHOESTRIN; SYNTHOFOLIN; SYNTOFOLIN; TAMPOVAGAN STILBOESTROL; TYLOSTERONE; VAGESTROL

Description: white crystalline powder or small plates; odorless

Use: a non-steroid synthetic estrogen used in biochemical research, in medicine (prevents spontaneous abortion) and veterinary medicine It is currently approved by the USFDA for the following uses in humans: (1) replacement therapy of estrogen deficiencies associated with menopause an other hormone-related conditions; (2) control of functional menstrual disorders; (3) relief or prevention of breast engorgement after child delivery; and (4) palliative therapy for cancer of the prostate and for cancer of the breast in postmenopausal women

Physical Properties

Freezing Point: 169 °C (336.2 °F)

Water Solubility: < 1 mg/mL at 20 C

Other Solubilities: DMSO: >=100 mg/ml at 21 °C; Methanol: Soluble; Acetone: >=100 mg/ml at 21 °C; Dioxane: Soluble; Chloroform: Soluble; Dilute alkali hydroxides: Soluble; Vegetable oils: Soluble; Ethyl Acetate: Soluble; Ether: Soluble.

Refraction Index: Alpha 1.594

Flash Point: Not available; probably combustible

RTECS Toxicity Data

Acute Oral: Woman TD_{Lo} Dose: 375 ug/kg/30D-I; Toxic Effects: Behavioral - Toxic psychosis; Behavioral - Excitement. Rat LD_{50} Dose: >3 gm/kg.

Acute Dermal: Human TD_{Lo} Route: Skin; Dose: 60 ug/kg/14D; Toxic Effects: Skin and appendages - Breast. Mouse LD_{Lo} Route: Subcutaneous Dose: 500 mg/kg.

Chronic (Multiple Dose) Oral: Rat Dose: 300 mg/kg/30D-I; Toxic Effects: Blood - Changes in serum composition; Biochemical - Transaminases; Biochemical - Other proteins. Rat Dose: 910 mg/kg/13W-I; Toxic Effects: Blood - Changes in serum composition; Biochemical - Transaminases; Biochemical - Other proteins. Rat Dose: 252 mg/kg/42D-I; Toxic Effects: Liver - Changes in liver weight; Endocrine - Changes in pituitary weight; DEATH - Changes in uterine weight.

Chronic (Multiple Dose) Dermal: Mouse Route: Subcutaneous; Dose: 2800 ug/kg/2W-I; Toxic Effects: Liver - Changes in Liver weight; Endocrine - Changes in spleen weight; Biochemical - Hepatic microsomal mixed oxidase(dealkylation, hyroxylation,etc).

Reproductive/Teratogenic: Man Route: Oral; Dose: 1065 mg/kg; Duration: male 30W prior to mating; Paternal Effects - Spermatogenesis; Testes, epididymis, sperm duct. Man Route: Oral; Dose: 521 mg/kg; Duration: male 2Y prior to mating; Paternal Effects - Testes, epididymis, sperm duct. Woman Route: Oral; Dose: 2500 ug/kg; Duration: female 5D prior to mating; Maternal Effects - Other effects on females. Woman Route: Oral; Dose: 5 mg/kg; Duration: female 5D prior to mating; Maternal Effects - Menstrual cycle changes or disorders; Effects on Fertility - Female fertility index. Woman Route: Oral; Dose: 1730 mg/kg; Duration: female 1-39W of pregnancy; Specific Developmental Abnormalities - Urogenital system; Effects on Newborn - Delayed effects. Woman Route: Oral; Dose: 147 mg/kg; Duration: female 7-34W of pregnancy; Effects on Newborn - Delayed effects. Woman Route: Oral; Dose: 21 mg/kg; Duration: female 13-15W of pregnancy Effects on Newborn - Delayed effects. Woman Route: Oral; Dose: 5 mg/kg; Duration: female 1-5D of pregnancy; Effects on Fertility - Female fertility index. Woman Route: Oral; Dose: 24 mg/kg; Duration: female 34W prior to mating Effects on Fertility - Other measures of fertility. Rat Route: Oral; Dose: 770 ug/kg; Duration: female 7D prior to mating; Effects on Fertility - Other measures of fertility. Rat Route: Oral; Dose: 473 ug/kg; Duration: female 1-22D of pregnancy; Maternal Effects - Postpartum; Effects on Newborn - Growth statistics. Rat Route: Oral; Dose: 100 ug/kg; Duration: male 10D prior to mating; Paternal Effects - Testes, epididymis, sperm duct; Prostate, seminal vessicle, Cowper's gland, accessory glands. Rat Route: Oral; Dose: 100 ug/kg; Duration: female 15D of pregnancy; Effects on Newborn - Biochemical and metabolic. Rat Route: Oral; Dose: 6300 ng/kg; Duration: female 7D prior to mating; Maternal Effects - Uterus, cervix, vagina. Rat Route: Oral; Dose: 100 ug/kg; Duration: female 5D of pregnancy; Effects on Fertility - Post-implantation mortality. Rat Route: Oral; Dose: 585 ug/kg; Duration: female 6-18D of pregnancy; Effects on Newborn - Viability index; Growth statistics. Rat Route: Oral; Dose: 405 mg/kg; Duration: female 10-18D of pregnancy Maternal Effects - Parturition; Effects on Newborn - Stillbirth. Monkey Route: Oral; Dose: 7200 ug/kg; Duration: female 18-23W of pregnancy Effects on Newborn - Delayed effects. Monkey Route: Oral; Dose: 1200 ug/kg; Duration: female 1-6D of pregnancy; Effects on Fertility - Female fertility index. Monkey Route: Oral; Dose: 1 mg/kg; Duration: female 1-6D of pregnancy; Effects on Fertility - Female fertility index. Monkey Route: Oral; Dose: 7 mg/kg; Duration: female 14-24W of pregnancy Effects on Fertility - Female fertility index.

Mutagenic: Human DNA Damage; Cell Type: fibroblast; Dose: 300 umol/L. Human DNA Damage; Cell Type: leukocyte; Dose: 20 umol/L. Human DNA Damage; Cell Type: liver; Dose: 18 mg/L. Human Unscheduled DNA Synthesis; Cell Type: liver; Dose: 5600 ug/L. Human Unscheduled DNA Synthesis; Cell Type: HeLa cell; Dose: 1 umol/L. Human DNA Inhibition; Cell Type: fibroblast; Dose: 100 umol/L. Human DNA Inhibition; Cell Type: lymphocyte; Dose: 10 umol/L. Human DNA Inhibition; Cell Type: HeLa cell; Dose: 100 umol/L. Human Mutations in Mammalian Somatic Cells; Cell Type: lymphocyte; Dose: 20 mg/L. Human Cytogenetic Analysis; Cell Type: lymphocyte; Dose: 100 ug/L. Human Sister Chromatid Exchenge; Cell Type: liver; Dose: 10 umol/L. Human Sister Chromatid Exchenge; Cell Type: fibroblast; Dose: 133 nmol/L. Human Sister Chromatid Exchenge; Cell Type: lymphocyte; Dose: 10 umol/L. Human Micronucleus Test; Cell Type: lymphocyte; Dose: 30 umol/L. Human Sex Chromosome Loss; Cell Type: lymphocyte; Dose: 6 mmol/L. Human Sex Chromosome Loss; Cell Type: fibroblast; Dose: 6 mg/L. Human Other Mutation Test Systems; Cell Type: other cell types; Dose: 20 umol/L. Human Other Mutation Test Systems; Cell Type: fibroblast; Dose: 6 mg/kg.

Tumorigenic: Man Route: Multiple routes; Dose: 25 mg/kg/2Y-C; Toxic Effects: Tumorigenic - Carcinogenic by RTECS criteria; Skin and appendages - Tumors. Man Route: Oral; Dose: 69 mg/kg/5Y-C; Toxic Effects: Tumorigenic - Neoplastic by RTECS criteria; Liver - Tumors. Man Route: Oral; Dose: 45990 ug/kg/3Y-C; Toxic Effects: Tumorigenic - Carcinogenic by RTECS criteria; Kidney, Ureter, and Bladder - Kidney tumors. Woman Route: Oral; Dose: 7655 ug/kg/4Y-C; Toxic Effects: Tumorigenic - Carcinogenic by RTECS criteria; Tumorigenic effects - Uterine tumors. Woman Route: Oral; Dose: 21 mg/kg (13-15W preg); Toxic Effects: Tumorigenic - Carcinogenic by RTECS criteria; Tumorigenic effects - Transplacental tumorigenesis; Tumorigenic effects - Other reproductive system tumors. Rat Route: Oral; Dose: 103 gm/kg/2Y-C; Toxic Effects: Tumorigenic - Carcinogenic by RTECS criteria; Liver - Tumors. Monkey Route: Implant; Dose: 48 mg/kg; Toxic Effects: Tumorigenic - Carcinogenic by RTECS criteria; Endocrine - Thyroid tumors.

Hazard Overviews

Health: Irritating to eyes/skin/respiratory tract. Toxic. Other Acute Effects: harmful if swallowed, inhaled, or absorbed through skin. Chronic Effects: may cause congenital malformation in the fetus; may alter genetic material; target organ: reproductive system. Carcinogen.

Fire: Will burn. Extinguishing agents: water spray; carbon dioxide, dry chemical powder or appropriate foam. Precautions: combustible liquid.

Reactivity: Incompatible with: strong oxidizing agents, strong bases, acid chlorides, acid anhydrides. Hazardous decomposition products: toxic fumes of: carbon monoxide, carbon dioxide.

Carcinogenicity: IARC - Group 1, Carcinogenic to humans; NIOSH - Not listed; NTP - Listed; ACGIH - Not listed; OSHA - Not listed; EPA - Not listed; MAK - Not listed

Primary Target Organs:

Eyes

Skin

Respiratory System

Repro-ductive

Environmental

Environmental Fate: If released to water, it may bioconcentrate in aquatic organisms and strongly adsorb to suspended solids. If released to the atmosphere, the vapor should rapidly oxidize primarily by reaction with ozone. The atmospheric half-life in the vapor phase at 298 °K has been estimated to be 11.40 minutes.

Environmental Physical Data

Octanol/Water Partition Coefficient: log K_{ow} = 5.07
Sorption Partition Coefficient: K_{oc} = 5.5×10^4
BCF: estimated at 4200

Regulations

RCRA 40CFR: Listed Hazardous Waste No. U089 Toxic Waste
CERCLA: 40CFR 302.4: Listed per RCRA Section 3001 RQ: 1 lb (0.454 kg)
SARA 40CFR 372.65: Not listed
SARA EHS 40CFR 355: Not listed
TSCA: Listed

Analytical Methods

Soil: SW846 8270C
Plasma: EPA 29

DIE9200 CAS #: 105-55-5

N,N'-DIETHYLTHIOUREA

RTECS: YS9800000
EINECS Number: 203-308-5
Molecular Formula: $C_5H_{12}N_2S$
Formula Weight: 132.22

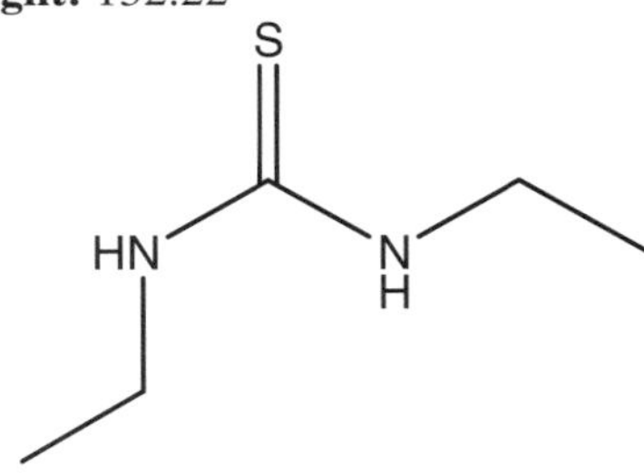

Chemical Structure

Synonyms: N,N'-DIETHYLTHIOCARBAMIDE; 1,3-DIETHYL-2-THIOUREA; 1,3-DIETHYLTHIOUREA; PENNZONE E; THIATE H; THIOUREA,N,N'-DIETHYL-; U 15030; UREA,1,3-DIETHYL-2-THIO-
Description: buff solid
Use: inhibitor of corrosion in metal pickling solutions; accelerator activator in elastomers

Physical Properties

Boiling Point: Decomposes
Freezing Point: 68 °C (154.4 °F) to 71 °C (159.8 °F)
Water Solubility: Slightly Soluble in Water
Other Solubilities: 95% Ethanol: >=100 mg/ml at 20 °C; Acetone: >=100 mg/ml at 20 °C; Benzene: Soluble; DMSO: >=100 mg/ml at 20 °C; Ether: Very Soluble; Ethyl Acetate: Soluble; Gasoline: Insoluble; Methanol: Soluble.
Flash Point: Not available; probably combustible

RTECS Toxicity Data

Acute Oral: Rat LD_{50} Dose: 316 mg/kg. Mouse LD_{Lo} Dose: 62 mg/kg.
Mutagenic: Mouse Mutations in Mammalian Somatic Cells; Cell Type: lymphocyte; Dose: 1500 mg/L.
Tumorigenic: Rat Route: Oral; Dose: 11 gm/kg/2Y-C; Toxic Effects: Tumorigenic - Carcinogenic by RTECS criteria; Endocrine - Thyroid tumors.

Hazard Overviews

Fire: Will burn.
Carcinogenicity: IARC - Not listed; NIOSH - Not listed; NTP - Not listed; ACGIH - Not listed; OSHA - Not listed; EPA - Not listed; MAK - Not listed

Environmental

Ecotoxicity: Fishes: creek chub 24h critical range 100-300 mg/l
Environmental Fate: If released to soil, biodegradation may be an important removal process. Because of the estimated low K_{oc} value in soil, it may rapidly leach in soil. The volatilization from moist soil is not expected to be an important removal process. Neither photolysis nor hydrolysis are expected to be important in water. Based on its low K_{oc} value, the compound will not remain adsorbed to particulate matter and sediment in water, but instead will be present in the aquatic phase. Based on its estimated low values for Henry's Law constant and bioconcentration factor, neither volatilization from water nor bioconcentration in aquatic organisms would be important. In the atmosphere, vapor phase would be removed from the atmosphere with an estimated half-life of 4 hours due to reaction with hydroxyl radicals. The primary removal processes for particulate from the atmosphere may be wet and dry deposition.

Environmental Physical Data

Henry's Law Constant: estimated at 6.85 $x10^{-8}$
Octanol/Water Partition Coefficient: log K_{OW} = 0.57
Sorption Partition Coefficient: K_{OC} = 49
BCF: estimated at 2

Regulations

RCRA 40CFR: Not listed
CERCLA: 40CFR 302.4: Not listed
SARA 40CFR 372.65: Not listed
SARA EHS 40CFR 355: Not listed
TSCA: Listed

DIF1000 CAS #: 35367-38-5

DIFLUBENZURON

RTECS: YS6200000
EINECS Number: 252-529-3
Molecular Formula: $C_{14}H_9ClF_2N_2O_2$
Formula Weight: 310.68
Synonyms: ASTONEX; BENZAMIDE,N-(((4-CHLOROPHENYL)AMINO)CARBONYL)-2,6-DIFLUORO-; (N-((4-CHLOROPHENYL)AMINO)CARBONYL)-2,6-DIFLUOROBENZAMIDE; N-(((4-CHLOROPHENYL)AMINO)CARBONYL)-2,6-DIFLUOROBENZAMIDE; N-((4-CHLOROPHENYL)AMINO)-2,6-DIFLUOROBENZAMIDE; N-(4-CHLOROPHENYLCARBAMOYL)-2,6-DIFLUOROBENZAMIDE; 1-(4-CHLOROPHENYL)-3-(2,6-DIFLUOROBENZOYL) UREA; 1-(4-CHLOROPHENYL)-3-(2,6-DIFLUOROBENZOYL)UREA; 1-(P-CHLOROPHENYL)-3-(2,6-DIFLUOROBENZOYL)UREA; DIFLURON; DIMILIN; DIMILIN G1; DIMILIN G4; DIMILIN ODC-45; DIMILIN WP-25; DU 112307; DUPHACID; ENT 29054; LARVAKIL; MICROMITE; OMS 1804; PDD 6040I; PH 60-40; PH-6040; PHILIPS-DUPHAR PH 60-40; TH 60-40; TH 6040; THOMPSON HAYWARD 6040; THOMPSON-HAYWARD TH6040
Description: colorless to white crystals or crystalline solid
Use: insect growth regulator; insecticide (larvicide); inhibits molting of larvae of mosquitoes, houseflies, stable flies, black flies, leaf miners, gypsy moths, codling moths, psyllids, pine processionary moths, boll weevils, velvet bean catepillars, cabbage-white caterpillars; to protect tree fruit (including citrus), horticultural crops, cotton, pome fruit, soya beans, brassicas, other field crops, mushroom cultivation & open areas; on; sciarid flies & phorid flies; incorporated in feed of cattle for control fly larvae in manure.

Physical Properties

Boiling Point: Decomposes
Freezing Point: 239 °C (462.2 °F)
Vapor Pressure: < 0.033 mPa at 50 °C
Water Solubility: 0.0000002%
Other Solubilities: in Acetone 6.5 g/l at 20 °C. In dimethylformamide 104, dioxane 20 (both in g/l at 25 °C); Moderately Soluble in polar organic solvents; very slightly Soluble in non-polar organic solvents.

RTECS Toxicity Data

Acute Oral: Rat LD_{50} Dose: 4640 mg/kg. Mouse LD_{50} Dose: 4640 mg/kg.
Acute Inhalation: Rat LC; Dose: >35 gm/m^3/6hr.
Acute Dermal: Rabbit LD_{50} Route: Skin; Dose: 2 gm/kg. Rat LD_{50} Route: Skin; Dose: >10 gm/kg. Rat LD_{50} Route: Subcutaneous Dose: >3400 mg/kg.
Mutagenic: Rabbit Other Mutation Test Systems; Cell Type: liver; Dose: 200 ug/L. Rabbit Other Mutation Test Systems; Cell Type: other cell types; Dose: 200 ug/L. Rabbit Other Mutation Test Systems; Cell Type: liver; Dose: 200 ug/L. Rabbit Other Mutation Test Systems; Cell Type: other cell types; Dose: 200 ug/L.

Hazard Overviews

Carcinogenicity: IARC - Not listed; NIOSH - Not listed; NTP - Not listed; ACGIH - Not listed; OSHA - Not listed; EPA - Not listed; MAK - Not listed

Environmental

Ecotoxicity: LC_{50} Rainbow trout 250 mg/l/96 hr (static bioassay) LC_{50} Mysidopsis bahia 2.1 ug/l/96 hr. /Conditions of bioassay not specified LC_{50} Coho salmon > 150 mg/l/96 hr. /Conditions of bioassay not specified LC_{50} Juvenile Rainbow trout > 150 mg/l/96 hr. /Conditions of bioassay not specified

Environmental Fate: If released to the atmosphere, it is expected to exist primarily in the particulate-phase and it will be susceptible to dry deposition. If released to water, biodegradation may be the primary fate process; half-lives averaged 2.4 days in freshwater sediment, 14 days in marine-sediment and 32 days in marine-water. Hydrolysis to p-chlorophenylurea may contribute to biodegradation at warm temperatures in alkaline aquatic systems. Adsorption to sediment (measured K_{oc} of 6,790) will be important. Volatilization and photolysis do not appear to be important. Bioconcentration in aquatic organisms may be important; however, one review suggests that it is not. In soil, adsorption will be important and leaching will not occur; biodegradation will be the primary fate process in soil with measured half-lives of 19-27 days. Furthermore, a complete degradation half-life of 10 days was reported in soil.

Environmental Physical Data

Henry's Law Constant: estimated at 1.23 $x10^{-9}$
Octanol/Water Partition Coefficient: log K_{ow} = 3.10
Sorption Partition Coefficient: K_{OC} = 1 $x10^{4}$
BCF: estimated at 1218

Regulations

RCRA 40CFR: Not listed
CERCLA: 40CFR 302.4: Not listed
SARA 40CFR 372.65: Listed
SARA EHS 40CFR 355: Not listed
TSCA: Listed

Analytical Methods

Food: AOAC 983.07

DIF2460 CAS #: 372-18-9

1,3-DIFLUOROBENZENE

RTECS: CZ5652000
EINECS Number: 206-746-5
Molecular Formula: $C_6H_4F_2$
Formula Weight: 114.10

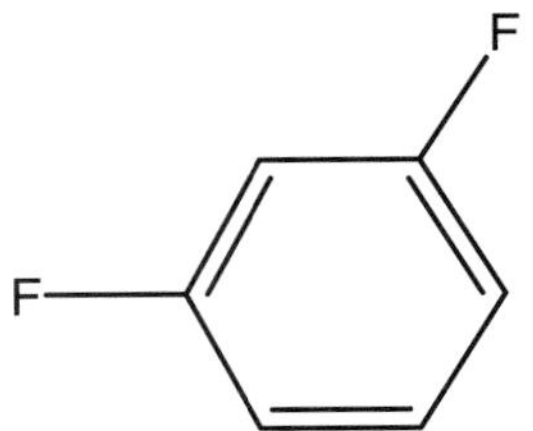

Chemical Structure

Synonyms: M-DIFLUOROBENZENE

Physical Properties

Boiling Point: 83 °C (181.4 °F)
Freezing Point: -59 °C (-74.2 °F)
Specific Gravity: 1.15720
Water Solubility: Insoluble
Other Solubilities: Acetone: Soluble; Benzene: Soluble
Refraction Index: 1.4374
Ionization Potential (eV): 9.33 +/-0.02

RTECS Toxicity Data

Acute Inhalation: Mouse LC_{50} Dose: 55 gm/m^3/2hr; Toxic Effects: Brain and coverings - Recordings from specific areas of CNS.

Hazard Overviews

Health: May cause irritation. Other Acute Effects: may be harmful by inhalation, ingestion, or skin absorption.

Fire: Hazards: vapor may travel considerable distance to source of ignition and flash back; container explosion may occur; emits toxic fumes. Extinguishing agents: carbon dioxide, dry chemical powder or appropriate foam. Precautions: combustible liquid.

Reactivity: Incompatible with: strong oxidizing agents. Hazardous decomposition products: toxic fumes of: carbon monoxide, carbon dioxide, hydrogen fluoride.

Carcinogenicity: IARC - Not listed; NIOSH - Not listed; NTP - Not listed; ACGIH - Not listed; OSHA - Not listed; EPA - Not listed; MAK - Not listed

Environmental

Regulations

RCRA 40CFR: Not listed
CERCLA: 40CFR 302.4: Not listed
SARA 40CFR 372.65: Not listed
SARA EHS 40CFR 355: Not listed
TSCA: Not listed

DIF5380 CAS #: 75-37-6

1,1-DIFLUOROETHANE

RTECS: KI1410000
DOT: UN1030; IMO2.0
EINECS Number: 200-866-1

Molecular Formula: $C_2H_4F_2$
Structured MF: CH_3CHF_2
Formula Weight: 66.05

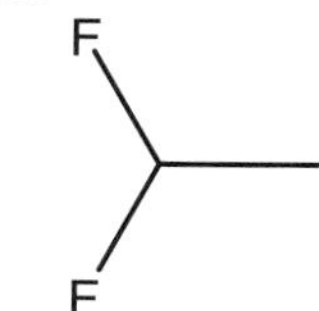

Chemical Structure

Synonyms: ALGOFRENE TYPE 67; DIFLUOROETHANE; DYMEL 152A; DYMEL 152; ETHANE,1,1-DIFLUORO-; ETHYLENE FLUORIDE; ETHYLIDENE DIFLUORIDE; ETHYLIDENE FLUORIDE; FC 152A; FREON 152; FREON 152A; GENETRON 152A; GENETRON 100; H-FC 152A; HALOCARBON 152A; R 152A; REFRIGERANT 152A
Description: colorless gas; odorless
Use: intermediate; refrigerant; aerosol propellant

Physical Properties

Boiling Point: -24.7 °C (-12 °F)
Freezing Point: -117 °C (-178.6 °F)
Specific Gravity: 0.95 at 20 °C
Vapor Density: 2.3 Air=1
Vapor Pressure: 4437.1 mm Hg at 25 °C
Water Solubility: Insoluble in Water
Other Solubilities: Very Soluble in Alcohol, Ether, Acetone, Benzene. Insoluble in Glycerol and glycol
Surface Tension: 11.25 dynes/cm at 20 °C
Refraction Index: 1.3011 at -72 °C
Critical Temperature: 114 °C
Critical Pressure: 652 psia
Ionization Potential (eV): 11.865 +/-0.030
LEL: 3.7% v/v
UEL: 18% v/v

RTECS Toxicity Data

Acute Inhalation: Mouse LC_{50} Dose: 977 gm/m^3/2hr. Rat LC_{Lo} Dose: 64000 ppm/4hr.
Mutagenic: Insects - D Melanogaster Sex Chromosome Loss; Route: Inhalation; Dose: 98 pph/10M.

Hazard Overviews

Flammable

Compressed Gas

4 1 0 —
Fire Diamond

Health: Irritating to respiratory tract. Simple asphyxiant which can displace available oxygen needed for breathing. Also Causes: narcosis, laryngeal spasm, edema, ventricular fibrillation, frostbite.
Fire: Flammable. Can form explosive mixtures in the air. Let fire burn if leak can't be stopped. For small fires use dry chemical or carbon dioxide. For large fires use water spray or fog. Containers may explode in heat of fire. Vapors may travel to an ignition source and flash back.
Reactivity: Stable. Hazardous polymerization cannot occur. Avoid: exposure to heat and ignition sources. Incompatible with: oxidizers; powdered or freshly heated aluminum and other light, divalent metals. Hazardous decomposition products: hydrogen fluoride; carbonyl fluoride.
Carcinogenicity: IARC - Not listed; NIOSH - Not listed; NTP - Not listed; ACGIH - Not listed; OSHA - Not listed; EPA - Not listed; MAK - Not listed
Primary Target Organs:

Eyes

Skin

Respiratory System

Cardio-vascular

Exposure Limits
AIHA WEEL: TWA: 1000 ppm.

Environmental

Environmental Fate: If released to the soil, it may rapidly volatilize from soil surfaces or leach through soil possibly into groundwater. If released to water, volatilization would be the dominant fate process based on a half-life of 2.4 hours from a model river. If released to the atmosphere, all is expected to exist in the vapor phase. In the troposphere, it reacts slowly with photochemically generated hydroxyl radicals (half-life of 472 days). This relatively slow half-life in the lower atmosphere suggests that some may gradually diffuse into the stratosphere (half-life of 20 years). In the stratosphere, it is expected to slowly photolyze and contribute to the catalytic removal of stratospheric ozone. However, no data are available it contributes to stratospheric ozone depletion. From its source of emissions, global atmospheric transport is expected to take place due to the stability of the chemical.
Cleanup/Disposal: Guide No. 115: Eliminate all ignition sources (no smoking, flares, sparks or flames in immediate area). All equipment used when handling the product must be grounded. Do not touch or walk through spilled material. Stop leak if you can do it without risk. If possible, turn leaking containers so that gas escapes rather than liquid. Use water spray to reduce vapors or divert vapor cloud drift. Do not direct water at spill or source of leak. Prevent spreading of vapors through sewers, ventilation systems and confined areas. Isolate area until gas has dispersed.

Environmental Physical Data

Henry's Law Constant: estimated at 2.042 x10^{-2}
Octanol/Water Partition Coefficient: log K_{ow} = 0.75
Sorption Partition Coefficient: K_{oc} = estimated at 51
BCF: estimated at 6.5
BOD: none

Regulations

RCRA 40CFR: Not listed
CERCLA: 40CFR 302.4: Not listed
SARA 40CFR 372.65: Not listed
SARA EHS 40CFR 355: Not listed
TSCA: Listed

DIF6110 CAS #: 624-72-6

1,2-DIFLUOROETHANE

RTECS: KI1410500
DOT: UN1030; IMO2.1
Molecular Formula: $C_2H_4F_2$
Formula Weight: 66.05
Synonyms: ETHANE,1,2-DIFLUORO-; ETHYLENE DIFLUORIDE; FC143; FLUOROCARBON FC143; FREON 152
Use: refrigeration & air conditioning

Physical Properties

Boiling Point: 30.7 °C (87 °F) at 760 mm Hg
Saturated Vapor Density: 2.443225862 kg/m^3
Vapor Pressure: 616.3 mm Hg at 25 °C
Other Solubilities: Soluble in Ether, Benzene, Chloroform

RTECS Toxicity Data

Mutagenic: Bacteria - S Typhimurium Mutations in Microorganisms; Dose: 50 pph/72H-C (-S9).

Hazard Overviews

Carcinogenicity: IARC - Not listed; NIOSH - Not listed; NTP - Not listed; ACGIH - Not listed; OSHA - Not listed; EPA - Not listed; MAK - Not listed

Environmental

Environmental Fate: The half-life for the reaction with photochemically produced hydroxyl radicals is estimated to be 0.3 years, indicating that it may be an important fate process in the atmosphere. If released to soil, it is expected to be very highly mobile. Rapid volatilization should occur from terrestrial surfaces. If released to water, volatilization is expected to occur rapidly and be the dominant fate process.
Cleanup/Disposal: Guide No. 115: Eliminate all ignition sources (no smoking, flares, sparks or flames in immediate area). All equipment used when handling the product must be grounded. Do not touch or walk through spilled material. Stop leak if you can do it without risk. If possible, turn leaking containers so that gas escapes rather than liquid. Use water spray to reduce vapors or divert vapor cloud drift. Do not direct water at spill or source of leak. Prevent spreading of vapors through sewers, ventilation systems and confined areas. Isolate area until gas has dispersed.

Environmental Physical Data

Henry's Law Constant: estimated at 0.387
Octanol/Water Partition Coefficient: log K_{ow} = 0.578
Sorption Partition Coefficient: K_{oc} = estimated at 49
BCF: estimated at 1.69

Regulations

RCRA 40CFR: Not listed
CERCLA: 40CFR 302.4: Not listed
SARA 40CFR 372.65: Not listed
SARA EHS 40CFR 355: Not listed
TSCA: Not listed

DIF6840 CAS #: 75-38-7

1,1-DIFLUOROETHENE

RTECS: KW0560000
DOT: UN1959; IMO2.1
EINECS Number: 200-867-7
Molecular Formula: $C_2H_2F_2$
Structured MF: $CH_2{=}CF_2$
Formula Weight: 64.038

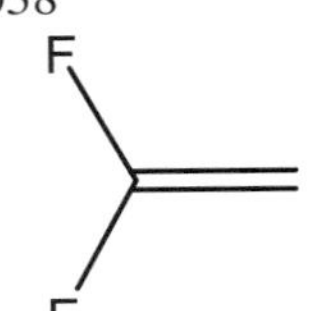

Chemical Structure

Synonyms: 1,1-DIFLUOROETHYLENE; DIFLUORO-1,1-ETHYLENE; ETHENE,1,1-DIFLUORO-; ETHYLENE,1,1-DIFLUORO-; GENETRON 1132A; HALOCARBON 1132A; R1132A; VDF; VINYLIDENE DIFLUORIDE; VINYLIDENE FLUORIDE
Description: colorless gas; faint ethereal odor
Use: in the manufacture of many polymers and copolymers; manufacture of poly vinylidene fluoride, and as a chemical intermediate in organic synthesis

Physical Properties

Boiling Point: -83 °C (-117 °F)
Freezing Point: -144 °C (-227.2 °F) at 1 atm
Vapor Density: 2.2 Air=1
Density: 0.617 g/cc at 24 °C (liquid)
Vapor Pressure: 3683 kPa at 21 °C
Water Solubility: Insoluble in Water
Other Solubilities: 95% Ethanol: Soluble; Ether: Soluble.
Surface Tension: 0.039942 N/m at melting point
Critical Temperature: 30.1 °C
Critical Pressure: 4434 kPa
Ionization Potential (eV): 10.29
Flash Point: Flammable
Autoignition Temperature: 640 °C
LEL: 5.5% v/v
UEL: 21.3% v/v

RTECS Toxicity Data

Acute Inhalation: Rat LC_{Lo} Dose: 128000 ppm/4hr; Toxic Effects: Lungs, Thorax, or Respiration - Other changes.
Mutagenic: Bacteria - S Typhimurium Mutations in Microorganisms; Dose: 50 pph/24H (-S9).
Tumorigenic: Rat Route: Oral; Dose: 1930 mg/kg/52W-I; Toxic Effects: Tumorigenic - Neoplastic by RTECS criteria; Sense organs and special senses - Tumors; Skin and appendages - Tumors.

Hazard Overviews

Flammable

Compressed Gas

Fire Diamond

Health: Simple asphyxiant which displaces oxygen needed for breathing. Stored as a compressed gas which can cause frostbite. Also Causes: slight intoxication, dizziness, headache, fatigue, difficulty breathing.

Fire: Flammable. Stop flow of gas. Use dry chemical or carbon dioxide to extinguish flame in order to shut off supply or repair leak. Use water spray to cool surroundings and prevent ignition of other materials.

Reactivity: Stable. Hazardous polymerization cannot occur. Avoid: exposure to heat and ignition sources. Incompatible with: hydrogen chloride when heated under pressure; aqueous potassium iodide. Hazardous decomposition products: fluoride; hydrogen fluoride.

Carcinogenicity: IARC - Group 3, Not classifiable as to carcinogenicity to humans; NIOSH - Not listed; NTP - Not listed; ACGIH - Not listed; OSHA - Not listed; EPA - Not listed; MAK - Class B, Justifiably suspected of having carcinogenic potential

Primary Target Organs:

Nervous System

Exposure Limits

NIOSH REL: TWA: 1 ppm. STEL: 5 ppm; use 1910.1017.

Environmental

Environmental Fate: When released to the atmosphere, it will degrade by reaction with photochemically produced hydroxyl radicals (estimated half-life of about 8 days). Reaction with atmospheric ozone will make a small contribution to its atmospheric degradation (estimated half-life of about 60 days). It exists as a gas under normal ambient conditions. If released to soil or water, most can be expected to volatilize to the atmosphere. Insufficient data are available to predict the relative importance or rate of biodegradation in soil or water conditions that preclude evaporation.

Cleanup/Disposal: Guide No. 116: Eliminate all ignition sources (no smoking, flares, sparks or flames in immediate area). All equipment used when handling the product must be grounded. Stop leak if you can do it without risk. Do not touch or walk through spilled material. Do not direct water at spill or source of leak. Use water spray to reduce vapors or divert vapor cloud drift. If possible, turn leaking containers so that gas escapes rather than liquid. Prevent entry into waterways, sewers, basements or confined areas. Isolate area until gas has dispersed.

Environmental Physical Data

Henry's Law Constant: estimated at 0.356

Octanol/Water Partition Coefficient: log K_{ow} = 1.24

Sorption Partition Coefficient: K_{oc} = estimated at 250 to 112

BCF: estimated at 33

Regulations

RCRA 40CFR: Not listed

CERCLA: 40CFR 302.4: Not listed

SARA 40CFR 372.65: Not listed

SARA EHS 40CFR 355: Not listed

TSCA: Listed

DIG1000 CAS #: 71-63-6

DIGITOXIN

RTECS: IH2275000

EINECS Number: 200-760-5

Molecular Formula: $C_{41}H_{64}O_{13}$

Formula Weight: 764.92

Chemical Structure

Synonyms: ACEDOXIN; ASTHENTHILO; CARD-20(22)-ENOLIDE,3-((O-2,6-DIDEOXY-BETA-D-RIBO-HEXOPYRANOSYL-(1-4)-O-2,6-DIDEOXY-BETA-D-RIBO-HEXOPYRANOSYL-(1-4)-2,6-DIDEOXY-BETA-D-RIBO-HEXOPYRANOSYL)OXY)-14-HYDROXY-,(3BETA,5BETA)-; CARDIDIGIN; CARDIGIN; CARDITALIN; CARDITOXIN; CORAMEDAN; CRISTAPURAT; CRYSTALLINE DIGITALIN; CRYSTODIGIN; (3BETA,5BETA)-3-((O-2,6-DIDEOXY-BETA-D-RIBOHEXOPYRANOSYL-(1-4)-O-2,6-DIDEOXY-BETA-D-RIBO-HEXOPYRANOSYL-(1-4)-2,6-DIDEOXY-BETA-D-RIBO-HEXOPYRANOSYL)OXY)-14-HYDROXYCARD-20(22)- ENOLIDE; DIGICOR; DIGILONG; DIGIMED; DIGIMERCK; DIGIPURAL; DIGISIDIN; DIGITALIN; DIGITALIN,CRYSTALLINE; DIGITALINE CRISTALLISEE; DIGITALINE NATIVELLE; DIGITALINUM VERUM; DIGITOKSIM; DIGITOPHYLLIN; DIGITOXIGENIN TRIDIGITOXOSIDE; DIGITOXIGENIN-TRIDIGITOXOSID; DIGITOXINUM; DIGITOXOSIDE; DIGITRIN; DITAVEN; GLUCODIGIN; LANATOXIN; MONO-DIGITOXID; MONODIGITOXOSIDE; MONO-GLYCOCARD; MYODIGIN; PURODIGIN; PURPURID; TARDIGAL; TRI-DIGITOXOSIDE; UNIDIGIN

Description: white or pale buff microcrystalline powder, leaflets, or very small elongated, rectangular plates; odorless

Use: medication: cardiotonic; medication (vet): cardiac tonic; cardiovascular drug (a cardiotonic glycoside); medication: treatment of low output congestive heart failure, management of atrial flutter, atrial fibrillation and paroxysmal atrial tachycardia

Physical Properties

Freezing Point: Anhydrous 256 °C (492.8 °F) to 257 °C (494.6 °F)
Water Solubility: 1 g/ 100L Water 20 °C
Other Solubilities: 1 g/40 ml Chloroform, 60 ml Alcohol, 400 ml Ethyl Acetate, at 20 °C. Soluble in Acetone, Amyl Alcohol, Pyridine. Sparingly Soluble in Ether, Petroleum Ether

RTECS Toxicity Data

Acute Oral: Man TD_{Lo} Dose: 71 ug/kg; Toxic Effects: Cardiac - Cardiomyopathy including infarction; Cardiac - Arrythmias (including changes in conduction). Woman TD_{Lo} Dose: 400 ug/kg; Toxic Effects: Cardiac - EKG changes not diagnostic of above; Gastrointestinal - Nausea or vomiting. Woman TD_{Lo} Dose: 300 ug/kg; Toxic Effects: Cardiac - Arrythmias (including changes in conduction). Infant TD_{Lo} Dose: 150 ug/kg; Toxic Effects: Cardiac - Arrythmias (including changes in conduction); Gastrointestinal - Nausea or vomiting. Man LD_{Lo} Dose: 286 ug/kg; Toxic Effects: Peripheral nerve and sensation - Pareshtesia; Cardiac - Pulse rate increased without fall in BP; Blood - Thrombocytopenia. Rat LD_{50} Dose: 23750 ug/kg.
Acute Dermal: Rat LD_{50} Route: Subcutaneous Dose: 16430 ug/kg. Mouse LD_{50} Route: Subcutaneous Dose: 22180 ug/kg; Toxic Effects: Brain and coverings - Recordings from specific areas of CNS; Behavioral - Convulsions or effect on seizure threshold; Cardiac - Other changes.
Chronic (Multiple Dose) Dermal: Rat Route: Subcutaneous; Dose: 175 mg/kg/35D-I; Toxic Effects: Endocrine - Changes in adrenal weight.
Reproductive/Teratogenic: Woman Route: Oral; Dose: 200 ug/kg; Duration: female 27-30W of pregnancy Effects on Newborn - Viability index; Other neonatal measures or effects.

Hazard Overviews

Poison

Health: May cause irritation. Poison. Other Acute Effects: may be fatal if inhaled, swallowed, or absorbed through skin; exposure can cause stomach pains; vomiting; diarrhea; target organs: heart;g.i. system; nerves.
Fire: Extinguishing agents: water spray; carbon dioxide, dry chemical powder or appropriate foam. Precautions: combustible liquid.
Reactivity: Incompatible with: strong oxidizing agents, strong acids. Hazardous decomposition products: toxic fumes of: carbon monoxide, carbon dioxide.
Carcinogenicity: IARC - Not listed; NIOSH - Not listed; NTP - Not listed; ACGIH - Not listed; OSHA - Not listed; EPA - Not listed; MAK - Not listed

Primary Target Organs:

Gastro-intestinal

Nervous System

Cardio-vascular

Environmental

Cleanup/Disposal: Guide No. 152: Do not touch damaged containers or spilled material unless wearing appropriate protective clothing. Stop leak if you can do it without risk. Prevent entry into waterways, sewers, basements or confined areas. Cover with plastic sheet to prevent spreading. Absorb or cover with dry earth, sand or other non-combustible material and transfer to containers. Do not get water inside containers.

Regulations

RCRA 40CFR: Not listed
CERCLA: 40CFR 302.4: Not listed
SARA 40CFR 372.65: Not listed TPQ: 100/10000 lb
SARA EHS 40CFR 355: Listed TPQ: 100 lb
TSCA: Listed

DIG3000 CAS #: 2238-07-5

DIGLYCIDYL ETHER

RTECS: KN2350000
EINECS Number: 218-802-6
Molecular Formula: $C_6H_{10}O_3$
Structured MF: $C_6H_{10}O_3$
Formula Weight: 130.16
Synonyms: BIS(2,3-EPOXYPROPYL) ETHER; DGE; DIALLYL ETHER DIOXIDE; DI(2,3-EPOXY)PROPYL ETHER; DI(2,3-EPOXYPROPYL) ETHER; DI(EPOXYPROPYL) ETHER; 2-EPOXYPROPYL ETHER; ETHER,BIS(2,3-EPOXYPROPYL); ETHER,DIGLYCIDYL; GLYCIDYL ETHER; NSC 54739; NSV 54739; OXIRANE,2,2'-(OXYBIS(METHYLENE))BIS-
Description: colorless liquid; strong, irritating odor
Use: reactive diluent for epoxy resins; chem intermediate; stabilizer of chlorinated org cmpd; textile-treating agent

Physical Properties

Boiling Point: 260 °C (500 °F) at 760 mm Hg
Specific Gravity: 1.1195 at 20 °C/4 °C
Vapor Density: 3.78 Air=1 at 25 °C
Saturated Vapor Density: 1.200495702 kg/m^3
Vapor Pressure: 0.09 mm Hg at 25 °C
Odor Threshold: 25 mg/m^3
Flash Point: 64 °C

RTECS Toxicity Data

Acute Oral: Rat LD_{50} Dose: 450 mg/kg; Toxic Effects: Brain and coverings - Recordings from specific areas of CNS; Behavioral - Somnolence (general depressed activity); Behavioral - Ataxia. Mouse LD_{50} Dose: 170 mg/kg; Toxic Effects: Brain and coverings - Recordings from specific areas of CNS; Behavioral - Somnolence (general depressed activity); Behavioral - Ataxia.

Acute Inhalation: Rat LC_{50} Dose: 200 ppm/4hr. Mouse LC_{50} Dose: 30 ppm/4hr; Toxic Effects: Lungs, Thorax, or Respiration - Other changes.

Acute Dermal: Rabbit LD_{Lo} Route: Skin; Dose: 1500 mg/kg; Toxic Effects: Sense organs and special senses - Lacrimation; Behavioral - Somnolence (general depressed activity); Skin and appendages - Dermatitis, irritative. Rat LD_{50} Route: Skin; Dose: 1 gm/kg.

Irritation Eye: Rabbit Standard Draize Test Dose: 113 mg; Reaction: severe. Rabbit Standard Draize Test Dose: 750 ug/24H; Reaction: severe.

Irritation Skin: Rabbit Standard Draize Test Dose: 20 mg/24H; Reaction: moderate. Rabbit Standard Draize Test Dose: 563 mg/3D; Reaction: severe.

Mutagenic: Bacteria - S Typhimurium Mutations in Microorganisms; Dose: 50 ug/plate (+/-S9).

Hazard Overviews

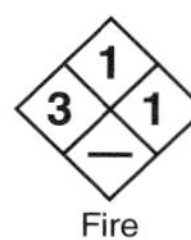

Fire Diamond

Health: Severe irritation to eyes/skin/respiratory tract. Toxic. Also Causes: skin burns; high concentrations: pulmonary edema, chemical pneumonitis; upon ingestion: GI tract irritation.

Fire: Use dry chemical, carbon dioxide, water spray, or regular foam.

Reactivity: Stable. Hazardous polymerization can occur in the presence of so-called curing agents, may polymerize at room temperature to form hard epoxy resins. Avoid: heat; ignition sources. Incompatible with: strong oxidizing sources. Hazardous decomposition products: acid smoke; fumes.

Carcinogenicity: IARC - Not listed; NIOSH - Listed as carcinogen; NTP - Not listed; ACGIH - Class A4, Not classifiable as a human carcinogen; OSHA - Not listed; EPA - Not listed; MAK - Class B, Justifiably suspected of having carcinogenic potential

Primary Target Organs:

Eyes

Skin

Respiratory System

Exposure Limits

OSHA PEL: STEL: 0.5 ppm; 2.8 mg/m^3.

OSHA PEL Vacated 1989 Limits: TWA: 0.1 ppm; 0.5 mg/m^3.

ACGIH TLV: TWA: 0.1 ppm; 0.53 mg/m^3.

NIOSH REL: TWA: 0.1 ppm; 0.5 mg/m^3.

NIOSH IDLH: 10 ppm.

DFG MAK: TWA: 0.1 ppm; 0.6 mg/m^3.

Respirator Recommendation

Exposure Range: >0.5 to <10 ppm Supplied Air, Constant Flow/Pressure Demand, Half Mask

Exposure Range: 10 to unlimited ppm Self-contained Breathing Apparatus, Pressure Demand, Full Face

Note: poor warning properties

Environmental

Regulations

RCRA 40CFR: Not listed

CERCLA: 40CFR 302.4: Not listed

SARA 40CFR 372.65: Not listed TPQ: 1000 lb

SARA EHS 40CFR 355: Listed TPQ: 1000 lb

TSCA: Listed

DIG5000 CAS #: 101-90-6

DIGLYCIDYL RESORCINOL ETHER

RTECS: VH1050000

EINECS Number: 202-987-5

Molecular Formula: $C_{12}H_{14}O_4$

Formula Weight: 222.2

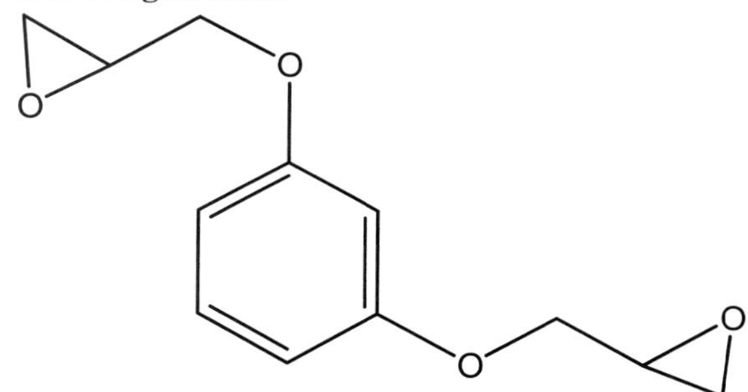

Chemical Structure

Synonyms: ARALDITE ERE 1359; BENZENE,M-BIS(2,3-EPOXYPROPOXY)-; META-BIS(2,3-EPOXYPROPOXY BENZENE); 1,3-BIS(2,3-EPOXYPROPOXY)BENZENE; M-BIS(2,3-EPOXYPROPOXY)BENZENE; M-BIS(GLYCIDYLOXY)BENZENE; META-BIS(GLYCIDYLOXY)BENZENE; DIGLYCIDYL ETHER OF RESORCINOL; 1,3-DIGLYCIDYLOXYBENZENE; DIGLYCIDYLRESORCINOL; ERE 1359; OXIRANE,2,2'-(1,3-PHENYLENEBIS(OXYMETHYLENE))BIS-; 2,2'-(1,3-PHENYLENEBIS(OXYMETHYLENE))BISOXIRANE; RDGE; RESORCINOL BIS(2,3-EPOXYPROPYL) ETHER; RESORCINOL BIS(2,3-EPOXYPROPYL)ETHER; RESORCINOL DIGLYCIDYL ETHER; RESORCINOL GLYCIDYL ETHER; RESORCINOL,DIGLYCIDYL-; RESORCINYL DIGLYCIDYL ETHER

Use: liquid epoxy resin and as a reactive diluent in the production of other epoxy resins; to cure polysulfide rubber; as a diluent to impart special properties to cured epoxy resins (e.g., as in aerospace applications)

Physical Properties

Boiling Point: 172 °C (342 °F) at 0.8 mm Hg

Freezing Point: 32 °C (89.6 °F) to 33 °C (91.4 °F)

Specific Gravity: 1.21 at 25 °C

Saturated Vapor Density: 1.62812559 kg/m^3

Density: 1.21 g/mL at 25 °C

Vapor Pressure: 40.7 mm Hg at 25 °C

Water Solubility: < 0.1 mg/mL at 18 C
Other Solubilities: 95% Ethanol: 10-50 mg/ml at 21 °C; Acetone: >=100 mg/ml at 21 °C; Benzene: miscible; Chloroform: miscible; DMSO: >=100 mg/ml at 21 °C; Methanol: miscible; Most organic resins: miscible.
Refraction Index: 1.541 at 25 °C/D
Flash Point: 176.667 °C Open Cup

RTECS Toxicity Data

Acute Oral: Rat LD_{50} Dose: 2570 mg/kg; Toxic Effects: Gastrointestinal - Ulceration or bleeding from duodenum. Mouse LD_{50} Dose: 980 mg/kg; Toxic Effects: Behavioral - Somnolence (general depressed activity).
Chronic (Multiple Dose) Oral: Rat Dose: 5320 mg/kg/14D-I; Toxic Effects: DEATH. Rat Dose: 13 gm/kg/13W-I; Toxic Effects: DEATH.
Irritation Skin: Rabbit Standard Draize Test Dose: 500 mg/24H; Reaction: moderate.
Mutagenic: Mouse Mutations in Mammalian Somatic Cells; Cell Type: lymphocyte; Dose: 100 ug/L. Hamster Cytogenetic Analysis; Cell Type: ovary; Dose: 8 mg/L.
Tumorigenic: Rat Route: Oral; Dose: 6180 mg/kg/2Y-I; Toxic Effects: Tumorigenic - Carcinogenic by RTECS criteria; Gastrointestinal - Tumors. Rat Route: Oral; Dose: 12875 mg/kg/2Y-I; Toxic Effects: Tumorigenic - Carcinogenic by RTECS criteria; Gastrointestinal - Tumors.

Hazard Overviews

Reactivity: Stable. Hazardous polymerization cannot occur. Avoid: high temperatures; ignition sources. Hazardous decomposition products: carbon oxide.
Carcinogenicity: IARC - Group 2B, Possibly carcinogenic to humans; NIOSH - Not listed; NTP - Class 2B, Reasonably anticipated to be a carcinogen, sufficient evidence of carcinogenicity from studies in experimental animals; ACGIH - Not listed; OSHA - Not listed; EPA - Not listed; MAK - Class A2, Unmistakably carcinogenic in animal experimentation only

Environmental

Cleanup/Disposal: Guide No. 171: Do not touch or walk through spilled material. Stop leak if you can do it without risk. Prevent dust cloud. Avoid inhalation of asbestos dust. Small Dry Spills: With clean shovel place material into clean, dry container and cover loosely; move containers from spill area. Small Spills: Take up with sand or other noncombustible absorbent material and place into containers for later disposal. Large Spills: Dike far ahead of liquid spill for later disposal. Cover powder spill with plastic sheet or tarp to minimize spreading. Prevent entry into waterways, sewers, basements or confined areas.

Regulations

RCRA 40CFR: Not listed
CERCLA: 40CFR 302.4: Not listed
SARA 40CFR 372.65: Listed
SARA EHS 40CFR 355: Not listed
TSCA: Listed

DIG7000 CAS #: 13561-08-5

2,6-DIGLYCIDYLPHENYL GLYCIDYL ETHER

RTECS: KN3150000
EINECS Number: 236-951-5
Molecular Formula: $C_{15}H_{18}O_4$
Formula Weight: 262.33
Synonyms: BIS(2,6-(2,3-EPOXYPROPYL))PHENYL GLYCIDYL ETHER; 2,6-DI(2,3-EPOXYPROPYL)PHENYL 2,3-EPOXYPROPYL; 2,6-DI(2,3-EPOXYPROPYL)PHENYL 2,3-EPOXYPROPYL ETHER; DIGLYCIDYLPHENYL GLYCIDYL ETHER; ETHER,2,6-BIS(2,3-EPOXYPROPYL)PHENYL 2,3-EPOXYPROPYL; OXIRANE,2,2'-(OXIRANYLMETHOXY)-1,3-PHENYLENE)BIS(METHYLENE))BIS-
Use: reactive diluent for epoxy resins

RTECS Toxicity Data

Acute Oral: Rat LD_{50} Dose: 1620 uL/kg.
Acute Dermal: Rabbit LD_{50} Route: Skin; Dose: 2520 uL/kg.

Hazard Overviews

Carcinogenicity: IARC - Not listed; NIOSH - Not listed; NTP - Not listed; ACGIH - Not listed; OSHA - Not listed; EPA - Not listed; MAK - Not listed

Environmental

Regulations

RCRA 40CFR: Not listed
CERCLA: 40CFR 302.4: Not listed
SARA 40CFR 372.65: Not listed
SARA EHS 40CFR 355: Not listed
TSCA: Listed

DIG9000 CAS #: 20830-75-5

DIGOXIN

RTECS: IH6125000
EINECS Number: 244-068-1
Molecular Formula: $C_{41}H_{64}O_{14}$
Formula Weight: 780.92

Chemical Structure

Synonyms: ACYGOXIN; CARD-20(22)-ENOLIDE,3-((O-2,6-DIDEOXY-BETA-D-RIBO-HEXOPYRANOSYL-(1-4)-O-2,6-DIDEOXY-BETA-D-RIBO-HEXOPYRANOSYL-(1-4)-2,6-DIDEOXY-BETA-D-RIBO-HEXOPYRANOSYL)OXY)-12,14-DIHYDROXY-,(3BETA,5BETA,12BETA)-; CARDOXIN; CHLOROFORMIC DIGITALIN; CORDIOXIL; DAVOXIN; (3BETA,5BETA,12BETA)-3-((O-2,6-DIDEOXY-BETA-D-RIBO-HEXAPYRANOSYL-(1-4)-2,6-DIDEOXY-BETA-D-RIBO-HEXOPYRANOSYL-(1-4)-2,6-DIDEOXY-BETA-D-RIBO-HEXOPYRANOSYL)OXY)-12 ,14-DIHYDROXYCARD-20(22)-ENOLIDE; DIGACIN; DIGITALIS GLYCOSIDE; DIGOKSYNA; DIGONIX; DIGOXIGENIN-TRIDIGITOXOSID; DIGOXINE; DILANACIN; DIXINA; DOKIM; DYNAMOS; HOMOLLE'S DIGITALIN; LANACORDIN; LANATILIN; LANICOR; LANOXICAPS; LANOXIN; LENOXICAPS; LENOXIN; LONGDIGOX; NEODIOXANIN; ROUGOXIN; SAROXIN; SK-DIGOXIN; STILLACOR; VANOXIN

Description: clear to white crystals or white crystalline powder; odorless

Use: medication: cardiotonic; medication (vet): cardiotonic; cardiovascular drug (a cardiotonic glycoside); medication: treatment of low output congestive heart failure, management of atrial flutter, atrial fibrillation and paroxysmal atrial tachycrdia

Physical Properties

Boiling Point: Decomposes at 230 °C (446 °F) to 265 °C (509 °F)

Freezing Point: Decomposes at 240 °C (464 °F)

Specific Gravity: 0.8062 at 20 °C

Saturated Vapor Density: 1.200409394 kg/m^3

Vapor Pressure: 0.01 mm Hg at 20 °C

Water Solubility: Insoluble in Water

Other Solubilities: Soluble in Pyridine or mixture of Chloroform & Alcohol; almost Insoluble in Ether, Acetone, Ethyl Acetate, Chloroform; more Soluble in hot 80% Alcohol than Gitoxin; Slightly Soluble in diluted Alcohol, and very slightly Soluble in 40% Propylene Glycol.

Refraction Index: 1.4420

Flash Point: 132 °C

RTECS Toxicity Data

Acute Oral: Man TD_{Lo} Dose: 333 ug/kg; Toxic Effects: Sense organs and special senses - Visual field changes; Cardiac - Pulse rate. Man TD_{Lo} Dose: 75 ug/kg; Toxic Effects: Cardiac - EKG changes not diagnostic of above; Cardiac - Pulse rate increased without fall in BP. Woman TD_{Lo} Dose: 500 ug/kg; Toxic Effects: Cardiac - EKG changes not diagnostic of above; Cardiac - Pulse rate increased without fall in BP; Gastrointestinal - Nausea or vomiting. Woman TD_{Lo} Dose: 165 ug/kg/33D-I; Toxic Effects: Behavioral - General anesthetic; Behavioral - Muscle contraction or spasticity. Woman TD_{Lo} Dose: 100 ug/kg; Toxic Effects: Behavioral - Anorexia (human); Cardiac - Arrythmias (including changes in conduction); Gastrointestinal - Nausea or vomiting. Child TD_{Lo} Dose: 127 ug/kg; Toxic Effects: Cardiac - Pulse rate. Child TD_{Lo} Dose: 54 ug/kg; Toxic Effects: Cardiac - Pulse rate; Nutritional and gross metabolic - Changes in potassium. Rat LD_{50} Dose: 28270 ug/kg.

Acute Dermal: Rat LD_{50} Route: Subcutaneous Dose: 8900 ng/kg. Mouse LD_{50} Route: Subcutaneous Dose: 8150 ng/kg.

Reproductive/Teratogenic: Domestic Animal Route: Intramuscular; Dose: 810 ug/kg; Duration: female 1-18W of pregnancy; Specific Developmental Abnormalities - Blood and lymphatic systems (including spleen and marrow).

Hazard Overviews

Poison

Health: Irritating. Poison. Other Acute Effects: may be fatal if inhaled, swallowed, or absorbed through skin; exposure can cause stomach pains; vomiting; diarrhea; target organs: heart, g.i. system, nerves.

Fire: Will burn. Extinguishing agents: water spray; carbon dioxide, dry chemical powder or appropriate foam. Precautions: combustible liquid.

Reactivity: Incompatible with: strong oxidizing agents, strong acids. Hazardous decomposition products: toxic fumes of: carbon monoxide, carbon dioxide.

Carcinogenicity: IARC - Not listed; NIOSH - Not listed; NTP - Not listed; ACGIH - Not listed; OSHA - Not listed; EPA - Not listed; MAK - Not listed

Primary Target Organs:

Eyes

Skin

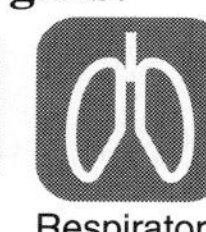

Respiratory System

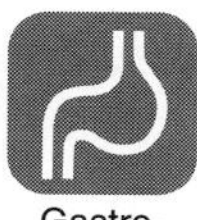

Gastro-intestinal

Nervous System

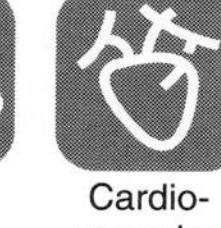

Cardio-vascular

Environmental

Regulations

RCRA 40CFR: Not listed

CERCLA: 40CFR 302.4: Not listed

SARA 40CFR 372.65: Not listed TPQ: 10/10000 lb

SARA EHS 40CFR 355: Listed TPQ: 10 lb

TSCA: Listed

DIH1000 **CAS #: 3648-21-3**

DIHEPTYL PHTHALATE

RTECS: TI1090000
EINECS Number: 222-885-4
Molecular Formula: $C_{22}H_{34}O_4$
Structured MF: $C_6H_4(COOC_7H_{15})_2$
Formula Weight: 362.45

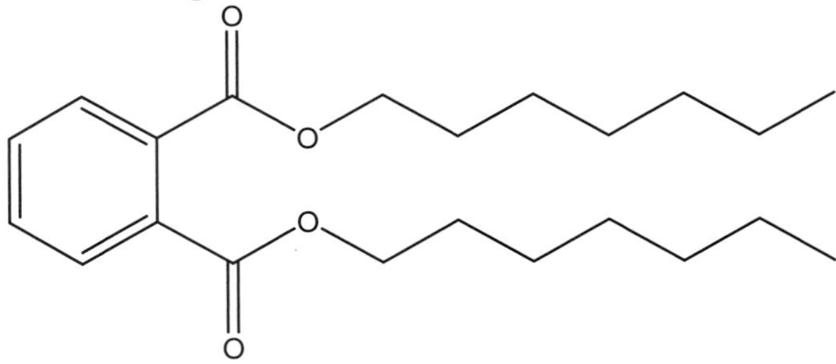

Chemical Structure

Synonyms: 1,2-BENZENEDICARBOXYLIC ACID,DIHEPTYL ESTER; 1,2-BENZENEDICARBOXYLIC ACID,DIHEPTYL ESTER (9CI); DIHEPTYL 1,2-BENZENEDICARBOXYLATE; DIHEPTYL ESTER OF PHTHALIC ACID; DI-N-HEPTYL PHTHALATE; HEPTYL PHTHALATE; PHTHALIC ACID,DIHEPTYL ESTER
Description: colorless liquid; practically odorless
Use: as a plasticizer for vinyl resins

Physical Properties

Boiling Point: 360 °C (680 °F) at 760 mm Hg
Specific Gravity: Liquid est. 1 at 20 °C
Vapor Pressure: Extremely low
Water Solubility: 0.01% in Water
Other Solubilities: Soluble in Benzene, Toluene, petrol, Kerosene, and mineral oils.
Surface Tension: 31.5 dynes/cm at 20 °C
Flash Point: 118 °C Closed Cup

RTECS Toxicity Data

Reproductive/Teratogenic: Mouse Route: Oral; Dose: 2500 mg/kg; Duration: female 9D of pregnancy; Specific Developmental Abnormalities - Skin and skin appendages.

Hazard Overviews

Health: Irritating to eyes/skin/respiratory tract. Harmful. Other Acute Effects: harmful if swallowed, inhaled, or absorbed through skin. Chronic Effects: overexposure may cause reproductive disorder(s) based on tests with laboratory animals.
Fire: Will burn. Hazards: emits toxic fumes. Extinguishing agents: water spray; carbon dioxide, dry chemical powder or appropriate foam. Precautions: combustible liquid.
Reactivity: Incompatible with: strong oxidizing agents. Hazardous decomposition products: toxic fumes of: carbon monoxide, carbon dioxide.
Carcinogenicity: IARC - Not listed; NIOSH - Not listed; NTP - Not listed; ACGIH - Not listed; OSHA - Not listed; EPA - Not listed; MAK - Not listed
Primary Target Organs:

Eyes

Skin

Respiratory System

Environmental

Environmental Fate: If released to soil, it may undergo microbial degradation, based on limited screening data. It is not expected to volatilize from the soil surface to the atmosphere, nor is it expected to leach through soil. If released to water, it may undergo microbial degradation. It is not expected to bioaccumulate in fish and aquatic organisms, nor is it expected to volatilize from water to the atmosphere. If released to the atmosphere, it is expected to be adsorbed to particulates. Destruction by the vapor phase reaction with photochemically produced hydroxyl radicals may occur.
Cleanup/Disposal: Guide No. 151: Do not touch damaged containers or spilled material unless wearing appropriate protective clothing. Stop leak if you can do it without risk. Prevent entry into waterways, sewers, basements or confined areas. Cover with plastic sheet to prevent spreading. Absorb or cover with dry earth, sand or other non-combustible material and transfer to containers. Do not get water inside containers.

Environmental Physical Data

Henry's Law Constant: 3.54×10^{-6}
Octanol/Water Partition Coefficient: log K_{ow} = > 2.12
BCF: none likely

Regulations

RCRA 40CFR: Not listed
CERCLA: 40CFR 302.4: Not listed
SARA 40CFR 372.65: Not listed
SARA EHS 40CFR 355: Not listed
TSCA: Listed

DIH1270 **CAS #: 109-31-9**

DIHEXYL AZELATE

RTECS: CM2100000
EINECS Number: 203-664-1
Molecular Formula: $C_{21}H_{40}O_4$
Formula Weight: 356.61

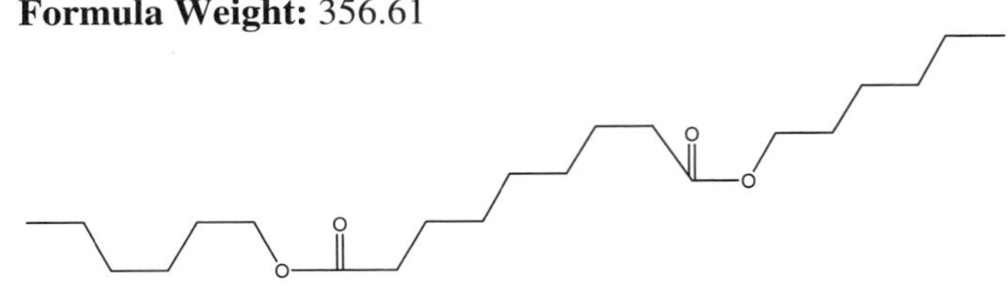

Chemical Structure

Synonyms: AZELAIC ACID,DIHEXYL ESTER; DI-N-HEXYL AZELATE; DI-N-HEXYLESTER KYSELINY AZELAOVE; DNHA; NONANEDIOIC ACID,DIHEXYL ESTER; PLASTOLEIN 9050; PLASTOLEIN 9051; PLASTOLEIN 9051 DHNZ

Description: liquid

Use: plasticizer for various resins, particularly polyvinyl chloride & vinyl copolymers; plasticizer for cellulose & vinyl resins; plasticizer for polypropylene food wrap

Physical Properties

Boiling Point: 216 °C (421 °F) at 5 mm Hg
Freezing Point: -10 °C (14 °F)
Specific Gravity: 0.927 at 20 °C
Flash Point: 204 °C
Autoignition Temperature: 238 °C

RTECS Toxicity Data

Acute Oral: Rat LD_{50} Dose: 16 mL/kg. Mouse LD_{Lo} Dose: 15 gm/kg.
Acute Dermal: Rabbit LD_{50} Route: Skin; Dose: >20 mL/kg.

Hazard Overviews

Health: Irritating to eyes/skin. Other Acute Effects: may be harmful by inhalation, ingestion, or skin absorption.
Fire: Will burn. Hazards: emits toxic fumes. Extinguishing agents: water spray; carbon dioxide, dry chemical powder or appropriate foam; do not direct a solid stream of water or foam into burning molten material; this may cause spattering and spread the fire. Precautions: combustible liquid.
Reactivity: Stable. Hazardous polymerization will not occur. Incompatible with: strong oxidizing agents, strong acids, strong bases. Hazardous decomposition products: toxic fumes of: carbon monoxide, carbon dioxide.
Carcinogenicity: IARC - Not listed; NIOSH - Not listed; NTP - Not listed; ACGIH - Not listed; OSHA - Not listed; EPA - Not listed; MAK - Not listed
Primary Target Organs:

Eyes

Skin

Environmental

Regulations

RCRA 40CFR: Not listed
CERCLA: 40CFR 302.4: Not listed
SARA 40CFR 372.65: Not listed
SARA EHS 40CFR 355: Not listed
TSCA: Listed

DIH1540 **CAS #: 84-75-3**

DIHEXYL PHTHALATE

RTECS: TI1100000
EINECS Number: 201-559-5
Molecular Formula: $C_{20}H_{30}O_4$
Structured MF: $C_6H_4(COO(CH_2)_5CH_3)_2$
Formula Weight: 334.50

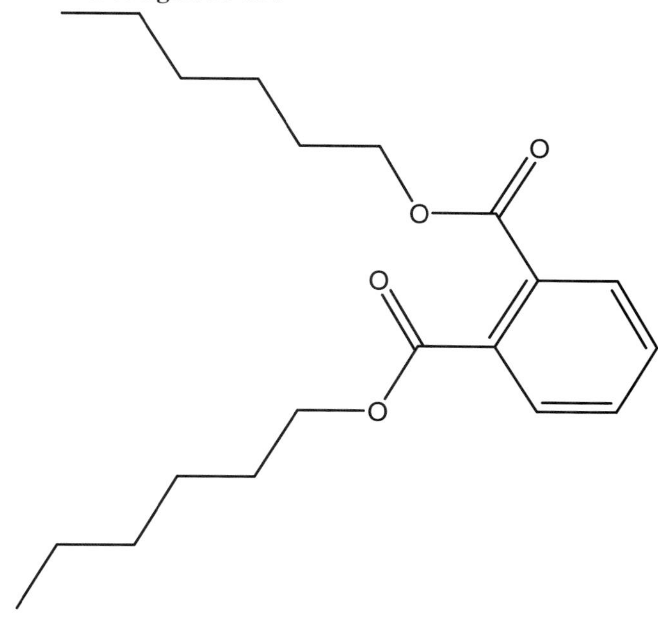

Chemical Structure

Synonyms: 1,2-BENZENEDICARBOXYLIC ACID,DIHEXYL ESTER; 1,2-BENZENEDIOIC ACID DIHEXYL ESTER; DI(2-ETHYLBUTYL) PHTHA}1-E; DIHEXYL 1,2-BENZENEDICARBOXYLATE; DIHEXYL ESTER PHTHALIC ACID; DI-N-HEXYL PHTHALATE; DIHEXYLESTER KYSELINY FTALOVE; HEXYL PHTHALATE,1,2-; PHTHALIC ACID,DIHEXYL ESTER; PHTHALIC ACID,DIHEXYL ETHER

Description: clear, oily liquid; slightly aromatic odor

Use: plasticizer for cellulose ester and vinyl plastics

Physical Properties

Boiling Point: 210 °C (410 °F) at 5 mm Hg
Freezing Point: -58 °C (-72.4 °F)
Specific Gravity: 0.995
Vapor Density: 11.5 Air=1
Vapor Pressure: 1.4 $x10^{-5}$ mm Hg
Water Solubility: Insoluble in Water
Other Solubilities: 95% Ethanol: Soluble; Acetone: Soluble; DMSO: Soluble.
Flash Point: 176.7 °C

RTECS Toxicity Data

Acute Oral: Rat LD_{50} Dose: 29600 mg/kg.
Acute Dermal: Rabbit LD_{50} Route: Skin; Dose: >20 mL/kg.
Chronic (Multiple Dose) Oral: Rat Dose: 25200 mg/kg/21D-C; Toxic Effects: Liver - Fatty liver degeneration; Liver - Changes in liver weight; Biochemical - Phosphatases. Bird Dose: 3862 mg/kg/30D-C; Toxic Effects: Behavioral - Food intake (animal); Biochemical - Lipids including transport.
Irritation Eye: Rabbit Standard Draize Test Dose: 500 mg/24H; Reaction: mild.
Irritation Skin: Rabbit Standard Draize Test Dose: 500 mg/24H; Reaction: mild.

Reproductive/Teratogenic: Rat Route: Oral; Dose: 9600 mg/kg; Duration: male 4D prior to mating; Paternal Effects - Testes, epididymis, sperm duct. Mouse Route: Oral; Dose: 13 gm/kg; Duration: male 7D prior to mating; Effects on Fertility - Mating Performance; Other measures of fertility; Effects on Newborn - Live birth index. Mouse Route: Oral; Dose: 7920 mg/kg; Duration: female 6-13D of pregnancy; Effects on Fertility - Litter size. Mouse Route: Oral; Dose: 181 gm/kg; Duration: male 18W prior to mating; Paternal Effects - Spermatogenesis; Testes, epididymis, sperm duct; Prostate, seminal vessicle, Cowper's gland, accessory glands. Mouse Route: Oral; Dose: 30 gm/kg; Duration: male 7D prior to mating; Paternal Effects - Spermatogenesis; Testes, epididymis, sperm duct; Effects on Fertility - Male fertility index. Mouse Route: Oral; Dose: 30 gm/kg; Duration: female 7D prior to mating; Effects on Fertility - Female fertility index.

Hazard Overviews

Fire: Will burn.
Carcinogenicity: IARC - Not listed; NIOSH - Not listed; NTP - Not listed; ACGIH - Not listed; OSHA - Not listed; EPA - Not listed; MAK - Not listed

Environmental

Environmental Fate: If released in soil, it would be expected to adsorb strongly to soil and/or biodegrade. If released into water, it will adsorb strongly to sediment and particulate matter in the water column and/or biodegrade. It may also be lost due to volatilization (half-life in a model river is 9.3 days). Considerable bioaccumulation in fish and aquatic organisms should occur. If released in air, it will most likely be in the form of an aerosol and be subject to gravitational settling. The vapor-phase ester should react with photochemically-produced hydroxyl radicals with an estimated half-life of 1.1 days.

Environmental Physical Data

Henry's Law Constant: estimated at 7.39 $x10^{-6}$
Octanol/Water Partition Coefficient: log K_{ow} = average 5.8
Sorption Partition Coefficient: K_{oc} = 9600
BCF: none likely

Regulations

RCRA 40CFR: Not listed
CERCLA: 40CFR 302.4: Not listed
SARA 40CFR 372.65: Not listed
SARA EHS 40CFR 355: Not listed
TSCA: Listed

DIH1810 CAS #: 143-16-8

DI-N-HEXYLAMINE

RTECS: IH6600000
EINECS Number: 205-588-4
Molecular Formula: $C_{12}H_{27}N$
Formula Weight: 185.40

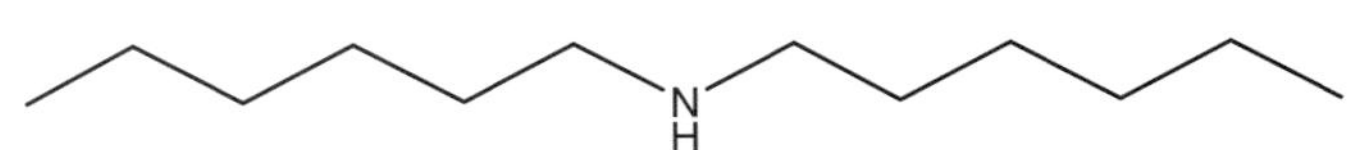

Chemical Structure

Synonyms: 1-HEXANAMINE,N-HEXYL-; N-HEXYL-1-HEXANAMINE
Description: colorless, clear liquid
Use: organic synthesis

Physical Properties

Boiling Point: 230 °C (446 °F)
Freezing Point: -13 °C (8.6 °F)
Specific Gravity: 0.778
Vapor Density: 6.4 Air=1
Density: 0.79 g/mL at 20 °C
Water Solubility: < 1 mg/mL at 26 C
Other Solubilities: 95% Ethanol: >=100 mg/ml at 26 °C; Acetone: >=100 mg/ml at 26 °C; DMSO: 5-10 mg/ml at 26 °C.
Refraction Index: 1.434
Flash Point: 95 °C

RTECS Toxicity Data

Acute Oral: Rat LD_{50} Dose: 380 mg/kg.
Acute Dermal: Rabbit LD_{50} Route: Skin; Dose: 170 mg/kg.
Irritation Eye: Rabbit Standard Draize Test Dose: 750 ug/24H; Reaction: severe.
Irritation Skin: Rabbit Open Draize Test Dose: 500 mg open; Reaction: mild.

Hazard Overviews

Poison

Corrosive

Health: Corrosive to eyes/skin/respiratory tract. Poison. Other Acute Effects: may be fatal if inhaled, swallowed, or absorbed through skin; readily absorbed through skin; inhalation may result in spasm; inflammation and edema of the larynx and bronchi; chemical pneumonitis; pulmonary edema; burning sensation; coughing; wheezing; laryngitis; shortness of breath; headache; nausea and vomiting.
Fire: Will burn. Hazards: emits toxic fumes. Extinguishing agents: carbon dioxide, dry chemical powder or appropriate foam. Precautions: combustible liquid.
Reactivity: Incompatible with: strong oxidizing agents, strong acids, acid chlorides, acid anhydrides. Hazardous decomposition products: toxic fumes of: carbon monoxide, carbon dioxide, nitrogen oxides.
Carcinogenicity: IARC - Not listed; NIOSH - Not listed; NTP - Not listed; ACGIH - Not listed; OSHA - Not listed; EPA - Not listed; MAK - Not listed

Primary Target Organs:

Eyes

Skin

Respiratory System

Environmental

Cleanup/Disposal: Guide No. 153: Eliminate all ignition sources (no smoking, flares, sparks or flames in immediate area). Do not touch damaged containers or spilled material unless wearing appropriate protective clothing. Stop leak if you can do it without risk. Prevent entry into waterways, sewers, basements or confined areas. Absorb or cover with dry earth, sand or other non-combustible material and transfer to containers. Do not get water inside containers.

Regulations

RCRA 40CFR: Not listed
CERCLA: 40CFR 302.4: Not listed
SARA 40CFR 372.65: Not listed
SARA EHS 40CFR 355: Not listed
TSCA: Listed

DIH2080 CAS #: 125-93-9

DIHYDROABIETIC ACID, TRIESTER WITH GLYCEROL

EINECS Number: 204-761-1
Molecular Formula: $C_{63}H_{98}O_6$
Formula Weight: 951.47
Synonyms: 1-PHENANTHRENECARBOXYLIC ACID,1,2,3,4,4A,4B,5,6,7,9,10,10A-DODECAHYDRO-1,4A-DIMETHYL-7-(1-METHYLETHYL)-, 1,2,3-PROPANETRIYL ESTER,(1R-(1ALPHA,4ABETA,4BALPHA,10AALPHA))-
Use: tackifying resin for adhesives-esp hot-melt adhesives; tackifying resin in chewing gum

Hazard Overviews

Carcinogenicity: IARC - Not listed; NIOSH - Not listed; NTP - Not listed; ACGIH - Not listed; OSHA - Not listed; EPA - Not listed; MAK - Not listed

Environmental

Regulations

RCRA 40CFR: Not listed
CERCLA: 40CFR 302.4: Not listed
SARA 40CFR 372.65: Not listed
SARA EHS 40CFR 355: Not listed
TSCA: Listed

DIH2350 CAS #: 583-39-1

DIHYDROBENZIMIDAZOLE THIONE

RTECS: DE1050000
EINECS Number: 209-502-6
Molecular Formula: $C_7H_6N_2S$
Formula Weight: 150.21

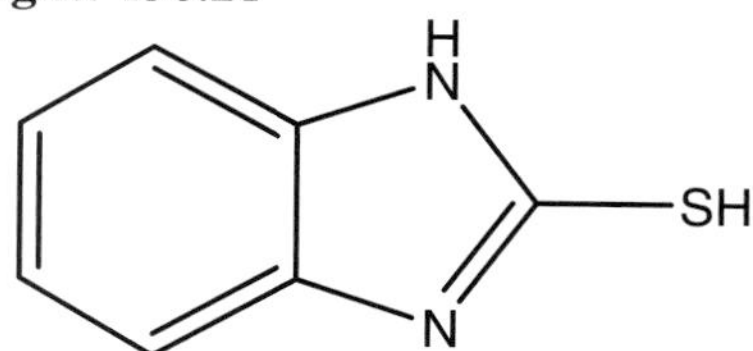

Chemical Structure

Synonyms: ANTIEGENE MB; ANTIOXIDANT MB; AOMB; ASM MB; 1H-BENZIMIDAZOLE-2-THIOL; 2-BENZIMIDAZOLETHIOL; 2-BENZIMIDAZOLINETHIONE; 2-BENZIMIDAZOLINTHION; 1,3-DIHYDRO-2H-BENZIMIDAZOLE-2-THIONE; 2-MERCAPTOBENZIMIDAZOLE; MERCAPTOBENZIMIDAZOLE; 2-MERCAPTOBENZOIMIDAZOLE; MERCAPTOBENZOIMIDAZOLE; 2-MERKAPTOBENZIMIDAZOL; MERKAPTOBENZIMIDAZOL; O-PHENYLENETHIOUREA; 2-THIOBENZIMIDAZOLE
Description: yellow to white crystals or cream colored powder; slight odor

Physical Properties

Freezing Point: 303 °C (577.4 °F) to 304 °C (579.2 °F)
Specific Gravity: 1.42
Vapor Pressure: 30.2 mm Hg at 61 °C
Water Solubility: < 1 mg/mL at 23.5 C
Other Solubilities: 95% Ethanol: 1-10 mg/ml at 23.5 °C; Acetone: 10-50 mg/ml at 23.5 °C; Absolute Ethanol: Soluble; DMSO: >=100 mg/ml at 23.5 °C; Methanol: Soluble.
Flash Point: Not available; probably combustible

RTECS Toxicity Data

Acute Oral: Mouse LD_{50} Dose: 750 mg/kg.
Chronic (Multiple Dose) Oral: Rat Dose: 1820 mg/kg/17W-I; Toxic Effects: Blood - Changes in serum composition.
Chronic (Multiple Dose) Inhalation: Rat Dose: 6200 ug/m^3/6H/13W-I; Toxic Effects: Endocrine - Changes in thymus weight; Endocrine - Changes in thyroid weight; Blood - Changes in erythrocite (RBC) cell count.
Irritation Eye: Rabbit Standard Draize Test Dose: 500 mg/24H; Reaction: mild.
Irritation Skin: Rabbit Standard Draize Test Dose: 500 mg/24H; Reaction: mild.
Reproductive/Teratogenic: Rat Route: Oral; Dose: 660 mg/kg; Duration: female 7-17D of pregnancy; Maternal Effects - Other effects on females. Rat Route: Oral; Dose: 110 mg/kg; Duration: female 7-17D of pregnancy; Maternal Effects - Other effects on females; Effects on Embryo or Fetus - Fetotoxicity; Specific Developmental Abnormalities - Musculoskeletal system. Rat Route: Oral; Dose: 330 mg/kg; Duration: female 7-17D of pregnancy; Specific

Developmental Abnormalities - Craniofacial (including nose and tongue); Urogenital system.

Hazard Overviews

Health: Irritating to eyes/skin/respiratory tract. Harmful. Other Acute Effects: may be harmful by inhalation, ingestion, or skin absorption.

Fire: Will burn. Hazards: emits toxic fumes. Extinguishing agents: water spray; carbon dioxide, dry chemical powder or appropriate foam. Precautions: combustible liquid.

Reactivity: Incompatible with: strong oxidizing agents, strong bases. Hazardous decomposition products: hydrogen cyanide, carbon monoxide, carbon dioxide, nitrogen oxides, sulfur oxides.

Carcinogenicity: IARC - Not listed; NIOSH - Not listed; NTP - Not listed; ACGIH - Not listed; OSHA - Not listed; EPA - Not listed; MAK - Not listed

Primary Target Organs:

Eyes

Skin

Respiratory System

Environmental

Regulations

RCRA 40CFR: Not listed
CERCLA: 40CFR 302.4: Not listed
SARA 40CFR 372.65: Not listed
SARA EHS 40CFR 355: Not listed
TSCA: Listed

DIH2620 CAS #: 119-84-6

3,4-DIHYDROCOUMARIN

RTECS: MW5775000
EINECS Number: 204-354-9
Molecular Formula: $C_9H_8O_2$
Formula Weight: 148.16

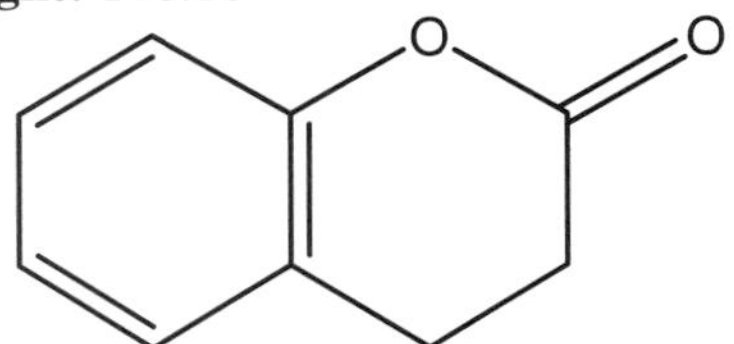

Chemical Structure

Synonyms: 1,2-BENZODIHYDROPYRONE; 2H-1-BENZOPYRAN-2-ONE,3,4-DIHYDRO-; 2H-1-BENZOPYRAN-2-ONE,3,4-DIHYDRO-(9CI); 2-CHROMANONE; CHROMAN,2-OXO-; COUMARIN,3,4,-DIHYDRO; COUMARIN,3,4-DIHYDRO-; 3,4-DIHYDRO-2H-1-BENZOPYRAN-2-ONE; DIHYDROCOUMARIN; HYDROCINNAMIC ACID,O-HYDROXY-,DELTA-LACTONE; HYDROCINNAMIC ACID,O-HYDROXY-,LACTONE; HYDROCOUMARIN; O-HYDROXYCINNAMIC ACID; MELILOTIC LACTONE; MELILOTIN; MELILOTINE; MELILOTOL; OXOCHROMAN

Description: white to light yellow, oily liquid; colorless crystals or leaflets; sweet - herbal odor
Use: flavoring agent; perfumery

Physical Properties

Boiling Point: 272 °C (522 °F)
Freezing Point: 25 °C (77 °F)
Specific Gravity: 1.169 at 18 °C
Saturated Vapor Density: 1.28693706 kg/m^3
Density: 1.169 g/mL at 18 °C
Vapor Pressure: 13.4 mm Hg at 25 °C
Water Solubility: Insoluble in Water
Other Solubilities: 95% Ethanol: >=100 mg/ml at 20 °C; Acetone: >=100 mg/ml at 20 °C; Chloroform: Soluble; DMSO: >=100 mg/ml at 20 °C.
Refraction Index: 1.5563 at 20 °C/D
Flash Point: 130 °C

RTECS Toxicity Data

Acute Oral: Rat LD_{50} Dose: 1460 mg/kg; Toxic Effects: Behavioral - Somnolence (general depressed activity). Guinea Pig LD_{50} Dose: 1760 mg/kg; Toxic Effects: Behavioral - Somnolence (general depressed activity).

Acute Dermal: Rabbit LD_{50} Route: Skin; Dose: >5 gm/kg.

Chronic (Multiple Dose) Oral: Rat Dose: 18 gm/kg/16D-I; Toxic Effects: DEATH. Rat Dose: 19500 mg/kg/13W-I; Toxic Effects: Liver - Other changes; Blood - Changes in erythrocite (RBC) cell count; Biochemical - True cholinesterase.

Irritation Skin: Rabbit Standard Draize Test Dose: 500 mg/24H; Reaction: moderate. Guinea Pig Standard Draize Test Dose: 1%/48H; Reaction: moderate.

Tumorigenic: Rat Route: Oral; Dose: 433 gm/kg/2Y-I; Toxic Effects: Tumorigenic - Neoplastic by RTECS criteria; Kidney, Ureter, and Bladder - Kidney tumors. Mouse Route: Oral; Dose: 144 gm/kg/2Y-I; Toxic Effects: Tumorigenic - Neoplastic by RTECS criteria; Liver - Tumors.

Hazard Overviews

Health: Irritating to eyes/skin/respiratory tract. Harmful. Other Acute Effects: may be harmful by inhalation, ingestion, or skin absorption.

Fire: Will burn. Hazards: emits toxic fumes. Extinguishing agents: water spray; carbon dioxide, dry chemical powder or appropriate foam. Precautions: combustible liquid.

Reactivity: Incompatible with: strong oxidizing agents. Hazardous decomposition products: toxic fumes of: carbon monoxide, carbon dioxide.

Carcinogenicity: IARC - Not listed; NIOSH - Not listed; NTP - Not listed; ACGIH - Not listed; OSHA - Not listed; EPA - Not listed; MAK - Not listed

Primary Target Organs:

Eyes

Skin

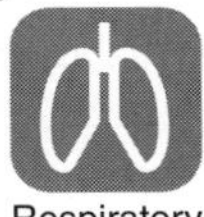
Respiratory System

Environmental

Regulations

RCRA 40CFR: Not listed
CERCLA: 40CFR 302.4: Not listed
SARA 40CFR 372.65: Not listed
SARA EHS 40CFR 355: Not listed
TSCA: Listed

DIH2890 **CAS #: 5234-68-4**

5,6-DIHYDRO-2-METHYL-N-PHENYL-1,4-OXATHIIN-3-CARBOXAMIDE

RTECS: RP4550000
EINECS Number: 226-031-1
Molecular Formula: $C_{12}H_{13}NO_2S$
Formula Weight: 235.31
Synonyms: CARBATHIIN; 5-CARBOXANILIDO-2,3-DIHYDRO-6-METHYL-1,4-OXATHIIN; CARBOXIN; CARBOXINE; CEREVAX; CEREVAX EXTRA; D 735; D-735; D735; DCMO; 2,3-DIHYDRO-5-CARBOXANILIDO-6-METHYL-1,4-OXATHIIN; 5,6-DIHYDRO-2-METHYL-3-CARBOXANILIDO-1,4-OXATHIIN; 2,3-DIHYDRO-6-METHYL-1,4-OXATHIIN-5-CARBOXANILIDE; 5,6-DIHYDRO-2-METHYL-1,4-OXATHI-IN-3-CARBOXANILIDE; 5,6-DIHYDRO-2-METHYL-1,4-OXATHIIN-3-CARBOXANILIDE; 5,6-DIHYDRO-2-METHYL-1,4-OXATHI-INE-3-CARBOXANILIDE; 2,3-DIHYDRO-6-METHYL-5-PHENYLCARBAMOYL-1,4-OXATHIIN; DMOC; DUAL MURGANIC RPB; ENHANCE; ENHANCE PLUS; F 735; FLO PRO V SEED PROTECTANT; GERMATE PLUS; KEMIKAR; KISVAX; MIST-O-MATIC MURGANIC; MURGANIC; 1,4-OXATHIIN-3-CARBOXAMIDE,5,6-DIHYDRO-2-METHYL-N-PHENYL; 1,4-OXATHIIN-3-CARBOXANILIDE,5,6-DIHYDRO-2-METHYL-; 1,4-OXATHIIN,2,3-DIHYDRO-5-CARBOXANILIDO-6-METHYL-; OXATIN; PRO-GRO; RTV VITAVAX; V 4X; VITAFLO 250; VITAFLOW; VITAVAX; VITAVAX 30-C; VITAVAX 735D; VITAVAX 75W; VITAVAX-200; VITAVAX-300; VITAVAX 100; VITAVAX 34; VITAVAX-THIRAM-LINDANE; VITAWAX
Description: white to off-white solid or crystals
Use: systemic fungicide for seed treatments of cereals against basidomycete type of fungi, eg, bunts & smuts; with other fungicides for other seed- & soil-borne seedling diseases; against rhizoctonia on cotton, groundnuts & vegetables; against smuts of small grains & rhizoctonia solani of sugar beets, loose smut on barley, wheat, & oats; against seedling diseases of barley, soya beans, vegetables, etc; & fairy rings on turf.

Physical Properties

Freezing Point: 93 °C (199.4 °F) to 95 °C (203 °F)
Saturated Vapor Density: 1.200000002 kg/m^3
Vapor Pressure: 1.8 $x10^{-7}$ mm Hg at 25 °C
Water Solubility: 170 mg/L at 25 °C
Other Solubilities: solubility (25 °C): 600 g/kg Acetone; 1500 g/kg dimethyl sulfoxide; 110 g/kg Ethanol; 210 g/kg Methanol.

RTECS Toxicity Data

Acute Oral: Rat LD_{50} Dose: 430 mg/kg; Toxic Effects: Behavioral - Somnolence (general depressed activity); Behavioral - Muscle contraction or spasticity. Mouse LD_{50} Dose: 3200 mg/kg.
Acute Inhalation: Rat LC_{50} Dose: >20 gm/m^3/1hr.
Acute Dermal: Rat LD_{50} Route: Skin; Dose: 1050 mg/kg.
Mutagenic: Rat Cytogenetic Analysis; Route: Intraperitoneal; Dose: 382 mg/kg/48H-C.

Hazard Overviews

Carcinogenicity: IARC - Not listed; NIOSH - Not listed; NTP - Not listed; ACGIH - Not listed; OSHA - Not listed; EPA - Not listed; MAK - Not listed

Environmental

Ecotoxicity: LC_{50} Rainbow trout 2 mg/l/96 hr /Conditions of bioassay not specified LC_{50} Mallard duck >4640 mg/kg diet/8 day
Environmental Fate: If released to soil, the expected mobility is medium to high based on estimated K_{oc} values of 80 and 259. If released to water, it will be essentially non-volatile. Aquatic bioconcentration and adsorption to sediment are not expected to be important fate processes based on estimated K_{oc} values of 80 and 259 and an estimated BCF value of 34. Several studies have shown that it biodegrades slowly. If released to the atmosphere, it will exist in both the vapor and particulate phases. In the vapor phase, it will degrade in the atmosphere by reaction with photochemically produced hydroxyl radicals with an estimated half-life of 2.4 hours. It will also degrade in the atmosphere by reaction with atmospheric ozone with an estimated half-life of 1 day. Physical removal from air can occur through wet and dry deposition.

Environmental Physical Data

Henry's Law Constant: estimated at 3.3 $x10^{-10}$
Sorption Partition Coefficient: estimated at 80
BCF: estimated at 34

Regulations

RCRA 40CFR: Not listed
CERCLA: 40CFR 302.4: Not listed
SARA 40CFR 372.65: Listed
SARA EHS 40CFR 355: Not listed
TSCA: Not listed

Analytical Methods

Soil: EPA PMD-CBX
Drinking Water: EPA 507, 525.2; AOAC 991.07; ASTM D5475
Food: FDA 212.1, 232.1, 232.4, 242.1
Plasma: FDA 211.1, 231.1, 252

DIH3160 **CAS #: 104-61-0**

DIHYDRO-5-PENTYL-2(3H)-FURANONE

RTECS: LU3675000
EINECS Number: 203-219-1
Molecular Formula: $C_9H_{16}O_2$
Formula Weight: 156.22

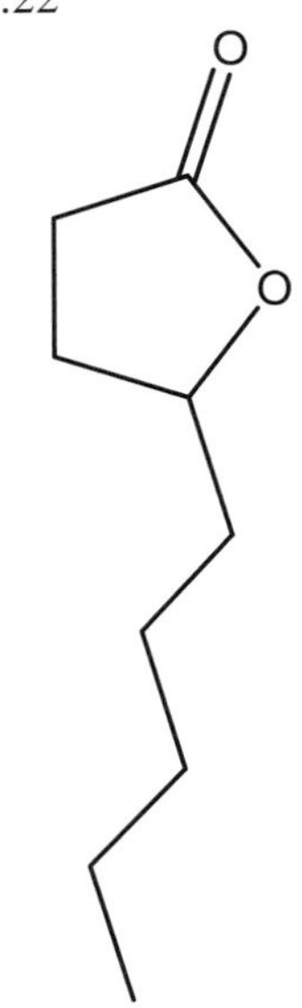

Chemical Structure

Synonyms: ALDEHYDE C-18; GAMMA-AMYL-GAMMA-BUTYROLACTONE; GAMMA-AMYLBUTYROLACTONE; GAMMA-N-AMYLBUTYROLACTONE; COCONUT ALDEHYDE; 2(3H)-FURANONE,DIHYDRO-5-PENTYL-; 4-HYDROXYNONANOIC ACID LACTONE; 4-HYDROXYNONANOIC ACID,GAMMA-LACTONE; GAMMA-NONALACTONE; 1,4-NONALOLIDE; NONANOIC ACID,4-HYDROXY-,GAMMA-LACTONE; GAMMA-NONANOLACTONE; 4-NONANOLIDE; GAMMA-NONANOLIDE; NONAN-1,4-OLIDE; 4-PENTYL-BUTANOLIDE; GAMMA-PENTYL-GAMMA-BUTYROLACTONE; PRUNOLIDE
Description: colorless to very pale, straw-yellow liquid; strong odor reminiscent of coconut
Use: in fragrances; as flavoring ingredient

Physical Properties

Boiling Point: 243 °C (469 °F)
Specific Gravity: 0.958 to 0.966 at 25/25 °C
Water Solubility: Insoluble in Water
Other Solubilities: Soluble in most fixed oils and mineral oil; practically Insoluble in Glycerol.
pH: Not more than 5
Refraction Index: 1.4460 to 1.4500 at 20 °C
Flash Point: > 100 °C

RTECS Toxicity Data

Acute Oral: Rat LD_{50} Dose: 6600 mg/kg. Guinea Pig LD_{50} Dose: 3440 mg/kg; Toxic Effects: Behavioral - Somnolence (general depressed activity); Gastrointestinal - Changes in structure or function of salivary glands.
Irritation Skin: Rabbit Standard Draize Test Dose: 100 mg/24H; Reaction: severe. Rabbit Standard Draize Test Dose: 500 mg/24H; Reaction: mild.
Mutagenic: Bacteria - B Subtilis DNA Repair; Dose: 20 mg/disc.

Hazard Overviews

Health: Severely irritating to eyes/skin/respiratory tract. Other Acute Effects: harmful if swallowed, inhaled, or absorbed through skin; high concentrations are extremely destructive to tissues of the mucous membranes and upper respiratory tract, eyes and skin; symptoms of exposure may include burning sensation, coughing, wheezing, laryngitis, shortness of breath, headache, nausea and vomiting.
Fire: Will burn. Extinguishing agents: water spray; carbon dioxide, dry chemical powder or appropriate foam. Precautions: combustible liquid.
Reactivity: Incompatible with: strong oxidizing agents, strong bases. Hazardous decomposition products: toxic fumes of: carbon monoxide, carbon dioxide.
Carcinogenicity: IARC - Not listed; NIOSH - Not listed; NTP - Not listed; ACGIH - Not listed; OSHA - Not listed; EPA - Not listed; MAK - Not listed
Primary Target Organs:

Eyes

Skin

Respiratory System

Environmental

Regulations

RCRA 40CFR: Not listed
CERCLA: 40CFR 302.4: Not listed
SARA 40CFR 372.65: Not listed
SARA EHS 40CFR 355: Not listed
TSCA: Listed

DIH3430 **CAS #: 473-55-2**

DIHYDROPINENE

EINECS Number: 207-467-1
Molecular Formula: $C_{10}H_{18}$
Formula Weight: 138.26

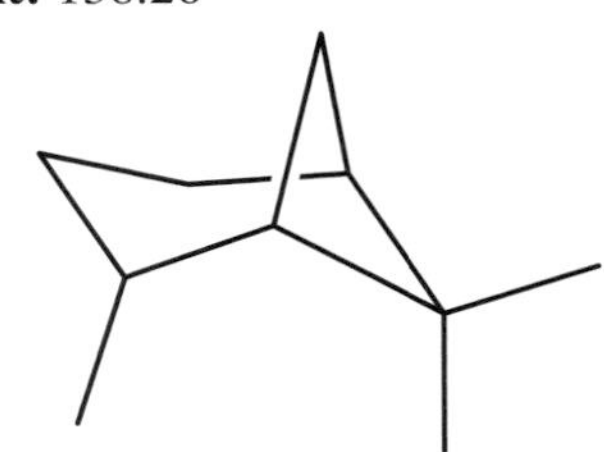

Chemical Structure

Synonyms: BICYCLO(3.1.1)HEPTANE,2,6,6-TRIMETHYL-; PINANE; 2,6,6-TRIMETHYLBICYCLO(3,1,1)HEPTANE; 2,7,7-TRIMETHYLBICYCLO(3.1.1)HEPTANE
Description: colorless, clear liquid; mild odor
Use: intermediate used in the manufacture of terpene alcohols

Physical Properties

Boiling Point: 151 °C (304 °F)
Specific Gravity: 0.8
Vapor Density: About 4.8 Air=1
Saturated Vapor Density: ~ 1.211897641 kg/m^3
Vapor Pressure: ~ 2 mm Hg at 20 °C
Water Solubility: < 1 mg/mL at 20 C
Other Solubilities: 95% Ethanol: >=100 mg/ml at 20 °C; Acetone: >=100 mg/ml at 20 °C; DMSO: 10-50 mg/ml at 20 °C.
Refraction Index: 1.4623
Flash Point: 40 °C
Autoignition Temperature: 273 °C
LEL: 0.7% v/v
UEL: 7.2% v/v at 160 °C

Hazard Overviews

Flammable

Fire: Flammable.
Carcinogenicity: IARC - Not listed; NIOSH - Not listed; NTP - Not listed; ACGIH - Not listed; OSHA - Not listed; EPA - Not listed; MAK - Not listed

Environmental

Cleanup/Disposal: Guide No. 128: Eliminate all ignition sources (no smoking, flares, sparks or flames in immediate area). All equipment used when handling the product must be grounded. Do not touch or walk through spilled material. Stop leak if you can do it without risk. Prevent entry into waterways, sewers, basements or confined areas. A vapor suppressing foam may be used to reduce vapors. Absorb or cover with dry earth, sand or other non-combustible material and transfer to containers. Use clean non-sparking tools to collect absorbed material. Large Spills: Dike far ahead of liquid spill for later disposal. Water spray may reduce vapor; but may not prevent ignition in closed spaces.

Regulations

RCRA 40CFR: Not listed
CERCLA: 40CFR 302.4: Not listed
SARA 40CFR 372.65: Not listed
SARA EHS 40CFR 355: Not listed
TSCA: Listed

DIH3700 **CAS #: 94-58-6**

DIHYDROSAFROLE

RTECS: DA6125000
EINECS Number: 202-344-9
Molecular Formula: $C_{10}H_{12}O_2$
Formula Weight: 164.22

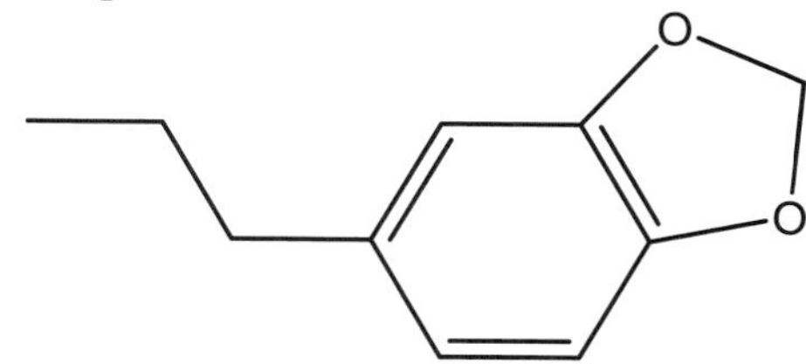

Chemical Structure

Synonyms: BENZENE,1,2-METHYLENEDIOXY-4-PROPYL-; 1,3-BENZODIOXOLE,5-PROPYL-; 2',3'-DIHYDROSAFROLE; 1,2-(METHYLENEDIOXY)-4-PROPYLBENZENE; 5-PROPYL-1,3-BENZODIOXOLE; 4-PROPYL-1,2-METHYLENEDIOXYBENZENE; SAFROLE,DIHYDRO-
Description: colorless, oily liquid
Use: fragrance in cosmetics (former use); flavoring agent in root beer (former use); chem int for piperonyl butoxide

Physical Properties

Boiling Point: 228 °C (442 °F)
Specific Gravity: 1.0695 at 20 °C
Other Solubilities: Soluble in Carbon Tetrachloride.
Refraction Index: 1.5187 at 25 °C/D

RTECS Toxicity Data

Acute Oral: Rat LD_{50} Dose: 2260 mg/kg. Mouse LD_{50} Dose: 3700 mg/kg.
Acute Dermal: Rabbit LD_{50} Route: Skin; Dose: >5 gm/kg.
Chronic (Multiple Dose) Oral: Rat Dose: 78750 mg/kg/15W-I; Toxic Effects: DEATH.
Irritation Skin: Rabbit Standard Draize Test Dose: 500 mg/24H; Reaction: mild.
Tumorigenic: Mouse Route: Oral; Dose: 101 gm/kg/81W-C; Toxic Effects: Tumorigenic - Carcinogenic by RTECS criteria; Gastrointestinal - Tumors; Liver - Tumors. Mouse Route: Oral; Dose: 163 gm/kg/81W-C; Toxic Effects: Tumorigenic - Carcinogenic by RTECS criteria; Lungs, Thorax, or Respiration - Tumors; Liver - Tumors.

Hazard Overviews

Carcinogenicity: IARC - Group 2B, Possibly carcinogenic to humans; NIOSH - Not listed; NTP - Not listed; ACGIH - Not listed; OSHA - Not listed; EPA - Not listed; MAK - Not listed

Environmental

Environmental Fate: If released to soil, it will have medium mobility. Volatilization may be important from moist soil surfaces. Insufficient data are available to determine the importance or rate of biodegradation in soil or water. It is not expected to hydrolyze in moist soil or in water since it lacks hydrolyzable functional groups. If released to water, it may adsorb to suspended solids and sediment. It may volatilize from water surfaces with estimated half-lives from a model river and model lake of 4 and 33 days, respectively. An estimated BCF value of 310 suggests that bioconcentration will be high in aquatic organisms. It may be susceptible to direct photolysis in surface water and in the atmosphere based upon its absorption of light at wavelengths >290 nm. If released to the atmosphere, it will exist mainly in the vapor-phase. The half-life for the vapor phase reaction with photochemically produced hydroxyl radicals has been estimated to be 6.7 hours. Particulate-phase may be physically removed from the air by wet and dry deposition.

Environmental Physical Data

Henry's Law Constant: estimated at 1.22×10^{-5}
Octanol/Water Partition Coefficient: log K_{ow} = 3.58
Sorption Partition Coefficient: K_{oc} = 300
BCF: estimated at 310

Regulations

RCRA 40CFR: Listed Hazardous Waste No. U090 Toxic Waste
CERCLA: 40CFR 302.4: Listed per RCRA Section 3001 RQ: 10 lb (4.535 kg)
SARA 40CFR 372.65: Listed
SARA EHS 40CFR 355: Not listed
TSCA: Listed

Analytical Methods

Soil: SW846 8270C

DIH3970 CAS #: 67-96-9

DIHYDROTACHYSTEROL

RTECS: WW0600000
EINECS Number: 200-672-7
Molecular Formula: $C_{28}H_{46}O$
Formula Weight: 398.74

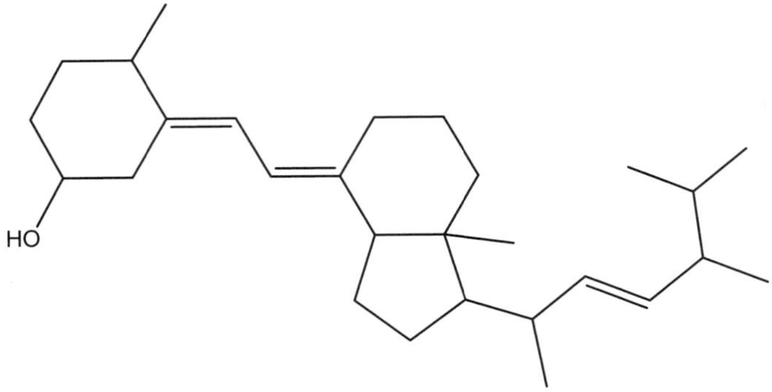

Chemical Structure

Synonyms: A.T. 10; ANTITANIL; ANTI-TETANY SUBSTANCE 10; AT 10; CALCAMINE; DHT2; DICHYSTROLUM; DIHYDRAL; DIHYDROTACHYSTEROL2; DYGRATYL; HYTAKEROL; PARTEROL; 9,10-SECOERGOSTA-5,7,22-TRIEN-3-OL,(3BETA,5E,7E,10ALPHA,22E)-; 9,10-SECOERGOSTA-5,7,22-TRIEN-3BETA-OL; 9,10-SECOERGOSTA-5,7,22-TRIEN-3-OL,(3-BETA,5E,7E,10-ALPHA,22E)-(9CI); TACHYSTEROL2,DIHYDRO-; TACHYSTEROL,DIHYDRO-; TACHYSTIN
Description: colorless or white crystals, crystalline powder, or needles; odorless
Use: medication: blood calcium regulator; medication (vet): in prevention of parturient paresis

Physical Properties

Freezing Point: 125 °C (257 °F) to 127 °C (260.6 °F)
Water Solubility: Insoluble in Water
Other Solubilities: Soluble in Alcohol; freely Soluble in Ether & Chloroform; Sparingly Soluble in vegetable oils.
RTECS Toxicity Data
Acute Oral: Mouse LD_{50} Dose: 288 mg/kg.
Reproductive/Teratogenic: Rat Route: Oral; Dose: 150 mg/kg; Duration: female 1-5D of pregnancy; Effects on Fertility - Pre-implantation mortality.

Hazard Overviews

Health: May cause irritation. Toxic. Other Acute Effects: harmful if swallowed; can cause weakness; headache; somnolence; nausea; vomiting; dry mouth; constipation; muscle pain; bone pain; metallic taste. The toxicological properties have not been thoroughly investigated.
Fire: Hazards: emits toxic fumes. Extinguishing agents: water spray; carbon dioxide, dry chemical powder or appropriate foam. Precautions: combustible liquid.
Reactivity: Stable. Hazardous polymerization will not occur. Hazardous decomposition products: carbon monoxide, carbon dioxide.
Carcinogenicity: IARC - Not listed; NIOSH - Not listed; NTP - Not listed; ACGIH - Not listed; OSHA - Not listed; EPA - Not listed; MAK - Not listed

Environmental

Regulations

RCRA 40CFR: Not listed
CERCLA: 40CFR 302.4: Not listed
SARA 40CFR 372.65: Not listed
SARA EHS 40CFR 355: Not listed
TSCA: Not listed

DIH4240 CAS #: 77-79-2

2,5-DIHYDROTHIOPHENE 1,1-DIOXIDE

RTECS: XM9100000
EINECS Number: 201-059-7
Molecular Formula: $C_4H_6O_2S$
Formula Weight: 118.16

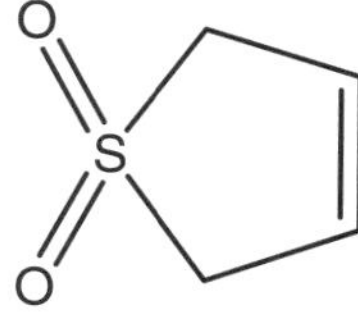

Chemical Structure

Synonyms: BUTADIENE SULFONE; 2,5-DIHYDROTHIOPHENE 1,1-DIOXIDE; 2,5-DIHYDROTHIOPHENE DIOXIDE; 2,5-DIHYDROTHIOPHENE SULFONE; 3-SULFOLENE; BETA-SULFOLENE; SULFOL-3-ENE; SULFOLENE; 1-THIA-3-CYCLOPENTENE 1,1-DIOXIDE; THIOPHENE,2,5-DIHYDRO-,1,1-DIOXIDE; THIOPHENE,2-5-DIHYDRO-1.1-DIOXIDE

Description: crystals; pungent odor

Use: specialty solvent in petroleum refining; manufacture of sulfolane which is also a solvent in petroleum refining 1

Physical Properties

Boiling Point: Decomposes

Freezing Point: 64 °C (147.2 °F) to 65.5 °C (149.9 °F)

Specific Gravity: 1.3 at 15/15 °C

Density: 1.314 g/mL at 70 °C

Water Solubility: Soluble in Water

Other Solubilities: 95% Ethanol: 10-50 mg/ml at 16 °C; Acetone: >=100 mg/ml at 16 °C; Absolute Ethanol: Soluble; Benzene: Soluble; Chloroform: Soluble; DMSO: >=100 mg/ml at 16 °C; Ether: Soluble; Organic solvents: Soluble; Toluene: Partially Soluble.

Ionization Potential (eV): 10.0 +/-0.7

Flash Point: 112 °C

RTECS Toxicity Data

Acute Oral: Rat LD_{50} Dose: 2830 mg/kg. Mouse LD_{50} Dose: 1600 mg/kg; Toxic Effects: Behavioral - Somnolence (general depressed activity); Behavioral - Convulsions or effect on seizure threshold; Gastrointestinal - Changes in structure or function of esophagus.

Acute Dermal: Rabbit LD_{50} Route: Skin; Dose: >5 gm/kg.

Hazard Overviews

Health: Severely irritating to eyes. Harmful. Other Acute Effects: may be harmful by inhalation, ingestion, or skin absorption.

Fire: Will burn. Hazards: emits toxic fumes. Extinguishing agents: water spray; carbon dioxide, dry chemical powder or appropriate foam. Precautions: combustible liquid.

Reactivity: Incompatible with: strong oxidizing agents. Hazardous decomposition products: toxic fumes of: carbon monoxide, carbon dioxide, sulfur oxides.

Carcinogenicity: IARC - Not listed; NIOSH - Not listed; NTP - Not listed; ACGIH - Not listed; OSHA - Not listed; EPA - Not listed; MAK - Not listed

Primary Target Organs:

Eyes

Environmental

Environmental Fate: If released to soil, it is expected to leach (estimated K_{oc} of 4). No data were located which suggest biodegradation is an important fate process in soil or water. If released to water, adsorption to sediment and bioconcentration in aquatic organisms are not expected to be environmentally important removal processes. A volatilization half-life of 225 days has been estimated for a model river (one meter deep). If released to the atmosphere, vapor-phase is expected to degrade by reaction with photochemically produced hydroxyl radicals (estimated half-life of 6.7 hours). Vapor phase may also be rapidly degraded by reaction with ozone in the troposphere; the half-life for this reaction in air can be estimated to be about 1.4 hours.

Environmental Physical Data

Henry's Law Constant: estimated at 4.3 $x10^{-6}$

Octanol/Water Partition Coefficient: log K_{ow} = -1.448

Sorption Partition Coefficient: K_{oc} = 4

BCF: estimated at 4.7

Regulations

RCRA 40CFR: Not listed

CERCLA: 40CFR 302.4: Not listed

SARA 40CFR 372.65: Not listed

SARA EHS 40CFR 355: Not listed

TSCA: Listed

DIH4510 CAS #: 147-47-7

1,2-DIHYDRO-2,2,4-TRIMETHYLQUINOLINE

RTECS: VB4900000

EINECS Number: 205-688-8

Molecular Formula: $C_{12}H_{15}N$

Formula Weight: 173.2

Synonyms: ACETONANIL; ACETONANYL; ACETONE ANIL; ACETONE ANIL (QUINOLINE DERIV.); AGERITE RESIN D; DHTMQ; FLECTOL A; FLECTOL H; FLECTOL PASTILLES; POLNOX R; QUINOLINE,1,2-DIHYDRO-2,2,4-TRIMETHYL-; TMQ; TRIMETHYL DIHYDROQUINOLINE; 2,2,4-TRIMETHYL-1,2-DIHYDROCHINOLIN; 2,2,4-TRIMETHYL-1,2-DIHYDROQUINOLINE; TRIMETHYL-1,2-DIHYDROQUINOLINE; 2,2,4-TRIMETHYL-1,2-DIHYDROQUINONE; VULKANOX HS/LG; VULKANOX HS/POWDER

Description: light tan powder; odorless

Use: as a rubber antioxidant

Physical Properties

Boiling Point: 255 °C (491 °F) to 260 °C (500 °F)

Freezing Point: 120 °C (248 °F)

Specific Gravity: 1.08 at 25 °C

Density: 1.03 g/mL at 25 °C

Water Solubility: < 0.1 mg/mL at 19 C

Other Solubilities: 95% Ethanol: >=100 mg/ml at 20.5 °C; Acetone: >=100 mg/ml at 20.5 °C; DMSO: >=100 mg/ml at 20.5 °C.

Refraction Index: 1.5895

Flash Point: 101 °C

RTECS Toxicity Data

Acute Oral: Rat LD_{50} Dose: 2 gm/kg; Toxic Effects: Behavioral - Somnolence (general depressed activity). Mouse LD_{50} Dose: 1450 mg/kg; Toxic Effects: Behavioral - Somnolence (general depressed activity).
Chronic (Multiple Dose) Oral: Rat Dose: 12 gm/kg/30D-I; Toxic Effects: Liver - Liver function tests impaired; DEATH.
Chronic (Multiple Dose) Inhalation: Rat Dose: 12500 ug/m^3/4H/21W-I; Toxic Effects: Kidney, Ureter, and Bladder - Other changes in urine composition; Blood - Changes in serum composition; Nutritional and gross metabolic - Weight loss or decreased weight gain.

Hazard Overviews

Health: Irritating to eyes/skin/respiratory tract. Harmful. Other Acute Effects: harmful if swallowed; may be harmful if inhaled, or absorbed through the skin.
Fire: Will burn. Hazards: emits toxic fumes. Extinguishing agents: water spray; carbon dioxide, dry chemical powder or appropriate foam. Precautions: combustible liquid.
Reactivity: Incompatible with: strong oxidizing agents. Hazardous decomposition products: toxic fumes of: carbon monoxide, carbon dioxide, nitrogen oxides.
Carcinogenicity: IARC - Not listed; NIOSH - Not listed; NTP - Not listed; ACGIH - Not listed; OSHA - Not listed; EPA - Not listed; MAK - Not listed
Primary Target Organs:

Eyes

Skin
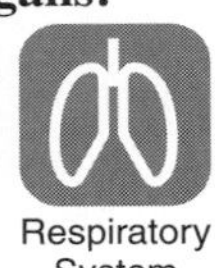
Respiratory System

Environmental

Regulations

RCRA 40CFR: Not listed
CERCLA: 40CFR 302.4: Not listed
SARA 40CFR 372.65: Not listed
SARA EHS 40CFR 355: Not listed
TSCA: Listed

DIH5320 CAS #: 490-78-8

2,6-DIHYDROXYACETOPHENONE

RTECS: AM7700000
EINECS Number: 207-716-4
Molecular Formula: $C_8H_8O_3$
Formula Weight: 152.16

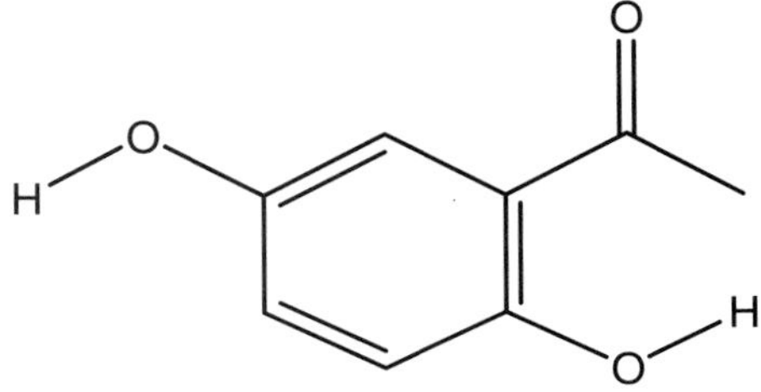

Chemical Structure

Synonyms: 2-ACETYLHYDROQUINONE; ACETYLHYDROQUINONE; 2',5'-DIHYDROXYACETOPHENONE; 2,5-DIHYDROXYACETOPHENONE; 1-(2,5-DIHYDROXYPHENYL)ETHANONE; ETHANONE,1-(2,5-DIHYDROXYPHENYL)-(9CI); QUINACETOPHENONE

Physical Properties

Freezing Point: 204 °C (399.2 °F) to 206 °C (402.8 °F)
Water Solubility: Insoluble
Other Solubilities: Ethanol: Soluble; Ether: Slightly soluble; Benzene: Slightly soluble

Hazard Overviews

Health: Irritating to eyes/skin/respiratory tract. Other Acute Effects: may be harmful by inhalation, ingestion, or skin absorption.
Fire: Extinguishing agents: water spray; carbon dioxide, dry chemical powder or appropriate foam. Precautions: combustible liquid.
Reactivity: Incompatible with: strong oxidizing agents, strong bases. Hazardous decomposition products: toxic fumes of: carbon monoxide, carbon dioxide.
Carcinogenicity: IARC - Not listed; NIOSH - Not listed; NTP - Not listed; ACGIH - Not listed; OSHA - Not listed; EPA - Not listed; MAK - Not listed
Primary Target Organs:

Eyes

Skin

Respiratory System

Environmental

Regulations

RCRA 40CFR: Not listed
CERCLA: 40CFR 302.4: Not listed
SARA 40CFR 372.65: Not listed
SARA EHS 40CFR 355: Not listed
TSCA: Listed

DIH5590 CAS #: 81-64-1

1,4-DIHYDROXY-9,10-ANTHRACENEDIONE

RTECS: CB6600000
EINECS Number: 201-368-7
Molecular Formula: $C_{14}H_8O_4$
Formula Weight: 240.2

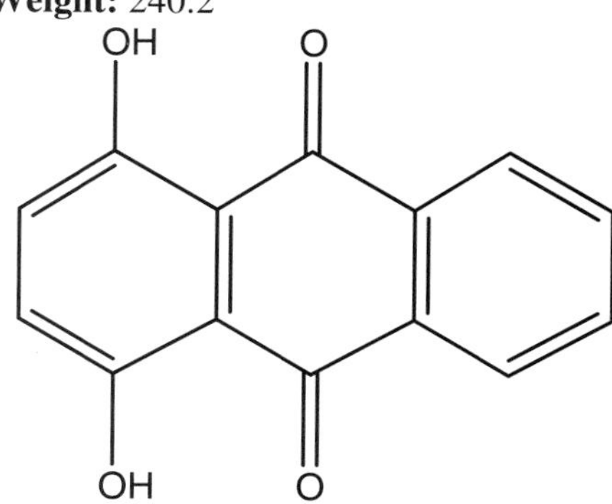

Chemical Structure

Synonyms: 9,10-ANTHRACENEDIONE,1,4-DIHYDROXY-; ANTHRAQUINONE,1,4-DIHYDROXY-; C.I. 58050; CHINIZARIN; 1,4-DIHYDROXYANTHRACHINON; 1,4-DIHYDROXY-9,10-ANTHRAQUINONE; 1,4-DIHYDROXYANTHRAQUINONE; 1,4-DIOXYANTHRAQUINONE; 1,4-DOA; QUINIZARIN; QUINIZARINE; SMOKE ORANGE R

Description: yellow to deep red crystals, needles, leaflets, plates

Use: antioxidant in synthetic lubricants; synthesis of experimental antitumor agents; fungicide for control of powdery mildew

Physical Properties

Freezing Point: 196 °C (384.8 °F)
Water Solubility: Soluble in Hot Water
Other Solubilities: Soluble in Ether; Soluble in aqueous alkalies & in Ammonia; 1 g Soluble in about 13 g of boiling Glacial Acetic Acid; Soluble in hot water & Benzene; Soluble in Potassium Hydroxide & Sulfuric acid.

RTECS Toxicity Data

Acute Oral: Rat LD_{50} Dose: >5 gm/kg. Mouse LD; Dose: >10 gm/kg.
Irritation Eye: Rabbit Standard Draize Test Dose: 500 mg/24H; Reaction: mild.
Mutagenic: Bacteria - S Typhimurium Mutations in Microorganisms; Dose: 100 ug/plate (+/-S9).

Hazard Overviews

Health: Irritating to eyes/skin/respiratory tract. Other Acute Effects: may be harmful by inhalation, ingestion, or skin absorption; may cause allergic skin reaction; sensitization by skin contact.

Fire: Extinguishing agents: water spray; carbon dioxide, dry chemical powder or appropriate foam. Precautions: combustible liquid.

Reactivity: Incompatible with: strong oxidizing agents, strong acids. Hazardous decomposition products: toxic fumes of: carbon monoxide, carbon dioxide.

Carcinogenicity: IARC - Not listed; NIOSH - Not listed; NTP - Not listed; ACGIH - Not listed; OSHA - Not listed; EPA - Not listed; MAK - Not listed

Primary Target Organs:

Eyes

Skin
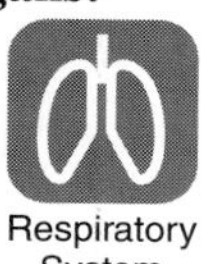
Respiratory System

Environmental

Regulations

RCRA 40CFR: Not listed
CERCLA: 40CFR 302.4: Not listed
SARA 40CFR 372.65: Not listed
SARA EHS 40CFR 355: Not listed
TSCA: Listed

DIH5860 CAS #: 72-48-0

1,2-DIHYDROXYANTHRAQUINONE

RTECS: CB6580000
EINECS Number: 200-782-5
Molecular Formula: $C_{14}H_8O_4$
Formula Weight: 240.22

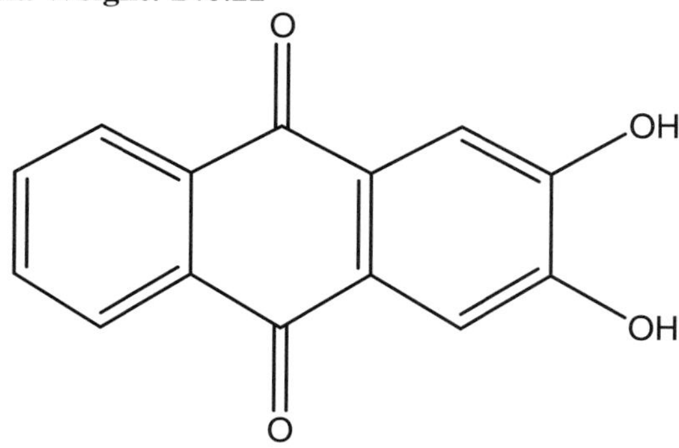

Chemical Structure

Synonyms: ALIZARIN; ALIZARIN B; ALIZARIN RED; ALIZARINA; ALIZARINE; ALIZARINE 3B; ALIZARINE B; ALIZARINE INDICATOR; ALIZARINE L PASTE; ALIZARINE LAKE RED 2P; ALIZARINE LAKE RED 3P; ALIZARINE LAKE RED IPX; ALIZARINE NAC; ALIZARINE PASTE 20% BLUISH; ALIZARINE RED; ALIZARINE RED B; ALIZARINE RED B2; ALIZARINE RED IP; ALIZARINE RED IPP; ALIZARINE RED L; ALIZARINPRIMEVEROSIDE; 9,10-ANTHRACENEDIONE,1,2-DIHYDROXY-; 1,2-ANTHRAQUINONEDIOL; C.I. 58000; C.I. MORDANT RED 11; C.I. PIGMENT RED 83; CERTIQUAL ALIZARINE; D AND C ORANGE NO. 15; DEEP CRIMSON MADDER 10821; 1,2-DIHYDROXYANTHRACHINON; 1,2-DIHYDROXY-9,10-ANTHRAQUINONE; ELJON MADDER; MITSUI ALIZARINE B; SANYO CARMINE L2B; TURKEY RED

Physical Properties

Boiling Point: 430 °C (806 °F)
Freezing Point: 287 °C (548.6 °F) to 289 °C (552.2 °F)
Specific Gravity: 0.95
Water Solubility: Slightly Soluble
Other Solubilities: Ethanol: Soluble; Ether: Soluble; Acetone: Soluble; Benzene: Soluble; Chloroform: Insoluble; Pyridine: Miscible; CS_2: Soluble; Methanol: Soluble
Flash Point: Combustible
Autoignition Temperature: 445 °C
RTECS Toxicity Data
Irritation Eye: Rabbit Standard Draize Test Dose: 500 mg/24H; Reaction: mild.
Mutagenic: Rat Unscheduled DNA Synthesis; Cell Type: liver; Dose: 10 mg/L. Bacteria - B Subtilis DNA Repair; Dose: 2 mg/disc.

Hazard Overviews

Health: Irritating to eyes/skin/respiratory tract. Other Acute Effects: may be harmful by inhalation, ingestion, or skin absorption.
Fire: Combustible. Extinguishing agents: water spray; carbon dioxide, dry chemical powder or appropriate foam. Precautions: combustible liquid.
Reactivity: Incompatible with: strong oxidizing agents, strong bases. Hazardous decomposition products: toxic fumes of: carbon monoxide, carbon dioxide.
Carcinogenicity: IARC - Not listed; NIOSH - Not listed; NTP - Not listed; ACGIH - Not listed; OSHA - Not listed; EPA - Not listed; MAK - Not listed
Primary Target Organs:

Eyes

Skin

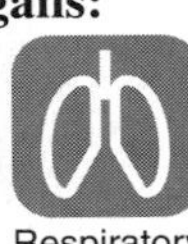
Respiratory System

Environmental

Regulations

RCRA 40CFR: Not listed
CERCLA: 40CFR 302.4: Not listed
SARA 40CFR 372.65: Not listed
SARA EHS 40CFR 355: Not listed
TSCA: Listed

DIH6400 **CAS #: 2373-98-0**

3,3-DIHYDROXYBENZIDINE

RTECS: DV4900000
Molecular Formula: $C_{12}H_{12}N_2O_2$
Formula Weight: 216.26
Synonyms: BENZIDINE,3,3'-DIHYDROXY-; M,M'-BIPHENOL,6,6'-DIAMINO-; (1,1'-BIPHENYL)-4,4'-DIAMINE,3,3'-DIHYDROXY-; (1,1'-BIPHENYL)-3,3'-DIOL,4,4'-DIAMINO-; 3,3'-BIPHENYLDIOL,4,4'-DIAMINO-; 6,6'-DIAMINO-M,M'-BIPHENOL; 4,4'-DIAMINO-(1,1'-BIPHENYL)-3,3'-DIOL; 4,4'-DIAMINO-3,3'-BIPHENYLDIOL; 3,3'-DIHYDROXYBENZIDINE; 3,3'-DIHYDROXY-4,4'-DIAMINOBIPHENYL; 3,3'-DIOXYBENZIDINE; 3,3'-DWUOKSYBENZYDYNA
Description: colorless crystals
Use: organic synthesis

Physical Properties

Freezing Point: 160 °C (320 °F)

RTECS Toxicity Data

Mutagenic: Bacteria - E Coli Phage Inhibition; Dose: 100 mmol/L.
Tumorigenic: Rat Route: Oral; Dose: 9950 mg/kg/52W-I; Toxic Effects: Tumorigenic - Neoplastic by RTECS criteria; Skin and appendages - Tumors. Rat Route: Subcutaneous; Dose: 5900 mg/kg/43W-I; Toxic Effects: Tumorigenic - Carcinogenic by RTECS criteria; Gastrointestinal - Tumors; Liver - Tumors.

Hazard Overviews

Carcinogenicity: IARC - Not listed; NIOSH - Not listed; NTP - Not listed; ACGIH - Not listed; OSHA - Not listed; EPA - Not listed; MAK - Not listed

Environmental

Regulations

RCRA 40CFR: Not listed
CERCLA: 40CFR 302.4: Not listed
SARA 40CFR 372.65: Not listed
SARA EHS 40CFR 355: Not listed
TSCA: Not listed

DIH6670 **CAS #: 131-56-6**

2,4-DIHYDROXYBENZOPHENONE

RTECS: DJ0700000
EINECS Number: 205-029-4
Molecular Formula: $C_{13}H_{10}O_3$
Formula Weight: 214.23

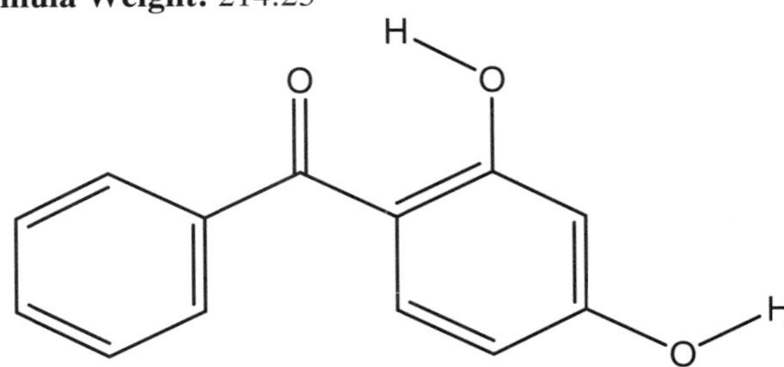

Chemical Structure

Synonyms: ADVASTAB 48; BENZOPHENONE,2,4-DIHYDROXY-; BENZORESORCINOL; DASTIB 263; 2,4-DIHYDROXYBENZOFENON; EASTMAN INHIBITOR DHPB; HHB; INHIBITOR DHBP; METHANONE,(2,4-DIHYDROXYPHENYL)PHENYL-; QUINSORB 010; RESBENZOPHENONE; SYNTASE 100; UF 1; UV 12; UVINUL 400; UVISTAT 12

Description: light-yellow, crystalline solid
Use: ultraviolet absorber in polymers; sunscreening agent; UV stabilizer for polymers

Physical Properties

Boiling Point: 194 °C (381 °F) at 1 mm Hg
Freezing Point: 142 °C (287.6 °F)
Density: 5.8 lb/gal at 20 °C
Water Solubility: Insoluble in Water
Other Solubilities: easily Soluble in Glacial Acetic Acid; scarecly Soluble in cold Benzene.

RTECS Toxicity Data

Acute Oral: Rat LD_{50} Dose: 8600 mg/kg; Toxic Effects: Behavioral - Somnolence (general depressed activity); Behavioral - Food intake (animal); Gastrointestinal - Hypermotility, diarrhea.
Chronic (Multiple Dose) Oral: Rat Dose: 54600 mg/kg/91D-I; Toxic Effects: Liver - Changes in liver weight; Blood - Pigmented or nucleated red Blood cells; Blood - Changes in erythrocite (RBC) cell count.
Irritation Eye: Rabbit Standard Draize Test Dose: 100 mg/24H; Reaction: moderate.

Hazard Overviews

Health: Irritating to eyes/skin/respiratory tract. Other Acute Effects: may be harmful by inhalation, ingestion, or skin absorption.
Fire: Extinguishing agents: water spray. Precautions: combustible liquid.
Reactivity: Incompatible with: strong oxidizing agents, strong bases. Hazardous decomposition products: toxic fumes of: carbon monoxide, carbon dioxide.
Carcinogenicity: IARC - Not listed; NIOSH - Not listed; NTP - Not listed; ACGIH - Not listed; OSHA - Not listed; EPA - Not listed; MAK - Not listed
Primary Target Organs:

Eyes

Skin

Respiratory System

Environmental

Regulations

RCRA 40CFR: Not listed
CERCLA: 40CFR 302.4: Not listed
SARA 40CFR 372.65: Not listed
SARA EHS 40CFR 355: Not listed
TSCA: Listed

DIH6940 CAS #: 110-64-5

1,4-DIHYDROXY-2-BUTENE

RTECS: EM4970000
EINECS Number: 203-787-0
Molecular Formula: $C_4H_8O_2$
Structured MF: $HOCH_2CH{=}CHCH_2OH$
Formula Weight: 88.12
Synonyms: AGRISYNTH B2D; 2-BUTENE-1,4-DIOL
Description: almost colorless liquid; odorless
Use: intermediate for alkyd resins, plasticizers, nylon, pharmaceuticals; cross linking agent for synthetic resins; mfr of fungicides; chem int for 1,4-butanediol; chem int for endosulfan (an insecticide); chem int for cis-1,4-bis(bromoacetoxy)-2-butene fungicide; chem int for pharmaceuticals, eg, pyridoxine & erythritol; comonomer-eg, for polyurethanes with toluene diisocyanate; chem int for org chems, eg, crotonaldehyde; reducing agent in chrome tanning baths

Physical Properties

Boiling Point: 141 °C (286 °F) to 149 °C (300 °F) at 20 mm Hg
Freezing Point: 7 °C (44.6 °F)
Specific Gravity: 1.067 to 1.074
Water Solubility: Very Soluble in Water
Other Solubilities: Very Soluble in Ethyl Alcohol, Acetone; Sparingly Soluble in Benzene
Refraction Index: 1.476 to 1.478 at 25 °C/D
Flash Point: 128 °C Open Cup

RTECS Toxicity Data

Acute Oral: Rat LD_{50} Dose: 1250 mg/kg.

Hazard Overviews

Health: Irritating to eyes/skin/respiratory tract. Harmful. Other Acute Effects: harmful if swallowed; may be harmful if inhaled; may be harmful if absorbed through the skin.
Fire: Will burn. Extinguishing agents: water spray; carbon dioxide, dry chemical powder or appropriate foam. Precautions: combustible liquid.
Reactivity: Incompatible with: strong oxidizing agents, strong reducing agents, acid chlorides, acid anhydrides, chloroformates, halogens, phosphorus, halides. Hazardous decomposition products: toxic fumes of: carbon monoxide, carbon dioxide.
Carcinogenicity: IARC - Not listed; NIOSH - Not listed; NTP - Not listed; ACGIH - Not listed; OSHA - Not listed; EPA - Not listed; MAK - Not listed
Primary Target Organs:

Eyes

Skin

Respiratory System

Environmental

Cleanup/Disposal: Guide No. 127: Eliminate all ignition sources (no smoking, flares, sparks or flames in immediate area). All equipment used when handling the product must be grounded. Do not touch or walk through spilled material. Stop leak if you can do it without risk. Prevent entry into waterways, sewers, basements or confined areas. A vapor suppressing foam may be used to reduce vapors. Absorb or cover with dry earth, sand or other non-combustible material and transfer to containers. Use clean non-sparking tools to collect absorbed material. Large Spills: Dike far ahead of liquid spill for later disposal. Water spray may reduce vapor; but may not prevent ignition in closed spaces.

Environmental Physical Data

BCF: no food chain concentration potential

Regulations

RCRA 40CFR: Not listed
CERCLA: 40CFR 302.4: Not listed
SARA 40CFR 372.65: Not listed
SARA EHS 40CFR 355: Not listed
TSCA: Listed

DIH7210 CAS #: 32222-06-3

1,2,5-DIHYDROXYCHOLECALCIFEROL

RTECS: FZ4645000
EINECS Number: 250-963-8
Molecular Formula: $C_{27}H_{44}O_3$
Formula Weight: 416.71

Chemical Structure

Synonyms: CALCITRIOL; CHOLECALCIFEROL,1A,25-DIHYDROXY-; 1,25-DHCC; 1,25-DIHYDROXYCHOLECALCIFEROL; 1-ALPHA,25-DIHYDROXYCHOLECALCIFEROL; 1A,25-DIHYDROXYCHOLECALCIFEROL; 1ALPHA,25-DIHYDROXYCHOLECALCIFEROL; 1,25-DIHYDROXYVITAMIN D3; 1-ALPHA,25-DIHYDROXYVITAMIN D3; 1ALPHA,25-DIHYDROXYVITAMIN D3; DIHYDROXYVITAMIN D3; RO 215535; ROCALTROL; (5Z,7E)-9,10-SECOCHESTA-5,7,10(19)-TRIENE-1-ALPHA,3-BETA,25-TRIOL; 9,10-SECOCHOLESTA-5,7,10(19)-TRIENE-1,3,25-TRIOL,(1ALPHA,3BETA,5Z,7E)-; 9,10SECO(5Z,7E)-5,7,10(19)-CHOLESTATRIENE-1ALPHA,3BETA,25-TRIOL; 9,10-SECOCHOLESTA-5,7,10(19)-TRIENE-1,3,25-TRIOL,(1-ALPHA,3-BETA,5Z,7E)-(9CI); SOLTRIOL

Description: colorless, crystalline solid

RTECS Toxicity Data

Acute Oral: Rat LD_{50} Dose: 620 ug/kg. Mouse LD_{50} Dose: 1350 ug/kg; Toxic Effects: Sense organs and special senses - Lacrimation; Behavioral - Ataxia; Nutritional and gross metabolic - Body temperature decrease.
Acute Dermal: Rat LD_{50} Route: Subcutaneous Dose: 66 ug/kg; Toxic Effects: Behavioral - Somnolence (general depressed activity); Behavioral - Ataxia; Nutritional and gross metabolic - Body temperature decrease. Mouse LD_{50} Route: Subcutaneous Dose: 145 ug/kg; Toxic Effects: Sense organs and special senses - Lacrimation; Behavioral - Ataxia; Nutritional and gross metabolic - Body temperature decrease.
Reproductive/Teratogenic: Rat Route: Oral; Dose: 55 ug/kg; Duration: female 7-17D of pregnancy; Effects on Embryo or Fetus - Fetotoxicity. Rat Route: Subcutaneous; Dose: 15 ug/kg; Duration: female 16-21D of pregnancy Effects on Embryo or Fetus - Fetotoxicity; Specific Developmental Abnormalities - Musculoskeletal system; Blood and lymphatic systems (including spleen and marrow).
Mutagenic: Human DNA Inhibition; Cell Type: leukocyte; Dose: 400 nmol/L/4H. Human DNA Inhibition; Cell Type: other cell types; Dose: 100 nmol/L/48H-C. Human Other Mutation Test Systems; Cell Type: other cell types; Dose: 50 nmol/L/96H-C.

Hazard Overviews

Poison

Health: Poison. Other Acute Effects: may be fatal if inhaled, swallowed, or absorbed through skin; lachrymator; symptoms of exposure may include burning sensation; coughing; wheezing; laryngitis; shortness of breath; headache; nausea; vomiting. Chronic Effects: overexposure may cause reproductive disorder(s) based on tests with laboratory animals; target organs: g.i. system, bones, kidneys, central nervous system.
Fire: Extinguishing agents: water spray; carbon dioxide, dry chemical powder or appropriate foam. Precautions: combustible liquid.
Reactivity: Stable. Hazardous polymerization will not occur. Incompatible with: strong oxidizing agents, acid chlorides, acid anhydrides, sensitive to light, may decompose on exposure to air and moisture. Hazardous decomposition products: toxic fumes of: carbon monoxide, carbon dioxide.
Carcinogenicity: IARC - Not listed; NIOSH - Not listed; NTP - Not listed; ACGIH - Not listed; OSHA - Not listed; EPA - Not listed; MAK - Not listed
Primary Target Organs:

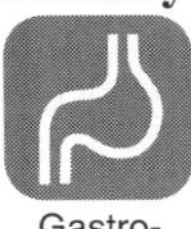
Gastro-intestinal

Nervous System

Kidneys

Bone

Environmental

Regulations

RCRA 40CFR: Not listed
CERCLA: 40CFR 302.4: Not listed
SARA 40CFR 372.65: Not listed
SARA EHS 40CFR 355: Not listed
TSCA: Not listed

DIH7480 CAS #: 2892-51-5

3,4-DIHYDROXY-3-CYCLOBUTENE-1,2-DIONE

RTECS: GU1800000
EINECS Number: 220-761-4
Molecular Formula: $C_4H_2O_4$
Formula Weight: 114.06

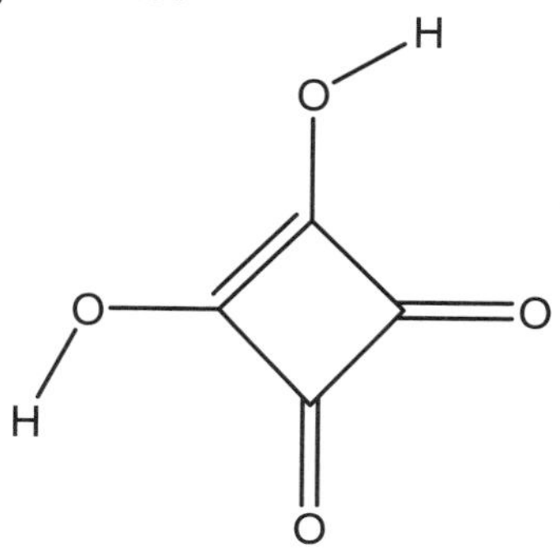

Chemical Structure

Synonyms: 3-CYCLOBUTENE-1,2-DIONE,3,4-DIHYDROXY-(9CI); CYCLOBUTENEDIONE,DIHYDROXY-(8CI); DIHYDROXYCYCLOBUTENEDIONE; QUADRATIC ACID; SQUARIC ACID

RTECS Toxicity Data

Tumorigenic: Mouse Route: Subcutaneous; Dose: 368 mg/kg/92W-I; Toxic Effects: Tumorigenic - Equivocal tumorigenic agent by RTECS criteria; Tumorigenic - Tumors at site of application.

Hazard Overviews

Health: Irritating to eyes/skin/respiratory tract. Other Acute Effects: may be harmful by inhalation, ingestion, or skin absorption; may cause allergic skin reaction.
Fire: Extinguishing agents: carbon dioxide, dry chemical powder or appropriate foam. Precautions: combustible liquid.
Reactivity: Incompatible with: strong oxidizing agents, strong bases. Hazardous decomposition products: toxic fumes of: carbon monoxide, carbon dioxide.
Carcinogenicity: IARC - Not listed; NIOSH - Not listed; NTP - Not listed; ACGIH - Not listed; OSHA - Not listed; EPA - Not listed; MAK - Not listed

Primary Target Organs:

Eyes

Skin

Respiratory System

Environmental

Regulations

RCRA 40CFR: Not listed
CERCLA: 40CFR 302.4: Not listed
SARA 40CFR 372.65: Not listed
SARA EHS 40CFR 355: Not listed
TSCA: Listed

DIH7750 CAS #: 2664-63-3

4,4-DIHYDROXYDIPHENYLSULFIDE

RTECS: SN0800000
EINECS Number: 220-197-9
Molecular Formula: $C_{12}H_{10}O_2S$
Formula Weight: 218.28

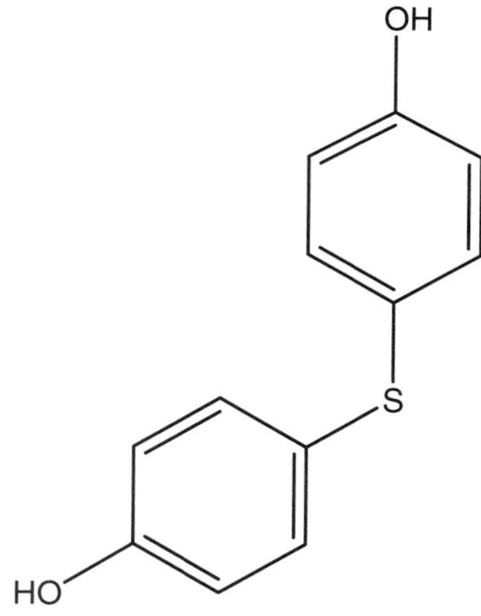

Chemical Structure

Synonyms: DFS; 4,4'-DIHYDROXYDIPHENYL SULFIDE; 4,4'-DIOXYDIPHENYLSULFIDE; SULFIDE,BIS(4-HYDROXYPHENYL); 4,4'-THIODIPHENOL

Physical Properties

Freezing Point: 151 °C (303.8 °F)
Water Solubility: Slightly Soluble
Other Solubilities: Ethanol: Slightly soluble; Ether: Slightly soluble; CS_2: Slightly soluble

RTECS Toxicity Data

Acute Oral: Rat LD_{50} Dose: 3362 mg/kg. Mouse LD_{50} Dose: 5500 mg/kg.
Acute Dermal: Rabbit LD_{50} Route: Skin; Dose: >10250 mg/kg.

Irritation Eye: Rabbit Standard Draize Test Dose: 100 mg; Reaction: severe.
Irritation Skin: Rabbit Standard Draize Test Dose: 500 mg; Reaction: mild.

Hazard Overviews

Corrosive

Health: Corrosive to eyes/skin/respiratory tract. Other Acute Effects: harmful if swallowed, inhaled, or absorbed through skin; inhalation may result in spasm, inflammation and edema of the larynx and bronchi, chemical pneumonitis and pulmonary edema; symptoms of exposure may include burning sensation; coughing; wheezing; laryngitis; shortness of breath; headache; nausea; vomiting.
Fire: Extinguishing agents: water spray; carbon dioxide, dry chemical powder or appropriate foam. Precautions: combustible liquid.
Reactivity: Incompatible with: strong oxidizing agents, strong bases. Hazardous decomposition products: toxic fumes of: carbon monoxide, carbon dioxide, sulfur oxides
Carcinogenicity: IARC - Not listed; NIOSH - Not listed; NTP - Not listed; ACGIH - Not listed; OSHA - Not listed; EPA - Not listed; MAK - Not listed
Primary Target Organs:

Eyes

Skin

Respiratory System

Environmental

Regulations

RCRA 40CFR: Not listed
CERCLA: 40CFR 302.4: Not listed
SARA 40CFR 372.65: Not listed
SARA EHS 40CFR 355: Not listed
TSCA: Listed

DIH8020 CAS #: 120-40-1

N,N-DI(2-HYDROXYETHY)LAURAMIDE

RTECS: JR1925000
EINECS Number: 204-393-1
Molecular Formula: $C_{16}H_{33}NO_3$
Structured MF: $CH_3(CH_2)_{10}CON(CH_2CH_2OH)_2$
Formula Weight: 287.50
Synonyms: N,N-BIS(2-HYDROXYETHYL)DODECANAMIDE; BIS(2-HYDROXYETHYL)LAURAMIDE; N,N-BIS(2-HYDROXYETHYL)LAURAMIDE; N,N-BIS(BETA-HYDROXYETHYL)LAURAMIDE; N,N-BIS(HYDROXYETHYL)LAURAMIDE; N,N-BIS(2-HYDROXYETHYL)LAUROYLAMIDE; N,N-BIS(2-HYDROXYETHYL)LAURYLAMIDE; CLINDROL 200L; CLINDROL 101CG; CLINDROL 203CG; CLINDROL 210CGN; CLINDROL 200 L; CLINDROL SUPERAMIDE 100L; COCO DIETHANOLAMIDE; COCONUT OIL AMIDE OF DIETHANOLAMINE; COMPERLAN LD; CONDENSATE PL; CRILLON L.D.E; CRILLON LDE; DIETHANOLLAURAMIDE; N,N-DIETHANOLLAURAMIDE; N,N-DIETHANOLLAURIC ACID AMIDE; N,N-DIETHYLOLLAURAMIDE; DODECANAMIDE,N,N-BIS(2-HYDROXYETHYL)-; EMID 6511; EMID 6541; ETHYLAN MLD; HETAMIDE ML; LAURAMIDE DEA; LAURIC ACID DIETHANOLAMIDE; LAURIC ACID DIETHANOLAMINE CONDENSATE; LAURIC DIETHANOLAMIDE; LAUROYL DIETHANOLAMIDE; LAUROYLDIETHANOLAMINE; LAURYL DIETHANOLAMIDE; LDA; LDE; MONAMID 150-LW; MONAMIDE 150LW; NINOL P-621; NINOL 4821; NINOL AA-62 EXTRA; NINOL AA62; ONYXOL 345; REWOMID DL 203/S; REWOMID DLMS; RICHAMIDE 6310; RICHAMIDE STD; ROLAMID CD; STANDAMID LD; STANDAMIDD LD; STEINAMID DL 203 S; STEPAN LDA; SUPER AMIDE L-9A; SUPER AMIDE L-9C; SYNOTOL L-60; UNAMIDE J-56; VARAMID ML 1; VARAMIDE ML 1
Description: amber waxy liquid
Use: cosmetic ingredient

Physical Properties

Boiling Point: 239 °C (462 °F) to 244 °C (471 °F)
Freezing Point: 38.7 °C (101.66 °F)
Density: About 8.1 lb/gal
Water Solubility: Insoluble in Water
Other Solubilities: 95% Ethanol: >=100 mg/ml at 23 °C; Acetone: <1 mg/ml at 23 °C; Absolute Ethanol: Soluble; DMSO: <1 mg/ml at 19 °C.
pH: 1% solution 9 to 11
Flash Point: Not available; probably combustible

RTECS Toxicity Data

Acute Oral: Rat LD_{50} Dose: 2700 mg/kg.

Hazard Overviews

Fire: Will burn.
Carcinogenicity: IARC - Not listed; NIOSH - Not listed; NTP - Not listed; ACGIH - Not listed; OSHA - Not listed; EPA - Not listed; MAK - Not listed

Environmental

Environmental Fate: If released to soil, the expected mobility is medium to high. If released to water, it will be essentially nonvolatile. Aquatic bioconcentration and adsorption to sediment are not expected to be important fate processes. In natural river water, it degraded 98% in 6 days (20 mg/l). If released to the atmosphere, it will exist primarily in the particulate phase. In the vapor phase, it will degrade in the atmosphere by reaction with photochemically produced hydroxyl radicals with an estimated half-life of approximately 8 hours. Physical removal from air can occur through wet and dry deposition.

Environmental Physical Data

Henry's Law Constant: estimated at 2.16×10^{-12}
Sorption Partition Coefficient: $K_{OC} = 52$
BCF: estimated at 29

Regulations

RCRA 40CFR: Not listed

CERCLA: 40CFR 302.4: Not listed
SARA 40CFR 372.65: Not listed
SARA EHS 40CFR 355: Not listed
TSCA: Listed

DIH8290 CAS #: 131-53-3

2,2-DIHYDROXY-4-METHOXY BENZOPHENONE

RTECS: DJ1049500
EINECS Number: 205-026-8
Molecular Formula: $C_{14}H_{12}O_4$
Formula Weight: 244.26

Chemical Structure

Synonyms: ADVASTAB 47; BENZOPHENONE-8; CYASORB; CYASORB UV 24; CYASORB UV 24 LIGHT ABSORBER; 2,2'-DIHYDROXY-4-METHOXYBENZOPHENONE; DIOXYBENZON; DIOXYBENZONE; (2-HYDROXY-4-METHOXYPHENYL)(2-HYDROXYPHENYL)METHANONE; METHANONE,(2-HYDROXY-4-METHOXYPHENYL)(2-HYDROXYPHENYL)-(9CI); SPECTRA-SORB UV 24; UF 2; UV 24
Description: yellow powder
Use: ultraviolet sunscreen

Physical Properties

Boiling Point: 170 °C (338 °F) to 175 °C (347 °F)
Other Solubilities: 95% Ethanol: Very Soluble.
Flash Point: Not available; probably combustible

RTECS Toxicity Data

Mutagenic: Mouse Mutations in Microorganisms; Cell Type: lymphocyte; Dose: 32 ug/plate (+S9). Bacteria - S Typhimurium Mutations in Microorganisms; Dose: 12500 ug/L (+S9). Bacteria - S Typhimurium Mutations in Microorganisms; Dose: 3 ug/plate (-S9).

Hazard Overviews

Health: Irritating to eyes/skin/respiratory tract. Other Acute Effects: may be harmful by inhalation, ingestion, or skin absorption.
Fire: Will burn. Hazards: emits toxic fumes. Extinguishing agents: water spray; carbon dioxide, dry chemical powder or appropriate foam. Precautions: combustible liquid.
Reactivity: Incompatible with: strong oxidizing agents. Hazardous decomposition products: toxic fumes of: carbon monoxide, carbon dioxide.

Carcinogenicity: IARC - Not listed; NIOSH - Not listed; NTP - Not listed; ACGIH - Not listed; OSHA - Not listed; EPA - Not listed; MAK - Not listed
Primary Target Organs:

Eyes

Skin

Respiratory System

Environmental

Regulations
RCRA 40CFR: Not listed
CERCLA: 40CFR 302.4: Not listed
SARA 40CFR 372.65: Not listed
SARA EHS 40CFR 355: Not listed
TSCA: Listed

DIH8560 CAS #: 2107-76-8

5,7-DIHYDROXY-4-METHYLCOUMARIN

EINECS Number: 218-289-9
Molecular Formula: $C_{10}H_8O_4$
Formula Weight: 192.18

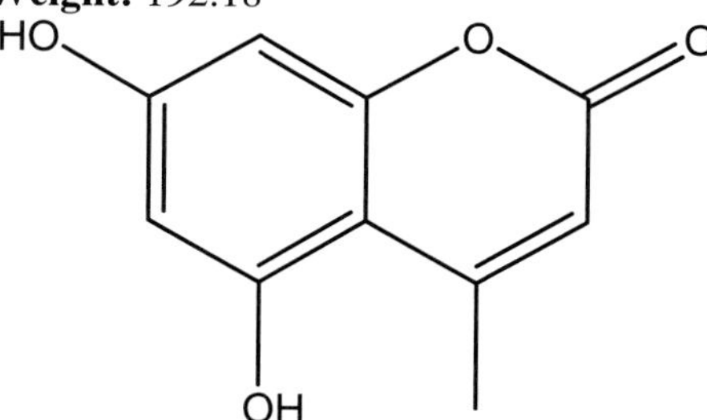

Chemical Structure

Synonyms: 2H-1-BENZOPYRAN-2-ONE,5,7-DIHYDROXY-4-METHYL-; COUMARIN,5,7-DIHYDROXY-4-METHYL-; 5,7-DIHYDROXY-4-METHYL-2H-1-BENZOPYRAN-2-ONE; 5-HYDROXY-4-METHYLUMBELLIFERONE; 4-METHYL-5,7-DIHYDROXYCOUMARIN
Description: yellow to white solid, needles, or leaves
Use: in suntan oils as a sunscreen; in wall paints as a whitening agent

Physical Properties

Freezing Point: 282 °C (539.6 °F) to 284 °C (543.2 °F)
Water Solubility: Slightly Soluble in Water
Other Solubilities: 95% Ethanol: Very Soluble in hot solvent; Benzene: Slightly Soluble; Chloroform: Slightly Soluble; Ether: Slightly Soluble; Toluene: Slightly Soluble.
Flash Point: Not available; probably combustible

Hazard Overviews

Fire: Will burn.

Carcinogenicity: IARC - Not listed; NIOSH - Not listed; NTP - Not listed; ACGIH - Not listed; OSHA - Not listed; EPA - Not listed; MAK - Not listed

Environmental

Regulations

RCRA 40CFR: Not listed
CERCLA: 40CFR 302.4: Not listed
SARA 40CFR 372.65: Not listed
SARA EHS 40CFR 355: Not listed
TSCA: Listed

DIH9100 CAS #: 66-22-8

2,4-DIHYDROXYPYRIMIDINE

RTECS: YQ8650000
EINECS Number: 200-621-9
Molecular Formula: $C_4H_4N_2O_2$
Formula Weight: 112.10

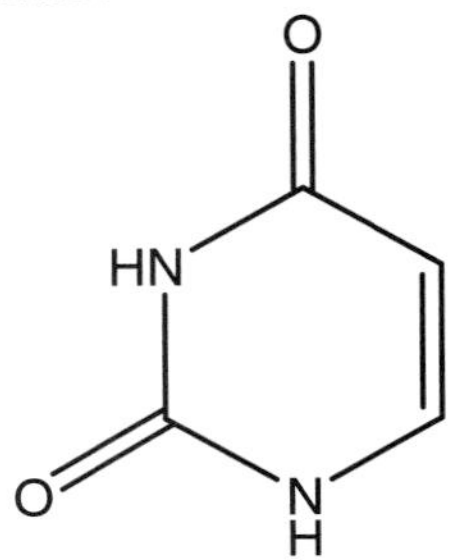

Chemical Structure

Synonyms: 2,4-DIOXOPYRIMIDINE; HYBAR X; PIROD; 2,4-PYRIMIDINEDIOL; 2,4-PYRIMIDINEDIONE; 2,4(1H,3H)-PYRIMIDINEDIONE (9CI); PYROD

Physical Properties

Freezing Point: 335 °C (635 °F)
Water Solubility: Slightly Soluble
Other Solubilities: Ethanol: Very Soluble; Ether: Very Soluble; dilute NH_3: Soluble
Ionization Potential (eV): 9.2 +/-0.5

RTECS Toxicity Data

Acute Oral: Rat LD_{50} Dose: >6 gm/kg. Mouse LD_{50} Dose: >8 gm/kg.

Reproductive/Teratogenic: Rat Route: Oral; Dose: 18 gm/kg; Duration: female 17-22D of pregnancy Specific Developmental Abnormalities - Blood and lymphatic systems (including spleen and marrow); Urogenital system; Effects on Newborn - Behavioral. Rat Route: Oral; Dose: 15400 mg/kg; Duration: female 7-17D of pregnancy; Effects on Embryo or Fetus - Fetotoxicity; Specific Developmental Abnormalities - Musculoskeletal system. Rat Route: Oral; Dose: 616 mg/kg; Duration: female 7-17D of pregnancy; Effects on Newborn - Sex ratio.

Mutagenic: Mouse Cytogenetic Analysis; Route: Intraperitoneal; Dose: 15 mg/kg. Bacteria - E Coli Phage Inhibition; Dose: 1 gm/L.

Tumorigenic: Rat Route: Oral; Dose: 131 gm/kg/2Y-C; Toxic Effects: Tumorigenic - Carcinogenic by RTECS criteria; Kidney, Ureter, and Bladder - Tumors; Kidney, Ureter, and Bladder - Kidney tumors. Rat Route: Oral; Dose: 378 gm/kg/30W-C; Toxic Effects: Tumorigenic - Equivocal tumorigenic agent by RTECS criteria; Kidney, Ureter, and Bladder - Tumors. Rat Route: Oral; Dose: 454 gm/kg/36W-C; Toxic Effects: Tumorigenic - Equivocal tumorigenic agent by RTECS criteria; Kidney, Ureter, and Bladder - Tumors. Rat Route: Oral; Dose: 235 gm/kg/20W-C; Toxic Effects: Tumorigenic - Equivocal tumorigenic agent by RTECS criteria; Kidney, Ureter, and Bladder - Tumors. Rat Route: Oral; Dose: 189 gm/kg/15W-C; Toxic Effects: Tumorigenic - Equivocal tumorigenic agent by RTECS criteria; Kidney, Ureter, and Bladder - Tumors.

Hazard Overviews

Health: May cause irritation. Other Acute Effects: may be harmful by inhalation, ingestion, or skin absorption.

Fire: Hazards: emits toxic fumes. Extinguishing agents: water spray; carbon dioxide, dry chemical powder or appropriate foam. Precautions: combustible liquid.

Reactivity: Incompatible with: strong oxidizing agents. Hazardous decomposition products: toxic fumes of: carbon monoxide, carbon dioxide, nitrogen oxides.

Carcinogenicity: IARC - Not listed; NIOSH - Not listed; NTP - Not listed; ACGIH - Not listed; OSHA - Not listed; EPA - Not listed; MAK - Not listed

Environmental

Environmental Physical Data

Octanol/Water Partition Coefficient: log K_{ow} = calculated at -1.07

Regulations

RCRA 40CFR: Not listed
CERCLA: 40CFR 302.4: Not listed
SARA 40CFR 372.65: Not listed
SARA EHS 40CFR 355: Not listed
TSCA: Listed

DII1200 CAS #: 83-73-8

DIIODOHYDROXYQUIN

RTECS: VC5775000
EINECS Number: 201-497-9
Molecular Formula: $C_9H_5I_2NO$
Formula Weight: 396.98

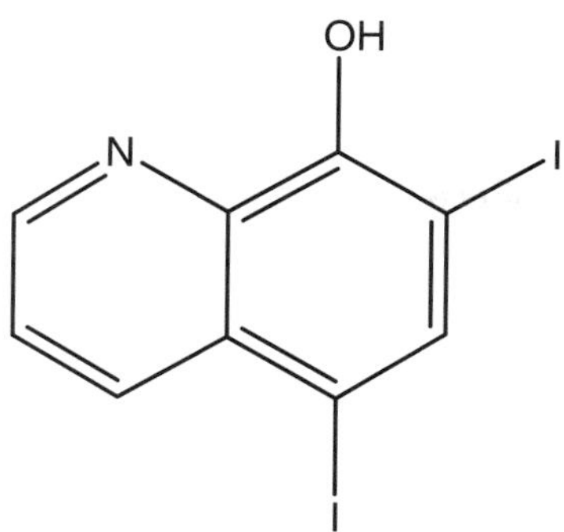

Chemical Structure

Synonyms: DIAMOEBIN; 5,7-DIIODO-8-HYDROXYQUINOLINE; DIIODOHYDROXYQUINOLINE; 5,7-DIIODO-OXINE; DIIODOQUIN; 5,7-DIIODO-8-QUINOLINOL; DINOLEINE; DIODOHYDROXYQUIN; DIODOQUIN; DIODOQUINE; 5,7-DIODO-8-QUINOLINOL; DIODOXYLIN; DI-QUINOL; DIREXIODE; DISOQUIN; DYODIN; EMBEQUIN; ENTERODIAMOEBIN; ENTEROSEPT; FLORAQUIN; FLUORAQUIN; 8-HYDROXY-5,7-DIIODOQUINOLINE; IODOQUINOL; IOQUIN; IOQUIN SUSPENSION; LANODOXIN; MOEBIQUIN; QUINADOME; 8-QUINOLINOL,5,7-DIIODO-; RAFAMEBIN; SEARLEQUIN; SEBAQUIN; SS 578; STANQUINATE; YODOXIN; ZOAQUIN

Description: light yellowish to tan, microcrystalline powder or crystals, medicinal grade is yellowish-brown powder; odorless or nearly so

Use: medication as antiamebic

Physical Properties

Freezing Point: 200 °C (392 °F) to 215 °C (419 °F)

Water Solubility: Almost Insoluble in Water

Other Solubilities: Sparingly Soluble in Alcohol, Ether, Acetone. Soluble in hot Pyridine & hot Dioxane

RTECS Toxicity Data

Acute Oral: Child TD_{Lo} Dose: 819 mg/kg/13D-I; Toxic Effects: Behavioral - Somnolence (general depressed activity); Behavioral - Convulsions or effect on seizure threshold; Blood - Oxidant related (GPD deficient) anemia. Child TD_{Lo} Dose: 120 gm/kg/2Y-I; Toxic Effects: Sense organs and special senses - Visual field changes.

Mutagenic: Mouse Micronucleus Test; Route: Oral; Dose: 80 mg/kg. Bacteria - E Coli DNA Repair; Dose: 260 nmol/plate.

Hazard Overviews

Health: Irritating to eyes/skin/respiratory tract. Harmful. Other Acute Effects: harmful if swallowed, inhaled, or absorbed through skin; may cause allergic skin reaction; may cause nervous system disturbances; target organs: eyes, nervous system. Chronic Effects: optic neuritis, optic atrophy and peripheral neuropathy have been reported on long term high dose usage; adverse reactions include fever, chills, headache, vertigo, thyroid enlargement, nausea, vomiting, diarrhea, abdominal cramps, pruritus ani, urticaria, and skin eruptions.

Fire: Hazards: emits toxic fumes. Extinguishing agents: water spray; carbon dioxide, dry chemical powder or appropriate foam. Precautions: combustible liquid.

Reactivity: Incompatible with: strong oxidizing agents, strong acids. Hazardous decomposition products: toxic fumes of: carbon monoxide, carbon dioxide, nitrogen oxides, hydrogen iodide.

Carcinogenicity: IARC - Not listed; NIOSH - Not listed; NTP - Not listed; ACGIH - Not listed; OSHA - Not listed; EPA - Not listed; MAK - Not listed

Primary Target Organs:

Eyes

Skin

Respiratory System

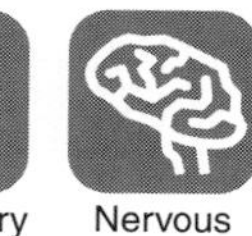
Nervous System

Environmental

Regulations

RCRA 40CFR: Not listed

CERCLA: 40CFR 302.4: Not listed

SARA 40CFR 372.65: Not listed

SARA EHS 40CFR 355: Not listed

TSCA: Listed

DII1600 CAS #: 1310-43-6

DIIRON MONOPHOSPHIDE

EINECS Number: 215-178-7

Molecular Formula: Fe_2P

Structured MF: Fe_2P

Formula Weight: 142.67

Synonyms: DI-IRON PHOSPHIDE; FERROUS PHOSPHIDE

Description: gray, hexagonal needles or blue-gray powder

Use: iron & steel mfr

Physical Properties

Freezing Point: 1420 °C (2588 °F)

Specific Gravity: 5.7

Water Solubility: Insoluble in Water

Other Solubilities: Soluble in mineral acids with the liberation of phosphene.

Hazard Overviews

Carcinogenicity: IARC - Not listed; NIOSH - Not listed; NTP - Not listed; ACGIH - Not listed; OSHA - Not listed; EPA - Not listed; MAK - Not listed

Exposure Limits

OSHA PEL Vacated 1989 Limits: STEL: 1 mg/m^3; as Fe salts.

ACGIH TLV: TWA: 1 mg/m^3; as Fe Salt.

NIOSH REL: TWA: 1 mg/m^3; as Fe salts.

Environmental

Regulations

RCRA 40CFR: Not listed

CERCLA: 40CFR 302.4: Not listed

SARA 40CFR 372.65: Not listed

SARA EHS 40CFR 355: Not listed
TSCA: Listed

DII1800 CAS #: 544-01-4

DI-ISOAMYL ETHER

EINECS Number: 208-857-4
Molecular Formula: $C_{10}H_{22}O$
Formula Weight: 158.28

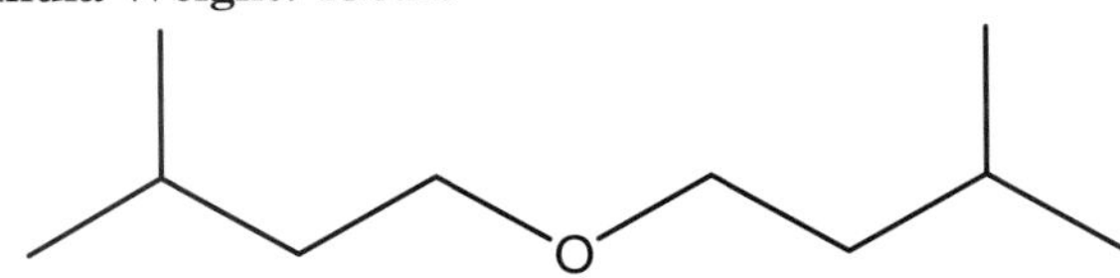

Chemical Structure

Synonyms: BUTANE,1,1'-OXYBIS(3-METHYL-; DIISOAMYL ETHER; DIISOPENTYL ETHER; ISOAMYL ETHER; ISOAMYL OXIDE; ISOPENTYL ETHER; 1,1'-OXYBIS(3-METHYLBUTANE)
Description: colorless liquid; pleasant, fruity odor
Use: solvent in Grignard reaction; solvent of odorous principles; mfr lacquers; regenerating rubber

Physical Properties

Boiling Point: 172 °C (342 °F)
Specific Gravity: 0.783 at 12 °C/4 °C
Saturated Vapor Density: 1.211089819 kg/m^3
Vapor Pressure: 0.210 kPa at 25 °C
Water Solubility: Insoluble in Water
Other Solubilities: miscible with Alcohol, Chloroform, Ether
Refraction Index: 1.408 at 20 °C/D
Flash Point: Combustible

Hazard Overviews

Health: Irritating. Other Acute Effects: may be harmful by inhalation, ingestion, or skin absorption.
Fire: Combustible. Hazards: emits toxic fumes; vapor may travel considerable distance to source of ignition and flash back. Extinguishing agents: carbon dioxide, dry chemical powder or appropriate foam; water spray. Precautions: combustible liquid.
Reactivity: Incompatible with: strong oxidizing agents. Hazardous decomposition products: toxic fumes of: carbon monoxide, carbon dioxide.
Carcinogenicity: IARC - Not listed; NIOSH - Not listed; NTP - Not listed; ACGIH - Not listed; OSHA - Not listed; EPA - Not listed; MAK - Not listed
Primary Target Organs:

Eyes

Skin

Respiratory System

Environmental

Environmental Fate: If released to soil, it will be subject to volatilization. It will be expected to exhibit moderate mobility in soil and, therefore, it may leach to groundwater. It will not be expected to hydrolyze in soil. If released to water, it will not be expected to significantly adsorb to sediment or suspended particulate matter, bioconcentrate in aquatic organisms, hydrolyze, directly photolyze, or photooxidize via reaction with photochemically produced hydroxyl radicals in the water, based upon estimated physical-chemical properties or analogies to other structurally related aliphatic ethers. In surface water it will be subject to rapid volatilization with estimated half-lives of 4.4 hr and 20 hr for volatilization from a river one meter deep flowing 1 m/sec with a wind velocity of 3 m/sec and a model pond. It may be resistant to biodegradation in environmental media based upon screening test data from studies using sewage inocula. Many ethers are known to be resistant to biodegradation. If released to the atmosphere, it will be expected to exist almost entirely in the vapor phase based on its vapor pressure. It will be susceptible to photooxidation via vapor phase reaction with photochemically produced hydroxyl radicals with a half-life of 18 hours estimated for this process. Direct photolysis will not be an important removal process since aliphatic ethers do not absorb light at wavelengths >290 nm.

Environmental Physical Data

Henry's Law Constant: calculated at 1.46 $x10^{-3}$
Sorption Partition Coefficient: K_{oc} = estimated at 237
BCF: estimated at 31

Regulations

RCRA 40CFR: Not listed
CERCLA: 40CFR 302.4: Not listed
SARA 40CFR 372.65: Not listed
SARA EHS 40CFR 355: Not listed
TSCA: Listed

DII2000 CAS #: 141-04-8

DIISOBUTYL ADIPATE

RTECS: AV1480000
EINECS Number: 205-450-3
Molecular Formula: $C_{14}H_{26}O_4$
Formula Weight: 258.36

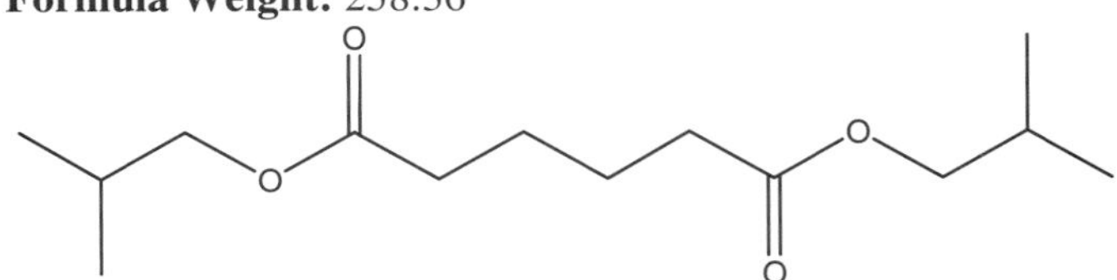

Chemical Structure

Synonyms: ADIPIC ACID BIS(2-METHYLPROPYL) ESTER; ADIPIC ACID,DIISOBUTYL ESTER; DIBA; FTAFLEX DIBA; HEXANEDIOIC ACID,BIS(2-METHYLPROPYL) ESTER; HEXANEDIOIC ACID,BIS(2-METHYLPROPYL) ESTER (9CI); ISOBUTYL ADIPATE

Description: colorless liquid; odorless
Use: plasticizer

Physical Properties

Boiling Point: 278 °C (532 °F) to 280 °C (536 °F)
Freezing Point: -20 °C (-4 °F)
Specific Gravity: 0.95 at 25 °C
Water Solubility: Insoluble in Water
Other Solubilities: Soluble in most organic solvents
Refraction Index: 1.4301
Flash Point: Combustible

RTECS Toxicity Data

Acute Oral: Guinea Pig LD_{50} Dose: 12300 uL/kg.
Reproductive/Teratogenic: Rat Route: Intraperitoneal; Dose: 1190 mg/kg; Duration: female 5-15D of pregnancy; Effects on Embryo or Fetus - Fetotoxicity. Rat Route: Intraperitoneal; Dose: 595 mg/kg; Duration: female 5-15D of pregnancy; Specific Developmental Abnormalities - Other developmental abnormalities.

Hazard Overviews

Health: May cause irritation to eyes/skin/respiratory tract. Other Acute Effects: may be harmful by inhalation, ingestion, or skin absorption.
Fire: Combustible. Hazards: emits toxic fumes. Extinguishing agents: carbon dioxide; dry chemical powder; foam and water spray are effective but may cause frothing. Precautions: combustible liquid.
Reactivity: Stable. Hazardous polymerization will not occur. Incompatible with: strong oxidizing agents, strong acids, strong bases. Hazardous decomposition products: toxic fumes of: carbon monoxide, carbon dioxide.
Carcinogenicity: IARC - Not listed; NIOSH - Not listed; NTP - Not listed; ACGIH - Not listed; OSHA - Not listed; EPA - Not listed; MAK - Not listed

Environmental

Regulations

RCRA 40CFR: Not listed
CERCLA: 40CFR 302.4: Not listed
SARA 40CFR 372.65: Not listed
SARA EHS 40CFR 355: Not listed
TSCA: Listed

DII2200 **CAS #: 108-82-7**

DIISOBUTYL CARBINOL

RTECS: MJ3325000
EINECS Number: 203-619-6
Molecular Formula: $C_9H_{20}O$
Structured MF: $[(CH_3)_2CHCH_2]_2CHOH$
Formula Weight: 144.26

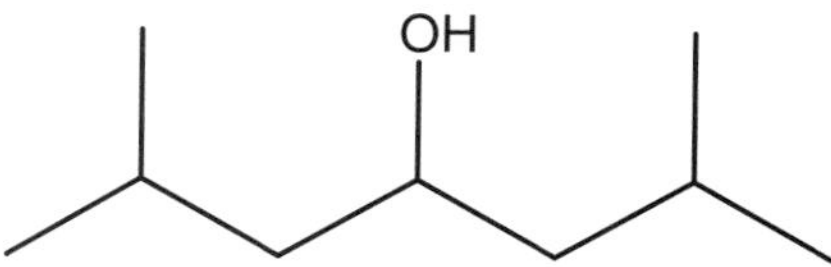

Chemical Structure

Synonyms: DIISOBUTYLCARBINOL; 2,6-DIMETHYL HEPTANOL-4; 2,6-DIMETHYL-4-HEPTANOL; 4-HEPTANOL,2,6-DIMETHYL-; 4-HYDROXY-2,6-DIMETHYL HEPTANE; SEC-NONYL ALCOHOL
Description: colorless liquid
Use: solvent for coating compositions of urea or melamine resins & for prepn of lubricant additives & plasticizers; surface-active agent; lubricant additives, rubber chemicals, flotation agents, antifoam agent; flavoring ingredient; reaction medium for hydrogen peroxide production; defoamer

Physical Properties

Boiling Point: 176 °C (349 °F) to 177 °C (351 °F) at 760 mm Hg
Freezing Point: < -65 °C (-85 °F)
Specific Gravity: 0.809 at 21 °C/4 °C
Vapor Density: 4.8 Air=1
Vapor Pressure: 0.06 psia
Water Solubility: Insoluble in Water
Other Solubilities: Soluble in Alcohol, Ether
Refraction Index: 1.423 at 21 °C
Flash Point: 72.222 °C Open Cup
Autoignition Temperature: 257 °C
LEL: 0.8% v/v
UEL: 6.1% v/v

RTECS Toxicity Data

Acute Oral: Rat LD_{50} Dose: 3560 mg/kg. Mouse LD_{50} Dose: 3530 mg/kg.
Acute Dermal: Rabbit LD_{50} Route: Skin; Dose: 4600 mg/kg.
Irritation Eye: Rabbit Standard Draize Test Dose: 500 mg/24H; Reaction: mild.
Irritation Skin: Rabbit Open Draize Test Dose: 10 mg/24H open; Reaction: mild. Rabbit Open Draize Test Dose: 500 mg open; Reaction: mild.

Hazard Overviews

Health: Irritating to eyes/skin. Other Acute Effects: may be harmful by inhalation, ingestion, or skin absorption.
Fire: Combustible. Hazards: emits toxic fumes. Extinguishing agents: carbon dioxide, dry chemical powder or appropriate foam; water spray. Precautions: combustible liquid.
Reactivity: Incompatible with: strong oxidizing agents. Hazardous decomposition products: toxic fumes of: carbon monoxide, carbon dioxide.
Carcinogenicity: IARC - Not listed; NIOSH - Not listed; NTP - Not listed; ACGIH - Not listed; OSHA - Not listed; EPA - Not listed; MAK - Not listed

Primary Target Organs:

Eyes

Skin

Environmental

Cleanup/Disposal: Guide No. 127: Eliminate all ignition sources (no smoking, flares, sparks or flames in immediate area). All equipment used when handling the product must be grounded. Do not touch or walk through spilled material. Stop leak if you can do it without risk. Prevent entry into waterways, sewers, basements or confined areas. A vapor suppressing foam may be used to reduce vapors. Absorb or cover with dry earth, sand or other non-combustible material and transfer to containers. Use clean non-sparking tools to collect absorbed material. Large Spills: Dike far ahead of liquid spill for later disposal. Water spray may reduce vapor; but may not prevent ignition in closed spaces.

Environmental Physical Data

BCF: no food chain concentration potential

Regulations

RCRA 40CFR: Not listed
CERCLA: 40CFR 302.4: Not listed
SARA 40CFR 372.65: Not listed
SARA EHS 40CFR 355: Not listed
TSCA: Listed

Analytical Methods

Water / Groundwater: ASTM D3695

DII2400 CAS #: 108-83-8

DIISOBUTYL KETONE

RTECS: MJ5775000
DOT: UN1157; IMO3.0
EINECS Number: 203-620-1
Molecular Formula: $C_9H_{18}O$
Structured MF: $((CH_3)_2CHCH_2)_2CO$
Formula Weight: 142.24

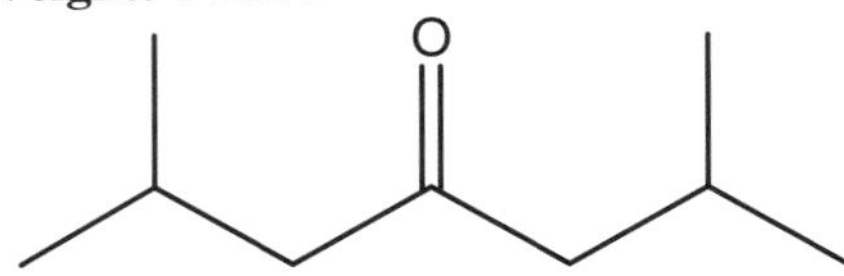

Chemical Structure

Synonyms: DIBK; DIISOBUTILCHETONE; DI-ISOBUTYLCETONE; DIISOBUTYLKETON; SEC-DIISOPROPYL ACETONE; SYM-DIISOPROPYL ACETONE; DIISOPROPYLACETONE; S-DIISOPROPYLACETONE; SYM-DIISOPROPYLACETONE; 2,6-DIMETHYL-HEPTAN-4-ON; 2,6-DIMETHYL-4-HEPTANONE; 2,6-DIMETHYLHEPTAN-4-ONE; 2,6-DIMETIL-EPTAN-4-ONE; 4-HEPTANONE,2,6-DIMETHYL-; ISOBUTYL KETONE; ISOVALERONE; VALERONE

Description: colorless liquid; peppermint odor

Use: solvent for nitrocellulose, rubber and synthetic resins, lacquers, coating compositions, roll-coating inks, stains and organic synthesis

Physical Properties

Boiling Point: 168 °C (334 °F)
Freezing Point: -41.5 °C (-42.7 °F)
Specific Gravity: 0.8053 at 20 °C/4 °C
Vapor Density: 4.9 Air=1
Density: 0.81 g/mL at 20 °C
Vapor Pressure: 2 mm Hg
Water Solubility: 0.05% by weight
Other Solubilities: 95% Ethanol: Very Soluble; Acetone: Very Soluble; Benzene: Very Soluble; Ether: Very Soluble.
Surface Tension: 23.92 dynes/cm at 22 °C
Odor Threshold: 0.660 to 1.8600 mg/m^3
Refraction Index: 1412 at 20 °C
Evaporation Rate: 30.8 Ether=1
Ionization Potential (eV): 9.04
Flash Point: 49 °C Tag Closed Cup
Autoignition Temperature: 396 °C
LEL: 0.8% v/v
UEL: 7.1% v/v at 93 °C

RTECS Toxicity Data

Acute Oral: Rat LD_{50} Dose: 5750 mg/kg. Mouse LD_{50} Dose: 1416 mg/kg.
Acute Inhalation: Rat LC_{Lo} Dose: 2000 ppm/4hr. Human TC_{Lo} Dose: 50 ppm; Toxic Effects: Sense organs and special senses - Other; Behavioral - Headache; Gastrointestinal - Nausea or vomiting.
Acute Dermal: Rabbit LD_{50} Route: Skin; Dose: 16 gm/kg.
Chronic (Multiple Dose) Oral: Rat Dose: 30 gm/kg/3W-I; Toxic Effects: Behavioral - Somnolence (general depressed activity); Liver - Changes in liver weight. Rat Dose: 130 gm/kg/13W-I; Toxic Effects: Liver - Changes in liver weight; Endocrine - Hypoglycemia; Nutritional and gross metabolic - Weight loss or decreased weight gain.
Chronic (Multiple Dose) Inhalation: Rat Dose: 98 ppm/6H/9D-I; Toxic Effects: Behavioral - Fluid intake; Kidney, Ureter, and Bladder - Urine volume increased; Kidney, Ureter, and Bladder - Other changes in urine composition. Rat Dose: 250 ppm/7H/6W-I; Toxic Effects: Liver - Changes in Liver weight; Kidney, Ureter, and Bladder - Changes in kidney weight.
Irritation Eye: Human Standard Draize Test Dose: 25 ppm/15M; Reaction: mild. Rabbit Standard Draize Test Dose: 500 mg; Reaction: mild.
Irritation Skin: Rabbit Open Draize Test Dose: 10 mg/24H open; Reaction: mild. Rabbit Open Draize Test Dose: 500 mg open; Reaction: mild.

Hazard Overviews

Flammable

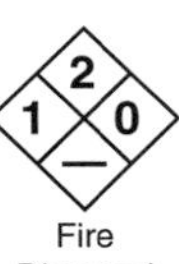

Fire Diamond

Health: Irritating to eyes/skin/respiratory tract. Also Causes: headache, dizziness, shortness of breath, nausea, fainting, narcosis, coma, defatting, skin cracking. Gl irritation upon ingestion. Chronic Effects: dermatitis.

Fire: Flammable. Can form explosive mixtures in the air. For small fires use dry chemical, carbon dioxide, or alcohol-resistant foam. For large fires use water spray, fog, or alcohol-resistant foam. Water may be ineffective. Containers may explode in heat of fire.

Reactivity: Stable. Hazardous polymerization cannot occur. Avoid: heat; ignition sources. Incompatible with: oxidizers; plastic; rubber; coatings. Hazardous decomposition products: acrid smoke; carbon oxide(s).

Carcinogenicity: IARC - Not listed; NIOSH - Listed as carcinogen; NTP - Not listed; ACGIH - Not listed; OSHA - Not listed; EPA - Not listed; MAK - Not listed

Primary Target Organs:

Eyes

Skin

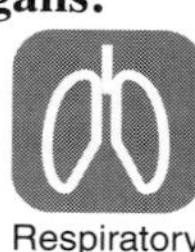
Respiratory System

Nervous System

Exposure Limits

OSHA PEL: TWA: 50 ppm; 290 mg/m^3.

OSHA PEL Vacated 1989 Limits: TWA: 25 ppm; 150 mg/m^3.

ACGIH TLV: TWA: 25 ppm; 145 mg/m^3.

NIOSH REL: TWA: 25 ppm; 150 mg/m^3.

NIOSH IDLH: 500 ppm.

DFG MAK: TWA: 50 ppm; 290 mg/m^3.

Respirator Recommendation

Exposure Range: >50 to <500 ppm Air Purifying, Negative Pressure, Half Mask

Exposure Range: 500 to unlimited ppm Self-contained Breathing Apparatus, Pressure Demand, Full Face

Cartridge Color: black

Environmental

Ecotoxicity: Aquatic toxicity: 65 ppm/24 hr/brine shrimp/TLm

Environmental Fate: If released to the atmosphere, it is degraded relatively rapidly by reaction with photochemically produced hydroxyl radicals (estimated half-life of 14.2 hours in air). The results of several biodegradation screening studies suggest that biodegradation will be an important fate process in water and soil. If released to water, volatilization may be important. Volatilization half-lives of 4.9 and 57 hours have been estimated for a model river (one meter deep) and a model environmental pond, respectively. If released to soil, it may leach significantly based upon an estimated K_{oc} value of 55. It can be expected to evaporate from dry surfaces.

Cleanup/Disposal: Guide No. 127: Eliminate all ignition sources (no smoking, flares, sparks or flames in immediate area). All equipment used when handling the product must be grounded. Do not touch or walk through spilled material. Stop leak if you can do it without risk. Prevent entry into waterways, sewers, basements or confined areas. A vapor suppressing foam may be used to reduce vapors. Absorb or cover with dry earth, sand or other non-combustible material and transfer to containers. Use clean non-sparking tools to collect absorbed material. Large Spills: Dike far ahead of liquid spill for later disposal. Water spray may reduce vapor; but may not prevent ignition in closed spaces.

Environmental Physical Data

Henry's Law Constant: estimated at 7.18 $x10^{-4}$

Sorption Partition Coefficient: K_{oc} = estimated at 155

BCF: estimated at 20

BOD: theoretical 4%, 5 days

Regulations

RCRA 40CFR: Not listed

CERCLA: 40CFR 302.4: Not listed

SARA 40CFR 372.65: Not listed

SARA EHS 40CFR 355: Not listed

TSCA: Listed

Analytical Methods

Air: ASTM D4490

Water / Groundwater: ASTM D3695

Indoor / Expired Air: NIOSH 1300

DII2600 **CAS #: 84-69-5**

DIISOBUTYL PHTHALATE

RTECS: TI1225000

EINECS Number: 201-553-2

Molecular Formula: $C_{16}H_{22}O_4$

Structured MF: o-$C_6H_4[COOCH_2CH(CH_3)_2]_2$

Formula Weight: 278.35

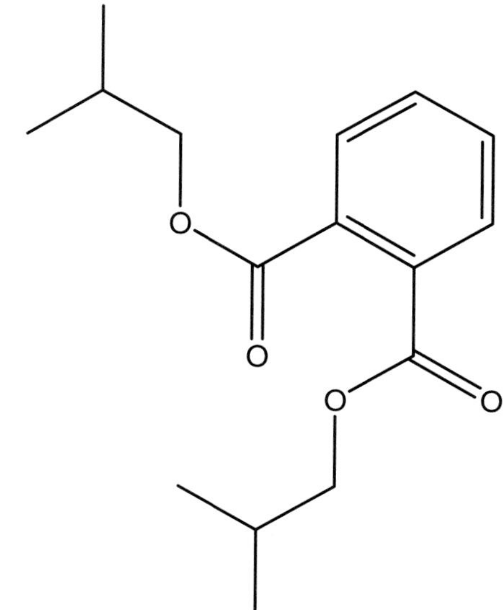

Chemical Structure

Synonyms: 1,2-BENZENEDICARBOXYLIC ACID,BIS(2-METHYLPROPYL) ESTER; DIBP; DI-ISO-BUTYL PHTHALATE; DIISOBUTYLESTER KYSELINY FTALOVE; HATCOL DIBP; HEXAPLAS M/1B; ISOBUTYL PHTHALATE; KODAFLEX DIBP; PALATINOL IC; PHTHALIC ACID,DIISOBUTYL ESTER

Description: liquid
Use: plasticizer for a wide variety of polymers

Physical Properties

Boiling Point: 295 °C (563 °F) to 298 °C (568 °F)
Freezing Point: -64 °C (-83.2 °F)
Specific Gravity: 1.049 at 15 °C
Vapor Density: 9.59 Air=1
Density: 1.04 g/mL at 20 °C
Water Solubility: Insoluble in Water
Other Solubilities: 95% Ethanol: Soluble; Acetone: Soluble; Benzene: Soluble; DMSO: Soluble; Ether: Soluble.
Refraction Index: 1.4900 at 25 °C/D
Flash Point: 185 °C Open Cup
Autoignition Temperature: 432 °C
LEL: 0.4% v/v

RTECS Toxicity Data

Acute Oral: Rat LD_{50} Dose: 15 gm/kg. Mouse LD_{50} Dose: 10 gm/kg; Toxic Effects: Behavioral - Muscle weakness; Behavioral - Coma; Lungs, Thorax, or Respiration - Respiratory stimulation.
Acute Dermal: Guinea Pig LD_{50} Route: Skin; Dose: 10 gm/kg.
Reproductive/Teratogenic: Rat Route: Oral; Dose: 8400 mg/kg; Duration: male 7D prior to mating; Paternal Effects - Testes, epididymis, sperm duct. Rat Route: Intraperitoneal; Dose: 1250 mg/kg; Duration: female 5-15D of pregnancy; Effects on Fertility - Post-implantation mortality. Rat Route: Intraperitoneal; Dose: 375 mg/kg; Duration: female 5-15D of pregnancy; Effects on Embryo or Fetus - Fetotoxicity; Specific Developmental Abnormalities - Musculoskeletal system.

Hazard Overviews

Health: May be irritating to eyes/skin/respiratory tract. Other Acute Effects: may be harmful by inhalation, ingestion, or skin absorption.
Fire: Will burn. Hazards: emits toxic fumes. Extinguishing agents: water spray; carbon dioxide, dry chemical powder or appropriate foam. Precautions: combustible liquid.
Reactivity: Stable. Hazardous polymerization will not occur. Incompatible with: strong oxidizing agents. Hazardous decomposition products: toxic fumes of: carbon monoxide, carbon dioxide.
Carcinogenicity: IARC - Not listed; NIOSH - Not listed; NTP - Not listed; ACGIH - Not listed; OSHA - Not listed; EPA - Not listed; MAK - Not listed

Environmental

Environmental Fate: If released to soil, it would be expected to adsorb strongly to soil and/or biodegrade. There is evidence that it may leach into groundwater and that there is considerable volatilization losses from plant surfaces. Similarly, if released to water it would be expected to adsorb to sediment and particulate matter in the water column and/or biodegrade. Volatilization from water should not be significant. It is expected to have a moderately large potential for bioconcentrating in fish. If released in air, it will exist partly in the vapor phase and partly in the form of an aerosol and be subject to gravitational settling. The vapor phase ester should react with photochemically-produced hydroxyl radicals with an estimated half-life of 1.8 days.
Cleanup/Disposal: Guide No. 171: Do not touch or walk through spilled material. Stop leak if you can do it without risk. Prevent dust cloud. Avoid inhalation of asbestos dust. Small Dry Spills: With clean shovel place material into clean, dry container and cover loosely; move containers from spill area. Small Spills: Take up with sand or other noncombustible absorbent material and place into containers for later disposal. Large Spills: Dike far ahead of liquid spill for later disposal. Cover powder spill with plastic sheet or tarp to minimize spreading. Prevent entry into waterways, sewers, basements or confined areas.

Environmental Physical Data

Henry's Law Constant: estimated at 1.41 $x10^{-6}$
Octanol/Water Partition Coefficient: log K_{ow} = 4.11
Sorption Partition Coefficient: K_{oc} = 4100
BCF: none likely

Regulations

RCRA 40CFR: Not listed
CERCLA: 40CFR 302.4: Not listed
SARA 40CFR 372.65: Not listed
SARA EHS 40CFR 355: Not listed
TSCA: Listed

Analytical Methods

Food: FDA 212.1, 232.1
Plasma: EPA 001; FDA 211.1, 231.1, 252

DII2800 CAS #: 1191-15-7

DIISOBUTYLALUMINUM HYDRIDE

RTECS: BD0710000
DOT: UN2220; UN2221
EINECS Number: 214-729-9
Molecular Formula: $C_8H_{19}Al$
Formula Weight: 142.25

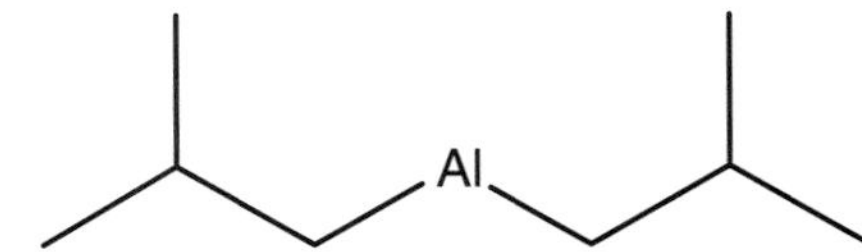

Chemical Structure

Synonyms: AL-ALCHILI; AL-DIISOBUTYL; ALUMINUM,DIISOBUTYLHYDRO-; ALUMINUM,HYDROBIS(2-METHYLPROPYL)-; ALUMINUM,HYDROBIS(2-METHYLPROPYL)-(9CI); ALUMINUM,HYDRODIISOBUTYL-; ALUMINUM,HYDRODIISOBUTYL-(8CI); BIS(ISOBUTYL)HYDROALUMINUM; DIBAL-H; DIISOBUTYLALUMINIUM HYDRIDE; DIISOBUTYLALUMINUM; DIISOBUTYLHYDROALUMINUM; HYDROBIS(2-METHYLPROPYL)ALUMINUM; HYDRODIISOBUTYLALUMINUM
Description: colorless liquid; characteristic musty odor

Use: reducing agent in pharmaceuticals; reducing agent for aldehydes, amides & olefins; polymerization catalyst for polyethylene; intermediate in organic phosphate insecticides, silicones, special saturated hydrocarbons, terpene alcohols, & organic acids

Physical Properties

Boiling Point: 105 °C (221 °F) at 0.2 mm Hg
Freezing Point: -80 °C (-112 °F)
Density: 0.798 g/mL
Bulk Density: 0.798
Other Solubilities: miscible with hydrocarbon solvents
Flash Point: Same as solvent

RTECS Toxicity Data

Acute Inhalation: Guinea Pig LC_{Lo} Dose: 70 gm/m^3/1hr; Toxic Effects: Lungs, Thorax, or Respiration - Other changes.

Hazard Overviews

Corrosive

Health: Corrosive to eyes/skin/respiratory tract. Toxic. Other Acute Effects: harmful if swallowed, inhaled, or absorbed through skin; inhalation may result in spasm, inflammation and edema of the larynx and bronchi, chemical pneumonitis and pulmonary edema; symptoms of exposure may include burning sensation; coughing; wheezing; laryngitis; shortness of breath; headache; nausea; vomiting; may cause nervous system disturbances. Chronic Effects: possible carcinogen; laboratory experiments have shown mutagenic effects; target organs: liver, pancreas, nerves.
Fire: Hazards: vapor may travel considerable distance to source of ignition and flash back; reacts with water to liberate flammable and/or explosive gas; container explosion may occur. Extinguishing agents: dry chemical powder; do not use water. Precautions: combustible liquid.
Reactivity: Hazardous polymerization will not occur. Incompatible with: alcohols, oxygen, oxidizing agents, acids, may decompose on exposure to air, reacts violently with water. Hazardous decomposition products: carbon monoxide, carbon dioxide, aluminum oxide, liberates flammable/explosive hydrogen gas, hydrogen chloride gas.
Carcinogenicity: IARC - Not listed; NIOSH - Not listed; NTP - Not listed; ACGIH - Not listed; OSHA - Not listed; EPA - Not listed; MAK - Not listed
Primary Target Organs:

Eyes

Skin

Respiratory System

Nervous System

Liver

Exposure Limits
OSHA PEL Vacated 1989 Limits: TWA: 2 mg/m^3; as Al soluble.
ACGIH TLV: TWA: 2 mg/m^3; as Al.
NIOSH REL: TWA: 2 mg/m^3; as Al soluble salts, alkyls.

Environmental

Regulations

RCRA 40CFR: Not listed
CERCLA: 40CFR 302.4: Not listed
SARA 40CFR 372.65: Not listed
SARA EHS 40CFR 355: Not listed
TSCA: Listed

DII3000 **CAS #: 25167-70-8**

DIISOBUTYLENE

EINECS Number: 246-690-9
Molecular Formula: C_8H_{16}
Structured MF: C_8H_{16}
Formula Weight: 112.21

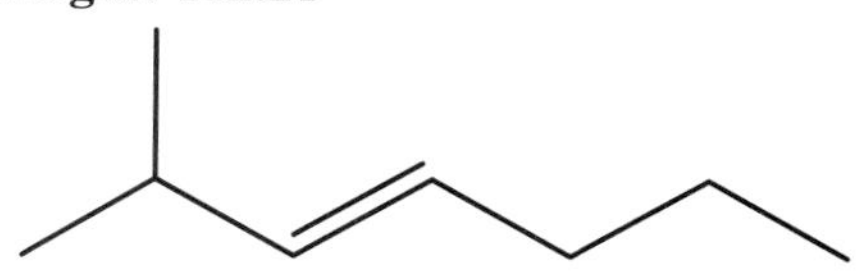

Chemical Structure

Description: colorless liquid

Physical Properties

Boiling Point: 102 °C (216 °F)
Freezing Point: -101 °C (-149.8 °F)
Specific Gravity: 0.723
Vapor Density: 3.99 Air=1
Vapor Pressure: Reid 1.6 psia
Water Solubility: Will float in slick on Water surface
Surface Tension: 20.7 dynes/cm at 20 °C
Critical Temperature: 287 °C
Critical Pressure: 380 psia
Flash Point: -29 °C
Autoignition Temperature: 391 °C
LEL: 0.8% v/v
UEL: 4.8% v/v

Hazard Overviews

Flammable

Fire: Flammable. Hazards: take precautionary measures against static discharges. Extinguishing agents: water spray; carbon dioxide, dry chemical powder or appropriate foam. Precautions: combustible liquid.
Carcinogenicity: IARC - Not listed; NIOSH - Not listed; NTP - Not listed; ACGIH - Not listed; OSHA - Not listed; EPA - Not listed; MAK - Not listed
Respirator Recommendation
Exposure Range: >600 to 30000 ppm Supplied Air, Constant Flow/Pressure Demand, Half Mask

Exposure Range: >30,000 to 600,000 ppm Supplied Air, Constant Flow/Pressure Demand, Full Face
Exposure Range: >600,000 to unlimited ppm Self-contained Breathing Apparatus, Pressure Demand, Full Face
Note: odor threshold unknown

Environmental

Cleanup/Disposal: Guide No. 127: Eliminate all ignition sources (no smoking, flares, sparks or flames in immediate area). All equipment used when handling the product must be grounded. Do not touch or walk through spilled material. Stop leak if you can do it without risk. Prevent entry into waterways, sewers, basements or confined areas. A vapor suppressing foam may be used to reduce vapors. Absorb or cover with dry earth, sand or other non-combustible material and transfer to containers. Use clean non-sparking tools to collect absorbed material. Large Spills: Dike far ahead of liquid spill for later disposal. Water spray may reduce vapor; but may not prevent ignition in closed spaces.

Regulations

RCRA 40CFR: Not listed
CERCLA: 40CFR 302.4: Not listed
SARA 40CFR 372.65: Not listed
SARA EHS 40CFR 355: Not listed
TSCA: Not listed

DII3600 CAS #: 51363-64-5

DIISODECYL PHENYL PHOSPHATE

EINECS Number: 257-153-3
Molecular Formula: $C_{26}H_{47}O_4P$
Formula Weight: 454.64
Synonyms: PHOSPHORIC ACID,DIISODECYL PHENYL ESTER
Use: plasticizer; hydraulic fluid

Hazard Overviews

Carcinogenicity: IARC - Not listed; NIOSH - Not listed; NTP - Not listed; ACGIH - Not listed; OSHA - Not listed; EPA - Not listed; MAK - Not listed

Environmental

Regulations

RCRA 40CFR: Not listed
CERCLA: 40CFR 302.4: Not listed
SARA 40CFR 372.65: Not listed
SARA EHS 40CFR 355: Not listed
TSCA: Listed

DII3800 CAS #: 26761-40-0

DIISODECYL PHTHALATE

RTECS: TI1270000
EINECS Number: 247-977-1
Molecular Formula: $C_{28}H_{46}O_4$
Structured MF: $C_{28}H_{46}O_4$
Formula Weight: 446.67

Chemical Structure

Synonyms: 1,2-BENZENEDICARBOXYLIC ACID,DIISODECYL ESTER; 1,2-BENZENEDICARBOXYLIC ACID,DIISODECYL ESTER (9CI); BIS(ISODECYL) PHTHALATE; BIS(ISODECYL)PHTHALATE; BIS(8-METHYLNONYL)PHTHALATE; DIDP; DIDP (PLASTICIZER); DI-ISODECYL PHTHALATE; PALATINOL Z; PHTHALIC ACID,BIS(8-METHYLNONYL) ESTER; PHTHALIC ACID,DIISODECYL ESTER; PLASTICIZED DDP; PX 120; SICOL 184; VESTINOL DZ
Description: clear viscous liquid; mild odor
Use: plasticizer for a wide variety of polymers

Physical Properties

Boiling Point: 250 °C (482 °F) to 257 °C (495 °F) at 4 mm Hg
Freezing Point: -50 °C (-58 °F)
Specific Gravity: 0.966 at 20 °C/20 °C
Density: 0.97 g/mL at 20 °C
Vapor Pressure: 1.1 mm Hg at 200 °C
Water Solubility: Insoluble
Other Solubilities: 95% Ethanol: Soluble; Acetone: Soluble; Benzene: Soluble; DMSO: Soluble; Ether: Soluble; Glycerol: Insoluble; Most organic solvents: Soluble; Some amines: Soluble.
Refraction Index: 1.483 at 25 °C/D
Flash Point: 232 °C Open Cup
Autoignition Temperature: 402 °C
LEL: 0.3% v/v

RTECS Toxicity Data

Acute Oral: Rat LD_{50} Dose: 64 gm/kg. Rabbit LD_{Lo} Dose: 22500 uL/kg.
Acute Inhalation: Rat LC; Dose: >130 mg/m^3/6hr. Mouse LC; Dose: >130 mg/m^3/6hr.
Acute Dermal: Rabbit LD_{50} Route: Skin; Dose: >3160 mg/kg.

Hazard Overviews

Fire Diamond

Fire: Will burn.
Carcinogenicity: IARC - Not listed; NIOSH - Not listed; NTP - Not listed; ACGIH - Not listed; OSHA - Not listed; EPA - Not listed; MAK - Not listed

Environmental

Environmental Fate: If released on land, it would be expected to adsorb strongly to soil and probably biodegrade. In water it will also absorb to sediment and particulate matter and very slowly biodegrade. Bioconcentration occurs in shellfish but it is not considered to bioaccumulate in fish. Depuration is rapid, occurring in a few days. In the air, it will be associated with aerosols and particulate matter. The vapor phase chemical will degrade by reaction with photochemically produced hydroxyl radicals (half-life 10.7 hr).

Environmental Physical Data

Henry's Law Constant: has not been measured
BCF: daphnia manga 116

Regulations

RCRA 40CFR: Not listed
CERCLA: 40CFR 302.4: Not listed
SARA 40CFR 372.65: Not listed
SARA EHS 40CFR 355: Not listed
TSCA: Listed

Analytical Methods

Plasma: EPA 001

DII4000 CAS #: 28553-12-0

DIISONONYL PHTHALATE

RTECS: CZ3850000
EINECS Number: 249-079-5
Molecular Formula: $C_{26}H_{42}O_4$
Formula Weight: 418.6

Chemical Structure

Synonyms: 1,2-BENZENEDICARBOXYLIC ACID,DIISONONYL ESTER; BIS(7-METHYLOCTYL) PHTHALATE; DINP; ENJ 2065; PALATINOL DN; PALATINOL N; PHTHALIC ACID,DIISONONYL ESTER; PHTHALISOCIZER DINP; SANSOCIZER DINP; VESTINOL NN; WITAMOL 150
Description: colorless, clear liquid; mild odor
Use: organic intermediate and plasticizer

Physical Properties

Vapor Pressure: 54 x10^{-5} torr
Water Solubility: 0.2 mg/L
Other Solubilities: 95% Ethanol: Soluble (>=10 mg/ml at 21 °C); Acetone: Soluble; Benzene: Soluble; DMSO: Soluble (>=10 mg/ml at 21 °C); Ether: Soluble; Most organic solvents: Soluble.
Flash Point: > 93.3 °C

Hazard Overviews

Health: Irritating. Other Acute Effects: may be harmful by inhalation, ingestion, or skin absorption. The toxicological properties have not been thoroughly investigated.
Fire: Will burn. Hazards: emits toxic fumes. Extinguishing agents: carbon dioxide, dry chemical powder or appropriate foam. Precautions: combustible liquid.
Reactivity: Stable. Hazardous polymerization will not occur. Hazardous decomposition products: toxic fumes of: carbon monoxide, carbon dioxide.
Carcinogenicity: IARC - Not listed; NIOSH - Not listed; NTP - Not listed; ACGIH - Not listed; OSHA - Not listed; EPA - Not listed; MAK - Not listed
Primary Target Organs:

Eyes

Skin

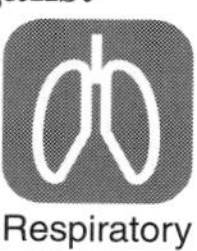
Respiratory System

Environmental

Environmental Fate: If released on land, it would be expected to adsorb strongly to soil and slowly biodegrade. In water it will also adsorb to sediment and particulate matter and very slowly biodegrade (approximately 1%/mo in freshwater/sediment systems). Moderate bioconcentration occurred in marine mussels and according to results for similar phthalate ester, should be higher in freshwater shellfish and should not bioaccumulate appreciably in fish. Depuration is rapid, occurring in a few days. In the air, it will be associated with aerosols and particulate matter. The vapor phase chemical will degrade by reaction with photochemically produced hydroxyl radicals (half-life 11.3 hr).

Environmental Physical Data

Henry's Law Constant: calculated at 1.5 x10^{-6}
BCF: coral 0.46

Regulations

RCRA 40CFR: Not listed
CERCLA: 40CFR 302.4: Not listed
SARA 40CFR 372.65: Not listed
SARA EHS 40CFR 355: Not listed
TSCA: Listed

Analytical Methods

Plasma: EPA 001

DII4200 CAS #: 1330-86-5

DIISOOCTYL ADIPATE

EINECS Number: 215-553-5
Molecular Formula: $C_{22}H_{42}O_4$

Formula Weight: 370.56

Synonyms: ADIPIC ACID,DIISOOCTYL ESTER; ADIPOL 10A; DI-ISO-OCTYL ADIPATE; DIMETHYL HEPTYLADIPATE; DIOA; HEXANEDIOIC ACID,DIISOOCTYL ESTER; ISOOCTYL ADIPATE; PX 208

Description: colorless or very pale amber liquid; slight aromatic odor

Use: low temp plasticizer for cellulose acetate butyrate resin, cellulose nitrate, ethyl cellulose, polymethyl methacrylate resins, polystyrene resins, polyvinyl chloride

Physical Properties

Boiling Point: 205 °C (401 °F) to 220 °C (428 °F) at 4 mm Hg

Freezing Point: -70 °C (-94 °F)

Specific Gravity: 0.93 at 20 °C

Vapor Pressure: < 0.12 mm Hg 150 °C

Flash Point: 210 °C

Hazard Overviews

Fire: Will burn.

Carcinogenicity: IARC - Not listed; NIOSH - Not listed; NTP - Not listed; ACGIH - Not listed; OSHA - Not listed; EPA - Not listed; MAK - Not listed

Environmental

Regulations

RCRA 40CFR: Not listed

CERCLA: 40CFR 302.4: Not listed

SARA 40CFR 372.65: Not listed

SARA EHS 40CFR 355: Not listed

TSCA: Listed

Analytical Methods

Plasma: EPA 001

DII4400 CAS #: 25168-26-7

2,4-DIISOOCTYL ESTERS

RTECS: AG8575000

DOT: UN2765; UN2766; UN2999; UN3000; NA2765; IMO3.2; IMO6.1

EINECS Number: 246-704-3

Molecular Formula: $C_{16}H_{22}Cl_2O_3$

Formula Weight: 333.3

Synonyms: ACETIC ACID,(2,4-DICHLOROPHENOXY)-,ISOOCTYL ESTER; AGROXONE 5; AMCHEM AMINE 2,4-D LOW VOLATILE ESTER WEED KILLER; AMOCO 2,4-D LV ESTER; CHIPMAN 2,4-D GRAN 20; CHIPMAN 2,4-D LOW VOLATILE ESTER 4L; CHIPMAN 2,4-D LOW VOLATILE ESTER 6L; CHIPMAN 2,4-D LOW VOLATILE ESTER 6; ORTHO 2,4-D LV ESTER 4; DED-WEED LV-69; DED-WEED SULV AMINE; DIAMOND SHAMROCK LO-VOL 4D; DIAMOND SHAMROCK LO-VOL 6D; 2,4-DICHLOROPHENOXYACETIC ACID ISOOCTYL ESTER; 2,4-DICHLOROPHENOXYACETIC ACID,ISOOCTYL ESTER; ESTASOL; ESTERON GE; ESTERON 6E HERBICIDE; FARMCO DLV-400 SPECIAL; FELCO LOW VOLATILE ESTER 600; FELCO LV 400 WEED KILLER; FS LV 400 WEED KILLER; GREEN CROSS ROADSIDE 2,4-D LOW VOLATILE WEED KILLER; GREEVER'S 2,4-D LOW VOLATILE ESTER WEED KILLER; HEDONAL; HENRY FIELD'S LAWN WEED KILLER; HERBATE ESTER 80; ISOOCTYL ALCOHOL (2,4-DICHLOROPHENOXY)ACETATE; ISOOCTYL ALCOHOL,(2,4-DICHLOROPHENOXY)ACETATE; ISOOCTYL 2,4-DICHLOROPHENOXYACETATE; 2,4-D ISOOCTYL ESTER; ISOOCTYL ESTER OF 2,4-DICHLOROPHENOXYACETIC ACID; ISOOKTYLESTER KYSELINY 2,4-DICHLORFENOXYOCTOVE; LINE RIDER LV-4D; LINE RIDER LV-6D; MILLER'S LO VOL 4# 2,4-D; MONSANTO 2,4-D GRANULAR; MONSANTO 2,4-D LOW VOLATILE ESTER; NACO LV-4D WEED KILLER; REED LV 2,4-D; REED LV 400 2,4-D; REED LV 600 2,4-D; RHODIA 2,4-D LOW VOLATILE ESTER 4L; SILVAPRON; SILVAPRON D; STULL'S LOW VOLATILE WEED KILLER; TOBACCO STATES BRAND ESTER 210; VISKO-RHAP 2,4-D LOW VOLATILE; WEEDTRINE-II

Description: dark liquid; fuel oil-like odor

Use: herbicide

Physical Properties

Boiling Point: 317 °C (603 °F)

Freezing Point: 12 °C (53.6 °F)

Specific Gravity: 1.152

Density: 1.025 g/cc at 25.2 °C

Vapor Pressure: 2 $x10^{-6}$ at 25 °C

Water Solubility: 0.07 mg/L at 20 °C

Other Solubilities: 95% Ethanol: >=100 mg/ml at 21 °C; Acetone: >=100 mg/ml at 21 °C; DMSO: 10-50 mg/ml at 21 °C.

Flash Point: > 79.444 °C Open Cup

RTECS Toxicity Data

Acute Oral: Rat LD_{50} Dose: 982 mg/kg.

Acute Dermal: Rabbit LD_{50} Route: Skin; Dose: >2 gm/kg.

Reproductive/Teratogenic: Rat Route: Oral; Dose: 1370 mg/kg; Duration: female 6-15D of pregnancy; Effects on Newborn - Live birth index. Rat Route: Oral; Dose: 302 mg/kg; Duration: female 9-12D of pregnancy; Specific Developmental Abnormalities - Musculoskeletal system. Rat Route: Oral; Dose: 188 mg/kg; Duration: female 6-15D of pregnancy; Specific Developmental Abnormalities - Homeostasis; Effects on Newborn - Growth statistics. Rat Route: Oral; Dose: 1131 mg/kg; Duration: female 6-15D of pregnancy; Effects on Fertility - Post-implantation mortality. Rat Route: Oral; Dose: 528 mg/kg; Duration: female 8-11D of pregnancy; Effects on Embryo or Fetus - Fetotoxicity.

Mutagenic: Human Sister Chromatid Exchenge; Cell Type: lymphocyte; Dose: 50 ug/L.

Tumorigenic: Mouse Route: Oral; Dose: 14 gm/kg/78W-I; Toxic Effects: Tumorigenic - Equivocal tumorigenic agent by RTECS criteria; Lungs, Thorax, or Respiration - Tumors; Liver - Tumors. Mouse Route: Subcutaneous; Dose: 21 mg/kg; Toxic Effects: Tumorigenic - Carcinogenic by RTECS criteria; Blood - Tumors.

Hazard Overviews

Fire: Combustible.

Carcinogenicity: IARC - Not listed; NIOSH - Not listed; NTP - Not listed; ACGIH - Not listed; OSHA - Not listed; EPA - Not listed; MAK - Not listed

Environmental

Ecotoxicity: LC_{50} Gammarus fasciatus mature 2.4 mg/l/96 hr static bioassay at 21 °C (95% confidence limit 1.9-3.0 mg/l) LC_{50} Gammarus lacustris 2400 ug/l/96 hr /Conditions of bioassay not specified LC_{50} Bluegill 8.8 - 66.3 mg/l/24 hr; 8.8 - 59.7 mg/l/48 hr /Conditions of bioassay not specified

Environmental Fate: If released on land it will be expected to undergo rapid hydrolysis, but the rate will be dependent on the pH. It will be expected to bind to soil and may be subject to microbially mediated hydrolysis. If released to water, the ester will be expected to extensively bind to sediments and significantly bioconcentrate. Hydrolysis may be significant at high pHs and may be enhanced by the presence of suspended sediments and variably affected by the presence of humic acids. Field studies and analogies to other 2,4-D esters suggest that evaporation from soil and water may be significant. A half-life of 6.1 hours has been estimated for evaporation from a river 1 m deep, flowing at 1 m/sec with a wind velocity of 3 m/sec. If released to the air the ester may be subject to direct photolysis and a half-life of 15.9 hours has been estimated for the reaction with photochemically produced hydroxyl radicals.

Cleanup/Disposal: Guide No. 131: Fully encapsulating, vapor protective clothing should be worn for spills and leaks with no fire. Eliminate all ignition sources (no smoking, flares, sparks or flames in immediate area). All equipment used when handling the product must be grounded. Do not touch or walk through spilled material. Stop leak if you can do it without risk. Prevent entry into waterways, sewers, basements or confined areas. A vapor suppressing foam may be used to reduce vapors. Small Spills: Absorb with earth, sand or other non-combustible material and transfer to containers for later disposal. Use clean non-sparking tools to collect absorbed material. Large Spills: Dike far ahead of liquid spill for later disposal. Water spray may reduce vapor; but may not prevent ignition in closed spaces. Guide No. 152: Do not touch damaged containers or spilled material unless wearing appropriate protective clothing. Stop leak if you can do it without risk. Prevent entry into waterways, sewers, basements or confined areas. Cover with plastic sheet to prevent spreading. Absorb or cover with dry earth, sand or other non-combustible material and transfer to containers. Do not get water inside containers.

Environmental Physical Data

Octanol/Water Partition Coefficient: log K_{ow} = 6.581

Sorption Partition Coefficient: K_{oc} = estimated at 2.5 $x10^4$ to 6.8 $x10^4$

BCF: estimated at 5.9 $x10^4$

Regulations

RCRA 40CFR: Not listed

CERCLA: 40CFR 302.4: Listed per CWA Section 311(b)(4) RQ: 100 lb (45.35 kg)

SARA 40CFR 372.65: Not listed

SARA EHS 40CFR 355: Not listed

TSCA: Not listed

Analytical Methods

Air: EPA TO-10

Soil: EPA PMD-TLC

Water / Groundwater: EPA 022

Food: FDA 212.1, 212.2, 232.1

Plasma: EPA 001; FDA 211.1, 231.1, 252

DII4600 CAS #: 1330-76-3

DIISOOCTYL MALEATE

EINECS Number: 215-547-2

Molecular Formula: $C_{20}H_{36}O_4$

Formula Weight: 340.51

Synonyms: 2-BUTENEDIOIC ACID (Z)-,DIISOOCTYL ESTER; DIOM; MALEIC ACID,DIISOOCTYL ESTER; RC COMONOMER DIOM

Use: comonomer for polymers with vinyl acetate for latex paint

Hazard Overviews

Carcinogenicity: IARC - Not listed; NIOSH - Not listed; NTP - Not listed; ACGIH - Not listed; OSHA - Not listed; EPA - Not listed; MAK - Not listed

Environmental

Regulations

RCRA 40CFR: Not listed

CERCLA: 40CFR 302.4: Not listed

SARA 40CFR 372.65: Not listed

SARA EHS 40CFR 355: Not listed

TSCA: Listed

DII4800 CAS #: 27215-10-7

DIISOOCTYL PHOSPHATE

RTECS: TC2400000

DOT: UN1902; IMO8.3

EINECS Number: 248-334-8

Molecular Formula: $C_{16}H_{35}O_4P$

Formula Weight: 322.48

Synonyms: DIISOOCTYL ACID PHOSPHATE; ISOOCTANOL,HYDROGEN PHOSPHATE; PHOSPHORIC ACID,DIISOOCTYL ESTER

Description: liquid

Use: in curing or urea-formaldehyde resin; rust inhibitors; mold lubricants; antistatic agent for non-cellulose fibers; leather tanning; bringing reclaimed rubber to workable viscosity; soldering flux; wetting agent for cutback asphalt; flameproofing plasticizer

Physical Properties

Specific Gravity: 1.02 at 31/4 °C

Water Solubility: Insoluble in Water

Other Solubilities: Not decomposed by air, water, or heat but will decompose if exposed to strong oxidizers.

Flash Point: Does not burn or burns with difficulty

Hazard Overviews

Corrosive

Fire: Noncombustible.
Carcinogenicity: IARC - Not listed; NIOSH - Not listed; NTP - Not listed; ACGIH - Not listed; OSHA - Not listed; EPA - Not listed; MAK - Not listed

Environmental

Cleanup/Disposal: Guide No. 153: Eliminate all ignition sources (no smoking, flares, sparks or flames in immediate area). Do not touch damaged containers or spilled material unless wearing appropriate protective clothing. Stop leak if you can do it without risk. Prevent entry into waterways, sewers, basements or confined areas. Absorb or cover with dry earth, sand or other non-combustible material and transfer to containers. Do not get water inside containers.

Regulations

RCRA 40CFR: Not listed
CERCLA: 40CFR 302.4: Not listed
SARA 40CFR 372.65: Not listed
SARA EHS 40CFR 355: Not listed
TSCA: Listed

DII5000 — CAS #: 27554-26-3

DIISOOCTYL PHTHALATE

RTECS: TI1300000
EINECS Number: 248-523-5
Molecular Formula: $C_{24}H_{38}O_4$
Structured MF: o-$C_6H_4[COO(CH_2)_5CH(CH_3)_2]_2$
Formula Weight: 390.54

Chemical Structure

Synonyms: 1,2-BENZENEDICARBOXYLIC ACID,DIISOOCTYL ESTER; CORFLEX 880; DIISOOCTYL 1,2-BENZENEDICARBOXYLATE; DI-ISO-OCTYL PHTHALATE; DIOP; FLEXOL PLASTICIZER DIOP; HEXAPLAS M/O; ISOOCTYL PHTHALATE; PHTHALIC ACID,BIS(6-METHYLHEPTYL)ESTER; PHTHALIC ACID,DIISOOCTYL ESTER
Description: nearly colorless, viscous liquid; mild odor
Use: plasticizer for poly vinyl chloride jackets for building wire; plasticizer for cellulosic & acrylate resins & synthetic rubber

Physical Properties

Boiling Point: 370 °C (698 °F)
Freezing Point: -4 °C (24.8 °F)
Specific Gravity: 0.986 at 20 °C
Vapor Density: 13.5 Air=1
Bulk Density: 8.20 lbs/gal at 20 °C
Vapor Pressure: 5.5×10^{-6} mm Hg
Water Solubility: Insoluble in Water
Other Solubilities: Soluble to various extents in many common organic solvents and oils.
Flash Point: 232 °C Closed Cup

RTECS Toxicity Data

Acute Oral: Rat LD_{50} Dose: 22 gm/kg. Mouse LD_{50} Dose: 2769 mg/kg.
Acute Dermal: Rabbit LD_{50} Route: Skin; Dose: 12600 uL/kg.
Irritation Skin: Rabbit Open Draize Test Dose: 500 mg open; Reaction: mild.

Hazard Overviews

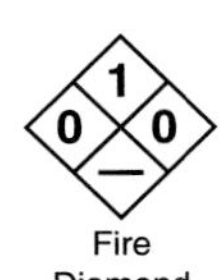

Fire Diamond

Health: Irritating to eyes/skin. Toxic. Other Acute Effects: harmful if swallowed, inhaled, or absorbed through skin. Chronic Effects: overexposure may cause reproductive disorder(s) based on tests with laboratory animals. Carcinogen.
Fire: Will burn. Extinguishing agents: water spray; carbon dioxide, dry chemical powder or appropriate foam. Precautions: combustible liquid.
Reactivity: Incompatible with: strong oxidizing agents. Hazardous decomposition products: toxic fumes of: carbon monoxide, carbon dioxide.
Carcinogenicity: IARC - Not listed; NIOSH - Not listed; NTP - Not listed; ACGIH - Not listed; OSHA - Not listed; EPA - Not listed; MAK - Not listed
Primary Target Organs:

Eyes Skin

Environmental

Environmental Fate: If released in soil, it would be expected to adsorb strongly to soil and slowly biodegrade. If released into water it will adsorb strongly to sediment and particulate matter in the water column. It may also be lost due to volatilization (half-life in a model river is 2.1 days), and biodegrade. Considerable bioaccumulation in fish and aquatic organisms may occur based upon its high estimated octanol/water partition coefficient. No experimental half-lives are available in soil or the aquatic environment. If released in air, it will most likely be in the form of an aerosol and be subject to gravitational settling. The vapor phase ester should react with photochemically produced hydroxyl radicals with an estimated half-life of 0.8 days.

Cleanup/Disposal: Guide No. 151: Do not touch damaged containers or spilled material unless wearing appropriate protective clothing. Stop leak if you can do it without risk. Prevent entry into waterways, sewers, basements or confined areas. Cover with plastic sheet to prevent spreading. Absorb or cover with dry earth, sand or other non-combustible material and transfer to containers. Do not get water inside containers.

Environmental Physical Data

Henry's Law Constant: estimated at 3.88 $x10^{-5}$

Octanol/Water Partition Coefficient: log K_{ow} = estimated at 3 to 4

Sorption Partition Coefficient: K_{oc} = 1.6 $x10^{4}$

BCF: likely to be high

Regulations

RCRA 40CFR: Not listed

CERCLA: 40CFR 302.4: Not listed

SARA 40CFR 372.65: Not listed

SARA EHS 40CFR 355: Not listed

TSCA: Listed

Analytical Methods

Food: FDA 212.1, 232.1

Plasma: EPA 001; FDA 252

DII5200 CAS #: 110-97-4

DIISOPROPANOLAMINE

RTECS: UB6600000

EINECS Number: 203-820-9

Molecular Formula: $C_6H_{15}NO_2$

Structured MF: $(CH_3CH(OH)CH_2)_2NH$

Formula Weight: 133.19

OH OH H N

Chemical Structure

Synonyms: BIS(2-HYDROXYPROPYL)AMINE; BIS(2-PROPANOL)AMINE; DIPA; DIPROPYL-2,2'-DIHYDROXY-AMINE; 1,1'-IMINOBIS(2-PROPANOL); 1,1'-IMINOBIS-2-PROPANOL; 1,1'-IMINODI-2-PROPANOL; 1,1'-IMINO-2-PROPANOL; 2-PROPANOL,1,1'-IMINOBIS-; 2-PROPANOL,1,1'-IMINOBIS-(9CI); 2-PROPANOL,1,1'-IMINODI-

Description: crystals; ammonia odor

Use: emulsifying agents for polishes, textile specialties, leather compounds, insecticides, cutting oils and water paints

Physical Properties

Boiling Point: 249 °C (480 °F) to 250 °C (482 °F) at 745 mm Hg

Freezing Point: 44.5 °C (112.1 °F) to 45.5 °C (113.9 °F)

Specific Gravity: 0.989 at 20 °C/4 °C

Density: 0.99 g/mL

Vapor Pressure: 0.0 psia

Water Solubility: Soluble in Water

Other Solubilities: 95% Ethanol: Soluble; Acetone: Soluble; Ether: Soluble.

Critical Temperature: 399 °C

Critical Pressure: 529 psia

Flash Point: 126.667 °C

Autoignition Temperature: 304 °C

LEL: 1.1% v/v

UEL: Estimated at 5.4% v/v

RTECS Toxicity Data

Acute Oral: Rat LD_{50} Dose: 4765 mg/kg; Toxic Effects: Behavioral - Somnolence (general depressed activity); Gastrointestinal - Hypermotility, diarrhea; Nutritional and gross metabolic - Body temperature decrease.

Acute Dermal: Mammal LD_{50} Route: Skin; Dose: >1 gm/kg.

Chronic (Multiple Dose) Oral: Rat Dose: 42 gm/kg/2W-C; Toxic Effects: Musculoskelital - Other changes; DEATH.

Irritation Eye: Rabbit Standard Draize Test Dose: 50 mg; Reaction: severe.

Irritation Skin: Rabbit Open Draize Test Dose: 500 mg open; Reaction: mild.

Hazard Overviews

Corrosive

Health: Corrosive to eyes/skin/respiratory tract. Other Acute Effects: harmful if swallowed, inhaled, or absorbed through skin; inhalation may result in spasm, inflammation and edema of the larynx and bronchi, chemical pneumonitis and pulmonary edema; symptoms of exposure may include burning sensation; coughing; wheezing; laryngitis; shortness of breath; headache; nausea; vomiting.

Fire: Will burn. Hazards: emits toxic fumes. Extinguishing agents: carbon dioxide, dry chemical powder or appropriate foam. Precautions: combustible liquid.

Reactivity: Incompatible with: strong oxidizing agents, strong acids. Hazardous decomposition products: toxic fumes of: carbon monoxide, carbon dioxide, nitrogen oxides.

Carcinogenicity: IARC - Not listed; NIOSH - Not listed; NTP - Not listed; ACGIH - Not listed; OSHA - Not listed; EPA - Not listed; MAK - Not listed

Primary Target Organs:

Eyes

Skin

Respiratory System

Environmental

Cleanup/Disposal: Guide No. 171: Do not touch or walk through spilled material. Stop leak if you can do it without risk. Prevent dust cloud. Avoid inhalation of asbestos dust. Small Dry Spills: With clean shovel place material into clean, dry container and cover loosely; move containers from spill area. Small Spills: Take up with sand or other noncombustible absorbent material and place into containers for later disposal. Large Spills: Dike far ahead of liquid spill for later disposal. Cover powder spill with plastic sheet or tarp to minimize spreading. Prevent entry into waterways, sewers, basements or confined areas.

Environmental Physical Data

Octanol/Water Partition Coefficient: log K_{ow} = measured at 0.79
BCF: no food chain concentration potential

Regulations

RCRA 40CFR: Not listed
CERCLA: 40CFR 302.4: Not listed
SARA 40CFR 372.65: Not listed
SARA EHS 40CFR 355: Not listed
TSCA: Listed

DII5800 CAS #: 107-56-2

O,O-DIISOPROPYL DITHIOPHOSPHATE

EINECS Number: 203-503-5
Molecular Formula: $C_6H_{15}O_2PS_2$
Formula Weight: 214.28
Synonyms: O,O-DIISOPROPYL DITHIOPHOSPHORIC ACID; O,O-DIISOPROPYL HYDROGEN DITHIOPHOSPHATE; O,O-DIISOPROPYL HYDROGEN PHOSPHORODITHIOATE; O,O-DIISOPROPYL PHOSPHORODITHIOATE; O,O-DIISOPROPYLPHOSPHORODITHIOIC ACID; ISOPROPYL AEROFLOAT; PHOSPHORODITHIOIC ACID,O,O-BIS(1-METHYLETHYL) ESTER; PHOSPHORODITHIOIC ACID,O,O-DIISOPROPYL ESTER

Hazard Overviews

Carcinogenicity: IARC - Not listed; NIOSH - Not listed; NTP - Not listed; ACGIH - Not listed; OSHA - Not listed; EPA - Not listed; MAK - Not listed

Environmental

Regulations

RCRA 40CFR: Not listed
CERCLA: 40CFR 302.4: Not listed
SARA 40CFR 372.65: Not listed
SARA EHS 40CFR 355: Not listed
TSCA: Listed

DII6000 CAS #: 94-11-1

2,4-DIISOPROPYL ESTER

RTECS: AG8750000
DOT: UN2765; UN2766; UN2999; UN3000; NA2765; IMO3.2; IMO6.1
EINECS Number: 202-305-6
Molecular Formula: $C_{11}H_{12}Cl_2O_3$
Structured MF: 2,4-$Cl_2C_6H_3OCH_2COOR$, where R=C_4H_9, C_3H_7 or $CH_2CH_2OC_4H_9$
Formula Weight: 263.12
Synonyms: ACETIC ACID,(2,4-DICHLOROPHENOXY)-,ISOPROPYL ESTER; ACETIC ACID,(2,4-DICHLOROPHENOXY)-,1-METHYLETHYL ESTER; ACETIC ACID,(2,4-DICHLOROPHENOXY)-,1-METHYLETHYL ESTER(9CI); AMCHEM WEED KILLER 650; BARBER'S WEED KILLER (ESTER FORMULATION); BRIDGEPORT SPOT WEED KILLER; CHEMICAL INSECTICIDE'S ISOPROPYL ESTER OF 2,4-D LIQUIDCONCENTRATE; CROP RIDER 3-34D-2; CROP RIDER 3.34D; 2,4-DICHLOROPHENOXYACETIC ACID ISOPROPYL ESTER; 2,4-DICHLOROPHENOXYACETIC ACID,ISOPROPYL ESTER; ESTERON 44; ISOPROPYL (2,4-DICHLOROPHENOXY)ACETATE; 2,4-D,ISOPROPYL ESTER; ISOPROPYL 2,4-D ESTER; ISOPROPYLESTER KYSELINY 2,4-DICHLORFENOXYOCTOVE; MONSANTO 2,4-D ISOPROPYL ESTER; NIAGARA ESTASOL; PARSONS 2,4-D WEED KILLER ISOPROPYL ESTER; SWIFT'S GOLD BEAR 44 ESTER; WEEDONE 128
Description: colorless liquid; may be a solid if pure; fuel oil-like odor
Use: herbicide for pasture & rangelands; agricultural use-eg, wheat, corn, grain sorghum, rice and other grains; industrial/commercial uses: lawns, turf, & aquatic use; other field crops-eg, nuts; component of herbicide for jungle defoliation (former use)

Physical Properties

Boiling Point: 130 °C (266 °F) at 1 mm Hg
Freezing Point: 24 °C (75.2 °F)
Specific Gravity: 1.255 to 1.27 at 25 °C/25 °C
Saturated Vapor Density: 1.200133844 kg/m^3
Vapor Pressure: 10.5 $x10^{-3}$ mm Hg at 25 °C
Water Solubility: Practically Insoluble in Water
Other Solubilities: Soluble in Alcohols and most oils.
Odor Threshold: Ester 0.02 ppm
Refraction Index: 1.5209
Flash Point: > 79.444 °C Open Cup

RTECS Toxicity Data

Acute Oral: Rat LD_{50} Dose: 700 mg/kg. Mouse LD_{50} Dose: 541 mg/kg; Toxic Effects: Gastrointestinal - Gastritis; Behavioral - Somnolence (general depressed activity); Liver - Fatty Liver degeneration.
Reproductive/Teratogenic: Mouse Route: Oral; Dose: 1326 mg/kg; Duration: female 7-15D of pregnancy; Effects on Embryo or Fetus - Fetotoxicity. Mouse Route: Subcutaneous; Dose: 846 mg/kg; Duration: female 6-14D of pregnancy; Effects on Embryo or Fetus - Extra embryonic structures; Fetotoxicity; Specific Developmental Abnormalities - Eye, ear. Mouse Route: Subcutaneous; Dose: 846 mg/kg;

Duration: female 6-14D of pregnancy; Specific Developmental Abnormalities - Craniofacial (including nose and tongue). Mouse Route: Subcutaneous; Dose: 414 mg/kg; Duration: female 6-14D of pregnancy; Effects on Fertility - Pre-implantation mortality.

Tumorigenic: Mouse Route: Oral; Dose: 12 gm/kg/78W-I; Toxic Effects: Tumorigenic - Equivocal tumorigenic agent by RTECS criteria; Lungs, Thorax, or Respiration - Tumors; Blood - Tumors.

Hazard Overviews

Fire: Combustible.

Carcinogenicity: IARC - Not listed; NIOSH - Not listed; NTP - Not listed; ACGIH - Not listed; OSHA - Not listed; EPA - Not listed; MAK - Not listed

Environmental

Ecotoxicity: LC_{50} Lepomis macrochirus (bluegill) 0.9 ppm/24 hr, 0.8 ppm/48 hr /Conditions of bioassay not specified

Environmental Fate: It is catalytically hydrolyzed in moist soils in less than a day. Hydrolysis in water is less rapid (half-life 23 days at neutral pH) but its removal may be accelerated in natural bodies of water by the presence of acclimated microorganisms, photodegradation, adsorption to sediment and particulate matter and favorable pHs. Hydrolysis is faster under alkaline conditions than under acid conditions but will be affected by humic acid concentration. It should not bioconcentrate appreciably in fish. It is very likely that the isopropyl ester will be transported long distances in the atmosphere. Its persistence in the atmosphere is unknown; however, the ester should be removed from air by rainout and possibly photodegradation.

Cleanup/Disposal: Guide No. 131: Fully encapsulating, vapor protective clothing should be worn for spills and leaks with no fire. Eliminate all ignition sources (no smoking, flares, sparks or flames in immediate area). All equipment used when handling the product must be grounded. Do not touch or walk through spilled material. Stop leak if you can do it without risk. Prevent entry into waterways, sewers, basements or confined areas. A vapor suppressing foam may be used to reduce vapors. Small Spills: Absorb with earth, sand or other non-combustible material and transfer to containers for later disposal. Use clean non-sparking tools to collect absorbed material. Large Spills: Dike far ahead of liquid spill for later disposal. Water spray may reduce vapor; but may not prevent ignition in closed spaces. Guide No. 152: Do not touch damaged containers or spilled material unless wearing appropriate protective clothing. Stop leak if you can do it without risk. Prevent entry into waterways, sewers, basements or confined areas. Cover with plastic sheet to prevent spreading. Absorb or cover with dry earth, sand or other non-combustible material and transfer to containers. Do not get water inside containers.

Environmental Physical Data

Sorption Partition Coefficient: K_{oc} = estimated at 531

BCF: no food chain concentration potential

Regulations

RCRA 40CFR: Not listed

CERCLA: 40CFR 302.4: Listed per CWA Section 311(b)(4) RQ: 100 lb (45.35 kg)

SARA 40CFR 372.65: Listed

SARA EHS 40CFR 355: Not listed

TSCA: Not listed

Analytical Methods

Air: EPA TO-10

Soil: EPA PMD-TLC

Water / Groundwater: EPA 022

Food: FDA 212.1, 232.1

Plasma: EPA 001; FDA 211.1, 231.1, 252

DII6200 **CAS #: 1445-75-6**

DIISOPROPYL METHYLPHOSPHONATE

RTECS: SZ9090000

EINECS Number: 215-896-0

Molecular Formula: $C_7H_{17}O_3P$

Formula Weight: 180.21

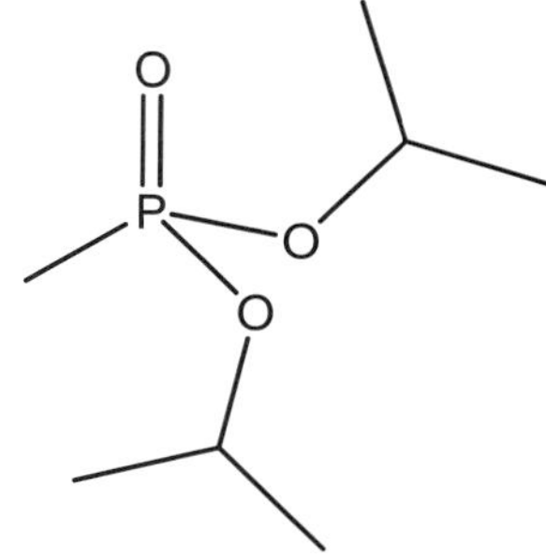

Chemical Structure

Synonyms: DIISOPROPYL METHANEPHOSPHONATE; DIMP; PHOSPHONIC ACID,METHYL-,BIS(1-METHYLETHYL) ESTER

Physical Properties

Boiling Point: 190 °C (374 °F)

Specific Gravity: 0.976

Vapor Pressure: 0.267 kPa at 70 °C

Water Solubility: 1 to 2 g/L at 25 °C

Surface Tension: 28.8 dynes/cm

Refraction Index: 1.4120

RTECS Toxicity Data

Acute Oral: Rat LD_{50} Dose: 826 mg/kg. Mouse LD_{50} Dose: 1041 mg/kg.

Hazard Overviews

Carcinogenicity: IARC - Not listed; NIOSH - Not listed; NTP - Not listed; ACGIH - Not listed; OSHA - Not listed; EPA - Class D, Not classifiable as to human carcinogenicity; MAK - Not listed

Environmental

Environmental Fate: Has a low vapor pressure and high solubility in water and will accordingly partition into water. If released in air, it will exist in the vapor phase and react with photochemically produced hydroxyl radicals with an estimated half-life of 5.2 hours. Due to its high water solubility, it would be susceptible to wash out by rain. If released in water or soil, it would be slowly lost by volatilization. The volatilization half-life in a model river is estimated to be 12.8 days and the half-life in a model lake is 98 days. Hydrolysis, biodegradation, and photolysis are not important environmental fate processes. When released to or deposited on soil, it will not adsorb strongly to soil and may be expected to leach. Volatilization appears to be the most important fate process. Volatilization is greater in moist soil than dry soil with 22% being lost in one study in 10 days. Biodegradation in soil is extremely slow; reported half-lives are 1 and 3 years, for acclimated and unacclimated soil, respectively. In the atmosphere, it would exist primarily in the vapor phase. It will react with photochemically produced hydroxyl radicals resulting in an estimated half-life of 5.2 hours. It does not absorb radiation >290 nm and therefore direct photolysis cannot occur.

Environmental Physical Data

Henry's Law Constant: estimated at 3.88×10^{-6}
Octanol/Water Partition Coefficient: log K_{ow} = 1.03
Sorption Partition Coefficient: log K_{oc} = estimated at 1.03
BCF: estimated at 5.1

Regulations

RCRA 40CFR: Not listed
CERCLA: 40CFR 302.4: Not listed
SARA 40CFR 372.65: Not listed
SARA EHS 40CFR 355: Not listed
TSCA: Listed

DII6400 **CAS #: 105-64-6**

DIISOPROPYL PEROXYDICARBONATE

RTECS: SD9800000
DOT: UN2133; UN2134
EINECS Number: 203-317-4
Molecular Formula: $C_8H_{14}O_6$
Structured MF: $C_3H_7OOCOOCOOC_3H_7$
Formula Weight: 206.22
Synonyms: DIISOPROPYL PERDICARBONATE; DIISOPROPYL PEROXYDIFORMATE; IPP; ISOPROPYL PERCARBONATE; ISOPROPYL PEROXYDICARBONATE; LUPEROX IPP; PEROXYDICARBONATE D'ISOPROPYLE; PEROXYDICARBONIC ACID,BIS(1-METHYLETHYL) ESTER; PEROXYDICARBONIC ACID,DIISOPROPYL ESTER
Description: white solid, liquid at room temp.
Use: low temp polymerization catalyst

Physical Properties

Boiling Point: Decomposes at 1 atm
Freezing Point: 8 °C (46.4 °F) to 10 °C (50 °F)
Specific Gravity: 1.08 at 15.5 °C/4 °C
Water Solubility: Almost Insoluble in Water
Other Solubilities: miscible with aliphatic & aromatic hydrocarbons, Esters, Ethers & chlorinated hydrocarbons
Refraction Index: 1.4034 at 20 °C/D
Flash Point: Not pertinent (combustible solid)

RTECS Toxicity Data

Acute Oral: Rat LD_{50} Dose: 2140 mg/kg.
Acute Dermal: Rabbit LD_{50} Route: Skin; Dose: 2025 mg/kg.
Irritation Eye: Rabbit Standard Draize Test Dose: 500 mg; Reaction: severe.

Hazard Overviews

Explosive

Flammable

Fire Diamond

Health: Irritating to eyes/skin/respiratory tract.
Fire: Flammable. Oxidizer. Liquid explosive when exposed to shock or friction. For small fires use dry chemical, carbon dioxide, water spray, or regular foam. Flood large fires with water from a distance. Fight fire from maximum distance.
Reactivity: Unstable, liquid is shock and friction sensitive. Hazardous polymerization cannot occur. Avoid: exposure to temperatures above 50 °F; crystallization. Incompatible with: strong alkalies; reducing agents; amines; organic matter; reactive monomers; 1 percent aniline; 1,2-diaminoethane or potassium iodide. Hazardous decomposition products: flammable gases; corrosive gases.
Carcinogenicity: IARC - Not listed; NIOSH - Not listed; NTP - Not listed; ACGIH - Not listed; OSHA - Not listed; EPA - Not listed; MAK - Not listed
Primary Target Organs:

Eyes

Skin

Respiratory System

Environmental

Cleanup/Disposal: Guide No. 148: Eliminate all ignition sources (no smoking, flares, sparks or flames in immediate area). Keep combustibles (wood, paper, oil, etc.) away from spilled material. Do not touch or walk through spilled material. Stop leak if you can do it without risk. Small Spills: Take up with inert, damp, noncombustible material using clean non-sparking tools and place into loosely covered plastic containers for later disposal. Large Spills: Dike far ahead of liquid spill for later disposal. Prevent entry into waterways, sewers, basements or confined areas. Do not clean-up or dispose of, except under supervision of a specialist.

Environmental Physical Data

BCF: no food chain concentration potential

Regulations

RCRA 40CFR: Not listed
CERCLA: 40CFR 302.4: Not listed
SARA 40CFR 372.65: Not listed
SARA EHS 40CFR 355: Not listed
TSCA: Listed

DII6600 CAS #: 108-18-9

DIISOPROPYLAMINE

RTECS: IM4025000
DOT: UN1158; IMO3.2
EINECS Number: 203-558-5
Molecular Formula: $C_6H_{15}N$
Structured MF: $[(CH_3)_2CH]_2NH$
Formula Weight: 101.19

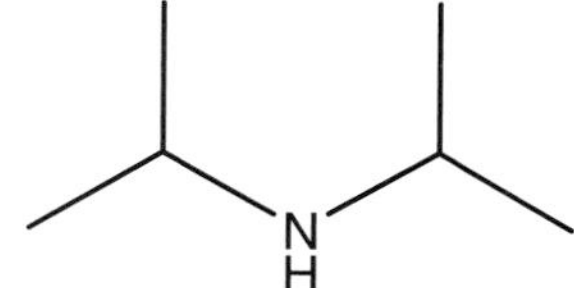

Chemical Structure

Synonyms: BIS(ISOPROPYL)AMINE; DIPA; N-(1-METHYLETHYL)-2-PROPANAMINE; 2-PROPANAMINE,N-(1-METHYLETHYL)-
Description: colorless liquid; ammonia-like odor
Use: catalyst for polymerization, solvent for extraction, intermediate for organic synthesis

Physical Properties

Boiling Point: 84 °C (183 °F) at 760 mm Hg
Freezing Point: -61 °C (-77.8 °F)
Specific Gravity: 0.7169 at 20 °C/4 °C
Vapor Density: 3.5 Air=1
Density: 0.722 g/mL at 22 °C
Bulk Density: 6.0 lbs/gal at 20 °C
Vapor Pressure: 70 mm Hg
Water Solubility: Miscible
Other Solubilities: 95% Ethanol: Soluble; Acetone: Soluble; aromatic hydrocarbons: Soluble; Benzene: Soluble; DMSO: Soluble; Ether: Soluble; Ethyl Acetate: Soluble; fatTY acids: Soluble.
Surface Tension: 20.04 4 dynes/cm at 16 °C
Odor Threshold: 1.8 ppm
pH: Strongly alkaline
Refraction Index: 1.3924 at 20 °C/D
Critical Temperature: 249.0 °C
Critical Pressure: 30 atm
Ionization Potential (eV): 7.73
Flash Point: -1 °C Open Cup
Autoignition Temperature: 316 °C
LEL: 1.1% v/v
UEL: 7.1% v/v

RTECS Toxicity Data

Acute Oral: Rat LD_{50} Dose: 770 mg/kg. Mouse LD_{50} Dose: 2120 mg/kg.
Acute Inhalation: Rat LC_{50} Dose: 4800 mg/m^3/2hr. Mouse LC_{50} Dose: 4200 mg/m^3/2hr.
Acute Dermal: Rabbit LD_{50} Route: Skin; Dose: >10 gm/kg. Guinea Pig LD_{Lo} Route: Subcutaneous Dose: 1400 mg/kg.
Chronic (Multiple Dose) Inhalation: Mouse Dose: 48300 ppb/6H/14D-I; Toxic Effects: Sense organs and special senses - Other. Rabbit Dose: 261 ppm/7H/54D-I; Toxic Effects: DEATH. Rabbit Dose: 597 ppm/7H/9D-I; Toxic Effects: DEATH.
Irritation Eye: Rabbit Standard Draize Test Dose: 750 ug open; Reaction: severe.
Irritation Skin: Rabbit Standard Draize Test Dose: 500 mg/24H; Reaction: mild.
Mutagenic: Bacteria - S Typhimurium Mutations in Microorganisms; Dose: 1 ug/plate (+/-S9).

Hazard Overviews

Flammable

Fire Diamond

Health: Irritating to eyes/respiratory tract. Also Causes: nausea, vomiting, headache, visual disturbances
Fire: Flammable. Can form explosive mixtures in the air. Use dry chemical, alcohol foam, or carbon dioxide to fight fires. Use a water spray to cool fire-exposed containers.
Reactivity: Stable. Hazardous polymerization cannot occur. Avoid: ignition sources. Incompatible with: strong oxidizers; strong acids. Hazardous decomposition products: oxides of nitrogen; carbon monoxide; other toxic gases.
Carcinogenicity: IARC - Not listed; NIOSH - Listed as carcinogen; NTP - Not listed; ACGIH - Not listed; OSHA - Not listed; EPA - Not listed; MAK - Not listed
Primary Target Organs:

Eyes

Skin

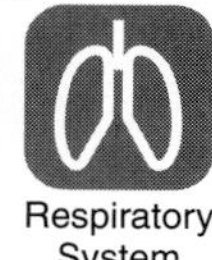

Respiratory System

Exposure Limits
OSHA PEL: TWA: 5 ppm; 20 mg/m^3; skin.
ACGIH TLV: TWA: 5 ppm; 21 mg/m^3.
NIOSH REL: TWA: 5 ppm; 20 mg/m^3.
NIOSH IDLH: 200 ppm.
Respirator Recommendation
Exposure Range: >5 to 50 ppm Air Purifying, Negative Pressure, Half Mask
Exposure Range: >50 to <200 ppm Air Purifying, Negative Pressure, Full Face
Exposure Range: 200 to unlimited ppm Self-contained Breathing Apparatus, Pressure Demand, Full Face
Cartridge Color: black

Environmental

Ecotoxicity: Aquatic toxicity: 60 ppm/24 hr/creek chub/lethal/fresh water 40-60 ppm/*/creek chub/critical range/ fresh water *Time period not specified.

Environmental Fate: If released to the atmosphere, the vapor phase is expected to degrade rapidly (estimated half- life of 4 hours) by reaction with photochemically produced hydroxyl radicals. If released to soil, it may volatilize from dry soil surfaces and hydrolyze in moist soils. Some may adsorb to soil. Biodegradation may be important in soil and water providing a sufficient acclimation period. High concentrations (> 50 ppm) will be toxic to microorganisms. If released to water, bioconcentration in aquatic organisms is not expected to be an important fate process. Adsorption to sediments is not expected to be an important transport process; however, some may partition from the water column to sediments and suspended material. Volatilization from water will be important; volatilization half- lives of 12 hours and 7 days have been estimated for a model river and model lake, respectively.

Cleanup/Disposal: Guide No. 132: Fully encapsulating, vapor protective clothing should be worn for spills and leaks with no fire. Eliminate all ignition sources (no smoking, flares, sparks or flames in immediate area). All equipment used when handling the product must be grounded. Do not touch or walk through spilled material. Stop leak if you can do it without risk. Prevent entry into waterways, sewers, basements or confined areas. A vapor suppressing foam may be used to reduce vapors. Absorb with earth, sand or other non-combustible material and transfer to containers (except for Hydrazine). Use clean non-sparking tools to collect absorbed material. Large Spills: Dike far ahead of liquid spill for later disposal. Water spray may reduce vapor; but may not prevent ignition in closed spaces.

Environmental Physical Data

Henry's Law Constant: 9.6×10^{-5}

Octanol/Water Partition Coefficient: log K_{ow} = calculated at 1.84

Sorption Partition Coefficient: K_{oc} = estimated at 190

BCF: estimated at 10

BOD: 1% BODT, 5 days

Regulations

RCRA 40CFR: Not listed

CERCLA: 40CFR 302.4: Not listed

SARA 40CFR 372.65: Not listed

SARA EHS 40CFR 355: Not listed

TSCA: Listed

Analytical Methods

Air: ASTM D4490

DII6800 CAS #: 96-80-0

DIISOPROPYLAMINOETHANOL

RTECS: KK5950000

EINECS Number: 202-536-2

Molecular Formula: $C_8H_{19}NO$

Formula Weight: 145.28

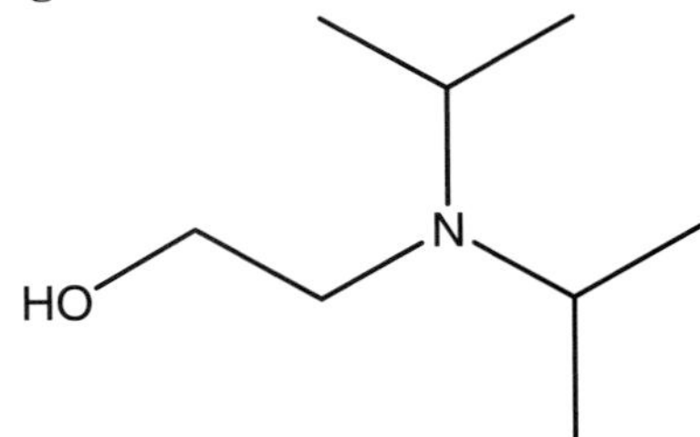

Chemical Structure

Synonyms: DIISOPROPYL ETHANOLAMINE; 2-DIISOPROPYLAMINOETHANOL; N,N-DIISOPROPYLETHANOLAMINE

Physical Properties

Boiling Point: 187 °C (369 °F)

Specific Gravity: 0.82600

Refraction Index: 1.4417

RTECS Toxicity Data

Acute Oral: Rat LD_{50} Dose: 860 mg/kg. Mouse LD_{50} Dose: 770 mg/kg.

Acute Inhalation: Rat LC_{50} Dose: 1965 mg/m^3/6hr. Mouse LC_{50} Dose: 1661 mg/m^3/6hr.

Acute Dermal: Rabbit LD_{50} Route: Skin; Dose: 450 uL/kg.

Irritation Eye: Rabbit Standard Draize Test Dose: 750 ug open; Reaction: severe.

Irritation Skin: Rabbit Open Draize Test Dose: 500 mg open; Reaction: mild.

Hazard Overviews

Corrosive

Health: Corrosive to eyes/skin/respiratory tract. Other Acute Effects: harmful by inhalation, in contact with skin and if swallowed; readily absorbed through skin; extremely destructive to tissue of the mucous membranes and upper respiratory tract, eyes and skin; inhalation may result in spasm, inflammation and edema of the larynx and bronchi, chemical pneumonitis and pulmonary edema; symptoms of exposure may include burning sensation, coughing, wheezing, laryngitis, shortness of breath, headache, nausea and vomiting.

Fire: Hazards: emits toxic fumes. Extinguishing agents: carbon dioxide, dry chemical powder or appropriate foam. Precautions: combustible liquid.

Carcinogenicity: IARC - Not listed; NIOSH - Not listed; NTP - Not listed; ACGIH - Not listed; OSHA - Not listed; EPA - Not listed; MAK - Not listed

Primary Target Organs:

Eyes

Skin

Respiratory System

Environmental

Ecotoxicity: Fishes: Pimephales promelas 4d LC_{50} 201 mg/l

Regulations

RCRA 40CFR: Not listed
CERCLA: 40CFR 302.4: Not listed
SARA 40CFR 372.65: Not listed
SARA EHS 40CFR 355: Not listed
TSCA: Listed

DII7000 CAS #: 16715-83-6

2-(DIISOPROPYLAMINO)ETHYL METHACRYLATE

EINECS Number: 240-772-8
Molecular Formula: $C_{12}H_{23}NO_2$
Formula Weight: 203.20
Synonyms: DIISOPROPYLAMINOETHYL METHACRYLATE; N,N-DIISOPROPYLAMINOETHYL METHACRYLATE; METHACRYLIC ACID,2-(DIISOPROPYLAMINO)ETHYL ESTER; 2-PROPENOIC ACID,2-METHYL-,2-(BIS(1-METHYLETHYL)AMINO)ETHYL ESTER
Use: comonomer for polymers used for adhesion in inks; comonomer for acrylic resins for floor waxes; chem int for monomeric quaternaries for water treatment

Hazard Overviews

Carcinogenicity: IARC - Not listed; NIOSH - Not listed; NTP - Not listed; ACGIH - Not listed; OSHA - Not listed; EPA - Not listed; MAK - Not listed

Environmental

Regulations

RCRA 40CFR: Not listed
CERCLA: 40CFR 302.4: Not listed
SARA 40CFR 372.65: Not listed
SARA EHS 40CFR 355: Not listed
TSCA: Listed

DII7200 CAS #: 577-55-9

1,2-DIISOPROPYLBENZENE

RTECS: CZ6350000
EINECS Number: 209-412-7
Molecular Formula: $C_{12}H_{18}$
Formula Weight: 162.27
Synonyms: O-DIISOPROPYLBENZENE
Description: colorless, clear liquid

Physical Properties

Boiling Point: 204 °C (399 °F) at 760 mm Hg
Freezing Point: -57 °C (-70.6 °F)
Specific Gravity: 0.8701 at 20 °C/4 °C
Vapor Pressure: 9.67 kPa at 125 °C
Water Solubility: Insoluble
Other Solubilities: Soluble in all proportions of Alcohol, Ether, Acetone and Benzene.
Refraction Index: 1.4960
Flash Point: 76.667 °C Open Cup
Autoignition Temperature: 449 °C

RTECS Toxicity Data

Acute Oral: Rat LD_{Lo} Dose: 5 gm/kg.

Hazard Overviews

Fire: Combustible.
Carcinogenicity: IARC - Not listed; NIOSH - Not listed; NTP - Not listed; ACGIH - Not listed; OSHA - Not listed; EPA - Not listed; MAK - Not listed

Environmental

Regulations

RCRA 40CFR: Not listed
CERCLA: 40CFR 302.4: Not listed
SARA 40CFR 372.65: Not listed
SARA EHS 40CFR 355: Not listed
TSCA: Listed

DII7400 CAS #: 99-62-7

1,3-DIISOPROPYLBENZENE

RTECS: CZ6334000
EINECS Number: 202-773-1
Molecular Formula: $C_{12}H_{18}$
Formula Weight: 162.27

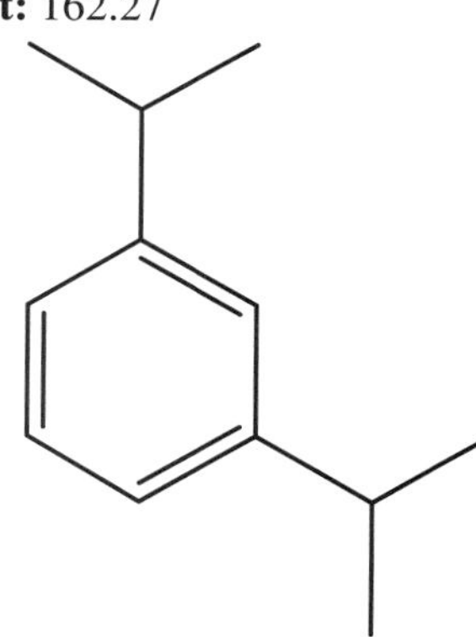

Chemical Structure

Synonyms: BENZENE,1,3-BIS(1-METHYLETHYL)-; BENZENE,1,3-BIS(1-METHYLETHYL)-(9CI); BENZENE,M-DIISOPROPYL-; 1,3-BIS(1-METHYLETHYL)BENZENE; M-DIISOPROPYLBENZENE; M-DIISOPROPYLBENZOL

Description: colorless liquid
Use: solvent; intermediate; chem int for resorcinol & acetone; solvent

Physical Properties

Boiling Point: 203.2 °C (398 °F) at 760 mm Hg
Freezing Point: -61 °C (-77.8 °F)
Specific Gravity: 0.8559 at 20 °C/4 °C
Vapor Pressure: 1 mm Hg at 34.7 °C
Water Solubility: Insoluble in Water
Other Solubilities: Soluble in all proportions of Alcohol, Ether, Acetone, Benzene.
Refraction Index: 1.4883
Autoignition Temperature: 449 °C

RTECS Toxicity Data

Acute Oral: Rat LD_{50} Dose: 7400 mg/kg. Mouse LD_{50} Dose: 3100 mg/kg.
Chronic (Multiple Dose) Oral: Rat Dose: 45500 ug/kg/26W-I; Toxic Effects: Behavioral - Muscle contraction or spasticity; Behavioral - Alteration of classical conditioning; Blood - Changes in erythrocite (RBC) cell count.
Chronic (Multiple Dose) Inhalation: Rat Dose: 1 gm/m^3/5H/22W-I; Toxic Effects: Liver - Liver function tests impaired; Blood - Changes in cell count (unspecified). Mouse Dose: 1 gm/m^3/5H/22W-I; Toxic Effects: Brain and coverings - Recordings from specific areas of CNS; Liver - Liver function tests impaired.
Reproductive/Teratogenic: Rat Route: Inhalation; Dose: 200 ug/m^3/5H Duration: female 22W prior to mating Maternal Effects - Menstrual cycle changes or disorders.

Hazard Overviews

Health: May cause irritation to eyes/skin. Other Acute Effects: may be harmful by inhalation, ingestion, or skin absorption.
Fire: Will burn. Extinguishing agents: water spray; carbon dioxide, dry chemical powder or appropriate foam. Precautions: combustible liquid.
Carcinogenicity: IARC - Not listed; NIOSH - Not listed; NTP - Not listed; ACGIH - Not listed; OSHA - Not listed; EPA - Not listed; MAK - Not listed

Environmental

Regulations

RCRA 40CFR: Not listed
CERCLA: 40CFR 302.4: Not listed
SARA 40CFR 372.65: Not listed
SARA EHS 40CFR 355: Not listed
TSCA: Listed

DII7600 **CAS #: 100-18-5**

1,4-DIISOPROPYLBENZENE

RTECS: CZ6360000
EINECS Number: 202-826-9
Molecular Formula: $C_{12}H_{18}$
Formula Weight: 162.29

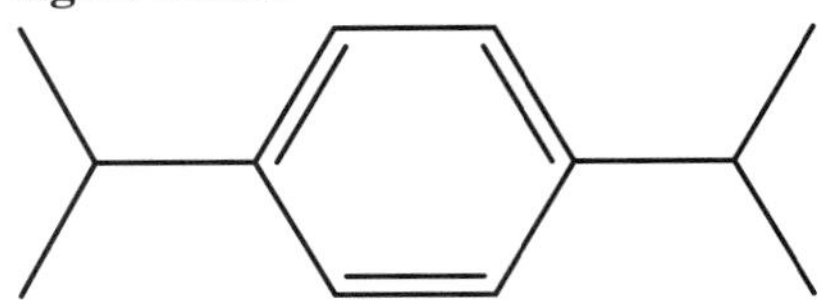

Chemical Structure

Synonyms: BENZENE,1,4-BIS(1-METHYLETHYL)-; BENZENE,1,4-BIS(1-METHYLETHYL)-(9CI); BENZENE,P-DIISOPROPYL-; 1,4-BIS(1-METHYLETHYL)BENZENE; P-DIISOPROPYLBENZENE; PARA-DIISOPROPYLBENZENE; P-DIISOPROPYLBENZOL
Description: white solid
Use: solvent; intermediate; chem int for hydroquinone & acetone

Physical Properties

Boiling Point: 210.3 °C (411 °F) at 760 mm Hg
Freezing Point: -17.1 °C (1.22 °F)
Specific Gravity: 0.8568 at 20 °C/4 °C
Vapor Pressure: 1 mm Hg at 40.0 °C
Water Solubility: Insoluble in Water
Other Solubilities: Soluble in all proportions of Alcohol, Ether, Acetone, Benzene.
Refraction Index: 1.4898
Ionization Potential (eV): 8.35

RTECS Toxicity Data

Acute Oral: Mouse LD_{50} Dose: 3400 mg/kg.

Hazard Overviews

Carcinogenicity: IARC - Not listed; NIOSH - Not listed; NTP - Not listed; ACGIH - Not listed; OSHA - Not listed; EPA - Not listed; MAK - Not listed

Environmental

Regulations

RCRA 40CFR: Not listed
CERCLA: 40CFR 302.4: Not listed
SARA 40CFR 372.65: Not listed
SARA EHS 40CFR 355: Not listed
TSCA: Listed

DII7800 **CAS #: 25321-09-9**

DIISOPROPYLBENZENE

RTECS: CZ6330000
EINECS Number: 246-835-6
Molecular Formula: $C_{12}H_{18}$
Structured MF: $C_6H_4[CH(CH_3)_2]_2$
Formula Weight: 162.30

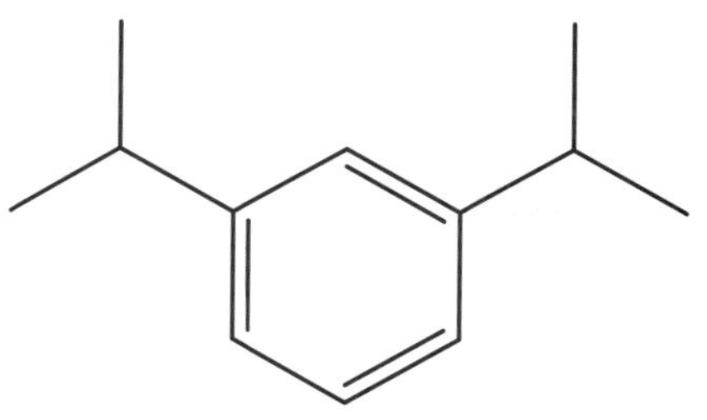

Chemical Structure

Synonyms: BENZENE,BIS(1-METHYLETHYL)-(9CI); BIS(1-METHYLETHYL)BENZENE
Description: amber, clear liquid
Use: blended into gasoline, diesel and other hydrocarbon fuels

Physical Properties

Boiling Point: 205 °C (401 °F)
Freezing Point: -63 °C (-81 °F) to -17 °C (1 °F)
Specific Gravity: 0.9 (Water=1)
Vapor Density: 5.6 Air=1
Vapor Pressure: 0.25 to 0.39 mm Hg at 25 °C
Flash Point: 77 °C Closed Cup
Autoignition Temperature: 449 °C
LEL: 0.9% v/v
UEL: 6.5% v/v

RTECS Toxicity Data

Acute Oral: Rat LD_{50} Dose: 6500 uL/kg.
Acute Inhalation: Rat LC_{Lo} Dose: 5300 mg/m^3/4hr. Mouse LC_{Lo} Dose: 5300 mg/m^3/2hr.
Acute Dermal: Rabbit LD_{50} Route: Skin; Dose: 16 mL/kg.
Irritation Eye: Rabbit Standard Draize Test Dose: 500 mg/24H; Reaction: mild.
Irritation Skin: Rabbit Standard Draize Test Dose: 100 mg/24H; Reaction: moderate.

Hazard Overviews

Fire: Combustible.
Carcinogenicity: IARC - Not listed; NIOSH - Not listed; NTP - Not listed; ACGIH - Not listed; OSHA - Not listed; EPA - Not listed; MAK - Not listed

Environmental

Environmental Fate: If released to the atmosphere, it will degrade by reaction with photochemically produced hydroxyl radicals (estimated half-life of 30 hr). If released to soil, it is not expected t leach based upon an estimated K_{oc} of 4000. If released to water, it may volatilize and partition to sediment. Insufficient data are available to predict the importance of biodegradation in soil or water.
Cleanup/Disposal: Guide No. 152: Do not touch damaged containers or spilled material unless wearing appropriate protective clothing. Stop leak if you can do it without risk. Prevent entry into waterways, sewers, basements or confined areas. Cover with plastic sheet to prevent spreading. Absorb or cover with dry earth, sand or other non-combustible material and transfer to containers. Do not get water inside containers.

Environmental Physical Data

Henry's Law Constant: estimated at 0.0204
Sorption Partition Coefficient: K_{oc} = 4000
BCF: 2.11 to 2.14

Regulations

RCRA 40CFR: Not listed
CERCLA: 40CFR 302.4: Not listed
SARA 40CFR 372.65: Not listed
SARA EHS 40CFR 355: Not listed
TSCA: Listed

DII8000 **CAS #: 26762-93-6**

DIISOPROPYLBENZENE HYDROPEROXIDE

RTECS: MX2440000
DOT: UN2171; IMO5.2
EINECS Number: 247-988-1
Molecular Formula: $C_{12}H_{18}O_2$
Structured MF: $(CH_3)_2CHC_6H_4C(CH_3)_2OOH+(CH_3)_2CHC_6H_4CH(CH_3)_2$
Formula Weight: 194.30
Synonyms: DIISOPROPYLBENZENE HYDROPEROXIDE,>72% IN SOLUTION; HYDROPEROXIDE,BIS(1-METHYLETHYL)PHENYL; HYDROPEROXIDE,DIISOPROPYLPHENYL
Description: colorless to pale yellow liquid; sharp, disagreeable odor
Use: chem int for hydroquinone (para-isomer); initiator for cold SBR polymerization; curing agent for polyester resins; chem int for resorcinol; strong oxidizing agent

Physical Properties

Freezing Point: < -9 °C (15.8 °F)
Specific Gravity: Liquid 0.956 at 15 °C
Water Solubility: May float or sink in water
Flash Point: 79.444 °C Closed Cup

Hazard Overviews

Corrosive

Fire: Combustible.
Carcinogenicity: IARC - Not listed; NIOSH - Not listed; NTP - Not listed; ACGIH - Not listed; OSHA - Not listed; EPA - Not listed; MAK - Not listed

Environmental

Cleanup/Disposal: Guide No. 145: Eliminate all ignition sources (no smoking, flares, sparks or flames in immediate area). Keep combustibles (wood, paper, oil, etc.) away from spilled material. Do not touch damaged containers or spilled material unless wearing appropriate protective clothing. Keep substance wet using water spray. Stop leak if you can do it without risk. Small Spills: Take up with inert, damp, noncombustible material using clean non-sparking tools and place into loosely covered plastic containers for later disposal. Large Spills: Wet down with water and dike for later disposal. Prevent entry into waterways, sewers, basements or confined areas.

Regulations

RCRA 40CFR: Not listed
CERCLA: 40CFR 302.4: Not listed
SARA 40CFR 372.65: Not listed
SARA EHS 40CFR 355: Not listed
TSCA: Listed

DII8200 CAS #: 95-29-4

N,N-DIISOPROPYL-2-BENZOTHIAZOLESULFENAMIDE

RTECS: DL6330000
EINECS Number: 202-407-0
Molecular Formula: $C_{13}H_{18}N_2S_2$
Formula Weight: 282.45
Synonyms: 2-BENZOTHIAZOLESULFENAMIDE,N,N-BIS(1-METHYLETHYL)-; 2-BENZOTHIAZOLESULFENAMIDE,N,N-DIISOPROPYL; DIBS; N,N-DIISOPROPYL BENZOTHIAZOLE-2-SULFENAMIDE; N,N-DIISOPROPYLBENZOTHIAZYL-2-SULFENAMIDE; DIPAC; DIPAK
Use: rubber accelerator

RTECS Toxicity Data

Acute Oral: Mouse LD_{50} Dose: 3892 mg/kg.
Chronic (Multiple Dose) Oral: Rabbit Dose: 1820 mg/kg/17W-I; Toxic Effects: Blood - Changes in serum composition.

Hazard Overviews

Carcinogenicity: IARC - Not listed; NIOSH - Not listed; NTP - Not listed; ACGIH - Not listed; OSHA - Not listed; EPA - Not listed; MAK - Not listed

Environmental

Regulations

RCRA 40CFR: Not listed
CERCLA: 40CFR 302.4: Not listed
SARA 40CFR 372.65: Not listed
SARA EHS 40CFR 355: Not listed
TSCA: Listed

DII8400 CAS #: 36876-13-8

DIISOPROPYLBIPHENYL

EINECS Number: 253-247-3
Molecular Formula: $C_{18}H_{22}$
Formula Weight: 238.38
Synonyms: 1,1'-BIPHENYL,AR,AR'-BIS(1-METHYLETHYL)-; DICUMYL
Use: solvent in the manufacture of carbonless copy paper; dielectric fluid in capacitors as a PCB replacement

Physical Properties

Other Solubilities: Soluble in most organic solvents.

Hazard Overviews

Carcinogenicity: IARC - Not listed; NIOSH - Not listed; NTP - Not listed; ACGIH - Not listed; OSHA - Not listed; EPA - Not listed; MAK - Not listed

Environmental

Octanol/Water Partition Coefficient: log K_{ow} = estimated at 7.3
BCF: flagfish 2896

Regulations

RCRA 40CFR: Not listed
CERCLA: 40CFR 302.4: Not listed
SARA 40CFR 372.65: Not listed
SARA EHS 40CFR 355: Not listed
TSCA: Listed

DII8600 CAS #: 55-91-4

DIISOPROPYLFLUOROPHOSPHATE

RTECS: TE5075000
EINECS Number: 200-247-6
Molecular Formula: $C_6H_{14}FO_3P$
Formula Weight: 184.15

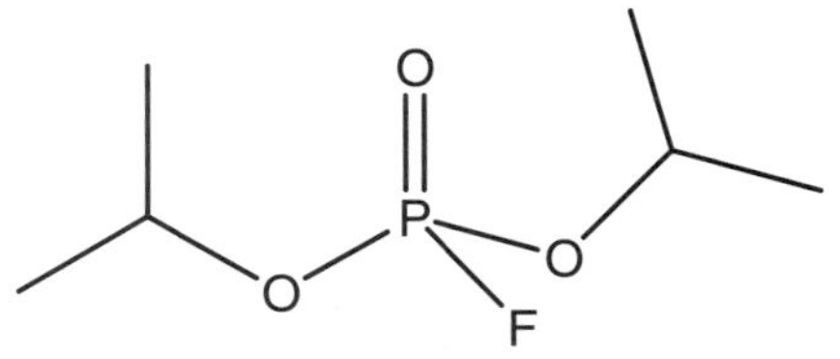

Chemical Structure

Synonyms: DFP; DIFLUPYL; DIFLUROPHATE; DIISOPROPOXYPHOSPHORYL FLUORIDE; DIISOPROPYL FLUOROPHOSPHATE; O,O-DIISOPROPYL FLUOROPHOSPHATE; DIISOPROPYL FLUOROPHOSPHONATE; DIISOPROPYL PHOSPHOFLUORIDATE; DIISOPROPYL PHOSPHOROFLUORIDATE; O,O'-DIISOPROPYL PHOSPHORYL FLUORIDE; DIISOPROPYLFLUORFOSFAT; DIISOPROPYLFLUOROPHOSPHORIC ACID ESTER; DIISOPROPYLFLUORPHOSPHORSAEUREESTER; DIISOPROPYLPHOSPHOROFLUORIDATE; DYFLOS; EA 1152; FLOROPRYL; FLUOPHOSPHORIC ACID,DIISOPROPYL ESTER; FLUORODIISOPROPYL PHOSPHATE; FLUOROPHOSPHORIC ACID,DIISOPROPYL ESTER; FLUOROPRYL; FLUOSTIGMINE; ISOFLUOROPHATE; ISOFLUROPHATE; ISOFLUROPHOSPHATE; ISOPROPYL FLUOPHOSPHATE; ISOPROPYL PHOSPHOROFLUORIDATE; NEOGLAUCIT; PF-3; PHOSPHOROFLUORIDIC ACID,BIS(1-METHYLETHYL) ESTER; PHOSPHOROFLUORIDIC ACID,DIISOPROPYL ESTER; T-1703; TL 466

Description: clear, colorless or faintly yellow oily liquid
Use: medication: cholinergic (ophthalmic); medication (vet): as a miotic; medication: treatment of primary open-angle glaucoma; in the treatment of aphkic glaucoma and accomodative estropia; insecticide (former use)

Physical Properties

Boiling Point: 62 °C (144 °F) at 9 mm Hg
Freezing Point: -82 °C (-115.6 °F)
Specific Gravity: 1.055
Vapor Density: 5.24 Air=1
Saturated Vapor Density: 1.204891026 kg/m^3
Vapor Pressure: 0.579 mm Hg at 20 °C
Water Solubility: 1.54% wt/wt 25 °C
Other Solubilities: Soluble in Ether; Soluble in Alcohol.
Refraction Index: 1.3830 at 25 °C/D

RTECS Toxicity Data

Acute Oral: Rat LD_{50} Dose: 5 mg/kg. Mouse LD_{50} Dose: 2 mg/kg.
Acute Inhalation: Rat LC_{50} Dose: 360 mg/m^3/10M; Toxic Effects: Behavioral - Muscle weakness; Lungs, Thorax, or Respiration - Dyspnea; Gastrointestinal - Changes in structure or function of salivary glands. Human TC_{Lo} Dose: 8200 ug/m^3/10M; Toxic Effects: Sense organs and special senses - Miosis (pupilliary dilation); Behavioral - Headache.
Acute Dermal: Rabbit LD_{50} Route: Subcutaneous Dose: 1 mg/kg. Monkey LD_{50} Route: Subcutaneous Dose: 1 mg/kg.
Chronic (Multiple Dose) Oral: Rat Dose: 16800 ug/kg/8W-C; Toxic Effects: Brain and coverings - Other degenerative changes; Blood - Other changes; Biochemical - True cholinesterase. Chicken Dose: 5 mg/kg/8W-I; Toxic Effects: Brain and coverings - Other degenerative changes; Behavioral - Ataxia; Biochemical - True cholinesterase.
Chronic (Multiple Dose) Dermal: Rat Route: Subcutaneous; Dose: 13 mg/kg/13D-I; Toxic Effects: Brain and coverings - Other degenerative changes; Biochemical - True cholinesterase. Chicken Route: Subcutaneous; Dose: 4 mg/kg/4D-I; Toxic Effects: Spinal Cord - Other degenerative changes; Peripheral Nerve and sensation - Structural change in nerve or sheath; Behavioral - Muscle weakness.
Reproductive/Teratogenic: Rat Route: Subcutaneous; Dose: 6 mg/kg; Duration: female 6-20D of pregnancy; Effects on Newborn - Stillbirth. Mouse Route: Intraperitoneal; Dose: 500 ug/kg; Duration: female 13D of pregnancy; Effects on Newborn - Biochemical and metabolic; Behavioral.

Hazard Overviews

Poison

Health: Poison. Other Acute Effects: danger: highly toxic cholinesterase inhibitor; may be fatal if inhaled, swallowed, or absorbed through skin; exposure can cause: coughing, chest pains, difficulty in breathing; nausea, dizziness and headache stomach pains, vomiting, diarrhea; convulsions; target organ: central nervous system.
Fire: Hazards: emits toxic fumes; container explosion may occur. Extinguishing agents: carbon dioxide, dry chemical powder or appropriate foam. Precautions: combustible liquid.
Reactivity: Incompatible with: strong bases, strong oxidizing agents, may decompose on exposure to moist air or water. Hazardous decomposition products: toxic fumes of: carbon monoxide, carbon dioxide, hydrogen fluoride; thermal decomposition may produce toxic fumes of phosphorus oxides and/or phosphine.
Carcinogenicity: IARC - Not listed; NIOSH - Not listed; NTP - Not listed; ACGIH - Not listed; OSHA - Not listed; EPA - Not listed; MAK - Not listed
Primary Target Organs:

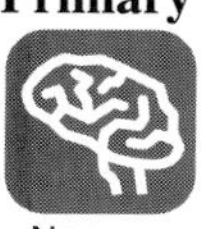
Nervous System

Environmental

Ecotoxicity: ET_{50} Squid 184 ppm/0.5 hr, toxic effect: cholinesterase inhibition.
Environmental Fate: In the atmosphere, it is estimated to have a half-life of 3.06 days. In moist soils at 25 °C, it should hydrolyze with a half-life of several days. Biodegradation may also be a significant removal mechanism. Volatilization from soil will be minimal except possibly on the surface of warm, dry soils. Due to its relatively high water solubility, it is expected to be highly mobile in soils. In aquatic systems it is not expected to bioconcentrate, sorb to sediments or volatilize to any significant degree. Hydrolysis and possibly biodegradation will be the most significant removal mechanisms. No information was found on aquatic photolysis.
Cleanup/Disposal: Guide No. 152: Do not touch damaged containers or spilled material unless wearing appropriate protective clothing. Stop leak if you can do it without risk. Prevent entry into waterways, sewers, basements or confined areas. Cover with plastic sheet to prevent spreading. Absorb or cover with dry earth, sand or other non-combustible material and transfer to containers. Do not get water inside containers.

Environmental Physical Data

Henry's Law Constant: estimated at 9.11 x10^{-6}
Octanol/Water Partition Coefficient: log K_{ow} = 1.92
Sorption Partition Coefficient: K_{oc} = estimated at 21.7
BCF: none likely

Regulations

RCRA 40CFR: Listed Hazardous Waste No. P043 Toxic Waste
CERCLA: 40CFR 302.4: Listed per RCRA Section 3001 RQ: 100 lb (45.35 kg)
SARA 40CFR 372.65: Not listed TPQ: 100 lb
SARA EHS 40CFR 355: Listed TPQ: 100 lb
TSCA: Listed

DII8800 CAS #: 38640-62-9

DIISOPROPYLNAPHTHALENE

RTECS: QJ1527000
EINECS Number: 254-052-6
Molecular Formula: $C_{16}H_{20}$
Formula Weight: 212.34
Synonyms: BIS(ISOPROPYL)NAPHTHALENE; BIS(1-METHYLETHYL)NAPHTHALENE; K 113; KMC 113; KMC-R 113; NAPHTHALENE,BIS(1-METHYLETHYL)-
Description: clear, yellowish-brown liquid; faint, sweet odor
Use: organic intermediate; solvent; chemical intermediate; substitute for polychlorinated biphenyls

Physical Properties

Boiling Point: 290 °C (554 °F) to 295 °C (563 °F)
Saturated Vapor Density: 1.200004991 kg/m^3
Density: 0.95 at 30 °C
Bulk Density: 7.9 lbs/gal
Vapor Pressure: 5 x10^{-4} mm Hg at 25 °C
Water Solubility: 0.11 mg/L
Flash Point: 140 °C Open Cup

RTECS Toxicity Data

Acute Oral: Mouse LD_{50} Dose: 3400 mg/kg; Toxic Effects: Behavioral - Somnolence (general depressed activity); Behavioral - Change in motor activity (specific assay); Blood - Hemorrhage.

Hazard Overviews

Fire: Will burn.
Carcinogenicity: IARC - Not listed; NIOSH - Not listed; NTP - Not listed; ACGIH - Not listed; OSHA - Not listed; EPA - Not listed; MAK - Not listed

Environmental

Ecotoxicity: LD_{50} Seriola quinqueradiata (yellowtail) IP approx 2 ml/kg
Environmental Fate: Should biodegrade in the environment, but hydrolysis should not be important. In sunlit environmental media, it will undergo direct photolysis. It is expected to be immobile in soil. It has the potential to bioconcentrate in aquatic systems. It may also partition from the water column to organic matter contained in sediments and suspended solids. Volatilization from environmental waters may be rapid. The volatilization half-lives from a model river and model pond, the latter considers the effect of adsorption, have been estimated to be 5.2 hr and 97.1 days, respectively. It is expected to exist entirely in the vapor phase in ambient air. In the atmosphere, reactions with photochemically produced hydroxyl radicals (half-life of 5.4 hr) is likely to be important.

Environmental Physical Data

Henry's Law Constant: calculated at 1.27 x10^{-3}
Octanol/Water Partition Coefficient: log K_{ow} = 4.90
Sorption Partition Coefficient: K_{oc} = estimated at 1.1 x10^4 to 1.47 x10^4
BCF: carp 203

Regulations

RCRA 40CFR: Not listed
CERCLA: 40CFR 302.4: Not listed
SARA 40CFR 372.65: Not listed
SARA EHS 40CFR 355: Not listed
TSCA: Listed

DII9000 CAS #: 2078-54-8

2,6-DIISOPROPYLPHENOL

RTECS: SL0810000
EINECS Number: 218-206-6
Molecular Formula: $C_{12}H_{18}O$
Formula Weight: 178.30

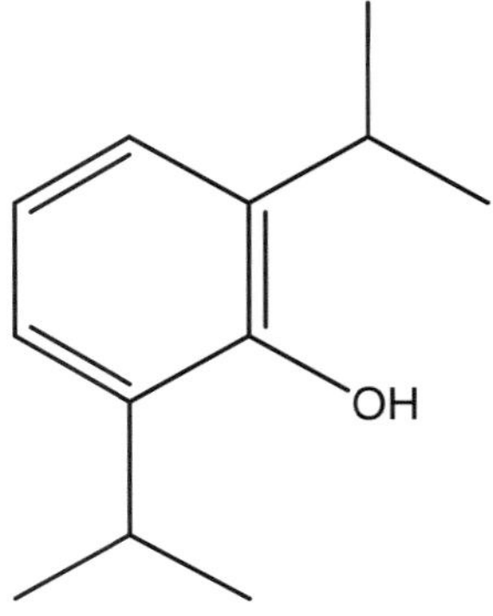

Chemical Structure

Synonyms: 2,6-BIS(1-METHYLETHYL)PHENOL; DIPRIVAN; ICI 35868; PHENOL,2,6-BIS(1-METHYLETHYL)-(9CI); PROPOFOL

Physical Properties

Boiling Point: 256 °C (492.8 °F) at 764 mm Hg
Freezing Point: 18 °C (64.4 °F)
Specific Gravity: 1.514
Water Solubility: Insoluble
Refraction Index: 1.5140

Hazard Overviews

Health: Irritating to eyes/skin/respiratory tract. Harmful. Other Acute Effects: harmful if swallowed; may be harmful if inhaled; may be harmful if absorbed through the skin; depending on the intensity and duration of exposure, effects may vary from mild irritation to severe destruction of tissue; prolonged contact can cause damage to the eyes severe irritation or burns.
Fire: Hazards: under fire conditions, material may decompose to form flammable and/or explosive mixtures in air. Extinguishing agents: water spray; carbon dioxide, dry chemical powder or appropriate foam. Precautions: combustible liquid.

Reactivity: Incompatible with: bases, acid chlorides, acid anhydrides, oxidizing agents. Hazardous decomposition products: toxic fumes of: carbon monoxide, carbon dioxide.
Carcinogenicity: IARC - Not listed; NIOSH - Not listed; NTP - Not listed; ACGIH - Not listed; OSHA - Not listed; EPA - Not listed; MAK - Not listed
Primary Target Organs:

Eyes

Skin

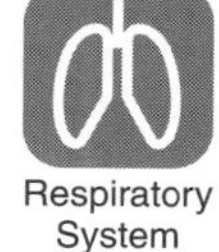
Respiratory System

Environmental

Regulations

RCRA 40CFR: Not listed
CERCLA: 40CFR 302.4: Not listed
SARA 40CFR 372.65: Not listed
SARA EHS 40CFR 355: Not listed
TSCA: Listed

DIL3000 CAS #: 105-74-8

DILAURYL PEROXIDE

RTECS: OF2625000
DOT: UN2124; UN2893; IMO5.2
EINECS Number: 203-326-3
Molecular Formula: $C_{24}H_{46}O_4$
Structured MF: $[CH_3(CH_2)_{10}COO]_2$
Formula Weight: 398.70

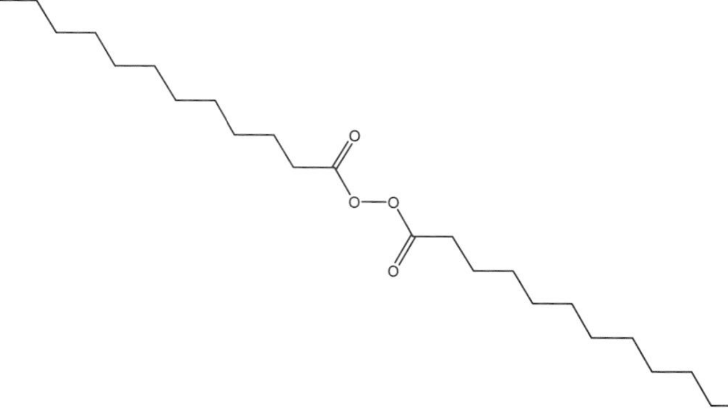

Chemical Structure

Synonyms: ALPEROX C; DILAUROYL PEROXIDE; DODECANOYL PEROXIDE; DYP-97F; LAUROX; LAUROYL PEROXIDE; LAURYDOL; LYP 97F; LYP 97; PEROXIDE,BIS(1-OXODODECYL); PEROXIDE,BIS(1-OXODODECYL)-; PEROXIDE,DIDODECANOYL; PEROXYDE DE LAUROYLE
Description: white powder, flakes, granules; faint soapy odor
Use: bleach for flour, vegetable oils, fats & waxes; polymerization agent in plastics ind; curing agent for rubber, unsaturated polyester resins, acrylates & styrenated alkyd resins; burn-out agent for acetate yarns; in cosmetics & pharmaceuticals ind; to catalyze polymerization of styrene; polymerization initiator-eg, for vinyl chloride polymers; cross-linking agent; composite material for dental fillings.

Physical Properties

Boiling Point: Decomposes
Freezing Point: 49 °C (120.2 °F)
Specific Gravity: Solid 0.91 at 25 °C (solid)
Water Solubility: Insoluble in Water
Other Solubilities: Soluble in oils and most organic solvents.
Flash Point: Not pertinent (oxidizing combustible solid)

RTECS Toxicity Data

Irritation Eye: Rabbit Standard Draize Test Dose: 500 mg/24H; Reaction: mild.
Tumorigenic: Mouse Route: Subcutaneous; Dose: 184 mg/kg/46W-I; Toxic Effects: Tumorigenic - Equivocal tumorigenic agent by RTECS criteria; Tumorigenic - Tumors at site of application.

Hazard Overviews

Explosive

Flammable

Fire Diamond

Health: Severely irritating to eyes/skin/respiratory tract. Also Causes: headache, stomach irritation.
Fire: Flammable. Oxidizer. Shock-sensitive. For small fires use dry chemical, water spray, carbon dioxide, or regular foam. For large fires flood area with water. Ignites organics and contains sufficient oxygen to support combustion in oxygen deficient atmospheres.
Reactivity: Unstable. Hazardous polymerization cannot occur. Avoid: adding dilauryl peroxide to resins before tertiary amine; adding to hot materials; placing in heated equipment; temperatures above 104 °F. Incompatible with: charcoal; thiocyanates; organic material; strong acids; alkalies; accelerators; reducers; heavy metal compounds; stabilizers. Hazardous decomposition products: carbon dioxide; lauric acid; n-docosane; n-undecane; undecyllaurate.
Carcinogenicity: IARC - Group 3, Not classifiable as to carcinogenicity to humans; NIOSH - Not listed; NTP - Not listed; ACGIH - Not listed; OSHA - Not listed; EPA - Not listed; MAK - Not listed
Primary Target Organs:

Eyes

Skin

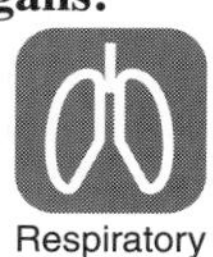
Respiratory System

Environmental

Cleanup/Disposal: Guide No. 145: Eliminate all ignition sources (no smoking, flares, sparks or flames in immediate area). Keep combustibles (wood, paper, oil, etc.) away from spilled material. Do not touch damaged containers or spilled material unless wearing appropriate protective clothing. Keep substance wet using water spray. Stop leak if you can do it without risk. Small Spills: Take up with inert, damp, noncombustible material using clean non-sparking tools and place into loosely covered plastic containers for later disposal. Large Spills: Wet down with water and dike for later disposal. Prevent entry into waterways, sewers, basements or confined areas.

Environmental Physical Data

BCF: no food chain concentration potential

Regulations

RCRA 40CFR: Not listed
CERCLA: 40CFR 302.4: Not listed
SARA 40CFR 372.65: Not listed
SARA EHS 40CFR 355: Not listed
TSCA: Listed

DIL6000 CAS #: 123-28-4

DILAURYL THIODIPROPIONATE

RTECS: UF8000000
EINECS Number: 204-614-1
Molecular Formula: $C_{30}H_{58}O_4S$
Formula Weight: 514.94

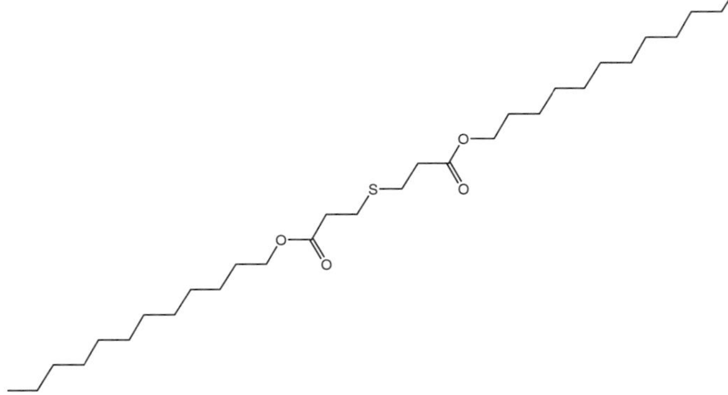

Chemical Structure

Synonyms: ADVASTAB 800; ANTIOXIDANT AS; ANTIOXIDANT LTDP; BIS(DODECYLOXYCARBONYLETHYL) SULFIDE; CARSTAB DLTDP; CYANOX LTDP; DIDODECYL 3,3'-THIODIPROPIONATE; DILAURYL 3,3'-THIODIPROPIONATE; DILAURYL BETA',BETA'-THIODIPROPIONATE; DILAURYL BETA,BETA'-THIODIPROPIONATE; DILAURYL BETA-THIODIPROPIONATE; DILAURYLESTER KYSELINY BETA',BETA'-THIODIPROPIONOVE; DLT; DLTDP; DLTP; DMPTP; IPOGNOX 89; IRGANOX PS 800; LAURYL 3,3'-THIODIPROPIONATE; LUSMIT; MILBAN F; NEGANOX DLTP; PLASTANOX LTDP; PLASTANOX LTDP ANTIOXIDANT; PROPANOIC ACID,3,3'-THIOBIS-,DIDODECYL ESTER; PROPIONIC ACID,3,3'-THIODI-,DIDODECYL ESTER; STABILIZER DLT; THIOBIS(DODECYL PROPIONATE); THIODIPROPIONIC ACID,DILAURYL ESTER; TYOX B
Description: white flakes; sweetish odor
Use: stabilization of food packaging films; antioxidant in edible fats; additive for high-pressure lubricants & greases; plasticizer & softening agent

Physical Properties

Freezing Point: 40 °C (104 °F)
Specific Gravity: 0.975 at 25 °C
Water Solubility: Insoluble in Water
Other Solubilities: Soluble in most organic solvents

RTECS Toxicity Data

Acute Oral: Rat LD_{50} Dose: >2500 mg/kg. Mouse LD_{50} Dose: >2 gm/kg.
Irritation Eye: Rabbit Standard Draize Test Dose: 500 mg/24H; Reaction: mild.

Hazard Overviews

Health: May be irritating to eyes/skin. Other Acute Effects: may be harmful by inhalation, ingestion, or skin absorption.
Fire: Hazards: emits toxic fumes; in powder form capable of creating a dust explosion. Extinguishing agents: water spray; carbon dioxide, dry chemical powder or appropriate foam. Precautions: combustible liquid.
Reactivity: Hazardous decomposition products: toxic fumes of: carbon monoxide, carbon dioxide, sulfur oxides.
Carcinogenicity: IARC - Not listed; NIOSH - Not listed; NTP - Not listed; ACGIH - Not listed; OSHA - Not listed; EPA - Not listed; MAK - Not listed

Environmental

Regulations

RCRA 40CFR: Not listed
CERCLA: 40CFR 302.4: Not listed
SARA 40CFR 372.65: Not listed
SARA EHS 40CFR 355: Not listed
TSCA: Listed

DIM1000 CAS #: 1165-48-6

DIMEFLINE

RTECS: LK8400000
EINECS Number: 214-616-4
Molecular Formula: $C_{20}H_{21}NO_3$
Formula Weight: 323.38
Synonyms: 4H-1-BENZOPYRAN-4-ONE,8-((DIMETHYLAMINO)METHYL)-7-METHOXY-3-METHYL-2-PHENYL-; DIMEFLIN; 8-(DIMETHYLAMINOMETHYL)-7-METHOXY-3-METHYLFLAVONE; 8-[(DIMETHYLAMINO)METHYL]-7-METHOXY-3-METHYLFLAVONE; 8-[(DIMETHYLAMINO)METHYL]-7-METHOXY-3-METHYL-2-PHENYL-4H-1-BENZOPYRAN-4-ONE; 8-DIMETHYLAMINOMETHYL-7-METHOXY-3-METHYL-2-PHENYLCHROMONE; 8-((DIMETHYLAMINO)METHYL)-7-METHOXY-3-METHYL-2-PHENYLFLAVONE; DW 62; FLAVONE,8-((DIMETHYLAMINO)METHYL)-7-METHOXY-3-METHYL-; FLAVONE,8-(DIMETHYLAMINOMETHYL)-7-METHOXY-3-METHYL-; MALIVAN; N-(7-METHOXY-3-METHYL-4-OXO-2-PHENYL-4H-CHROMEN-8-YL)METHYL-N,N-DIMETHYLAMINE; REANIMIL
Description: crystals
Use: respiratory stimulant

Physical Properties

Boiling Point: Decomposes at 213 °C (415 °F) to 214 °C (417.2 °F)
Freezing Point: Decomposes at 213 °C (415.4 °F) to 214 °C (417.2 °F)
Other Solubilities: Chloroform: Soluble

RTECS Toxicity Data

Acute Oral: Rat LD_{50} Dose: 40 mg/kg; Toxic Effects: Behavioral - Convulsions or effect on seizure threshold; Behavioral - Excitment; Behavioral - Rigidity. Mouse LD_{50} Dose: 12 mg/kg; Toxic Effects: Behavioral - Convulsions or effect on seizure threshold; Behavioral - Excitment; Behavioral - Rigidity.
Acute Dermal: Mouse LD_{50} Route: Subcutaneous Dose: 4 mg/kg; Toxic Effects: Behavioral - Convulsions or effect on seizure threshold; Behavioral - Excitment; Behavioral - Rigidity. Hamster LD_{50} Route: Subcutaneous Dose: 12 mg/kg; Toxic Effects: Behavioral - Convulsions or effect on seizure threshold; Behavioral - Excitment; Behavioral - Rigidity.

Hazard Overviews

Carcinogenicity: IARC - Not listed; NIOSH - Not listed; NTP - Not listed; ACGIH - Not listed; OSHA - Not listed; EPA - Not listed; MAK - Not listed

Environmental

Regulations

RCRA 40CFR: Not listed
CERCLA: 40CFR 302.4: Not listed
SARA 40CFR 372.65: Not listed
SARA EHS 40CFR 355: Not listed
TSCA: Not listed

DIM1050 CAS #: 115-26-4

DIMEFOX

RTECS: TD4025000
EINECS Number: 204-076-8
Molecular Formula: $C_4H_{12}FN_2OP$
Formula Weight: 154.13
Synonyms: BFP; BFPO; BIS(DIMETHYLAMIDO)FLUOROPHOSPHATE; BIS(DIMETHYLAMIDO)-PHOSPHORYL FLUORIDE; BIS(DIMETHYLAMIDO)PHOSPHORYL FLUORIDE; BIS(DIMETHYLAMINO)FLUOROPHOSPHATE; BISDIMETHYLAMINOFLUOROPHOSPHINE OXIDE; CR 409; DIFO; DMF; ENT 19,109; FLUOPHOSPHORIC ACID DI(DIMETHYLAMIDE); FLUORURE DE N,N,N',N'-TETRAMETHYLE PHOSPHORO-DIAMIDE; HANANE; PESTOX 14; PESTOX IV; PESTOX XIV; PHOSPHINE OXIDE,BIS(DIMETHYLAMINO)FLUORO-; PHOSPHORODIAMIDIC FLUORIDE,TETRAMETHYL-; S-14; T-2002; TERRA-SYSTAM; TERRA-SYTAM; TERRASYTUM; TETRA SYTAM; N,N,N',N'-TETRAMETHYL-DIAMIDO-FLUOR-PHOSPHIN-OXID; N,N,N',N'-TETRAMETHYL-DIAMIDO-FOSFORZUUR-FLUORIDE; TETRAMETHYLDIAMIDOPHOSPHORIC FLUORIDE; N,N,N',N'-TETRAMETHYL-DIAMIDO-PHOSPHORSAEURE-FLUORID; N,N,N',N'-TETRAMETHYLPHOSPHORODIAMIDIC FLUORIDE; N,N,N,N-TETRAMETHYLPHOSPHORODIAMIDIC FLUORIDE; TETRAMETHYLPHOSPHORODIAMIDIC FLUORIDE; N,N,N',N'-TETRAMETIL-FOSFORODIAMMIDO-FLUORURO; TL 792; WACKER S 14/10
Description: colorless liquid; fishy odor
Use: systemic acaricide; systemic insecticide

Physical Properties

Boiling Point: 86 °C (187 °F) at 15 mm Hg
Specific Gravity: 1.1151 at 20 °C/4 °C
Saturated Vapor Density: 1.202452639 kg/m^3
Vapor Pressure: 0.36 mm Hg at 25 °C
Water Solubility: Freely Soluble in Water
Other Solubilities: miscible with most organic solvents.
Refraction Index: 1.4267 at 20 °C/D

RTECS Toxicity Data

Acute Oral: Rat LD_{50} Dose: 1 mg/kg; Toxic Effects: Behavioral - Coma; Lungs, Thorax, or Respiration - Respiratory stimulation; Gastrointestinal - Changes in structure or function of salivary glands. Monkey LD_{50} Dose: 2 mg/kg.
Acute Dermal: Rabbit LD_{50} Route: Subcutaneous Dose: 6 mg/kg; Toxic Effects: Behavioral - Coma; Lungs, Thorax, or Respiration - Respiratory stimulation; Gastrointestinal - Changes in structure or function of salivary glands. Rat LD_{50} Route: Skin; Dose: 2 mg/kg. Rat LD_{50} Route: Subcutaneous Dose: 300 ug/kg.

Hazard Overviews

Carcinogenicity: IARC - Not listed; NIOSH - Not listed; NTP - Not listed; ACGIH - Not listed; OSHA - Not listed; EPA - Not listed; MAK - Not listed

Environmental

Regulations

RCRA 40CFR: Not listed
CERCLA: 40CFR 302.4: Not listed
SARA 40CFR 372.65: Not listed TPQ: 500 lb
SARA EHS 40CFR 355: Listed TPQ: 500 lb
TSCA: Not listed

DIM1100 CAS #: 523-87-5

DIMENHYDRINATE

RTECS: XH5082000
EINECS Number: 208-350-8
Molecular Formula: $C_{24}H_{28}ClN_5O_3$
Formula Weight: 469.96

Chemical Structure

Synonyms: AMOSYT; ANAUTINE; ANDRAMINE; ANTEMIN; AVIOMARIN; O-BENZHYDRYLDIMETHYLAMINOETHANOL 8-CHLOROTHEOPHYLLINATE; BENZHYDRYL-BETA-DIMETHYLAMINOETHYLETHER 8-CHLOROTHEOPHYLLINE; 2-(BENZHYDRYLOXY)-N,N-DIMETHYLETHYLAMINE 8-CHLOROTHEOPHTHEOPHYLLINATE; 2-(BENZHYDRYLOXY)-N,N-DIMETHYLETHYLAMINE CMPD WITH 8 CHLOROTHEOPHYLLINE; 2-(BENZHYDRYLOXY)-N,N-DIMETHYLETHYLAMINE COMPD. WITH 8-CHLOROTHEOPHYLLINE; 2-(BENZHYDRYLOXY)-N,N-DIMETHYLETHYLAMINE8-CHLOROTHEOPHYLLINATE; CHLORANAUTINE; 8-CHLORO-3,7-DIHYDRO-1,3-DIMETHYL-1H-PURINE-2,6-DIONE CMPD WITH; 8-CHLOROTHEOPHYLLINE CMPD WITH 2-(DIPHENYLMETHOXY)-N,N-DIMETHYLAMINE (1:1); 8-CHLOROTHEOPHYLLINE,COMPD. WITH 2-(DIPHENYLMETHOXY)-N,N-DIMETHYLETHYLAMINE (1:1); DIAMARIN; BETA-DIMETHYLAMINOETHYL BENZHYDRYL ETHER 1,3-DIMETHYL-8-CHLOROXANTHINE; BETA-DIMETHYLAMINOETHYL BENZHYDRYL ETHER1,3-DIMETHYL-8-CHLOROXANTHINE; N,N-DIMETHYL-2-DIPHENYLMETHOXYETHYLAMINE8-CHLOROTHEOPHYLLINATE; DIPHENHYDRAMINE 8-CHLOROTHEOPHYLLINATE; DIPHENHYDRAMINE 8-CHLOROTHEOPHYLLINE; DIPHENHYDRINATE; 2-(DIPHENYLMETHOXY)-N,N-DIMETHYLETHANAMINE (1:1); 2-(DIPHENYLMETHOXY)-N,N-DIMETHYLETHYLAMINE8-CHLOROTHEOPHYLLINATE; 2-(DIPHENYLMETHOXY)-N,N-DIMETHYLETHYLANIME CMPD WITH 8-CHLOROTHEOPHYLLINE (1:; DRAMAMIN; DRAMAMINE; DRAMARIN; DRAMYL; DROMYL; ELDODRAM; EMEDYL; EMES; EPHA; ETHYLAMINE 2-(DIPHENYLMETHOXY)-N,N-DIMETHYL-,COMPD WITH8-CHLOROTHEOPHYLLINE (1:1); ETHYLAMINE,N,N-DIMETHYL-2-(DIPHENYLMETHOXY)-,COMPD.WITH 8-CHLOROTHEOPHYLLINE; FASTON; GRAVINOL; GRAVOL; MENHYDRINATE; NEO-NAVIGAN; NOVAMIN; NOVAMINE; PERMITAL; 1H-PYRINE-2,6-DIONE,8-CHLORO-3,7-DIHYDRO-1,3-DIMETHYL-,COMPD WITH 2-(DIPHENYLMETHOXY)-N,N-DIMETHYLETHANAMINE(1:1); REIDAMINE; REISE-ENGLETTEN; REMOVINE; SUPREMAL; TEODRAMIN; THEOPHYLLINE,8-CHLORO-,COMPD WITH2-(DIPHENYLMETHOXY)-N,N-DIMETHYETHYLAMINE (1:1); TRAVELIN; TRAVELMIN; VOMEX A; XAMAMINA; XAMAMINE

Description: white crystalline powder; odorless

Use: antinauseant; (vet)has been used as antiemetic; antihistamine

Physical Properties

Freezing Point: 102 °C (215.6 °F) to 107 °C (224.6 °F)

Water Solubility: About 3 mg/mL

Other Solubilities: 95% Ethanol: Soluble (>=10 mg/ml at 22 °C); Benzene: Soluble; DMSO: Soluble (>=10 mg/ml at 22 °C); Ether: Very slightly Soluble.

pH: Saturated solution 6.8 to 7.3

Flash Point: Not available; probably nonflammable

RTECS Toxicity Data

Acute Oral: Man TD_{Lo} Dose: 11400 ug/kg; Toxic Effects: Behavioral - Hallucinations, distorted perceptions. Infant TD_{Lo} Dose: 114 mg/kg; Toxic Effects: Behavioral - Convulsions or effect on seizure threshold; Behavioral - Coma; Cardiac - Arrythmias (including changes in conduction). Woman LD_{Lo} Dose: 100 mg/kg; Toxic Effects: Brain and coverings - Increased intracranial pressure; Behavioral - Convulsions or effect on seizure threshold; Cardiac - Arrythmias (including changes in conduction). Rat LD_{50} Dose: 1320 mg/kg.

Mutagenic: Bacteria - S Typhimurium Mutations in Microorganisms; Dose: 333 ug/plate (+S9).

Hazard Overviews

Health: Irritating to eyes/skin/respiratory tract. Toxic. Other Acute Effects: harmful if swallowed, inhaled, or absorbed through skin; exposure to high concentrations may cause hallucinations and distorted perceptions; adverse reaction symptoms: sedation; sleepiness; dizziness; disturbed coordination; epigastric distress; urticaria; anaphylactic shock; photosensitivity; chills; hypotension; headache; palpitations; tachycardia; extrasystoles; hemolytic anemia; agranulocytosis; thrombocytopenia; fatigue; confusion; excitation; nervousness; tremor; irritability; insomnia; euphoria; paraesthesia; blurred vision; diplopia; vertigo; tinnitus; acute labyrinthitis; neuritis; convulsions; anorexia; nausea; vomiting; diarrhea; constipation; urinary frequency; difficult urination; urinary retention; early menses; target organs: eyes; ears; central nervous system. Chronic Effects: may cause congenital malformation in the fetus. Possible human carcinogen.

Fire: Noncombustible. Extinguishing agents: water spray; carbon dioxide, dry chemical powder or appropriate foam. Precautions: combustible liquid.

Reactivity: Stable. Hazardous polymerization will not occur. Hazardous decomposition products: toxic fumes of: carbon monoxide, carbon dioxide, nitrogen oxides, hydrogen chloride gas.

Carcinogenicity: IARC - Not listed; NIOSH - Not listed; NTP - Not listed; ACGIH - Not listed; OSHA - Not listed; EPA - Not listed; MAK - Not listed

Primary Target Organs:

Eyes

Skin

Respiratory System

Nervous System

Environmental

Regulations

RCRA 40CFR: Not listed

CERCLA: 40CFR 302.4: Not listed

SARA 40CFR 372.65: Not listed

SARA EHS 40CFR 355: Not listed

TSCA: Listed

DIM1150 **CAS #: 59-52-9**

DIMERCAPROL

RTECS: UB2625000
EINECS Number: 200-433-7
Molecular Formula: $C_3H_8OS_2$
Formula Weight: 124.23

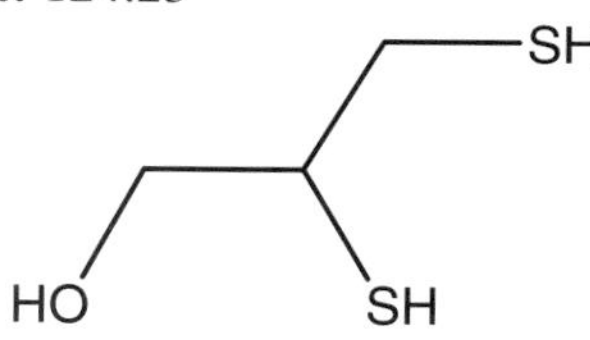

Chemical Structure

Synonyms: ANTOXOL; BAL; BRITISH ANTI-LEWISITE; BRITISH ANTILEWISITE; DICAPTOL; DIMERCAPROL PROPANOL; DIMERCAPTOL; 2,3-DIMERCAPTOL-1-PROPANOL; 1,2-DIMERCAPTO-3-PROPANOL; 2,3-DIMERCAPTOPROPAN-1-OL; 2,3-DIMERCAPTOPROPANOL; DIMERCAPTOPROPANOL; DIMERKAPROL; DIMERSOL; DITHIOGLYCERINE; 1,2-DITHIOGLYCEROL; ALPHA,BETA-DITHIOGLYCEROL; DITHIOGLYCEROL; 2,3-DITHIOPROPANOL; DMP; GLYCEROL,1,2-DITHIO-; 3-HYDROXY-1,2-PROPANEDITHIOL; PANOBAL; 1-PROPANOL,2,3-DIMERCAPTO-; SULFACTIN

Description: colorless, viscous oily liquid; pungent offensive odor of mercaptans

Use: antidote to arsenic, gold and mercury poisoning; chelating agent; detoxicant for heavy metal poisoning

Physical Properties

Boiling Point: 120 °C (248 °F) at 15 mm Hg
Freezing Point: 77 °C (170.6 °F)
Specific Gravity: 1.2385 at 25 °C/4 °C
Vapor Density: 4.3 Air=1
Saturated Vapor Density: 1.225509887 kg/m^3
Vapor Pressure: 4.92 mm Hg at 25 °C
Water Solubility: 8.7 g dissolves in 100 ml Water (decomposes)
Other Solubilities: 95% Ethanol: >=100 mg/ml at 20 °C; Acetone: >=100 mg/ml at 20 °C; DMSO: >=100 mg/ml at 20 °C; Ether: Soluble; Oils: Soluble.
Refraction Index: 1.5720 at 25 °C/D
Flash Point: > 112 °C

RTECS Toxicity Data

Acute Oral: Mouse LD_{50} Dose: 217 mg/kg.
Acute Dermal: Rat LD_{50} Route: Subcutaneous Dose: 2 gm/kg.
Reproductive/Teratogenic: Mouse Route: Subcutaneous; Dose: 100 mg/kg; Duration: female 11D of pregnancy; Effects on Embryo or Fetus - Fetal death. Mouse Route: Subcutaneous; Dose: 100 mg/kg; Duration: female 12D of pregnancy; Specific Developmental Abnormalities - Musculoskeletal system. Mouse Route: Subcutaneous; Dose: 200 mg/kg; Duration: female 12-13D of pregnancy Specific Developmental Abnormalities - Craniofacial (including nose and tongue).

Hazard Overviews

Health: Irritating to eyes/skin/respiratory tract. Toxic. Other Acute Effects: harmful if swallowed, inhaled, or absorbed through skin; exposure can cause nausea; headache; vomiting; target organ: heart.

Fire: Will burn. Hazards: emits toxic fumes. Extinguishing agents: carbon dioxide, dry chemical powder or appropriate foam. Precautions: combustible liquid.

Reactivity: Incompatible with: bases, oxidizing agents, reducing agents, alkali metals. Hazardous decomposition products: toxic fumes of: carbon monoxide, carbon dioxide, sulfur oxides, hydrogen sulfide gas.

Carcinogenicity: IARC - Not listed; NIOSH - Not listed; NTP - Not listed; ACGIH - Not listed; OSHA - Not listed; EPA - Not listed; MAK - Not listed

Primary Target Organs:

Eyes

Skin

Respiratory System

Cardio-vascular

Environmental

Cleanup/Disposal: Guide No. 171: Do not touch or walk through spilled material. Stop leak if you can do it without risk. Prevent dust cloud. Avoid inhalation of asbestos dust. Small Dry Spills: With clean shovel place material into clean, dry container and cover loosely; move containers from spill area. Small Spills: Take up with sand or other noncombustible absorbent material and place into containers for later disposal. Large Spills: Dike far ahead of liquid spill for later disposal. Cover powder spill with plastic sheet or tarp to minimize spreading. Prevent entry into waterways, sewers, basements or confined areas.

Regulations

RCRA 40CFR: Not listed
CERCLA: 40CFR 302.4: Not listed
SARA 40CFR 372.65: Not listed
SARA EHS 40CFR 355: Not listed
TSCA: Listed

DIM1200 **CAS #: 1072-71-5**

2,5-DIMERCAPTO-1,3,4-THIADIAZOLE

RTECS: XI3850000
EINECS Number: 214-014-1
Molecular Formula: $C_2H_2N_2S_3$
Formula Weight: 150.24

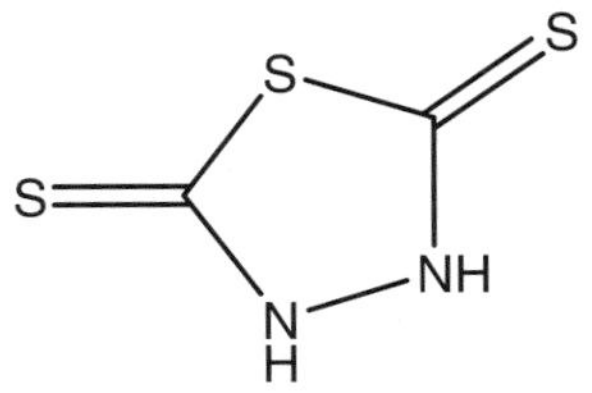

Chemical Structure

Synonyms: BISMUTHIOL I; 2,5-DIMERCAPTOTHIADIAZOLE; PY 61H; 1,3,4-THIADIAZOL-DITHIOL-(2,5)

Physical Properties

Freezing Point: 162 °C (323.6 °F)
Water Solubility: 0.03%

RTECS Toxicity Data

Acute Oral: Quail LD_{50} Dose: >316 mg/kg.
Irritation Eye: Rabbit Standard Draize Test Dose: 100 mg; Reaction: severe.

Hazard Overviews

Health: Irritating to eyes/skin/respiratory tract. Other Acute Effects: may be harmful by inhalation, ingestion, or skin absorption; exposure can cause nausea; headache; vomiting.
Fire: Hazards: emits toxic fumes. Extinguishing agents: water spray; carbon dioxide, dry chemical powder or appropriate foam. Precautions: combustible liquid.
Reactivity: Incompatible with: strong oxidizing agents. Hazardous decomposition products: toxic fumes of: carbon monoxide, carbon dioxide, nitrogen oxides, sulfur oxides.
Carcinogenicity: IARC - Not listed; NIOSH - Not listed; NTP - Not listed; ACGIH - Not listed; OSHA - Not listed; EPA - Not listed; MAK - Not listed
Primary Target Organs:

Eyes

Skin
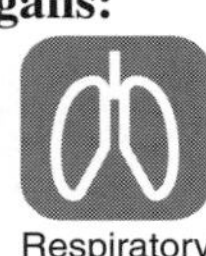
Respiratory System

Environmental

Regulations

RCRA 40CFR: Not listed
CERCLA: 40CFR 302.4: Not listed
SARA 40CFR 372.65: Not listed
SARA EHS 40CFR 355: Not listed
TSCA: Listed

DIM1250 CAS #: 9006-65-9

DIMETHICONE

Molecular Formula: Unknown
Description: colorless oil
Use: ointment base & topical drug ingredient skin protectant; as lubricants, water repellents; defoaming agent in ready-to-serve foods; constituent of antacid preparations

Physical Properties

Water Solubility: Immiscible with Water
Other Solubilities: miscible with Chloroform, Ether

Hazard Overviews

Carcinogenicity: IARC - Not listed; NIOSH - Not listed; NTP - Not listed; ACGIH - Not listed; OSHA - Not listed; EPA - Not listed; MAK - Not listed

Environmental

Regulations

RCRA 40CFR: Not listed
CERCLA: 40CFR 302.4: Not listed
SARA 40CFR 372.65: Not listed
SARA EHS 40CFR 355: Not listed
TSCA: Not listed

DIM1300 CAS #: 55290-64-7

DIMETHIPIN

RTECS: JO5090000
EINECS Number: 259-572-7
Molecular Formula: $C_6H_{10}O_4S_2$
Formula Weight: 210.28
Synonyms: 2,3-DIHYDRO-5,6-DIMETHYL-1,4-DITHIIN 1,1,4,4-TETROXIDE; 1,4-DITHIIN,2,3-DIHYDRO-5,6-DIMETHYL-,1,1,4,4-TETRAOXIDE; HARVADE; N 252; OXIDIMETHIIN; TETRATHIIN; TETRATHIIN (DESICCANT); UBI-N 252

Physical Properties

Freezing Point: 167 °C (332.6 °F) to 169 °C (336.2 °F)
Water Solubility: 3 g/l 25 °C
Other Solubilities: 180 g/kg acetone, 10 g/kg xylene

RTECS Toxicity Data

Acute Oral: Rat LD_{50} Dose: 1150 mg/kg. Mouse LD_{50} Dose: 440 mg/kg.
Acute Inhalation: Rat LC_{50} Dose: >20 gm/m^3/1hr.
Acute Dermal: Rabbit LD_{50} Route: Skin; Dose: 8 gm/kg.

Hazard Overviews

Carcinogenicity: IARC - Not listed; NIOSH - Not listed; NTP - Not listed; ACGIH - Not listed; OSHA - Not listed; EPA - Class C, Possible human carcinogen; MAK - Not listed

Environmental

Regulations

RCRA 40CFR: Not listed
CERCLA: 40CFR 302.4: Not listed
SARA 40CFR 372.65: Listed

SARA EHS 40CFR 355: Not listed
TSCA: Not listed

DIM1350 CAS #: 60-51-5

DIMETHOATE

RTECS: TE1750000
EINECS Number: 200-480-3
Molecular Formula: $C_5H_{12}NO_3PS_2$
Structured MF: $(CH_3O)_2PSSCH_2CONHCH_3$
Formula Weight: 229.28
Synonyms: AADIMETHOAL; AC-12880; AC-18682; ACETIC ACID,O,O-DIMETHYLDITHIOPHOSPHORYL-,N-MONOMETHYLAMIDE SALT; AMERICAN CYANAMID 12,880; AMERICAN CYANAMID 12880; BI-58; BI 58 EC; 8014 BIS HC; CEKUTHOATE; CL 12880; CYGON; CYGON 2-E; CYGON 4E; CYGON 400; CYGON INSECTICIDE; DAPHENE; DE-FEND; DEFEND; DEMOS-L40; DEVIGON; DIMATE 267; DIMET; DIMETATE; DIMETHOAAT; DIMETHOAT; DIMETHOAT TECH 95%; DIMETHOAT TECHNISCH 95%; DIMETHOATE-267; DIMETHOGEN; O,O-DIMETHYL S-(2-(METHYLAMINO)-2-OXOETHYL)PHOSPHORODITHIOATE; O,O-DIMETHYL S-(N-METHYLCARBAMOYLMETHYL) DITHIOPHOSPHATE; O,O-DIMETHYL METHYLCARBAMOYLMETHYL PHOSPHORODITHIOATE; O,O-DIMETHYL S-METHYLCARBAMOYLMETHYL PHOSPHORODITHIOATE; O,O-DIMETHYL S-(N-METHYLCARBAMOYLMETHYL)PHOSPHORODITHIOATE; O,O-DIMETHYL S-(N-METHYLCARBAMYLMETHYL)THIOTHIONOPHOSPHATE; O,O-DIMETHYLDITHIOPHOSPHORYLACETIC ACID,N-MONOMETHYLAMIDE SALT; O,O-DIMETHYLDITHIOPHOSPHORYLACETIC ACID,N-MONOMETHYLAMIDESALT; O,O-DIMETHYL-DITHIOPHOSPHORYLESSIGSAEURE MONOMETHYLAMID; O,O-DIMETHYL-S-(N-METHYL-CARBAMOYL)-METHYL-DITHIOFOSFAAT; (O,O-DIMETHYL-S-(N-METHYL-CARBAMOYL-METHYL)-DITHIOPHOSPHAT); O,O-DIMETHYL-S-(N-MONOMETHYL)-CARBAMYL METHYLDITHIOPHOSPHATE; O,O-DIMETHYL-S-(2-OXO-3-AZA-BUTYL)-DITHIOPHOSPHAT; O,O-DIMETIL-S-(N-METIL-CARBAMOIL-METIL)-DITIOFOSFATO; DIMETON; DIMEVUR; DITHIOPHOSPHATE DE O,O-DIMETHYLE ET DE S(-N-METHYLCARBAMOYL-METHYLE); DITHIOPHOSPHATE DE O,O-DIMETHYLE ET DES(-N-METHYLCARBAMOYL-METHYLE); EI-12880; END 24650; ENT 24,650; ENT 24650; EPA PESTICIDE CODE 035001; EXPERIMENTAL INSECTICIDE 12,880; FERKETHION; FIP; FORTION NM; FOSFAMID; FOSFATOX R; FOSFOTOX; FOSFOTOX R; FOSFOTOX R 35; FOSTION; FOSTION M M; FOSTION MM; L-395; LURGO; S-METHYLCARBAMOYLMETHYL O,O-DIMETHYL PHOSPHORODITHIOATE; N-MONOMETHYLAMIDE OF O,O-DIMETHYLDITHIOPHOSPHORYLACETICACID; NC-262; OMS 111; OMS 94; PEI 75; PERFECTHION; PERFEKTHION; PHOSPHAMID; PHOSPHAMIDE; PHOSPHORODITHIOIC ACID O,O-DIMETHYL ESTER,ESTER WITH2-MERCAPTO-N-METHYLACETAMIDE; PHOSPHORODITHIOIC ACID,O,O-DIMETHYL ESTER,S-ESTER WITH2-MERCAPTO-N-METHYLACETAMIDE; PHOSPHORODITHIOIC ACID,O,O-DIMETHYL-S-(2-(METHYLAMINO)-2-OXOETHYL) ESTER (9CI); PHOSPHORODITHIOIC ACID,O,O-DIMETHYLS-(2-(METHYLAMINO)-2-OXOETHYL) ESTER; RACUSAN; REBELATE; ROGODAN; ROGOR; ROGOR 20L; ROGOR 40; ROGOR L; ROGOR P; ROXION; ROXION U.A; ROXION UA; SALUT; SEVIGOR; SINORATOX; SISTEMIN; SOLUT; SYSTEMIN; SYSTOATE; TARA; TARA 909; TRIMETION
Description: colorless crystals; camphor-like mercaptan odor
Use: broad range contact and systemic acaricide and insecticide; in the control of bots in livestock

Physical Properties

Boiling Point: 107 °C (225 °F) at 0.05 mm Hg
Freezing Point: 49 °C (120.2 °F)
Specific Gravity: 1.277 at 65 °C
Density: 1.281 g/mL at 50 °C
Vapor Pressure: 1.1 mPa at 25 °C
Water Solubility: 1 to 10 mg/mL at 24 °C
Other Solubilities: Very Soluble in Ethanol, Chloroform, Acetone; Slightly Soluble in Diethyl Ether; Insoluble in Petroleum Ether; Slightly Soluble in aromatic hydrocarbons; Soluble in Cyclohexanone; low solubility in Xylene, Hexane.
Refraction Index: 1.5334 at 65 °C/D
Flash Point: 107 °C Closed Cup

RTECS Toxicity Data

Acute Oral: Man TD_{Lo} Dose: 286 mg/kg; Toxic Effects: Peripheral nerve and sensation - Fasciculations; Behavioral - Coma; Lungs, Thorax, or Respiration - Dyspnea. Man TD_{Lo} Dose: 300 mg/kg; Toxic Effects: Behavioral - Coma; Cardiac - Pulse rate; Vascular - BP lowering not characterized in autonomic section. Man TD_{Lo} Dose: 357 mg/kg; Toxic Effects: Brain and coverings - Other degenerative changes; Cardiac - Cardiomyopathy including infarction; Kidney, Ureter, and Bladder - Chgs in tubules (inc acute renal failure, acute tubular necrosis. Human LD_{50} Dose: 30 mg/kg. Rat LD_{50} Dose: 60 mg/kg.
Acute Dermal: Rabbit LD_{50} Route: Skin; Dose: 1 gm/kg; Toxic Effects: Behavioral - Excitment. Rat LD_{50} Route: Skin; Dose: 353 mg/kg. Rat LD_{50} Route: Subcutaneous Dose: 350 mg/kg; Toxic Effects: Sense organs and special senses - Lacrimation; Lungs, Thorax, or Respiration - Dyspnea; Gastrointestinal - Changes in structure or function of salivary glands.
Chronic (Multiple Dose) Oral: Rat Dose: 630 mg/kg/28D-I; Toxic Effects: Blood - Changes in serum composition; Biochemical - Multiple enzyme effects. Rat Dose: 1092 mg/kg/52W-C; Toxic Effects: Blood - Other changes; Biochemical - True cholinesterase. Rat Dose: 300 mg/kg/6W-I; Toxic Effects: Brain and coverings - Changes in surface EEG; Peripheral Nerve and sensation - Recording from peripheral motor nerve; Biochemical - True cholinesterase. Rat Dose: 3330 mg/kg/90D-C; Toxic Effects: Sense organs and special senses - Other; Nutritional and gross metabolic - Weight loss or decreased weight gain; Biochemical - True cholinesterase. Rat Dose: 3852 mg/kg/26W-I; Toxic Effects: Lungs, Thorax, or Respiration - Structural or functional change in trachea or bronchi; Liver - Fatty liver degeneration; Kidney, Ureter, and Bladder - Changes in both tubules and glomeruli.
Chronic (Multiple Dose) Inhalation: Rat Dose: 4500 ug/m^3/13W-I; Toxic Effects: Kidney, Ureter, and Bladder - Other changes in urine composition; Blood - Changes in leukocyte (WBC) cell count; Biochemical - True cholinesterase. Rat Dose: 257 ug/m^3/24H/15W-I; Toxic Effects: Kidney, Ureter, and Bladder - Other changes in urine

composition; Blood - Changes in serum composition; Biochemical - True cholinesterase.

Reproductive/Teratogenic: Rat Route: Oral; Dose: 120 mg/kg; Duration: female 6-15D of pregnancy; Specific Developmental Abnormalities - Musculoskeletal system. Mouse Route: Oral; Dose: 1050 mg/kg; Duration: multigenerations; Effects on Fertility - Female fertility index; Other measures of fertility; Effects on Newborn - Weaning or lactation index. Mouse Route: Oral; Dose: 220 mg/kg; Duration: female 6-16D of pregnancy; Effects on Embryo or Fetus - Fetotoxicity. Mouse Route: Oral; Dose: 440 mg/kg; Duration: female 6-16D of pregnancy; Specific Developmental Abnormalities - Musculoskeletal system.

Mutagenic: Human Unscheduled DNA Synthesis; Cell Type: fibroblast; Dose: 100 umol/L. Human Cytogenetic Analysis; Cell Type: lymphocyte; Dose: 44 umol/L. Human Sister Chromatid Exchenge; Cell Type: lymphocyte; Dose: 2 mg/L. Human Sister Chromatid Exchenge; Cell Type: embryo; Dose: 229 mg/L.

Tumorigenic: Rat Route: Oral; Dose: 256 mg/kg/4W-I; Toxic Effects: Tumorigenic - Carcinogenic by RTECS criteria; Liver - Tumors; Blood - Tumors. Rat Route: Intramuscular; Dose: 176 mg/kg/6W-I; Toxic Effects: Tumorigenic - Carcinogenic by RTECS criteria; Liver - Tumors; Blood - Tumors.

Hazard Overviews

Fire: Will burn.

Carcinogenicity: IARC - Not listed; NIOSH - Not listed; NTP - Not listed; ACGIH - Not listed; OSHA - Not listed; EPA - Not listed; MAK - Not listed

Environmental

Ecotoxicity: The estimated 48-hr and 72-hr TLm values for zebrafish Brachydanio rerio embryos, exposed to dimethoate, were 940 mg/l and 259 mg/l, respectively LC_{50} Red Crayfish 1.0 mg/l (48-hr) /From table LC_{50} Ring-necked pheasants 332 mg/l in 5-day diet (95% confidence limit 293-376 mg/l), age 10 days LD_{50} Honey bees 0.9 mg/bee LD_{50} Redwinged blackbird oral 6.60-17.8 mg/kg Acute oral LD_{50} for farm animals (mg/kg bw): >50 for horse; 80 for sheep; and 70 for cattle /from table LC_{50} Pteronarcys californica (stoneflies) 0.043 mg/l/96 hr (95% confidence limit 0.036-0.051 mg/l), second year class, temp 21 °C. Static bioassay without aeration, pH 7.2-7.5, water hardness 40-50 mg/l as calcium carbonate LC_{50} Salmo gairdneri (rainbow trout) 20.0 ppm/24 hr /Conditions of bioassay not given EC_{50} Skeletonema costatum (marine algae), effect: decreased dry weight is 9.5 mg/l/96 hr. /Conditions of bioassay not specified LC_{50} Gammarus lacustris (scuds) 0.20 mg/l/96 hr (95% confidence limit 0.15-0.27 mg/l), mature, temp 21 °C. Static bioassay without aeration, pH 7.2-7.5, water hardness 40-50 mg/l as calcium carbonate and alkalinity of 30-35 mg/l LC_{50} Lapomis macrochirus (bluegill) 6.0 mg/l/96 hr, wt 0.3 g, temp 24 °C. Static bioassay without aeration, pH 7.2-7.5, water hardness 40-50 mg/l as calcium carbonate and alkalinity of 30-35 mg/l

Environmental Fate: If released to soil, it will not adsorb to the soil and will be subject to leaching. Evaporation from dry soil surfaces, and biodegradation in soil may be important removal mechanisms from soil. Soil half-lives of approximately 4 and 2.5 days were reported during drought and moderate rainfall conditions. However, a half-life of 122 days has also been measured in soil. Based on these half-lives and the fact that hydrolysis may be important in water under basic conditions, it may be susceptible to hydrolysis in moist, basic soils. In water, adsorption to sediment and bioconcentration in fish are not expected to be important transport processes. Photolysis and evaporation are not expected to be important fate processes of dimethoate in water. It may be subject to biodegradation in natural waters based on a half-life of 8 weeks for degradation in raw river water. Hydrolysis may be important based on estimated aqueous hydrolysis half-lives of 118 and 3.7 day at pH 7 and pH 9, respectively. In the atmosphere, it may exist in the vapor- and particulate-phases. Degradation of vapor-phase by reaction with photochemically produced hydroxyl radicals (estimated half-life of 5 days) will be important. Particulate-phase may be removed from air via dry and wet deposition.

Cleanup/Disposal: Guide No. 152: Do not touch damaged containers or spilled material unless wearing appropriate protective clothing. Stop leak if you can do it without risk. Prevent entry into waterways, sewers, basements or confined areas. Cover with plastic sheet to prevent spreading. Absorb or cover with dry earth, sand or other non-combustible material and transfer to containers. Do not get water inside containers.

Environmental Physical Data

Henry's Law Constant: estimated at 1.05×10^{-10}

Octanol/Water Partition Coefficient: log K_{ow} = 0.78

Sorption Partition Coefficient: K_{oc} = 18 to 36

BCF: carp 1.1 to 2.4

Regulations

RCRA 40CFR: Listed Hazardous Waste No. P044 Toxic Waste

CERCLA: 40CFR 302.4: Listed per RCRA Section 3001 RQ: 10 lb (4.535 kg)

SARA 40CFR 372.65: Listed TPQ: 500/10000 lb

SARA EHS 40CFR 355: Listed TPQ: 10 lb

TSCA: Listed

Analytical Methods

Soil: EPA PMD-DME, PMD-TLC; SW846 3640A, 8141, 8141A, 8270B, 8270C, 8321, 8321A

Water / Groundwater: EPA P-005-1, 1657, 022; SW846 8141, 8141A

Food: FDA 212.1, 232.1, 232.3, 232.4, 242.1

Plasma: EPA 001, 027, 29; FDA 211.1, 231.1, 252

DIM1400 CAS #: 828-00-2

DIMETHOXANE

RTECS: AH1350000

EINECS Number: 212-579-9

Molecular Formula: $C_8H_{14}O_4$

Structured MF: $CH_3COOC_4H_5O_2(CH_3)_2$
Formula Weight: 174.19

Chemical Structure

Synonyms: ACETIC ACID,2,6-DIMETHYL-M-DIOXAN-4-YL ESTER; ACETIC ACID,ESTER WITH 2,6-DIMETHYL-M-DIOXAN-4-OL; ACETOMETHOXAN; ACETOMETHOXANE; 6-ACETOXY-2,4-DIMETHYL-1,3-DIOXANE; 6-ACETOXY-2,4-DIMETHYL-M-DIOXANE; 6-ACETOXY-2,4-DIMETHYL-META-DIOXANE; DDOA; 2,4-DIMETHYL-6-ACETOXY-1,3-DIOXANE; 2,6-DIMETHYL-1,3-DIOXAN-4-OL ACETATE; 2,6-DIMETHYL-M-DIOXAN-4-OL ACETATE; 2,6-DIMETHYL-META-DIOXAN-4-OL ACETATE; 2,4-DIMETHYL-6-M-DIOXANYL ACETATE; 2,6-DIMETHYL-M-DIOXAN-4-YL ACETATE; 2,6-DIMETHYL-META-DIOXAN-4-YL ACETATE; 2,6-DIMETHYL-M-DIOXAN-4-YL ESTER ACETIC ACID; 1,3-DIOXAN-4-OL,2,6-DIMETHYL-,ACETATE; 1,3-DIOXAN-4-OL,2-6-DIMETHYL-,ACETATE; M-DIOXAN-4-OL,2,6-DIMETHYL-,ACETATE; DIOXIN; DIOXIN (BACTERICIDE); DIOXIN (BACTERICIDE) (OBS.); G1V GARD DXN; GIV GARD DXN; GIV GARD DXN-CO

Description: yellow to light amber, clear liquid; mustard-like odor

Use: microbicide; preservative for cutting oils, resins, oils, emulsions, waterbased paints, cosmetics, inks, dyes, textile chemicals, fabric softeners, adhesives and antistatic lubricants; and petrol additive

Physical Properties

Boiling Point: 74 °C (165 °F) to 75 °C (167 °F) at 6 mm Hg
Freezing Point: < 25 °C (77 °F)
Specific Gravity: 1.0655 at 20 °C/4 °C
Vapor Density: > 1 Air=1
Density: 1.068 to 1.076 g/mL at 25 °C
Water Solubility: Miscible with Water
Other Solubilities: 95% Ethanol: >=100 mg/ml at 22 °C; Acetone: >=100 mg/ml at 22 °C; Corn oil: miscible; DMSO: >=100 mg/ml at 22 °C; Many organic solvents: miscible.
Odor Threshold: 1.0 ppm
Refraction Index: 1.430 at 20 °C/D
Evaporation Rate: > 1 Butyl Acetate=1
Flash Point: 61 °C

RTECS Toxicity Data

Acute Oral: Rat LD_{50} Dose: 1930 mg/kg. Mouse LD_{Lo} Dose: 2800 mg/kg.

Chronic (Multiple Dose) Oral: Rat Dose: 12 gm/kg/16D-I; Toxic Effects: Gastrointestinal - Other changes; Liver - Changes in liver weight. Rat Dose: 32500 mg/kg/13W-I; Toxic Effects: Brain and coverings - Changes in brain weight; Gastrointestinal - Other changes; Kidney, Ureter, and Bladder - Changes in kidney weight. Rat Dose: 24 gm/kg/16D-I; Toxic Effects: Gastrointestinal - Other changes; Liver - Changes in liver weight; DEATH.

Mutagenic: Hamster Cytogenetic Analysis; Cell Type: ovary; Dose: 20200 ug/L. Hamster Sister Chromatid Exchenge; Cell Type: ovary; Dose: 3660 ug/L.

Tumorigenic: Rat Route: Oral; Dose: 948 gm/kg/88W-I; Toxic Effects: Tumorigenic - Carcinogenic by RTECS criteria; Liver - Tumors; Blood - Leukemia. Mouse Route: Oral; Dose: 25750 mg/kg/2Y-C; Toxic Effects: Tumorigenic - Equivocal tumorigenic agent by RTECS criteria; Gastrointestinal - Tumors.

Hazard Overviews

Health: Irritating to eyes/skin/respiratory tract. Harmful. Other Acute Effects: may be harmful by inhalation, ingestion, or skin absorption. Chronic Effects: possible carcinogen; target organs: liver, bladder.

Fire: Combustible. Hazards: emits toxic fumes. Extinguishing agents: water spray; carbon dioxide, dry chemical powder or appropriate foam; use water spray to cool fire-exposed containers. Precautions: combustible liquid.

Reactivity: Stable. Hazardous polymerization will not occur. Incompatible with: strong oxidizing agents, strong acids, strong bases, readily hydrolyzed. Hazardous decomposition products: toxic fumes of: carbon monoxide, carbon dioxide.

Carcinogenicity: IARC - Group 3, Not classifiable as to carcinogenicity to humans; NIOSH - Not listed; NTP - Not listed; ACGIH - Not listed; OSHA - Not listed; EPA - Not listed; MAK - Not listed

Primary Target Organs:

Eyes

Skin

Respiratory System

Liver

Environmental

Environmental Fate: If released on soil, it would readily leach. Chemicals containing acetal and ester groups are susceptible to chemical hydrolysis, the hydrolysis of the acetal ring being favored under acidic conditions and the ester under alkaline conditions. No hydrolysis rate data are available. Based on limited data, it may also biodegrade but no rate data are available. If released in water, it would be expected to hydrolyze and possibly biodegrade, rate data in water are also lacking. Volatilization from water, adsorption to sediment, or bioconcentration in aquatic organisms are not expected to be important fate processes in water. In the atmosphere, it will react with photochemically-produced hydroxyl radicals, resulting in an estimated atmospheric half-life of 7.9 hr. Since it is miscible in water, it should also be washed out by rain.

Environmental Physical Data

Henry's Law Constant: estimated at 1.24×10^{-7}
Octanol/Water Partition Coefficient: **log K_{ow} = 0.49**
Sorption Partition Coefficient: **log K_{oc} =** estimated at 0.56
BCF: calculated at 1.4

Regulations

RCRA 40CFR: Not listed
CERCLA: 40CFR 302.4: Not listed
SARA 40CFR 372.65: Not listed
SARA EHS 40CFR 355: Not listed
TSCA: Listed

DIM1450 CAS #: 91-10-1

2,6-DIMETHOXY PHENOL

RTECS: SL0900000
EINECS Number: 202-041-1
Molecular Formula: $C_8H_{10}O_3$
Formula Weight: 154.18

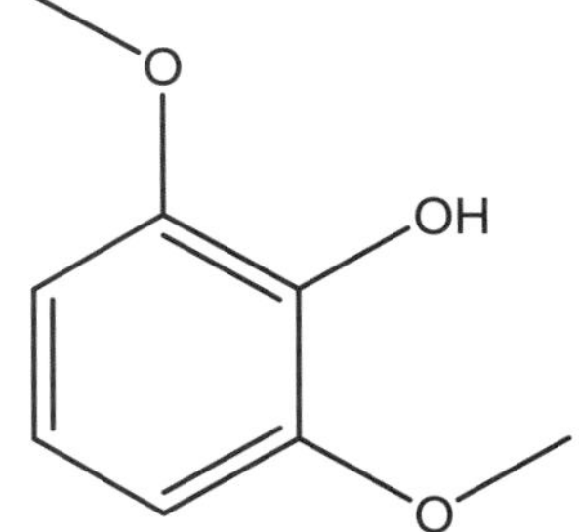

Chemical Structure

Synonyms: ALDRICH; 2,6-DIMETHOXYPHENOL; 1,3-DIMETHYL PYROGALLATE; 2,6-DWUMETOKSYFENOL; PYROGALLOL 1,3-DIMETHYL ETHER; PYROGALLOL DIMETHYLETHER; SYRINGOL

Physical Properties

Boiling Point: 261 °C (501.8 °F)
Freezing Point: 55 °C (131 °F)
Water Solubility: 0.02%
Other Solubilities: Ether: Very Soluble; Ethanol: Very Soluble

RTECS Toxicity Data

Acute Oral: Rat LD_{50} Dose: 550 mg/kg; Toxic Effects: Behavioral - General anesthetic. Mouse LD_{50} Dose: 2500 mg/kg.
Chronic (Multiple Dose) Oral: Rat Dose: 14828 mg/kg/36D-I; Toxic Effects: Nutritional and gross metabolic - Weight loss or decreased weight gain; DEATH.

Hazard Overviews

Health: Irritating to eyes/skin/respiratory tract. Harmful. Other Acute Effects: harmful if swallowed; may be harmful if inhaled or absorbed through the skin; depending on the intensity and duration of exposure, effects may vary from mild irritation to severe destruction of tissue; prolonged contact can cause damage to the eyes, severe irritation or burns.
Fire: Extinguishing agents: carbon dioxide, dry chemical powder or appropriate foam. Precautions: combustible liquid.
Carcinogenicity: IARC - Not listed; NIOSH - Not listed; NTP - Not listed; ACGIH - Not listed; OSHA - Not listed; EPA - Not listed; MAK - Not listed
Primary Target Organs:

Eyes

Skin

Respiratory System

Environmental

Regulations

RCRA 40CFR: Not listed
CERCLA: 40CFR 302.4: Not listed
SARA 40CFR 372.65: Not listed
SARA EHS 40CFR 355: Not listed
TSCA: Listed

DIM1500 CAS #: 2735-04-8

2,4-DIMETHOXYANILINE

RTECS: BX4200000
EINECS Number: 220-355-7
Molecular Formula: $C_8H_{11}NO_2$
Formula Weight: 153.18

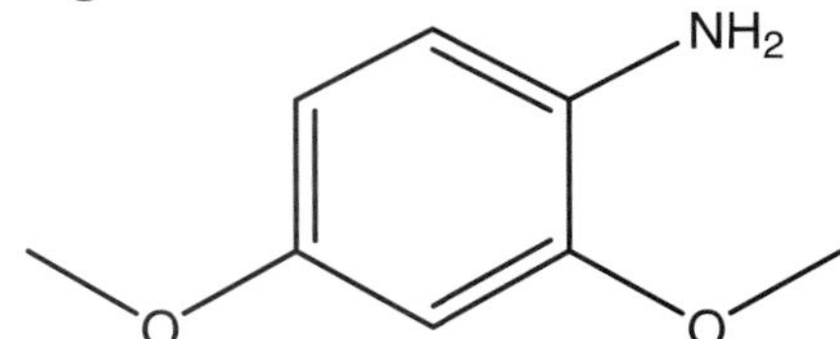

Chemical Structure

Synonyms: ANILINE,2,4-DIMETHOXY-; BENZENAMINE,2,4-DIMETHOXY-
Description: plates
Use: toxic, hazardous decomposition products

Physical Properties

Boiling Point: 262 °C (503.6 °F)
Freezing Point: 33.5 °C (92.3 °F)
Water Solubility: Slightly Soluble in Water
Other Solubilities: 95% Ethanol: <1 mg/ml at 17 °C; Acetone: 10-50 mg/ml at 17 °C; Benzene: Soluble; DMSO: 50-100 mg/ml at 17 °C; Ether: Soluble.

RTECS Toxicity Data

Acute Oral: Rat LD_{50} Dose: 464 mg/kg. Mouse LD_{50} Dose: 1 gm/kg.
Mutagenic: Rat Morphological Transformation; Cell Type: embryo; Dose: 55 ug/plate. Bacteria - S Typhimurium Mutations in Microorganisms; Dose: 10 ug/plate (-S9).

Hazard Overviews

Health: Irritating to eyes/skin/respiratory tract. Toxic. Other Acute Effects: harmful if swallowed, inhaled, or absorbed through skin; absorption into the body leads to the formation of methemoglobin which in sufficient concentration causes cyanosis; onset may be delayed 2 to 4 hours or longer.

Fire: Hazards: emits toxic fumes. Extinguishing agents: water spray; carbon dioxide, dry chemical powder or appropriate foam. Precautions: combustible liquid.

Reactivity: Incompatible with: acids, acid chlorides, acid anhydrides, chloroformates, strong oxidizing agents, may discolor on exposure to light. Hazardous decomposition products: toxic fumes of: carbon monoxide, carbon dioxide, nitrogen oxides.

Carcinogenicity: IARC - Not listed; NIOSH - Not listed; NTP - Not listed; ACGIH - Not listed; OSHA - Not listed; EPA - Not listed; MAK - Not listed

Primary Target Organs:

Eyes

Skin

Respiratory System

Environmental

Regulations

RCRA 40CFR: Not listed
CERCLA: 40CFR 302.4: Not listed
SARA 40CFR 372.65: Not listed
SARA EHS 40CFR 355: Not listed
TSCA: Listed

DIM1600 CAS #: 119-90-4

3,3'-DIMETHOXYBENZIDINE

RTECS: DD0875000
EINECS Number: 204-355-4
Molecular Formula: $C_{14}H_{16}N_2O_2$
Structured MF: $[C_6H_3(OCH_3)NH_2]_2$
Formula Weight: 244.29

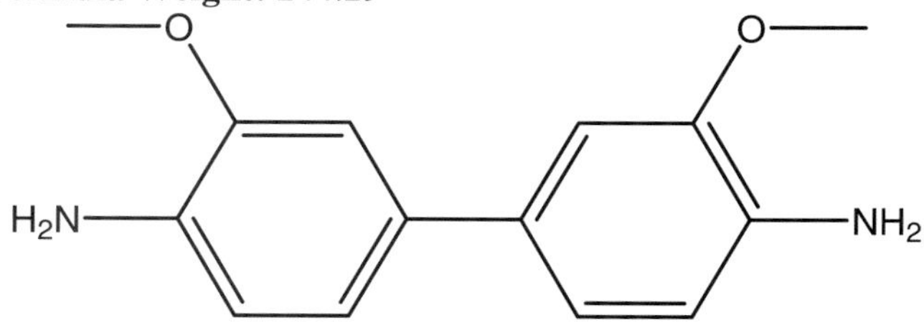

Chemical Structure

Synonyms: ACETAMINE DIAZO BLACK RD; ACETAMINE DIAZO NAVY RD; AMACEL DEVELOPED NAVY SD; AZOENE FAST BLUE BASE; AZOENE FAST BLUE SALT; AZOFIX BLUE B SALT; AZOGENE FAST BLUE B; AZOGENE FAST BLUE B SALT; BENZIDENE,3,3'-DIMETHOXY-; BIANISIDINE; (1,1'-BIPHENYL)-4,4'-DIAMINE,3,3'-DIMETHOXY-; BIPHENYL,4,4'DIAMINO-3,3'DIMETHOXY; BLUE BASE IRGA B; BLUE BASE NB; BLUE BN BASE; BLUE BN SALT; BLUE SALT NB; BRENTAMINE FAST BLUE B BASE; BRENTAMINE FAST BLUE B SALT; C.I. 24110; C.I. 37235; C.I. AZOIC DIAZO COMPONENT 48; C.I. AZOIC DIAZO COMPONENT 48,FAST BLUE B SALT; C.I. DISPERSE BLACK 6; CELLITAZOL B; CELLITAZOL BN; CIBACETE DIAZO NAVY BLUE 2B; DIACEL NAVY DC; DIACELLITON FAST GREY G; 4,4'-DIAMINO-3,3'-BIPHENYLDIOL DIMETHYL ETHER; 4,4'-DIAMINO-3,3'-DIMETHOXYBIPHENYL; 4,4'-DIAMINO-3,3'-DIMETHOXYDIPHENYL; DI-P-AMINO-DI-M-METHOXYDIPHENYL; DI-P-AMINODI-M-METHOXYDIPHENYL; DI-PARA-AMINODI-META-METHOXYDIPHENYL; O-DIANISIDIN; O-DIANISIDINA; 3,3'-DIANISIDINE; DIANISIDINE; O,O'-DIANISIDINE; O-DIANISIDINE; DIATO BLUE BASE B; DIATO BLUE SALT B; DIAZO FAST BLUE B; 3,3'-DIMETHOXYBENZIDIN; 3,3'-DIMETHOXY-[1,1'-BIPHENYL]-4,4'-DIAMINE; 3,3'-DIMETHOXYBIPHENYL-4,4'-DIAMINE; 3,3'-DIMETHOXY-4,4'-DIAMINOBIPHENYL; 3,3'-DIMETOSSIBENZODINA; DMOB; FAST BLUE B BASE; FAST BLUE BN SALT; FAST BLUE DS SALT; FAST BLUE DSC BASE; FAST BLUE SALT B; FAST BLUE SALT BN; HILTONIL FAST BLUE B BASE; HILTOSAL FAST BLUE B SALT; HINDASOL BLUE B SALT; KAKO BLUE B SALT; KAYAKU BLUE B BASE; KAYAKU BLUE B SALT; LAKE BLUE B BASE; MEISEI TERYL DIAZO BLUE HR; MITSUI BLUE B BASE; MITSUI BLUE B SALT; NAPHTHANIL BLUE B BASE; NATASOL BLUE B SALT; NEUTROSEL NAVY BN; SANYO FAST BLUE B SALT; SETACYL DIAZO NAVY R; SPECTROLENE BLUE B

Description: colorless crystals, leaflets, or needles that turn a violet color on standing

Use: chemical intermediate for the production of azo dyes and o-Dianisidine diisocyanate; dye for leather, paper, plastics, rubber and textiles; and in the detection of metals, thiocyanates, and nitrites

Physical Properties

Freezing Point: 137 °C (278.6 °F)
Vapor Density: 8.5 Air=1
Saturated Vapor Density: 1.2 kg/m^3
Vapor Pressure: 8.8 x10^{-9} mm Hg at 25 °C
Water Solubility: Practically Insoluble in Water
Other Solubilities: 95% Ethanol: <1 mg/ml at 20 °C; Acetone: 5-10 mg/ml at 20 °C; Benzene: Soluble; Chloroform: Soluble; DMSO: >=100 mg/ml at 20 °C; Ether: Soluble; Most lipids: Probably Soluble; Most organic solvents: Probably Soluble.
pH: Weak base
Flash Point: 206 °C Closed Cup

RTECS Toxicity Data

Acute Oral: Rat LD_{50} Dose: 1920 mg/kg. Dog LD_{Lo} Dose: 600 mg/kg.

Mutagenic: Human Unscheduled DNA Synthesis; Cell Type: HeLa cell; Dose: 100 nmol/L. Rat Unscheduled DNA Synthesis; Cell Type: liver; Dose: 500 umol/L.

Tumorigenic: Rat Route: Oral; Dose: 12 gm/kg/56W-I; Toxic Effects: Tumorigenic - Equivocal tumorigenic agent by RTECS criteria; Sense organs and special senses - Tumors; Skin and appendages - Tumors. Hamster Route: Oral; Dose: 588 gm/kg/70W-C; Toxic Effects: Tumorigenic - Equivocal tumorigenic agent by RTECS criteria; Kidney, Ureter, and Bladder - Tumors.

Hazard Overviews

Health: Irritating to eyes/skin/respiratory tract. Toxic. Other Acute Effects: harmful if swallowed, inhaled, or absorbed through skin; causes sneezing; target organ: bladder. Chronic Effects: may alter genetic material; may cause heritable genetic damage; target organ: bladder. Carcinogen.

Fire: Will burn. Hazards: emits toxic fumes. Extinguishing agents: water spray; carbon dioxide, dry chemical powder or appropriate foam. Precautions: combustible liquid.

Reactivity: Incompatible with: strong oxidizing agents. Hazardous decomposition products: toxic fumes of: carbon monoxide, carbon dioxide, nitrogen oxides.

Carcinogenicity: IARC - Group 2B, Possibly carcinogenic to humans; NIOSH - Listed as carcinogen; NTP - Class 2B, Reasonably anticipated to be a carcinogen, sufficient evidence of carcinogenicity from studies in experimental animals; ACGIH - Not listed; OSHA - Not listed; EPA - Not listed; MAK - Class A2, Unmistakably carcinogenic in animal experimentation only

Primary Target Organs:

Eyes

Skin

Respiratory System

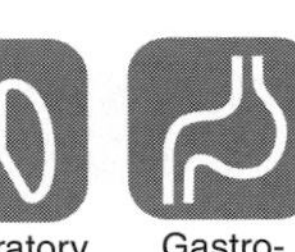
Gastro-intestinal

Environmental

Environmental Fate: No information on fate was found; however, its fate may be predicted based on information on the fate of benzidine and other aromatic amines. If it is released to the soil it may sorb to the soil with the amount of adsorption dependent on the pH of the soil, adsorption increasing with decreasing pH. It may also react with natural substances in the soil like cations such as Fe(III), clay minerals, and aromatic amines are known to form covalent bonds with humic materials. No information on the biodegradation was found; however, it may be subject to biodegradation as benzidine was 79% degraded in 4 weeks in silty clay loam. If released to water it may readily degrade due to reaction with hydroxyl radicals, redox reactions with naturally occurring cations, and perhaps photodegradation. As such, degradation should occur more rapidly in humic waters because of the presence of reactive radicals, cations and molecules and the fact that aromatic amines are known to form covalent bonds with humic materials. Its adsorption to sediments will be pH dependent, with adsorption increasing as pH decreases due to the increase in the amount of ionized cation species present. It will not bioconcentrate in aquatic organisms or hydrolyze. No information on the biodegradation in natural waters was found; limited data from laboratory screening tests suggest that it may be susceptible to biodegradation. Since it absorbs light above 290 nm, it may be susceptible to direct photolysis near the surface of natural waters. If released to the atmosphere, it may be subject to direct photolysis due to its absorption of light greater than 290 nm. It should exist almost entirely in the particulate phase in the ambient atmosphere; the estimated vapor phase half-life in the atmosphere is 2 hours as a result of reaction with photochemically produced hydroxyl radicals.

Environmental Physical Data

Henry's Law Constant: estimated at 4.7×10^{-11}
Octanol/Water Partition Coefficient: log K_{ow} = 1.808
Sorption Partition Coefficient: K_{oc} = estimated at 48.8
BCF: estimated at 100

Regulations

RCRA 40CFR: Listed Hazardous Waste No. U091 Toxic Waste
CERCLA: 40CFR 302.4: Listed per RCRA Section 3001 RQ: 100 lb (45.35 kg)
SARA 40CFR 372.65: Listed
SARA EHS 40CFR 355: Not listed
TSCA: Listed

Analytical Methods

Soil: SW846 8270B, 8270C
Water / Groundwater: EPA 1625, 553; SW846 8325
Drinking Water: EPA 553
Plasma: EPA 29

DIM1650 **CAS #: 20325-40-0**

3,3'-DIMETHOXYBENZIDINE DIHYDROCHLORIDE

RTECS: DD1050000
EINECS Number: 243-737-5
Molecular Formula: $C_{14}H_{18}Cl_2N_2O_2$
Formula Weight: 317.24

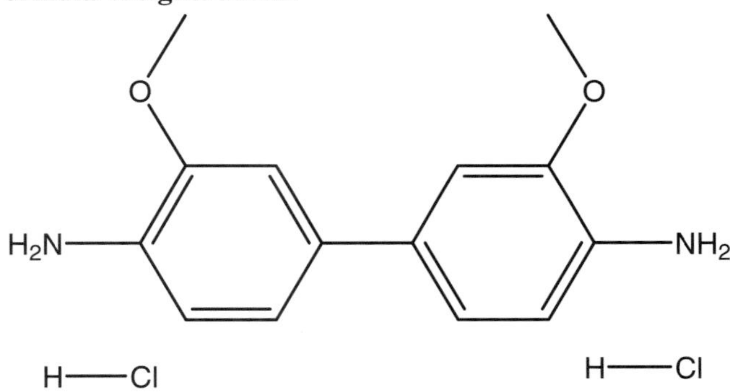

Chemical Structure

Synonyms: (1,1'-BIPHENYL)-4,4'-DIAMINE,3,3-DIMETHOXY-,DIHYDROCHLORIDE (9CI); C.I. DISPERSE BLACK 6 DIHYDROCHLORIDE; DIANISIDINE DIHYDROCHLORIDE; O-DIANISIDINE DIHYDROCHLORIDE; 3,3'-DIMETHOXY-(1,1-BIPHENYL)-4,4'-DIAMINE DIHYDROCHLORIDE; 3,3'-DIMETHOXY-4,4'-DIAMINOBIPHENYL DIHYDROCHLORIDE
Description: off-white powder

Physical Properties

Freezing Point: 268 °C (514.4 °F)
Water Solubility: 10 to 50 mg/mL at 20 °C

Other Solubilities: 95% Ethanol: <1 mg/ml at 20 °C; Acetone: <1 mg/ml at 20 °C; DMSO: 1-5 mg/ml at 20 °C.
Flash Point: Not available; probably combustible

RTECS Toxicity Data

Mutagenic: Mouse Mutations in Microorganisms; Cell Type: lymphocyte; Dose: 50 mg/L (+S9). Mouse Mutations in Mammalian Somatic Cells; Cell Type: lymphocyte; Dose: 33 mg/L.
Tumorigenic: Rat Route: Oral; Dose: 1040 mg/kg/1Y-I; Toxic Effects: Tumorigenic - Carcinogenic by RTECS criteria; Sense organs and special senses - Tumors; Tumorigenic effects - Testicular tumors. Rat Route: Oral; Dose: 11 gm/kg/51W-I; Toxic Effects: Tumorigenic - Carcinogenic by RTECS criteria; Sense organs and special senses - Tumors; Skin and appendages - Tumors. Rat Route: Oral; Dose: 6497 mg/kg/91W-C; Toxic Effects: Tumorigenic - Carcinogenic by RTECS criteria; Gastrointestinal - Tumors; Liver - Tumors.

Hazard Overviews

Health: Irritating to eyes/skin/respiratory tract. Toxic. Other Acute Effects: harmful if swallowed, inhaled, or absorbed through skin; causes sneezing. Chronic Effects: may alter genetic material; target organ: bladder. Carcinogen.
Fire: Will burn. Hazards: emits toxic fumes. Extinguishing agents: water spray; carbon dioxide, dry chemical powder or appropriate foam. Precautions: combustible liquid.
Reactivity: Incompatible with: strong oxidizing agents. Hazardous decomposition products: toxic fumes of: carbon monoxide, carbon dioxide, nitrogen oxides, hydrogen chloride gas.
Carcinogenicity: IARC - Not listed; NIOSH - Not listed; NTP - Listed; ACGIH - Not listed; OSHA - Not listed; EPA - Not listed; MAK - Not listed
Primary Target Organs:

Eyes Skin Respiratory System

Environmental

Regulations

RCRA 40CFR: Not listed
CERCLA: 40CFR 302.4: Not listed
SARA 40CFR 372.65: Listed
SARA EHS 40CFR 355: Not listed
TSCA: Listed

DIM1700 CAS #: 111984-09-9

3,3'-DIMETHOXYBENZIDINE HYDROCHLORIDE

Molecular Formula: $C_{14}H_{17}N_2O_2Cl$
Formula Weight: 280.75

Hazard Overviews

Carcinogenicity: IARC - Not listed; NIOSH - Not listed; NTP - Not listed; ACGIH - Not listed; OSHA - Not listed; EPA - Not listed; MAK - Not listed

Environmental

Regulations

RCRA 40CFR: Not listed
CERCLA: 40CFR 302.4: Not listed
SARA 40CFR 372.65: Listed
SARA EHS 40CFR 355: Not listed
TSCA: Not listed

DIM1750 CAS #: 91-93-0

3,3-DIMETHOXYBENZIDINE-4,4-DIISOCYANATE

RTECS: NQ8800000
Molecular Formula: $C_{16}H_{12}N_2O_4$
Structured MF: $[OCN(CH_3O)C_6H_3]_2$
Formula Weight: 296.30

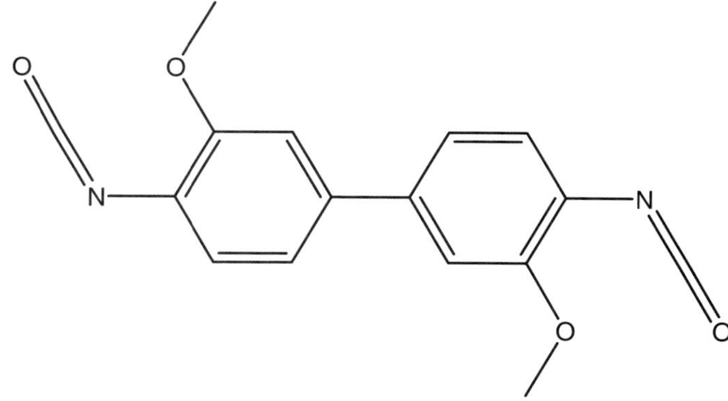

Chemical Structure

Synonyms: 1,1'-BIPHENYL, 4,4'-DIISOCYANATO-3,3'-DIMETHOXY-; DIANISIDINE DIISOCYANATE; 4,4'-DIISOCYANATO-3,3'-DIMETHOXY-1,1'-BIPHENYL; 3,3'-DIMETHOXYBENZIDINE-4,4'-DIISOCYANATE; 3,3'-DIMETHOXY-4,4'-BIPHENYL DIISOCYANATE; 3,3'-DIMETHOXY-4,4'-BIPHENYLENE DIISOCYANATE; 3,3'-DIMETHOXY-4,4'-BIPHENYLENEIISOCYANATE; 3,3'-DIMETHOXY-4,4'-BIPHENYLYLENE ISOCYANATE; 3,3'-DIMETHOXY-4,4'-BIPHENYLYLENE ISOCYANIC ACID ESTER; 3,3'-DIMETHOXY-4,4'-DIPHENYLYL ISOCYANATE; ISOCYANIC ACID, 3,3'-DIMETHOXY-4,4'-BIPHENYLENE ESTER; ISOCYANIC ACID, 3,3'-DIMETHOXY-4,4'-BIPHENYLYLENE ESTER
Description: gray to brown powder
Use: in polymers and adhesive systems, as a high-strength backbone or cross-linking intermediate, as a component of polyurethane elastomers and possibly in coating, gaskets and shock absorbers

Physical Properties

Freezing Point: 112 °C (233.6 °F)

Water Solubility: < 1 mg/mL at 17.5 C
Other Solubilities: 95% Ethanol: <1 mg/ml at 17.5 °C; Acetone: <1 mg/ml at 17.5 °C; DMSO: <1 mg/ml at 17.5 °C; Esters: Soluble; Ketones: Soluble; Methanol: <1 mg/ml at 19 °C; Toluene: <1 mg/ml at 19 °C.
Flash Point: Not available; probably combustible

RTECS Toxicity Data

Mutagenic: Rat Morphological Transformation; Cell Type: embryo; Dose: 81 ug/plate. Bacteria - S Typhimurium Mutations in Microorganisms; Dose: 3300 ng/plate (+S9), 3 ug/plate (-S9).
Tumorigenic: Rat Route: Oral; Dose: 565 gm/kg/78W-I; Toxic Effects: Tumorigenic - Carcinogenic by RTECS criteria; Blood - Leukemia; Skin and appendages - Tumors. Rat Route: Oral; Dose: 1200 gm/kg/78W-I; Toxic Effects: Tumorigenic - Carcinogenic by RTECS criteria; Blood - Leukemia; Skin and appendages - Tumors.

Hazard Overviews

Corrosive

Health: Corrosive to eyes/skin/respiratory tract. Toxic. Other Acute Effects: toxic by inhalation, in contact with skin and if swallowed; burning sensation; coughing; wheezing; laryngitis; shortness of breath; headache; nausea and vomiting; allergic reactions in certain sensitive individuals following prolonged or repeated exposure; lung irritation; chest pain and edema which may be fatal. Chronic Effects: nervous system disturbances; development of inflammatory and ulcerous lesions of the penis, prepuce and scrotum in animals; increased toxic effects with exposure to and/or consumption of alcohol; target organs: blood, skin, lymphatic, system, female reproductive system, brain, liver, kidneys, bladder. Possible human carcinogen. Causes reproductive effects in animals.
Fire: Will burn. Hazards: emits toxic fumes. Extinguishing agents: water spray; carbon dioxide, dry chemical powder or appropriate foam. Precautions: combustible liquid.
Carcinogenicity: IARC - Group 3, Not classifiable as to carcinogenicity to humans; NIOSH - Not listed; NTP - Not listed; ACGIH - Not listed; OSHA - Not listed; EPA - Not listed; MAK - Not listed
Primary Target Organs:

Eyes

Skin

Respiratory System

Liver

Kidneys

Blood

Environmental

Regulations

RCRA 40CFR: Not listed
CERCLA: 40CFR 302.4: Listed as Compound per CWA Section 307(a) per CAA Section 112
SARA 40CFR 372.65: Listed
SARA EHS 40CFR 355: Not listed
TSCA: Not listed

DIM1800	CAS #: 534-15-6

1,1-DIMETHOXYETHANE

RTECS: AB2825000
DOT: UN2377; IMO3.2
EINECS Number: 208-589-8
Molecular Formula: $C_4H_{10}O_2$
Formula Weight: 90.14

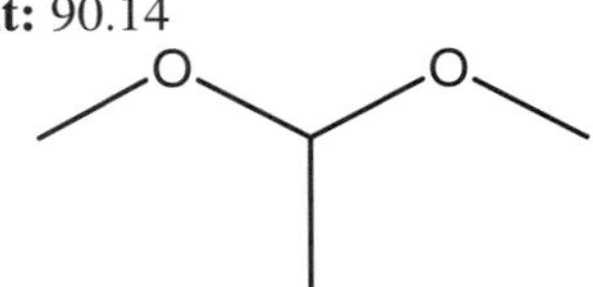

Chemical Structure

Synonyms: ACETALDEHYDE METHYL ACETAL; ACETALDEHYDE,DIMETHYL ACETAL; DIMETHYL ACETAL; DIMETHYL ALDEHYDE; DIMETHYLACETAL; ETHANE,1,1-DIMETHOXY-; ETHYLIDENE DIMETHYL ETHER; METHYL FORMYL
Description: colorless mobile liquid; green vegetable odor
Use: in medicine; organic synthesis; flavoring ingredient fruit flavors, apple, whiskey, orange; chemical intermediate for organic synthesis; fragrance & flavoring agent

Physical Properties

Boiling Point: 64.5 °C (148 °F)
Freezing Point: -113.2 °C (-171.76 °F)
Specific Gravity: 0.85015 at 20 °C/4 °C
Saturated Vapor Density: 1.403060254 kg/m^3
Vapor Pressure: 61 mm Hg at 20 °C
Water Solubility: Miscible with Water
Other Solubilities: miscible with Alcohol, Chloroform, Ether.
Refraction Index: 1.3668 at 20 °C/D
Ionization Potential (eV): 9.65 +/-0.2
Flash Point: 26.667 °C

RTECS Toxicity Data

Acute Oral: Rat LD_{50} Dose: 6500 mg/kg. Rabbit LD_{50} Dose: 4507 mg/kg.
Acute Inhalation: Rat LC_{50} Dose: 3000 ppm/4hr; Toxic Effects: Behavioral - General anesthetic.
Acute Dermal: Rabbit LD_{50} Route: Skin; Dose: 20 gm/kg.
Irritation Eye: Rabbit Standard Draize Test Dose: 100 mg/24H; Reaction: moderate.
Irritation Skin: Rabbit Open Draize Test Dose: 10 mg/24H open; Reaction: mild.

Hazard Overviews

Flammable

Health: Irritating to eyes/skin/respiratory tract. Other Acute Effects: may be harmful by inhalation, ingestion, or skin absorption.
Fire: Flammable. Hazards: emits toxic fumes; vapor may travel considerable distance to source of ignition and flash back; container explosion may occur. Extinguishing agents: water spray; carbon dioxide, dry chemical powder or appropriate foam. Precautions: combustible liquid.
Reactivity: Incompatible with: strong oxidizing agents, strong acids. Hazardous decomposition products: toxic fumes of: carbon monoxide, carbon dioxide.
Carcinogenicity: IARC - Not listed; NIOSH - Not listed; NTP - Not listed; ACGIH - Not listed; OSHA - Not listed; EPA - Not listed; MAK - Not listed
Primary Target Organs:

Eyes

Skin

Respiratory System

Environmental

Environmental Fate: If released to soil, it will have very high mobility. Volatilization may be important from moist and dry soil surfaces. According to one study it shows a slight ability to support growth for an organism isolated from soil enriched with triethylene glycol as a carbon source. If released to water, it would not adsorb to suspended solids and sediment. It would volatilize from water surfaces with estimated half-lives for a model river and model lake of 15 hours and 7.5 days, respectively. An estimated BCF value of 0.86 suggests that it will not bioconcentrate in aquatic organisms. If released to the atmosphere, it will exist in the vapor phase. Vapor-phase is degraded in the atmosphere by reaction with photochemically produced hydroxyl radicals with an estimated half-life of about 43 hours.
Cleanup/Disposal: Guide No. 127: Eliminate all ignition sources (no smoking, flares, sparks or flames in immediate area). All equipment used when handling the product must be grounded. Do not touch or walk through spilled material. Stop leak if you can do it without risk. Prevent entry into waterways, sewers, basements or confined areas. A vapor suppressing foam may be used to reduce vapors. Absorb or cover with dry earth, sand or other non-combustible material and transfer to containers. Use clean non-sparking tools to collect absorbed material. Large Spills: Dike far ahead of liquid spill for later disposal. Water spray may reduce vapor; but may not prevent ignition in closed spaces.

Environmental Physical Data

Henry's Law Constant: estimated at 6.7×10^{-5}
Octanol/Water Partition Coefficient: log K_{ow} = 0.22
Sorption Partition Coefficient: log K_{oc} = 1
BCF: estimated at 0.86

Regulations

RCRA 40CFR: Not listed
CERCLA: 40CFR 302.4: Not listed
SARA 40CFR 372.65: Not listed
SARA EHS 40CFR 355: Not listed
TSCA: Listed

DIM1850 CAS #: 110-71-4

1,2-DIMETHOXYETHANE

RTECS: KI1451000
DOT: UN2252; IMO3.2
EINECS Number: 203-794-9
Molecular Formula: $C_4H_{10}O_2$
Structured MF: $CH_3OCH_2CH_2OCH_3$
Formula Weight: 90.12

Chemical Structure

Synonyms: ANSUL ETHER 12'; ANSUL ETHER 121; ALPHA,BETA-DIMETHOXYETHANE; DIMETHOXYETHANE; DIMETHYL CELLOSOLVE; DIMETHYL CELLOSOLVE SOLVENT; DIMETHYLCELLOSOLVE; 2,5-DIOXAHEXANE; DME; DME (SOLVENT); EGDME; ETHANE,1,2-DIMETHOXY-; 1,2-ETHANEDIOL,DIMETHYL ETHER; ETHYLENE DIMETHYL ETHER; ETHYLENE GLYCOL DIMETHYL ETHER; GDME; GLYCOL DIMETHYL ETHER; GLYME; MONOETHYLENE GLYCOL DIMETHYL ETHER; MONOGLYME
Description: colorless to water-white liquid; sharp ethereal odor
Use: solvent; to facilitate the formation of alkali metal-hydrocarbon adducts and is used in the Reformatsky reaction with methyl gamma-bromocrotonate

Physical Properties

Boiling Point: 82 °C (180 °F) to 83 °C (181 °F) at 760 mm Hg
Freezing Point: Either -71 °C (-95.8 °F) to -50 °C (-58 °F)
Specific Gravity: 0.86285 at 20 °C/4 °C
Vapor Density: 3.1 Air=1
Density: 0.8683 g/mL at 20 °C
Vapor Pressure: 40 mm Hg at 10.7 °C
Water Solubility: Miscible with Water
Other Solubilities: 95% Ethanol: >=100 mg/ml at 22 °C; Acetone: >=100 mg/ml at 22 °C; Alcohol: miscible; Benzene: Soluble; DMSO: >=100 mg/ml at 22 °C; Ether: Soluble; Hydrocarbons: Soluble; Petroleum Ether: Soluble.
Surface Tension: 0.02461 N/m at 20 °C
pH: 8.2
Refraction Index: 1.3813 at 20 °C/D
Evaporation Rate: 4.99 Butyl Acetate=1
Critical Temperature: 263 °C
Critical Pressure: 561 psia
Ionization Potential (eV): 9.2
Flash Point: -2 °C Closed Cup

Autoignition Temperature: 202 °C
LEL: 1.6% v/v
UEL: 10.4% v/v

RTECS Toxicity Data

Acute Oral: Rat LD_{50} Dose: >3200 mg/kg. Mouse LD_{50} Dose: 3200 mg/kg.

Acute Inhalation: Rat LC_{Lo} Dose: 63 gm/m^3/6hr; Toxic Effects: Behavioral - Somnolence (general depressed activity); Behavioral - Irritability.

Chronic (Multiple Dose) Inhalation: Rat Dose: 4000 ppm/4H/2W-I; Toxic Effects: Behavioral - Change in psychophysiological tests; Lungs, Thorax, or Respiration - Other changes; DEATH.

Reproductive/Teratogenic: Rat Route: Oral; Dose: 660 mg/kg; Duration: female 8-18D of pregnancy; Effects on Fertility - Post-implantation mortality; Specific Developmental Abnormalities - Musculoskeletal system; Homeostasis. Rat Route: Oral; Dose: 660 mg/kg; Duration: female 8-18D of pregnancy; Effects on Newborn - Stillbirth; Live birth index; Viability index. Rat Route: Oral; Dose: 660 mg/kg; Duration: female 8-18D of pregnancy; Effects on Newborn - Growth statistics.

Hazard Overviews

Flammable

Explosive

Health: Irritating. Harmful. Other Acute Effects: harmful if swallowed, inhaled, or absorbed through skin. Chronic Effects: overexposure may cause reproductive disorder(s) based on tests with laboratory animals; target organs: liver, kidneys, blood, central nervous system, female reproductive system, male reproductive system.

Fire: Flammable. Hazards: vapor may travel considerable distance to source of ignition and flash back; container explosion may occur; forms explosive mixtures in air. Extinguishing agents: carbon dioxide, dry chemical powder or appropriate foam; water may be effective for cooling, but may not effect extinguishment. Precautions: combustible liquid.

Reactivity: Incompatible with: oxidizing agents, strong acids. Hazardous decomposition products: toxic fumes of: carbon monoxide, carbon dioxide.

Carcinogenicity: IARC - Not listed; NIOSH - Not listed; NTP - Not listed; ACGIH - Not listed; OSHA - Not listed; EPA - Not listed; MAK - Not listed

Primary Target Organs:

Eyes

Skin

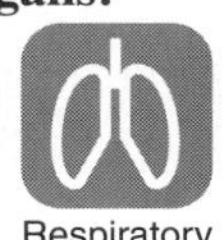
Respiratory System

Liver

Kidneys

Blood

Environmental

Environmental Fate: If released to the atmosphere, it will mainly exist in the vapor phase in the ambient atmosphere based on a measured vapor pressure of 48 mm Hg at 20 °C. Vapor-phase is degraded in the atmosphere by reaction with photochemically produced hydroxyl radicals with an estimated half-life of about 25 hours. An estimated K_{oc} of 18 suggests that it will have very high mobility in soil. Volatilization may be possible from dry soil surfaces, based on its vapor pressure, but will be very slow from moist soil surfaces given an estimated Henry's Law constant of 1.1 x10^{-6} atm-cu m/mole. Limited data, based on pure culture studies, suggest that this compound may be resistant to biodegradation in both soil and water. It is not expected to adsorb to suspended matter in the water column based on its K_{oc} value. It may volatilize slowly from water surfaces given an estimated Henry's Law constant of 1.1 x10^{-6} atm-cu m/mole. Estimated half-lives for a model river and model lake are 33 and 240 days, respectively. Bioconcentration in aquatic organisms will be low based on an estimated BCF value of 0.4.

Cleanup/Disposal: Guide No. 127: Eliminate all ignition sources (no smoking, flares, sparks or flames in immediate area). All equipment used when handling the product must be grounded. Do not touch or walk through spilled material. Stop leak if you can do it without risk. Prevent entry into waterways, sewers, basements or confined areas. A vapor suppressing foam may be used to reduce vapors. Absorb or cover with dry earth, sand or other non-combustible material and transfer to containers. Use clean non-sparking tools to collect absorbed material. Large Spills: Dike far ahead of liquid spill for later disposal. Water spray may reduce vapor; but may not prevent ignition in closed spaces.

Environmental Physical Data

Henry's Law Constant: estimated at 1.1 x10^{-6}
Octanol/Water Partition Coefficient: log K_{ow} = -0.21
Sorption Partition Coefficient: K_{oc} = estimated at 18
BCF: estimated at 0.4

Regulations

RCRA 40CFR: Not listed
CERCLA: 40CFR 302.4: Not listed
SARA 40CFR 372.65: Not listed
SARA EHS 40CFR 355: Not listed
TSCA: Listed

DIM1900 CAS #: 122-07-6

2,2-DIMETHOXYETHYLMETHYLLAMINE

EINECS Number: 204-520-0
Molecular Formula: $C_5H_{13}NO_2$
Formula Weight: 119.17

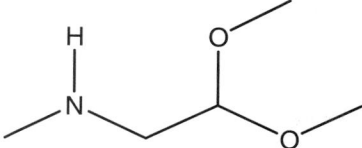

Chemical Structure

Synonyms: ACETALDEHYDE,(METHYLAMINO)-,DIMETHYL ACETAL; ETHANAMINE,2,2-DIMETHOXY-N-METHYL-; 2-(METHYLAMINO)ACETALDEHYDE DIMETHYL ACETAL; N-METHYLAMINOACETALDEHYDE DIMETHYL ACETAL

Physical Properties

Boiling Point: 140 °C (284 °F)
Specific Gravity: 0.92800
Refraction Index: 1.4115

Hazard Overviews

Health: Irritating to eyes/skin/respiratory tract. Other Acute Effects: harmful if swallowed, inhaled, or absorbed through skin; lachrymator; exposure may cause burning sensation; coughing; wheezing; laryngitis; shortness of breath; headache; nausea; vomiting.

Fire: Hazards: vapor may travel considerable distance to source of ignition and flash back; container explosion may occur; forms explosive mixtures in air. Extinguishing agents: carbon dioxide, dry chemical powder or appropriate foam; water may be effective for cooling, but may not effect extinguishment. Precautions: combustible liquid.

Reactivity: Incompatible with: oxidizing agents, acids. Hazardous decomposition products: toxic fumes of: carbon monoxide, carbon dioxide, nitrogen oxides.

Carcinogenicity: IARC - Not listed; NIOSH - Not listed; NTP - Not listed; ACGIH - Not listed; OSHA - Not listed; EPA - Not listed; MAK - Not listed

Primary Target Organs:

Eyes

Skin

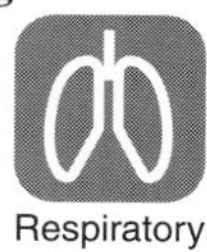
Respiratory System

Environmental

Regulations

RCRA 40CFR: Not listed
CERCLA: 40CFR 302.4: Not listed
SARA 40CFR 372.65: Not listed
SARA EHS 40CFR 355: Not listed
TSCA: Listed

DIM2050 CAS #: 70-38-2

DIMETHRIN

RTECS: GZ1455000
Molecular Formula: $C_{19}H_{26}O_2$
Formula Weight: 286.39

Synonyms: CHRYSANTHEMUMIC ACID,2,4-DIMETHYLBENZYL ESTER; CYCLOPROPANECARBOXYLIC ACID,2,2-DIMETHYL-3-(2-METHYL-1-PROPENYL)-,(2,4-DIMETHYLPHENYL)METHYL ESTER; DIMETHRINE; 2,4-DIMETHYLBENZYL 2,2-DIMETHYL-3-(2-METHYLPROPENYL)CYCLOPROPANECARBOXYLATE; 2,4-DIMETHYLBENZYL ESTER OF CIS,TRANS-CHRYSANTHEMUMIC ACID; 2,4-DIMETHYLBENZYL-(I)-CIS-TRANS-CHRYSANTHEMUMATE; 2,4-DIMETHYLBENZYLCHRYSANTHEMUMATE; 2,4-DIMETHYLBENZYL(1RS)-CIS,TRANS-2,2-DIMETHYL-3-(2-METHYLPROP-1-ENYL) CYCLOPROPANECARBOXYLATE; 2,4-DIMETHYLBENZYLESTER KYSELINY CHRYSANTHEMOVE; (2,4-DIMETHYLPHENYL)METHYL 2,2-DIMETHYL-3-(2-METHYL-1-PROPENYL)CYCLOPROPANECARBOXYLATE; DIMETRIN; ENT-21170; SHA 034101

Use: insecticide; insecticide for use in ponds and swamps as a mosquito larvicide

Physical Properties

Boiling Point: 175 °C (347 °F)
Specific Gravity: 0.98
Water Solubility: Insoluble in Water
Other Solubilities: Soluble in petroleum hydrocarbons, aromatic petroleum derivatives, alcohols, and methylene chloride; decomposed by strong alkali

RTECS Toxicity Data

Acute Oral: Rat LD_{50} Dose: 40 gm/kg. Mouse LD_{50} Dose: 10 gm/kg; Toxic Effects: Automatic Nervous System - Other (direct) parasympathomimetic; Behavioral - Convulsions or effect on seizure threshold; Blood - Other changes.

Acute Dermal: Rabbit LD; Route: Skin; Dose: >4900 mg/kg.

Chronic (Multiple Dose) Oral: Rat Dose: 182 gm/kg/52W-C; Toxic Effects: Liver - Changes in liver weight. Rat Dose: 57 gm/kg/16W-C; Toxic Effects: Liver - Changes in liver weight; Kidney, Ureter, and Bladder - Changes in kidney weight; Nutritional and gross metabolic - Weight loss or decreased weight gain.

Hazard Overviews

Carcinogenicity: IARC - Not listed; NIOSH - Not listed; NTP - Not listed; ACGIH - Not listed; OSHA - Not listed; EPA - Not listed; MAK - Not listed

Environmental

Ecotoxicity: LC_{50} Perca flavescens 28 ug/l/96 hr. Static bioassay without aeration, pH 7.2-7.5, water hardness 40-50 mg/l as calcium carbonate and alkalinity of 30-35 mg/l.

Environmental Fate: If released on land, it is expected to adsorb strongly to the soil surface. Photolysis may occur on the soil surface. Its degradation in soil is unknown. If released in water, it is expected to adsorb strongly to sediment and particulate matter in the water column. It would be lost by volatilization having an estimated half-life of 1.0 day in a model river. Photolysis may occur in the surface layers of water. If released on land, dimethrin is expected to adsorb strongly to the soil surface. Photolysis may occur on the soil surface. Its degradation in soil in unknown. If released in water, it is expected to adsorb strongly to sediment and particulate matter in the water column. It would be lost by volatilization having an estimated half-life of 1.0 day in a model river. Photolysis may occur in the surface layers of water. While it is estimated to strongly bioconcentrate in aquatic organism, this may not be realized since it may metabolize. If released into the atmosphere as a spray, it would be removed by gravitational settling. Vapor phase reacts with photochemically-produced hydroxyl radicals and ozone resulting in an estimated atmospheric half-life of 25 minutes.

Environmental Physical Data

Henry's Law Constant: estimated at 7.61 x10^{-5}
Octanol/Water Partition Coefficient: log K_{ow} = 6.57
Sorption Partition Coefficient: K_{oc} = estimated at 4.48
BCF: calculated at 5.8 x10^4

Regulations

RCRA 40CFR: Not listed
CERCLA: 40CFR 302.4: Not listed
SARA 40CFR 372.65: Not listed
SARA EHS 40CFR 355: Not listed
TSCA: Not listed

DIM2100 CAS #: 792-74-5

DIMETHYL 4,4'-BIPHENYLDICARBOXYLATE

EINECS Number: 212-341-4
Molecular Formula: $C_{16}H_{14}O_4$
Formula Weight: 270.29

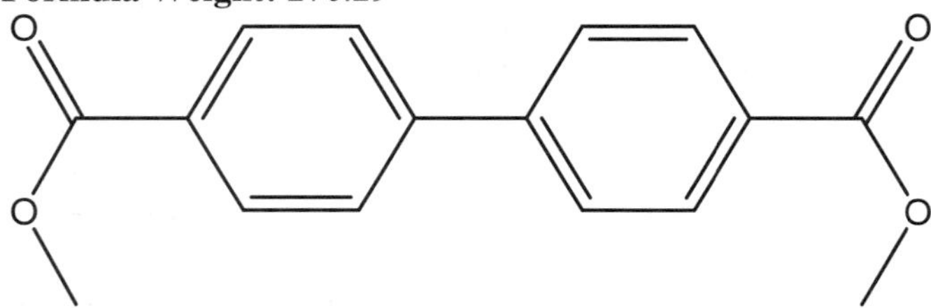

Chemical Structure

Synonyms: (1,1'-BIPHENYL)-4,4'-DICARBOXYLIC ACID,DIMETHYL ESTER; 4,4'-BIPHENYLDICARBOXYLIC ACID,DIMETHYL ESTER
Use: in prepn of insect growth regulators

Hazard Overviews

Health: Irritating to eyes/skin/respiratory tract. Harmful. Other Acute Effects: harmful if swallowed, inhaled, or absorbed through skin.
Fire: Hazards: emits toxic fumes; under fire conditions, material may decompose to form flammable and/or explosive mixtures in air. Extinguishing agents: water spray; carbon dioxide, dry chemical powder or appropriate foam. Precautions: combustible liquid.
Reactivity: Stable. Hazardous polymerization will not occur. Incompatible with: strong oxidizing agents, acids, bases. Hazardous decomposition products: toxic fumes of: carbon monoxide, carbon dioxide.
Carcinogenicity: IARC - Not listed; NIOSH - Not listed; NTP - Not listed; ACGIH - Not listed; OSHA - Not listed; EPA - Not listed; MAK - Not listed
Primary Target Organs:

Eyes

Skin

Respiratory System

Environmental

Regulations

RCRA 40CFR: Not listed
CERCLA: 40CFR 302.4: Not listed
SARA 40CFR 372.65: Not listed
SARA EHS 40CFR 355: Not listed
TSCA: Listed

DIM2150 CAS #: 1467-79-4

DIMETHYL CYANAMIDE

RTECS: GS6475000
EINECS Number: 215-991-7
Molecular Formula: $C_3H_6N_2$
Formula Weight: 70.11

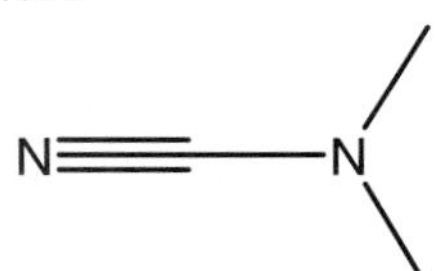

Chemical Structure

Synonyms: CARBAMIC ACID NITRILE; CARBONIC ACID DIETHYLAMIDE; CYANAMIDE,DIMETHYL-; N-CYANODIMETHYLAMINE; N-CYANO-N-METHYLMETHANAMINE; DIMETHYLCYANAMIDE; N,N-DIMETHYLCYANAMIDE; DIMETHYLKYANAMID
Description: colorless, mobile liquid
Use: organic intermediate; solvent

Physical Properties

Boiling Point: 163.5 °C (326 °F) at 760 mm Hg
Freezing Point: -41 °C (-41.8 °F)
Specific Gravity: 0.876
Vapor Density: 2.42 Air=1
Density: 0.877 at 30
Vapor Pressure: 40 mm Hg at 80 °C
Water Solubility: Very Soluble in Water
Other Solubilities: 95% Ethanol: Soluble; Acetone: Soluble; Ether: Soluble.
Refraction Index: 1.4089 at 19 °C/D
Ionization Potential (eV): 9.0
Flash Point: 71 °C Closed Cup

RTECS Toxicity Data

Acute Oral: Rat LD_{50} Dose: 146 mg/kg; Toxic Effects: Behavioral - Convulsions or effect on seizure threshold; Behavioral - Muscle weakness; Lungs, Thorax, or Respiration - Dyspnea. Mouse LD_{50} Dose: 73 mg/kg; Toxic Effects: Behavioral - Convulsions or effect on seizure threshold; Behavioral - Muscle weakness; Lungs, Thorax, or Respiration - Dyspnea.

Acute Inhalation: Rat LC_{50} Dose: 2500 mg/m^3; Toxic Effects: Behavioral - Convulsions or effect on seizure threshold; Behavioral - Muscle weakness; Lungs, Thorax, or Respiration - Dyspnea. Mouse LC_{50} Dose: 2800 mg/m^3; Toxic Effects: Behavioral - Convulsions or effect on seizure threshold; Behavioral - Muscle weakness; Lungs, Thorax, or Respiration - Dyspnea.

Acute Dermal: Mouse LD_{50} Route: Skin; Dose: 125 mg/kg. Guinea Pig LD_{Lo} Route: Skin; Dose: 5 gm/kg.

Hazard Overviews

Poison

Fire Diamond

Health: Severely irritating to eyes/skin/respiratory tract. Poison. Other Acute Effects: may be fatal if inhaled, swallowed, or absorbed through skin; symptoms of exposure may include burning sensation; coughing; wheezing; laryngitis; shortness of breath; headache; nausea; vomiting.

Fire: Combustible. Hazards: emits toxic fumes. Extinguishing agents: water spray; carbon dioxide, dry chemical powder or appropriate foam. Precautions: combustible liquid.

Reactivity: Incompatible with: strong acids, strong bases, strong oxidizing agents, strong reducing agents, may decompose on exposure to moist air or water. Hazardous decomposition products: thermal decomposition may produce carbon monoxide, carbon dioxide, and nitrogen oxides.

Carcinogenicity: IARC - Not listed; NIOSH - Not listed; NTP - Not listed; ACGIH - Not listed; OSHA - Not listed; EPA - Not listed; MAK - Not listed

Primary Target Organs:

Eyes

Skin

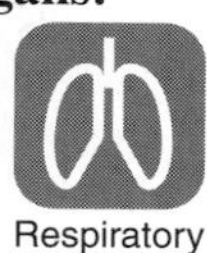
Respiratory System

Environmental

Cleanup/Disposal: Guide No. 153: Eliminate all ignition sources (no smoking, flares, sparks or flames in immediate area). Do not touch damaged containers or spilled material unless wearing appropriate protective clothing. Stop leak if you can do it without risk. Prevent entry into waterways, sewers, basements or confined areas. Absorb or cover with dry earth, sand or other non-combustible material and transfer to containers. Do not get water inside containers.

Regulations

RCRA 40CFR: Not listed

CERCLA: 40CFR 302.4: Listed as Compound per CWA Section 307(a) per CAA Section 112

SARA 40CFR 372.65: Not listed

SARA EHS 40CFR 355: Not listed

TSCA: Listed

DIM2400 **CAS #: 756-80-9**

O,O-DIMETHYL DITHIOPHOSPHATE

RTECS: TE0525000

EINECS Number: 212-053-9

Molecular Formula: $C_2H_7O_2PS_2$

Formula Weight: 158.18

Synonyms: O,O-DIMETHYL DITHIOPHOSPHORIC ACID; O,O'-DIMETHYL HYDROGEN DITHIOPHOSPHATE; O,O-DIMETHYL HYDROGEN DITHIOPHOSPHATE; DIMETHYL PHOSPHODITHIONATE; DIMETHYL PHOSPHORODITHIOATE; O,O-DIMETHYL PHOSPHORODITHIOATE; O,O-DIMETHYL PHOSPHORODITHIOIC ACID; O,O-DIMETHYLDITHIOPHOSPHATE; DIMETHYLDITHIOPHOSPHORIC ACID; O,O-DIMETHYLPHOSPHORODITHIOATE; KWAS DWUMETYLO-DWUTIOFOSFOROWY; KYSELINA O,O-DIMETHYLDITHIOFOSFORCNA; METHYL PHOSPHORODITHIOATE (6CI,7CI); PHOSPHORODITHIOIC ACID,O,O-DIMETHYL ESTER

Use: chem int for malathion (insecticide), azinphos-methyl (insecticide), phosmet (insecticide), methidathion (insecticide)

RTECS Toxicity Data

Acute Oral: Rat LD_{50} Dose: 1 gm/kg. Mouse LD_{50} Dose: 1550 mg/kg.

Irritation Eye: Rabbit Standard Draize Test Dose: 250 ug/24H; Reaction: severe.

Irritation Skin: Rabbit Standard Draize Test Dose: 20 mg/24H; Reaction: moderate.

Reproductive/Teratogenic: Rat Route: Inhalation; Dose: 161 mg/m^3/6H Duration: male 11W prior to mating; Effects on Fertility - Male fertility index.

Hazard Overviews

Carcinogenicity: IARC - Not listed; NIOSH - Not listed; NTP - Not listed; ACGIH - Not listed; OSHA - Not listed; EPA - Not listed; MAK - Not listed

Environmental

Regulations

RCRA 40CFR: Not listed

CERCLA: 40CFR 302.4: Not listed

SARA 40CFR 372.65: Not listed

SARA EHS 40CFR 355: Not listed

TSCA: Listed

DIM2450 **CAS #: 115-10-6**

DIMETHYL ETHER

RTECS: PM4780000
DOT: UN1033; IMO2
EINECS Number: 204-065-8
Molecular Formula: C_2H_6O
Structured MF: CH_3OCH_3
Formula Weight: 46.07

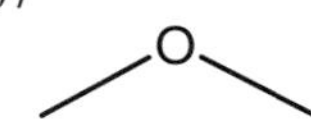

Chemical Structure

Synonyms: DIMETHYL OXIDE; ETHER,DIMETHYL; ETHER,METHYL; METHANE,OXYBIS-; METHOXY METHANE; METHOXYMETHANE; METHYL ETHER; METHYL OXIDE; OXYBISMETHANE; WOOD ETHER

Description: colorless compressed gas; etheral odor

Use: in refrigeration; cooling medium, solvent, extractant, propellant for aerosol, catalyst and stabilizer for polymerization

Physical Properties

Boiling Point: -24 °C (-11 °F)
Freezing Point: -138.5 °C (-217.3 °F)
Specific Gravity: 1.617
Vapor Density: 1.6 Air=1
Vapor Pressure: 6.8 kPa at 200 °K
Water Solubility: 1 vol Water takes up 37 vol gas
Other Solubilities: Acetone: Soluble; Alcohol: Soluble; Benzene: Slightly Soluble; Chloroform: Soluble; Ether: Soluble.
Surface Tension: 21 dynes/cm at 40 °C
Refraction Index: 1.3441 at -42.5 °C/D
Evaporation Rate: > 1 Butyl Acetate=1
Critical Temperature: 127 °C
Critical Pressure: 52.6 atm
Ionization Potential (eV): 10.025 +/-0.025
Flash Point: Not pertinent (flammable gas)
Autoignition Temperature: 350 °C
LEL: 3.4% v/v
UEL: 26.7% v/v

RTECS Toxicity Data

Acute Inhalation: Rat LC_{50} Dose: 308 gm/m^3; Toxic Effects: Behavioral - General anesthetic.

Chronic (Multiple Dose) Inhalation: Rat Dose: 2 pph/6H/30W-I; Toxic Effects: Liver - Changes in Liver weight; Blood - Changes in serum composition.

Hazard Overviews

Flammable

Compressed Gas

Fire Diamond

Health: Irritating to eyes/skin/respiratory tract. Also Causes: intoxication, incoordination, blurred vision, headache, dizziness, unconsciousness, severe frostbite, irritated mucous membranes and eyes.

Fire: Flammable. Can form explosive mixtures in the air. Stop flow of gas. If gas is burning, control but do not extinguish flame until gas flow is controlled. Use carbon dioxide or dry chemical to extinguish flame. Use abundant water spray to disperse vapors, to cool fire-exposed containers, and to protect workers shutting off gas.

Reactivity: Stable. Hazardous polymerization cannot occur. Avoid: heat; ignition sources; hot surfaces; contact with oxygen at room temperature, in sunlight, or for long periods of time. Incompatible with: oxidizing agents; aluminum hydride and lithium aluminum hydride. Hazardous decomposition products: toxic carbon monoxide; toxic carcinogenic formaldehyde.

Carcinogenicity: IARC - Not listed; NIOSH - Not listed; NTP - Not listed; ACGIH - Not listed; OSHA - Not listed; EPA - Not listed; MAK - Not listed

Primary Target Organs:

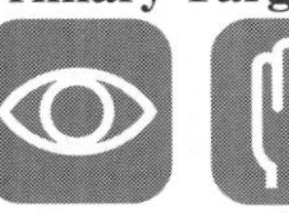
Eyes

Skin

Mucous Membranes

Nervous System

Exposure Limits
DFG MAK: TWA: 1000 ppm; 1910 mg/m^3.
AIHA WEEL: TWA: 1000 ppm.

Respirator Recommendation
Exposure Range: >500 to 25000 ppm Supplied Air, Constant Flow/Pressure Demand, Half Mask
Exposure Range: >25,000 to 500,000 ppm Supplied Air, Constant Flow/Pressure Demand, Full Face
Exposure Range: >500,000 to unlimited ppm Self-contained Breathing Apparatus, Pressure Demand, Full Face
Note: odor threshold unknown

Environmental

Environmental Fate: If released to soil, it will be subject to volatilization. It will be expected to exhibit very high mobility in soil and, therefore, it may leach to groundwater. If released to water, it will not be expected to significantly adsorb to sediment or suspended particulate matter, bioconcentrate in aquatic organisms, directly photolyze, or photooxidize via reaction with photochemically produced hydroxyl radicals in the water, based upon estimated physical-chemical properties or analogies to other structurally related aliphatic ethers. It will not be expected to hydrolyze in water or soil. In surface water it will be subject to rapid volatilization with estimated half-lives for volatilization of 2.6 hr and 30 hr from a river one meter deep flowing 1 m/sec with a wind velocity of 3 m/sec and a model pond, respectively. It may be resistant to biodegradation in environmental media based upon screening test data for the structurally related diethyl ether from studies using activated sludge or sewage inocula. Many ethers are known to be resistant to biodegradation. If released to the atmosphere, it will be expected to exist almost entirely in the vapor phase based upon its vapor pressure. It will be susceptible to photooxidation via vapor phase reaction with photochemically produced hydroxyl radicals with an estimated half-life of 5.4 days for this process. It also will be susceptible to photooxidation via vapor phase reaction with nitrate radicals in nighttime air with an estimated half-life of greater than or equal to 22 days for this process. Direct photolysis will not be an important removal process since it does not absorb light at wavelengths >290 nm.

Cleanup/Disposal: Guide No. 115: Eliminate all ignition sources (no smoking, flares, sparks or flames in immediate area). All equipment used when handling the product must be grounded. Do not touch or walk through spilled material. Stop leak if you can do it without risk. If possible, turn leaking containers so that gas escapes rather than liquid. Use water spray to reduce vapors or divert vapor cloud drift. Do not direct water at spill or source of leak. Prevent spreading of vapors through sewers, ventilation systems and confined areas. Isolate area until gas has dispersed.

Environmental Physical Data

Henry's Law Constant: 9.78×10^{-4}
Sorption Partition Coefficient: K_{OC} = estimated at 14
BCF: estimated at 1.7

Regulations

RCRA 40CFR: Not listed
CERCLA: 40CFR 302.4: Not listed
SARA 40CFR 372.65: Not listed
SARA EHS 40CFR 355: Not listed
TSCA: Listed

Analytical Methods

Air: ASTM D4490

DIM2500 CAS #: 16090-49-6

DIMETHYL GERMANIUM SULFIDE

Molecular Formula: C_2H_6GeS
Formula Weight: 134.73
Synonyms: GERMANE,DIMETHYLTHIOXO-; GERMANIUM SULFIDE,DIMETHYL-

Hazard Overviews

Carcinogenicity: IARC - Not listed; NIOSH - Not listed; NTP - Not listed; ACGIH - Not listed; OSHA - Not listed; EPA - Not listed; MAK - Not listed

Environmental

Regulations

RCRA 40CFR: Not listed
CERCLA: 40CFR 302.4: Not listed
SARA 40CFR 372.65: Not listed
SARA EHS 40CFR 355: Not listed
TSCA: Not listed

DIM2550 CAS #: 1119-40-0

DIMETHYL GLUTARATE

EINECS Number: 214-277-2
Molecular Formula: $C_7H_{12}O_4$
Structured MF: $CH_3O_2OC(CH_2)_3CO_2CH_3$
Formula Weight: 160.17

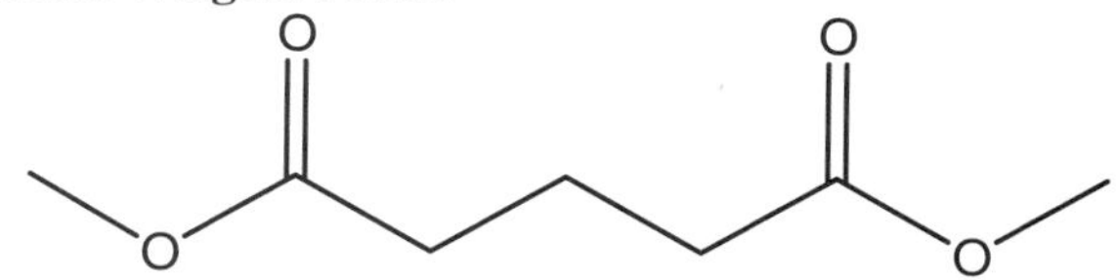

Chemical Structure

Synonyms: DIMETHYL PENTANEDIOATE; GLUTARIC ACID,DIMETHYL ESTER; METHYL GLUTARATE; PENTANEDIOIC ACID,DIMETHYL ESTER
Description: liquid; faint agreeable odor
Use: chem int for epichlorohydrin-polyamide resins for paper; chem int for polyester resins (coatings, plasticizers); chem int for polyamide resins

Physical Properties

Boiling Point: 213.5 °C (416 °F) to 214 °C (417 °F) at 752 mm Hg
Freezing Point: -42.5 °C (-44.5 °F)
Specific Gravity: 1.0876 at 20 °C/4 °C
Vapor Density: 5.52 Air=1
Other Solubilities: Very Soluble in Alcohol, Ether
Refraction Index: 1.4242 at 20 °C/D
Flash Point: 103 °C Closed Cup

Hazard Overviews

Health: Irritating. Other Acute Effects: may be harmful by inhalation, ingestion, or skin absorption.
Fire: Will burn. Extinguishing agents: carbon dioxide, dry chemical powder or appropriate foam. Precautions: combustible liquid.
Reactivity: Incompatible with: acids, bases, oxidizing agents, reducing agents. Hazardous decomposition products: toxic fumes of: carbon monoxide, carbon dioxide.
Carcinogenicity: IARC - Not listed; NIOSH - Not listed; NTP - Not listed; ACGIH - Not listed; OSHA - Not listed; EPA - Not listed; MAK - Not listed
Primary Target Organs:

Eyes

Skin

Respiratory System

Environmental

Cleanup/Disposal: Guide No. 154: Eliminate all ignition sources (no smoking, flares, sparks or flames in immediate area). Do not touch damaged containers or spilled material unless wearing appropriate protective clothing. Stop leak if

you can do it without risk. Prevent entry into waterways, sewers, basements or confined areas. Absorb or cover with dry earth, sand or other non-combustible material and transfer to containers. Do not get water inside containers.

Regulations

RCRA 40CFR: Not listed
CERCLA: 40CFR 302.4: Not listed
SARA 40CFR 372.65: Not listed
SARA EHS 40CFR 355: Not listed
TSCA: Listed

DIM2600 CAS #: 94-60-0

DIMETHYL HEXAHYDROTEREPHTHALATE

EINECS Number: 202-347-5
Molecular Formula: $C_{10}H_{16}O_4$
Formula Weight: 200.2

Chemical Structure

Synonyms: 1,4-CYCLOHEXANEDICARBOXYLIC ACID,DIMETHYL ESTER; 1,4-CYCLOHEXANEDICARBOXYLIC DIMETHYL ESTER; DIMETHYL 1,4-CYCLOHEXANEDICARBOXYLATE
Description: partially crystalline solid
Use: plasticizer

Physical Properties

Boiling Point: 265 °C (509 °F)
Specific Gravity: 1.102
Other Solubilities: In all proportions in most organic solvents
Refraction Index: 1.4580 at 20 °C

Hazard Overviews

Health: May cause irritation to eyes and skin. Other Acute Effects: may be harmful by inhalation, ingestion, or skin absorption.
Fire: Extinguishing agents: water spray; carbon dioxide, dry chemical powder or appropriate foam. Precautions: combustible liquid.
Carcinogenicity: IARC - Not listed; NIOSH - Not listed; NTP - Not listed; ACGIH - Not listed; OSHA - Not listed; EPA - Not listed; MAK - Not listed

Environmental

Regulations

RCRA 40CFR: Not listed
CERCLA: 40CFR 302.4: Not listed
SARA 40CFR 372.65: Not listed
SARA EHS 40CFR 355: Not listed
TSCA: Listed

DIM2650 CAS #: 868-85-9

DIMETHYL HYDROGEN PHOSPHITE

RTECS: SZ7710000
EINECS Number: 212-783-8
Molecular Formula: $C_2H_7O_3P$
Structured MF: $(CH_3O)_2P(O)H$
Formula Weight: 110.05

Chemical Structure

Synonyms: BIS(HYDROXYMETHYL)PHOSPHINE OXIDE; DIMETHOXYPHOSPHINE OXIDE; DIMETHYL ACID PHOSPHITE; DIMETHYL PHOSPHITE; DIMETHYL PHOSPHONATE; DIMETHYL PHOSPHOROUS ACID; DIMETHYLACID PHOSPHITE; DIMETHYLESTER KYSELINI FOSFORITE; DIMETHYLESTER KYSELINY FOSFORITE; DIMETHYLFOSFIT; DIMETHYLFOSFONAT; DIMETHYLHYDROGEN PHOSPHITE; DIMETHYLHYDROGENPHOSPHITE; DMHP; HYDROGEN DIMETHYL PHOSPHITE; METHYL PHOSPHONATE; METHYL PHOSPHONATE ((MEO)2HPO); PHOSPHONIC ACID,DIMETHYL ESTER; PHOSPHOROUS ACID DIMETHYL ESTER
Description: colorless , mobile liquid; mild odor
Use: as a flame retardant; a methylating agent for n-heterocyclic compounds and s-methylate thioamides; a hydrophosphorylating reagent for alkenes and alkynes

Physical Properties

Boiling Point: 72 °C (162 °F) to 73 °C (163 °F) at 25 mm Hg
Specific Gravity: 1.2 at 20 °C/4 °C
Vapor Density: 3.79 Air=1
Saturated Vapor Density: 1.204412886 kg/m^3
Vapor Pressure: 1 mm Hg at 25 °C
Water Solubility: Soluble in Water
Other Solubilities: 95% Ethanol: >=100 mg/ml at 19.5 °C; Acetone: >=100 mg/ml at 19.5 °C; Alcohol: miscible; Common organic solvents: miscible; DMSO: >=100 mg/ml at 19.5 °C; Ether: miscible; Pyridine: >10%; Pyrimidine: Soluble.
pH: Acid
Refraction Index: 1.400 at 25 °C/D
Ionization Potential (eV): 10.53
Flash Point: 29 °C Closed Cup

RTECS Toxicity Data

Acute Oral: Rat LD_{50} Dose: 3040 mg/kg. Mouse LD_{50} Dose: 1831 mg/kg; Toxic Effects: Automatic Nervous System - Other (direct) parasympathomimetic; Lungs, Thorax, or Respiration - Respiratory stimulation; Gastrointestinal - Changes in structure or function of salivary glands.
Acute Inhalation: Rat LC_{50} Dose: >20 gm/m^3.
Acute Dermal: Rabbit LD_{50} Route: Skin; Dose: 2400 mg/kg. Rat LD_{50} Route: Subcutaneous Dose: 2970 mg/kg.
Chronic (Multiple Dose) Oral: Rat Dose: 7500 mg/kg/15D-I; Toxic Effects: Behavioral - Somnolence (general depressed activity); DEATH. Rat Dose: 26 gm/kg/13W-I; Toxic Effects: Sense organs and special senses - Corneal damage; Nutritional and gross metabolic - Weight loss or decreased weight gain; DEATH. Rat Dose: 30 gm/kg/15D-I; Toxic Effects: DEATH. Rat Dose: 97500 mg/kg/13W-I; Toxic Effects: Lungs, Thorax, or Respiration - Chronic pulmonary edema; DEATH.
Irritation Eye: Rabbit Standard Draize Test Dose: 20 mg/24H; Reaction: moderate.
Irritation Skin: Rabbit Standard Draize Test Dose: 500 mg/24H; Reaction: mild.
Mutagenic: Mouse Mutations in Microorganisms; Cell Type: lymphocyte; Dose: 1700 mg/L (+S9). Mouse Mutations in Mammalian Somatic Cells; Cell Type: lymphocyte; Dose: 2100 mg/L.
Tumorigenic: Rat Route: Oral; Dose: 103 gm/kg/2Y-I; Toxic Effects: Tumorigenic - Carcinogenic by RTECS criteria; Lungs, Thorax, or Respiration - Tumors; Gastrointestinal - Tumors.

Hazard Overviews

Flammable

Health: Irritating to eyes/skin/respiratory tract. Harmful. Other Acute Effects: may be harmful by inhalation, ingestion, or skin absorption; exposure can cause nausea; headache; vomiting. Chronic Effects: possible carcinogen; target organs: eyes, kidneys, nerves.
Fire: Flammable. Hazards: emits toxic fumes; vapor may travel considerable distance to source of ignition and flash back. Extinguishing agents: water spray; carbon dioxide, dry chemical powder or appropriate foam. Precautions: combustible liquid.
Reactivity: Incompatible with: strong oxidizing agents, strong bases, strong acids, acid chlorides, may decompose on exposure to moist air or water. Hazardous decomposition products: toxic fumes of: carbon monoxide, carbon dioxide; thermal decomposition may produce toxic fumes of phosphorus oxides and/or phosphine.
Carcinogenicity: IARC - Group 3, Not classifiable as to carcinogenicity to humans; NIOSH - Not listed; NTP - Not listed; ACGIH - Not listed; OSHA - Not listed; EPA - Not listed; MAK - Not listed

Primary Target Organs:

Eyes

Skin

Respiratory System

Nervous System

Kidneys

Environmental

Environmental Fate: In soil, it will probably hydrolyze and leach to groundwater (where it will completely hydrolyze); it will not volatilize significantly. In water, it will hydrolyze but will not evaporate readily nor sorb to sediments or biota. It is not known if it biodegrades, although it may photodegrade since it appears to be photosensitive. The fate in air is not known, although photolysis may occur.
Cleanup/Disposal: Guide No. 128: Eliminate all ignition sources (no smoking, flares, sparks or flames in immediate area). All equipment used when handling the product must be grounded. Do not touch or walk through spilled material. Stop leak if you can do it without risk. Prevent entry into waterways, sewers, basements or confined areas. A vapor suppressing foam may be used to reduce vapors. Absorb or cover with dry earth, sand or other non-combustible material and transfer to containers. Use clean non-sparking tools to collect absorbed material. Large Spills: Dike far ahead of liquid spill for later disposal. Water spray may reduce vapor; but may not prevent ignition in closed spaces.

Environmental Physical Data

Octanol/Water Partition Coefficient: log K_{ow} = expected to be low

Regulations

RCRA 40CFR: Not listed
CERCLA: 40CFR 302.4: Not listed
SARA 40CFR 372.65: Not listed
SARA EHS 40CFR 355: Not listed
TSCA: Listed

DIM2800 **CAS #: 1459-93-4**

DIMETHYL ISOPHTHALATE

RTECS: NT2540000
EINECS Number: 215-951-9
Molecular Formula: $C_{10}H_{10}O_4$
Formula Weight: 194.19

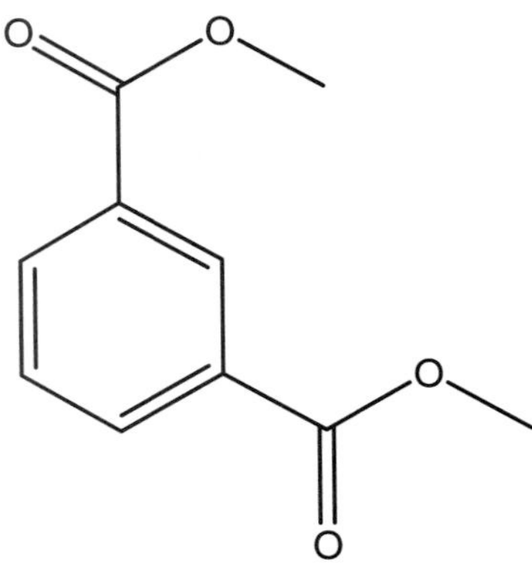

Chemical Structure

Synonyms: 1,3-BENZENEDICARBOXYLIC ACID,DIMETHYL ESTER; DIMETHYL 1,3-BENZENEDICARBOXYLATE; DIMETHYL M-PHTHALATE; DIMETHYLESTER KYSELINY ISOFTALOVE; DIMETHYLESTER KYSELINY TEREFTALOVE; ISOPHTHALIC ACID,DIMETHYL ESTER; METHYL 3-CARBOMETHOXYBENZOATE; METHYL ISOPHTHALATE

Description: solid or needles

Use: plasticizer; chem intermed for polyester resins

Physical Properties

Boiling Point: 282 °C (540 °F)
Freezing Point: 67 °C (152.6 °F) to 68 °C (154.4 °F)
Specific Gravity: 1.194 at 20 °C/4 °C
Saturated Vapor Density: 1.200012592 kg/m^3
Vapor Pressure: 1.4 x10^{-3} torr at 25 °C
Water Solubility: Estimated 820 ppm at 25 °C
Refraction Index: 1.5168 at 20 °C/D

RTECS Toxicity Data

Irritation Eye: Rabbit Standard Draize Test Dose: 500 mg/24H; Reaction: mild.

Hazard Overviews

Health: Irritating to eyes/skin/respiratory tract. Other Acute Effects: may be harmful by inhalation, ingestion, or skin absorption.

Fire: Extinguishing agents: water spray; carbon dioxide, dry chemical powder or appropriate foam. Precautions: combustible liquid.

Reactivity: Incompatible with: strong oxidizing agents. Hazardous decomposition products: toxic fumes of: carbon monoxide, carbon dioxide.

Carcinogenicity: IARC - Not listed; NIOSH - Not listed; NTP - Not listed; ACGIH - Not listed; OSHA - Not listed; EPA - Not listed; MAK - Not listed

Primary Target Organs:

Eyes

Skin

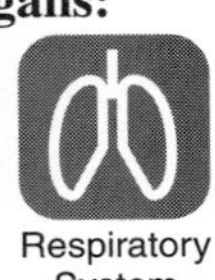
Respiratory System

Environmental

Environmental Fate: If released to the atmosphere, it should react with photochemically produced hydroxy radicals with an estimated half life of 25 days. By analogy to the structurally similar compound dimethyl phthalate, it should be a candidate for direct photochemical degradation. If released to soil, it would be expected to display moderate to high mobility. By analogy to the structurally similar compound, dimethyl phthalate, biodegradation should be an important process. Volatilization from soil is not a likely fate process. If released to water, it should biodegrade, should not adsorb to soil, sediment or suspended matter, and should not bioaccumulate in aquatic organisms. Degradation from direct photolysis may ensue, and hydrolysis should not be an important process in waters at a pH less than 9. The estimated volatilization half life for a model river 1 m deep, flowing at 1 m/sec, and a wind velocity of 3 m/sec is 115 days, and thus volatilization from water to the atmosphere should not be an important fate process.

Environmental Physical Data

Henry's Law Constant: 4.43 x10^{-7}
Octanol/Water Partition Coefficient: log K_{ow} = 2.35
Sorption Partition Coefficient: K_{oc} = 109 to 450
BCF: estimated at 14 to 36

Regulations

RCRA 40CFR: Not listed
CERCLA: 40CFR 302.4: Not listed
SARA 40CFR 372.65: Not listed
SARA EHS 40CFR 355: Not listed
TSCA: Listed

Analytical Methods

Plasma: EPA 001

DIM2850 CAS #: 624-48-6

DIMETHYL MALEATE

RTECS: EM6300000
EINECS Number: 210-848-5
Molecular Formula: $C_6H_8O_4$
Formula Weight: 144.14

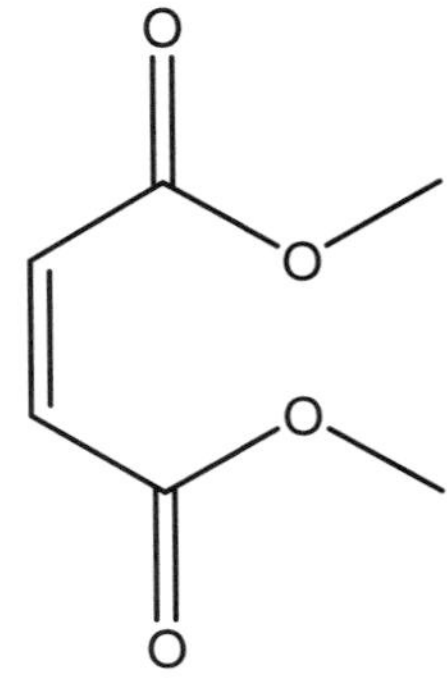

Chemical Structure

Synonyms: DIMETHYLESTER KYSELINY MALEINOVE; MALEIC ACID,DIMETHYL ESTER; METHYL MALEATE; SIPOMER DMM

Physical Properties

Boiling Point: 205 °C (401 °F)
Freezing Point: -19 °C (-2.2 °F)
Specific Gravity: 1.15
Vapor Density: 4.97 Air=1
Vapor Pressure: 9.14 kPa at 125 °C
Water Solubility: Slightly
Other Solubilities: Ether: Soluble; Carbon Tetrachloride: Soluble; ligroin: Insoluble
Refraction Index: 1.4416

RTECS Toxicity Data

Acute Oral: Rat LD_{50} Dose: 1410 mg/kg.
Acute Inhalation: Mouse LC; Dose: >1500 mg/m^3/10M.
Acute Dermal: Rabbit LD_{50} Route: Skin; Dose: 530 mg/kg; Toxic Effects: Skin and appendages - Primary irritation.
Chronic (Multiple Dose) Dermal: Rat Route: Skin; Dose: 40 mg/kg/28D-I; Toxic Effects: Behavioral - Food intake (animal); Skin and appendages - Dermatitis, other; Nutritional and gross metabolic - Weight loss or decreased weight gain.
Irritation Eye: Rabbit Standard Draize Test Dose: 500 mg/24H; Reaction: mild. Rabbit Standard Draize Test Dose: 500 mg; Reaction: severe.

Hazard Overviews

Health: Irritating to eyes/skin/respiratory tract. Toxic. Other Acute Effects: harmful if swallowed, inhaled, or absorbed through skin; readily absorbed through skin.
Fire: Extinguishing agents: carbon dioxide, dry chemical powder or appropriate foam. Precautions: combustible liquid.
Reactivity: Incompatible with: acids, bases, oxidizing agents, reducing agents. Hazardous decomposition products: toxic fumes of: carbon monoxide, carbon dioxide.
Carcinogenicity: IARC - Not listed; NIOSH - Not listed; NTP - Not listed; ACGIH - Not listed; OSHA - Not listed; EPA - Not listed; MAK - Not listed
Primary Target Organs:

Eyes

Skin

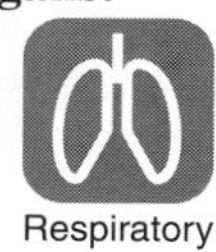
Respiratory System

Environmental

Regulations

RCRA 40CFR: Not listed
CERCLA: 40CFR 302.4: Not listed
SARA 40CFR 372.65: Not listed
SARA EHS 40CFR 355: Not listed
TSCA: Listed

DIM2900 CAS #: 756-79-6

DIMETHYL METHYLPHOSPHONATE

RTECS: SZ9120000
EINECS Number: 212-052-3
Molecular Formula: $C_3H_9O_3P$
Structured MF: $CH_3PO(OCH_3)_2$
Formula Weight: 124.08

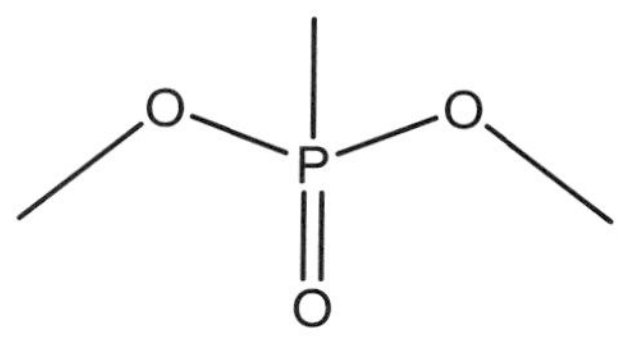

Chemical Structure

Synonyms: DIMETHOXYMETHYLPHOSPHINE OXIDE; DIMETHYL METHANEPHOSPHONATE; DMMP; FYROL DMMP; METHANEPHOSPHONIC ACID DIMETHYL ESTER; METHYL PHOSPHONIC ACID,DIMETHYL ESTER; METHYLPHOSPHONIC ACID DIMETHYL ESTER; PHOSPHONIC ACID,METHYL-,DIMETHYL ESTER; PYROL DMMP
Description: colorless, clear liquid; pleasant odor
Use: gasoline additive, hydraulic fluid additive, heavy metal extractor, solvent, stimulant for nerve gas agents, additive flame retardant in plastics, extractant for rare earths and synthetic intermediate; forms an anion which may be alkylated or acylated; to methylate nucleotide bases and to convert carboxylic acids into methyl esters

Physical Properties

Boiling Point: 181 °C (358 °F) at 754 mm Hg
Freezing Point: 69 °C (156.2 °F)
Specific Gravity: 1.15 at 20 °C/4 °C
Saturated Vapor Density: 1.206212123 kg/m^3
Density: 1.145 g/mL
Vapor Pressure: 1.2 mm Hg at 25 °C
Water Solubility: Soluble in Water
Other Solubilities: 95% Ethanol: >=100 mg/ml at 21 °C; Acetone: >=100 mg/ml at 21 °C; DMSO: >=100 mg/ml at 21 °C.
Refraction Index: 1.4099 at 30 °C/D
Flash Point: 43 °C

RTECS Toxicity Data

Acute Oral: Rat LD_{50} Dose: 8210 mg/kg; Toxic Effects: Behavioral - Muscle weakness. Mouse LD_{50} Dose: >6810 mg/kg.
Chronic (Multiple Dose) Oral: Rat Dose: 75 gm/kg/15D-C; Toxic Effects: DEATH. Rat Dose: 130 gm/kg/13W-I; Toxic Effects: Liver - Changes in liver weight; DEATH. Rat Dose: 150 gm/kg/15D-C; Toxic Effects: Gastrointestinal - Gastritis; Gastrointestinal - Other changes; DEATH. Rat Dose: 260 gm/kg/13W-I; Toxic Effects: DEATH.
Reproductive/Teratogenic: Rat Route: Oral; Dose: 63 gm/kg; Duration: male 63D prior to mating; Paternal Effects -

Spermatogenesis; Effects on Fertility - Male fertility index. Rat Route: Oral; Dose: 126 gm/kg; Duration: male 63D prior to mating; Paternal Effects - Testes, epididymis, sperm duct; Prostate, seminal vessicle, Cowper's gland, accessory glands. Rat Route: Oral; Dose: 15750 mg/kg; Duration: male 63D prior to mating; Effects on Fertility - Post-implantation mortality. Rat Route: Oral; Dose: 61250 mg/kg; Duration: male 7W prior to mating; Paternal Effects - Spermatogenesis; Testes, epididymis, sperm duct.

Mutagenic: Rat Sperm Morphology; Route: Oral; Dose: 61250 mg/kg/7W-I. Mouse Dominant Lethal Test; Route: Oral; Dose: 65 gm/kg/13W-C.

Tumorigenic: Rat Route: Oral; Dose: 515 gm/kg/2Y-C; Toxic Effects: Tumorigenic - Carcinogenic by RTECS criteria; Kidney, Ureter, and Bladder - Kidney tumors.

Hazard Overviews

Flammable

Health: Irritating to eyes/skin/respiratory tract. Harmful. Other Acute Effects: harmful if swallowed, inhaled, or absorbed through skin; exposure can cause nausea; headache; vomiting. Chronic Effects: Possible carcinogen.

Fire: Flammable. Hazards: emits toxic fumes; vapor may travel considerable distance to source of ignition and flash back. Extinguishing agents: water spray; carbon dioxide, dry chemical powder or appropriate foam. Precautions: combustible liquid.

Reactivity: Incompatible with: strong oxidizing agents, strong bases. Hazardous decomposition products: toxic fumes of: carbon monoxide, carbon dioxide; thermal decomposition may produce toxic fumes of phosphorus oxides and/or phosphine.

Carcinogenicity: IARC - Not listed; NIOSH - Not listed; NTP - Not listed; ACGIH - Not listed; OSHA - Not listed; EPA - Not listed; MAK - Not listed

Primary Target Organs:

Eyes

Skin

Respiratory System

Environmental

Environmental Fate: Will hydrolyze to the half ester and methanol with an estimated half-life of 13.2 years at 20 °C. Half-lives in muddy water range from 7-210 days, depending on the initial concentration and temperature. Half-lives in soil range from 0.2-60 days with an average half-life of 12.4 days. No information was found on volatilization rates, photolysis, or biodegradation.

Cleanup/Disposal: Guide No. 128: Eliminate all ignition sources (no smoking, flares, sparks or flames in immediate area). All equipment used when handling the product must be grounded. Do not touch or walk through spilled material. Stop leak if you can do it without risk. Prevent entry into waterways, sewers, basements or confined areas. A vapor suppressing foam may be used to reduce vapors. Absorb or cover with dry earth, sand or other non-combustible material and transfer to containers. Use clean non-sparking tools to collect absorbed material. Large Spills: Dike far ahead of liquid spill for later disposal. Water spray may reduce vapor; but may not prevent ignition in closed spaces.

Environmental Physical Data

BCF: none likely

Regulations

RCRA 40CFR: Not listed
CERCLA: 40CFR 302.4: Not listed
SARA 40CFR 372.65: Not listed
SARA EHS 40CFR 355: Not listed
TSCA: Listed

DIM3050 CAS #: 141-91-3

2,6-DIMETHYL MORPHOLINE

RTECS: QE1750000
EINECS Number: 205-509-3
Molecular Formula: $C_6H_{13}NO$
Formula Weight: 115.20

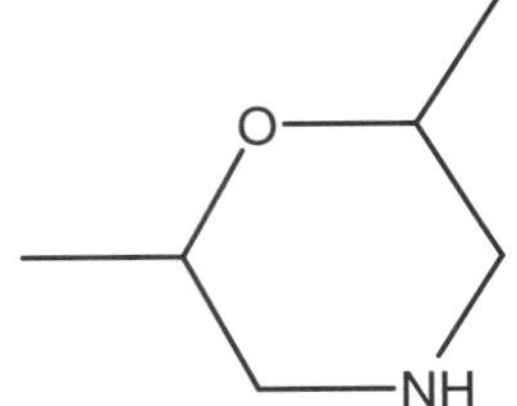

Chemical Structure

Synonyms: 2,6-DIMETHYLMORFOLIN; 2,6-DIMETHYLMORPHOLINE; 2,6-DIMETHYL-2,3,5,6-TETRAHYDRO-4H-1,4-OXAZINE; MORPHOLINE,2,6-DIMETHYL-

Description: liquid

Use: corrosion inhibitors, stabilizers for chlorinated solvents, rubless flow polishes, rubber accelerators, germicides and textile finishing agents

Physical Properties

Boiling Point: 146.6 °C (296 °F) at 760 mm Hg
Freezing Point: -85 °C (-121 °F)
Specific Gravity: 0.9346 at 20 °C/20 °C
Density: 0.93 g/mL at 20 °C
Water Solubility: Soluble in all Proportions
Other Solubilities: 95% Ethanol: Very Soluble; Acetone: Soluble; Benzene: Very Soluble; DMSO: Soluble.
Refraction Index: 1.4460 at 20 °C/D
Flash Point: 44 °C Open Cup

RTECS Toxicity Data

Acute Oral: Rat LD_{50} Dose: 2830 mg/kg.

Acute Dermal: Rabbit LD_{50} Route: Skin; Dose: 710 uL/kg; Toxic Effects: Skin and appendages - Primary irritation.
Irritation Eye: Rabbit Standard Draize Test Dose: 2 mg/24H; Reaction: severe.
Irritation Skin: Rabbit Standard Draize Test Dose: 500 mg/24H; Reaction: mild. Rabbit Open Draize Test Dose: 10 mg/24H open; Reaction: mild.
Mutagenic: Bacteria - S Typhimurium Mutations in Microorganisms; Dose: 6666 ug/plate (+S9), 3333 ug/plate (-S9).

Hazard Overviews

Flammable

Health: Severely irritating to eyes/skin/respiratory tract. Toxic. Other Acute Effects: harmful if swallowed, inhaled, or absorbed through skin; readily absorbed through skin.
Fire: Flammable. Hazards: emits toxic fumes; vapor may travel considerable distance to source of ignition and flash back. Extinguishing agents: water spray; carbon dioxide, dry chemical powder or appropriate foam. Precautions: combustible liquid.
Reactivity: Incompatible with: strong oxidizing agents, strong acids. Hazardous decomposition products: toxic fumes of: carbon monoxide, carbon dioxide, nitrogen oxides.
Carcinogenicity: IARC - Not listed; NIOSH - Not listed; NTP - Not listed; ACGIH - Not listed; OSHA - Not listed; EPA - Not listed; MAK - Not listed
Primary Target Organs:

Eyes

Skin

Respiratory System

Environmental

Cleanup/Disposal: Guide No. 131: Fully encapsulating, vapor protective clothing should be worn for spills and leaks with no fire. Eliminate all ignition sources (no smoking, flares, sparks or flames in immediate area). All equipment used when handling the product must be grounded. Do not touch or walk through spilled material. Stop leak if you can do it without risk. Prevent entry into waterways, sewers, basements or confined areas. A vapor suppressing foam may be used to reduce vapors. Small Spills: Absorb with earth, sand or other non-combustible material and transfer to containers for later disposal. Use clean non-sparking tools to collect absorbed material. Large Spills: Dike far ahead of liquid spill for later disposal. Water spray may reduce vapor; but may not prevent ignition in closed spaces.

Regulations

RCRA 40CFR: Not listed
CERCLA: 40CFR 302.4: Not listed
SARA 40CFR 372.65: Not listed
SARA EHS 40CFR 355: Not listed
TSCA: Listed

DIM3100 **CAS #: 597-25-1**

DIMETHYL MORPHOLINOPHOSPHORAMIDATE

RTECS: SZ9660000
Molecular Formula: $C_6H_{14}NO_4P$
Formula Weight: 195.18
Synonyms: DIMETHYL MORPHOLINOPHOSPHONATE; DIMETHYL 4-MORPHOLINYLPHOSPHONATE; DIMETHYLMORPHOLINOPHOSPHONATE; DMMPA; HC 1717; MORPHOLINOPHOSPHONIC ACID DIMETHYL ESTER; 4-MORPHOLINYLPHOSPHONIC ACID DIMETHYL ESTER; PHOSPHONIC ACID,MORPHOLINO-,DIMETHYL ESTER; PHOSPHONIC ACID,4-MORPHOLINYL-,DIMETHYL ESTER; PHOSPHONIC ACID,4-MORPHOLINYL-,DIMETHYL ESTER (9CI)
Description: colorless, clear, slightly viscous liquid
Use: as a PRC stimulant for chemical agents in studying effectiveness in chemical defense procedures; a SRI proposed anticholinesterase stimulant (in military training)

Physical Properties

Boiling Point: Decomposes at 205 °C (401 °F)
Specific Gravity: 1.2237 at 25/22 °C
Saturated Vapor Density: 1.212667078 kg/m^3
Vapor Pressure: 1.4 mm Hg at 23 °C
Water Solubility: >= 100 mg/mL at 18 C
Other Solubilities: 95% Ethanol: >=100 mg/ml at 18 °C; Acetone: >=100 mg/ml at 18 °C; DMSO: >=100 mg/ml at 18 °C.
Refraction Index: 1.4537
Flash Point: > 93 °C

RTECS Toxicity Data

Acute Oral: Rat LD_{50} Dose: 5910 mg/kg. Mouse LD_{50} Dose: 3300 mg/kg.
Chronic (Multiple Dose) Oral: Rat Dose: 35 gm/kg/14D-I; Toxic Effects: DEATH. Rat Dose: 52 gm/kg/13W-I; Toxic Effects: Liver - Changes in liver weight; DEATH. Rat Dose: 28 gm/kg/14D-I; Toxic Effects: DEATH. Rat Dose: 26 gm/kg/13W-I; Toxic Effects: DEATH.
Mutagenic: Mouse Mutations in Mammalian Somatic Cells; Cell Type: lymphocyte; Dose: 2200 mg/L. Hamster Cytogenetic Analysis; Cell Type: ovary; Dose: 3 gm/L.
Tumorigenic: Rat Route: Oral; Dose: 309 gm/kg/2Y-I; Toxic Effects: Tumorigenic - Carcinogenic by RTECS criteria; Blood - Leukemia.

Hazard Overviews

Fire: Will burn.
Carcinogenicity: IARC - Not listed; NIOSH - Not listed; NTP - Not listed; ACGIH - Not listed; OSHA - Not listed; EPA - Not listed; MAK - Not listed

Environmental

Regulations

RCRA 40CFR: Not listed
CERCLA: 40CFR 302.4: Not listed
SARA 40CFR 372.65: Not listed
SARA EHS 40CFR 355: Not listed
TSCA: Not listed

DIM3150 CAS #: 112-75-4

DIMETHYL MYRISTAMINE

EINECS Number: 204-002-4
Molecular Formula: $C_{16}H_{35}N$
Formula Weight: 241.2

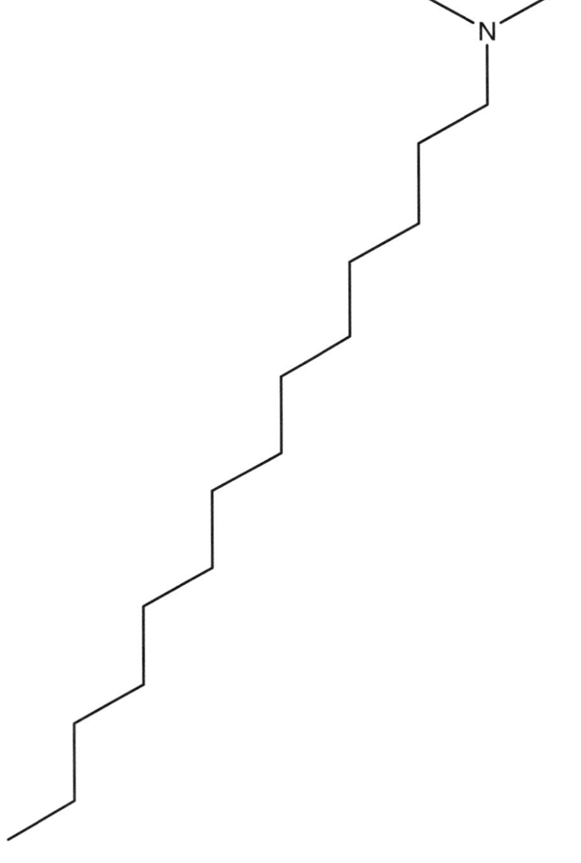

Chemical Structure

Synonyms: ARMEEN DM 14D; DIMETHYLMYRISTAMINE; DIMETHYLMYRISTYLAMINE; N,N-DIMETHYLMYRISTYLAMINE; DIMETHYLTETRADECYLAMINE; N,N-DIMETHYL-N-TETRADECYLAMINE; N,N-DIMETHYLTETRADECYLAMINE; MYRISTYLDIMETHYLAMINE; 1-TETRADECANAMINE,N,N-DIMETHYL-; TETRADECYLAMINE,N,N-DIMETHYL-; TETRADECYLDIMETHYLAMINE
Description: liquid
Use: cationic detergent; corrosion inhibitor

Hazard Overviews

Carcinogenicity: IARC - Not listed; NIOSH - Not listed; NTP - Not listed; ACGIH - Not listed; OSHA - Not listed; EPA - Not listed; MAK - Not listed

Environmental

Regulations

RCRA 40CFR: Not listed
CERCLA: 40CFR 302.4: Not listed
SARA 40CFR 372.65: Not listed
SARA EHS 40CFR 355: Not listed
TSCA: Listed

DIM3250 CAS #: 2524-03-0

DIMETHYL PHOSPHOROCHLORIDOTHIOATE

RTECS: TD1830000
DOT: UN2267; NA2267; IMO8.0
EINECS Number: 219-754-9
Molecular Formula: $C_2H_6ClO_2PS$
Formula Weight: 160.56

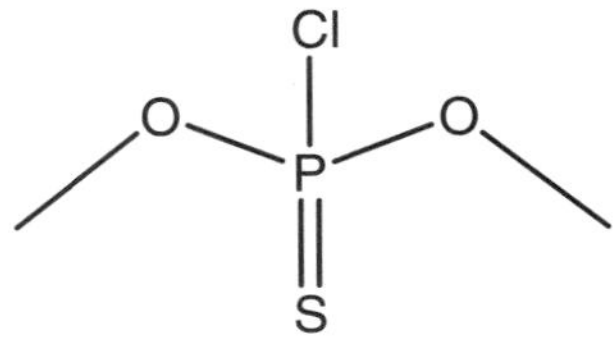

Chemical Structure

Synonyms: CHLORODIMETHOXYPHOSPHINE SULFIDE; DIMETHOXY THIOPHOSPHONYL CHLORIDE; DIMETHYL CHLOROTHIONOPHOSPHATE; O,O-DIMETHYL CHLOROTHIONOPHOSPHATE; DIMETHYL CHLOROTHIOPHOSPHATE; O,O-DIMETHYL CHLOROTHIOPHOSPHATE; O,O-DIMETHYL PHOSPHOROCHLOROTHIOATE; O,O-DIMETHYL PHOSPHOROTHIONOCHLORIDATE; DIMETHYL THIONOPHOSPHOROCHLORIDATE; O,O-DIMETHYL THIONOPHOSPHOROCHLORIDATE; O,O-DIMETHYL THIOPHOSPHORIC ACID CHLORIDE; DIMETHYL THIOPHOSPHOROCHLORIDATE; DIMETHYL THIOPHOSPHORYL CHLORIDE; DIMETHYLCHLORTHIOFOSFAT; O,O-DIMETHYLESTER KYSELINY CHLORTHIOFOSFORECNE; O,O-DIMETHYLPHOSPHOROCHLORIDOTHIOATE; O,O-DIMETHYLTHIONOPHOSPHORYL CHLORIDE; DMPCT; METHYL PCT; PHOSPHONOTHIOIC ACID,CHLORO-,O,O-DIMETHYL ESTER; PHOSPHOROCHLORIDOTHIOIC ACID,O,O-DIMETHYL ESTER
Description: colorless to light amber liquid
Use: intermediate for insecticides, pesticides, fungicides; oil & gasoline additives; plasticizers; corrosion inhibitors; flame retardants; flotation agents; chem int for methyl parathion, abate, famphur, fenthion, monitor, ronnel & sumithion; intermediate in the production of pesticides such as methyl parathion, temephos, and fenitrothion

Physical Properties

Boiling Point: 66 °C (151 °F) to 67 °C (153 °F) at 16 mm Hg
Specific Gravity: 1.320 at 25 °C
Vapor Pressure: 4 mm Hg at 40 °C
Water Solubility: Insoluble in Water
Other Solubilities: Soluble in Alcohol, Benzene, Acetone, Carbon Tetrachloride, Chloroform, Ethyl Acetate; Slightly Soluble in Hexane.
Refraction Index: 1.4795 at 25 °C/D

RTECS Toxicity Data

Acute Oral: Rat LD_{Lo} Dose: 1 gm/kg. Mouse LD_{50} Dose: 1800 mg/kg.
Acute Inhalation: Rat LC_{50} Dose: 340 mg/m^3/4hr. Mouse LC_{50} Dose: 320 mg/m^3/2hr.
Acute Dermal: Rabbit LD_{Lo} Route: Skin; Dose: 750 mg/kg.

Hazard Overviews

Poison

Corrosive

Health: Corrosive to eyes/skin/respiratory tract. Poison. Other Acute Effects: may be fatal if inhaled; highly toxic cholinesterase inhibitor; harmful if swallowed or absorbed through skin; inhalation may result in spasm, inflammation and edema of the larynx and bronchi, chemical pneumonitis and pulmonary edema; symptoms of exposure may include burning sensation; coughing; wheezing; laryngitis; shortness of breath; headache; nausea; vomiting; target organ: lungs.
Fire: Hazards: water hydrolyzes material liberating acidic gas which in contact with metal surfaces can generate flammable and/or explosive hydrogen gas; emits toxic fumes; container explosion may occur. Extinguishing agents: carbon dioxide, dry chemical powder or appropriate foam; do not use water. Precautions: combustible liquid.
Reactivity: Incompatible with: strong bases, strong oxidizing agents, may decompose on exposure to moist air or water. Hazardous decomposition products: toxic fumes of: carbon monoxide, carbon dioxide, hydrogen chloride gas; thermal decomposition may produce toxic fumes of phosphorus oxides and/or phosphine; sulfur oxides.
Carcinogenicity: IARC - Not listed; NIOSH - Not listed; NTP - Not listed; ACGIH - Not listed; OSHA - Not listed; EPA - Not listed; MAK - Not listed
Primary Target Organs:

Eyes

Skin

Respiratory System

Environmental

Environmental Fate: If released to the atmosphere, it will degrade rapidly in the vapor-phase by reaction with photochemically produced hydroxyl radicals (estimated half-life of 6.5 hours). If released to moist soil or water, it is expected to hydrolyze.
Cleanup/Disposal: Guide No. 156: Eliminate all ignition sources (no smoking, flares, sparks or flames in immediate area). All equipment used when handling the product must be grounded. Do not touch damaged containers or spilled material unless wearing appropriate protective clothing. Stop leak if you can do it without risk. A vapor suppressing foam may be used to reduce vapors. For chlorosilanes, use AFFF alcohol-resistant medium expansion foam to reduce vapors. Do not get water on spilled substance or inside containers. Use water spray to reduce vapors or divert vapor cloud drift. Prevent entry into waterways, sewers, basements or confined areas. Small Spills: Cover with dry earth, dry sand, or other non-combustible material followed with plastic sheet to minimize spreading or contact with rain. Use clean non-sparking tools to collect material and place it into loosely covered plastic containers for later disposal.

Environmental Physical Data

Henry's Law Constant: estimated at 0.00152
Sorption Partition Coefficient: K_{oc} = estimated at 10.5
BCF: not significant

Regulations

RCRA 40CFR: Not listed
CERCLA: 40CFR 302.4: Not listed
SARA 40CFR 372.65: Listed TPQ: 500 lb
SARA EHS 40CFR 355: Listed TPQ: 500 lb
TSCA: Listed

DIM3300 **CAS #: 131-11-3**

DIMETHYL PHTHALATE

RTECS: TI1575000
EINECS Number: 205-011-6
Molecular Formula: $C_{10}H_{10}O_4$
Structured MF: C_6H_4-1,2-$(CO_2CH_3)_2$
Formula Weight: 194.19

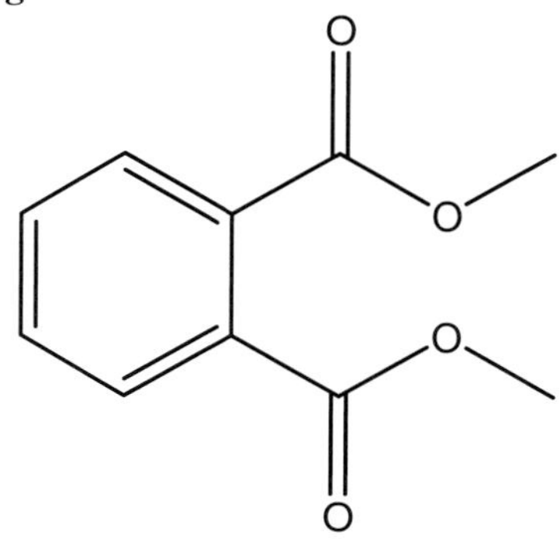

Chemical Structure

Synonyms: AVOLIN; 1,2-BENZENEDICARBOXYLIC ACID,DIMETHYL ESTER; DIMETHYL 1,2-BENZENDICARBOXYLATE; DIMETHYL 1,2-BENZENEDICARBOXYLATE; DIMETHYL BENZENE-O-DICARBOXYLATE; DIMETHYL

BENZENEORTHODICARBOXYLATE; DIMETHYL ESTER OF 1,2-BENZENEDICARBOXYLIC ACID; DIMETHYL O-PHTHALATE; O-DIMETHYL PHTHALATE; DIMETHYL-1,2-BENZENEDICARBOXYLATE; DIMETHYLESTER KYSELINY FTALOVE; DMF (INSECT REPELLANT); DMF (INSECT REPELLENT); DMP; ENT 262; EPA PESTICIDE CHEMICAL CODE 028002; FERMINE; METHYL PHTHALATE; MIPAX; NTM; PALATINOL M; PHTHALATE ACID,DIMETHYL ESTER; PHTHALIC ACID METHYL ESTER; PHTHALIC ACID,DIMETHYL ESTER; PHTHALSAEUREDIMETHYLESTER; REPEFTAL; SOLVANOM; SOLVARONE; UNIMOLL DM

Description: clear, colorless oily liquid; slight ester odor

Use: solvent and plasticizer for cellulose acetate and cellulose acetatebutyrate compositions; film; insect repellent

Physical Properties

Boiling Point: 283.7 °C (543 °F) at 760 mm Hg

Freezing Point: 5.5 °C (41.9 °F)

Specific Gravity: 1.196 at 15.6 °C/15.6 °C

Vapor Density: 6.69 Air=1

Bulk Density: 9.93 lbs/gal at 68 °F

Vapor Pressure: 1 mm Hg at 100.3 °C

Water Solubility: 0.4% by weight

Other Solubilities: 95% Ethanol: >=100 mg/ml at 20 °C; Acetone: >=100 mg/ml at 20 °C; Benzene: Soluble; Chloroform: Soluble; DMSO: >=100 mg/ml at 20 °C; Ether: Soluble; Mineral oil: Soluble; Paraffin hydrocarbons: Insoluble; Petroleum Ether: Insoluble.

Refraction Index: 1.5168 at 20 °C/D

Evaporation Rate: < 1 Butyl Acetate=1

Ionization Potential (eV): 9.64

Flash Point: 146.111 °C

Autoignition Temperature: 491 °C

LEL: 0.9% v/v

RTECS Toxicity Data

Acute Oral: Rat LD_{50} Dose: 6800 mg/kg; Toxic Effects: Behavioral - Somnolence (general depressed activity); Behavioral - Withdrawal; Nutritional and gross metabolic - Weight loss or decreased weight gain. Mouse LD_{50} Dose: 6800 mg/kg; Toxic Effects: Behavioral - Somnolence (general depressed activity); Behavioral - Withdrawal; Nutritional and gross metabolic - Weight loss or decreased weight gain.

Acute Inhalation: Cat LC_{Lo} Dose: 9300 mg/m^3/6.5hr.

Acute Dermal: Rabbit LD_{50} Route: Skin; Dose: >20 mL/kg. Rat LD_{50} Route: Skin; Dose: >4800 mg/kg.

Chronic (Multiple Dose) Inhalation: Rat Dose: 1840 ug/kg/4H/17W-I; Toxic Effects: Liver - Other changes; Kidney, Ureter, and Bladder - Other changes; Blood - Other changes.

Chronic (Multiple Dose) Dermal: Rabbit Route: Skin; Dose: 360 mL/kg/13W-I; Toxic Effects: DEATH.

Reproductive/Teratogenic: Rat Route: Intraperitoneal; Dose: 338 mg/kg; Duration: female 5-15D of pregnancy; Effects on Fertility - Post-implantation mortality; Effects on Embryo or Fetus - Fetotoxicity; Specific Developmental Abnormalities - Musculoskeletal system. Rat Route: Intraperitoneal; Dose: 338 mg/kg; Duration: female 5-15D of pregnancy; Specific Developmental Abnormalities - Eye, ear; Other developmental abnormalities. Rat Route: Intraperitoneal; Dose: 1125 mg/kg; Duration: female 5-15D of pregnancy; Effects on Embryo or Fetus - Fetal death. Rat Route: Intraperitoneal; Dose: 1500 mg/kg; Duration: female 3-9D of pregnancy; Effects on Fertility - Litter size.

Mutagenic: Rat Cytogenetic Analysis; Route: Skin; Dose: 25 gm/kg/4W-I. Bacteria - S Typhimurium Mutations in Microorganisms; Dose: 200 ug/plate (+S9).

Hazard Overviews

Fire Diamond

Health: Irritating to eyes/respiratory tract. Also Causes: gastrointestinal irritation, vomiting, diarrhea, and CNS depression.

Fire: Will burn. Use agent suitable for surrounding fire. Water or foam may cause frothing.

Reactivity: Stable. Hazardous polymerization cannot occur. Avoid: heat; ignition sources. Incompatible with: nitrates; strong oxidizers; strong acids; strong alkalis. Hazardous decomposition products: carbon oxide(s).

Carcinogenicity: IARC - Not listed; NIOSH - Not listed; NTP - Not listed; ACGIH - Not listed; OSHA - Not listed; EPA - Class D, Not classifiable as to human carcinogenicity; MAK - Not listed

Primary Target Organs:

Eyes

Respiratory System

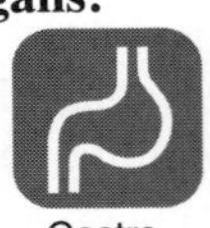
Gastro-intestinal

Nervous System

Exposure Limits

OSHA PEL: TWA: 5 mg/m^3.

ACGIH TLV: TWA: 5 mg/m^3.

NIOSH REL: TWA: 5 mg/m^3.

NIOSH IDLH: 2000 mg/m^3.

Respirator Recommendation

Exposure Range: >5 to 50 mg/m^3 Air Purifying, Negative Pressure, Half Mask

Exposure Range: >50 to 500 mg/m^3 Air Purifying, Negative Pressure, Full Face

Exposure Range: >500 to 5000 mg/m^3 Supplied Air, Constant Flow/Pressure Demand, Full Face Self-contained Breathing Apparatus, Pressure Demand, Full Face

Cartridge Color: black with dust/mist prefilter (use P100 or consult supervisor for appropriate dust/mist prefilter)

Environmental

Ecotoxicity: EC_{50} Skeletonema costatum (alga) 26,100 ug/l/96 hr effect: Chlorophyll a /Conditions of bioassay not specified

Environmental Fate: Its primary loss mechanism appears to be biodegradation. Half-lives of 8-11 days and 0.2 days has been determined in river water, but no half-lives are available for soil or groundwater although it is utilized by soil microorganisms and degrades under anaerobic conditions. Little adsorption to soil or sediment will occur. It will not bioconcentrate in fish. If emitted into the atmosphere it will most likely be as an aerosol and it will be subject to rainout and gravitational settling. Photodegradation by hydroxyl radicals will also occur (estimated half-life 23.8 hr

Cleanup/Disposal: Guide No. 171: Do not touch or walk through spilled material. Stop leak if you can do it without risk. Prevent dust cloud. Avoid inhalation of asbestos dust. Small Dry Spills: With clean shovel place material into clean, dry container and cover loosely; move containers from spill area. Small Spills: Take up with sand or other noncombustible absorbent material and place into containers for later disposal. Large Spills: Dike far ahead of liquid spill for later disposal. Cover powder spill with plastic sheet or tarp to minimize spreading. Prevent entry into waterways, sewers, basements or confined areas.

Environmental Physical Data

Henry's Law Constant: 5.3 x10^{-5}

Octanol/Water Partition Coefficient: log K_{ow} = calculated at 2.12

Sorption Partition Coefficient: K_{oc} = estimated at 44 to 160

BCF: brown shrimp 4.7

Regulations

RCRA 40CFR: Listed Hazardous Waste No. U102 Toxic Waste

CERCLA: 40CFR 302.4: Listed per RCRA Section 3001 per CWA Section 307(a) RQ: 5000 lb (2268 kg)

SARA 40CFR 372.65: Listed

SARA EHS 40CFR 355: Not listed

TSCA: Listed

Analytical Methods

Soil: CLP LC_SV, MC_SVOA, OHC; EPA 16, 1625, PMD-DNE; SW846 3640A, 8060, 8061, 8061A, 8250A, 8270B, 8270C, 8410

Water / Groundwater: EPA 1625, 606, 625, 625-S, S-002-1; APHA 6410-B; USGS O3118

Drinking Water: EPA 506, 525.1, 525.2

Food: FDA 212.1, 232.1

Plasma: EPA 001, 29; FDA 211.1, 231.1, 252

DIM3400 **CAS #: 106-65-0**

DIMETHYL SUCCINATE

RTECS: WM7675000

EINECS Number: 203-419-9

Molecular Formula: $C_6H_{10}O_4$

Structured MF: $CH_3OCOCH_2CH_2COOCH_3$

Formula Weight: 146.14

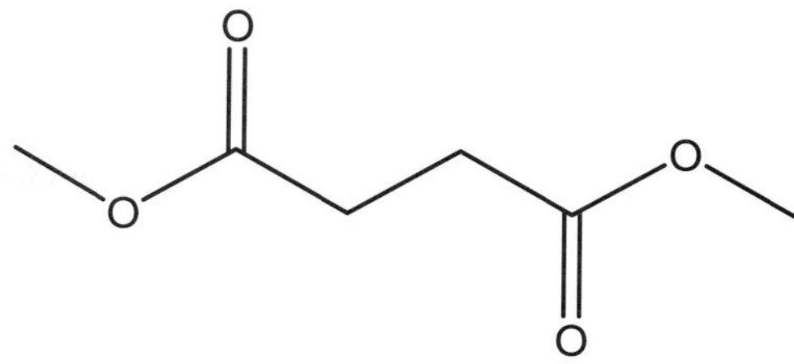

Chemical Structure

Synonyms: BUTANEDIOIC ACID,DIMETHYL ESTER; DIMETHYL BUTANEDIOATE; METHYL SUCCINATE; SUCCINIC ACID,DIMETHYL ESTER

Description: colorless liquid, solidifies when cold; pleasant, ethereal, winey odor

Use: manufacture of lacquers, dyes, esters for perfumes and succinates and in photography

Physical Properties

Boiling Point: 195.3 °C (384 °F) at 760 mm Hg

Freezing Point: 19.5 °C (67.1 °F)

Specific Gravity: 1.1202 at 18 °C/4 °C

Vapor Density: 5.04 Air=1

Saturated Vapor Density: 1.201913358 kg/m^3

Density: 1.121 at 20 °C

Vapor Pressure: 0.3 mm Hg at 20 °C

Water Solubility: 1 parts in 120 parts Water

Other Solubilities: 95% Ethanol: >=100 mg/ml at 23 °C; Acetone: >=100 mg/ml at 23 °C; Alcohol: Soluble; DMSO: >=100 mg/ml at 23 °C; Ether: Soluble.

Refraction Index: 1.4197 at 20 °C/D

Evaporation Rate: < 0.1 Butyl Acetate=1

Flash Point: 85 °C Closed Cup

Autoignition Temperature: 365 °C

LEL: 1.0% v/v

UEL: 8.5% v/v

RTECS Toxicity Data

Acute Oral: Rat LD_{50} Dose: >5 gm/kg.

Acute Dermal: Rabbit LD_{50} Route: Skin; Dose: >5 gm/kg.

Hazard Overviews

Fire: Combustible.

Carcinogenicity: IARC - Not listed; NIOSH - Not listed; NTP - Not listed; ACGIH - Not listed; OSHA - Not listed; EPA - Not listed; MAK - Not listed

Environmental

Cleanup/Disposal: Guide No. 127: Eliminate all ignition sources (no smoking, flares, sparks or flames in immediate area). All equipment used when handling the product must be grounded. Do not touch or walk through spilled material. Stop leak if you can do it without risk. Prevent entry into waterways, sewers, basements or confined areas. A vapor suppressing foam may be used to reduce vapors. Absorb or cover with dry earth, sand or other non-combustible material and transfer to containers. Use clean non-sparking tools to collect absorbed material. Large Spills: Dike far ahead of

liquid spill for later disposal. Water spray may reduce vapor; but may not prevent ignition in closed spaces.

Regulations

RCRA 40CFR: Not listed
CERCLA: 40CFR 302.4: Not listed
SARA 40CFR 372.65: Not listed
SARA EHS 40CFR 355: Not listed
TSCA: Listed

DIM3450 **CAS #: 77-78-1**

DIMETHYL SULFATE

RTECS: WS8225000
DOT: UN1595; IMO6.1
EINECS Number: 201-058-1
Molecular Formula: $C_2H_6O_4S$
Structured MF: $(CH_3)_2SO_4$
Formula Weight: 126.13

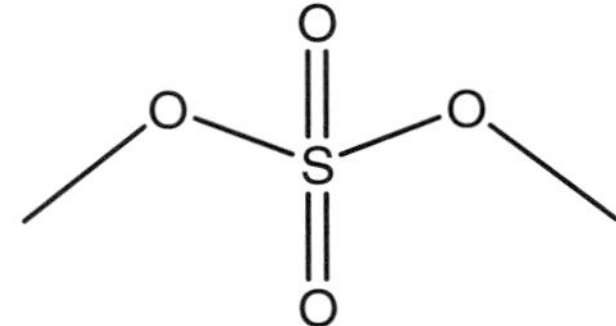

Chemical Structure

Synonyms: DIMETHYL ESTER OF SULFURIC ACID; DIMETHYL MONOSULFATE; DIMETHYL SULPHATE; DIMETHYLESTER KYSELINY SIROVE; DIMETHYLSULFAAT; DIMETHYLSULFAT; DIMETHYLSULFATE; DIMETILSOLFATO; DMS; DMS (METHYL SULFATE); DMS(METHYL SULFATE); DWUMETYLOWY SIARCZAN; METHYL SULFATE; METHYLE (SULFATE DE); SULFATE DE DIMETHYLE; SULFATE DE METHYLE; SULFATE DIMETHYLIQUE; SULFATO DE DIMETILO; SULFURIC ACID,DIMETHYL ESTER
Description: colorless liquid; faint onion-like odor
Use: formerly war gas; solvent in mfr of dyes, colors, perfumes, pharmaceuticals; solvent for separation of mineral oils; analysis of auto fluids; methylating agent for amines & phenols, polyurethane-based adhesives; methylating & sulfating agent for agrichems, fabric softeners, dyes, synthetic drugs, & intermediates; alkylating agent to convert compounds eg, thiols to corresponding methyl derivatives; sulfating & sulfonating agent.

Physical Properties

Boiling Point: Decomposes about 188 °C (370 °F)
Freezing Point: -31.7 °C (-25.06 °F)
Specific Gravity: 1.3283 at 20 °C
Vapor Density: 4.35 Air=1
Saturated Vapor Density: 1.202644192 kg/m^3
Vapor Pressure: 0.5 mm Hg at 20 °C
Water Solubility: 2.8 g/100ml at 18 °C
Other Solubilities: Soluble in Ether, dioxane, Acetone, aromatic hydrocarbons; Sparingly Soluble in Carbon Disulfide, aliphatic hydrocarbons; Soluble in Alcohol.
Surface Tension: Estimated at 20 dynes/cm
pH: Expected that dilute solutions of dimethyl sulfate would be acid
Refraction Index: 1.3874 at 20 °C/D
Flash Point: 115.6 °C Open Cup
Autoignition Temperature: 188 °C
LEL: 3.6% v/v
UEL: 23% v/v

RTECS Toxicity Data

Acute Oral: Rat LD_{50} Dose: 205 mg/kg. Mouse LD_{50} Dose: 140 mg/kg.
Acute Inhalation: Rat LC_{50} Dose: 45 mg/m^3/4hr; Toxic Effects: Lungs, Thorax, or Respiration - Dyspnea; Lungs, Thorax, or Respiration - Cyanosis; Blood - Hemorrhage. Human LC_{Lo} Dose: 97 ppm/10M.
Acute Dermal: Rabbit LD_{Lo} Route: Subcutaneous Dose: 53 mg/kg; Toxic Effects: Behavioral - Muscle contraction or spasticity; Skin and appendages - Dermatitis, other. Rat LD_{50} Route: Subcutaneous Dose: 100 mg/kg.
Irritation Eye: Rabbit Standard Draize Test Dose: 50 ug/24H; Reaction: severe. Rabbit Nonstandard Exposure Dose: 100 mg/4S rinse; Reaction: severe.
Irritation Skin: Rabbit Open Draize Test Dose: 10 mg/24H open; Reaction: severe.
Reproductive/Teratogenic: Rat Route: Intravenous; Dose: 100 mg/kg; Duration: female 15D of pregnancy; Effects on Embryo or Fetus - Fetal death.
Mutagenic: Human DNA Damage; Cell Type: lymphocyte; Dose: 1 mmol/L. Human DNA Damage; Cell Type: fibroblast; Dose: 100 umol/L. Human Unscheduled DNA Synthesis; Cell Type: lymphocyte; Dose: 10 umol/L. Human Unscheduled DNA Synthesis; Cell Type: fibroblast; Dose: 50 umol/L. Human Unscheduled DNA Synthesis; Cell Type: HeLa cell; Dose: 400 umol/L. Human DNA Inhibition; Cell Type: HeLa cell; Dose: 2 mmol/L. Human DNA Inhibition; Cell Type: fibroblast; Dose: 75 umol/L. Human DNA Adduct; Cell Type: HeLa cell; Dose: 5 umol/L. Human Cytogenetic Analysis; Route: Unreported; Dose: 2 mg/m^3. Human Sister Chromatid Exchenge; Cell Type: fibroblast; Dose: 1 umol/L.
Tumorigenic: Rat Route: Inhalation; Dose: 17 mg/m^3/19W-I; Toxic Effects: Tumorigenic - Equivocal tumorigenic agent by RTECS criteria; Sense organs and special senses - Tumors; Blood - Lymphomax including Hodgkin's disease. Rat Route: Subcutaneous; Dose: 50 mg/kg; Toxic Effects: Tumorigenic - Equivocal tumorigenic agent by RTECS criteria; Tumorigenic - Tumors at site of application.

Hazard Overviews

Corrosive

Fire Diamond

Health: Irritating to eyes/skin/respiratory tract. Also causes: headache; giddiness; photophobia; painful coughing,severe pulmonary damage; prostration; convulsions; delerium;

paralysis; coma, and delayed damage to the kidneys/liver/heart. Chronic Effects: Cancer.

Fire: Combustible. Use foam, dry chemical, or carbon dioxide. Use water spray to cool fire-exposed containers. Dimethyl sulfate is not readily combustible and is not likely to cause a fire at room temperature.

Reactivity: Stable. Hazardous polymerization cannot occur. Avoid: sources of ignition such as high temperatures and open flame. Incompatible with: dimethyl sulfate; concentrated ammonia; metal salts; ammonium hydroxide; sodium azide. Hazardous decomposition products: carbon monoxide; sulfur oxides.

Carcinogenicity: IARC - Group 2A, Probably carcinogenic to humans; NIOSH - Listed as carcinogen; NTP - Class 2B, Reasonably anticipated to be a carcinogen, sufficient evidence of carcinogenicity from studies in experimental animals; ACGIH - Class A3, Animal carcinogen; OSHA - Not listed; EPA - Class B2, Probable human carcinogen based on animal studies; MAK - Class A2, Unmistakably carcinogenic in animal experimentation only

Primary Target Organs:

Eyes

Skin

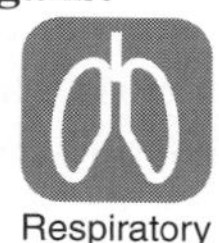
Respiratory System

Nervous System

Liver

Kidneys

Exposure Limits

OSHA PEL: TWA: 1 ppm; 5 mg/m^3.

OSHA PEL Vacated 1989 Limits: TWA: 0.1 ppm; 0.5 mg/m^3.

ACGIH TLV: TWA: 0.1 ppm; 0.52 mg/m^3.

NIOSH REL: TWA: 0.1 ppm; 0.5 mg/m^3.

NIOSH IDLH: 7 ppm.

Respirator Recommendation

Exposure Range: >1 to <7 ppm Supplied Air, Constant Flow/Pressure Demand, Half Mask

Exposure Range: 7 to unlimited ppm Self-contained Breathing Apparatus, Pressure Demand, Full Face

Note: poor warning properties

Environmental

Ecotoxicity: LC_{50} Lepomis machrochirus (bluegill sunfish) 7.5 g/cu m (7.5 ppm)/96 hr /Conditions of bioassay not specified LC_{50} Menidia beryllina (tidewater silverside) 15 g/cu m (15 ppm)/96 hr /Conditions of bioassay not specified

Environmental Fate: If released in water, it will hydrolyze (half-life 1.2 hr). If released on land, it will probably hydrolyze in damp soil or groundwater. In the atmosphere, it will be subject to gravitational settling and wash out by rain as well as reaction with photochemically produced hydroxyl radicals (estimated half-life 6.5 days). It would not be expected to bioconcentrate in fish.

Cleanup/Disposal: Guide No. 156: Eliminate all ignition sources (no smoking, flares, sparks or flames in immediate area). All equipment used when handling the product must be grounded. Do not touch damaged containers or spilled material unless wearing appropriate protective clothing. Stop leak if you can do it without risk. A vapor suppressing foam may be used to reduce vapors. For chlorosilanes, use AFFF alcohol-resistant medium expansion foam to reduce vapors. Do not get water on spilled substance or inside containers. Use water spray to reduce vapors or divert vapor cloud drift. Prevent entry into waterways, sewers, basements or confined areas. Small Spills: Cover with dry earth, dry sand, or other non-combustible material followed with plastic sheet to minimize spreading or contact with rain. Use clean non-sparking tools to collect material and place it into loosely covered plastic containers for later disposal.

Environmental Physical Data

Henry's Law Constant: calculated at 4.03 x10^{-6}

Octanol/Water Partition Coefficient: log K_{ow} = estimated at 0.032

Sorption Partition Coefficient: K_{oc} = estimated at 15.6

BCF: none expected

Regulations

RCRA 40CFR: Listed Hazardous Waste No. U103 Toxic Waste

CERCLA: 40CFR 302.4: Listed per RCRA Section 3001 RQ: 100 lb (45.35 kg)

SARA 40CFR 372.65: Listed TPQ: 500 lb

SARA EHS 40CFR 355: Listed TPQ: 100 lb

TSCA: Listed

Analytical Methods

Air: ASTM D4490; Canada 1RM-6

Indoor / Expired Air: NIOSH 2524

DIM3500 CAS #: 75-18-3

DIMETHYL SULFIDE

RTECS: PV5075000

DOT: UN1164; IMO3.1

EINECS Number: 200-846-2

Molecular Formula: C_2H_6S

Structured MF: $(CH_3)_2S$

Formula Weight: 62.13

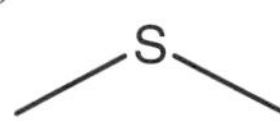

Chemical Structure

Synonyms: DIMETHYL MONOSULFIDE; DIMETHYL SULPHIDE; DIMETHYL THIOETHER; DIMETHYLSULFID; DMS; EXACT-S; METHANE,THIOBIS-; METHANETHIOMETHANE; METHYL MONOSULFIDE; METHYL SULPHIDE; METHYL THIOETHER; METHYLTHIOMETHANE; SULFURE DE METHYLE; 2-THIAPROPANE; THIOBIS(METHANE); THIOBISMETHANE; 2-THIOPROPANE

Description: colorless liquid; unpleasant odor of wild radish or cabbage-like odor

Use: gas odorant; catalyst impregnator; solvent for anhydrous mineral salts; flavoring ingredient in foods & beverages; chem int for the solvent dimethyl sulfoxide

Physical Properties

Boiling Point: 37.3 °C (99 °F)
Freezing Point: -98.27 °C (-144.886 °F)
Specific Gravity: 0.8483 at 20 °C/4 °C
Vapor Density: 2.14 Air=1
Vapor Pressure: 15 mm Hg
Water Solubility: Insoluble in Water
Other Solubilities: Soluble in alcohol and ether
Surface Tension: Estimated at 30 dynes/cm
Odor Threshold: 0.001 ppm
Refraction Index: 1.4355 at 20 °C/D
Critical Temperature: 229 °C
Critical Pressure: 826 psia
Ionization Potential (eV): 8.69 +/-0.01
Flash Point: -48.333 °C
Autoignition Temperature: 206 °C
LEL: 2.2% v/v
UEL: 19.7% v/v

RTECS Toxicity Data

Acute Oral: Rat LD_{50} Dose: 3300 mg/kg; Toxic Effects: Behavioral - General anesthetic; Behavioral - Change in motor activity (specific assay); Behavioral - Irritability. Mouse LD_{50} Dose: 3700 mg/kg; Toxic Effects: Behavioral - Change in motor activity (specific assay); Behavioral - Antipsychotic; Nutritional and gross metabolic - Other changes.
Acute Inhalation: Rat LC_{50} Dose: 40250 ppm; Toxic Effects: Lungs, Thorax, or Respiration - Other changes; Gastrointestinal - Hypermotility, diarrhea; Kidney, ureter, and Bladder - Urine volume increased. Mouse LC_{50} Dose: 31620 ug/m^3.
Acute Dermal: Rabbit LD_{50} Route: Skin; Dose: >5 gm/kg.
Chronic (Multiple Dose) Oral: Rat Dose: 21 gm/kg/70D-I; Toxic Effects: Biochemical - Catalases. Rat Dose: 24500 mg/kg/14W-I; Toxic Effects: Gastrointestinal - Other changes; Endocrine - Changes in thyroid weight.
Irritation Eye: Rabbit Standard Draize Test Dose: 250 ug/24H; Reaction: severe.
Irritation Skin: Rabbit Standard Draize Test Dose: 500 mg/24H; Reaction: mild.

Hazard Overviews

Corrosive

Flammable

Health: Severely irritating to eyes; irritating to skin/respiratory tract. Harmful. Other Acute Effects: harmful if swallowed; may be harmful if inhaled; may be harmful if absorbed through the skin; nausea; headache; vomiting.
Fire: Flammable. Hazards: vapor may travel considerable distance to source of ignition and flash back; container explosion may occur; forms explosive mixtures in air; emits toxic fumes. Extinguishing agents: carbon dioxide, dry chemical powder or appropriate foam; water may be effective for cooling, but may not effect extinguishment. Precautions: combustible liquid.
Reactivity: Incompatible with: strong oxidizing agents. Hazardous decomposition products: toxic fumes of: carbon monoxide, carbon dioxide, sulfur oxides, hydrogen sulfide gas.
Carcinogenicity: IARC - Not listed; NIOSH - Not listed; NTP - Not listed; ACGIH - Not listed; OSHA - Not listed; EPA - Not listed; MAK - Not listed
Primary Target Organs:

Eyes

Skin

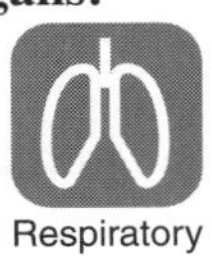
Respiratory System

Environmental

Environmental Fate: If released to soil or water, volatilization is expected to be the dominant fate process (half-life from a model river 2.7 hours). Partial removal by microbial degradation under both aerobic and anaerobic conditions is also possible. At water surfaces partial removal by sensitized photolysis may also take place. Chemical hydrolysis, reaction with photochemically generated alkylperoxy radicals, bioaccumulation in aquatic organisms, and adsorption to suspended solids and sediments are not expected to be important fate processes in water. If released to the atmosphere, nighttime reaction with gaseous nitrate radicals may be the dominant removal mechanism. In the atmosphere over the continental US the half-life for this reaction is estimated to be on the order of several hours in clean air and one hour in moderately polluted air. This reaction is not expected to be an important fate process in remote marine atmospheres due to the low concentration of nitrate radicals in these areas. During daylight hours, it is expected to react with either photochemically generated hydroxyl radicals or atomic oxygen. The overall atmospheric half-life based on these two reactions has been estimated to be 2.2 days.
Cleanup/Disposal: Guide No. 130: Eliminate all ignition sources (no smoking, flares, sparks or flames in immediate area). All equipment used when handling the product must be grounded. Do not touch or walk through spilled material. Stop leak if you can do it without risk. Prevent entry into waterways, sewers, basements or confined areas. A vapor suppressing foam may be used to reduce vapors. Absorb or cover with dry earth, sand or other non-combustible material and transfer to containers. Use clean non-sparking tools to collect absorbed material. Large Spills: Dike far ahead of liquid spill for later disposal. Water spray may reduce vapor; but may not prevent ignition in closed spaces.

Environmental Physical Data

Henry's Law Constant: 1.69 x10^{-3}
Octanol/Water Partition Coefficient: log K_{ow} = 0.84
Sorption Partition Coefficient: K_{oc} = 19 to 68
BCF: estimated at 2 to 3
BOD: ~ 1 lb/lb

Regulations

RCRA 40CFR: Not listed
CERCLA: 40CFR 302.4: Not listed
SARA 40CFR 372.65: Not listed
SARA EHS 40CFR 355: Not listed
TSCA: Listed

Analytical Methods

Air: ASTM D4490

DIM3550 CAS #: 67-68-5

DIMETHYL SULFOXIDE

RTECS: PV6210000
EINECS Number: 200-664-3
Molecular Formula: C_2H_6OS
Structured MF: $(CH_3)_2SO$
Formula Weight: 78.13

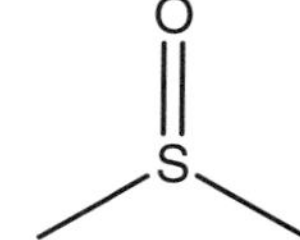

Chemical Structure

Synonyms: A 10846; DELTAN; DEMASORB; DEMAVET; DEMESO; DEMSODROX; DERMASORB; DIMETHYL SULPHOXIDE; DIMEXIDE; DIPIRARTRIL-TROPICO; DMS-70; DMS-90; DMS 70; DMS 90; DMSO; DOLICUR; DOLIGUR; DOMOSO; DROMISOL; DURASORB; GAMASOL 90; HYADUR; INFILTRINA; M 176; METHANE,SULFINYLBIS-; METHYL SULFOXIDE; METHYLSULFINYLMETHANE; NSC-763; RIMSO-5; RIMSO-50; SOMIPRONT; SQ 9453; SULFINYLBIS(METHANE); SULFINYLBISMETHANE; SYNTEXAN; TOPSYM; TOPSYM (RESCINDED)

Description: colorless liquid; odorless

Use: in antifreeze, hydraulic fluid, paint and varnish removers and potentially as a pharmaceutical solvent; as a clearing solution for examination of microscopic fungus, as a solvent, as a treatment for interstitial cystitis, as an analytical reagent, in the manufacture of synthetic fibers, industrial cleaners and pesticides, in the preservation of cells at low temperatures and in the diffusion of drugs into the bloodstream by topical applications; in medicine as an anti-inflammatory agent; in veterinary medicine; in plant pathology and nutrition, as a metal-complexing agent and as an analgesic

Physical Properties

Boiling Point: 189 °C (372 °F)
Freezing Point: 18.45 °C (65.21 °F)
Specific Gravity: 1.1 at 20 °C/4 °C
Vapor Density: 2.71
Saturated Vapor Density: 1.201123481 kg/m^3
Density: 1.1 g/mL at 20 °C
Vapor Pressure: 0.42 mm Hg at 20 °C
Water Solubility: Soluble in Water
Other Solubilities: 95% Ethanol: >=100 mg/ml at 20 °C; Acetone: >=100 mg/ml at 20 °C; Alcohol: Soluble; Benzene: Soluble; Chloroform: Soluble; DMSO: >=100 mg/ml at 20 °C; Ether: Soluble; Most organic solvents (except straight chain hydrocarbons): Soluble.
Refraction Index: 1.4795 at 20 °C/D
Evaporation Rate: > 1 Ether=1
Ionization Potential (eV): 9.10 +/-0.2
Flash Point: 95 °C Open Cup
Autoignition Temperature: 215 °C
LEL: 2.6% v/v
UEL: 42% v/v

RTECS Toxicity Data

Acute Oral: Rat LD_{50} Dose: 14500 mg/kg; Toxic Effects: Sense organs and special senses - Hemorrage; Sense organs and special senses - Conjunctive irritation. Mouse LD_{50} Dose: 7920 mg/kg.

Acute Dermal: Rat LD_{50} Route: Skin; Dose: 40 gm/kg. Rat LD_{50} Route: Subcutaneous Dose: 12 gm/kg; Toxic Effects: Behavioral - Change in motor activity (specific assay); Lungs, Thorax, or Respiration - Dyspnea.

Chronic (Multiple Dose) Oral: Rat Dose: 1070 gm/kg/13W-I; Toxic Effects: Blood - Other changes; Nutritional and gross metabolic - Weight loss or decreased weight gain; DEATH. Monkey Dose: 4864 gm/kg/78W-I; Toxic Effects: DEATH.

Chronic (Multiple Dose) Dermal: Dog Route: Skin; Dose: 389 gm/kg/17W-C; Toxic Effects: Sense organs and special senses - Changes in refraction; Sense organs and special senses - Other. Pig Route: Skin; Dose: 4698 mL/kg/58W-I; Toxic Effects: Sense organs and special senses - Changes in refraction; Behavioral - Fluid intake.

Irritation Eye: Rabbit Standard Draize Test Dose: 500 mg/24H; Reaction: mild.

Irritation Skin: Rabbit Standard Draize Test Dose: 500 mg/24H; Reaction: mild. Rabbit Open Draize Test Dose: 10 mg/24H open; Reaction: mild.

Reproductive/Teratogenic: Rat Route: Subcutaneous; Dose: 30750 mg/kg; Duration: female 8-10D of pregnancy; Effects on Fertility - Post-implantation mortality; Litter size. Rat Route: Intraperitoneal; Dose: 56 gm/kg; Duration: female 6-12D of pregnancy; Effects on Fertility - Abortion. Rat Route: Intraperitoneal; Dose: 6600 mg/kg; Duration: female 7-15D of pregnancy; Effects on Fertility - Post-implantation mortality.

Mutagenic: Human Other Mutation Test Systems; Cell Type: lymphocyte; Dose: 140 mmol/L. Rat Cytogenetic Analysis; Route: Intraperitoneal; Dose: 25 gm/kg/5D.

Tumorigenic: Rat Route: Oral; Dose: 59 gm/kg/81W-I; Toxic Effects: Tumorigenic - Equivocal tumorigenic agent by RTECS criteria; Skin and appendages - Tumors. Rat Route: Subcutaneous; Dose: 220 gm/kg/82W-I; Toxic Effects: Tumorigenic - Equivocal tumorigenic agent by RTECS criteria; Skin and appendages - Tumors.

Hazard Overviews

Fire Diamond

Health: Irritating to eyes/skin. Also Causes: shortness of breath, bronchial asthma, headaches, dizziness, sedation. Chronic Effects: teratogenic effects (based on animal studies), contact dermatitis.

Fire: Will burn. Can form explosive mixtures in the air. Use water spray, alcohol foam, carbon dioxide, or dry chemical.

Reactivity: Stable. Hazardous polymerization cannot occur. Avoid: contact of with skin. Incompatible with: strong oxidizing agents (potassium permanganate; nitric acid); nonmetal chlorides (silicon chloride; phosphorous trichlorate; thionyl chloride); iodine pentafluoride; diborane; sodium hydride; silver flouride; methyl bromide; tolyl chloride; perchloric acid; periodic acid. Hazardous decomposition products: sulfur dioxide; methyl mercaptan; formaldehyde.

Carcinogenicity: IARC - Not listed; NIOSH - Not listed; NTP - Not listed; ACGIH - Not listed; OSHA - Not listed; EPA - Not listed; MAK - Not listed

Primary Target Organs:

Eyes

Skin

Respiratory System

Nervous System

Environmental

Ecotoxicity: Aquatic toxicity: 33,500 ppm/48 hr/bluegill/TLm/fresh water

Environmental Fate: If released on land, it would be expected to leach in the soil. Some volatilization from dry soil and surfaces may be expected. It is fairly resistant to biodegradation based upon screening tests. It may be reduced by some reducing agents that may occur in soil. If sprayed on land as an aerosol, it will be lost primarily by volatilization. If released in water, it should disproportionate to dimethyl sulfide and dimethyl sulfone. This disproportionation is catalyzed by light and there is some evidence that catalysts or photosensitizers in natural water may affect the rate. It may also be reduced by reducing agents that may occur in natural waters. In water, volatilization, adsorption to sediment, and bioconcentration in fish should not be important. In the atmosphere, it will exist primarily in the vapor phase and be removed by both wet and dry deposition. It will react with photochemically-produced hydroxyl radicals with a half-life of about 7 hr.

Cleanup/Disposal: Guide No. 171: Do not touch or walk through spilled material. Stop leak if you can do it without risk. Prevent dust cloud. Avoid inhalation of asbestos dust. Small Dry Spills: With clean shovel place material into clean, dry container and cover loosely; move containers from spill area. Small Spills: Take up with sand or other noncombustible absorbent material and place into containers for later disposal. Large Spills: Dike far ahead of liquid spill for later disposal. Cover powder spill with plastic sheet or tarp to minimize spreading. Prevent entry into waterways, sewers, basements or confined areas.

Environmental Physical Data

Henry's Law Constant: 7.77×10^{-9}

Octanol/Water Partition Coefficient: log K_{ow} = -2.03

BCF: no food chain concentration potential

Regulations

RCRA 40CFR: Not listed

CERCLA: 40CFR 302.4: Not listed

SARA 40CFR 372.65: Not listed

SARA EHS 40CFR 355: Not listed

TSCA: Listed

Analytical Methods

Other: EPA 1666, 1671

DIM3600 CAS #: 120-61-6

DIMETHYL TEREPHTHALATE

RTECS: WZ1225000

EINECS Number: 204-411-8

Molecular Formula: $C_{10}H_{10}O_4$

Structured MF: $CH_3OCOC_6H_4COOCH_3$

Formula Weight: 194.19

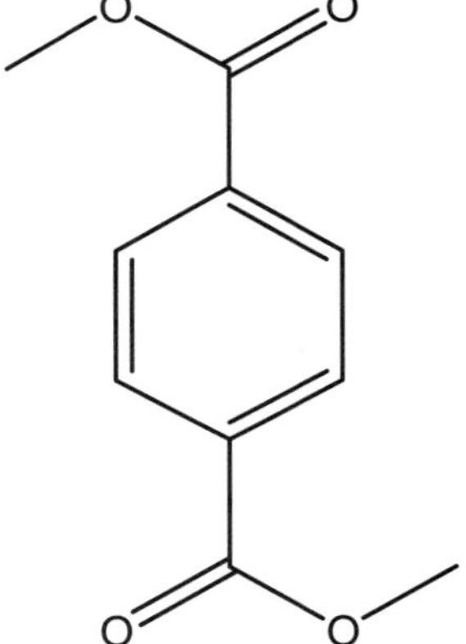

Chemical Structure

Synonyms: 1,4-BENZENEDICARBOXYLIC ACID,DIMETHYL ESTER; 1,4-BENZENEDICARBOXYLIC ACID,DIMETHYL ESTER (9CI); DI-ME TEREPHTHALATE; DIMETHYL 1,4-BENZENEDICARBOXYLATE; DIMETHYL ESTER OF 1,4-BENZENEDICARBOXYLIC ACID; DIMETHYL P-PHTHALATE; DIMETHYLESTER KYSELINY ISOFTALOVE; DIMETHYLESTER KYSELINY TEREFTALOVE; DIMETHYLTEREPHTHALATE; DMT; METHYL 4-CARBOMETHOXYBENZOATE; TEREPHTHALIC ACID METHYL ESTER; TEREPHTHALIC ACID,DIMETHYL ESTER

Description: white powder, crystalline solid; slight ester odor

Use: herbicide intermediate; a component of paints adhesives, printing inks and coatings; in the production of polyester resins, films and fibers and in the production of polyethylene terephthalate (PET) fibers for home furnishings, tire cord, ropes and clothing; in analytical chemistry

Physical Properties

Boiling Point: 300 °C (572 °F)
Freezing Point: 140 °C (284 °F)
Specific Gravity: 1.362
Vapor Density: 6.7 Air=1
Vapor Pressure: 16 mm Hg at 100 °C
Water Solubility: < 1 mg/mL at 13 C
Other Solubilities: 95% Ethanol: <1 mg/ml at 13 °C; Acetone: <1 mg/ml at 13 °C; Alcohol (hot): Soluble; DMSO: <1 mg/ml at 13 °C; Ether: Soluble; Toluene: 10-50 mg/ml at 23 °C.
Surface Tension: 2.5332 x10^{-2} N/m at melting point
Flash Point: 146 °C Open Cup
Autoignition Temperature: 570 °C
LEL: 33 g/m^3

RTECS Toxicity Data

Acute Oral: Rat LD_{50} Dose: >3200 mg/kg.
Acute Dermal: Guinea Pig LD_{50} Route: Skin; Dose: >5 gm/kg.
Chronic (Multiple Dose) Inhalation: Rat Dose: 1 mg/m^3/2H/22W-I; Toxic Effects: Sense organs and special senses - Other; Vascular - BP lowering not characterized in autonomic section; Nutritional and gross metabolic - Weight loss or decreased weight gain.
Irritation Eye: Rabbit Standard Draize Test Dose: 500 mg/24H; Reaction: mild.
Mutagenic: Mouse Micronucleus Test; Route: Intraperitoneal; Dose: 200 umol/kg.

Hazard Overviews

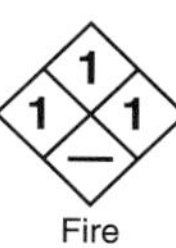

Fire Diamond

Health: Mildly irritating to eyes/skin/respiratory tract.
Fire: Will burn. May form explosive dust-air mixtures. Use water spray, dry chemical, carbon dioxide, or alcohol foam.
Reactivity: Stable. Hazardous polymerization cannot occur. Avoid: dust; ignition sources; oxidizing agents. Incompatible with: strong oxidizing agents; strong bases. Hazardous decomposition products: toxic gases.
Carcinogenicity: IARC - Not listed; NIOSH - Not listed; NTP - Not listed; ACGIH - Not listed; OSHA - Not listed; EPA - Not listed; MAK - Not listed
Primary Target Organs:

Eyes

Skin

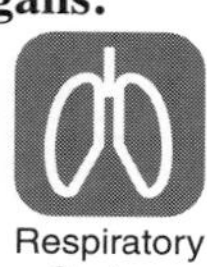
Respiratory System

Exposure Limits
AIHA WEEL: TWA: 10 mg/m^3total; OTHER: 5 mg/m^3respirable.
Respirator Recommendation
Exposure Range: >5 to 50 mg/m^3 Air Purifying, Negative Pressure, Half Mask
Exposure Range: >50 to 500 mg/m^3 Air Purifying, Negative Pressure, Full Face
Exposure Range: >500 to 5000 mg/m^3 Supplied Air, Constant Flow/Pressure Demand, Full Face
Exposure Range: >5000 to unlimited mg/m^3 Self-contained Breathing Apparatus, Pressure Demand, Full Face
Cartridge Color: dust/mist filter (use P100 or consult supervisor for appropriate dust/mist filter)

Environmental

Environmental Fate: If released to soil, it will have medium to very high soil mobility. Volatilization is expected to occur from both moist and dry soils. Biodegradation is expected in both soils and aquatic conditions based on several biodegradation studies. If released to water, hydrolysis may be an important fate process based on estimated base-catalyzed hydrolysis half-lives of 26.4 days and 264 days for pH's of 8 and 7, respectively. Bioconcentration and adsorption to sediment are not expected to be very important fate process in aquatic environments. It will essentially non-volatile from water surfaces. If released to air, vapor-phase will react with photochemically-produced hydroxyl radicals with a half-life of 3-28 days. Particulate may be removed from the atmosphere by both wet and dry deposition.

Environmental Physical Data

Henry's Law Constant: estimated at 6.1 x10^{-6}
Octanol/Water Partition Coefficient: log K_{ow} = 2.25
Sorption Partition Coefficient: K_{oc} = estimated at 36
BCF: estimated at 30
BOD: 84% BODT, 3 days

Regulations

RCRA 40CFR: Not listed
CERCLA: 40CFR 302.4: Not listed
SARA 40CFR 372.65: Not listed
SARA EHS 40CFR 355: Not listed
TSCA: Listed

Analytical Methods

Water / Groundwater: EPA 6

DIM3650 **CAS #: 1861-32-1**

DIMETHYL TETRACHLOROTEREPHTHALATE

RTECS: WZ1500000
EINECS Number: 217-464-7
Molecular Formula: $C_{10}H_6Cl_4O_4$
Formula Weight: 332.0

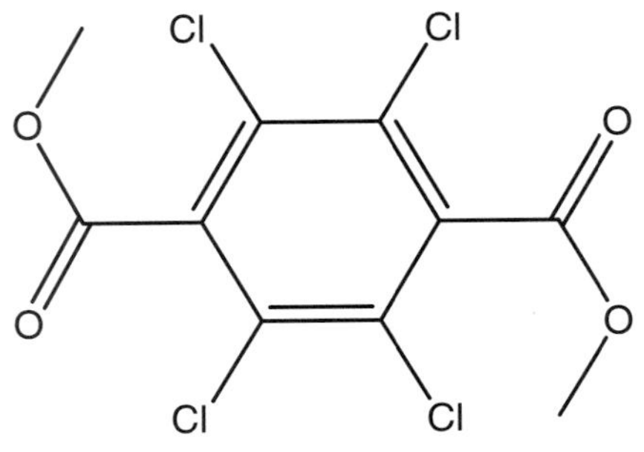

Chemical Structure

Synonyms: 1,4-BENZENEDICARBOXYLIC ACID,2,3,5,6-TETRACHLORO-,DIMETHYL ESTER; CHLOROTHAL; CHLORTHAL-DIMETHYL; CHLORTHAL-METHYL; DAC 893; DACTHAL; DACTHALOR; DAKTAL; DCPA; DIMETHYL 2,3,5,6-TETRACHLOROBENZENE-1,4-DICARBOXYLATE; DIMETHYL 2,3,5,6-TETRACHLOROTEREPHTHALATE; DIMETHYLESTER KYSELINY TETRACHLORTEREFTALOVE; FATAL; RID; TEREPHTHALIC ACID,TETRACHLORO-,DIMETHYL ESTER; 2,3,5,6-TETRACHLOROTEREPHTHALIC ACID DIMETHYL ESTER; TETRACHLOROTEREPHTHALIC ACID DIMETHYL ESTER; 2,3,5,6-TETRACHLOROTEREPHTHALIC ACID,DIMETHYL ETHER; 2,3,5,6-TETRACHLORPHTHALSAURE-DIMETHYLESTER

Description: colorless crystals; essentially odorless

Use: selective non-systemic herbicide for pre-emergence control of annual grasses & broad-leaved weeds; presently approved for on turf, ornamentals, strawberries, & agronomic crops incl cotton, soybeans, & field beans; effective against smooth & hairy crabgrass, witchgrass, green & yellow foxtails, fall panicum, etc; against carpet weedpurslane & common chickweed; tolerated by crop plants; extraordinarily toxic to dodder, for dodder control in seed alfalfa.

Physical Properties

Boiling Point: Decomposes at 360 °C (680 °F) to 370 °C (698 °F)

Freezing Point: 155 °C (311 °F) to 156 °C (312.8 °F)

Specific Gravity: Dtahal W-75 1.7

Saturated Vapor Density: 1.200000041 kg/m^3

Vapor Pressure: 2.5 $x10^{-6}$ mm Hg at 25 °C

Water Solubility: < 5% at 0 °C

Other Solubilities: Solubility (g/kg solvent at 25 °C): 120 g in dioxan, 250 g in Benzene, 170 g in Toluene, 140 g in Xylene, 100 g in Acetone, 70 g in Carbon Tetrachloride; 7% at 25 °C in Carbon Tetrachloride.

RTECS Toxicity Data

Acute Oral: Rat LD_{50} Dose: 3 gm/kg.

Acute Inhalation: Rat LC_{50} Dose: >5700 mg/m^3/4hr.

Acute Dermal: Rabbit LD_{50} Route: Skin; Dose: 10 gm/kg.

Reproductive/Teratogenic: Rat Route: Oral; Dose: 700 mg/kg; Duration: female 10W prior to mating Effects on Fertility - Pre-implantation mortality; Post-implantation mortality; Effects on Embryo or Fetus - Fetotoxicity. Rat Route: Oral; Dose: 700 mg/kg; Duration: male 10W prior to mating; Effects on Fertility - Post-implantation mortality; Effects on Newborn - Behavioral.

Hazard Overviews

Carcinogenicity: IARC - Not listed; NIOSH - Not listed; NTP - Not listed; ACGIH - Not listed; OSHA - Not listed; EPA - Not listed; MAK - Not listed

Environmental

Ecotoxicity: Fishes: Lepomis macrochirus 48h LC_{50} 700 mg/l

Environmental Fate: If released to the atmosphere, it will exist in both the vapor phase and the particulate phase based on an experimental vapor pressure of 2.5 $x10^{-6}$ mm Hg. In the vapor phase, it should react slowly with hydroxyl radicals with an estimated half-life of 36 days. Particulate phase may be removed physically from air by wet and dry deposition. It is expected to be essentially immobile in soil based on an estimated K_{oc} of 5900; it may bind strongly to organic matter or clay in the soil. This compound will biodegrade with products of monomethyl tetrachloroterephthalate and tetrachloroterephthalic acid. Half-lives ranging from 11 to 289 days depending on the soil temperature and moisture conditions (optimal conditions, 20-30 °C, 0.2 kg H2O/kg soil) have been reported. Volatilization from moist soil surfaces is also possible. During a 21 day period, 36-52% of the total measured loss from soil was accounted for by volatilization and 26% by breakdown in soil. Photodegradation on soil surfaces may occur with a half-life of 5 hours (reaction products include monomethyl tetrachloroterephthalate, tetrachloroterephthalic acid, and 1,2,4,5-tetrachlorobenzene). In water, it should bind strongly to particulate matter and sediment in the water column based on an estimated K_{oc} value of 5900. Biodegradation may occur based on evidence from several soil studies. It should bioconcentrate in aquatic organisms based on an estimated BCF value of 1300; this compound has been detected in fish at several locations.

Environmental Physical Data

Henry's Law Constant: 2.18 $x10^{-6}$

Octanol/Water Partition Coefficient: log K_{ow} = 4.40

Sorption Partition Coefficient: K_{oc} = 5900

BCF: estimated at 1300

Regulations

RCRA 40CFR: Not listed

CERCLA: 40CFR 302.4: Not listed

SARA 40CFR 372.65: Not listed

SARA EHS 40CFR 355: Not listed

TSCA: Not listed

Analytical Methods

Soil: SW846 8081

Water / Groundwater: EPA 608.2, 022

Drinking Water: EPA 508, 508.1, 515.2, 525.2; AOAC 990.06

Food: FDA 212.1, 212.2, 232.1, 232.4, 242.1; AOAC 970.05, 970.06

Indoor / Expired Air: ASTM D4861

Plasma: EPA 001; FDA 211.1, 231.1, 252

Other: EPA 1656

DIM3700 CAS #: 544-97-8

DIMETHYL ZINC

DOT: UN1370; IMO4.2
EINECS Number: 208-884-1
Molecular Formula: C_2H_6Zn
Structured MF: $(CH_3)_2Zn$
Formula Weight: 95.45

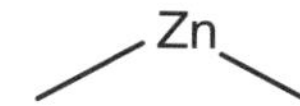

Chemical Structure

Synonyms: DIMETHYLZINC; METHYLZINC; ZINC METHYL; ZINC,DIMETHYL-
Description: colorless mobile liquid; peculiar, garlic like odor
Use: in ter-butyl alcohol synthesis; as methylating agent in methyltitanium trichloride

Physical Properties

Boiling Point: 46 °C (115 °F)
Freezing Point: -40 °C (-40 °F)
Specific Gravity: 1.386 at 10.5 °C/4 °C
Vapor Pressure: Saturated vapor pressure 6.014 lb/sq in at 70 °C
Water Solubility: Decomposes in Cold Water
Other Solubilities: Soluble in Ether, miscible with hydrocarbons.
Surface Tension: 18 dynes/cm
Ionization Potential (eV): 9.00 +/-0.3
Flash Point: Ignites spontaneously
Autoignition Temperature: < -18 °C

Hazard Overviews

Corrosive

Flammable

Fire Diamond

Health: Corrosive to eyes/skin/respiratory tract. Toxic. Other Acute Effects: harmful if swallowed, inhaled, or absorbed through skin; inhalation may result in spasm, inflammation and edema of the larynx and bronchi, chemical pneumonitis and pulmonary edema; symptoms of exposure may include burning sensation; coughing; wheezing; laryngitis; shortness of breath; headache; nausea; vomiting. Chronic Effects: may cause nervous system disturbances; inhalation studies have demonstrated the development of inflammatory and ulcerous lesions of the penis, prepuce and scrotum in animals; target organs: liver, kidneys, bladder, brain.
Fire: Flammable. Hazards: vapor may travel considerable distance to source of ignition and flash back; reacts with water to liberate flammable and/or explosive gas; emits toxic fumes; catches fire if exposed to air. Extinguishing agents: carbon dioxide, dry chemical powder or appropriate foam; water may be effective for cooling, but may not effect extinguishment. Precautions: combustible liquid.
Reactivity: Stable. Hazardous polymerization will not occur. Incompatible with: oxidizing agents, protect from moisture. Hazardous decomposition products: carbon monoxide, carbon dioxide.
Carcinogenicity: IARC - Not listed; NIOSH - Not listed; NTP - Not listed; ACGIH - Not listed; OSHA - Not listed; EPA - Not listed; MAK - Not listed
Primary Target Organs:

Eyes

Skin

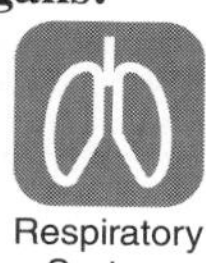
Respiratory System

Nervous System

Liver

Kidneys

Environmental

Cleanup/Disposal: Guide No. 135: Fully encapsulating, vapor protective clothing should be worn for spills and leak with no fire. Eliminate all ignition sources (no smoking, flares, sparks or flames in immediate area). Do not touch or walk through spilled material. Stop leak if you can do it without risk. Small Spills: Cover with dry earth, dry sand, or other non-combustible material followed with plastic sheet to minimize spreading or contact with rain. Use clean non-sparking tools to collect material and place it into loosely covered plastic containers for later disposal. Prevent entry into waterways, sewers, basements or confined areas.

Environmental Physical Data

BOD: none

Regulations

RCRA 40CFR: Not listed
CERCLA: 40CFR 302.4: Listed as Compound per CWA Section 307(a)
SARA 40CFR 372.65: Listed as Compound
SARA EHS 40CFR 355: Not listed
TSCA: Listed

DIM3750 CAS #: 127-19-5

N,N-DIMETHYLACETAMIDE

RTECS: AB7700000
EINECS Number: 204-826-4
Molecular Formula: C_4H_9NO
Structured MF: $CH_3CON(CH_3)_2$
Formula Weight: 87.12

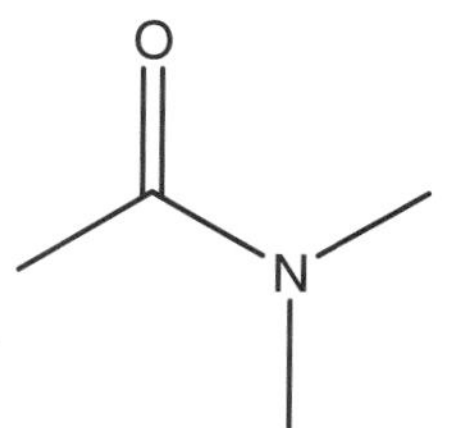

Chemical Structure

Synonyms: ACETAMIDE,N,N-DIMETHYL-; ACETDIMETHYLAMIDE; ACETIC ACID DIMETHYLAMIDE; ACETIC ACID,DIMETHYLAMIDE; ACETIMETHYLAMIDE; ACETYLDIMETHYLAMINE; DIMETHYL ACETAMIDE; N,N-DIMETHYL ACETAMIDE; DIMETHYLACETAMID; DIMETHYLACETAMIDE; DIMETHYLACETONE AMIDE; DIMETHYLAMID KYSELINY OCTOVE; DIMETHYLAMIDE ACETATE; DIMETHYLAMIDE ACETATE,DMA; N,N-DIMETHYLETHANAMIDE; DMA; DMAC; NSC 3138; U-5954

Description: colorless liquid; ammonia odor

Use: dipolar aprotic solvent for many organic reactions and industrial applications

Physical Properties

Boiling Point: 165 °C (329 °F) at 758 mm Hg
Freezing Point: -20 °C (-4 °F)
Specific Gravity: 0.9366 at 25 °C/4 °C
Vapor Density: 3 Air=1
Density: 0.937 g/mL at 20 °C
Vapor Pressure: 2 mm Hg
Water Solubility: Soluble in Water
Other Solubilities: 95% Ethanol: >=100 mg/ml at 22 °C; Acetone: >=100 mg/ml at 22 °C; DMSO: >=100 mg/ml at 22 °C; Ether: Soluble.
Surface Tension: 34 dynes/cm at 20 °C
Odor Threshold: 163.8 mg/m^3
Refraction Index: 1.4230 at 20 °C/D
Critical Temperature: 385 °C
Critical Pressure: 39.7 atm
Ionization Potential (eV): 8.81
Flash Point: 70 °C Open Cup
Autoignition Temperature: 490 °C
LEL: 1.8% v/v
UEL: 11.5% v/v at 320 °F

RTECS Toxicity Data

Acute Oral: Rat LD_{50} Dose: 4300 mg/kg. Mouse LD_{50} Dose: 4620 mg/kg.

Acute Inhalation: Rat LC_{50} Dose: 2475 ppm/1hr; Toxic Effects: Nutritional and gross metabolic - Weight loss or decreased weight gain. Mouse LC_{50} Dose: 7200 mg/m^3; Toxic Effects: Behavioral - Ataxia; Lungs, Thorax, or Respiration - Other changes; Liver - Other changes.

Acute Dermal: Rabbit LD_{50} Route: Skin; Dose: 2240 mg/kg; Toxic Effects: Skin and appendages - Primary irritation. Rat LD_{50} Route: Skin; Dose: >2 gm/kg.

Chronic (Multiple Dose) Oral: Rat Dose: 364 mg/kg/26W-I; Toxic Effects: Liver - Other changes; Biochemical - Other esterases; Biochemical - Transaminases. Rat Dose: 4500 mg/kg/90D-C; Toxic Effects: Blood - Changes in erythrocite (RBC) cell count; Blood - Changes in leukocyte (WBC) cell count.

Chronic (Multiple Dose) Inhalation: Rat Dose: 103 ppm/6H/26W-I; Toxic Effects: Liver - Other changes. Rat Dose: 300 ppm/6H/12D-I; Toxic Effects: Blood - Changes in serum composition; Nutritional and gross metabolic - Weight loss or decreased weight gain; Biochemical - Lipids including transport. Rat Dose: 120 ppm/6H/64D-I; Toxic Effects: Liver - Changes in Liver weight. Rat Dose: 100 ppm/6H/2Y-I; Toxic Effects: Liver - Other changes.

Chronic (Multiple Dose) Dermal: Rabbit Route: Skin; Dose: 8 gm/kg/4D-I; Toxic Effects: Liver - Other changes; Nutritional and gross metabolic - Weight loss or decreased weight gain; DEATH. Dog Route: Skin; Dose: 113 gm/kg/6W-I; Toxic Effects: Liver - Fatty Liver degeneration; DEATH.

Irritation Eye: Rabbit Standard Draize Test Dose: 100 mg; Reaction: mild.

Irritation Skin: Rabbit Open Draize Test Dose: 10 mg/24H open; Reaction: mild.

Reproductive/Teratogenic: Rat Route: Oral; Dose: 5600 mg/kg; Duration: female 6-19D of pregnancy; Effects on Fertility - Post-implantation mortality; Effects on Embryo or Fetus - Fetotoxicity; Specific Developmental Abnormalities - Craniofacial (including nose and tongue). Rat Route: Oral; Dose: 5600 mg/kg; Duration: female 6-19D of pregnancy; Specific Developmental Abnormalities - Musculoskeletal system; Cardiovascular (circulatory) system; Homeostasis. Rat Route: Inhalation; Dose: 288 ppm/6H; Duration: male 10D prior to mating; Paternal Effects - Testes, epididymis, sperm duct. Rat Route: Inhalation; Dose: 281 ppm/6H; Duration: female 6-15D of pregnancy; Effects on Embryo or Fetus - Fetotoxicity. Rat Route: Inhalation; Dose: 300 ppm/6H; Duration: female 10W prior to mating Effects on Newborn - Growth statistics. Rat Route: Inhalation; Dose: 282 ppm/6H; Duration: female 6-15D of pregnancy; Effects on Embryo or Fetus - Fetotoxicity.

Mutagenic: Mouse DNA Inhibition; Route: Unreported; Dose: 4400 mg/kg. Hamster Sister Chromatid Exchenge; Cell Type: ovary; Dose: 10 gm/L.

Hazard Overviews

Fire Diamond

Health: Irritating to eyes/skin/respiratory tract. Also Causes: headache, dizziness, drowsiness, delusions, auditory/visual hallucinations, disorientation, CNS depression, lethargy, sweating, weakness. Chronic Effects: Liver enlargement, elevated cholesterol, reproductive effects, liver damage, teratogenic effects.

Fire: Combustible. Use carbon dioxide, dry chemical, or alcohol foam. Do not scatter with a high pressure water stream. Vapors may travel to an ignition source and flash back.

Reactivity: Stable. Hazardous polymerization cannot occur. Avoid: skin contact; heat; ignition sources. Incompatible with: strong oxidizing agents; such as iron; carbon tetrachloride; benzene hexachloride. Hazardous decomposition products: ammonia; carbon monoxide; carbon dioxide.

Carcinogenicity: IARC - Not listed; NIOSH - Listed as carcinogen; NTP - Not listed; ACGIH - Not listed; OSHA - Not listed; EPA - Not listed; MAK - Not listed

Primary Target Organs:

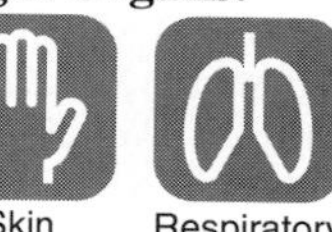

Eyes Skin Respiratory System Nervous System Liver

Exposure Limits
OSHA PEL: TWA: 10 ppm; 35 mg/m^3.
ACGIH TLV: TWA: 10 ppm; 36 mg/m^3.
NIOSH REL: TWA: 10 ppm; 35 mg/m^3.
NIOSH IDLH: 300 ppm.
DFG MAK: TWA: 10 ppm; 35 mg/m^3.

Respirator Recommendation
Exposure Range: >10 to <300 ppm Supplied Air, Constant Flow/Pressure Demand, Half Mask
Exposure Range: 300 to unlimited ppm Self-contained Breathing Apparatus, Pressure Demand, Full Face
Note: poor warning properties

Environmental

Ecotoxicity: LC_{50} Pimephales promelas (fathead minnow) 1.50 g/l/96 hr (confidence limit 1.21- 1.86 g/l), flow-through bioassay with measured concentrations, 23.1 °C, dissolved oxygen 6.6 mg/l, hardness 45.0 mg/l calcium carbonate, alkalinity 43.5 mg/l calcium carbonate, and pH 7.7

Environmental Fate: If released to soil, it will display very high mobility. An experimental Henry's Law constant indicates that it will not volatilize from moist soil to the atmosphere; volatilization from dry soil to the atmosphere should be slow because of its low vapor pressure. It is stable to hydrolysis except under strongly acidic or basic conditions, and it is not expected to hydrolyze in soil. If released to water, it will not bioconcentrate in fish and aquatic organisms nor will it adsorb to sediment and suspended organic matter or volatilize from water. The estimated half-life for volatilization from a model river is 2800 days. It will be partially dissociated in basic waters. If released to the atmosphere, it may undergo a rapid gas-phase reaction with photochemically produced hydroxyl radicals. An estimated half-life for this process is 6.1 hours. It may also undergo atmospheric removal by wet deposition processes.

Cleanup/Disposal: Guide No. 171: Do not touch or walk through spilled material. Stop leak if you can do it without risk. Prevent dust cloud. Avoid inhalation of asbestos dust. Small Dry Spills: With clean shovel place material into clean, dry container and cover loosely; move containers from spill area. Small Spills: Take up with sand or other noncombustible absorbent material and place into containers for later disposal. Large Spills: Dike far ahead of liquid spill for later disposal. Cover powder spill with plastic sheet or tarp to minimize spreading. Prevent entry into waterways, sewers, basements or confined areas.

Environmental Physical Data

Henry's Law Constant: 1.22 x10^{-8}
Octanol/Water Partition Coefficient: log K_{ow} = -0.77
Sorption Partition Coefficient: K_{oc} = calculated at 9.1
BCF: calculated at 0.15

Regulations

RCRA 40CFR: Not listed
CERCLA: 40CFR 302.4: Not listed
SARA 40CFR 372.65: Not listed
SARA EHS 40CFR 355: Not listed
TSCA: Listed

Analytical Methods

Air: ASTM D4490
Food: AOAC 971.07
Indoor / Expired Air: NIOSH 2004
Other: EPA 1665

DIM3800 CAS #: 762-42-5

DIMETHYLACETYLENEDICARB-OXYLATE

RTECS: ES0175000
EINECS Number: 212-098-4
Molecular Formula: $C_6H_6O_4$
Formula Weight: 142.12

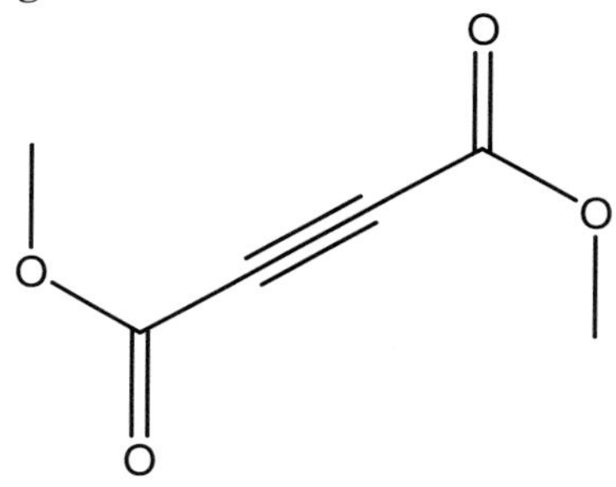

Chemical Structure

Synonyms: ACETYLENEDICARBOXYLIC ACID,DIMETHYL ESTER; 1,2-BIS(METHOXYCARBONYL)ETHYNE; DI(CARBOMETHOXY)ACETYLENE; DIMETHYL ACETYLENEDICARBOXYLIC ACID; DIMETHYL ETHYNEDICARBOXYLATE; METHYL ACETYLENEDICARBOXYLATE

Physical Properties

Boiling Point: 205 °C (401 °F)
Specific Gravity: 1.156
Vapor Density: 4.9 Air=1
Vapor Pressure: 0.13
Water Solubility: Negligible
Other Solubilities: Ethanol: Soluble; Ether: Soluble; Carbon Tetrachloride: Soluble
Refraction Index: 1.4434

RTECS Toxicity Data

Acute Oral: Rat LD_{Lo} Dose: 50 mg/kg. Mouse LD_{50} Dose: 550 mg/kg; Toxic Effects: Behavioral - Somnolence (general depressed activity); Behavioral - Ataxia; Behavioral - Coma.

Hazard Overviews

Corrosive

Health: Corrosive to eyes/skin/respiratory tract. Other Acute Effects: harmful if swallowed, inhaled, or absorbed through skin; lachrymator; inhalation may result in spasm, inflammation and edema of the larynx and bronchi, chemical pneumonitis and pulmonary edema; symptoms of exposure may include burning sensation; coughing; wheezing; laryngitis; shortness of breath; exposure can cause nausea, headache, vomiting, cyanosis, convulsions.
Fire: Extinguishing agents: water spray; carbon dioxide, dry chemical powder or appropriate foam. Precautions: combustible liquid.
Reactivity: Incompatible with: strong oxidizing agents, strong acids, strong bases, strong reducing agents. Hazardous decomposition products: toxic fumes of: carbon monoxide, carbon dioxide.
Carcinogenicity: IARC - Not listed; NIOSH - Not listed; NTP - Not listed; ACGIH - Not listed; OSHA - Not listed; EPA - Not listed; MAK - Not listed
Primary Target Organs:

Eyes

Skin

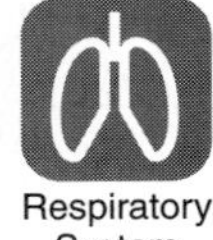
Respiratory System

Environmental

Regulations

RCRA 40CFR: Not listed
CERCLA: 40CFR 302.4: Not listed
SARA 40CFR 372.65: Not listed
SARA EHS 40CFR 355: Not listed
TSCA: Listed

DIM3850 CAS #: 2008-39-1

2,4-DIMETHYLAMINE

RTECS: AG8400000
DOT: UN2765; UN2766; UN2999; UN3000; IMO3.2; IMO6.1
EINECS Number: 217-915-8
Molecular Formula: $C_{10}H_{13}Cl_2NO_3$
Formula Weight: 266.1
Synonyms: ACETIC ACID,(2,4-DICHLOROPHENOXY)-,CMPD WITHDIMETHYLAMINE (1:1); ACETIC ACID,(2,4-DICHLOROPHENOXY)-,CMPD WITHN-METHYLMETHANAMINE (1:1); ACETIC ACID,(2,4-DICHLOROPHENOXY)-COMPD. WITH N-METHYL MEHTDNAMINE (1:1); ACETIC ACID,(2,4'DICHLOROPHENOXY)-,COMPD. WITH N-METHYLMETHANAMINE (1:1) (9CI); AMINOL; AMINOL (HERBICIDE); AMINOPRELIK 39; BANVEL-720; BANVEL 3 LIQUID HERBICIDE; BARBER'S WEED KILLER (AMINE FORMULATION); BEST 4 SERVIS BRAND LAWN WEED KILLER; BLADEX G; BLITZ 64; BRABANT 2,4-D AMINE; CHIPMAN 2,4-D AMINE NO 4; CHIPMAN LAWN WEEDKILLER; CHIPMAN MECOPROP + 2,4-D WEEDKILLER LIQUID; CO-OP PREMIUM LAWN WEED KILLER; D 50; 2,4-D ACETATE; 2,4-D AMINE; 2,4-D AMINE SALT; 2,4-D DIMETHYLAMINE SALT; 2,4-D DMA; 2,4-D DNA; D 50 (PESTICIDE); DED-WEED SULV; DEFY; DEMISE; DESORMONE; DIAMOND SHAMROCK AMINE 6D; (2,4-DICHLOROPHENOXY)ACETIC ACID DIMETHYLAMINE; (2,4-DICHLOROPHENOXY)ACETIC ACID DIMETHYLAMINE SALT; DIKAMIN D; DIMETHYLAMINE SALT OF 2,4-D; DIMETHYLAMINE,(2,4-DICHLOROPHENOXY) ACETATE; DIMETHYLAMINE,(2,4-DICHLOROPHENOXY)ACETATE; DIMETHYLAMMONIUM (2,4-DICHLOROPHENOXY)ACETATE; DIMETHYLAMMONIUM 2,4-DICHLOROPHENOXYACETATE; DMA-4; DOW DMA-4; DU PONT LAWN WEED KILLER; DU PONT TURF FOOD WITH WEED KILLER; DU PONT WEED KILLER NO 2; FLORO TOX 2,4-D AMINE WEED KILLER; FORMULA 40; FS AMINE 400 WEED KILLER; GREEN CROSS KILLEX SPOT WEEDER PRESSURIZED SPRAY; GREEN CROSS POISON IVY KILLER; HERBITEX; HORMIN; LIQUID CLEARIT VEGETATION KILLER; LIQUID WONDER WEEDER; MANCO KILL-WEED; MARQUETTE HERBITEX PLUS; MECOTURF PLUS 2,4-D LIQUID WEEDKILLER; MONOSAN; MONSANTO 2,4-D AMINE; MORSELECT; NORKEM 40T; PACIFIC COOPERATIVES P 2,4-D AMINE WEED KILLER; PARSONS 2,4-D WEED KILLER; PARSONS 2,4-D WEED KILLER NO 40; PFIZER; PHORDENE; REED AMINE 400; SHIRWEED 500; SPRAYGRAZE; ORTHO SUPER WEED-B-GON SPRAY; SURE DEATH 2,4-D AMINE WEEDKILLER; TECHNE 2,4-D AMINE WEED KILLER; U-46 D-FLUID; VIGORO DANDELIONS KILLER; WEEDAR 64; WEEDAR 96; WEED-RHAP A-4D; WILSON'S MULTI-WEEDER; ZEHRUNG 2,4-D SELECTIVE AMINE WEED KILLER
Description: white solid; odorless
Use: herbicide

Physical Properties

Freezing Point: 85 °C (185 °F) to 87 °C (188.6 °F)
Saturated Vapor Density: 1.200000001 kg/m^3
Vapor Pressure: 8 x10^{-8} mm Hg at 25 °C
Water Solubility: 300 g/100 g of Water at 20 °C
Other Solubilities: Soluble in Methyl, Ethyl, and Isopropyl Alcohols, Acetone; Insoluble in Kerosene & Diesel Oil.
Flash Point: Nonflammable

RTECS Toxicity Data

Mutagenic: Human Cytogenetic Analysis; Cell Type: lymphocyte; Dose: 500 umol/L. Mouse Cytogenetic Analysis; Route: Oral; Dose: 14 mg/kg/15D-C.

Hazard Overviews

Fire: Noncombustible.
Carcinogenicity: IARC - Not listed; NIOSH - Not listed; NTP - Not listed; ACGIH - Not listed; OSHA - Not listed; EPA - Not listed; MAK - Not listed

Environmental

Ecotoxicity: LC_{50} Phasianus colchicus (ring-necked pheasants) oral more than 5000 ppm in 5-day diet LC_{50} Lepomis macrochirus (Bluegill) 166-542 mg/l/24 hr; 166-458 mg/l/48 hr /Conditions of bioassay not specified LC_{50} Oncorhynchus kitsuchi (coho salmon) yearling > 200 mg/l/96 hr /Conditions of bioassay not specified
Environmental Fate: In environmental fate studies the disappearance of the free acid, 2,4-D rather than the

dimethylamine salt is generally monitored. If applied to crops, much of the herbicide is metabolized by the foliage. The measured half-life in soil is 4-23 days in cooler climates and less in warmer climates. If released in water, biodegradation will be the primary degradative process and half-lives range from 0.5-11 days. It is metabolized in fish and bioconcentration is not expected to be appreciable. Adsorption to sediment should not be significant. If released in air, it will be in the form of an aerosol and be subject to gravitational settling. Photolysis may occur and the vapor phase chemical, which is only expected to be a minor form, is estimated to be attacked by photochemically produced hydroxyl radicals (half-life 23.6 hr.).

Cleanup/Disposal: Guide No. 131: Fully encapsulating, vapor protective clothing should be worn for spills and leaks with no fire. Eliminate all ignition sources (no smoking, flares, sparks or flames in immediate area). All equipment used when handling the product must be grounded. Do not touch or walk through spilled material. Stop leak if you can do it without risk. Prevent entry into waterways, sewers, basements or confined areas. A vapor suppressing foam may be used to reduce vapors. Small Spills: Absorb with earth, sand or other non-combustible material and transfer to containers for later disposal. Use clean non-sparking tools to collect absorbed material. Large Spills: Dike far ahead of liquid spill for later disposal. Water spray may reduce vapor; but may not prevent ignition in closed spaces. Guide No. 152: Do not touch damaged containers or spilled material unless wearing appropriate protective clothing. Stop leak if you can do it without risk. Prevent entry into waterways, sewers, basements or confined areas. Cover with plastic sheet to prevent spreading. Absorb or cover with dry earth, sand or other non-combustible material and transfer to containers. Do not get water inside containers.

Environmental Physical Data

Henry's Law Constant: will be low
Sorption Partition Coefficient: K_{OC} = 72 to 136
BCF: catfish and bluegill 0.1 to 0.47

Regulations

RCRA 40CFR: Not listed
CERCLA: 40CFR 302.4: Not listed
SARA 40CFR 372.65: Not listed
SARA EHS 40CFR 355: Not listed
TSCA: Not listed

DIM3900 CAS #: 124-40-3

DIMETHYLAMINE

RTECS: IP8750000
DOT: UN1032; UN1160; IMO2.1; IMO3.2
EINECS Number: 204-697-4
Molecular Formula: C_2H_7N
Structured MF: $(CH_3)_2NH$
Formula Weight: 45.08

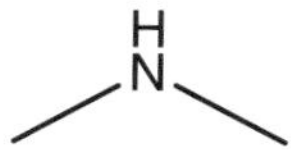

Chemical Structure

Synonyms: N,N-DIMETHYLAMINE; DIMETHYLAMINE (ANHYDROUS); DIMETHYLAMINE,ANHYDROUS; DMA; METHANAMINE,N-METHYL-; METHANAMINE,N-METHYL-(9CI); N-METHYLMETHANAMINE
Description: colorless gas at room temp., liquid below 44 F(6.7 C); strong fish odor
Use: rubber vulcanization accelerator, tanning agent, propellant, detergent soaps, pharmaceutical

Physical Properties

Boiling Point: 6.8 °C (44 °F)
Freezing Point: -92.2 °C (-133.96 °F)
Specific Gravity: Liquid 0.6804 at 0 °C/4 °C
Vapor Density: 1.6 Air=1
Density: 0.937 g/cc at 17 °C
Bulk Density: ~ 7.8 lbs/gal
Vapor Pressure: 1.7 mm Hg
Water Solubility: Very Soluble in Water
Other Solubilities: Ether: Soluble; Ethanol: Soluble.
Surface Tension: 18.1 dynes/cm
Odor Threshold: 0.0486 mg/m^3
pH: Aqueous solution is highly alkaline
Refraction Index: 1.350 at 17 °C/D
Evaporation Rate: > 1 Butyl Acetate=1
Critical Temperature: 164 °C
Critical Pressure: 5.340 mPa
Ionization Potential (eV): 8.24
Flash Point: -6.667 °C Closed Cup
Autoignition Temperature: 400 °C
LEL: 2.8% v/v
UEL: 14.4% v/v

RTECS Toxicity Data

Acute Oral: Rat LD_{50} Dose: 698 mg/kg; Toxic Effects: Behavioral - Excitment; Behavioral - Muscle weakness; Gastrointestinal - Ulceration or bleeding from stomach. Mouse LD_{50} Dose: 316 mg/kg; Toxic Effects: Behavioral - Excitment; Behavioral - Muscle weakness; Gastrointestinal - Ulceration or bleeding from stomach.
Acute Inhalation: Rat LC_{50} Dose: 4540 ppm/6hr; Toxic Effects: Sense organs and special senses - Other; Sense organs and special senses - Other; Lungs, Thorax, or Respiration - Dyspnea. Mouse LC_{50} Dose: 4725 ppm/2hr.
Chronic (Multiple Dose) Oral: Rat Dose: 85 mg/kg/35W-I; Toxic Effects: Behavioral - Alteration of classical conditioning; Liver - Changes in liver weight. Rat Dose: 37128 gm/kg/39W-C; Toxic Effects: Liver - Changes in liver weight; Biochemical - Phosphatases; Biochemical - Hepatic microsomal mixed oxidase(dealkylation, hyroxylation,etc).
Chronic (Multiple Dose) Inhalation: Rat Dose: 175 ppm/6H/1Y-I; Toxic Effects: Blood - Changes in serum composition; Nutritional and gross metabolic - Weight loss or decreased weight gain; Biochemical - Phosphatases. Rat

Dose: 175 ppm/6H/9D-I; Toxic Effects: Sense organs and special senses - Other. Rat Dose: 175 ppm/6H/2Y-I; Toxic Effects: Sense organs and special senses - Other; Nutritional and gross metabolic - Weight loss or decreased weight gain.
Mutagenic: Rat Cytogenetic Analysis; Route: Inhalation; Dose: 50 ug/m^3. Hamster Cytogenetic Analysis; Cell Type: ovary; Dose: 10 mmol/L.

Hazard Overviews

Corrosive

Flammable

Fire Diamond

Health: Corrosive to eyes/skin/respiratory tract. Also Causes: coughing, sneezing, difficulty breathing, nausea, pulmonary edema, conjunctivitis, corneal damage, intense pain, permanent corneal opacity, blindness, frostbite, necrosis, burns of the gastrointestinal tract, frostbite. Chronic Effects: dermatitis, conjunctivitis, bronchitis, liver injury.
Fire: Flammable. Can form explosive mixtures in the air. Stop flow of gas before extinguishing fire. Use water spray, dry chemical, or alcohol resistant foam. Use water spray to cool fire-exposed containers. Fight fire from maximum distance.
Reactivity: Stable. Hazardous polymerization cannot occur. Avoid: heat; ignition sources. Incompatible with: oxidizing materials; acids; sources of halogens; mercury; hypochlorite; acraldehyde; maleic anhydride; aluminum; brass; copper; zinc; plastics; rubber; coatings. Hazardous decomposition products: carbon monoxide; carbon dioxide; hydrocarbons; amine vapors; toxic oxides of nitrogen.
Carcinogenicity: IARC - Not listed; NIOSH - Not listed; NTP - Not listed; ACGIH - Class A4, Not classifiable as a human carcinogen; OSHA - Not listed; EPA - Not listed; MAK - Not listed
Primary Target Organs:

Eyes

Skin

Respiratory System

Exposure Limits
OSHA PEL: TWA: 10 ppm; 18 mg/m^3.
NIOSH REL: TWA: 10 ppm; 18 mg/m^3.
NIOSH IDLH: 500 ppm.
DFG MAK: TWA: 2 ppm; 4 mg/m^3.
Respirator Recommendation
Exposure Range: >10 to 100 ppm Air Purifying, Negative Pressure, Half Mask
Exposure Range: >100 to <500 ppm Air Purifying, Negative Pressure, Full Face
Exposure Range: 500 to unlimited ppm Self-contained Breathing Apparatus, Pressure Demand, Full Face
Cartridge Color: green

Environmental

Ecotoxicity: Aquatic toxicity: 50 ppm/24 hr/chub/died/fresh water >100 ppm/48 hr/shrimp/LC_{50}/salt water
Environmental Fate: If released on land, it should volatilize and leach into the soil. The chemical would be expected to biodegrade in several weeks. If released into water, it will readily biodegrade (half-life 1.5 days) and also be removed by volatilization (half-life 35 hr in a model river). Adsorption to sediment and bioconcentration in aquatic organisms will not be appreciable. In the atmosphere, it will react with photochemically produced hydroxyl radicals and degrade with a 5.9 hr half-life. Degradation will be faster in polluted atmospheres. It will also be scavenged by rain. It is a precursor of dimethylnitrosamine. The latter is formed in the atmosphere in the presence of nitrogen oxides and in lake water, sewage, and soil in the presence of nitrite ions.
Cleanup/Disposal: Guide No. 118: Eliminate all ignition sources (no smoking, flares, sparks or flames in immediate area). All equipment used when handling the product must be grounded. Fully encapsulating, vapor protective clothing should be worn for spills and leaks with no fire. Do not touch or walk through spilled material. Stop leak if you can do it without risk. If possible, turn leaking containers so that gas escapes rather than liquid. Use water spray to reduce vapors or divert vapor cloud drift. Do not direct water at spill or source of leak. Guide No. 129: Eliminate all ignition sources (no smoking, flares, sparks or flames in immediate area). All equipment used when handling the product must be grounded. Do not touch or walk through spilled material. Stop leak if you can do it without risk. Prevent entry into waterways, sewers, basements or confined areas. A vapor suppressing foam may be used to reduce vapors. Absorb or cover with dry earth, sand or other non-combustible material and transfer to containers. Use clean non-sparking tools to collect absorbed material. Large Spills: Dike far ahead of liquid spill for later disposal. Water spray may reduce vapor; but may not prevent ignition in closed spaces.

Environmental Physical Data

Henry's Law Constant: 7.24 x10^{-4}
Octanol/Water Partition Coefficient: log K_{ow} = -0.38
Sorption Partition Coefficient: K_{oc} = 434.9
BCF: estimated at 0.30
BOD: 0%, 5 days

Regulations

RCRA 40CFR: Listed Hazardous Waste No. U092 Ignitable Waste
CERCLA: 40CFR 302.4: Listed per CWA Section 311(b)(4) per RCRA Section 3001 RQ: 1000 lb (453.5 kg)
SARA 40CFR 372.65: Listed
SARA EHS 40CFR 355: Not listed
TSCA: Listed

Analytical Methods

Air: ASTM D4490
Indoor / Expired Air: NIOSH 2010
Other: EPA 1666, 1671

DIM3950 CAS #: 2300-66-5

DIMETHYLAMINE DICAMBA

RTECS: DG7506170
EINECS Number: 218-951-7
Molecular Formula: $C_{10}H_{13}Cl_2NO_3$
Formula Weight: 266.14
Synonyms: O-ANISIC ACID,3,6-DICHLORO-,COMPD. WITH DIMETHYLAMINE(1:1); BANEX; BANVEL; BANVEL 4S; BANVEL D; DIANATE; DICAMBA AMINE; DICAMBA DIMETHYLAMINE SALT; 3,6-DICHLORO-2-METHOXYBENZOIC ACID COMPD. WITH N-METHYLMETHANAMINE (1:1); DIMETHYLAMINE SALT OF DICAMBA

RTECS Toxicity Data

Acute Oral: Rat LD_{50} Dose: 2629 mg/kg.
Acute Dermal: Rabbit LD_{50} Route: Skin; Dose: >2 gm/kg.

Hazard Overviews

Carcinogenicity: IARC - Not listed; NIOSH - Not listed; NTP - Not listed; ACGIH - Not listed; OSHA - Not listed; EPA - Not listed; MAK - Not listed

Environmental

Regulations

RCRA 40CFR: Not listed
CERCLA: 40CFR 302.4: Not listed
SARA 40CFR 372.65: Listed
SARA EHS 40CFR 355: Not listed
TSCA: Not listed

Analytical Methods

Soil: EPA PMD-TLC

DIM4000 CAS #: 60-11-7

4-DIMETHYLAMINOAZOBENZENE

RTECS: BX7350000
EINECS Number: 200-455-7
Molecular Formula: $C_{14}H_{15}N_3$
Structured MF: $C_6H_5N{=}NC_6H_4N(CH_3)_2$
Formula Weight: 225.28

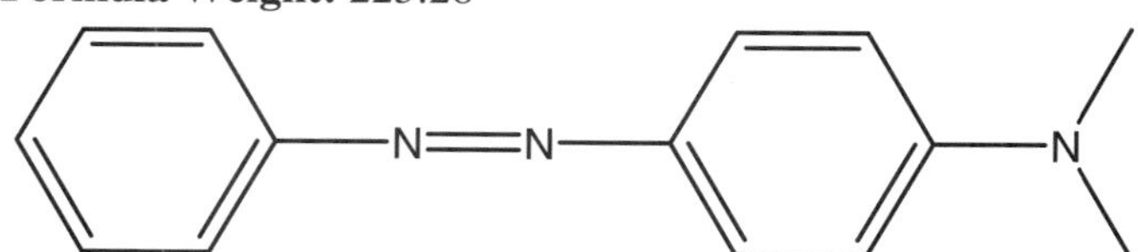

Chemical Structure

Synonyms: ATUL FAST YELLOW R; AZOBENZENE,P-DIMETHYLAMINO-; BENZENAMINE,N,N-DIMETHYL-4-(PHENYLAZO)-; BENZENAMINE,N,N-DIMETHYL-4-(PHENYLAZO)-(9CI); BENZENEAZODIMETHYLANILINE; BRILLIANT FAST OIL YELLOW; BRILLIANT FAST SPIRIT YELLOW; BRILLIANT FAST YELLOW; BRILLIANT OIL YELLOW; BUTTER YELLOW; C.I. 11020; C.I. SOLVENT YELLOW 2; CERASINE YELLOW GG; DAB; DAB (CARCINOGEN); DIMETHYL YELLOW; DIMETHYL YELLOW ANALAR; DIMETHYL YELLOW N,N-DIMETHYLANILINE; P-DIMETHYLAMINOAZOBENZEN; 4-(N,N-DIMETHYLAMINO)AZOBENZENE; DIMETHYLAMINOAZOBENZENE; N,N-DIMETHYL-4-AMINOAZOBENZENE; N,N-DIMETHYL-P-AMINOAZOBENZENE; P-(DIMETHYLAMINO)AZOBENZENE; P-DIMETHYLAMINOAZOBENZENE; 4-DIMETHYLAMINOAZOBENZOL; DIMETHYLAMINOAZOBENZOL; P-DIMETHYLAMINO-AZOBENZOL; 4-DIMETHYLAMINOPHENYLAZOBENZENE; N,N-DIMETHYL-P-AZOANILINE; N,N-DIMETHYL-P-(PHENYLAZO)ANILINE; N,N-DIMETHYL-P-PHENYLAZOANILINE; N,N-DIMETHYL-4-(PHENYLAZO)BENZAMINE; N,N-DIMETHYL-4-(PHENYLAZO)BENZENAMINE; DMAB; ENIAL YELLOW 2G; FAST OIL YELLOW B; FAST YELLOW; FAT YELLOW; FAT YELLOW A; FAT YELLOW AD OO; FAT YELLOW ES; FAT YELLOW ES EXTRA; FAT YELLOW EXTRA CONC; FAT YELLOW R; FAT YELLOW R (8186); GRASAL BRILLIANT YELLOW; IKETON YELLOW EXTRA; JAUNE DE BEURRE; METHYL YELLOW; OIL YELLOW; OIL YELLOW 2G; OIL YELLOW G-2; OIL YELLOW 20; OIL YELLOW 2625; OIL YELLOW 7463; OIL YELLOW BB; OIL YELLOW D; OIL YELLOW DN; OIL YELLOW FF; OIL YELLOW FN; OIL YELLOW G; OIL YELLOW GG; OIL YELLOW GR; OIL YELLOW II; OIL YELLOW N; OIL YELLOW PEL; OIL YELLOW S; OLEAL YELLOW 2G; ORGANOL YELLOW ADM; ORIENT OIL YELLOW GG; P.D.A.B; PETROL YELLOW WT; 4-(PHENYLAZO)-N,N-DIMETHYLANILINE; RESINOL YELLOW GR; RESOFORM YELLOW GGA; SILOTRAS YELLOW T2G; SOMALIA YELLOW A; STEAR YELLOW JB; SUDAN GG; SUDAN YELLOW; SUDAN YELLOW GG; SUDAN YELLOW GGA; TOYO OIL YELLOW G; WAXOLINE YELLOW AD; WAXOLINE YELLOW ADS; YELLOW G SOLUBLE IN GREASE; ZLUT MASELNA; ZLUT ROZPOUSTEDLOVA 2
Description: yellow solid, crystalline leaflets; odorless
Use: dye (for polishes and other wax products, polystyrene, petrol and soap); indicator in volumetric analysis (yellow @ pH 4.0; red @ pH 2.9); for the determination of free HCl in gastric juice; spot test identification of peroxidized fats; and formerly as a food dye

Physical Properties

Boiling Point: Sublimes
Freezing Point: 114 °C (237.2 °F) to 117 °C (242.6 °F)
Vapor Density: 7.78 Air=1
Vapor Pressure: 3.3×10^{-7} mm Hg
Water Solubility: 0.001% by weight
Other Solubilities: Very Soluble in Pyridine.
Flash Point: Not available; probably combustible

RTECS Toxicity Data

Acute Oral: Rat LD_{50} Dose: 200 mg/kg. Mouse LD_{50} Dose: 300 mg/kg.
Reproductive/Teratogenic: Mouse Route: Subcutaneous; Dose: 200 mg/kg; Duration: female 10D of pregnancy; Specific Developmental Abnormalities - Musculoskeletal system. Mouse Route: Intraperitoneal; Dose: 3 gm/kg; Duration: male 5D prior to mating; Paternal Effects - Spermatogenesis.
Mutagenic: Human Unscheduled DNA Synthesis; Cell Type: HeLa cell; Dose: 10 nmol/L. Human Unscheduled DNA Synthesis; Cell Type: fibroblast; Dose: 4 mg/L. Human DNA Inhibition; Cell Type: HeLa cell; Dose: 100 umol/L.

Tumorigenic: Rat Route: Oral; Dose: 5426 mg/kg/17W-C; Toxic Effects: Tumorigenic - Carcinogenic by RTECS criteria; Liver - Tumors. Rat Route: Oral; Dose: 2600 mg/kg/13W-C; Toxic Effects: Tumorigenic - Neoplastic by RTECS criteria; Liver - Tumors. Rat Route: Oral; Dose: 13 gm/kg/53W-C; Toxic Effects: Tumorigenic - Neoplastic by RTECS criteria; Liver - Tumors. Rat Route: Oral; Dose: 1920 mg/kg/14W-C; Toxic Effects: Tumorigenic - Equivocal tumorigenic agent by RTECS criteria; Liver - Tumors. Rat Route: Oral; Dose: 1800 mg/kg/14W-C; Toxic Effects: Tumorigenic - Equivocal tumorigenic agent by RTECS criteria; Liver - Tumors. Rat Route: Oral; Dose: 2331 mg/kg/7W-I; Toxic Effects: Tumorigenic - Equivocal tumorigenic agent by RTECS criteria; Liver - Tumors. Rat Route: Oral; Dose: 17200 mg/kg/17W-C; Toxic Effects: Tumorigenic - Neoplastic by RTECS criteria; Liver - Tumors. Rat Route: Oral; Dose: 8316 mg/kg/33W-C; Toxic Effects: Tumorigenic - Neoplastic by RTECS criteria; Liver - Tumors. Rat Route: Oral; Dose: 3990 mg/kg/19W-C; Toxic Effects: Tumorigenic - Neoplastic by RTECS criteria; Lungs, Thorax, or Respiration - Consolidation; Liver - Tumors. Rat Route: Oral; Dose: 800 mg/kg/64D-C; Toxic Effects: Tumorigenic - Equivocal tumorigenic agent by RTECS criteria; Liver - Tumors.

Hazard Overviews

Fire Diamond

Health: Irritating to skin. Toxic. Also Causes: weakness, dizziness, feeling of euphoria, shortness of breath, irregular muscular action. Chronic Effects: dermatitis, may cause cancer.

Fire: Noncombustible. Use extinguishing agents suitable for surrounding fire.

Reactivity: Stable. Hazardous polymerization cannot occur. Avoid: dust generation. Hazardous decomposition products: nitrogen oxide gases.

Carcinogenicity: IARC - Group 2B, Possibly carcinogenic to humans; NIOSH - Listed as carcinogen; NTP - Class 2B, Reasonably anticipated to be a carcinogen, sufficient evidence of carcinogenicity from studies in experimental animals; ACGIH - Not listed; OSHA - Listed as a carcinogen; EPA - Not listed; MAK - Not listed

Primary Target Organs:

Skin

Respiratory System

Nervous System

Liver

Respirator Recommendation

Exposure Range: unlimited Self-contained Breathing Apparatus, Pressure Demand, Full Face Air Purifying, Negative Pressure, Full Face

Note: TLV not established

Environmental

Environmental Fate: If it is released to soil it may bind to the soil based on an estimated K_{oc} of 7390 and therefore should not leach to the groundwater. However, since it has a pKa of 3.226 at 25 °C, it exists partially as a cation and the extent of its adsorption to soils and sediments should be affected by the pH of the medium. It should not hydrolyze in soils. No information was found on its biodegradation in soils. If it is released to water it may bioconcentrate in aquatic organisms, adsorb to sediment, and may be subject to direct photolysis. It should not hydrolyze or evaporate from water. Based on a laboratory screening test using an inoculum from settled domestic wastewater, it may be subject to biodegradation. If it is released to the atmosphere, it may be subject to direct photolysis and the estimated vapor phase half-life in the atmosphere is 7.04 hr as a result of photochemically produced hydroxyl radicals adding to the aromatic rings; however, it may exist primarily adsorbed onto particulate matter due to its very low vapor pressure.

Cleanup/Disposal: Guide No. 154: Eliminate all ignition sources (no smoking, flares, sparks or flames in immediate area). Do not touch damaged containers or spilled material unless wearing appropriate protective clothing. Stop leak if you can do it without risk. Prevent entry into waterways, sewers, basements or confined areas. Absorb or cover with dry earth, sand or other non-combustible material and transfer to containers. Do not get water inside containers.

Environmental Physical Data

Henry's Law Constant: calculated at 7.1 $x10^{-9}$
Octanol/Water Partition Coefficient: log K_{ow} = 4.58
Sorption Partition Coefficient: K_{oc} = estimated at 7390
BCF: estimated at 1780

Regulations

RCRA 40CFR: Listed Hazardous Waste No. U093 Toxic Waste
CERCLA: 40CFR 302.4: Listed per RCRA Section 3001 RQ: 10 lb (4.535 kg)
SARA 40CFR 372.65: Listed
SARA EHS 40CFR 355: Not listed
TSCA: Listed

Analytical Methods

Soil: SW846 3640A, 8250A, 8270B, 8270C
Water / Groundwater: EPA 1625
Plasma: EPA 29

DIM4050 CAS #: 23103-98-2

2-(DIMETHYLAMINO)-5,6-DIMETHYL-4-PYRIMIDINYLDIMETHYLCARBAM

RTECS: EZ9100000
EINECS Number: 245-430-1
Molecular Formula: $C_{11}H_{18}N_4O_2$
Formula Weight: 238.33

Synonyms: ABOL; AFICIDA; APHOX; DIMETHYLCARBAMIC ACID 2-(DIMETHYLAMINO)-5,6-DIMETHYL-4-PYRIMIDINYL ESTER; 5,6-DIMETHYL-2-DIMETHYLAMINO-4-PYRIMIDINYLDIMETHYLCARBAMATE; 5,6-DWUMETYLO-2-DWUMETYLOAMINO-4-PIRIMIDYNYLODWUKARBAMINIAN; ENT-27766; FERNOS; PIRIMICARB; PIRIMOR; PIRIMOR 50 DP; PIRIMOR G; PIRIMOR GRANULATE; PP 062; PP062; PRIMICARBE; PYRIMOR; RAPID

Physical Properties

Freezing Point: 90.5 °C (194.9 °F)
Water Solubility: 2.7 g/l 25 °C
Other Solubilities: 4.0 g/l acetone, 2.5 g/l ethanol, 2.9 g/l xylene, 3.3 g/l chloroform

RTECS Toxicity Data

Acute Oral: Rat LD_{50} Dose: 100 mg/kg. Mouse LD_{50} Dose: 68 mg/kg.
Acute Dermal: Rabbit LD_{Lo} Route: Skin; Dose: 900 mg/kg. Rat LD_{50} Route: Skin; Dose: >500 mg/kg.
Mutagenic: Human Cytogenetic Analysis; Cell Type: lymphocyte; Dose: 10 mg/L. Mouse Cytogenetic Analysis; Route: Oral; Dose: 2 mg/kg.

Hazard Overviews

Carcinogenicity: IARC - Not listed; NIOSH - Not listed; NTP - Not listed; ACGIH - Not listed; OSHA - Not listed; EPA - Not listed; MAK - Not listed

Environmental

Regulations

RCRA 40CFR: Not listed
CERCLA: 40CFR 302.4: Not listed
SARA 40CFR 372.65: Not listed
SARA EHS 40CFR 355: Not listed
TSCA: Not listed

Analytical Methods

Soil: EPA PMD-PJB
Food: FDA 232.4, 242.1; AOAC 982.08

DIM4100 CAS #: 108-01-0

2-DIMETHYLAMINOETHANOL

RTECS: KK6125000
EINECS Number: 203-542-8
Molecular Formula: $C_4H_{11}NO$
Structured MF: $(CH_3)_2NCH_2CH_2OH$
Formula Weight: 89.14

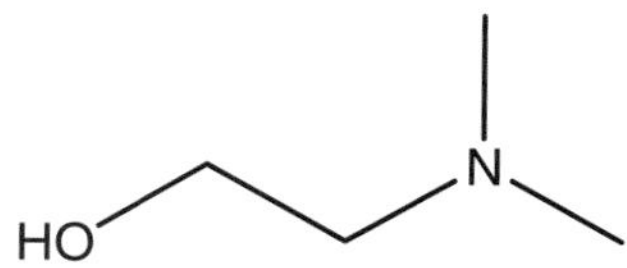

Chemical Structure

Synonyms: AMIETOL M 21; BIMANOL; DEANOL; DIMETHYLAETHANOLAMIN; DIMETHYLAMINOAETHANOL; (DIMETHYLAMINO)ETHANOL; 2-(DIMETHYLAMINO)ETHANOL; 2-(N,N-DIMETHYLAMINO)ETHANOL; BETA-(DIMETHYLAMINO)ETHANOL; BETA-DIMETHYLAMINOETHANOL; DIMETHYLAMINOETHANOL; N,N-DIMETHYL-2-AMINOETHANOL; N,N-DIMETHYLAMINOETHANOL; N-DIMETHYLAMINOETHANOL; BETA-DIMETHYLAMINOETHYL ALCOHOL; DIMETHYLETHANOLAMINE; N,N-DIMETHYLETHANOLAMINE; DIMETHYL(2-HYDROXYETHYL)AMINE; DIMETHYL(HYDROXYETHYL)AMINE; N,N-DIMETHYL(2-HYDROXYETHYL)AMINE; N,N-DIMETHYL-2-HYDROXYETHYLAMINE; N,N-DIMETHYL-N-(2-HYDROXYETHYL)AMINE; N,N-DIMETHYL-N-(BETA-HYDROXYETHYL)AMINE; DIMETHYLMONOETHANOLAMINE; DMAE; 2-DWUMETYLOAMINOETANOLU; ETHANOL,2-DIMETHYLAMINO-; (2-HYDROXYETHYL)DIMETHYLAMINE; BETA-HYDROXYETHYLDIMETHYLAMINE; N-(2-HYDROXYETHYL)DIMETHYLAMINE; KALPUR P; LIPARON; NORCHOLINE; PROPAMINE A; TEXACAT DME
Description: colorless liquid; amine odor
Use: antidepressant (CNS stimulant); intermediate in the synthesis of dyestuffs; textile auxiliaries; pharmaceuticals; corrosion inhibitors; curing epoxy, amine and polyamide resins; emulsifier; paints and coatings

Physical Properties

Boiling Point: 135 °C (275 °F) at 758 mm Hg
Freezing Point: -59 °C (-74.2 °F)
Specific Gravity: 0.8866 at 20 °C/4 °C
Vapor Density: 3.03 Air=1
Saturated Vapor Density: 1.225540399 kg/m^3
Bulk Density: 7.4 lbs/gal at 20 °C
Vapor Pressure: 7.8 mm Hg at 22 °C
Water Solubility: Miscible with Water
Other Solubilities: 95% Ethanol: >=100 mg/ml at 23 °C; Acetone: >=100 mg/ml at 23 °C; Benzene: miscible; DMSO: >=100 mg/ml at 23 °C; Ether: miscible; Many organic solvents: Soluble.
Surface Tension: 27.1 dynes/cm at 24.5 °C
Odor Threshold: Detection 0.015 ppm
pH: 12.1
Refraction Index: 1.43 at 20 °C/D
Evaporation Rate: < 1 Butyl Acetate=1
Critical Temperature: 300 °C
Critical Pressure: 600 psia
Ionization Potential (eV): 8.2
Flash Point: 40.556 °C Open Cup
Autoignition Temperature: 295 °C
LEL: 1.6% v/v
UEL: 11.9% v/v

RTECS Toxicity Data

Acute Oral: Rat LD_{50} Dose: 2 gm/kg.
Acute Inhalation: Rat LC_{50} Dose: 1641 ppm/4hr; Toxic Effects: Sense organs and special senses - Lacrimation; Behavioral - Ataxia; Lungs, Thorax, or Respiration - Dyspnea. Mouse LC_{50} Dose: 3250 mg/m^3; Toxic Effects: Brain and coverings - Recordings from specific areas of CNS; Sense organs and special senses - Conjunctive irritation; Behavioral - Convulsions or effect on seizure threshold.

Acute Dermal: Rabbit LD_{50} Route: Skin; Dose: 1370 uL/kg. Mouse LD_{50} Route: Subcutaneous Dose: 961 mg/kg; Toxic Effects: Behavioral - Tremor; Behavioral - Coma; Lungs, Thorax, or Respiration - Other changes.
Chronic (Multiple Dose) Inhalation: Rat Dose: 670 mg/m^3/4H/22W-I; Toxic Effects: Endocrine - Changes in adrenal weight; Blood - Other changes; Nutritional and gross metabolic - Weight loss or decreased weight gain. Rat Dose: 288 ppm/6H/2W-I; Toxic Effects: Sense organs and special senses - Corneal damage; Behavioral - Food intake (animal); DEATH. Rat Dose: 76 ppm/6H/13W-I; Toxic Effects: Nutritional and gross metabolic - Weight loss or decreased weight gain.
Irritation Eye: Rabbit Standard Draize Test Dose: 750 ug open; Reaction: severe.
Irritation Skin: Rabbit Open Draize Test Dose: 445 mg open; Reaction: mild.

Hazard Overviews

Corrosive

Flammable

Fire Diamond

Health: Corrosive to eyes/skin/respiratory tract. Also Causes: asthmatic symptoms, permanent eye injury.
Fire: Flammable. For small fires use dry chemical, carbon dioxide, water spray, or standard foam. For large fires use water spray, fog, or standard foam. Containers may explode in heat of fire.
Reactivity: Stable. Hazardous polymerization cannot occur. Avoid: heat; ignition sources. Incompatible with: oxidizing agents; strong acids; copper; copper alloys; zinc; galvanized steel; zinc alloys; cellulose nitrate. Hazardous decomposition products: carbon monoxide; nitrogen oxides.
Carcinogenicity: IARC - Not listed; NIOSH - Not listed; NTP - Not listed; ACGIH - Not listed; OSHA - Not listed; EPA - Not listed; MAK - Not listed
Primary Target Organs:

Eyes

Skin

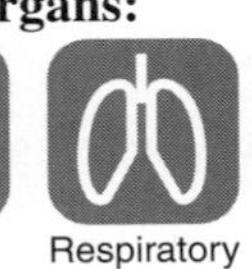
Respiratory System

Mucous Membranes

Environmental

Ecotoxicity: Aquatic toxicity: 10-100 ppm/96 hr/finfish/TLm
Cleanup/Disposal: Guide No. 132: Fully encapsulating, vapor protective clothing should be worn for spills and leaks with no fire. Eliminate all ignition sources (no smoking, flares, sparks or flames in immediate area). All equipment used when handling the product must be grounded. Do not touch or walk through spilled material. Stop leak if you can do it without risk. Prevent entry into waterways, sewers, basements or confined areas. A vapor suppressing foam may be used to reduce vapors. Absorb with earth, sand or other non-combustible material and transfer to containers (except for Hydrazine). Use clean non-sparking tools to collect absorbed material. Large Spills: Dike far ahead of liquid spill for later disposal. Water spray may reduce vapor; but may not prevent ignition in closed spaces.

Regulations

RCRA 40CFR: Not listed
CERCLA: 40CFR 302.4: Not listed
SARA 40CFR 372.65: Not listed
SARA EHS 40CFR 355: Not listed
TSCA: Listed

DIM4200 CAS #: 2867-47-2

N,N-DIMETHYLAMINOETHYL METHACRYLATE

RTECS: OZ4200000
EINECS Number: 220-688-8
Molecular Formula: $C_8H_{15}NO_2$
Formula Weight: 157.22

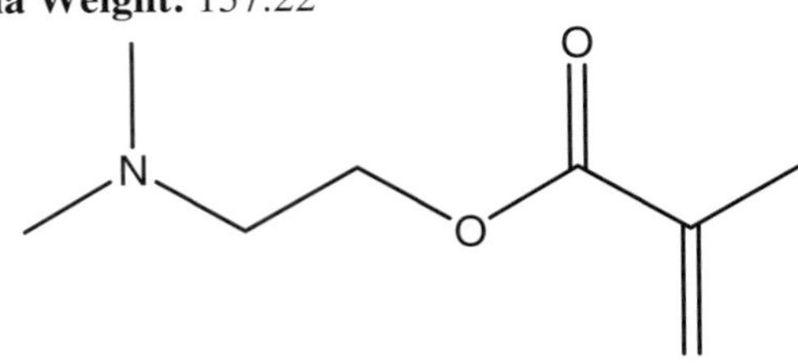

Chemical Structure

Synonyms: AGEFLEX FM-1; (DIMETHYLAMINO)ETHYL METHACRYLATE; 2-(DIMETHYLAMINO)ETHYL METHACRYLATE; 2-(N,N-DIMETHYLAMINO)ETHYL METHACRYLATE; 2-DIMETHYLAMINOETHYL METHACRYLATE; BETA-(N,N-DIMETHYLAMINO)ETHYL METHACRYLATE; BETA-DIMETHYLAMINOETHYL METHACRYLATE; DIMETHYLAMINOETHYL METHACRYLATE; 2-DIMETHYLAMINOETHYLESTER KYSELINY METHAKRYLOVE; N,N-DIMETHYLETHANOLAMINE METHACRYLATE; ETHANOL,2-(DIMETHYLAMINO)-,METHACRYLATE; METHACRYLIC ACID,2-(DIMETHYLAMINO)ETHYL ESTER; 2-PROPENOIC ACID,2-METHYL-,2-(DIMETHYLAMINO)ETHYL ESTER
Description: liquid
Use: in binders for coatings, textile chems, dispersing agents for nonaqueous systems, antistatic agents, stabilizers for chlorinated polymers, ion exchange resins, emulsifying agents, cationic precipitating agents; in dental materials, antithrombogenic prosthetics, experimental in slow-release capsules, in cosmetics; comonomer for for water treatment, for acrylic polymers for architectural paints, for viscosity index improvers, for adhesives, for homopolymer; intermed for acetate & nitrate salts.

Physical Properties

Boiling Point: 182 °C (360 °F) 190 °C (374 °F)
Freezing Point: About -30 °C (-22 °F)
Vapor Density: 5.4 Air=1
Density: 0.933 g/cu cm at 25 °C
Water Solubility: Soluble in Water
Refraction Index: 1.4376 at 25 °C/D

Flash Point: 73.9 °C Tag Open Cup

RTECS Toxicity Data

Acute Oral: Rat LD_{50} Dose: 1751 mg/kg.
Acute Inhalation: Rat LC_{50} Dose: 620 mg/m^3/4hr. Mouse LC_{50} Dose: 1800 mg/m^3/2hr.

Hazard Overviews

Corrosive

Fire Diamond

Health: Corrosive to eyes/skin/respiratory tract. Harmful. Other Acute Effects: harmful if swallowed, inhaled, or absorbed through skin; lachrymator; inhalation may result in spasm, inflammation and edema of the larynx and bronchi, chemical pneumonitis and pulmonary edema; symptoms of exposure may include burning sensation; coughing; wheezing; laryngitis; shortness of breath; headache; nausea; vomiting; prolonged or repeated exposure may cause allergic reactions in certain sensitive individuals.
Fire: Combustible. Hazards: emits toxic fumes; container explosion may occur. Extinguishing agents: water spray; carbon dioxide, dry chemical powder or appropriate foam. Precautions: combustible liquid.
Reactivity: Light sensitive. Incompatible with: strong acids, strong bases, strong oxidizing agents, strong reducing agents, store away from heat and direct sunlight, may decompose on exposure to moist air or water, may undergo autopolymerization, light sensitive. Hazardous decomposition products: thermal decomposition may produce carbon monoxide, carbon dioxide, and nitrogen oxides.
Carcinogenicity: IARC - Not listed; NIOSH - Not listed; NTP - Not listed; ACGIH - Not listed; OSHA - Not listed; EPA - Not listed; MAK - Not listed
Primary Target Organs:

Eyes

Skin

Respiratory System

Environmental

Regulations

RCRA 40CFR: Not listed
CERCLA: 40CFR 302.4: Not listed
SARA 40CFR 372.65: Not listed
SARA EHS 40CFR 355: Not listed
TSCA: Listed

DIM4250 CAS #: 30558-43-1

2-(DIMETHYLAMINO)-N-HYDROXY-2-OXO-ETHANIMIDOTHIOIC ACID

EINECS Number: 250-239-1
Molecular Formula: $C_5H_{10}N_2O_2S$
Formula Weight: 162.23

Hazard Overviews

Carcinogenicity: IARC - Not listed; NIOSH - Not listed; NTP - Not listed; ACGIH - Not listed; OSHA - Not listed; EPA - Not listed; MAK - Not listed

Environmental

Regulations

RCRA 40CFR: Listed Hazardous Waste No. U394 Toxic Waste
CERCLA: 40CFR 302.4: Listed per RCRA Section 3001 RQ: 1 lb (0.454 kg)
SARA 40CFR 372.65: Not listed
SARA EHS 40CFR 355: Not listed
TSCA: Listed

DIM4300 CAS #: 51-82-1

DIMETHYLAMINOMETHYL N,N-DIMETHYLDITHIOCARBAMATE

RTECS: FA6000000
EINECS Number: 200-126-8
Molecular Formula: $C_6H_{14}N_2S_2$
Formula Weight: 178.34
Synonyms: CARBAMIC ACID,DIMETHYLDITHIO-,(DIMETHYLAMINO)METHYLESTER; CARBAMIC ACID,DITHIO-,N,N-DIMETHYL-,DIMETHYLAMINOMETHYLESTER; CARBAMODITHIOIC ACID,DIMETHYL-,(DIMETHYLAMINO)METHYLESTER; N,N-DIMETHYLDITHIOCARBAMIC ACID DIMETHYLAMINOMETHYL ESTER; N,N-DIMETHYL-DITHIOCARBAMINSAEURE-DIMETHYLAMINOMETHYL-ESTER

Hazard Overviews

Carcinogenicity: IARC - Not listed; NIOSH - Not listed; NTP - Not listed; ACGIH - Not listed; OSHA - Not listed; EPA - Not listed; MAK - Not listed

Environmental

Regulations

RCRA 40CFR: Not listed
CERCLA: 40CFR 302.4: Not listed
SARA 40CFR 372.65: Not listed
SARA EHS 40CFR 355: Not listed
TSCA: Listed

DIM4350 CAS #: 1738-25-6

3-(DIMETHYLAMINO)-PROPIONITRILE

RTECS: UG1575000
EINECS Number: 217-090-4
Molecular Formula: $C_5H_{10}N_2$

Structured MF: $(CH_3)_2NCH_2CH_2CN$
Formula Weight: 98.17

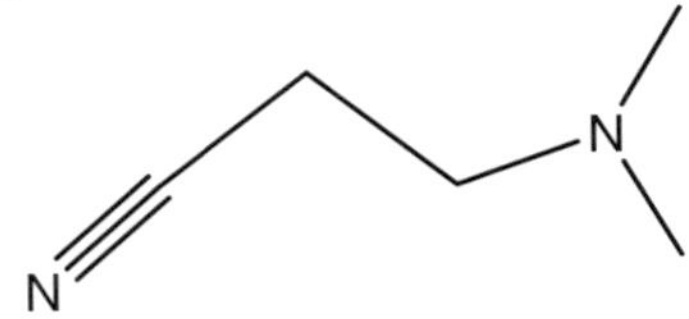

Chemical Structure

Synonyms: 3-DIMETHYLAMINOPROPANNITRIL; 3-(DIMETHYLAMINO)PROPIONITRILE; 3-(N,N-DIMETHYLAMINO)PROPIONITRILE; BETA-(DIMETHYLAMINO)PROPIONITRILE; BETA-DIMETHYLAMINOPROPIONITRILE; BETA-N-DIMETHYLAMINOPROPIONITRILE; DIMETHYLAMINOPROPIONITRILE; N,N-DIMETHYLAMINO-3-PROPIONITRILE; DMAPN; PROPANENITRILE,3-(DIMETHYLAMINO)-; PROPIONITRILE,3-(DIMETHYLAMINO)-
Description: colorless, mobile fluid
Use: catalyst in polyurethane foam manufacture

Physical Properties

Boiling Point: 172 °C (342 °F)
Freezing Point: -44.2 °C (-47.56 °F)
Specific Gravity: 0.8617 at 30 °C
Vapor Density: 3.4 Air=1
Vapor Pressure: 10 mm Hg at 57 °C
Water Solubility: Miscible with Water
Other Solubilities: miscible with Alcohol, and other solvents
Flash Point: 63.889 °C Closed Cup

RTECS Toxicity Data

Acute Oral: Rat LD_{50} Dose: 2600 mg/kg. Mouse LD_{50} Dose: 1500 mg/kg.
Acute Dermal: Rabbit LD_{50} Route: Skin; Dose: 1410 mg/kg.
Irritation Eye: Rabbit Standard Draize Test Dose: 20 mg/24H; Reaction: moderate.
Irritation Skin: Rabbit Standard Draize Test Dose: 500 mg/24H; Reaction: mild.

Hazard Overviews

Corrosive

Health: Corrosive to eyes/skin/respiratory tract. Toxic. Other Acute Effects: may be fatal if inhaled, swallowed, or absorbed through skin; inhalation may result in spasm, inflammation and edema of the larynx and bronchi, chemical pneumonitis and pulmonary edema; symptoms of exposure may include burning sensation; coughing; wheezing; laryngitis; shortness of breath; headache; nausea; vomiting; may cause nervous system disturbances. Chronic Effects: may cause reproductive disorders; target organs: central nervous system, bladder, male reproductive system.

Fire: Combustible. Hazards: emits toxic fumes. Extinguishing agents: carbon dioxide, dry chemical powder or appropriate foam; water spray. Precautions: combustible liquid.
Reactivity: Incompatible with: acids, acid chlorides, acid anhydrides, strong oxidizing agents, carbon dioxide. Hazardous decomposition products: toxic fumes of: carbon monoxide, carbon dioxide, nitrogen oxides, hydrogen cyanide.
Carcinogenicity: IARC - Not listed; NIOSH - Listed as carcinogen; NTP - Not listed; ACGIH - Not listed; OSHA - Not listed; EPA - Not listed; MAK - Not listed
Primary Target Organs:

Eyes

Skin

Respiratory System

Nervous System
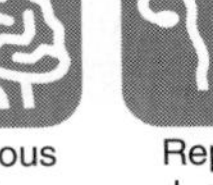
Repro-ductive

Environmental

Regulations

RCRA 40CFR: Not listed
CERCLA: 40CFR 302.4: Not listed
SARA 40CFR 372.65: Not listed
SARA EHS 40CFR 355: Not listed
TSCA: Listed

DIM4400 **CAS #: 1628-58-6**

2-P-(DIMETHYLAMINOSTYRYL)-BENZOTHIAZOLE

RTECS: DL4200000
EINECS Number: 216-622-2
Molecular Formula: $C_{17}H_{16}N_2S$
Formula Weight: 280.41

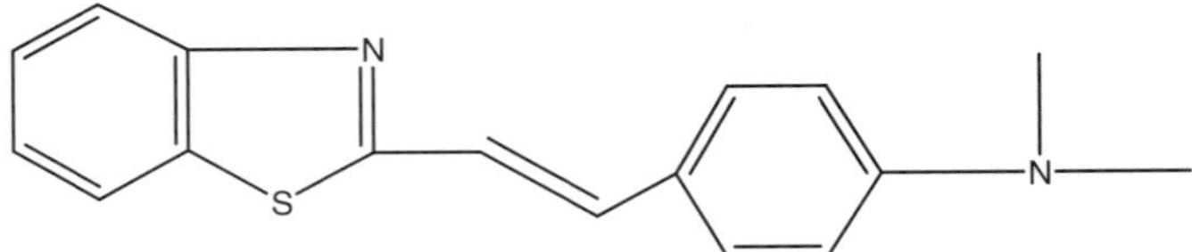

Chemical Structure

Synonyms: BENZENAMINE,4-(2-(2-BENZOTHIAZOLYL)ETHENYL)-N,N-DIMETHYL-; BENZOTHIAZOLE,2-(P-(DIMETHYLAMINO)STYRYL)-; 2-(4-DIMETHYLAMINOSTYRYL)BENZOTHIAZOLE; 2-(P-(DIMETHYLAMINO)STYRYL)BENZOTHIAZOLE
Use: no evidence of commercial use in us

RTECS Toxicity Data

Tumorigenic: Rat Route: Oral; Dose: 35 gm/kg/1Y-I; Toxic Effects: Tumorigenic - Neoplastic by RTECS criteria; Skin and appendages - Tumors; Tumorigenic effects - Testicular tumors.

Hazard Overviews

Carcinogenicity: IARC - Not listed; NIOSH - Not listed; NTP - Not listed; ACGIH - Not listed; OSHA - Not listed; EPA - Not listed; MAK - Not listed

Environmental

Regulations

RCRA 40CFR: Not listed
CERCLA: 40CFR 302.4: Not listed
SARA 40CFR 372.65: Not listed
SARA EHS 40CFR 355: Not listed
TSCA: Listed

DIM4450 CAS #: 897-55-2

4-(4-DIMETHYLAMINOSTYRYL)QUIN-OLINE

RTECS: VB7000000
Molecular Formula: $C_{19}H_{18}N_2$
Formula Weight: 274.39
Synonyms: 4M20; BENZENAMINE,N,N-DIMETHYL-4-(2-(4-QUINOLINYL)ETHENYL)-; 2-(4-N,N-DIMETHYLAMINOSTYRYL)QUINOLINE; 4-(4-DIMETHYLAMINOSTYRYL)QUINOLINE; 4-(P-(DIMETHYLAMINO)STYRYL)QUINOLINE; 4-(P-DIMETHYLAMINOSTYRYL)QUINOLINE; NSC-10482; NSC-4236; QUINOLINE,4-(P-(DIMETHYLAMINO)STYRYL)-
Use: radiomimetic drug

RTECS Toxicity Data

Tumorigenic: Mouse Route: Intravenous; Dose: 100 mg/kg; Toxic Effects: Tumorigenic - Neoplastic by RTECS criteria; Liver - Tumors; Tumorigenic effects - Ovarian tumors.
Carcinogenicity: IARC - Not listed; NIOSH - Not listed; NTP - Not listed; ACGIH - Not listed; OSHA - Not listed; EPA - Not listed; MAK - Not listed

Environmental

Regulations

RCRA 40CFR: Not listed
CERCLA: 40CFR 302.4: Not listed
SARA 40CFR 372.65: Not listed
SARA EHS 40CFR 355: Not listed
TSCA: Listed

DIM4500 CAS #: 121-69-7

N,N-DIMETHYLANILINE

RTECS: BX4725000
DOT: UN2253; IMO6.1
EINECS Number: 204-493-5
Molecular Formula: $C_8H_{11}N$
Formula Weight: 121.18

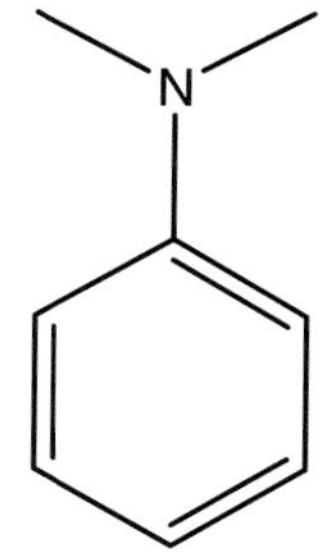

Chemical Structure

Synonyms: ANILINE,N,N-DIMETHYL-; BENZENAMINE,N,N-DIMETHYL-; (DIMETHYLAMINO)BENZENE; DIMETHYLANILINE; N,N-DIMETHYLBENZENAMINE; N,N-DIMETHYLBENZENEAMINE; DIMETHYLPHENYLAMINE; N,N-DIMETHYLPHENYLAMINE; (DIMETYLAMINO)BENZENE; DWUMETYLOANILINA; VERSNELLER NL 63/10
Description: clear, light yellow to brown liquid; fish-like or amine odor
Use: in the manufacture of basic dyes, vanillin and Michler's ketone; as reagent for methanol, methyl furfural, H2O2, nitrate, alcohol and formaldehyde; as a stabilizer and with MBTH in a colorimetric peroxidase determination; reagent in a sensitive procedure using p-anisidene-N,N-dimethylaniline for the catalytic determination of microamounts of ferric and ferrous ions in as little as 10 E-7 M

Physical Properties

Boiling Point: 192 °C (378 °F) to 194 °C (381 °F)
Freezing Point: 2.45 °C (36.41 °F)
Specific Gravity: 0.956 at 20 °C/4 °C
Vapor Density: 4.17 Air=1
Vapor Pressure: 1 mm Hg at 29.5 °C
Water Solubility: 2% by weight
Other Solubilities: 95% Ethanol: >=100 mg/ml at 20 °C; Acetone: >=100 mg/ml at 20 °C; Benzene: Soluble; Chloroform: Soluble; DMSO: >=100 mg/ml at 20 °C; Ether: Soluble.
Refraction Index: 1.5582 at 20 °C/D
Ionization Potential (eV): 7.14
Flash Point: 61 °C
Autoignition Temperature: 371 °C

RTECS Toxicity Data

Acute Oral: Human LD_{Lo} Dose: 50 mg/kg; Toxic Effects: Gastrointestinal - Nausea or vomiting; Gastrointestinal - Other changes. Rat LD_{50} Dose: 1410 mg/kg.
Acute Inhalation: Rat LC_{Lo} Dose: 250 mg/m^3/4hr; Toxic Effects: Behavioral - Somnolence (general depressed activity); Behavioral - Excitement.
Acute Dermal: Rabbit LD_{50} Route: Skin; Dose: 1770 mg/kg. Rat LD_{Lo} Route: Subcutaneous Dose: 100 mg/kg.
Chronic (Multiple Dose) Oral: Rat Dose: 32500 mg/kg/13W-I; Toxic Effects: Kidney, Ureter, and Bladder - Other changes; Blood - Changes in spleen; Nutritional and gross metabolic - Weight loss or decreased weight gain. Rat Dose: 16250 mg/kg/13W-I; Toxic Effects: Liver - Other changes;

metabolic - Weight loss or decreased weight gain. Rat Dose: 16250 mg/kg/13W-I; Toxic Effects: Liver - Other changes; Blood - Changes in bone marrow not included above; Blood - Changes in spleen.

Chronic (Multiple Dose) Inhalation: Rat Dose: 10700 ug/m^3/5H/17W-I; Toxic Effects: Blood - Pigmented or nucleated red Blood cells; Blood - Methemoglobinemia-Carboxyhemoglobin; Biochemical - Catalases. Rat Dose: 300 ug/m^3/24H/14W-C; Toxic Effects: Behavioral - Muscle contraction or spasticity; Blood - Methemoglobinemia-Carboxyhemoglobin; Blood - Changes in erythrocite (RBC) cell count.

Irritation Eye: Rabbit Standard Draize Test Dose: 20 mg/24H; Reaction: moderate.

Irritation Skin: Rabbit Standard Draize Test Dose: 500 mg/24H; Reaction: mild. Rabbit Open Draize Test Dose: 10 mg/24H open; Reaction: mild.

Mutagenic: Rat DNA Damage; Route: Intraperitoneal; Dose: 485 mg/kg. Mouse DNA Damage; Route: Intraperitoneal; Dose: 485 mg/kg.

Tumorigenic: Rat Route: Oral; Dose: 15450 mg/kg/2Y-C; Toxic Effects: Tumorigenic - Equivocal tumorigenic agent by RTECS criteria; Endocrine - Tumors. Mouse Route: Oral; Dose: 15450 mg/kg/2Y-C; Toxic Effects: Tumorigenic - Equivocal tumorigenic agent by RTECS criteria; Endocrine - Tumors.

Hazard Overviews

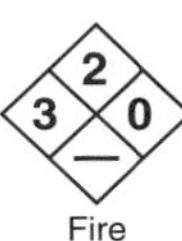

Fire Diamond

Health: Irritating to eyes/skin. Also Causes: CNS depression, methemoglobinemia, drop in blood pressure, tachycardia.

Fire: Combustible. Use water fog, dry chemical, carbon dioxide, or foams. Use water sprays to cool fire-exposed containers. Never direct solid streams of water into pools of burning liquid since this tends to scatter and spread the flames.

Reactivity: Stable. Hazardous polymerization cannot occur. Avoid: heat; ignition sources. Incompatible with: strong acids; strong oxidizing agents; benzoyl peroxide. Hazardous decomposition products: toxic carbon monoxide; toxic aniline vapors.

Carcinogenicity: IARC - Group 3, Not classifiable as to carcinogenicity to humans; NIOSH - Not listed; NTP - Not listed; ACGIH - Class A4, Not classifiable as a human carcinogen; OSHA - Not listed; EPA - Not listed; MAK - Class B, Justifiably suspected of having carcinogenic potential

Primary Target Organs:

Eyes

Skin

Nervous System

Cardio-vascular

Blood

Exposure Limits

OSHA PEL: TWA: 5 ppm; 25 mg/m^3; skin.

OSHA PEL Vacated 1989 Limits: TWA: 5 ppm; 25 mg/m^3; STEL: 10 ppm; 50 mg/m^3.

ACGIH TLV: TWA: 5 ppm; 25 mg/m^3; STEL: 10 ppm; 50 mg/m^3.

NIOSH IDLH: 100 ppm.

DFG MAK: TWA: 5 ppm; 25 mg/m^3.

Respirator Recommendation

Exposure Range: >5 to 50 ppm Air Purifying, Negative Pressure, Half Mask

Exposure Range: >50 to <100 ppm Air Purifying, Negative Pressure, Full Face

Exposure Range: 100 to unlimited ppm Self-contained Breathing Apparatus, Pressure Demand, Full Face

Cartridge Color: black

Environmental

Ecotoxicity: LC_{50} Pimephales promelas (fathead minnow) 52.6 g/l/96 hr (confidence limit is not relevant), flow-through bioassay with measured concentrations, 25.3 °C, dissolved oxygen 5.9 mg/l, hardness 43.5 mg/l calcium carbonate, alkalinity 46.0 mg/l calcium carbonate

Environmental Fate: If released to the atmosphere, it will degrade rapidly by reaction with hydroxyl radicals (estimated half-life of 2.6 hr) and ozone (estimated half-life of 29.4 hr). If released to soil, it may be susceptible to moderate to high leaching based on an estimated K_{oc} range of 80 to 430. However, a 241-day field study demonstrated that the component of waste sludge did not leach significantly in soil. If released to water, reaction with sunlight produced singlet oxygen, which has an estimated half-life of 9.6 hr, may be the major degradation process. Direct photolysis may also contributed to aquatic degradation. The results of various screening studies suggest that it is not readily biodegradable in unacclimated media; however acclimated inocula are apparently capable of degrading the compound.

Cleanup/Disposal: Guide No. 153: Eliminate all ignition sources (no smoking, flares, sparks or flames in immediate area). Do not touch damaged containers or spilled material unless wearing appropriate protective clothing. Stop leak if you can do it without risk. Prevent entry into waterways, sewers, basements or confined areas. Absorb or cover with dry earth, sand or other non-combustible material and transfer to containers. Do not get water inside containers.

Environmental Physical Data

Octanol/Water Partition Coefficient: log K_{ow} = 2.31

Sorption Partition Coefficient: K_{oc} = 80 to 430

BCF: not significant

Regulations

RCRA 40CFR: Not listed

CERCLA: 40CFR 302.4: Listed per CAA Section 112 RQ: 100 lb (45.35 kg)

SARA 40CFR 372.65: Listed

SARA EHS 40CFR 355: Not listed

TSCA: Listed

Analytical Methods

Air: ASTM D4490

Indoor / Expired Air: NIOSH 2002

Other: EPA 1665

DIM4550 **CAS #: 781-43-1**

9,10-DIMETHYLANTHRACENE

RTECS: CA9685000
EINECS Number: 212-308-4
Molecular Formula: $C_{16}H_{14}$
Formula Weight: 206.29

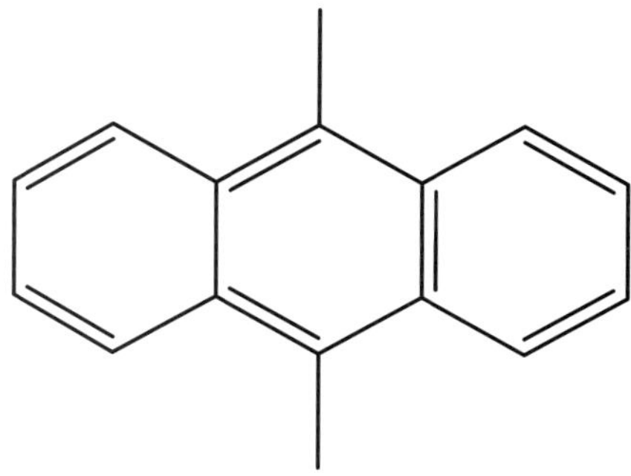

Chemical Structure

Synonyms: ANTHRACENE,9,10-DIMETHYL-

Physical Properties

Boiling Point: 360 °C (680 °F)
Freezing Point: 182 °C (359.6 °F) to 184 °C (363.2 °F)

RTECS Toxicity Data

Mutagenic: Human Sister Chromatid Exchenge; Cell Type: lymphocyte; Dose: 100 umol/L. Hamster Morphological Transformation; Cell Type: kidney; Dose: 80 ug/L.
Tumorigenic: Mouse Route: Skin; Dose: 40 mg/kg/20D-I; Toxic Effects: Tumorigenic - Carcinogenic by RTECS criteria; Skin and appendages - Tumors. Mouse Route: Skin; Dose: 1100 mg/kg/46W-I; Toxic Effects: Tumorigenic - Equivocal tumorigenic agent by RTECS criteria; Lungs, Thorax, or Respiration - Tumors; Skin and appendages - Tumors.

Hazard Overviews

Health: Irritating. Harmful. Other Acute Effects: may be harmful by inhalation, ingestion, or skin absorption. Chronic Effects: laboratory experiments have shown mutagenic effect.
Fire: Hazards: emits toxic fumes. Extinguishing agents: water spray; carbon dioxide, dry chemical powder or appropriate foam. Precautions: combustible liquid.
Reactivity: Incompatible with: strong oxidizing agents. Hazardous decomposition products: toxic fumes of: carbon monoxide, carbon dioxide.
Carcinogenicity: IARC - Not listed; NIOSH - Not listed; NTP - Not listed; ACGIH - Not listed; OSHA - Not listed; EPA - Not listed; MAK - Not listed

Primary Target Organs:

Eyes

Skin

Respiratory System

Environmental

Environmental Fate: If released to soil, it is expected to be immobile and it may photolyze on soil surfaces. Volatilization and hydrolysis are not expected to be environmentally relevant fate processes in either soil and water. Insufficient data are available to predict the significance of biodegradation. If released to water, it may adsorb strongly to suspended solids and sediments or it may photolyze in near surface waters (estimated midday, midsummer half-life 0.35 hours). It is not expected to bioaccumulate significantly in fish with microsomal oxidase, but it may bioaccumulate significantly in other aquatic organisms. If released to the atmosphere, it is expected to exist in both vapor and particulate form. This compound may react with photochemically generated hydroxyl radicals (estimated vapor phase half-life 2 hours), it may undergo direct photolysis, or it may be subject to dry deposition. Adsorption onto particulate matter may retard direct photolysis and reaction with hydroxyl radicals.

Environmental Physical Data

Henry's Law Constant: estimated at 8.5×10^{-4}
Octanol/Water Partition Coefficient: log K_{ow} = 5.79
Sorption Partition Coefficient: K_{oc} = 3.8×10^{5} to 4.03×10^{5}
BCF: fish 100 to 1000

Regulations

RCRA 40CFR: Not listed
CERCLA: 40CFR 302.4: Not listed
SARA 40CFR 372.65: Not listed
SARA EHS 40CFR 355: Not listed
TSCA: Not listed

DIM4600 **CAS #: 75-60-5**

DIMETHYLARSENIC ACID

RTECS: CH7525000
DOT: UN1572; IMO6.1
EINECS Number: 200-883-4
Molecular Formula: $C_2H_7AsO_2$
Structured MF: $(CH_3)_2AsOOH$
Formula Weight: 138.01

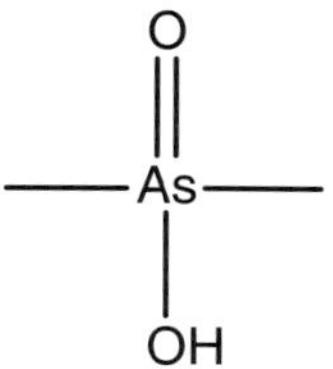

Chemical Structure

Synonyms: ACIDE CACODYLIQUE; ACIDE DIMETHYLARSENIQUE; ACIDE DIMETHYLARSINIQUE; AGENT BLUE; ALKARGEN; ANSAR; ANSAR 138; ARSAN; ARSINE OXIDE,HYDROXYDIMETHYL-; ARSINIC ACID,DIMETHYL-(9CI); BOLLS-EYE; CACODYLIC ACID; CHEXMATE; COTTON AIDE HC; DILIC; DIMETHYLARSINIC ACID; DMAA; ERASE; HYDROXYDIMETHYL ARSINE OXIDE; HYDROXYDIMETHYLARSINE OXIDE; KYSELINA KAKODYLOVA; MONCIDE; MONTAR; PHYTAR; PHYTAR 138; PHYTAR 560; PHYTAR 600; RAD-E-CAT 25; RAD-E-CATE; RAD-E-CATE 25; SALVO; SILVISAR; SILVISAR 510; SYLVICOR

Description: colorless to white triclinic crystals, water solutions may be dyed blue; odorless

Use: chemical synthesis, herbicide, defoliant

Physical Properties

Boiling Point: > 200 °C (392 °F) at 1 atm
Freezing Point: 195 °C (383 °F) to 196 °C (384.8 °F)
Specific Gravity: Estimated > 1.1 at 20 °C/4 °C
Vapor Density: 7.2 Air=1
Saturated Vapor Density: 4.879829401 kg/m^3
Density: > 1.1 g/mL
Vapor Pressure: 620 mm Hg at 20 °C
Water Solubility: 1 parts in 0.5 parts Water
Other Solubilities: 95% Ethanol: Very Soluble; Ether: Insoluble.
Flash Point: Nonflammable

RTECS Toxicity Data

Acute Oral: Rat LD_{50} Dose: 644 mg/kg.

Acute Inhalation: Rat LC_{Lo} Dose: >2600 mg/m^3/2hr.

Chronic (Multiple Dose) Oral: Rat Dose: 2260 mg/kg/4W-I; Toxic Effects: Kidney, Ureter, and Bladder - Chgs in tubules (inc acute renal failure, acute tubular necrosis; Nutritional and gross metabolic - Weight loss or decreased weight gain; DEATH.

Reproductive/Teratogenic: Rat Route: Oral; Dose: 400 mg/kg; Duration: female 7-16D of pregnancy; Effects on Embryo or Fetus - Fetotoxicity; Specific Developmental Abnormalities - Musculoskeletal system. Rat Route: Oral; Dose: 300 mg/kg; Duration: female 7-16D of pregnancy; Specific Developmental Abnormalities - Craniofacial (including nose and tongue). Rat Route: Oral; Dose: 500 mg/kg; Duration: female 7-16D of pregnancy; Effects on Embryo or Fetus - Fetal death. Rat Route: Oral; Dose: 400 mg/kg; Duration: female 6-15D of pregnancy; Effects on Embryo or Fetus - Fetotoxicity.

Mutagenic: Mouse Mutations in Mammalian Somatic Cells; Cell Type: lymphocyte; Dose: 1080 mg/L. Mouse Micronucleus Test; Route: Intraperitoneal; Dose: 7900 mg/kg/24H.

Tumorigenic: Mouse Route: Subcutaneous; Dose: 464 mg/kg; Toxic Effects: Tumorigenic - Equivocal tumorigenic agent by RTECS criteria; Lungs, Thorax, or Respiration - Tumors; Blood - Tumors.

Hazard Overviews

Health: Irritating to eyes/skin/respiratory tract. Toxic. Other Acute Effects: toxic by inhalation and if swallowed; may be harmful if absorbed through the skin; burning and dryness of the oral and nasal cavities; metallic taste; drowsiness; loss of appetite; tremors; convulsions; respiratory arrest; garlic odor to breath; perspiration; muscle spasms; irritation of the gastrointestinal tract; nausea; vomiting; diarrhea; shock and death; target organs: kidneys, g.i. system. Chronic Effects: alteration of genetic material; heritable genetic damage; possible risk of congenital malformation in the fetus; target organs: kidneys, g.i. system, heart, brain, skin, bone marrow, nerves, liver. Known human carcinogen. Possibly causes reproductive effects in humans.

Fire: Noncombustible. Hazards: emits toxic fumes. Extinguishing agents: water spray; carbon dioxide, dry chemical powder or appropriate foam. Precautions: combustible liquid.

Reactivity: Stable. Hazardous polymerization will not occur. Incompatible with: strong oxidizing agents, strong bases. Hazardous decomposition products: toxic fumes of: carbon monoxide, carbon dioxide, arsenic oxides.

Carcinogenicity: IARC - Not listed; NIOSH - Not listed; NTP - Not listed; ACGIH - Not listed; OSHA - Not listed; EPA - Class D, Not classifiable as to human carcinogenicity; MAK - Not listed

Primary Target Organs:

Eyes

Skin

Respiratory System

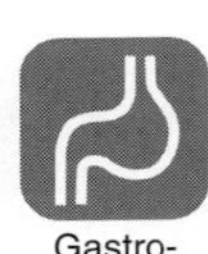
Gastro-intestinal

Kidneys

Exposure Limits
OSHA PEL: TWA: 0.5 mg/m^3; as As.
ACGIH TLV: TWA: 0.01 mg/m^3; as As.
NIOSH REL: STEL: 0.002 mg/m^3; Ceiling (15 min) as As.

Environmental

Ecotoxicity: Aquatic toxicity: 100 ppm/96 hr/scud/not toxic

Environmental Fate: If released to the atmosphere, it will exist in the particulate phase where it can be physically removed by wet and dry deposition. If released to soil and water, it will degrade primarily through biodegradation. The rate of microbial degradation will probably depend upon microbial adaptation. Adsorption in soil and water has been found to depend on clay content, iron oxide content, and pH; adsorption increases with increasing clay, iron oxide content and pH. Adsorption to sediment may be an important environmental sink in the aquatic environment.

Cleanup/Disposal: Guide No. 151: Do not touch damaged containers or spilled material unless wearing appropriate protective clothing. Stop leak if you can do it without risk. Prevent entry into waterways, sewers, basements or confined areas. Cover with plastic sheet to prevent spreading. Absorb or cover with dry earth, sand or other non-combustible material and transfer to containers. Do not get water inside containers.

Environmental Physical Data

BCF: mosquito fish 21

Regulations

RCRA 40CFR: Listed Hazardous Waste No. U136 Toxic Waste
CERCLA: 40CFR 302.4: Listed per RCRA Section 3001 RQ: 1 lb (0.454 kg)
SARA 40CFR 372.65: Listed as Compound
SARA EHS 40CFR 355: Not listed
TSCA: Listed

Analytical Methods

Indoor / Expired Air: NIOSH 5022

DIM4650 CAS #: 57-97-6

7,12-DIMETHYLBENZ(A)ANTHRACENE

RTECS: CW3850000
EINECS Number: 200-359-5
Molecular Formula: $C_{20}H_{16}$
Formula Weight: 256.33

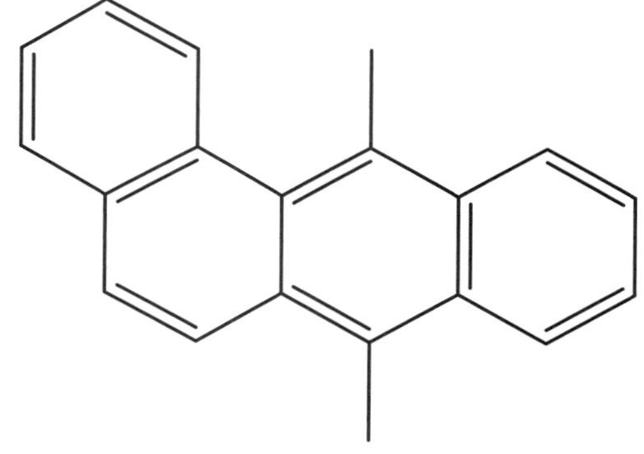

Chemical Structure

Synonyms: BENZ(A)ANTHRACENE,7,12-DIMETHYL-; DBA; 7,12-DIMETHYL-1,2-BENZANTHRACENE; 7,12-DIMETHYLBENZANTHRACENE; 9,10-DIMETHYL-1,2-BENZANTHRACENE; 9,10-DIMETHYL-BENZANTHRACENE; 9,10-DIMETHYLBENZ(A)ANTHRACENE; DIMETHYLBENZ(A)ANTHRACENE; DIMETHYLBENZANTHRACENE; 7,12-DIMETHYLBENZANTHRANCENE; 9,10-DIMETHYL-1,2-BENZANTHRAZEN; DIMETHYLBENZANTHRENE; 7,12-DIMETHYLBENZO(A)ANTHRACENE; 1,4-DIMETHYL-2,3-BENZPHENANTHRENE; 7,12-DMBA; DMBA
Description: faint greenish-yellow tinge plates or leaflets; odorless
Use: as a research chemical for testing antineoplastic drugs by inducing malignant tumors

Physical Properties

Freezing Point: 122 °C (251.6 °F) to 123 °C (253.4 °F)
Water Solubility: Insoluble in Water
Other Solubilities: 95% Ethanol: <1 mg/ml at 18 °C; Acetone: 10-50 mg/ml at 18 °C; Acid: Soluble; Alcohol: Slightly Soluble; Benzene: Soluble; Carbon Disulfide: Soluble; DMSO: 10-50 mg/ml at 18 °C; Toluene: Soluble.
Ionization Potential (eV): 7.10 +/-1.0
Flash Point: 86.111 °C Tag Closed Cup

RTECS Toxicity Data

Acute Oral: Rat LD_{50} Dose: 327 mg/kg. Mouse LD_{50} Dose: 340 mg/kg.
Acute Dermal: Guinea Pig LD_{Lo} Route: Subcutaneous Dose: 20 mg/kg.
Irritation Skin: Mouse Standard Draize Test Dose: 64 ug; Reaction: mild.
Reproductive/Teratogenic: Rat Route: Oral; Dose: 672 mg/kg; Duration: female 15D prior to mating Effects on Newborn - Live birth index; Germ cell effects. Rat Route: Intravenous; Dose: 20 mg/kg; Duration: female 9D of pregnancy; Specific Developmental Abnormalities - Central nervous system; Eye, ear; Craniofacial (including nose and tongue). Rat Route: Intravenous; Dose: 20 mg/kg; Duration: female 13D of pregnancy; Specific Developmental Abnormalities - Musculoskeletal system. Rat Route: Intravenous; Dose: 20 mg/kg; Duration: female 5D of pregnancy; Effects on Embryo or Fetus - Fetal death. Rat Route: Intravenous; Dose: 25 mg/kg; Duration: male 1D prior to mating; Effects on Embryo or Fetus - Fetal death. Rat Route: Intravenous; Dose: 31 mg/kg; Duration: male 1D prior to mating; Paternal Effects - Spermatogenesis; Testes, epididymis, sperm duct. Rat Route: Intravenous; Dose: 10 mg/kg; Duration: male 1D prior to mating; Paternal Effects - Testes, epididymis, sperm duct.
Mutagenic: Human DNA Damage; Cell Type: lymphocyte; Dose: 500 nmol/L. Human Unscheduled DNA Synthesis; Cell Type: other cell types; Dose: 10 mg/L. Human Unscheduled DNA Synthesis; Cell Type: HeLa cell; Dose: 100 nmol/L. Human Unscheduled DNA Synthesis; Cell Type: other cell types; Dose: 1 mg/L. Human Unscheduled DNA Synthesis; Cell Type: lymphocyte; Dose: 5 umol/L. Human Unscheduled DNA Synthesis; Cell Type: liver; Dose: 1 mmol/L. Human DNA Inhibition; Cell Type: HeLa cell; Dose: 1 umol/L. Human DNA Inhibition; Cell Type: fibroblast; Dose: 1 mg/L. Human DNA Adduct; Cell Type: embryo; Dose: 200 nmol/L. Human DNA Adduct; Cell Type: HeLa cell; Dose: 5 umol/L. Human Mutations in Mammalian Somatic Cells; Cell Type: other cell types; Dose: 100 nmol/L. Human Mutations in Mammalian Somatic Cells; Cell Type: fibroblast; Dose: 12500 ug/L. Human Cytogenetic Analysis; Cell Type: lymphocyte; Dose: 10 umol/L. Human Cytogenetic Analysis; Cell Type: leukocyte; Dose: 50 mmol/L. Human Sister Chromatid Exchenge; Cell Type: lymphocyte; Dose: 100 nmol/L. Human Sister Chromatid Exchenge; Cell Type: liver; Dose: 1 mmol/L. Human Micronucleus Test; Cell Type: lymphocyte; Dose: 50 umol/L. Human Morphological Transformation; Cell Type: mammary gland; Dose: 500 ug/L.
Tumorigenic: Rat Route: Oral; Dose: 37500 ug/kg (14-20D preg); Toxic Effects: Tumorigenic - Carcinogenic by RTECS criteria; Brain and coverings - Tumors; Tumorigenic effects - Transplacental tumorigenesis. Rat Route: Oral; Dose: 15 mg/kg; Toxic Effects: Tumorigenic - Carcinogenic by RTECS criteria; Skin and appendages - Tumors. Rat Route:

Oral; Dose: 150 mg/kg; Toxic Effects: Tumorigenic - Carcinogenic by RTECS criteria; Skin and appendages - Tumors. Rat Route: Oral; Dose: 10 mg/kg; Toxic Effects: Tumorigenic - Carcinogenic by RTECS criteria; Skin and appendages - Tumors. Rat Route: Oral; Dose: 37500 ug/kg; Toxic Effects: Tumorigenic - Carcinogenic by RTECS criteria; Skin and appendages - Tumors. Monkey Route: Skin; Dose: 1600 mg/kg/65W-I; Toxic Effects: Tumorigenic - Carcinogenic by RTECS criteria; Skin and appendages - Tumors.

Hazard Overviews

Health: May cause irritation. Toxic. Other Acute Effects: harmful if swallowed, inhaled, or absorbed through skin. Chronic Effects: carcinogen; may alter genetic material; may cause heritable genetic damage.

Fire: Combustible. Extinguishing agents: water spray; carbon dioxide, dry chemical powder or appropriate foam. Precautions: combustible liquid.

Reactivity: Incompatible with: strong oxidizing agents. Hazardous decomposition products: toxic fumes of: carbon monoxide, carbon dioxide.

Carcinogenicity: IARC - Not listed; NIOSH - Not listed; NTP - Not listed; ACGIH - Not listed; OSHA - Not listed; EPA - Not listed; MAK - Not listed

Environmental

Environmental Fate: If released into air or water it will be partially associated with particulate matter. In the atmosphere, it will be subject to photolysis and photooxidation although no rates could be found in the literature. The particulate-bound compound may be transported moderate distances before settling out. In the water, it will sorb strongly to sediment and the unadsorbed in surface layers of water will photodegrade. Its slow desorption from sediment and particulate matter may maintain a low concentration of the chemical for long periods of time. Based upon physical properties, it should bioconcentrate in fish. In soil, the strongly adsorbed material will remain in the upper few centimeters of soil. Its fate there is unknown.

Cleanup/Disposal: Guide No. 171: Do not touch or walk through spilled material. Stop leak if you can do it without risk. Prevent dust cloud. Avoid inhalation of asbestos dust. Small Dry Spills: With clean shovel place material into clean, dry container and cover loosely; move containers from spill area. Small Spills: Take up with sand or other noncombustible absorbent material and place into containers for later disposal. Large Spills: Dike far ahead of liquid spill for later disposal. Cover powder spill with plastic sheet or tarp to minimize spreading. Prevent entry into waterways, sewers, basements or confined areas.

Environmental Physical Data

Octanol/Water Partition Coefficient: log K_{ow} = 5.65
Sorption Partition Coefficient: K_{oc} = sediments 2.357 x10^5
BCF: clams 3.13

Regulations

RCRA 40CFR: Listed Hazardous Waste No. U094 Toxic Waste
CERCLA: 40CFR 302.4: Listed per RCRA Section 3001 RQ: 1 lb (0.454 kg)
SARA 40CFR 372.65: Listed
SARA EHS 40CFR 355: Not listed
TSCA: Listed

Analytical Methods

Soil: SW846 3640A, 8250A, 8270B, 8270C
Water / Groundwater: EPA 1625; ASTM D4763
Plasma: EPA 29

DIM4700 CAS #: 119-93-7

3,3'-DIMETHYLBENZIDINE

RTECS: DD1225000
EINECS Number: 204-358-0
Molecular Formula: $C_{14}H_{16}N_2$
Structured MF: $(C_6H_3(CH_3)NH_2)_2$
Formula Weight: 212.28

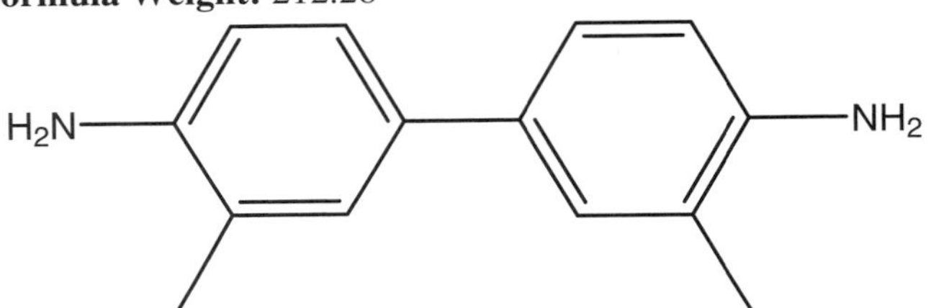

Chemical Structure

Synonyms: BENZIDINE,3,3'-DIMETHYL-; BIANISIDINE; (1,1'-BIPHENYL)-4,4'-DIAMINE,3,3'-DIMETHYL-; (1,1'-BIPHENYL)-4,4'-DIAMINE-3,3'-DIMETHYL-; 4,4'-BI-O-TOLUIDINE; C.I. 37230; C.I. AZOIC DIAZO COMPONENT 113; 4,4'-DIAMINO-3,3'-DIMETHYLBIPHENYL; 4,4'-DIAMINO-3,3'-DIMETHYLDIPHENYL; DIAMINODITOLYL; DIAMINOTOLYL; 3,3'-DIMETHYLBENZIDIN; 3,3'-DIMETHYL-(1,1'-BIPHENYL)-4,4'-DIAMINE; 3,3'-DIMETHYL-4,4'-BIPHENYLDIAMINE; 3,3'-DIMETHYLBIPHENYL-4,4'-DIAMINE; 3,3'-DIMETHYL-4,4'-DIAMINOBIPHENYL; 3,3'-DIMETHYL-4,4'-DIPHENYLDIAMINE; 3,3'-DIMETHYLDIPHENYL-4,4'-DIAMINE; 4,4'-DI-O-TOLUIDINE; DMB; FAST DARK BLUE BASE R; ORTHOTOLIDINE; 2-TOLIDIN; O-TOLIDIN; 2-TOLIDINA; 2-TOLIDINE; 3,3'-TOLIDINE; O,O'-TOLIDINE; O-TOLIDINE; TOLIDINE

Description: white, red crystals, crystalline powder, glistening plates

Use: manufacture of dyes, as a very sensitive reagent for gold (1:10 million detectable) and for free chlorine in water, and in the production of polyurethane-based high-strength elastomers, coating and rigid plastics

Physical Properties

Boiling Point: 300 °C (572 °F)
Freezing Point: 129 °C (264.2 °F) to 131 °C (267.8 °F)
Specific Gravity: 1
Water Solubility: Slightly Soluble in Water

Other Solubilities: 95% Ethanol: <1 mg/ml at 19 °C; Acetone: >=100 mg/ml at 19 °C; Acetic Acid: Soluble; Alcohol: Soluble; DMSO: >=100 mg/ml at 19 °C; Dilute acids: Soluble; Ether: Soluble.
Flash Point: Not available; probably combustible

RTECS Toxicity Data

Acute Oral: Dog LD_{Lo} Dose: 600 mg/kg.
Mutagenic: Human Unscheduled DNA Synthesis; Cell Type: HeLa cell; Dose: 1 umol/L. Rat Unscheduled DNA Synthesis; Cell Type: liver; Dose: 1 umol/L.
Tumorigenic: Rat Route: Oral; Dose: 4500 mg/kg/27D-I; Toxic Effects: Tumorigenic - Carcinogenic by RTECS criteria; Skin and appendages - Tumors. Rat Route: Subcutaneous; Dose: 1650 mg/kg/33W-I; Toxic Effects: Tumorigenic - Equivocal tumorigenic agent by RTECS criteria; Skin and appendages - Tumors. Rat Route: Subcutaneous; Dose: 9 gm/kg/51W-I; Toxic Effects: Tumorigenic - Equivocal tumorigenic agent by RTECS criteria; Sense organs and special senses - Tumors. Rat Route: Subcutaneous; Dose: 5040 mg/kg/56W-I; Toxic Effects: Tumorigenic - Equivocal tumorigenic agent by RTECS criteria; Sense organs and special senses - Tumors; Endocrine - Tumors. Rat Route: Subcutaneous; Dose: 3240 mg/kg/34W-I; Toxic Effects: Tumorigenic - Equivocal tumorigenic agent by RTECS criteria; Endocrine - Tumors; Skin and appendages - Tumors.

Hazard Overviews

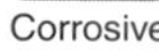
Corrosive

Fire Diamond

Health: Corrosive to skin/eyes. Toxic. Also Causes: methemoglobin, cyanosis, nasal irritation, headache, difficult breathing, dizziness, mental confusion, weakness, convulsions, coma, redness, swelling, blisters, necrosis, swelling eyelids, sensitivity to light, diarrhea. Chronic Effects: possible cancer hazard.
Fire: Will burn. Use dry chemical, carbon dioxide, fog, water spray, or regular foam.
Reactivity: Stable. Hazardous polymerization cannot occur. Avoid: light; heat; ignition sources. Hazardous decomposition products: carbon dioxide; toxic nitrogen oxide fumes.
Carcinogenicity: IARC - Group 2B, Possibly carcinogenic to humans; NIOSH - Listed as carcinogen; NTP - Class 2B, Reasonably anticipated to be a carcinogen, sufficient evidence of carcinogenicity from studies in experimental animals; ACGIH - Class A3, Animal carcinogen; OSHA - Not listed; EPA - Not listed; MAK - Class A2, Unmistakably carcinogenic in animal experimentation only
Primary Target Organs:

Eyes

Skin

Nervous System

Blood

Exposure Limits
NIOSH REL: STEL: 0.02 mg/m^3; 60-minute, skin.
Respirator Recommendation
Exposure Range: unlimited Self-contained Breathing Apparatus, Pressure Demand, Full Face
Note: TLV not established

Environmental

Environmental Fate: Released to soil it will have a moderate tendency to sorb to organic matter But is expected to covalently bond to humic material like other aromatic amines. Thus leaching should not be rapid in soils. Based upon screening studies biodegradation may be an important degradative process while hydrolysis will not be important. No information was found about volatilization. In water, it will have a low tendency to sorb but should covalently bond to sediments. Bioconcentration in aquatic organisms should not be important, because of its low octanol/water partition coefficient. Biodegradation will probably be significant based on screening studies, while hydrolysis will not. No information was found on photolysis or evaporation. An estimated half-life for vapor phase in the atmosphere is 4.0 hours based upon reaction with photochemically generated hydroxyl radicals.

Environmental Physical Data

Octanol/Water Partition Coefficient: log K_{ow} = 2.34
Sorption Partition Coefficient: K_{oc} = 2.34
BCF: calculated at 35

Regulations

RCRA 40CFR: Listed Hazardous Waste No. U095 Toxic Waste
CERCLA: 40CFR 302.4: Listed per RCRA Section 3001 RQ: 10 lb (4.535 kg)
SARA 40CFR 372.65: Listed
SARA EHS 40CFR 355: Not listed
TSCA: Listed

Analytical Methods

Soil: SW846 8270B, 8270C
Water / Groundwater: EPA 553
Drinking Water: EPA 553
Plasma: EPA 29

DIM4750 **CAS #: 612-82-8**

3,3'-DIMETHYLBENZIDINE DIHYDROCHLORIDE

RTECS: DD1226000
EINECS Number: 210-322-5
Molecular Formula: $C_{14}H_{18}Cl_2N_2$
Formula Weight: 285.24

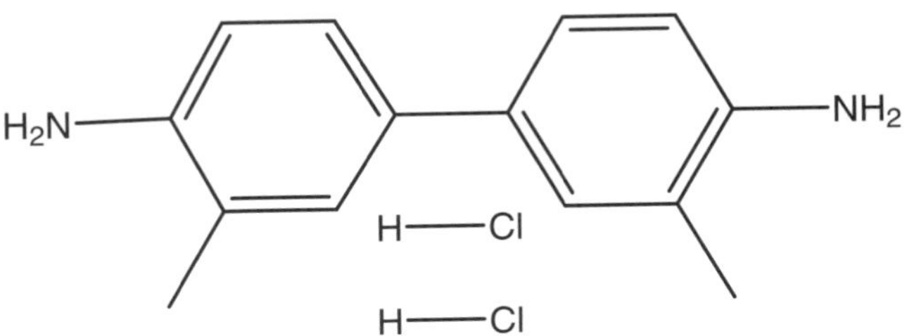

Chemical Structure

Synonyms: 4,4'-DIAMINO-3,3'-DIMETHYLBIPHENYL DIHYDROCHLORIDE; 3,3'-DIMETHYLBIPHENYL-4,4'-BIPHENYLDIAMINE DIHYDROCHLORIDE; 2,3'-DIMETHYLBIPHENYL-4,4'-DIAMINE DIHYDROCHLORIDE; O-TOLIDINE DIHYDROCHLORIDE
Description: light tan powder

Physical Properties

Water Solubility: 10 to 50 mg/mL at 22 °C
Other Solubilities: 95% Ethanol: <1 mg/ml at 15 °C; Acetone: <1 mg/ml at 22 °C; DMSO: 10-50 mg/ml at 22 °C.
Flash Point: Not available; probably combustible

RTECS Toxicity Data

Chronic (Multiple Dose) Oral: Rat Dose: 1120 mg/kg/14D-C; Toxic Effects: Endocrine - Other changes; Nutritional and gross metabolic - Weight loss or decreased weight gain; DEATH. Rat Dose: 18200 mg/kg/13W-C; Toxic Effects: Endocrine - Changes in thymus weight; Blood - Changes in bone marrow not included above; Blood - Changes in erythrocite (RBC) cell count.
Mutagenic: Hamster Cytogenetic Analysis; Cell Type: ovary; Dose: 10 mg/L. Hamster Sister Chromatid Exchenge; Cell Type: ovary; Dose: 20 mg/L.
Tumorigenic: Rat Route: Oral; Dose: 1820 mg/kg/65W-C; Toxic Effects: Tumorigenic - Carcinogenic by RTECS criteria; Liver - Tumors; Skin and appendages - Tumors. Mouse Route: Oral; Dose: 15288 mg/kg/78W-C; Toxic Effects: Tumorigenic - Carcinogenic by RTECS criteria; Lungs, Thorax, or Respiration - Tumors.

Hazard Overviews

Health: Irritating to eyes/skin/respiratory tract. Toxic. Other Acute Effects: harmful if swallowed, inhaled, or absorbed through skin. Chronic Effects: may alter genetic material; target organs: liver, kidneys, g.i. system, bone marrow. Carcinogen.
Fire: Will burn. Hazards: emits toxic fumes. Extinguishing agents: water spray; carbon dioxide, dry chemical powder or appropriate foam. Precautions: combustible liquid.
Reactivity: Incompatible with: strong oxidizing agents, light sensitive, air sensitive. Hazardous decomposition products: toxic fumes of: carbon monoxide, carbon dioxide, nitrogen oxides, hydrogen chloride gas.
Carcinogenicity: IARC - Not listed; NIOSH - Not listed; NTP - Not listed; ACGIH - Not listed; OSHA - Not listed; EPA - Not listed; MAK - Not listed

Primary Target Organs:

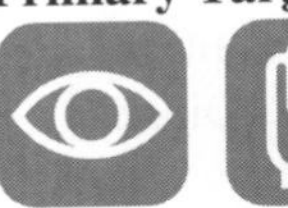
Eyes

Skin
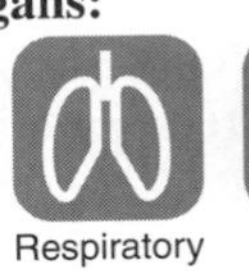
Respiratory System
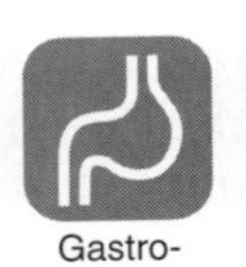
Gastro-intestinal

Liver

Kidneys

Environmental

Regulations

RCRA 40CFR: Not listed
CERCLA: 40CFR 302.4: Not listed
SARA 40CFR 372.65: Listed
SARA EHS 40CFR 355: Not listed
TSCA: Listed

Analytical Methods

Water / Groundwater: SW846 8325

DIM4800 **CAS #: 41766-75-0**

3,3'-DIMETHYLBENZIDINE DIHYDROFLUORIDE

Molecular Formula: $C_{14}H_{18}N_2F_2$
Formula Weight: 252.31

Hazard Overviews

Carcinogenicity: IARC - Not listed; NIOSH - Not listed; NTP - Not listed; ACGIH - Not listed; OSHA - Not listed; EPA - Not listed; MAK - Not listed

Environmental

Regulations

RCRA 40CFR: Not listed
CERCLA: 40CFR 302.4: Not listed
SARA 40CFR 372.65: Listed
SARA EHS 40CFR 355: Not listed
TSCA: Not listed

DIM4850 **CAS #: 103-83-3**

N,N-DIMETHYLBENZYLAMINE

RTECS: DP4500000
EINECS Number: 203-149-1
Molecular Formula: $C_9H_{13}N$
Structured MF: $C_6H_5CH_2N(CH_3)_2$
Formula Weight: 135.21

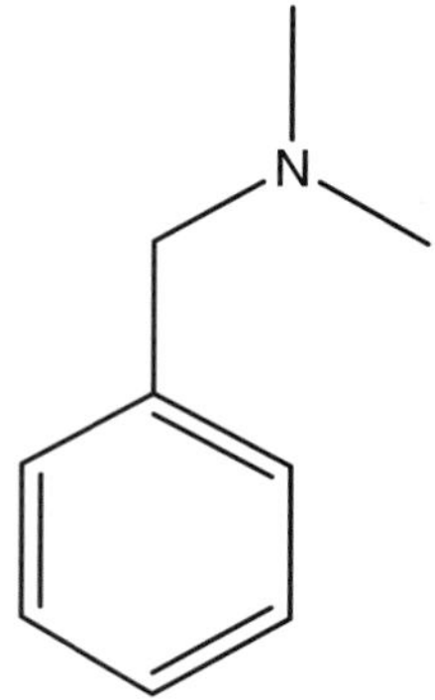

Chemical Structure

Synonyms: ARALDITE ACCELERATOR 062; BDMA; BENZENEMETHANAMINE,N,N-DIMETHYL-(9CI); BENZYL-N,N-DIMETHYLAMINE; BENZYLDIMETHYLAMINE; N-BENZYLDIMETHYLAMINE; N,N-DIMETHYLBENZENEMETHANAMINE; DIMETHYLBENZYLAMINE; N-(PHENYLMETHYL)DIMETHYLAMINE; SUMINE 2015

Description: clear, colorless to slight yellow liquid; fish or ammonia odor

Physical Properties

Boiling Point: ~ 175 °C (347 °F)
Freezing Point: -75 °C (-103 °F)
Specific Gravity: Liquid 0.915 at 10 °C
Vapor Density: Estimate 4.66 Air=1
Water Solubility: Slightly Soluble
Other Solubilities: Ethanol: Miscible; Ether: Miscible
pH: > 7
Refraction Index: 1.5011
Flash Point: 130 °C

RTECS Toxicity Data

Acute Oral: Rat LD_{50} Dose: 265 mg/kg.

Acute Inhalation: Rat LC_{Lo} Dose: 1200 mg/m^3/2hr; Toxic Effects: Behavioral - Somnolence (general depressed activity); Behavioral - Ataxia; Lungs, Thorax, or Respiration - Dyspnea. Mouse LC_{Lo} Dose: 1200 mg/m^3/2hr; Toxic Effects: Behavioral - Somnolence (general depressed activity); Behavioral - Ataxia; Lungs, Thorax, or Respiration - Dyspnea.

Acute Dermal: Rabbit LD_{50} Route: Skin; Dose: 1660 mg/kg; Toxic Effects: Behavioral - Tremor; Behavioral - Excitment.

Chronic (Multiple Dose) Inhalation: Rat Dose: 30 mg/m^3/4H/26W-I; Toxic Effects: Brain and coverings - Recordings from specific areas of CNS; Kidney, Ureter, and Bladder - Other changes in urine composition.

Irritation Eye: Rabbit Standard Draize Test Dose: 5 mg; Reaction: severe.

Irritation Skin: Rabbit Standard Draize Test Dose: 500 mg/4H; Reaction: severe.

Hazard Overviews

Corrosive

Fire Diamond

Health: Irritating to eyes/skin/respiratory tract. Also Causes: respiratory stresses. Chronic Effects: skin sensitization.

Fire: Combustible. Use dry chemical, carbon dioxide, water fog, or foams to put out benzyldimethylamine fires. Use water sprays to cool fire-exposed containers.

Reactivity: Stable. Hazardous polymerization cannot occur. Avoid: heat; ignition sources. Incompatible with: nitric acid; sulfuric acid; hydrochloric acid; benzyldimethylamine; strong oxidizing agents. Hazardous decomposition products: carbon monoxide; ammonia; hydrogen cyanide; oxides of carbon; oxides of nitrogen.

Carcinogenicity: IARC - Not listed; NIOSH - Not listed; NTP - Not listed; ACGIH - Not listed; OSHA - Not listed; EPA - Not listed; MAK - Not listed

Primary Target Organs:

Eyes

Skin

Respiratory System

Environmental

Cleanup/Disposal: Guide No. 153: Eliminate all ignition sources (no smoking, flares, sparks or flames in immediate area). Do not touch damaged containers or spilled material unless wearing appropriate protective clothing. Stop leak if you can do it without risk. Prevent entry into waterways, sewers, basements or confined areas. Absorb or cover with dry earth, sand or other non-combustible material and transfer to containers. Do not get water inside containers.

Environmental Physical Data

BCF: no food chain concentration potential

Regulations

RCRA 40CFR: Not listed
CERCLA: 40CFR 302.4: Not listed
SARA 40CFR 372.65: Not listed
SARA EHS 40CFR 355: Not listed
TSCA: Listed

DIM4900 **CAS #: 75-83-2**

2,2-DIMETHYLBUTANE

RTECS: EJ9300000
DOT: UN1208; IMO3.1
EINECS Number: 200-906-8
Molecular Formula: C_6H_{14}
Structured MF: $CH_3CH_2C(CH_3)_3$
Formula Weight: 86.18

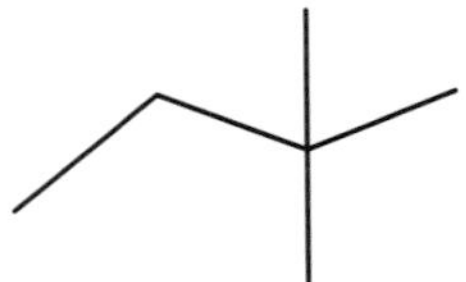

Chemical Structure

Synonyms: BUTANE,2,2-DIMETHYL-; NEOHEXANE
Description: colorless liquid; mild gasoline odor
Use: as an intermediate for agricultural chemicals; a component of high-octane motor and aviation fuels

Physical Properties

Boiling Point: 49.7 °C (121 °F) at 760 mm Hg
Freezing Point: -99.9 °C (-147.82 °F)
Specific Gravity: 0.6485 at 20 °C/4 °C
Vapor Density: 3 Air=1
Density: 0.6444 g/mL at 25 °C
Vapor Pressure: 400 mm Hg at 31.0 °C
Water Solubility: < 1 mg/mL at 22 C
Other Solubilities: 95% Ethanol: >=100 mg/ml at 22 °C; Acetone: >=100 mg/ml at 22 °C; Alcohol: Soluble; Benzene: >10%; DMSO: 10-50 mg/ml at 23 °C; Ether: Soluble.
Surface Tension: 16.3 dynes/cm at 20 °C
Refraction Index: 1.3688 at 20 °C/D
Evaporation Rate: < 1 Ether=1
Critical Temperature: 216 °C
Critical Pressure: 447 psia
Ionization Potential (eV): 9.79
Flash Point: -48 °C Closed Cup
Autoignition Temperature: 405 °C
LEL: 1.2% v/v
UEL: 7.0% v/v

Hazard Overviews

Flammable

Health: Irritating to eyes/skin/respiratory tract. Other Acute Effects: harmful if inhaled or swallowed.
Fire: Flammable. Hazards: vapor may travel considerable distance to source of ignition and flash back; extremely flammable; container explosion may occur; forms explosive mixtures in air. Extinguishing agents: carbon dioxide, dry chemical powder or appropriate foam. Precautions: combustible liquid.
Reactivity: Incompatible with: oxidizing agents. Hazardous decomposition products: toxic fumes of: carbon monoxide, carbon dioxide.
Carcinogenicity: IARC - Not listed; NIOSH - Not listed; NTP - Not listed; ACGIH - Not listed; OSHA - Not listed; EPA - Not listed; MAK - Not listed

Primary Target Organs:

Eyes

Skin

Respiratory System

Exposure Limits
ACGIH TLV: TWA: 500 ppm; 1760 mg/m^3; STEL: 1000 ppm; 3500 mg/m^3.
NIOSH REL: TWA: 100 ppm; 350 mg/m^3. STEL: 510 ppm; 1800 mg/m^3; Ceiling, 15 min.

Environmental

Cleanup/Disposal: Guide No. 128: Eliminate all ignition sources (no smoking, flares, sparks or flames in immediate area). All equipment used when handling the product must be grounded. Do not touch or walk through spilled material. Stop leak if you can do it without risk. Prevent entry into waterways, sewers, basements or confined areas. A vapor suppressing foam may be used to reduce vapors. Absorb or cover with dry earth, sand or other non-combustible material and transfer to containers. Use clean non-sparking tools to collect absorbed material. Large Spills: Dike far ahead of liquid spill for later disposal. Water spray may reduce vapor; but may not prevent ignition in closed spaces.

Environmental Physical Data

BCF: no food chain concentration potential

Regulations

RCRA 40CFR: Not listed
CERCLA: 40CFR 302.4: Not listed
SARA 40CFR 372.65: Not listed
SARA EHS 40CFR 355: Not listed
TSCA: Listed

DIM4950 CAS #: 79-29-8

2,3-DIMETHYLBUTANE

RTECS: EJ9350000
DOT: UN2457; IMO3.1
EINECS Number: 201-193-6
Molecular Formula: C_6H_{14}
Structured MF: $(CH_3)_2CHCH(CH_3)_2$
Formula Weight: 86.18

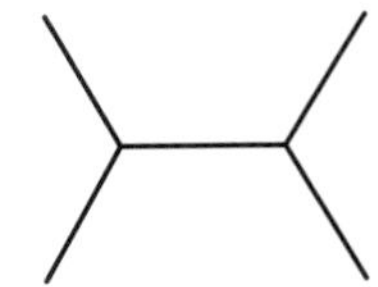

Chemical Structure

Synonyms: BIISOPROPYL; BUTANE,2,3-DIMETHYL-; DIISOPROPYL; 1,1,2,2-TETRAMETHYLETHANE
Description: colorless liquid; mild gasoline-like odor
Use: component of high octane fuel; in organic synthesis; as a solvent for vegetable oils, glues, coatings and paints, in

gasoline, as an intermediate for chemicals and in rubber solvents and petroleum ether

Physical Properties

Boiling Point: 57.9 °C (136 °F) at 760 mm Hg
Freezing Point: -128.8 °C (-199.84 °F)
Specific Gravity: 0.6616 at 20 °C/4 °C
Vapor Density: 3 Air=1
Saturated Vapor Density: 1.93161343 kg/m^3
Density: 0.66164 g/mL at 20 °C
Vapor Pressure: 235 mm Hg at 25 °C
Water Solubility: < 1 mg/mL at 23.5 C
Other Solubilities: 95% Ethanol: >=100 mg/ml at 23.5 °C; Acetone: >=100 mg/ml at 23.5 °C; Alcohol: Very Soluble; Benzene: >10%; DMSO: 10-50 mg/ml at 23.5 °C; Ether: >10%.
Surface Tension: 2.3229 x10^{-2} N/m at melting point
Refraction Index: 1.375
Evaporation Rate: < 1 Ethylether=1
Critical Temperature: 227 °C
Critical Pressure: 3.1268 x10^6 Pa
Ionization Potential (eV): 9.79
Flash Point: -29 °C Closed Cup
Autoignition Temperature: 405 °C
LEL: 1.2% v/v
UEL: 7.0% v/v

RTECS Toxicity Data

Chronic (Multiple Dose) Oral: Rat Dose: 10 gm/kg/4W-I; Toxic Effects: Kidney, Ureter, and Bladder - Chgs in tubules (inc acute renal failure, acute tubular necrosis; Biochemical - Peptidases; DEATH.

Hazard Overviews

Flammable

Health: Irritating to eyes/skin/respiratory tract. Other Acute Effects: harmful if inhaled or swallowed.
Fire: Flammable. Hazards: vapor may travel considerable distance to source of ignition and flash back; extremely flammable; container explosion may occur; forms explosive mixtures in air. Extinguishing agents: carbon dioxide, dry chemical powder or appropriate foam. Precautions: combustible liquid.
Reactivity: Incompatible with: oxidizing agents. Hazardous decomposition products: toxic fumes of: carbon monoxide, carbon dioxide.
Carcinogenicity: IARC - Not listed; NIOSH - Not listed; NTP - Not listed; ACGIH - Not listed; OSHA - Not listed; EPA - Not listed; MAK - Not listed
Primary Target Organs:

Eyes

Skin

Respiratory System

Exposure Limits
ACGIH TLV: TWA: 500 ppm; 1760 mg/m^3; STEL: 1000 ppm; 3500 mg/m^3.
NIOSH REL: TWA: 100 ppm; 350 mg/m^3. STEL: 510 ppm; 1800 mg/m^3; Ceiling, 15 min.

Environmental

Environmental Fate: If released to soil, it will have low mobility. Volatilization may be important from moist and dry soil surfaces. Insufficient date are available to determine the rate or importance of biodegradation in soil or water. If released to water, it would adsorb to suspended solids and sediment. It would volatilize from water surfaces with estimated half-lives for a model river and model lake of 2.7 hours and 3.7 days, respectively. An estimated BCF value of 230 suggests that it will bioconcentrate in aquatic organisms. If released to the atmosphere, it will exist as a vapor. Vapor-phase is degraded in the atmosphere by reaction with photochemically produced hydroxyl radicals with an estimated half-life of about 5.4 days. Vapor-phase is also degraded in the atmosphere by reaction with nitrate radicals with an estimated half-life of about 334 days.
Cleanup/Disposal: Guide No. 128: Eliminate all ignition sources (no smoking, flares, sparks or flames in immediate area). All equipment used when handling the product must be grounded. Do not touch or walk through spilled material. Stop leak if you can do it without risk. Prevent entry into waterways, sewers, basements or confined areas. A vapor suppressing foam may be used to reduce vapors. Absorb or cover with dry earth, sand or other non-combustible material and transfer to containers. Use clean non-sparking tools to collect absorbed material. Large Spills: Dike far ahead of liquid spill for later disposal. Water spray may reduce vapor; but may not prevent ignition in closed spaces.

Environmental Physical Data

Henry's Law Constant: estimated at 1.2
Octanol/Water Partition Coefficient: log K_{ow} = 3.42
Sorption Partition Coefficient: K_{oc} = estimated at 1700
BCF: estimated at 230

Regulations

RCRA 40CFR: Not listed
CERCLA: 40CFR 302.4: Not listed
SARA 40CFR 372.65: Not listed
SARA EHS 40CFR 355: Not listed
TSCA: Listed

DIM5000 **CAS #: 75-97-8**

3,3-DIMETHYL-2-BUTANONE

RTECS: EL7700000
EINECS Number: 200-920-4
Molecular Formula: $C_6H_{12}O$
Formula Weight: 100.16

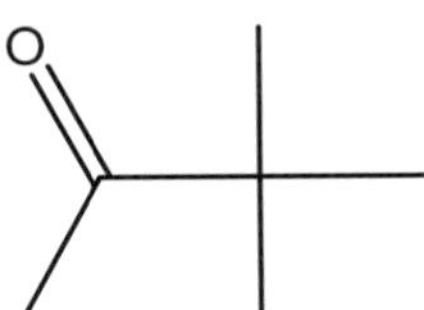

Chemical Structure

Synonyms: 2-BUTANONE,3,3-DIMETHYL-; T-BUTYL METHYL KETONE; TERT-BUTYL METHYL KETONE; 2,2-DIMETHYL-3-BUTANONE; 2,2-DIMETHYLBUTANONE; 3,3-DIMETHYLBUTANONE; KETONE,T-BUTYL METHYL; METHYL T-BUTYL KETONE; METHYL TERT-BUTYL KETONE; METHYLTERT-BUTYL KETONE; PINACOLIN; PINACOLINE; PINACOLONE; PINAKOLIN; 1,1,1-TRIMETHYLACETONE

Description: colorless liquid; peppermint or camphor-like odor

Use: for fungicides, eg triademefon; synthesis of herbicide metribuzin

Physical Properties

Boiling Point: 106.1 °C (223 °F)
Freezing Point: -52.5 °C (-62.5 °F)
Specific Gravity: 0.7229 at 25 °C/25 °C
Saturated Vapor Density: Calculated 1.32204392 kg/m^3
Vapor Pressure: Calculated 31.5 mm Hg at 25 °C
Water Solubility: 2.44% at 15 °C
Other Solubilities: Soluble in Alcohol, Ether, Acetone
Surface Tension: 2.9879 $x10^{-2}$ N/m at melting point
Refraction Index: 1.3952 at 25 °C/D
Critical Temperature: 291 °C
Critical Pressure: 3.32 $x10^6$ Pa
Ionization Potential (eV): 9.12 +/-0.2
Flash Point: 23 °C

RTECS Toxicity Data

Acute Oral: Rat LD_{50} Dose: 610 mg/kg. Mouse LD_{50} Dose: 1625 mg/kg.
Acute Inhalation: Mouse LC_{50} Dose: 5700 mg/m^3.

Hazard Overviews

Flammable

Health: May cause irritation. Other Acute Effects: may be harmful by inhalation, ingestion, or skin absorption.

Fire: Flammable. Hazards: vapor may travel considerable distance to source of ignition and flash back; emits toxic fumes. Extinguishing agents: carbon dioxide, dry chemical powder or appropriate foam; water spray. Precautions: combustible liquid.

Reactivity: Incompatible with: strong oxidizing agents. Hazardous decomposition products: toxic fumes of: carbon monoxide, carbon dioxide.

Carcinogenicity: IARC - Not listed; NIOSH - Not listed; NTP - Not listed; ACGIH - Not listed; OSHA - Not listed; EPA - Not listed; MAK - Not listed

Environmental

Ecotoxicity: LC_{50} Pimephales promelas (fathead minnow) 87 mg/l 96 hr flow-through bioassay, wt 0.12 g, water hardness 45.5 mg/l $CaCO_3$, temp: 25 + or - 1 °C, pH 7.5, dissolved oxygen greater than 60% of

Environmental Fate: If released to the atmosphere, it will degrade by reaction with photochemically produced hydroxyl radicals (estimated half-life of about 13 days). Physical removal from air through wet deposition may occur. If released to water, volatilization may be an important fate process. Volatilization half-lives of 7 and 79 hours have been estimated for a model river (1 m deep) and a model pond (2 m deep) respectively. If released to soil, estimated K_{oc} values of 10-20 suggest that it will be highly mobile. With a relatively high vapor pressure of 31.5 mm Hg at 25 °C, it can be expected to evaporate from terrestrial surfaces to the atmosphere. Insufficient data are available to predict the relative importance or rate of biodegradation in soil or water.

Environmental Physical Data

Henry's Law Constant: estimated at 0.000218
Sorption Partition Coefficient: K_{OC} = 20
BCF: estimated at 2

Regulations

RCRA 40CFR: Not listed
CERCLA: 40CFR 302.4: Not listed
SARA 40CFR 372.65: Not listed
SARA EHS 40CFR 355: Not listed
TSCA: Listed

DIM5050 CAS #: 793-24-8

N-(1,3-DIMETHYLBUTYL)-N'-PHENYL-1,4-BENZENEDIAMINE

RTECS: ST0900000
EINECS Number: 212-344-0
Molecular Formula: $C_{18}H_{24}N_2$
Formula Weight: 268.44

Synonyms: ANTIOXIDANT 4-20; ANTIOXIDANT 4020; ANTIOXIDANT CD 13; ANTOZITE 67E; ANTOZITE 67F; ANTOZITE 67; 1,4-BENZENEDIAMINE,N-(1,3-DIMETHYLBUTYL)-N'-PHENYL-; CD 13; DBDA; DIAFEN 13; N-(1,3-DIMETHYLBUTYL)-N'-PHENYL-P-PHENYLENEDIAMINE; DMBPD; DUSANTOX 6PPD; FLEXZONE 7F; FLEXZONE 7L; NOCRANE 6C; NOCRANE 7 L; OZONON 6C; PERMANAX 120; PERMANAX 6PPD; N-PHENYL-N'-(1,3-DIMETHYL BUTYL)-PARA-PHENYLENEDIAMINE; P-PHENYLENEDIAMINE,N-(1,3-DIMETHYLBUTYL)-N'-PHENYL-; SANTOFLEX 13; UOP 562; UOP 588; VULKANOX 4020; WINGSTAY 300

Description: dark, violet solid

Use: antioxidant, antiozonant & polymer stabilizer; antiozonant & antioxidant for butadiene rubber, nitrile-butadiene rubber, styrene-butadiene rubber, synthetic isoprene rubber

Physical Properties

Boiling Point: 260 °C (500 °F)
Freezing Point: 50 °C (122 °F)
Specific Gravity: 1.07
Water Solubility: < 1 mg/mL at 17 C
Other Solubilities: 95% Ethanol: >=100 mg/ml at 20.5 °C; Acetone: >=100 mg/ml at 20.5 °C; DMSO: >=100 mg/ml at 20.5 at °C.
Flash Point: 204.444 °C Open Cup

RTECS Toxicity Data

Acute Oral: Rat LD_{50} Dose: 3580 mg/kg; Toxic Effects: Sense organs and special senses - Other; Behavioral - Food intake (animal); Gastrointestinal - Hypermotility, diarrhea.
Acute Dermal: Rabbit LD_{50} Route: Skin; Dose: >7940 mg/kg; Toxic Effects: Behavioral - Food intake (animal); Behavioral - Change in motor activity (specific assay).

Hazard Overviews

Fire: Will burn.
Carcinogenicity: IARC - Not listed; NIOSH - Not listed; NTP - Not listed; ACGIH - Not listed; OSHA - Not listed; EPA - Not listed; MAK - Not listed

Environmental

Regulations

RCRA 40CFR: Not listed
CERCLA: 40CFR 302.4: Not listed
SARA 40CFR 372.65: Not listed
SARA EHS 40CFR 355: Not listed
TSCA: Listed

DIM5100 CAS #: 79-44-7

DIMETHYLCARBAMOYL CHLORIDE

RTECS: FD4200000
DOT: UN2262; IMO8.0
EINECS Number: 201-208-6
Molecular Formula: C_3H_6ClNO
Formula Weight: 107.55

Chemical Structure

Synonyms: CARBAMIC ACID,DIMETHYL-(9CI); CARBAMIC CHLORIDE,DIMETHYL-; CARBAMOYL CHLORIDE,DIMETHYL-; CARBAMYL CHLORIDE,N,N-DIMETHYL-; CHLORID KYSELINY DIMETHYLKARBAMINOVE; CHLOROFORMIC ACID DIMETHYLAMIDE; DDC; DIMETHYL CARBAMIC CHLORIDE; DIMETHYL CARBAMOYL CHLORIDE; DIMETHYL CARBAMYL CHLORIDE; DIMETHYLAMID KYSELINY CHLORMRAVENCI; (DIMETHYLAMINO)CARBONYL CHLORIDE; N,N-DIMETHYLAMINOCARBONYL CHLORIDE; DIMETHYLCARBAMIC ACID CHLORIDE; N,N-DIMETHYLCARBAMIC ACID CHLORIDE; DIMETHYLCARBAMIC CHLORIDE; DIMETHYLCARBAMIDOYL CHLORIDE; N,N-DIMETHYLCARBAMIDOYL CHLORIDE; N,N-DIMETHYLCARBAMOYL CHLORIDE; DIMETHYLCARBAMYL CHLORIDE; N,N-DIMETHYLCARBAMYL CHLORIDE; N-N-DIMETHYLCARBAMYL CHLORIDE; DIMETHYLCHLOROFORMAMIDE; DIMETHYLKARBAMOYLCHLORID; DMCC; TL 389
Description: clear, colorless liquid
Use: chemical intermediate in the production of drugs and pesticides

Physical Properties

Boiling Point: 165 °C (329 °F) to 167 °C (333 °F)
Freezing Point: -33 °C (-27.4 °F)
Specific Gravity: 1.1678 at 20 °C
Vapor Density: 3.73
Saturated Vapor Density: 1.210649156 kg/m^3
Vapor Pressure: 2.49 mm Hg at 25 °C
Water Solubility: Reacts
Other Solubilities: 95% Ethanol: >=100 mg/ml at 21 °C; Acetone: >=100 mg/ml at 21 °C; DMSO: >=100 mg/ml at 21 °C.
Refraction Index: 1.4529 at 20 °C/D
Flash Point: 68 °C

RTECS Toxicity Data

Acute Oral: Rat LD_{50} Dose: 1 gm/kg; Toxic Effects: Behavioral - Somnolence (general depressed activity).
Acute Inhalation: Rat LC_{50} Dose: 180 ppm/6hr. Mouse LC_{Lo} Dose: 1000 mg/m^3/10M.
Mutagenic: Human Unscheduled DNA Synthesis; Cell Type: fibroblast; Dose: 32 ug/L. Mouse Mutations in Microorganisms; Cell Type: lymphocyte; Dose: 187 mg/L (+S9).
Tumorigenic: Rat Route: Inhalation; Dose: 1 ppm/6H/6W-I; Toxic Effects: Tumorigenic - Carcinogenic by RTECS criteria; Sense organs and special senses - Tumors. Rat Route: Inhalation; Dose: 1 ppm/6H/6W-I; Toxic Effects: Tumorigenic - Equivocal tumorigenic agent by RTECS criteria; Lungs, Thorax, or Respiration - Tumors. Rat Route: Inhalation; Dose: 1 ppm; Toxic Effects: Tumorigenic - Equivocal tumorigenic agent by RTECS criteria; Sense organs and special senses - Tumors.

Hazard Overviews

Corrosive

Health: Corrosive to eyes/skin/respiratory tract. Toxic. Other Acute Effects: harmful if swallowed, inhaled, or absorbed through skin; extremely destructive to tissue of the mucous membranes and upper respiratory tract, eyes and skin; inhalation may result in spasm, inflammation and edema of the larynx and bronchi, chemical pneumonitis and pulmonary edema; symptoms of exposure may include burning sensation, coughing, wheezing, laryngitis, shortness of breath, headache, nausea and vomiting. Chronic Effects: may cause heritable genetic damage. Carcinogen.

Fire: Combustible. Hazards: emits toxic fumes. Extinguishing agents: carbon dioxide, dry chemical powder or appropriate foam. Precautions: combustible liquid.

Reactivity: Incompatible with: strong oxidizing agents, strong bases, may decompose on exposure to moist air or water. Hazardous decomposition products: toxic fumes of: carbon monoxide, carbon dioxide, nitrogen oxides, hydrogen chloride gas.

Carcinogenicity: IARC - Group 2A, Probably carcinogenic to humans; NIOSH - Listed as carcinogen; NTP - Class 2B, Reasonably anticipated to be a carcinogen, sufficient evidence of carcinogenicity from studies in experimental animals; ACGIH - Class A3, Animal carcinogen; OSHA - Not listed; EPA - Not listed; MAK - Class A2, Unmistakably carcinogenic in animal experimentation only

Primary Target Organs:

Eyes

Skin

Respiratory System

Respirator Recommendation

Exposure Range: unlimited Self-contained Breathing Apparatus, Pressure Demand, Full Face Supplied Air, Constant Flow/Pressure Demand, Full Face

Note: TLV not established

Environmental

Environmental Fate: If released to soil, it will be expected to rapidly hydrolyze if the soil is moist, based upon the observed rapid hydrolysis in aqueous solution. Since it rapidly hydrolyzes, adsorption to and volatilization from moist soil are not expected to be significant processes, although no data specifically regarding the fate in soil were located. Biodegradation is not expected to be an important process relative to hydrolysis, although no data concerning biodegradation were located. If released to water, it will be expected to rapidly hydrolyze, based upon a half-life of 54 sec at 14 °C calculated from a measured hydrolysis rate constant. Since it rapidly hydrolyzes, bioconcentration, volatilization, and adsorption to sediment and suspended solids are not expected to be significant processes. No data were located concerning biodegradation, but it probably abiotically degrades via hydrolysis significantly faster than via biodegradation. If released to the atmosphere, it will be expected to exist almost entirely in the vapor phase based upon its vapor pressure. It will be susceptible to photooxidation via vapor phase reaction with photochemically produced hydroxyl radicals with a half-life of 6 hours estimated for this process.

Cleanup/Disposal: Guide No. 156: Eliminate all ignition sources (no smoking, flares, sparks or flames in immediate area). All equipment used when handling the product must be grounded. Do not touch damaged containers or spilled material unless wearing appropriate protective clothing. Stop leak if you can do it without risk. A vapor suppressing foam may be used to reduce vapors. For chlorosilanes, use AFFF alcohol-resistant medium expansion foam to reduce vapors. Do not get water on spilled substance or inside containers. Use water spray to reduce vapors or divert vapor cloud drift. Prevent entry into waterways, sewers, basements or confined areas. Small Spills: Cover with dry earth, dry sand, or other non-combustible material followed with plastic sheet to minimize spreading or contact with rain. Use clean non-sparking tools to collect material and place it into loosely covered plastic containers for later disposal.

Environmental Physical Data

BCF: not significant

Regulations

RCRA 40CFR: Listed Hazardous Waste No. U097 Toxic Waste

CERCLA: 40CFR 302.4: Listed per RCRA Section 3001 RQ: 1 lb (0.454 kg)

SARA 40CFR 372.65: Listed

SARA EHS 40CFR 355: Not listed

TSCA: Listed

DIM5150 CAS #: 88-04-0

3,5-DIMETHYL-4-CHLOROPHENOL

RTECS: ZE6850000

EINECS Number: 201-793-8

Molecular Formula: C_8H_9ClO

Formula Weight: 156.62

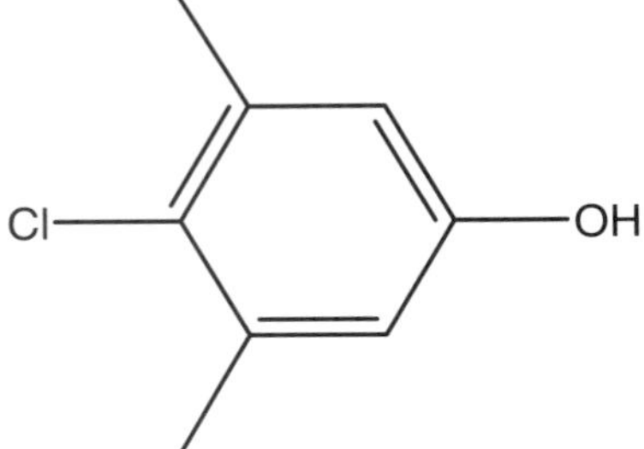

Chemical Structure

Synonyms: BENZYTOL; 4-CHLORO-3,5-DIMETHYLPHENOL; 2-CHLORO-5-HYDROXY-1,3-DIMETHYLBENZENE; 4-CHLORO-1-HYDROXY-3,5-DIMETHYLBENZENE; 2-CHLORO-5-HYDROXY-M-XYLENE; 2-CHLORO-M-XYLENOL; 4-CHLORO-3,5-XYLENOL; CHLORO-XYLENOL; P-CHLORO-M-XYLENOL; DESSON; DETTOL; ESPADOL; HUSEPT EXTRA; NIPACIDE MX; OTTASEPT; OTTASEPT EXTRA; PCMX; PHENOL,4-CHLORO-3,5-DIMETHYL-; RBA 777

Physical Properties

Boiling Point: 246 °C (474.8 °F)

Freezing Point: 115 °C (239 °F)

Water Solubility: 0.3 g/l

Other Solubilities: Ethanol: Soluble; Ether: Soluble; Benzene: Slightly soluble; Petroleum Ether: Slightly soluble

RTECS Toxicity Data

Acute Oral: Rat LD_{50} Dose: 3830 mg/kg; Toxic Effects: Lungs, Thorax, or Respiration - Other changes; Gastrointestinal - Other changes; Kidney, Ureter, and Bladder - Other changes. Mouse LD_{50} Dose: 1 gm/kg; Toxic Effects: Behavioral - Altered sleep time (including change in righting reflex); Behavioral - Ataxia; Gastrointestinal - Other changes.

Irritation Eye: Rabbit Standard Draize Test Dose: 100 mg; Reaction: moderate.

Reproductive/Teratogenic: Rat Route: Oral; Dose: 17100 mg/kg; Duration: female 1-19D of pregnancy; Effects on Embryo or Fetus - Fetotoxicity; Specific Developmental Abnormalities - Musculoskeletal system.

Mutagenic: Bacteria - E Coli Other Mutation Test Systems; Dose: 100 umol/L.

Hazard Overviews

Health: Irritating to eyes/skin/respiratory tract. Harmful. Other Acute Effects: harmful if swallowed; may be harmful if inhaled or absorbed through the skin; depending on the intensity and duration of exposure, effects may vary from mild irritation to severe destruction of tissue; prolonged contact can cause damage to the eyes, severe irritation, or burns; prolonged or repeated exposure may cause allergic reactions in certain sensitive individuals.

Fire: Hazards: emits toxic fumes. Extinguishing agents: water spray; carbon dioxide, dry chemical powder or appropriate foam. Precautions: combustible liquid.

Reactivity: Incompatible with: strong oxidizing agents, strong bases. Hazardous decomposition products: toxic fumes of: carbon monoxide, carbon dioxide, hydrogen chloride gas.

Carcinogenicity: IARC - Not listed; NIOSH - Not listed; NTP - Not listed; ACGIH - Not listed; OSHA - Not listed; EPA - Not listed; MAK - Not listed

Primary Target Organs:

Eyes

Skin

Respiratory System

Environmental

Regulations

RCRA 40CFR: Not listed

CERCLA: 40CFR 302.4: Listed as Compound per CWA Section 307(a)

SARA 40CFR 372.65: Listed as Compound

SARA EHS 40CFR 355: Not listed

TSCA: Listed

Analytical Methods

Soil: EPA PMD-PFH

DIM5250 CAS #: 98-94-2

N,N-DIMETHYLCYCLOHEXYLAMINE

RTECS: GX1198000
EINECS Number: 202-715-5
Molecular Formula: $C_8H_{17}N$
Structured MF: $(CH_3)_2NC_6H_{11}$
Formula Weight: 127.26

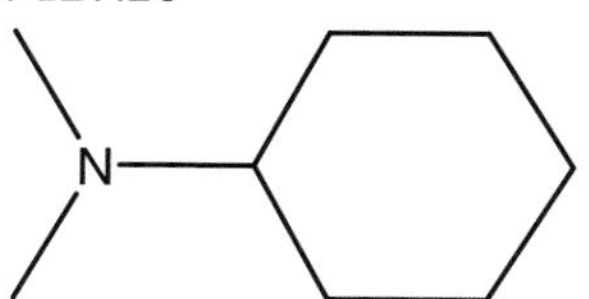

Chemical Structure

Synonyms: CYCLOHEXANAMINE,N,N-DIMETHYL-; CYCLOHEXANAMINE,N,N-DIMETHYL-(9CI); CYCLOHEXYLAMINE,N,N-DIMETHYL-; CYCLOHEXYLDIMETHYLAMINE; N-CYCLOHEXYLDIMETHYLAMINE; (DIMETHYLAMINO)CYCLOHEXANE; N,N-DIMETHYLAMINOCYCLOHEXANE; N,N-DIMETHYLCYCLOHEXANAMINE; DIMETHYLCYCLOHEXYLAMINE; N,N-DIMETHYL-N-CYCLOHEXYLAMINE; POLYCAT 8

Description: water-white liquid; musky ammonia odor

Use: catalyst for polyurethane foams; intermediate for rubber accelerators; treatment of textiles; chemical intermediate for rubber accelerators & dyes; catalyst for polyurethane foams; component of textile treatments

Physical Properties

Boiling Point: 162 °C (324 °F) at 1 atm
Freezing Point: < -77 °C (-106.6 °F)
Specific Gravity: 0.849 at 20 °C/20 °C
Vapor Density: Estimate 4.4 Air=1
Water Solubility: Partly Soluble in Water
Other Solubilities: miscible with Alcohol, Benzene, Acetone
Ionization Potential (eV): 7.5
Flash Point: 43.333 °C Open Cup

RTECS Toxicity Data

Acute Oral: Rat LD_{50} Dose: 348 mg/kg. Mouse LD_{50} Dose: 320 mg/kg; Toxic Effects: Behavioral - Muscle contraction or spasticity.

Acute Inhalation: Rat LC_{50} Dose: 1889 mg/m^3/2hr; Toxic Effects: Behavioral - Muscle contraction or spasticity. Mouse LC_{50} Dose: 1100 mg/m^3/2hr; Toxic Effects: Behavioral - Muscle contraction or spasticity.

Hazard Overviews

Corrosive

Flammable

Health: Corrosive to eyes/skin/respiratory tract. Harmful. Other Acute Effects: harmful by inhalation, in contact with skin and if swallowed; extremely destructive to tissue of the mucous membranes and upper respiratory tract, eyes and skin; inhalation may result in spasm, inflammation and edema of the larynx and bronchi, chemical pneumonitis and pulmonary edema; symptoms of exposure may include burning sensation, coughing, wheezing, laryngitis, shortness of breath, headache, nausea and vomiting.

Fire: Flammable. Hazards: emits toxic fumes. Extinguishing agents: carbon dioxide, dry chemical powder or appropriate foam. Precautions: combustible liquid.

Carcinogenicity: IARC - Not listed; NIOSH - Not listed; NTP - Not listed; ACGIH - Not listed; OSHA - Not listed; EPA - Not listed; MAK - Not listed

Primary Target Organs:

Eyes

Skin

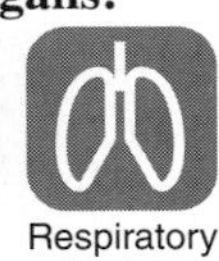
Respiratory System

Environmental

Cleanup/Disposal: Guide No. 152: Do not touch damaged containers or spilled material unless wearing appropriate protective clothing. Stop leak if you can do it without risk. Prevent entry into waterways, sewers, basements or confined areas. Cover with plastic sheet to prevent spreading. Absorb or cover with dry earth, sand or other non-combustible material and transfer to containers. Do not get water inside containers.

Regulations

RCRA 40CFR: Not listed
CERCLA: 40CFR 302.4: Not listed
SARA 40CFR 372.65: Not listed
SARA EHS 40CFR 355: Not listed
TSCA: Listed

DIM5300 CAS #: 3081-20-7

DIMETHYLCYCLOPINACOXYSILANE

EINECS Number: 221-376-4

Physical Properties

Boiling Point: 64 °C (147 °F)
Specific Gravity: 0.91
Vapor Density: > 1 Air=1
Water Solubility: Reacts
Other Solubilities: Soluble in Toluene
Evaporation Rate: < 1 Butyl Acetate=1

Hazard Overviews

Carcinogenicity: IARC - Not listed; NIOSH - Not listed; NTP - Not listed; ACGIH - Not listed; OSHA - Not listed; EPA - Not listed; MAK - Not listed

Environmental

Regulations

RCRA 40CFR: Not listed
CERCLA: 40CFR 302.4: Not listed
SARA 40CFR 372.65: Not listed
SARA EHS 40CFR 355: Not listed
TSCA: Not listed

DIM5400 CAS #: 78-63-7

2,5-DIMETHYLDI(T-BUTYLPEROXY)HEXANE

RTECS: MO1835000
EINECS Number: 201-128-1
Molecular Formula: $C_{16}H_{34}O_4$
Formula Weight: 290.50

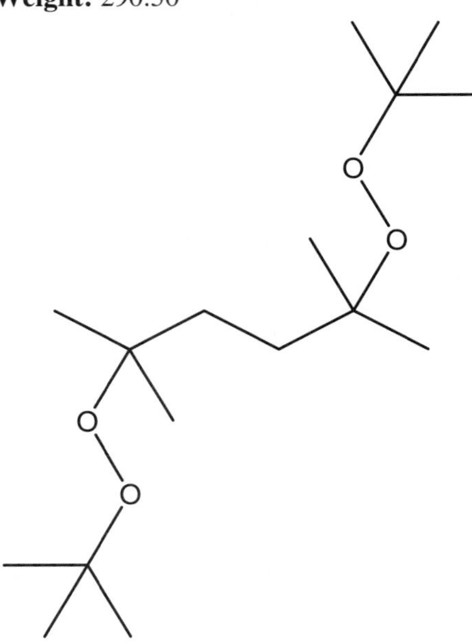

Chemical Structure

Synonyms: 2,5-DIMETHYL-2,5-DI(T-BUTYLPEROXY)HEXANE; PEROXIDE,(1,1,4,4-TETRAMETHYL-1,4-BUTANEDIYL)BIS((1,1-DIMETHYLETHYL); PEROXIDE,(1,1,4,4-TETRAMETHYLTETRAMETHYLENE)BIS(TERT-BUTYL; (1,1,4,4-TETRAMETHYL-1,4-BUTANEDIYL)BIS[(1,1-DIMETHYLETHYL) PEROXIDE; TRIGONOX 101-101/45; VAROX

Description: yellow, clear liquid

Use: catalyst in polyethylene cross linking; styrene polymerization; polyester resins

Physical Properties

Boiling Point: 50 °C (122 °F) to 52 °C (126 °F)
Freezing Point: 8 °C (46.4 °F)
Water Solubility: < 1 mg/mL at 23 C
Other Solubilities: 95% Ethanol: >=100 mg/ml at 23 °C; Acetone: >=100 mg/ml at 23 °C; DMSO: 1-5 mg/ml at 23 °C.
Flash Point: 85 °C

RTECS Toxicity Data

Acute Oral: Rat LD_{50} Dose: >3200 mg/kg.

Hazard Overviews

Health: Irritating to eyes/skin/respiratory tract. Other Acute Effects: may be harmful by inhalation, ingestion, or skin absorption.

Fire: Combustible. Hazards: emits toxic fumes; may accelerate combustion; contact with other material may cause fire. Extinguishing agents: water spray; carbon dioxide, dry chemical powder or appropriate foam. Precautions: combustible liquid.

Reactivity: Unstable. Incompatible with: strong oxidizing agents, acids, reducing agents, organic materials, finely powdered metals. Hazardous decomposition products: toxic fumes of: carbon monoxide, carbon dioxide.

Carcinogenicity: IARC - Not listed; NIOSH - Not listed; NTP - Not listed; ACGIH - Not listed; OSHA - Not listed; EPA - Not listed; MAK - Not listed

Primary Target Organs:

Eyes

Skin

Respiratory System

Environmental

Cleanup/Disposal: Guide No. 145: Eliminate all ignition sources (no smoking, flares, sparks or flames in immediate area). Keep combustibles (wood, paper, oil, etc.) away from spilled material. Do not touch damaged containers or spilled material unless wearing appropriate protective clothing. Keep substance wet using water spray. Stop leak if you can do it without risk. Small Spills: Take up with inert, damp, noncombustible material using clean non-sparking tools and place into loosely covered plastic containers for later disposal. Large Spills: Wet down with water and dike for later disposal. Prevent entry into waterways, sewers, basements or confined areas.

Regulations

RCRA 40CFR: Not listed
CERCLA: 40CFR 302.4: Not listed
SARA 40CFR 372.65: Not listed
SARA EHS 40CFR 355: Not listed
TSCA: Listed

DIM5450 CAS #: 75-78-5

DIMETHYLDICHLOROSILANE

RTECS: VV3150000
DOT: UN1162; IMO3.2
EINECS Number: 200-901-0
Molecular Formula: $C_2H_6Cl_2Si$
Structured MF: $(CH_3)_2SiCl_2$
Formula Weight: 129.O7

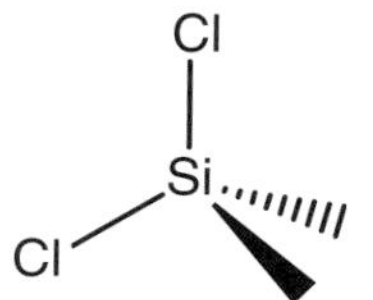

Chemical Structure

Synonyms: DICHLORODIMETHYLSILANE; DICHLORODIMETHYLSILICON; DIMETHYL-DICHLORSILAN; INERTON DW-DMC; SILANE,DICHLORODIMETHYL-

Description: colorless liquid; sharp, irritating hydrochloric acid odor

Use: in ethchlorvynol assay; int for silicone products; high-purity derivatization reagent for gas chromatography

Physical Properties

Boiling Point: 70.5 °C (159 °F)
Freezing Point: < -70 °C (-94 °F)
Specific Gravity: 1.07
Vapor Density: 4.45 Air=1
Vapor Pressure: 5.73 kPa at 0 °C
Water Solubility: Reacts with Water
Other Solubilities: Soluble in Benzene & Ether
Surface Tension: 20.1 dynes/cm at 20 °C
pH: Acid
Refraction Index: 1.4023
Ionization Potential (eV): 10.7
Flash Point: -8.889 °C Open Cup
Autoignition Temperature: > 399 °C
LEL: 1.4% v/v
UEL: 9.5% v/v

RTECS Toxicity Data

Acute Oral: Rat LD_{50} Dose: 5660 uL/kg; Toxic Effects: Behavioral - Somnolence (general depressed activity); Lungs, Thorax, or Respiration - Dyspnea; Liver - Other changes.

Acute Inhalation: Rat LC_{50} Dose: 930 ppm/4hr. Mouse LC_{50} Dose: 300 mg/m^3/2hr.

Irritation Eye: Rabbit Standard Draize Test Dose: 5 mg/24H; Reaction: severe.

Irritation Skin: Rabbit Standard Draize Test Dose: 20 mg/24H; Reaction: moderate.

Hazard Overviews

Corrosive

Flammable

Fire Diamond

Health: Corrosive, causes severe burns to eyes/skin/respiratory tract. Also Causes: difficulty breathing, chest pain, pulmonary edema, low blood oxygen, shock, circulatory collapse, kidney failure, blindnesss.

Fire: Flammable. Reacts with water producing excessive heat and hydrochloric acid. Use dry chemical or carbon dioxide. Do not use water. Can form explosive mixtures in air.

Reactivity: Stable. Hazardous polymerization cannot occur. Avoid: exposure to heat and ignition sources; water; oxidizers. Incompatible with: oxidizers; water/moisture. Hazardous decomposition products: chlorine; hydrogen chloride; phosgene gas.

Carcinogenicity: IARC - Not listed; NIOSH - Not listed; NTP - Not listed; ACGIH - Not listed; OSHA - Not listed; EPA - Not listed; MAK - Not listed

Primary Target Organs:

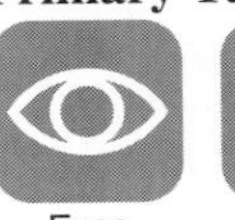

Eyes Skin Respiratory System

Exposure Limits

AIHA WEEL: STEL: 1 ppm ceiling.

Environmental

Cleanup/Disposal: Guide No. 155: Eliminate all ignition sources (no smoking, flares, sparks or flames in immediate area). All equipment used when handling the product must be grounded. Do not touch damaged containers or spilled material unless wearing appropriate protective clothing. Stop leak if you can do it without risk. A vapor suppressing foam may be used to reduce vapors. For chlorosilanes, use AFFF alcohol-resistant medium expansion foam to reduce vapors. Do not get water on spilled substance or inside containers. Use water spray to reduce vapors or divert vapor cloud drift. Prevent entry into waterways, sewers, basements or confined areas. Small Spills: Cover with dry earth, dry sand, or other non-combustible material followed with plastic sheet to minimize spreading or contact with rain. Use clean non-sparking tools to collect material and place it into loosely covered plastic containers for later disposal.

Environmental Physical Data

BCF: none

Regulations

RCRA 40CFR: Not listed
CERCLA: 40CFR 302.4: Not listed
SARA 40CFR 372.65: Listed TPQ: 500 lb
SARA EHS 40CFR 355: Listed TPQ: 500 lb
TSCA: Listed

DIM5500 CAS #: 78-62-6

DIMETHYLDIETHOXYISLANE

RTECS: VV3590000
EINECS Number: 201-127-6
Molecular Formula: $C_6H_{16}O_2Si$
Formula Weight: 148.31

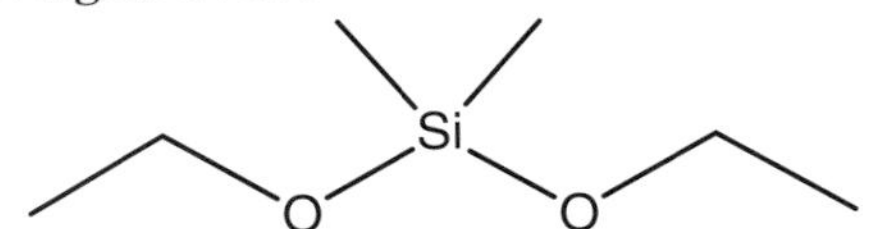

Chemical Structure

Synonyms: DIETHOXYDIMETHYLSILANE; DIMETHYL-DIETHOXYSILAN; DIMETHYLDIETHOXYSILANE

Physical Properties

Boiling Point: 113 °C (235 °F)
Freezing Point: -97 °C (-142.6 °F)
Specific Gravity: 0.848
Vapor Density: < 1 Air=1
Vapor Pressure: 0.071 kPa at -25 °C
Water Solubility: Reacts
Other Solubilities: Soluble in Ethanol & Toluene
Refraction Index: 1.3805
Evaporation Rate: > 1 Butyl Acetate=1

RTECS Toxicity Data

Acute Oral: Rat LD_{50} Dose: 9280 mg/kg.
Acute Inhalation: Rat LC_{50} Dose: 20 gm/m^3/8hr.
Chronic (Multiple Dose) Inhalation: Rat Dose: 4000 ppm/7H/10D-I; Toxic Effects: DEATH.
Irritation Eye: Rabbit Standard Draize Test Dose: 500 mg/24H; Reaction: mild.
Irritation Skin: Rabbit Standard Draize Test Dose: 500 mg/24H; Reaction: mild.

Hazard Overviews

Health: Irritating to eyes/skin/respiratory tract. Other Acute Effects: may be harmful by inhalation, ingestion, or skin absorption; prolonged exposure can cause: nausea, dizziness and headache; gastrointestinal disturbances.

Fire: Hazards: vapor may travel considerable distance to source of ignition and flash back; container explosion may occur; emits toxic fumes. Extinguishing agents: carbon dioxide, dry chemical powder or appropriate foam. Precautions: combustible liquid.

Reactivity: Incompatible with: strong oxidizing agents, strong acids, may decompose on exposure to moist air or water. Hazardous decomposition products: toxic fumes of: carbon monoxide, carbon dioxide, silicon oxide.

Carcinogenicity: IARC - Not listed; NIOSH - Not listed; NTP - Not listed; ACGIH - Not listed; OSHA - Not listed; EPA - Not listed; MAK - Not listed

Primary Target Organs:

Eyes Skin Respiratory System

Environmental

Regulations

RCRA 40CFR: Not listed
CERCLA: 40CFR 302.4: Not listed
SARA 40CFR 372.65: Not listed
SARA EHS 40CFR 355: Not listed
TSCA: Listed

DIM5550 CAS #: 6026-42-2

DIMETHYLDIISOPROPYLAMINO-SILANE

EINECS Number: 227-885-8

Physical Properties

Boiling Point: 157.8 °C (316 °F)
Specific Gravity: < 1
Vapor Density: > 1 Air=1
Water Solubility: Reacts
Evaporation Rate: < 1 Butyl Acetate=1

Hazard Overviews

Carcinogenicity: IARC - Not listed; NIOSH - Not listed; NTP - Not listed; ACGIH - Not listed; OSHA - Not listed; EPA - Not listed; MAK - Not listed

Environmental

Regulations

RCRA 40CFR: Not listed
CERCLA: 40CFR 302.4: Not listed
SARA 40CFR 372.65: Not listed
SARA EHS 40CFR 355: Not listed
TSCA: Not listed

DIM5650 CAS #: 107-64-2

DIMETHYLDIOCTADECYLAMMONIUM CHLORIDE

RTECS: BQ1923000
EINECS Number: 203-508-2
Molecular Formula: $C_{38}H_{80}ClN$
Formula Weight: 586.64

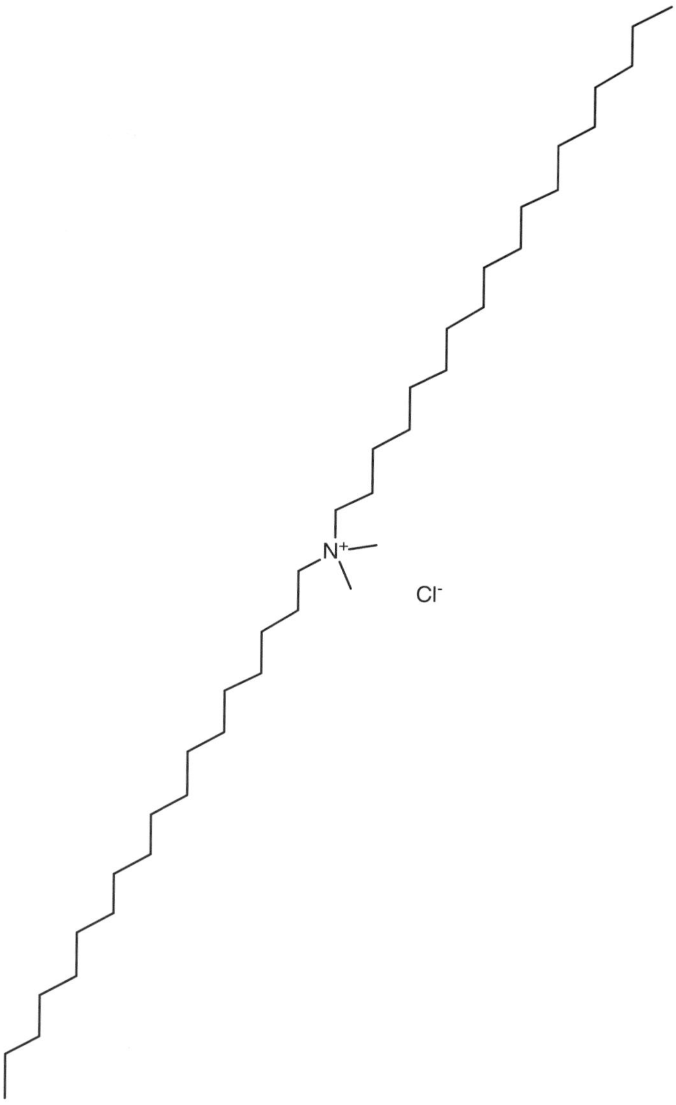

N^+ Cl^-

Chemical Structure

Synonyms: ALIQUAT 207; AMMONIUM,DIMETHYLDIOCTADECYL-,CHLORIDE; AROSURF TA 100; ARQUAD R 40; DIMETHYLDISTEARYLAMMONIUM CHLORIDE; N,N-DIMETHYL-N-OCTADECYL-1-OCTADECANAMINIUM CHLORIDE; DIOCTADECYLDIMETHYLAMMONIUM CHLORIDE; N,N-DIOCTADECYL-N,N-DIMETHYLAMMONIUM CHLORIDE; DISTEARYL DIMETHYLAMMONIUM CHLORIDE; DISTEARYLDIMETHYLAMMONIUM CHLORIDE; GENAMIN DSAC; KD 83; 1-OCTADECANAMINIUM,N,N-DIMETHYL-N-OCTADECYL-,CHLORIDE; 1-OCTADECANAMINIUM,N,N-DIMETHYL-N-OCTADECYL-,CHLORIDE(9CI); Q-D 86P; QUATERNIUM-5; QUATERNIUM 5; TALOFLOC; VARISOFT 100

Use: surfactant; laundry softener; in hair rinse formulations; fabric softener for retail, commercial & textile mills; cationic surfactant in cosmetic products-eg, creme rinses

RTECS Toxicity Data

Acute Oral: Rat LD_{50} Dose: 11300 mg/kg; Toxic Effects: Behavioral - Somnolence (general depressed activity);

Gastrointestinal - Hypermotility, diarrhea; Skin and appendages - Hair.

Hazard Overviews

Health: May be irritating to eyes/skin/respiratory tract. Other Acute Effects: may be harmful by inhalation, ingestion, or skin absorption.

Fire: Hazards: emits toxic fumes. Extinguishing agents: water spray; carbon dioxide, dry chemical powder or appropriate foam. Precautions: combustible liquid.

Reactivity: Stable. Hazardous polymerization will not occur. Incompatible with: strong oxidizing agents. Hazardous decomposition products: toxic fumes of: carbon monoxide, carbon dioxide, nitrogen oxides, hydrogen chloride gas.

Carcinogenicity: IARC - Not listed; NIOSH - Not listed; NTP - Not listed; ACGIH - Not listed; OSHA - Not listed; EPA - Not listed; MAK - Not listed

Environmental

Environmental Physical Data

BCF: bluegill lab water 260

Regulations

RCRA 40CFR: Not listed
CERCLA: 40CFR 302.4: Not listed
SARA 40CFR 372.65: Not listed
SARA EHS 40CFR 355: Not listed
TSCA: Listed

DIM5850 CAS #: 112-18-5

N,N-DIMETHYL-1-DODECANAMINE

RTECS: JR6600000
EINECS Number: 203-943-8
Molecular Formula: $C_{14}H_{31}N$
Formula Weight: 213.46

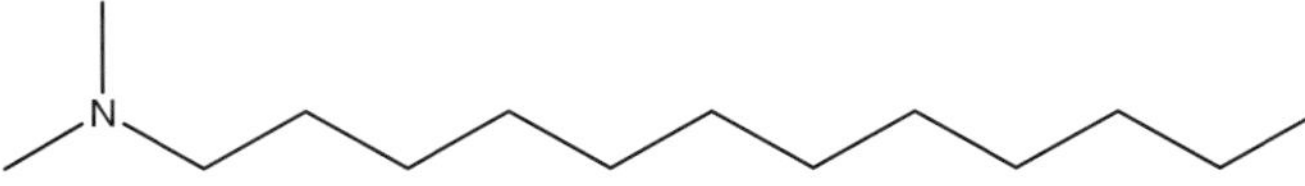

Chemical Structure

Synonyms: ADMA-2; ADMA 2; ANTIOXIDANT DDA; ARMEEN DM 12D; ARMEEN DM-12D; BARLENE 125; DDA; DDA (CORROSION INHIBITOR); N,N-DIMETHYL-N-DODECYLAMINE; N,N-DIMETHYLDODECYLAMINE; N,N-DIMETHYLLAURYLAMINE; 1-DODECANAMINE,N,N-DIMETHYL-; I-DODECANAMINE,N,N-DIMETHYL; DODECYLAMINE,N,N-DIMETHYL-; DODECYLDIMETHYLAMINE; N-DODECYLDIMETHYLAMINE; FARMIN DM 20; IPL; N-LAURLYDIMETHYLAMINE; LAURYLDIMETHYLAMINE; N-LAURYLDIMETHYLAMINE; MONOLAURYL DIMETHYLAMINE; RC 5629

Description: liquid

Use: corrosion inhibitor; acid-stable emulsifier

Physical Properties

Other Solubilities: 95% Ethanol; Acetone; DMSO.

Flash Point: Not available; probably combustible

RTECS Toxicity Data

Irritation Eye: Rabbit Standard Draize Test Dose: 50 ug/24H; Reaction: severe.

Irritation Skin: Rabbit Standard Draize Test Dose: 2 mg/24H; Reaction: severe.

Hazard Overviews

Corrosive

Health: Corrosive to eyes/skin/respiratory tract. Other Acute Effects: harmful if swallowed, inhaled, or absorbed through skin; inhalation may result in spasm, inflammation and edema of the larynx and bronchi, chemical pneumonitis and pulmonary edema; symptoms of exposure may include burning sensation; coughing; wheezing; laryngitis; shortness of breath; headache; nausea; vomiting.

Fire: Will burn. Hazards: emits toxic fumes. Extinguishing agents: carbon dioxide, dry chemical powder or appropriate foam. Precautions: combustible liquid.

Reactivity: Incompatible with: strong oxidizing agents, strong acids. Hazardous decomposition products: toxic fumes of: carbon monoxide, carbon dioxide, nitrogen oxides.

Carcinogenicity: IARC - Not listed; NIOSH - Not listed; NTP - Not listed; ACGIH - Not listed; OSHA - Not listed; EPA - Not listed; MAK - Not listed

Primary Target Organs:

Eyes

Skin

Respiratory System

Environmental

Regulations

RCRA 40CFR: Not listed
CERCLA: 40CFR 302.4: Not listed
SARA 40CFR 372.65: Not listed
SARA EHS 40CFR 355: Not listed
TSCA: Listed

DIM5900 CAS #: 1643-20-5

N,N-DIMETHYL-N-DODECYLAMINE OXIDE

RTECS: JR6650000
EINECS Number: 216-700-6
Molecular Formula: $C_{14}H_{31}NO$
Formula Weight: 229.46

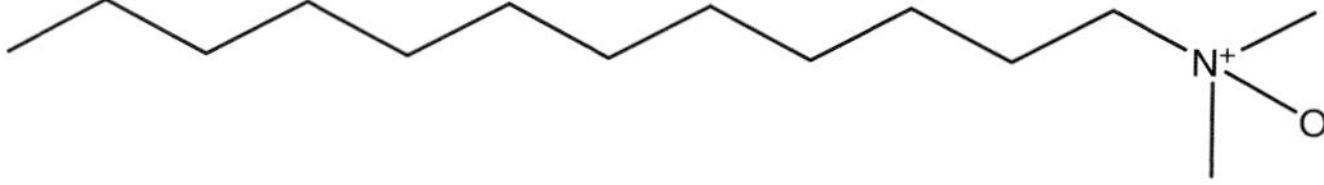

Chemical Structure

Synonyms: AMMONYX AO; AMMONYX LO; AMONYX AO; AROMOX DMCD; AROMOX DMMC-W; CONCO XAL; DDNO; DIMETHYLAURYLAMINE OXIDE; DIMETHYLDODECYLAMINE N-OXIDE; DIMETHYLDODECYLAMINE OXIDE; N,N-DIMETHYLDODECYLAMINE OXIDE; N,N-DIMETHYL-DODECYLAMINOXID; DIMETHYLLAURYLAMINE OXIDE; N,N-DIMETYLDODECYLAMINE OXIDE; 1-DODECANAMINE,N,N-DIMETHL-,N-OXIDE; DODECYCLDIMETHYLAMINE OXIDE; DODECYLAMINE,N,N-DIMETHYL-,N-OXIDE; DODECYLDIMETHYLAMINE OXIDE; N-DODECYLDIMETHYLAMINE OXIDE; EMPIGEN OB; LAURAMINE OXIDE; LAURYLDIMETHYLAMINE N-OXIDE; LAURYLDIMETHYLAMINE OXIDE; N-LAURYL-N,N-DIMETHYLAMINE OXIDE; N-LAURYLDIMETHYLAMINE N-OXIDE

Use: detergents

Physical Properties

Freezing Point: 130 °C (266 °F) to 131 °C (267.8 °F)

Flash Point: Not available; probably nonflammable

RTECS Toxicity Data

Irritation Eye: Rabbit Standard Draize Test Dose: 50 ug/24H; Reaction: severe.

Irritation Skin: Rabbit Standard Draize Test Dose: 2 mg/24H; Reaction: severe.

Hazard Overviews

Corrosive

Health: Corrosive to eyes/skin/respiratory tract. Other Acute Effects: harmful if swallowed, inhaled, or absorbed through skin; inhalation may result in spasm, inflammation and edema of the larynx and bronchi, chemical pneumonitis and pulmonary edema; symptoms of exposure may include burning sensation; coughing; wheezing; laryngitis; shortness of breath; headache; nausea; vomiting.

Fire: Noncombustible. Hazards: emits toxic fumes. Extinguishing agents: carbon dioxide, dry chemical powder or appropriate foam; water spray. Precautions: combustible liquid.

Reactivity: Stable. Hazardous polymerization will not occur. Incompatible with: strong oxidizing agents. Hazardous decomposition products: toxic fumes of: carbon monoxide, carbon dioxide, nitrogen oxides, ammonia.

Carcinogenicity: IARC - Not listed; NIOSH - Not listed; NTP - Not listed; ACGIH - Not listed; OSHA - Not listed; EPA - Not listed; MAK - Not listed

Primary Target Organs:

Eyes

Skin

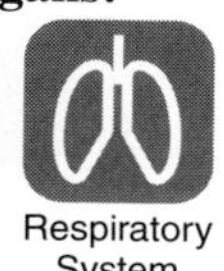

Respiratory System

Environmental

Cleanup/Disposal: Guide No. 171: Do not touch or walk through spilled material. Stop leak if you can do it without risk. Prevent dust cloud. Avoid inhalation of asbestos dust. Small Dry Spills: With clean shovel place material into clean, dry container and cover loosely; move containers from spill area. Small Spills: Take up with sand or other noncombustible absorbent material and place into containers for later disposal. Large Spills: Dike far ahead of liquid spill for later disposal. Cover powder spill with plastic sheet or tarp to minimize spreading. Prevent entry into waterways, sewers, basements or confined areas.

Regulations

RCRA 40CFR: Not listed

CERCLA: 40CFR 302.4: Not listed

SARA 40CFR 372.65: Not listed

SARA EHS 40CFR 355: Not listed

TSCA: Listed

DIM6000 CAS #: 598-56-1

N,N-DIMETHYLETHYLAMINE

EINECS Number: 209-940-8

Molecular Formula: $C_4H_{11}N$

Formula Weight: 73.14

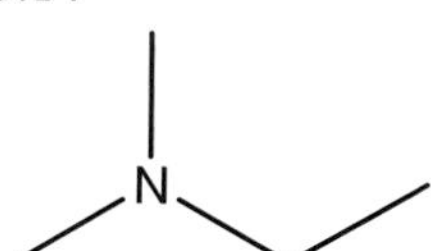

Chemical Structure

Synonyms: ETHANAMINE,N,N-DIMETHYL-; ETHYLAMINE,N,N-DIMETHYL-; N-ETHYLDIMETHYLAMINE; METHANAMINE,N-ETHYL-N-METHYL-

Use: curing catalyst for resin bonding of sand cores

Physical Properties

Boiling Point: 36.5 °C (97.7 °F)

Freezing Point: -140 °C (-220 °F)

Specific Gravity: 0.67500

Refraction Index: 1.3705

Ionization Potential (eV): 7.74 +/-0.05

Hazard Overviews

Corrosive

Health: Corrosive to eyes/skin/respiratory tract. Other Acute Effects: harmful if swallowed, inhaled, or absorbed through skin; inhalation may result in spasm, inflammation and edema of the larynx and bronchi, chemical pneumonitis and pulmonary edema; prolonged contact can cause severe irritation or burns; damage to the eyes.
Fire: Hazards: vapor may travel considerable distance to source of ignition and flash back; container explosion may occur; emits toxic fumes; forms explosive mixtures in air. Extinguishing agents: carbon dioxide, dry chemical powder or appropriate foam; water may be effective for cooling, but may not effect extinguishment. Precautions: combustible liquid.
Reactivity: Incompatible with: acids, acid chlorides, acid anhydrides, strong oxidizing agents, carbon dioxide. Hazardous decomposition products: thermal decomposition may produce carbon monoxide, carbon dioxide, and nitrogen oxides.
Carcinogenicity: IARC - Not listed; NIOSH - Not listed; NTP - Not listed; ACGIH - Not listed; OSHA - Not listed; EPA - Not listed; MAK - Not listed
Primary Target Organs:

Eyes

Skin
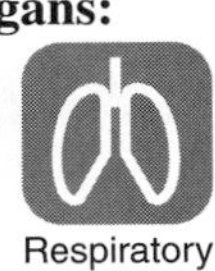
Respiratory System

Exposure Limits
DFG MAK: TWA: 25 ppm; 75 mg/m^3.

Environmental

Regulations

RCRA 40CFR: Not listed
CERCLA: 40CFR 302.4: Not listed
SARA 40CFR 372.65: Not listed
SARA EHS 40CFR 355: Not listed
TSCA: Listed

DIM6050 CAS #: 88-60-8

2-(1,1-DIMETHYLETHYL)-5-METHYLPHENOL

RTECS: SK1578180
EINECS Number: 201-842-3
Molecular Formula: $C_{11}H_{16}O$
Formula Weight: 164.25

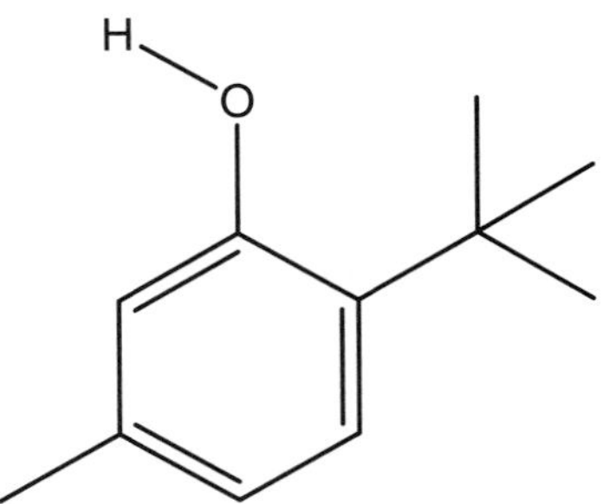

Chemical Structure

Synonyms: BENZENE,1-TERT-BUTYL-2-HYDROXY-4-METHYL-; M-CRESOL,6-TERT-BUTYL-; PHENOL,2-(1,1-DIMETHYLETHYL)-5-METHYL-
Use: chem int for musk ambrette, a fixative in perfumes; chem int for bridged phenol compds (rubber antioxidants); chem int for phenolic antioxidants used in thermoplastics; int for reaction product with sulfur dichloride; antioxidant & stabilizer for coumarone-indene resins; antioxidant & stabilizer for polyvinyl ethers; stabilizer for trichloro- & perchloroethylene

Physical Properties

Boiling Point: 224 °C (435 °F) at 760 mm Hg
Freezing Point: 46 °C (114.8 °F)
Specific Gravity: 0.922 at 80 °C/4 °C
Water Solubility: Insoluble in Water
Other Solubilities: Soluble in Alcohol, Ether, Acetone, organic solvents
Refraction Index: 1.5250 at 20 °C

RTECS Toxicity Data

Acute Oral: Mouse LD_{50} Dose: 1080 mg/kg.

Hazard Overviews

Health: Severely irritating. Harmful. Other Acute Effects: harmful if swallowed, inhaled, or absorbed through skin; high concentrations are extremely destructive to tissues of the mucous membranes, upper respiratory tract, eyes and skin; symptoms of exposure may include burning sensation, coughing, wheezing, laryngitis, shortness of breath, headache, nausea and vomiting.
Fire: Hazards: emits toxic fumes. Extinguishing agents: water spray; carbon dioxide, dry chemical powder or appropriate foam. Precautions: combustible liquid.
Reactivity: Incompatible with: strong oxidizing agents. Hazardous decomposition products: toxic fumes of: carbon monoxide, carbon dioxide.
Carcinogenicity: IARC - Not listed; NIOSH - Not listed; NTP - Not listed; ACGIH - Not listed; OSHA - Not listed; EPA - Not listed; MAK - Not listed

Regulations
RCRA 40CFR: Not listed
CERCLA: 40CFR 302.4: Listed per CAA Section 112 RQ: 100 lb (45.35 kg)
SARA 40CFR 372.65: Listed
SARA EHS 40CFR 355: Not listed
TSCA: Listed

Analytical Methods
Air: ASTM D4490
Soil: EPA 1625
Water / Groundwater: EPA 1625
Indoor / Expired Air: NIOSH 2004
Other: EPA 1665

DIM6200 CAS #: 110-03-2
2,5-DIMETHYL-2,5-HEXANEDIOL

EINECS Number: 203-731-5
Molecular Formula: $C_8H_{18}O_2$
Formula Weight: 146.23

Chemical Structure

Synonyms: 2,5-DIMETHYLHEXANE-2,5-DIOL; DIMETHYLHEXANEDIOL; 2,5-HEXANEDIOL,2,5-DIMETHYL-; NSC 5595; 1,1,4,4-TETRAMETHYL-1,4-BUTANEDIOL
Description: white crystals, prisms, or flakes
Use: chemical intermediate

Physical Properties
Boiling Point: 118 °C (244 °F) at 15 mm Hg
Freezing Point: 92 °C (197.6 °F)
Density: 0.898 g/mL at 20 °C
Water Solubility: Soluble in Water
Other Solubilities: Soluble in Acetone; Insoluble in Carbon Tetrachloride, Kerosene.

Hazard Overviews
Health: Irritating to eyes/skin. Other Acute Effects: may be harmful by inhalation, ingestion, or skin absorption.
Fire: Extinguishing agents: water spray; carbon dioxide, dry chemical powder or appropriate foam. Precautions: combustible liquid.
Reactivity: Incompatible with: strong acids, strong oxidizing agents, strong reducing agents, acid chlorides, acid anhydrides. Hazardous decomposition products: toxic fumes of: carbon monoxide, carbon dioxide.

Carcinogenicity: IARC - Not listed; NIOSH - Not listed; NTP - Not listed; ACGIH - Not listed; OSHA - Not listed; EPA - Not listed; MAK - Not listed
Primary Target Organs:

Eyes

Skin

Environmental

Regulations
RCRA 40CFR: Not listed
CERCLA: 40CFR 302.4: Not listed
SARA 40CFR 372.65: Not listed
SARA EHS 40CFR 355: Not listed
TSCA: Listed

DIM6250 CAS #: 142-30-3
2,5-DIMETHYL-3-HEXYNE-2,5-DIOL

EINECS Number: 205-533-4
Molecular Formula: $C_8H_{14}O_2$
Formula Weight: 142.20

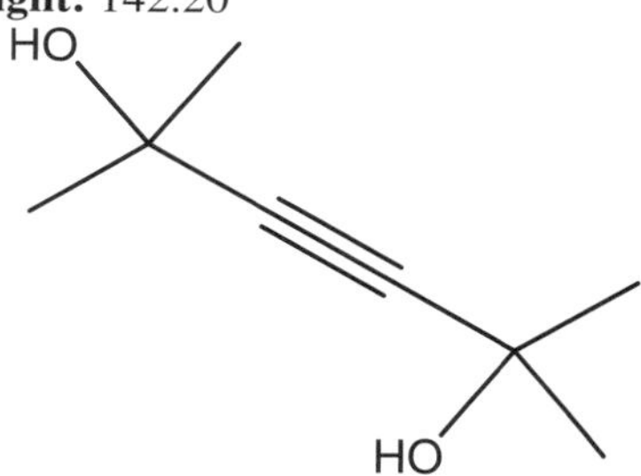

Chemical Structure

Synonyms: ACETYLENEPINACOL; D 43; DIMETHYLHEXYNEDIOL; 3-HEXYNE-2,5-DIOL,2,5-DIMETHYL-; KEMITRACIN-50; TETRAMETHYLBUTYNEDIOL
Description: white crystals
Use: wire-drawing lubricant; antifoaming agent; coupling agent in resin coatings; chemical intermediate; chem int for fragrances & peroxide catalysts; surface-active agent; metal treating chemical

Physical Properties
Boiling Point: 205 °C (401 °F) at 759 mm Hg
Freezing Point: 95 °C (203 °F)
Specific Gravity: 0.949 at 20 °C/20 °C
Water Solubility: Soluble in Water
Other Solubilities: Slightly Soluble in Carbon Tetrachloride, Naphtha; Very Soluble in Ethyl Acetate.

Hazard Overviews
Health: Irritating to eyes/skin. Harmful. Other Acute Effects: may be harmful by inhalation, ingestion, or skin absorption; prolonged or repeated exposure may cause allergic reactions

in certain sensitive individuals; may cause sensitization by inhalation and skin contact.

Fire: Hazards: in powder form capable of creating a dust explosion. Extinguishing agents: water spray; carbon dioxide, dry chemical powder or appropriate foam. Precautions: combustible liquid.

Reactivity: Incompatible with: strong acids, strong oxidizing agents, strong reducing agents, acid chlorides, acid anhydrides, strong bases. Hazardous decomposition products: toxic fumes of: carbon monoxide, carbon dioxide.

Carcinogenicity: IARC - Not listed; NIOSH - Not listed; NTP - Not listed; ACGIH - Not listed; OSHA - Not listed; EPA - Not listed; MAK - Not listed

Primary Target Organs:

Eyes

Skin

Environmental

Regulations

RCRA 40CFR: Not listed
CERCLA: 40CFR 302.4: Not listed
SARA 40CFR 372.65: Not listed
SARA EHS 40CFR 355: Not listed
TSCA: Listed

DIM6350 CAS #: 77-71-4

5,5-DIMETHYLHYDANTOIN

RTECS: MU0977000
EINECS Number: 201-051-3
Molecular Formula: $C_5H_8N_2O_2$
Formula Weight: 128.15

Chemical Structure

Synonyms: DANTOIN DMH; DIMETHYLHYDANTOIN; 5,5-DIMETHYL-2,4-IMIDAZOLIDINEDIONE; DM HYDANTOIN; DMH; HYDANTOIN,5,5-DIMETHYL-; 2,4-IMIDAZOLIDINEDIONE,5,5-DIMETHYL-; NSC 8652; T10

Description: white, crystalline solid

Use: synthesis; preparation of water-soluble polymers; in industry when weather resistance, high temperature resistance, good color, or high filler loading is required

Physical Properties

Freezing Point: 178 °C (352.4 °F)
Water Solubility: Soluble in Water
Other Solubilities: Soluble in Alcohol, and Ether

RTECS Toxicity Data

Acute Oral: Rat LD_{50} Dose: 7800 mg/kg; Toxic Effects: Behavioral - Somnolence (general depressed activity); Behavioral - Withdrawal; Lungs, Thorax, or Respiration - Dyspnea. Rabbit LD_{50} Dose: 12660 mg/kg; Toxic Effects: Behavioral - Somnolence (general depressed activity); Behavioral - Rigidity; Lungs, Thorax, or Respiration - Dyspnea.

Acute Dermal: Mouse LD_{50} Route: Subcutaneous Dose: 2800 mg/kg.

Chronic (Multiple Dose) Oral: Rat Dose: 2464 mg/kg/22W-I; Toxic Effects: Liver - Other changes; Biochemical - Catalases.

Hazard Overviews

Health: May cause irritation to eyes/skin. Other Acute Effects: may be harmful by inhalation, ingestion, or skin absorption; can cause CNS depression.

Fire: Hazards: emits toxic fumes. Extinguishing agents: water spray; carbon dioxide, dry chemical powder or appropriate foam. Precautions: combustible liquid.

Reactivity: Incompatible with: strong oxidizing agents. Hazardous decomposition products: toxic fumes of: carbon monoxide, carbon dioxide, nitrogen oxides.

Carcinogenicity: IARC - Not listed; NIOSH - Not listed; NTP - Not listed; ACGIH - Not listed; OSHA - Not listed; EPA - Not listed; MAK - Not listed

Environmental

Ecotoxicity: Fishes: Pimephales promelas 4d LC_{50} 16,500 mg/l

Regulations

RCRA 40CFR: Not listed
CERCLA: 40CFR 302.4: Not listed
SARA 40CFR 372.65: Not listed
SARA EHS 40CFR 355: Not listed
TSCA: Listed

DIM6400 CAS #: 57-14-7

1,1-DIMETHYLHYDRAZINE

RTECS: MV2450000
DOT: UN1163; IMO3.2
EINECS Number: 200-316-0
Molecular Formula: $C_2H_8N_2$
Structured MF: $(CH_3)_2NNH_2$
Formula Weight: 60.10

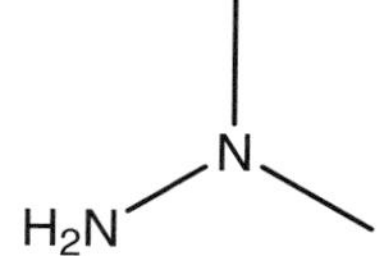

Chemical Structure

Synonyms: ASYM DIMETHYLHYDRAZINE; ASYMMETRIC DIMETHYLHYDRAZINE; ASYMMETRIC-DIMETHYLHYDRAZINE; DIMAZIN; DIMAZINE; AS-DIMETHYL HYDRAZINE; 1,1-DIMETHYLHYDRAZIN; AS-DIMETHYLHYDRAZINE; DIMETHYLHYDRAZINE; N,N-DIMETHYLHYDRAZINE; U-DIMETHYLHYDRAZINE; DIMETHYLHYDRAZINE UNSYMMETRICAL; DIMETHYLHYDRAZINE,UNSYMMETRICAL; N,N-DIMETILIDRAZINA; DMH; HYDRAZINE,1,1-DIMETHYL-; NIESYMETRYCZNA DWU METYLOHYDRAZYNA; UDMH; UNS-DIMETHYLHYDRAZINE; UNSYM-DIMETHYLHYDRAZINE; UNSYMMETRICAL DIMETHYLHYDRAZINE; UNSYMMETRICAL-DIMETHYLHYDRAZINE

Description: colorless liquid; fishy or ammonia-like odor

Use: a chemical intermediate in the manufacture of plant growth inhibitors and aminimides, as a protecting group for carbonyl compounds, as a reagent for conversion of aldehydes to nitriles, as a component of jet and rocket fuels, in chemical synthesis, as a stabilizer for organic peroxide fuel additives, as an absorbent for acid gases and in photography

Physical Properties

Boiling Point: 63.9 °C (147 °F)
Freezing Point: -58 °C (-72.4 °F)
Specific Gravity: 0.7914 at 22 °C/4 °C
Vapor Density: 1.94 Air=1
Saturated Vapor Density: 1.465890867 kg/m^3
Density: 0.7914 g/mL at 22 °C
Vapor Pressure: 20.93 kPa at 25 °C
Water Solubility: Miscible
Other Solubilities: 95% Ethanol: >=100 mg/ml at 24 °C; Acetone: >=100 mg/ml at 24 °C; Alcohol: Soluble; DMSO: >=100 mg/ml at 24 °C; Dimethyl formamide: miscible; Ether: miscible; Hydrocarbons: miscible; Petroleum Ether: miscible.
Surface Tension: 28 dynes/cm
Odor Threshold: 12.0 to 20 mg/m^3
pH: Strongly alkaline liquid
Refraction Index: 1.40753 at 22.3 °C/D
Critical Temperature: 249 °C
Critical Pressure: 53.5 atm
Ionization Potential (eV): 8.05
Flash Point: -15 °C Closed Cup
Autoignition Temperature: 249 °C
LEL: 2% v/v
UEL: 95% v/v

RTECS Toxicity Data

Acute Oral: Rat LD_{50} Dose: 122 mg/kg. Mouse LD_{50} Dose: 265 mg/kg.

Acute Inhalation: Rat LC_{50} Dose: 252 ppm/4hr; Toxic Effects: Sense organs and special senses - Other; Behavioral - Convulsions or effect on seizure threshold; Lungs, Thorax, or Respiration - Dyspnea. Mouse LC_{50} Dose: 172 ppm/4hr; Toxic Effects: Sense organs and special senses - Other; Behavioral - Convulsions or effect on seizure threshold; Lungs, Thorax, or Respiration - Dyspnea.

Acute Dermal: Rabbit LD_{50} Route: Skin; Dose: 1060 mg/kg. Guinea Pig LD_{50} Route: Skin; Dose: 1329 mg/kg; Toxic Effects: Behavioral - Convulsions or effect on seizure threshold.

Chronic (Multiple Dose) Inhalation: Rat Dose: 34 gm/m^3/13M/40D; Toxic Effects: Lungs, Thorax, or Respiration - Chronic pulmonary edema.

Reproductive/Teratogenic: Rat Route: Intraperitoneal; Dose: 600 mg/kg; Duration: female 6-15D of pregnancy; Effects on Fertility - Post-implantation mortality; Effects on Embryo or Fetus - Fetotoxicity.

Mutagenic: Human DNA Damage; Cell Type: fibroblast; Dose: 300 umol/L. Human Morphological Transformation; Cell Type: other cell types; Dose: 50 mg/L. Human Morphological Transformation; Cell Type: fibroblast; Dose: 167 umol/L.

Tumorigenic: Rat Route: Oral; Dose: 150 mg/kg/7W-I; Toxic Effects: Tumorigenic - Equivocal tumorigenic agent by RTECS criteria; Sense organs and special senses - Tumors; Gastrointestinal - Colon tumors. Rat Route: Oral; Dose: 300 mg/kg/14W-I; Toxic Effects: Tumorigenic - Equivocal tumorigenic agent by RTECS criteria; Sense organs and special senses - Tumors; Gastrointestinal - Colon tumors.

Hazard Overviews

Corrosive

Flammable

Fire Diamond

Health: Corrosive to eyes/skin/respiratory tract. Toxic. Also Causes: difficulty breathing, CNS depression, liver injury, hypoxia. Chronic Effects: hemolytic anemia and convulsions; cancer.

Fire: Flammable. Can form explosive mixtures in the air. Use a dry chemical, alcohol foam, or carbon dioxide. Water may be ineffective in extinguishing a fire involving 1,1-dimethylhydrazine. Use water spray to cool fire-exposed tanks or containers.

Reactivity: Stable, in dark, cold conditions. Hazardous polymerization cannot occur. Avoid: ignition sources. Incompatible with: strong oxidizing agents; halogens; ferric oxide (iron rust). Hazardous decomposition products: hydrogen gas; ammonia; dimethylamine; hydrazoic acid.

Carcinogenicity: IARC - Group 2B, Possibly carcinogenic to humans; NIOSH - Listed as carcinogen; NTP - Class 2B, Reasonably anticipated to be a carcinogen, sufficient evidence of carcinogenicity from studies in experimental animals; ACGIH - Class A3, Animal carcinogen; OSHA - Not listed; EPA - Not listed; MAK - Class A2, Unmistakably carcinogenic in animal experimentation only

Primary Target Organs:

Eyes

Skin

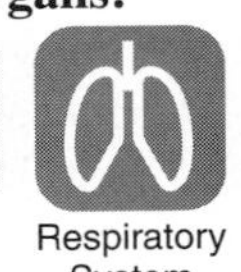
Respiratory System

Nervous System

Liver

Blood

Exposure Limits

OSHA PEL: TWA: 0.5 ppm; 1 mg/m^3.

ACGIH TLV: TWA: 0.01 ppm; 0.025 mg/m^3.

NIOSH REL: STEL: 0.06 ppm; 0.15 mg/m^3; 2-hour.

NIOSH IDLH: 15 ppm.

Respirator Recommendation

Exposure Range: >0.5 to <15 ppm Supplied Air, Constant Flow/Pressure Demand, Half Mask

Exposure Range: 15 to unlimited ppm Self-contained Breathing Apparatus, Pressure Demand, Full Face

Note: poor warning properties

Environmental

Ecotoxicity: LC_{50} Ictalurus punctatus (channel catfish) 11.35 mg/l/96 hr. /Conditions of bioassay not specified LC_{50} Daphnia 38 mg/l/24 hr. /Conditions of bioassay not specified

Environmental Fate: Release to soil will result in some degradation and adsorption to soils containing clay and organic carbon. According to limited data, it is not expected to leach or volatilize from soils containing clay or organic content. However, a low K_{oc} value suggests it will generally be highly mobile in most soils. Volatilization and leaching to groundwater may result upon release to sandy soil. Biodegradation is not expected to be significant. Release to water is expected to result in oxidation at a rate directly proportional to the pH. Bioconcentration is not expected to be important in aquatic systems. Volatilization from the water surface and adsorption to sediments containing organic carbon may occur. Biodegradation is not expected to be significant. Release to the atmosphere is expected to result in rapid degradation by ozone or nitrogen dioxide.

Cleanup/Disposal: Guide No. 131: Fully encapsulating, vapor protective clothing should be worn for spills and leaks with no fire. Eliminate all ignition sources (no smoking, flares, sparks or flames in immediate area). All equipment used when handling the product must be grounded. Do not touch or walk through spilled material. Stop leak if you can do it without risk. Prevent entry into waterways, sewers, basements or confined areas. A vapor suppressing foam may be used to reduce vapors. Small Spills: Absorb with earth, sand or other non-combustible material and transfer to containers for later disposal. Use clean non-sparking tools to collect absorbed material. Large Spills: Dike far ahead of liquid spill for later disposal. Water spray may reduce vapor; but may not prevent ignition in closed spaces.

Environmental Physical Data

Henry's Law Constant: cannot be calculated

Octanol/Water Partition Coefficient: log K_{ow} = -1.37

Sorption Partition Coefficient: K_{oc} = 4

BCF: calculated at -1.37

Regulations

RCRA 40CFR: Listed Hazardous Waste No. U098 Toxic Waste

CERCLA: 40CFR 302.4: Listed per RCRA Section 3001 RQ: 10 lb (4.535 kg)

SARA 40CFR 372.65: Listed TPQ: 1000 lb

SARA EHS 40CFR 355: Listed TPQ: 10 lb

TSCA: Listed

Analytical Methods

Air: ASTM D4490

Indoor / Expired Air: NIOSH 3515

DIM6450 CAS #: 540-73-8

1,2-DIMETHYLHYDRAZINE

RTECS: MV2625000

DOT: UN2382; IMO3.1

Molecular Formula: $C_2H_8N_2$

Structured MF: $(CH_3)HN{=}NHCH_3$

Formula Weight: 60.10

Synonyms: 1,2-DIMETHYLHYDRAZIN; 1,2-DIMETHYL-HYDRAZINE; N,N'-DIMETHYLHYDRAZINE; SYM-DIMETHYLHYDRAZINE; DIMETHYLHYDRAZINE,SYMMETRICAL; 1,2-DMH; DMH; HYDRAZINE,1,2-DIMETHYL-; HYDRAZOMETHANE; SDMH; SYMETRYCZNA DWUMETYLOHYDRAZYNA; SYMMETRICAL-DIMETHYLHYDRAZINE

Description: colorless, clear liquid; fishy or ammoniacal odor

Use: evaluated as high energy rocket fuel; research chemical

Physical Properties

Boiling Point: 81 °C (178 °F) at 753 mm Hg

Freezing Point: -9 °C (15.8 °F)

Specific Gravity: 0.8274 at 20 °C/4 °C

Vapor Density: Estimate 2.07 Air=1

Vapor Pressure: 68 mm Hg at 24.46 °C

Water Solubility: > 10%

Other Solubilities: >10% in Ethanol; >10% in ethyl Ether.

pH: Alkaline

Refraction Index: 1.4209 at 20 °C/D

Flash Point: < 23 °C Closed Cup

RTECS Toxicity Data

Acute Oral: Rat LD_{50} Dose: 100 mg/kg. Mouse LD_{50} Dose: 36 mg/kg.

Acute Inhalation: Rat LC_{Lo} Dose: 280 ppm/4hr; Toxic Effects: Sense organs and special senses - Other; Behavioral - Convulsions or effect on seizure threshold; Lungs, Thorax, or Respiration - Dyspnea.

Acute Dermal: Rat LD_{50} Route: Subcutaneous Dose: 220 mg/kg. Mouse LD_{50} Route: Subcutaneous Dose: 24 mg/kg.

Reproductive/Teratogenic: Rat Route: Intraperitoneal; Dose: 100 mg/kg; Duration: female 6-15D of pregnancy; Effects on Embryo or Fetus - Fetotoxicity.

Mutagenic: Human Unscheduled DNA Synthesis; Cell Type: lung; Dose: 100 uL/L. Human DNA Adduct; Cell Type: lung;

Dose: 1290 umol/L. Human DNA Adduct; Cell Type: fibroblast; Dose: 20 umol. Human Morphological Transformation; Cell Type: fibroblast; Dose: 230 umol/L.

Tumorigenic: Rat Route: Oral; Dose: 120 mg/kg/4W-I; Toxic Effects: Tumorigenic - Equivocal tumorigenic agent by RTECS criteria; Gastrointestinal - Tumors; Sense organs and special senses - Other. Monkey Route: Subcutaneous; Dose: 400 mg/kg/2Y-I; Toxic Effects: Tumorigenic - Equivocal tumorigenic agent by RTECS criteria; Gastrointestinal - Colon tumors. Monkey Route: Subcutaneous; Dose: 608 mg/kg/2Y-I; Toxic Effects: Tumorigenic - Equivocal tumorigenic agent by RTECS criteria; Tumorigenic effects - Uterine tumors.

Hazard Overviews

Corrosive

Flammable

Fire: Flammable.

Carcinogenicity: IARC - Group 2B, Possibly carcinogenic to humans; NIOSH - Not listed; NTP - Not listed; ACGIH - Not listed; OSHA - Not listed; EPA - Not listed; MAK - Class A2, Unmistakably carcinogenic in animal experimentation only

Environmental

Environmental Fate: If released to the atmosphere, it will degrade rapidly by reaction with photochemically produced hydroxyl radicals (estimated half-life of 3.1 hours). Reaction with atmospheric ozone may also be significant. If released to soil containing traces of heavy metal ions, it may degrade rapidly. In the presence of traces of heavy metal ions (which act as catalysts), it is rapidly dehydrogenated to azomethane. Degradation in soil may also occur through reaction with oxygen. It is miscible in water and may therefore leach significantly in soil. Evaporation from dry surfaces is likely to occur. If released to water, it can degrade through direct reaction with oxygen and by reaction with heavy metal ions. Insufficient data are available to predict the relative importance of biodegradation in soil or water.

Cleanup/Disposal: Guide No. 131: Fully encapsulating, vapor protective clothing should be worn for spills and leaks with no fire. Eliminate all ignition sources (no smoking, flares, sparks or flames in immediate area). All equipment used when handling the product must be grounded. Do not touch or walk through spilled material. Stop leak if you can do it without risk. Prevent entry into waterways, sewers, basements or confined areas. A vapor suppressing foam may be used to reduce vapors. Small Spills: Absorb with earth, sand or other non-combustible material and transfer to containers for later disposal. Use clean non-sparking tools to collect absorbed material. Large Spills: Dike far ahead of liquid spill for later disposal. Water spray may reduce vapor; but may not prevent ignition in closed spaces.

Environmental Physical Data

Octanol/Water Partition Coefficient: log K_{ow} = -1.37

Sorption Partition Coefficient: K_{oc} = 4

BCF: calculated at -1.27

Regulations

RCRA 40CFR: Listed Hazardous Waste No. U099 Toxic Waste

CERCLA: 40CFR 302.4: Listed per RCRA Section 3001 RQ: 1 lb (0.454 kg)

SARA 40CFR 372.65: Not listed

SARA EHS 40CFR 355: Not listed

TSCA: Not listed

DIM6600 CAS #: 2999-74-8

DIMETHYLMAGNESIUM

Molecular Formula: C_2H_6Mg

Formula Weight: 54.38

Synonyms: MAGNESIUM,DIMETHYL-

Description: white solid crystals

Physical Properties

Freezing Point: Stable to 240 °C (464 °F)

Other Solubilities: Very Slightly Soluble in Ether

Flash Point: Inflammable

Hazard Overviews

Fire: Noncombustible.

Carcinogenicity: IARC - Not listed; NIOSH - Not listed; NTP - Not listed; ACGIH - Not listed; OSHA - Not listed; EPA - Not listed; MAK - Not listed

Environmental

Regulations

RCRA 40CFR: Not listed

CERCLA: 40CFR 302.4: Not listed

SARA 40CFR 372.65: Not listed

SARA EHS 40CFR 355: Not listed

TSCA: Listed

DIM6650 CAS #: 766-39-2

2,3-DIMETHYLMALEIC ANHYDRIDE

RTECS: ON4025000

EINECS Number: 212-165-8

Molecular Formula: $C_6H_6O_3$

Formula Weight: 126.12

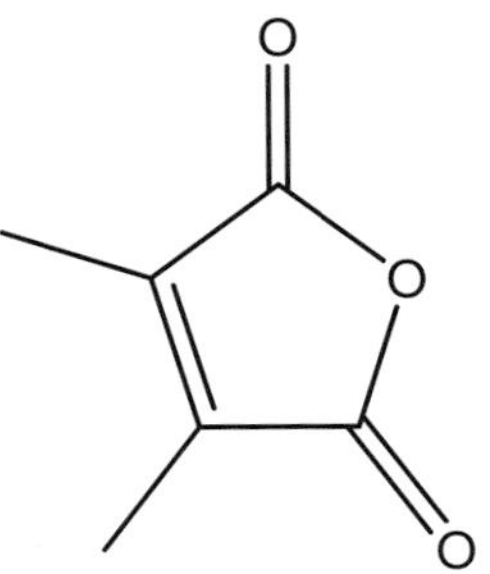

Chemical Structure

Synonyms: DIMETHYLMALEIC ACID ANHYDRIDE; ALPHA,BETA-DIMETHYLMALEIC ANHYDRIDE; DIMETHYLMALEIC ANHYDRIDE; 2,5-FURANDIONE,3,4-DIMETHYL-; MALEIC ANHYDRIDE,DIMETHYL-; PYROCINCHONIC ANHYDRIDE
Use: research chem (chem int for org synth; protein reagent)

Physical Properties

Boiling Point: 223 °C (433 °F)
Freezing Point: 96 °C (204.8 °F)
Specific Gravity: 1.107 at 100 °C/4 °C
Water Solubility: Slightly Soluble in Water
Other Solubilities: Very Soluble in Alcohol, Ether, Benzene, Chloroform

RTECS Toxicity Data

Tumorigenic: Rat Route: Subcutaneous; Dose: 2600 mg/kg/65W-I; Toxic Effects: Tumorigenic - Equivocal tumorigenic agent by RTECS criteria; Tumorigenic - Tumors at site of application.

Hazard Overviews

Health: Irritating to eyes/skin/respiratory tract. Other Acute Effects: may be harmful by inhalation, ingestion, or skin absorption.
Fire: Extinguishing agents: water spray; carbon dioxide, dry chemical powder or appropriate foam. Precautions: combustible liquid.
Reactivity: Incompatible with: strong oxidizing agents, strong acids, strong bases, may decompose on exposure to moist air or water. Hazardous decomposition products: toxic fumes of: carbon monoxide, carbon dioxide.
Carcinogenicity: IARC - Not listed; NIOSH - Not listed; NTP - Not listed; ACGIH - Not listed; OSHA - Not listed; EPA - Not listed; MAK - Not listed
Primary Target Organs:

Eyes

Skin

Respiratory System

Environmental

Regulations

RCRA 40CFR: Not listed
CERCLA: 40CFR 302.4: Not listed
SARA 40CFR 372.65: Not listed
SARA EHS 40CFR 355: Not listed
TSCA: Listed

DIM6850 CAS #: 575-43-9

1,6-DIMETHYLNAPHTHALENE

EINECS Number: 209-385-1
Molecular Formula: $C_{12}H_{12}$
Formula Weight: 156.23

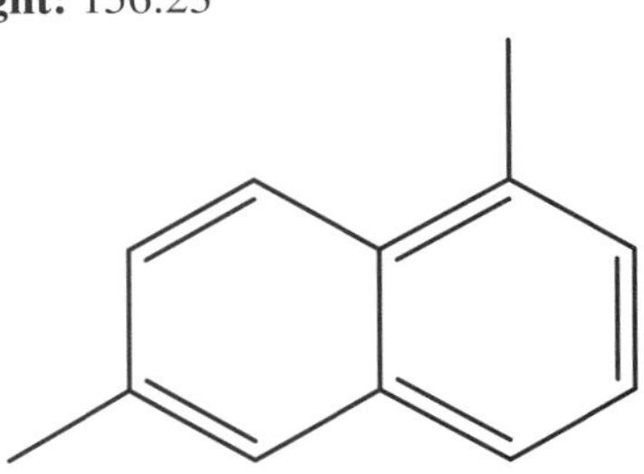

Chemical Structure

Synonyms: NAPHTHALENE,1,6-DIMETHYL-
Description: pale yellow, clear liquid

Physical Properties

Boiling Point: 264 °C (507 °F)
Freezing Point: -16.9 °C (1.58 °F)
Specific Gravity: 1.0021 at 20 °C
Saturated Vapor Density: 1.200101137 kg/m^3
Density: 1.0021 g/mL at 20 °C
Vapor Pressure: 1.46 x10^{-2} mm Hg at 25 °C
Water Solubility: < 1 mg/mL at 20.5 C
Other Solubilities: 95% Ethanol: >=100 mg/ml at 20.5 °C; Acetone: >=100 mg/ml at 20.5 °C; Benzene: >10%; DMSO: >=100 mg/ml at 20.5 °C; Ether: >10%.
Odor Threshold: 0.0428 mg/m^3
Refraction Index: 1.61656 at 20 °C/D
Flash Point: > 110 °C

Hazard Overviews

Health: Irritating to eyes/skin. Acute Effects: may be harmful by inhalation, ingestion, or skin absorption.
Fire: Will burn. Extinguishing agents: water spray; carbon dioxide, dry chemical powder or appropriate foam. Precautions: combustible liquid.
Reactivity: Incompatible with: strong oxidizing agents. Hazardous decomposition products: toxic fumes of: carbon monoxide, carbon dioxide.
Carcinogenicity: IARC - Not listed; NIOSH - Not listed; NTP - Not listed; ACGIH - Not listed; OSHA - Not listed; EPA - Not listed; MAK - Not listed
Primary Target Organs:

Eyes

Skin

Environmental

Environmental Fate: Should biodegrade in the environment. However, hydrolysis should not be important. In sunlit environmental media, it may undergo direct photolysis. It is expected to be slightly mobile to immobile in soil. It has the potential to bioconcentrate in aquatic systems. It may also partition from the water column to organic matter contained in sediments and suspended solids. Volatilization from environmental waters may be important. The volatilization half-lives from a model river and a model pond, the latter considers the effect of adsorption, have been estimated to be 6.2 hr and 79.3 days, respectively. It is expected to exist entirely in the vapor phase in ambient air. In the atmosphere, reaction with photochemically produced hydroxyl radicals (estimated half-life of 5.1 hr) is likely to be important.

Environmental Physical Data

Henry's Law Constant: estimated at 4.25 $x10^{-4}$
Octanol/Water Partition Coefficient: log K_{ow} = 4.61
Sorption Partition Coefficient: K_{oc} = 4625 to 7670
BCF: calculated at 2.82 to 3.77

Regulations

RCRA 40CFR: Not listed
CERCLA: 40CFR 302.4: Not listed
SARA 40CFR 372.65: Not listed
SARA EHS 40CFR 355: Not listed
TSCA: Not listed

DIM6950 CAS #: 138-89-6

N,N-DIMETHYL-P-NITROSOANILINE

RTECS: BX7175000
EINECS Number: 205-343-1
Molecular Formula: $C_8H_{10}N_2O$
Formula Weight: 150.18

Chemical Structure

Synonyms: ACCELERENE; ACCELERINE; ANILINE,N,N-DIMETHYL-4-NITRO-; ANILINE,N,N-DIMETHYL-P-NITROSO-; BENZENAMINE,N,N-DIMETHYL-4-NITROSO-; BENZENAMINE,N,N-DIMETHYL-4-NITROSO-(9CI); 4-(DIMETHYLAMINO)NITROSOBENZENE; P-(DIMETHYLAMINO)NITROSOBENZENE; P-(N,N-DIMETHYLAMINO)NITROSOBENZENE; N,N-DIMETHYL-P-NITROSO ANILINE; N,N-DIMETHYL-4-NITROSOANILINE; N,N-DIMETHYL-PARA-NITROSOANILINE; N,N-DIMETHYL-4-NITROSOBENZENAMINE; DIMETHYL(P-NITROSOPHENYL)AMINE; NDMA; NITROSOANILINE,N,N-DIMETHYL-P-; 4-NITROSO-N,N-DIMETHYLANILINE; 4-NITROSODIMETHYLANILINE; P-NITROSO-N,N-DIMETHYLANILINE; P-NITROSODIMETHYLANILINE; PARA-NITROSODIMETHYLANILINE; PARANITROSODIMETHYLANILIDE; ULTRA BRILLIANT BLUE P

Description: greenish-yellow solid, plates, or leaflets
Use: production of methylene blue, vulcanization accelerator, manufacture of organic compounds, in printing fabrics

Physical Properties

Freezing Point: 92.5 °C (198.5 °F) to 93.5 °C (200.3 °F)
Specific Gravity: 1.145 at 20 °C
Density: 1.145 g/mL at 20 °C
Water Solubility: Insoluble in Water
Other Solubilities: 95% Ethanol: Soluble; Acetone: Soluble; DMSO: Soluble; Ether: Soluble.
Flash Point: Not available; flammable

RTECS Toxicity Data

Acute Oral: Mammal LD_{Lo} Dose: 650 mg/kg; Toxic Effects: Behavioral - Somnolence (general depressed activity); Behavioral - Convulsions or effect on seizure threshold; Kidney, Ureter, and Bladder - Hematuria.
Mutagenic: Bacteria - S Typhimurium Mutations in Microorganisms; Dose: 10 ug/plate (+S9), 33 ug/plate (-S9).
Tumorigenic: Rat Route: Oral; Dose: 7300 mg/kg/1Y-C; Toxic Effects: Tumorigenic - Equivocal tumorigenic agent by RTECS criteria; Gastrointestinal - Tumors; Blood - Lymphomax including Hodgkin's disease.

Hazard Overviews

Flammable

Health: May cause irritation. Toxic. Other Acute Effects: harmful if swallowed, inhaled, or absorbed through skin.
Fire: Flammable. Hazards: emits toxic fumes; in powder form capable of creating a dust explosion. Extinguishing agents: water spray; carbon dioxide, dry chemical powder or appropriate foam. Precautions: combustible liquid.
Reactivity: Incompatible with: strong oxidizing agents, strong acids, strong reducing agents. Hazardous decomposition products: toxic fumes of: carbon monoxide, carbon dioxide, nitrogen oxides.
Carcinogenicity: IARC - Not listed; NIOSH - Not listed; NTP - Not listed; ACGIH - Not listed; OSHA - Not listed; EPA - Not listed; MAK - Not listed

Environmental

Cleanup/Disposal: Guide No. 135: Fully encapsulating, vapor protective clothing should be worn for spills and leak with no fire. Eliminate all ignition sources (no smoking, flares, sparks or flames in immediate area). Do not touch or walk through spilled material. Stop leak if you can do it without risk. Small Spills: Cover with dry earth, dry sand, or other non-combustible material followed with plastic sheet to minimize spreading or contact with rain. Use clean non-sparking tools to collect material and place it into loosely covered plastic containers for later disposal. Prevent entry into waterways, sewers, basements or confined areas.

Regulations

RCRA 40CFR: Not listed
CERCLA: 40CFR 302.4: Not listed
SARA 40CFR 372.65: Not listed
SARA EHS 40CFR 355: Not listed
TSCA: Listed

DIM7000 CAS #: 1854-26-8

DIMETHYLOLDIHYDROXYETHYLENE UREA

RTECS: NJ0607000
EINECS Number: 217-451-6
Molecular Formula: $C_5H_{10}N_2O_5$
Formula Weight: 178.14
Synonyms: ARKOFIX; ARKOFIX NG; 1,3-BIS(HYDROXYMETHYL)-4,5-DIHYDROXY-2-IMIDAZOLINONE; CASSURIT LR; DEPREMOL G; (4,5-DIHYDROXY-1,3-BIS(HYDROXYMETHYL)-2-IMIDAZOLIDINONE; 4,5-DIHYDROXY-1,3-BIS(HYDROXYMETHYL)-2-IMIDAZOLIDINONE; DIHYDROXYDIMETHYLOLETHYLENEUREA; DIHYDROXYETHYLENEDIMETHYLOLUREA; DIMETHYLOGLYOXAMONOUREINE; N,N'-DIMETHYLOL-4,5-DIHYDROXYETHYLENEUREA; DIMETHYLOLDIHYDROXYIMIDAZOLIDINONE; DIMETHYLOLGLYGOXALUREA; DIMETHYLOLGLYOXALUREA; DMDHEU; FIRMATEX RK; FIXAPRET CP; FIXAPRET CP 40; FIXAPRET CPK; FIXAPRET CPN; FIXAPRET CPNS; HYLITE LF; 2-IMIDAZOLIDINONE,4,5-DIHYDROXY-1,3-BIS(HYDROXYMETHYL)-; KNITTEX LE; LE; NEUPERM 6FN; NEUPERM GFN; NS 11; PERMAFRESH 113B; PERMAFRESH 183; PERMAFRESH LF; PERMAFRESH LH; PERMAFRESH LKS; PERMAFURESSHU LKS; PROTOCOL C; PROX DW; READPRET KPN; REAPRET KPN; SARCOSET 6M; SARCOSET GM; SUMITEX FSK; SUMITEX NS; SUMITEX NS 1 SPE; SUMITEX NS 2; SUMITEX NS 1SPE; VERAPRET DH; VERAPRET DKH; WNM
Description: pale yellow, clear liquid
Use: in wrinkle resistant textile finishes

Physical Properties

Water Solubility: >= 100 mg/mL at 20 C
Other Solubilities: 95% Ethanol: May decompose; Acetone: May decompose; DMSO: >=100 mg/ml at 20 °C.
Flash Point: Not available; probably combustible

RTECS Toxicity Data

Irritation Eye: Rabbit Standard Draize Test Dose: 500 mg/24H; Reaction: mild.
Irritation Skin: Rabbit Standard Draize Test Dose: 500 mg/24H; Reaction: severe.
Mutagenic: Insects - D Melanogaster Specific Locus Test; Route: Oral; Dose: 60 ppb. Bacteria - S Typhimurium Mutations in Microorganisms; Dose: 3333 ug/plate (+/-S9).

Hazard Overviews

Fire: Will burn.
Carcinogenicity: IARC - Not listed; NIOSH - Not listed; NTP - Not listed; ACGIH - Not listed; OSHA - Not listed; EPA - Not listed; MAK - Not listed

Environmental

Regulations

RCRA 40CFR: Not listed
CERCLA: 40CFR 302.4: Not listed
SARA 40CFR 372.65: Not listed
SARA EHS 40CFR 355: Not listed
TSCA: Listed

DIM7050 CAS #: 10143-23-4

2,3-DIMETHYL-1-PENTANOL

RTECS: SA6650000
Molecular Formula: $C_7H_{16}O$
Structured MF: $HOCH_2CHCH_3CHCH_3CH_2CH_3$
Formula Weight: 116.23
Synonyms: 2,3-DIMETHYLPENTANOL; 1-PENTANOL,2,3-DIMETHYL-; 1-PENTANOL,2,3-DIMETHYL-,(3S)-(-)-
Description: clear, colorless liquid

Physical Properties

Boiling Point: 155 °C (311 °F) to 156 °C (313 °F)
Density: 0.839 g/cu cm at 23.3 °C
Vapor Pressure: 1 mm Hg at 50 °C
Water Solubility: 1 to 5 mg/mL at 22.5 °C
Other Solubilities: 95% Ethanol: >=100 mg/ml at 22 °C; Acetone: >=100 mg/ml at 22 °C; DMSO: >=100 mg/ml at 22 °C.
Flash Point: 61.1 °C

RTECS Toxicity Data

Acute Oral: Rat LD_{50} Dose: 2380 mg/kg.
Acute Dermal: Rabbit LD_{50} Route: Skin; Dose: 2500 mg/kg; Toxic Effects: Skin and appendages - Primary irritation.
Irritation Skin: Rabbit Open Draize Test Dose: 10 mg/24H open; Reaction: mild.

Hazard Overviews

Fire: Combustible.
Carcinogenicity: IARC - Not listed; NIOSH - Not listed; NTP - Not listed; ACGIH - Not listed; OSHA - Not listed; EPA - Not listed; MAK - Not listed

Environmental

Regulations

RCRA 40CFR: Not listed
CERCLA: 40CFR 302.4: Not listed
SARA 40CFR 372.65: Not listed
SARA EHS 40CFR 355: Not listed
TSCA: Not listed

DIM7100 CAS #: 122-09-8

A,A-DIMETHYLPHENETHYLAMINE

RTECS: SH4025000
EINECS Number: 204-522-1
Molecular Formula: $C_{10}H_{15}N$
Formula Weight: 149.23
Synonyms: 2-AMINO-2-METHYL-1-PHENYLPROPANE; BENZEETHANAMINE,ALPHA,ALPHA-DIMETHYL (9CI); BENZENEETHANAMINE,ALPHA,ALPHA-DIMETHYL-; ALPHA-BENZYLISOPROPYLAMINE; ALPHA,ALPHA-DIMETHYLBENZENEETHANAMINE; ALPHA,ALPHA-DIMETHYLPHENETHYLAMINE; 1,1-DIMETHYL-2-PHENYLETHYLAMINE; ALPHA,ALPHA-DIMETHYL-BETA-PHENYLETHYLAMINE; DUROMINE; ETHANAMINE,1,1-DIMETHYL-2-PHENYL-; IONAMIN; LINYL; LIPOPILL; LONAMIN; MG 18370; MG 18570; MIRAPRONT; NORMEPHENTERMINE; OBERMINE; OMNIBEX; PHENETHYLAMINE,ALPHA,ALPHA-DIMETHYL-; PHENTERMINE; PHENTROL; PHENTROL 2; PHENTROL 3; PHENTROL 4; 2-PHENYL-TERT-BUTYLAMINE; PHENYL-TERT-BUTYLAMINE; WILPO
Description: colorless, mobile, oily liquid; characteristic odor
Use: medication

Physical Properties

Boiling Point: 205 °C (401 °F) at 750 mm Hg
Freezing Point: 198 °C (388.4 °F)
Water Solubility: Slightly Soluble in Water
Other Solubilities: Very Soluble in Chloroform; Insoluble in Ether; Very Slightly Soluble in Benzene & Acetone; Soluble in Ethanol.

RTECS Toxicity Data

Acute Oral: Man TD_{Lo} Dose: 1429 ug/kg; Toxic Effects: Automatic Nervous System - Sympathomimetic. Mouse LD_{50} Dose: 105 mg/kg.
Mutagenic: Mold - A Nidulans Sex Chromosome Loss; Dose: 1 mg/L.

Hazard Overviews

Carcinogenicity: IARC - Not listed; NIOSH - Not listed; NTP - Not listed; ACGIH - Not listed; OSHA - Not listed; EPA - Not listed; MAK - Not listed

Environmental

Regulations

RCRA 40CFR: Listed Hazardous Waste No. P046 Toxic Waste
CERCLA: 40CFR 302.4: Listed per RCRA Section 3001 RQ: 5000 lb (2268 kg)
SARA 40CFR 372.65: Not listed
SARA EHS 40CFR 355: Not listed
TSCA: Listed

Analytical Methods

Soil: SW846 8270B, 8270C
Plasma: EPA 29

DIM7150 CAS #: 526-75-0

2,3-DIMETHYLPHENOL

RTECS: ZE5500000
DOT: UN2261; IMO6.1
EINECS Number: 208-395-3
Molecular Formula: $C_8H_{10}O$
Structured MF: $(CH_3)_2C_6H_3OH$
Formula Weight: 122.18

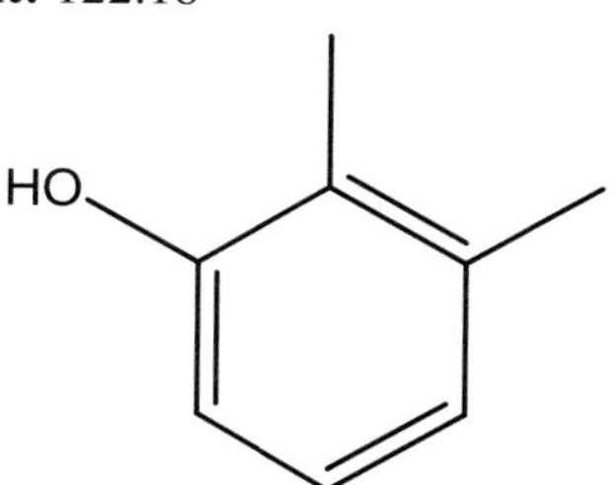

Chemical Structure

Synonyms: BENZENE,1,2-DIMETHYL-3-HYDROXY-; 1-HYDROXY-2,3-DIMETHYLBENZENE; 3-HYDROXY-O-XYLENE; PHENOL,2,3-DIMETHYL-; VIC-O-XYLENOL; XYELLENOL 100; 2,3-XYLENOL; O-3-XYLENOL; O-XYLENOL
Description: needles
Use: for preparation of coal tar disinfectants; manufacture of artificial resins; manufacture of solvents, pharmaceuticals, insecticides, fungicides, plasticizers and rubber chemicals; additive to lubricants and gasolines; wetting agents; and dyestuffs

Physical Properties

Boiling Point: 218 °C (424 °F)
Freezing Point: 75 °C (167 °F)
Specific Gravity: Commercial 1.02 to 1.03 at 15 °C
Vapor Pressure: 1 mm Hg at 56.0 °C
Water Solubility: < 1 mg/mL at 23 C
Other Solubilities: Very Soluble in Benzene, Chloroform.
Odor Threshold: Detection in Water 0.5 mg/l
Refraction Index: 1.5420 at 20 °C
Flash Point: Not available; probably combustible

RTECS Toxicity Data

Acute Inhalation: Rat LC; Dose: >85500 ug/m^3/4hr.

Hazard Overviews

Corrosive

Health: Corrosive to eyes/skin/respiratory tract. Toxic. Other Acute Effects: harmful if swallowed, inhaled, or absorbed through skin; material is extremely destructive to tissue of the mucous membranes and upper respiratory tract, eyes and skin; inhalation may result in spasm, inflammation and edema of the larynx and bronchi, chemical pneumonitis and pulmonary edema; symptoms of exposure may include burning sensation, coughing, wheezing, laryngitis, shortness of breath, headache, nausea and vomiting.

Fire: Will burn. Hazards: under fire conditions, material may decompose to form flammable and/or explosive mixtures in air. Extinguishing agents: water spray; carbon dioxide, dry chemical powder or appropriate foam. Precautions: combustible liquid.

Reactivity: Incompatible with: bases, acid chlorides, acid anhydrides, oxidizing agents, corrodes steel, brass, copper, copper alloys. Hazardous decomposition products: toxic fumes of: carbon monoxide, carbon dioxide.

Carcinogenicity: IARC - Not listed; NIOSH - Not listed; NTP - Not listed; ACGIH - Not listed; OSHA - Not listed; EPA - Not listed; MAK - Not listed

Primary Target Organs:

Eyes

Skin

Respiratory System

Environmental

Ecotoxicity: LC_{50} Daphnia magna (cladoceran) 16.0 mg/l/48 hr /Static bioassay

Environmental Fate: If released into water, it would be expected to rapidly biodegrade with a half life on the order of a few hours to a few days at ambient temperature, and in humic waters reaction with alkyl peroxy radicals should occur with a similar rate constant. The solubility in water combined with its low volatility should preclude volatilization from water into air. The low K_{oc} (estimated range 41.2-1452) and the high rate of aquatic degradation suggest that sorption onto sediment should not be an important process. Similarly, high mobility in soil should be considered. Atmospheric degradation by photochemically produced hydroxyl radicals should be a rapid more rapid in urban areas through the reaction with nitrate radicals. Rain washout would also be an effective method of atmospheric removal.

Cleanup/Disposal: Guide No. 153: Eliminate all ignition sources (no smoking, flares, sparks or flames in immediate area). Do not touch damaged containers or spilled material unless wearing appropriate protective clothing. Stop leak if you can do it without risk. Prevent entry into waterways, sewers, basements or confined areas. Absorb or cover with dry earth, sand or other non-combustible material and transfer to containers. Do not get water inside containers.

Environmental Physical Data

Henry's Law Constant: 9.3×10^{-6}

Octanol/Water Partition Coefficient: log K_{ow} = 2.35

Sorption Partition Coefficient: K_{oc} = estimated at 41.2 to 452

BCF: estimated at 36

Regulations

RCRA 40CFR: Not listed

CERCLA: 40CFR 302.4: Not listed

SARA 40CFR 372.65: Not listed

SARA EHS 40CFR 355: Not listed

TSCA: Listed

Analytical Methods

Soil: SW846 8041

DIM7200 CAS #: 105-67-9

2,4-DIMETHYLPHENOL

RTECS: ZE5600000

DOT: UN2261; IMO6.1

EINECS Number: 203-321-6

Molecular Formula: $C_8H_{10}O$

Structured MF: $(CH_3)_2C_6H_3OH$

Formula Weight: 122.16

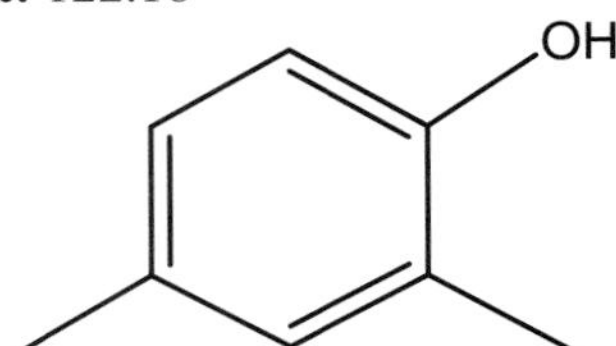

Chemical Structure

Synonyms: ASYM-O-XYLENOL; BACTICIN; BENZENE,2,4-DIMETHYL-1-HYDROXY-; BULK LYSOL BRAND DISINFECTANT; 4,6-DIMETHYLPHENOL; 2,4-DMP; DU COR CONCENTRATED FLY INSECTICIDE; EPA PESTICIDE CHEMICAL CODE 086804; GABLE-TITE DAR CREOSOTE (CREOLA); GABLE-TITE LIGHT CREOSOTE (CREOLA); GALLEX; 1-HYDROXY-2,4-DIMETHYLBENZENE; 4-HYDROXY-1,3-DIMETHYLBENZENE; 4-HYDROXY-M-XYLENE; LYSOL BRAND DISINFECTANT; PHENOL,2,4-DIMETHYL-; 2,4-XYLENOL; AS-M-XYLENOL; M-4-XYLENOL; M-XYLENOL

Description: colorless crystals or needles

Use: in the manufacture of a wide range of commercial products for industry and agriculture; in the manufacture of phenolic antioxidants, disinfectants, solvents, pharmaceuticals, insecticides fungicides, plasticizers, rubber chemicals, polyphenylene oxide, dyestuffs, wetting agents, artificial resins and plastics; in the preparation of coal tar disinfectants; an additive or constituent of lubricants, gasolines and cresylic acid

Physical Properties

Boiling Point: 211.5 °C (413 °F) at 766 mm Hg

Freezing Point: 25.4 °C (77.72 °F) to 26 °C (78.8 °F)

Specific Gravity: 0.965 at 20 °C/4 °C

Saturated Vapor Density: 1.200314986 kg/m^3

Density: 0.965 g/mL at 20 °C

Vapor Pressure: 0.0621 mm Hg at 20 °C

Water Solubility: < 1 mg/mL at 18 C

Other Solubilities: 95% Ethanol: >=100 mg/ml at 18 °C; Acetone: >=100 mg/ml at 18 °C; Alkalies: Very Soluble; Benzene: Freely Soluble; Chloroform: Freely Soluble; DMSO: >=100 mg/ml at 18 °C; Ether: Freely Soluble; Sodium hydroxide solutions: Soluble.

Odor Threshold: 0.0005 to 0.4 mg/m^3

Refraction Index: 1.5420 at 14 °C/D

Ionization Potential (eV): 8.0 +/-0.2
Flash Point: > 112 °C Closed Cup

RTECS Toxicity Data

Acute Oral: Rat LD_{50} Dose: 3200 mg/kg. Mouse LD_{50} Dose: 809 mg/kg.
Acute Inhalation: Rat LC; Dose: >30 mg/m^3.
Acute Dermal: Rat LD_{50} Route: Skin; Dose: 1040 mg/kg.
Chronic (Multiple Dose) Oral: Rat Dose: 6 gm/kg/10D-I; Toxic Effects: Liver - Changes in liver weight; Blood - Changes in serum composition; Blood - Changes in leukocyte (WBC) cell count. Rat Dose: 48600 mg/kg/90D-I; Toxic Effects: Liver - Changes in liver weight; Blood - Changes in serum composition; Nutritional and gross metabolic - Weight loss or decreased weight gain.
Tumorigenic: Mouse Route: Skin; Dose: 16 gm/kg/39W-I; Toxic Effects: Tumorigenic - Carcinogenic by RTECS criteria; Skin and appendages - Tumors.

Hazard Overviews

Fire: Will burn.
Carcinogenicity: IARC - Not listed; NIOSH - Not listed; NTP - Not listed; ACGIH - Not listed; OSHA - Not listed; EPA - Not listed; MAK - Not listed

Environmental

Ecotoxicity: LC_{50} Lemina minor (duckweed) 292,800 ug/l/48 hr /Static, unmeasured bioassay LC_{50} Pimephales promelas (fathead minnow) 17 mg/l/96 hr /Flow through bioassay TLm Carassius carassius (crucian carp) 30 mg/l/24 hr /Conditions of bioassay not specified LC_{50} Daphnia magna (cladoceran) 2,120 ug/l/48 hr /Static, unmeasured bioassay TLm Salvelinus (trout embryo) 28 mg/l/24 hr /Conditions of bioassay not specified
Environmental Fate: When released in water it will degrade principally due to biodegradation with a half-life of hours to days at ambient temperature. In humic waters oxidation by alkyl peroxy radicals may also be important. Adsorption to sediment and particulate matter in the water column will only be moderate and bioconcentration in fish should not be significant. If spilled on soil, it will probably adsorb moderately to the soil and biodegrade in several days. In the atmosphere, it will degrade during daylight hours by reaction with photochemically produced hydroxyl radicals (half-life 8 hr). At night it will probably degrade very rapidly by reaction with nitrate radicals. Washout by rain will also be an effective removal process.
Cleanup/Disposal: Guide No. 153: Eliminate all ignition sources (no smoking, flares, sparks or flames in immediate area). Do not touch damaged containers or spilled material unless wearing appropriate protective clothing. Stop leak if you can do it without risk. Prevent entry into waterways, sewers, basements or confined areas. Absorb or cover with dry earth, sand or other non-combustible material and transfer to containers. Do not get water inside containers.

Environmental Physical Data

Henry's Law Constant: estimated at 6 x10^{-7}
Octanol/Water Partition Coefficient: log K_{ow} = 2.30
Sorption Partition Coefficient: K_{oc} = estimated at 425
BCF: fish 1.8 to 2.18

Regulations

RCRA 40CFR: Listed Hazardous Waste No. U101 Toxic Waste
CERCLA: 40CFR 302.4: Listed per RCRA Section 3001 per CWA Section 307(a) RQ: 100 lb (45.35 kg)
SARA 40CFR 372.65: Listed
SARA EHS 40CFR 355: Not listed
TSCA: Listed

Analytical Methods

Air: EPA 0020
Soil: CLP LC_SV, MC_SVOA, OHC; EPA 16, 1625; SW846 1311, 3630B, 3640A, 3650A, 3650B, 8040A, 8041, 8250A, 8270B, 8270C
Water / Groundwater: EPA S-001-1, S-002-1, 1625, 604, 625, 625-S, 6; APHA 6410-B, 6420-BA, 6420-BB, 6420-C; ASTM D4763; USGS O3117
Plasma: EPA 29
Other: EPA O-005-1

DIM7250 CAS #: 95-87-4

2,5-DIMETHYLPHENOL

RTECS: ZE5775000
DOT: UN2261; IMO6.1
EINECS Number: 202-461-5
Molecular Formula: $C_8H_{10}O$
Structured MF: $(CH_3)_2C_6H_3OH$
Formula Weight: 122.16

OH

Chemical Structure

Synonyms: 1,4-DIMETHYL-2-HYDROXYBENZENE; 3,6-DIMETHYLPHENOL; 2,5-DMP; 1-HYDROXY-2,5-DIMETHYLBENZENE; HYDROXY-P-XYLENE; 6-METHYL-M-CRESOL; PHENOL,2,5-DIMETHYL-; 1,2,5-XYLENOL; 2,5-XYLENOL; P-XYLENOL
Description: crystals, needles, or prisms
Use: for preparation of coal tar disinfectants; manufacture of artificial resins, solvents, pharmaceuticals, insecticides, fungicides, plasticizers and rubber chemicals; additive to lubricants and gasolines; wetting agents; and dyestuffs

Physical Properties

Boiling Point: 211.5 °C (413 °F) at 762 mm Hg
Freezing Point: 74.5 °C (166.1 °F)
Specific Gravity: 0.965 at 80 °C
Density: 0.965 g/mL at 20 °C

Vapor Pressure: 10 mm Hg at 91.3 °C
Water Solubility: < 1 mg/mL at 20 C
Other Solubilities: Very Soluble in Benzene.
Odor Threshold: 0.0005 mg/m^3
Flash Point: Not available; probably combustible

RTECS Toxicity Data

Acute Oral: Rat LD_{50} Dose: 444 mg/kg. Mouse LD_{50} Dose: 383 mg/kg; Toxic Effects: Behavioral - Ataxia; Behavioral - Muscle contraction or spasticity; Lungs, Thorax, or Respiration - Dyspnea.
Tumorigenic: Mouse Route: Skin; Dose: 4000 mg/kg/20W-I; Toxic Effects: Tumorigenic - Equivocal tumorigenic agent by RTECS criteria; Skin and appendages - Tumors.

Hazard Overviews

Corrosive

Health: Corrosive to eyes/skin/respiratory tract. Harmful. Other Acute Effects: harmful by inhalation, in contact with skin and if swallowed; readily absorbed through skin; material is extremely destructive to tissue of the mucous membranes and upper respiratory tract, eyes and skin; inhalation may result in spasm, inflammation and edema of the larynx and bronchi, chemical pneumonitis and pulmonary edema; symptoms of exposure may include burning sensation, coughing, wheezing, laryngitis, shortness of breath, headache, nausea and vomiting.
Fire: Will burn. Hazards: under fire conditions, material may decompose to form flammable and/or explosive mixtures in air. Extinguishing agents: water spray; carbon dioxide, dry chemical powder or appropriate foam. Precautions: combustible liquid.
Carcinogenicity: IARC - Not listed; NIOSH - Not listed; NTP - Not listed; ACGIH - Not listed; OSHA - Not listed; EPA - Not listed; MAK - Not listed
Primary Target Organs:

Eyes

Skin

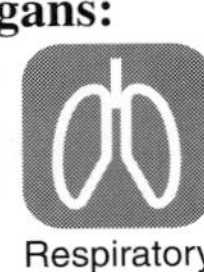

Respiratory System

Environmental

Ecotoxicity: LC_{50} Salmo gairdneri (rainbow trout) 3.2-5.6 mg/l/96 hr /Static bioassay LC_{50} Daphnia magna (cladoceran) 10.0 mg/1/48 hr /Static bioassay
Environmental Fate: If released into water, it would be expected to rapidly biodegrade with a half-life on the order of a few hours to a few days at ambient temperature, and in humic waters reaction with alkyl peroxy radicals should occur with a similar rate constant. The solubility in water combined with its low volatility should preclude volatilization from water into air. The low K_{oc} (estimated range 41.2 - 452) and the high rate of aquatic degradation suggest that sorption onto sediment should not be an important process. Similarly, high mobility through soil should be expected. Atmospheric degradation by photochemically produced hydroxyl radicals should be a facile process in sunlight (estimated half-life = 3.3 hr), and degradation at night should be more rapid in urban areas through the reaction with nitrate radicals. Rain washout would also be an effective method of atmospheric removal.
Cleanup/Disposal: Guide No. 153: Eliminate all ignition sources (no smoking, flares, sparks or flames in immediate area). Do not touch damaged containers or spilled material unless wearing appropriate protective clothing. Stop leak if you can do it without risk. Prevent entry into waterways, sewers, basements or confined areas. Absorb or cover with dry earth, sand or other non-combustible material and transfer to containers. Do not get water inside containers.

Environmental Physical Data

Henry's Law Constant: 1.05 x10^{-6}
Octanol/Water Partition Coefficient: log K_{ow} = 2.33
Sorption Partition Coefficient: K_{oc} = estimated at 41.2 to 452
BCF: estimated at 35

Regulations

RCRA 40CFR: Not listed
CERCLA: 40CFR 302.4: Not listed
SARA 40CFR 372.65: Not listed
SARA EHS 40CFR 355: Not listed
TSCA: Listed

Analytical Methods

Soil: SW846 8041

DIM7300 CAS #: 576-26-1

2,6-DIMETHYLPHENOL

RTECS: ZE6125000
DOT: UN2261; IMO6.1
EINECS Number: 209-400-1
Molecular Formula: $C_8H_{10}O$
Structured MF: $(CH_3)_2C_6H_3OH$
Formula Weight: 122.16

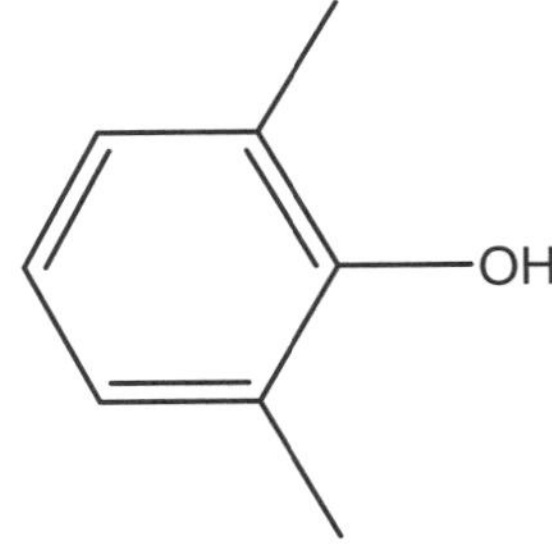

Chemical Structure

Synonyms: 2,6-DMP; 1-HYDROXY-2,6-DIMETHYLBENZENE; 2-HYDROXY-M-XYLENE; PHENOL,2,6-DIMETHYL-; VIC-M-XYLENOL; 2,6-XYLENOL; M-2-XYLENOL; XYLENOL 235
Description: white crystaline solid; sweet, tar-like odor

Use: for preparation of coal tar disinfectants; manufacture of artificial resins; manufacture of solvents, pharmaceuticals, insecticides, fungicides, plasticizers and rubber chemicals; additive to lubricants and gasolines; wetting agents; dyestuffs; and manufacture of polyphenylene oxide

Physical Properties

Boiling Point: 203 °C (397 °F)
Freezing Point: 49 °C (120.2 °F)
Saturated Vapor Density: 1.200723031 kg/m^3
Density: 1.01 g/mL at 20 °C
Vapor Pressure: 0.019 kPa at 25 °C
Water Solubility: < 1 mg/mL at 20 C
Other Solubilities: Very Soluble in Benzene, Chloroform.
Odor Threshold: 0.0002 mg/m^3
Ionization Potential (eV): 8.05 +/-0.2
Flash Point: 73 °C Closed Cup

RTECS Toxicity Data

Acute Oral: Rat LD_{50} Dose: 296 mg/kg. Mouse LD_{50} Dose: 450 mg/kg; Toxic Effects: Behavioral - Somnolence (general depressed activity); Behavioral - Convulsions or effect on seizure threshold; Lungs, Thorax, or Respiration - Other changes.
Acute Inhalation: Mammal LC; Dose: >270 mg/m^3.
Acute Dermal: Rabbit LD_{50} Route: Skin; Dose: 1 gm/kg. Rat LD_{50} Route: Skin; Dose: 2325 mg/kg; Toxic Effects: Behavioral - Somnolence (general depressed activity); Behavioral - Convulsions or effect on seizure threshold; Lungs, Thorax, or Respiration - Other changes.
Chronic (Multiple Dose) Oral: Rat Dose: 2065 mg/kg/10W-I; Toxic Effects: Lungs, Thorax, or Respiration - Changes in lung weight; Liver - Changes in liver weight; Endocrine - Changes in spleen weight. Rat Dose: 1470 mg/kg/35W-I; Toxic Effects: Liver - Other changes; Blood - Changes in spleen; Blood - Changes in serum composition.
Chronic (Multiple Dose) Inhalation: Rat Dose: 22 mg/m^3/4H/19W-I; Toxic Effects: Liver - Liver function tests impaired; Kidney, Ureter, and Bladder - Other changes in urine composition; Blood - Changes in serum composition. Rat Dose: 1 mg/m^3/24H/6D-C; Toxic Effects: Brain and coverings - Recordings from specific areas of CNS; Blood - Changes in serum composition; Biochemical - True cholinesterase.
Tumorigenic: Mouse Route: Skin; Dose: 4000 mg/kg/20W-I; Toxic Effects: Tumorigenic - Equivocal tumorigenic agent by RTECS criteria; Skin and appendages - Tumors.

Hazard Overviews

Corrosive

Fire Diamond

Health: Corrosive to skin/respiratory tract. Toxic. Also Causes: headache, dizziness, nausea, difficulty breathing, hypotension, arrhythmias, coma. Chronic Effects: liver and kidney damage.
Fire: Combustible. For small fires use dry chemical, carbon dioxide, water spray, or standard foam. For large fires use water spray, fog, or standard foam. Do not scatter fire with solid stream of water. Container may explode in heat of fire.
Reactivity: Stable. Hazardous polymerization cannot occur. Avoid: heat; ignition sources. Incompatible with: strong oxidizing agents. Hazardous decomposition products: toxic vapors and gases; carbon monoxide.
Carcinogenicity: IARC - Not listed; NIOSH - Not listed; NTP - Not listed; ACGIH - Not listed; OSHA - Not listed; EPA - Not listed; MAK - Not listed
Primary Target Organs:

Eyes

Skin

Mucous Membranes
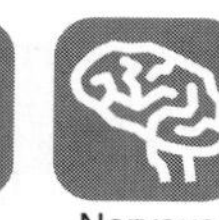
Nervous System

Liver

Kidneys

Environmental

Ecotoxicity: LC_{50} Pimephales promelas (fathead minnow) > 27 mg/l/96 hr /freshwater, flow through bioassay LC_{50} Daphnia magna (cladoceran) 11.2 mg/1/48 hr /Static bioassay LC_{100} Terahymena pyriformis (ciliate) 2.66 mmole/l/24 hr /Conditions of bioassay not specified
Environmental Fate: If released into water, it would be expected to rapidly biodegrade with a half life on the order of a few hours to a few days at ambient temperature, and in humic waters reaction with alkyl peroxy radicals should occur with a similar rate constant. The solubility in water combined with its low volatility should preclude volatilization from water into air. The low K_{oc} (estimated range 37-457) and the high rate of aquatic degradation suggest that sorption onto sediment should not be an important process. Similarly, high mobility through soil should be expected. Atmospheric degradation by photochemically produced hydroxyl radicals should be a rapid process in sunlight (estimated half-life = 7.1 hr), and degradation at night should be more rapid in urban areas through the reaction with nitrate radicals. Rain washout would also be an effective method of atmospheric removal.
Cleanup/Disposal: Guide No. 153: Eliminate all ignition sources (no smoking, flares, sparks or flames in immediate area). Do not touch damaged containers or spilled material unless wearing appropriate protective clothing. Stop leak if you can do it without risk. Prevent entry into waterways, sewers, basements or confined areas. Absorb or cover with dry earth, sand or other non-combustible material and transfer to containers. Do not get water inside containers.

Environmental Physical Data

Henry's Law Constant: 4.90 x10^{-6}
Octanol/Water Partition Coefficient: log K_{ow} = 2.36
Sorption Partition Coefficient: K_{oc} = estimated at 37 to 458
BCF: estimated at 37

Regulations

RCRA 40CFR: Not listed
CERCLA: 40CFR 302.4: Not listed
SARA 40CFR 372.65: Listed
SARA EHS 40CFR 355: Not listed
TSCA: Listed

Analytical Methods

Soil: SW846 8041

DIM7350 CAS #: 95-65-8

3,4-DIMETHYLPHENOL

RTECS: ZE6300000
DOT: UN2261; IMO6.1
EINECS Number: 202-439-5
Molecular Formula: $C_8H_{10}O$
Structured MF: $(CH_3)_2C_6H_3OH$
Formula Weight: 122.17

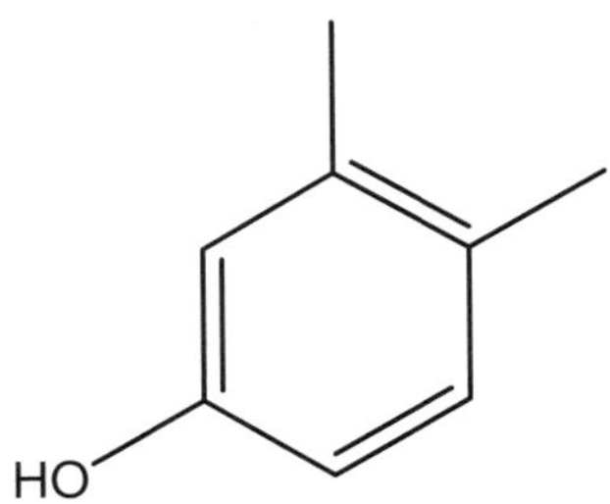

Chemical Structure

Synonyms: 4,5-DIMETHYLPHENOL; 3,4-DMP; 1-HYDROXY-3,4-DIMETHYLBENZENE; 4-HYDROXY-1,2-DIMETHYLBENZENE; 4-HYDROXY-O-XYLENE; PHENOL,3,4-DIMETHYL-; 1,3,4-XYLENOL; 3,4-XYLENOL; AS-O-XYLENOL; O-4-XYLENOL
Description: needles or prisms
Use: preparation of coal tar disinfectants; manufacture of artificial resins; manufacture of solvents, pharmaceuticals, insecticides, fungicides, plasticizers and rubber chemicals; additive to lubricants and gasolines; wetting agents; and dyestuffs

Physical Properties

Boiling Point: 225 °C (437 °F)
Freezing Point: 62.5 °C (144.5 °F)
Specific Gravity: 0.983 at 20 °C/4 °C
Density: 0.983 g/mL at 80 °C
Vapor Pressure: 10 mm Hg at 107.7 °C
Water Solubility: < 1 mg/mL at 20 C
Other Solubilities: Very Soluble in Benzene, Chloroform.
Odor Threshold: 0.003 mg/m^3
Flash Point: Not available; probably combustible

RTECS Toxicity Data

Acute Oral: Rat LD_{50} Dose: 727 mg/kg. Mouse LD_{50} Dose: 400 mg/kg; Toxic Effects: Behavioral - Ataxia; Behavioral - Muscle contraction or spasticity; Lungs, Thorax, or Respiration - Dyspnea.

Chronic (Multiple Dose) Oral: Rat Dose: 5075 mg/kg/10W-I; Toxic Effects: Liver - Changes in liver weight; Endocrine - Changes in spleen weight; Nutritional and gross metabolic - Weight loss or decreased weight gain.

Tumorigenic: Mouse Route: Skin; Dose: 4000 mg/kg/20W-I; Toxic Effects: Tumorigenic - Equivocal tumorigenic agent by RTECS criteria; Skin and appendages - Tumors.

Hazard Overviews

Corrosive

Health: Corrosive to eyes/skin/respiratory tract. Harmful. Other Acute Effects: extremely destructive to tissue of the mucous membranes and upper respiratory tract, eyes and skin; harmful if swallowed, inhaled, or absorbed through skin; inhalation may result in spasm, inflammation and edema of the larynx and bronchi, chemical pneumonitis and pulmonary edema; symptoms of exposure may include burning sensation, coughing, wheezing, laryngitis, shortness of breath, headache, nausea and vomiting.

Fire: Will burn. Hazards: under fire conditions, material may decompose to form flammable and/or explosive mixtures in air. Extinguishing agents: water spray; carbon dioxide, dry chemical powder or appropriate foam. Precautions: combustible liquid.

Carcinogenicity: IARC - Not listed; NIOSH - Not listed; NTP - Not listed; ACGIH - Not listed; OSHA - Not listed; EPA - Not listed; MAK - Not listed

Primary Target Organs:

Eyes

Skin

Respiratory System

Environmental

Ecotoxicity: TLm Carassius carassius (Crucian carp) 21 mg/l/24 hr /Conditions of bioassay not specified TLm Rutilus rutilus (roach) 16 mg/l/24 hr /Conditions of bioassay not specified LC_{50} Pimephales promelas (fathead minnows) 14 mg/l/72 hr /Static bioassay in Lake Superior Water at 18-22 °C TLm Salvelinus (trout embryos) 7 mg/l/24 hr /Conditions of bioassay not specified

Environmental Fate: If released into water, it would be expected to rapidly biodegrade with a half-life on the order of a few hours to a few days at ambient temperature and in humic waters reaction with alkyl peroxy radicals should occur with a similar rate constant. The solubility in water combined with its low volatility should preclude volatilization from water into air. The low K_{oc} (estimated range 40-389) and the high rate of aquatic degradation suggest that sorption onto sediment should not be an important process. Similarly, high mobility through soil should be expected. Atmospheric degradation by photochemically produced hydroxyl radicals should be a rapid process in sunlight (estimated half-life = 3.3 hr), and degradation at night should be more rapid in urban areas through the reaction with nitrate radicals. Rain washout would also be an effective method of atmospheric removal.

Cleanup/Disposal: Guide No. 153: Eliminate all ignition sources (no smoking, flares, sparks or flames in immediate

area). Do not touch damaged containers or spilled material unless wearing appropriate protective clothing. Stop leak if you can do it without risk. Prevent entry into waterways, sewers, basements or confined areas. Absorb or cover with dry earth, sand or other non-combustible material and transfer to containers. Do not get water inside containers.

Environmental Physical Data

Henry's Law Constant: 4.41 x10^{-7}
Octanol/Water Partition Coefficient: log K_{ow} = 2.23
Sorption Partition Coefficient: K_{oc} = estimated at 39.8 to 389
BCF: estimated at 29

Regulations

RCRA 40CFR: Not listed
CERCLA: 40CFR 302.4: Not listed
SARA 40CFR 372.65: Not listed
SARA EHS 40CFR 355: Not listed
TSCA: Listed

Analytical Methods

Soil: SW846 8041

DIM7400 CAS #: 108-68-9

3,5-DIMETHYLPHENOL

RTECS: ZE6475000
DOT: UN2261; IMO6.1
EINECS Number: 203-606-5
Molecular Formula: $C_8H_{10}O$
Structured MF: $(CH_3)_2C_6H_3OH$
Formula Weight: 122.16

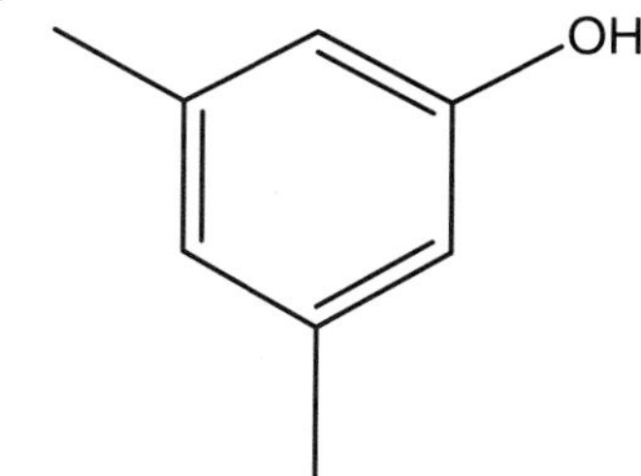

Chemical Structure

Synonyms: BENZENE,1,3-DIMETHYL-5-HYDROXY-; 3,5-DMP; 1-HYDROXY-3,5-DIMETHYLBENZENE; 5-HYDROXY-M-XYLENE; PHENOL,3,5-DIMETHYL-; 1,3,5-XYLENOL; 3,5-XYLENOL; M-5-XYLENOL; SYM-M-XYLENOL; XYLENOL 100; XYLENOL 200
Description: white crystals or needles
Use: for preparation of coal tar disinfectants; manufacture of artificial resins; manufacture of solvents, pharmaceuticals, insecticides, fungicides, plasticizers and rubber chemicals; additive to lubricants and gasolines; wetting agents; and dyestuffs

Physical Properties

Boiling Point: 219.5 °C (427 °F)
Freezing Point: 64 °C (147.2 °F)
Specific Gravity: 0.968 at 20 °C/4 °C
Density: 0.96 g/mL at 20 °C
Vapor Pressure: 1 mm Hg at 62.0 °C
Water Solubility: Slightly Soluble in Water
Other Solubilities: Very Soluble in Benzene, Chloroform, Ether.
Odor Threshold: 4.10 mg/m^3
Flash Point: Not available; probably combustible

RTECS Toxicity Data

Acute Oral: Rat LD_{50} Dose: 608 mg/kg. Mouse LD_{50} Dose: 477 mg/kg; Toxic Effects: Behavioral - Ataxia; Behavioral - Muscle contraction or spasticity; Lungs, Thorax, or Respiration - Dyspnea.
Acute Inhalation: Rat LC; Dose: >4 mg/m^3.
Irritation Eye: Rabbit Standard Draize Test Dose: 250 ug; Reaction: severe.
Tumorigenic: Mouse Route: Skin; Dose: 4000 mg/kg/20W-I; Toxic Effects: Tumorigenic - Equivocal tumorigenic agent by RTECS criteria; Skin and appendages - Tumors.

Hazard Overviews

Corrosive

Health: Corrosive to eyes/skin/respiratory tract. Other Acute Effects: harmful if swallowed, inhaled, or absorbed through skin; inhalation may result in spasm, inflammation and edema of the larynx and bronchi, chemical pneumonitis and pulmonary edema; symptoms of exposure may include burning sensation, coughing, wheezing, laryngitis, shortness of breath, headache, nausea and vomiting.
Fire: Will burn. Hazards: under fire conditions, material may decompose to form flammable and/or explosive mixtures in air. Extinguishing agents: water spray; carbon dioxide, dry chemical powder or appropriate foam. Precautions: combustible liquid.
Reactivity: Incompatible with: bases, acid chlorides, acid anhydrides, oxidizing agents, corrodes steel, brass, copper, copper alloys. Hazardous decomposition products: toxic fumes of: carbon monoxide, carbon dioxide.
Carcinogenicity: IARC - Not listed; NIOSH - Not listed; NTP - Not listed; ACGIH - Not listed; OSHA - Not listed; EPA - Not listed; MAK - Not listed
Primary Target Organs:

Eyes

Skin

Respiratory System

Environmental

Ecotoxicity: TLm Carassius carassius (Crucian carp) 53 mg/l/24 hr /Conditions of bioassay not specified TLm Tinca tinca (tench) 52 mg/l/24 hr /Conditions of bioassay not specified LC_{100} Tetrahymena pyriformis (protozoa ciliate) 2.3 mmole/l/24 hr /Conditions of bioassay not specified TLm

Salvelinus (trout embryos) 50 mg/l/24 hr /Conditions of bioassay not specified

Environmental Fate: If released into water, it would be expected to rapidly biodegrade with a half-life on the order of a few hours to a few days at ambient temperature, and in humic waters reaction with alkyl peroxy radicals should occur with a similar rate constant. The solubility in water combined with its low volatility should preclude volatilization from water into air. The low K_{oc} (estimated range 41-452) and the high rate of aquatic degradation suggest that sorption onto sediment should not be an important process. Similarly, high mobility through soil should be expected. Atmospheric degradation by photochemically produced hydroxyl radicals should be a facile process in sunlight (estimated half-life = 1.9 hr), and degradation at night should be more rapid in urban areas through the reaction with nitrate radicals. Rain washout would also be an effective method of atmospheric removal.

Cleanup/Disposal: Guide No. 153: Eliminate all ignition sources (no smoking, flares, sparks or flames in immediate area). Do not touch damaged containers or spilled material unless wearing appropriate protective clothing. Stop leak if you can do it without risk. Prevent entry into waterways, sewers, basements or confined areas. Absorb or cover with dry earth, sand or other non-combustible material and transfer to containers. Do not get water inside containers.

Environmental Physical Data

Henry's Law Constant: 3.70 $\times 10^{-6}$
Octanol/Water Partition Coefficient: log K_{ow} = 2.35
Sorption Partition Coefficient: K_{oc} = soils 460 to 1400
BCF: estimated at 36
BOD: sewage seed 0.82 lb/lb, 5 days

Regulations

RCRA 40CFR: Not listed
CERCLA: 40CFR 302.4: Not listed
SARA 40CFR 372.65: Not listed
SARA EHS 40CFR 355: Not listed
TSCA: Listed

Analytical Methods

Water / Groundwater: ASTM D4763

DIM7450 CAS #: 99-98-9

DIMETHYL-P-PHENYLENEDIAMINE

RTECS: ST0874000
EINECS Number: 202-807-5
Molecular Formula: $C_8H_{12}N_2$
Structured MF: $(CH_3)_2NC_6H_4NH_2$
Formula Weight: 136.22

Chemical Structure

Synonyms: 4-AMINO-N,N-DIMETHYLANILINE; P-AMINO-N,N-DIMETHYLANILINE; P-AMINODIMETHYLANILINE; 1,4-BENZENEDIAMINE,N,N-DIMETHYL-; 1,4-BENZENEDIAMINE,N,N-DIMETHYL-(9CI); C.I. 76075; 4-(DIMETHYLAMINO)ANILINE; 4-DIMETHYLAMINOANILINE; DIMETHYLAMINOANILINE; P-(DIMETHYLAMINO)ANILINE; 4-(DIMETHYLAMINO)BENZENAMINE; 4-(DIMETHYLAMINO)PHENYLAMINE; P-DIMETHYLAMINOPHENYLAMINE; N,N-DIMETHYL-1,4-BENZENEDIAMINE; N,N-DIMETHYL-P-FENYLENDIAMIN; DIMETHYL-PARAPHENYLENEDIAMINE; DIMETHYL-PARA-PHENYLENEDIAMINE; N,N-DIMETHYL-1,4-PHENYLENEDIAMINE; N,N-DIMETHYL-P-PHENYLENEDIAMINE; DMPD; P-PHENYLENEDIAMINE,N,N-DIMETHYL-

Description: reddish-violet crystals; colorless, asbestos-like, needles

Use: as a base for production of methylene blue, as a photodeveloper, as a reagent for cellulose, as a reagent for the detection of hydrogen sulfide, in organic synthesis and as a reagent for certain bacteria; in microscopy and in tests for acetone, uric acid thallic salts, oxydases, lignin, ozone, H2O2, hydrogen sulfide and bromine

Physical Properties

Boiling Point: 262 °C (504 °F)
Freezing Point: 53 °C (127.4 °F)
Specific Gravity: 1.036 at 20 °C/4 °C
Vapor Density: 4.69 Air=1
Water Solubility: < 1 mg/mL at 20 C
Other Solubilities: 95% Ethanol: >=100 mg/ml at 20 °C; Acetone: >=100 mg/ml at 20 °C; Alcohol: Soluble; Benzene: Soluble; Chloroform: Soluble; DMSO: >=100 mg/ml at 20 °C; Ether: Soluble.
Flash Point: 90 °C

RTECS Toxicity Data

Acute Oral: Rat LD_{50} Dose: 50 mg/kg; Toxic Effects: Behavioral - Tremor; Behavioral - Convulsions or effect on seizure threshold. Mouse LD_{50} Dose: 30 mg/kg; Toxic Effects: Behavioral - Tremor; Behavioral - Convulsions or effect on seizure threshold.

Acute Inhalation: Rabbit LC_{Lo} Dose: 500 ppb; Toxic Effects: Behavioral - Convulsions or effect on seizure threshold; Behavioral - Excitment; Lungs, Thorax, or Respiration - Dyspnea. Guinea Pig LC_{Lo} Dose: 240 ppb; Toxic Effects: Behavioral - Convulsions or effect on seizure threshold; Behavioral - Excitment; Lungs, Thorax, or Respiration - Dyspnea.

Acute Dermal: Human TD_{Lo} Route: Skin; Dose: 14 ug/kg; Toxic Effects: Blood - Hemorrhage; Skin and appendages - Primary irritation; Skin and appendages - Dermatitis, allergic. Rabbit LD_{Lo} Route: Skin; Dose: 60 mg/kg; Toxic Effects:

Behavioral - General anesthetic; Behavioral - Convulsions or effect on seizure threshold; Lungs, Thorax, or Respiration - Cyanosis. Rabbit LD_{Lo} Route: Subcutaneous Dose: 60 mg/kg; Toxic Effects: Behavioral - Somnolence (general depressed activity); Behavioral - Convulsions or effect on seizure threshold; Behavioral - Coma.

Chronic (Multiple Dose) Oral: Rat Dose: 649 mg/kg/11D-C; Toxic Effects: Liver - Changes in liver weight; Endocrine - Other changes; Nutritional and gross metabolic - Weight loss or decreased weight gain.

Mutagenic: Rat Unscheduled DNA Synthesis; Cell Type: liver; Dose: 100 umol/L. Hamster Unscheduled DNA Synthesis; Cell Type: liver; Dose: 100 umol/L.

Hazard Overviews

Poison

Corrosive

Health: Irritating to eyes/skin/respiratory tract. Poison. Other Acute Effects: may be fatal if inhaled, swallowed, or absorbed through skin; absorption into the body leads to the formation of methemoglobin which in sufficient concentration causes cyanosis; onset may be delayed 2 to 4 hours or longer.

Fire: Combustible. Hazards: emits toxic fumes. Extinguishing agents: water spray; carbon dioxide, dry chemical powder or appropriate foam. Precautions: combustible liquid.

Carcinogenicity: IARC - Not listed; NIOSH - Not listed; NTP - Not listed; ACGIH - Not listed; OSHA - Not listed; EPA - Not listed; MAK - Not listed

Primary Target Organs:

Eyes

Skin

Respiratory System

Environmental

Environmental Fate: If released to the atmosphere, it will degrade rapidly (estimated half-life of 1.368 hr) by reaction with photochemically produced hydroxyl radicals. If released to soil it may be susceptible to high amounts of leaching in soil based on an estimated K_{oc} range of 10.8 to 19.1. However it may undergo a covalent chemical bonding with humic materials which can result in its chemical alteration to a latent form and prevent leaching. If released to water, covalent bonding with humic materials in the water column and sediments may result in significant partitioning from the water column. In the water column it may be susceptible to significant photooxidation via hydroxyl and peroxy radicals and singlet molecular oxygen. Direct photolysis may also contribute to aquatic degradation. The lack of data makes statements regarding degradation impractical.

Cleanup/Disposal: Guide No. 154: Eliminate all ignition sources (no smoking, flares, sparks or flames in immediate area). Do not touch damaged containers or spilled material unless wearing appropriate protective clothing. Stop leak if you can do it without risk. Prevent entry into waterways, sewers, basements or confined areas. Absorb or cover with dry earth, sand or other non-combustible material and transfer to containers. Do not get water inside containers.

Environmental Physical Data

Henry's Law Constant: estimated at 1.94 $x10^{-8}$
Octanol/Water Partition Coefficient: log K_{ow} = 1.11
Sorption Partition Coefficient: K_{oc} = 10.8 to 19.1
BCF: estimated at 2.33 to 4.11

Regulations

RCRA 40CFR: Not listed
CERCLA: 40CFR 302.4: Not listed
SARA 40CFR 372.65: Not listed TPQ: 10/10000 lb
SARA EHS 40CFR 355: Listed TPQ: 10 lb
TSCA: Listed

DIM7700 CAS #: 463-82-1

2,2-DIMETHYLPROPANE

RTECS: TY1190000
EINECS Number: 207-343-7
Molecular Formula: C_5H_{12}
Formula Weight: 72.15
Synonyms: 2,2-DIMETHYLPROPANE,OTHER THAN PENTANE AND ISOPENTANE; NEOPENTANE; TERT-PENTANE; PROPANE,2,2-DIMETHYL-; TETRAMETHYLMETHANE; 1,1,1-TRIMETHYLETHANE
Description: colorless gas, liquid
Use: research chemical; in butyl rubber

Physical Properties

Boiling Point: 9.5 °C (49 °F)
Freezing Point: -16.55 °C (2.21 °F)
Specific Gravity: Liquid 0.613 at 0 °C/0 °C
Vapor Pressure: 0.7 kPa at -73 °C
Water Solubility: Insoluble in Water
Other Solubilities: Soluble in Alcohol, Ether.
Refraction Index: 1.3476 at 6 °C/D
Ionization Potential (eV): =< 10.21 +/-0.04
Flash Point: < -7 °C
Autoignition Temperature: 450 °C
LEL: 1.4% v/v
UEL: 7.5% v/v

RTECS Toxicity Data

Acute Inhalation: Mouse LC_{Lo} Dose: 1097 gm/m^3/2hr; Toxic Effects: Behavioral - Excitement.

Hazard Overviews

Flammable

Fire Diamond

Health: Irritating to eyes/skin. Also Causes: CNS depression. Chronic Effects: dermatitis.

Fire: Flammable. Can form explosive mixtures in air. For small fires use dry chemical or carbon dioxide. For large fires use water spray or fog. Containers may explode in heat of fire. Vapors may travel to an ignition source and flash back. Fight fire from maximum distance.

Reactivity: Stable. Hazardous polymerization cannot occur. Avoid: heat; ignition sources; oxidizers. Incompatible with: oxidizers. Hazardous decomposition products: acrid smoke; irritating gas.

Carcinogenicity: IARC - Not listed; NIOSH - Not listed; NTP - Not listed; ACGIH - Not listed; OSHA - Not listed; EPA - Not listed; MAK - Not listed

Primary Target Organs:

Eyes

Skin

Nervous System

Environmental

Regulations

RCRA 40CFR: Not listed
CERCLA: 40CFR 302.4: Not listed
SARA 40CFR 372.65: Not listed
SARA EHS 40CFR 355: Not listed
TSCA: Listed

Analytical Methods

Air: ASTM D2820

DIM7750 CAS #: 109-55-7

N,N-DIMETHYL-1,3-PROPANEDIAMINE

RTECS: TX7525000
EINECS Number: 203-680-9
Molecular Formula: $C_5H_{14}N_2$
Formula Weight: 102.21

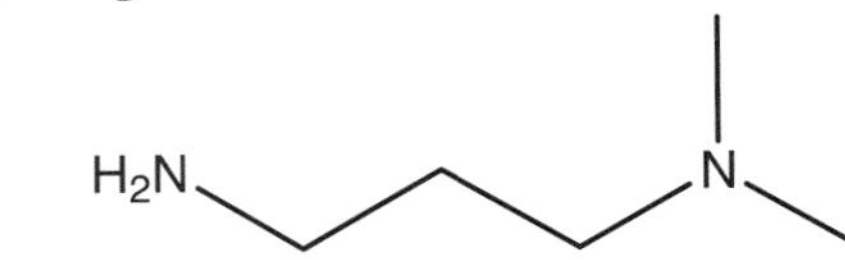

Chemical Structure

Synonyms: 1-AMINO-3-(DIMETHYLAMINO)PROPANE; 1-AMINO-3-DIMETHYLAMINOPROPANE; 3-AMINO-1-(DIMETHYLAMINO)PROPANE; 3-(DIMETHYLAMINE)PROPYLAMINE; 1-DIMETHYLAMINO-3-AMINOPROPANE; 3-(DIMETHYLAMINO)-1-PROPANAMINE; 3-(DIMETHYLAMINO)-1-PROPYLAMINE; 3-(DIMETHYLAMINO)PROPYLAMINE; 3-(N,N-DIMETHYLAMINO)PROPYLAMINE; 3-DIMETHYLAMINOPROPYLAMINE; DIMETHYLAMINOPROPYLAMINE; GAMMA-DIMETHYLAMINOPROPYLAMINE; N,N-DIMETHYL-N-(3-AMINOPROPYL)AMINE; N,N-DIMETHYL-1,3-DIAMINOPROPANE; N,N-DIMETHYL-1,3-PROPYLENEDIAMINE; N,N-DIMETHYLPROPYLENEDIAMINE; N,N-DIMETHYLTRIMETHYLENEDIAMINE; 1,3-PROPANEDIAMINE,N,N-DIMETHYL-; PROPYLAMINE,3-(N,N-DIMETHYLAMINO)-

Description: colorless liquid

Use: curing agent for epoxy resins; organic intermediate

Physical Properties

Boiling Point: 123 °C (253 °F)
Freezing Point: -70 °C (-94 °F)
Specific Gravity: 0.81 at 30 °C
Vapor Density: 3.52 Air=1
Vapor Pressure: 10 mm Hg at 30 °C
Water Solubility: Soluble in Water
Other Solubilities: Organic solvents: Soluble.
pH: 1:100 solution 12
Refraction Index: 1.4328 at 25 °C/D
Flash Point: 35 °C Closed Cup

RTECS Toxicity Data

Acute Oral: Rat LD_{50} Dose: 1870 mg/kg.
Acute Dermal: Rabbit LD_{50} Route: Skin; Dose: 600 uL/kg.
Irritation Eye: Rabbit Standard Draize Test Dose: 5 mg; Reaction: moderate.

Hazard Overviews

Corrosive

Flammable

Health: Corrosive to eyes/skin/respiratory tract. Toxic. Other Acute Effects: harmful if swallowed, inhaled, or absorbed through skin; readily absorbed through skin; lachrymator; inhalation may result in spasm, inflammation and edema of the larynx and bronchi, chemical pneumonitis and pulmonary edema.

Fire: Flammable. Hazards: vapor may travel considerable distance to source of ignition and flash back; container explosion may occur; emits toxic fumes; forms explosive mixtures in air. Extinguishing agents: carbon dioxide, dry chemical powder or appropriate foam; water may be effective for cooling, but may not effect extinguishment. Precautions: combustible liquid.

Reactivity: Incompatible with: acids, acid chlorides, acid anhydrides, strong oxidizing agents, carbon dioxide. Hazardous decomposition products: thermal decomposition may produce carbon monoxide, carbon dioxide, and nitrogen oxides.

Carcinogenicity: IARC - Not listed; NIOSH - Not listed; NTP - Not listed; ACGIH - Not listed; OSHA - Not listed; EPA - Not listed; MAK - Not listed

Primary Target Organs:

Eyes

Skin

Respiratory System

Environmental

Cleanup/Disposal: Guide No. 128: Eliminate all ignition sources (no smoking, flares, sparks or flames in immediate area). All equipment used when handling the product must be grounded. Do not touch or walk through spilled material. Stop leak if you can do it without risk. Prevent entry into waterways, sewers, basements or confined areas. A vapor suppressing foam may be used to reduce vapors. Absorb or cover with dry earth, sand or other non-combustible material and transfer to containers. Use clean non-sparking tools to collect absorbed material. Large Spills: Dike far ahead of liquid spill for later disposal. Water spray may reduce vapor; but may not prevent ignition in closed spaces.

Regulations

RCRA 40CFR: Not listed
CERCLA: 40CFR 302.4: Not listed
SARA 40CFR 372.65: Not listed
SARA EHS 40CFR 355: Not listed
TSCA: Listed

DIM7800 CAS #: 75-98-9

2,2-DIMETHYLPROPANOIC ACID

RTECS: TO7700000
EINECS Number: 200-922-5
Molecular Formula: $C_5H_{10}O_2$
Structured MF: $(CH_3)_3CCOOH$
Formula Weight: 102.13

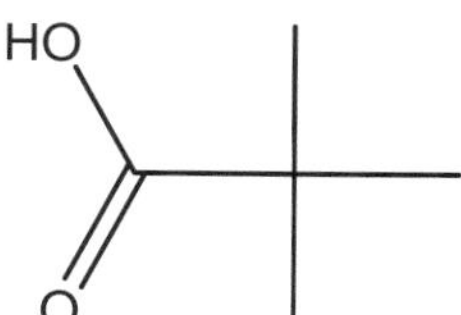

Chemical Structure

Synonyms: ACETIC ACID,TRIMETHYL-; 2,2-DIMETHYLPROPIONIC ACID; ALPHA,ALPHA-DIMETHYLPROPIONIC ACID; KYSELINA 2,2-DIMETHYLPROPIONOVA; KYSELINA PIVALOVA; NEOPENTANOIC ACID; TERT-PENTANOIC ACID; PIVALIC ACID; PROPANOIC ACID,2,2-DIMETHYL-; PROPIONIC ACID,2,2-DIMETHYL-; TRIMETHYLACETIC ACID; VERSATIC 5
Description: colored crystals
Use: intermediate, as a replacement for some natural materials; in pharmaceuticals

Physical Properties

Boiling Point: 163.8 °C (327 °F)
Freezing Point: 35.5 °C (95.9 °F)
Specific Gravity: 0.905 at 50 °C
Vapor Density: Estimate 3.5 Air=1
Water Solubility: 1 g dissolves in 40 ml Water
Other Solubilities: freely Soluble in Alcohol, Ether
pH: Acid
Refraction Index: 1.3931 at 36.5 °C/D
Ionization Potential (eV): 10.08
Flash Point: 64 °C Closed Cup

RTECS Toxicity Data

Acute Oral: Rat LD_{50} Dose: 900 mg/kg.
Acute Dermal: Rat LD_{50} Route: Skin; Dose: 1900 mg/kg.
Tumorigenic: Mouse Route: Skin; Dose: 188 mg/kg/47W-I; Toxic Effects: Tumorigenic - Equivocal tumorigenic agent by RTECS criteria; Lungs, Thorax, or Respiration - Tumors; Skin and appendages - Tumors.

Hazard Overviews

Corrosive

Health: Corrosive to eyes/skin/respiratory tract. Other Acute Effects: harmful if swallowed, inhaled, or absorbed through skin; extremely destructive to tissue of the mucous membranes and upper respiratory tract, eyes and skin; inhalation may result in spasm, inflammation and edema of the larynx and bronchi; chemical pneumonitis; pulmonary edema; burning sensation; coughing; wheezing; laryngitis; shortness of breath; headache; nausea; vomiting.
Fire: Combustible. Hazards: in powder form capable of creating a dust explosion. Extinguishing agents: carbon dioxide, dry chemical powder or appropriate foam; water spray. Precautions: combustible liquid.
Reactivity: Incompatible with: bases, oxidizing agents, reducing agents. Hazardous decomposition products: toxic fumes of: carbon monoxide, carbon dioxide.
Carcinogenicity: IARC - Not listed; NIOSH - Not listed; NTP - Not listed; ACGIH - Not listed; OSHA - Not listed; EPA - Not listed; MAK - Not listed
Primary Target Organs:

Eyes

Skin
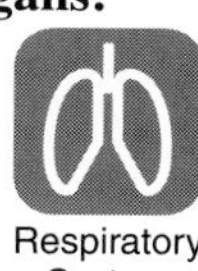
Respiratory System

Environmental

Ecotoxicity: LD_{50} Goldfish 400 to 375 mg/l/(24 to 96 hr) at pH 5 /Conditions of bioassay not specified
Environmental Fate: If released to the environment, under environmental conditions (pH 5-9) it will predominantly be in the form of the 2,2-dimethylpropanoic acid ion (pKa of 5.031). No data are available to determine whether adsorption to soil and sediment is an important fate process for the ion. If released to soil, biodegradation may be an important fate process. If released to water, volatilization, aquatic oxidation with hydroxyl radicals, and hydrolysis are not expected to be important fate processes. Biodegradation in water may be important, but no rate data are available. If released to the ambient atmosphere vapor-phase is expected to moderately degrade by reaction with photochemically produced hydroxyl radicals (estimated half-life of 27 days).

Cleanup/Disposal: Guide No. 154: Eliminate all ignition sources (no smoking, flares, sparks or flames in immediate area). Do not touch damaged containers or spilled material unless wearing appropriate protective clothing. Stop leak if you can do it without risk. Prevent entry into waterways, sewers, basements or confined areas. Absorb or cover with dry earth, sand or other non-combustible material and transfer to containers. Do not get water inside containers.

Environmental Physical Data

Henry's Law Constant: 1.28×10^{-6}
Sorption Partition Coefficient: K_{OC} = estimated at 818
BCF: estimated at 111
BOD: 24% BODT, 5 days

Regulations

RCRA 40CFR: Not listed
CERCLA: 40CFR 302.4: Not listed
SARA 40CFR 372.65: Not listed
SARA EHS 40CFR 355: Not listed
TSCA: Listed

DIM7850 CAS #: 80-46-6

4-(1,1-DIMETHYLPROPYL)PHENOL

RTECS: SM6825000
EINECS Number: 201-280-9
Molecular Formula: $C_{11}H_{16}O$
Formula Weight: 164.24

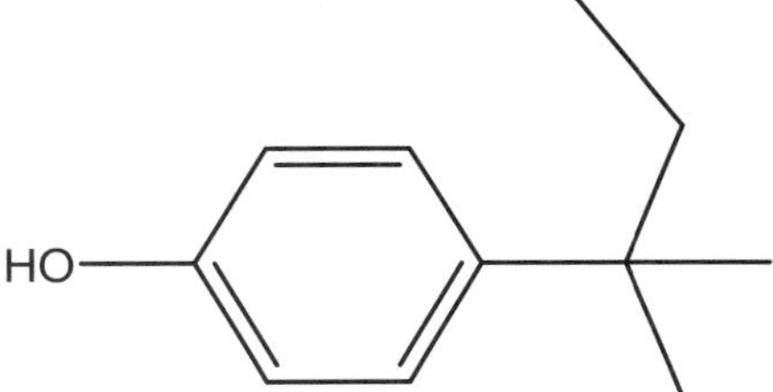

Chemical Structure

Synonyms: AMILFENOL; AMILPHENOL; AMYL PHENOL 4T; 4-T-AMYLPHENOL; 4-TERT-AMYLPHENOL; P-TERT-AMYLPHENOL; PARA-TERT-AMYLPHENOL; P-(1,1-DIMETHYLPROPYL)PHENOL; P-(ALPHA,ALPHA-DIMETHYLPROPYL)PHENOL; 1-HYDROXY-4(2-METHYL-2-BUTYL)BENZENE; 2-METHYL-2-P-HYDROXYPHENYLBUTANE; PENTAPHEN; 4-TERT-PENTYLPHENOL; P-T-PENTYLPHENOL; P-TERT-PENTYLPHENOL; PHENOL,4-(1,1-DIMETHYLPROPYL)-; PHENOL,P-TERT-PENTYL-; PTAP; UCAR AMYL PHENOL 4T
Description: white crystals
Use: manufacture of oil soluble resins; intermediate for organic mercury germicide pesticides and chemicals used in rubber and petroleum industries

Physical Properties

Boiling Point: 262.5 °C (505 °F)
Freezing Point: 94 °C (201.2 °F) to 95 °C (203 °F)
Specific Gravity: 0.962 at 20 °C/4 °C
Vapor Pressure: 1.7928×10^{-2} Pa at melting point
Water Solubility: Practically Insoluble in Water
Other Solubilities: 95% Ethanol: >=100 mg/ml at 22 °C; Acetone: >=100 mg/ml at 22 °C; Benzene: Soluble; DMSO: >=100 mg/ml at 22 °C; Ether: Soluble.
Surface Tension: Predicted 6.4552×10^{-2} N/m at melting point
Flash Point: 111 °C Open Cup

RTECS Toxicity Data

Acute Oral: Rat LD_{50} Dose: 1830 mg/kg.
Acute Dermal: Rabbit LD_{50} Route: Skin; Dose: 2 gm/kg.
Irritation Eye: Rabbit Standard Draize Test Dose: 1%; Reaction: severe. Rabbit Standard Draize Test Dose: 500 mg; Reaction: severe.

Hazard Overviews

Corrosive

Health: Corrosive to eyes/skin/respiratory tract. Harmful. Other Acute Effects: harmful if swallowed or absorbed through skin; may be harmful if inhaled; extremely destructive to tissue of the mucous membranes and upper respiratory tract, eyes and skin; inhalation may result in spasm, inflammation and edema of the larynx and bronchi, chemical pneumonitis and pulmonary edema; symptoms of exposure may include burning sensation, coughing, wheezing, laryngitis, shortness of breath, headache, nausea and vomiting; depigmentation of skin and hair.
Fire: Will burn. Hazards: emits toxic fumes. Extinguishing agents: carbon dioxide, dry chemical powder or appropriate foam. Precautions: combustible liquid.
Carcinogenicity: IARC - Not listed; NIOSH - Not listed; NTP - Not listed; ACGIH - Not listed; OSHA - Not listed; EPA - Not listed; MAK - Not listed
Primary Target Organs:

Eyes

Skin

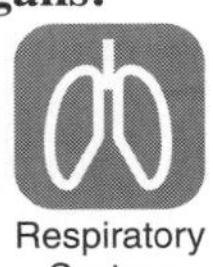
Respiratory System

Environmental

Ecotoxicity: LC_{50} Pimephales promelas (fathead minnow) 2.50 (1.87-3.34) mg/l 72 & 96 hr, wt 102 mg, flow-through bioassay, dissolved oxygen 7.4 (4.6-8.8) mg/l, water hardness 44.9 (42.4-46.6) mg/l as $CaCO_3$, pH 6.9-7.7, alkalinity 42.9 (39.6-61.4) mg/l $CaCO_3$
Environmental Fate: Insufficient data are available to predict the relative importance or rate of biodegradation in soil or water. In the vapor phase it will degrade in the atmosphere by reaction with photochemically produced hydroxyl radicals with an estimated half-life of approximately 9 hours. Reaction with atmospheric nitrate radicals may be important.
Cleanup/Disposal: Guide No. 171: Do not touch or walk through spilled material. Stop leak if you can do it without risk. Prevent dust cloud. Avoid inhalation of asbestos dust.

Small Dry Spills: With clean shovel place material into clean, dry container and cover loosely; move containers from spill area. Small Spills: Take up with sand or other noncombustible absorbent material and place into containers for later disposal. Large Spills: Dike far ahead of liquid spill for later disposal. Cover powder spill with plastic sheet or tarp to minimize spreading. Prevent entry into waterways, sewers, basements or confined areas.

Environmental Physical Data

Henry's Law Constant: estimated at 2.03 $x10^{-6}$
Octanol/Water Partition Coefficient: log K_{ow} = 4.03
Sorption Partition Coefficient: K_{oc} = 3800
BCF: estimated at 34

Regulations

RCRA 40CFR: Not listed
CERCLA: 40CFR 302.4: Not listed
SARA 40CFR 372.65: Not listed
SARA EHS 40CFR 355: Not listed
TSCA: Listed

Analytical Methods

Soil: EPA PMD-PFH

DIM7900 CAS #: 5910-89-4

2,3-DIMETHYLPYRAZINE

RTECS: UQ2625000
EINECS Number: 227-630-0
Molecular Formula: $C_6H_8N_2$
Formula Weight: 108.16

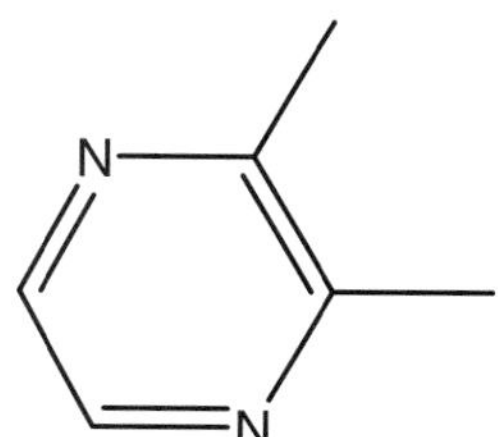

Chemical Structure

Physical Properties

Boiling Point: 156 °C (312.8 °F)
Freezing Point: 11 °C (51.8 °F) to 13 °C (55.4 °F)
Specific Gravity: 1.02810
Water Solubility: Miscible
Other Solubilities: Ethanol: Soluble; Ether: Soluble

RTECS Toxicity Data

Acute Oral: Rat LD_{50} Dose: 613 mg/kg.

Hazard Overviews

Health: Irritating to eyes/skin/respiratory tract. Harmful. Other Acute Effects: harmful if swallowed; may be harmful if inhaled; may be harmful if absorbed through the skin.

Fire: Hazards: emits toxic fumes. Extinguishing agents: carbon dioxide, dry chemical powder or appropriate foam; water spray. Precautions: combustible liquid.
Reactivity: Incompatible with: strong oxidizing agents, strong acids. Hazardous decomposition products: toxic fumes of: carbon monoxide, carbon dioxide, nitrogen oxides.
Carcinogenicity: IARC - Not listed; NIOSH - Not listed; NTP - Not listed; ACGIH - Not listed; OSHA - Not listed; EPA - Not listed; MAK - Not listed
Primary Target Organs:

Eyes

Skin
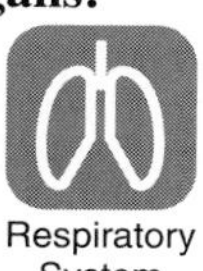
Respiratory System

Environmental

Regulations

RCRA 40CFR: Not listed
CERCLA: 40CFR 302.4: Not listed
SARA 40CFR 372.65: Not listed
SARA EHS 40CFR 355: Not listed
TSCA: Listed

DIM7950 CAS #: 123-32-0

2,5-DIMETHYLPYRAZINE

RTECS: UQ2800000
EINECS Number: 204-618-3
Molecular Formula: $C_6H_8N_2$
Formula Weight: 108.16

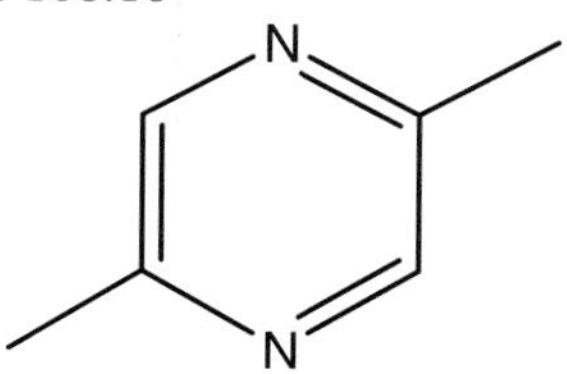

Chemical Structure

Physical Properties

Boiling Point: 155 °C (311 °F)
Freezing Point: 15 °C (59 °F)
Specific Gravity: 0.98870
Water Solubility: Miscible
Other Solubilities: Ethanol: Miscible; Ether: Miscible; Acetone: Soluble; Chloroform: Soluble
Refraction Index: 1.4980

RTECS Toxicity Data

Acute Oral: Rat LD_{50} Dose: 1020 mg/kg.
Mutagenic: Hamster Cytogenetic Analysis; Cell Type: ovary; Dose: 2500 mg/L. Yeast - S Cerevisiae Mutations in Microorganisms; Dose: 33800 mg/L (-S9).

Hazard Overviews

Health: Irritating to eyes/skin/respiratory tract. Harmful. Other Acute Effects: harmful if swallowed; may be harmful if inhaled or absorbed through the skin.

Fire: Hazards: emits toxic fumes. Extinguishing agents: water spray; carbon dioxide, dry chemical powder or appropriate foam. Precautions: combustible liquid.

Reactivity: Incompatible with: strong oxidizing agents, protect from moisture, strong acids. Hazardous decomposition products: toxic fumes of: carbon monoxide, carbon dioxide, nitrogen oxides.

Carcinogenicity: IARC - Not listed; NIOSH - Not listed; NTP - Not listed; ACGIH - Not listed; OSHA - Not listed; EPA - Not listed; MAK - Not listed

Primary Target Organs:

Eyes

Skin

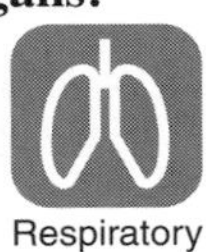
Respiratory System

Environmental

Regulations

RCRA 40CFR: Not listed
CERCLA: 40CFR 302.4: Not listed
SARA 40CFR 372.65: Not listed
SARA EHS 40CFR 355: Not listed
TSCA: Listed

DIM8050 CAS #: 15679-24-0

2,7-DIMETHYLPYRENE

EINECS Number: 239-762-6
Molecular Formula: $C_{18}H_{14}$
Formula Weight: 230.31
Synonyms: PYRENE,2,7-DIMETHYL-

Physical Properties

Freezing Point: 230 °C (446 °F)

Hazard Overviews

Carcinogenicity: IARC - Not listed; NIOSH - Not listed; NTP - Not listed; ACGIH - Not listed; OSHA - Not listed; EPA - Not listed; MAK - Not listed

Environmental

Regulations

RCRA 40CFR: Not listed
CERCLA: 40CFR 302.4: Not listed
SARA 40CFR 372.65: Not listed
SARA EHS 40CFR 355: Not listed
TSCA: Not listed

DIM8150 CAS #: 124-28-7

DIMETHYLSTEARYLAMINE

RTECS: RG4200000
EINECS Number: 204-694-8
Molecular Formula: $C_{20}H_{43}N$
Formula Weight: 297.55

Chemical Structure

Synonyms: ARMEEN DM 18D; DIMANTINE; N,N-DIMETHYL-1-OCTADECANAMINE; N,N-DIMETHYLOCTADECANAMINE; DIMETHYL-N-OCTADECYLAMINE; DIMETHYLOCTADECYLAMINE; N,N-DIMETHYL-N-OCTADECYLAMINE; N,N-DIMETHYLOCTADECYLAMINE; N,N-DIMETHYLOKTADECYLAMIN; DIMETHYLSTEARAMINE; N,N-DIMETHYLSTEARYLAMINE; DYMANTHINE; KEMAMINE 9902D; 1-OCTADECANAMINE,N,N-DIMETHYL-; OCTADECYLAMINE,N,N-DIMETHYL-; OCTADECYLDIMETHYLAMINE; STEARYLDIMETHYLAMINE

Use: chem int for n,n-dimethyl-n-octadecylamine oxide; chem int for benzyl dimethyl octadecyl ammonium chloride; corrosion inhibitor for condensing systems; anthelmintic agent (hydrochloride salt)

Physical Properties

Freezing Point: 22.89 °C (73.202 °F)

RTECS Toxicity Data

Irritation Eye: Rabbit Standard Draize Test Dose: 20 mg/24H; Reaction: moderate.

Irritation Skin: Rabbit Standard Draize Test Dose: 20 mg/24H; Reaction: moderate.

Hazard Overviews

Carcinogenicity: IARC - Not listed; NIOSH - Not listed; NTP - Not listed; ACGIH - Not listed; OSHA - Not listed; EPA - Not listed; MAK - Not listed

Environmental

Regulations

RCRA 40CFR: Not listed
CERCLA: 40CFR 302.4: Not listed
SARA 40CFR 372.65: Not listed
SARA EHS 40CFR 355: Not listed
TSCA: Listed

DIM8250 CAS #: 26636-01-1

DIMETHYLTIN BIS(ISOOCTYL MERCAPTOACETATE)

RTECS: WH6721000

DOT: UN2788; UN3146; IMO6.1
EINECS Number: 247-862-6
Molecular Formula: $C_{22}H_{44}O_4S_2Sn$
Formula Weight: 555.47
Synonyms: ACETIC ACID,2,2'-((DIMETHYLSTANNYLENE)BIS(THIO))BIS-,DIISOOCTYL ESTER; ACETIC ACID,2,2'-((DIMETHYLSTANNYLENE)BIS(THIO))BIS-,DIISOOCTYL ESTER (9CI); ADVASTAB TM 181S; ADVASTAB TM 181; BIS((((ISOOCTYLOXY)CARBONYL)METHYL)THIO)DIMETHYLTIN; DIISOOCTYL ((DIMETHYLSTANNYLENE)DITHIO)DIACETATE; DIMETHYLTIN S,S'-BIS(ISOOCTYL MERCAPTOACETATE); DIMETHYLTIN BIS(ISOOCTYL THIOGLYCOLATE); DIMETHYLTINBIS(ISOOCTYLMERCAPTOACETATE); DIMETHYLTIN-BIS(ISOOCTYLTHIOGLYCOLATE); DIMETHYLZINN-S,S'-BIS(ISOOCTYLTHIOGLYCOLAT); STANNANE,BIS(ISOOCTYLOXYCARBONYLMETHYLTHIO)DIMETHYL-; T 40 (ESTER); TIN,BIS((CARBOXYMETHYL)THIO)DIMETHYL-,DIISOOCTYL ESTER(7CI); TIN,DIMETHYL-,BIS(ISOOCTYLTHIOGLYCOLLATE); TM 181S
Use: heat stabilizer for pvc resins; catalyst for mfr of polyurethane foam; in germany approved for food packaging; dialkyltin compounds best general-purpose stabilizers for polyvinyl chloride, esp if colorlessness & transparency required; catalysts for high resiliency foam in automotive seating; esterification catalysts for mfr of org esters in plasticizers, lubricants, & heat-transfer fluids.

RTECS Toxicity Data

Acute Oral: Rat LD_{50} Dose: 604 mg/kg; Toxic Effects: Behavioral - Ataxia; Lungs, Thorax, or Respiration - Respiratory depression.

Hazard Overviews

Carcinogenicity: IARC - Not listed; NIOSH - Not listed; NTP - Not listed; ACGIH - Class A4, Not classifiable as a human carcinogen; OSHA - Not listed; EPA - Not listed; MAK - Not listed
Exposure Limits
OSHA PEL: TWA: 0.1 mg/m^3; as Sn.
OSHA PEL Vacated 1989 Limits: TWA: 0.1 mg/m^3.
ACGIH TLV: TWA: 2 mg/m^3; see NIOSH.
NIOSH REL: TWA: 2 mg/m^3; inorganic (.1 if organic).
DFG MAK: TWA: 2 mg/m^3; See NIOSH.

Environmental

Cleanup/Disposal: Guide No. 153: Eliminate all ignition sources (no smoking, flares, sparks or flames in immediate area). Do not touch damaged containers or spilled material unless wearing appropriate protective clothing. Stop leak if you can do it without risk. Prevent entry into waterways, sewers, basements or confined areas. Absorb or cover with dry earth, sand or other non-combustible material and transfer to containers. Do not get water inside containers.

Environmental Physical Data

BCF: concentrates few mg/kg

Regulations

RCRA 40CFR: Not listed
CERCLA: 40CFR 302.4: Not listed
SARA 40CFR 372.65: Not listed
SARA EHS 40CFR 355: Not listed
TSCA: Listed

DIM8300 **CAS #: 68928-76-7**

DIMETHYLTIN NEODECANOATE

EINECS Number: 273-028-6

Physical Properties

Specific Gravity: 1.136
Water Solubility: Negligible

Hazard Overviews

Carcinogenicity: IARC - Not listed; NIOSH - Not listed; NTP - Not listed; ACGIH - Not listed; OSHA - Not listed; EPA - Not listed; MAK - Not listed

Environmental

Regulations

RCRA 40CFR: Not listed
CERCLA: 40CFR 302.4: Not listed
SARA 40CFR 372.65: Not listed
SARA EHS 40CFR 355: Not listed
TSCA: Not listed

DIM8350 **CAS #: 2223-82-7**

2,2-DIMETHYLTRIMETHYLENE ACRYLATE

RTECS: AS8925000
EINECS Number: 218-741-5
Molecular Formula: $C_{11}H_{16}O_4$
Formula Weight: 212.27

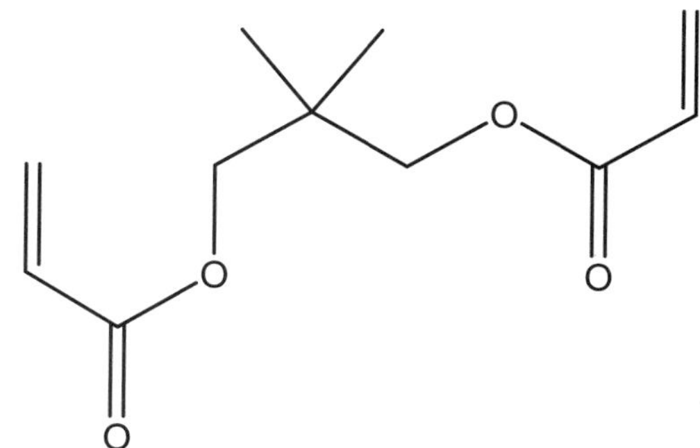

Chemical Structure

Synonyms: ACRYLIC ACID,2,2-DIMETHYL-1,3-PROPANEDIOL DIESTER; ACRYLIC ACID,2,2-DIMETHYLTRIMETHYLENE ESTER; DIMETHYLOLPROPANE DIACRYLATE; 2,2-DIMETHYLPROPANE-1,3-DIOL DIACRYLATE; NEOPENTYL GLYCOL DIACRYLATE; NEOPENTYLGLYCOL DIACRYLATE; 1,3-PROPANEDIOL,2,2-DIMETHYL-,DIACRYLATE; 2-PROPENOIC ACID,2,2-DIMETHYL-1,3-

PROPANEDIYL ESTER; 2-PROPENOIC ACID,2,2-DIMETHYL-1,3-PROPANEDIYL ESTER (9CI); SR 247; VISCOAT 247
Use: crosslinking agent for UV curable polymers for inks and coatings

Physical Properties

Other Solubilities: Reacts readily with electrophilic, free-radical, and nucleophilic agent
Refraction Index: 1.4542

RTECS Toxicity Data

Acute Oral: Rat LD_{50} Dose: 5190 uL/kg.
Acute Dermal: Rabbit LD_{50} Route: Skin; Dose: 283 uL/kg.
Irritation Skin: Rabbit Open Draize Test Dose: 500 mg open; Reaction: severe.
Tumorigenic: Mouse Route: Skin; Dose: 46800 mg/kg/28W-I; Toxic Effects: Tumorigenic - Carcinogenic by RTECS criteria; Skin and appendages - Tumors; Tumorigenic - Tumors at site of application.

Hazard Overviews

Health: Severely irritating to eyes; irritating to skin/respiratory tract. Harmful. Other Acute Effects: harmful if swallowed, inhaled, or absorbed through skin; may cause allergic skin reaction. Chronic Effects: Possible human carcinogen.
Fire: Hazards: emits toxic fumes; closed containers may rupture and explode during runaway polymerization; water may be effective for cooling, but may not effect extinguishment. Extinguishing agents: carbon dioxide, dry chemical powder or appropriate foam. Precautions: combustible liquid.
Reactivity: Incompatible with: heat, strong oxidizing agents, direct sunlight, inert gases, free radical initiators. Hazardous decomposition products: toxic fumes of: carbon monoxide, carbon dioxide.
Carcinogenicity: IARC - Not listed; NIOSH - Not listed; NTP - Not listed; ACGIH - Not listed; OSHA - Not listed; EPA - Not listed; MAK - Not listed
Primary Target Organs:

Eyes

Skin

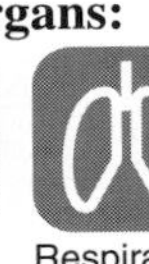
Respiratory System

Environmental

Regulations

RCRA 40CFR: Not listed
CERCLA: 40CFR 302.4: Not listed
SARA 40CFR 372.65: Not listed
SARA EHS 40CFR 355: Not listed
TSCA: Listed

DIM8400 **CAS #: 598-94-7**

1,1-DIMETHYLUREA

RTECS: YS9867985
EINECS Number: 209-957-0
Molecular Formula: $C_3H_8N_2O$
Formula Weight: 88.13

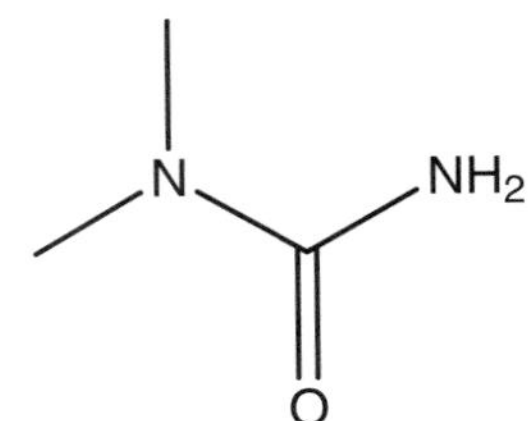

Chemical Structure

Synonyms: ASYM-DIMETHYLUREA; N,N-DIMETHYLHARNSTOFF; N,N-DIMETHYLUREA; UREA,1,1-DIMETHYL-; UREA,N,N-DIMETHYL-
Description: monoclinic prisms

Physical Properties

Freezing Point: 182 °C (359.6 °F)
Specific Gravity: 1.255
Water Solubility: Soluble in Water
Other Solubilities: Slightly Soluble in Alcohol
Ionization Potential (eV): =< 8.96

Hazard Overviews

Health: Irritating to eyes/skin/respiratory tract. Other Acute Effects: may be harmful by inhalation, ingestion, or skin absorption.
Fire: Extinguishing agents: water spray; carbon dioxide, dry chemical powder or appropriate foam. Precautions: combustible liquid.
Reactivity: Incompatible with: strong oxidizing agents. Hazardous decomposition products: toxic fumes of: carbon monoxide, carbon dioxide, nitrogen oxides.
Carcinogenicity: IARC - Not listed; NIOSH - Not listed; NTP - Not listed; ACGIH - Not listed; OSHA - Not listed; EPA - Not listed; MAK - Not listed
Primary Target Organs:

Eyes

Skin

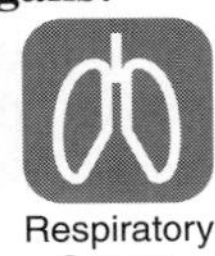
Respiratory System

Environmental

Regulations

RCRA 40CFR: Not listed
CERCLA: 40CFR 302.4: Not listed
SARA 40CFR 372.65: Not listed
SARA EHS 40CFR 355: Not listed
TSCA: Listed

DIM8450 CAS #: 96-31-1

1,3-DIMETHYLUREA

RTECS: YS9868000
EINECS Number: 202-498-7
Molecular Formula: $C_3H_8N_2O$
Formula Weight: 88.13

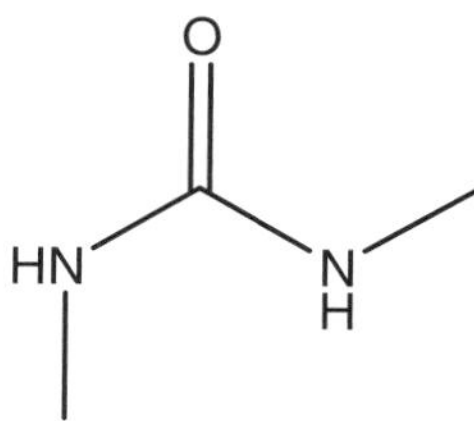

Chemical Structure

Synonyms: N,N'-DIMETHYLHARNSTOFF; N,N'-DIMETHYLUREA; SYM-DIMETHYLUREA; SYMMETRIC DIMETHYLUREA; UREA,1,3-DIMETHYL-; UREA,N,N'-DIMETHYL-
Description: colorless prisms
Use: intermediate in the synthesis of drugs

Physical Properties

Boiling Point: 268 °C (514 °F) to 270 °C (518 °F)
Freezing Point: 108 °C (226.4 °F)
Specific Gravity: 1.142
Density: 1.142 g/mL
Water Solubility: Very Soluble in Water
Other Solubilities: 95% Ethanol: >=100 mg/ml at 21 °C; Acetone: >=100 mg/ml at 21 °C; DMSO: >=100 mg/ml at 21 °C; Ether: Insoluble.
Ionization Potential (eV): =< 9.23
Flash Point: Not available; probably combustible

RTECS Toxicity Data

Reproductive/Teratogenic: Rat Route: Oral; Dose: 2 gm/kg; Duration: female 12D of pregnancy; Effects on Embryo or Fetus - Fetotoxicity. Mouse Route: Oral; Dose: 2 gm/kg; Duration: female 10D of pregnancy; Effects on Fertility - Post-implantation mortality; Effects on Embryo or Fetus - Fetotoxicity; Specific Developmental Abnormalities - Craniofacial (including nose and tongue). Mouse Route: Oral; Dose: 2 gm/kg; Duration: female 10D of pregnancy; Specific Developmental Abnormalities - Musculoskeletal system.
Mutagenic: Human DNA Inhibition; Cell Type: lymphocyte; Dose: 40 mmol/L. Protozoa - C Reinhardi Mutations in Microorganisms; Dose: 400 mmol/L (-S9).

Hazard Overviews

Health: May cause irritation. Other Acute Effects: may be harmful by inhalation, ingestion, or skin absorption; ingestion may cause moderate gastric upset with possible kidney and bladder involvement. Chronic Effects: an animal study indicates teratogenicity when administered to mice at high levels.
Fire: Will burn. Hazards: emits toxic fumes. Extinguishing agents: water spray; carbon dioxide, dry chemical powder or appropriate foam. Precautions: combustible liquid.
Carcinogenicity: IARC - Not listed; NIOSH - Not listed; NTP - Not listed; ACGIH - Not listed; OSHA - Not listed; EPA - Not listed; MAK - Not listed

Environmental

Ecotoxicity: Algae: Scenedesmus subspicatus 72h EC_{20} 121 mg/l Daphnia magna Straus 24-48h EC_0 500 mg/l Leuciscus idus 96h LC_{50} 10,000

Environmental Physical Data

Octanol/Water Partition Coefficient: log K_{ow} = 0.78

Regulations

RCRA 40CFR: Not listed
CERCLA: 40CFR 302.4: Not listed
SARA 40CFR 372.65: Not listed
SARA EHS 40CFR 355: Not listed
TSCA: Listed

DIM8500 CAS #: 513-37-1

DIMETHYLVINYL CHLORIDE

RTECS: UC8045000
EINECS Number: 208-158-4
Molecular Formula: C_4H_7Cl
Structured MF: $(CH_3)_2C{=}CHCl$
Formula Weight: 90.55

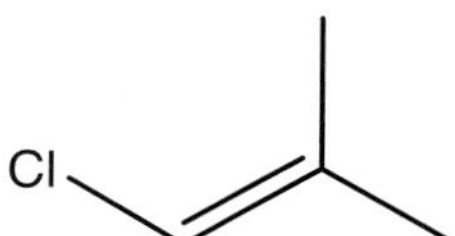

Chemical Structure

Synonyms: 1-CHLOROISOBUTENE; 1-CHLOROISOBUTYLENE; ALPHA-CHLOROISOBUTYLENE; 1-CHLORO-2-METHYL-1-PROPENE; 1-CHLORO-2-METHYLPROPENE; 2,2-DIMETHYLVINYL CHLORIDE; BETA,BETA-DIMETHYLVINYL CHLORIDE; DIMETHYLVINYLCHLORIDE; ISOCROTYL CHLORIDE; 2-METHYL-1-CHLOROPROPENE; 2-METHYL-1-PROPENYL CHLORIDE; 1-PROPENE,1-CHLORO-2-METHYL-; PROPENE,1-CHLORO-2-METHYL-
Description: clear, colorless liquid
Use: organic synthesis

Physical Properties

Boiling Point: 68 °C (154 °F) at 754 mm Hg
Specific Gravity: 0.9186 at 20 °C/4 °C
Saturated Vapor Density: 1.777743376 kg/m^3
Vapor Pressure: 172.4 mm Hg at 25.5 °C
Water Solubility: 1 to 5 mg/mL at 20 °C

Other Solubilities: 95% Ethanol: >=100 mg/ml at 20 °C; Acetone: >=100 mg/ml at 20 °C; Acid: Soluble; Chloroform: Soluble; DMSO: >=100 mg/ml at 20 °C; Ether: miscible.
Refraction Index: 1.4221 at 20 °C
Flash Point: -1 °C

RTECS Toxicity Data

Acute Oral: Rat LD_{50} Dose: 4465 mg/kg; Toxic Effects: Sense organs and special senses - Other; Behavioral - Muscle weakness; Lungs, Thorax, or Respiration - Dyspnea. Mouse LD_{50} Dose: 3160 mg/kg; Toxic Effects: Behavioral - Somnolence (general depressed activity); Lungs, Thorax, or Respiration - Dyspnea; Gastrointestinal - Hypermotility, diarrhea.
Acute Inhalation: Rat LC_{50} Dose: 400 mg/m^3/4hr. Mouse LC_{50} Dose: 181 gm/m^3; Toxic Effects: Behavioral - General anesthetic; Lungs, Thorax, or Respiration - Respiratory depression; Lungs, Thorax, or Respiration - Other changes.
Chronic (Multiple Dose) Oral: Rat Dose: 17500 mg/kg/14D-I; Toxic Effects: Behavioral - Muscle weakness; Musculoskelital - Other changes; DEATH. Rat Dose: 48750 mg/kg/13W-I; Toxic Effects: Behavioral - Somnolence (general depressed activity); Gastrointestinal - Necrotic changes; DEATH.
Mutagenic: Mouse Mutations in Mammalian Somatic Cells; Cell Type: lymphocyte; Dose: 400 ug/L. Hamster Sister Chromatid Exchenge; Cell Type: ovary; Dose: 500 mg/L.
Tumorigenic: Rat Route: Oral; Dose: 51500 mg/kg/2Y-I; Toxic Effects: Tumorigenic - Carcinogenic by RTECS criteria; Sense organs and special senses - Tumors. Mouse Route: Oral; Dose: 51 gm/kg/2Y-I; Toxic Effects: Tumorigenic - Carcinogenic by RTECS criteria; Gastrointestinal - Tumors. Mouse Route: Oral; Dose: 102 gm/kg/2Y-I; Toxic Effects: Tumorigenic - Neoplastic by RTECS criteria; Skin and appendages - Tumors.

Hazard Overviews

Flammable

Health: Irritating to eyes/skin/respiratory tract. Toxic. Other Acute Effects: harmful if swallowed, inhaled, or absorbed through skin; symptoms of exposure may include burning sensation, coughing, wheezing, laryngitis, shortness of breath, headache, nausea and vomiting. Chronic Effects: Carcinogen.
Fire: Flammable. Hazards: vapor may travel considerable distance to source of ignition and flash back; container explosion may occur; forms explosive mixtures in air; emits toxic fumes. Extinguishing agents: carbon dioxide, dry chemical powder or appropriate foam; water may be effective for cooling, but may not effect extinguishment. Precautions: combustible liquid.
Reactivity: Incompatible with: strong oxidizing agents, strong bases. Hazardous decomposition products: toxic fumes of: carbon monoxide, carbon dioxide, hydrogen chloride gas.

Carcinogenicity: IARC - Group 2B, Possibly carcinogenic to humans; NIOSH - Not listed; NTP - Listed; ACGIH - Not listed; OSHA - Not listed; EPA - Not listed; MAK - Not listed
Primary Target Organs:

Eyes

Skin

Respiratory System

Environmental

Environmental Fate: If released to water, volatilization will be rapid with estimated half-lives of 2.9 hours and 3.8 days from a model environmental river and a model lake, respectively. Adsorption to sediment and bioconcentration are not likely fate processes. Insufficient information is available to determine the rate or importance of biodegradation in soil or water. If released to the atmosphere, it will exist in the vapor phase. Vapor-phase will degrade in the atmosphere by reaction with photochemically produced hydroxyl radicals with an estimated half-life of approximately 21 hours and by reaction with atmospheric ozone with an estimated half-life of 26 hours. Removal of atmospheric compound may occur through wet deposition. If released to soil, it is expected to have high mobility based on estimated K_{oc} values of 68 to 98. Volatilization is expected from both moist and dry soils.
Cleanup/Disposal: Guide No. 128: Eliminate all ignition sources (no smoking, flares, sparks or flames in immediate area). All equipment used when handling the product must be grounded. Do not touch or walk through spilled material. Stop leak if you can do it without risk. Prevent entry into waterways, sewers, basements or confined areas. A vapor suppressing foam may be used to reduce vapors. Absorb or cover with dry earth, sand or other non-combustible material and transfer to containers. Use clean non-sparking tools to collect absorbed material. Large Spills: Dike far ahead of liquid spill for later disposal. Water spray may reduce vapor; but may not prevent ignition in closed spaces.

Environmental Physical Data

Henry's Law Constant: estimated at 0.019
Sorption Partition Coefficient: K_{oc} = 68
BCF: estimated at 12

Regulations

RCRA 40CFR: Not listed
CERCLA: 40CFR 302.4: Not listed
SARA 40CFR 372.65: Not listed
SARA EHS 40CFR 355: Not listed
TSCA: Not listed

DIM8650 **CAS #: 644-64-4**

DIMETILAN

RTECS: EZ9084000
EINECS Number: 211-420-0
Molecular Formula: $C_{10}H_{16}N_4O_3$

Formula Weight: 240.27

Synonyms: CARBAMIC ACID,DIMETHYL-,1-((DIMETHYLAMINO)CARBONYL)-5-METHYL-1H-PYRAZOL-3-YL ESTER; CARBAMIC ACID,DIMETHYL-,ESTER WITH3-HYDROXY-N,N,5-TRIMETHYLPYRAZOLE-1-CARBOXAMIDE; CARBAMIC ACID,DIMETHYL-,5-METHYL-1H-PYRAZOL-3-YL ESTER; DIMETHYL 2-CARBAMYL-3-METHYLPYRAZOLYLDIMETHYLCARBAMATE; DIMETHYLCARBAMIC ACID 1-((DIMETHYLAMINO)CARBONYL)-5-METHYL-1H-PYRAZOL-3-YL ESTER; DIMETHYLCARBAMIC ACID ESTER WITH 3-HYDROXY-N,N,5-TRIMETHYLPYRAZOLE-1-CARBOXAMIDE; DIMETHYLCARBAMIC ACID ESTER WITH3-HYDROXY-N,N,5-TRIMETHYLPYRAZOLE-1-CARBOXAMIDE; DIMETHYLCARBAMIC ACID1-((DIMETHYLAMINO)CARBONYL)-5-METHYL-1H-PYRAZOL-3-YL ESTER; 1-DIMETHYLCARBAMOYL-5-METHYL-3-PYRAZOLYL DIMETHYLCARBAMATE; 1-DIMETHYLCARBAMOYL-5-METHYLPYRAZOL-3-YL DIMETHYLCARBAMATE; 2-DIMETHYLCARBAMOYL-3-METHYL-5-PYRAZOLYL DIMETHYLCARBAMATE; 2-DIMETHYLCARBAMOYL-3-METHYLPYRAZOLYL-(5)-N,N-DIMETHYLCARBAMAT; 2-(N,N-DIMETHYLCARBAMYL)-3-METHYLPYRAZOLYL-5 N,N-DIMETHYLCARBAMATE; 2-(N,N-DIMETHYLCARBAMYL)-3-METHYLPYRAZOLYL-5N,N-DIMETHYLCARBAMATE; DIMETILANE; ENT 25595-X; ENT 25,922; FLY BANDS; G 22870; G-22870; GEIGY 22870; GEIGY GS-13332; GS-13332; 5-METHYL-1H-PYRAZOL-3-YL DIMETHYLCARBAMATE; OMS 479; PYRAZOLE-1-CARBOXAMIDE,3-HYDROXY-N,N,5-TRIMETHYL-,DIMETHYL CARBAMATE (ESTER); PYRAZOLE-1-CARBOXAMIDE,3-HYDROXY-N,N,5-TRIMETHYL-,DIMETHYLCARBAMATE (ESTER); SNIP; SNIP FLY; SNIP FLY BANDS

Description: colorless solid

Use: formerly an insecticide for insect control on livestock

Physical Properties

Boiling Point: 200 °C (392 °F) to 210 °C (410 °F) at 13 mm Hg

Freezing Point: 68 °C (154.4 °F)

Specific Gravity: 1.065

Saturated Vapor Density: 1.20000115 kg/m^3

Vapor Pressure: 1 x10^{-4} mm Hg at 20 °C

Water Solubility: Soluble in Water

Other Solubilities: Readily Soluble in Chlorobenzene.

Flash Point: About 1.67 °C Open Cup

RTECS Toxicity Data

Acute Oral: Rat LD_{50} Dose: 25 mg/kg. Mouse LD_{50} Dose: 60 mg/kg.

Acute Dermal: Rat LD_{50} Route: Skin; Dose: 600 mg/kg.

Hazard Overviews

Flammable

Fire: Flammable.

Carcinogenicity: IARC - Not listed; NIOSH - Not listed; NTP - Not listed; ACGIH - Not listed; OSHA - Not listed; EPA - Not listed; MAK - Not listed

Environmental

Regulations

RCRA 40CFR: Listed Hazardous Waste No. P191 Toxic Waste

CERCLA: 40CFR 302.4: Listed per RCRA Section 3001 RQ: 1 lb (0.454 kg)

SARA 40CFR 372.65: Not listed TPQ: 500/10000 lb

SARA EHS 40CFR 355: Listed TPQ: 1 lb

TSCA: Not listed

DIM8700 CAS #: 80-06-8

DIMITE

RTECS: DC7875000

DOT: UN2761; IMO6.1

EINECS Number: 201-246-3

Molecular Formula: $C_{14}H_{12}Cl_2O$

Formula Weight: 267.16

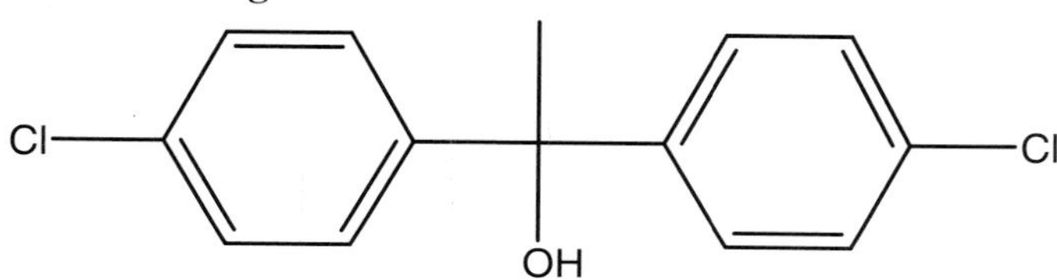

Chemical Structure

Synonyms: BCPE; BENZENEMETHANOL,4-CHLORO-ALPHA-(4-CHLOROPHENYL)-ALPHA-METHYL-; BENZHYDROL,4,4'-DICHLORO-ALPHA-METHYL-; 1,1-BIS(4-CHLOROPHENYL)ETHANOL; 1,1-BIS(P-CHLOROPHENYL)ETHANOL; BIS(P-CHLOROPHENYL)METHYL CARBINOL; 1,1-BIS(P-CHLOROPHENYL)METHYLCARBINOL; 1,1-BIS(4-CHLORPHENYL)-AETHANOL; BPE; CHLORFENETHOL; 4-CHLORO-ALPHA-(4-CHLOROPHENYL)-ALPHA-METHYLBENZENEMENTHANOL; 4-CHLORO-ALPHA-(4-CHLOROPHENYL)-ALPHA-METHYLBENZENEMETHANOL; DCPC; DCPE; DICHLORODIPHENYLETHANOL; P,P'-DICHLORODIPHENYLMETHYLCARBINOL; 4,4'-DICHLORO-(METHYL BENZHYDROL); 4,4'-DICHLORO-ALPHA-METHYLBENZHYDROL; 4,4'-DICHLORO-ALPHA-METHYLBENZOHYDROL; DI(P-CHLOROPHENYL) METHYLCARBINOL; DI-(P-CHLOROPHENYL) METHYLCARBINOL; DI-(P-CHLOROPHENYL)-ETHANOL; DIMIT; DMC; ENT 9,624; ETHANOL,1,1-BIS(P-CHLOROPHENYL)-; QIKRON

Description: white solid or colorless crystals

Use: former uses: non-systemic acaricide, insecticide for control of spider mites in fruit, vegetables, vines, hops, cotton, and ornamental cultivation

Physical Properties

Freezing Point: 69 °C (156.2 °F) to 69.5 °C (157.1 °F)

Vapor Pressure: Low

Water Solubility: Insoluble in Water

Other Solubilities: Soluble in organic solvents; 4.3 g in 100 ml Petroleum Ether at 25-30 °C; 125 g in 100 ml Ethanol at 25-30 °C; 110 g in 100 ml Toluene at 25-30 °C.

Flash Point: Does not burn or burns with difficulty

RTECS Toxicity Data

Acute Oral: Rat LD_{50} Dose: 500 mg/kg. Mouse LD_{Lo} Dose: 750 mg/kg; Toxic Effects: Behavioral - Change in motor activity (specific assay).
Acute Dermal: Rat LD_{50} Route: Skin; Dose: 10 gm/kg.

Hazard Overviews

Health: May be irritating to eyes/skin. Toxic. Other Acute Effects: harmful if inhaled or swallowed.
Fire: Noncombustible. Hazards: emits toxic fumes. Extinguishing agents: water spray; carbon dioxide, dry chemical powder or appropriate foam. Precautions: combustible liquid.
Carcinogenicity: IARC - Not listed; NIOSH - Not listed; NTP - Not listed; ACGIH - Not listed; OSHA - Not listed; EPA - Not listed; MAK - Not listed

Environmental

Ecotoxicity: LC_{50} Fathead minnow 1.4 mg/l/96 hr at 18 °C, wt 1.2 g (95% confidence limit 1.0 - 2.1 mg/l) LC_{50} Channel catfish 0.9 mg/l/96 hr at 18 °C, wt 1.4 g (95% confidence limit 0.7 - 1.2 mg/l
Cleanup/Disposal: Guide No. 151: Do not touch damaged containers or spilled material unless wearing appropriate protective clothing. Stop leak if you can do it without risk. Prevent entry into waterways, sewers, basements or confined areas. Cover with plastic sheet to prevent spreading. Absorb or cover with dry earth, sand or other non-combustible material and transfer to containers. Do not get water inside containers.

Regulations

RCRA 40CFR: Not listed
CERCLA: 40CFR 302.4: Not listed
SARA 40CFR 372.65: Not listed
SARA EHS 40CFR 355: Not listed
TSCA: Not listed

Analytical Methods

Food: FDA 232.4, 242.1
Plasma: EPA 028

DIN1000 CAS #: 29091-05-2

DINITRAMINE

RTECS: XS9990000
EINECS Number: 249-419-2
Molecular Formula: $C_{11}H_{13}F_3N_4O_4$
Formula Weight: 322.2
Synonyms: 1,3-BENZENEDIAMINE,N(SUP 3),N(SUP 3)-DIETHYL-2,4-DINITRO-6-(TRIFLUOROMETHYL)-; COBEKO; COBEX; COBEX (HERBICIDE); COBEXO; DIETHAMINE; 3-DIETHYLAMINO-2,4-DINITRO-6-TRIFLUOROMETHYLANILINE; N(3),N(3)-DIETHYL-2,4-DINITRO-6-(TRIFLUOROMETHYL)-1,3-BENZENEDIAMINE; N(SUP 3),N(SUP 3)-DIETHYL-2,4-DINITRO-6-(TRIFLUOROMETHYL)-1,3-BENZENEDIAMINE; N(1),N(1)-DIETHYL-2,6-DINITRO-4-TRIFLUOROMETHYL-M-PHENYLENEDIAMINE; N(4),N(4)-DIETHYL-ALPHA,ALPHA,ALPHA-TRIFLUORO-3,5-DINITROTOLUENE-2,4-DIAMINE; N(SUP 4),N(SUP 4)-DIETHYL-ALPHA,ALPHA,ALPHA-TRIFLUORO-3,5-DINITROTOLUENE-2,4-DIAMINE; DINITROAMINE; TOLUENE-2,4-DIAMINE,N(SUP 4),N(SUP 4)-DIETHYL-ALPHA,ALPHA,ALPHA-TRIFLUORO-3,5-DINITRO-; USB-3584
Description: yellow crystals
Use: former use selective pre-plant soil incorporated control of many annual grass and broad leaf weeds in cotton, soya beans, groundnuts, peas, beans, safflowers, sunflowers, carrots, turnips, fennel, chickory, and in transported tomatoes, capsicums, aubergines, and brassicas

Physical Properties

Freezing Point: 98 °C (208.4 °F) to 99 °C (210.2 °F)
Density: 1.5 g/cc at 25 °C
Vapor Pressure: 0.479 mPa at 25 °C
Water Solubility: 1 mg/L at 20 °C
Other Solubilities: In Acetone 1040, Chloroform 670, Benzene 473, Xylene 227, Ethanol 107, hexane 14 (all in g/l at 20 °C).
Flash Point: Combustible

RTECS Toxicity Data

Acute Oral: Rat LD_{50} Dose: 3 gm/kg. Duck LD_{50} Dose: >10 gm/kg.
Acute Dermal: Rabbit LD_{50} Route: Skin; Dose: 2 gm/kg.

Hazard Overviews

Fire: Combustible.
Carcinogenicity: IARC - Not listed; NIOSH - Not listed; NTP - Not listed; ACGIH - Not listed; OSHA - Not listed; EPA - Not listed; MAK - Not listed

Environmental

Ecotoxicity: LC_{50} Salmo trutta (Brown trout) 590 ug/l/96 hr at 12 °C, 0.7 g. Static bioassay without aeration, pH 7.2-7.5, water hardness 40-50 mg/l as calcium carbonate and alkalinity of 30-35 mg/l LC_{50} Oncorhynchus kisutcho (Coho salmon) 600 ug/l/96 hr at 12 °C, 0.9 g. Static bioassay without aeration, pH 7.2-7.5, water hardness 40-50 mg/l as calcium carbonate and alkalinity of 30-35 mg/l

Regulations

RCRA 40CFR: Not listed
CERCLA: 40CFR 302.4: Not listed
SARA 40CFR 372.65: Not listed
SARA EHS 40CFR 355: Not listed
TSCA: Not listed

Analytical Methods

Soil: SW846 8091
Plasma: FDA 211.1, 231.1

DIN1190 CAS #: 7617-57-4

1,2-DINITRCSOBENZENE

RTECS: CZ7875000
Molecular Formula: $C_6H_4N_2O_2$
Formula Weight: 136.12

Synonyms: BENZENE,1,2-DINITROSO-; BENZENE,O-DINITROSO-; 1,2-DINITROSOBENZENE

Use: crosslinking agent for natural & butyl rubber; monomer for polydinitrosobenzene-vulcanizing accelerator

RTECS Toxicity Data

Acute Oral: Rat LD_{Lo} Dose: 250 mg/kg.

Hazard Overviews

Carcinogenicity: IARC - Not listed; NIOSH - Not listed; NTP - Not listed; ACGIH - Not listed; OSHA - Not listed; EPA - Not listed; MAK - Not listed

Environmental

Regulations

RCRA 40CFR: Not listed
CERCLA: 40CFR 302.4: Not listed
SARA 40CFR 372.65: Not listed
SARA EHS 40CFR 355: Not listed
TSCA: Not listed

DIN1570 CAS #: 97-02-9

2,4-DINITROANILINE

RTECS: BX9100000
DOT: UN1596; IMO6.1
EINECS Number: 202-553-5
Molecular Formula: $C_6H_5N_3O_4$
Structured MF: $NH_2C_6H_3(NO_2)_2$
Formula Weight: 183.12

Chemical Structure

Synonyms: 1-AMINO-2,4-DINITROBENZENE; ANILINE,2,4-DINITRO-; BENZENAMINE,2,4-DINITRO-; BENZENAMINE,2,4-DINITRO-(9CI); 2,4-DINITRANILINE; 2,4-DINITROAMINOBENZENE; 2,4-DINITROANILIN; 2,4-DINITROANILINA; 2,4-DINITROBENZENAMINE; 2,4-DINITROPHENYLAMINE; DNA

Description: yellow to green-yellow needle-like crystals; musty odor

Use: intermediate for azo pigments; toner pigment in printing inks; corrosion inhibitor

Physical Properties

Boiling Point: 56.7 °C (134 °F)
Freezing Point: 187.5 °C (369.5 °F) to 188 °C (370.4 °F)
Vapor Density: 6.31 Air=1
Saturated Vapor Density: 1.200000005 kg/m^3
Density: 1.615 g/mL at 14 °C
Vapor Pressure: 5.94 x10^{-7} mm Hg at 25 °C
Water Solubility: < 0.1 mg/mL at 23 C
Other Solubilities: 95% Ethanol: <1 mg/ml at 21.5 °C; Acetone: 50-100 mg/ml at 21.5 °C; DMSO: >=100 mg/ml at 21.5 °C; Hydrochloric Acid (hot): Soluble.
Flash Point: 223.889 °C Closed Cup

RTECS Toxicity Data

Acute Oral: Rat LD_{50} Dose: 285 mg/kg. Mouse LD_{50} Dose: 370 mg/kg.

Irritation Eye: Rabbit Standard Draize Test Dose: 500 mg/24H; Reaction: mild.

Reproductive/Teratogenic: Rat Route: Inhalation; Dose: 1100 ug/m^3/4 Duration: female 1-7D of pregnancy; Effects on Embryo or Fetus - Fetotoxicity. Rat Route: Inhalation; Dose: 17 mg/m^3/4H; Duration: female 1-7D of pregnancy; Effects on Fertility - Pre-implantation mortality. Rat Route: Inhalation; Dose: 17 mg/m^3/4H; Duration: female 1-7D of pregnancy; Effects on Embryo or Fetus - Fetal death.

Mutagenic: Rat Unscheduled DNA Synthesis; Cell Type: liver; Dose: 50 umol/L. Bacteria - S Typhimurium Mutations in Microorganisms; Dose: 10 ug/plate (+S9). Bacteria - S Typhimurium Mutations in Microorganisms; Dose: 2 ug/plate (-S9).

Hazard Overviews

Explosive

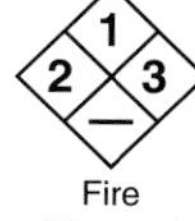

Fire Diamond

Health: Irritating to eyes/skin/respiratory tract. Toxic. Also Causes: skin burns upon prolonged exposure, headache, nausea, stupor; high exposure: methemoglobinemia, CNS depression.

Fire: Will burn. Explodes when heated. Fight fire with alcohol foam, polymer foam, dry chemical, or carbon dioxide. Water or regular foam may cause frothing.

Reactivity: Unstable, can detonate if heated under confinement. Hazardous polymerization cannot occur. Avoid: heat; ignition sources. Incompatible with: oxidizers; charcoal; chlorine and hydrochloric acid; strong acids; acid chlorides; acid anhydrides. Hazardous decomposition products: toxic carbon oxide; nitrogen oxide gases.

Carcinogenicity: IARC - Not listed; NIOSH - Not listed; NTP - Not listed; ACGIH - Not listed; OSHA - Not listed; EPA - Not listed; MAK - Not listed

Primary Target Organs:

Eyes

Skin

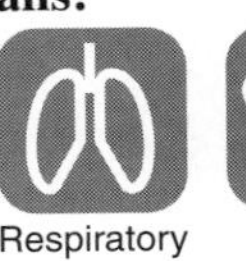
Respiratory System

Nervous System

Blood

Environmental

Ecotoxicity: LC_{50} Pimephales promelas (fathead minnow) 14.2 mg/l/96 hr (Confidence limit 13.5 to 15.0 mg/l). Affected fish lost schooling behavior and swam near the tank surface with half being hyperactive and half hypoactive. They had increased respiration and hemorrhaging, were darkly colored, and lost equilibrium prior to death

Environmental Fate: If released to the atmosphere, it is expected to degrade relatively rapidly (estimated half-life of 17.7 hr) by reaction with photochemically produced hydroxyl radicals. If released to soil it may undergo a covalent chemical bonding with humic materials which can result in its chemical alteration to a latent form and prevent leaching. In the absence of covalent bonding, moderate leaching may be possible. If released to water, covalent bonding with humic materials in the water column and sediments may result in partitioning from the water column to sediments. By analogy to aromatic amine chemical class, in the water column it may be susceptible to photooxidation via hydroxyl and peroxy radicals. One screening study suggests that biodegradation in water will not be significant; however, insufficient data are available to assess the relative importance of biodegradation in soil or water.

Cleanup/Disposal: Guide No. 153: Eliminate all ignition sources (no smoking, flares, sparks or flames in immediate area). Do not touch damaged containers or spilled material unless wearing appropriate protective clothing. Stop leak if you can do it without risk. Prevent entry into waterways, sewers, basements or confined areas. Absorb or cover with dry earth, sand or other non-combustible material and transfer to containers. Do not get water inside containers.

Environmental Physical Data

Henry's Law Constant: estimated at 1.51×10^{-10}

Octanol/Water Partition Coefficient: log K_{ow} = 1.84

Sorption Partition Coefficient: K_{oc} = estimated at 240

BCF: estimated at 15

Regulations

RCRA 40CFR: Not listed

CERCLA: 40CFR 302.4: Not listed

SARA 40CFR 372.65: Not listed

SARA EHS 40CFR 355: Not listed

TSCA: Listed

Analytical Methods

Soil: SW846 8131

DIN1760 CAS #: 99-65-0

1,3-DINITROBENZENE

RTECS: CZ7350000

DOT: UN1597; IMO6.1

EINECS Number: 202-776-8

Molecular Formula: $C_6H_4N_2O_4$

Structured MF: $C_6H_4(NO_2)_2$

Formula Weight: 168.12

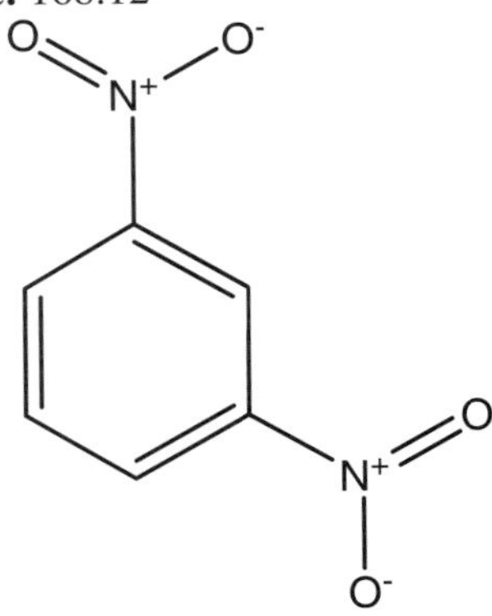

Chemical Structure

Synonyms: BENZENE,1,3-DINITRO-; BENZENE,M-DINITRO-; BINITROBENZENE; 2,4-DINITROBENZENE; DINITROBENZENE; M-DINITROBENZENE; META-DINITROBENZENE; 1,3-DINITROBENZOL; M-DNB; DWUNITROBENZEN; NSC-7189

Use: organic synthesis, dye intermediates

Physical Properties

Boiling Point: 300 °C (572 °F) to 303 °C (577 °F)

Freezing Point: 89 °C (192.2 °F) to 90 °C (194 °F)

Specific Gravity: 1.575 at 18 °C/4 °C

Vapor Density: 5.8 Air=1 at boiling point

Saturated Vapor Density: 1.200029532 kg/m^3

Vapor Pressure: 5.13×10^{-6} atm

Water Solubility: 1 g dissolves in 2000 ml Cold Water

Other Solubilities: Acetone: Very Soluble; Alcohol: Very Soluble; Benzene: Very Soluble; Chloroform: Soluble; Ether: Soluble; Methanol: Soluble; Pyrimidene: Very Soluble; Toluene: Soluble.

Ionization Potential (eV): 10.43

Flash Point: 150 °C Closed Cup

RTECS Toxicity Data

Acute Oral: Human LD_{Lo} Dose: 28 mg/kg. Rat LD_{50} Dose: 59500 ug/kg; Toxic Effects: Behavioral - Somnolence (general depressed activity); Lungs, Thorax, or Respiration - Dyspnea; Skin and appendages - Hair.

Acute Dermal: Man TD_{Lo} Route: Skin; Dose: 4 mg/kg/2D-I; Toxic Effects: Behavioral - Change in motor activity (specific assay); Lungs, Thorax, or Respiration - Cyanosis. Rabbit LD_{50} Route: Skin; Dose: 1900 mg/kg.

Chronic (Multiple Dose) Oral: Rat Dose: 156 mg/kg/90D-I; Toxic Effects: Endocrine - Changes in spleen weight; Blood -

Methemoglobinemia-Carboxyhemoglobin. Rat Dose: 265 mg/kg/8W-C; Toxic Effects: Blood - Pigmented or nucleated red Blood cells; Nutritional and gross metabolic - Weight loss or decreased weight gain; DEATH - Changes in testicular weight. Rat Dose: 296 mg/kg/16W-C; Toxic Effects: Endocrine - Changes in spleen weight; Nutritional and gross metabolic - Weight loss or decreased weight gain; DEATH - Changes in testicular weight.

Reproductive/Teratogenic: Rat Route: Oral; Dose: 24 mg/kg; Duration: male 1D prior to mating; Paternal Effects - Spermatogenesis; Testes, epididymis, sperm duct. Rat Route: Oral; Dose: 33600 ug/kg; Duration: female 16W prior to mating Maternal Effects - Ovaries, fallopian tubes. Rat Route: Oral; Dose: 48 mg/kg; Duration: male 1D prior to mating; Effects on Fertility - Male fertility index. Rat Route: Oral; Dose: 90 mg/kg; Duration: male 12W prior to mating; Paternal Effects - Spermatogenesis.

Mutagenic: Bacteria - S Typhimurium Mutations in Microorganisms; Dose: 3300 ng/plate (+S9). Bacteria - S Typhimurium Mutations in Microorganisms; Dose: 50 ug/plate (-S9).

Hazard Overviews

Reactivity: Unstable, shock and friction sensitive. Hazardous polymerization cannot occur. Avoid: shock; friction; heating. Incompatible with: strong oxidizers; caustics; metals; nitric acid; tetranitromethane; steam. Hazardous decomposition products: carbon dioxide; toxic nitrogen oxides.

Carcinogenicity: IARC - Not listed; NIOSH - Not listed; NTP - Not listed; ACGIH - Not listed; OSHA - Not listed; EPA - Class D, Not classifiable as to human carcinogenicity; MAK - Class B, Justifiably suspected of having carcinogenic potential

Exposure Limits

OSHA PEL: TWA: 1 mg/m^3; skin.

ACGIH TLV: TWA: 0.15 ppm; 1 mg/m^3.

NIOSH REL: TWA: 1 mg/m^3.

NIOSH IDLH: 50 mg/m^3.

Respirator Recommendation

Exposure Range: >1 to 10 mg/m^3 Air Purifying, Negative Pressure, Half Mask

Exposure Range: >10 to <50 mg/m^3 Air Purifying, Negative Pressure, Full Face

Exposure Range: 50 to unlimited mg/m^3 Self-contained Breathing Apparatus, Pressure Demand, Full Face

Cartridge Color: black with dust/mist prefilter (use P100 or consult supervisor for appropriate dust/mist prefilter)

Environmental

Ecotoxicity: Aquatic toxicity: 8-10 mg/l/6 hr/minnows/min. lethal dose/ fresh water

Environmental Fate: Release is expected to result in adsorption to clay but adsorption to other soils is expected to be weak and leaching may occur. Volatilization from soil surfaces may occur but is expected to be slow. Reduction to aromatic amines may occur under anaerobic conditions. Release to water may result in biodegradation and slow volatilization from the water surface. Direct photolysis may occur based on its absorption of UV light greater than 290 nm. Bioconcentration and hydrolysis are not expected to be significant fate process. Release to the atmosphere is expected to result in the reaction with photochemically generated hydroxyl radicals with an estimated half-life of 14.15 hr. Direct photolysis may also occur.

Cleanup/Disposal: Guide No. 152: Do not touch damaged containers or spilled material unless wearing appropriate protective clothing. Stop leak if you can do it without risk. Prevent entry into waterways, sewers, basements or confined areas. Cover with plastic sheet to prevent spreading. Absorb or cover with dry earth, sand or other non-combustible material and transfer to containers. Do not get water inside containers.

Environmental Physical Data

Henry's Law Constant: estimated at 2.33 $x10^{-6}$

Octanol/Water Partition Coefficient: log K_{ow} = 1.49

Sorption Partition Coefficient: log K_{oc} = estimated at 1.39

BCF: trout muscle 0.93

Regulations

RCRA 40CFR: Not listed

CERCLA: 40CFR 302.4: Listed per CWA Section 311(b)(4) RQ: 100 lb (45.35 kg)

SARA 40CFR 372.65: Listed

SARA EHS 40CFR 355: Not listed

TSCA: Listed

Analytical Methods

Soil: SW846 3640A, 8091, 8270B, 8270C, 8330

Water / Groundwater: EPA 1625

Plasma: EPA 29

DIN1950 CAS #: 25154-54-5

DINITROBENZENE

RTECS: CZ7340000

DOT: UN1597; IMO6.1

EINECS Number: 246-673-6

Molecular Formula: $C_6H_4N_2O_4$

Structured MF: $C_6H_4(NO_2)_2$

Formula Weight: 168

Synonyms: BENZENE,DINITRO-; DINITROBENZENE,LIQUID OR SOLID; DINITROBENZOL

Description: white white to pale yellow crystalline solid

Use: organic synthesis; dyes; camphor substitute in cellulose nitrate dinitrobenzene

Physical Properties

Boiling Point: 301 °C (574 °F)

Freezing Point: 90 °C (194 °F)

Specific Gravity: 1.58

Vapor Pressure: < 1 mm Hg

Water Solubility: Insoluble in Water

Flash Point: 150 °C Closed Cup

Hazard Overviews

Explosive

Health: Corrosive to eyes/skin/respiratory tract. Toxic. Also Causes: headache, nausea, vomiting, dizziness, yellowish color of eyes/hair/skin, difficulty breathing, general weakness, cyanosis; and possible progression to convulsions, coma, necrosis, or severe eye damage. Chronic Effects: anemia, paresthesis in the feet/ankles/hands, and visual reduction.

Fire: Explosive when exposed to shock or friction, or when heated. For small fires use dry chemical, carbon dioxide, water spray, or regular foam. For large fires use water spray, fog, or regular foam. Apply water as fog in flooding amounts since solid streams of water may be ineffective. Use care when applying water, fog, or foam as they may cause frothing. Will burn.

Carcinogenicity: IARC - Not listed; NIOSH - Not listed; NTP - Not listed; ACGIH - Not listed; OSHA - Not listed; EPA - Not listed; MAK - Not listed

Primary Target Organs:

Eyes

Skin

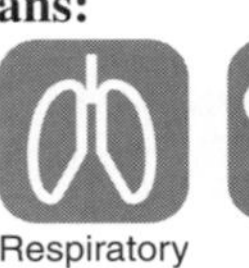
Respiratory System

Nervous System

Liver

Blood

Environmental

Ecotoxicity: Aquatic toxicity: 8 to 10 mg/l/6-hour/minimum lethal dose/Minnows/fresh water

Cleanup/Disposal: Guide No. 152: Do not touch damaged containers or spilled material unless wearing appropriate protective clothing. Stop leak if you can do it without risk. Prevent entry into waterways, sewers, basements or confined areas. Cover with plastic sheet to prevent spreading. Absorb or cover with dry earth, sand or other non-combustible material and transfer to containers. Do not get water inside containers.

Regulations

RCRA 40CFR: Not listed
CERCLA: 40CFR 302.4: Listed per CWA Section 311(b)(4) RQ: 100 lb (45.35 kg)
SARA 40CFR 372.65: Not listed
SARA EHS 40CFR 355: Not listed
TSCA: Not listed

Analytical Methods

Soil: SW846 3650B, 8090

DIN2140 CAS #: 528-29-0

O-DINITROBENZENE

RTECS: CZ7450000
DOT: UN1597; IMO6.1
EINECS Number: 208-431-8
Molecular Formula: $C_6H_4N_2O_4$
Structured MF: $C_6H_4(NO_2)_2$
Formula Weight: 168.12

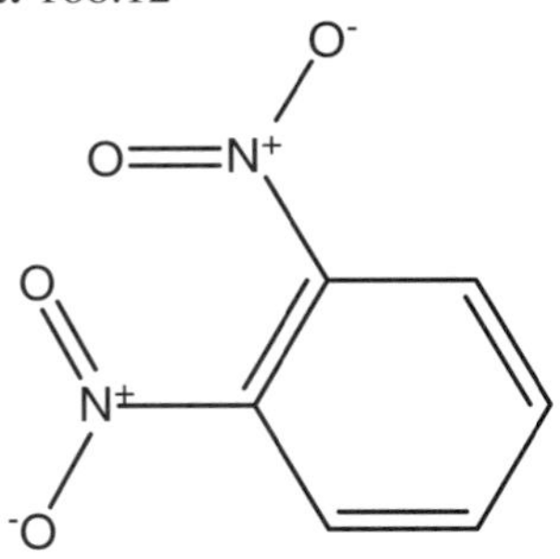

Chemical Structure

Synonyms: BENZENE,1,2-DINITRO-; BENZENE,O-DINITRO-; BENZENE,1,2-DINITRO-(9CI); 1,2-DINITROBENZENE; ORTHO-DINITROBENZENE; NSC 60682

Use: synthesis of dyestuffs, dyestuff intermediates, explosives, and celluloid production; in organic synthesis; as a camphor substitute in cellulose nitrate; manufacture of dyes, & medicines

Physical Properties

Boiling Point: 319 °C (606 °F)
Freezing Point: 118 °C (244.4 °F)
Vapor Density: 5.8 Air=1
Density: 1.565 g/mL at 17 °C
Water Solubility: 0.05% by weight
Other Solubilities: Soluble in Alcohol, Benzene, Chloroform
Refraction Index: 1.565
Ionization Potential (eV): 10.71
Flash Point: 150 °C Closed Cup

RTECS Toxicity Data

Mutagenic: Bacteria - S Typhimurium Mutations in Microorganisms; Dose: 250 ug/plate (+/-S9).

Hazard Overviews

Reactivity: Unstable, highly shock- and friction-sensitive. Hazardous polymerization cannot occur. Avoid: shock; friction; heating. Incompatible with: strong oxidizers; caustics; tin; zinc; nitric acid (all isomers); tetranitromethane (meta); steam. Hazardous decomposition products: carbon dioxide; nitrogen oxides.

Carcinogenicity: IARC - Not listed; NIOSH - Not listed; NTP - Not listed; ACGIH - Not listed; OSHA - Not listed; EPA - Class D, Not classifiable as to human carcinogenicity; MAK - Not listed

Exposure Limits
OSHA PEL: TWA: 1 mg/m^3; skin.
NIOSH REL: TWA: 1 mg/m^3.
NIOSH IDLH: 50 mg/m^3.

Respirator Recommendation
Exposure Range: >1 to 10 mg/m^3 Air Purifying, Negative Pressure, Half Mask

Exposure Range: >10 to <50 mg/m^3 Air Purifying, Negative Pressure, Full Face

Exposure Range: 50 to unlimited mg/m^3 Self-contained Breathing Apparatus, Pressure Demand, Full Face

Cartridge Color: black with dust/mist prefilter (use P100 or consult supervisor for appropriate dust/mist prefilter)

Environmental

Ecotoxicity: Aquatic toxicity: 8-10 ppm/6 hr/minnow/minimum lethal dose/hard water/23 °C; Food chain concentration potential: May have accumulative effects

Environmental Fate: Release to soil is expected to result in adsorption to clay but adsorption to other soils is expected to be weak and from the latter leaching to groundwater may occur. Volatilization from soil surfaces may occur but is expected to be slow. Reduction to aromatic amines may occur under anaerobic conditions. Release to water may result in biodegradation and slow volatilization from the water surface. Direct photolysis may occur. Bioconcentration, and hydrolysis are not expected to be significant. Release to the atmosphere is expected to result in the reaction with photochemically generated hydroxyl radicals with an estimated half-life of 14.15 hr. Direct photolysis may also occur.

Cleanup/Disposal: Guide No. 152: Do not touch damaged containers or spilled material unless wearing appropriate protective clothing. Stop leak if you can do it without risk. Prevent entry into waterways, sewers, basements or confined areas. Cover with plastic sheet to prevent spreading. Absorb or cover with dry earth, sand or other non-combustible material and transfer to containers. Do not get water inside containers.

Environmental Physical Data

Henry's Law Constant: 2.33 x10^{-6}

Octanol/Water Partition Coefficient: log K_{OW} = 1.58

Sorption Partition Coefficient: log K_{OC} = estimated at 1.47

BCF: trout muscle 0.98

Regulations

RCRA 40CFR: Not listed

CERCLA: 40CFR 302.4: Listed per CWA Section 311(b)(4) RQ: 100 lb (45.35 kg)

SARA 40CFR 372.65: Listed

SARA EHS 40CFR 355: Not listed

TSCA: Not listed

Analytical Methods

Soil: SW846 8270B, 8270C

Plasma: EPA 29

DIN2330 CAS #: 100-25-4

P-DINITROBENZENE

RTECS: CZ7525000

DOT: UN1597; IMO6.1

EINECS Number: 202-833-7

Molecular Formula: $C_6H_4N_2O_4$

Structured MF: $C_6H_4(NO_2)_2$

Formula Weight: 168.12

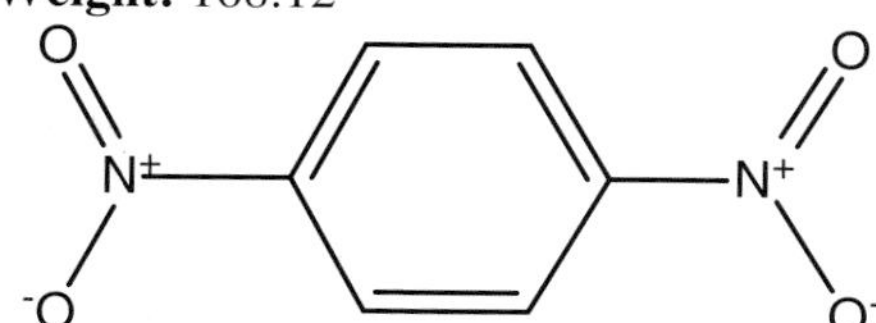

Chemical Structure

Synonyms: BENZENE,1,4-DINITRO-; BENZENE,P-DINITRO-; 1,4-DINITROBENZENE; DINITROBENZENE,PARA-; PARA-DINITROBENZENE; DITHANE A-4; NSC 3809

Use: organic synthesis; dyes; camphor substitute in cellulose nitrate; to reduce aniline, which has wide application in the manufacture of dyes, & medicines

Physical Properties

Boiling Point: 299 °C (570 °F)

Freezing Point: 173 °C (343.4 °F) to 174 °C (345.2 °F)

Specific Gravity: 1.625 at 18 °C/4 °C

Vapor Density: 5.8 Air=1 at boiling point

Saturated Vapor Density: < 1.207574592 kg/m^3

Vapor Pressure: < 1 mm Hg at 20 °C

Water Solubility: 0.01% by weight

Other Solubilities: 1 g dissolves in 300 ml Alcohol; sparingly Soluble in Chloroform, and Ethyl Acetate.

Ionization Potential (eV): 10.50

Flash Point: 150 °C

RTECS Toxicity Data

Acute Oral: Cat LD_{Lo} Dose: 29 mg/kg.

Mutagenic: Bacteria - S Typhimurium Mutations in Microorganisms; Dose: 5 ug/plate (+S9), 25 ug/plate (-S9).

Hazard Overviews

Reactivity: Unstable, shock and friction sensitive. Hazardous polymerization cannot occur. Avoid: shock; friction; heating. Incompatible with: strong oxidizers; caustics; metals; nitric acid; tetranitromethane; steam. Hazardous decomposition products: carbon dioxide; toxic nitrogen oxides.

Carcinogenicity: IARC - Not listed; NIOSH - Not listed; NTP - Not listed; ACGIH - Not listed; OSHA - Not listed; EPA - Class D, Not classifiable as to human carcinogenicity; MAK - Class B, Justifiably suspected of having carcinogenic potential

Exposure Limits

OSHA PEL: TWA: 1 mg/m^3; skin.

NIOSH REL: TWA: 1 mg/m^3.

NIOSH IDLH: 50 mg/m^3.

Respirator Recommendation

Exposure Range: >1 to 10 mg/m^3 Air Purifying, Negative Pressure, Half Mask

Exposure Range: >10 to <50 mg/m^3 Air Purifying, Negative Pressure, Full Face

Exposure Range: 50 to unlimited mg/m^3 Self-contained Breathing Apparatus, Pressure Demand, Full Face

Cartridge Color: black with dust/mist prefilter (use P100 or consult supervisor for appropriate dust/mist prefilter)

Environmental

Ecotoxicity: LC_{50} Pimephales promelas (fathead minnow) 0.603 (0.581-0.627) mg/l 96 hr, wt 164 mg, flow-through bioassay, dissolved oxygen 7.4 (4.6-8.8) mg/l, water hardness 44.9 (42.4-46.6) mg/l as $CaCO_3$, pH 6.9-7.7, alkalinity 42.9 (39.6-61.4) mg/l $CaCO_3$

Environmental Fate: Released to soil may result in adsorption to clay but adsorption to other soils is expected to be weak; therefore, leaching in soils may occur. Volatilization from soil surfaces may occur but is expected to be slow. Reduction to aromatic amines may occur under anaerobic conditions. Release to water may result in biodegradation and slow volatilization from water surfaces. Direct photolysis may occur based upon its absorption of UV light >290 nm. Bioconcentration and hydrolysis are not expected to be significant. Release to the atmosphere is expected to result in the reaction with photochemically generated hydroxyl radicals with an estimated half-life of 14.15 hr. Direct photolysis in the atmosphere may also occur.

Cleanup/Disposal: Guide No. 152: Do not touch damaged containers or spilled material unless wearing appropriate protective clothing. Stop leak if you can do it without risk. Prevent entry into waterways, sewers, basements or confined areas. Cover with plastic sheet to prevent spreading. Absorb or cover with dry earth, sand or other non-combustible material and transfer to containers. Do not get water inside containers.

Environmental Physical Data

Henry's Law Constant: 2.33 $x10^{-6}$
Octanol/Water Partition Coefficient: log K_{ow} = 1.46
Sorption Partition Coefficient: log K_{oc} = estimated at 1.36
BCF: trout muscle 0.91

Regulations

RCRA 40CFR: Not listed
CERCLA: 40CFR 302.4: Listed per CWA Section 311(b)(4) RQ: 100 lb (45.35 kg)
SARA 40CFR 372.65: Listed
SARA EHS 40CFR 355: Not listed
TSCA: Listed

Analytical Methods

Soil: SW846 8091, 8270B, 8270C
Water / Groundwater: EPA 1625
Plasma: EPA 29

DIN2900 CAS #: 1528-74-1

4,4-DINITROBIPHENYL

RTECS: DV4000000
EINECS Number: 216-210-2
Molecular Formula: $C_{12}H_8N_2O_4$
Formula Weight: 244.22
Synonyms: 1,1'-BIPHENYL,4,4'-DINITRO-; BIPHENYL,4,4'-DINITRO-; 4,4'-DINITROBIFENYL; 4,4'-DINITROBIPHENYL; 4,4'-DINITRODIPHENYL
Description: needles

Physical Properties

Freezing Point: 240 °C (464 °F) to 243 °C (469.4 °F)
Water Solubility: Insoluble in Water
Other Solubilities: Soluble in hot Alcohol, Benzene & Acetic Acid

RTECS Toxicity Data

Irritation Eye: Rabbit Standard Draize Test Dose: 500 mg/24H; Reaction: mild.

Mutagenic: Bacteria - S Typhimurium Mutations in Microorganisms; Dose: 2500 ng/plate (+S9), 5 ug/plate (-S9). Bacteria - S Typhimurium Other Mutation Test Systems; Dose: 1 mg/L.

Tumorigenic: Rat Route: Oral; Dose: 950 mg/kg/W-I; Toxic Effects: Tumorigenic - Equivocal tumorigenic agent by RTECS criteria; Sense organs and special senses - Tumors; Skin and appendages - Tumors.

Hazard Overviews

Carcinogenicity: IARC - Not listed; NIOSH - Not listed; NTP - Not listed; ACGIH - Not listed; OSHA - Not listed; EPA - Not listed; MAK - Not listed

Environmental

Regulations

RCRA 40CFR: Not listed
CERCLA: 40CFR 302.4: Listed as Compound per CWA Section 307(a)
SARA 40CFR 372.65: Not listed
SARA EHS 40CFR 355: Not listed
TSCA: Listed

DIN3090 CAS #: 88-85-7

DINITROBUTYL PHENOL

RTECS: SJ9800000
DOT: UN2779; UN2780; UN3013; UN3014; IMO3.2; IMO6.1
EINECS Number: 201-861-7
Molecular Formula: $C_{10}H_{12}N_2O_5$
Structured MF: $C_6H_2OH(NO_2)_2C_4H_9$
Formula Weight: 240.24

Chemical Structure

Synonyms: AATOX; ARETIT; BASANITE; BLAARTOX; BNP 20; BNP 30; BUTAPHEN; BUTAPHENE; 2-SEC-BUTYL-4,6-DINITROPHENOL; CALDON; CHEMOX; CHEMOX P.E; CHEMOX GENERAL; CHEMOX PE; DBNF; DESICOIL; DIBUTOX; DIBUTOX 20CE; DINITRALL; 4,6-DINITRO-2-SEC BUTYLFENOL; DINITRO-ORTHO-SEC-BUTYL PHENOL; 4,6-DINITRO-2-SEC.BUTYLFENOL; 2,4-DINITRO-6-SEC-BUTYLPHENOL; 4,6-DINITRO-2-SEC-BUTYLPHENOL; 4,6-DINITRO-O-SEC-BUTYLPHENOL; DINITROBUTYLPHENOL; 2,4-DINITRO-6-(1-METHYL-PROPYL)PHENOL; 2,4-DINITRO-6-(1-METHYLPROPYL)PHENOL; 4,6-DINITRO-2-(1-METHYL-N-PROPYL)PHENOL; 4,6-DINITRO-2-(1-METHYL-PROPYL)PHENOL; DINOSEB; DINOSEBE; DN 289; DNBP; DNOSBP; DNSBP; DOW GENERAL; DOW GENERAL WEED KILLER; DOW SELECTIVE WEED KILLER; DYTOP; ELGETOL; ELGETOL 318; ENT 1,122; GEBUTOX; GEBUTOX; KNOX-WEED; HEL-FIRE; HIVERTOX; KILOSEB; KNOX-WEED; LADOB; LASEB; 6-(1-METHYL-PROPYL)-2,4-DINITROFENOL; 2-(1-METHYLPROPYL)-4,6-DINITROPHENOL; 6-(1-METIL-PROPIL)-2,4-DINITRO-FENOLO; NITROPONE; NITROPONE C; PHENOL,2-SEC-BUTYL-4,6-DINITRO-; PHENOL,2-(1-METHYLPROPYL)-4,6-DINITRO-; PHENOTAN; PNOSBP; PREMERGE; PREMERGE 3; SINOX GENERAL; SPARIC; SPURGE; SUBITEX; UNICORP DNBP; UNICROP DNBP; VERTAC DINITRO WEED KILLER; VERTAC GENERAL WEED KILLER; VERTAC SELECTIVE WEED KILLER; WSX-8365

Description: yellow crystals; orange solid; pungent odor

Use: miticide former use; insecticide or ovicide but must be used in the dormant growth season or as a salt form to reduce toxicity; herbicide for preemergence treatment former use

Physical Properties

Freezing Point: 38 °C (100.4 °F) to 42 °C (107.6 °F)
Specific Gravity: 1.2647 at 45 °C/4 °C
Vapor Density: 7.90 Air=1
Vapor Pressure: 1 mm Hg at 151.1 °C
Water Solubility: 0.0052 g/100 g Water
Other Solubilities: 95% Ethanol: Soluble; Ether: Soluble.
pH: Acidic Phenol
Refraction Index: 1.5707 at 45 °C/D
Flash Point: 15.6 °C Closed Cup

RTECS Toxicity Data

Acute Oral: Rat LD_{50} Dose: 25 mg/kg; Toxic Effects: Behavioral - Somnolence (general depressed activity); Behavioral - Convulsions or effect on seizure threshold; Lungs, Thorax, or Respiration - Respiratory stimulation. Mouse LD_{50} Dose: 16 mg/kg.

Acute Inhalation: Cat LC_{Lo} Dose: 45 mg/m^3/3hr.

Acute Dermal: Rabbit LD_{50} Route: Skin; Dose: 80 mg/kg. Rat LD_{50} Route: Skin; Dose: 80 mg/kg. Rat LD_{50} Route: Subcutaneous Dose: 20368 ug/kg.

Chronic (Multiple Dose) Oral: Rat Dose: 72800 ug/kg/26W-I; Toxic Effects: Endocrine - Other changes; Blood - Other changes; Nutritional and gross metabolic - Weight loss or decreased weight gain.

Irritation Eye: Rabbit Standard Draize Test Dose: 50 ug/24H; Reaction: severe.

Reproductive/Teratogenic: Rat Route: Oral; Dose: 1201 mg/kg; Duration: male 77D prior to mating; Effects on Fertility - Male fertility index. Rat Route: Oral; Dose: 820 mg/kg; Duration: male 8W prior to mating; Paternal Effects - Testes, epididymis, sperm duct; Effects on Fertility - Female fertility index; Male fertility index. Rat Route: Oral; Dose: 820 mg/kg; Duration: male 8W prior to mating; Effects on Newborn - Viability index; Weaning or lactation index; Growth statistics. Rat Route: Oral; Dose: 100 mg/kg; Duration: female 6-15D of pregnancy; Specific Developmental Abnormalities - Eye, ear. Rat Route: Oral; Dose: 78 mg/kg; Duration: male 20D prior to mating; Paternal Effects - Spermatogenesis. Rat Route: Oral; Dose: 150 mg/kg; Duration: female 6-15D of pregnancy; Effects on Embryo or Fetus - Fetotoxicity.

Mutagenic: Yeast - S Cerevisiae Gene Conversion; Dose: 185 ppm.

Tumorigenic: Mouse Route: Oral; Dose: 764 mg/kg/78W-I; Toxic Effects: Tumorigenic - Equivocal tumorigenic agent by RTECS criteria; Lungs, Thorax, or Respiration - Tumors; Liver - Tumors.

Hazard Overviews

Poison

Corrosive

Flammable

Health: Irritating to eyes/skin. Poison. Other Acute Effects: may be fatal if inhaled, swallowed, or absorbed through skin; causes severe eye irritation; absorption into the body leads to the formation of methemoglobin which in sufficient concentration causes cyanosis; onset may be delayed 2 to 4 hours or longer. Chronic Effects: teratogen; may cause harm to the unborn child; target organs: kidneys, liver. Lab experiments have shown mutagenic effects.

Fire: Flammable. Extinguishing agents: carbon dioxide, dry chemical powder or appropriate foam. Precautions: combustible liquid.

Reactivity: Stable. Hazardous polymerization will not occur. Hazardous decomposition products: thermal decomposition may produce carbon monoxide, carbon dioxide, and nitrogen oxides.

Carcinogenicity: IARC - Not listed; NIOSH - Not listed; NTP - Not listed; ACGIH - Not listed; OSHA - Not listed; EPA - Class D, Not classifiable as to human carcinogenicity; MAK - Not listed

Primary Target Organs:

Eyes

Skin

Liver

Kidneys

Environmental

Ecotoxicity: LC_{50} Japanese quail oral 409 ppm, 5 day diet, age 14 days (95% confidence limit 356-470 ppm) LC_{50} yearling Coho salmon 100 ug/l/96 hr /conditions of bioassay not specified LD_{50} Pheasant oral 26.4 mg/kg, 12 males, age 4 mo (95% confidence limit 21.0-33.3 mg/kg) LC_{50} Ictalurus punctatus (channel catfish) 0.028-0.058 mg/l/96 hr under standard conditions at 12 °C in soft, reconstituted water

Environmental Fate: Release to soil is expected to result in biodegradation and it will only weakly adsorb to soils and should, therefore, leach to groundwater. However, it may bind more strongly to clay soils, especially at acidic pH. Photolytic degradation from soil surface may be important. It may photodegrade in surface water with a half-life of 14-18 days. Hydrolysis in water may not be important. It is unlikely to undergo significant biodegradation in most natural waters. Volatilization from water is expected to be slow and bioconcentration is expected to be insignificant. The half-life for the reaction of vapor phase with photochemically generated hydroxyl radicals in the atmosphere was estimated to be 14.1 days. Wet deposition may remove some of the compound from air.

Cleanup/Disposal: Guide No. 131: Fully encapsulating, vapor protective clothing should be worn for spills and leaks with no fire. Eliminate all ignition sources (no smoking, flares, sparks or flames in immediate area). All equipment used when handling the product must be grounded. Do not touch or walk through spilled material. Stop leak if you can do it without risk. Prevent entry into waterways, sewers, basements or confined areas. A vapor suppressing foam may be used to reduce vapors. Small Spills: Absorb with earth, sand or other non-combustible material and transfer to containers for later disposal. Use clean non-sparking tools to collect absorbed material. Large Spills: Dike far ahead of liquid spill for later disposal. Water spray may reduce vapor; but may not prevent ignition in closed spaces. Guide No. 153: Eliminate all ignition sources (no smoking, flares, sparks or flames in immediate area). Do not touch damaged containers or spilled material unless wearing appropriate protective clothing. Stop leak if you can do it without risk. Prevent entry into waterways, sewers, basements or confined areas. Absorb or cover with dry earth, sand or other non-combustible material and transfer to containers. Do not get water inside containers.

Environmental Physical Data

Henry's Law Constant: 5.04 $\times 10^{-4}$
Sorption Partition Coefficient: K_{oc} = 124
BCF: estimated at 68

Regulations

RCRA 40CFR: Listed Hazardous Waste No. P020 Toxic Waste
CERCLA: 40CFR 302.4: Listed per RCRA Section 3001 RQ: 1000 lb (453.5 kg)
SARA 40CFR 372.65: Listed TPQ: 100/10000 lb
SARA EHS 40CFR 355: Listed TPQ: 1,000 lb
TSCA: Listed

Analytical Methods

Soil: EPA PMD-DOG; SW846 3640A, 8040A, 8150B, 8151, 8270B, 8270C, 8321, 8321A
Water / Groundwater: EPA P-008-1, 1658, 615, X_89_176
Drinking Water: EPA 515.1, 515.2, 555
Plasma: EPA 29; FDA 231.1, 252

DIN3280 CAS #: 5388-62-5

2,6-DINITRO-4-CHLOROANILINE

RTECS: BX9250000
EINECS Number: 226-381-5
Molecular Formula: $C_6H_4ClN_3O_4$
Formula Weight: 217.57

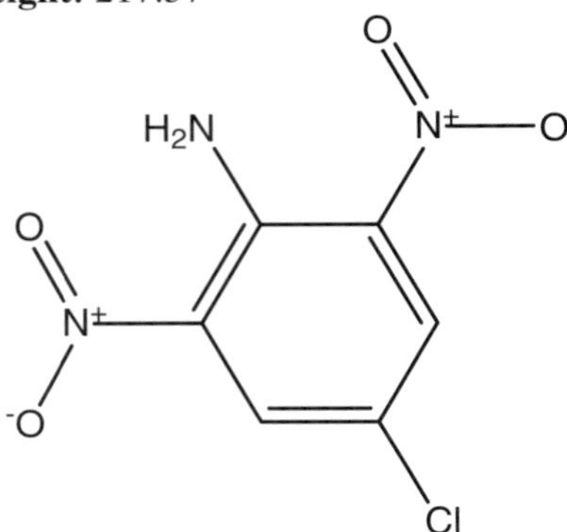

Chemical Structure

Synonyms: ANILINE,4-CHLORO-2,6-DINITRO-; BENZENAMINE,4-CHLORO-2,6-DINITRO-; 4-CHLORO-2,6-DINITROANILINE
Description: orange-yellow needles

Physical Properties

Freezing Point: 147 °C (296.6 °F)
Other Solubilities: Soluble in hot Alcohol

RTECS Toxicity Data

Acute Oral: Rat LD_{50} Dose: 400 mg/kg.

Hazard Overviews

Health: Irritating to eyes/skin/respiratory tract. Toxic. Other Acute Effects: may be harmful by inhalation, ingestion, or skin absorption.

Fire: Hazards: emits toxic fumes. Extinguishing agents: water spray; carbon dioxide, dry chemical powder or appropriate foam. Precautions: combustible liquid.

Reactivity: Incompatible with: strong oxidizing agents. Hazardous decomposition products: toxic fumes of: carbon monoxide, carbon dioxide, nitrogen oxides, hydrogen chloride gas.

Carcinogenicity: IARC - Not listed; NIOSH - Not listed; NTP - Not listed; ACGIH - Not listed; OSHA - Not listed; EPA - Not listed; MAK - Not listed

Primary Target Organs:

Eyes

Skin

Respiratory System

Environmental

Regulations

RCRA 40CFR: Not listed
CERCLA: 40CFR 302.4: Not listed
SARA 40CFR 372.65: Not listed
SARA EHS 40CFR 355: Not listed
TSCA: Listed

DIN3470 CAS #: 97-00-7

DINITROCHLOROBENZENE

RTECS: CZ0525000
DOT: UN1577; IMO6.1
EINECS Number: 202-551-4
Molecular Formula: $C_6H_3ClN_2O_4$
Formula Weight: 202.56

Chemical Structure

Synonyms: BENZENE,1-CHLORO-2,4-DINITRO-; CDNB; 1-CHLOOR-2,4-DINITROBENZEEN; 1-CHLOR-2,4-DINITROBENZENE; 1-CHLORO-2,4-DINITROBENZEEN; 1-CHLORO-2,4-DINITROBENZENE; 4-CHLORO-1,3-DINITROBENZENE; 6-CHLORO-1,3-DINITROBENZENE; CHLORODINITROBENZENE; 1-CHLORO-2,4-DINITROBENZOL; 1-CLORO-2,4-DINITROBENZENE; 1,3-DINITRO-4-CHLOROBENZENE; 2,4-DINITRO-1-CHLOROBENZENE; 2,4-DINITROCHLOROBENZENE; DINITROCHLOROBENZOL; 2,4-DINITROPHENYL CHLORIDE; DNCB
Description: yellow rhombic crystals; almond-like odor
Use: reagent for the detection & determination of nicotinic acid, nicotinamide and other pyridine cmpd; dyes; organic synthesis; algicide in coolant water of air conditioning systems

Physical Properties

Boiling Point: 315 °C (599 °F)
Freezing Point: 54 °C (129.2 °F)
Specific Gravity: 1.7
Vapor Density: 6.98 Air=1
Water Solubility: Sparingly Soluble in Water
Other Solubilities: Soluble in Ether, Benzene & Carbon Disulfide Sparingly Soluble in cold, freely Soluble in hot Alcohol
Refraction Index: 1.5857
LEL: 2.0% v/v
UEL: 22% v/v

RTECS Toxicity Data

Acute Oral: Rat LD_{50} Dose: 780 mg/kg.
Acute Dermal: Rabbit LD_{50} Route: Skin; Dose: 130 mg/kg; Toxic Effects: Skin and appendages - Primary irritation.
Chronic (Multiple Dose) Oral: Rat Dose: 2340 mg/kg/30D-I; Toxic Effects: Brain and coverings - Other degenerative changes; Liver - Other changes.
Chronic (Multiple Dose) Inhalation: Rat Dose: 200 ug/m^3/4H/17W-I; Toxic Effects: Blood - Changes in serum composition; Biochemical - Other proteins.
Irritation Eye: Rabbit Standard Draize Test Dose: 50 ug/24H; Reaction: severe.
Irritation Skin: Rabbit Standard Draize Test Dose: 2 mg/24H; Reaction: severe.
Mutagenic: Rat DNA Damage; Cell Type: liver; Dose: 5 umol/L. Mouse DNA Damage; Route: Intraperitoneal; Dose: 30 mg/kg.

Hazard Overviews

Poison

Explosive

Fire Diamond

Health: Irritating to eyes/skin/respiratory tract. Poison. Also Causes: methemoglobinemia, itching skin, vesicles, exfoliation. Chronic Effects: optic nerve inflammation, blurry vision, leg pains, burning of feet.
Fire: Combustible and can explode upon shock or friction. For small fires, use dry chemical, carbon dioxide, water spray, or regular foam. For large fires, use water spray, fog, or regular foam. Be aware that water or foam may cause frothing. Fight from maximum distance.
Reactivity: Unstable, shock sensitive and can explode from friction. Hazardous polymerization cannot occur. Avoid: heat; ignition sources; friction. Incompatible with: ammonia at 170 °C/40 bar; hydrazine hydrate; hydrazine sulfate; oxidizers; alkaline materials. Hazardous decomposition products: carbon dioxide; nitrogen oxides.
Carcinogenicity: IARC - Not listed; NIOSH - Not listed; NTP - Not listed; ACGIH - Not listed; OSHA - Not listed; EPA - Not listed; MAK - Not listed

Primary Target Organs:

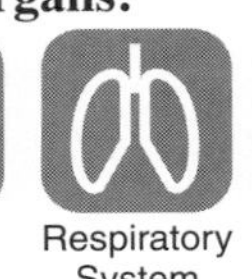

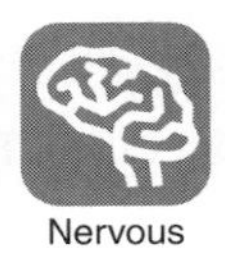

Environmental

Environmental Fate: If released to water, some may partition to sediment. Volatilization from water is not expected to be an environmentally significant fate process (volatilization half-life of approximately 166 days from a model river). Data are available suggesting that it may not biodegrade in water systems; however, data are limited. In soil, its K_{oc} of about 1,390 suggests low soil mobility. No data are available on other potential degradation processes in soil or water. If released to the atmosphere, vapor is degraded slowly by reaction with photochemically produced hydroxyl radicals (estimated half-life of 2 years in air).

Cleanup/Disposal: Guide No. 153: Eliminate all ignition sources (no smoking, flares, sparks or flames in immediate area). Do not touch damaged containers or spilled material unless wearing appropriate protective clothing. Stop leak if you can do it without risk. Prevent entry into waterways, sewers, basements or confined areas. Absorb or cover with dry earth, sand or other non-combustible material and transfer to containers. Do not get water inside containers.

Environmental Physical Data

Henry's Law Constant: estimated at 3.15 $x10^{-7}$

Octanol/Water Partition Coefficient: log K_{ow} = calculated at 2.1

Sorption Partition Coefficient: K_{OC} = 1390

BCF: estimated at 191

Regulations

RCRA 40CFR: Not listed

CERCLA: 40CFR 302.4: Listed as Compound per CWA Section 307(a)

SARA 40CFR 372.65: Not listed

SARA EHS 40CFR 355: Not listed

TSCA: Listed

Analytical Methods

Soil: SW846 8091

Water / Groundwater: EPA 646

DIN3660 CAS #: 534-52-1

DINITRO-O-CRESOL

RTECS: GO9625000

DOT: UN1598; IMO6.1

EINECS Number: 208-601-1

Molecular Formula: $C_7H_6N_2O_5$

Structured MF: $C_6H_2(CH_3)OH(NO_2)_2$

Formula Weight: 198.13

Chemical Structure

Synonyms: ANTINONIN; ANTINONNIN; ARBOROL; C.I. 10310; CAPSINE; CHEMSECT DNOC; O-CRESOL,4,6,-DINITRO-; DEGRASSAN; DEKRYSIL; DETAL; DETOL; DILLEX; DINITRO; 2,4-DINITRO-O-CRESOL; 3,5-DINITRO-O-CRESOL; 4,6-DINITRO-O-CRESOL; 4,6-DINITROCRESOL; DINITROCRESOL; 4,6-DINITRO-O-CRESOLO; DINITRO-O-CRESOL,SOLID OR SOLUTION; DINITRODENDTROXAL; 3,5-DINITRO-2-HYDROXYTOLUENE; 4,6-DINITRO-O-KRESOL; 4,6-DINITROKRESOL; DINITROL; DINITROMETHYL CYCLOHEXYLTRIENOL; 4,6-DINITRO-2-METHYL PHENOL; 2,4-DINITRO-6-METHYLPHENOL; DINITROSOL; DINOC; DINOK; DINURANIA; DITROSOL; DN; DN-DRY MIX NO. 2; DNOC; DNOK; DWUNITRO-O-KREZOL; EFFUSAN; EFFUSAN 3436; ELGETOL; ELGETOL 30; ELIPOL; ENT 154; EXTRAR; HEDOLIT; HEDOLITE; K III; K IV; KRENITE; KREOZAN; KRESAMONE; KRESONITE-E; KREZOTOL 50; LE DINITROCRESOL-4,6; LIPAN; 6-METHYL-2,4-DINITROCRESOL; 2-METHYL-4,6-DINITROPHENOL; NITRADOR; NITROFAN; ORANZ VIKTORIA; PHENOL,2-METHYL-4,6-DINITRO-(9CI); PROKARBOL; RAFEX; RAFEX 35; RAFEX 5; RAPHALOX; RAPHATOX; SANDOLIN; SANDOLIN A; SELINON; SINOX; TOLUENE,3,5-DINITRO-2-HYDROXY-; TRIFOCIDE; TRIFRINA; WINTERWASH; ZAHLREICHE BEZEICHNUNGEN

Description: yellow prisms or needles; odorless

Use: dormant ovicidal spray for fruit trees (highly phytotoxic and cannot be used successfully on actively growing plants); Herbicide; insecticide

Physical Properties

Boiling Point: 312 °C (594 °F)

Freezing Point: 87.5 °C (189.5 °F)

Vapor Density: 6.8 Air=1

Density: 6.82

Vapor Pressure: 0.00005 mm Hg

Water Solubility: 0.013% at 15 °C

Other Solubilities: Acetone: Soluble; Alcohol: Soluble; Ether: Soluble; Petroleum Ether: Slightly Soluble.

Flash Point: Noncombustible Solid

Autoignition Temperature: 340 °C

RTECS Toxicity Data

Acute Oral: Man TD_{Lo} Dose: 7500 ug/kg/7D; Toxic Effects: Behavioral - Somnolence (general depressed activity); Behavioral - Headache. Rat LD_{50} Dose: 7 mg/kg.

Acute Inhalation: Cat LC_{Lo} Dose: 40 mg/m^3/4hr; Toxic Effects: Behavioral - Somnolence (general depressed activity); Behavioral - Muscle weakness; Lungs, Thorax, or Respiration - Dyspnea. Human TC_{Lo} Dose: 1 mg/m^3; Toxic

Effects: Brain and coverings - Recordings from specific areas of CNS; Cardiac - Other changes; Gastrointestinal - Other changes.

Acute Dermal: Child LD_{Lo} Route: Skin; Dose: 500 mg/kg; Toxic Effects: Behavioral - Coma; Behavioral - Headache; Gastrointestinal - Nausea or vomiting. Rabbit LD_{50} Route: Skin; Dose: 1 gm/kg.

Chronic (Multiple Dose) Oral: Rat Dose: 45500 ug/kg/26W-I; Toxic Effects: Liver - Other changes; Blood - Other changes; Biochemical - Phosphatases.

Chronic (Multiple Dose) Inhalation: Cat Dose: 2 mg/m^3/4H/4W-I; Toxic Effects: Endocrine - Hyperglycemia; Blood - Changes in leukocyte (WBC) cell count; DEATH.

Irritation Eye: Rabbit Standard Draize Test Dose: 20 mg/24H; Reaction: moderate.

Irritation Skin: Rabbit Standard Draize Test Dose: 105 mg/9D-I; Reaction: mild.

Mutagenic: Rat Cytogenetic Analysis; Route: Oral; Dose: 15 mg/kg. Insects - D Melanogaster Sex Chromosome Loss; Route: Oral; Dose: 250 umol/L.

Hazard Overviews

Poison

Corrosive

Explosive

Health: Irritating to eyes/skin/respiratory tract. Poison. Other Acute Effects: may be fatal if inhaled, swallowed, or absorbed through skin; absorption into the body leads to the formation of methemoglobin which in sufficient concentration causes cyanosis; onset may be delayed 2 to 4 hours or longer; symptoms of exposure may include burning sensation, coughing, wheezing, laryngitis, shortness of breath, headache, nausea and vomiting.

Fire: Noncombustible. Hazards: in powder form capable of creating a dust explosion; may explode when heated; may be shock-sensitive. Extinguishing agents: water spray; carbon dioxide, dry chemical powder or appropriate foam. Precautions: combustible liquid.

Reactivity: Incompatible with: oxidizing agents, reducing agents, strong bases. Hazardous decomposition products: thermal decomposition may produce carbon monoxide, carbon dioxide, and nitrogen oxides.

Carcinogenicity: IARC - Not listed; NIOSH - Listed as carcinogen; NTP - Not listed; ACGIH - Not listed; OSHA - Not listed; EPA - Not listed; MAK - Not listed

Primary Target Organs:

Eyes

Skin

Respiratory System

Exposure Limits

OSHA PEL: TWA: 0.2 mg/m^3; skin.
ACGIH TLV: TWA: 0.2 mg/m^3.
NIOSH REL: TWA: 0.2 mg/m^3.
NIOSH IDLH: 5 mg/m^3.
DFG MAK: TWA: 0.2 mg/m^3.

Respirator Recommendation

Exposure Range: >0.2 to <5 mg/m^3 Air Purifying, Negative Pressure, Half Mask

Exposure Range: 5 to unlimited mg/m^3 Self-contained Breathing Apparatus, Pressure Demand, Full Face

Cartridge Color: dust/mist filter (use P100 or consult supervisor for appropriate dust/mist filter)

Environmental

Ecotoxicity: LC_{50} Coturnix >75,000 ppm, no overt signs of toxicity to 5,000 ppm LC_{50} Gammarus fasciatus (scud) 1100 ug/l/96 hr at 21 °C (95% confidence limit 730-1600 ug/l), mature /static bioassay LC_{50} Salmo gairdneri (rainbow trout) 66 ug/l/96 hr at 13 °C (95% confidence limit 37-117 ug/l), wt 1.2 g /static bioassay EC_{50} Daphnia pulex (daphnid) 145 ug/l/48 hr at 21 °C (95% confidence limit 100-210 ug/l), first instar /static bioassay LD_{50} Mallard oral 22.7 mg/kg (95% confidence limit 18.0-28.5 mg/kg), 5-7 mo old males LC_{50} Pteronarcys 320 ug/l/96 hr at 15 °C (95% confidence limit 230-450 ug/l), SECOND YR CLASS /static bioassay

Environmental Fate: If released to soil, it will usually disappear within a few weeks to 2 months when applied at normal pesticide rates. Biodegradation is probably the main removal process from agricultural soils. Estimated K_{oc} values (225-590) suggest that it will have medium to low soil mobility. If released to water, direct photolysis may occur since it absorbs light in the environmentally important range of the spectrum; The half-life for photooxidation via peroxy radicals has been estimated to be 58 days. The significance of biodegradation in natural waters cannot be predicted with certainty from the available data; The results of one screening study suggest that concentration may be an important factor in determining the ability of microbes to biotransform it. Aquatic hydrolysis, volatilization, bioconcentration, and adsorption to sediments are not expected to be important fate processes. If released to air, it may exist in both the vapor and adsorbed (to particulates) phases. In the vapor- phase, it will react rapidly with photochemically produced hydroxyl radicals at an estimated half-life rate of 8 hours. Particulate-phase will be susceptible to wet and dry deposition.

Cleanup/Disposal: Guide No. 153: Eliminate all ignition sources (no smoking, flares, sparks or flames in immediate area). Do not touch damaged containers or spilled material unless wearing appropriate protective clothing. Stop leak if you can do it without risk. Prevent entry into waterways, sewers, basements or confined areas. Absorb or cover with dry earth, sand or other non-combustible material and transfer to containers. Do not get water inside containers.

Environmental Physical Data

Henry's Law Constant: estimated at 4.8 x10^{-11}
Octanol/Water Partition Coefficient: log K_{ow} = 2.564
Sorption Partition Coefficient: K_{oc} = 225 to 590
BCF: estimated at 52

Regulations

RCRA 40CFR: Listed Hazardous Waste No. P047 Toxic Waste

CERCLA: 40CFR 302.4: Listed per RCRA Section 3001 per CWA Section 307(a) RQ: 10 lb (4.535 kg)
SARA 40CFR 372.65: Listed
SARA EHS 40CFR 355: Listed TPQ: 10 lb
TSCA: Listed

Analytical Methods

Soil: CLP MC_SVOA, OHC; EPA 16, 1625; SW846 3640A, 3650A, 3650B, 8040A, 8250A, 8270B, 8270C, 8410
Water / Groundwater: EPA 1625, 604, 625, 625-S; APHA 6410-B, 6420-BA, 6420-BB, 6420-C; USGS O3117
Plasma: EPA 29; FDA 231.1, 252
Other: EPA O-005-1

DIN3850 CAS #: 131-89-5

2,4-DINITRO-6-CYCLOHEXYLPHENOL

RTECS: SK6650000
DOT: UN2588; IMO6.1
EINECS Number: 205-042-5
Molecular Formula: $C_{12}H_{14}N_2O_5$
Structured MF: $C_{12}H_{14}N_2O_5$
Formula Weight: 266.25
Synonyms: 6-CICLOESIL-2,4-DINITR-FENOLO; 2-CYCLOHEXYL-4,6-DINITROFENOL; 2-CYCLOHEXYL-4,6-DINITROPHENOL; 6-CYCLOHEXYL-2,4-DINITROPHENOL; DINEX; 2,4-DINITRO-6-CYCLOHEXYL PHENOL; 4,6-DINITRO-O-CYCLOHEXYLPHENOL; DINITRO-O-CYCLOHEXYLPHENOL; DINITRO-ORTHO-CYCLOHEXYLPHENOL; DINITROCYCLOHEXYLPHENOL; DN; DN-111; DN 1; DN DRY MIX NO 1; DN DRY MIX NO. 1; DN DUST NO. 12; DN (PESTICIDE); DNOCHP; DOWSPRAY 17; DRY MIX NO 1; DRY MIX NO. 1; ENT 157; PEDINEX; PHENOL,2-CYCLOHEXYL-4,6-DINITRO-; PHENOL,6-CYCLOHEXYL-2,4-DINITRO-; SN 46
Description: crystalline solid
Use: insecticide, especially in control of citrus red mite; molluscide; acaricide for control of citrus red mites

Physical Properties

Freezing Point: 106.5 °C (223.7 °F) to 107.5 °C (225.5 °F)
Vapor Density: 9.2 Air=1
Water Solubility: Very Slightly Soluble in Water
Other Solubilities: Soluble in Alcohol.
Flash Point: Flammable

RTECS Toxicity Data

Acute Oral: Rat LD_{50} Dose: 65 mg/kg. Mouse LD_{50} Dose: 50 mg/kg.
Acute Dermal: Rabbit LD_{Lo} Route: Subcutaneous Dose: 40 mg/kg; Toxic Effects: Behavioral - Convulsions or effect on seizure threshold; Lungs, Thorax, or Respiration - Other changes; Nutritional and gross metabolic - Body temperature decrease. Mouse LD_{Lo} Route: Subcutaneous Dose: 30 mg/kg; Toxic Effects: Lungs, Thorax, or Respiration - Other changes.
Chronic (Multiple Dose) Oral: Mouse Dose: 4200 mg/kg/2W-C; Toxic Effects: DEATH.
Irritation Skin: Rabbit Standard Draize Test Dose: 105 mg/9D-I; Reaction: moderate.

Hazard Overviews

Flammable

Fire: Flammable.
Carcinogenicity: IARC - Not listed; NIOSH - Not listed; NTP - Not listed; ACGIH - Not listed; OSHA - Not listed; EPA - Not listed; MAK - Not listed

Environmental

Environmental Fate: Release to the soil will result in strong adsorption to soils so little leaching to groundwater is expected. Direct photolysis at the soil surface may be significant. Biodegradation may occur. Volatilization is not expected to be significant. Release to water will result in bioconcentration, adsorption to sediments and possibly direct photolysis (based on its absorption of light of greater than 290 nm). Volatilization is expected to be slow. As a weak acid, ionization will occur in alkaline waters and the ionized species is not expected to bioconcentrate or volatilize. Release to air may result in direct photolysis and adsorption of some to particulate matter is expected. The half-life of the reaction of vapor phase with photochemically generated hydroxyl radicals in the atmosphere was estimated to be 6.8 h.
Cleanup/Disposal: Guide No. 151: Do not touch damaged containers or spilled material unless wearing appropriate protective clothing. Stop leak if you can do it without risk. Prevent entry into waterways, sewers, basements or confined areas. Cover with plastic sheet to prevent spreading. Absorb or cover with dry earth, sand or other non-combustible material and transfer to containers. Do not get water inside containers.

Environmental Physical Data

Octanol/Water Partition Coefficient: log K_{ow} = 4.65
Sorption Partition Coefficient: K_{oc} = 3.9
BCF: estimated at 2141

Regulations

RCRA 40CFR: Listed Hazardous Waste No. P034 Toxic Waste
CERCLA: 40CFR 302.4: Listed per RCRA Section 3001 RQ: 100 lb (45.35 kg)
SARA 40CFR 372.65: Not listed
SARA EHS 40CFR 355: Not listed
TSCA: Not listed

Analytical Methods

Soil: SW846 8270B, 8270C
Plasma: EPA 29

DIN4040 **CAS #: 4097-33-0**

2,6-DINITRO-4-OCTYLPHENOL

Molecular Formula: $C_{14}H_{20}N_2O_5$
Formula Weight: 296.32
Synonyms: PHENOL,2,6-DINITRO-4-OCTYL-

Hazard Overviews

Carcinogenicity: IARC - Not listed; NIOSH - Not listed; NTP - Not listed; ACGIH - Not listed; OSHA - Not listed; EPA - Not listed; MAK - Not listed

Environmental

Regulations

RCRA 40CFR: Not listed
CERCLA: 40CFR 302.4: Listed as Compound per CWA Section 307(a)
SARA 40CFR 372.65: Not listed
SARA EHS 40CFR 355: Not listed
TSCA: Not listed

DIN4230 **CAS #: 66-56-8**

2,3-DINITROPHENOL

RTECS: SL2700000
EINECS Number: 200-628-7
Molecular Formula: $C_6H_4N_2O_5$
Formula Weight: 184.12

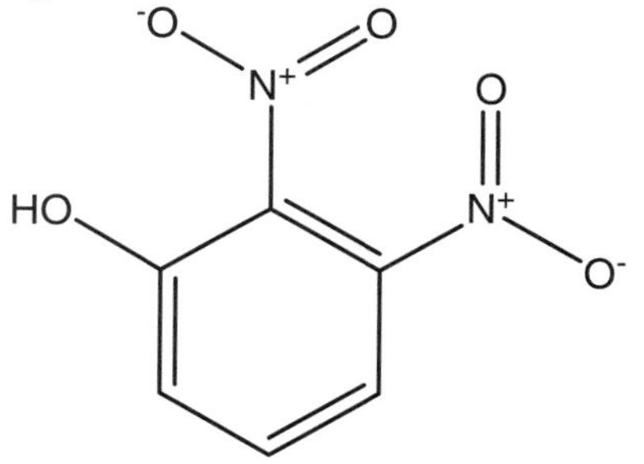

Chemical Structure

Synonyms: 2,3-DINITROFENOL

Physical Properties

Boiling Point: Sublimes
Freezing Point: 145 °C (293 °F)
Specific Gravity: 1.68 at 20 °C
Saturated Vapor Density: 1.242228675 kg/m^3
Vapor Pressure: $1x10^{-5}$mm Hg at 20 °C
Water Solubility: Slightly Soluble in Cold Water
Other Solubilities: Soluble in Acetone, Benzene, Carbon Tetrachloride, Chloroform, ethyl Alcohol, ethyl Ether and Pyridine
Flash Point: Varies depending on isomer

RTECS Toxicity Data

Mutagenic: Bacteria - S Typhimurium Mutations in Microorganisms; Dose: 250 ug/plate (+/-S9).

Hazard Overviews

Reactivity: Stable, with at least 15% water, explosive if dry and exposed to heat, flame, or shock.. Hazardous polymerization cannot occur. Avoid: exposure to heat; flame; shock and allowing to become dry. Incompatible with: steam; strong oxidizers; alkalies; ammonia. Hazardous decomposition products: carbon dioxide; nitrogen oxides.
Carcinogenicity: IARC - Not listed; NIOSH - Not listed; NTP - Not listed; ACGIH - Not listed; OSHA - Not listed; EPA - Not listed; MAK - Not listed

Environmental

Regulations

RCRA 40CFR: Not listed
CERCLA: 40CFR 302.4: Listed as Compound per CWA Section 307(a)
SARA 40CFR 372.65: Not listed
SARA EHS 40CFR 355: Not listed
TSCA: Not listed

DIN4420 **CAS #: 51-28-5**

2,4-DINITROPHENOL

RTECS: SL2800000
DOT: UN0076; UN1320; UN1599; IMO1.1D; IMO4.1; IMO6.1
EINECS Number: 200-087-7
Molecular Formula: $C_6H_4N_2O_5$
Structured MF: $(NO_2)_2C_6H_3OH$
Formula Weight: 184.11

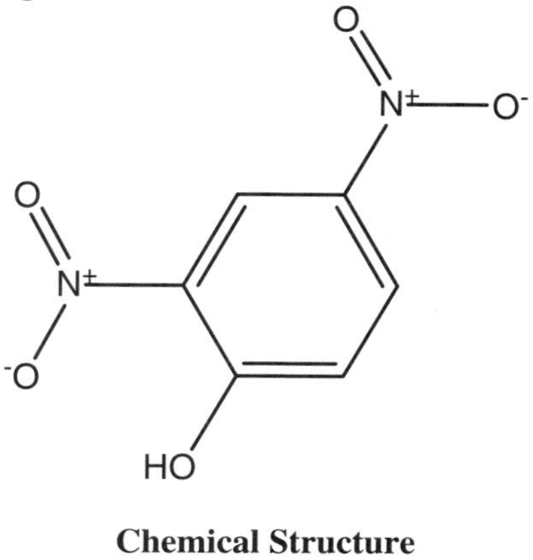

Chemical Structure

Synonyms: ALDIFEN; CAMELLO MOSQUITO COILS; CHEMOX PE; COBRA SALTS (IMPREGNA SALTS); 2,4-DINITROFENOL; DINITROFENOLO; ALPHA-DINITROPHENOL; DINOFAN; 2,4-DNP; DNP; FENOSYL CARBON N; FENOXYL; FENOXYL CARBON N; 1-HYDROXY-2,4-DINITROBENZENE; MAROXOL-50; NITRO KLEENUP; NITROPHEN; NITROPHENE; NSC 1532; OSMOPLASTIC-R; OSMOTOX-PLUS; PHENOL,2,4-DINITRO; PHENOL,2,4-DINITRO-; PHENOL,ALPHA-DINITRO-; SHIRAKIKU BRAND MOSQUITO COILS;

SOLFO BLACK B; SOLFO BLACK BB; SOLFO BLACK G; SOLFO BLACK SB; SOLFO BLACK 2B SUPRA; TERTROSULPHUR BLACK PB; TERTROSULPHUR PBR

Description: yellow crystals

Use: in the manufacture of dyes, diaminophenol, wood preservatives, insecticides, explosives, herbicides, photographic developers, picric acid and picramic acid; a reagent for the detection of potassium and ammonium ions and in chemical synthesis; useful tool in biochemical research; formerly used as a metabolic stimulator to aid in weight reduction

Physical Properties

Boiling Point: Sublimes
Freezing Point: 112 °C (233.6 °F) to 114 °C (237.2 °F)
Specific Gravity: 1.683
Vapor Density: 6.35
Saturated Vapor Density: 1.200000169 kg/m^3
Density: 1.683 g/mL at 24 °C
Vapor Pressure: 2 x10^{-5} mm Hg at 25 °C
Water Solubility: 5600 mg/L in Water at 18 °C
Other Solubilities: 15.55 g/100 g in Ethyl Acetate at 15 °C; 35.90 g/100 g in Acetone at 15 °C; 5.39 g/100 g in Chloroform at 15 °C; 20.08 g/100 g in Pyridine at 15 °C; 0.423 g/100 g in Carbon Tetrachloride at 15 °C.
pH: 2.6
Flash Point: Flammable

RTECS Toxicity Data

Acute Oral: Human LD_{Lo} Dose: 36 mg/kg; Toxic Effects: Behavioral - Coma; Cardiac - Change in rate; Nutritional and gross metabolic - Body temperature increase. Rat LD_{50} Dose: 30 mg/kg.

Acute Inhalation: Dog LC_{Lo} Dose: 300 mg/m^3/30M.

Acute Dermal: Rabbit LD_{Lo} Route: Subcutaneous Dose: 20 mg/kg. Rat LD_{50} Route: Subcutaneous Dose: 25 mg/kg.

Chronic (Multiple Dose) Oral: Rat Dose: 6510 ug/kg/30W-I; Toxic Effects: Behavioral - Alteration of classical conditioning. Rabbit Dose: 126 ug/kg/30W-I; Toxic Effects: Blood - Other changes.

Irritation Skin: Rabbit Standard Draize Test Dose: 300 mg/4W-I; Reaction: mild.

Reproductive/Teratogenic: Rat Route: Oral; Dose: 2040 mg/kg; Duration: female 8D prior to mating; Effects on Newborn - Stillbirth; Weaning or lactation index. Mouse Route: Intraperitoneal; Dose: 40800 ug/kg; Duration: female 10-12D of pregnancy Effects on Embryo or Fetus - Fetotoxicity.

Mutagenic: Mouse Cytogenetic Analysis; Route: Intraperitoneal; Dose: 10 gm/kg. Hamster DNA Inhibition; Cell Type: lung; Dose: 7 mmol/L.

Hazard Overviews

Poison

Explosive

Flammable

Fire Diamond

Health: Irritating to eyes/skin/respiratory tract. Poison. Also Causes: headache, coughing, profuse sweating, thirst, fatigue, fever, rapid pulse, difficult breathing, anxiety, confusion, convulsions, loss of consciousness, pulmonary edema, possible circulatory or respiratory collapse, irritation, redness, swelling, blisters, burning pain, watering of eyes, inflammation of lids, burning of mouth, throat, salivation, dizziness, nausea, vomiting, bright yellow stools. Liver and kidney damage may develop after exposure. Chronic Effects: Hypersensitivity.

Fire: Explosive when heated, or exposed to shock, and flammable. Flood fires with water. If water is unavailable, use dry chemical or dirt. Fight fire from maximum distance.

Reactivity: Stable.with at least 15% water; explosive when dry and subject to heat, flame, or shock. Hazardous polymerization cannot occur. Avoid: exposure to heat; flame; shock and allowing to become dry. Incompatible with: steam; strong oxidizers; the 2,4-isomer forms explosive salts with alkalies and ammonia. Hazardous decomposition products: carbon dioxide; nitrogen oxides.

Carcinogenicity: IARC - Not listed; NIOSH - Not listed; NTP - Not listed; ACGIH - Not listed; OSHA - Not listed; EPA - Not listed; MAK - Not listed

Primary Target Organs:

Eyes

Skin

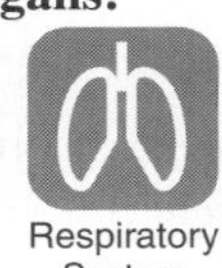
Respiratory System

Nervous System

Liver

Kidneys

Environmental

Ecotoxicity: LC_{50} Herring embryo 5,500 ug/l/96 hr. /Conditions of bioassay not specified LC_{50} Skeletonema costatum (alga) 98,700 ug/l/96 hr /Conditions of bioassay not specified LC_{50} Cyprinodon variegatus (marine sheepshead minnow) 5,500-29,400 ug/l/96 hr /Conditions of bioassay not specified Chlorella vulgaris (algae) exposed to 9,200 ug/l of 2,4-dinitrophenol had 70% growth inhibition in 80 hr. /From table LC_{50} Lepomis macrochirus (bluegill) 620 ug/l/96 hr /Conditions of bioassay not specified LC_{50} Daphnia magna (freshwater cladoceran) 4,090-4,710 ug/l/96 hr /Conditions of bioassay not specified LC_{50} Salmo salar (Atlantic salmon) 700 ug/l/96 hr. /Conditions of bioassay not specified

Environmental Fate: If released to soil it is expected to be highly mobile, although there is a possibility that some of this compound will adsorb to some clay minerals. May inhibit microbial growth of some aerobic microbes, but there are other microorganisms which may degrade this compound in the environment; possible biotransformation mechanisms include the reduction of the nitro group, hydroxylation of the aromatic ring and displacement of the nitro group; possible biodegradation products include, 2-amino-4-nitrophenol, 4-amino-2-nitrophenol, and nitrite. It is not expected to volatilize significantly from wet or dry soil surfaces. If released to water, it is expected to react with alkylperoxy radicals (calculated half-life 58 days) and it has the potential to photolyze due to absorption of UV light wavelengths > 290 nm it is not expected to bioaccumulate in aquatic organisms or volatilize and aerobic biodegradation appears to be slow. This compound is not expected to adsorb significantly to suspended solids or sediments, although there is a possibility that some of this compound will adsorb to clay minerals. If released to air, it is expected to exist in both vapor and particulate form. It may photolyze, it may be physically removed by settling or washout in precipitation or it may react with photochemically generated hydroxyl radicals (calculated vapor-phase half-life 14 hours).

Cleanup/Disposal: Guide No. 113: Eliminate all ignition sources (no smoking, flares, sparks or flames in immediate area). All equipment used when handling the product must be grounded. Do not touch or walk through spilled material. Small Spills: Flush area with flooding quantities of water. Large Spills: Wet down with water and dike for later disposal. Keep wetted product wet by slowly adding flooding quantities of water. Guide No. 153: Eliminate all ignition sources (no smoking, flares, sparks or flames in immediate area). Do not touch damaged containers or spilled material unless wearing appropriate protective clothing. Stop leak if you can do it without risk. Prevent entry into waterways, sewers, basements or confined areas. Absorb or cover with dry earth, sand or other non-combustible material and transfer to containers. Do not get water inside containers.

Environmental Physical Data

Henry's Law Constant: calculated at 8 x10^{-10}

Octanol/Water Partition Coefficient: log K_{ow} = 1.54

Sorption Partition Coefficient: K_{oc} = estimated at 36 to 164

BCF: calculated at 8.1

BOD: pure bacterial < 1 lb/lb, .94 days

Regulations

RCRA 40CFR: Listed Hazardous Waste No. P048 Toxic Waste

CERCLA: 40CFR 302.4: Listed per CWA Section 311(b)(4) per RCRA Section 3001 per CWA Section 307(a) RQ: 10 lb (4.535 kg)

SARA 40CFR 372.65: Listed

SARA EHS 40CFR 355: Not listed

TSCA: Listed

Analytical Methods

Soil: CLP MC_SVOA, OHC; EPA 16, 1625; SW846 3640A, 8040A, 8250A, 8270B, 8270C, 8410

Water / Groundwater: EPA 1625, 604, 625, 625-S, 6; APHA 6410-B, 6420-BA, 6420-BB, 6420-C; USGS O3117

Plasma: EPA 29

Other: EPA O-005-1

DIN4610 CAS #: 329-71-5

2,5-DINITROPHENOL

RTECS: SL2900000

DOT: UN0076; UN1320; UN1599; IMO1.1D IMO4.1; IMO6.1

EINECS Number: 206-348-1

Molecular Formula: $C_6H_4N_2O_5$

Structured MF: $(NO_2)_2C_6H_3 \cdot OH$

Formula Weight: 184

OH
O
O
N+
N+
-O
O-

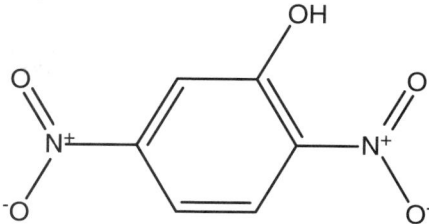

Chemical Structure

Synonyms: 2,5-DINITROFENOL; GAMMA-DINITROPHENOL; 2,5-DNP; PHENOL,2,5-DINITRO-; PHENOL,GAMMA-DINITRO-

Use: in manufacture of dyes and organic chemicals, as an indicator; as a reagent for the detection of potassium and ammonium ions

Physical Properties

Freezing Point: Either 104 °C (219.2 °F) to 108 °C (226.4 °F)

Specific Gravity: 1.68

Vapor Density: 6.35. Air=1

Water Solubility: Slightly Soluble in Cold Water

Other Solubilities: Slightly Soluble in cold Alcohol; Soluble in hot Alcohol, fixed alkali hydroxides.

Odor Threshold: Detection 2.4 mg/l

pH: 4

Flash Point: Combustible though it may require some effort to ignite

RTECS Toxicity Data

Mutagenic: Bacteria - S Typhimurium Mutations in Microorganisms; Dose: 100 ug/plate (+/-S9).

Hazard Overviews

Reactivity: Stable. Hazardous polymerization cannot occur. Avoid: heat; flame; shock; allowing to become dry. Incompatible with: strong oxidizers; alkalies; ammonia. Hazardous decomposition products: carbon dioxide; toxic nitrogen oxides.

Carcinogenicity: IARC - Not listed; NIOSH - Not listed; NTP - Not listed; ACGIH - Not listed; OSHA - Not listed; EPA - Not listed; MAK - Not listed

Environmental

Ecotoxicity: Aquatic toxicity: 100 ppm/fish/critical concentration 30 ppm/Minnow/toxic threshold in oxygen utilization

Environmental Fate: If released to soil, it appears to be rather persistent. It is expected to be highly mobile, although there is a possibility that some of this compound will adsorb to clay minerals. It appears to be resistant to aerobic biodegradation and is not expected to volatilize from wet or dry soil surfaces. If released to water, it is expected to react with alkylperoxy radicals (half-life on the order of 2 months) and it has the potential to photolyze due to absorption of UV light wavelengths > 290 nm. It is not expected to undergo aerobic biodegradation, bioaccumulate in aquatic organisms or volatilize significantly. This compound is not expected to adsorb significantly to suspended solids or sediments, although there is a possibility that some of this compound will adsorb to clay minerals. If released to the atmosphere, it is expected to exist in both vapor and particulate form. It may directly photolyze, it may be physically removed by settling or washout in precipitation, or it may react in the vapor phase with photochemically generated hydroxyl radicals (calculated half-life 14 days).

Cleanup/Disposal: Guide No. 113: Eliminate all ignition sources (no smoking, flares, sparks or flames in immediate area). All equipment used when handling the product must be grounded. Do not touch or walk through spilled material. Small Spills: Flush area with flooding quantities of water. Large Spills: Wet down with water and dike for later disposal. Keep wetted product wet by slowly adding flooding quantities of water. Guide No. 153: Eliminate all ignition sources (no smoking, flares, sparks or flames in immediate area). Do not touch damaged containers or spilled material unless wearing appropriate protective clothing. Stop leak if you can do it without risk. Prevent entry into waterways, sewers, basements or confined areas. Absorb or cover with dry earth, sand or other non-combustible material and transfer to containers. Do not get water inside containers.

Environmental Physical Data

Henry's Law Constant: calculated at 4×10^{-11}

Octanol/Water Partition Coefficient: log K_{ow} = 1.75

Sorption Partition Coefficient: K_{oc} = estimated at 213

BCF: estimated at 13

Regulations

RCRA 40CFR: Not listed

CERCLA: 40CFR 302.4: Listed per CWA Section 311(b)(4) RQ: 10 lb (4.535 kg)

SARA 40CFR 372.65: Not listed

SARA EHS 40CFR 355: Not listed

TSCA: Not listed

Analytical Methods

Soil: SW846 8041

DIN4800 CAS #: 573-56-8

2,6-DINITROPHENOL

RTECS: SL2975000

DOT: UN0076; UN1320; UN1599; IMO1.1; IMO4.1; IMO6.1

EINECS Number: 209-357-9

Molecular Formula: $C_6H_4N_2O_5$

Structured MF: $C_6H_3OH(NO_2)_2$

Formula Weight: 184.12

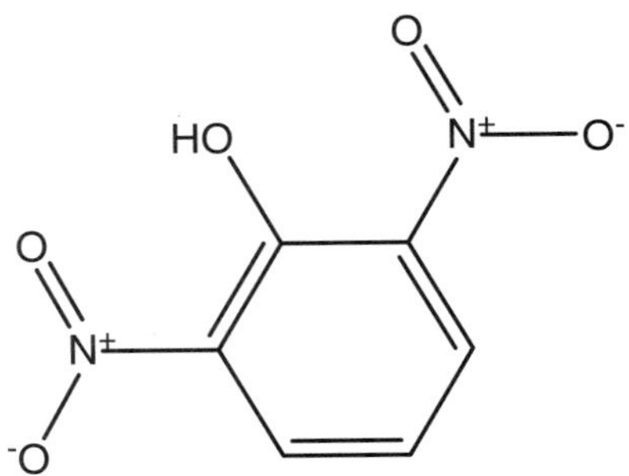

Chemical Structure

Synonyms: 2,6-DINITROFENOL; BETA-DINITROPHENOL; PHENOL,2,6-DINITRO-; PHENOL,BETA-DINITRO-

Use: in manufacture of dyes and organic chemicals, as indicator; dyes, especially sulfur colors; picric acid; picramic acid; preservation of lumber; manufacture of the photographic developer diaminophenol hydrochloride; explosives manufacture

Physical Properties

Specific Gravity: Estimated 1.68

Vapor Density: 6.35

Water Solubility: Slightly Soluble in Cold Water

Other Solubilities: Soluble in Acetone, Benzene, or Pyrimidine.

Odor Threshold: Detection 10 mg/l

pH: 2

Flash Point: Combustible though it may require some effort to ignite

Hazard Overviews

Reactivity: Stable. Hazardous polymerization cannot occur. Avoid: heat; flame; allowing dinitrophenol to become dry. Incompatible with: steam; strong oxidizers. Hazardous decomposition products: carbon dioxide; toxic nitrogen oxides.

Carcinogenicity: IARC - Not listed; NIOSH - Not listed; NTP - Not listed; ACGIH - Not listed; OSHA - Not listed; EPA - Not listed; MAK - Not listed

Environmental

Ecotoxicity: Aquatic toxicity: 46.3-51.6 ppm/48 hr/bluegill/TLm/20 °C; Waterfowl toxicity: 25 ppm

Environmental Fate: If released to soil, it is expected to be highly mobile, although there is a possibility that some of this compound will adsorb to clay minerals. It appears to be resistant to aerobic biodegradation and is not expected to volatilize significantly from wet or dry soil surfaces. If released to water, it is expected to react with alkylperoxy radicals (half-life on the order of 2 months) and it has the potential to photolyze due to absorption of UV light wavelengths > 290 nm. It is not expected to undergo aerobic biodegradation, bioaccumulate in aquatic organisms or volatilize significantly. This compound is not expected to adsorb significantly to suspended solids or sediments, although there is a possibility that some of this compound will adsorb to clay minerals. If released to the atmosphere, it is expected to exist in both vapor and particulate form. It may undergo direct photolysis, it may be physically removed by settling or washout in precipitation, or it may react in the vapor phase with photochemically generated hydroxyl radicals (half-life 14 days).

Cleanup/Disposal: Guide No. 113: Eliminate all ignition sources (no smoking, flares, sparks or flames in immediate area). All equipment used when handling the product must be grounded. Do not touch or walk through spilled material. Small Spills: Flush area with flooding quantities of water. Large Spills: Wet down with water and dike for later disposal. Keep wetted product wet by slowly adding flooding quantities of water. Guide No. 153: Eliminate all ignition sources (no smoking, flares, sparks or flames in immediate area). Do not touch damaged containers or spilled material unless wearing appropriate protective clothing. Stop leak if you can do it without risk. Prevent entry into waterways, sewers, basements or confined areas. Absorb or cover with dry earth, sand or other non-combustible material and transfer to containers. Do not get water inside containers.

Environmental Physical Data

Henry's Law Constant: calculated at 4×10^{-11}
Octanol/Water Partition Coefficient: log K_{ow} = 1.37
Sorption Partition Coefficient: K_{oc} = estimated at 133
BCF: estimated at 6

Regulations

RCRA 40CFR: Not listed
CERCLA: 40CFR 302.4: Listed per CWA Section 311(b)(4) RQ: 10 lb (4.535 kg)
SARA 40CFR 372.65: Not listed
SARA EHS 40CFR 355: Not listed
TSCA: Not listed

DIN4990 CAS #: 577-71-9

3,4-DINITROPHENOL

RTECS: SL3000000
EINECS Number: 209-415-3
Molecular Formula: $C_6H_4NO_3$
Formula Weight: 184.1

Chemical Structure

Synonyms: 3,4-DINITROFENOL

Physical Properties

Boiling Point: Sublimes
Freezing Point: 134 °C (273.2 °F)
Specific Gravity: 1.68 at 20 °C
Saturated Vapor Density: 1.24222323 kg/m^3
Vapor Pressure: 1×10^{-5}mm Hg at 20 °C
Water Solubility: Slightly Soluble in Cold Water
Other Solubilities: Soluble in Acetone, Benzene, Carbon Tetrachloride, Chloroform, ethyl Alcohol, ethyl Ether and Pyridine
Flash Point: Varies depending on isomer

RTECS Toxicity Data

Mutagenic: Bacteria - S Typhimurium Mutations in Microorganisms; Dose: 100 ug/plate (+/-S9).

Hazard Overviews

Reactivity: Stable. Hazardous polymerization cannot occur. Avoid: heat; flame; shock; allowing to become dry. Incompatible with: steam; strong oxidizers. Hazardous decomposition products: carbon dioxide; toxic nitrogen oxides.
Carcinogenicity: IARC - Not listed; NIOSH - Not listed; NTP - Not listed; ACGIH - Not listed; OSHA - Not listed; EPA - Not listed; MAK - Not listed

Environmental

Regulations

RCRA 40CFR: Not listed
CERCLA: 40CFR 302.4: Listed as Compound per CWA Section 307(a)
SARA 40CFR 372.65: Not listed
SARA EHS 40CFR 355: Not listed
TSCA: Not listed

DIN5180 CAS #: 586-11-8

3,5-DINITROPHENOL

RTECS: SL3050000
Molecular Formula: $C_6H_4NO_3$
Formula Weight: 184.1

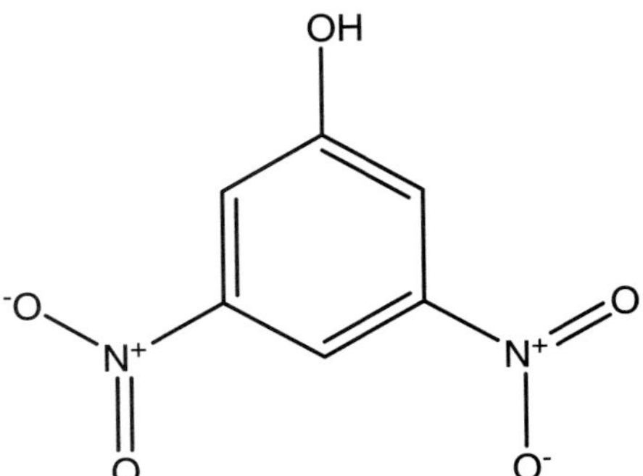

Chemical Structure

Synonyms: 3,5-DINITROFENOL

Physical Properties

Boiling Point: Sublimes
Freezing Point: 125 °C (257 °F)
Specific Gravity: 1.68 at 20 °C
Saturated Vapor Density: 1.24222323 kg/m^3
Vapor Pressure: $1x10^{-5}$mm Hg at 20 °C
Water Solubility: Slightly Soluble in Cold Water
Other Solubilities: Soluble in Acetone, Benzene, Carbon Tetrachloride, Chloroform, ethyl Alcohol, ethyl Ether and Pyridine
Flash Point: Varies depending on isomer

Hazard Overviews

Reactivity: Stable with at least 15% water by weight, but becomes explosive when dry and subjected to shock, heat, or flame. Hazardous polymerization cannot occur. Avoid: heat; flame; allowing to become dry. Incompatible with: steam; strong oxidizers. Hazardous decomposition products: carbon dioxide; toxic nitrogen oxides.
Carcinogenicity: IARC - Not listed; NIOSH - Not listed; NTP - Not listed; ACGIH - Not listed; OSHA - Not listed; EPA - Not listed; MAK - Not listed

Environmental

Regulations

RCRA 40CFR: Not listed
CERCLA: 40CFR 302.4: Listed as Compound per CWA Section 307(a)
SARA 40CFR 372.65: Not listed
SARA EHS 40CFR 355: Not listed
TSCA: Not listed

DIN5370 CAS #: 25550-58-7

DINITROPHENOL

RTECS: SL2625000
DOT: UN0076; UN1320; UN1599; IMO1.1D; IMO4.1; IMO6.1
EINECS Number: 247-096-2
Molecular Formula: $C_6H_4N_2O_5$
Formula Weight: 184.1
Synonyms: PHENOL,DINITRO-
Use: dyes, especially sulfur colors; picric acid; picramic acid; preservation of lumber; manufacture of the photographic developer diaminophenol hydrochloride; explosives manufacture; weed control; mfr of dyes, diaminophenol, etc; wood preservative; insecticide; indicator; reagent for the detection of potassium and ammonium ions

Physical Properties

Boiling Point: Sublimes
Freezing Point: 114 °C (237.2 °F) to 115 °C (239 °F)
Specific Gravity: 1.68
Vapor Density: 6.35
Vapor Pressure: Order of $1x10^{-5}$mm Hg or less
Water Solubility: Slightly Soluble in Cold Water
Other Solubilities: Soluble in Chloroform.
Flash Point: Combustible though it may require some effort to ignite

RTECS Toxicity Data

Acute Oral: Rat LD_{Lo} Dose: 30 mg/kg.

Hazard Overviews

Reactivity: Stable with at least 15% water, but, becomes explosive when dry and subjected to shock, heat, or flame. Hazardous polymerization cannot occur. Avoid: exposure to heat; flame; shock and allowing to become dry. Incompatible with: steam; strong oxidizers; alkalies and ammonia (the 2,4-isomer forms explosive salts). Hazardous decomposition products: carbon dioxide; toxic nitrogen oxides.
Carcinogenicity: IARC - Not listed; NIOSH - Not listed; NTP - Not listed; ACGIH - Not listed; OSHA - Not listed; EPA - Not listed; MAK - Not listed

Environmental

Environmental Fate: If released to soil, they are expected to be highly mobile, although there is a possibility that some absorption to clay minerals will take place. In general, they appear to be resistant to aerobic biodegradation and are not expected to volatilize significantly from soil. If released to water, they may undergo direct photolysis due to absorption of UV light wavelengths > 290 nm or they may react with alkylperoxy radicals (half-life on the order of 2 months). They are not expected to adsorb to suspended organic solids or sediments, although there is a possibility that some adsorption to clay minerals will take place. They are not expected to bioaccumulate in aquatic organisms or volatilize significantly from water. If released to the atmosphere, they are expected to exist in both vapor and particulate form. They may photolyze, they may be physically removed by settling or washout in precipitation, or they may react with photochemically generated hydroxyl radical (estimated vapor-phase half-life 14 hours).
Cleanup/Disposal: Guide No. 113: Eliminate all ignition sources (no smoking, flares, sparks or flames in immediate area). All equipment used when handling the product must be grounded. Do not touch or walk through spilled material.

Small Spills: Flush area with flooding quantities of water. Large Spills: Wet down with water and dike for later disposal. Keep wetted product wet by slowly adding flooding quantities of water. Guide No. 153: Eliminate all ignition sources (no smoking, flares, sparks or flames in immediate area). Do not touch damaged containers or spilled material unless wearing appropriate protective clothing. Stop leak if you can do it without risk. Prevent entry into waterways, sewers, basements or confined areas. Absorb or cover with dry earth, sand or other non-combustible material and transfer to containers. Do not get water inside containers.

Environmental Physical Data

Henry's Law Constant: calculated at 1 x10^{-11}
Octanol/Water Partition Coefficient: log K_{ow} = 1.54
Sorption Partition Coefficient: K_{oc} = estimated at 160 to 460
BCF: none expected

Regulations

RCRA 40CFR: Not listed
CERCLA: 40CFR 302.4: Listed per CWA Section 311(b)(4) RQ: 10 lb (4.535 kg)
SARA 40CFR 372.65: Not listed
SARA EHS 40CFR 355: Not listed
TSCA: Not listed

DIN5560 CAS #: 17977-09-2

2,2-DINITROPROPYL ACRYLATE

EINECS Number: 241-897-0
Molecular Formula: $C_6H_8N_2O_6$
Formula Weight: 204.14
Synonyms: ACRYLIC ACID,2,2-DINITROPROPYL ESTER; 2,2-DINITROPROPYL 2-PROPENOATE; 2-PROPENOIC ACID,2,2-DINITROPROPYL ESTER
Use: monomer for polymers used in rocket propellants

Physical Properties

Other Solubilities: Reacts readily with electrophilic, free-radical, and nucleophilic agent

Hazard Overviews

Carcinogenicity: IARC - Not listed; NIOSH - Not listed; NTP - Not listed; ACGIH - Not listed; OSHA - Not listed; EPA - Not listed; MAK - Not listed

Environmental

Regulations

RCRA 40CFR: Not listed
CERCLA: 40CFR 302.4: Not listed
SARA 40CFR 372.65: Not listed
SARA EHS 40CFR 355: Not listed
TSCA: Listed

DIN5940 CAS #: 101-25-7

DINITROSOPENTAMETHYLENETETR AMINE

RTECS: XA5250000
EINECS Number: 202-928-3
Molecular Formula: $C_5H_{10}N_6O_2$
Formula Weight: 186.2
Synonyms: ACETO DNPT 100; ACETO DNPT 40; ACETO DNPT 80; CELLMIC A 80; CELMIKE A; CHEMPOR N 90; CHEMPOR PC 65; CHKHZ 18; DINITROSOPENTAMETHENETETRAMINE; DI-N-NITROSOPENTAMETHYLENE TETRAMINE; N,N-DINITROSOPENTAMETHYLENE TETRAMINE; 3,4-DI-N-NITROSOPENTAMETHYLENETETRAMINE; 3,7-DI-N-NITROSOPENTAMETHYLENETETRAMINE; N(1),N(3)-DINITROSOPENTAMETHYLENETETRAMINE; N(SUP 1),N(SUP 3)-DINITROSOPENTAMETHYLENETETRAMINE; N,N'-DINITROSOPENTAMETHYLENETETRAMINE; N,N-DINITROSOPENTAMETHYLENETETRAMINE; N,N'-DINITROSOPENTAMETHYLENETETRAMINE,NOT >82% WITHPHLEGMATIZER; 3,7-DINITROSO-1,3,5,7-TETRAAZABICYCLO(3.3.1)NONANE; 3,7-DINITROSO-1,3,5,7-TETRAAZABICYCLO-(3,3,1)-NONANE; DIPENTAX; DNPM; DNPMT; DNPT; 1,5-ENDOMETHYLENE-3,7-DINITROSO-1,3,5,7-TETRAAZACYCLOOCTANE; KHEMPOR N 90; 1,5-METHYLENE-3,7-DINITROSO-1,3,5,7-TETRAAZACYCLOOCTANE; MICROPOR; MIKROFOR N; NSC 73599; OPEX; OPEX 80; OPEX 93; PENTAMETHYLENETETRAMINE,DINITROSO-; POROFOR CHKHC-18; POROFOR CHKHZ-18; POROFOR DNO/F; POROPHOR B; POROTOR DNO/F; 1,3,5,7-TETRAAZABICYCLO(3.3.1)NONANE,3,7-DINITROSO-; UNICEL 100; UNICEL NDX; UNICEL-ND; VULCACEL B-40; VULCACEL BN; VULCACEL BN 94
Description: light cream-colored powder; light-yellow needles
Use: blowing agent for natural & synthetic unicellular rubber, polyvinyl chloride plastisols, for epoxy, polyester & silicone resins

Physical Properties

Boiling Point: Decomposes at 190 °C (374 °F) to 200 °C (392 °F)
Freezing Point: Decomposes at 190 °C (374 °F) to 200 °C (392 °F)
Water Solubility: About 1%
Other Solubilities: Somewhat Soluble in Methyl Ethyl Ketone & Acetonitrile; Slightly Soluble in Methanol, Ethanol; Slightly Soluble in Benzene, Ether, Acetone; Readily Soluble in Dimethyl Sulfoxide.

RTECS Toxicity Data

Acute Oral: Rat LD_{50} Dose: 940 mg/kg; Toxic Effects: Behavioral - Tremor; Behavioral - Convulsions or effect on seizure threshold; Lungs, Thorax, or Respiration - Other changes.
Acute Dermal: Rat LD_{50} Route: Subcutaneous Dose: 220 mg/kg; Toxic Effects: Behavioral - Tremor; Behavioral - Convulsions or effect on seizure threshold; Lungs, Thorax, or Respiration - Other changes. Mouse LD_{50} Route: Subcutaneous Dose: 140 mg/kg; Toxic Effects: Behavioral -

Tremor; Behavioral - Convulsions or effect on seizure threshold; Lungs, Thorax, or Respiration - Other changes.
Mutagenic: Rat Cytogenetic Analysis; Cell Type: liver; Dose: 1 mg/L. Mouse Mutations in Mammalian Somatic Cells; Cell Type: lymphocyte; Dose: 175 mg/L.

Hazard Overviews

Carcinogenicity: IARC - Group 3, Not classifiable as to carcinogenicity to humans; NIOSH - Not listed; NTP - Not listed; ACGIH - Not listed; OSHA - Not listed; EPA - Not listed; MAK - Not listed

Environmental

Regulations

RCRA 40CFR: Not listed
CERCLA: 40CFR 302.4: Listed as Compound per CWA Section 307(a)
SARA 40CFR 372.65: Not listed
SARA EHS 40CFR 355: Not listed
TSCA: Listed

DIN6130 CAS #: 140-79-4

1,4-DINITROSOPIPERAZINE

RTECS: TL6300000
EINECS Number: 205-434-6
Molecular Formula: $C_4H_8N_4O_2$
Formula Weight: 144.14
Synonyms: DINITROSOPIPERAZIN; DINITROSOPIPERAZINE; N,N'-DINITROSOPIPERAZINE; DNPZ; NSC 339; PIPERAZINE,1,4-DINITROSO-
Description: pale yellow plates
Use: research chemical

Physical Properties

Freezing Point: 158 °C (316.4 °F)
Water Solubility: Slightly Soluble in Water
Other Solubilities: Slightly Soluble in Acetone Very Soluble in hot Alcohol

RTECS Toxicity Data

Acute Oral: Rat LD_{50} Dose: 160 mg/kg.
Acute Dermal: Rat LD_{50} Route: Subcutaneous Dose: 160 mg/kg.
Reproductive/Teratogenic: Mouse Route: Oral; Dose: 140 mg/kg; Duration: female 15-21D of pregnancy Effects on Newborn - Live birth index. Mouse Route: Oral; Dose: 400 mg/kg; Duration: female 20D after birth; Effects on Newborn - Weaning or lactation index.
Mutagenic: Human DNA Damage; Cell Type: lung; Dose: 100 umol/plate. Human Sister Chromatid Exchenge; Cell Type: lymphocyte; Dose: 10 mmol/L.
Tumorigenic: Rat Route: Oral; Dose: 1040 mg/kg/1Y-I; Toxic Effects: Tumorigenic - Carcinogenic by RTECS criteria; Gastrointestinal - Tumors; Liver - Tumors. Rat Route: Oral; Dose: 1800 mg/kg/64W-C; Toxic Effects: Tumorigenic - Equivocal tumorigenic agent by RTECS criteria; Gastrointestinal - Tumors. Rat Route: Oral; Dose: 2250 mg/kg/50W-I; Toxic Effects: Tumorigenic - Equivocal tumorigenic agent by RTECS criteria; Sense organs and special senses - Tumors; Liver - Tumors. Rat Route: Oral; Dose: 560 mg/kg/10W-C; Toxic Effects: Tumorigenic - Equivocal tumorigenic agent by RTECS criteria; Sense organs and special senses - Tumors. Rat Route: Oral; Dose: 300 mg/kg; Toxic Effects: Tumorigenic - Equivocal tumorigenic agent by RTECS criteria; Skin and appendages - Tumors. Rat Route: Oral; Dose: 1120 mg/kg/20W-C; Toxic Effects: Tumorigenic - Equivocal tumorigenic agent by RTECS criteria; Sense organs and special senses - Tumors. Rat Route: Oral; Dose: 1680 mg/kg/30W-C; Toxic Effects: Tumorigenic - Equivocal tumorigenic agent by RTECS criteria; Sense organs and special senses - Tumors.

Hazard Overviews

Carcinogenicity: IARC - Not listed; NIOSH - Not listed; NTP - Not listed; ACGIH - Not listed; OSHA - Not listed; EPA - Not listed; MAK - Not listed

Environmental

Regulations

RCRA 40CFR: Not listed
CERCLA: 40CFR 302.4: Not listed
SARA 40CFR 372.65: Not listed
SARA EHS 40CFR 355: Not listed
TSCA: Listed

DIN6320 CAS #: 128-42-7

4,4'-DINITRO-2,2'-STILBENEDISULFONIC ACID

RTECS: WJ6625000
EINECS Number: 204-885-6
Molecular Formula: $C_{14}H_{10}N_2O_{10}S_2$
Formula Weight: 430.38
Synonyms: BENZENESULFONIC ACID,2,2'-(1,2-ETHENEDIYL)BIS(5-NITRO-; DINITROSTILBENEDISULFONIC ACID; DNS; KYSELINA 4,4'-DINITROSTILBEN-2,2'-DISULFONOVA; 2,2'-STILBENEDISULFONIC ACID,4,4'-DINITRO-
Use: chem int for yellow azo-azoxy dyes used on cotton; chem int for orange & red dyes for cellulosic fibers; chem int for its diamine; chem int for brighteners

RTECS Toxicity Data

Acute Oral: Rat LD_{50} Dose: 12600 mg/kg. Mouse LD_{50} Dose: 47 gm/kg; Toxic Effects: Liver - Liver function tests impaired; Kidney, Ureter, and Bladder - Renal function tests depressed.
Irritation Eye: Rabbit Standard Draize Test Dose: 500 mg/24H; Reaction: mild.
Irritation Skin: Rabbit Standard Draize Test Dose: 500 mg/24H; Reaction: mild.

Hazard Overviews

Carcinogenicity: IARC - Not listed; NIOSH - Not listed; NTP - Not listed; ACGIH - Not listed; OSHA - Not listed; EPA - Not listed; MAK - Not listed

Environmental

Regulations

RCRA 40CFR: Not listed
CERCLA: 40CFR 302.4: Not listed
SARA 40CFR 372.65: Not listed
SARA EHS 40CFR 355: Not listed
TSCA: Listed

DIN6510 CAS #: 148-01-6

3,5-DINITRO-O-TOLUAMIDE

RTECS: XS4200000
EINECS Number: 205-706-4
Molecular Formula: $C_8H_7N_3O_5$
Structured MF: $(NO_2)_2C_6H_2(CH_3)CONH_2$
Formula Weight: 225.18
Synonyms: BENZAMIDE,2-METHYL-3,5-DINITRO-; COCCIDINE A; COCCIDOT; D.O.T; DINITOLMID; DINITOLMIDE; 2-METHYL-3,5-DINITROBENZAMIDE; ZOALENE; ZOAMIX
Description: yellowish, crystalline solid

Physical Properties

Freezing Point: 177.22 °C (350.996 °F)
Water Solubility: Slight
Other Solubilities: Soluble in acetone, acetonitrile, dioxane, and dimethylformamide
Flash Point: Noncombustible Solid

RTECS Toxicity Data

Acute Oral: Rat LD_{50} Dose: 600 mg/kg.
Mutagenic: Bacteria - B Subtilis DNA Repair; Dose: 1 mg/disc. Bacteria - S Typhimurium Mutations in Microorganisms; Dose: 500 ug/plate (+/-S9).

Hazard Overviews

Fire: Noncombustible.
Carcinogenicity: IARC - Not listed; NIOSH - Not listed; NTP - Not listed; ACGIH - Class A4, Not classifiable as a human carcinogen; OSHA - Not listed; EPA - Not listed; MAK - Not listed
Exposure Limits
OSHA PEL Vacated 1989 Limits: TWA: 5 mg/m^3.
ACGIH TLV: TWA: 5 mg/m^3.
NIOSH REL: TWA: 5 mg/m^3.
Respirator Recommendation
Exposure Range: >5 to 50 mg/m^3 Air Purifying, Negative Pressure, Half Mask
Exposure Range: >50 to 500 mg/m^3 Air Purifying, Negative Pressure, Full Face
Exposure Range: >500 to 5000 mg/m^3 Supplied Air, Constant Flow/Pressure Demand, Full Face
Exposure Range: >5000 to unlimited mg/m^3 Self-contained Breathing Apparatus, Pressure Demand, Full Face
Cartridge Color: dust/mist filter (use P100 or consult supervisor for appropriate dust/mist filter)

Environmental

Regulations

RCRA 40CFR: Not listed
CERCLA: 40CFR 302.4: Not listed
SARA 40CFR 372.65: Not listed
SARA EHS 40CFR 355: Not listed
TSCA: Not listed

DIN6700 CAS #: 602-01-7

2,3-DINITROTOLUENE

RTECS: XT1400000
DOT: UN1600; UN2038; IMO6.1
EINECS Number: 210-013-5
Molecular Formula: $C_7H_6N_2O_4$
Formula Weight: 182.14

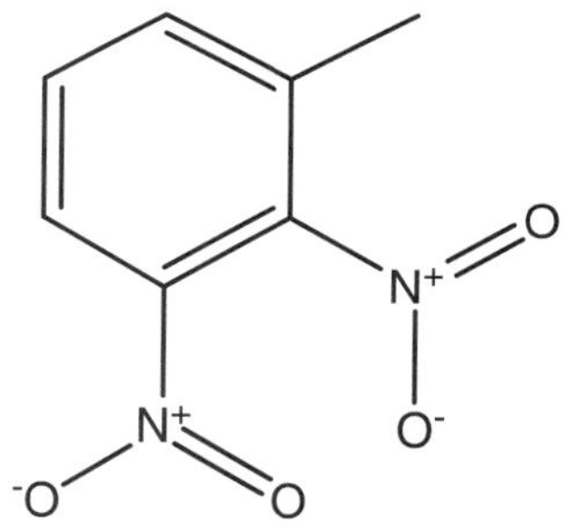

Chemical Structure

Synonyms: BENZENE,1-METHYL-2,3-DINITRO-; BENZENE,1-METHYL-2,3-DINITRO-(9CI); 2,3-DNT; 1-METHYL-2,3-DINITROBENZENE; TOLUENE,2,3-DINITRO-
Description: yellow crystals

Physical Properties

Freezing Point: 63 °C (145.4 °F)
Water Solubility: Insoluble
Other Solubilities: Ethanol: Soluble; Ether: Soluble; Chloroform: Slightly soluble
Flash Point: Combustible material

RTECS Toxicity Data

Acute Oral: Rat LD_{50} Dose: 911 mg/kg; Toxic Effects: Behavioral - Somnolence (general depressed activity). Mouse LD_{50} Dose: 1072 mg/kg.
Irritation Skin: Rabbit Standard Draize Test Dose: 500 mg/24H; Reaction: mild.
Mutagenic: Rat DNA Damage; Cell Type: liver; Dose: 300 umol/L. Bacteria - S Typhimurium Mutations in

Microorganisms; Dose: 50 ug/plate (+S9). Bacteria - S Typhimurium Mutations in Microorganisms; Dose: 1 mg/plate (-S9).

Hazard Overviews

Poison

Fire Diamond

Health: Irritating. Poison. Acute Effects: harmful if swallowed, inhaled, or absorbed through skin; may cause sensitization by inhalation and skin contact; readily absorbed through skin; absorption into the body leads to the formation of methemoglobin which in sufficient concentration causes cyanosis; onset may be delayed 2 to 4 hours or longer; symptoms of exposure may include burning sensation; coughing; wheezing; laryngitis; shortness of breath; headache; nausea; vomiting; target organs: blood, liver, spleen.

Fire: Combustible. Hazards: in powder form capable of creating a dust explosion; may explode when heated; may be shock-sensitive. Extinguishing agents: water spray; carbon dioxide, dry chemical powder or appropriate foam. Precautions: combustible liquid.

Reactivity: Incompatible with: oxidizing agents, reducing agents, strong bases. Hazardous decomposition products: thermal decomposition may produce carbon monoxide, carbon dioxide, and nitrogen oxides.

Carcinogenicity: IARC - Not listed; NIOSH - Not listed; NTP - Not listed; ACGIH - Not listed; OSHA - Not listed; EPA - Not listed; MAK - Not listed

Primary Target Organs:

Eyes

Skin

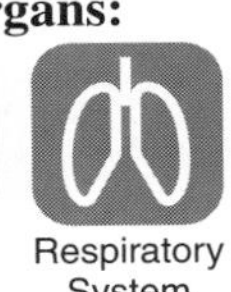
Respiratory System

Liver

Blood

Environmental

Ecotoxicity: The log of the 48 hr immobilization concentration (IC_{50}) of Daphnia magna was 1.49 umol/l for 2,3-dinitrotoluene in static tests. Using a semi-static procedure, the log of the 21 day IC_{50} was 0.99 umol/l for 2,3-dinitrotoluene LC_{50} Cyprinodon variegatus (sheepshead minnow) 2280 ug/l/96 hr /Static, unmeasured EC_{50} Selenastrum capricornutum (alga) 1370 ug/l/96 hr, Cell numbers /Static, unmeasured EC_{50} Skeletonema costatum (alga) 370 ug/l/96 hr, Cell numbers /Static, unmeasured LC_{50} Lepomis macrochirus (bluegill) 330 ug/l/96 hr /Static, unmeasured LC_{50} Mysidopsis bahia (mysid shrimp) 590 ug/l/96 hr /Static, unmeasured

Environmental Fate: If released to soil, it will be expected to have moderate mobility and may leach to groundwater. No information on biodegradation in soil was found; however, biodegradation may be an important fate process in soil based on a study in which it was shown to be cometabolized in natural surface water to which yeast extract was added. It will not hydrolyze in soils. If it is released to water, it will not be expected to bioconcentrate in aquatic organisms but may moderately sorb to sediment. It will not hydrolyze in water and evaporation from water is not expected to be an important transport process although biodegradation may be. It may be susceptible to photolysis based on the behavior of 2,4-dinitrotoluene which had sunlight photolysis half-lives of 43 hr in distilled water and 2.7, 9.6, and 3.7 hr in river, bay, and pond waters, respectively. If it is released to the atmosphere, it may be susceptible to direct photolysis based on the behavior of 2,4-dinitrotoluene in water. The estimated vapor phase half-life in the atmosphere is 8 hr as a result of addition of photochemically produced hydroxyl radicals to the aromatic ring.

Cleanup/Disposal: Guide No. 152: Do not touch damaged containers or spilled material unless wearing appropriate protective clothing. Stop leak if you can do it without risk. Prevent entry into waterways, sewers, basements or confined areas. Cover with plastic sheet to prevent spreading. Absorb or cover with dry earth, sand or other non-combustible material and transfer to containers. Do not get water inside containers.

Environmental Physical Data

Henry's Law Constant: 8.79 $\times 10^{-8}$
Octanol/Water Partition Coefficient: log K_{ow} = 2.0
Sorption Partition Coefficient: K_{OC} = estimated at 290
BCF: estimated at 19

Regulations

RCRA 40CFR: Not listed
CERCLA: 40CFR 302.4: Not listed
SARA 40CFR 372.65: Not listed
SARA EHS 40CFR 355: Not listed
TSCA: Listed

DIN6890 CAS #: 121-14-2

2,4-DINITROTOLUENE

RTECS: XT1575000
DOT: UN1600; UN2038; IMO6.1
EINECS Number: 204-450-0
Molecular Formula: $C_7H_6N_2O_4$
Structured MF: $CH_3C_6H_3(NO_2)_2$
Formula Weight: 182.14

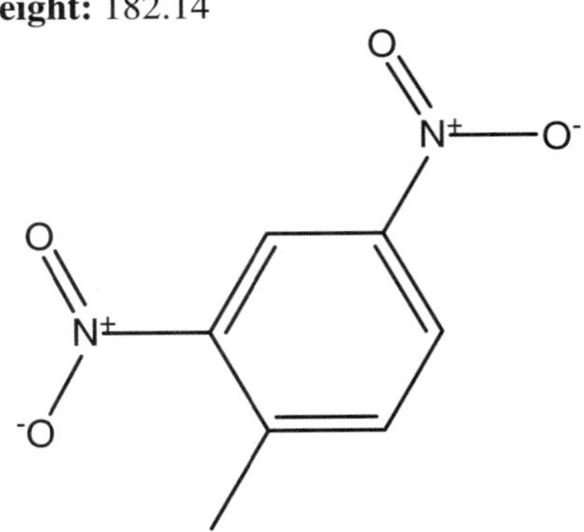

Chemical Structure

Synonyms: BENZENE,1-METHYL-2,4-DINITRO-; DINITROTOLUENE; 2,4-DINITROTOLUOL; DINITROTOLUOL; 2,4-DNT; DNT; 1-METHYL-2,4-DINITROBENZENE; 4-METHYL-1,3-DINITROBENZENE; TOLUENE,2,4-DINITRO-
Description: orange to yellow solid, liquid; slight odor
Use: organic synthesis; toluidines; dyes; explosives

Physical Properties

Boiling Point: 300 °C (572 °F)
Freezing Point: 71 °C (159.8 °F)
Specific Gravity: 1.379 at 20 °C
Vapor Density: 6.27 Air=1
Saturated Vapor Density: 1.208337931 kg/m^3
Density: 1.3208 g/mL at 71 °C
Vapor Pressure: 1 mm Hg at 20 °C
Water Solubility: 300 ppm at 20 °C
Other Solubilities: Ethanol, at 15 °C, 30.46 g/l; Diethyl Ether, at 22 °C, 94 g/l; Carbon Disulfide, at 17 °C, 21.9 g/l.
Refraction Index: 1.442
Flash Point: 206.667 °C Closed Cup

RTECS Toxicity Data

Acute Oral: Rat LD_{50} Dose: 268 mg/kg. Mouse LD_{50} Dose: 790 mg/kg.
Acute Dermal: Cat LD_{Lo} Route: Subcutaneous Dose: 25 mg/kg.
Chronic (Multiple Dose) Oral: Rat Dose: 13195 mg/kg/13W-C; Toxic Effects: Behavioral - Food intake (animal); DEATH. Rat Dose: 24752 mg/kg/2Y-C; Toxic Effects: Blood - Normocytic anemia; Blood - Changes in other cell count. Rat Dose: 12410 mg/kg/1Y-C; Toxic Effects: Blood - Normocytic anemia; Nutritional and gross metabolic - Weight loss or decreased weight gain; DEATH - Changes in testicular weight. Rat Dose: 76 gm/kg/26W-C; Toxic Effects: Liver - Changes in liver weight; Blood - Changes in serum composition; DEATH - Changes in testicular weight.
Irritation Skin: Rabbit Standard Draize Test Dose: 500 mg/24H; Reaction: mild.
Reproductive/Teratogenic: Rat Route: Oral; Dose: 3094 mg/kg; Duration: male 13W prior to mating; Effects on Fertility - Pre-implantation mortality; Other measures of fertility. Rat Route: Oral; Dose: 8463 mg/kg; Duration: male 13W prior to mating; Paternal Effects - Spermatogenesis.
Mutagenic: Rat DNA Damage; Cell Type: liver; Dose: 3 mmol/L. Rat Unscheduled DNA Synthesis; Cell Type: other cell types; Dose: 10 umol/L. Rat Unscheduled DNA Synthesis; Route: Oral; Dose: 35 mg/kg.
Tumorigenic: Rat Route: Oral; Dose: 2620 mg/kg/78W-C; Toxic Effects: Tumorigenic - Neoplastic by RTECS criteria; Skin and appendages - Tumors. Rat Route: Oral; Dose: 5460 mg/kg/78W-C; Toxic Effects: Tumorigenic - Neoplastic by RTECS criteria; Skin and appendages - Tumors. Rat Route: Oral; Dose: 28 gm/kg/2Y-C; Toxic Effects: Tumorigenic - Carcinogenic by RTECS criteria; Liver - Tumors; Nutritional and gross metabolic - Weight loss or decreased weight gain. Rat Route: Oral; Dose: 12775 mg/kg/Y-C; Toxic Effects: Tumorigenic - Equivocal tumorigenic agent by RTECS criteria; Liver - Tumors.

Hazard Overviews

Explosive

Fire Diamond

Health: Irritating to eyes/skin/respiratory tract. Toxic. Also Causes: cyanosis, headache, dizziness, weakness, nausea, vomiting, dyspnea, drowsiness, unconsciousness, stomach cramps, diarrhea, skin ulceration, necrosis, and severe eye damage. Chronic Effects: anemia, affects the liver's drug-metabolizing enzymes, and a liver mutagen and carcinogen (rodents).
Fire: Explosive and will burn. For small fires use dry chemical, carbon dioxide, water spray, or regular foam. For large fires use water spray, fog, or regular foam. Fight fire from maximum distance.
Reactivity: Unstable, explosive when exposed to heat/friction/contamination; slightly sensitive to impact shock. Hazardous polymerization cannot occur. Avoid: heat; friction. Incompatible with: organic materials; strong reducing agents; metallic hydrides; strong oxidizing agents; caustics; metals; plastics; rubber; coatings. Hazardous decomposition products: toxic nitrogen oxides.
Carcinogenicity: IARC - Group 2B, Possibly carcinogenic to humans; NIOSH - Not listed; NTP - Not listed; ACGIH - Not listed; OSHA - Not listed; EPA - Not listed; MAK - Not listed
Primary Target Organs:

Eyes

Skin

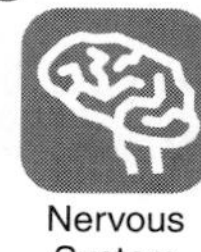
Nervous System

Liver

Cardio-vascular

Blood

Exposure Limits
OSHA PEL: TWA: 1.5 mg/m^3; skin.

Environmental

Ecotoxicity: LC_{50} Pimephales promelas (fathead minnow) 24.3 mg/l/96 hr (confidence limit 23.0 to 25.6 mg/l). Affected fish lost schooling behavior, were hypoactive and underreactive to external stimuli, swam near the tank surface and had increased respiration Toxicity threshold (the concentration decreasing Uronema parduczi population decreased by 5%) was < 1 mg/l for 2,4-dinitrotoluene
Environmental Fate: In soil, it will be slightly mobile (estimated K_{oc} = 282). Based on aqueous biodegradation tests, it may biodegrade in both aerobic and anaerobic zones of soil. In water it will not bioconcentrate significantly (experimental BCF=204) and will have a slight tendency to partition to suspended and sediment organic matter (log Kow = 1.98). Volatilization from water will not be significant. Photolysis will probably be the most important removal process in water. Photolytic half-lives in river, bay and pond waters were 2.7, 9.6 and 3.7 hr, respectively. One source gave a theoretical half-life of 11 hr in natural waters while another estimated the

half-life as 1.7 days in the Rhine River. This reaction was found to be accelerated in the presence of humic material. The importance of biodegradation in natural waters is unknown, although a number of conflicting screening test results are available. In the atmosphere, it is estimated to have a half-life of 71 days. It has been detected in drinking water, seawater, river water, and in wastewater from 2,4,6-trinitrotoluene production.

Cleanup/Disposal: Guide No. 152: Do not touch damaged containers or spilled material unless wearing appropriate protective clothing. Stop leak if you can do it without risk. Prevent entry into waterways, sewers, basements or confined areas. Cover with plastic sheet to prevent spreading. Absorb or cover with dry earth, sand or other non-combustible material and transfer to containers. Do not get water inside containers.

Environmental Physical Data

Henry's Law Constant: calculated at 8.79 x10^{-8}
Octanol/Water Partition Coefficient: log K_{ow} = 1.98
Sorption Partition Coefficient: K_{oc} = 282
BCF: guppy 204

Regulations

RCRA 40CFR: Listed Hazardous Waste No. U105 Toxic Waste
CERCLA: 40CFR 302.4: Listed per CWA Section 311(b)(4) per RCRA Section 3001 per CWA Section 307(a) RQ: 10 lb (4.535 kg)
SARA 40CFR 372.65: Listed
SARA EHS 40CFR 355: Not listed
TSCA: Listed

Analytical Methods

Soil: CLP LC_SV, MC_SVOA, OHC; EPA 16, 1625; SW846 3640A, 3650A, 3650B, 8090, 8091, 8250A, 8270B, 8270C, 8275, 8330, 8410; DOE OP130R
Water / Groundwater: EPA 1625, 609, 625, 625-S, 6; AOAC 986.22; APHA 6410-B; USGS O3118
Drinking Water: EPA 525.2
Plasma: EPA 29

DIN7080 CAS #: 619-15-8

2,5-DINITROTOLUENE

RTECS: XT1750000
DOT: UN1600; UN2038; IMO6.1
EINECS Number: 210-581-4
Molecular Formula: $C_7H_6N_2O_4$
Formula Weight: 182.14
Synonyms: BENZENE,2-METHYL-1,4-DINITRO-; BENZENE,2-METHYL-1,4-DINITRO-(9CI); 2,5-DNT; 2-METHYL-1,4-DINITROBENZENE; TOLUENE,2,5-DINITRO-
Description: needles

Physical Properties

Freezing Point: 52.5 °C (126.5 °F)
Specific Gravity: 1.282 at 111 °C
Other Solubilities: Soluble in Alcohol, Benzene
Flash Point: Combustible material

RTECS Toxicity Data

Acute Oral: Rat LD_{50} Dose: 517 mg/kg; Toxic Effects: Behavioral - Somnolence (general depressed activity). Mouse LD_{50} Dose: 652 mg/kg; Toxic Effects: Behavioral - Somnolence (general depressed activity).
Irritation Skin: Rabbit Standard Draize Test Dose: 500 mg/24H; Reaction: moderate.
Mutagenic: Bacteria - S Typhimurium Mutations in Microorganisms; Dose: 10 ug/plate (+/-S9).

Hazard Overviews

Fire Diamond

Fire: Combustible.
Carcinogenicity: IARC - Not listed; NIOSH - Not listed; NTP - Not listed; ACGIH - Not listed; OSHA - Not listed; EPA - Not listed; MAK - Not listed

Environmental

Ecotoxicity: Crustaceans: water flea 48h LC_{50} 3.1 mg/l
Environmental Fate: If released to soil it will be expected to have moderate mobility and may leach to groundwater. No information on biodegradation in soil was found; however, biodegradation may be an important fate process in soil based on a study in which it was shown to be cometabolized in natural surface water to which yeast extract was added. It will not hydrolyze in soils. If released in to water, it will not be expected to bioconcentrate in aquatic organisms but may moderately sorb to sediment. It will not hydrolyze in water, evaporation from water is not expected to be an important transport process although biodegradation may be. It may be susceptible to photolysis based on the behavior of 2,4-dinitrotoluene which had photolysis half-lives of 43 hr in distilled water and 2.7, 9.6, and 3.7 hr in river, bay, and pond waters, respectively. If it is released to the atmosphere, may be susceptible to direct photolysis based on the behavior of 2,4-dinitrotoluene in water. The estimated vapor phase half-life in the atmosphere is 8 hr as a result of addition of photochemically produced hydroxyl radicals to the aromatic ring.
Cleanup/Disposal: Guide No. 152: Do not touch damaged containers or spilled material unless wearing appropriate protective clothing. Stop leak if you can do it without risk. Prevent entry into waterways, sewers, basements or confined areas. Cover with plastic sheet to prevent spreading. Absorb or cover with dry earth, sand or other non-combustible material and transfer to containers. Do not get water inside containers.

Environmental Physical Data

Henry's Law Constant: 8.79 x10^{-8}
Octanol/Water Partition Coefficient: log K_{ow} = 1.997
Sorption Partition Coefficient: K_{oc} = estimated at 290

BCF: estimated at 19

Regulations

RCRA 40CFR: Not listed
CERCLA: 40CFR 302.4: Not listed
SARA 40CFR 372.65: Not listed
SARA EHS 40CFR 355: Not listed
TSCA: Listed

DIN7270 CAS #: 606-20-2

2,6-DINITROTOLUENE

RTECS: XT1925000
DOT: UN1600; UN2038; IMO6.1
EINECS Number: 210-106-0
Molecular Formula: $C_7H_6N_2O_4$
Structured MF: $(NO_2)_2C_6H_3CH_3$
Formula Weight: 182.14

Chemical Structure

Synonyms: BENZENE,2-METHYL-1,3-DINITRO-; BENZENE,2-METHYL-1,3-DINITRO-(9CI); 2,6-DNT; 1-METHYL-2,6-DINITROBENZENE; 2-METHYL-1,3-DINITROBENZENE
Description: yellow to red solid; slight odor
Use: manufacture of dyes, explosives (TNT), urethane polymers, flexible and rigid foams, and elastomers; in organic synthesis

Physical Properties

Boiling Point: 285 °C (545 °F)
Freezing Point: 66 °C (150.8 °F)
Specific Gravity: 1.2833 at 111 °C
Vapor Density: 6.28 Air=1
Density: 1.2833 g/mL at 111 °C
Vapor Pressure: 5.67 $x10^{-4}$ mm Hg
Water Solubility: < 1 mg/mL at 18 C
Other Solubilities: 95% Ethanol: 10-50 mg/ml at 18 °C; Acetone: >=100 mg/ml at 18 °C; Alcohol: Soluble; Benzene: Very Soluble; DMSO: >=100 mg/ml at 18 °C; Ethanol: >10%; Ether: Soluble.
Odor Threshold: 0.1 ppm in water
Refraction Index: 1.479
Flash Point: 206.667 °C Closed Cup

RTECS Toxicity Data

Acute Oral: Rat LD_{50} Dose: 177 mg/kg. Mouse LD_{50} Dose: 621 mg/kg; Toxic Effects: Behavioral - Somnolence (general depressed activity).
Chronic (Multiple Dose) Oral: Rat Dose: 13500 mg/kg/90D-I; Toxic Effects: Blood - Normocytic anemia; Blood - Methemoglobinemia-Carboxyhemoglobin; DEATH. Rat Dose: 27 gm/kg/90D-C; Toxic Effects: Blood - Normocytic anemia; Blood - Methemoglobinemia-Carboxyhemoglobin; DEATH. Rat Dose: 9 gm/kg/90D-C; Toxic Effects: Blood - Normocytic anemia; Blood - Methemoglobinemia-Carboxyhemoglobin; DEATH.
Irritation Skin: Rabbit Standard Draize Test Dose: 500 mg/24H; Reaction: mild.
Mutagenic: Rat DNA Damage; Cell Type: liver; Dose: 3 mmol/L. Rat Unscheduled DNA Synthesis; Cell Type: other cell types; Dose: 10 umol/L. Rat Unscheduled DNA Synthesis; Route: Oral; Dose: 5 mg/kg.
Tumorigenic: Rat Route: Oral; Dose: 2555 mg/kg/1Y-C; Toxic Effects: Tumorigenic - Carcinogenic by RTECS criteria; Liver - Tumors. Rat Route: Oral; Dose: 5110 mg/kg/1Y-C; Toxic Effects: Tumorigenic - Equivocal tumorigenic agent by RTECS criteria; Liver - Tumors.

Hazard Overviews

Poison

Corrosive

Fire Diamond

Health: Corrosive to eyes/skin/respiratory tract. Poison. Other Acute Effects: harmful if swallowed, inhaled, or absorbed through skin; may cause sensitization by inhalation and skin contact; readily absorbed through skin; neurological hazard; inhalation may result in spasm, inflammation and edema of the larynx and bronchi, chemical pneumonitis and pulmonary edema; symptoms of exposure may include burning sensation; coughing; wheezing; laryngitis; shortness of breath; headache; nausea; vomiting; may cause allergic respiratory and skin reactions; nervous system disturbances. Chronic Effects: may cause reproductive disorders; data from animal studies have shown reduced sperm formation, testicular atrophy in exposed dogs, rats and mice, and nonfunctioning ovaries in mice; target organs: blood, liver, spleen. Carcinogen.
Fire: Will burn. Hazards: in powder form capable of creating a dust explosion; may explode when heated; may be shock-sensitive. Extinguishing agents: water spray; carbon dioxide, dry chemical powder or appropriate foam. Precautions: combustible liquid.
Reactivity: Incompatible with: oxidizing agents, reducing agents, strong bases. Hazardous decomposition products: thermal decomposition may produce carbon monoxide, carbon dioxide, and nitrogen oxides.
Carcinogenicity: IARC - Group 2B, Possibly carcinogenic to humans; NIOSH - Not listed; NTP - Not listed; ACGIH - Not listed; OSHA - Not listed; EPA - Not listed; MAK - Not listed

Primary Target Organs:

Eyes

Skin

Respiratory System

Liver

Blood

Environmental

Ecotoxicity: Aquatic toxicity: 10-100 ppm/96 hr/fin fish/TLm

Environmental Fate: If released to soil, it is expected to biodegrade. It should be fairly mobile, based on experiments in sandy loam and sandy slit loam soil. If released to water, it will readily biodegrade. Photooxidation should be rapid in surface layers of water. It should not adsorb appreciably to sediments or suspended solids. Volatilization from water or soil will not be significant. In the atmosphere, it will not react with photochemically produce hydroxyl radicals. Its half-life is estimated to be 47 days.

Cleanup/Disposal: Guide No. 152: Do not touch damaged containers or spilled material unless wearing appropriate protective clothing. Stop leak if you can do it without risk. Prevent entry into waterways, sewers, basements or confined areas. Cover with plastic sheet to prevent spreading. Absorb or cover with dry earth, sand or other non-combustible material and transfer to containers. Do not get water inside containers.

Environmental Physical Data

Henry's Law Constant: estimated at 9.26×10^{-8}

Octanol/Water Partition Coefficient: log K_{ow} = 1.72

Sorption Partition Coefficient: log K_{oc} = 1.72

BCF: algal biomass 5225

BOD: not pertinent

Regulations

RCRA 40CFR: Listed Hazardous Waste No. U106 Toxic Waste

CERCLA: 40CFR 302.4: Listed per CWA Section 311(b)(4) per RCRA Section 3001 per CWA Section 307(a) RQ: 100 lb (45.35 kg)

SARA 40CFR 372.65: Listed

SARA EHS 40CFR 355: Not listed

TSCA: Listed

Analytical Methods

Soil: CLP MC_SVOA, OHC; EPA 16, 1625; SW846 3640A, 8090, 8091, 8250A, 8270B, 8270C, 8330, 8410

Water / Groundwater: EPA 1625, 609, 625, 625-S, 6; APHA 6410-B; USGS O3118

Drinking Water: EPA 525.2

Plasma: EPA 29

DIN7460 **CAS #: 610-39-9**

3,4-DINITROTOLUENE

RTECS: XT2100000

DOT: UN1600; UN2038; IMO6.1

EINECS Number: 210-222-1

Molecular Formula: $C_7H_6N_2O_4$

Structured MF: $(NO_2)_2C_6H_3{\bullet}CH_3$

Formula Weight: 182.14

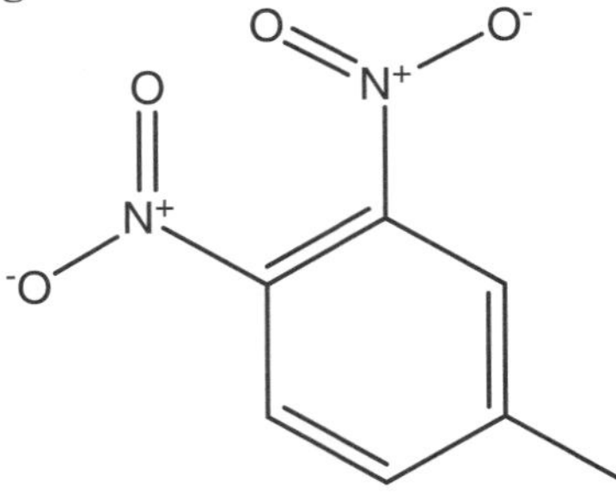

Chemical Structure

Synonyms: BENZENE,4-METHYL-1,2-DINITRO-; BENZENE,4-METHYL-1,2-DINITRO-(9CI); 3,4-DNT; 4-METHYL-1,2-DINITROBENZENE; TOLUENE,3,4-DINITRO-

Description: yellow to red solid or needles; slight odor

Use: organic synthesis; toluidines; dyes; explosives; in mfr of toluene diamine

Physical Properties

Boiling Point: Decomposes at 1 atm

Freezing Point: 58.3 °C (136.94 °F)

Specific Gravity: 1.2594 at 111 °C

Vapor Density: 6.28

Vapor Pressure: 1 mm Hg

Water Solubility: Insoluble

Other Solubilities: Soluble in Alcohol

Flash Point: 206.667 °C Closed Cup

RTECS Toxicity Data

Acute Oral: Rat LD_{50} Dose: 807 mg/kg. Mouse LD_{50} Dose: 747 mg/kg; Toxic Effects: Behavioral - Somnolence (general depressed activity).

Irritation Skin: Rabbit Standard Draize Test Dose: 500 mg/24H; Reaction: mild.

Mutagenic: Rat DNA Damage; Cell Type: liver; Dose: 300 umol/L. Bacteria - S Typhimurium Mutations in Microorganisms; Dose: 1 mg/plate (+S9). Bacteria - S Typhimurium Mutations in Microorganisms; Dose: 10 ug/plate (-S9).

Hazard Overviews

Poison

Fire Diamond

Health: Irritating. Poison. Acute Effects: may be fatal if inhaled, swallowed, or absorbed through skin; may cause sensitization by inhalation and skin contact; readily absorbed through skin; absorption into the body leads to the formation of methemoglobin which in sufficient concentration causes cyanosis; onset may be delayed 2 to 4 hours or longer; target organs: blood, liver, spleen. Chronic Effects: damage to the liver; blood effects.

Fire: Will burn. Hazards: container explosion may occur; may explode when heated. Extinguishing agents: water spray; carbon dioxide; dry chemical powder. Precautions: combustible liquid.
Reactivity: Incompatible with: heat, oxidizing agents, reducing agents, strong bases. Hazardous decomposition products: carbon monoxide, carbon dioxide, nitrogen oxides.
Carcinogenicity: IARC - Not listed; NIOSH - Not listed; NTP - Not listed; ACGIH - Not listed; OSHA - Not listed; EPA - Not listed; MAK - Not listed
Primary Target Organs:

Eyes

Skin
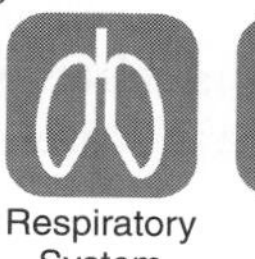
Respiratory System

Liver

Blood

Environmental

Ecotoxicity: The log of the 48 hr immobilization concentration (IC_{50}) of Daphnia magna was 1.49 umol/l for 3,4-dinitrotoluene in static tests. Using a semi-static procedure, the log of the 21 day IC_{50} was 0.78 umol/l for 3,4-dinitrotoluene The log of the concentration of 3,4-dinitrotoluene causing a 50% decrease of Photobacterium phosphoreum bioluminescence after 15 min of exposure was 1.58 umol/l
Environmental Fate: If released to soil, it will be expected to have moderate mobility and may leach to groundwater. No information on biodegradation in soil was found; however, biodegradation may be an important fate process in soil based on a study in which it was shown to be cometabolized in natural surface water to which yeast extract was added. It will not hydrolyze in soils. If it is released to water, it will not be expected to bioconcentrate in aquatic organisms but may moderately sorb to sediment. It will not hydrolyze in water and evaporation from water is not expected to be an important transport process although biodegradation may be. It may be susceptible to photolysis based on the behavior of 2,4-dinitrotoluene which had sunlight photolysis half-lives of 43 hr in distilled water and 2.7, 9.6, and 3.7 hr in river, bay, and pond waters, respectively. If it is released to the atmosphere, it may be susceptible to direct photolysis based on the behavior of 2,4-DNT in water. The estimated vapor phase half-life in the atmosphere is 8 hr as a result of addition of photochemically produced hydroxyl radicals to the aromatic ring.
Cleanup/Disposal: Guide No. 152: Do not touch damaged containers or spilled material unless wearing appropriate protective clothing. Stop leak if you can do it without risk. Prevent entry into waterways, sewers, basements or confined areas. Cover with plastic sheet to prevent spreading. Absorb or cover with dry earth, sand or other non-combustible material and transfer to containers. Do not get water inside containers.

Environmental Physical Data

Henry's Law Constant: 8.79×10^{-8}
Octanol/Water Partition Coefficient: log K_{ow} = estimated at 1.997
BCF: estimated at 19

Regulations

RCRA 40CFR: Not listed
CERCLA: 40CFR 302.4: Listed per CWA Section 311(b)(4) per CWA Section 307(a) RQ: 10 lb (4.535 kg)
SARA 40CFR 372.65: Not listed
SARA EHS 40CFR 355: Not listed
TSCA: Listed

DIN7650 CAS #: 618-85-9

3,5-DINITROTOLUENE

RTECS: XT2150000
DOT: UN1600; UN2038; IMO6.1
EINECS Number: 210-566-2
Molecular Formula: $C_7H_6N_2O_4$
Formula Weight: 182.14
Synonyms: BENZENE,1-METHYL-3,5-DINITRO-; BENZENE,1-METHYL-3,5-DINITRO-(9CI); 3,5-DNT; 1-METHYL-3,5-DINITROBENZENE; TOLUENE,3,5-DINITRO-
Description: yellowish to reddish needles
Use: intermediates for production of toluene diisocyanate; organic synthesis; toluidines; dyes; explosives

Physical Properties

Freezing Point: 93 °C (199.4 °F)
Specific Gravity: 1.277
Water Solubility: Slightly Soluble
Other Solubilities: > 10% in ethyl Ether; > 10% in Ethanol; > 10% in Chloroform.

RTECS Toxicity Data

Acute Oral: Rat LD_{50} Dose: 216 mg/kg; Toxic Effects: Behavioral - Somnolence (general depressed activity). Mouse LD_{50} Dose: 607 mg/kg; Toxic Effects: Behavioral - Somnolence (general depressed activity).
Mutagenic: Bacteria - S Typhimurium Mutations in Microorganisms; Dose: 100 ug/plate (+S9), 500 ug/plate (-S9).

Hazard Overviews

Fire Diamond

Carcinogenicity: IARC - Group 3, Not classifiable as to carcinogenicity to humans; NIOSH - Not listed; NTP - Not listed; ACGIH - Not listed; OSHA - Not listed; EPA - Not listed; MAK - Not listed

Environmental

Environmental Fate: If released to water, it may photolyze (half-life of 28 days). Some may also partition to sediment. Volatilization from water is not expected to be an environmentally significant fate process (volatilization half-life of approximately 125 days from a model river). In soil, its K_{oc} of about 414 suggests moderate soil mobility. No data are available on other potential degradation processes in soil or water. If released to the atmosphere, vapor phase is degraded slowly by reaction with photochemically produced hydroxyl radicals (estimated half-life of 87 days in air).

Cleanup/Disposal: Guide No. 152: Do not touch damaged containers or spilled material unless wearing appropriate protective clothing. Stop leak if you can do it without risk. Prevent entry into waterways, sewers, basements or confined areas. Cover with plastic sheet to prevent spreading. Absorb or cover with dry earth, sand or other non-combustible material and transfer to containers. Do not get water inside containers.

Environmental Physical Data

Henry's Law Constant: estimated at 3.97×10^{-7}
Octanol/Water Partition Coefficient: log K_{ow} = 2.28
Sorption Partition Coefficient: K_{oc} = 414
BCF: estimated at 32

Regulations

RCRA 40CFR: Not listed
CERCLA: 40CFR 302.4: Not listed
SARA 40CFR 372.65: Not listed
SARA EHS 40CFR 355: Not listed
TSCA: Not listed

DIN7840 CAS #: 25321-14-6

DINITROTOLUENE

RTECS: XT1300000
DOT: UN1600; UN2038; IMO6.1
EINECS Number: 246-836-1
Molecular Formula: $C_7H_6N_2O_4$
Structured MF: $CH_3C_6H_3(NO_2)_2$
Formula Weight: 182.14
Synonyms: BENZENE,METHYLDINITRO-; BINITROTOLUENE; DINITROPHENYLMETHANE; DINITROTOLUENE (2,4 AND 2,6 MIX); DINITROTOLUENES,LIQUID OR SOLID; DINITROTOLUOL; DNT; METHYLDINITROBENZENE; TDNT; TOLUENE,AR,AR-DINITRO; TOLUENE,AR,AR-DINITRO-; TOLUENE,DINITRO-
Use: organic synthesis; toluidines; dyes; explosives

Physical Properties

Boiling Point: 300 °C (572 °F)
Freezing Point: 70 °C (158 °F)
Vapor Density: 6.3 Air=1
Density: 1.3208 g/mL
Vapor Pressure: 1 mm Hg
Water Solubility: Insoluble in Water
Other Solubilities: 95% Ethanol: Soluble; Acetone: Soluble; DMSO: Soluble; Ether: Soluble.
Flash Point: 206.667 °C Closed Cup
Autoignition Temperature: 280 °C

RTECS Toxicity Data

Acute Oral: Mouse LD_{50} Dose: 750 mg/kg; Toxic Effects: Behavioral - Convulsions or effect on seizure threshold; Lungs, Thorax, or Respiration - Dyspnea. Guinea Pig LD_{50} Dose: 1300 mg/kg; Toxic Effects: Brain and coverings - Other degenerative changes; Lungs, Thorax, or Respiration - Dyspnea; Lungs, Thorax, or Respiration - Respiratory stimulation.

Reproductive/Teratogenic: Rat Route: Oral; Dose: 1050 mg/kg; Duration: female 7-20D of pregnancy; Specific Developmental Abnormalities - Blood and lymphatic systems (including spleen and marrow); Effects on Newborn - Delayed effects. Rat Route: Oral; Dose: 196 mg/kg; Duration: female 7-20D of pregnancy; Effects on Newborn - Behavioral. Rat Route: Oral; Dose: 2100 mg/kg; Duration: female 7-20D of pregnancy; Effects on Fertility - Post-implantation mortality. Rat Route: Oral; Dose: 1400 mg/kg; Duration: female 7-20D of pregnancy; Specific Developmental Abnormalities - Blood and lymphatic systems (including spleen and marrow).

Mutagenic: Rat Unscheduled DNA Synthesis; Route: Oral; Dose: 35 mg/kg. Rat Morphological Transformation; Route: Oral; Dose: 294 mg/kg/3W-C.

Tumorigenic: Rat Route: Oral; Dose: 12775 mg/kg/Y-C; Toxic Effects: Tumorigenic - Equivocal tumorigenic agent by RTECS criteria; Liver - Tumors.

Hazard Overviews

Reactivity: Unstable, explosive when exposed to heat, friction, or contamination; slightly sensitive to impact shock. Hazardous polymerization cannot occur. Avoid: exposure to heat; friction. Incompatible with: organic materials; strong reducing agents (sodium sulphide; zinc powder; sodium hyposulphite; metallic hydrides); strong oxidizing agents (bichromates; peroxides; chlorates); caustics; metals (tin and zinc); corrodes some forms of plastics, rubber, and coatings. Hazardous decomposition products: carbon dioxide; toxic nitrogen oxides.

Carcinogenicity: IARC - Not listed; NIOSH - Not listed; NTP - Not listed; ACGIH - Class A2, Suspected human carcinogen; OSHA - Not listed; EPA - Not listed; MAK - Not listed

Exposure Limits
ACGIH TLV: TWA: 0.15 mg/m^3.
NIOSH IDLH: 50 mg/m^3.

Respirator Recommendation
Exposure Range: >1.5 to <50 mg/m^3 Supplied Air, Constant Flow/Pressure Demand, Half Mask
Exposure Range: 50 to unlimited mg/m^3 Self-contained Breathing Apparatus, Pressure Demand, Full Face
Note: odor threshold unknown

Environmental

Environmental Fate: If released to soil, they are expected to have moderate mobility and may leach to groundwater. No information on biodegradation in soil was found; however, biodegradation may be an important fate process in soil based on the metabolism of all the isomers in surface water to which yeast extract was added. They should not hydrolyze in soils. If released to water, they should not bioconcentrate in aquatic organisms but may moderately sorb to sediment. The Henry's Law Constant is so low that evaporation from water is not expected to be an important transport process. Biodegradation may be an important fate process in water. They may be susceptible to photolysis based on the behavior of 2,4-dinitrotoluene which had photolytic half-lives of 43 hr in distilled water and 2.7, 9.6, and 3.7 hr in river, bay, and pond waters, respectively, and the behavior of 2,6-DNT in sunlit water where it had a half-life of 12 minutes due to an indirect photoreaction. If released to the atmosphere, they may be susceptible to direct photolysis. The estimated vapor phase half-life in the atmosphere is 8 hr as a result of addition of photochemically produced hydroxyl radicals to the aromatic ring.

Cleanup/Disposal: Guide No. 152: Do not touch damaged containers or spilled material unless wearing appropriate protective clothing. Stop leak if you can do it without risk. Prevent entry into waterways, sewers, basements or confined areas. Cover with plastic sheet to prevent spreading. Absorb or cover with dry earth, sand or other non-combustible material and transfer to containers. Do not get water inside containers.

Environmental Physical Data

Henry's Law Constant: 8.79 x10^{-8}

Octanol/Water Partition Coefficient: log K_{ow} = estimated at 1.997

Sorption Partition Coefficient: K_{oc} = estimated at 201 to 290

BCF: calculated at 5225

Regulations

RCRA 40CFR: Not listed

CERCLA: 40CFR 302.4: Listed per CWA Section 311(b)(4) per CWA Section 307(a) RQ: 10 lb (4.535 kg)

SARA 40CFR 372.65: Listed

SARA EHS 40CFR 355: Not listed

TSCA: Listed

DIN8030 CAS #: 39300-45-3

DINOCAP

RTECS: GQ5775000

DOT: UN1663; UN3013

EINECS Number: 254-408-0

Molecular Formula: $C_{18}H_{24}N_2O_6$

Formula Weight: 364.39

Synonyms: ARATHANE; 2-BUTENOIC ACID 2-(1-METHYLHEPTYL)-4,6-DINITROPHENYL ESTER; 2-BUTENOIC ACID,2-(OR 4)-ISOOCTYL-4,6(OR 2,6)-DINITROPHENYL ESTER (9CI); 2-BUTENOIC ACID,2(OR 4)-ISOOCTYL-4,6(OR2,6)-DINITROPHENYL ESTER; CAPRANE; 2-CAPRYL-4,6-DINITROPHENYL CROTONATE; CAPRYLDINITROPHENYL CROTONATE; CARATAN; CARATHANE; CR 1639; CROTONATE DE 2,4-DINITRO 6-(1-METHYL-HEPTYL)-PHENYLE; CROTONIC ACID 2,4-DINITRO-6-(1-METHYLHEPTYL)PHENYL ESTER; CROTONIC ACID 2,4-DINITRO-6-(2-OCTYL)PHENYL ESTER; CROTONIC ACID 2-(1-METHYLHEPTYL)-4,6-DINITROPHENYL ESTER; CROTOTHANE; DINITRO METHYLHEPTYPHENYL CROTONATE; 4,6-DINITRO-2-(2-CAPRYL)PHENYL CROTONATE; 4,6-DINITRO-2-CAPRYLPHENYL CROTONATE; DINITROCAPRYLPHENYL CROTONATE; 2,4-DINITRO-6-(1-METHYLHEPTYL)PHENYL CROTONATE; 4,6-DINITRO-2-(1-METHYLHEPTYL)PHENYL CROTONATE; DINITRO(1-METHYLHEPTYL)PHENYL CROTONATE; 2,4-DINITRO-6-(1-METHYLHEPTYL)-PHENYLCROTONAT; 2,4-DINITRO-6-(2-OCTYL)PHENYL CROTONATE; DINOKAP; DINOKAPU; DPC; DPC (PESTICIDE); ENT 24727; ISOCOTHANE; KARATAN; KARATHANE; KARATHANE 25; KARATHANE FN 57; KARATHANE WD; (6-(1-METHYL-HEPTYL)-2,4-DINITRO-FENYL)-CROTONAAT; 2-(1-METHYLHEPTYL)-4,6-DINITROFENYLESTER KYSELINYKROTONOVE; 2-(1-METHYLHEPTYL)-4,6-DINITROPHENYL CROTONATE; (6-(1-METHYL-HEPTYL)-2,3-DINITRO-PHENYL)-CROTONAT; (6-(1-METIL-EPITL)-2,4-DINITRO-FENIL)-CROTONATO; MILDEX; PHENOL,2-(1-METHYLHEPTYL)-4,6-DINITRO-,CROTONATE (ESTER)

Description: dark brown liquid

Use: non-systemic acaricide and contact fungicide recommended to control powdery mildews on various fruits, grape vines and ornamentals; seed treatments (former use)

Physical Properties

Boiling Point: 138 °C (280 °F) to 140 °C (284 °F) at 0.05 mm Hg

Saturated Vapor Density: 1.200000001 kg/m^3

Density: 1.1 g/cu cm at 20 °C

Vapor Pressure: 4 x10^{-8} mm Hg at 20-25 °C

Water Solubility: Sparingly Soluble in Water

Other Solubilities: Soluble in most organic solvents (e.g., Acetone, Methanol, heptane).

RTECS Toxicity Data

Acute Oral: Rat LD_{50} Dose: 1102 mg/kg; Toxic Effects: Liver - Other changes; Kidney, Ureter, and Bladder - Other changes. Mouse LD_{50} Dose: 49500 ug/kg.

Acute Inhalation: Rat LC_{50} Dose: 360 mg/m^3/4hr.

Acute Dermal: Rabbit LD_{50} Route: Skin; Dose: 9400 mg/kg.

Reproductive/Teratogenic: Mouse Route: Oral; Dose: 120 mg/kg; Duration: female 7-16D of pregnancy; Effects on Newborn - Growth statistics; Behavioral. Mouse Route: Oral; Dose: 250 mg/kg; Duration: female 7-16D of pregnancy; Effects on Fertility - Post-implantation mortality; Effects on Newborn - Viability index. Mouse Route: Oral; Dose: 50 mg/kg; Duration: female 7-16D of pregnancy; Effects on Embryo or Fetus - Fetotoxicity.

Mutagenic: Bacteria - S Typhimurium Mutations in Microorganisms; Dose: 500 ug/plate (+S9). Yeast - S Cerevisiae Mutations in Microorganisms; Dose: 5 ppm (-S9).

Hazard Overviews

Carcinogenicity: IARC - Not listed; NIOSH - Not listed; NTP - Not listed; ACGIH - Not listed; OSHA - Not listed; EPA - Not listed; MAK - Not listed

Environmental

Ecotoxicity: LC_{50} Coturnix japonica (Japanese quail), 14 days old, oral (5 day ad libitum in diet) 790 ppm (95% confidence intervals 662-934 ppm) LC_{50} Salmo gairdneri (rainbow trout) 15 ug/l at 13 °C (95% confidence limit 14-16 ug/l), wt 2 g.

Static bioassay without aeration, pH 7.2-7.5, water hardness 40-50 mg/l as calcium carbonate and alkalinity of 30-35 mg/l LC_{50} Gammarus fasciatus (Lake trout) 75 ug/l at 15 °C (95% confidence limit 57-99 ug/l), mature. Static bioassay without aeration, pH 7.2-7.5, water hardness 272 ppm calcium carbonate and alkalinity of 30-35 mg/l

Environmental Fate: If released in soil, it would tightly bind to soil. Its persistence in soil under normal use is reported to be 2-4 weeks. In one study, 32% mineralization occurred in 98 days. If released in water, it would bind tightly to sediment and particulate matter in the water column. It is not expected to volatilize. In alkaline water, it may undergo chemical hydrolysis resulting in estimated half-lives of 129 and 12.9 days at pH 8 and 9, respectively. It is stable in acidic media. It would be expected to bioconcentrate in aquatic organisms. If sprayed into the atmosphere, it would exist primarily as an aerosol and be removed by gravitational settling. Vapor-phase reacts with photochemically- produced hydroxyl radicals and ozone in the atmosphere, with a resulting half-life of about a day.

Cleanup/Disposal: Guide No. 131: Fully encapsulating, vapor protective clothing should be worn for spills and leaks with no fire. Eliminate all ignition sources (no smoking, flares, sparks or flames in immediate area). All equipment used when handling the product must be grounded. Do not touch or walk through spilled material. Stop leak if you can do it without risk. Prevent entry into waterways, sewers, basements or confined areas. A vapor suppressing foam may be used to reduce vapors. Small Spills: Absorb with earth, sand or other non-combustible material and transfer to containers for later disposal. Use clean non-sparking tools to collect absorbed material. Large Spills: Dike far ahead of liquid spill for later disposal. Water spray may reduce vapor; but may not prevent ignition in closed spaces. Guide No. 153: Eliminate all ignition sources (no smoking, flares, sparks or flames in immediate area). Do not touch damaged containers or spilled material unless wearing appropriate protective clothing. Stop leak if you can do it without risk. Prevent entry into waterways, sewers, basements or confined areas. Absorb or cover with dry earth, sand or other non-combustible material and transfer to containers. Do not get water inside containers.

Environmental Physical Data

Henry's Law Constant: calculated at 4.79 x10^{-9}
Octanol/Water Partition Coefficient: log K_{ow} = 34,4000
Sorption Partition Coefficient: K_{oc} = estimated at 4.4 x10^{4}
BCF: calculated at 1700

Regulations

RCRA 40CFR: Not listed
CERCLA: 40CFR 302.4: Not listed
SARA 40CFR 372.65: Listed
SARA EHS 40CFR 355: Not listed
TSCA: Not listed

Analytical Methods

Soil: EPA PMD-TLC; SW846 8270B, 8270C
Water / Groundwater: EPA 646
Food: FDA 212.1, 232.1, 232.4, 242.1
Plasma: EPA 001, 028, 29; FDA 211.1, 231.1, 252

DIN8220 CAS #: 151-32-6

DI-N-NONYL ADIPATE

EINECS Number: 205-789-7
Molecular Formula: $C_{24}H_{46}O_4$
Formula Weight: 288.26
Synonyms: ADIMOLL DN; ADIPIC ACID,DINONYL ESTER; BISOFLEX DNA; DINONYL ADIPATE; HEXANEDIOIC ACID,DINONYL ESTER; PLASTOMOLL NA
Description: colorless liquid
Use: low temp plasticizer for cellulose acetate butyrate, ethyl cellulose resins, polymethyl methacrylate resins, polystyrene resins, polyvinyl chloride resins, vinyl chloride-acetate resins

Physical Properties

Boiling Point: 201 °C (394 °F) to 210 °C (410 °F) at 1 mm Hg
Specific Gravity: 0.926 at 25 °C
Refraction Index: 1.4523 at 20 °C/D

Hazard Overviews

Carcinogenicity: IARC - Not listed; NIOSH - Not listed; NTP - Not listed; ACGIH - Not listed; OSHA - Not listed; EPA - Not listed; MAK - Not listed

Environmental

Regulations

RCRA 40CFR: Not listed
CERCLA: 40CFR 302.4: Not listed
SARA 40CFR 372.65: Not listed
SARA EHS 40CFR 355: Not listed
TSCA: Listed

DIN8410 CAS #: 84-76-4

DINONYL PHTHALATE

RTECS: TI1800000
EINECS Number: 201-560-0
Molecular Formula: $C_{26}H_{42}O_4$
Structured MF: $C_6H_4(COOC_9H_{19})_2$
Formula Weight: 418.62

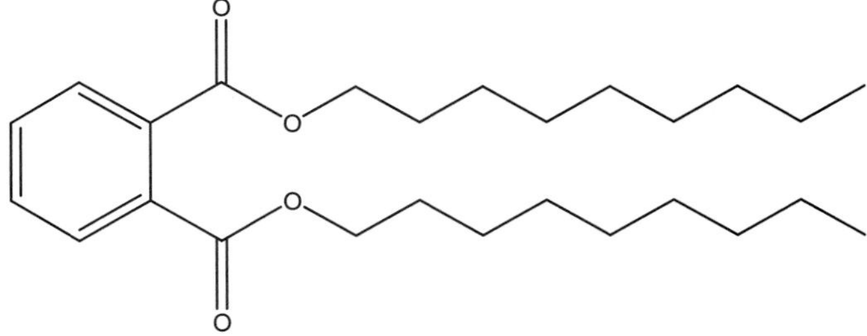

Chemical Structure

Synonyms: 1,2-BENZENEDICARBOXYLIC ACID,DINONYL ACID; 1,2-BENZENEDICARBOXYLIC ACID,DINONYL ESTER; BISOFLEX 91; BISOFLEX DNP; BISOLFLEX 91; DINONYL 1,2-BENZENEDICARBOXYLATE; DI-N-NONYL PHTHALATE; DI-N-NONYLPHTHALATE (DNNP); DITRIMETHYLHEXYL PHTHALATE; NONYL PHTHALATE; PHTHALIC ACID,DINONYL ESTER; UNIMOLL DN

Description: colorless liquid; odorless

Use: making vinyl mixes which have to withstand heat and resist migration and extraction by detergents; in the technology of plastisols and coating pastes; as stationary liquid phase in chromatography

Physical Properties

Boiling Point: 413 °C (775 °F)
Specific Gravity: 0.972 at 20 °C
Vapor Density: 14.44 Air=1
Vapor Pressure: 1 mm Hg at 205 °C
Water Solubility: Insoluble in Water
Other Solubilities: Soluble to various extents in many common organic solvents and oils
Refraction Index: 1.4871 at 20 °C/D
Flash Point: 216 °C Closed Cup

RTECS Toxicity Data

Acute Oral: Mouse LD_{50} Dose: 21500 mg/kg; Toxic Effects: Behavioral - Somnolence (general depressed activity); Behavioral - Withdrawal; Nutritional and gross metabolic - Weight loss or decreased weight gain. Guinea Pig LD_{50} Dose: 21500 mg/kg; Toxic Effects: Behavioral - Somnolence (general depressed activity); Behavioral - Withdrawal; Nutritional and gross metabolic - Weight loss or decreased weight gain.

Hazard Overviews

Fire: Will burn.
Carcinogenicity: IARC - Not listed; NIOSH - Not listed; NTP - Not listed; ACGIH - Not listed; OSHA - Not listed; EPA - Not listed; MAK - Not listed

Environmental

Environmental Fate: There are limited screening data indicating that if released to soil, it may undergo microbial degradation. It is not expected to volatilize from the soil surface to the atmosphere, nor is it expected to leach through soil. If released to water, it may undergo microbial degradation. It is not expected to bioaccumulate in fish and aquatic organisms, nor is it expected to volatilize from water to the atmosphere. If released to the atmosphere, it is expected to be adsorbed to particulates. Destruction by the vapor phase reaction with photochemically produced hydroxyl radicals may occur.

Environmental Physical Data

Henry's Law Constant: 1.41 $\times 10^{-5}$
Octanol/Water Partition Coefficient: log K_{ow} = > 2.12
BCF: none likely

Regulations

RCRA 40CFR: Not listed
CERCLA: 40CFR 302.4: Not listed
SARA 40CFR 372.65: Not listed
SARA EHS 40CFR 355: Not listed
TSCA: Listed

Analytical Methods

Plasma: EPA 001

DIN8600 CAS #: 137-99-5

2,4-DINONYLPHENOL

EINECS Number: 205-310-1
Molecular Formula: $C_{24}H_{42}O$
Formula Weight: 346.58
Synonyms: PHENOL,2,4-DINONYL-
Description: colorless liquid
Use: in prepn of lubricating oil additives, resins, plasticizers, surface active agents nonylphenol; solvent dinonyl phenol; in stabilizers, petroleum demulsifiers, fungicides, antioxidant for rubber nonylphenol; chem int for ethoxylated nonionic oil-soluble emulsifier; chem int for ethoxylated phosphated anionic surfactant; chem int for ethylene oxide adducts; comonomer for polymer with formaldehyde

Physical Properties

Water Solubility: Insoluble in Water
Other Solubilities: Soluble in organic solvents.

Hazard Overviews

Carcinogenicity: IARC - Not listed; NIOSH - Not listed; NTP - Not listed; ACGIH - Not listed; OSHA - Not listed; EPA - Not listed; MAK - Not listed

Environmental

Regulations

RCRA 40CFR: Not listed
CERCLA: 40CFR 302.4: Not listed
SARA 40CFR 372.65: Not listed
SARA EHS 40CFR 355: Not listed
TSCA: Listed

DIN8790 CAS #: 544-40-1

DINORMAL BUTYL SULFIDE

RTECS: ER6417000
EINECS Number: 208-870-5
Molecular Formula: $C_8H_{18}S$
Formula Weight: 146.32

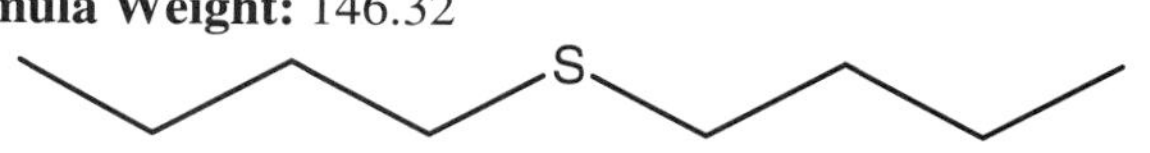

Chemical Structure

Synonyms: BUTYL MONOSULFIDE; N-BUTYL-SULFIDE; BUTYLTHIOBUTANE; N-DIBUTYL SULFIDE; DIBUTYL SULPHIDE; DIBUTYL THIOETHER; DI-N-BUTYLSULFIDE; 5-THIANONANE; THIANONANE-5

Physical Properties

Boiling Point: 189 °C (372.2 °F)
Freezing Point: -80 °C (-112 °F)
Specific Gravity: 0.83860
Vapor Pressure: 15.3 kPa at 125 °C
Water Solubility: Insoluble
Other Solubilities: Ether: Very Soluble; Ethanol: Very Soluble; Chloroform: Very Soluble
Refraction Index: 1.4530
Ionization Potential (eV): 8.2

RTECS Toxicity Data

Acute Oral: Rat LD_{50} Dose: 2220 mg/kg.
Acute Dermal: Rabbit LD_{50} Route: Skin; Dose: >5 gm/kg.
Irritation Skin: Rabbit Standard Draize Test Dose: 500 mg/24H; Reaction: moderate.

Hazard Overviews

Health: Irritating to eyes/skin/respiratory tract. Other Acute Effects: may be harmful by inhalation, ingestion, or skin absorption; exposure can cause nausea; headache; vomiting.
Fire: Hazards: emits toxic fumes. Extinguishing agents: water spray; carbon dioxide, dry chemical powder or appropriate foam. Precautions: combustible liquid.
Reactivity: Incompatible with: strong bases, strong oxidizing agents. Hazardous decomposition products: toxic fumes of: carbon monoxide, carbon dioxide, sulfur oxides, hydrogen sulfide gas.
Carcinogenicity: IARC - Not listed; NIOSH - Not listed; NTP - Not listed; ACGIH - Not listed; OSHA - Not listed; EPA - Not listed; MAK - Not listed
Primary Target Organs:

Eyes

Skin

Respiratory System

Environmental

Ecotoxicity: Fishes: Pimephales promelas 48h LC_{50} 3.5 mg/l

Environmental Physical Data

Octanol/Water Partition Coefficient: log K_{ow} = 4.02

Regulations

RCRA 40CFR: Not listed
CERCLA: 40CFR 302.4: Not listed
SARA 40CFR 372.65: Not listed
SARA EHS 40CFR 355: Not listed
TSCA: Listed

DIN8980 CAS #: 1420-07-1

DINOTERB

RTECS: SK0160000
DOT: UN1599; IMO6.1
EINECS Number: 215-813-8
Molecular Formula: $C_{10}H_{12}N_2O_5$
Formula Weight: 240.24

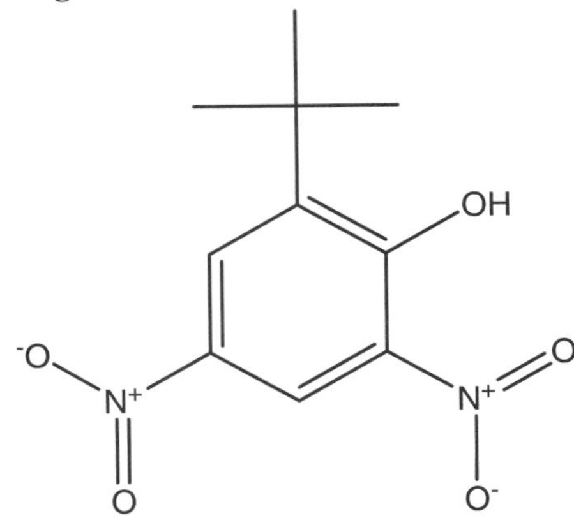

Chemical Structure

Synonyms: 2-TERT-BUTYL-4,6-DINITROPHENOL; 2-(1,1-DIMETHYLETHYL)-4,6-DINITROPHENOL; 2,4-DINITRO-6-TERT-BUTYLPHENOL; DINOTERBE; DNTBP; HERBOGIL; PHENOL,2-TERT-BUTYL-4,6-DINITRO-; PHENOL,O-T-BUTYL-4,6-DINITRO-; PHENOL,2-(1,1-DIMETHYLETHYL)-4,6-DINITRO-
Description: yellow solid or crystals; phenol-like odor
Use: contact herbicide for post-emergence control of annual weeds in cereals, lucerne; for pre-emergence control of annual weeds in beans, peas, & potatoes; pre-harvest desiccant for potatoes & leguminous seed crops; to control varieties of insects & acari in fruit trees & grapevines; maize; dinoterb acetate was better tolerated by plants than dinoterb; 2,4-dinitro-6-tert-butylphenol, rodenticide effective against mice & rats.

Physical Properties

Freezing Point: 125.5 °C (257.9 °F) to 126.5 °C (259.7 °F)
Vapor Pressure: 1.5 $x10^{-6}$ mbar at 20 °C
Water Solubility: Practically Insoluble in Water
Other Solubilities: Soluble in aqueous alkalis

RTECS Toxicity Data

Acute Oral: Rat LD_{50} Dose: 26 mg/kg. Mouse LD_{50} Dose: 19500 ug/kg.
Acute Dermal: Guinea Pig LD_{50} Route: Skin; Dose: 150 mg/kg.

Hazard Overviews

Carcinogenicity: IARC - Not listed; NIOSH - Not listed; NTP - Not listed; ACGIH - Not listed; OSHA - Not listed; EPA - Not listed; MAK - Not listed

Environmental

Cleanup/Disposal: Guide No. 153: Eliminate all ignition sources (no smoking, flares, sparks or flames in immediate area). Do not touch damaged containers or spilled material unless wearing appropriate protective clothing. Stop leak if you can do it without risk. Prevent entry into waterways, sewers, basements or confined areas. Absorb or cover with dry earth, sand or other non-combustible material and transfer to containers. Do not get water inside containers.

Environmental Physical Data

Octanol/Water Partition Coefficient: log K_{ow} = 3.51
BCF: none likely

Regulations

RCRA 40CFR: Not listed
CERCLA: 40CFR 302.4: Not listed
SARA 40CFR 372.65: Not listed TPQ: 500/10000 lb
SARA EHS 40CFR 355: Listed TPQ: 500 lb
TSCA: Not listed

DIO1000 CAS #: 123-79-5

DI-N-OCTYL ADIPATE

EINECS Number: 204-652-9
Molecular Formula: $C_{22}H_{42}O_4$
Formula Weight: 370.00
Synonyms: ADIMOLL DO; ADIPIC ACID,DIOCTYL ESTER; BIS-(2-ETHYLHEXYL)HEXANEDIOATE; DICAPRYLYL ADIPATE; DIOCTYL ADIPATE; HEXANEDIOIC ACID,DIOCTYL ESTER; OCTYL ADIPATE
Description: colorless or very pale amber liquid; slight, aromatic odor
Use: primary type of plasticizer for use with resins at low temp; plasticizer for synthetic rubbers, nitrocellulose, & ethyl cellulose

Physical Properties

Boiling Point: 214 °C (417 °F) at 5 mm Hg
Freezing Point: -70 °C (-94 °F)
Specific Gravity: 0.924 to 0.93 at 20 °C/20 °C
Water Solubility: Insoluble in Water
Other Solubilities: Insoluble or Very Slightly Soluble in Glycerine & glycols; Soluble in most organic solvents
Flash Point: 206 °C Open Cup
Autoignition Temperature: 377 °C
LEL: 0.4% v/v

Hazard Overviews

Fire: Will burn.
Carcinogenicity: IARC - Not listed; NIOSH - Not listed; NTP - Not listed; ACGIH - Not listed; OSHA - Not listed; EPA - Not listed; MAK - Not listed

Environmental

Environmental Fate: If released to air, it will degrade relatively rapidly by reaction with photochemically produced hydroxyl radicals (estimated half-life of 17 hr). If released to soil or water, it is expected to biodegrade. An estimated K_{oc} value of 5000 suggests that it will be relatively immobile in soil and may partition from the water column to sediment in the aquatic environment.

Environmental Physical Data

Henry's Law Constant: estimated at 4.34 $x10^{-7}$
Octanol/Water Partition Coefficient: log K_{ow} = 6.11
Sorption Partition Coefficient: K_{oc} = estimated at 5000
BCF: carp 27

Regulations

RCRA 40CFR: Not listed
CERCLA: 40CFR 302.4: Not listed
SARA 40CFR 372.65: Not listed
SARA EHS 40CFR 355: Not listed
TSCA: Listed

DIO1670 CAS #: 19102-74-0

DIOCTYL PEROXIDE

Molecular Formula: $C_{16}H_{34}O_2$
Formula Weight: 258.50
Synonyms: CAPRYLYL PEROXIDE; OCTANE,1,1'-DIOXYBIS-; OCTYL PEROXIDE

Hazard Overviews

Carcinogenicity: IARC - Not listed; NIOSH - Not listed; NTP - Not listed; ACGIH - Not listed; OSHA - Not listed; EPA - Not listed; MAK - Not listed

Environmental

Regulations

RCRA 40CFR: Not listed
CERCLA: 40CFR 302.4: Not listed
SARA 40CFR 372.65: Not listed
SARA EHS 40CFR 355: Not listed
TSCA: Not listed

DIO2340 CAS #: 1639-66-3

DI-N-OCTYL SODIUM SULFOSUCCINATE

RTECS: WN0620000
EINECS Number: 216-684-0
Molecular Formula: $C_{20}H_{38}NaO_7S$
Formula Weight: 445.20
Synonyms: BU-CERUMEN; BUTANEDIOIC ACID,SULFO-,1,4-DIOCTYL ESTER,SODIUM SALT; BUTANEDIOIC ACID,SULFO-,1,4-DIOCTYL ESTER,SODIUM SALT(9CI); DIOKTYLESTER SULFOJANTARANU SODNEHO; ELFANOL 883; MONAWET MO-70; MONAWET MO-84 R2W; MONAWET MO-70 RP; SODIUM DI-N-OCTYL SULFOSUCCINATE; SOLBALEITE; SUCCINIC ACID,SULFO-,1,4-DIOCTYL ESTER,SODIUM SALT; SUCCINIC ACID,SULFO-,DIOCTYL ESTER,SODIUM SALT; SULFOSUCCINIC ACID 1,4-DIOCTYL ESTER SODIUM SALT

Use: stool softener

RTECS Toxicity Data

Acute Oral: Rat LD_{50} Dose: 1900 mg/kg. Mouse LD_{50} Dose: 4800 mg/kg.

Hazard Overviews

Carcinogenicity: IARC - Not listed; NIOSH - Not listed; NTP - Not listed; ACGIH - Not listed; OSHA - Not listed; EPA - Not listed; MAK - Not listed

Environmental

Regulations

RCRA 40CFR: Not listed
CERCLA: 40CFR 302.4: Not listed
SARA 40CFR 372.65: Not listed
SARA EHS 40CFR 355: Not listed
TSCA: Listed

DIO3010 CAS #: 2373-23-1

DIOCTYL SULFOSUCCINATE

RTECS: WN0600000
Molecular Formula: $C_{20}H_{38}O_7S$
Formula Weight: 422.58
Synonyms: BUTANEDIOIC ACID,SULFO-,1,4-DIOCTYL ESTER; BUTANEDIOIC ACID,SULFO-,1,4-DIOCTYL ESTER (9CI); 1,4-DIOCTYL SULFOBUTANEDIOATE; EMPIMIN OT; SUCCINIC ACID,SULFO-,1,4-DIOCTYL ESTER
Use: chemical intermediate; surface acting agent used to prolong pyrethroid insecticide activity; in sludge removal

Physical Properties

Acute Oral: Rat LD_{50} Dose: 1900 mg/kg.

Hazard Overviews

Carcinogenicity: IARC - Not listed; NIOSH - Not listed; NTP - Not listed; ACGIH - Not listed; OSHA - Not listed; EPA - Not listed; MAK - Not listed

Environmental

Regulations

RCRA 40CFR: Not listed
CERCLA: 40CFR 302.4: Not listed
SARA 40CFR 372.65: Not listed
SARA EHS 40CFR 355: Not listed
TSCA: Not listed

DIO3680 CAS #: 101-67-7

4,4'-DIOCTYLDIPHENYLAMINE

EINECS Number: 202-965-5
Molecular Formula: $C_{28}H_{43}N$
Structured MF: $CH_3(CH_2)_7C_6H_4NHC_6H_4(CH_2)_7CH_3$
Formula Weight: 393.72
Synonyms: BENZENAMINE,4-OCTYL-N-(4-OCTYLPHENYL)-; BIS(P-OCTYLPHENYL)AMINE; DI-N-OCTYL DIPHENYLAMINE; P,P'-DIOCTYLDIPHENYLAMINE; P,P-DIOCTYLDIPHENYLAMINE; 4,4'-DIOCTYLPHENYLAMINE; DIPHENYLAMINE,4,4'-DIOCTYL-; 4-OCTYL-N-(4-OCTYLPHENYL)BENZENAMINE
Description: light tan powder; slight amine odor
Use: antioxidant for petroleum-based & synthetic lubricants & plastics; antioxidant for rubbers & plastics; antioxidant for petroleum-based & synthetic lubricants

Physical Properties

Freezing Point: 80 °C (176 °F) to 90 °C (194 °F)
Specific Gravity: 0.99
Vapor Density: 1.01 Air=1
Water Solubility: Insoluble in Water
Other Solubilities: 95% Ethanol: <1 mg/ml at 21 °C; Acetone: >=100 mg/ml at 21 °C; DMSO: <1 mg/ml at 21 °C.
Flash Point: 213 °C

Hazard Overviews

Fire: Will burn.
Carcinogenicity: IARC - Not listed; NIOSH - Not listed; NTP - Not listed; ACGIH - Not listed; OSHA - Not listed; EPA - Not listed; MAK - Not listed

Environmental

Regulations

RCRA 40CFR: Not listed
CERCLA: 40CFR 302.4: Not listed
SARA 40CFR 372.65: Not listed
SARA EHS 40CFR 355: Not listed
TSCA: Listed

DIO4350 CAS #: 103-96-8

DI-2-OCTYL-P-PHENYLENEDIAMINE

EINECS Number: 203-162-2
Molecular Formula: $C_{22}H_{40}N_2$
Formula Weight: 324.57

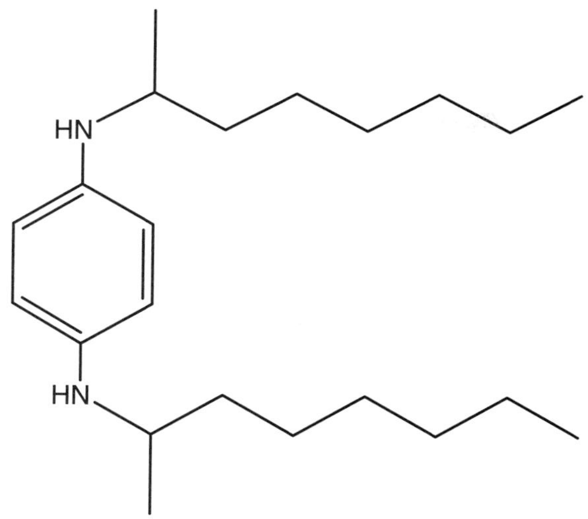

Chemical Structure

Synonyms: ANTOZITE 1; 1,4-BENZENEDIAMINE,N,N'-BIS(1-METHYLHEPTYL)-; N,N'-BIS(1-METHYLHEPTYL)-1,4-BENZENEDIAMINE; N,N'-BIS(1-METHYLHEPTYL)-P-PHENYLENEDIAMINE; N,N'-BIS(2-OCTYL)-P-PHENYLENEDIAMINE; N,N'-DI(1-METHYLHEPTYL)-P-PHENYLENEDIAMINE; N,N'-DI(2-OCTYL)-P-PHENYLENEDIAMINE; N,N'-DI(2-OCTYL)-PARA-PHENYLENEDIAMINE; ELASTOZONE 30; P-PHENYLENEDIAMINE,N,N'-BIS(1-METHYLHEPTYL)-; SANTOFLEX 217; TENEMENE 30; UOP 288
Use: rubber antioxidant; antiozonant for natural, isoprene, & butadiene rubbers; antiozonant for styrene-butadiene rubber

Physical Properties

Flash Point: 215.556 °C Open Cup

Hazard Overviews

Fire: Will burn.
Carcinogenicity: IARC - Not listed; NIOSH - Not listed; NTP - Not listed; ACGIH - Not listed; OSHA - Not listed; EPA - Not listed; MAK - Not listed

Environmental

Regulations

RCRA 40CFR: Not listed
CERCLA: 40CFR 302.4: Not listed
SARA 40CFR 372.65: Not listed
SARA EHS 40CFR 355: Not listed
TSCA: Listed

DIO5020 **CAS #: 25088-57-7**

DIOLEYL HYDROGEN PHOSPHITE

EINECS Number: 246-608-1
Molecular Formula: $C_{36}H_{71}O_3P$
Formula Weight: 582.94
Synonyms: DIOLEYL PHOSPHITE; DIOLEYL PHOSPHONATE; PHOSPHONIC ACID,DI-9-OCTADECENYL ESTER (Z,Z)-
Use: antioxidant

Hazard Overviews

Carcinogenicity: IARC - Not listed; NIOSH - Not listed; NTP - Not listed; ACGIH - Not listed; OSHA - Not listed; EPA - Not listed; MAK - Not listed

Environmental

Regulations

RCRA 40CFR: Not listed
CERCLA: 40CFR 302.4: Not listed
SARA 40CFR 372.65: Not listed
SARA EHS 40CFR 355: Not listed
TSCA: Not listed

DIO5690 **CAS #: 3329-91-7**

DIOSCORINE

RTECS: JG5787500
Molecular Formula: $C_{13}H_{19}NO_2$
Formula Weight: 221.29
Synonyms: SPIRO(2-AZABICYCLO(2.2.2)OCTANE-5,2'-(2H)PYRAN)-6'(3'H)-ONE,2,4'-DIMETHYL-,(1R-(1ALPHA,4ALPHA,5ALPHA))-
Description: greenish-yellow prisms

Physical Properties

Freezing Point: 54 °C (129.2 °F) to 55 °C (131 °F)
Water Solubility: Soluble in Water
Other Solubilities: Soluble in Alcohol, Acetone, Chloroform; Slightly Soluble in Ether, Benzene, Petroleum Ether

Hazard Overviews

Carcinogenicity: IARC - Not listed; NIOSH - Not listed; NTP - Not listed; ACGIH - Not listed; OSHA - Not listed; EPA - Not listed; MAK - Not listed

Environmental

Regulations

RCRA 40CFR: Not listed
CERCLA: 40CFR 302.4: Not listed
SARA 40CFR 372.65: Not listed
SARA EHS 40CFR 355: Not listed
TSCA: Not listed

DIO6360 **CAS #: 505-22-6**

1,3-DIOXANE

RTECS: JG8224000
DOT: UN1165; IMO3.2
EINECS Number: 208-005-1
Molecular Formula: $C_4H_8O_2$
Formula Weight: 88.11

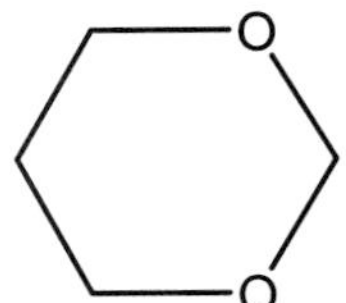

Chemical Structure

Synonyms: 1,3-DIOXACYCLOHEXANE; M-DIOXAN; M-DIOXANE; META-DIOXANE; M-DIOXIN,DIHYDRO-; 1,3-PROPANEDIOL FORMAL; TRIMETHYLENE GLYCOL METHYLENE ETHER; TRIMETHYLENE METHYLENE DIOXIDE
Description: colorless, clear liquid; alcohol like odor
Use: monomer for acetyl resins (no evidence of current use)

Physical Properties

Boiling Point: 105 °C (221 °F) at 755 mm Hg
Freezing Point: -42 °C (-43.6 °F)
Specific Gravity: 1.0342 at 20 °C/4 °C
Density: 1.032 g/mL
Vapor Pressure: 1.16 kPa at 0 °C
Water Solubility: Soluble in Water
Other Solubilities: 95% Ethanol: >=100 mg/ml at 20 °C; Acetone: >=100 mg/ml at 20 °C; Alcohol: >10%; Benzene: Soluble; DMSO: >=100 mg/ml at 20 °C; Ether: >10%.
Refraction Index: 1.4165 at 20 °C/D
Ionization Potential (eV): 9.8 +/-0.2
Flash Point: 2 °C
LEL: 2% v/v
UEL: 22% v/v

RTECS Toxicity Data

Mutagenic: Hamster Cytogenetic Analysis; Cell Type: ovary; Dose: 1300 mg/L. Hamster Sister Chromatid Exchenge; Cell Type: ovary; Dose: 2080 mg/L.

Hazard Overviews

Flammable

Health: Irritating to eyes/skin/respiratory tract. Other Acute Effects: may be harmful by inhalation, ingestion, or skin absorption.
Fire: Flammable. Hazards: vapor may travel considerable distance to source of ignition and flash back; emits toxic fumes. Extinguishing agents: carbon dioxide, dry chemical powder or appropriate foam; water spray. Precautions: combustible liquid.
Reactivity: Incompatible with: strong oxidizing agents. Hazardous decomposition products: toxic fumes of: carbon monoxide, carbon dioxide.
Carcinogenicity: IARC - Not listed; NIOSH - Not listed; NTP - Not listed; ACGIH - Not listed; OSHA - Not listed; EPA - Not listed; MAK - Not listed

Primary Target Organs:

Eyes

Skin

Respiratory System

Environmental

Environmental Fate: Limited prediction of environmental fate is based entirely on physical properties, chemical structure and analogy to similar compounds since experimental data are not available. If released to the atmosphere, it is expected to exist in the gas phase where it will be degraded relatively rapidly (estimated half-life of 2 days) by reaction with photochemically formed hydroxyl radicals. If released to soil, leaching may be possible since it is miscible in water. If released to water, volatilization is expected to be slow based on estimated half-lives of 7.2 days, 77.5 days, and 35.5 months from a shallow model river, a model environmental pond, and Lake Zurich, respectively. Aquatic hydrolysis, bioconcentration, and adsorption to sediment are not expected to be important. No data are available pertaining to biodegradation in the environment
Cleanup/Disposal: Guide No. 127: Eliminate all ignition sources (no smoking, flares, sparks or flames in immediate area). All equipment used when handling the product must be grounded. Do not touch or walk through spilled material. Stop leak if you can do it without risk. Prevent entry into waterways, sewers, basements or confined areas. A vapor suppressing foam may be used to reduce vapors. Absorb or cover with dry earth, sand or other non-combustible material and transfer to containers. Use clean non-sparking tools to collect absorbed material. Large Spills: Dike far ahead of liquid spill for later disposal. Water spray may reduce vapor; but may not prevent ignition in closed spaces.

Environmental Physical Data

Henry's Law Constant: 4.88 $\times 10^{-6}$
Octanol/Water Partition Coefficient: log K_{ow} = -0.419
BCF: estimated at 0.3

Regulations

RCRA 40CFR: Not listed
CERCLA: 40CFR 302.4: Not listed
SARA 40CFR 372.65: Not listed
SARA EHS 40CFR 355: Not listed
TSCA: Not listed

DIO7030 CAS #: 123-91-1

1,4-DIOXANE

RTECS: JG8225000
DOT: UN1165; IMO3.2
EINECS Number: 204-661-8
Molecular Formula: $C_4H_8O_2$
Formula Weight: 88.10

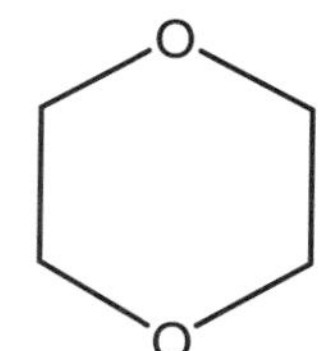

Chemical Structure

Synonyms: 1,4-DIETHYLENE DIOXIDE; DIETHYLENE DIOXIDE; DIETHYLENE ETHER; DI(ETHYLENE OXIDE); DIETHYLENE OXIDE; 1,4-DIETHYLENEDIOXIDE; DIOKAN; DIOKSAN; DIOSSANO-1,4; DIOXAAN-1,4; 1,4-DIOXACYCLOHEXANE; 1,4-DIOXAN; DIOXAN; DIOXAN-1,4; P-DIOXAN; PARA-DIOXAN; DIOXANE; DIOXANE-1,4; P-DIOXANE; PARA-DIOXANE; DIOXANNE; 1,4-DIOXAN,TETRAHYDRO-; P-DIOXIN,TETRAHYDRO-; DIOXYETHYLENE ETHER; ETHYLENE GLYCOL ETHYLENE ETHER; GLYCOL ETHYLENE ETHER; GLYCOLETHYLENETHER; TETRAHYDRO-1,4-DIOXIN; TETRAHYDRO-P-DIOXIN; TETRAHYDRO-PARA-DIOXIN

Description: colorless liquid; ethereal odor

Use: solvent for cellulose acetate, ethyl cellulose, benzyl cellulose, resins, oils, waxes, dyes, many organic as well as some inorganic compounds, lacquers, paints, varnishes, paint and varnish removers, cleaning and detergent preparations, cements, cosmetics, deodorants fumigants, fats, greases, mineral oil, polyvinyl polymers, plastics, adhesive sealants, pharmaceuticals, rubber chemicals, surface coatings and electrical, agricultural and biochemical intermediates; in the preparation of histological slides, as a wetting and dispersing agent in textile processing, in dye baths, in stain and printing compositions, in emulsions, in polishing compounds, as a stabilizer for chlorinated solvents and as a scintillation counter

Physical Properties

Boiling Point: 101.1 °C (214 °F) at 760 mm Hg
Freezing Point: 11.8 °C (53.24 °F)
Specific Gravity: 1.0337 at 20 °C/4 °C
Vapor Density: 3.03 Air=1
Saturated Vapor Density: 1.293315789 kg/m^3
Density: 1.03 g/mL
Vapor Pressure: 29 mm Hg at 20 °C
Water Solubility: Miscible with Water
Other Solubilities: miscible with aromatic hydrocarbons and oils.
Surface Tension: 36.9 dynes/cm at 25 °C
Odor Threshold: 620 mg/m^3
Refraction Index: 1.4175 at 20 °C/D
Evaporation Rate: 2.7 Butyl Acetate=1
Critical Temperature: 312 °C
Critical Pressure: 50.7 atm
Ionization Potential (eV): 9.13
Flash Point: 13 °C
Autoignition Temperature: 180 °C
LEL: 2.0% v/v
UEL: 22% v/v

RTECS Toxicity Data

Acute Oral: Rat LD_{50} Dose: 7120 mg/kg. Mouse LD_{50} Dose: 5300 mg/kg.

Acute Inhalation: Rat LC_{50} Dose: 46 gm/m^3/2hr; Toxic Effects: Sense organs and special senses - Other. Human LC_{Lo} Dose: 470 ppm/3D; Toxic Effects: Brain and coverings - Other degenerative changes; Lungs, Thorax, or Respiration - Other changes; Liver - Other changes. Human TC_{Lo} Dose: 470 ppm; Toxic Effects: Behavioral - Convulsions or effect on seizure threshold; Vascular - BP elevation not characterized in autonomic section; Gastrointestinal - Other changes. Human TC_{Lo} Dose: 5500 ppm/1M; Toxic Effects: Sense organs and special senses - Lacrimation; Sense organs and special senses - Conjunctive irritation; Lungs, Thorax, or Respiration - Other changes.

Acute Dermal: Rabbit LD_{50} Route: Skin; Dose: 7600 uL/kg.

Chronic (Multiple Dose) Inhalation: Rat Dose: 6000 ppm/4H/2W-I; Toxic Effects: Behavioral - Change in psychophysiological tests. Rat Dose: 20500 ug/m^3/13W-I; Toxic Effects: Kidney, Ureter, and Bladder - Proteinuria; Nutritional and gross metabolic - Changes in chlorine; Biochemical - Transaminases.

Irritation Eye: Rabbit Standard Draize Test Dose: 100 mg/24H; Reaction: moderate. Rabbit Standard Draize Test Dose: 100 mg; Reaction: severe.

Irritation Skin: Rabbit Open Draize Test Dose: 515 mg open; Reaction: mild.

Reproductive/Teratogenic: Rat Route: Oral; Dose: 10 gm/kg; Duration: female 6-15D of pregnancy; Effects on Embryo or Fetus - Fetotoxicity; Specific Developmental Abnormalities - Musculoskeletal system.

Mutagenic: Human DNA Inhibition; Cell Type: HeLa cell; Dose: 400 mmol/L. Rat DNA Damage; Route: Oral; Dose: 2550 mg/kg. Rat DNA Damage; Cell Type: liver; Dose: 300 umol/L.

Tumorigenic: Rat Route: Oral; Dose: 185 gm/kg/2Y-C; Toxic Effects: Tumorigenic - Carcinogenic by RTECS criteria; Sense organs and special senses - Tumors; Liver - Tumors. Rat Route: Inhalation; Dose: 111 ppm/7H/2Y-I; Toxic Effects: Tumorigenic - Equivocal tumorigenic agent by RTECS criteria; Blood - Tumors; Blood - Lymphomax including Hodgkin's disease. Rat Route: Oral; Dose: 416 gm/kg/57W-C; Toxic Effects: Tumorigenic - Equivocal tumorigenic agent by RTECS criteria; Sense organs and special senses - Tumors; Liver - Tumors. Rat Route: Oral; Dose: 408 gm/kg/2Y-C; Toxic Effects: Tumorigenic - Carcinogenic by RTECS criteria; Sense organs and special senses - Tumors; Liver - Tumors. Rat Route: Oral; Dose: 416 gm/kg/57W-C; Toxic Effects: Tumorigenic - Carcinogenic by RTECS criteria; Sense organs and special senses - Tumors; Liver - Tumors. Rat Route: Oral; Dose: 528 gm/kg/63W-I; Toxic Effects: Tumorigenic - Equivocal tumorigenic agent by RTECS criteria; Liver - Tumors; Kidney, Ureter, and Bladder - Kidney tumors.

Hazard Overviews

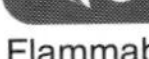
Flammable

Fire Diamond

Health: Irritating to the eyes/skin/respiratory tract. Also Causes: stomach pain, anorexia, vomiting, kidney and liver damage, and edema of the lungs and brain. Chronic Effects: possible cancer hazard.

Fire: Flammable. Can form explosive mixtures in the air. Use water as fog, dry chemical, carbon dioxide, or alcohol-resistant foam. Dioxane slowly reacts with air to form explosive peroxides. Use water spray to cool fire-exposed containers.

Reactivity: Stable in closed container under nitrogen gas blanket. Hazardous polymerization can occur, dry material reacts slowly with atmospheric oxygen to form explosive peroxides. Avoid: heat; ignition sources; always test for explosive peroxides. Incompatible with: silver perchlorate; hydrogen and raney nickel; oxidizing agents. Hazardous decomposition products: toxic gases.

Carcinogenicity: IARC - Group 2B, Possibly carcinogenic to humans; NIOSH - Listed as carcinogen; NTP - Class 2B, Reasonably anticipated to be a carcinogen, sufficient evidence of carcinogenicity from studies in experimental animals; ACGIH - Not listed; OSHA - Not listed; EPA - Class B2, Probable human carcinogen based on animal studies; MAK - Class B, Justifiably suspected of having carcinogenic potential

Primary Target Organs:

Eyes

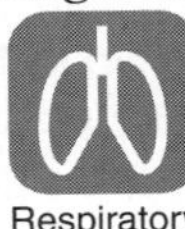
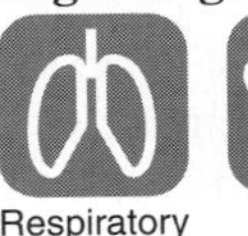
Respiratory System

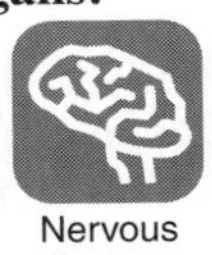
Nervous System

Liver

Kidneys

Exposure Limits

OSHA PEL: TWA: 100 ppm; 360 mg/m^3.

OSHA PEL Vacated 1989 Limits: TWA: 25 ppm; 90 mg/m^3.

ACGIH TLV: TWA: 25 ppm; 90 mg/m^3.

NIOSH REL: STEL: 1 ppm; 3.6 mg/m^3; 30-minute.

NIOSH IDLH: 500 ppm.

DFG MAK: TWA: 50 ppm; 180 mg/m^3.

Respirator Recommendation

Exposure Range: >100 to <500 ppm Supplied Air, Constant Flow/Pressure Demand, Half Mask

Exposure Range: 500 to unlimited ppm Self-contained Breathing Apparatus, Pressure Demand, Full Face

Note: poor warning properties

Environmental

Ecotoxicity: LC_{50} Lepomis macrochirus 10,000 ppm/96 hr in a static bioassay in fresh water at 23 °C, mild aeration applied after 24 hr. LC_{50} Menidia beryllina 6,700 ppm/96 hr in a static bioassay in synthetic seawater at 23 °C, mild aeration applied after 24 hr

Environmental Fate: When released to water, it is not expected to hydrolyze and may volatilize, although its infinite water solubility precludes estimating the volatilization half-life. Based on its infinite water solubility and low estimated soil sorption partition coefficient, when released to soil it is expected to leach to groundwater. It is not expected to bioconcentrate in fish or biodegrade in soil or water. Material which enters the atmosphere is expected to degrade fairly quickly. After 3.4 hr, 50% mixed with NO and subjected to environmental UV radiation had degraded. A half-life of 6.69 hr was estimated for the reaction with atmospheric hydroxyl radicals. The expected products of this reaction are aldehydes and ketones.

Cleanup/Disposal: Guide No. 127: Eliminate all ignition sources (no smoking, flares, sparks or flames in immediate area). All equipment used when handling the product must be grounded. Do not touch or walk through spilled material. Stop leak if you can do it without risk. Prevent entry into waterways, sewers, basements or confined areas. A vapor suppressing foam may be used to reduce vapors. Absorb or cover with dry earth, sand or other non-combustible material and transfer to containers. Use clean non-sparking tools to collect absorbed material. Large Spills: Dike far ahead of liquid spill for later disposal. Water spray may reduce vapor; but may not prevent ignition in closed spaces.

Environmental Physical Data

Octanol/Water Partition Coefficient: log K_{ow} = -0.27

Sorption Partition Coefficient: log K_{oc} = estimated at 1.23

BCF: not significant

BOD: theoretical 0%, 10 days

Regulations

RCRA 40CFR: Listed Hazardous Waste No. U108 Toxic Waste

CERCLA: 40CFR 302.4: Listed per RCRA Section 3001 RQ: 100 lb (45.35 kg)

SARA 40CFR 372.65: Listed

SARA EHS 40CFR 355: Not listed

TSCA: Listed

Analytical Methods

Air: ASTM D3686, D3687, D4490

Soil: SW846 5031, 8015B, 8240B, 8260A, 8260B

Water / Groundwater: ASTM D3695

Indoor / Expired Air: NIOSH 1602; EPA IP-1B

Plasma: EPA 29

DIO7700 CAS #: 78-34-2

DIOXATHION

RTECS: TE3350000

EINECS Number: 201-107-7

Molecular Formula: $C_{12}H_{26}O_6P_2S_4$

Formula Weight: 456.54

Synonyms: AC 528; 2,3-BIS(DIETHOXYPHOSPHINOTHIOYLTHIO)-1,4-DIOXANE; BIS(DITHIOPHOSPHATE DE O,O-DIETHYLE) DE S,S'-(1,4-DIOXANNE-2,3-DIYLE); CO-NAV; DELNATEX; DELNAV; DELTIC; 1,4-DIOSSAN-2,3-DIYL-BIS(O,O-DIETIL-DITIOFOSFATO); 1,4-DIOXAAN-2,3-DIYL-BIS(O,O-DIETHYL-DITHIOFOSFAAT); 2,3-P-DIOXAN-S,S'-BIS(O,O-DIAETHYLDITHIOPHOSPHAT); 2,3-P-

DIOXANDITHIOL S,S-BIS(O,O-DIETHYLPHOSPHORODITHIOATE); 1,4-DIOXAN-2,3-DIYL BIS(O,O-DIETHYLPHOSPHOROTHIOLOTHIONATE); 1,4-DIOXAN-2,3-DIYL O,O,O',O'-TETRAETHYL DI(PHOSPHOROMITHIOATE); 1,4-DIOXAN-2,3-DIYL O,O,O',O',-TETRAETHYLDI(PHOSPHORODITHIOATE); 1,4-DIOXAN-2,3-DIYL O,O,O',O'-TETRAETHYLDI(PHOSPHOROMITHIOATE); 1,4-DIOXAN-2,3-DIYL-BIS(O,O-DIAETHYL-DITHIOPHOSPHAT); 2,3-P-DIOXANE S,S-BIS(O,O-DIETHYLPHOSPHORODITHIOATE); DIOXANE PHOSPHATE; 2,3-P-DIOXANEDITHIOL S,S-BIS(O,O-DIETHYLPHOSPHORODITHIOATE); P-DIOXANE-2,3-DITHIOL,S,S-DIESTER WITH O,O-DIETHYL PHOSHORODITHIOATE; P-DIOXANE-2,3-DITHIOL,S,S-DIESTER WITH O,O-DIETHYLPHOSPHORODITHIOATE; P-DIOXANE-2,3-DIYL ETHYL PHOSPHORODITHIOATE; 2,3-P-DIOXANETHIOL S,S-BIS(O,O-DIETHYL PHOSPHORO-DITHIOATE); DIOXATION; DIOXOTHION; ENT 22,897; ENT 22879; ENT 22897; HERCULES 528; HERCULES AC528; KAVADEL; NAVADEL; PHOSPHORODITHIOIC ACID S,S'-1,4-DIOXANE-2,3-DIYL O,O,O',O'-TETRAETHYL ESTER; PHOSPHORODITHIOIC ACID S,S'-P-DIOXANE-2,3-DIYL O,O,O',O'-TETRAETHYL ESTER; PHOSPHORODITHIOIC ACID,S,S'-1,4-DIOXANE-2,3-DIYL O,O,O',O'-TETRAETHYL ESTER (9CI); PHOSPHORODITHIOIC ACID,S,S'-1,4-DIOXANE-2,3-DIYLO,O,O',O'-TETRAETHYL ESTER; PHOSPHORODITHIOIC ACID,S,S'-P-DIOXANE-2,3-DIYLO,O,O',O'-TETRAETHYL ESTER; RUPHOS

Description: tan, brown liquid

Use: as an insecticide, miticide and acaricide

Physical Properties

Boiling Point: 60 °C (140 °F) to 68 °C (154 °F)
Freezing Point: -20 °C (-4 °F)
Specific Gravity: 1.257 at 26 °C/4 °C
Saturated Vapor Density: 1.614349111 kg/m^3
Density: 1.24 to 1.26 g/mL at 25 °C
Bulk Density: 1.257
Vapor Pressure: 17.8 mm Hg at 25 °C
Water Solubility: Practically Insoluble in Water
Other Solubilities: 10 g/kg Hexane, Kerosene; Soluble in aromatic hydrocarbons, Ethers, Esters, & Ketones; Soluble in Ethanol, Benzene, Acetone; Slightly Soluble in petroleum oils.
Refraction Index: 1.5420 at 20 °C/D
Flash Point: Nonflammable

RTECS Toxicity Data

Acute Oral: Human TD_{Lo} Dose: 9 mg/kg/60D; Toxic Effects: Biochemical - True cholinesterase. Rat LD_{50} Dose: 20 mg/kg.

Acute Inhalation: Rat LC_{50} Dose: 1398 mg/m^3/1hr; Toxic Effects: Sense organs and special senses - Lacrimation; Behavioral - Convulsions or effect on seizure threshold; Gastrointestinal - Hypermotility, diarrhea. Mouse LC_{50} Dose: 340 mg/m^3/1hr; Toxic Effects: Sense organs and special senses - Lacrimation; Behavioral - Convulsions or effect on seizure threshold; Gastrointestinal - Hypermotility, diarrhea.

Acute Dermal: Rabbit LD_{50} Route: Skin; Dose: 85 mg/kg. Rat LD_{50} Route: Skin; Dose: 63 mg/kg; Toxic Effects: Sense organs and special senses - Lacrimation; Behavioral - Convulsions or effect on seizure threshold; Gastrointestinal - Hypermotility, diarrhea. Rat LD_{50} Route: Subcutaneous Dose: 95 mg/kg.

Chronic (Multiple Dose) Oral: Rat Dose: 70980 ug/kg/13W-C; Toxic Effects: Biochemical - True cholinesterase. Dog Dose: 9600 ug/kg/12D-I; Toxic Effects: Biochemical - True cholinesterase.

Mutagenic: Bacteria - S Typhimurium Mutations in Microorganisms; Dose: 6667 ug/plate (+S9), 10 mg/plate (-S9).

Hazard Overviews

Poison

Fire Diamond

Health: Mildly irritating to eyes/skin. Poison. Also Causes: CNS effects, conjunctivitis, headache, nausea, sweating, abdominal cramps, blurred vision, confusion, drowsiness, bronchorrhea, bronchospasm, pulmonary edema, convulsions, death.

Fire: Noncombustible. Use agent suitable for surrounding fire. Hazardous combustion products may include carbon oxide(s), sulfur oxide(s), phosphorus oxide(s).

Reactivity: Stable. Hazardous polymerization cannot occur. Avoid: exposure to heat. Incompatible with: alkalis; tin. Hazardous decomposition products: carbon; phosphorus; and sulfur oxides.

Carcinogenicity: IARC - Not listed; NIOSH - Not listed; NTP - Not listed; ACGIH - Class A4, Not classifiable as a human carcinogen; OSHA - Not listed; EPA - Not listed; MAK - Not listed

Primary Target Organs:

Eyes

Skin

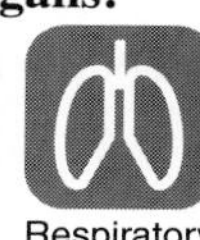
Respiratory System

Nervous System

Blood

Exposure Limits
OSHA PEL Vacated 1989 Limits: TWA: 0.2 mg/m^3.
ACGIH TLV: TWA: 0.2 mg/m^3.
NIOSH REL: TWA: 0.2 mg/m^3.

Respirator Recommendation
Exposure Range: >0.2 to 2 mg/m^3 Air Purifying, Negative Pressure, Half Mask
Exposure Range: >2 to 20 mg/m^3 Air Purifying, Negative Pressure, Full Face
Exposure Range: >20 to 200 mg/m^3 Supplied Air, Constant Flow/Pressure Demand, Full Face
Exposure Range: >200 to unlimited mg/m^3 Self-contained Breathing Apparatus, Pressure Demand, Full Face
Cartridge Color: black with magenta (P100)

Environmental

Ecotoxicity: EC_{50} Daphnia magna 0.35 ug/l/48 hr at 2 °C, first instar (95% confidence limit 0.25-0.49 ug/L). Static bioassay without aeration, pH 7.2-7.5, water hardness 40-50 mg/l as calcium carbonate and alkalinity of 30-35 mg/l LC_{50} Mallard oral approx 3600 ppm in 5-day LC_{50} Gammarus fasciatus 8.6 ug/l/96 hr at 15 °C, mature (95% confidence limit 5.4-13.8 ug/l). Static bioassay without aeration, pH 7.2-7.5, water hardness 40-50 mg/l as calcium carbonate and alkalinity of 30-35 mg/l LD_{50} Phasianus colchicus (pheasant) oral 240 mg/kg, 3-7 months old males (95% confidence limit 190-302 mg/kg) Aquatic Toxicity: TLm: Bluegill 14 ug/l/48

Environmental Fate: If released in soil, it will degrade with a half-life ranging from 15 to 55 days in different soil types with dissipation being more rapid in drier, sandy soils. If released into water, it would adsorb moderately to sediment and particulate matter in the water column. Moderate bioconcentration in aquatic organisms would also be anticipated. Otherwise its fate in water is unknown. It would probably be released into the atmosphere in the form of an aerosol and be subject to gravitational settling. Otherwise its fate in the atmosphere is unknown.

Cleanup/Disposal: Guide No. 152: Do not touch damaged containers or spilled material unless wearing appropriate protective clothing. Stop leak if you can do it without risk. Prevent entry into waterways, sewers, basements or confined areas. Cover with plastic sheet to prevent spreading. Absorb or cover with dry earth, sand or other non-combustible material and transfer to containers. Do not get water inside containers.

Environmental Physical Data

Octanol/Water Partition Coefficient: log K_{ow} = 2.99
Sorption Partition Coefficient: K_{oc} = 1000
BCF: calculated at 110

Regulations

RCRA 40CFR: Not listed
CERCLA: 40CFR 302.4: Not listed
SARA 40CFR 372.65: Not listed TPQ: 500 lb
SARA EHS 40CFR 355: Listed TPQ: 500 lb
TSCA: Not listed

Analytical Methods

Soil: EPA PMD-TLC; SW846 8141, 8141A, 8270C
Water / Groundwater: EPA P-005-1, 1657, 614.1, 022; SW846 8141, 8141A
Food: FDA 212.1, 232.1, 232.3, 232.4, 242.1
Plasma: EPA 001, 027, 028; FDA 211.1, 231.1, 252

DIO8370 CAS #: 646-06-0

1,3-DIOXOLANE

RTECS: JH6760000
DOT: UN1166; IMO3.2
EINECS Number: 211-463-5
Molecular Formula: $C_3H_6O_2$
Formula Weight: 74.09

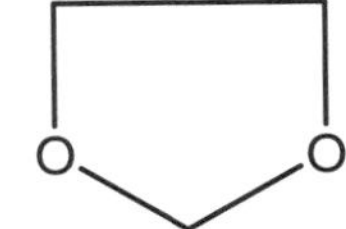

Chemical Structure

Synonyms: 1,3-DIOXACYCLOPENTANE; 1,3-DIOXOLAN; DIOXOLAN; DIOXOLANE; 1,3-DIOXOLE,DIHYDRO-; ETHYLENE GLYCOL FORMAL; FORMAL GLYCOL; GLYCOL FORMAL; GLYCOL METHYLENE ETHER; GLYCOLFORMAL

Description: water-white liquid

Use: low-boiling solvent and extractant for oils, fats, waxes, dyes, and cellulose derivatives; crosslinking agent for phenolic novolak resins

Physical Properties

Boiling Point: 78 °C (172 °F) at 765 mm Hg
Freezing Point: -95 °C (-139 °F)
Specific Gravity: 1.06 at 20 °C/4 °C
Vapor Density: 2.6 Air=1
Saturated Vapor Density: 1.371849365 kg/m^3
Bulk Density: 8.2 lbs/gal at 20 °C
Vapor Pressure: 70 mm Hg at 20 °C
Water Solubility: Soluble in Water
Other Solubilities: Soluble in Alcohol, Ether, Acetone
Refraction Index: 1.3974 at 20 °C/D
Ionization Potential (eV): 9.9 +/-0.1
Flash Point: 1.67 °C Open Cup

RTECS Toxicity Data

Acute Oral: Rat LD_{50} Dose: 3 gm/kg. Mouse LD_{50} Dose: 3200 mg/kg.

Acute Inhalation: Rat LC_{50} Dose: 20650 mg/m^3/4hr. Mouse LC_{50} Dose: 10500 mg/m^3/2hr.

Acute Dermal: Rabbit LD_{50} Route: Skin; Dose: 8480 uL/kg.

Chronic (Multiple Dose) Inhalation: Rat Dose: 1360 mg/m^3/4H/22W-I; Toxic Effects: Kidney, Ureter, and Bladder - Urine volume increased; Blood - Other changes. Rat Dose: 105000 mg/m^3/50D-I; Toxic Effects: Brain and coverings - Recordings from specific areas of CNS.

Irritation Eye: Rabbit Standard Draize Test Dose: 750 ug open; Reaction: severe.

Irritation Skin: Rabbit Open Draize Test Dose: 530 mg open; Reaction: mild.

Mutagenic: Rat DNA Damage; Route: Intraperitoneal; Dose: 290 mg/kg.

Hazard Overviews

Flammable

Health: Irritating to eyes/skin/respiratory tract. Harmful. Other Acute Effects: may be harmful by inhalation, ingestion, or skin absorption; target organ: blood.

Fire: Flammable. Hazards: vapor may travel considerable distance to source of ignition and flash back; under fire conditions, material may decompose to form flammable and/or explosive mixtures in air. Extinguishing agents: carbon dioxide, dry chemical powder or appropriate foam; do not use water. Precautions: combustible liquid.

Reactivity: Incompatible with: acids, strong oxidizing agents, protect from moisture. Hazardous decomposition products: toxic fumes of: carbon monoxide, carbon dioxide.

Carcinogenicity: IARC - Not listed; NIOSH - Not listed; NTP - Not listed; ACGIH - Not listed; OSHA - Not listed; EPA - Not listed; MAK - Not listed

Primary Target Organs:

Eyes

Skin

Respiratory System

Blood

Environmental

Environmental Fate: If released to soil, it is expected to leach (estimated K_{oc} of 15) and volatilize from dry soil surfaces. No data were located which suggest biodegradation is an important fate process in soil or water. If released to water, hydrolysis, aquatic oxidation with photochemically produced hydroxyl radicals, adsorption to sediment and bioconcentration in aquatic organisms are not expected to be environmentally important removal processes. Volatilization half-lives of 34 hours and 15 days have been estimated for a model river (one meter deep) and a model environmental pond, respectively. Its complete water solubility suggests that it may be susceptible to long distance transport in aquatic environments. If released to the atmosphere, it is expected to exist in the vapor phase. Vapor-phase is expected to degrade by reaction with photochemically produced hydroxyl radicals (estimated half-life of 1.1 days). Based on its complete water solubility, removal from air via wet deposition may occur.

Cleanup/Disposal: Guide No. 127: Eliminate all ignition sources (no smoking, flares, sparks or flames in immediate area). All equipment used when handling the product must be grounded. Do not touch or walk through spilled material. Stop leak if you can do it without risk. Prevent entry into waterways, sewers, basements or confined areas. A vapor suppressing foam may be used to reduce vapors. Absorb or cover with dry earth, sand or other non-combustible material and transfer to containers. Use clean non-sparking tools to collect absorbed material. Large Spills: Dike far ahead of liquid spill for later disposal. Water spray may reduce vapor; but may not prevent ignition in closed spaces.

Environmental Physical Data

Henry's Law Constant: 2.4×10^{-5}

Octanol/Water Partition Coefficient: log K_{ow} = -0.37

Sorption Partition Coefficient: K_{oc} = 15

BCF: estimated at 0.3

Regulations

RCRA 40CFR: Not listed

CERCLA: 40CFR 302.4: Not listed

SARA 40CFR 372.65: Not listed

SARA EHS 40CFR 355: Not listed

TSCA: Listed

DIO9040 CAS #: 96-49-1

1,3-DIOXOLAN-2-ONE

RTECS: FF9550000

EINECS Number: 202-510-0

Molecular Formula: $C_3H_4O_3$

Formula Weight: 88.06

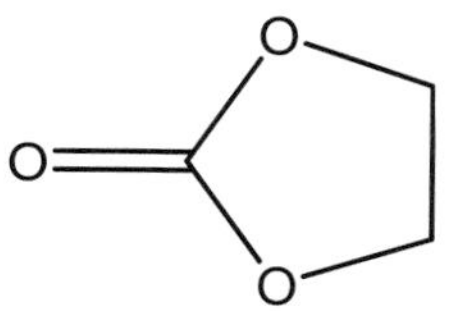

Chemical Structure

Synonyms: CYCLIC ETHYLENE CARBONATE; DIOXOLONE-2; ETHYLENE CARBONATE; ETHYLENE CARBONIC ACID; ETHYLENE GLYCOL CARBONATE; ETHYLENE GLYCOL,CYCLIC CARBONATE; ETHYLENESTER KYSELINY UHLICITE; GLYCOL CARBONATE

Description: monoclinic plates; colorless solid or liquid; odorless

Use: solvent for polymers & resins, plasticizer, intermediate for pharmaceuticals, rubber chemicals, textile finishing agents, hydroxyethylation reactions

Physical Properties

Boiling Point: 248 °C (478 °F)

Freezing Point: 39 °C (102.2 °F) to 40 °C (104 °F)

Specific Gravity: 1.3214 at 39 °C/4 °C

Saturated Vapor Density: 1.200072375 kg/m^3

Vapor Pressure: 0.003 kPa at 25 °C

Water Solubility: Soluble in Water

Other Solubilities: Soluble in n-butanol and Carbon Tetrachloride, Ethyl Acetate.

Refraction Index: 1.4148

Ionization Potential (eV): 10.40 +/-1.0

Flash Point: 143 °C Open Cup

RTECS Toxicity Data

Acute Oral: Rat LD_{50} Dose: 10 gm/kg.

Acute Dermal: Rabbit LD_{50} Route: Skin; Dose: >3 gm/kg.

Irritation Skin: Rabbit Open Draize Test Dose: 660 mg open; Reaction: mild.

Hazard Overviews

Health: Irritating to eyes; may cause skin irritation. Harmful. Other Acute Effects: may be harmful by inhalation, ingestion, or skin absorption.

Fire: Will burn. Extinguishing agents: water spray; carbon dioxide, dry chemical powder or appropriate foam. Precautions: combustible liquid.

Carcinogenicity: IARC - Not listed; NIOSH - Not listed; NTP - Not listed; ACGIH - Not listed; OSHA - Not listed; EPA - Not listed; MAK - Not listed

Primary Target Organs:

Eyes

Environmental

Regulations

RCRA 40CFR: Not listed
CERCLA: 40CFR 302.4: Not listed
SARA 40CFR 372.65: Not listed
SARA EHS 40CFR 355: Not listed
TSCA: Listed

DIP1000 CAS #: 126-58-9

DIPENTAERYTHRITOL

EINECS Number: 204-794-1
Molecular Formula: $C_{10}H_{22}O_7$
Formula Weight: 254.26

OH HO HO O OH OH HO

Chemical Structure

Synonyms: BIS(PENTAERYTHRITOL); DIPENTEK; 1,3-PROPANEDIOL,2,2'-(OXYBIS(METHYLENE))BIS(2-(HYDROXYMETHYL)-
Description: off-white, freeflowing crystalline powder
Use: in paints & coatings

Physical Properties

Freezing Point: 212 °C (413.6 °F) to 220 °C (428 °F)
Specific Gravity: 1.33 at 25 °C/4 °C
Water Solubility: Non-hygroscopic
Other Solubilities: Solubility at 25 °C in: Methanol 0.25 g/100 g Ethanol 0.07 g/100 g Acetone < 0.1 g 100 g Benzene < 0.1 g/100 g.

Hazard Overviews

Health: May be irritating to eyes/skin. Other Acute Effects: may be harmful by inhalation, ingestion, or skin absorption.
Fire: Hazards: emits toxic fumes. Extinguishing agents: water spray; carbon dioxide, dry chemical powder or appropriate foam. Precautions: combustible liquid.
Reactivity: Incompatible with: strong oxidizing agents. Hazardous decomposition products: toxic fumes of: carbon monoxide, carbon dioxide.
Carcinogenicity: IARC - Not listed; NIOSH - Not listed; NTP - Not listed; ACGIH - Not listed; OSHA - Not listed; EPA - Not listed; MAK - Not listed

Environmental

Regulations

RCRA 40CFR: Not listed
CERCLA: 40CFR 302.4: Not listed
SARA 40CFR 372.65: Not listed
SARA EHS 40CFR 355: Not listed
TSCA: Listed

DIP1180 CAS #: 79-74-3

2,5-DI-T-PENTYLHYDROQUINONE

RTECS: MX6300000
EINECS Number: 201-222-2
Molecular Formula: $C_{16}H_{26}O_2$
Formula Weight: 250.4

OH HO

Chemical Structure

Synonyms: 1,4-BENZENEDIOL,2,5-BIS(1,1-DIMETHYLPROPYL)-; 2,5-BIS(1,1-DIMETHYLPROPYL)-1,4-BENZENEDIOL; 2,5-BIS(1,1-DIMETHYLPROPYL)HYDROQUINONE; DAHQ; 2,5-DI-TERT-AMYLBENZENE-1,4-DIOL; 2,5-DI(TERT-AMYL)HYDROQUINONE; 2,5-DI-T-AMYLHYDROQUINONE; 2,5-DI-TERT-AMYLHYDROQUINONE; 2,5-DI-TERT-PENTYLBENZENE-1,4-DIOL; 2,5-DI(TERT-PENTYL)HYDROQUINONE; 2,5-DI-TERT-PENTYLHYDROQUINONE; HYDROQUINONE,2,5-DI-TERT-AMYL-; HYDROQUINONE,2,5-DI-T-PENTYL-; HYDROQUINONE,2,5-DI-TERT-PENTYL-; SANTOUAR A; SANTOVAR A
Description: crystals or buff powder
Use: antioxidant for uncured rubber & for unsaturated resins & oils; food packaging; polymerization inhibitor; as staining protector in rubber

Physical Properties

Freezing Point: 179.4 °C (354.92 °F) to 180.4 °C (356.72 °F)
Specific Gravity: 1.05 at 25 °C
Water Solubility: Slightly Soluble in Water
Other Solubilities: Soluble in Alcohol & Benzene

RTECS Toxicity Data

Acute Oral: Rat LD_{50} Dose: 2 gm/kg. Rabbit LD_{50} Dose: 2 gm/kg.

Hazard Overviews

Health: Irritating to eyes/skin/respiratory tract. Other Acute Effects: may be harmful by inhalation, ingestion, or skin absorption; depending on the intensity and duration of exposure, effects may vary from mild irritation to severe destruction of tissue; prolonged contact can cause: damage to the lungs, severe irritation or burns.

Fire: Hazards: emits toxic fumes; in powder form capable of creating a dust explosion. Extinguishing agents: water spray; carbon dioxide, dry chemical powder or appropriate foam. Precautions: combustible liquid.

Reactivity: Stable. Hazardous polymerization will not occur. Incompatible with: strong oxidizing agents, strong bases. Hazardous decomposition products: toxic fumes of: carbon monoxide, carbon dioxide.

Carcinogenicity: IARC - Not listed; NIOSH - Not listed; NTP - Not listed; ACGIH - Not listed; OSHA - Not listed; EPA - Not listed; MAK - Not listed

Primary Target Organs:

Eyes

Skin

Respiratory System

Environmental

Regulations

RCRA 40CFR: Not listed
CERCLA: 40CFR 302.4: Not listed
SARA 40CFR 372.65: Not listed
SARA EHS 40CFR 355: Not listed
TSCA: Listed

DIP1360 CAS #: 120-95-6

2,4-DI-T-PENTYLPHENOL

RTECS: SL3500000
EINECS Number: 204-439-0
Molecular Formula: $C_{16}H_{26}O$
Formula Weight: 234.39

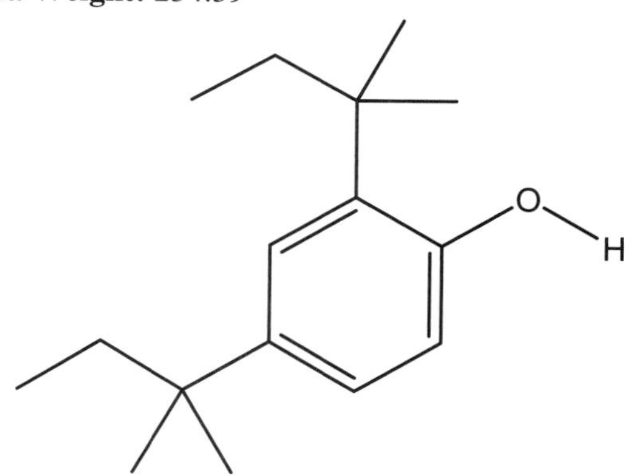

Chemical Structure

Synonyms: 2,4-DI-TERT-AMYLPHENOL; DI-TERT-AMYLPHENOL; 2,4-DI-TERT-PENTYLPHENOL; PHENOL,2,4-BIS(1,1-DIMETHYLPROPYL)-; PHENOL,2,4-DI-TERT-PENTYL-; PRODOX 156

Use: chem int for condensate with diazotized o-nitroaniline, dipentylphenoxyalkyl acids, dipentylphenoxyalkyl & aryl acid chlorides, dipentylphenoxyalkyl & aryl amides, dipentylphenoxyalkyl nitriles

Physical Properties

Boiling Point: 169 °C (336.2 °F)
Freezing Point: 26 °C (78.8 °F)

RTECS Toxicity Data

Acute Oral: Rat LD_{50} Dose: 330 mg/kg.
Irritation Eye: Rabbit Standard Draize Test Dose: 100 mg; Reaction: moderate.

Hazard Overviews

Corrosive

Health: Corrosive causes burns to eyes/skin/respiratory tract. Toxic. Other Acute Effects: harmful if swallowed, inhaled, or absorbed through skin; inhalation may result in spasm; inflammation and edema of the larynx and bronchi; chemical pneumonitis; pulmonary edema; exposure may cause burning sensation; coughing; wheezing; laryngitis; shortness of breath; headache; nausea; vomiting.

Fire: Hazards: emits toxic fumes. Extinguishing agents: water spray; carbon dioxide, dry chemical powder or appropriate foam. Precautions: combustible liquid.

Reactivity: Incompatible with: strong oxidizing agents. Hazardous decomposition products: toxic fumes of: carbon monoxide, carbon dioxide.

Carcinogenicity: IARC - Not listed; NIOSH - Not listed; NTP - Not listed; ACGIH - Not listed; OSHA - Not listed; EPA - Not listed; MAK - Not listed

Primary Target Organs:

Eyes

Skin

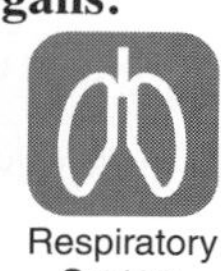
Respiratory System

Environmental

Regulations

RCRA 40CFR: Not listed
CERCLA: 40CFR 302.4: Not listed
SARA 40CFR 372.65: Not listed
SARA EHS 40CFR 355: Not listed
TSCA: Listed

DIP1540 **CAS #: 82-66-6**

DIPHENADIONE

RTECS: NK5600000
EINECS Number: 201-434-5
Molecular Formula: $C_{23}H_{16}O_3$
Formula Weight: 340.36
Synonyms: CONTRAX-D; DIDANDIN; DIDION; DIPAXIN; DIPHACIN; DIPHACINONE; DIPHACINS; DIPHENACIN; DIPHENANDIONE; 2-DIPHENYL ACETYL-1,3-INDANEDIONE; 2-DIPHENYLACETYL-1,3-DIKETOHYDRINDENE; 2-DIPHENYL-ACETYL-INDAN-1,3-DION; 2-(DIPHENYLACETYL)-1,3-INDANDIONE; 2-(DIPHENYLACETYL)INDAN-1,3-DIONE; 2-DIPHENYLACETYL-1,3-INDANDIONE; 2-DIPHENYLACETYL-1,3-INDANEDIONE; 2-(DIPHENYLACETYL)-1H-INDENE-1,3(2H)-DIONE; 2-(DIPHENYLACETYL)-1H-INDENE-2,3(2H)DIONE; GOLD CREST; 1,3-INDANDIONE,2-(DIPHENYLACETYL)-; 1,3-INDANDIONE,2-DIPHENYLACETYL-; KILL-KO RAT KILLER; ORAGULANT; P.C.Q; PARAKAKES; PCQ; PID; PROMAR; PROMAR PCQ; RAMIK; RATINDAN 1; RODENT CAKE; SOLVAN; U 1363; U-1363
Description: pale yellow crystalline powder or crystals; odorless
Use: medication: anticoagulant; rodenticide, used to control mice, rats, prairie dogs ground squirrels, voles and other rodents; medication (vet): anticoagulant, atherosclerotic; baticide

Physical Properties

Freezing Point: 146 °C (294.8 °F) to 147 °C (296.6 °F)
Density: 1.281 at 25 °C
Vapor Pressure: Technical 13.7 nPa at 25 °C
Water Solubility: Practically Insoluble in Water
Other Solubilities: Soluble in Methyl Cyanide, Cyclohexane; Soluble in Ether, Glacial Acetic Acid; Soluble in Acetone, Acetic Acid; In Chloroform 204, Toluene 73, Xylene 50, Ethanol 2.1, heptane 1.8 (all in g/kg); Soluble in alkalis with the formation of salts.
pH: Acid reaction
Refraction Index: Alpha 1.670

RTECS Toxicity Data

Acute Oral: Rat LD_{50} Dose: 1500 ug/kg. Mouse LD_{50} Dose: 28300 ug/kg.
Acute Inhalation: Rat LC_{50} Dose: 2 gm/m^3/4hr.
Acute Dermal: Rat LD_{50} Route: Skin; Dose: 200 mg/kg.

Hazard Overviews

Carcinogenicity: IARC - Not listed; NIOSH - Not listed; NTP - Not listed; ACGIH - Not listed; OSHA - Not listed; EPA - Not listed; MAK - Not listed

Environmental

Ecotoxicity: LD_{50} Anas platyrhynchos (Mallard duck) oral 3158 mg/kg LC_{50} Salmo gairdneri (Rainbow trout) 2.8 mg/l/96 hr /Conditions of bioassay not specified

Regulations

RCRA 40CFR: Not listed
CERCLA: 40CFR 302.4: Not listed
SARA 40CFR 372.65: Not listed TPQ: 10/10000 lb
SARA EHS 40CFR 355: Listed TPQ: 10 lb
TSCA: Not listed

Analytical Methods

Soil: EPA PMD-DOZ

DIP1720 **CAS #: 957-51-7**

DIPHENAMID

RTECS: AB8050000
EINECS Number: 213-482-4
Molecular Formula: $C_{16}H_{17}NO$
Formula Weight: 239.34

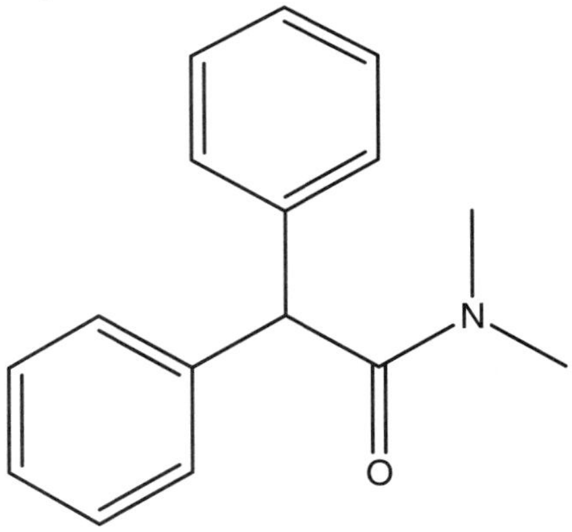

Chemical Structure

Synonyms: 80W; ACETAMIDE,N,N-DIMETHYL-2,2-DIPHENYL-; BENZENEACETAMIDE,N,N-DIMETHYL-ALPHA-PHENYL-; BENZENEACETAMIDE,N,N-DIMETHYL-ALPHA-PHENYL-(9CI); DIAMIDE; DIF 4; DIFENAMID; DIFENAMIDE; DIHERBID; DIMETHYLAMID KYSELINY DIFENYLOCTOVE; N,N-DIMETHYL-ALPHA,ALPHA-DIPHENYL ACETAMIDE; N,N,DIMETHYL-2,2-DIPHENYLACETAMIDE; N,N-DIMETHYL-2,2-DIPHENYLACETAMIDE; N,N-DIMETHYL-ALPHA,ALPHA-DIPHENYLACETAMIDE; N,N-DIMETHYLDIPHENYLACETAMIDE; N,N-DIMETHYL-',' -DIPHENYLACETMIDE; N,N-ALPHA-DIMETHYLPHENYLBENZENEACETAMIDE; N,N-DIMETHYL-ALPHA-PHENYLBENZENEACETAMIDE; DIMID; DIMIT; DIPHENAMIDE; DIPHENYLAMIDE; 2,2-DIPHENYL-N,N-DIMETHYLACETAMIDE; DYMID; ENIDE; ENIDE 50W; ENIDE 50; ENT-28567; FDN; FENAM; L 34314; L-34314; LILLY 34,314; RIDEON; TREFMID; U 4513; ZARUR
Description: colorless to white; crystals or prisms; no appreciable odor
Use: selective preemergence herbicide for control of annual broadleaf weeds & grasses; against carpetweed, chickweed, knotweed, lambsquarters, pigweed, purslane, & smartweed; in cotton, potatoes, sweetpotatoes, tomatoes, vegetables, capsicums, okra, soya beans, groundnuts, tobacco, pome fruit, stone fruit, citrus fruit, bush fruit, strawberries, forestry nurseries, ornamental plants, shrubs, & trees; plant growth regulator.

Physical Properties

Freezing Point: 134.5 °C (274.1 °F) to 135.5 °C (275.9 °F)
Specific Gravity: 1.17 at 23.3 °C
Vapor Pressure: Negligible at 20 °C
Water Solubility: 0.026 g/100ml at 27 °C
Other Solubilities: Moderately Soluble in polar organic solvents.
Flash Point: Nonflammable

RTECS Toxicity Data

Acute Oral: Rat LD_{50} Dose: 685 mg/kg. Monkey LD_{50} Dose: 1 gm/kg.
Acute Dermal: Rat LD_{50} Route: Skin; Dose: >6320 mg/kg. Mouse LD_{50} Route: Subcutaneous Dose: 800 mg/kg.
Chronic (Multiple Dose) Oral: Rat Dose: 7280 mg/kg/26W-I; Toxic Effects: Blood - Pigmented or nucleated red Blood cells; Blood - Changes in serum composition; Biochemical - Cytochrome oxidases (including oxidative phosphorylation).
Mutagenic: Mouse Cytogenetic Analysis; Route: Unreported; Dose: 10 mg/kg.

Hazard Overviews

Health: Irritating to eyes/skin/respiratory tract. Harmful. Other Acute Effects: harmful if swallowed; may be harmful if inhaled; may be harmful if absorbed through the skin.
Fire: Noncombustible. Hazards: emits toxic fumes. Extinguishing agents: water spray; carbon dioxide, dry chemical powder or appropriate foam. Precautions: combustible liquid.
Carcinogenicity: IARC - Not listed; NIOSH - Not listed; NTP - Not listed; ACGIH - Not listed; OSHA - Not listed; EPA - Not listed; MAK - Not listed
Primary Target Organs:

Eyes

Skin

Respiratory System

Environmental

Ecotoxicity: LC_{50} Pimephales promelas (fathead minnow) 48 mg/l/96 hr (95% confidence limit 38-60 mg/l) at 18 °C, wt 0.9 g. static bioassay without aeration, pH 7.2-7.5, water hardness 40-50 mg/l as calcium carbonate and alkalinity of 30-35 mg/l LC_{50} Asellus more than 100 mg/l/96 hr at 15 °C, mature. static bioassay without aeration, pH 7.2-7.5, water hardness 40-50 mg/l as calcium carbonate and alkalinity of 30-35 mg/l EC_{50} Daphnia magna 58 mg/l/48 hr, immobilization (95% confidence limit 43-79 mg/l) at 21 °C, 1st instar. static bioassay without aeration, pH 7.2-7.5, water hardness 40-50 mg/l as calcium carbonate and alkalinity of 30-35 mg/l EC_{50} Cypridopsis vidua 51 mg/l/48 hr, immobilization (95% confidence limit 37-71 mg/l) at 21 °C, mature. static bioassay without aeration, pH 7.2-7.5, water hardness 40-50 mg/l as calcium carbonate and alkalinity of 30-35 mg/l LC_{50} Palaemonetes 32 mg/l/96 hr (95% confidence limit 29-35 mg/l) at 21 °C, mature. static bioassay without aeration, pH 7.2-7.5, water hardness 40-50 mg/l as calcium carbonate and alkalinity of 30-35 mg/l
Environmental Fate: If released to soil, loss will occur primarily due to biodegradation. Loss from soil due to volatilization and photolysis should not be important. It is expected to have a moderate mobility in soils. Depending on soil characteristics and rainfall, the persistence in soil may be 3-8 months. If released to water, the major process for loss is probably biodegradation. Hydrolysis, photolysis, bioconcentration, and volatilization should not be important processes in water. It will be found completely adsorbed to particulate matter in the atmosphere and may be removed by dry and wet deposition. Partial removal will also occur as a result of dry and wet deposition.

Environmental Physical Data

Henry's Law Constant: estimated at 2.42 $x10^{-11}$
Octanol/Water Partition Coefficient: log K_{ow} = 2.17
Sorption Partition Coefficient: K_{oc} = estimated at 210
BCF: aquatic organism 27

Regulations

RCRA 40CFR: Not listed
CERCLA: 40CFR 302.4: Not listed
SARA 40CFR 372.65: Listed
SARA EHS 40CFR 355: Not listed
TSCA: Not listed

Analytical Methods

Soil: EPA PMD-DPA
Water / Groundwater: EPA 645
Drinking Water: EPA 507, 525.2; AOAC 991.07; ASTM D5475
Food: FDA 212.1, 232.1, 232.3, 232.4, 242.1
Plasma: EPA 001; FDA 252

DIP1900 **CAS #: 58-73-1**

DIPHENHYDRAMINE

RTECS: KR6825000
EINECS Number: 200-396-7
Molecular Formula: $C_{17}H_{21}NO$
Formula Weight: 255.35
Synonyms: ALLEDRYL; ALLERGAN; ALLERGICAL; ALLERGIN; ALLERGINA; ALLERGIVAL; AMIDRYL; ANTISTOMINUM; ANTOMIN; AUTOMIN; BAGODRYL; BARAMINE; BENA; BENACHLOR; BENADRIN; BENAPON; BENODIN; BENODINE; BENYLAN; BENZANTINE; BENZHYDRAMINE; BENZHYDRAMINUM; BENZHYDRIL; BENZHYDROAMINA; O-BENZHYDRYLDIMETHYLAMINOETHANOL; N-(BENZHYDRYLOKSY-ETYLO)DWUMETYLOAMINA; 2-(BENZHYDRYLOXY),-N,N-DIMETHYLETHYLAMINE; 2-(BENZHYDRYLOXY)-N,N-DIMETHYLETHYLAMINE; 2-(BENZOHYDRYLOXY)-N,N-DIMETHYLETHYLAMINE; BETRAMIN; DABYLEN; DEBENDRIN; DERMISTINA; DERMODRIN; DESENTOL; DIABENYL; DIABYLEN; DIBENDRIN; DIBONDRIN; DIFEDRYL; DIFENHYDRAMIN; DIFENIDRAMINA; DIHIDRAL; DIMEDROL; BETA-DIMETHYLAMINO-AETHYL-BENZHYDRYL-AETHER; BETA-DIMETHYLAMINOETHANOL DIPHENYLMETHYL ETHER; ALPHA-(2-DIMETHYLAMINOETHOXY)DIPHENYLMETHANE; BETA-

DIMETHYLAMINOETHYLBENZHYDRYL ETHER; BETA-DIMETHYLAMINOETHYLBENZHYDRYLETHER; DIPHANTINE; DIPHENYLHYDRAMINE; 2(DIPHENYLMETHOXY)-N,N-DIMETHYLETHYLAMINE; 2-(DIPHENYLMETHOXY)-N,N-DIMETHYLETHYLAMINE; DRYLISTAN; DYLAMON; ETANAUTINE; ETHANAMINE,2-(DIPHENYLMETHOXY)-N,N-DIMETHYL-; ETHYLAMINE,N,N-DIMETHYL-2-(DIPHENYLMETHOXY)-; HISTAXIN; HYADRINE; IBIODRAL; MEDIDRYL; MEPHADRYL; NAUSEN; NOVAMINA; PM 255; PROBEDRYL; RESTAMIN; RIGIDYL; S51; SYNTEDRIL

Description: white crystalline powder or crystals; odorless

Use: antihistaminic; vet as antihistaminic, in anti-motion sickness hydrochloride; antihistaminic in human & veterinary medicine; antiemetic for motion sickness; sedative; antiparkinsonism drug

Physical Properties

Boiling Point: 150 °C (302 °F) to 165 °C (329 °F) at 2.0 mm Hg

Water Solubility: 1 g/1 mL Water

Other Solubilities: 1 g dissolves in: 2 ml Alcohol, 50 ml Acetone, 2 ml Chloroform; Very Slightly Soluble in Benzene, Ether

pH: 1% aqueous solution about 5.5

RTECS Toxicity Data

Acute Oral: Human TD_{Lo} Dose: 714 ug/kg; Toxic Effects: Behavioral - Somnolence (general depressed activity); Behavioral - Alteration of operant conditioning; Behavioral - Change in psychophysiological tests. Rat LD_{50} Dose: 390 mg/kg; Toxic Effects: Behavioral - Convulsions or effect on seizure threshold; Behavioral - Change in motor activity (specific assay).

Acute Dermal: Rat LD_{50} Route: Subcutaneous Dose: 474 mg/kg. Mouse LD_{50} Route: Subcutaneous Dose: 50 mg/kg.

Reproductive/Teratogenic: Rat Route: Subcutaneous; Dose: 440 mg/kg; Duration: female 1-22D of pregnancy; Maternal Effects - Other effects on females. Rat Route: Subcutaneous; Dose: 440 mg/kg; Duration: female 1-22D of pregnancy; Effects on Newborn - Growth statistics.

Mutagenic: Human DNA Inhibition; Cell Type: fibroblast; Dose: 12500 ug/L. Human Cytogenetic Analysis; Cell Type: fibroblast; Dose: 100 mg/L. Human Other Mutation Test Systems; Cell Type: fibroblast; Dose: 12500 ug/L.

Hazard Overviews

Carcinogenicity: IARC - Not listed; NIOSH - Not listed; NTP - Not listed; ACGIH - Not listed; OSHA - Not listed; EPA - Not listed; MAK - Not listed

Environmental

Regulations

RCRA 40CFR: Not listed
CERCLA: 40CFR 302.4: Not listed
SARA 40CFR 372.65: Not listed
SARA EHS 40CFR 355: Not listed
TSCA: Listed

DIP2080 **CAS #: 972-02-1**

DIPHENIDOL

RTECS: TM4970000
EINECS Number: 213-540-9
Molecular Formula: $C_{21}H_{27}NO$
Formula Weight: 309.43

Synonyms: AVOMOL; BENZHYDROL,ALPHA-(3-PIPERIDINOPROPYL)-; CEPHADOL; DIFENIDOL; DIPHENADOL; ALPHA,ALPHA-DIPHENYL-1-PIPERIDINEBUTANOL; DIPHENYL(3-(1-PIPERIDYL)PROPYL)CARBINOL; NOMETIC; 1-PIPERIDINEBUTANOL,ALPHA,ALPHA-DIPHENYL-; SK&F NO 478-A; SK&F NO. 478-A; SKF 478; VONTROL

Description: white crystalline powder or needles; odorless

Use: medication: antiemetic; medication: useful in mgmnt of nausea & vomiting assoc with infectious diseases, malignancies, radiation sickness, general anesthesia, & treatment with antineoplastic agents

Physical Properties

Freezing Point: 104 °C (219.2 °F) to 105 °C (221 °F)

Water Solubility: Insoluble in Water

Other Solubilities: freely Soluble in Methanol; Soluble in water & Chloroform; Practically Insoluble in Ether, Benzene, & Petroleum Ether

RTECS Toxicity Data

Acute Oral: Rat LD_{50} Dose: 815 mg/kg. Mouse LD_{50} Dose: 450 mg/kg.

Acute Dermal: Rat LD_{50} Route: Subcutaneous Dose: 304 mg/kg.

Hazard Overviews

Carcinogenicity: IARC - Not listed; NIOSH - Not listed; NTP - Not listed; ACGIH - Not listed; OSHA - Not listed; EPA - Not listed; MAK - Not listed

Environmental

Regulations

RCRA 40CFR: Not listed
CERCLA: 40CFR 302.4: Not listed
SARA 40CFR 372.65: Not listed
SARA EHS 40CFR 355: Not listed
TSCA: Not listed

DIP2260 **CAS #: 915-30-0**

DIPHENOXYLATE

RTECS: NS5290000
EINECS Number: 213-020-1
Molecular Formula: $C_{30}H_{32}N_2O_2$
Formula Weight: 452.57

Synonyms: 1-(3-CYANO-3,3-DIPHENYLPROPYL)-4-PHENYL-ISONIPECOTIC ACIDETHYL ESTER; 1-(3-CYANO-3,3-

DIPHENYLPROPYL)-4-PHENYLISONIPECOTIC ACIDETHYL ESTER; 1-(3-CYANO-3,3-DIPHENYLPROPYL)-4-PHENYL-4-PIPERIDINECARBOXYLIC ACIDETHYL ESTER; 2,2-DIPHENYL-4-(4-CARBETHOXY-4-PHENYLPIPERIDINO)BUTYRONITRILE; ETHYL 1-(3-CYANO-3,3-DIPHENYLPROPYL)-4-PHENYLISONIPECOTATE; ETHYL1-(3-CYANO-3,3-DIPHENYLPROPYL)-4-PHENYL-4-PIPERIDINECARBOXYLATE; ISONIPECOTIC ACID,1-(3-CYANO-3,3-DIPHENYLPROPYL)-4-PHENYL-,ETHYL ESTER; 4-PIPERIDINECARBOXYLIC ACID,1-(3-CYANO-3,3-DIPHENYLPROPYL)-4-PHENYL-,ETHYL ESTER; 4-PIPERIDINECARBOXYLIC ACID,1-(3-CYANO-3,3-DIPHENYLPROPYL)-4-PHENYL-,ETHYL ESTER (9CI); R-1132

Description: white crystalline powder or crystals; odorless

Physical Properties

Freezing Point: Crystals at 220.5 °C (428.9 °F) to 222 °C (431.6 °F)

Water Solubility: 0.8 mg/l at 25 °C

Other Solubilities: Practically Insoluble in Ether; slightly Soluble in isopropanol; sparingly Soluble in Acetone.

pH: Saturated solution about 3.3

RTECS Toxicity Data

Acute Oral: Rat LD_{50} Dose: 221 mg/kg; Toxic Effects: Gastrointestinal - Decreased motility or constipation. Mouse LD_{50} Dose: 337 mg/kg.

Hazard Overviews

Health: Toxic. Other Acute Effects: may be fatal if swallowed; may cause respiratory depression; constipation; anorexia; nausea; vomiting; headache; drowsiness; depression; skin effects. Chronic Effects: prolonged or repeated exposure can lead to habituation or addiction.

Fire: Extinguishing agents: water spray; carbon dioxide, dry chemical powder or appropriate foam. Precautions: combustible liquid.

Carcinogenicity: IARC - Not listed; NIOSH - Not listed; NTP - Not listed; ACGIH - Not listed; OSHA - Not listed; EPA - Not listed; MAK - Not listed

Environmental

Regulations

RCRA 40CFR: Not listed

CERCLA: 40CFR 302.4: Not listed

SARA 40CFR 372.65: Not listed

SARA EHS 40CFR 355: Not listed

TSCA: Not listed

DIP2440 **CAS #: 102-09-0**

DIPHENYL CARBONATE

RTECS: FG0500000

EINECS Number: 203-005-8

Molecular Formula: $C_{13}H_{10}O_3$

Formula Weight: 214.22

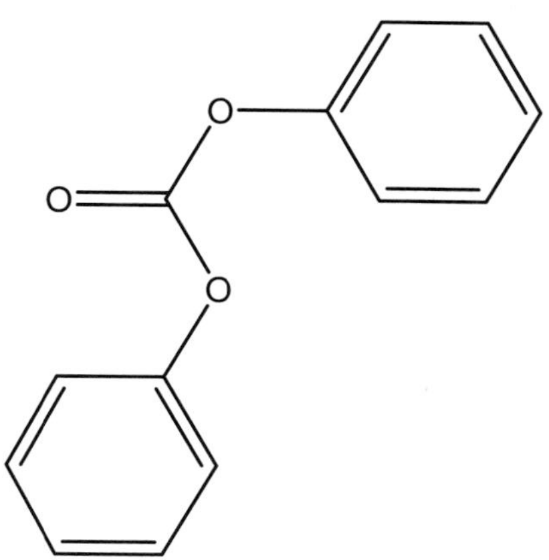

Chemical Structure

Synonyms: CARBONIC ACID,DIPHENYL ESTER; PHENYL CARBONATE

Description: white, crystalline solid or lustrous needles

Use: in molten state as solvent for nitrocellulose; plasticizer & solvent; synthesis of polycarbonate resins

Physical Properties

Boiling Point: 302 °C (576 °F) to 306 °C (583 °F)

Freezing Point: 80 °C (176 °F) to 81 °C (177.8 °F)

Specific Gravity: 1.1215 at 87 °C/4 °C

Water Solubility: Practically Insoluble in Water

Other Solubilities: Soluble in Acetone, Carbon Tetrachloride, other organic solvents.

Ionization Potential (eV): 9.01 +/-2.0

RTECS Toxicity Data

Tumorigenic: Mouse Route: Oral; Dose: 28 gm/kg/78W-I; Toxic Effects: Tumorigenic - Equivocal tumorigenic agent by RTECS criteria; Lungs, Thorax, or Respiration - Tumors; Liver - Tumors. Mouse Route: Subcutaneous; Dose: 1000 mg/kg; Toxic Effects: Tumorigenic - Neoplastic by RTECS criteria; Liver - Tumors.

Hazard Overviews

Health: May cause irritation to eyes/skin/respiratory tract. Harmful. Other Acute Effects: harmful if swallowed or absorbed through skin; may be harmful if inhaled; prolonged exposure can cause lung irritation or dermatitis; prolonged or repeated exposure may cause allergic reactions in certain sensitive individuals.

Fire: Hazards: emits toxic fumes. Extinguishing agents: water spray; carbon dioxide, dry chemical powder or appropriate foam. Precautions: combustible liquid.

Reactivity: Stable. Hazardous polymerization will not occur. Incompatible with: strong oxidizing agents. Hazardous decomposition products: toxic fumes of: carbon monoxide, carbon dioxide.

Carcinogenicity: IARC - Not listed; NIOSH - Not listed; NTP - Not listed; ACGIH - Not listed; OSHA - Not listed; EPA - Not listed; MAK - Not listed

Environmental

Regulations

RCRA 40CFR: Not listed
CERCLA: 40CFR 302.4: Not listed
SARA 40CFR 372.65: Not listed
SARA EHS 40CFR 355: Not listed
TSCA: Listed

DIP2620 CAS #: 26444-49-5

DIPHENYL CRESYL PHOSPHATE

RTECS: TC5520000
EINECS Number: 247-693-8
Molecular Formula: $C_{19}H_{17}O_4P$
Structured MF: $(CH_3C_6H_4)_2PO_4$
Formula Weight: 340.33
Synonyms: CRESOL DIPHENYL PHOSPHATE; CRESYL DIPHENYL PHOSPHATE; CRESYL PHENYL PHOSPHATE; CRESYLDIPHENYL PHOSPHATE; DIPHENYL CRESOL PHOSPHATE; DIPHENYL TOLYL ESTER PHOSPHORIC ACID; DIPHENYL TOLYL PHOSPHATE; DISFLAMOLL DPK; KRONITEX CDP; METHYL PHENYL DIPHENYL PHOSPHATE; METHYLPHENYL DIPHENYL PHOSPHATE; MONOCRESYL DIPHENYL PHOSPHATE; MONOCRESYL DIPHENYLPHOSPHATE; PHOSFLEX 112; PHOSPHORIC ACID,CRESYL DIPHENYL ESTER; PHOSPHORIC ACID,DIPHENYL TOLYL ESTER; PHOSPHORIC ACID,METHYLPHENYL DIPHENYL; PHOSPHORIC ACID,METHYLPHENYL DIPHENYL ESTER; PHOSPHORIC ACID,METHYLPHENYL DIPHENYL ESTER (9CI); PHOSPHORIC ACID,METHYLPHENYL DIPHENYL ESTER(9CI); SANTICIZER 140; TOLYL DIPHENYL PHOSPHATE
Description: colorless liquid; slight odor
Use: plasticizer; extreme-pressure lubricant; hydraulic fluid; gasoline additive; food packaging

Physical Properties

Boiling Point: 390 °C (734 °F) at 760 mm Hg
Freezing Point: -38 °C (-36.4 °F)
Specific Gravity: 1.208 at 25 °C
Vapor Density: 11.7 Air=1
Saturated Vapor Density: 1.201695082 kg/m^3
Density: 1.2 g/mL
Vapor Pressure: 0.1 mm Hg at 25 °C
Water Solubility: Insoluble in Water
Other Solubilities: Acetone: Soluble; Benzene: Soluble; Ether: Soluble; Glycerol: Insoluble.
Flash Point: 232 °C Closed Cup

RTECS Toxicity Data

Acute Oral: Rat LD_{50} Dose: 6400 mg/kg. Chicken LD_{50} Dose: >10 gm/kg.
Acute Dermal: Rabbit LD_{Lo} Route: Skin; Dose: 3160 mg/kg; Toxic Effects: Behavioral - Muscle weakness; Liver - Other changes; Kidney, Ureter, and Bladder - Other changes.

Hazard Overviews

Fire Diamond

Health: Severely irritating to eyes/skin. Also Causes: burns, kidney, liver, nervous system and cardiovascular disturbances, peeling skin, localized anesthesia, ochronosis, hypersensitvity, nausea, or vomiting.
Fire: Will burn. Use dry chemical, carbon dioxide, fog, or a water mist. The use of foam and heavy water sprays is discouraged because they may cause frothing.
Reactivity: Stable. Hazardous polymerization cannot occur. Incompatible with: wet, alkaline conditions; some plastics and elastomers (vinyl-based resins and natural rubbers). Hazardous decomposition products: carbon dioxide; toxic fumes of phosphorus oxide.
Carcinogenicity: IARC - Not listed; NIOSH - Not listed; NTP - Not listed; ACGIH - Not listed; OSHA - Not listed; EPA - Not listed; MAK - Not listed
Primary Target Organs:

Eyes

Skin
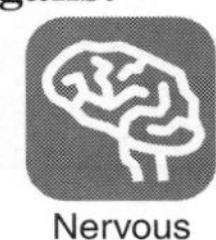
Nervous System

Liver

Kidneys

Cardio-vascular

Environmental

Regulations

RCRA 40CFR: Not listed
CERCLA: 40CFR 302.4: Not listed
SARA 40CFR 372.65: Not listed
SARA EHS 40CFR 355: Not listed
TSCA: Listed

Analytical Methods

Plasma: EPA 001

DIP2800 CAS #: 101-84-8

DIPHENYL ETHER

RTECS: KN8970000
EINECS Number: 202-981-2
Molecular Formula: $C_{12}H_{10}O$
Structured MF: $C_6H_5OC_6H_5$
Formula Weight: 170.20

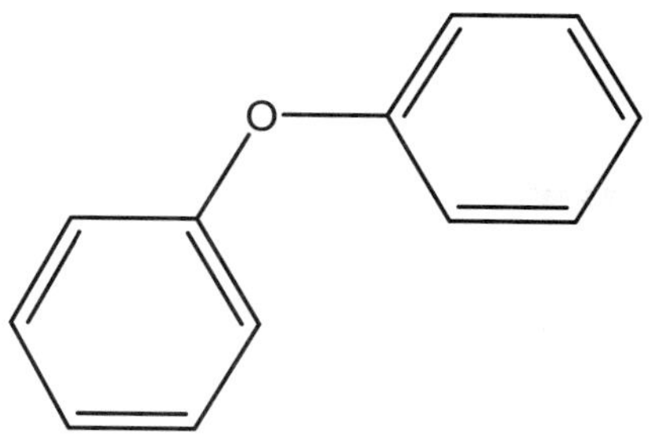

Chemical Structure

Synonyms: BENZENE,1,1'-OXYBIS-; BENZENE,PHENOXY-; BIPHENYL OXIDE; DIPHENYL OXIDE; ETHER,DIPHENYL; GERANIUM CRYSTALS; 1,1'-OXYBISBENZENE; OXYDIPHENYL; PHENOXY BENZENE; PHENOXYBENZENE; PHENYL ETHER; PHENYL OXIDE

Description: colorless crystals, liquid above 82 F (28 C); geranium-like odor

Use: heat-transfer medium, perfuming soaps

Physical Properties

Boiling Point: 259 °C (498 °F)
Freezing Point: 28 °C (82.4 °F)
Specific Gravity: Liquid 1.075 at 20 °C
Vapor Density: 5.86 Air=1
Saturated Vapor Density: 1.200172976 kg/m^3
Vapor Pressure: 0.0225 mm Hg at 25 °C
Water Solubility: Insoluble in Water
Other Solubilities: 95% Ethanol: Soluble; Benzene: Soluble; Ether: Soluble.
Surface Tension: 40.05 dynes/cm at 20 °C
Odor Threshold: 0.1 ppm
Refraction Index: 1.5787 at 25 °C/D
Critical Temperature: 490 °C
Critical Pressure: 3.1300 x10^6 Pa
Ionization Potential (eV): 8.09
Flash Point: 115 °C Closed Cup
Autoignition Temperature: 618 °C
LEL: 0.7% v/v
UEL: 6.0% v/v

RTECS Toxicity Data

Acute Oral: Rat LD_{50} Dose: 3370 mg/kg.
Irritation Skin: Rabbit Standard Draize Test Dose: 500 mg/24H; Reaction: mild.

Hazard Overviews

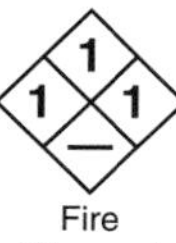

Fire Diamond

Health: Mildly irritating to eyes/skin/respiratory tract. Also Causes: nausea, burning of eyes, reddening.

Fire: Will burn. May form explosive dust-air mixtures. Forms explosive peroxides on standing. Use dry chemical or carbon dioxide. Water or foam may cause frothing.

Reactivity: Stable. Hazardous polymerization cannot occur. Avoid: heat; ignition sources. Incompatible with: strong oxidizers; chlorosulfonic acid. Hazardous decomposition products: carbon oxides.

Carcinogenicity: IARC - Not listed; NIOSH - Listed as carcinogen; NTP - Not listed; ACGIH - Not listed; OSHA - Not listed; EPA - Not listed; MAK - Not listed

Primary Target Organs:

Eyes

Skin

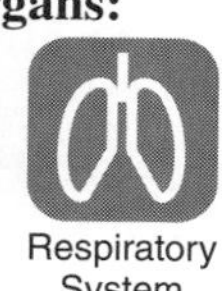
Respiratory System

Exposure Limits
OSHA PEL: TWA: 1 ppm; 7 mg/m^3; vapor.
ACGIH TLV: TWA: 1 ppm; 7 mg/m^3; STEL: 2 ppm; 14 mg/m^3.
NIOSH REL: TWA: 1 ppm; 7 mg/m^3.
NIOSH IDLH: 100 ppm.
DFG MAK: TWA: 1 ppm; 7 mg/m^3.

Environmental

Ecotoxicity: LC_{50} Pimephales promelas (fathead minnow) 4.0 mg/l 96 hr flow-through bioassay, wt 0.12 g, water hardness 45.5 mg/l $CaCO_3$, temp: 25 + or - 1 °C, pH 7.5, dissolved oxygen greater than 60% of saturation

Environmental Fate: If released to the atmosphere, it will degrade readily in the vapor phase by reaction with photochemically produced hydroxyl radicals (estimated half-life of about 20 hr). If released to soil or water, it will biodegrade relatively slowly based upon the results of two biological screening studies. Volatilization from terrestrial surfaces and water to air is expected to be an important transport process. A K_{oc} value of 1950 suggests that it will have low mobility in soil and may partition to sediment in water.

Cleanup/Disposal: Guide No. 127: Eliminate all ignition sources (no smoking, flares, sparks or flames in immediate area). All equipment used when handling the product must be grounded. Do not touch or walk through spilled material. Stop leak if you can do it without risk. Prevent entry into waterways, sewers, basements or confined areas. A vapor suppressing foam may be used to reduce vapors. Absorb or cover with dry earth, sand or other non-combustible material and transfer to containers. Use clean non-sparking tools to collect absorbed material. Large Spills: Dike far ahead of liquid spill for later disposal. Water spray may reduce vapor; but may not prevent ignition in closed spaces.

Environmental Physical Data

Henry's Law Constant: estimated at 0.00028
Octanol/Water Partition Coefficient: log K_{ow} = 4.21
Sorption Partition Coefficient: K_{oc} = 1950
BCF: rainbow trout 195

Regulations

RCRA 40CFR: Not listed
CERCLA: 40CFR 302.4: Not listed
SARA 40CFR 372.65: Not listed
SARA EHS 40CFR 355: Not listed

TSCA: Listed

Analytical Methods

Air: ASTM D3686, D3687
Soil: SW846 3640A
Water / Groundwater: EPA 1625; ASTM D4763
Indoor / Expired Air: NIOSH 1617, 2013

DIP2980 CAS #: 115-88-8

DIPHENYL OCTYL PHOSPHATE

EINECS Number: 204-113-8
Molecular Formula: $C_{20}H_{27}O_4P$
Formula Weight: 362.21
Synonyms: DISFLAMOLL DPO; OCTYL DIPHENYL PHOSPHATE; PHOSPHORIC ACID,OCTYL DIPHENYL ESTER
Use: flame-retardant plasticizer; plasticizer for polyvinyl chloride; hydraulic fluid in aircraft

Physical Properties

Water Solubility: 0.14 mg/L

Hazard Overviews

Carcinogenicity: IARC - Not listed; NIOSH - Not listed; NTP - Not listed; ACGIH - Not listed; OSHA - Not listed; EPA - Not listed; MAK - Not listed

Environmental

Environmental Fate: If released to the atmosphere, it will exist in both the vapor phase and in the particulate phase based on an estimated vapor pressure of 1.24×10^{-7} mm Hg. In the vapor phase will be degraded by hydroxyl radicals with an estimated half-life of 10.3 hours. Particulate phase may be removed physically from air by wet and dry deposition. In soil, it is expected to be immobile. Hydrolysis may be an important fate process, particularly under alkaline soil conditions. In water, hydrolysis is expected to be a major fate process despite its low water solubility, especially if the water is alkaline. An estimated Henry's Law constant of 2.5×10^{-7} atm-cu m/mole indicates that this compound will not volatilize from water surfaces. BCF values of 1900 and 41000 suggest that it may bioconcentrate in aquatic organisms. However, rapid depuration rates have been reported for phosphate esters.

Environmental Physical Data

Henry's Law Constant: estimated at 2.5×10^{-7}
Octanol/Water Partition Coefficient: log K_{ow} = 6.37
Sorption Partition Coefficient: K_{oc} = 1.3×10^4 to 7×10^4
BCF: estimated at 4.1×10^4

Regulations

RCRA 40CFR: Not listed
CERCLA: 40CFR 302.4: Not listed
SARA 40CFR 372.65: Not listed
SARA EHS 40CFR 355: Not listed
TSCA: Not listed

DIP3160 CAS #: 71449-78-0

DIPHENYL 4-THIOPHENOXYPHENYL

Physical Properties

Specific Gravity: 0.97
Water Solubility: Insoluble

Hazard Overviews

Carcinogenicity: IARC - Not listed; NIOSH - Not listed; NTP - Not listed; ACGIH - Not listed; OSHA - Not listed; EPA - Not listed; MAK - Not listed

Environmental

Regulations

RCRA 40CFR: Not listed
CERCLA: 40CFR 302.4: Not listed
SARA 40CFR 372.65: Not listed
SARA EHS 40CFR 355: Not listed
TSCA: Not listed

DIP3340 CAS #: 102-08-9

DIPHENYL THIOUREA

RTECS: FE1225000
EINECS Number: 203-004-2
Molecular Formula: $C_{13}H_{12}N_2S$
Formula Weight: 228.32

HN
S
NH

Chemical Structure

Synonyms: A 1; DFT; 1,3-DIFENYLTHIOMOCOVINA; N,N'-DIPHENYLTHIOCARAMIDE; N,N'-DIPHENYLTHIOCARBAMIDE; S-DIPHENYLTHIOCARBAMIDE; 1,3-DIPHENYL-2-THIOUREA; 1,3-DIPHENYLTHIOUREA; DIPHENYLTHIOUREA; N,N'-DIPHENYLTHIOUREA; SYM-DIPHENYL-THIOUREA; SYM-DIPHENYLTHIOUREA; 2-FENYLOTIOMOCZNIK; NOCCELER C; RHENOCURE CA; STABILISATOR C; SULFOCARBANILIDE; THIOCARBANILIDE; THIOKARBANILID; THIOUREA,N,N'-DIPHENYL-; THIOUREA,SYM-DIPHENYL-; THIOUREA,N,N'-DIPHENYL-(9CI); UREA,1,3-DIPHENYL-2-THIO-; VULKACIT CA

Description: white to faint gray powder or crystalline leaflets
Use: vulcanizing accelerator; sulfur dyes; pharmaceuticals; flotation agent; acid inhibitor; intermediate for organic synthesis

Physical Properties

Boiling Point: Decomposes
Freezing Point: 154 °C (309.2 °F)
Specific Gravity: 1.32
Density: 1.32 g/mL
Water Solubility: Almost Insoluble in Water
Other Solubilities: 95% Ethanol: Very Soluble; Chloroform: Very Soluble; Ether: Very Soluble; olive oil: Very Soluble.
Flash Point: Not available; probably combustible

RTECS Toxicity Data

Acute Oral: Rat LD_{50} Dose: 50 mg/kg. Rabbit LD_{Lo} Dose: 1500 mg/kg; Toxic Effects: Behavioral - Ataxia; Lungs, Thorax, or Respiration - Cyanosis; Gastrointestinal - Hypermotility, diarrhea.
Reproductive/Teratogenic: Rat Route: Oral; Dose: 1500 mg/kg; Duration: female 6-20D of pregnancy; Effects on Embryo or Fetus - Fetotoxicity. Rat Route: Oral; Dose: 3 gm/kg; Duration: female 6-20D of pregnancy; Specific Developmental Abnormalities - Craniofacial (including nose and tongue); Musculoskeletal system. Rat Route: Oral; Dose: 3 gm/kg; Duration: female 6-20D of pregnancy; Effects on Fertility - Post-implantation mortality.

Hazard Overviews

Health: May cause irritation to eyes/skin. Other Acute Effects: may be harmful by inhalation, ingestion, or skin absorption.
Fire: Will burn. Hazards: emits toxic fumes. Extinguishing agents: water spray; carbon dioxide, dry chemical powder or appropriate foam. Precautions: combustible liquid.
Reactivity: Incompatible with: strong oxidizing agents. Hazardous decomposition products: toxic fumes of: carbon monoxide, carbon dioxide, nitrogen oxides, sulfur oxides.
Carcinogenicity: IARC - Not listed; NIOSH - Not listed; NTP - Not listed; ACGIH - Not listed; OSHA - Not listed; EPA - Not listed; MAK - Not listed

Environmental

Regulations

RCRA 40CFR: Not listed
CERCLA: 40CFR 302.4: Not listed
SARA 40CFR 372.65: Not listed
SARA EHS 40CFR 355: Not listed
TSCA: Listed

DIP3520 CAS #: 122-39-4

N,N-DIPHENYLAMINE

RTECS: JJ7800000
EINECS Number: 204-539-4
Molecular Formula: $C_{12}H_{11}N$
Structured MF: $(C_6H_5)_2NH$
Formula Weight: 169.22

Chemical Structure

Synonyms: ANILINE,N-PHENYL-; ANILINOBENZENE; BENZENAMINE,N-PHENYL-; BENZENAMINE,N-PHENYL-(9CI); BENZENE,ANILINO-; BENZENE,(PHENYLAMINO)-; BIG DIPPER; C.I. 10355; DECCOSCALD 282; DFA; DIFENYLAMIN; DIPHENYLAMINE; DPA; N-FENYLANILIN; NO SCALD; NO-SCALD; NO-SCALD DPA 283; N-PHENYL ANILINE; (PHENYLAMINO)BENZENE; N-PHENYLANILINE; N-PHENYLBENZENAMINE; N-PHENYLBENZENEAMINE; SCALDIP; SHIELD DPA
Description: colorless crystalline solid; floral odor
Use: in rubber antioxidants and accelerators, in solid rocket propellants, as an insecticide, in the manufacture of dyes, in pharmaceuticals, in veterinary medicine (used topically in anti-screwworm mixtures and in tests for nitrate or nitrite poisoning), in storage preservation of apples, as a stabilizer for nitrocellulose explosives and cellulose, and in analytical chemistry for the detection of NO3, ClO3 and other strong oxidizer It is used in image formation in films and to protect rice from the effects o thiolcarbamate herbicides; controls weatherfleck in tobacco and inhibits algae formation; active against body lice, chiggers and houseflies

Physical Properties

Boiling Point: 302 °C (576 °F)
Freezing Point: 53 °C (127.4 °F) to 54 °C (129.2 °F)
Specific Gravity: 1.16
Vapor Density: 5.82 Air=1
Density: 1.16 g/mL
Vapor Pressure: 1 mm Hg at 108.3 °C
Water Solubility: 0.03% by weight
Other Solubilities: Very Soluble in Ethyl Acetate, Carbon Tetrachloride, Acetone, Pyridine; Soluble in Petroleum Ether.
Surface Tension: 39.3 dynes/cm at 60 °C
Odor Threshold: 0.05 ppm
Evaporation Rate: < 1 Butyl Acetate=1
Ionization Potential (eV): 7.40
Flash Point: 153 °C Open Cup
Autoignition Temperature: 634 °C

RTECS Toxicity Data

Acute Oral: Rat LD_{50} Dose: 2 gm/kg; Toxic Effects: Lungs, Thorax, or Respiration - Cyanosis. Mouse LD_{50} Dose: 1750 mg/kg; Toxic Effects: Lungs, Thorax, or Respiration - Cyanosis.
Chronic (Multiple Dose) Oral: Rat Dose: 9660 mg/kg/92W-C; Toxic Effects: Blood - Other changes. Rat Dose: 28 gm/kg/28D-I; Toxic Effects: Kidney, Ureter, and Bladder -

Chgs in tubules (inc acute renal failure, acute tubular necrosis; Blood - Changes in bone marrow not included above; Nutritional and gross metabolic - Weight loss or decreased weight gain.

Reproductive/Teratogenic: Rat Route: Oral; Dose: 7500 mg/kg; Duration: female 17-22D of pregnancy Specific Developmental Abnormalities - Urogenital system.

Hazard Overviews

Explosive

Fire Diamond

Health: Irritating to eyes/skin/respiratory tract. Also Causes: bladder symptoms, rapid heart beat, high BP, fatigue, anorexia, cyanosis, vomiting, emaciation. Chronic Effects: skin rashes, loss of appetite, diarrhea, low body temperature, weight loss.

Fire: Will burn. May form explosive dust-air mixtures. Use dry chemical, carbon dioxide, alcohol foam, or water spray. Do not create dusty conditions in the fire area.

Reactivity: Stable. Hazardous polymerization cannot occur. Avoid: excessive heat; ignition sources. Incompatible with: oxidizing materials; hexachloromelamine; trichloromelamine. Hazardous decomposition products: carbon monoxide; oxides of nitrogen.

Carcinogenicity: IARC - Not listed; NIOSH - Not listed; NTP - Not listed; ACGIH - Class A4, Not classifiable as a human carcinogen; OSHA - Not listed; EPA - Not listed; MAK - Not listed

Primary Target Organs:

Eyes

Skin

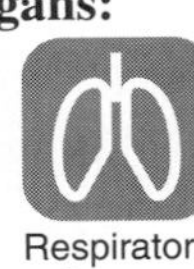
Respiratory System

Cardio-vascular

Blood

Exposure Limits

OSHA PEL Vacated 1989 Limits: TWA: 10 mg/m^3.

ACGIH TLV: TWA: 10 mg/m^3.

Respirator Recommendation

Exposure Range: >10 to 100 mg/m^3 Air Purifying, Negative Pressure, Half Mask

Exposure Range: >100 to 1000 mg/m^3 Air Purifying, Negative Pressure, Full Face

Exposure Range: >1000 to 10,000 mg/m^3 Supplied Air, Constant Flow/Pressure Demand, Full Face

Exposure Range: >10,000 to unlimited mg/m^3 Self-contained Breathing Apparatus, Pressure Demand, Full Face

Cartridge Color: black with dust/mist prefilter (use P100 or consult supervisor for appropriate dust/mist prefilter)

Environmental

Cleanup/Disposal: Guide No. 129: Eliminate all ignition sources (no smoking, flares, sparks or flames in immediate area). All equipment used when handling the product must be grounded. Do not touch or walk through spilled material. Stop leak if you can do it without risk. Prevent entry into waterways, sewers, basements or confined areas. A vapor suppressing foam may be used to reduce vapors. Absorb or cover with dry earth, sand or other non-combustible material and transfer to containers. Use clean non-sparking tools to collect absorbed material. Large Spills: Dike far ahead of liquid spill for later disposal. Water spray may reduce vapor; but may not prevent ignition in closed spaces.

Environmental Physical Data

Octanol/Water Partition Coefficient: log K_{ow} = 3.22 to 3.5

BCF: no food chain concentration potential

Regulations

RCRA 40CFR: Not listed

CERCLA: 40CFR 302.4: Not listed

SARA 40CFR 372.65: Listed

SARA EHS 40CFR 355: Not listed

TSCA: Listed

Analytical Methods

Soil: EPA PMD-DPF; SW846 3640A, 8250A, 8270B, 8270C, 8275

Water / Groundwater: EPA 1625, 620; APHA 6040-B; ASTM D4763

Food: FDA 212.1, 232.1, 232.4, 242.1

Plasma: FDA 252, 29

DIP3700 CAS #: 587-84-8

DIPHENYLAMINE SULFATE

RTECS: JJ9285000

EINECS Number: 209-605-6

Molecular Formula: $C_{12}H_{13}NO_4S$

Formula Weight: 267.32

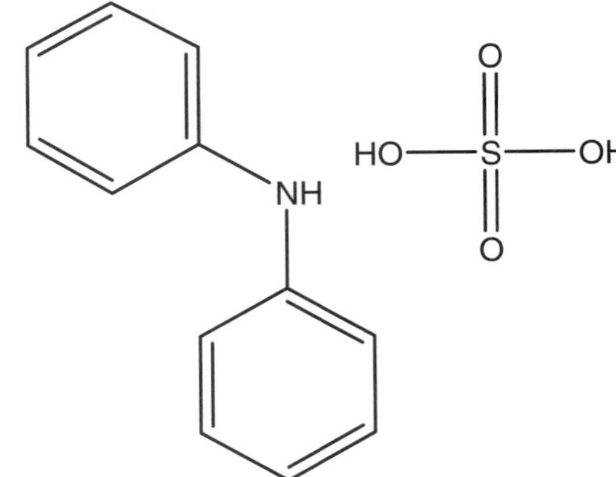

Chemical Structure

Synonyms: BENZENAMINE,N-PHENYL-,SULFATE (1:1); DIPHENYLAMINE,HYDROGEN SULFATE; DIPHENYLAMINE,SULFATE

Description: white to yellowish powder

Use: mfr dyes; stabilizing nitrocellulose explosives & celluloid; in analytical chemistry for detection of nitrates, chlorates & other oxidizing substances; medication (vet): topically in anti-screwworm mixtures; formerly in tests for nitrate or nitrite poisoning

Physical Properties

Freezing Point: 125 °C (257 °F)
Water Solubility: Practically Insoluble in Water
Other Solubilities: Soluble in Alcohol, Sulfuric acid

Hazard Overviews

Carcinogenicity: IARC - Not listed; NIOSH - Not listed; NTP - Not listed; ACGIH - Not listed; OSHA - Not listed; EPA - Not listed; MAK - Not listed

Environmental

Regulations

RCRA 40CFR: Not listed
CERCLA: 40CFR 302.4: Not listed
SARA 40CFR 372.65: Not listed
SARA EHS 40CFR 355: Not listed
TSCA: Listed

DIP4060 **CAS #: 103-29-7**

1,2-DIPHENYLETHANE

RTECS: DT4375000
EINECS Number: 203-096-4
Molecular Formula: $C_{14}H_{14}$
Formula Weight: 182.28

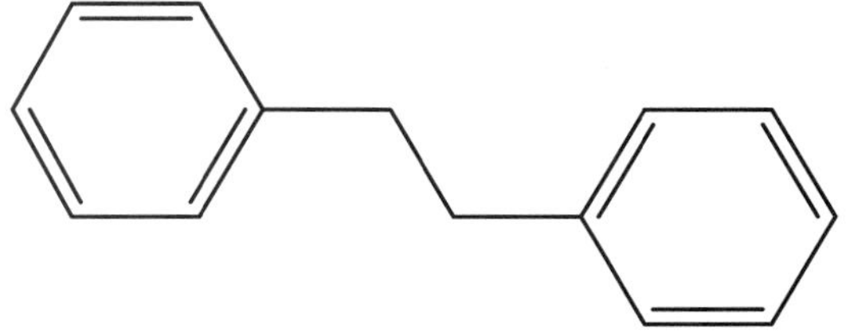

Chemical Structure

Synonyms: DIBENZYL; ETHANE,1,2-DIPHENYL-

Physical Properties

Boiling Point: 284 °C (543.2 °F)
Freezing Point: 50 °C (122 °F) to 53 °C (127.4 °F)
Specific Gravity: 0.97800
Water Solubility: Insoluble
Other Solubilities: Ethanol: Soluble; Ether: Soluble; CS_2: Soluble; liq SO{\{dn Slightly soluble} Soluble; liq NH_3 Insoluble
Refraction Index: 1.5476

RTECS Toxicity Data

Acute Oral: Rat LD_{50} Dose: 4518 mg/kg; Toxic Effects: Sense organs and special senses - Lacrimation; Behavioral - Change in motor activity (specific assay); Gastrointestinal - Hypermotility, diarrhea.
Acute Dermal: Rabbit LD_{50} Route: Skin; Dose: >5 gm/kg.

Hazard Overviews

Health: May be irritating to eyes/skin. Other Acute Effects: may be harmful by inhalation, ingestion, or skin absorption.
Fire: Hazards: emits toxic fumes. Extinguishing agents: water spray; carbon dioxide, dry chemical powder or appropriate foam. Precautions: combustible liquid.
Reactivity: Incompatible with: strong oxidizing agents. Hazardous decomposition products: toxic fumes of: carbon monoxide, carbon dioxide.
Carcinogenicity: IARC - Not listed; NIOSH - Not listed; NTP - Not listed; ACGIH - Not listed; OSHA - Not listed; EPA - Not listed; MAK - Not listed

Environmental

Environmental Physical Data

Octanol/Water Partition Coefficient: log K_{ow} = 4.79 to 4.82

Regulations

RCRA 40CFR: Not listed
CERCLA: 40CFR 302.4: Not listed
SARA 40CFR 372.65: Not listed
SARA EHS 40CFR 355: Not listed
TSCA: Listed

DIP4240 **CAS #: 645-49-8**

CIS-1,2-DIPHENYLETHYLENE

EINECS Number: 211-445-7
Molecular Formula: $C_{14}H_{12}$
Formula Weight: 180.24

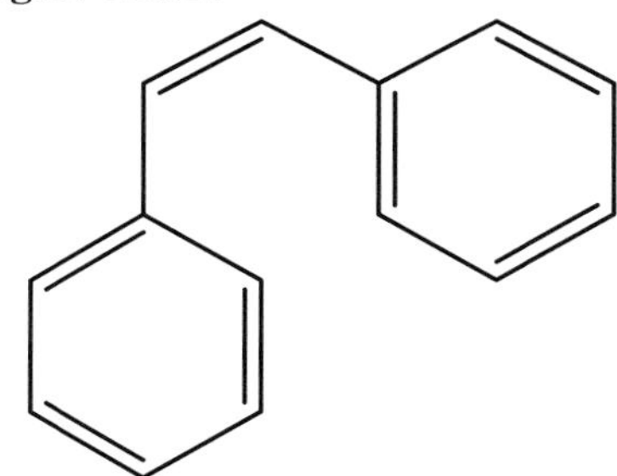

Chemical Structure

Synonyms: BENZENE,1,1'-(1,2-ETHENEDIYL)BIS-,(Z)-; CIS-DIPHENYLETHENE; ISOSTILBENE; (Z)-STILBENE; CIS-STILBENE; STILBENE,(Z)-
Description: yellow oily liquid
Use: manufacture of dyes and optical bleaches; crystals are used as phosphor and scintillators

Physical Properties

Boiling Point: 135 °C (275 °F) at 10 mm Hg
Freezing Point: 1 °C (33.8 °F)
Specific Gravity: 1.01430
Vapor Pressure: 0.199 kPa at 100 °C
Water Solubility: Insoluble

Other Solubilities: 95% Ethanol: Very Soluble; Benzene: Soluble.
Refraction Index: 1.6188 at 25 °C/D
Ionization Potential (eV): 7.80 +/-0.4
Flash Point: Not available; probably combustible

Hazard Overviews

Health: Irritating to eyes/skin. Other Acute Effects: may be harmful by inhalation, ingestion, or skin absorption.
Fire: Will burn. Extinguishing agents: water spray; carbon dioxide, dry chemical powder or appropriate foam. Precautions: combustible liquid.
Reactivity: Incompatible with: strong oxidizing agents. Hazardous decomposition products: toxic fumes of: carbon monoxide, carbon dioxide.
Carcinogenicity: IARC - Not listed; NIOSH - Not listed; NTP - Not listed; ACGIH - Not listed; OSHA - Not listed; EPA - Not listed; MAK - Not listed
Primary Target Organs:

Eyes

Skin

Environmental

Regulations

RCRA 40CFR: Not listed
CERCLA: 40CFR 302.4: Not listed
SARA 40CFR 372.65: Not listed
SARA EHS 40CFR 355: Not listed
TSCA: Not listed

DIP4420 CAS #: 103-30-0

TRANS-1,2-DIPHENYLETHYLENE

RTECS: WJ4926500
EINECS Number: 203-098-5
Molecular Formula: $C_{14}H_{12}$
Formula Weight: 180.24

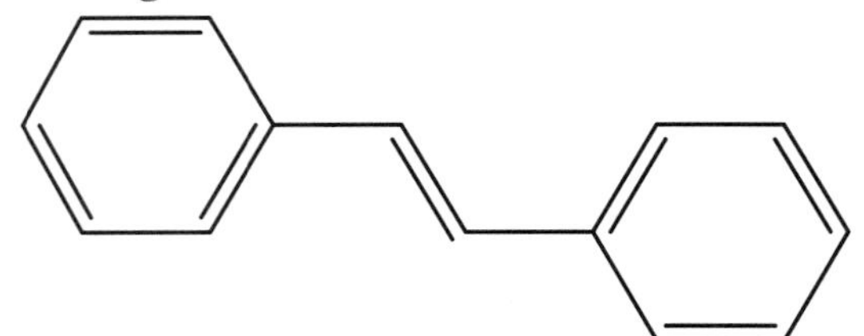

Chemical Structure

Synonyms: BENZENE,1,1'-(1,2-ETHENEDIYL)BIS-,(E)-; BENZENE,1,1'-(1,2-ETHENEDIYL)BIS-,(E)-(9CI); TRANS-BIBENZAL; TRANS-BIBENZYLIDENE; TRANS-1,2-DIPHENYLETHENE; TRANS-DIPHENYLETHENE; (E)-1,2-DIPHENYLETHYLENE; TRANS-ALPHA,BETA-DIPHENYLETHYLENE; TRANS-1,1'-(1,2-ETHENDIYL)-BIS(BENZENE); (E)-1,1'-(1,2-ETHENEDIYL)BISBENZENE; (E)-STILBENE; STILBENE,(E)-; TRANS-STILBENE; TOLULYNE; TOLUYLENE
Description: colorless or slightly yellow crystals
Use: manufacture of dyes and optical bleaches; crystals are used as phosphor and scintillators

Physical Properties

Boiling Point: 306 °C (583 °F) to 307 °C (585 °F) at 760 mm Hg
Freezing Point: 124 °C (255.2 °F)
Specific Gravity: 0.9707
Vapor Pressure: 0.784 kPa at 150 °C
Water Solubility: Practically Insoluble in Water
Other Solubilities: 95% Ethanol: Soluble; Benzene: Very Soluble; Ether: Very Soluble.
Refraction Index: 1.6264 at 17 °C
Ionization Potential (eV): 7.70 +/-0.7
Flash Point: Not available; probably combustible

RTECS Toxicity Data

Acute Oral: Mouse LD_{50} Dose: 920 mg/kg.

Hazard Overviews

Health: May be irritating to eyes/skin. Harmful. Other Acute Effects: harmful if swallowed; may be harmful if inhaled; may be harmful if absorbed through the skin.
Fire: Will burn. Hazards: emits toxic fumes. Extinguishing agents: water spray; carbon dioxide, dry chemical powder or appropriate foam. Precautions: combustible liquid.
Reactivity: Incompatible with: strong oxidizing agents. Hazardous decomposition products: toxic fumes of: carbon monoxide, carbon dioxide.
Carcinogenicity: IARC - Not listed; NIOSH - Not listed; NTP - Not listed; ACGIH - Not listed; OSHA - Not listed; EPA - Not listed; MAK - Not listed

Environmental

Regulations

RCRA 40CFR: Not listed
CERCLA: 40CFR 302.4: Not listed
SARA 40CFR 372.65: Not listed
SARA EHS 40CFR 355: Not listed
TSCA: Listed

DIP4600 CAS #: 1241-94-7

DIPHENYL-2-ETHYLHEXYL PHOSPHATE

RTECS: TC6125000
EINECS Number: 214-987-2
Molecular Formula: $C_{20}H_{27}O_4P$
Formula Weight: 362.41
Synonyms: DIPHENYL 2-ETHYLHEXYL PHOSPHATE; DISFLAMOLL DPO; ETHYL HEXYL DIPHENYLPHOSPHATE; 2-ETHYLHEXYL DIPHENYL PHOSPHATE; 2-ETHYLHEXYL DIPHENYL PHOSPHORATE; 2-ETHYLHEXYL

DIPHENYLPHOSPHATE; (2-ETHYLHEXYL)-DIFENYLFOSFAT; 1-HEXANOL,2-ETHYL-,ESTER WITH DIPHENYL PHOSPHATE; OCTICIZER; PHOSPHORIC ACID,2-ETHYLHEXYL DIPHENYL ESTER; SANTICIZER 141

Description: liquid

Use: plasticizer; fire retardant

Physical Properties

Boiling Point: 375 °C (707 °F)

Freezing Point: -30 °C (-22 °F)

Specific Gravity: 1.09000

Saturated Vapor Density: 1.200001142 kg/m^3

Vapor Pressure: 6.29 $x10^{-5}$ mm Hg at 25 °C

Water Solubility: 1.9 mg/L at 25 °C

Other Solubilities: 95% Ethanol: Soluble; Acetone: Soluble; Benzene: Soluble; Ether: Soluble.

Refraction Index: 1.510 at 25 °C/D

Flash Point: Probably combustible

RTECS Toxicity Data

Acute Oral: Rat LD_{Lo} Dose: 10 gm/kg. Rabbit LD_{50} Dose: 218 mg/kg.

Acute Dermal: Rabbit LD_{50} Route: Skin; Dose: >7900 mg/kg.

Hazard Overviews

Fire: Combustible.

Carcinogenicity: IARC - Not listed; NIOSH - Not listed; NTP - Not listed; ACGIH - Not listed; OSHA - Not listed; EPA - Not listed; MAK - Not listed

Environmental

Environmental Fate: If released to soil, calculated soil adsorption coefficients indicate that it may display slight to moderate mobility in soil. It may slowly volatilize from moist soil to the atmosphere, although adsorption to soil may attenuate the rate of this process. It will not volatilize from dry soil to the atmosphere. It has the potential to biodegrade in soil under aerobic conditions. If released to water, it is expected to biodegrade under aerobic conditions and it may slowly biodegrade in sediments under both aerobic and anaerobic conditions. It may slowly volatilize from water to the atmosphere, although adsorption to sediment may attenuate the rate of this process. The estimated half-life for volatilization from a model river is 4.7 days. Outdoor experiments performed in small ponds indicate that it will adsorb to sediment and suspended organic matter, will not bioconcentrate in fish and aquatic organisms and will only slowly volatilize to the atmosphere. If released to the atmosphere, it is unlikely to undergo direct photochemical degradation.

Environmental Physical Data

Henry's Law Constant: 1.58 $x10^{-5}$

Octanol/Water Partition Coefficient: log K_{ow} = extrapolated 5.730

Sorption Partition Coefficient: K_{oc} = calculated at 430 to 3100

BCF: % in minnows 1 to 2

Regulations

RCRA 40CFR: Not listed

CERCLA: 40CFR 302.4: Not listed

SARA 40CFR 372.65: Not listed

SARA EHS 40CFR 355: Not listed

TSCA: Listed

Analytical Methods

Plasma: EPA 001

DIP4780 CAS #: 102-06-7

N,N'-DIPHENYLGUANIDINE

RTECS: MF0875000

EINECS Number: 203-002-1

Molecular Formula: $C_{13}H_{13}N_3$

Structured MF: HN=C($NHC_6H_5)_2$

Formula Weight: 211.29

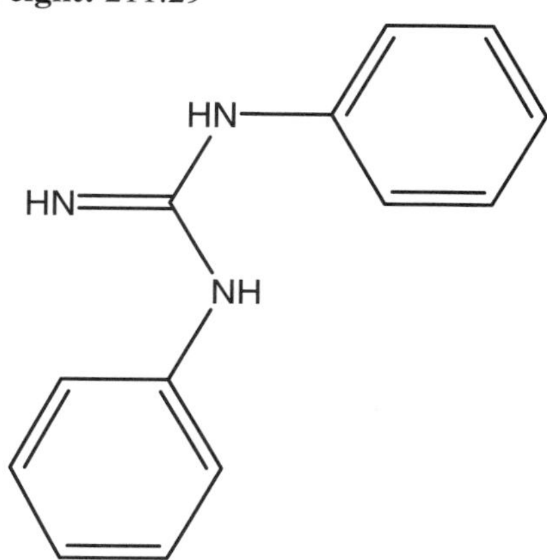

Chemical Structure

Synonyms: ACCELERATOR D; DENAX; DENAX DPG; DFG; 1,3-DIFENYLGUANID; 1,3-DIPHENYLGUANIDINE; DIPHENYLGUANIDINE; SYM-DIPHENYLGUANIDINE; DPG; DPG ACCELERATOR; DWUFENYLOGUANIDYNA; GUANIDINE,1,3-DIPHENYL-; GUANIDINE,N,N'-DIPHENYL-; MELANILINE; NOCCELER D; VULCACID D; VULCAFOR DPG; VULKACIT D/C; VULKACIT D; VULKACITE D; VULKAZIT

Description: white powder or monoclinic needles; slight odor

Use: primary standard for acids; accelerator for the vulcanization of rubber; and complexing agent in the detection of metals and organic bases

Physical Properties

Boiling Point: 170 °C (338 °F)

Freezing Point: 150 °C (302 °F)

Specific Gravity: 1

Density: 1.15 g/mL at 25 °C

Vapor Pressure: Negligible

Water Solubility: Sparingly Soluble in Water

pH: Aqueous solution is strongly alkaline

Flash Point: 170 °C

RTECS Toxicity Data

Acute Oral: Rat LD_{50} Dose: 323 mg/kg. Mouse LD_{50} Dose: 150 mg/kg; Toxic Effects: Peripheral nerve and sensation - Flaccid paralysis without anesthesia; Behavioral - Ataxia.

Acute Dermal: Rabbit LD_{50} Route: Skin; Dose: >794 mg/kg. Guinea Pig LD_{Lo} Route: Subcutaneous Dose: 200 mg/kg; Toxic Effects: Behavioral - Altered sleep time (including change in righting reflex); Behavioral - Tetany.

Chronic (Multiple Dose) Oral: Rat Dose: 3808 mg/kg/17W-I; Toxic Effects: Blood - Pigmented or nucleated red Blood cells; Blood - Changes in erythrocite (RBC) cell count; Biochemical - Catalases.

Reproductive/Teratogenic: Mouse Route: Oral; Dose: 76 mg/kg; Duration: female 1-19D of pregnancy; Specific Developmental Abnormalities - Musculoskeletal system. Mouse Route: Oral; Dose: 190 mg/kg; Duration: female 1-19D of pregnancy; Effects on Fertility - Pre-implantation mortality. Mouse Route: Oral; Dose: 56 mg/kg; Duration: male 7D prior to mating; Effects on Fertility - Male fertility index; Pre-implantation mortality. Mouse Route: Oral; Dose: 196 mg/kg; Duration: male 49D prior to mating; Paternal Effects - Spermatogenesis. Mouse Route: Oral; Dose: 140 mg/kg; Duration: male 35D prior to mating; Paternal Effects - Testes, epididymis, sperm duct.

Mutagenic: Mouse Body Fluid Assay; Indicator Organism: Bacteria - S Typhimurium; Dose: 36 ng/kg. Bacteria - S Typhimurium Mutations in Microorganisms; Dose: 360 ng/plate (+S9). Bacteria - S Typhimurium Mutations in Microorganisms; Dose: 33 ug/plate (-S9).

Hazard Overviews

Health: Irritating to eyes/skin/respiratory tract. Toxic. Other Acute Effects: harmful if swallowed, inhaled, or absorbed through skin; prolonged or repeated exposure may cause allergic reactions in certain sensitive individuals.

Fire: Will burn. Hazards: emits toxic fumes. Extinguishing agents: water spray; carbon dioxide, dry chemical powder or appropriate foam. Precautions: combustible liquid.

Reactivity: Incompatible with: strong oxidizing agents. Hazardous decomposition products: toxic fumes of: carbon monoxide, carbon dioxide, nitrogen oxides.

Carcinogenicity: IARC - Not listed; NIOSH - Not listed; NTP - Not listed; ACGIH - Not listed; OSHA - Not listed; EPA - Not listed; MAK - Not listed

Primary Target Organs:

Eyes

Skin

Respiratory System

Environmental

Cleanup/Disposal: Guide No. 171: Do not touch or walk through spilled material. Stop leak if you can do it without risk. Prevent dust cloud. Avoid inhalation of asbestos dust. Small Dry Spills: With clean shovel place material into clean, dry container and cover loosely; move containers from spill area. Small Spills: Take up with sand or other noncombustible absorbent material and place into containers for later disposal. Large Spills: Dike far ahead of liquid spill for later disposal. Cover powder spill with plastic sheet or tarp to minimize spreading. Prevent entry into waterways, sewers, basements or confined areas.

Regulations

RCRA 40CFR: Not listed
CERCLA: 40CFR 302.4: Not listed
SARA 40CFR 372.65: Not listed
SARA EHS 40CFR 355: Not listed
TSCA: Listed

DIP4960 CAS #: 530-50-7

1,1-DIPHENYLHYDRAZINE

RTECS: MV3450000
EINECS Number: 208-483-1
Molecular Formula: $C_{12}H_{12}N_2$
Formula Weight: 184.23
Synonyms: ALPHA,ALPHA-DIPHENYLHYDRAZINE; N,N-DIPHENYLHYDRAZINE; HYDRAZINE,1,1-DIPHENYL; HYDROZOBENZENE
Description: yellow crystals
Use: reagent for determination of arabinose & lactose

Physical Properties

Boiling Point: 220 °C (428 °F) at 40 mm Hg
Freezing Point: Either 34.5 °C (94.1 °F) to 44 °C (111.2 °F)
Specific Gravity: 1.19 at 16 °C/4 °C
Vapor Pressure: 1 mm Hg at 126 °C
Water Solubility: Slightly Soluble in Water
Other Solubilities: >10% in Benzene; >10% in ethyl Ether; >10% in Ethanol; >10% in Chloroform.

RTECS Toxicity Data

Mutagenic: Bacteria - S Typhimurium Mutations in Microorganisms; Dose: 130 nmol/plate (-S9).

Hazard Overviews

Carcinogenicity: IARC - Not listed; NIOSH - Not listed; NTP - Not listed; ACGIH - Not listed; OSHA - Not listed; EPA - Not listed; MAK - Not listed

Environmental

Environmental Fate: If released to soil or water, it may be subject to photolysis. Volatilization is not expected to be a significant fate process. This compound is expected to bioaccumulate slightly in aquatic organisms; however, bioaccumulation is not expected to be a significant fate process. If released to air, this compound is expected to exist partly in the vapor phase and partly in particulate form (adsorbed to airborne particulate matter). Both phases may be subject to photolysis. Vapor may also react with photochemically generated hydroxyl radicals (half life 3 hours) and particles may also be subject to dry deposition.

Environmental Physical Data

Henry's Law Constant: estimated at 2.91 x10^{-8}
Octanol/Water Partition Coefficient: log K_{ow} = 3.01
Sorption Partition Coefficient: K_{oc} = estimated at 1034
BCF: estimated at 114

Regulations

RCRA 40CFR: Not listed
CERCLA: 40CFR 302.4: Listed as Compound per CWA Section 307(a)
SARA 40CFR 372.65: Not listed
SARA EHS 40CFR 355: Not listed
TSCA: Not listed

DIP5140 CAS #: 122-66-7

1,2-DIPHENYLHYDRAZINE

RTECS: MW2625000
EINECS Number: 204-563-5
Molecular Formula: $C_{12}H_{12}N_2$
Structured MF: $C_6H_5NHNHC_6H_5$
Formula Weight: 184.24

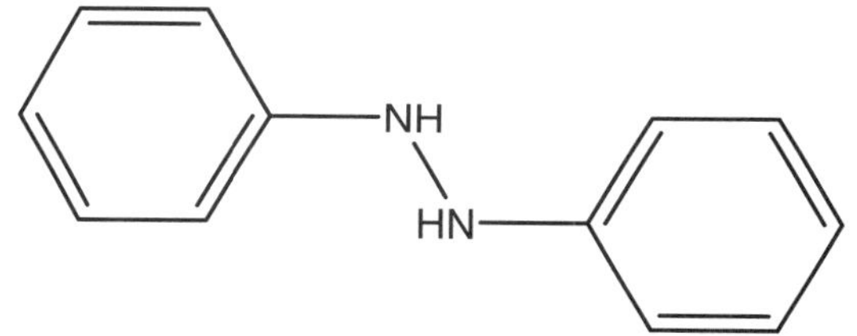

Chemical Structure

Synonyms: BENZENE,1,1'-HYDRAZOBIS-; BENZENE,HYDRAZODI-; N,N'-BIANILINE; (SYM)-DIPHENYLHYDRAZINE; N,N'-DIPHENYLHYDRAZINE; 1,2-DIPHENYLHYDRAZINE (9CI); DPH; HYDRAZINE,1,2-DIPHENYL-; HYDRAZOBENZEN; HYDRAZOBENZENE; 1,1'-HYDRAZODIBENZENE; HYDRAZODIBENZENE; SYMMETRICAL DIPHENYL HYDRAZINE
Description: tablets
Use: a precursor in the manufacture of benzidine and an intermediate in the production of dyes; in the synthesis of phenylbutazone, a potent anti-inflammatory (antiarthritic) drug

Physical Properties

Boiling Point: Decomposes
Freezing Point: 131 °C (267.8 °F)
Specific Gravity: 1.158 at 16 °C/4 °C
Density: 1.158 g/mL at 16 °C
Vapor Pressure: 1 torr at 103 °C
Water Solubility: Slightly Soluble in Water
Other Solubilities: 95% Ethanol: 10-50 mg/ml at 27 °C; Acetone: 50-100 mg/ml at 27 °C; Acetic Acid: Insoluble; Acetylene: Insoluble; Alcohol: Very Soluble; Benzene: Slightly Soluble; DMSO: >=100 mg/ml at 27 °C; Ethanol: >10%.
Flash Point: Not available; probably combustible

RTECS Toxicity Data

Mutagenic: Mouse DNA Inhibition; Route: Intraperitoneal; Dose: 100 mg/kg. Hamster Cytogenetic Analysis; Cell Type: ovary; Dose: 14 mg/L.
Tumorigenic: Rat Route: Oral; Dose: 2620 mg/kg/78W-C; Toxic Effects: Tumorigenic - Carcinogenic by RTECS criteria; Liver - Tumors; Skin and appendages - Tumors. Rat Route: Oral; Dose: 9820 mg/kg/78W-C; Toxic Effects: Tumorigenic - Carcinogenic by RTECS criteria; Liver - Tumors; Skin and appendages - Tumors. Rat Route: Oral; Dose: 36 gm/kg/53W-I; Toxic Effects: Tumorigenic - Equivocal tumorigenic agent by RTECS criteria; Skin and appendages - Tumors; Tumorigenic effects - Uterine tumors.

Hazard Overviews

Explosive

Health: May cause irritation to eyes/skin/respiratory tract. Toxic. Other Acute Effects: harmful if swallowed, inhaled, or absorbed through skin. Chronic Effects: may alter genetic material; may cause heritable genetic damage. Carcinogen.
Fire: Will burn. Hazards: emits toxic fumes. Extinguishing agents: water spray; carbon dioxide, dry chemical powder or appropriate foam. Precautions: combustible liquid.
Reactivity: Incompatible with: strong oxidizing agents, strong acids, acid chlorides, acid anhydrides. Hazardous decomposition products: toxic fumes of: carbon monoxide, carbon dioxide, nitrogen oxides.
Carcinogenicity: IARC - Not listed; NIOSH - Not listed; NTP - Listed; ACGIH - Not listed; OSHA - Not listed; EPA - Class B2, Probable human carcinogen based on animal studies; MAK - Not listed

Environmental

Ecotoxicity: LC_{50} Lepomis macrochirus (bluegill) 0.27 mg/l/96 hr (static bioassay, total water hardness 28-44 mg/l calcium carbonate, pH 6.7-7.4, oxygen concentration 5.3-7.0 mg/l) LC_{50} Daphnia magna (cladoceran) static method 4,100 ug/l/96-hr
Environmental Fate: It is in rapid redox equilibrium with azobenzene and under most conditions it will be quickly converted to it (half life a few minutes) by air, cations such as copper(II) etc. Degradation will also result from photolysis and possibly biodegradation but these processes would not compete with oxidation under most conditions. It will also adsorb moderately to soil, sediment and particulate matter which may increase the oxidation rate by facilitating contact with environmentally prevalent cations which catalyze its oxidation.

Environmental Physical Data

Octanol/Water Partition Coefficient: log K_{ow} = 2.94
Sorption Partition Coefficient: K_{oc} = 947
BCF: calculated at 2.00

Regulations

RCRA 40CFR: Listed Hazardous Waste No. U109 Toxic Waste
CERCLA: 40CFR 302.4: Listed per RCRA Section 3001 per CWA Section 307(a) RQ: 10 lb (4.535 kg)
SARA 40CFR 372.65: Listed
SARA EHS 40CFR 355: Not listed
TSCA: Listed

Analytical Methods

Soil: SW846 3640A, 8250A, 8270B, 8270C; EPA 16
Water / Groundwater: EPA 1625, 625-S, 6; APHA 6040-B
Plasma: EPA 29

DIP5320 CAS #: 101-68-8

DIPHENYLMETHYL DIISOCYANATE

RTECS: NQ9350000
DOT: UN2489; IMO6.1
EINECS Number: 202-966-0
Molecular Formula: $C_{15}H_{10}N_2O_2$
Structured MF: $CH_2(C_6H_4NCO)_2$
Formula Weight: 250.27

Chemical Structure

Synonyms: BENZENE,1,1'-METHYLENEBIS(4-ISOCYANATO-; BENZENE,1,1'-METHYLENEBIS(4-ISOCYANATO)-(9CI); BENZENE,1,1'-METHYLENEBIS(4-ISOCYANATO-(9CI); BIS(1,4-ISOCYANATOPHENYL)METHANE; BIS(4-ISOCYANATOPHENYL)METHANE; BIS(P-ISOCYANATOPHENYL)METHANE; BIS(PARA-ISOCYANATOPHENYL)METHANE; CARADATE 30; DESMODUR 44; DIFENIL-METAN-DIISOCIANATO; DIFENYLMETHAAN-DISSOCYANAAT; 4-4'-DIISOCYANATE DE DIPHENYLMETHANE; 4,4'-DIISOCYANATODIPHENYLMETHANE; 4,4'DIISOCYANATODIPHENYLMETHANE; DI-(4-ISOCYANATOPHENYL)METHANE; DIPHENYL METHANE DIISOCYANATE; DIPHENYLMETHAN-4,4'-DIISOCYANAT; 4,4'-DIPHENYLMETHANE DIISOCYANATE; DIPHENYLMETHANE 4,4'-DIISOCYANATE; DIPHENYLMETHANE DIISOCYANATE; P,P'-DIPHENYLMETHANE DIISOCYANATE; PARA,PARA'-DIPHENYLMETHANE DIISOCYANATE; DIPHENYLMETHANE-4,4'-DIISOCYANATE; HYLENE M50; ISOCYANIC ACID,METHYLENEDI-P-PHENYLENE ESTER; ISONATE; ISONATE 125M; ISONATE 125 MF; ISONATE 125MF; MDI; METHYLENE BIS(4-PHENYL ISOCYANATE); METHYLENE BISPHENYL ISOCYANATE; METHYLENE BIS(4-PHENYLISOCYANATE); METHYLENE DI-P-PHENYLENE ESTER OF ISOCYANIC ACID; METHYLENE DI-P-PHENYLENE ISOCYANATE; 1,1'-METHYLENEBIS(4-ISOCYANATOBENZENE); 1,1-METHYLENEBIS(4-ISOCYANATOBENZENE); METHYLENEBIS(4-ISOCYANATOBENZENE); 4,4'-METHYLENEBIS(PHENYL ISOCYANATE); METHYLENEBIS(4-PHENYL ISOCYANATE); METHYLENEBIS(P-PHENYL ISOCYANATE); METHYLENEBIS(PARA-PHENYL ISOCYANATE); P,P'-METHYLENEBIS(PHENYL ISOCYANATE); PARA,PARA'-METHYLENEBIS(PHENYL ISOCYANATE); METHYLENEBIS(4-PHENYLENE ISOCYANATE); METHYLENEBIS(P-PHENYLENE ISOCYANATE); METHYLENEBIS(PARA-PHENYLENE ISOCYANATE); 4,4'-METHYLENEDIPHENYL DIISOCYANATE; METHYLENEDIPHENYL DIISOCYANATE; 4,4'-METHYLENEDIPHENYL ISOCYANATE; 4,4'-METHYLENEDIPHENYLDIISOCYANATE; 4,4'-METHYLENEDI-P-PHENYLENE DIISOCYANATE; METHYLENEDI-P-PHENYLENE DIISOCYANATE; METHYLENEDI-PARA-PHENYLENE DIISOCYANATE; 4,4'-METHYLENEDIPHENYLENE ISOCYANATE; METHYLENEDI-P-PHENYLENE ISOCYANATE; METHYLENEDI-PARA-PHENYLENE ISOCYANATE; NACCONATE 300; NOCCONATE 300; RUBINATE 44

Description: white, light yellow flakes, fused solid, liquid above 99 F (37.2 C); odorless

Use: preparation of polyurethane resin and spandex fibers, bonding rubber to rayon

Physical Properties

Boiling Point: 196 °C (385 °F) at 5 mm Hg
Freezing Point: 37 °C (98.6 °F)
Specific Gravity: 1.197 at 70 °C
Density: 1.2 g/mL at 20 °C
Vapor Pressure: 0.001 mm Hg at 40 °C
Water Solubility: 0.2% by weight
Other Solubilities: 95% Ethanol: reacts violently; Acetone: Soluble; Benzene: Soluble; DMSO: Soluble; Kerosene: Soluble; Nitrobenzene: Soluble.
Refraction Index: 1.5906 at 50 °C/D
Flash Point: 202.222 °C Open Cup

RTECS Toxicity Data

Acute Oral: Rat LD_{50} Dose: 9200 mg/kg; Toxic Effects: Behavioral - Somnolence (general depressed activity); Behavioral - Ataxia; Nutritional and gross metabolic - Body temperature decrease. Mouse LD_{50} Dose: 2200 mg/kg.

Acute Inhalation: Rat LC_{50} Dose: 178 mg/m^3. Human TC_{Lo} Dose: 130 ppb/30M; Toxic Effects: Immunological including allergic - Increased immune response; Nutritional and gross metabolic - Body temperature increase.

Irritation Eye: Rabbit Standard Draize Test Dose: 100 mg; Reaction: moderate.

Mutagenic: Human DNA Damage; Route: Inhalation; Dose: 20 ppb/15M-C. Human Cytogenetic Analysis; Cell Type: lymphocyte; Dose: 540 mg/L. Human Sister Chromatid Exchenge; Cell Type: lymphocyte; Dose: 2170 mg/L.

Hazard Overviews

Fire Diamond

Health: Severely irritating to eyes/skin/respiratory tract. Also Causes: coughing, chest pain, bluish face and lips, wheezing, headache, nausea, diarrhea, dizziness, fatigue, bronchitis, pulmonary edema, itching, eczema, lid inflammation, corneal burns, eye lesions, painful swallowing, abdominal and stomach pain, state of shock, convulsions, sensitization. Chronic Effects: asthmatic bronchitis.

Fire: Will burn. Use dry chemical, waterspray, fog, or regular foam. Container may explode in heat of fire.
Reactivity: Stable. Hazardous polymerization cannot occur. Avoid: heat; ignition sources. Incompatible with: alcohols; strong alkalis; acids. Hazardous decomposition products: carbon dioxide; nitrogen oxides; toxic isocyanate vapors.
Carcinogenicity: IARC - Not listed; NIOSH - Not listed; NTP - Not listed; ACGIH - Not listed; OSHA - Not listed; EPA - Not listed; MAK - Class B, Justifiably suspected of having carcinogenic potential
Primary Target Organs:

Eyes

Skin

Respiratory System

Exposure Limits
OSHA PEL: STEL: 0.02 ppm; 0.2 mg/m^3.
NIOSH REL: TWA: 0.005 ppm; 0.05 mg/m^3. STEL: 0.02 ppm; 0.2 mg/m^3; 10-minute.
NIOSH IDLH: 75 mg/m^3.
DFG MAK: TWA: 0.005 ppm; 0.05 mg/m^3.
Respirator Recommendation
Exposure Range: >0.2 to 10 mg/m^3 Supplied Air, Constant Flow/Pressure Demand, Half Mask
Exposure Range: >10 to <75 mg/m^3 Supplied Air, Constant Flow/Pressure Demand, Full Face
Exposure Range: 75 to unlimited mg/m^3 Self-contained Breathing Apparatus, Pressure Demand, Full Face
Note: odor threshold unknown

Environmental

Environmental Fate: If released to the atmosphere, it will exist primarily in the vapor-phase where it will degrade by reaction with photochemically produced hydroxyl radicals (estimated half-life of 32 hours). Reaction with atmospheric water vapor may increase the atmospheric degradation rate several fold. It reacts readily with water. If released to water or moist soil in a spill situation, agglomerations react with water to form a hard crust of inert, water-insoluble material comprised of polyureas; the polyurea crusts can entrap and prevent further contact with water, thereby increasing the persistence time.
Cleanup/Disposal: Guide No. 156: Eliminate all ignition sources (no smoking, flares, sparks or flames in immediate area). All equipment used when handling the product must be grounded. Do not touch damaged containers or spilled material unless wearing appropriate protective clothing. Stop leak if you can do it without risk. A vapor suppressing foam may be used to reduce vapors. For chlorosilanes, use AFFF alcohol-resistant medium expansion foam to reduce vapors. Do not get water on spilled substance or inside containers. Use water spray to reduce vapors or divert vapor cloud drift. Prevent entry into waterways, sewers, basements or confined areas. Small Spills: Cover with dry earth, dry sand, or other non-combustible material followed with plastic sheet to minimize spreading or contact with rain. Use clean non-sparking tools to collect material and place it into loosely covered plastic containers for later disposal.

Environmental Physical Data
BCF: not significant

Regulations
RCRA 40CFR: Not listed
CERCLA: 40CFR 302.4: Listed per CAA Section 112 RQ: 5000 lb (2268 kg)
SARA 40CFR 372.65: Listed
SARA EHS 40CFR 355: Not listed
TSCA: Listed

Analytical Methods
Indoor / Expired Air: NIOSH 5521

DIP5500 CAS #: 144-79-6

DIPHENYLMETHYLCHLOROSILANE

EINECS Number: 205-639-0
Molecular Formula: $C_{13}H_{13}ClSi$
Formula Weight: 232.78

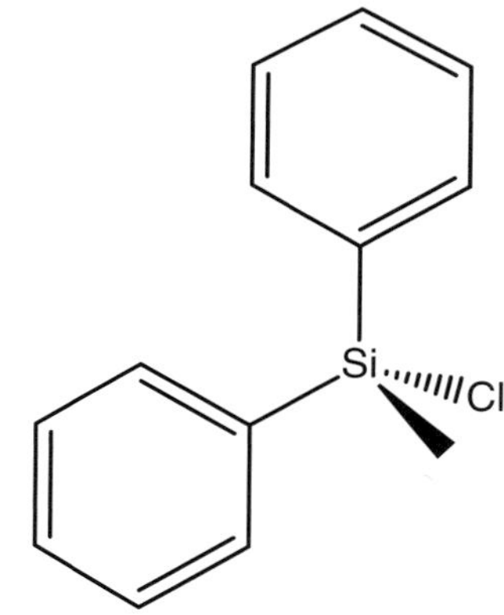

Chemical Structure

Physical Properties

Boiling Point: 295 °C (563 °F)
Specific Gravity: 1.128
Vapor Density: > 1 Air=1
Water Solubility: Reacts
Odor Threshold: 10.0 ppm
Refraction Index: 1.5742
Evaporation Rate: < 1 Butyl Acetate=1

Hazard Overviews

Corrosive

Health: Corrosive to eyes/skin/respiratory tract. Other Acute Effects: harmful if swallowed, inhaled, or absorbed through skin; inhalation may result in spasm; inflammation and edema of the larynx and bronchi; chemical pneumonitis and pulmonary edema; burning sensation; coughing; wheezing; laryngitis; shortness of breath; headache; nausea; vomiting.

Fire: Hazards: emits toxic fumes. Extinguishing agents: water spray; carbon dioxide, dry chemical powder or appropriate foam. Precautions: combustible liquid.

Reactivity: Incompatible with: strong oxidizing agents, may decompose on exposure to moist air or water. Hazardous decomposition products: toxic fumes of: carbon monoxide, carbon dioxide, silicon oxide, hydrogen chloride gas.

Carcinogenicity: IARC - Not listed; NIOSH - Not listed; NTP - Not listed; ACGIH - Not listed; OSHA - Not listed; EPA - Not listed; MAK - Not listed

Primary Target Organs:

Eyes

Skin

Respiratory System

Environmental

Regulations

RCRA 40CFR: Not listed
CERCLA: 40CFR 302.4: Not listed
SARA 40CFR 372.65: Not listed
SARA EHS 40CFR 355: Not listed
TSCA: Not listed

DIP5680 CAS #: 92-71-7

2,5-DIPHENYLOXAZOLE

RTECS: RP6825000
EINECS Number: 202-181-3
Molecular Formula: $C_{15}H_{11}NO$
Formula Weight: 221.27

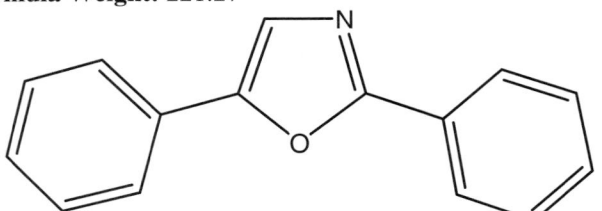

Chemical Structure

Physical Properties

Boiling Point: 360 °C (680 °F)
Freezing Point: 72 °C (161.6 °F) to 73 °C (163.4 °F)
Specific Gravity: 1.09400
Water Solubility: Insoluble
Other Solubilities: Ethanol: Very Soluble; Ether: Very Soluble; Chloroform: Slightly soluble
Refraction Index: 1.6231

Hazard Overviews

Health: May be irritating to eyes/skin. Other Acute Effects: may be harmful by inhalation, ingestion, or skin absorption.

Fire: Hazards: emits toxic fumes. Extinguishing agents: water spray; carbon dioxide, dry chemical powder or appropriate foam. Precautions: combustible liquid.

Carcinogenicity: IARC - Not listed; NIOSH - Not listed; NTP - Not listed; ACGIH - Not listed; OSHA - Not listed; EPA - Not listed; MAK - Not listed

Environmental

Regulations

RCRA 40CFR: Not listed
CERCLA: 40CFR 302.4: Not listed
SARA 40CFR 372.65: Not listed
SARA EHS 40CFR 355: Not listed
TSCA: Listed

DIP6040 CAS #: 74-31-7

N,N-DIPHENYL-P-PHENYLENEDIAMINE

RTECS: ST2275000
EINECS Number: 200-806-4
Molecular Formula: $C_{18}H_{16}N_2$
Structured MF: $(C_6H_5NH)_2C_6H_4$
Formula Weight: 260.32

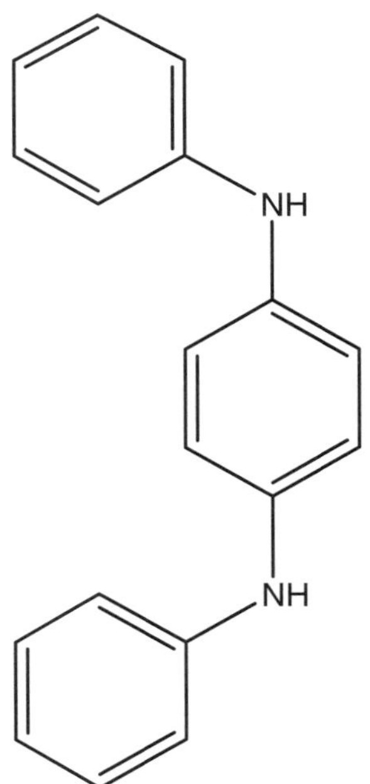

Chemical Structure

Synonyms: AGERITE; AGERITE DPPD; ALTOFANE DIP; ANTAGE DP; ANTIGENE P; ANTIOXIDANT H; 1,4-BENZENEDIAMINE,N,N'-DIPHENYL-; 1,4-BIS(PHENYLAMINO)BENZENE; P-BIS(PHENYLAMINO)BENZENE; DFFD; DIAFEN; DIAFEN FF; 1,4-

DIANILINOBENZENE; N,N'-DIFENYL-P-FENYLENDIAMIN; N,N'-DIPHENYL-1,4-BENZENEDIAMINE; N,N-DIPHENYL-1,4-BENZENEDIAMINE; N,N'-DIPHENYL-1,4-DIAMINOBENZENE; 4,4'-DIPHENYL-P-PHENYLENEDIAMINE; DIPHENYL-P-PHENYLENEDIAMINE; N,N'-DIPHENYL-1,4-PHENYLENEDIAMINE; N,N'-DIPHENYL-P-PHENYLENEDIAMINE; DPPD; EKALAND DPPD; FLEXAMINE G; JZF; NAUGARD J; NOCRAC DP; NONFLEX H; NONOX DPPD; PERMANAX 18; PERMANAX DPPD; 4-PHENYLAMINODIPHENYLAMINE; P-PHENYLAMINODIPHENYLAMINE; P-PHENYLENEDIPHENYLDIAMINE; STABILIZER DPPD

Description: colorless leaflets or greenish-brown; gray powder

Use: as an antioxidant for rubber, petroleum oils and feedstuffs; as a stabilizer, a polymerization inhibitor, to retard copper degradation and as an intermediate for dyes, drugs, plastics an detergents

Physical Properties

Boiling Point: 220 °C (428 °F) to 225 °C (437 °F) at 0.5 mm Hg

Freezing Point: 150 °C (302 °F) to 151 °C (303.8 °F)

Specific Gravity: 1.2

Vapor Density: 9 Air=1

Density: 1.2 g/mL

Water Solubility: Almost Insoluble in Water

Other Solubilities: DMSO: 50-100 mg/ml at 20 °C; Alcohol: >10%; Benzene: >10%; Chloroform: Soluble; Ether: Soluble; Ethyl Acetate: Soluble; Glacial Acetic Acid: Soluble; Ethyl Alcohol: >10%.

Refraction Index: 1.557

Flash Point: 204 °C

RTECS Toxicity Data

Acute Oral: Rat LD_{50} Dose: 2370 mg/kg. Mouse LD_{50} Dose: 18 gm/kg.

Chronic (Multiple Dose) Oral: Rat Dose: 182 gm/kg/2Y-C; Toxic Effects: Liver - Changes in liver weight; Kidney, Ureter, and Bladder - Chgs in tubules (inc acute renal failure, acute tubular necrosis; Blood - Changes in serum composition.

Irritation Eye: Rabbit Standard Draize Test Dose: 500 mg/24H; Reaction: mild.

Reproductive/Teratogenic: Rat Route: Oral; Dose: 450 mg/kg; Duration: female 14D prior to mating Maternal Effects - Parturition; Effects on Newborn - Stillbirth. Rat Route: Oral; Dose: 2500 mg/kg; Duration: female 1-22D of pregnancy; Effects on Fertility - Post-implantation mortality.

Mutagenic: Hamster Mutations in Mammalian Somatic Cells; Cell Type: lung; Dose: 30 mg/L. Hamster Cytogenetic Analysis; Cell Type: lung; Dose: 1800 ug/L.

Tumorigenic: Mouse Route: Subcutaneous; Dose: 1000 mg/kg; Toxic Effects: Tumorigenic - Equivocal tumorigenic agent by RTECS criteria; Lungs, Thorax, or Respiration - Tumors; Blood - Tumors.

Hazard Overviews

Health: Irritating to eyes/skin/respiratory tract. Other Acute Effects: may be harmful by inhalation, ingestion, or skin absorption.

Fire: Will burn. Hazards: emits toxic fumes. Extinguishing agents: water spray; carbon dioxide, dry chemical powder or appropriate foam. Precautions: combustible liquid.

Reactivity: Incompatible with: strong oxidizing agents, strong acids. Hazardous decomposition products: toxic fumes of: carbon monoxide, carbon dioxide, nitrogen oxides.

Carcinogenicity: IARC - Not listed; NIOSH - Not listed; NTP - Not listed; ACGIH - Not listed; OSHA - Not listed; EPA - Not listed; MAK - Not listed

Primary Target Organs:

Eyes

Skin

Respiratory System

Environmental

Regulations

RCRA 40CFR: Not listed

CERCLA: 40CFR 302.4: Not listed

SARA 40CFR 372.65: Not listed

SARA EHS 40CFR 355: Not listed

TSCA: Listed

DIP6400 CAS #: 136-35-6

1,3-DIPHENYL-1-TRIAZENE

RTECS: XY2625000

EINECS Number: 205-240-1

Molecular Formula: $C_{12}H_{11}N_3$

Formula Weight: 197.11

Chemical Structure

Synonyms: ANILINE,N-(PHENYLAZO)-; ANILINOAZOBENZENE; BENZENEAZOANILIDE; BENZENEAZOANILINE; CELLOFOR; DAAB; ALPHA-DIAZOAMIDOBENZOL; DIAZOAMINOBENZEN; DIAZOAMINOBENZENE; P-DIAZOAMINOBENZENE; DIAZOAMINOBENZOL; DIAZOBENZENEANILIDE; 1,3-DIPHENYLTRIAZENE; N-(PHENYLAZO)ANILINE; 1-TRIAZENE,1,3-DIPHENYL-; TRIAZENE,1,3-DIPHENYL-

Description: golden-yellow, small crystals

Use: organic synthesis, dyes, insecticide

Physical Properties

Boiling Point: Explodes at 150 °C (302 °F)

Freezing Point: 98 °C (208.4 °F)

Density: 6.8

Water Solubility: Insoluble in Water

Other Solubilities: 95% Ethanol: 10-50 mg/ml at 18 °C; Acetone: >100 mg/ml at 18 °C; Benzene: Very Soluble; DMSO: >=100 mg/ml at 18 °C; Ether: Very Soluble.
Flash Point: Not available; probably combustible

RTECS Toxicity Data

Mutagenic: Bacteria - S Typhimurium Mutations in Microorganisms; Dose: 300 ng/plate (-S9).
Tumorigenic: Mouse Route: Oral; Dose: 1480 mg/kg/59D-C; Toxic Effects: Tumorigenic - Equivocal tumorigenic agent by RTECS criteria; Gastrointestinal - Tumors. Mouse Route: Skin; Dose: 30 gm/kg/46W-I; Toxic Effects: Tumorigenic - Equivocal tumorigenic agent by RTECS criteria; Skin and appendages - Tumors; Tumorigenic - Tumors at site of application.

Hazard Overviews

Health: Irritating to eyes/skin/respiratory tract. Toxic. Other Acute Effects: harmful if swallowed, inhaled, or absorbed through skin.
Fire: Will burn. Hazards: emits toxic fumes; container explosion may occur. Extinguishing agents: water spray; carbon dioxide, dry chemical powder or appropriate foam. Precautions: combustible liquid.
Reactivity: Incompatible with: strong oxidizing agents, store away from heat and direct sunlight. Hazardous decomposition products: toxic fumes of: carbon monoxide, carbon dioxide, nitrogen oxides.
Carcinogenicity: IARC - Not listed; NIOSH - Not listed; NTP - Not listed; ACGIH - Not listed; OSHA - Not listed; EPA - Not listed; MAK - Not listed
Primary Target Organs:

Eyes

Skin

Respiratory System

Environmental

Regulations

RCRA 40CFR: Not listed
CERCLA: 40CFR 302.4: Not listed
SARA 40CFR 372.65: Not listed
SARA EHS 40CFR 355: Not listed
TSCA: Listed

DIP6760 CAS #: 503-38-8

DIPHOSGENE

RTECS: LQ7350000
EINECS Number: 207-965-9
Molecular Formula: $C_2Cl_4O_2$
Structured MF: $ClCO_2CCl_3$
Formula Weight: 197.83

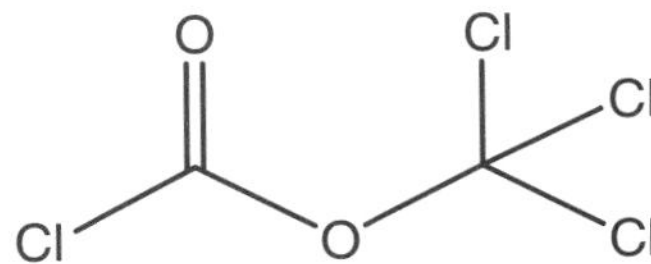

Chemical Structure

Synonyms: CARBONOCHLORIDIC ACID TRICHLOROMETHYL ESTER; CARBONOCHLORIDIC ACID,TRICHLOROMETHYL ESTER; DIFOSGEN; DIPHOSGEN; FORMIC ACID,CHLORO-,TRICHLOROMETHYL ESTER; METHANOL,TRICHLORO-,CHLOROFORMATE; PERCHLOROMETHYL FORMATE; SUPERPALITE; TRICHLORMETHYLESTER KYSELINY CHLORMRAVENCI; TRICHLOROMETHYL CHLOROFORMATE
Description: colorless liquid; odor similar to newmown hay
Use: organic synthesis; military poison gas; in synthesis of isocyanides

Physical Properties

Boiling Point: 128 °C (262 °F) at 760 mm Hg
Freezing Point: -57 °C (-70.6 °F)
Specific Gravity: 1.6525 at 14 °C
Water Solubility: Insoluble in Water
Other Solubilities: Soluble in Benzene, Alcohol, and Ether.
Refraction Index: 1.4566 at 22 °C/D

RTECS Toxicity Data

Acute Inhalation: Rabbit LC_{Lo} Dose: 900 mg/m^3/15M; Toxic Effects: Lungs, Thorax, or Respiration - Ciliary function changes; Lungs, Thorax, or Respiration - Acute pulmonary edema; Lungs, Thorax, or Respiration - Other changes.

Hazard Overviews

Poison

Corrosive

Health: Corrosive to eyes/skin/respiratory tract. Poison. Other Acute Effects: may be fatal if inhaled, swallowed, or absorbed through skin; material is extremely destructive to tissue of the mucous membranes and upper respiratory tract, eyes and skin; inhalation may result in spasm, inflammation and edema of the larynx and bronchi, chemical pneumonitis and pulmonary edema; symptoms of exposure may include burning sensation, coughing, wheezing, laryngitis, shortness of breath, headache, nausea and vomiting.
Fire: Hazards: emits toxic fumes. Extinguishing agents: water spray; carbon dioxide, dry chemical powder or appropriate foam. Precautions: combustible liquid.
Reactivity: Incompatible with: strong oxidizing agents, bases, sensitive to heat. Hazardous decomposition products: toxic fumes of: carbon monoxide, carbon dioxide, hydrogen chloride gas, phosgene gas.
Carcinogenicity: IARC - Not listed; NIOSH - Not listed; NTP - Not listed; ACGIH - Not listed; OSHA - Not listed; EPA - Not listed; MAK - Not listed

Primary Target Organs:

Eyes

Skin

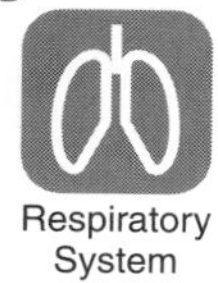
Respiratory System

Environmental

Regulations

RCRA 40CFR: Not listed
CERCLA: 40CFR 302.4: Not listed
SARA 40CFR 372.65: Not listed
SARA EHS 40CFR 355: Not listed
TSCA: Listed

DIP7120 CAS #: 2164-07-0

DIPOTASSIUM ENDOTHALL

RTECS: RN8223400
EINECS Number: 218-498-5
Molecular Formula: $C_8H_8K_2O_5$
Formula Weight: 262.36
Synonyms: ENDOTHALL DIPOTASSIUM SALT

Hazard Overviews

Carcinogenicity: IARC - Not listed; NIOSH - Not listed; NTP - Not listed; ACGIH - Not listed; OSHA - Not listed; EPA - Not listed; MAK - Not listed

Environmental

Regulations

RCRA 40CFR: Not listed
CERCLA: 40CFR 302.4: Not listed
SARA 40CFR 372.65: Listed
SARA EHS 40CFR 355: Not listed
TSCA: Listed

DIP7300 CAS #: 7758-11-4

DIPOTASSIUM PHOSPHATE

EINECS Number: 231-834-5
Molecular Formula: $H_3K_2O_4P$
Formula Weight: 174.18

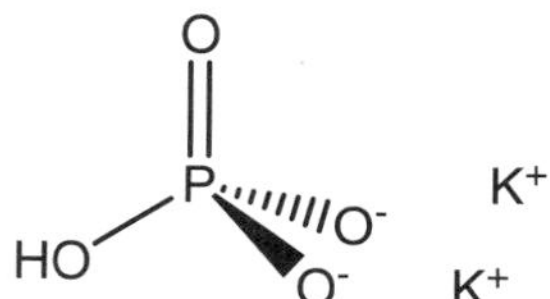

Chemical Structure

Synonyms: DIBASIC POTASSIUM PHOSPHATE; DIKALIUM PHOSPHATE; DIPOTASSIUM HYDROGEN PHOSPHATE; DIPOTASSIUM MONOHYDROGEN PHOSPHATE; DIPOTASSIUM MONOPHOSPHATE; DIPOTASSIUM ORTHOPHOSPHATE; DIPOTASSIUM-O-PHOSPHATE; DKP; HYDROGEN DIPOTASSIUM PHOSPHATE; PHOSPHORIC ACID,DIPOTASSIUM SALT; POTASSIUM DIBASIC PHOSPHATE; POTASSIUM HYDROGEN PHOSPHATE; POTASSIUM MONOHYDROGEN PHOSPHATE; POTASSIUM MONOPHOSPHATE; POTASSIUM ORTHOPHOSPHATE,MONO-H; POTASSIUM PHOSPHATE,DIBASIC
Description: colorless to white granules, crystals, or powder
Use: buffering agent in anti-freeze; nutrient in the culturing of antibiotics; saline bulk cathartic; sequestrant; emulsifier in specified cheeses; corrosion inhibitor in ethylene glycol antifreeze; buffering agent in coffee creamers; fertilizer (home & garden); paper processing agent; humectant; pharmaceuticals; in foods as buffer, yeast food; laboratory reagent

Physical Properties

Freezing Point: Decomposes
Water Solubility: Very Soluble in Water
Other Solubilities: Very Soluble in Alcohol.
pH: 1% solution 8.8

Hazard Overviews

Health: Irritating to eyes/skin/respiratory tract. Other Acute Effects: may be harmful by inhalation, ingestion, or skin absorption.
Fire: Hazards: emits toxic fumes. Extinguishing agents: water spray; carbon dioxide, dry chemical powder or appropriate foam. Precautions: combustible liquid.
Reactivity: Incompatible with: strong oxidizing agents, protect from moisture. Hazardous decomposition products: nature of decomposition products not known;.
Carcinogenicity: IARC - Not listed; NIOSH - Not listed; NTP - Not listed; ACGIH - Not listed; OSHA - Not listed; EPA - Not listed; MAK - Not listed
Primary Target Organs:

Eyes

Skin

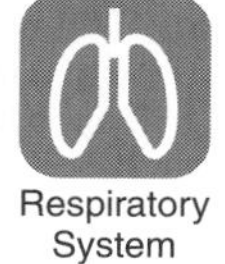
Respiratory System

Environmental

Regulations

RCRA 40CFR: Not listed
CERCLA: 40CFR 302.4: Not listed
SARA 40CFR 372.65: Not listed
SARA EHS 40CFR 355: Not listed
TSCA: Listed

DIP7480 CAS #: 136-45-8

DIPROPYL ISOCINCHOMERONATE

RTECS: US8000000
EINECS Number: 205-245-9
Molecular Formula: $C_{13}H_{17}NO_4$
Formula Weight: 251.31

Synonyms: DIPROPYL 2,5-PYRIDINEDICARBOXYLATE; DIPROPYLESTER KYSELINY PYRIDIN-2,5-DIKARBOXYLOVE; DI-N-PROPYL-ISOCINCHOMERONATE; ENT 17595; ISOCINCHOMERONYL DIPROPYLESTER; MGK R-326; MGK 326; MGK REPELLENT-326; PYRIDIN-2,5-DICARBONSAEURE-DI-N-PROPYLESTER; R-326; REPPER 333

RTECS Toxicity Data

Acute Oral: Rat LD_{50} Dose: 5230 mg/kg. Mouse LD_{50} Dose: 1600 mg/kg.
Acute Dermal: Rabbit LD_{50} Route: Skin; Dose: 9500 mg/kg. Rat LD_{50} Route: Skin; Dose: 9400 mg/kg.

Hazard Overviews

Carcinogenicity: IARC - Not listed; NIOSH - Not listed; NTP - Not listed; ACGIH - Not listed; OSHA - Not listed; EPA - Not listed; MAK - Not listed

Environmental

Regulations

RCRA 40CFR: Not listed
CERCLA: 40CFR 302.4: Not listed
SARA 40CFR 372.65: Listed
SARA EHS 40CFR 355: Not listed
TSCA: Not listed

Analytical Methods

Water / Groundwater: EPA 633.1

DIP7660 CAS #: 123-19-3

DIPROPYL KETONE

RTECS: MJ5600000
EINECS Number: 204-608-9
Molecular Formula: $C_7H_{14}O$
Structured MF: $(CH_3CH_2CH_2)_2CO$
Formula Weight: 114.21

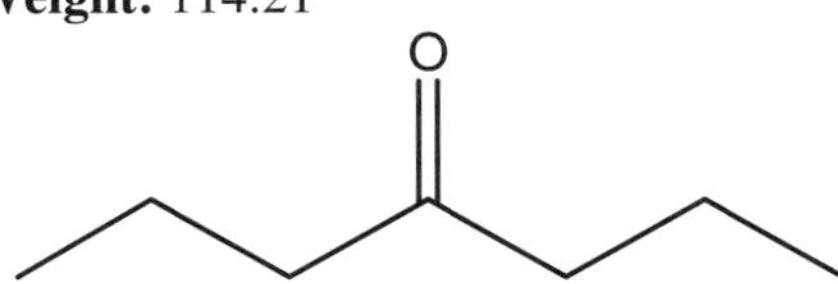

Chemical Structure

Synonyms: BUTYRONE; DPK; GBL; 4-HEPTANONE; HEPTAN-4-ONE; PROPYL KETONE
Description: colorless liquid; pleasant, penetrating odor

Physical Properties

Boiling Point: 143.89 °C (291 °F)
Freezing Point: -32.78 °C (-27.004 °F)
Specific Gravity: 0.82
Vapor Density: 3.93
Bulk Density: 6.79 lbs/gal at 20 °C
Vapor Pressure: 5 mm Hg
Water Solubility: Insoluble
Other Solubilities: miscible with Alcohol and Ether
Surface Tension: 25.2 dynes/cm
Refraction Index: 1.4069
Ionization Potential (eV): 9.10
Flash Point: 49 °C

RTECS Toxicity Data

Acute Oral: Rat LD_{50} Dose: 3730 uL/kg.
Acute Inhalation: Rat LC_{50} Dose: 2690 ppm/6hr; Toxic Effects: Behavioral - Sleep; Behavioral - Somnolence (general depressed activity); Lungs, Thorax, or Respiration - Respiratory depression.
Acute Dermal: Rabbit LD_{50} Route: Skin; Dose: 5660 uL/kg.
Chronic (Multiple Dose) Oral: Rat Dose: 30 gm/kg/3W-I; Toxic Effects: Behavioral - Somnolence (general depressed activity); Liver - Changes in liver weight; Nutritional and gross metabolic - Weight loss or decreased weight gain. Rat Dose: 130 gm/kg/13W-I; Toxic Effects: Kidney, Ureter, and Bladder - Changes in kidney weight; Endocrine - Hypoglycemia; Nutritional and gross metabolic - Weight loss or decreased weight gain.
Irritation Eye: Rabbit Standard Draize Test Dose: 500 mg/24H; Reaction: mild.
Irritation Skin: Rabbit Standard Draize Test Dose: 500 mg/24H; Reaction: mild.

Hazard Overviews

Flammable

Fire Diamond

Health: Severely irritating to eyes; irritating to skin/respiratory tract. Also Causes: CNS depression, aspiration pneumonitis, dermatitis, corneal damage. Chronic: liver damage, reduced blood glucose.
Fire: Flammable. For small fires use dry chemical, carbon dioxide, water spray, or alcohol-resistant foam. For large fires use water spray, fog, or alcohol-resistant foam.
Reactivity: Stable. Hazardous polymerization cannot occur. Avoid: heat; ignition sources; contact with oxidizing materials. Incompatible with: oxidizing materials. Hazardous decomposition products: carbon dioxide; acrid smoke, fumes.
Carcinogenicity: IARC - Not listed; NIOSH - Listed as carcinogen; NTP - Not listed; ACGIH - Not listed; OSHA - Not listed; EPA - Not listed; MAK - Not listed
Primary Target Organs:

Eyes

Skin

Respiratory System

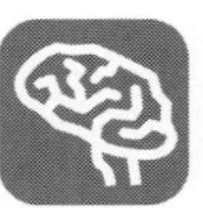

Nervous System

Liver

Exposure Limits
OSHA PEL Vacated 1989 Limits: TWA: 50 ppm; 235 mg/m^3.
ACGIH TLV: TWA: 50 ppm; 223 mg/m^3.
NIOSH REL: TWA: 50 ppm; 235 mg/m^3.

Respirator Recommendation
Exposure Range: >50 to 2500 ppm Supplied Air, Constant Flow/Pressure Demand, Half Mask
Exposure Range: >2500 to 50,000 ppm Supplied Air, Constant Flow/Pressure Demand, Full Face
Exposure Range: >50,000 to unlimited ppm Self-contained Breathing Apparatus, Pressure Demand, Full Face
Note: odor theshold unknown

Environmental

Regulations
RCRA 40CFR: Not listed
CERCLA: 40CFR 302.4: Not listed
SARA 40CFR 372.65: Not listed
SARA EHS 40CFR 355: Not listed
TSCA: Listed

DIP7840 CAS #: 598-03-8

DIPROPYL SULFONE

EINECS Number: 209-913-0
Molecular Formula: $C_6H_{14}O_2S$
Formula Weight: 150.2

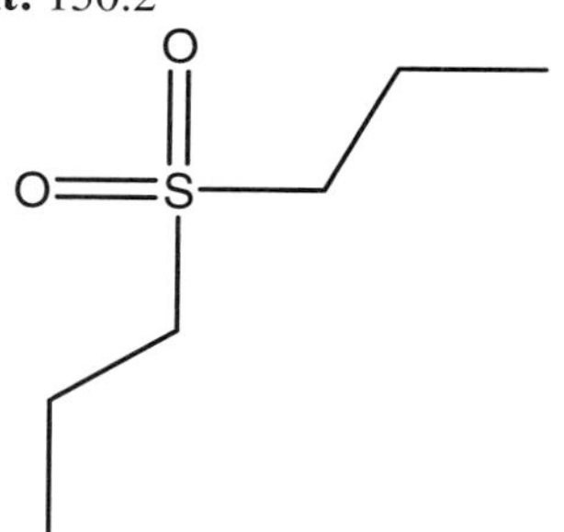

Chemical Structure

Physical Properties
Freezing Point: 30 °C (86 °F)
Specific Gravity: 1.02780
Water Solubility: Slightly Soluble
Other Solubilities: Ethanol: Soluble; Ether: Soluble
Refraction Index: 1.4456

Hazard Overviews
Health: Irritating to eyes/skin. Other Acute Effects: may be harmful by inhalation, ingestion, or skin absorption.
Fire: Hazards: emits toxic fumes. Extinguishing agents: water spray; carbon dioxide, dry chemical powder or appropriate foam. Precautions: combustible liquid.
Reactivity: Incompatible with: strong oxidizing agents. Hazardous decomposition products: toxic fumes of: carbon monoxide, carbon dioxide, sulfur oxides.
Carcinogenicity: IARC - Not listed; NIOSH - Not listed; NTP - Not listed; ACGIH - Not listed; OSHA - Not listed; EPA - Not listed; MAK - Not listed

Primary Target Organs:

Eyes

Skin

Environmental

Regulations
RCRA 40CFR: Not listed
CERCLA: 40CFR 302.4: Not listed
SARA 40CFR 372.65: Not listed
SARA EHS 40CFR 355: Not listed
TSCA: Listed

DIP8020 CAS #: 142-84-7

DIPROPYLAMINE

RTECS: JL9200000
DOT: UN2383; IMO3.2
EINECS Number: 205-565-9
Molecular Formula: $C_6H_{15}N$
Structured MF: $(CH_3CH_2CH_2)_2NH$
Formula Weight: 101.19

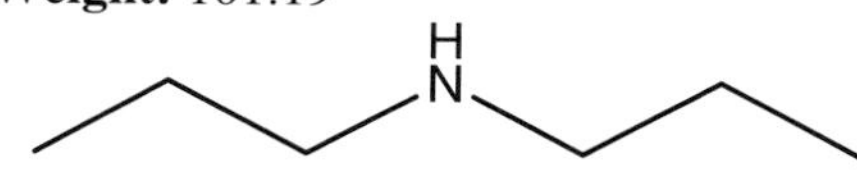

Chemical Structure

Synonyms: DI-N-PROPYLAMINE; N-DIPROPYLAMINE; 1-PROPANAMINE,N-PROPYL-; N-PROPYL-1-PROPANAMINE
Description: colorless to water-white liquid; ammonia odor
Use: intermediate in organic synthesis

Physical Properties
Boiling Point: 109 °C (228 °F) to 110 °C (230 °F)
Freezing Point: -39.6 °C (-39.28 °F)
Specific Gravity: 0.738 at 20 °C/4 °C
Vapor Density: 3.5 Air=1
Density: 0.74 g/mL at 20 °C
Bulk Density: 6.1 lbs/gal
Vapor Pressure: Saturated vapor pressure 0.433 lb/sq in at 70 °C
Water Solubility: Soluble >= 10 mg/mL
Other Solubilities: 95% Ethanol: Soluble (>=10 mg/ml); Acetone: Very Soluble; Benzene: Very Soluble; Ether: Very Soluble.
Surface Tension: 6.58 dynes/cm at 20 °C
Odor Threshold: 0.4140 to 0.8250 mg/m^3
Refraction Index: 1.4055 at 20 °C/D
Critical Temperature: 277 °C
Critical Pressure: 456 psia
Ionization Potential (eV): 7.84 +/-0.1
Flash Point: 17 °C Open Cup
Autoignition Temperature: 299 °C

RTECS Toxicity Data

Acute Oral: Rat LD_{50} Dose: 460 mg/kg.
Acute Inhalation: Rat LC_{50} Dose: 4400 mg/m^3/4hr. Mouse LC_{50} Dose: 3070 mg/m^3/2hr.
Acute Dermal: Rabbit LD_{50} Route: Skin; Dose: 1250 mg/kg; Toxic Effects: Skin and appendages - Primary irritation. Rabbit LD_{50} Route: Subcutaneous Dose: 1250 mg/kg.
Chronic (Multiple Dose) Oral: Rat Dose: 910 mg/kg/26W-I; Toxic Effects: Biochemical - Catalases. Rabbit Dose: 910 mg/kg/26W-I; Toxic Effects: Biochemical - Catalases.

Hazard Overviews

Corrosive

Flammable

Health: Corrosive to eyes/skin/respiratory tract. Harmful. Other Acute Effects: harmful if swallowed, inhaled, or absorbed through skin; causes burns; material is extremely destructive to tissue of the mucous membranes and upper respiratory tract, eyes and skin; inhalation may result in spasm, inflammation and edema of the larynx and bronchi; chemical pneumonitis; pulmonary edema; burning sensation; coughing; wheezing; laryngitis; shortness of breath; headache; nausea and vomiting; coughing; chest pains; difficulty in breathing.
Fire: Flammable. Hazards: vapor may travel considerable distance to source of ignition and flash back. Extinguishing agents: carbon dioxide, dry chemical powder or appropriate foam. Precautions: combustible liquid.
Reactivity: Incompatible with: acids, oxidizing agents. Hazardous decomposition products: thermal decomposition may produce carbon monoxide, carbon dioxide, and nitrogen oxides.
Carcinogenicity: IARC - Not listed; NIOSH - Not listed; NTP - Not listed; ACGIH - Not listed; OSHA - Not listed; EPA - Not listed; MAK - Not listed
Primary Target Organs:

Eyes

Skin

Respiratory System

Environmental

Ecotoxicity: Aquatic toxicity: 20-60 ppm/24 hr/creek chub/critical range/fresh water
Environmental Fate: If released to the atmosphere, it is rapidly degraded (estimated half-life of 4.6 hr) by reaction with photochemically produced hydroxyl radicals. If released to water, it is physically removed by volatilization. Volatilization half-lives of 0.83 and 9.5 days have been estimated for a shallow (1 m deep) model river and an environmental pond, respectively. If released to soil, it is expected to be moderately to very mobile and easily leached based upon estimated K_{oc} values of 15-393. Evaporation from dry soil is likely to occur. A single screening study has demonstrated that it is readily biodegraded by activated sludge inocula.
Cleanup/Disposal: Guide No. 132: Fully encapsulating, vapor protective clothing should be worn for spills and leaks with no fire. Eliminate all ignition sources (no smoking, flares, sparks or flames in immediate area). All equipment used when handling the product must be grounded. Do not touch or walk through spilled material. Stop leak if you can do it without risk. Prevent entry into waterways, sewers, basements or confined areas. A vapor suppressing foam may be used to reduce vapors. Absorb with earth, sand or other non-combustible material and transfer to containers (except for Hydrazine). Use clean non-sparking tools to collect absorbed material. Large Spills: Dike far ahead of liquid spill for later disposal. Water spray may reduce vapor; but may not prevent ignition in closed spaces.

Environmental Physical Data

Henry's Law Constant: 5.1 x10^{-5}
Octanol/Water Partition Coefficient: log K_{ow} = 1.67
Sorption Partition Coefficient: K_{oc} = estimated at 15 to 193
BCF: estimated at 1.04

Regulations

RCRA 40CFR: Listed Hazardous Waste No. U110 Ignitable Waste
CERCLA: 40CFR 302.4: Listed per RCRA Section 3001 RQ: 5000 lb (2268 kg)
SARA 40CFR 372.65: Not listed
SARA EHS 40CFR 355: Not listed
TSCA: Listed

DIP8200 **CAS #: 52888-80-9**

DIPROPYLCARBAMOTHIOIC ACID, S-(PHENYLMETHYL) ESTER

RTECS: FD3835000
EINECS Number: 401-730-6
Molecular Formula: $C_{14}H_{21}NOS$
Formula Weight: 251.42
Synonyms: ARKADE; S-BENZYL DIPROPYL THIOLCARBAMATE; S-BENZYL DIPROPYLTHIOCARBAMATE; BENZYL DIPROPYLTHIOLCARBAMATE; BOXER; CARBAMIC ACID,DIPROPYLTHIO-,S-BENZYL ESTER (7CI); S-(PHENYLMETHYL) DIPROPYLCARBAMOTHIOATE; PROSULFOCARB; R 15574; SC 0574

RTECS Toxicity Data

Acute Oral: Rat LD_{50} Dose: 1820 mg/kg.
Acute Inhalation: Rat LC_{50} Dose: >4700 mg/m^3/4hr.
Acute Dermal: Rabbit LD_{50} Route: Skin; Dose: >2 gm/kg.

Hazard Overviews

Carcinogenicity: IARC - Not listed; NIOSH - Not listed; NTP - Not listed; ACGIH - Not listed; OSHA - Not listed; EPA - Not listed; MAK - Not listed

Environmental

Regulations

RCRA 40CFR: Listed Hazardous Waste No. U387 Toxic Waste
CERCLA: 40CFR 302.4: Listed per RCRA Section 3001 RQ: 1 lb (0.454 kg)
SARA 40CFR 372.65: Not listed
SARA EHS 40CFR 355: Not listed
TSCA: Not listed

DIP8380 **CAS #: 25265-71-8**

DIPROPYLENE GLYCOL

RTECS: UB8765000
EINECS Number: 246-770-3
Molecular Formula: $C_6H_{14}O_3$
Formula Weight: 134.18
Synonyms: OXYBISPROPANOL; PROPANOL,OXYBIS-
Description: colorless liquid; odorless
Use: in polyester and alkyd resins, reinforced plastics plasticizers, solvents, fragrances, lacquer, paint, printing ink and shellac varnish; in the manufacture of nitrocellulose solvent

Physical Properties

Boiling Point: 231.9 °C (449 °F) at 760 mm Hg
Freezing Point: -40 °C (-40 °F)
Specific Gravity: 1.0252 at 20 °C/20 °C
Vapor Density: 4.63 Air=1
Saturated Vapor Density: < 1.200057267 kg/m^3
Density: 1.02 to 1.03 g/mL
Vapor Pressure: < 0.01 mm Hg at 20 °C
Water Solubility: Miscible with Water
Other Solubilities: 95% Ethanol: >=100 mg/ml at 24 °C; Acetone: >=100 mg/ml at 24 °C; Alcohol: Soluble; Chloroform: miscible; DMSO: >=100 mg/ml at 24 °C; Ether: miscible; Methanol: Soluble; Toluene: Soluble.
Refraction Index: 1.439 at 25 °C
Critical Temperature: 382 °C
Critical Pressure: 36 atm
Flash Point: 121.111 to 137.778 °C Open Cup
Autoignition Temperature: 305 °C
LEL: 2.2% v/v
UEL: 12.6% v/v

RTECS Toxicity Data

Acute Oral: Rat LD_{50} Dose: 14850 mg/kg. Mammal LD_{50} Dose: 15 gm/kg.
Acute Dermal: Rabbit LD_{50} Route: Skin; Dose: >20 mL/kg.
Reproductive/Teratogenic: Rat Route: Oral; Dose: 50 gm/kg; Duration: female 6-15D of pregnancy; Maternal Effects - Other effects on females.

Hazard Overviews

Fire Diamond

Fire: Will burn.
Carcinogenicity: IARC - Not listed; NIOSH - Not listed; NTP - Not listed; ACGIH - Not listed; OSHA - Not listed; EPA - Not listed; MAK - Not listed

Environmental

Environmental Fate: It is not expected to undergo hydrolysis or direct photolysis in the environment. The miscibility in water suggests that volatilization, adsorption and bioconcentration are not important fate processes. This is supported by an estimated Henry's Law constant of 3.58 x10^{-9} atm-cu m/mole at 25 °C which indicates that volatilization from natural waters and moist soil should be extremely slow. A low estimated log BCF suggests it should not bioconcentrate among aquatic organisms. A low K_{oc} indicates it should not partition from the water column to organic matter contained in sediments and suspended solids, and it should be highly mobile in soil. Although aerobic biodegradation screening test data suggests that the rate should be slow, biodegradation may still be an important removal mechanism from aerobic soil and water. In the atmosphere, it is expected to exist almost entirely in the vapor phase and reactions with photochemically produced hydroxyl radicals should be important (estimated half-life of 13 hours). Physical removal from air by precipitation and dissolution in clouds may occur; however, its short atmospheric residence time suggests that wet deposition is of limited importance.
Cleanup/Disposal: Guide No. 171: Do not touch or walk through spilled material. Stop leak if you can do it without risk. Prevent dust cloud. Avoid inhalation of asbestos dust. Small Dry Spills: With clean shovel place material into clean, dry container and cover loosely; move containers from spill area. Small Spills: Take up with sand or other noncombustible absorbent material and place into containers for later disposal. Large Spills: Dike far ahead of liquid spill for later disposal. Cover powder spill with plastic sheet or tarp to minimize spreading. Prevent entry into waterways, sewers, basements or confined areas.

Environmental Physical Data

Henry's Law Constant: calculated at 3.58 x10^{-9}
Octanol/Water Partition Coefficient: log K_{ow} = -1.07
Sorption Partition Coefficient: K_{oc} = 6
BCF: calculated at -1.04

Regulations

RCRA 40CFR: Not listed
CERCLA: 40CFR 302.4: Not listed
SARA 40CFR 372.65: Not listed
SARA EHS 40CFR 355: Not listed
TSCA: Listed

DIP8560 **CAS #: 1320-18-9**

2,4-DIPROPYLENE GLYCOL BUTYL ETHER ESTER

RTECS: AG8886000
DOT: UN2765; UN2766; UN2999; UN3000; NA2765; IMO3.2; IMO6.1
Molecular Formula: $C_{15}H_{20}Cl_2O_5$
Formula Weight: 351
Synonyms: ACETIC ACID,(2,4-DICHLOROPHENOXY)-,BUTOXY PROPYLENEDERIV; ACETIC ACID,2,4-DICHLOROPHENOXY-,BUTOXYPROPYL ESTER; 2,4-D PGBE; 2,4-D,PROPYLENE GLYCOL BUTYL ETHER ESTER; 2,4-DICHLOROPHENOXY,2-BUTOXYMETHYLETHYL ESTER; DUPONT LAWN WEEDER; ESTERON TEN-TEN; ESTERON 99 WEED KILLER; ESTERON 99 WEED KILLER CONCENTRATE; VERTON 2D; VERTON 4D
Description: fuel oil-like odor

Physical Properties

Saturated Vapor Density: 1.242076225 kg/m^3
Vapor Pressure: 2.4 mm Hg 25 °C
Water Solubility: Generally Immiscible or Insoluble in Water
Other Solubilities: Soluble in non-polar organic solvents such as Hexane, Benzene, Acetone, and Alcohols.
Flash Point: > 79.444 °C Open Cup

Hazard Overviews

Fire: Combustible.
Carcinogenicity: IARC - Not listed; NIOSH - Not listed; NTP - Not listed; ACGIH - Not listed; OSHA - Not listed; EPA - Not listed; MAK - Not listed

Environmental

Ecotoxicity: LC_{50} Pteronarcys californica (stoneflies) second yr class 2.6 (95% confidence interval 1.8-3.8) mg/l/96 hr at 10 °C; water hardness 272 ppm $CaCO_3$ /pH not specified; static bioassay LC_{50} Lepomis macrochirus (bluegill, 1.0 g) 0.6 (95% confidence interval 0.4-0.7) mg/l/96 hr at 18 °C /pH not specified; static bioassay LC_{50} Palaemonetes kadiakensis (grass shrimp) mature 0.4 (95% confidence interval 0.09-1.4) mg/l/96 hr at 21 °C, water hardness 272 ppm $CaCO_3$ /pH not specified; static bioassay LC_{50} Simocephalus serrulatus (daphnids) first instar 4.9 (95% confidence interval 4.0-6.7) mg/l/96 hr at 15 °C /pH not specified; static bioassay
Cleanup/Disposal: Guide No. 131: Fully encapsulating, vapor protective clothing should be worn for spills and leaks with no fire. Eliminate all ignition sources (no smoking, flares, sparks or flames in immediate area). All equipment used when handling the product must be grounded. Do not touch or walk through spilled material. Stop leak if you can do it without risk. Prevent entry into waterways, sewers, basements or confined areas. A vapor suppressing foam may be used to reduce vapors. Small Spills: Absorb with earth, sand or other non-combustible material and transfer to containers for later disposal. Use clean non-sparking tools to collect absorbed material. Large Spills: Dike far ahead of liquid spill for later disposal. Water spray may reduce vapor; but may not prevent ignition in closed spaces. Guide No. 152: Do not touch damaged containers or spilled material unless wearing appropriate protective clothing. Stop leak if you can do it without risk. Prevent entry into waterways, sewers, basements or confined areas. Cover with plastic sheet to prevent spreading. Absorb or cover with dry earth, sand or other non-combustible material and transfer to containers. Do not get water inside containers.

Regulations

RCRA 40CFR: Not listed
CERCLA: 40CFR 302.4: Listed per CWA Section 311(b)(4) RQ: 100 lb (45.35 kg)
SARA 40CFR 372.65: Listed
SARA EHS 40CFR 355: Not listed
TSCA: Not listed

Analytical Methods

Soil: EPA PMD-TLC

DIP8740 **CAS #: 34590-94-8**

DIPROPYLENE GLYCOL MONOMETHYL ETHER

RTECS: JM1575000
EINECS Number: 252-104-2
Molecular Formula: $C_7H_{16}O_3$
Structured MF: $CH_3OC_3H_6OC_3H_6OH$
Formula Weight: 148.2

O O OH

Chemical Structure

Synonyms: ARCOSOLV; 1,4-DIMETHYL-3,6-DIOXA-1-HEPTANOL; DIPROPYLENE GLYCOL METHYL ETHER; DOWANOL 50B; DOWANOL-50B; DOWANOL DPM; 1-(2-METHOXYISOPROPOXY)-2-PROPANOL; PPG-2 METHYL ETHER; PROPANOL,(2-METHOXYMETHYLETHOXY)-; UCAR SOLVENT 2LM
Description: colorless liquid; mild, pleasant, ethereal odor
Use: in mfr of various cosmetics; solvent in hard-surface liquid household cleaners & water-based surface coatings; coupling agent in water-based polishes; solvent for nitrocellulose & other synthetic resins; solvent in hydraulic brake fluids

Physical Properties

Boiling Point: 190 °C (374 °F)
Freezing Point: -80 °C (-112 °F)
Specific Gravity: 0.95 at 25/4 °C
Vapor Density: 5.11 Air=1
Vapor Pressure: 0.4 mm Hg at 26 °C

Water Solubility: Miscible with Water
Other Solubilities: completely miscible with Acetone, Ethanol, Benzene, Carbon Tetrachloride, Ether, Methanol, Monochlorbenzene & Petroleum Ether.
Surface Tension: 28.8 dynes/cm
Odor Threshold: 210 to 6000 mg/m^3
Refraction Index: 1.419 at 25 °C
Flash Point: 85 °C Closed Cup
LEL: 1.1% v/v
UEL: 3.0% v/v

RTECS Toxicity Data

Acute Oral: Rat LD_{50} Dose: 5135 mg/kg; Toxic Effects: Behavioral - Sleep. Dog LD_{50} Dose: 7500 mg/kg; Toxic Effects: Lungs, Thorax, or Respiration - Other changes.
Acute Dermal: Rabbit LD_{50} Route: Skin; Dose: 9500 mg/kg.
Irritation Eye: Human Standard Draize Test Dose: 8 mg; Reaction: mild. Rabbit Standard Draize Test Dose: 238 mg; Reaction: mild. Rabbit Standard Draize Test Dose: 500 mg/24H; Reaction: mild.
Irritation Skin: Rabbit Open Draize Test Dose: 500 mg open; Reaction: mild.

Hazard Overviews

Fire Diamond

Health: Irritating to eyes/skin/respiratory tract. Also Causes: headache, weakness.
Fire: Combustible. Use water fog, dry chemical, alcohol foam, or carbon dioxide. Water or foam may cause frothing. Use water spray to cool fire-exposed tanks/containers.
Reactivity: Stable. Hazardous polymerization cannot occur. Avoid: contact with heat; sparks; open flame; possible ignition sources. Incompatible with: strong oxidizing agents. Hazardous decomposition products: carbon dioxide; carbon monoxide.
Carcinogenicity: IARC - Not listed; NIOSH - Listed as carcinogen; NTP - Not listed; ACGIH - Not listed; OSHA - Not listed; EPA - Not listed; MAK - Not listed
Primary Target Organs:

Eyes

Respiratory System

Nervous System

Exposure Limits
OSHA PEL: TWA: 100 ppm; 600 mg/m^3.
OSHA PEL Vacated 1989 Limits: TWA: 100 ppm; 600 mg/m^3; STEL: 1500 ppm; 900 mg/m^3.
ACGIH TLV: TWA: 100 ppm; 606 mg/m^3; STEL: 150 ppm; 909 mg/m^3.
NIOSH REL: TWA: 100 ppm; 600 mg/m^3. STEL: 150 ppm; 900 mg/m^3; skin.
NIOSH IDLH: 600 ppm.
DFG MAK: TWA: 50 ppm; 300 mg/m^3.

Respirator Recommendation
Exposure Range: >100 to <600 ppm Supplied Air, Constant Flow/Pressure Demand, Half Mask
Exposure Range: 600 to unlimited ppm Self-contained Breathing Apparatus, Pressure Demand, Full Face
Note: poor warning properties

Environmental

Environmental Fate: Because of its high solubility and low vapor pressure, it would be expected to partition to the aquatic phase of the environment. In water, it would not be expected to sorb to sediments or to bioconcentrate. The main degradation mechanism in water is, in all likelihood, biodegradation, while photolysis and hydrolysis are probably insignificant. Evaporative transfer from water to the atmosphere is expected to be minimal. However, in the atmosphere it is estimated to have a half-life of approximately 3.4 hours. Besides photochemical reactions, it may be removed from the atmosphere by washout. In soil, it will be highly mobile, and hence, leach to groundwater. In moist soil, as in aquatic systems, biodegradation will probably be the primary removal mechanism. However, compound which is on the surface of dry soil may evaporate.
Cleanup/Disposal: Guide No. 127: Eliminate all ignition sources (no smoking, flares, sparks or flames in immediate area). All equipment used when handling the product must be grounded. Do not touch or walk through spilled material. Stop leak if you can do it without risk. Prevent entry into waterways, sewers, basements or confined areas. A vapor suppressing foam may be used to reduce vapors. Absorb or cover with dry earth, sand or other non-combustible material and transfer to containers. Use clean non-sparking tools to collect absorbed material. Large Spills: Dike far ahead of liquid spill for later disposal. Water spray may reduce vapor; but may not prevent ignition in closed spaces.

Environmental Physical Data

BOD: 0% BODT, 5 days

Regulations

RCRA 40CFR: Not listed
CERCLA: 40CFR 302.4: Not listed
SARA 40CFR 372.65: Not listed
SARA EHS 40CFR 355: Not listed
TSCA: Listed

DIP8920 **CAS #: 621-64-7**

DI-N-PROPYLNITROSAMINE

RTECS: JL9700000
EINECS Number: 210-698-0
Molecular Formula: $C_6H_{14}N_2O$
Formula Weight: 130.22

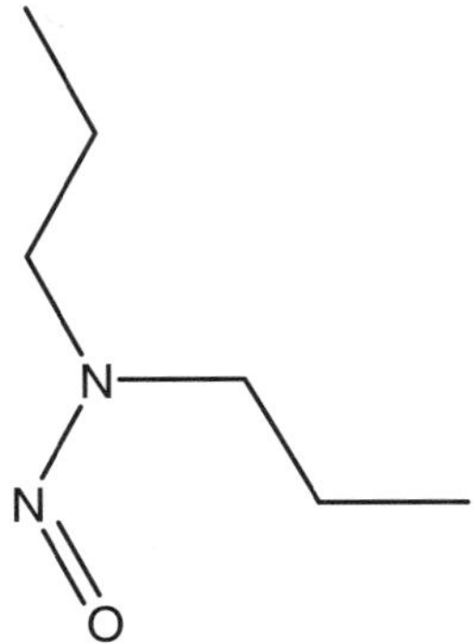

Chemical Structure

Synonyms: DIPROPYL NITROSAMINE; DIPROPYLAMINE,N-NITROSO-; DIPROPYLNITROSAMINE; N,N-DI-N-PROPYLNITROSAMINE; N,N-DIPROPYLNITROSAMINE; DPN; DPNA; NDPA; N-NITROSO-N-DIPROPYLAMINE; N-NITROSODI-N-PROPYLAMINE; N-NITROSODIPROPYLAMINE; NITROSODIPROPYLAMINE; N-NITROSO-N-PROPYL-1-PROPANAMINE; NITROUS DIPROPYLAMIDE; 1-PROPANAMINE,N-NITROSO-N-PROPYL-; PROPANAMINE,N-NITROSO-N-PROPYL-; PROPYLAMINE,N-NITROSO-N-DI-

Description: yellow liquid

Use: research chemical

Physical Properties

Boiling Point: 206 °C (403 °F)

Specific Gravity: 0.916 at 20 °C/4 °C

Saturated Vapor Density: 1.200473952 kg/m^3

Vapor Pressure: 0.086 mm Hg at 20 °C

Water Solubility: 1% at 0 °C

Other Solubilities: Alcohol: Very Soluble; Ether: Very Soluble.

Refraction Index: 1.4437 at 20 °C/D

RTECS Toxicity Data

Acute Oral: Rat LD_{50} Dose: 480 mg/kg. Hamster LD_{50} Dose: >400 mg/kg.

Acute Dermal: Rat LD_{50} Route: Subcutaneous Dose: 487 mg/kg; Toxic Effects: Lungs, Thorax, or Respiration - Other changes; Gastrointestinal - Ulceration or bleeding from stomach. Mouse LD_{50} Route: Subcutaneous Dose: 689 mg/kg.

Mutagenic: Human DNA Damage; Cell Type: liver; Dose: 3200 umol/L. Human Unscheduled DNA Synthesis; Cell Type: liver; Dose: 1800 umol/L. Human Unscheduled DNA Synthesis; Cell Type: HeLa cell; Dose: 100 umol/L.

Tumorigenic: Rat Route: Oral; Dose: 660 mg/kg/60W-I; Toxic Effects: Tumorigenic - Carcinogenic by RTECS criteria; Sense organs and special senses - Tumors; Liver - Tumors. Rat Route: Oral; Dose: 1880 mg/kg/30W-I; Toxic Effects: Tumorigenic - Equivocal tumorigenic agent by RTECS criteria; Gastrointestinal - Tumors; Liver - Tumors. Rat Route: Oral; Dose: 1350 mg/kg/30W-I; Toxic Effects: Tumorigenic - Equivocal tumorigenic agent by RTECS criteria; Gastrointestinal - Tumors; Liver - Tumors. Rat Route: Oral; Dose: 1150 mg/kg/30W-C; Toxic Effects: Tumorigenic - Equivocal tumorigenic agent by RTECS criteria; Gastrointestinal - Tumors. Rat Route: Oral; Dose: 1056 mg/kg/30W-I; Toxic Effects: Tumorigenic - Carcinogenic by RTECS criteria; Sense organs and special senses - Tumors; Liver - Tumors. Rat Route: Oral; Dose: 675 mg/kg/30W-C; Toxic Effects: Tumorigenic - Equivocal tumorigenic agent by RTECS criteria; Lungs, Thorax, or Respiration - Tumors; Gastrointestinal - Tumors.

Hazard Overviews

Health: Irritating to eyes/skin/respiratory tract. Toxic. Other Acute Effects: harmful if swallowed, inhaled, or absorbed through skin. Chronic Effects: mutagen; target organs: liver, kidneys, throat, lungs. Carcinogen. The toxicological properties have not been thoroughly investigated.

Fire: Extinguishing agents: water spray; carbon dioxide, dry chemical powder or appropriate foam. Precautions: combustible liquid.

Reactivity: Stable. Hazardous polymerization will not occur. Incompatible with: avoid contact with copper salts, mercury salts, strong mineral acids. Hazardous decomposition products: toxic fumes of: carbon monoxide, carbon dioxide, nitrogen oxides.

Carcinogenicity: IARC - Group 2B, Possibly carcinogenic to humans; NIOSH - Not listed; NTP - Class 2B, Reasonably anticipated to be a carcinogen, sufficient evidence of carcinogenicity from studies in experimental animals; ACGIH - Not listed; OSHA - Not listed; EPA - Class B2, Probable human carcinogen based on animal studies; MAK - Class A2, Unmistakably carcinogenic in animal experimentation only

Primary Target Organs:

Eyes

Skin

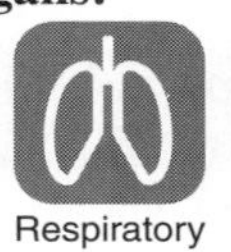
Respiratory System

Mucous Membranes

Liver

Kidneys

Environmental

Environmental Fate: If released to the surface of soils, it will rapidly volatilize. In soil, it is not expected to sorb strongly to organic matter and thus may leach into groundwater. In general, it is not expected to be persistent in soils because of removal by volatilization and biodegradation. If released to water, it will have a slight tendency to sorb to suspended organic matter, biota, and sediments. Bioconcentration will not be significant. Volatilization from water will probably not be significant except possibly from shallow rivers. Photolysis may be the most important degradative process in water. The rate of biodegradation cannot be assessed and hydrolysis is probably not important. In the atmosphere, it probably will rapidly photolyze

Environmental Physical Data

Henry's Law Constant: estimated at 1.4 x10^{-6}

Octanol/Water Partition Coefficient: log K_{ow} = 1.360

Sorption Partition Coefficient: K_{oc} = 131

BCF: calculated at 100

Regulations

RCRA 40CFR: Listed Hazardous Waste No. U111 Toxic Waste
CERCLA: 40CFR 302.4: Listed per RCRA Section 3001 per CWA Section 307(a) RQ: 10 lb (4.535 kg)
SARA 40CFR 372.65: Listed
SARA EHS 40CFR 355: Not listed
TSCA: Listed

Analytical Methods

Air: EPA 0020
Soil: CLP MC_SVOA, OHC; EPA 16, 1625; SW846 3640A, 8070, 8070A, 8250A, 8270B, 8270C, 8410
Water / Groundwater: EPA 1625, 611, 625, 625-S, 6; APHA 6040-B, 6410-B; USGS O3118
Indoor / Expired Air: NIOSH 2522
Plasma: EPA 29

DIP9100 CAS #: 2253-43-2

DI-N-PROPYLPHOSPHORODITHIOIC ACID

EINECS Number: 218-847-1
Molecular Formula: $C_6H_{15}O_2PS_2$
Formula Weight: 214.27
Synonyms: DIPROPYL DITHIOPHOSPHATE; O,O-DIPROPYL DITHIOPHOSPHATE; DIPROPYL PHOSPHORODITHIOATE; O,O-DIPROPYL PHOSPHORODITHIOATE; O,O-DIPROPYL PHOSPHORODITHIOTIC ACID; O,O-DIPROPYLDITHIOPHOSPHORIC ACID; PHOSPHORODITHIOIC ACID,O,O-DIPROPYL ESTER
Use: chem int for the insecticide, aspon (former use)

Hazard Overviews

Carcinogenicity: IARC - Not listed; NIOSH - Not listed; NTP - Not listed; ACGIH - Not listed; OSHA - Not listed; EPA - Not listed; MAK - Not listed

Environmental

Regulations

RCRA 40CFR: Not listed
CERCLA: 40CFR 302.4: Not listed
SARA 40CFR 372.65: Not listed
SARA EHS 40CFR 355: Not listed
TSCA: Listed

DIQ3000 CAS #: 85-00-7

DIQUAT

RTECS: JM5690000
DOT: NA2781
EINECS Number: 201-579-4
Molecular Formula: $C_{12}H_{12}Br_2N_2$
Formula Weight: 344.07

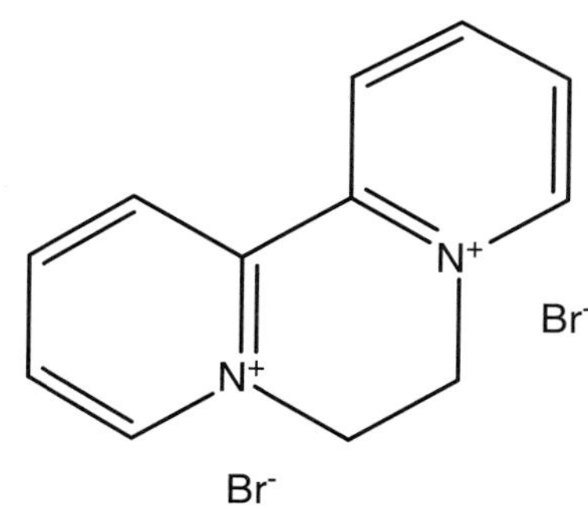

Chemical Structure

Synonyms: 1,1'-AETHYLEN-2,2'-BIPYRIDINIUM-DIBROMID; 1,1'-AETHYLEN-2,2'-BIPYRIDIUM-DIBROMID; AQUACIDE; CLEANSWEEP; DEIQUAT; DEIQUAT DIBROMIDE; DEXTRONE; DIBROMIDE; 9,10-DIHYDRO-8A,10,-DIAZONIAPHENANTHRENE DIBROMIDE; 9,10-DIHYDRO-8A,10A-DIAZONIAPHENANTHRENE DIBROMIDE; 9,10-DIHYDRO-8A,10A-DIAZONIAPHENANTHRENE(1,1'-ETHYLENE-2,2'-BIPYRIDYLIUM)-; 9,10-DIHYDRO-8A,10A-DIAZONIAPHENANTHRENE(1,1'-ETHYLENE-2,2'-BIPYRIDYLIUM)DIBROMIDE; 5,6-DIHYDRO-DIPYRIDO(1,2A;2,1C)PYRAZINIUM DIBROMIDE; 26,7-DIHYDROPYRIDO(1,2-A:2',1'C)PYRAZINEDIUM DIBROMIDE; 6,7-DIHYDROPYRIDO(1,2-A:2',1'-C)PYRAZINEDIUM DIBROMIDE; DIPYRIDO(1,2-A:2',1'-C)PYRAZINEDIIUM,6,7-DIHYDRO-,DIBROMIDE; DIQUAT DIBROMIDE; 1,1-ETHYLENE 2,2-DIPYRIDYLIUM DIBROMIDE; ETHYLENE DIPYRIDYLIUM DIBROMIDE; 1,1'-ETHYLENE-2,2'-BIPYRIDINIUM DIBROMIDE; 1,1'-ETHYLENE-2,2'-BIPYRIDYLIUM DIBROMIDE; 1,1'-ETHYLENE-2,2'-BIPYRIDYLLIUM DIBROMIDE; 1,1'-ETHYLENE-2,2'-DIPYRIDYLIUM DIBROMIDE; 1,1-ETHYLENE-2,2-DIPYRIDYLIUM DIBROMIDE; FB/2; PREEGLONE; REGLON; REGLON DIBROMIDE; REGLONE; WEEDTRINE-D
Description: colorless to yellow crystals; odorless
Use: contact herbicide, desiccant

Physical Properties

Boiling Point: Decomposes
Freezing Point: 335 °C (635 °F) to 340 °C (644 °F)
Specific Gravity: 1.22 to 1.27 at 20 °C/20 °C
Density: 1.22 to 1.27 g/mL at 20 °C
Vapor Pressure: < $1x10^{-5}$mbar at 20 °C
Water Solubility: 70 g/100ml at 20 °C
Other Solubilities: 95% Ethanol: Slightly Soluble; Hydroxylic solvents: Slightly Soluble; Non-polar organic solvents: Insoluble.
Flash Point: Not available; probably combustible

RTECS Toxicity Data

Acute Oral: Rat LD_{50} Dose: 120 mg/kg; Toxic Effects: Lungs, Thorax, or Respiration - Other changes. Domestic Animal LD_{50} Dose: 30 mg/kg.
Acute Dermal: Rabbit LD_{50} Route: Skin; Dose: >500 mg/kg. Rat LD_{50} Route: Skin; Dose: 433 mg/kg. Rat LD_{50} Route: Subcutaneous Dose: 20 mg/kg.
Irritation Eye: Rabbit Standard Draize Test Dose: 10 mg; Reaction: mild.
Irritation Skin: Rabbit Standard Draize Test Dose: 400 mg/kg/20D; Reaction: mild.

Reproductive/Teratogenic: Rat Route: Intravenous; Dose: 15 mg/kg; Duration: female 17D of pregnancy; Effects on Fertility - Post-implantation mortality; Effects on Embryo or Fetus - Fetal death. Rat Route: Intraperitoneal; Dose: 7 mg/kg; Duration: female 7D of pregnancy; Effects on Embryo or Fetus - Fetal death; Specific Developmental Abnormalities - Musculoskeletal system.

Mutagenic: Human Unscheduled DNA Synthesis; Cell Type: other cell types; Dose: 20 mg/L. Human Unscheduled DNA Synthesis; Cell Type: fibroblast; Dose: 1 umol/L. Human DNA Inhibition; Cell Type: lymphocyte; Dose: 500 mg/L.

Hazard Overviews

Fire: Will burn.

Carcinogenicity: IARC - Not listed; NIOSH - Not listed; NTP - Not listed; ACGIH - Class A4, Not classifiable as a human carcinogen; OSHA - Not listed; EPA - Not listed; MAK - Not listed

Exposure Limits

OSHA PEL Vacated 1989 Limits: TWA: 0.5 mg/m^3.

ACGIH TLV: TWA: 0.5 mg/m^3; Inhalable.

Environmental

Ecotoxicity: LD_{50} Mallard oral 564 mg/kg (95% confidence limit 324-982 mg/kg) 3-4 mo old males 96-hr median lethal concentration values for herbicides to yearling Coho salmon determined material was moderately toxic in freshwater (96-hr LC_{50} 10-30 mg/l) LC_{50} Japanese quail oral 1346 ppm (95% confidence limit 1178-1540 ppm), 14 days LC_{50} brown trout yearlings 570 mg/l 48-hr LC_{50} Mallard oral more than 5000 ppm (no mortality to 2500 ppm, 33% at 5000 ppm) Lowest observed concentration Salmo gairdneri (rainbow trout) >17 mg/l LC_{50} Gammarus fasciatus (Scud) more than 100 mg/l/96 hr, age mature, water temp 15 °C LC_{50} Stizostedionvitreum 2100 ug/l/96 hr

Environmental Fate: In the atmosphere, it will exist mainly as an aerosol and be subject to photolysis (half-life approx. 2 days) and gravitational settling. It will tightly bind to the upper layers of soil where it may remain for long periods of time. The binding to some soils is considered to be irreversible and makes it unavailable for biodegradation and photodegradation. It is removed rapidly from aquatic systems, principally by adsorption. If adsorption is initially to weeds, biodegradation to soluble or volatile products occurs in several weeks. When sorbed to sediment, little or no degradation probably occurs. In any case, it disappears from the water in 2-4 weeks. Little or no bioconcentration in fish will occur.

Cleanup/Disposal: Guide No. 151: Do not touch damaged containers or spilled material unless wearing appropriate protective clothing. Stop leak if you can do it without risk. Prevent entry into waterways, sewers, basements or confined areas. Cover with plastic sheet to prevent spreading. Absorb or cover with dry earth, sand or other non-combustible material and transfer to containers. Do not get water inside containers.

Environmental Physical Data

Octanol/Water Partition Coefficient: log K_{ow} = -3.05

BCF: none expected

Regulations

RCRA 40CFR: Not listed

CERCLA: 40CFR 302.4: Listed per CWA Section 311(b)(4) RQ: 1000 lb (453.5 kg)

SARA 40CFR 372.65: Not listed

SARA EHS 40CFR 355: Not listed

TSCA: Not listed

Analytical Methods

Soil: EPA PMD-DQT

Water / Groundwater: ASTM D4763

DIR1000 CAS #: 2429-74-5

DIRECT BLUE 15

RTECS: QJ6420000

EINECS Number: 219-385-3

Molecular Formula: $C_{34}H_{28}N_6Na_4O_{16}S_4$

Formula Weight: 996.88

Chemical Structure

Synonyms: AIREDALE BLUE D; AIZEN DIRECT SKY BLUE 5B; AIZEN DIRECT SKY BLUE 5BH; AMANIL SKY BLUE; ATLANTIC SKY BLUE A; ATUL DIRECT SKY BLUE; AZINE SKY BLUE 5B; BELAMINE SKY BLUE A; BENZANIL SKY BLUE; BENZO SKY BLUE A-CF; BENZO SKY BLUE S; BIS(AZO))BIS(5-AMINO-4-HYDROXY-,TETRASODIUM SALT; C.I. 24400; C.I. DIRECT BLUE 15; C.I. DIRECT BLUE 15,TETRASODIUM SALT; CARTASOL BLUE 2GF; CHLORAMINE SKY BLUE 4B; CHLORAMINE SKY BLUE A; CHROME LEATHER PURE BLUE; CRESOTINE PURE BLUE; DIACOTTON SKY BLUE 5B; DIAMINE BLUE; DIAMINE BLUE 6B; DIAMINE SKY BLUE; DIAMINE SKY BLUE CI; DIAPHTAMINE PURE BLUE; DIAZOL PURE BLUE 4B; 3,3'-((3,3'-DIMETHOXY-4,4'-BIPHENYLYLENE)BIS(AZO))BIS(5-AMINO-4-HYDROXY-; DIPHENYL BRILLIANT BLUE; DIPHENYL SKY BLUE 6B; DIRECT BLUE 10G; DIRECT BLUE HH; DIRECT PURE BLUE; DIRECT PURE BLUE M; DIRECT PURE BLUE N; DIRECT SKY BLUE; DIRECT SKY BLUE 5B; DIRECT SKY BLUE A; ENIANIL PURE BLUE AN; FENAMIN SKY BLUE; HISPAMIN SKY BLUE 3B; KAYAFECT BLUE Y; KAYAKU DIRECT SKY BLUE 5B; MITSUI DIRECT SKY BLUE 5B; MODR PRIMA 15; NAPHTAMINE BLUE 10G; 2,7-NAPHTHALENEDISULFONIC ACID,3,3'-((3,3'-DIMETHOXY(1,1'-BIPHENYL)-4,4'-DIYL)BIS(AZO))BIS(5-AMINO-4-HYDROXY-

,TETRASODIU///; 2,7-NAPHTHALENEDISULFONIC ACID,3,3'-((3,3'-DIMETHOXY-4,4'-BIPHENYLYLENE)-; 2,7-NAPHTHALENEDISULFONIC ACID),TETRASODIUM SALT; NIAGARA BLUE 4B; NIAGARA SKY BLUE; NIPPON DIRECT SKY BLUE; NIPPON SKY BLUE; NITSUI DIRECT SKY BLUE 5B; NITTO DIRECT SKY BLUE 5B; OXAMINE SKY BLUE 5B; PAPER BLUE S; PHENAMINE SKY BLUE A; PONTACYL SKY BLUE 4BX; PONTAMINE SKY BLUE 5BX; SHIKISO DIRECT SKY BLUE 5B; SKY BLUE 4B; SKY BLUE 5B; TERTRODIRECT BLUE F; VONDACEL BLUE HH

Description: dark blue to deep purple, microcrystalline powder

Use: dye used in the textile industry to dye both natural and synthetic fibers; as a biological stain and as a tint for cinematograph films

Physical Properties

Freezing Point: > 300 °C (572 °F)

Water Solubility: 10 to 50 mg/mL at 20 °C

Other Solubilities: 95% Ethanol: <1 mg/ml at 20 °C; Acetone: <1 mg/ml at 20 °C; DMSO: <1 mg/ml at 20 °C; Most organic solvents: Insoluble.

Flash Point: Not available; probably combustible

RTECS Toxicity Data

Reproductive/Teratogenic: Rat Route: Intraperitoneal; Dose: 70 mg/kg; Duration: female 8D of pregnancy; Effects on Fertility - Post-implantation mortality. Rat Route: Intraperitoneal; Dose: 200 mg/kg; Duration: female 8D of pregnancy; Specific Developmental Abnormalities - Central nervous system; Eye, ear. Rat Route: Intraperitoneal; Dose: 140 mg/kg; Duration: female 8D of pregnancy; Effects on Embryo or Fetus - Other effects to embryo or fetus.

Mutagenic: Bacteria - S Typhimurium Mutations in Microorganisms; Dose: 500 ug/plate (+S9). Bacteria - S Typhimurium Mutations in Microorganisms; Dose: 100 ug/plate (-S9).

Hazard Overviews

Health: Irritating. Toxic. Other Acute Effects: harmful if swallowed, inhaled, or absorbed through skin. Chronic Effects: possible teratogen. Carcinogen.

Fire: Will burn. Hazards: emits toxic fumes. Extinguishing agents: water spray; carbon dioxide, dry chemical powder or appropriate foam. Precautions: combustible liquid.

Reactivity: Stable. Hazardous polymerization will not occur. Incompatible with: strong oxidizing agents. Hazardous decomposition products: toxic fumes of: carbon monoxide, carbon dioxide, nitrogen oxides, sulfur oxides.

Carcinogenicity: IARC - Group 2B, Possibly carcinogenic to humans; NIOSH - Not listed; NTP - Not listed; ACGIH - Not listed; OSHA - Not listed; EPA - Not listed; MAK - Not listed

Primary Target Organs:

Eyes

Skin

Respiratory System

Environmental

Regulations

RCRA 40CFR: Not listed

CERCLA: 40CFR 302.4: Not listed

SARA 40CFR 372.65: Not listed

SARA EHS 40CFR 355: Not listed

TSCA: Listed

DIR3000 CAS #: 28407-37-6

DIRECT BLUE 218

RTECS: GS2169000

EINECS Number: 249-008-8

Molecular Formula: $C_{32}H_{16}Cu_2N_6Na_4O_{16}S_4$

Formula Weight: 1087.84

Synonyms: AMANIL SUPRA BLUE 9GL; ANAMIL SUPRA BLUE 9GL; C.I. 24401; C.I. DIRECT BLUE 218; COPPER, (MU-((TETRAHYDROGEN3,3'-((3,3'-DIHYDROXY-4,4'-BIPHENYLENE)BIS(AZO))BIS(5-AMINO-4-HYDROXY-2,7-NAPHTHALENEDISULFONATO))(4-)))DI-, TETRASODIUM SALT; CUPRATE(4-),(MU-((3,3'-((3,3'-DIHYDROXY(1,1'-BIPHENYL)-4,4'-DIYL)BIS(AZO))BIS-(5-AMINO-4-HYDROXY-2,7-NAPHTHALENEDISULFONATO))(8-)))DI-, TETRASODIUMSALT; (MU-((3,3'-((3,3'-DIHYDROXY(1,1'-BIPHENYL)-4,4'-DIYL)BIS(AZO))BIS(5-AMINO-; FAST BLUE 7GLN; FASTUSOL BLUE 9GLP; 4-HYDROXY-2,7-NAPHTHALENEDISULFONATO))(8-))DICUPRATE(4-),TETRASODIUM; PONTAMINE BOND BLUE B; PONTAMINE FAST BLUE 7GLN; SOLANTINE BLUE 10GL

Description: dark blue to deep purple amorphous powder

Use: dyestuff, intermediate, biological stain

Physical Properties

Water Solubility: 10 to 50 mg/mL at 17 °C

Other Solubilities: 95% Ethanol: <1 mg/ml at 24 °C; Acetone: <1 mg/ml at 17 °C; Acids: Soluble; Alkalies: Soluble; Carbon Tetrachloride: Insoluble; Chloroform: Insoluble; DMSO: <1 mg/ml at 24 °C.

Flash Point: Not available; probably combustible

RTECS Toxicity Data

Acute Oral: Rat LD_{50} Dose: 3290 mg/kg. Rabbit LD_{Lo} Dose: 3920 mg/kg.

Acute Dermal: Rabbit LD_{Lo} Route: Skin; Dose: 8 gm/kg.

Mutagenic: Bacteria - S Typhimurium Mutations in Microorganisms; Dose: 1 mg/plate (+S9), 125 ug/plate (-S9).

Tumorigenic: Rat Route: Oral; Dose: 438 gm/kg/2Y-C; Toxic Effects: Tumorigenic - Neoplastic by RTECS criteria; Lungs, Thorax, or Respiration - Tumors.

Hazard Overviews

Fire: Will burn.

Carcinogenicity: IARC - Not listed; NIOSH - Not listed; NTP - Not listed; ACGIH - Not listed; OSHA - Not listed; EPA - Not listed; MAK - Not listed

Environmental

Regulations

RCRA 40CFR: Not listed
CERCLA: 40CFR 302.4: Listed as Compound per CWA Section 307(a)
SARA 40CFR 372.65: Listed
SARA EHS 40CFR 355: Not listed
TSCA: Listed

DIR6000 CAS #: 2602-46-2

DIRECT BLUE 6

RTECS: QJ6400000
EINECS Number: 220-012-1
Molecular Formula: $C_{32}H_{20}N_6Na_4O_{14}S_4$
Formula Weight: 936.8
Synonyms: AIREDALE BLUE 2BD; AIZEN DIRECT BLUE 2BH; AMANIL BLUE 2BX; ATLANTIC BLUE 2B; ATUL DIRECT BLUE 2B; AZOCARD BLUE 2B; AZOMINE BLUE 2B; BELAMINE BLUE 2B; BENCIDAL BLUE 2B; BENZANIL BLUE 2B; BENZO BLUE BBA-CF; BENZO BLUE BBN-CF; BENZO BLUE GS; 3,3'-((4,4'-BIPHENYLYLENE)BIS(AZO))BIS(5-AMINO-4-HYDROXY)-2,7-NAPHTHALENE-; BLUE 2B; BLUE 2B SALT; BRASILAMINA BLUE 2B; BRAZILAMINA BLUE 2B; C.I. 22610; C.I. DIRECT BLUE 6; C.I. DIRECT BLUE 6,TETRASODIUM SALT; CALCOMINE BLUE 2B; CHLORAMINE BLUE 2B; CHLORAZOL BLUE B; CHLORAZOL BLUE BP; CHROME LEATHER BLUE 2B; CRESOTINE BLUE 2B; DIACOTTON BLUE BB; DIAMINE BLUE 2B; DIAMINE BLUE BB; DIAPHTAMINE BLUE BB; DIAZINE BLUE 2B; DIAZOL BLUE 2B; DIPHENYL BLUE 2B; DIPHENYL BLUE KF; DIPHENYL BLUE M2B; DIRECT BLUE 2B; DIRECT BLUE A; DIRECT BLUE BB; DIRECT BLUE GS; DIRECT BLUE K; DIRECT BLUE M2B; DISULFONIC ACID TETRASODIUM SALT; ENIANIL BLUE 2BN; FENAMIN BLUE 2B; FIXANOL BLUE 2B; HISPAMIN BLUE 2B; INDIGO BLUE 2B; KAYAKU DIRECT; KAYAKU DIRECT BLUE BB; MITSUI DIRECT BLUE 2BN; MODR PRIMA 6; NAPHTAMINE BLUE 2B; 2,7-NAPHTHALENEDISULFONIC ACID,3,3'-((1,1'-BIPHENYL)-4,4'-DIYLBIS(AZO))BIS(5-AMINO-4-HYDROXY-,TETRASODIUM SALT; NIAGARA BLUE 2B; NIPPON BLUE BB; PARAMINE BLUE 2B; PHENAMINE BLUE BB; PHENO BLUE 2B; PONTAMINE BLUE BB; SODIUM DIPHENYL-4,4'-BIS-AZO-2; SODIUM DIPHENYL-4,4'-BIS-AZO-2"-8"-AMINO-1"-NAPHTHOL-3",6"-DISULPHONATE; TERTRODIRECT BLUE 2B; VONDACEL BLUE 2B
Description: blue violet solid
Use: dye for fabric, leather, cotton, cellulosic materials, paper, silk, wool and nylon fibers; to stain biological materials and to produce aqueous writing inks; reportedly used in hair dyes."

Physical Properties

Boiling Point: Decomposes
Freezing Point: Decomposes
Water Solubility: Soluble in Water
Other Solubilities: 95% Ethanol: <1 mg/ml at 20.0 °C; Acetone: <1 mg/ml at 20.0 °C; Cellosolve: Slightly Soluble; DMSO: 1-5 mg/ml at 20.0 °C; Ethanol: Slightly Soluble; Ethylene Glycol monoethyl Ether: Slightly Soluble; Other organic solvents: Insoluble.
Flash Point: Not available; probably combustible

RTECS Toxicity Data

Acute Oral: Rat LD_{50} Dose: 8760 mg/kg. Rabbit LD_{Lo} Dose: 4 gm/kg.
Acute Dermal: Rabbit LD_{Lo} Route: Skin; Dose: 2 gm/kg.
Reproductive/Teratogenic: Rat Route: Intraperitoneal; Dose: 200 mg/kg; Duration: female 8D of pregnancy; Specific Developmental Abnormalities - Central nervous system; Eye, ear; Effects on Embryo or Fetus - Fetotoxicity.
Mutagenic: Rat DNA Adduct; Route: Intraperitoneal; Dose: 61200 ug/kg. Bacteria - S Typhimurium Mutations in Microorganisms; Dose: 100 ug/plate (+S9). Bacteria - S Typhimurium Mutations in Microorganisms; Dose: 100 nmol/plate (-S9).

Hazard Overviews

Fire: Will burn.
Carcinogenicity: IARC - Not listed; NIOSH - Not listed; NTP - Listed; ACGIH - Not listed; OSHA - Not listed; EPA - Not listed; MAK - Not listed

Environmental

Regulations

RCRA 40CFR: Not listed
CERCLA: 40CFR 302.4: Not listed
SARA 40CFR 372.65: Listed
SARA EHS 40CFR 355: Not listed
TSCA: Listed

Analytical Methods

Indoor / Expired Air: NIOSH 5013

DIR9000 CAS #: 2429-82-5

DIRECT BROWN 2

EINECS Number: 219-391-6
Molecular Formula: $C_{29}H_{21}N_5Na_2O_7S$
Formula Weight: 629.58
Synonyms: AIREDALE BROWN MD; AIZEN DIRECT BROWN MH; AMANIL BROWN MR; ATLANTIC BROWN M; ATUL DIRECT BROWN MR; AZINE BROWN M; AZOCARD BROWN M; AZOMINE BROWN M; BELAMINE FAST BROWN M; BENCIDAL FAST BROWN M; BENZANIL BROWN M; BENZANOL BROWN M; BENZO BROWN M; BENZOIC ACID,5-((4'-((7-AMINO-1-HYDROXY-3-SULFO-2-NAPHTHALENYL)AZO)(1,1'-BIPHENYL)-4-YL)AZO)-2-HYDROXY-,DISODIUM SALT; BENZOIC ACID,5-((4'-((7-AMIONO-1-HYDROXY-3-SULFO-2-NAPHTHALENYL)AZO)(1,1'-; BIPHENYL)-4-YL)AZO)-2-HYDROXY-,DISODIUM SALT; BRASILAMINA FAST BROWN 3RA; BROWN M; C.I. 22311; C.I. DIRECT BROWN 2; C.I. DIRECT BROWN 2,DISODIUM SALT; CALCOMINE BROWN MCW; CHLORAMINE BROWN M; CHLORAMINE BROWN 2ME; CHLORAZOL BROWN M; CHROME LEATHER BROWN M; COLUMBIA BROWN M; CRESOTINE BROWN RC; CUTAMINE BROWN CM; DIACOTTON BROWN M; DIAMINE BROWN; DIAMINE BROWN M; DIAMINE BROWN MBA-CF; DIAPHTAMINE BROWN M; DIAZINE BROWN; DIAZINE BROWN M; DIAZO BROWN MC; DIAZOL BROWN M; DIPHENYL BROWN; DIPHENYL BROWN V; DIPHENYL FAST BROWN MD; DIRECT BROWN 3RB; DIRECT FAST BROWN; DIRECT FAST BROWN M; ENIANIL FAST BROWN M; ERIE FAST BROWN 3RB; FENAMIN BROWN M; HISPAMIN FAST BROWN 3R2B;

JAPANOL BROWN M; KAYAKU DIRECT BROWN M; MAHOGANY EMBL; MITSUI DIRECT BROWN M; NAPHTAMINE BROWN DC; PARAMINE FAST BROWN M; PHENAMINE BROWN MB; UNION BROWN M

Description: reddish-brown solid or dark brown powder

Use: paper and leather dye

Physical Properties

Water Solubility: 1 to 10 mg/mL at 20 °C

Other Solubilities: 95% Ethanol: <1 mg/ml at 20 °C; Acetone: <1 mg/ml at 23 °C; DMSO: 1-10 mg/ml at 20 °C; Most organic solvents: Insoluble.

Flash Point: Not available; probably combustible

Hazard Overviews

Fire: Will burn.

Carcinogenicity: IARC - Not listed; NIOSH - Not listed; NTP - Not listed; ACGIH - Not listed; OSHA - Not listed; EPA - Not listed; MAK - Not listed

Environmental

Regulations

RCRA 40CFR: Not listed

CERCLA: 40CFR 302.4: Not listed

SARA 40CFR 372.65: Not listed

SARA EHS 40CFR 355: Not listed

TSCA: Listed

DIS1000 CAS #: 7778-43-0

DISODIUM ARSENITE

RTECS: CG0875000

DOT: UN1685; IMO6.1

EINECS Number: 231-902-4

Molecular Formula: $AsH_3Na_2O_4$

Formula Weight: 185.91

Synonyms: ARSENIC ACID,DISODIUM SALT; DISODIUM ARSENATE; DISODIUM ARSENIC ACID; DISODIUM HYDROGEN ARSENATE; DISODIUM HYDROGEN ORTHOARSENATE; DISODIUM MONOHYDROGEN ARSENATE; SHAUGHNESSY 013505; SODIUM ACID ARSENATE; SODIUM ARSENATE; SODIUM ARSENATE DIBASIC; SODIUM ARSENATE DIBASIC,ANHYDROUS; SODIUM ARSENATE,DIBASIC; SODIUM BIARSENATE

Description: colorless, clear powder or crystals; odorless

Use: all registered products for nonwood that contain inorganic arsenicals sodium arsenate cancelled & applications denied; references: 53 fr 5524 (2/24/88); 53 fr 24787 (6/30/88); med: formerly antimalarial; dermatologic; poison on fly-papers; formerly toxicant in ant syrups; insecticide former use; agent in mfr of wood preservatives-wolman & boliden salts; med: (vet): has been in parasitism (internally & externally), for nonparasitic skin & blood diseases, in rheumatism, asthma & heaves, & alternative.

Physical Properties

Freezing Point: 57 °C (134.6 °F)

Specific Gravity: 1.87

Water Solubility: Very Soluble in Water

Other Solubilities: Insoluble in Ether.

pH: Aqueous solution is alkaline

RTECS Toxicity Data

Mutagenic: Human DNA Inhibition; Cell Type: other cell types; Dose: 10 umol/L. Human DNA Inhibition; Cell Type: lymphocyte; Dose: 100 nmol/L. Human Cytogenetic Analysis; Cell Type: other cell types; Dose: 50 umol/L. Human Sister Chromatid Exchenge; Cell Type: lymphocyte; Dose: 2500 ug/L. Human Other Mutation Test Systems; Cell Type: lymphocyte; Dose: 100 nmol/L.

Hazard Overviews

Carcinogenicity: IARC - Group 1, Carcinogenic to humans; NIOSH - Listed as carcinogen; NTP - Class 1, Known to be a carcinogen; ACGIH - Class A1, Confirmed human carcinogen; OSHA - Listed as a carcinogen; EPA - Class A, Human carcinogen; MAK - Class A1, Capable of inducing malignant tumors as shown by experience with humans

Exposure Limits

OSHA PEL: TWA: 0.01 mg/m^3; as As inorganic.

ACGIH TLV: TWA: 0.01 mg/m^3; as As.

NIOSH REL: STEL: 0.002 mg/m^3; ceiling (15 min) as As.

Respirator Recommendation

Exposure Range: >0.01 to <5 mg/m^3 Supplied Air, Constant Flow/Pressure Demand, Full Face

Exposure Range: 5 to unlimited mg/m^3 Self-contained Breathing Apparatus, Pressure Demand, Full Face

Note: as arsenic, inorganic compounds; refer to 29CFR 1910.1018 for more specific respirator recommendations

Environmental

Ecotoxicity: LC_{50} Drosophila melanogaster (fruit fly) 0.54 mM/7 days

Cleanup/Disposal: Guide No. 151: Do not touch damaged containers or spilled material unless wearing appropriate protective clothing. Stop leak if you can do it without risk. Prevent entry into waterways, sewers, basements or confined areas. Cover with plastic sheet to prevent spreading. Absorb or cover with dry earth, sand or other non-combustible material and transfer to containers. Do not get water inside containers.

Regulations

RCRA 40CFR: Not listed

CERCLA: 40CFR 302.4: Listed as Compound per CWA Section 307(a) per CAA Section 112

SARA 40CFR 372.65: Listed as Compound

SARA EHS 40CFR 355: Not listed

TSCA: Listed

DIS1530 CAS #: 62-33-9

DISODIUM CALCIUM EDTA

RTECS: EV7700000

EINECS Number: 200-529-9

Molecular Formula: $C_{10}H_{12}CaN_2Na_2O_8$
Formula Weight: 374.28

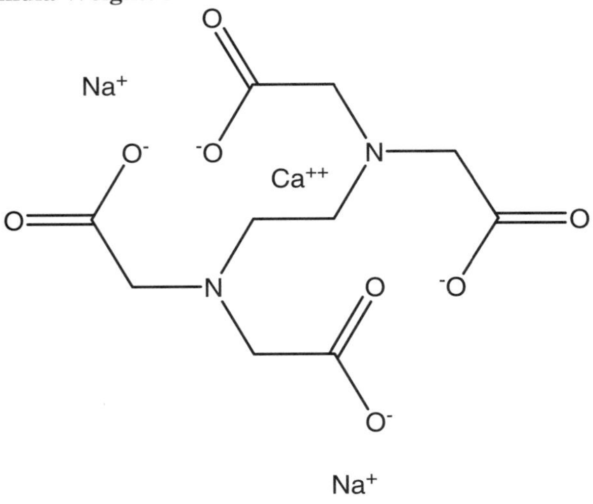

Chemical Structure

Synonyms: ACETIC ACID,(ETHYLENEDINITRILO)TETRA-,CALCIUM DISODIUMSALT; ADSORBONAC; ANTALLIN; CALCIATE(2-),((N,N'-1,2-ETHANEDIYLBIS(N-(CARBOXYMETHYL)GLYCINATO))(4-)-N,N',O,O',O(N),O(N'))-,DISODIUM,(OC-6-21)-; CALCIATE(2-),((ETHYLENEDINITRILO)TETRAACETATO)-,DISODIUM; CALCITETRACEMATE DISODIUM; CALCIUM DISODIUM EDATHAMIL; CALCIUM DISODIUM EDETATE; CALCIUM DISODIUM EDTA; CALCIUM DISODIUM ETHYLENDIAMINETETRAACETATE; CALCIUM DISODIUM ETHYLENEDIAMINETETRAACETATE; CALCIUM DISODIUM (ETHYLENEDINITRILO)TETRAACETATE; CALCIUM DISODIUM VERSENATE; CALCIUM EDTA; CALCIUM TITRIPLEX; CHELATON; DISODIUM CALCIUM ETHYLENEDIAMINETETRAACETATE; EDATHAMIL CALCIUM DISODIUM; EDETAMIN; EDETAMINE; EDETATE CALCIUM; EDETIC ACID CALCIUM DISODIUM SALT; EDTA CALCIUM DISODIUM SALT; EDTACAL; ETHYLENEDIAMINETETRAACETIC ACID CALCIUM DISODIUM SALT; ETHYLENEDIAMINETETRAACETIC ACID CALCIUM DISODIUM SALT(1:1:2); ETHYLENEDIAMINETETRAACETIC ACID,CALCIUM DISODIUM CHELATE; (ETHYLENEDINITRILO)TETRAACETIC ACID CALCIUM DISODIUM SALT; GLYCINE,N,N'-1,2-ETHANEDIYLBIS(N-(CARBOXYMETHYL)-,CALCIUM SODIUM SALT (1:1:2); LEDCLAIR; MONOCALCIUM DISODIUM EDTA; MOSATIL; RIKELATE CALCIUM; SODIUM CALCIUM EDETATE; SORMETAL; TETACIN; TETACIN-CALCIUM; TETAZINE; VERSENE; VERSENE CA
Description: powder or crystals; odorless
Use: chelating agent for lead tetrahydrate; sequestering agent; vet: anticoagulant; treatment of cardiac arrhythmias, hypercalcemia; antigushing agent in fermented malt beverages; analytical reagent; sequestering agent in food; in pharmaceuticals

Physical Properties

Density: Apparent 6.9 lb/gal
Water Solubility: 103 g/100 mL
Other Solubilities: Insoluble in organic solvents
pH: 7

RTECS Toxicity Data

Acute Oral: Rat LD_{50} Dose: 10 gm/kg. Mouse LD_{50} Dose: 10 gm/kg.

Hazard Overviews

Health: Acute Effects: may be harmful by inhalation, ingestion, or skin absorption.
Fire: Extinguishing agents: carbon dioxide, dry chemical powder or appropriate foam. Precautions: combustible liquid.
Reactivity: Stable. Hazardous polymerization will not occur. Incompatible with: strong oxidizing agents. Hazardous decomposition products: thermal decomposition may produce carbon monoxide, carbon dioxide, and nitrogen oxides.
Carcinogenicity: IARC - Not listed; NIOSH - Not listed; NTP - Not listed; ACGIH - Not listed; OSHA - Not listed; EPA - Not listed; MAK - Not listed

Environmental

Ecotoxicity: Fishes: bluegill (Lepomis macrochirus) 96h LC_{50},S 2,340 mg/l

Regulations

RCRA 40CFR: Not listed
CERCLA: 40CFR 302.4: Not listed
SARA 40CFR 372.65: Not listed
SARA EHS 40CFR 355: Not listed
TSCA: Listed

DIS2060 CAS #: 138-93-2

DISODIUM CYANODITHIOIMIDOCARBONATE

RTECS: FD3405000
EINECS Number: 205-346-8
Molecular Formula: $C_2H_2N_2Na_2S_2$
Formula Weight: 164.16
Synonyms: BUSANAT 586; CYANODITHIOIMIDOCARBONIC ACID DISODIUM SALT; IMIDOCARBONIC ACID,CYANODITHIO-,DISODIUM SALT

Hazard Overviews

Carcinogenicity: IARC - Not listed; NIOSH - Not listed; NTP - Not listed; ACGIH - Not listed; OSHA - Not listed; EPA - Not listed; MAK - Not listed

Environmental

Regulations

RCRA 40CFR: Not listed
CERCLA: 40CFR 302.4: Listed as Compound per CWA Section 307(a) per CAA Section 112
SARA 40CFR 372.65: Listed
SARA EHS 40CFR 355: Not listed
TSCA: Not listed

Analytical Methods

Water / Groundwater: EPA 630.1

DIS2590 CAS #: 7782-95-8

DISODIUM DIHYDROGEN HYPOPHOSPHATE

EINECS Number: 231-972-6
Molecular Formula: $H_4Na_2O_6P_2$
Formula Weight: 205.96
Synonyms: DISODIUM DIHYDROGEN SUBPHOSPHATE; DISODIUM HYPOPHOSPHATE; HYPOPHOSPHORIC ACID,DISODIUM SALT; SODIUM ACID HYPOPHOSPHATE; SODIUM DIHYDROGEN HYPOPHOSPHATE
Description: colorless, monoclinic crystals

Physical Properties

Freezing Point: Hexahydrate 250 °C (482 °F)
Density: Hexahydrate 1.849
Water Solubility: 2 g/100 cc
Other Solubilities: Soluble in dilute Sulfuric acid, Ammonium Hydroxide; Insoluble in Ethanol.
Refraction Index: 1.468 to 1.504

Hazard Overviews

Carcinogenicity: IARC - Not listed; NIOSH - Not listed; NTP - Not listed; ACGIH - Not listed; OSHA - Not listed; EPA - Not listed; MAK - Not listed

Environmental

Regulations

RCRA 40CFR: Not listed
CERCLA: 40CFR 302.4: Not listed
SARA 40CFR 372.65: Not listed
SARA EHS 40CFR 355: Not listed
TSCA: Listed

DIS3120 CAS #: 139-33-3

DISODIUM EDTA

RTECS: AH4375000
EINECS Number: 205-358-3
Molecular Formula: $C_{10}H_{14}N_2Na_2O_8$
Formula Weight: 336.24

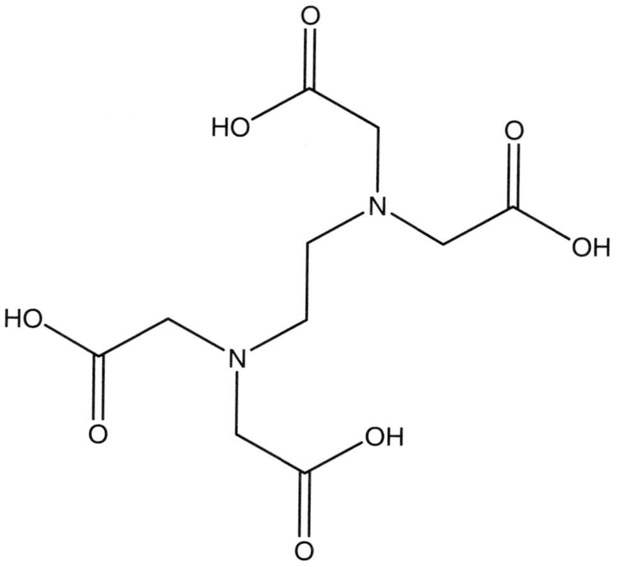

Chemical Structure

Synonyms: CHELADRATE; CHELAPLEX III; CHELATON III; COMPLEXON III; DINATRIUM ETHYLENDIAMINTETRAACETAT; D'E.D.T.A. DISODIQUE; DISODIUM DIACID ETHYLENEDIAMINETETRAACETATE; DISODIUM DIHYDROGEN ETHYLENEDIAMINETETRAACETATE; DISODIUM DIHYDROGEN(ETHYLENEDINITRILO)TETRAACETATE; DISODIUM EDATHAMIL; DISODIUM EDETATE; DISODIUM ETHYLENEDIAMINETETRAACETATE; DISODIUM ETHYLENEDIAMINETETRAACETIC ACID; DISODIUM (ETHYLENEDINITRILO)TETRAACETATE; DISODIUM (ETHYLENEDINITRILO)TETRAACETIC ACID; DISODIUM SALT OF EDTA; DISODIUM SEQUESTRENE; DISODIUM TETRACEMATE; DISODIUM VERSENATE; DISODIUM VERSENE; DOTITE 2NA; EDATHAMIL DISODIUM; EDETATE DISODIUM; EDETIC ACID DISODIUM SALT; EDTA DISODIUM; ENDRATE DISODIUM; N,N'-1,2-ETHANEDIYLBIS(N-(CARBOXYMETHYL)GLYCINE) DISODIUMSALT; ETHYLENEBIS(IMINODIACETIC ACID) DISODIUM SALT; ETHYLENEDIAMINETETRAACETATE,DISODIUM SALT; ETHYLENEDIAMINETETRAACETIC ACID,DISODIUM SALT; (ETHYLENEDINITRILO)-TETRAACETIC ACID DISODIUM SALT; F 1; F 1 (COMPLEXON); GLYCINE,N,N'-1,2-ETHANEDIYLBIS(N-(CARBOXYMETHYL)-,DISODIUM SALT (9CI); KIRESUTO B; KOMPLEXON III; METAQUEST B; PERMA KLEER 50 CRYSTALS DISODIUM SALT; PERMA KLEER DI CRYSTALS; SELEKTON B 2; SEQUESTRENE SODIUM 2; SODIUM VERSENATE; TETRACEMATE DISODIUM; TITRIPLEX III; TRILON BD; TRIPLEX III; VERESENE DISODIUM SALT; VERSENE SODIUM 2

Physical Properties

Bulk Density: 6.5 lbs/gal
Water Solubility: Freely Soluble
pH: 5% solution 4 to 6

RTECS Toxicity Data

Acute Oral: Rat LD_{50} Dose: 2 gm/kg. Mouse LD_{50} Dose: 2050 mg/kg.
Chronic (Multiple Dose) Oral: Rat Dose: 540 gm/kg/90D-I; Toxic Effects: Nutritional and gross metabolic - Weight loss or decreased weight gain; DEATH. Rabbit Dose: 15 gm/kg/30D-I; Toxic Effects: Liver - Other changes; Kidney,

Ureter, and Bladder - Other changes; Endocrine - Other changes.

Reproductive/Teratogenic: Rat Route: Oral; Dose: 31429 mg/kg; Duration: female 1-22D of pregnancy; Effects on Fertility - Female fertility index. Rat Route: Oral; Dose: 12857 mg/kg; Duration: female 7-15D of pregnancy; Effects on Fertility - Litter size; Effects on Embryo or Fetus - Fetotoxicity; Fetal death. Rat Route: Oral; Dose: 12857 mg/kg; Duration: female 7-15D of pregnancy; Specific Developmental Abnormalities - Central nervous system; Eye, ear; Craniofacial (including nose and tongue). Rat Route: Oral; Dose: 12857 mg/kg; Duration: female 7-15D of pregnancy; Specific Developmental Abnormalities - Musculoskeletal system.

Mutagenic: Grasshopper Cytogenetic Analysis; Route: Parenteral; Dose: 1 mmol/L.

Hazard Overviews

Health: May cause irritation to eyes/skin/respiratory tract. Harmful. Other Acute Effects: harmful if swallowed; may be harmful if inhaled or absorbed through the skin.

Fire: Extinguishing agents: noncombustible; use extinguishing media appropriate to surrounding fire conditions. Precautions: combustible liquid.

Reactivity: Stable. Hazardous polymerization will not occur. Incompatible with: strong oxidizing agents. Hazardous decomposition products: toxic fumes of: carbon monoxide, carbon dioxide, nitrogen oxides, sodium/sodium oxides.

Carcinogenicity: IARC - Not listed; NIOSH - Not listed; NTP - Not listed; ACGIH - Not listed; OSHA - Not listed; EPA - Not listed; MAK - Not listed

Environmental

Regulations

RCRA 40CFR: Not listed
CERCLA: 40CFR 302.4: Not listed
SARA 40CFR 372.65: Not listed
SARA EHS 40CFR 355: Not listed
TSCA: Listed

DIS3650 CAS #: 144-33-2

DISODIUM HYDROGEN CITRATE

RTECS: GE7580000
EINECS Number: 205-623-3
Molecular Formula: $C_6H_8Na_2O_7$
Formula Weight: 236.08

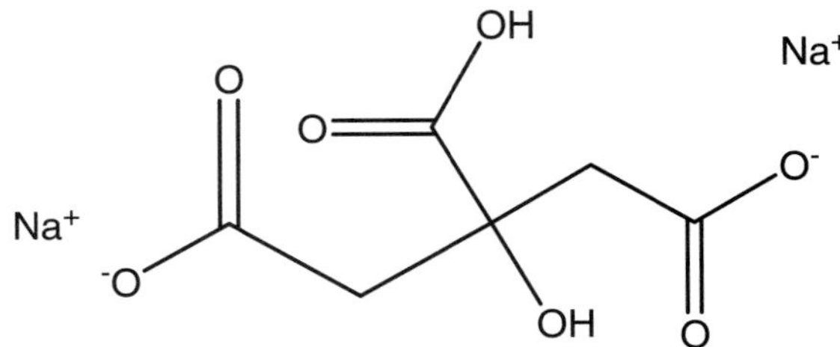

Chemical Structure

Synonyms: ALKACITRON; CITRALKA; CITRIC ACID,DISODIUM SALT; DISODIUM CITRATE; DISODIUM MONOHYDROGEN CITRATE; NATRIUM CITRICUM; 1,2,3-PROPANETRICARBOXYLIC ACID,2-HYDROXY-,DISODIUM SALT; 1,2,3-PROPANETRICARBOXYLIC ACID,2-HYDROXY-,DISODIUM SALT(9CI); SODIUM CITRATE; SODIUM CITRATE,ACID

Description: white powder

Use: anticoagulant in transfusions

Physical Properties

Freezing Point: 240 °C (464 °F)
Water Solubility: 1 g/ <2mL Water
pH: 3% wt/vol solution in water 4.9 to 5.2

RTECS Toxicity Data

Acute Dermal: Mouse LD_{50} Route: Subcutaneous Dose: 2580 mg/kg.

Hazard Overviews

Health: Irritating to eyes/skin. Other Acute Effects: may be harmful by inhalation, ingestion, or skin absorption.

Fire: Hazards: emits toxic fumes. Extinguishing agents: water spray; carbon dioxide, dry chemical powder or appropriate foam. Precautions: combustible liquid.

Reactivity: Hazardous polymerization will not occur. Incompatible with: strong oxidizing agents, protect from moisture. Hazardous decomposition products: toxic fumes of: carbon monoxide, carbon dioxide.

Carcinogenicity: IARC - Not listed; NIOSH - Not listed; NTP - Not listed; ACGIH - Not listed; OSHA - Not listed; EPA - Not listed; MAK - Not listed

Primary Target Organs:

Eyes

Skin

Environmental

Regulations

RCRA 40CFR: Not listed
CERCLA: 40CFR 302.4: Not listed
SARA 40CFR 372.65: Not listed
SARA EHS 40CFR 355: Not listed
TSCA: Listed

DIS4180 CAS #: 3655-00-3

DISODIUM LAURIMINODIPROPIONATE

EINECS Number: 222-899-0
Molecular Formula: $C_{18}H_{35}NNa_2O_4$
Formula Weight: 375.12
Synonyms: BETA-ALANINE,N-(2-CARBOXYETHYL)-N-DODECYL-,DISODIUM SALT; DERIPHAT 160; DISODIUM 3,3'-(DODECYLIMINO)BIS(PROPIONATE); DISODIUM BETA,BETA'-(LAURYLIMINO)DIPROPIONATE; DISODIUM N-LAURYL-BETA,BETA'-IMINODIPROPIONATE; DISODIUM N-LAURYL-BETA-IMINODIPROPIONATE; N-LAURYL-BETA-IMINODIPROPIONATE SODIUM SALT; PROPIONIC ACID,3,3'-(DODECYLIMINO)DI-,DISODIUM SALT; SODIUM N-DODECYLIMINODIPROPIONATE; SODIUM N-LAURYL-BETA-IMINODIPROPIONATE
Use: surfactants in cosmetic & detergent formulations

Hazard Overviews

Carcinogenicity: IARC - Not listed; NIOSH - Not listed; NTP - Not listed; ACGIH - Not listed; OSHA - Not listed; EPA - Not listed; MAK - Not listed

Environmental

Regulations

RCRA 40CFR: Not listed
CERCLA: 40CFR 302.4: Not listed
SARA 40CFR 372.65: Not listed
SARA EHS 40CFR 355: Not listed
TSCA: Listed

DIS4710 CAS #: 144-21-8

DISODIUM METHANEARSONATE

RTECS: PA2275000
EINECS Number: 205-620-7
Molecular Formula: $CH_5AsNa_2O_3$
Formula Weight: 139.96
Synonyms: ANSAR 184; ANSAR 8100; ANSAR DSMA LIQUID; ARRHENAL; ARSINYL; ARSONIC ACID,METHYL-,DISODIUM SALT; ARSYNAL; CACODYL NEW; CHIPCO CRAB KLEEN; CLOUT; CRAB-E-RAD; CRAB-3-RAD 100; CRALO-E-RAD; DAL-E-RAD 100; DIARSEN; DIMET; DINATE; DISODIUM METHANEARSENATE; DISODIUM METHYL ARSONATE; DISODIUM METHYLARSENATE; DISODIUM METHYLARSONATE; DISODIUM MONOMETHYLARSONATE; DISOMAR; DISOMEAR; DI-TAC; DMA; DMA 100; DREXEL DSMA LIQUID; DSMA; DSMA LIQUID; JON-TROL; MAA SODIUM SALT; METHANEARSONIC ACID,DISODIUM SALT; METHAR; METHAR 30; METHARSAN; METHARSINAT; METHYLARSONAT DISODNY; NAMATE; NEOASYCODILE; SODAR; SODIUM METHANEARSONATE; SODIUM METHARSONATE; SODIUM METHYLARSONATE; SOMAR; STENOSINE; TONARSAN; TONARSEN; TONARSIN; VERSAR DSMA LQ; WEED BROOM; WEED-HOE; WEED-E-RAD; WEED-E-RAD 360; WEED-E-RAD DMA POWDER
Description: colorless to white crystalline solid
Use: herbicide; selective pre-emergence & post-emergent contact herbicide with systemic properties; to control grass weeds in cotton, for turf treatment (eg, against crabgrass), & on uncropped land; in zoysia, bluegrass, & bermuda grass lawn; particularly effective against perennial weed johnsongrass; in mfr of pharmaceuticals.

Physical Properties

Freezing Point: 132 °C (269.6 °F) to 139 °C (282.2 °F)
Density: Hydrate 1
Water Solubility: Hygroscopic
Other Solubilities: Soluble in Alcohol.
Flash Point: Nonflammable

RTECS Toxicity Data

Acute Oral: Rat LD_{50} Dose: 821 mg/kg. Mouse LD_{50} Dose: 1150 mg/kg.
Acute Inhalation: Rat LC_{50} Dose: >22100 mg/m^3/4hr.
Acute Dermal: Rabbit LD_{50} Route: Skin; Dose: 10 gm/kg.
Reproductive/Teratogenic: Hamster Route: Intraperitoneal; Dose: 600 mg/kg; Duration: female 9D of pregnancy; Effects on Embryo or Fetus - Other effects to embryo or fetus. Hamster Route: Intraperitoneal; Dose: 500 mg/kg; Duration: female 12D of pregnancy; Effects on Embryo or Fetus - Fetotoxicity; Fetal death.

Hazard Overviews

Fire: Noncombustible.
Carcinogenicity: IARC - Not listed; NIOSH - Not listed; NTP - Not listed; ACGIH - Not listed; OSHA - Not listed; EPA - Not listed; MAK - Not listed
Exposure Limits
OSHA PEL: TWA: 0.5 mg/m^3; as As.
ACGIH TLV: TWA: 0.01 mg/m^3; as As.
NIOSH REL: STEL: 0.002 mg/m^3; Ceiling (15 min) as As.

Environmental

Ecotoxicity: LC_{50} Bluegill sunfish (lepomis macrochirus) > 1000 ppm/48 hr/fresh water/conditions of bioassay not specified

Regulations

RCRA 40CFR: Not listed
CERCLA: 40CFR 302.4: Listed as Compound per CWA Section 307(a) per CAA Section 112
SARA 40CFR 372.65: Listed as Compound
SARA EHS 40CFR 355: Not listed
TSCA: Not listed

DIS5770 CAS #: 7558-79-4

DISODIUM PHOSPHATE

RTECS: WC4500000
EINECS Number: 231-448-7
Molecular Formula: $H_3Na_2O_4P$
Structured MF: Na_2HPO_4

Formula Weight: 141.98

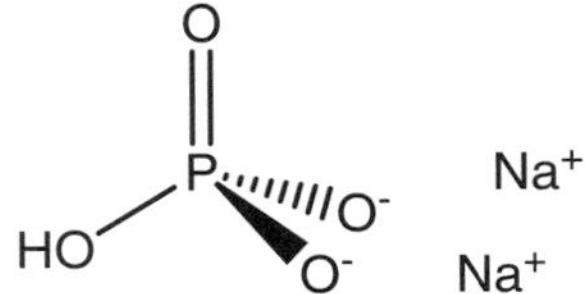

Chemical Structure

Synonyms: ACETEST; ANHYDROUS SODIUM ACID PHOSPHATE; DIBASIC SODIUM PHOSPHATE; DISODIUM ACID ORTHOPHOSPHATE; DISODIUM ACID PHOSPHATE; DISODIUM HYDROGEN PHOSPHATE; DISODIUM HYDROPHOSPHATE; DISODIUM MONOHYDROGEN PHOSPHATE; DISODIUM ORTHOPHOSPHATE; DISODIUM PHOSPHORIC ACID; DSP; EXCICCATED SODIUM PHOSPHATE; EXSICCATED SODIUM PHOSPHATE; NATRIUMPHOSPHAT; PHOSPHATE OF SODA; PHOSPHORIC ACID,DISODIUM SALT; SECONDARY SODIUM PHOSPHATE; SODA PHOSPHATE; SODIUM HYDROGEN PHOSPHATE; SODIUM MONOHYDROGEN PHOSPHATE; SODIUM MONOHYDROGEN PHOSPHATE (2:1:1); SODIUM ORTHOPHOSPHATE,SECONDARY; SODIUM PHOSPHATE; SODIUM PHOSPHATE,DIBASIC; SODIUM PHOSPHATE,DIBASIC HEPTAHYDRATE

Description: colorless, white crystals, powder; odorless

Use: sequestrant, buffer and emulsifier in foods; mordant in dyeing; used to weight silk; in tanning; in the manufacture of enamels, ceramics, detergents, and boiler compounds; as fireproofing agent; in soldering and brazing as reagent and buffer in analytical chemistry; fertilizers; pharmaceuticals

Physical Properties

Freezing Point: Loses H_20 at 92.5 °C (198.5 °F)
Specific Gravity: 1.5
Water Solubility: 14 lb Soluble in 100 gal Water
Other Solubilities: 95% Ethanol: <1 mg/ml at 20 °C; Acetone: <1 mg/ml at 20 °C; DMSO: <1 mg/ml at 20 °C.
pH: 1% aqueous solution at 25 °C 9.1
Refraction Index: Alpha 1.441
Flash Point: Nonflammable

RTECS Toxicity Data

Acute Oral: Rat LD_{50} Dose: 17 gm/kg.
Acute Dermal: Rat LD_{Lo} Route: Subcutaneous Dose: 1 gm/kg.
Irritation Eye: Rabbit Standard Draize Test Dose: 500 mg/24H; Reaction: mild.
Irritation Skin: Rabbit Standard Draize Test Dose: 500 mg/24H; Reaction: mild.

Hazard Overviews

Fire Diamond

Health: Irritating to eyes/skin/respiratory tract. Also Causes: coughing, difficulty in breathing, nausea.
Fire: Noncombustible. Use agent suitable for surrounding fire.
Reactivity: Stable. Hazardous polymerization cannot occur. Avoid: exposure to air. Incompatible with: alkaloids; lead acetate; antipyrine; chloral hydrate; resorcinol; pyrogallol. Hazardous decomposition products: fumes of phosphorus oxides; sodium oxide.
Carcinogenicity: IARC - Not listed; NIOSH - Not listed; NTP - Not listed; ACGIH - Not listed; OSHA - Not listed; EPA - Not listed; MAK - Not listed
Primary Target Organs:

Eyes

Skin

Respiratory System

Environmental

Cleanup/Disposal: Guide No. 171: Do not touch or walk through spilled material. Stop leak if you can do it without risk. Prevent dust cloud. Avoid inhalation of asbestos dust. Small Dry Spills: With clean shovel place material into clean, dry container and cover loosely; move containers from spill area. Small Spills: Take up with sand or other noncombustible absorbent material and place into containers for later disposal. Large Spills: Dike far ahead of liquid spill for later disposal. Cover powder spill with plastic sheet or tarp to minimize spreading. Prevent entry into waterways, sewers, basements or confined areas.

Environmental Physical Data

BOD: none

Regulations

RCRA 40CFR: Not listed
CERCLA: 40CFR 302.4: Listed per CWA Section 311(b)(4) RQ: 5000 lb (2268 kg)
SARA 40CFR 372.65: Not listed
SARA EHS 40CFR 355: Not listed
TSCA: Listed

DIS6300 CAS #: 7758-16-9

DISODIUM PYROPHOSPHATE

RTECS: UX6475000
EINECS Number: 231-835-0
Molecular Formula: $H_4Na_2O_7P_2$
Formula Weight: 221.97

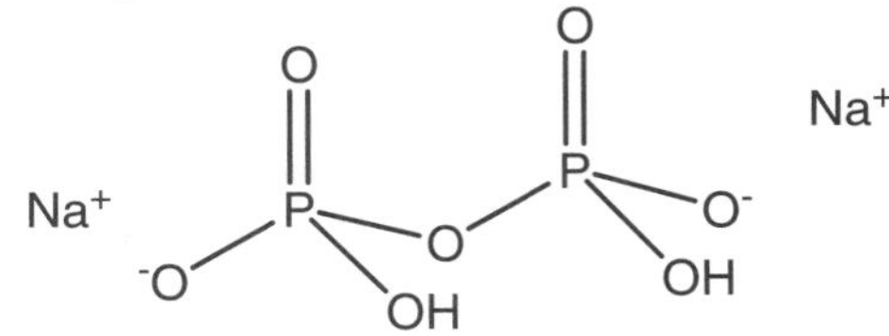

Chemical Structure

Synonyms: DINATRIUMPYROPHOSPHAT; DIPHOSPHORIC ACID,DISODIUM SALT; DISODIUM ACID PYROPHOSPHATE; DISODIUM DIHYDROGEN DIPHOSPHATE; DISODIUM DIHYDROGEN PYROPHOSPHATE; DISODIUM DIPHOSPHATE; PYROPHOSPHORIC ACID,DISODIUM SALT; SAPP; SODIUM ACID PYROPHOSPHATE; SODIUM PYROPHOSPHATE; SODIUM PYROPHOSPHATE,ACID

Description: white, fused masses or powder

Use: food acidulant; sequestrant; cleaning of food & dairy equipment; leavening agent in baking powder, in prepared mixes (food grade); in oil well drilling muds; in detergent builders; in water treatment; electroplating; metal cleaning & phosphatizing; buffer; peptizing agent in cheese & meat products; frozen desserts

Physical Properties

Boiling Point: Decomposes at 220 °C (428 °F)
Freezing Point: Decomposes at 220 °C (428 °F)
Density: Hexahydrate 1.86
Water Solubility: Soluble in Water
Refraction Index: 1.4599 to 1.469

RTECS Toxicity Data

Acute Oral: Mouse LD_{50} Dose: 2650 mg/kg.
Acute Dermal: Rabbit LD; Route: Skin; Dose: >300 mg/kg. Mouse LD_{50} Route: Subcutaneous Dose: 480 mg/kg.

Hazard Overviews

Health: Irritating to eyes/skin/respiratory tract. Other Acute Effects: may be harmful by inhalation, ingestion, or skin absorption.

Fire: Hazards: emits toxic fumes. Extinguishing agents: carbon dioxide, dry chemical powder or appropriate foam. Precautions: combustible liquid.

Reactivity: Incompatible with: strong oxidizing agents. Hazardous decomposition products: thermal decomposition may produce toxic fumes of phosphorus oxides and/or phosphine.

Carcinogenicity: IARC - Not listed; NIOSH - Not listed; NTP - Not listed; ACGIH - Not listed; OSHA - Not listed; EPA - Not listed; MAK - Not listed

Primary Target Organs:

Eyes

Skin

Respiratory System

Environmental

Regulations

RCRA 40CFR: Not listed
CERCLA: 40CFR 302.4: Not listed
SARA 40CFR 372.65: Not listed
SARA EHS 40CFR 355: Not listed
TSCA: Listed

DIS6830 CAS #: 2475-45-8

DISPERSE BLUE 1

RTECS: CB0540000
EINECS Number: 219-603-7
Molecular Formula: $C_{14}H_{12}N_4O_2$
Formula Weight: 268.28

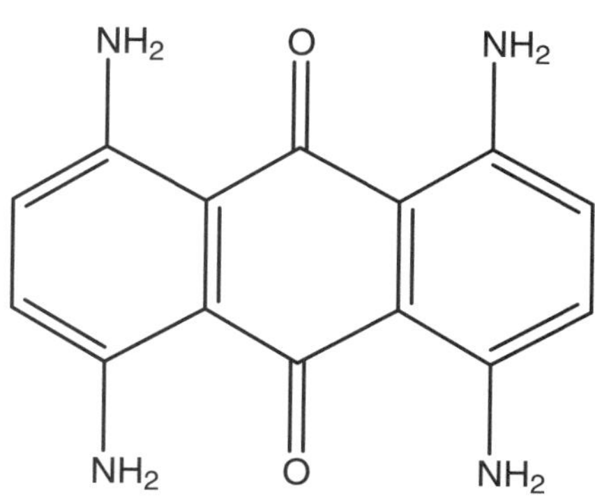

Chemical Structure

Synonyms: ACETATE BLUE 6; ACETATE BLUE G; ACETOQUINONE BLUE L; ACETOQUINONE BLUE R; ACETYLON FAST BLUE G; AMACEL BLUE GG; AMACEL PURE BLUE B; 9,10-ANTHRACENEDIONE,1,4,5,8-TETRAAMINO-; ANTHRAQUINONE,1,4,5,8-TETRAAMINO-; ANTHRAQUINONE,1,4,5,8-TETRAMINO-; ARTISIL BLUE SAP; ARTISIL BLUE SAP CONC; BRASILAZET BLUE GR; C.I. 64500; C.I. DISPERSE BLUE 1; C.I. SOLVENT BLUE 18; CELANTHRENE PURE BLUE BRS; CELLITON BLUE BB-CF; CELLITON BLUE EXTRA; CELLITON BLUE G; CELLITON BLUE GA-CF; CIBACET SAPPHIRE BLUE G; CILLA BLUE EXTRA; DIACELLITION FAST BLUE R; DIACELLITON FAST BLUE R; DISPERSE BLUE NO 1; DISPERSE BLUE NO. 1; DRACET SAPPHIRE BLUE G; DURANOL BRILLIANT BLUE CB; FENACET BLUE G; GRASOL BLUE 2GS; KAYALON FAST BLUE BR; MICROSETILE BLUE EB; MIKETON FAST BLUE; MIKETON FAST BLUE B; NACELAN BLUE G; NEOSETILE BLUE EB; NYLOQUINONE BLUE 2J; ORACET SAPPHIRE BLUE G; PERLITON BLUE B; SERINYL BLUE 2G; SERINYL BLUE 3G; SERINYL BLUE 3GN; SETACYL BLUE 2GS; SETACYL BLUE 2GS II; SUPRACET BRILLIANT BLUE 2GN; SUPRACET DEEP BLUE R; 1,4,5,8-TETRAAMINO-9,10-ANTHRAQUINONE; 1,4,5,8-TETRAAMINOANTHRAQUINONE; 1,4,5,8-TETRAMINOANTHRAQUINONE

Description: blue-black microcrystalline powder

Use: dye for cellulose acetate, nylon, polyester and other synthetic fibers, thermoplastics and wool sheepskins; alcoholic and ester solvents; vinyl and alkyl resin and cellulose ester lacquers; cellulose acetate and polystyrene molding powders

Physical Properties

Freezing Point: 332 °C (629.6 °F)
Saturated Vapor Density: 1.2 kg/m^3
Vapor Pressure: 1.8×10^{-8} mm Hg at 25 °C
Water Solubility: 30 ug/l in Water at 25 °C
Other Solubilities: DMSO: <1 mg/ml at 24 °C; 95% Ethanol: <1 mg/ml at 14 °C; Acetone: <1 mg/ml at 20 °C; Toluene: <1 mg/ml at 22 °C; Benzene: Slightly Soluble; Linseed oil: Soluble.
Flash Point: Not available; probably combustible

RTECS Toxicity Data

Acute Oral: Rat LD; Dose: >3 gm/kg; Toxic Effects: Kidney, Ureter, and Bladder - Other changes in urine composition. Mouse LD; Dose: >2 gm/kg; Toxic Effects: Kidney, Ureter, and Bladder - Other changes in urine composition.

Mutagenic: Bacteria - S Typhimurium Mutations in Microorganisms; Dose: 33 ug/plate (+S9). Bacteria - S Typhimurium Mutations in Microorganisms; Dose: 100 ug/plate (-S9).

Tumorigenic: Rat Route: Oral; Dose: 90125 mg/kg/2Y-C; Toxic Effects: Tumorigenic - Carcinogenic by RTECS criteria; Kidney, Ureter, and Bladder - Tumors; Blood - Leukemia. Rat Route: Oral; Dose: 180 gm/kg/2Y-C; Toxic Effects: Tumorigenic - Carcinogenic by RTECS criteria; Endocrine - Tumors; Tumorigenic effects - Testicular tumors.

Hazard Overviews

Health: Irritating to eyes/skin. Toxic. Other Acute Effects: harmful if swallowed, inhaled, or absorbed through skin. Chronic Effects: target organs: bladder, liver, lungs. Carcinogen.

Fire: Will burn. Hazards: emits toxic fumes. Extinguishing agents: water spray; carbon dioxide, dry chemical powder or appropriate foam. Precautions: combustible liquid.

Reactivity: Incompatible with: strong oxidizing agents. Hazardous decomposition products: toxic fumes of: carbon monoxide, carbon dioxide, nitrogen oxides.

Carcinogenicity: IARC - Group 2B, Possibly carcinogenic to humans; NIOSH - Not listed; NTP - Not listed; ACGIH - Not listed; OSHA - Not listed; EPA - Not listed; MAK - Not listed

Primary Target Organs:

Eyes

Skin

Respiratory System

Liver

Environmental

Environmental Fate: If released to the atmosphere, it will exist mainly in the particulate phase where it will be removed through both wet and dry deposition. Based on estimated K_{oc} values of 1000 and 30000, it should have low to no mobility in soil. Volatilization from either moist or dry soil surfaces is not expected. It is expected to bind to particulate matter and sediment in the water column based on its estimated K_{oc} values. It is not expected to volatilize from water surfaces based on an estimated Henry's Law constant of 2.1 $x10^{-7}$ atm-cu m/mol. It may bioconcentrate in aquatic organisms given estimated BCF values of 110 and 4500.

Environmental Physical Data

Henry's Law Constant: estimated at 2.1 $x10^{-7}$
Octanol/Water Partition Coefficient: log K_{ow} = 2.98
Sorption Partition Coefficient: K_{oc} = 1000
BCF: estimated at 110

Regulations

RCRA 40CFR: Not listed
CERCLA: 40CFR 302.4: Not listed
SARA 40CFR 372.65: Not listed
SARA EHS 40CFR 355: Not listed
TSCA: Listed

DIS7360	CAS #: 693-36-7

DISTEARYL THIODIPROPIONATE

RTECS: UF8010000
EINECS Number: 211-750-5
Molecular Formula: $C_{42}H_{82}O_4S$
Formula Weight: 683.18

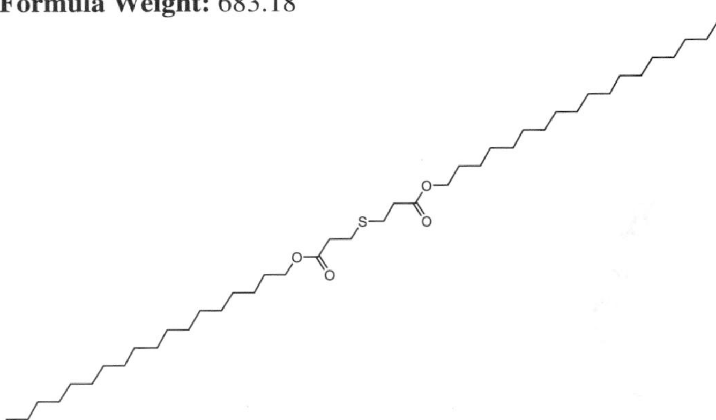

Chemical Structure

Synonyms: ADVASTAB 802; ADVASTAB PS 802; ANTIOK S; ARBESTAB DSTDP; CYANOX STDP; CYANOX-STDP; 3,3'-DIOCTADECYL THIODIPROPIONATE; DIOCTADECYL 3,3'-THIODIPROPIONATE; DIOCTADECYL THIODIPROPIONATE; DISTEARYL 3,3'-THIODIPROPIONATE; DISTEARYL BETA,BETA'-THIODIPROPIONATE; DISTEARYL BETA-THIODIPROPIONATE; DISTEARYL THIOPROPIONATE; DSTDP; DSTP; HOSTANOX SE 2; HOSTANOX VP-SE 2; IRGANOX PS 802; LUSMIT SS; NAUGARD DSTDP; PLASTANOX STDP; PLASTANOX STDP ANTIOXIDANT; PROPANOIC ACID,3,3'-THIOBIS-,DIOCTADECYL ESTER; PROPANOIC ACID,3,3'-THIOBIS-,DIOCTADECYL ESTER (9CI); PROPIONIC ACID,3,3'-THIODI-,DIOCTADECYL ESTER; PS 802; SEENOX DS; SUMILIZER TPS; THIODIPROPIONIC ACID,DISTEARYL ESTER; YOSHINOX DSTDP

Description: white flakes

Use: with primary antioxidants for vegetable oils; placticizer; softening agent; antioxidant (as synergist) for polypropylene, polyethylene, fats, thermoplastics & rubbers; preservative in animal feed

Physical Properties

Water Solubility: Insoluble in Water
Other Solubilities: Very Soluble in Benzene & olefin polymers.

RTECS Toxicity Data

Acute Oral: Rat LD_{50} Dose: >2500 mg/kg. Mouse LD_{50} Dose: >2 gm/kg.

Hazard Overviews

Health: Irritating to eyes/skin/respiratory tract. Other Acute Effects: may be harmful by inhalation, ingestion, or skin absorption.

Fire: Hazards: emits toxic fumes; in powder form capable of creating a dust explosion. Extinguishing agents: water spray;

carbon dioxide, dry chemical powder or appropriate foam. Precautions: combustible liquid.

Reactivity: Stable. Hazardous polymerization will not occur. Incompatible with: strong oxidizing agents. Hazardous decomposition products: toxic fumes of: carbon monoxide, carbon dioxide, sulfur oxides.

Carcinogenicity: IARC - Not listed; NIOSH - Not listed; NTP - Not listed; ACGIH - Not listed; OSHA - Not listed; EPA - Not listed; MAK - Not listed

Primary Target Organs:

Eyes

Skin

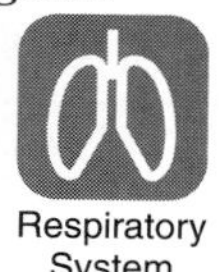
Respiratory System

Environmental

Regulations

RCRA 40CFR: Not listed
CERCLA: 40CFR 302.4: Not listed
SARA 40CFR 372.65: Not listed
SARA EHS 40CFR 355: Not listed
TSCA: Listed

DIS7890 **CAS #: 97-77-8**

DISULFIRAM

RTECS: JO1225000
EINECS Number: 202-607-8
Molecular Formula: $C_{10}H_{20}N_2S_4$
Structured MF: $[(C_2H_5)_2NCS_2]_2$
Formula Weight: 296.54

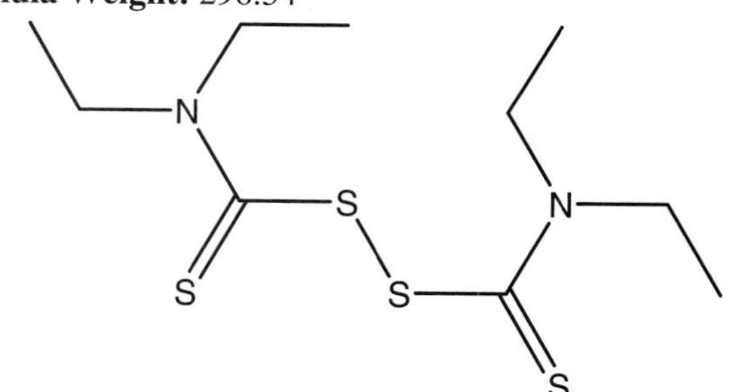

Chemical Structure

Synonyms: ABSTENSIL; ABSTENSYL; ABSTINIL; ABSTINYL; ALCOPHOBIN; ALK-AUBS; ANTABUS; ANTABUSE; ANTADIX; ANTAENYL; ANTAETHAN; ANTAETHYL; ANTAETIL; ANTALCOL; ANTETAN; ANTETHYL; ANTETIL; ANTEYL; ANTIAETHAN; ANTIETANOL; ANTI-ETHYL; ANTIETIL; ANTIKOL; ANTIVITIUM; AVERSAN; AVERZAN; BIS((DIETHYLAMINO)THIOXOMETHYL)DISULPHIDE; BIS(DIETHYLTHIOCARBAMOYL) DISULFIDE; BIS(N,N-DIETHYLTHIOCARBAMOYL) DISULFIDE; BIS(N,N-DIETHYLTHIOCARBAMOYL)DISULFIDE; BIS(DIETHYLTHIOCARBAMOYL)DISULPHIDE; BIS(N,N-DIETHYLTHIOCARBAMOYL)DISULPHIDE; BONIBAL; CONTRALIN; CONTRAPOT; CRONETAL; DICUPRAL; DISETIL; DISULFAN; DISULFURAM; DISULPHURAM; 1,1'-DITHIOBIS(N,N-DIETHYLTHIOFORMAMIDE); DUPON 4472; DUPONT FUNGICIDE 4472; EKAGOM TEDS; ENT 27,340; EPHORRAN; ESPENAL; ESPERAL; ETABUS; ETHYL THIRAM; ETHYL THIUDAD; ETHYL THIURAD; ETHYL TUADS; ETHYL TUEX; ETHYLDITHIOURAME; ETHYLDITHIURAME; EXHORAN; EXHORRAN; FORMAMIDE,1,1'-DITHIOBIS(N,N-DIETHYLTHIO-; HOCA; KROTENAL; NOCBIN; NOXAL; NSC 190940; REFUSAL; RO-SULFIRAM; RO-SULFRAM-500; STOPAETHYL; STOPETHYL; STOPETYL; TATD; TENURID; TENUTEX; TETD; TETIDIS; TETRADIN; TETRADINE; TETRAETHYLTHIOPEROXYDICARBONIC DIAMIDE; TETRAETHYLTHIRAM DISULPHIDE; TETRAETHYLTHIURAM; TETRAETHYLTHIURAM DISULFIDE; N,N,N',N',-TETRAETHYLTHIURAM DISULPHIDE; N,N,N',N'-TETRAETHYLTHIURAM DISULPHIDE; TETRAETHYLTHIURAM DISULPHIDE; TETRAETIL; TETURAM; TETURAMIN; THIOPEROXYDICARBONIC DIAMIDE (((H2N)C(S))2S2),TETRAETHYL-; THIOSAN; THIOSCABIN; THIRERANIDE; THIURAM DISULFIDE,TETRAETHYL-; THIURAM E; THIURANIDE; TILLRAM; TIURAM; TTD; TTS

Description: white, yellow, light gray crystals, powder; barely detectable odor

Use: as a fungicide, seed disinfectant, bactericide, insecticide, rubber accelerator and vulcanizer; a deterrent to alcohol use

Physical Properties

Boiling Point: 117 °C (243 °F) at 17 mm Hg
Freezing Point: 70 °C (158 °F)
Specific Gravity: 1.3
Density: 1.27 g/mL at 20 °C
Vapor Pressure: ~ 0 mm Hg at 20 °C
Water Solubility: 0.02 g/100 ml at 0 °C
Other Solubilities: Slightly Soluble in light petroleum.
pH: Solution obtained by shaking 1 g with 30 mL of water 6
Flash Point: Not available; probably combustible

RTECS Toxicity Data

Acute Oral: Man TD_{Lo} Dose: 150 mg/kg/6W-I; Toxic Effects: Musculoskelital - Joints. Woman TD_{Lo} Dose: 140 mg/kg/2W; Toxic Effects: Behavioral - Hallucinations, distorted perceptions; Behavioral - Convulsions or effect on seizure threshold. Woman TD_{Lo} Dose: 90 mg/kg/18D-I; Toxic Effects: Liver - Jaundice, other or unclassified. Child TD_{Lo} Dose: 150 mg/kg; Toxic Effects: Behavioral - Somnolence (general depressed activity); Behavioral - Hallucinations, distorted perceptions; Gastrointestinal - Nausea or vomiting. Human LD_{Lo} Dose: 160 mg/kg; Toxic Effects: Behavioral - Coma; Gastrointestinal - Ulceration or bleeding from large intestine; Liver - Hepatitis, fibrous (cirrhosis, post-necrotic scarring). Man LD_{Lo} Dose: 150 mg/kg/6W-I; Toxic Effects: Liver - Hepatitis (hepatocellular necrosis), diffuse. Rat LD_{50} Dose: 500 mg/kg.

Acute Dermal: Rat LD; Route: Subcutaneous Dose: >4 gm/kg.

Chronic (Multiple Dose) Oral: Rat Dose: 7 gm/kg/7D-I; Toxic Effects: Automatic Nervous System - Other (direct) parasympathomimetic; Gastrointestinal - Ulceration or bleeding from stomach; Blood - Changes in spleen. Rat Dose: 6750 mg/kg/6W-I; Toxic Effects: Brain and coverings - Other degenerative changes; Nutritional and gross metabolic - Changes in other metals. Rat Dose: 7770 mg/kg/3W-I; Toxic Effects: Peripheral Nerve and sensation - Recording from peripheral motor nerve; Nutritional and gross metabolic - Weight loss or decreased weight gain. Rat Dose: 6545 mg/kg/13W-I; Toxic Effects: Automatic Nervous System - Parasympatholytic; Gastrointestinal - Contraction (isolated tissue); Nutritional and gross metabolic - Weight loss or decreased weight gain. Rat Dose: 720 mg/kg/60D-I; Toxic Effects: Brain and coverings - Other degenerative changes; Biochemical - Phosphatases. Rat Dose: 44640 mg/kg/2Y-C; Toxic Effects: Nutritional and gross metabolic - Weight loss or decreased weight gain; DEATH.

Irritation Eye: Rabbit Standard Draize Test Dose: 100 mg; Reaction: mild.

Reproductive/Teratogenic: Rat Route: Oral; Dose: 5 gm/kg; Duration: female 3-12D of pregnancy; Effects on Fertility - Post-implantation mortality; Effects on Embryo or Fetus - Fetotoxicity. Rat Route: Oral; Dose: 310 mg/kg; Duration: male 31D prior to mating; Paternal Effects - Testes, epididymis, sperm duct.

Mutagenic: Human DNA Inhibition; Cell Type: HeLa cell; Dose: 100 umol/L. Rat Morphological Transformation; Route: Oral; Dose: 12600 mg/kg/6W.

Tumorigenic: Mouse Route: Oral; Dose: 35 gm/kg/78W-I; Toxic Effects: Tumorigenic - Neoplastic by RTECS criteria; Liver - Tumors. Mouse Route: Subcutaneous; Dose: 1000 mg/kg; Toxic Effects: Tumorigenic - Neoplastic by RTECS criteria; Lungs, Thorax, or Respiration - Tumors; Blood - Tumors.

Hazard Overviews

Fire Diamond

Health: Toxic. Also Causes: metallic/garlic taste in mouth, fever, ataxia, nausea, diarrhea, tachypnea, hypertension, lethargy, encephalopathy, coma, changes in kidneys, hematuria, ketonuria, skin eruptions. Chronic Effects: liver damage, joint inflammation, mutagen, reproductive hazard, carcinogen.

Fire: Noncombustible. Use agent suitable for surrounding fire.

Reactivity: Stable. Hazardous polymerization cannot occur. Avoid: generation of dusts. Hazardous decomposition products: carbon dioxide.

Carcinogenicity: IARC - Group 3, Not classifiable as to carcinogenicity to humans; NIOSH - Not listed; NTP - Not listed; ACGIH - Class A4, Not classifiable as a human carcinogen; OSHA - Not listed; EPA - Not listed; MAK - Not listed

Primary Target Organs:

Skin

Nervous System

Exposure Limits

OSHA PEL Vacated 1989 Limits: TWA: 2 mg/m^3.

ACGIH TLV: TWA: 2 mg/m^3.

NIOSH REL: TWA: 2 mg/m^3.

DFG MAK: TWA: 2 mg/m^3.

Respirator Recommendation

Exposure Range: >2 to 20 mg/m^3 Air Purifying, Negative Pressure, Half Mask

Exposure Range: >20 to 200 mg/m^3 Air Purifying, Negative Pressure, Full Face

Exposure Range: >200 to 2000 mg/m^3 Supplied Air, Constant Flow/Pressure Demand, Full Face

Exposure Range: >2000 to unlimited mg/m^3 Self-contained Breathing Apparatus, Pressure Demand, Full Face

Cartridge Color: black with magenta (P100)

Environmental

Ecotoxicity: Crustaceans: Daphnia magna 48h LC_{50} 0.12; Fishes: Poecilia reticulata 96h LC_{50} 0.32

Environmental Fate: Based on data for, thiram, its half-life in water should be approximately 5 days at pH 7. It should also photodegrade in air, surface water, or on the soil surface. It is expected to adsorb strongly to soil and sediment. Its biodegradability is unknown. Bioconcentration in aquatic organisms should be low to moderate. Volatilization will not be significant and it will not partition to the atmosphere. If released in the atmosphere as an aerosol or attached to particulate matter, it will be removed by gravitational settling. Vapor-phase would react with photochemically-produced hydroxyl radicals (half-life 0.045 days).

Environmental Physical Data

Henry's Law Constant: < 1.46 x10^{-8}

Octanol/Water Partition Coefficient: log K_{ow} = 3.88

Sorption Partition Coefficient: K_{oc} = 237

BCF: calculated at 136 to 523

Regulations

RCRA 40CFR: Listed Hazardous Waste No. U403 Toxic Waste

CERCLA: 40CFR 302.4: Listed per RCRA Section 3001 RQ: 1 lb (0.454 kg)

SARA 40CFR 372.65: Not listed

SARA EHS 40CFR 355: Not listed

TSCA: Listed

DIS8420 CAS #: 82-48-4

1,8-DISULFOANTHRAQUINONE

RTECS: CB0554000

EINECS Number: 201-426-1

Molecular Formula: $C_{14}H_8O_8S_2$

Formula Weight: 368.34

Synonyms: 1,8-ANTHRAQUINONEDISULFONIC ACID

Physical Properties

Freezing Point: 293 °C (559.4 °F)

Water Solubility: Very Soluble

Other Solubilities: Ethanol: Very Soluble

DIT1000 CAS #: 3347-22-6

DITHIANON

RTECS: QL0700000
EINECS Number: 222-098-6
Molecular Formula: $C_{14}H_4N_2O_2S_2$
Formula Weight: 296.33
Synonyms: DELAN; DELAN (FUNGICIDE); DELAN-COL; 2,3-DICARBONITRILO-1,4-DIATHIAANTHRACHINON; 2,3-DICYANO-1,4-DITHIA-ANTHRAQUINONE; 2,3-DICYANO-1,4-DITHIAANTHRAQUINONE; 5,10-DIHYDRO-5,10-DIOXONAPHTHO(2,3-B)-P-DITHIIN-2,3-DICARBONITRILE; 5,10-DIHYDRO-5,10-DIOXONAPHTHO-(2,3-B)-1,4-DITHIIN-2,3-DICARBONITRILE; 5,10-DIHYDRO-5,10-DIOXONAPHTHO-(2,3B)-P-DITHIIN-2,3-DICARBONITRILE; 5,10-DIHYDROXY-5,10-DIOXONAPHTHO-(2,3,B)-P-DITHIIN-2,3-DICARBONITRILE; 2,3-DINITRILO-1,4-DITHIA-9,10-ANTHRAQUINONE; 2,3-DINITRILO-1,4-DITHIA-ANTHRAQUINONE; 2,3-DINITRILO-1,4-DITHIAANTHRAQUINONE; 2,3-DINITRILO-1,4-DITHIOANTHRACHINON; 1,4-DITHIAANTHRAQUINONE-2,3-DICARBONITRILE; 1,4-DITHIAANTHRAQUINONE-2,3-DINITRILE; DITHIANONE; IT 931; MV 119A; NAPHTHO(2,3-B)-1,4-DITHIIN-2,3-DICARBONITRILE,5,10-DIHYDRO-5,10-DIOXO-; NAPHTHO(2,3-B)-P-DITHIIN-2,3-DICARBONITRILE,5,10-DIHYDRO-5,10-DIOXO-; STAUFFER MV-119A; THYNON
Description: gray-brown needles or crystals; odorless
Use: fungicide against foliar diseases of small fruit , eg, pome & stone fruit, ornamentals, & citrus; not against powdery mildews; fungicide seed dressing; against venturia spp (apple, cherry, & pear), microthyriella rubi (apple), coccomyces hiemalis & stigmina carpophila (cherry), monilia spp, taphrina deformans & tranzschelia discolor (apricot & peach), plasmopara viticola (grape), etc.

Physical Properties

Freezing Point: 220 °C (428 °F)
Vapor Pressure: < 6.7×10^{-7} mbar at 20 °C
Water Solubility: Insoluble in Water
Other Solubilities: Soluble in Acetone.

RTECS Toxicity Data

Acute Oral: Rat LD_{50} Dose: 638 mg/kg. Mouse LD_{50} Dose: 1110 mg/kg.
Acute Inhalation: Rat LC_{50} Dose: 3 gm/m^3/4hr.
Acute Dermal: Rat LD_{50} Route: Skin; Dose: >2 gm/kg.
Mutagenic: Mouse Morphological Transformation; Cell Type: fibroblast; Dose: 25 mg/L.

Hazard Overviews

Carcinogenicity: IARC - Not listed; NIOSH - Not listed; NTP - Not listed; ACGIH - Not listed; OSHA - Not listed; EPA - Not listed; MAK - Not listed

Environmental

Ecotoxicity: LC_{50} Channel catfish 130 ug/l/96 hr at 18 °C (95% confidence limit 120-140 ug/l), wt 1.6 g

Regulations

RCRA 40CFR: Not listed
CERCLA: 40CFR 302.4: Not listed
SARA 40CFR 372.65: Not listed
SARA EHS 40CFR 355: Not listed
TSCA: Not listed

Analytical Methods

Food: FDA 232.4, 242.1

DIT2330 CAS #: 514-73-8

DITHIAZANINE IODIDE

RTECS: DL7060000
EINECS Number: 208-186-7
Molecular Formula: $C_{23}H_{23}IN_2S_2$
Formula Weight: 518.50

Chemical Structure

Synonyms: ABMINTHIC; ANELMID; ANGUIFUGAN; BENZOTHIAZOLIUM,3-ETHYL-2-(5-(3-ETHYL-2(3H)-BENZOTHIAZOLYLIDENE)-1,3-PENTADIENYL)-,IODIDE; COMPOUND 01748; DEJO; DELVEX; DESELMINE; 3,3'-DIETHYLPENTAMETHINETHIACYANINE IODIDE; 3,3'-DIETHYLTHIADICARBOCYANINE IODIDE; DIETHYLTHIADICARBOCYANINE IODIDE; DILOMBRIN; DILOMBRINE; DITHIAZANIN IODIDE; DITHIAZINE; DITHIAZININE; DIZAN; EASTMAN 7663; 3-ETHYL-2-(5-(3-ETHYL-2-BENZOTHIAZOLINYLIDENE)-1,3-PENTADIENYL)BENZOTHIAZOLIUM IODIDE; 3-ETHYL-2-(5-(3-ETHYL-2-(3H)-BENZOTHIAZOLYLIDENE)-1,3-PENTADIENYL) BENZOTHIAZOLIUM IODIDE; 3-ETHYL-2-(5-(3-ETHYLBENZOTHIAZOL-2(3H)-YLIDENE)PENTA-1,3-DIENYL)BENZOTHIAZOLIUMIODIDE; L-01748; NECTOCYD; NETOCYD; NK 136; OMNI-PASSIN; OSSIURENE; PARTEL; TELMICID; TELMID; TELMIDE; VERCIDON
Description: dark greenish crystalline powder or blue violet powder
Use: sensitizer for photographic emulsions; medication: anthelmintic (nematodes); medication (vet): anthelmintic for heartworm microfilariae, hookworms, ascarids, whipworms, strongyloides stercoralis, and spirocerca lupi in dogs

Physical Properties

Freezing Point: Decomposes at ~ 248 °C (478.4 °F)
Water Solubility: Practically Insoluble in Water
Other Solubilities: Practically Insoluble in Ether; Very slightly Soluble in Alcohol and methyl Alcohol; Can be solubilized with polyvinylpyrrolidone.

Hazard Overviews

Poison

Health: Irritating to eyes/skin/respiratory tract. Poison. Other Acute Effects: may be fatal if inhaled, swallowed, or absorbed through skin.
Fire: Hazards: emits toxic fumes. Extinguishing agents: water spray; carbon dioxide, dry chemical powder or appropriate foam. Precautions: combustible liquid.
Reactivity: Incompatible with: strong oxidizing agents, sensitive to light. Hazardous decomposition products: toxic fumes of: carbon monoxide, carbon dioxide, nitrogen oxides, sulfur oxides, hydrogen iodide.
Carcinogenicity: IARC - Not listed; NIOSH - Not listed; NTP - Not listed; ACGIH - Not listed; OSHA - Not listed; EPA - Not listed; MAK - Not listed
Primary Target Organs:

Eyes

Skin

Respiratory System

Environmental

Regulations

RCRA 40CFR: Not listed
CERCLA: 40CFR 302.4: Not listed
SARA 40CFR 372.65: Not listed TPQ: 500/10000 lb
SARA EHS 40CFR 355: Listed TPQ: 500 lb
TSCA: Listed

DIT3660 CAS #: 142-46-1

2,5-DITHIOBIUREA

RTECS: EC1460000
EINECS Number: 205-537-6
Molecular Formula: $C_2H_6N_4S_2$
Formula Weight: 150.24

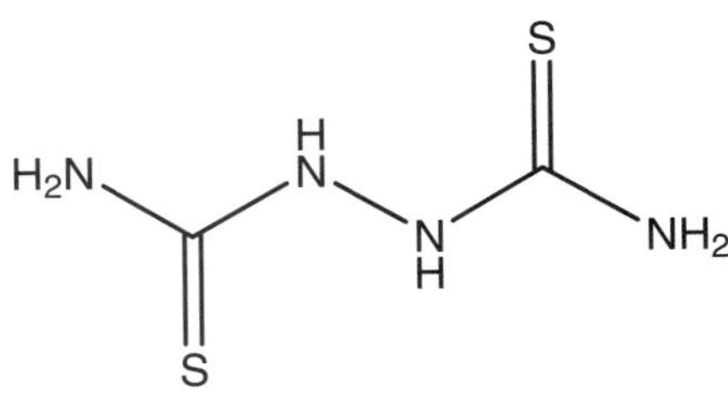

Chemical Structure

Synonyms: BISTHIOCARBAMYL HYDRAZINE; N,N'-BISTHIOCARBAMYL HYDRAZINE; BIS(THIOUREA); BITHIOUREA; BIUREA,2,5-DITHIO-; DITHIOBIUREA; DITHIOCARBAMOYLHYDRAZINE; 2,5-DITHIODIUREA; 2,5-DITHIOUREA; DITHIOUREA; 1,2-HYDRAZINEDICARBOTHIOAMIDE
Description: brown powder
Use: organic synthesis

Physical Properties

Freezing Point: 200 °C (392 °F) to 203 °C (397.4 °F)
Water Solubility: < 0.1 mg/mL 23 C
Other Solubilities: 95% Ethanol: Insoluble (<1 mg/ml at 23 °C); DMSO: Soluble (>=10 mg/ml).
Flash Point: Not available; probably combustible

RTECS Toxicity Data

Acute Oral: Rat LD; Dose: >500 mg/kg.
Mutagenic: Hamster Sister Chromatid Exchenge; Cell Type: ovary; Dose: 145 mg/L.
Tumorigenic: Mouse Route: Oral; Dose: 655 gm/kg/78W-C; Toxic Effects: Tumorigenic - Equivocal tumorigenic agent by RTECS criteria; Liver - Tumors. Mouse Route: Oral; Dose: 1310 gm/kg/78W-C; Toxic Effects: Tumorigenic - Equivocal tumorigenic agent by RTECS criteria; Liver - Tumors.

Hazard Overviews

Health: Irritating to eyes/skin/respiratory tract. Other Acute Effects: may be harmful by inhalation, ingestion, or skin absorption.
Fire: Will burn. Hazards: emits toxic fumes. Extinguishing agents: water spray; carbon dioxide, dry chemical powder or appropriate foam. Precautions: combustible liquid.
Reactivity: Incompatible with: strong oxidizing agents. Hazardous decomposition products: toxic fumes of: carbon monoxide, carbon dioxide, nitrogen oxides, sulfur oxides.
Carcinogenicity: IARC - Not listed; NIOSH - Not listed; NTP - Not listed; ACGIH - Not listed; OSHA - Not listed; EPA - Not listed; MAK - Not listed
Primary Target Organs:

Eyes

Skin
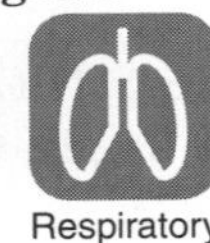
Respiratory System

Environmental

Regulations

RCRA 40CFR: Not listed

CERCLA: 40CFR 302.4: Not listed
SARA 40CFR 372.65: Not listed
SARA EHS 40CFR 355: Not listed
TSCA: Listed

DIT4990 **CAS #: 541-53-7**

2,4-DITHIOBIURET

RTECS: EC1575000
EINECS Number: 208-784-8
Molecular Formula: $C_2H_5N_3S_2$
Formula Weight: 135.20

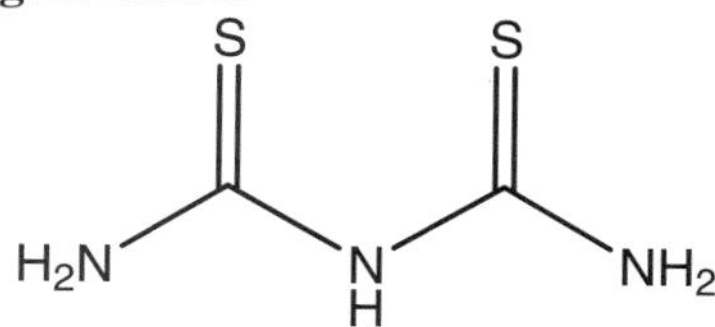

Chemical Structure

Synonyms: ALLOPHANIMIDIC ACID,DITHIO-; BIURET,2,4-DITHIO-; BIURET,DITHIO-; DITHIOBIURET; DTB; IMIDODICARBONIMIDOTHIOIC DIAMIDE; IMIDODICARBONODITHIOIC DIAMIDE; THIOIMIDODICARBONIC DIAMIDE; THIOIMIDODICARBONIC DIAMIDE (((H2N)C(S))2NH); UREA,2-THIO-1-(THIOCARBAMOYL)-
Description: colorless crystals
Use: plasticizer; rubber accelerator; int in resin mfr; in making insecticides & rodenticides; to delay wilting of flowers

Physical Properties

Boiling Point: Decomposes
Freezing Point: 181 °C (358 °F)
Density: 1.522 g/mL at 30 °C
Water Solubility: 0.27 g/100ml at 27 °C
Other Solubilities: 2.2 g/100 g Ethanol, 16 g/100 g Acetone, about 34 g/100 g Cellosolve in 1%Sodium Hydroxide= 3.6 g/100 g in 5% Sodium Hydroxide= 16 g/100 g in 10% Sodium Hydroxide= 29 g/100 g Soluble in alkali.
pH: Saturated aqueous solution at 30 °C 5.8

RTECS Toxicity Data

Acute Oral: Rat LD_{50} Dose: 5 mg/kg.
Acute Dermal: Rabbit LD_{Lo} Route: Subcutaneous Dose: 100 mg/kg; Toxic Effects: Peripheral Nerve and sensation - Flaccid paralysis without anesthesia; Lungs, Thorax, or Respiration - Cyanosis; Endocrine - Hyperglycemia.
Chronic (Multiple Dose) Dermal: Rabbit Route: Subcutaneous; Dose: 84 mg/kg/12D-I; Toxic Effects: Peripheral Nerve and sensation - Flaccid paralysis without anesthesia; Behavioral - Food intake (animal); Kidney, Ureter, and Bladder - Other changes.

Hazard Overviews

Carcinogenicity: IARC - Not listed; NIOSH - Not listed; NTP - Not listed; ACGIH - Not listed; OSHA - Not listed; EPA - Not listed; MAK - Not listed

Environmental

Environmental Fate: If released to soil, it will be expected to hydrolyze relatively rapidly if the soil is moist, based upon a calculated half-life for hydrolysis in water of 4.1 days at 25 °C. Based upon this relatively rapid hydrolysis, adsorption to and volatilization from moist soil are not expected to be significant processes, although no data specifically regarding fate in soil were located. If released to water, it will be expected to hydrolyze relatively rapidly, based upon a measured hydrolysis rate constant of 7.1 $x10^{-3}$ hr-1 in aqueous solution at 25 °C which corresponds to a half-life of 4.1 days. Based upon this relatively rapid hydrolysis, bioconcentration, volatilization, and adsorption to sediment and suspended solids are not expected to be significant processes. No data were located concerning biodegradation, but it probably abiotically hydrolyzes significantly faster than it biodegrades. If released to the atmosphere, it will be susceptible to photooxidation via vapor phase reaction with photochemically produced hydroxyl radicals. An atmospheric half-life of 3.8 hours at an atmospheric concentration of 5 $x10^5$ hydroxyl radicals per cu cm has been calculated for this process based upon an estimated rate constant.

Environmental Physical Data

BCF: not significant

Regulations

RCRA 40CFR: Listed Hazardous Waste No. P049 Toxic Waste
CERCLA: 40CFR 302.4: Listed per RCRA Section 3001 RQ: 100 lb (45.35 kg)
SARA 40CFR 372.65: Listed TPQ: 100/10000 lb
SARA EHS 40CFR 355: Listed TPQ: 100 lb
TSCA: Listed

DIT6320 **CAS #: 103-34-4**

4,4'-DITHIODIMORPHOLINE

RTECS: QE3325000
EINECS Number: 203-103-0
Molecular Formula: $C_8H_{16}N_2O_2S_2$
Formula Weight: 236.38

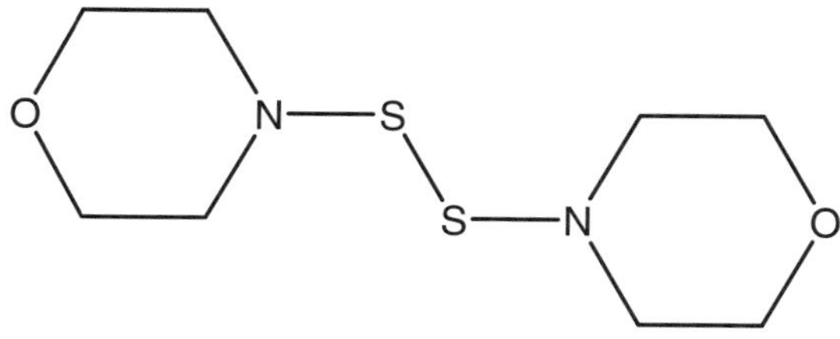

Chemical Structure

Synonyms: ACCEL R; N,N'-BISMORPHOLINE DISULFIDE; BISMORPHOLINO DISULFIDE; DEOVULC M; DIMORPHOLINE DISULFIDE; DIMORPHOLINE N,N'-DISULFIDE; DIMORPHOLINO DISULFIDE; DISULFIDE,DIMORPHOLINO-; 4,4'-DITHIOBIS(MORPHOLINE); DITHIOBISMORPHOLINE; N,N'-DITHIODIMORFOLIN; N,N'-DITHIODIMORPHOLINE; N,N-DITHIODIMORPHOLINE; 4,4'-DITHIOMORPHOLINE; MORPHOLINE DISULFIDE; MORPHOLINE,N,N'-DISULFIDE-; MORPHOLINE,4,4'-DITHIOBIS-; MORPHOLINE,4,4'-DITHIODI-; MORPHOLINODISULFIDE; SANFEL R; SULFASAN; SULFASAN R; VANAX A; VULNOC

Description: crystals; gray to tan powder

Use: staining protector in rubber; fungicide; organosulfur accelerator; accelerator for natural & synthetic rubbers; vulcanizer for natural & synthetic rubbers; curing agent for poly(fluoroalkoxyphosphazenes)

Physical Properties

Freezing Point: 125 °C (257 °F)
Specific Gravity: 1.36 at 25 °C
Other Solubilities: Chloroform: Soluble
Flash Point: Minimum 121.111 °C

RTECS Toxicity Data

Acute Oral: Rat LD_{50} Dose: 4300 mg/kg. Mouse LD_{50} Dose: 1660 mg/kg.

Acute Inhalation: Mouse LC_{50} Dose: 1624 mg/m^3.

Chronic (Multiple Dose) Oral: Rat Dose: 5375 mg/kg/43W-I; Toxic Effects: Brain and coverings - Recordings from specific areas of CNS; Liver - Liver function tests impaired; Biochemical - True cholinesterase.

Chronic (Multiple Dose) Inhalation: Rat Dose: 200 mg/m^3/24H/17W-C; Toxic Effects: Sense organs and special senses - Conjunctive irritation; Skin and appendages - Dermatitis, irritative.

Mutagenic: Bacteria - B Subtilis DNA Repair; Dose: 1 mg/disc. Bacteria - S Typhimurium Mutations in Microorganisms; Dose: 100 ug/plate (-S9).

Hazard Overviews

Health: Irritating to eyes/skin/respiratory tract. Other Acute Effects: may be harmful by inhalation, ingestion, or skin absorption; exposure can cause: nausea, headache and vomiting.

Fire: Will burn. Hazards: emits toxic fumes. Extinguishing agents: water spray; carbon dioxide, dry chemical powder or appropriate foam. Precautions: combustible liquid.

Reactivity: Incompatible with: strong oxidizing agents. Hazardous decomposition products: toxic fumes of: carbon monoxide, carbon dioxide, nitrogen oxides, sulfur oxides.

Carcinogenicity: IARC - Not listed; NIOSH - Not listed; NTP - Not listed; ACGIH - Not listed; OSHA - Not listed; EPA - Not listed; MAK - Not listed

Primary Target Organs:

Eyes

Skin

Respiratory System

Environmental

Regulations

RCRA 40CFR: Not listed
CERCLA: 40CFR 302.4: Not listed
SARA 40CFR 372.65: Not listed
SARA EHS 40CFR 355: Not listed
TSCA: Listed

DIT7650 **CAS #: 27157-94-4**

O,O-DITOLYL PHOSPHORODITHIOATE

EINECS Number: 248-273-7
Molecular Formula: $C_{14}H_{15}O_2PS_2$
Formula Weight: 310.38

Synonyms: CRESYL AEROFLOAT; PHOSPHORODITHIOIC ACID,O,O-BIS(METHYLPHENYL) ESTER; PHOSPHORODITHIOIC ACID,O,O-DITOLYL ESTER

Use: flotation collector; collector in refinement processing for silver sulfide, copper sulfide, lead sulfide, & zinc sulfide ores

Hazard Overviews

Carcinogenicity: IARC - Not listed; NIOSH - Not listed; NTP - Not listed; ACGIH - Not listed; OSHA - Not listed; EPA - Not listed; MAK - Not listed

Environmental

Regulations

RCRA 40CFR: Not listed
CERCLA: 40CFR 302.4: Not listed
SARA 40CFR 372.65: Not listed
SARA EHS 40CFR 355: Not listed
TSCA: Listed

DIT8980 **CAS #: 119-06-2**

DITRIDECYL PHTHALATE

RTECS: TI1950000
EINECS Number: 204-294-3
Molecular Formula: $C_{34}H_{58}O_4$
Structured MF: $C_6H_4(COOC_{13}H_{27})_2$
Formula Weight: 530.92

Synonyms: 1,2-BENZENEDICARBOXYLIC ACID,DITRIDECYL ESTER; BIS(TRIDECYL) PHTHALATE; DITRIDECYL 1,2-BENZENEDICARBOXYLATE; DI-(TRIDECYL) PHTHALATE; DI-(TRIDECYL)PHTHALATE; DTDP; JAYFLEX DTDP; NUOPLAZ; PHTHALIC ACID,DITRIDECYL ESTER; POLYCIZER 962-BPA; STAFLEX DTDP; 1-TRIDECANOL,PHTHALATE; TRUFLEX DTDP

Description: colorless liquid; nearly odorless

Use: plasticizer

Physical Properties

Boiling Point: > 285 °C (545 °F) at 5 mm Hg

Freezing Point: < -37 °C (-35 °F)
Specific Gravity: 0.951 at 20 °C/20 °C
Vapor Density: 18.3 Air=1
Density: 0.95 g/mL at 20 °C
Vapor Pressure: Extremely low
Water Solubility: 0.34 mg/L at 24 °C
Other Solubilities: DMSO: Soluble.
Flash Point: 243.3 °C Open Cup

RTECS Toxicity Data

Acute Oral: Rat LD_{50} Dose: >64 mL/kg.
Irritation Skin: Rabbit Open Draize Test Dose: 10 mg/24H open; Reaction: mild.

Hazard Overviews

Fire: Will burn.
Carcinogenicity: IARC - Not listed; NIOSH - Not listed; NTP - Not listed; ACGIH - Not listed; OSHA - Not listed; EPA - Not listed; MAK - Not listed

Environmental

Environmental Fate: It is expected to adsorb to sediment and suspended organic matter, and bioaccumulate in fish and aquatic organisms. Volatilization from water to the atmosphere may be a significant fate process, although the rate of volatilization may be decreased considerably by adsorption to sediment and suspended matter. The volatilization half-life from a model pond which takes into account adsorption processes, is over 200 days. Hydrolysis is not expected to occur. There are limited data indicating that if released to soil, it may be susceptible to biodegradation. In the atmosphere, it is expected to be adsorbed to particulates. Destruction by the vapor phase reaction with photochemically produced hydroxyl radicals may occur.
Cleanup/Disposal: Guide No. 171: Do not touch or walk through spilled material. Stop leak if you can do it without risk. Prevent dust cloud. Avoid inhalation of asbestos dust. Small Dry Spills: With clean shovel place material into clean, dry container and cover loosely; move containers from spill area. Small Spills: Take up with sand or other noncombustible absorbent material and place into containers for later disposal. Large Spills: Dike far ahead of liquid spill for later disposal. Cover powder spill with plastic sheet or tarp to minimize spreading. Prevent entry into waterways, sewers, basements or confined areas.

Environmental Physical Data

Henry's Law Constant: 2.23 x10^{-4}
Octanol/Water Partition Coefficient: log K_{ow} = > 2.12
Sorption Partition Coefficient: K_{oc} = calculated at 7900
BCF: calculated at 1140

Regulations

RCRA 40CFR: Not listed
CERCLA: 40CFR 302.4: Not listed
SARA 40CFR 372.65: Not listed
SARA EHS 40CFR 355: Not listed
TSCA: Listed

DIU3000 CAS #: 3648-20-2

DIUNDECYL PHTHALATE

RTECS: TI1980000
EINECS Number: 222-884-9
Molecular Formula: $C_{30}H_{50}O_4$
Structured MF: $C_6H_4(COOC_{11}H_{23})_2$
Formula Weight: 474.72
Synonyms: 1,2-BENZENEDICARBOXYLIC ACID,DIUNDECYL ESTER; 1,2-BENZENEDICARBOXYLIC ACID,DIUNDECYL ESTER (9CI); DI-N-UNDECYL PHTHALATE; DUP; PHTHALIC ACID,DIUNDECYL ESTER; SANTICIZER 711
Description: amber to golden clear liquid; odorless
Use: plasticizer

Physical Properties

Vapor Density: 15.3 Air=1
Water Solubility: Insoluble < 1 mg/mL at 21 C
Other Solubilities: 95% Ethanol: Soluble (>=10 mg/ml at 21 °C); DMSO: Slightly Soluble (1-10 mg/ml at 21 °C).
Flash Point: > 93.3 °C

RTECS Toxicity Data

Acute Oral: Rat LD; Dose: >20 gm/kg.
Acute Inhalation: Rat LC; Dose: >6040 mg/m^3/6hr; Toxic Effects: Lungs, Thorax, or Respiration - Acute pulmonary edema; Nutritional and gross metabolic - Weight loss or decreased weight gain.
Acute Dermal: Rabbit LD; Route: Skin; Dose: >10 gm/kg.
Irritation Eye: Rabbit Standard Draize Test Dose: 100 mg; Reaction: mild.

Hazard Overviews

Fire: Will burn.
Carcinogenicity: IARC - Not listed; NIOSH - Not listed; NTP - Not listed; ACGIH - Not listed; OSHA - Not listed; EPA - Not listed; MAK - Not listed

Environmental

Environmental Fate: If released on land, it would be expected to adsorb strongly to soil and should biodegrade. In water it will also adsorb to sediment and particulate matter and very slowly biodegrade. Bioconcentration occurs in shellfish but to a lesser extent in fish. However depuration is rapid, occurring in a few days. In the air, it will be associated with aerosols and particulate matter. Any vapor phase chemical should degrade by reaction with photochemically produced hydroxyl radicals (estimated half-life 10.2 hr).

Environmental Physical Data

BCF: may be expected to bioconcentrate

Regulations

RCRA 40CFR: Not listed
CERCLA: 40CFR 302.4: Not listed
SARA 40CFR 372.65: Not listed
SARA EHS 40CFR 355: Not listed

TSCA: Listed

DIU6000 **CAS #: 330-54-1**

DIURON

RTECS: YS8925000
EINECS Number: 206-354-4
Molecular Formula: $C_9H_{10}Cl_2N_2O$
Structured MF: $C_6H_3Cl_2NHCON(CH_3)_2$
Formula Weight: 233.10

Chemical Structure

Synonyms: AF 101; CEKIURON; CRISURON; DAILON; DCMU; DIATER; 3-(3,4-DICHLOOR-FENYL)-1,1-DIMETHYLUREUM; DICHLORFENIDIM; 3-(3,4-DICHLOROPHENOL)-1,1-DIMETHYLUREA; 1-(3,4-DICHLOROPHENYL)-3,3-DIMETHYLUREA; 3-(3,4-DICHLOROPHENYL)-1,1-DIMETHYLUREA; N'-(3,4-DICHLOROPHENYL)-N,N-DIMETHYLUREA; N-(3,4-DICHLOROPHENYL)-N',N'-DIMETHYLUREA; 1-(3,4-DICHLOROPHENYL)-3,3-DIMETHYLUREE; 3-(3,4-DICHLOR-PHENYL)-1,1-DIMETHYL-HARNSTOFF; 3-(3,4-DICLORO-FENYL)-1,1-DIMETIL-UREA; 1,1-DIMETHYL-3-(3,4-DICHLOROPHENYL)UREA; DI-ON; DIREX; DIREX 4L; DIRUROL; DIUMATE; DIUREX; DIURON 4L; DMU; DP HARDENER 95; DREXEL; DREXEL DIURON 4L; DURAN; DYNEX; FARMCO DIURON; HERBATOX; HW 920; KARAMEX; KARMEX; KARMEX D; KARMEX DIURON HERBICIDE; KARMEX DW; LUCENIT; MARMER; SUP'R FLO; TELVAR; TELVAR DIURON WEED KILLER; TIGREX; UNIDRON; UREA,3-(3,4-DICHLOROPHENYL)-1,1-DIMETHYL-; UREA,N'-(3,4-DICHLOROPHENYL)-N,N-DIMETHYL-; UROX D; VONDURON
Description: white crystals; odorless
Use: pre-emergence selective & total weed control herbicide; for general weed control on non-crop areas; for subsequent annual maintenance to prevent re-infestation by seedlings; against emerging & young broadleaf & grass weeds well mosses selectively; on asparagus, citrus, cotton, pineapple, sugarcane, temperate tree, & bush fruits, alfalfa, wheat & vineyards; for indust area weed control, in higher rainfall climates.

Physical Properties

Boiling Point: Decomposes at 180 °C (356 °F) to 190 °C (374 °F)
Freezing Point: 158 °C (316.4 °F) to 159 °C (318.2 °F)
Specific Gravity: 1.48 Water=1 at 4 °C
Vapor Density: 8.04 Air=1
Vapor Pressure: mm Hg
Water Solubility: 42 ppm in Water at 25 °C
Other Solubilities: Very low in hydrocarbon solvents; 53 g/kg in Acetone at 27 °C; 1200 ppm Soluble in Benzene at 27 °C; 1400 ppm Soluble in Butyl Stearate at 27 °C; 900 ppm Soluble in Cottonseed Oil (refined) at 27 °C.
Flash Point: Not flammable

RTECS Toxicity Data

Acute Oral: Rat LD_{50} Dose: 1017 mg/kg; Toxic Effects: Behavioral - General anesthetic; Behavioral - Ataxia.
Acute Dermal: Rat LD_{50} Route: Skin; Dose: >5 gm/kg.
Chronic (Multiple Dose) Oral: Rat Dose: 52500 ug/kg/30D-C; Toxic Effects: Blood - Changes in serum composition; Liver - Changes in liver weight; Biochemical - Transaminases. Rat Dose: 13620 mg/kg/61W-C; Toxic Effects: Blood - Methemoglobinemia-Carboxyhemoglobin; Blood - Changes in erythrocite (RBC) cell count; Blood - Changes in leukocyte (WBC) cell count.
Chronic (Multiple Dose) Inhalation: Rat Dose: 90 mg/m^3/4H/60D-I; Toxic Effects: Kidney, Ureter, and Bladder - Proteinuria; Blood - Changes in spleen; Endocrine - Changes in thyroid weight.
Reproductive/Teratogenic: Rat Route: Oral; Dose: 360 mg/kg; Duration: female 6-15D of pregnancy; Effects on Embryo or Fetus - Fetotoxicity. Rat Route: Oral; Dose: 1250 mg/kg; Duration: female 6-15D of pregnancy; Specific Developmental Abnormalities - Musculoskeletal system.
Mutagenic: Mouse DNA Inhibition; Route: Oral; Dose: 1 gm/kg. Bacteria - S Typhimurium Mutations in Microorganisms; Dose: 3 ug/plate (-S9).
Tumorigenic: Mouse Route: Oral; Dose: 153 gm/kg/78W-I; Toxic Effects: Tumorigenic - Equivocal tumorigenic agent by RTECS criteria; Lungs, Thorax, or Respiration - Tumors; Blood - Tumors.

Hazard Overviews

Fire Diamond

Health: Irritating to eyes/skin/respiratory tract. Also Causes: possible chest discomfort, cough, nausea, vomiting, diarrhea. Chronic Effects: sclerodermatous reaction, anemia and methemoglobinemia have been reported in animals.
Fire: Will burn, but does not ignite readily. Use agent suitable for surrounding fire. Dust may ignite when dispersed in air, especially confined spaces.
Reactivity: Stable. Hazardous polymerization cannot occur. Avoid: excessive heat; acids; alkalis. Incompatible with: strong acids; alkalis. Hazardous decomposition products: dimethylamine; 2,4-dichlorophenyl isocyanate; chlorine; carbon oxide(s); nitrogen.
Carcinogenicity: IARC - Not listed; NIOSH - Not listed; NTP - Not listed; ACGIH - Class A4, Not classifiable as a human carcinogen; OSHA - Not listed; EPA - Not listed; MAK - Not listed

Primary Target Organs:

Eyes

Skin

Respiratory System

Exposure Limits
OSHA PEL Vacated 1989 Limits: TWA: 10 mg/m^3.
ACGIH TLV: TWA: 10 mg/m^3.
NIOSH REL: TWA: 10 mg/m^3.
Respirator Recommendation
Exposure Range: >10 to 500 mg/m^3 Supplied Air, Constant Flow/Pressure Demand, Half Mask
Exposure Range: >500 to 10,000 mg/m^3 Supplied Air, Constant Flow/Pressure Demand, Full Face
Exposure Range: >10,000 to unlimited mg/m^3 Self-contained Breathing Apparatus, Pressure Demand, Full Face
Note: odor threshold unknown

Environmental

Ecotoxicity: LC_{50} Bobwhite quail oral (5 days) 1730 ppm (95% confidence limit 1482-2035 ppm), age 9 days LC_{50} Cutthroat trout 1.4 mg/l/96 hr (95% confidence limit 1.1-1.9 mg/l), at 10 °C, wt 0.3 g /. static bioassay without aeration, pH 7.2-7.5, water hardness 40-50 mg/l as calcium carbonate and alkalinity of 30-35 mg/l LC_{50} Pteronarcys (insect) 1.2 ug/l/48 hr Agmenellum quadroplicatum (blue-green alge) (strain PR6) lethal at 2 $x10^{-6}$ M (no specific isomer) LC_{50} Gammarus lacustris (crustacean) 160 ug/l/96 hr /Conditions of bioassay not specified LC_{50} Oncorhynchus kisutch (fish), 16000 ug/l/48 hr /Conditions of bioassay not specified LC_{50} Rainbow trout 4.3 ppm/48 hr /Conditions of bioassay not specified

Environmental Fate: It is a strongly adsorbed, highly persistent chemical and if released in soil will remain in the upper 5-10 cm of soil and have a half-life of about 330 days. If released into water it will adsorb to the sediment where it will slowly biodegrade after acclimation. The major product of the 6-7 degradation compounds that were isolated was 3,4-dichloroaniline and this metabolite may be further metabolized to an azobenzene derivative. In clear surface layers of water, sunlight irradiation will degrade it in a matter of days. Slow biodegradation may also occur. Bioconcentration in fish is not appreciable. It will degrade in air probably within a half-life in the vapor phase of 5.8 days due to reaction with hydroxyl radical.

Environmental Physical Data

Octanol/Water Partition Coefficient: log K_{ow} = 2.77
Sorption Partition Coefficient: K_{oc} = 400
BCF: estimated at 1.88

Regulations

RCRA 40CFR: Not listed
CERCLA: 40CFR 302.4: Listed per CWA Section 311(b)(4) RQ: 100 lb (45.35 kg)
SARA 40CFR 372.65: Listed
SARA EHS 40CFR 355: Not listed
TSCA: Listed

Analytical Methods

Soil: EPA PMD-TLC; SW846 8321A
Water / Groundwater: EPA 553, 632; SW846 8325
Drinking Water: EPA 553; AOAC 992.14
Food: FDA 212.1, 212.2, 232.1, 232.4, 242.1
Indoor / Expired Air: ASTM D4861
Plasma: EPA 001, 027; FDA 211.1, 231.1, 242.4, 252

DIV5000 **CAS #: 1314-34-7**

DIVANADIUM TRIOXIDE

RTECS: YW3050000
DOT: UN2860; IMO6.1
EINECS Number: 215-230-9
Molecular Formula: O_3V_2
Structured MF: V_2O_5
Formula Weight: 149.88

O^{--} V^{+++} O^{--} V^{+++} O^{--}

Chemical Structure

Synonyms: VANADIC OXIDE; VANADIUM OXIDE; VANADIUM(3+) OXIDE; VANADIUM OXIDE,SESQUI; VANADIUM SESQUIOXIDE; VANADIUM TRIOXIDE; VANADIUM TRIOXIDE,NONFUSED FORM
Description: black powder
Use: catalyst in making ethanol from ethylene; catalyst in sulfuric & nitric acid manufacture; medication

Physical Properties

Boiling Point: Decomposes at 1750 °C (3182 °F)
Freezing Point: 1940 °C (3524 °F)
Specific Gravity: 4.87 at 18 °C/4 °C
Vapor Pressure: ~ 0 mm Hg
Water Solubility: Insoluble in Water
Other Solubilities: Soluble in Nitric acid; Hydrogen Fluoride; alkali.
Flash Point: Not flammable

RTECS Toxicity Data

Acute Oral: Mouse LD_{50} Dose: 130 mg/kg.
Acute Dermal: Mouse LD_{50} Route: Subcutaneous Dose: 130 mg/kg.
Mutagenic: Hamster Cytogenetic Analysis; Cell Type: ovary; Dose: 18 mg/L. Hamster Sister Chromatid Exchenge; Cell Type: ovary; Dose: 1470 ug/L.

Hazard Overviews

Health: Irritating to eyes/skin/respiratory tract. Toxic. Other Acute Effects: harmful if swallowed, inhaled, or absorbed through skin.
Fire: Noncombustible. Hazards: emits toxic fumes. Extinguishing agents: noncombustible; use extinguishing

media appropriate to surrounding fire conditions. Precautions: combustible liquid.
Reactivity: Incompatible with: may decompose on exposure to air.
Carcinogenicity: IARC - Not listed; NIOSH - Not listed; NTP - Not listed; ACGIH - Not listed; OSHA - Not listed; EPA - Not listed; MAK - Not listed
Primary Target Organs:

Eyes

Skin

Respiratory System

Environmental

Cleanup/Disposal: Guide No. 154: Eliminate all ignition sources (no smoking, flares, sparks or flames in immediate area). Do not touch damaged containers or spilled material unless wearing appropriate protective clothing. Stop leak if you can do it without risk. Prevent entry into waterways, sewers, basements or confined areas. Absorb or cover with dry earth, sand or other non-combustible material and transfer to containers. Do not get water inside containers.

Regulations

RCRA 40CFR: Not listed
CERCLA: 40CFR 302.4: Not listed
SARA 40CFR 372.65: Not listed
SARA EHS 40CFR 355: Not listed
TSCA: Listed

Analytical Methods

Indoor / Expired Air: NIOSH 7504

DOC3000 CAS #: 112-85-6
N-DOCOSANOIC ACID

EINECS Number: 204-010-8
Molecular Formula: $C_{22}H_{44}O_2$
Formula Weight: 340.57

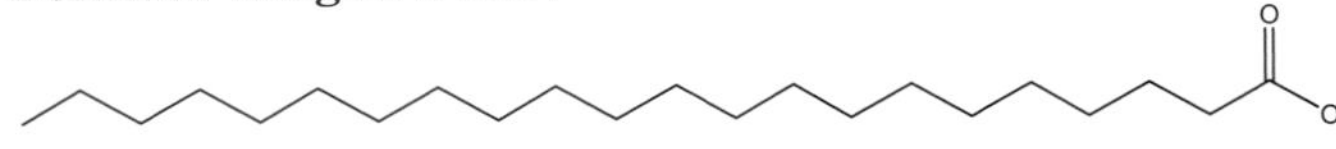

Chemical Structure

Synonyms: BEHENIC ACID; 1-DOCOSANOIC ACID; DOCOSANOIC ACID; DOCOSOIC ACID; GLYCON B-70; HYDROFOL 2022-55; HYDROFOL ACID 560
Description: waxy solid
Use: in cosmetics; waxes; plasticizers; chemicals; stabilizers; chem int for lithium docosanoate (lithium behenate); chem int for silver docosanoate (silver behenate); chem int for docosanamide (behenamide); chem int for docosylamine (behenylamine) and derivatives; chem int for other metal salts, eg, sodium and iron(+3); chem int for higher alkyl esters, eg, octadecyl ester

Physical Properties

Boiling Point: 306 °C (583 °F) at 60 mm Hg
Freezing Point: 79.95 °C (175.91 °F)
Specific Gravity: 0.8221 at 100 °C/4 °C
Vapor Pressure: 0.001 kPa at 150 °C
Water Solubility: Slightly Soluble in Water
Other Solubilities: 0.218 g are Soluble in 100 ml of 91.5% Ethanol at 25 °C; 0.116 g are Soluble in 100 ml of 86.2% Ethanol at 25 °C; 0.011 g are Soluble in 100 ml of 63.07% Ethanol at 25 °C; 0.1922 g are Soluble in 100 g of Ether at 16 °C.
pH: Neutralization value: 164.73
Refraction Index: 1.4270 at 100 °C/D

Hazard Overviews

Health: Irritating to eyes/skin/respiratory tract. Other Acute Effects: may be harmful by inhalation, ingestion, or skin absorption.
Fire: Hazards: in powder form capable of creating a dust explosion. Extinguishing agents: carbon dioxide, dry chemical powder or appropriate foam; water spray. Precautions: combustible liquid.
Reactivity: Incompatible with: bases, oxidizing agents, reducing agents. Hazardous decomposition products: toxic fumes of: carbon monoxide, carbon dioxide.
Carcinogenicity: IARC - Not listed; NIOSH - Not listed; NTP - Not listed; ACGIH - Not listed; OSHA - Not listed; EPA - Not listed; MAK - Not listed
Primary Target Organs:

Eyes

Skin

Respiratory System

Environmental

Regulations

RCRA 40CFR: Not listed
CERCLA: 40CFR 302.4: Not listed
SARA 40CFR 372.65: Not listed
SARA EHS 40CFR 355: Not listed
TSCA: Listed

DOC6000 CAS #: 661-19-8
1-DOCOSANOL

RTECS: JR1315000
EINECS Number: 211-546-6
Molecular Formula: $C_{22}H_{46}O$
Formula Weight: 326.61

OH

Chemical Structure

Synonyms: BEHENIC ALCOHOL; BEHENYL ALCOHOL; N-DOCOSANOL; DOCOSYL ALCOHOL; IK 2; LIDAVOL; TADENAN
Description: colorless waxy solid
Use: synthetic fibers; lubricants; evaporation retardant on water surfaces

Physical Properties

Boiling Point: 180 °C (356 °F) at 0.22 mm Hg
Freezing Point: 71 °C (159.8 °F)
Water Solubility: Slightly Soluble in Water
Other Solubilities: Slightly Soluble in Acetone; Very Soluble in Alcohol, Methanol, hot Acetone, hot petroleum; Soluble in Chloroform

RTECS Toxicity Data

Acute Dermal: Mouse LD_{50} Route: Subcutaneous Dose: >800 mg/kg.

Hazard Overviews

Health: Irritating to eyes/skin. Other Acute Effects: may be harmful by inhalation, ingestion, or skin absorption.
Fire: Hazards: emits toxic fumes. Extinguishing agents: water spray; carbon dioxide, dry chemical powder or appropriate foam. Precautions: combustible liquid.
Reactivity: Incompatible with: strong oxidizing agents. Hazardous decomposition products: toxic fumes of: carbon monoxide, carbon dioxide.
Carcinogenicity: IARC - Not listed; NIOSH - Not listed; NTP - Not listed; ACGIH - Not listed; OSHA - Not listed; EPA - Not listed; MAK - Not listed
Primary Target Organs:

Eyes

Skin

Environmental

Regulations

RCRA 40CFR: Not listed
CERCLA: 40CFR 302.4: Not listed
SARA 40CFR 372.65: Not listed
SARA EHS 40CFR 355: Not listed
TSCA: Listed

DOD1000 **CAS #: 14979-34-1**

DODECACHLORODICYCLOPENTADI ENE

Molecular Formula: $C_{10}Cl_{12}$
Formula Weight: 545.55
Synonyms: HEXACHLOROCYCLOPENTADIENE CYCLIC DIMER; 4,7-METHANO-1H-INDENE,1,1,2,3,3A,4,5,6,7,7A,8,8-DODECACHLORO-3A,4,7,7A-TETRAHYDRO-; 4,7-METHANOINDENE,1,1,2,3,3A,4,5,6,7,7A,8,8-DODECACHLORO-3A,4,7,7A-TETRAHYDRO-

Hazard Overviews

Carcinogenicity: IARC - Not listed; NIOSH - Not listed; NTP - Not listed; ACGIH - Not listed; OSHA - Not listed; EPA - Not listed; MAK - Not listed

Environmental

Regulations

RCRA 40CFR: Not listed
CERCLA: 40CFR 302.4: Not listed
SARA 40CFR 372.65: Not listed
SARA EHS 40CFR 355: Not listed
TSCA: Not listed

DOD1360 **CAS #: 713-95-1**

DELTA-DODECALACTONE

RTECS: UQ0850000
EINECS Number: 211-932-4
Molecular Formula: $C_{12}H_{22}O_2$
Formula Weight: 198.31

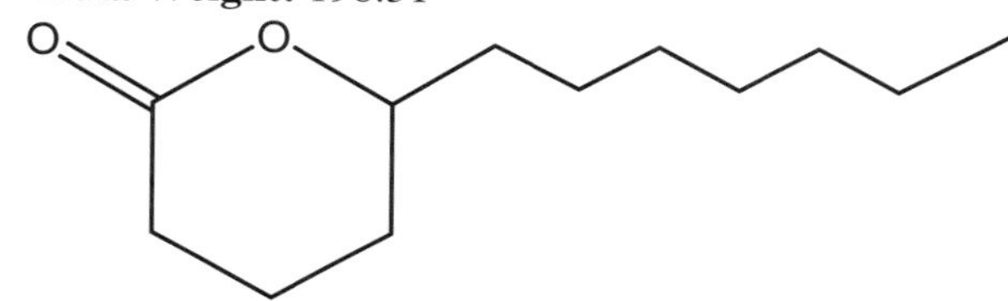

Chemical Structure

Synonyms: DODECANOIC ACID,5-HYDROXY-,DELTA-LACTONE; DELTA-HEPTYL-DELTA-VALEROLACTONE; N-HEPTYL-DELTA-VALEROLACTONE; 5-HYDROXYDODECANOIC ACID DELTA-LACTONE; 5-HYDROXYDODECANOIC ACID LACTONE; 5-HYDROXYDODECANOIC ACID,DELTA-LACTONE; 2H-PYRAN-2-ONE,6-HEPTYLTETRAHYDRO-
Description: colorless to very pale straw-yellow viscous liquid; powerful, fresh-fruit, oily odor
Use: flavor chemical; food additive, flavoring

Physical Properties

Water Solubility: Insoluble in Water
Other Solubilities: Soluble in Ethanol; poorly Soluble in Propylene Glycol.

RTECS Toxicity Data

Acute Oral: Rat LD_{50} Dose: >5 gm/kg.
Acute Dermal: Rabbit LD_{50} Route: Skin; Dose: >5 gm/kg.
Irritation Skin: Rabbit Standard Draize Test Dose: 500 mg/24H; Reaction: moderate.

Hazard Overviews

Health: Irritating to eyes/skin/respiratory tract. Other Acute Effects: may be harmful by inhalation, ingestion, or skin absorption.

Fire: Hazards: emits toxic fumes. Extinguishing agents: water spray; carbon dioxide, dry chemical powder or appropriate foam. Precautions: combustible liquid.

Reactivity: Incompatible with: strong oxidizing agents. Hazardous decomposition products: toxic fumes of: carbon monoxide, carbon dioxide.

Carcinogenicity: IARC - Not listed; NIOSH - Not listed; NTP - Not listed; ACGIH - Not listed; OSHA - Not listed; EPA - Not listed; MAK - Not listed

Primary Target Organs:

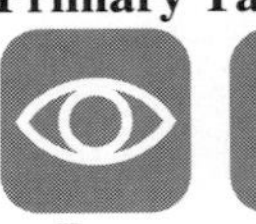
Eyes

Skin

Respiratory System

Environmental

Regulations

RCRA 40CFR: Not listed
CERCLA: 40CFR 302.4: Not listed
SARA 40CFR 372.65: Not listed
SARA EHS 40CFR 355: Not listed
TSCA: Listed

DOD1720 CAS #: 102-87-4

1-DODECANAMINE,N,N-DIDODECYL

EINECS Number: 203-063-4
Molecular Formula: $C_{36}H_{75}N$
Formula Weight: 522

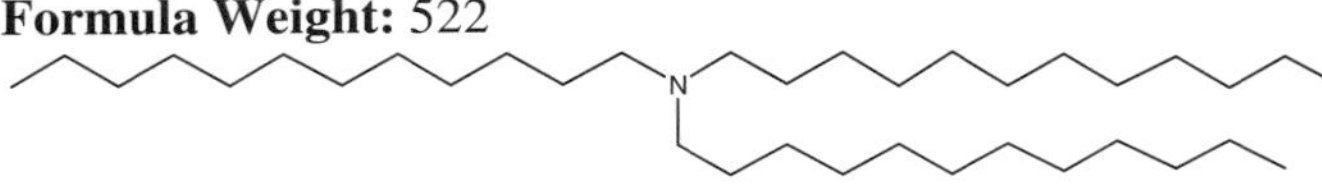

Chemical Structure

Physical Properties

Boiling Point: > 218 °C (424 °F)
Freezing Point: 8.9 °C (48.02 °F)
Specific Gravity: 0.8
Vapor Density: > 1 Air=1
Odor Threshold: 46.8 ppm
pH: 9
Evaporation Rate: < 1 Butyl Acetate=1

Hazard Overviews

Health: Irritating to eyes/skin/respiratory tract. Other Acute Effects: may be harmful by inhalation, ingestion, or skin absorption.

Fire: Hazards: emits toxic fumes. Extinguishing agents: water spray; carbon dioxide, dry chemical powder or appropriate foam. Precautions: combustible liquid.

Reactivity: Incompatible with: strong oxidizing agents, sensitive to air. Hazardous decomposition products: toxic fumes of: carbon monoxide, carbon dioxide, nitrogen oxides.

Carcinogenicity: IARC - Not listed; NIOSH - Not listed; NTP - Not listed; ACGIH - Not listed; OSHA - Not listed; EPA - Not listed; MAK - Not listed

Primary Target Organs:

Eyes

Skin

Respiratory System

Environmental

Regulations

RCRA 40CFR: Not listed
CERCLA: 40CFR 302.4: Not listed
SARA 40CFR 372.65: Not listed
SARA EHS 40CFR 355: Not listed
TSCA: Not listed

DOD2080 CAS #: 112-40-3

DODECANE

RTECS: JR2125000
EINECS Number: 203-967-9
Molecular Formula: $C_{12}H_{26}$
Formula Weight: 170.34

Chemical Structure

Synonyms: ADAKANE 12; BIHEXYL; DIHEXYL; N-DODECAN; N-DODECANE; DUODECANE

Description: colorless liquid; hydrocarbon odor

Use: organic synthesis; solvent; hydrocarbon standard; jet fuel research; manufacture of paraffin products; rubber industry; paper processing industry; distillation chaser

Physical Properties

Boiling Point: 216.3 °C (421 °F) at 760 mm Hg
Freezing Point: -9.6 °C (14.72 °F)
Specific Gravity: 0.7487 at 20 °C/4 °C
Vapor Density: 5.96 Air=1
Vapor Pressure: 1 mm Hg at 47.8 °C
Water Solubility: < 1 mg/mL at 25 C
Other Solubilities: 95% Ethanol: >=100 mg/ml at 25 °C; Acetone: >=100 mg/ml at 25 °C; Chloroform: Soluble; DMSO: <1 mg/ml at 25 °C; Ether: Very Soluble.
Odor Threshold: 37 mg/m^3
Refraction Index: 1.4216 at 20 °C
Critical Temperature: 386 °C
Critical Pressure: 17.9 atm
Flash Point: 73.89 °C
Autoignition Temperature: 203 °C
LEL: 0.6% v/v

RTECS Toxicity Data

Tumorigenic: Mouse Route: Skin; Dose: 11 gm/kg/22W-I; Toxic Effects: Tumorigenic - Equivocal tumorigenic agent by RTECS criteria; Skin and appendages - Tumors; Tumorigenic - Tumors at site of application.

Hazard Overviews

Fire Diamond

Health: Irritating to eyes/skin/respiratory tract. Also Causes: narcotic effect; defatting agent, drying, cracking, and possible dermatitis of the skin; gastrointestinal irritation; aspiration hazard.

Fire: Combustible. Use dry chemical, carbon dioxide, or foam. If the leak or spill has caught fire, use water spray to cool fire-exposed containers. If it can be done without risk, remove material from the fire area. Use smothering techniques to extinguish the fire.

Reactivity: Stable. Hazardous polymerization cannot occur. Avoid: heat; ignition sources; strong oxidizers. Incompatible with: strong oxidizers. Hazardous decomposition products: carbon oxides.

Carcinogenicity: IARC - Not listed; NIOSH - Not listed; NTP - Not listed; ACGIH - Not listed; OSHA - Not listed; EPA - Not listed; MAK - Not listed

Primary Target Organs:

Eyes

Skin

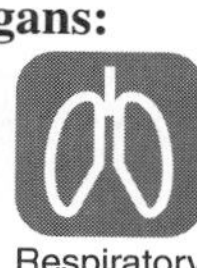
Respiratory System

Nervous System

Environmental

Environmental Fate: Releases into water will decrease in concentration (half-life 0.5-4 days) due to adsorption to sediment and particulate matter in the water column, biodegradation, and possibly volatilization, particularly from oil slicks. Released on land it will be retained in the upper layers of soil and biodegrade within several months, especially if microbial populations are acclimated. In the atmosphere, it is most likely associated with particulate matter and will be subject to gravitational settling. The free compound will slowly photodegrade. It did not bioconcentrate in the one species of fish studied but does bioconcentrate in algae and mussels.

Environmental Physical Data

Sorption Partition Coefficient: K_{oc} = estimated at 6.7 x10^4

BCF: golden orfes 1.72

Regulations

RCRA 40CFR: Not listed

CERCLA: 40CFR 302.4: Not listed

SARA 40CFR 372.65: Not listed

SARA EHS 40CFR 355: Not listed

TSCA: Listed

Analytical Methods

Soil: EPA 1625

Water / Groundwater: EPA 1625

DOD2440 CAS #: 693-23-2

DODECANEDIOIC ACID

EINECS Number: 211-746-3

Molecular Formula: $C_{12}H_{22}O_4$

Formula Weight: 230.31

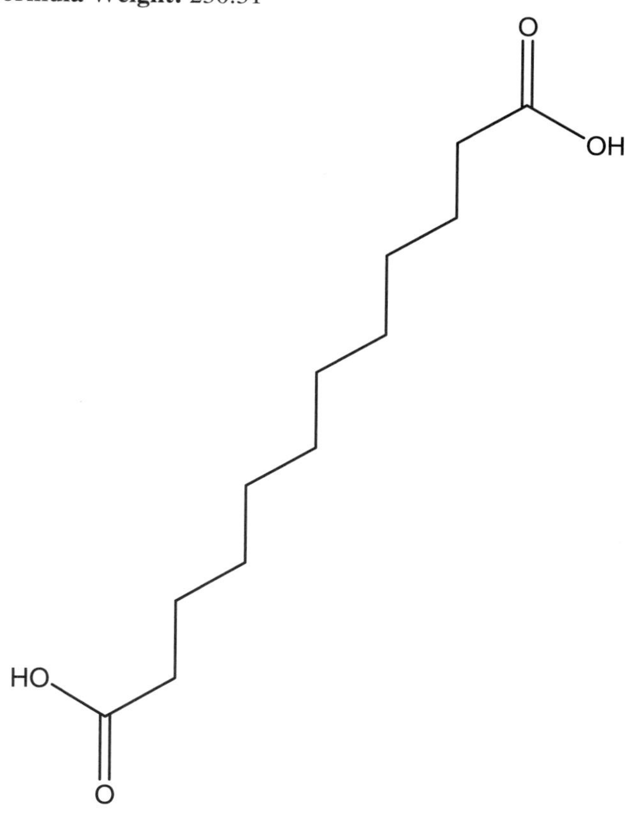

Chemical Structure

Synonyms: DECAMETHYLENEDICARBOXYLIC ACID; 1,10-DECANEDICARBOXYLIC ACID; 1,10-DICARBOXYDECANE; 1,12-DODECANEDIOIC ACID

Use: intermediate

Physical Properties

Boiling Point: 222 °C (431.6 °F)

Freezing Point: 130 °C (266 °F) to 133 °C (271.4 °F)

Specific Gravity: 1.15000

Water Solubility: Slightly Soluble in Hot Water

Other Solubilities: Soluble in hot Toluene, Alcohol, hot Acetic Acid

Hazard Overviews

Health: Irritating to eyes/skin/respiratory tract. Other Acute Effects: may be harmful by inhalation, ingestion, or skin absorption.

Fire: Hazards: in powder form capable of creating a dust explosion. Extinguishing agents: water spray; carbon dioxide, dry chemical powder or appropriate foam. Precautions: combustible liquid.
Reactivity: Incompatible with: strong oxidizing agents, reducing agents. Hazardous decomposition products: toxic fumes of: carbon monoxide, carbon dioxide.
Carcinogenicity: IARC - Not listed; NIOSH - Not listed; NTP - Not listed; ACGIH - Not listed; OSHA - Not listed; EPA - Not listed; MAK - Not listed
Primary Target Organs:

Eyes

Skin

Respiratory System

Environmental

Regulations

RCRA 40CFR: Not listed
CERCLA: 40CFR 302.4: Not listed
SARA 40CFR 372.65: Not listed
SARA EHS 40CFR 355: Not listed
TSCA: Listed

DOD2800 CAS #: 112-55-0

1-DODECANETHIOL

RTECS: JR3155000
EINECS Number: 203-984-1
Molecular Formula: $C_{12}H_{26}S$
Structured MF: $CH_3(CH_2)_{11}SH$
Formula Weight: 202.41

SH

Chemical Structure

Synonyms: 1-DODECANETHIOL; N-DODECANETHIOL; 1-DODECYL MERCAPTAN; DODECYL MERCAPTAN; M-DODECYL MERCAPTAN; N-DODECYL MERCAPTAN; LAURYL MERCAPTAN; M-LAURYL MERCAPTAN; N-LAURYL MERCAPTAN; LAURYL MERCAPTIDE; 1-MERCAPTODODECANE; PENNFLOAT M; PENNFLOAT S
Description: colorless to water-white to pale yellow oily liquid, solid below 15 deg F; mild, skunk-like odor
Use: manufacture of synthetic rubber and plastics; synthesis of pharmaceuticals; in insecticides and fungicides; nonionic detergent

Physical Properties

Boiling Point: 142 °C (288 °F) to 145 °C (293 °F) at 15 mm Hg
Freezing Point: -7 °C (19.4 °F)
Specific Gravity: 0.845 at 20 °C/20 °C
Vapor Density: Estimated > 1 Air=1
Vapor Pressure: Estimated < 1
Water Solubility: Insoluble in Water
Other Solubilities: 95% Ethanol: >10 mg/ml at 21 °C; DMSO: <1 mg/ml at 21 °C; Ethyl Acetate: Soluble; Gasoline: Soluble; Methanol: Soluble.
Surface Tension: Estimated at 30 dynes/cm
Odor Threshold: 4 mg/m^3
Refraction Index: 1.4589 at 20 °C/D
Flash Point: 128 °C Open Cup

RTECS Toxicity Data

Mutagenic: Rat Cytogenetic Analysis; Route: Inhalation; Dose: 5020 ug/m^3/16W.

Hazard Overviews

Health: Severely irritating to eyes, irritating to skin/respiratory tract. Other Acute Effects: may be harmful by inhalation, ingestion, or skin absorption; risk of serious damage to eyes; may cause allergic skin reaction; exposure can cause nausea; headache; vomiting.
Fire: Will burn. Hazards: emits toxic fumes. Extinguishing agents: carbon dioxide, dry chemical powder or appropriate foam; water may be effective for cooling, but may not effect extinguishment. Precautions: combustible liquid.
Reactivity: Stable. Hazardous polymerization will not occur. Incompatible with: bases, oxidizing agents, reducing agents, alkali metals. Hazardous decomposition products: toxic fumes of: carbon monoxide, carbon dioxide, sulfur oxides, hydrogen sulfide gas.
Carcinogenicity: IARC - Not listed; NIOSH - Listed as carcinogen; NTP - Not listed; ACGIH - Not listed; OSHA - Not listed; EPA - Not listed; MAK - Not listed
Primary Target Organs:

Eyes

Skin

Respiratory System

Exposure Limits
NIOSH REL: STEL: 0.5 ppm; 4.1 mg/m^3; 15-minute.

Environmental

Cleanup/Disposal: Guide No. 131: Fully encapsulating, vapor protective clothing should be worn for spills and leaks with no fire. Eliminate all ignition sources (no smoking, flares, sparks or flames in immediate area). All equipment used when handling the product must be grounded. Do not touch or walk through spilled material. Stop leak if you can do it without risk. Prevent entry into waterways, sewers, basements or confined areas. A vapor suppressing foam may be used to reduce vapors. Small Spills: Absorb with earth, sand or other non-combustible material and transfer to containers for later disposal. Use clean non-sparking tools to collect absorbed material. Large Spills: Dike far ahead of liquid spill for later disposal. Water spray may reduce vapor; but may not prevent ignition in closed spaces.

Regulations

RCRA 40CFR: Not listed
CERCLA: 40CFR 302.4: Not listed

SARA 40CFR 372.65: Not listed
SARA EHS 40CFR 355: Not listed
TSCA: Listed

DOD3160 CAS #: 143-07-7

DODECANOIC ACID

RTECS: OE9800000
EINECS Number: 205-582-1
Molecular Formula: $C_{12}H_{24}O_2$
Structured MF: $CH_3(CH_2)_{10}CO_2H$
Formula Weight: 200.32

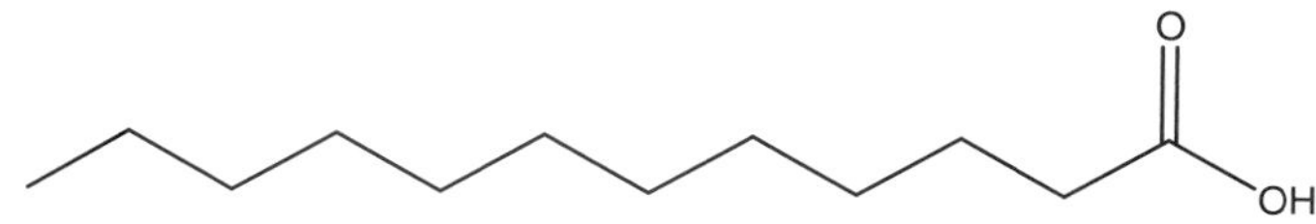

Chemical Structure

Synonyms: C-1297; N-DODECANOIC ACID; DODECOIC ACID; DUODECYCLIC ACID; DUODECYLIC ACID; HYDROFOL ACID 1255; HYDROFOL ACID 1295; HYSTRENE 9512; LAURIC ACID; LAUROSTEARIC ACID; NEO-FAT 12-43; NEO-FAT 12; NINOL AA62 EXTRA; 1-UNDECANECARBOXYLIC ACID; UNDECANE-1-CARBOXYLIC ACID; WECOLINE 1295
Description: colorless solid; white, crystalline powder; characteristic, odor like oil of bay
Use: alkylate resins; wetting agents; soaps; detergents; cosmetics; insecticides; food additives; raw material for esters; surfactants

Physical Properties

Boiling Point: 298.9 °C (570 °F)
Freezing Point: 44.2 °C (111.56 °F)
Specific Gravity: 0.883 at 20 °C/4 °C
Vapor Density: 6.91 Air=1
Density: 0.883 g/mL
Vapor Pressure: 1 mm Hg at 121 °C
Water Solubility: Insoluble in Water
Other Solubilities: 95% Ethanol: >=100 mg/ml at 21 °C; Acetone: >=100 mg/ml at 21 °C; Benzene: Soluble; DMSO: >=100 mg/ml at 21 °C; Ether: Soluble; Methanol: Very Soluble; Petroleum Ether: Soluble; Propyl Alcohol: Soluble.
Surface Tension: 26.6 mN/m at 70 °C
Odor Threshold: 01 mg/m^3
pH: Acid
Refraction Index: 1.4304
Flash Point: > 110 °C Closed Cup

RTECS Toxicity Data

Acute Oral: Rat LD_{50} Dose: 12 gm/kg.
Chronic (Multiple Dose) Oral: Rat Dose: 333 gm/kg/13W-C; Toxic Effects: Behavioral - Food intake (animal); Behavioral - Fluid intake.
Irritation Eye: Rabbit Standard Draize Test Dose: 100 mg; Reaction: mild.
Irritation Skin: Rabbit Standard Draize Test Dose: 500 mg; Reaction: mild.
Mutagenic: Yeast - S Cerevisiae Cytogenetic Analysis; Dose: 10 mg/L.
Tumorigenic: Mouse Route: Skin; Dose: 108 gm/kg/15W-I; Toxic Effects: Tumorigenic - Neoplastic by RTECS criteria; Tumorigenic - Active as anti-cancer agent.

Hazard Overviews

Health: Irritating to eyes/skin/respiratory tract. Other Acute Effects: may be harmful by inhalation, ingestion, or skin absorption.
Fire: Will burn. Hazards: in powder form capable of creating a dust explosion. Extinguishing agents: carbon dioxide, dry chemical powder or appropriate foam; water spray. Precautions: combustible liquid.
Reactivity: Incompatible with: bases, oxidizing agents, reducing agents. Hazardous decomposition products: toxic fumes of: carbon monoxide, carbon dioxide.
Carcinogenicity: IARC - Not listed; NIOSH - Not listed; NTP - Not listed; ACGIH - Not listed; OSHA - Not listed; EPA - Not listed; MAK - Not listed
Primary Target Organs:

Eyes

Skin

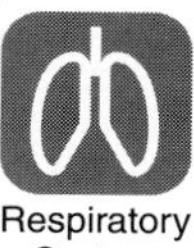
Respiratory System

Environmental

Ecotoxicity: Fishes: Red killifish (Oryzias latipes): 96h LC_{50}:in seawater: no mortality at saturation conc
Environmental Fate: If released to the atmosphere, it will degrade by reaction with photochemically produced hydroxyl radicals (estimated half-life of about 1.21 days). It can be physically removed from air by rainfall. If released to soil or water, it is expected to biodegrade; biodegradation screening studies have demonstrated that it biodegrades readily. In water and soil, some adsorption to sediments and soil may take place; however, adsorption is expected to vary with pH.

Environmental Physical Data

Henry's Law Constant: estimated at 9.3×10^{-6}
Octanol/Water Partition Coefficient: log K_{ow} = 4.2
Sorption Partition Coefficient: K_{oc} = 300
BCF: estimated at 255 to 916
BOD: 6.1% BODT, 1 days

Regulations

RCRA 40CFR: Not listed
CERCLA: 40CFR 302.4: Not listed
SARA 40CFR 372.65: Not listed
SARA EHS 40CFR 355: Not listed
TSCA: Listed

DOD3520 CAS #: 27342-88-7

DODECANOL

Molecular Formula: $C_{12}H_{26}O$
Formula Weight: 186.34
Synonyms: ALCOHOL C-12; N-DODECANOL; DODECYL ALCOHOL; LAURYL ALCOHOL
Description: crystalline solid
Use: manufacture of detergents; manufacture of sulfonic acid esters which are used as wetting agents; intermediate for production of thiodipropionate, lauryl chloride, specialty ethoxylates and sulfate defoamers; lube additives; pharmaceuticals; rubbers; textiles; perfumes; foaming agents

Physical Properties

Boiling Point: 259 °C (498 °F)
Freezing Point: 24 °C (75.2 °F)
Specific Gravity: 0.8309 at 24 °C/4 °C
Vapor Density: 6.43 Air=1
Vapor Pressure: 1 mm Hg at 91 °C
Water Solubility: Insoluble in Water
Other Solubilities: Soluble in Alcohol & Ether

Hazard Overviews

Carcinogenicity: IARC - Not listed; NIOSH - Not listed; NTP - Not listed; ACGIH - Not listed; OSHA - Not listed; EPA - Not listed; MAK - Not listed

Environmental

Regulations

RCRA 40CFR: Not listed
CERCLA: 40CFR 302.4: Not listed
SARA 40CFR 372.65: Not listed
SARA EHS 40CFR 355: Not listed
TSCA: Listed

DOD3880 CAS #: 112-16-3

DODECANOYL CHLORIDE

EINECS Number: 203-941-7
Molecular Formula: $C_{12}H_{23}ClO$
Formula Weight: 218.77

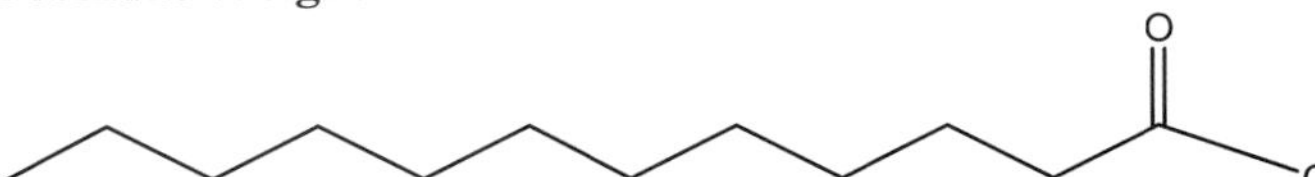

Chemical Structure

Synonyms: DODECANOIC ACID,CHLORIDE; N-DODECANOYL CHLORIDE; LAURIC ACID CHLORIDE; LAUROYL CHLORIDE
Description: water-white liquid
Use: surfactant; polymerization initiator; antienzyme agent; foamer; in synthesis of lauroyl peroxide, sodium n-lauroyl sarcosinate and other sarcosinates

Physical Properties

Boiling Point: 145 °C (293 °F) at 18 mm Hg
Freezing Point: -17 °C (1.4 °F)
Specific Gravity: 0.91690
Water Solubility: Decomposes in Water
Other Solubilities: Soluble in Ether
Refraction Index: 1.4458 at 20 °C/D

Hazard Overviews

Corrosive

Health: Corrosive to eyes/skin/respiratory tract. Other Acute Effects: harmful if swallowed, inhaled, or absorbed through skin; lachrymator; inhalation may result in spasm, inflammation and edema of the larynx and bronchi, chemical pneumonitis and pulmonary edema; symptoms of exposure may include burning sensation; coughing; wheezing; laryngitis; shortness of breath; headache; nausea; vomiting.
Fire: Hazards: water hydrolyzes material liberating acidic gas which in contact with metal surfaces can generate flammable and/or explosive hydrogen gas; emits toxic fumes. Extinguishing agents: carbon dioxide, dry chemical powder or appropriate foam. Precautions: combustible liquid.
Reactivity: Incompatible with: strong oxidizing agents, strong bases, amines, do not allow contact with water. Hazardous decomposition products: toxic fumes of: carbon monoxide, carbon dioxide, hydrogen chloride gas.
Carcinogenicity: IARC - Not listed; NIOSH - Not listed; NTP - Not listed; ACGIH - Not listed; OSHA - Not listed; EPA - Not listed; MAK - Not listed
Primary Target Organs:

Eyes

Skin

Respiratory System

Environmental

Regulations

RCRA 40CFR: Not listed
CERCLA: 40CFR 302.4: Not listed
SARA 40CFR 372.65: Not listed
SARA EHS 40CFR 355: Not listed
TSCA: Listed

DOD4240 CAS #: 112-41-4

1-DODECENE

EINECS Number: 203-968-4
Molecular Formula: $C_{12}H_{24}$
Structured MF: $CH_3(CH_2)_9CH{=}CH_2$
Formula Weight: 168.33

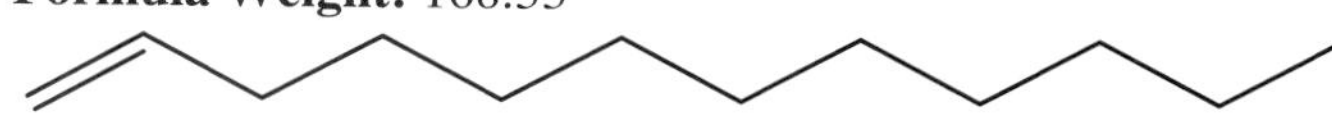

Chemical Structure

Synonyms: ADACENE 12; ALPHA-DODECENE; N-DODEC-1-ENE; ALPHA-DODECYLENE
Description: colorless liquid; mild, pleasant odor
Use: chem int for n-dodecyl dimethyl amine; in flavors, perfumes, medicines, oils, dyes and resins

Physical Properties

Boiling Point: 213.4 °C (416 °F) at 760 mm Hg
Freezing Point: -35.2 °C (-31.36 °F)
Specific Gravity: 0.7584 at 20 °C/4 °C
Vapor Pressure: 5 mm Hg at 74 °C
Water Solubility: Insoluble in Water
Other Solubilities: Soluble in Alcohol, Benzene, Ether, Acetone
Surface Tension: 26.6 dyne/cm at 20 °C
Refraction Index: 1.43002 at 20 °C
Autoignition Temperature: 255 °C

Hazard Overviews

Health: Irritating to eyes/skin/respiratory tract. Harmful. Other Acute Effects: harmful if inhaled or swallowed.
Fire: Will burn. Extinguishing agents: carbon dioxide, dry chemical powder or appropriate foam; water spray. Precautions: combustible liquid.
Reactivity: Incompatible with: bases, oxidizing agents, reducing agents. Hazardous decomposition products: toxic fumes of: carbon monoxide, carbon dioxide.
Carcinogenicity: IARC - Not listed; NIOSH - Not listed; NTP - Not listed; ACGIH - Not listed; OSHA - Not listed; EPA - Not listed; MAK - Not listed
Primary Target Organs:

Eyes

Skin

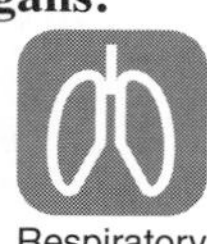
Respiratory System

Environmental

Environmental Fate: If released to soil, it is expected to adsorb strongly to soil and be essentially immobile. It may slowly volatilize from dry soil and rapidly volatilize from moist soil, although strong adsorption may attenuate the rate of this process. Pure culture studies indicate that it has the potential to biodegrade under aerobic conditions in soil or water, but the rate of biodegradation in unknown. If released to water, it is expected to significantly bioconcentrate in fish and aquatic organisms and it is also expected to significantly adsorb to sediment and suspended organic matter. It may rapidly volatilize from water to the atmosphere; its half-life for volatilization from a model river is 3.8 hr. Strong adsorption to sediment may attenuate the rate of this process. The half-life for volatilization from a model pond, which takes into account adsorptive processes, is 17 months. If released to the atmosphere, it is expected to undergo gas-phase oxidation with photochemically produced hydroxyl radicals and ozone. Estimated half-lives for these processes are 10 hr and 23 hr, respectively.

Environmental Physical Data

Henry's Law Constant: 4.25
Octanol/Water Partition Coefficient: log K_{ow} = 6.5
Sorption Partition Coefficient: K_{oc} = calculated at 8.16 $\times 10^4$
BCF: calculated at 5.1 $\times 10^4$

Regulations

RCRA 40CFR: Not listed
CERCLA: 40CFR 302.4: Not listed
SARA 40CFR 372.65: Not listed
SARA EHS 40CFR 355: Not listed
TSCA: Listed

DOD4600 CAS #: 25378-22-7

DODECENE

EINECS Number: 246-922-9
Molecular Formula: $C_{12}H_{24}$
Formula Weight: 168.33
Synonyms: 1-DODECENE; ALPHA-DODECYLENE; DODECYLENE; PROPENE,POLYMERS,TETRAMER; 1-PROPENE,TETRAMER; PROPENE,TETRAMER; PROPYLENE TETRAMER; TETRAPROPYLENE
Description: colorless liquid
Use: as flavor additive; perfume; medicines; oils; dyes; resins

Physical Properties

Boiling Point: 213.4 °C (416 °F) at 760 mm Hg
Freezing Point: -35.23 °C (-31.414 °F)
Specific Gravity: 0.7584 at 20 °C/4 °C
Vapor Density: 5.81
Vapor Pressure: 1 mm Hg at 47.2 °C
Water Solubility: Insoluble in Water
Other Solubilities: Soluble in Alcohol, Ether, Acetone, Petroleum Ether.
Refraction Index: 1.4300 at 20 °C
Flash Point: Alpha < 100 °C
Autoignition Temperature: Alpha dodecylene 255 °C

Hazard Overviews

Fire Diamond

Fire: Combustible.
Carcinogenicity: IARC - Not listed; NIOSH - Not listed; NTP - Not listed; ACGIH - Not listed; OSHA - Not listed; EPA - Not listed; MAK - Not listed

Environmental

Regulations

RCRA 40CFR: Not listed
CERCLA: 40CFR 302.4: Not listed
SARA 40CFR 372.65: Not listed
SARA EHS 40CFR 355: Not listed
TSCA: Listed

DOD4960 CAS #: 25377-73-5

DODECENYLSUCCINIC ANHYDRIDE

RTECS: WN1225000
EINECS Number: 246-917-1
Molecular Formula: $C_{16}H_{26}O_3$
Formula Weight: 281.44

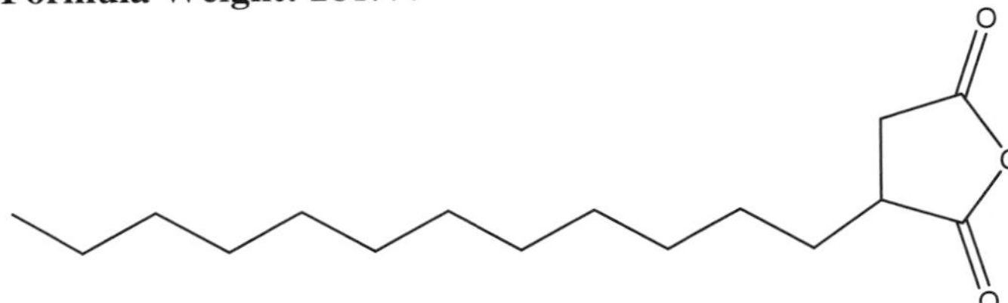

Chemical Structure

Synonyms: DDS; DDS A; 2,5-FURANDIONE,3-(DODECENYL)DIHYDRO-; SUCCINIC ANHYDRIDE,DODECENYL-
Description: light-yellow, clear, viscous oil
Use: alkyd, epoxy, & other resins; anticorrosion agents; plasticizers; wetting agents for bituminous compd, & vulcanizable products

Physical Properties

Boiling Point: 180 °C (356 °F) to 182 °C (359.6 °F) at 5 mm Hg
Density: 1.002 at 75 °C
Flash Point: 177 °C Cleveland Open Cup

RTECS Toxicity Data

Acute Inhalation: Rat LC_{Lo} Dose: 1220 mg/m^3/4hr.

Hazard Overviews

Health: Irritating to eyes/skin/respiratory tract. Other Acute Effects: may be harmful by inhalation, ingestion, or skin absorption; prolonged or repeated exposure may cause allergic reactions in certain sensitive individuals; exposure can cause blurred vision; dermatitis; may cause sneezing.
Fire: Will burn. Hazards: emits toxic fumes. Extinguishing agents: water spray; carbon dioxide, dry chemical powder or appropriate foam. Precautions: combustible liquid.
Reactivity: Stable. Hazardous polymerization will not occur. Incompatible with: strong oxidizing agents, amines, bases, may decompose on exposure to moist air or water. Hazardous decomposition products: toxic fumes of: carbon monoxide, carbon dioxide.
Carcinogenicity: IARC - Not listed; NIOSH - Not listed; NTP - Not listed; ACGIH - Not listed; OSHA - Not listed; EPA - Not listed; MAK - Not listed
Primary Target Organs:

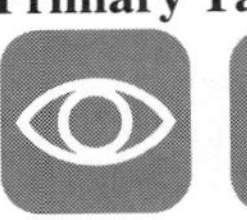
Eyes

Skin

Respiratory System

Environmental

Regulations

RCRA 40CFR: Not listed
CERCLA: 40CFR 302.4: Not listed
SARA 40CFR 372.65: Not listed
SARA EHS 40CFR 355: Not listed
TSCA: Listed

DOD5320 CAS #: 27176-87-0

DODECYL BENZENE SULFONIC ACID

RTECS: DB6600000
EINECS Number: 248-289-4
Molecular Formula: $C_{18}H_{30}O_3S$
Structured MF: $CH_3(CH_2)_{11}C_6H_{45}O_3H$
Formula Weight: 326.54

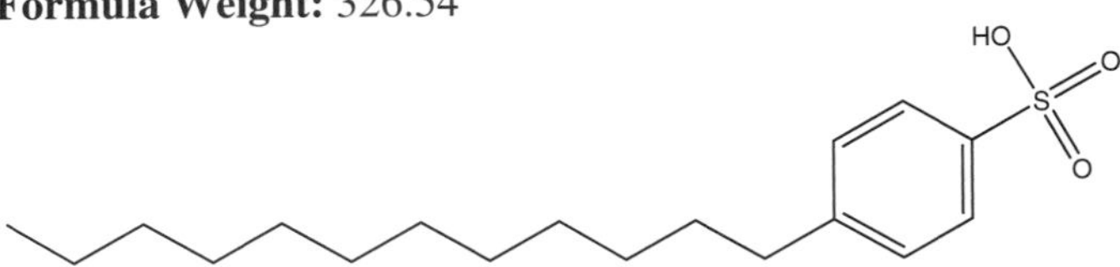

Chemical Structure

Synonyms: BENZENESULFONIC ACID,DODECYL-; BIO-SOFT S 100; CALSOFT LAS 99; DDBSA; DOBANIC ACID 83; DOBANIC ACID JN; DODECYL BENZENESULFONATE; DODECYLBENZENESULFONIC ACID; N-DODECYLBENZENESULFONIC ACID; DODECYLBENZENESULPHONIC ACID; E 7256; ELFAN WA SULPHONIC ACID; LAURYLBENZENESULFONIC ACID; MARLON AS 3; NACCONOL 98SA; NANSA 1042P; NANSA SSA; RICHONIC ACID B; SULFRAMIN ACID 1298; WITCO 1298 SULFONIC ACID
Description: colorless to light yellow to brown liquid; bland odor
Use: dodecylbenzenesulfonic acid impregnation medium for thin layer chromatography, absorbent in gas chromatography

for polychlorinated biphenyls, etc; incr stability of triforine acaricidal & fungicidal agents (preferably n-dodecylbenzenesulfonic acid or tetrapropylene benzene sulfonic acid); dodecylbenzenesulfonate in separation of iron (iii) hydroxide flox; electronic cleaning chems, pickling baths, & detergent mfr; neutralizing agents for acids & caustics.

Physical Properties

Boiling Point: > 204.5 °C (400 °F)
Specific Gravity: 1 at 25 °C
Water Solubility: Soluble in Water
Odor Threshold: 200 mg/m^3
Flash Point: 148.889 °C Open Cup

RTECS Toxicity Data

Acute Oral: Rat LD_{50} Dose: 650 mg/kg.

Hazard Overviews

Fire: Will burn.
Carcinogenicity: IARC - Not listed; NIOSH - Not listed; NTP - Not listed; ACGIH - Not listed; OSHA - Not listed; EPA - Not listed; MAK - Not listed

Environmental

Ecotoxicity: Aquatic toxicity: 4.2 to 5.6 ppm/96-hour/Fathead minnows; Bluegills/TLm/soft water; 3.5 to 4.4 ppm/96-hour/Fathead minnows/TLm/hard water; 5 to 15 ppm/guppy/lethal concentration; 0.05 ppm reduced fertility rate of common mussel; 5 ppm reduced percentage of fertilized eggs of fish
Cleanup/Disposal: Guide No. 171: Do not touch or walk through spilled material. Stop leak if you can do it without risk. Prevent dust cloud. Avoid inhalation of asbestos dust. Small Dry Spills: With clean shovel place material into clean, dry container and cover loosely; move containers from spill area. Small Spills: Take up with sand or other noncombustible absorbent material and place into containers for later disposal. Large Spills: Dike far ahead of liquid spill for later disposal. Cover powder spill with plastic sheet or tarp to minimize spreading. Prevent entry into waterways, sewers, basements or confined areas.

Environmental Physical Data

BOD: activated < 1 lb/lb, 5 days

Regulations

RCRA 40CFR: Not listed
CERCLA: 40CFR 302.4: Listed per CWA Section 311(b)(4) RQ: 1000 lb (453.5 kg)
SARA 40CFR 372.65: Not listed
SARA EHS 40CFR 355: Not listed
TSCA: Listed

DOD5680 CAS #: 2461-18-9

DODECYL GLYCIDYL ETHER

EINECS Number: 219-554-1
Molecular Formula: $C_{15}H_{30}O_2$
Formula Weight: 242.45

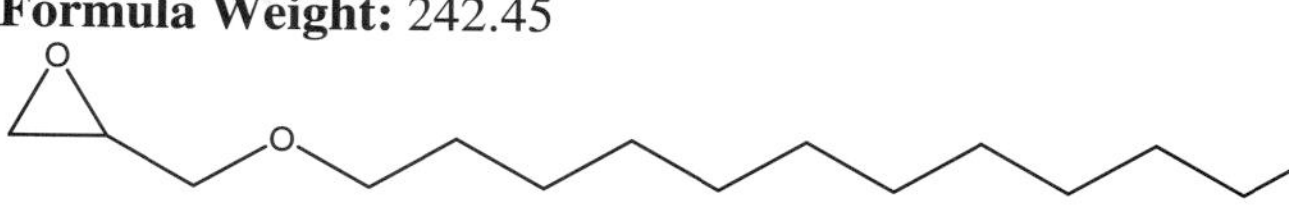

Chemical Structure

Synonyms: N-DODECYL GLYCIDYL ETHER; ((DODECYLOXY)METHYL)OXIRANE; ETHER,DODECYL 2,3-EPOXYPROPYL; LAURYL GLYCIDYL ETHER; OXIRANE,((DODECYLOXY)METHYL)-; PROPANE,1-(DODECYLOXY)-2,3-EPOXY-
Description: colorless, clear viscous liquid
Use: preparation of cosmetics & pharmaceuticals; preparation of water-in-oil emulsifier; secondary heat stabilizer for polymers (eg, PVC); reactive diluent for epoxy resins

Physical Properties

Water Solubility: < 1 mg/mL at 19 C
Other Solubilities: 95% Ethanol: >=100 mg/ml at 19 °C; Acetone: >=100 mg/ml at 19 °C; DMSO: 10-50 mg/ml at 19 °C.
Flash Point: Not available; probably combustible

Hazard Overviews

Fire: Will burn.
Carcinogenicity: IARC - Not listed; NIOSH - Not listed; NTP - Not listed; ACGIH - Not listed; OSHA - Not listed; EPA - Not listed; MAK - Not listed

Environmental

Regulations

RCRA 40CFR: Not listed
CERCLA: 40CFR 302.4: Not listed
SARA 40CFR 372.65: Not listed
SARA EHS 40CFR 355: Not listed
TSCA: Listed

DOD6040 CAS #: 142-90-5

N-DODECYL METHACRYLATE

RTECS: OZ4300000
EINECS Number: 205-570-6
Molecular Formula: $C_{16}H_{30}O_2$
Structured MF: $CH_2{=}CCH_3CO_2(CH_2)_{11}CH_3$
Formula Weight: 254.41

Chemical Structure

Synonyms: ACRYLIC ACID,2-METHYL-,DODECYL ESTER; AGEFLEX FM 246; DODECYL METHACRYLATE; DODECYL 2-METHYL-2-PROPENOATE; EPA PESTICIDE CODE 053101; LAURYL METHACRYLATE; LAURYLESTER KYSELINY METHAKRYLOVE; METAZENE; METHACRYLIC ACID DODECYL ESTER; METHACRYLIC ACID,LAURYL ESTER; 2-PROPENOIC ACID,2-METHYL-,DODECYL ESTER

Description: liquid

Use: polymerizable monomer for plastics, molding powders, solvent coatings, adhesives, oil additives; emulsions for textile, leather, and paper finishing; as a deodorant to mask methyl sulfide odors in industry; to delay volatilization of insecticides; monomer for viscosity index improvers for lubricating oil; monomer for pour-point depressants for distillate fuels; used in dentistry as restorative materials, adhesives, prosthetic devices

Physical Properties

Boiling Point: 272 °C (522 °F) to 344 °C (651 °F)
Freezing Point: -20 °C (-4 °F)
Specific Gravity: 0.868 at 20 °C/20 °C
Vapor Density: Estimate 8.8 Air=1
Bulk Density: 0.868 g/ml
Water Solubility: Insoluble in Water
Refraction Index: 1.444 at 25 °C
Flash Point: 132 °C Cleveland Open Cup

RTECS Toxicity Data

Irritation Eye: Rabbit Standard Draize Test Dose: 500 mg/24H; Reaction: mild.
Irritation Skin: Rabbit Standard Draize Test Dose: 500 mg/24H; Reaction: mild.

Hazard Overviews

Health: Irritating to eyes/skin/respiratory tract. Other Acute Effects: may be harmful by inhalation, ingestion, or skin absorption; vapor or mist is irritating.
Fire: Will burn. Hazards: emits toxic fumes. Extinguishing agents: water spray; carbon dioxide, dry chemical powder or appropriate foam. Precautions: combustible liquid.
Reactivity: Incompatible with: strong oxidizing agents. Hazardous decomposition products: toxic fumes of: carbon monoxide, carbon dioxide.
Carcinogenicity: IARC - Not listed; NIOSH - Not listed; NTP - Not listed; ACGIH - Not listed; OSHA - Not listed; EPA - Not listed; MAK - Not listed
Primary Target Organs:

Eyes

Skin

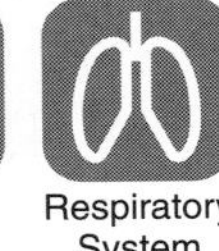
Respiratory System

Environmental

Regulations

RCRA 40CFR: Not listed
CERCLA: 40CFR 302.4: Not listed
SARA 40CFR 372.65: Not listed
SARA EHS 40CFR 355: Not listed
TSCA: Listed

DOD6400 CAS #: 151-41-7

DODECYL SULFATE

RTECS: WT0700000
EINECS Number: 205-791-8
Molecular Formula: $C_{12}H_{26}O_4S$
Formula Weight: 266.44
Synonyms: DODECANSULFONIC ACID,HYDROXY-; DODECYL HYDROGEN SULFATE; N-DODECYL SULFATE; DODECYLSULFURIC ACID; LAURYL SULFATE; LAURYL SULFURIC ACID; LAURYL SULPHATE; MONODODECYL HYDROGEN SULFATE; SULFURIC ACID MONOLAURYL ESTER; SULFURIC ACID,MONODODECYL ESTER
Use: in shampoos; in treatment of waste water; chem int for salts, incl sodium & ammonium, used as surfactant

RTECS Toxicity Data

Acute Oral: Rat LD_{50} Dose: 1300 mg/kg.

Hazard Overviews

Carcinogenicity: IARC - Not listed; NIOSH - Not listed; NTP - Not listed; ACGIH - Not listed; OSHA - Not listed; EPA - Not listed; MAK - Not listed

Environmental

Regulations

RCRA 40CFR: Not listed
CERCLA: 40CFR 302.4: Not listed
SARA 40CFR 372.65: Not listed
SARA EHS 40CFR 355: Not listed
TSCA: Listed

DOD6760 CAS #: 124-22-1

DODECYLAMINE

RTECS: JR6475000
EINECS Number: 204-690-6
Molecular Formula: $C_{12}H_{27}N$
Formula Weight: 185.36

H_2N

Chemical Structure

Synonyms: ALAMINE 4; AMINE BB; 1-AMINODODECANE; ARMEEN 12D; 1-DODECANAMINE; 1-DODECANAMINE (9CI); 1-

DODECYLAMINE; N-DODECYLAMINE; KEMAMINE P690; LAURINAMINE; LAURYLAMINE; N-LAURYLAMINE; MONODODECYLAMINE; NISSAN AMINE BB

Description: oil; amine odor

Use: chem int for the fungicide, dodecylguanidine acetate; chem int for cationic surfactants

Physical Properties

Boiling Point: 259 °C (498 °F) at 760 mm Hg
Freezing Point: 28.3 °C (82.94 °F)
Specific Gravity: 0.8015 at 20 °C/4 °C
Water Solubility: Slightly Soluble in Water
Other Solubilities: Soluble in all proportions in Alcohol, Ether, Benzene, Chloroform, Carbon Tetrachloride
Refraction Index: 1.4421 at 20 °C/D

RTECS Toxicity Data

Acute Oral: Rat LD_{50} Dose: 1020 mg/kg; Toxic Effects: Gastrointestinal - Other changes. Mouse LD_{50} Dose: 1160 mg/kg; Toxic Effects: Gastrointestinal - Other changes.
Irritation Eye: Rabbit Standard Draize Test Dose: 50 ug/24H; Reaction: severe.
Irritation Skin: Rabbit Standard Draize Test Dose: 750 ug/24H; Reaction: severe.

Hazard Overviews

Corrosive

Health: Corrosive to eyes/skin/respiratory tract. Harmful. Other Acute Effects: harmful if swallowed, inhaled, or absorbed through skin; material is extremely destructive to tissue of the mucous membranes and upper respiratory tract, eyes and skin; inhalation may result in spasm, inflammation and edema of the larynx and bronchi, chemical pneumonitis and pulmonary edema; may cause burning sensation; coughing; wheezing; laryngitis; shortness of breath; headache; nausea; vomiting.
Fire: Hazards: emits toxic fumes. Extinguishing agents: water spray; carbon dioxide, dry chemical powder or appropriate foam. Precautions: combustible liquid.
Reactivity: Incompatible with: acids, acid chlorides, acid anhydrides, oxidizing agents. Hazardous decomposition products: thermal decomposition may produce carbon monoxide, carbon dioxide, and nitrogen oxides.
Carcinogenicity: IARC - Not listed; NIOSH - Not listed; NTP - Not listed; ACGIH - Not listed; OSHA - Not listed; EPA - Not listed; MAK - Not listed
Primary Target Organs:

Eyes

Skin

Respiratory System

Environmental

Regulations

RCRA 40CFR: Not listed
CERCLA: 40CFR 302.4: Not listed
SARA 40CFR 372.65: Not listed
SARA EHS 40CFR 355: Not listed
TSCA: Listed

DOD7480 **CAS #: 123-01-3**

DODECYLBENZENE

RTECS: CZ9540000
EINECS Number: 204-591-8
Molecular Formula: $C_{18}H_{30}$
Structured MF: $C_6H_5(CH_2)_{11}CH_3$
Formula Weight: 246.44

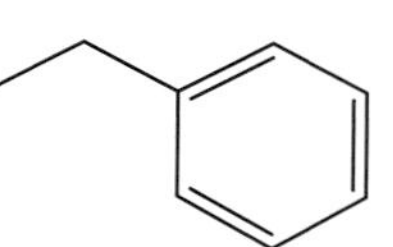

Chemical Structure

Synonyms: ALKYLATE P 1; BENZENE,DODECYL-; DETERGENT ALKYLATE; DETERGENT ALKYLATE NO 2; DODECANE,1-PHENYL-; N-DODECYLBENZENE; DODECYLBENZENE (LINEAR); LAURYLBENZENE; MARLICAN; NALKYLENE 500; PHENYLDODECAN; 1-PHENYLDODECANE; UCANE ALKYLATE 12
Description: colorless liquid; weak oily odor
Use: detergents of ABS or LAS type; a vehicle & solvent for polynuclear aromatics in carcinogenesis studies; to produce aryl alkyl sulfonates in the soap & detergent industries

Physical Properties

Boiling Point: 290 °C (554 °F) to 410 °C (770 °F)
Freezing Point: < -60 °C (-76 °F)
Specific Gravity: 0.9 at 20 °C/4 °C
Saturated Vapor Density: < 1.211838838 kg/m^3
Density: 8.47
Vapor Pressure: < 1 mm Hg at 24 °C
Water Solubility: Insoluble in Water
Other Solubilities: 95% Ethanol: >=100 mg/ml at 25 °C; Acetone: >=100 mg/ml at 25 °C; DMSO: 1-5 mg/ml at 25 °C; Ether: Soluble.
Surface Tension: 30.12 dynes/cm at 20 °C
Refraction Index: 1.482
Flash Point: 140.6 °C

RTECS Toxicity Data

Acute Oral: Rat LD; Dose: >10 mL/kg; Toxic Effects: Liver - Other changes.

Hazard Overviews

Health: May cause irritation to eyes/skin. Other Acute Effects: may be harmful by inhalation, ingestion, or skin absorption.

Fire: Will burn. Extinguishing agents: water spray; carbon dioxide, dry chemical powder or appropriate foam. Precautions: combustible liquid.

Reactivity: Incompatible with: strong oxidizing agents. Hazardous decomposition products: toxic fumes of: carbon monoxide, carbon dioxide.

Carcinogenicity: IARC - Not listed; NIOSH - Not listed; NTP - Not listed; ACGIH - Not listed; OSHA - Not listed; EPA - Not listed; MAK - Not listed

Environmental

Regulations

RCRA 40CFR: Not listed
CERCLA: 40CFR 302.4: Not listed
SARA 40CFR 372.65: Not listed
SARA EHS 40CFR 355: Not listed
TSCA: Listed

Analytical Methods

Water / Groundwater: ASTM D4763

DOD8200 CAS #: 27193-86-8

DODECYLPHENOL

RTECS: SL3675000
EINECS Number: 248-312-8
Molecular Formula: $C_{18}H_{30}O$
Structured MF: $C_6H_4OHC_{12}H_{25}$
Formula Weight: 262.48
Synonyms: T-DET; DODECYLPHENOL (MIXED ISOMERS); PHENOL,DODECYL-; PHENOL,DODECYL-,MIXED ISOMERS
Description: straw-colored liquid; phenolic odor
Use: chemical intermediate for metal salts used as lube oil additives & for ethoxylated dodecylphenol; solvent; intermediate for surface-active agents; in resins, fungicides, bactericides, dyes, pharmaceuticals, adhesives, and rubber chemicals; coupling agent in emulsifer blends

Physical Properties

Boiling Point: 314 °C (597 °F) to 334 °C (633 °F)
Specific Gravity: 0.94 at 20 °C/20 °C
Vapor Density: 9.04 Air=1
Saturated Vapor Density: 1.200000881 kg/m^3
Vapor Pressure: 6.93 x10^{-5} mm Hg at 25 °C
Water Solubility: Soluble in Water
Other Solubilities: Soluble in organic solvents
Flash Point: 163 °C Open Cup

RTECS Toxicity Data

Acute Oral: Rat LD_{50} Dose: 2100 mg/kg; Toxic Effects: Gastrointestinal - Hypermotility, diarrhea; Gastrointestinal - Peritonitis; Nutritional and gross metabolic - Weight loss or decreased weight gain.

Acute Dermal: Rabbit LD_{50} Route: Skin; Dose: 5 gm/kg; Toxic Effects: Skin and appendages - Primary irritation.

Hazard Overviews

Fire Diamond

Fire: Will burn.

Carcinogenicity: IARC - Not listed; NIOSH - Not listed; NTP - Not listed; ACGIH - Not listed; OSHA - Not listed; EPA - Not listed; MAK - Not listed

Environmental

Ecotoxicity: Lethal threshold in salmon (salmo salar) was 0.15 mg/l. Both static and flow-through tests were applied

Environmental Fate: If released to air, it will degrade relatively rapidly by reaction with photochemically produced hydroxyl radicals (estimated half-life of 3.7-6.65 hr). Night-time reaction with nitrate radicals can also remove it from the atmosphere. If released to soil, it is not expected to leach based upon an estimated K_{oc} of 90,000. If released to water, it may partition from the water column to sediment and suspended material. Abiotic degradation in water may occur through photo-oxidation but rates are not available. An estimated BCF of 59,000 suggests bioconcentration may be important. Volatilization can physically remove it from water; however, the importance of volatilization depends upon the degree of adsorption to sediment. Insufficient data are available to predict the importance of biodegradation in water or soil.

Cleanup/Disposal: Guide No. 153: Eliminate all ignition sources (no smoking, flares, sparks or flames in immediate area). Do not touch damaged containers or spilled material unless wearing appropriate protective clothing. Stop leak if you can do it without risk. Prevent entry into waterways, sewers, basements or confined areas. Absorb or cover with dry earth, sand or other non-combustible material and transfer to containers. Do not get water inside containers.

Environmental Physical Data

Henry's Law Constant: estimated at 3.46 x10^{-5}
Octanol/Water Partition Coefficient: log K_{ow} = 5.64
Sorption Partition Coefficient: K_{oc} = 9 x10^4
BCF: bioconcentration factor 6000

Regulations

RCRA 40CFR: Not listed
CERCLA: 40CFR 302.4: Not listed
SARA 40CFR 372.65: Not listed
SARA EHS 40CFR 355: Not listed
TSCA: Listed

DOD8920 CAS #: 4484-72-4

DODECYLTRICHLOROSILANE

RTECS: VV3940000
DOT: UN1771; IMO8.0
EINECS Number: 224-769-9

Molecular Formula: $C_{12}H_{25}Cl_3Si$
Structured MF: $CH_3(CH_2)_{11}SiCl_3$
Formula Weight: 303.81

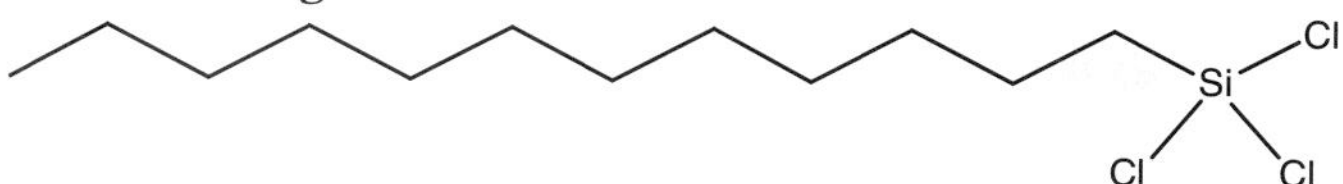

Chemical Structure

Synonyms: SILANE,DODECYLTRICHLORO-; SILANE,TRICHLORODODECYL-; TRICHLORODODECYLSILANE
Description: colorless to yellow liquid; sharp pungent odor, like hydrochloric acid
Use: intermediate for silicones

Physical Properties

Boiling Point: 288 °C (550 °F)
Specific Gravity: 1.026 at 25 °C/25 °C
Water Solubility: Reacts with water
pH: Acid
Refraction Index: 1.4251 at 25 °C/D
Flash Point: > 65.556 °C Open Cup

Hazard Overviews

Corrosive

Health: Corrosive to eyes/skin/respiratory tract. Other Acute Effects: harmful if swallowed, inhaled, or absorbed through skin; inhalation may result in spasm, inflammation and edema of the larynx and bronchi, chemical pneumonitis and pulmonary edema; symptoms of exposure may include burning sensation; coughing; wheezing; laryngitis; shortness of breath; headache; nausea; vomiting.
Fire: Combustible. Hazards: emits toxic fumes. Extinguishing agents: carbon dioxide, dry chemical powder or appropriate foam; do not use water. Precautions: combustible liquid.
Reactivity: Incompatible with: strong oxidizing agents, strong bases, alcohols, may decompose on exposure to moist air or water. Hazardous decomposition products: toxic fumes of: carbon monoxide, carbon dioxide, hydrogen chloride gas, silicon oxide.
Carcinogenicity: IARC - Not listed; NIOSH - Not listed; NTP - Not listed; ACGIH - Not listed; OSHA - Not listed; EPA - Not listed; MAK - Not listed
Primary Target Organs:

Eyes

Skin

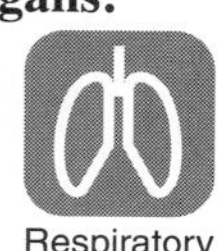
Respiratory System

Environmental

Cleanup/Disposal: Guide No. 156: Eliminate all ignition sources (no smoking, flares, sparks or flames in immediate area). All equipment used when handling the product must be grounded. Do not touch damaged containers or spilled material unless wearing appropriate protective clothing. Stop leak if you can do it without risk. A vapor suppressing foam may be used to reduce vapors. For chlorosilanes, use AFFF alcohol-resistant medium expansion foam to reduce vapors. Do not get water on spilled substance or inside containers. Use water spray to reduce vapors or divert vapor cloud drift. Prevent entry into waterways, sewers, basements or confined areas. Small Spills: Cover with dry earth, dry sand, or other non-combustible material followed with plastic sheet to minimize spreading or contact with rain. Use clean non-sparking tools to collect material and place it into loosely covered plastic containers for later disposal.

Environmental Physical Data

BCF: no food chain concentration potential

Regulations

RCRA 40CFR: Not listed
CERCLA: 40CFR 302.4: Not listed
SARA 40CFR 372.65: Not listed
SARA EHS 40CFR 355: Not listed
TSCA: Listed

DOP5000 CAS #: 51-61-6

DOPAMINE

RTECS: UX1088000
EINECS Number: 200-110-0
Molecular Formula: $C_8H_{11}NO_2$
Formula Weight: 153.18
Synonyms: 4-(2-AMINOETHYL)-1,2-BENZENEDIOL; 4-(2-AMINOETHYL)PYROCATECHOL; ASL 279; 1,2-BENZENEDIOL,4-(2-AMINOETHYL)-; 3,4-DIHYDROXYPHENETHYLAMINE; ALPHA-(3,4-DIHYDROXYPHENYL)-BETA-AMINOETHANE; 2-(3,4-DIHYDROXYPHENYL)ETHYLAMINE; 3,4-DIHYDROXYPHENYLETHYLAMINE; DOPAMIN; HYDROXYTYRAMIN; 3-HYDROXYTYRAMINE; OXYTYRAMINE; PYROCATECHOL,4-(2-AMINOETHYL)-
Description: stout prisms; odorless
Use: medication: dopaminergic agent

Physical Properties

Boiling Point: Decomposes at 210 °C (410 °F) to 214 °C (417.2 °F)
Freezing Point: Decomposes at 210 °C (410 °F) to 214 °C (417.2 °F)
Water Solubility: Freely Soluble in Water
Other Solubilities: Soluble in Methanol, in hot 95% Ethanol, in aqueous solution of alkali hydroxides; Practically Insoluble in Petroleum Ether, Ether, Benzene, Chloroform, Toluene

RTECS Toxicity Data

Reproductive/Teratogenic: Woman Route: Intravenous; Dose: 120 ug/kg; Duration: female 2H after birth; Maternal Effects - Other effects on females. Rat Route: Subcutaneous; Dose: 300 mg/kg; Duration: female 30D prior to mating Maternal Effects - Ovaries, fallopian tubes. Rat Route: Subcutaneous; Dose: 100 mg/kg; Duration: female 10-14D of pregnancy Effects on Newborn - Weaning or lactation index; Growth statistics; Other postnatal measures or effects. Rat Route: Subcutaneous; Dose: 100 mg/kg; Duration: female 10-14D of pregnancy Specific Developmental Abnormalities - Eye, ear.

Mutagenic: Human DNA Damage; Cell Type: fibroblast; Dose: 50 mg/L. Human DNA Inhibition; Cell Type: other cell types; Dose: 3800 nmol/L.

Hazard Overviews

Carcinogenicity: IARC - Not listed; NIOSH - Not listed; NTP - Not listed; ACGIH - Not listed; OSHA - Not listed; EPA - Not listed; MAK - Not listed

Environmental

Regulations

RCRA 40CFR: Not listed
CERCLA: 40CFR 302.4: Not listed
SARA 40CFR 372.65: Not listed
SARA EHS 40CFR 355: Not listed
TSCA: Not listed

DOX1000 CAS #: 309-29-5

DOXAPRAM

RTECS: UY5769500
EINECS Number: 206-216-3
Molecular Formula: $C_{24}H_{30}N_2O_2$
Formula Weight: 378.50
Synonyms: AHR 619; DOPRAM; DOPREAM; 1-ETHYL-4-(2-MORPHOLINOETHYL)-3,3-DIPHENYL-2-PYRROLIDINONE; 2-PYRROLIDINONE,1-ETHYL-4-(2-MORPHOLINOETHYL)-3,3-DIPHENYL-; 2-PYRROLIDINONE,1-ETHYL-4-(2-(4-MORPHOLINYL)ETHYL)-3,3-DIPHENYL-; 2-PYRROLIDINONE,1-ETHYL-4-(2-(4-MORPHOLINYL)ETHYL)-3,3-DIPHENYL-(9CI)
Description: white to off-white crystalline powder or crystals
Use: medication: respiratory stimulant

Physical Properties

Freezing Point: 123 °C (253.4 °F) to 124 °C (255.2 °F)
Water Solubility: Soluble in Water
Other Solubilities: Sparingly Soluble in Alcohol; Slightly Soluble in Chloroform; Practically Insoluble in Ether

RTECS Toxicity Data

Reproductive/Teratogenic: Domestic Animal Route: Intravenous; Dose: 1800 ug/kg; Duration: female 16W of pregnancy; Effects on Embryo or Fetus - Other effects to embryo or fetus.

Hazard Overviews

Carcinogenicity: IARC - Not listed; NIOSH - Not listed; NTP - Not listed; ACGIH - Not listed; OSHA - Not listed; EPA - Not listed; MAK - Not listed

Environmental

Regulations

RCRA 40CFR: Not listed
CERCLA: 40CFR 302.4: Not listed
SARA 40CFR 372.65: Not listed
SARA EHS 40CFR 355: Not listed
TSCA: Not listed

DOX3000 CAS #: 1668-19-5

DOXEPIN

RTECS: HQ4300000
Molecular Formula: $C_{19}H_{21}NO$
Formula Weight: 279.37
Synonyms: DIBENZ(B,E)OXEPIN-DELTA(11(6H),GAMMA)-PROPYLAMINE,N,N-DIMETHYL-; 11-(3-DIMETHYLAMINO-PROPYLIDEN)-6,11-DIHYDRO-DIBENZ(B,E)OXIPIN; 11-(3-(DIMETHYLAMINO)PROPYLIDENE)-6H-DIBENZ(B,E)0XEPINE; 11-(3-DIMETHYLAMINOPROPYLIDENE)-6,11-DIHYDRODIBENZ(B,E)OXIPIN; N,N-DIMETHYLDIBENZ(B,E)OXEPIN-DELTA(11(6H),GAMMA)-PROPYLAMINE; DOXEPINE; P 3693A; 1-PROPANAMINE,3-DIBENZ(B,E)OXEPIN-11(6H)-YLIDENE-N,N-DIMETHYL-
Description: oily liquid; odorless
Use: antidepressant; vet: antipruritic

Physical Properties

Boiling Point: 154 °C (309 °F) to 157 °C (315 °F) at 0.03 mm Hg
Freezing Point: 184 °C (363.2 °F) to 186 °C (366.8 °F)
Water Solubility: 1 g/mL
Other Solubilities: 1 g Soluble in 2 ml Alcohol; 10 ml Chloroform

RTECS Toxicity Data

Acute Oral: Human LD_{Lo} Dose: 60 mg/kg; Toxic Effects: Lungs, Thorax, or Respiration - Bronchiectasis. Rat LD_{50} Dose: 147 mg/kg; Toxic Effects: Automatic Nervous System - Parasympatholytic; Behavioral - Somnolence (general depressed activity).

Hazard Overviews

Carcinogenicity: IARC - Not listed; NIOSH - Not listed; NTP - Not listed; ACGIH - Not listed; OSHA - Not listed; EPA - Not listed; MAK - Not listed

Environmental

Regulations

RCRA 40CFR: Not listed
CERCLA: 40CFR 302.4: Not listed
SARA 40CFR 372.65: Not listed
SARA EHS 40CFR 355: Not listed
TSCA: Not listed

DOX6000 CAS #: 564-25-0

DOXYCYCLINE

RTECS: QI8650000
EINECS Number: 209-271-1
Molecular Formula: $C_{22}H_{24}N_2O_8$
Formula Weight: 462.46
Synonyms: AZUDOXAT; 6-DEOXY-5-HYDROXYTETRACYCLINE; ALPHA-6-DEOXY-5-HYDROXYTETRACYCLINE; 6-ALPHA-DEOXY-5-OXYTETRACYCLINE; ALPHA-6-DEOXYOXYTETRACYCLINE; ALPHA-6-DEOXYOXYTETRACYCLINE MONOHYDRATE; 6-DEOXYTETRACYCLINE; 4-(DIMETHYLAMINO)-1,4,4A,5,5A,6,11,12A-OCTAHYDRO-3,5,10,12,12A-PENTAHYDROXY-6-METHYL-1,11-DIOXO-2-NAPHTHACENECARBOXAMIDE MONOHYDRATE; DORYX; DOXICICLINA; DOXITARD; DOXY-CAPS; ALPHA-DOXYCYCLINE; DOXY-PUREN; DOXY-TABS; DOXYTETRACYCLINE; GS-3065; 5-HYDROXY-ALPHA-6-DEOXYTETRACYCLINE; 5-HYDROXY-ALPHA-6-DEOXYTETRACYCLINE MONOHYDRATE; INVESTIN; LIVIATIN; 2-NAPHTHACENECARBOXAMIDE,4-(DIMETHYLAMINO)-1,4,4A,5,5A,6,11,12A-OCTAHYDRO-3,5,10,12,12A-PENTAHYDROXY-6-METHYL-1,11-DIOXO-; 2-NAPHTHACENECARBOXAMIDE,4-(DIMETHYLAMINO)-1,4,4A,5,5A,6,11,12A-OCTAHYDRO-3,5,10,12,12A-PENTAHYDROXY-6-METHYL,1,11-DIOXO-,(45-(4ALPHA,4AALPHA,5ALPHA,5AALPHA,6ALPHA,12AALPHA))-; NORDOX; OXYTETRACYCLINE,6-DEOXY-; SPANOR; ALPHA-6-TETRACYCLINE MONOHYDRATE; VIBRAMYCIN; VIBRA-TABS; VIBRAVENOS
Description: yellow, crystalline powder
Use: medication: antibacterial

Physical Properties

Water Solubility: Very Slightly Soluble in Water
Other Solubilities: Slightly Soluble in Alcohol. /Doxycycline hyclate/; Sparingly Soluble in Alcohol. /Doxycycline monohydrate/.

RTECS Toxicity Data

Acute Oral: Woman TD_{Lo} Dose: 68 mg/kg/24D-I; Toxic Effects: Musculoskelital - Joints; Immunological including allergic - Othr immed. (hmrl): urticaria, allergic rhinitis, serum sickness; Nutritional and gross metabolic - Body temperature increase. Rat LD_{50} Dose: >2 gm/kg.
Reproductive/Teratogenic: Mouse Route: Oral; Dose: 750 mg/kg; Duration: female 2D of pregnancy; Effects on Embryo or Fetus - Fetotoxicity.
Mutagenic: Human DNA Inhibition; Cell Type: lymphocyte; Dose: 3750 ug/L.

Hazard Overviews

Health: Irritating to eyes/skin/respiratory tract. Harmful. Other Acute Effects: harmful if swallowed, inhaled, or absorbed through skin; can cause phototoxic reactions; gastrointestinal disturbances; turn teeth yellow; reduce mineralization. Chronic Effects: overexposure may cause reproductive disorder(s) based on tests with laboratory animals; possible risk of harm to the unborn child; target organs: teeth, bones.
Fire: Hazards: emits toxic fumes. Extinguishing agents: carbon dioxide, dry chemical powder or appropriate foam; water spray. Precautions: combustible liquid.
Reactivity: Stable. Hazardous polymerization will not occur. Hazardous decomposition products: toxic fumes of: carbon monoxide, carbon dioxide, nitrogen oxides.
Carcinogenicity: IARC - Not listed; NIOSH - Not listed; NTP - Not listed; ACGIH - Not listed; OSHA - Not listed; EPA - Not listed; MAK - Not listed
Primary Target Organs:

Eyes

Skin

Respiratory System

Bone

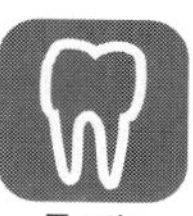
Teeth

Environmental

Regulations

RCRA 40CFR: Not listed
CERCLA: 40CFR 302.4: Not listed
SARA 40CFR 372.65: Not listed
SARA EHS 40CFR 355: Not listed
TSCA: Not listed

DOX9000 CAS #: 469-21-6

DOXYLAMINE

RTECS: US9250000
EINECS Number: 207-414-2
Molecular Formula: $C_{17}H_{22}N_2O$
Formula Weight: 270.38
Synonyms: 2-[ALPHA-(2-DIMETHYLAMINOETHOXY)-ALPHA-METHYLBENZYL] PYRIDINE; 2-(ALPHA-(2-(DIMETHYLAMINO)ETHOXY)-ALPHA-METHYLBENZYL)PYRIDINE; 2-[ALPHA-(2-DIMETHYLAMINOETHOXY)-ALPHA-METHYLBENZYL]PYRIDINE; 2-DIMETHYLAMINOETHOXYPHENYL-METHYL-2-PICOLINE; 2-DIMETHYLAMINOETHOXYPHENYLMETHYL-2-PICOLINE; N,N-DIMETHYL-2-[1-PHENYL-1-(2-PYRIDINYL)ETHOXY]ETHANAMINE; ETHANAMINE,N,N-DIMETHYL-2-(1-PHENYL-1-(2-PYRIDINYL)ETHOXY)-; ETHANAMINE,N,N-DIMETHYL-2-(1-PHENYL-1-(2-PYRIDINYL)ETHOXY)-(9CI); PHENYL-2-PYRIDYLMETHYL-BETA-N,N-DIMETHYLAMINOETHYL ETHER; PYRIDINE,2-(ALPHA-(2-(DIMETHYLAMINO)ETHOXY)-ALPHA-METHYLBENZYL)-
Description: liquid; characteristic odor
Use: antihistaminic

Physical Properties

Boiling Point: 137 °C (279 °F) to 141 °C (286 °F) at 0.5 mm Hg
Freezing Point: Crystals at 101 °C (213.8 °F) to 104 °C (219.2 °F)
Water Solubility: 1 g dissolves in 1 mL Water
Other Solubilities: 1 g dissolves in 2 ml Alcohol; Slightly Soluble in Benzene & Ether; 1 g dissolves in 2 ml Chloroform.
pH: 1% aqueous solution 4.9 to 5.1

Refraction Index: Calculated Alpha 1.525

RTECS Toxicity Data

Acute Oral: Mouse LD_{50} Dose: 470 mg/kg; Toxic Effects: Behavioral - Convulsions or effect on seizure threshold. Rabbit LD_{50} Dose: 250 mg/kg; Toxic Effects: Behavioral - Convulsions or effect on seizure threshold; Behavioral - Rigidity.

Acute Dermal: Rat LD_{50} Route: Subcutaneous Dose: 440 mg/kg; Toxic Effects: Behavioral - Convulsions or effect on seizure threshold. Mouse LD_{50} Route: Subcutaneous Dose: 460 mg/kg; Toxic Effects: Behavioral - Convulsions or effect on seizure threshold.

Hazard Overviews

Carcinogenicity: IARC - Not listed; NIOSH - Not listed; NTP - Not listed; ACGIH - Not listed; OSHA - Not listed; EPA - Not listed; MAK - Not listed

Environmental

Regulations

RCRA 40CFR: Not listed
CERCLA: 40CFR 302.4: Not listed
SARA 40CFR 372.65: Not listed
SARA EHS 40CFR 355: Not listed
TSCA: Not listed

DRO3000 CAS #: 58-19-5

DROMOSTANOLONE

RTECS: BV8063500
EINECS Number: 200-367-9
Molecular Formula: $C_{20}H_{32}O_2$
Formula Weight: 304.18

Synonyms: 5ALPHA-ANDROSTAN-3-ONE,17BETA-HYDROXY-2ALPHA-METHYL-; ANDROSTAN-3-ONE,17-HYDROXY-2-METHYL-,(2-ALPHA,5-ALPHA,17-BETA)-; ANDROSTAN-3-ONE,17-HYDROXY-2-METHYL-,(2ALPHA,5ALPHA,17BETA)-; DIHYDRO-2-ALPHA-METHYLTESTOSTERONE; DIHYDRO-2ALPHA-METHYLTESTOSTERONE; DROSTANOLONE; 17-BETA-HYDROXY-2-ALPHA-METHYL-5-ALPHA-ANDROSTAN-3-ONE; MEDROSTERON; MEDROTESTRON; METHOLONE; 2-ALPHA-METHYLDIHYDROTESTOSTERONE; 2ALPHA-METHYLDIHYDROTESTOSTERONE; 2-ALPHA-METHYL-17-BETA-HYDROXY-5-ALPHA-ANDROSTAN-3-ONE

Description: white to creamy white, crystalline powder or crystals; odorless or has faint odor

Use: medication: antineoplastic propionate; medication: restricted to treatment of metastatic carcinoma of breast; may improve anemias

Physical Properties

Freezing Point: Crystals at 126 °C (258.8 °F) to 130 °C (266 °F)
Water Solubility: Practically Insoluble in Water
Other Solubilities: Soluble in Chloroform, Ether, & Methanol

RTECS Toxicity Data

Reproductive/Teratogenic: Rabbit Route: Subcutaneous; Dose: 30 mg/kg; Duration: female 1-3D of pregnancy; Effects on Fertility - Pre-implantation mortality. Mouse Route: Subcutaneous; Dose: 2 mg/kg; Duration: female 1D prior to mating; Maternal Effects - Uterus, cervix, vagina.

Hazard Overviews

Carcinogenicity: IARC - Not listed; NIOSH - Not listed; NTP - Not listed; ACGIH - Not listed; OSHA - Not listed; EPA - Not listed; MAK - Not listed

Environmental

Regulations

RCRA 40CFR: Not listed
CERCLA: 40CFR 302.4: Not listed
SARA 40CFR 372.65: Not listed
SARA EHS 40CFR 355: Not listed
TSCA: Not listed

DRO6000 CAS #: 548-73-2

DROPERIDOL

RTECS: DE2100000
EINECS Number: 208-957-8
Molecular Formula: $C_{22}H_{22}FN_3O_2$
Formula Weight: 379.44

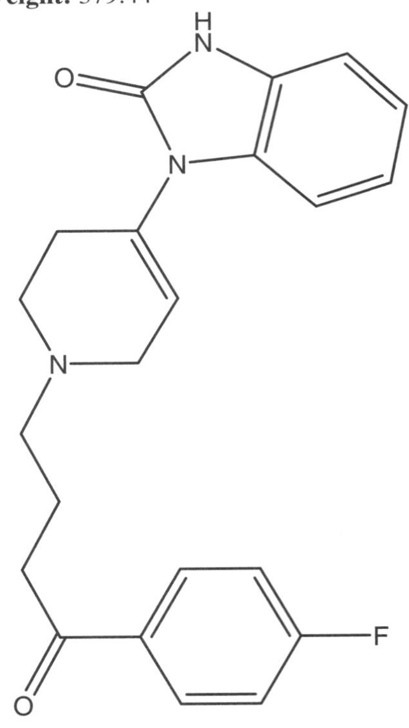

Chemical Structure

Synonyms: 2-BENZIMIDAZOLINONE,1-(1-(3-(P-FLUOROBENZOYL)PROPYL)-1,2,3,6-TETRAHYDRO-4-PYRIDYL)-;

2H-BENZIMIDAZOL-2-ONE,1-(1-(4-(4-FLUOROPHENYL)-4-OXOBUTYL)-1,2,3,6-TETRAHYDRO-4-PYRIDINYL)-1,3-DIHYDRO-; DEHIDROBENZPERIDOL; DEHYDROBENZPERIDOL; DEIDROBENZPERIDOLO; DHBP; DIHIDROBENZPERIDOL; DRIDOL; DROLEPTAN; 1-(1-(3-(P-FLUOROBENZOYL)PROPYL)-1,2,3,6-TETRAHYDRO-4-PYRIDYL)-2-BENZIMIDAZOLINONE; 1-(1-(4-(P-FLUOROPHENYL)-4-OXOBUTYL)-1,2,3,6-TETRAHYDRO-4-PYRIDYL)-2-BENZIMIDAZOLINONE; 1-(1-(4-(P-FLUOROPHENYL-4-OXOBUTYL)-1,2,3,6-TETRAHYDRO-4-PYRIDYL)-2-BENZIMIDAZOLINONE; HALKAN; INAPPIN; INAPSIN; INAPSINE; INNOVAN; INOPSIN; INOVAL; LEPTANAL; LEPTOFEN; MCN-JR-4749; MCN-JR 4749; PROPERIDOL; R 4749; R4749; SINTODRIL; SINTOSIAN; THALAMONAL; VETKALM

Description: white to light tan, amorphous or microcrystalline powder; odorless

Use: medication: tranquilizer; medication (vet): tranquilizer

Physical Properties

Freezing Point: 144 °C (291.2 °F) to 148 °C (298.4 °F)

Water Solubility: 1

Other Solubilities: 1 g Soluble in 5.5 ml Chloroform, 600 ml Alcohol, Slightly Soluble in Ether

RTECS Toxicity Data

Acute Oral: Human TD_{Lo} Dose: 223 ug/kg/12D-I; Toxic Effects: Behavioral - Wakefulness; Behavioral - Tremor; Behavioral - Muscle weakness. Rat LD_{50} Dose: 750 mg/kg.

Acute Dermal: Rat LD_{50} Route: Subcutaneous Dose: >100 mg/kg. Mouse LD_{50} Route: Subcutaneous Dose: 125 mg/kg.

Hazard Overviews

Health: Irritating. Harmful. Other Acute Effects: harmful if swallowed; may be harmful if inhaled; may be harmful if absorbed through the skin; causes drowsiness; mild hypotension; target organ: nerves

Fire: Hazards: emits toxic fumes. Extinguishing agents: water spray; carbon dioxide, dry chemical powder or appropriate foam. Precautions: combustible liquid.

Reactivity: Stable. Hazardous polymerization will not occur. Hazardous decomposition products: toxic fumes of: carbon monoxide, carbon dioxide, nitrogen oxides, sulfur oxides.

Carcinogenicity: IARC - Not listed; NIOSH - Not listed; NTP - Not listed; ACGIH - Not listed; OSHA - Not listed; EPA - Not listed; MAK - Not listed

Primary Target Organs:

Eyes

Skin

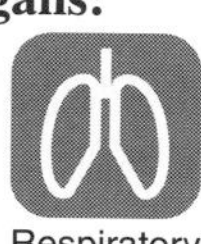
Respiratory System

Nervous System

Environmental

Regulations

RCRA 40CFR: Not listed

CERCLA: 40CFR 302.4: Not listed

SARA 40CFR 372.65: Not listed

SARA EHS 40CFR 355: Not listed

TSCA: Not listed

DYD5000 **CAS #: 152-62-5**

DYDROGESTERONE

RTECS: TU3671000

EINECS Number: 205-806-8

Molecular Formula: $C_{21}H_{28}O_2$

Formula Weight: 312.44

Synonyms: 6-DEHYDRO-9BETA,10ALPHA-PROGESTERONE; 6-DEHYDRO-RETRO-PROGESTERONE; 6-DEHYDRORETROPROGESTERONE; DIPHASTON; DUFASTON; DUPHASTON; DUVARON; GESTATRON; GYNOREST; HYDROGESTERONE; HYDROGESTRONE; 10ALPHA-ISOPREGNENONE; ISOPREGNENONE; 9BETA,10ALPHA-PREGNA-4,6-DIENE-3,20-DIONE; PREGNA-4,6-DIENE-3,20-DIONE,(9BETA,10ALPHA)-; PREGNA-4,6-DIENE-3,20-DIONE,(9-BETA,10-ALPHA)-(9CI); PRODEL; RETRO-6-DEHYDROPROGESTERONE; RETRONE; DELTA(6)-RETROPROGESTERONE; DELTA(SUP 6)-RETROPROGESTERONE; RETROPROGESTERONE,6-DEHYDRO-; TEROLUT

Description: crystals

Use: medication as progestin

Physical Properties

Freezing Point: 169 °C (336.2 °F)

RTECS Toxicity Data

Acute Oral: Rat LD_{50} Dose: >4600 mg/kg. Mouse LD_{50} Dose: >7200 mg/kg.

Reproductive/Teratogenic: Rat Route: Subcutaneous; Dose: 70 mg/kg; Duration: female 14D prior to mating Maternal Effects - Ovaries, fallopian tubes; Effects on Fertility - Other measures of fertility. Rabbit Route: Oral; Dose: 1600 mg/kg; Duration: female 13-28D of pregnancy Specific Developmental Abnormalities - Urogenital system.

Hazard Overviews

Carcinogenicity: IARC - Not listed; NIOSH - Not listed; NTP - Not listed; ACGIH - Not listed; OSHA - Not listed; EPA - Not listed; MAK - Not listed

Environmental

Regulations

RCRA 40CFR: Not listed

CERCLA: 40CFR 302.4: Not listed

SARA 40CFR 372.65: Not listed

SARA EHS 40CFR 355: Not listed

TSCA: Not listed

DYP5000 **CAS #: 479-18-5**

DYPHYLLINE

RTECS: XH5100000

EINECS Number: 207-526-1

Molecular Formula: $C_{10}H_{14}N_4O_4$

Formula Weight: 254.25

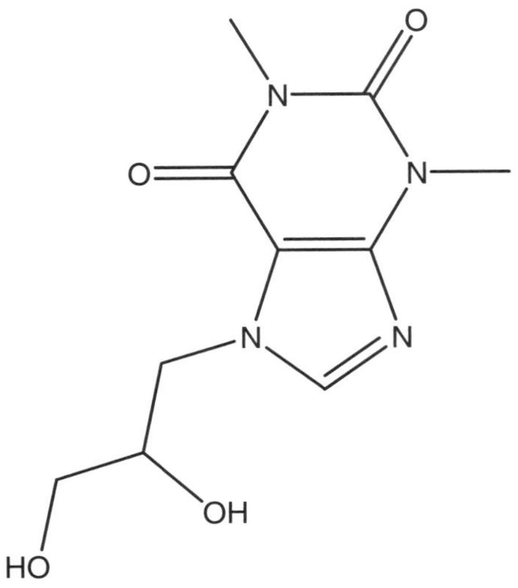

Chemical Structure

Synonyms: AFI-PHYLLIN; AFI-PHYLLINE; ARISTOPHYLLIN; ASTMAMASIT; ASTROPHYLLIN; CIRCAIN; CIRCAIR; CORONAL; CORONARIN; CORPHYLLIN; COR-THEOPHYLLINE; DIHYDROXYPROPYL THEOPHYLLINE; DIHYDROXYPROPYL THEOPYLIN; 7-(2,3-DIHYDROXYPROPYL)-3,7-DIHYDRO-1,3-DIMETHYL-1H-PURINE-2,6-DIONE; 7-(2,3-DIHYDROXYPROPYL)-1,3-DIMETHYLXANTHINE; 7-(2,3-DIHYDROXYPROPYL)-THEOPHYLLINE; 7-(2,3-DIHYDROXYPROPYL)THEOPHYLLINE; 7-(BETA,GAMMA-DIHYDROXYPROPYL)THEOPHYLLINE; (1,2-DIHYDROXY-3-PROPYL)THIOPHYLLIN; DILOR; 1,3-DIMETHYL-7-(2,3-DIHYDROXYPROPYL)XANTHINE; 7-(2,3-DIOXYPROPYL)THEOPHYLLINE; DIPHYLLIN; DIPROFILLIN; DIPROFILLINE; DIPROPHYLLIN; DIPROPHYLLINE; DIPROPHYLLINEE; GLYFYLLIN; GLYPHYLLIN; GLYPHYLLINE; HIDROXITEOFILLINA; HIPHYLLIN; HYPHYLLINE; LUFYLLIN; NEOPHYLLIN; NEOPHYLLIN M; NEOPHYLLINE; NEOSTENORASAN; NEOSTENOVASAN; NEOTHYLLINE; NEOTILINA; NEO-VASOPHYLLINE; NEUFIL; NEUTRAFIL; NEUTRAFILLINA; NEUTRAPHYLLIN; NEUTRAPHYLLINE; NEUTROXANTINA; PROPYLPHYLLIN; PROPYPHYLLIN; PROTHEOPHYLLINE; PURIFILIN; 1H-PURINE-2,6-DIONE,7(2,3-DIHYDROXYPROPYL)-3,7-DIHYDRO-1,3-DIMETHYL-; 1H-PURINE-2,6-DIONE,7-(2,3-DIHYDROXYPROPYL)-3,7-DIHYDRO-1,3-DIMETHYL; SIBEPHYLLIN; SIBEPHYLLINE; SILBEPHYLLINE; SOLUFILIN; SOLUFYLLIN; SOLUPHYLLIN; SYNTHOPHYLLINE; TEFILAN; TESFEN; THEAL; THEAL AMPULES; THEFYLAN; THEOPHYLLINE,7-(2,3-DIHYDROXYPROPYL)-

Description: white amorphous solid or crystals

Use: medication: smooth muscle relaxant

Physical Properties

Freezing Point: 158 °C (316.4 °F)

Water Solubility: Freely Soluble in Water

Other Solubilities: Soluble in Alcohol: 2 g/100 ml; in Chloroform: 1 g/100 ml.

pH: 1% aqueous solution 6.6 to 7.3

RTECS Toxicity Data

Acute Oral: Rat LD; Dose: >400 mg/kg. Mouse LD_{50} Dose: 1954 mg/kg.

Acute Dermal: Rat LD_{50} Route: Subcutaneous Dose: 1253 mg/kg. Mouse LD_{50} Route: Subcutaneous Dose: 1052 mg/kg.

Hazard Overviews

Health: May cause irritation. Harmful. Other Acute Effects: harmful if swallowed; may be harmful if inhaled or absorbed through the skin.

Fire: Hazards: emits toxic fumes. Extinguishing agents: carbon dioxide; dry chemical powder; water spray. Precautions: combustible liquid.

Reactivity: Incompatible with: strong oxidizing agents. Hazardous decomposition products: thermal decomposition may produce carbon monoxide, carbon dioxide, and nitrogen oxides.

Carcinogenicity: IARC - Not listed; NIOSH - Not listed; NTP - Not listed; ACGIH - Not listed; OSHA - Not listed; EPA - Not listed; MAK - Not listed

Environmental

Regulations

RCRA 40CFR: Not listed

CERCLA: 40CFR 302.4: Not listed

SARA 40CFR 372.65: Not listed

SARA EHS 40CFR 355: Not listed

TSCA: Listed

DYR5000 CAS #: 101-05-3

DYRENE

RTECS: XY7175000

DOT: UN2763; UN2764; UN2997; UN2998; IMO3.2; IMO6.1

EINECS Number: 202-910-5

Molecular Formula: $C_9H_5Cl_3N_4$

Formula Weight: 275.51

Synonyms: ANILAZIN; ANILAZINE; ANIYALINE; B-622; BORTRYSAN; 2-(2-CHLORANILIN)-4,6-DICHLOR-1,3,5-TRIAZIN; (O-CHLOROANILINO)DICHLOROTRIAZINE; 2-CHLORO-N-(4,6-DICHLORO-1,3,5-TRIAZIN-2-YL)ANILINE; 2,4-DICHLORO-6-O-CHLORANILINO-S-TRIAZINE; 2,4-DICHLORO-6-(2-CHLOROANILINO)-1,3,5-TRIAZINE; 2,4-DICHLORO-6-(O-CHLOROANILINO)-S-TRIAZINE; 4,6-DICHLORO-N-(2-CHLOROPHENYL)-1,3,5-TRIAZIN-2-AMINE; DIREX; DIREZ; DYRENE 50W; ENT 26,058; KEMATE; TRIASYN; TRIAZIN; 1,3,5-TRIAZIN-2-AMINE,4,6-DICHLORO-N-(2-CHLOROPHENYL)-; TRIAZINE; TRIAZINE (PESTICIDE); S-TRIAZINE,2,4-DICHLORO-6-(O-CHLOROANILINO)-; ZINOCHLOR

Description: white to tan crystals

Use: as a foliage fungicide in the control of early and late blights of potatoes and tomatoes; to control anthracnos in cucurbits, leaf spot diseases (Alternaria, Cercospora and Septoria) in many crops, glume blotch of wheat, Helminthosporium esp. in wheat and barley, brown patch, dollar spot, snow mold and other diseases of turf, Botrytis, Colletotrichum and Pyrenophora esp. on many crops; on vegetables, ornamentals, berry fruit, melons, watermelons, coffee and tobacco; as a bactericide and wood preservative

Physical Properties

Freezing Point: 159 °C (318.2 °F) to 160 °C (320 °F)

Specific Gravity: 1.8 at 20 °C
Vapor Pressure: 820 nPa at 20 °C
Water Solubility: < 1 mg/mL at 21 C
Other Solubilities: 5 g/100 ml Toluene, 4 g/100 ml Xylene, & 10 g/100 ml Acetone, each at 30 °C.
Flash Point: Not available; probably combustible

RTECS Toxicity Data

Acute Oral: Rat LD_{50} Dose: 2700 mg/kg. Mouse LD_{50} Dose: 6020 mg/kg.
Acute Inhalation: Rat LC_{50} Dose: >228 mg/m^3/1hr.
Acute Dermal: Rabbit LD_{50} Route: Skin; Dose: >9400 mg/kg. Rat LD_{50} Route: Skin; Dose: >5 gm/kg.
Irritation Skin: Man Standard Draize Test Dose: 0.1%; Reaction: moderate. Rabbit Standard Draize Test Dose: 500 mg; Reaction: severe.
Mutagenic: Human Mutations in Mammalian Somatic Cells; Cell Type: lymphocyte; Dose: 5 mg/L. Rat Morphological Transformation; Cell Type: embryo; Dose: 990 ng/plate.

Hazard Overviews

Fire: Will burn.
Carcinogenicity: IARC - Not listed; NIOSH - Not listed; NTP - Not listed; ACGIH - Not listed; OSHA - Not listed; EPA - Not listed; MAK - Not listed

Environmental

Ecotoxicity: LC_{50} Lepomis microlophus (redear sunfish) less than 140 ug/l/96 hr at 18 °C, wt 2.5 g. Static bioassay without aeration, pH 7.2-7.5, water hardness 40-50 mg/l as calcium carbonate and alkalinity of 30-35 mg/l LD_{50} Anas platyrhynchos (mallard duck) oral above 2000 mg/kg, male, 3-4 months old LD_{50} oral Agelaius phoeniceus (redwinged blackbird) 100 mg/kg; avian repellency value: R50 = 40.0 mg/kg; repellency toxicity index (hazard factor) = -0.4 LC_{50} Gammarus fasciatus (scuds) 0.27 ug/l/96 hr at 15 °C, mature (95% confidence limit 0.21-0.35 ug/l). Static bioassay without aeration, pH 7.2-7.5, water hardness 40-50 mg/l as calcium carbonate and alkalinity of 30-35 mg/l
Environmental Fate: When applied to soil, it binds strongly to the organic components of soil. After several days only a few percent appears in soil extracts. However, in one experiment, little was mineralized after 110 days. Therefore it is not clear to what extent degradation occurs compared with being chemically bound to soil components. The recommended field half-life value is 1 day. Under ordinary agricultural use, it would enter water associated with soil in runoff or as a result of inadvertent spraying over water. It will strongly adsorb to sediment and particulate matter in the water column and hydrolyze. At 22 °C, the hydrolytic half-lives are 33 days and 22 hr at pH 7 and 9, respectively. It will be nonvolatile from water. It is estimated to have a moderate potential for bioconcentration in aquatic organisms. It should exist predominantly as an aerosol in the atmosphere and will be removed from the air by gravitational settling. Vapor-phase reacts with photochemically-produced hydroxyl radicals, resulting in an estimated atmospheric half-life of 9.0 hr.

Cleanup/Disposal: Guide No. 131: Fully encapsulating, vapor protective clothing should be worn for spills and leaks with no fire. Eliminate all ignition sources (no smoking, flares, sparks or flames in immediate area). All equipment used when handling the product must be grounded. Do not touch or walk through spilled material. Stop leak if you can do it without risk. Prevent entry into waterways, sewers, basements or confined areas. A vapor suppressing foam may be used to reduce vapors. Small Spills: Absorb with earth, sand or other non-combustible material and transfer to containers for later disposal. Use clean non-sparking tools to collect absorbed material. Large Spills: Dike far ahead of liquid spill for later disposal. Water spray may reduce vapor; but may not prevent ignition in closed spaces. Guide No. 151: Do not touch damaged containers or spilled material unless wearing appropriate protective clothing. Stop leak if you can do it without risk. Prevent entry into waterways, sewers, basements or confined areas. Cover with plastic sheet to prevent spreading. Absorb or cover with dry earth, sand or other non-combustible material and transfer to containers. Do not get water inside containers.

Environmental Physical Data

Henry's Law Constant: calculated at 3.4 $x10^{-7}$
Octanol/Water Partition Coefficient: log K_{ow} = 3.88
Sorption Partition Coefficient: K_{oc} = estimated at 1000
BCF: calculated at 523

Regulations

RCRA 40CFR: Not listed
CERCLA: 40CFR 302.4: Not listed
SARA 40CFR 372.65: Listed
SARA EHS 40CFR 355: Not listed
TSCA: Not listed

Analytical Methods

Soil: EPA PMD-TLC; SW846 8270B, 8270C
Water / Groundwater: EPA 022
Food: FDA 212.1, 232.1, 232.4, 242.1; AOAC 988.04
Plasma: EPA 001, 027, 028, 29; FDA 211.1, 231.1, 252

EAS5000 CAS #: 8006-87-9

EAST INDIAN SANDALWOOD OIL

RTECS: RJ3697000
Molecular Formula: Unknown
Formula Weight: N/A
Synonyms: ARHEOL; OIL OF SANTAL; OILS,SANDALWOOD; SANDAL OIL; SANDALWOOD OIL; SANDALWOOD OIL,EAST INDIAN; SANTAL OIL; SANTALWOOD OIL
Description: pale yellow, somewhat viscid liquid; strong, persistent, warm, woody odor characteristic sandalwood
Use: medication: former urinary anti-infective; medication: (vet): formerly as a urinary antiseptic; perfumery; flavoring; fragrance in cosmetics, soaps, and detergents; flavoring agent in foods, beverages, and chewing gum

Physical Properties

Specific Gravity: 0.965 to 0.98 at 25 °C/25 °C
Water Solubility: Very Slightly Soluble in Water
Other Solubilities: Soluble in fixed oils; Insoluble in Glycerin.
pH: 0.5 to 8.0
Refraction Index: 1.500 to 1.510 at 20 °C/D

RTECS Toxicity Data

Acute Oral: Rat LD_{50} Dose: 5580 mg/kg.
Acute Dermal: Rabbit LD_{50} Route: Skin; Dose: >5 gm/kg.
Irritation Skin: Mouse Standard Draize Test Dose: 100%; Reaction: mild.

Hazard Overviews

Carcinogenicity: IARC - Not listed; NIOSH - Not listed; NTP - Not listed; ACGIH - Not listed; OSHA - Not listed; EPA - Not listed; MAK - Not listed

Environmental

Regulations

RCRA 40CFR: Not listed
CERCLA: 40CFR 302.4: Not listed
SARA 40CFR 372.65: Not listed
SARA EHS 40CFR 355: Not listed
TSCA: Listed

ECH5000 CAS #: 520-68-3

ECHIMIDINE

Molecular Formula: $C_{20}H_{31}NO_7$
Formula Weight: 397.52
Synonyms: 2-BUTENOIC ACID,2-METHYL-,7-((2,3-DIHYDROXY-2-(1-HYDROXYETHYL)-3-METHYL-1-OXOBUTOXY)METHYL) - 2,3,5,7A-TETRAHYDRO-1H-PYRROLIZIN-1-YLESTER

Hazard Overviews

Carcinogenicity: IARC - Not listed; NIOSH - Not listed; NTP - Not listed; ACGIH - Not listed; OSHA - Not listed; EPA - Not listed; MAK - Not listed

Environmental

Regulations

RCRA 40CFR: Not listed
CERCLA: 40CFR 302.4: Not listed
SARA 40CFR 372.65: Not listed
SARA EHS 40CFR 355: Not listed
TSCA: Not listed

EGO5000 CAS #: 59204-74-9

EGOMAKETONE

RTECS: JX2580000
Molecular Formula: $C_{10}H_{12}O_2$
Formula Weight: 164.22
Synonyms: EGOMACETONE; 3-PENTEN-1-ONE,1-(3-FURANYL)-4-METHYL-
Use: food flavoring; in soyo folk medicine; in biochemical & physiological studies; essential oil of perilla frutescens

Hazard Overviews

Carcinogenicity: IARC - Not listed; NIOSH - Not listed; NTP - Not listed; ACGIH - Not listed; OSHA - Not listed; EPA - Not listed; MAK - Not listed

Environmental

Regulations

RCRA 40CFR: Not listed
CERCLA: 40CFR 302.4: Not listed
SARA 40CFR 372.65: Not listed
SARA EHS 40CFR 355: Not listed
TSCA: Not listed

EIC5000 CAS #: 629-96-9

1-EICOSANOL

EINECS Number: 211-119-4
Molecular Formula: $C_{20}H_{42}O$
Formula Weight: 298.56

OH

Chemical Structure

Synonyms: ARACHIC ALCOHOL; ARACHIDIC ALCOHOL; ARACHIDYL ALCOHOL; N-1-EICOSANOL; N-EICOSANOL; EICOSYL ALCOHOL; PRI-N-EICOSYL ALCOHOL

Description: white, wax-like solid or crystals
Use: lubricants; rubber; plastics; textiles; research

Physical Properties

Boiling Point: 369 °C (696 °F) at 760 mm Hg
Freezing Point: 72.5 °C (162.5 °F) to 73 °C (163.4 °F)
Specific Gravity: 0.8405 at 20 °C/4 °C
Water Solubility: Insoluble in Water
Other Solubilities: Very Soluble in Acetone; Slightly Soluble in Alcohol; Soluble in hot Benzene, petroleum
Refraction Index: 1.4550 at 20 °C/D (supercooled)

Hazard Overviews

Health: Irritating. Other Acute Effects: may be harmful by inhalation, ingestion, or skin absorption.
Fire: Extinguishing agents: carbon dioxide, dry chemical powder or appropriate foam; water spray. Precautions: combustible liquid.
Reactivity: Incompatible with: strong oxidizing agents. Hazardous decomposition products: toxic fumes of: carbon monoxide, carbon dioxide.
Carcinogenicity: IARC - Not listed; NIOSH - Not listed; NTP - Not listed; ACGIH - Not listed; OSHA - Not listed; EPA - Not listed; MAK - Not listed
Primary Target Organs:

Eyes

Skin

Respiratory System

Environmental

Regulations

RCRA 40CFR: Not listed
CERCLA: 40CFR 302.4: Not listed
SARA 40CFR 372.65: Not listed
SARA EHS 40CFR 355: Not listed
TSCA: Listed

ELA5000 CAS #: 23315-05-1

ELAIOMYCIN

RTECS: EL5075000
Molecular Formula: $C_{13}H_{26}N_2O_3$
Formula Weight: 258.35
Synonyms: 2-BUTANOL,4-METHOXY-3-(1-OCTENYL-ONN-AZOXY)-,(E,Z)-(2S,3S)-; 2-BUTANOL,4-METHOXY-3-(1-OCTENYL-ONN-AZOXY)-,(S-(R*,R*-(E,Z)))-; 2-BUTANOL,4-METHOXY-3-(1-OCTENYL-ONN-AZOXY)-,D-THREO-; (2S,3S)-4-METHOXY-3-(1'-CIS-OCTENYL-CIS-AZOXY)-2-BUTANOL; (E,Z)-(2S,3S)-4-METHOXY-3-(1-OCTENYL)-ONN-AZOXY)-2-BUTANOL; (E,Z)-(2S,3S)-4-METHOXY-3-(1-OCTENYL-ONN-AZOXY)-2-BUTANOL; D-THREO-4-METHOXY-3-(1-OCTENYLAZOXY)-2-BUTANOL; D-THREO-METHOXY-3-(1-OCTENYL-ONN-AZOXY)-2-BUTANOL
Description: pale yellow oil

Physical Properties

Boiling Point: Acetate 84 °C (183 °F) to 90 °C (194 °F)
Water Solubility: Sparingly Soluble in Water
Other Solubilities: Soluble in practically all common organic solvents
Refraction Index: 1.4798 at 25 °C/D

RTECS Toxicity Data

Acute Dermal: Mouse LD_{50} Route: Subcutaneous Dose: 50 mg/kg.
Tumorigenic: Rat Route: Oral; Dose: 35 mg/kg; Toxic Effects: Tumorigenic - Equivocal tumorigenic agent by RTECS criteria; Liver - Tumors; Kidney, Ureter, and Bladder - Kidney tumors.

Hazard Overviews

Carcinogenicity: IARC - Not listed; NIOSH - Not listed; NTP - Not listed; ACGIH - Not listed; OSHA - Not listed; EPA - Not listed; MAK - Not listed

Environmental

Regulations

RCRA 40CFR: Not listed
CERCLA: 40CFR 302.4: Not listed
SARA 40CFR 372.65: Not listed
SARA EHS 40CFR 355: Not listed
TSCA: Not listed

ELE5000 CAS #: 8012-95-1

ELECTRICAL INSULATING OIL

RTECS: PY8030000
EINECS Number: 232-384-2
Molecular Formula: Unspecified or Variable
Formula Weight: Varies
Synonyms: ADEPSINE OIL; ALBOLINE; BALNEOL; BAYOL 55; BAYOL F; BLANDLUBE; BLANDOL WHITE MINERAL OIL; CARNEA 21; CLEARTECK; (COMPONENT OF) AGORAL; (COMPONENT OF) ALPHA KERI; (COMPONENT OF) KERI LOTION; CRYSTOL 325; CRYSTOSOL; DRAKEOL; ERVOL; FILTRAWHITE; FLEXON 845; FONOLINE; FRIGOL; GLORIA; GLYMOL; HEAVY LIQUID PETROLATUM; HEAVY MINERAL OIL; HEAVY MINERAL OIL MIST; HEVYTECK; HYDROCARBON OILS; IRGAWAX 361; KAYDOL; KONDREMUL; KREMOL; LIGNITE OIL; LIQUID PARAFFIN; LIQUID PETROLATUM; LIQUID VASELINE; MAGIESOL 44; MINERAL OILS; MOLOL; NEO-CULTOL; NUJOL; OIL MIST; OIL MIST,MINERAL; PARAFFIN OIL; PARAFFIN OIL MIST; PARAFFIN OILS; PARAFFINS; PAROL; PAROLEINE; PENETECK; PENRECO; PERFECTA; PETROGALAR; PETROLATUM,LIQ; PETROLATUM,LIQUID; PRIMOL; PRIMOL 355; PRIMOL D; PROTOPET; SAXOL; SHELLFLEX 371N; SUNPAR 150; TECH PET F; TRIONA B; ULTROL 7; USP MINERAL OIL; UVASOL; WHITE MINERAL OIL; WHITE MINERAL OIL MIST; WHITE OILS
Description: nearly colorless liquid; slight petroleum odor
Use: lubricant, as an ingredient in various ointment bases, as an emollient to the skin in irritant conditions, as a moisturizer, as a laxative, as a cathartic, as an ocular lubricant and as an ingredient in various pharmaceutical preparations; in

cosmetic products, to remove crusts and for chronic constipation; when sterilized, an aseptic dressing and as a lubricant for catheters and surgical instruments

Physical Properties

Boiling Point: 360 °C (680 °F)
Freezing Point: -17.78 °C (-0.004 °F)
Specific Gravity: 0.875 to 0.905 heavy
Density: 0.83 to 0.86 g/mL (light)
Vapor Pressure: < 0.5 mm Hg
Water Solubility: Insoluble in Water
Other Solubilities: miscible with most fixed oils; not miscible with Castor Oil; Soluble in volatile oils.
Surface Tension: 27 dynes/cm at 20 °C
Flash Point: 193 °C Open Cup
Autoignition Temperature: 260 to 371 °C

RTECS Toxicity Data

Acute Oral: Mouse LD_{50} Dose: 22 gm/kg.
Irritation Eye: Rabbit Standard Draize Test Dose: 500 mg; Reaction: moderate.
Irritation Skin: Rabbit Standard Draize Test Dose: 100 mg/24H; Reaction: mild. Guinea Pig Standard Draize Test Dose: 100 mg/24H; Reaction: mild.
Tumorigenic: Man Route: Inhalation; Dose: 5 mg/m^3/5Y-I; Toxic Effects: Tumorigenic - Carcinogenic by RTECS criteria; Gastrointestinal - Tumors; Tumorigenic effects - Testicular tumors. Mouse Route: Skin; Dose: 332 gm/kg/20W-I; Toxic Effects: Tumorigenic - Equivocal tumorigenic agent by RTECS criteria; Skin and appendages - Tumors.

Hazard Overviews

Fire Diamond

Health: Irritating to eyes/skin/respiratory tract. Also Causes: cough, difficult breathing; ingestion: diarrhea, abdominal pain. Chronic Effects: dermatitis, folliculitis, oil acne, skin cancer (animal data).
Fire: Will burn. Use carbon dioxide, foam, dry chemical, or water fog, or spray. Use water spray to cool fire-exposed containers. Use dry chemical or carbon dioxide fire in electrical equipment.
Reactivity: Stable. Hazardous polymerization cannot occur. Avoid: heat; ignition sources; strong oxidizing agents. Incompatible with: strong oxidizing agents (concentrated oxygen; chlorine; nitric acid). Hazardous decomposition products: carbon monoxide.
Carcinogenicity: IARC - Not listed; NIOSH - Listed as carcinogen; NTP - Not listed; ACGIH - Not listed; OSHA - Not listed; EPA - Not listed; MAK - Not listed

Primary Target Organs:

Eyes

Skin

Respiratory System

Exposure Limits
OSHA PEL: TWA: 5 mg/m^3.
ACGIH TLV: TWA: 5 mg/m^3; STEL: 10 mg/m^3.
NIOSH REL: TWA: 5 mg/m^3. STEL: 10 mg/m^3.
NIOSH IDLH: 2500 mg/m^3.
Respirator Recommendation
Exposure Range: >5 to 50 ppm Air Purifying, Negative Pressure, Half Mask
Exposure Range: >50 to 500 ppm Air Purifying, Negative Pressure, Full Face
Exposure Range: >500 to <2500 ppm Supplied Air, Constant Flow/Pressure Demand, Full Face
Exposure Range: 2500 to unlimited ppm Self-contained Breathing Apparatus, Pressure Demand, Full Face
Cartridge Color: dust/mist filter (use P100 or consult supervisor for appropriate dust/mist filter)

Environmental

Cleanup/Disposal: Guide No. 171: Do not touch or walk through spilled material. Stop leak if you can do it without risk. Prevent dust cloud. Avoid inhalation of asbestos dust. Small Dry Spills: With clean shovel place material into clean, dry container and cover loosely; move containers from spill area. Small Spills: Take up with sand or other noncombustible absorbent material and place into containers for later disposal. Large Spills: Dike far ahead of liquid spill for later disposal. Cover powder spill with plastic sheet or tarp to minimize spreading. Prevent entry into waterways, sewers, basements or confined areas.

Regulations

RCRA 40CFR: Not listed
CERCLA: 40CFR 302.4: Not listed
SARA 40CFR 372.65: Not listed
SARA EHS 40CFR 355: Not listed
TSCA: Listed

Analytical Methods

Air: EPA 0010; ASTM D3416

EME3000 CAS #: 483-18-1

EMETINE

RTECS: DK1750000
EINECS Number: 207-592-1
Molecular Formula: $C_{29}H_{40}N_2O_4$
Formula Weight: 480.63
Synonyms: 2H-BENZO(A)QUINOLIZINE,3-ETHYL-1,3,4,6,7,11B-HEXAHYDRO-9,10-DIMETHOXY-2-((1,2,3,4-TETRAHYDRO-6,7-DIMET/; CEPHAELINE METHYL ETHER; CEPHALINE-O-METHYL ETHER; EMETAN,6',7',10,11-TETRAMETHOXY-; EMETIN; (-)-

EMETINE; METHYL CEPHAELINE; NSC-33669; NSC 33669; 6',7',10,11-TETRAMETHOXYEMETAN

Description: white amorphous powder

Use: medication: antiamebic agent; medication (vet): as an antiamebic and in lung worm infection

Physical Properties

Freezing Point: 74 °C (165.2 °F)

Water Solubility: 1 g/100 mL Water

Other Solubilities: Sparingly Soluble in Petroleum Ether, in solution of Potassium or Sodium Hydroxide; freely Soluble in Methanol, Ethyl Acetate.

pH: Strong alkaline reaction

RTECS Toxicity Data

Acute Dermal: Man TD_{Lo} Route: Subcutaneous Dose: 10 mg/kg/10D; Toxic Effects: Behavioral - Muscle weakness; Cardiac - Arrythmias (including changes in conduction); Gastrointestinal - Other changes. Rabbit LD_{Lo} Route: Subcutaneous Dose: 30 mg/kg.

Irritation Eye: Rabbit Standard Draize Test Dose: 2000 ppm; Reaction: severe.

Irritation Skin: Rabbit Standard Draize Test Dose: 1 pph/24H; Reaction: moderate.

Mutagenic: Mouse DNA Inhibition; Route: Intraperitoneal; Dose: 25 mg/kg. Other Microorganisms Mutations in Microorganisms; Dose: 1250 mg/L (-S9).

Hazard Overviews

Carcinogenicity: IARC - Not listed; NIOSH - Not listed; NTP - Not listed; ACGIH - Not listed; OSHA - Not listed; EPA - Not listed; MAK - Not listed

Environmental

Regulations

RCRA 40CFR: Not listed

CERCLA: 40CFR 302.4: Not listed

SARA 40CFR 372.65: Not listed

SARA EHS 40CFR 355: Not listed

TSCA: Not listed

EME6000 CAS #: 316-42-7

EMETINE, DIHYDROCHLORIDE

RTECS: JY5250000

EINECS Number: 206-259-8

Molecular Formula: $C_{29}H_{42}Cl_2N_2O_4$

Formula Weight: 553.63

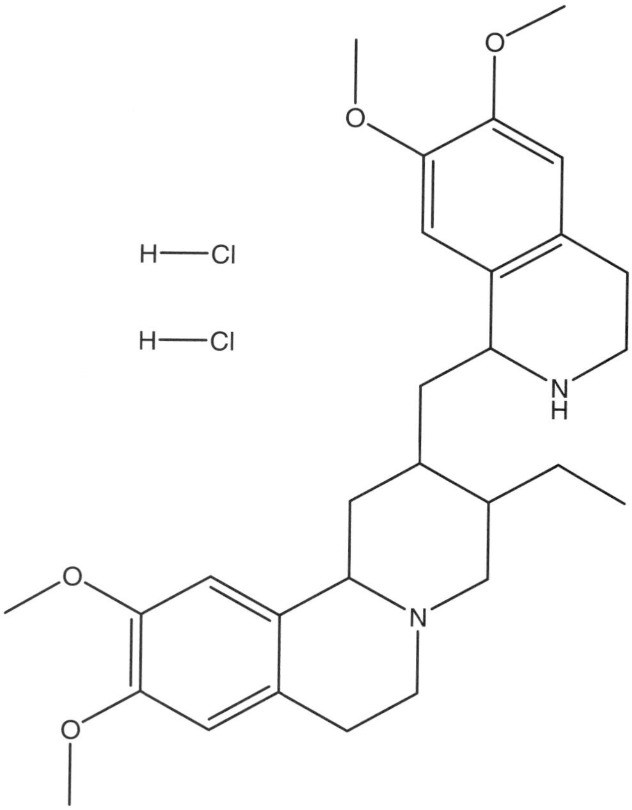

H—Cl

H—Cl

Chemical Structure

Synonyms: AMEBICIDE; EMETAN,6',7',10,11-TETRAMETHOXY-,DIHYDROCHLORIDE (9CI); (-)-EMETINE DIHYDROCHLORIDE; EMETINE DIHYDROCHLORIDE; L-EMETINE DIHYDROCHLORIDE; (-)-EMETINE HYDROCHLORIDE; L-EMETINE HYDROCHLORIDE; EMETINE,HYDROCHLORIDE; NSC-33669; PURUM; 6',7',10,11-TETRAMETHOXYEMETAN DIHYDROCHLORIDE

Description: white powder

Use: antiamebic; an antiamebic and for lungworm infection

Physical Properties

Freezing Point: Decomposes at 235 °C (455 °F) to 270 °C (518 °F)

Water Solubility: Soluble >= 10 mg/mL at 21 C

Other Solubilities: 95% Ethanol: Soluble (>=10 mg/ml at 21 °C); DMSO: Soluble (>=10 mg/ml at 21 °C); Ether: Insoluble.

Flash Point: Probably combustible

RTECS Toxicity Data

Acute Oral: Rat LD_{50} Dose: 12 ug/kg. Mouse LD_{50} Dose: 15 ug/kg.

Acute Dermal: Man LD_{Lo} Route: Subcutaneous Dose: 25 mg/kg/20D-I; Toxic Effects: Gastrointestinal - Hypermotility, diarrhea; Gastrointestinal - Nausea or vomiting; Lungs, Thorax, or Respiration - Dyspnea. Woman LD_{Lo} Route: Subcutaneous Dose: 2770 ug/kg/4D-I Toxic Effects: Behavioral - Hallucinations, distorted perceptions; Gastrointestinal - Hypermotility, diarrhea. Monkey LD_{Lo} Route: Subcutaneous Dose: 13886 ug/kg.

Chronic (Multiple Dose) Dermal: Rat Route: Subcutaneous; Dose: 35 mg/kg/7W-I; Toxic Effects: Cardiac - EKG changes not diagnostic of above; Nutritional and gross metabolic - Weight loss or decreased weight gain.

Hazard Overviews

Poison

Health: Irritating to eyes/skin/respiratory tract. Poison. Other Acute Effects: may be fatal if inhaled, swallowed, or absorbed through skin. Chronic Effects: when ingested produces nausea; vomiting; injection produces pain; muscle stiffness; necrosis; abscess formation; other adverse effects include nausea; vomiting; diarrhea; dizziness; headache; urticaria; purpuric skin rash; material accumulates mainly in the liver but appreciable levels are also found in the kidney, spleen and lungs persisting for several months in these tissues; prolonged adminstration or large doses may cause lesions of the heart, kidneys, liver, gastro-intestinal tract and skeletal muscle; severe degenerative myocarditis may occur giving rise to sudden cardiac failure and death; other serious cardiovascular effects include precordial pain, dyspnea, tachycardia and hypotension; EKG changes include t-wave inversion, prolongation of the q-t interval and st elevation.
Fire: Combustible. Hazards: emits toxic fumes. Extinguishing agents: carbon dioxide; dry chemical powder; water spray. Precautions: combustible liquid.
Reactivity: Incompatible with: strong oxidizing agents. Hazardous decomposition products: thermal decomposition may produce carbon monoxide, carbon dioxide, and nitrogen oxides; hydrogen chloride gas.
Carcinogenicity: IARC - Not listed; NIOSH - Not listed; NTP - Not listed; ACGIH - Not listed; OSHA - Not listed; EPA - Not listed; MAK - Not listed
Primary Target Organs:

Eyes

Skin

Respiratory System

Environmental

Cleanup/Disposal: Guide No. 154: Eliminate all ignition sources (no smoking, flares, sparks or flames in immediate area). Do not touch damaged containers or spilled material unless wearing appropriate protective clothing. Stop leak if you can do it without risk. Prevent entry into waterways, sewers, basements or confined areas. Absorb or cover with dry earth, sand or other non-combustible material and transfer to containers. Do not get water inside containers.

Regulations

RCRA 40CFR: Not listed
CERCLA: 40CFR 302.4: Not listed
SARA 40CFR 372.65: Not listed TPQ: 1/10000 lb
SARA EHS 40CFR 355: Listed TPQ: 1 lb
TSCA: Listed

ENA5000 CAS #: 75847-73-3

ENALAPRIL

Molecular Formula: $C_{20}H_{28}N_2O_5$
Formula Weight: 376.45
Synonyms: 1-(N-((S)-1-CARBOXY-3-PHENYLPROPYL)-L-ALANYL)-L-PROLINE1'-ETHYL ESTER; (S)-1-(N-(1-(ETHOXYCARBONYL)-3-PHENYLPROPYL)-L-ALANYL)-L-PROLINE
Description: white to off-white crystalline powder
Use: medication: antihypertensive; medication: antihypertensive agent and agent for treatment of heart failure

Physical Properties

Freezing Point: 143 °C (289.4 °F) to 144.5 °C (292.1 °F)
Water Solubility: 25 mg/mL
Other Solubilities: Solubility (g/ml): Alcohol 0.08, Methanol 0.20; Sparingly Soluble in Methanol. /Enalaprilat/; Solubilities of 80 mg/ml in Alcohol at room temperature. /Enalapril maleate/.
pH: 1% water 2.6

Hazard Overviews

Carcinogenicity: IARC - Not listed; NIOSH - Not listed; NTP - Not listed; ACGIH - Not listed; OSHA - Not listed; EPA - Not listed; MAK - Not listed

Environmental

Regulations

RCRA 40CFR: Not listed
CERCLA: 40CFR 302.4: Not listed
SARA 40CFR 372.65: Not listed
SARA EHS 40CFR 355: Not listed
TSCA: Not listed

END1000 CAS #: 959-98-8

A-ENDOSULFAN

RTECS: RB9275100
Molecular Formula: $C_9H_6Cl_6O_3S$
Formula Weight: 406.91
Synonyms: ALPHA-BENZOEPIN; A-ENDOSULFAN-ALPHA; ALPHA-ENDOSULFAN; ENDOSULFAN 1; ENDOSULFAN A; ALPHA-THIODAN; BETA-THIONEX
Description: brown crystals
Use: insecticide

Physical Properties

Freezing Point: 108 °C (226.4 °F) to 110 °C (230 °F)
Water Solubility: Insoluble
Other Solubilities: 95% Ethanol: Soluble; Acetone: Soluble; Benzene: Soluble; Ether: Soluble.

RTECS Toxicity Data

Acute Oral: Rat LD_{50} Dose: 76 mg/kg.

Hazard Overviews

Carcinogenicity: IARC - Not listed; NIOSH - Not listed; NTP - Not listed; ACGIH - Not listed; OSHA - Not listed; EPA - Not listed; MAK - Not listed

Environmental

Cleanup/Disposal: Guide No. 151: Do not touch damaged containers or spilled material unless wearing appropriate protective clothing. Stop leak if you can do it without risk. Prevent entry into waterways, sewers, basements or confined areas. Cover with plastic sheet to prevent spreading. Absorb or cover with dry earth, sand or other non-combustible material and transfer to containers. Do not get water inside containers.

Regulations

RCRA 40CFR: Not listed
CERCLA: 40CFR 302.4: Listed per CWA Section 307(a) RQ: 1 lb (0.454 kg)
SARA 40CFR 372.65: Not listed
SARA EHS 40CFR 355: Not listed
TSCA: Not listed

Analytical Methods

Soil: CLP LC_PEST, MC_PEST, OHC; SW846 3630B, 3640A, 8080A, 8081, 8081A, 8250A, 8270B, 8270C; EPA 16, 3, P-002-1, P-011-1; USGS O5104, O7104
Water / Groundwater: EPA P-003-1, P-004-1, 608, 617, 625, 625-S, 680; APHA 6410-B, 6630-B, 6630-C, 6630-D; USGS O3104
Drinking Water: EPA 508, 508.1, 525.2; AOAC 990.06
Food: FDA 212.1, 212.2, 232.1, 232.4, 242.1; EPA 4; USGS O9104
Plasma: EPA 001, 003, 027, 028, 29; FDA 211.1, 231.1, 252, 253
Other: EPA P-009-1, 1656

END2140 CAS #: 33213-65-9

B-ENDOSULFAN

RTECS: RB9875200
Molecular Formula: $C_9H_6Cl_6O_3S$
Formula Weight: 406.91
Synonyms: BETA-BENZOEPIN; B-ENDOSULFAN-BETA; BETA-ENDOSULFAN; ENDOSULFAN 2; ENDOSULFAN B; GENERAL WEED KILLER; BETA-THIODAN; ALPHA-THIONEX
Description: brown crystals
Use: insecticide

Physical Properties

Freezing Point: 208 °C (406.4 °F) to 210 °C (410 °F)
Water Solubility: Insoluble
Other Solubilities: 95% Ethanol: Soluble; Acetone: Soluble; Benzene: Soluble; Ether: Soluble.

RTECS Toxicity Data

Acute Oral: Rat LD_{50} Dose: 240 mg/kg.

Hazard Overviews

Carcinogenicity: IARC - Not listed; NIOSH - Not listed; NTP - Not listed; ACGIH - Not listed; OSHA - Not listed; EPA - Not listed; MAK - Not listed

Environmental

Cleanup/Disposal: Guide No. 151: Do not touch damaged containers or spilled material unless wearing appropriate protective clothing. Stop leak if you can do it without risk. Prevent entry into waterways, sewers, basements or confined areas. Cover with plastic sheet to prevent spreading. Absorb or cover with dry earth, sand or other non-combustible material and transfer to containers. Do not get water inside containers.

Regulations

RCRA 40CFR: Not listed
CERCLA: 40CFR 302.4: Listed per CWA Section 307(a) RQ: 1 lb (0.454 kg)
SARA 40CFR 372.65: Not listed
SARA EHS 40CFR 355: Not listed
TSCA: Not listed

Analytical Methods

Soil: CLP LC_PEST, MC_PEST, OHC; SW846 3630B, 3640A, 8080A, 8081, 8081A, 8250A, 8270B, 8270C; EPA 16, 3, P-002-1, P-011-1; USGS O5104, O7104
Water / Groundwater: EPA P-003-1, P-004-1, 608, 617, 625, 625-S, 680; APHA 6410-B, 6630-B, 6630-C, 6630-D
Drinking Water: EPA 508, 508.1, 525.2; AOAC 990.06
Food: FDA 212.1, 212.2, 232.1, 232.4, 242.1; EPA 4
Plasma: EPA 001, 003, 028, 29; FDA 211.1, 231.1, 252, 253
Other: EPA P-009-1, 1656

END3280 CAS #: 115-29-7

ENDOSULFAN

RTECS: RB9275000
DOT: NA2761
EINECS Number: 204-079-4
Molecular Formula: $C_9H_6Cl_6O_3S$
Structured MF: $C_9H_6Cl_6O_3S$
Formula Weight: 406.95
Synonyms: BENZODIOXATHIEPIN-3-OXIDE; BENZOEPIN; BEOSIT; BIO 5,462; BIO 5,642; CHLORTHIEPIN; CRISULFAN; CYCLODAN; DEVISULPHAN; 2-DIMETHANOL; ENDOCEL; ENDOSOL; ENDOSULPHAN; ENSURE; ENT 23,979; EPA PESTICIDE CODE 079401; FMC 5462; GOLDENLEAF TOBACCO SPRAY; 1,2,3,4,7,7-HEXACHLOROBICYCLO(2.2.1)HEPTEN-5,6-BIOXYMETHYLENE SULFITE; 1,2,3,4,7,7-HEXACHLOROBICYCLO(2.2.1)HEPTEN-5,6-BIOXYMETHYLENESULFITE; 1,2,3,4,7,7-HEXACHLOROBICYCLO-2,2,1-HEPTEN-5,6-BISOXYMETHYLENE SULFITE; ALPHA,BETA-1,2,3,4,7,7-HEXACHLOROBICYCLO)2.2.1)-2-HEPTENE-5,6-BISOXY-; 1,2,3,4,7,7-HEXACHLOROBICYCLO(2.2.1)-2-HEPTENE-5,6-BISOXYMETHYLENE SULFITE; ALPHA,BETA-1,2,3,4,7,7-

HEXACHLOROBICYCLO(2.2.1)-2-HEPTENE-5,6-BISOXYMETHYLENE SULFITE; 6,7,8,9,10-HEXACHLORO-1,5,5A,6,9,9A-HEXACHLORO-6,9-METHANO-2,4,3 -BENZO-DIOXATHIEPIN-3-OXIDE; 6,7,8,9,10,10-HEXACHLORO-1,5,5A,6,9,9A-HEXAHYDRO-6,9-METHANO-2,4,3-; HEXACHLOROHEXAHYDROMETHANO 2,4,3-BENZODIOXATHIEPIN-3-OXIDE; 6,7,8,9,10,10-HEXACHLORO-1,5,5A,6,9,9A-HEXAHYDRO-6,9-METHANO-2,4,3-BENZO(E)DIOXATHIEPIN 3-OXIDE; 6,7,8,9,10,10-HEXACHLORO-1,5,5A,6,9,9A-HEXAHYDRO-6,9-METHANO-2,4,3-BENZODIOXATHIEPIN-3-OXIDE; 1,4,5,6,7,7-HEXACHLORO-5-NORBORNENE-2,3-DIMETHANOL CYCLIC SULFITE; 1,4,5,6,7,7-HEXACHLORO-5-NORBORNENE-2,3-DIMETHANOL CYCLICSULFITE; 1,4,5,6,7,7-HEXACHLORO-5-NORBORNENE-2,3-DIMETHANOL,CYCLIC SULFATE; 1,4,5,6,7,7-HEXACHLORO-8,9,10-TRINORBORN-5-EN-2,3-YLENEDIMETHYL SULPHITE; HILDAN; HOE 2,671; HOE 2671; INSECTOPHENE; KOP-THIODAN; MALIX; 6,9-METHANO-2,4,3-BENZODIOXATHIEPIN,6,7,8,9,10,10-HEXACHLORO-1,5,5A,6,9,9A-HEXAHYDRO-,3-OXIDE; METHYLENE SULFITE; NIA 5462; NIAGARA 5,462; NIAGRA 5462; 5-NORBORNENE-2,3-DIMETHANOL,1,4,5,6,7,7-HEXACHLORO-,CYCLIC SULFITE; OMS 570; PFF THIODAN 4E; RASAYANSULFAN; SD-4314; SULFUROUS ACID,CYCLIC ESTER WITH 1,4,5,6,7,7-HEXACHLORO-5-NORBORNENE-; SULFUROUS ACID,CYCLIC ESTER WITH 1,4,5,6,7,7-HEXACHLORO-5-NORBORNENE-2,3-DIMETHANOL; THIFOR; THIMUL; THIODAN; THIODAN 35; THIODAN DUST INSECTICIDE; THIODAN 4EC; THIODAN 4EC INSECTICIDE; THIODAN 4E INSECTICIDE LIQUID; THIODAN 50 W; THIODAN 50 WP; THIODAN 50 WP INSECTICIDE; THIOFOR; THIOMUL; THIONATE; THIONEX; THIOSULFAN; THIOSULFAN TIONEL; THIOTOX; THIOTOX (INSECTICIDE); TIONEL; TIONEX; TIOVEL

Description: tan semi-waxy solid; slight sulfur dioxide odor
Use: insecticide; pesticide

Physical Properties

Boiling Point: 106 °C (223 °F) at 0.7 mm Hg
Freezing Point: 106 °C (222.8 °F)
Specific Gravity: 1.735 at 20 °C
Vapor Density: 14 Air=1
Vapor Pressure: 0.009 mm Hg at 80 °C
Water Solubility: Practically Insoluble in Water
Other Solubilities: 95% Ethanol: 10-50 mg/ml at 23 °C; Acetone: >=100 mg/ml at 23 °C; Chloroform: Soluble; DMSO: 10-50 mg/ml at 23 °C; Kerosene: Soluble; Most organic solvents: Soluble; Xylene: Soluble.
pH: Tap water 7.2
Flash Point: Not available; probably combustible

RTECS Toxicity Data

Acute Oral: Man TD_{Lo} Dose: 86 mg/kg; Toxic Effects: Behavioral - Convulsions or effect on seizure threshold; Behavioral - Coma; Lungs, Thorax, or Respiration - Cyanosis. Man LD_{Lo} Dose: 418 uL/kg; Toxic Effects: Sense organs and special senses - Other; Behavioral - Convulsions or effect on seizure threshold; Cardiac - Pulse rate. Rat LD_{50} Dose: 18 mg/kg.

Acute Inhalation: Rat LC_{50} Dose: 80 mg/m^3/4hr. Cat LC_{50} Dose: 90 mg/m^3/4hr.

Acute Dermal: Rabbit LD_{50} Route: Skin; Dose: 90 mg/kg; Toxic Effects: Sense organs and special senses - Other; Skin and appendages - Dermatitis, irritative. Rabbit LD_{50} Route: Subcutaneous Dose: 360 mg/kg; Toxic Effects: Behavioral - Change in motor activity (specific assay); Gastrointestinal - Nausea or vomiting; Gastrointestinal - Contraction (isolated tissue).

Chronic (Multiple Dose) Oral: Rat Dose: 150 mg/kg/15D-I; Toxic Effects: Liver - Changes in liver weight; Kidney, Ureter, and Bladder - Changes in kidney weight; DEATH - Changes in testicular weight. Rat Dose: 450 mg/kg/60D-I; Toxic Effects: Lungs, Thorax, or Respiration - Changes in lung weight; Liver - Changes in liver weight; DEATH. Rat Dose: 840 mg/kg/7W-I; Toxic Effects: Endocrine - Hyperglycemia; Blood - Other changes; Nutritional and gross metabolic - Changes in calcium. Rat Dose: 185 mg/kg/22W-C; Toxic Effects: Endocrine - Changes in spleen weight; Immunological including allergic - Decrease in humoral immune response. Rat Dose: 45 mg/kg/30D-I; Toxic Effects: Liver - Other changes; Kidney, Ureter, and Bladder - Changes in kidney weight; Blood - Changes in serum composition. Rat Dose: 126 mg/kg/6W-C; Toxic Effects: Liver - Changes in liver weight; Immunological including allergic - Decrease in humoral immune response.

Reproductive/Teratogenic: Rat Route: Oral; Dose: 45 mg/kg; Duration: female 6-14D of pregnancy; Effects on Fertility - Post-implantation mortality; Specific Developmental Abnormalities - Musculoskeletal system. Rat Route: Oral; Dose: 600 mg/kg; Duration: male 60D prior to mating; Paternal Effects - Testes, epididymis, sperm duct; Other effects on male.

Mutagenic: Human Sister Chromatid Exchenge; Cell Type: lymphocyte; Dose: 1 umol/L. Mouse Mutations in Mammalian Somatic Cells; Cell Type: lymphocyte; Dose: 18600 ug/L.

Tumorigenic: Mouse Route: Oral; Dose: 330 mg/kg/78W-I; Toxic Effects: Tumorigenic - Neoplastic by RTECS criteria; Lungs, Thorax, or Respiration - Tumors. Mouse Route: Subcutaneous; Dose: 2 mg/kg; Toxic Effects: Tumorigenic - Equivocal tumorigenic agent by RTECS criteria; Lungs, Thorax, or Respiration - Tumors; Liver - Tumors.

Hazard Overviews

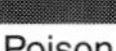
Poison

Fire Diamond

Health: Poison. Chronic Effects: headache, weakness, dizziness, abdominal pain, a feeling of warmness, loss of appetite, confusion, fainting, foaming at the mouth, convulsions, and death

Fire: Noncombustible. Use agent suitable for surrounding fire. Possible hazardous combustion products include sulfur oxide(s) and chlorine gas.

Reactivity: Stable. Hazardous polymerization cannot occur. Avoid: elevated temperatures Incompatible with: acids; alkalis; iron. Hazardous decomposition products: sulfur oxides; chlorine.

Carcinogenicity: IARC - Not listed; NIOSH - Not listed; NTP - Not listed; ACGIH - Class A4, Not classifiable as a human carcinogen; OSHA - Not listed; EPA - Not listed; MAK - Not listed

Primary Target Organs:

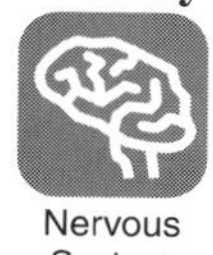
Nervous System

Exposure Limits

OSHA PEL: TWA: 0.1 mg/m^3; skin.

OSHA PEL Vacated 1989 Limits: TWA: 0.1 mg/m^3.

ACGIH TLV: TWA: 0.1 mg/m^3.

NIOSH REL: TWA: 0.1 mg/m^3.

Respirator Recommendation

Exposure Range: >0.1 to 5 mg/m^3 Supplied Air, Constant Flow/Pressure Demand, Half Mask

Exposure Range: >5 to 10 mg/m^3 Supplied Air, Constant Flow/Pressure Demand, Full Face

Exposure Range: >10 to unlimited mg/m^3 Self-contained Breathing Apparatus, Pressure Demand, Full Face

Note: odor threshold unknown

Environmental

Ecotoxicity: LC_{50} Bobwhite quail oral 805 ppm (95% confidence limit 690-939 ppm), age 9 days LC_{50} Corethra plumicornis 200 ug/l/24 hr /From table; Conditions of bioassay not specified TL_{50} Palaemon macrodactylus 3.4 (1.8-6.5) ug/l/96 hr /in a flow through bioassay LC_{100} Bufo bufo (true toad) 15 ug/l/24 hr /Conditions of bioassay not specified LC_{50} Lepomis macrochirus (Bluegill) 1.2 ug/l/96 hr at 18 °C (95% confidence limit 0.9-1.7 ug/l), wt 1.0 g, static bioassay LC_{50} Mugil cephalis (mullet) 0.38 ug/l/96 hr in a flow through bioassay LC_{50} Pteronarcys californica (Stone flies) 2.3 ug/l/96 hr at 15 °C (95% confidence limit 1.6-3.3 ug/l), second year class, static bioassay LC_{50} Ischura sp (insect) naiads 235 ug/l/24 hr at 8 °C; 120 ug/l/48 hr at 8 °C; 62 ug/l/120 hr at 8 °C

Environmental Fate: Release to soil will most likely result in biodegradation and in hydrolysis, especially under alkaline conditions. On the soil surface it may photolyze. Volatilization and leaching are not expected to be significant due to the high estimated soil-sorption coefficients of the isomers. When released to water, isomers are expected to hydrolyze readily under alkaline conditions, and more slowly at neutral and acidic pH values (alpha half-lives = 35.4 and 150.6 days for pH 7 and 5.5, respectively; beta half-lives = 37.5 and 187.3 days for pH 7 and 5.5 respectively). Volatilization and biodegradation are also expected to be significant. Photolysis and oxidation may also be important. Bioconcentration is expected to be significant. Released to the atmosphere it will react with photochemically generated hydroxyl radicals with an estimated half-life of 1.23 hr. Bioconcentration is expected to be significant.

Cleanup/Disposal: Guide No. 151: Do not touch damaged containers or spilled material unless wearing appropriate protective clothing. Stop leak if you can do it without risk. Prevent entry into waterways, sewers, basements or confined areas. Cover with plastic sheet to prevent spreading. Absorb or cover with dry earth, sand or other non-combustible material and transfer to containers. Do not get water inside containers.

Environmental Physical Data

Henry's Law Constant: 1.04 x10^{-5}

Octanol/Water Partition Coefficient: log K_{ow} = 3.55

Sorption Partition Coefficient: K_{oc} = estimated at 3.46

BCF: grass shrimp concentration factors 81 to 245

Regulations

RCRA 40CFR: Listed Hazardous Waste No. P050 Toxic Waste

CERCLA: 40CFR 302.4: Listed per CWA Section 311(b)(4) per RCRA Section 3001 per CWA Section 307(a) RQ: 1 lb (0.454 kg)

SARA 40CFR 372.65: Not listed TPQ: 10/10000 lb

SARA EHS 40CFR 355: Listed TPQ: 1 lb

TSCA: Not listed

Analytical Methods

Soil: EPA PMD-TLC

Water / Groundwater: EPA 022; ASTM D3086

Food: FDA 212.1, 212.2, 232.1, 232.4, 242.1; AOAC 976.23, 983.08

Plasma: FDA 231.1, 252

END4420 **CAS #: 1031-07-8**

ENDOSULFAN SULFATE

RTECS: RB9150000

Molecular Formula: $C_9H_6Cl_6O_4S$

Formula Weight: 422.95

Synonyms: 6,7,8,9,10,10-HEXACHLORO-1,5,5A,6,9,9A-HEXAHYDRO-6,9-METHANO-2,4,3-BENZODIOXATHIEPIN-3,3-DIOXIDE; 6,9-METHANO-2,4,3-BENZODIOXATHIEPIN,6,7,8,9,10,10-HEXACHLORO-1,5,5A,6,9,9A-HEXAHYDRO-,3,3-DIOXIDE

Use: not commercially produced

Physical Properties

Freezing Point: 181 °C (357.8 °F)

Saturated Vapor Density: 1.200000214 kg/m^3

Vapor Pressure: 1 x10^{-5} mm Hg at 25 °C

Water Solubility: 0.22 mg/L tap Water (pH 7.2) at 22 °C

RTECS Toxicity Data

Acute Oral: Rat LD_{50} Dose: 18 mg/kg.

Hazard Overviews

Carcinogenicity: IARC - Not listed; NIOSH - Not listed; NTP - Not listed; ACGIH - Not listed; OSHA - Not listed; EPA - Not listed; MAK - Not listed

Environmental

Ecotoxicity: LC_{50} Aedes aegypti (mosquito) 150 ug/l/48 hr LC_{50} Daphnia magna (water flea) 140 ug/l/48 hr /Conditions of bioassay not specified LC_{50} Lebistes reticulatus (guppy) 1.6 ug/l/48 hr /Conditions of bioassay not specified

Environmental Fate: If released to the soil, it will be expected to bind to the soil, and will not be expected to leach to the groundwater. No information about hydrolysis in soils was found; however, this may be an important fate process based on reported hydrolysis half-lives for endosulfan isomers of 35.4-37.5 days at pH 7.0 and 150.6-187.3 days at pH 5.5. Biodegradation may be an important fate process with a half-life of 11 weeks incubated with mixed cultures from a sandy loam soil. If released to water, it will be expected to adsorb to the sediment and may bioconcentrate in aquatic organisms. Photolysis may not be an important fate process based on the stability of thin films exposed to light >300 nm. It was not degraded in standard screening tests using settled domestic wastewater as inoculum. However, it has been reported to be biodegraded when exposed to mixed cultures from a sandy loam soil. Hydrolysis in water may be an important fate process based on reported hydrolysis half-lives for endosulfan isomers of 35.4-37.5 days at pH 7.0 and 150.6-187.3 days at pH 5.5. Evaporation from water may be an important transport process based on an estimated half-life of 43 hr for evaporation from a river 1 m deep, flowing 1 m/sec with a wind velocity of 3 m/sec. Evaporation from lakes and deeper streams and rivers will be slower and adsorption to sediments will slow evaporation. If released to the atmosphere, it will react with hydroxyl radicals with a resulting estimated vapor phase half-life in the atmosphere of 1.23 hr. Photolysis may not be an important fate process based on the stability of thin films to light >300 nm.

Cleanup/Disposal: Guide No. 151: Do not touch damaged containers or spilled material unless wearing appropriate protective clothing. Stop leak if you can do it without risk. Prevent entry into waterways, sewers, basements or confined areas. Cover with plastic sheet to prevent spreading. Absorb or cover with dry earth, sand or other non-combustible material and transfer to containers. Do not get water inside containers.

Environmental Physical Data

Henry's Law Constant: calculated at 4.7×10^{-4}

Octanol/Water Partition Coefficient: log K_{ow} = 3.66

Sorption Partition Coefficient: K_{oc} = estimated at 2330 to 1.42×10^{4}

BCF: fish 935 to 1741

Regulations

RCRA 40CFR: Not listed

CERCLA: 40CFR 302.4: Listed per CWA Section 307(a) RQ: 1 lb (0.454 kg)

SARA 40CFR 372.65: Not listed

SARA EHS 40CFR 355: Not listed

TSCA: Not listed

Analytical Methods

Soil: CLP LC_PEST, MC_PEST, OHC; SW846 3630B, 3640A, 8080A, 8081, 8081A, 8250A, 8270B, 8270C; EPA 16, 3, P-002-1, P-011-1; USGS O5104, O7104

Water / Groundwater: EPA P-003-1, P-004-1, 608, 617, 625, 625-S, 680; APHA 6410-B, 6630-C, 6630-D

Drinking Water: EPA 508, 508.1, 525.2; AOAC 990.06

Food: FDA 212.1, 212.2, 232.1, 232.4, 242.1; EPA 4; AOAC 976.23

Plasma: EPA 001, 028, 29; FDA 211.1, 231.1, 252, 253

Other: EPA 1656

END5560 CAS #: 145-73-3

ENDOTHALL

RTECS: RN7875000

EINECS Number: 205-660-5

Molecular Formula: $C_8H_{10}O_5$

Structured MF: $O(COONa)_2$

Formula Weight: 186.18

Synonyms: ACCELERATE; AQUATHOL; AQUATHOL PLUS; AQUATHOL PLUS GRANULAR; 1,2-BENZENEDICARBOXYLIC ACID,HEXAHYDRO-3,6-ENDO-OXY-; 1,2-CYCLOHEXANEDICARBOXYLIC ACID,3,6-ENDO-EPOXY-; DES-I-CATE; 3,6-ENDOOXOHEXAHYDROPHTHALIC ACID; 3,6-ENDOOXYPHTHALIC ACID,HEXAHYDRO-; ENDOTHAL; ENDOTHAL TECHNICAL; ENDOTHALL TURF HERBICIDE; ENDOTHALL WEED KILLER; 3,6-ENDOXOHEXAHYDROPHTHALIC ACID; 3,6-EPOXYCYCLOHEXANE-1,2-DICARBOXYLIC ACID; HERBICIDE 273; HERBON PENNOUT; HYDOUT; HYDROTHOL; 7-OXABICYCLO(2.2.1)HEPTANE-2,3-DICARBOXYLIC ACID; PHTHALIC ACID,HEXAHYDRO-3,6-ENDO-OXY-; TRI-ENDOTHAL

Description: crystalline, white solid; odorless

Use: pre- & post-emergence herbicide, defoliant, desiccant, aquatic algicide & growth regulator; desiccant on lucerne & potato, for defoliation of cotton & to control algae & aquatic weeds; for control of weeds in red beet, spinach & sugar beet, turf, hops sucker suppression; alfalfa, clover desiccants; cotton harvest aids; potato vine killers; can be in combination with propham.

Physical Properties

Freezing Point: Decomposes at ~ 144 °C (291.2 °F)

Specific Gravity: 1.431

Vapor Pressure: Very low at room temperature

Water Solubility: 10 g/100g at 20 °C

Other Solubilities: solubility at 20 °C, g/100 g: Acetone 7.0, Benzene 0.01, Dioxane 7.6, Ether 0.1, Isopropanol 1.7, Methanol 28.0.

Flash Point: Nonflammable

RTECS Toxicity Data

Acute Oral: Man LD_{Lo} Dose: 571 uL/kg; Toxic Effects: Cardiac - Other changes; Kidney, Ureter, and Bladder - Urine volume decreased; Nutritional and gross metabolic - Metabolic acidosis. Rat LD_{50} Dose: 38 mg/kg.

Acute Dermal: Rat LD_{50} Route: Skin; Dose: >1 gm/kg.

Hazard Overviews

Fire: Noncombustible.

Carcinogenicity: IARC - Not listed; NIOSH - Not listed; NTP - Not listed; ACGIH - Not listed; OSHA - Not listed; EPA - Not listed; MAK - Not listed

Environmental

Ecotoxicity: LC_{50} Salmo gairdneri (Rainbow trout) 0.14 mg/l/96 hr at 13 °C (95% confidence limit 0.08-0.24 mg/l), wt 1.2 g / static bioassay without aeration, pH 7.2-7.5, water hardness 40-50 mg/l as calcium carbonate LC_{50} Pteronarcys (Stoneflies) 0.05 mg/l/96 hr at 15 °C, second yr class / static bioassay without aeration, pH 7.2-7.5, water hardness 40-50 mg/l as calcium carbonate and alkalinity of 30-35 mg/l LC_{50} Morone saxatilis (striped bass fingerlings) 710 ppm/96 hr LC_{50} Oncorhunchus kisutch (Coho salmon) more than 100 mg/l/96 hr at 13 °C, wt 1.4 g / static bioassay without aeration, pH 7.2-7.5, water hardness 40-50 mg/l as calcium carbonate and alkalinity of 30-35 mg/l LC_{50} Chinook salmon 5 mg/l/14 days

Environmental Fate: If released to soil, it is expected to rapidly biodegrade under aerobic conditions. The half-life in soil is reported to be 4 to 9 days. It should be highly mobile in soil; however, rapid degradation would limit the extent of leaching. Chemical hydrolysis and volatilization are not expected to be significant. If released to water, it should rapidly biodegrade under aerobic conditions (half-life approx. 1 week or less) and biodegrade more slowly under anaerobic conditions. Glutamic acid is a major biotransformation product under aerobic conditions. It is not expected to oxidize, chemically hydrolyze, photolyze, volatilize, bioaccumulate or adsorb to suspended solids or sediments in water. If released to the atmosphere, it is expected to exist predominantly on particles and should either settle out or wash out in precipitation. It is not expected to chemically react or photolyze in the atmosphere.

Environmental Physical Data

Sorption Partition Coefficient: K_{oc} = sediments < 2
BCF: bluegill 1

Regulations

RCRA 40CFR: Listed Hazardous Waste No. P088 Toxic Waste
CERCLA: 40CFR 302.4: Listed per RCRA Section 3001 RQ: 1000 lb (453.5 kg)
SARA 40CFR 372.65: Not listed
SARA EHS 40CFR 355: Not listed
TSCA: Not listed

Analytical Methods

Soil: EPA PMD-ENB
Drinking Water: EPA 548, 548.1

END6700 **CAS #: 2778-04-3**

ENDOTHION

RTECS: TF8225000
DOT: UN2783
EINECS Number: 220-472-3
Molecular Formula: $C_9H_{13}O_6PS$
Formula Weight: 280.25

Synonyms: AC-18,737; O,O-DIMETHYL S-(5-METHOXY-4-OXO-4H-PYRAN-2-YL)PHOSPHOROTHIOATE; O,O-DIMETHYL S-(5-METHOXYPYRONYL-2-METHYL) THIOPHOSPHATE; O,O-DIMETHYL PHOSPHOROTHIOATE S-ESTER WITH2-(MERCAPTOMETHYL)-5-METHOXY-4H-PYRAN-4-ONE; O,O-DIMETHYL-S-((5-METHOXY-PYRON-2-YL)-METHYL)-THIOLPHOSPHAT; ENDOCID; ENDOCIDE; ENT 24,653; EXOTHION; FMC 5767; 5-METHOXY-2-(DIMETHOXYPHOSPHINYLTHIOMETHYL) PYRONE-4; 5-METHOXY-2-(DIMETHOXYPHOSPHINYLTHIOMETHYL)PYRONE-4; S-((5-METHOXY-4-OXO-4H-PYRAN-2-YL)METHYL) O,O-DIMETHYLPHOSPHOROTHIOATE; S-5-METHOXY-4-OXOPYRAN-2-YLMETHYL DIMETHYLPHOSPHOROTHIOATE; S-(5-METHOXY-4-PYRON-2-YLMETHYL) DIMETHYL PHOSPHOROTHIOATE; S-(5-METHOXY-4-PYRON-2-YLMETHYL) DIMETHYLPHOSPHOROTHIOLATE; S-5-METHOXY-4-PYRON-2-YLMETHYL OO-DIMETHYLPHOSPHOROTHIOATE; S-((5-METHOXY-4H-PYRON-2-YL)-METHYL)-O,O-DIMETHYL-MONOTHIOFOSFAAT; S-((5-METHOXY-4H-PYRON-2-YL)-METHYL)-O,O-DIMETHYL-MONOTHIOPHOSPHAT; S-((5-METOSSI-4H-PIRON-2-IL)-METIL)-O,O-DIMETIL-MONOTIOFOSFATO; NIA-5767; NIAGARA 5767; PHOSPHATE 100; PHOSPHOPYRON; PHOSPHOPYRONE; PHOSPHOROTHIOIC ACID,O,O-DIMETHYL ESTER,S-ESTER WITH2-(MERCAPTOMETHYL)-5-METHOXY-4H-PYRAN-4-ONE; PHOSPHOROTHIOIC ACID,S-((5-METHOXY-4-OXO-4H-PYRAN-2-YL)METHYL) O,O-DIMETHYLESTER; 7175 RP; THIOPHOSPHATE DE O,O-DIMETHYLE ET DE S-((5-METHOXY-4-PYRONYL)-METHYLE); THIOPHOSPHATE DE O,O-DIMETHYLE ET DES-((5-METHOXY-4-PYRONYL)-METHYLE)

Description: white crystals; slight odor

Use: systemic aphicide, acaricide; for the control of sap-feeding insects and mites of orchard, field, and market garden crops

Physical Properties

Freezing Point: 96 °C (204.8 °F)
Water Solubility: 150 g Soluble in 100 ml Water at 20 °C
Other Solubilities: Soluble in Acetone, Benzene; Insoluble in Ether or Carbon Tetrachloride; Soluble in Ethanol.

RTECS Toxicity Data

Acute Oral: Rat LD_{50} Dose: 23 mg/kg.

Hazard Overviews

Fire Diamond

Carcinogenicity: IARC - Not listed; NIOSH - Not listed; NTP - Not listed; ACGIH - Not listed; OSHA - Not listed; EPA - Not listed; MAK - Not listed

Environmental

Cleanup/Disposal: Guide No. 152: Do not touch damaged containers or spilled material unless wearing appropriate protective clothing. Stop leak if you can do it without risk. Prevent entry into waterways, sewers, basements or confined areas. Cover with plastic sheet to prevent spreading. Absorb or cover with dry earth, sand or other non-combustible material and transfer to containers. Do not get water inside containers.

Regulations

RCRA 40CFR: Not listed
CERCLA: 40CFR 302.4: Not listed
SARA 40CFR 372.65: Not listed TPQ: 500/10000 lb
SARA EHS 40CFR 355: Listed TPQ: 500 lb
TSCA: Not listed

END7840 CAS #: 72-20-8

ENDRIN

RTECS: IO1575000
DOT: UN2761; UN2995; UN2996; NA2761; IMO6.1
EINECS Number: 200-775-7
Molecular Formula: $C_{12}H_8Cl_6O$
Formula Weight: 380.93
Synonyms: COMPOUND 269; 5,8-DIMETHANONAPHTHALENE; 1,4:5,8-DIMETHANONAPHTHALENE,1,2,3,4,10,10-HEXACHLORO-6,7-EPOXY-1,4,4A,5,6,7,8,8A-OCTAHYDRO-ENDO,ENDO-; 2,7:3,6-DIMETHANONAPHTH(2,3-B)OXIRENE,3,4,5,6,9,9-HEXACHLORO-1A,2,2A,3,6,6A,7,7A-OCTAHYDRO-,(1AALPHA,2BETA,2ABETA,3ALPHA,6ALPHA,6ABETA,7BETA,7AALPHA)-; EN 57; ENDREX; ENDRICOL; ENDRINE; ENT 17,251; ENT 17251; EXPERIMENTAL INSECTICIDE 269; 3,4,5,6,9,9-HEXACHLORO-1AALPHA,2BETA,2ABETA,3ALPHA,6ALPHA,6ABETA,7BETA,7AALPHA-OCTAHYDRO-2,7:3,6-DIMETHANONAPHTH[2,3-B]OXIRENE; 1,2,3,4,10,10-HEXACHLORO-6,7-EPOXY-1,4,4A,5,6,7,8,8A-OCTAHYDRO-1 ,4-ENDO,ENDO-5,8-DIMETHANONAPHTHALENE; 1,2,3,4,10,10-HEXACHLORO-6,7-EPOXY,1,4,4A,5,6,7,8,8A-OCTAHYDRO-1,4-ENDO-5,8-DIMETHANONAPHTHALENE; 1,2,3,4,10,10-HEXACHLORO-6,7-EPOXY-1,4,4A,5,6,7,8,8A-OCTAHYDRO-1,4-ENDO,ENDO; 1,2,3,4,10,10-HEXACHLORO-6,7-EPOXY-1,4,4A,5,6,7,8,8A-OCTAHYDRO-1,4-ENDO,ENDO-5,8-DIMETHANONAPHTHALENE; 1,2,3,4,10,10-HEXACHLORO-6,7-EPOXY-1,4,4A,5,6,7,8,8A-OCTAHYDRO-ENDO,ENDO-1,4:5,8-DIMETHANONAPHTHALENE; HEXACHLOROEPOXYOCTAHYDRO-ENDO,ENDO-DIMETHANONAPHTHALENE; 3,4,5,6,9,9,-HEXACHLORO-1A,2,2A,3,6,6A,7,7A-OCTAHYDRO-2,7:3,6-DIMETHANO-; 3,4,5,6,9,9-HEXACHLORO-1A,2,2A,3,6,6A,7,7A-OCTAHYDRO-2,7:3,6-DIMETHANO-; (1A ALPHA,2BETA,2A BETA,3ALPHA,6ALPHA,6A BETA,7BETA,7A ALPHA)-3,4,5,6,9,9-HEXACHLORO-1A,2,2A,3,6,6A,7,7A-OCTAHYDRO-2,7:3,6-DIMETHANONAPHTH[2,3-B]OXIRENE; 3,4,5,6,9,9-HEXACHLORO-1A,2,2A,3,6,6A,7,7A-OCTAHYDRO-2,7:3,6-DIMETHANONAPHTH(2,3-B)OXIRENE; 3,4,5,6,9,9-HEXACHLORO-1A,2,2A,3,6,6A,7,7A-OCTAHYDRO-2,7:3,6-DIMETHANONAPHTH[2,3-B]OXIRENE; (1R,4S,4AS,5S,6S,7R,8R,8AR)-1,2,3,4,10,10-HEXACHLORO-1,4,4A,5,6,7,8,8A-OCTAHYDRO-6,7-EPOXY-1,4:5,8-DIMETHANONAPHTHALENE; 1,2,3,4,10,10-HEXACHLORO-1R,4S,4AS,5S,6,7R,8R,8AR-OCTAHYDRO-6,7-EPOXY-1,4:5,8-DIMETHANONAPHTHALENE; HEXADRIN; LATKA 269; MENDRIN; NAPH[2,3-B]OXIRENE; NAPHTH(2,3-B)OXIRENE; NENDRIN; OKTANEX; OMS 197; SD 3419
Description: colorless to white to tan, crystalline solid; odorless or mild, chemical odor
Use: as an insecticide, avicide, rodenticide and pesticide

Physical Properties

Boiling Point: Decomposes
Freezing Point: < 392 °C (737.6 °F)
Specific Gravity: 1.7 at 20 °C
Saturated Vapor Density: 1.200000004 kg/m^3
Vapor Pressure: 2 $x10^{-7}$ mm Hg at 25 °C
Water Solubility: Insoluble
Other Solubilities: Insoluble in Methanol.
Odor Threshold: 1.80 $x10^{-2}$ ppm
Flash Point: Nonflammable solid or combustible solution
Autoignition Temperature: 454 to 482 °C
LEL: 1.1% v/v
UEL: 7% v/v

RTECS Toxicity Data

Acute Oral: Man LD_{Lo} Dose: 171 mg/kg; Toxic Effects: Behavioral - Convulsions or effect on seizure threshold; Kidney, Ureter, and Bladder - Chgs in tubules (inc acute renal failure, acute tubular necrosis; Blood - Thrombocytopenia. Rat LD_{50} Dose: 3 mg/kg. Monkey LD_{50} Dose: 3 mg/kg.
Acute Dermal: Rabbit LD_{50} Route: Skin; Dose: 60 mg/kg. Rat LD_{50} Route: Skin; Dose: 12 mg/kg.
Chronic (Multiple Dose) Oral: Rat Dose: 1975 mg/kg/79D-I; Toxic Effects: Sense organs and special senses - Hemorrage; Behavioral - Excitment; DEATH. Rat Dose: 219 mg/kg/2Y-C; Toxic Effects: Liver - Changes in liver weight; Nutritional and gross metabolic - Other changes.
Chronic (Multiple Dose) Inhalation: Mouse Dose: 360 ppb/7H/22W-I; Toxic Effects: DEATH. Rabbit Dose: 360 ppb/7H/24W-I; Toxic Effects: DEATH.
Chronic (Multiple Dose) Dermal: Rabbit Route: Skin; Dose: 1400 mg/kg/14W-I; Toxic Effects: DEATH.
Reproductive/Teratogenic: Rat Route: Oral; Dose: 2320 ug/kg; Duration: female 4D prior to mating; Effects on Fertility - Pre-implantation mortality; Litter size; Specific Developmental Abnormalities - Musculoskeletal system. Rat Route: Intratesticular; Dose: 10 mg/kg; Duration: male 10D prior to mating; Paternal Effects - Spermatogenesis.
Mutagenic: Rat Cytogenetic Analysis; Route: Parenteral; Dose: 1 mg/kg. Rat Sperm Morphology; Route: Parenteral; Dose: 10 mg/kg/10D-C.

Hazard Overviews

Explosive

Carcinogenicity: IARC - Group 3, Not classifiable as to carcinogenicity to humans; NIOSH - Not listed; NTP - Not listed; ACGIH - Class A4, Not classifiable as a human carcinogen; OSHA - Not listed; EPA - Class D, Not classifiable as to human carcinogenicity; MAK - Not listed
Exposure Limits
OSHA PEL: TWA: 0.1 mg/m^3; skin.
ACGIH TLV: TWA: 0.1 mg/m^3.
NIOSH REL: TWA: 0.1 mg/m^3.
NIOSH IDLH: 2 mg/m^3.
DFG MAK: TWA: 0.1 mg/m^3.
Respirator Recommendation
Exposure Range: >0.1 to <2 mg/m^3 Air Purifying, Negative Pressure, Half Mask

Exposure Range: 2 to unlimited mg/m^3 Self-contained Breathing Apparatus, Pressure Demand, Full Face
Cartridge Color: black with magenta (P100)

Environmental

Ecotoxicity: LD_{50} Callipepla Californica (California quail) female oral 1.19 mg/kg (95% Confidence limit 0.857-1.65 mg/kg) 9-10 mo old LC_{50} Bobwhite quail oral 14 ppm (95% confidence limit 11-24 ppm), age 17 days LC_{50} Perca flavescens (Yellow perch) 0.15 ug/l/96 hr at 12 °C (95% confidence limit 0.12-0.18 ug/l) fingerling / /flow-through toxicity test LD_{50} Tympanuchus phasianellus (Sharp-tailed grouse) female oral 1.06 mg/kg (95% Confidence limit) LC_{50} Asellus brevicaudus (Sowbug) 1.5 ug/l/96 hr at 15 °C (95% confidence limit 0.9-3.7 ug/l), mature. static bioassay without aeration, pH 7.2-7.5, water hardness 40-50 mg/l as calcium carbonate and alkalinity of 30-35 mg/l LC_{50} Orconectes nais (Crayfish) 3.2 ug/l/96 hr at 21 °C (95% confidence limit 1.6-7.5 ug/l), early instar. static bioassay without aeration, pH 7.2-7.5, water hardness 40-50 mg/l as calcium carbonate and alkalinity of 30-35 mg/l LC_{50} Claassenia sabulosa (Stone fly) 0.08 ug/l/96 hr at 15 °C (95% confidence limit 0.06-0.09 ug/l), 2nd year class. static bioassay without aeration, pH 7.2-7.5, water hardness 40-50 mg/l as calcium carbonate and alkalinity of 30-35 mg/l LC_{50} Ophiocephalus punctatus (fish) 0.033 ppm/96 hr /Conditions of bioassay not specified LD_{50} Odocoileus Hemionus hemionus (Mule deer) female oral 6.25-12.5 mg/kg, 10 mo old LC_{50} Pagurus longicarpus (Hermit crab) 12 ppb/96 hr, static lab bioassay LC_{50} Mephales promela (Fathead minnow) 1.8 ug/l/96 hr at 18 °C (95% confidence limit 1.0-3.0 ug/l), wt 1.2 g. static bioassay without aeration, pH 7.2-7.5, water hardness 40-50 mg/l as calcium carbonate and alkalinity of 30-35 mg/l
Environmental Fate: It is very persistent, but it is known to photodegrade to delta-ketoendrin (half-life 7 days - June). Released to soil it will persist for long periods (up to 14 years or more), will reach the air either through very slow evaporation or adsorption on dust particles, will not leach to groundwater, and will reach surface water with surface runoff. Once endrin reaches surface waters it will adsorb strongly to sediments, bioconcentrate in fish, and photodegrade. Biodegradation will not be an important process. Fate in the atmosphere is unknown, but it probably will be primarily associated with particulate matter and be removed mainly by rainout and dry deposition.
Cleanup/Disposal: Guide No. 131: Fully encapsulating, vapor protective clothing should be worn for spills and leaks with no fire. Eliminate all ignition sources (no smoking, flares, sparks or flames in immediate area). All equipment used when handling the product must be grounded. Do not touch or walk through spilled material. Stop leak if you can do it without risk. Prevent entry into waterways, sewers, basements or confined areas. A vapor suppressing foam may be used to reduce vapors. Small Spills: Absorb with earth, sand or other non-combustible material and transfer to containers for later disposal. Use clean non-sparking tools to collect absorbed material. Large Spills: Dike far ahead of liquid spill for later disposal. Water spray may reduce vapor; but may not prevent ignition in closed spaces. Guide No. 151: Do not touch damaged containers or spilled material unless wearing appropriate protective clothing. Stop leak if you can do it without risk. Prevent entry into waterways, sewers, basements or confined areas. Cover with plastic sheet to prevent spreading. Absorb or cover with dry earth, sand or other non-combustible material and transfer to containers. Do not get water inside containers.

Environmental Physical Data

Henry's Law Constant: estimated at 4×10^{-7}
Octanol/Water Partition Coefficient: log K_{ow} = calculated at 5.6
Sorption Partition Coefficient: K_{oc} = estimated at 3.4×10^{4}
BCF: snails 1150

Regulations

RCRA 40CFR: Listed Hazardous Waste No. P051 Toxic Waste
CERCLA: 40CFR 302.4: Listed per CWA Section 311(b)(4) per RCRA Section 3001 per CWA Section 307(a) RQ: 1 lb (0.454 kg)
SARA 40CFR 372.65: Not listed TPQ: 500/10000 lb
SARA EHS 40CFR 355: Listed TPQ: 1 lb
TSCA: Not listed

Analytical Methods

Air: EPA TO-10
Soil: CLP LC_PEST, MC_PEST, OHC; EPA PMD-TLC, 16, 3, 025, P-002-1, P-011-1; SW846 3630B, 3640A, 8080A, 8081, 8081A, 8250A, 8270B, 8270C; DOE OP130R; USGS O5104, O7104
Water / Groundwater: EPA P-003-1, P-004-1, 608, 617, 625, 625-S, 680, 022; APHA 6410-B, 6630-B, 6630-C, 6630-D; ASTM D3086; USGS O3104
Drinking Water: EPA 505, 508, 508.1, 525.1, 525.2; AOAC 990.06; ASTM D5175
Food: FDA 212.1, 212.2, 232.1, 232.4, 242.1; EPA 4, XENO; AOAC 961.05, 970.52, 972.05; USGS O9104
Indoor / Expired Air: NIOSH 5519
Plasma: EPA 001, 003, 004, 027, 028, 29; FDA 211.1, 231.1, 251.1, 252, 253
Other: EPA P-009-1, 1656

END8980 **CAS #: 7421-93-4**

ENDRIN ALDEHYDE

DOT: UN2761; UN2995; UN2996; IMO6.1
Molecular Formula: $C_{12}H_8Cl_6O$
Formula Weight: 380.89
Synonyms: 1,2,4-METHENECYCLOPENTA(C,D)PENTALENE-R-CARBOXALDEHYDE,2,2A,3,3,4,7-HEXACHLORODECAHYDRO; 1,2,4-METHENOCYCLOPENTA(CD)PENTALENE-5-CARBOXALDEHYDE,2,2A,3,3,4,7-HEXACHLORODECAHYDRO-,(1ALPHA,2BETA,2ABETA,4BETA,4ABETA,5BETA,6ABETA,6BBETA,7R*)-; SD 7442
Use: not commercially used but occurs as an impurity of endrin

Physical Properties

Freezing Point: 235 °C (455 °F)
Saturated Vapor Density: 1.200000004 kg/m^3
Vapor Pressure: 2 x10^{-7} torr at 25 °C
Water Solubility: 0.25 ppm at 25 °C
Flash Point: Will not burn

Hazard Overviews

Fire Diamond

Fire: Noncombustible.
Carcinogenicity: IARC - Not listed; NIOSH - Not listed; NTP - Not listed; ACGIH - Not listed; OSHA - Not listed; EPA - Not listed; MAK - Not listed

Environmental

Environmental Fate: If released to soil, it is not expected to leach in most soil types based on estimated K_{oc} values of 8500 to 45000. Additional data were not available to predict the environmental fate in soil. If released to water, adsorption to sediments and bioconcentration appears to be important fate processes. Aquatic hydrolysis, oxidation (via peroxy radicals or singlet oxygen), or volatilization are not expected to be significant. No data are available to predict biodegradability. If released to the atmosphere, it may exist predominantly in the adsorbed-phase. Particulates will be subject to wet and dry deposition.
Cleanup/Disposal: Guide No. 131: Fully encapsulating, vapor protective clothing should be worn for spills and leaks with no fire. Eliminate all ignition sources (no smoking, flares, sparks or flames in immediate area). All equipment used when handling the product must be grounded. Do not touch or walk through spilled material. Stop leak if you can do it without risk. Prevent entry into waterways, sewers, basements or confined areas. A vapor suppressing foam may be used to reduce vapors. Small Spills: Absorb with earth, sand or other non-combustible material and transfer to containers for later disposal. Use clean non-sparking tools to collect absorbed material. Large Spills: Dike far ahead of liquid spill for later disposal. Water spray may reduce vapor; but may not prevent ignition in closed spaces. Guide No. 151: Do not touch damaged containers or spilled material unless wearing appropriate protective clothing. Stop leak if you can do it without risk. Prevent entry into waterways, sewers, basements or confined areas. Cover with plastic sheet to prevent spreading. Absorb or cover with dry earth, sand or other non-combustible material and transfer to containers. Do not get water inside containers.

Environmental Physical Data

Henry's Law Constant: 2.9 x10^{-9}
Octanol/Water Partition Coefficient: log K_{ow} = 5.34
Sorption Partition Coefficient: K_{oc} = estimated at 8500 to 4.5 x10^4
BCF: calculated at 2200

Regulations

RCRA 40CFR: Not listed
CERCLA: 40CFR 302.4: Listed per CWA Section 307(a) RQ: 1 lb (0.454 kg)
SARA 40CFR 372.65: Not listed
SARA EHS 40CFR 355: Not listed
TSCA: Not listed

Analytical Methods

Air: EPA TO-10
Soil: CLP LC_PEST, MC_PEST; SW846 3630B, 3640A, 8080A, 8081, 8081A, 8250A, 8270B, 8270C; EPA 16, 3
Water / Groundwater: EPA 608, 617, 625, 680, 6; APHA 6410-B, 6630-C, 6630-D
Drinking Water: EPA 508, 508.1, 525.2; AOAC 990.06
Food: FDA 212.1, 232.1; EPA 4
Plasma: EPA 001, 027, 028, 29; FDA 211.1, 231.1, 252
Other: EPA P-009-1, 1656

EPH5000 CAS #: 299-42-3

(L)-EPHEDRINE

RTECS: KB0700000
EINECS Number: 206-080-5
Molecular Formula: $C_{10}H_{15}NO$
Formula Weight: 165.23

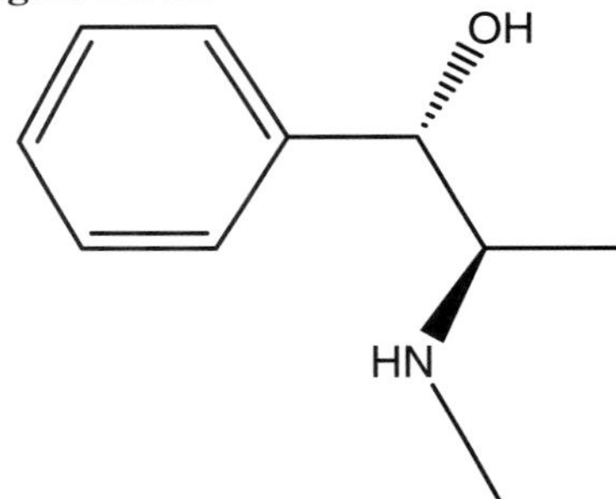

Chemical Structure

Synonyms: BENZENEMETHANOL,ALPHA-(1-(METHYLAMINO)ETHYL)-,(R-(R*,S*))-; BIOPHEDRIN; ECIPHIN; EFEDRIN; EPHEDRAL; EPHEDRATE; EPHEDREMAL; EPHEDRIN; (-)-EPHEDRINE; EPHEDRINE; EPHEDRINE,(-)-; L(-)-EPHEDRINE; L-EPHEDRINE; EPHEDRITAL; EPHEDROL; EPHEDROSAN; EPHEDROTAL; EPHEDSOL; EPHENDRONAL; EPHOXAMIN; FEDRIN; ALPHA-HYDROXY-BETA-METHYL AMINE PROPYLBENZENE; 1-HYDROXY-2-METHYLAMINO-1-PHENYLPROPANE; ALPHA-HYDROXY-BETA-METHYLAMINOPROPYLBENZENE; ISOFEDROL; KRATEDYN; MANADRIN; MANDRIN; ALPHA-[1-(METHYLAMINO)ETHYL]BENZENEMETHANOL; (-)-ALPHA-(1-METHYLAMINOETHYL)BENZYL ALCOHOL; 1-ALPHA-(1-METHYLAMINOETHYL)BENZYL ALCOHOL; ALPHA[1-(METHYLAMINO)ETHYL]BENZYL ALCOHOL; L-ALPHA-(1-METHYLAMINOETHYL)BENZYL ALCOHOL; 1-2-METHYLAMINO-1-PHENYLPROPANOL; 2-METHYLAMINO-1-PHENYL-1-PROPANOL; L-2-METHYLAMINO-1-PHENYLPROPANOL; NASOL; NOREPHEDRINE,N-METHYL-; 1-PHENYL-1-HYDROXY-2-METHYLAMINOPROPANE; 1-PHENYL-2-

METHYLAMINOPROPANOL; SANEDRINE; 1-SEDRIN; VENCIPON; ZEPHROL

Description: waxy solid, crystals or granules

Use: medication: adrenergic (bronchodilator); vet: sympathomimetic to counteract hypotension assoc with anesthesia; as mydriatic; in allergic reactions & as CNS stimulant; bronchodilator for treatment of asthma; nasal decongestant for treatment of allergies

Physical Properties

Boiling Point: 255 °C (491 °F)
Freezing Point: 34 °C (93.2 °F)
Specific Gravity: 1.00850
Water Solubility: 1 g dissolves in about 20 ml Water
Other Solubilities: 1 g dissolves in 14 ml Alcohol; Practically Insoluble in Ether, Chloroform /L-form HCL/; 1 g dissolves in 95 ml Alcohol; freely Soluble in hot Alcohol /L-form Sulfate/.
pH: Aqueous solution 1 in 200 is 10.8

RTECS Toxicity Data

Acute Oral: Rat LD_{50} Dose: 600 mg/kg. Mouse LD_{50} Dose: 689 mg/kg.

Acute Dermal: Rabbit LD_{Lo} Route: Subcutaneous Dose: 320 mg/kg; Toxic Effects: Cardiac - Other changes; Lungs, Thorax, or Respiration - Other changes. Rat LD_{50} Route: Subcutaneous Dose: 300 mg/kg.

Reproductive/Teratogenic: Rat Route: Intraperitoneal; Dose: 50 mg/kg; Duration: female 9D of pregnancy; Specific Developmental Abnormalities - Cardiovascular (circulatory) system.

Hazard Overviews

Health: Irritating to eyes/skin/respiratory tract. Toxic. Other Acute Effects: harmful if swallowed, inhaled, or absorbed through skin. Chronic Effects: causes rapid pulse, rise in blood pressure and other actions similar to epinephrine; has been known to cause allergic sensitization; a sympathomimetic agent with direct and indirect effects on adrenergic receptors; target organs: nerves, heart.

Fire: Hazards: emits toxic fumes. Extinguishing agents: water spray; carbon dioxide, dry chemical powder or appropriate foam. Precautions: combustible liquid.

Reactivity: Incompatible with: strong oxidizing agents, strong acids, acid chlorides, acid anhydrides, may discolor on exposure to light. Hazardous decomposition products: toxic fumes of: carbon monoxide, carbon dioxide, nitrogen oxides.

Carcinogenicity: IARC - Not listed; NIOSH - Not listed; NTP - Not listed; ACGIH - Not listed; OSHA - Not listed; EPA - Not listed; MAK - Not listed

Primary Target Organs:

Eyes

Skin

Respiratory System

Nervous System

Cardio-vascular

Environmental

Regulations

RCRA 40CFR: Not listed
CERCLA: 40CFR 302.4: Not listed
SARA 40CFR 372.65: Not listed
SARA EHS 40CFR 355: Not listed
TSCA: Listed

EPI3000 CAS #: 3132-64-7

A-EPIBROMOHYDRIN

RTECS: TX4115000
DOT: UN2558; IMO6.1
EINECS Number: 221-525-3
Molecular Formula: C_3H_5BrO
Formula Weight: 136.98

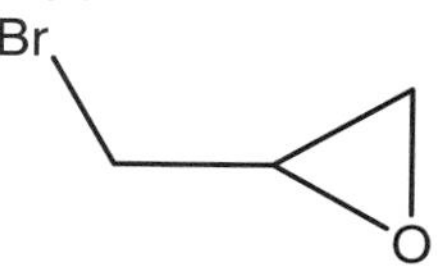

Chemical Structure

Synonyms: 1-BROMO-2,3-EPOXYPROPANE; 3-BROMO-1,2-EPOXYPROPANE; BROMOMETHYL OXIRANE; (BROMOMETHYL)ETHYLENE OXIDE; 2-BROMOMETHYLOXIRAN; (BROMOMETHYL)OXIRANE; 2-(BROMOMETHYL)OXIRANE; 3-BROMOPROPYLENE OXIDE; EPIBROMHIDRINA; EPIBROMHYDRIN; EPIBROMHYDRINE; EPIBROMOHYDRIN; EPIBROMOHYDRINE; 1,2-EPOXY-3-BROMOPROPANE; OXIRANE,(BROMOMETHYL)-; PROPANE,1-BROMO-2,3-EPOXY-; PROPANE,3-BROMO-1,2-EPOXY-

Description: yellow, clear liquid

Use: sporicide; flame retardant

Physical Properties

Boiling Point: 134.6 °C (274 °F)
Freezing Point: -40 °C (-40 °F)
Specific Gravity: 1.601
Saturated Vapor Density: Estimated 1.211758258 kg/m^3
Vapor Pressure: Estimated 2 mm Hg at 25 °C
Water Solubility: < 1 mg/mL at 22 C
Other Solubilities: 95% Ethanol: >=100 mg/ml at 20 °C; Acetone: >=100 mg/ml at 20 °C; Benzene: Soluble; Chloroform: Soluble; DMSO: >=100 mg/ml at 20 °C; Ether: Soluble.
Refraction Index: 1.482
Flash Point: < 22 °C

RTECS Toxicity Data

Mutagenic: Human DNA Inhibition; Cell Type: HeLa cell; Dose: 1500 umol/L. Hamster Sister Chromatid Exchenge; Cell Type: lung; Dose: 50 umol/L.

Hazard Overviews

Poison

Corrosive

Flammable

Health: Corrosive to eyes/skin/respiratory tract. Poison. Other Acute Effects: may be fatal if inhaled, swallowed, or absorbed through skin; lachrymator; inhalation may result in spasm, inflammation and edema of the larynx and bronchi, chemical pneumonitis and pulmonary edema; symptoms of exposure may include burning sensation; coughing; wheezing; laryngitis; shortness of breath; headache; nausea; vomiting; prolonged exposure can cause damage to the lungs, liver, and kidneys. Chronic Effects: laboratory experiments have shown mutagenic effects.
Fire: Flammable. Hazards: container explosion may occur; under fire conditions, material may decompose to form flammable and/or explosive mixtures in air; emits toxic fumes. Extinguishing agents: water spray; carbon dioxide, dry chemical powder or appropriate foam. Precautions: combustible liquid.
Reactivity: Incompatible with: acids, bases, oxidizing agents, sodium, zinc, aluminum, magnesium and their alloys. Hazardous decomposition products: toxic fumes of: carbon monoxide, carbon dioxide, hydrogen bromide gas.
Carcinogenicity: IARC - Not listed; NIOSH - Not listed; NTP - Not listed; ACGIH - Not listed; OSHA - Not listed; EPA - Not listed; MAK - Not listed
Primary Target Organs:

Eyes

Skin

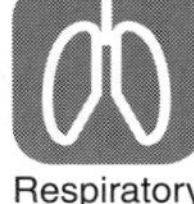
Respiratory System

Environmental

Environmental Fate: Vapor-phase in the ambient atmosphere is expected to degrade by reaction with photochemically produced hydroxyl radicals (estimated half-life of 28 days). If released to soil, adsorption will not be important and leaching can occur. Hydrolysis is expected to be the primary fate process in moist soil and water based on a measured half-life of 16 days in water at 25 °C and pH 7. Hydrolysis related products depend upon the ionic content of the environmental media. 1-Bromo-2,3-propanediol will be a possible product; however, anions such as chloride will yield 1-bromo-3-chloro-2-propanol. If released to water, adsorption from the water column to sediment and suspended materials and bioconcentration in aquatic organisms are not expected to be important aquatic fate mechanisms. A volatilization half-life of 2.5 days has been estimated for a model river (one meter deep) indicating moderate volatilization from shallow rivers.
Cleanup/Disposal: Guide No. 131: Fully encapsulating, vapor protective clothing should be worn for spills and leaks with no fire. Eliminate all ignition sources (no smoking, flares, sparks or flames in immediate area). All equipment used when handling the product must be grounded. Do not touch or walk through spilled material. Stop leak if you can do it without risk. Prevent entry into waterways, sewers, basements or confined areas. A vapor suppressing foam may be used to reduce vapors. Small Spills: Absorb with earth, sand or other non-combustible material and transfer to containers for later disposal. Use clean non-sparking tools to collect absorbed material. Large Spills: Dike far ahead of liquid spill for later disposal. Water spray may reduce vapor; but may not prevent ignition in closed spaces.

Environmental Physical Data

Henry's Law Constant: estimated at 1.84×10^{-5}
Octanol/Water Partition Coefficient: log K_{ow} = estimated at -0.07
Sorption Partition Coefficient: K_{oc} = 22
BCF: estimated at 0.5

Regulations

RCRA 40CFR: Not listed
CERCLA: 40CFR 302.4: Not listed
SARA 40CFR 372.65: Not listed
SARA EHS 40CFR 355: Not listed
TSCA: Listed

EPI6000 CAS #: 106-89-8

EPICHLOROHYDRIN

RTECS: TX4900000
DOT: UN2023; IMO6.1
EINECS Number: 203-439-8
Molecular Formula: C_3H_5ClO
Formula Weight: 92.53

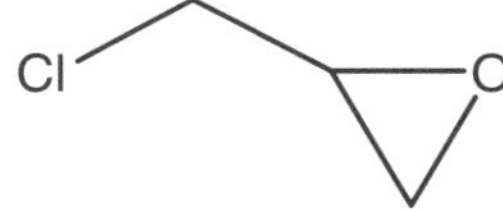

Chemical Structure

Synonyms: 1-CHLOOR-2,3-EPOXY-PROPAAN; 1-CHLOR-2,3-EPOXY-PROPAN; 1-CHLORO-2,3-EPOXYPROPANE; 3-CHLORO-1,2-EPOXYPROPANE; 1-CHLORO-2,3-EPOXYPROPONE; (CHLOROMETHYL)ETHYLENE OXIDE; 2-(CHLOROMETHYL)OXIRANE; CHLOROMETHYLOXIRANE; 3-CHLORO-1,2-PROPANE OXIDE; 3-CHLORO-1,2-PROPYLENE OXIDE; CHLOROPROPYLENE OXIDE; GAMMA-CHLOROPROPYLENE OXIDE; 1-CLORO-2,3-EPOSSIPROPANO; ECH; EPICHLOORHYDRINE; EPICHLORHYDRIN; EPICHLORHYDRINE; (DL)-ALPHA-EPICHLOROHYDRIN; ALPHA-EPICHLOROHYDRIN; EPI-CHLOROHYDRIN; EPICHLOROHYDRYNA; EPICHLOROPHYDRIN; ALPHA-EPICHOROHYDRIN; EPICLORIDRINA; 1,2-EPOXY-3-CHLOROPROPANE; 2,3-EPOXYPROPYL CHLORIDE; GLYCEROL EPICHLORHYDRIN; GLYCEROL EPICHLOROHYDRIN; GLYCIDYL CHLORIDE; OXIRANE,(CHLOROMETHYL)-; OXIRANE,2-(CHLOROMETHYL); SKEKHG
Description: colorless liquid; chloroform-like odor
Use: manufacture of epoxy resins, glycerol and various other intermediates; a solvent for natural and synthetic resins, gums, cellulose esters and ethers, paints, varnishes, nail

enamels and lacquers, and cements for celluloid; in surface-active agents pharmaceuticals, insecticides, agricultural chemicals, textile chemicals, coatings, adhesives, ion-exchange resins, plasticizers, glycidyl esters, ethymyl-ethylenic alcohol and fatty acid derivatives; a stabilizer in chlorine-containing materials and an intermediate in the preparation of condensates with polyfunctional substances

Physical Properties

Boiling Point: 116.5 °C (242 °F)
Freezing Point: -48 °C (-54.4 °F)
Specific Gravity: 1.1801 at 20 °C/4 °C
Vapor Density: 3.29 Air=1
Density: 1.182 g/cu cm at 18.5 °C
Bulk Density: 9.78 lbs/gal
Vapor Pressure: 10 mm Hg at 16.6 °C
Water Solubility: 7% by weight
Other Solubilities: miscible with Alcohol, Ether, Chloroform; miscible with Trichloroethylene, Carbon Tetrachloride; Soluble in Alcohol, Ether and Benzene.
Surface Tension: 37.0 dynes/cm
Odor Threshold: Perception 0.3 mg/m^3
Refraction Index: 1.44195 at 11.6 °C/D
Evaporation Rate: 1.35 Butyl Acetate=1
Ionization Potential (eV): 10.60
Flash Point: 33.889 °C Closed Cup
Autoignition Temperature: 416 °C
LEL: 3.8% v/v
UEL: 21% v/v

RTECS Toxicity Data

Acute Oral: Rat LD_{50} Dose: 90 mg/kg. Mouse LD_{50} Dose: 195 mg/kg; Toxic Effects: Behavioral - Somnolence (general depressed activity); Behavioral - Tremor; Behavioral - Ataxia.
Acute Inhalation: Rat LC_{50} Dose: 250 ppm/8hr. Human TC_{Lo} Dose: 40 ppm/2hr; Toxic Effects: Lungs, Thorax, or Respiration - Other changes. Human TC_{Lo} Dose: 20 ppm; Toxic Effects: Sense organs and special senses - Other; Sense organs and special senses - Other.
Acute Dermal: Rabbit LD_{50} Route: Skin; Dose: 515 mg/kg. Rat LD_{50} Route: Subcutaneous Dose: 150 mg/kg; Toxic Effects: Kidney, Ureter, and Bladder - Urine volume increased.
Chronic (Multiple Dose) Oral: Rat Dose: 1218 mg/kg/9W-I; Toxic Effects: Liver - Changes in liver weight; Endocrine - Changes in adrenal weight; Biochemical - Other proteins. Rat Dose: 910 mg/kg/26W-I; Toxic Effects: Behavioral - Alteration of classical conditioning.
Chronic (Multiple Dose) Inhalation: Rat Dose: 20 mg/m^3/24H/14W-C; Toxic Effects: Brain and coverings - Other degenerative changes; Behavioral - Alteration of classical conditioning; Blood - Other changes. Rat Dose: 170 mg/m^3/3H/17W-C; Toxic Effects: Lungs, Thorax, or Respiration - Structural or functional change in trachea or bronchi; Kidney, Ureter, and Bladder - Chgs in tubules (inc acute renal failure, acute tubular necrosis; DEATH.
Irritation Eye: Rabbit Standard Draize Test Dose: 100 mg/24H; Reaction: moderate. Rabbit Standard Draize Test Dose: 100 mg; Reaction: severe.
Reproductive/Teratogenic: Rat Route: Oral; Dose: 180 mg/kg; Duration: male 12D prior to mating; Effects on Fertility - Male fertility index. Rat Route: Oral; Dose: 25 mg/kg; Duration: male 1D prior to mating; Paternal Effects - Spermatogenesis. Rat Route: Oral; Dose: 1050 mg/kg; Duration: male 21D prior to mating; Paternal Effects - Spermatogenesis; Testes, epididymis, sperm duct; Effects on Fertility - Male fertility index. Rat Route: Oral; Dose: 288 mg/kg; Duration: male 23D prior to mating; Effects on Fertility - Pre-implantation mortality. Rat Route: Inhalation; Dose: 50 ppm/6H; Duration: male 50D prior to mating; Effects on Fertility - Male fertility index.
Mutagenic: Human Unscheduled DNA Synthesis; Cell Type: fibroblast; Dose: 32 ug/L. Human Unscheduled DNA Synthesis; Cell Type: lymphocyte; Dose: 500 umol/L. Human DNA Inhibition; Cell Type: HeLa cell; Dose: 2700 umol/L. Human DNA Inhibition; Cell Type: lymphocyte; Dose: 5 mmol/L. Human Mutations in Mammalian Somatic Cells; Cell Type: other cell types; Dose: 500 umol/L. Human Cytogenetic Analysis; Cell Type: lymphocyte; Dose: 10 umol/L/24H. Human Cytogenetic Analysis; Cell Type: leukocyte; Dose: 1 umol/L. Human Sister Chromatid Exchenge; Cell Type: lymphocyte; Dose: 10 nmol/L.
Tumorigenic: Rat Route: Oral; Dose: 60 gm/kg/81W-I; Toxic Effects: Tumorigenic - Carcinogenic by RTECS criteria; Gastrointestinal - Tumors. Rat Route: Inhalation; Dose: 100 ppm/6H/30D-C; Toxic Effects: Tumorigenic - Carcinogenic by RTECS criteria; Sense organs and special senses - Tumors; Lungs, Thorax, or Respiration - Acute pulmonary edema. Rat Route: Inhalation; Dose: 100 ppm/6H/6W-I; Toxic Effects: Tumorigenic - Equivocal tumorigenic agent by RTECS criteria; Lungs, Thorax, or Respiration - Tumors. Rat Route: Inhalation; Dose: 100 ppm; Toxic Effects: Tumorigenic - Equivocal tumorigenic agent by RTECS criteria; Sense organs and special senses - Tumors. Rat Route: Inhalation; Dose: 30 ppm/6H/57W-I; Toxic Effects: Tumorigenic - Equivocal tumorigenic agent by RTECS criteria; Sense organs and special senses - Tumors; Lungs, Thorax, or Respiration - Chronic pulmonary edema. Rat Route: Oral; Dose: 36 gm/kg/81W-I; Toxic Effects: Tumorigenic - Equivocal tumorigenic agent by RTECS criteria; Gastrointestinal - Tumors. Rat Route: Oral; Dose: 85050 mg/kg/81W-C; Toxic Effects: Tumorigenic - Neoplastic by RTECS criteria; Gastrointestinal - Tumors. Rat Route: Oral; Dose: 42525 mg/kg/81W-C; Toxic Effects: Tumorigenic - Equivocal tumorigenic agent by RTECS criteria; Gastrointestinal - Tumors. Rat Route: Oral; Dose: 5150 mg/kg/2Y-I; Toxic Effects: Tumorigenic - Equivocal tumorigenic agent by RTECS criteria; Gastrointestinal - Tumors; Endocrine - Tumors.

Hazard Overviews

Corrosive

Flammable

Fire Diamond

Health: Corrosive to eyes/skin/respiratory tract. Toxic. Also Causes: nausea, abdominal discomfort, pain in liver region, difficult breathing, cyanosis, nasal irritation, chemical pneumonitis, blistering, dermatitis, liver and kidney damage. Chronic Effects: CNS depression, sensitization.

Fire: Flammable. Can form explosive mixtures in the air. Alcohol foam is recommended extinguishing agent. Water spray may be used to cool fire-exposed containers, to disperse vapors, to flush away spills from exposure, and to dilute spills to nonflammable mixtures. Do not get water in containers. Avoid use of dry chemical if the fire occurs in a confined vent container.

Reactivity: Unstable. Hazardous polymerization can occur in the presence of strong acids, caustic alkalies, aluminum, aluminum chloride, iron (III) chloride, zinc, and curing agents such as ethylenediamine. Incompatible with: active hydrogen atoms; strong oxidizers; strong acids; caustics; mercaptans; zinc; aluminum; chlorides of iron; potassium tert-butoxide; trichloroethylene; isopropylamine; aniline. Hazardous decomposition products: phosgene gas; carbon dioxide; carbon monoxide; hydrogen chloride; chlorine.

Carcinogenicity: IARC - Group 2A, Probably carcinogenic to humans; NIOSH - Listed as carcinogen; NTP - Class 2B, Reasonably anticipated to be a carcinogen, sufficient evidence of carcinogenicity from studies in experimental animals; ACGIH - Not listed; OSHA - Not listed; EPA - Class B2, Probable human carcinogen based on animal studies; MAK - Class A2, Unmistakably carcinogenic in animal experimentation only

Primary Target Organs:

Eyes

Skin

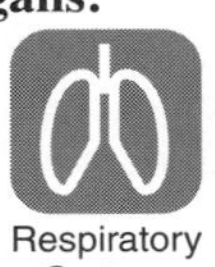
Respiratory System

Liver

Kidneys

Exposure Limits

OSHA PEL: TWA: 5 ppm; 19 mg/m^3; skin.

OSHA PEL Vacated 1989 Limits: TWA: 2 ppm; 8 mg/m^3.

ACGIH TLV: TWA: 2 ppm; 7.6 mg/m^3.

NIOSH IDLH: 75 ppm.

Respirator Recommendation

Exposure Range: >5 to <75 ppm Supplied Air, Constant Flow/Pressure Demand, Half Mask

Exposure Range: 75 to unlimited ppm Self-contained Breathing Apparatus, Pressure Demand, Full Face

Note: poor warning properties

Environmental

Ecotoxicity: Toxicity threshold (cell multiplication) inhibition test: Scenedesmus quadricauda (green algae) 5.4 mg/l TLm Sheephead minnow 11.8 ppm/ 96 hr /Conditions of bioassay not specified Toxicity threshold (cell multiplication) test: Entosiphon sulcatum (protozoa) 35 mg/l LC_{50} Menidia beryllina 18 ppm/96 hr, static bioassay in synthetic seawater at 23 °C, mild aeration applied after 24 hr LC_{50} Lepomis macrochirus (goldfish) 42 ppm/96 hr at 23 °C, static bioassay in freshwater at 23 °C, mild aeration applied after 24 hr

Environmental Fate: If released into water it will be lost primarily by evaporation (half-life 29 hr in a typical river) and hydrolysis (half-life 8.2 days). It will neither adsorb appreciably to sediment nor bioconcentrate in fish. If spilled on land, it will evaporate and leach into the groundwater where it will hydrolyze. Biodegradation and chemical reactions with ions and reactive species may accelerate its loss in soil and water but data from field studies are lacking. In the atmosphere, it will degrade by reaction with photochemically produced hydroxyl radicals (est. half-life 4 days).

Cleanup/Disposal: Guide No. 131: Fully encapsulating, vapor protective clothing should be worn for spills and leaks with no fire. Eliminate all ignition sources (no smoking, flares, sparks or flames in immediate area). All equipment used when handling the product must be grounded. Do not touch or walk through spilled material. Stop leak if you can do it without risk. Prevent entry into waterways, sewers, basements or confined areas. A vapor suppressing foam may be used to reduce vapors. Small Spills: Absorb with earth, sand or other non-combustible material and transfer to containers for later disposal. Use clean non-sparking tools to collect absorbed material. Large Spills: Dike far ahead of liquid spill for later disposal. Water spray may reduce vapor; but may not prevent ignition in closed spaces.

Environmental Physical Data

Henry's Law Constant: 3.8 x10^{-5}

Octanol/Water Partition Coefficient: log K_{ow} = 0.22 to 0.30

Sorption Partition Coefficient: K_{oc} = estimated at 123

BCF: estimated at 0.66

Regulations

RCRA 40CFR: Listed Hazardous Waste No. U041 Toxic Waste

CERCLA: 40CFR 302.4: Listed per CWA Section 311(b)(4) per RCRA Section 3001 RQ: 100 lb (45.35 kg)

SARA 40CFR 372.65: Listed TPQ: 1000 lb

SARA EHS 40CFR 355: Listed TPQ: 100 lb

TSCA: Listed

Analytical Methods

Air: ASTM D4490

Soil: SW846 8010B, 8240B, 8260A, 8260B

Water / Groundwater: ASTM D3695

Indoor / Expired Air: NIOSH 1010; EPA IP-1B

Plasma: EPA 29

EPN5000 CAS #: 2104-64-5

EPN

RTECS: TB1925000

DOT: UN2783; UN2784; UN3017; UN3018; IMO3.2; IMO6.1
EINECS Number: 218-276-8
Molecular Formula: $C_{14}H_{14}NO_4PS$
Structured MF: $C_2H_5O(C_6H_5)P(S)OC_6H_4NO_2$
Formula Weight: 323.31
Synonyms: O-AETHYL-O-(4-NITRO-PHENYL)-PHENYL-MONOTHIOPHOSPHONAT; BENZENEPHOSPHONIC ACID,THIONO-,ETHYL-P-NITROPHENYL ESTER; BENZENEPHOSPHOTHIONIC ACID,ETHYL 4-NITRO-PHENYL ESTER; BENZENEPHOSPHOTHIONIC ACID,ETHYL-4-NITRO PHENYLESTER; ENT 17,798; ENT 17 298; EPN-300; EPN 300; ETHOXY-(((4-NITROPHENOXY)PHENYL)PHOSPHINE) SULFIDE; ETHOXY-4-NITROPHENOXYPHENYLPHOSPHINE SULFIDE; ETHYL P-NITROPHENYL BENZENETHIONOPHOSPHONATE; O-ETHYL O-(4-NITROPHENYL) BENZENETHIONOPHOSPHONATE; ETHYL P-NITROPHENYL BENZENETHIOPHOSPHATE; ETHYL P-NITROPHENYL BENZENETHIOPHOSPHONATE; O-ETHYL O-P-NITROPHENYL BENZENETHIOPHOSPHONATE; ETHYL P-NITROPHENYL PHENYLPHOSPHONOTHIOATE; O-ETHYL O-(4-NITROPHENYL) PHENYLPHOSPHONOTHIOATE; O-ETHYL O-(P-NITROPHENYL) PHENYLPHOSPHONOTHIOATE; O-ETHYL O-4-NITROPHENYL PHENYL-PHOSPHONOTHIOATE; O-ETHYL O-4-NITROPHENYL PHENYLPHOSPHONOTHIOATE; O-ETHYL O-P-NITROPHENYL PHENYLPHOSPHONOTHIOATE; O-ETHYL O-P-NITROPHENYL PHENYLPHOSPHONOTHIOLATE; O-ETHYL O-(P-NITROPHENYL) PHENYLPHOSPHONOTHIONATE; O-ETHYL O-P-NITROPHENYL PHENYLPHOSPHOROTHIOATE; ETHYL P-NITROPHENYL THIONOBENZENEPHOSPHATE; ETHYL P-NITROPHENYL THIONOBENZENEPHOSPHONATE; O-ETHYL O-(4-NITROPHENYL)BENZENETHIONOPHOSPHONATE; ETHYL (P-NITROPHENYL)THIONOBENZENEPHOSPHONATE; O-ETHYL PHENYL (P-NITROPHENYL) THIOPHOSPHONATE; O-ETHYL PHENYL P-NITROPHENYL THIOPHOSPHONATE; O-ETHYL PHENYLPHOSPHONOTHIOIC ACID O-(4-NITROPHENYL) ESTER; O-ETHYL-O-P-NITROFENYLESTER KYSELINY FENYLTHIOFOSFONOVE; O-ETHYL-O-((4-NITRO-FENYL)-FENYL)-MONOTHIOFOSFONAAT; O-ETHYL-O,P-NITROPHENYL PHENYLPHOSPHOROTHIOATE; O-ETHYL-O-P-NITROPHENYLPHENYLPHOSPHONOTHIOATE; ETHYL-P-NITROPHENYLTHIONOBENZENEPHOSPHATE; O-ETIL-O-((4-NITRO-FENIL)-FENIL)-MONOTIOFOSFONATO; KASUTOP DUST; MEIDON 15 DUST; O-(4-NITROPHENYL) O-ETHYL PHENYL THIOPHOSPHONATE; OMS 219; PHENOL,P-NITRO-,O-ESTER WITH O-ETHYL PHENYLPHOSPHONOTHIOATE; PHENYLPHOSPHONOTHIOATE,O-ETHYL-O-P-NITROPHENYL-; PHENYLPHOSPHONOTHIOIC ACID O-ETHYL O-(4-NITROPHENYL) ESTER; PHENYLPHOSPHONOTHIOIC ACID O-ETHYL O-P-NITROPHENYL ESTER; PHENYLTHIOPHOSPHONATE DE O-ETHYLE ET O-4-NITROPHENYLE; PHOSPHONOTHIOIC ACID,PHENYL-,O-ETHYL O-(4-NITROPHENYL)ESTER; PHOSPHONOTHIOIC ACID,PHENYL-,O-ETHYL O-(P-NITROPHENYL)ESTER; PIN; SANTOX; TSUMAPHOS
Description: light yellow to white crystalline powder or light yellow oily liquid; brown liquid above 97 deg F; aromatic odor
Use: non-systemic insecticide & acaricide (former use); for cotton;; rice, vegetables, tobacco, fruit, nuts; effective against lepidopterous & other leaf-eating larvae, apple flea weevil, plum curculio, coddling moth, soil insects; aphids, mites, scale insects, european corn borer, mosquito larvae, boll weevil, pink bollworm, codling moth, plum curculio, etc.

Physical Properties

Boiling Point: 215 °C (419 °F) at 5 mm Hg
Freezing Point: 36 °C (96.8 °F)
Specific Gravity: 1.268 at 25 °C
Vapor Pressure: 0.126 mPa at 25 °C
Water Solubility: Insoluble
Other Solubilities: Soluble in Ether.
Refraction Index: 1.6021 at 25 °C/D
Flash Point: Noncombustible Solid

Hazard Overviews

Fire: Noncombustible.
Carcinogenicity: IARC - Not listed; NIOSH - Not listed; NTP - Not listed; ACGIH - Class A4, Not classifiable as a human carcinogen; OSHA - Not listed; EPA - Not listed; MAK - Not listed
Exposure Limits
OSHA PEL: TWA: 0.5 mg/m^3; skin.
ACGIH TLV: TWA: 0.1 mg/m^3.
NIOSH REL: TWA: 0.5 mg/m^3.
NIOSH IDLH: 5 mg/m^3.
DFG MAK: TWA: 0.5 mg/m^3.
Respirator Recommendation
Exposure Range: >0.5 to <5 mg/m^3 Air Purifying, Negative Pressure, Half Mask
Exposure Range: 5 to unlimited mg/m^3 Self-contained Breathing Apparatus, Pressure Demand, Full Face
Cartridge Color: black with magenta (P100)

Environmental

Ecotoxicity: LD_{50} Coturnix japonica Japaneses quail female oral 5.25 mg/kg (95% confidence limits 3.79-7.28 mg/kg) LC_{50} Coturnix japonica Japanese quail, 14 days old, oral (5-day ad libitum in diet) 437 ppm (confidence intervals 302-632 ppm LC_{50} Cyprinodon variegatus (Sheepshead minnow) 188.9 ug/l/96 hr /conditions of bioassay not specified LD_{50} Anas platyrhynchos (Mallard duck) oral 3.08 mg/kg (95% confidence limit 2.38-4.00), 3 mo old females LC_{50} Penaeus duorarum (Pink shrimp) 0.29 ug/l/96 hr /conditions of bioassay not specified LC_{50} Lagodon rhomboides (Pinfish) 18.3 ug/l/96 hr /conditions of bioassay not specified LC_{50} Palaemonetes 0.6 ug/l/96 hr in hard water (272 ppm $CaCO_3$) at 21 °C (95% confidence limit 0.4-0.8 ug/l) mature STage. static bioassay without aeration, pH 7.2-7.5, water hardness 40-50 mg/l as calcium carbonate and alkalinity of 30-35 mg/l
Environmental Fate: If released to soil or water, breakdown is expected to proceed primarily through hydrolysis and oxidation to phenylphosphonic acid. Other degradation products are EPN-oxon, desethyl EPN-oxon, O-ethyl S-methylphenylphosphonothiolate, p-nitrophenol, O-ethyl O-methylphenylphosphonate, and O-ethylphenylphosphonate. The half-life in soil under field conditions has been found to range from about 2 weeks to 1 month. It is expected to be relatively immobile in soil. This compound is not expected to volatilize significantly from dry soil surfaces. In water, adsorption to suspended solids and sediments may be an important fate process. Significant bioaccumulation in aquatic organisms is expected only when continuous exposure to this compound occurs. Volatilization should not be a significant fate process. Based on a vapor pressure of 9.45 $x10^{-7}$ mm Hg

at 25 °C, it is expected to exist partly in the vapor phase and partly in the particulate phase in the atmosphere. Vapor is expected to react rapidly with photochemically generated hydroxyl radicals (half-life 5 hours). Particulate phase may be removed from the atmosphere by wet or dry deposition. In addition, it has the potential to undergo direct photolysis.

Cleanup/Disposal: Guide No. 131: Fully encapsulating, vapor protective clothing should be worn for spills and leaks with no fire. Eliminate all ignition sources (no smoking, flares, sparks or flames in immediate area). All equipment used when handling the product must be grounded. Do not touch or walk through spilled material. Stop leak if you can do it without risk. Prevent entry into waterways, sewers, basements or confined areas. A vapor suppressing foam may be used to reduce vapors. Small Spills: Absorb with earth, sand or other non-combustible material and transfer to containers for later disposal. Use clean non-sparking tools to collect absorbed material. Large Spills: Dike far ahead of liquid spill for later disposal. Water spray may reduce vapor; but may not prevent ignition in closed spaces. Guide No. 152: Do not touch damaged containers or spilled material unless wearing appropriate protective clothing. Stop leak if you can do it without risk. Prevent entry into waterways, sewers, basements or confined areas. Cover with plastic sheet to prevent spreading. Absorb or cover with dry earth, sand or other non-combustible material and transfer to containers. Do not get water inside containers.

Environmental Physical Data

Henry's Law Constant: estimated at 1.3 $x10^{-7}$
Octanol/Water Partition Coefficient: log K_{ow} = 3.85
Sorption Partition Coefficient: K_{oc} = estimated at 2340 to 2960
BCF: topmouth gundgeon 2346

Regulations

RCRA 40CFR: Not listed
CERCLA: 40CFR 302.4: Not listed
SARA 40CFR 372.65: Not listed TPQ: 100/10000 lb
SARA EHS 40CFR 355: Listed TPQ: 100 lb
TSCA: Not listed

Analytical Methods

Soil: EPA PMD-TLC; SW846 8141, 8141A, 8270B, 8270C
Water / Groundwater: EPA P-005-1, 1657, 614.1, 022; SW846 8141, 8141A
Food: FDA 212.1, 232.1, 232.3, 232.4, 242.1; AOAC 974.22
Indoor / Expired Air: NIOSH 5012
Plasma: EPA 001, 027, 028, 29; FDA 211.1, 231.1, 252

EPO1730 CAS #: 57608-57-8

EPOXY RESIN ERR-0100

RTECS: KD4385000

Physical Properties

Boiling Point: Varies
Specific Gravity: Varies
Vapor Pressure: Low
Water Solubility: Varies
Flash Point: Generally > 93.3 °C

RTECS Toxicity Data

Acute Oral: Rat LD_{50} Dose: >16 gm/kg.
Acute Dermal: Rabbit LD_{50} Route: Skin; Dose: >8 gm/kg.
Irritation Skin: Rabbit Standard Draize Test Dose: 4500 mg/3D-I; Reaction: mild.

Hazard Overviews

Reactivity: Stable. Polymerization can occur exothermically when in contact with curing agents (organic amines; acid anhydrides; polyamides). Avoid: exposure to heat and ignition sources. Incompatible with: oxidizing agents; strong acids; mercaptans; bases; bulk contact with curing agents. Hazardous decomposition products: carbon oxides; various hydrocarbons.

Carcinogenicity: IARC - Not listed; NIOSH - Not listed; NTP - Not listed; ACGIH - Not listed; OSHA - Not listed; EPA - Not listed; MAK - Not listed

Environmental

Regulations

RCRA 40CFR: Not listed
CERCLA: 40CFR 302.4: Not listed
SARA 40CFR 372.65: Not listed
SARA EHS 40CFR 355: Not listed
TSCA: Not listed

EPO2460 CAS #: 930-22-3

3,4-EPOXY-1-BUTENE

RTECS: EM7350000
EINECS Number: 213-210-4
Molecular Formula: C_4H_6O
Formula Weight: 70.09

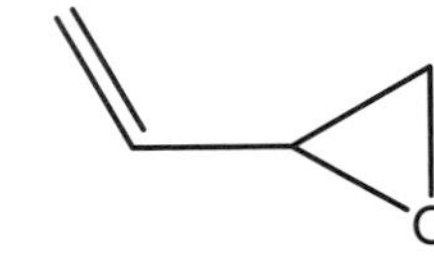

Chemical Structure

Synonyms: BUTADIENE EPOXIDE; BUTADIENE MONOEPOXIDE; BUTADIENE MONO-OXIDE; BUTADIENE MONOOXIDE; BUTADIENE MONOXIDE; 1,3-BUTADIENE OXIDE; BUTADIENE OXIDE; 1-BUTENE,3,4-EPOXY-; 1,2-EPOXY-3-BUTENE; 1,2-EPOXYBUTENE; 1,2-EPOXYBUTENE-3; 3,4-EPOXYBUTENE; ETHENYLOXIRANE; 1,2-OXIDO-3-BUTENE; OXIRANE,ETHENYL-; VINYLETHYLENE OXIDE; VINYLOXIRANE
Description: light yellow, clear liquid
Use: comonomer with propylene oxide for polyethers; reactive diluent for epoxy resins

Physical Properties

Boiling Point: 70 °C (158 °F) at 760 mm Hg
Freezing Point: -135 °C (-211 °F)
Specific Gravity: 0.9006 at 0 °C
Vapor Density: 2.4 Air=1
Density: 0.9006 g/mL at 0 °C
Water Solubility: 10 to 50 mg/mL at 21 °C
Other Solubilities: 95% Ethanol: >=100 mg/ml at 21 °C; Acetone: >=100 mg/ml at 21 °C; Benzene: Soluble; DMSO: >=100 mg/ml at 21 °C; Ether: Soluble.
Refraction Index: 1.4168
Ionization Potential (eV): 9.52
Flash Point: < -50 °C
Autoignition Temperature: 430 °C

RTECS Toxicity Data

Mutagenic: Human Mutations in Mammalian Somatic Cells; Cell Type: lymphocyte; Dose: 150 umol/L. Human Sister Chromatid Exchenge; Cell Type: lymphocyte; Dose: 25 umol/L.
Tumorigenic: Mouse Route: Skin; Dose: 492 gm/kg/41W-I; Toxic Effects: Tumorigenic - Equivocal tumorigenic agent by RTECS criteria; Skin and appendages - Tumors; Tumorigenic - Tumors at site of application.

Hazard Overviews

Fire Diamond

Health: Irritating to eyes/skin. Harmful. Other Acute Effects: harmful if swallowed, inhaled, or absorbed through skin; possible risk of irreversible effects. Chronic Effects: possible carcinogen.
Fire: Flammable. Hazards: vapor may travel considerable distance to source of ignition and flash back; container explosion may occur. Extinguishing agents: water spray; carbon dioxide, dry chemical powder or appropriate foam. Precautions: combustible liquid.
Carcinogenicity: IARC - Not listed; NIOSH - Not listed; NTP - Not listed; ACGIH - Not listed; OSHA - Not listed; EPA - Not listed; MAK - Not listed
Primary Target Organs:

Eyes

Skin

Environmental

Cleanup/Disposal: Guide No. 128: Eliminate all ignition sources (no smoking, flares, sparks or flames in immediate area). All equipment used when handling the product must be grounded. Do not touch or walk through spilled material. Stop leak if you can do it without risk. Prevent entry into waterways, sewers, basements or confined areas. A vapor suppressing foam may be used to reduce vapors. Absorb or cover with dry earth, sand or other non-combustible material and transfer to containers. Use clean non-sparking tools to collect absorbed material. Large Spills: Dike far ahead of liquid spill for later disposal. Water spray may reduce vapor; but may not prevent ignition in closed spaces.

Regulations

RCRA 40CFR: Not listed
CERCLA: 40CFR 302.4: Not listed
SARA 40CFR 372.65: Not listed
SARA EHS 40CFR 355: Not listed
TSCA: Listed

EPO3190 CAS #: 286-20-4

1,2-EPOXYCYCLOHEXANE

RTECS: RN7175000
EINECS Number: 206-007-7
Molecular Formula: $C_6H_{10}O$
Formula Weight: 98.16

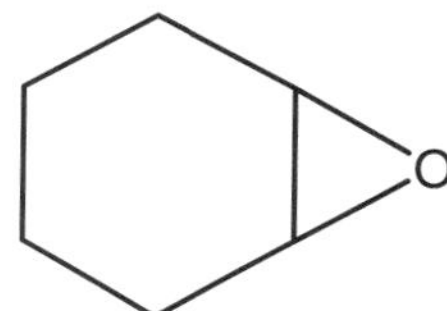

Chemical Structure

Synonyms: CCHO; CYCLOHEXANE OXIDE; CYCLOHEXANE,1,2-EPOXY-; CYCLOHEXENE EPOXIDE; 1,2-CYCLOHEXENE OXIDE; CYCLOHEXENE 1-OXIDE; CYCLOHEXENE OXIDE; CYCLOHEXENE,OXIDE; CYCLOHEXENE-1-OXIDE; CYCLOHEXYLENE OXIDE; 7-OXABICYCLO(4.1.0)HEPTANE; TETRAMETHYLENEOXIRANE
Description: colorless liquid; strong odor
Use: chemical intermediate; chem int for epoxy resins & photoreactive polymers

Physical Properties

Boiling Point: 129 °C (264 °F) to 130 °C (266 °F)
Freezing Point: > -10 °C (14 °F)
Specific Gravity: 0.967 at 25 °C/4 °C
Water Solubility: Insoluble in Water
Other Solubilities: Very Soluble in Benzene; Soluble in Chloroform.
Refraction Index: 1.4503 at 25 °C/D
Ionization Potential (eV): 9.82
Flash Point: 81 °C Closed Cup

RTECS Toxicity Data

Acute Oral: Rat LD_{50} Dose: 1090 uL/kg.
Acute Inhalation: Rat LC_{Lo} Dose: 2000 ppm/4hr.
Acute Dermal: Rabbit LD_{50} Route: Skin; Dose: 630 uL/kg.
Mutagenic: Hamster Mutations in Mammalian Somatic Cells; Cell Type: lung; Dose: 5 mmol/L. Hamster Sister Chromatid Exchenge; Cell Type: lung; Dose: 5 mmol/L.

Hazard Overviews

Health: Irritating to eyes/skin. Toxic. Other Acute Effects: harmful if swallowed, inhaled, or absorbed through skin.

Fire: Combustible. Hazards: vapor may travel considerable distance to source of ignition and flash back; container explosion may occur; emits toxic fumes. Extinguishing agents: carbon dioxide, dry chemical powder or appropriate foam. Precautions: combustible liquid.

Reactivity: Incompatible with: acids, bases, oxidizing agents. Hazardous decomposition products: carbon monoxide, carbon dioxide.

Carcinogenicity: IARC - Not listed; NIOSH - Not listed; NTP - Not listed; ACGIH - Not listed; OSHA - Not listed; EPA - Not listed; MAK - Not listed

Primary Target Organs:

Environmental

Regulations

RCRA 40CFR: Not listed
CERCLA: 40CFR 302.4: Not listed
SARA 40CFR 372.65: Not listed
SARA EHS 40CFR 355: Not listed
TSCA: Listed

EPO3920 **CAS #: 2386-87-0**

3,4-EPOXYCYCLOHEXYLMETHYL-3,4-EPOXYCYCLOHEXANE CARBOXYLATE

RTECS: RN7750000
EINECS Number: 219-207-4
Molecular Formula: $C_{14}H_{20}O_4$
Formula Weight: 252.34

Chemical Structure

Synonyms: CHISSONOX 221 MONOMER; 3,4-EPOXYCYCLOHEXANECARBOXYLIC ACID(3,4-EPOXYCYCLOHEXYLMETHYL) ESTER; 3,4-EPOXYCYCLOHEXANEMETHYL 3,4-EPOXYCYCLOHEXANECARBOXYLATE; 3,4-EPOXYCYCLOHEXYLMETHYL 3,4-EPOXYCYCLOHEXANE CARBOXYLATE; (3,4-EPOXYCYCLOHEXYL)METHYL 3,4-EPOXYCYCLOHEXYLCARBOXYLATE; 3,4-EPOXYCYCLOHEXYLMETHYL3',4'-EPOXYCYCLOHEXANECARBOXYLATE; ERL-4221; 7-OXABICYCLO(4.1.0)HEPTANE-3-CARBOXYLIC ACID,7-OXABICYCLO(4.1.0)HEPT-3-YLMETHYL ESTER; UT 632

Use: stabilizer for organophosphorus insecticides; reactive diluent for epoxy resins

RTECS Toxicity Data

Acute Oral: Rat LD_{50} Dose: 4490 mg/kg.
Acute Dermal: Rabbit LD_{50} Route: Skin; Dose: 20 mL/kg.

Hazard Overviews

Health: Irritating to eyes/skin/respiratory tract. Toxic. Other Acute Effects: may be harmful by inhalation, ingestion, or skin absorption; may cause allergic skin reaction; exposure can cause gastrointestinal disturbances nausea; vomiting; diarrhea. Chronic Effects: may alter genetic material.

Fire: Hazards: container explosion may occur; emits toxic fumes. Extinguishing agents: carbon dioxide, dry chemical powder or appropriate foam. Precautions: combustible liquid.

Reactivity: Stable. Hazardous polymerization may occur. Incompatible with: amines, acids, strong bases. Hazardous decomposition products: toxic fumes of: carbon monoxide, carbon dioxide.

Carcinogenicity: IARC - Not listed; NIOSH - Not listed; NTP - Not listed; ACGIH - Not listed; OSHA - Not listed; EPA - Not listed; MAK - Not listed

Primary Target Organs:

Environmental

Regulations

RCRA 40CFR: Not listed
CERCLA: 40CFR 302.4: Not listed
SARA 40CFR 372.65: Not listed
SARA EHS 40CFR 355: Not listed
TSCA: Listed

EPO4650 **CAS #: 285-67-6**

1,2-EPOXYCYCLOPENTANE

RTECS: RN8935000
EINECS Number: 206-005-6
Molecular Formula: C_5H_8O
Formula Weight: 84.13

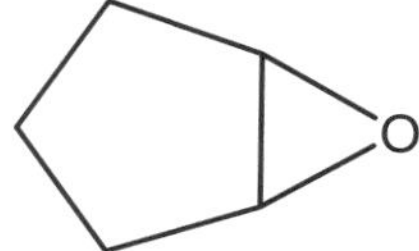

Chemical Structure

Synonyms: CYCLOPENTANE OXIDE; CYCLOPENTANE,1,2-EPOXY-; CYCLOPENTENE EPOXIDE; CYCLOPENTENE OXIDE; CYCLOPENTENE,OXIDE; CYCLOPENTENEOXIDE; 6-OXABICYCLO(3.1.0)HEXANE
Use: research chemical

Physical Properties

Boiling Point: 102 °C (215.6 °F)
Specific Gravity: 0.96400
Refraction Index: 1.4336

RTECS Toxicity Data

Mutagenic: Hamster Sister Chromatid Exchenge; Cell Type: lung; Dose: 40 mmol/L. Bacteria - K Kpneumoniae Mutations in Microorganisms; Dose: 2 mmol/L (-S9). Bacteria - S Typhimurium Mutations in Microorganisms; Dose: 15 umol/plate (+S9).

Hazard Overviews

Health: Irritating to eyes/skin/respiratory tract. Other Acute Effects: may be harmful by inhalation, ingestion, or skin absorption.
Fire: Hazards: vapor may travel considerable distance to source of ignition and flash back; container explosion may occur; emits toxic fumes. Extinguishing agents: carbon dioxide, dry chemical powder or appropriate foam. Precautions: combustible liquid.
Reactivity: Incompatible with: acids, bases, oxidizing agents. Hazardous decomposition products: toxic fumes of: carbon monoxide, carbon dioxide.
Carcinogenicity: IARC - Not listed; NIOSH - Not listed; NTP - Not listed; ACGIH - Not listed; OSHA - Not listed; EPA - Not listed; MAK - Not listed
Primary Target Organs:

Eyes

Skin

Respiratory System

Environmental

Regulations

RCRA 40CFR: Not listed
CERCLA: 40CFR 302.4: Not listed
SARA 40CFR 372.65: Not listed
SARA EHS 40CFR 355: Not listed
TSCA: Listed

EPO5380 CAS #: 2855-19-8

1,2-EPOXYDODECANE

RTECS: JR2450000
EINECS Number: 220-667-3
Molecular Formula: $C_{12}H_{24}O$
Formula Weight: 184.36

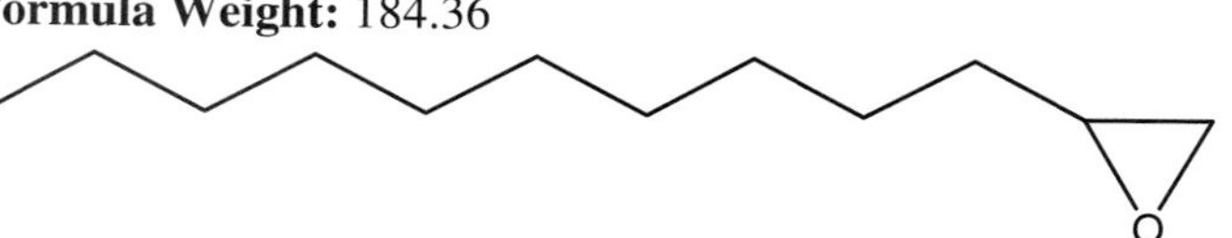

Chemical Structure

Synonyms: DECYL OXIRANE; DODECANE,1,2-EPOXY-; DODECENE EPOXIDE; 1,2-DODECENE OXIDE; 1-DODECENE OXIDE; ALPHA-DODECENE OXIDE; 1,2-EPOXY-N-DODECANE; 1,2-EPOXYDODEKAN; NEDOX 1200; OXIRANE,DECYL-
Description: colorless, clear liquid
Use: secondary heat stabilizer for polymers

Physical Properties

Boiling Point: 124 °C (255 °F) to 125 °C (257 °F)
Specific Gravity: 0.844
Saturated Vapor Density: 1.252867514 kg/m^3
Vapor Pressure: 6.25 mm Hg at 25 °C
Water Solubility: < 1 mg/mL at 23 C
Other Solubilities: 95% Ethanol: >=100 mg/ml at 23 °C; Acetone: >=100 mg/ml at 23 °C; DMSO: 50-100 mg/ml at 23 °C.
Refraction Index: 1.4355
Flash Point: 44 °C

Hazard Overviews

Flammable

Health: Irritating to eyes/skin/respiratory tract. Other Acute Effects: may be harmful by inhalation, ingestion, or skin absorption.
Fire: Flammable. Hazards: emits toxic fumes; vapor may travel considerable distance to source of ignition and flash back. Extinguishing agents: water spray; carbon dioxide, dry chemical powder or appropriate foam. Precautions: combustible liquid.
Reactivity: Incompatible with: strong oxidizing agents, strong acids, strong bases, may decompose on exposure to moist air or water. Hazardous decomposition products: toxic fumes of: carbon monoxide, carbon dioxide.
Carcinogenicity: IARC - Not listed; NIOSH - Not listed; NTP - Not listed; ACGIH - Not listed; OSHA - Not listed; EPA - Not listed; MAK - Not listed

Primary Target Organs:

Environmental

Cleanup/Disposal: Guide No. 128: Eliminate all ignition sources (no smoking, flares, sparks or flames in immediate area). All equipment used when handling the product must be grounded. Do not touch or walk through spilled material. Stop leak if you can do it without risk. Prevent entry into waterways, sewers, basements or confined areas. A vapor suppressing foam may be used to reduce vapors. Absorb or cover with dry earth, sand or other non-combustible material and transfer to containers. Use clean non-sparking tools to collect absorbed material. Large Spills: Dike far ahead of liquid spill for later disposal. Water spray may reduce vapor; but may not prevent ignition in closed spaces.

Regulations

RCRA 40CFR: Not listed
CERCLA: 40CFR 302.4: Not listed
SARA 40CFR 372.65: Not listed
SARA EHS 40CFR 355: Not listed
TSCA: Listed

EPO6110 **CAS #: 7320-37-8**

1,2-EPOXYHEXADECANE

RTECS: ML9450000
EINECS Number: 230-786-2
Molecular Formula: $C_{16}H_{32}O$
Formula Weight: 240.48

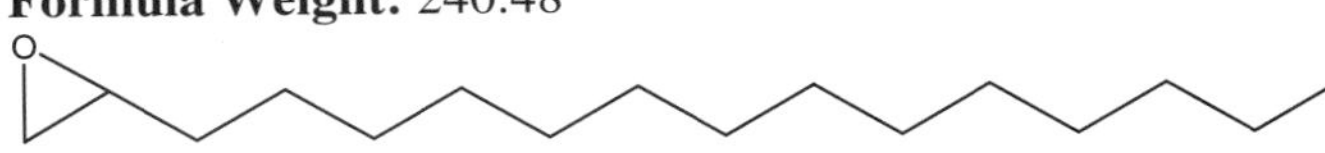

Chemical Structure

Synonyms: HEXADECANE,1,2-EPOXY-; 1,2-HEXADECENE EPOXIDE; HEXADECENE EPOXIDE; HEXADECYLENE OXIDE; OXIRANE,TETRADECYL-; TETRADECYL OXIRANE
Description: white waxy solid or colorless, clear liquid or white powder; faint pleasant odor
Use: organic synthesis

Physical Properties

Boiling Point: 270 °C (518 °F) to 275 °C (527 °F)
Freezing Point: 25 °C (77 °F) to 27 °C (80.6 °F)
Specific Gravity: 0.849 at 20/4 °C
Density: 0.85 g/mL
Water Solubility: Reacts
Other Solubilities: 95% Ethanol: Reaction; Acetone: 50-100 mg/ml at 18 °C; DMSO: <1 mg/ml at 18 °C.
Flash Point: Not available; probably combustible

RTECS Toxicity Data

Mutagenic: Mouse Mutations in Microorganisms; Cell Type: lymphocyte; Dose: 24 mg/L (+S9). Mouse Mutations in Mammalian Somatic Cells; Cell Type: lymphocyte; Dose: 18 mg/L.
Tumorigenic: Mouse Route: Skin; Dose: 53 gm/kg/44W-I; Toxic Effects: Tumorigenic - Equivocal tumorigenic agent by RTECS criteria; Skin and appendages - Tumors; Tumorigenic - Tumors at site of application.

Hazard Overviews

Health: Irritating to eyes/skin/respiratory tract. Toxic. Other Acute Effects: harmful if swallowed, inhaled, or absorbed through skin; exposure may cause nausea; vomiting; diarrhea. Chronic Effects: Carcinogen.
Fire: Will burn. Hazards: emits toxic fumes. Extinguishing agents: water spray; carbon dioxide, dry chemical powder or appropriate foam. Precautions: combustible liquid.
Reactivity: Incompatible with: strong oxidizing agents, strong acids, strong bases, amines, may decompose on exposure to moist air or water. Hazardous decomposition products: toxic fumes of: carbon monoxide, carbon dioxide.
Carcinogenicity: IARC - Not listed; NIOSH - Not listed; NTP - Not listed; ACGIH - Not listed; OSHA - Not listed; EPA - Not listed; MAK - Not listed
Primary Target Organs:

Environmental

Regulations

RCRA 40CFR: Not listed
CERCLA: 40CFR 302.4: Not listed
SARA 40CFR 372.65: Not listed
SARA EHS 40CFR 355: Not listed
TSCA: Listed

EPO6840 **CAS #: 470-67-7**

1,4-EPOXY-P-MENTHANE

RTECS: OS9274000
EINECS Number: 207-428-9
Molecular Formula: $C_{10}H_{18}O$
Formula Weight: 154.26
Synonyms: 1,4-CINEOL; 1,4-CINEOLE; P-CINEOLE; ISOCINEOLE; 1-METHYL-4-(1-METHYLETHYL)-7-OXABICYCLO(2.2.1)HEPTANE; 7-OXABICYCLO(2.2.1)HEPTANE,1-ISOPROPYL-4-METHYL-(6CI); 7-OXABICYCLO(2.2.1)HEPTANE,1-METHYL-4-(1-METHYLETHYL)-; 7-OXABICYCLO(2.2.1)HEPTANE,1-METHYL-4-(1-METHYLETHYL)-(9CI)
Description: colorless liquid; minty lime oil odor

Use: in pharmaceuticals (cough syrups, expectorants); perfumery; used in essential oil imitations, mint, toothpaste flavors; fragrance and flavoring agent

Physical Properties

Boiling Point: 173 °C (343 °F) to 174 °C (345 °F)
Freezing Point: 1 °C (33.8 °F)
Specific Gravity: 0.8997 at 20 °C
Water Solubility: Slightly Soluble in Water
Other Solubilities: Soluble in Benzene, Petroleum Ether. Soluble in all proportions in Ether, Alcohol
Refraction Index: 1.4562 at 20 °C/D

RTECS Toxicity Data

Acute Oral: Rat LD_{50} Dose: 3100 mg/kg.
Acute Dermal: Rabbit LD_{50} Route: Skin; Dose: >5 gm/kg.

Hazard Overviews

Health: Irritating to eyes/skin/respiratory tract. Harmful. Other Acute Effects: harmful if swallowed, inhaled, or absorbed through skin.
Fire: Hazards: emits toxic fumes. Extinguishing agents: water spray; carbon dioxide, dry chemical powder or appropriate foam. Precautions: combustible liquid.
Reactivity: Incompatible with: strong oxidizing agents. Hazardous decomposition products: toxic fumes of: carbon monoxide, carbon dioxide.
Carcinogenicity: IARC - Not listed; NIOSH - Not listed; NTP - Not listed; ACGIH - Not listed; OSHA - Not listed; EPA - Not listed; MAK - Not listed
Primary Target Organs:

Eyes

Skin

Respiratory System

Environmental

Regulations

RCRA 40CFR: Not listed
CERCLA: 40CFR 302.4: Not listed
SARA 40CFR 372.65: Not listed
SARA EHS 40CFR 355: Not listed
TSCA: Listed

EPO7570 CAS #: 930-37-0

1,2-EPOXY-3-METHOXYPROPANE

RTECS: TZ3530000
EINECS Number: 213-216-7
Molecular Formula: $C_4H_8O_2$
Formula Weight: 88.12
Synonyms: GLYCIDOL METHYL ETHER; GLYCIDYL METHYL ETHER; 3-METHOXY-1,2-EPOXYPROPANE; (METHOXYMETHYL)OXIRANE; METHOXYMETHYLOXIRANE; 3-METHOXYPROPYLENE OXIDE; METHYL GLYCIDYL ETHER; OXIRANE,(METHOXYMETHYL)-; OXIRANE,(METHOXYMETHYL)-(9CI); PROPANE,1,2-EPOXY-3-METHOXY-
Description: colorless, clear liquid
Use: research chemical

Physical Properties

Boiling Point: 113 °C (235.4 °F)
Specific Gravity: 0.98900
Water Solubility: >= 100 mg/mL at 20.5 C
Other Solubilities: 95% Ethanol: >=100 mg/ml at 20.5 °C; Acetone: >=100 mg/ml at 20.5 °C; DMSO: >=100 mg/ml at 20.5 °C.
Refraction Index: 1.4320
Flash Point: < 20.6 °C

RTECS Toxicity Data

Mutagenic: Hamster Sister Chromatid Exchenge; Cell Type: lung; Dose: 1250 umol/L. Bacteria - K Kpneumoniae Mutations in Microorganisms; Dose: 200 umol/L (-S9). Bacteria - S Typhimurium Mutations in Microorganisms; Dose: 100 ug/plate (+S9), 33 ug/plate (-S9).

Hazard Overviews

Flammable

Health: Irritating to eyes/skin/respiratory tract. Other Acute Effects: may be harmful by inhalation, ingestion, or skin absorption.
Fire: Flammable. Hazards: vapor may travel considerable distance to source of ignition and flash back; emits toxic fumes. Extinguishing agents: water spray; carbon dioxide, dry chemical powder or appropriate foam. Precautions: combustible liquid.
Carcinogenicity: IARC - Not listed; NIOSH - Not listed; NTP - Not listed; ACGIH - Not listed; OSHA - Not listed; EPA - Not listed; MAK - Not listed
Primary Target Organs:

Eyes

Skin

Respiratory System

Environmental

Regulations

RCRA 40CFR: Not listed
CERCLA: 40CFR 302.4: Not listed
SARA 40CFR 372.65: Not listed
SARA EHS 40CFR 355: Not listed
TSCA: Listed

EPO8300 CAS #: 503-30-0

1,3-EPOXYPROPANE

RTECS: RQ6825000
EINECS Number: 207-964-3
Molecular Formula: C_3H_6O
Formula Weight: 58.08

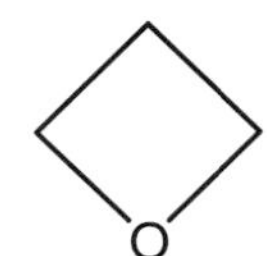

Chemical Structure

Synonyms: CYCLOOXABUTANE; OXACYCLOBUTANE; OXETAN; OXETANE; ALPHA,GAMMA-PROPANE OXIDE; PROPANE,1,3-EPOXY-; 1,3-PROPYLENE OXIDE; TRIMETHYLENE OXIDE; TRIMETHYLENOXID
Description: oil; agreeable aromatic odor
Use: chem int for epoxy resins & 3-substituted propanols

Physical Properties

Boiling Point: 48 °C (118 °F) at 750 mm Hg
Freezing Point: -97 °C (-142.6 °F)
Specific Gravity: 0.893 at 25 °C/4 °C
Water Solubility: Soluble in all Proportions
Other Solubilities: 95% Ethanol: >=100 mg/ml at 20 °C; Acetone: >=100 mg/ml at 20 °C; DMSO: >=100 mg/ml at 20 °C; Ether: Soluble.
Refraction Index: 1.3895 at 25 °C/D
Ionization Potential (eV): 9.668 +/-0.005
Flash Point: -28 °C

RTECS Toxicity Data

Acute Dermal: Rat LD_{50} Route: Subcutaneous Dose: 500 mg/kg.
Mutagenic: Bacteria - S Typhimurium Mutations in Microorganisms; Dose: 3333 ug/plate (-S9).
Tumorigenic: Rat Route: Subcutaneous; Dose: 2240 mg/kg/56W-I; Toxic Effects: Tumorigenic - Equivocal tumorigenic agent by RTECS criteria; Tumorigenic - Tumors at site of application.

Hazard Overviews

Flammable

Health: May cause irritation. Other Acute Effects: may be harmful by inhalation, ingestion, or skin absorption.
Fire: Flammable. Hazards: vapor may travel considerable distance to source of ignition and flash back; container explosion may occur; forms explosive mixtures in air. Extinguishing agents: carbon dioxide, dry chemical powder or appropriate foam; water may be effective for cooling, but may not effect extinguishment. Precautions: combustible liquid.
Reactivity: Incompatible with: oxidizing agents, strong acids. Hazardous decomposition products: toxic fumes of: carbon monoxide, carbon dioxide.
Carcinogenicity: IARC - Not listed; NIOSH - Not listed; NTP - Not listed; ACGIH - Not listed; OSHA - Not listed; EPA - Not listed; MAK - Not listed

Environmental

Cleanup/Disposal: Guide No. 127: Eliminate all ignition sources (no smoking, flares, sparks or flames in immediate area). All equipment used when handling the product must be grounded. Do not touch or walk through spilled material. Stop leak if you can do it without risk. Prevent entry into waterways, sewers, basements or confined areas. A vapor suppressing foam may be used to reduce vapors. Absorb or cover with dry earth, sand or other non-combustible material and transfer to containers. Use clean non-sparking tools to collect absorbed material. Large Spills: Dike far ahead of liquid spill for later disposal. Water spray may reduce vapor; but may not prevent ignition in closed spaces.

Regulations

RCRA 40CFR: Not listed
CERCLA: 40CFR 302.4: Not listed
SARA 40CFR 372.65: Not listed
SARA EHS 40CFR 355: Not listed
TSCA: Listed

EPO9030 CAS #: 5455-98-1

N-(2,3-EPOXYPROPYL)PHTHALIMIDE

RTECS: TI4950000
EINECS Number: 226-710-2
Molecular Formula: $C_{11}H_9NO_3$
Formula Weight: 203.20

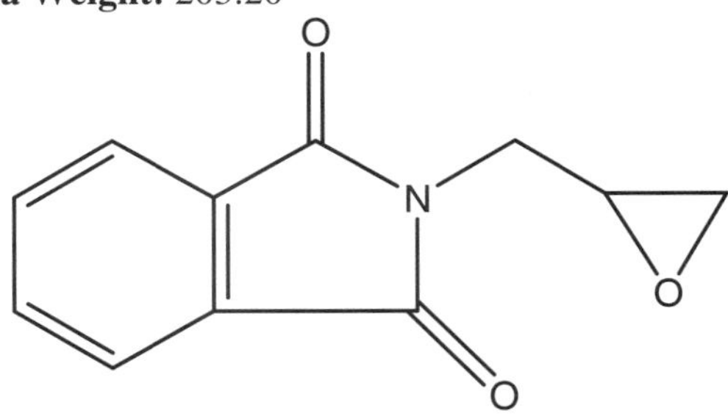

Chemical Structure

Synonyms: 2,3-EPOXYPROPYLPHTHALIMIDE; N-(2,3-EPOXYPROPYL)-PHTHALIMIDE; N-GLYCIDYLPHTHALIMIDE; 1H-ISOINDOLE-1,3(2H)-DIONE,2-(OXIRANYLMETHYL)-; 1H-ISOINDOLE-1,3(2H)-DIONE,2-(OXIRANYLMETHYL)-(9CI); 2-(OXIRANYLMETHYL)-1H-ISOINDOLE-1,3(2H)-DIONE; PHTHALIMIDE,N-(2,3-EPOXYPROPYL)-
Description: white powder

Physical Properties

Other Solubilities: reacts rapidly with Amino, Hydroxyl, & Carboxyl groups.

RTECS Toxicity Data

Acute Oral: Rat LD_{50} Dose: 4700 mg/kg; Toxic Effects: Behavioral - Somnolence (general depressed activity); Behavioral - Ataxia; Gastrointestinal - Changes in structure or function of salivary glands.

Acute Inhalation: Rat LC_{Lo} Dose: 4400 mg/m^3/4hr; Toxic Effects: Sense organs and special senses - Other; Lungs, Thorax, or Respiration - Other changes.

Irritation Eye: Rabbit Standard Draize Test Dose: 100 mg; Reaction: severe.

Irritation Skin: Rabbit Standard Draize Test Dose: 500 mg/24H; Reaction: mild.

Mutagenic: Hamster Mutations in Mammalian Somatic Cells; Cell Type: ovary; Dose: 10 mg/L. Hamster Sister Chromatid Exchenge; Cell Type: lung; Dose: 78 umol/L.

Hazard Overviews

Health: Irritating to eyes/skin/respiratory tract. Harmful. Other Acute Effects: may be harmful by inhalation, ingestion, or skin absorption. Chronic Effects: laboratory experiments have shown mutagenic effects.

Fire: Hazards: emits toxic fumes. Extinguishing agents: water spray; carbon dioxide, dry chemical powder or appropriate foam. Precautions: combustible liquid.

Reactivity: Incompatible with: acids, bases, oxidizing agents. Hazardous decomposition products: carbon monoxide, carbon dioxide, nitrogen oxides.

Carcinogenicity: IARC - Not listed; NIOSH - Not listed; NTP - Not listed; ACGIH - Not listed; OSHA - Not listed; EPA - Not listed; MAK - Not listed

Primary Target Organs:

Eyes

Skin

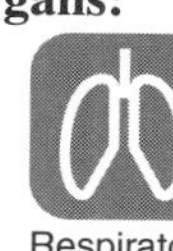
Respiratory System

Environmental

Regulations

RCRA 40CFR: Not listed
CERCLA: 40CFR 302.4: Not listed
SARA 40CFR 372.65: Not listed
SARA EHS 40CFR 355: Not listed
TSCA: Listed

ERB5000 CAS #: 136-25-4

ERBON

RTECS: UF1400000
Molecular Formula: $C_{11}H_9Cl_5O_3$
Formula Weight: 366.45

Synonyms: BARON; 2,2-DICHLOROPROPIONIC ACID 2-(2,4,5-TRICHLOROPHENOXY)ETHYLESTER; 2,2-DICHLOROPROPIONIC ACID,2-(2,4,5-TRICHLOROPHENOXY)ETHYL ESTER; ERBN; ETHANOL,2-(2,4,5-TRICHLOROPHENOXY)-,2,2-DICHLOROPROPIONATE; NOVEGE; NOVON; PENTANATE; PROPANOIC ACID,2,2-DICHLORO-,2-(2,4,5-TRICHLOROPHENOXY)ETHYL ESTER; PROPIONIC ACID,2,2-DICHLORO-,2-(2,4,5-TRICHLOROPHENOXY)ETHYL ESTER; 2-(2,4,5-TRICHLORFENOXY)ETHYLESTER KYSELINY 2,2-DICHLORPROPIONOVE; 2-(2,4,5-TRICHLOROPHENOXY)ETHYL 2,2-DICHLOROPROPIONATE; 2,4,5-TRICHLOROPHENOXYETHYL-ALPHA,ALPHA-DICHLOROPROPIONATE

Description: white solid or crystals

Use: non-selective, translocatable herbicide

Physical Properties

Boiling Point: 161 °C (322 °F) to 164 °C (327 °F)
Freezing Point: 49 °C (120.2 °F) to 50 °C (122 °F)
Specific Gravity: 1.47 at 20 °C/20 °C
Water Solubility: Insoluble in Water
Other Solubilities: Soluble in Acetone, Alcohol, Kerosene, Xylene
Flash Point: Nonflammable

RTECS Toxicity Data

Acute Oral: Rat LD_{50} Dose: 1 gm/kg. Mouse LD_{50} Dose: 912 mg/kg.

Chronic (Multiple Dose) Oral: Rat Dose: 227 mg/kg/26/W-I; Toxic Effects: Behavioral - Alteration of classical conditioning; Liver - Liver function tests impaired. Rabbit Dose: 227 mg/kg/26/W-I; Toxic Effects: Liver - Liver function tests impaired.

Hazard Overviews

Fire: Noncombustible.

Carcinogenicity: IARC - Not listed; NIOSH - Not listed; NTP - Not listed; ACGIH - Not listed; OSHA - Not listed; EPA - Not listed; MAK - Not listed

Environmental

Regulations

RCRA 40CFR: Not listed
CERCLA: 40CFR 302.4: Not listed
SARA 40CFR 372.65: Not listed
SARA EHS 40CFR 355: Not listed
TSCA: Not listed

Analytical Methods

Soil: EPA PMD-CPH

ERG1000 CAS #: 60-79-7

ERGONOVINE

RTECS: KE5075000
EINECS Number: 200-485-0
Molecular Formula: $C_{19}H_{23}N_3O_2$
Formula Weight: 325.39

Chemical Structure

Synonyms: BASERGIN; 9,10-DIDEHYDRO-N-(2-HYDROXY-1-METHYLETHYL)-6-METHYLERGOLINE-8BETA(S)-CARBOXAMIDE; 9,10-DIDEHYDRO-N-(ALPHA-(HYDROXYMETHYL)ETHYL)-6-METHYLERGOLINE-8-BETA-CARBOXAMIDE; 9,10-DIHYDRO-N-(ALPHA-(HYDROXYMETHYL)ETHYL)-6-METHYLERGOLINE-8-BETA-CARBOXAMIDE; ERGOATETRINE; ERGOBASINE; ERGOKLININE; ERGOLINE-8-CARBOXAMIDE,9,10-DIDEHYDRO-N-(2-HYDROXY-1-METHYLETHYL)-6-METHYL-,(8BETA(S))-; ERGOLINE-8BETA-CARBOXAMIDE,9,10-DIDEHYDRO-N-((S)-2-HYDROXY-1-METHYLETHYL)-6-METHYL-; ERGOMETRIN; ERGOMETRINE; ERGOSTETRINE; ERGOTOCINE; ERGOTRATE; ERMETRINE; N-(1-(HYDROXYMETHYL)ETHYL)-D-LYSERGAMIDE; N-(ALPHA-(HYDROXYMETHYL)ETHYL)-D-LYSERGAMIDE; N-(1-(HYDROXYMETHYL)ETHYL)-D-LYSERGOMIDE; N-(1-HYDROXYMETHYL)ETHYL)-D-LYSERGOMIDE; N-(ALPHA-(HYDROXYMETHYL)ETHYL)-D-LYSERGOMIDE; LYSERGAMIDE,N-((S)-2-HYDROXY-1-METHYLETHYL)-; D-LYSERGIC ACID 1-HYDROXYMETHYLETHYLAMIDE; D-LYSERGIC ACID L-2-PROPANOLAMIDE; LYSERGIC ACID PROPANOLAMIDE; D-LYSERGIC ACID-L,2-PROPANOLAMIDE; D-LYSERGIC ACID-L-PROPANOLAMIDE; L-LYSERGIC-L(BETA-HYDROXYISOPROPYLAMIDE); MARGONOVINE; NEOFEMERGEN; SECACORNIN; SECOMETRIN

Description: fine needles or solvated crystals; odorless

Use: oxytocic ergonovine & salts; vet: oxytocic maleate

Physical Properties

Freezing Point: 162 °C (323.6 °F)

Water Solubility: 1 g/36 mL Water

Other Solubilities: 1 g Soluble 120 ml Alcohol; nearly Insoluble in Ether, Chloroform; Slightly Soluble in Hydrobromide.

RTECS Toxicity Data

Acute Dermal: Rat LD_{Lo} Route: Subcutaneous Dose: 500 mg/kg; Toxic Effects: Behavioral - Convulsions or effect on seizure threshold; Lungs, Thorax, or Respiration - Respiratory stimulation; Nutritional and gross metabolic - Body temperature increase.

Hazard Overviews

Poison

Health: Poison. Other Acute Effects: may be fatal if inhaled, swallowed, or absorbed through skin; exposure can cause nausea; dizziness; headache; stomach pains; vomiting; diarrhea; other symptoms include shortness of breathe; coronary spasm; angina. Chronic Effects: an oxytocic, in pregnant women overexposure may result in abortion or fetal harm; target organs: female reproductive system; vascular system.

Fire: Hazards: emits toxic fumes. Extinguishing agents: carbon dioxide, dry chemical powder or appropriate foam. Precautions: combustible liquid.

Reactivity: Stable. Hazardous polymerization will not occur. Hazardous decomposition products: toxic fumes of: carbon monoxide, carbon dioxide, nitrogen oxides.

Carcinogenicity: IARC - Not listed; NIOSH - Not listed; NTP - Not listed; ACGIH - Not listed; OSHA - Not listed; EPA - Not listed; MAK - Not listed

Primary Target Organs:

Blood

Repro-ductive

Environmental

Regulations

RCRA 40CFR: Not listed

CERCLA: 40CFR 302.4: Not listed

SARA 40CFR 372.65: Not listed

SARA EHS 40CFR 355: Not listed

TSCA: Not listed

ERG3000 CAS #: 57-87-4

ERGOSTEROL

EINECS Number: 200-352-7

Molecular Formula: $C_{28}H_{44}O$

Formula Weight: 396.63

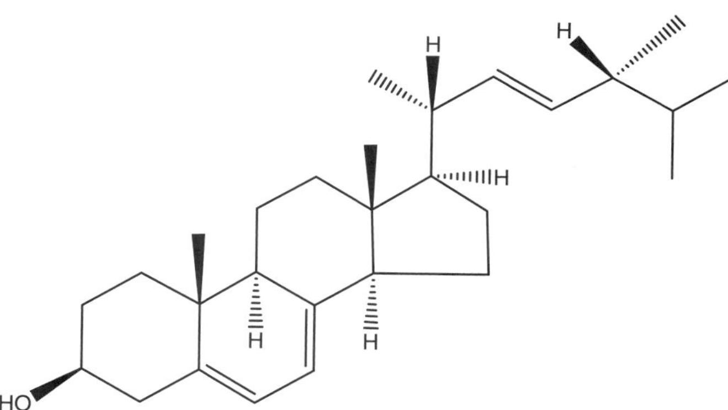

Chemical Structure

Synonyms: DELTA-5,7,22-ERGOSTATRIEN-3BETA-OL; ERGOSTA-5,7,22-TRIEN-3-OL,(3BETA,22E)-; ERGOSTA-5,7,22-TRIEN-3BETA-OL; ERGOSTA-5:6,7:8,22:23-TRIEN-3-OL; ERGOSTERIN; PROVITAMIN D; PROVITAMIN D2
Description: colorless crystals, needles, or plates
Use: antirachitic vitamin; chem int for vitamin D2

Physical Properties

Boiling Point: 250 °C (482 °F) at 0.01 mm Hg
Freezing Point: 168 °C (334.4 °F)
Water Solubility: Practically Insoluble in Water
Other Solubilities: 1 g dissolves in 45 ml boiling Alcohol, in 70 ml Ether, in 39 ml boiling Ether, in 31 ml Chloroform; Slightly Soluble in Petroleum Ether; Soluble in Benzene, Acetic Acid.

Hazard Overviews

Poison

Health: Poison. Other Acute Effects: may be fatal if swallowed. Chronic Effects: possible teratogen; overexposure may cause reproductive disorders based on tests with laboratory animals. The toxicological properties have not been thoroughly investigated.
Fire: Extinguishing agents: carbon dioxide, dry chemical powder or appropriate foam. Precautions: combustible liquid.
Reactivity: Stable. Hazardous polymerization will not occur. Incompatible with: acids, strong oxidizing agents. Hazardous decomposition products: toxic fumes of: carbon monoxide, carbon dioxide.
Carcinogenicity: IARC - Not listed; NIOSH - Not listed; NTP - Not listed; ACGIH - Not listed; OSHA - Not listed; EPA - Not listed; MAK - Not listed

Environmental

Regulations

RCRA 40CFR: Not listed
CERCLA: 40CFR 302.4: Not listed
SARA 40CFR 372.65: Not listed
SARA EHS 40CFR 355: Not listed
TSCA: Not listed

ERG7000 CAS #: 113-15-5

ERGOTAMINE

RTECS: KE7700000
EINECS Number: 204-023-9
Molecular Formula: $C_{33}H_{35}N_5O_5$
Formula Weight: 581.65
Synonyms: ERGOTAMAN-3',6',18-TRIONE,12'-HYDROXY-2'-METHYL-5'-(PHENYLMETHYL)-,(5'ALPHA)-; ERGOTAMIN; 12'-HYDROXY-2'-METHYL-5'ALPHA-(PHENYLMETHYL)ERGOTAMAN-3',6',18-TRIONE
Description: needles, plates, or elongated prisms
Use: vasoconstrictor (specific for migraine); vet: as oxytocic tartrate

Physical Properties

Boiling Point: Decomposes at 200 °C (392 °F)
Freezing Point: Decomposes at 213 °C (415.4 °F) to 214 °C (417.2 °F)
Water Solubility: Moderately Soluble in Water
Other Solubilities: Soluble in about 70 parts Methanol, 150 parts Acetone, 300 parts Alcohol; freely Soluble in Chloroform, Pyridine, Glacial Acetic Acid; moderately Soluble in Ethyl Acetate; Slightly Soluble in Benzene; almost Insoluble in Petroleum Ether.

RTECS Toxicity Data

Acute Dermal: Human TD_{Lo} Route: Subcutaneous Dose: 171 ug/kg/17D-I Toxic Effects: Musculoskelital - Changes in teeth and supporting structures; Skin and appendages - Corrosive. Infant TD_{Lo} Route: Subcutaneous Dose: 4400 ug/kg/4D-I Toxic Effects: Vascular - Regional or general arteriolar constriction.

Hazard Overviews

Carcinogenicity: IARC - Not listed; NIOSH - Not listed; NTP - Not listed; ACGIH - Not listed; OSHA - Not listed; EPA - Not listed; MAK - Not listed

Environmental

Regulations

RCRA 40CFR: Not listed
CERCLA: 40CFR 302.4: Not listed
SARA 40CFR 372.65: Not listed
SARA EHS 40CFR 355: Not listed
TSCA: Not listed

ERG9000 CAS #: 379-79-3

ERGOTAMINE TARTRATE

RTECS: KE8225000
EINECS Number: 206-835-9

Molecular Formula: $C_{70}H_{76}N_{10}O_{16}$
Formula Weight: 1313.56

Chemical Structure

Synonyms: ERGAM; ERGATE; ERGOMAR; ERGOSTAT; ERGOTAMINE BITARTRATE; ERGOTARTRATE; ETIN; EXMIGRA; FEMERGIN; GOTAMINE; GOTAMINE TARTRATE; GYNERGEN; LINGRAINE; LINGRAN; NEO-ERGOTIN; RIGETAMIN; SECAGYN; SECUPAN; TARTRATE
Description: pale beige powder
Use: as a vasoconstrictor, specifically for migraines; as an oxytocic

Physical Properties

Freezing Point: Decomposes at 203 °C (397.4 °F)
Water Solubility: < 1 mg/mL at 20 C
Other Solubilities: 95% Ethanol: <1 mg/ml at 20 °C; Acetone: 1-10 mg/ml at 18 °C; DMSO: >=100 mg/ml at 18 °C.
Flash Point: Not available; probably combustible

RTECS Toxicity Data

Acute Oral: Human TD_{Lo} Dose: 3700 ug/kg/26W-I; Toxic Effects: Behavioral - Hallucinations, distorted perceptions; Behavioral - Convulsions or effect on seizure threshold; Gastrointestinal - Nausea or vomiting. Man TD_{Lo} Dose: 214 ug/kg; Toxic Effects: Behavioral - Muscle contraction or spasticity. Woman TD_{Lo} Dose: 700 ug/kg/14D-I; Toxic Effects: Behavioral - Convulsions or effect on seizure threshold; Vascular - Regional or general arteriolar constriction. Woman TD_{Lo} Dose: 11 mg/kg/13W-I; Toxic Effects: Vascular - BP elevation not characterized in autonomic section.
Acute Dermal: Cat LD_{50} Route: Subcutaneous Dose: 11 mg/kg.
Reproductive/Teratogenic: Rat Route: Oral; Dose: 100 mg/kg; Duration: female 6-15D of pregnancy; Effects on Fertility - Pre-implantation mortality; Post-implantation mortality; Effects on Embryo or Fetus - Fetotoxicity. Rat Route: Oral; Dose: 100 mg/kg; Duration: female 6-15D of pregnancy; Specific Developmental Abnormalities - Musculoskeletal system. Rat Route: Oral; Dose: 60 mg/kg; Duration: female 5-16D of pregnancy; Effects on Embryo or Fetus - Fetotoxicity. Rat Route: Oral; Dose: 300 mg/kg; Duration: female 5-16D of pregnancy; Effects on Fertility - Other measures of fertility.
Mutagenic: Human Cytogenetic Analysis; Cell Type: lymphocyte; Dose: 100 ug/L. Mouse Cytogenetic Analysis; Route: Intraperitoneal; Dose: 200 mg/kg/24H-I.

Hazard Overviews

Poison

Health: Poison. Other Acute Effects: may be fatal if inhaled, swallowed, or absorbed through skin; exposure can cause nausea; dizziness; headache; stomach pains; vomiting; diarrhea; thirst; changes in blood pressure and heart rate; tingling or numbness in the extremities; confusion. Chronic Effects: an oxytocic; in pregnant women overexposure may result in abortion or fetal harm; other effects include peripheral circulatory disturbances and gangrene; reproductive disorder(s) based on tests with laboratory animals; laboratory experiments have shown mutagenic effects; target organs: female reproductive system, vascular system.
Fire: Will burn. Hazards: emits toxic fumes. Extinguishing agents: carbon dioxide, dry chemical powder or appropriate foam. Precautions: combustible liquid.
Reactivity: Stable. Hazardous polymerization will not occur. Hazardous decomposition products: toxic fumes of: carbon monoxide, carbon dioxide, nitrogen oxides.
Carcinogenicity: IARC - Not listed; NIOSH - Not listed; NTP - Not listed; ACGIH - Not listed; OSHA - Not listed; EPA - Not listed; MAK - Not listed
Primary Target Organs:

Blood

Repro-ductive

Environmental

Cleanup/Disposal: Guide No. 154: Eliminate all ignition sources (no smoking, flares, sparks or flames in immediate area). Do not touch damaged containers or spilled material unless wearing appropriate protective clothing. Stop leak if you can do it without risk. Prevent entry into waterways, sewers, basements or confined areas. Absorb or cover with dry earth, sand or other non-combustible material and transfer to containers. Do not get water inside containers.

Regulations

RCRA 40CFR: Not listed
CERCLA: 40CFR 302.4: Not listed

SARA 40CFR 372.65: Not listed TPQ: 500/10000 lb
SARA EHS 40CFR 355: Listed TPQ: 500 lb
TSCA: Listed

ERI3000 CAS #: 8007-27-0

ERIGERON OIL

Molecular Formula: Unknown
Synonyms: FLEABANE OIL; OIL OF CANADA FLEABANE; OIL OF ERIGERON; OIL OF FLEABANE; OILS,FLEABANE
Description: colorless to pale-yellow liquid; slightly pungent, herbaceous odor
Use: in perfumery; flavoring for foods

Hazard Overviews

Carcinogenicity: IARC - Not listed; NIOSH - Not listed; NTP - Not listed; ACGIH - Not listed; OSHA - Not listed; EPA - Not listed; MAK - Not listed

Environmental

Regulations

RCRA 40CFR: Not listed
CERCLA: 40CFR 302.4: Not listed
SARA 40CFR 372.65: Not listed
SARA EHS 40CFR 355: Not listed
TSCA: Listed

ERU3000 CAS #: 112-84-5

ERUCAMIDE

EINECS Number: 204-009-2
Molecular Formula: $C_{22}H_{43}NO$
Formula Weight: 337.37

O N H H

Chemical Structure

Synonyms: 13-DOCOSENAMIDE,(Z)-; 13-DOCOSENAMIDE,CIS-; ERUCIC ACID AMIDE; ERUCYL AMIDE; ERUCYLAMIDE
Description: solid
Use: foam stabilizer; solvent for waxes & resins; emulsions; antiblock agent for polyethylene; as adherent

Physical Properties

Freezing Point: 75 °C (167 °F) to 80 °C (176 °F)
Other Solubilities: Soluble in Isopropanol Slightly Soluble in Alcohol, Acetone

Hazard Overviews

Health: Irritating to eyes/skin/respiratory tract. Other Acute Effects: may be harmful by inhalation, ingestion, or skin absorption; target organs: heart, liver. The toxicological properties have not been thoroughly investigated.
Fire: Hazards: emits toxic fumes. Extinguishing agents: water spray; carbon dioxide, dry chemical powder or appropriate foam. Precautions: combustible liquid.
Reactivity: Stable. Hazardous polymerization will not occur. Incompatible with: strong oxidizing agents. Hazardous decomposition products: toxic fumes of: carbon monoxide, carbon dioxide, nitrogen oxides.
Carcinogenicity: IARC - Not listed; NIOSH - Not listed; NTP - Not listed; ACGIH - Not listed; OSHA - Not listed; EPA - Not listed; MAK - Not listed
Primary Target Organs:

Eyes

Skin

Respiratory System

Liver

Cardio-vascular

Environmental

Environmental Fate: Hydrolysis in water and soil should not be important. Based on sunlight absorption characteristics of chromophoric groups present in it, photolysis in water and soil surfaces may be important, but the rates of photolysis are unknown. It was found to be readily assimilated by fungi but its biodegradation in soil and water is unknown. The estimated low vapor pressure indicates volatilization from soil may not be important. The estimated Henry's Law constant suggests that volatilization from water should be negligible. Based upon an estimated K_{oc} value, it may have very low mobility in soil and should remain strongly sorbed to suspended solids and sediments in water. The estimated log BCF indicates bioconcentration in aquatic organisms should be important. The estimated vapor pressure suggests that it should be present both in the vapor and particulate phase in air. If present in the atmosphere in the vapor form, it should react with photochemically produced hydroxyl radicals and ozone with an estimated half-life of 3.2 hours and 2.1 hours, respectively. Particulate salt may be partially removed from the atmosphere by dry deposition.

Environmental Physical Data

Henry's Law Constant: estimated at 2.844×10^{-6}
Sorption Partition Coefficient: K_{oc} = estimated at 4.02

BCF: 6310

Regulations

RCRA 40CFR: Not listed
CERCLA: 40CFR 302.4: Not listed
SARA 40CFR 372.65: Not listed
SARA EHS 40CFR 355: Not listed
TSCA: Listed

ERU6000 CAS #: 112-86-7

ERUCIC ACID

EINECS Number: 204-011-3
Molecular Formula: $C_{22}H_{42}O_2$
Formula Weight: 338.56

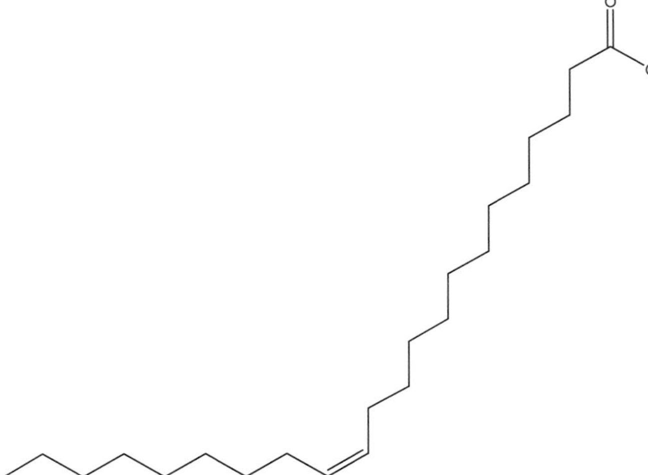

Chemical Structure

Synonyms: 13-DOCOSENOIC ACID (CIS); 13-DOCOSENOIC ACID,(Z)-; CIS-13-DOCOSENOIC ACID; DELTA 13:14-DOCOSENOIC ACID; DELTA(13)-CIS-DOCOSENOIC ACID; Z-13-DOCOSENOIC ACID

Description: needles

Use: ingredient in appetite depressants; lubricant additive; chem int for eurcamide, stearyleurcamide & dibasic acids; polyethylene film additive; water resistant nylon additive

Physical Properties

Boiling Point: 265 °C (509 °F) at 15 mm Hg
Freezing Point: 33.8 °C (92.84 °F)
Specific Gravity: 0.86 at 55 °C/4 °C
Water Solubility: Insoluble in Water
Other Solubilities: about 175 g dissolve in 100 ml Ethanol; about 160 g dissolve in 100 ml Methanol; Very Soluble in Ether
Refraction Index: 1.4534 at 45 °C/D

Hazard Overviews

Health: Irritating to eyes/skin/respiratory tract. Other Acute Effects: may be harmful by inhalation, ingestion, or skin absorption; target organs: heart, liver. The toxicological properties have not been thoroughly investigated.

Fire: Hazards: emits toxic fumes. Extinguishing agents: water spray; carbon dioxide, dry chemical powder or appropriate foam. Precautions: combustible liquid.

Reactivity: Stable. Hazardous polymerization will not occur. Incompatible with: strong oxidizing agents. Hazardous decomposition products: toxic fumes of: carbon monoxide, carbon dioxide, nitrogen oxides.

Carcinogenicity: IARC - Not listed; NIOSH - Not listed; NTP - Not listed; ACGIH - Not listed; OSHA - Not listed; EPA - Not listed; MAK - Not listed

Primary Target Organs:

Eyes

Skin

Respiratory System

Liver

Cardio-vascular

Environmental

Regulations

RCRA 40CFR: Not listed
CERCLA: 40CFR 302.4: Not listed
SARA 40CFR 372.65: Not listed
SARA EHS 40CFR 355: Not listed
TSCA: Listed

ERY1000 CAS #: 114-07-8

ERYTHROMYCIN

RTECS: KF4375000
EINECS Number: 204-040-1
Molecular Formula: $C_{37}H_{67}NO_{13}$
Formula Weight: 733.92

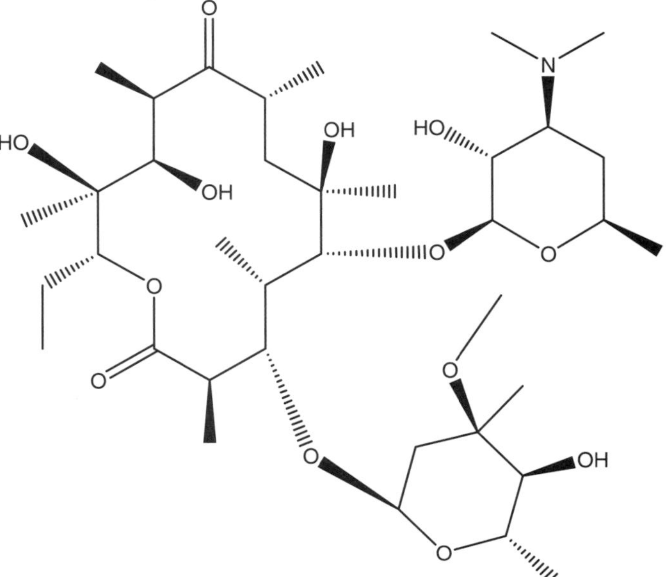

Chemical Structure

Synonyms: ABBOTICIN; ABOMACETIN; DOTYCIN; EM; EMU; ERYCIN; ERYCINUM; ERYTHROCIN; ERYTHROGRAN; ERYTHROGUENT; ERYTHROMAST 36; ERYTHROMID; ERYTHROMYCIN A; ERYTHROMYCIN SODIUM LAURYL SULFATE; ERYTHROMYCIN,COMPD WITH MONODODECYL SULFATE,SODIUM SALT; ILOTYCIN; E-MYCIN; OXACYCLOTETRADECANE-2,10-DIONE,4-((2,6-DIDEOXY-3-C-METHYL-3-O-METHYL-ALPHA-L-RIBO-HEXOPYRANOSYL)-OXY)-14-ETHYL-7,12 ,13-TR///; PANTOMICINA; PENTADECANOIC ACID,3-((2,6-DIDEOXY-3-C-METHYL-3-O-METHYL-ALPHA-L-RIBO-HEXOPYRANOSYL)-OXY)-6,11,12,13-TETRAHYDROXY-2 ,4,6,8///; PROPIOCINE; RETCIN; ROBIMYCIN; SULFURIC ACID,MONODODECYL ESTER,SODIUM SALT,COMPD WITHERYTHROMYCIN; TORLAMICINA

Description: white or slightly yellow crystals or powder; odorless or practically odorless

Use: medication: macrolide antibiotic

Physical Properties

Freezing Point: 135 °C (275 °F) to 140 °C (284 °F)

Water Solubility: About 2 mg/mL

Other Solubilities: freely Soluble in Alcohol, Acetone, Chloroform, Acetonitrile, Ethyl Acetate; moderately Soluble in Ether, Ethylene Dichloride, Amyl Acetate.

pH: Saturated solution 8 to 10.5

Refraction Index: 1.4683 at 25 °C/D

RTECS Toxicity Data

Acute Oral: Child TD_{Lo} Dose: 10 mg/kg/1D-I; Toxic Effects: Behavioral - Somnolence (general depressed activity); Nutritional and gross metabolic - Body temperature decrease. Rat LD_{50} Dose: 4600 mg/kg.

Acute Dermal: Rat LD_{Lo} Route: Subcutaneous Dose: 427 mg/kg. Mouse LD_{50} Route: Subcutaneous Dose: 1800 mg/kg.

Chronic (Multiple Dose) Oral: Rat Dose: 33600 mg/kg/6W-I; Toxic Effects: Nutritional and gross metabolic - Weight loss or decreased weight gain; DEATH.

Reproductive/Teratogenic: Rat Route: Oral; Dose: 6 gm/kg; Duration: female 10-15D of pregnancy Effects on Embryo or Fetus - Fetotoxicity. Rat Route: Subcutaneous; Dose: 50 mg/kg; Duration: female 6-10D of pregnancy; Effects on Fertility - Litter size; Effects on Embryo or Fetus - Fetotoxicity; Fetal death.

Mutagenic: Bacteria - E Coli DNA Repair; Dose: 600 ug/disc.

Hazard Overviews

Health: Irritating to eyes/skin/respiratory tract. Harmful. Other Acute Effects: may be harmful by inhalation, ingestion, or skin absorption; may cause allergic reaction; gastrointestinal disturbances; overexposure to high concentrations may cause reversible deafness; target organs: liver, gastrointestinal system, ears.

Fire: Extinguishing agents: carbon dioxide, dry chemical powder or appropriate foam. Precautions: combustible liquid.

Reactivity: Stable. Hazardous polymerization will not occur. Hazardous decomposition products: carbon monoxide, carbon dioxide, nitrogen oxides.

Carcinogenicity: IARC - Not listed; NIOSH - Not listed; NTP - Not listed; ACGIH - Not listed; OSHA - Not listed; EPA - Not listed; MAK - Not listed

Primary Target Organs:

Eyes

Skin

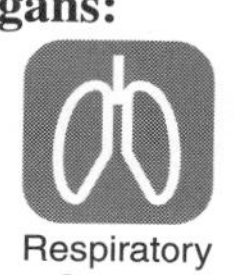
Respiratory System

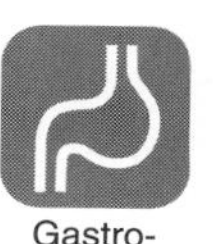
Gastro-intestinal

Liver

Exposure Limits

AIHA WEEL: TWA: 3 mg/m^3.

Respirator Recommendation

Exposure Range: >3 to 30 mg/m^3 Air Purifying, Negative Pressure, Half Mask

Exposure Range: >30 to 300 mg/m^3 Air Purifying, Negative Pressure, Full Face

Exposure Range: >300 to 3000 mg/m^3 Supplied Air, Constant Flow/Pressure Demand, Full Face

Exposure Range: >3000 to unlimited mg/m^3 Self-contained Breathing Apparatus, Pressure Demand, Full Face

Cartridge Color: dust/mist filter (use P100 or consult supervisor for appropriate dust/mist filter)

Environmental

Regulations

RCRA 40CFR: Not listed

CERCLA: 40CFR 302.4: Not listed

SARA 40CFR 372.65: Not listed

SARA EHS 40CFR 355: Not listed

TSCA: Not listed

ERY3000 **CAS #: 1264-62-6**

ERYTHROMYCIN ETHYL SUCCINATE

RTECS: WM9800000

EINECS Number: 215-033-8

Molecular Formula: $C_{43}H_{75}NO_{16}$

Formula Weight: 862.19

Chemical Structure

Synonyms: ERYTHROCIN ETHYL SUCCINATE; ERYTHROMYCIN,2'-(ETHYL BUTANEDIOATE); ERYTHROMYCIN,ETHYL SUCCINATE; ERYTHROMYCIN,MONO(ETHYL SUCCINATE); ERYTHROMYCIN,MONO(ETHYL SUCCINATE) (ESTER); ERYTHROPED; PEDIAMYCIN; SUCCINIC ACID,MONOETHYL ESTER,MONOESTER WITHERYTHROMYCIN

Description: white or slightly yellow, crystalline powder; odorless or practically so

Use: antimicrobial agent

Physical Properties

Water Solubility: Very Slightly Soluble in Water

Other Solubilities: freely Soluble in Alcohol, Soluble in Polyethylene Glycol.

Refraction Index: Alpha 1.490

RTECS Toxicity Data

Acute Oral: Man TD_{Lo} Dose: 5714 ug/kg; Toxic Effects: Behavioral - Ataxia. Child TD_{Lo} Dose: 420 mg/kg/7D-I; Toxic Effects: Gastrointestinal - Nausea or vomiting; Liver - Other changes.

Acute Dermal: Rat LD_{50} Route: Subcutaneous Dose: 4167 mg/kg; Toxic Effects: Gastrointestinal - Hypermotility, diarrhea.

Hazard Overviews

Health: Harmful. Other Acute Effects: may be harmful by inhalation, ingestion, or skin absorption; may cause allergic reaction; exposure can cause stomach pains; vomiting; diarrhea.

Fire: Hazards: emits toxic fumes. Extinguishing agents: water spray; carbon dioxide, dry chemical powder or appropriate foam. Precautions: combustible liquid.

Reactivity: Stable. Hazardous polymerization will not occur. Hazardous decomposition products: thermal decomposition may produce carbon monoxide, carbon dioxide, and nitrogen oxides.

Carcinogenicity: IARC - Not listed; NIOSH - Not listed; NTP - Not listed; ACGIH - Not listed; OSHA - Not listed; EPA - Not listed; MAK - Not listed

Environmental

Regulations

RCRA 40CFR: Not listed
CERCLA: 40CFR 302.4: Not listed
SARA 40CFR 372.65: Not listed
SARA EHS 40CFR 355: Not listed
TSCA: Not listed

ERY5000 CAS #: 643-22-1

ERYTHROMYCIN STEARATE

RTECS: KF5785000
EINECS Number: 211-396-1
Molecular Formula: $C_{55}H_{103}NO_{15}$
Formula Weight: 1018.59

Synonyms: ABBOTICINE; BRISTAMYCIN; DOWMYCIN E; ERATREX; ERYPAR; ERYTHROCIN STEARATE; ERYTHROMYCIN STEARATE (SALT); ERYTHROMYCIN STEARIC ACID SALT; ERYTHROMYCIN,OCTADECANOATE (SALT); ERYTHROMYCIN,STEARATE (SALT); ETHRIL; GALLIMYCIN; MEBERYT; OE 7; PANTOMICINA; PFIZER-E; QIDMYCIN; SK-ERYTHROMYCIN

Description: white to slightly yellow crystals or powder; practically odorless

Use: an antibiotic effective against infections caused by Gram-positive bacteria, including some beta-hemolytic streptococci, pneumococci and staphylococci; used in antibiotic treatment of upper and lower respiratory tract infections caused by chlamydia trachomatis and intestinal amebiasis; used in treatment of syphillis in patients allergic to penicillin; and may be effective in treating Legionnaire's disease

Physical Properties

Freezing Point: 100 °C (212 °F) to 104 °C (219.2 °F)
Water Solubility: Practically Insoluble in Water
Other Solubilities: 95% Ethanol: >=100 mg/ml at 22 °C; Acetone: 50-100 mg/ml at 22 °C; DMSO: 1-5 mg/ml at 22 °C; Ether: Soluble.
pH: Solution is alkaline
Flash Point: Not available; probably combustible

RTECS Toxicity Data

Acute Oral: Mouse LD_{50} Dose: 3112 mg/kg.

Acute Dermal: Mouse LD_{50} Route: Subcutaneous Dose: >2500 mg/kg.

Chronic (Multiple Dose) Oral: Rat Dose: 13037 mg/kg/14D-C; Toxic Effects: Behavioral - Food intake (animal); Gastrointestinal - Other changes; Nutritional and gross metabolic - Weight loss or decreased weight gain. Rat Dose: 71 gm/kg/14D-C; Toxic Effects: Behavioral - Food intake (animal); Gastrointestinal - Other changes; DEATH.

Hazard Overviews

Health: Irritating to eyes/skin/respiratory tract. Harmful. Other Acute Effects: may be harmful by inhalation, ingestion, or skin absorption; may cause allergic reaction; overexposure to high concentrations may cause reversible deafness; prolonged or repeated exposure may result in superinfection; target organ: liver.
Fire: Will burn. Extinguishing agents: carbon dioxide, dry chemical powder or appropriate foam. Precautions: combustible liquid.
Reactivity: Stable. Hazardous polymerization will not occur. Hazardous decomposition products: carbon monoxide, carbon dioxide, nitrogen oxides.
Carcinogenicity: IARC - Not listed; NIOSH - Not listed; NTP - Not listed; ACGIH - Not listed; OSHA - Not listed; EPA - Not listed; MAK - Not listed
Primary Target Organs:

Eyes

Skin

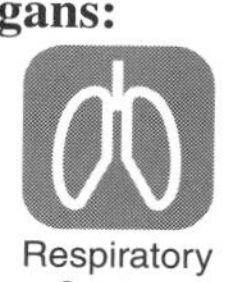
Respiratory System

Liver

Environmental

Regulations

RCRA 40CFR: Not listed
CERCLA: 40CFR 302.4: Not listed
SARA 40CFR 372.65: Not listed
SARA EHS 40CFR 355: Not listed
TSCA: Not listed

ERY7000 CAS #: 23451-24-3

ERYTHROPHLAMINE HYDROCHLORIDE

RTECS: TP8670000
Molecular Formula: $C_{25}H_{40}ClNO_6$
Formula Weight: 486.11
Synonyms: 1-PHENANTHRENECARBOXYLIC ACID,7-(2-(2-(DIMETHYLAMINO)ETHOXY)-2-OXOETHYLIDENE)TETRADECAHYDRO-2-HYDROXY-1,4A,8-TRIMETHYL-9-OXO-, METHYL ESTER, HYDROCHLORIDE,(1R-(1ALPHA,2ALPHA,4AALPHA,4BBETA,7E,8BETA,8AALPHA,10ABETA))-; 8-BETA-PODOCARPAN-16-OIC ACID,13-(CARBOXYMETHYLENE)-3-BETA-HYDROXY-14-METHYL -7-OXO-,13-(2-DIMETHYLAMINO)ETHYL) 16-METHYL ESTER, HYDROCHLORIDE; 8BETA-PODOCARPAN-16-OIC ACID,13-(CARBOXYMETHYLENE)-3BETA-HYDROXY-14-METHYL-7-OXO-,13-(2-(DIMETHYLAMINO) ETHYL)METHYL ESTER HYDROCHLORIDE
Description: crystals

Physical Properties

Freezing Point: Crystals at 149 °C (300.2 °F) to 150 °C (302 °F)

Hazard Overviews

Carcinogenicity: IARC - Not listed; NIOSH - Not listed; NTP - Not listed; ACGIH - Not listed; OSHA - Not listed; EPA - Not listed; MAK - Not listed

Environmental

Regulations

RCRA 40CFR: Not listed
CERCLA: 40CFR 302.4: Not listed
SARA 40CFR 372.65: Not listed
SARA EHS 40CFR 355: Not listed
TSCA: Not listed

ERY9000 CAS #: 36150-73-9

ERYTHROPHLEINE

RTECS: SF7325000
Molecular Formula: $C_{24}H_{39}NO_5$
Formula Weight: 421.64
Synonyms: NORCASSAMIDINE; 1-PHENANTHRENECARBOXYLIC ACID,TETRADECAHYDRO-9-HYDROXY-7-(2-(2-METHYLAMINO)ETHOXY)-2-OXOETHYLIDENE)-1,4A,8-TRIMETHYL-,METHYL ESTER; 1-PHENANTHRENECARBOXYLIC ACID,TETRADECAHYDRO-9-HYDROXY-1,4A,8-TRIMETHYL-7-(2-(2-(METHYLAMINO)ETHOXY)-2-OXOETHYLIDENE)-, METHYL ESTER,(1S-(1ALPHA,4AALPHA,4BBETA,7E,8BETA,8AALPHA,9ALPHA,10ABETA))-; TETRADECAHYDRO-9-HYDROXY-1,4A,8-TRIMETHYL-7-(2-(2-(METHYLAMINO)ETHOXY)-2-OXOETHYLIDENE)-1-PHENANTHRENECARBOXYLIC ACID METHYL ESTER
Description: solid
Use: medication: cardiotonic agent

Physical Properties

Freezing Point: 115 °C (239 °F)
Water Solubility: Soluble in Water
Other Solubilities: Soluble in Alcohol

Hazard Overviews

Carcinogenicity: IARC - Not listed; NIOSH - Not listed; NTP - Not listed; ACGIH - Not listed; OSHA - Not listed; EPA - Not listed; MAK - Not listed

Environmental

Regulations

RCRA 40CFR: Not listed
CERCLA: 40CFR 302.4: Not listed
SARA 40CFR 372.65: Not listed
SARA EHS 40CFR 355: Not listed
TSCA: Not listed

ESF5000 CAS #: 66230-04-4

ESFENVALERATE

RTECS: CY1576367
Molecular Formula: $C_{25}H_{22}ClNO_3$
Formula Weight: 419.9
Synonyms: A ALPHA; ASANA; ASANA XL; (S)-ALPHA-CYANO-3-PHENOXYBENZYL(S)-2-(4-CHLOROPHENYL)-3-METHYLBUTYRATE; [S-(R*,R*]-CYANO(3-PHENOXYPHENYL)METHYL4-CHLORO-2-(1-METHYLETHYL)BENZENEACETATE; FENVALERATE ALPHA; FENVALERATE A ALPHA; HALMARK; OMS 3023; S 1844; S 5602 A ALPHA; S-1844; S-5602ALPHA; S-5620A ALPHA; SUMI-ALPHA; SUMI-ALFA; SUMICIDIN A ALPHA
Description: viscous yellow or brown liquid or white crystalline solid
Use: control of a wide range of insect pests on cotton, maize, groundnuts, soya beans, sugar cane, sunflowers, sorghum, fruit trees, vegetables, ornamentals, and noncrop land; insecticide

Physical Properties

Boiling Point: 151 °C (304 °F) to 167 °C (333 °F)
Freezing Point: 59 °C (138.2 °F) to 60.2 °C (140.36 °F)
Specific Gravity: 1.175 at 25 °C/25 °C
Saturated Vapor Density: 7.108009667 kg/m^3
Vapor Pressure: 0.037 mPa at 25 °C
Water Solubility: < 1 mg/L at 20 °C
Other Solubilities: Solubility (1%): acetonitrile >60, Chloroform >60, DMF >60, DMSO >60, Ethyl Acetate >60, Acetone >60, ethyl cellosolve 40-50, n-hexane 1-5, Kerosene <1, Methanol 7-10, alpha-methylnaphthalene 50-60, Xylene >60.
Flash Point: Burns with difficulty

RTECS Toxicity Data

Acute Oral: Rat LD_{50} Dose: 325 mg/kg.
Acute Dermal: Rabbit LD_{50} Route: Skin; Dose: >2 gm/kg. Rat LD_{50} Route: Skin; Dose: >5 gm/kg.

Hazard Overviews

Fire: Will burn.
Carcinogenicity: IARC - Group 3, Not classifiable as to carcinogenicity to humans; NIOSH - Not listed; NTP - Not listed; ACGIH - Not listed; OSHA - Not listed; EPA - Not listed; MAK - Not listed

Environmental

Environmental Fate: If released to the atmosphere, it will degrade rapidly in the vapor phase by reaction with photochemically produced hydroxyl radicals (estimated half-life of 10 hr). If released to moist soil or water, it can degrade through aqueous hydrolysis. It hydrolyzes in water with estimated half-lives of 16.3 days, 1.63 days and 3.9 hours at pH 7, 8 and 9, respectively, at 25 °C. Photodegradation may contribute to degradation on soil surfaces and waters exposed to sunlight. It has been shown to partition from the water column to sediment and macrophytes; sediment will probably be the major reservoir in aquatic systems. Leaching in soil is not expected to be important.

Environmental Physical Data

Henry's Law Constant: estimated at 1.19×10^{-7}
Octanol/Water Partition Coefficient: log K_{ow} = 6.77
Sorption Partition Coefficient: K_{oc} = estimated at 5.0
BCF: fish 1400

Regulations

RCRA 40CFR: Not listed
CERCLA: 40CFR 302.4: Not listed
SARA 40CFR 372.65: Not listed
SARA EHS 40CFR 355: Not listed
TSCA: Not listed

Analytical Methods

Food: FDA 212.1, 232.1, 232.4, 242.1
Plasma: FDA 211.1, 231.1, 252

ESM5000 CAS #: 103598-03-4

ESMOLOL

Synonyms: BREVIBLOC
Description: white or off white crystalline powder
Use: medication: beta-adrenergic blocking drug

Physical Properties

Water Solubility: 650 mg/L at room temperature
Other Solubilities: Solubility in Alcohol: 350 mg/l at room temperature.

Hazard Overviews

Carcinogenicity: IARC - Not listed; NIOSH - Not listed; NTP - Not listed; ACGIH - Not listed; OSHA - Not listed; EPA - Not listed; MAK - Not listed

Environmental

Regulations

RCRA 40CFR: Not listed
CERCLA: 40CFR 302.4: Not listed
SARA 40CFR 372.65: Not listed
SARA EHS 40CFR 355: Not listed
TSCA: Not listed

EST5000 CAS #: 53-16-7

ESTRONE

RTECS: KG8575000
EINECS Number: 200-164-5
Molecular Formula: $C_{18}H_{22}O_2$
Formula Weight: 270.36

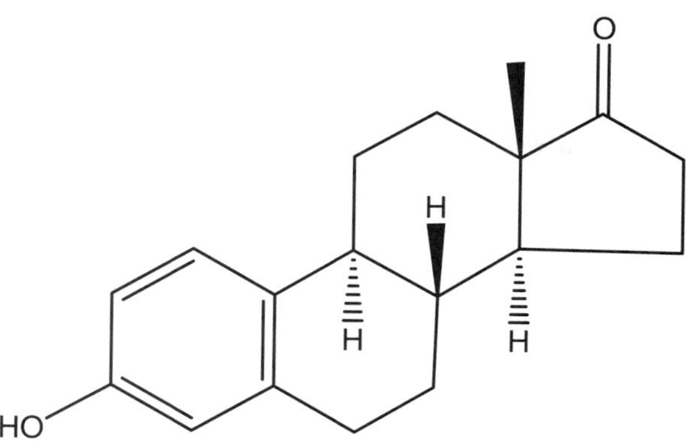

Chemical Structure

Synonyms: AQUACRINE; CRINOVARYL; CRISTALLOVAR; CRYSTOGEN; DESTRONE; DISYNFORMON; E(SUB 1); ENDOFOLLICULINA; ESTERONE; 1,3,5(10)-ESTRATRIEN-3-OL-17-ONE; 1,3,5-ESTRATRIEN-3-OL-17-ONE; DELTA-1,3,5-ESTRATRIEN-3-BETA-OL-17-ONE; ESTRA-1,3,5(10)-TRIEN-17-ONE,3-HYDROXY-; ESTROL; ESTRON; ESTRONA; ESTRONE-A; ESTROVARIN; ESTRUGENONE; ESTRUSOL; FEMESTRONE INJECTION; FEMIDYN; FOLIKRIN; FOLIPEX; FOLISAN; FOLLESTRINE; FOLLESTROL; FOLLICULAR HORMONE; FOLLICULIN; FOLLICULINE; FOLLICULINE BENZOATE; FOLLICUNODIS; FOLLIDRIN; FOLLIDRIN (TABLETS); GLANDUBOLIN; HIESTRONE; HORMOFOLLIN; HORMOVARINE; 3-HYDROXYESTRA-1,3,5(10)-TRIEN-17-ONE; 3-HYDROXY-17-KETO-ESTRA-1,3,5-TRIENE; 3-HYDROXY-17-KETO-OESTRA-1,3,5-TRIENE; 3-HYDROXY-1,3,5(10)-OESTRA-TRIEN-17-ONE; 3-HYDROXY-1,3,5(10)-OESTRATRIEN-17-ONE; 3-HYDROXY-OESTRA-1,3,5(10)-TRIEN-17-ONE; KESTRONE; KETODESTRIN; KETOHYDROXY-ESTRATRIENE; KETOHYDROXYESTRIN; KETOHYDROXYOESTRIN; KOLPON; MENAGEN; MENFORMON; 1,3,5(10)-OESTRATRIEN-3-OL-17-ONE; 1,3,5-OESTRATRIEN-3-OL-17-ONE; DELTA-1,3,5-OESTRATRIEN-3-BETA-OL-17-ONE; OESTRIN; OESTROFORM; OESTRONE; OESTROPEROS; OVEX [TABLETS]; OVIFOLLIN; PERLATAN; SOLLICULIN; THEELIN; THELESTRIN; THELYKININ; THYNESTRON; TOKOKIN; UNDEN; UNDEN (PHARMACEUTICAL); WYNESTRON

Description: white to creamy white, crystalline powder; odorless

Use: medication (vet): estrogenic hormone therapy; in prepn of commercial 19-nor steroids; medication: estrogen

Physical Properties

Freezing Point: 254.5 °C (490.1 °F) to 256 °C (492.8 °F)
Specific Gravity: Beta 1.236
Water Solubility: Insoluble in Water
Other Solubilities: Soluble in hot Acetone, secondary Pyrimidine; Slightly Soluble in Ether, Benzene, hot Chloroform, hot Alcohol; 1 g in 50 ml boiling Alcohol, 145 ml boiling Benzene; Soluble in Pyridine; Soluble in dioxane and Pyrimidine.

RTECS Toxicity Data

Reproductive/Teratogenic: Man Route: Implant; Dose: 1586 ug/kg; Duration: male 60D prior to mating; Paternal Effects - Spermatogenesis; Impotence. Rat Route: Oral; Dose: 100 ug/kg; Duration: female 1D prior to mating; Maternal Effects - Uterus, cervix, vagina. Rat Route: Oral; Dose: 560 ug/kg; Duration: male 14D prior to mating; Paternal Effects - Prostate, seminal vessicle, Cowper's gland, accessory glands. Rat Route: Oral; Dose: 2240 ug/kg; Duration: male 14D prior to mating; Paternal Effects - Testes, epididymis, sperm duct. Rat Route: Oral; Dose: 4945 ng/kg; Duration: female 1D prior to mating; Maternal Effects - Uterus, cervix, vagina. Monkey Route: Oral; Dose: 6250 ug/kg; Duration: female 5D prior to mating; Maternal Effects - Uterus, cervix, vagina.

Mutagenic: Rat DNA Adduct; Route: Oral; Dose: 870 nmol/kg. Rat Cytogenetic Analysis; Route: Intraperitoneal; Dose: 10 mg/kg.

Tumorigenic: Rat Route: Implant; Dose: 16 mg/kg; Toxic Effects: Tumorigenic - Equivocal tumorigenic agent by RTECS criteria; Skin and appendages - Tumors. Rat Route: Implant; Dose: 80 mg/kg; Toxic Effects: Tumorigenic - Equivocal tumorigenic agent by RTECS criteria; Skin and appendages - Tumors.

Hazard Overviews

Health: May be irritating to eyes/skin/respiratory tract. Toxic. Other Acute Effects: may be harmful by inhalation, ingestion, or skin absorption. Chronic Effects: may impair fertility; may cause harm to the unborn child; may cause congenital malformation in the fetus; may cause reproductive disorders; sufficient evidence for carcinogenicity in experimental animals; studies in humans strongly suggest an increased incidence of endometrial carcinoma; target organs: female reproductive system, male reproductive system, cardiovascular system, liver. Possible human carcinogen.

Fire: Hazards: emits toxic fumes. Extinguishing agents: water spray; carbon dioxide, dry chemical powder or appropriate foam. Precautions: combustible liquid.

Reactivity: Incompatible with: strong oxidizing agents. Hazardous decomposition products: toxic fumes of: carbon monoxide, carbon dioxide.

Carcinogenicity: IARC - Not listed; NIOSH - Not listed; NTP - Listed; ACGIH - Not listed; OSHA - Not listed; EPA - Not listed; MAK - Not listed

Primary Target Organs:

Liver

Cardio-vascular

Repro-ductive

Environmental

Regulations

RCRA 40CFR: Not listed
CERCLA: 40CFR 302.4: Not listed
SARA 40CFR 372.65: Not listed
SARA EHS 40CFR 355: Not listed
TSCA: Listed

ETH1000 **CAS #: 58-54-8**

ETHACRYNIC ACID

RTECS: AG6600000

EINECS Number: 200-384-1
Molecular Formula: $C_{13}H_{12}Cl_2O_4$
Formula Weight: 303.15

Chemical Structure

Synonyms: ACETIC ACID,(2,3-DICHLORO-4-(2-METHYLENEBUTYRYL)PHENOXY)-; ACETIC ACID,(2,3-DICHLORO-4-(2-METHYLENE-1-OXOBUTYL)PHENOXY)-(9CI); CRINURYL; 2,3-DICHLORO-4-(2-METHYLENEBUTYRL)PHENOXY ACETIC ACID; 2,3-DICHLORO-4-(2-METHYLENEBUTYRYL)PHENOXY ACETIC ACID; (2,3-DICHLORO-4-(2-METHYLENEBUTYRYL)PHENOXY)ACETIC ACID; (2,3-DICHLORO-4-(2-METHYLENE-1-OXOBUTYL)PHENOXY)-ACETIC ACID; (2,3-DICHLORO-4-(2-METHYLENE-1-OXOBUTYL)PHENOXY)ACETIC ACID; EDECRIL; EDECRIN; EDECRINA; ENDECRIL; ETACRINIC ACID; ETAKRINIC ACID; HIDROMEDIN; HYDROMEDIN; KYSELINA 4-(2-(1-BUTENYL)KARBONYL)-2,3-DICHLORFENOXYOCTOVA; KYSELINA ETHAKRYNOVA; METHYLENEBUTYRL PHENOXYACETIC ACID; (4-(2-METHYLENEBUTYRL)2,3-DICHLOROPHENOXY)ACETIC ACID; METHYLENEBUTYRYL PHENOXYACETIC ACID; (4-(2-METHYLENEBUTYRYL)-2,3-DICHLOROPHENOXY)ACETIC ACID; MINGIT; MK-595; OTACRIL; REDMAX; REOMAX; TALADREN; UREGIT
Description: crystals
Use: diuretic, suggested diuretic in dogs

Physical Properties

Freezing Point: 121 °C (249.8 °F) to 122 °C (251.6 °F)
Water Solubility: Sparingly Soluble in Water
Other Solubilities: 95% Ethanol: >=100 mg/ml at 24 °C; Acetone: >=100 mg/ml at 24 °C; DMSO: >=100 mg/ml at 24 °C.
Flash Point: Not available; probably combustible

RTECS Toxicity Data

Acute Oral: Man TD_{Lo} Dose: 3 mg/kg; Toxic Effects: Sense organs and special senses - Change in acuity; Kidney, Ureter, and Bladder - Urine volume increased. Woman TD_{Lo} Dose: 4 mg/kg; Toxic Effects: Sense organs and special senses - Change in acuity. Rat LD_{50} Dose: 1 gm/kg.
Acute Dermal: Mouse LD_{50} Route: Subcutaneous Dose: 189 mg/kg.

Hazard Overviews

Health: Irritating to eyes/skin/respiratory tract. Toxic. Other Acute Effects: harmful if swallowed, inhaled, or absorbed through skin. Chronic Effects: prolonged exposure can cause gastrointestinal disturbances; stomach pains; vomiting; diarrhea; target organ: kidneys.
Fire: Will burn. Hazards: emits toxic fumes. Extinguishing agents: water spray; carbon dioxide, dry chemical powder or appropriate foam. Precautions: combustible liquid.
Reactivity: Stable. Hazardous polymerization will not occur. Incompatible with: bases, ammonia. Hazardous decomposition products: toxic fumes of: carbon monoxide, carbon dioxide, hydrogen chloride gas.
Carcinogenicity: IARC - Not listed; NIOSH - Not listed; NTP - Not listed; ACGIH - Not listed; OSHA - Not listed; EPA - Not listed; MAK - Not listed
Primary Target Organs:

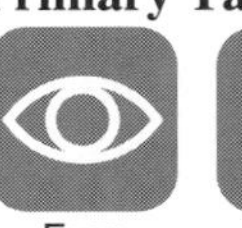
Eyes

Skin

Respiratory System

Kidneys

Environmental

Regulations

RCRA 40CFR: Not listed
CERCLA: 40CFR 302.4: Not listed
SARA 40CFR 372.65: Not listed
SARA EHS 40CFR 355: Not listed
TSCA: Not listed

ETH1050 CAS #: 74-55-5

ETHAMBUTOL

RTECS: EL3640000
EINECS Number: 200-810-6
Molecular Formula: $C_{10}H_{24}N_2O_2$
Formula Weight: 204.31
Synonyms: (+)-N,N'-BIS(1-(HYDROXYMETHYL)PROPYL)ETHYLENEDIAMINE; D,N,N'-BIS(1-HYDROXYMETHYLPROPYL)ETHYLENEDIAMINE; D-N,N'-BIS(1-HYDROXYMETHYLPROPYL)ETHYLENEDIAMINE; 1-BUTANOL,2,2'-(1,2-ETHANEDIYLDIIMINO)BIS-,(R-(R*,R*))-; 1-BUTANOL,2,2'-(ETHYLENEDIIMINO)DI-,(+)-; DADIBUTOL; DIAMBUTOL; EMB; D-ETHAMBUTOL; -2,2'-(1,2-ETHANEDIYLDIIMINO)BIS-1-BUTANOL; D-2,2'-(ETHYLENEDIIMINO)BIS(1-BUTANOL); (+)-2,2'-(ETHYLENEDIIMINO)DI-1-BUTANOL; D-2,2'-(ETHYLENEDIIMINO)DI-1-BUTANOL; TIBUTOL
Description: white, crystalline powder; essentially odorless
Use: antibacterial (tuberculostatic)

Physical Properties

Freezing Point: 87.5 °C (189.5 °F) to 88.8 °C (191.84 °F)
Water Solubility: Soluble in Water
Other Solubilities: Soluble in Propylene Glycol; Slightly Soluble in Alcohol; Insoluble in Chloroform.

RTECS Toxicity Data

Acute Oral: Man TD_{Lo} Dose: 600 mg/kg; Toxic Effects: Peripheral nerve and sensation - Structural change in nerve or sheath; Behavioral - Change in motor activity (specific assay); Kidney, Ureter, and Bladder - Other changes in urine composition. Man TD_{Lo} Dose: 1200 mg/kg; Toxic Effects:

Skin and appendages - Dermatitis, other; Immunological including allergic - Anaphylaxis. Woman TD_{Lo} Dose: 900 mg/kg/60D-I; Toxic Effects: Liver - Jaundice, cholestatic.
Acute Dermal: Mammal LD_{50} Route: Subcutaneous Dose: 890 mg/kg; Toxic Effects: Lungs, Thorax, or Respiration - Other changes.

Hazard Overviews

Carcinogenicity: IARC - Not listed; NIOSH - Not listed; NTP - Not listed; ACGIH - Not listed; OSHA - Not listed; EPA - Not listed; MAK - Not listed

Environmental

Regulations

RCRA 40CFR: Not listed
CERCLA: 40CFR 302.4: Not listed
SARA 40CFR 372.65: Not listed
SARA EHS 40CFR 355: Not listed
TSCA: Not listed

ETH1100 CAS #: 74-84-0

ETHANE

RTECS: KH3800000
DOT: UN1035; UN1961; IMO2.1
EINECS Number: 200-814-8
Molecular Formula: C_2H_6
Structured MF: C_2H_6
Formula Weight: 30.08

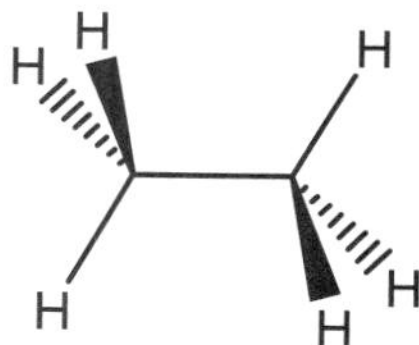

Chemical Structure

Synonyms: BIMETHYL; DIMETHYL; ETHYL HYDRIDE; METHYLMETHANE
Description: colorless gas; odorless
Use: in the prodn of ethylene by high temp thermal cracking; as a feedstock in the prodn of vinyl chloride; in the synthesis of chlorinated hydrocarbons; refrigerant; as fuel gas

Physical Properties

Boiling Point: -88 °C (-126 °F)
Freezing Point: -172 °C (-277.6 °F)
Specific Gravity: 0.446 at 0 °C/4 °C
Vapor Density: 1.04 Air=1
Vapor Pressure: Calculated 31459 mm Hg at 25 °C
Water Solubility: 60.2 ppm at 25 °C
Other Solubilities: 46 ml/100 ml Alcohol at 4 °C; > 10% in Benzene; Soluble in Ether.
Surface Tension: Estimated at 45 dynes/cm
Odor Threshold: 184.500 to 1105.7700 ppm
Refraction Index: 1.0377 at 0 °C/D & 546 MM HG
Critical Temperature: 32 °C
Critical Pressure: 48.2 atm
Ionization Potential (eV): 11.56 +/-0.02
Flash Point: -135 °C
Autoignition Temperature: 472 °C
LEL: 3.0% v/v
UEL: 12.5% v/v

Hazard Overviews

Health: A compressed gas which may cause frostbite. A simple asphyxiant which can displace available oxygen. Also Causes: rapid respiration and air hunger, diminished mental alertness, impaired muscular coordination, faulty judgment, depression of all sensations, rapid fatigue, emotional instability, nausea, vomiting, prostration, loss of consciousness, convulsions, deep coma, and death.
Fire: Flammable. Can form explosive mixtures in air. Stop flow of ethane gas before extinguishing fire. Carbon dioxide, dry chemical, or a halogenated compound can be used to extinguish flames to permit immediate access to shutoff valves. Keep surrounding area cool with direct water spray. Protect personnel attempting to stop leak with water spray Treat any fire that involves it as an emergency.
Reactivity: Stable. Hazardous polymerization cannot occur. Avoid: ignition sources (open flame; mechanical or electrical sparks; uninsulated sources of heat; lighted tobacco products; steam lines). Incompatible with: chlorine; chlorine dioxide. Hazardous decomposition products: carbon monoxide.
Carcinogenicity: IARC - Not listed; NIOSH - Not listed; NTP - Not listed; ACGIH - Not listed; OSHA - Not listed; EPA - Not listed; MAK - Not listed
Primary Target Organs:

Respirator Recommendation
Exposure Range: unlimited Self-contained Breathing Apparatus, Pressure Demand, Full Face
Note: asphyxiant, respirator needed only if atmosphere is oxygen deficient

Environmental

Environmental Fate: Photolysis, hydrolysis and bioconcentration are not expected to be important environmental fate processes. Biodegradation may occur in soil and water; however, volatilization is expected to be the dominant fate process. To a lesser extent, adsorption may also occur. A K_{oc} range of 230 to 460 indicates a medium mobility class in soil. In aquatic systems, it may partition from the water column to organic matter contained in sediments and suspended materials. A Henry's Law constant of 5.01 $x10^{-1}$

atm-cu m/mole at 25 °C suggests extremely rapid volatilization from environmental waters. The volatilization half lives from a model river and a model pond, the latter considers the effect of adsorption, have been estimated to be 1.5 hr and 1.9 days, respectively. It is expected to exist entirely in the vapor phase in ambient air. Reactions with photochemically produced hydroxyl radicals in the atmosphere have been shown to occur (average half life of 52 days or greater). Data also suggests that nighttime reactions with radical species and nitrogen oxides may contribute to atmospheric transformation.

Cleanup/Disposal: Guide No. 115: Eliminate all ignition sources (no smoking, flares, sparks or flames in immediate area). All equipment used when handling the product must be grounded. Do not touch or walk through spilled material. Stop leak if you can do it without risk. If possible, turn leaking containers so that gas escapes rather than liquid. Use water spray to reduce vapors or divert vapor cloud drift. Do not direct water at spill or source of leak. Prevent spreading of vapors through sewers, ventilation systems and confined areas. Isolate area until gas has dispersed.

Environmental Physical Data

Henry's Law Constant: calculated at 5.01 $x10^{-1}$
Octanol/Water Partition Coefficient: log K_{ow} = 1.81
Sorption Partition Coefficient: K_{oc} = estimated at 230 to 460
BCF: estimated at 1.15
BOD: none

Regulations

RCRA 40CFR: Not listed
CERCLA: 40CFR 302.4: Not listed
SARA 40CFR 372.65: Not listed
SARA EHS 40CFR 355: Not listed
TSCA: Listed

Analytical Methods

Air: ASTM D2820

ETH1150 CAS #: 79-40-3

ETHANEDITHIOAMIDE

RTECS: RP1575000
EINECS Number: 201-203-9
Molecular Formula: $C_2H_4N_2S_2$
Formula Weight: 120.2

S, S, H_2N, NH_2

Chemical Structure

Synonyms: DITHIOOXAMIDE; DITHIOXAMIDE; HYDRORUBEANIC ACID; OXALDIIMIDIC ACID,DITHIO-; OXALIC ACID,DITHIONO-,DIAMIDE; OXAMIDE,DITHIO-; RUBEAN; RUBEANE; RUBEANIC ACID; RVK

Description: red crystals or orange-red crystalline powder
Use: chemical intermediate; reagent for copper, cobalt, & nickel; as stabilizer of ascorbic acid soln

Physical Properties

Boiling Point: Decomposes at 200 °C (392 °F)
Freezing Point: Decomposes at 200 °C (392 °F)
Water Solubility: Slightly Soluble in Water
Other Solubilities: Soluble in concentrated Sulfuric acid; Soluble in Acetone & Chloroform.

RTECS Toxicity Data

Acute Oral: Rat LD_{Lo} Dose: 500 mg/kg. Mouse LD_{50} Dose: 350 mg/kg; Toxic Effects: Behavioral - Somnolence (general depressed activity); Behavioral - Convulsions or effect on seizure threshold; Behavioral - Change in motor activity (specific assay).

Hazard Overviews

Health: Irritating to eyes/skin/respiratory tract. Toxic. Other Acute Effects: harmful if swallowed, inhaled, or absorbed through skin.
Fire: Extinguishing agents: water spray; carbon dioxide, dry chemical powder or appropriate foam. Precautions: combustible liquid.
Reactivity: Incompatible with: strong oxidizing agents. Hazardous decomposition products: toxic fumes of: carbon monoxide, carbon dioxide, nitrogen oxides, sulfur oxides.
Carcinogenicity: IARC - Not listed; NIOSH - Not listed; NTP - Not listed; ACGIH - Not listed; OSHA - Not listed; EPA - Not listed; MAK - Not listed
Primary Target Organs:

Eyes

Skin

Respiratory System

Environmental

Regulations

RCRA 40CFR: Not listed
CERCLA: 40CFR 302.4: Not listed
SARA 40CFR 372.65: Not listed
SARA EHS 40CFR 355: Not listed
TSCA: Listed

ETH1200 CAS #: 540-63-6

1,2-ETHANEDITHIOL

RTECS: KI3325000
EINECS Number: 208-752-3
Molecular Formula: $C_2H_6S_2$
Formula Weight: 94.20

Chemical Structure

Synonyms: 1,2-DIMERCAPTOETHANE; DITHIOETHYLENEGLYCOL; DITHIOGLYCOL; 1,2-DITHIOL ETHANE; ETHYL HYDROPERSULFIDE; A-ETHYLENE DIMERCAPTAN; ALPHA-ETHYLENE DIMERCAPTAN; ETHYLENE DIMERCAPTAN; S-ETHYLENE DIMERCAPTAN; ETHYLENE DITHIOGLYCOL; ETHYLENE GLYCOL,DITHIO-; ETHYLENE MERCAPTAN; ETHYLENEDIMERCAPTAN; ETHYLENEDITHIOL

Description: liquid

Physical Properties

Boiling Point: 146 °C (295 °F) at 760 mm Hg
Freezing Point: -41.2 °C (-42.16 °F)
Specific Gravity: 1.123 at 23.5 °C
Water Solubility: Insoluble in Water
Other Solubilities: Soluble in Ethanol, Ether, Acetone; Soluble in oxygenated solvents.
Refraction Index: 1.5589 at 20 °C/D

RTECS Toxicity Data

Acute Oral: Mouse LD_{50} Dose: 342 mg/kg.

Hazard Overviews

Health: Irritating to eyes/skin/respiratory tract. Toxic. Other Acute Effects: harmful if swallowed, inhaled, or absorbed through skin; exposure can cause nausea, headache and vomiting.

Fire: Hazards: emits toxic fumes. Extinguishing agents: carbon dioxide, dry chemical powder or appropriate foam. Precautions: combustible liquid.

Reactivity: Incompatible with: bases, oxidizing agents, reducing agents, alkali metals. Hazardous decomposition products: toxic fumes of: carbon monoxide, carbon dioxide, sulfur oxides, hydrogen sulfide gas.

Carcinogenicity: IARC - Not listed; NIOSH - Not listed; NTP - Not listed; ACGIH - Not listed; OSHA - Not listed; EPA - Not listed; MAK - Not listed

Primary Target Organs:

Eyes

Skin

Respiratory System

Environmental

Regulations

RCRA 40CFR: Not listed
CERCLA: 40CFR 302.4: Not listed
SARA 40CFR 372.65: Not listed
SARA EHS 40CFR 355: Not listed
TSCA: Listed

ETH1250 **CAS #: 62-55-5**

ETHANETHIOAMIDE

RTECS: AC8925000
DOT: UN1035
EINECS Number: 200-541-4
Molecular Formula: C_2H_5NS
Formula Weight: 75.14

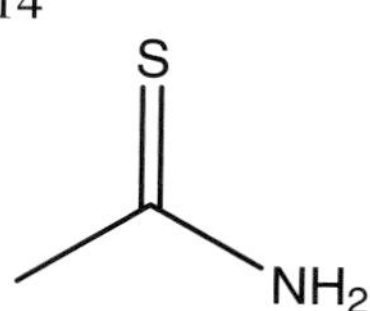

Chemical Structure

Synonyms: ACETAMIDE,THIO-; ACETIC ACID,THIONO-,AMIDE; ACETOTHIOAMIDE; TAA; THIACETAMIDE; THIOACETAMIDE; THIOACETIMIDIC ACID

Description: colorless crystals, leaflets or plates; slight odor of mercaptans

Use: substitute for hydrogen sulfide in the laboratory

Physical Properties

Freezing Point: 113 °C (235.4 °F) to 114 °C (237.2 °F)
Water Solubility: 16.3 g/100ml at 25 °C
Other Solubilities: 95% Ethanol: <1 mg/ml at 20 °C; Acetone: 50-100 mg/ml 20 °C; Benzene: Slightly Soluble; DMSO: 5-10 mg/ml at 20 °C; Ether: Slightly Soluble.
Odor Threshold: in 50% Ethylene glycol 94.7 ppm
Ionization Potential (eV): 8.36 +/-0.2
Flash Point: Not available; probably combustible

RTECS Toxicity Data

Acute Oral: Rat LD_{50} Dose: 301 mg/kg.

Acute Dermal: Mouse LD_{Lo} Route: Subcutaneous Dose: 2 gm/kg.

Chronic (Multiple Dose) Oral: Rat Dose: 933 mg/kg/10W-C; Toxic Effects: Liver - Other changes; Nutritional and gross metabolic - Weight loss or decreased weight gain. Rat Dose: 240 mg/kg/4D-I; Toxic Effects: Biochemical - Phosphatases; Biochemical - Other hydrolases. Rat Dose: 1890 mg/kg/4W-C; Toxic Effects: Blood - Changes in serum composition; Biochemical - Phosphatases; Biochemical - Hepatic microsomal mixed oxidase(dealkylation, hyroxylation,etc). Rat Dose: 3648 mg/kg/27W-C; Toxic Effects: Liver - Other changes.

Reproductive/Teratogenic: Rat Route: Intraperitoneal; Dose: 1 gm/kg; Duration: female 7D of pregnancy; Effects on Fertility - Post-implantation mortality. Rat Route: Intraperitoneal; Dose: 150 mg/kg; Duration: female 9-11D of pregnancy; Effects on Embryo or Fetus - Cytological changes (inc. somatic cell genetic material); Specific Developmental Abnormalities - Hepatobiliary system.

Mutagenic: Human DNA Inhibition; Cell Type: HeLa cell; Dose: 150 mmol/L. Monkey Cytogenetic Analysis; Cell Type: kidney; Dose: 50 mg/L. Monkey Micronucleus Test;

Cell Type: kidney; Dose: 100 mg/L. Human Other Mutation Test Systems; Cell Type: fibroblast; Dose: 100 mg/L. Monkey Other Mutation Test Systems; Cell Type: kidney; Dose: 25 mg/L/48H.

Tumorigenic: Rat Route: Oral; Dose: 7350 mg/kg/40W-C; Toxic Effects: Tumorigenic - Carcinogenic by RTECS criteria; Liver - Tumors. Rat Route: Oral; Dose: 6000 mg/kg/43W-C; Toxic Effects: Tumorigenic - Equivocal tumorigenic agent by RTECS criteria; Lungs, Thorax, or Respiration - Tumors; Liver - Tumors. Rat Route: Oral; Dose: 7200 mg/kg/51W-C; Toxic Effects: Tumorigenic - Equivocal tumorigenic agent by RTECS criteria; Liver - Tumors. Rat Route: Oral; Dose: 1008 mg/kg/9W-C; Toxic Effects: Tumorigenic - Equivocal tumorigenic agent by RTECS criteria; Liver - Tumors. Rat Route: Oral; Dose: 9900 mg/kg/71W-C; Toxic Effects: Tumorigenic - Equivocal tumorigenic agent by RTECS criteria; Lungs, Thorax, or Respiration - Tumors; Liver - Tumors. Rat Route: Oral; Dose: 1600 mg/kg/12W-C; Toxic Effects: Tumorigenic - Equivocal tumorigenic agent by RTECS criteria; Liver - Tumors. Rat Route: Oral; Dose: 5140 mg/kg/47W-C; Toxic Effects: Tumorigenic - Equivocal tumorigenic agent by RTECS criteria; Liver - Tumors; Tumorigenic effects - Prostate tumors. Rat Route: Oral; Dose: 7665 mg/kg/1Y-C; Toxic Effects: Tumorigenic - Equivocal tumorigenic agent by RTECS criteria; Liver - Tumors. Rat Route: Oral; Dose: 4320 mg/kg/34W-C; Toxic Effects: Tumorigenic - Equivocal tumorigenic agent by RTECS criteria; Liver - Tumors.

Hazard Overviews

Health: Irritating to eyes/skin/respiratory tract. Toxic. Other Acute Effects: harmful if swallowed, inhaled, or absorbed through skin. Chronic Effects: target organs: thyroid; liver. Probable carcinogen.

Fire: Will burn. Hazards: emits toxic fumes. Extinguishing agents: water spray; carbon dioxide, dry chemical powder or appropriate foam. Precautions: combustible liquid.

Reactivity: Stable. Hazardous polymerization will not occur. Incompatible with: strong oxidizing agents, strong acids, strong bases. Hazardous decomposition products: toxic fumes of: carbon monoxide, carbon dioxide, nitrogen oxides, sulfur oxides, hydrogen sulfide gas.

Carcinogenicity: IARC - Group 2B, Possibly carcinogenic to humans; NIOSH - Not listed; NTP - Listed; ACGIH - Not listed; OSHA - Not listed; EPA - Not listed; MAK - Not listed

Primary Target Organs:

Eyes

Skin

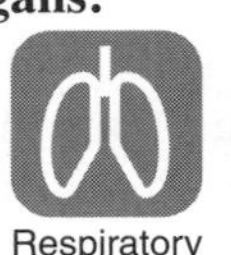
Respiratory System

Liver

Glandular System

Environmental

Environmental Fate: In the event that it is released on land it might leach into groundwater since it is poorly adsorbed by soil. It is stable in water but could degrade by photolysis. It would not be expected to volatilize, adsorb to sediment or bioconcentrate in fish. Its fate in groundwater is unknown.

Cleanup/Disposal: Guide No. 115: Eliminate all ignition sources (no smoking, flares, sparks or flames in immediate area). All equipment used when handling the product must be grounded. Do not touch or walk through spilled material. Stop leak if you can do it without risk. If possible, turn leaking containers so that gas escapes rather than liquid. Use water spray to reduce vapors or divert vapor cloud drift. Do not direct water at spill or source of leak. Prevent spreading of vapors through sewers, ventilation systems and confined areas. Isolate area until gas has dispersed.

Environmental Physical Data

Octanol/Water Partition Coefficient: log K_{ow} = calculated at -0.46 to 0.36

BCF: none likely

Regulations

RCRA 40CFR: Listed Hazardous Waste No. U218 Toxic Waste

CERCLA: 40CFR 302.4: Listed per RCRA Section 3001 RQ: 10 lb (4.535 kg)

SARA 40CFR 372.65: Listed

SARA EHS 40CFR 355: Not listed

TSCA: Listed

Analytical Methods

Soil: EPA 1625

Water / Groundwater: EPA 1625

ETH1300 CAS #: 64-17-5

ETHANOL

RTECS: KQ6300000

DOT: UN1170; IMO3.2; IMO3.3

EINECS Number: 200-578-6

Molecular Formula: C_2H_6O

Structured MF: CH_3CH_2OH

Formula Weight: 46.07

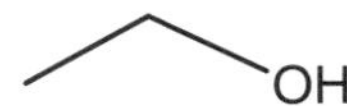

Chemical Structure

Synonyms: ABSOLUTE ETHANOL; AETHANOL; AETHYLALKOHOL; ALCOHOL; ALCOHOL ANHYDROUS; ALCOHOL DEHYDRATED; ALCOHOL,ANHYDROUS; ALCOHOL,ETHYL; ALCOOL ETHYLIQUE; ALCOOL ETILICO; ALGRAIN; ALKOHOL; ALKOHOLU ETYLOWEGO; ANHYDROL; COLOGNE SPIRIT; COLOGNE SPIRITS (ALCOHOL); ETANOLO; ETHANOL 200 PROOF; ETHANOL SOLUTION; ETHYL ALCOHOL; ETHYL ALCOHOL ANHYDROUS; ETHYL HYDRATE; ETHYL HYDROXIDE; ETHYLALCOHOL; ETOH; ETYLOWY ALKOHOL; FERMENTATION ALCOHOL; GRAIN ALCOHOL; JAYSOL; JAYSOL S; METHYLCARBINOL; MOLASSES ALCOHOL; POTATO ALCOHOL; SD ALCOHOL 23-HYDROGEN; SPIRIT; SPIRITS OF WINE; SPIRT; TECSOL; TECSOL C

Description: colorless liquid; ethanol odor

Use: alcoholic beverages; solvent in laboratory and industry (for resins, fats, fatty acids, oils, and hydrocarbons); extraction medium; antiseptic; sedative; manufacture of perfumes, pharmaceuticals (rubbing compounds, lotions, tonics, and colognes), denatured alcohol, acetaldehyde, acetic acid, ethylacetate, ethylene, 2-ethyl hexanol, nitrocellulose, ethylchloride, ether, butadiene, ethylene dibromide, lacquers, plastics and plasticizers, cosmetic rubber and rubber accelerators, aerosols, mouthwash products, soaps and cleaning preparations, polishes, dyes, adhesives, inks, preservatives, pesticides and explosives; gasoline additive/substitute; elastomers; surface coatings; antifreeze; yeast growth medium; organic synthesis; in veterinary medicine a an antiseptic, to destroy nerve tissue and as a solvent and dehydrating agent

Physical Properties

Boiling Point: 78.5 °C (173 °F)
Freezing Point: -114.1 °C (-173.38 °F)
Specific Gravity: 0.789 at 20 °C/4 °C
Vapor Density: 1.59 Air=1
Saturated Vapor Density: 1.255113485 kg/m^3
Density: 0.79 g/mL at 20 °C
Vapor Pressure: 59.3 mm Hg at 25 °C
Water Solubility: > 10% in Water
Other Solubilities: 95% Ethanol: >=100 mg/ml at 23 °C; Acetone: >=100 mg/ml at 23 °C; Benzene: Soluble; Chloroform: miscible; DMSO: >=100 mg/ml at 23 °C; Ether: miscible; Most organic solvents: miscible; Methanol: miscible.
Odor Threshold: 10 ppm
Refraction Index: 1.361 at 20 °C/D
Evaporation Rate: 2.7 Butyl Acetate=1
Critical Temperature: 243 °C
Critical Pressure: 63 atm
Ionization Potential (eV): 9.51
Flash Point: 13 °C Closed Cup
Autoignition Temperature: 363 °C
LEL: 3.3% v/v
UEL: 19% v/v

RTECS Toxicity Data

Acute Oral: Man TD_{Lo} Dose: 3371 uL/kg; Toxic Effects: Behavioral - Altered sleep time (including change in righting reflex); Behavioral - Excitment; Behavioral - Coma. Man TD_{Lo} Dose: 700 mg/kg; Toxic Effects: Behavioral - Change in psychophysiological tests. Man TD_{Lo} Dose: 50 mg/kg; Toxic Effects: Gastrointestinal - Alteration in gastric secretion; Gastrointestinal - Other changes. Man TD_{Lo} Dose: 1430 ug/kg; Toxic Effects: Behavioral - Change in motor activity (specific assay); Behavioral - Ataxia; Behavioral - Antipsychotic. Woman TD_{Lo} Dose: 256 gm/kg/12W; Toxic Effects: Behavioral - Hallucinations, distorted perceptions; Endocrine - Effect on menstrual cycle. Child TD_{Lo} Dose: 14400 mg/kg/30M-I; Toxic Effects: Behavioral - Coma; Lungs, Thorax, or Respiration - Dyspnea; Gastrointestinal - Nausea or vomiting. Human LD_{Lo} Dose: 1400 mg/kg; Toxic Effects: Behavioral - Sleep; Behavioral - Headache; Gastrointestinal - Nausea or vomiting. Child LD_{Lo} Dose: 2 gm/kg; Toxic Effects: Lungs, Thorax, or Respiration - Other changes; Liver - Fatty Liver degeneration; Blood - Other changes. Rat LD_{50} Dose: 7060 mg/kg; Toxic Effects: Lungs, Thorax, or Respiration - Other changes.
Acute Inhalation: Rat LC_{50} Dose: 20000 ppm/10hr. Mouse LC_{50} Dose: 39 gm/m^3/4hr.
Acute Dermal: Infant LD_{Lo} Route: Subcutaneous Dose: 19440 mg/kg; Toxic Effects: Behavioral - Convulsions or effect on seizure threshold; Behavioral - Coma; Nutritional and gross metabolic - Body temperature decrease. Rabbit LD_{Lo} Route: Skin; Dose: 20 gm/kg.
Chronic (Multiple Dose) Oral: Rat Dose: 1825 gm/kg/1Y-C; Toxic Effects: Liver - Fatty liver degeneration; Liver - Changes in liver weight; Nutritional and gross metabolic - Weight loss or decreased weight gain. Rat Dose: 19 gm/kg/21D-C; Toxic Effects: Biochemical - Hepatic microsomal mixed oxidase(dealkylation, hyroxylation,etc). Rat Dose: 280 gm/kg/5W-I; Toxic Effects: Cardiac - Changes in heart weight; Blood - Changes in serum composition; DEATH - Changes in testicular weight. Rat Dose: 851 gm/kg/10W-C; Toxic Effects: Biochemical - Hepatic microsomal mixed oxidase(dealkylation, hyroxylation,etc); Biochemical - Dehydrogenases. Monkey Dose: 210 gm/kg/5W-I; Toxic Effects: Liver - Fatty liver degeneration.
Irritation Eye: Rabbit Standard Draize Test Dose: 500 mg/24H; Reaction: mild. Rabbit Standard Draize Test Dose: 500 mg; Reaction: severe.
Irritation Skin: Rabbit Standard Draize Test Dose: 20 mg/24H; Reaction: moderate. Rabbit Open Draize Test Dose: 400 mg open; Reaction: mild.
Reproductive/Teratogenic: Woman Route: Oral; Dose: 41 gm/kg; Duration: female 41W of pregnancy; Effects on Newborn - Apgor score (human only); Other neonatal measures or effects; Drug dependence. Woman Route: Oral; Dose: 250 mg/kg; Duration: female 37W of pregnancy; Effects on Embryo or Fetus - Other effects to embryo or fetus. Woman Route: Intravenous; Dose: 8 gm/kg; Duration: female 32W of pregnancy; Effects on Newborn - Apgor score (human only); Other neonatal measures or effects. Woman Route: Intrauterine; Dose: 200 mg/kg; Duration: female 5D prior to mating; Effects on Fertility - Female fertility index. Rat Route: Oral; Dose: 4 gm/kg; Duration: female 13D of pregnancy; Effects on Embryo or Fetus - Cytological changes (inc. somatic cell genetic material). Rat Route: Oral; Dose: 322 gm/kg; Duration: male 35D prior to mating; Paternal Effects - Spermatogenesis; Testes, epididymis, sperm duct. Rat Route: Oral; Dose: 12 gm/kg; Duration: female 9-12D of pregnancy; Effects on Embryo or Fetus - Fetotoxicity. Rat Route: Oral; Dose: 132 gm/kg; Duration: female 1-22D of pregnancy; Maternal Effects - Parturition; Effects on Newborn - Growth statistics; Behavioral. Rat Route: Oral; Dose: 24 gm/kg; Duration: female 14-16D of pregnancy Specific Developmental Abnormalities - Central nervous system; Other developmental abnormalities. Rat Route: Oral; Dose: 354 gm/kg; Duration: female 10D after birth; Effects on Newborn - Biochemical and metabolic. Rat Route: Oral; Dose: 4 gm/kg; Duration: female 6-15D of pregnancy;

Specific Developmental Abnormalities - Eye, ear; Urogenital system. Rat Route: Oral; Dose: 35295 mg/kg; Duration: female 1-15D of pregnancy; Effects on Fertility - Female fertility index; Pre-implantation mortality; Effects on Fertility - Post-implantation mortality. Rat Route: Oral; Dose: 44 gm/kg; Duration: female 7-17D of pregnancy; Specific Developmental Abnormalities - Musculoskeletal system; Urogenital system. Rat Route: Inhalation; Dose: 20000 ppm/7H; Duration: female 1-22D of pregnancy; Specific Developmental Abnormalities - Other developmental abnormalities. Monkey Route: Oral; Dose: 78 gm/kg; Duration: female 4-23W of pregnancy; Effects on Fertility - Abortion. Monkey Route: Oral; Dose: 400 mg/kg; Duration: female 2-21W of pregnancy; Effects on Newborn - Growth statistics. Monkey Route: Oral; Dose: 206 gm/kg; Duration: female 90D prior to mating Maternal Effects - Menstrual cycle changes or disorders. Monkey Route: Oral; Dose: 32400 mg/kg; Duration: female 2-19W of pregnancy; Effects on Embryo or Fetus - Fetotoxicity; Specific Developmental Abnormalities - Central nervous system; Craniofacial (including nose and tongue). Monkey Route: Oral; Dose: 43200 mg/kg; Duration: female 1-24W of pregnancy; Effects on Embryo or Fetus - Extra embryonic structures.

Mutagenic: Human DNA Inhibition; Cell Type: lymphocyte; Dose: 220 mmol/L. Human Cytogenetic Analysis; Cell Type: lymphocyte; Dose: 1160 gm/L. Human Cytogenetic Analysis; Cell Type: fibroblast; Dose: 12000 ppm. Human Cytogenetic Analysis; Cell Type: leukocyte; Dose: 1 pph/72H-C. Human Sister Chromatid Exchenge; Cell Type: lymphocyte; Dose: 500 ppm/72H-C.

Tumorigenic: Mouse Route: Oral; Dose: 320 mg/kg/50W-I; Toxic Effects: Tumorigenic - Equivocal tumorigenic agent by RTECS criteria; Liver - Tumors; Blood - Lymphomax including Hodgkin's disease. Mouse Route: Oral; Dose: 400 gm/kg/57W-I; Toxic Effects: Tumorigenic - Equivocal tumorigenic agent by RTECS criteria; Gastrointestinal - Tumors.

Hazard Overviews

Flammable

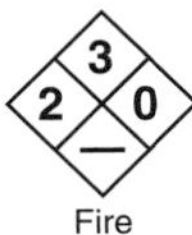

Fire Diamond

Health: Irritating to eyes/skin/respiratory tract. Also Causes: headache, nervousness, dizziness, tremors, fatigue, nausea, unconsciousness, sleep disorders, hallucinations, distorted perceptions, convulsions, motor activity changes, loss of muscular coordination, vomiting, teratogenic effects. Chronic Effects: skin defatting, dermatitis, somnolence, lack of concentration.

Fire: Flammable. Can form explosive mixtures in air. Use carbon dioxide, dry chemical or alcohol foam. Use water spray to cool fire-exposed containers, to reduce fire intensity, to suppress vapors, or to dilute and flush spilled liquid. Water may be ineffective.

Reactivity: Stable, but absorbs water rapidly from air. Hazardous polymerization cannot occur. Avoid: heat and ignition sources. Incompatible with: strong oxidizing agents; perchlorates; calcium hypochlorite; chlorine oxides; mercuric nitrate; hydrogen peroxide; nitric acid; bromine pentafluoride; uranyl perchlorate; nitrosyl perchlorate; chromyl chloride; permanganic acid; uranium hexafluoride; iodine heptafluoride; disulfuryl difluoride; acetyl chloride; acetyl bromide; manganese perchlorate and 2,2-dimethoxypropane; cyanuric acid and water; dichloromethane and sulfuric acid and nitrate or nitrite; tetrachlorisilane and water; phosphorous (III) oxide; platinum; disulfuric acid and nitric acid; potassium tert-butoxide and acids; acetic anhydride and sodium hydrogen sulfate; silver nitrate; ammonia and silver nitrate; magnesium perchlorate; sodium; nitric acid and silver; silver (I) oxide and ammonia or hydrazine. Hazardous decomposition products: oxides of carbon.

Carcinogenicity: IARC - Not listed; NIOSH - Not listed; NTP - Not listed; ACGIH - Class A4, Not classifiable as a human carcinogen; OSHA - Not listed; EPA - Not listed; MAK - Not listed

Primary Target Organs:

Eyes

Skin

Respiratory System

Nervous System

Exposure Limits

OSHA PEL: TWA: 1000 ppm; 1900 mg/m^3.
ACGIH TLV: TWA: 1000 ppm; 1880 mg/m^3.
NIOSH REL: TWA: 1000 ppm; 1900 mg/m^3.
NIOSH IDLH: 3300 ppm; LEL.
DFG MAK: TWA: 1000 ppm; 1900 mg/m^3.

Respirator Recommendation

Exposure Range: >1000 to <3300 ppm Supplied Air, Constant Flow/Pressure Demand, Half Mask

Exposure Range: 3300 to unlimited ppm Self-contained Breathing Apparatus, Pressure Demand, Full Face

Environmental

Ecotoxicity: EC_{50} Pimephales promelas (fathead minnows) 12.9 g/l/96 hr; age 30 days old, water hardness 47.3 mg/l ($CaCO_3$), temp 24.3 °C, pH 7.60, dissolved oxygen 6.8 mg/l, alkalinity 43.7 mg/l ($CaCO_3$); tank vol: 6.3 l; additions: 3.81 vol/day /Flow-through bioassay LC_{50} Pimephales promelas (fathead minnows) 14.2 g/l/96 hr (95% confidence limit 13.4-15.1 g/l); age 30 days old, water hardness 47.5 mg/l ($CaCO_3$), temp 24.0 °C, pH 7.56, dissolved oxygen 6.6 mg/l, alkalinity 40.9 mg/l ($CaCO_3$); tank vol: 6.3 l Toxicity Threshold (Cell Multiplication Inhibition Test) Entosiphon sulcatum (protozoa) 65 mg/l LC_{50} Salmo gairdneri (rainbow trout) 13000 mg/l/96 hr at 12 °C (95% confidence limit 12000-16000 mg/l), wt 0.8 g /static bioassay Toxicity Threshold (Cell Multiplication Inhibition Test) Microcystis aeruginosa (algae) 1450 mg/l LC_{50} Palaemonetes > 250 mg/l/96 hr at 21 °C, mature /static bioassay

Environmental Fate: When spilled on land it is apt to volatilize, biodegrade, and leach into the ground water, but no data on the rates of these processes could be found. Its fate in ground water is unknown. When released into water it will volatilize and probably biodegrade. It would not be expected to adsorb to sediment or bioconcentrate in fish. Although no data on its biodegradation in natural waters could be found, laboratory tests suggest that it may readily biodegrade and its detection in water systems may be due in part to its extensive use in industry with possible relatively steady and large levels of discharges. When released to the atmosphere it will photodegrade in hours (polluted urban atmosphere) to an estimated range of 4 to 6 days in less polluted areas. Rainout should be significant.

Cleanup/Disposal: Guide No. 127: Eliminate all ignition sources (no smoking, flares, sparks or flames in immediate area). All equipment used when handling the product must be grounded. Do not touch or walk through spilled material. Stop leak if you can do it without risk. Prevent entry into waterways, sewers, basements or confined areas. A vapor suppressing foam may be used to reduce vapors. Absorb or cover with dry earth, sand or other non-combustible material and transfer to containers. Use clean non-sparking tools to collect absorbed material. Large Spills: Dike far ahead of liquid spill for later disposal. Water spray may reduce vapor; but may not prevent ignition in closed spaces.

Environmental Physical Data

Henry's Law Constant: 5×10^{-6}

Octanol/Water Partition Coefficient: log K_{ow} = -0.31

BCF: none likely

BOD: 125%, 5 days

Regulations

RCRA 40CFR: Not listed

CERCLA: 40CFR 302.4: Not listed

SARA 40CFR 372.65: Not listed

SARA EHS 40CFR 355: Not listed

TSCA: Listed

Analytical Methods

Air: ASTM D3686, D3687, D4490

Soil: EPA 80APP-F1, 80APP-F2; SW846 5031, 5032, 8015A, 8015B, 8240B, 8260A, 8260B

Water / Groundwater: ASTM D3695

Indoor / Expired Air: NIOSH 1400

Plasma: EPA 29; NIOSH 8002

Other: EPA 1666, 1671

ETH1400 CAS #: 113-18-8

ETHCHLORVYNOL

RTECS: SB4725000

Molecular Formula: C_7H_9ClO

Formula Weight: 144.61

Synonyms: A 71; AETHYL-CHLORVYNOL; ALVINOL; ARVYNOL; 1-CHLORO-3-ETHYL-1-PENTEN-4-YL-3-OL; 1-CHLORO-3-ETHYL-1-PENTEN-4-YN-3-OL; 5-CHLORO-3-ETHYLPENT-1-YN-4-EN-3-OL; 1-CHLORO-3-ETHYLPENT-1-4-YN-3-OL; BETA-CHLOROVINYL ETHYL ETHYNYL CARBINOL; 3-(BETA-CHLOROVINYL)-1-PENTYN-3-OL; BETA-CHLORVINYL ETHYL ETHYNYL CARBINOL; ETCHLORVINOLO; ETHCHLOROVYNOL; ETHCHLORVINOL; ETHCHLORVINYL; ETHCLORVYNOL; ETHOCHLORVYNOL; ETHYCHLORVYNOL; ETHYL BETA-CHLOROVINYL ETHYNYL CARBINOL; NORMONSON; NORMOSAN; NORMOSON; NOSTEL; 1-PENTEN-4-YN-3-OL,1-CHLORO-3-ETHYL-; PLACIDIL; PLACIDYL; ROERIDORM; SERENESIL; SERENIL; SERENSIL

Description: colorless to yellow liquid; pungent aromatic odor

Use: medication: sedative; hypnotic; medication: anticonvulsant; muscle relaxant

Physical Properties

Boiling Point: 173 °C (343 °F) to 174 °C (345 °F) at 760 mm Hg

Specific Gravity: 1.065 to 1.07 at 25 °C/4 °C

Water Solubility: Immiscible with Water

Other Solubilities: miscible with most organic solvents

Refraction Index: 1.4675 to 1.4800 at 25 °C/D

RTECS Toxicity Data

Acute Oral: Woman TD_{Lo} Dose: 10 mg/kg; Toxic Effects: Behavioral - General anesthetic. Woman TD_{Lo} Dose: 15 mg/kg/2D; Toxic Effects: Blood - Thrombocytopenia.

Acute Dermal: Rat LD_{50} Route: Subcutaneous Dose: 200 mg/kg. Mouse LD_{50} Route: Subcutaneous Dose: 240 mg/kg.

Chronic (Multiple Dose) Oral: Dog Dose: 11550 mg/kg/77W-I; Toxic Effects: Behavioral - Ataxia.

Reproductive/Teratogenic: Rat Route: Oral; Dose: 1680 mg/kg; Duration: female 1-21D of pregnancy; Effects on Newborn - Growth statistics; Biochemical and metabolic. Rat Route: Oral; Dose: 400 mg/kg; Duration: female 1-20D of pregnancy; Specific Developmental Abnormalities - Urogenital system; Effects on Newborn - Behavioral.

Hazard Overviews

Carcinogenicity: IARC - Not listed; NIOSH - Not listed; NTP - Not listed; ACGIH - Not listed; OSHA - Not listed; EPA - Not listed; MAK - Not listed

Environmental

Regulations

RCRA 40CFR: Not listed

CERCLA: 40CFR 302.4: Not listed

SARA 40CFR 372.65: Not listed

SARA EHS 40CFR 355: Not listed

TSCA: Not listed

ETH1500 CAS #: 2622-21-1

ETHENYLCYCLOHEXENE

RTECS: GW6649500

EINECS Number: 220-070-8

Molecular Formula: C_8H_{12}

Formula Weight: 108.19

Synonyms: CYCLOHEXENE,1-ETHENYL-; CYCLOHEXENE,1-VINYL-; 1-ETHENYLCYCLOHEXENE; 1-VINYL-1-CYCLOHEXENE; 1-VINYLCYCLOHEXENE; 2-VINYLCYCLOHEXENE

Use: research chemical; chem int (possible use)

Physical Properties

Boiling Point: 145 °C (293 °F) at 760 mm Hg
Specific Gravity: 0.8623 at 15 °C/4 °C
Water Solubility: Insoluble in Water
Other Solubilities: Soluble in Ether & Benzene; Very Soluble in Methyl Alcohol
Refraction Index: 1.4915

RTECS Toxicity Data

Acute Inhalation: Rat LC_{50} Dose: 10500 mg/m^3/4hr.

Hazard Overviews

Carcinogenicity: IARC - Not listed; NIOSH - Not listed; NTP - Not listed; ACGIH - Not listed; OSHA - Not listed; EPA - Not listed; MAK - Not listed

Environmental

Regulations

RCRA 40CFR: Not listed
CERCLA: 40CFR 302.4: Not listed
SARA 40CFR 372.65: Not listed
SARA EHS 40CFR 355: Not listed
TSCA: Listed

ETH1550 CAS #: 16672-87-0

ETHEPHON

RTECS: SZ7100000
EINECS Number: 240-718-3
Molecular Formula: $C_2H_6ClO_3P$
Formula Weight: 144.49

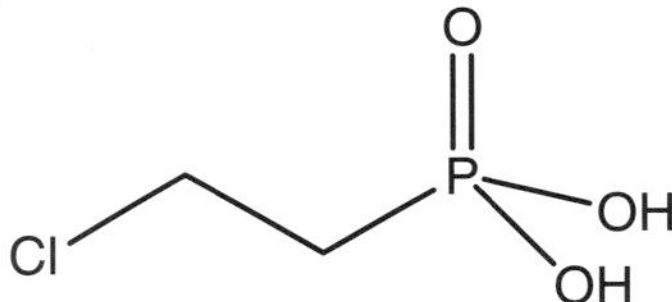

Chemical Structure

Synonyms: ACIDE CHLORO-2-ETHYL-PHOSPHONIQUE; AMCHEM 68-250; BROMOFLOR; CAMPOSAN; CEP; 2-CEPA; CEPA; CEPHA; CEPHA 10LS; CERONE; CHIPCO FLOREL PRO; 2-CHLORAETHYL-PHOSPHONSAEURE; CHLORETHEPHON; (2-CHLOROEHTYL)PHOSPHONIC ACID; 2-CHLOROETHANEPHOSPHONIC ACID; 2-(CHLOROETHYL)PHOSPHONIC ACID; 2-CHLOROETHYLPHOSPHONIC ACID; CHLOROETHYLPHOSPHONIC ACID; ETHEFON; ETHEL; ETHEPON; ETHEVERSE; ETHREL; ETHREL C; FLORDIMEX; FLOREL; G 996; GAGRO; KAMPOSAN; PHOSPHONIC ACID,(2-CHLOROETHYL)-; PREP; ROLL-FRUCT; TERPAL; TOMATHREL
Description: white waxy solid or needles
Use: plant growth regulator for barley, blackberries, guava, macadamia nuts, ornamentals, peppers, wheat, etc; produces ethylene responses in (fruit ripening, abscission, flower induction, breaking apical dominance); stimulator of latex flow in rubber; color enhancer; to promote ripening in cranberries, morello cherries, etc; to loosen fruit in currants, gooseberries, etc; to increase flower bud development in apple trees; to prevent lodging in flax; to improve sturdiness of onion seed crop.

Physical Properties

Freezing Point: 74 °C (165.2 °F) to 75 °C (167 °F)
Specific Gravity: 1.58
Saturated Vapor Density: < 1.671756817 kg/m^3
Vapor Pressure: < 0.01 mPa at 20 °C
Water Solubility: Freely Soluble in Water
Other Solubilities: Very Slightly Soluble in aromatic solvents; Very Soluble in Alcohol; Soluble in short-chain Alcohols; sparingly Soluble in nonpolar organic solvents; Insoluble in Kerosene, Diesel Oil.
Flash Point: Nonflammable

RTECS Toxicity Data

Acute Oral: Rat LD_{50} Dose: 3400 mg/kg; Toxic Effects: Biochemical - Other enzymes. Mouse LD_{50} Dose: 2850 mg/kg.
Acute Inhalation: Rat LC_{50} Dose: 90 mg/m^3/4hr.
Acute Dermal: Rabbit LD_{50} Route: Skin; Dose: 5730 mg/kg.

Hazard Overviews

Poison

Health: Irritating to eyes/skin/respiratory tract. Poison. Other Acute Effects: may be fatal if inhaled; may be harmful if swallowed; may be harmful if absorbed through the skin.
Fire: Noncombustible. Hazards: emits toxic fumes. Extinguishing agents: water spray; carbon dioxide, dry chemical powder or appropriate foam. Precautions: combustible liquid.
Reactivity: Incompatible with: strong oxidizing agents, strong bases, protect from moisture. Hazardous decomposition products: toxic fumes of: carbon monoxide, carbon dioxide, hydrogen chloride gas; thermal decomposition may produce toxic fumes of phosphorus oxides and/or phosphine.
Carcinogenicity: IARC - Not listed; NIOSH - Not listed; NTP - Not listed; ACGIH - Not listed; OSHA - Not listed; EPA - Not listed; MAK - Not listed
Primary Target Organs:

Eyes

Skin

Respiratory System

Environmental

Ecotoxicity: LC_{50} Lepomis macrochirus (Bluegil) 300 mg/l/96 hr /Conditions of bioassay not specified LD_{50} Bobwhite quail oral 1000 ppm

Regulations

RCRA 40CFR: Not listed
CERCLA: 40CFR 302.4: Not listed
SARA 40CFR 372.65: Not listed
SARA EHS 40CFR 355: Not listed
TSCA: Not listed

Analytical Methods

Food: FDA 212.1, 232.1, 232.4, 242.1
Plasma: FDA 211.1, 231.1, 252

ETH1600 CAS #: 60-29-7

ETHER

RTECS: KI5775000
DOT: UN1155; IMO3.1
EINECS Number: 200-467-2
Molecular Formula: $C_4H_{10}O$
Structured MF: $C_2H_5OC_2H_5$
Formula Weight: 74.12

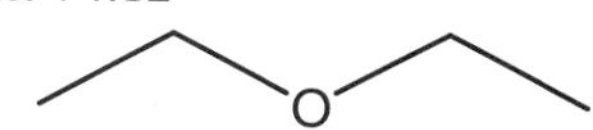

Chemical Structure

Synonyms: AETHER; ANAESTHETIC ETHER; ANESTHESIA ETHER; ANESTHETIC ETHER; DIAETHYLAETHER; DIETHYL ETHER; DIETHYL OXIDE; DWUETYLOWY ETER; ETERE ETILICO; ETHANE,1,1'-OXYBIS-; ETHER ETHYLIQUE; ETHER,ETHYL; ETHOXYETHANE; ETHYL ETHER; ETHYL OXIDE; 3-OXAPENTANE; 1,1'-OXYBISETHANE; OXYDE D'ETHYLE; PRONARCOL; SOLVENT ETHER; SULFURIC ETHER

Description: colorless, water white liquid; sweet, aromatic, odor

Use: as a solvent for waxes, fats, oils, perfumes, alkaloids, gums, resins, hydrocarbons, raw rubber and smokeless powder; solvent for nitrocellulose when mixed with alcohol; as an important reagent in organic synthesis, especially in Grignard and Wurtz type reactions; in the manufacture of gun powder, as a primer for gasoline engines, in the manufacture of ethylene, as an alcohol denaturant, as a refrigerant, in diesel fuels, in dry cleaning and as an anesthetic (via inhalation);for cleaning the skin before surgical operations and for removing adhesive plaster from the skin; in veterinary medicine as an inhalation anesthetic; for colic (orally) and as a stimulant (subcutaneously)

Physical Properties

Boiling Point: 34.6 °C (94 °F) at 760 mm Hg
Freezing Point: -116.3 °C (-177.34 °F)
Specific Gravity: 0.7134 at 20 °C/4 °C
Vapor Density: 2.55 Air=1
Density: 0.7134 g/mL at 20 °C
Bulk Density: 6 lbs/gal at 20 °C
Vapor Pressure: Extremely low
Water Solubility: 8.43% by weight in Water at 15 °C
Other Solubilities: miscible with lower alphatic alcohols, Benzene, Chloroform, Petroleum Ether, fat solvents, many oils; Soluble in concentrated Hydrochloric Acid; Soluble in Acetone; Soluble in solvent Naphtha.
Surface Tension: 17.0 dynes/cm at 20 °C
Odor Threshold: 0.99 to 3.0 mg/m^3
Refraction Index: 1.35555 at 15 °C/D
Evaporation Rate: 37.5 Butyl Acetate=1
Critical Temperature: 192.7 °C
Critical Pressure: 35.6 atm
Ionization Potential (eV): 9.53
Flash Point: -45 °C Closed Cup
Autoignition Temperature: 180 to 190 °C
LEL: 1.9% v/v
UEL: 36% v/v

RTECS Toxicity Data

Acute Oral: Man LD_{Lo} Dose: 260 mg/kg. Rat LD_{50} Dose: 1215 mg/kg.

Acute Inhalation: Mouse LC_{50} Dose: 31000 ppm/30M; Toxic Effects: Behavioral - Convulsions or effect on seizure threshold. Human TC_{Lo} Dose: 200 ppm; Toxic Effects: Sense organs and special senses - Other.

Acute Dermal: Rabbit LD; Route: Skin; Dose: >20 mL/kg. Mouse LD_{Lo} Route: Subcutaneous Dose: 8 mg/kg.

Chronic (Multiple Dose) Inhalation: Mouse Dose: 1000 ppm/24H/5W-C; Toxic Effects: Liver - Other changes; Nutritional and gross metabolic - Weight loss or decreased weight gain. Guinea Pig Dose: 1 pph/24H/5W-C; Toxic Effects: Nutritional and gross metabolic - Weight loss or decreased weight gain; DEATH.

Irritation Eye: Rabbit Standard Draize Test Dose: 100 mg; Reaction: moderate.

Irritation Skin: Rabbit Open Draize Test Dose: 360 mg open; Reaction: mild. Guinea Pig Standard Draize Test Dose: 50 mg/24H; Reaction: severe.

Mutagenic: Mouse DNA Inhibition; Cell Type: embryo; Dose: 2850 mg/L. Hamster Other Mutation Test Systems; Cell Type: fibroblast; Dose: 1 pph.

Hazard Overviews

Flammable

Fire Diamond

Health: Irritating to eyes/skin/respiratory tract. Also Causes: fatigue, giddiness, and unconsciousness. Chronic Effects: dermatitis, anorexia, headache, drowsiness, and psychological disturbances.

Fire: Highly flammable. Can form explosive mixtures in the air. Vapor may cause flash fire. Use water as fog, dry chemical, carbon dioxide, or alcohol-resistant foam. Can float on water surface and may re-ignite. Use a water spray to cool fire-exposed containers and to flush spills away from sources of ignition. Use a blanketing effect to smother flames on the surface of the burning ether.

Reactivity: Unstable, prolonged exposure to air and sunlight causes the formation of unstable peroxides that can explode if distilled or heated. Hazardous polymerization cannot occur. Avoid: ignition sources; old ether that has been exposed to air or sunlight. Incompatible with: acetyl peroxides; air; bromoazide; chlorine; chlorine trifluoride; chromic anhydride; chromyl chloride; lithium aluminum hydride; nitrosyl perchlorate; ozone; perchloric acid; permanganates; sulfuric acid; potassium peroxide; sodium peroxide; triethyl aluminum; trimethyl aluminum; liquid oxygen; nitric acid; nitrates. Hazardous decomposition products: carbon monoxide.

Carcinogenicity: IARC - Not listed; NIOSH - Not listed; NTP - Not listed; ACGIH - Not listed; OSHA - Not listed; EPA - Not listed; MAK - Not listed

Primary Target Organs:

Eyes

Skin

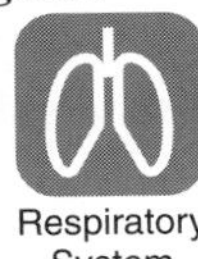
Respiratory System

Nervous System

Exposure Limits

OSHA PEL: TWA: 400 ppm; 1200 mg/m^3.

OSHA PEL Vacated 1989 Limits: TWA: 400 ppm; 1200 mg/m^3; STEL: 500 ppm; 1500 mg/m^3.

ACGIH TLV: TWA: 400 ppm; 1210 mg/m^3.

NIOSH IDLH: 1900 ppm; LEL.

DFG MAK: TWA: 400 ppm; 1200 mg/m^3.

Respirator Recommendation

Exposure Range: >400 to 1000 ppm Air Purifying, Negative Pressure, Half Mask

Exposure Range: >1000 to <1900 ppm Supplied Air, Constant Flow/Pressure Demand, Half Mask

Exposure Range: 1900 to unlimited ppm Self-contained Breathing Apparatus, Pressure Demand, Full Face

Cartridge Color: black

Environmental

Ecotoxicity: LC_{50} Poecilia reticulata (guppy) 2138 ppm/14 day /Conditions of bioassay not specified LC_{50} Pimephales promelas (fathead minnow) 2560 mg/l/96 hr /Conditions of bioassay not specified

Environmental Fate: If is released to soil, it will be subject to volatilization. It will be expected to exhibit high mobility in soil and, therefore, it may leach to groundwater. If released to water, it will not be expected to significantly adsorb to sediment or suspended particulate matter, bioconcentrate in aquatic organisms or hydrolyze, based upon estimated physical-chemical properties or analogies to other structurally related aliphatic ethers. It will not significantly photooxidize via reaction with photochemically produced hydroxyl radicals in the water. In surface water it will be subject to rapid volatilization with estimated half-lives of 3.1 hr and 1.5 days for volatilization from a river one meter deep flowing 1 m/sec with a wind velocity of 3 m/sec and a model pond respectively. Data from aqueous screening studies suggest that it is resistant to biodegradation in environmental media. It will not be expected to hydrolyze in water or soil. If released to the atmosphere, it will be expected to exist almost entirely in the vapor phase based upon a reported vapor pressure of 537 mm Hg at 25 °C. It will be susceptible to photooxidation via vapor phase reaction with photochemically produced hydroxyl radicals with a half-life of 29 hr estimated for this process. Direct photolysis will not be an important removal process since it does not absorb light at wavelengths >290.

Cleanup/Disposal: Guide No. 127: Eliminate all ignition sources (no smoking, flares, sparks or flames in immediate area). All equipment used when handling the product must be grounded. Do not touch or walk through spilled material. Stop leak if you can do it without risk. Prevent entry into waterways, sewers, basements or confined areas. A vapor suppressing foam may be used to reduce vapors. Absorb or cover with dry earth, sand or other non-combustible material and transfer to containers. Use clean non-sparking tools to collect absorbed material. Large Spills: Dike far ahead of liquid spill for later disposal. Water spray may reduce vapor; but may not prevent ignition in closed spaces.

Environmental Physical Data

Henry's Law Constant: 1.23 x10^{-3}

Octanol/Water Partition Coefficient: log K_{ow} = 0.89

Sorption Partition Coefficient: K_{oc} = estimated at 73

BCF: estimated at 2.8

BOD: 3%, 5 days

Regulations

RCRA 40CFR: Listed Hazardous Waste No. U117 Ignitable Waste

CERCLA: 40CFR 302.4: Listed per RCRA Section 3001 RQ: 100 lb (45.35 kg)

SARA 40CFR 372.65: Not listed

SARA EHS 40CFR 355: Not listed

TSCA: Listed

Analytical Methods

Air: EPA VA-006-1, VA-008-1; ASTM D3686, D3687, D4490

Soil: EPA 1624; SW846 1311, 8015A, 8015B, 8260A, 8260B

Water / Groundwater: ASTM D3695

Drinking Water: EPA 524.2

Indoor / Expired Air: NIOSH 1610

Plasma: EPA 29

ETH1700 **CAS #: 126-52-3**

ETHINAMATE

RTECS: GV9275000

EINECS Number: 204-789-4

Molecular Formula: $C_9H_{13}NO_2$

Formula Weight: 167.20

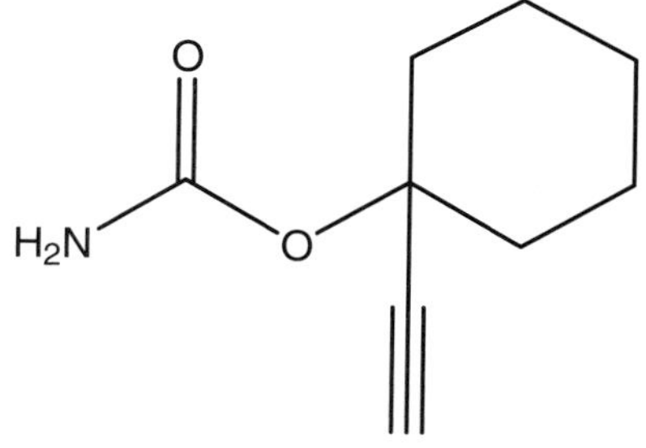

Chemical Structure

Synonyms: AETHINYL-CYCLOHEXYL-CARBAMAT; CARBAMATE DE L'ETHINYLCYCLOHEXANOL; CARBAMIC ACID 1-ETHYNYLCYCLOHEXYL ESTER; CARBAMIC ACID,1-ETHYNYLCYCLOHEXYL ESTER; CYCLOHEXANOL,1-ETHYNYL-,CARBAMATE; ETHINAMAT; 1-ETHINYLCYCLOHEXYL CARBAMATE; 1-ETHINYLCYCLOHEXYL CARBONATE; ETHIONAMIDE; 1-ETHYNYLCYCLOHEXANOL CARBAMATE; 1-ETHYNYLCYCLOHEXYL CARBAMATE; ETHYNYLCYCLOHEXYL CARBAMATE; ETINAMATE; VALAMIN; VALAMINA; VALAMINETTA; VALAMINETTAE; VALAMINETTEN; VALMID; VALMIDATE; VOLAMIN

Description: white powder or needles; essentially odorless

Use: medication: hypnotic agent

Physical Properties

Boiling Point: 118 °C (244 °F) to 122 °C (252 °F) at 3 mm Hg

Freezing Point: 96 °C (204.8 °F) to 98 °C (208.4 °F)

Water Solubility: 1 g Soluble in about 500 ml Water

Other Solubilities: freely Soluble in Alcohol, Chloroform, & Ether; Solubleat 25 °C; Ethanol 35.0%, hexane 2.0%, sesame oil 0.7%, 1,2-Propylene Glycol 22%; 1 in 3 part of Alcohol, 1 in 50 part of light petroleum, and 1 in 140 part of sesame oil.

pH: Saturated solution about 6.5

Refraction Index: Alpha 1.53

RTECS Toxicity Data

Acute Oral: Human LD_{Lo} Dose: 57 mg/kg. Rat LD_{50} Dose: 331 mg/kg; Toxic Effects: Behavioral - General anesthetic; Behavioral - Somnolence (general depressed activity).

Acute Dermal: Rat LD_{50} Route: Subcutaneous Dose: 390 mg/kg.

Reproductive/Teratogenic: Mouse Route: Intraperitoneal; Dose: 400 mg/kg; Duration: female 10D of pregnancy; Effects on Embryo or Fetus - Fetal death; Specific Developmental Abnormalities - Other developmental abnormalities.

Hazard Overviews

Health: Toxic. Other Acute Effects: harmful if swallowed, inhaled, or absorbed through skin; exposure can cause nausea; vomiting; drowsiness; dizziness. Chronic Effects: prolonged or repeated exposure can lead to habituation or addiction.

Fire: Extinguishing agents: carbon dioxide, dry chemical powder or appropriate foam. Precautions: combustible liquid.

Reactivity: Stable. Hazardous polymerization will not occur. Hazardous decomposition products: toxic fumes of: carbon monoxide, carbon dioxide, nitrogen oxides.

Carcinogenicity: IARC - Not listed; NIOSH - Not listed; NTP - Not listed; ACGIH - Not listed; OSHA - Not listed; EPA - Not listed; MAK - Not listed

Environmental

Regulations

RCRA 40CFR: Not listed

CERCLA: 40CFR 302.4: Not listed

SARA 40CFR 372.65: Not listed

SARA EHS 40CFR 355: Not listed

TSCA: Listed

ETH1800 CAS #: 563-12-2

ETHION

RTECS: TE4550000

DOT: UN3018; NA2783; IMO6.1

EINECS Number: 209-242-3

Molecular Formula: $C_9H_{22}O_4P_2S_4$

Structured MF: $[(C_2H_5O)_2PSS]_2CH_2$

Formula Weight: 384.48

Synonyms: AC 3422; BIS(S-(DIETHOXYPHOSPHINOTHIOYL)MERCAPTO)METHANE; DIETHION; EMBATHION; ENT 24,105; ETHANOX; ETHIOL 100; ETHODAN; ETHOPAZ; ETHYL METHYLENE PHOSPHORODITHIOATE; FMC-1240; FOSFATOX E; FOSFONO 50; HYLEMOX; KWIT; METHYLEEN-S,S'-BIS(O,O-DIETHYL-DITHIOFOSFAAT); S,S'-METHYLEN-BIS(O,O-DIAETHYL-DITHIOPHOSPHAT); S,S'-METHYLENE O,O,O',O'-TETRAETHYL PHOSPHORODITHIOATE; METHYLENE-S,S'-BIS(O,O-DIAETHYL-DITHIOPHOSPHAT); NIA 1240; NIAGARA 1240; NIALATE; PHOSPHORODITHIOIC ACID,O,O-DIETHYL ESTER,S,S-DIESTERWITH METHANEDITHIOL; PHOSPHORODITHIOIC ACID,O,O-DIETYL ESTER,S,S-DIESTER WITHMETHANEDITHIOL; PHOSPHORODITHIOIC ACID,S,S'-METHYLENEO,O,O',O'-TETRAETHYL ESTER; PHOSPHOTOX E; RHODOCIDE; RODOCID; RP 8167; O,O,O',O'-TETRAAETHYL-BIS(DITHIOPHOSPHAT); TETRAETHYL S,S'-METHYLENE BIS(PHOSPHOROTHIOLOTHIONATE); O,O,O',O'-TETRAETHYL S,S'-METHYLENE DI(PHOSPHORODITHIOATE); O,O,O,O-TETRAETHYL S,S'-METHYLENEBIS(DITHIOPHOSPHATE); O,O,O',O'-TETRAETHYL S,S'-METHYLENEBISPHOSPHORDITHIOATE; O,O,O',O'-TETRAETHYL-S,S'-METHYLENEBISPHOSPHORODITHIOATE; VEGFRU FOSMITE

Description: colorless to amber liquid; odorless (when pure) or disagreeable odor (technical grade)

Use: non-systemic insecticide & acaricide on food, fiber, & ornamental crops, eg, citrus fruit, apples, nuts, other fruit, & cotton; aphids, mites, scales, thrips, leafhoppers, maggots, & foliar feeding larvae cattle dip for ticks & treatment for buffalo flies;apples, to kill eggs & scales; in control of boophilus.

Physical Properties

Boiling Point: Decomposes > 150 °C (302 °F)

Freezing Point: -13 °C (8.6 °F) to -12 °C (10.4 °F)

Specific Gravity: 1.22 at 20 °C/4 °C
Vapor Density: 13.26 Air=1
Saturated Vapor Density: 1.200000029 kg/m^3
Vapor Pressure: 1.5 x10^{-6} mm Hg at 25 °C
Water Solubility: Slightly Soluble in Water
Other Solubilities: Soluble in petroleum oils, most organic solvents; Soluble in Methyl Alcohol & Ethyl Alcohol; miscible with aromatics.
Odor Threshold: Emulsifiable 0.6 mg/l
Refraction Index: 1.5490 at 20 °C/D
Flash Point: 176 °C

RTECS Toxicity Data

Acute Oral: Human TD_{Lo} Dose: 100 ug/kg; Toxic Effects: Biochemical - True cholinesterase. Infant TD_{Lo} Dose: 15700 ug/kg; Toxic Effects: Peripheral nerve and sensation - Flaccid paralysis without anesthesia; Behavioral - Change in motor activity (specific assay); Nutritional and gross metabolic - Body temperature increase. Rat LD_{50} Dose: 13 mg/kg.
Acute Inhalation: Rat LC_{50} Dose: 864 mg/m^3.
Acute Dermal: Rat LD_{50} Route: Skin; Dose: 62 mg/kg. Guinea Pig LD_{50} Route: Skin; Dose: 915 mg/kg.
Chronic (Multiple Dose) Oral: Dog Dose: 675 mg/kg/90D-C; Toxic Effects: Blood - Other changes; Nutritional and gross metabolic - Weight loss or decreased weight gain; Biochemical - True cholinesterase. Dog Dose: 182 mg/kg/1Y-C; Toxic Effects: Blood - Other changes; Biochemical - True cholinesterase.

Hazard Overviews

Poison

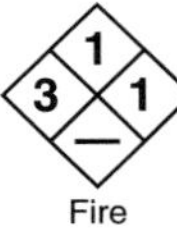

Fire Diamond

Health: Irritating to eyes/respiratory tract. Poison. Also Causes: abdominal cramps, nausea, salivation, cough, sweating, anxiety, confusion, muscle twitching, depression, dizziness, rapid heart beat, high blood pressure, eye effects, death. Chronic Effects: acetylcholine build-up
Fire: Will burn. Use dry chemical, carbon dioxide, water spray, or regular foam.
Reactivity: Unstable, slowly oxidizes on exposure to air. Hazardous polymerization cannot occur. Avoid: heat; alkalis; acids. Incompatible with: acids; alkalis. Hazardous decomposition products: carbon; sulfur; phosphorus oxide(s).
Carcinogenicity: IARC - Not listed; NIOSH - Not listed; NTP - Not listed; ACGIH - Not listed; OSHA - Not listed; EPA - Not listed; MAK - Not listed
Primary Target Organs:

Eyes

Skin

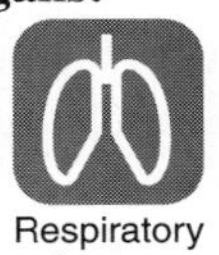
Respiratory System

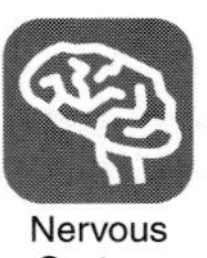
Nervous System

Cardio-vascular

Exposure Limits
OSHA PEL Vacated 1989 Limits: TWA: 0.4 mg/m^3.
ACGIH TLV: TWA: 0.4 mg/m^3.

Respirator Recommendation
Exposure Range: >0.4 to 4 mg/m^3 Air Purifying, Negative Pressure, Half Mask
Exposure Range: >4 to 40 mg/m^3 Air Purifying, Negative Pressure, Full Face
Exposure Range: >40 to 400 mg/m^3 Supplied Air, Constant Flow/Pressure Demand, Full Face
Exposure Range: >400 to unlimited mg/m^3 Self-contained Breathing Apparatus, Pressure Demand, Full Face
Cartridge Color: black with magenta (P100)

Environmental

Ecotoxicity: LC_{50} Fathead minnow 720 ug/l/96 hr at 18 °C, fingerling LC_{50} Gammarus fasciatus 1.8 ug/l/96 hr at 21 °C (95% confidence limit 1.3- 2.4 ug/l), mature / static bioassay LC_{50} Micropterus salmoides 150 ug/l/96 hr /Conditions of bioassay not specified LC_{50} Ring-necked pheasant greater than 5000 ppm in 5-day diet, age 10 days LC_{50} Palaemonetes kadiakensis 5.6 ug/l/96 hr at 15 °C (95% confidence limit 3.2-9.8 ug/l), mature / static bioassay
Environmental Fate: If released to soil, microbial degradation may be an important process for removal based on degradation studies with sterile versus nonsterile soils. It is susceptible to slow aqueous hydrolysis at acidic and neutral conditions with the hydrolysis becoming relatively rapid at pHs approaching 9 and above; therefore, hydrolysis may be the dominant transformation mechanism in very alkaline moist soils. Based on measured K_{oc} values of 6451 to 15435, it is not expected to leach in soil. The persistence half-life in various soils under laboratory conditions has been found to vary from 1.3 to 8 weeks. Greenhouse and field studies with an organic soil have found half-lives of 16 to 49 weeks and significant carry-over from one planting season to the next. If released to water, microbial degradation may not be important in relation to abiotic processes based on degradation studies with natural waters versus sterilized natural waters. The hydrolysis half-lives at 25 °C are 63, 58, 25, and 8.4 weeks at pHs 5, 6, 7, and 8, respectively, with a half-life of 1 day at pH 10 and 30 °C. It may be susceptible to some photooxidation in sunlit natural water. The high measured K_{oc} values suggest that adsorption to sediment will be an important transport process. Based on a measured log Kow of 5.073, bioconcentration is potentially significant. Volatilization from water is not expected to be important, with the possible exception of rapidly moving shallow streams. The persistence half-life in three different natural waters under laboratory conditions was found to vary from 4 to 22 weeks; it was suggested that the faster rate of degradation in one of the natural waters was due to the presence of dissolved ions which caused an unidentified catalytic effect. If released to the atmosphere, it will exist in both the vapor-phase and adsorbed-phase, although the adsorbed-phase may dominate. In the vapor-phase, it will react rapidly with photochemically produced hydroxyl radicals at an estimated half-life rate of 6.95 hours. In the adsorbed-phase it will be subject to wet and dry deposition.
Cleanup/Disposal: Guide No. 152: Do not touch damaged containers or spilled material unless wearing appropriate

protective clothing. Stop leak if you can do it without risk. Prevent entry into waterways, sewers, basements or confined areas. Cover with plastic sheet to prevent spreading. Absorb or cover with dry earth, sand or other non-combustible material and transfer to containers. Do not get water inside containers.

Environmental Physical Data

Henry's Law Constant: estimated at 6.9 x10^{-7}
Octanol/Water Partition Coefficient: log K_{ow} = 5.073
Sorption Partition Coefficient: K_{oc} = 1.54 x10^{4}
BCF: estimated at 4220

Regulations

RCRA 40CFR: Not listed
CERCLA: 40CFR 302.4: Listed per CWA Section 311(b)(4) RQ: 10 lb (4.535 kg)
SARA 40CFR 372.65: Not listed TPQ: 1000 lb
SARA EHS 40CFR 355: Listed TPQ: 10 lb
TSCA: Not listed

Analytical Methods

Soil: EPA PMD-ETI, PMD-TLC, 025; SW846 8270B, 8270C; USGS O5104, O7104
Water / Groundwater: EPA P-005-1, 1657, 614, 614.1, 022; USGS O3104
Drinking Water: AOAC 991.07; ASTM D5475
Food: FDA 212.1, 232.1, 232.2, 232.3, 232.4, 242.1; AOAC 968.24, 970.52, 979.04
Indoor / Expired Air: NIOSH 5600
Plasma: EPA 001, 003, 027, 028, 29; FDA 211.1, 231.1, 252

ETH1850 CAS #: 535-32-0

(D)-ETHIONINE

RTECS: ES6825100
EINECS Number: 208-612-1
Molecular Formula: $C_6H_{13}NO_2S$
Formula Weight: 163.23

Chemical Structure

Synonyms: D-2-AMINO-4-(ETHYLTHIO)BUTYRIC ACID; BUTYRIC ACID,2-AMINO-4-(ETHYLTHIO)-,D-; D-ETHIONINE; D-HOMOCYSTEINE,S-ETHYL-

Physical Properties

Freezing Point: Decomposes at 278 °C (532.4 °F)

Hazard Overviews

Health: May cause irritation. Toxic. Other Acute Effects: harmful if swallowed, inhaled, or absorbed through skin. Chronic Effects: may alter genetic material; target organ: liver. Carcinogen.
Fire: Hazards: emits toxic fumes. Extinguishing agents: water spray; carbon dioxide, dry chemical powder or appropriate foam. Precautions: combustible liquid.
Reactivity: Incompatible with: strong oxidizing agents. Hazardous decomposition products: toxic fumes of: carbon monoxide, carbon dioxide.
Carcinogenicity: IARC - Not listed; NIOSH - Not listed; NTP - Not listed; ACGIH - Not listed; OSHA - Not listed; EPA - Not listed; MAK - Not listed
Primary Target Organs:

Liver

Environmental

Regulations

RCRA 40CFR: Not listed
CERCLA: 40CFR 302.4: Not listed
SARA 40CFR 372.65: Not listed
SARA EHS 40CFR 355: Not listed
TSCA: Not listed

ETH1900 CAS #: 13073-35-3

(L)-ETHIONINE

RTECS: ES6825300
EINECS Number: 235-966-4
Molecular Formula: $C_6H_{13}NO_2S$
Formula Weight: 162.23

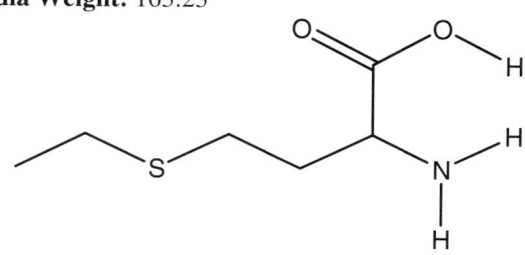

Chemical Structure

Synonyms: ALPHA-AMINO-GAMMA-(ETHYLMERCAPTO)BUTYRIC ACID; 2-AMINO-4-(ETHYLTHIO)BUTYRIC ACID; L-2-AMINO-4-(ETHYLTHIO)BUTYRIC ACID; BUTANOIC ACID,2-AMINO-4-(ETHYLTHIO)-,; BUTYRIC ACID,2-AMINO-4-(ETHYLENE)-,L-; ETH; ETHIONINE; L-ETHIONINE; L-S-ETHYLHOMOCYSTEINE; HOMOCYSTEINE S-ETHYL ETHER; HOMOCYSTEINE,S-ETHYL-; L-HOMOCYSTEINE,S-ETHYL-
Description: crystals
Use: exptl carcinogen; biological research chemical

Physical Properties

Freezing Point: Decomposes at 272 °C (521.6 °F) to 274 °C (525.2 °F)

RTECS Toxicity Data

Mutagenic: Human DNA Inhibition; Cell Type: lymphocyte; Dose: 4 mmol/L. Human Sister Chromatid Exchenge; Cell Type: lymphocyte; Dose: 2 mmol/L.
Tumorigenic: Mouse Route: Oral; Dose: 44100 mg/kg/2Y-C; Toxic Effects: Tumorigenic - Carcinogenic by RTECS criteria; Lungs, Thorax, or Respiration - Tumors; Liver - Tumors.

Hazard Overviews

Health: Irritating to eyes/skin/respiratory tract. Toxic. Other Acute Effects: harmful if swallowed, inhaled, or absorbed through skin. Chronic Effects: may alter genetic material; target organs: liver. Carcinogen.
Fire: Hazards: emits toxic fumes. Extinguishing agents: water spray; carbon dioxide, dry chemical powder or appropriate foam. Precautions: combustible liquid.
Reactivity: Incompatible with: strong oxidizing agents. Hazardous decomposition products: toxic fumes of: carbon monoxide, carbon dioxide.
Carcinogenicity: IARC - Not listed; NIOSH - Not listed; NTP - Not listed; ACGIH - Not listed; OSHA - Not listed; EPA - Not listed; MAK - Not listed
Primary Target Organs:

Eyes

Skin

Respiratory System

Liver

Environmental

Regulations

RCRA 40CFR: Not listed
CERCLA: 40CFR 302.4: Not listed
SARA 40CFR 372.65: Not listed
SARA EHS 40CFR 355: Not listed
TSCA: Listed

ETH1950 CAS #: 116-01-8

ETHOATE-METHYL

RTECS: TE0960000
EINECS Number: 204-121-1
Molecular Formula: $C_6H_{14}NO_3PS_2$
Formula Weight: 243.30
Synonyms: AC 18706; AMERICAN CYANAMID 18706; B 77; B/77; CL 18706; DIMETHOATE-ETHYL; O,O-DIMETHYL S-(N-ETHYLCARBAMOYLMETHYL) DITHIOPHOSPHATE; O,O-DIMETHYL S-(N-ETHYLCARBAMOYLMETHYL) PHOSPHORODITHIOATE; DWUTIOFOSFORAN S-N-ETYLOKARBAMYLOMETYLO-O,O-DWUMETYLOWY; EI-18706; ENT-25,506; ENT-25506; ETHOATE METHYL; N-ETHYLAMIDE OF O,O-DIMETHYL DITHIOPHOSPHORYLACETIC ACID; S-(2-(ETHYLAMINO)-2-OXOETHYL) O,O-DIMETHYLPHOSPHORODITHIOATE; S-(N-ETHYLCARBAMOYLMETHYL) DIMETHYL PHOSPHORODITHIOATE; S-ETHYLCARBAMOYLMETHYL O,O-DIMETHYL PHOSPHORODITHIOATE; ETOAT METYLOWY; FITIOS; FITIOS B/77; N-MONOETHYLAMIDE OF O,O-DIMETHYLDITHIOPHOSPHORYLACETICACID; OMS 252; PHOSHOROTHIOIC ACID,S-(2-(ETHYLAMINO)-2-OXOETHYL) O,O-DIMETHYL ESTER; PHOSPHORODITHIOIC ACID,O,O-DIMETHYL ESTER,S-ESTER WITHN-ETHYL-2-MERCAPTOACETAMIDE; PHOSPHORODITHIOIC ACID,S-(2-(ETHYLAMINO)-2-OXOETHYL)O,O-DIMETHYL ESTER; VEL 88
Description: white crystalline solid; slightly aromatic odor
Use: systemic insecticide & acaricide

Physical Properties

Freezing Point: 65.5 °C (149.9 °F) to 66.7 °C (152.06 °F)
Specific Gravity: 1.164 at 70 °C
Water Solubility: 8.5 g/L at 26 °C
Other Solubilities: 0.95% in Olive Oil, 630 g/kg in Benzene, 83O g/kg in Chloroform, 430 g/kg in Diethyl Ether, 6O g/kg in Xylene; Very Soluble in Acetone, Ethanol
Refraction Index: 1.5225 at 70 °C/D

RTECS Toxicity Data

Acute Oral: Rat LD_{50} Dose: 125 mg/kg. Mouse LD_{50} Dose: 350 mg/kg.
Acute Dermal: Rat LD_{50} Route: Skin; Dose: 2 gm/kg.
Chronic (Multiple Dose) Oral: Rat Dose: 675 mg/kg/9W-I; Toxic Effects: Liver - Changes in liver weight; Biochemical - True cholinesterase; DEATH.

Hazard Overviews

Carcinogenicity: IARC - Not listed; NIOSH - Not listed; NTP - Not listed; ACGIH - Not listed; OSHA - Not listed; EPA - Not listed; MAK - Not listed

Environmental

Regulations

RCRA 40CFR: Not listed
CERCLA: 40CFR 302.4: Not listed
SARA 40CFR 372.65: Not listed
SARA EHS 40CFR 355: Not listed
TSCA: Not listed

ETH2000 CAS #: 77-15-6

ETHOHEPTAZINE

RTECS: CM2800000
EINECS Number: 201-007-3
Molecular Formula: $C_{16}H_{23}NO_2$
Formula Weight: 261.35
Synonyms: AETHOHEPTAZIN; 1H-AZEPINE-4-CARBOXYLIC ACID,HEXAHYDRO-1-METHYL-4-PHENYL-,ETHYL ESTER; AZEPINE-4-CARBOXYLIC ACID,HEXAHYDRO-1-METHYL-4-PHENYL-,ETHYL ESTER; 4-CARBETHOXY-1-METHYL-4-PHENYLAZACYCLOHEPTANE; 4-CARBETHOXY-1-METHYL-4-

PHENYLHEXAMETHYLENIMINE; ETHYL HEPTAZINE; ETHYL HEXAHYDRO-1-METHYL-4-PHENYL-AZEPINE-4-CARBOXYLATE; HEXAHYDRO-1-METHYL-4-PHENYL-4-AZEPINECARBOXYLIC ACID ETHYLESTER; 1-METHYL-4-CARBETHOXY-4-PHENYLHEXAMETHYLENEIMINE; 1-METHYL-4-CARBETHOXY-4-PHENYLHEXAMETHYLENIMINE; WY 401; ZACTANE

Description: liquid

Use: medication as analgesic

Physical Properties

Boiling Point: 127 °C (261 °F) to 129 °C (264 °F) at 0.5 mm Hg

Specific Gravity: 1.038 at 26 °C/4 °C

Water Solubility: 1 in 100 solution

pH: 1 in 100 solution 3.5 to 4.5

Refraction Index: 1.5210 at 26 °C/D

RTECS Toxicity Data

Acute Oral: Rat LD_{50} Dose: 355 mg/kg. Mouse LD_{50} Dose: 318 mg/kg.

Hazard Overviews

Carcinogenicity: IARC - Not listed; NIOSH - Not listed; NTP - Not listed; ACGIH - Not listed; OSHA - Not listed; EPA - Not listed; MAK - Not listed

Environmental

Regulations

RCRA 40CFR: Not listed

CERCLA: 40CFR 302.4: Not listed

SARA 40CFR 372.65: Not listed

SARA EHS 40CFR 355: Not listed

TSCA: Not listed

ETH2050 CAS #: 13194-48-4

ETHOPROPHOS

RTECS: TE4025000

DOT: UN2783; UN2784; UN3017; UN3018; IMO3.2; IMO6.1

EINECS Number: 236-152-1

Molecular Formula: $C_8H_{19}O_2PS_2$

Structured MF: $CH_3CH_2OP(O)(SCH_2CH_2CH_3)_2$

Formula Weight: 242.36

Synonyms: ENT 27,318; ETHOPROP; O-ETHYL S,S-DIPROPYL DITHIOPHOSPHATE; O-ETHYL S,S-DIPROPYL PHOSPHORODITHIOATE; O-ETHYL S,S-DIPROPYLPHOSPHORODITHIOATE; O-ETHYL-S,S-DIPROPYL PHOSPHORODITHIONATE; JOLT; MOBIL V-C 9-104; MOCAP; PHOSPHORODITHIOIC ACID O-ETHYL S,S-DIPROPYL ESTER; PHOSPHORODITHIOIC ACID,O-ETHYL S,S-DIPROPYL ESTER; PROPHOS; ROVOKIL; V-C 9-104; V-C CHEMICAL V-C 9-104; VC 9-104; VC9-104; VIRGINIA-CAROLINA VC 9-104

Description: pale yellow, clear liquid; strong mercaptan-like odor

Use: formerly non-systemic, non-fumigant nematicide & soil insecticide; against plant-parasitic nematodes & soil insects in ornamentals, potatoes, citrus, tobacco, pineapples, sugar cane, turf, etc; for wireworm, rootworm, cutworm, root borer, fleabeetle larvae control; on corn; potato, peanuts; banana, plaintain, beans (snap, lima), cabbage, corn,cucumber, etc.

Physical Properties

Boiling Point: 86 °C (187 °F) to 91 °C (196 °F) at 0.2 mm Hg

Freezing Point: 20 °C (68 °F)

Specific Gravity: 1.094 at 20 °C/4 °C

Saturated Vapor Density: 1.200004414 kg/m^3

Vapor Pressure: 3.8 $x10^{-4}$ mm Hg at 20-25 °C

Water Solubility: Slightly Soluble in Water

Other Solubilities: Soluble in Acetone, Ethanol, Hexane, Kerosene, Xylene; >300 g/kg /in/ cyclohexane, 1,2-dichloroethane, Diethyl Ether, Ethyl Acetate, petroleum spirit.

Flash Point: Does not burn or burns with difficulty

RTECS Toxicity Data

Acute Oral: Rat LD_{50} Dose: 34 mg/kg. Rabbit LD_{50} Dose: 55 mg/kg.

Acute Dermal: Rabbit LD_{50} Route: Skin; Dose: 2400 ug/kg. Rat LD_{50} Route: Skin; Dose: 60 mg/kg.

Hazard Overviews

Fire: Noncombustible.

Carcinogenicity: IARC - Not listed; NIOSH - Not listed; NTP - Not listed; ACGIH - Not listed; OSHA - Not listed; EPA - Not listed; MAK - Not listed

Environmental

Ecotoxicity: LC_{50} Lepomis macrochirus (Bluegill) 2.07 mg/l/96 hr /Conditions of bioassay not specified LD_{50} Hen oral 5.6 mg/kg LC_{50} Colinis virginianus (Bobwhite) 33 ppm in 5 day diet (95% confidence limit 27-40 ppm), age 14 days

Environmental Fate: If released to the atmosphere, it will degrade rapidly in the vapor phase by reaction with photochemically produced hydroxyl radicals (half-life of about 5.7 hr). When released to soil and water, it will degrade primarily through biodegradation. Screening studies have indicated that biodegradation is the major degradation process in soil. Aqueous hydrolysis may become an important process in alkaline water and soils; hydrolysis is not important at pH 7 or less, but it is reported to become rapid at pH 9 and above. Adsorption studies have shown that it is moderately to highly mobile in soil. The US Dept of Agriculture's Pesticide Properties Database lists a soil half-life of 25 days, but it may range from 3-56 days. Due to microbial adaptation, it degrades more rapidly in soils that have had prior exposure.

Cleanup/Disposal: Guide No. 131: Fully encapsulating, vapor protective clothing should be worn for spills and leaks with no fire. Eliminate all ignition sources (no smoking, flares, sparks or flames in immediate area). All equipment used when handling the product must be grounded. Do not touch or walk through spilled material. Stop leak if you can do it without risk. Prevent entry into waterways, sewers, basements or confined areas. A vapor suppressing foam may be used to reduce vapors. Small Spills: Absorb with earth, sand or other non-combustible material and transfer to containers for later disposal. Use clean non-sparking tools to collect absorbed material. Large Spills: Dike far ahead of liquid spill for later disposal. Water spray may reduce vapor; but may not prevent ignition in closed spaces. Guide No. 152: Do not touch damaged containers or spilled material unless wearing appropriate protective clothing. Stop leak if you can do it without risk. Prevent entry into waterways, sewers, basements or confined areas. Cover with plastic sheet to prevent spreading. Absorb or cover with dry earth, sand or other non-combustible material and transfer to containers. Do not get water inside containers.

Environmental Physical Data

Henry's Law Constant: estimated at 1.62×10^{-7}
Sorption Partition Coefficient: K_{oc} = calculated at 115
BCF: sheepshead 4 to 17

Regulations

RCRA 40CFR: Not listed
CERCLA: 40CFR 302.4: Not listed
SARA 40CFR 372.65: Listed TPQ: 1000 lb
SARA EHS 40CFR 355: Listed TPQ: 1000 lb
TSCA: Not listed

Analytical Methods

Soil: SW846 8140, 8141, 8141A
Water / Groundwater: EPA 1657, 622, 022; SW846 8141, 8141A
Drinking Water: EPA 507, 525.2; AOAC 991.07; ASTM D5475
Food: FDA 212.1, 232.1, 232.3, 232.4, 242.1
Indoor / Expired Air: NIOSH 5600
Plasma: EPA 001; FDA 211.1, 231.1, 252

ETH2100 CAS #: 77-67-8

ETHOSUXIMIDE

RTECS: WN2800000
EINECS Number: 201-048-7
Molecular Formula: $C_7H_{11}NO_2$
Formula Weight: 141.17

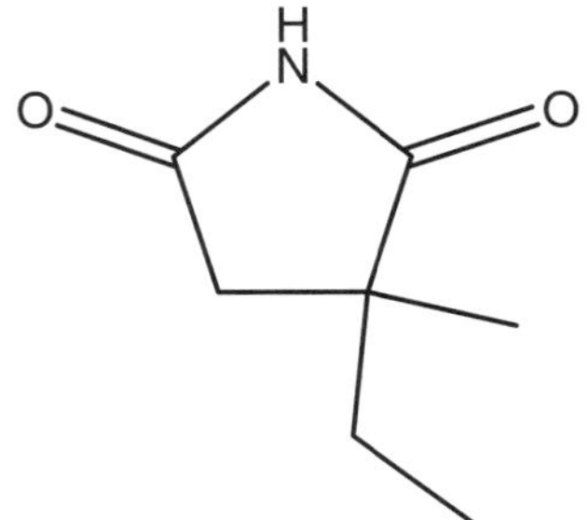

Chemical Structure

Synonyms: AETHOSUXIMIDE; ASAMID; ATYSMAL; C.I. 366; CAPITUS; EMESIDE; EPILEO PETIT MAL; ETHOSUCCIMIDE; ETHOSUCCINIMIDE; ETHOSUXIDE; 3-ETHYL-3-METHYL-2,5-PYRROLIDINE-DIONE; 3-ETHYL-3-METHYLPYRROLIDINE-2,5-DIONE; 2-ETHYL-2-METHYLSUCCINIMIDE; 3-ETHYL-3-METHYLSUCCINIMIDE; ALPHA-ETHYL-ALPHA-METHYLSUCCINIMIDE; ETHYMAL; ETOMAL; ETOSUXIMID; ETOSUXIMIDA; H 940; H-490; MESENTOL; 3-METHYL-3-ETHYLPYRROLIDINE-2,5-DIONE; ALPHA-METHYL-ALPHA-ETHYLSUCCINIMIDE; GAMMA-METHYL-GAMMA-ETHYL-SUCCINIMIDE; PEMAL; PEMALIN; PENTINIMID; PETINIMID; PETNIDAN; PIKNOLEPSIN; PM 671; PYKNOLEPSINUM; 2,5-PYRROLIDINEDIONE,3-ETHYL-3-METHYL-; RONTON; SIMATIN; SIMATIN(E); SUCCIMAL; SUCCIMITIN; SUCCINIMIDE,2-ETHYL-2-METHYL-; SUXILEP; SUXIMAL; SUXIN; SUXINUTIN; THETAMID; THILOPEMAL; ZARAONDAN; ZARODAN; ZARONDAN; ZARONDAN-SAFT; ZARONTIN; ZARTALIN
Description: white to off-white crystalline powder or waxy solid; characteristic odor
Use: medication: anticonvulsant to control petit mal epileptic seizures

Physical Properties

Freezing Point: 65 °C (149 °F)
Water Solubility: Freely Soluble in Water
Other Solubilities: Very Slightly Soluble in solvent Hexane.

RTECS Toxicity Data

Acute Oral: Mouse LD_{50} Dose: 1530 mg/kg; Toxic Effects: Behavioral - Altered sleep time (including change in righting reflex); Behavioral - Ataxia; Lungs, Thorax, or Respiration - Dyspnea.
Acute Dermal: Mouse LD_{50} Route: Subcutaneous Dose: 1810 mg/kg.
Reproductive/Teratogenic: Rat Route: Oral; Dose: 33300 ug/kg; Duration: female 6-14D of pregnancy; Effects on Embryo or Fetus - Extra embryonic structures; Fetotoxicity; Specific Developmental Abnormalities - Eye, ear. Rat Route: Oral; Dose: 66600 ug/kg; Duration: female 6-14D of pregnancy; Effects on Fertility - Post-implantation mortality. Rat Route: Oral; Dose: 166 mg/kg; Duration: female 6-14D of pregnancy; Specific Developmental Abnormalities - Other developmental abnormalities. Rat Route: Oral; Dose: 2250 mg/kg; Duration: female 9-17D of pregnancy; Effects on Embryo or Fetus - Fetotoxicity; Fetal death.
Mutagenic: Human Cytogenetic Analysis; Cell Type: lymphocyte; Dose: 30 mg/L.

Hazard Overviews

Health: May cause irritation. Harmful. Other Acute Effects: harmful if swallowed; exposure can cause gastrointestinal disturbances; target organs: central nervous system, blood.
Fire: Hazards: emits toxic fumes. Extinguishing agents: carbon dioxide, dry chemical powder or appropriate foam. Precautions: combustible liquid.
Reactivity: Stable. Hazardous polymerization will not occur. Hazardous decomposition products: thermal decomposition may produce carbon monoxide, carbon dioxide, and nitrogen oxides.
Carcinogenicity: IARC - Not listed; NIOSH - Not listed; NTP - Not listed; ACGIH - Not listed; OSHA - Not listed; EPA - Not listed; MAK - Not listed

Primary Target Organs:

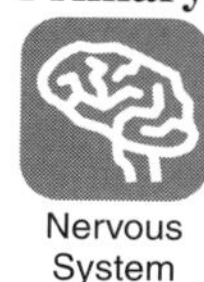

Environmental

Regulations

RCRA 40CFR: Not listed
CERCLA: 40CFR 302.4: Not listed
SARA 40CFR 372.65: Not listed
SARA EHS 40CFR 355: Not listed
TSCA: Not listed

ETH2150 CAS #: 1197-37-1

4-ETHOXY-1,2-BENZENEDIAMINE

EINECS Number: 214-825-0
Molecular Formula: $C_8H_{12}N_2O$
Formula Weight: 152.20
Synonyms: 1,2-BENZENEDIAMINE,4-ETHOXY-; O-PHENYLENEDIAMINE,4-ETHOXY-

Physical Properties

Boiling Point: 295 °C (563 °F)
Freezing Point: 72 °C (161.6 °F)
Water Solubility: Very Soluble
Other Solubilities: Ethanol: Soluble; Ether: Soluble; Chloroform: Soluble

Hazard Overviews

Carcinogenicity: IARC - Not listed; NIOSH - Not listed; NTP - Not listed; ACGIH - Not listed; OSHA - Not listed; EPA - Not listed; MAK - Not listed

Environmental

Regulations

RCRA 40CFR: Not listed
CERCLA: 40CFR 302.4: Not listed
SARA 40CFR 372.65: Not listed
SARA EHS 40CFR 355: Not listed
TSCA: Listed

ETH2200 CAS #: 67801-06-3

4-ETHOXY-1,3-BENZENEDIAMINE DIHYDROCHLORIDE

RTECS: ST2480000
EINECS Number: 267-125-2
Molecular Formula: $C_8H_{14}Cl_2N_2O$
Formula Weight: 236.11

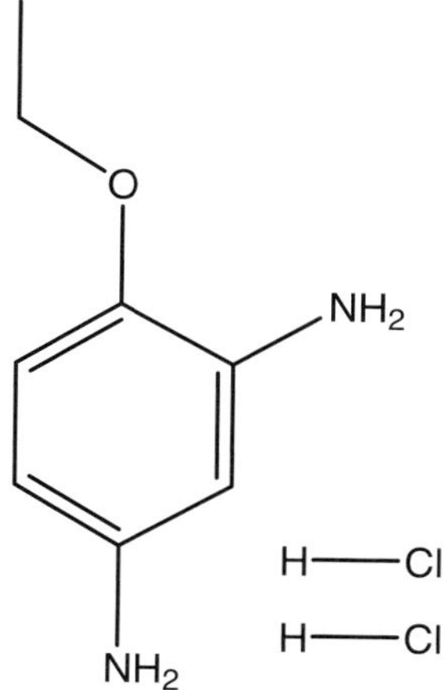

Chemical Structure

Synonyms: 1,3-BENZENEDIAMINE,4-ETHOXY-,DIHYDROCHLORIDE; 1,3-BENZENEDIAMINE,4-ETHOXY-,DIHYDROCHLORIDE (9CI); 2,4-DIAMINOETHOXYBENZENE DIHYDROCHLORIDE

RTECS Toxicity Data

Mutagenic: Bacteria - S Typhimurium Mutations in Microorganisms; Dose: 10 ug/plate (-S9).

Hazard Overviews

Carcinogenicity: IARC - Not listed; NIOSH - Not listed; NTP - Not listed; ACGIH - Not listed; OSHA - Not listed; EPA - Not listed; MAK - Not listed

Environmental

Regulations

RCRA 40CFR: Not listed
CERCLA: 40CFR 302.4: Listed as Compound per CWA Section 307(a)
SARA 40CFR 372.65: Not listed
SARA EHS 40CFR 355: Not listed
TSCA: Listed

ETH2250 CAS #: 68015-98-5

4-ETHOXY-1,3-BENZENEDIAMINE SULFATE

EINECS Number: 268-164-8
Molecular Formula: $C_8H_{14}N_2O_5S$
Formula Weight: 250.166
Synonyms: 1,3-BENZENEDIAMINE,4-ETHOXY-,SULFATE (1:1)

Hazard Overviews

Carcinogenicity: IARC - Not listed; NIOSH - Not listed; NTP - Not listed; ACGIH - Not listed; OSHA - Not listed; EPA - Not listed; MAK - Not listed

Environmental

Regulations

RCRA 40CFR: Not listed
CERCLA: 40CFR 302.4: Not listed
SARA 40CFR 372.65: Not listed
SARA EHS 40CFR 355: Not listed
TSCA: Listed

ETH2300 CAS #: 103-75-3

ETHOXYDIHYDROPYRAN

RTECS: UP8925000
EINECS Number: 203-141-8
Molecular Formula: $C_7H_{12}O_2$
Structured MF: $OCH{=}{=}CHCH_2CH_2CHOC_2H_5$
Formula Weight: 128.19

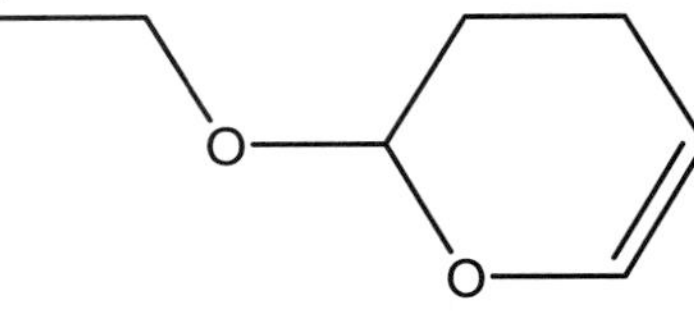

Chemical Structure

Synonyms: 2-ETHOXY-2,3-DIHYDRO(4H)PYRAN; 2-ETHOXY-2,3-DIHYDRO-GAMMA-PYRAN; 2-ETHOXY-3,4-DIHYDRO-1,2-PYRAN; 2-ETHOXY-3,4-DIHYDRO-2H-PYRAN; 2-ETHOXY-3,4-DIHYDROPYRAN; 2-ETHOXYDIHYDROPYRAN,IN PREGNANCY DIAGNOSIS; 2-ETHOXYPYRAN; 2H-PYRAN,2-ETHOXY-3,4-DIHYDRO-
Description: liquid
Use: stabilizer; intermediate

Physical Properties

Boiling Point: 143 °C (289 °F)
Freezing Point: -100 °C (-148 °F)
Specific Gravity: 0.97 at 20 °C/20 °C
Water Solubility: Very Slightly Soluble in Water
Surface Tension: Estimated at 25 dynes/cm
Refraction Index: 1.4394
Flash Point: 43.889 °C Open Cup

RTECS Toxicity Data

Acute Oral: Rat LD_{50} Dose: 6160 mg/kg.
Acute Inhalation: Rat LC_{Lo} Dose: 8000 ppm/4hr.
Acute Dermal: Rabbit LD_{50} Route: Skin; Dose: 3560 uL/kg.
Irritation Eye: Rabbit Standard Draize Test Dose: 50 mg; Reaction: moderate.
Irritation Skin: Rabbit Open Draize Test Dose: 10 mg/24H open; Reaction: mild.

Hazard Overviews

Flammable

Health: Irritating to eyes/skin/respiratory tract. Other Acute Effects: may be harmful by inhalation, ingestion, or skin absorption.
Fire: Flammable. Hazards: vapor may travel considerable distance to source of ignition and flash back; container explosion may occur; forms explosive mixtures in air. Extinguishing agents: carbon dioxide, dry chemical powder or appropriate foam; water may be effective for cooling, but may not effect extinguishment. Precautions: combustible liquid.
Reactivity: Incompatible with: oxidizing agents, strong acids, alcohols. Hazardous decomposition products: toxic fumes of: carbon monoxide, carbon dioxide.
Carcinogenicity: IARC - Not listed; NIOSH - Not listed; NTP - Not listed; ACGIH - Not listed; OSHA - Not listed; EPA - Not listed; MAK - Not listed
Primary Target Organs:

Eyes

Skin

Respiratory System

Environmental

Cleanup/Disposal: Guide No. 128: Eliminate all ignition sources (no smoking, flares, sparks or flames in immediate area). All equipment used when handling the product must be grounded. Do not touch or walk through spilled material. Stop leak if you can do it without risk. Prevent entry into waterways, sewers, basements or confined areas. A vapor suppressing foam may be used to reduce vapors. Absorb or cover with dry earth, sand or other non-combustible material and transfer to containers. Use clean non-sparking tools to collect absorbed material. Large Spills: Dike far ahead of liquid spill for later disposal. Water spray may reduce vapor; but may not prevent ignition in closed spaces.

Regulations

RCRA 40CFR: Not listed
CERCLA: 40CFR 302.4: Not listed
SARA 40CFR 372.65: Not listed
SARA EHS 40CFR 355: Not listed
TSCA: Listed

ETH2350 CAS #: 29718-44-3

ETHOXY-1-DODECANOL

Molecular Formula: $C_{14}H_{30}O_2$
Formula Weight: 230.39
Synonyms: 1-DODECANOL,ETHOXY-; DODECYL ALCOHOL (ETHOXYLATED); DODECYL ALCOHOL,ETHOXYLATED

Use: no evidence of use in US

Hazard Overviews

Carcinogenicity: IARC - Not listed; NIOSH - Not listed; NTP - Not listed; ACGIH - Not listed; OSHA - Not listed; EPA - Not listed; MAK - Not listed

Environmental

Regulations

RCRA 40CFR: Not listed
CERCLA: 40CFR 302.4: Not listed
SARA 40CFR 372.65: Not listed
SARA EHS 40CFR 355: Not listed
TSCA: Not listed

ETH2400 CAS #: 110-80-5

2-ETHOXYETHANOL

RTECS: KK8050000
DOT: UN1171; IMO3.3
EINECS Number: 203-804-1
Molecular Formula: $C_4H_{10}O_2$
Formula Weight: 90.12

Chemical Structure

Synonyms: ATHYLENGLYKOL-MONOATHYLATHER; CELLOSOLVE; CELLOSOLVE SOLVENT; CELOSOLV; DOWANOL 8; DOWANOL EE; EGEE; EKTASOLVE EE; EMKANOL; ETER MONOETILICO DEL ETILENGLICOL; ETHANOL,2-ETHOXY-; ETHER MONOETHYLIQUE DE L'ETHYLENE-GLYCOL; BETA-ETHOXYETHANOL; ETHYL CELLOSOLVE; ETHYL ETHYLENE GLYCOL; ETHYL GLYCOL; ETHYLENE GLYCOL ETHYL ETHER; ETHYLENE GLYCOL MONOETHYL ETHER; ETOKSYETYLOWY ALKOHOL; GLYCOL ETHER EE; GLYCOL ETHYL ETHER; GLYCOL MONOETHYL ETHER; HYDROXY ETHER; JEFFERSOL EE; OXITOL; POLY-SOLV EE; SOLVULOSE
Description: colorless liquid; sweet, ether-like odor
Use: solvent for nitrocellulose, lacquers, and dopes; in varnish removers, cleansing solutions, dye baths; finishing leather; water pigments and dye solution; increasing stability of emulsions; solvent for natural and synthetic resins; mutual solvent for formulation of soluble oils; dyeing and printing textiles; anti-icing additive for aviation fuels; and antifreeze for explosives

Physical Properties

Boiling Point: 135 °C (275 °F) at 760 mm Hg
Freezing Point: -90 °C (-130 °F)
Specific Gravity: 0.931 at 20 °C/20 °C
Vapor Density: 3.1 Air=1
Saturated Vapor Density: 1.212645517 kg/m^3
Density: 0.925 g/cu cm at 23.9 °C
Vapor Pressure: 3.8 mm Hg at 20 °C
Water Solubility: Miscible
Other Solubilities: 95% Ethanol: >=100 mg/ml at 18 °C; Acetone: >=100 mg/ml at 18 °C; DMSO: >=100 mg/ml at 18 °C; Ether: miscible; Liquid esters: miscible.
Surface Tension: 28.2 dynes/cm at 25 °C
Odor Threshold: > 225 ppm
Refraction Index: 1.408 at 20 °C/D
Evaporation Rate: 0.32 Butyl Acetate=1
Ionization Potential (eV): 9.6
Flash Point: 44 °C Closed Cup
Autoignition Temperature: 235 °C
LEL: 1.7% v/v
UEL: 15.6% v/v

RTECS Toxicity Data

Acute Oral: Rat LD_{50} Dose: 2125 mg/kg; Toxic Effects: Behavioral - Somnolence (general depressed activity); Behavioral - Withdrawal; Lungs, Thorax, or Respiration - Respiratory depression. Mouse LD_{50} Dose: 2451 mg/kg.
Acute Inhalation: Rat LC_{50} Dose: 2000 ppm/7hr. Mouse LC_{50} Dose: 1820 ppm/7hr; Toxic Effects: Behavioral - Analgesia; Lungs, Thorax, or Respiration - Dyspnea; Kidney, ureter, and Bladder - Hematuria.
Acute Dermal: Rabbit LD_{50} Route: Skin; Dose: 3300 mg/kg. Rabbit LD_{50} Route: Subcutaneous Dose: 2 gm/kg.
Chronic (Multiple Dose) Oral: Rat Dose: 26452 mg/kg/6W-I; Toxic Effects: Behavioral - Food intake (animal); Blood - Pigmented or nucleated red Blood cells; Nutritional and gross metabolic - Weight loss or decreased weight gain. Rat Dose: 25 gm/kg/2W-I; Toxic Effects: DEATH. Rat Dose: 18655 mg/kg/13W-C; Toxic Effects: Endocrine - Changes in thymus weight; Nutritional and gross metabolic - Weight loss or decreased weight gain; DEATH - Changes in testicular weight.
Chronic (Multiple Dose) Inhalation: Rat Dose: 4000 ppm/4H/2W-I; Toxic Effects: Nutritional and gross metabolic - Weight loss or decreased weight gain. Rat Dose: 1370 mg/m^3/7H/5W-I; Toxic Effects: Blood - Changes in spleen.
Irritation Eye: Rabbit Standard Draize Test Dose: 50 mg; Reaction: moderate. Rabbit Standard Draize Test Dose: 500 mg/24H; Reaction: mild.
Irritation Skin: Rabbit Open Draize Test Dose: 500 mg open; Reaction: mild.
Reproductive/Teratogenic: Rat Route: Oral; Dose: 500 mg/kg; Duration: male 5D prior to mating; Effects on Fertility - Other measures of fertility. Rat Route: Oral; Dose: 7820 mg/kg; Duration: female 1-21D of pregnancy; Effects on Fertility - Post-implantation mortality; Abortion. Rat Route: Oral; Dose: 600 mg/kg; Duration: female 10-12D of pregnancy Effects on Embryo or Fetus - Fetotoxicity; Specific Developmental Abnormalities - Musculoskeletal system; Cardiovascular (circulatory) system. Rat Route: Oral; Dose: 1800 mg/kg; Duration: female 7-15D of pregnancy; Effects on Embryo or Fetus - Fetal death. Rat Route: Oral; Dose: 4500 mg/kg; Duration: male 6W prior to mating; Paternal Effects - Spermatogenesis. Rat Route: Inhalation; Dose: 10 ppm/6H; Duration: female 6-15D of pregnancy; Effects on Fertility - Litter size; Specific Developmental Abnormalities - Musculoskeletal system. Rat Route:

Inhalation; Dose: 200 ppm/7H; Duration: female 1-19D of pregnancy; Effects on Embryo or Fetus - Fetotoxicity; Specific Developmental Abnormalities - Musculoskeletal system; Cardiovascular (circulatory) system. Rat Route: Inhalation; Dose: 600 ppm/7H; Duration: female 7-13D of pregnancy; Effects on Fertility - Post-implantation mortality; Effects on Embryo or Fetus - Fetal death. Rat Route: Inhalation; Dose: 100 ppm/7H; Duration: female 14-20D of pregnancy Maternal Effects - Parturition; Effects on Newborn - Behavioral.

Mutagenic: Rat Sperm Morphology; Route: Oral; Dose: 23400 mg/kg/5W-I. Hamster Cytogenetic Analysis; Cell Type: ovary; Dose: 6830 mg/L.

Hazard Overviews

Flammable

Fire Diamond

Health: Irritating to eyes/skin. Also Causes: dizziness, cyanosis, pulmonary edema, albuminuria, hyperbilirubinemia, oligospermia, unconsciousness, gynecological disorders, fetal abnormalities, kidney and liver damage, GI hemorrhages, anemia, leukopenia, hematuria, conjunctiva, cyanosis, tonic-clonic seizures, acetone on breath, acidosis. Chronic Effects: albuminuria, hyperbilirubinemia, possible reproductive effects/birth defects.

Fire: Flammable. Can form explosive mixtures in air. Vapors may travel to ignition source and flash back. For small fires use dry chemical, carbon dioxide, water spray, or alcohol-resistant foam. For large fires use water spray, fog, or alcohol-resistant foam. Containers may explode in heat of fire. Forms highly explosive peroxide when exposed to air.

Reactivity: Unstable, forms highly explosive peroxides when exposed to air. Hazardous polymerization cannot occur. Avoid: heat; ignition sources. Incompatible with: oxidizers; hydrogen peroxide and toluene and polyacrylamide gel. Hazardous decomposition products: carbon oxides.

Carcinogenicity: IARC - Not listed; NIOSH - Not listed; NTP - Not listed; ACGIH - Not listed; OSHA - Not listed; EPA - Not listed; MAK - Not listed

Primary Target Organs:

Eyes

Skin

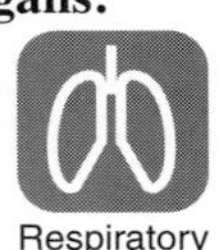
Respiratory System

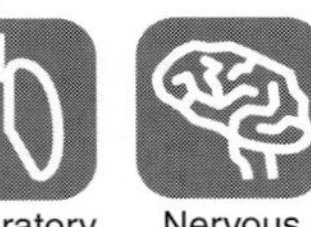
Nervous System

Liver

Kidneys

Exposure Limits

OSHA PEL: TWA: 200 ppm; 740 mg/m^3; skin.

ACGIH TLV: TWA: 5 ppm; 18 mg/m^3.

NIOSH REL: TWA: 0.5 ppm; 1.8 mg/m^3.

NIOSH IDLH: 500 ppm.

DFG MAK: TWA: 5 ppm; 19 mg/m^3.

Respirator Recommendation

Exposure Range: >200 to <500 ppm Air Purifying, Negative Pressure, Half Mask

Exposure Range: 500 to unlimited ppm Self-contained Breathing Apparatus, Pressure Demand, Full Face

Cartridge Color: black

Environmental

Ecotoxicity: LC_{50} Menidia beryllina >10,000 ppm/96 hr (static bioassay in synthetic seawater at 23 °C, mild aeration applied after 24 hr) LD_{50} Goldfish >5000 mg/l/24 hr (modified ASTM D 1345) LC_{50} Lepomis macrochirus >10,000 ppm/96 hr (static bioassay in fresh water at 23 °C, mild aeration applied after 24 hr)

Environmental Fate: Release to the soil is expected to result in volatilization from the soil surface and leaching to groundwater. Biodegradation is expected to be significant. Release to water will result in volatilization from the water surface and biodegradation. Minimal adsorption to sediments is expected and bioconcentration is not expected to be significant. Release to the atmosphere is expected to result in rapid degradation by nitrogen dioxides and the estimated half-life for the reaction between the vapor phase and photochemically generated hydroxyl radicals is 11.41 hr.

Cleanup/Disposal: Guide No. 127: Eliminate all ignition sources (no smoking, flares, sparks or flames in immediate area). All equipment used when handling the product must be grounded. Do not touch or walk through spilled material. Stop leak if you can do it without risk. Prevent entry into waterways, sewers, basements or confined areas. A vapor suppressing foam may be used to reduce vapors. Absorb or cover with dry earth, sand or other non-combustible material and transfer to containers. Use clean non-sparking tools to collect absorbed material. Large Spills: Dike far ahead of liquid spill for later disposal. Water spray may reduce vapor; but may not prevent ignition in closed spaces.

Environmental Physical Data

Henry's Law Constant: 0.05

Octanol/Water Partition Coefficient: log K_{ow} = -0.1

Sorption Partition Coefficient: log K_{oc} = 1.32

BCF: no food chain concentration potential

BOD: 1.58 lb/lb, 5 days

Regulations

RCRA 40CFR: Listed Hazardous Waste No. U359 Toxic Waste

CERCLA: 40CFR 302.4: Listed per RCRA Section 3001 RQ: 1000 lb (453.5 kg)

SARA 40CFR 372.65: Listed

SARA EHS 40CFR 355: Not listed

TSCA: Listed

Analytical Methods

Air: ASTM D4490

Indoor / Expired Air: NIOSH 1403

ETH2450 CAS #: 111-15-9

2-ETHOXYETHYL ACETATE

RTECS: KK8225000

DOT: UN1172; IMO3.3

EINECS Number: 203-839-2

Molecular Formula: $C_6H_{12}O_3$
Structured MF: $CH_3COOCH_2CH_2OC_2H_5$
Formula Weight: 132.16

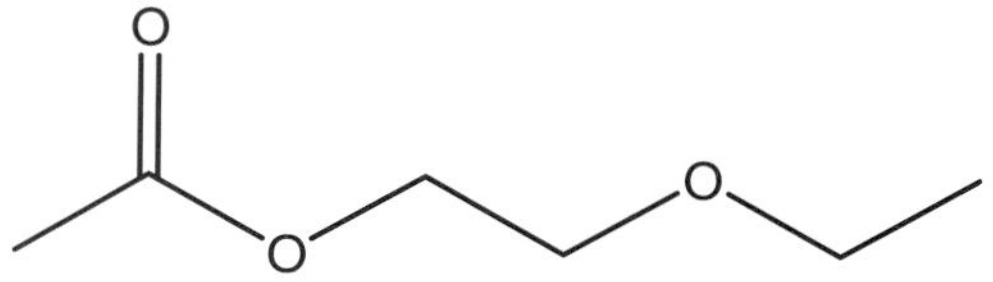

Chemical Structure

Synonyms: ACETATE DE CELLOSOLVE; ACETATE DE L'ETHER MONOETHYLIQUE DE L'ETHYLENE-GLYCOL; ACETATE D'ETHYLGLYCOL; ACETATO DI CELLOSOLVE; ACETIC ACID,2-ETHOXYETHYL ESTER; 1-ACETOXY-2-ETHOXYETHANE; 2-AETHOXY-AETHYLACETAT; AETHYLENGLYKOLAETHERACETAT; CELLOSOLVE ACETATE; CELOSOLVACETAT; CSAC; EGEEA; EKTASOLVE EE ACETATE SOLVENT; ETHANOL,2-ETHOXY-,ACETATE; ETHOXY ACETATE; 2-ETHOXYETHANOL ACETATE; 2-ETHOXYETHANOL,ESTER WITH ACETIC ACID; BETA-ETHOXYETHYL ACETATE; ETHOXYETHYL ACETATE; 2-ETHOXY-ETHYLACETAAT; 2-ETHOXYETHYLACETATE; 2-ETHOXYETHYLE,ACETATE DE; 2-ETHOXYETHYLESTER KYSELINY OCTOVE; ETHYL CELLOSOLVE ACETAAT; ETHYL CELLOSOLVE ACETATE; ETHYL GLYCOL ACETATE; ETHYLENE GLYCOL ETHYL ETHER ACETATE; ETHYLENE GLYCOL MONOETHYL ETHER ACETATE; ETHYLENE GLYCOL MONOETHYL ETHER ACETYLATED; ETHYLGLYCOL ACETATE; ETHYLGLYKOLACETAT; 2-ETOSSIETIL-ACETATO; GLYCOL ETHER EE ACETATE; GLYCOL MONOETHYL ETHER ACETATE; OCTAN ETOKSYETLU; OCTAN ETOKSYETYLU; OXITOL ACETATE; OXYTOL ACETATE; POLY-SOLV EE ACETATE

Description: colorless liquid; unpleasant odor

Use: in automobile lacquers to retard evaporation and impart high gloss; solvent for nitrocellulose, oils and resins; retards "blushing" in lacquers, varnish removers; wood stains; textiles; leather and cosmetic ingredient

Physical Properties

Boiling Point: 156 °C (313 °F)
Freezing Point: -61.7 °C (-79.06 °F)
Specific Gravity: 0.975 at 20 °C/20 °C
Vapor Density: 4.72 Air=1
Saturated Vapor Density: 1.206740036 kg/m^3
Density: 0.975 g/mL at 20 °C
Vapor Pressure: 1.2 mm Hg at 20 °C
Water Solubility: 1 parts in about 6 parts Water
Other Solubilities: 95% Ethanol: >=100 mg/ml at 19 °C; Acetone: >=100 mg/ml at 19 °C; Aromatic hydrocarbons: miscible; Corn oil: Very Soluble; DMSO: >=100 mg/ml at 19 °C; Ether: Soluble.
Surface Tension: 31.8 dynes/cm at 25 °C
Odor Threshold: 0.056 ppm
Refraction Index: 1.4054 at 20 °C/D
Evaporation Rate: 0.21 Butyl Acetate=1
Critical Temperature: 334 °C
Critical Pressure: 440 psia
Flash Point: 52 °C Closed Cup
Autoignition Temperature: 379 °C
LEL: 1.24% v/v
UEL: 12.7% v/v

RTECS Toxicity Data

Acute Oral: Rat LD_{50} Dose: 2700 mg/kg. Rabbit LD_{50} Dose: 1950 mg/kg.

Acute Inhalation: Rat LC_{50} Dose: 12100 mg/m^3/8hr. Rabbit LC_{50} Dose: >2000 ppm/4hr; Toxic Effects: Kidney, ureter, and Bladder - Hematuria.

Acute Dermal: Rabbit LD_{50} Route: Skin; Dose: 10500 uL/kg. Mouse LD_{Lo} Route: Subcutaneous Dose: 5 gm/kg.

Irritation Eye: Rabbit Standard Draize Test Dose: 40 mg; Reaction: moderate.

Irritation Skin: Rabbit Open Draize Test Dose: 490 mg open; Reaction: mild.

Reproductive/Teratogenic: Rat Route: Inhalation; Dose: 600 ppm/7H; Duration: female 7-15D of pregnancy; Effects on Embryo or Fetus - Fetal death. Rat Route: Inhalation; Dose: 200 ppm/6H; Duration: female 6-15D of pregnancy; Effects on Embryo or Fetus - Fetotoxicity; Specific Developmental Abnormalities - Respiratory system. Rat Route: Inhalation; Dose: 300 ppm/6H; Duration: female 6-15D of pregnancy; Specific Developmental Abnormalities - Cardiovascular (circulatory) system; Urogenital system. Rat Route: Inhalation; Dose: 50 mg/kg/4H; Duration: female 1-19D of pregnancy; Effects on Fertility - Pre-implantation mortality; Post-implantation mortality. Rat Route: Inhalation; Dose: 50 ppm/6H; Duration: female 6-15D of pregnancy; Specific Developmental Abnormalities - Musculoskeletal system.

Hazard Overviews

Fire Diamond

Health: Irritating to eyes/skin/respiratory tract. Also Causes: CNS effects, kidney/blood damage. Chronic Effects: personality changes, may cause reproductive effects/birth defects based on animal data.

Fire: Flammable. Can form explosive mixtures in the air. Use dry chemical, carbon dioxide, or alcohol-resistant foam. Fire may be difficult to extinguish and may re-ignite.

Reactivity: Stable. Hazardous polymerization cannot occur. Avoid: heat; ignition sources; distillation which may cause explosion. Incompatible with: oxidizers; nitrates; strong acids; bases. Hazardous decomposition products: acrid smoke; irritating fumes.

Carcinogenicity: IARC - Not listed; NIOSH - Listed as carcinogen; NTP - Not listed; ACGIH - Not listed; OSHA - Not listed; EPA - Not listed; MAK - Not listed

Primary Target Organs:

Eyes

Skin

Respiratory System

Nervous System

Kidneys

Blood

Exposure Limits

OSHA PEL: TWA: 100 ppm; 540 mg/m^3; skin.
ACGIH TLV: TWA: 5 ppm; 27 mg/m^3.
NIOSH REL: TWA: 0.5 ppm; 2.7 mg/m^3.

NIOSH IDLH: 500 ppm.
DFG MAK: TWA: 5 ppm; 27 mg/m^3.
Respirator Recommendation
Exposure Range: >100 to <500 ppm Air Purifying, Negative Pressure, Half Mask
Exposure Range: 500 to unlimited ppm Self-contained Breathing Apparatus, Pressure Demand, Full Face
Cartridge Color: black

Environmental

Ecotoxicity: LC_{50} Pimephales promelas (fathead minnow) 42.2 (40.7-43.6) mg/l 96 hr, wt 148 mg, flow-through bioassay, dissolved oxygen 7.4 (4.6-8.8) mg/l, water hardness 44.9 (42.4-46.6) mg/l as $CaCO_3$, pH 6.9-7.7, alkalinity 42.9 (39.6-61.4) mg/l $CaCO_3$, temp: 26.4 °C
Environmental Fate: If released to the atmosphere, it will degrade primarily by reaction with photochemically produced hydroxyl radicals (estimated half-life of 1.2 days). If released to soil or water, it is expected to degrade via biodegradation. Several biodegradation screening tests have demonstrated that it is readily biodegradable. It may leach readily in soils based upon its high water solubility.
Cleanup/Disposal: Guide No. 129: Eliminate all ignition sources (no smoking, flares, sparks or flames in immediate area). All equipment used when handling the product must be grounded. Do not touch or walk through spilled material. Stop leak if you can do it without risk. Prevent entry into waterways, sewers, basements or confined areas. A vapor suppressing foam may be used to reduce vapors. Absorb or cover with dry earth, sand or other non-combustible material and transfer to containers. Use clean non-sparking tools to collect absorbed material. Large Spills: Dike far ahead of liquid spill for later disposal. Water spray may reduce vapor; but may not prevent ignition in closed spaces.

Environmental Physical Data

Henry's Law Constant: estimated at 1.25 $x10^{-6}$
Octanol/Water Partition Coefficient: log K_{ow} = measured at 0.24
Sorption Partition Coefficient: K_{oc} = estimated at 5
BCF: calculated at 0.6
BOD: theoretical 36%, 5 days

Regulations

RCRA 40CFR: Not listed
CERCLA: 40CFR 302.4: Not listed
SARA 40CFR 372.65: Not listed
SARA EHS 40CFR 355: Not listed
TSCA: Listed

Analytical Methods

Air: ASTM D3686, D3687, D4490
Indoor / Expired Air: NIOSH 1450
Plasma: EPA 001

ETH2500 CAS #: 106-74-1

2-ETHOXYETHYL ACRYLATE

RTECS: AS9800000
EINECS Number: 203-429-3
Molecular Formula: $C_7H_{12}O_3$
Formula Weight: 144.17

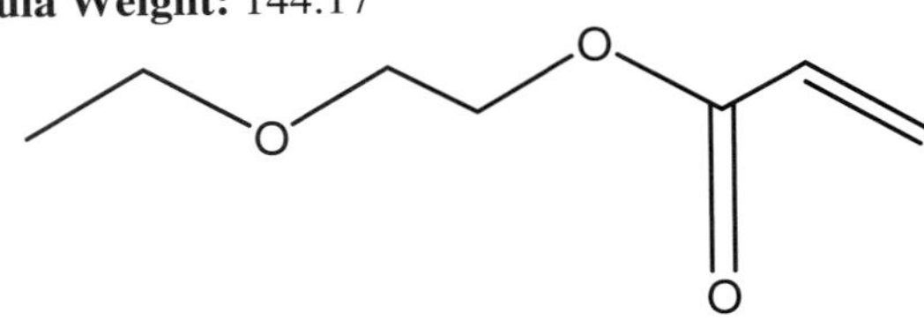

Chemical Structure

Synonyms: ACRYLIC ACID,2-ETHOXYETHANOL ESTER; ACRYLIC ACID,2-ETHOXYETHYL ESTER; CELLOSOLVE ACRYLATE; ETHANOL,2-ETHOXY-,ACRYLATE; ETHOXYETHYL ACRYLATE; 2-ETHOXYETHYLESTER KYSELINY AKRYLOVE; 2-ETHOXYETHYL-2-PROPENOATE; ETHYLENE GLYCOL MONOETHYL ETHER ACRYLATE; ETHYLENE GLYCOL MONOETHYL ETHER PROPENOATE; 2-PROPENOIC ACID,2-ETHOXYETHYL ESTER; 2-PROPENOIC ACID-2-ETHOXYETHYL ESTER
Use: monomer for specialty elastomers-eg, adhesives, laminates; in elastomers

Physical Properties

Boiling Point: 174 °C (345 °F)
Freezing Point: -47 °C (-52.6 °F)
Specific Gravity: 0.983 at 20 °C
Other Solubilities: Reacts readily with electrophilic, free-radical, and nucleophilic agent
Refraction Index: 1.4274

RTECS Toxicity Data

Acute Oral: Rat LD_{50} Dose: 1070 mg/kg.
Acute Inhalation: Rat LC_{Lo} Dose: 500 ppm/4hr.
Acute Dermal: Rabbit LD_{50} Route: Skin; Dose: 1010 uL/kg.
Irritation Eye: Rabbit Standard Draize Test Dose: 20 mg open; Reaction: severe.
Irritation Skin: Rabbit Open Draize Test Dose: 10 mg/24H open; Reaction: mild. Rabbit Open Draize Test Dose: 500 mg open; Reaction: mild.

Hazard Overviews

Carcinogenicity: IARC - Not listed; NIOSH - Not listed; NTP - Not listed; ACGIH - Not listed; OSHA - Not listed; EPA - Not listed; MAK - Not listed

Environmental

Regulations

RCRA 40CFR: Not listed
CERCLA: 40CFR 302.4: Not listed
SARA 40CFR 372.65: Not listed
SARA EHS 40CFR 355: Not listed
TSCA: Listed

ETH2550 CAS #: 26983-51-7

ETHOXYETHYL HYDROXY MERCURY

Hazard Overviews

Carcinogenicity: IARC - Not listed; NIOSH - Not listed; NTP - Not listed; ACGIH - Not listed; OSHA - Not listed; EPA - Not listed; MAK - Not listed

Environmental

Regulations

RCRA 40CFR: Not listed
CERCLA: 40CFR 302.4: Not listed
SARA 40CFR 372.65: Not listed
SARA EHS 40CFR 355: Not listed
TSCA: Not listed

ETH2600 CAS #: 62-44-2

N-(4-ETHOXYPHENYL)ACETAMIDE

RTECS: AM4375000
EINECS Number: 200-533-0
Molecular Formula: $C_{10}H_{13}NO_2$
Formula Weight: 179.21

Chemical Structure

Synonyms: ACETAMIDE,N-(4-ETHOXYPHENYL)-; ACETAMIDE,N-(4-ETHOXYPHENYL)-(9CI); 1-ACETAMIDO-4-ETHOXYBENZENE; ACETANILIDE,4'-ETHOXY-; ACETIC ACID,AMIDE,N(4-ETHOXYPHENYL)-; ACETO-PARA-PHENALIDE; P-ACETOPHENETIDE; PARA-ACETOPHENETIDE; ACETO-PARA-PHENETIDIDE; P-ACETOPHENETIDIDE; PARA-ACETOPHENETIDIDE; ACETOPHENETIDIN; ACETO-4-PHENETIDINE; ACETOPHENETIDINE; P-ACETOPHENETIDINE; PARA-ACETOPHENETIDINE; ACETOPHENETIN; ACET-P-PHENALIDE; ACET-PARA-PHENALIDE; ACET-P-PHENETIDIN; ACET-PARA-PHENETIDIN; ACETPHENETIDIN; P-ACETPHENETIDIN; PARA-ACETPHENETIDIN; ACETYLPHENETIDIN; N-ACETYL-P-PHENETIDINE; N-ACETYL-PARA-PHENETIDINE; ACHROCIDIN; ANAPAC; ASA COMPOUND; BROMO SELTZER; BUFF-A-COMP; CITRA-FORT; CLISTANOL; CODEMPIRAL; COMMOTIONAL; CONTRADOL; CONTRADOULEUR; CORICIDIN; CORIFORTE; CORYBAN-D; DAPRISAL; DARVON COMPOUND; DASIKON; DASIN; DASIN CH; DOLOSTOP; EDRISAL; EMPIRAL; EMPIRIN COMPOUND; EMPRAZIL; EMPRAZIL-C; EPRAGEN; PARA-ETHOXY-ACETANILID; 4'-ETHOXYACETANILIDE; 4'ETHOXYACETANILIDE; 4-ETHOXYACETANILIDE; P-ETHOXYACETANILIDE; PARA-ETHOXYACETANILIDE; P-ETHOXYANILID KYSELINY OCTOVE; N-PARA-ETHOXYPHENYL-ACETAMIDE; N-PARA-ETHOXYPHENYLACETAMIDE; FENACETIN; FENACETINA; FENEDINA; FENIA; FENIDINA; FENINA; FIORINAL; FORTACYL; GELONIDA; GEWODIN; HELVAGIT; HJORTON'S POWDER; HOCOPHEN; KAFA; KALMIN; MALEX; MELABON; MELAFORTE; NORGESIC; PAMPRIN; PARACETOPHENETIDIN; PARACETOPHENTIDIN; PARAMETTE; PARATODOL; PERCODAN; PERTONAL; PHENACET; PARA-PHENACETIN; PHENACETIN; PHENACETINE; PHENACETINUM; PHENACITIN; PHENACON; PHENAPHEN; PHENAPHEN PLUS; PHENAZETIN; PHENAZETINA; PHENEDINA; P-PHENETIDINE,N-ACETYL; P-PHENETIDINE,N-ACETYL-; PHENIDIN; PHENIN; PHENODYNE; PYRAPHEN; PYRROXATE; QUADRONAL; ROBAXISAL-PH; SALGYDAL; SANALGINE; SARIDON; SERANEX; SINEDAL; SINUBID; SINUTAB; STELLACYL; SUPER ANAHIST; SUPRALGIN; SYNALGOS-DC; SYNALOGOS; TACOL; TERRACYDIN; TETRACYDIN; THEPHORIN A-C; TREUPEL; VEGANINE; VIDEN; WIGRAINE; XARIL; ZACTIRIN COMPOUND

Description: white, glistening crystals, usually scales, or fine white, crystalline powder; odorless

Use: analgesic; antipyretic

Physical Properties

Boiling Point: Decomposes
Freezing Point: 134 °C (273.2 °F) to 135 °C (275 °F)
Saturated Vapor Density: 1.200025844 kg/m^3
Vapor Pressure: 3.16×10^{-3} mm Hg at 25 °C
Water Solubility: 1 g dissolves in 1310 ml Cold Water
Other Solubilities: Very Soluble in Pyrimidine; Soluble in Acetone; Slightly Soluble in Benzene.
pH: Saturated solution is neutral
Refraction Index: 1.571
Flash Point: Not available; probably combustible

RTECS Toxicity Data

Acute Oral: Rat LD_{50} Dose: 3600 mg/kg. Mouse LD_{50} Dose: 866 mg/kg.

Acute Inhalation: Mouse LC_{50} Dose: 33900 mg/m^3.

Acute Dermal: Rabbit LD_{50} Route: Subcutaneous Dose: 1 gm/kg; Toxic Effects: Behavioral - Convulsions or effect on seizure threshold. Mouse LD_{50} Route: Subcutaneous Dose: 1625 mg/kg.

Chronic (Multiple Dose) Oral: Rat Dose: 10 gm/kg/4W-I; Toxic Effects: Blood - Methemoglobinemia-Carboxyhemoglobin; Blood - Changes in other cell count; Blood - Changes in erythrocite (RBC) cell count. Rat Dose: 16 gm/kg/43W-C; Toxic Effects: Kidney, Ureter, and Bladder - Other changes in urine composition.

Chronic (Multiple Dose) Inhalation: Rat Dose: 190 mg/m^3/4H/30D-I; Toxic Effects: Behavioral - Alteration of classical conditioning; Liver - Liver function tests impaired.

Rat Dose: 15 mg/m^3/4H/17W-I; Toxic Effects: Brain and coverings - Recordings from specific areas of CNS; Kidney, Ureter, and Bladder - Proteinuria; Blood - Changes in serum composition.

Reproductive/Teratogenic: Rat Route: Oral; Dose: 50336 mg/kg; Duration: male 17W prior to mating; Effects on Fertility - Male fertility index. Rat Route: Oral; Dose: 24 gm/kg; Duration: female 1-20D of pregnancy; Effects on Fertility - Pre-implantation mortality; Effects on Embryo or Fetus - Fetotoxicity. Rat Route: Oral; Dose: 6 gm/kg; Duration: female 1-20D of pregnancy; Specific Developmental Abnormalities - Musculoskeletal system. Rat Route: Oral; Dose: 62920 mg/kg; Duration: male 22W prior to mating; Paternal Effects - Testes, epididymis, sperm duct. Rat Route: Oral; Dose: 57200 mg/kg; Duration: male 20W prior to mating; Paternal Effects - Spermatogenesis.

Mutagenic: Human Mutations in Microorganisms; Cell Type: lymphocyte; Dose: 1500 mg/L (+S9). Rat DNA Damage; Route: Intraperitoneal; Dose: 165 mg/kg. Rat DNA Damage; Route: Oral; Dose: 82500 ug/kg.

Tumorigenic: Human Route: Oral; Dose: 7300 mg/kg/Y-C; Toxic Effects: Tumorigenic - Carcinogenic by RTECS criteria; Kidney, Ureter, and Bladder - Kidney tumors. Human Route: Oral; Dose: 28 gm/kg/28Y-I; Toxic Effects: Tumorigenic - Carcinogenic by RTECS criteria; Kidney, Ureter, and Bladder - Tumors. Man Route: Oral; Dose: 57 gm/kg/47Y-I; Toxic Effects: Tumorigenic - Carcinogenic by RTECS criteria; Kidney, Ureter, and Bladder - Kidney tumors. Man Route: Oral; Dose: 27 gm/kg/10Y-I; Toxic Effects: Tumorigenic - Carcinogenic by RTECS criteria; Kidney, Ureter, and Bladder - Kidney tumors. Man Route: Oral; Dose: 126 gm/kg/25Y-I; Toxic Effects: Tumorigenic - Carcinogenic by RTECS criteria; Kidney, Ureter, and Bladder - Tumors. Woman Route: Oral; Dose: 80 gm/kg/63Y-I; Toxic Effects: Tumorigenic - Carcinogenic by RTECS criteria; Kidney, Ureter, and Bladder - Kidney tumors. Woman Route: Oral; Dose: 140 gm/kg/13Y-I; Toxic Effects: Tumorigenic - Carcinogenic by RTECS criteria; Kidney, Ureter, and Bladder - Kidney tumors. Rat Route: Oral; Dose: 572 gm/kg/60W-C; Toxic Effects: Tumorigenic - Carcinogenic by RTECS criteria; Sense organs and special senses - Tumors; Kidney, Ureter, and Bladder - Tumors. Rat Route: Oral; Dose: 9450 mg/kg/45W-C; Toxic Effects: Tumorigenic - Equivocal tumorigenic agent by RTECS criteria; Liver - Tumors. Rat Route: Oral; Dose: 206 gm/kg/2Y-C; Toxic Effects: Tumorigenic - Equivocal tumorigenic agent by RTECS criteria; Sense organs and special senses - Tumors; Skin and appendages - Tumors.

Hazard Overviews

Health: May cause irritation. Toxic. Other Acute Effects: harmful if swallowed, inhaled, or absorbed through skin; absorption into the body leads to the formation of methemoglobin which in sufficient concentration causes cyanosis; onset may be delayed 2 to 4 hours or longer. Chronic Effects: carcinogen; may alter genetic material; may cause heritable genetic damage; target organs: blood; kidneys; liver.

Fire: Will burn. Hazards: emits toxic fumes. Extinguishing agents: water spray; carbon dioxide, dry chemical powder or appropriate foam. Precautions: combustible liquid.

Reactivity: Incompatible with: strong oxidizing agents, strong acids, strong bases, strong reducing agents. Hazardous decomposition products: toxic fumes of: carbon monoxide, carbon dioxide, nitrogen oxides.

Carcinogenicity: IARC - Group 2A, Probably carcinogenic to humans; NIOSH - Not listed; NTP - Listed; ACGIH - Not listed; OSHA - Not listed; EPA - Not listed; MAK - Not listed

Primary Target Organs:

Liver

Kidneys

Blood

Environmental

Environmental Fate: If released to soil, it is expected to have moderately high mobility. If released to water, chemical hydrolysis, chemical oxidation, bioaccumulation in aquatic organism, and adsorption to suspended solids and sediments are not expected to be significant fate processes. The volatilization half-life from a model river has been estimated to exist to be 35 days. However, due to the lack of data on other fate processes the relative importance of the volatilization is not known. If released to the atmosphere, it is expected to exist almost entirely in the vapor phase. The dominant removal mechanism is expected to be reaction with photochemically generated hydroxyl radicals (half-life 5 hours).

Environmental Physical Data

Henry's Law Constant: 1.4 x10^{-6}
Octanol/Water Partition Coefficient: log K_{ow} = 1.58
Sorption Partition Coefficient: K_{oc} = estimated at 139 to 172
BCF: estimated at 9 to 18

Regulations

RCRA 40CFR: Listed Hazardous Waste No. U187 Toxic Waste
CERCLA: 40CFR 302.4: Listed per RCRA Section 3001 RQ: 100 lb (45.35 kg)
SARA 40CFR 372.65: Not listed
SARA EHS 40CFR 355: Not listed
TSCA: Listed

Analytical Methods

Soil: SW846 3640A, 8250A, 8270B, 8270C
Water / Groundwater: EPA 1625
Plasma: EPA 29

ETH2650 CAS #: 91-53-2

ETHOXYQUIN

RTECS: VB8225000
EINECS Number: 202-075-7

Molecular Formula: $C_{14}H_{19}NO$
Formula Weight: 217.30

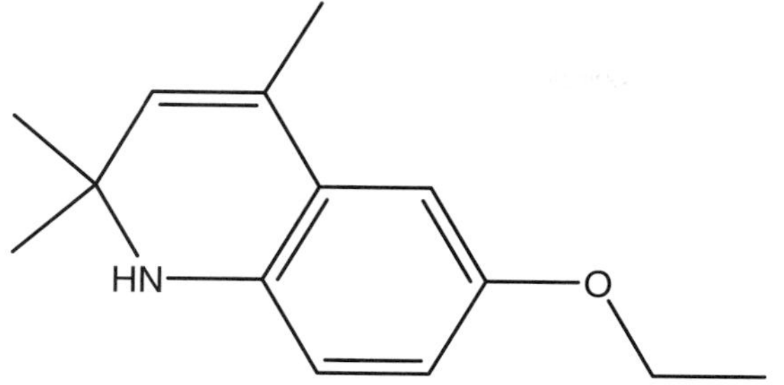

Chemical Structure

Synonyms: AMEA 100; ANTAGE AW; ANTIOXIDANT EC; ANTOX; ARIES ANTOX; DAWE'S NUTRIGARD; 1,2-DIHYDRO-6-ETHOXY-2,2,4-TRIMETHYLQUINOLINE; 1,2-DIHYDRO-2,2,4-TRIMETHYL-6-ETHOXYQUINOLINE; EMQ; EQ; ETHOXYCHIN; 6-ETHOXY-1,2-DIHYDRO-2,2,4-TRIMETHYLQUINOLINE; ETHOXYQUINE; 6-ETHOXY-2,2,4-TRIMETHYL-1,2-DIHYDROQUINOLINE; NIFLEX; NIFLEX D; NIX-SCALD; NOCRAC AW; PERMANAX 103; QUINOL ED; QUINOLINE,6-ETHOXY-1,2-DIHYDRO-2,2,4-TRIMETHYL-; SANTOFLEX A; SANTOFLEX AW; SANTOQUIN; SANTOQUINE; STOP-SCALD; 2,2,4-TRIMETHYL-6-ETHOXY-1,2-DIHYDROQUINOLINE

Description: yellow liquid; mercaptan-like odor

Use: as a food additive; as an antidegradation agent for rubber, insecticide, antioxidant, flex-cracking inhibitor, post-harvest dip to prevent scald on apples and pears, fungicide and plant growth regulator

Physical Properties

Boiling Point: 123 °C (253 °F) to 125 °C (257 °F) at 2 mm Hg
Freezing Point: About
Specific Gravity: 1.029 to 1.031 at 25 °C/25 °C
Vapor Density: 7.48 Air=1
Density: 1.03 g/mL at 25 °C
Vapor Pressure: 0.0000182 mm Hg at 0 °C
Water Solubility: < 1 mg/mL at 20 C
Other Solubilities: 95% Ethanol: >=100 mg/ml at 20 °C; Acetone: >=100 mg/ml at 20 °C; DMSO: >=100 mg/ml at 20 °C; Fats: Soluble; Oils: Soluble; Organic solvents: Soluble.
Refraction Index: 1.569 to 1.672 at 25 °C/D
Flash Point: 107 °C

RTECS Toxicity Data

Acute Oral: Rat LD_{50} Dose: 800 mg/kg. Mouse LD_{50} Dose: 1730 mg/kg.

Chronic (Multiple Dose) Oral: Rat Dose: 63 gm/kg/30W-I; Toxic Effects: Kidney, Ureter, and Bladder - Changes in kidney weight; Nutritional and gross metabolic - Weight loss or decreased weight gain. Rat Dose: 122 mg/kg/58W-C; Toxic Effects: Kidney, Ureter, and Bladder - Other changes. Rat Dose: 12 gm/kg/29W-C; Toxic Effects: Liver - Changes in liver weight; Kidney, Ureter, and Bladder - Chgs in tubules (inc acute renal failure, acute tubular necrosis; Kidney, Ureter, and Bladder - Changes in kidney weight.

Mutagenic: Rat Unscheduled DNA Synthesis; Route: Oral; Dose: 13 gm-kg/4W-C. Bacteria - S Typhimurium Mutations in Microorganisms; Dose: 200 ug/plate (-S9).

Hazard Overviews

Health: Irritating to eyes/skin. Harmful. Other Acute Effects: harmful if swallowed; may be harmful if inhaled or absorbed through the skin; may cause allergic skin reaction. Chronic Effects: may produce reversible liver damage; target organ: liver.

Fire: Will burn. Hazards: emits toxic fumes. Extinguishing agents: carbon dioxide, dry chemical powder or appropriate foam. Precautions: combustible liquid.

Carcinogenicity: IARC - Not listed; NIOSH - Not listed; NTP - Not listed; ACGIH - Not listed; OSHA - Not listed; EPA - Not listed; MAK - Not listed

Primary Target Organs:

Eyes

Skin

Liver

Environmental

Regulations

RCRA 40CFR: Not listed
CERCLA: 40CFR 302.4: Not listed
SARA 40CFR 372.65: Not listed
SARA EHS 40CFR 355: Not listed
TSCA: Listed

Analytical Methods

Food: FDA 212.1, 232.1, 232.4, 242.1
Plasma: FDA 211.1, 231.1, 252

ETH2700 **CAS #: 452-35-7**

ETHOXZOLAMIDE

RTECS: DL6390000
EINECS Number: 207-199-5
Molecular Formula: $C_9H_{10}N_2O_3S_2$
Formula Weight: 258.33

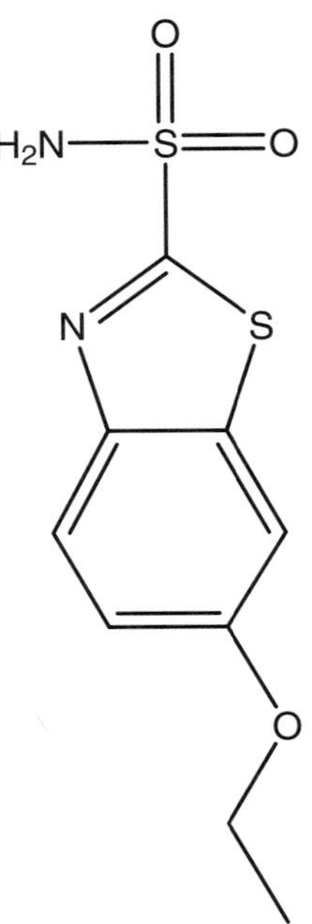

Chemical Structure

Synonyms: 2-BENZOTHIAZOLESULFONAMIDE,6-ETHOXY-; CARDRASE; DIURETIC C; ETHOXAZOLAMIDE; 6-ETHOXY-2-BENZOTHIAZOLESULFONAMIDE; ETHOXYZOLAMIDE; ETOXZOLAMIDE; GLAUCOTENSIL; MINGORAL; U-4191

Description: white or slightly yellow crystalline powder; odorless

Use: carbonic anhydrase inhibitor; diuretic to reduce intraocular pressure

Physical Properties

Freezing Point: 188 °C (370.4 °F) to 190.5 °C (374.9 °F)

Water Solubility: Practically Insoluble in Water

Other Solubilities: Slightly Soluble in Alcohol, Acetone, Chloroform & Ether

RTECS Toxicity Data

Reproductive/Teratogenic: Rat Route: Oral; Dose: 3300 mg/kg; Duration: female 1-22D of pregnancy; Specific Developmental Abnormalities - Musculoskeletal system; Effects on Embryo or Fetus - Fetal death; Effects on Fertility - Post-implantation mortality.

Hazard Overviews

Health: Irritating to eyes/skin/respiratory tract. Harmful. Other Acute Effects: may be harmful by inhalation, ingestion, or skin absorption; target organ: kidneys. Chronic Effects: may cause reproductive effects based on tests with laboratory animals.

Fire: Hazards: emits toxic fumes. Extinguishing agents: water spray; carbon dioxide, dry chemical powder or appropriate foam. Precautions: combustible liquid.

Reactivity: Incompatible with: strong oxidizing agents. Hazardous decomposition products: toxic fumes of: carbon monoxide, carbon dioxide, nitrogen oxides, sulfur oxides.

Carcinogenicity: IARC - Not listed; NIOSH - Not listed; NTP - Not listed; ACGIH - Not listed; OSHA - Not listed; EPA - Not listed; MAK - Not listed

Primary Target Organs:

Eyes

Skin

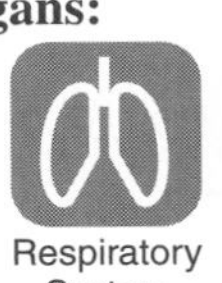
Respiratory System

Kidneys

Environmental

Environmental Physical Data

Octanol/Water Partition Coefficient: log K_{ow} = 2.01

Regulations

RCRA 40CFR: Not listed

CERCLA: 40CFR 302.4: Not listed

SARA 40CFR 372.65: Not listed

SARA EHS 40CFR 355: Not listed

TSCA: Not listed

ETH2750 CAS #: 141-78-6

ETHYL ACETATE

RTECS: AH5425000

DOT: UN1173; IMO3.2

EINECS Number: 205-500-4

Molecular Formula: $C_4H_8O_2$

Structured MF: $CH_3COOCH_2CH_3$

Formula Weight: 88.10

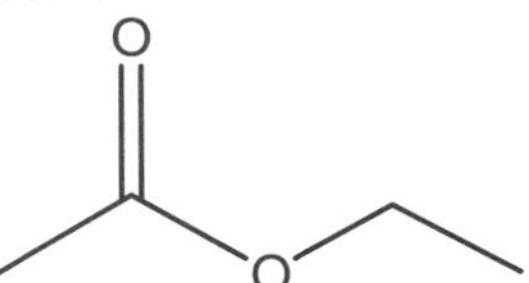

Chemical Structure

Synonyms: ACETATE D'ETHYLE; ACETATO DE ETILO; ACETIC ACID ETHYL ESTER; ACETIC ACID,ETHYL ESTER; ACETIC ETHER; ACETIDIN; ACETOXYETHANE; AETHYLACETAT; EPA PESTICIDE CHEMICAL CODE 044003; ESSIGESTER; ETHYL ACETIC ESTER; ETHYL ESTER; ETHYL ESTER OF ACETIC ACID; ETHYL ETHANOATE; ETHYLACETAAT; ETHYLE (ACETATE D'); ETHYLESTER KYSELINY OCTOVE; ETILE (ACETATO DI); OCTAN ETYLU; VINEGAR NAPHTHA

Description: colorless liquid; fruity odor

Use: in artificial fruit essences and as a solvent for varnishes, airplane dopes, coatings, plastics, lacquers, nitrocellulose, smokeless powders, organic synthesis and pharmaceuticals; as an insect fumigant and in the manufacture of artificial leather, photographic film and plates, artificial silks, perfumes and cleaning textiles; a synthetic flavoring substance and adjuvant

Physical Properties

Boiling Point: 77 °C (171 °F)

Freezing Point: -83 °C (-117.4 °F)

Specific Gravity: 0.902 at 20 °C/4 °C
Vapor Density: 3.04 Air=1
Saturated Vapor Density: 1.434898367 kg/m^3
Density: 0.8945 g/mL
Bulk Density: 0.8945 gr/ml
Vapor Pressure: 73 mm Hg at 20 °C
Water Solubility: 50 to 100 mg/mL at 21 °C
Other Solubilities: 95% Ethanol: >=100 mg/ml at 21 °C; Acetone: >=100 mg/ml at 21 °C; Alcohol: Soluble; Benzene: Soluble; Chloroform: Soluble; DMSO: >=100 mg/ml at 21 °C; Ether: Soluble; Most organic solvents: miscible.
Surface Tension: 24 dynes/cm
Odor Threshold: 0.0196 to 655 mg/m^3
Refraction Index: 1.3719 at 20 °C
Evaporation Rate: 4.5 Butyl Acetate=1
Critical Temperature: 250 °C
Critical Pressure: 3.8 mn/sq m
Ionization Potential (eV): 10.01
Flash Point: 7.2 °C Open Cup
Autoignition Temperature: 427 °C
LEL: 2.2% v/v
UEL: 9% v/v

RTECS Toxicity Data

Acute Oral: Rat LD_{50} Dose: 5620 mg/kg. Mouse LD_{50} Dose: 4100 mg/kg; Toxic Effects: Behavioral - Somnolence (general depressed activity); Behavioral - Change in motor activity (specific assay); Behavioral - Coma.
Acute Inhalation: Rat LC_{50} Dose: 200 gm/m^3; Toxic Effects: Behavioral - Somnolence (general depressed activity); Lungs, Thorax, or Respiration - Acute pulmonary edema; Gastrointestinal - Changes in structure or function of salivary glands. Human TC_{Lo} Dose: 400 ppm; Toxic Effects: Sense organs and special senses - Other; Sense organs and special senses - Conjunctive irritation; Lungs, Thorax, or Respiration - Other changes.
Acute Dermal: Rabbit LD_{50} Route: Skin; Dose: >20 mL/kg. Rat LD_{Lo} Route: Subcutaneous Dose: 5 gm/kg.
Mutagenic: Hamster Cytogenetic Analysis; Cell Type: fibroblast; Dose: 9 gm/L. Yeast - S Cerevisiae Sex Chromosome Loss; Dose: 24400 ppm.

Hazard Overviews

Flammable

Fire Diamond

Health: Irritating to eyes/skin/respiratory tract. Also Causes (high concent): possible CNS depression (weakness, drowsiness, unconsciousness). Chronic Effects: dermatitis, conjunctival irritation with corneal clouding.
Fire: Flammable. Can form explosive mixtures in the air. For small fires use dry chemical, carbon dioxide, water spray, or alcohol-resistant foam. For large fires use water spray, fog, or alcohol-resistant foam.
Reactivity: Unstable, is slowly decomposed by moisture. Hazardous polymerization cannot occur. Avoid: heat; ignition sources. Incompatible with: nitrates; strong oxidizers; acids; alkalis; ignites on contact with potassium tert-butoxide; potentially explosive reaction with lithium tetrahydroaluminate; violent reaction with chlorosulfonic acid. Hazardous decomposition products: carbon oxide(s).
Carcinogenicity: IARC - Not listed; NIOSH - Not listed; NTP - Not listed; ACGIH - Not listed; OSHA - Not listed; EPA - Not listed; MAK - Not listed
Primary Target Organs:

Eyes

Skin

Respiratory System

Exposure Limits
OSHA PEL: TWA: 400 ppm; 1400 mg/m^3.
ACGIH TLV: TWA: 400 ppm; 1440 mg/m^3.
NIOSH REL: TWA: 400 ppm; 1400 mg/m^3.
NIOSH IDLH: 2000 ppm; LEL.
DFG MAK: TWA: 400 ppm; 1400 mg/m^3.
Respirator Recommendation
Exposure Range: >400 to 1000 ppm Air Purifying, Negative Pressure, Half Mask
Exposure Range: >1000 to <2000 ppm Supplied Air, Constant Flow/Pressure Demand, Half Mask
Exposure Range: 2000 to unlimited ppm Self-contained Breathing Apparatus, Pressure Demand, Full Face
Cartridge Color: black

Environmental

Ecotoxicity: LC_{50} Heteropneustes fossilis (common indian catfish) 212.5 ppm/96 hr /conditions of bioassay not specified
Environmental Fate: If released into water it will be lost primarily by evaporation (half-life 10 hr in a typical river) and biodegradation. While readily biodegraded in most aquatic tests, the rate in natural waters is unknown. Bioconcentration in fish will be insignificant. If released on land, it will partially evaporate and partially leach into the ground. Biodegradation will probably occur both in soil and groundwater, however, experimental data are lacking. In the atmosphere, it will react with photochemically produced hydroxyl radicals (half-life 8.3 days). A few percent an hour will disappear under photochemical smog situations.
Cleanup/Disposal: Guide No. 129: Eliminate all ignition sources (no smoking, flares, sparks or flames in immediate area). All equipment used when handling the product must be grounded. Do not touch or walk through spilled material. Stop leak if you can do it without risk. Prevent entry into waterways, sewers, basements or confined areas. A vapor suppressing foam may be used to reduce vapors. Absorb or cover with dry earth, sand or other non-combustible material and transfer to containers. Use clean non-sparking tools to collect absorbed material. Large Spills: Dike far ahead of liquid spill for later disposal. Water spray may reduce vapor; but may not prevent ignition in closed spaces.

Environmental Physical Data

Henry's Law Constant: 0
Octanol/Water Partition Coefficient: log K_{ow} = 0.73

BCF: not significant
BOD: theoretical 66%, 5 days

Regulations

RCRA 40CFR: Listed Hazardous Waste No. U112 Ignitable Waste
CERCLA: 40CFR 302.4: Listed per RCRA Section 3001 RQ: 5000 lb (2268 kg)
SARA 40CFR 372.65: Not listed
SARA EHS 40CFR 355: Not listed
TSCA: Listed

Analytical Methods

Air: ASTM D3686, D3687, D4490
Soil: SW846 5031, 8015B, 8260A, 8260B
Water / Groundwater: ASTM D3695
Food: AOAC 975.06
Indoor / Expired Air: NIOSH 1457
Plasma: EPA 29
Other: EPA 1666

ETH2800 CAS #: 141-97-9

ETHYL ACETOACETATE

RTECS: AK5250000
EINECS Number: 205-516-1
Molecular Formula: $C_6H_{10}O_3$
Structured MF: $CH_3COCH_2COOC_2H_5$
Formula Weight: 130.14

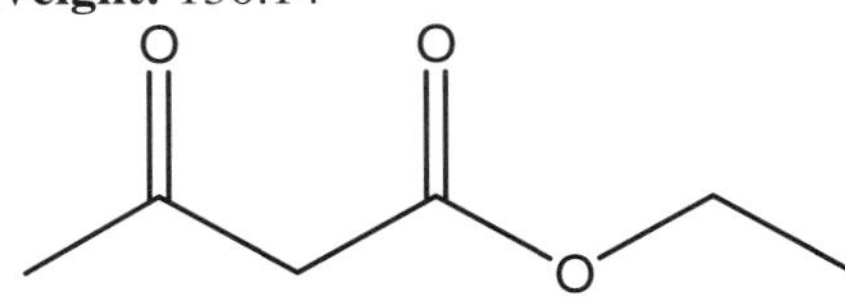

Chemical Structure

Synonyms: ACETOACETIC ACID ETHYL ESTER; ACETOACETIC ACID,ETHYL ESTER; ACETOACETIC ESTER; ACETOCTAN ETHYLNATY; ACTIVE ACETYL ACETATE; ACTIVE ACETYLACETATE; BUTANOIC ACID,3-OXO-,ETHYL ESTER; DIACETIC ESTER; DIACETIC ETHER; EAA; 1-ETHOXYBUTANE-1,3-DIONE; ETHYL ACETYL ACETATE; ETHYL ACETYLACETATE; ETHYL ACETYLACETONATE; ETHYL BETA-KETOBUTYRATE; ETHYL 3-OXOBUTANOATE; ETHYL 3-OXOBUTYRATE; ETHYLACETACETAT; ETHYLESTER KYSELINY ACETOCTOVE; ETHYL-3-OXOBUTANOATE; 3-OXOBUTANOIC ACID ETHYL ESTER
Description: clear liquid; fruity odor
Use: flavoring agent in food; chemical intermediate for yellow pigments in paints, lacquers, inks, photographic & vat dyes; amino acids; methionine; antimalarial agents (atabrine); thiamine (vitamin b1); analgesics; antipyrene; aminopyrene

Physical Properties

Boiling Point: 180.8 °C (357 °F) at 760 mm Hg
Freezing Point: -45 °C (-49 °F)
Specific Gravity: 1.0282 at 20 °C/4 °C
Vapor Density: 4.48 Air=1
Saturated Vapor Density: 1.203924826 kg/m^3
Vapor Pressure: 0.095 kPa at 25 °C
Water Solubility: 1 parts in about 35 parts Water
Other Solubilities: Soluble in Benzene, & Chloroform; miscible with Ether, & Acetone.
Surface Tension: 32.5 dynes/cm at 20 °C
Refraction Index: 1.4171
Evaporation Rate: 0.1 Butyl Acetate=1
Flash Point: 84.4 °C Closed Cup
Autoignition Temperature: 295 °C
LEL: 1.4% v/v
UEL: 9.5% v/v at 176 °C

RTECS Toxicity Data

Acute Oral: Rat LD_{50} Dose: 3980 mg/kg. Mouse LD_{50} Dose: 5105 mg/kg.
Acute Dermal: Rabbit LD; Route: Skin; Dose: >20 mL/kg.
Irritation Eye: Rabbit Standard Draize Test Dose: 100 mg/24H; Reaction: moderate. Rabbit Standard Draize Test Dose: 100 mg; Reaction: severe.
Irritation Skin: Rabbit Open Draize Test Dose: 510 mg open; Reaction: mild.
Mutagenic: Bacteria - B Subtilis DNA Repair; Dose: 21 mg/disc. Bacteria - E Coli Mutations in Microorganisms; Dose: 200 ug/plate (-S9).

Hazard Overviews

Fire Diamond

Health: Irritating to eyes/skin/respiratory tract.
Fire: Combustible. Use dry chemical, alcohol foam, or carbon dioxide to fight fires involving ethyl acetoacetate. Use a water spray to cool fire-exposed tanks or containers. Do not use a solid stream of water because this may scatter and spread the fire.
Reactivity: Stable. Hazardous polymerization cannot occur. Avoid: sources of ignition. Incompatible with: zinc; tribromopentyl alcohol.. Hazardous decomposition products: carbon dioxide; carbon monoxide.
Carcinogenicity: IARC - Not listed; NIOSH - Not listed; NTP - Not listed; ACGIH - Not listed; OSHA - Not listed; EPA - Not listed; MAK - Not listed
Primary Target Organs:

Eyes

Skin

Respiratory System

Mucous Membranes

Environmental

Cleanup/Disposal: Guide No. 130: Eliminate all ignition sources (no smoking, flares, sparks or flames in immediate area). All equipment used when handling the product must be grounded. Do not touch or walk through spilled material. Stop leak if you can do it without risk. Prevent entry into waterways, sewers, basements or confined areas. A vapor suppressing foam may be used to reduce vapors. Absorb or cover with dry earth, sand or other non-combustible material and transfer to containers. Use clean non-sparking tools to collect absorbed material. Large Spills: Dike far ahead of liquid spill for later disposal. Water spray may reduce vapor; but may not prevent ignition in closed spaces.

Environmental Physical Data

Octanol/Water Partition Coefficient: log K_{ow} = calculated at 0.27
BCF: no food chain concentration potential

Regulations

RCRA 40CFR: Not listed
CERCLA: 40CFR 302.4: Not listed
SARA 40CFR 372.65: Not listed
SARA EHS 40CFR 355: Not listed
TSCA: Listed

Analytical Methods

Water / Groundwater: ASTM D3695

ETH2850 CAS #: 140-88-5

ETHYL ACRYLATE

RTECS: AT0700000
DOT: UN1917; IMO3.2
EINECS Number: 205-438-8
Molecular Formula: $C_5H_8O_2$
Structured MF: $H_2C{=}CHCOOCH_2CH_3$
Formula Weight: 100.11

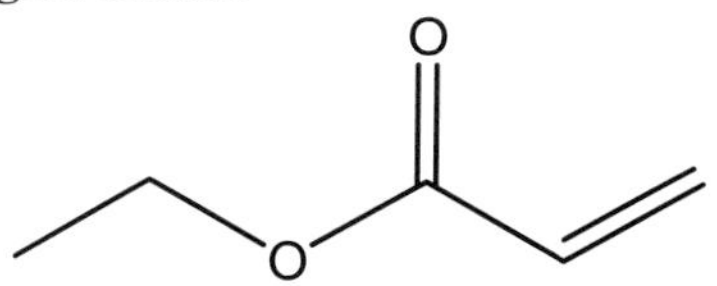

Chemical Structure

Synonyms: ACRYLATE D'ETHYLE; ACRYLIC ACID,ETHYL ESTER; ACRYLIC ACID,ETHYL ESTER (INHIBITED); ACRYLSAEUREAETHYLESTER; AETHYLACRYLAT; AKRYLANEM ETYLU; CARBOSET 511; ETHOXYCARBONYLETHYLENE; ETHYL ACRYLATE (INHIBITED); ETHYL ACRYLATE,INHIBITED; ETHYL ESTER OF ACRYLIC ACID; ETHYL 2-PROPENOATE; ETHYL PROPENOATE; ETHYLACRYLAAT; ETHYLAKRYLAT; ETHYLESTER KYSELINY AKRYLOVE; ETIL ACRILATO; ETILACRILATULUI; 2-PROPENOIC ACID ETHYL ESTER; 2-PROPENOIC ACID,ETHYL ESTER; PROPENOIC ACID,ETHYL ESTER (INHIBITED)
Description: colorless liquid; sharp, acrid odor
Use: manufacturing of acrylic resins for use in paint formulations, industrial coatings and latexes, in the manufacturing of plastics such as ethylene ethyl acrylate and in the manufacturing of polyacrylate elastomers and acrylic rubber; in the forming of denture materials, water emulsion vehicle for paints, textiles and paper coatings, leather finishes, resins or adhesives and lends flexibility to hard films

Physical Properties

Boiling Point: 99.4 °C (211 °F)
Freezing Point: -71.2 °C (-96.16 °F)
Specific Gravity: 0.9234 at 20 °C/4 °C
Vapor Density: 3.45 Air=1
Saturated Vapor Density: 1.313440454 kg/m^3
Bulk Density: 7.6 lbs/gal at 20 °C
Vapor Pressure: 29.3 mm Hg at 20 °C
Water Solubility: 2% by weight
Other Solubilities: 95% Ethanol: >=100 mg/ml at 21 °C; Acetone: >=100 mg/ml at 21 °C; Alcohol: Soluble; Chloroform: Soluble; DMSO: >=100 mg/ml at 21 °C; Ether: Soluble.
Surface Tension: 25 dynes/cm
Odor Threshold: 0.0008 to 32.0000 mg/m^3
pH: 0.0008% may cause it to be slightly acid
Refraction Index: 1.404 at 20 °C/D
Evaporation Rate: 3.3 Butyl Acetate=1
Critical Temperature: 279 °C
Critical Pressure: 544 psia
Ionization Potential (eV): 10.30
Flash Point: 10 °C Open Cup
Autoignition Temperature: 372 °C
LEL: 1.4% v/v
UEL: 14% v/v

RTECS Toxicity Data

Acute Oral: Rat LD_{50} Dose: 800 mg/kg. Mouse LD_{50} Dose: 1799 mg/kg.
Acute Inhalation: Rat LC_{50} Dose: 1414 ppm/4hr; Toxic Effects: Sense organs and special senses - Other; Lungs, Thorax, or Respiration - Dyspnea; Gastrointestinal - Changes in structure or function of salivary glands. Human TC_{Lo} Dose: 50 ppm; Toxic Effects: Sense organs and special senses - Other; Sense organs and special senses - Other; Lungs, Thorax, or Respiration - Other changes.
Acute Dermal: Rabbit LD_{50} Route: Skin; Dose: 500 uL/kg; Toxic Effects: Lungs, Thorax, or Respiration - Acute pulmonary edema; Liver - Other changes; Skin and appendages - Dermatitis, other. Rat LD_{Lo} Route: Skin; Dose: 1800 mg/kg.
Chronic (Multiple Dose) Oral: Rat Dose: 109 gm/kg/2Y-C; Toxic Effects: Behavioral - Food intake (animal); Behavioral - Fluid intake; Nutritional and gross metabolic - Weight loss or decreased weight gain. Rat Dose: 11200 mg/kg/14D-I; Toxic Effects: Gastrointestinal - Ulceration or bleeding from stomach. Rat Dose: 800 mg/kg/4D-I; Toxic Effects: Gastrointestinal - Ulceration or bleeding from stomach; Gastrointestinal - Other changes.
Chronic (Multiple Dose) Inhalation: Rat Dose: 75 ppm/6H/2Y-I; Toxic Effects: Sense organs and special senses - Other; Nutritional and gross metabolic - Weight loss or decreased weight gain. Rat Dose: 300 ppm/7H/30D-I; Toxic Effects: Kidney, Ureter, and Bladder - Changes in kidney weight; Nutritional and gross metabolic - Weight loss or decreased weight gain; DEATH. Rat Dose: 75 ppm/6H/26W-I; Toxic Effects: Sense organs and special senses - Other.
Irritation Eye: Rabbit Standard Draize Test Dose: 45 mg; Reaction: mild.

Irritation Skin: Rabbit Standard Draize Test Dose: 10 mg/24H; Reaction: mild. Rabbit Open Draize Test Dose: 500 mg open; Reaction: mild.
Mutagenic: Mouse Mutations in Microorganisms; Cell Type: lymphocyte; Dose: 20 mg/L (+S9). Mouse Mutations in Mammalian Somatic Cells; Cell Type: lymphocyte; Dose: 20 mg/L.
Tumorigenic: Rat Route: Oral; Dose: 51500 mg/kg/2Y-I; Toxic Effects: Tumorigenic - Carcinogenic by RTECS criteria; Gastrointestinal - Tumors. Mouse Route: Oral; Dose: 103 gm/kg/2Y-I; Toxic Effects: Tumorigenic - Carcinogenic by RTECS criteria; Gastrointestinal - Tumors.

Hazard Overviews

Flammable

Fire Diamond

Health: Severely irritating to eyes/skin/respiratory tract. Also Causes: dizziness, nausea, vomiting, headache. Chronic Effects: Skin sensitization, possible cancer hazard.
Fire: Flammable. May polymerize explosively if uninhibited. Use dry chemical, alcohol foam, or carbon dioxide. Water may be ineffective in extinguishing a fire involving ethyl acrylate. Use water spray to cool fire-exposed tanks/containers.
Reactivity: Stable. Hazardous polymerization can occur if the inhibitor levels become low or if it is exposed to heat, sparks, or open flame. Avoid: heat; sparks; open flame; free radical polymerization initiators. Incompatible with: oxidizing agents; initiators of polymerization; acids; strong alkalies; atmospheric moisture. Hazardous decomposition products: carbon monoxide; carbon dioxide.
Carcinogenicity: IARC - Group 2B, Possibly carcinogenic to humans; NIOSH - Listed as carcinogen; NTP - Class 2B, Reasonably anticipated to be a carcinogen, sufficient evidence of carcinogenicity from studies in experimental animals; ACGIH - Class A4, Not classifiable as a human carcinogen; OSHA - Not listed; EPA - Not listed; MAK - Not listed
Primary Target Organs:

Eyes

Skin

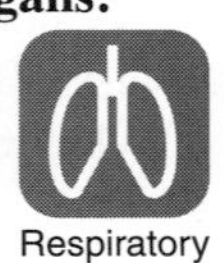
Respiratory System

Mucous Membranes

Exposure Limits
OSHA PEL: TWA: 25 ppm; 100 mg/m^3; skin.
OSHA PEL Vacated 1989 Limits: TWA: 5 ppm; 20 mg/m^3; STEL: 25 ppm; 100 mg/m^3.
ACGIH TLV: TWA: 5 ppm; 20 mg/m^3; STEL: 15 ppm; 61 mg/m^3.
NIOSH IDLH: 300 ppm.
DFG MAK: TWA: 5 ppm; 20 mg/m^3.
Respirator Recommendation
Exposure Range: >25 to 250 ppm Air Purifying, Negative Pressure, Half Mask
Exposure Range: >250 to <300 ppm Air Purifying, Negative Pressure, Full Face
Exposure Range: 300 to unlimited ppm Self-contained Breathing Apparatus, Pressure Demand, Full Face
Cartridge Color: black

Environmental

Ecotoxicity: Bacteria: Pseudomonas putida: inhibition of cell multiplication starts at 270 mg/l Algae: Microcystis aeruginosa: inhibition of cell multiplication starts at 14 mg/l
Environmental Fate: If concentrated compound is spilled into water or onto land, it is likely to undergo polymerization to an insoluble resin. Low concentrations released on land should volatilize, leach into the ground, and probably biodegrade. If low concentrations are released into water, it will volatilize (half-life 7.4 hr in a model river) and probably biodegrade. Adsorption to sediment, bioconcentration in aquatic organisms, and hydrolysis (except if the pH is > 9) will not be important. In the atmosphere, vapor will be attacked by ozone and photochemically produced hydroxyl radicals (half-life 6.5 hr) and scavenged by rain.
Cleanup/Disposal: Guide No. 129: Eliminate all ignition sources (no smoking, flares, sparks or flames in immediate area). All equipment used when handling the product must be grounded. Do not touch or walk through spilled material. Stop leak if you can do it without risk. Prevent entry into waterways, sewers, basements or confined areas. A vapor suppressing foam may be used to reduce vapors. Absorb or cover with dry earth, sand or other non-combustible material and transfer to containers. Use clean non-sparking tools to collect absorbed material. Large Spills: Dike far ahead of liquid spill for later disposal. Water spray may reduce vapor; but may not prevent ignition in closed spaces.

Environmental Physical Data

Henry's Law Constant: calculated at 3.05 x10^{-4}
Octanol/Water Partition Coefficient: log K_{ow} = 1.32
Sorption Partition Coefficient: K_{oc} = 22
BCF: calculated at 6
BOD: theoretical 66%, 5 days

Regulations

RCRA 40CFR: Listed Hazardous Waste No. U113 Ignitable Waste
CERCLA: 40CFR 302.4: Listed per RCRA Section 3001 RQ: 1000 lb (453.5 kg)
SARA 40CFR 372.65: Listed
SARA EHS 40CFR 355: Not listed
TSCA: Listed

Analytical Methods

Air: ASTM D3686, D3687, D4490
Water / Groundwater: ASTM D3695
Indoor / Expired Air: NIOSH 1450

ETH2900 **CAS #: 541-85-5**

ETHYL AMYL KETONE

RTECS: MJ7350000
EINECS Number: 208-793-7
Molecular Formula: $C_8H_{16}O$
Structured MF: $C_2H_5COCH_2CH(CH_3)CH_2CH_3$
Formula Weight: 128.24

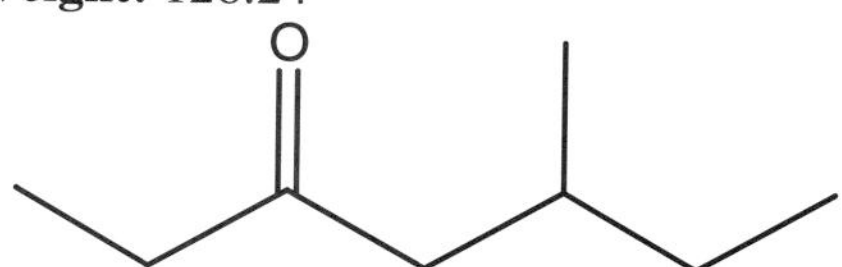

Chemical Structure

Synonyms: AMYL ETHYL KETONE; EAK; ETHYL SEC-AMYL KETONE; ETHYL ISOAMYL KETONE; 3-METHYL-5-HEPTANONE; 5-METHYL-3-HEPTANONE
Description: colorless liquid; mild fruity odor

Physical Properties

Boiling Point: 157.22 °C (315 °F)
Freezing Point: -56.67 °C (-70.006 °F)
Specific Gravity: 0.82
Vapor Density: 4.42 Air=1
Bulk Density: 83 lbs/gal
Vapor Pressure: 2 mm Hg
Water Solubility: Insoluble
Other Solubilities: Soluble in 4 volumes of 60 % Alcohol and many organic solvents
Surface Tension: 24.6 dynes/cm at 20 °C
Odor Threshold: 6 ppm
Refraction Index: 1.416
Evaporation Rate: 0.3 Butyl Acetate=1
Flash Point: 59 °C

RTECS Toxicity Data

Acute Oral: Rat LD_{50} Dose: 3500 mg/kg. Mouse LD_{50} Dose: 3800 mg/kg.
Acute Inhalation: Rat LC_{Lo} Dose: 3484 ppm/8hr. Mouse LC_{Lo} Dose: 3484 ppm/4hr.
Acute Dermal: Rabbit LD_{50} Route: Skin; Dose: >16 gm/kg.
Irritation Skin: Rabbit Standard Draize Test Dose: 500 mg; Reaction: mild.

Hazard Overviews

Flammable

Fire Diamond

Health: Irritating to eyes/skin/respiratory tract. Also Causes: CNS depression with headache, nausea, and narcosis. Chronic Effects: dryness and cracking of skin.
Fire: Flammable. For small fires use dry chemical, carbon dioxide, alcohol-resistant foam, or water spray. For large fires use fog, alcohol-resistant foam or water spray. Do not scatter material with a solid high-pressure water stream.
Reactivity: Stable. Hazardous polymerization cannot occur. Avoid: heat; ignition sources. Incompatible with: oxidizing agents. Hazardous decomposition products: carbon oxides; acrid smoke.
Carcinogenicity: IARC - Not listed; NIOSH - Listed as carcinogen; NTP - Not listed; ACGIH - Not listed; OSHA - Not listed; EPA - Not listed; MAK - Not listed
Primary Target Organs:

Eyes

Skin

Respiratory System

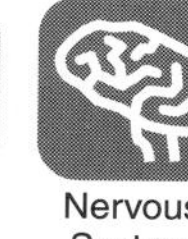

Nervous System

Exposure Limits
OSHA PEL: TWA: 25 ppm; 130 mg/m^3.
ACGIH TLV: TWA: 25 ppm; 131 mg/m^3.
NIOSH REL: TWA: 25 ppm; 130 mg/m^3.
NIOSH IDLH: 100 ppm.
Respirator Recommendation
Exposure Range: >25 to <100 ppm Air Purifying, Negative Pressure, Half Mask
Exposure Range: 100 to unlimited ppm Self-contained Breathing Apparatus, Pressure Demand, Full Face
Cartridge Color: black

Environmental

Cleanup/Disposal: Guide No. 127: Eliminate all ignition sources (no smoking, flares, sparks or flames in immediate area). All equipment used when handling the product must be grounded. Do not touch or walk through spilled material. Stop leak if you can do it without risk. Prevent entry into waterways, sewers, basements or confined areas. A vapor suppressing foam may be used to reduce vapors. Absorb or cover with dry earth, sand or other non-combustible material and transfer to containers. Use clean non-sparking tools to collect absorbed material. Large Spills: Dike far ahead of liquid spill for later disposal. Water spray may reduce vapor; but may not prevent ignition in closed spaces.

Regulations

RCRA 40CFR: Not listed
CERCLA: 40CFR 302.4: Not listed
SARA 40CFR 372.65: Not listed
SARA EHS 40CFR 355: Not listed
TSCA: Listed

Analytical Methods

Indoor / Expired Air: NIOSH 1301

ETH2950 **CAS #: 87-25-2**

ETHYL ANTHRANILATE

RTECS: DG2448000
EINECS Number: 201-735-1
Molecular Formula: $C_9H_{11}NO_2$
Formula Weight: 165.19

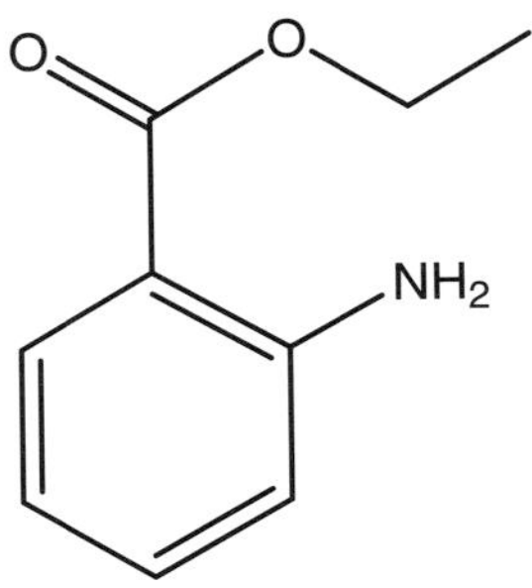

Chemical Structure

Synonyms: 2-AMINOBENZOIC ACID,ETHYL ESTER; AMINOBENZOIC ACID,ETHYL ESTER; O-AMINOBENZOIC ACID,ETHYL ESTER; ANTHRANILIC ACID,ETHYL ESTER; BENZOIC ACID,2-AMINO-,ETHYL ESTER; 2-CARBOETHOXYANILINE; O-(ETHOXYCARBONYL)ANILINE; ETHYL 2-AMINOBENZOATE; ETHYL O-AMINOBENZOATE; ETHYL-ORTHO-AMINOBENZOATE

Description: colorless liquid; faint orange-flowers odor

Use: therapeutic category (humans); topical anesthetic; therapeutic category (veterinary); perfumery; flavoring

Physical Properties

Boiling Point: 268 °C (514 °F) at 760 mm Hg

Freezing Point: 13 °C (55.4 °F)

Specific Gravity: 1.1174 at 20 °C/4 °C

Density: 1.12 g/mL

Water Solubility: Insoluble in Water

Other Solubilities: 95% Ethanol: Soluble; DMSO: Insoluble; Ether: Soluble.

Refraction Index: 1.5646 at 20 °C/D

Flash Point: Not available; probably combustible

RTECS Toxicity Data

Acute Oral: Rat LD_{50} Dose: 3750 mg/kg.

Irritation Skin: Rabbit Standard Draize Test Dose: 500 mg/24H; Reaction: moderate.

Hazard Overviews

Health: Irritating to eyes/skin/respiratory tract. Other Acute Effects: may be harmful by inhalation, ingestion, or skin absorption.

Fire: Will burn. Extinguishing agents: carbon dioxide, dry chemical powder or appropriate foam. Precautions: combustible liquid.

Reactivity: Incompatible with: acids, bases, oxidizing agents, reducing agents. Hazardous decomposition products: toxic fumes of: carbon monoxide, carbon dioxide, nitrogen oxides.

Carcinogenicity: IARC - Not listed; NIOSH - Not listed; NTP - Not listed; ACGIH - Not listed; OSHA - Not listed; EPA - Not listed; MAK - Not listed

Primary Target Organs:

Eyes Skin Respiratory System

Environmental

Regulations

RCRA 40CFR: Not listed

CERCLA: 40CFR 302.4: Not listed

SARA 40CFR 372.65: Not listed

SARA EHS 40CFR 355: Not listed

TSCA: Listed

ETH3050 CAS #: 100-41-4

ETHYL BENZENE

RTECS: DA0700000

DOT: UN1175; IMO3.2

EINECS Number: 202-849-4

Molecular Formula: C_8H_{10}

Structured MF: $C_6H_5 \bullet C_2H_5$

Formula Weight: 106.16

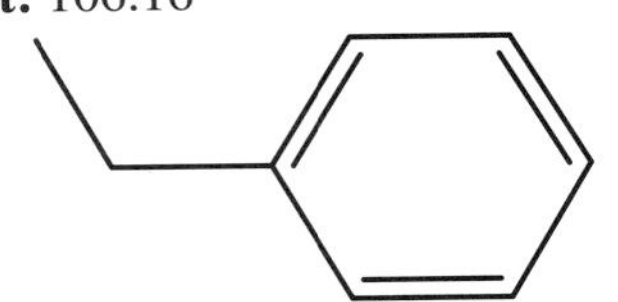

Chemical Structure

Synonyms: AETHYLBENZOL; BENZENE,ETHYL-; EB; ETHYLBENZEEN; ETHYLBENZOL; ETILBENZENE; ETYLOBENZEN; PHENYLETHANE

Description: colorless liquid; pungent odor

Use: intermediate in production of styrene; organic synthesis; solvent; dilutant; antiknock agent; acetophenone manufacture; asphalt constituent; naphtha constituent

Physical Properties

Boiling Point: 136.2 °C (277 °F) at 760 mm Hg

Freezing Point: -95 °C (-139 °F)

Specific Gravity: 0.867 at 20 °C/4 °C

Vapor Density: 3.66 Air=1

Saturated Vapor Density: 1.242010889 kg/m^3

Density: 0.866 g/cu cm at 21 °C

Bulk Density: 7.21 lbs/gal at 25 °C

Vapor Pressure: 10 mm Hg at 25.90 °C

Water Solubility: 0.01% by weight

Other Solubilities: 95% Ethanol: 1-10 mg/ml at 23 °C; Acetone: >=100 mg/ml at 23 °C; Ammonia: Insoluble; Benzene: Soluble; Carbon Tetrachloride: Soluble; DMSO: >=100 mg/ml at 23 °C; Ether: Soluble; Most organic solvents: Soluble; Sulfur dioxide: Soluble.

Surface Tension: 35.4 dynes/cm

Odor Threshold: 8.7 to 870.0 mg/m^3

Refraction Index: 1.4959 at 20 °C/D

Evaporation Rate: ~ 94 Ether=1

Critical Temperature: 173 °C

Critical Pressure: 523 psia

Ionization Potential (eV): 8.76

Flash Point: 12.8 °C Closed Cup
Autoignition Temperature: 432 °C
LEL: 1.6% v/v
UEL: 7% v/v

RTECS Toxicity Data

Acute Oral: Rat LD_{50} Dose: 3500 mg/kg; Toxic Effects: Liver - Other changes; Kidney, Ureter, and Bladder - Other changes.

Acute Inhalation: Rat LC_{Lo} Dose: 4000 ppm/4hr. Human TC_{Lo} Dose: 100 ppm/8hr; Toxic Effects: Sense organs and special senses - Other; Behavioral - Sleep; Lungs, Thorax, or Respiration - Other changes.

Acute Dermal: Rabbit LD_{50} Route: Skin; Dose: 17800 uL/kg.

Chronic (Multiple Dose) Inhalation: Rat Dose: 740 ppm/6H/92D-I; Toxic Effects: Lungs, Thorax, or Respiration - Changes in lung weight; Liver - Changes in Liver weight; Kidney, Ureter, and Bladder - Changes in kidney weight. Rat Dose: 782 ppm/6H/4W-I; Toxic Effects: Liver - Changes in Liver weight; Blood - Changes in leukocyte (WBC) cell count; Blood - Changes in platelet cell count.

Irritation Eye: Rabbit Standard Draize Test Dose: 500 mg; Reaction: severe.

Irritation Skin: Rabbit Open Draize Test Dose: 15 mg/24H open; Reaction: mild.

Reproductive/Teratogenic: Rat Route: Inhalation; Dose: 97 ppm/7H; Duration: female 15D prior to mating Effects on Fertility - Female fertility index. Rat Route: Inhalation; Dose: 985 ppm/7H; Duration: female 1-19D of pregnancy; Effects on Embryo or Fetus - Fetotoxicity. Rat Route: Inhalation; Dose: 96 ppm/7H; Duration: female 1-19D of pregnancy; Specific Developmental Abnormalities - Musculoskeletal system. Rat Route: Inhalation; Dose: 600 mg/m^3/24 Duration: female 7-15D of pregnancy; Effects on Fertility - Post-implantation mortality; Effects on Embryo or Fetus - Fetal death; Specific Developmental Abnormalities - Musculoskeletal system. Rat Route: Inhalation; Dose: 2400 mg/m^3/2 Duration: female 7-15D of pregnancy; Effects on Embryo or Fetus - Fetotoxicity.

Mutagenic: Human Sister Chromatid Exchenge; Cell Type: lymphocyte; Dose: 10 mmol/L. Mouse Mutations in Mammalian Somatic Cells; Cell Type: lymphocyte; Dose: 80 mg/L.

Hazard Overviews

Flammable

Fire Diamond

Health: Irritating to eyes/skin/respiratory tract. Also Causes: chest constriction, vertigo, narcosis, cramps, respiratory paralysis. Chronic Effects: fatigue, sleepiness, headache, blood disorders, lymphocytosis.

Fire: Flammable. Can form explosive mixtures in the air. For small fires, use dry chemical, carbon dioxide, or alcohol-resistant foam. For large fires, use fog or alcohol-resistant foam. Use water only if other agents are unavailable; Floats on water and may travel to an ignition source and spread fire.

Reactivity: Stable. Hazardous polymerization cannot occur. Avoid: heat. Incompatible with: oxidizers. Hazardous decomposition products: acrid smoke; irritating fumes.

Carcinogenicity: IARC - Not listed; NIOSH - Not listed; NTP - Not listed; ACGIH - Not listed; OSHA - Not listed; EPA - Class D, Not classifiable as to human carcinogenicity; MAK - Not listed

Primary Target Organs:

Eyes

Skin

Respiratory System

Nervous System

Blood

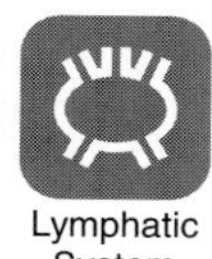
Lymphatic System

Exposure Limits

OSHA PEL: TWA: 100 ppm; 435 mg/m^3.

OSHA PEL Vacated 1989 Limits: TWA: 100 ppm; 435 mg/m^3; STEL: 125 ppm; 545 mg/m^3.

ACGIH TLV: TWA: 100 ppm; 434 mg/m^3; STEL: 125 ppm; 543 mg/m^3.

NIOSH REL: TWA: 100 ppm; 435 mg/m^3. STEL: 125 ppm; 545 mg/m^3.

NIOSH IDLH: 800 ppm; LEL.

DFG MAK: TWA: 100 ppm; 440 mg/m^3.

Respirator Recommendation

Exposure Range: >100 to <800 ppm Air Purifying, Negative Pressure, Half Mask

Exposure Range: 800 to unlimited ppm Self-contained Breathing Apparatus, Pressure Demand, Full Face

Cartridge Color: black

Environmental

Ecotoxicity: LC_{50} Cyprinodon variegatus (sheepshead minnow) 275 mg/l 96 hr in a static unmeasured bioassay LC_{50} Pimephales promelas (fathead minnow) 12.1 mg/l/96 hr (confidence limit 11.5 - 12.7 mg/l), flow-through bioassay with measured concentrations, 26.1 °C, dissolved oxygen 7.0 mg/l, hardness 45.6 mg/l calcium carbonate, alkalinity 43.0 mg/l Toxicity threshold (cell multiplication inhibition test): Pseudomonas putida (bacteria) 12 mg/l LC_{50} Palaemonetes pugio (grass shrimp, adult) 14,400 ug/l/24 hr in a static unmeasured bioassay LC_{50} Palaemonetes pugio (grass shrimp, larva) 10,200 ug/l/24 hr in a static unmeasured bioassay Toxicity threshold (cell multiplication inhibition test): Microcystis aeruginosa (algae) 33 mg/l; Scenedesmus quadricauda (green algae) > 160 mg/l

Environmental Fate: If released to the atmosphere, it exist predominantly in the vapor phase based on its vapor pressure where it will photochemically degrade by reaction with hydroxyl radicals (half-life 0.5 to 2 days) and partially return to earth in rain. It will not be subject to direct photolysis. Releases into water will decrease in concentration by evaporation and biodegradation. The time for this decrease and the primary loss processes will depend on the season, and the turbulence and microbial populations in the particular body of water. Representative half-lives are several days to 2 weeks. Some may be adsorbed by sediment but significant bioconcentration in fish is not expected to occur based upon its octanol/water partition coefficient. It is only adsorbed moderately by soil. It will not significantly hydrolyze in water or soil.

Cleanup/Disposal: Guide No. 129: Eliminate all ignition sources (no smoking, flares, sparks or flames in immediate area). All equipment used when handling the product must be grounded. Do not touch or walk through spilled material. Stop leak if you can do it without risk. Prevent entry into waterways, sewers, basements or confined areas. A vapor suppressing foam may be used to reduce vapors. Absorb or cover with dry earth, sand or other non-combustible material and transfer to containers. Use clean non-sparking tools to collect absorbed material. Large Spills: Dike far ahead of liquid spill for later disposal. Water spray may reduce vapor; but may not prevent ignition in closed spaces.

Environmental Physical Data

Henry's Law Constant: 8.44×10^{-3}
Octanol/Water Partition Coefficient: log K_{ow} = 3.15
Sorption Partition Coefficient: K_{oc} = 164
BCF: goldfish 1.9
BOD: theoretical 2.8%, 5 days

Regulations

RCRA 40CFR: Not listed
CERCLA: 40CFR 302.4: Listed per CWA Section 311(b)(4) per CWA Section 307(a) RQ: 1000 lb (453.5 kg)
SARA 40CFR 372.65: Listed
SARA EHS 40CFR 355: Not listed
TSCA: Listed

Analytical Methods

Air: EPA OA-002-1, VA-001-1, VA-003-1, VA-005-1, VA-006-1, VA-008-1, VG-006-1, VG-007-1, VG-011-1, TO-1, TO-14; ASTM D3686, D3687, D4490
Soil: CLP LC_VOA, MC_VOA, OHC; EPA 7, 1624, VG-008-1, VG-010-1, VS-001-1, VS-002-1, VW-010-1; SW846 1311, 5021, 5032, 5041, 5041A, 8020A, 8021A, 8240B, 8260A, 8260B; DOE OS040, OS060
Water / Groundwater: EPA 602, 624, 624-S, VW-001-1, VW-002-1, VW-003-1, VW-004-1, VW-008-1, VW-014-1; APHA 6040-B, 6210-B, 6210-D, 6220-B, 6230-D; ASTM D3695, D3871, D4763; USGS O3115
Drinking Water: EPA 502.2, 503.1, 524.1, 524.2; APHA 6210-C, 6220-C
Food: EPA 5
Indoor / Expired Air: NIOSH 1501; EPA IP-1A, IP-1B
Plasma: EPA 29
Other: EPA VS-006-1, VW-011-1

ETH3100 CAS #: 74-96-4

ETHYL BROMIDE

RTECS: KH6475000
DOT: UN1891; IMO6.1
EINECS Number: 200-825-8
Molecular Formula: C_2H_5Br
Structured MF: CH_3CH_2Br
Formula Weight: 108.98

Br

Chemical Structure

Synonyms: BROMIC ETHER; BROMOETHANE; BROMURE D'ETHYLE; ETHANE,BROMO-; ETYLU BROMEK; HALON 2001; HYDROBROMIC ETHER; MONOBROMOETHANE
Description: colorless liquid; ethereal odor
Use: as an ethylating agent in organic synthesis, in medicine as an anesthetic, as a refrigerant, as a solvent, as a grain and fruit fumigant and in gasoline; formerly used as a topical and inhalation anesthetic

Physical Properties

Boiling Point: 38.2 °C (101 °F)
Freezing Point: -119 °C (-182.2 °F)
Specific Gravity: 1.4612 at 20 °C/4 °C
Vapor Density: 3.76 Air=1
Saturated Vapor Density: 3.233611252 kg/m^3
Density: 1.4708 g/mL at 15 °C
Bulk Density: 12 to 12.1 lbs/gal
Vapor Pressure: 467 mm Hg at 25 °C
Water Solubility: 1.067 g/100 g Water at 0 °C
Other Solubilities: 95% Ethanol: >=100 mg/ml at 16 °C; Acetone: >=100 mg/ml at 16 °C; Alcohol: miscible; Chloroform: miscible; DMSO: >=100 mg/ml at 16 °C; Ether: Soluble; Organic solvents: miscible.
Surface Tension: 24.15 dyne/cm at 20 °C
Odor Threshold: 890 mg/m^3
Refraction Index: 1.42386 at 25 °C
Ionization Potential (eV): 10.29
Flash Point: -20 °C Closed Cup
Autoignition Temperature: 511 °C
LEL: 6.8% v/v
UEL: 8% v/v

RTECS Toxicity Data

Acute Oral: Rat LD_{50} Dose: 1350 mg/kg.
Acute Inhalation: Rat LC_{50} Dose: 26980 ppm/1hr. Mouse LC_{50} Dose: 16230 ppm/1hr; Toxic Effects: Behavioral - Somnolence (general depressed activity); Gastrointestinal - Hypermotility, diarrhea.
Chronic (Multiple Dose) Inhalation: Rat Dose: 2 pph/6H/14D-I; Toxic Effects: DEATH. Rat Dose: 1600 ppm/6H/14W-I; Toxic Effects: Blood - Changes in spleen; DEATH.
Mutagenic: Hamster Sister Chromatid Exchenge; Cell Type: ovary; Dose: 100 mg/L. Bacteria - S Typhimurium Mutations in Microorganisms; Dose: 75 ng/plate (+S9), 50 ng/plate (-S9).
Tumorigenic: Rat Route: Inhalation; Dose: 100 ppm/6H/2Y-I; Toxic Effects: Tumorigenic - Equivocal tumorigenic agent by RTECS criteria; Brain and coverings - Tumors. Mouse Route: Inhalation; Dose: 200 ppm/6H/2Y-I; Toxic Effects:

Tumorigenic - Carcinogenic by RTECS criteria; Tumorigenic effects - Uterine tumors.

Hazard Overviews

Flammable

Fire Diamond

Health: Irritating to eyes/skin/respiratory tract. Also Causes: drowsiness, slurred speech, and incoordination.

Fire: Flammable. Can form explosive mixtures in the air. Use dry chemical, foam, or carbon dioxide. Water may be ineffective in extinguishing fire. Use water spray to cool fire-exposed tanks/container.

Reactivity: Stable. Hazardous polymerization cannot occur. Avoid: contact with some forms of plastics, rubber, and coatings. Incompatible with: strong oxidizing agents; chemically active metals (sodium; potassium; calcium; powdered aluminum; zinc; magnesium). Hazardous decomposition products: carbon dioxide; carbon monoxide; toxic fumes of bromine.

Carcinogenicity: IARC - Group 3, Not classifiable as to carcinogenicity to humans; NIOSH - Not listed; NTP - Not listed; ACGIH - Class A3, Animal carcinogen; OSHA - Not listed; EPA - Not listed; MAK - Class A2, Unmistakably carcinogenic in animal experimentation only

Primary Target Organs:

Eyes

Skin

Respiratory System

Liver

Kidneys

Cardio-vascular

Exposure Limits

OSHA PEL: TWA: 200 ppm; 890 mg/m^3.

OSHA PEL Vacated 1989 Limits: TWA: 200 ppm; 890 mg/m^3; STEL: 250 ppm; 1110 mg/m^3.

ACGIH TLV: TWA: 5 ppm; 22 mg/m^3.

NIOSH IDLH: 2000 ppm.

Respirator Recommendation

Exposure Range: >200 to 1000 ppm Air Purifying, Negative Pressure, Half Mask

Exposure Range: >1000 to <2000 ppm Supplied Air, Constant Flow/Pressure Demand, Half Mask

Exposure Range: 2000 to unlimited ppm Self-contained Breathing Apparatus, Pressure Demand, Full Face

Cartridge Color: black

Environmental

Environmental Fate: If released to the atmosphere, it will degrade relatively slowly by reaction with photochemically produced hydroxyl radicals (estimated half-life of 48 days). If released to water, it will be removed through hydrolysis and volatilization. The aqueous hydrolysis half-lives at 20 and 25 °C are 40 and 30 days, respectively. The volatilization half-lives from a model environmental river (1 meter deep) and model pond have been estimated to be 3.2 hr and 38.2 hr, respectively. If released to soil, it will be susceptible to hydrolysis under wet soil conditions. Its detection in landfill leachate demonstrates that environmental leaching can occur. A relatively high vapor pressure of 467 mm Hg at 25 °C and Henry's Law of 7.49 x10^{-3} atm-cu meter/mole indicate that evaporation from moist and dry soils may occur. Biodegradation is expected to be an important fate process in both water and soil.

Cleanup/Disposal: Guide No. 131: Fully encapsulating, vapor protective clothing should be worn for spills and leaks with no fire. Eliminate all ignition sources (no smoking, flares, sparks or flames in immediate area). All equipment used when handling the product must be grounded. Do not touch or walk through spilled material. Stop leak if you can do it without risk. Prevent entry into waterways, sewers, basements or confined areas. A vapor suppressing foam may be used to reduce vapors. Small Spills: Absorb with earth, sand or other non-combustible material and transfer to containers for later disposal. Use clean non-sparking tools to collect absorbed material. Large Spills: Dike far ahead of liquid spill for later disposal. Water spray may reduce vapor; but may not prevent ignition in closed spaces.

Environmental Physical Data

Henry's Law Constant: estimated at 7.49 x10^{-3}

Octanol/Water Partition Coefficient: log K_{ow} = 1.61

Sorption Partition Coefficient: K_{oc} = estimated at 29

BCF: estimated at 3.7

Regulations

RCRA 40CFR: Not listed

CERCLA: 40CFR 302.4: Not listed

SARA 40CFR 372.65: Not listed

SARA EHS 40CFR 355: Not listed

TSCA: Listed

Analytical Methods

Air: ASTM D3686, D3687, D4490

Indoor / Expired Air: NIOSH 1011; EPA IP-1B

ETH3150 CAS #: 105-36-2

ETHYL 2-BROMOACETATE

RTECS: AF6000000

EINECS Number: 203-290-9

Molecular Formula: $C_4H_7BrO_2$

Formula Weight: 167.01

Chemical Structure

Synonyms: ACETIC ACID,BROMO-,ETHYL ESTER; ANTOL; BROMOACETIC ACID ETHYL ESTER; BROMOACETIC ACID,ETHYL ESTER; ETHOXYCARBONYLMETHYL BROMIDE; ETHYL BROMACETATE; ETHYL ALPHA-BROMOACETATE; ETHYL BROMOACETATE; ETHYL MONOBROMOACETATE

Description: clear colorless liquid

Use: tear gas; chem int & reactant in synth of an analgesia; chem int, eg, for pharmaceuticals & beta-keto acids; lacrimator or warning agent for poisonous, odorless gases

Physical Properties

Boiling Point: 168 °C (334 °F) to 169 °C (336 °F) at 760 mm Hg
Freezing Point: -13.8 °C (7.16 °F)
Specific Gravity: 1.5059 at 20 °C/20 °C
Vapor Density: 5.8 Air=1
Water Solubility: Insoluble in Water
Other Solubilities: Soluble in Benzene.
Refraction Index: 1.4489 at 20 °C
Flash Point: 48 °C

RTECS Toxicity Data

Tumorigenic: Mouse Route: Subcutaneous; Dose: 252 mg/kg/63W-I; Toxic Effects: Tumorigenic - Neoplastic by RTECS criteria; Tumorigenic - Tumors at site of application.

Hazard Overviews

Corrosive

Flammable

Health: Corrosive to eyes/skin/respiratory tract. Other Acute Effects: harmful if swallowed, inhaled, or absorbed through skin; inhalation may result in spasm, inflammation and edema of the larynx and bronchi, chemical pneumonitis and pulmonary edema; symptoms of exposure may include burning sensation, coughing, wheezing, laryngitis, shortness of breath, headache, nausea and vomiting.
Fire: Flammable. Hazards: emits toxic fumes. Extinguishing agents: carbon dioxide, dry chemical powder or appropriate foam. Precautions: combustible liquid.
Reactivity: Incompatible with: acids, bases, oxidizing agents, reducing agents. Hazardous decomposition products: toxic fumes of: carbon monoxide, carbon dioxide, hydrogen bromide gas.
Carcinogenicity: IARC - Not listed; NIOSH - Not listed; NTP - Not listed; ACGIH - Not listed; OSHA - Not listed; EPA - Not listed; MAK - Not listed
Primary Target Organs:

Eyes

Skin

Respiratory System

Environmental

Regulations

RCRA 40CFR: Not listed
CERCLA: 40CFR 302.4: Not listed
SARA 40CFR 372.65: Not listed
SARA EHS 40CFR 355: Not listed
TSCA: Listed

ETH3200 CAS #: 97-95-0

ETHYL BUTANOL

RTECS: EL3850000
EINECS Number: 202-621-4
Molecular Formula: $C_6H_{14}O$
Structured MF: $(C_2H_5)_2CHCH_2OH$
Formula Weight: 102.18

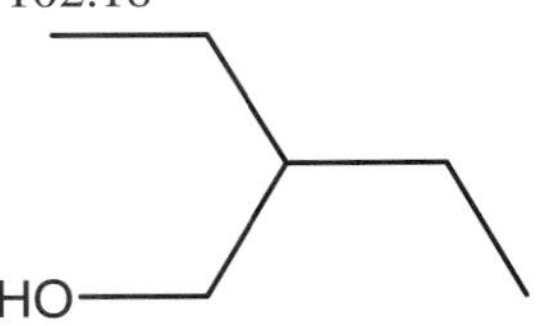

Chemical Structure

Synonyms: 1-BUTANOL,2-ETHYL-; 2-ETHYL-1-BUTANOL; 2-ETHYLBUTANOL; 2-ETHYLBUTANOL-1; 2-ETHYLBUTYL ALCOHOL; SEC-HEXYL ALCOHOL; 3-METHYLOLPENTANE; 3-PENTYLCARBINOL; SEC-PENTYLCARBINOL; PSEUDOHEXYL ALCOHOL
Description: colorless liquid; mild, nonresidual odor
Use: solvent for oils, resins, waxes, dyes; syn of perfumes, drugs; flavoring; solvent for printing inks; component in lacquers; mfr of surface active agents, synthetic lubricants, and lubricant additives

Physical Properties

Boiling Point: 146.27 °C (295 °F) at 760 mm Hg
Freezing Point: < -15 °C (5 °F)
Specific Gravity: 0.8326 at 20 °C/4 °C
Vapor Density: 0.01 Air=1
Saturated Vapor Density: 1.203585953 kg/m^3
Bulk Density: 6.93 lbs/gal at 20 °C
Vapor Pressure: 0.9 mm Hg at 20 °C
Water Solubility: Slightly Soluble in Water
Other Solubilities: miscible with most organic solvents.
Surface Tension: 24.3 dynes/cm at 25 °C
Refraction Index: 1.4220 at 20 °C/D
Flash Point: 58.333 °C Open Cup
Autoignition Temperature: 304 °C
LEL: 1.9% v/v
UEL: 8.8% v/v

RTECS Toxicity Data

Acute Oral: Rat LD_{50} Dose: 1850 mg/kg. Rabbit LD_{50} Dose: 1200 mg/kg.
Acute Dermal: Rabbit LD_{50} Route: Skin; Dose: 1260 uL/kg.
Irritation Eye: Rabbit Standard Draize Test Dose: 250 ug open; Reaction: severe. Rabbit Standard Draize Test Dose: 250 ug/24H; Reaction: severe.
Irritation Skin: Rabbit Standard Draize Test Dose: 500 mg/24H; Reaction: mild. Rabbit Open Draize Test Dose: 415 mg open; Reaction: mild.

Hazard Overviews

Flammable

Health: Severely irritating to eyes; may irritate skin. Harmful. Other Acute Effects: harmful if swallowed or absorbed through skin; may be harmful if inhaled; exposure can cause damage to the eyes.

Fire: Flammable. Extinguishing agents: carbon dioxide, dry chemical powder or appropriate foam; water may be effective for cooling, but may not effect extinguishment. Precautions: combustible liquid.

Carcinogenicity: IARC - Not listed; NIOSH - Not listed; NTP - Not listed; ACGIH - Not listed; OSHA - Not listed; EPA - Not listed; MAK - Not listed

Primary Target Organs:

Eyes

Environmental

Cleanup/Disposal: Guide No. 153: Eliminate all ignition sources (no smoking, flares, sparks or flames in immediate area). Do not touch damaged containers or spilled material unless wearing appropriate protective clothing. Stop leak if you can do it without risk. Prevent entry into waterways, sewers, basements or confined areas. Absorb or cover with dry earth, sand or other non-combustible material and transfer to containers. Do not get water inside containers.

Environmental Physical Data

BCF: no food chain concentration potential

Regulations

RCRA 40CFR: Not listed
CERCLA: 40CFR 302.4: Not listed
SARA 40CFR 372.65: Not listed
SARA EHS 40CFR 355: Not listed
TSCA: Listed

Analytical Methods

Water / Groundwater: ASTM D3695

ETH3250 **CAS #: 617-79-8**

2-ETHYL BUTYLAMINE

RTECS: EO4900000
EINECS Number: 210-530-6
Molecular Formula: $C_6H_{15}N$
Structured MF: $C_2H_5NHC_4H_9$
Formula Weight: 101.22
Synonyms: 1-AMINO-2-ETHYLBUTANE; 1-BUTANAMINE,2-ETHYL-; 2-ETHYLBUTYLAMINE

Physical Properties

Boiling Point: 125 °C (257 °F)
Specific Gravity: 0.776 at 20 °C/20 °C
Vapor Density: 3.49 Air=1
Water Solubility: Partly Soluble
Surface Tension: Estimated at 21 dynes/cm
Refraction Index: 1.407
Critical Temperature: Estimated at 296.5 °C
Critical Pressure: Estimated at 440.9 psia
Flash Point: 21.111 °C

RTECS Toxicity Data

Acute Oral: Rat LD_{50} Dose: 390 mg/kg.
Acute Inhalation: Rat LC_{50} Dose: 500 ppm/4hr.
Acute Dermal: Rabbit LD_{50} Route: Skin; Dose: 2 mL/kg.
Irritation Eye: Rabbit Standard Draize Test Dose: 250 ug open; Reaction: severe.
Irritation Skin: Rabbit Standard Draize Test Dose: 5 mg/24H; Reaction: severe.

Hazard Overviews

Corrosive

Flammable

Health: Corrosive to eyes/skin/respiratory tract. Other Acute Effects: harmful if swallowed, inhaled, or absorbed through skin; inhalation may result in spasm, inflammation and edema of the larynx and bronchi, chemical pneumonitis and pulmonary edema; prolonged contact can cause severe irritation or burns.

Fire: Flammable. Hazards: vapor may travel considerable distance to source of ignition and flash back; container explosion may occur; emits toxic fumes; forms explosive mixtures in air. Extinguishing agents: carbon dioxide, dry chemical powder or appropriate foam; water may be effective for cooling, but may not effect extinguishment. Precautions: combustible liquid.

Reactivity: Incompatible with: acids, acid chlorides, acid anhydrides, strong oxidizing agents, carbon dioxide. Hazardous decomposition products: thermal decomposition may produce carbon monoxide, carbon dioxide, and nitrogen oxides.

Carcinogenicity: IARC - Not listed; NIOSH - Not listed; NTP - Not listed; ACGIH - Not listed; OSHA - Not listed; EPA - Not listed; MAK - Not listed

Primary Target Organs:

Eyes

Skin

Respiratory System

Environmental

Cleanup/Disposal: Guide No. 132: Fully encapsulating, vapor protective clothing should be worn for spills and leaks with no fire. Eliminate all ignition sources (no smoking, flares,

sparks or flames in immediate area). All equipment used when handling the product must be grounded. Do not touch or walk through spilled material. Stop leak if you can do it without risk. Prevent entry into waterways, sewers, basements or confined areas. A vapor suppressing foam may be used to reduce vapors. Absorb with earth, sand or other non-combustible material and transfer to containers (except for Hydrazine). Use clean non-sparking tools to collect absorbed material. Large Spills: Dike far ahead of liquid spill for later disposal. Water spray may reduce vapor; but may not prevent ignition in closed spaces.

Regulations

RCRA 40CFR: Not listed
CERCLA: 40CFR 302.4: Not listed
SARA 40CFR 372.65: Not listed
SARA EHS 40CFR 355: Not listed
TSCA: Not listed

ETH3300 CAS #: 105-54-4

ETHYL N-BUTYRATE

RTECS: ET1660000
DOT: UN1180; IMO3.3
EINECS Number: 203-306-4
Molecular Formula: $C_6H_{12}O_2$
Structured MF: $CH_3CH_2CH_2COOC_2H_5$
Formula Weight: 116.16

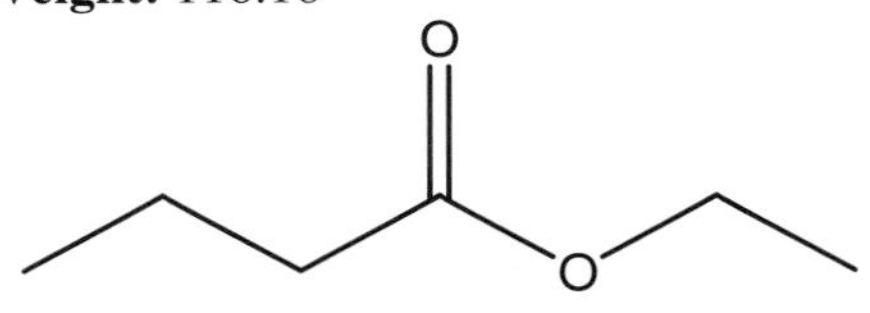

Chemical Structure

Synonyms: BUTANOIC ACID ETHYL ESTER; BUTANOIC ACID,ETHYL ESTER; BUTYRIC ACID,ETHYL ESTER; BUTYRIC ETHER; ETHYL BUTANOATE; ETHYL BUTYRATE
Description: colorless liquid; fruity odor with pineapple undernote
Use: in mfr of artificial rum; the so-called pineapple oil; as flavor for foods, beverages, & chewing gums; fragrance ingredient for soap, perfume, creams, & lotions; solvent for cellulosic lacquers

Physical Properties

Boiling Point: 120 °C (248 °F) to 121 °C (250 °F)
Freezing Point: -100.8 °C (-149.44 °F)
Specific Gravity: 0.879 at 20 °C/4 °C
Vapor Density: 4 Air=1
Saturated Vapor Density: 1.27156276 kg/m^3
Vapor Pressure: 2.01 kPa at 25 °C
Water Solubility: 1 parts in 150 parts Water
Other Solubilities: miscible with Alcohol, Ether
Surface Tension: 24.5 dynes/cm at 20 °C
Odor Threshold: 0.015 ppm
pH: Not more than 1
Refraction Index: 1.4000 at 20 °C/D
Critical Temperature: 293 °C
Critical Pressure: 460 psia
Flash Point: 29 °C Open Cup
Autoignition Temperature: 463 °C

RTECS Toxicity Data

Acute Oral: Rat LD_{50} Dose: 13 gm/kg; Toxic Effects: Behavioral - Somnolence (general depressed activity); Behavioral - Coma. Rabbit LD_{50} Dose: 5228 mg/kg.
Irritation Skin: Rabbit Standard Draize Test Dose: 500 mg/24H; Reaction: moderate.

Hazard Overviews

Flammable

Fire: Flammable.
Carcinogenicity: IARC - Not listed; NIOSH - Not listed; NTP - Not listed; ACGIH - Not listed; OSHA - Not listed; EPA - Not listed; MAK - Not listed

Environmental

Environmental Fate: If released to soil, it will not be expected to strongly adsorb to soil but will be expected to exhibit very high mobility in the soil based upon an estimated K_{oc} of 41. It may be susceptible to volatilization from near-surface soils based upon the predicted rapid volatilization rates from surface water and relatively high vapor pressure. Hydrolysis will not be expected to be an important removal process except in alkaline soils based upon hydrolysis rates measured in water. It is not known whether it will be subject to biodegradation in soil. If released to water, it will not be expected to strongly adsorb to sediment or suspended particulate matter based upon an estimated K_{oc} of 41 or to bioconcentrate in aquatic organisms based upon an estimated BCF of 12. It is not known whether it will be subject to biodegradation in natural waters. It will be subject to rapid volatilization from surface water based upon an estimated half-life of 5.5 hr for volatilization from a model river (1 meter deep flowing 1 m/sec with a wind speed of 3/m sec) calculated using an estimated Henry's Law constant. The volatilization half-life from a model pond has been estimated to be 65 hr. Indirect photooxidation with hydroxyl radicals in water is not expected to be an important removal process. It should not directly photolysis in water. Hydrolysis will not be expected to be an important removal process except in alkaline waters based upon hydrolysis half-lives of 101 years, 6.3 years, 229 days, and 23 days at pH 5,7,8, and 9, respectively, calculated using measured base and acid hydrolysis rate constants. If released to the atmosphere, it will be expected to exist almost entirely in the vapor phase. It will be susceptible to rapid reaction with photochemically produced hydroxyl radicals in the atmosphere based upon a half-life of 6 days which was calculated from a measured rate constant. It should not directly photolyze in the atmosphere.

Cleanup/Disposal: Guide No. 129: Eliminate all ignition sources (no smoking, flares, sparks or flames in immediate area). All equipment used when handling the product must be grounded. Do not touch or walk through spilled material. Stop leak if you can do it without risk. Prevent entry into waterways, sewers, basements or confined areas. A vapor suppressing foam may be used to reduce vapors. Absorb or cover with dry earth, sand or other non-combustible material and transfer to containers. Use clean non-sparking tools to collect absorbed material. Large Spills: Dike far ahead of liquid spill for later disposal. Water spray may reduce vapor; but may not prevent ignition in closed spaces.

Environmental Physical Data

Henry's Law Constant: calculated at 3.99×10^{-4}
Octanol/Water Partition Coefficient: log K_{ow} = 1.73
Sorption Partition Coefficient: K_{oc} = 41
BCF: estimated at 12

Regulations

RCRA 40CFR: Not listed
CERCLA: 40CFR 302.4: Not listed
SARA 40CFR 372.65: Not listed
SARA EHS 40CFR 355: Not listed
TSCA: Listed

ETH3350 **CAS #: 75-00-3**

ETHYL CHLORIDE

RTECS: KH7525000
DOT: UN1037; IMO2.3
EINECS Number: 200-830-5
Molecular Formula: C_2H_5Cl
Structured MF: H_3CCH_2Cl
Formula Weight: 64.52

Chemical Structure

Synonyms: AETHYLCHLORID; AETHYLCHLORIDE; AETHYLIS; AETHYLIS CHLORIDUM; ANODYNON; CHELEN; CHLOORETHAAN; CHLORENE; CHLORETHYL; CHLORIDUM; CHLOROAETHAN; CHLOROETHANE; CHLORURE D'ETHYLE; CHLORYL; CHLORYL ANESTHETIC; CLORENE; CLORETILO; CLOROETANO; CLORURO DI ETILE; DUBLOFIX; ETHANE,CHLORO-; ETHER CHLORATUS; ETHER CHLORIDUM; ETHER HYDROCHLORIC; ETHER MURIATIC; ETYLU CHLOREK; HYDROCHLORIC ETHER; KELENE; MONOCHLORETHANE; MONOCHLOROETHANE; MURIATIC ETHER; NARCOTILE

Description: colorless gas at room temp., liquid when compressed; ethereal, slightly pungent odor

Use: in the manufacture of tetraethyl lead, ethyl cellulose sulfanol and trional; as an anesthetic; in organic synthesis; as an alkylating agent; in refrigeration; as an analytical reagent; as a solvent for phosphorous, sulfur, fats, oils, resins and waxes; in insecticides; and as an ethylating agent in the manufacture of dyes and drugs

Physical Properties

Boiling Point: 12.3 °C (54 °F) at 760 mm Hg
Freezing Point: -138.7 °C (-217.66 °F)
Specific Gravity: 0.9214 at 0 °C/4 °C
Vapor Density: 2.22 Air=1
Vapor Pressure: 1000 mm Hg at 20 °C
Water Solubility: 0.574 g/100 ml in Water at 20 °C
Other Solubilities: 95% Ethanol: 48 g/100 mL at 21 °C.
Surface Tension: Estimated at 40 dynes/cm
Odor Threshold: Recognition 10 to 12 mg/m^3
Refraction Index: 1.3676 at 20 °C/D
Evaporation Rate: Much higher than 1 Butyl Acetate=1
Critical Temperature: 187 °C
Critical Pressure: 52 atm
Ionization Potential (eV): 10.97
Flash Point: -50 °C Closed Cup
Autoignition Temperature: 519 °C
LEL: 3.8% v/v
UEL: 15.4% v/v

RTECS Toxicity Data

Acute Inhalation: Rat LC_{50} Dose: 152 gm/m^3/2hr; Toxic Effects: Brain and coverings - Other degenerative changes; Behavioral - General anesthetic; Blood - Hemorrhage. Mouse LC_{50} Dose: 146 gm/m^3/2hr.

Chronic (Multiple Dose) Inhalation: Rat Dose: 19000 ppm/6H/13W-I; Toxic Effects: Liver - Changes in Liver weight. Rat Dose: 14 gm/m^3/2H/60D-I; Toxic Effects: Brain and coverings - Recordings from specific areas of CNS. Rat Dose: 570 mg/m^3/4H/26W-I; Toxic Effects: Liver - Liver function tests impaired; Nutritional and gross metabolic - Weight loss or decreased weight gain.

Reproductive/Teratogenic: Rat Route: Inhalation; Dose: 60 mg/m^3/4H; Duration: male 26W prior to mating; Paternal Effects - Spermatogenesis.

Mutagenic: Hamster Mutations in Mammalian Somatic Cells; Cell Type: ovary; Dose: 2340 mg/L. Bacteria - E Coli Mutations in Microorganisms; Dose: 1 pph (+/-S9). Bacteria - S Typhimurium Mutations in Microorganisms; Dose: 10 ug/plate (+/-S9).

Tumorigenic: Rat Route: Inhalation; Dose: 15000 ppm/6H/2Y-I; Toxic Effects: Tumorigenic - Equivocal tumorigenic agent by RTECS criteria; Skin and appendages - Tumors. Mouse Route: Inhalation; Dose: 15000 ppm/6H/2Y-I; Toxic Effects: Tumorigenic - Carcinogenic by RTECS criteria; Liver - Tumors; Tumorigenic effects - Uterine tumors.

Hazard Overviews

Flammable

Compressed Gas

Fire Diamond

Health: Irritating to eyes/skin/respiratory tract. Also Causes: CNS depression, cardiac arrythmias, abdominal cramps, frostbite.

Fire: Flammable. Explosive when heated. Can form explosive mixtures in the air. For small fires use dry chemical, water spray, or regular foam. For large fires use water spray, fog, or regular foam.

Reactivity: Stable. Hazardous polymerization cannot occur. Avoid: heat and ignition sources. Incompatible with: oxidizing materials; water/steam; sodium; potassium; calcium; powdered aluminum; powdered copper; powdered magnesium, powdered zinc. Hazardous decomposition products: phosgene; hydrogen chloride gas.

Carcinogenicity: IARC - Group 3, Not classifiable as to carcinogenicity to humans; NIOSH - Not listed; NTP - Not listed; ACGIH - Class A3, Animal carcinogen; OSHA - Not listed; EPA - Not listed; MAK - Class B, Justifiably suspected of having carcinogenic potential

Primary Target Organs:

Eyes

Skin

Respiratory System

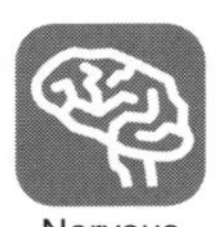
Nervous System

Cardio-vascular

Exposure Limits

OSHA PEL: TWA: 1000 ppm; 2600 mg/m^3.

ACGIH TLV: TWA: 100 ppm; 264 mg/m^3.

NIOSH IDLH: 3800 ppm; LEL.

Respirator Recommendation

Exposure Range: >1000 to <3800 ppm Supplied Air, Constant Flow/Pressure Demand, Half Mask

Exposure Range: 3800 to unlimited ppm Self-contained Breathing Apparatus, Pressure Demand, Full Face

Note: odor threshold unknown

Environmental

Environmental Fate: If released to the atmosphere, the dominant environmental fate process will be reaction with photochemically generated hydroxyl radicals, estimated half-life of about 40 days. If released to surface water, volatilization will be the dominant process as half-lives ranging from 1.1-5.6 days have been predicted for representative bodies of water. In groundwater, where volatilization may not be able to occur, hydrolysis may be the most important removal mechanism. The hydrolysis half-life has been estimated to be 38 days at 25 °C. Very limited biodegradation data suggest that it may be biodegradable, but insufficient data are available to estimate the relative importance of biodegradation in the environment. Aquatic bioconcentration, adsorption, direct photolysis, and oxidation are not important. If released to soil, it will evaporate rapidly where release to air is possible. It is susceptible to significant leaching.

Cleanup/Disposal: Guide No. 115: Eliminate all ignition sources (no smoking, flares, sparks or flames in immediate area). All equipment used when handling the product must be grounded. Do not touch or walk through spilled material. Stop leak if you can do it without risk. If possible, turn leaking containers so that gas escapes rather than liquid. Use water spray to reduce vapors or divert vapor cloud drift. Do not direct water at spill or source of leak. Prevent spreading of vapors through sewers, ventilation systems and confined areas. Isolate area until gas has dispersed.

Environmental Physical Data

Henry's Law Constant: estimated at 0.00848

Octanol/Water Partition Coefficient: log K_{ow} = 1.43

Sorption Partition Coefficient: K_{oc} = estimated at 143

BCF: none

Regulations

RCRA 40CFR: Not listed

CERCLA: 40CFR 302.4: Listed per CWA Section 307(a) RQ: 100 lb (45.35 kg)

SARA 40CFR 372.65: Listed

SARA EHS 40CFR 355: Not listed

TSCA: Listed

Analytical Methods

Air: EPA OA-002-1, VG-007-1, VG-011-1, TO-14; ASTM D4490

Soil: CLP LC_VOA, MC_VOA, OHC; EPA 7, 1624; SW846 5021, 5032, 5041, 5041A, 8010B, 8021A, 8240B, 8260A, 8260B

Water / Groundwater: EPA 601, 624, 624-S, VW-014-1; APHA 6210-B, 6210-D, 6230-B, 6230-D; USGS O3115

Drinking Water: EPA 502.1, 502.2, 524.1, 524.2; APHA 6210-C, 6230-C

Food: EPA 5

Indoor / Expired Air: NIOSH 2519; EPA IP-1A

Plasma: EPA 29

ETH3400 CAS #: 105-39-5

ETHYL CHLOROACETATE

RTECS: AF9110000

DOT: UN1181

EINECS Number: 203-294-0

Molecular Formula: $C_4H_7ClO_2$

Structured MF: $ClCH_2COOC_2H_5$

Formula Weight: 122.56

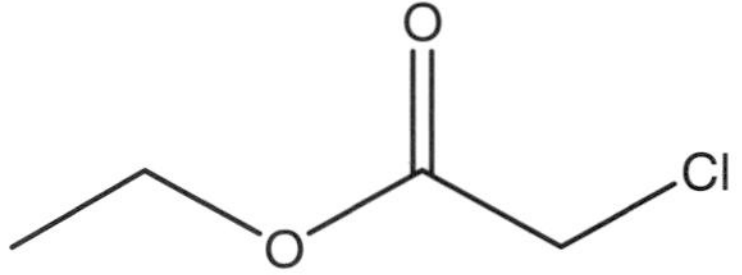

Chemical Structure

Synonyms: ACETIC ACID,CHLORO-,ETHYL ESTER; CHLOROACETIC ACID,ETHYL ESTER; ETHYL CHLORACETATE; ETHYL ALPHA-CHLOROACETATE; ETHYL CHLOROETHANOATE; ETHYL MONOCHLORACETATE; ETHYL MONOCHLOROACETATE; ETHYLESTER KYSELINY CHLOROCTOVE

Description: water-white liquid; fruity, pungent odor

Use: org synthesis; vat dyestuff; chem int for sodium fluoroacetate, a rodenticide & other derivatives, eg, dl-aspartic acid; chem int for benazolin, a herbicide

Physical Properties

Boiling Point: 144 °C (291 °F) to 146 °C (295 °F)
Freezing Point: -21 °C (-5.8 °F)
Specific Gravity: 1.1498 at 20 °C/4 °C
Vapor Density: 4.3 Air=1
Vapor Pressure: 0.121 kPa at 0 °C
Water Solubility: Insoluble in Water
Other Solubilities: Soluble in Benzene.
Surface Tension: Estimated at 24 dynes/cm
Refraction Index: 1.4227 at 20 °C/D
Flash Point: 54 °C

RTECS Toxicity Data

Acute Oral: Rat LD_{50} Dose: 235 mg/kg. Mouse LD_{50} Dose: 350 mg/kg.
Acute Dermal: Rabbit LD_{50} Route: Skin; Dose: 230 mg/kg. Mouse LD_{50} Route: Subcutaneous Dose: 250 mg/kg; Toxic Effects: Peripheral Nerve and sensation - Flaccid paralysis without anesthesia; Lungs, Thorax, or Respiration - Respiratory stimulation; Lungs, Thorax, or Respiration - Other changes.
Irritation Eye: Rabbit Standard Draize Test Dose: 250 ug/24H; Reaction: severe. Rabbit Standard Draize Test Dose: 250 ug; Reaction: severe.
Tumorigenic: Mouse Route: Intraperitoneal; Dose: 2940 mg/kg/8W-I; Toxic Effects: Tumorigenic - Neoplastic by RTECS criteria; Lungs, Thorax, or Respiration - Tumors.

Hazard Overviews

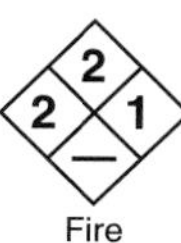

Fire Diamond

Health: Irritating to eyes/respiratory tract. Also Causes: it is among the halogenated acid esters which are capable of producing pulmonary edema.
Fire: Combustible. Use dry chemical, carbon dioxide, or regular foam. If water spray is used, protect against the generation of toxic fumes. Container may explode in heat of fire.
Reactivity: Stable. Hazardous polymerization cannot occur. Avoid: heat; ignition sources. Incompatible with: oxidizers; water; steam; alkalis; sodium cyanide. Hazardous decomposition products: toxic chlorine; hydrogen chloride; phosgene gases.
Carcinogenicity: IARC - Not listed; NIOSH - Not listed; NTP - Not listed; ACGIH - Not listed; OSHA - Not listed; EPA - Not listed; MAK - Not listed
Primary Target Organs:

Eyes

Respiratory System

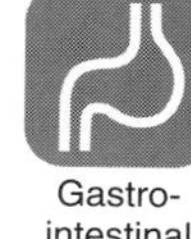
Gastro-intestinal

Environmental

Environmental Fate: If released to the soil, it will not readily evaporate but will leach to groundwater if complete hydrolysis does not occur first. If released to water it is not likely to volatilize extensively or adsorb strongly to sediments. The ester will slowly hydrolyze in water. Entry to the atmosphere is unlikely, but any that does enter should degrade with an estimated half-life of 14.95 days. No information on the biodegradation in the soil or water was available. It is unknown whether it will photolyze at environmental wavelengths.
Cleanup/Disposal: Guide No. 155: Eliminate all ignition sources (no smoking, flares, sparks or flames in immediate area). All equipment used when handling the product must be grounded. Do not touch damaged containers or spilled material unless wearing appropriate protective clothing. Stop leak if you can do it without risk. A vapor suppressing foam may be used to reduce vapors. For chlorosilanes, use AFFF alcohol-resistant medium expansion foam to reduce vapors. Do not get water on spilled substance or inside containers. Use water spray to reduce vapors or divert vapor cloud drift. Prevent entry into waterways, sewers, basements or confined areas. Small Spills: Cover with dry earth, dry sand, or other non-combustible material followed with plastic sheet to minimize spreading or contact with rain. Use clean non-sparking tools to collect material and place it into loosely covered plastic containers for later disposal.

Environmental Physical Data

BCF: none likely

Regulations

RCRA 40CFR: Not listed
CERCLA: 40CFR 302.4: Not listed
SARA 40CFR 372.65: Not listed
SARA EHS 40CFR 355: Not listed
TSCA: Listed

ETH3450 CAS #: 541-41-3

ETHYL CHLOROFORMATE

RTECS: LQ6125000
DOT: UN1182; IMO3.2; IMO6.1
EINECS Number: 208-778-5
Molecular Formula: $C_3H_5ClO_2$
Structured MF: $ClCOOC_2H_5$
Formula Weight: 108.53

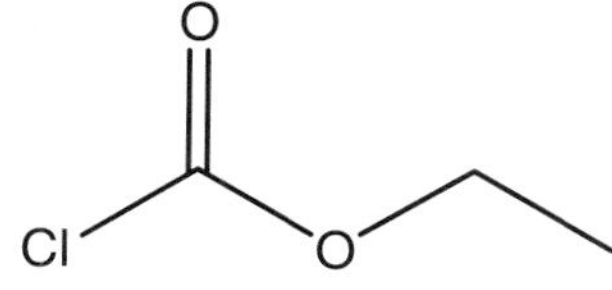

Chemical Structure

Synonyms: CARBONOCHLORIDIC ACID,ETHYL ESTER; CATHYL CHLORIDE; CHLORAMEISENSAEUREAETHYLESTER;

CHLOROCARBONATE D'ETHYLE; CHLOROCARBONIC ACID ETHYL ESTER; CHLOROFORMIC ACID ETHYL ESTER; CLHORAMEISENSAEUREAETHYLESTER; CLOROFORMIATO DE ETILO; ECF; ETHOXYCARBONYL CHLORIDE; ETHYL CARBONOCHLORIDATE; ETHYL CHLOROCARBONATE; ETHYL CHLOROMETHANOATE; ETHYLCHLOORFORMIAAT; ETHYLE,CHLOROFORMIAT D'; ETHYLESTER KYSELINY CHLORMRAVENCI; ETIL CLOROCARBONATO; ETIL CLOROFORMIATO; FORMIC ACID,CHLORO-,ETHYL ESTER; TL 423

Description: water-white liquid; sharp, like hydrochloric acid, odor

Use: organic synthesis; in the manufacture of ore flotation agents by reaction with various alkyl xanthates

Physical Properties

Boiling Point: 95 °C (203 °F) at 760 mm Hg
Freezing Point: -80.6 °C (-113.08 °F)
Specific Gravity: 1.1403 at 20 °C/4 °C
Vapor Density: 3.7 Air=1
Vapor Pressure: 40 mm Hg
Water Solubility: Practically Insoluble in Water
Other Solubilities: miscible with Alcohol, Benzene, Chloroform, Ether.
Surface Tension: 27.5 dynes/cm at 15 °C
pH: Acid
Refraction Index: 1.3947 at 20 °C/D
Flash Point: 16 °C Closed Cup
Autoignition Temperature: 500 °C

RTECS Toxicity Data

Acute Oral: Rat LD_{50} Dose: 270 mg/kg.
Acute Inhalation: Rat LC_{50} Dose: 145 ppm/1hr. Mouse LC_{Lo} Dose: 2260 mg/m^3/10M.
Acute Dermal: Rabbit LD_{50} Route: Skin; Dose: 7120 mg/kg.

Hazard Overviews

Poison

Corrosive

Flammable

Health: Corrosive to eyes/skin/respiratory tract. Poison. Other Acute Effects: may be fatal if inhaled, swallowed, or absorbed through skin; lachrymator; inhalation may result in spasm, inflammation and edema of the larynx and bronchi, chemical pneumonitis and pulmonary edema; symptoms of exposure may include burning sensation; coughing; wheezing; laryngitis; shortness of breath; headache; nausea; vomiting; exposure can cause pulmonary edema; effects may be delayed. Chronic Effects: target organ: lungs.

Fire: Flammable. Hazards: emits toxic fumes; vapor may travel considerable distance to source of ignition and flash back; container explosion may occur. Extinguishing agents: carbon dioxide, dry chemical powder or appropriate foam; do not use water. Precautions: combustible liquid.

Reactivity: Stable. Hazardous polymerization will not occur. Incompatible with: acids, bases, alcohols, amines, iron and iron salts, metals, rust activated carbon, do not allow contact with water. Hazardous decomposition products: toxic fumes of: carbon monoxide, carbon dioxide, hydrogen chloride gas, phosgene gas, ethanol.

Carcinogenicity: IARC - Not listed; NIOSH - Not listed; NTP - Not listed; ACGIH - Not listed; OSHA - Not listed; EPA - Not listed; MAK - Not listed

Primary Target Organs:

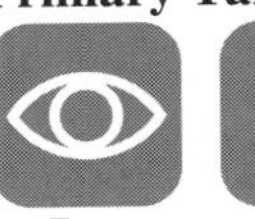
Eyes

Skin

Respiratory System

Environmental

Cleanup/Disposal: Guide No. 155: Eliminate all ignition sources (no smoking, flares, sparks or flames in immediate area). All equipment used when handling the product must be grounded. Do not touch damaged containers or spilled material unless wearing appropriate protective clothing. Stop leak if you can do it without risk. A vapor suppressing foam may be used to reduce vapors. For chlorosilanes, use AFFF alcohol-resistant medium expansion foam to reduce vapors. Do not get water on spilled substance or inside containers. Use water spray to reduce vapors or divert vapor cloud drift. Prevent entry into waterways, sewers, basements or confined areas. Small Spills: Cover with dry earth, dry sand, or other non-combustible material followed with plastic sheet to minimize spreading or contact with rain. Use clean non-sparking tools to collect material and place it into loosely covered plastic containers for later disposal.

Environmental Physical Data

BCF: no food chain concentration potential

Regulations

RCRA 40CFR: Not listed
CERCLA: 40CFR 302.4: Not listed
SARA 40CFR 372.65: Listed
SARA EHS 40CFR 355: Not listed
TSCA: Listed

Analytical Methods

Air: ASTM D4490

ETH3500 CAS #: 2941-64-2

S-ETHYL CHLOROTHIOLFORMATE

EINECS Number: 220-928-1
Molecular Formula: $C_3H_5C_1OS$
Structured MF: $CH_3CH_2SC(O)Cl$
Formula Weight: 124.59

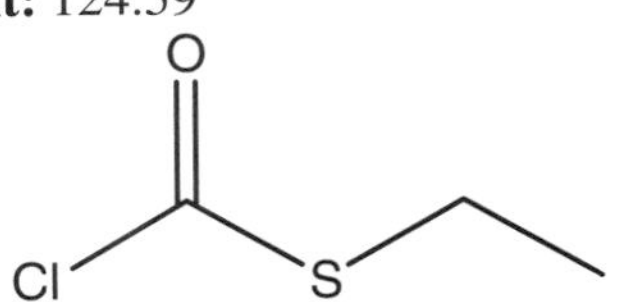

Chemical Structure

Synonyms: CARBONOCHLORIDOTHIOIC ACID,S-ETHYL ESTER; CHLOROTHIOFORMIC ACID ETHYL ESTER; S-ETHYL CARBONOCHLORIDOTHIOATE; S-ETHYL CHLOROTHIOCARBONATE; ETHYL CHLOROTHIOFORMATE; S-ETHYL CHLOROTHIOFORMATE; ETHYL CHLOROTHIOLFORMATE; ETHYL CHLOROTHIOLOFORMATE; ETHYL THIOCHLOROFORMATE; S-ETHYL THIOCHLOROFORMATE; FORMIC ACID,CHLOROTHIO-,S-ETHYL ESTER
Description: dark purple liquid
Use: synthesis of S-esters

Physical Properties

Boiling Point: 132 °C (270 °F)
Specific Gravity: 1.19 at 20/4 °C
Saturated Vapor Density: 1.77770363 kg/m^3
Density: 1.195 g/mL
Vapor Pressure: 111 mm Hg at 25 °C
Water Solubility: < 1 mg/mL at 19 C
Other Solubilities: 95% Ethanol: >=100 mg/ml at 19 °C; Acetone: >=100 mg/ml at 19 °C; DMSO: >=100 mg/ml at 19 °C.
Refraction Index: 1.482
Flash Point: 30 °C

Hazard Overviews

Corrosive

Flammable

Health: Corrosive to eyes/skin/respiratory tract. Harmful. Other Acute Effects: harmful if swallowed, inhaled, or absorbed through skin; lachrymator; inhalation may result in spasm, inflammation and edema of the larynx and bronchi, chemical pneumonitis and pulmonary edema; symptoms of exposure may include burning sensation; coughing; wheezing; laryngitis; shortness of breath; headache; nausea; vomiting.
Fire: Flammable. Hazards: vapor may travel considerable distance to source of ignition and flash back; emits toxic fumes. Extinguishing agents: carbon dioxide, dry chemical powder or appropriate foam. Precautions: combustible liquid.
Reactivity: Incompatible with: strong oxidizing agents, strong bases, may decompose on exposure to moist air or water. Hazardous decomposition products: toxic fumes of: carbon monoxide, carbon dioxide, sulfur oxides, hydrogen chloride gas.
Carcinogenicity: IARC - Not listed; NIOSH - Not listed; NTP - Not listed; ACGIH - Not listed; OSHA - Not listed; EPA - Not listed; MAK - Not listed
Primary Target Organs:

Eyes

Skin

Respiratory System

Environmental

Cleanup/Disposal: Guide No. 155: Eliminate all ignition sources (no smoking, flares, sparks or flames in immediate area). All equipment used when handling the product must be grounded. Do not touch damaged containers or spilled material unless wearing appropriate protective clothing. Stop leak if you can do it without risk. A vapor suppressing foam may be used to reduce vapors. For chlorosilanes, use AFFF alcohol-resistant medium expansion foam to reduce vapors. Do not get water on spilled substance or inside containers. Use water spray to reduce vapors or divert vapor cloud drift. Prevent entry into waterways, sewers, basements or confined areas. Small Spills: Cover with dry earth, dry sand, or other non-combustible material followed with plastic sheet to minimize spreading or contact with rain. Use clean non-sparking tools to collect material and place it into loosely covered plastic containers for later disposal.

Regulations

RCRA 40CFR: Not listed
CERCLA: 40CFR 302.4: Not listed
SARA 40CFR 372.65: Not listed
SARA EHS 40CFR 355: Not listed
TSCA: Listed

ETH3550 CAS #: 105-56-6

ETHYL CYANOACETATE

RTECS: AG4110000
EINECS Number: 203-309-0
Molecular Formula: $C_5H_7NO_2$
Formula Weight: 113.12

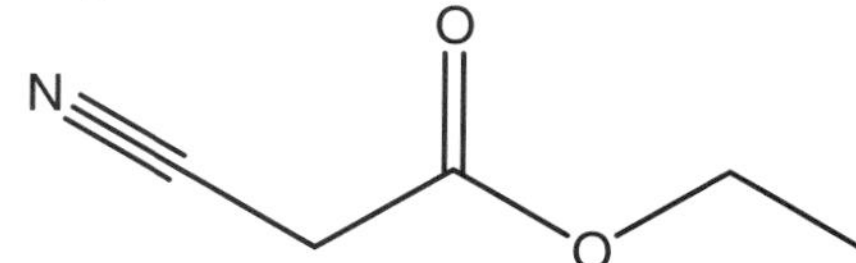

Chemical Structure

Synonyms: ACETIC ACID,CYANO-,ETHYL ESTER; CYANACETATE ETHYLE; CYANOACETIC ACID ETHYL ESTER; CYANOACETIC ESTER; ESTERE CIANOACETICO; ETHYL CYANACETATE; ETHYL CYANOETHANOATE; ETHYLESTER KYSELINY KYANOCTOVE; MALONIC ACID ETHYL ESTER NITRILE
Description: colorless to pale straw-colored liquid; slight, pleasant odor
Use: organic synthesis; pharmaceuticals; dyes

Physical Properties

Boiling Point: 206 °C (403 °F) at 760 mm Hg
Freezing Point: -22 °C (-7.6 °F)
Specific Gravity: 1.056 at 25 °C/4 °C
Vapor Pressure: 0.177 kPa at 75 °C
Water Solubility: Insoluble in Water
Other Solubilities: Soluble in Ammonia, aqueous solutions of alkalies. miscible with Alcohol, Ether
Refraction Index: 1.41793 at 20.5 °C/D
Flash Point: 110 °C

RTECS Toxicity Data

Acute Oral: Rat LD_{Lo} Dose: 400 mg/kg.

Acute Dermal: Rabbit LD_{Lo} Route: Subcutaneous Dose: 1410 mg/kg; Toxic Effects: Peripheral Nerve and sensation - Spastic parapysis with or without sensory change; Lungs, Thorax, or Respiration - Respiratory stimulation. Guinea Pig LD_{50} Route: Subcutaneous Dose: 1115 mg/kg; Toxic Effects: Behavioral - Tremor; Behavioral - Muscle contraction or spasticity; Gastrointestinal - Nausea or vomiting.

Hazard Overviews

Fire: Will burn.

Carcinogenicity: IARC - Not listed; NIOSH - Not listed; NTP - Not listed; ACGIH - Not listed; OSHA - Not listed; EPA - Not listed; MAK - Not listed

Environmental

Regulations

RCRA 40CFR: Not listed

CERCLA: 40CFR 302.4: Listed as Compound per CWA Section 307(a) per CAA Section 112

SARA 40CFR 372.65: Not listed

SARA EHS 40CFR 355: Not listed

TSCA: Listed

ETH3600 CAS #: 5459-93-8

N-ETHYL CYCLOHEXYLAMINE

RTECS: GX1225000

EINECS Number: 226-733-8

Molecular Formula: $C_8H_{17}N$

Structured MF: $C_2H_5NHC_6H_{11}$

Formula Weight: 127.23

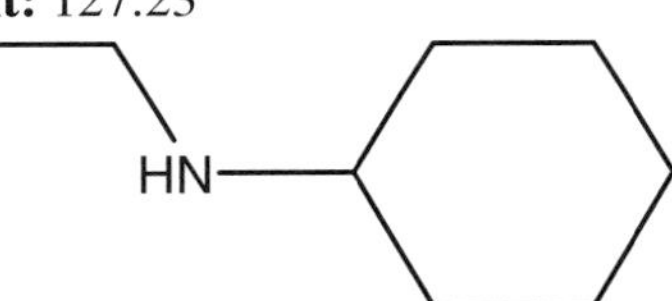

Chemical Structure

Synonyms: ACCELERATOR HX; CYCLOHEXANAMINE,N-ETHYL-; CYCLOHEXANE,(ETHYLAMINO)-; CYCLOHEXYLAMINE,N-ETHYL-; N-CYCLOHEXYLETHYLAMINE; N-ETHYLCYCLOHEXANAMINE; N-ETHYL(CYCLOHEXYL)AMINE; N-ETHYLCYCLOHEXYLAMINE; VULKACIT HX

Physical Properties

Boiling Point: 164 °C (327 °F) at 760 mm Hg

Freezing Point: -45 °C (-49 °F)

Specific Gravity: 0.868 at 0 °C/0 °C

Vapor Density: 4.4 Air=1

Water Solubility: Slightly Soluble in Water

Other Solubilities: Soluble in all proportions in Alcohol, Ether

Surface Tension: 29.52 dynes/cm at 20 °C

Refraction Index: 1.4198 at 20 °C

Critical Temperature: 358 °C

Critical Pressure: 446.91 psia

Flash Point: 30 °C Open Cup

Autoignition Temperature: 285 °C

RTECS Toxicity Data

Acute Oral: Rat LD_{50} Dose: 590 mg/kg.

Acute Inhalation: Rat LC_{Lo} Dose: 500 ppm/4hr.

Acute Dermal: Rabbit LD_{50} Route: Skin; Dose: 750 mg/kg; Toxic Effects: Skin and appendages - Primary irritation.

Irritation Eye: Rabbit Standard Draize Test Dose: 250 ug/24H; Reaction: severe.

Irritation Skin: Rabbit Standard Draize Test Dose: 5 mg/24H; Reaction: severe.

Hazard Overviews

Corrosive

Flammable

Fire Diamond

Health: Corrosive to eyes/skin/respiratory tract. Other Acute Effects: harmful if swallowed, inhaled, or absorbed through skin; inhalation may result in spasm, inflammation and edema of the larynx and bronchi, chemical pneumonitis and pulmonary edema; symptoms of exposure may include burning sensation; coughing; wheezing; laryngitis; shortness of breath; headache; nausea; vomiting.

Fire: Flammable. Hazards: emits toxic fumes; vapor may travel considerable distance to source of ignition and flash back. Extinguishing agents: carbon dioxide, dry chemical powder or appropriate foam. Precautions: combustible liquid.

Reactivity: Incompatible with: strong oxidizing agents, strong acids, absorbs carbon dioxide from air. Hazardous decomposition products: toxic fumes of: carbon monoxide, carbon dioxide, nitrogen oxides.

Carcinogenicity: IARC - Not listed; NIOSH - Not listed; NTP - Not listed; ACGIH - Not listed; OSHA - Not listed; EPA - Not listed; MAK - Not listed

Primary Target Organs:

Eyes

Skin

Respiratory System

Environmental

Cleanup/Disposal: Guide No. 132: Fully encapsulating, vapor protective clothing should be worn for spills and leaks with no fire. Eliminate all ignition sources (no smoking, flares, sparks or flames in immediate area). All equipment used when handling the product must be grounded. Do not touch or walk through spilled material. Stop leak if you can do it without risk. Prevent entry into waterways, sewers, basements or confined areas. A vapor suppressing foam may be used to reduce vapors. Absorb with earth, sand or other non-combustible material and transfer to containers (except for

Hydrazine). Use clean non-sparking tools to collect absorbed material. Large Spills: Dike far ahead of liquid spill for later disposal. Water spray may reduce vapor; but may not prevent ignition in closed spaces.

Regulations

RCRA 40CFR: Not listed
CERCLA: 40CFR 302.4: Not listed
SARA 40CFR 372.65: Not listed
SARA EHS 40CFR 355: Not listed
TSCA: Listed

ETH3650 CAS #: 10287-53-3

ETHYL 4-DIMETHYLAMINO BENZOATE

EINECS Number: 233-634-3
Molecular Formula: $C_{11}H_{15}NO_2$
Formula Weight: 193.25

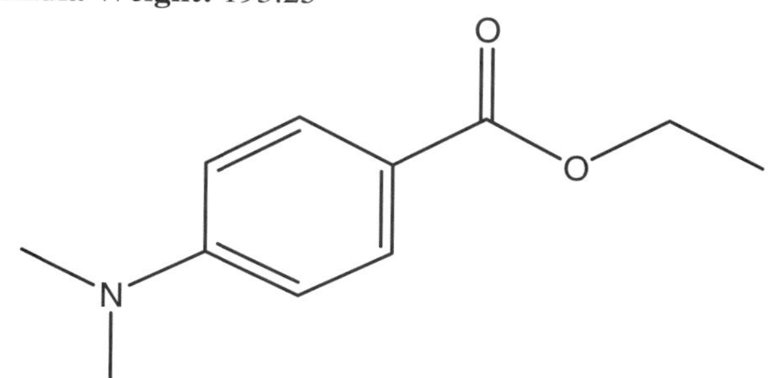

Chemical Structure

Physical Properties

Boiling Point: > 250 °C (482 °F)
Freezing Point: 67 °C (152.6 °F)
Specific Gravity: 1.00990

Hazard Overviews

Health: May cause irritation. Other Acute Effects: may be harmful by inhalation, ingestion, or skin absorption.
Fire: Hazards: emits toxic fumes. Extinguishing agents: carbon dioxide, dry chemical powder or appropriate foam. Precautions: combustible liquid.
Reactivity: Incompatible with: acids, bases, oxidizing agents, reducing agents. Hazardous decomposition products: toxic fumes of: carbon monoxide, carbon dioxide, nitrogen oxides.
Carcinogenicity: IARC - Not listed; NIOSH - Not listed; NTP - Not listed; ACGIH - Not listed; OSHA - Not listed; EPA - Not listed; MAK - Not listed

Environmental

Regulations

RCRA 40CFR: Not listed
CERCLA: 40CFR 302.4: Not listed
SARA 40CFR 372.65: Not listed
SARA EHS 40CFR 355: Not listed
TSCA: Not listed

ETH3700 CAS #: 759-94-4

ETHYL DIPROPYLTHIOCARBAMATE

RTECS: FA4550000
EINECS Number: 212-073-8
Molecular Formula: $C_9H_{19}NOS$
Formula Weight: 189.35
Synonyms: S-AETHYL-N,N-DIPROPYLTHIOLCARBAMAT; ALIROX; CARBAMIC ACID,DIPROPYLTHIO-,S-ETHYL ESTER; CARBAMOTHIOIC ACID,DIPROPYL-,S-ETHYL ESTER; CARBAMOTHIOIC ACID,DIPROPYL-,S-ETHYL ESTER (9CI); DIPROPYLCARBAMOTHIOIC ACID S-ETHYL ESTER; DIPROPYLTHIOCARBAMIC ACID S-ETHYL ESTER; N,N-DIPROPYLTHIOCARBAMIC ACID S-ETHYL ESTER; N,N-DIPROPYLTHIOCARBAMIC ACID S ETHYL ESTER; DIPROPYLTHIOCARBAMIC ACID,S-ETHYL ESTER; EPTAM; EPTAM 6E; EPTC; ERADICANE; S-ETHYL DIPROPYLCARBAMOTHIOATE; ETHYL N,N-DI-N-PROPYLTHIOCARBAMATE; S-ETHYL DIPROPYLTHIOCARBAMATE; S-ETHYL N,N-DI-N-PROPYLTHIOCARBAMATE; ETHYL DI-N-PROPYLTHIOLCARBAMATE; ETHYL N,N-DI-N-PROPYLTHIOLCARBAMATE; S-ETHYL-N,N-DIPROPYLTHIOCARBAMATE; S-ETHYLDIPROPYLTHIOCARBAMATE; ETHYL-N,N-DIPROPYLTHIOLCARBAMATE; FDA 1541; GENEP EPTC; NIPTAN; R 1608; R-1608; STAUFFER R 1608; TORBIN; WITOX
Description: pale-dark yellow clear liquid; aromatic odor
Use: selective herbicide

Physical Properties

Boiling Point: 232 °C (450 °F) at 760 mm Hg
Specific Gravity: 0.955 at 30 °C
Saturated Vapor Density: 1.200296837 kg/m^3
Density: 0.95 g/mL at 30 °C
Vapor Pressure: 0.034 mm Hg at 25 °C
Water Solubility: < 0.1 mg/mL at 22.5 C
Other Solubilities: 95% Ethanol: Soluble (>=10 mg/ml at 30 °C); Benzene: miscible; DMSO: Soluble (>=10 mg/ml at 30 °C); Isopropanol: miscible; Methanol: miscible; Toluene: miscible; Xylene: miscible.
Refraction Index: 1.4750 at 30 °C/D
Flash Point: 116 °C Open Cup

RTECS Toxicity Data

Acute Oral: Rat LD_{50} Dose: 916 mg/kg. Mouse LD_{50} Dose: 750 mg/kg.
Acute Inhalation: Rat LC_{Lo} Dose: 200 mg/m^3/4hr; Toxic Effects: Sense organs and special senses - Lacrimation; Behavioral - Somnolence (general depressed activity); Gastrointestinal - Changes in structure or function of salivary glands. Human TC_{Lo} Dose: 135 mg/m^3/90M; Toxic Effects: Behavioral - Headache; Behavioral - Withdrawal; Gastrointestinal - Nausea or vomiting.
Acute Dermal: Rabbit LD_{50} Route: Skin; Dose: 1460 mg/kg. Rat LD_{50} Route: Skin; Dose: 3200 mg/kg.
Mutagenic: Bacteria - E Coli Mutations in Microorganisms; Dose: 5 mg/L (-S9).

Hazard Overviews

Fire: Will burn.
Carcinogenicity: IARC - Not listed; NIOSH - Not listed; NTP - Not listed; ACGIH - Not listed; OSHA - Not listed; EPA - Not listed; MAK - Not listed

Environmental

Ecotoxicity: LC_{50} Salvelinus namaycush (lake trout) 16.2 mg/l/96 hr at 10 °C (95% confidence limit 14.8-17.7 mg/l), wt 0.9 g, static bioassay without aeration, pH 7.2-7.5, water hardness 40-50 mg/l as calcium carbonate LC_{50} Gammarus fasciatus 23 mg/l/96 hr /Conditions of bioassay not specified LC_{50} ASELLUS 23 mg/l/96 hr at 15 °C (95% confidence limit 15-36 mg/l), mature, static bioassay without aeration, pH 7.2-7.5, water hardness 40-50 mg/l as calcium carbonate and alkalinity of 30-35 mg/l
Environmental Fate: If released to soil, microbial degradation is reported to be a major removal process. Furthermore, volatilization from soil surfaces will be an important fate process (estimated half-life of 3.4 hours from a soil surface) if applied when the surface is wet at the time of application and the herbicide is not immediately incorporated into the soil. It has moderate-high mobility in soil (K_{oc}'s of 109-283); its detection in groundwater demonstrates that it can leach. The half-life in soil typically ranges from 6 to 35 days; the US Dept of Agriculture's Pesticide Properties Database lists a soil half-life of 6 days. If released to water, it may degrade through microbial degradation. Volatilization will be important (estimated volatilization half-lives of 2.4 and 82 days from a model river and environmental pond, respectively). Hydrolysis and bioconcentration will not be important. If released to the atmosphere, it will exist almost entirely in the vapor phase; vapor phase readily degrades by reaction with photochemically produced hydroxyl radicals (estimated half-life of 14 hours). Physical removal from the atmosphere occurs through wet deposition.

Environmental Physical Data

Henry's Law Constant: estimated at 2.26×10^{-5}
Sorption Partition Coefficient: $K_{oc} = 200$
BCF: fish 12 to 22

Regulations

RCRA 40CFR: Listed Hazardous Waste No. U390 Toxic Waste
CERCLA: 40CFR 302.4: Listed per RCRA Section 3001 RQ: 1 lb (0.454 kg)
SARA 40CFR 372.65: Listed
SARA EHS 40CFR 355: Not listed
TSCA: Listed

Analytical Methods

Soil: EPA PMD-EPT
Water / Groundwater: EPA 634
Drinking Water: EPA 507, 525.2
Food: FDA 212.1, 232.1; AOAC 974.05
Plasma: EPA 001; FDA 252

ETH3750 **CAS #: 121-39-1**

ETHYL 2,3-EPOXY-3-PHENYLPROPIONATE

RTECS: MB4970000
EINECS Number: 204-467-3
Molecular Formula: $C_{11}H_{12}O_3$
Formula Weight: 192.23

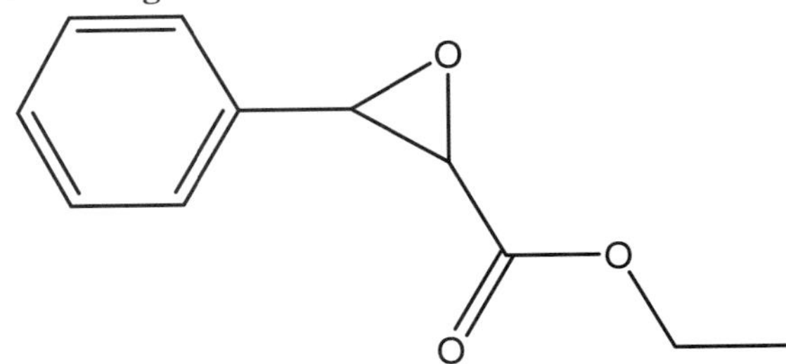

Chemical Structure

Synonyms: ETHYL ALPHA,BETA-EPOXYHYDROCINNAMATE; ETHYL ALPHA,BETA-EPOXY-ALPHA-PHENYLPROPIONATE; ETHYL 3-PHENYLGLYCIDATE; ETHYL PHENYLGLYCIDATE; ETHYL 3-PHENYLOXIRANECARBOXYLATE; GLYCIDIC ACID,3-PHENYL-,ETHYL ESTER; OXIRANECARBOXYLIC ACID,3-PHENYL-,ETHYL ESTER; OXIRANECARBOXYLIC ACID,3-PHENYL-,ETHYL ESTER (9CI); 3-OXO-3-PHENYLPROPANOIC ACID ETHYL ESTER; 3-PHENYLGLYCIDIC ACID,ETHYL ESTER; 3-PHENYL-OXIRANECARBOXYLIC ACID ETHYL ESTER
Description: colorless to pale-yellow liquid; strong fruity odor suggestive of strawberry
Use: as fragrance in soaps, detergents, creams, lotions, perfume; in berry & cherry food flavorings; flavoring agent in gelatins, puddings & baked goods; flavoring agent in candy, ice cream & beverages

Physical Properties

Boiling Point: 265 °C (509 °F) to 270 °C (518 °F)
Freezing Point: <
Specific Gravity: 1.12 to 1.128 at 25/25 °C
Water Solubility: Slightly Soluble 1 to 10 mg/mL
Other Solubilities: 95% Ethanol: Soluble (>=10 mg/ml); Ether: Soluble.
pH: Not more than 1
Refraction Index: 1.519 to 1.5230 at 20 °C
Flash Point: 100 °C

RTECS Toxicity Data

Acute Oral: Rat LD_{50} Dose: 2300 mg/kg.
Mutagenic: Mouse Morphological Transformation; Cell Type: fibroblast; Dose: 103 mg/L. Hamster Mutations in Microorganisms; Cell Type: ovary; Dose: 103 mg/L (+S9).

Hazard Overviews

Health: Irritating to eyes/skin/respiratory tract. Harmful. Other Acute Effects: harmful if swallowed, inhaled, or absorbed through skin.

Fire: Will burn. Hazards: emits toxic fumes. Extinguishing agents: water spray; carbon dioxide, dry chemical powder or appropriate foam. Precautions: combustible liquid.
Reactivity: Incompatible with: strong oxidizing agents, may decompose on exposure to moist air or water. Hazardous decomposition products: toxic fumes of: carbon monoxide, carbon dioxide.
Carcinogenicity: IARC - Not listed; NIOSH - Not listed; NTP - Not listed; ACGIH - Not listed; OSHA - Not listed; EPA - Not listed; MAK - Not listed
Primary Target Organs:

Eyes

Skin

Respiratory System

Environmental

Regulations

RCRA 40CFR: Not listed
CERCLA: 40CFR 302.4: Not listed
SARA 40CFR 372.65: Not listed
SARA EHS 40CFR 355: Not listed
TSCA: Listed

ETH3800 **CAS #: 763-69-9**

ETHYL ETHOXYPROPIONATE

RTECS: UF3325000
EINECS Number: 212-112-9
Molecular Formula: $C_7H_{14}O_3$
Structured MF: $C_2H_5OOCCH_2CH_2OC_2H_5$
Formula Weight: 146.21

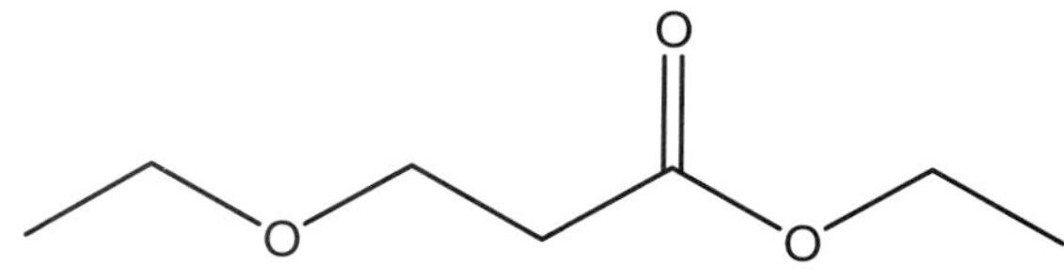

Chemical Structure

Synonyms: ETHOXYPROPIONIC ACID,ETHYL ESTER; ETHYL BETA-ETHOXYPROPIONATE; ETHYLESTER KYSELINY 3-ETHOXYPROPIONOVE
Description: water-white, liquid; esteric odor

Physical Properties

Boiling Point: 170 °C (338 °F) at 1 atm
Freezing Point: -100 °C (-148 °F)
Specific Gravity: 0.95
Vapor Density: 5 Air=1
Water Solubility: Floats on water
Refraction Index: 1.4065
Flash Point: 59 °C Closed Cup
LEL: 05% v/v

RTECS Toxicity Data

Acute Oral: Rat LD_{50} Dose: 5 gm/kg.
Acute Dermal: Rabbit LD_{50} Route: Skin; Dose: 10 mL/kg.
Irritation Eye: Rabbit Standard Draize Test Dose: 500 mg/24H; Reaction: mild.
Irritation Skin: Rabbit Standard Draize Test Dose: 500 mg/24H; Reaction: mild. Rabbit Open Draize Test Dose: 10 mg/24H open; Reaction: mild.

Hazard Overviews

Flammable

Health: Irritating. Other Acute Effects: may be harmful by inhalation, ingestion, or skin absorption.
Fire: Flammable. Hazards: emits toxic fumes. Extinguishing agents: carbon dioxide, dry chemical powder or appropriate foam; water spray. Precautions: combustible liquid.
Reactivity: Incompatible with: strong oxidizing agents, strong bases. Hazardous decomposition products: toxic fumes of: carbon monoxide, carbon dioxide.
Carcinogenicity: IARC - Not listed; NIOSH - Not listed; NTP - Not listed; ACGIH - Not listed; OSHA - Not listed; EPA - Not listed; MAK - Not listed
Primary Target Organs:

Eyes

Skin

Respiratory System

Environmental

Cleanup/Disposal: Guide No. 154: Eliminate all ignition sources (no smoking, flares, sparks or flames in immediate area). Do not touch damaged containers or spilled material unless wearing appropriate protective clothing. Stop leak if you can do it without risk. Prevent entry into waterways, sewers, basements or confined areas. Absorb or cover with dry earth, sand or other non-combustible material and transfer to containers. Do not get water inside containers.

Regulations

RCRA 40CFR: Not listed
CERCLA: 40CFR 302.4: Not listed
SARA 40CFR 372.65: Not listed
SARA EHS 40CFR 355: Not listed
TSCA: Listed

ETH3850 **CAS #: 1497-68-3**

O-ETHYL ETHYLTHIOPHOSPHONYL CHLORIDE

EINECS Number: 216-095-9
Molecular Formula: $C_4H_{10}ClOPS$

Formula Weight: 172.60
Synonyms: ETHANEPHOSPHONOCHLORIDOTHIOIC ACID,O-ETHYL ESTER; O-ETHYL ETHANEPHOSPHONOTHIONOCHLORIDATE; O-ETHYL ETHYLPHOSPHONOCHLORIDOTHIOATE; ETHYLTHIONOPHOSPHONIC ACID O-ETHYL ESTER CHLORIDE; PHOSPHONOCHLORIDOTHIOIC ACID,ETHYL-,O-ETHYL ESTER
Use: reagent to prepare organophosphate insecticides & cyanodithioimidocarbonate herbicides

Hazard Overviews

Carcinogenicity: IARC - Not listed; NIOSH - Not listed; NTP - Not listed; ACGIH - Not listed; OSHA - Not listed; EPA - Not listed; MAK - Not listed

Environmental

Regulations

RCRA 40CFR: Not listed
CERCLA: 40CFR 302.4: Not listed
SARA 40CFR 372.65: Not listed
SARA EHS 40CFR 355: Not listed
TSCA: Listed

ETH3900 CAS #: 459-72-3

ETHYL FLUOROACETATE

RTECS: AH7175000
EINECS Number: 207-297-8
Molecular Formula: $C_4H_7FO_2$
Formula Weight: 106.11

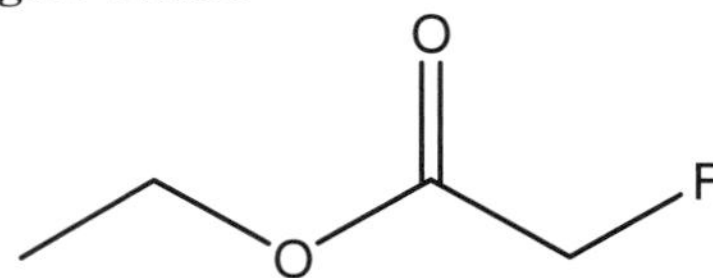

Chemical Structure

Synonyms: ACETIC ACID,FLUORO-,ETHYL ESTER; ETHYLESTER KYSELINY FLUOROCTOVE
Description: liquid; odor of ethyl acetate

Physical Properties

Boiling Point: 121.6 °C (251 °F) at 758 mm Hg
Specific Gravity: 1.0926 at 20.5 °C
Water Solubility: Soluble in Water
Refraction Index: 1.3767 at 20.5 °C/D

Hazard Overviews

Poison

Corrosive

Health: Corrosive to eyes/skin/respiratory tract. Poison. Other Acute Effects: may be fatal if inhaled, swallowed, or absorbed through skin; material is extremely destructive to tissue of the mucous membranes and upper respiratory tract, eyes and skin; inhalation may result in spasm, inflammation and edema of the larynx and bronchi, chemical pneumonitis and pulmonary edema; symptoms of exposure may include burning sensation, coughing, wheezing, laryngitis, shortness of breath, headache, nausea and vomiting; target organs: central nervous system, heart.
Fire: Hazards: emits toxic fumes; vapor may travel considerable distance to source of ignition and flash back. Extinguishing agents: carbon dioxide, dry chemical powder or appropriate foam. Precautions: combustible liquid.
Reactivity: Incompatible with: acids, bases, oxidizing agents, reducing agents. Hazardous decomposition products: toxic fumes of: carbon monoxide, carbon dioxide, hydrogen fluoride.
Carcinogenicity: IARC - Not listed; NIOSH - Not listed; NTP - Not listed; ACGIH - Not listed; OSHA - Not listed; EPA - Not listed; MAK - Not listed
Primary Target Organs:

Eyes

Skin

Respiratory System

Nervous System

Cardio-vascular

Environmental

Regulations

RCRA 40CFR: Not listed
CERCLA: 40CFR 302.4: Not listed
SARA 40CFR 372.65: Not listed
SARA EHS 40CFR 355: Not listed
TSCA: Listed

ETH3950 CAS #: 109-94-4

ETHYL FORMATE

RTECS: LQ8400000
DOT: UN1190; IMO3.1
EINECS Number: 203-721-0
Molecular Formula: $C_3H_6O_2$
Structured MF: CH_3CH_2OCHO
Formula Weight: 74.09

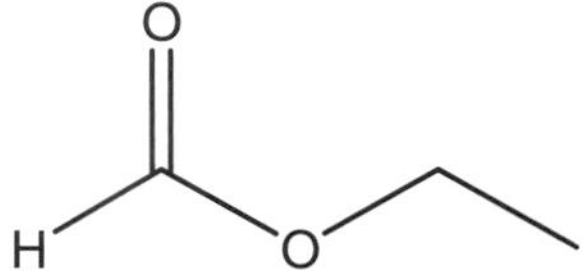

Chemical Structure

Synonyms: AETHYLFORMIAT; AREGINAL; ETHYL ESTER OF FORMIC ACID; ETHYL FORMIC ESTER; ETHYL METHANOATE; ETHYLE (FORMIATE D'); ETHYLESTER KYSELINY MRAVENCI; ETHYLFORMIAAT; ETILE (FORMIATO DI); FORMIC ACID,ETHYL ESTER; FORMIC ETHER; MROWCZAN ETYLU
Description: colorless liquid; fruity odor

Use: solvent for oils & greases, nitrocellulose, cellulose acetate (in lacquer ind & artificial silk mfr), celluloid (in mfr of safety glass; in shoe ind); flavor for lemonade & essences; mfr of artificial rum & arrac; fungicide & larvicide; in org synthesis; acetone substitute; in synthetic resins; in prodn of estazolam, anticonvulsant drug; fumigant for dried fruits; in synthesis of synthetic sex hormones.

Physical Properties

Boiling Point: 54.5 °C (130 °F) at 760 mm Hg
Freezing Point: -80.5 °C (-112.9 °F)
Specific Gravity: 0.9168 at 20 °C
Vapor Density: 2.6 Air=1
Vapor Pressure: 200 mm Hg
Water Solubility: 9 parts/100 at 18 °C
Other Solubilities: Soluble in all proportions in Alcohol, Ether; Very Soluble in Acetone; Soluble in most organic solvents; 88,250 mg/L at 25 °C.
Surface Tension: Estimated at 28 dynes/cm
Odor Threshold: 19 ppm
Refraction Index: 1.3598 at 20 °C/D
Critical Temperature: 235 °C
Critical Pressure: 686 psia
Ionization Potential (eV): 10.61
Flash Point: -20 °C Closed Cup
Autoignition Temperature: 455 °C
LEL: 2.8% v/v
UEL: 16% v/v

RTECS Toxicity Data

Acute Oral: Rat LD_{50} Dose: 1850 mg/kg; Toxic Effects: Behavioral - Somnolence (general depressed activity); Lungs, Thorax, or Respiration - Dyspnea. Rabbit LD_{50} Dose: 2075 mg/kg.
Acute Inhalation: Rat LC_{Lo} Dose: 8000 ppm/4hr.
Acute Dermal: Rabbit LD_{50} Route: Skin; Dose: >20 mL/kg. Rabbit LD_{Lo} Route: Subcutaneous Dose: 1 gm/kg.
Irritation Eye: Rabbit Standard Draize Test Dose: 100 mg/24H; Reaction: moderate.
Irritation Skin: Rabbit Open Draize Test Dose: 460 mg open; Reaction: mild.
Tumorigenic: Mouse Route: Skin; Dose: 110 gm/kg/9W-I; Toxic Effects: Tumorigenic - Equivocal tumorigenic agent by RTECS criteria; Skin and appendages - Tumors.

Hazard Overviews

Flammable

Fire Diamond

Health: Irritating to eyes/skin/respiratory tract. Also Causes: pulmonary edema, CNS depression (exposure to high concentrations); esophageal burns.
Fire: Flammable. Can form explosive mixtures in the air. Use dry chemical, carbon dioxide, water spray, fog, or alcohol-resistant foam. Container may explode in heat of fire. Vapors may travel to an ignition source and flash back.

Reactivity: Stable. Hazardous polymerization cannot occur. Avoid: heat; ignition sources. Incompatible with: nitrates; strong oxidizers; strong alkalis; strong acids. Hazardous decomposition products: acrid smoke.
Carcinogenicity: IARC - Not listed; NIOSH - Listed as carcinogen; NTP - Not listed; ACGIH - Not listed; OSHA - Not listed; EPA - Not listed; MAK - Not listed
Primary Target Organs:

Eyes

Skin

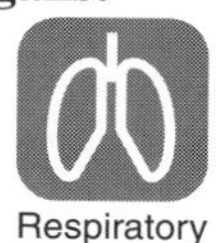
Respiratory System

Exposure Limits
OSHA PEL: TWA: 100 ppm; 300 mg/m^3.
ACGIH TLV: TWA: 100 ppm; 303 mg/m^3.
NIOSH REL: TWA: 100 ppm; 300 mg/m^3.
NIOSH IDLH: 1500 ppm.
DFG MAK: TWA: 100 ppm; 300 mg/m^3.
Respirator Recommendation
Exposure Range: >100 to 1000 ppm Air Purifying, Negative Pressure, Half Mask
Exposure Range: >1000 to <1500 ppm Supplied Air, Constant Flow/Pressure Demand, Half Mask
Exposure Range: 1500 to unlimited ppm Self-contained Breathing Apparatus, Pressure Demand, Full Face
Cartridge Color: black

Environmental

Environmental Fate: If released to soil, it would be expected to volatilize from the soil surface and readily leach. It would be expected to readily biodegrade and chemically hydrolyze, especially in alkaline soil. It has a relatively high Henry's Law constant and therefore, will readily volatilize. A volatilization half-life of 4.5 hours was predicted from a model river. It would not be expected to sorb to sediment or bioconcentrate in aquatic organisms. Hydrolysis will be an important fate process, especially in alkaline waters (half-life is 3.1 days and 7.5 hr at pH 7 and 8, respectively). Based on limited data from screening tests supplemented by predictions of mathematical models and data on similar chemicals, it will rapidly biodegrade. In the atmosphere, it will react with photochemically-produced hydroxyl radicals, with an estimated half-life of 11 days. It is fairly soluble in water and should be readily washed out of the air by rain.
Cleanup/Disposal: Guide No. 129: Eliminate all ignition sources (no smoking, flares, sparks or flames in immediate area). All equipment used when handling the product must be grounded. Do not touch or walk through spilled material. Stop leak if you can do it without risk. Prevent entry into waterways, sewers, basements or confined areas. A vapor suppressing foam may be used to reduce vapors. Absorb or cover with dry earth, sand or other non-combustible material and transfer to containers. Use clean non-sparking tools to collect absorbed material. Large Spills: Dike far ahead of liquid spill for later disposal. Water spray may reduce vapor; but may not prevent ignition in closed spaces.

Environmental Physical Data

Henry's Law Constant: 1.8
Octanol/Water Partition Coefficient: log K_{ow} = 0.23
Sorption Partition Coefficient: K_{oc} = estimated at 8
BCF: estimated at 0.88
BOD: 30%, 10 days

Regulations

RCRA 40CFR: Not listed
CERCLA: 40CFR 302.4: Not listed
SARA 40CFR 372.65: Not listed
SARA EHS 40CFR 355: Not listed
TSCA: Listed

Analytical Methods

Air: ASTM D3686, D3687
Indoor / Expired Air: NIOSH 1452

ETH4000 CAS #: 4016-11-9

ETHYL GLYCIDYL ETHER

RTECS: TZ3200000
EINECS Number: 223-671-3
Molecular Formula: $C_5H_{10}O_2$
Formula Weight: 102.15

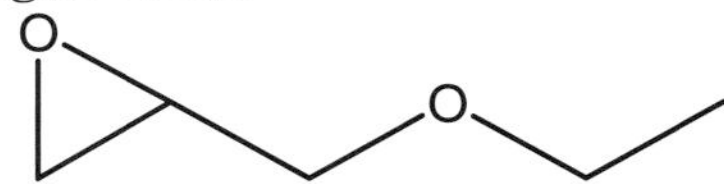

Chemical Structure

Synonyms: 1,2-EPOXY-3-ETHOXY-PROPANE; 1,2-EPOXY-3-ETHOXYPROPANE; 3-ETHOXY-1,2-EPOXYPROPANE; (ETHOXYMETHYL)OXIRANE; OXIRANE,(ETHOXYMETHYL)-; OXIRANE,(ETHOXYMETHYL)-(9CI); PROPANE,1,2-EPOXY-3-ETHOXY-
Use: no evidence of current commercial use in US

Physical Properties

Boiling Point: 128 °C (262.4 °F)
Specific Gravity: 0.97
Water Solubility: Soluble
Refraction Index: 1.4320

RTECS Toxicity Data

Mutagenic: Bacteria - K Kpneumoniae Mutations in Microorganisms; Dose: 500 umol/L (-S9). Bacteria - S Typhimurium Mutations in Microorganisms; Dose: 8 mmol/L (+S9).

Hazard Overviews

Carcinogenicity: IARC - Not listed; NIOSH - Not listed; NTP - Not listed; ACGIH - Not listed; OSHA - Not listed; EPA - Not listed; MAK - Not listed

Environmental

Regulations

RCRA 40CFR: Not listed
CERCLA: 40CFR 302.4: Not listed
SARA 40CFR 372.65: Not listed
SARA EHS 40CFR 355: Not listed
TSCA: Listed

ETH4050 CAS #: 52738-99-5

2-ETHYL HEXALDEHYDE

EINECS Number: 258-146-8
Molecular Formula: $C_6H_{11}O$
Structured MF: C_4H_9CHCHO
Formula Weight: 99.15
Description: colorless liquid

Physical Properties

Boiling Point: 163.4 °C (326 °F)
Freezing Point: -100 °C (-148 °F)
Specific Gravity: 0.823
Water Solubility: Will float in slick on Water surface
Flash Point: 51.67 °C

Hazard Overviews

Flammable

Fire: Flammable.
Carcinogenicity: IARC - Not listed; NIOSH - Not listed; NTP - Not listed; ACGIH - Not listed; OSHA - Not listed; EPA - Not listed; MAK - Not listed

Environmental

Regulations

RCRA 40CFR: Not listed
CERCLA: 40CFR 302.4: Not listed
SARA 40CFR 372.65: Not listed
SARA EHS 40CFR 355: Not listed
TSCA: Not listed

ETH4200 CAS #: 120-47-8

ETHYL 4-HYDROXYBENZOATE

RTECS: DH2190000
EINECS Number: 204-399-4
Molecular Formula: $C_9H_{10}O_3$
Formula Weight: 166.17

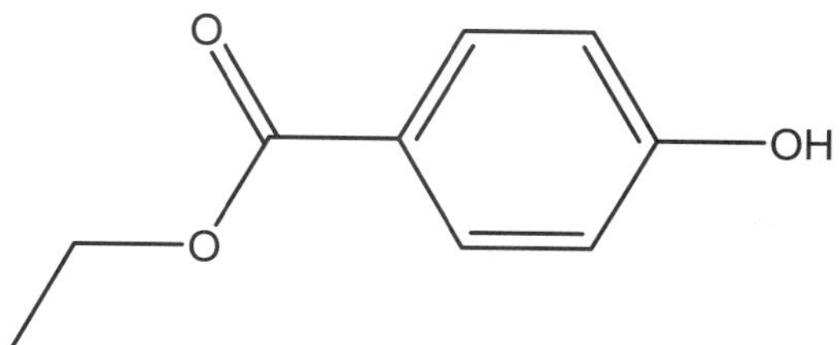

Chemical Structure

Synonyms: ASEPTOFORM E; BONOMOLD OE; CARBETHOXYPHENOL; P-CARBETHOXYPHENOL; EASEPTOL; ESTER ETYLOWYKWASU P-HYDROKSYBENZOESOWEGO; ETHYL BUTEX; ETHYL P-HYDROXYBENZOATE; ETHYL P-OXYBENZOATE; ETHYL PARABEN; ETHYL PARASEPT; ETHYLESTER KYSELINY P-HYDROXYBENZOOVE; P-HYDROXYBENZOATE ETHYL ESTER; 4-HYDROXYBENZOIC ACID ETHYL ESTER; P-HYDROXYBENZOIC ACID ETHYL ESTER; P-HYDROXYBENZOIC ETHYL ESTER; MYCOCTEN; NAPAGIN A; NIPAGIN A; NIPAGINA A; NIPAZIN A; P-OXYBENZOESAEUREAETHYLESTER; SOLBROL A; TEGOSEPT E

Description: small, colorless crystals or white powder; odorless

Use: preservative for pharmaceuticals; preservative for adhesives in food packaging; preservative for antiseptic creams & ointments, other galenicals, cosmetic preparations, adhesives

Physical Properties

Boiling Point: Decomposes at 297 °C (567 °F) to 298 °C (568 °F)

Freezing Point: 116 °C (240.8 °F)

Water Solubility: Slightly Soluble

Other Solubilities: 25 g/100 g of Propylene Glycol at 25 °C; at 25 °C: 0.5 g/100 g Glycerin; 1 g/100 g Peanut Oil; Soluble in Carbon Disulfide; Slightly Soluble in Petroleum Ether, Chloroform.

RTECS Toxicity Data

Acute Oral: Rat LD; Dose: >200 mg/kg. Mouse LD_{50} Dose: 3 gm/kg.

Chronic (Multiple Dose) Oral: Rat Dose: 504 gm/kg/12W-C; Toxic Effects: Nutritional and gross metabolic - Weight loss or decreased weight gain; DEATH.

Reproductive/Teratogenic: Rat Route: Oral; Dose: 45600 mg/kg; Duration: female 8-15D of pregnancy; Specific Developmental Abnormalities - Other developmental abnormalities.

Mutagenic: Hamster Cytogenetic Analysis; Cell Type: fibroblast; Dose: 250 mg/L. Hamster Cytogenetic Analysis; Cell Type: lung; Dose: 440 mg/L.

Hazard Overviews

Health: Irritating to eyes/skin/respiratory tract. Other Acute Effects: may be harmful by inhalation, ingestion, or skin absorption.

Fire: Hazards: emits toxic fumes. Extinguishing agents: water spray; carbon dioxide, dry chemical powder or appropriate foam. Precautions: combustible liquid.

Reactivity: Incompatible with: strong oxidizing agents, strong bases. Hazardous decomposition products: toxic fumes of: carbon monoxide, carbon dioxide.

Carcinogenicity: IARC - Not listed; NIOSH - Not listed; NTP - Not listed; ACGIH - Not listed; OSHA - Not listed; EPA - Not listed; MAK - Not listed

Primary Target Organs:

Eyes

Skin

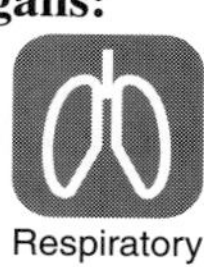

Respiratory System

Environmental

Regulations

RCRA 40CFR: Not listed

CERCLA: 40CFR 302.4: Not listed

SARA 40CFR 372.65: Not listed

SARA EHS 40CFR 355: Not listed

TSCA: Listed

ETH4250 **CAS #: 624-42-0**

ETHYL ISOAMYL KETONE

EINECS Number: 210-844-3

Molecular Formula: $C_8H_{16}O$

Formula Weight: 128.21

Synonyms: 3-HEPTANONE,6-METHYL-; 6-METHYL-3-HEPTANONE

Physical Properties

Boiling Point: 163 °C (325 °F) at 734 mm Hg

Specific Gravity: 0.8304 at 20 °C

Water Solubility: Insoluble in Water

Other Solubilities: > 10% in Benzene; > 10% in Ether; > 10% in Ethanol.

Refraction Index: 1.4209 at 20 °C/D

Hazard Overviews

Carcinogenicity: IARC - Not listed; NIOSH - Not listed; NTP - Not listed; ACGIH - Not listed; OSHA - Not listed; EPA - Not listed; MAK - Not listed

Environmental

Regulations

RCRA 40CFR: Not listed

CERCLA: 40CFR 302.4: Not listed

SARA 40CFR 372.65: Not listed

SARA EHS 40CFR 355: Not listed

TSCA: Not listed

ETH4300 **CAS #: 109-90-0**

ETHYL ISOCYANATE

RTECS: NQ8825000
DOT: UN2481
EINECS Number: 203-717-9
Molecular Formula: C_3H_5NO
Formula Weight: 71.09

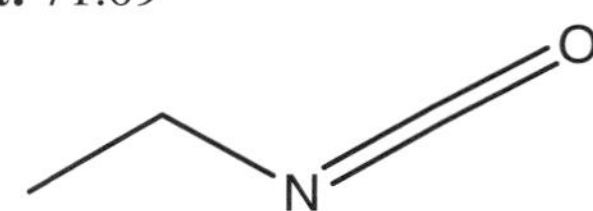

Chemical Structure

Synonyms: ETHANE,ISOCYANATO-; ISOCYANATOETHANE; ISOCYANATOETHENE; ISOCYANIC ACID,ETHYL ESTER
Description: liquid
Use: pharmaceutical and pesticide intermediate

Physical Properties

Boiling Point: 60 °C (140 °F)
Specific Gravity: 0.9031 at 20 °C/4 °C
Water Solubility: Insoluble
Other Solubilities: Soluble in chlorinated and aromatic hydrocarbons.
Refraction Index: 1.3808 at 20 °C/D
Ionization Potential (eV): 10.1
Flash Point: 76.6 °C Cleveland Open Cup

RTECS Toxicity Data

Mutagenic: Bacteria - S Typhimurium Mutations in Microorganisms; Dose: 50 ug/plate (+S9).

Hazard Overviews

Poison

Corrosive

Health: Severely irritating to eyes/skin/respiratory tract. Poison. Other Acute Effects: may be fatal if inhaled, swallowed, or absorbed through skin; severe lachrymator; possible sensitizer; inhalation may result in spasm, inflammation and edema of the larynx and bronchi, chemical pneumonitis and pulmonary edema; repeated exposure may cause asthma; prolonged or repeated exposure may cause allergic reactions in certain sensitive individuals; prolonged contact can cause nausea; dizziness; headache; lung irritation; chest pain and edema which may be fatal.
Fire: Combustible. Hazards: vapor may travel considerable distance to source of ignition and flash back; emits toxic fumes. Extinguishing agents: carbon dioxide; dry chemical powder. Precautions: combustible liquid.
Reactivity: Incompatible with: water, alcohols, strong bases, amines, acids, strong oxidizing agents, heat. Hazardous decomposition products: thermal decomposition may produce carbon monoxide, carbon dioxide, and nitrogen oxides; hydrogen cyanide.
Carcinogenicity: IARC - Not listed; NIOSH - Not listed; NTP - Not listed; ACGIH - Not listed; OSHA - Not listed; EPA - Not listed; MAK - Not listed
Primary Target Organs:

Eyes

Skin

Respiratory System

Environmental

Cleanup/Disposal: Guide No. 155: Eliminate all ignition sources (no smoking, flares, sparks or flames in immediate area). All equipment used when handling the product must be grounded. Do not touch damaged containers or spilled material unless wearing appropriate protective clothing. Stop leak if you can do it without risk. A vapor suppressing foam may be used to reduce vapors. For chlorosilanes, use AFFF alcohol-resistant medium expansion foam to reduce vapors. Do not get water on spilled substance or inside containers. Use water spray to reduce vapors or divert vapor cloud drift. Prevent entry into waterways, sewers, basements or confined areas. Small Spills: Cover with dry earth, dry sand, or other non-combustible material followed with plastic sheet to minimize spreading or contact with rain. Use clean non-sparking tools to collect material and place it into loosely covered plastic containers for later disposal.

Regulations

RCRA 40CFR: Not listed
CERCLA: 40CFR 302.4: Not listed
SARA 40CFR 372.65: Not listed
SARA EHS 40CFR 355: Not listed
TSCA: Listed

ETH4350 **CAS #: 97-64-3**

ETHYL LACTATE

RTECS: OD5075000
DOT: UN1192; IMO3.3
EINECS Number: 202-598-0
Molecular Formula: $C_5H_{10}O_3$
Structured MF: $CH_3CHOHCOOC_2H_5$
Formula Weight: 118.13

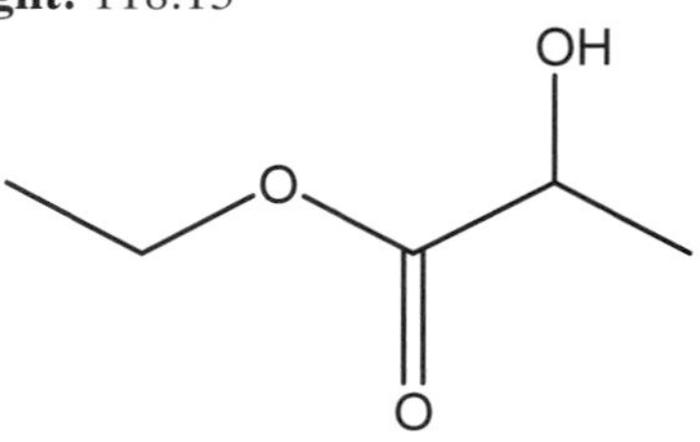

Chemical Structure

Synonyms: ACTYLOL; ACYTOL; ETHYL 2-HYDROXYPROPIONATE; ETHYL ALPHA-HYDROXYPROPIONATE; ETHYLESTER KYSELINY MLECNE; 2-HYDROXYPROPANOIC ACID ETHYL ESTER; LACTATE D'ETHYLE; LACTIC ACID,ETHYL ESTER; PROPANOIC ACID,2-HYDROXY-,ETHYL ESTER; SOLACTOL

Description: colorless liquid; characteristic odor

Use: solvent for basic dyes, hard copals; in lacquer industry; & manufacture of safety glass; stencil sheets; preservative; solvent for nitrocellulose, cellulose acetate, cellulose ethers and other resins; gelatinant for varnishes; flavoring agent in beverages and various foods; to reduce alkalinity in cosmetic prepn

Physical Properties

Boiling Point: 154 °C (309 °F)
Freezing Point: -25 °C (-13 °F)
Specific Gravity: 1.042 at 14 °C/4 °C
Vapor Density: 4.1 Air=1
Bulk Density: 8.55 lbs/gal at 20 °C
Vapor Pressure: 5 mm Hg at 30 °C
Water Solubility: Miscible with Water
Other Solubilities: miscible with Gasoline.
Surface Tension: 29.20 dynes/cm at 20 °C
Odor Threshold: In alcoholic beverages 14 ppm
Refraction Index: 1.4229 at 20 °C
Flash Point: 46.1 °C Closed Cup
Autoignition Temperature: 400 °C
LEL: 1.5% v/v
UEL: 11.4% v/v

RTECS Toxicity Data

Acute Oral: Rat LD_{50} Dose: >5 gm/kg. Mouse LD_{50} Dose: 2500 mg/kg.

Acute Dermal: Rabbit LD_{50} Route: Skin; Dose: >5 gm/kg. Mouse LD_{50} Route: Subcutaneous Dose: 2500 mg/kg.

Hazard Overviews

Flammable

Health: May cause irritation. Other Acute Effects: may be harmful by inhalation, ingestion, or skin absorption.

Fire: Flammable. Hazards: emits toxic fumes. Extinguishing agents: carbon dioxide, dry chemical powder or appropriate foam. Precautions: combustible liquid.

Carcinogenicity: IARC - Not listed; NIOSH - Not listed; NTP - Not listed; ACGIH - Not listed; OSHA - Not listed; EPA - Not listed; MAK - Not listed

Environmental

Cleanup/Disposal: Guide No. 129: Eliminate all ignition sources (no smoking, flares, sparks or flames in immediate area). All equipment used when handling the product must be grounded. Do not touch or walk through spilled material. Stop leak if you can do it without risk. Prevent entry into waterways, sewers, basements or confined areas. A vapor suppressing foam may be used to reduce vapors. Absorb or cover with dry earth, sand or other non-combustible material and transfer to containers. Use clean non-sparking tools to collect absorbed material. Large Spills: Dike far ahead of liquid spill for later disposal. Water spray may reduce vapor; but may not prevent ignition in closed spaces.

Environmental Physical Data

BCF: no food chain concentration potential

Regulations

RCRA 40CFR: Not listed
CERCLA: 40CFR 302.4: Not listed
SARA 40CFR 372.65: Not listed
SARA EHS 40CFR 355: Not listed
TSCA: Listed

ETH4400 CAS #: 75-08-1

ETHYL MERCAPTAN

RTECS: KI9625000
DOT: UN2363; IMO3.1
EINECS Number: 200-837-3
Molecular Formula: C_2H_6S
Structured MF: CH_3CH_2SH
Formula Weight: 62.13

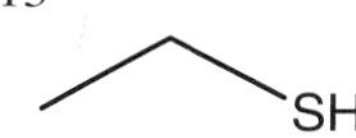

Chemical Structure

Synonyms: AETHANETHIOL; AETHYLMERCAPTAN; ETANTIOLO; ETHAANTHIOL; ETHANETHIOL; ETHYL HYDROSULFIDE; ETHYL SULFHYDRATE; ETHYL THIOALCOHOL; ETHYLMERCAPTAAN; ETHYLMERKAPTAN; ETILMERCAPTANO; LPG ETHYL MERCAPTAN 1010; MERCAPTOETHANE; THIOETHANOL; THIOETHYL ALCOHOL

Description: colorless liquid, gas; penetrating "skunk-like" garlic or leek odor

Use: intermediate & starting material in mfr of plastics, insecticides, antioxidants; odorant for natural gas; int in prodn of acaricides, defoliants, pharmaceuticals, & adhesives

Physical Properties

Boiling Point: 35 °C (95 °F) at 760 mm Hg
Freezing Point: -144.4 °C (-227.92 °F)
Specific Gravity: 0.83907 at 20 °C/4 °C
Vapor Density: 2.14 Air=1
Saturated Vapor Density: 1.997284574 kg/m^3
Vapor Pressure: 442 mm Hg at 20 °C
Water Solubility: 6.76 g/L at 20 °C
Other Solubilities: Soluble in Acetone; dilute alkali.
Surface Tension: Estimated at 25 dynes/cm
Odor Threshold: 0.0003 ppm
Refraction Index: 1.431 at 20 °C/D
Critical Temperature: 225.5 °C
Critical Pressure: 54.2 atm
Ionization Potential (eV): 9.29
Flash Point: -48.333 °C Closed Cup

Autoignition Temperature: 299 °C
LEL: 2.8% v/v
UEL: 18% v/v

RTECS Toxicity Data

Acute Oral: Rat LD_{50} Dose: 682 mg/kg; Toxic Effects: Behavioral - Muscle weakness; Behavioral - Ataxia; Lungs, Thorax, or Respiration - Cyanosis.
Acute Inhalation: Rat LC_{50} Dose: 4420 ppm/4hr; Toxic Effects: Peripheral Nerve and sensation - Spastic parapysis with or without sensory change; Behavioral - Excitment; Lungs, Thorax, or Respiration - Cyanosis. Mouse LC_{50} Dose: 2770 ppm/4hr; Toxic Effects: Behavioral - Excitment; Behavioral - Change in motor activity (specific assay); Lungs, Thorax, or Respiration - Cyanosis.
Irritation Eye: Rabbit Standard Draize Test Dose: 100 mg/24H; Reaction: moderate.
Irritation Skin: Rabbit Standard Draize Test Dose: 500 mg/24H; Reaction: mild.

Hazard Overviews

Flammable

Fire Diamond

Health: Irritating to eyes/skin/respiratory tract. Also Causes: headache, nausea, restlessness, incoordination, skeletal muscle paralysis, cyanosis, respiratory depression. Chronic Effects: Changes in smell/taste.
Fire: Flammable. Can form explosive mixtures in the air. For small fires use dry chemical, carbon dioxide, or regular foam. For large fires use fog or regular foam. Water should not be used unless other agents are unavailable because it can produce toxic/flammable vapors.
Reactivity: Stable. Hazardous polymerization cannot occur. Avoid: exposure to heat, flames, and sparks; contact with oxidizing materials, acids, water. Incompatible with: dry bleach; calcium oxychloride; oxidizing materials; water/steam. Hazardous decomposition products: highly toxic fumes of sulfur oxides.
Carcinogenicity: IARC - Not listed; NIOSH - Listed as carcinogen; NTP - Not listed; ACGIH - Not listed; OSHA - Not listed; EPA - Not listed; MAK - Not listed
Primary Target Organs:

Eyes

Skin

Respiratory System

Mucous Membranes

Nervous System

Exposure Limits
OSHA PEL: STEL: 10 ppm; 25 mg/m^3.
OSHA PEL Vacated 1989 Limits: TWA: 0.5 ppm; 1 mg/m^3.
ACGIH TLV: TWA: 0.5 ppm; 1.3 mg/m^3.
NIOSH REL: STEL: 0.5 ppm; 1.3 mg/m^3; 15-minute.
NIOSH IDLH: 500 ppm.
DFG MAK: TWA: 0.5 ppm; 1 mg/m^3.

Respirator Recommendation
Exposure Range: >10 to 100 ppm Air Purifying, Negative Pressure, Half Mask
Exposure Range: >100 to <500 ppm Air Purifying, Negative Pressure, Full Face
Exposure Range: 500 to unlimited ppm Self-contained Breathing Apparatus, Pressure Demand, Full Face
Cartridge Color: black

Environmental

Environmental Fate: If released to the atmosphere, it will exist primarily in the vapor-phase where it will degrade readily by reaction with photochemically produced hydroxyl radicals (estimated half-life of 8 hours) and night-time nitrate radicals (estimated half-life of 1 hr). If released to surface water, volatilization may be important. Volatilization half-lives of 2.5 and 29 hours can be estimated for a model river (1 meter deep) and model pond (2 meters deep), respectively. If released to soil, it may be transported by leaching, gas penetration, and surface evaporation. Insufficient data are available to assess the relative importance of biodegradation in soil or water.
Cleanup/Disposal: Guide No. 130: Eliminate all ignition sources (no smoking, flares, sparks or flames in immediate area). All equipment used when handling the product must be grounded. Do not touch or walk through spilled material. Stop leak if you can do it without risk. Prevent entry into waterways, sewers, basements or confined areas. A vapor suppressing foam may be used to reduce vapors. Absorb or cover with dry earth, sand or other non-combustible material and transfer to containers. Use clean non-sparking tools to collect absorbed material. Large Spills: Dike far ahead of liquid spill for later disposal. Water spray may reduce vapor; but may not prevent ignition in closed spaces.

Environmental Physical Data

Henry's Law Constant: 0.0045
Sorption Partition Coefficient: K_{oc} = estimated at 34
BCF: estimated at 4

Regulations

RCRA 40CFR: Not listed
CERCLA: 40CFR 302.4: Not listed
SARA 40CFR 372.65: Not listed
SARA EHS 40CFR 355: Not listed
TSCA: Listed

Analytical Methods

Air: EPA VA-006-1, VA-008-1; ASTM D4490
Indoor / Expired Air: NIOSH 2542

ETH4450 **CAS #: 97-63-2**

ETHYL METHACRYLATE

RTECS: OZ4550000
DOT: UN2277; IMO3.2
EINECS Number: 202-597-5
Molecular Formula: $C_6H_{10}O_2$

Structured MF: $CH_2=CCH_3COOC_2H_5$
Formula Weight: 114.14

Chemical Structure

Synonyms: ETHYL 2-METHACRYLATE; ETHYL ALPHA-METHYL ACRYLATE; ETHYL 2-METHYLACRYLATE; ETHYL ALPHA-METHYLACRYLATE; ETHYL 2-METHYL-2-PROPENOATE; ETHYLESTER KYSELINY METHAKRYLOVE; METHACRYLIC ACID,ETHYL ESTER; 2-METHYL-2-PROPENOIC ACID,ETHYL ESTER; 2-PROPENOIC ACID,2-METHYL-,ETHYL ESTER; RHOPLEX AC-33; RHOPLEX AC-33 (ROHM AND HAAS)

Description: colorless, liquid; acrid odor

Use: Widely known as "Plexiglass"(in the polymer form), this compound is used to make polymers, which in turn are used for building, automotive, aerospace an furniture industries; by dentists as dental plates, artificial teeth and orthopedic cement; as a chemical intermediate in organic synthesis

Physical Properties

Boiling Point: 117 °C (243 °F) at 760 mm Hg
Freezing Point: > -75 °C (-103 °F)
Specific Gravity: 0.9135 at 20 °C/4 °C
Vapor Density: 3.9 Air=1
Saturated Vapor Density: 1.264898004 kg/m^3
Density: 0.917 g/mL
Vapor Pressure: 14 torr at 20 °C
Water Solubility: 5 to 10 mg/mL at 20 °C
Other Solubilities: 95% Ethanol: >=100 mg/ml at 20 °C; Acetone: >=100 mg/ml at 20 °C; DMSO: >=100 mg/ml at 20 °C; Ether: Very Soluble.
Refraction Index: 1.4147 at 20 °C/D
Flash Point: 35 °C Cleveland Open Cup
Autoignition Temperature: 393 °C
LEL: 1.8% v/v
UEL: Saturation% v/v

RTECS Toxicity Data

Acute Oral: Rat LD_{50} Dose: 14800 mg/kg; Toxic Effects: Behavioral - Muscle weakness; Behavioral - Coma; Lungs, Thorax, or Respiration - Respiratory depression. Mouse LD_{50} Dose: 7836 mg/kg.

Acute Inhalation: Rat LC_{50} Dose: 8300 ppm/4hr; Toxic Effects: Sense organs and special senses - Other; Sense organs and special senses - Other; Lungs, Thorax, or Respiration - Dyspnea.

Acute Dermal: Rabbit LD; Route: Skin; Dose: >10 mL/kg. Rat LD_{Lo} Route: Subcutaneous Dose: 25 gm/kg; Toxic Effects: Behavioral - Muscle weakness; Behavioral - Coma; Lungs, Thorax, or Respiration - Respiratory depression.

Reproductive/Teratogenic: Rat Route: Intraperitoneal; Dose: 735 mg/kg; Duration: female 5-15D of pregnancy; Effects on Embryo or Fetus - Fetotoxicity. Rat Route: Intraperitoneal; Dose: 366 mg/kg; Duration: female 5-15D of pregnancy; Effects on Fertility - Post-implantation mortality; Specific Developmental Abnormalities - Other developmental abnormalities.

Mutagenic: Mouse Mutations in Mammalian Somatic Cells; Cell Type: lymphocyte; Dose: 1450 mg/L. Mouse Cytogenetic Analysis; Cell Type: lymphocyte; Dose: 900 mg/L.

Hazard Overviews

Flammable

Explosive

Health: Irritating to eyes/skin/respiratory tract. Other Acute Effects: harmful if swallowed, inhaled, or absorbed through skin; lachrymator; symptoms of exposure may include burning sensation, coughing, wheezing, laryngitis, shortness of breath, headache, nausea and vomiting; may cause allergic reaction.

Fire: Flammable. Hazards: emits toxic fumes; vapor may travel considerable distance to source of ignition and flash back; forms explosive mixtures in air. Extinguishing agents: water spray; carbon dioxide, dry chemical powder or appropriate foam. Precautions: combustible liquid.

Carcinogenicity: IARC - Not listed; NIOSH - Not listed; NTP - Not listed; ACGIH - Not listed; OSHA - Not listed; EPA - Not listed; MAK - Not listed

Environmental

Environmental Fate: If released to the atmosphere, reaction with photochemically produced hydroxyl radicals should ensue with an estimated half-life of 8 hours. Rain wash-out may be a significant fate process, and direct photochemical degradation is not indicated. If released to water, hydrolysis should be extremely slow under neutral conditions, yet may become important in basic waters. It should be capable of undergoing microbial degradation. If released to soil, it should display moderate to high mobility and should undergo microbial degradation. Volatilization from both water and soil should be an important fate process.

Cleanup/Disposal: Guide No. 129: Eliminate all ignition sources (no smoking, flares, sparks or flames in immediate area). All equipment used when handling the product must be grounded. Do not touch or walk through spilled material. Stop leak if you can do it without risk. Prevent entry into waterways, sewers, basements or confined areas. A vapor suppressing foam may be used to reduce vapors. Absorb or cover with dry earth, sand or other non-combustible material and transfer to containers. Use clean non-sparking tools to collect absorbed material. Large Spills: Dike far ahead of liquid spill for later disposal. Water spray may reduce vapor; but may not prevent ignition in closed spaces.

Environmental Physical Data

Henry's Law Constant: 2.8 $x10^{-4}$
Octanol/Water Partition Coefficient: log K_{ow} = 1.94

Sorption Partition Coefficient: K_{oc} = estimated at 271
BCF: estimated at 18

Regulations

RCRA 40CFR: Listed Hazardous Waste No. U118 Toxic Waste
CERCLA: 40CFR 302.4: Listed per RCRA Section 3001 RQ: 1000 lb (453.5 kg)
SARA 40CFR 372.65: Not listed
SARA EHS 40CFR 355: Not listed
TSCA: Listed

Analytical Methods

Soil: SW846 3640A, 5032, 8240B, 8260A, 8260B
Drinking Water: EPA 524.2
Plasma: EPA 29

ETH4500 CAS #: 62-50-0

ETHYL METHANESULFONATE

RTECS: PB2100000
EINECS Number: 200-536-7
Molecular Formula: $C_3H_8O_3S$
Structured MF: $CH_3SO_3C_2H_5$
Formula Weight: 124.15

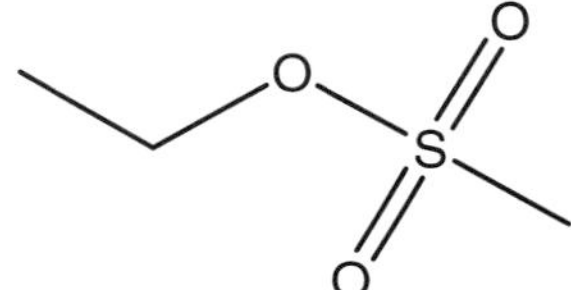

Chemical Structure

Synonyms: EMS; ENT 26396; ETHYL ESTER OF METHANESULFONIC ACID; ETHYL ESTER OF METHANESULPHONIC ACID; ETHYL ESTER OF METHYLSULFONIC ACID; ETHYL ESTER OF METHYLSULPHONIC ACID; ETHYL MESYLATE; ETHYL METHANESULPHONATE; ETHYL METHANSULFONATE; ETHYL METHANSULPHONATE; ETHYLESTER KYSELINY METHANSULFONOVE; ETHYLMETHANESULFONATE; ETHYLMETHANESULPHONATE; HALF-MYLERAN; METHANESULPHONIC ACID ETHYL ESTER; METHANESULPHONIC ACID,ETHYL ESTER; METHYLSULFONIC ACID ETHYL ESTER; METHYLSULFONIC ACID,ETHYL ESTER; NSC 26805
Description: colorless liquid
Use: considered for use as a reversible male chemosterilant for insects and mammalian pests and as a possible human male contraceptive; as a research tool and model compound to study mutagenesis and carcinogenesis; has mutagenic potential for use in plant breeding

Physical Properties

Boiling Point: 213 °C (415 °F) to 213.5 °C (416 °F) at 761 mm Hg
Specific Gravity: 1.1452 at 22 °C/4 °C
Saturated Vapor Density: 1.20169923 kg/m^3
Vapor Pressure: 0.328 mm Hg at 25 °C
Water Solubility: 50 to 100 mg/mL at 27 °C
Other Solubilities: 95% Ethanol: >=100 mg/ml at 27 °C; Acetone: >=100 mg/ml at 27 °C; DMSO: >=100 mg/ml at 27 °C.
Refraction Index: 1.418
Flash Point: 100 °C

RTECS Toxicity Data

Acute Oral: Mouse LD_{50} Dose: 470 mg/kg.
Reproductive/Teratogenic: Rat Route: Oral; Dose: 500 mg/kg; Duration: male 5D prior to mating; Effects on Fertility - Male fertility index; Pre-implantation mortality. Rat Route: Intravenous; Dose: 62 mg/kg; Duration: female 18D of pregnancy; Effects on Embryo or Fetus - Other effects to embryo or fetus.
Mutagenic: Human DNA Damage; Cell Type: other cell types; Dose: 3 mmol/L. Human DNA Damage; Cell Type: fibroblast; Dose: 50 mmol/L. Human DNA Repair; Cell Type: fibroblast; Dose: 20 umol/L. Human Unscheduled DNA Synthesis; Cell Type: other cell types; Dose: 500 mg/L. Human Unscheduled DNA Synthesis; Cell Type: fibroblast; Dose: 1 mmol/L/3H. Human Unscheduled DNA Synthesis; Cell Type: HeLa cell; Dose: 1 mmol/L. Human Unscheduled DNA Synthesis; Cell Type: lung; Dose: 400 mg/L. Human Unscheduled DNA Synthesis; Cell Type: other cell types; Dose: 500 mg/L. Human Unscheduled DNA Synthesis; Cell Type: other cell types; Dose: 1 mmol/L. Human DNA Inhibition; Cell Type: HeLa cell; Dose: 10 mmol/L/30M-C. Human DNA Adduct; Cell Type: HeLa cell; Dose: 5 umol/L. Human Mutations in Mammalian Somatic Cells; Cell Type: lymphocyte; Dose: 50 umol/L. Human Mutations in Mammalian Somatic Cells; Cell Type: fibroblast; Dose: 500 mg/L. Human Mutations in Mammalian Somatic Cells; Cell Type: other cell types; Dose: 150 mg/L. Human Cytogenetic Analysis; Cell Type: fibroblast; Dose: 1 umol/L. Human Cytogenetic Analysis; Cell Type: lymphocyte; Dose: 250 mg/L. Human Sister Chromatid Exchenge; Cell Type: other cell types; Dose: 3 mmol/L. Human Sister Chromatid Exchenge; Cell Type: lymphocyte; Dose: 500 umol/L. Human Sister Chromatid Exchenge; Cell Type: fibroblast; Dose: 1 umol/L. Human Sister Chromatid Exchenge; Cell Type: lung; Dose: 60 mg/L. Human Morphological Transformation; Cell Type: other cell types; Dose: 10 mg/L. Human Other Mutation Test Systems; Cell Type: lymphocyte; Dose: 400 umol/L.
Tumorigenic: Rat Route: Oral; Dose: 1043 gm/kg/12W-C; Toxic Effects: Tumorigenic - Carcinogenic by RTECS criteria; Skin and appendages - Tumors. Rat Route: Oral; Dose: 3353 mg/kg/13W-C; Toxic Effects: Tumorigenic - Carcinogenic by RTECS criteria; Skin and appendages - Tumors. Rat Route: Oral; Dose: 1050 gm/kg/12W-C; Toxic Effects: Tumorigenic - Carcinogenic by RTECS criteria; Skin and appendages - Tumors.

Hazard Overviews

Corrosive

Health: May cause irritation. Toxic. Other Acute Effects: harmful if swallowed, inhaled, or absorbed through skin. Chronic Effects: carcinogen; may alter genetic material; may cause heritable genetic damage; target organs: lungs; kidneys.
Fire: Will burn. Hazards: emits toxic fumes. Extinguishing agents: water spray; carbon dioxide, dry chemical powder or appropriate foam. Precautions: combustible liquid.
Reactivity: Incompatible with: strong oxidizing agents, strong bases. Hazardous decomposition products: toxic fumes of: carbon monoxide, carbon dioxide, sulfur oxides.
Carcinogenicity: IARC - Group 2B, Possibly carcinogenic to humans; NIOSH - Not listed; NTP - Listed; ACGIH - Not listed; OSHA - Not listed; EPA - Not listed; MAK - Not listed
Primary Target Organs:

Respiratory System

Kidneys

Environmental

Environmental Fate: If released to water, it will hydrolyze (half-life 96 hours at 20 °C). Direct photolysis, reaction with alkylperoxy radicals and singlet oxygen, bioaccumulation in aquatic organisms, and adsorption to suspended solids and sediments in water are not expected to be important fate processes. If released to moist soil, it is expected to hydrolyze as fast, if not faster, than in water. Mobility is expected to be extremely limited. If released to dry soil, this compound is expected to volatilize fairly rapidly. If released to the atmosphere, it is expected to exist almost entirely in the vapor phase. This compound may be removed from the atmosphere by reaction with photochemically generated hydroxyl radicals (estimated half-life 30 days) or by wet deposition.
Cleanup/Disposal: Guide No. 171: Do not touch or walk through spilled material. Stop leak if you can do it without risk. Prevent dust cloud. Avoid inhalation of asbestos dust. Small Dry Spills: With clean shovel place material into clean, dry container and cover loosely; move containers from spill area. Small Spills: Take up with sand or other noncombustible absorbent material and place into containers for later disposal. Large Spills: Dike far ahead of liquid spill for later disposal. Cover powder spill with plastic sheet or tarp to minimize spreading. Prevent entry into waterways, sewers, basements or confined areas.

Environmental Physical Data

Octanol/Water Partition Coefficient: log K_{ow} = 0.09
Sorption Partition Coefficient: K_{oc} = estimated at 27
BCF: estimated at 1

Regulations

RCRA 40CFR: Listed Hazardous Waste No. U119 Toxic Waste
CERCLA: 40CFR 302.4: Listed per RCRA Section 3001 RQ: 1 lb (0.454 kg)
SARA 40CFR 372.65: Not listed
SARA EHS 40CFR 355: Not listed
TSCA: Listed

Analytical Methods

Soil: SW846 3640A, 8250A, 8270B, 8270C
Water / Groundwater: EPA 1625
Plasma: EPA 29

ETH4550 **CAS #: 540-67-0**

ETHYL METHYL ETHER

RTECS: KO0260000
DOT: UN1039; IMO2.0
Molecular Formula: C_3H_8O
Formula Weight: 60.10
Synonyms: ETHANE,METHOXY-; ETHER,ETHYL METHYL; METHANE,ETHOXY; METHANE,ETHOXY-; METHOXYETHANE; METHYL ETHYL ETHER
Description: colorless liquid
Use: solvent (aerosol); medication: anesthetic agent former use

Physical Properties

Boiling Point: 10.8 °C (51 °F) at 760 mm Hg
Freezing Point: -113 °C (-171.4 °F)
Specific Gravity: 0.7252 at 0 °C/0 °C
Water Solubility: Soluble in Water
Other Solubilities: Soluble in all proportions in Alcohol, Ether.
Refraction Index: 1.3420 at 4 °C/D
Critical Temperature: 164.7 °C
Critical Pressure: 43.4 atm
Ionization Potential (eV): 9.72 +/-0.07
Flash Point: Flammable, dangerous fire risk
Autoignition Temperature: 190 °C
LEL: 2% v/v
UEL: 10.1% v/v

RTECS Toxicity Data

Acute Inhalation: Mouse LC_{50} Dose: 1082 gm/m^3/15M; Toxic Effects: Behavioral - General anesthetic.

Hazard Overviews

Flammable

Explosive

Fire: Flammable.
Carcinogenicity: IARC - Not listed; NIOSH - Not listed; NTP - Not listed; ACGIH - Not listed; OSHA - Not listed; EPA - Not listed; MAK - Not listed

Environmental

Environmental Fate: If released to soil, it will be subject to volatilization. It will be expected to exhibit very high mobility in soil and, therefore, it may leach to groundwater. It will not be expected to hydrolyze in soil. If released to water, it will not be expected to significantly adsorb to sediment or suspended particulate matter, bioconcentrate in aquatic organisms, hydrolyze, directly photolyze, or photooxidize via reaction with photochemically produced hydroxyl radicals in the water, based upon estimated physical-chemical properties or analogies to other structurally related aliphatic ethers. In surface water, it will be subject to rapid volatilization with estimated half-lives of 4.6 and 2.2 days for volatilization from a river one meter deep flowing 1 m/sec with a wind velocity of 3 m/sec and a model pond, respectively. It may be resistant to biodegradation in environmental media based upon screening test data for the structurally related diethyl ether from studies using activated sludge or sewage inocula. Many ethers are known to be resistant to biodegradation. If released to the atmosphere, it will be expected to exist almost entirely in the vapor phase based on its vapor pressure. It will be susceptible to photooxidation via vapor phase reaction with photochemically produced hydroxyl radicals with an estimated half-life of 2.6 days for this process. Direct photolysis will not be an important removal process since aliphatic ethers do not absorb light at wavelengths >290 nm.

Cleanup/Disposal: Guide No. 115: Eliminate all ignition sources (no smoking, flares, sparks or flames in immediate area). All equipment used when handling the product must be grounded. Do not touch or walk through spilled material. Stop leak if you can do it without risk. If possible, turn leaking containers so that gas escapes rather than liquid. Use water spray to reduce vapors or divert vapor cloud drift. Do not direct water at spill or source of leak. Prevent spreading of vapors through sewers, ventilation systems and confined areas. Isolate area until gas has dispersed.

Environmental Physical Data

Henry's Law Constant: 2.93×10^{-4}
Octanol/Water Partition Coefficient: log K_{ow} = 0.341
Sorption Partition Coefficient: K_{oc} = estimated at 37
BCF: estimated at 1.1

Regulations

RCRA 40CFR: Not listed
CERCLA: 40CFR 302.4: Not listed
SARA 40CFR 372.65: Not listed
SARA EHS 40CFR 355: Not listed
TSCA: Not listed

ETH4600 CAS #: 80-40-0

ETHYL P-METHYLBENZENESULFONATE

RTECS: XT6825000
EINECS Number: 201-276-7

Molecular Formula: $C_9H_{12}O_3S$
Formula Weight: 200.25

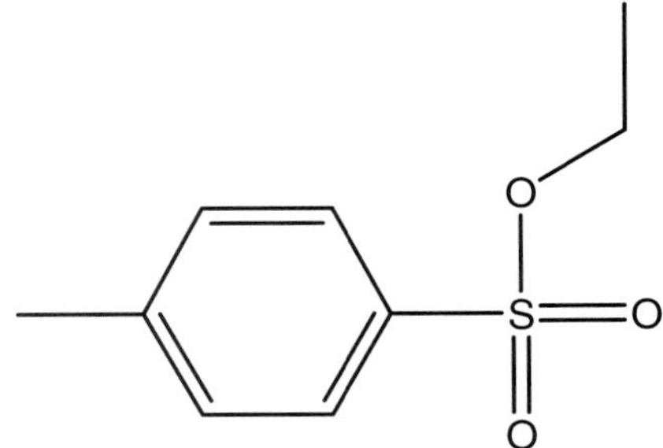

Chemical Structure

Synonyms: BENZENESULFONIC ACID,4-METHYL-,ETHYL ESTER; ETHYL P-METHYL BENZENESULFONATE; ETHYL PTS; ETHYL P-TOLUENESULFONATE; ETHYL P-TOSYLATE; ETHYL TOSYLATE; ETHYLESTER KYSELINY P-TOLUENSULFONOVE; ETHYL-P-TOLUENESULFONATE; P-TOLUENESULFONIC ACID,ETHYL ESTER; P-TOLUOLSULFONSAEURE AETHYL ESTER
Description: monoclinic crystals or prisms
Use: plasticizer for cellulose acetate; ethylating agent

Physical Properties

Boiling Point: 173 °C (343 °F) at 15 mm Hg
Freezing Point: 33 °C (91.4 °F)
Specific Gravity: 1.17
Water Solubility: Insoluble in Water
Other Solubilities: Soluble in hot Ethyl Acetate.
Flash Point: 157.778 °C

RTECS Toxicity Data

Acute Dermal: Rat LD_{50} Route: Subcutaneous Dose: 500 mg/kg.

Mutagenic: Hamster Morphological Transformation; Cell Type: embryo; Dose: 500 ug/L. Insects - D Melanogaster Sex Chromosome Loss; Route: Oral; Dose: 1 mmol/L. Insects - D Melanogaster Sex Chromosome Loss; Route: Parenteral; Dose: 10 mmol/L.

Tumorigenic: Rat Route: Subcutaneous; Dose: 50 mg/kg; Toxic Effects: Tumorigenic - Equivocal tumorigenic agent by RTECS criteria; Tumorigenic - Tumors at site of application. Rat Route: Subcutaneous; Dose: 3250 mg/kg/65W-I; Toxic Effects: Tumorigenic - Equivocal tumorigenic agent by RTECS criteria; Tumorigenic - Tumors at site of application.

Hazard Overviews

Health: Irritating to eyes/skin/respiratory tract. Harmful. Other Acute Effects: may be harmful by inhalation, ingestion, or skin absorption. Chronic Effects: possible risk of irreversible effects; possible mutagen; laboratory experiments have shown mutagenic effects.

Fire: Will burn. Hazards: emits toxic fumes. Extinguishing agents: water spray; carbon dioxide, dry chemical powder or appropriate foam. Precautions: combustible liquid.

Carcinogenicity: IARC - Not listed; NIOSH - Not listed; NTP - Not listed; ACGIH - Not listed; OSHA - Not listed; EPA - Not listed; MAK - Not listed

Primary Target Organs:

Eyes

Skin

Respiratory System

Environmental

Environmental Physical Data

Octanol/Water Partition Coefficient: log K_{ow} = 1.81

Regulations

RCRA 40CFR: Not listed
CERCLA: 40CFR 302.4: Not listed
SARA 40CFR 372.65: Not listed
SARA EHS 40CFR 355: Not listed
TSCA: Listed

ETH4650 CAS #: 105-40-8

ETHYL N-METHYLCARBAMATE

RTECS: FC2625000
EINECS Number: 203-295-6
Molecular Formula: $C_4H_9NO_2$
Formula Weight: 103.12

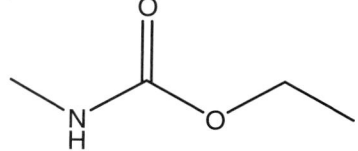

Chemical Structure

Synonyms: CARBAMIC ACID,METHYL-,ETHYL ESTER; ETHYL METHYLAMINOFORMATE; ETHYL METHYLCARBAMATE; ETHYLESTER KYSELINY METHYLKARBAMINOVE; ETHYL-N-METHYL CARBAMATE; N-METHYL URETHAN; METHYL URETHANE; METHYLCARBAMIC ACID,ETHYL ESTER; N-METHYLURETHAN; METHYLURETHANE; N-METHYLURETHANE
Description: colorless, clear liquid
Use: research chemical

Physical Properties

Boiling Point: 170 °C (338 °F) at 760 mm Hg
Specific Gravity: 1.0115 at 20 °C/4 °C
Density: 1.011 g/cc at 19.4 °C
Vapor Pressure: 1 mm Hg at 26.5 °C
Water Solubility: Very Soluble in Water
Other Solubilities: 95% Ethanol: >=100 mg/ml at 20 °C; Acetone: >=100 mg/ml at 20 °C; DMSO: >=100 mg/ml at 20 °C.
Refraction Index: 1.4183 at 20 °C/D
Flash Point: 73.06 °C

RTECS Toxicity Data

Acute Dermal: Mouse LD_{50} Route: Subcutaneous Dose: 1360 mg/kg; Toxic Effects: Behavioral - Somnolence (general depressed activity).
Reproductive/Teratogenic: Hamster Route: Intraperitoneal; Dose: 495 mg/kg; Duration: female 8D of pregnancy; Specific Developmental Abnormalities - Central nervous system; Body wall; Musculoskeletal system. Hamster Route: Intraperitoneal; Dose: 495 mg/kg; Duration: female 8D of pregnancy; Effects on Embryo or Fetus - Fetotoxicity.
Mutagenic: Mouse Mutations in Microorganisms; Cell Type: lymphocyte; Dose: 5 mmol/L (+S9).
Tumorigenic: Mouse Route: Skin; Dose: 10 gm/kg/1W-I; Toxic Effects: Tumorigenic - Equivocal tumorigenic agent by RTECS criteria; Skin and appendages - Tumors. Mouse Route: Intraperitoneal; Dose: 6500 mg/kg/13W-I; Toxic Effects: Tumorigenic - Equivocal tumorigenic agent by RTECS criteria; Lungs, Thorax, or Respiration - Tumors.

Hazard Overviews

Fire: Combustible.
Carcinogenicity: IARC - Not listed; NIOSH - Not listed; NTP - Not listed; ACGIH - Not listed; OSHA - Not listed; EPA - Not listed; MAK - Not listed

Environmental

Regulations

RCRA 40CFR: Not listed
CERCLA: 40CFR 302.4: Not listed
SARA 40CFR 372.65: Not listed
SARA EHS 40CFR 355: Not listed
TSCA: Listed

ETH4700 CAS #: 77-83-8

ETHYL METHYLPHENYLGLYCIDATE

RTECS: MW5250000
EINECS Number: 201-061-8
Molecular Formula: $C_{12}H_{14}O_3$
Formula Weight: 206.26

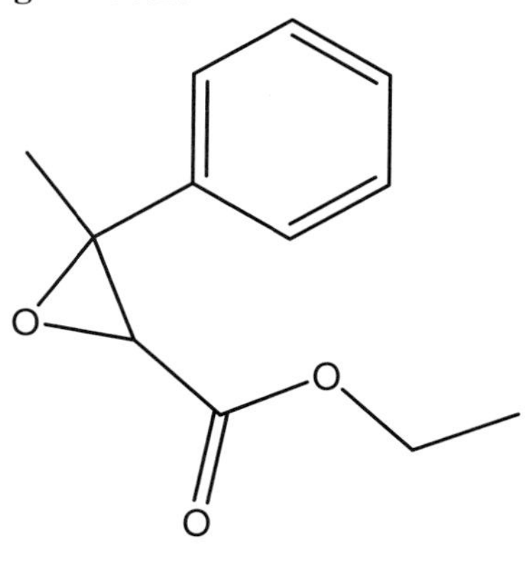

Chemical Structure

Synonyms: ALDEHYDE C-16; C-16 ALDEHYDE; ALDEHYDE C16; BUTANOIC ACID,2,3-EPOXY-3-PHENYL-,ETHYL ESTER; EMPG; ALPHA,BETA-EPOXY-BETA-METHYLHYDROCINNAMIC ACID ETHYL ESTER; ALPHA,BETA-EPOXY-BETA-METHYLHYDROCINNAMIC ACID,ETHYL ESTER; ALPHA,BETA-EPOXY-BETA-METHYLHYDROCINNAMIC ACID,ETHYLESTER; ETHYL ALPHA,BETA-EPOXY-BETA-METHYLHYDROCINNAMATE; ETHYL 2,3-EPOXY-3-METHYL-3-PHENYLPROPIONATE; ETHYL ALPHA,BETA-EPOXY-BETA-METHYLPHENYLPROPIONATE; ETHYL ESTER OF 2,3-EPOXY-3-PHENYLBUTANOIC ACID; ETHYL 3-METHYL-3-PHENYLGLYCIDATE; ETHYL BETA-METHYLPHENYLGLYCIDATE; FRAESEOL; FRAESOL; HYDROCINNAMIC ACID,ALPHA,BETA-EPOXY-BETA-METHYL-,ETHYLESTER; 3-METHYL-3-PHENYLGLYCIDIC ACID ETHYL ESTER; OXIRANECARBOXYLIC ACID,3-METHYL-3-PHENYL-,ETHYL ESTER; STRAWBERRY ALDEHYDE

Description: colorless to pale-yellow liquid; strong fruity odor

Use: synthetic flavoring substance and adjuvant; perfumery

Physical Properties

Boiling Point: 272 °C (522 °F) to 275 °C (527 °F)
Specific Gravity: 1.0442 at 20 °C
Density: 1.1 to 1.12 at 20 °C
Water Solubility: Insoluble < 1 mg/mL at 20 C
Other Solubilities: 95% Ethanol: Soluble (>=10 mg/ml at 20 °C); DMSO: Soluble (>=10 mg/ml at 20 °C).
Refraction Index: 1.5182 at 20 °C
Flash Point: 134 °C

RTECS Toxicity Data

Acute Oral: Rat LD_{50} Dose: 5470 mg/kg; Toxic Effects: Sense organs and special senses - Other; Behavioral - Somnolence (general depressed activity); Skin and appendages - Hair. Guinea Pig LD_{50} Dose: 4050 mg/kg; Toxic Effects: Behavioral - Somnolence (general depressed activity); Gastrointestinal - Other changes.

Mutagenic: Hamster Cytogenetic Analysis; Cell Type: ovary; Dose: 50 mg/L. Hamster Sister Chromatid Exchenge; Cell Type: ovary; Dose: 16 mg/L.

Hazard Overviews

Health: May be irritating to eyes/skin/respiratory tract. Other Acute Effects: may be harmful by inhalation, ingestion, or skin absorption. Chronic Effects: target organs: nerves, testes.

Fire: Will burn. Extinguishing agents: water spray; carbon dioxide, dry chemical powder or appropriate foam. Precautions: combustible liquid.

Reactivity: Stable. Hazardous polymerization will not occur. Incompatible with: strong oxidizing agents. Hazardous decomposition products: toxic fumes of: carbon monoxide, carbon dioxide.

Carcinogenicity: IARC - Not listed; NIOSH - Not listed; NTP - Not listed; ACGIH - Not listed; OSHA - Not listed; EPA - Not listed; MAK - Not listed

Primary Target Organs:

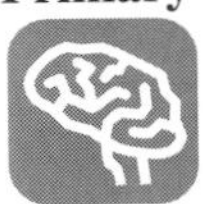
Nervous System

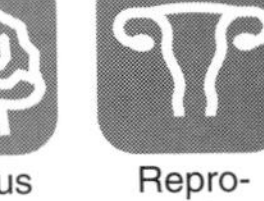
Repro-ductive

Environmental

Regulations

RCRA 40CFR: Not listed
CERCLA: 40CFR 302.4: Not listed
SARA 40CFR 372.65: Not listed
SARA EHS 40CFR 355: Not listed
TSCA: Listed

ETH4750 CAS #: 625-58-1

ETHYL NITRATE

RTECS: QU7900000
DOT: NA1993
EINECS Number: 210-903-3
Molecular Formula: $C_2H_5NO_3$
Formula Weight: 91.07
Synonyms: ETHYLESTER KYSELINY DUSICNE; NITRIC ACID,ETHYL ESTER; NITRIC ETHER
Description: colorless liquid; pleasant odor
Use: in org synthesis of dyes; chem int for drugs, perfumes; component of rocket propellants

Physical Properties

Boiling Point: 87.2 °C (189 °F) at 762 mm Hg
Freezing Point: -94.6 °C (-138.28 °F)
Specific Gravity: 1.1084 at 20 °C/4 °C
Vapor Density: 3.1 Air=1
Vapor Pressure: 2.17 kPa at 0 °C
Water Solubility: 1 g/100 mL at °C
Other Solubilities: Soluble in Alcohol and Ether
Refraction Index: 1.3852
Ionization Potential (eV): 11.22
Flash Point: 10 °C
LEL: 4.0% v/v

RTECS Toxicity Data

Mutagenic: Insects - D Melanogaster Sex Chromosome Loss; Route: Inhalation; Dose: 1200 ppm/3D. Bacteria - S Typhimurium Mutations in Microorganisms; Dose: 10 umol/plate (+/-S9).

Hazard Overviews

Flammable

Fire Diamond

Fire: Flammable.

Carcinogenicity: IARC - Not listed; NIOSH - Not listed; NTP - Not listed; ACGIH - Not listed; OSHA - Not listed; EPA - Not listed; MAK - Not listed

Environmental

Cleanup/Disposal: Guide No. 128: Eliminate all ignition sources (no smoking, flares, sparks or flames in immediate

area). All equipment used when handling the product must be grounded. Do not touch or walk through spilled material. Stop leak if you can do it without risk. Prevent entry into waterways, sewers, basements or confined areas. A vapor suppressing foam may be used to reduce vapors. Absorb or cover with dry earth, sand or other non-combustible material and transfer to containers. Use clean non-sparking tools to collect absorbed material. Large Spills: Dike far ahead of liquid spill for later disposal. Water spray may reduce vapor; but may not prevent ignition in closed spaces.

Regulations

RCRA 40CFR: Not listed
CERCLA: 40CFR 302.4: Not listed
SARA 40CFR 372.65: Not listed
SARA EHS 40CFR 355: Not listed
TSCA: Listed

ETH4800 CAS #: 109-95-5

ETHYL NITRITE

RTECS: RA0810000
DOT: UN1194; IMO3.1
EINECS Number: 203-722-6
Molecular Formula: $C_2H_5NO_2$
Structured MF: C_2H_5ONO
Formula Weight: 75.07

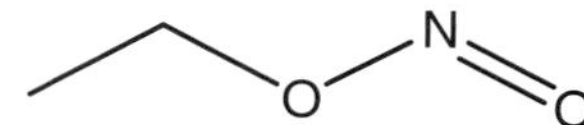

Chemical Structure

Synonyms: ETHYL NITRITE SOLUTIONS; ETHYLESTER KYSELINY DUSITE; HYPONITROUS ETHER; NITROSYL ETHOXIDE; NITROUS ACID ETHYL ESTER; NITROUS ACID,ETHYL ESTER; NITROUS ETHER; NITROUS ETHYL ETHER; SPIRIT OF ETHYL NITRITE; SWEET SPIRIT OF NITER; SWEET SPIRIT OF NITRE
Description: colorless to yellowish liquid; sweet, rum-like odor
Use: for preparing spirit nitrous ether; chem intermediate; flavor in foods & beverages; organic reactions; flavors useful in rum, brandy, fruit flavors

Physical Properties

Boiling Point: 17 °C (63 °F)
Freezing Point: -50 °C (-58 °F)
Specific Gravity: 0.9 at 15 °C/15 °C
Vapor Density: 2.6 Air=1
Water Solubility: Slightly Soluble in Water
Other Solubilities: miscible with Alcohol, Ether.
Surface Tension: 35 dynes/cm
Refraction Index: 1.3418 at 10 °C/D
Ionization Potential (eV): 10.53 +/-0.01
Flash Point: -35 °C Closed Cup
Autoignition Temperature: Decomposes at 90 °C
LEL: 4.0% v/v
UEL: 50% v/v

RTECS Toxicity Data

Acute Inhalation: Mouse LC; Dose: >200 mg/m^3/15M. Rat LC_{50} Dose: 160 ppm/4hr; Toxic Effects: Behavioral - Altered sleep time (including change in righting reflex); Behavioral - Excitment; Lungs, Thorax, or Respiration - Cyanosis.

Hazard Overviews

Explosive

Flammable

Fire Diamond

Health: Severely irritating to eyes/skin; corrosive depending on concentration/duration. Toxic. Also Causes: headache, tachycardia, hypotension, methemoglobinemia, unconsciousness.
Fire: Explosive when heated. Flammable. Use dry chemical, foam, carbon dioxide, or water spray. Fight fire from maximum distance. Use water spray to cool fire-exposed containers. Do not get water inside containers.
Reactivity: Unstable, decomposes on exposure to air, light, or moisture; on standing, it gradually decomposes. Hazardous polymerization cannot occur. Avoid: heat; ignition sources; acids; acid fumes. Incompatible with: broad range of materials. Hazardous decomposition products: nitrogen oxides.
Carcinogenicity: IARC - Not listed; NIOSH - Not listed; NTP - Not listed; ACGIH - Not listed; OSHA - Not listed; EPA - Not listed; MAK - Not listed
Primary Target Organs:

Eyes

Skin

Blood

Environmental

Cleanup/Disposal: Guide No. 131: Fully encapsulating, vapor protective clothing should be worn for spills and leaks with no fire. Eliminate all ignition sources (no smoking, flares, sparks or flames in immediate area). All equipment used when handling the product must be grounded. Do not touch or walk through spilled material. Stop leak if you can do it without risk. Prevent entry into waterways, sewers, basements or confined areas. A vapor suppressing foam may be used to reduce vapors. Small Spills: Absorb with earth, sand or other non-combustible material and transfer to containers for later disposal. Use clean non-sparking tools to collect absorbed material. Large Spills: Dike far ahead of liquid spill for later disposal. Water spray may reduce vapor; but may not prevent ignition in closed spaces.

Environmental Physical Data

BCF: no food chain concentration potential

Regulations

RCRA 40CFR: Not listed
CERCLA: 40CFR 302.4: Not listed
SARA 40CFR 372.65: Not listed

SARA EHS 40CFR 355: Not listed
TSCA: Listed

ETH4850 CAS #: 99-77-4

ETHYL 4-NITROBENZOATE

RTECS: DH5600000
EINECS Number: 202-786-2
Molecular Formula: $C_9H_9NO_4$
Formula Weight: 195.19

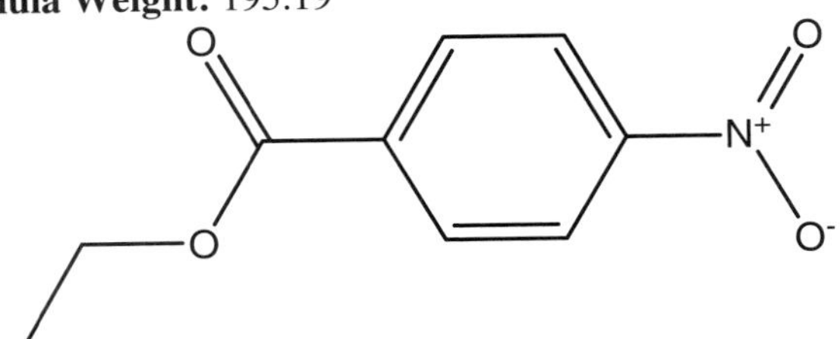

Chemical Structure

Synonyms: ETHYL P-NITROBENZOATE; P-NITROBENZOIC ACID,ETHYL ESTER

Physical Properties

Boiling Point: 186.3 °C (367.34 °F)
Freezing Point: 57 °C (134.6 °F)
Specific Gravity: 0.36
Water Solubility: Negligible
Other Solubilities: Soluble in Ethylbenzene
Evaporation Rate: < 1 Butyl Acetate=1

Hazard Overviews

Health: May cause irritation. Other Acute Effects: may be harmful by inhalation, ingestion, or skin absorption.
Fire: Hazards: emits toxic fumes. Extinguishing agents: water spray; carbon dioxide, dry chemical powder or appropriate foam. Precautions: combustible liquid.
Carcinogenicity: IARC - Not listed; NIOSH - Not listed; NTP - Not listed; ACGIH - Not listed; OSHA - Not listed; EPA - Not listed; MAK - Not listed

Environmental

Regulations

RCRA 40CFR: Not listed
CERCLA: 40CFR 302.4: Not listed
SARA 40CFR 372.65: Not listed
SARA EHS 40CFR 355: Not listed
TSCA: Listed

ETH4900 CAS #: 122-51-0

ETHYL ORTHOFORMATE

RTECS: RM6475000
EINECS Number: 204-550-4
Molecular Formula: $C_7H_{16}O_3$
Formula Weight: 148.23

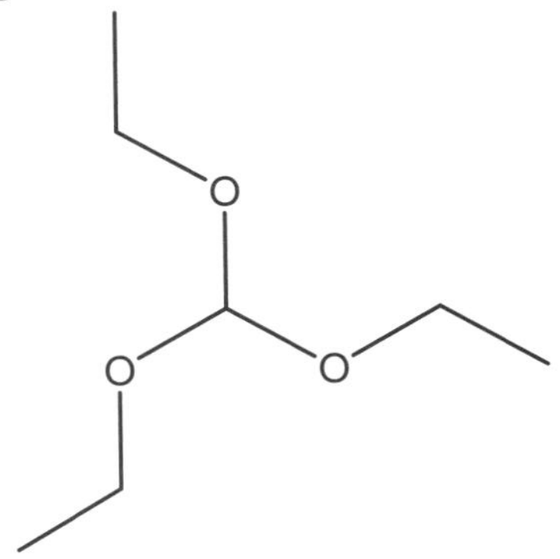

Chemical Structure

Synonyms: AETHON; ETHONE; ETHYLESTER KYSELINY ORTHOMRAVENCI; METHANE,TRIETHOXY-; 1,1',1'-(METHYLIDYNETRIS(OXY))TRIS(ETHANE); ORTHOFORMIC ACID,ETHYL ESTER; ORTHOMRAVENCAN ETHYLNATY; TRIETHOXYMETHANE; TRIETHYL ORTHOFORMATE; TRIETHYLESTER KYSELINY ORTHOMRAVENCI

Physical Properties

Boiling Point: 146 °C (294.8 °F)
Freezing Point: -76 °C (-104.8 °F)
Specific Gravity: 0.89090
Water Solubility: Decomposes 1.3 g/l
Other Solubilities: Ethanol: Soluble; Ether: Soluble
Refraction Index: 1.3922
Flash Point: 30 °C Closed Cup

RTECS Toxicity Data

Acute Oral: Rat LD_{50} Dose: 7060 mg/kg.
Acute Inhalation: Rat LC_{Lo} Dose: 4000 ppm/8hr.
Acute Dermal: Rabbit LD_{50} Route: Skin; Dose: 20 mL/kg. Guinea Pig LD_{50} Route: Skin; Dose: >10 mL/kg.
Irritation Eye: Rabbit Standard Draize Test Dose: 100 mg/24H; Reaction: moderate.
Irritation Skin: Rabbit Standard Draize Test Dose: 500 mg/24H; Reaction: mild. Rabbit Open Draize Test Dose: 10 mg/24H open; Reaction: mild.

Hazard Overviews

Flammable

Health: Irritating to eyes/skin/respiratory tract. Other Acute Effects: may be harmful by inhalation, ingestion, or skin absorption.
Fire: Flammable. Hazards: vapor may travel considerable distance to source of ignition and flash back; under fire conditions, material may decompose to form flammable and/or explosive mixtures in air. Extinguishing agents: carbon

dioxide, dry chemical powder or appropriate foam; do not use water. Precautions: combustible liquid.

Reactivity: Incompatible with: acids, strong oxidizing agents, protect from moisture. Hazardous decomposition products: toxic fumes of: carbon monoxide, carbon dioxide.

Carcinogenicity: IARC - Not listed; NIOSH - Not listed; NTP - Not listed; ACGIH - Not listed; OSHA - Not listed; EPA - Not listed; MAK - Not listed

Primary Target Organs:

Eyes

Skin

Respiratory System

Environmental

Regulations

RCRA 40CFR: Not listed
CERCLA: 40CFR 302.4: Not listed
SARA 40CFR 372.65: Not listed
SARA EHS 40CFR 355: Not listed
TSCA: Listed

ETH5000 CAS #: 101-99-5

ETHYL N-PHENYLCARBAMATE

RTECS: FD8925000
EINECS Number: 202-995-9
Molecular Formula: $C_9H_{11}NO_2$
Formula Weight: 165.21

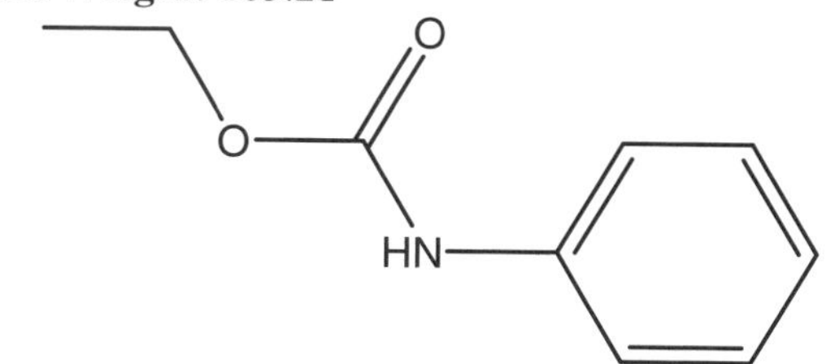

Chemical Structure

Synonyms: CARBAMIC ACID,PHENYL-,ETHYL ESTER; CARBAMIC ACID,PHENYL-,ETHYL ESTER (9CI); CARBANILIC ACID,ETHYL ESTER; EPC (THE PLANT REGULATOR); ETHANOL,CARBANILATE (7CI); N-(ETHOXYCARBONYL)ANILINE; ETHYL CARBANILATE; ETHYL PHENYLCARBAMATE; ETHYL N-PHENYLURETHAN; ETHYL N-PHENYLURETHANE; ETHYLESTER KYSELINY KARBANILOVE; ETHYL-N-PHENYLCARBAMATE; EUPHORIN; KEIMSTOP; PHENYLETHYL CARBAMATE; PHENYLURETHAN; PHENYLURETHAN(E); N-PHENYLURETHANE; PHENYLURETHANE; URETHAN,PHENYL; URETHAN,PHENYL-

Description: white acicular crystals; plates; aromatic odor

Use: fungicide; research chem (exptl neoplastic agent, citrus fungicide); plant growth regulator (former use)

Physical Properties

Boiling Point: 238 °C (460 °F)
Freezing Point: 52 °C (125.6 °F) to 53 °C (127.4 °F)
Specific Gravity: 1.1064 at 20 °C/4 °C
Water Solubility: Slightly Soluble in Water
Other Solubilities: freely Soluble in Alcohol, Ether
Refraction Index: 1.1064 at 20 °C/D

RTECS Toxicity Data

Acute Dermal: Mouse LD_{Lo} Route: Subcutaneous Dose: 1 gm/kg; Toxic Effects: Behavioral - Somnolence (general depressed activity).

Tumorigenic: Mouse Route: Skin; Dose: 20 gm/kg/2W-I; Toxic Effects: Tumorigenic - Neoplastic by RTECS criteria; Skin and appendages - Tumors. Mouse Route: Skin; Dose: 72 gm/kg/15W-I; Toxic Effects: Tumorigenic - Neoplastic by RTECS criteria; Skin and appendages - Tumors.

Hazard Overviews

Health: May cause irritation. Other Acute Effects: may be harmful by inhalation, ingestion, or skin absorption. The toxicological properties have not been thoroughly investigated.

Fire: Hazards: emits toxic fumes. Extinguishing agents: carbon dioxide, dry chemical powder or appropriate foam. Precautions: combustible liquid.

Reactivity: Stable. Hazardous polymerization will not occur. Hazardous decomposition products: toxic fumes of: carbon monoxide, carbon dioxide, nitrogen oxides.

Carcinogenicity: IARC - Not listed; NIOSH - Not listed; NTP - Not listed; ACGIH - Not listed; OSHA - Not listed; EPA - Not listed; MAK - Not listed

Environmental

Regulations

RCRA 40CFR: Not listed
CERCLA: 40CFR 302.4: Not listed
SARA 40CFR 372.65: Not listed
SARA EHS 40CFR 355: Not listed
TSCA: Listed

ETH5050 CAS #: 1498-51-7

ETHYL PHOSPHORODICHLORIDATE

RTECS: TD4390000
DOT: NA1760
EINECS Number: 216-099-0
Molecular Formula: $C_2H_5Cl_2O_2P$
Structured MF: $Cl_2(OC_2H_5)PO$
Formula Weight: 162.94

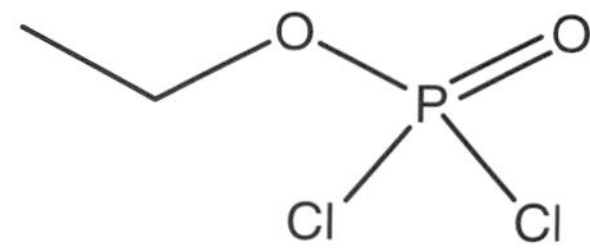

Chemical Structure

Synonyms: DICHLOROETHOXYPHOSPHINE OXIDE; DICHLOROPHOSPHORIC ACID,ETHYL ESTER; ETHYL

DICHLOROPHOSPHATE; ETHYLPHOSPHORIC ACID DICHLORIDE; PHOSPHORODICHLORIDIC ACID,ETHYL ESTER

Description: colorless liquid

Use: intermediate in the manufacture of the pesticide ethoprop

Physical Properties

Boiling Point: 167 °C (333 °F)
Specific Gravity: 1.35 at 19 °C
Water Solubility: Reacts with water
Surface Tension: 32.2 dynes/cm
Refraction Index: 1.4338

Hazard Overviews

Corrosive

Health: Corrosive to eyes/skin/respiratory tract. Other Acute Effects: harmful if swallowed, inhaled, or absorbed through skin; inhalation may result in spasm, inflammation and edema of the larynx and bronchi, chemical pneumonitis and pulmonary edema; symptoms of exposure may include burning sensation; coughing; wheezing; laryngitis; shortness of breath; headache; nausea; vomiting.

Fire: Hazards: emits toxic fumes; water hydrolyzes material liberating acidic gas which in contact with metal surfaces can generate flammable and/or explosive hydrogen gas. Extinguishing agents: carbon dioxide, dry chemical powder or appropriate foam; do not use water. Precautions: combustible liquid.

Reactivity: Incompatible with: strong oxidizing agents, strong bases, may decompose on exposure to moist air or water. Hazardous decomposition products: toxic fumes of: carbon monoxide, carbon dioxide, hydrogen chloride gas; thermal decomposition may produce toxic fumes of phosphorus oxides and/or phosphine.

Carcinogenicity: IARC - Not listed; NIOSH - Not listed; NTP - Not listed; ACGIH - Not listed; OSHA - Not listed; EPA - Not listed; MAK - Not listed

Primary Target Organs:

Eyes

Skin

Respiratory System

Environmental

Cleanup/Disposal: Guide No. 154: Eliminate all ignition sources (no smoking, flares, sparks or flames in immediate area). Do not touch damaged containers or spilled material unless wearing appropriate protective clothing. Stop leak if you can do it without risk. Prevent entry into waterways, sewers, basements or confined areas. Absorb or cover with dry earth, sand or other non-combustible material and transfer to containers. Do not get water inside containers.

Regulations

RCRA 40CFR: Not listed
CERCLA: 40CFR 302.4: Not listed
SARA 40CFR 372.65: Not listed
SARA EHS 40CFR 355: Not listed
TSCA: Listed

ETH5100 CAS #: 105-37-3

ETHYL PROPIONATE

RTECS: UF3675000
DOT: UN1195
EINECS Number: 203-291-4
Molecular Formula: $C_5H_{10}O_2$
Structured MF: $C_2H_5COOC_2H_5$
Formula Weight: 102.15

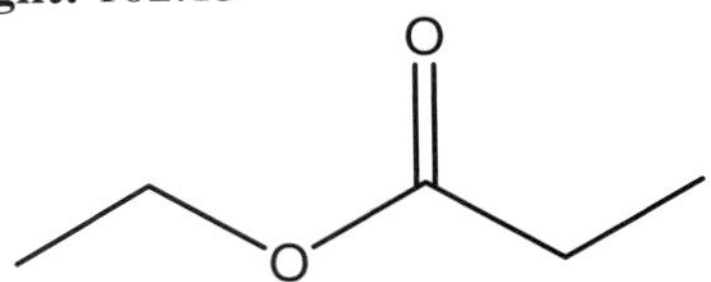

Chemical Structure

Synonyms: ETHYL PROPANOATE; ETHYLESTER KYSELINY PROPIONOVE; PROPANOIC ACID,ETHYL ESTER; PROPIONATE D'ETHYLE; PROPIONIC ACID,ETHYL ESTER; PROPIONIC ESTER; PROPIONIC ETHER

Description: colorless liquid; reminiscent of rum and pineapple odor

Use: solvent for various natural & synthetic resins; cutting agent for pyroxylin; fragrance for chewing gum, baked goods & candy; solvent for cellulose ethers & esters; chem int for pyrimethamine

Physical Properties

Boiling Point: 99 °C (210 °F)
Freezing Point: -73 °C (-99.4 °F)
Specific Gravity: 0.891 at 20 °C/4 °C
Vapor Density: 3.52 Air=1
Saturated Vapor Density: 1.348506178 kg/m^3
Vapor Pressure: 4.97 kPa at 25 °C
Water Solubility: 1 parts in about 60 parts Water
Other Solubilities: Soluble in most organic solvents.
Odor Threshold: Recognition 0.3 to 0.5 mg/m^3
Refraction Index: 1.3844 at 20 °C/D
Ionization Potential (eV): 10.00 +/-0.5
Flash Point: 12 °C Closed Cup
Autoignition Temperature: 440 °C
LEL: 1.9% v/v
UEL: 11% v/v

RTECS Toxicity Data

Acute Oral: Rat LD_{50} Dose: 8732 mg/kg; Toxic Effects: Behavioral - Somnolence (general depressed activity); Lungs, Thorax, or Respiration - Other changes; Gastrointestinal - Other changes.

Acute Dermal: Rabbit LD_{Lo} Route: Skin; Dose: 14256 mg/kg; Toxic Effects: Lungs, Thorax, or Respiration - Other changes.

Irritation Eye: Rabbit Standard Draize Test Dose: 100 mg; Reaction: moderate.
Irritation Skin: Rabbit Standard Draize Test Dose: 500 mg/24H; Reaction: moderate.

Hazard Overviews

Flammable

Health: Irritating to eyes/skin/respiratory tract. Other Acute Effects: may be harmful by inhalation, ingestion, or skin absorption; prolonged exposure can cause: narcotic effect gastrointestinal disturbances coughing, chest pains, difficulty in breathing.
Fire: Flammable. Hazards: vapor may travel considerable distance to source of ignition and flash back; container explosion may occur. Extinguishing agents: carbon dioxide, dry chemical powder or appropriate foam; water may be effective for cooling, but may not effect extinguishment. Precautions: combustible liquid.
Reactivity: Stable. Incompatible with: oxidizing agents, bases, acids, moisture, heat. Hazardous decomposition products: toxic fumes of: carbon monoxide, carbon dioxide.
Carcinogenicity: IARC - Not listed; NIOSH - Not listed; NTP - Not listed; ACGIH - Not listed; OSHA - Not listed; EPA - Not listed; MAK - Not listed
Primary Target Organs:

Eyes

Skin

Respiratory System

Environmental

Ecotoxicity: Toxicity threshold (cell multiplication inhibition test): green algae (Scenedesmus quadricauda): 7d EC_0 14 mg/l
Cleanup/Disposal: Guide No. 129: Eliminate all ignition sources (no smoking, flares, sparks or flames in immediate area). All equipment used when handling the product must be grounded. Do not touch or walk through spilled material. Stop leak if you can do it without risk. Prevent entry into waterways, sewers, basements or confined areas. A vapor suppressing foam may be used to reduce vapors. Absorb or cover with dry earth, sand or other non-combustible material and transfer to containers. Use clean non-sparking tools to collect absorbed material. Large Spills: Dike far ahead of liquid spill for later disposal. Water spray may reduce vapor; but may not prevent ignition in closed spaces.

Environmental Physical Data

Octanol/Water Partition Coefficient: log K_{ow} = 1.21

Regulations

RCRA 40CFR: Not listed
CERCLA: 40CFR 302.4: Not listed
SARA 40CFR 372.65: Not listed
SARA EHS 40CFR 355: Not listed
TSCA: Listed

ETH5150 CAS #: 78-10-4

ETHYL SILICATE

RTECS: VV9450000
DOT: UN1292; IMO3.3
EINECS Number: 201-083-8
Molecular Formula: $C_8H_{20}O_4Si$
Structured MF: $(C_2H_5)_4SiO_4$
Formula Weight: 208.30

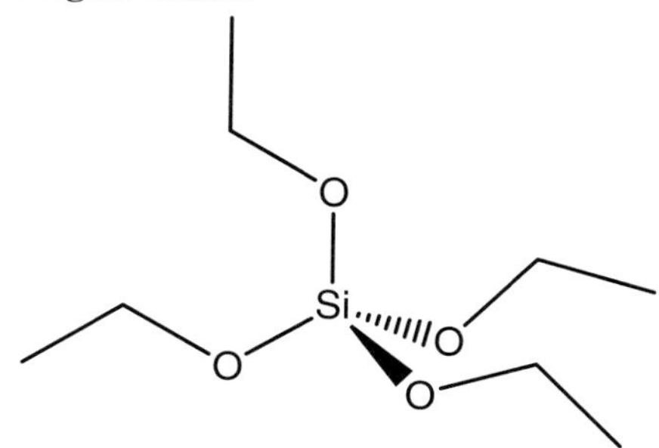

Chemical Structure

Synonyms: DYNASIL A; ES 100; ES 28; ES 28 (ESTER); ETHYL ORTHOSILICATE; ETHYL SILICATE (CONDENSED); ETYLU KRZEMIAN; ORTHOSILICIC ACID,TETRAETHYL ESTER; SILANE,TETRAETHOXY-; SILICATE D'ETHYLE; SILICATE TETRAETHYLIQUE; SILICIC ACID,TETRAETHYL ESTER; SILICON ETHOXIDE; SILICON TETRAETHOXIDE; SILIKAN L; TEOS; TETRAETHOXYSILANE; TETRAETHOXYSILICON; TETRAETHYL ORTHOSILICATE; TETRAETHYL SILICATE; TETRAETHYL-O-SILICATE; TETRAETHYLSILIKAT
Description: colorless liquid; alcohol-like odor
Use: in weatherproof & acid-proof mortars & cements; for hardening stone, arresting decay & disintegration; in lost wax process for casting of alloys; refractory bricks, other molded objects, heat-resistant & chem-resistant paints, protective coatings, lacquers; bonding agent intermed; source of silicon oxide; in ceramics, glass frosting, & low heat glasses; in prodn of dental investments; investment castings for copper alloys; in prodn of inorganic zinc resistant coatings.

Physical Properties

Boiling Point: 168.8 °C (336 °F) at 760 mm Hg
Freezing Point: -82.5 °C (-116.5 °F)
Specific Gravity: 0.933 at 20 °C/4 °C
Vapor Density: 7.22 Air=1
Saturated Vapor Density: 1.20976225 kg/m^3
Bulk Density: 7.8 lbs/gal at 20 °C
Vapor Pressure: 1 mm Hg at 20 °C
Water Solubility: Practically Insoluble in Water
Other Solubilities: Soluble in Ether; Slightly Soluble in Benzene; miscible with Alcohol.
Surface Tension: 22.8 dynes/cm at 20 °C
Odor Threshold: 85 ppm
Refraction Index: 1.3818 at 25 °C/D
Ionization Potential (eV): 9.77
Flash Point: 51.667 °C Open Cup

LEL: 1.3% v/v
UEL: 23% v/v

RTECS Toxicity Data

Acute Oral: Rat LD_{50} Dose: 6270 mg/kg.
Acute Inhalation: Rat LC_{Lo} Dose: 1000 ppm/4hr. Mouse LC_{Lo} Dose: 1 gm/m^3/2hr; Toxic Effects: Behavioral - General anesthetic; Lungs, Thorax, or Respiration - Structural or functional change in trachea or bronchi; Kidney, ureter, and Bladder - Chgs in tubules (inc acute renal failure).
Acute Dermal: Rabbit LD_{50} Route: Skin; Dose: 6300 uL/kg.
Chronic (Multiple Dose) Inhalation: Rat Dose: 1000 ppm/7H/3D-I; Toxic Effects: Kidney, Ureter, and Bladder - Chgs in tubules (inc acute renal failure, acute tubular necrosis; Kidney, Ureter, and Bladder - Hematuria; DEATH. Mouse Dose: 200 ppm/6H/4W-I; Toxic Effects: Sense organs and special senses - Other; Kidney, Ureter, and Bladder - Chgs in tubules (inc acute renal failure, acute tubular necrosis; Nutritional and gross metabolic - Weight loss or decreased weight gain.
Irritation Eye: Rabbit Standard Draize Test Dose: 100 mg; Reaction: mild. Rabbit Standard Draize Test Dose: 500 mg/24H; Reaction: mild.
Irritation Skin: Rabbit Standard Draize Test Dose: 500 mg/24H; Reaction: moderate.

Hazard Overviews

Flammable

Fire Diamond

Health: Severely irritating to eyes. Irritating to skin/respiratory tract. Also Causes: shortness of breath, wheezing, contact dermatitis and redness due to ethyl silicate's defatting solvent effect.
Fire: Flammable. Can form explosive mixtures in the air. Use carbon dioxide, water sprays, foams or dry chemicals.
Reactivity: Stable. Hazardous polymerization cannot occur. Avoid: exposure to heat and ignition sources such as open flames; lighted tobacco products; electrical/mechanical sparks. Incompatible with: strong oxidizing agents. Hazardous decomposition products: acrid smoke and fumes; carbon dioxide; toxic carbon monoxide; silicon compounds.
Carcinogenicity: IARC - Not listed; NIOSH - Listed as carcinogen; NTP - Not listed; ACGIH - Not listed; OSHA - Not listed; EPA - Not listed; MAK - Not listed
Primary Target Organs:

Eyes

Skin

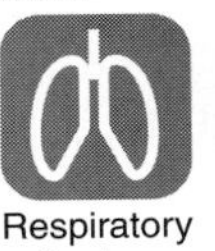
Respiratory System

Mucous Membranes

Exposure Limits
OSHA PEL: TWA: 100 ppm; 850 mg/m^3.
OSHA PEL Vacated 1989 Limits: TWA: 10 ppm; 85 mg/m^3.
ACGIH TLV: TWA: 10 ppm; 85 mg/m^3.
NIOSH REL: TWA: 10 ppm; 85 mg/m^3.
NIOSH IDLH: 700 ppm.
DFG MAK: TWA: 20 ppm; 170 mg/m^3.
Respirator Recommendation
Exposure Range: >100 to <700 ppm Air Purifying, Negative Pressure, Half Mask
Exposure Range: 700 to unlimited ppm Self-contained Breathing Apparatus, Pressure Demand, Full Face
Cartridge Color: black

Environmental

Cleanup/Disposal: Guide No. 132: Fully encapsulating, vapor protective clothing should be worn for spills and leaks with no fire. Eliminate all ignition sources (no smoking, flares, sparks or flames in immediate area). All equipment used when handling the product must be grounded. Do not touch or walk through spilled material. Stop leak if you can do it without risk. Prevent entry into waterways, sewers, basements or confined areas. A vapor suppressing foam may be used to reduce vapors. Absorb with earth, sand or other non-combustible material and transfer to containers (except for Hydrazine). Use clean non-sparking tools to collect absorbed material. Large Spills: Dike far ahead of liquid spill for later disposal. Water spray may reduce vapor; but may not prevent ignition in closed spaces.

Environmental Physical Data

BCF: no food chain concentration potential

Regulations

RCRA 40CFR: Not listed
CERCLA: 40CFR 302.4: Not listed
SARA 40CFR 372.65: Not listed
SARA EHS 40CFR 355: Not listed
TSCA: Listed

ETH5250 CAS #: 20941-65-5

ETHYL TELLURAC

RTECS: WY2950000
EINECS Number: 244-121-9
Molecular Formula: $C_{20}H_{40}N_4S_8Te$
Formula Weight: 720.6
Synonyms: CARBAMODITHIOIC ACID,DIETHYL-,TETRAKIS(ANHYDROSULFIDE)WITH THIOTELLURIC ACID; CARBAMODITHIOIC ACID,DIETHYL-,TETRAKIS(ANHYDROSULFIDE)WITH THIOTELLURIC ACID (H4TES4); DIETHYLDITHIO CARBAMIC ACID TELLURIUM SALT; DIETHYLDITHIOCARBAMIC ACID TELLURIUM SALT; ETHYLTELLURAC; NOCCELER TTTE; TELLURAC; TELLURIUM BIS(DIETHYLDITHIOCARBAMATE); TELLURIUM DIETHYLDITHIOCARBAMATE; TELLURIUM(IV) DIETHYL DITHIOCARBAMATE; TELLURIUM,TETRAKIS(DIETHYLCARBAMODITHIOATEO-S,S')-,(DD-8-111"1"1'1'1'"1"')-; TELLURIUM,TETRAKIS(DIETHYLDITHIOCARBAMATE)-; TELLURIUM,TETRAKIS(DIETHYLDITHIOCARBAMATO)-; TETRAKIS(DIETHYLCARBAMODITHIOATO-S,S')TELLURIUM; TETRAKIS(DIETHYLDITHIOCARBAMATO)TELLURIUM
Description: yellow to orange-yellow powder
Use: a primary or secondary rubber vulcanization accelerator

Physical Properties

Freezing Point: 108 °C (226.4 °F) to 118 °C (244.4 °F)
Density: 1.44 g/mL
Water Solubility: Insoluble in Water
Other Solubilities: 95% Ethanol: <1 mg/ml at 30 °C; Acetone: 1-5 mg/ml at 16 °C; Benzene: Soluble; Carbon Disulfide: Soluble; Chloroform: Soluble; DMSO: 1-10 mg/ml at 30 °C; Gasoline: Slightly Soluble.
Flash Point: Not available; probably combustible

RTECS Toxicity Data

Tumorigenic: Mouse Route: Oral; Dose: 113 gm/kg/107W-C; Toxic Effects: Tumorigenic - Equivocal tumorigenic agent by RTECS criteria; Sense organs and special senses - Tumors.

Hazard Overviews

Fire: Will burn.
Carcinogenicity: IARC - Group 3, Not classifiable as to carcinogenicity to humans; NIOSH - Not listed; NTP - Not listed; ACGIH - Not listed; OSHA - Not listed; EPA - Not listed; MAK - Not listed
Exposure Limits
OSHA PEL: TWA: 0.1 mg/m^3; as Te.
ACGIH TLV: TWA: 0.1 mg/m^3; as Te.
NIOSH REL: TWA: 0.1 mg/m^3; as Te.
DFG MAK: TWA: 0.1 mg/m^3; as Se.

Environmental

Regulations

RCRA 40CFR: Not listed
CERCLA: 40CFR 302.4: Not listed
SARA 40CFR 372.65: Not listed
SARA EHS 40CFR 355: Not listed
TSCA: Listed

ETH5300 CAS #: 542-90-5

ETHYL THIOCYANATE

RTECS: XK9900000
EINECS Number: 208-833-3
Molecular Formula: C_3H_5NS
Formula Weight: 87.14

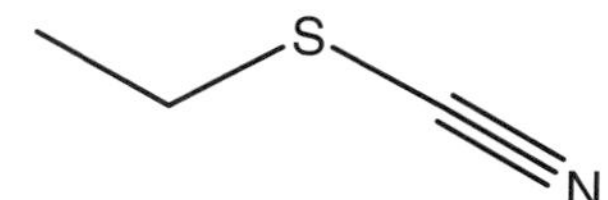

Chemical Structure

Synonyms: AETHYLRHODANID; ETHANE,THIOCYANATO-; ETHYL RHODANATE; ETHYL SULFOCYANATE; ETHYLTHIOKYANAT; THIOCYANIC ACID,ETHYL ESTER
Description: volatile liquid
Use: agricultural insecticide; sometimes employed as insecticidal fumigants; on the skin as delousing preparations and as livestock sprays

Physical Properties

Boiling Point: 146 °C (295 °F)
Freezing Point: -85.5 °C (-121.9 °F)
Specific Gravity: 1.0071 at 22 °C/4 °C
Vapor Pressure: 8.34 kPa at 75 °C
Water Solubility: Insoluble in Water
Other Solubilities: miscible with Alcohol, Ether.
Refraction Index: 1.4684 at 15 °C/D

RTECS Toxicity Data

Acute Oral: Rat LD_{Lo} Dose: 201 mg/kg; Toxic Effects: Behavioral - Somnolence (general depressed activity); Lungs, Thorax, or Respiration - Dyspnea; Lungs, Thorax, or Respiration - Cyanosis. Cat LD_{Lo} Dose: 10 mg/kg; Toxic Effects: Behavioral - Somnolence (general depressed activity); Lungs, Thorax, or Respiration - Other changes; Gastrointestinal - Other changes.
Acute Dermal: Rabbit LD_{Lo} Route: Subcutaneous Dose: 15 mg/kg; Toxic Effects: Behavioral - Convulsions or effect on seizure threshold; Behavioral - Ataxia; Behavioral - Muscle contraction or spasticity. Rat LD_{Lo} Route: Skin; Dose: 2512 mg/kg; Toxic Effects: Behavioral - Somnolence (general depressed activity). Rat LD_{Lo} Route: Subcutaneous Dose: 40 mg/kg; Toxic Effects: Behavioral - Somnolence (general depressed activity); Lungs, Thorax, or Respiration - Dyspnea; Lungs, Thorax, or Respiration - Cyanosis.
Chronic (Multiple Dose) Oral: Rat Dose: 804 mg/kg/2W-I; Toxic Effects: Lungs, Thorax, or Respiration - Other changes; Liver - Other changes; Kidney, Ureter, and Bladder - Changes in both tubules and glomeruli.

Hazard Overviews

Poison

Health: Severely irritating. Poison. Other Acute Effects: may be fatal if inhaled, swallowed, or absorbed through skin; high concentrations are extremely destructive to tissues of the mucous membranes and upper respiratory tract, eyes and skin; symptoms of exposure may include burning sensation, coughing, wheezing, laryngitis, shortness of breath, headache, nausea and vomiting.
Fire: Hazards: emits toxic fumes. Extinguishing agents: water spray; carbon dioxide, dry chemical powder or appropriate foam. Precautions: combustible liquid.
Reactivity: Incompatible with: strong oxidizing agents, strong bases, strong acids, may decompose on exposure to moist air or water. Hazardous decomposition products: toxic fumes of: carbon monoxide, carbon dioxide, nitrogen oxides, sulfur oxides.
Carcinogenicity: IARC - Not listed; NIOSH - Not listed; NTP - Not listed; ACGIH - Not listed; OSHA - Not listed; EPA - Not listed; MAK - Not listed

Primary Target Organs:

Environmental

Regulations

RCRA 40CFR: Not listed
CERCLA: 40CFR 302.4: Listed as Compound per CWA Section 307(a) per CAA Section 112
SARA 40CFR 372.65: Not listed TPQ: 10000 lb
SARA EHS 40CFR 355: Listed TPQ: 10000 lb
TSCA: Listed

ETH5350 CAS #: 121-32-4

ETHYL VANILLIN

RTECS: CU6125000
EINECS Number: 204-464-7
Molecular Formula: $C_9H_{10}O_3$
Formula Weight: 166.18

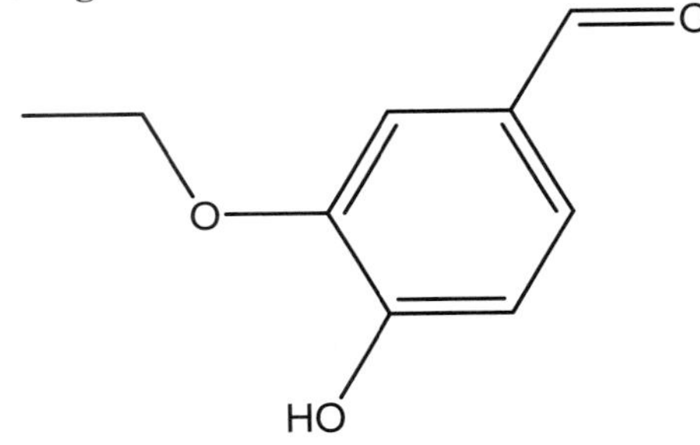

Chemical Structure

Synonyms: BENZALDEHYDE,3-ETHOXY-4-HYDROXY-; BOUBONAL; BOURBONAL; ETHAVAN; ETHOVAN; 3-ETHOXY-4-HYDROXYBENZALDEHYDE; ETHYLPROTAL; ETHYLPROTOCATECHUIC ALDEHYDE; ETHYLVANILLIN; 4-HYDROXY-3-ETHOXYBENZALDEHYDE; PROTOCATECHUIC ALDEHYDE ETHYL ETHER; QUANTROVANIL; VANILLAL; VANILLIN,ETHYL-; VANIROM; VANIROME
Description: white or slightly yellowish crystals or colorless flakes; intense vanilla odor
Use: in flavoring and perfumery

Physical Properties

Boiling Point: 285 °C (545 °F)
Freezing Point: 77 °C (170.6 °F) to 78 °C (172.4 °F)
Water Solubility: 1.3% at 50 °C
Other Solubilities: 95% Ethanol: Soluble in hot solvent; Benzene: Soluble; Chloroform: Soluble; Ether: Soluble; Glycerol: Soluble; Propylene Glycol: Soluble.
pH: Solutions are acid
Refraction Index: Alpha 1.485
Flash Point: Low, combustible

RTECS Toxicity Data

Acute Oral: Rat LD_{50} Dose: 1590 mg/kg. Rabbit LD_{Lo} Dose: 3 gm/kg.
Acute Dermal: Rat LD_{Lo} Route: Subcutaneous Dose: 1800 mg/kg.
Chronic (Multiple Dose) Oral: Rat Dose: 4480 mg/kg/70D-I; Toxic Effects: Cardiac - Other changes; Blood - Changes in spleen; Nutritional and gross metabolic - Weight loss or decreased weight gain.
Irritation Skin: Human Standard Draize Test Dose: 10 mg/48H; Reaction: mild.
Mutagenic: Human Sister Chromatid Exchenge; Cell Type: lymphocyte; Dose: 1 mmol/L. Hamster Cytogenetic Analysis; Cell Type: fibroblast; Dose: 250 mg/L.

Hazard Overviews

Health: Irritating to eyes/skin/respiratory tract. Other Acute Effects: harmful if swallowed; may be harmful if inhaled or absorbed through the skin.
Fire: Will burn. Hazards: emits toxic fumes. Extinguishing agents: water spray; carbon dioxide, dry chemical powder or appropriate foam. Precautions: combustible liquid.
Reactivity: Incompatible with: strong oxidizing agents, strong bases, sensitive to light. Hazardous decomposition products: toxic fumes of: carbon monoxide, carbon dioxide.
Carcinogenicity: IARC - Not listed; NIOSH - Not listed; NTP - Not listed; ACGIH - Not listed; OSHA - Not listed; EPA - Not listed; MAK - Not listed
Primary Target Organs:

Environmental

Cleanup/Disposal: Guide No. 171: Do not touch or walk through spilled material. Stop leak if you can do it without risk. Prevent dust cloud. Avoid inhalation of asbestos dust. Small Dry Spills: With clean shovel place material into clean, dry container and cover loosely; move containers from spill area. Small Spills: Take up with sand or other noncombustible absorbent material and place into containers for later disposal. Large Spills: Dike far ahead of liquid spill for later disposal. Cover powder spill with plastic sheet or tarp to minimize spreading. Prevent entry into waterways, sewers, basements or confined areas.

Regulations

RCRA 40CFR: Not listed
CERCLA: 40CFR 302.4: Not listed
SARA 40CFR 372.65: Not listed
SARA EHS 40CFR 355: Not listed
TSCA: Listed

ETH5400 **CAS #: 12075-68-2**

ETHYLALUMINUM SESQUICHLORIDE

RTECS: BD1950000
DOT: UN1925; UN2220; UN2221; UN3052; IMO4.2
EINECS Number: 235-137-7
Molecular Formula: $C_6H_{15}Al_2Cl_3$
Structured MF: $(C_2H_5)_3Al_2Cl_3$
Formula Weight: 247.5

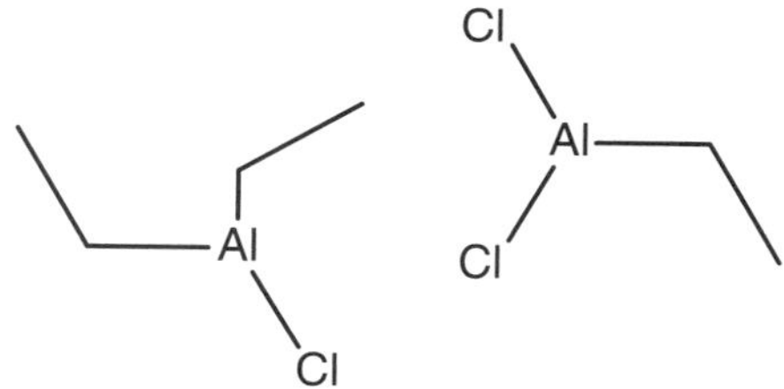

Chemical Structure

Synonyms: ALUMINUM,CHLORODIETHYL-,MIXT WITH DICHLOROETHYLALUMINUM; ALUMINUM,DICHLOROETHYL-,MIXT WITH CHLORODIETHYLALUMINUM; ALUMINUM,TRICHLOROTRIETHYLDI-; EASC; ETHYL ALUMINUM SESQUICHLORIDE; SESQUIETHYLALUMINUM CHLORIDE; TRICHLOROTRIETHYLDIALUMINIUM; TRICHLOROTRIETHYLDIALUMINUM; TRIETHYLALUMINUM SESQUICHLORIDE; TRIETHYLDIALUMINUM TRICHLORIDE; TRIETHYLTRICHLORODIALUMINUM
Description: yellow, clear liquid
Use: catalyst for olefin polymerization; aromatic hydrogenation; intermediate

Physical Properties

Boiling Point: 204 °C (399 °F)
Freezing Point: -20 °C (-4 °F)
Vapor Density: 8.49 Air=1
Density: 1.092 g/cu cm at 25 °C
Vapor Pressure: 0.30 mm Hg
Water Solubility: Reacts Violently with Water
Other Solubilities: Soluble in toluene
Surface Tension: Estimated at 32 dynes/cm
Flash Point: -20 °C Closed Cup

Hazard Overviews

Corrosive

Flammable

Health: Corrosive to eyes/skin/respiratory tract. Toxic. Other Acute Effects: harmful if swallowed, inhaled, or absorbed through skin; inhalation may result in spasm, inflammation and edema of the larynx and bronchi, chemical pneumonitis and pulmonary edema; symptoms of exposure may include burning sensation; coughing; wheezing; laryngitis; shortness of breath; headache; nausea; vomiting. Chronic Effects: may cause nervous system disturbances; inhalation studies have demonstrated the development of inflammatory and ulcerous lesions of the penis, prepuce and scrotum in animals; target organs: brain, liver, kidneys, bladder.
Fire: Flammable. Hazards: pyrophoric; catches fire if exposed to air; container explosion may occur; reacts with water to liberate flammable and/or explosive gas. Extinguishing agents: dry chemical powder; do not use water. Precautions: combustible liquid.
Reactivity: Incompatible with: alcohols, oxygen, oxidizing agents, acids, bases, reacts violently with water. Hazardous decomposition products: toxic fumes of: hydrogen chloride gas, aluminum oxide, carbon monoxide, carbon dioxide.
Carcinogenicity: IARC - Not listed; NIOSH - Not listed; NTP - Not listed; ACGIH - Not listed; OSHA - Not listed; EPA - Not listed; MAK - Not listed
Primary Target Organs:

Eyes

Skin

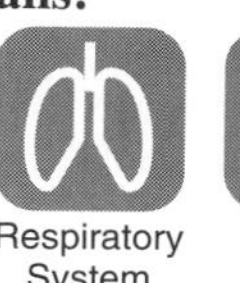
Respiratory System

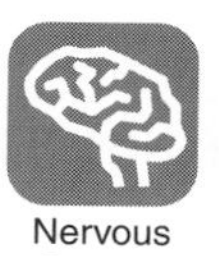
Nervous System

Liver

Kidneys

Exposure Limits
OSHA PEL Vacated 1989 Limits: TWA: 2 mg/m^3; as Al soluble.
ACGIH TLV: TWA: 2 mg/m^3; as Al.
NIOSH REL: TWA: 2 mg/m^3; as Al soluble salts, alkyls.

Environmental

Cleanup/Disposal: Guide No. 135: Fully encapsulating, vapor protective clothing should be worn for spills and leak with no fire. Eliminate all ignition sources (no smoking, flares, sparks or flames in immediate area). Do not touch or walk through spilled material. Stop leak if you can do it without risk. Small Spills: Cover with dry earth, dry sand, or other non-combustible material followed with plastic sheet to minimize spreading or contact with rain. Use clean non-sparking tools to collect material and place it into loosely covered plastic containers for later disposal. Prevent entry into waterways, sewers, basements or confined areas.

Environmental Physical Data

BOD: none

Regulations

RCRA 40CFR: Not listed
CERCLA: 40CFR 302.4: Not listed
SARA 40CFR 372.65: Not listed
SARA EHS 40CFR 355: Not listed
TSCA: Listed

ETH5450 **CAS #: 578-54-1**

2-ETHYLANILINE

RTECS: BX9800000
EINECS Number: 209-424-2
Molecular Formula: $C_8H_{11}N$
Formula Weight: 121.20

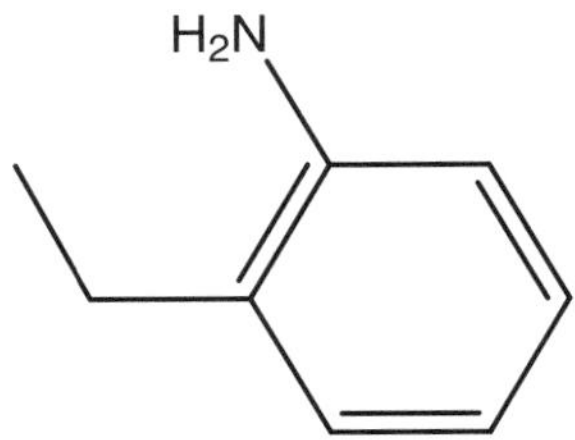

Chemical Structure

Synonyms: O-AMINOETHYLBENZENE; ANILINE,2-ETHYL-; ANILINE,O-ETHYL-; ANILINE,O-ETHYL-(8CI); BENZENAMINE,2-ETHYL-; BENZENAMINE,2-ETHYL-(9CI); 2-ETHYL ANILINE; O-ETHYLANILINE; 2-ETHYLBENZENAMINE

Description: brown liquid

Use: intermediate for pharmaceuticals, dyestuffs, pesticides, and other products

Physical Properties

Boiling Point: 214 °C (417 °F)
Freezing Point: -44 °C (-47.2 °F)
Specific Gravity: 0.982 at 20 °C
Water Solubility: Insoluble in Water
Other Solubilities: Soluble in Alcohol & Toluene
Refraction Index: 1.5584
Flash Point: 85 °C Open Cup

RTECS Toxicity Data

Acute Oral: Rat LD_{50} Dose: 1260 mg/kg; Toxic Effects: Behavioral - Somnolence (general depressed activity); Lungs, Thorax, or Respiration - Cyanosis; Blood - Changes in spleen.

Hazard Overviews

Health: Irritating to eyes/skin/respiratory tract. Toxic. Other Acute Effects: harmful if swallowed, inhaled, or absorbed through skin; causes severe eye irritation; absorption into the body leads to the formation of methemoglobin which in sufficient concentration causes cyanosis; onset may be delayed 2 to 4 hours or longer.

Fire: Combustible. Hazards: emits toxic fumes. Extinguishing agents: water spray; carbon dioxide, dry chemical powder or appropriate foam. Precautions: combustible liquid.

Reactivity: Incompatible with: acids, acid chlorides, acid anhydrides, chloroformates, strong oxidizing agents. Hazardous decomposition products: toxic fumes of: carbon monoxide, carbon dioxide, nitrogen oxides.

Carcinogenicity: IARC - Not listed; NIOSH - Not listed; NTP - Not listed; ACGIH - Not listed; OSHA - Not listed; EPA - Not listed; MAK - Not listed

Primary Target Organs:

Eyes

Skin

Respiratory System

Environmental

Ecotoxicity: Green Algae: Scenedesmus subspicatus growth inhib. IC_{10}: 6.5 mg/l

Regulations

RCRA 40CFR: Not listed
CERCLA: 40CFR 302.4: Not listed
SARA 40CFR 372.65: Not listed
SARA EHS 40CFR 355: Not listed
TSCA: Listed

ETH5500 CAS #: 103-69-5

N-ETHYLANILINE

RTECS: BX9780000
EINECS Number: 203-135-5
Molecular Formula: $C_8H_{11}N$
Formula Weight: 121.18

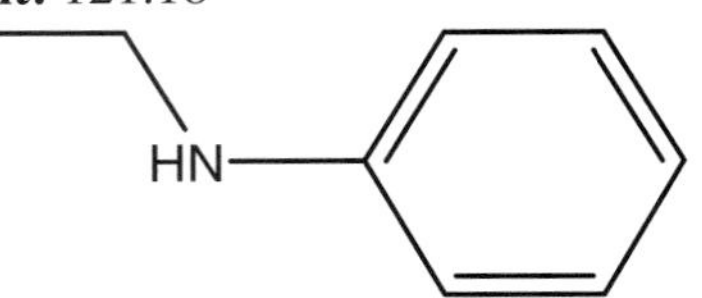

Chemical Structure

Synonyms: AETHYLANILIN; ANILINE,N-ETHYL-; ANILINOETHANE; ANILINOETHANOL; BENZENAMINE,N-ETHYL-; BENZENAMINE,N-ETHYL-(9CI); ETHYL PHENYLAMINE; N-ETHYLAMINOBENZENE; ETHYLANILINE; N-ETHYLBENZENAMINE; N-ETHYLBENZENAMINO; N-ETHYLBENZENEAMINE; N-ETHYLBENZENEAMINO; N-ETHYLOMINOBENZENE; ETHYLPHENYLAMINE; N-ETHYL-N-PHENYLAMINE

Description: colorless clear to straw-colored, yellow-brown oily liquid; aniline-like odor

Use: organic synthesis

Physical Properties

Boiling Point: 204.5 °C (400 °F)
Freezing Point: -63.5 °C (-82.3 °F)
Specific Gravity: 0.9625 at 20 °C/4 °C
Vapor Density: 4.2 Air=1
Vapor Pressure: 1 mm Hg at 38.5 °C
Water Solubility: Insoluble in Water
Other Solubilities: 95% Ethanol: >=100 mg/ml at 22 °C; Acetone: >=100 mg/ml at 22 °C; Benzene: Soluble; DMSO: >=100 mg/ml at 22 °C; Ether: Soluble.
Refraction Index: 1.5559
Ionization Potential (eV): =< 7.67 +/-1.0
Flash Point: 85 °C Open Cup

RTECS Toxicity Data

Acute Inhalation: Rat LC_{50} Dose: >1130 mg/m^3/4hr.
Acute Dermal: Rat LD_{50} Route: Skin; Dose: 4700 mg/kg.

Hazard Overviews

Poison

Health: Irritating to eyes/skin/respiratory tract. Poison. Other Acute Effects: harmful if swallowed, inhaled, or absorbed through skin.
Fire: Combustible. Hazards: emits toxic fumes. Extinguishing agents: carbon dioxide, dry chemical powder or appropriate foam; water spray. Precautions: combustible liquid.
Reactivity: Incompatible with: acids, acid chlorides, acid anhydrides, strong oxidizing agents, carbon dioxide, air sensitive, light sensitive. Hazardous decomposition products: thermal decomposition may produce carbon monoxide, carbon dioxide, and nitrogen oxides.
Carcinogenicity: IARC - Not listed; NIOSH - Not listed; NTP - Not listed; ACGIH - Not listed; OSHA - Not listed; EPA - Not listed; MAK - Not listed
Primary Target Organs:

Eyes

Skin

Respiratory System

Environmental

Ecotoxicity: Algae: Scenedesmus subspicatus 72h EC_{10} 74 mg/l; Crustaceans: Daphnia magna Straus 24h EC_0 12 mg/l
Cleanup/Disposal: Guide No. 153: Eliminate all ignition sources (no smoking, flares, sparks or flames in immediate area). Do not touch damaged containers or spilled material unless wearing appropriate protective clothing. Stop leak if you can do it without risk. Prevent entry into waterways, sewers, basements or confined areas. Absorb or cover with dry earth, sand or other non-combustible material and transfer to containers. Do not get water inside containers.

Environmental Physical Data

Octanol/Water Partition Coefficient: log K_{ow} = measured at 2.01

Regulations

RCRA 40CFR: Not listed
CERCLA: 40CFR 302.4: Not listed
SARA 40CFR 372.65: Not listed
SARA EHS 40CFR 355: Not listed
TSCA: Listed

ETH5550 CAS #: 98-69-1

4-ETHYLBENZENESULFONIC ACID

EINECS Number: 202-693-7
Molecular Formula: $C_8H_{10}O_3S$
Formula Weight: 186.24

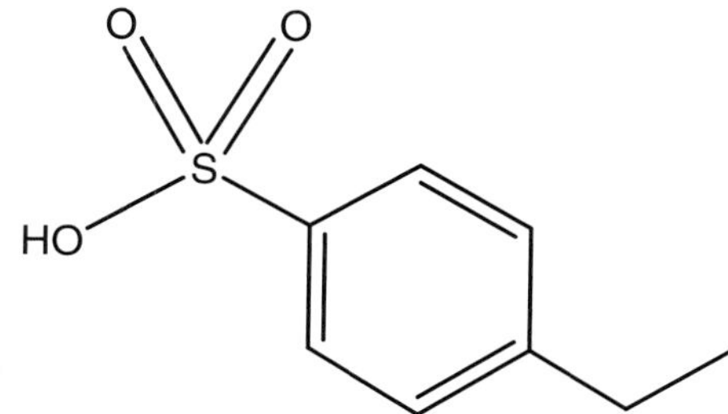

Chemical Structure

Synonyms: BENZENESULFONIC ACID,4-ETHYL-; BENZENESULFONIC ACID,P-ETHYL-; P-ETHYLBENZENESULFONIC ACID; PHENYLETHANE-P-SULFONATE
Use: in wastewater treatment

Physical Properties

Specific Gravity: 1.229
Flash Point: > 110 °C

Hazard Overviews

Corrosive

Health: Corrosive to eyes/skin/respiratory tract. Other Acute Effects: harmful if swallowed, inhaled, or absorbed through skin; extremely destructive to tissue of the mucous membranes and upper respiratory tract, eyes and skin; inhalation may result in spasm, inflammation and edema of the larynx and bronchi, chemical pneumonitis and pulmonary edema; symptoms of exposure may include burning sensation, coughing, wheezing, laryngitis, shortness of breath, headache, nausea and vomiting.
Fire: Will burn. Hazards: emits toxic fumes. Extinguishing agents: carbon dioxide, dry chemical powder or appropriate foam. Precautions: combustible liquid.
Carcinogenicity: IARC - Not listed; NIOSH - Not listed; NTP - Not listed; ACGIH - Not listed; OSHA - Not listed; EPA - Not listed; MAK - Not listed
Primary Target Organs:

Eyes

Skin

Respiratory System

Environmental

Regulations

RCRA 40CFR: Not listed
CERCLA: 40CFR 302.4: Not listed
SARA 40CFR 372.65: Not listed
SARA EHS 40CFR 355: Not listed
TSCA: Listed

ETH5600 CAS #: 92-59-1

N-ETHYL-N-BENZYLANILINE

EINECS Number: 202-169-8
Molecular Formula: $C_{15}H_{17}N$
Formula Weight: 211.29

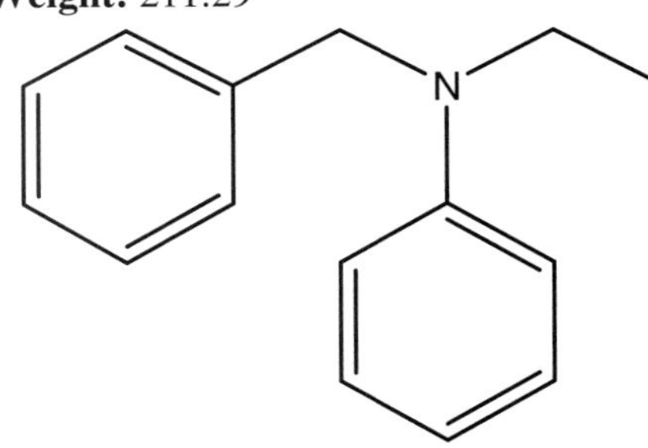

Chemical Structure

Synonyms: AMINE,BENZYL ETHYL PHENYL; BENZENEMETHANAMINE,N-ETHYL-N-PHENYL-; BENZYLAMINE,N-ETHYL-N-PHENYL-; BENZYLETHYL PHENYLAMINE; N-BENZYL-N-ETHYLANILINE; BENZYLETHYLPHENYLAMINE; ETHYL BENZYLANILINE; ETHYLBENZYLANILINE; N-ETHYL-N-PHENYLBENZENEMETHANAMINE; N-ETHYL-N-PHENYLBENZYLAMINE; PHENYLETHYLBENZYLAMINE
Description: colorless to light yellow, oily liquid or pale yellow crystals
Use: intermediate for triphenylmethane dyestuff synthesis; in organic synthesis and as a photographic intermediate

Physical Properties

Boiling Point: 287 °C (549 °F) at 710 mm Hg
Freezing Point: 34 °C (93.2 °F)
Specific Gravity: 1.034 at 19 °C/4 °C
Density: 7.28
Water Solubility: Practically Insoluble in Water
Other Solubilities: 95% Ethanol: >=100 mg/ml at 22 °C; Acetone: >=100 mg/ml at 22 °C; DMSO: >=100 mg/ml at 22 °C; Ether: Soluble.
Refraction Index: 1.5938 at 23 °C/D
Flash Point: 140 °C Open Cup
Autoignition Temperature: < 500 °C

Hazard Overviews

Health: Irritating to eyes/skin/respiratory tract. Harmful. Other Acute Effects: may be harmful by inhalation, ingestion, or skin absorption.
Fire: Will burn. Hazards: emits toxic fumes. Extinguishing agents: water spray; carbon dioxide, dry chemical powder or appropriate foam. Precautions: combustible liquid.
Carcinogenicity: IARC - Not listed; NIOSH - Not listed; NTP - Not listed; ACGIH - Not listed; OSHA - Not listed; EPA - Not listed; MAK - Not listed

Primary Target Organs:

Eyes

Skin

Respiratory System

Environmental

Regulations

RCRA 40CFR: Not listed
CERCLA: 40CFR 302.4: Not listed
SARA 40CFR 372.65: Not listed
SARA EHS 40CFR 355: Not listed
TSCA: Listed

ETH5650 CAS #: 538-07-8

ETHYLBIS(2-CHLOROETHYL)AMINE

RTECS: YE1225000
Molecular Formula: $C_6H_{13}Cl_2N$
Formula Weight: 170.08
Synonyms: BIS(2-CHLOROETHYL)ETHYLAMINE; 2-CHLORO-N-(2-CHLOROETHYL)-N-ETHYLETHANAMINE; 2,2'-DICHLOROTRIETHYLAMINE; ETHANAMINE,2-CHLORO-N-(2-CHLOROETHYL)-N-ETHYL-; ETHYL-S; ETHYLBIS(BETA-CHLOROETHYL)AMINE; HN1; TL 1149; TL 329; TRIETHYLAMINE,2,2'-DICHLORO-
Description: liquid; faint fishy, amine odor

Physical Properties

Boiling Point: Trans 48 °C (118 °F) to 49 °C (119.3 °F)
Freezing Point: -34 °C (-29.2 °F)
Specific Gravity: 1.0861 at 23 °C/ 4 °C
Saturated Vapor Density: 2.582635209 kg/m^3
Vapor Pressure: 180 to 265 mm Hg at 20 °C
Water Solubility: Practically Insoluble in Water
Other Solubilities: miscible with many organic solvents
Refraction Index: 1.4653 at 25 °C/D
Flash Point: About 18 to 21 °C

RTECS Toxicity Data

Acute Oral: Rat LD_{50} Dose: 2500 ug/kg.
Acute Dermal: Rabbit LD_{50} Route: Skin; Dose: 15 mg/kg. Rat LD_{50} Route: Skin; Dose: 17 mg/kg.

Hazard Overviews

Flammable

Fire: Flammable.
Carcinogenicity: IARC - Not listed; NIOSH - Not listed; NTP - Not listed; ACGIH - Not listed; OSHA - Not listed; EPA - Not listed; MAK - Not listed

Environmental

Regulations
RCRA 40CFR: Not listed
CERCLA: 40CFR 302.4: Not listed
SARA 40CFR 372.65: Not listed TPQ: 500 lb
SARA EHS 40CFR 355: Listed TPQ: 500 lb
TSCA: Listed

ETH5700 CAS #: 3953-10-4

2-ETHYLBUTYL ACRYLATE

RTECS: AT0300000
EINECS Number: 223-546-3
Molecular Formula: $C_9H_{16}O_2$
Formula Weight: 156.25
Synonyms: ACRYLIC ACID,2-ETHYLBUTYL ESTER; 2-ETHYLBUTYLACRYLATE; 2-ETHYLBUTYLESTER KYSELINY AKRYLOVE; 2-PROPENOIC ACID,2-ETHYLBUTYL ESTER; 2-PROPENOIC ACID,2-ETHYLBUTYL ESTER (9CI)
Use: monomer

Physical Properties
Boiling Point: 82 °C (180 °F)
Freezing Point: -56.7 °C (-70.06 °F)
Specific Gravity: 0.8964
Vapor Density: 5.4
Saturated Vapor Density: 1.211778131 kg/m^3
Vapor Pressure: 1.7 mm Hg at 20 °C
Water Solubility: Insoluble in Water
Other Solubilities: Reacts readily with electrophilic, free-radical, and nucleophilic agent
Flash Point: 52 °C Open Cup

RTECS Toxicity Data
Acute Oral: Rat LD_{50} Dose: 6490 mg/kg.
Acute Dermal: Rabbit LD_{50} Route: Skin; Dose: 5500 uL/kg.
Irritation Eye: Rabbit Standard Draize Test Dose: 500 mg/24H; Reaction: mild.
Irritation Skin: Rabbit Standard Draize Test Dose: 20 mg/24H; Reaction: moderate. Rabbit Open Draize Test Dose: 10 mg/24H open; Reaction: severe.

Hazard Overviews

Flammable

Fire Diamond

Fire: Flammable.
Carcinogenicity: IARC - Not listed; NIOSH - Not listed; NTP - Not listed; ACGIH - Not listed; OSHA - Not listed; EPA - Not listed; MAK - Not listed

Environmental

Regulations
RCRA 40CFR: Not listed
CERCLA: 40CFR 302.4: Not listed
SARA 40CFR 372.65: Not listed
SARA EHS 40CFR 355: Not listed
TSCA: Listed

ETH5750 CAS #: 32900-06-4

ETHYLBUTYL THIOUREA

EINECS Number: 251-285-5
Molecular Formula: $C_7H_{16}N_2S$
Formula Weight: 150.19

Hazard Overviews
Carcinogenicity: IARC - Not listed; NIOSH - Not listed; NTP - Not listed; ACGIH - Not listed; OSHA - Not listed; EPA - Not listed; MAK - Not listed

Environmental

Regulations
RCRA 40CFR: Not listed
CERCLA: 40CFR 302.4: Not listed
SARA 40CFR 372.65: Not listed
SARA EHS 40CFR 355: Not listed
TSCA: Listed

ETH5800 CAS #: 88-09-5

2-ETHYLBUTYRIC ACID

RTECS: ET1400000
EINECS Number: 201-796-4
Molecular Formula: $C_6H_{12}O_2$
Formula Weight: 116.18

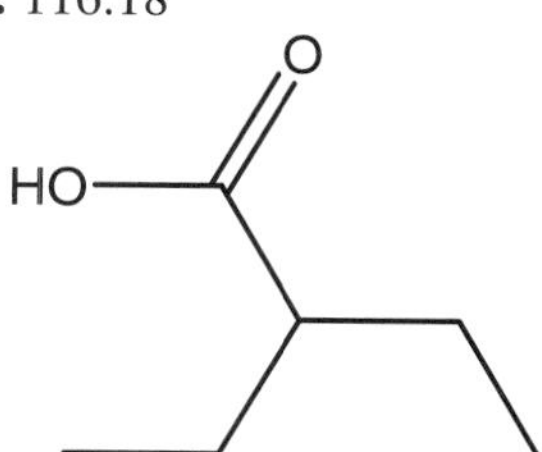

Chemical Structure

Synonyms: ACETIC ACID,DIETHYL-; DIETHYLACETIC ACID; 2-ETHYL BUTANOIC ACID; ALPHA-ETHYLBUTYRIC ACID; KYSELINA DIETHYLOCTOVA; 3-PENTANECARBOXYLIC ACID

Physical Properties
Boiling Point: 99 °C (210.2 °F) to 101 °C (213.8 °F) at 18 mm Hg

Freezing Point: -14 °C (6.8 °F)
Specific Gravity: 0.92390
Bulk Density: 7.7 lbs/gal
Vapor Pressure: 2.41 kPa at 100 °C
Water Solubility: 18 g/l
Other Solubilities: Ethanol: Miscible; Ether: Miscible; Carbon Tetrachloride: Slightly soluble
Refraction Index: 1.4132
Flash Point: 98 °C

RTECS Toxicity Data

Acute Oral: Rat LD_{50} Dose: 2200 mg/kg.
Acute Dermal: Rabbit LD_{50} Route: Skin; Dose: 520 uL/kg.
Irritation Eye: Rabbit Standard Draize Test Dose: 250 ug open; Reaction: severe.
Irritation Skin: Rabbit Standard Draize Test Dose: 500 mg/24H; Reaction: mild. Rabbit Open Draize Test Dose: 10 mg/24H open; Reaction: mild.

Hazard Overviews

Corrosive

Health: Corrosive to eyes/skin/respiratory system. Harmful. Other Acute Effects: harmful if swallowed, inhaled, or absorbed through skin; inhalation may result in spasm, inflammation and edema of the larynx and bronchi, chemical pneumonitis and pulmonary edema; symptoms of exposure may include burning sensation, coughing, wheezing, laryngitis, shortness of breath, headache, nausea and vomiting.
Fire: Will burn. Extinguishing agents: carbon dioxide, dry chemical powder or appropriate foam; water spray. Precautions: combustible liquid.
Reactivity: Incompatible with: bases, oxidizing agents, reducing agents. Hazardous decomposition products: toxic fumes of: carbon monoxide, carbon dioxide.
Carcinogenicity: IARC - Not listed; NIOSH - Not listed; NTP - Not listed; ACGIH - Not listed; OSHA - Not listed; EPA - Not listed; MAK - Not listed
Primary Target Organs:

Eyes

Skin

Respiratory System

Environmental

Regulations

RCRA 40CFR: Not listed
CERCLA: 40CFR 302.4: Not listed
SARA 40CFR 372.65: Not listed
SARA EHS 40CFR 355: Not listed
TSCA: Listed

ETH5850 **CAS #: 9004-57-3**

ETHYLCELLULOSE

RTECS: FJ5950500
Molecular Formula: Unknown
Synonyms: AMPACET E/C; AQUACOAT; AQUACOAT ECD 30; AQUACOAT ECD 30FMC; CELLULOSE ETHYL; CELLULOSE ETHYLATE; CELLULOSE,ETHYL ETHER; CELLULOSE,TRIETHYL ETHER; ET 100 (CELLULOSE DERIVATIVE); ETHOCEL; ETHOCEL N-200; ETHOCEL 150; ETHOCEL 890; ETHOCEL E 50; ETHOCEL E 7; ETHOCEL E50; ETHOCEL E7; ETHOCEL MED; ETHOCEL N 10; ETHOCEL N 200; ETHOCEL N 7; ETHOCEL N10; ETHOCEL N200; ETHOCEL N7; ETHOCEL STD; ETS; ETS (POLYSACCHARIDE); G 200; G 50; G 50 (POLYSACCHARIDE); N 5; NIXON E/C; SPT 50 CPS; SPT 50CPS; T 100; T 100 (POLYSACCHARIDE); TRIETHYL CELLULOSE
Description: white to light tan powder or needles, available commercially as clear film or sheet; odorless
Use: mfr of plastics, lacquers, furniture trim, flashlight cases, adhesives, & wire insulation; in packaging of films, paper, textiles, etc; pigment-grinding bases, toughening agent for plastics; proximity fuses; vitamin prepn; casing for rocket propellants; food additive; surface coating; thermoplastic for molding powders & sheeting; in adhesives & textile printing inks; in sustained-controlled release; taste masking; moisture barrier.

Physical Properties

Freezing Point: 240 °C (464 °F) to 255 °C (491 °F)
Specific Gravity: 1.07 to 1.18
Water Solubility: Negligible
Other Solubilities: freely Soluble in mixture of aromatic hydrocarbons with Alcohol.
pH: Aqueous solution is neutral

RTECS Toxicity Data

Acute Oral: Rat LD_{50} Dose: >5 gm/kg.
Acute Dermal: Rabbit LD_{50} Route: Skin; Dose: >5 gm/kg.
Irritation Skin: Rabbit Standard Draize Test Dose: 500 mg/24H; Reaction: mild.

Hazard Overviews

Health: Irritating to eyes/skin/respiratory tract. Other Acute Effects: may be harmful by inhalation, ingestion, or skin absorption.
Fire: Hazards: in powder form capable of creating a dust explosion; emits toxic fumes; static charges generated by emptying package in or near flammable vapors may cause flash fire. Extinguishing agents: water spray; carbon dioxide, dry chemical powder or appropriate foam. Precautions: combustible liquid.
Carcinogenicity: IARC - Not listed; NIOSH - Not listed; NTP - Not listed; ACGIH - Not listed; OSHA - Not listed; EPA - Not listed; MAK - Not listed

Primary Target Organs:

Eyes

Skin

Respiratory System

Environmental

Regulations

RCRA 40CFR: Not listed
CERCLA: 40CFR 302.4: Not listed
SARA 40CFR 372.65: Not listed
SARA EHS 40CFR 355: Not listed
TSCA: Listed

ETH5900 CAS #: 92-49-9

N-ETHYL-N-(2-CHLOROETHYL)ANILINE

RTECS: BX1300000
EINECS Number: 202-159-3
Molecular Formula: $C_{10}H_{14}ClN$
Formula Weight: 183.7
Synonyms: ANILINE,4-CHLORO-N,N-DIETHYL-; ANILINE,N-(2-CHLOROETHYL)-N-ETHYL-; BENZENAMINE,N-(2-CHLOROETHYL)-N-ETHYL-; BENZENAMINE,N-(2-CHLOROETHYL)-N-ETHYL-(9CI); N-(2-CHLOROETHYL)-N-ETHYLANILINE; N-(2-CHLOROETHYL)-N-ETHYLBENZENAMINE; EMERY 5770; ETHYL(CHLOROETHYL)ANILINE
Description: needles

Physical Properties

Boiling Point: 251 °C (484 °F) to 253 °C (487 °F)
Freezing Point: 45.5 °C (113.9 °F) to 46.5 °C (115.7 °F)
Other Solubilities: Soluble in hot Alcohol

RTECS Toxicity Data

Acute Oral: Rat LD_{50} Dose: 616 mg/kg.
Acute Dermal: Rabbit LD_{50} Route: Skin; Dose: 200 mg/kg; Toxic Effects: Skin and appendages - Corrosive.

Hazard Overviews

Poison

Health: Severely irritating. Poison. Other Acute Effects: may be fatal if inhaled, swallowed, or absorbed through skin; high concentrations are extremely destructive to tissues of the mucous membranes, upper respiratory tract, eyes and skin; symptoms of exposure may include burning sensation, coughing, wheezing, laryngitis, shortness of breath, headache, nausea and vomiting.

Fire: Hazards: emits toxic fumes. Extinguishing agents: water spray; carbon dioxide, dry chemical powder or appropriate foam. Precautions: combustible liquid.
Carcinogenicity: IARC - Not listed; NIOSH - Not listed; NTP - Not listed; ACGIH - Not listed; OSHA - Not listed; EPA - Not listed; MAK - Not listed
Primary Target Organs:

Eyes

Skin

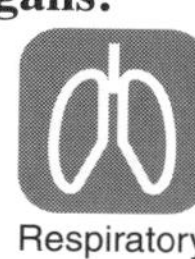
Respiratory System

Environmental

Regulations

RCRA 40CFR: Not listed
CERCLA: 40CFR 302.4: Not listed
SARA 40CFR 372.65: Not listed
SARA EHS 40CFR 355: Not listed
TSCA: Listed

ETH5950 CAS #: 598-14-1

ETHYLDICHLOROARSINE

RTECS: CH3500000
DOT: UN1892; IMO6.1
EINECS Number: 209-919-3
Molecular Formula: $C_2H_5AsCl_2$
Formula Weight: 174.89
Synonyms: ARSENIC DICHLOROETHANE; ARSINE,DICHLOROETHYL-; ARSONOUS DICHLORIDE,ETHYL-; ARSONOUS DICHLORIDE,ETHYL-(9CI); DICHLOROETHYLARSINE; DICK; ED; ETHYLARSONOUS DICHLORIDE; ETHYLDICHLORARSINE; TL 214
Description: colorless, mobile liquid; biting, irritant fruit-like odor
Use: military poison gas

Physical Properties

Boiling Point: Decomposes at 156 °C (313 °F)
Freezing Point: -65 °C (-85 °F)
Specific Gravity: 1.742 at 14 °C
Vapor Density: 6 Air=1.29
Saturated Vapor Density: 1.218189915 kg/m^3
Vapor Pressure: 2.29 mm Hg at 21.5 °C
Water Solubility: Decomposes in Water
Other Solubilities: Soluble in alcohol, benzene, ether
Flash Point: Does not burn or burns with difficulty

RTECS Toxicity Data

Acute Inhalation: Human LC_{Lo} Dose: 14 ppm/30M. Cat LC_{Lo} Dose: 50 mg/m^3/40M; Toxic Effects: Sense organs and special senses - Lacrimation; Lungs, Thorax, or Respiration - Structural or functional change in trachea or bronchi.
Acute Dermal: Mouse LD_{Lo} Route: Skin; Dose: 20 mg/kg. Cat LD_{Lo} Route: Subcutaneous Dose: 1 mg/kg; Toxic Effects:

Behavioral - Excitment; Lungs, Thorax, or Respiration - Acute pulmonary edema; Gastrointestinal - Changes in structure or function of salivary glands.

Hazard Overviews

Corrosive

Fire: Noncombustible.
Carcinogenicity: IARC - Not listed; NIOSH - Not listed; NTP - Not listed; ACGIH - Not listed; OSHA - Not listed; EPA - Not listed; MAK - Not listed
Exposure Limits
OSHA PEL: TWA: 0.5 mg/m^3; as As.
ACGIH TLV: TWA: 0.01 mg/m^3; as As.
NIOSH REL: STEL: 0.002 mg/m^3; Ceiling (15 min) as As.

Environmental

Cleanup/Disposal: Guide No. 151: Do not touch damaged containers or spilled material unless wearing appropriate protective clothing. Stop leak if you can do it without risk. Prevent entry into waterways, sewers, basements or confined areas. Cover with plastic sheet to prevent spreading. Absorb or cover with dry earth, sand or other non-combustible material and transfer to containers. Do not get water inside containers.

Regulations

RCRA 40CFR: Not listed
CERCLA: 40CFR 302.4: Listed as Compound per CWA Section 307(a) per CAA Section 112
SARA 40CFR 372.65: Listed as Compound
SARA EHS 40CFR 355: Not listed
TSCA: Listed

ETH6000 CAS #: 1789-58-8

ETHYLDICHLOROSILANE

RTECS: VV3230000
EINECS Number: 217-255-0
Molecular Formula: $C_2H_6Cl_2Si$
Structured MF: $C_2H_5SiHCl_2$
Formula Weight: 129.07

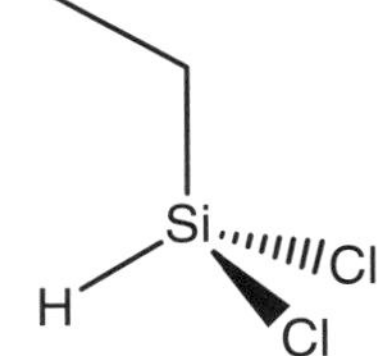

Chemical Structure

Synonyms: DICHLOROETHYLSILANE; MONOETHYLDICHLOROSILANE; SILANE,DICHLOROETHYL-
Description: colorless liquid; sharp, irritating odor
Use: chem int for silicones

Physical Properties

Boiling Point: 75.5 °C (168 °F)
Specific Gravity: 1.1
Vapor Density: 4.45 Air=1
Water Solubility: Water Soluble
Surface Tension: 21.7 dynes/cm at 20 °C
pH: Acid
Flash Point: -1.111 °C
LEL: 2.9% v/v

Hazard Overviews

Poison

Corrosive

Flammable

Fire Diamond

Health: Corrosive to eyes/skin/respiratory tract. Poison. Other Acute Effects: may be fatal if inhaled, swallowed, or absorbed through skin; inhalation may result in spasm, inflammation and edema of the larynx and bronchi, chemical pneumonitis and pulmonary edema; symptoms of exposure may include burning sensation; coughing; wheezing; laryngitis; shortness of breath; headache; nausea; vomiting; may cause allergic respiratory reaction.
Fire: Flammable. Hazards: vapor may travel considerable distance to source of ignition and flash back; emits toxic fumes. Extinguishing agents: carbon dioxide, dry chemical powder or appropriate foam. Precautions: combustible liquid.
Reactivity: Incompatible with: strong oxidizing agents, reacts violently with water. Hazardous decomposition products: toxic fumes of: carbon monoxide, carbon dioxide, hydrogen chloride gas, silicon oxide.
Carcinogenicity: IARC - Not listed; NIOSH - Not listed; NTP - Not listed; ACGIH - Not listed; OSHA - Not listed; EPA - Not listed; MAK - Not listed
Primary Target Organs:

Eyes

Skin

Respiratory System

Environmental

Cleanup/Disposal: Guide No. 171: Do not touch or walk through spilled material. Stop leak if you can do it without risk. Prevent dust cloud. Avoid inhalation of asbestos dust. Small Dry Spills: With clean shovel place material into clean, dry container and cover loosely; move containers from spill area. Small Spills: Take up with sand or other noncombustible absorbent material and place into containers for later disposal. Large Spills: Dike far ahead of liquid spill for later disposal. Cover powder spill with plastic sheet or tarp to minimize spreading. Prevent entry into waterways, sewers, basements or confined areas.

Environmental Physical Data

BCF: no food chain concentration potential

Regulations
RCRA 40CFR: Not listed
CERCLA: 40CFR 302.4: Not listed
SARA 40CFR 372.65: Not listed
SARA EHS 40CFR 355: Not listed
TSCA: Listed

ETH6050 CAS #: 74-85-1
ETHYLENE

RTECS: KU5340000
DOT: UN1038; UN1962; IMO2.1
EINECS Number: 200-815-3
Molecular Formula: C_2H_4
Structured MF: CH_2=CH_2
Formula Weight: 28.05

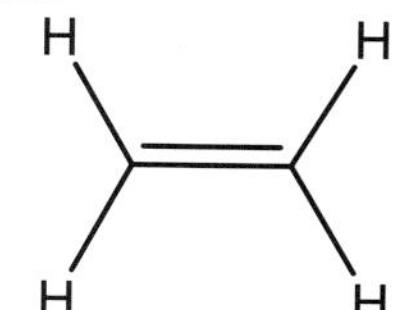

Chemical Structure

Synonyms: ACETENE; ATHYLEN; BICARBURRETTED HYDROGEN; ELAYL; ETHENE; ETILENO; LIQUID ETHYENE; LIQUID ETHYLENE; OLEFIANT GAS
Description: colorless gas; sweet odor
Use: oxyethylene welding & cutting metals; mfr mustard gas & other organics; refrigerant; intermed for ethylbenzene, ethyl alcohol, acetaldehyde, linear primary alcohols, vinyl acetate monomer, & anesthetics; monomer for polyethylene; cooling medium; solvent; mfr of tetraethyl lead; plant growth regulator; compressed gas to initiate degreening & ripening of bananas, to induce flowering of pineapples; med: inhalation anesthetic.

Physical Properties
Boiling Point: -102.4 °C (-152 °F) at 700 mm Hg
Freezing Point: -169 °C (-272.2 °F)
Vapor Density: 0.978 Air=1
Density: 567.37 kg/cu m (-103.8 °C)
Vapor Pressure: 4040 kPa at -1.5 °C
Water Solubility: 1 vol dissolves in about 4 vol Water at 0 °C
Other Solubilities: Soluble in Acetone, Benzene; 131 mg/l at 20 °C; 256 cu cm/l at 0 °C.
Surface Tension: 16 dynes/cm at 04 °C
Odor Threshold: Detection 2.60 x10^2 ppm
Refraction Index: 1.363 at 100 °C/D
Evaporation Rate: > 1 Butyl Acetate=1
Critical Temperature: 9.6 °C
Critical Pressure: 50.7 atm
Ionization Potential (eV): 10.5138 +/-0.0006
Flash Point: ~ -136 °C Closed Cup
Autoignition Temperature: 490 °C
LEL: 2.7% v/v
UEL: 36% v/v

RTECS Toxicity Data
Acute Inhalation: Mammal LC_{Lo} Dose: 950000 ppm/5M.

Hazard Overviews

Flammable

Compressed Gas

Fire Diamond

Health: A simple asphyxiant which can displace available oxygen needed for breathing. Also Causes: frostbite, rapid respiration, diminished alertness, impaired coordination, faulty judgement, rapid fatigue, unconsciousness, convulsions, coma, death.
Fire: Extremely flammable. Can form explosive mixtures in air. Extinguish by shutting off source of gas. Use water sprays to cool fire-exposed containers and to protect personnel. Treat any fire situation involving rapidly escaping and burning ethylene gas as an emergency.
Reactivity: Stable. Hazardous polymerization cannot occur. Avoid: ignition sources (open flame; lighted cigarettes or pipes; uninsulated heating elements; electrical or mechanical sparks); accidental or uncontrollably rapid release from high-pressure cylinders, tank cars, or pipelines. Incompatible with: aluminum chloride; benzoyl peroxide and carbon tetrachloride; bromotrichloro-methane; carbon tetrachloride; chlorine; chlorine dioxide; nitrogen dioxide; nitromethane and aluminum chloride; ozone. Hazardous decomposition products: carbon dioxide; carbon monoxide.
Carcinogenicity: IARC - Group 3, Not classifiable as to carcinogenicity to humans; NIOSH - Not listed; NTP - Not listed; ACGIH - Class A4, Not classifiable as a human carcinogen; OSHA - Not listed; EPA - Not listed; MAK - Class B, Justifiably suspected of having carcinogenic potential
Primary Target Organs:

Eyes

Skin

Nervous System

Respirator Recommendation
Exposure Range: unlimited Self-contained Breathing Apparatus, Pressure Demand, Full Face
Note: asphyxiant, respirator needed only if atmosphere is oxygen deficient

Environmental
Ecotoxicity: Aquatic toxicity: 22 ppm/1 hr/sunfish/killed/fresh water
Environmental Fate: In the atmosphere, gas-phase may be degraded by ozone (half-life of 6.5 days), nitrate radicals (half-life of 190 days), or photochemically-produced hydroxyl radicals (estimated half-life of 1.9 days). Hydrolysis, bioconcentration, adsorption, and biodegradation are not expected to be important fate processes in soil or aquatic ecosystems. In soil and water, it may oxidize to

ethylene oxide. The high vapor pressure suggests that the gas may permeate through soil and sediment. Volatilization is expected to be the primary environmental fate process in soil and water. Volatilization half-lives from a model river and a model environmental lake have been estimated to be 1.6 and 50 hr, respectively.

Cleanup/Disposal: Guide No. 115: Eliminate all ignition sources (no smoking, flares, sparks or flames in immediate area). All equipment used when handling the product must be grounded. Do not touch or walk through spilled material. Stop leak if you can do it without risk. If possible, turn leaking containers so that gas escapes rather than liquid. Use water spray to reduce vapors or divert vapor cloud drift. Do not direct water at spill or source of leak. Prevent spreading of vapors through sewers, ventilation systems and confined areas. Isolate area until gas has dispersed. Guide No. 116: Eliminate all ignition sources (no smoking, flares, sparks or flames in immediate area). All equipment used when handling the product must be grounded. Stop leak if you can do it without risk. Do not touch or walk through spilled material. Do not direct water at spill or source of leak. Use water spray to reduce vapors or divert vapor cloud drift. If possible, turn leaking containers so that gas escapes rather than liquid. Prevent entry into waterways, sewers, basements or confined areas. Isolate area until gas has dispersed.

Environmental Physical Data

Henry's Law Constant: 0.228
Octanol/Water Partition Coefficient: log K_{ow} = 1.13
Sorption Partition Coefficient: K_{oc} = estimated at 300
BCF: none

Regulations

RCRA 40CFR: Not listed
CERCLA: 40CFR 302.4: Not listed
SARA 40CFR 372.65: Listed
SARA EHS 40CFR 355: Not listed
TSCA: Listed

Analytical Methods

Air: ASTM D2820, D4490
Water / Groundwater: USGS O3114

ETH6100 CAS #: 107-07-3

ETHYLENE CHLOROHYDRIN

RTECS: KK0875000
DOT: UN1135; IMO6.1
EINECS Number: 203-459-7
Molecular Formula: C_2H_5ClO
Structured MF: $ClCH_2CH_2OH$
Formula Weight: 80.51

Chemical Structure

Synonyms: AETHYLENECHLORHYDRIN; 2-CHLOORETHANOL; 2-CHLORAETHANOL; 2-CHLORETHANOL; 2-CHLORO-1-ETHANOL; 2-CHLOROETHANOL; BETA-CHLOROETHANOL; CHLOROETHANOL; DELTA-CHLOROETHANOL; 2-CHLOROETHYL ALCOHOL; BETA-CHLOROETHYL ALCOHOL; CHLOROETHYLOWY ALKOHOL; 2-CLOROETANOLO; ETHANOL,2-CHLORO-; ETHENE,CHLOROHYDRIN; ETHYLCHLOROHYDRIN; ETHYLEEN-CHLOORHYDRINE; ETHYLENE CHLORHYDRIN; ETHYLENE GLYCOL,CHLOROHYDRIN; GLICOL MONOCLORIDRINA; GLYCOL CHLOROHYDRIN; GLYCOL MONOCHLOROHYDRIN; GLYCOLMONOCHLOORHYDRINE; GLYCOMONOCHLORHYDRIN; 2-HYDROXYETHYL CHLORIDE; BETA-HYDROXYETHYL CHLORIDE; MONOCHLORHYDRINE DU GLYCOL; 2-MONOCHLOROETHANOL

Description: colorless liquid; faint ethereal odor

Use: solvent for cellulose acetate, ethylcellulose cellulose esters, resins and waxes; in the manufacture of insecticides, to speed the germination of seeds, treating sweet potatoes before planting, to activate sprouting of dormant potatoes, as a cleaning solvent, in the introduction of hydroxyethyl group in organic synthesis, in the manufacture of ethylene oxide and ethylene glycol and in the synthesis of amines, carbitols, indigo, malonic acid and novocain; for the separation of butadiene from hydrocarbon mixtures, in dewaxing and removing cycloalkanes from mineral oil, in the refining of rosin, in the manufacture of certain pesticides and in the extraction of pure lignin; in the dyeing and cleaning industry to remove tar spots, as a cleaning agent for machines and as a solvent in fabric dyeing

Physical Properties

Boiling Point: 128 °C (262 °F) to 130 °C (266 °F) at 760 mm Hg
Freezing Point: -67.5 °C (-89.5 °F)
Specific Gravity: 1.197 at 20 °C/4 °C
Vapor Density: 2.78 Air=1
Saturated Vapor Density: 1.213742232 kg/m^3
Density: 1.2 g/cu cm at 21.1 °C
Bulk Density: 10.0 lbs/gal at 20 °C
Vapor Pressure: 4.9 mm Hg at 20 °C
Water Solubility: Soluble in all Proportions
Other Solubilities: 95% Ethanol: >=100 mg/ml at 19 °C; Acetone: >=100 mg/ml at 19 °C; Alcohol: miscible; DMSO: >=100 mg/ml at 19 °C; Ethanol: >10%; Most organic liquids: Soluble.
Odor Threshold: 0.4 ppm
Refraction Index: 1.4419 at 20 °C/D
Ionization Potential (eV): 10.90
Flash Point: 40 °C Open Cup
Autoignition Temperature: 425 °C
LEL: 4.9% v/v
UEL: 15.9% v/v

RTECS Toxicity Data

Acute Oral: Rat LD_{50} Dose: 71 mg/kg; Toxic Effects: Behavioral - Somnolence (general depressed activity); Lungs, Thorax, or Respiration - Dyspnea. Mouse LD_{50} Dose: 81 mg/kg.
Acute Inhalation: Rat LC_{50} Dose: 290 mg/m^3; Toxic Effects: Behavioral - Somnolence (general depressed activity); Lungs,

Thorax, or Respiration - Dyspnea. Man LC_{Lo} Dose: 305 ppm/2hr; Toxic Effects: Liver - Other changes; Kidney, ureter, and Bladder - Other changes.

Acute Dermal: Rabbit LD_{50} Route: Skin; Dose: 67 mg/kg. Rabbit LD_{50} Route: Subcutaneous Dose: 100 mg/kg;

Chronic (Multiple Dose) Oral: Rat Dose: 5670 mg/kg/12W-I; Toxic Effects: Liver - Changes in liver weight; Endocrine - Changes in pituitary weight; DEATH. Rat Dose: 91 mg/kg/26W-I; Toxic Effects: Gastrointestinal - Changes in structure or function of exocrine pancreas.

Chronic (Multiple Dose) Inhalation: Rat Dose: 3400 mg/m^3/15M/11D-I; Toxic Effects: Liver - Fatty Liver degeneration; Blood - Hemorrhage; DEATH. Rat Dose: 1 mg/m^3/4H/17W-I; Toxic Effects: Blood - Changes in serum composition; Blood - Other changes; Biochemical - Transaminases.

Chronic (Multiple Dose) Dermal: Rat Route: Skin; Dose: 6188 mg/kg/14D-I; Toxic Effects: DEATH. Rat Route: Skin; Dose: 32500 mg/kg/13W-I; Toxic Effects: Lungs, Thorax, or Respiration - Chronic pulmonary edema; DEATH.

Irritation Eye: Rabbit Standard Draize Test Dose: 2 mg; Reaction: severe. Rabbit Standard Draize Test Dose: 33 mg; Reaction: moderate. Rabbit Standard Draize Test Dose: 9 mg/6H; Reaction: moderate.

Irritation Skin: Rabbit Standard Draize Test Dose: 200 mg/2H; Reaction: mild.

Reproductive/Teratogenic: Mouse Route: Oral; Dose: 1100 mg/kg; Duration: female 6-16D of pregnancy; Effects on Embryo or Fetus - Fetotoxicity; Specific Developmental Abnormalities - Hepatobiliary system.

Mutagenic: Rat Cytogenetic Analysis; Route: Inhalation; Dose: 1 mg/m^3. Rat Other Mutation Test Systems; Route: Oral; Dose: 20 mg/L. Rat Other Mutation Test Systems; Cell Type: liver; Dose: 12 gm/L. Rat Other Mutation Test Systems; Route: Inhalation; Dose: 1 mg/m^3.

Hazard Overviews

Poison

Flammable

Fire Diamond

Health: Irritating to eyes/skin/respiratory tract. Poison. Also Causes: pulmonary edema, respiratory failure, incoordination, confusion, vertigo, weakness, nausea, vomiting, seizures, low blood pressure, collapse, unconsciousness, nausea, headache, delirium, coma, corneal burns, urine may show red blood cells, albumin, casts. Chronic Effects: CNS and liver damage.

Fire: Flammable. Let burn if leak can't be stopped. Use dry chemical, alcohol resistant foam, or carbon dioxide. Only use water when other agents are unavailable because ethylene chlorohydrin is water reactive. If water must be used, protect against toxic fumes. Fight fire from maximum distance. If water spray is used to cool fire-exposed containers, use flooding quantities of water because solid streams of water may be ineffective. Do not get water inside containers.

Reactivity: Stable. Hazardous polymerization cannot occur. Avoid: heat; ignition sources; water; steam. Incompatible with: oxidizing materials; acids; bases; ethylene diamine; chlorosulfonic acid; sodium hydroxide; water; steam. Hazardous decomposition products: toxic hydrogen chloride; phosgene fumes.

Carcinogenicity: IARC - Not listed; NIOSH - Not listed; NTP - Not listed; ACGIH - Class A4, Not classifiable as a human carcinogen; OSHA - Not listed; EPA - Not listed; MAK - Not listed

Primary Target Organs:

Eyes

Skin

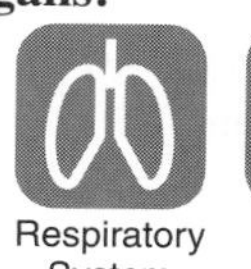
Respiratory System

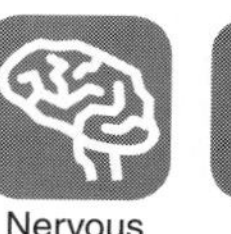
Nervous System

Liver

Kidneys

Exposure Limits

OSHA PEL: TWA: 5 ppm; 16 mg/m^3.

OSHA PEL Vacated 1989 Limits: STEL: 1 ppm; 3 mg/m^3; Ceiling.

NIOSH REL: STEL: 1 ppm; 3 mg/m^3; skin.

NIOSH IDLH: 7 ppm.

DFG MAK: TWA: 1 ppm; 3 mg/m^3.

Respirator Recommendation

Exposure Range: >5 to <7 ppm Supplied Air, Constant Flow/Pressure Demand, Half Mask

Exposure Range: 7 to unlimited ppm Self-contained Breathing Apparatus, Pressure Demand, Full Face

Note: odor threshold unknown

Environmental

Ecotoxicity: LC_{50} Pimephales promelas (fathead minnow) 83.7 mg/l/96 hr (95% confidence limit 75.0-93.4 mg/l), flow-through bioassay with measured concentrations, 24.0 °C, dissolved oxygen 7.1 mg/l, hardness 52.3 mg/l calcium carbonate, alkalinity 44.2 mg/l calcium carbonate, and pH 7.22.

Environmental Fate: If released on soil, it would be expected to leach because of its estimated very low adsorptivity to soil and may biodegrade. It should not volatilize from moist soil, but may evaporate from dry soil and other surfaces. Fate in the aquatic environment is not clear. Based on data from screening tests, it may biodegrade. It would not be expected to volatilize from surface waters, adsorb to sediment or bioconcentrate in aquatic organisms. It reacts with photochemically-produced hydroxyl radicals in the atmosphere, as a result of which its half-life in the atmosphere will be approximately 11.5 days. It is soluble in water and would be washed out of the air by rain.

Cleanup/Disposal: Guide No. 131: Fully encapsulating, vapor protective clothing should be worn for spills and leaks with no fire. Eliminate all ignition sources (no smoking, flares, sparks or flames in immediate area). All equipment used when handling the product must be grounded. Do not touch or walk through spilled material. Stop leak if you can do it without risk. Prevent entry into waterways, sewers, basements or confined areas. A vapor suppressing foam may be used to reduce vapors. Small Spills: Absorb with earth, sand or other non-combustible material and transfer to containers for later disposal. Use clean non-sparking tools to collect absorbed material. Large Spills: Dike far ahead of liquid spill for later disposal. Water spray may reduce vapor; but may not prevent ignition in closed spaces.

Environmental Physical Data

Henry's Law Constant: estimated at 1.04 x10^{-7}
Octanol/Water Partition Coefficient: log K_{ow} = 0.03
Sorption Partition Coefficient: log K_{oc} = estimated at 1.33
BCF: calculated at 0.62
BOD: 0.50 lb/lb, 10 days

Regulations

RCRA 40CFR: Not listed
CERCLA: 40CFR 302.4: Not listed
SARA 40CFR 372.65: Not listed TPQ: 500 lb
SARA EHS 40CFR 355: Listed TPQ: 500 lb
TSCA: Listed

Analytical Methods

Soil: SW846 8240B, 8260A, 8260B, 8430
Water / Groundwater: ASTM D3695
Indoor / Expired Air: NIOSH 2513
Plasma: EPA 29

ETH6150 CAS #: 109-78-4

ETHYLENE CYANOHYDRIN

RTECS: MU5250000
EINECS Number: 203-704-8
Molecular Formula: C_3H_5NO
Structured MF: $HOCH_2CH_2CN$
Formula Weight: 71.08

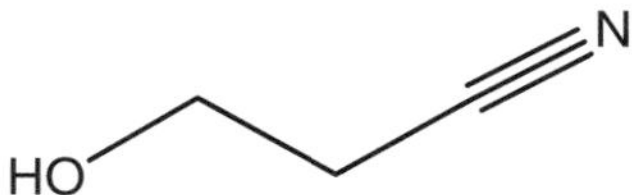

Chemical Structure

Synonyms: AETHYLENCYANHYDRIN; 2-CYANOETHANOL; BETA-CYANOETHANOL; 2-CYANOETHANOL ALCOHOL; 2-CYANOETHYL ALCOHOL; GLYCOL CYANOHYDRIN; BETA-HPN; HYDRACRYLONITRILE; 2-HYDROXYCYANOETHANE; 2-HYDROXYETHYL CYANIDE; 2-HYDROXYETHYLKYANID; 3-HYDROXYPROPANENITRILE; 3-HYDROXYPROPANOIC ACID,NITRILE; 3-HYDROXYPROPIONITRILE; BETA-HYDROXYPROPIONITRILE; METHANOLACETONITRILE; PROPANENITRILE,3-HYDROXY-; PROPANOIC ACID,3-HYDROXY,NITRILE; PROPIONITRILE,3-HYDROXY-
Description: water-white to straw-colored liquid;; practically odorless
Use: solvent for certain cellulose esters and inorganic salts; also as an organic intermediate for acrylates

Physical Properties

Boiling Point: 230 °C (446 °F) at 760 mm Hg
Freezing Point: -46 °C (-50.8 °F)
Specific Gravity: 1.0588 at 0 °C
Vapor Density: 2.5 Air=1
Saturated Vapor Density: 1.200183289 kg/m^3
Density: 1.0588 g/mL
Vapor Pressure: 0.08 mm Hg at 25 °C
Water Solubility: Soluble in all Proportions
Other Solubilities: Insoluble in Benzene, Carbon Tetrachloride & Naphtha; 2.3% (wt/wt) in Ether at 15 °C; miscible with Acetone, Methyl Ethyl Ketone; Insoluble in Petroleum Ether; Soluble in Chloroform.
Refraction Index: 1.4240 at 20 °C/D
Critical Temperature: 429 °C
Critical Pressure: 720 psia
Flash Point: 129.4 °C Open Cup
Autoignition Temperature: 505 °C
LEL: 2.3% v/v
UEL: 12.1% v/v

RTECS Toxicity Data

Acute Oral: Rat LD_{50} Dose: 3200 mg/kg. Mouse LD_{50} Dose: 1800 mg/kg.
Acute Inhalation: Mouse LC_{33} Dose: 300 mg/m^3/2hr.
Acute Dermal: Rabbit LD_{50} Route: Skin; Dose: 5 mL/kg.
Chronic (Multiple Dose) Oral: Rat Dose: 8100 mg/kg/90D-C; Toxic Effects: Brain and coverings - Changes in brain weight; Cardiac - Changes in heart weight.
Irritation Eye: Rabbit Standard Draize Test Dose: 500 mg; Reaction: moderate.
Irritation Skin: Rabbit Open Draize Test Dose: 520 mg open; Reaction: mild.

Hazard Overviews

Health: Irritating to eyes/skin/respiratory tract. Harmful. Other Acute Effects: may be harmful by inhalation, ingestion, or skin absorption; repeated exposure can cause damage to the kidneys.
Fire: Will burn. Hazards: emits toxic fumes. Extinguishing agents: carbon dioxide, dry chemical powder or appropriate foam; do not use water. Precautions: combustible liquid.
Reactivity: Incompatible with: acids, bases, oxidizing agents, moisture, heat do not allow contact with water, avoid contact with acid. Hazardous decomposition products: toxic fumes of: hydrogen cyanide, carbon monoxide, carbon dioxide, nitrogen oxides.
Carcinogenicity: IARC - Not listed; NIOSH - Not listed; NTP - Not listed; ACGIH - Not listed; OSHA - Not listed; EPA - Not listed; MAK - Not listed
Primary Target Organs:

Eyes

Skin

Respiratory System

Environmental

Cleanup/Disposal: Guide No. 130: Eliminate all ignition sources (no smoking, flares, sparks or flames in immediate area). All equipment used when handling the product must be grounded. Do not touch or walk through spilled material. Stop leak if you can do it without risk. Prevent entry into waterways, sewers, basements or confined areas. A vapor suppressing foam may be used to reduce vapors. Absorb or cover with dry earth, sand or other non-combustible material and transfer to containers. Use clean non-sparking tools to collect absorbed material. Large Spills: Dike far ahead of liquid spill for later disposal. Water spray may reduce vapor; but may not prevent ignition in closed spaces.

Environmental Physical Data

BCF: no food chain concentration potential

Regulations

RCRA 40CFR: Not listed
CERCLA: 40CFR 302.4: Listed as Compound per CWA Section 307(a) per CAA Section 112
SARA 40CFR 372.65: Not listed
SARA EHS 40CFR 355: Not listed
TSCA: Listed

ETH6200 CAS #: 110-30-5

N,N-ETHYLENE DISTEARYLAMIDE

EINECS Number: 203-755-6
Molecular Formula: $C_{38}H_{76}N_2O_2$
Formula Weight: 593.04

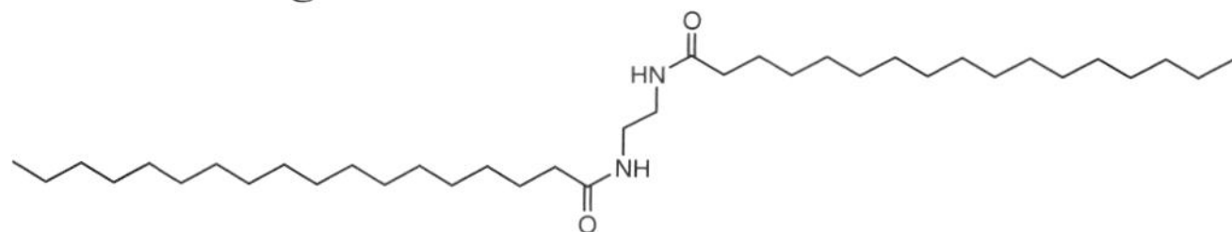

Chemical Structure

Synonyms: ABRIL WAX 10DS; ACRAWAX CT; ACROWAX C; ADVAWACHS 280; ADVAWAX; ADVAWAX 275; ADVAWAX 280; ARMOWAX EBS-P; 1,2-BIS(OCTADECANAMIDO)ETHANE; CARLISLE 280; CARLISLE WAX 280; CHEMETRON 100; N,N'-DISTEAROYLETHYLENEDIAMINE; N,N'-ETHYLENEBIS(STEARAMIDE); N,N'-ETHYLENEDISTEARAMIDE; KEMAMIDE W 40; LUBROL EA; MICROTOMIC 280; OCTADECANAMIDE,N,N'-1,2-ETHANEDIYLBIS-; OCTADECANAMIDE,N,N'-ETHYLENEBIS-; PLASTFLOW; WAX C
Use: synthetic wax; in lacquers, varnishes; as a plastic lubricant and antiblocking agent; as extender or substitue for carnauba wax; in bath oil bars; internal lubricant for ABS & PVC plastics; detackifier & mold release agent for rubber compounding; defoamer for steam boilers & textile & paper indust; heat stabilizer for plastics

Physical Properties

Freezing Point: High melting point

Hazard Overviews

Health: Irritating to eyes/skin/respiratory tract. Other Acute Effects: may be harmful by inhalation, ingestion, or skin absorption.
Fire: Hazards: emits toxic fumes. Extinguishing agents: water spray; dry chemical powder; appropriate foam. Precautions: combustible liquid.
Reactivity: Stable. Hazardous polymerization will not occur. Incompatible with: strong oxidizing agents. Hazardous decomposition products: toxic fumes of: carbon monoxide, carbon dioxide, nitrogen oxides.
Carcinogenicity: IARC - Not listed; NIOSH - Not listed; NTP - Not listed; ACGIH - Not listed; OSHA - Not listed; EPA - Not listed; MAK - Not listed
Primary Target Organs:

Eyes

Skin

Respiratory System

Environmental

Regulations

RCRA 40CFR: Not listed
CERCLA: 40CFR 302.4: Not listed
SARA 40CFR 372.65: Not listed
SARA EHS 40CFR 355: Not listed
TSCA: Listed

ETH6250 CAS #: 371-62-0

ETHYLENE FLUOROHYDRIN

RTECS: KL1575000
EINECS Number: 206-740-2
Molecular Formula: C_2H_5FO
Formula Weight: 64.07

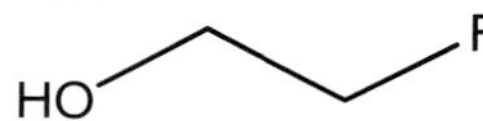

Chemical Structure

Synonyms: ETHANOL,2-FLUORO-; 2-FLUOROETHANOL; BETA-FLUOROETHANOL; TL 741
Use: rodenticide and insecticide

Physical Properties

Boiling Point: 104 °C (218 °F)
Freezing Point: -26 °C (-16 °F)
Specific Gravity: 1.104 at 20 °C
Vapor Pressure: 0.563 kPa at 0 °C
Water Solubility: Miscible with Water
Other Solubilities: Ethanol: Miscible; Ether: Miscible; Acetone: Very Soluble; Chloroform: Slightly soluble
Refraction Index: 1.3647
Ionization Potential (eV): 10.66

RTECS Toxicity Data

Acute Oral: Rat LD_{50} Dose: 5 mg/kg.
Acute Inhalation: Rat LC_{50} Dose: 200 mg/m^3/10M. Monkey LC_{50} Dose: 1500 mg/m^3/10M; Toxic Effects: Behavioral - Coma; Lungs, Thorax, or Respiration - Respiratory stimulation; Gastrointestinal - Changes in structure or function of salivary glands.

Acute Dermal: Mouse LD_{50} Route: Subcutaneous Dose: 15 mg/kg.

Hazard Overviews

Poison

Health: Irritating to eyes/skin/respiratory tract. Poison. Other Acute Effects: may be fatal if inhaled, swallowed, or absorbed through skin.
Fire: Hazards: vapor may travel considerable distance to source of ignition and flash back; container explosion may occur; emits toxic fumes. Extinguishing agents: carbon dioxide, dry chemical powder or appropriate foam; water spray. Precautions: combustible liquid.
Reactivity: Incompatible with: strong acids, strong oxidizing agents, strong reducing agents, acid chlorides, acid anhydrides, phosphorus, halides. Hazardous decomposition products: toxic fumes of: carbon monoxide, carbon dioxide, hydrogen fluoride.
Carcinogenicity: IARC - Not listed; NIOSH - Not listed; NTP - Not listed; ACGIH - Not listed; OSHA - Not listed; EPA - Not listed; MAK - Not listed
Primary Target Organs:

Eyes

Skin
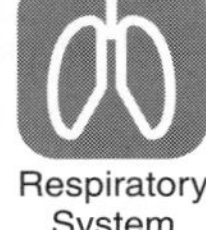
Respiratory System

Environmental

Regulations

RCRA 40CFR: Not listed
CERCLA: 40CFR 302.4: Not listed
SARA 40CFR 372.65: Not listed TPQ: 10 lb
SARA EHS 40CFR 355: Listed TPQ: 10 lb
TSCA: Listed

ETH6300 CAS #: 107-21-1

ETHYLENE GLYCOL

RTECS: KW2975000
DOT: UN1153
EINECS Number: 203-473-3
Molecular Formula: $C_2H_6O_2$
Structured MF: $HOCH_2CH_2OH$
Formula Weight: 62.07

HO ⁄\⁄ OH

Chemical Structure

Synonyms: ATHYLENGLYKOL; 1,2-DIHYDROXYETHANE; DOWTHERM; DOWTHERM SR 1; 1,2-ETHANDIOL; 1,2-ETHANEDIOL; ETHANE-1,2-DIOL; ETHYLENE ALCOHOL; ETHYLENE DIHYDRATE; FRIDEX; GLYCOL; GLYCOL ALCOHOL; 2-HYDROXYETHANOL; LUTROL-9; M.E.G; MACROGOL 400 BPC; MEG; MONOETHYLENE GLYCOL; NORKOOL; RAMP; TESCOL; UCAR 17
Description: colorless liquid; odorless
Use: in antifreeze, in hydraulic brake fluids, as an industrial humectant, as an ingredient of electrolytic condensers, as a solve in the paint and plastics industries, in the formulation of printers inks, stamp pad inks and ball point pen inks, as a softening agent for cellophane a as a stabilizer for soybean foam used to extinguish oil and gasoline fires; in the synthesis of safety explosives, glyoxal, unsaturated estertype alkyd resins, plasticizers, elastomers, synthetic fibers and synthetic waxes; in asphalt emulsion, as a heat transfer agent and as ingredient for deicing airport runways

Physical Properties

Boiling Point: 197.6 °C (388 °F) at 760 mm Hg
Freezing Point: -13 °C (8.6 °F)
Specific Gravity: 1.1135 at 20 °C/4 °C
Vapor Density: 2.14 Air=1
Saturated Vapor Density: 1.200108033 kg/m^3
Density: 1.1155 g/mL at 20 °C
Bulk Density: 9.31 lbs/gal at 15 °C
Vapor Pressure: 0.06 mm Hg at 20 °C
Water Solubility: Miscible with Water
Other Solubilities: Slightly Soluble in Ether (1:200); Practically Insoluble in Benzene, chlorinated hydrocarbons, Petroleum Ether, oils.
Surface Tension: 48.4 dynes/cm at 20 °C
Odor Threshold: 3 at 20 °C
Refraction Index: 1.43312 at 15 °C/D
Evaporation Rate: < 0.01 Butyl Acetate=1
Ionization Potential (eV): 10.16
Flash Point: 111 °C Closed Cup
Autoignition Temperature: 398 °C
LEL: 3.2% v/v

RTECS Toxicity Data

Acute Oral: Man TD_{Lo} Dose: 15 gm/kg; Toxic Effects: Peripheral nerve and sensation - Sensory change involving peripheral nerve; Gastrointestinal - Ulceration or bleeding from small intestine; Kidney, Ureter, and Bladder - Renal function tests depressed. Man TD_{Lo} Dose: 1195 mg/kg; Toxic Effects: Peripheral nerve and sensation - Sensory change involving peripheral nerve; Kidney, Ureter, and Bladder - Renal function tests depressed. Man TD_{Lo} Dose: 16 gm/kg; Toxic Effects: Behavioral - Coma; Kidney, Ureter, and Bladder - Renal function tests depressed; Nutritional and gross metabolic - Metabolic acidosis. Child TD_{Lo} Dose: 5500 mg/kg; Toxic Effects: Behavioral - General anesthetic; Lungs, Thorax, or Respiration - Respiratory stimulation; Kidney, Ureter, and Bladder - Other changes. Human LD_{Lo} Dose: 786 mg/kg; Toxic Effects: Behavioral - Convulsions or effect on seizure threshold; Behavioral - Coma; Gastrointestinal - Hypermotility, diarrhea. Human LD_{Lo} Dose: 398 mg/kg; Toxic Effects: Behavioral - Headache; Gastrointestinal -

Nausea or vomiting; Liver - Other changes. Rat LD_{50} Dose: 4700 mg/kg.

Acute Inhalation: Rat LC_{50} Dose: 10876 mg/kg. Human TC_{Lo} Dose: 10000 mg/m^3; Toxic Effects: Sense organs and special senses - Lacrimation; Lungs, Thorax, or Respiration - Cough; Lungs, Thorax, or Respiration - Other changes.

Acute Dermal: Rabbit LD_{50} Route: Skin; Dose: 9530 uL/kg. Rat LD_{50} Route: Subcutaneous Dose: 2800 mg/kg.

Chronic (Multiple Dose) Oral: Rat Dose: 438 gm/kg/2Y-C; Toxic Effects: Behavioral - Fluid intake; Kidney, Ureter, and Bladder - Chgs in tubules (inc acute renal failure, acute tubular necrosis; DEATH. Rat Dose: 85230 mg/kg/90D-C; Toxic Effects: Kidney, Ureter, and Bladder - Changes in kidney weight; Blood - Changes in serum composition; Nutritional and gross metabolic - Changes in phosphorus. Rat Dose: 12 mL/kg/6D-I; Toxic Effects: Nutritional and gross metabolic - Changes in calcium; Nutritional and gross metabolic - Changes in phosphorus; Biochemical - Phosphatases. Rat Dose: 26150 mg/kg/10D-C; Toxic Effects: Kidney, Ureter, and Bladder - Chgs in tubules (inc acute renal failure, acute tubular necrosis; Blood - Changes in serum composition; Biochemical - Dehydrogenases. Monkey Dose: 5751 gm/kg/22W-I; Toxic Effects: Kidney, Ureter, and Bladder - Chgs in tubules (inc acute renal failure, acute tubular necrosis.

Chronic (Multiple Dose) Inhalation: Rat Dose: 8400 ug/m^3/24H/12W-C; Toxic Effects: Brain and coverings - Changes in brain weight; Kidney, Ureter, and Bladder - Urine volume decreased; Nutritional and gross metabolic - Weight loss or decreased weight gain. Rat Dose: 12 mg/m^3/8H/90D-I; Toxic Effects: Sense organs and special senses - Corneal damage; DEATH. Rat Dose: 1 mg/m^3/32W-I; Toxic Effects: Lungs, Thorax, or Respiration - Other changes; Liver - Liver function tests impaired; Kidney, Ureter, and Bladder - Chgs in tubules (inc acute renal failure, acute tubular necrosis.

Irritation Eye: Rabbit Standard Draize Test Dose: 100 mg/1H; Reaction: mild. Rabbit Standard Draize Test Dose: 1440 mg/6H; Reaction: moderate. Rabbit Standard Draize Test Dose: 500 mg/24H; Reaction: mild.

Irritation Skin: Rabbit Open Draize Test Dose: 555 mg open; Reaction: mild.

Reproductive/Teratogenic: Rat Route: Oral; Dose: 50 gm/kg; Duration: female 6-15D of pregnancy; Specific Developmental Abnormalities - Skin and skin appendages; Musculoskeletal system; Blood and lymphatic systems (including spleen and marrow). Rat Route: Oral; Dose: 8580 mg/kg; Duration: female 6-15D of pregnancy; Effects on Embryo or Fetus - Fetotoxicity; Specific Developmental Abnormalities - Musculoskeletal system. Rat Route: Oral; Dose: 12500 mg/kg; Duration: female 6-15D of pregnancy; Specific Developmental Abnormalities - Craniofacial (including nose and tongue); Musculoskeletal system. Rat Route: Oral; Dose: 25 gm/kg; Duration: female 6-15D of pregnancy; Maternal Effects - Uterus, cervix, vagina; Effects on Fertility - Litter size; Effects on Embryo or Fetus - Fetotoxicity. Rat Route: Oral; Dose: 50 gm/kg; Duration: female 6-15D of pregnancy; Effects on Fertility - Post-implantation mortality. Rat Route: Inhalation; Dose: 2500 mg/m^3/6 Duration: female 6-15D of pregnancy; Maternal Effects - Other effects on females; Specific Developmental Abnormalities - Musculoskeletal system; Other developmental abnormalities.

Mutagenic: Human DNA Inhibition; Cell Type: lymphocyte; Dose: 320 mmol/L. Rat Cytogenetic Analysis; Route: Oral; Dose: 1200 mg/kg.

Hazard Overviews

Fire Diamond

Health: Causes: (when heated or misted): throat irritation, headache, eye effects, coma; CNS, CVS, pulmonary, kidney, and liver effects. Chronic Effects: severe kidney problems, may cause birth defects based on animal data.

Fire: Use water as fog, dry chemical, carbon dioxide or alcohol-resistant foam.

Reactivity: Stable. Hazardous polymerization cannot occur. Avoid: heat; ignition sources; water. Incompatible with: chlorosulfonic acid; sulfuric acid; oleum; sodium hydroxide; phosphorus pentasulfide; silvered-copper wire; dimethyl terephthalate; titanium butoxide; chromium trioxide; potassium permanga-nate; sodium peroxide; ammonium dichromate; silver chlorate; sodium chloride; uranyl nitrate. Hazardous decomposition products: carbon dioxide; acrid smoke; irritating vapors.

Carcinogenicity: IARC - Not listed; NIOSH - Not listed; NTP - Not listed; ACGIH - Class A4, Not classifiable as a human carcinogen; OSHA - Not listed; EPA - Not listed; MAK - Not listed

Primary Target Organs:

Eyes

Respiratory System

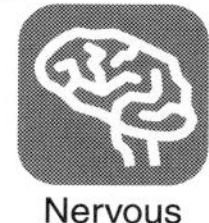
Nervous System

Liver

Kidneys

Cardio-vascular

Exposure Limits

OSHA PEL Vacated 1989 Limits: STEL: 50 ppm; 125 mg/m^3; Ceiling.

DFG MAK: TWA: 10 ppm; 26 mg/m^3.

Respirator Recommendation

Exposure Range: >39.4 to 394 ppm Air Purifying, Negative Pressure, Half Mask

Exposure Range: >394 to 1000 ppm Air Purifying, Negative Pressure, Full Face

Exposure Range: >1000 to 1970 ppm Supplied Air, Constant Flow/Pressure Demand, Half Mask

Exposure Range: >1970 to 39,400 ppm Supplied Air, Constant Flow/Pressure Demand, Full Face

Exposure Range: >39,400 to unlimited ppm Self-contained Breathing Apparatus, Pressure Demand, Full Face

Cartridge Color: black with dust/mist prefilter (use P100 or consult supervisor for appropriate dust/mist prefilter)

Environmental

Ecotoxicity: Toxicity threshold (cell multiplication inhibition test): Algae (Microcystis aeruginosa) 2,000 mg/l; Green algae (Scenedesmus quandricauda) > 10,000 mg/l LC_{50} Guppies (Poecilia reticulata) 49,300 ppm/7 days /Conditions of bioassay not specified LC_{50} Brown shrimp (Crangon crangon) given > 100 mg/l/48 hr aerated salt water LC_{50} Goldfish given 5000 mg/l/24 hr at 20 °C static conditions Toxicity threshold (cell multiplication inhibition test): Protozoa (Entosiphon sulcatum) and (Uronema parduczi Chatton-Lwoff) >10,000 mg/l LD_{50} Carassius auratus (goldfish) >5,000 mg/l/24 hr modified ASTM D

Environmental Fate: It will readily biodegrade in water (half-life ca. 3 days). No data are available that report its fate in soils; however, by analogy to its fate in water, biodegradation is probably fast and the dominate removal mechanism. Should it leach into the groundwater, biodegradation may occur. In water it is not expected to bioconcentrate in aquatic organisms adsorb to sediments or volatilize. If it volatilizes, it will react in the atmosphere with hydroxyl radical with a half-life of about 1 day.

Cleanup/Disposal: Guide No. 127: Eliminate all ignition sources (no smoking, flares, sparks or flames in immediate area). All equipment used when handling the product must be grounded. Do not touch or walk through spilled material. Stop leak if you can do it without risk. Prevent entry into waterways, sewers, basements or confined areas. A vapor suppressing foam may be used to reduce vapors. Absorb or cover with dry earth, sand or other non-combustible material and transfer to containers. Use clean non-sparking tools to collect absorbed material. Large Spills: Dike far ahead of liquid spill for later disposal. Water spray may reduce vapor; but may not prevent ignition in closed spaces.

Environmental Physical Data

Henry's Law Constant: estimated at 2.34×10^{-10}
Octanol/Water Partition Coefficient: log K_{ow} = -1.36
BCF: fish 10
BOD: theoretical 12.5%, 5 days

Regulations

RCRA 40CFR: Not listed
CERCLA: 40CFR 302.4: Listed per CAA Section 112 RQ: 5000 lb (2268 kg)
SARA 40CFR 372.65: Listed
SARA EHS 40CFR 355: Not listed
TSCA: Listed

Analytical Methods

Soil: SW846 8430
Water / Groundwater: ASTM D3695
Drinking Water: EPA 17
Other: EPA 1666, 1671

ETH6350 CAS #: 111-55-7

ETHYLENE GLYCOL DIACETATE

RTECS: KW4025000
EINECS Number: 203-881-1
Molecular Formula: $C_6H_{10}O_4$
Structured MF: $(CH_2OOCCH_3)_2$
Formula Weight: 146.14

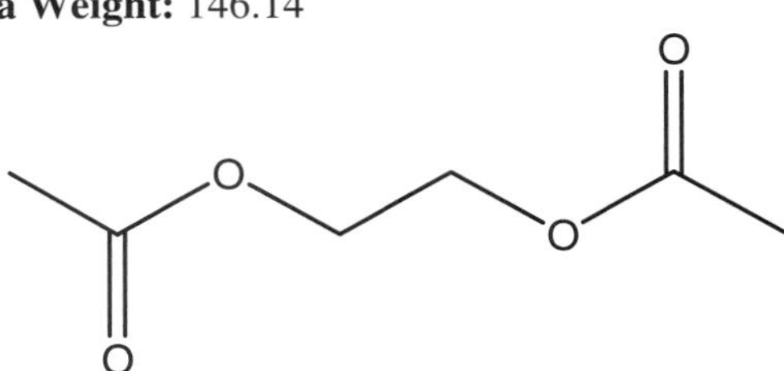

Chemical Structure

Synonyms: 1,2-DIACETOXYETHANE; 1,2-ETHANEDIOL DIACETATE; ETHANEDIOL DIACETATE; 1,2-ETHANEDIOL,DIACETATE; ETHYLENE ACETATE; ETHYLENE DIACETATE; ETHYLENE DIETHANOATE; ETHYLENE GLYCOL ACETATE; ETHYLENE GLYCOL,DIACETATE; GLYCOL DIACETATE

Description: colorless liquid; slight odor of ethyl acetate

Use: solvent for cellulose esters and ethers; resins, lacquers, printing inks, perfume fixative, nondiscoloring plasticizer for ethyl benzyl cellulose

Physical Properties

Boiling Point: 190 °C (374 °F) to 191 °C (376 °F)
Freezing Point: -31 °C (-23.8 °F)
Specific Gravity: 1.104
Vapor Density: 5.04 Air=1
Saturated Vapor Density: 1.202551143 kg/m^3
Bulk Density: 9.2 lbs/gal at 20 °C
Vapor Pressure: 0.4 mm Hg at 20 °C
Water Solubility: >= 100 mg/mL at 20 C
Other Solubilities: 95% Ethanol: >=100 mg/ml at 20 °C; Acetone: >=100 mg/ml at 20 °C; Benzene: Soluble; DMSO: >=100 mg/ml at 20 °C; Diethyl Ether: miscible; Ether: Soluble.
Surface Tension: Estimated at 20 dynes/cm
Odor Threshold: Absolute perception limit 0.093 ppm
Refraction Index: 1.415 at 20 °C/D
Flash Point: 88 °C Closed Cup
Autoignition Temperature: 482 °C
LEL: 1.6% v/v

RTECS Toxicity Data

Acute Oral: Rat LD_{50} Dose: 6850 mg/kg. Guinea Pig LD_{50} Dose: 4940 mg/kg; Toxic Effects: Behavioral - General anesthetic; Gastrointestinal - Other changes; Kidney, Ureter, and Bladder - Other changes.

Acute Dermal: Rabbit LD_{50} Route: Skin; Dose: 8480 uL/kg.

Chronic (Multiple Dose) Inhalation: Rat Dose: 200 mg/m^3/4H/17W-I; Toxic Effects: Liver - Other changes; Blood - Other changes; Biochemical - Transaminases.
Irritation Eye: Rabbit Standard Draize Test Dose: 500 mg; Reaction: mild.

Hazard Overviews

Health: Irritating to eyes/skin/respiratory tract. Other Acute Effects: may be harmful by inhalation, ingestion, or skin absorption.
Fire: Combustible. Extinguishing agents: water spray; carbon dioxide, dry chemical powder or appropriate foam. Precautions: combustible liquid.
Reactivity: Incompatible with: strong oxidizing agents, strong acids, strong bases, strong reducing agents. Hazardous decomposition products: toxic fumes of: carbon monoxide, carbon dioxide.
Carcinogenicity: IARC - Not listed; NIOSH - Not listed; NTP - Not listed; ACGIH - Not listed; OSHA - Not listed; EPA - Not listed; MAK - Not listed
Primary Target Organs:

Eyes

Skin

Respiratory System

Environmental

Ecotoxicity: LC_{50} Lepomis macrochirus (bluegill sunfish) 90 ppm/96 hr, static bioassay in fresh water at 23 °C: mild aeration applied after 24 hr LC_{50} Menidia beryllina (tidewater silverside) 78 ppm/96 hr, static bioassay in synthetic seawater at 23 °C: mild aeration applied after 24 hr
Cleanup/Disposal: Guide No. 128: Eliminate all ignition sources (no smoking, flares, sparks or flames in immediate area). All equipment used when handling the product must be grounded. Do not touch or walk through spilled material. Stop leak if you can do it without risk. Prevent entry into waterways, sewers, basements or confined areas. A vapor suppressing foam may be used to reduce vapors. Absorb or cover with dry earth, sand or other non-combustible material and transfer to containers. Use clean non-sparking tools to collect absorbed material. Large Spills: Dike far ahead of liquid spill for later disposal. Water spray may reduce vapor; but may not prevent ignition in closed spaces.

Environmental Physical Data

Octanol/Water Partition Coefficient: log K_{ow} = calculated at 0.10 to 0.38
BCF: no food chain concentration potential

Regulations

RCRA 40CFR: Not listed
CERCLA: 40CFR 302.4: Not listed
SARA 40CFR 372.65: Not listed
SARA EHS 40CFR 355: Not listed
TSCA: Listed

Analytical Methods

Water / Groundwater: ASTM D3695

ETH6400 CAS #: 629-14-1

ETHYLENE GLYCOL DIETHYL ETHER

RTECS: KI1225000
DOT: UN1153; IMO3.3
EINECS Number: 211-076-1
Molecular Formula: $C_6H_{14}O_2$
Structured MF: $CH_3CH_2OCH_2CH_2OCH_2CH_3$
Formula Weight: 118.18

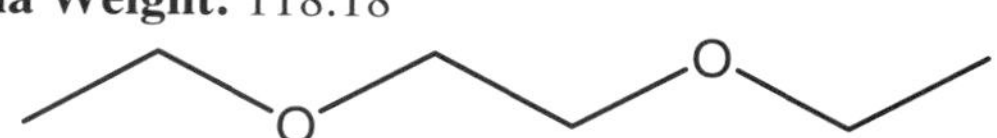

Chemical Structure

Synonyms: 1,2-DIETHOXYETHANE; DIETHYL CELLOSOLVE; DIETHYLETHER ETHYLENGLYKOLU; ETHANE,1,2-DIETHOXY-; 1,2-ETHANEDIOL,DIETHYL ETHER; 2-ETHOXYETHYL ETHYL ETHER; ETHYL GLYME; GLYME-1
Description: colorless liquid; sweetish odor
Use: solvent for ester gum, shellac, some resins and some oils; as a solvent and diluent for detergents and in organic synthesis as a reaction medium

Physical Properties

Boiling Point: 123.5 °C (254 °F) at 760 mm Hg
Freezing Point: -74 °C (-101.2 °F)
Specific Gravity: 0.8484 at 20 °C
Vapor Density: 4.07 Air=1
Saturated Vapor Density: 1.245642033 kg/m^3
Density: 0.8484 g/mL at 20 °C
Vapor Pressure: 9.4 mm Hg at 20 °C
Water Solubility: 0.02%
Other Solubilities: 95% Ethanol: >=100 mg/ml at 19 °C; Acetone: >=100 mg/ml at 19 °C; Benzene: >10%; DMSO: >=100 mg/ml at 19 °C; Ether: >10%.
Surface Tension: Estimated at 26 dynes/cm at 20 °C
Refraction Index: 1.3860 at 20 °C/D
Flash Point: 35 °C Open Cup
Autoignition Temperature: 208 °C

RTECS Toxicity Data

Acute Oral: Rat LD_{50} Dose: 4390 mg/kg; Toxic Effects: Behavioral - General anesthetic; Gastrointestinal - Other changes; Kidney, Ureter, and Bladder - Other changes. Guinea Pig LD_{50} Dose: 2440 mg/kg; Toxic Effects: Behavioral - General anesthetic; Gastrointestinal - Other changes; Kidney, Ureter, and Bladder - Other changes.
Acute Inhalation: Rat LC_{Lo} Dose: 8000 ppm/4hr.
Irritation Eye: Rabbit Standard Draize Test Dose: 100 mg; Reaction: severe.
Reproductive/Teratogenic: Rabbit Route: Oral; Dose: 1400 mg/kg; Duration: female 6-19D of pregnancy; Effects on Fertility - Post-implantation mortality; Effects on Embryo or Fetus - Fetotoxicity; Specific Developmental Abnormalities - Musculoskeletal system. Mouse Route: Oral; Dose: 23640

mg/kg; Duration: female 7-14D of pregnancy; Effects on Fertility - Post-implantation mortality. Mouse Route: Oral; Dose: 23640 mg/kg; Duration: female 6-13D of pregnancy; Effects on Fertility - Litter size; Effects on Newborn - Live birth index; Growth statistics. Mouse Route: Oral; Dose: 5000 mg/kg; Duration: female 6-15D of pregnancy; Specific Developmental Abnormalities - Craniofacial (including nose and tongue); Musculoskeletal system. Mouse Route: Oral; Dose: 10 gm/kg; Duration: female 6-15D of pregnancy; Effects on Embryo or Fetus - Fetotoxicity; Fetal death.

Hazard Overviews

Flammable

Fire Diamond

Health: Irritating to eyes/respiratory tract. Also Causes: transitory injury of the cornea, CNS depression, anemia, nausea, vomiting, headache, diarrhea, dyspnea, kidney damage, hemoglobinurie, hematuria, and red blood cell hemolysis, abdominal and lumbar pain, polyuria, oliguria, anuria, or pathological lesions in the brain, lungs, liver, and heart. Chronic Effects: liver, kidney, and nervous system damage, reproductive effects.

Fire: Flammable. For small fires use dry chemical, carbon dioxide, or alcohol-resistant foam. For large fires use fog, or alcohol-resistant foam. Water may be ineffective; use in absence of other extinguishing media.

Reactivity: Stable. Hazardous polymerization cannot occur. Avoid: excessive vapor generation. Incompatible with: oxidizing materials. Hazardous decomposition products: carbon dioxide.

Carcinogenicity: IARC - Not listed; NIOSH - Not listed; NTP - Not listed; ACGIH - Not listed; OSHA - Not listed; EPA - Not listed; MAK - Not listed

Primary Target Organs:

Eyes

Mucous Membranes

Nervous System

Liver

Kidneys

Blood

Environmental

Cleanup/Disposal: Guide No. 127: Eliminate all ignition sources (no smoking, flares, sparks or flames in immediate area). All equipment used when handling the product must be grounded. Do not touch or walk through spilled material. Stop leak if you can do it without risk. Prevent entry into waterways, sewers, basements or confined areas. A vapor suppressing foam may be used to reduce vapors. Absorb or cover with dry earth, sand or other non-combustible material and transfer to containers. Use clean non-sparking tools to collect absorbed material. Large Spills: Dike far ahead of liquid spill for later disposal. Water spray may reduce vapor; but may not prevent ignition in closed spaces.

Environmental Physical Data

BOD: 0.10 lb/lb, 5 days

Regulations

RCRA 40CFR: Not listed
CERCLA: 40CFR 302.4: Not listed
SARA 40CFR 372.65: Not listed
SARA EHS 40CFR 355: Not listed
TSCA: Listed

ETH6450 **CAS #: 628-96-6**

ETHYLENE GLYCOL DINITRATE

RTECS: KW5600000
EINECS Number: 211-063-0
Molecular Formula: $C_2H_4N_2O_6$
Structured MF: $O_2NOCH_2CH_2ONO_2$
Formula Weight: 152.06
Synonyms: DINITRATE D'ETHYLENE GLYCOL; DINITRATO DE ETILENGLICOL; DINITROGLICOL; DINITROGLYCOL; EGDN; 1,2-ETHANEDIOL DINITRATE; ETHANEDIOL DINITRATE; 1,2-ETHANEDIOL,DINITRATE; ETHYLENE DINITRATE; ETHYLENE GLYCOL,DINITRATE; ETHYLENE NITRATE; ETHYLENGLYKOLDINITRAT; GLYCOL DINITRATE; GLYCOL (DINITRATE DE); GLYCOLDINITRAAT; GLYKOLDINITRAT; NITROGLYCOL; NITROGLYKOL
Description: yellowish liquid
Use: to lower the freezing point of dynamite mixture and increase its stability

Physical Properties

Boiling Point: 197 °C (387 °F) to 200 °C (392 °F) at 760 mm Hg
Freezing Point: -22.3 °C (-8.14 °F)
Specific Gravity: 1.4918 at 20 °C/4 °C
Vapor Density: 5.24 Air=1
Saturated Vapor Density: 1.200482413 kg/m^3
Vapor Pressure: 0.072 mm Hg at 25 °C
Water Solubility: Insoluble
Other Solubilities: Soluble in Carbon Tetrachloride, Benzene, Toluene, Acetone; Soluble in dilute alkali; miscible with most organic solvents.
Refraction Index: 1.499 at 15 °C
Flash Point: 215 °C Closed Cup

RTECS Toxicity Data

Acute Dermal: Rabbit LD_{Lo} Route: Subcutaneous Dose: 300 mg/kg; Toxic Effects: Vascular - Other changes; Blood - Normocytic anemia. Cat LD_{Lo} Route: Subcutaneous Dose: 75 mg/kg; Toxic Effects: Blood - Normocytic anemia.

Hazard Overviews

Explosive

Fire Diamond

Health: Causes: throbbing headache, immediate drop in blood pressure, dizziness, nausea, vomiting, abdominal pain, CNS depression, tachycardia, diastolic hypertension, reduced pulse pressure.

Fire: Explosive. Use water spray, alcohol foam, dry chemical, or carbon dioxide to fight fires from a remote or explosion-proof location. Use water spray to cool fire-exposed containers.

Reactivity: Stable. Hazards polymerization cannot occur. Avoid: sources of ignition; contact with any material until compatibility is known. Incompatible with: sources of ignition. Hazardous decomposition products: oxides of nitrogen; carbon monoxide.

Carcinogenicity: IARC - Not listed; NIOSH - Listed as carcinogen; NTP - Not listed; ACGIH - Not listed; OSHA - Not listed; EPA - Not listed; MAK - Not listed

Primary Target Organs:

Gastro-intestinal

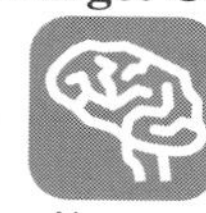
Nervous System

Cardio-vascular

Blood

Exposure Limits

OSHA PEL: STEL: 0.2 ppm; 1 mg/m^3; skin.

OSHA PEL Vacated 1989 Limits: STEL: 0.1 mg/m^3.

ACGIH TLV: TWA: 0.05 ppm; 0.31 mg/m^3.

NIOSH REL: STEL: 0.1 mg/m^3; skin.

NIOSH IDLH: 75 mg/m^3.

DFG MAK: TWA: 0.05 ppm; 0.3 mg/m^3.

Respirator Recommendation

Exposure Range: >0.1 to 5 mg/m^3 Supplied Air, Constant Flow/Pressure Demand, Half Mask

Exposure Range: >5 to <75 mg/m^3 Supplied Air, Constant Flow/Pressure Demand, Full Face

Exposure Range: 75 to unlimited mg/m^3 Self-contained Breathing Apparatus, Pressure Demand, Full Face

Note: odor threshold unknown

Environmental

Ecotoxicity: Fishes: perch 3-4d LOLC 5 mg/l

Environmental Physical Data

Octanol/Water Partition Coefficient: log K_{ow} = 1.16

Regulations

RCRA 40CFR: Not listed

CERCLA: 40CFR 302.4: Not listed

SARA 40CFR 372.65: Not listed

SARA EHS 40CFR 355: Not listed

TSCA: Listed

Analytical Methods

Air: ASTM D4490

ETH6550 CAS #: 111-76-2

ETHYLENE GLYCOL MONOBUTYL ETHER

RTECS: KJ8575000

DOT: UN2369; IMO6.1

EINECS Number: 203-905-0

Molecular Formula: $C_6H_{14}O_2$

Structured MF: $CH_3(CH_2)_3OCH_2CH_2OH$

Formula Weight: 118.20

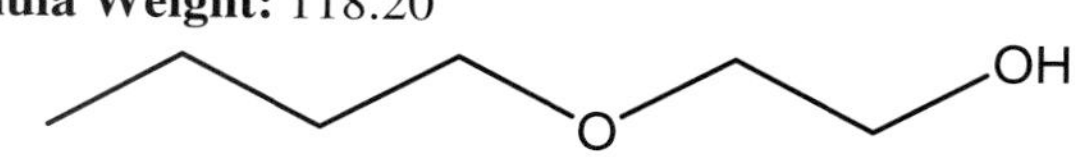

Chemical Structure

Synonyms: BUCS; BUTOKSYETYLOWY ALKOHOL; 2-BUTOSSI-ETANOLO; 2-BUTOXY-AETHANOL; 2-BUTOXY-1-ETHANOL; 2-BUTOXYETHANOL; BETA-BUTOXYETHANOL; BUTOXYETHANOL; N-BUTOXYETHANOL; BUTYL CELLOSOLVE; BUTYL CELLU-SOL; O-BUTYL ETHYLENE GLYCOL; BUTYL GLYCOL; BUTYL OXITOL; BUTYLCELOSOLV; BUTYLGLYCOL; CHIMEC NR; DOWANOL EB; EGBE; EKTASOLVE EB; EPA PESTICIDE CHEMICAL CODE 011501; ETER MONOBUTILICO DEL ETILENGLICOL; ETHANOL,2-BUTOXY-; ETHER MONOBUTYLIQUE DE L'ETHYLENEGLYCOL; ETHYLENE GLYCOL N-BUTYL; ETHYLENE GLYCOL BUTYL ETHER; ETHYLENE GLYCOL N-BUTYL ETHER; ETHYLENE GLYCOL,MONOBUTYL ETHER; GAFCOL EB; GLYCOL BUTYL ETHER; GLYCOL ETHER EB; GLYCOL ETHER EB ACETATE; GLYCOL MONOBUTYL ETHER; JEFFERSOL EB; MONOBUTYL ETHER OF ETHYLENE GLYCOL; MONOBUTYL ETHYLENE GLYCOL ETHER; MONOBUTYL GLYCOL ETHER; 3-OXA-1-HEPTANOL; POLY-SOLV EB

Description: clear liquid; pleasant odor

Use: solvent for nitrocellulose, resins, grease, oil and albumin; dry cleaning; spray lacquers; quick-drying lacquers; varnishes; enamels; varnish removers; textiles (preventing spotting in printing or dyeing); emulsifier for petroleum

Physical Properties

Boiling Point: 171 °C (340 °F) to 172 °C (342 °F)

Freezing Point: -70 °C (-94 °F)

Specific Gravity: 0.9012 at 20 °C/4 °C

Vapor Density: 4.07 Air=1

Saturated Vapor Density: 1.203691034 kg/m^3

Vapor Pressure: 0.76 mm Hg at 20 °C

Water Solubility: 1 parts in 20 parts Water

Other Solubilities: 95% Ethanol: >=100 mg/ml at 22 °C; Acetone: >=100 mg/ml at 22 °C; DMSO: >=100 mg/ml at 22 °C; Ether: Soluble; Mineral oil: Soluble; Most organic solvents: Soluble.

Surface Tension: 27.4 dynes/cm at 25 °C

Odor Threshold: 0.10 ppm

Refraction Index: 1.4198 at 20 °C/D

Evaporation Rate: 0.1

Critical Temperature: 368 °C

Critical Pressure: 470 psia

Ionization Potential (eV): 10.00

Flash Point: 61.667 °C Closed Cup

Autoignition Temperature: 238 °C

LEL: 1.1% v/v

UEL: 12.7% v/v at 135 °C

RTECS Toxicity Data

Acute Oral: Woman TD_{Lo} Dose: 600 mg/kg; Toxic Effects: Behavioral - Coma; Lungs, Thorax, or Respiration - Dyspnea; Nutritional and gross metabolic - Metabolic acidosis. Woman

TD_{Lo} Dose: 7813 uL/kg; Toxic Effects: Behavioral - Coma; Vascular - BP lowering not characterized in autonomic section; Nutritional and gross metabolic - Metabolic acidosis. Rat LD_{50} Dose: 470 mg/kg.

Acute Inhalation: Rat LC_{50} Dose: 450 ppm/4hr; Toxic Effects: Behavioral - Ataxia; Nutritional and gross metabolic - Weight loss or decreased weight gain. Human TC_{Lo} Dose: 195 ppm/8hr; Toxic Effects: Gastrointestinal - Nausea or vomiting. Human TC_{Lo} Dose: 100 ppm; Toxic Effects: Sense organs and special senses - Other; Sense organs and special senses - Other; Lungs, Thorax, or Respiration - Other changes.

Acute Dermal: Rabbit LD_{50} Route: Skin; Dose: 220 mg/kg. Mouse LD_{Lo} Route: Subcutaneous Dose: 500 mg/kg.

Chronic (Multiple Dose) Oral: Rat Dose: 139 gm/kg/90D-C; Toxic Effects: Liver - Changes in liver weight; Kidney, Ureter, and Bladder - Changes in kidney weight; Nutritional and gross metabolic - Weight loss or decreased weight gain. Rat Dose: 9324 mg/m3/21D-C; Toxic Effects: Behavioral - Fluid intake; Nutritional and gross metabolic - Weight loss or decreased weight gain. Rat Dose: 1500 mg/kg/12D-I; Toxic Effects: Blood - Pigmented or nucleated red Blood cells; Blood - Other changes; Blood - Changes in erythrocite (RBC) cell count. Rat Dose: 13290 mg/kg/6W-I; Toxic Effects: Liver - Changes in liver weight; Blood - Changes in erythrocite (RBC) cell count; Biochemical - Phosphatases.

Chronic (Multiple Dose) Inhalation: Rat Dose: 1540 mg/m^3/7H/5W-I; Toxic Effects: Blood - Changes in erythrocite (RBC) cell count. Rat Dose: 432 ppm/7H/30D-I; Toxic Effects: Kidney, Ureter, and Bladder - Hematuria; Blood - Other changes; DEATH. Rat Dose: 245 ppm/6H/9D-I; Toxic Effects: Liver - Changes in Liver weight; Blood - Pigmented or nucleated red Blood cells; Blood - Changes in erythrocite (RBC) cell count. Rat Dose: 10 mg/m^3/24H/13W-C; Toxic Effects: Endocrine - Hypoglycemia; Blood - Changes in erythrocite (RBC) cell count; Biochemical - Transaminases.

Irritation Eye: Rabbit Standard Draize Test Dose: 100 mg/24H; Reaction: moderate. Rabbit Standard Draize Test Dose: 100 mg; Reaction: severe.

Irritation Skin: Rabbit Open Draize Test Dose: 500 mg open; Reaction: mild.

Reproductive/Teratogenic: Rat Route: Oral; Dose: 6279 mg/kg; Duration: male 13W prior to mating; Paternal Effects - Spermatogenesis. Rat Route: Inhalation; Dose: 200 ppm/6H; Duration: female 6-15D of pregnancy; Maternal Effects - Uterus, cervix, vagina; Effects on Fertility - Post-implantation mortality; Litter size. Rat Route: Inhalation; Dose: 25 ppm/6H; Duration: female 6-15D of pregnancy; Specific Developmental Abnormalities - Musculoskeletal system. Rat Route: Inhalation; Dose: 12 mg/kg/4H; Duration: female 1-19D of pregnancy; Effects on Fertility - Post-implantation mortality.

Mutagenic: Bacteria - S Typhimurium Mutations in Microorganisms; Dose: 19 umol/plate (-S9), 76 umol/plate (+S9).

Hazard Overviews

Fire Diamond

Health: Irritating to eyes/skin/respiratory tract. Toxic. Also Causes: nausea, headache; high concentration: hemolytic anemia, unconsciousness, liver and kidney damage. Chronic Effects: may cause reproductive and birth defects.

Fire: Combustible. Can form explosive mixtures in the air. Use water as fog, dry chemical, carbon dioxide, or alcohol-resistant foam. Do not use a solid stream of water since the stream scatters and spreads the fire. Use water spray to cool fire-exposed tanks or containers.

Reactivity: Stable. Hazardous polymerization cannot occur. Avoid: plastics; rubber; coatings; auto-oxidation may produce peroxides. Incompatible with: oxidizing materials; heat; flame; metallic aluminum . Hazardous decomposition products: toxic vapors; gases.

Carcinogenicity: IARC - Not listed; NIOSH - Listed as carcinogen; NTP - Not listed; ACGIH - Not listed; OSHA - Not listed; EPA - Not listed; MAK - Not listed

Primary Target Organs:

Eyes

Skin

Respiratory System

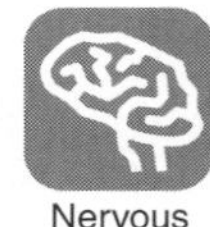
Nervous System

Liver

Kidneys

Exposure Limits

OSHA PEL: TWA: 50 ppm; 240 mg/m^3; skin.

OSHA PEL Vacated 1989 Limits: TWA: 25 ppm; 120 mg/m^3.

ACGIH TLV: TWA: 25 ppm; 121 mg/m^3.

NIOSH REL: TWA: 5 ppm; 24 mg/m^3.

NIOSH IDLH: 700 ppm.

DFG MAK: TWA: 20 ppm; 100 mg/m^3.

Respirator Recommendation

Exposure Range: >50 to 500 ppm Air Purifying, Negative Pressure, Half Mask

Exposure Range: >500 to <700 ppm Air Purifying, Negative Pressure, Full Face

Exposure Range: 700 to unlimited ppm Self-contained Breathing Apparatus, Pressure Demand, Full Face

Cartridge Color: black

Note: use dust/mist prefilter if mist is present

Environmental

Ecotoxicity: LC_{50} Poecilia reticulata (guppy) 983 ppm/7 day. /Conditions of bioassay not specified LC_{50} Crangon crangon (brown shrimp) 800 mg/l/48 hr (range: 600-1000 mg/l). /Conditions of bioassay not specified LC_{50} Menidia beryllina 1250 ppm/96 hr (static bioassay in synthetic seawater at 23 °C, mild aeration applied after 24 hr)

Environmental Fate: It is not expected to undergo hydrolysis or direct photolysis in the environment. The complete miscibility in water suggests that volatilization, adsorption and bioconcentration are not important fate processes. This is supported by an estimated Henry's Law constant of 2.08 $x10^{-8}$ atm-cu m/mole at 25 °C which indicates that volatilization from environmental waters and moist soil should be extremely slow. A low estimated log BCF suggests it should not bioconcentrate among aquatic organisms. A low K_{oc} indicates it should not partition from the water column to organic matter contained in sediments and suspended solids, and it should be highly mobile in soil. Limited monitoring data has shown it can leach to ground water. Aqueous screening test data indicate that biodegradation is likely to be the most important removal mechanism from aerobic soil and water. In the atmosphere, it is expected to exist almost entirely in the vapor phase and reactions with photochemically produced hydroxyl radicals should be important (estimated half-life of 17 hours). Physical removal from air by precipitation and dissolution in clouds may occur; however, its short atmospheric residence time suggests that wet deposition is of limited importance.

Cleanup/Disposal: Guide No. 152: Do not touch damaged containers or spilled material unless wearing appropriate protective clothing. Stop leak if you can do it without risk. Prevent entry into waterways, sewers, basements or confined areas. Cover with plastic sheet to prevent spreading. Absorb or cover with dry earth, sand or other non-combustible material and transfer to containers. Do not get water inside containers.

Environmental Physical Data

Henry's Law Constant: calculated at 2.08 $x10^{-8}$
Octanol/Water Partition Coefficient: log K_{ow} = 0.83
Sorption Partition Coefficient: K_{oc} = 67
BCF: calculated at 0.40
BOD: theoretical 26%, 5 days

Regulations

RCRA 40CFR: Not listed
CERCLA: 40CFR 302.4: Not listed
SARA 40CFR 372.65: Not listed
SARA EHS 40CFR 355: Not listed
TSCA: Listed

Analytical Methods

Air: ASTM D3686, D3687, D4490
Indoor / Expired Air: NIOSH 1403

ETH6600 **CAS #: 112-07-2**

ETHYLENE GLYCOL MONOBUTYL ETHER ACETATE

RTECS: KJ8925000
EINECS Number: 203-933-3
Molecular Formula: $C_8H_{16}O_3$
Structured MF: $C_4H_9O(CH_2)_2OCOCH_3$
Formula Weight: 160.24

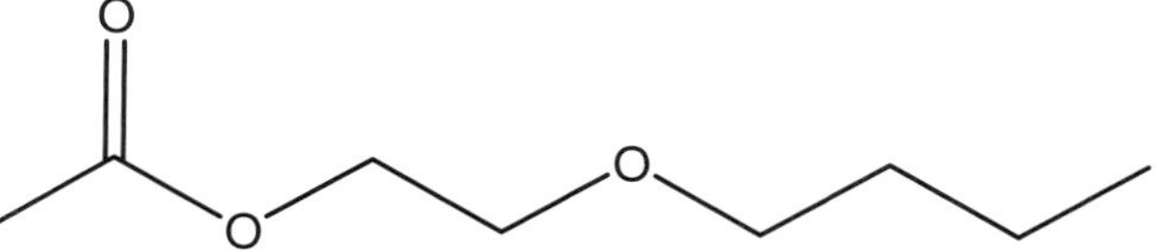

Chemical Structure

Synonyms: ACETIC ACID, 2-BUTOXYETHYL ESTER; 2-BUTOXYETHANOL ACETATE; 2-BUTOXYETHANOL,ACETATE; 2-BUTOXYETHYL ACETATE; BUTOXYETHYL ACETATE; 2-BUTOXYETHYLESTER KYSELINY OCTOVE; BUTYL CELLOSOLVE ACETATE; BUTYL GLYCOL ACETATE; BUTYLCELOSOLVACETAT; BUTYLGLYCOL ACETATE; EGBEA; EKTASOLVE EB; EKTASOLVE EB ACETATE; ETHANOL, 2-BUTOXY-,ACETATE; ETHYLENE GLYCOL BUTYL ETHER ACETATE; GLYCOL MONOBUTYL ETHER ACETATE
Description: colorless liquid; fruity odor
Use: high-boiling solvent for nitrocellulose lacquers, epoxy resins, multicolor lacquers; film coalescing aid for polyvinyl acetate latex

Physical Properties

Boiling Point: 192.3 °C (378 °F)
Freezing Point: -64.5 °C (-84.1 °F)
Specific Gravity: 0.9422 at 20 °C/20 °C
Vapor Density: 5.4 Air=1
Vapor Pressure: 0.3 mm Hg
Water Solubility: 1.1 g/100 g at 20 °C
Other Solubilities: Soluble in hydrocarbons & organic solvents.
Surface Tension: Estimated at 26 dynes/cm
Refraction Index: 1.4200 at 20 °C
Flash Point: 71 °C Closed Cup
Autoignition Temperature: 341 °C
LEL: 0.88% v/v
UEL: 8.54% v/v at 135 °C

RTECS Toxicity Data

Acute Oral: Rat LD_{50} Dose: 2400 mg/kg; Toxic Effects: Kidney, Ureter, and Bladder - Hematuria; Kidney, Ureter, and Bladder - Other changes in urine composition. Mouse LD_{50} Dose: 3200 mg/kg.
Acute Dermal: Rabbit LD_{50} Route: Skin; Dose: 1500 mg/kg; Toxic Effects: Kidney, Ureter, and Bladder - Hematuria; Kidney, Ureter, and Bladder - Other changes in urine composition; Blood - Normocytic anemia.
Irritation Eye: Rabbit Standard Draize Test Dose: 500 mg/24H; Reaction: mild.
Irritation Skin: Rabbit Open Draize Test Dose: 500 mg open; Reaction: mild.

Hazard Overviews

Fire Diamond

Health: Mildly irritating to skin/respiratory tract. Also Causes: gastrointestinal irritation upon ingestion.

Fire: Combustible. To fight fire, use alcohol foam.
Reactivity: Stable. Hazardous polymerization cannot occur. Incompatible with: oxidizers; nitrates. Hazardous decomposition products: acrid smoke; irritating fumes.
Carcinogenicity: IARC - Not listed; NIOSH - Listed as carcinogen; NTP - Not listed; ACGIH - Not listed; OSHA - Not listed; EPA - Not listed; MAK - Not listed
Primary Target Organs:

Skin

Respiratory System

Exposure Limits
NIOSH REL: TWA: 5 ppm; 33 mg/m^3.
DFG MAK: TWA: 20 ppm; 135 mg/m^3.

Environmental

Environmental Fate: If released to the atmosphere, it will degrade primarily by reaction with photochemically produced hydroxyl radicals (estimated half-life of 11.8 hr). If released to soil or water, it is expected to degrade via biodegradation. One biodegradation screening study has demonstrated that it is readily biodegradable. It may leach readily in soils based upon an estimated K_{oc} of 26. The importance of leaching may be lessened if rapid biodegradation occurs.
Cleanup/Disposal: Guide No. 153: Eliminate all ignition sources (no smoking, flares, sparks or flames in immediate area). Do not touch damaged containers or spilled material unless wearing appropriate protective clothing. Stop leak if you can do it without risk. Prevent entry into waterways, sewers, basements or confined areas. Absorb or cover with dry earth, sand or other non-combustible material and transfer to containers. Do not get water inside containers.

Environmental Physical Data

Henry's Law Constant: estimated at 7.19 x10^{-6}
Octanol/Water Partition Coefficient: log K_{ow} = measured at 1.51
Sorption Partition Coefficient: K_{oc} = estimated at 26
BCF: calculated at 3.2

Regulations

RCRA 40CFR: Not listed
CERCLA: 40CFR 302.4: Not listed
SARA 40CFR 372.65: Not listed
SARA EHS 40CFR 355: Not listed
TSCA: Listed

ETH6650 CAS #: 122-99-6

ETHYLENE GLYCOL MONOPHENYL ETHER

RTECS: KM0350000
EINECS Number: 204-589-7
Molecular Formula: $C_8H_{10}O_2$
Structured MF: $C_6H_5OCH_2CH_2OH$
Formula Weight: 138.16

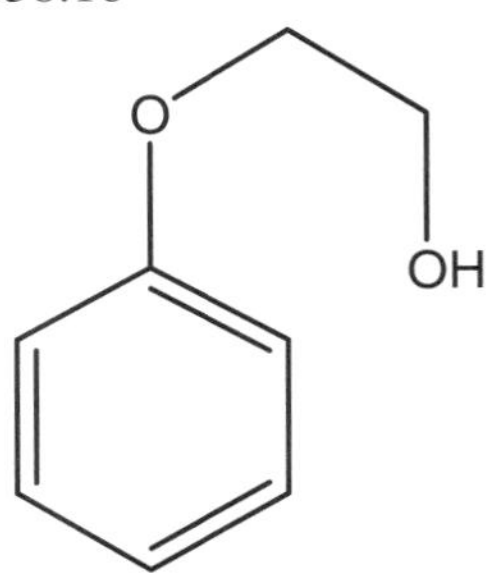

Chemical Structure

Synonyms: AROSOL; DOWANOL EP; DOWANOL EPH; EMERESSENCE 1160; EMERY 6705; ETHANOL,2-PHENOXY-; ETHYLENE GLYCOL PHENYL ETHER; 2-FENOXYETHANOL; FENYL-CELLOSOLVE; FENYLCELOSOLV; GLYCOL MONOPHENYL ETHER; 2-HYDROXYETHYL PHENYL ETHER; BETA-HYDROXYETHYL PHENYL ETHER; 1-HYDROXY-2-PHENOXYETHANE; PHENOXETHOL; PHENOXETOL; 2-PHENOXYETHANOL; BETA-PHENOXYETHANOL; PHENOXYETHANOL; 2-PHENOXYETHYL ALCOHOL; BETA-PHENOXYETHYL ALCOHOL; PHENOXYETHYL ALCOHOL; PHENOXYTOL; PHENYL CELLOSOLVE; PHENYLMONOGLYCOL ETHER; PLASTIAZAN-41; ROSE ETHER
Description: colorless, oily liquid; faint aromatic odor
Use: fixative for perfumes, a bactericide (in conjunction with quaternary ammonium compounds), a insect repellent, a topical antiseptic, a solvent for cellulose acetate, dyes, inks and resins, in organic synthesis of plasticizers, in germicides, in pharmaceuticals, in cosmetics and in preservatives

Physical Properties

Boiling Point: 245.2 °C (473 °F) at 760 mm Hg
Freezing Point: 14 °C (57.2 °F)
Specific Gravity: 1.1094 at 20 °C/20 °C
Vapor Density: 4.77 Air=1
Saturated Vapor Density: < 1.200059434 kg/m^3
Density: 1.105 to 1.11 g/mL at 20 °C
Bulk Density: 9.2 lbs/gal
Vapor Pressure: < 0.01 mm Hg at 20 °C
Water Solubility: 2.67 g/100 ml Water
Other Solubilities: 95% Ethanol: >=100 mg/ml at 20 °C; Acetone: >=100 mg/ml at 20 °C; Arachis oil: 1 in 50 at 20 °C; DMSO: >=100 mg/ml at 20 °C; Ether: Soluble; Glycerol: miscible; Olive oil: 1 in 50 at 20 °C; Sodium hydroxide solutions: Freely Soluble.
Surface Tension: 42 dynes/cm
Refraction Index: 1.534 at 20 °C/D
Flash Point: 121.111 °C Open Cup

RTECS Toxicity Data

Acute Oral: Rat LD_{50} Dose: 1260 mg/kg; Toxic Effects: Behavioral - General anesthetic; Gastrointestinal - Other changes; Kidney, Ureter, and Bladder - Other changes.
Acute Dermal: Rabbit LD_{50} Route: Skin; Dose: 5 mL/kg. Rat LD_{50} Route: Skin; Dose: 14422 mg/kg; Toxic Effects: Lungs, Thorax, or Respiration - Acute pulmonary edema.

Chronic (Multiple Dose) Oral: Rat Dose: 182 gm/kg/13W-I; Toxic Effects: Kidney, Ureter, and Bladder - Changes in kidney weight; Nutritional and gross metabolic - Weight loss or decreased weight gain; DEATH.
Chronic (Multiple Dose) Dermal: Rabbit Route: Skin; Dose: 14 gm/kg/14D-I; Toxic Effects: Blood - Other changes; DEATH.
Irritation Eye: Rabbit Standard Draize Test Dose: 250 ug/24H; Reaction: severe. Rabbit Standard Draize Test Dose: 6 mg; Reaction: moderate.
Irritation Skin: Rabbit Standard Draize Test Dose: 500 mg/24H; Reaction: mild.
Mutagenic: Bacteria - E Coli DNA Inhibition; Dose: 2000 ppm. Bacteria - E Coli Other Mutation Test Systems; Dose: 2000 ppm.

Hazard Overviews

Health: Severely irritating to eyes; irritating to skin/respiratory tract. Toxic. Other Acute Effects: harmful if swallowed, inhaled, or absorbed through skin. Chronic Effects: may cause reproductive disorders; may impair fertility.
Fire: Will burn. Hazards: emits toxic fumes. Extinguishing agents: water spray; carbon dioxide, dry chemical powder or appropriate foam. Precautions: combustible liquid.
Reactivity: Incompatible with: strong oxidizing agents, acid chlorides, acid anhydrides. Hazardous decomposition products: toxic fumes of: carbon monoxide, carbon dioxide.
Carcinogenicity: IARC - Not listed; NIOSH - Not listed; NTP - Not listed; ACGIH - Not listed; OSHA - Not listed; EPA - Not listed; MAK - Not listed
Primary Target Organs:

Eyes

Skin

Respiratory System

Environmental

Ecotoxicity: EC_{50} Pimephales promelas (fathead minnow) 344 mg/l/96 hr (confidence limit 337 - 352 mg/l), flow-through bioassay with measured concentrations, 26.6 °C, dissolved oxygen 6.0 mg/l, hardness 45.0 mg/l calcium carbonate, alkalinity 42.0 mg/l calcium carbonate, and pH 7.62. Effect: loss of equilibrium LC_{50} Pimephales promelas (fathead minnow) 344 mg/l/96 hr (confidence limit 337 - 352 mg/l), flow-through bioassay with measured concentrations, 26.6 °C, dissolved oxygen 6.0 mg/l, hardness 45.0 mg/l calcium carbonate, alkalinity 42. mg/l calcium carbonate, and pH 7.62
Environmental Fate: If released to air, it will degrade relatively rapidly by reaction with photochemically produced hydroxyl radicals (estimated half-life of 111.6 hr). Physical removal from air via wet deposition is possible since it is relatively soluble in water. If released to soil or water, it is expected to degrade through biodegradation. Leaching in soil is possible. It will evaporate slowly from terrestrial surfaces.
Cleanup/Disposal: Guide No. 171: Do not touch or walk through spilled material. Stop leak if you can do it without risk. Prevent dust cloud. Avoid inhalation of asbestos dust. Small Dry Spills: With clean shovel place material into clean, dry container and cover loosely; move containers from spill area. Small Spills: Take up with sand or other noncombustible absorbent material and place into containers for later disposal. Large Spills: Dike far ahead of liquid spill for later disposal. Cover powder spill with plastic sheet or tarp to minimize spreading. Prevent entry into waterways, sewers, basements or confined areas.

Environmental Physical Data

Henry's Law Constant: estimated at 2.0×10^{-7}
Octanol/Water Partition Coefficient: log K_{ow} = 1.16
Sorption Partition Coefficient: K_{oc} = estimated at 16
BCF: estimated at 2
BOD: 21% BODT, 5 days

Regulations

RCRA 40CFR: Not listed
CERCLA: 40CFR 302.4: Not listed
SARA 40CFR 372.65: Not listed
SARA EHS 40CFR 355: Not listed
TSCA: Listed

ETH6700 **CAS #: 2807-30-9**

ETHYLENE GLYCOL MONOPROPYL ETHER

RTECS: KM2800000
EINECS Number: 220-548-6
Molecular Formula: $C_5H_{12}O_2$
Structured MF: $CH_3CH_2CH_2OCH_2CH_2OH$
Formula Weight: 104.15

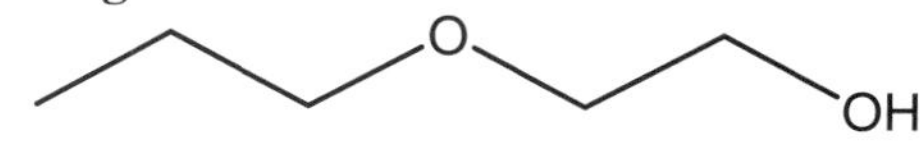

Chemical Structure

Synonyms: EKTASOLVE EP; ETHANOL 2-PROPOXY; ETHYLENE GLYCOL MONO PROPYL ETHER; ETHYLENE GLYCOL MONO-N-PROPYL ETHER; MONOPROPYL ETHER OF ETHYLENE GLYCOL; 2-PROPOXYETHANOL; PROPYL CELLOSOLVE; N-PROPYL OXITOL GLYCOL
Description: volatile liquid; mild ethereal odor

Physical Properties

Boiling Point: 150 °C (302 °F) to 152 °C (306 °F) at 760 mm Hg
Specific Gravity: 0.9112 at 20 °C/4 °C
Vapor Density: 3.6 Air=1
Saturated Vapor Density: 1.211865789 kg/m^3
Vapor Pressure: 2.9 mm Hg at 25 °C
Water Solubility: Soluble in Water
Other Solubilities: Soluble in Alcohol, Ether
Refraction Index: 1.412 at 25 °C
Flash Point: 51.667 °C Open Cup
LEL: 1.26% v/v
UEL: 15.8% v/v at 127 °C

RTECS Toxicity Data

Acute Oral: Rat LD_{50} Dose: 3089 mg/kg; Toxic Effects: Behavioral - Tremor; Lungs, Thorax, or Respiration - Dyspnea; Nutritional and gross metabolic - Weight loss or decreased weight gain. Mouse LD_{50} Dose: 1774 mg/kg; Toxic Effects: Behavioral - Somnolence (general depressed activity); Behavioral - Tremor; Lungs, Thorax, or Respiration - Dyspnea.

Acute Inhalation: Mouse LC_{50} Dose: 1530 ppm/7hr; Toxic Effects: Behavioral - Analgesia; Lungs, Thorax, or Respiration - Dyspnea; Kidney, ureter, and Bladder - Hematuria. Rat LC_{Lo} Dose: 2000 ppm/4hr.

Acute Dermal: Rabbit LD_{50} Route: Skin; Dose: 960 uL/kg. Guinea Pig LD_{50} Route: Skin; Dose: 1 gm/kg; Toxic Effects: Behavioral - Tremor; Lungs, Thorax, or Respiration - Dyspnea; Nutritional and gross metabolic - Weight loss or decreased weight gain.

Irritation Eye: Rabbit Standard Draize Test Dose: 100 mg; Reaction: severe. Rabbit Standard Draize Test Dose: 750 ug/24H; Reaction: severe.

Irritation Skin: Rabbit Standard Draize Test Dose: 500 mg/24H; Reaction: mild. Guinea Pig Standard Draize Test Dose: 500 mg; Reaction: mild.

Reproductive/Teratogenic: Rat Route: Inhalation; Dose: 100 ppm/6H; Duration: female 6-15D of pregnancy; Specific Developmental Abnormalities - Musculoskeletal system. Mouse Route: Oral; Dose: 16 gm/kg; Duration: female 7-14D of pregnancy; Effects on Newborn - Stillbirth.

Hazard Overviews

Flammable

Health: Severely irritating to eyes; irritating to skin/respiratory tract. Toxic. Other Acute Effects: harmful if swallowed, inhaled, or absorbed through skin; readily absorbed through skin; target organs: blood; lungs, liver; kidneys.

Fire: Flammable. Hazards: emits toxic fumes. Extinguishing agents: carbon dioxide, dry chemical powder or appropriate foam; water spray. Precautions: combustible liquid.

Reactivity: Stable. Hazardous polymerization will not occur. Incompatible with: strong oxidizing agents, may decompose on exposure to light, sensitive to air, sensitive to heat. Hazardous decomposition products: toxic fumes of: carbon monoxide, carbon dioxide.

Carcinogenicity: IARC - Not listed; NIOSH - Not listed; NTP - Not listed; ACGIH - Not listed; OSHA - Not listed; EPA - Not listed; MAK - Not listed

Primary Target Organs:

Eyes

Skin

Respiratory System

Liver

Kidneys

Blood

Environmental

Cleanup/Disposal: Guide No. 128: Eliminate all ignition sources (no smoking, flares, sparks or flames in immediate area). All equipment used when handling the product must be grounded. Do not touch or walk through spilled material. Stop leak if you can do it without risk. Prevent entry into waterways, sewers, basements or confined areas. A vapor suppressing foam may be used to reduce vapors. Absorb or cover with dry earth, sand or other non-combustible material and transfer to containers. Use clean non-sparking tools to collect absorbed material. Large Spills: Dike far ahead of liquid spill for later disposal. Water spray may reduce vapor; but may not prevent ignition in closed spaces.

Regulations

RCRA 40CFR: Not listed
CERCLA: 40CFR 302.4: Not listed
SARA 40CFR 372.65: Not listed
SARA EHS 40CFR 355: Not listed
TSCA: Listed

ETH6750 CAS #: 97-90-5

ETHYLENE METHACRYLATE

RTECS: OZ4400000
EINECS Number: 202-617-2
Molecular Formula: $C_{10}H_{14}O_4$
Formula Weight: 198.22

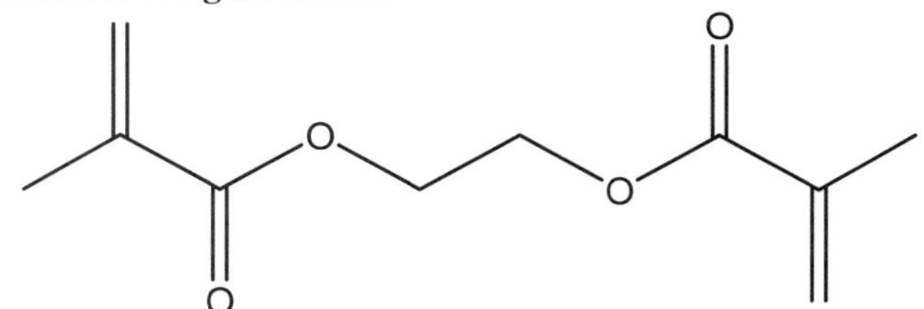

Chemical Structure

Synonyms: AGEFLEX EGDM; 1,2-BIS(METHACRYLOYLOXY)ETHANE; DIGLYCOL DIMETHACRYLATE; ETHANEDIOL DIMETHACRYLATE; ETHYLDIOL METACRYLATE; ETHYLDIOL METHACRYLATE; ETHYLENE GLYCOL BIS(METHACRYLATE); ETHYLENE GLYCOL DIMETHACRYLATE; GLYCOL DIMETHACRYLATE; METHACRYLIC ACID,ETHYLENE ESTER; 2-PROPENOIC ACID,2-METHYL-,1,2-ETHANEDIYL ESTER; 2-PROPENOIC ACID,2-METHYL-,1,2-ETHANEDIYL ESTER (9CI); SARTOMER SR 206; SR 206

Use: in polymers for dental restorative resins; in synthetic wound dressings; rate-controlling barrier in slow release capsules; in acrylic nail prepn; crosslinking agent for thermoset & UV curable acrylics; crosslinking agent for polymers used in contact lenses; crosslinking aid in peroxide curing of epdm & epm rubbers; as curing agent

Physical Properties

Boiling Point: 260 °C (500 °F)
Freezing Point: -40 °C (-40 °F)

Specific Gravity: 1.055 at 20 °C/20 °C
Other Solubilities: > 10% in Ethanol; > 10% in ligroin.
Refraction Index: 1.4532 at 25 °C

RTECS Toxicity Data

Acute Oral: Rat LD_{50} Dose: 3300 mg/kg. Mouse LD_{50} Dose: 2 gm/kg.
Mutagenic: Mouse Mutations in Mammalian Somatic Cells; Cell Type: lymphocyte; Dose: 5820 umol/L.

Hazard Overviews

Health: Irritating to eyes/skin/respiratory tract. Harmful. Other Acute Effects: harmful if swallowed, inhaled, or absorbed through skin; lachrymator; symptoms of exposure may include burning sensation, coughing, wheezing, laryngitis, shortness of breath, headache, nausea and vomiting.
Fire: Hazards: emits toxic fumes. Extinguishing agents: water spray; carbon dioxide, dry chemical powder or appropriate foam. Precautions: combustible liquid.
Carcinogenicity: IARC - Not listed; NIOSH - Not listed; NTP - Not listed; ACGIH - Not listed; OSHA - Not listed; EPA - Not listed; MAK - Not listed
Primary Target Organs:

Eyes

Skin

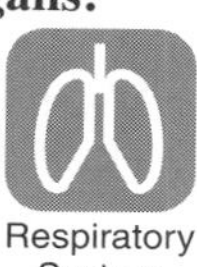
Respiratory System

Environmental

Regulations

RCRA 40CFR: Not listed
CERCLA: 40CFR 302.4: Not listed
SARA 40CFR 372.65: Not listed
SARA EHS 40CFR 355: Not listed
TSCA: Listed

ETH6800 CAS #: 75-21-8

ETHYLENE OXIDE

RTECS: KX2450000
DOT: UN1040; IMO2.3
EINECS Number: 200-849-9
Molecular Formula: C_2H_4O
Formula Weight: 44.06

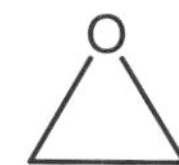

Chemical Structure

Synonyms: AETHYLENOXID; AMPROLENE; ANPROLENE; ANPROLINE; DIHYDROOXIRENE; DIMETHYLENE OXIDE; E O; E.O; ENT-26263; EPA PESTICIDE CHEMICAL CODE 042301; 1,2-EPOXY ETHANE; 1,2-EPOXYAETHAN; 1,2-EPOXYETHANE; EPOXYETHANE; ETHENE OXIDE; ETHOX; ETHYLEENOXIDE; ETHYLENE OXIDE,WITH NITROGEN; ETHYLENE (OXYDE D'); ETILENE (OSSIDO DI); ETO; ETYLENU TLENEK; FEMA NO 2433; MERPOL; OXACYCLOPROPANE; OXANE; ALPHA,BETA-OXIDOETHANE; OXIDOETHANE; OXIRAAN; OXIRAN; OXIRANE; OXIRENE,DIHYDRO-; OXYFUME; OXYFUME 12; STERILIZING GAS ETHYLENE OXIDE 100%; T-GAS
Description: colorless gas, liquid above 51 F (11 C); sweet, ether-like odor
Use: as an intermediate in the production of ethylene glycol, as an intermediate for polyethylene terephthalate polyester fiber and film production, and in the manufacture of non-ionic surface-active agent diethylene glycol, triethylene glycol, ethanolamines, chorine and chorine chloride, and other organic chemicals; as a fungicide for treatment by fumigation of books; dental, pharmaceutical, medical and scientific equipment and supplies (glass, metals, plastics, rubber or textiles), drugs, leather, motor oil, paper, soil, bedding for experimental animals, clothing, furs, furniture and transportation vehicles (jet aircraft, buses and railroad passenger cars); to sterilize foodstuffs such as spices, cocoa, flour, dried egg powder, desiccated coconut, dried fruits and dehydrated vegetables; to accelerate the maturing of tobacco leaves; as a rocket propellant, to sterilize surgical instruments, as a starting material for the manufacture of acrylonitrile, as a petroleum demulsifier and as an industrial sterilant (e.g. medical plastic tubing)

Physical Properties

Boiling Point: 10.7 °C (51 °F) at 760 mm Hg
Freezing Point: -111 °C (-167.8 °F)
Specific Gravity: 0.882 at 10 °C/10 °C
Vapor Density: 1.49
Density: 0.869 g/mL at 20 °C
Bulk Density: 7.25 lbs/gal at 20 °C
Vapor Pressure: Calculated 1314 mm Hg at 25 °C
Water Solubility: Miscible in all proportions
Other Solubilities: 95% Ethanol: miscible; Acetone: miscible; Alcohol: Soluble; Benzene: miscible; Carbon Tetrachloride: miscible; Ether: miscible; Methanol: miscible; Organic solvents: miscible.
Surface Tension: Liquid 24.3 dyne/cm at 20 °C
Odor Threshold: 520 to 1400 mg/m^3
Refraction Index: 1.3597 at 7 °C/D
Critical Temperature: 196 °C
Critical Pressure: 7.1941 $x10^6$ Pa
Ionization Potential (eV): 10.56
Flash Point: < -17.778 °C Open Cup
Autoignition Temperature: 429 °C
LEL: 3% v/v
UEL: 100% v/v

RTECS Toxicity Data

Acute Oral: Rat LD_{50} Dose: 72 mg/kg. Guinea Pig LD_{50} Dose: 270 mg/kg.
Acute Inhalation: Rat LC_{50} Dose: 800 ppm/4hr; Toxic Effects: Lungs, Thorax, or Respiration - Other changes; Liver - Other changes; Kidney, ureter, and Bladder - Other changes. Human TC_{Lo} Dose: 12500 ppm/10S; Toxic Effects: Sense organs and special senses - Other. Woman TC_{Lo} Dose: 500 ppm/2M; Toxic Effects: Behavioral - Convulsions or effect on seizure

threshold; Gastrointestinal - Nausea or vomiting; Lungs, Thorax, or Respiration - Other changes.

Acute Dermal: Rat LD_{50} Route: Subcutaneous Dose: 187 mg/kg. Cat LD_{Lo} Route: Subcutaneous Dose: 100 mg/kg.

Chronic (Multiple Dose) Inhalation: Rat Dose: 406 ppm/6H/6W-I; Toxic Effects: Nutritional and gross metabolic - Weight loss or decreased weight gain; DEATH. Rat Dose: 300 ug/m^3/24H/83D-C; Toxic Effects: Behavioral - Muscle contraction or spasticity; Blood - Other changes; Nutritional and gross metabolic - Changes in chlorine. Monkey Dose: 100 ppm/7H/2Y-I; Toxic Effects: Sense organs and special senses - Other.

Irritation Eye: Rabbit Standard Draize Test Dose: 18 mg/6H; Reaction: moderate.

Reproductive/Teratogenic: Rat Route: Inhalation; Dose: 100 ppm/6H; Duration: female 6-15D of pregnancy; Effects on Embryo or Fetus - Fetotoxicity. Rat Route: Inhalation; Dose: 100 ppm/6H; Duration: male 12W prior to mating; Effects on Newborn - Live birth index. Rat Route: Inhalation; Dose: 3600 ug/m^3/2 Duration: male 60D prior to mating; Paternal Effects - Testes, epididymis, sperm duct; Effects on Fertility - Pre-implantation mortality. Rat Route: Inhalation; Dose: 100 ppm/6H; Duration: female 12W prior to mating Effects on Fertility - Pre-implantation mortality; Effects on Newborn - Live birth index. Rat Route: Inhalation; Dose: 150 ppm/7H; Duration: female 7-16D of pregnancy; Effects on Embryo or Fetus - Fetotoxicity; Specific Developmental Abnormalities - Craniofacial (including nose and tongue); Musculoskeletal system. Monkey Route: Inhalation; Dose: 50 ppm/7H; Duration: male 96W prior to mating; Paternal Effects - Spermatogenesis. Monkey Route: Inhalation; Dose: 50 ppm/7H; Duration: male 2Y prior to mating; Paternal Effects - Spermatogenesis.

Mutagenic: Human DNA Damage; Cell Type: fibroblast; Dose: 5 mmol/L. Human Unscheduled DNA Synthesis; Cell Type: leukocyte; Dose: 4 mmol/L. Human Mutations in Mammalian Somatic Cells; Cell Type: fibroblast; Dose: 5 mmol/L. Human Cytogenetic Analysis; Route: Inhalation; Dose: 5 ppm/Y. Monkey Cytogenetic Analysis; Route: Inhalation; Dose: 100 ppm/7H/2Y-I. Human Sister Chromatid Exchenge; Cell Type: lymphocyte; Dose: 4 pph. Human Sister Chromatid Exchenge; Cell Type: fibroblast; Dose: 36 ppm/24H. Human Sister Chromatid Exchenge; Cell Type: lymphocyte; Dose: 10 mg/L. Human Sister Chromatid Exchenge; Route: Inhalation; Dose: 380 ppb/6H/4Y-I. Human Sister Chromatid Exchenge; Route: Inhalation; Dose: 60 ppm/12W-I. Monkey Sister Chromatid Exchenge; Route: Inhalation; Dose: 50 ppm/7H/2Y-I.

Tumorigenic: Rat Route: Oral; Dose: 1186 mg/kg/2Y-I; Toxic Effects: Tumorigenic - Carcinogenic by RTECS criteria; Gastrointestinal - Tumors; Liver - Tumors. Rat Route: Inhalation; Dose: 33 ppm/6H/2Y-I; Toxic Effects: Tumorigenic - Carcinogenic by RTECS criteria; Brain and coverings - Tumors; Blood - Leukemia. Rat Route: Oral; Dose: 5112 mg/kg/2Y-I; Toxic Effects: Tumorigenic - Carcinogenic by RTECS criteria; Gastrointestinal - Tumors; Liver - Tumors. Rat Route: Inhalation; Dose: 50 ppm/7H/2Y-I; Toxic Effects: Tumorigenic - Carcinogenic by RTECS criteria; Blood - Tumors. Rat Route: Inhalation; Dose: 33 ppm/6H/2Y-I; Toxic Effects: Tumorigenic - Equivocal tumorigenic agent by RTECS criteria; Brain and coverings - Tumors.

Hazard Overviews

Corrosive

Explosive

Flammable

Fire Diamond

Health: Corrosive to eyes/skin/respiratory tract. Toxic. Also Causes: headache, vomiting, incoordination. Chronic Effects: memory loss, alkylated hemoglobin; possible reproductive/cancer hazard.

Fire: Extremely flammable! Forms explosive mixtures in air. For small fires, use dry chemical, carbon dioxide, or alcohol-resistant foam. For large fires, use fog, or alcohol-resistant foam. Also beware of explosive decomposition. Fight fire from maximum distance. Use water spray only when other agents are unavailable because EtO must be very dilute before it can be completely extinguished. This dilution process is too time consuming for an urgent fire situation. Let fire burn unless leak can be stopped immediately.

Reactivity: Unstable, undergoes slow polymerization. Polymerization can occur with: ammonia; alkali hydroxides; amines; metallic potassium; acids; covalent halides (aluminum chloride; iron chloride; tin chloride; aluminum oxide; iron oxide; rust). Avoid: exposure to heat and ignition sources. Incompatible with: glycerol at 392 °F (200 °C); oxidizers; alcohols; mercaptans; magnesium perchlorate; alkane thiols; trimethylamine; copper; bromoethane; leather, rubber, and many plastics. Hazardous decomposition products: carbon dioxide; acrid smoke; irritating fumes.

Carcinogenicity: IARC - Group 2A, Probably carcinogenic to humans; NIOSH - Listed as carcinogen; NTP - Class 2A, Reasonably anticipated to be a carcinogen, limited evidence of carcinogenicity from studies in humans; ACGIH - Class A2, Suspected human carcinogen; OSHA - Listed as a carcinogen; EPA - Not listed; MAK - Class A2, Unmistakably carcinogenic in animal experimentation only

Primary Target Organs:

Eyes

Skin

Respiratory System

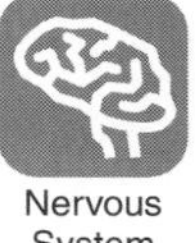
Nervous System

Liver

Kidneys

Exposure Limits

ACGIH TLV: TWA: 1 ppm; 0.18 mg/m^3.

NIOSH IDLH: 800 ppm.

Respirator Recommendation

Exposure Range: >1 to 50 ppm Supplied Air, Constant Flow/Pressure Demand, Half Mask

Exposure Range: >50 to <800 ppm Supplied Air, Constant Flow/Pressure Demand, Full Face

Exposure Range: 800 to unlimited ppm Self-contained Breathing Apparatus, Pressure Demand, Full Face

Note: poor warning properties

Environmental

Ecotoxicity: LC_{50} Goldfish 90 mg/l/24 hr modified ASTM D

Environmental Fate: Once in the atmosphere it will degrade very slowly by reaction with hydroxyl-radicals (estimated half-life 211 days). Releases into water will be removed by volatilization, hydrolysis and to a lesser extent, biodegradation. The volatilization half-lives for its removal from a model river and model lake are 5.9 hr and 3.8 days, respectively. It will not adsorb strongly to soil or bioconcentrate in fish.

Cleanup/Disposal: Guide No. 119: Eliminate all ignition sources (no smoking, flares, sparks or flames in immediate area). All equipment used when handling the product must be grounded. Fully encapsulating, vapor protective clothing should be worn for spills and leaks with no fire. Do not touch or walk through spilled material. Stop leak if you can do it without risk. Do not direct water at spill or source of leak. Use water spray to reduce vapors or divert vapor cloud drift. For chlorosilanes, use AFFF alcohol-resistant medium expansion foam to reduce vapors. If possible, turn leaking containers so that gas escapes rather than liquid. Prevent entry into waterways, sewers, basements or confined areas. Isolate area until gas has dispersed.

Environmental Physical Data

Henry's Law Constant: estimated at 1.48 $x10^{-4}$

Octanol/Water Partition Coefficient: log K_{ow} = -0.3

Sorption Partition Coefficient: K_{oc} = 16

BCF: none likely

Regulations

RCRA 40CFR: Listed Hazardous Waste No. U115 Toxic Waste Ignitable Waste

CERCLA: 40CFR 302.4: Listed per RCRA Section 3001 RQ: 10 lb (4.535 kg)

SARA 40CFR 372.65: Listed TPQ: 1000 lb

SARA EHS 40CFR 355: Listed TPQ: 10 lb

TSCA: Listed

Analytical Methods

Air: ASTM D4490

Soil: SW846 8015B, 8240B, 8260A, 8260B

Water / Groundwater: ASTM D3695

Indoor / Expired Air: NIOSH 1614, 3702; ASTM D4413

Plasma: EPA 29

ETH6850 CAS #: 420-12-2

ETHYLENE SULFIDE

RTECS: KX3500000

EINECS Number: 206-993-9

Molecular Formula: C_2H_4S

Formula Weight: 60.12

Chemical Structure

Synonyms: AETHYLENSULFID; 2,3-DIHYDROTHIIRENE; ETHYLENE EPISULFIDE; ETHYLENE EPISULPHIDE; ETHYLENE SULPHIDE; THIACYCLOPROPANE; THIIRAN; THIIRANE; THIIRENE,2,3-DIHYDRO-

Description: colorless liquid

Use: chemical intermediate; research chem for use as a biocide and chem int

Physical Properties

Boiling Point: 55 °C (131 °F) to 56 °C (133 °F)

Freezing Point: -109 °C (-164.2 °F)

Specific Gravity: 1.0368 at 0 °C/4 °C

Water Solubility: Immiscible with Water

Other Solubilities: miscible with Alcohol, Ether Soluble in Acetone, Chloroform

Refraction Index: 1.4935

Ionization Potential (eV): 9.05 +/-0.01

RTECS Toxicity Data

Acute Oral: Rat LD_{50} Dose: 178 mg/kg; Toxic Effects: Behavioral - General anesthetic; Lungs, Thorax, or Respiration - Structural or functional change in trachea or bronchi; Lungs, Thorax, or Respiration - Respiratory depression. Mouse LD_{50} Dose: 35600 ug/kg; Toxic Effects: Behavioral - Muscle weakness; Behavioral - Ataxia; Behavioral - Muscle contraction or spasticity.

Acute Inhalation: Rat LC_{50} Dose: 690 ppm/6hr; Toxic Effects: Lungs, Thorax, or Respiration - Fibrosis, focal (pneumoconiosis); Lungs, Thorax, or Respiration - Acute pulmonary edema; Blood - Hemorrhage. Mouse LC_{50} Dose: 1400 mg/m^3; Toxic Effects: Behavioral - Muscle weakness; Behavioral - Ataxia; Behavioral - Muscle contraction or spasticity.

Acute Dermal: Rat LD_{50} Route: Subcutaneous Dose: 90 mg/kg.

Irritation Eye: Rabbit Standard Draize Test Dose: 100 mg; Reaction: severe.

Tumorigenic: Rat Route: Subcutaneous; Dose: 400 mg/kg/50W-I; Toxic Effects: Tumorigenic - Equivocal tumorigenic agent by RTECS criteria; Tumorigenic - Tumors at site of application.

Hazard Overviews

Poison

Health: Severely irritating to eyes; irritating to skin. Poison. Other Acute Effects: may be fatal if inhaled, swallowed, or absorbed through skin; exposure can cause nausea, headache and vomiting. Chronic Effects: Possible carcinogen.

Fire: Hazards: vapor may travel considerable distance to source of ignition and flash back; container explosion may occur; forms explosive mixtures in air; emits toxic fumes. Extinguishing agents: carbon dioxide, dry chemical powder or appropriate foam; water may be effective for cooling, but may not effect extinguishment. Precautions: combustible liquid.

Reactivity: Incompatible with: strong bases, strong oxidizing agents. Hazardous decomposition products: toxic fumes of: carbon monoxide, carbon dioxide, sulfur oxides, hydrogen sulfide gas.

Carcinogenicity: IARC - Group 3, Not classifiable as to carcinogenicity to humans; NIOSH - Not listed; NTP - Not listed; ACGIH - Not listed; OSHA - Not listed; EPA - Not listed; MAK - Not listed

Primary Target Organs:

Eyes

Skin

Environmental

Regulations

RCRA 40CFR: Not listed
CERCLA: 40CFR 302.4: Not listed
SARA 40CFR 372.65: Not listed
SARA EHS 40CFR 355: Not listed
TSCA: Listed

ETH6900 CAS #: 96-45-7

ETHYLENE THIOUREA

RTECS: NI9625000
EINECS Number: 202-506-9
Molecular Formula: $C_3H_6N_2S$
Formula Weight: 102.17

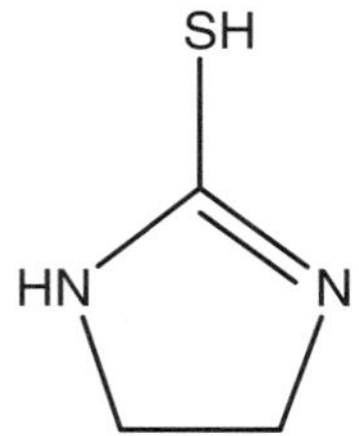

Chemical Structure

Synonyms: AKROCHEM ETU-22; APEROCHEM ETU-22; 4,5-DIHYDROIMIDAZOLE-2(3H)-THIONE; 4,5-DIHYDRO-2-MERCAPTOIMIDAZOLE; L'ETHYLENE THIOUREE; 1,3-ETHYLENE-2-THIOUREA; 1,3-ETHYLENETHIOUREA; ETHYLENETHIOUREA; N,N'-ETHYLENETHIOUREA; N,N-ETHYLENE-THIOUREA; N,N-ETHYLENETHIOUREA; ETU; 2-IMADAZOLINE-2-THIOL; IMIDAZOLE-2(3H)-THIONE,4,5-DIHYDRO-; 2-IMIDAZOLIDINE THIONE; 2-IMIDAZOLIDINE-2-THIONE; 2-IMIDAZOLIDINETHIONE; IMIDAZOLIDINETHIONE; IMIDAZOLINE,2-MERCAPTO-; 2-IMIDAZOLINE-2-THIOL; IMIDAZOLINE-2-THIOL; IMIDAZOLINE-2(3H)-THIONE; IMIDIZOLIDENETHIONE; 2-MERCAPTO-4,5-DIHYDROIMIDAZOLE; 2-MERCAPTO-2-IMIDAZOLINE; 2-MERCAPTOIMIDAZOLINE; MERCAPTOIMIDAZOLINE; MERCAZIN I; 2-MERKAPTOIMIDAZOLIN; NA-22; NA-22-D; NOCCELER 22; PENNAC CRA; RHENOGRAN ETU; RHODANIN S 62; ROBAC 22; RODANIN S-62; SANCELLER 22; SODIUM-22 NEOPRENE ACCELERATOR; SOXINOL 22; TETRAHYDRO-2H-IMIDAZOLE-2-THIONE; 2-THIOL-DIHYDROGLYOXALINE; 2-THIOLDIHYDROGLYOXALINE; THIOUREA,N,N'-(1,2-ETHANEDIYL)-; UREA,1,3-ETHYLENE-2-THIO-; VULKACIT NPV/C; VULKACIT NPV/C2; WARECURE C

Description: white to pale green crystals; faint amine odor

Use: as a polymer vulcanizing and curing agent; an accelerator in the curing of polychloroprene (neoprene) and other elastomers; in electroplating baths, as an intermediate for anti-oxidants, in insecticides, dyes, pharmaceuticals and synthetic resins

Physical Properties

Boiling Point: 230 °C (446 °F) to 312.78 °C (595 °F)
Freezing Point: 200 °C (392 °F) to 203 °C (397.4 °F)
Saturated Vapor Density: < 1.203983848 kg/m^3
Vapor Pressure: < 1 mm Hg at 20 °C
Water Solubility: 1 to 5 mg/mL at 18 °C
Other Solubilities: Slightly Soluble at room temperature in Acetic Acid, Naphtha.
Ionization Potential (eV): 8.15
Flash Point: 252 °C

RTECS Toxicity Data

Acute Oral: Rat LD_{50} Dose: 1832 mg/kg. Mouse LD_{50} Dose: 3 gm/kg.

Chronic (Multiple Dose) Oral: Rat Dose: 297 mg/kg/28D-C; Toxic Effects: Endocrine - Other changes; Blood - Changes in serum composition. Rat Dose: 764 mg/kg/90D-C; Toxic Effects: Endocrine - Other changes; Endocrine - Changes in thyroid weight; Blood - Changes in serum composition. Rat Dose: 220 mg/kg/7W-C; Toxic Effects: Behavioral - Food intake (animal); Endocrine - Changes in thyroid weight; Nutritional and gross metabolic - Weight loss or decreased weight gain. Rat Dose: 3600 mg/kg/17W-C; Toxic Effects: Endocrine - Other changes; Endocrine - Changes in thyroid weight; Nutritional and gross metabolic - Weight loss or decreased weight gain.

Irritation Eye: Rabbit Standard Draize Test Dose: 500 mg/24H; Reaction: mild.

Reproductive/Teratogenic: Rat Route: Oral; Dose: 60 mg/kg; Duration: female 13D of pregnancy; Specific Developmental Abnormalities - Urogenital system. Rat Route: Oral; Dose: 100 mg/kg; Duration: female 13D of pregnancy; Specific Developmental Abnormalities - Gastrointestinal system. Rat Route: Oral; Dose: 60 mg/kg; Duration: female 13D of pregnancy; Specific Developmental Abnormalities - Craniofacial (including nose and tongue). Rat Route: Oral; Dose: 10 mg/kg; Duration: female 17D of pregnancy; Effects on Newborn - Viability index; Weaning or lactation index. Rat Route: Oral; Dose: 50 mg/kg; Duration: female 12D of pregnancy; Effects on Embryo or Fetus - Fetotoxicity; Specific Developmental Abnormalities - Musculoskeletal system. Rat Route: Oral; Dose: 30 mg/kg; Duration: female

15D of pregnancy; Specific Developmental Abnormalities - Central nervous system; Effects on Newborn - Weaning or lactation index. Rat Route: Inhalation; Dose: 120 mg/m^3/3H Duration: female 7-14D of pregnancy; Effects on Embryo or Fetus - Fetotoxicity; Fetal death. Rat Route: Inhalation; Dose: 27200 ug/m^3/ Duration: female 7-14D of pregnancy; Specific Developmental Abnormalities - Musculoskeletal system. Rat Route: Oral; Dose: 60 mg/kg; Duration: female 13D of pregnancy; Specific Developmental Abnormalities - Urogenital system. Rat Route: Oral; Dose: 60 mg/kg; Duration: female 13D of pregnancy; Specific Developmental Abnormalities - Craniofacial (including nose and tongue).

Mutagenic: Rat Cytogenetic Analysis; Route: Parenteral; Dose: 200 mg/kg. Rat Cytogenetic Analysis; Route: Oral; Dose: 200 mg/kg.

Tumorigenic: Rat Route: Oral; Dose: 10920 mg/kg/2Y-C; Toxic Effects: Tumorigenic - Carcinogenic by RTECS criteria; Endocrine - Thyroid tumors. Rat Route: Oral; Dose: 5306 mg/kg/77W-C; Toxic Effects: Tumorigenic - Carcinogenic by RTECS criteria; Lungs, Thorax, or Respiration - Tumors; Endocrine - Thyroid tumors. Rat Route: Oral; Dose: 11466 mg/kg/78W-C; Toxic Effects: Tumorigenic - Carcinogenic by RTECS criteria; Endocrine - Thyroid tumors. Rat Route: Oral; Dose: 44 gm/kg/2Y-C; Toxic Effects: Tumorigenic - Carcinogenic by RTECS criteria; Endocrine - Thyroid tumors; Tumorigenic effects - Testicular tumors. Rat Route: Oral; Dose: 146 gm/kg/2Y-C; Toxic Effects: Tumorigenic - Carcinogenic by RTECS criteria; Endocrine - Thyroid tumors. Rat Route: Oral; Dose: 9125 mg/kg/2Y-C; Toxic Effects: Tumorigenic - Carcinogenic by RTECS criteria; Endocrine - Tumors; Endocrine - Thyroid tumors. Rat Route: Oral; Dose: 5470 mg/kg/26W-C; Toxic Effects: Tumorigenic - Equivocal tumorigenic agent by RTECS criteria; Liver - Tumors.

Hazard Overviews

Fire Diamond

Health: Chronic Effects: reduced thyroid activity, yellow/ivory complexion, thickening/drying of the skin, enlarged tongue, dry, brittle hair. Possible cancer hazard. May cause teratogenic (harmful to fetus) effects.

Fire: Will burn. Use carbon dioxide, dry chemical, water fog, or foams.

Reactivity: Stable. Hazardous polymerization cannot occur. Avoid: direct contact; inhalation of dust particles. Incompatible with: alkaline hypochlorite; ultraviolet radiation. Hazardous decomposition products: oxides of sulfur; oxides of nitrogen; carbon monoxide.

Carcinogenicity: IARC - Group 2B, Possibly carcinogenic to humans; NIOSH - Listed as carcinogen; NTP - Class 2B, Reasonably anticipated to be a carcinogen, sufficient evidence of carcinogenicity from studies in experimental animals; ACGIH - Not listed; OSHA - Not listed; EPA - Not listed; MAK - Class B, Justifiably suspected of having carcinogenic potential

Primary Target Organs:

Eyes

Skin

Repro-ductive

Glandular System

Environmental

Ecotoxicity: Crustaceans: Daphnia magna 48h LC_{50} 26 mg/l; Fishes: creek chub critical range 24h 6,000-8,000 mg/l

Environmental Fate: If released to soil, it is readily degraded with half-lives of less than one week observed under field conditions. Although soil microorganisms can degrade it, degradation in soil may be accomplished by chemical conversion to ethyleneurea followed by microbial attack. It is expected to be highly mobile in most wet soils with the exception of soils with high organic content. Strong adsorption and/or degradation has been observed when moist soil is dried. Small quantities may volatilize from plant and soil surfaces. If released to water, it is susceptible to significant photooxidation in the presence of sunlight and appropriate photosensitizers which are present in many natural waters. Observed photooxidation products include glycine sulfate and ethyleneurea. It is not expected to significantly hydrolyze, volatilize, bioconcentrate, directly photolyze or adsorb to sediments in water. Sufficient data are not available to predict the significance of biodegradation in natural water systems. If released to air in the vapor phase, it is expected to react with photochemically produced hydroxyl radicals with an estimated half-life of 8.55 days.

Environmental Physical Data

Henry's Law Constant: 0.308×10^{-9}

Octanol/Water Partition Coefficient: log K_{ow} = -0.66

Sorption Partition Coefficient: K_{oc} = estimated at 19

BCF: estimated at 0.73

Regulations

RCRA 40CFR: Listed Hazardous Waste No. U116 Toxic Waste

CERCLA: 40CFR 302.4: Listed per RCRA Section 3001 RQ: 10 lb (4.535 kg)

SARA 40CFR 372.65: Listed

SARA EHS 40CFR 355: Not listed

TSCA: Listed

Analytical Methods

Soil: EPA PMD-MAU

Water / Groundwater: EPA 1625, 553

Drinking Water: EPA 509, 553; AOAC 992.31

Food: FDA 232.4, 242.1; AOAC 978.16

Indoor / Expired Air: NIOSH 5011

ETH6950 CAS #: 111-54-6

ETHYLENEBISDITHIOCARBAMIC ACID

DOT: UN2771; UN2772; UN3005; UN3006; IMO3.2; IMO6.1
Molecular Formula: $C_4H_8N_2S_4$
Formula Weight: 212.38
Synonyms: CARBAMIC ACID,ETHYLENEBIS(DITHIO-; CARBAMODITHIOIC ACID,1,2-ETHANEDIYLBIS-; 1,2-ETHANEDICARBAMIC ACID,TETRATHIO-; ETHYLENEBIS(DITHIOCARBAMIC ACID)
Use: as fungicides

Hazard Overviews

Carcinogenicity: IARC - Not listed; NIOSH - Not listed; NTP - Not listed; ACGIH - Not listed; OSHA - Not listed; EPA - Not listed; MAK - Not listed

Environmental

Cleanup/Disposal: Guide No. 131: Fully encapsulating, vapor protective clothing should be worn for spills and leaks with no fire. Eliminate all ignition sources (no smoking, flares, sparks or flames in immediate area). All equipment used when handling the product must be grounded. Do not touch or walk through spilled material. Stop leak if you can do it without risk. Prevent entry into waterways, sewers, basements or confined areas. A vapor suppressing foam may be used to reduce vapors. Small Spills: Absorb with earth, sand or other non-combustible material and transfer to containers for later disposal. Use clean non-sparking tools to collect absorbed material. Large Spills: Dike far ahead of liquid spill for later disposal. Water spray may reduce vapor; but may not prevent ignition in closed spaces. Guide No. 151: Do not touch damaged containers or spilled material unless wearing appropriate protective clothing. Stop leak if you can do it without risk. Prevent entry into waterways, sewers, basements or confined areas. Cover with plastic sheet to prevent spreading. Absorb or cover with dry earth, sand or other non-combustible material and transfer to containers. Do not get water inside containers.

Regulations

RCRA 40CFR: Listed Hazardous Waste No. U114 Toxic Waste
CERCLA: 40CFR 302.4: Listed per RCRA Section 3001 RQ: 5000 lb (2268 kg)
SARA 40CFR 372.65: Not listed
SARA EHS 40CFR 355: Not listed
TSCA: Not listed

ETH7000 CAS #: 107-15-3

ETHYLENEDIAMINE

RTECS: KH8575000
DOT: UN1604; IMO8.2
EINECS Number: 203-468-6
Molecular Formula: $C_2H_8N_2$
Structured MF: $H_2NCH_2CH_2NH_2$
Formula Weight: 60.10

Chemical Structure

Synonyms: AETHALDIAMIN; AETHYLENEDIAMIN; ALGICODE 106L; AMERSTAT 274; BETA-AMINOETHYLAMINE; 1,2-DIAMINOAETHAN; 1,2-DIAMINO-ETHAAN; 1,2-DIAMINOETHANE; 1,2-DIAMINO-ETHANO; DIMETHYLENEDIAMINE; EPA PESTICIDE CHEMICAL CODE 004205; 1,2-ETHANEDIAMINE; ETHYLEENDIAMINE; ETHYLENDIAMINE; 1,2-ETHYLENEDIAMINE; ETHYLENE-DIAMINE; ETHYLENEDIAMINE (ANHYDROUS)
Description: clear, colorless liquid; ammonia odor
Use: stabilizer for rubber latex, inhibitor in antifreeze solutions, in the manufacture of chelating agents and fungicides, as an emulsifier and in neutralizing oils, textile lubricants, rubber accelerators, dyes, organic synthesis, adhesives, pharmaceuticals and in veterinary medicine as a urinary acidifier; as a food additive in human food; solvent for albumin, shellac, casein and sulfur; in medicine

Physical Properties

Boiling Point: 116 °C (241 °F) to 117 °C (243 °F)
Freezing Point: 8.5 °C (47.3 °F)
Specific Gravity: 0.898 at 25 °C/4 °C
Vapor Density: 2.07 Air=1
Saturated Vapor Density: 1.218118149 kg/m^3
Bulk Density: 7.50 lbs/gal at 20 °C
Vapor Pressure: 10.7 mm Hg at 20 °C
Water Solubility: Miscible
Other Solubilities: 95% Ethanol: >=100 mg/ml at 17 °C; Acetone: >=100 mg/ml at 17 °C; Alcohol: Soluble; Benzene: Insoluble; DMSO: >=100 mg/ml at 17 °C; Ether: Slightly Soluble.
Odor Threshold: 2.50 to 28 mg/m^3
pH: 25% solution at 25 °C 11.9
Refraction Index: 1.4568 at 20 °C/D
Evaporation Rate: 0.91 Butyl Acetate=1
Critical Temperature: 320 °C
Critical Pressure: 941 psia
Ionization Potential (eV): 8.60
Flash Point: 34 °C Closed Cup
Autoignition Temperature: Anhydrous 76% 385 °C
LEL: 4.2% v/v
UEL: Anhydrous 76% 14.4% v/v

RTECS Toxicity Data

Acute Oral: Rat LD_{50} Dose: 500 mg/kg. Guinea Pig LD_{50} Dose: 470 mg/kg; Toxic Effects: Behavioral - Somnolence (general depressed activity); Gastrointestinal - Other changes; Kidney, Ureter, and Bladder - Other changes.

Acute Inhalation: Rat LC_{Lo} Dose: 4000 ppm/8hr. Human TC_{Lo} Dose: 200 ppm; Toxic Effects: Peripheral Nerve and sensation - Pareshtesia.

Acute Dermal: Rabbit LD_{50} Route: Skin; Dose: 730 uL/kg. Rabbit LD_{Lo} Route: Subcutaneous Dose: 500 mg/kg.

Chronic (Multiple Dose) Oral: Rat Dose: 3500 mg/kg/7D-C; Toxic Effects: Liver - Changes in liver weight; Kidney, Ureter, and Bladder - Changes in kidney weight; Nutritional and gross metabolic - Weight loss or decreased weight gain.

Chronic (Multiple Dose) Inhalation: Rat Dose: 700 ug/m^3/17W-I; Toxic Effects: Brain and coverings - Other degenerative changes; Blood - Changes in spleen; Biochemical - Phosphatases.

Irritation Eye: Rabbit Standard Draize Test Dose: 750 ug/24H; Reaction: severe. Rabbit Standard Draize Test Dose: 750 ug; Reaction: severe.

Irritation Skin: Rabbit Open Draize Test Dose: 10 mg/24H open; Reaction: severe. Rabbit Open Draize Test Dose: 450 mg open; Reaction: moderate.

Reproductive/Teratogenic: Mouse Route: Oral; Dose: 3200 mg/kg; Duration: female 6-13D of pregnancy; Effects on Newborn - Growth statistics.

Mutagenic: Bacteria - S Typhimurium Mutations in Microorganisms; Dose: 33 ug/plate (+S9), 1 mg/plate (-S9).

Hazard Overviews

Corrosive

Flammable

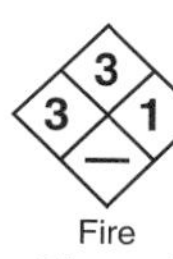

Fire Diamond

Health: Corrosive to eyes/skin/respiratory tract. Also Causes: dermatitis, coughing, wheezing, laryngitis, nausea, pulmonary edema. Chronic Effects: dermatitis, asthma.

Fire: Flammable. Can form explosive mixtures in the air. For small fires use dry chemical, carbon dioxide, water spray, fog, or alcohol-resistant foam. For large fires, use water spray, fog, or alcohol-resistant foam. Solid streams of water may spread fire.

Reactivity: Stable. Hazardous polymerization cannot occur. Avoid: heat; ignition sources; strong alkalinity. Incompatible with: copper and its alloys; acetic acid; acetic anhydride; acrylic acid; acrylonitrile; allyl chloride; beta-propiolactone; carbon disulfide; carbon tetrachloride; cellulose nitrate; chlorosulfonic acid; diisopropylperoxydicarbonate; hydrochloric acid; epichlorohydrin, oleum; mesityl oxide; nitric acid; nitromethane; silver perchlorate; sulfuric acid; vinyl acetate. Hazardous decomposition products: carbon; nitrogen oxides; ammonia.

Carcinogenicity: IARC - Not listed; NIOSH - Not listed; NTP - Not listed; ACGIH - Class A4, Not classifiable as a human carcinogen; OSHA - Not listed; EPA - Class D, Not classifiable as to human carcinogenicity; MAK - Not listed

Primary Target Organs:

Eyes

Skin

Respiratory System

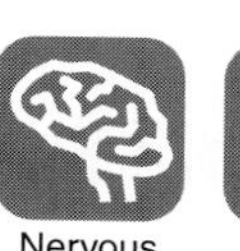

Nervous System

Kidneys

Exposure Limits

OSHA PEL: TWA: 10 ppm; 25 mg/m^3.

ACGIH TLV: TWA: 0.5 ppm; 0.88 mg/m^3.

NIOSH REL: TWA: 10 ppm; 25 mg/m^3.

NIOSH IDLH: 1000 ppm.

DFG MAK: TWA: 10 ppm; 25 mg/m^3.

Respirator Recommendation

Exposure Range: >10 to 100 ppm Air Purifying, Negative Pressure, Half Mask

Exposure Range: >100 to <1000 ppm Air Purifying, Negative Pressure, Full Face

Exposure Range: 1000 to unlimited ppm Self-contained Breathing Apparatus, Pressure Demand, Full Face

Cartridge Color: black

Environmental

Ecotoxicity: Aquatic toxicity: 60 ppm/24 hr/chub/killed/fresh water

Cleanup/Disposal: Guide No. 132: Fully encapsulating, vapor protective clothing should be worn for spills and leaks with no fire. Eliminate all ignition sources (no smoking, flares, sparks or flames in immediate area). All equipment used when handling the product must be grounded. Do not touch or walk through spilled material. Stop leak if you can do it without risk. Prevent entry into waterways, sewers, basements or confined areas. A vapor suppressing foam may be used to reduce vapors. Absorb with earth, sand or other non-combustible material and transfer to containers (except for Hydrazine). Use clean non-sparking tools to collect absorbed material. Large Spills: Dike far ahead of liquid spill for later disposal. Water spray may reduce vapor; but may not prevent ignition in closed spaces.

Environmental Physical Data

Octanol/Water Partition Coefficient: log K_{ow} = 0.05

BCF: no food chain concentration potential

BOD: theoretical 75%, 5 days

Regulations

RCRA 40CFR: Not listed

CERCLA: 40CFR 302.4: Listed per CWA Section 311(b)(4) RQ: 5000 lb (2268 kg)

SARA 40CFR 372.65: Not listed TPQ: 10000 lb

SARA EHS 40CFR 355: Listed TPQ: 5,000 lb

TSCA: Listed

Analytical Methods

Air: ASTM D4490

Indoor / Expired Air: NIOSH 2540

ETH7050 **CAS #: 24464-90-2**

ETHYLENEDIAMINE CARBAMATE

Molecular Formula: $C_{(x+2)}H_{(3x+8)}N_{(x+2)}O_{(2x)}$
Formula Weight: Varies
Synonyms: CARBAMIC ACID,COMPD WITH 1,2-ETHANEDIAMINE; CARBAMIC ACID,COMPD WITH ETHYLENEDIAMINE

Hazard Overviews

Carcinogenicity: IARC - Not listed; NIOSH - Not listed; NTP - Not listed; ACGIH - Not listed; OSHA - Not listed; EPA - Not listed; MAK - Not listed

Environmental

Regulations

RCRA 40CFR: Not listed
CERCLA: 40CFR 302.4: Not listed
SARA 40CFR 372.65: Not listed
SARA EHS 40CFR 355: Not listed
TSCA: Not listed

ETH7150 **CAS #: 60-00-4**

ETHYLENEDIAMINE TETRAACETIC ACID

RTECS: AH4025000
DOT: NA9117
EINECS Number: 200-449-4
Molecular Formula: $C_{10}H_{16}N_2O_8$
Structured MF: $(HOOCCH_2)_2NCH_2CH_2N(CH_2COOH)_2$
Formula Weight: 292.28

Chemical Structure

Synonyms: ACETIC ACID,2,2',2'',2'''-(1,2-ETHANEDIYLDINITRILO)TETRAKIS-; ACETIC ACID,(ETHYLENEDINITRILO)TETRA-; ACIDE ETHYLENEDIAMINETETRACETIQUE; CELON A; CELON ATH; CHEELOX; CHEELOX BF ACID; CHEMCOLOX 340; CLEWAT TAA; COMPLEXON II; 3,6-DIAZAOCTANEDIOIC ACID,3,6-BIS(CARBOXYMETHYL)-; EDATHAMIL; EDETIC; EDETIC ACID; EDTA; EDTA ACID; EDTA (CHELATING AGENT); ENDRATE; EPA PESTICIDE CHEMICAL CODE 039101; N,N'-1,2-ETHANEDIYLBIS(N-(CARBOXYMETHYL)GLYCINE); ETHYLENEBISIMINODIACETIC ACID; ETHYLENEDIAMINETETRAACETATE; ETHYLENEDIAMINE-N,N,N',N'-TETRAACETIC ACID; ETHYLENEDIAMINETETRAACETIC ACID; ETHYLENEDINITRILOTETRAACETIC ACID; GLUMA CLEANSER; GLYCINE,N,N'-1,2-ETHANEDIYLBIS(N-(CARBOXYMETHYL)-; GLYCINE,N,N'-1,2-ETHANEDIYLBIS(N-(CARBOXYMETHYL)-(9CI); HAMP-ENE ACID; HAVIDOTE; ICRF 185; KOMPLEXON II; KYSELINA ETHYLENDIAMINTETRAOCTOVA; METAQUEST A; NERVANAID B ACID; NULLAPON B ACID; NULLAPON BF ACID; PERMA KLEER 50 ACID; QUESTEX 4H; SEQ 100; SEQUESTRENE AA; SEQUESTRIC ACID; SEQUESTROL; TETRINE ACID; TITRIPLEX; TRICON BW; TRILON BS; TRILON BW; UNIVERSNE ACID; VERSENE; VERSENE ACID; VINKEIL 100; WARKEELATE ACID; YD 30
Description: white crystaline powder; odorless
Use: important metal chelating agent (in food, pharmaceuticals, cooling water, detergents, indust germicides, etc); antioxidant in foods; intermed for micronutrients; bleaching agent in film processing; etching agent in metal finishin; med: in blood anticoagulants, reducing blood cholesterol, to treat skin lesions in chromium-exposed workers, etc; (vet): for heavy metal poisoning; agricultural chem sprays, decontamination of radioactive surfaces, in anal chemistry (eluting agent in ion exchange, spectrophotometric titration, etc); in textiles dyeing, clarification of liq.

Physical Properties

Freezing Point: Decomposes at 240 °C (464 °F)
Water Solubility: 0.5 g/L in Water at 25 °C
Other Solubilities: Soluble in solutions of alkali hydroxides; Insoluble in common organic solvents.
Flash Point: Nonflammable

RTECS Toxicity Data

Acute Oral: Mouse LD_{50} Dose: 30 mg/kg.
Reproductive/Teratogenic: Rat Route: Oral; Dose: 7632 mg/kg; Duration: female 7-14D of pregnancy; Specific Developmental Abnormalities - Eye, ear; Craniofacial (including nose and tongue); Musculoskeletal system. Rat Route: Oral; Dose: 7632 mg/kg; Duration: female 7-14D of pregnancy; Specific Developmental Abnormalities - Cardiovascular (circulatory) system; Respiratory system; Urogenital system. Rat Route: Oral; Dose: 7632 mg/kg; Duration: female 7-14D of pregnancy; Effects on Fertility - Post-implantation mortality; Effects on Embryo or Fetus - Fetotoxicity.
Mutagenic: Rat DNA Inhibition; Cell Type: other cell types; Dose: 600 umol/L. Rabbit DNA Inhibition; Cell Type: kidney; Dose: 250 umol/L.

Hazard Overviews

Fire Diamond

Health: Mildly irritating to eyes/skin/respiratory tract.
Fire: Noncombustible. Use agent suitable for surrounding fire.

Reactivity: Stable. Hazardous polymerization cannot occur. Hazardous decomposition products: fumes of nitrogen oxides.
Carcinogenicity: IARC - Not listed; NIOSH - Not listed; NTP - Not listed; ACGIH - Not listed; OSHA - Not listed; EPA - Not listed; MAK - Not listed
Primary Target Organs:

Eyes

Skin

Respiratory System

Environmental

Ecotoxicity: Toxicity Threshold (Cell Multiplication Inhibition Test) Scenedesmus quadricauda (green algae) 11 mg/l /Conditions of bioassay not specified LC_{50} (Lepomis macrochirus) bluegill 159 mg/l/96 hr; 96 hr no adverse effect level: 100 mg/l /Static bioassay Toxicity Threshold (Cell Multiplication Inhibition Test) Uronema parduczi Chatton-Lwoff (protozoa) 17 mg/l /Conditions of bioassay not specified
Environmental Fate: If released to soil, it is expected to complex with trace metals and alkaline earth metals present in the soil, thereby causing an increase in the total solubility of the metals. It may eventually predominate as the Fe(III) chelate in acidic soils and as the Ca chelate in alkaline soils. Biodegradation in aerobic soils is the dominant removal mechanism, although biodegradation in anaerobic soils is negligible. In various soils common values for mineralization of 2-4 ppm are 13-45% after 15 weeks and 65-70% after 45 weeks. The compound and its chelates are expected to leach readily through soil and significant volatilization from soil is not expected. If released to water, it is expected to complex with trace metals and alkaline earth metals. Biodegradation is expected to take place relatively slowly under aerobic conditions and to be negligible under anaerobic conditions. Cometabolism has been suggested as the mechanism for biodegradation. Compounds identified as possible biodegradation products of the ammonium ferric chelate are as follows: ethylenediamine triacetic acid (ED3A), iminodiacetic acid (IDA), N,N-ethylenediamine diacetic acid (N,N-EDDA), N,N'-EDDA, ethylenediamine monoacetic acid (EDMA), nitrilotriacetic acid (NTA) and glycine. It may react with photochemically generated hydroxyl radicals (half-life 229 days) and it may photodegrade. The following photodegradation products of Fe(III)-EDTA have been identified: carbon monoxide, formaldehyde, ED3A, N,N-EDDA, N,N'-EDDA, IDA, EDMA and glycine. It is not expected to bioaccumulate in aquatic organisms, adsorb to suspended solids or sediments or volatilize from water surfaces. If released to the atmosphere, it should adsorb to particulate matter and appears to have the potential to photolyze.
Cleanup/Disposal: Guide No. 171: Do not touch or walk through spilled material. Stop leak if you can do it without risk. Prevent dust cloud. Avoid inhalation of asbestos dust. Small Dry Spills: With clean shovel place material into clean, dry container and cover loosely; move containers from spill area. Small Spills: Take up with sand or other noncombustible absorbent material and place into containers for later disposal. Large Spills: Dike far ahead of liquid spill for later disposal. Cover powder spill with plastic sheet or tarp to minimize spreading. Prevent entry into waterways, sewers, basements or confined areas.

Environmental Physical Data

BCF: bluegills 2
BOD: 1%, 5 days

Regulations

RCRA 40CFR: Not listed
CERCLA: 40CFR 302.4: Listed per CWA Section 311(b)(4) RQ: 5000 lb (2268 kg)
SARA 40CFR 372.65: Not listed
SARA EHS 40CFR 355: Not listed
TSCA: Listed

Analytical Methods

Soil: DOE OM510R
Water / Groundwater: ASTM D3113

ETH7200 CAS #: 120-93-4

ETHYLENEUREA

RTECS: NJ0570000
EINECS Number: 204-436-4
Molecular Formula: $C_3H_6N_2O$
Formula Weight: 86.10

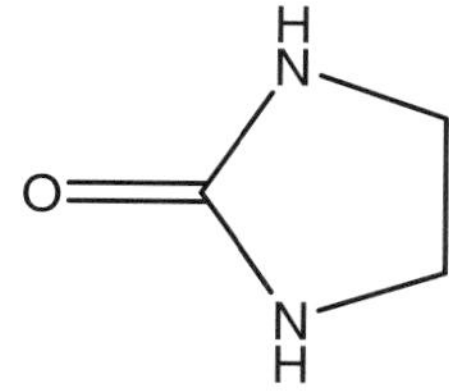

Chemical Structure

Synonyms: ETHYLENE UREA; N,N'-ETHYLENEUREA; 2-IMIDAZOLIDINONE; 2-IMIDAZOLIDONE; 2-OXOMIDAZOLIDINE; SD 6073; UREA,N,N'-(1,2-ETHANEDIYL)-; UREA,1,3-ETHYLENE-
Description: white lumpy powder or needles
Use: mfr high polymers; in formulation of plasticizers, lacquers, & adhesives; insecticide; chem int for textile & leather finishing agents

Physical Properties

Boiling Point: 100 °C (212 °F)
Freezing Point: 131 °C (267.8 °F)
Water Solubility: Very Soluble in Water
Other Solubilities: Slightly Soluble in Chloroform.
Ionization Potential (eV): 8.9

RTECS Toxicity Data

Reproductive/Teratogenic: Rat Route: Intraperitoneal; Dose: 15 mg/kg; Duration: male 1D prior to mating; Effects on Fertility - Male fertility index.

Tumorigenic: Mouse Route: Subcutaneous; Dose: 1000 mg/kg; Toxic Effects: Tumorigenic - Carcinogenic by RTECS criteria; Blood - Tumors.

Hazard Overviews

Health: May cause irritation to eyes/skin/respiratory tract. Harmful. Other Acute Effects: may be harmful by inhalation, ingestion, or skin absorption. Chronic Effects: danger of cumulative effects. Possible human carcinogen. The toxicological properties have not been thoroughly investigated.

Fire: Extinguishing agents: water spray; carbon dioxide, dry chemical powder or appropriate foam. Precautions: combustible liquid.

Reactivity: Stable. Hazardous polymerization will not occur. Hazardous decomposition products: toxic fumes of: carbon monoxide, carbon dioxide, nitrogen oxides.

Carcinogenicity: IARC - Not listed; NIOSH - Not listed; NTP - Not listed; ACGIH - Not listed; OSHA - Not listed; EPA - Not listed; MAK - Not listed

Environmental

Ecotoxicity: Crustaceans: Daphnia magna 48h LC_{50} 5,600 mg/l; Fishes: Poecilia reticulata 96h LC_{50} 13,000 mg/l

Environmental Physical Data

Octanol/Water Partition Coefficient: log K_{ow} = -0.96

Regulations

RCRA 40CFR: Not listed
CERCLA: 40CFR 302.4: Not listed
SARA 40CFR 372.65: Not listed
SARA EHS 40CFR 355: Not listed
TSCA: Listed

ETH7250 CAS #: 151-56-4

ETHYLENIMINE

RTECS: KX5075000
DOT: UN1185; IMO3.2
EINECS Number: 205-793-9
Molecular Formula: C_2H_5N
Structured MF: $CH_2{=}CH_2{=}NH$
Formula Weight: 43.08

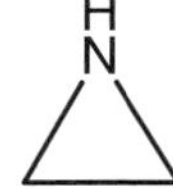

Chemical Structure

Synonyms: AETHYLENIMIN; AMINOETHYLENE; AZACYCLOPROPANE; AZIRAN; AZIRANE; AZIRIDIN; AZIRIDINE; 1H-AZIRINE,DIHYDRO-; DIHYDROAZIRENE; DIHYDRO-1H-AZIRINE; DIHYDROAZIRINE; DIMETHYLENEIMINE; DIMETHYLENIMINE; EI; ENT-50324; ETHYLEENIMINE; ETHYLENEIMINE; ETHYLENEIMINE,INHIBITED; ETHYLIMINE; ETILENIMINA; TL 337; VINYLAMINE

Description: colorless liquid; intense ammonia odor

Use: polymerization products in paper ind; in adhesives, binders, petro refining chem, fuels & lubricants, coating resins, varnishes, lacquers, agricultural chems, cosmetics, ion exchange resins, photographic chems, & surfacants; monomer for polyethyleneimine; comonomer for polymers (eg, with ethylenediamine); intermed for n-(2-hydroxyethyl)ethyleneimine; mfr of taurine, triethylmelamine; in textile ind for incr wet strength, flameproofing, etc; pharmaceuticals; control of insect pests by chemosterilization; sporacidal action.

Physical Properties

Boiling Point: 56 °C (133 °F) to 57 °C (135 °F) at 760 mm Hg
Freezing Point: -71.5 °C (-96.7 °F)
Specific Gravity: 0.8321 at 24 °C/4 °C
Vapor Density: 1.5 Air=1
Vapor Pressure: 160 mm Hg
Water Solubility: Miscible
Other Solubilities: Soluble in Alcohol, Ether, Acetate, Benzene.
Surface Tension: 34.5 dynes/cm at 20 °C
Odor Threshold: Perception 2 ppm
pH: Strongly alkaline
Refraction Index: 1.4123
Ionization Potential (eV): 9.20
Flash Point: -11 °C
Autoignition Temperature: 322 °C
LEL: 3.6% v/v
UEL: 46% v/v

RTECS Toxicity Data

Acute Oral: Rat LD_{50} Dose: 15 mg/kg; Toxic Effects: Behavioral - General anesthetic; Gastrointestinal - Other changes; Kidney, Ureter, and Bladder - Other changes.

Acute Inhalation: Rat LC_{50} Dose: 100 mg/m^3/2hr. Mouse LC_{50} Dose: 400 mg/m^3/2hr; Toxic Effects: Behavioral - Muscle weakness; Lungs, Thorax, or Respiration - Other changes; Nutritional and gross metabolic - Body temperature decrease.

Acute Dermal: Rabbit LD_{Lo} Route: Subcutaneous Dose: 50 mg/kg. Guinea Pig LD_{50} Route: Skin; Dose: 14 mg/kg;

Chronic (Multiple Dose) Inhalation: Rat Dose: 700 ug/m^3/4H/17W-I; Toxic Effects: Kidney, Ureter, and Bladder - Proteinuria; Blood - Changes in other cell count. Rat Dose: 10 mg/m^3/4H/6W-I; Toxic Effects: Behavioral - Muscle weakness; Nutritional and gross metabolic - Weight loss or decreased weight gain.

Irritation Eye: Rabbit Standard Draize Test Dose: 2 mg; Reaction: severe.

Reproductive/Teratogenic: Rat Route: Inhalation; Dose: 800 ug/m^3/24 Duration: male 1D prior to mating; Paternal Effects - Testes, epididymis, sperm duct. Rat Route: Inhalation; Dose: 19 mg/m^3; Duration: male 6W prior to mating; Paternal Effects - Testes, epididymis, sperm duct; Maternal Effects - Ovaries, fallopian tubes; Effects on Embryo or Fetus - Fetal death. Rat Route: Inhalation; Dose: 19 mg/m^3; Duration: male 6W prior to mating; Effects on Newborn - Live birth index; Behavioral. Rat Route: Inhalation; Dose: 700 ug/m^3;

Duration: male 17W prior to mating; Effects on Newborn - Growth statistics; Delayed effects.
Mutagenic: Human DNA Inhibition; Cell Type: HeLa cell; Dose: 500 umol/L. Human Cytogenetic Analysis; Cell Type: lymphocyte; Dose: 1 umol/L/1H. Human Cytogenetic Analysis; Cell Type: leukocyte; Dose: 10 umol/L/1H. Human Cytogenetic Analysis; Cell Type: lung; Dose: 10 umol/L.
Tumorigenic: Rat Route: Subcutaneous; Dose: 20 mg/kg/67D-I; Toxic Effects: Tumorigenic - Neoplastic by RTECS criteria; Tumorigenic - Tumors at site of application. Rat Route: Subcutaneous; Dose: 10 mg/kg/8W-I; Toxic Effects: Tumorigenic - Equivocal tumorigenic agent by RTECS criteria; Tumorigenic - Tumors at site of application.

Hazard Overviews

Poison

Corrosive

Explosive

Flammable
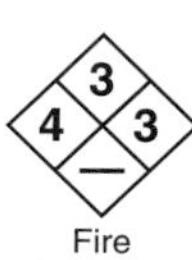

Fire Diamond

Health: Corrosive, causes severe burns to eyes/skin/respiratory tract. Poison. Also Causes: dizziness, headache, pulmonary edema, nausea, vomiting, kidney damage. Chronic Effects: headache, weariness, pneumonia, CNS, and heart effects.
Fire: Flammable. May polymerize explosively. Use alcohol foam or dry chemical. Fight advanced or large fires from a remote or explosion-proof location. Use a water spray to cool fire-exposed containers.
Reactivity: Stable, but requires a caustic such as sodium hydroxide as an inhibitor. Hazardous polymerization can occur if there is an insufficient concentration of inhibitor, elevated temperature, or contact with catalysts. Avoid: heating or concentration. Incompatible with: acetic anhydride; acrolein; allyl chloride; carbon disulfide; chlorine; epichlorohydrin; glyoxal; oleum; beta-propiolactone; silver; sodium hypochlorite; vinyl acetate; acids. Hazardous decomposition products: nitrogen; carbon monoxide.
Carcinogenicity: IARC - Group 3, Not classifiable as to carcinogenicity to humans; NIOSH - Listed as carcinogen; NTP - Not listed; ACGIH - Class A3, Animal carcinogen; OSHA - Listed as a carcinogen; EPA - Not listed; MAK - Class A2, Unmistakably carcinogenic in animal experimentation only
Primary Target Organs:

Eyes

Skin
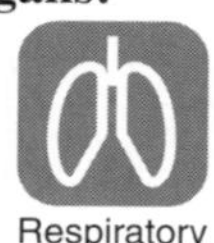
Respiratory System

Nervous System

Liver

Kidneys

Exposure Limits
NIOSH IDLH: 100 ppm.
Respirator Recommendation
Exposure Range: >0.5 to 25 ppm Supplied Air, Constant Flow/Pressure Demand, Half Mask
Exposure Range: >25 to <100 ppm Supplied Air, Constant Flow/Pressure Demand, Full Face
Exposure Range: 100 to unlimited ppm Self-contained Breathing Apparatus, Pressure Demand, Full Face

Note: poor warning properties

Environmental

Ecotoxicity: Toxicity threshold (cell multiplication test) Microcystis aeruginosa (algae) 0.12 mg/l /Time not specified Toxicity threshold (cell multiplication test) Entosiphon sulcatum (protozoa) 4.3 mg/l /Time not specified
Environmental Fate: It is a reactive molecule but there is no data on its fate in environmental media. In the atmosphere, it should react with hydroxyl radicals (estimated half-life 1.5 days). If released in water, it will hydrolyze at neutral pH in about 5 months but it is apt to be lost much faster by evaporation or chemical reactions with metal ions. While it should rapidly evaporate from soil, it may also leach into the soil or complex with metal ions in the soil. It would not be expected to bioconcentrate in fish. No monitoring data could be found.
Cleanup/Disposal: Guide No. 131: Fully encapsulating, vapor protective clothing should be worn for spills and leaks with no fire. Eliminate all ignition sources (no smoking, flares, sparks or flames in immediate area). All equipment used when handling the product must be grounded. Do not touch or walk through spilled material. Stop leak if you can do it without risk. Prevent entry into waterways, sewers, basements or confined areas. A vapor suppressing foam may be used to reduce vapors. Small Spills: Absorb with earth, sand or other non-combustible material and transfer to containers for later disposal. Use clean non-sparking tools to collect absorbed material. Large Spills: Dike far ahead of liquid spill for later disposal. Water spray may reduce vapor; but may not prevent ignition in closed spaces.

Environmental Physical Data

Henry's Law Constant: 1.1 $x10^{-5}$
BCF: no tendency

Regulations

RCRA 40CFR: Listed Hazardous Waste No. P054 Toxic Waste
CERCLA: 40CFR 302.4: Listed per RCRA Section 3001 RQ: 1 lb (0.454 kg)
SARA 40CFR 372.65: Listed TPQ: 500 lb
SARA EHS 40CFR 355: Listed TPQ: 1 lb
TSCA: Listed

Analytical Methods

Air: ASTM D4490
Indoor / Expired Air: NIOSH 3514

ETH7300 CAS #: 965-90-2

ETHYLESTRENOL

RTECS: RC8961100
EINECS Number: 213-523-6
Molecular Formula: $C_{20}H_{32}O$
Formula Weight: 288.46

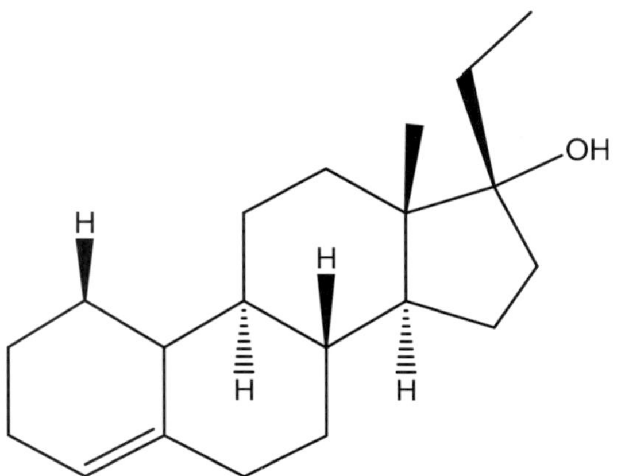

Chemical Structure

Synonyms: DURABOLIN-O; DURABORAL; 17ALPHA-ETHYLESTR-4-EN-17BETA-OL; 17ALPHA-ETHYL-17BETA-HYDROXY-4-ESTRENE; ETHYLNANDROL; 17BETA-HYDROXY-17ALPHA-ETHYL-19-NOR-4-ANDROSTENE; MAXIBALIN; MAXIBOLIN; NEODURABOLIN; 19-NOR-17ALPHA-PREGN-4-EN-17-OL; 19-NORPREGN-4-EN-17-OL,(17ALPHA)-; 19-NORPREGN-4-EN-17-OL,(17-ALPHA)-(9CI); ORABOLIN; ORG-483; ORGABOLIN; ORGABORAL

Description: white to creamy white crystals; odorless

Use: medication: anabolic steroid

Physical Properties

Freezing Point: 76 °C (168.8 °F) to 78 °C (172.4 °F)

Water Solubility: Practically Insoluble in Water

Other Solubilities: freely Soluble in Alcohol. Soluble in Chloroform, Acetone, Methanol.

RTECS Toxicity Data

Acute Oral: Mouse LD_{50} Dose: >666700 ug/kg.

Reproductive/Teratogenic: Rat Route: Oral; Dose: 1 gm/kg; Duration: female 2-6D of pregnancy; Effects on Fertility - Post-implantation mortality; Abortion. Rat Route: Oral; Dose: 1 gm/kg; Duration: female 15-19D of pregnancy Effects on Newborn - Stillbirth. Rat Route: Oral; Dose: 20 mg/kg; Duration: female 17-20D of pregnancy Specific Developmental Abnormalities - Urogenital system. Rat Route: Oral; Dose: 15 mg/kg; Duration: male 15D prior to mating; Paternal Effects - Prostate, seminal vessicle, Cowper's gland, accessory glands.

Hazard Overviews

Carcinogenicity: IARC - Not listed; NIOSH - Not listed; NTP - Not listed; ACGIH - Not listed; OSHA - Not listed; EPA - Not listed; MAK - Not listed

Environmental

Regulations

RCRA 40CFR: Not listed

CERCLA: 40CFR 302.4: Not listed

SARA 40CFR 372.65: Not listed

SARA EHS 40CFR 355: Not listed

TSCA: Not listed

ETH7350 CAS #: 123-05-7

2-ETHYLHEXALDEHYDE

RTECS: MN7525000

DOT: UN1191; IMO3.3

EINECS Number: 204-596-5

Molecular Formula: $C_8H_{16}O$

Structured MF: C_4H_9CHCHO

Formula Weight: 128.22

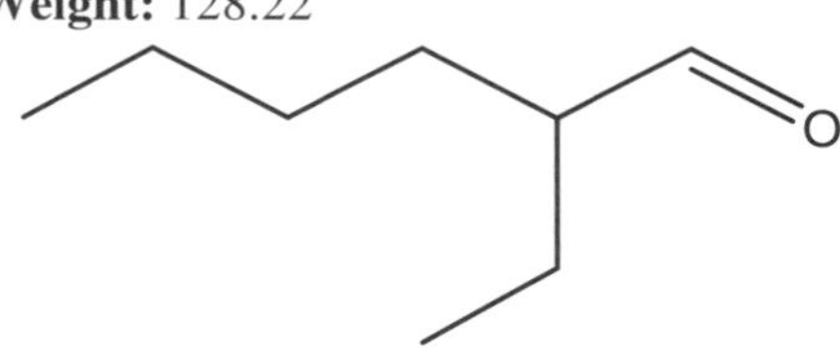

Chemical Structure

Synonyms: BUTYL ETHYL ACETALDEHYDE; BUTYLETHYLACETALDEHYDE; ETHYL BUTYLACETALDEHYDE; ETHYLBUTYLACETALDEHYDE; 2-ETHYLCAPROALDEHYDE; ALPHA-ETHYLCAPROALDEHYDE; ETHYLHEXALDEHYDE; 2-ETHYLHEXANAL; 2-ETHYLHEXYLALDEHYDE; 3-FORMYLHEPTANE; HEXANAL,2-ETHYL-; OCTYL ALDEHYDE; BETA-PROPYL-ALPHA-ETHYLACROLEIN

Description: colorless liquid; mild odor

Use: insecticide; organic synthesis (intermediate); warning agent and leak detector

Physical Properties

Boiling Point: 163 °C (325 °F) at 760 mm Hg

Freezing Point: < -100 °C (-148 °F)

Specific Gravity: 0.854 at 20 °C

Vapor Density: 4.42 Air=1

Saturated Vapor Density: 1.20972392 kg/m^3

Density: 0.8205 g/mL at 20 °C

Vapor Pressure: 1.8 mm Hg at 20 °C

Water Solubility: < 1 mg/mL at 21 C

Other Solubilities: 95% Ethanol: >=100 mg/ml at 21 °C; Acetone: >=100 mg/ml at 21 °C; DMSO: >=100 mg/ml at 21 °C.

Refraction Index: 1.4142 at 20 °C/D

Flash Point: 44 °C Closed Cup

Autoignition Temperature: 190 °C

LEL: 0.85% v/v

UEL: 7.2% v/v

RTECS Toxicity Data

Acute Oral: Rat LD_{50} Dose: 3078 mg/kg; Toxic Effects: Behavioral - Muscle weakness. Mouse LD_{50} Dose: 3550 mg/kg.

Acute Inhalation: Rat LC_{Lo} Dose: 4000 ppm/4hr.

Acute Dermal: Rabbit LD_{50} Route: Skin; Dose: 5040 uL/kg. Rat LD_{50} Route: Skin; Dose: >20 mL/kg; Toxic Effects: Skin and appendages - Dermatitis, other.

Chronic (Multiple Dose) Oral: Rat Dose: 25200 mg/kg/3W-C; Toxic Effects: Blood - Changes in serum composition.
Irritation Eye: Rabbit Standard Draize Test Dose: 500 mg/24H; Reaction: mild.
Irritation Skin: Rabbit Standard Draize Test Dose: 20 mg/24H; Reaction: moderate. Rabbit Open Draize Test Dose: 10 mg/24H open; Reaction: severe.

Hazard Overviews

Flammable

Health: Irritating to eyes/skin/respiratory tract. Other Acute Effects: may be harmful by inhalation, ingestion, or skin absorption; target organ: liver.
Fire: Flammable. Hazards: emits toxic fumes; vapor may travel considerable distance to source of ignition and flash back. Extinguishing agents: water spray; carbon dioxide, dry chemical powder or appropriate foam. Precautions: combustible liquid.
Reactivity: Incompatible with: strong oxidizing agents, strong bases. Hazardous decomposition products: toxic fumes of: carbon monoxide, carbon dioxide.
Carcinogenicity: IARC - Not listed; NIOSH - Not listed; NTP - Not listed; ACGIH - Not listed; OSHA - Not listed; EPA - Not listed; MAK - Not listed
Primary Target Organs:

Eyes

Skin

Respiratory System

Liver

Environmental

Environmental Fate: If released to soil, it will have moderate mobility. Volatilization may be important from moist and dry soil surfaces. If released to water, it may adsorb to suspended solids and sediment to a limited extent. It may volatilize from water surfaces with estimated half-lives for a model river and model lake of about 4.6 hours and 5 days, respectively. An estimated BCF value of 21 suggests that it will not bioconcentrate in aquatic organisms. Insufficient data are available to determine the rate of biodegradation in soil or water. If released to the atmosphere, it will exist in the vapor phase. Vapor-phase is degraded in the atmosphere by reaction with photochemically produced hydroxyl radicals with an estimated half-life of about 11 hours.
Cleanup/Disposal: Guide No. 129: Eliminate all ignition sources (no smoking, flares, sparks or flames in immediate area). All equipment used when handling the product must be grounded. Do not touch or walk through spilled material. Stop leak if you can do it without risk. Prevent entry into waterways, sewers, basements or confined areas. A vapor suppressing foam may be used to reduce vapors. Absorb or cover with dry earth, sand or other non-combustible material and transfer to containers. Use clean non-sparking tools to collect absorbed material. Large Spills: Dike far ahead of liquid spill for later disposal. Water spray may reduce vapor; but may not prevent ignition in closed spaces.

Environmental Physical Data

Henry's Law Constant: estimated at 7.6 $x10^{-4}$
Sorption Partition Coefficient: K_{oc} = estimated at 160
BCF: estimated at 21

Regulations

RCRA 40CFR: Not listed
CERCLA: 40CFR 302.4: Not listed
SARA 40CFR 372.65: Not listed
SARA EHS 40CFR 355: Not listed
TSCA: Listed

Analytical Methods

Water / Groundwater: ASTM D3695

ETH7400 CAS #: 94-96-2

2-ETHYL-1,3-HEXANEDIOL

RTECS: MO2625000
EINECS Number: 202-377-9
Molecular Formula: $C_8H_{18}O_2$
Formula Weight: 146.22

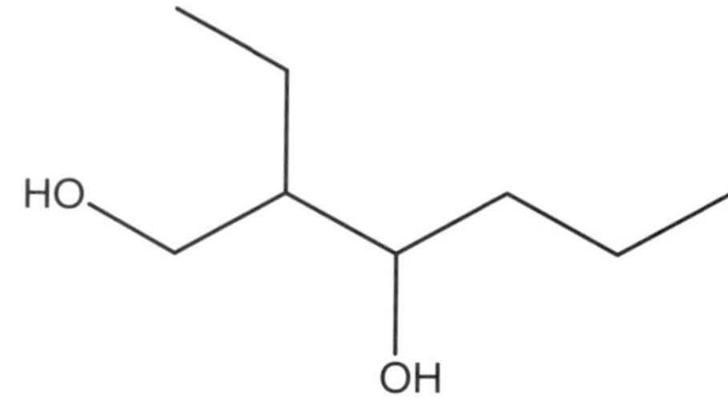

Chemical Structure

Synonyms: 34496; 6-12 INSECT REPELLENT; 6-12-INSECT REPELLENT; CARBIDE 6-12; COMPOUND 6-12 INSECT REPELLANT; COMPOUND 6-12 INSECT REPELLENT; ENT 375; EPA PESTICIDE CHEMICAL CODE 041001; ETHOHEXADIOL; ETHYL HEXANEDIOL; ETHYL HEXYLENE GLYCOL; 2-ETHYLHEXANE-1,3-DIOL; 2-ETHYLHEXANEDIOL-1,3; ETHYLHEXANEDIOL; 2-ETHYL-1,3-HEXYLENE GLYCOL; 2-ETHYL-3-PROPYL-1,3-PROPANEDIOL; 1,3-HEXANEDIOL,2-ETHYL-; 3-HYDROXYMETHYL-N-HEPTAN-4-OL; LATKA 612; OCTYLENE GLYCOL; REPELLANT 612; REPELLENT 612; RUTGERS 6-12; RUTGERS 612
Description: colorless, slightly oily liquid; odorless
Use: solvent for resins & inks; plasticizer; chem intermediate in prodn of polyurethane resins; insect repellent, cosmetics; vehicle and solvent in printing inks, medicine, chelating agent for boric acid

Physical Properties

Boiling Point: 244.2 °C (472 °F) at 760 mm Hg
Freezing Point: -40 °C (-40 °F)
Specific Gravity: 0.9325 at 22 °C/4 °C
Vapor Density: 5.03
Saturated Vapor Density: < 1.200063822 kg/m^3
Bulk Density: 7.8 lbs/gal at 20 °C

Vapor Pressure: < 0.01 mm Hg at 20 °C
Water Solubility: 0.6% at 0 °C
Other Solubilities: miscible with Chloroform, Diethyl Ether.
Refraction Index: 1.4530 at 22 °C/D
Evaporation Rate: Negligible Butyl Acetate=1
Flash Point: 110 °C Open Cup
Autoignition Temperature: 335 °C

RTECS Toxicity Data

Acute Oral: Rat LD_{50} Dose: 1400 mg/kg. Mouse LD_{50} Dose: 1900 mg/kg.
Acute Inhalation: Rat LC; Dose: >4800 ppm/8hr.
Acute Dermal: Rabbit LD_{50} Route: Skin; Dose: 2 gm/kg. Guinea Pig LD_{50} Route: Skin; Dose: >10 mL/kg.
Chronic (Multiple Dose) Dermal: Rabbit Route: Skin; Dose: 180 mL/kg/13W-I; Toxic Effects: Skin and appendages - Dermatitis, other; DEATH.
Irritation Eye: Rabbit Standard Draize Test Dose: 20 mg; Reaction: severe.
Irritation Skin: Rabbit Standard Draize Test Dose: 500 mg; Reaction: mild.
Reproductive/Teratogenic: Rat Route: Oral; Dose: 20 gm/kg; Duration: female 6-15D of pregnancy; Effects on Fertility - Post-implantation mortality; Effects on Embryo or Fetus - Fetotoxicity; Specific Developmental Abnormalities - Musculoskeletal system. Rat Route: Oral; Dose: 20 gm/kg; Duration: female 6-15D of pregnancy; Specific Developmental Abnormalities - Other developmental abnormalities.

Hazard Overviews

Health: Irritating to eyes/skin. Harmful. Other Acute Effects: harmful by inhalation, in contact with skin and if swallowed; do not breathe fumes. Chronic Effects: possible risk of irreversible effects; possible teratogen; overexposure may cause reproductive disorders.
Fire: Will burn. Extinguishing agents: water spray; carbon dioxide, dry chemical powder or appropriate foam. Precautions: combustible liquid.
Carcinogenicity: IARC - Not listed; NIOSH - Not listed; NTP - Not listed; ACGIH - Not listed; OSHA - Not listed; EPA - Not listed; MAK - Not listed
Primary Target Organs:

Eyes

Skin

Environmental

Regulations

RCRA 40CFR: Not listed
CERCLA: 40CFR 302.4: Not listed
SARA 40CFR 372.65: Not listed
SARA EHS 40CFR 355: Not listed
TSCA: Listed

Analytical Methods

Soil: EPA PMD-EUX

ETH7450 CAS #: 149-57-5

2-ETHYLHEXANOIC ACID

RTECS: MO7700000
EINECS Number: 205-743-6
Molecular Formula: $C_8H_{16}O_2$
Structured MF: $CH_3(CH_2)_3CHC_2H_5CO_2H$
Formula Weight: 144.22

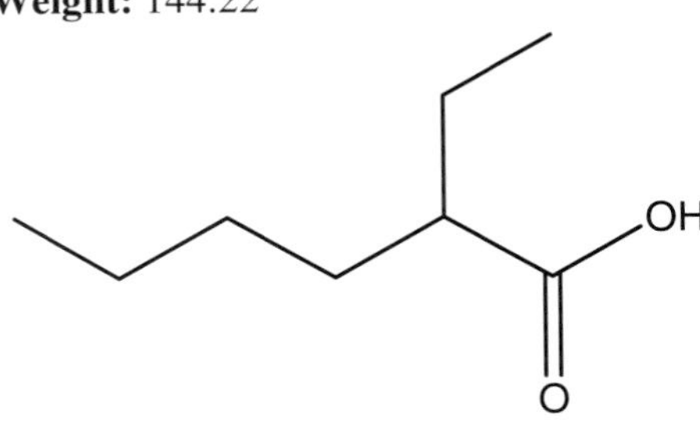

Chemical Structure

Synonyms: 2-BUTYLBUTANOIC ACID; BUTYLETHYLACETIC ACID; 2-ETHYLCAPROIC ACID; ALPHA-ETHYLCAPROIC ACID; ALPHA-ETHYLHEXANOIC ACID; ETHYLHEXANOIC ACID; 2-ETHYLHEXOIC ACID; ETHYLHEXOIC ACID; 3-HEPTANECARBOXYLIC ACID; HEXANOIC ACID,2-ETHYL-; HEXONIC ACID,2-ETHYL-; KYSELINA 2-ETHYLKAPRONOVA; KYSELINA HEPTAN-3-KARBOXYLOVA
Description: clear liquid
Use: in the manufacture of detergents, also as a drying agent for paints and varnishes and as a gelling agent for hydrocarbons

Physical Properties

Boiling Point: 228 °C (442 °F) at 755 mm Hg
Freezing Point: -83 °C (-117.4 °F)
Specific Gravity: 0.9031 at 25 °C/4 °C
Vapor Density: 4.98 Air=1
Water Solubility: Soluble in Hot Water
Other Solubilities: 95% Ethanol: >=100 mg/ml at 19 °C; Acetone: >=100 mg/ml at 19 °C; DMSO: >=100 mg/ml at 19 °C.
pH: Acid
Refraction Index: 1.4241 at 20 °C/D
Flash Point: 118 °C Open Cup
Autoignition Temperature: 371 °C
LEL: 0.8% v/v
UEL: 6% v/v

RTECS Toxicity Data

Acute Oral: Rat LD_{50} Dose: 3 gm/kg.
Acute Dermal: Rabbit LD_{50} Route: Skin; Dose: 1260 uL/kg. Guinea Pig LD_{50} Route: Skin; Dose: 6300 uL/kg.
Chronic (Multiple Dose) Oral: Rat Dose: 25200 mg/kg/3W-C; Toxic Effects: Blood - Changes in serum composition.

Irritation Eye: Rabbit Standard Draize Test Dose: 20 mg; Reaction: severe.
Irritation Skin: Rabbit Open Draize Test Dose: 450 mg open; Reaction: mild.
Reproductive/Teratogenic: Rat Route: Oral; Dose: 1803 mg/kg; Duration: female 12D of pregnancy; Specific Developmental Abnormalities - Musculoskeletal system; Cardiovascular (circulatory) system; Urogenital system. Rat Route: Oral; Dose: 5 gm/kg; Duration: female 6-15D of pregnancy; Effects on Embryo or Fetus - Fetotoxicity.
Mutagenic: Human Sister Chromatid Exchenge; Cell Type: lymphocyte; Dose: 630 umol/L.

Hazard Overviews

Corrosive

Health: Corrosive to eyes/skin/respiratory tract. Other Acute Effects: harmful if swallowed, inhaled, or absorbed through skin; inhalation may result in spasm; inflammation and edema of the larynx and bronchi; chemical pneumonitis and pulmonary edema; burning sensation; coughing; wheezing; laryngitis; shortness of breath; headache; nausea; vomiting; target organ: liver.
Fire: Will burn. Extinguishing agents: carbon dioxide, dry chemical powder or appropriate foam; water spray. Precautions: combustible liquid.
Reactivity: Incompatible with: bases, oxidizing agents, reducing agents. Hazardous decomposition products: toxic fumes of: carbon monoxide, carbon dioxide.
Carcinogenicity: IARC - Not listed; NIOSH - Not listed; NTP - Not listed; ACGIH - Not listed; OSHA - Not listed; EPA - Not listed; MAK - Not listed
Primary Target Organs:

Eyes

Skin

Respiratory System

Liver

Environmental

Cleanup/Disposal: Guide No. 171: Do not touch or walk through spilled material. Stop leak if you can do it without risk. Prevent dust cloud. Avoid inhalation of asbestos dust. Small Dry Spills: With clean shovel place material into clean, dry container and cover loosely; move containers from spill area. Small Spills: Take up with sand or other noncombustible absorbent material and place into containers for later disposal. Large Spills: Dike far ahead of liquid spill for later disposal. Cover powder spill with plastic sheet or tarp to minimize spreading. Prevent entry into waterways, sewers, basements or confined areas.

Regulations

RCRA 40CFR: Not listed
CERCLA: 40CFR 302.4: Not listed
SARA 40CFR 372.65: Not listed
SARA EHS 40CFR 355: Not listed
TSCA: Listed

ETH7550 CAS #: 104-76-7

2-ETHYLHEXANOL

RTECS: MP0350000
EINECS Number: 203-234-3
Molecular Formula: $C_8H_{18}O$
Structured MF: $CH_3(CH_2)_3CH(C_2H_5)CH_2OH$
Formula Weight: 130.22

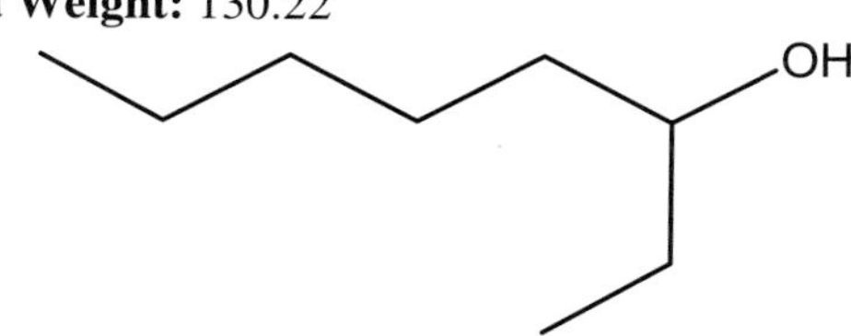

Chemical Structure

Synonyms: 2-AETHYLHEXANOL; 2-ETHYL-1-HEXANOL; ETHYLHEXANOL; 2-ETHYLHEXYL ALCOHOL; 1-HEXANOL,2-ETHYL-; OCTYL ALCOHOL
Description: colorless liquid; sweet floral odor reminiscent of rose
Use: as a solvent for dyes, resins, oils, antifoaming agents and nitrocellulose; as a wetting agent, in organic synthesis, in paints lacquers, baking finishes, plasticizers, inks, rubber, paper, lubricants, photography and dry cleaning, as a plasticizer for PVC resins, in textile finishing compounds and in mercerizing textiles

Physical Properties

Boiling Point: 184.34 °C (364 °F)
Freezing Point: -76 °C (-104.8 °F)
Specific Gravity: 0.8344 at 20 °C/20 °C
Vapor Density: 4.49 Air=1
Saturated Vapor Density: 1.200275554 kg/m^3
Density: 0.831 g/cu cm at 20 °C
Bulk Density: 6.9 lbs/gal at 20 °C
Vapor Pressure: 0.05 mm Hg at 20 °C
Water Solubility: 1 parts in about 720 parts Water
Other Solubilities: 95% Ethanol: >=100 mg/ml at 18 °C; Acetone: >=100 mg/ml at 18 °C; Alcohol: Soluble; Benzene: Soluble; Chloroform: Soluble; DMSO: >=100 mg/ml at 18 °C; Ether: Soluble; Organic solvents: Soluble.
Surface Tension: 27.6 dynes/cm at 20 °C
Refraction Index: 1.4300 at 20 °C/D
Critical Temperature: 339.8 °C
Critical Pressure: 2.76 Pa
Flash Point: 81 °C
Autoignition Temperature: 231 °C
LEL: 0.88% v/v
UEL: 9.7% v/v

RTECS Toxicity Data

Acute Oral: Rat LD_{50} Dose: 3730 mg/kg; Toxic Effects: Brain and coverings - Recordings from specific areas of CNS; Behavioral - Somnolence (general depressed activity); Lungs, Thorax, or Respiration - Dyspnea. Mouse LD_{50} Dose: 2500 mg/kg.
Acute Dermal: Rabbit LD_{50} Route: Skin; Dose: 1970 mg/kg. Rat LD_{50} Route: Subcutaneous Dose: 650 mg/kg.
Chronic (Multiple Dose) Oral: Rat Dose: 9 gm/kg/11D-I; Toxic Effects: Gastrointestinal - Ulceration or bleeding from stomach; Liver - Changes in liver weight; Kidney, Ureter, and Bladder - Changes in kidney weight. Rat Dose: 16250 mg/kg/13W-I; Toxic Effects: Gastrointestinal - Other changes; Liver - Changes in liver weight; Kidney, Ureter, and Bladder - Changes in kidney weight. Rat Dose: 14 gm/kg/17D-I; Toxic Effects: Liver - Changes in liver weight; Biochemical - Phosphatases; Biochemical - Other hydrolases. Rat Dose: 25200 mg/kg/3W-C; Toxic Effects: Blood - Changes in serum composition. Rat Dose: 260 gm/kg/2Y-I; Toxic Effects: DEATH.
Irritation Eye: Rabbit Standard Draize Test Dose: 20 mg/24H; Reaction: moderate. Rabbit Standard Draize Test Dose: 20 mg; Reaction: severe.
Irritation Skin: Rabbit Standard Draize Test Dose: 500 mg/24H; Reaction: moderate. Rabbit Open Draize Test Dose: 415 mg open; Reaction: mild.
Reproductive/Teratogenic: Rat Route: Oral; Dose: 1628 mg/kg; Duration: female 12D of pregnancy; Specific Developmental Abnormalities - Musculoskeletal system; Urogenital system. Mouse Route: Oral; Dose: 12 gm/kg; Duration: female 7-14D of pregnancy; Effects on Newborn - Live birth index; Viability index; Growth statistics. Mouse Route: Oral; Dose: 12200 mg/kg; Duration: female 6-13D of pregnancy; Effects on Fertility - Litter size; Effects on Newborn - Live birth index; Growth statistics.
Mutagenic: Bacteria - S Typhimurium Mutations in Microorganisms; Dose: 500 umol/L (+S9).

Hazard Overviews

Fire Diamond

Health: Irritating to eyes/skin/respiratory tract. Also Causes: headache, weakness, nausea, giddiness, ataxia, CNS depression. Chronic Effects: skin redness/sensitization, liver/kidney damage (in animals).
Fire: Combustible. Can form explosive mixtures in the air. For small fires use dry chemical, carbon dioxide, water spray, or alcohol-resistant foam. For large fires use water spray, fog, or alcohol-resistant foam. Do not use a solid stream of water.
Reactivity: Stable. Hazardous polymerization cannot occur. Avoid: heat; ignition sources. Incompatible with: oxidizing materials. Hazardous decomposition products: carbon dioxide; acrid smoke; vapors.

Carcinogenicity: IARC - Not listed; NIOSH - Not listed; NTP - Not listed; ACGIH - Not listed; OSHA - Not listed; EPA - Not listed; MAK - Not listed
Primary Target Organs:

Eyes

Skin

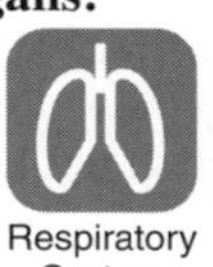
Respiratory System

Nervous System

Environmental

Ecotoxicity: Aquatic toxicity: 19 ppm/24 hr/brine shrimp/TLm
Environmental Fate: If released in soil, it would be expected to leach and readily biodegrade after some acclimation. Some volatilization may occur from both moist and dry soil. If released in water, it will volatilize; the estimated half-life in a model river is 1.7 days. It is readily biodegradable in screening tests and one river die-away test and should biodegrade. Adsorption to sediment and bioconcentration in fish is not expected to be important. In the atmosphere, it will occur as a vapor and react with photochemically produced hydroxyl radicals. Its estimated half-life in the atmosphere is 1.2 days. Since it is moderately water soluble, it may also be washed out of the atmosphere by rain.
Cleanup/Disposal: Guide No. 171: Do not touch or walk through spilled material. Stop leak if you can do it without risk. Prevent dust cloud. Avoid inhalation of asbestos dust. Small Dry Spills: With clean shovel place material into clean, dry container and cover loosely; move containers from spill area. Small Spills: Take up with sand or other noncombustible absorbent material and place into containers for later disposal. Large Spills: Dike far ahead of liquid spill for later disposal. Cover powder spill with plastic sheet or tarp to minimize spreading. Prevent entry into waterways, sewers, basements or confined areas.

Environmental Physical Data

Henry's Law Constant: calculated at 2.65×10^{-5}
Sorption Partition Coefficient: K_{oc} = estimated at 105
BCF: estimated at 13
BOD: 88%, 5 days

Regulations

RCRA 40CFR: Not listed
CERCLA: 40CFR 302.4: Not listed
SARA 40CFR 372.65: Not listed
SARA EHS 40CFR 355: Not listed
TSCA: Listed

Analytical Methods

Water / Groundwater: ASTM D3695

ETH7600 CAS #: 103-09-3

2-ETHYLHEXYL ACETATE

RTECS: AH5600000
EINECS Number: 203-079-1

Molecular Formula: $C_{10}H_{20}O_2$
Structured MF: $CH_3COOCH_2CH(CH_2H_5)C_4H_9$
Formula Weight: 172.27

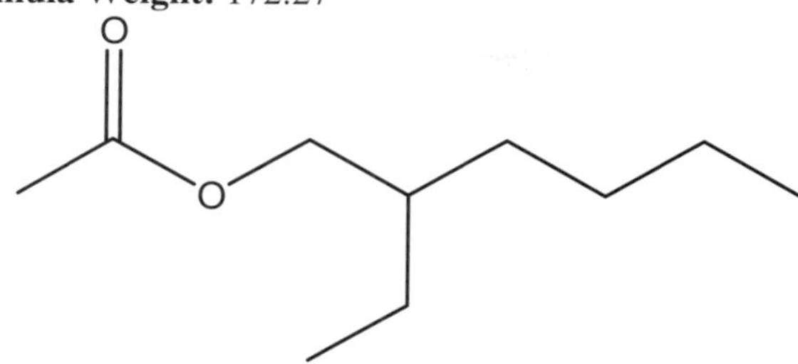

Chemical Structure

Synonyms: ACETIC ACID ALPHA-ETHYLHEXYL ESTER; ACETIC ACID,2-ETHYLHEXYL ESTER; 2-ETHYL-1-HEXANOL ACETATE; 2-ETHYLHEXANYL ACETATE; 2-ETHYL-1-HEXYL ACETATE; BETA-ETHYLHEXYL ACETATE; 2-ETHYLHEXYL ETHANOATE; 2-ETHYLHEXYLESTER KYSELINY OCTOVE; OCTYL ACETATE
Description: water-white liquid
Use: solvent for nitrocellulose, some resins, waxes & oils; baking finishes

Physical Properties

Boiling Point: 199 °C (390 °F)
Freezing Point: -93 °C (-135.4 °F)
Specific Gravity: 0.873 at 20 °C/20 °C
Vapor Density: 5.93 Air=1
Saturated Vapor Density: 1.203120218 kg/m^3
Vapor Pressure: 0.4 mm Hg at 20 °C
Water Solubility: Very Slightly Soluble in Water
Other Solubilities: Soluble in Ether.
Odor Threshold: 0.51 to 1.47 mg/m^3
Refraction Index: 1.4204 at 20 °C
Flash Point: 13 °C Closed Cup
Autoignition Temperature: 268 °C
LEL: 0.76% v/v
UEL: 8.14% v/v

RTECS Toxicity Data

Acute Oral: Rat LD_{50} Dose: 3 gm/kg.
Acute Dermal: Rabbit LD; Route: Skin; Dose: >20 mL/kg. Guinea Pig LD_{50} Route: Skin; Dose: >20 mL/kg.
Irritation Eye: Rabbit Standard Draize Test Dose: 250 ug/24H; Reaction: severe. Rabbit Standard Draize Test Dose: 500 mg; Reaction: mild.
Irritation Skin: Rabbit Open Draize Test Dose: 500 mg open; Reaction: mild.

Hazard Overviews

Flammable

Fire: Flammable.

Carcinogenicity: IARC - Not listed; NIOSH - Not listed; NTP - Not listed; ACGIH - Not listed; OSHA - Not listed; EPA - Not listed; MAK - Not listed

Environmental

Environmental Fate: If released to soil, it is expected to display slight mobility. Volatilization is expected to occur from both moist and dry soils. Hydrolysis in soil is not expected to be a significant process except in highly basic soils with a pH >9. If released to water, it is expected to rapidly volatilize to the atmosphere. The half-life for volatilization from a model river is 4.2 h. It may adsorb to sediment and suspended organic matter and it may bioconcentrate in fish and aquatic organisms. Hydrolysis in aquatic systems is not expected to be a significant process except under basic conditions of pH >9. In the atmosphere, it is expected to undergo a gas-phase reaction with photochemically produced hydroxyl radicals, with an estimated half-life on the order of 1.5 days. It may undergo atmospheric removal by wet deposition processes.
Cleanup/Disposal: Guide No. 128: Eliminate all ignition sources (no smoking, flares, sparks or flames in immediate area). All equipment used when handling the product must be grounded. Do not touch or walk through spilled material. Stop leak if you can do it without risk. Prevent entry into waterways, sewers, basements or confined areas. A vapor suppressing foam may be used to reduce vapors. Absorb or cover with dry earth, sand or other non-combustible material and transfer to containers. Use clean non-sparking tools to collect absorbed material. Large Spills: Dike far ahead of liquid spill for later disposal. Water spray may reduce vapor; but may not prevent ignition in closed spaces.

Environmental Physical Data

Henry's Law Constant: 2.81 $x10^{-3}$
Octanol/Water Partition Coefficient: log K_{ow} = 3.72
Sorption Partition Coefficient: K_{oc} = calculated at 2515
BCF: calculated at 395

Regulations

RCRA 40CFR: Not listed
CERCLA: 40CFR 302.4: Not listed
SARA 40CFR 372.65: Not listed
SARA EHS 40CFR 355: Not listed
TSCA: Listed

Analytical Methods

Water / Groundwater: ASTM D3695

ETH7650 **CAS #: 103-11-7**

2-ETHYLHEXYL ACRYLATE

RTECS: AT0855000
EINECS Number: 203-080-7
Molecular Formula: $C_{11}H_{20}O_2$
Structured MF: $CH_2{=}CH{\bullet}CO{\bullet}O{\bullet}CH_2CHC_2H_5(CH_2)_3CH_3$
Formula Weight: 184.28

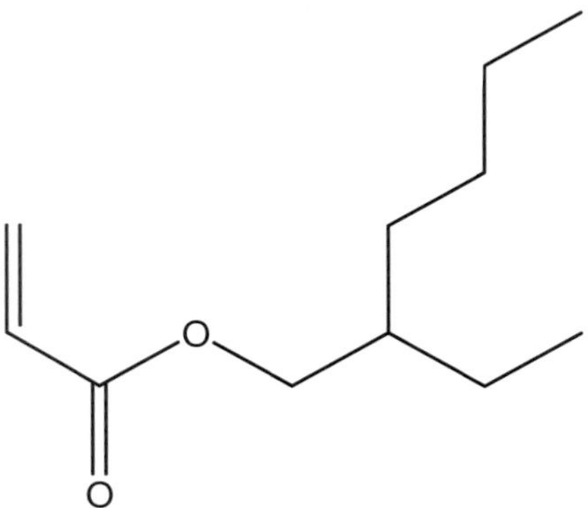

Chemical Structure

Synonyms: ACRYLIC ACID,2-ETHYLHEXYL ESTER; ACRYLIC ACID-2-ETHYLHEXYL ESTER; 2-ETHYLHEXYL 2-PROPENOATE; 2-ETHYLHEXYLESTER KYSELINY AKRYLOVE; 1-HEXANOL,2-ETHYL-,ACRYLATE; OCTYL ACRYLATE; 2-PROPENOIC ACID,2-ETHYLHEXYL ESTER; 2-PROPENOIC ACID-2-ETHYLHEXYL ESTER; 2-PROPENOIC ACID,2-ETHYLHEXYL ESTER (9CI)

Description: colorless liquid; sharp, objectionable odor

Use: monomer for plastics, protective coating, paper treatment, water-based paints

Physical Properties

Boiling Point: 214 °C (417 °F) to 218 °C (424 °F)
Freezing Point: -90 °C (-130 °F)
Specific Gravity: 0.8869 at 20 °C/20 °C
Vapor Density: 6.35 Air=1
Vapor Pressure: 1 mm Hg at 50.0 °C
Water Solubility: 0.01 g/100 g Water
Other Solubilities: 95% Ethanol: >= 100 mg/ml at 19 °C; Acetone: >= 100 mg/ml at 19 °C; DMSO: 10-50 mg/ml at 19 °C.
Surface Tension: Estimated at 26 dynes/cm
Refraction Index: 1.4332 at 25 °C
Flash Point: 82.222 °C Open Cup
Autoignition Temperature: 252 °C
LEL: 0.8% v/v
UEL: Inhibited 6.4% v/v

RTECS Toxicity Data

Acute Oral: Rat LD_{50} Dose: 6500 uL/kg. Mouse LD_{50} Dose: 4400 mg/kg; Toxic Effects: Behavioral - Excitement.

Acute Inhalation: Mammal LC_{50} Dose: >450 mg/m^3; Toxic Effects: Behavioral - Somnolence (general depressed activity); Nutritional and gross metabolic - Weight loss or decreased weight gain. Mouse LC_{Lo} Dose: 600 mg/m^3; Toxic Effects: Behavioral - Excitement.

Acute Dermal: Rabbit LD_{50} Route: Skin; Dose: 8480 uL/kg. Rat LD; Route: Skin; Dose: >12 gm/kg.

Chronic (Multiple Dose) Oral: Rat Dose: 4020 mg/kg/30D-I; Toxic Effects: Brain and coverings - Recordings from specific areas of CNS; Biochemical - True cholinesterase; Biochemical - Catalases. Rat Dose: 1785 mg/kg/26W-I; Toxic Effects: Liver - Other changes; Blood - Other changes; Biochemical - True cholinesterase.

Chronic (Multiple Dose) Inhalation: Rat Dose: 103 mg/m^3/19W-I; Toxic Effects: Brain and coverings - Recordings from specific areas of CNS; Endocrine - Changes in thyroid weight; Biochemical - Catalases.

Irritation Eye: Rabbit Standard Draize Test Dose: 5 mg; Reaction: severe. Rabbit Standard Draize Test Dose: 500 mg/24H; Reaction: mild.

Irritation Skin: Rabbit Standard Draize Test Dose: 20 mg/24H; Reaction: moderate. Rabbit Open Draize Test Dose: 10 mg/24H open; Reaction: severe.

Tumorigenic: Mouse Route: Skin; Dose: 187 gm/kg/78W-I; Toxic Effects: Tumorigenic - Carcinogenic by RTECS criteria; Skin and appendages - Tumors; Tumorigenic - Tumors at site of application. Mouse Route: Skin; Dose: 240 gm/kg/2Y-C; Toxic Effects: Tumorigenic - Neoplastic by RTECS criteria; Skin and appendages - Tumors; Tumorigenic - Tumors at site of application.

Hazard Overviews

Fire Diamond

Health: Irritating to eyes/skin/respiratory tract. Also Causes: pulmonary edema, headache, nausea, vomiting, drowsiness, convulsions, severe eye injury, convulsions, abdominal discomfort, vomiting, diarrhea. Chronic Effects: sensitization, respiratory tract damage.

Fire: Combustible. Use water spray, dry chemical, foam, or carbon dioxide. Water spray may be ineffective in extinguishing fire. Water or foam may cause frothing. Use water spray to cool fire-exposed containers.

Reactivity: Unstable. Hazardous polymerization can occur, by elevated temperatures, peroxides, oxidizers; usually contains inhibitors to prevent polymerization. Avoid: heat; ignition sources; sunlight; x-ray; ultraviolet radiation; strong bases; oxygen-free atmospheres. Incompatible with: strong acids; alkalies; amines; halogens; halogen compounds; oxidizing materials; polymerization initiators; peroxides. Hazardous decomposition products: carbon monoxide; carbon dioxide; various organic compounds.

Carcinogenicity: IARC - Group 3, Not classifiable as to carcinogenicity to humans; NIOSH - Not listed; NTP - Not listed; ACGIH - Not listed; OSHA - Not listed; EPA - Not listed; MAK - Not listed

Primary Target Organs:

Eyes

Skin

Respiratory System

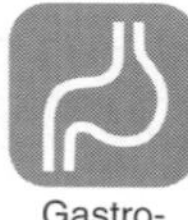
Gastro-intestinal

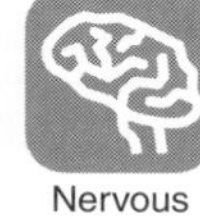
Nervous System

Environmental

Ecotoxicity: Aquatic toxicity: 72 ppm/24 hr/ brine shrimp/TLm

Environmental Fate: If released to soil, it will be expected to exhibit moderate mobility in soil and, therefore, it may leach to groundwater. It may hydrolyze, especially in alkaline soils based upon hydrolysis data for the structurally similar ethyl acrylate. It may biodegrade in soil based upon the biodegradability of butyl acrylate in aqueous screening tests. It may volatilize from near surface soil and other surfaces. If released to water, it will not be expected to adsorb to sediment or suspended particulate matter or bioconcentrate in aquatic organisms. Hydrolysis may be a significant process especially in alkaline waters based upon hydrolysis data for the structurally similar ethyl acrylate. It may directly photolyze in sunlight based upon the slight absorption of light at wavelengths > 290 nm by ethyl acrylate and other acrylate esters. It may biodegrade in natural waters based upon the biodegradability of butyl acrylate and ethyl acrylate in aqueous screening tests. It will significantly volatilize from water with an estimated half-life of 7.3 hr for volatilization from a model river. The volatilization half-life from a model pond, which considers the effect of adsorption, has been estimated to be 2.7 days. If released to the atmosphere, it will be expected to exist almost entirely in the vapor phase based upon a reported vapor pressure of 0.146 mm Hg at 23 °C. It will be susceptible to photooxidation via vapor phase reaction with photochemically produced hydroxyl radicals and ozone. An atmospheric half-life of 10.3 hours at an atmospheric concentration of 5 $x10^5$ hydroxyl radicals per cu cm and 7 $x10^{11}$ ozone molecules per cu cm has been estimated for this process.

Cleanup/Disposal: Guide No. 171: Do not touch or walk through spilled material. Stop leak if you can do it without risk. Prevent dust cloud. Avoid inhalation of asbestos dust. Small Dry Spills: With clean shovel place material into clean, dry container and cover loosely; move containers from spill area. Small Spills: Take up with sand or other noncombustible absorbent material and place into containers for later disposal. Large Spills: Dike far ahead of liquid spill for later disposal. Cover powder spill with plastic sheet or tarp to minimize spreading. Prevent entry into waterways, sewers, basements or confined areas.

Environmental Physical Data

Henry's Law Constant: calculated at 3.54 $x10^{-4}$
Sorption Partition Coefficient: K_{oc} = estimated at 363
BCF: estimated at 46
BOD: theoretical 9%, 5 days

Regulations

RCRA 40CFR: Not listed
CERCLA: 40CFR 302.4: Not listed
SARA 40CFR 372.65: Not listed
SARA EHS 40CFR 355: Not listed
TSCA: Listed

Analytical Methods

Water / Groundwater: ASTM D3695

ETH7700 CAS #: 85-69-8

2-ETHYLHEXYL BUTYL PHTHALATE

EINECS Number: 201-623-2
Molecular Formula: $C_{20}H_{30}O_4$
Formula Weight: 334.46
Synonyms: 1,2-BENZENEDICARBOXYLIC ACID,BUTYL 2-ETHYLHEXYL ESTER; BUTYL 2-ETHYLHEXYL PHTHALATE; PHTHALIC ACID,BUTYL 2-ETHYLHEXYL ESTER
Description: colorless or very slightly colored liquid

Physical Properties

Vapor Pressure: Extremely low
Water Solubility: 1 mg/L at 20 °C
Other Solubilities: Soluble to various extents in many common organic solvents and oils.

Hazard Overviews

Carcinogenicity: IARC - Not listed; NIOSH - Not listed; NTP - Not listed; ACGIH - Not listed; OSHA - Not listed; EPA - Not listed; MAK - Not listed

Environmental

Environmental Fate: If released to water, it is expected to undergo microbial degradation based, on limited biodegradation data. Adsorption to sediment and suspended organic matter, as well as bioaccumulation in fish and aquatic organisms is expected to occur. Volatilization from water to the atmosphere is expected to be a slow process. Hydrolysis is not expected to be significant. If released to soil, it may undergo biodegradation under aerobic conditions, based on limited biodegradation data. In the atmosphere, it is expected to be adsorbed to particulates. Destruction by the vapor phase reaction with photochemically produced hydroxyl radicals may occur.

Environmental Physical Data

Henry's Law Constant: 2.13 $x10^{-6}$
Octanol/Water Partition Coefficient: log K_{ow} = > 2.12
Sorption Partition Coefficient: K_{oc} = calculated at 4370
BCF: none likely

Regulations

RCRA 40CFR: Not listed
CERCLA: 40CFR 302.4: Not listed
SARA 40CFR 372.65: Not listed
SARA EHS 40CFR 355: Not listed
TSCA: Listed

ETH7750 CAS #: 1070-03-7

2-ETHYLHEXYL DIHYDROGEN PHOSPHATE

EINECS Number: 213-967-0
Molecular Formula: $C_8H_{19}O_4P$
Formula Weight: 218.33
Synonyms: MONO(2-ETHYLHEXYL)PHOSPHATE; PHOSPHORIC ACID,MONO(2-ETHYLHEXYL) ESTER
Use: surface-active agent (hydrotrope, penetrant, wetting agent); recovery of alum; for phosphorus removal in wastewater treatment; extractant; acidic extractant for metals

Hazard Overviews

Carcinogenicity: IARC - Not listed; NIOSH - Not listed; NTP - Not listed; ACGIH - Not listed; OSHA - Not listed; EPA - Not listed; MAK - Not listed

Environmental

Regulations

RCRA 40CFR: Not listed
CERCLA: 40CFR 302.4: Not listed
SARA 40CFR 372.65: Not listed
SARA EHS 40CFR 355: Not listed
TSCA: Listed

ETH7800 CAS #: 141-38-8

2-ETHYLHEXYL EPOXYSTEARATE

RTECS: RG1600000
EINECS Number: 205-482-8
Molecular Formula: $C_{26}H_{50}O_3$
Formula Weight: 410.76
Synonyms: 9,10-EPOXYOCTADECANOIC ACID,2-ETHYLHEXYL ESTER; 9,10-EPOXYSTEARIC ACID,2-ETHYLHEXYL ESTER; 2-ETHYLHEXYL 9,10-EPOXYOCTADECANOATE; OCTADECANOIC ACID,9,10-EPOXY-,2-ETHYLHEXYL ESTER; OXIRANEOCTANOIC ACID,3-OCTYL-,2-ETHYLHEXYL ESTER; OXIRANOCTANOICACID,3-OCTYL,2-ETHYLHEXYL ESTER
Description: amber, clear viscous liquid
Use: plasticizer, heat & light stabilizer for polyvinyl chloride

Physical Properties

Saturated Vapor Density: 1.308084501 kg/m^3
Density: 0.0894 g/cu cm at 25 °C
Vapor Pressure: 5.2 mm Hg at 25.0 °C
Water Solubility: < 1 mg/mL at 20 C
Other Solubilities: 95% Ethanol: >=100 mg/ml at 20 °C; Acetone: >=100 mg/ml at 20 °C; DMSO: <1 mg/ml at 20 °C.
Flash Point: > 93.33 °C

RTECS Toxicity Data

Acute Oral: Rat LD_{50} Dose: 30800 mg/kg.
Irritation Skin: Rabbit Open Draize Test Dose: 10 mg/24H open; Reaction: mild.

Hazard Overviews

Fire: Will burn.
Carcinogenicity: IARC - Not listed; NIOSH - Not listed; NTP - Not listed; ACGIH - Not listed; OSHA - Not listed; EPA - Not listed; MAK - Not listed

Environmental

Regulations

RCRA 40CFR: Not listed
CERCLA: 40CFR 302.4: Not listed
SARA 40CFR 372.65: Not listed
SARA EHS 40CFR 355: Not listed
TSCA: Listed

ETH7850 CAS #: 688-84-6

2-ETHYLHEXYL METHACRYLATE

RTECS: OZ4630000
EINECS Number: 211-708-6
Molecular Formula: $C_{12}H_{22}O_2$
Formula Weight: 198.30

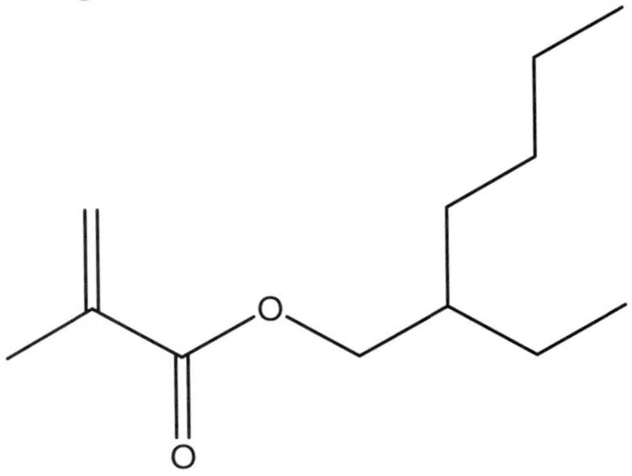

Chemical Structure

Synonyms: 2-ETHYLHEXYL METHACRYATE; 2-ETHYL-1-HEXYL METHACRYLATE; METHACRYLATE,2-ETHYLISOHEXY; METHACRYLIC ACID,2-ETHYLHEXYL ESTER; 2-PROPENOIC ACID,2-METHYL-,2-ETHYLHEXYL ESTER
Use: comonomer for acrylic polymers used in latex paints; comonomer for acrylic polymers used in lacquers; monomer for polymers used as viscosity index improvers

Physical Properties

Boiling Point: 113 °C (235 °F)
Specific Gravity: 0.88000
Vapor Density: 6.8 Air=1
Saturated Vapor Density: < 1.209217786 kg/m^3
Vapor Pressure: < 1 mm Hg at 20 °C
Refraction Index: 1.4380 at 20 °C

Hazard Overviews

Health: Irritating to eyes/skin/respiratory tract. Other Acute Effects: harmful if swallowed, inhaled, or absorbed through skin; lachrymator; symptoms of exposure may include burning sensation; coughing; wheezing; laryngitis; shortness of breath; headache; nausea; vomiting.
Fire: Hazards: emits toxic fumes. Extinguishing agents: water spray; carbon dioxide, dry chemical powder or appropriate foam. Precautions: combustible liquid.
Reactivity: Incompatible with: strong oxidizing agents, may polymerize on exposure to light, heat-sensitive, strong bases, strong acids. Hazardous decomposition products: toxic fumes of: carbon monoxide, carbon dioxide.
Carcinogenicity: IARC - Not listed; NIOSH - Not listed; NTP - Not listed; ACGIH - Not listed; OSHA - Not listed; EPA - Not listed; MAK - Not listed

Primary Target Organs:

Environmental

Regulations

RCRA 40CFR: Not listed
CERCLA: 40CFR 302.4: Not listed
SARA 40CFR 372.65: Not listed
SARA EHS 40CFR 355: Not listed
TSCA: Listed

ETH7950 **CAS #: 31044-12-9**

2-ETHYLHEXYL SODIUM PHOSPHATE

EINECS Number: 250-443-0
Molecular Formula: $C_8H_{19}Na_xO_4P$
Formula Weight: Varies
Synonyms: PHOSPHORIC ACID,MONO(2-ETHYLHEXYL) ESTER,SODIUM SALT
Use: surface active agent (detergent, dispersant); corrosion inhibitor; as lubricating oil additive

Hazard Overviews

Carcinogenicity: IARC - Not listed; NIOSH - Not listed; NTP - Not listed; ACGIH - Not listed; OSHA - Not listed; EPA - Not listed; MAK - Not listed

Environmental

Regulations

RCRA 40CFR: Not listed
CERCLA: 40CFR 302.4: Not listed
SARA 40CFR 372.65: Not listed
SARA EHS 40CFR 355: Not listed
TSCA: Listed

ETH8000 **CAS #: 21245-02-3**

2-ETHYLHEXYL-P-DIMETHYLAMINO

EINECS Number: 244-289-3
Molecular Formula: $C_{17}H_{27}NO_2$
Formula Weight: 277.41

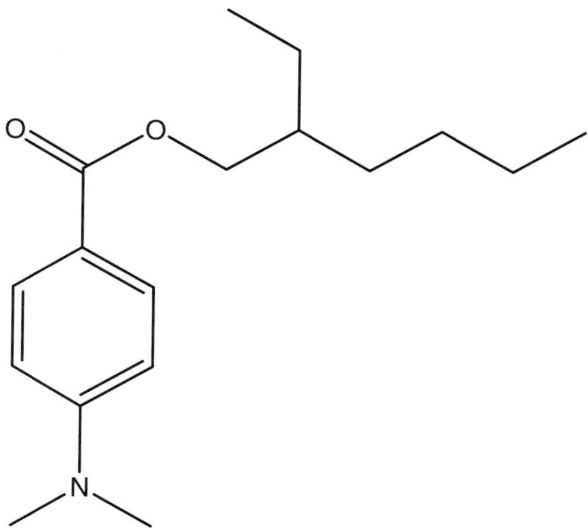

Chemical Structure

Physical Properties

Boiling Point: 365 °C (689 °F)
Specific Gravity: 0.995
Water Solubility: < 2
Evaporation Rate: < 1 Butyl Acetate=1

Hazard Overviews

Health: Irritating to eyes/skin/respiratory tract. Other Acute Effects: may be harmful by inhalation, ingestion, or skin absorption.
Fire: Hazards: emits toxic fumes. Extinguishing agents: water spray; carbon dioxide, dry chemical powder or appropriate foam. Precautions: combustible liquid.
Reactivity: Stable. Hazardous polymerization will not occur. Incompatible with: strong oxidizing agents, strong acids. Hazardous decomposition products: toxic fumes of: carbon monoxide, carbon dioxide, nitrogen oxides.
Carcinogenicity: IARC - Not listed; NIOSH - Not listed; NTP - Not listed; ACGIH - Not listed; OSHA - Not listed; EPA - Not listed; MAK - Not listed
Primary Target Organs:

Environmental

Regulations

RCRA 40CFR: Not listed
CERCLA: 40CFR 302.4: Not listed
SARA 40CFR 372.65: Not listed
SARA EHS 40CFR 355: Not listed
TSCA: Not listed

ETH8100 **CAS #: 496-03-7**

2-ETHYL-3-HYDROXYHEXANAL

EINECS Number: 207-812-6
Molecular Formula: $C_8H_{16}O_2$
Formula Weight: 144.22
Synonyms: BUTYRALDOL; HEXANAL,2-ETHYL-3-HYDROXY-; 3-HYDROXY-2-ETHYLHEXANAL

Hazard Overviews

Carcinogenicity: IARC - Not listed; NIOSH - Not listed; NTP - Not listed; ACGIH - Not listed; OSHA - Not listed; EPA - Not listed; MAK - Not listed

Environmental

Regulations

RCRA 40CFR: Not listed
CERCLA: 40CFR 302.4: Not listed
SARA 40CFR 372.65: Not listed
SARA EHS 40CFR 355: Not listed
TSCA: Not listed

ETH8150 **CAS #: 75-34-3**

ETHYLIDENE DICHLORIDE

RTECS: KI0175000
DOT: UN2362; IMO3.2
EINECS Number: 200-863-5
Molecular Formula: $C_2H_4Cl_2$
Structured MF: Cl_2CHCH_3
Formula Weight: 98.97

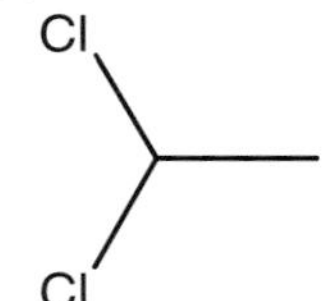

Chemical Structure

Synonyms: AETHYLIDENCHLORID; ASYMMETRICAL DICHLOROETHANE; CHLORINATED HYDROCHLORIC ETHER; CHLORURE D'ETHYLIDENE; CLORURO DI ETILIDENE; 1,1-DICHLOORETHAAN; 1,1-DICHLORAETHAN; 1,1-DICHLORETHANE; 1,1-DICHLOROETHANE; ALPHA ALPHA DICHLOROETHANE; DICHLOROETHANE,1,1-; 1,1-DICLOROETANO; ETHANE,1,1-DICHLORO-; ETHYLIDENE CHLORIDE; 1,1-ETHYLIDENE DICHLORIDE
Description: colorless liquid; chloroform odor
Use: in vinyl chloride manufacturing, chlorinated solve intermediate, coupling agent in antiknock gasoline, metal degreasing agent, organic synthesis, ore flotation, manufacturing of H3CCCl3, chemical intermediate, an industrial solvent, in fumigant formulations, extraction solvent, as a dewaxer of mineral oils, extractant for heat-sensitive substances, manufacturing of high vacuum rubber and silicon grease and as a paint, varnish an finish remover; formerly used as an anesthetic

Physical Properties

Boiling Point: 57.3 °C (135 °F)
Freezing Point: -96.9 °C (-142.42 °F)
Specific Gravity: 1.175 at 20 °C/4 °C
Vapor Density: 3.44 Air=1
Saturated Vapor Density: 2.091450817 kg/m^3
Density: 1.256 g/cc at 21 °C
Vapor Pressure: 234 torr at 25 °C
Water Solubility: 0.5 g/100 ml Water at 20 °C
Other Solubilities: 95% Ethanol: >=100 mg/ml at 20 °C; Acetone: >=100 mg/ml at 20 °C; Alcohol: Soluble; Benzene: Soluble; DMSO: >=100 mg/ml at 20 °C; Ether: Soluble; Fixed and volatile oils: Soluble; Most common solvents: miscible.
Surface Tension: 24.75 dynes/cm
Odor Threshold: 445.5 to 810 mg/m^3
Refraction Index: 1.4167 at 20 °C
Evaporation Rate: 11.6 Butyl Acetate=1
Critical Temperature: 128 °C
Critical Pressure: 734.8 psia
Ionization Potential (eV): 11.06
Flash Point: 13.889 °C Open Cup
Autoignition Temperature: 458 °C
LEL: 5.6% v/v
UEL: 11.4% v/v

RTECS Toxicity Data

Acute Oral: Rat LD_{50} Dose: 725 mg/kg.
Acute Inhalation: Rat LC_{50} Dose: 13000 ppm/4hr.
Chronic (Multiple Dose) Oral: Rat Dose: 240 gm/kg/12W-I; Toxic Effects: Brain and coverings - Recordings from specific areas of CNS; DEATH.
Reproductive/Teratogenic: Rat Route: Inhalation; Dose: 6000 ppm/7H; Duration: female 6-15D of pregnancy; Specific Developmental Abnormalities - Musculoskeletal system.
Mutagenic: Rat Unscheduled DNA Synthesis; Cell Type: liver; Dose: 13 mmol/L. Mold - A Nidulans Sex Chromosome Loss; Dose: 2000 ppm.
Tumorigenic: Mouse Route: Oral; Dose: 185 gm/kg/78W-I; Toxic Effects: Tumorigenic - Equivocal tumorigenic agent by RTECS criteria; Tumorigenic effects - Uterine tumors. Mouse Route: Oral; Dose: 1300 gm/kg/78W-I; Toxic Effects: Tumorigenic - Equivocal tumorigenic agent by RTECS criteria; Tumorigenic effects - Uterine tumors.

Hazard Overviews

Flammable

Fire Diamond

Health: Irritating to eyes/respiratory tract. Also Causes: headache, dizziness, coughing, staggering, disturbed vision, irregular heartbeat, unconsciousness, narcosis, coma, pulmonary edema, defatting, redness, swelling, watering eyes, lid inflammation, death due to cardiac or respiratory failure Chronic Effects: rash, scaliness, neurological effects. Flammable.

Fire: Flammable. Can form explosive mixtures in the air. For small fires use dry chemical, carbon dioxide, or alcohol-resistant foam. For large fires use fog or alcohol-resistant foam. Water may be ineffective unless used as a blanket. Fight fire from maximum distance.

Reactivity: Stable. Hazardous polymerization cannot occur. Avoid: exposure to heat and ignition sources. Incompatible with: strong oxidizers; caustics; some forms of plastics, rubber, and coatings. Hazardous decomposition products: carbon dioxide; irritating hydrogen chloride and toxic phosgene fumes.

Carcinogenicity: IARC - Not listed; NIOSH - Not listed; NTP - Not listed; ACGIH - Class A4, Not classifiable as a human carcinogen; OSHA - Not listed; EPA - Class C, Possible human carcinogen; MAK - Not listed

Primary Target Organs:

Eyes

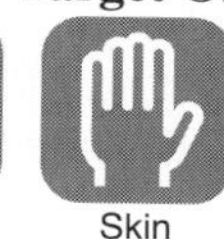
Skin

Respiratory System

Nervous System

Cardio-vascular

Exposure Limits

OSHA PEL: TWA: 100 ppm; 400 mg/m^3.

ACGIH TLV: TWA: 100 ppm; 405 mg/m^3.

NIOSH REL: TWA: 100 ppm; 400 mg/m^3.

NIOSH IDLH: 3000 ppm.

DFG MAK: TWA: 100 ppm; 400 mg/m^3.

Respirator Recommendation

Exposure Range: >100 to <3000 ppm Supplied Air, Constant Flow/Pressure Demand, Half Mask

Exposure Range: 3000 to unlimited ppm Self-contained Breathing Apparatus, Pressure Demand, Full Face

Note: poor warning properties

Environmental

Ecotoxicity: LC_{50} Lepomis macrochirus (bluegill) 550 ppm/96 hr, static bioassay in fresh water at 23 °C; mild aeration applied after 24 hr (no specific isomer LC_{50} Poecilia reticulata (guppies) 202 ppm/7 days. /Conditions of bioassay not specified TLm Lagodon rhomboides (pinperch) 160 mg/l/24 hr. /Conditions of bioassay not specified TLm Artemia salina (brine shrimp) 320 mg/l/24 hr. /Conditions of bioassay not specified

Environmental Fate: If released on land, it will rapidly volatilize, although it may also leach into groundwater where its fate is unknown. Bioconcentration in aquatic organisms will not be important. If released in water it will be removed by volatilization with a half-life of 6-9 days, 5-8 days, and 24-32 hr, respectively in a typical pond, lake, or river. In the atmosphere, it will degrade (half-life 62 days) by reaction with photochemically produced hydroxyl radicals, and it will be scavenged by rain.

Cleanup/Disposal: Guide No. 130: Eliminate all ignition sources (no smoking, flares, sparks or flames in immediate area). All equipment used when handling the product must be grounded. Do not touch or walk through spilled material. Stop leak if you can do it without risk. Prevent entry into waterways, sewers, basements or confined areas. A vapor suppressing foam may be used to reduce vapors. Absorb or cover with dry earth, sand or other non-combustible material and transfer to containers. Use clean non-sparking tools to collect absorbed material. Large Spills: Dike far ahead of liquid spill for later disposal. Water spray may reduce vapor; but may not prevent ignition in closed spaces.

Environmental Physical Data

Octanol/Water Partition Coefficient: log K_{ow} = 1.9

Sorption Partition Coefficient: K_{oc} = estimated at 43

BCF: estimated at 1.3

BOD: 0.05 g/g, 10 days

Regulations

RCRA 40CFR: Listed Hazardous Waste No. U076 Toxic Waste

CERCLA: 40CFR 302.4: Listed per RCRA Section 3001 per CWA Section 307(a) RQ: 1000 lb (453.5 kg)

SARA 40CFR 372.65: Listed

SARA EHS 40CFR 355: Not listed

TSCA: Listed

Analytical Methods

Air: EPA 0031, OA-002-1, VA-001-1, VA-005-1, VG-006-1, VG-007-1, VG-011-1, TO-14; ASTM D3686, D3687, D4490

Soil: CLP LC_VOA, MC_VOA, OHC; EPA 7, 1624, VG-008-1, VS-001-1, VS-002-1; SW846 5021, 5032, 5041, 5041A, 8010B, 8021A, 8240B, 8260A, 8260B

Water / Groundwater: EPA 601, 624, 624-S, VW-001-1, VW-002-1, VW-003-1, VW-008-1; APHA 6210-B, 6210-D, 6230-B, 6230-D; USGS O3115

Drinking Water: EPA 502.1, 502.2, 524.1, 524.2; APHA 6210-C, 6230-C

Food: EPA 5

Indoor / Expired Air: NIOSH 1003; EPA IP-1A, IP-1B

Plasma: EPA 29

Other: EPA VS-006-1, VW-012-1, VW-013-1

ETH8200 **CAS #: 16219-75-3**

ETHYLIDENE NORBORNENE

RTECS: RB9450000

EINECS Number: 240-347-7

Molecular Formula: C_9H_{12}

Structured MF: C_9H_{12}

Formula Weight: 120.21

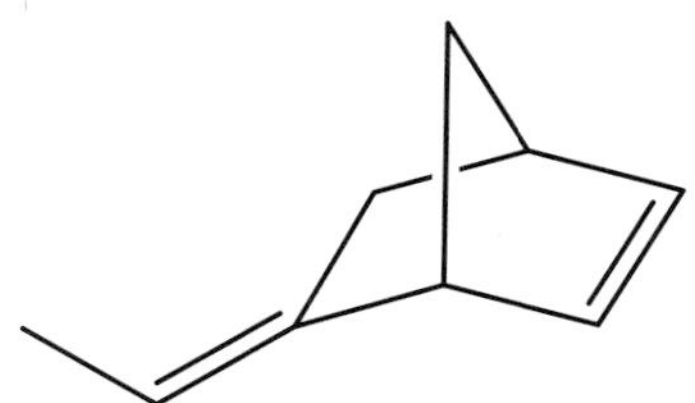

Chemical Structure

Synonyms: BICYCLO(2.2.1)HEPT-2-ENE,5-ETHYLIDENE-; ENB; 5-ETHYLIDENEBICYCLO(2,2,1)HEP-2-ENE; 5-ETHYLIDENEBICYCLO(2,2,1)HEPT-2-ENE; 5-ETHYLIDENEBICYCLO(2.2.1)-HEPT-2-ENE; 5-ETHYLIDENEBICYCLO(2.2.1)HEPT-2-ENE; 5-ETHYLIDENE-2-NORBORNENE; ETHYLIDENENORBORNENE; 2-NORBORNENE,5-ETHYLIDENE-

Description: colorless liquid; slight odor of turpentine

Use: as the third monomer in EPDM elastomers

Physical Properties

Boiling Point: 67 °C (153 °F) at 50 mm Hg

Freezing Point: -80 °C (-112 °F)

Specific Gravity: 0.8958 at 20 °C

Vapor Density: 4.1 Air=1

Saturated Vapor Density: 1.220857459 kg/m^3

Vapor Pressure: 4.2 mm Hg at 20 °C

Other Solubilities: 95% Ethanol: >=100 mg/ml at 18 °C; Acetone: >=100 mg/ml at 18 °C; DMSO: 50-100 mg/ml at 18 °C.

Odor Threshold: 0.007 to 0.014 ppm

Flash Point: 38.333 °C Open Cup

LEL: 0.9% v/v

UEL: 6.4% v/v

RTECS Toxicity Data

Acute Oral: Rat LD_{50} Dose: 2830 uL/kg. Mouse LD_{50} Dose: 3250 mg/kg.

Acute Inhalation: Rat LC_{50} Dose: 1246 ppm/4hr; Toxic Effects: Behavioral - Muscle weakness. Human TC_{Lo} Dose: 6 ppm/30M; Toxic Effects: Sense organs and special senses - Other; Sense organs and special senses - Conjunctive irritation; Sense organs and special senses - Change in function.

Acute Dermal: Rabbit LD_{50} Route: Skin; Dose: 9170 uL/kg.

Irritation Skin: Rabbit Open Draize Test Dose: 445 mg open; Reaction: mild.

Hazard Overviews

Fire Diamond

Health: Irritating to eyes/skin/respiratory tract. Also Causes: difficulty breathing, headache, confusion, reddening of skin (upon prolonged contact).

Fire: Combustible. Use dry chemical, carbon dioxide, fog, or alcohol foam. Water may be ineffective as an extinguishing agent, but should be used to cool fire-exposed containers.

Reactivity: Unstable, reacts with oxygen; stabilized with tert-butyl catechol. Hazardous polymerization cannot occur. Avoid: vapor generation; contact with oxygen. Incompatible with: oxygen. Hazardous decomposition products: carbon dioxide.

Carcinogenicity: IARC - Not listed; NIOSH - Listed as carcinogen; NTP - Not listed; ACGIH - Not listed; OSHA - Not listed; EPA - Not listed; MAK - Not listed

Primary Target Organs:

Eyes

Skin

Respiratory System

Exposure Limits

OSHA PEL Vacated 1989 Limits: STEL: 5 ppm; 25 mg/m^3; Ceiling.

ACGIH TLV: STEL: 5 ppm; 25 mg/m^3; Ceiling.

Respirator Recommendation

Exposure Range: >5 to 50 ppm Air Purifying, Negative Pressure, Half Mask

Exposure Range: >50 to 500 ppm Air Purifying, Negative Pressure, Full Face

Exposure Range: >500 to 5000 ppm Supplied Air, Constant Flow/Pressure Demand, Full Face

Exposure Range: >5000 to unlimited ppm Self-contained Breathing Apparatus, Pressure Demand, Full Face

Cartridge Color: black

Environmental

Cleanup/Disposal: Guide No. 128: Eliminate all ignition sources (no smoking, flares, sparks or flames in immediate area). All equipment used when handling the product must be grounded. Do not touch or walk through spilled material. Stop leak if you can do it without risk. Prevent entry into waterways, sewers, basements or confined areas. A vapor suppressing foam may be used to reduce vapors. Absorb or cover with dry earth, sand or other non-combustible material and transfer to containers. Use clean non-sparking tools to collect absorbed material. Large Spills: Dike far ahead of liquid spill for later disposal. Water spray may reduce vapor; but may not prevent ignition in closed spaces.

Environmental Physical Data

BCF: no food chain concentration potential

Regulations

RCRA 40CFR: Not listed

CERCLA: 40CFR 302.4: Not listed

SARA 40CFR 372.65: Not listed

SARA EHS 40CFR 355: Not listed

TSCA: Listed

ETH8250 CAS #: 2235-25-8

ETHYLMERCURIC PHOSPHATE

RTECS: OW3750000
EINECS Number: 218-790-2
Molecular Formula: $C_2H_7O_4HgP$
Formula Weight: 326.64
Synonyms: EMP; ETHYL MERCURIC PHOSPHATE; ETHYL MERCURY PHOSPHATE; ETHYLMERCURY PHOSPHATE; ETHYLMERKURIDIHYDROGENFOSFAT; GRANOSAN M; LIGNASAN; N. I. CERESAN; NEW IMPROVED CERESAN; NEW IMPROVED GRANOSAN; RUBERON; SOILSIN
Description: white powder; garlic-like odor
Use: seed fungicide (former use); timber preservative

Physical Properties

Water Solubility: Soluble in Water
Other Solubilities: quantitatively decomposed by strong acids.

RTECS Toxicity Data

Acute Oral: Human LD_{Lo} Dose: 8614 ug/kg/13W. Rat LD_{50} Dose: 48 mg/kg; Toxic Effects: Behavioral - Altered sleep time (including change in righting reflex); Behavioral - Somnolence (general depressed activity); Behavioral - Ataxia.
Acute Dermal: Mouse LD_{50} Route: Subcutaneous Dose: 76 mg/kg.
Reproductive/Teratogenic: Mouse Route: Subcutaneous; Dose: 40 mg/kg; Duration: female 10D of pregnancy; Effects on Embryo or Fetus - Cytological changes (inc. somatic cell genetic material).

Hazard Overviews

Carcinogenicity: IARC - Not listed; NIOSH - Not listed; NTP - Not listed; ACGIH - Not listed; OSHA - Not listed; EPA - Not listed; MAK - Not listed
Exposure Limits
OSHA PEL: TWA: 0.01 mg/m^3; as Hg.
ACGIH TLV: TWA: .01 mg/m^3; STEL: .03 mg/m^3.
NIOSH REL: TWA: .01 mg/m^3. STEL: .04 mg/m^3; Ceiling, Alkyl Compound.

Environmental

Regulations

RCRA 40CFR: Not listed
CERCLA: 40CFR 302.4: Not listed
SARA 40CFR 372.65: Not listed
SARA EHS 40CFR 355: Not listed
TSCA: Not listed

ETH8300 CAS #: 102-27-2

N-ETHYL-3-METHYLANILINE

RTECS: CY0440000
EINECS Number: 203-019-4
Molecular Formula: $C_9H_{13}N$
Formula Weight: 135.21

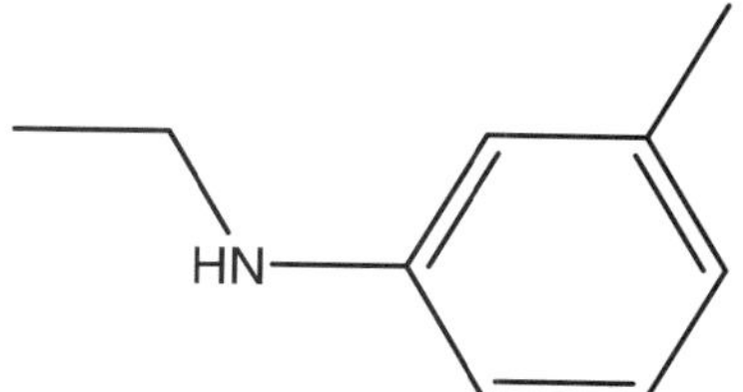

Chemical Structure

Synonyms: BENZENAMINE,N-ETHYL-3-METHYL-; N-ETHYL-3-METHYLBENZENAMINE; N-ETHYL-M-TOLUIDINE; M-METHYL-N-ETHYLANILINE; TOLUENE,3(ETHYLAMINO)-; M-TOLUIDINE,N-ETHYL-
Use: dye and photographic chemical intermediate

Physical Properties

Boiling Point: 221 °C (430 °F)
Specific Gravity: 0.92630
Other Solubilities: Soluble in Alcohol, Ether
Refraction Index: 1.5451 at 20 °C/D
Flash Point: Combustible

RTECS Toxicity Data

Acute Oral: Rat LD_{50} Dose: 580 mg/kg. Mouse LD_{50} Dose: 280 mg/kg.

Hazard Overviews

Poison

Health: Irritating to eyes/skin/respiratory tract. Poison. Other Acute Effects: may be fatal if inhaled, swallowed, or absorbed through skin; absorption into the body leads to the formation of methemoglobin which in sufficient concentration causes cyanosis; onset may be delayed 2 to 4 hours or longer.
Fire: Combustible. Hazards: emits toxic fumes. Extinguishing agents: water spray; carbon dioxide, dry chemical powder or appropriate foam. Precautions: combustible liquid.
Reactivity: Incompatible with: acids, acid chlorides, acid anhydrides, chloroformates, strong oxidizing agents. Hazardous decomposition products: toxic fumes of: carbon monoxide, carbon dioxide, nitrogen oxides.
Carcinogenicity: IARC - Not listed; NIOSH - Not listed; NTP - Not listed; ACGIH - Not listed; OSHA - Not listed; EPA - Not listed; MAK - Not listed

Primary Target Organs:

Eyes

Skin

Respiratory System

Environmental

Regulations

RCRA 40CFR: Not listed
CERCLA: 40CFR 302.4: Not listed
SARA 40CFR 372.65: Not listed
SARA EHS 40CFR 355: Not listed
TSCA: Listed

ETH8350 CAS #: 56046-62-9

N-(2-(ETHYL(3-METHYL-4-NITROSOPHENYL)AMINO)ETHYL)-METHANES

EINECS Number: 259-958-5
Molecular Formula: $C_{12}H_{19}N_3O_3S$
Formula Weight: 285.35
Synonyms: METHANESULFONAMIDE,N-(2-(ETHYL(3-METHYL-4-NITROSOPHENYL)AMINO)ETHYL)-
Use: captive chemical intermediate

Hazard Overviews

Carcinogenicity: IARC - Not listed; NIOSH - Not listed; NTP - Not listed; ACGIH - Not listed; OSHA - Not listed; EPA - Not listed; MAK - Not listed

Environmental

Environmental Fate: If released to soil, it should have low mobility. Volatilization should not be important from moist or dry soil surfaces. Insufficient data are available to determine the rate or importance of biodegradation in soil or water. If released to water, it should adsorb to suspended solids and sediment. It will be essentially non-volatile from water surfaces. An estimated BCF value of 24 suggests that it will not bioconcentrate in aquatic ESULFONA organisms. If released to the atmosphere, it will exist as both a vapor and particulate. Vapor-phase is degraded in the atmosphere by reaction with photochemically produced hydroxyl radicals with an estimated half-life of about 1.7 hours. Particulate-phase may be physically removed from the air by wet and dry deposition.

Environmental Physical Data

Henry's Law Constant: estimated at 1×10^{-10}
Octanol/Water Partition Coefficient: log K_{ow} = 2.13
Sorption Partition Coefficient: K_{oc} = 1950
BCF: estimated at 24

Regulations

RCRA 40CFR: Not listed
CERCLA: 40CFR 302.4: Not listed
SARA 40CFR 372.65: Not listed
SARA EHS 40CFR 355: Not listed
TSCA: Listed

ETH8400 CAS #: 100-74-3

N-ETHYLMORPHOLINE

RTECS: QE4025000
EINECS Number: 202-885-0
Molecular Formula: $C_6H_{13}NO$
Structured MF: $C_4H_8ONCH_2CH_3$
Formula Weight: 115.18

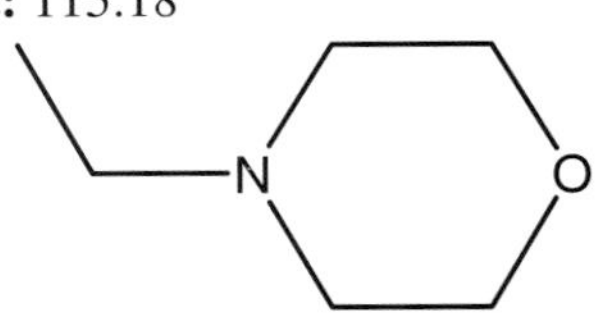

Chemical Structure

Synonyms: N-ETHYLMORFOLIN; 4-ETHYLMORPHOLINE; ETHYLMORPHOLINE; MORPHOLINE,4-ETHYL-; NEM
Description: colorless liquid; ammonia odor
Use: intermediate for dyes, pharmaceuticals, rubber accelerators and emulsifying agents; solvent for fats, fatty oils, dyes, resins and oils; catalyst in making polyurethane foams

Physical Properties

Boiling Point: 138 °C (280 °F) to 139 °C (282 °F) at 763 mm Hg
Freezing Point: -62.78 °C (-81.004 °F)
Specific Gravity: 0.8996 at 20 °C/4 °C
Vapor Density: 4 Air=1
Saturated Vapor Density: 1.228622396 kg/m^3
Vapor Pressure: 6.1 mm Hg at 20 °C
Water Solubility: Soluble in all Proportions
Other Solubilities: 95% Ethanol: >=100 mg/ml at 20 °C; Acetone: >=100 mg/ml at 20 °C; Benzene: Soluble; DMSO: >=100 mg/ml at 20 °C; Ether: Soluble.
Odor Threshold: 25 ppm
Refraction Index: 1.4400 at 20 °C/D
Flash Point: 32 °C Open Cup
LEL: 1.0% v/v
UEL: 9.8% v/v

RTECS Toxicity Data

Acute Oral: Rat LD_{50} Dose: 1780 mg/kg. Mouse LD_{50} Dose: 1200 mg/kg.
Acute Inhalation: Mouse LC_{50} Dose: 18000 mg/m^3/2hr. Rat LC_{Lo} Dose: 2000 ppm/4hr.
Irritation Eye: Rabbit Standard Draize Test Dose: 2 mg open; Reaction: severe.
Irritation Skin: Rabbit Open Draize Test Dose: 453 mg open; Reaction: mild.

Mutagenic: Bacteria - S Typhimurium Mutations in Microorganisms; Dose: 6667 ug/plate (-S9).

Hazard Overviews

Flammable

Fire Diamond

Health: Severe irritation to eyes/respiratory tract. Mildly irritating to skin. Also Causes: drowsiness, olfactory fatique, corneal edema, blurred vision, blue-gray vision, GI tract irritation upon ingestion.

Fire: Flammable. Can form explosive mixtures in the air. Fight fire with dry chemical, carbon dioxide, water spray, or foam. Fight fire from maximum distance.

Reactivity: Stable. Hazardous polymerization cannot occur. Avoid: heat; ignition sources. Incompatible with: strong acids; oxidizers. Hazardous decomposition products: carbon oxide; nitrogen oxide; ammonia gas.

Carcinogenicity: IARC - Not listed; NIOSH - Listed as carcinogen; NTP - Not listed; ACGIH - Not listed; OSHA - Not listed; EPA - Not listed; MAK - Not listed

Primary Target Organs:

Eyes

Respiratory System

Exposure Limits

OSHA PEL: TWA: 20 ppm; 94 mg/m^3; skin.

OSHA PEL Vacated 1989 Limits: TWA: 5 ppm; 23 mg/m^3.

ACGIH TLV: TWA: 5 ppm; 24 mg/m^3.

NIOSH REL: TWA: 5 ppm; 23 mg/m^3.

NIOSH IDLH: 100 ppm.

Respirator Recommendation

Exposure Range: >20 to <100 ppm Air Purifying, Negative Pressure, Half Mask

Exposure Range: 100 to unlimited ppm Self-contained Breathing Apparatus, Pressure Demand, Full Face

Cartridge Color: black

Environmental

Environmental Fate: If released to water, it will be essentially non-volatile. Insufficient data are available to predict the relative importance or rate of biodegradation in soil or water. If released to the atmosphere, it will exist primarily in the vapor phase. Vapor-phase will degrade in the atmosphere by reaction with photochemically produced hydroxyl radicals with an estimated half-life of approximately 2.5 hours. Removal of atmospheric material may occur through wet deposition. If released to soil, it is expected to have very high mobility based on an estimated K_{oc} value of 12. Volatilization is not expected from moist soils, but will be important from dry soils.

Cleanup/Disposal: Guide No. 128: Eliminate all ignition sources (no smoking, flares, sparks or flames in immediate area). All equipment used when handling the product must be grounded. Do not touch or walk through spilled material. Stop leak if you can do it without risk. Prevent entry into waterways, sewers, basements or confined areas. A vapor suppressing foam may be used to reduce vapors. Absorb or cover with dry earth, sand or other non-combustible material and transfer to containers. Use clean non-sparking tools to collect absorbed material. Large Spills: Dike far ahead of liquid spill for later disposal. Water spray may reduce vapor; but may not prevent ignition in closed spaces.

Environmental Physical Data

Henry's Law Constant: estimated at 2.47 x10^{-8}

Sorption Partition Coefficient: K_{oc} = estimated at 12

BCF: not significant

Regulations

RCRA 40CFR: Not listed

CERCLA: 40CFR 302.4: Not listed

SARA 40CFR 372.65: Not listed

SARA EHS 40CFR 355: Not listed

TSCA: Listed

Analytical Methods

Air: ASTM D4490

Water / Groundwater: ASTM D3695

ETH8450 CAS #: 32976-88-8

ETHYLNITROSOBIURET

RTECS: EC1750000

Molecular Formula: $C_4H_8N_4O_3$

Formula Weight: 160.16

Synonyms: BIURET,1-ETHYL-1-NITROSO-; ENBU; N-ETHYL-N-NITROSOBIURET; IMIDODICARBONIC DIAMIDE,N-ETHYL-N-NITROSO-; N-NITROSO-N-ETHYLBIURET

RTECS Toxicity Data

Acute Oral: Rat LD_{50} Dose: 1050 mg/kg.

Reproductive/Teratogenic: Rat Route: Oral; Dose: 300 mg/kg; Duration: female 14D of pregnancy; Effects on Embryo or Fetus - Fetal death.

Tumorigenic: Rat Route: Oral; Dose: 400 mg/kg; Toxic Effects: Tumorigenic - Equivocal tumorigenic agent by RTECS criteria; Gastrointestinal - Tumors. Rat Route: Oral; Dose: 100 mg/kg (22D preg); Toxic Effects: Tumorigenic - Carcinogenic by RTECS criteria; Tumorigenic effects - Transplacental tumorigenesis; Brain and coverings - Tumors. Rat Route: Oral; Dose: 100 mg/kg (15D preg); Toxic Effects: Tumorigenic - Equivocal tumorigenic agent by RTECS criteria; Tumorigenic effects - Transplacental tumorigenesis; Brain and coverings - Tumors.

Hazard Overviews

Carcinogenicity: IARC - Not listed; NIOSH - Not listed; NTP - Not listed; ACGIH - Not listed; OSHA - Not listed; EPA - Not listed; MAK - Not listed

Environmental

Regulations

RCRA 40CFR: Not listed
CERCLA: 40CFR 302.4: Not listed
SARA 40CFR 372.65: Not listed
SARA EHS 40CFR 355: Not listed
TSCA: Not listed

ETH8500 CAS #: 759-73-9

N-ETHYL-N-NITROSOUREA

RTECS: YT3150000
EINECS Number: 212-072-2
Molecular Formula: $C_3H_7N_3O_2$
Structured MF: $H_2NCON(NO)CH_2CH_3$
Formula Weight: 117.13

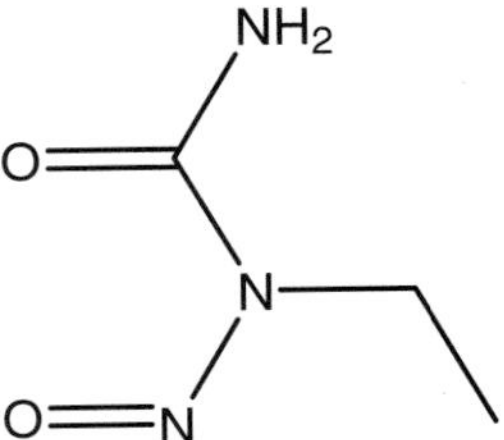

Chemical Structure

Synonyms: AENH; AETHYLNITROSO-HARNSTOFF; ANH; CARBAMIDE,N-ETHYL-N-NITROSO-; ENU; N-ETHYL-N-NITROSOCARBAMIDE; 1-ETHYL-1-NITROSOMOCOVINA; 1-ETHYL-1-NITROSOUREA; ETHYLNITROSOUREA; N-ETHYL-N-NITROSO-UREA; N-ETHYLNITROSOUREA; NEU; N-NITROSO-N-ETHYLUREA; NITROSOETHYLUREA; NSC 45403; UREA,1-ETHYL-1-NITROSO-; UREA,N-ETHYL-N-NITROSO-

Description: yellow-pink crystals

Use: primarily for tumor induction and related research in experimental animals; as a research mutagen, as an ethylating agent and in the synthesis of diazoethane

Physical Properties

Freezing Point: Decomposes at 103 °C (217.4 °F) to 104 °C (219.2 °F)

Water Solubility: 1.3% at 0 °C

Other Solubilities: 95% Ethanol: >=100 mg/ml at 21 °C; Acetone: >=100 mg/ml at 21 °C; Chloroform: >10%; DMSO: >=100 mg/ml at 21 °C; Non-polar organic solvents: Insoluble; Polar organic solvents: Soluble.

Flash Point: Not available; probably combustible

RTECS Toxicity Data

Acute Oral: Rat LD_{50} Dose: 300 mg/kg. Mouse LD_{50} Dose: 960 mg/kg.

Acute Dermal: Rat LD_{50} Route: Subcutaneous Dose: 240 mg/kg.

Reproductive/Teratogenic: Rat Route: Oral; Dose: 440 mg/kg; Duration: female 7-17D of pregnancy; Maternal Effects - Uterus, cervix, vagina; Effects on Embryo or Fetus - Fetotoxicity. Rat Route: Oral; Dose: 125 mg/kg; Duration: female 7-16D of pregnancy; Effects on Fertility - Post-implantation mortality; Effects on Embryo or Fetus - Fetal death; Specific Developmental Abnormalities - Other developmental abnormalities. Rat Route: Oral; Dose: 150 mg/kg; Duration: female 7-16D of pregnancy; Effects on Fertility - Litter size; Specific Developmental Abnormalities - Urogenital system. Rat Route: Oral; Dose: 100 mg/kg; Duration: female 7-16D of pregnancy; Specific Developmental Abnormalities - Central nervous system; Eye, ear. Rat Route: Oral; Dose: 75 mg/kg; Duration: female 7-16D of pregnancy; Effects on Embryo or Fetus - Extra embryonic structures; Fetotoxicity.

Mutagenic: Human DNA Damage; Cell Type: other cell types; Dose: 1 mmol/L. Human DNA Damage; Cell Type: liver; Dose: 1800 umol/L. Human DNA Damage; Cell Type: fibroblast; Dose: 1500 ug/L. Human DNA Damage; Cell Type: HeLa cell; Dose: 20 mg. Human DNA Damage; Cell Type: lymphocyte; Dose: 1 mmol/L. Human Unscheduled DNA Synthesis; Cell Type: liver; Dose: 320 umol/L. Human Unscheduled DNA Synthesis; Cell Type: HeLa cell; Dose: 1 mmol/L. Human Unscheduled DNA Synthesis; Cell Type: other cell types; Dose: 25 umol/L. Human Unscheduled DNA Synthesis; Cell Type: other cell types; Dose: 350 umol/L. Human DNA Inhibition; Cell Type: HeLa cell; Dose: 100 umol/L. Human Mutations in Mammalian Somatic Cells; Cell Type: fibroblast; Dose: 2 mmol/L. Human Mutations in Mammalian Somatic Cells; Cell Type: lymphocyte; Dose: 100 umol/L. Human Cytogenetic Analysis; Cell Type: lung; Dose: 200 ug/L. Human Cytogenetic Analysis; Cell Type: fibroblast; Dose: 100 umol/L/1H. Human Cytogenetic Analysis; Cell Type: lymphocyte; Dose: 25 mg/L. Monkey Cytogenetic Analysis; Route: Subcutaneous; Dose: 500 umol/kg. Human Sister Chromatid Exchenge; Cell Type: lung; Dose: 10 ug/L. Human Sister Chromatid Exchenge; Cell Type: other cell types; Dose: 200 umol/L. Human Sister Chromatid Exchenge; Cell Type: fibroblast; Dose: 100 umol/L.

Tumorigenic: Rat Route: Oral; Dose: 10 mg/kg (19D preg); Toxic Effects: Tumorigenic - Carcinogenic by RTECS criteria; Tumorigenic effects - Transplacental tumorigenesis; Brain and coverings - Tumors. Rat Route: Oral; Dose: 374 mg/kg/2W-C; Toxic Effects: Tumorigenic - Carcinogenic by RTECS criteria; Blood - Leukemia. Rat Route: Oral; Dose: 180 mg/kg; Toxic Effects: Tumorigenic - Equivocal tumorigenic agent by RTECS criteria; Cardiac - Tumors; Gastrointestinal - Tumors. Rat Route: Oral; Dose: 520 mg/kg/1Y-I; Toxic Effects: Tumorigenic - Equivocal tumorigenic agent by RTECS criteria; Gastrointestinal - Tumors; Blood - Leukemia. Monkey Route: Intravenous; Dose: 610 mg/kg/2Y-I; Toxic Effects: Tumorigenic - Equivocal tumorigenic agent by RTECS criteria; Vascular - Tumors; Endocrine - Thyroid tumors.

Hazard Overviews

Health: Toxic. Other Acute Effects: harmful if swallowed, inhaled, or absorbed through skin; exposure can cause severe damage to haematopoietic, lymphoid and other tissue. Chronic Effects: teratogen; mutagen; suspected of causing cancers of the lung, nasal sinuses, brain, esophagus, stomach, liver, bladder and kidney.

Fire: Will burn. Hazards: emits toxic fumes. Extinguishing agents: carbon dioxide, dry chemical powder or appropriate foam. Precautions: combustible liquid.

Reactivity: Stable. Hazardous polymerization will not occur. Hazardous decomposition products: toxic fumes of: carbon monoxide, carbon dioxide, nitrogen oxides.

Carcinogenicity: IARC - Group 2A, Probably carcinogenic to humans; NIOSH - Not listed; NTP - Listed; ACGIH - Not listed; OSHA - Not listed; EPA - Not listed; MAK - Not listed

Environmental

Environmental Fate: If it is released to soil, it is not expected to sorb to the soil and may leach to groundwater. It may hydrolyze to diazoethane in moist soils depending on the pH; with reported half-lives in water solution of 31, 1.5, 0.1 and 0.05 hr at pH 6.0, 7.0, 8.0, and 9.0, respectively. Little information was located on the photodegradation; however, based on the susceptibility of other N-nitroso compounds to photolysis, it may readily photodegrade in sunlight. Whether photodegradation would take place on soil is unknown. No information on the biodegradation in soils or waters was found. If released to water, it will not be expected to sorb to sediments or bioconcentrate in aquatic organisms. It will likely hydrolyze and may photodegrade, but it will not be expected to significantly volatilize. If released to the atmosphere, it may be susceptible to direct photolysis based on the behavior of other N-nitroso compounds in sunlight. The estimated vapor phase half-life in the atmosphere is 14.68 days as a result of reaction with photochemically produced hydroxyl radicals. Rainout may be an important physical removal process due to high water solubility.

Cleanup/Disposal: Guide No. 171: Do not touch or walk through spilled material. Stop leak if you can do it without risk. Prevent dust cloud. Avoid inhalation of asbestos dust. Small Dry Spills: With clean shovel place material into clean, dry container and cover loosely; move containers from spill area. Small Spills: Take up with sand or other noncombustible absorbent material and place into containers for later disposal. Large Spills: Dike far ahead of liquid spill for later disposal. Cover powder spill with plastic sheet or tarp to minimize spreading. Prevent entry into waterways, sewers, basements or confined areas.

Environmental Physical Data

Henry's Law Constant: estimated at 1.3×10^{-10}

Octanol/Water Partition Coefficient: $\log K_{ow} = 0.23$

Sorption Partition Coefficient: K_{oc} = estimated at 9.45

BCF: estimated at 0.88

Regulations

RCRA 40CFR: Listed Hazardous Waste No. U176 Toxic Waste

CERCLA: 40CFR 302.4: Listed per RCRA Section 3001 RQ: 1 lb (0.454 kg)

SARA 40CFR 372.65: Listed

SARA EHS 40CFR 355: Not listed

TSCA: Listed

ETH8550 CAS #: 90-00-6

2-ETHYLPHENOL

RTECS: SL4025000

EINECS Number: 201-958-4

Molecular Formula: $C_8H_{10}O$

Structured MF: $CH_3CH_2C_6H_4OH$

Formula Weight: 122.18

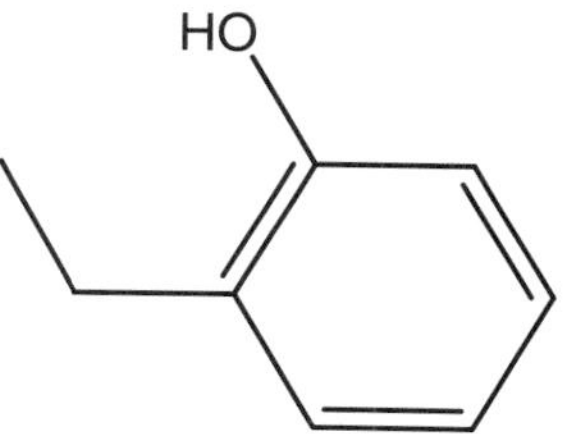

Chemical Structure

Synonyms: BENZENE,1-ETHYL-2-HYDROXY-; 1-ETHYL-2-HYDROXYBENZENE; O-ETHYLPHENOL; FLOROL; PHENOL,2-ETHYL-; PHENOL,O-ETHYL-; PHLOROL

Description: crystals or colorless liquid; phenol odor

Use: starting material for photochemicals

Physical Properties

Boiling Point: 204.5 °C (400 °F)

Freezing Point: 18 °C (64.4 °F)

Specific Gravity: 1.0146 at 25 °C

Vapor Density: 4.21 Air=1

Saturated Vapor Density: 1.200776218 kg/m^3

Density: 1.01459 g/mL at 25 °C

Vapor Pressure: 0.153 mm Hg at 25 °C

Water Solubility: Practically Insoluble in Water

Other Solubilities: 95% Ethanol: >=100 mg/ml at 22 °C; Acetone: >=100 mg/ml at 22 °C; Acetic Acid: Soluble; Benzene: Soluble; DMSO: >=100 mg/ml at 22 °C; Ether: Soluble.

Odor Threshold: 0.3 mg/l

pH: Acid

Refraction Index: 1.5372

Flash Point: 78 °C Closed Cup

RTECS Toxicity Data

Acute Oral: Mouse LD_{50} Dose: 600 mg/kg.
Tumorigenic: Mouse Route: Skin; Dose: 3100 mg/kg/12W-I; Toxic Effects: Tumorigenic - Neoplastic by RTECS criteria; Skin and appendages - Tumors.

Hazard Overviews

Health: Irritating to eyes/skin/respiratory tract. Harmful. Other Acute Effects: harmful if swallowed, inhaled, or absorbed through skin; depending on the intensity and duration of exposure, effects may vary from mild irritation to severe destruction of tissue; prolonged contact can cause damage to the eyes, severe irritation or burns.
Fire: Combustible. Hazards: emits toxic fumes. Extinguishing agents: carbon dioxide, dry chemical powder or appropriate foam. Precautions: combustible liquid.
Reactivity: Incompatible with: acid chlorides, acid anhydrides, oxidizing agents. Hazardous decomposition products: toxic fumes of: carbon monoxide, carbon dioxide.
Carcinogenicity: IARC - Not listed; NIOSH - Not listed; NTP - Not listed; ACGIH - Not listed; OSHA - Not listed; EPA - Not listed; MAK - Not listed
Primary Target Organs:

Eyes

Skin

Respiratory System

Environmental

Ecotoxicity: Toxicity to microorganisms: biodegradation inhib. EC_{50}: 200 mg/l.
Environmental Fate: If released to the atmosphere, it will mainly exist in the vapor phase based on a measured vapor pressure of 0.153 mm Hg at 25 °C. Vapor-phase is degraded in the atmosphere by reaction with photochemically produced hydroxyl radicals with an estimated half-life of about 9 hours. An estimated K_{oc} of 530 suggests that it will have moderate to low mobility in soil. Volatilization from dry and moist soil surfaces is possible, but should not be a major fate process for this compound based on values for vapor pressure and Henry's Law constant, respectively. Limited data suggests that this compound may biodegrade in both soil and water under aerobic and anaerobic conditions. In water, it may adsorb to suspended matter in the water column based on its K_{oc} value. It may volatilize from water surfaces given an estimated Henry's Law constant of 4.6 x10^{-6} atm-cu m/mole. Estimated half-lives for a model river and model lake are 9 and 68 days, respectively. Bioconcentration in aquatic organisms is moderate, not high based on an estimated BCF value of 44.
Cleanup/Disposal: Guide No. 171: Do not touch or walk through spilled material. Stop leak if you can do it without risk. Prevent dust cloud. Avoid inhalation of asbestos dust. Small Dry Spills: With clean shovel place material into clean, dry container and cover loosely; move containers from spill area. Small Spills: Take up with sand or other noncombustible absorbent material and place into containers for later disposal. Large Spills: Dike far ahead of liquid spill for later disposal. Cover powder spill with plastic sheet or tarp to minimize spreading. Prevent entry into waterways, sewers, basements or confined areas.

Environmental Physical Data

Henry's Law Constant: estimated at 4.6 x10^{-6}
Octanol/Water Partition Coefficient: log K_{ow} = 2.47
Sorption Partition Coefficient: K_{oc} = estimated at 530
BCF: estimated at 44

Regulations

RCRA 40CFR: Not listed
CERCLA: 40CFR 302.4: Not listed
SARA 40CFR 372.65: Not listed
SARA EHS 40CFR 355: Not listed
TSCA: Listed

ETH8600 CAS #: 620-17-7

3-ETHYLPHENOL

EINECS Number: 210-627-3
Molecular Formula: $C_8H_{10}O$
Formula Weight: 122.17

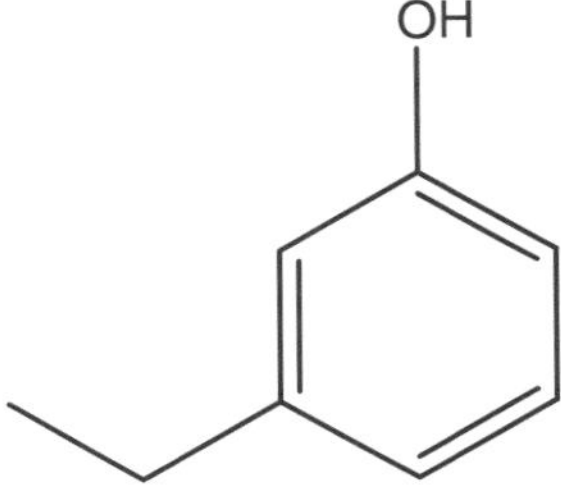

Chemical Structure

Synonyms: BENZENE,1-ETHYL-3-HYDROXY-; 1-ETHYL-3-HYDROXYBENZENE; M-ETHYLPHENOL; META-ETHYLPHENOL; PHENOL,3-ETHYL-; PHENOL,M-ETHYL-
Description: colorless liquid
Use: starting material for photochemicals; production of phenolic resins

Physical Properties

Boiling Point: 218.4 °C (425 °F) at 760 mm Hg
Freezing Point: -4 °C (24.8 °F)
Specific Gravity: 1.0283 at 20 °C/4 °C
Saturated Vapor Density: 1.200253639 kg/m^3
Vapor Pressure: 0.05 mm Hg at 25 °C
Water Solubility: Slightly Soluble in Water
Other Solubilities: Slightly Soluble in Chloroform Very Soluble in Alcohol, Ether

Hazard Overviews

Health: Irritating to eyes/skin/respiratory tract. Harmful. Other Acute Effects: harmful if swallowed, inhaled, or absorbed through skin.

Fire: Hazards: emits toxic fumes. Extinguishing agents: carbon dioxide, dry chemical powder or appropriate foam. Precautions: combustible liquid.

Reactivity: Incompatible with: acid chlorides, acid anhydrides, oxidizing agents. Hazardous decomposition products: toxic fumes of: carbon monoxide, carbon dioxide.

Carcinogenicity: IARC - Not listed; NIOSH - Not listed; NTP - Not listed; ACGIH - Not listed; OSHA - Not listed; EPA - Not listed; MAK - Not listed

Primary Target Organs:

Environmental

Ecotoxicity: Toxicity to microorganisms: biodegradation inhib. EC_{50}: 200 mg/l

Environmental Fate: If released to the atmosphere, will mainly exist in the vapor phase based on an experimental vapor pressure of 0.05 mm Hg at 25 °C. Vapor-phase is degraded in the atmosphere by reaction with photochemically produced hydroxyl radicals with an estimated half-life of about 5 hours. An estimated K_{oc} of 480 suggests that it will have moderate mobility in soil. Volatilization from dry and moist soil surfaces is possible, but should not be a major fate process for this compound. Based on limited data, this compound may biodegrade in both soil and water. A system where water was passed through contaminated soil (initial concentration = 73 ug/L) and then through an upflow aerated column was capable of 93% removal in 37 days. In water, it may adsorb to suspended matter in the water column based on its K_{oc} value. May volatilize from water surfaces given an estimated Henry's Law constant of 1.1 $x10^{-6}$ atm-cu m/mole. Estimated half-lives for a model river and model lake are 37 and 274 days, respectively. Bioconcentration in aquatic organisms may occur based on an estimated BCF value of 40.

Environmental Physical Data

Henry's Law Constant: estimated at 1.1 $x10^{-6}$

Octanol/Water Partition Coefficient: log K_{ow} = 2.40

Sorption Partition Coefficient: K_{oc} = estimated at 480

BCF: estimated at 40

Regulations

RCRA 40CFR: Not listed

CERCLA: 40CFR 302.4: Not listed

SARA 40CFR 372.65: Not listed

SARA EHS 40CFR 355: Not listed

TSCA: Listed

ETH8650 **CAS #: 123-07-9**

4-ETHYLPHENOL

RTECS: SL4040000

EINECS Number: 204-598-6

Molecular Formula: $C_8H_{10}O$

Formula Weight: 122.17

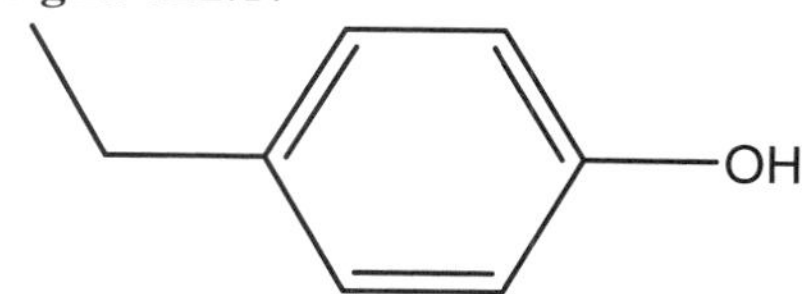

Chemical Structure

Synonyms: 1-ETHYL-4-HYDROXYBENZENE; PARA-ETHYLPHENOL; 4-HYDROXYPHENYLETHANE; PHENOL,4-ETHYL-; PHENOL,P-ETHYL-

Description: colorless or white needles; tend to yellow on exposure to light; powerful woody-phenolic, yet somewhat sweet medicinal, odor

Use: synthetic flavor; for prodn of phenolic resins; for low-temperature tar & in varnish ind; pure 4-ethylphenol starting material for prodn of 4-vinylphenol & of antioxidants (eg, 2,6-di-t-butyl-4-ethylphenol & 2,2'-methylenebis(6-t-butyl-4-ethylphenol)), which are ysed in rubber & polymers; intermed for pharmaceuticals & dyes.

Physical Properties

Boiling Point: 217.9 °C (424 °F)

Freezing Point: 46 °C (114.8 °F)

Specific Gravity: 1.011 at 20 °C

Vapor Pressure: 1.12 kPa at 100 °C

Water Solubility: Slightly Soluble in Water

Other Solubilities: Soluble in Alcohol, Ether, & Benzene. Soluble in Carbon Disulfide, & Acetone

Refraction Index: 1.5239 at 25 °C

Ionization Potential (eV): 7.84 +/-0.2

Flash Point: 104 °C Open Cup

Hazard Overviews

Health: Irritating to eyes/skin/respiratory tract. Other Acute Effects: harmful if swallowed, inhaled, or absorbed through skin; depending on the intensity and duration of exposure, effects may vary from mild irritation to severe destruction of tissue; prolonged contact can cause: damage to the eyes; severe irritation; burns.

Fire: Will burn. Hazards: emits toxic fumes. Extinguishing agents: carbon dioxide, dry chemical powder or appropriate foam. Precautions: combustible liquid.

Reactivity: Incompatible with: acid chlorides, acid anhydrides, oxidizing agents. Hazardous decomposition products: toxic fumes of: carbon monoxide, carbon dioxide.

Carcinogenicity: IARC - Not listed; NIOSH - Not listed; NTP - Not listed; ACGIH - Not listed; OSHA - Not listed; EPA - Not listed; MAK - Not listed

Primary Target Organs:

Respiratory System

Environmental

Ecotoxicity: Toxicity to microorganisms: biodegradation inhib. EC_{50} 250 mg/l

Environmental Fate: If released to the atmosphere, it will mainly exist in the vapor phase based on a experimental vapor pressure of 0.0372 mm Hg at 25 °C. Vapor-phase is degraded in the atmosphere by reaction with photochemically produced hydroxyl radicals with an estimated half-life of about 9 hours. An estimated K_{oc} of 600 suggests that it will have low mobility in soil. Volatilization from dry and moist soil surfaces is possible, but should not be a major fate process for this compound. Based on limited data, this compound may biodegrade in both soil and water. A system where water was passed through contaminated soil (initial concentration=4000 ug/L) and then through an upflow aerated column was capable of 76% removal in 37 days. In water, it may adsorb to suspended matter in the water column based on its K_{oc} value. It may volatilize from water surfaces given an estimated Henry's Law constant of 1.2 x10^{-6} atm-cu m/mole. Estimated half-lives for a model river and model lake are 33 and 245 days, respectively. Bioconcentration in aquatic organisms may occur based on an estimated BCF value of 54.

Environmental Physical Data

Henry's Law Constant: estimated at 1.2 x10^{-6}
Octanol/Water Partition Coefficient: log K_{ow} = 2.58
Sorption Partition Coefficient: K_{oc} = estimated at 600
BCF: estimated at 54

Regulations

RCRA 40CFR: Not listed
CERCLA: 40CFR 302.4: Not listed
SARA 40CFR 372.65: Not listed
SARA EHS 40CFR 355: Not listed
TSCA: Listed

ETH8700 CAS #: 1125-27-5

ETHYLPHENYLDICHLOROSILANE

RTECS: VV3270000
EINECS Number: 214-407-8
Molecular Formula: $C_8H_{10}Cl_2Si$
Structured MF: $(C_2H_5)(C_6H_5)SiCl_2$
Formula Weight: 205.17
Synonyms: DICHLOROETHYLPHENYLSILANE; ETHYL PHENYL DICHLOROSILANE; SILANE,DICHLOROETHYLPHENYL-
Description: colorless liquid; sharp odor like hydrochloric acid

Physical Properties

Boiling Point: > 149 °C (300 °F) at 1 atm
Specific Gravity: Liquid 1.159 at 15 °C
Vapor Density: 7.07 Air=1
Water Solubility: Reacts with water
Surface Tension: Estimated at 25 dynes/cm
pH: Acid
Flash Point: > 65.556 °C Open Cup

RTECS Toxicity Data

Acute Inhalation: Rat LC; Dose: >10 mg/m^3/2hr; Toxic Effects: Behavioral - Somnolence (general depressed activity); Lungs, Thorax, or Respiration - Dyspnea. Mouse LC; Dose: >10 mg/m^3/2hr; Toxic Effects: Behavioral - Somnolence (general depressed activity); Lungs, Thorax, or Respiration - Dyspnea.

Chronic (Multiple Dose) Inhalation: Rat Dose: 10 mg/m^3/3H/35W-I; Toxic Effects: Lungs, Thorax, or Respiration - Emphysema; Lungs, Thorax, or Respiration - Other changes; Kidney, Ureter, and Bladder - Other changes.

Hazard Overviews

Fire: Combustible.
Carcinogenicity: IARC - Not listed; NIOSH - Not listed; NTP - Not listed; ACGIH - Not listed; OSHA - Not listed; EPA - Not listed; MAK - Not listed

Environmental

Cleanup/Disposal: Guide No. 171: Do not touch or walk through spilled material. Stop leak if you can do it without risk. Prevent dust cloud. Avoid inhalation of asbestos dust. Small Dry Spills: With clean shovel place material into clean, dry container and cover loosely; move containers from spill area. Small Spills: Take up with sand or other noncombustible absorbent material and place into containers for later disposal. Large Spills: Dike far ahead of liquid spill for later disposal. Cover powder spill with plastic sheet or tarp to minimize spreading. Prevent entry into waterways, sewers, basements or confined areas.

Environmental Physical Data

BCF: no food chain concentration potential

Regulations

RCRA 40CFR: Not listed
CERCLA: 40CFR 302.4: Not listed
SARA 40CFR 372.65: Not listed
SARA EHS 40CFR 355: Not listed
TSCA: Listed

ETH8800 CAS #: 645-62-5

2-ETHYL-3-PROPYLACROLEIN

RTECS: MP6300000
EINECS Number: 211-448-3
Molecular Formula: $C_8H_{14}O$
Structured MF: $CH_3(CH_2)_2CH{=}CCHO$
Formula Weight: 126.22

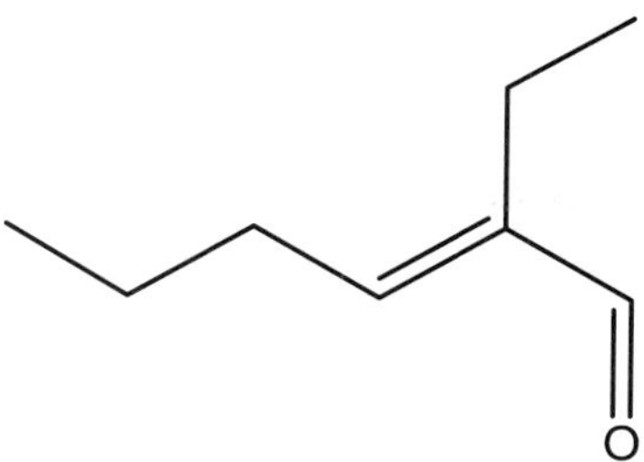

Chemical Structure

Synonyms: ACROLEIN,2-ETHYL-3-PROPYL-; 2-ETHYL-2-HEXEN-1-AL; 2-ETHYL-2-HEXENAL; 2-ETHYLHEXENAL; ALPHA-ETHYL-2-HEXENAL; 2-ETHYL-3-PROPYL ACROLEIN; ALPHA-ETHYL-BETA-N-PROPYLACROLEIN; ALPHA-ETHYL-BETA-PROPYLACROLEIN; 2-ETHYL-3-PROPYLACRYLALDEHYDE; 2-HEXENAL,2-ETHYL-

Description: colorless to yellow liquid; sharp, powerful irritating odor

Use: insecticide, intermediate for organic synthesis, leakage indicator

Physical Properties

Boiling Point: 175 °C (347 °F)
Freezing Point: 100 °C (212 °F)
Specific Gravity: 0.8518
Vapor Density: 4.4 Air=1
Saturated Vapor Density: 1.205293285 kg/m^3
Bulk Density: 7.1 lbs/gal at 20 °C
Vapor Pressure: 1 mm Hg at 20 °C
Water Solubility: 0.07 g/100 ml
Other Solubilities: 95% Ethanol: >=100 mg/ml at 21 °C; Acetone: >=100 mg/ml at 21 °C; DMSO: >=100 mg/ml at 21 °C.
Surface Tension: 28.2 dynes/cm at 20 °C
Flash Point: 68.333 °C Open Cup
Autoignition Temperature: 200 °C

RTECS Toxicity Data

Acute Oral: Rat LD_{50} Dose: 3 gm/kg.
Acute Dermal: Guinea Pig LD_{50} Route: Skin; Dose: >20 mL/kg.
Irritation Eye: Rabbit Standard Draize Test Dose: 500 mg; Reaction: severe.

Hazard Overviews

Fire Diamond

Fire: Combustible.
Carcinogenicity: IARC - Not listed; NIOSH - Not listed; NTP - Not listed; ACGIH - Not listed; OSHA - Not listed; EPA - Not listed; MAK - Not listed

Environmental

Cleanup/Disposal: Guide No. 171: Do not touch or walk through spilled material. Stop leak if you can do it without risk. Prevent dust cloud. Avoid inhalation of asbestos dust. Small Dry Spills: With clean shovel place material into clean, dry container and cover loosely; move containers from spill area. Small Spills: Take up with sand or other noncombustible absorbent material and place into containers for later disposal. Large Spills: Dike far ahead of liquid spill for later disposal. Cover powder spill with plastic sheet or tarp to minimize spreading. Prevent entry into waterways, sewers, basements or confined areas.

Environmental Physical Data

BCF: no food chain concentration potential
BOD: theoretical 52%, 10 days

Regulations

RCRA 40CFR: Not listed
CERCLA: 40CFR 302.4: Not listed
SARA 40CFR 372.65: Not listed
SARA EHS 40CFR 355: Not listed
TSCA: Listed

ETH8850 **CAS #: 672-04-8**

M-(1-ETHYLPROPYL)PHENYL METHYLCARBAMATE

Molecular Formula: $C_{13}H_{19}NO_2$
Formula Weight: 221.30
Synonyms: BUX; CARBAMIC ACID,METHYL-,M-(1-ETHYLPROPYL)PHENYL ESTER; PHENOL,3-(1-ETHYLPROPYL)-,METHYLCARBAMATE
Description: colorless to yellow, low melting solid
Use: formerly in soil insecticide bux for corn rootworms, water weevils, & on other field crops

Physical Properties

Boiling Point: 125 °C (257 °F) to 130 °C (266 °F)
Freezing Point: 40 °C (104 °F) to 45 °C (113 °F)
Saturated Vapor Density: 1.200000314 kg/m^3
Density: 1.031 at 25 °C
Vapor Pressure: 3 x10^{-5} mm Hg at 30 °C
Water Solubility: < 0%
Other Solubilities: Soluble in most solvents, but less in aliphatic hydrocarbons.

Hazard Overviews

Carcinogenicity: IARC - Not listed; NIOSH - Not listed; NTP - Not listed; ACGIH - Not listed; OSHA - Not listed; EPA - Not listed; MAK - Not listed

Environmental

Ecotoxicity: LC_{50} goldfish 0.56 mg/l/96 hr LC_{50} Mallards oral more than 5000 ppm in 5-day diet (12% mortality at 1000 ppm, 38% at 5000 ppm) LC_{50} trout 0.064 mg/l/96 hr

Regulations

RCRA 40CFR: Not listed
CERCLA: 40CFR 302.4: Not listed

SARA 40CFR 372.65: Not listed
SARA EHS 40CFR 355: Not listed
TSCA: Not listed

ETH9000 CAS #: 1420-55-9

ETHYLTHICPERAZINE

RTECS: SP0280000
EINECS Number: 215-819-0
Molecular Formula: $C_{22}H_{29}N_3S_2$
Formula Weight: 399.62
Synonyms: 3-ETHYLMERCAPTO-10-(1'-METHYLPIPERAZINYL-4'-PROPYL)PHENOTHIAZINE; ETHYLTHIOPERAZINE; 10H-PHENOTHIAZINE,2-(ETHYLTHIO)-10-(3-(4-METHYL-1-PIPERAZINYL)PROPYL)-; PHENOTHIAZINE,2-(ETHYLTHIO)-10-(3-(4-METHYL-1-PIPERAZINYL)PROPYL)-; THEITHYLPERAZINE; THIETHYLPERAZINE
Description: crystals; slight odor
Use: medication: antiemetic agent

Physical Properties

Boiling Point: 227 °C (441 °F) at 0.01 mm Hg
Water Solubility: Freely Soluble in Water
Other Solubilities: Poorly Soluble in Methanol, absolute Ethanol; Very poorly Soluble in Benzene, Ether, Chloroform
pH: 1 in 100 solution 2.8 to 3.8

RTECS Toxicity Data

Acute Oral: Mouse LD_{50} Dose: 680 mg/kg; Toxic Effects: Behavioral - Antipsychotic.

Hazard Overviews

Carcinogenicity: IARC - Not listed; NIOSH - Not listed; NTP - Not listed; ACGIH - Not listed; OSHA - Not listed; EPA - Not listed; MAK - Not listed

Environmental

Regulations

RCRA 40CFR: Not listed
CERCLA: 40CFR 302.4: Not listed
SARA 40CFR 372.65: Not listed
SARA EHS 40CFR 355: Not listed
TSCA: Not listed

ETH9050 CAS #: 110-77-0

2-(ETHYLTHIO)ETHANOL

RTECS: KL1225000
EINECS Number: 203-802-0
Molecular Formula: $C_4H_{10}OS$
Formula Weight: 106.18

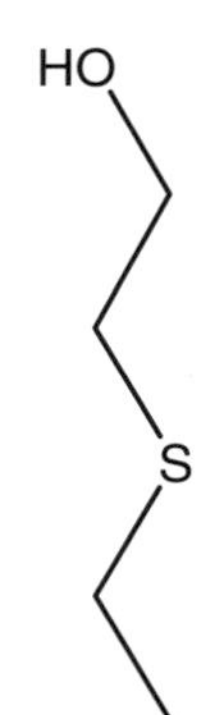

Chemical Structure

Synonyms: ETHANOL,2-(ETHYLTHIO)-; ETHYL 2-HYDROXYETHYL SULFIDE; ETHYL BETA-HYDROXYETHYL SULFIDE; ETHYL 2-HYDROXYETHYL THIOETHER; ETHYL THIOETHANOL; BETA-ETHYLMERKAPTOETHANOL; BETA-ETHYLTHIOETHANOL; BETA-HYDROXYDIETHYL SULFIDE; 2-HYDROXYETHYL ETHYL SULFIDE
Description: pale straw liquid
Use: int for pesticides, lubricating and cutting oil additives, flotation agents, plasticizers; in synthesis; chem int for pesticides, eg, disulfoton

Physical Properties

Boiling Point: 184.5 °C (364 °F) at 760 mm Hg
Freezing Point: About -100 °C (-148 °F)
Specific Gravity: 1.0166 at 20 °C/4 °C
Water Solubility: Slightly Soluble in Water
Other Solubilities: Soluble in Alcohol, Very Soluble in Acetone.
Refraction Index: 1.4867 at 20 °C/D

RTECS Toxicity Data

Irritation Eye: Rabbit Standard Draize Test Dose: 750 ug/24H; Reaction: severe.
Irritation Skin: Rabbit Standard Draize Test Dose: 2 mg/24H; Reaction: severe.

Hazard Overviews

Corrosive

Health: Corrosive to eyes/skin/respiratory tract. Other Acute Effects: harmful if swallowed, inhaled, or absorbed through skin; inhalation may result in spasm, inflammation and edema of the larynx and bronchi, chemical pneumonitis and pulmonary edema; symptoms of exposure may include burning sensation; coughing; wheezing; laryngitis; shortness of breath; headache; nausea; vomiting.
Fire: Hazards: emits toxic fumes. Extinguishing agents: water spray; carbon dioxide, dry chemical powder or appropriate foam. Precautions: combustible liquid.
Reactivity: Incompatible with: strong oxidizing agents. Hazardous decomposition products: toxic fumes of: carbon

monoxide, carbon dioxide, sulfur oxides, hydrogen sulfide gas.

Carcinogenicity: IARC - Not listed; NIOSH - Not listed; NTP - Not listed; ACGIH - Not listed; OSHA - Not listed; EPA - Not listed; MAK - Not listed

Primary Target Organs:

Eyes

Skin

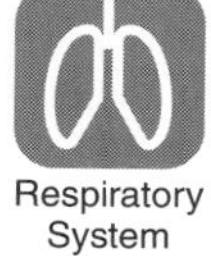
Respiratory System

Environmental

Regulations

RCRA 40CFR: Not listed
CERCLA: 40CFR 302.4: Not listed
SARA 40CFR 372.65: Not listed
SARA EHS 40CFR 355: Not listed
TSCA: Listed

ETH9100 CAS #: 1077-56-1

N-ETHYL-O-TOLUENESULFONAMIDE

EINECS Number: 214-073-3
Molecular Formula: $C_9H_{13}NO_2S$
Formula Weight: 199.27
Synonyms: BENZENESULFONAMIDE,N-ETHYL-2-METHYL-; O-TOLUENESULFONAMIDE,N-ETHYL-
Use: plasticizer for polyamides, cellulose acetate, gasoline-resistant coatings

Hazard Overviews

Carcinogenicity: IARC - Not listed; NIOSH - Not listed; NTP - Not listed; ACGIH - Not listed; OSHA - Not listed; EPA - Not listed; MAK - Not listed

Environmental

Regulations

RCRA 40CFR: Not listed
CERCLA: 40CFR 302.4: Not listed
SARA 40CFR 372.65: Not listed
SARA EHS 40CFR 355: Not listed
TSCA: Listed

ETH9150 CAS #: 1762-26-1

ETHYLTRIMETHYLLEAD

EINECS Number: 217-169-3
Molecular Formula: $C_5H_{14}Pb$
Formula Weight: 281.4
Synonyms: ETHYLTRIMETHYLPLUMBANE; PLUMBANE,ETHYLTRIMETHYL-; TRIMETHYLETHYLLEAD
Use: gasoline additive

Hazard Overviews

Carcinogenicity: IARC - Not listed; NIOSH - Not listed; NTP - Not listed; ACGIH - Not listed; OSHA - Not listed; EPA - Not listed; MAK - Not listed

Exposure Limits
OSHA PEL: STEL: 0.05 mg/m^3; as Pb inorganic.
ACGIH TLV: TWA: 0.05 mg/m^3; as Pb inorganic.
NIOSH REL: TWA: 0.1 mg/m^3; < inorganic; blood Pb<.06mg/100g.
DFG MAK: TWA: 0.1 mg/m^3; as Pb inorganic.

Environmental

Regulations

RCRA 40CFR: Not listed
CERCLA: 40CFR 302.4: Listed as Compound per CWA Section 307(a) per CAA Section 112
SARA 40CFR 372.65: Listed as Compound
SARA EHS 40CFR 355: Not listed
TSCA: Listed

ETH9200 CAS #: 151-01-9

ETHYLXANTHATE

RTECS: FG1370000
EINECS Number: 205-780-8
Molecular Formula: $C_3H_6OS_2$
Formula Weight: 122.21
Synonyms: CARBONIC ACID,DITHIO-,O-ETHYL ESTER; CARBONODITHIOIC ACID,O-ETHYL ESTER; ETHOXYDITHIOFORMIC ACID; O-ETHYL DITHIOCARBAMATE; O-ETHYL DITHIOCARBONATE; ETHYL XANTHATE; ETHYL XANTHOGENATE; ETHYLXANTHIC ACID; ETHYLXANTHOGENIC ACID; XANTHATE; XANTHIC ACID,ETHYL-; XANTHOGENIC ACID; XANTHOGENIC ACID,ETHYL-
Description: liquid

Physical Properties

Boiling Point: Decomposes at 25 °C (77 °F)
Freezing Point: -53 °C (-63.4 °F)
Water Solubility: Slightly Soluble in Water

RTECS Toxicity Data

Mutagenic: Rat Cytogenetic Analysis; Route: Inhalation; Dose: 13050 ug/m^3/16W-I.

Hazard Overviews

Carcinogenicity: IARC - Not listed; NIOSH - Not listed; NTP - Not listed; ACGIH - Not listed; OSHA - Not listed; EPA - Not listed; MAK - Not listed

Environmental

Regulations

RCRA 40CFR: Not listed
CERCLA: 40CFR 302.4: Not listed
SARA 40CFR 372.65: Not listed

SARA EHS 40CFR 355: Not listed
TSCA: Listed

ETH9250 CAS #: 78-27-3

1-ETHYNYL-1-CYCLOHEXANOL

RTECS: GV9100000
EINECS Number: 201-100-9
Molecular Formula: $C_8H_{12}O$
Formula Weight: 124.20

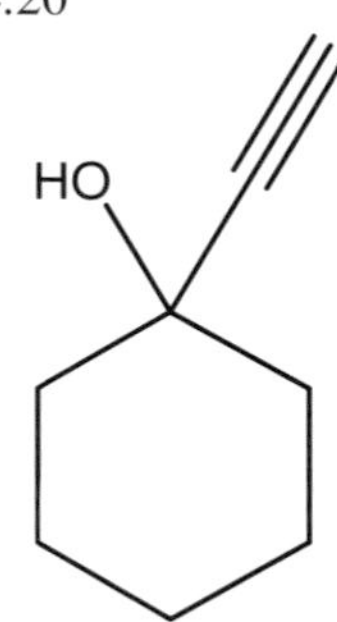

Chemical Structure

Synonyms: 1-ETHYNYLCYCLOHEXAN-1-OL; 1-ETHYNYLCYCLOHEXANOL; NSC 8194

Physical Properties

Boiling Point: 180 °C (356 °F)
Freezing Point: 33 °C (91.4 °F)
Specific Gravity: 0.98730
Water Solubility: 1 g/l
Other Solubilities: Ethanol: Soluble; Benzene: Soluble; Chloroform: Slightly soluble; Petroleum Ether: Soluble
pH: 7
Refraction Index: 1.4822

RTECS Toxicity Data

Acute Oral: Rat LD_{50} Dose: 600 mg/kg.
Acute Dermal: Rabbit LD_{50} Route: Skin; Dose: 1 gm/kg; Toxic Effects: Skin and appendages - Primary irritation.
Irritation Skin: Rabbit Open Draize Test Dose: 10 mg/24H open; Reaction: mild.

Hazard Overviews

Health: May be irritating to eyes/skin/respiratory tract. Toxic. Other Acute Effects: harmful if swallowed or absorbed through skin; may be harmful if inhaled.
Fire: Extinguishing agents: water spray; carbon dioxide, dry chemical powder or appropriate foam. Precautions: combustible liquid.
Reactivity: Stable. Hazardous polymerization will not occur. Incompatible with: oxidizing agents. Hazardous decomposition products: toxic fumes of: carbon monoxide, carbon dioxide.

Carcinogenicity: IARC - Not listed; NIOSH - Not listed; NTP - Not listed; ACGIH - Not listed; OSHA - Not listed; EPA - Not listed; MAK - Not listed

Environmental

Regulations

RCRA 40CFR: Not listed
CERCLA: 40CFR 302.4: Not listed
SARA 40CFR 372.65: Not listed
SARA EHS 40CFR 355: Not listed
TSCA: Listed

ETO5000 CAS #: 33419-42-0

ETOPOSIDE

RTECS: KC0190000
EINECS Number: 251-509-1
Molecular Formula: $C_{29}H_{32}O_{13}$
Formula Weight: 588.58

Chemical Structure

Synonyms: DEMETHYL-EPIPODOPHYLLOTOXIN ETHYLIDENE GLUCOSIDE; 4'-DEMETHYLEPIPODOPHYLLOTOXIN 9-(4,6-O-ETHYLIDENE-BETA-D-GLUCOPYRANOSIDE); 4'-DEMETHYLEPIPODOPHYLLOTOXIN ETHYLIDENE-BETA-D-GLUCOSIDE; 4-DEMETHYLEPIPODOPHYLLOTOXIN BETA-D-ETHYLIDENEGLUCOSIDE; 4'-DEMETHYLEPIPODOPHYLLOTOXIN9-(4,6-O-ETHYLIDENE-BETA-D-GLUCOPYRANOSIDE); 4-DEMETHYL-EPIPODOPHYLLOTOXIN-BETA-D-ETHYLIDENE-GLUCOSIDE; 4'-O-DEMETHYL-1-O-(4,6-O-ETHYLIDENE-BETA-D-GLUCOPYRANOSYL)EPIPODOPHYLLOTOXIN; EPE; EPEC; EPIPODOPHYLLOTOXIN,4'-DEMETHYL-,9-(4,6-O-ETHYLIDENE-BETA-D-GLUCOPYRANOSIDE); EPIPODOPHYLLOTOXIN,4'-DEMETHYL-,4,6-O-ETHYLIDENE-BETA-D-GLUCOPYRANOSIDE (8CI); 9-((4,6-O-ETHYLIDINE-BETA-D-GLUCOPYRANOSYL)OSY)-5,8,8A,9-TETRAHYDRO-5-(4- HYDROXY-3,4-DIMETHYLOXYPHENYL)FURO(3',4":6,7) NAPTHO-(2,3-D)-1,3-

DIOXOL-6 (5AH)-ONE; LASTET; NK 171; NSC-141540; NSC 141540; VEPESID; VEPESID J; VP-16; VP 16213

Description: yellow-brown crystalline powder or crystals

Use: medication: antineoplastic agent with antimitotic properties

Physical Properties

Freezing Point: 236 °C (456.8 °F) to 251 °C (483.8 °F)

Water Solubility: About 0 mg/mL

Other Solubilities: Soluble in Alcohol: approx 0.76 mg/ml

RTECS Toxicity Data

Acute Oral: Human TD_{Lo} Dose: 16 mg/kg/5D-I; Toxic Effects: Blood - Agranulocytosis; Blood - Aplastic anemia; Blood - Changes in bone marrow not included above.

Mutagenic: Human DNA Damage; Cell Type: lymphocyte; Dose: 1 umol/L. Human DNA Damage; Cell Type: other cell types; Dose: 2 umol/L. Human DNA Damage; Cell Type: HeLa cell; Dose: 1 umol/L. Human DNA Damage; Cell Type: leukocyte; Dose: 500 nmol/L. Human DNA Damage; Cell Type: mammary gland; Dose: 24 umol/L. Human DNA Damage; Cell Type: other cell types; Dose: 100 nmol/L/1H. Human DNA Inhibition; Cell Type: other cell types; Dose: 10 umol/L. Human DNA Inhibition; Cell Type: HeLa cell; Dose: 25 umol/L. Human DNA Inhibition; Cell Type: leukocyte; Dose: 500 nmol/L. Human Cytogenetic Analysis; Cell Type: lymphocyte; Dose: 100 ug/L. Human Sister Chromatid Exchenge; Cell Type: lymphocyte; Dose: 25 ug/L. Human Micronucleus Test; Cell Type: lymphocyte; Dose: 50 nmol/L. Human Other Mutation Test Systems; Cell Type: HeLa cell; Dose: 20 umol/L. Human Other Mutation Test Systems; Cell Type: lymphocyte; Dose: 5 mg/L.

Hazard Overviews

Health: Toxic. Other Acute Effects: harmful if swallowed, inhaled, or absorbed through skin; exposure can cause gastrointestinal disturbances; target organ: bone marrow. Chronic Effects: teratogen; mutagen. Possible human carcinogen.

Fire: Hazards: emits toxic fumes. Extinguishing agents: water spray; carbon dioxide, dry chemical powder or appropriate foam. Precautions: combustible liquid.

Reactivity: Stable. Hazardous polymerization will not occur. Hazardous decomposition products: toxic fumes of: carbon monoxide, carbon dioxide.

Carcinogenicity: IARC - Not listed; NIOSH - Not listed; NTP - Not listed; ACGIH - Not listed; OSHA - Not listed; EPA - Not listed; MAK - Not listed

Environmental

Regulations

RCRA 40CFR: Not listed

CERCLA: 40CFR 302.4: Not listed

SARA 40CFR 372.65: Not listed

SARA EHS 40CFR 355: Not listed

TSCA: Not listed

ETR5000 CAS #: 38260-54-7

ETRIMFOS

RTECS: TF8350000

DOT: UN2783; UN2784; UN3017; UN3018; IMO3.2; IMO6.1

EINECS Number: 253-855-9

Molecular Formula: $C_{10}H_{17}N_2O_4PS$

Formula Weight: 292.29

Synonyms: O,O-DIMETHYLO-(2-ETHYL-4-ETHOXYPYRIMIDINYL)-6-THIONOPHOSPHATE; EKAMET; EKAMET G; EKAMET ULV; ENT 29126; O-(6-ETHOXY-2-ETHYL-4-PYRIMIDINYL) O,O-DIMETHYLPHOSPHOROTHIOATE; O-6-ETHOXY-2-ETHYLPYRIMIDIN-4-YL O,O-DIMETHYLPHOSPHOROTHIOATE; O-(6-ETHOXY-2-ETHYL-4-PYRIMIDINYL)-PHOSPHOROTHIOIC ACIDO,O-DIMETHYL ESTER; OMS 1806; SAN 197; SAN 197 I; SATISFAR; SATISFAR DP-2; SATISFAR LS 3; SATISFAR LS 5

Description: colorless oil; slight characteristic odor

Use: broad-range non-systemic insecticide; for control of lepidoptera, coleoptera, diptera, & hemiptera on fruit trees (incl citrus), olives, potatoes, vegetables, maize, lucerne, ornamentals, tobacco, & rice; for control of mites, lice, etc in stored products; all purpose insecticide for gardens; controls scales in fruit trees, citrus & olives, grape berry moths in grapes, european corn borer, colorado potato beetle, pests of tobacco; granular for pyralidae in paddy rice.

Physical Properties

Freezing Point: -3.35 °C (25.97 °F)

Specific Gravity: 1.195 at 20 °C

Saturated Vapor Density: 1.200000932 kg/m^3

Vapor Pressure: 6.5 $x10^{-5}$ mm Hg at 20 °C

Water Solubility: 40 mg/L at 23 °C

Other Solubilities: completely miscible with Acetone, Chloroform, dimethyl sulfoxide, Ethanol, hexane, Methanol, Xylene.

RTECS Toxicity Data

Acute Oral: Rat LD_{50} Dose: 1800 mg/kg. Mouse LD_{50} Dose: 437 mg/kg.

Acute Dermal: Rabbit LD_{50} Route: Skin; Dose: >500 mg/kg. Rat LD_{50} Route: Skin; Dose: >2 gm/kg.

Hazard Overviews

Carcinogenicity: IARC - Not listed; NIOSH - Not listed; NTP - Not listed; ACGIH - Not listed; OSHA - Not listed; EPA - Not listed; MAK - Not listed

Environmental

Ecotoxicity: LC_{50} Rainbow trout 24 ug/l/96 hr /Conditions of bioassay not specified LC_{50} Quail dietary 740 mg/kg/8 day /Conditions of bioassay not specified

Environmental Fate: If released to the atmosphere, it will degrade rapidly in the vapor-phase by reaction with photochemically produced hydroxyl radicals (half-life of 4.8 hr). If released to soil or water, it will degrade readily through hydrolysis and biodegradation. It hydrolyzes in water with a half-life of 14-16 days at pH6-pH9 and 25 °C. It is rapidly metabolized in plants, animals and soil to 6-ethoxy-2-ethylpyrimidin-4-ol (EEHP) and other derivatives. Its half-life in soil is typically 3-8 days. It is expected to have a low to moderate leaching potential in soil.

Cleanup/Disposal: Guide No. 131: Fully encapsulating, vapor protective clothing should be worn for spills and leaks with no fire. Eliminate all ignition sources (no smoking, flares, sparks or flames in immediate area). All equipment used when handling the product must be grounded. Do not touch or walk through spilled material. Stop leak if you can do it without risk. Prevent entry into waterways, sewers, basements or confined areas. A vapor suppressing foam may be used to reduce vapors. Small Spills: Absorb with earth, sand or other non-combustible material and transfer to containers for later disposal. Use clean non-sparking tools to collect absorbed material. Large Spills: Dike far ahead of liquid spill for later disposal. Water spray may reduce vapor; but may not prevent ignition in closed spaces. Guide No. 152: Do not touch damaged containers or spilled material unless wearing appropriate protective clothing. Stop leak if you can do it without risk. Prevent entry into waterways, sewers, basements or confined areas. Cover with plastic sheet to prevent spreading. Absorb or cover with dry earth, sand or other non-combustible material and transfer to containers. Do not get water inside containers.

Environmental Physical Data

Henry's Law Constant: estimated at 6.2 x10^{-7}
Sorption Partition Coefficient: K_{oc} = estimated at 570
BCF: estimated at 77

Regulations

RCRA 40CFR: Not listed
CERCLA: 40CFR 302.4: Not listed
SARA 40CFR 372.65: Not listed
SARA EHS 40CFR 355: Not listed
TSCA: Not listed

Analytical Methods

Food: FDA 212.1, 232.1, 232.3, 232.4, 242.1
Plasma: FDA 211.1, 231.1, 252

EUG3000 **CAS #: 97-53-0**

EUGENOL

RTECS: SJ4375000
EINECS Number: 202-589-1
Molecular Formula: $C_{10}H_{12}O_2$
Structured MF: $H_2C=CHCH_2C_6H_3(OCH_3)OH$
Formula Weight: 164.20

OH
O

Chemical Structure

Synonyms: 4-ALLYLCATECHOL-2-METHYL ETHER; 4-ALLYLGUAIACOL; 5-ALLYLGUAIACOL; ALLYLGUAIACOL; P-ALLYLGUAIACOL; 1-ALLYL-4-HYDROXY-3-METHOXYBENZENE; 4-ALLYL-1-HYDROXY-2-METHOXYBENZENE; 4-ALLYL-2-METHOXYPHENOL; CARYOPHYLLIC ACID; EUGENIC ACID; 1,3,4-EUGENOL; P-EUGENOL; FA 100; 1-HYDROXY-4-ALLYL-2-METHOXYBENZENE; 1-HYDROXY-2-METHOXY-4-ALLYLBENZENE; 4-HYDROXY-3-METHOXYALLYLBENZENE; 1-HYDROXY-2-METHOXY-4-PROP-2-ENYLBENZENE; 1-HYDROXY-2-METHOXY-4-PROPENYLBENZENE; 2-METHOXY-4-ALLYLPHENOL; 2-METHOXY-1-HYDROXY-4-ALLYLBENZENE; 2-METHOXY-4-(2-PROPEN-1-YL)PHENOL; 2-METHOXY-4-(2-PROPENYL)PHENOL; 2-METHOXY-4-PROP-2-ENYLPHENOL; 2-METOKSY-4-ALLILOFENOL; PHENOL,4-ALLYL-2-METHOXY-; PHENOL,2-METHOXY-4-(2-PROPENYL)-; SYNTHETIC EUGENOL

Description: colorless or pale yellow liquid; crystals; odor of cloves

Use: in dentistry as a flavoring agent and mild rubefacient in dentifrices, as an obtundent for hypersensitive dentine, caries or exposed pulp, in dental cement preparations, as an analgesic and anesthetic and, mixed with zinc oxide, as a temporary dental filling; in perfumery, in the production of isoeugenol for the manufacture of vanillin, a an insect attractant, in the flavor industry, in essential oils and in medicine (local antiseptic and analgesic); in as an international standard in the GC determination of thymol in biological samples

Physical Properties

Boiling Point: 253.2 °C (488 °F) at 760 mm Hg
Freezing Point: -9.2 °C (15.44 °F) to -9.1 °C (15.62 °F)
Specific Gravity: 1.0664 at 20 °C/4 °C
Vapor Density: > 1 Air=1
Density: 1.064 to 1.068 g/mL
Vapor Pressure: 1 mm Hg at 78.4 °C
Water Solubility: Practically Insoluble in Water
Other Solubilities: freely Soluble in Benzene, Chloroform, Ether, hot Alcohol
Refraction Index: 1.5416 at 19.5 °C/D
Evaporation Rate: < 1 Butyl Acetate=1
Flash Point: ~ 104 °C

RTECS Toxicity Data

Acute Oral: Rat LD_{50} Dose: 1930 mg/kg; Toxic Effects: Behavioral - Altered sleep time (including change in righting

reflex); Behavioral - Coma; Lungs, Thorax, or Respiration - Dyspnea. Mouse LD_{50} Dose: 3 gm/kg; Toxic Effects: Behavioral - Somnolence (general depressed activity).
Acute Inhalation: Rat LC; Dose: >2580 mg/m^3/4hr; Toxic Effects: Behavioral - Somnolence (general depressed activity).
Acute Dermal: Rat LD_{Lo} Route: Subcutaneous Dose: 5 gm/kg; Toxic Effects: Behavioral - Convulsions or effect on seizure threshold; Behavioral - Muscle weakness; Kidney, Ureter, and Bladder - Other changes.
Chronic (Multiple Dose) Oral: Rat Dose: 68 gm/kg/34D-I; Toxic Effects: Liver - Other changes; DEATH. Rat Dose: 70 gm/kg/14D-C; Toxic Effects: DEATH.
Irritation Skin: Human Standard Draize Test Dose: 40 mg/48H; Reaction: mild. Man Standard Draize Test Dose: 16 mg/48H; Reaction: moderate. Rabbit Standard Draize Test Dose: 100 mg/24H; Reaction: severe.
Mutagenic: Mouse Mutations in Mammalian Somatic Cells; Cell Type: lymphocyte; Dose: 21300 ug/L. Mouse Micronucleus Test; Route: Intraperitoneal; Dose: 296 mg/kg/24H. Mouse Micronucleus Test; Route: Oral; Dose: 29589 mg/kg/24H.
Tumorigenic: Mouse Route: Oral; Dose: 37080 mg/kg/2Y-I; Toxic Effects: Tumorigenic - Equivocal tumorigenic agent by RTECS criteria; Liver - Tumors.

Hazard Overviews

Corrosive

Health: Irritating to eyes/skin/respiratory tract. Harmful. Other Acute Effects: harmful by inhalation, in contact with skin and if swallowed; may cause allergic reaction. Chronic Effects: may cause sensitization by inhalation and skin contact; possible risk of irreversible effects; target organs: kidneys, central nervous system, lungs and gastrointestinal system. Possible carcinogen.
Fire: Will burn. Hazards: emits toxic fumes. Extinguishing agents: water spray; carbon dioxide, dry chemical powder or appropriate foam. Precautions: combustible liquid.
Carcinogenicity: IARC - Group 3, Not classifiable as to carcinogenicity to humans; NIOSH - Not listed; NTP - Not listed; ACGIH - Not listed; OSHA - Not listed; EPA - Not listed; MAK - Not listed
Primary Target Organs:

Eyes

Skin

Respiratory System
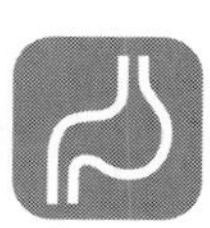
Gastro-intestinal

Nervous System

Kidneys

Environmental

Cleanup/Disposal: Guide No. 171: Do not touch or walk through spilled material. Stop leak if you can do it without risk. Prevent dust cloud. Avoid inhalation of asbestos dust. Small Dry Spills: With clean shovel place material into clean, dry container and cover loosely; move containers from spill area. Small Spills: Take up with sand or other noncombustible absorbent material and place into containers for later disposal. Large Spills: Dike far ahead of liquid spill for later disposal. Cover powder spill with plastic sheet or tarp to minimize spreading. Prevent entry into waterways, sewers, basements or confined areas.

Environmental Physical Data

Octanol/Water Partition Coefficient: log K_{ow} = 2.31

Regulations

RCRA 40CFR: Not listed
CERCLA: 40CFR 302.4: Not listed
SARA 40CFR 372.65: Not listed
SARA EHS 40CFR 355: Not listed
TSCA: Listed

EUG6000 CAS #: 10031-96-6

EUGENYL FORMATE

RTECS: SJ4551000
EINECS Number: 233-099-6
Molecular Formula: $C_{11}H_{12}O_3$
Formula Weight: 192.22
Synonyms: 4-ALLYL-2-METHOXYPHENOL FORMATE; 4-ALLYL-2-METHOXYPHENYL FORMATE; EUGENOL FORMATE; PHENOL,4-ALLYL-2-METHOXY-,FORMATE; PHENOL,2-METHOXY-4-(2-PROPENYL)-,FORMATE; 4-(2-PROPENYL)-2-METHOXYPHENYL FORMATE
Description: odor reminiscent of orris
Use: condiment

Physical Properties

Boiling Point: 270 °C (518 °F)
Specific Gravity: 1.105 to 1.109 at 15 °C
Other Solubilities: Soluble 1:2-3.5 in 80% Alcohol
pH: Maximum 0.9
Refraction Index: 1.524 to 1.5265 at 20 °C

RTECS Toxicity Data

Acute Oral: Rat LD_{50} Dose: 3400 mg/kg.
Acute Dermal: Rabbit LD_{50} Route: Skin; Dose: >5 gm/kg.
Irritation Skin: Rabbit Standard Draize Test Dose: 500 mg/24H; Reaction: mild.

Hazard Overviews

Corrosive

Carcinogenicity: IARC - Not listed; NIOSH - Not listed; NTP - Not listed; ACGIH - Not listed; OSHA - Not listed; EPA - Not listed; MAK - Not listed

Environmental

Regulations

RCRA 40CFR: Not listed

CERCLA: 40CFR 302.4: Not listed
SARA 40CFR 372.65: Not listed
SARA EHS 40CFR 355: Not listed
TSCA: Not listed

EXA5000 **CAS #: 33791-58-1**

3A,4,5,6,7,7A-11EXAHYDRO-4,7-METHANOINDENYL ACRYLATE

EINECS Number: 251-678-1
Molecular Formula: $C_{13}H_{16}O_2$
Formula Weight: 204.27
Synonyms: ACRYLIC ACID,3A,4,5,6,7,7A-HEXAHYDRO-4,7-METHANOINDENYLESTER; 2-PROPEONIC ACID,3A,4,5,6,7,7A-HEXANYDRO-4,7-METHANO-1H-INDENYL ESTER

Physical Properties

Other Solubilities: Reacts readily with electrophilic, free-radical, and nucleophilic agent

Hazard Overviews

Health: Irritating to eyes/skin/respiratory tract. Other Acute Effects: may be harmful by inhalation, ingestion, or skin absorption; exposure can cause nausea; dizziness; headache.
Fire: Hazards: emits toxic fumes; closed containers may rupture and explode during runaway polymerization. Extinguishing agents: carbon dioxide, dry chemical powder or appropriate foam; water or foam may cause frothing. Precautions: combustible liquid.
Reactivity: Unstable. Incompatible with: polymerizing initiators, peroxides, strong oxidizing agents, copper, copper alloys ,steel, iron and iron salts, rust, strong bases, sensitive to heat, sensitive to light. Hazardous decomposition products: toxic fumes of: carbon monoxide, carbon dioxide.
Carcinogenicity: IARC - Not listed; NIOSH - Not listed; NTP - Not listed; ACGIH - Not listed; OSHA - Not listed; EPA - Not listed; MAK - Not listed
Primary Target Organs:

Eyes

Skin

Respiratory System

Environmental

Regulations

RCRA 40CFR: Not listed
CERCLA: 40CFR 302.4: Not listed
SARA 40CFR 372.65: Not listed
SARA EHS 40CFR 355: Not listed
TSCA: Listed

FAM5000 **CAS #: 52-85-7**

FAMPHUR

RTECS: TF7650000
DOT: UN2783; UN2784; UN3017; UN3018; IMO3.2; IMO6.1
EINECS Number: 200-154-0
Molecular Formula: $C_{10}H_{16}NO_5PS_2$
Formula Weight: 325.36
Synonyms: AC 38023; AMERICAN CYANAMID-38023; AMERICAN CYANAMID 38,023; AMERICAN CYANAMID CL-38,023; BO-ANA; CL-38023; CL 38023; O,O-DIMETHYL O-(P-(N,N-DIMETHYLSULFAMOYL) PHENYL)PHOSPHOROTHIOATE; O,O-DIMETHYL O-(P-(DIMETHYLSULFAMOYL)PHENYL)PHOSPHOROTHIOATE; O,O-DIMETHYL O-(P-(N,N-DIMETHYLSULFAMOYL)PHENYL)PHOSPHOROTHIOATE; O-(4-((DIMETHYLAMINO)SULFONYL)PHENYL) O,O-DIMETHYLPHOSPHOROTHIOATE; O-4-DIMETHYLSULFAMOYLPHENYL O,O-DIMETHYL PHOSPHOROTHIOATE; DOVIP; ENT 25,644; EPA PESTICIDE CODE 059901; FAMFUR; FAMOFOS; FAMOPHOS; FAMOPHOS WARBEX; FAMPHOS; FANFOS; P-HYDROXY-N,N-DIMETHYLBENZENESULFONAMIDE ESTER WITHPHOSPHOROTHIOIC ACID O,O-DIMETHYL ESTER; PHOSPHOROTHIOIC ACID,O,O-DIMETHYL ESTER,O-ESTER WITHP-HYDROXY-N,N-DIMETHYLBENZENE-SULFONAMIDE; PHOSPHOROTHIOIC ACID,O-(4-((DIMETHYLAMINO)SULFONYL)PHENYL) O,O-DIMETHYL ESTER; VARBEX; WARBEX; WARBEXOL
Description: colorless crystalline powder
Use: vet: insecticide (grubicide); vet: insecticide for control of lice infestations; insecticide for control of cattle grub and lice infestations for grub control

Physical Properties

Freezing Point: 52.5 °C (126.5 °F) to 53.5 °C (128.3 °F)
Water Solubility: Sparingly Soluble in Water
Other Solubilities: In aqueous isopropanol (45%) 23 g/kg (20 °C) in Xylene 300 g/kg (5 °C); Soluble in Acetone, carbon tetrachloride, Chloroform, cyclohexanone, dichloromethane, Toluene; sparingly Soluble in aliphatic hydrocarbons
Flash Point: Does not burn or burns with difficulty

RTECS Toxicity Data

Acute Oral: Rat LD_{50} Dose: 28 mg/kg. Mouse LD_{50} Dose: 9500 ug/kg.
Acute Dermal: Rabbit LD_{50} Route: Skin; Dose: 1460 mg/kg. Rat LD_{50} Route: Skin; Dose: 400 mg/kg.
Chronic (Multiple Dose) Oral: Rat Dose: 13500 ug/kg/90D-C; Toxic Effects: Biochemical - True cholinesterase.

Hazard Overviews

Fire: Noncombustible.
Carcinogenicity: IARC - Not listed; NIOSH - Not listed; NTP - Not listed; ACGIH - Not listed; OSHA - Not listed; EPA - Not listed; MAK - Not listed

Environmental

Ecotoxicity: LC_{50} Anas platyrhynchos (Mallard duck) oral approx 35 ppm in 5-day diet, age 10 days LD_{50} Agelaius phoeniceus (Red-winged blackbird) oral 1.78 mg/kg (95% confidence limit 1.0-3.2 mg/kg LD_{50} Anas platyrhynchos (Mallard duck) oral 9.87 mg/kg (95% confidence limit 5.88-16.6 mg/kg), 3-4 mo old
Environmental Fate: If released to water, it is expected to undergo slow hydrolysis; an experimentally determined half-life of 115 days has been determined under neutral conditions. Hydrolysis under basic conditions will occur at a more rapid rate, with a calculated half-life value of 60 days at pH 10. It may adsorb to sediment and suspended organic matter, but it is not expected to bioconcentrate in fish and aquatic organisms. If released to soil, slow hydrolysis may occur in this media.
Cleanup/Disposal: Guide No. 131: Fully encapsulating, vapor protective clothing should be worn for spills and leaks with no fire. Eliminate all ignition sources (no smoking, flares, sparks or flames in immediate area). All equipment used when handling the product must be grounded. Do not touch or walk through spilled material. Stop leak if you can do it without risk. Prevent entry into waterways, sewers, basements or confined areas. A vapor suppressing foam may be used to reduce vapors. Small Spills: Absorb with earth, sand or other non-combustible material and transfer to containers for later disposal. Use clean non-sparking tools to collect absorbed material. Large Spills: Dike far ahead of liquid spill for later disposal. Water spray may reduce vapor; but may not prevent ignition in closed spaces. Guide No. 152: Do not touch damaged containers or spilled material unless wearing appropriate protective clothing. Stop leak if you can do it without risk. Prevent entry into waterways, sewers, basements or confined areas. Cover with plastic sheet to prevent spreading. Absorb or cover with dry earth, sand or other non-combustible material and transfer to containers. Do not get water inside containers.

Environmental Physical Data

Octanol/Water Partition Coefficient: log K_{ow} = 2.28
Sorption Partition Coefficient: K_{oc} = 419
BCF: not expected to be significant

Regulations

RCRA 40CFR: Listed Hazardous Waste No. P097 Toxic Waste
CERCLA: 40CFR 302.4: Listed per RCRA Section 3001 RQ: 1000 lb (453.5 kg)
SARA 40CFR 372.65: Listed
SARA EHS 40CFR 355: Not listed
TSCA: Listed

Analytical Methods

Soil: SW846 8141, 8141A, 8270B, 8270C, 8321, 8321A
Water / Groundwater: EPA 1657, 622.1; SW846 8141, 8141A

Food: FDA 212.1, 232.1, 232.3, 232.4, 242.1
Plasma: EPA 001, 29; FDA 252

FAR5000 CAS #: 4602-84-0

FARNESOL

RTECS: JR4979000
EINECS Number: 225-004-1
Molecular Formula: $C_{15}H_{26}O$
Formula Weight: 222.41

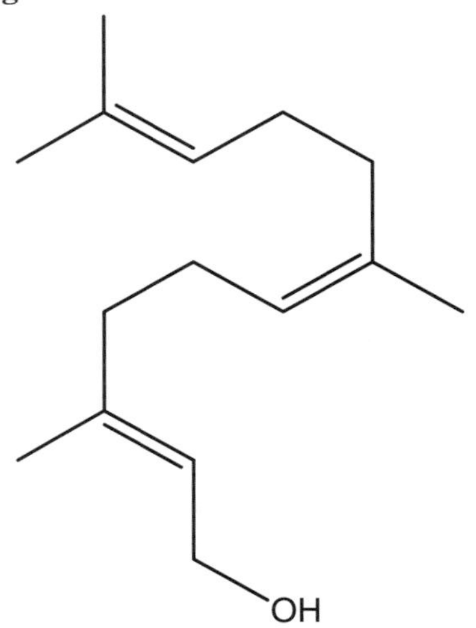

Chemical Structure

Synonyms: 2,6,10-DODECATRIEN-1-OL,3,7,11-TRIMETHYL-; FARNESYL ALCOHOL; 2,6,10-TRIMETHYL-2,6,10-DODECATRIEN-12-OL; 3,7,11-TRIMETHYL-2,6,10-DODECATRIEN-1-OL
Description: colorless liquid; flowery odor or weak citrus-lime odor
Use: in perfumery, to emphasize the odor of sweet floral perfumes, such as lilac and cyclamen

Physical Properties

Boiling Point: 110 °C (230 °F) to 113 °C (235 °F)
Specific Gravity: 0.8871 at 20 °C/4 °C
Water Solubility: Insoluble in Water
Other Solubilities: Acetone: Soluble; Alcohol: Very Soluble; Ether: Soluble; Organic solvents: Soluble.
Refraction Index: 1.4891

RTECS Toxicity Data

Acute Oral: Rat LD_{50} Dose: 6 gm/kg; Toxic Effects: Behavioral - Somnolence (general depressed activity); Behavioral - Antipsychotic. Mouse LD_{50} Dose: 7400 mg/kg; Toxic Effects: Behavioral - Somnolence (general depressed activity); Behavioral - Antipsychotic.
Mutagenic: Other Insects DNA Inhibition; Cell Type: ovary; Dose: 100 umol/L. Other Insects DNA Inhibition; Cell Type: ovary; Dose: 100 umol/L.

Hazard Overviews

Health: Irritating. Other Acute Effects: may be harmful by inhalation, ingestion, or skin absorption.
Fire: Extinguishing agents: water spray; carbon dioxide, dry chemical powder or appropriate foam. Precautions: combustible liquid.
Reactivity: Incompatible with: strong oxidizing agents. Hazardous decomposition products: toxic fumes of: carbon monoxide, carbon dioxide.
Carcinogenicity: IARC - Not listed; NIOSH - Not listed; NTP - Not listed; ACGIH - Not listed; OSHA - Not listed; EPA - Not listed; MAK - Not listed
Primary Target Organs:

Eyes

Skin

Respiratory System

Environmental

Regulations

RCRA 40CFR: Not listed
CERCLA: 40CFR 302.4: Not listed
SARA 40CFR 372.65: Not listed
SARA EHS 40CFR 355: Not listed
TSCA: Listed

FEN1000 CAS #: 22224-92-6

FENAMIPHOS

RTECS: TB3675000
DOT: UN2783; UN2784; UN3017; UN3018; IMO3.2; IMO6.1
EINECS Number: 244-848-1
Molecular Formula: $C_{13}H_{22}NO_3PS$
Structured MF: $C_{13}H_{22}NO_3PS$
Formula Weight: 303.4
Synonyms: O-AETHYL-O-(3-METHYL-4-METHYLTHIOPHENYL)-ISOPROPYLAMIDO-PHOSPHORSAEUREESTER; BAY 68138; BAY SRA 3886; ENT 27572; EPA PESTICIDE CHEMICAL CODE 100601; ESTER; ETHYL 3-METHYL-4-(METHYLTHIO)PHENYLISOPROPYLPHOSHORAMIDATE; ETHYL 3-METHYL-4-(METHYLTHIO)PHENYL(1-METHYLETHYL)PHOSPHORAMIDATE; ETHYL 3-METHYL-4-(METHYLTHIO)PHENYL-(1-METHYLETHYL)PHOSPHORAMIDATE; ETHYL 4-(METHYLTHIO)-M-TOLYL ISOPROPYLPHOSPHORAMIDATE; ETHYL 4-METHYLTHIO-M-TOLYL ISOPROPYLPHOSPHORAMIDATE; ETHYL-3-METHYL-4-(METHYLTHIO)PHENYL(1-METHYLETHYL)PHOSPHORAMIDATE; ETHYL-4-(METHYLTHIO)-M-TOLYL ISOPROPYLPHOSPHORAMIDATE; ISOPROPYLAMINO-O-ETHYL-(4-METHYLMERCAPTO-3-METHYLPHENYL)PHOSPHATE; ISOPROPYLPHOSPHORAMIDIC ACID,4-(METHYLTHIO)-M-TOLYL ETHYL ESTER; METHAPHENAMIPHOS; 1-(METHYLETHYL)-ETHYL 3-METHYL-4-(METHYLTHIO)PHENYLPHOSPHORAMIDATE; 1-(METHYLETHYL)-ETHYL-3-METHYL-4-(METHYLTHIO)PHENYLPHOSPHORAMIDATE; (1-METHYLETHYL)-PHOSPHORAMIDIC ACID,ETHYL(3-METHYL-4-(METHYLTHIO)PHENYL); NEMACUR; NEMACUR P; NSC 195106;

PHENAMIPHOS; PHOSPHORAMIDIC ACID,ISOPROPYL-,ETHYL4-(METHYLTHIO)-M-TOLYL ESTER; PHOSPHORAMIDIC ACID,(1-METHYLETHYL)-,ETHYL (3-METHYL-4-(METHYLTHIO)PHENYL) ESTER; PHOSPHORAMIDIC ACID,(1-METHYLETHYL)-,ETHYL3-METHYL-4-(METHYLTHIO)PHENYL ESTER; SRA 3886

Description: colorless to off-white to tan, waxy solid or crystals

Use: systemic agricultural nematocide; selective insecticide

Physical Properties

Boiling Point: Decomposes

Freezing Point: 49.2 °C (120.56 °F)

Saturated Vapor Density: 1.200000015 kg/m^3

Density: 1.15 g/cu cm at 20 °C

Vapor Pressure: 1 x10^{-6} mm Hg at 25 °C

Water Solubility: 700 mg/L at 20 °C

Other Solubilities: 95% Ethanol: >=100 mg/ml at 21 °C; Acetone: >=100 mg/ml at 21 °C; DMSO: >=100 mg/ml at 21 °C.

Flash Point: Not available; probably combustible

RTECS Toxicity Data

Acute Oral: Rat LD$_{50}$ Dose: 8 mg/kg. Mouse LD$_{50}$ Dose: 22700 ug/kg.

Acute Inhalation: Rat LC$_{50}$ Dose: 91 mg/m^3/4hr.

Acute Dermal: Rat LD$_{50}$ Route: Skin; Dose: 80 mg/kg.

Hazard Overviews

Fire: Will burn.

Carcinogenicity: IARC - Not listed; NIOSH - Not listed; NTP - Not listed; ACGIH - Class A4, Not classifiable as a human carcinogen; OSHA - Not listed; EPA - Not listed; MAK - Not listed

Exposure Limits

OSHA PEL Vacated 1989 Limits: TWA: 0.1 mg/m^3.

ACGIH TLV: TWA: 0.1 mg/m^3.

NIOSH REL: TWA: 0.1 mg/m^3.

Respirator Recommendation

Exposure Range: >0.1 to 1 mg/m^3 Air Purifying, Negative Pressure, Half Mask

Exposure Range: >1 to 10 mg/m^3 Air Purifying, Negative Pressure, Full Face

Exposure Range: >10 to 100 mg/m^3 Supplied Air, Constant Flow/Pressure Demand, Full Face

Exposure Range: >100 to unlimited mg/m^3 Self-contained Breathing Apparatus, Pressure Demand, Full Face

Cartridge Color: black w/magenta

Environmental

Environmental Fate: If released to the atmosphere, it should exist in both the vapor and particulate phases based on an experimental vapor pressure of 1 x10^{-6} mm Hg at 25 °C. Vapor-phase is degraded in the atmosphere by reaction with photochemically produced hydroxyl radicals with an estimated half-life of about 2 hours. Particulate-phase may be physically removed from the air by wet and dry deposition. Measured K$_{oc}$ values ranging from 94-370 suggest that it will have moderate to high mobility in soil. It rapidly degrades in soil through a combination of biodegradation, thiooxidation, and photolysis on the soil surface. The main transformation product of these processes is fenamiphos sulfoxide. Half-lives of the parent from days to weeks are measured in untreated soils; half-lives of one to a few days are measured in previously treated soils due to microbial acclimation. In natural water, it had a half-life of 1.8 days. The main compound formed was fenamiphos sulfoxide. It was converted to fenamiphos sulfoxide through thiooxidation (over a period of 16 hours in water) and via photolysis in aqueous solution (complete degradation required 90 minutes). Volatilization from water surfaces is not expected based on an estimated Henry's Law constant of 9.7 x10^{-11} atm-cu m/mole. This compound may bioconcentrate in aquatic organisms given an estimated BCF of 170; a measured BCF of 468 in earthworms was reported.

Cleanup/Disposal: Guide No. 131: Fully encapsulating, vapor protective clothing should be worn for spills and leaks with no fire. Eliminate all ignition sources (no smoking, flares, sparks or flames in immediate area). All equipment used when handling the product must be grounded. Do not touch or walk through spilled material. Stop leak if you can do it without risk. Prevent entry into waterways, sewers, basements or confined areas. A vapor suppressing foam may be used to reduce vapors. Small Spills: Absorb with earth, sand or other non-combustible material and transfer to containers for later disposal. Use clean non-sparking tools to collect absorbed material. Large Spills: Dike far ahead of liquid spill for later disposal. Water spray may reduce vapor; but may not prevent ignition in closed spaces. Guide No. 152: Do not touch damaged containers or spilled material unless wearing appropriate protective clothing. Stop leak if you can do it without risk. Prevent entry into waterways, sewers, basements or confined areas. Cover with plastic sheet to prevent spreading. Absorb or cover with dry earth, sand or other non-combustible material and transfer to containers. Do not get water inside containers.

Environmental Physical Data

Henry's Law Constant: estimated at 1.2 x10^{-9}

Octanol/Water Partition Coefficient: log K$_{ow}$ = 3.23

Sorption Partition Coefficient: K$_{oc}$ = 94 to 370

BCF: estimated at 170

Regulations

RCRA 40CFR: Not listed

CERCLA: 40CFR 302.4: Not listed

SARA 40CFR 372.65: Not listed TPQ: 10/10000 lb

SARA EHS 40CFR 355: Listed TPQ: 10 lb

TSCA: Not listed

Analytical Methods

Soil: EPA PMD-TLC

Drinking Water: EPA 507, 525.2; AOAC 991.07; ASTM D5475

Food: FDA 212.1, 232.1, 232.3, 232.4, 242.1

Indoor / Expired Air: NIOSH 5600

Plasma: FDA 211.1, 231.1, 252

FEN1570 **CAS #: 60168-88-9**

FENARIMOL (A-(2-CHLOROPHENYL)-ALPHA-4-CHLOROPHENYL)-

RTECS: UV9279400
EINECS Number: 262-095-7
Molecular Formula: $C_{17}H_{12}Cl_2N_2O$
Formula Weight: 331.21
Synonyms: BLOC; (2-CHLOROPHENYL)-ALPHA-(4-CHLOROPHENYL)-5-PYRIMIDINEMETHANOL; ALPHA-(2-CHLOROPHENYL)-ALPHA-(4-CHLOROPHENYL)-5-PYRIMIDINEMETHANOL; EL 222; FENARIMOL; RIMIDIN; RUBIGAN

Physical Properties

Freezing Point: 117 °C (242.6 °F) to 119 °C (246.2 °F)
Other Solubilities: Slightly soluble in hexane

RTECS Toxicity Data

Acute Oral: Rat LD_{50} Dose: 2500 mg/kg. Mouse LD_{50} Dose: 4500 mg/kg.
Acute Dermal: Rabbit LD_{50} Route: Skin; Dose: >2 gm/kg.
Reproductive/Teratogenic: Rat Route: Oral; Dose: 980 mg/kg; Duration: male 4D prior to mating; Effects on Fertility - Male fertility index.
Mutagenic: Mouse Cytogenetic Analysis; Route: Oral; Dose: 450 mg/kg. Mold - A Nidulans Sex Chromosome Loss; Dose: 6 mg/L.

Hazard Overviews

Carcinogenicity: IARC - Not listed; NIOSH - Not listed; NTP - Not listed; ACGIH - Not listed; OSHA - Not listed; EPA - Not listed; MAK - Not listed

Environmental

Regulations

RCRA 40CFR: Not listed
CERCLA: 40CFR 302.4: Listed as Compound per CWA Section 307(a)
SARA 40CFR 372.65: Listed
SARA EHS 40CFR 355: Not listed
TSCA: Listed

Analytical Methods

Soil: EPA PMD-FBR
Water / Groundwater: EPA 633.1
Drinking Water: EPA 507, 525.2; AOAC 991.07; ASTM D5475
Food: FDA 212.1, 232.1, 232.4, 242.1
Plasma: FDA 211.1, 231.1, 252
Other: EPA 1656

FEN2140 **CAS #: 13356-08-6**

FENBUTATIN OXIDE

RTECS: JN8770000
DOT: UN2786; UN2787; UN3019; UN3020; IMO3.2; IMO6.1
EINECS Number: 236-407-7
Molecular Formula: $C_{60}H_{37}OSn_2$
Formula Weight: 1052.66
Synonyms: BENDEX; BIS(TRIS(BETA,BETA-DIMETHYLPHENETHYL)TIN)OXIDE; BIS(TRIS(2-METHYL-2-PHENYLPROPYL)TIN)OXIDE; DI(TRI-(2,2-DIMETHYL-2-PHENYLETHYL)TIN)OXIDE; ENT 27738; FENBUTATIN OXIDE; FENBUTATIN-OXYDE; FENYLBUTATIN OXIDE; FENYLBUTYLSTANNIUM OXIDE; HEXAKIS; HEXAKIS (2-METHYL-2-PHENYLPROPYL)-DISTANNOXANE; HEXAKIS(BETA,BETA-DIMETHYLPHENETHYL)-DISTANNOXANE; HEXAKIS(2-METHYL-2-PHENYLPROPYL)DISTANNOXANE; OSDARAN; SD 14114; SHELL SD-14114; TORQUE; VENDEX
Description: white crytalline powder; mild odor
Use: control of the mobile stage of a wide range of phytophagus mites on citrus, glasshouse crops, ornamentally top fruit, vegetable and vines

Physical Properties

Boiling Point: 235 °C (455 °F) to 240 °C (464 °F) at 0.05 mm Hg
Freezing Point: 138 °C (280.4 °F) to 139 °C (282.2 °F)
Water Solubility: 0.005 mg/L at 23 °C
Other Solubilities: Very slightly Soluble in aliphatic hydrocarbons and mineral oils.
Flash Point: Does not burn or burns with difficulty

RTECS Toxicity Data

Acute Oral: Rat LD_{50} Dose: 2630 mg/kg. Mouse LD_{50} Dose: 1450 mg/kg.
Acute Dermal: Rabbit LD_{50} Route: Skin; Dose: >2 gm/kg. Rat LD_{50} Route: Skin; Dose: 1 gm/kg.

Hazard Overviews

Fire: Noncombustible.
Carcinogenicity: IARC - Not listed; NIOSH - Not listed; NTP - Not listed; ACGIH - Class A4, Not classifiable as a human carcinogen; OSHA - Not listed; EPA - Not listed; MAK - Not listed

Exposure Limits
OSHA PEL: TWA: 0.1 mg/m^3; as Sn.
OSHA PEL Vacated 1989 Limits: TWA: 0.1 mg/m^3.
ACGIH TLV: TWA: 2 mg/m^3; see NIOSH.
NIOSH REL: TWA: 2 mg/m^3; inorganic (.1 if organic).
DFG MAK: TWA: 2 mg/m^3; See NIOSH.

Environmental

Ecotoxicity: LC_{50} Rainbow trout 0.27 mg active ingredient /l (as WP formulation) 48 hr LD_{50} Bee acute oral >0.1
Environmental Fate: May hydrolyze chemically in water and in moist soil to tris(2-methyl-2-phenylpropyl)tin hydroxide,

but the hydroxide could be slowly reconverted to the parent compound at ambient temperature. Photolysis and biodegradation in soil and water may not be important. Both estimation method and experimental results show that adsorption to soil and suspended solids and sediments in water may be important. It should be essentially immobile in soil. The apparent dissipation half-life in test soils due to irreversible adsorption ranged from 8 to 30 days. An estimated bioconcentration factor indicates bioconcentration in aquatic organisms may also be important.

Cleanup/Disposal: Guide No. 131: Fully encapsulating, vapor protective clothing should be worn for spills and leaks with no fire. Eliminate all ignition sources (no smoking, flares, sparks or flames in immediate area). All equipment used when handling the product must be grounded. Do not touch or walk through spilled material. Stop leak if you can do it without risk. Prevent entry into waterways, sewers, basements or confined areas. A vapor suppressing foam may be used to reduce vapors. Small Spills: Absorb with earth, sand or other non-combustible material and transfer to containers for later disposal. Use clean non-sparking tools to collect absorbed material. Large Spills: Dike far ahead of liquid spill for later disposal. Water spray may reduce vapor; but may not prevent ignition in closed spaces. Guide No. 153: Eliminate all ignition sources (no smoking, flares, sparks or flames in immediate area). Do not touch damaged containers or spilled material unless wearing appropriate protective clothing. Stop leak if you can do it without risk. Prevent entry into waterways, sewers, basements or confined areas. Absorb or cover with dry earth, sand or other non-combustible material and transfer to containers. Do not get water inside containers.

Environmental Physical Data

Sorption Partition Coefficient: K_{oc} = estimated at 4.91
BCF: estimated at 4.09

Regulations

RCRA 40CFR: Not listed
CERCLA: 40CFR 302.4: Not listed
SARA 40CFR 372.65: Listed
SARA EHS 40CFR 355: Not listed
TSCA: Listed

FEN2710 CAS #: 458-24-2

FENFLURAMINE

RTECS: SH6820000
EINECS Number: 207-276-3
Molecular Formula: $C_{12}H_{16}F_3N$
Formula Weight: 231.27
Synonyms: BENZENEETHANAMINE,N-ETHYL-ALPHA-METHYL-3-(TRIFLUOROMETHYL)-; 2-ETHYLAMINO-1-(3-TRIFLUOROMETHYLPHENYL)PROPANE; N-ETHYL-ALPHA-METHYL-3-(TRIFLUOROMETHYL)BENZENEETHANAMINE; N-ETHYL-ALPHA-METHYL-3-TRIFLUOROMETHYLPHENETHYLAMINE; N-ETHYL-ALPHA-METHYL-M-(TRIFLUOROMETHYL)PHENETHYLAMINE; PHENETHYLAMINE,N-ETHYL-ALPHA-METHYL-M-(TRIFLUOROMETHYL)-; S 768; 3-(TRIFLUOROMETHYL)-N-ETHYL-ALPHA-METHYLPHENETHYLAMINE; 1-(META-TRIFLUOROMETHYL-PHENYL)-2 ETHYLAMINOPROPANE
Description: white to off-white amorphous powder or crystals; characteristic odor
Use: medication: anorexic; an anorexigenic agent in human medicine and used experimentally as a means of controlling weight in broiler hens

Physical Properties

Boiling Point: 108 °C (226 °F) to 112 °C (234 °F)
Water Solubility: Sparingly Soluble in Water
Other Solubilities: Sparingly Soluble in Ethanol.

RTECS Toxicity Data

Acute Oral: Man TD_{Lo} Dose: 4286 ug/kg; Toxic Effects: Automatic Nervous System - Sympathomimetic; Behavioral - Hallucinations, distorted perceptions. Human LD_{Lo} Dose: 50 mg/kg; Toxic Effects: Behavioral - Convulsions or effect on seizure threshold; Behavioral - Muscle contraction or spasticity; Cardiac - Other changes. Rat LD_{50} Dose: 130 mg/kg.
Reproductive/Teratogenic: Rat Route: Oral; Dose: 280 mg/kg; Duration: female 7-20D of pregnancy; Effects on Newborn - Behavioral.

Hazard Overviews

Carcinogenicity: IARC - Not listed; NIOSH - Not listed; NTP - Not listed; ACGIH - Not listed; OSHA - Not listed; EPA - Not listed; MAK - Not listed

Environmental

Regulations

RCRA 40CFR: Not listed
CERCLA: 40CFR 302.4: Not listed
SARA 40CFR 372.65: Not listed
SARA EHS 40CFR 355: Not listed
TSCA: Not listed

FEN3280 CAS #: 122-14-5

FENITROTHION

RTECS: TG0350000
EINECS Number: 204-524-2
Molecular Formula: $C_9H_{12}NO_5PS$
Formula Weight: 277.25
Synonyms: AC-47300; ACCOTHION; ACEOTHION; AGRIA 1050; AGRIYA 1050; AGROTHION; AKOTION; AMERICAN CYANAMID CL-47,300; ARBOGAL; BAY 41831; BAYER 41831; BAYER S 5660; CEKUTROTHION; CL 47300; CP 47114; M-CRESOL,4-NITRO-,O-ESTER WITH O,O-DIMETHYLPHOSPHOROTHIOATE; CYFEN; CYTEL; CYTEN; O,O-DIMETHYL O-(3-METHYL-4 NITROPHENYL) PHOSPHOROTHIONATE; O,O-DIMETHYL O-(3-METHYL) PHOSPHOROTHIOATE; O,O-DIMETHYL O-(3-METHYL-4-NITROPHENYL) PHOSPHOROTHIOATE; DIMETHYL 3-METHYL-4-NITROPHENYL PHOSPHOROTHIONATE; O,O-DIMETHYL O-(3-METHYL-4-NITROPHENYL) PHOSPHOROTHIONATE; O,O-DIMETHYL O-(3-METHYL-4-NITROPHENYL) THIOPHOSPHATE;

O,O-DIMETHYL O-(4-NITRO-3-METHYLPHENYL)THIOPHOSPHATE; O,O-DIMETHYL O-4-NITRO-M-TOLYL PHOSPHOROTHIOATE; DIMETHYL 4-NITRO-M-TOLYL PHOSPHOROTHIONATE; O,O-DIMETHYL-O-(3-METHYL-4-NITROFENYL)-MONOTHIOFOSFAAT; O,O-DIMETHYL-O-(3-METHYL-4-NITRO-PHENYL)-MONOTHIOPHOSPHAT; O,O-DIMETHYL-O-(4-NITRO-5-METHYLPHENYL)-THIONOPHOSPHAT; O,O-DIMETIL-O-(3-METIL-4-NITROFENIL) FOSFOROTIOATO; O,O-DIMETIL-O-(3-METIL-4-NITRO-FENIL)-MONOTIOFOSFATO; DYBAR; EI 47300; ENT 25,715; FALITHION; FENITION; FENITOX; FENITROTION; FENTROTHIONE; FOLITHION; FOLITHION EC 50; H-35-F 87 (BVM); 8057HC; INSECTIGAS F; KOTION; MACBAR; MEP; MEP (PESTICIDE); METATHIO E-50; METATHION; METATHION E 50; METATHIONE; METATHIONINE; METATHIONINE E50; METATION; METATION E50; 3-METHYL-4-NITROPHENYL DIMETHYL PHOSPHOROTHIOATE; METHYLNITROPHOS; MGLAWIK F; MONSANTO CP 47114; NITROPHOS; NOVATHION; NUVANOL; OLEOSUMIFENE; OMS 43; OVADOFOS; OWADOFOS; PENNWALT C-4852; PHENITROTHION; PHOSPHOROTHIOIC ACID,O,O-DIMETHYL O-(3-METHYL-4-NITROPHENYL) ESTER; PHOSPHOROTHIOIC ACID,O,O-DIMETHYL O-(4-NITRO-M-TOLYL)ESTER; PHOSPHOROTHIOIC ACID,O,O-DIMETHYLO-(3-METHYL-4-NITROPHENYL) ESTER; S 112A; S 5660; S-1102A; SUMIFENE; SUMITHIAN; SUMITHION; SUMITOMO S-1102A; SUPER SUMITHION; THIOPHOSPHATE DE O,O-DIMETHYLE ET DE O-(3-METHYL-4-NITROPHENYLE); THIOPHOSPHATE DE O,O-DIMETHYLE ET DEO-(3-METHYL-4-NITROPHENYLE); TIONOFOSFORAN O,O-DWUMETYLO-O-(3-METYLO)-4-NITROFENYLOWY; VERTHION

Description: yellow-brown oily liquid

Use: contact & stomach insecticide; acaricide; for controlling chewing & sucking insects on cereals, cotton, orchard fruits, rice, vegetables, and forest fly, mosquito & cockroach residual contact sprays for farms & in public health programs

Physical Properties

Boiling Point: 118 °C (244 °F) at 0.05 mm Hg
Specific Gravity: 1.32 to 1.34 at 25 °C/25 °C
Saturated Vapor Density: 1.20000073 kg/m^3
Vapor Pressure: 5.4 x10^{-5} mm Hg at 20 °C
Water Solubility: 0.0002%
Other Solubilities: Readily Soluble in dichloromethane, 2-propanol, Toluene, hardly Soluble in n-hexane.
Refraction Index: 1.5528 at 25 °C/D

RTECS Toxicity Data

Acute Oral: Woman TD$_{Lo}$ Dose: 800 mg/kg; Toxic Effects: Gastrointestinal - Hypermotility, diarrhea; Gastrointestinal - Nausea or vomiting; Lungs, Thorax, or Respiration - Dyspnea. Man LD$_{Lo}$ Dose: 429 uL/kg; Toxic Effects: Behavioral - Coma; Lungs, Thorax, or Respiration - Respiratory depression; Gastrointestinal - Other changes. Rat LD$_{50}$ Dose: 250 mg/kg.

Acute Inhalation: Rat LC$_{50}$ Dose: 378 mg/m^3/4hr; Toxic Effects: Behavioral - Tremor; Behavioral - Muscle contraction or spasticity; Lungs, Thorax, or Respiration - Dyspnea.

Acute Dermal: Rabbit LD$_{50}$ Route: Skin; Dose: 1250 mg/kg. Rat LD$_{50}$ Route: Skin; Dose: 1002 mg/kg. Rat LD$_{50}$ Route: Subcutaneous Dose: 1300 mg/kg.

Chronic (Multiple Dose) Oral: Rat Dose: 400 mg/kg/40D-I; Toxic Effects: DEATH. Rat Dose: 760 mg/kg/12W-I; Toxic Effects: Blood - Other changes; Biochemical - True cholinesterase. Rat Dose: 406 mg/kg/28D-C; Toxic Effects: Endocrine - Hyperglycemia; Endocrine - Changes in adrenal weight; Blood - Other changes. Rat Dose: 456 mg/kg/1Y-C; Toxic Effects: Cardiac - Changes in heart weight; DEATH - Changes in uterine weight.

Chronic (Multiple Dose) Inhalation: Rat Dose: 1 mg/m^3/4H/13W-I; Toxic Effects: Biochemical - True cholinesterase. Rat Dose: 17 mg/m^3/17W-I; Toxic Effects: Biochemical - True cholinesterase. Rat Dose: 15 mg/m^3/2H/4W-I; Toxic Effects: Brain and coverings - Other degenerative changes; Blood - Other changes; Biochemical - True cholinesterase. Rat Dose: 57 mg/m^3/2H/30D-I; Toxic Effects: Biochemical - True cholinesterase; Biochemical - Other esterases; DEATH.

Chronic (Multiple Dose) Dermal: Rabbit Route: Skin; Dose: 10500 mg/kg/21D-I; Toxic Effects: Behavioral - Sleep; Gastrointestinal - Hypermotility, diarrhea; DEATH.

Reproductive/Teratogenic: Rat Route: Oral; Dose: 90 mg/kg; Duration: female 7-15D of pregnancy; Effects on Newborn - Behavioral; Other postnatal measures or effects.

Mutagenic: Human Cytogenetic Analysis; Cell Type: leukocyte; Dose: 40 mg/L. Human Cytogenetic Analysis; Cell Type: lymphocyte; Dose: 40 mg/L. Human Sister Chromatid Exchenge; Cell Type: embryo; Dose: 55400 ug/L.

Hazard Overviews

Explosive

Carcinogenicity: IARC - Not listed; NIOSH - Not listed; NTP - Not listed; ACGIH - Not listed; OSHA - Not listed; EPA - Not listed; MAK - Not listed

Environmental

Ecotoxicity: LD$_{50}$ Bobwhite quail oral 23.6 mg/kg (95% confidence limit 12.6-43.5 mg/kg), 5 mo old females LC$_{50}$ Atlantic salmon 1.7 mg/l/96 hr, water 12 °C (95% confidence limit 1.5-2.0 mg/l), wt 0.4 g. Static bioassay without aeration, pH 7.2-7.5, water hardness 40-50 mg/l as calcium carbonate and alkalinity of 30-35 mg/l LC$_{50}$ Gammarus fasciatus 0.003 mg/l/96 hr, hard water 15 °C, mature. Static bioassay without aeration, pH 7.2-7.5, water hardness 40-50 mg/l as calcium carbonate and alkalinity of 30-35 mg/l LD$_{50}$ Mule deer oral greater than 727 mg/kg LC$_{50}$ Daphnia 0.0092 ppm/3 hr. /Conditions of bioassay not specified LC$_{50}$ Brook trout 2.0 mg/l/96 hr, water 12 °C (95% confidence limit 1.6-2.5 mg/l), wt 0.4 g. Static bioassay without aeration, pH 7.2-7.5, water hardness 40-50 mg/l as calcium carbonate and alkalinity of 30-35 mg/l

Environmental Fate: Abiotic hydrolysis in soil and water should not be important. The photolysis in clear water and soil surfaces should be important and the photolysis half-life has been estimated to be about 1 day. Biodegradation may be important in water and in soil and biodegradation is expected to be faster under anaerobic than aerobic conditions. The volatilization loss from soil and water should not be important, but the volatilization from surface slicks following application has been reported to be very fast. Based on experimentally determined soil sorption coefficients, it should show medium to low mobility in soil and it should moderately to strongly adsorb to suspended solids and sediments in water. The experimentally determined BCF values suggest that it should moderately bioconcentrate in aquatic organisms. Based on an estimated rate constant, vapor should react with photochemically produced hydroxyl radicals with a half-life of 6.2 days. The removal of atmospheric vapor phase by direct photolysis may also be important, but the photolysis rates under sunlight irradiation is not known. Partial removal from the atmosphere by dry deposition of spray droplets or particulate is also possible.

Environmental Physical Data

Henry's Law Constant: 6.56×10^{-7}
Octanol/Water Partition Coefficient: log K_{ow} = 3.38
Sorption Partition Coefficient: K_{oc} = soils 424
BCF: topmouth gundgeon 246

Regulations

RCRA 40CFR: Not listed
CERCLA: 40CFR 302.4: Not listed
SARA 40CFR 372.65: Not listed
SARA EHS 40CFR 355: Listed
TSCA: Not listed

Analytical Methods

Soil: SW846 8141A
Water / Groundwater: EPA 622.1, 022; SW846 8141, 8141A
Food: FDA 212.1, 232.1, 232.3, 232.4, 242.1; AOAC 985.07, 989.02
Plasma: EPA 001, 028; FDA 211.1, 231.1, 252

FEN3850 CAS #: 8006-84-6

FENNEL OIL

RTECS: LJ2550000
Molecular Formula: Unknown
Formula Weight: N/A
Synonyms: BITTER FENCHEL; BITTER FENNEL OIL; FENCHEL OEL; FENOUIL AMER; FENOUIL DOUX; FINOCCHIO; FINOCCHIO AMARO; FINOCCHIO DOLCE; FLORENCE FENNEL; HELLER FENCHEL; HINOJO; HINOJO AMARGO; OIL OF FENNEL; OILS,FENNEL; OILS,FOENICULUM VULGARE; SWEET FENNEL OIL
Description: colorless to pale yellow oil; odor of fennel
Use: to cover taste of unpleasant medicines; essential oil used as source of anethole; flavoring cattle feed; in compounded oils for sweets & pastry; for liqueur applications; medication: formerly used as carminative

Physical Properties

Specific Gravity: 0.953 to 0.973 at 25 deg/25 °C
Water Solubility: Slightly Soluble in Water
Other Solubilities: Soluble in 1 vol 90% or in 8 volumes 80% Alcohol; Very Soluble in Chloroform, Ether
Refraction Index: 1.5280 to 1.5380 at 20 °C/D

RTECS Toxicity Data

Acute Oral: Rat LD_{50} Dose: 3120 mg/kg. Mouse LD_{50} Dose: 3100 mg/kg.
Acute Dermal: Rabbit LD_{50} Route: Skin; Dose: >5 gm/kg.
Irritation Skin: Rabbit Standard Draize Test Dose: 500 mg/24H; Reaction: moderate. Mouse Standard Draize Test Dose: 100%; Reaction: severe.
Mutagenic: Bacteria - S Typhimurium Mutations in Microorganisms; Dose: 2500 ug/plate (-S9).

Hazard Overviews

Carcinogenicity: IARC - Not listed; NIOSH - Not listed; NTP - Not listed; ACGIH - Not listed; OSHA - Not listed; EPA - Not listed; MAK - Not listed

Environmental

Regulations

RCRA 40CFR: Not listed
CERCLA: 40CFR 302.4: Not listed
SARA 40CFR 372.65: Not listed
SARA EHS 40CFR 355: Not listed
TSCA: Listed

FEN4420 CAS #: 31879-05-7

FENOPROFEN

RTECS: MU6646000
EINECS Number: 250-850-3
Molecular Formula: $C_{15}H_{14}O_3$
Formula Weight: 242.28
Synonyms: BENZENEACETIC ACID,ALPHA-METHYL-3-PHENOXY-,(+-)-; ALPHA-DL-2-(3-PHENOXYPHENYL)PROPIONIC ACID; DL-2-(3-PHENOXYPHENYL)PROPIONIC ACID; HYDRATROPIC ACID,M-PHENOXY-,(+-)-; LILLY 53838; ALPHA-METHYL-3-PHENOXYBENZENEACETIC ACID; (+-)-M-PHENOXYHYDROTROPIC ACID; (+-)-2-(3-PHENOXYPHENYL)PROPIONIC ACID
Description: viscous oil
Use: medication: anti-inflammatory; analgesic

Physical Properties

Boiling Point: 168 °C (334 °F) to 171 °C (340 °F) at 0.11 mm Hg
Refraction Index: 1.5742 at 25 °C/D

RTECS Toxicity Data

Acute Oral: Human TD_{Lo} Dose: 4286 ug/kg; Toxic Effects: Endocrine - Evidence of thyroid hyperfunction. Woman TD_{Lo} Dose: 1092 mg/kg/13W-I; Toxic Effects: Kidney, Ureter, and Bladder - Chgs in tubules (inc acute renal failure, acute tubular necrosis; Kidney, Ureter, and Bladder - Interstitial nephritis; Kidney, Ureter, and Bladder - Other changes. Woman TD_{Lo} Dose: 9386 mg/kg/1Y-I; Toxic Effects: Kidney, Ureter, and Bladder - Interstitial nephritis.

Hazard Overviews

Carcinogenicity: IARC - Not listed; NIOSH - Not listed; NTP - Not listed; ACGIH - Not listed; OSHA - Not listed; EPA - Not listed; MAK - Not listed

Environmental

Regulations

RCRA 40CFR: Not listed
CERCLA: 40CFR 302.4: Not listed
SARA 40CFR 372.65: Not listed
SARA EHS 40CFR 355: Not listed
TSCA: Not listed

FEN4990 CAS #: 66441-23-4

FENOXAPROP ETHYL

RTECS: UA2454000
EINECS Number: 266-362-9
Molecular Formula: $C_{18}H_{16}ClNO_5$
Formula Weight: 291.38
Synonyms: ACCLAIM; (+/-)-2-(4-((6-CHLORO-2-BENZOXAZOLYL)OXY)PHENOXY)PROPANOIC ACID ETHYL ESTER; DEPON; ETHYL (D+)-2-(4-(6-CHLOR-2-BENZOXAZOLYLOXY)PHENOXY)PROPANOATE; ETHYL (+-)-2-(4-(6-CHLORO-2-BENZOXAZOLYLOXY)PHENOXY)PROPANOATE; EXCEL; FENOXAPROP-ETHYL; FURORE; HOE-A 25-01; HOE 33171; OPTION; PUMA; WHIP
Description: colorless solid
Use: herbicide to control grassy weeds in broadleaved crops and established turfgrass; controls emerged annual and perennial grass weeds in such crops as beans, beets, cotton, groundnuts, potatoes, soyabeans and vegetables

Physical Properties

Boiling Point: 200 °C (392 °F) at 100 pa
Freezing Point: 84 °C (183.2 °F) to 85 °C (185 °F)
Vapor Pressure: 0.19 $x10^{-5}$ Pa at 20 °C
Water Solubility: 1 mg/L at 25 °C
Other Solubilities: Solubility at 20 deg c: >0.5% in hexane, >1% in cyclohexane, Ethanol, 1-octanol, >20%in Ethyl Acetate, >30% in Toluene, >50% in Acetone.

RTECS Toxicity Data

Acute Oral: Rat LD_{50} Dose: 2357 mg/kg. Mouse LD_{50} Dose: 4670 mg/kg.
Acute Inhalation: Rat LC_{50} Dose: >510 mg/m^3.
Acute Dermal: Rabbit LD_{50} Route: Skin; Dose: >2 gm/kg. Rat LD_{50} Route: Skin; Dose: >2 gm/kg.

Hazard Overviews

Carcinogenicity: IARC - Not listed; NIOSH - Not listed; NTP - Not listed; ACGIH - Not listed; OSHA - Not listed; EPA - Not listed; MAK - Not listed

Environmental

Ecotoxicity: LD_{50} Japnese quail oral > 5000 mg/kg

Environmental Physical Data

Henry's Law Constant: 3 $x10^{-7}$

Regulations

RCRA 40CFR: Not listed
CERCLA: 40CFR 302.4: Not listed
SARA 40CFR 372.65: Listed
SARA EHS 40CFR 355: Not listed
TSCA: Not listed

Analytical Methods

Food: FDA 212.1, 232.1, 232.3, 232.4, 242.1
Plasma: FDA 211.1, 231.1, 252

FEN5560 CAS #: 72490-01-8

FENOXYCARB

RTECS: FD0423000
DOT: UN2757; UN2758; UN2991; UN2992; IMO3.2; IMO6.1
EINECS Number: 276-696-7
Molecular Formula: $C_{17}H_{19}NO_4$
Formula Weight: 301.3
Synonyms: ABG 6215; ETHYL 2-(4-PHENOXYPHENOXY)ETHYL CARBAMATE; ETHYL (2-(4-PHENOXYPHENOXY)ETHYL)CARBAMATE; ETHYL 2-(4-PHENOXYPHENOXY)ETHYLCARBAMATE; ETHYL(2-(P-PHENOXY)ETHYL)CARBAMATE; ETHYL(2-(4-PHENOXYPHENOXY)ETHYL)CARBAMATE; INSEGAR; LOGIC; (2-(4-PHENOXYPHENOXY)ETHYL)CARBAMIC ACID ETHYL ESTER; PICTYL; PYCTYL; RO 13-5223; TORUS; VARIKILL
Description: colorless crystals
Use: control of lepidoptera, scale insects, and psyllids on fruit (including citrus), cotton, and ornamentals; coleoptera and lepidopera in stored products; control of cockroaches, fleas, mosquito larvae, and fire ants in public health situations; control of platynota idaeusalis

Physical Properties

Freezing Point: 53 °C (127.4 °F) to 54 °C (129.2 °F)
Saturated Vapor Density: 1.389090785 kg/m^3
Vapor Pressure: 0.0017 mPa at 25 °C
Water Solubility: 6 mg/kg at 20 °C
Other Solubilities: Solubility (20 °C) 5 g/kg hexane > 250 g/kg most organic solvents.

RTECS Toxicity Data

Acute Oral: Rat LD_{50} Dose: 16800 mg/kg. Mouse LD_{50} Dose: >5 gm/kg.
Acute Inhalation: Rat LC_{50} Dose: >480 mg/m^3.
Acute Dermal: Rat LD_{50} Route: Skin; Dose: >2 gm/kg.

Hazard Overviews

Carcinogenicity: IARC - Not listed; NIOSH - Not listed; NTP - Not listed; ACGIH - Not listed; OSHA - Not listed; EPA - Not listed; MAK - Not listed

Environmental

Ecotoxicity: LC_{50} Rainbow trout ca 2 mg/l/96 hr /Conditions of bioassay not specified LD_{50} Japanese quail oral > 7000 mg/kg

Environmental Fate: If released to soil, it will rapidly biodegrade. The hydrolysis and photolysis in soil will not be important. The mobility in soil is expected to be low. If released to water, it will degrade rapidly due to biodegradation. Hydrolysis and photolysis of this insecticide do not appear to be important in water. The bioconcentration factor in bluegill sunfish has been experimentally determined to be 20 which suggests that bioconcentration in aquatic organisms is not an important fate process. However, low water solubility would suggest that bioconcentration may be important in aquatic organisms that can not metabolize it. If released to air, vapor-phase may react with photochemically-produced hydroxyl radicals with an estimated half-life of 2.9 hours. Particulate may be removed from the atmosphere by dry deposition.

Cleanup/Disposal: Guide No. 131: Fully encapsulating, vapor protective clothing should be worn for spills and leaks with no fire. Eliminate all ignition sources (no smoking, flares, sparks or flames in immediate area). All equipment used when handling the product must be grounded. Do not touch or walk through spilled material. Stop leak if you can do it without risk. Prevent entry into waterways, sewers, basements or confined areas. A vapor suppressing foam may be used to reduce vapors. Small Spills: Absorb with earth, sand or other non-combustible material and transfer to containers for later disposal. Use clean non-sparking tools to collect absorbed material. Large Spills: Dike far ahead of liquid spill for later disposal. Water spray may reduce vapor; but may not prevent ignition in closed spaces. Guide No. 151: Do not touch damaged containers or spilled material unless wearing appropriate protective clothing. Stop leak if you can do it without risk. Prevent entry into waterways, sewers, basements or confined areas. Cover with plastic sheet to prevent spreading. Absorb or cover with dry earth, sand or other non-combustible material and transfer to containers. Do not get water inside containers.

Environmental Physical Data

Henry's Law Constant: estimated at 8.42×10^{-7}

Sorption Partition Coefficient: $K_{OC} = 1622$

BCF: bluegill 21

Regulations

RCRA 40CFR: Not listed

CERCLA: 40CFR 302.4: Not listed

SARA 40CFR 372.65: Listed

SARA EHS 40CFR 355: Not listed

TSCA: Not listed

Analytical Methods

Food: FDA 232.3, 232.4, 242.1

FEN6130 CAS #: 64257-84-7

FENPROPATHRIN

RTECS: GZ2090500

Molecular Formula: $C_{22}H_{23}NO_3$

Formula Weight: 349.43

Synonyms: (R,S)-ALPHA-CYANO-3-PHENOXYBENZYL 2,2,3,3-TETRAMETHYLCYCLOPROPANECARBOXYLATE; (RS)-ALPHA-CYANO-3-PHENOXYBENZYL2,2,3,3-TETRAMETHYLCYCLOPROPANECARBOXYLATE; ALPHA-CYANO-3-PHENOXYBENZYL2,2,3,3-TETRAMETHYLCYCLOPROPANECARBOXYLATE; CYANO(3-PHENOXYPHENYL)METHYL2,2,3,3-TETRAMETHYLCYCLOPROPANECARBOXYLATE; DANITOL; ORTHO DANITOL; FENPROPANATE; FENPROPATHRINE; MEOTHRIN; RODY; S 3206; SD 41706; (+-)-2,2,3,3-TETRAMETHYLCYCLOPROPANECARBOXYLIC ACIDCYANO(3-PHENOXYPHENYL)METHYL ESTER; 2,2,3,3-TETRAMETHYLCYCLOPROPANECARBOXYLIC ACIDCYANO(3-PHENOXYPHENYL)METHYL ESTER; WL 41706

Description: pale yellow to yellow brown liquid or solid

Use: insecticide, acaracide; control of mites (except rust mites) & insects (eg whiteflies, lepidopterous larvae, leaf miners, leafworms, bollworms, leafrollers, armyworms, cabbageworms, cabbage looper, aphids, tortrixies, psyllas, bugs, fruit moths, tuberworms, cutworms, budworms, diamondback moth, mosquito bugs) on pome fruit, citrus fruit, vines, vegetables, ornamentals (incl trees), cotton, field crops, & glasshouse crops (cucurbits, tomatoes, ornamentals).

Physical Properties

Freezing Point: 45 °C (113 °F) to 50 °C (122 °F)

Specific Gravity: 1.15 at 25 °C

Vapor Pressure: 0.73 mPa at 20 °C

Water Solubility: 0.33 mg/L at 25 °C

Other Solubilities: In Xylene 1000, cyclohexanone 1000, Methanol 337 (all in g/kg at 25 °C)

Refraction Index: 1.5283 at 26 °C/D

Flash Point: Burns with difficulty

Hazard Overviews

Fire: Will burn.

Carcinogenicity: IARC - Not listed; NIOSH - Not listed; NTP - Not listed; ACGIH - Not listed; OSHA - Not listed; EPA - Not listed; MAK - Not listed

Environmental

Ecotoxicity: LC_{50} Bluegill sunfish 1.95 ug/l/48 hr /Conditions of bioassay not specified LD_{50} Mallard ducks oral 1089 mg/kg LC_{50} Bobwhite quail dietary >10,000 mg/kg diet/8 day

Regulations

RCRA 40CFR: Not listed

CERCLA: 40CFR 302.4: Not listed

SARA 40CFR 372.65: Not listed

SARA EHS 40CFR 355: Not listed

TSCA: Not listed

FEN6700 **CAS #: 115-90-2**

FENSULFOTHION

RTECS: TF3850000
DOT: UN2783
EINECS Number: 204-114-3
Molecular Formula: $C_{11}H_{17}O_4PS_2$
Structured MF: $C_{11}H_{17}O_4PS_2$ $((C_2H_5O)_2P(S)OC_6H_4S(O)CH_3)$
Formula Weight: 308.35
Synonyms: BAY 25141; BAYER 25141; CHEMAGRO 25141; DACONIT; DASANIT; O,O-DIAETHYL-O-4-METHYLSULFINYL-PHENYL-MONOTHIOPHOSPHAT; O,O-DIAETHYL-O-(4-METHYLSULFINYL-PHENYL)-THIONOPHOSPHAT; O,O-DIETHYL O-(4-(METHYLSULFINYL)PHENYL) PHOSPHOROTHIOATE; O,O-DIETHYL O-(P-(METHYLSULFINYL)PHENYL) PHOSPHOROTHIOATE; DIETHYL P-METHYLSULFINYLPHENYL THIOPHOSPHATE; O,O-DIETHYL O-P-(METHYLSULFINYL)PHENYL THIOPHOSPHATE; P,O-DIETHYL O-P-(METHYLSULFINYL)PHENYL THIOPHOSPHATE; O,O-DIETHYL O-(P-METHYLSULFINYL)PHENYL)PHOSPHOROTHIOATE; ENT 24,945; OMS 37; PHENOL,P-(METHYLSULFINYL)-,O-ESTER WITH O,O-DIETHYLPHOSPHOROTHIOATE; PHOSPHOROTHIOIC ACID,O,O-DIETHYLO-(4-(METHYLSULFINYL)PHENYL) ESTER; PHOSPHOROTHIOIC ACID,O,O-DIETHYLO-(P-(METHYLSULFINYL)PHENYL) ESTER; S 767; TERRACUR P; VUAGT 108; VUAGT 96
Description: yellowish-brown oily liquid
Use: insecticide against soil insects; mosquito larvicide; insecticide & nematocide principally for tobacco, corn, and various other crops

Physical Properties

Boiling Point: 138 °C (280 °F) to 141 °C (286 °F) at 0.01 mm Hg
Specific Gravity: 1.202 at 20 °C/4 °C
Water Solubility: 2 g/L at 25 °C
Other Solubilities: Soluble in most organic solventsexcept aliphatics
Refraction Index: 1.540 at 25 °C/D
Flash Point: Combustible Liquid

RTECS Toxicity Data

Acute Oral: Rat LD_{50} Dose: 2200 ug/kg. Guinea Pig LD_{50} Dose: 9 mg/kg; Toxic Effects: Biochemical - True cholinesterase.
Acute Inhalation: Rat LC_{50} Dose: 113 mg/m^3/1hr.
Acute Dermal: Rat LD_{50} Route: Skin; Dose: 3 mg/kg. Duck LD_{50} Route: Skin; Dose: 3 mg/kg.

Hazard Overviews

Poison

Fire Diamond

Health: Irritating to eyes/skin. Poison. Also Causes: salivation, watering eyes, nausea, vomiting, diarrhea, pinpoint pupils, blurred vision, hypertension, cardiac arrythmias, muscle cramps, weakness, agitation, headache, giddiness, dizziness, incoordination, confusion, delirium, seizures, difficulty breathing, respiratory paralysis.
Fire: Noncombustible. Use agent suitable for surrounding fire.
Reactivity: Unstable, readily oxidizes in air to form sulfone and isomerizes to the S-ethyl isomer. Hazardous polymerization cannot occur. Avoid: exposure to alkalis. Incompatible with: alkalis. Hazardous decomposition products: carbon oxide(s); phosphorus oxide(s); sulfur oxide(s).
Carcinogenicity: IARC - Not listed; NIOSH - Not listed; NTP - Not listed; ACGIH - Class A4, Not classifiable as a human carcinogen; OSHA - Not listed; EPA - Not listed; MAK - Not listed
Primary Target Organs:

Skin

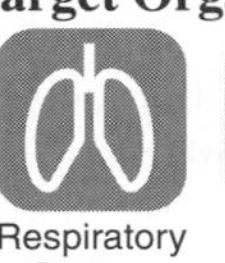
Respiratory System

Nervous System

Cardio-vascular

Blood

Exposure Limits
OSHA PEL Vacated 1989 Limits: TWA: 0.1 mg/m^3.
ACGIH TLV: TWA: 0.1 mg/m^3.
NIOSH REL: TWA: 0.1 mg/m^3.
Respirator Recommendation
Exposure Range: >0.1 to 5 mg/m^3 Supplied Air, Constant Flow/Pressure Demand, Half Mask
Exposure Range: >5 to 100 mg/m^3 Supplied Air, Constant Flow/Pressure Demand, Full Face
Exposure Range: >100 to unlimited mg/m^3 Self-contained Breathing Apparatus, Pressure Demand, Full Face
Note: odor threshold unknown

Environmental

Ecotoxicity: LC_{50} Bobwhite 35 ppm in 5-day diet (95% confidence limit 29-43 ppm), age 10 mo LC_{50} (96 hr): for Rainbow trout 8.8 mg/l; for Golden orfe 6.8 mg/l LD_{50} Mallard oral 0.749 mg/kg (95% confidence limit 0.595-0.944 mg/kg), 5-7 mo old females LC_{50} Gammarus fasciatus 0.1 mg/l at 15 °C (95% confidence limit 0.007-0.014 ppm), age mature LD_{50} Sharp-tailed grouse oral 0.500-1.00 mg/kg, males
Environmental Fate: If released to soil, it will be transformed by microbial degradation and chemical hydrolysis. Adaptation of soil microbes has been demonstrated to be important which significantly shortens the persistence upon annual reapplication. The estimated K_{oc} values of 77 and 105 suggest that it may be susceptible to significant leaching in many soil types. A correlation between increased adsorption with increasing soil organic content has been demonstrated. In soils of very high organic content (65%), such as onion fields in which it has pesticide use, it has been shown to be retained in the upper soil layers. Volatilization from soil is not expected to be significant. The initial persistence half-life in soil which has not received previous treatment has been observed to vary from less than one week to several weeks. If released to the aquatic environment, it will hydrolyze; the measured hydrolysis half-life in pure water at 25 °C is 58-87 days over the pH range of 4.5 to 8.0. Microbial degradation in

natural water may also be important, especially under anaerobic conditions. It is not expected to volatilize, bioconcentrate in aquatic organisms, or adsorb to sediments significantly in natural waters. If released to the atmosphere, vapor-phase will react rapidly with a resultant estimated half-life of 7.03 hours at 25 °C in atmosphere with a typical average concentration of 8 x10^5 hydroxyl radicals/cu m. However, it will not react with ozone. If released to air in solid-phase, such as dusts or mists generated by pesticide application, it is likely to be removed from the air by dry deposit. Localized removal by rainfall washout may be possible.

Cleanup/Disposal: Guide No. 152: Do not touch damaged containers or spilled material unless wearing appropriate protective clothing. Stop leak if you can do it without risk. Prevent entry into waterways, sewers, basements or confined areas. Cover with plastic sheet to prevent spreading. Absorb or cover with dry earth, sand or other non-combustible material and transfer to containers. Do not get water inside containers.

Environmental Physical Data

Henry's Law Constant: estimated at 1.8 x10^{-10}
Octanol/Water Partition Coefficient: log K_{ow} = 2.23
Sorption Partition Coefficient: K_{oc} = estimated at 77
BCF: estimated at 1.46

Regulations

RCRA 40CFR: Not listed
CERCLA: 40CFR 302.4: Not listed
SARA 40CFR 372.65: Not listed TPQ: 500 lb
SARA EHS 40CFR 355: Listed TPQ: 500 lb
TSCA: Not listed

Analytical Methods

Soil: EPA PMD-TLC; SW846 8140, 8141, 8141A, 8270B, 8270C, 8321, 8321A
Water / Groundwater: EPA 1657, 622; SW846 8141, 8141A
Food: FDA 212.1, 232.1, 232.3, 232.4, 242.1; AOAC 983.09, 986.07
Plasma: EPA 001, 028, 29; FDA 211.1, 231.1, 252

FEN7270 **CAS #: 437-38-7**

FENTANYL

RTECS: UE5550000
EINECS Number: 207-113-6
Molecular Formula: $C_{22}H_{28}N_2O$
Formula Weight: 336.46
Synonyms: FENTANEST; FENTANIL; FENTANILA; FENTANYLUM; LEPTANAL; PENTANYL; 1-PHENETHYL-4-(N-PHENYLPROPIONAMIDO)PIPERIDINE; N-(1-PHENETHYL-4-PIPERIDINYL)-N-PHENYLPROPIONAMIDE; N-(1-PHENETHYL-4-PIPERIDYL)PROPIONANILIDE; 1-PHENETHYL-4-N-PROPIONYLANILINOPIPERIDINE; N-PHENETHYL-4-(N-PROPIONYLANILINO)PIPERIDINE; PHENTANYL; N-PHENYL-N-[1-(2-PHENYLETHYL)-4-PIPERIDINYL]PROPANAMIDE; PROPANAMIDE,N-PHENYL-N-(1-(2-PHENYLETHYL)-4-PIPERIDINYL)-; PROPANAMIDE,N-PHENYL-N-(1-(2-PHENYLETHYL)-4-PIPERIDINYL)-(9CI); PROPIONANILIDE,N-(1-PHENETHYL-4-PIPERIDYL)-; R 4263; SENTONIL
Description: crystals; odorless
Use: medication: synthetic anesthetic/analgesic; medication (vet): analgesic/tranquilization

Physical Properties

Freezing Point: 83 °C (181.4 °F) to 84 °C (183.2 °F)
Other Solubilities: Soluble in Methanol; sparingly Soluble in Chloroform.

RTECS Toxicity Data

Acute Oral: Rat LD_{50} Dose: 18 mg/kg. Mouse LD_{50} Dose: 368 mg/kg.
Acute Dermal: Rat LD_{50} Route: Subcutaneous Dose: 1500 ug/kg. Mouse LD_{50} Route: Subcutaneous Dose: 62 mg/kg.

Hazard Overviews

Poison

Health: Poison. Other Acute Effects: may be fatal if inhaled, swallowed, or absorbed through skin; may cause allergic reaction; exposure can cause light-headedness, dizziness, sedation, nausea, vomiting, sweating, respiratory depression, apnea, circulatory depression, respiratory arrest, shock and cardiac arrest. Chronic Effects: prolonged or repeated exposure can lead to habituation or addiction; target organs: central nervous system, heart.
Fire: Hazards: emits toxic fumes. Extinguishing agents: carbon dioxide, dry chemical powder or appropriate foam; water spray. Precautions: combustible liquid.
Reactivity: Stable. Hazardous polymerization will not occur. Hazardous decomposition products: toxic fumes of: carbon monoxide, carbon dioxide, nitrogen oxides.
Carcinogenicity: IARC - Not listed; NIOSH - Not listed; NTP - Not listed; ACGIH - Not listed; OSHA - Not listed; EPA - Not listed; MAK - Not listed
Primary Target Organs:

Nervous System

Cardio-vascular

Environmental

Regulations

RCRA 40CFR: Not listed
CERCLA: 40CFR 302.4: Not listed
SARA 40CFR 372.65: Not listed
SARA EHS 40CFR 355: Not listed
TSCA: Not listed

FEN7840 CAS #: 55-38-9

FENTHION

RTECS: TF9625000
DOT: UN2783; UN2784; UN3017; UN3018; IMO3.2; IMO6.1
EINECS Number: 200-231-9
Molecular Formula: $C_{10}H_{15}O_3PS_2$
Structured MF: $(CH_3O)_2PSOC_6H_3CH_3SCH_3$
Formula Weight: 278.34
Synonyms: B 29493; BAY 29493; BAY-BASSA; BAYCID; BAYER S-1752; BAYER 29493; BAYER 29 493; BAYER 9007; BAYTEX; M-CRESOL,4-(METHYLTHIO)-,O-ESTER WITH O,O-DIMETHYL PHOSPHOROTHIOATE; M-CRESOL,4-(METHYLTHIO)-,O-ESTER WITH O,O-DIMETHYLPHOSPHOROTHIOATE; DALF; O,O-DIMETHYL O-4-(METHYLMERCAPTO)-3-METHYLPHENYLPHOSPHOROTHIOATE; O,O-DIMETHYL)-(4-METHYLMERCAPTO-3-METHYLPHENYL)THIONOPHOSPHATE; O,O-DIMETHYL O-(3-METHYL-4-METHYLMERCAPTOPHENYL)PHOSPHOROTHIOATE; O,O-DIMETHYL O-3-METHYL-4-METHYLTHIOPHENYL PHOSPHOROTHIOATE; O,O-DIMETHYL O-(3-METHYL-4-METHYLTHIOPHENYL) THIOPHOSPHATE; O,O-DIMETHYL O-3-METHYL-4-METHYLTHIOPHENYLPHOSPHOROTHIOATE; O,O-DIMETHYL O-[3-METHYL-4-(METHYLTHIO)PHENYL]PHOSPHOROTHIOATE; O,O-DIMETHYL O-(4-METHYLTHIO-3-METHYLPHENYL) THIOPHOSPHATE; O,O-DIMETHYL O-(4-METHYLTHIO-3-METHYLPHENYL)PHOSPHOROTHIOATE; O,O-DIMETHYL O-[4-(METHYLTHIO)PHENYL] PHOSPHOROTHIOATE; O,O-DIMETHYL O-(4-(METHYLTHIO)-M-TOLYL) PHOSPHOROTHIOATE; O,O-DIMETHYL O-4-METHYLTHIO-M-TOLYL PHOSPHOROTHIOATE; O,O-DIMETHYL-O-4-(METHYLMERCAPTO)-3-METHYLPHENYL PHOSPHOROTHIOATE; O,O-DIMETHYL-O-4-(METHYLMERCAPTO)-3-METHYLPHENYL THIOPHOSPHATE; O,O-DIMETHYL-O-4-(METHYLMERCAPTO)-3-METHYLPHENYLPHOSPHOROTHIOATE; O,O-DIMETHYL-O-4-(METHYLMERCAPTO)-3-METHYLPHENYLTHIOPHOSPHATE; O,O-DIMETHYL-O-(3-METHYL-4-METHYLMERCAPTOPHENYL)PHOSPHOROTHIOATE; O,O-DIMETHYL-O-(3-METHYL-4-METHYLTHIO-FENYL)-MONOTHIOFOSFAAT; O,O-DIMETHYL-O-3-METHYL-4-METHYLTHIOPHENYL PHOSPHOROTHIOATE; O,O-DIMETHYL-O-(3-METHYL-4-METHYLTHIOPHENYL)-MONOTHIOPHOSPHAT; O,O-DIMETHYL-O-(3-METHYL-4-METHYLTHIO)PHENYL)PHOSPHOROTHIOATE; O,O-DIMETHYL-O-(3-METHYL-4-METHYLTHIO-PHENYL)-THIONOPHOSPHAT; O,O-DIMETHYL-O-(4-(METHYLTHIO)-M-TOLYL)PHOSPHOROTHIOATE; O,O-DIMETHYL-(O-(4-METHYLTHIO)-M-TOLYL)PHOSPHOROTHIOIC ACID ESTER; O,O-DIMETHYLO-(3-METHYL-4-METHYLTHIO-FENYL)-MONOTHIOFOSFAAT; O,O-DIMETIL-O-(3-METIL-4-METILTIO-FENIL)-MONOTIOFOSFATO; DMTP; ENT 25,540; ENTEX; FENTHION 4E; FIGURON; HEBAYCID; LEBAYCID; MERCAPTOPHOS; 4-METHYLMERCAPTO-3-METHYLPHENYL DIMETHYL THIOPHOSPHATE; MPP; OMS 2; PHENTHION; PHOSPHOROTHIOIC ACID O,O-DIMETHYL O-(3-METHYL-4-(METHYLTHIO)PHENYL) ESTER; PHOSPHOROTHIOIC ACID O,O-DIMETHYLO-[3-METHYL-4-(METHYLTHIO)PHENYL] ESTER; PHOSPHOROTHIOIC ACID,O,O-DIMETHYL-,O-(4-METHYLTHIO)-M-TOLYL ESTER; PHOSPHOROTHIOIC ACID,O,O-DIMETHYLO-(3-METHYL-4-(METHYLTHIO)PHENYL) ESTER; QUELETON; QUELETOX; S 1752; SPOTTON; TALODEX; THIOPHOSPHATE DE O,O-DIMETHYLE ET DE O-(3-METHYL-4-METHYLTHIOPHENYLE); THIOPHOSPHATE DE O,O-DIMETHYLE ET DEO-(3-METHYL-4-METHYLTHIOPHENYLE); TIGUVON
Description: yellow, tan oily liquid; garlic odor
Use: insecticide, acaricide

Physical Properties

Boiling Point: 87 °C (189 °F) at 0.01 mm Hg
Freezing Point: 7 °C (44.6 °F)
Specific Gravity: 1.25 at 20 °C/4 °C
Density: 1.25 g/mL
Vapor Pressure: 4 mPa at 20 °C
Water Solubility: 0.006% by weight
Other Solubilities: 95% Ethanol: Soluble (>=10 mg/ml at 22.5 °C); Acetone: Soluble; DMSO: Soluble (>=10 mg/ml at 22.5 °C); Ether: Soluble.
Refraction Index: 1.5698 at 20 °C/D
Flash Point: Not available; probably combustible

RTECS Toxicity Data

Acute Oral: Man TD_{Lo} Dose: 643 uL/kg; Toxic Effects: Peripheral nerve and sensation - Fasciculations; Behavioral - Ataxia; Cardiac - Pulse rate. Man TD_{Lo} Dose: 257 mg/kg; Toxic Effects: Cardiac - Pulse rate increased without fall in BP; Gastrointestinal - Hypermotility, diarrhea; Gastrointestinal - Nausea or vomiting. Woman TD_{Lo} Dose: 1200 uL/kg; Toxic Effects: Behavioral - Coma; Cardiac - Pulse rate; Gastrointestinal - Hypermotility, diarrhea. Woman TD_{Lo} Dose: 525 mg/kg; Toxic Effects: Sense organs and special senses - Miosis (pupilliary dilation); Behavioral - Muscle contraction or spasticity; Lungs, Thorax, or Respiration - Respiratory depression. Rat LD_{50} Dose: 180 mg/kg.
Acute Inhalation: Rat LC_{50} Dose: 800 mg/m^3/4hr. Mouse LC_{Lo} Dose: 1 gm/m^3/2hr.
Acute Dermal: Rat LD_{50} Route: Skin; Dose: 330 mg/kg; Toxic Effects: Behavioral - Tremor; Behavioral - Convulsions or effect on seizure threshold; Behavioral - Excitment. Mouse LD_{50} Route: Skin; Dose: 500 mg/kg. Mouse LD_{50} Route: Subcutaneous Dose: 144 mg/kg.
Chronic (Multiple Dose) Oral: Rat Dose: 172 mg/kg/6D-I; Toxic Effects: Peripheral Nerve and sensation - Structural change in nerve or sheath; Blood - Changes in bone marrow not included above; Biochemical - True cholinesterase. Bird Dose: 7500 ug/kg/5D-C; Toxic Effects: Behavioral - Food intake (animal); Nutritional and gross metabolic - Weight loss or decreased weight gain; DEATH.
Chronic (Multiple Dose) Dermal: Rat Route: Skin; Dose: 380 mg/kg/19D-I; Toxic Effects: Brain and coverings - Other degenerative changes; Blood - Changes in serum composition; Biochemical - True cholinesterase. Rat Route: Skin; Dose: 3225 mg/kg/43W-I; Toxic Effects: Brain and coverings - Other degenerative changes. Rat Route: Skin; Dose: 365 mg/kg/5D-I; Toxic Effects: DEATH. Rat Route: Subcutaneous; Dose: 5200 mg/kg/1Y-I; Toxic Effects: Sense organs and special senses - Retinal changes (pigmentary deposition, retinitis, other); Liver - Other changes; Biochemical - True cholinesterase. Rat Route: Subcutaneous; Dose: 260 mg/kg/13W-I; Toxic Effects: Sense organs and special senses - Retinal changes (pigmentary deposition, retinitis, other).

Reproductive/Teratogenic: Mouse Route: Oral; Dose: 1050 mg/kg; Duration: multigenerations; Effects on Fertility - Female fertility index; Effects on Newborn - Viability index; Weaning or lactation index. Mouse Route: Intraperitoneal; Dose: 40 mg/kg; Duration: female 11D of pregnancy; Effects on Embryo or Fetus - Fetotoxicity; Specific Developmental Abnormalities - Craniofacial (including nose and tongue); Musculoskeletal system.

Mutagenic: Hamster Sister Chromatid Exchenge; Cell Type: lung; Dose: 40 mg/L. Bacteria - S Typhimurium Mutations in Microorganisms; Dose: 333 ug/plate (-S9).

Tumorigenic: Mouse Route: Oral; Dose: 1730 mg/kg/103W-C; Toxic Effects: Tumorigenic - Equivocal tumorigenic agent by RTECS criteria; Skin and appendages - Tumors.

Hazard Overviews

Fire Diamond

Health: Toxic. Also Causes: nausea, vomiting, abdominal cramps, diarrhea, dizziness, fatigue, blurred vision, tearing; twitching, loss of muscle coordination, slurred speech, confusion, difficulty breathing, convulsions, coma, and death. Chest tightness, runny nose, and delayed pulmonary edema are common in inhalation exposures.

Fire: Noncombustible. Use extinguishing agents suitable for surrounding fire.

Reactivity: Stable. Hazardous polymerization cannot occur. Avoid: excessive heat. Incompatible with: alkalis (> pH 9); oxidizers. Hazardous decomposition products: sulfur and phosphorus oxides.

Carcinogenicity: IARC - Not listed; NIOSH - Not listed; NTP - Not listed; ACGIH - Class A4, Not classifiable as a human carcinogen; OSHA - Not listed; EPA - Not listed; MAK - Not listed

Primary Target Organs:

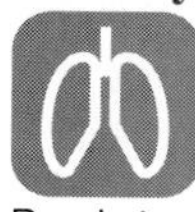
Respiratory System

Nervous System

Exposure Limits

OSHA PEL Vacated 1989 Limits: TWA: 0.2 mg/m^3.
ACGIH TLV: TWA: 0.2 mg/m^3.
DFG MAK: TWA: 0.2 mg/m^3.

Respirator Recommendation

Exposure Range: >0.2 to 10 mg/m^3 Supplied Air, Constant Flow/Pressure Demand, Half Mask

Exposure Range: >10 to 200 mg/m^3 Supplied Air, Constant Flow/Pressure Demand, Full Face

Exposure Range: >200 to unlimited mg/m^3 Self-contained Breathing Apparatus, Pressure Demand, Full Face

Note: poor warning properties

Environmental

Ecotoxicity: LD_{50} Bobwhite 30 ppm in 5 day diet (95% confidence limit 21-41 mg/kg) 10 day old LD_{50} Zenaida macroura (Mourning dove) oral 2.50 mg/kg (95% confidence limit 1.25-5.00 mg/kg) LD_{50} Anas platyrhynchos (Mallard duck) male oral 5.94 mg/kg (95% confidence limit 4.28-8.23 mg/kg) 4 mo old LC_{50} Gammarus lacustris (scud) 8.4 ug/l/96 hr (95% confidence limit 5.0-12.0 ug/l), mature. Static bioassay without aeration, pH 7.2-7.5, water hardness 40-50 mg/l as calcium carbonate and alkalinity of 30-35 mg/l LC_{50} Orconectes nais (crayfish) 50 ug/l (95% confidence limit 35-90 ug/l), early instar. Static bioassay without aeration, pH 7.2-7.5, water hardness 40-50 mg/l as calcium carbonate and alkalinity of 30-35 mg/l LC_{50} Pteronarcys californica (stonefly) 4.5 ug/l/96 hr (95% confidence limit 3.3-6.7 ug/l), second yr class. Static bioassay without aeration, pH 7.2-7.5, water hardness 40-50 mg/l as calcium carbonate and alkalinity of 30-35 mg/l LC_{50} Palaemonetes kadiakensis (glass shrimp) 10 ug/l/96 hr (95% confidence limit 7.1-14.0 ug/l), mature. Static bioassay without aeration, pH 7.2-7.5, water hardness 40-50 mg/l as calcium carbonate and alkalinity of 30-35 mg/l LC_{50} Oncorhynchus kisutch (coho salmon) 1320 ug/l/96 hr (95% confidence limit 1020-1680 ug/l), wt 0.4 g. Static bioassay without aeration, pH 7.2-7.5, water hardness 40-50 mg/l as calcium carbonate and alkalinity of 30-35 mg/l LC_{50} Salmo gairdneri (rainbow steelhead trout) 930 ug/l/96 hr (95% confidence limit 750-1150 ug/l), wt 1 g. Static bioassay without aeration, pH 7.2-7.5, water hardness 40-50 mg/l as calcium carbonate and alkalinity of 30-35 mg/l

Environmental Fate: If released to the atmosphere, it will degrade rapidly in the vapor-phase by reaction with photochemically produced hydroxyl radicals (half-life of about 5 hr). When released to soil or water, it will degrade through photodegradation and biodegradation; in the presence of sunlight, photodegradation will likely dominate. The persistence half life in water under field conditions has been reported to range from 2.9 to 21.1 days for various ocean, river, swamp, lake and canal waters. However, it may be more persistent in some environments, such as salt marsh sediments (below several mm deep), where light and oxygen are limited. The US Dept of Agriculture's Pesticide Properties Database lists a soil half-life of 34 days. It is expected to have very low soil mobility.

Cleanup/Disposal: Guide No. 131: Fully encapsulating, vapor protective clothing should be worn for spills and leaks with no fire. Eliminate all ignition sources (no smoking, flares, sparks or flames in immediate area). All equipment used when handling the product must be grounded. Do not touch or walk through spilled material. Stop leak if you can do it without risk. Prevent entry into waterways, sewers, basements or confined areas. A vapor suppressing foam may be used to reduce vapors. Small Spills: Absorb with earth, sand or other non-combustible material and transfer to containers for later disposal. Use clean non-sparking tools to collect absorbed material. Large Spills: Dike far ahead of liquid spill for later disposal. Water spray may reduce vapor; but may not prevent ignition in closed spaces. Guide No. 152: Do not touch damaged containers or spilled material unless wearing appropriate protective clothing. Stop leak if you can do it without risk. Prevent entry into waterways, sewers, basements or confined areas. Cover with plastic sheet to prevent spreading. Absorb or cover with dry earth, sand or other non-combustible material and transfer to containers. Do not get water inside containers.

Environmental Physical Data

Henry's Law Constant: estimated at 1.46 x10^{-6}
Octanol/Water Partition Coefficient: log K_{ow} = 4.09
Sorption Partition Coefficient: K_{oc} = estimated at 1400 to 4000
BCF: guppies 1.66 x10^{4}

Regulations

RCRA 40CFR: Not listed
CERCLA: 40CFR 302.4: Not listed
SARA 40CFR 372.65: Listed
SARA EHS 40CFR 355: Not listed
TSCA: Not listed

Analytical Methods

Soil: EPA PMD-TLC; SW846 8140, 8141, 8141A, 8270B, 8270C
Water / Groundwater: EPA P-005-1, 1657, 622, 022; SW846 8141, 8141A
Food: FDA 212.1, 232.1, 232.3, 232.4, 242.1
Plasma: EPA 001, 29; FDA 211.1, 231.1, 252

FEN8410 **CAS #: 101-42-8**

FENURON

RTECS: YT1450000
EINECS Number: 202-941-4
Molecular Formula: $C_9H_{12}N_2O$
Structured MF: $C_6H_5NHCON(CH_3)_2$
Formula Weight: 164.21

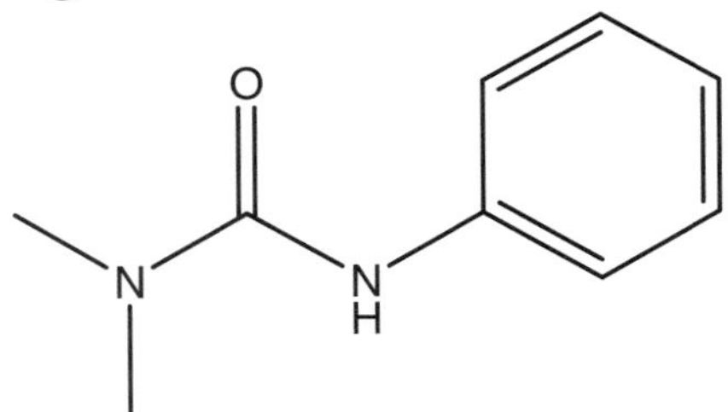

Chemical Structure

Synonyms: BEET-KLEEN; CROPTEX CHROME; CROPTEX RUBY; DIBAR; 1,1-DIMETHYL-3-PHENYLUREA; 1,2-DIMETHYL-3-PHENYLUREA; N,N-DIMETHYL-N'-PHENYLUREA; N,N-DIMETHYL-N-PHENYLUREA; DYBAR; ELECTRUM; FALISILVAN; FENIDIN; FENULON; HERBON YELLOW; OMICURE 94; PDU; 1-PHENYL-3,3-DIMETHYLUREA; 3(PHENYL)-1,1-DIMETHYLUREA; 3-PHENYL-1,1-DIMETHYLUREA; N-PHENYL-N',N'-DIMETHYLUREA; PREMALOX; PUD (HERBICIDE); QUINTEX; RED
Description: colorless to white, crystalline solid
Use: weed and brush killer; herbicide; control of woody plants and deep rooted perennial weeds, particularly on non crop land

Physical Properties

Freezing Point: 133 °C (271.4 °F) to 134 °C (273.2 °F)
Specific Gravity: 1.08 at 20 °C/20 °C
Vapor Pressure: 1.6 x10^{-4} mm Hg at 60 °C
Water Solubility: 3850 ppm at 25 °C
Other Solubilities: sparingly Soluble in hydrocarbons; In Ethanol 108.8, diethylether 5.5, Acetone 80.2, Benzene 3.1, Chloroform 125, n-hexane 0.2, groundnut oil 1.0 (all in g/kg at 20-25 °C).

RTECS Toxicity Data

Acute Oral: Rat LD_{50} Dose: 6400 mg/kg. Mouse LD_{50} Dose: 4700 mg/kg.
Chronic (Multiple Dose) Oral: Rat Dose: 13886 mg/kg/61W-C; Toxic Effects: Blood - Methemoglobinemia-Carboxyhemoglobin; Blood - Other changes. Guinea Pig Dose: 19264 mg/kg/43W-I; Toxic Effects: Kidney, Ureter, and Bladder - Other changes in urine composition; Endocrine - Evidence of thyroid hypofunction; Blood - Other changes.
Mutagenic: Mouse DNA Inhibition; Route: Oral; Dose: 500 mg/kg.

Hazard Overviews

Health: May be irritating to eyes/skin. Other Acute Effects: may be harmful by inhalation, ingestion, or skin absorption; absorption into the body leads to the formation of methemoglobin which in sufficient concentration causes cyanosis; onset may be delayed 2 to 4 hours or longer.
Fire: Hazards: emits toxic fumes. Extinguishing agents: water spray; carbon dioxide, dry chemical powder or appropriate foam. Precautions: combustible liquid.
Reactivity: Incompatible with: strong oxidizing agents. Hazardous decomposition products: toxic fumes of: carbon monoxide, carbon dioxide, nitrogen oxides.
Carcinogenicity: IARC - Not listed; NIOSH - Not listed; NTP - Not listed; ACGIH - Not listed; OSHA - Not listed; EPA - Not listed; MAK - Not listed

Environmental

Ecotoxicity: LC_{50} Guppy 610 mg/l/48 hr /Conditions of bioassay not specified
Environmental Fate: If released to the atmosphere, degradation of vapor-phase by reaction with photochemically produced hydroxyl radicals (estimated half-life of 2.3 hours) will be important. Photolysis may be important and particulate-phase will be removed from air via dry deposition. If released to soil, biodegradation will be the primary fate process. A complete degradation half-life of 2.2-4.5 months was measured in soil. Photolysis may be important on soil surfaces exposed to sunlight. Adsorption to soil will not be important. In water, volatilization, adsorption to sediment, hydrolysis and bioconcentration in aquatic organisms will not be important. Biodegradation is expected to be the primary fate process and some photodegradation may occur.

Environmental Physical Data

Henry's Law Constant: estimated at 9.71 $x10^{-10}$
Octanol/Water Partition Coefficient: log K_{ow} = 0.87
Sorption Partition Coefficient: K_{oc} = 27 to 43
BCF: estimated at 1 to 6

Regulations

RCRA 40CFR: Not listed
CERCLA: 40CFR 302.4: Not listed
SARA 40CFR 372.65: Not listed
SARA EHS 40CFR 355: Not listed
TSCA: Listed

Analytical Methods

Soil: SW846 8321A
Water / Groundwater: EPA 621, 632
Plasma: FDA 242.4

FEN8980 CAS #: 51630-58-1

FENVALERATE

RTECS: CY1576350
EINECS Number: 257-326-3
Molecular Formula: $C_{25}H_{22}ClNO_3$
Formula Weight: 419.92
Synonyms: AQMATRINE; BALMARK; BELMARK; 4-CHLORO-ALPHA-(1-METHYLETHYL)BENZENEACETIC ACIDCYANO(3-PHENOXYPHENYL)METHYL ESTER; ALPHA-CYANO-3-PHENOXYBENZYL 2-(4-CHLOROPHENYL)ISOVALERATE; ALPHA-CYANO-3-PHENOXYBENZYL2-(4-CHLOROPHENYL)-3-METHYLBUTYRATE; ALPHA-CYANO-3-PHENOXY-BENZYLALPHA-(4-CHLOROPHENYL)ISOVALERATE; ALPHA-CYANO-3-PHENOXY-BENZYLALPHA-ISOPROPYL-4-CHLOROPHENYLACETATE; (+)ALPHA-CYANO-3-PHENOXYBENZYL-(+)-ALPHA-(4-CHLOROPHENYL)ISOVALERATE; (RS)-ALPHA-CYANO-3-PHENOXYBENZYL(RS)-2-(4-CHLOROPHENYL)-3-METHYLBUTYRATE; ALPHA-CYANO-3-PHENOXYBENZYL-2-(4-CHLOROPHENYL)-3-METHYLBUTYRATE; ALPHA-CYANO-3-PHENOXYBENZYLISOPROPYL-4-CHLOROPHENYLACETATE; CYANO(3-PHENOXYPHENYL)METHYL 4-CHLORO-ALPHA-(1-METHYLETHYL)BENZENEACETATE; CYANO(3-PHENOXYPHENYL)METHYL4-CHLORO-ALPHA-(1-METHYLETHYL)BENZENE ACETATE; CYANO(3-PHENOXYPHENYL)METHYL4-CHLORO-ALPHA-(1-METHYLETHYL)BENZENEACETATE; (CYANO(3-PHENOXYPHENYL)METHYL-4-CHLORO-ALPHA-(1-METHYLETHYL)PHENYLACETATE); ECTRIN; EPA SHAUGHNESSY CODE: 109301; EVERCIDE 2362; FENKILL; FENVAL; FENVALERATE; GOLD CREST TRIBUTE; INSECTRAL; OMS-2000; PHENVALERATE; PYDRIN; PYRIDIN; S 5602; S-5602; SANMARTON; SD-43775; SD 43775; SUMIBAC; SUMICIDIN; SUMIFLEECE; SUMIFLY; SUMIPOWER; SUMITICK; SUMKIDIN; TIRADE; WL 43775; WL43775
Description: yellow, clear viscous liquid; mild chemical odor
Use: highly active contact insecticide effective against wide range of pests, incl strains resistant to organochlorine, organophosphorus, & carbamate insecticides; controls insects that attack leaves or fruits on crops, incl cotton, fruit, vegetables; control of chewing, sucking, & boring insects (particularly lepidoptera, diptera, orthoptera, hemiptera, & coleoptera) in fruit, vines, olives, hops,in public health & animal husbandry, controlling flies, ear ticks; med: (vet) ectoparasiticide.

Physical Properties

Boiling Point: Decomposes
Specific Gravity: 1.17 at 23 °C/4 °C
Saturated Vapor Density: 1.2 kg/m^3
Vapor Pressure: 1.1 $x10^{-8}$ mm Hg at 25 °C
Water Solubility: < 1 mg/L at 20 °C
Other Solubilities: Readily Soluble in most organic solvents, in Acetone, Ethanol, Chloroform, cyclohexanone, Xylene, all > 1 kg/kg at 23 °C.
Refraction Index: 1.5533 at 20 °C/D
Flash Point: Burns with difficulty

RTECS Toxicity Data

Mutagenic: Human Cytogenetic Analysis; Cell Type: lymphocyte; Dose: 2 mg/L. Human Sister Chromatid Exchenge; Cell Type: lymphocyte; Dose: 10 mg/L.

Hazard Overviews

Fire: Will burn.
Carcinogenicity: IARC - Group 3, Not classifiable as to carcinogenicity to humans; NIOSH - Not listed; NTP - Not listed; ACGIH - Not listed; OSHA - Not listed; EPA - Not listed; MAK - Not listed

Environmental

Ecotoxicity: LC_{50} Rainbow trout formulated product 21.0 ppb active ingredient/24 hr (static test) LC_{50} Rainbow trout 0.0036 mg/l/96 hr /Conditions of bioassay not specified LC_{50} Salmo salar (Atlantic salmon) 1 ug/l/96 hr, juvenile /Conditions of bioassay not specified LC_{50} Mallard dietary 5500 ppm /Conditions of bioassay not specified
Environmental Fate: If released to the atmosphere, it will degrade rapidly in the vapor phase by reaction with photochemically produced hydroxyl radicals (estimated half-life of 10 hr). If released to soil or water, it can degrade through biodegradation, photodegradation and aqueous hydrolysis. Screening studies have suggested that biodegradation is the primary route of degradation. Photodegradation may become the major degradation route on terrestrial surfaces (soil, plants, etc.) or shallow waters exposed to sunlight. Aqueous hydrolysis may become important when the medium pH exceeds 8. It is not expected to leach in soil. In aquatic ecosystems, it is expected to partition from the water column to sediment and suspended matter. It is reported to have half-lives of 1 to 18 days on soil surfaces, 15 days to 3 months within soil systems, 8 to 14 days on plants, and 4 to 15 days in natural water.

Environmental Physical Data

Henry's Law Constant: estimated at 1.19 $x10^{-7}$
Octanol/Water Partition Coefficient: log K_{ow} = 4.42
Sorption Partition Coefficient: K_{oc} = sediments 4.04 to 4.22
BCF: eastern oysters 4700

Regulations

RCRA 40CFR: Not listed

CERCLA: 40CFR 302.4: Listed as Compound per CWA Section 307(a) per CAA Section 112
SARA 40CFR 372.65: Listed
SARA EHS 40CFR 355: Not listed
TSCA: Not listed

Analytical Methods

Water / Groundwater: EPA 1660
Food: FDA 212.1, 232.1, 232.4, 242.1
Indoor / Expired Air: ASTM D4861
Plasma: FDA 211.1, 231.1, 252

FER1000 CAS #: 14484-64-1

FERBAM

RTECS: NO8750000
EINECS Number: 238-484-2
Molecular Formula: $C_9H_{18}FeN_3S_6$
Structured MF: $((CH_3)_2NCS_2)_3Fe$
Formula Weight: 416.50
Synonyms: AAFERTIS; AAFERZIMAG; BERCEMA FERTAM 50; CARBAMIC ACID,DIMETHYLDITHIO-,IRON SALT; CARBAMIC ACID,DIMETHYLDITHIO-,IRON(3+)SALT; CARBAMIC ACID,DIMETHYLELITHIA,IRON SALT; CARBANATE; CORMATE; DIMETHYL-CARBAMODITHIOIC ACID IRON(3+) SALT; DIMETHYLCARBAMODITHIOIC ACID,IRON COMPLEX; DIMETHYLCARBAMODITHIOIC ACID,IRON(3+) SALT; DIMETHYLDITHIOCARBAMIC ACID,FERRIC SALT; DIMETHYLDITHIOCARBAMIC ACID,IRON SALT; DIMETHYLDITHIOCARBAMIC ACID,IRON(3+) SALT; EISENDIMETHYLDITHIOCARBAMAT; EISEN(III)-TRIS(N,N-DIMETHYLDITHIOCARBAMAT); ENT 14,689; F 40; FERBAM 50; FERBAME; FERBAM,IRON SALT; FERBAN; FERBECK; FERBERK; FERMACIDE; FERMATE; FERMATE FERBAM FUNGICIDE; FERMOCIDE; FERRADOW; FERRIC DIMETHYL DITHIOCARBAMATE; FERRIC DIMETHYLDITHIOATE; FERRIC DIMETHYLDITHIOCARBAMATE; FUKLASIN; FUKLASIN ULTRA; FUKLAZIN; HEXAFERB; HOKMATE; IRON DIMETHYLDITHIOCARBAMATE; IRON TRIS(DIMETHYLDITHIOCARBAMATE); IRON(III) DIMETHYLDITHIOCARBAMATE; IRON,TRIS(DIMETHYLCARBAMODITHIOATO-S,S')-,(OC-6-11)-; IRON,TRIS(DIMETHYLDITHIOCARBAMATO)-; KARBAM BLACK; KNOCKMATE; LIROMATE; NIACIDE; (OC-6-11)-TRIS(DIMETHYLCARBAMODITHIOATO-S,S')IRON; STAUFFER FERBAM; SUP'R-FLO FERBAM FLOWABLE; TRIFUNGOL; TRIMANZONE; TRIS(DIMETHYLCARBAMODITHIOATE-S,S')IRON; TRIS(DIMETHYLCARBAMODITHIOATO-S,S')IRON; TRIS(DIMETHYLCARBAMODTHIOATO-S,S')IRON; TRIS(N,N-DIMETHYLDITHIOCARBAMATO) IRON(111); TRIS(DIMETHYLDITHIOCARBAMATO)IRON; TRIS(N,N-DIMETHYLDITHIOCARBAMATO)IRON(III); VANCIDE FE95
Description: dark brown to black solid or fluffy powder; odorless
Use: fungicide used on fruit, nuts, vegetables, ornamental crops, in household applications; protection of foliage against fungal pathogens incl taphrina deformans of peaches; fungicide for apple scab, cedar apple rust, peachleaf curl, tobacco blue mold, cranberry diseases; protective fungicide to other crops

Physical Properties

Boiling Point: Decomposes
Freezing Point: Decomposes at > 180 °C (356 °F)
Vapor Pressure: Negligible at room temperature
Water Solubility: 120 ppm in Water
Other Solubilities: Soluble in Acetone, Chloroform, Pyridine, Acetonitrile
pH: 5
Ionization Potential (eV): 7.72
Flash Point: Decomposition products are flammable

RTECS Toxicity Data

Acute Oral: Rat LD_{50} Dose: 1130 mg/kg. Mouse LD_{50} Dose: 3400 mg/kg; Toxic Effects: Behavioral - Somnolence (general depressed activity); Behavioral - Excitment; Behavioral - Ataxia.
Chronic (Multiple Dose) Oral: Rat Dose: 45 gm/kg/30D-C; Toxic Effects: Nutritional and gross metabolic - Weight loss or decreased weight gain; DEATH. Rat Dose: 91250 mg/kg/2Y-C; Toxic Effects: Brain and coverings - Other degenerative changes; DEATH.
Reproductive/Teratogenic: Rat Route: Oral; Dose: 1140 mg/kg; Duration: female 6-15D of pregnancy; Effects on Fertility - Litter size; Effects on Embryo or Fetus - Fetotoxicity. Rat Route: Oral; Dose: 536 mg/kg; Duration: female 16-22D of pregnancy Effects on Newborn - Growth statistics. Rat Route: Oral; Dose: 3479 mg/kg; Duration: female 16-22D of pregnancy Effects on Newborn - Viability index; Weaning or lactation index.
Mutagenic: Mouse Sperm Morphology; Route: Oral; Dose: 5 gm/kg/5D-C. Bacteria - B Subtilis DNA Repair; Dose: 2 ug/disc.
Tumorigenic: Mouse Route: Oral; Dose: 3500 mg/kg/78W-I; Toxic Effects: Tumorigenic - Equivocal tumorigenic agent by RTECS criteria; Lungs, Thorax, or Respiration - Tumors; Liver - Tumors. Mouse Route: Subcutaneous; Dose: 100 mg/kg; Toxic Effects: Tumorigenic - Carcinogenic by RTECS criteria; Blood - Tumors.

Hazard Overviews

Flammable

Fire: Flammable.
Carcinogenicity: IARC - Group 3, Not classifiable as to carcinogenicity to humans; NIOSH - Not listed; NTP - Not listed; ACGIH - Class A4, Not classifiable as a human carcinogen; OSHA - Not listed; EPA - Not listed; MAK - Not listed
Exposure Limits
OSHA PEL: TWA: 15 mg/m^3; total dust.
OSHA PEL Vacated 1989 Limits: TWA: 10 mg/m^3.
ACGIH TLV: TWA: 10 mg/m^3.
NIOSH REL: TWA: 10 mg/m^3; as Fe (Salts).
NIOSH IDLH: 800 mg/m^3.
DFG MAK: TWA: 15 mg/m^3.

Respirator Recommendation

Exposure Range: >15 to 150 mg/m^3 Air Purifying, Negative Pressure, Half Mask

Exposure Range: >150 to <800 mg/m^3 Air Purifying, Negative Pressure, Full Face

Exposure Range: 800 to unlimited mg/m^3 Self-contained Breathing Apparatus, Pressure Demand, Full Face

Cartridge Color: dust/mist filter (use P100 or consult supervisor for appropriate dust/mist filter)

Environmental

Environmental Fate: If released to the atmosphere, it will exist as an aerosol and will be removed from the air physically through dry and wet deposition. In soil, it is expected to have moderate mobility. Under acidic conditions, it will almost completely degrade within 4 to 5 weeks; under neutral conditions, it will require 14 to 15 weeks to degrade. It will biodegrade in soil with a half life of 8 weeks. It is not expected to volatilize from water surfaces. It may hydrolyze under acidic conditions. This compound is not expected to bioconcentrate in aquatic organisms.

Environmental Physical Data

Sorption Partition Coefficient: K_{oc} = 300

BCF: calculated at 40

Regulations

RCRA 40CFR: Listed Hazardous Waste No. U396 Toxic Waste

CERCLA: 40CFR 302.4: Listed per RCRA Section 3001 RQ: 1 lb (0.454 kg)

SARA 40CFR 372.65: Listed

SARA EHS 40CFR 355: Not listed

TSCA: Listed

Analytical Methods

Water / Groundwater: EPA 630, 630.1

Food: AOAC 965.15

FER1310 CAS #: 1185-57-5

FERRIC AMMONIUM CITRATE

RTECS: GE7540000

EINECS Number: 214-686-6

Molecular Formula: $C_6H_{(3x+8)}Fe_xN_xO_7$

Structured MF: Mixture of $FeC_6H_5O_7$, $(NH_4)2HC_6H_5O_7$ and water of hydration

Formula Weight: 709.44

HO O
O
HO
OH
HO O
Fe
NH_3

Chemical Structure

Synonyms: AMMONIUM FERRIC CITRATE; AMMONIUM IRON (III) CITRATE; AMMONIUM IRON(III) CITRATE; CITRIC ACID,AMMONIUM IRON(3+) SALT; FAC; IRON AMMONIUM CITRATE; 1,2,3-PROPANETRICARBOXYLIC ACID,2-HYDROXY-,AMMONIUMIRON(3+) SALT; 1,2,3-PROPANETRICARBOXYLIC ACID,2-HYDROXY-,AMMONIUMIRON(3+) SALT (9CI)

Description: brown form - reddish-brown granules, garnet-red transparent scales, or brownish-yellow powder; green form - green transparent, scales, pearls, granules, or powder; odorless or slight ammonia odor

Use: for blueprints; in photography; feed additive; ferric ammonium citrate, med: hematinic, source of nutrient iron, particularly for milk fortification, (vet) iron deficiency anemia, astringents for skin disorders; drug colorant (exempted from fda certifications); water-soluble anticaking agent at very low concentrations for salt for food use.

Physical Properties

Boiling Point: Decomposes at 1 atm

Specific Gravity: 1.8 at 20 °C/4 °C

Water Solubility: Extremely Soluble in Water

Other Solubilities: Practically Insoluble in Alcohol

Flash Point: Nonflammable

Hazard Overviews

Health: Irritating to eyes/skin/respiratory tract. Other Acute Effects: may be harmful by inhalation, ingestion, or skin absorption; overdose may have a corrosive effect on the gastrointestinal mucosa and be followed by necrosis, perforation and stricture formation; several hours may elapse before symptoms that can include epigastric pain, diarrhea, vomiting, nausea and hematemesis occur; after apparent recovery a person may experience metabolic acidosis, convulsions and coma hours or days later; further complications may develop leading to acute liver necrosis that can result in death due to hepatic coma.

Fire: Noncombustible. Hazards: emits toxic fumes. Extinguishing agents: water spray; carbon dioxide, dry chemical powder or appropriate foam. Precautions: combustible liquid.

Reactivity: Stable. Hazardous polymerization will not occur. Incompatible with: strong oxidizing agents, may decompose on exposure to light. Hazardous decomposition products: toxic fumes of: carbon monoxide, carbon dioxide, nitrogen oxides, ammonia, iron and iron salts.

Carcinogenicity: IARC - Not listed; NIOSH - Not listed; NTP - Not listed; ACGIH - Not listed; OSHA - Not listed; EPA - Not listed; MAK - Not listed

Primary Target Organs:

Exposure Limits

OSHA PEL Vacated 1989 Limits: STEL: 1 mg/m^3; as Fe salts.

ACGIH TLV: TWA: 1 mg/m^3; as Fe Salt.

NIOSH REL: TWA: 1 mg/m^3; as Fe salts.

Respirator Recommendation

Exposure Range: >1 to 10 mg/m^3 Air Purifying, Negative Pressure, Half Mask

Exposure Range: >10 to 100 mg/m^3 Air Purifying, Negative Pressure, Full Face

Exposure Range: >100 to 1000 mg/m^3 Supplied Air, Constant Flow/Pressure Demand, Full Face

Exposure Range: >1000 to unlimited mg/m^3 Self-contained Breathing Apparatus, Pressure Demand, Full Face

Cartridge Color: dust/mist filter (use P100 or consult supervisor for appropriate dust/mist filter)

Note: as iron salts, soluble

Environmental

Environmental Physical Data

BCF: no food chain concentration potential

BOD: sewage 0.40 lb/lb, 5 days

Regulations

RCRA 40CFR: Not listed

CERCLA: 40CFR 302.4: Listed per CWA Section 311(b)(4) RQ: 1000 lb (453.5 kg)

SARA 40CFR 372.65: Not listed

SARA EHS 40CFR 355: Not listed

TSCA: Listed

FER1620 CAS #: 14221-47-7

FERRIC AMMONIUM OXALATE

RTECS: LJ8932000

DOT: NA9119

EINECS Number: 238-090-0

Molecular Formula: $C_6H_{12}FeN_3O_{12}$

Formula Weight: 374.06

Synonyms: AMMONIUM FERRIC OXALATE; AMMONIUM FERRIOXALATE; AMMONIUM TRIOXALATOFERRATE(III); FERRATE(3-),TRIS(ETHANEDIOATO(2-)-O,O')-,TRIAMMONIUM,(OC-6-11)-; FERRATE(3-),TRIS(ETHANEDIOATO(2-)-O,O')-,TRIAMMONIUM,(OC-6-11)-(9CI); FERRATE(3-),TRIS(OXALATO)-,TRIAMMONIUM; TRIAMMONIUM TRIS-[ETHANEDIOATO(2-)-O,O'] FERRATE(3-); TRIAMMONIUM TRIS-(ETHANEDIOATO(2-)-O,O')FERRATE(3-1); TRIAMMONIUM TRIS(OXALATO)FERRATE(3-)

Description: yellowish green to bright-green, crystals or solid; slight burnt sugar odor

Use: in photography, blueprints; coloring of aluminum & aluminum alloys

Physical Properties

Boiling Point: Decomposes at 160 °C (320 °F) to 170 °C (338 °F)

Freezing Point: Decomposes at 160 °C (320 °F) to 170 °C (338 °F)

Density: Hydrate 1.78 at 17 °C

Water Solubility: Very Soluble in Water

RTECS Toxicity Data

Irritation Eye: Rabbit Standard Draize Test Dose: 100 mg; Reaction: moderate. Rabbit Nonstandard Exposure Dose: 100 mg/4S rinse; Reaction: mild.

Irritation Skin: Rabbit Standard Draize Test Dose: 500 mg; Reaction: moderate.

Hazard Overviews

Carcinogenicity: IARC - Not listed; NIOSH - Not listed; NTP - Not listed; ACGIH - Not listed; OSHA - Not listed; EPA - Not listed; MAK - Not listed

Exposure Limits

OSHA PEL Vacated 1989 Limits: STEL: 1 mg/m^3; as Fe salts.

ACGIH TLV: TWA: 1 mg/m^3; as Fe Salt.

NIOSH REL: TWA: 1 mg/m^3; as Fe salts.

Respirator Recommendation

Exposure Range: >1 to 10 mg/m^3 Air Purifying, Negative Pressure, Half Mask

Exposure Range: >10 to 100 mg/m^3 Air Purifying, Negative Pressure, Full Face

Exposure Range: >100 to 1000 mg/m^3 Supplied Air, Constant Flow/Pressure Demand, Full Face

Exposure Range: >1000 to unlimited mg/m^3 Self-contained Breathing Apparatus, Pressure Demand, Full Face

Cartridge Color: dust/mist filter (use P100 or consult supervisor for appropriate dust/mist filter)

Note: as iron salts, soluble

Environmental

Cleanup/Disposal: Guide No. 171: Do not touch or walk through spilled material. Stop leak if you can do it without risk. Prevent dust cloud. Avoid inhalation of asbestos dust. Small Dry Spills: With clean shovel place material into clean, dry container and cover loosely; move containers from spill area. Small Spills: Take up with sand or other noncombustible absorbent material and place into containers for later disposal. Large Spills: Dike far ahead of liquid spill for later disposal. Cover powder spill with plastic sheet or tarp to minimize spreading. Prevent entry into waterways, sewers, basements or confined areas.

Regulations

RCRA 40CFR: Not listed
CERCLA: 40CFR 302.4: Not listed
SARA 40CFR 372.65: Not listed
SARA EHS 40CFR 355: Not listed
TSCA: Listed

FER2240 CAS #: 2944-67-4

FERRIC AMMONIUM OXALATE TRIHYDRATE

EINECS Number: 220-952-2
Molecular Formula: $C_6H_{12}FeN_3O_{12}$
Structured MF: $Fe(NH_4)_3(C_2O_4)_3 \cdot 3H_2O$
Formula Weight: 428
Description: yellowish-green powder; light burnt-sugar odor

Physical Properties

Boiling Point: Decomposes at 1 atm
Specific Gravity: 1.78
Water Solubility: 4.27 x10^5 ppm at 0 °C
Flash Point: Nonflammable

Hazard Overviews

Fire: Noncombustible.

Carcinogenicity: IARC - Not listed; NIOSH - Not listed; NTP - Not listed; ACGIH - Not listed; OSHA - Not listed; EPA - Not listed; MAK - Not listed

Environmental

Cleanup/Disposal: Guide No. 171: Do not touch or walk through spilled material. Stop leak if you can do it without risk. Prevent dust cloud. Avoid inhalation of asbestos dust. Small Dry Spills: With clean shovel place material into clean, dry container and cover loosely; move containers from spill area. Small Spills: Take up with sand or other noncombustible absorbent material and place into containers for later disposal. Large Spills: Dike far ahead of liquid spill for later disposal. Cover powder spill with plastic sheet or tarp to minimize spreading. Prevent entry into waterways, sewers, basements or confined areas.

Environmental Physical Data

BCF: no food chain concentration potential
BOD: sewage 0.14 lb/lb, 5 days

Regulations

RCRA 40CFR: Not listed
CERCLA: 40CFR 302.4: Not listed
SARA 40CFR 372.65: Not listed
SARA EHS 40CFR 355: Not listed
TSCA: Not listed

FER2550 CAS #: 10102-49-5

FERRIC ARSENATE

RTECS: NO4585000
DOT: UN1606; IMO6.1
EINECS Number: 233-274-7
Molecular Formula: $AsFeH_3O_4$
Formula Weight: 194.77
Synonyms: ARSENATE OF IRON,FERRIC; ARSENIC ACID (H3ASO4),IRON(3+) SALT (1:1); IRON ARSENATE; IRON(III) ARSENATE (1:1)
Description: green or brown powder
Use: insecticide

Physical Properties

Boiling Point: Decomposes
Freezing Point: Decomposes
Density: 3.18
Water Solubility: Solubility is Negligble
Other Solubilities: Insoluble in Nitric acid; Soluble in Hydrochloric Acid.
Refraction Index: 1.765 to 1.797
Flash Point: Nonflammable

Hazard Overviews

Fire: Noncombustible.

Carcinogenicity: IARC - Group 1, Carcinogenic to humans; NIOSH - Listed as carcinogen; NTP - Class 1, Known to be a carcinogen; ACGIH - Class A1, Confirmed human carcinogen; OSHA - Listed as a carcinogen; EPA - Class A, Human carcinogen; MAK - Class A1, Capable of inducing malignant tumors as shown by experience with humans

Exposure Limits
OSHA PEL Vacated 1989 Limits: STEL: 1 mg/m^3; as Fe (salts).
ACGIH TLV: TWA: 1 mg/m^3; as Fe Salt.
NIOSH REL: TWA: 1 mg/m^3; as Fe (Salts).
Respirator Recommendation
Exposure Range: >0.01 to <5 mg/m^3 Supplied Air, Constant Flow/Pressure Demand, Full Face
Exposure Range: 5 to unlimited mg/m^3 Self-contained Breathing Apparatus, Pressure Demand, Full Face
Note: as arsenic, inorganic compounds; refer to 29CFR 1910.1018 for more specific respirator recommendations

Environmental

Cleanup/Disposal: Guide No. 151: Do not touch damaged containers or spilled material unless wearing appropriate protective clothing. Stop leak if you can do it without risk. Prevent entry into waterways, sewers, basements or confined areas. Cover with plastic sheet to prevent spreading. Absorb or cover with dry earth, sand or other non-combustible material and transfer to containers. Do not get water inside containers.

Environmental Physical Data

BCF: biomagnification does not occur

Regulations

RCRA 40CFR: Not listed
CERCLA: 40CFR 302.4: Listed as Compound per CWA Section 307(a) per CAA Section 112
SARA 40CFR 372.65: Listed as Compound
SARA EHS 40CFR 355: Not listed
TSCA: Not listed

FER2860 CAS #: 7705-08-0

FERRIC CHLORIDE

RTECS: LJ9100000
DOT: UN1773; UN2582; IMO8.0
EINECS Number: 231-729-4
Molecular Formula: Cl_3Fe
Structured MF: $FeCl_3$
Formula Weight: 162.22

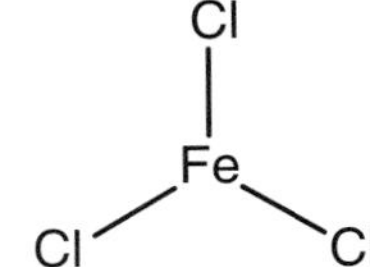

Chemical Structure

Synonyms: CHLORURE FERRIQUE; CHLORURE PERRIQUE; FERRIC CHLORIDE,SOLID; FERRIC CHLORIDE,SOLID,ANHYDROUS; FERRIC CHLORIDE,SOLUTION; FLORES MARTIS; IRON CHLORIDE; IRON CHLORIDE,SOLID; IRON (III) CHLORIDE; IRON PERCHLORIDE; IRON SESQUICHLORIDE,SOLID; IRON TRICHLORIDE; IRON(III) CHLORIDE; NATURAL MOLYSITE; PERCHLORURE DE FER

Description: greenish-black liquid; slight hydrochloric acid odor
Use: treatment of sewage and industrial wastes; etching agent for engraving, photography, and printed circuitry; catalyst; mordant; oxidizing, chlorinating and condensing agent, disinfectant, pigment, medicine; feed additive; water purification

Physical Properties

Boiling Point: About 316 °C (601 °F)
Freezing Point: About 300 °C (572 °F)
Specific Gravity: 2.898 at 25 °C/4 °C
Density: 2.9 g/mL at 25 °C
Vapor Pressure: 1 mm Hg at 194.0 °C
Water Solubility: 74.4 g/100g at 0 °C
Other Solubilities: 95% Ethanol: 10-50 mg/ml at 20 °C;: Glycerol: Soluble; Acetone: >=100 mg/ml at 20 °C; DMSO: 50-100 mg/ml at 20 °C; Ether: Very Soluble.
pH: 0.1 molar aqueous solution 2.0
Flash Point: Nonflammable

RTECS Toxicity Data

Acute Oral: Rat LD_{50} Dose: 450 mg/kg. Mouse LD_{50} Dose: 895 mg/kg.
Reproductive/Teratogenic: Rat Route: Intravaginal; Dose: 29 mg/kg; Duration: female 1D prior to mating; Effects on Fertility - Pre-implantation mortality. Rat Route: Intratesticular; Dose: 12976 ug/kg; Duration: male 1D prior to mating; Paternal Effects - Spermatogenesis; Testes, epididymis, sperm duct.
Mutagenic: Bacteria - E Coli Phage Inhibition; Dose: 41 ng/well. Bacteria - E Coli Other Mutation Test Systems; Dose: 500 nmol/tube.

Hazard Overviews

Corrosive

Fire Diamond

Health: Corrosive, causes severe burns to eyes/skin/respiratory system. Also Causes: local brown discoloration of the conjunctiva, abdominal pain, vomiting, diarrhea, rapid/weak pulse, low blood pressure, cyanosis, secondary shock, coma. Chronic Effects: mottling of lungs.
Fire: Noncombustible. Use agent suitable for surrounding fire. Reacts with water to produce toxic and corrosive fumes.
Reactivity: Stable, in closed, airtight, moisture-proof containers. Hazardous polymerization cannot occur. Avoid: excessive temperature; accidental exposure to any form of moisture (water, steam, or moist air). Incompatible with: allyl chloride; potassium; sodium. Hazardous decomposition products: ferrous chloride; chlorine gas; hydrogen chloride gas.
Carcinogenicity: IARC - Not listed; NIOSH - Not listed; NTP - Not listed; ACGIH - Not listed; OSHA - Not listed; EPA - Not listed; MAK - Not listed

Primary Target Organs:

Eyes

Skin

Respiratory System

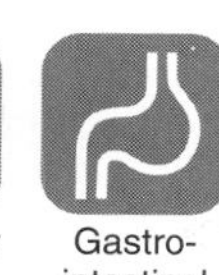
Gastro-intestinal

Exposure Limits

OSHA PEL Vacated 1989 Limits: STEL: 1 mg/m^3; as Fe salts.

ACGIH TLV: TWA: 1 mg/m^3; as Fe Salt.

NIOSH REL: TWA: 1 mg/m^3; as Fe salts.

Respirator Recommendation

Exposure Range: >1 to 10 mg/m^3 Air Purifying, Negative Pressure, Half Mask

Exposure Range: >10 to 100 mg/m^3 Air Purifying, Negative Pressure, Full Face

Exposure Range: >100 to 1000 mg/m^3 Supplied Air, Constant Flow/Pressure Demand, Full Face

Exposure Range: >1000 to unlimited mg/m^3 Self-contained Breathing Apparatus, Pressure Demand, Full Face

Cartridge Color: dust/mist filter (use P100 or consult supervisor for appropriate dust/mist filter)

Note: as iron salts, soluble

Environmental

Ecotoxicity: TLm Daphnia 15 ppm/96 hr fresh water /Conditions of bioassay not specified

Cleanup/Disposal: Guide No. 154: Eliminate all ignition sources (no smoking, flares, sparks or flames in immediate area). Do not touch damaged containers or spilled material unless wearing appropriate protective clothing. Stop leak if you can do it without risk. Prevent entry into waterways, sewers, basements or confined areas. Absorb or cover with dry earth, sand or other non-combustible material and transfer to containers. Do not get water inside containers. Guide No. 157: Eliminate all ignition sources (no smoking, flares, sparks or flames in immediate area). All equipment used when handling the product must be grounded. Do not touch damaged containers or spilled material unless wearing appropriate protective clothing. Stop leak if you can do it without risk. A vapor suppressing foam may be used to reduce vapors. Do not get water inside containers. Use water spray to reduce vapors or divert vapor cloud drift. Prevent entry into waterways, sewers, basements or confined areas. Small Spills: Cover with dry earth, dry sand, or other non-combustible material followed with plastic sheet to minimize spreading or contact with rain. Use clean non-sparking tools to collect material and place it into loosely covered plastic containers for later disposal.

Environmental Physical Data

BCF: no food chain concentration potential

BOD: none

Regulations

RCRA 40CFR: Not listed

CERCLA: 40CFR 302.4: Listed per CWA Section 311(b)(4) RQ: 1000 lb (453.5 kg)

SARA 40CFR 372.65: Not listed

SARA EHS 40CFR 355: Not listed

TSCA: Listed

FER3170 CAS #: 10025-77-1

FERRIC CHLORIDE, HEXAHYDRATE

RTECS: NO5425000

Molecular Formula: $Cl_3FeH_{12}O_6$

Formula Weight: 270.32

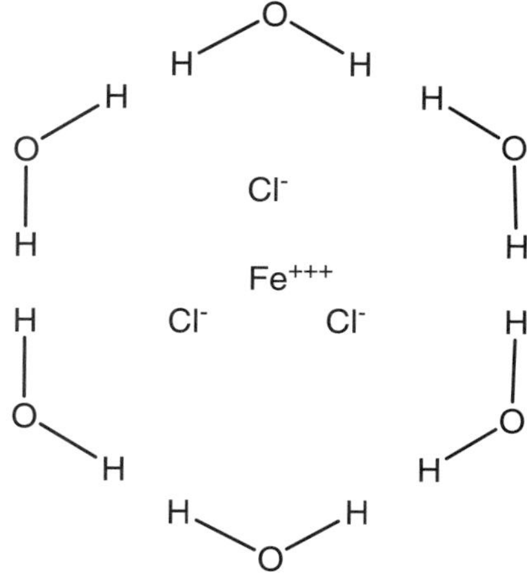

Chemical Structure

Synonyms: FERRIC TRICHLORIDE HEXAHYDRATE; IRON (III),CHLORIDE,HEXAHYDRATE; IRON TRICHLORIDE HEXAHYDRATE

Physical Properties

Boiling Point: 280 °C (536 °F)

Freezing Point: 37 °C (98.6 °F)

Specific Gravity: 1.82

Water Solubility: Soluble

Other Solubilities: Soluble in alcohol, acetone, ether

Odor Threshold: 10.0 ppm

RTECS Toxicity Data

Acute Oral: Rat LD_{Lo} Dose: 900 mg/kg.

Mutagenic: Human DNA Inhibition; Cell Type: lymphocyte; Dose: 4800 umol/L. Rat Cytogenetic Analysis; Cell Type: Ascites tumor; Dose: 500 mg/kg.

Hazard Overviews

Carcinogenicity: IARC - Not listed; NIOSH - Not listed; NTP - Not listed; ACGIH - Not listed; OSHA - Not listed; EPA - Not listed; MAK - Not listed

Exposure Limits

OSHA PEL Vacated 1989 Limits: STEL: 1 mg/m^3; as Fe salts.

ACGIH TLV: TWA: 1 mg/m^3; as Fe Salt.

NIOSH REL: TWA: 1 mg/m^3; as Fe salts.

Respirator Recommendation

Exposure Range: >1 to 10 mg/m^3 Air Purifying, Negative Pressure, Half Mask

Exposure Range: >10 to 100 mg/m^3 Air Purifying, Negative Pressure, Full Face
Exposure Range: >100 to 1000 mg/m^3 Supplied Air, Constant Flow/Pressure Demand, Full Face
Exposure Range: >1000 to unlimited mg/m^3 Self-contained Breathing Apparatus, Pressure Demand, Full Face
Cartridge Color: dust/mist filter (use P100 or consult supervisor for appropriate dust/mist filter)
Note: as iron salts, soluble

Environmental

Regulations

RCRA 40CFR: Not listed
CERCLA: 40CFR 302.4: Not listed
SARA 40CFR 372.65: Not listed
SARA EHS 40CFR 355: Not listed
TSCA: Not listed

FER3480 CAS #: 7783-50-8

FERRIC FLUORIDE

RTECS: NO6865000
EINECS Number: 232-002-4
Molecular Formula: F_3Fe
Structured MF: FeF_3
Formula Weight: 112.85

F^-

Fe^{+++}

F^- F^-

Chemical Structure

Synonyms: FERRIC TRIFLUORIDE; IRON FLUORIDE; IRON TRIFLUORIDE
Description: green crystals
Use: as catalyst in organic reactions; ceramics (porcelain, pottery); fluorinating agent; medication: as astringents in the treatment of some skin disorders

Physical Properties

Boiling Point: Sublimes at > 1000 °C (1832 °F)
Freezing Point: Sublimes at > 1000 °C (1832 °F)
Specific Gravity: 3.52
Water Solubility: Very Slightly Soluble in Water
Other Solubilities: freely Soluble in dilute Hydrogen Fluoride Practically Insoluble in Alcohol, Ether, Benzene 0.008 g/100 g liquid Hydrogen Fluoride.
Ionization Potential (eV): 12.5 +/-0.3
Flash Point: Not flammable

Hazard Overviews

Corrosive

Health: Corrosive to eyes/skin/respiratory tract. Other Acute Effects: harmful if swallowed, inhaled, or absorbed through skin; inhalation may result in spasm, inflammation and edema of the larynx and bronchi, chemical pneumonitis and pulmonary edema; symptoms of exposure may include burning sensation; coughing; wheezing; laryngitis; shortness of breath; headache; nausea; vomiting; symptoms of overexposure may include salivation; abdominal pain; fever; labored breathing; can reduce serum calcium levels possibly causing fatal hypocalcemia;. Chronic Effects: prolonged exposure to dusts, vapors or mists results in perforation of the nasal septum; excessive calcification of the bones, ligaments and tendons.
Fire: Noncombustible. Hazards: emits toxic fumes. Extinguishing agents: dry chemical powder. Precautions: combustible liquid.
Reactivity: Incompatible with: strong bases. Hazardous decomposition products: hydrogen fluoride.
Carcinogenicity: IARC - Not listed; NIOSH - Not listed; NTP - Not listed; ACGIH - Not listed; OSHA - Not listed; EPA - Not listed; MAK - Not listed
Primary Target Organs:

Eyes

Skin
Respiratory System

Exposure Limits
OSHA PEL: TWA: 2.5 mg/m^3; as F.
OSHA PEL Vacated 1989 Limits: STEL: 1 mg/m^3; as Fe (salts).
ACGIH TLV: TWA: 1 mg/m^3; as Fe Salt.
NIOSH REL: TWA: 1 mg/m^3; as Fe (Salts).

Environmental

Ecotoxicity: Aquatic toxicity: 419 mg(F)/l/96-hour/TLm/Mosquito fish 120 mg(F)/l/4 days/lethal concentration/goldfish
Cleanup/Disposal: Guide No. 157: Eliminate all ignition sources (no smoking, flares, sparks or flames in immediate area). All equipment used when handling the product must be grounded. Do not touch damaged containers or spilled material unless wearing appropriate protective clothing. Stop leak if you can do it without risk. A vapor suppressing foam may be used to reduce vapors. Do not get water inside containers. Use water spray to reduce vapors or divert vapor cloud drift. Prevent entry into waterways, sewers, basements or confined areas. Small Spills: Cover with dry earth, dry sand, or other non-combustible material followed with plastic sheet to minimize spreading or contact with rain. Use clean non-sparking tools to collect material and place it into loosely covered plastic containers for later disposal.

Regulations
RCRA 40CFR: Not listed
CERCLA: 40CFR 302.4: Listed per CWA Section 311(b)(4) RQ: 100 lb (45.35 kg)
SARA 40CFR 372.65: Not listed
SARA EHS 40CFR 355: Not listed
TSCA: Listed

FER3790 CAS #: 10421-48-4
FERRIC NITRATE

RTECS: QU8915000
DOT: UN1466; IMO5.1
EINECS Number: 233-899-5
Molecular Formula: $FeH_3N_3O_9$
Structured MF: $Fe(NO_3)_3 \cdot 6H_2O$
Formula Weight: 241.87
Synonyms: FERRIC NITRATE,NONAHYDRATE; IRON (III) NITRATE,ANHYDROUS; IRON NITRATE; IRON TRINITRATE; NITRIC ACID,IRON(3+) SALT
Description: pale violet, grayish-white crystals
Use: mordant in dyeing; weighting silks; tanning; corrosion inhibitor; reagent in anal chemistry; medication: as astringent in the treatment of some skin disorders

Physical Properties
Boiling Point: Decomposes at 1 atm
Freezing Point: Nonahydrate 47 °C (116.6 °F)
Specific Gravity: 1.68
Water Solubility: Freely Soluble in Water
Other Solubilities: Freely Soluble in Alcohol, Acetone; slightly Soluble in cold concentrated nitric acid.
Flash Point: Nonflammable

RTECS Toxicity Data
Mutagenic: Bacteria - E Coli Other Mutation Test Systems; Dose: 1250 nmol/tube.

Hazard Overviews

Fire Diamond

Health: Irritating to eyes/skin/respiratory tract. Also Causes: by ingestion: vomiting, abdominal pain, headache, shallow breathing, shock. Chronic Effects: liver damage.
Fire: Combustible. Strong oxidizer capable of igniting other combustibles. Fight fire with water only. Do not use dry chemical or carbon dioxide.
Reactivity: Stable. Hazardous polymerization cannot occur. Avoid: excessive dust generation; contact with reducing materials; combustibles; storage in metal containers. Incompatible with: aluminum; cyanides; phosphorus; acetylene gas; reducing materials (stannous chloride; thiocyanates; sodium hypophosphite; potassium ferrocyanide; tin; magnesium); combustibles (wood, paper, oil). Hazardous decomposition products: nitrogen oxides; nitric acid vapors.
Carcinogenicity: IARC - Not listed; NIOSH - Not listed; NTP - Not listed; ACGIH - Not listed; OSHA - Not listed; EPA - Not listed; MAK - Not listed
Primary Target Organs:

Eyes

Skin
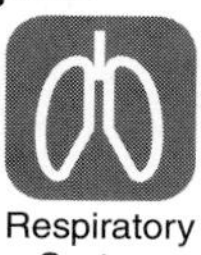
Respiratory System
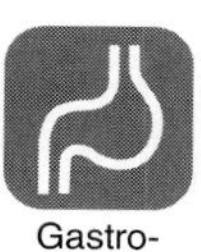
Gastro-intestinal

Liver

Exposure Limits
OSHA PEL Vacated 1989 Limits: STEL: 1 mg/m^3; as Fe salts.
ACGIH TLV: TWA: 1 mg/m^3; as Fe Salt.
NIOSH REL: TWA: 1 mg/m^3; as Fe salts.
Respirator Recommendation
Exposure Range: >1 to 10 mg/m^3 Air Purifying, Negative Pressure, Half Mask
Exposure Range: >10 to 100 mg/m^3 Air Purifying, Negative Pressure, Full Face
Exposure Range: >100 to 1000 mg/m^3 Supplied Air, Constant Flow/Pressure Demand, Full Face
Exposure Range: >1000 to unlimited mg/m^3 Self-contained Breathing Apparatus, Pressure Demand, Full Face
Cartridge Color: dust/mist filter (use P100 or consult supervisor for appropriate dust/mist filter)
Note: as iron salts, soluble

Environmental
Cleanup/Disposal: Guide No. 140: Keep combustibles (wood, paper, oil, etc.) away from spilled material. Do not touch damaged containers or spilled material unless wearing appropriate protective clothing. Stop leak if you can do it without risk. Do not get water inside containers. Small Dry Spills: With clean shovel place material into clean, dry container and cover loosely; move containers from spill area. Small Liquid Spills: Use a non-combustible material like vermiculite, sand or earth to soak up the product and place into a container for later disposal. Large Spills: Dike far ahead of liquid spill for later disposal. Following product recovery, flush area with water.

Environmental Physical Data
BCF: no food chain concentration potential
BOD: none

Regulations
RCRA 40CFR: Not listed
CERCLA: 40CFR 302.4: Listed per CWA Section 311(b)(4) RQ: 1000 lb (453.5 kg)
SARA 40CFR 372.65: Not listed
SARA EHS 40CFR 355: Not listed
TSCA: Listed

FER4100 **CAS #: 1309-37-1**

FERRIC OXIDE

RTECS: NO7400000
EINECS Number: 215-168-2
Molecular Formula: Fe_2O_3
Structured MF: Fe_2O_3
Formula Weight: 159.70

O^{--} Fe^{+++}

O^{--}

Fe^{+++} O^{--}

Chemical Structure

Synonyms: ANCHRED STANDARD; ANHYDROUS IRON OXIDE; ANHYDROUS OXIDE OF IRON; ARMENIAN BOLE; BAUXITE RESIDUE; BLACK OXIDE OF IRON; BLENDED RED OXIDES OF IRON; BURNT SIENNA; BURNT UMBER; BURNTISLAND RED; C.I. 77491; C.I. PIGMENT RED 101; C.I. PIGMENT RED 102; C.I. PIGMENT RED 101 AND 102; CALCOTONE RED; CAPUT MORTUUM; COLCOTHAR; COLLOIDAL FERRIC OXIDE; CROCUS MARTIS ADSTRINGENS; DEANOX; DEANOX DNX PIGMENTS; EISENOXYD; ENGLISH RED; FERRIC OXIDE (COLLOIDAL); FERRUGO; INDIAN RED; IRON (III) OXIDE; IRON OXIDE; IRON OXIDE PIGMENTS; IRON OXIDE RED; IRON RED; IRON SESQUIOXIDE; IRON(III) OXIDE IRON(III)OXIDE; JEWELER'S ROUGE; LEVANOX RED 130A; LIGHT RED; MANUFACTURED IRON OXIDES; MARS BROWN; MARS RED; NATURAL HEMATITE; NATURAL IRON OXIDES; NATURAL RED OXIDE; OCHRE; PRUSSIAN BROWN; QUICK ROUGE; RADDLE; 11554 RED; RED IRON OXIDE; RED OCHRE; RED OXIDE; RED OXIDE D3452; RED OXIDE D6984; RED OXIDE OF IRON; ROUGE; RUBIGO; SIENNA; SPECULAR IRON; STONE RED; SUPRA; SYNTHETIC IRON OXIDE; VENETIAN RED; VITRIOL RED; VOGEL'S IRON RED; YELLOW FERRIC OXIDE; YELLOW OXIDE OF IRON
Description: reddish brown dust, fume
Use: polishing agent for glass, precious metals, diamonds; in magnets, magnetic tapes; catalyst; pigment, mfr of ferrites for electronic uses; metallurgy; gas purification; paint & rubber pigment; in thermite; mordant; lab rgnt; catalyst (p-hydrogen); feed additive; memory cores for computers; in cement; fine red powder of ferric oxide, dental abrasive; to make high quality ferrites & similar ceramic materials lightfast; uv blocking red pigment for paints, wood varnishes, fiber dying, etc.

Physical Properties

Freezing Point: 565 °C (1049 °F)
Specific Gravity: 5.24
Vapor Pressure: ~ 0 mm Hg
Water Solubility: Insoluble in Water
Other Solubilities: Soluble in acids.
Refraction Index: 3.01
Flash Point: Noncombustible

RTECS Toxicity Data

Acute Dermal: Dog LD_{Lo} Route: Subcutaneous Dose: 30 mg/kg.
Chronic (Multiple Dose) Inhalation: Rat Dose: 500 ug/m^3/24H/61D-C; Toxic Effects: Brain and coverings - Other degenerative changes; Blood - Changes in serum composition; Biochemical - True cholinesterase.
Tumorigenic: Rat Route: Subcutaneous; Dose: 135 mg/kg; Toxic Effects: Tumorigenic - Equivocal tumorigenic agent by RTECS criteria; Tumorigenic - Tumors at site of application.

Hazard Overviews

Fire Diamond

Health: Irritating to eyes/skin/respiratory tract. Chronic Effects: siderosis.
Fire: Noncombustible. Use extinguishing agents suitable for surrounding fire.
Reactivity: Stable. Hazardous polymerization cannot occur. Avoid: dust generation; fumes. Incompatible with: hydrazine; calcium hypochlorite; performic acid; bromine pentafluoride.
Carcinogenicity: IARC - Group 3, Not classifiable as to carcinogenicity to humans; NIOSH - Not listed; NTP - Not listed; ACGIH - Class A4, Not classifiable as a human carcinogen; OSHA - Not listed; EPA - Not listed; MAK - Not listed
Primary Target Organs:

Eyes

Skin

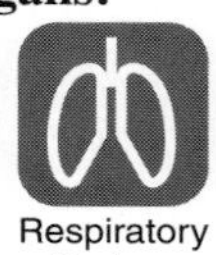
Respiratory System

Mucous Membranes

Exposure Limits
OSHA PEL: TWA: 15 mg/m^3; total dust.
OSHA PEL Vacated 1989 Limits: STEL: 1 mg/m^3; as Fe (salts).
ACGIH TLV: TWA: 5 mg/m^3; as Fe.
NIOSH REL: TWA: 5 mg/m^3; as Fe (Salts).
NIOSH IDLH: 2500 mg/m^3; as Fe.
DFG MAK: TWA: 6 mg/m^3.
Respirator Recommendation
Exposure Range: >10 to 100 mg/m^3 Air Purifying, Negative Pressure, Half Mask
Exposure Range: >100 to 1000 mg/m^3 Air Purifying, Negative Pressure, Full Face
Exposure Range: >1000 to <2500 mg/m^3 Supplied Air, Constant Flow/Pressure Demand, Full Face
Exposure Range: 2500 to unlimited mg/m^3 Self-contained Breathing Apparatus, Pressure Demand, Full Face
Cartridge Color: magenta (P100)

Environmental

Regulations

RCRA 40CFR: Not listed
CERCLA: 40CFR 302.4: Not listed
SARA 40CFR 372.65: Not listed
SARA EHS 40CFR 355: Not listed
TSCA: Listed

FER4410 **CAS #: 10045-86-0**

FERRIC PHOSPHATE

EINECS Number: 233-149-7
Molecular Formula: FeH_3O_4P
Formula Weight: 150.83

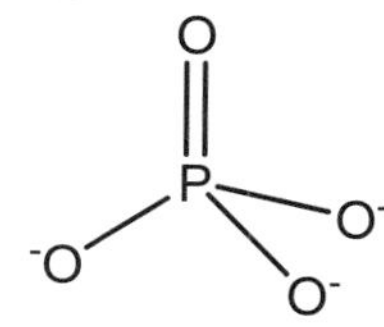

Fe^{+++}

Chemical Structure

Synonyms: FERRIC ORTHOPHOSPHATE; PHOSPHORIC ACID,IRON(3+) SALT (1:1)
Description: white, grayish-white, or light pink, crystals or powder
Use: food and feed supplement, particularly in bread enrichment; as fertilizer

Physical Properties

Freezing Point: Loses water at 140 °C (284 °F)
Density: Dihydrate 2.87
Water Solubility: Practically Insoluble in Water
Other Solubilities: readily Soluble in Hydrochloric Acid; slowly Soluble in Nitric acid; Soluble in Sulfuric acid.

Hazard Overviews

Carcinogenicity: IARC - Not listed; NIOSH - Not listed; NTP - Not listed; ACGIH - Not listed; OSHA - Not listed; EPA - Not listed; MAK - Not listed
Exposure Limits
OSHA PEL Vacated 1989 Limits: STEL: 1 mg/m^3; as Fe salts.
ACGIH TLV: TWA: 1 mg/m^3; as Fe Salt.
NIOSH REL: TWA: 1 mg/m^3; as Fe salts.
Respirator Recommendation
Exposure Range: >1 to 10 mg/m^3 Air Purifying, Negative Pressure, Half Mask
Exposure Range: >10 to 100 mg/m^3 Air Purifying, Negative Pressure, Full Face
Exposure Range: >100 to 1000 mg/m^3 Supplied Air, Constant Flow/Pressure Demand, Full Face
Exposure Range: >1000 to unlimited mg/m^3 Self-contained Breathing Apparatus, Pressure Demand, Full Face
Cartridge Color: dust/mist filter (use P100 or consult supervisor for appropriate dust/mist filter)
Note: as iron salts, soluble

Environmental

Regulations
RCRA 40CFR: Not listed
CERCLA: 40CFR 302.4: Not listed
SARA 40CFR 372.65: Not listed
SARA EHS 40CFR 355: Not listed
TSCA: Listed

FER4720 **CAS #: 10058-44-3**

FERRIC PYROPHOSPHATE

EINECS Number: 233-190-0
Molecular Formula: $Fe_4H_{12}O_{21}P_6$
Formula Weight: 745.25

Fe

Chemical Structure

Synonyms: DIPHOSPHORIC ACID,IRON(3+) SALT (3:4); IRON PYROPHOSPHATE; PYROPHOSPHORIC ACID,IRON(3+) SALT (3:4)
Description: yellowish-white powder
Use: catalyst; in fireproofing of synthetic fibers; in corrosion-preventing pigments; hematinic agent; nutrient &/or dietary supplement food additive; trace mineral added to animal feeds

Physical Properties

Water Solubility: Practically Insoluble in Water
Other Solubilities: Soluble in acid, alkali Citrate; Soluble in Mineral acids; Practically Insoluble in Acetic Acid

Hazard Overviews

Health: May cause irritation to eyes/skin/respiratory tract. Other Acute Effects: may be harmful by inhalation, ingestion, or skin absorption. The toxicological properties have not been thoroughly investigated.
Fire: Hazards: emits toxic fumes. Extinguishing agents: water spray; carbon dioxide, dry chemical powder or appropriate foam. Precautions: combustible liquid.
Reactivity: Stable. Hazardous polymerization will not occur. Hazardous decomposition products: toxic fumes of: carbon monoxide, carbon dioxide, phosphorus, iron and iron salts, phosphorous oxides.
Carcinogenicity: IARC - Not listed; NIOSH - Not listed; NTP - Not listed; ACGIH - Not listed; OSHA - Not listed; EPA - Not listed; MAK - Not listed

Exposure Limits
OSHA PEL Vacated 1989 Limits: STEL: 1 mg/m^3; as Fe salts.
ACGIH TLV: TWA: 1 mg/m^3; as Fe Salt.
NIOSH REL: TWA: 1 mg/m^3; as Fe salts.
Respirator Recommendation
Exposure Range: >1 to 10 mg/m^3 Air Purifying, Negative Pressure, Half Mask
Exposure Range: >10 to 100 mg/m^3 Air Purifying, Negative Pressure, Full Face
Exposure Range: >100 to 1000 mg/m^3 Supplied Air, Constant Flow/Pressure Demand, Full Face
Exposure Range: >1000 to unlimited mg/m^3 Self-contained Breathing Apparatus, Pressure Demand, Full Face
Cartridge Color: dust/mist filter (use P100 or consult supervisor for appropriate dust/mist filter)
Note: as iron salts, soluble

Environmental

Regulations

RCRA 40CFR: Not listed
CERCLA: 40CFR 302.4: Not listed
SARA 40CFR 372.65: Not listed
SARA EHS 40CFR 355: Not listed
TSCA: Listed

FER5030 **CAS #: 10028-22-5**

FERRIC SULFATE

RTECS: NO8505000
EINECS Number: 233-072-9
Molecular Formula: $Fe_2H_6O_{12}S_3$
Structured MF: $Fe_2(SO_4)_3$
Formula Weight: 399.88

Chemical Structure

Synonyms: DIIRON TRISULFATE; IRON PERSULFATE; IRON SESQUISULFATE; IRON SULFATE; IRON SULFATE (2:3); IRON(3+) SULFATE; IRON TERSULFATE; IRON(III) SULFATE; SULFURIC ACID,IRON(3+) SALT (3:2)
Description: grayish-white powder or yellow crystals; odorless
Use: in prepn of iron alums, other iron salts & pigments; in etching aluminum, pickling stainless steel & copper; mordant in textile dyeing & calico printing; polymerization catalyst; fertilizer; coagulant in water & sewage treatment; oxidant in coal conversion process (desulfurization); pigments, rgnt, disinfectant, chelated iron products, intermed; med: ferric salts, skin astringents.

Physical Properties

Boiling Point: Decomposes at 1 atm
Freezing Point: Decomposes at 480 °C (896 °F)
Specific Gravity: 3.097 at 18 °C
Water Solubility: Slowly Soluble in Water
Other Solubilities: Insoluble in sulfuric acid, Ammonia.
Refraction Index: 1.814
Flash Point: Nonflammable

RTECS Toxicity Data

Mutagenic: Bacteria - E Coli Other Mutation Test Systems; Dose: 250 nmol/tube.

Hazard Overviews

Corrosive

Health: Irritating to eyes/skin/respiratory tract. Other Acute Effects: may be harmful by inhalation, ingestion, or skin absorption. Chronic Effects: overdose may have a corrosive effect on the gastrointestinal mucosa and be followed by necrosis, perforation and stricture formation; several hours may elapse before symptoms that can include epigastric pain, diarrhea, vomiting, nausea, and hematemesis; after apparent recovery a person may experience metabolic acidosis; convulsions; and coma hours or days later; further complications may develop leading to acute liver necrosis that can result in death due to hepatic coma; toxic when injected into experimental animals.
Fire: Noncombustible. Hazards: emits toxic fumes. Extinguishing agents: noncombustible; use extinguishing media appropriate to surrounding fire conditions. Precautions: combustible liquid.
Reactivity: Stable. Hazardous polymerization will not occur. Incompatible with: strong oxidizing agents, moisture sensitive, light sensitive. Hazardous decomposition products: toxic fumes of: sulfur oxides.
Carcinogenicity: IARC - Not listed; NIOSH - Not listed; NTP - Not listed; ACGIH - Not listed; OSHA - Not listed; EPA - Not listed; MAK - Not listed
Primary Target Organs:

Eyes

Skin

Respiratory System

Exposure Limits
OSHA PEL Vacated 1989 Limits: STEL: 1 mg/m^3; as Fe salts.
ACGIH TLV: TWA: 1 mg/m^3; as Fe Salt.
NIOSH REL: TWA: 1 mg/m^3; as Fe salts.

Respirator Recommendation

Exposure Range: >1 to 10 mg/m^3 Air Purifying, Negative Pressure, Half Mask

Exposure Range: >10 to 100 mg/m^3 Air Purifying, Negative Pressure, Full Face

Exposure Range: >100 to 1000 mg/m^3 Supplied Air, Constant Flow/Pressure Demand, Full Face

Exposure Range: >1000 to unlimited mg/m^3 Self-contained Breathing Apparatus, Pressure Demand, Full Face

Cartridge Color: dust/mist filter (use P100 or consult supervisor for appropriate dust/mist filter)

Note: as iron salts, soluble

Environmental

Ecotoxicity: Aquatic toxicity: 0.716 ppm/12-24 hr/shiners, carp, suckers/killed/fresh water 133 ppm/48 hr/mosquito fish/TLm/fresh water

Cleanup/Disposal: Guide No. 140: Keep combustibles (wood, paper, oil, etc.) away from spilled material. Do not touch damaged containers or spilled material unless wearing appropriate protective clothing. Stop leak if you can do it without risk. Do not get water inside containers. Small Dry Spills: With clean shovel place material into clean, dry container and cover loosely; move containers from spill area. Small Liquid Spills: Use a non-combustible material like vermiculite, sand or earth to soak up the product and place into a container for later disposal. Large Spills: Dike far ahead of liquid spill for later disposal. Following product recovery, flush area with water.

Environmental Physical Data

BCF: no food chain concentration potential

BOD: none

Regulations

RCRA 40CFR: Not listed

CERCLA: 40CFR 302.4: Listed per CWA Section 311(b)(4) RQ: 1000 lb (453.5 kg)

SARA 40CFR 372.65: Not listed

SARA EHS 40CFR 355: Not listed

TSCA: Listed

FER5340 CAS #: 10045-89-3

FERROUS AMMONIUM SULFATE

RTECS: WS5890000

EINECS Number: 233-151-8

Molecular Formula: $FeH_{10}N_2O_8S_2$

Structured MF: $Fe(SO_4)\bullet(NH_4)_2SO_4\bullet6H_2O$

Formula Weight: 286.09

Fe^{++} NH_4^+ NH_4^+ $^-O-SO_2-O^-$ $^-O-SO_2-O^-$

Chemical Structure

Synonyms: AMMONIUM FERROUS SULFATE; AMMONIUM IRON SULFATE; AMMONIUM IRON SULFATE (2:2:1); DIAMMONIUM FERROUS SULFATE; DIAMMONIUM IRON SULFATE; FERROUS DIAMMONIUM DISULFATE; MOHR'S SALT; SULFURIC ACID,AMMONIUM IRON(2+) SALT (2:2:1)

Description: light green crystals

Use: in photography; as analytical standard; as polymerization catalyst; in dosimeters; analytical chemistry; metallurgy; in iron-plating baths; medication: for prevention and treatment of iron deficiency

Physical Properties

Boiling Point: Decomposes at 100 °C (212 °F) to 110 °C (230 °F)

Freezing Point: Decomposes at 100 °C (212 °F) to 110 °C (230 °F)

Specific Gravity: 1.86

Water Solubility: Soluble in Water

Other Solubilities: Insoluble in Ethanol.

Refraction Index: 1.487 to 1.4999

Flash Point: Nonflammable

Hazard Overviews

Fire Diamond

Health: Irritating to eyes/skin/respiratory tract. Also Causes: upon ingestion of large amounts: vomiting, abdominal pain, bloody diarrhea, hematemesis, lethargy, shock, acidosis, cyanosis, fever, liver injury.

Fire: Use dry chemical, carbon dioxide, Halon, water spray, fog, or standard foam.

Reactivity: Stable, but slowly oxidizes in air. Hazardous polymerization cannot occur. Avoid: heat; light; moisture. Hazardous decomposition products: ammonia fumes; oxides of nitrogen.

Carcinogenicity: IARC - Not listed; NIOSH - Not listed; NTP - Not listed; ACGIH - Not listed; OSHA - Not listed; EPA - Not listed; MAK - Not listed

Primary Target Organs:

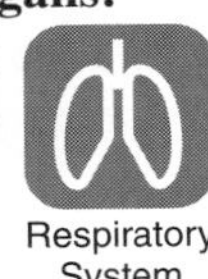

Exposure Limits

OSHA PEL Vacated 1989 Limits: STEL: 1 mg/m^3; as Fe salts.

ACGIH TLV: TWA: 1 mg/m^3; as Fe Salt.

NIOSH REL: TWA: 1 mg/m^3; as Fe salts.

Respirator Recommendation

Exposure Range: >1 to 10 mg/m^3 Air Purifying, Negative Pressure, Half Mask

Exposure Range: >10 to 100 mg/m^3 Air Purifying, Negative Pressure, Full Face

Exposure Range: >100 to 1000 mg/m^3 Supplied Air, Constant Flow/Pressure Demand, Full Face

Exposure Range: >1000 to unlimited mg/m^3 Self-contained Breathing Apparatus, Pressure Demand, Full Face

Cartridge Color: dust/mist filter (use P100 or consult supervisor for appropriate dust/mist filter)

Note: as iron salts, soluble

Environmental

Cleanup/Disposal: Guide No. 141: Keep combustibles (wood, paper, oil, etc.) away from spilled material. Do not touch damaged containers or spilled material unless wearing appropriate protective clothing. Stop leak if you can do it without risk. Small Dry Spills: With clean shovel place material into clean, dry container and cover loosely; move containers from spill area. Large Spills: Dike far ahead of spill for later disposal.

Environmental Physical Data

BCF: no food chain concentration potential

Regulations

RCRA 40CFR: Not listed

CERCLA: 40CFR 302.4: Listed per CWA Section 311(b)(4) RQ: 1000 lb (453.5 kg)

SARA 40CFR 372.65: Not listed

SARA EHS 40CFR 355: Not listed

TSCA: Listed

FER5650 CAS #: 10102-50-8

FERROUS ARSENATE

RTECS: NO4580000

DOT: UN1608; IMO6.1

EINECS Number: 233-275-2

Molecular Formula: $As_2Fe_3H_6O_8$

Formula Weight: 445.39

Synonyms: ARSENATE OF IRON,FERROUS; ARSENIC ACID (H3ASO4),IRON(2+) SALT (2:3); IRON ARSENATE; IRON(II) ARSENATE (3:2)

Description: green powder

Use: insecticide hexahydrate; vet: formerly as a hematinic tonic for cats & dogs in convalescence from surgery or chronic debilitating diseases; formerly used as anthelmintic for lambs

Physical Properties

Water Solubility: Insoluble in Water

Other Solubilities: Soluble in acids; Soluble in dilute Hydrochloric Acid; Slightly Soluble in NH_4OH.

Hazard Overviews

Carcinogenicity: IARC - Group 1, Carcinogenic to humans; NIOSH - Listed as carcinogen; NTP - Class 1, Known to be a carcinogen; ACGIH - Class A1, Confirmed human carcinogen; OSHA - Listed as a carcinogen; EPA - Class A, Human carcinogen; MAK - Class A1, Capable of inducing malignant tumors as shown by experience with humans

Exposure Limits

OSHA PEL Vacated 1989 Limits: STEL: 1 mg/m^3; as Fe (salts).

ACGIH TLV: TWA: 1 mg/m^3; as Fe Salt.

NIOSH REL: TWA: 1 mg/m^3; as Fe (Salts).

Respirator Recommendation

Exposure Range: >0.01 to <5 mg/m^3 Supplied Air, Constant Flow/Pressure Demand, Full Face

Exposure Range: 5 to unlimited mg/m^3 Self-contained Breathing Apparatus, Pressure Demand, Full Face

Note: as arsenic, inorganic compounds; refer to 29CFR 1910.1018 for more specific respirator recommendations

Environmental

Cleanup/Disposal: Guide No. 151: Do not touch damaged containers or spilled material unless wearing appropriate protective clothing. Stop leak if you can do it without risk. Prevent entry into waterways, sewers, basements or confined areas. Cover with plastic sheet to prevent spreading. Absorb or cover with dry earth, sand or other non-combustible material and transfer to containers. Do not get water inside containers.

Environmental Physical Data

BCF: biomagnification does not occur

Regulations

RCRA 40CFR: Not listed

CERCLA: 40CFR 302.4: Listed as Compound per CWA Section 307(a) per CAA Section 112

SARA 40CFR 372.65: Listed as Compound

SARA EHS 40CFR 355: Not listed

TSCA: Not listed

FER5960 CAS #: 563-71-3

FERROUS CARBONATE

EINECS Number: 209-259-6

Molecular Formula: CH_2FeO_3

Formula Weight: 117.87

Synonyms: BLAUD'S MASS; CARBONIC ACID,IRON(2+) SALT (1:1); IRON (2+) CARBONATE; IRON CARBONATE; IRON II CARBONATE
Use: hematinic; vet: nutrient source of iron; vet: trace mineral added to animal feeds; trace mineral added to foods

Physical Properties

Freezing Point: Decomposes
Specific Gravity: 3.8
Water Solubility: 0.0067 g Soluble in 1 Liter of Water at 25 °C
Other Solubilities: Soluble in acid.

Hazard Overviews

Carcinogenicity: IARC - Not listed; NIOSH - Not listed; NTP - Not listed; ACGIH - Not listed; OSHA - Not listed; EPA - Not listed; MAK - Not listed
Exposure Limits
OSHA PEL Vacated 1989 Limits: STEL: 1 mg/m^3; as Fe salts.
ACGIH TLV: TWA: 1 mg/m^3; as Fe Salt.
NIOSH REL: TWA: 1 mg/m^3; as Fe salts.
Respirator Recommendation
Exposure Range: >1 to 10 mg/m^3 Air Purifying, Negative Pressure, Half Mask
Exposure Range: >10 to 100 mg/m^3 Air Purifying, Negative Pressure, Full Face
Exposure Range: >100 to 1000 mg/m^3 Supplied Air, Constant Flow/Pressure Demand, Full Face
Exposure Range: >1000 to unlimited mg/m^3 Self-contained Breathing Apparatus, Pressure Demand, Full Face
Cartridge Color: dust/mist filter (use P100 or consult supervisor for appropriate dust/mist filter)
Note: as iron salts, soluble

Environmental

Regulations

RCRA 40CFR: Not listed
CERCLA: 40CFR 302.4: Not listed
SARA 40CFR 372.65: Not listed
SARA EHS 40CFR 355: Not listed
TSCA: Listed

FER6270 CAS #: 7758-94-3

FERROUS CHLORIDE

RTECS: NO5400000
DOT: NA1759; NA1760
EINECS Number: 231-843-4
Molecular Formula: Cl_2Fe
Structured MF: $FeCl_2$
Formula Weight: 126.76

Fe^{++}

Cl^- Cl^-

Chemical Structure

Synonyms: FERRO 66; FERROUS DICHLORIDE; IRON CHLORIDE; IRON(2+) CHLORIDE; IRON DICHLORIDE; IRON PROTOCHLORIDE; IRON(II) CHLORIDE; IRON(II) CHLORIDE (1:2); NATURAL LAWRENCITE
Description: white, pale green crystals; odorless
Use: in metallurgy; reducing agent; in pharmaceutical prepns; mordant in dyeing; mfr of ferric chloride; sewage treatment; hot bath in iron plating baths to yield more ductile deposit; coagulants (ferric & ferrous chloride) in dewatering sludge; med: ferrous salts available for prevention & treatment of iron deficiency.

Physical Properties

Boiling Point: 1023 °C (1873 °F)
Freezing Point: 674 °C (1245.2 °F)
Specific Gravity: 3.16 at 25 °C/4 °C
Vapor Pressure: 10 mm Hg at 700 °C
Water Solubility: Freely Soluble in Water
Other Solubilities: Soluble in 100 cc Alcohol: 64.4 g at 10 °C, 105.7 g at 100 °C.
pH: Acid
Flash Point: Nonflammable

RTECS Toxicity Data

Acute Oral: Rat LD_{50} Dose: 450 mg/kg.
Mutagenic: Hamster Morphological Transformation; Cell Type: embryo; Dose: 2500 umol/L. Bacteria - E Coli Phage Inhibition; Dose: 50 ng/well.

Hazard Overviews

Fire Diamond

Health: Mildly irritating to eyes/skin/respiratory tract. Also Causes: upon ingestion: gastrointestinal irritation.
Fire: Noncombustible. Use agent suitable for surrounding fire.
Reactivity: Stable, in closed, airtight, moisture-proof containers. Hazardous polymerization cannot occur. Avoid: exposure to any form of moisture (water, steam, or moist air). Incompatible with: ethylene oxide; potassium; sodium. Hazardous decomposition products: chloride fumes.
Carcinogenicity: IARC - Not listed; NIOSH - Not listed; NTP - Not listed; ACGIH - Not listed; OSHA - Not listed; EPA - Not listed; MAK - Not listed
Primary Target Organs:

Eyes

Skin

Respiratory System

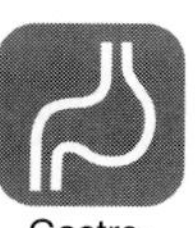
Gastro-intestinal

Exposure Limits
OSHA PEL Vacated 1989 Limits: STEL: 1 mg/m^3; as Fe salts.
ACGIH TLV: TWA: 1 mg/m^3; as Fe Salt.
NIOSH REL: TWA: 1 mg/m^3; as Fe salts.

Respirator Recommendation

Exposure Range: >1 to 10 mg/m^3 Air Purifying, Negative Pressure, Half Mask

Exposure Range: >10 to 100 mg/m^3 Air Purifying, Negative Pressure, Full Face

Exposure Range: >100 to 1000 mg/m^3 Supplied Air, Constant Flow/Pressure Demand, Full Face

Exposure Range: >1000 to unlimited mg/m^3 Self-contained Breathing Apparatus, Pressure Demand, Full Face

Cartridge Color: dust/mist filter (use P100 or consult supervisor for appropriate dust/mist filter)

Note: as iron salts, soluble

Environmental

Ecotoxicity: Aquatic toxicity: <38 ppm/64 hr/daphnia/toxic/fresh water

Cleanup/Disposal: Guide No. 154: Eliminate all ignition sources (no smoking, flares, sparks or flames in immediate area). Do not touch damaged containers or spilled material unless wearing appropriate protective clothing. Stop leak if you can do it without risk. Prevent entry into waterways, sewers, basements or confined areas. Absorb or cover with dry earth, sand or other non-combustible material and transfer to containers. Do not get water inside containers.

Environmental Physical Data

BCF: no food chain concentration potential

Regulations

RCRA 40CFR: Not listed

CERCLA: 40CFR 302.4: Listed per CWA Section 311(b)(4) RQ: 100 lb (45.35 kg)

SARA 40CFR 372.65: Not listed

SARA EHS 40CFR 355: Not listed

TSCA: Listed

FER6580 CAS #: 23383-11-1

FERROUS CITRATE

EINECS Number: 245-625-1

Molecular Formula: $C_6H_8Fe_xO_7$

Formula Weight: Varies

Synonyms: CITRIC ACID,IRON(2+) SALT; 1,2,3-PROPANETRICARBOXYLIC ACID,2-HYDROXY-,IRON(2+) SALT

Description: very slightly colored powder or white crystals, or reddish-brown scales

Use: hematinic; in ferrokinetic studies of iron absorption & turnover rates

Physical Properties

Freezing Point: Decomposes at 350 °C (662 °F)

Water Solubility: Practically Insoluble in Water

Other Solubilities: Soluble in Ammonium Hydroxide; Practically Insoluble in Alcohol, Acetone

Hazard Overviews

Carcinogenicity: IARC - Not listed; NIOSH - Not listed; NTP - Not listed; ACGIH - Not listed; OSHA - Not listed; EPA - Not listed; MAK - Not listed

Exposure Limits

OSHA PEL Vacated 1989 Limits: STEL: 1 mg/m^3; as Fe salts.

ACGIH TLV: TWA: 1 mg/m^3; as Fe Salt.

NIOSH REL: TWA: 1 mg/m^3; as Fe salts.

Respirator Recommendation

Exposure Range: >1 to 10 mg/m^3 Air Purifying, Negative Pressure, Half Mask

Exposure Range: >10 to 100 mg/m^3 Air Purifying, Negative Pressure, Full Face

Exposure Range: >100 to 1000 mg/m^3 Supplied Air, Constant Flow/Pressure Demand, Full Face

Exposure Range: >1000 to unlimited mg/m^3 Self-contained Breathing Apparatus, Pressure Demand, Full Face

Cartridge Color: dust/mist filter (use P100 or consult supervisor for appropriate dust/mist filter)

Note: as iron salts, soluble

Environmental

Regulations

RCRA 40CFR: Not listed

CERCLA: 40CFR 302.4: Not listed

SARA 40CFR 372.65: Not listed

SARA EHS 40CFR 355: Not listed

TSCA: Not listed

FER6890 CAS #: 299-29-6

FERROUS GLUCONATE

RTECS: LZ5150000

EINECS Number: 206-076-3

Molecular Formula: $C_{12}H_{22}FeO_{14}$

Formula Weight: 446.16

Chemical Structure

Synonyms: BIOFERGATE; CEREVON; ENTRON; FENTON; FERAVOL; FERGON; FERGON PREPARATIONS; FERLUCON; FEROX; FERRIN 55; FERRO-AGEPHA; FERROGLYCONICUM; FERRONICUM; FERROSE; FERRUM POLON; GLUCOFERRON; GLUCO-FERRUM; GLUCOMAX; GLUCONIC ACID IRON SALT; GLUCONIC ACID,IRON(2+) SALT (2:1),D-; GLUFERATE; IROMIN; IROMON (GADOR); IRON GLUCONATE; IRON-II GLUCONATE; IROX; IROX (GADOR); NIONATE; RAY-GLUCIRON

Description: slight odor of caramel
Use: hematinic; nutrient or dietary supplement; trace mineral added to animal feed; as coloring and flavoring in foods; a food colorant for coloring ripe olives

Physical Properties

Freezing Point: dihydrate 98 °C (208.4 °F)
Water Solubility: Soluble in Water
Other Solubilities: Soluble in Glycerin.
pH: Acid
Refraction Index: Alpha 1.545
Flash Point: Combustible

RTECS Toxicity Data

Acute Oral: Child TD_{Lo} Dose: 162 mg/kg; Toxic Effects: Gastrointestinal - Hypermotility, diarrhea; Gastrointestinal - Nausea or vomiting. Rat LD_{50} Dose: 2237 mg/kg; Toxic Effects: Gastrointestinal - Ulceration or bleeding from stomach.
Tumorigenic: Mouse Route: Subcutaneous; Dose: 2600 mg/kg/13W-I; Toxic Effects: Tumorigenic - Equivocal tumorigenic agent by RTECS criteria; Blood - Lymphomax including Hodgkin's disease; Tumorigenic effects - Ovarian tumors.

Hazard Overviews

Corrosive

Health: Data not available.
Fire: Combustible. Hazards: hazardous polymerization will not occur. Extinguishing agents: water spray; carbon dioxide, dry chemical powder or appropriate foam. Precautions: combustible liquid.
Reactivity: Hazardous polymerization will not occur.
Carcinogenicity: IARC - Not listed; NIOSH - Not listed; NTP - Not listed; ACGIH - Not listed; OSHA - Not listed; EPA - Not listed; MAK - Not listed
Exposure Limits
OSHA PEL Vacated 1989 Limits: STEL: 1 mg/m^3; as Fe salts.
ACGIH TLV: TWA: 1 mg/m^3; as Fe Salt.
NIOSH REL: TWA: 1 mg/m^3; as Fe salts.
Respirator Recommendation
Exposure Range: >1 to 10 mg/m^3 Air Purifying, Negative Pressure, Half Mask
Exposure Range: >10 to 100 mg/m^3 Air Purifying, Negative Pressure, Full Face
Exposure Range: >100 to 1000 mg/m^3 Supplied Air, Constant Flow/Pressure Demand, Full Face
Exposure Range: >1000 to unlimited mg/m^3 Self-contained Breathing Apparatus, Pressure Demand, Full Face
Cartridge Color: dust/mist filter (use P100 or consult supervisor for appropriate dust/mist filter)
Note: as iron salts, soluble

Environmental

Regulations

RCRA 40CFR: Not listed
CERCLA: 40CFR 302.4: Not listed
SARA 40CFR 372.65: Not listed
SARA EHS 40CFR 355: Not listed
TSCA: Listed

FER7200 **CAS #: 5905-52-2**

FERROUS LACTATE

RTECS: OD5525000
EINECS Number: 227-608-0
Molecular Formula: $C_6H_{12}FeO_6$
Formula Weight: 233.99

HO–CH(CH3)–C(=O)–O–Fe–O–C(=O)–CH(CH3)–OH

Chemical Structure

Synonyms: IRON LACTATE; IRON(2) LACTATE; IRON(2+) LACTATE; LACTIC ACID,IRON(2+) SALT (2:1); PROPANOIC ACID,2-HYDROXY-,IRON(2+) SALT (2:1)
Description: greenish-white powder or crystalline masses; slight characteristic odor
Use: hematinic; vet: hematinic; food additive & dietary supplement

Physical Properties

Freezing Point: Decomposes
Water Solubility: 2 g/100 mL
Other Solubilities: Soluble in alkali Citrates, Slightly Soluble in Alcohol; Insoluble in Ether
Flash Point: Combustible

RTECS Toxicity Data

Acute Oral: Mouse LD_{50} Dose: 147 mg/kg; Toxic Effects: Behavioral - Somnolence (general depressed activity); Behavioral - Convulsions or effect on seizure threshold; Lungs, Thorax, or Respiration - Dyspnea.
Tumorigenic: Mouse Route: Subcutaneous; Dose: 4200 mg/kg/21W-I; Toxic Effects: Tumorigenic - Equivocal tumorigenic agent by RTECS criteria; Tumorigenic - Tumors at site of application.

Hazard Overviews

Health: Toxic. Other Acute Effects: toxic if swallowed; do not breathe dust; avoid contact with skin and eyes.
Fire: Combustible. Hazards: hazardous polymerization will not occur. Extinguishing agents: water spray; carbon dioxide,

dry chemical powder or appropriate foam. Precautions: combustible liquid.

Reactivity: Hazardous polymerization will not occur.

Carcinogenicity: IARC - Not listed; NIOSH - Not listed; NTP - Not listed; ACGIH - Not listed; OSHA - Not listed; EPA - Not listed; MAK - Not listed

Exposure Limits

OSHA PEL Vacated 1989 Limits: STEL: 1 mg/m^3; as Fe salts.

ACGIH TLV: TWA: 1 mg/m^3; as Fe Salt.

NIOSH REL: TWA: 1 mg/m^3; as Fe salts.

Respirator Recommendation

Exposure Range: >1 to 10 mg/m^3 Air Purifying, Negative Pressure, Half Mask

Exposure Range: >10 to 100 mg/m^3 Air Purifying, Negative Pressure, Full Face

Exposure Range: >100 to 1000 mg/m^3 Supplied Air, Constant Flow/Pressure Demand, Full Face

Exposure Range: >1000 to unlimited mg/m^3 Self-contained Breathing Apparatus, Pressure Demand, Full Face

Cartridge Color: dust/mist filter (use P100 or consult supervisor for appropriate dust/mist filter)

Note: as iron salts, soluble

Environmental

Regulations

RCRA 40CFR: Not listed

CERCLA: 40CFR 302.4: Not listed

SARA 40CFR 372.65: Not listed

SARA EHS 40CFR 355: Not listed

TSCA: Not listed

FER7510 **CAS #: 516-03-0**

FERROUS OXALATE

EINECS Number: 208-217-4

Molecular Formula: C_2FeO_4

Formula Weight: 143.87

Synonyms: ETHANEDIOIC ACID,IRON(2+) SALT (1:1); FERROX; IRON OXALATE; IRON(2+) OXALATE; IRON PROTOXALATE; IRON(II) OXALATE; OXALIC ACID,IRON(2+) SALT (1:1)

Description: pale yellow, crystalline powder; odorless

Use: photographic developer for silver bromide-gelatin plates; for decorative glassware; to impact greenish-brown tint to optical glass (sunglasses, windshields, railroad car windows); pigment for plastics, paints, lacquers

Physical Properties

Freezing Point: Decomposes at 190 °C (374 °F)

Density: Dihydrate 2.28

Water Solubility: Slightly Soluble in Water

Other Solubilities: Soluble in dilute Mineral acids.

Hazard Overviews

Carcinogenicity: IARC - Not listed; NIOSH - Not listed; NTP - Not listed; ACGIH - Not listed; OSHA - Not listed; EPA - Not listed; MAK - Not listed

Exposure Limits

OSHA PEL Vacated 1989 Limits: STEL: 1 mg/m^3; as Fe salts.

ACGIH TLV: TWA: 1 mg/m^3; as Fe Salt.

NIOSH REL: TWA: 1 mg/m^3; as Fe salts.

Respirator Recommendation

Exposure Range: >1 to 10 mg/m^3 Air Purifying, Negative Pressure, Half Mask

Exposure Range: >10 to 100 mg/m^3 Air Purifying, Negative Pressure, Full Face

Exposure Range: >100 to 1000 mg/m^3 Supplied Air, Constant Flow/Pressure Demand, Full Face

Exposure Range: >1000 to unlimited mg/m^3 Self-contained Breathing Apparatus, Pressure Demand, Full Face

Cartridge Color: dust/mist filter (use P100 or consult supervisor for appropriate dust/mist filter)

Note: as iron salts, soluble

Environmental

Regulations

RCRA 40CFR: Not listed

CERCLA: 40CFR 302.4: Not listed

SARA 40CFR 372.65: Not listed

SARA EHS 40CFR 355: Not listed

TSCA: Listed

FER7820 **CAS #: 1345-25-1**

FERROUS OXIDE

EINECS Number: 215-721-8

Molecular Formula: FeO

Structured MF: FeO

Formula Weight: 71.85

Fe=O

Chemical Structure

Synonyms: C.I. 77489; IRON MONOOXIDE; IRON MONOXIDE; NATURAL WUESTITE

Description: jet-black powder or crystals

Use: in mfr of green, heat-absorbing glass; in steel mfr; in enamels; catalyst; to provide color, hiding power, &/or reinforcement in paints, coatings, construction materials, plastics, etc; catalysts in chem processing; micronutrient in animal feeds & fertilizers; in mfr of ferrites (ceramic magnets) for computer memory cores, hi-fi speakers, color tv, & other electrical & electronic products; magnetic pigments in audio & video tapes, cassettes, printing inks.

Physical Properties

Freezing Point: 1360 °C (2480 °F)

Specific Gravity: 5.7
Water Solubility: Practically Insoluble in Water
Other Solubilities: Insoluble in Alcohol.
Refraction Index: 2.23
Ionization Potential (eV): 8.9 +/-0.2
Flash Point: Spontaneously flammable

Hazard Overviews

Flammable

Health: Irritating to eyes/skin. Other Acute Effects: harmful if swallowed, inhaled, or absorbed through skin; overdose may have a corrosive effect on the gastrointestinal mucosa and be followed by necrosis, perforation and stricture formation; several hours may elapse before symptoms that can include epigastric pain, diarrhea, vomiting, nausea and hematemesis occur; after apparent recovery a person may experience metabolic acidosis, convulsions and coma hours or days later; further complications may develop leading to acute liver necrosis that can result in death due to hepatic coma.
Fire: Flammable. Hazards: emits toxic fumes. Extinguishing agents: carbon dioxide; dry chemical powder. Precautions: combustible liquid.
Reactivity: Incompatible with: strong oxidizing agents, sensitive to air.
Carcinogenicity: IARC - Not listed; NIOSH - Not listed; NTP - Not listed; ACGIH - Not listed; OSHA - Not listed; EPA - Not listed; MAK - Not listed
Primary Target Organs:

Eyes

Skin

Exposure Limits
OSHA PEL Vacated 1989 Limits: STEL: 1 mg/m^3; as Fe salts.
ACGIH TLV: TWA: 1 mg/m^3; as Fe Salt.
NIOSH REL: TWA: 1 mg/m^3; as Fe salts.
Respirator Recommendation
Exposure Range: >1 to 10 mg/m^3 Air Purifying, Negative Pressure, Half Mask
Exposure Range: >10 to 100 mg/m^3 Air Purifying, Negative Pressure, Full Face
Exposure Range: >100 to 1000 mg/m^3 Supplied Air, Constant Flow/Pressure Demand, Full Face
Exposure Range: >1000 to unlimited mg/m^3 Self-contained Breathing Apparatus, Pressure Demand, Full Face
Cartridge Color: dust/mist filter (use P100 or consult supervisor for appropriate dust/mist filter)
Note: as iron salts, soluble

Environmental

Regulations

RCRA 40CFR: Not listed
CERCLA: 40CFR 302.4: Not listed
SARA 40CFR 372.65: Not listed
SARA EHS 40CFR 355: Not listed
TSCA: Listed

FER8130 CAS #: 7720-78-7

FERROUS SULFATE

RTECS: NO8500000
EINECS Number: 231-753-5
Molecular Formula: FeH_2O_4S
Structured MF: $FeSO_4 \cdot 7H_2O$
Formula Weight: 151.91
Synonyms: COPPERAS; DURETTER; DUROFERON; EXSICCATED FERROUS SULFATE; EXSICCATED FERROUS SULPHATE; FEOSOL; FEOSPAN; FER-IN-SOL; FERO-GRADUMET; FERRALYN; FERRO-GRADUMET; FERROSULFAT; FERROSULFATE; FERRO-THERON; FERROUS SULFATE (1:1); FERROUS SULPHATE; FERSOLATE; GREEN VITRIOL; IRON (II) SULFATE (1:1); IRON MONOSULFATE; IRON PROTOSULFATE; IRON SULFATE; IRON SULFATE (1:1); IRON(2+) SULFATE; IRON(2+) SULFATE (1:1); IRON VITRIOL; IRON(II) SULFATE; IROSPAN; MOL-IRON; SAL CHALYBIS; SLOW FE; SLOW-FE; SULFERROUS; SULFURIC ACID IRON SALT (1:1); SULFURIC ACID,IRON(2+) SALT (1:1)
Use: in mfr of iron, iron cmpd, other sulfates; in iron electroplating; in fertilizer; in radiation dosimeters; reducing agent in chem processes; wood preservative; weed killer; in prevention of chlorosis in plants; in other pesticides, writing ink, process engraving & lithography; dye for leather; in etching aluminum, water treatment, anal chem (for nitrates); polymerization catalyst; med: hematinic, (vet) in iron deficiency; for manure to control fly breeding & deodorant for cesspool-like areas.

Physical Properties

Boiling Point: 300 °C (572 °F)
Freezing Point: 64 °C (147.2 °F)
Specific Gravity: 2.99 to 3.08
Water Solubility: Soluble in Water
Other Solubilities: Practically Insoluble in Alcohol /Ferrous Sulfate Heptahydrate/; Insoluble in Alcohol /Ferrous sulfate pentahydrate/.
pH: Aqueous solution is acid in reaction
Refraction Index: 1.471 to 1.486
Flash Point: Nonflammable

RTECS Toxicity Data

Acute Oral: Woman TD_{Lo} Dose: 10560 ug/kg; Toxic Effects: Gastrointestinal - Ulceration or bleeding from stomach. Woman TD_{Lo} Dose: 600 mg/kg; Toxic Effects: Behavioral - Somnolence (general depressed activity); Behavioral - Aggression; Gastrointestinal - Nausea or vomiting. Child TD_{Lo} Dose: 750 mg/kg; Toxic Effects: Cardiac - Pulse rate; Vascular - BP lowering not characterized in autonomic section; Gastrointestinal - Nausea or vomiting. Child TD_{Lo} Dose: 20 mg/kg; Toxic Effects: Brain and coverings - Recordings from specific areas of CNS; Behavioral -

Somnolence (general depressed activity); Behavioral - Coma. Child TD_{Lo} Dose: 150 mg/kg; Toxic Effects: Behavioral - Coma; Gastrointestinal - Hypermotility, diarrhea; Gastrointestinal - Nausea or vomiting. Woman LD_{Lo} Dose: 60 mg/kg; Toxic Effects: Behavioral - Somnolence (general depressed activity); Gastrointestinal - Nausea or vomiting; Kidney, Ureter, and Bladder - Chgs in tubules (inc acute renal failure, acute tubular necrosis. Infant LD_{Lo} Dose: 4400 ug/kg; Toxic Effects: Behavioral - Coma; Gastrointestinal - Nausea or vomiting; Kidney, Ureter, and Bladder - Chgs in tubules (inc acute renal failure, acute tubular necrosis. Child LD_{Lo} Dose: 390 mg/kg; Toxic Effects: Behavioral - Somnolence (general depressed activity); Behavioral - Anorexia (human); Nutritional and gross metabolic - Body temperature increase. Rat LD_{50} Dose: 319 mg/kg.

Acute Dermal: Rat LD_{50} Route: Subcutaneous Dose: 155 mg/kg. Mouse LD_{50} Route: Subcutaneous Dose: 60300 ug/kg.

Reproductive/Teratogenic: Rat Route: Oral; Dose: 7200 mg/kg; Duration: female 9-14D of pregnancy; Effects on Embryo or Fetus - Fetal death. Rat Route: Intratesticular; Dose: 12153 ug/kg; Duration: male 1D prior to mating; Paternal Effects - Testes, epididymis, sperm duct.

Mutagenic: Hamster Cytogenetic Analysis; Cell Type: fibroblast; Dose: 1250 mg/L. Hamster Cytogenetic Analysis; Cell Type: ovary; Dose: 5 mmol/L.

Tumorigenic: Mouse Route: Subcutaneous; Dose: 1600 mg/kg/16W-I; Toxic Effects: Tumorigenic - Equivocal tumorigenic agent by RTECS criteria; Tumorigenic - Tumors at site of application.

Hazard Overviews

Reactivity: Stable. Hazardous polymerization cannot occur. Avoid: exposure to water; alkalies; sunlight Incompatible with: alkalis; soluble carbonates; gold and silver salts; lead acetate; lime water; potassium iodide; potassium, sodium tartate; sodium borate; tannin. Hazardous decomposition products: sulfur oxides fumes.

Carcinogenicity: IARC - Not listed; NIOSH - Not listed; NTP - Not listed; ACGIH - Not listed; OSHA - Not listed; EPA - Not listed; MAK - Not listed

Exposure Limits

OSHA PEL Vacated 1989 Limits: STEL: 1 mg/m^3; as Fe salts.

ACGIH TLV: TWA: 1 mg/m^3; as Fe Salt.

NIOSH REL: TWA: 1 mg/m^3; as Fe salts.

Respirator Recommendation

Exposure Range: >1 to 10 mg/m^3 Air Purifying, Negative Pressure, Half Mask

Exposure Range: >10 to 100 mg/m^3 Air Purifying, Negative Pressure, Full Face

Exposure Range: >100 to 1000 mg/m^3 Supplied Air, Constant Flow/Pressure Demand, Full Face

Exposure Range: >1000 to unlimited mg/m^3 Self-contained Breathing Apparatus, Pressure Demand, Full Face

Cartridge Color: dust/mist filter (use P100 or consult supervisor for appropriate dust/mist filter)

Note: as iron salts, soluble

Environmental

Ecotoxicity: LC_{50} Shrimp 56 ppm /48 hr/ salt water /Conditions of bioassay not specified

Cleanup/Disposal: Guide No. 140: Keep combustibles (wood, paper, oil, etc.) away from spilled material. Do not touch damaged containers or spilled material unless wearing appropriate protective clothing. Stop leak if you can do it without risk. Do not get water inside containers. Small Dry Spills: With clean shovel place material into clean, dry container and cover loosely; move containers from spill area. Small Liquid Spills: Use a non-combustible material like vermiculite, sand or earth to soak up the product and place into a container for later disposal. Large Spills: Dike far ahead of liquid spill for later disposal. Following product recovery, flush area with water.

Environmental Physical Data

BCF: no food chain concentration potential

Regulations

RCRA 40CFR: Not listed

CERCLA: 40CFR 302.4: Listed per CWA Section 311(b)(4) RQ: 1000 lb (453.5 kg)

SARA 40CFR 372.65: Not listed

SARA EHS 40CFR 355: Not listed

TSCA: Listed

FER8440 CAS #: 7782-63-0

FERROUS SULFATE HEPTAHYDRATE

RTECS: NO8510000

Molecular Formula: $FeH_{14}O_{11}S$

Formula Weight: 278.05

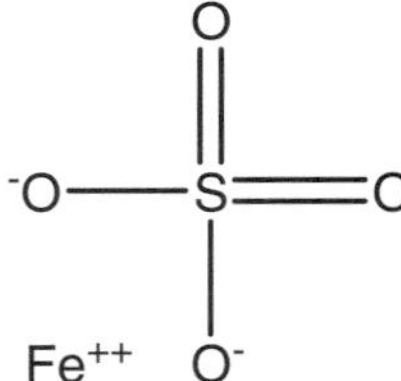

Chemical Structure

Synonyms: FESOFOR; FESOTYME; GREEN VITROL; HAEMOFORT; IRON PROTOSULFATE; IRON SULFATE HEPTAHYDRATE; IRON VITROL; IRONATE; IROSUL; MOL-IRON; PRESFERSUL

Description: blue-green crystals, granules; odorless

Physical Properties

Freezing Point: 64 °C (147.2 °F)

Specific Gravity: 1.89

Water Solubility: 16 g/100 g Cold Water

pH: 10% solution 3.7

Flash Point: Noncombustible

RTECS Toxicity Data

Acute Oral: Rat LD_{Lo} Dose: 1389 mg/kg. Mouse LD_{50} Dose: 1520 mg/kg.

Acute Dermal: Rabbit LD_{Lo} Route: Subcutaneous Dose: 279 mg/kg.

Mutagenic: Mouse Micronucleus Test; Cell Type: other cell types; Dose: 500 umol/L. Bacteria - E Coli Mutations in Microorganisms; Dose: 30 umol/L (-S9).

Hazard Overviews

Fire Diamond

Health: Irritating to eyes/skin/respiratory tract. Also Causes: ingestion can cause abdominal pain, constipation, and diarrhea. Chronic Effects: liver, spleen, and lymph damage; mottled lung appearance.

Fire: Noncombustible. Use extinguishing agents suitable for surrounding fire.

Reactivity: Stable. Hazardous polymerization cannot occur. Avoid: exposure to moisture; alkalies; sunlight. Incompatible with: alkalis; soluble carbonates; gold and silver salts; lead acetate; lime water; potassium iodide; potassium; sodium tartate; sodium borate; tannin. Hazardous decomposition products: toxic sulfur oxides fumes.

Carcinogenicity: IARC - Not listed; NIOSH - Not listed; NTP - Not listed; ACGIH - Not listed; OSHA - Not listed; EPA - Not listed; MAK - Not listed

Primary Target Organs:

Eyes

Skin
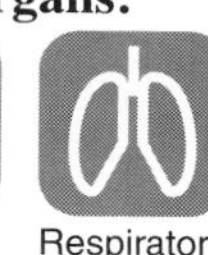
Respiratory System
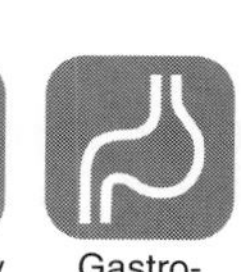
Gastro-intestinal

Liver

Lymphatic System

Exposure Limits

OSHA PEL Vacated 1989 Limits: STEL: 1 mg/m^3; as Fe salts.

ACGIH TLV: TWA: 1 mg/m^3; as Fe Salt.

NIOSH REL: TWA: 1 mg/m^3; as Fe salts.

Respirator Recommendation

Exposure Range: >1 to 10 mg/m^3 Air Purifying, Negative Pressure, Half Mask

Exposure Range: >10 to 100 mg/m^3 Air Purifying, Negative Pressure, Full Face

Exposure Range: >100 to 1000 mg/m^3 Supplied Air, Constant Flow/Pressure Demand, Full Face

Exposure Range: >1000 to unlimited mg/m^3 Self-contained Breathing Apparatus, Pressure Demand, Full Face

Cartridge Color: dust/mist filter (use P100 or consult supervisor for appropriate dust/mist filter)

Note: as iron salts, soluble

Environmental

Regulations

RCRA 40CFR: Not listed

CERCLA: 40CFR 302.4: Listed per CWA Section 311(b)(4) RQ: 1000 lb (453.5 kg)

SARA 40CFR 372.65: Not listed

SARA EHS 40CFR 355: Not listed

TSCA: Not listed

FER8750 **CAS #: 1317-37-9**

FERROUS SULFIDE

EINECS Number: 215-268-6

Molecular Formula: FeS

Structured MF: FeS

Formula Weight: 87.92

Fe═S

Chemical Structure

Synonyms: BLACK IRON SULFIDE; C.I. 77540; FERROUS MONOSULFIDE; IRON MONOSULFIDE; IRON PROTOSULFIDE; IRON SULFURET

Description: colorless hexagonal crystals when pure, usually gray to brownish-black lumps, rods or granular powder

Use: lab source of hydrogen sulfide; in the ceramic industry; paint pigment; in anodes; in lubricant coatings; treatment of exhaust gases & heavy metal pollution; pigment in glass containers; pigment in hair dyes & ceramics (former uses)

Physical Properties

Boiling Point: Decomposes

Freezing Point: 1194 °C (2181.2 °F)

Specific Gravity: 4.84

Water Solubility: 0.00062 g/100 cc Water at 18 °C

Other Solubilities: Insoluble in Nitric acid. Soluble in acids with evolution of Hydrogen Sulfide.

Hazard Overviews

Health: Irritating to eyes/skin/respiratory tract. Other Acute Effects: may be harmful by inhalation, ingestion, or skin absorption.

Fire: Hazards: emits toxic fumes. Extinguishing agents: water spray; carbon dioxide, dry chemical powder or appropriate foam. Precautions: combustible liquid.

Reactivity: Incompatible with: strong oxidizing agents, moisture sensitive, air sensitive, strong acids. Hazardous decomposition products: sulfur oxides, hydrogen sulfide gas.

Carcinogenicity: IARC - Not listed; NIOSH - Not listed; NTP - Not listed; ACGIH - Not listed; OSHA - Not listed; EPA - Not listed; MAK - Not listed

Primary Target Organs:

Eyes

Skin

Respiratory System

Exposure Limits

OSHA PEL Vacated 1989 Limits: STEL: 1 mg/m^3; as Fe salts.

ACGIH TLV: TWA: 1 mg/m^3; as Fe Salt.
NIOSH REL: TWA: 1 mg/m^3; as Fe salts.
Respirator Recommendation
Exposure Range: >1 to 10 mg/m^3 Air Purifying, Negative Pressure, Half Mask
Exposure Range: >10 to 100 mg/m^3 Air Purifying, Negative Pressure, Full Face
Exposure Range: >100 to 1000 mg/m^3 Supplied Air, Constant Flow/Pressure Demand, Full Face
Exposure Range: >1000 to unlimited mg/m^3 Self-contained Breathing Apparatus, Pressure Demand, Full Face
Cartridge Color: dust/mist filter (use P100 or consult supervisor for appropriate dust/mist filter)
Note: as iron salts, soluble

Environmental

Regulations
RCRA 40CFR: Not listed
CERCLA: 40CFR 302.4: Not listed
SARA 40CFR 372.65: Not listed
SARA EHS 40CFR 355: Not listed
TSCA: Listed

FER9060 CAS #: 12604-58-9

FERROVANADIUM DUST

RTECS: LK2900000
Molecular Formula: FeV
Structured MF: FeV
Formula Weight: 106.8
Synonyms: FERRO V; FERROVANADIUM
Description: gray, black dust; odorless

Physical Properties
Freezing Point: 1480 °C (2696 °F) to 1520 °C (2768 °F)
Vapor Pressure: ~ 0 mm Hg
Water Solubility: Insoluble
Flash Point: Noncombustible

Hazard Overviews

Explosive

Fire Diamond

Health: Irritating to eyes/respiratory tract. Chronic Effects: in animals causes: pulmonary changes consisting of bronchitis and perialveolitis (inflammation of the lungs).
Fire: Can be ignited by heat or flame. May form explosive dust-air mixtures. Use dry chemical, carbon dioxide, water spray, fog, or regular foam.
Reactivity: Stable. Hazardous polymerization cannot occur. Avoid: excessive dust generation; exposure to heat or flame; contact with strong oxidizers. Incompatible with: strong oxidizers. Hazardous decomposition products: iron and vanadium oxides.

Carcinogenicity: IARC - Not listed; NIOSH - Listed as carcinogen; NTP - Not listed; ACGIH - Not listed; OSHA - Not listed; EPA - Not listed; MAK - Not listed
Primary Target Organs:

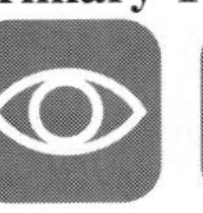
Eyes

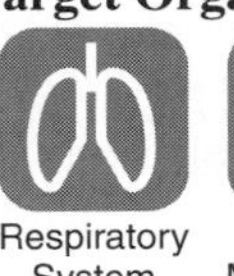
Respiratory System

Mucous Membranes

Exposure Limits
OSHA PEL: TWA: 1 mg/m^3.
OSHA PEL Vacated 1989 Limits: TWA: 1 mg/m^3; STEL: 3 mg/m^3.
ACGIH TLV: TWA: 1 mg/m^3; STEL: 3 mg/m^3.
NIOSH REL: TWA: 1 mg/m^3; as Fe (Salts). STEL: 3 mg/m^3.
NIOSH IDLH: 500 mg/m^3.
DFG MAK: TWA: 1 mg/m^3.
Respirator Recommendation
Exposure Range: >1 to 10 mg/m^3 Air Purifying, Negative Pressure, Half Mask
Exposure Range: >10 to 100 mg/m^3 Air Purifying, Negative Pressure, Full Face
Exposure Range: >100 to <500 mg/m^3 Supplied Air, Constant Flow/Pressure Demand, Full Face
Exposure Range: 500 to unlimited mg/m^3 Self-contained Breathing Apparatus, Pressure Demand, Full Face
Cartridge Color: dust/mist filter (use P100 or consult supervisor for appropriate dust/mist filter)

Environmental

Regulations
RCRA 40CFR: Not listed
CERCLA: 40CFR 302.4: Not listed
SARA 40CFR 372.65: Not listed
SARA EHS 40CFR 355: Not listed
TSCA: Not listed

FIB6000 CAS #: 65997-17-3

FIBROUS GLASS (SILICATE BASE)

EINECS Number: 266-046-0
Molecular Formula: Unspecified or Variable
Description: colorless fibers; odorless

Physical Properties
Freezing Point: 1000 °C (1832 °F) to 2000 °C (3632 °F)
Specific Gravity: About 2.5
Water Solubility: Negligible
Flash Point: Noncombustible

Hazard Overviews

Fire Diamond

Health: Irritating to the eyes/skin/respiratory tract. Chronic exposure may result in dermatitis and bronchitis. A suspect cancer hazard.

Fire: Noncombustible. Use extinguishing agents suitable for surrounding fire.

Reactivity: Stable. Hazardous polymerization cannot occur. Incompatible with: hot, strong alkaline solutions; hydrofluoric, fluosilicic, and phosphoric acids. Hazardous decomposition products: carbon dioxide; carbon monoxide; traces of hydrogen cyanide.

Carcinogenicity: IARC - Not listed; NIOSH - Not listed; NTP - Not listed; ACGIH - Not listed; OSHA - Not listed; EPA - Not listed; MAK - Not listed

Primary Target Organs:

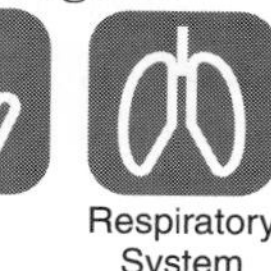

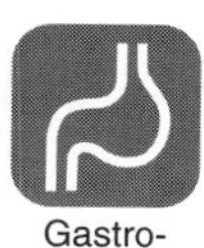

Environmental

Regulations

RCRA 40CFR: Not listed
CERCLA: 40CFR 302.4: Not listed
SARA 40CFR 372.65: Not listed
SARA EHS 40CFR 355: Not listed
TSCA: Listed

FIN5000 CAS #: 98319-26-7

FINASTERIDE

RTECS: CL5245000
Molecular Formula: $C_{23}H_{36}N_2O_2$
Formula Weight: 372.55

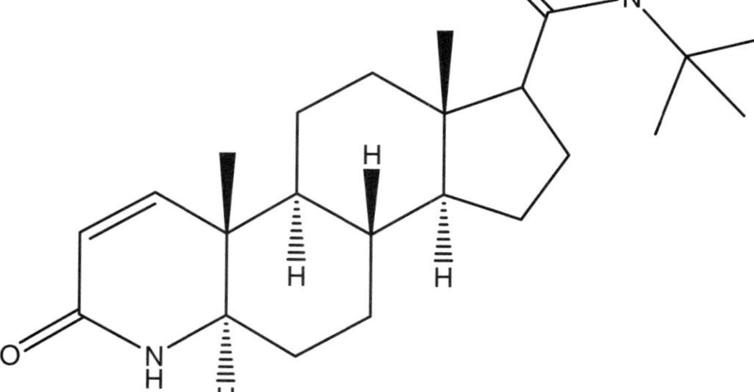

Chemical Structure

Synonyms: 4-AZAANDROST-1-ENE-17-CARBOXAMIDE,N-(1,1-DIMETHYLETHYL)-3-OXO-,(5ALPHA,17BETA)-; 17 BETA-(N-TERT-BUTYLCARBAMOYL)-4-AZA-5ALPHA-ANDROST-1-EN-3-ONE; N-TERT-BUTYL-3-OXO-4-AZA-5ALPHA-ANDROST-1-ENE-17BETA-CARBOXAMIDE; (5 ALPHA,17BETA)-(1,1-DIMETHYLETHYL)-3-OXO-4-AZAANDROST-1-ENE-17-CARBOXAMIDE; L-652,931; N-(2-METHYL-2-PROPYL)-3-OXO-4-AZA-5-ALPHA-ANDROST-1-ENE-17-BETA-CARBOXAMIDE; MK-0906; MK-906; PROSCAR

Description: white to off-white crystalline solid

Use: medication: treatment of benign prostatic hypertrophy

Physical Properties

Freezing Point: 252 °C (485.6 °F) to 254 °C (489.2 °F)

Water Solubility: Very Slightly Soluble in Water

Other Solubilities: Freely Soluble in Chloroform, DMSO, Ethanol, Methanol, n-propanol; sparingly Soluble in Propylene Glycol, Polyethylene Glycol 400; very slightly Soluble in 0.1N hydrogen chloride, 0.1N sodium hydroxide.

RTECS Toxicity Data

Acute Oral: Rat LD_{50} Dose: 418 mg/kg; Toxic Effects: Behavioral - Somnolence (general depressed activity); Gastrointestinal - Ulceration or bleeding from stomach; Lungs, Thorax, or Respiration - Respiratory depression; Mouse LD_{50} Dose: 486 mg/kg; Toxic Effects: Behavioral - Somnolence (general depressed activity); Behavioral - Ataxia; Lungs, Thorax, or Respiration - Respiratory depression.

Acute Dermal: Rat LD_{50} Route: Subcutaneous Dose: >2 gm/kg; Toxic Effects: Skin and appendages - Dermatitis, other.

Hazard Overviews

Health: Irritating to eyes/skin/respiratory tract. Toxic. Other Acute Effects: harmful if swallowed; may be harmful if inhaled; may be harmful if inhaled. Chronic Effects: may cause congenital malformation in the fetus; may cause reproductive disorders; target organ: male reproductive system.

Fire: Hazards: emits toxic fumes. Extinguishing agents: water spray; carbon dioxide, dry chemical powder or appropriate foam. Precautions: combustible liquid.

Carcinogenicity: IARC - Not listed; NIOSH - Not listed; NTP - Not listed; ACGIH - Not listed; OSHA - Not listed; EPA - Not listed; MAK - Not listed

Primary Target Organs:

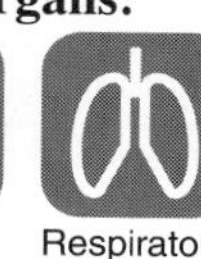

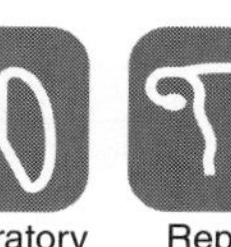

Environmental

Regulations

RCRA 40CFR: Not listed
CERCLA: 40CFR 302.4: Not listed
SARA 40CFR 372.65: Not listed
SARA EHS 40CFR 355: Not listed
TSCA: Not listed

FLO3000 CAS #: 519-95-9

FLORANTYRONE

EINECS Number: 208-279-2
Molecular Formula: $C_{20}H_{14}O_3$
Formula Weight: 302.31
Synonyms: ANCHOL; BILYN; CISTOPLEX; FLORANTYRON; FLUOCHOL; 8-FLUORANTHENEBUTYRIC ACID,GAMMA-OXO-; BETA-(8-FLUORANTHOYL)PROPIONIC ACID; BETA-(8-FLUORANTHYLOYL)PROPIONIC ACID; FLUORANTYRONE; 8-FLUORATHENEBUTANOIC ACID,GAMMA-OXO-; IDROBIL; IDROEPAR; GAMMA-OXO-8-FLUORANTHENEBUTANOIC ACID; GAMMA-OXO-8-FLUORANTHENEBUTYRIC ACID; SC 1674; ZANCHOL
Description: fine platelets
Use: medication: hydrocholeretic agent

Physical Properties

Freezing Point: 195 °C (383 °F)
Other Solubilities: Soluble in Methanol, Ethanol,aqueous solution of Sodium Carbonate.

Hazard Overviews

Carcinogenicity: IARC - Not listed; NIOSH - Not listed; NTP - Not listed; ACGIH - Not listed; OSHA - Not listed; EPA - Not listed; MAK - Not listed

Environmental

Regulations

RCRA 40CFR: Not listed
CERCLA: 40CFR 302.4: Not listed
SARA 40CFR 372.65: Not listed
SARA EHS 40CFR 355: Not listed
TSCA: Not listed

FLO6000 CAS #: 50-91-9

FLOXURIDINE

RTECS: YU7525000
EINECS Number: 200-072-5
Molecular Formula: $C_9H_{11}FN_2O_5$
Formula Weight: 246.21

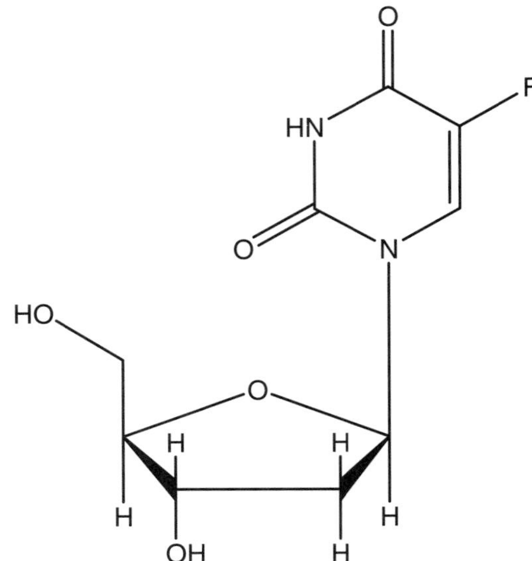

Chemical Structure

Synonyms: 2'-DEOXY-5-FLUOROURIDINE; DEOXYFLUOROURIDINE; 1-(2-DEOXY-BETA-D-RIBOFURANOSYL)-5-FLUOROURACIL; 1-BETA-D-2'-DEOXYRIBOFURANOSYL-5-FLUOROURACIL; 1-BETA-D-2'-DEOXYRIBOFURANOSYL-5-FLUROURACIL; FDUR; FLOXURIDIN; 5-FLUOR-1-(BETA-2'-DEOXYRIBOFURANOSYL)PYRIMIDIN-2,4(1H,3H)-DION; 5-FLUORO-2'-DEOXY-BETA-URIDINE; 5-FLUORO-2'-DEOXYURIDINE; 5-FLUORO-2-DEOXYURIDINE; 5-FLUORODEOXYURIDINE; BETA-5-FLUORO-2'-DEOXYURIDINE; FLUORODEOXYURIDINE; 5-FLUOROURACIL 2'-DEOXYRIBOSIDE; 5-FLUOROURACIL DEOXYRIBOSIDE; FLUORURIDINE DEOXYRIBOSE; 5-FUDR; FUDR; NSC-27640; NSC 27640; RO 5-0360; URIDINE,2'-DEOXY-5-FLUORO-
Description: white to off-white solid or crystals; odorless
Use: antiviral

Physical Properties

Freezing Point: 150 °C (302 °F) to 151 °C (303.8 °F)
Water Solubility: 1 g in about 2 ml Water
Other Solubilities: 1 g in about 4.5 ml Isopropanol, about 7.5 ml Methanol, about 12 ml Alcohol, about 27 ml Acetone Insoluble in Chloroform, Ether & Benzene
Flash Point: Not available; probably combustible

RTECS Toxicity Data

Acute Oral: Rat LD_{50} Dose: 215 mg/kg. Mouse LD_{50} Dose: 147 mg/kg.
Reproductive/Teratogenic: Rat Route: Intravenous; Dose: 50 mg/kg; Duration: female 11D of pregnancy; Specific Developmental Abnormalities - Musculoskeletal system. Rat Route: Intraperitoneal; Dose: 100 mg/kg; Duration: female 12D of pregnancy; Effects on Fertility - Post-implantation mortality; Specific Developmental Abnormalities - Craniofacial (including nose and tongue). Rat Route: Intraperitoneal; Dose: 100 mg/kg; Duration: female 12D of pregnancy; Specific Developmental Abnormalities - Craniofacial (including nose and tongue); Other developmental abnormalities. Rat Route: Intraperitoneal; Dose: 25 mg/kg; Duration: female 12D of pregnancy; Specific Developmental Abnormalities - Musculoskeletal

system. Rat Route: Intraperitoneal; Dose: 25 mg/kg; Duration: female 10D of pregnancy; Specific Developmental Abnormalities - Central nervous system; Eye, ear; Musculoskeletal system.

Mutagenic: Human DNA Damage; Cell Type: other cell types; Dose: 24 mg/L. Human DNA Damage; Cell Type: other cell types; Dose: 1 mmol/L. Human Unscheduled DNA Synthesis; Cell Type: HeLa cell; Dose: 100 nmol/L. Human DNA Inhibition; Cell Type: mammary gland; Dose: 100 umol/L. Human DNA Inhibition; Cell Type: lymphocyte; Dose: 10 umol/L. Human DNA Inhibition; Cell Type: other cell types; Dose: 500 ug/L. Human DNA Inhibition; Cell Type: HeLa cell; Dose: 10 umol/L. Human DNA Inhibition; Cell Type: lymphocyte; Dose: 25 mg/L. Human DNA Inhibition; Cell Type: other cell types; Dose: 1 umol/L. Human DNA Adduct; Cell Type: leukocyte; Dose: 500 nmol/L. Human Cytogenetic Analysis; Cell Type: leukocyte; Dose: 200 umol/L/24H. Human Cytogenetic Analysis; Cell Type: lymphocyte; Dose: 200 nmol/L. Human Cytogenetic Analysis; Cell Type: leukocyte; Dose: 50 mg/L. Human Other Mutation Test Systems; Cell Type: other cell types; Dose: 1 mg/L. Human Other Mutation Test Systems; Cell Type: other cell types; Dose: 1 umol/L. Human Other Mutation Test Systems; Cell Type: other cell types; Dose: 10 mg/L. Human Other Mutation Test Systems; Cell Type: HeLa cell; Dose: 5 mg/L. Human Other Mutation Test Systems; Cell Type: mammary gland; Dose: 100 umol/L.

Hazard Overviews

Health: May cause irritation. Toxic. Other Acute Effects: harmful if swallowed, inhaled, or absorbed through skin. Chronic Effects: possible teratogen; possible mutagen; laboratory experiments have shown mutagenic effects; overexposure may cause reproductive disorder(s) based on tests with laboratory animals; target organs: bone marrow, g.i. system, vascular system, immune system, heart.

Fire: Will burn. Extinguishing agents: water spray; carbon dioxide, dry chemical powder or appropriate foam. Precautions: combustible liquid.

Reactivity: Incompatible with: strong oxidizing agents, strong bases. Hazardous decomposition products: toxic fumes of: carbon monoxide, carbon dioxide, nitrogen oxides, hydrogen fluoride.

Carcinogenicity: IARC - Not listed; NIOSH - Not listed; NTP - Not listed; ACGIH - Not listed; OSHA - Not listed; EPA - Not listed; MAK - Not listed

Primary Target Organs:

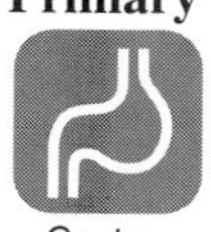
Gastro-intestinal

Cardio-vascular

Bone

Lymphatic System

Environmental

Environmental Physical Data

Octanol/Water Partition Coefficient: log K_{ow} = -1.16

Regulations

RCRA 40CFR: Not listed
CERCLA: 40CFR 302.4: Not listed
SARA 40CFR 372.65: Not listed
SARA EHS 40CFR 355: Not listed
TSCA: Not listed

FLU1000 CAS #: 69806-50-4

FLUAZIFOP-BUTYL

RTECS: UA3000000
EINECS Number: 274-125-6
Molecular Formula: $C_{19}H_{20}F_3NO_4$
Formula Weight: 383.40
Synonyms: BUTYL 2-(4-(5-TRIFLUOROMETHYL-2-PYRIDINYLOXY)PHENOXY)PROPANOATE; FUSILADE; HACHE UNO SUPER; HALOKON; IH 773B; ONECIDE; ONECIDE EC; PP 009; PROPIONIC ACID,2-(P-((5-(TRIFLUOROMETHYL)-2-PYRIDYL)OXY)PHENOXY)-,BUTYL ESTER; SL-236; TF 1169; TS-7236

Physical Properties

Boiling Point: 165 °C (329 °F) at 0.02 mm Hg
Freezing Point: 13 °C (55.4 °F)
Specific Gravity: 1.21000
Water Solubility: 1 mg/l
Other Solubilities: Miscible in acetone, cyclohexanone, hexane, methanol, dichloromethane, xylene, 24 g/l propylene glycol

RTECS Toxicity Data

Acute Oral: Rat LD_{50} Dose: 2910 mg/kg. Mouse LD_{50} Dose: 1490 mg/kg.
Acute Inhalation: Rat LC_{50} Dose: >5240 mg/m^3.
Acute Dermal: Rat LD_{50} Route: Skin; Dose: >6050 mg/kg. Rat LD_{50} Route: Subcutaneous Dose: >5 gm/kg.

Hazard Overviews

Carcinogenicity: IARC - Not listed; NIOSH - Not listed; NTP - Not listed; ACGIH - Not listed; OSHA - Not listed; EPA - Not listed; MAK - Not listed

Environmental

Regulations

RCRA 40CFR: Not listed
CERCLA: 40CFR 302.4: Not listed
SARA 40CFR 372.65: Listed
SARA EHS 40CFR 355: Not listed
TSCA: Not listed

Analytical Methods

Food: FDA 212.1, 232.1, 232.3, 232.4, 242.1; AOAC 984.08
Plasma: FDA 211.1, 231.1, 252

FLU1210 CAS #: 33245-39-5

FLUCHLORALIN

RTECS: XU5160000
EINECS Number: 251-426-0
Molecular Formula: $C_{12}H_{13}ClF_3N_3O_4$
Formula Weight: 355.70
Synonyms: BAS 392-H; BAS 3921H; BAS 3924H; BAS 392H; BAS 3920; BAS 3921; BAS 3922; BAS 392 04 H; BASALIN; BENZENAMINE,N-(2-CHLOROETHYL)-2,6-DINITRO-N-PROPYL-4-(TRIFLUOROMETHYL)-; N-(2-CHLOROETHYL)-2,6-DINITRO-N-PROPYL-4-(TRIFLUOROMETHYL)ANILINE; N-(2-CHLOROETHYL)-2,6-DINITRO-N-PROPYL-4-(TRIFLUOROMETHYL)BENZENAMIDE; N-(2-CHLOROETHYL)-ALPHA,ALPHA,ALPHA-TRIFLUORO-2,6-DINITRO-N-PROPYL-P-TOLUIDINE; N-PROPYL-N-(2-CHLOROETHYL)-2,6-DINITRO-4-TRIFLUOROMETHYLANILINE; N-PROPYL-N-(2-CHLOROETHYL)-ALPHA,ALPHA,ALPHA-TRIFLUORO-2,6-DINITRO-P-TOLUIDINE; P-TOLUIDINE,N-(2-CHLOROETHYL)-2,6-DINITRO-N-PROPYL-ALPHA,ALPHA,ALPHA-TRIFLUORO-; P-TOLUIDINE,N-(2-CHLOROETHYL)-ALPHA,ALPHA,ALPHA-TRIFLUORO-2,6-DINITRO-N-PROPYL-
Description: orange-yellow crystalline solid; faint, medicinal odor
Use: herbicide (former use)

Physical Properties

Freezing Point: 42 °C (107.6 °F) to 43 °C (109.4 °F)
Saturated Vapor Density: 1.200000498 kg/m^3
Vapor Pressure: 0.28 x10^{-4} mm Hg at 20 °C
Water Solubility: 10 ppm
Other Solubilities: g substance per 100 g of solvent at 20 °C: Acetone, Benzene, Chloroform, Ether; Ethyl Acetate more than 100.0 g; Cyclohexane 25.1 g, Ethanol 17.7 g; Olive Oil approximately 26.0 g.
Flash Point: 30 °C

RTECS Toxicity Data

Acute Oral: Rat LD_{50} Dose: 2940 mg/kg. Mouse LD_{50} Dose: 730 mg/kg.
Acute Inhalation: Rat LC_{50} Dose: 8400 mg/m^3/4hr.
Acute Dermal: Rabbit LD_{50} Route: Skin; Dose: >10 gm/kg. Rat LD_{50} Route: Skin; Dose: >4 gm/kg.
Mutagenic: Human Cytogenetic Analysis; Cell Type: lymphocyte; Dose: 2500 ug/L. Human Micronucleus Test; Cell Type: lymphocyte; Dose: 2500 ug/L.

Hazard Overviews

Flammable

Fire: Flammable.
Carcinogenicity: IARC - Not listed; NIOSH - Not listed; NTP - Not listed; ACGIH - Not listed; OSHA - Not listed; EPA - Not listed; MAK - Not listed

Environmental

Ecotoxicity: Fishes: bluegill sunfish 96h LC_{50} 16 mg/l rainbow trout 96h LC_{50} 12 mg/l
Environmental Fate: If released to the atmosphere, it will degrade rapidly in the vapor-phase by reaction with photochemically produced hydroxyl radicals (half-life of about 5 hr). If applied to soil surfaces, it can dissipate rapidly through photodecomposition and volatilization; half-lives of 1.5-4.0 days have been observed following surface application; however, immediate soil incorporation greatly increases soil persistence. Various persistence studies have reported soil half-lives ranging from several weeks to several months. Within soil systems, biodegradation is the probable route of disappearance, but it appears to be much slower than surface photolysis or volatilization. Under field conditions, it has not been found to leach. If released to water, it can degrade through photodecomposition and biodegradation. Volatilization may contribute to its loss from water.

Environmental Physical Data

Henry's Law Constant: 5.63 x10^{-5}
Sorption Partition Coefficient: K_{OC} = 3600
BCF: estimated at 250

Regulations

RCRA 40CFR: Not listed
CERCLA: 40CFR 302.4: Not listed
SARA 40CFR 372.65: Not listed
SARA EHS 40CFR 355: Not listed
TSCA: Not listed

Analytical Methods

Soil: SW846 8270B, 8270C
Water / Groundwater: EPA 646
Food: FDA 212.1, 212.2, 232.1, 232.4, 242.1
Plasma: EPA 29; FDA 211.1, 231.1, 252

FLU1420 CAS #: 70124-77-5

FLUCYTHRINATE

RTECS: CY1578620
EINECS Number: 274-322-7
Molecular Formula: $C_{26}H_{23}F_2NO_4$
Formula Weight: 451.48
Synonyms: AC-222705; AC 222705; (RS)-ALPHA-CYANO-3-PHENOXYBENZYL(S)-2-(4-DIFLUOROMETHOXYPHENYL)-3-METHYLBUTYRATE; (+ -)-CYANO-(3-PHENOXYPHENYL)METHYL(+)-4-(DIFLUOROMETHOXY)-ALPHA-(METHYLETHYL) BENZENEACETATE; (+-)-CYANO(3-PHENOXYPHENYL)METHYL(+)-4-(DIFLUOROMETHOXY)-ALPHA-(1-METHYLETHYL)BENZENEACETATE; (RS)-CYANO-(3-PHENOXYPHENYL)METHYL(S)-4-(DIFLUOROMETHOXY)-ALPHA-(1-METHYLETHYL)-BENZENEACETATE; CYBOLT; CYTHRIN; 4-(DIFLUOROMETHOXY)-ALPHA-(1-METHYLETHYL)BENZENEACETICACID CYANO(3-PHENOXY-PHENYL)METHYL ESTER; FUNCHIONG JUJR; GUARDIAN; PAY-OFF; STOCK GUARD
Description: viscous liquid; faint ester-like odor

Use: insecticide; for control of wide range of insect pests, particularly on cotton (bollworms, leafworms, sucking insects, whiteflies, beetles, etc), & fruit trees (lepidoptera, homoptera, coleoptera, etc); on vines, strawberries, citrus fruit, bananas, vegetables, soya beans, cereals, sugar beet, sunflowers, tobacco, ornamentals, etc; efficacy at higher temperatures greater than for several other pyrethroid insecticides; med: (vet) ectoparasiticide.

Physical Properties

Boiling Point: 108 °C (226 °F) at 0.35 mm Hg
Specific Gravity: 1.189 at 22 °C
Saturated Vapor Density: 1.40709099 kg/m^3
Vapor Pressure: 0.0012 mPa at 25 °C
Water Solubility: 1 mg/L at 21 °C
Other Solubilities: in corn oil 560, cottonseed oil 300, soya bean oil 300, hexane 90 (all in g/l at 21 °C).
Refraction Index: 1.541 at 25 °C/D
Flash Point: Burns with difficulty

RTECS Toxicity Data

Acute Oral: Rat LD_{50} Dose: 67 mg/kg. Mouse LD_{50} Dose: 76 mg/kg.
Acute Inhalation: Rat LC_{50} Dose: 4850 mg/m^3/4hr.
Acute Dermal: Rabbit LD_{50} Route: Skin; Dose: >1 gm/kg.

Hazard Overviews

Fire: Will burn.
Carcinogenicity: IARC - Not listed; NIOSH - Not listed; NTP - Not listed; ACGIH - Not listed; OSHA - Not listed; EPA - Not listed; MAK - Not listed

Environmental

Ecotoxicity: LC_{50} Rainbow trout 0.32 ug/l/96 hr /Conditions of bioassay not specified LD_{50} Mallard duck oral >2510 mg/kg LC_{50} Bob white quail dietary 3443 mg/kg diet/8 day LD_{50} Honey bee 0.078 ug/bee (topical application as a dust)
Environmental Fate: If released to the atmosphere, it will degrade rapidly in the vapor phase by reaction with photochemically produced hydroxyl radicals (estimated half-life of 6 hr). Particulate phase will be removed physically from air by wet and dry deposition. If released to soil or water, it can degrade through biodegradation, hydrolysis and photodegradation. Various persistence studies have suggested that it will have environmental half-lives of roughly one week to several weeks in soil or natural water. Estimated log K_{oc} values of 4.31-4.75 suggest that partitioning from water to sediment and suspended matter will occur; also these log K_{oc} values suggest that leaching should not occur in most soils. Persistence in sediment is expected to be longer than in the water-phase.

Environmental Physical Data

Henry's Law Constant: estimated at 8.61 $x10^{-8}$
Octanol/Water Partition Coefficient: log K_{ow} = 6.20
Sorption Partition Coefficient: log K_{oc} = estimated at 4.75
BCF: estimated at 3 $x10^4$

Regulations

RCRA 40CFR: Not listed
CERCLA: 40CFR 302.4: Not listed
SARA 40CFR 372.65: Not listed
SARA EHS 40CFR 355: Not listed
TSCA: Not listed

Analytical Methods

Food: FDA 212.2, 232.4, 242.1

FLU1630 CAS #: 2022-85-7

FLUCYTOSINE

RTECS: HA6040000
EINECS Number: 217-968-7
Molecular Formula: $C_4H_4FN_3O$
Formula Weight: 129.09

NH2 F N HO N

Chemical Structure

Synonyms: ALCOBON; 4-AMINO-5-FLUORO-2(1H)-PYRIMIDINONE; ANCOBON; ANCOTIL; CYTOSINE,5-FLUORO-; 5-FC; FLUCYTOSIN; FLUOCYTOSINE; 5-FLUOROCYSTOSINE; 5-FLUOROCYTOSIN; 5-FLUOROCYTOSINE; 2-HYDROXY-4-AMINO-5-FLUOROPYRIMIDINE; 2(1H)-PYRIMIDINONE,4-AMINO-5-FLUORO-; RO 2-9915
Description: white crystalline solid; odorless
Use: antifungal

Physical Properties

Boiling Point: Decomposes at 295 °C (563 °F) to 297 °C (566.6 °F)
Freezing Point: Decomposes at 295 °C (563 °F) to 297 °C (566.6 °F)
Water Solubility: 1.5 g Soluble in 100 ml Water at 25 °C
Other Solubilities: 1 g in about 12 ml 0.1 N HCL.

RTECS Toxicity Data

Acute Oral: Rat LD_{50} Dose: >15 gm/kg. Mouse LD_{50} Dose: >15 gm/kg.
Acute Dermal: Rat LD_{50} Route: Subcutaneous Dose: 3336 mg/kg. Mouse LD_{50} Route: Subcutaneous Dose: 1 gm/kg.
Mutagenic: Yeast - S Cerevisiae Mutations in Microorganisms; Dose: 8 mg/L (-S9). Other Microorganisms DNA Inhibition; Dose: 100 mg/L.

Hazard Overviews

Health: Irritating. Harmful. Other Acute Effects: harmful if swallowed, inhaled, or absorbed through skin. Chronic

Effects: overexposure may cause reproductive disorder(s) based on tests with laboratory animals; possible teratogen; target organ: bone marrow.

Fire: Hazards: emits toxic fumes. Extinguishing agents: water spray; carbon dioxide, dry chemical powder or appropriate foam. Precautions: combustible liquid.

Reactivity: Incompatible with: strong oxidizing agents, may discolor on exposure to light. Hazardous decomposition products: toxic fumes of: carbon monoxide, carbon dioxide, nitrogen oxides, hydrogen fluoride.

Carcinogenicity: IARC - Not listed; NIOSH - Not listed; NTP - Not listed; ACGIH - Not listed; OSHA - Not listed; EPA - Not listed; MAK - Not listed

Primary Target Organs:

Eyes

Skin

Respiratory System

Bone

Environmental

Regulations

RCRA 40CFR: Not listed
CERCLA: 40CFR 302.4: Not listed
SARA 40CFR 372.65: Not listed
SARA EHS 40CFR 355: Not listed
TSCA: Not listed

FLU1840 CAS #: 127-31-1

FLUDROCORTISONE

RTECS: TU5025000
EINECS Number: 204-833-2
Molecular Formula: $C_{21}H_{29}FO_5$
Formula Weight: 380.46

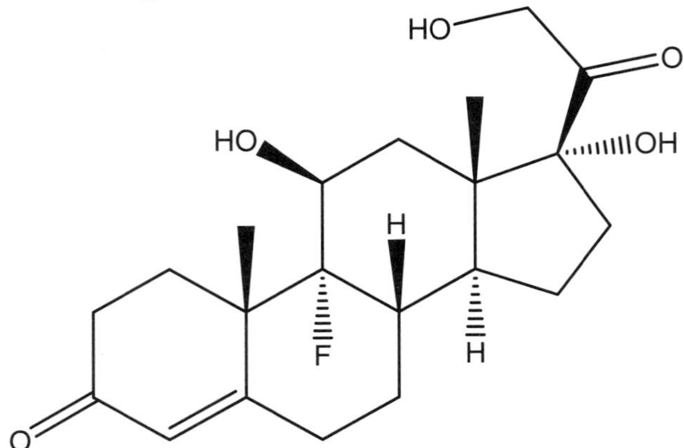

Chemical Structure

Synonyms: ALFLORONE; ASTONIN-H; F-COL; F-CORTEF; FLORINEF; 9ALPHA-FLUDROCORTISONE; FLUDROCORTONE; FLUDRONE; FLUDRONEF; FLUOHYDRISONE; FLUOHYDROCORTISONE; 9-ALPHA-FLUOROCORTISOL; 9-FLUOROCORTISOL; 9ALPHA-FLUOROCORTISOL; FLUOROCORTISOL; FLUOROCORTISONE; 9-ALPHA-FLUOROHYDROCORTISONE; 9-FLUOROHYDROCORTISONE; 9ALPHA-FLUOROHYDROCORTISONE; 9-ALPHA-FLUORO-17-HYDROXYCORTICOSTERONE; 9ALPHA-FLUORO-17-HYDROXYCORTICOSTERONE; 9-ALPHA-FLUORO-11-BETA,17-ALPHA,21-TRIHYDROXY-4-PREGNENE-3,20-DIONE; 9-FLUORO-11-BETA,17,21-TRIHYDROXYPREGN-4-ENE-3,20-DIONE; PREGN-4-ENE-3,20-DIONE,9-FLUORO-11,17,21-TRIHYDROXY-,(11BETA)-; PREGN-4-ENE-3,20-DIONE,9-FLUORO-11BETA,17,21-TRIHYDROXY-; U 5963

Description: crystals; odorless or practically odorless

Use: medication: anti-inflammatory agent; vet: synthetic adrenocortical steroid; medication: mineralocorticoid

Physical Properties

Boiling Point: Decomposes at 260 °C (500 °F) to 262 °C (503.6 °F)

Freezing Point: Decomposes at 260 °C (500 °F) to 262 °C (503.6 °F)

Water Solubility: 0.14 mg/mL at 0 °C

Other Solubilities: Slightly Soluble in Ether; Sparingly Soluble in Alcohol & Chloroform.

RTECS Toxicity Data

Reproductive/Teratogenic: Rat Route: Parenteral; Dose: 8750 ug/kg; Duration: female 7D prior to mating; Maternal Effects - Uterus, cervix, vagina. Rabbit Route: Oral; Dose: 1 mg/kg; Duration: female 13-16D of pregnancy Effects on Embryo or Fetus - Fetal death. Rabbit Route: Oral; Dose: 2 mg/kg; Duration: female 13-16D of pregnancy Specific Developmental Abnormalities - Craniofacial (including nose and tongue).

Hazard Overviews

Health: Irritating to eyes/skin/respiratory tract. Harmful. Other Acute Effects: may be harmful by inhalation, ingestion, or skin absorption; exposure can cause infection of the throat and mouth; hoarseness; disturbances of electrolyte balance manifested by retention of sodium and water resulting in odema and hypertension; exposure to high concentrations can cause moon face; hirsutism; buffalo hump; flushing; increased bruising; acne; sometimes leading to a fully developed Cushing's syndrome; exposure can cause cardia edema; increased blood volume; hypertension; cardiac arrhythmias; enlargement of the heart; headaches; weakness of the extremities. Chronic Effects: possible risk of congenital malformation in the fetus; target organs: adrenal gland, kidneys, thyroid, pituitary, female reproductive system.

Fire: Hazards: emits toxic fumes. Extinguishing agents: noncombustible; use extinguishing media appropriate to surrounding fire conditions. Precautions: combustible liquid.

Reactivity: Stable. Hazardous polymerization will not occur. Incompatible with: strong oxidizing agents. Hazardous decomposition products: toxic fumes of: carbon monoxide, carbon dioxide, hydrogen fluoride.

Carcinogenicity: IARC - Not listed; NIOSH - Not listed; NTP - Not listed; ACGIH - Not listed; OSHA - Not listed; EPA - Not listed; MAK - Not listed

Primary Target Organs:

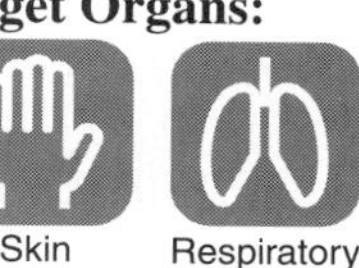

Eyes Skin Respiratory System Kidneys Repro-ductive Glandular System

Environmental

Regulations

RCRA 40CFR: Not listed
CERCLA: 40CFR 302.4: Not listed
SARA 40CFR 372.65: Not listed
SARA EHS 40CFR 355: Not listed
TSCA: Not listed

FLU2050 CAS #: 4301-50-2

FLUENETIL

RTECS: DU8335000
Molecular Formula: $C_{16}H_{15}FO_2$
Formula Weight: 258.31
Synonyms: ACETIC ACID,4-BIPHENYLYL-,2-FLUOROETHYL ESTER; 4-BIFENILACETATO DE 2-FLUOROETILO; 2-(4-BIFENIL)ACETO DE 2-FLUOROETIL; 2-(4-BIFENILIL)-ACETO DI 2-FLUOROETILE; (1,1'-BIPHENYL)-4-ACETIC ACID,2-FLUOROETHYL ESTER; 4-BIPHENYLACETIC ACID,2-FLUOROETHYL ESTER; 2-(4-BIPHENYLYL) ACETATE DE 2-FLUOROETHYLE; BIPHENYLYL-4,2-ACETATE DE (FLUORO-2-ETHYLE); EPA PESTICIDE CHEMICAL CODE 462200; FLUENETHYL; FLUENTHYL; FLUENYL; 2-FLUOR-ETHYL-4-BIPHENYLACETAT; 2-FLUORETHYL-4-BIPHENYLACETAT; 2-FLUORETHYLESTER KYSELINY XENYLOCTOVE; 2-FLUOROETHYL (1,1'-BIPHENYL)-4-ACETATE; 2-FLUOROETHYL 4-BIPHENYLACETATE; 2-FLUOROETHYL BIPHENYL-4-YLACETATE; 2-FLUOROETHYL-4-BIFENYLACETAAT; LABROL EC; LAMBROL; M 2060; MYTROL; TH 367-1

RTECS Toxicity Data

Acute Oral: Rat LD_{50} Dose: 6 mg/kg.
Acute Dermal: Rat LD_{Lo} Route: Skin; Dose: 4 mg/kg.

Hazard Overviews

Carcinogenicity: IARC - Not listed; NIOSH - Not listed; NTP - Not listed; ACGIH - Not listed; OSHA - Not listed; EPA - Not listed; MAK - Not listed

Environmental

Regulations

RCRA 40CFR: Not listed
CERCLA: 40CFR 302.4: Not listed
SARA 40CFR 372.65: Not listed TPQ: 100/10000 lb
SARA EHS 40CFR 355: Listed TPQ: 100 lb
TSCA: Not listed

FLU2260 CAS #: 148-56-1

FLUMETHAZIDE

RTECS: DK9800000
EINECS Number: 205-717-4
Molecular Formula: $C_8H_6F_3N_3O_4S_2$
Formula Weight: 329.28
Synonyms: ADEMIL; ADEMOL; FLUDEMIL; FLUMETHIAZID; FLUMETHIAZIDE; FLUMETIAZID; ROUTRAX; 7-SULFAMOYL-6-TRIFLUOROMETHYL-2H-1,2,4-BENZOTHIADIAZINE1,1-DIOXIDE; 6-(TRIFLUOROMETHYL)-2H-1,2,4-BENZOTHIADIAZINE-7-SULFONAMIDE 1,1-DIOXIDE; 6-(TRIFLUOROMETHYL)-1,2,4-BENZO-THIADIAZINE-7-SULFONAMIDE1,1-DIOXIDE; 6-(TRIFLUOROMETHYL)-1,4,2-BENZOTHIADIAZINE-7-SULFONAMIDO 1,1-DIOXIDE; 6-TRIFLUOROMETHYL-7-SULFAMOYL-4H-1,4,2-BENZOTHIADIAZINE 1,1-DIOXIDE; 6-TRIFLUOROMETHYL-7-SULFAMOYL-4H-1,2,4-BENZOTHIADIAZINE1,1-DIOXIDE; 6-TRIFLUOROMETHYL-7-SULFAMYL-1,2,4-BENZOTHIADIAZINE1,1-DIOXIDE; 6-TRIFLUOROMETHYL-7-SULFAMYL-1,2,4-BENZOTHIADIAZINE-1,1-DIOXIDE; TRIFLUOROMETHYLTHIAZIDE
Description: crystals
Use: carbonic anhydrase inhibitor

Physical Properties

Boiling Point: Decomposes at 305.4 °C (582 °F) to 307.8 °C (586 °F)
Freezing Point: Decomposes at 305.4 °C (581.72 °F) to 307.8 °C (586.04 °F)
Water Solubility: Sparingly Soluble in Water
Other Solubilities: Practically Insoluble in Ethyl Acetate, Methyl Ethyl Ketone, Benzene, Toluene; Soluble in Methanol, Ethanol, Dimethyl Formamide

RTECS Toxicity Data

Acute Oral: Mouse LD_{50} Dose: >8 gm/kg; Toxic Effects: Peripheral nerve and sensation - Spastic parapysis with or without sensory change; Behavioral - Convulsions or effect on seizure threshold; Lungs, Thorax, or Respiration - Other changes.

Hazard Overviews

Carcinogenicity: IARC - Not listed; NIOSH - Not listed; NTP - Not listed; ACGIH - Not listed; OSHA - Not listed; EPA - Not listed; MAK - Not listed

Environmental

Regulations

RCRA 40CFR: Not listed
CERCLA: 40CFR 302.4: Not listed
SARA 40CFR 372.65: Not listed
SARA EHS 40CFR 355: Not listed
TSCA: Not listed

FLU2470 CAS #: 67-73-2

FLUOCINOLONE ACETONIDE

RTECS: TU3830000
EINECS Number: 200-668-5
Molecular Formula: $C_{24}H_{30}F_2O_6$
Formula Weight: 452.50

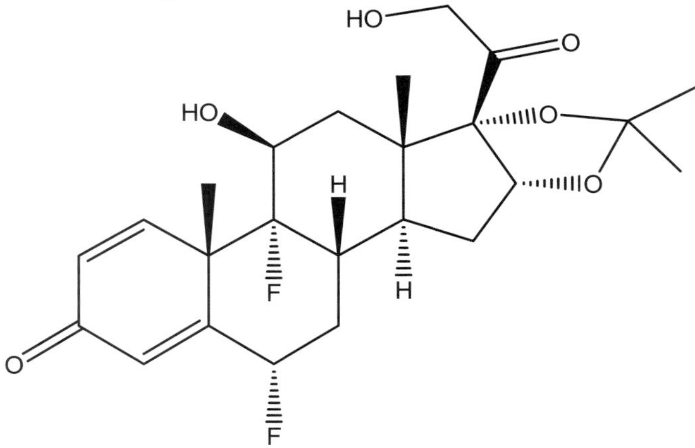

Chemical Structure

Synonyms: DERMALAR; 6-ALPHA,9-ALPHA-DIFLUORO-16-ALPHA-HYDROXYPREDNISOLONE 16,17-ACETONIDE; 6ALPHA,9ALPHA-DIFLUORO-16ALPHA-HYDROXYPREDNISOLONE16,17-ACETONIDE; FLUCINAR; FLUCORT; FLUOCINOLONE 16,17-ACETONIDE; 6ALPHA-FLUOROTRIAMCINOLONE ACETONIDE; FLUOVITIF; JELLIN; LOCALYN; LOCALYN SYNTEX; PERCUTINA; PREGNA-1,4-DIENE-3,20-DIONE,6,9-DIFLUORO-11,21-DIHYDROXY-16,17-((1-METHYLETHYLIDENE)BIS(OXY))-,(6ALPHA,11BETA,16ALPHA)-; PREGNA-1,4-DIENE-3,20-DIONE,6ALPHA,9-DIFLUORO-11BETA,16ALPHA,17,21-TETRAHYDROXY-,CYCLIC 16,17-ACETAL WITH ACETONE; RADIOCIN; RS-1401 AT; SINALAR; SYNALAR; SYNAMOL; SYNANDONE; SYNANDRONE; SYNSAC
Description: white, crystalline powder; odorless
Use: medication: glucocorticoid; anti-inflammatory agent

Physical Properties

Boiling Point: Decomposes
Freezing Point: 266 °C (510.8 °F)
Water Solubility: Insoluble in Water
Other Solubilities: Soluble in Alcohol, Acetone, & Methanol; Slightly Soluble in Chloroform

RTECS Toxicity Data

Acute Oral: Rat LD_{50} Dose: >4 gm/kg. Mouse LD_{50} Dose: >4 gm/kg.
Acute Dermal: Rat LD_{50} Route: Subcutaneous Dose: 108 mg/kg. Mouse LD_{50} Route: Subcutaneous Dose: 200 mg/kg.
Chronic (Multiple Dose) Dermal: Rat Route: Subcutaneous; Dose: 1750 ug/kg/35D-I; Toxic Effects: Nutritional and gross metabolic - Weight loss or decreased weight gain; DEATH.
Reproductive/Teratogenic: Rat Route: Subcutaneous; Dose: 1 mg/kg; Duration: female 1-20D of pregnancy; Effects on Fertility - Litter size; Effects on Embryo or Fetus - Fetotoxicity. Rabbit Route: Subcutaneous; Dose: 1200 ug/kg; Duration: female 8-31D of pregnancy; Effects on Newborn - Stillbirth. Rabbit Route: Subcutaneous; Dose: 1625 ug/kg; Duration: female 6-18D of pregnancy; Effects on Fertility - Abortion.
Mutagenic: Human Other Mutation Test Systems; Cell Type: lymphocyte; Dose: 500 nmol/L. Human Other Mutation Test Systems; Route: Skin; Dose: 2000 ppm.

Hazard Overviews

Health: Irritating to eyes/skin/respiratory tract. Harmful. Other Acute Effects: harmful if swallowed, inhaled, or absorbed through skin; possible risk of irreversible effects. Chronic Effects: overexposure may cause reproductive disorder(s) based on tests with laboratory animals; laboratory experiments have shown mutagenic effects
Fire: Hazards: emits toxic fumes. Extinguishing agents: water spray; carbon dioxide, dry chemical powder or appropriate foam. Precautions: combustible liquid.
Reactivity: Incompatible with: strong oxidizing agents. Hazardous decomposition products: toxic fumes of: carbon monoxide, carbon dioxide hydrogen fluoride.
Carcinogenicity: IARC - Not listed; NIOSH - Not listed; NTP - Not listed; ACGIH - Not listed; OSHA - Not listed; EPA - Not listed; MAK - Not listed
Primary Target Organs:

Eyes

Skin

Respiratory System

Environmental

Regulations

RCRA 40CFR: Not listed
CERCLA: 40CFR 302.4: Not listed
SARA 40CFR 372.65: Not listed
SARA EHS 40CFR 355: Not listed
TSCA: Not listed

FLU2680 CAS #: 2164-17-2

FLUOMETURON

RTECS: YT1575000
DOT: UN2767; UN2768; UN3001; UN3002; IMO3.2; IMO6.1
EINECS Number: 218-500-4
Molecular Formula: $C_{10}H_{11}F_3N_2O$
Structured MF: $CF_3C_6H_4NHCON(CH_3)_2$
Formula Weight: 232.21

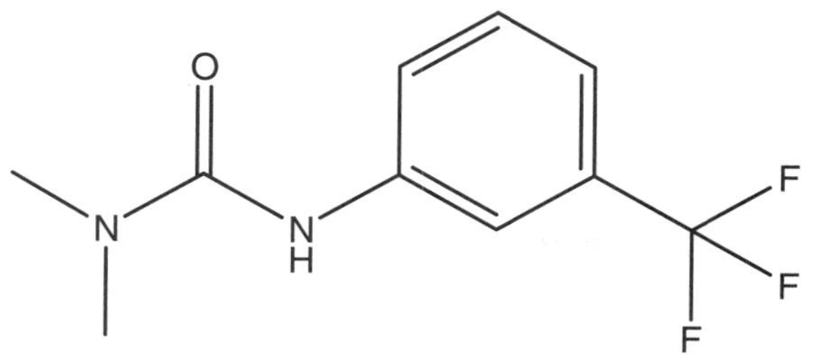

Chemical Structure

Synonyms: C 2059; CIBA 2059; COTOGARD; COTORAN; COTORAN MULTI 50WP; COTORON; COTTONEX; 1,1-DIMETHYL-3-(3-TRIFLUOROMETHYLPHENYL)UREA; N,N-DIMETHYL-N'-(3-(TRIFLUOROMETHYL)PHENYL)UREA; 1,1-DIMETHYL-3-(ALPHA,ALPHA,ALPHA-TRIFLUORO-M-TOLYL) UREA; 1,1-DIMETHYL-3-(ALPHA,ALPHA,ALPHA-TRIFLUORO-M-TOLYL)UREA; HERBICIDE C-2059; HIGALCOTON; KOTORAN; LANEX; PAKHTARAN; 3-(5-TRIFLUORMETHYLPHENYL)-1,1-DIMETHYLHARNSTOFF; 3-(M-TRIFLUOROMETHLYLPHENYL)-1,1-DIMETHYLUREA; 3-(M-TRIFLUOROMETHYL PHENYL)-1,1-DIMETHYLUREA; 3-(3-TRIFLUOROMETHYLPHENYL)-1,1-DIMETHYLUREA; 3-(M-TRIFLUOROMETHYLPHENYL)-1,1-DIMETHYLUREA; N'-(3 TRIFLUOROMETHYLPHENYL)-N,N-DIMETHYLUREA; N-(3-TRIFLUOROMETHYL)PHENYL)-1,1-DIMETHYLUREA; N-(3-TRIFLUOROMETHYLPHENYL)-N'-N'-DIMETHYLUREA; N-(3-TRIFLUOROMETHYLPHENYL)-N,N'DIMETHYLUREA; N-(3-TRIFLUOROMETHYLPHENYL)-N,N-DIMETHYLUREA; N-(M-TRIFLUOROMETHYLPHENYL)-N',N'-DIMETHYLUREA; N-(META-TRIFLUOROMETHYLPHENYL)-N,N'-DIMETHYLUREA; UREA,N,N-DIMETHYL-N'-(3-(TRIFLUOROMETHYL)PHENYL)-; UREA,1,1-DIMETHYL-3-(ALPHA,ALPHA,ALPHA-TRIFLUORO-M-TOLYL)-

Description: colorless to white crystals; odorless

Use: herbicide, especially used on cotton; absorbed through roots to control broadleaf and grass weeds

Physical Properties

Freezing Point: 163 °C (325.4 °F) to 1664.5 °C (3028.1 °F)
Saturated Vapor Density: 1.200000006 kg/m^3
Density: 1.39 g/cu cm at 20 °C
Vapor Pressure: 5 x10^{-7} mm Hg at 20 °C
Water Solubility: 80 ppm at 25 °C
Other Solubilities: 95% Ethanol: 50-100 mg/ml at 21.5 °C; Acetone: >=100 mg/ml at 21.5 °C; DMSO: >=100 mg/ml at 21.5 °C; Dichloromethane: 23 g/mL; Dimethylformamide: Soluble; Hexane: 170 g/mL; Isopropanol: Soluble; Methanol: 110 g/mL; Octan-1-ol: 22 g/mL.
Flash Point: Not available; probably combustible

RTECS Toxicity Data

Acute Oral: Rat LD_{50} Dose: 1450 mg/kg. Mouse LD_{50} Dose: 900 mg/kg; Toxic Effects: Behavioral - Somnolence (general depressed activity); Behavioral - Coma; Lungs, Thorax, or Respiration - Respiratory stimulation.

Acute Dermal: Rat LD_{50} Route: Skin; Dose: >2 gm/kg.

Mutagenic: Rat Morphological Transformation; Cell Type: embryo; Dose: 56 ug/plate. Mouse DNA Inhibition; Route: Oral; Dose: 1 gm/kg.

Hazard Overviews

Fire: Will burn.

Carcinogenicity: IARC - Group 3, Not classifiable as to carcinogenicity to humans; NIOSH - Not listed; NTP - Not listed; ACGIH - Not listed; OSHA - Not listed; EPA - Not listed; MAK - Not listed

Environmental

Ecotoxicity: LC_{50} Ictalurus punctatus (Channel catfish) 0.6 mg/l/96 hr at 18 °C (95% confidence interval 0.3-1.3 mg/l), wt 0.8 g. Static bioassay without aeration, pH 7.2-7.5, water hardness 40-50 mg/l as calcium carbonate and alkalinity of 30-35 mg/l LD_{50} Anas platyrhynchos (Mallard) oral >2000 mg/kg

Environmental Fate: In soil, it is transformed in the environment primarily by biodegradation. The half-life of disappearance from soil is about 85 days. It is highly to moderately mobile in soil. Both biodegradation and photolysis are expected to be the major pathways for the loss in water. Bioconcentration in aquatic organisms should not be important. Reaction of vapor phase with photochemically produced hydroxyl radicals may be an important fate process in the atmosphere (estimated half-life 4.1 hours). Removal can also occur as a result of dry and wet deposition.

Cleanup/Disposal: Guide No. 131: Fully encapsulating, vapor protective clothing should be worn for spills and leaks with no fire. Eliminate all ignition sources (no smoking, flares, sparks or flames in immediate area). All equipment used when handling the product must be grounded. Do not touch or walk through spilled material. Stop leak if you can do it without risk. Prevent entry into waterways, sewers, basements or confined areas. A vapor suppressing foam may be used to reduce vapors. Small Spills: Absorb with earth, sand or other non-combustible material and transfer to containers for later disposal. Use clean non-sparking tools to collect absorbed material. Large Spills: Dike far ahead of liquid spill for later disposal. Water spray may reduce vapor; but may not prevent ignition in closed spaces. Guide No. 151: Do not touch damaged containers or spilled material unless wearing appropriate protective clothing. Stop leak if you can do it without risk. Prevent entry into waterways, sewers, basements or confined areas. Cover with plastic sheet to prevent spreading. Absorb or cover with dry earth, sand or other non-combustible material and transfer to containers. Do not get water inside containers.

Environmental Physical Data

Henry's Law Constant: estimated at 1.45 x10^{-9}
Octanol/Water Partition Coefficient: log K_{ow} = 2.20
Sorption Partition Coefficient: log K_{oc} = 1.46 to 2.24
BCF: estimated at 1.65

Regulations

RCRA 40CFR: Not listed
CERCLA: 40CFR 302.4: Not listed
SARA 40CFR 372.65: Listed
SARA EHS 40CFR 355: Not listed
TSCA: Listed

Analytical Methods
Soil: SW846 8321A
Water / Groundwater: EPA 632
Drinking Water: AOAC 992.14
Food: AOAC 977.07
Indoor / Expired Air: ASTM D4861
Plasma: FDA 242.4

FLU2890 CAS #: 405-30-1
FLUORBENSIDE

RTECS: WQ2975200
Molecular Formula: $C_{13}H_{10}ClFS$
Formula Weight: 252.74
Synonyms: BENZENE,1-CHLORO-4-(((4-FLUOROPHENYL)THIO)METHYL)-; 4-CHLOROBENZYL 4'-FLUOROPHENYL SULFIDE; P-CHLOROBENZYL P-FLUOROPHENYL SULFIDE; 4-CHLOROBENZYL 4'-FLUOROPHENYL SULPHIDE; P-CHLOROBENZYL P-FLUOROPHENYL SULPHIDE; 4-CHLOROBENZYL 4-FLUORPHENYL SULPHIDE; 1-CHLORO-4-(((4-FLUOROPHENYL)THIO)METHYL)BENZENE; FLUOROPARACIDE; FLUOROSULFACIDE; FLUOROSULPHACIDE; FLUORPARACIDE; FLUORSULPHACIDE; HRS 942; RD 2454; SULFIDE,P-CHLOROBENZYL P-FLUOROPHENYL
Description: crystals
Use: acaricide

Physical Properties
Freezing Point: 36 °C (96.8 °F)
Water Solubility: Insoluble in Water
Other Solubilities: Soluble in Acetone & oils Soluble in Acetone & oils

RTECS Toxicity Data
Acute Oral: Rat LD_{50} Dose: 3 gm/kg.

Hazard Overviews
Carcinogenicity: IARC - Not listed; NIOSH - Not listed; NTP - Not listed; ACGIH - Not listed; OSHA - Not listed; EPA - Not listed; MAK - Not listed

Environmental

Regulations
RCRA 40CFR: Not listed
CERCLA: 40CFR 302.4: Not listed
SARA 40CFR 372.65: Not listed
SARA EHS 40CFR 355: Not listed
TSCA: Not listed

FLU3100 CAS #: 86-73-7
FLUORENE

RTECS: LL5670000
EINECS Number: 201-695-5
Molecular Formula: $C_{13}H_{10}$
Formula Weight: 166.21

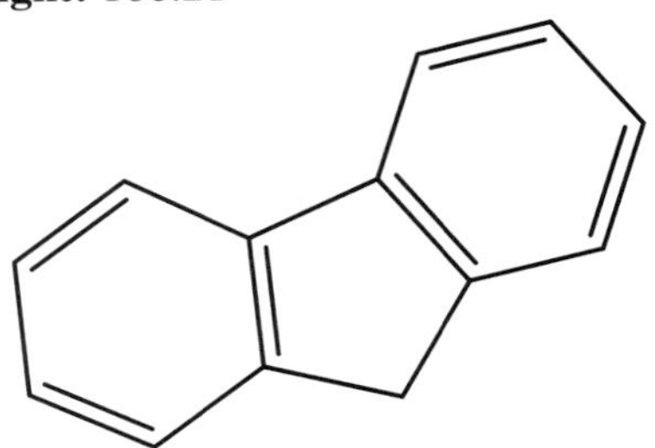

Chemical Structure

Synonyms: 2,3-BENZINDENE; O-BIPHENYLENEMETHANE; O-BIPHENYLMETHANE; ALPHA-DIPHENYLENEMETHANE; DIPHENYLENEMETHANE; ALPHA-DIPHENYLENEMETHANE-9H-FLUORENE; 9H-FLUORENE; METHANE,DIPHENYLENE-; 2,2'-METHYLENEBIPHENYL
Description: dazzling white leaflets or flakes , or small, white, crystalline plates, fluorescent when impure
Use: resinous products; dyestuffs

Physical Properties
Boiling Point: 295 °C (563 °F)
Freezing Point: 116 °C (240.8 °F) to 117 °C (242.6 °F)
Specific Gravity: 1.203 at 0 °C/4 °C
Saturated Vapor Density: 1.200002391 kg/m^3
Vapor Pressure: 3.2 $x10^{-4}$ mm Hg at 20 °C
Water Solubility: Insoluble 1.7 mg/kg
Other Solubilities: 95% Ethanol: Soluble in hot; Acetone: Soluble; Benzene: Soluble; Carbon desulfide: Soluble; Carbon Tetrachloride: Soluble; Ether: Soluble; Pyridine: Soluble; Toluene: Soluble.
Ionization Potential (eV): 7.89 +/-0.2

RTECS Toxicity Data
Mutagenic: Mouse DNA Damage; Cell Type: lymphocyte; Dose: 150 umol/L. Mouse Mutations in Microorganisms; Cell Type: lymphocyte; Dose: 19500 nmol/L (+S9).

Hazard Overviews
Health: May be irritating to eyes/skin. Other Acute Effects: may be harmful by inhalation, ingestion, or skin absorption. The toxicological properties have not been thoroughly investigated.
Fire: Extinguishing agents: water spray; carbon dioxide, dry chemical powder or appropriate foam. Precautions: combustible liquid.
Reactivity: Incompatible with: strong oxidizing agents. Hazardous decomposition products: toxic fumes of: carbon monoxide, carbon dioxide.
Carcinogenicity: IARC - Group 3, Not classifiable as to carcinogenicity to humans; NIOSH - Not listed; NTP - Not listed; ACGIH - Not listed; OSHA - Not listed; EPA - Class D, Not classifiable as to human carcinogenicity; MAK - Not listed

Environmental

Ecotoxicity: TLm Neanthes arenaceodentata 1.0 ppm/96 hr at 22 °C in a static bioassay, seawater

Environmental Fate: If released to the atmosphere, it will exist primarily in the vapor phase where it will degrade readily by photochemically produced hydroxyl radicals (estimated half-life of 29 hr). Particulate phase can be removed from air physically via wet and dry deposition; it has been detected in rain, snow and fog samples. Some particulate phase can be stable to photo-oxidation which will permit its long range global transport. If released to soil or water, it will biodegrade readily (aerobically) in the presence of acclimated microbes; microbial adaptation is an important fate process. Biodegradation can be slow in pristine soils or waters (or under conditions of limited oxygen). Strong adsorption to soil and water sediment is an important transport process; it has been detected in numerous, widespread sediment samples. The half-life of fluorene in soil has been reported to range from 2 to 64 days.

Environmental Physical Data

Henry's Law Constant: 0.0001
Octanol/Water Partition Coefficient: log K_{ow} = 4.18 to 4.38
Sorption Partition Coefficient: log K_{oc} = log 3.70 to 4.21
BCF: calculated at 3.17

Regulations

RCRA 40CFR: Not listed
CERCLA: 40CFR 302.4: Listed per CWA Section 307(a) RQ: 5000 lb (2268 kg)
SARA 40CFR 372.65: Not listed
SARA EHS 40CFR 355: Not listed
TSCA: Listed

Analytical Methods

Air: EPA TO-13; California 429
Soil: CLP LC_SV, MC_SVOA, OHC; EPA 16, 1625, PAH-005, PAH-007, PAH-011, PAH-012, S-004-1; SW846 1311, 3611A, 3630B, 3640A, 8100, 8250A, 8270B, 8270C, 8275, 8275A, 8310, 8410; DOE OS050
Water / Groundwater: EPA PAH-002, PAH-006, S-002-1, 1625, 610, 625, 625-S, 6; APHA 6040-B, 6410-B, 6440-B, 6440-C; ASTM D4657; USGS O3113, O3118
Drinking Water: EPA 525.1, 525.2, 550, 550.1
Indoor / Expired Air: NIOSH 5506, 5515; EPA IP-7-A, IP-7-B
Plasma: EPA 29
Other: EPA PAH-009

FLU3310 CAS #: 53-94-1

N-FLUORENE-2-YL-HYDROXYLAMINE

RTECS: NC3570000
Molecular Formula: $C_{13}H_{11}NO$
Formula Weight: 197.25

Synonyms: 9H-FLUOREN-2-AMINE,N-HYDROXY-; 2-FLUORENYLHYDROXYLAMINE; N-2-FLUORENYLHYDROXYLAMINE; N-HYDROXY-2-AMINOFLUORENE; N-HYDROXY-2-FLUORENAMINE; HYDROXYLAMINE,N-FLUOREN-2-YL-

RTECS Toxicity Data

Mutagenic: Human Unscheduled DNA Synthesis; Cell Type: other cell types; Dose: 100 nmol/L. Rat DNA Damage; Cell Type: lymphocyte; Dose: 500 ug/L. Rat DNA Damage; Cell Type: liver; Dose: 40 umol/L.

Tumorigenic: Rat Route: Subcutaneous; Dose: 420 mg/kg/9W-I; Toxic Effects: Tumorigenic - Neoplastic by RTECS criteria; Skin and appendages - Tumors; Tumorigenic - Tumors at site of application. Guinea Pig Route: Intraperitoneal; Dose: 1600 mg/kg/17W-I; Toxic Effects: Tumorigenic - Carcinogenic by RTECS criteria; Tumorigenic - Tumors at site of application.

Hazard Overviews

Carcinogenicity: IARC - Not listed; NIOSH - Not listed; NTP - Not listed; ACGIH - Not listed; OSHA - Not listed; EPA - Not listed; MAK - Not listed

Environmental

Regulations

RCRA 40CFR: Not listed
CERCLA: 40CFR 302.4: Not listed
SARA 40CFR 372.65: Not listed
SARA EHS 40CFR 355: Not listed
TSCA: Not listed

FLU3520 CAS #: 486-25-9

FLUORENONE

RTECS: LL8925000
EINECS Number: 207-630-7
Molecular Formula: $C_{13}H_8O$
Formula Weight: 180.22

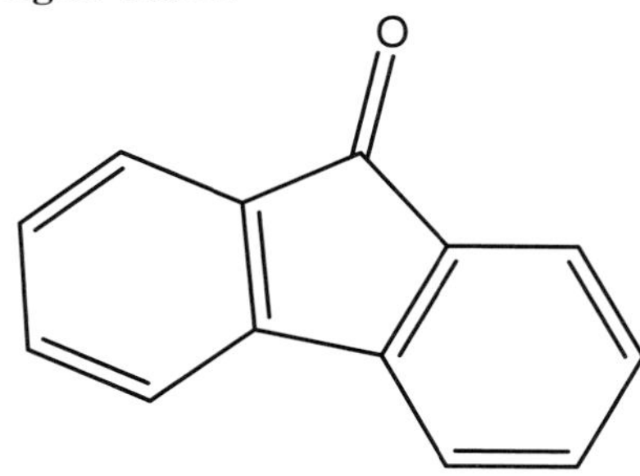

Chemical Structure

Synonyms: DIPHENYLENE KETONE; 9-FLUORENONE; 9H-FLUOREN-9-ONE; FLUOREN-9-ONE; 9-OXOFLUORENE
Description: yellow rhombic bipyramidal crystals

Use: intermediate; reagent (Oppenauer oxidation); chemical intermediate in the synthesis of: 2-methyl-3-phenylbenzyl alcohol

Physical Properties

Boiling Point: 341.5 °C (647 °F) at 760 mm Hg
Freezing Point: 84 °C (183.2 °F)
Specific Gravity: 1.13 at 99 °C/4 °C
Water Solubility: Insoluble in Water
Other Solubilities: Soluble in oxygenated and aromatic solvents.
Refraction Index: 1.6369 at 99 °C/D

RTECS Toxicity Data

Tumorigenic: Rat Route: Subcutaneous; Dose: 300 mg/kg/26W-I; Toxic Effects: Tumorigenic - Equivocal tumorigenic agent by RTECS criteria; Skin and appendages - Tumors.

Hazard Overviews

Health: May cause irritation to eyes/skin. Other Acute Effects: may be harmful by inhalation, ingestion, or skin absorption.
Fire: Hazards: emits toxic fumes. Extinguishing agents: water spray; carbon dioxide, dry chemical powder or appropriate foam. Precautions: combustible liquid.
Reactivity: Incompatible with: strong oxidizing agents. Hazardous decomposition products: toxic fumes of: carbon monoxide, carbon dioxide.
Carcinogenicity: IARC - Not listed; NIOSH - Not listed; NTP - Not listed; ACGIH - Not listed; OSHA - Not listed; EPA - Not listed; MAK - Not listed

Environmental

Ecotoxicity: Fishes: FWA 2A: bluegill: 96h LC_{50},S 474 mg/l
Environmental Fate: If released to the atmosphere, it will exist in both the vapor and particulate phases in the ambient atmosphere based on an estimated vapor pressure of 5.7×10^{-5} mm Hg at 25 °C. Vapor-phase is degraded in the atmosphere by reaction with photochemically produced hydroxyl radicals with an estimated half-life of about 3 days. An estimated K_{oc} of 2300 suggests that it will have only slight mobility in soil. Volatilization from dry and moist soil surfaces should not be a major fate process for this compound. Based on limited data, this compound may biodegrade in both soil and water. Groundwater, taken from a gasoline contaminated aquifer, was used to inoculate samples containing it as the sole carbon source; complete degradation was observed by day 11-15. In water, it is expected to adsorb to sediment and suspended matter based on its K_{oc} value. It may volatilize slowly from water surfaces given an estimated Henry's Law constant of 6.8×10^{-7} atm-cu m/mole. Estimated half-lives for a model river and model lake are 73 and 530 days, respectively. Bioconcentration in aquatic organisms may occur based on an estimated BCF value of 310; it has been detected in samples of catfish and snails.

Environmental Physical Data

Henry's Law Constant: estimated at 6.8×10^{-7}
Octanol/Water Partition Coefficient: log K_{ow} = 3.58
Sorption Partition Coefficient: K_{oc} = estimated at 2300
BCF: estimated at 310

Regulations

RCRA 40CFR: Not listed
CERCLA: 40CFR 302.4: Not listed
SARA 40CFR 372.65: Not listed
SARA EHS 40CFR 355: Not listed
TSCA: Listed

FLU3730 — CAS #: 2321-07-5

FLUORESCEIN

RTECS: LM5075000
EINECS Number: 219-031-8
Molecular Formula: $C_{20}H_{12}O_5$
Formula Weight: 332.30

HO OH O O O

Chemical Structure

Synonyms: BENZOIC ACID,2-(6-HYDROXY-3-OXO-3H-XANTHEN-9-YL); BENZOIC ACID,O-(6-HYDROXY-3-OXO-3H-XANTHEN-9-YL)-; C.I. 45350; C.I. 45350:1; C.I. 45350A; C.I. ACID YELLOW 73; C.I. 45350 (FREE ACID); C.I. SOLVENT YELLOW 94; 9-(O-CARBOXYPHENYL)-6-HYDROXY-3-ISOXANTHENONE; 9-(O-CARBOXYPHENYL)-6-HYDROXY-3H-XANTHEN-3-ONE; D AND C YELLOW NO 7; D&C YELLOW NO. 7; D&C YELLOW NO. 7; 3',4'-DEHYDROXYFLUORAN; 3',6'-DIHYDROXYFLUORAN; DIHYDROXYFLUORANE; 3',6'-DIHYDROXYSPIRO(ISOBENZOFURAN-1(3H),9'(9H)-XANTHEN)-3-ONE; 3,6-DIHYDROXYSPIRO(XANTHENE-9,3'-PHTHALIDE); DIRESORCINOLPHTHALEIN; FLUORAN,3',6'-DIHYDROXY-; 3',6'-FLUORANDIOL; 3,6-FLUORANDIOL; FLUORESCEIN RED; FLUORESCEINE; HIDACID FLUORESCEIN; O-(6-HYDROXY-3-OXO-3H-XANTHEN-9-YL)BENZOIC ACID; RESORCINOLPHTHALEIN; SOAP YELLOW F; SPIRO(ISOBENZOFURAN-1(3H),9'-(9H)XANTHEN)-3-ONE,3',6'-DIHYDROXY-; 3H-XANTHEN-3-ONE,9-(O-CARBOXYPHENYL)-6-HYDROXY-; 11712 YELLOW; 11712 YELLOW9-(O-CARBOXYPHENYL)-6-HYDROXY-3H-XANTHEN-3-ONE; ZLUT KYSELA 73
Description: yellowish-red to red powder or prisms
Use: water flow tracer for examining subterranean waters, dyeing silk and wool, diagnostic aid(corneal trauma indicator), indicator and reagent for bromine, dye in drugs and cosmetics

Physical Properties

Freezing Point: 314 °C (597.2 °F)

Water Solubility: Insoluble in Water
Other Solubilities: 95% Ethanol: Soluble; Acetone: Soluble; Benzene: Insoluble; Dilute acids: Soluble; Chloroform: Insoluble; Dilute alkalis: Soluble; Glacial Acetic Acid: Soluble; Ether: Slightly Soluble; Methanol: Soluble; Petroleum Ether: Insoluble
Flash Point: Not available; probably combustible

RTECS Toxicity Data

Mutagenic: Bacteria - E Coli DNA Adduct; Dose: 15 umol/L.

Hazard Overviews

Health: Irritating to eyes/skin/respiratory tract. Other Acute Effects: may be harmful by inhalation, ingestion, or skin absorption; target organ: kidneys.
Fire: Will burn. Hazards: emits toxic fumes. Extinguishing agents: water spray; carbon dioxide, dry chemical powder or appropriate foam. Precautions: combustible liquid.
Reactivity: Incompatible with: strong oxidizing agents. Hazardous decomposition products: toxic fumes of: carbon monoxide, carbon dioxide.
Carcinogenicity: IARC - Not listed; NIOSH - Not listed; NTP - Not listed; ACGIH - Not listed; OSHA - Not listed; EPA - Not listed; MAK - Not listed
Primary Target Organs:

Eyes

Skin

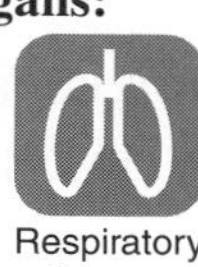
Respiratory System

Kidneys

Environmental

Regulations

RCRA 40CFR: Not listed
CERCLA: 40CFR 302.4: Not listed
SARA 40CFR 372.65: Not listed
SARA EHS 40CFR 355: Not listed
TSCA: Listed

FLU3940 **CAS #: 7782-41-4**

FLUORINE

RTECS: LM6475000
DOT: UN1045; IMO2.3
EINECS Number: 231-954-8
Molecular Formula: F_2
Structured MF: F_2
Formula Weight: 37.99
Synonyms: BIFLUORIDEN; FLUOR; FLUOR (DUTCH,FRENCH,GERMAN,POLISH); FLUORINE-19; FLUORINE,COMPRESSED; FLUORO; FLUORURES ACIDE; FLUORURI ACIDI; SAEURE FLUORIDE
Description: pale yellow, greenish-yellow gas; pungent, sharp, irritating odor
Use: mfr fluorochemicals & plastics; rocket propellant; chem intermed sulfur hexafluoride, chlorine trifluoride, bromine trifluoride uranium hexafluoride, molybedenum hexafluoride, perchloryl fluoride, oxygen difluoride

Physical Properties

Boiling Point: -188.13 °C (-307 °F)
Freezing Point: -219.61 °C (-363.298 °F)
Specific Gravity: 1.5127 at -188.13 °C (liquid
Vapor Density: 1.695 Air=1.29
Saturated Vapor Density: 1.200489474 kg/m^3
Vapor Pressure: 1 mm Hg at -223.0 °C
Water Solubility: Reacts
Other Solubilities: Reacts vigorously with most oxidizable substances at room temperature, frequently with ignition
Odor Threshold: 6.0 mg/m^3
Refraction Index: 1.0002
Critical Temperature: -129 °C
Critical Pressure: 55 atm
Ionization Potential (eV): 15.70
Flash Point: Nonflammable

RTECS Toxicity Data

Acute Inhalation: Rat LC_{50} Dose: 185 ppm/1hr; Toxic Effects: Sense organs and special senses - Conjunctive irritation; Lungs, Thorax, or Respiration - Dyspnea; Nutritional and gross metabolic - Weight loss or decreased weight gain. Mouse LC_{50} Dose: 150 ppm/1hr; Toxic Effects: Sense organs and special senses - Iritis; Lungs, Thorax, or Respiration - Dyspnea; Nutritional and gross metabolic - Weight loss or decreased weight gain.
Irritation Eye: Human Standard Draize Test Dose: 25 ppm/5M; Reaction: mild.

Hazard Overviews

Poison

Corrosive

Compressed Gas

Fire Diamond

Health: Corrosive to eyes/skin/respiratory tract. Poison. Also Causes: pulmonary edema, gastrointestinal problems, severe burns, frostbite. Chronic Effects: osteosclerosis.
Fire: Noncombustible. However, it is a strong oxidizer capable of igniting combustibles. Use agent suitable for surrounding fire. Do not use water to extinguish fire. Extinguish fire only if flow can be stopped. Contain fire and let burn, if possible.
Reactivity: Unstable, reacts vigorously with a wide range of both organic and inorganic compounds. Hazardous polymerization cannot occur. Avoid: heat; ignition sources; mixing with out establishing chemical compatibility. Incompatible with: water; most organic matter; hydrogen-containing molecules; alkali metals and alkaline earths; oxides of sulfur, nitrogen, and phosphorus; hydrogen; hydrogen and oxygen; halocarbons; hydrocarbons; nitric acid; sulfur dioxide; carbon monoxide; stainless steel; sodium acetate; sodium bromate; silver cyanide; cyanoguanidine; perchloric acid; potassium; potassium hydroxide; acetonitrile and chlorine; ice; alkanes and oxygen; ceramic materials;

acetylene; covalent halides; halogens; dicyanogen; hydrogen halide gases or concentrated solutions; metal acetylides and carbides; metal cyano complexes; metals; metal salts; metal oxides; metal borides; metal iodides; metal silicides; nickel (IV) oxide; non-metals; non-metal oxides; oxygenated organic compounds; oxygen and polymers; sulfides; xenon and catalysts; boron nitride; hexalithium disilicide and heat; nitrogenous bases; gallic acid; cesium heptafluoride propoxide; seleninyl fluoride; hydrogen sulfide; sodium dicyanamides. Hazardous decomposition products: toxic fluoride fumes.

Carcinogenicity: IARC - Not listed; NIOSH - Not listed; NTP - Not listed; ACGIH - Not listed; OSHA - Not listed; EPA - Not listed; MAK - Not listed

Primary Target Organs:

Eyes

Skin

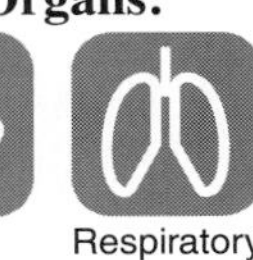
Respiratory System

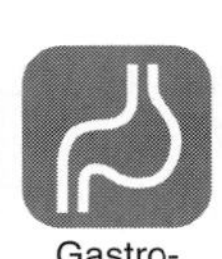
Gastro-intestinal

Bone

Exposure Limits

OSHA PEL: TWA: 0.1 ppm; 0.2 mg/m^3.

ACGIH TLV: TWA: 1 ppm; 1.6 mg/m^3; STEL: 2 ppm; 3.1 mg/m^3.

NIOSH REL: TWA: 0.1 ppm; 0.2 mg/m^3.

NIOSH IDLH: 25 ppm.

DFG MAK: TWA: 0.1 ppm; 0.2 mg/m^3.

Respirator Recommendation

Exposure Range: >0.1 to 5 ppm Supplied Air, Constant Flow/Pressure Demand, Half Mask

Exposure Range: >5 to <25 ppm Supplied Air, Constant Flow/Pressure Demand, Full Face

Exposure Range: 25 to unlimited ppm Self-contained Breathing Apparatus, Pressure Demand, Full Face

Environmental

Ecotoxicity: Aquatic toxicity: 2.3 ppm/*/trout/TLm/fresh water *Time period not specified

Cleanup/Disposal: Guide No. 124: Fully encapsulating, vapor protective clothing should be worn for spills and leaks with no fire. Do not touch or walk through spilled material. Keep combustibles (wood, paper, oil, etc.) away from spilled material. Stop leak if you can do it without risk. Use water spray to reduce vapors or divert vapor cloud drift. Do not direct water at spill or source of leak. If possible, turn leaking containers so that gas escapes rather than liquid. Prevent entry into waterways, sewers, basements or confined areas. Isolate area until gas has dispersed. Ventilate the area.

Environmental Physical Data

BCF: no food chain concentration potential

BOD: none

Regulations

RCRA 40CFR: Listed Hazardous Waste No. P056 Toxic Waste

CERCLA: 40CFR 302.4: Listed per RCRA Section 3001 RQ: 10 lb (4.535 kg)

SARA 40CFR 372.65: Listed TPQ: 500 lb

SARA EHS 40CFR 355: Listed TPQ: 10 lb

TSCA: Listed

Analytical Methods

Food: AOAC 921.04, 929.04, 933.03, 945.05

Other: AOAC 961.16

FLU4150 CAS #: 640-19-7

FLUOROACETAMIDE

RTECS: AC1225000

EINECS Number: 211-363-1

Molecular Formula: C_2H_4FNO

Formula Weight: 77.06

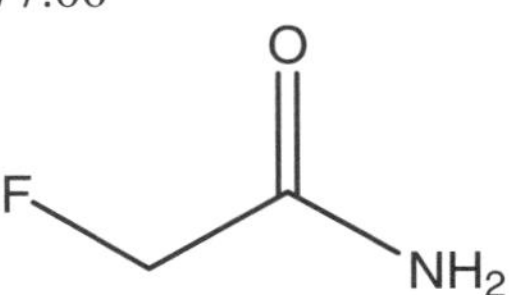

Chemical Structure

Synonyms: 1081; ACETAMIDE,2-FLUORO-; AFL 1081; AMID KYSELINY FLUOROCTOVE; BARAN; COMPOUND 1081; FAA; FLUORAKIL 100; 2-FLUOROACETAMIDE; FLUOROACETIC ACID AMIDE; FLUTRITEX 1; FUSSOL; MEGATOX; MONOFLUOROACETAMIDE; NAVRON; RODEX; YANOCK

Description: colorless crystalline powder

Use: rodenticide; insecticide proposed mainly for use on fruits to combat scale in sects, aphids and mites

Physical Properties

Boiling Point: Sublimes

Freezing Point: Sublimes

Water Solubility: Freely Soluble in Water

Other Solubilities: moderately Soluble in Ethanol; Sparingly Soluble in aliphatic & aromatic hydrocarbons.

RTECS Toxicity Data

Acute Oral: Human LD_{Lo} Dose: 2 mg/kg. Rat LD_{50} Dose: 5750 ug/kg.

Acute Inhalation: Mouse LC_{50} Dose: 550 mg/m^3.

Acute Dermal: Rat LD_{50} Route: Skin; Dose: 80 mg/kg. Mouse LD_{50} Route: Skin; Dose: 34 mg/kg. Mouse LD_{50} Route: Subcutaneous Dose: 34 mg/kg; Toxic Effects: Behavioral - Convulsions or effect on seizure threshold; Behavioral - Excitment; Skin and appendages - Hair.

Reproductive/Teratogenic: Rat Route: Oral; Dose: 90 mg/kg; Duration: male 30D prior to mating; Paternal Effects - Testes, epididymis, sperm duct.

Mutagenic: Rat Cytogenetic Analysis; Route: Intraperitoneal; Dose: 4 mg/kg. Rat Sperm Morphology; Route: Intraperitoneal; Dose: 4 mg/kg.

Hazard Overviews

Poison

Health: Irritating to eyes/skin/respiratory tract. Poison. Other Acute Effects: may be fatal if inhaled, swallowed, or absorbed through skin; target organs: heart, nerves. Chronic Effects: may cause reproductive disorders; possible teratogen; metabolized to the toxic fluoroacetate; symptoms of fluoroacetate poisoning include nausea; vomiting, cardiac irregularities, convulsions, cyanosis and death resulting from ventricular fibrillation or respiratory failure.
Fire: Hazards: emits toxic fumes. Extinguishing agents: water spray; carbon dioxide, dry chemical powder or appropriate foam. Precautions: combustible liquid.
Reactivity: Incompatible with: strong oxidizing agents, strong acids, strong bases, strong reducing agents. Hazardous decomposition products: toxic fumes of: carbon monoxide, carbon dioxide, nitrogen oxides, hydrogen fluoride.
Carcinogenicity: IARC - Not listed; NIOSH - Not listed; NTP - Not listed; ACGIH - Not listed; OSHA - Not listed; EPA - Not listed; MAK - Not listed
Primary Target Organs:

Eyes

Skin

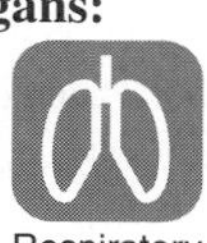
Respiratory System

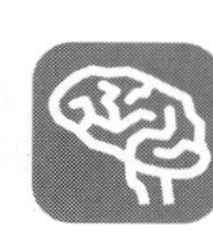
Nervous System

Cardio-vascular

Environmental

Environmental Fate: If released to soil, microbial degradation may be its primary mode of decomposition based upon the results of a single laboratory study. Its high water solubility and estimated K_{oc} of 6.4 suggest that it may leach in soil. If released to water, biodegradation may be the most important fate process. Aqueous hydrolysis is very slow (half-life of 2.4 years at 25 °C and pH 7). Aquatic volatilization, adsorption to sediment, and bioconcentration are not expected to be important. If released to the atmosphere, it will degrade relatively rapidly in the vapor-phase by reaction with photochemically formed hydroxyl radicals (estimated half-life of 19.2 hours).
Cleanup/Disposal: Guide No. 154: Eliminate all ignition sources (no smoking, flares, sparks or flames in immediate area). Do not touch damaged containers or spilled material unless wearing appropriate protective clothing. Stop leak if you can do it without risk. Prevent entry into waterways, sewers, basements or confined areas. Absorb or cover with dry earth, sand or other non-combustible material and transfer to containers. Do not get water inside containers.

Environmental Physical Data

Henry's Law Constant: estimated at 2.23×10^{-8}
Octanol/Water Partition Coefficient: log K_{ow} = -1.05
Sorption Partition Coefficient: K_{oc} = estimated at 6.4
BCF: estimated at 0.1

Regulations

RCRA 40CFR: Listed Hazardous Waste No. P057 Toxic Waste
CERCLA: 40CFR 302.4: Listed per RCRA Section 3001 RQ: 100 lb (45.35 kg)
SARA 40CFR 372.65: Not listed TPQ: 100/10000 lb
SARA EHS 40CFR 355: Listed TPQ: 100 lb
TSCA: Listed

FLU4360 **CAS #: 144-49-0**

FLUOROACETIC ACID

RTECS: AH5950000
DOT: UN2642; IMO6.1
EINECS Number: 205-631-7
Molecular Formula: $C_2H_3FO_2$
Formula Weight: 78.04
Synonyms: ACETIC ACID,FLUORO-; ACIDE-MONOFLUORACETIQUE; ACIDO MONOFLUORACETIO; ACIDO MONOFLUOROACETIO; CYMONIC ACID; FAA; FLUOROACETATE; 2-FLUOROACETIC ACID; ALPHA-FLUOROACETIC ACID; FLUOROETHANOIC ACID; GIFBLAAR POISON; HFA; MFA; MONOFLUORAZIJNZUUR; MONOFLUORESSIGSAURE; MONOFLUOROACETATE; MONOFLUOROACETIC ACID
Description: colorless solid
Use: rodenticide fluoroacetic acid, sodium salt

Physical Properties

Boiling Point: 165 °C (329 °F) at 760 mm Hg
Freezing Point: 35.2 °C (95.36 °F)
Specific Gravity: 1.3693 at 36 °C
Water Solubility: Soluble
Other Solubilities: Soluble in Alcohol

RTECS Toxicity Data

Acute Oral: Rat LD_{50} Dose: 4680 ug/kg. Mouse LD_{50} Dose: 7 mg/kg.
Acute Dermal: Rat LD_{50} Route: Subcutaneous Dose: 5 mg/kg; Toxic Effects: Behavioral - Somnolence (general depressed activity); Behavioral - Convulsions or effect on seizure threshold; Lungs, Thorax, or Respiration - Dyspnea. Mouse LD_{50} Route: Subcutaneous Dose: 281 mg/kg.

Hazard Overviews

Poison

Corrosive

Fire Diamond

Health: Corrosive to eyes/skin. Also Causes: hyperactivity, muscular spasms of hands or feet, convulsions, respiratory depression, irregular heartbeat, ventricular fibrillation, cardiac arrest. Immediate medical attention is vital!
Fire: Noncombustible. Use dry chemical, carbon dioxide, water spray, or foam. Fire produces irritating and toxic gases.

Reactivity: Stable. Hazardous polymerization cannot occur. Incompatible with: alkaline metals. Hazardous decomposition products: fluoride; sodium oxide.
Carcinogenicity: IARC - Not listed; NIOSH - Not listed; NTP - Not listed; ACGIH - Not listed; OSHA - Not listed; EPA - Not listed; MAK - Not listed
Primary Target Organs:

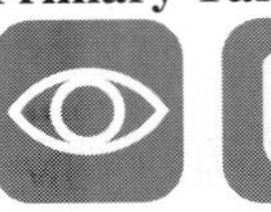
Eyes

Skin

Mucous Membranes

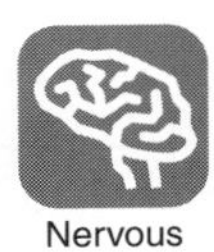
Nervous System

Cardio-vascular

Environmental

Ecotoxicity: Toxicity threshold (cell multiplication inhibition test): Algae (Microcystis aeruginosa): 8d EC_0 0.0004 mg/l
Environmental Fate: If released to the atmosphere, it is expected to exist almost entirely in the vapor-phase where it will degrade slowly by reaction with photochemically produced hydroxyl radicals (estimated half-life of 162 days). Physical removal from air can occur via rainfall and other wet deposition processes. If released to soil or water, it will probably degrade via biodegradation. It will exist predominantly in the ionized form in water and moist soil. The sorption characteristics of ionized fluoroacetic acid are not known.
Cleanup/Disposal: Guide No. 156: Eliminate all ignition sources (no smoking, flares, sparks or flames in immediate area). All equipment used when handling the product must be grounded. Do not touch damaged containers or spilled material unless wearing appropriate protective clothing. Stop leak if you can do it without risk. A vapor suppressing foam may be used to reduce vapors. For chlorosilanes, use AFFF alcohol-resistant medium expansion foam to reduce vapors. Do not get water on spilled substance or inside containers. Use water spray to reduce vapors or divert vapor cloud drift. Prevent entry into waterways, sewers, basements or confined areas. Small Spills: Cover with dry earth, dry sand, or other non-combustible material followed with plastic sheet to minimize spreading or contact with rain. Use clean non-sparking tools to collect material and place it into loosely covered plastic containers for later disposal.

Environmental Physical Data

Henry's Law Constant: estimated at 1.09 $x10^{-6}$
Octanol/Water Partition Coefficient: log K_{ow} = estimated at -0.061
Sorption Partition Coefficient: log K_{oc} = -0.61
BCF: estimated at 0.5

Regulations

RCRA 40CFR: Not listed
CERCLA: 40CFR 302.4: Not listed
SARA 40CFR 372.65: Not listed TPQ: 10/10000 lb
SARA EHS 40CFR 355: Listed TPQ: 10 lb
TSCA: Listed

Analytical Methods

Food: AOAC 949.09, 949.10

FLU4990 CAS #: 359-06-8

FLUOROACETYL CHLORIDE

RTECS: AO6825000
EINECS Number: 206-623-6
Molecular Formula: C_2H_2ClFO
Formula Weight: 96.49

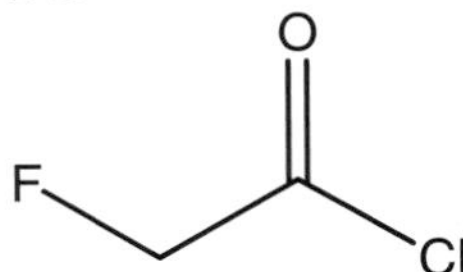

Chemical Structure

Synonyms: ACETYL CHLORIDE,FLUORO-; CHLORID KYSELINY FLUOROCTOVE; TL 670

RTECS Toxicity Data

Acute Inhalation: Mouse LC_{Lo} Dose: 200 mg/m^3/10M. Guinea Pig LC_{Lo} Dose: 100 mg/m^3/10M.

Hazard Overviews

Carcinogenicity: IARC - Not listed; NIOSH - Not listed; NTP - Not listed; ACGIH - Class A4, Not classifiable as a human carcinogen; OSHA - Not listed; EPA - Not listed; MAK - Not listed
Exposure Limits
OSHA PEL: TWA: 2.5 mg/m^3; as F.
ACGIH TLV: TWA: 2.5 mg/m^3; as F.
NIOSH REL: TWA: 2.5 mg/m^3; as F.
DFG MAK: TWA: 2.5 mg/m^3; as F.

Environmental

Regulations

RCRA 40CFR: Not listed
CERCLA: 40CFR 302.4: Not listed
SARA 40CFR 372.65: Not listed TPQ: 10 lb
SARA EHS 40CFR 355: Listed TPQ: 10 lb
TSCA: Listed

FLU5200 CAS #: 371-40-4

4-FLUOROANILINE

RTECS: BY1575000
DOT: UN2941; UN2944; IMO6.1
EINECS Number: 206-735-5
Molecular Formula: C_6H_6FN
Structured MF: 4-$FC_6H_4NH_2$
Formula Weight: 111.12

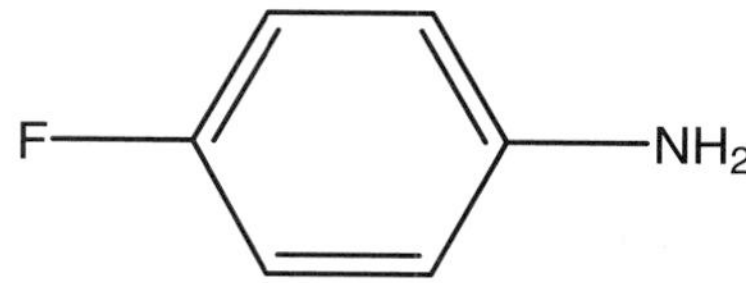

Chemical Structure

Synonyms: ANILINE,P-FLUORO-; BENZENAMINE,4-FLUORO-; BENZENAMINE,4-FLUORO-(9CI); 4-FLUORANILIN; P-FLUOROANILINE; PARA-FLUOROANILINE; 4-FLUOROBENZENAMINE; P-FLUOROPHENYLAMINE

Description: pale yellow liquid; mild sweet odor

Use: intermediate in mfr of herbicides & plant growth regulators; dye intermediate; prepn of para-fluorophenol

Physical Properties

Boiling Point: 188 °C (370 °F)
Freezing Point: -1.9 °C (28.58 °F)
Specific Gravity: 1.1725 at 20 °C/4 °C
Saturated Vapor Density: 1.204471143 kg/m^3
Vapor Pressure: 1 mm Hg at 25 °C
Water Solubility: Very Slightly Soluble in Water
Other Solubilities: Soluble in Alcohol, Ether.
Refraction Index: 1.51954 at 20 °C/D
Ionization Potential (eV): =< 8.18
Flash Point: 74 °C Closed Cup

RTECS Toxicity Data

Acute Oral: Rat LD_{50} Dose: 417 mg/kg.
Irritation Eye: Rabbit Standard Draize Test Dose: 250 ug/24H; Reaction: severe.
Irritation Skin: Rabbit Standard Draize Test Dose: 2 mg/24H; Reaction: severe.
Mutagenic: Bacteria - S Typhimurium Mutations in Microorganisms; Dose: 1 umol/plate (-S9).

Hazard Overviews

Poison

Corrosive

Health: Corrosive to eyes/skin/respiratory tract. Poison. Other Acute Effects: may be fatal if inhaled, swallowed, or absorbed through skin; material is extremely destructive to tissue of the mucous membranes and upper respiratory tract, eyes and skin; inhalation may result in spasm, inflammation and edema of the larynx and bronchi, chemical pneumonitis and pulmonary edema; symptoms of exposure may include burning sensation, coughing, wheezing, laryngitis, shortness of breath, headache, nausea and vomiting; absorption into the body leads to the formation of methemoglobin which in sufficient concentration causes cyanosis; onset may be delayed 2 to 4 hours or longer.

Fire: Combustible. Hazards: emits toxic fumes. Extinguishing agents: water spray; carbon dioxide, dry chemical powder or appropriate foam. Precautions: combustible liquid.

Reactivity: Incompatible with: acids, acid chlorides, acid anhydrides, chloroformates, strong oxidizing agents. Hazardous decomposition products: toxic fumes of: carbon monoxide, carbon dioxide, nitrogen oxides, hydrogen fluoride.

Carcinogenicity: IARC - Not listed; NIOSH - Not listed; NTP - Not listed; ACGIH - Not listed; OSHA - Not listed; EPA - Not listed; MAK - Not listed

Primary Target Organs:

Eyes

Skin

Respiratory System

Environmental

Environmental Fate: If released to the atmosphere, it should degrade rapidly in the vapor-phase (estimated half-life of 6.5 hr) by reaction with photochemically produced hydroxyl radicals. If released to soil or water, covalent bonding to humic materials may be important. The covalent bonding process may represent a mechanism by which it may be converted to a latent form in the biosphere for herbicides(1) and p-fluoroanile(2). If covalent bonding does not occur, significant leaching in soil may occur and sorption to sediments in water will not be important. When existing in a dissolved state in the water column, it is probably susceptible to significant degradation in sunlight via reaction with photochemically produced oxidants such as hydroxyl and peroxy radicals.

Cleanup/Disposal: Guide No. 153: Eliminate all ignition sources (no smoking, flares, sparks or flames in immediate area). Do not touch damaged containers or spilled material unless wearing appropriate protective clothing. Stop leak if you can do it without risk. Prevent entry into waterways, sewers, basements or confined areas. Absorb or cover with dry earth, sand or other non-combustible material and transfer to containers. Do not get water inside containers.

Environmental Physical Data

Henry's Law Constant: estimated at 1.17×10^{-5}
Octanol/Water Partition Coefficient: log K_{ow} = 1.15
Sorption Partition Coefficient: K_{oc} = estimated at 10 to 100
BCF: estimated at 4.4

Regulations

RCRA 40CFR: Not listed
CERCLA: 40CFR 302.4: Not listed
SARA 40CFR 372.65: Not listed
SARA EHS 40CFR 355: Not listed
TSCA: Listed

FLU5410 **CAS #: 393-52-2**

2-FLUOROBENZOYL CHLORIDE

RTECS: DM6640000
EINECS Number: 206-887-2
Molecular Formula: C_7H_4ClFO

Formula Weight: 158.57

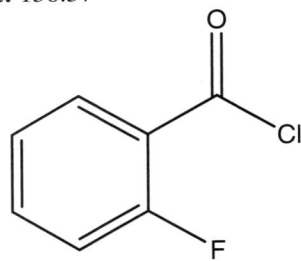

Chemical Structure

Synonyms: BENZOYL CHLORIDE,2-FLUORO-; BENZOYL CHLORIDE,O-FLUORO-; O-FLUOROBENZENE CARBONYL CHLORIDE; O-FLUOROBENZOYL CHLORIDE

Description: colorless liquid

Use: organic synthesis

Physical Properties

Boiling Point: 206 °C (403 °F)

Freezing Point: 4 °C (39.2 °F)

Density: 1.304 g/mL at 20 °C

Water Solubility: Decomposes

Other Solubilities: 95% Ethanol: Decomposes; Benzene: Very Soluble; Carbon Disulfide: Very Soluble; DMSO: Soluble; Ether: Very Soluble; Oils: Very Soluble.

Refraction Index: 1.5365

Flash Point: 82 °C

RTECS Toxicity Data

Mutagenic: Bacteria - S Typhimurium Mutations in Microorganisms; Dose: 100 ug/plate (-S9).

Hazard Overviews

Corrosive

Health: Corrosive to eyes/skin/respiratory tract. Harmful. Other Acute Effects: harmful if swallowed, inhaled, or absorbed through skin; material is extremely destructive to tissue of the mucous membranes and upper respiratory tract, eyes and skin; inhalation may result in spasm, inflammation and edema of the larynx and bronchi, chemical pneumonitis and pulmonary edema; symptoms of exposure may include burning sensation, coughing, wheezing, laryngitis, shortness of breath, headache, nausea and vomiting.

Fire: Combustible. Hazards: water hydrolyzes material liberating acidic gas which in contact with metal surfaces can generate flammable and/or explosive hydrogen gas; container explosion may occur; emits toxic fumes. Extinguishing agents: carbon dioxide; dry chemical powder; do not use water. Precautions: combustible liquid.

Reactivity: Incompatible with: water, alcohols, oxidizing agents, strong bases. Hazardous decomposition products: carbon monoxide, carbon dioxide, hydrogen chloride gas, phosgene gas, hydrogen fluoride.

Carcinogenicity: IARC - Not listed; NIOSH - Not listed; NTP - Not listed; ACGIH - Class A4, Not classifiable as a human carcinogen; OSHA - Not listed; EPA - Not listed; MAK - Not listed

Primary Target Organs:

Eyes

Skin

Respiratory System

Exposure Limits

OSHA PEL: TWA: 2.5 mg/m^3; as F.

ACGIH TLV: TWA: 2.5 mg/m^3; as F.

NIOSH REL: TWA: 2.5 mg/m^3; as F.

DFG MAK: TWA: 2.5 mg/m^3; as F.

Environmental

Cleanup/Disposal: Guide No. 171: Do not touch or walk through spilled material. Stop leak if you can do it without risk. Prevent dust cloud. Avoid inhalation of asbestos dust. Small Dry Spills: With clean shovel place material into clean, dry container and cover loosely; move containers from spill area. Small Spills: Take up with sand or other noncombustible absorbent material and place into containers for later disposal. Large Spills: Dike far ahead of liquid spill for later disposal. Cover powder spill with plastic sheet or tarp to minimize spreading. Prevent entry into waterways, sewers, basements or confined areas.

Regulations

RCRA 40CFR: Not listed

CERCLA: 40CFR 302.4: Not listed

SARA 40CFR 372.65: Not listed

SARA EHS 40CFR 355: Not listed

TSCA: Listed

FLU5620 CAS #: 16872-11-0

FLUOROBORIC ACID

RTECS: ED2685000

DOT: UN1775; IMO8.2

EINECS Number: 240-898-3

Molecular Formula: BF_4H

Structured MF: BF_4•H

Formula Weight: 87.82

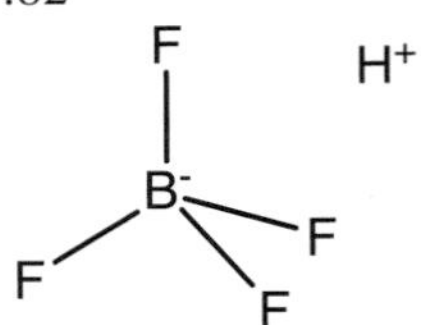

Chemical Structure

Synonyms: BORATE(1-),TETRAFLUORO-,HYDROGEN; BOROFLUORIC ACID; FLUOBORIC ACID; HYDROFLUOBORIC ACID; HYDROGEN TETRAFLUOROBORATE; HYDROGEN TETRAFLUOROBORATE(1-); TETRAFLUOROBORIC ACID

Description: colorless liquid; pungent odor

Use: catalyst for preparing acetals, esterifying cellulose; to clean metal surfaces before welding; to brighten aluminum; rgnt for sodium in presence of magnesium & potassium ions; in electrolytes for plating metals eg, chromium, iron, nickel, copper, silver, & lead, etc; for making stabilized diazo salts (diazonium & tetrazonium fluoborates); retards fermentation; stripping soln for removal of solder & plated metals; in bright dipping soln for sn-pb alloys in printed circuits, etc.

Physical Properties

Boiling Point: Decomposes at 130 °C (266 °F)
Specific Gravity: About 1.84
Water Solubility: Soluble in Hot Water
Other Solubilities: miscible with Alcohol.
Surface Tension: 48% Solution 65.3 dynes/cm
pH: Strong acid
Refraction Index: 20 at 20 °C/D

Hazard Overviews

Corrosive

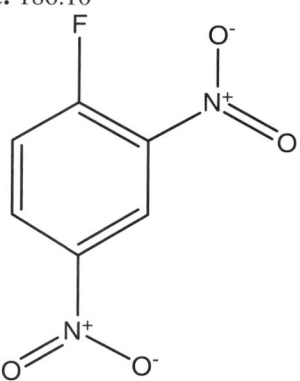

Fire Diamond

Health: Corrosive to eyes/skin/respiratory tract. Also Causes: pulmonary edema, blindness, burns of the mucous membranes and GI tract.

Fire: Noncombustible. Use agent suitable for surrounding fire. Cool fire-exposed containers with water spray. This material can react with metals to produce flammable and explosive hydrogen gas.

Reactivity: Stable. Hazardous polymerization cannot occur. Avoid: exposure to metals. Incompatible with: alkalis; acetic anhydride; heaxafluoroisopropylideneaminolithium; metals. Hazardous decomposition products: hydrogen fluoride; boron.

Carcinogenicity: IARC - Not listed; NIOSH - Not listed; NTP - Not listed; ACGIH - Not listed; OSHA - Not listed; EPA - Not listed; MAK - Not listed

Primary Target Organs:

Eyes

Skin

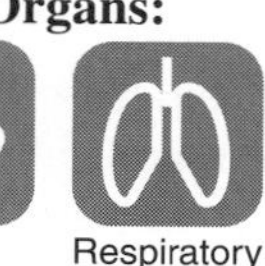
Respiratory System

Mucous Membranes

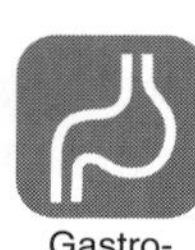
Gastro-intestinal

Exposure Limits
OSHA PEL: TWA: 2.5 mg/m^3; as F.

Environmental

Cleanup/Disposal: Guide No. 154: Eliminate all ignition sources (no smoking, flares, sparks or flames in immediate area). Do not touch damaged containers or spilled material unless wearing appropriate protective clothing. Stop leak if you can do it without risk. Prevent entry into waterways, sewers, basements or confined areas. Absorb or cover with dry earth, sand or other non-combustible material and transfer to containers. Do not get water inside containers.

Regulations

RCRA 40CFR: Not listed
CERCLA: 40CFR 302.4: Not listed
SARA 40CFR 372.65: Not listed
SARA EHS 40CFR 355: Not listed
TSCA: Listed

FLU5830 CAS #: 70-34-8

1-FLUORO-2,4-DINITROBENZENE

RTECS: CZ7800000
EINECS Number: 200-734-3
Molecular Formula: $C_6H_3FN_2O_4$
Formula Weight: 186.10

Chemical Structure

Synonyms: BENZENE,2,4-DINITRO-1-FLUORO-; BENZENE,1-FLUORO-2,4-DINITRO-; 1,3-DINITRO-4-FLUOROBENZENE; 2,4-DINITRO-1-FLUOROBENZENE; 2,4-DINITROFLUOROBENZENE; 2,4-DNFB; DNFB; DNP-F; FDNB; 1,2,4-FLUORODINITROBENZENE; SANGER'S REAGENT

Description: pale yellow crystals

Use: an alkylating agent; for labeling terminal amino groups of polypeptides and for characterizing amines and amino acids

Physical Properties

Boiling Point: 296 °C (565 °F)
Freezing Point: 25.8 °C (78.44 °F)
Specific Gravity: 1.4718 at 84 °C
Density: 1.4718 g/mL at 84 °C
Water Solubility: 0.1 to 1 mg/mL at 19.7 °C
Other Solubilities: 95% Ethanol: >=100 mg/ml at 22 °C; Acetone: >=100 mg/ml at 22 °C; Benzene: Soluble; DMSO: >=100 mg/ml at 22 °C; Ether: Soluble.
Refraction Index: 1.569
Flash Point: > 112 °C

RTECS Toxicity Data

Acute Oral: Rat LD_{Lo} Dose: 50 mg/kg.
Acute Dermal: Mouse LD_{Lo} Route: Skin; Dose: 100 mg/kg.
Mutagenic: Hamster Morphological Transformation; Cell Type: kidney; Dose: 80 ug/L. Bacteria - E Coli DNA Repair; Dose: 10 mg/well/16H.

Hazard Overviews

Poison

Health: Irritating to eyes/skin/respiratory tract. Poison. Other Acute Effects: may be fatal if inhaled, swallowed, or absorbed through skin; may cause allergic respiratory and skin reactions; causes blisters on contact with skin. Chronic Effects: carcinogen; may alter genetic material; may cause heritable genetic damage.

Fire: Will burn. Hazards: emits toxic fumes. Extinguishing agents: water spray; carbon dioxide, dry chemical powder or appropriate foam. Precautions: combustible liquid.

Reactivity: Incompatible with: strong bases, strong oxidizing agents. Hazardous decomposition products: toxic fumes of: carbon monoxide, carbon dioxide, nitrogen oxides, hydrogen fluoride.

Carcinogenicity: IARC - Not listed; NIOSH - Not listed; NTP - Not listed; ACGIH - Not listed; OSHA - Not listed; EPA - Not listed; MAK - Not listed

Primary Target Organs:

Eyes

Skin

Respiratory System

Environmental

Cleanup/Disposal: Guide No. 171: Do not touch or walk through spilled material. Stop leak if you can do it without risk. Prevent dust cloud. Avoid inhalation of asbestos dust. Small Dry Spills: With clean shovel place material into clean, dry container and cover loosely; move containers from spill area. Small Spills: Take up with sand or other noncombustible absorbent material and place into containers for later disposal. Large Spills: Dike far ahead of liquid spill for later disposal. Cover powder spill with plastic sheet or tarp to minimize spreading. Prevent entry into waterways, sewers, basements or confined areas.

Regulations

RCRA 40CFR: Not listed
CERCLA: 40CFR 302.4: Not listed
SARA 40CFR 372.65: Not listed
SARA EHS 40CFR 355: Not listed
TSCA: Listed

FLU6250 **CAS #: 350-46-9**

1-FLUORO-4-NITROBENZENE

RTECS: DA1400000
EINECS Number: 206-502-8
Molecular Formula: $C_6H_4FNO_2$
Formula Weight: 141.10

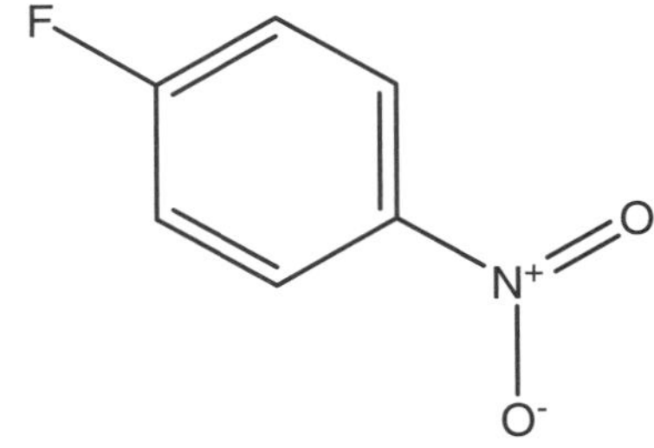

Chemical Structure

Synonyms: BENZENE,1-FLUORO-4-NITRO-; 4-FLUORO-1-NITROBENZENE; 4-FLUORONITROBENZENE; P-FLUORONITROBENZENE; 1-NITRO-4-FLUOROBENZENE; 4-NITROFLUOROBENZENE; P-NITROFLUOROBENZENE

Description: yellow needles

Use: a component of hair dyes

Physical Properties

Boiling Point: 206 °C (403 °F) to 207 °C (405 °F)
Freezing Point: 27 °C (80.6 °F)
Specific Gravity: 1.33 at 20 °C/4 °C
Water Solubility: Insoluble in Water
Other Solubilities: Soluble in Alcohol, Ether
Refraction Index: 1.5316
Ionization Potential (eV): 9.90

RTECS Toxicity Data

Acute Oral: Rat LD_{Lo} Dose: 250 mg/kg.

Mutagenic: Bacteria - S Typhimurium Mutations in Microorganisms; Dose: 160 nL/plate (+S9).

Hazard Overviews

Health: Irritating. Toxic. Other Acute Effects: harmful if swallowed, inhaled, or absorbed through skin; absorption into the body leads to the formation of methemoglobin which in sufficient concentration causes cyanosis; onset may be delayed 2 to 4 hours or longer.

Fire: Hazards: emits toxic fumes. Extinguishing agents: water spray; carbon dioxide, dry chemical powder or appropriate foam. Precautions: combustible liquid.

Reactivity: Incompatible with: strong bases, strong oxidizing agents. Hazardous decomposition products: toxic fumes of: carbon monoxide, carbon dioxide, nitrogen oxides, hydrogen fluoride.

Carcinogenicity: IARC - Not listed; NIOSH - Not listed; NTP - Not listed; ACGIH - Not listed; OSHA - Not listed; EPA - Not listed; MAK - Not listed

Primary Target Organs:

Eyes

Skin

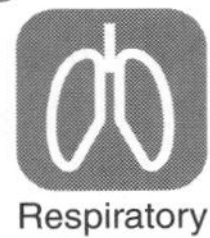
Respiratory System

Environmental

Ecotoxicity: LC_{50} Pimephales promelas (fathead minnow) 28.4 mg/l 96 hr flow-through bioassay, wt 0.12 g, water hardness

45.5 mg/l $CaCO_3$, temp: 25 +/- 1 °C, pH 7.5, dissolved oxygen greater than 60% of saturation

Environmental Physical Data

Octanol/Water Partition Coefficient: log K_{ow} = 1.80

Regulations

RCRA 40CFR: Not listed
CERCLA: 40CFR 302.4: Not listed
SARA 40CFR 372.65: Not listed
SARA EHS 40CFR 355: Not listed
TSCA: Listed

FLU6460 CAS #: 371-41-5

4-FLUOROPHENOL

RTECS: SL4550000
EINECS Number: 206-736-0
Molecular Formula: C_6H_5FO
Formula Weight: 112.11

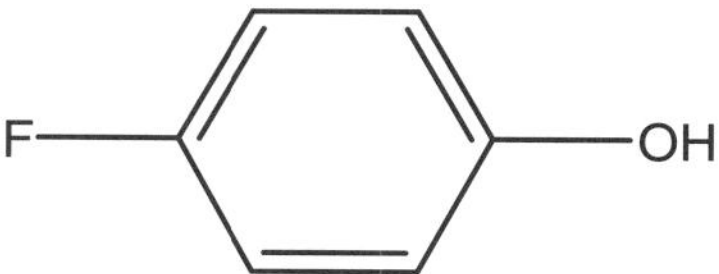

Chemical Structure

Physical Properties

Boiling Point: 185 °C (365 °F) to 188 °C (370.4 °F)
Freezing Point: 46 °C (114.8 °F) to 48 °C (118.4 °F)
Specific Gravity: 1.18890
Water Solubility: 60 g/l at 0 °C
Other Solubilities: Acetone: Soluble; Petroleum Ether: Soluble
Ionization Potential (eV): 8.5

RTECS Toxicity Data

Reproductive/Teratogenic: Rat Route: Oral; Dose: 1 gm/kg; Duration: female 11D of pregnancy; Effects on Fertility - Litter size.
Tumorigenic: Mouse Route: Skin; Dose: 10 gm/kg/25W-I; Toxic Effects: Tumorigenic - Carcinogenic by RTECS criteria; Skin and appendages - Tumors.

Hazard Overviews

Poison

Corrosive

Health: Corrosive to eyes/skin/respiratory tract. Poison. Other Acute Effects: may be fatal if inhaled, swallowed, or absorbed through skin; material is extremely destructive to tissue of the mucous membranes and upper respiratory tract, eyes and skin; inhalation may result in spasm, inflammation and edema of the larynx and bronchi, chemical pneumonitis and pulmonary edema; symptoms of exposure may include burning sensation, coughing, wheezing, laryngitis, shortness of breath, headache, nausea and vomiting; exposure can cause stomach pains, vomiting, diarrhea, dizziness irregular breathing and weakness.
Fire: Hazards: emits toxic fumes. Extinguishing agents: carbon dioxide, dry chemical powder or appropriate foam. Precautions: combustible liquid.
Reactivity: Incompatible with: acid chlorides, acid anhydrides, oxidizing agents. Hazardous decomposition products: toxic fumes of: carbon monoxide, carbon dioxide, hydrogen fluoride.
Carcinogenicity: IARC - Not listed; NIOSH - Not listed; NTP - Not listed; ACGIH - Not listed; OSHA - Not listed; EPA - Not listed; MAK - Not listed
Primary Target Organs:

Eyes

Skin

Respiratory System

Environmental

Regulations

RCRA 40CFR: Not listed
CERCLA: 40CFR 302.4: Not listed
SARA 40CFR 372.65: Not listed
SARA EHS 40CFR 355: Not listed
TSCA: Listed

FLU6670 CAS #: 459-22-3

4-FLUOROPHENYLACETONITRILE

RTECS: AM0210000
EINECS Number: 207-286-8
Molecular Formula: C_8H_6FN
Formula Weight: 135.15

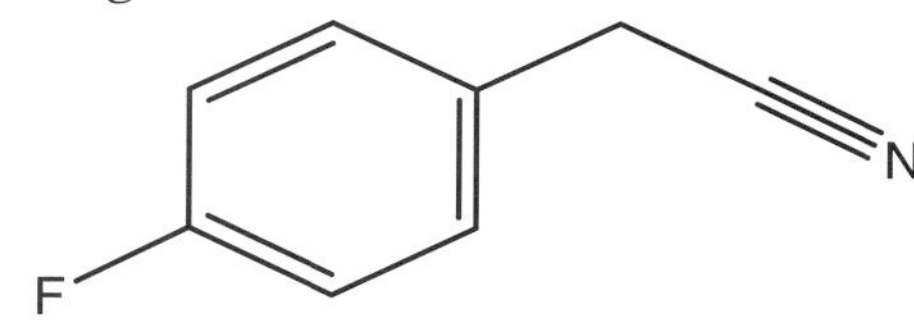

Chemical Structure

Synonyms: BENZENEACETONITRILE,4-FLUORO-; 4-FLUOROBENZENEACETONITRILE; P-FLUOROBENZYL CYANIDE; 4-FLUOROBENZYLCYANIDE; P-FLUOROPHENYLACETONITRILE

Physical Properties

Boiling Point: 119 °C (246.2 °F) to 120 °C (248 °F) at 18 mm Hg
Freezing Point: 86 °C (186.8 °F)
Specific Gravity: 1.13900
Refraction Index: 1.5002

Hazard Overviews

Health: Irritating to eyes/skin/respiratory tract. Harmful. Other Acute Effects: may be harmful by inhalation, ingestion, or skin absorption.

Fire: Hazards: emits toxic fumes. Extinguishing agents: water spray; carbon dioxide, dry chemical powder or appropriate foam. Precautions: combustible liquid.

Reactivity: Incompatible with: strong oxidizing agents. Hazardous decomposition products: toxic fumes of: carbon monoxide, carbon dioxide, nitrogen oxides, hydrogen fluoride.

Carcinogenicity: IARC - Not listed; NIOSH - Not listed; NTP - Not listed; ACGIH - Not listed; OSHA - Not listed; EPA - Not listed; MAK - Not listed

Primary Target Organs:

Eyes

Skin

Respiratory System

Environmental

Regulations

RCRA 40CFR: Not listed
CERCLA: 40CFR 302.4: Not listed
SARA 40CFR 372.65: Not listed
SARA EHS 40CFR 355: Not listed
TSCA: Listed

FLU7300 CAS #: 372-48-5

2-FLUOROPYRIDINE

RTECS: UT3900000
EINECS Number: 206-757-5
Molecular Formula: C_5H_4FN
Formula Weight: 97.10

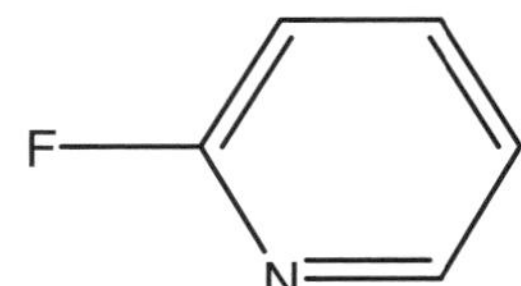

Chemical Structure

Physical Properties

Boiling Point: 126 °C (258.8 °F) at 753 mm Hg
Specific Gravity: 1.12800
Refraction Index: 1.4574
Ionization Potential (eV): 9.4

RTECS Toxicity Data

Mutagenic: Bacteria - S Typhimurium Mutations in Microorganisms; Dose: 5 mg/plate (-S9).

Hazard Overviews

Health: Irritating to eyes/skin/respiratory tract. Other Acute Effects: may be harmful by inhalation, ingestion, or skin absorption.

Fire: Hazards: vapor may travel considerable distance to source of ignition and flash back; container explosion may occur; emits toxic fumes; forms explosive mixtures in air. Extinguishing agents: carbon dioxide, dry chemical powder or appropriate foam; water may be effective for cooling, but may not effect extinguishment. Precautions: combustible liquid.

Reactivity: Incompatible with: acids, acid chlorides, acid anhydrides, strong oxidizing agents, carbon dioxide. Hazardous decomposition products: thermal decomposition may produce carbon monoxide, carbon dioxide, and nitrogen oxides; hydrogen fluoride.

Carcinogenicity: IARC - Not listed; NIOSH - Not listed; NTP - Not listed; ACGIH - Not listed; OSHA - Not listed; EPA - Not listed; MAK - Not listed

Primary Target Organs:

Eyes

Skin

Respiratory System

Environmental

Regulations

RCRA 40CFR: Not listed
CERCLA: 40CFR 302.4: Not listed
SARA 40CFR 372.65: Not listed
SARA EHS 40CFR 355: Not listed
TSCA: Not listed

FLU7510 CAS #: 51-21-8

FLUOROURACIL

RTECS: YR0350000
EINECS Number: 200-085-6
Molecular Formula: $C_4H_3FN_2O_2$
Formula Weight: 130.08

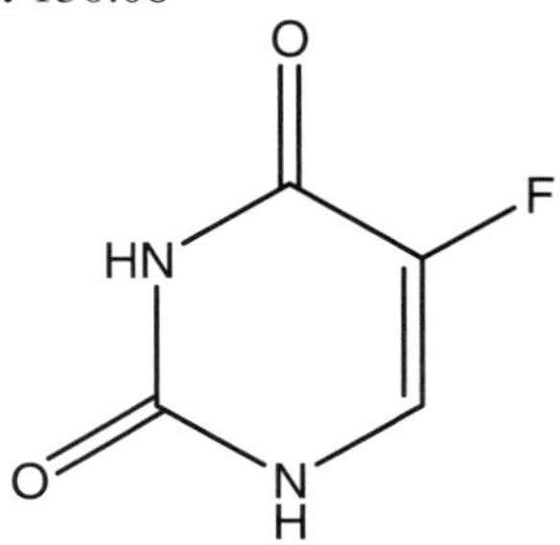

Chemical Structure

Synonyms: ADRUCIL; ARUMEL; CARZONAL; CINCO FU; 2,4-DIOXO-5-FLUOROPYRIMIDINE; EFFLUDERM (FREE BASE); EFUDEX; EFUDIX; EFURIX; 5-FARACIL; 5-FLUORACIL; 5-FLUOR-

2,4-DIHYDROXYPYRIMIDIN; FLUOROBLASTIN; FLUOROPLEX; FLUOROPLEX,FLURACIL; 5-FLUOROPYRIMIDIN-2,4-DIOL; 5-FLUORO-2,4(1H,3H)-PYRIMIDINEDIONE; 5-FLUORO-2,4-(1H,3H)-PYRIMIDINEDIONE; 5-FLUORO-2,4-PYRIMIDINEDIONE; 5-FLUOROPYRIMIDINE-2,4-(1H,3H)-DIONE; 5-FLUOROPYRIMIDINE-2,4-DIONE; 5-FLUOROURACIL; FLUORO-URACILE; FLUORO-URACILO; FLUOROURACILUM; 5-FLUOR-2,4-PYRIMIDINDIOL; 5-FLUOR-2,4(1H,3H)-PYRIMIDINDION; 5-FLUORURACIL; FLURACIL; FLURACILUM; FLURI; FLURIL; FLURO URACIL; FLUROBLASTIN; FT-207; FTORURACIL; 5-FU; FU; NSC-19893; NSC 19893; PHTHORURACIL; 2,4(1H,3H)-PYRIMIDINEDIONE,5-FLUORO-; QUEROPLEX; RO 2-9757; TIMAZIN; U-8953; ULUP; URACIL,5-FLUORO-

Description: white to practically white crystalline powder; practically odorless

Use: an antineoplastic, used alone or in conjunction with radiotherapy in the palliative treatment of cancers of the gastrointestinal tract, breast and respiratory tract; used topically in the treatment of solar or actinic keratoses and other skin tumors including Bowen's disease and superficial basal cell carcinomas

Physical Properties

Freezing Point: Decomposes at 282 °C (539.6 °F)

Water Solubility: 1 g in 80 ml Water

Other Solubilities: 1 g/100 ml of Propylene Glycol.

pH: Commercial solution 9.0

Flash Point: Not available; probably combustible

RTECS Toxicity Data

Acute Oral: Human TD_{Lo} Dose: 450 mg/kg/30D; Toxic Effects: Gastrointestinal - Other changes; Blood - Changes in bone marrow not included above; Tumorigenic - Active as anti-cancer agent. Rat LD_{50} Dose: 230 mg/kg.

Acute Dermal: Rat LD_{50} Route: Subcutaneous Dose: 217 mg/kg. Mouse LD_{50} Route: Subcutaneous Dose: 169 mg/kg.

Chronic (Multiple Dose) Oral: Rat Dose: 1092 mg/kg/30D-I; Toxic Effects: Endocrine - Changes in thymus weight; Blood - Normocytic anemia; Biochemical - Transaminases. Rat Dose: 6552 mg/kg/26W-I; Toxic Effects: Endocrine - Changes in thymus weight; Liver - Changes in liver weight; Blood - Changes in erythrocite (RBC) cell count. Rat Dose: 700 mg/kg/40W-I; Toxic Effects: Nutritional and gross metabolic - Weight loss or decreased weight gain; Blood - Changes in serum composition; Biochemical - Transaminases.

Chronic (Multiple Dose) Dermal: Rat Route: Subcutaneous; Dose: 21 mg/kg/7D-I; Toxic Effects: Blood - Changes in bone marrow not included above; Blood - Changes in leukocyte (WBC) cell count; Blood - Changes in platelet cell count. Mouse Route: Subcutaneous; Dose: 300 mg/kg/60D-I; Toxic Effects: Endocrine - Changes in spleen weight; DEATH; DEATH - Changes in testicular weight.

Reproductive/Teratogenic: Woman Route: Intravenous; Dose: 150 mg/kg; Duration: female 20-31W of pregnancy Effects on Newborn - Other neonatal measures or effects. Woman Route: Intravenous; Dose: 240 mg/kg; Duration: female 11-14W of pregnancy Specific Developmental Abnormalities - Musculoskeletal system. Rat Route: Oral; Dose: 35 mg/kg; Duration: female 7-13D of pregnancy; Specific Developmental Abnormalities - Musculoskeletal system. Rat Route: Oral; Dose: 175 mg/kg; Duration: female 7-13D of pregnancy; Effects on Fertility - Litter size; Effects on Embryo or Fetus - Fetotoxicity; Fetal death. Rat Route: Oral; Dose: 700 mg/kg; Duration: female 35D prior to mating Maternal Effects - Ovaries, fallopian tubes; Uterus, cervix, vagina.

Mutagenic: Human DNA Damage; Cell Type: other cell types; Dose: 13 mg/L. Human DNA Damage; Cell Type: other cell types; Dose: 1 mmol/L. Human Unscheduled DNA Synthesis; Cell Type: other cell types; Dose: 20 mg/L. Human Unscheduled DNA Synthesis; Cell Type: other cell types; Dose: 1 mmol/L. Human DNA Inhibition; Cell Type: other cell types; Dose: 200 mg/L. Human DNA Inhibition; Cell Type: other cell types; Dose: 1 umol/L. Human DNA Inhibition; Cell Type: ovary; Dose: 1 mg/L. Human DNA Adduct; Cell Type: leukocyte; Dose: 2600 nmol/L. Human Sister Chromatid Exchenge; Cell Type: lymphocyte; Dose: 100 pmol/L. Human Body Fluid Assay; Indicator Organism: Bacteria - S Typhimurium; Dose: 7 mg/kg. Human Other Mutation Test Systems; Cell Type: other cell types; Dose: 5 umol/L/6H. Human Other Mutation Test Systems; Cell Type: leukocyte; Dose: 2600 nmol/L. Human Other Mutation Test Systems; Cell Type: other cell types; Dose: 1 mmol/L. Human Other Mutation Test Systems; Cell Type: other cell types; Dose: 1 umol/L. Human Other Mutation Test Systems; Route: Skin; Dose: 1 pph. Human Other Mutation Test Systems; Cell Type: other cell types; Dose: 1 mg/L. Human Other Mutation Test Systems; Cell Type: HeLa cell; Dose: 50 mg/L.

Tumorigenic: Mouse Route: Intraperitoneal; Dose: 1500 mg/kg/50W-I; Toxic Effects: Tumorigenic - Carcinogenic by RTECS criteria; Lungs, Thorax, or Respiration - Tumors; Blood - Tumors.

Hazard Overviews

Health: Irritating to eyes/skin/respiratory tract. Toxic. Other Acute Effects: harmful if swallowed; may be harmful if inhaled; may be harmful if absorbed through the skin; causes photosensitivity; exposure to light can result in allergic reactions resulting in dermatologic lesions, which can vary from sunburnlike responses to edematous, vesiculated lesions or bullae; may cause nausea, vomiting, anorexia, diarrhea, stomatitis, fever, malaise, weakness, headache, depression, skin rash, erythema, bone marrow depression, bleeding syndrome and renal impairment; deaths have occured. Chronic Effects: may alter genetic material; may cause congenital malformation in the fetus; may cause reproductive disorders; target organs: heart, bone marrow, immune system, blood.

Fire: Will burn. Hazards: emits toxic fumes. Extinguishing agents: carbon dioxide, dry chemical powder or appropriate foam. Precautions: combustible liquid.

Reactivity: Stable. Hazardous polymerization will not occur. Incompatible with: strong oxidizing agents, strong bases. Hazardous decomposition products: toxic fumes of: carbon monoxide, carbon dioxide, nitrogen oxides, hydrogen fluoride.

Carcinogenicity: IARC - Group 3, Not classifiable as to carcinogenicity to humans; NIOSH - Not listed; NTP - Not listed; ACGIH - Not listed; OSHA - Not listed; EPA - Not listed; MAK - Not listed

Primary Target Organs:

Eyes

Skin

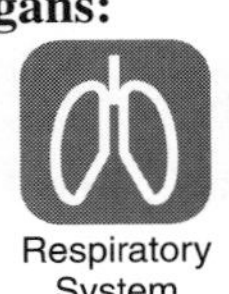
Respiratory System

Cardio-vascular

Blood

Bone

Environmental

Regulations

RCRA 40CFR: Not listed
CERCLA: 40CFR 302.4: Not listed
SARA 40CFR 372.65: Listed TPQ: 500/10000 lb
SARA EHS 40CFR 355: Listed TPQ: 500 lb
TSCA: Listed

FLU7720 CAS #: 7789-21-1

FLUOSULFONIC ACID

RTECS: LP0715000
EINECS Number: 232-149-4
Molecular Formula: FHO_3S
Structured MF: FSO_3H
Formula Weight: 100.07

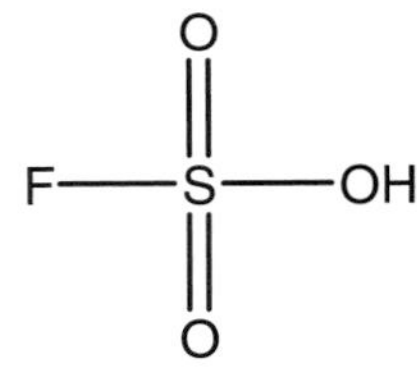

Chemical Structure

Synonyms: FLUOROSULFONIC ACID; FLUOROSULFURIC ACID; FLUOSULFURIC ACID; MONOFLUOROSULFURIC ACID
Description: colorless to reddish-brown liquid; choking odor
Use: fluorinating agent; catalyst in alkylation, acylation, polymerization & condensation reactions; in hydrofluorination of olefins; in prodn of substituted pyridines; in prodn of petroleum products; prepn of magic acid; in electropolishing

Physical Properties

Boiling Point: 163 °C (325 °F) at 760 mm Hg
Freezing Point: -89 °C (-128.2 °F)
Specific Gravity: 1.726 at 25 °C/4 °C
Water Solubility: Salts hydrolyzed by Water
Other Solubilities: Soluble in Nitrobenzene
pH: Considerably more acid than concentrated H2SO4
Flash Point: Nonflammable

Hazard Overviews

Poison

Corrosive

Health: Corrosive to eyes/skin/respiratory tract. Poison. Other Acute Effects: may be fatal if inhaled, swallowed, or absorbed through skin; inhalation may result in spasm; inflammation and edema of the larynx and bronchi; chemical pneumonitis; pulmonary edema; exposure may cause burning sensation; coughing; wheezing; laryngitis; shortness of breath; headache; nausea; vomiting.
Fire: Noncombustible. Hazards: water hydrolyzes material liberating acidic gas which in contact with metal surfaces can generate flammable and/or explosive hydrogen gas; emits toxic fumes. Extinguishing agents: noncombustible; use extinguishing media appropriate to surrounding fire conditions; do not use water. Precautions: combustible liquid.
Reactivity: Incompatible with: strong bases, alcohols, do not allow water to enter container because of violent reaction, avoid contact with metals. Hazardous decomposition products: toxic fumes of: hydrogen fluoride, sulfur oxides.
Carcinogenicity: IARC - Not listed; NIOSH - Not listed; NTP - Not listed; ACGIH - Not listed; OSHA - Not listed; EPA - Not listed; MAK - Not listed

Primary Target Organs:

Eyes

Skin

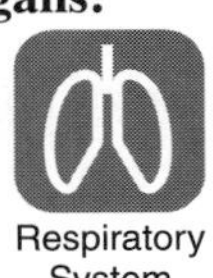
Respiratory System

Exposure Limits
OSHA PEL: TWA: 2.5 mg/m^3; as F.

Environmental

Cleanup/Disposal: Guide No. 144: Eliminate all ignition sources (no smoking, flares, sparks or flames in immediate area). Do not touch damaged containers or spilled material unless wearing appropriate protective clothing. Stop leak if you can do it without risk. Use water spray to reduce vapors or divert vapor cloud drift. Do not get water on spilled substance or inside containers. Large Spills: Do not clean-up or dispose of, except under supervision of a specialist.

Environmental Physical Data

BCF: no food chain concentration potential

Regulations

RCRA 40CFR: Not listed
CERCLA: 40CFR 302.4: Not listed
SARA 40CFR 372.65: Not listed
SARA EHS 40CFR 355: Not listed
TSCA: Listed

FLU7930 CAS #: 54910-89-3

FLUOXETINE

RTECS: DA8326855
Molecular Formula: $C_{17}H_{18}F_3NO$
Formula Weight: 309.33
Synonyms: BENZENEPROPAMINE,N-METHYL-GAMMA-(4-(TRIFLUOROMETHYL)PHENOXY)-(+ -)-; FLUOXETINA; (+) -N-METHYL-3-PHENYL-3-(ALPHA,ALPHA,ALPHA-TRIFLUORO-P-TOLYL)OXY)PROPYLAMINE; (-) -N-METHYL-3-PHENYL-3-(ALPHA,ALPHA,ALPHA-TRIFLUORO-P-TOLYL)OXY)PROPYLAMINE; N-METHYL-3-(P-TRIFLUOROMETHYL PHENOXY)-3-PHENYLPROPYLAMINE; (+)-N-METHYL-GAMMA-(4-(TRIFLUOROMETHYL)PHENOXY)BENZENEPROPANAMINE; (+-)-N-METHYL-GAMMA-(4-(TRIFLUOROMETHYL)PHENOXY)BENZENEPROPANAMINE; (-)-N-METHYL-GAMMA-(4-(TRIFLUOROMETHYL)PHENOXY)BENZENEPROPANAMINE; PROZAC; PULVULES
Description: white to off-white crystalline solid or crystals
Use: medication

Physical Properties

Freezing Point: 179 °C (354.2 °F) to 182 °C (359.6 °F)
Water Solubility: 50 mg/mL

RTECS Toxicity Data

Acute Oral: Woman TD_{Lo} Dose: 72 mg/kg/26W-I; Toxic Effects: Behavioral - Somnolence (general depressed activity). Rat LD_{50} Dose: 825 mg/kg.

Hazard Overviews

Carcinogenicity: IARC - Not listed; NIOSH - Not listed; NTP - Not listed; ACGIH - Not listed; OSHA - Not listed; EPA - Not listed; MAK - Not listed

Environmental

Regulations

RCRA 40CFR: Not listed
CERCLA: 40CFR 302.4: Not listed
SARA 40CFR 372.65: Not listed
SARA EHS 40CFR 355: Not listed
TSCA: Not listed

FLU8140 CAS #: 76-43-7

FLUOXYMESTERONE

RTECS: BV8390000
EINECS Number: 200-961-8
Molecular Formula: $C_{20}H_{29}FO_3$
Formula Weight: 336.45

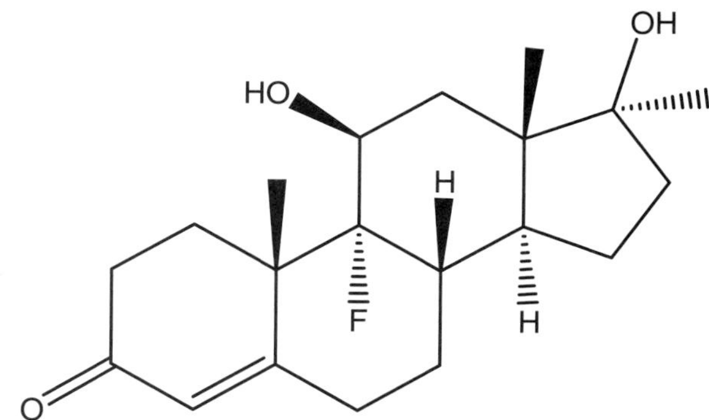

Chemical Structure

Synonyms: ANDROFLUORENE; ANDROFLUORONE; ANDROST-4-EN-3-ONE,9-FLUORO-11,17-DIHYDROXY-17-METHYL-,(11BETA,17BETA)-; ANDROST-4-EN-3-ONE,9-FLUORO-11BETA,17BETA-DIHYDROXY-17-METHYL-; ANDROST-4-EN-3-ONE,9-FLUORO-11,17-DIHYDROXY-17-METHYL-,(11-BETA,17-BETA)-(9CI); ANDROSTEROLO; 11BETA,17BETA-DIHYDROXY-9ALPHA-FLUORO-17ALPHA-METHYL-4-ANDROSTEN-3-ONE; 11-BETA,17-BETA-DIHYDROXY-9-ALPHA-FLUORO-17-ALPHA-METHYL-4-ANDROSTER-3-ONE; FLUORO-9-ALPHA DIHYDROXY-11-BETA,17-BETA METHYL-17-ALPHAANDROSTENE-4 ONE-3; 9-ALPHA-FLUORO-11-BETA,17-BETA-DIHYDROXY-17-ALPHA-METHYL-4-ANDROSTENE-3-ONE; 9-FLUORO-11-BETA,17-BETA-DIHYDROXY-17-METHYLANDROST-4-EN-3-ONE; 9-ALPHA-FLUORO-11-BETA-HYDROXY-17-METHYLTESTOSTERONE; 9-ALPHA-FLUORO-17-ALPHA-METHYL-11-BETA,17-DIHYDROXY-4-ANDROSTEN-3-ONE; FLUOTESTIN; FLUOXIMESTERONE; FLUOXYMESTRONE; FLUSTERON; FLUTESTOS; HALOTESTIN; 17-ALPHA-METHYL-9-ALPHA-FLUORO-11-BETA-HYDROXYTESTERONE; NEO-ORMONAL; NSC-12165; ORA TESTRYL; ORALSTERONE; ORATESTIN; ORA-TESTRYL; TESTORAL; U 6040; ULTANDREN; ULTANDRENE
Description: white solid; white or practically white crystalline powder; odorless
Use: androgen; antineoplastic agent in cancer chemotherapy

Physical Properties

Boiling Point: Decomposes at 270 °C (518 °F)
Freezing Point: About 240 °C (464 °F)
Water Solubility: Practically Insoluble in Water
Other Solubilities: Soluble in Pyridine; slightly Soluble in Acetone, Chloroform; sparingly Soluble in Methanol; practically Insoluble in Ether, Benzene, hexanes.

RTECS Toxicity Data

Acute Oral: Human TD_{Lo} Dose: 400 ug/kg; Toxic Effects: Lungs, Thorax, or Respiration - Other changes; Skin and appendages - Dermatitis, other; Biochemical - Transaminases.
Chronic (Multiple Dose) Oral: Rat Dose: 437500 ug/kg/35D-C; Toxic Effects: Blood - Changes in leukocyte (WBC) cell count; Endocrine - Adrenal cortex hypoplasia; Endocrine - Changes in thymus weight.
Reproductive/Teratogenic: Man Route: Oral; Dose: 12 mg/kg; Duration: male 84D prior to mating; Paternal Effects - Spermatogenesis. Rat Route: Oral; Dose: 800 mg/kg; Duration: female 3-10D of pregnancy; Effects on Fertility - Abortion. Rat Route: Oral; Dose: 500 mg/kg; Duration: female 3-7D of pregnancy; Effects on Fertility - Litter size.

Rat Route: Oral; Dose: 8 mg/kg; Duration: male 10D prior to mating; Paternal Effects - Prostate, seminal vessicle, Cowper's gland, accessory glands. Rat Route: Oral; Dose: 438 mg/kg; Duration: male 35D prior to mating; Paternal Effects - Testes, epididymis, sperm duct.

Mutagenic: Human Cytogenetic Analysis; Cell Type: lymphocyte; Dose: 1 mg/L. Human Sister Chromatid Exchenge; Cell Type: lymphocyte; Dose: 1 mg/L.

Hazard Overviews

Health: Irritating to eyes/skin/respiratory tract. Harmful. Other Acute Effects harmful if swallowed, inhaled, or absorbed through skin. Chronic Effects: development of hepatic adenomas, hepatocellular carcinoma, peliosis hepatis, cholestatic hepatitis and jaundice, associated with prolonged use of 17-alpha-alkylandrogen compounds; in women, virilization; in pregnant women, virilization of external genitalia of the female fetus; suppression of blood clotting factors; hirsutism; male pattern baldness; seborrhea; acne; hypersensitivity (including skin manifestations and anaphylactoid reactions); retention of sodium, water, chloride, calcium, potassium and inorganic phosphates; possible risk of harm to the unborn child; target organs: liver, reproductive system. Possibly causes reproductive effects in humans. The toxicological properties have not been thoroughly investigated.

Fire: Hazards: emits toxic fumes. Extinguishing agents: water spray; carbon dioxide, dry chemical powder or appropriate foam. Precautions: combustible liquid.

Reactivity: Stable. Hazardous polymerization will not occur. Incompatible with: strong oxidizing agents. Hazardous decomposition products: toxic fumes of: carbon monoxide, carbon dioxide, hydrogen fluoride.

Carcinogenicity: IARC - Not listed; NIOSH - Not listed; NTP - Not listed; ACGIH - Not listed; OSHA - Not listed; EPA - Not listed; MAK - Not listed

Primary Target Organs:

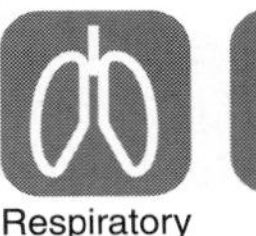

Eyes | Skin | Respiratory System | Liver | Reproductive

Environmental

Environmental Physical Data

Octanol/Water Partition Coefficient: log K_{ow} = 2.38

Regulations

RCRA 40CFR: Not listed
CERCLA: 40CFR 302.4: Not listed
SARA 40CFR 372.65: Not listed
SARA EHS 40CFR 355: Not listed
TSCA: Not listed

FLU8350 CAS #: 69-23-8

FLUPHENAZINE

RTECS: TL9730000
EINECS Number: 200-702-9
Molecular Formula: $C_{22}H_{26}F_3N_3OS$
Formula Weight: 437.52

Synonyms: ANATENSOL; DAPOTUM; ELINOL; FLUORFENAZINE; FLUOROPHENAZINE; FLUORPHENAZINE; 10-(3-(2-HYDROXYETHYL)PIPERAZINOPROPYL)-2-(TRIFLUOROMETHYL)PHENOTHIAZINE; 1-(2-HYDROXYETHYL)-4-(3-(TRIFLUOROMETHYL-10-PHENOTHIAZINYL)PROPYL)-PIPERAZINE; LYOGEN; MODECATE; MODITEN (TABL OR ELIXIR); OMCA; PACINOL; PERMITIL; PHENOTHIAZINE,10-(3-(4-(2-HYDROXYETHYL)-1-PIPERAZINYL)PROPYL)-2-(TRIFLUOROMETHYL)-; PHENOTHIAZINE,10-(3-(4-(2-HYDROXYETHYL)-1-PIPERAZINYL)PROPYL-2-(TRIFLUOROMETHYL); 1-PIPERAZINEETHANOL,4-(3-(2-(TRIFLUOROMETHYL)-10H-PHENOTHIAZIN-10-YL)PROPYL)-; PROLIXIN; PROLIXINE; S94; SEVINOL; SIQUALINE; SIQUALON; SQ 4918; TENSOFIN; TRANCIN; TRIFLUMETHAZINE; 2-(TRIFLUOROMETHYL)-10-(3-(1-(BETA-HYDROXYETHYL)-4-PIPERAZINYL)PROPYL)PHENOTHIAZINE; 4-(3-(2-(TRIFLUOROMETHYL)-10H-PHENOTHIAZIN-10-YL)PROPYL)-1-PIPERAZINEETHANOL; 4-(3-(-TRIFLUOROMETHYL-10-PHENOTHIAZYL)-PROPYL)-1-PIPERAZINEETHANOL; 4-(3-(2-TRIFLUOROMETHYL-10-PHENOTHIAZYL)-PROPYL)-1-PIPERAZINEETHANOL; VALAMINA; VESPAZINE

Description: dark brown viscous oil; characteristic odor
Use: medication

Physical Properties

Boiling Point: 250 °C (482 °F) to 252 °C (486 °F) at 0.3 mm Hg
Water Solubility: Insoluble in Water
Other Solubilities: Soluble in Alcohol, Acetone, Benzene, & Ether

RTECS Toxicity Data

Acute Oral: Mouse LD_{50} Dose: 220 mg/kg.

Acute Dermal: Rat LD_{50} Route: Subcutaneous Dose: 640 mg/kg; Toxic Effects: Sense organs and special senses - Ptosis; Automatic Nervous System - Central sympatholytic; Behavioral - Alteration of classical conditioning.

Reproductive/Teratogenic: Rat Route: Oral; Dose: 62 mg/kg; Duration: female 9D prior to mating; Effects on Fertility - Litter size. Rat Route: Oral; Dose: 18 mg/kg; Duration: female 9D prior to mating; Effects on Fertility - Mating Performance.

Hazard Overviews

Carcinogenicity: IARC - Not listed; NIOSH - Not listed; NTP - Not listed; ACGIH - Not listed; OSHA - Not listed; EPA - Not listed; MAK - Not listed

Environmental

Regulations

RCRA 40CFR: Not listed
CERCLA: 40CFR 302.4: Not listed

SARA 40CFR 372.65: Not listed
SARA EHS 40CFR 355: Not listed
TSCA: Not listed

FLU8560 CAS #: 53-34-9

FLUPREDNISOLONE

EINECS Number: 200-170-8
Molecular Formula: $C_{21}H_{27}FO_5$
Formula Weight: 378.45
Synonyms: ALPHADROL; ETADROL; FI 6150; 6ALPHA-FLUORO-1-DEHYDROHYDROCORTISONE; 6ALPHA-FLUOROPREDNISOLONE; 6ALPHA-FLUORO-1,4-PREGNADIENE-11BETA,17ALPHA,21-TRIOL-3,20-DIONE; 6ALPHA-FLUORO-11BETA,17ALPHA,21-TRIHYDROXYPREGNA-1,4-DIENE-3,20-DIONE; PREGNA-1,4-DIENE-3,20-DIONE,6-FLUORO-11,17,21-TRIHYDROXY-,(6ALPHA,11BETA)-; PREGNA-1,4-DIENE-3,20-DIONE,6ALPHA-FLUORO-11BETA,17,21-TRIHYDROXY-; U 7800; U-7800; VLADICORT
Description: white to off-white powder or crystals; odorless
Use: medication: glucocorticoid; anti-inflammatory agent

Physical Properties

Freezing Point: 208 °C (406.4 °F) to 213 °C (415.4 °F)
Water Solubility: Practically Insoluble in Water
Other Solubilities: Slightly Soluble in Chloroform, Ether, Ethylene Dichloride; Very Slightly Soluble in Benzene.

Hazard Overviews

Carcinogenicity: IARC - Not listed; NIOSH - Not listed; NTP - Not listed; ACGIH - Not listed; OSHA - Not listed; EPA - Not listed; MAK - Not listed

Environmental

Regulations

RCRA 40CFR: Not listed
CERCLA: 40CFR 302.4: Not listed
SARA 40CFR 372.65: Not listed
SARA EHS 40CFR 355: Not listed
TSCA: Not listed

FLU8770 CAS #: 17617-23-1

FLURAZEPAM

RTECS: DF2368050
EINECS Number: 241-591-7
Molecular Formula: $C_{21}H_{23}ClFN_3O$
Formula Weight: 387.89
Synonyms: 2H-1,4-BENZODIAZEPIN-2-ONE,7-CHLORO-1-(2-(DIETHYLAMINO)ETHYL)-5-(2-FLUOROPHENYL)-1,3-DIHYDRO-; 2H-1,4-BENZODIAZEPIN-2-ONE,7-CHLORO-1-(2-(DIETHYLAMINO)ETHYL)-5-(O-FLUOROPHENYL)-1,3-DIHYDRO-; 2H-1,4-BENZODIAZEPIN-2-ONE,1,3-DIHYDRO-7-CHLORO-1-(2-(DIETHYLAMINO)ETHYL)-5-(O-FLUOROPHENYL)-; 7-CHLORO-1-(2-(DIETHYLAMINO)ETHYL)-5-(2-FLUOROPHENYL)-1H-1,4-BENZODIAZEPIN-2(3H)-ONE; 7-CHLORO-1-(2-(DIETHYLAMINO)ETHYL)-5-(2-FLUOROPHENYL)-1,3-DIHYDRO-2H-1,4-BENZODIAZEPIN-2-ONE; DALMANE; DALMANE-R; RO 56901/3; RO-5-6901/3
Description: white rods; slight odor to odorless
Use: medication; in the treatment of anxiety and nervousness

Physical Properties

Freezing Point: 77 °C (170.6 °F) to 82 °C (179.6 °F)
Water Solubility: Moderately Hydroscopic
Other Solubilities: freely Soluble in Alcohol; Slightly Soluble in Chloroform

RTECS Toxicity Data

Acute Oral: Human TD_{Lo} Dose: 380 ug/kg; Toxic Effects: Lungs, Thorax, or Respiration - Dyspnea; Lungs, Thorax, or Respiration - Other changes. Rat LD_{50} Dose: 980 mg/kg; Toxic Effects: Behavioral - Somnolence (general depressed activity); Behavioral - Change in motor activity (specific assay); Lungs, Thorax, or Respiration - Respiratory stimulation.
Acute Dermal: Rat LD_{50} Route: Subcutaneous Dose: >3844 mg/kg; Toxic Effects: Behavioral - Somnolence (general depressed activity); Behavioral - Change in motor activity (specific assay); Lungs, Thorax, or Respiration - Respiratory stimulation.

Hazard Overviews

Carcinogenicity: IARC - Not listed; NIOSH - Not listed; NTP - Not listed; ACGIH - Not listed; OSHA - Not listed; EPA - Not listed; MAK - Not listed

Environmental

Regulations

RCRA 40CFR: Not listed
CERCLA: 40CFR 302.4: Not listed
SARA 40CFR 372.65: Not listed
SARA EHS 40CFR 355: Not listed
TSCA: Not listed

FLU8980 CAS #: 69409-94-5

FLUVALINATE

RTECS: YV9397100
Molecular Formula: $C_{26}H_{22}ClF_3N_2O_2$
Formula Weight: 502.93
Synonyms: N-(2-CHLORO-4-(TRIFLUOROMETHYL)PHENYL)-DL-VALINE CYANO(3-PHENOXYPHENYL)METHYL ESTER; N-[2-CHLORO-4-(TRIFLUOROMETHYL)-PHENYL]-DL-VALINECYANO(3-PHENOXYPHENYL)METHYL ESTER; (RS)-ALPHA-CYANO-3-PHENOXYBENZYL-2-[2-CHLORO-4-(TRIFLUOROMETHYL)ANILINO-3-METHYL-BUTANOATE]; (RS)-ALPHA-CYANO-3-PHENOXYBENZYLN-[2-CHLORO-ALPHA,ALPHA,ALPHA-TRIFLUORO-P-TOLYL)-D-VALINATE; CYANO(3-PHENOXYPHENYL)METHYLN-[2-CHLORO-4-(TRIFLUOROMETHYL)PHENYL]-D-VALINATE; KARTAN; KLARTAN; MARVIK; MAVRIK; MAVRIK AQUAFLOW; MAVRIK HR; SPUR; ZR-3210; ZR 3210
Description: yellow-amber liquid

Use: insecticide; control of a wide range of insects (including lepidoptera, aphids, thrips, leafhoppers, whiteflies) and spider mites on indoor and outdoor ornamentals, apples, pears, peaches, vines, cereals, vegetables, cotton, tobacco, and turf

Physical Properties

Boiling Point: > 450 °C (842 °F)
Specific Gravity: 1.29 at 25 °C
Saturated Vapor Density: 1.200000003 kg/m^3
Vapor Pressure: 1 $x10^{-7}$ mm Hg at 25 °C
Water Solubility: 2 ppb
Other Solubilities: Soluble in aromatic hydrocarbons, Alcohols, diethylether, and dichloromethane.
Refraction Index: 1.549 at 20 °C/D
Flash Point: Burns with difficulty

RTECS Toxicity Data

Acute Oral: Rat LD_{50} Dose: 261 mg/kg. Mouse LD_{50} Dose: 2042 mg/kg.
Acute Inhalation: Rat LC_{50} Dose: 439 mg/m^3/4hr.
Acute Dermal: Rat LD_{50} Route: Skin; Dose: >20 gm/kg. Rat LD_{50} Route: Subcutaneous Dose: >2 gm/kg.

Hazard Overviews

Fire: Will burn.
Carcinogenicity: IARC - Not listed; NIOSH - Not listed; NTP - Not listed; ACGIH - Not listed; OSHA - Not listed; EPA - Not listed; MAK - Not listed

Environmental

Ecotoxicity: LC_{50} Mallard duck dietary >5620 mg/kg diet/8 day LC_{50} Bluegill sunfish 0.00089 mg/l/96 hr /Conditions of bioassay not specified LD_{50} Bobwhite quail oral >2510 mg/kg
Environmental Fate: If released to the atmosphere, it will degrade rapidly in the vapor-phase by reaction with photochemically produced hydroxyl radicals (estimated half-life of 3.9 hr). Particulate-phase will be removed physically from air by wet and dry deposition. If released to soil or water, it can degrade through biodegradation, hydrolysis and photodegradation. Laboratory and field studies have measured aerobic soil half-lives of 6-8 days and an anaerobic soil half-life of 15 days. It hydrolyzes in water with half-lives of 30 days at pH 3 and pH 6 and 1-2 hr at pH 9 and 25 °C. The photodegradation half-life (as a thin-film or as an aqueous solution in a 100 ml Erlenmeyer flask) has been measured to be about 1 day of sunlight exposure. It is expected to partition from water to sediment and suspended matter in aquatic media. Results of laboratory and field studies have indicated that it will not leach in soil.

Environmental Physical Data

Henry's Law Constant: estimated at 1.45 $x10^{-8}$
Sorption Partition Coefficient: log K_{OC} = estimated at log 4.9
BCF: estimated at 1.2 $x10^4$

Regulations

RCRA 40CFR: Not listed
CERCLA: 40CFR 302.4: Not listed
SARA 40CFR 372.65: Listed
SARA EHS 40CFR 355: Not listed
TSCA: Not listed

Analytical Methods

Food: FDA 212.1, 212.2, 232.1, 232.4, 242.1
Plasma: FDA 252

FOL3000 CAS #: 59-30-3

FOLIC ACID

RTECS: LP5425000
EINECS Number: 200-419-0
Molecular Formula: $C_{19}H_{19}N_7O_6$
Formula Weight: 441.45

Chemical Structure

Synonyms: ACFOL; ACIFOLIC; 2-AMINO-6-((P-((1,3-DICARBOXYPROPYL)CARBAMOYL)ANILINO)METHYL)-4-PTERIDINOL; N-(4-(((2-AMINO-1,4-DIHYDRO-4-OXO-6-PTERIDINYL)METHYL)AMINO)BENZOYL)-L-; N-(P-(((2-AMINO-4-HYDROXY-6-PTERIDINYL)METHYL)AMINO)BENZOYL)-L-GLUTAMIC ACID; ANTIANEMIA FACTOR; CIFOLIC; CYTOFOL; FACTOR U; FOLACID; FOLACIN; FOLAEMIN; FOLASIC; FOLATE; FOLBAL; FOLCIDIN; FOLCYSTEINE; FOLDINE; FOLETTES; FOLIAMIN; FOLICO; FOLINA; FOLIPAC; FOLSAN; FOLSAURE; FOLSAV; FOLUITE; FOLVITE; GLUTAMIC ACID; L-GLUTAMIC ACID,N-(4-(((2-AMINO-1,4-DIHYDRO-4-OXO-6-PTERIDINYL)METHYL)AMINO)BENZO-YL)-; GLUTAMIC ACID,N-(P-(((2-AMINO-4-HYDROXY-6-PTERIDINYL)METHYL)AMINO)BENZOYL)-,L-; GLUTAMIC ACID,N-(P-(((2-AMINO-4-HYDROXYPYRIMIDO(4,5-B)PYRAZIN-6-YL)METHYL)AMINO)BENZOYL)-,L; GLUTAMIC ACID,PTEROYL-,L-; INCAFOLIC; KYSELINA LISTOVA; LIVER LACTOBACILLUS CASEI FACTOR; MILLAFOL; NIFOLIN; NOVOFOLACID; NSC 3073; PGA; PTEGLU; 4-PTERIDINOL,2-AMINO-6-((P-((1,3-DICARBOXYPROPYL)CARBAMOYL)ANILINO)METHYL)-; L-PTEROYLGLUTAMIC ACID; PTEROYL-L-GLUTAMIC ACID; PTEROYLGLUTAMIC ACID; PTEROYL-L-MONOGLUTAMIC ACID; PTEROYLMONOGLUTAMIC ACID; VITAMIN B11; VITAMIN BC; VITAMIN BE; VITAMIN M
Description: yellowish-orange crystals or platelets; ordorless or almost ordorless
Use: in medicine, in nutrition and as a food additive; in hematopoietic vitamins for cat therapy; in the treatment of megaloblastic anemias to stimulate erythrocyte, lymphocyte and platelet production, and in the treatment of pernicious anemia; correcting the folate deficiency associated with the use of anticonvulsant; in veterinary medicine as a nutrient factor (as a dietary requirement for poultry)

Physical Properties

Freezing Point: Decomposes at 250 °C (482 °F)
Water Solubility: Very Slightly Soluble in Cold Water
Other Solubilities: Practically Insoluble Alcohol, Acetone, Chloroform, and Ether; Soluble in dilute solutions of alkali hydroxides and carbonates, yielding a clear orange-brown solution; Soluble in Hydrochloric Acid and in sulphuric acid.
pH: Suspension of 1 g in 10 ml 4 to 4.8
Flash Point: Not available; probably combustible

RTECS Toxicity Data

Acute Oral: Mouse LD_{50} Dose: 10 gm/kg.
Acute Dermal: Mouse LD_{Lo} Route: Subcutaneous Dose: 200 mg/kg; Toxic Effects: Kidney, Ureter, and Bladder - Other changes; Blood - Changes in spleen.
Reproductive/Teratogenic: Rat Route: Parenteral; Dose: 150 mg/kg; Duration: female 10D of pregnancy; Effects on Embryo or Fetus - Cytological changes (inc. somatic cell genetic material).
Mutagenic: Rat Unscheduled DNA Synthesis; Route: Intraperitoneal; Dose: 150 mg/kg. Rat Cytogenetic Analysis; Cell Type: lymphocyte; Dose: 10 ug/L.

Hazard Overviews

Health: May cause irritation. Acute Effects: may be harmful by inhalation, ingestion, or skin absorption.
Fire: Will burn. Extinguishing agents: water spray; carbon dioxide, dry chemical powder or appropriate foam. Precautions: combustible liquid.
Reactivity: Incompatible with: strong oxidizing agents, may decompose on exposure to light. Hazardous decomposition products: toxic fumes of: carbon monoxide, carbon dioxide, nitrogen oxides.
Carcinogenicity: IARC - Not listed; NIOSH - Not listed; NTP - Not listed; ACGIH - Not listed; OSHA - Not listed; EPA - Not listed; MAK - Not listed

Environmental

Environmental Fate: If released to soil, photolysis on soil surfaces and adsorption to soil may be important fate processes. No data were located which suggest biodegradation is an important fate process in soil or water. If released to water, volatilization, bioconcentration in aquatic organisms, and reaction with photochemically produced OH radicals are not expected to be important aquatic fate process. In near surface water, it may be susceptible to photolysis. Adsorption to sediment and suspended materials may be important in aquatic systems. If released to the atmosphere, vapor-phase is expected to degrade rapidly by reaction with photochemically produced hydroxyl radicals (estimated half-life of 1.5 hours).

Environmental Physical Data

Henry's Law Constant: $< 1 \times 10^{-10}$
Sorption Partition Coefficient: $K_{oc} = 3370$
BCF: estimated at 474

Regulations

RCRA 40CFR: Not listed
CERCLA: 40CFR 302.4: Not listed
SARA 40CFR 372.65: Not listed
SARA EHS 40CFR 355: Not listed
TSCA: Listed

FOL6000 CAS #: 133-07-3

FOLPET

RTECS: TI5685000
EINECS Number: 205-088-6
Molecular Formula: $C_9H_4Cl_3NO_2S$
Formula Weight: 296.58

O N S Cl Cl Cl O

Chemical Structure

Synonyms: ACRYPTAN; COSAN I; ENT 26539; FALTAN; FALTEX; FOLNIT; FOLPAN; FOLPEL; FTALAN; FUNGITROL; FUNGITROL 11; FUNGITROL II; INTERCIDE TMP; 1H-ISOINDOLE-1,3(2H)-DIONE,2-((TRICHLOROMETHYL)THIO)-; MURPHY'S ROSE FUNGICIDE; ORTHO PHALTAN 50W; ORTHOPHALTAN; PHALTAN; PHALTANE; PHTHALIMIDE,N-((TRICHLOROMETHYL)THIO)-; PHTHALTAN; SPOLACID; THIOPHAL; TRICHLORMETHYLTHIOIMID KYSELINY FTALOVE; N-(TRICHLOR-METHYLTHIO)-PHTHALAMID; N-(TRICHLORMETHYLTHIO)PHTHALIMIDE; N-(TRICHLOROMETHANESULPHENYL)PHTHALIMIDE; N-(TRICHLOROMETHYLMERCAPTO)PHTHALIMIDE; N-(TRICHLOROMETHYLMERCPATO)PHTHALIMIDE; 2-((TRICHLOROMETHYL)THIO)-1H-ISOINDOLE-1,3(2H)-DIONE; N-((TRICHLOROMETHYL)THIO)PHTHALIMIDE; N-(TRICHLOROMETHYLTHIO)PHTHALIMIDE; TRICHLOROMETHYLTHIOPHTHALIMIDE; TROYSAN ANTI-MILDEW O; VINICOLL
Description: light colored powder; colorless to white crystals
Use: broad spectrum contact fungicide in paints & plastics; against fairly wide range of fungi that cause plant diseases; inhibits growth (does not eradicate established fungi); for foliage application against venturia species, downy mildews, spot & scab diseases, rot, and blights; in vegetables, eg, leeks, fruit, eg, citrus, vines, ornamentals, etc; seed & plant bed treatment; fungicide-bactericide for vinyls, paints, & enamels.

Physical Properties

Freezing Point: 177 °C (350.6 °F)
Water Solubility: 1 mg/L at room temperature
Other Solubilities: Slightly Soluble in Ethanol & Acetone; 3-4% in aliphatic ketones and 0.1-1% in hydrocarbons; In

Chloroform 87, Benzene 22, isopropanol 12.5 (all in g/l at 20 °C).

RTECS Toxicity Data

Acute Oral: Rat LD_{50} Dose: 7540 mg/kg. Mouse LD_{50} Dose: 1546 mg/kg.

Acute Inhalation: Rat LC_{50} Dose: >5 gm/m^3/2hr. Mouse LC_{50} Dose: >6 gm/m^3/2hr.

Acute Dermal: Rabbit LD_{50} Route: Skin; Dose: >22600 mg/kg. Mouse LD_{Lo} Route: Skin; Dose: 5 gm/kg.

Reproductive/Teratogenic: Rat Route: Oral; Dose: 500 mg/kg; Duration: male 5D prior to mating; Effects on Embryo or Fetus - Fetal death. Rat Route: Intraperitoneal; Dose: 500 mg/kg; Duration: male 5D prior to mating; Effects on Embryo or Fetus - Fetal death.

Mutagenic: Human DNA Damage; Cell Type: HeLa cell; Dose: 1300 nmol/L. Human Unscheduled DNA Synthesis; Cell Type: fibroblast; Dose: 100 umol/L. Human DNA Inhibition; Cell Type: lymphocyte; Dose: 5 mg/L. Human DNA Inhibition; Cell Type: HeLa cell; Dose: 1300 nmol/L. Human Cytogenetic Analysis; Cell Type: lymphocyte; Dose: 15 ug/L/15M.

Tumorigenic: Mouse Route: Oral; Dose: 437 gm/kg/2Y-C; Toxic Effects: Tumorigenic - Carcinogenic by RTECS criteria; Gastrointestinal - Tumors. Mouse Route: Subcutaneous; Dose: 1000 mg/kg; Toxic Effects: Tumorigenic - Equivocal tumorigenic agent by RTECS criteria; Lungs, Thorax, or Respiration - Tumors; Blood - Tumors.

Hazard Overviews

Carcinogenicity: IARC - Not listed; NIOSH - Not listed; NTP - Not listed; ACGIH - Not listed; OSHA - Not listed; EPA - Class B2, Probable human carcinogen based on animal studies; MAK - Not listed

Environmental

Ecotoxicity: LC_{50} Salmo gairdneri (rainbow trout) 39 ug/l/96 hr at 12 °C, wt 1.5 g (95% confidence limit 18-85 ug/l). Static bioassay without aeration, pH 7.2-7.5, water hardness 40-50 mg/l as calcium carbonate and alkalinity of 30-35 mg/l LD_{50} Anas platyrhynchos (mallard) male oral more than 2000 mg/kg, 3-4 month old LC_{50} Gammarus fasciatus (scud) 2500 ug/l/96 hr at 15 °C, mature (95% confidence limit 1994-3134 ug/l). Static bioassay without aeration, pH 7.2-7.5, water hardness 40-50 mg/l as calcium carbonate and alkalinity of 30-35 mg/l LC_{50} Oncorhynchus kisutch (coho salmon) 106 ug/l/96 hr at 12 °C, wt 1.0 g (95% confidence limit 82-137 ug/l). Static bioassay without aeration, pH 7.2-7.5, water hardness 40-50 mg/l as calcium carbonate and alkalinity of 30-35 mg/l LC_{50} Micropterus dolomieui (smallmouth bass) 91 ug/l/96 hr at 12 °C, fingerling (95% confidence limit73-113 ug/l). Static bioassay without aeration, pH 7.2-7.5, water hardness 40-50 mg/l as calcium carbonate and alkalinity of 30-35 mg/l

Environmental Fate: If released to soil, it is not expected to be mobile and, therefore, very little leaching in soil is expected. It is expected to degrade in the ambient atmosphere by reaction with photochemically produced hydroxyl radicals with a half-life of 0.37 day. If released to water, neither volatilization nor bioconcentration are expected to be important fate processes. In alkaline aqueous media, it undergoes rapid chemical hydrolysis. Absorption to sediments will be important.

Environmental Physical Data

Henry's Law Constant: 3.8 x10^{-6}

Octanol/Water Partition Coefficient: log K_{ow} = 2.85

Sorption Partition Coefficient: K_{oc} = estimated at 1848

BCF: estimated at 86

Regulations

RCRA 40CFR: Not listed

CERCLA: 40CFR 302.4: Not listed

SARA 40CFR 372.65: Listed

SARA EHS 40CFR 355: Not listed

TSCA: Listed

Analytical Methods

Air: EPA TO-10

Soil: EPA PMD-TLC

Water / Groundwater: EPA 022

Food: FDA 212.1, 212.2, 232.1, 232.3, 232.4, 242.1; AOAC 977.03

Indoor / Expired Air: EPA IP-8; ASTM D4861

Plasma: EPA 001, 028; FDA 211.1, 231.1, 252

FOM5000 CAS #: 72178-02-0

FOMESAFEN

RTECS: CV2475000

EINECS Number: 276-439-9

Molecular Formula: $C_{15}H_{10}ClF_3N_2O_6S$

Formula Weight: 438.78

Synonyms: 5-(2-CHLORO-4-(TRIFLUOROMETHYL)PHENOXY)-N-(METHYLSULFONYL)-2-NITROBENZAMIDE; FLEX; FOMESAFENE; PP021; REFLEX

Physical Properties

Freezing Point: 220 °C (428 °F) to 221 °C (429.8 °F)

Specific Gravity: 1.28000

Water Solubility: 50 mg/l 20 °C

Other Solubilities: 300 g/l acetone, 150 g/l cyclohexanone, 10 g/l dichloromethane, 1.9 g/l xylene

RTECS Toxicity Data

Acute Oral: Rat LD_{50} Dose: 1250 mg/kg. Duck LD_{50} Dose: >5 gm/kg.

Acute Dermal: Rabbit LD_{50} Route: Skin; Dose: >1 gm/kg.

Hazard Overviews

Carcinogenicity: IARC - Not listed; NIOSH - Not listed; NTP - Not listed; ACGIH - Not listed; OSHA - Not listed; EPA - Class C, Possible human carcinogen; MAK - Not listed

Environmental

Regulations

RCRA 40CFR: Not listed
CERCLA: 40CFR 302.4: Not listed
SARA 40CFR 372.65: Listed
SARA EHS 40CFR 355: Not listed
TSCA: Not listed

FON5000 CAS #: 944-22-9

FONOFOS

RTECS: TA5950000
DOT: UN2783; UN2784; UN3017; UN3018; IMO6.1
EINECS Number: 213-408-0
Molecular Formula: $C_{10}H_{15}OPS_2$
Structured MF: $C_{10}H_{15}OPS_2$
Formula Weight: 246.32
Synonyms: O-AETHYL-S-PHENYL-AETHYL-DITHIOPHOSPHONAT; DIFONATE; DIFONATUL; DOUBLEDOWN; DYFONAT; DYFONATE; DYFONATE TILLAM 1-4E; DYPHONATE; ENT 25,796; ENT 25796; O-ETHYL S-PHENYL ETHYLDITHIOPHOSPHONATE; (+/-)-O-ETHYL S-PHENYL ETHYLPHOSPHONODITHIOATE; O-ETHYL S-PHENYL (RS)-ETHYLPHOSPHONODITHIOATE; O-ETHYL S-PHENYL ETHYLPHOSPHONODITHIOATE; O-ETHYL S-PHENYL ETHYLPHOSPHONOTHIOLOTHIONATE; O-ETHYL-S-PHENYL ETHYLPHOSPHONODITHIOATE; O-ETHYL-S-PHENYL ETHYLPHOSPHOROTHIOATE; ETHYLPHOSPHONODITHIOIC ACID O-ETHYL S-PHENYL ESTER; FONOPHOS; N 2790; N-2790; OMS 410; PHENYL ETHYLDITHIOPHOSPHONATE; PHOSPHONODITHIOIC ACID,ETHYL-,O-ETHYL S-PHENYL ESTER; STAUFFER N 2790
Description: colorless to light-yellow liquid; aromatic pungent mercaptan odor
Use: soil insecticide in controlling soil insects eg, corn rootworms (diabrotica species), wireworms (elateridae), garden symphylan (scutigerella immaculata), root maggots (hylemya species), crickets (gryllidae) & other soil, foliar pests; in cereals, maize, sorghum, vegetables, ornamentals, fruit (incl citrus & bananas), vines, olives, potatoes, sugar beet, sugar cane, groundnuts, tobacco, & turf; soil fumigant.

Physical Properties

Boiling Point: 130 °C (266 °F) at 0.1 mm Hg
Specific Gravity: 1.16 at 25 °C/25 °C
Vapor Pressure: 28 mPa at 25 °C
Water Solubility: 0.001% by weight
Other Solubilities: miscible with Acetone, Ethanol, Kerosene, 4-Methylpentan-2-one, Xylene
Refraction Index: 1.5883 at 30 °C/D
Flash Point: > 94 °C

RTECS Toxicity Data

Acute Oral: Rat LD_{50} Dose: 3 mg/kg. Dog LD_{50} Dose: 3 mg/kg.
Acute Inhalation: Rat LC_{50} Dose: 1900 mg/m^3/1hr.
Acute Dermal: Rabbit LD_{50} Route: Skin; Dose: 25 mg/kg. Rat LD_{50} Route: Skin; Dose: 147 mg/kg.

Hazard Overviews

Fire: Will burn.
Carcinogenicity: IARC - Not listed; NIOSH - Not listed; NTP - Not listed; ACGIH - Class A4, Not classifiable as a human carcinogen; OSHA - Not listed; EPA - Not listed; MAK - Not listed
Exposure Limits
OSHA PEL Vacated 1989 Limits: TWA: 0.1 mg/m^3.
ACGIH TLV: TWA: 0.1 mg/m^3.
NIOSH REL: TWA: 0.1 mg/m^3.
Respirator Recommendation
Exposure Range: >0.1 to 5 mg/m^3 Supplied Air, Constant Flow/Pressure Demand, Half Mask
Exposure Range: >5 to 100 mg/m^3 Supplied Air, Constant Flow/Pressure Demand, Full Face
Exposure Range: >100 to unlimited mg/m^3 Self-contained Breathing Apparatus, Pressure Demand, Full Face
Note: odor threshold unknown

Environmental

Ecotoxicity: LD_{50} Anas platyrhynchos (Mallard) oral 16.9 mg/kg (95% confidence limit 13.4-21.3 mg/l), males 3-4 mo of age LD_{50} Lepomis macrolophus (Bluegill) 0.007 mg/l/96 hr at 24 °C (95% confidence limit 0.005-0.009 mg/l), wt 1.0 g. Static bioassay without aeration, pH 7.2-7.5, water hardness 40-50 mg/l as calcium carbonate and alkalinity of 30-35 mg/l
Environmental Fate: If released to soil, loss will occur primarily due to biodegradation. Loss from soil due to volatilization and hydrolysis may not be important. With the exception of sandy soils, it is expected to have a low mobility in most soils. The half-life in soil is about 40 days. The major pathway for the loss from water may be adsorption to suspended solids and sediment. Minor loss may occur as a result of biodegradation and photolysis. Some bioconcentration may occur in aquatic organisms. Reaction with photochemically produced hydroxyl radicals may be one of the most important loss processes in the atmosphere. The half-life due to this reaction has been estimated to be 1.4 hours. Partial removal will also occur as a result of dry and wet deposition.
Cleanup/Disposal: Guide No. 131: Fully encapsulating, vapor protective clothing should be worn for spills and leaks with no fire. Eliminate all ignition sources (no smoking, flares, sparks or flames in immediate area). All equipment used when handling the product must be grounded. Do not touch or walk through spilled material. Stop leak if you can do it without risk. Prevent entry into waterways, sewers, basements or confined areas. A vapor suppressing foam may be used to reduce vapors. Small Spills: Absorb with earth, sand or other non-combustible material and transfer to containers for later disposal. Use clean non-sparking tools to collect absorbed material. Large Spills: Dike far ahead of liquid spill for later disposal. Water spray may reduce vapor; but may not prevent ignition in closed spaces. Guide No. 152: Do not touch damaged containers or spilled material unless wearing appropriate protective clothing. Stop leak if you can do it without risk. Prevent entry into waterways, sewers, basements

or confined areas. Cover with plastic sheet to prevent spreading. Absorb or cover with dry earth, sand or other non-combustible material and transfer to containers. Do not get water inside containers.

Environmental Physical Data

Henry's Law Constant: estimated at 5.4×10^{-6}
Octanol/Water Partition Coefficient: log K_{ow} = 3.94
Sorption Partition Coefficient: log K_{oc} = 1.83 to 3.71
BCF: estimated at 2.11

Regulations

RCRA 40CFR: Not listed
CERCLA: 40CFR 302.4: Not listed
SARA 40CFR 372.65: Not listed TPQ: 500 lb
SARA EHS 40CFR 355: Listed TPQ: 500 lb
TSCA: Not listed

Analytical Methods

Soil: SW846 8141, 8141A
Water / Groundwater: EPA 622.1, 022; SW846 8141, 8141A
Food: FDA 212.1, 212.2, 232.1, 232.3, 232.4, 242.1
Indoor / Expired Air: NIOSH 5600
Plasma: EPA 001, 028; FDA 211.1, 231.1, 252

FOR1000 CAS #: 50-00-0

FORMALDEHYDE

RTECS: LP8925000
DOT: UN1198; UN2209; IMO3.2; IMO9.3
EINECS Number: 200-001-8
Molecular Formula: CH_2O
Structured MF: $H_2C=O$
Formula Weight: 30.03

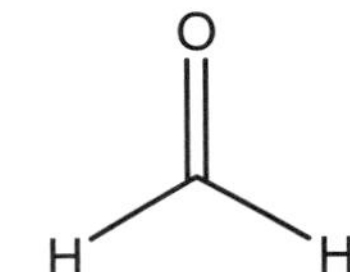

Chemical Structure

Synonyms: ALDEHYD MRAVENCI; ALDEHYDE FORMIQUE; ALDEIDE FORMICA; BFV; DORMOL; FA; FANNOFORM; FORMALDEHYD; FORMALDEHYDE SOLUTION; FORMALDEHYDE,GAS; FORMALIN; FORMALIN 40; FORMALINA; FORMALINE; FORMALIN-LOESUNGEN; FORMALITH; FORMIC ALDEHYDE; FORMOL; FYDE; HCHO; HOCH; IVALON; KARSAN; LYSOFORM; METHALDEHYDE; METHANAL; METHANAL FORMALIN; METHYL ALDEHYDE; METHYLENE GLYCOL; METHYLENE OXIDE; MORBICID; OPLOSSINGEN; OXOMETHANE; OXYMETHYLENE; PARAFORM; POLYOXYMETHYLENE GLYCOLS; SUPERLYSOFORM

Description: colorless liquid; pungent, suffocating odor

Use: as a preservative, disinfectant and antiseptic, in embalming solutions and in the manufacture of phenolic resins, artificial sil cellulose esters, dyes, urea, thiourea, melamine resins, organic chemicals, glass mirrors and explosives; in improving fastness of dyes on fabrics, in tanning and preserving hides, in mordanting and waterproofing fabrics, as a germicide and fungicide for vegetables and other plants, in destroying flies and other insects, in preserving and coagulating rubber latex and to prevent mildew and spelt in wheat and rot in oats; to rend casein, albumin, and gelatin insoluble, in chemical analysis, as a tissue fixative, as a component of particle board and plywood and in the manufacture of Pentaerythritol, Hexamethylenetetramine and 1,4-Butanediol; in ceiling and wall insulation, in resins used to wrinkle-proof fabrics, in photography for hardening gelatin plates and papers, for toning gelatin-chloride papers and for chrome printing and developing; an intermediate in drug manufacture and a pesticide intermediate

Physical Properties

Boiling Point: -19.5 °C (-3 °F)
Freezing Point: -92 °C (-133.6 °F)
Specific Gravity: 1.067 (Air=1)
Vapor Density: About 1.075 for gas
Vapor Pressure: 10 mm Hg at -88.0 °C
Water Solubility: 55% at 0 °C
Other Solubilities: 95% Ethanol: >=100 mg/ml at 20.5 °C; Acetone: >=100 mg/ml at 20.5 °C; Benzene: Soluble; Chloroform: Immiscible; DMSO: >=100 mg/ml at 20.5 °C; Ether: Soluble; Most organic solvents: Soluble; Petroleum Ether: Insoluble.
Odor Threshold: 0.5 to 1 ppm
pH: 2.8 to 4
Refraction Index: solution 1.3746 at 20 °C/D
Critical Temperature: 137.2 to 141 °C
Critical Pressure: 6.78 mPa
Ionization Potential (eV): 10.88
Flash Point: 40% solution 60 °C
Autoignition Temperature: Formaldehyde gas 300 °C
LEL: 7.0% v/v
UEL: Solution 73% v/v

RTECS Toxicity Data

Mutagenic: Human DNA Damage; Cell Type: fibroblast; Dose: 100 umol/L. Human DNA Damage; Cell Type: lung; Dose: 100 umol/L. Human DNA Damage; Cell Type: other cell types; Dose: 100 umol/L. Human Unscheduled DNA Synthesis; Cell Type: HeLa cell; Dose: 10 nmol/L. Human DNA Inhibition; Cell Type: other cell types; Dose: 210 umol/L. Human DNA Inhibition; Cell Type: HeLa cell; Dose: 400 umol/L. Human Mutations in Mammalian Somatic Cells; Cell Type: lymphocyte; Dose: 130 umol/L. Human Cytogenetic Analysis; Cell Type: lymphocyte; Dose: 10 mg/L. Human Cytogenetic Analysis; Cell Type: fibroblast; Dose: 2 mmol/L. Human Sister Chromatid Exchenge; Cell Type: lymphocyte; Dose: 125 umol/L. Human Other Mutation Test Systems; Cell Type: other cell types; Dose: 210 umol/L. Human Other Mutation Test Systems; Cell Type: lymphocyte; Dose: 10 mg/L.

Hazard Overviews

Corrosive

Flammable

Fire Diamond: 2 / 3 / 0 / —

Health: Corrosive to eye/skin/respiratory tract. Toxic. Also Causes: can sensitize skin. Chronic Effects: dermatitis, kidney damage. Suspect cancer hazard. Possible reproductive/teratogenic hazard (based on animal data).

Fire: Flammable. Can form explosive mixtures in the air. Use water spray, dry chemical, carbon dioxide, or alcohol-resistant foam.

Reactivity: Unstable, slowly oxidizes in air. Polymerization readily occurs. Avoid: exposure to heat and ignition sources. Incompatible with: carbon steel; hydrogen chloride; hydrogen peroxide; magnesium carbonate hydroxide; nitromethane; peroxyformic acid; phenol; potassium permanganate; performic acid; perchloric acid/aniline; nitrogen dioxide (explosive at 180 °C). Hazardous decomposition products: carbon dioxide; acrid smoke; toxic vapors.

Carcinogenicity: IARC - Group 2A, Probably carcinogenic to humans; NIOSH - Listed as carcinogen; NTP - Class 2B, Reasonably anticipated to be a carcinogen, sufficient evidence of carcinogenicity from studies in experimental animals; ACGIH - Class A2, Suspected human carcinogen; OSHA - Listed as a carcinogen; EPA - Class B1, Probable human carcinogen based on epidemiologic studies; MAK - Class B, Justifiably suspected of having carcinogenic potential

Primary Target Organs:

Eyes

Skin

Respiratory System

Kidneys

Exposure Limits

OSHA PEL: TWA: 0.75 ppm; STEL: 2 ppm.

ACGIH TLV: STEL: 0.3 ppm; 0.37 mg/m^3; Ceiling.

NIOSH REL: TWA: 0.016 ppm. STEL: 0.1 ppm; 15 mg/m^3; 15-minute.

NIOSH IDLH: 20 ppm.

DFG MAK: TWA: 0.5 ppm; 0.6 mg/m^3.

Respirator Recommendation

Exposure Range: >0.75 to 7.5 ppm Air Purifying, Negative Pressure, Half Mask

Exposure Range: >7.5 to <20 ppm Air Purifying, Negative Pressure, Full Face

Exposure Range: 20 to unlimited ppm Self-contained Breathing Apparatus, Pressure Demand, Full Face

Cartridge Color: olive (must be approved)

Note: replace filters every 3 hrs or end of shift whichever is sooner

Environmental

Ecotoxicity: LC_{50} Striped bass larvae 10 mg/l/48-96 hr; static bioassay LC_{50} Atlantic salmon (Salmo salar) 173 ul/l/96 hr flow-through LC_{50} Pimephales promelas (fathead minnow) 24.1 mg/l/96 hr (confidence limit 22.6 - 25.7 mg/l), flow-through bioassay with measured concentrations, 21.7 °C, dissolved oxygen 7.4 mg/l, hardness 50.8 mg/l calcium carbonate LC_{50} Channel catfish (Ictalurus punctatus) 65.8 ul/l/96 hr flow-through

Environmental Fate: Removed by direct photolysis and oxidation by photochemically produced hydroxyl radicals (half-life a few hours). Additional quantities are removed by dry deposition, rain or by dissolving in the ocean and other surface waters. In the aqueous compartment biodegradation takes place in a few days.

Cleanup/Disposal: Guide No. 132: Fully encapsulating, vapor protective clothing should be worn for spills and leaks with no fire. Eliminate all ignition sources (no smoking, flares, sparks or flames in immediate area). All equipment used when handling the product must be grounded. Do not touch or walk through spilled material. Stop leak if you can do it without risk. Prevent entry into waterways, sewers, basements or confined areas. A vapor suppressing foam may be used to reduce vapors. Absorb with earth, sand or other non-combustible material and transfer to containers (except for Hydrazine). Use clean non-sparking tools to collect absorbed material. Large Spills: Dike far ahead of liquid spill for later disposal. Water spray may reduce vapor; but may not prevent ignition in closed spaces.

Environmental Physical Data

Henry's Law Constant: 3.27 x10^{-7}

Octanol/Water Partition Coefficient: log K_{ow} = 0.35

BCF: none

BOD: theoretical 37%, 5 days

Regulations

RCRA 40CFR: Listed Hazardous Waste No. U122 Toxic Waste

CERCLA: 40CFR 302.4: Listed per CWA Section 311(b)(4) per RCRA Section 3001 RQ: 100 lb (45.35 kg)

SARA 40CFR 372.65: Listed TPQ: 500 lb

SARA EHS 40CFR 355: Listed TPQ: 100 lb

TSCA: Listed

Analytical Methods

Air: EPA TO-11, TO-5, 0011, 0011A, 0100, 8520; ASTM D4490; SW846 8315, 8315A, 8520

Water / Groundwater: EPA 6

Drinking Water: EPA 554

Food: AOAC 897.01, 898.01

Indoor / Expired Air: NIOSH 2539, 2541, 3500, 5700; EPA IP-6A, IP-6B, IP-6C, TO-14C, 0100; ASTM D5014

Other: EPA 1667

FOR2330 **CAS #: 75-12-7**

FORMAMIDE

RTECS: LQ0525000

EINECS Number: 200-842-0

Molecular Formula: CH_3NO

Structured MF: $HCONH_2$

Formula Weight: 45.04

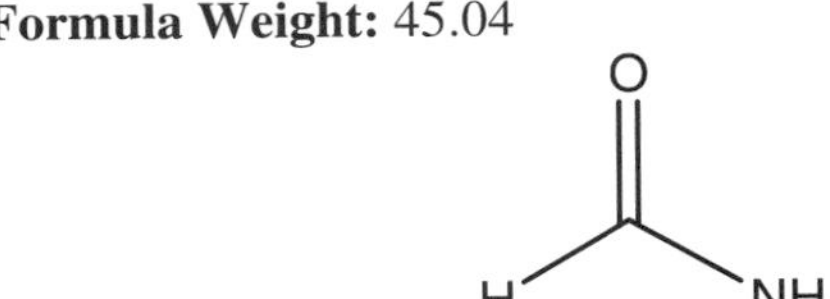

Chemical Structure

Synonyms: AMID KYSELINY MRAVENCI; CARBAMALDEHYDE; FORMIC ACID,AMIDE; FORMIMIDIC ACID; METHANAMIDE; METHANOIC ACID,AMIDE

Description: clear liquid; odorless or slight ammonia odor

Use: in the manufacture of formates and hydrocyanic acid It is also used as a softener for paper, gums and animal glues, as an ionizing and pharmaceutical solvent and as a versatile synthetic reagent

Physical Properties

Boiling Point: 210.5 °C (411 °F) at 760 mm Hg
Freezing Point: 2.55 °C (36.59 °F)
Specific Gravity: 1.1334 at 20 °C/4 °C
Vapor Density: 1.56 Air=1
Density: 1.1334 g/cu cm at 20 °C
Vapor Pressure: 29.7 mm Hg at 129.4 °C
Water Solubility: Soluble in Water
Other Solubilities: miscible with Methanol, Phenol, Dioxane, Acetic Acid, Ethylene Glycol, USP Glycerol.
Surface Tension: 58.35 dynes/cm
Odor Threshold: 150 mg/m^3
pH: 0.5 molar aqueous solution 7.1
Refraction Index: 1.44911 at 15 °C/D
Evaporation Rate: < 1 Butyl Acetate=1
Ionization Potential (eV): 10.20
Flash Point: 154.444 °C Open Cup
Autoignition Temperature: 154 °C

RTECS Toxicity Data

Acute Oral: Rat LD_{50} Dose: 5577 mg/kg; Toxic Effects: Automatic Nervous System - Other (direct) parasympathomimetic; Behavioral - Ataxia; Kidney, Ureter, and Bladder - Incontinence. Mouse LD_{50} Dose: 3150 mg/kg.

Acute Dermal: Rabbit LD_{Lo} Route: Skin; Dose: 6 gm/kg. Rat LD_{50} Route: Subcutaneous Dose: >4 gm/kg.

Chronic (Multiple Dose) Oral: Rat Dose: 910 mg/kg/26W-I; Toxic Effects: Brain and coverings - Recordings from specific areas of CNS; Liver - Liver function tests impaired; Kidney, Ureter, and Bladder - Proteinuria.

Chronic (Multiple Dose) Inhalation: Rat Dose: 1500 ppm/6H/2W-I; Toxic Effects: Blood - Changes in leukocyte (WBC) cell count; Blood - Changes in platelet cell count; Nutritional and gross metabolic - Weight loss or decreased weight gain.

Irritation Eye: Rabbit Standard Draize Test Dose: 100 mg; Reaction: severe.

Reproductive/Teratogenic: Rat Route: Oral; Dose: 2 gm/kg; Duration: female 7D of pregnancy; Effects on Fertility - Post-implantation mortality; Effects on Embryo or Fetus - Fetotoxicity. Rat Route: Oral; Dose: 7980 mg/kg; Duration: female 7-12D of pregnancy; Specific Developmental Abnormalities - Craniofacial (including nose and tongue); Musculoskeletal system.

Mutagenic: Non-mammalian species Cytogenetic Analysis; Cell Type: other cell types; Dose: 500 mmol/L. Non-mammalian species Other Mutation Test Systems; Cell Type: other cell types; Dose: 500 mmol/L.

Hazard Overviews

Fire Diamond

Health: Irritating to eyes/skin/respiratory tract. May be absorbed through the skin in toxic amounts.

Fire: Will burn. Use water spray, alcohol foam, carbon dioxide, or dry chemical. Fires may produce toxic gases such as carbon monoxide and ammonia.

Reactivity: Stable. Hazardous polymerization cannot occur. Avoid: sources of ignition; excessive heat. Incompatible with: strong oxidizing agents; acids; bases. Hazardous decomposition products: carbon monoxide; ammonia gas.

Carcinogenicity: IARC - Not listed; NIOSH - Listed as carcinogen; NTP - Not listed; ACGIH - Not listed; OSHA - Not listed; EPA - Not listed; MAK - Not listed

Primary Target Organs:

Eyes

Skin

Respiratory System

Mucous Membranes

Exposure Limits

OSHA PEL Vacated 1989 Limits: TWA: 20 ppm; 30 mg/m^3; STEL: 30 ppm; 45 mg/m^3.

ACGIH TLV: TWA: 10 ppm; 18 mg/m^3.

NIOSH REL: TWA: 10 ppm; 15 mg/m^3.

Respirator Recommendation

Exposure Range: >10 to 500 ppm Supplied Air, Constant Flow/Pressure Demand, Half Mask

Exposure Range: >500 to 10,000 ppm Supplied Air, Constant Flow/Pressure Demand, Full Face

Exposure Range: >10,000 to unlimited ppm Self-contained Breathing Apparatus, Pressure Demand, Full Face

Note: poor warning properties

Environmental

Environmental Fate: If released to the atmosphere, vapor phase is degraded relatively rapidly by reaction with photochemically produced hydroxyl radicals (estimated half-life of 2.1 hours in air). Based on the complete water solubility, removal by rainout may be important. Several biodegradation screening studies have observed significant biodegradation; therefore, if released to soil, it may be subject to biodegradation. Based on an estimated K_{oc} value of 3.6, it can be expected to leach significantly. Based on a low vapor pressure, it may volatilize from dry surface soils; however, volatilization from moist soil surfaces is not a significant fate process. If released to water, biodegradation is expected to be the most important fate process. The chemical structure suggests that it may be susceptible to environmental hydrolysis; although, no rate data are available that indicate hydrolysis is an important aquatic fate process. Volatilization, adsorption to sediment and suspended materials, and bioconcentration (estimated BCF of 0.0417) are not expected to be significant fate processes in aquatic environments.

Cleanup/Disposal: Guide No. 171: Do not touch or walk through spilled material. Stop leak if you can do it without risk. Prevent dust cloud. Avoid inhalation of asbestos dust. Small Dry Spills: With clean shovel place material into clean, dry container and cover loosely; move containers from spill area. Small Spills: Take up with sand or other noncombustible absorbent material and place into containers for later disposal. Large Spills: Dike far ahead of liquid spill for later disposal. Cover powder spill with plastic sheet or tarp to minimize spreading. Prevent entry into waterways, sewers, basements or confined areas.

Environmental Physical Data

Henry's Law Constant: 1.39×10^{-9}
Octanol/Water Partition Coefficient: log K_{ow} = -1.51
Sorption Partition Coefficient: K_{oc} = estimated at 3.6
BCF: estimated at 0.042
BOD: 0%, 5 days

Regulations

RCRA 40CFR: Not listed
CERCLA: 40CFR 302.4: Not listed
SARA 40CFR 372.65: Not listed
SARA EHS 40CFR 355: Not listed
TSCA: Listed

Analytical Methods

Other: EPA 1666, 1671

FOR3660 CAS #: 23422-53-9

FORMETANATE HYDROCHLORIDE

RTECS: FC2514000
EINECS Number: 245-656-0
Molecular Formula: $C_{11}H_{16}ClN_3O_2$
Formula Weight: 257.8
Synonyms: CARBAMIC ACID,METHYL-,ESTER WITHN'-(M-HYDROXYPHENYL)-N,N-DIMETHYLFORMAMIDINE,MONOHYDROCHLORIDE; CARZOL; CARZOL SP; DICARZOL; M-[((DIMETHYLAMINO) METHYLENE) AMINO] PHENYLMETHYLCARBAMATE HYDROCHLORIDE; M-(((DIMETHYLAMINO)METHYLENE)AMINO)PHENYL METHYL CARBAMATE,HYDROCHLORIDE; M-(((DIMETHYLAMINO)METHYLENE)AMINO)PHENYL METHYLCARBAMATEHYDROCHLORIDE; 3-DIMETHYLAMINOMETHYLENEIMINOPHENYL-N-METHYLCARBAMATE,HYDROCHLORIDE; N,N-DIMETHYL-N'-(((METHYLAMINO)CARBONYL)OXY)PHENYLMETHANIMIDAMIDE MONOHYDROCHLORIDE; N,N-DIMETHYL-N'-(((METHYLAMINO)CARBONYL)OXY)PHENYLMETHANIMIDAMIDE MONOHYDROCHLORIDE; ENT-27566; ENT 27566; EP-332; EP 332; FORMETANATE MONOHYDROCHLORIDE; METHANIMIDAMIDE,N,N-DIMETHYL-N'-(3-(((METHYLAMINO)CARBONYL)OXY)PHENYL)-,MONOHYDROCHLORIDE; MORTON EP 332; NOR-AM EP 332; SCHERING 36056; SN 36056
Description: colorless crystals or white powder; faint odor
Use: insecticide & acaricide on deciduous fruit & nuts & alfalfa; active against the mobile stages of fruit tree spider mites and is recommended for use in greenhouses

Physical Properties

Boiling Point: Decomposes at 200 °C (392 °F) to 202 °C (396 °F)
Freezing Point: Decomposes at 200 °C (392 °F) to 202 °C (395.6 °F)
Vapor Pressure: Practically zero at room temperature
Water Solubility: 1%
Other Solubilities: Benzene <0.1%, Acetone <0.1%, Chloroform 0.2%, Methanol about 25%, hexane <0.1%.

RTECS Toxicity Data

Acute Oral: Rat LD_{50} Dose: 20 mg/kg. Mouse LD_{50} Dose: 18 mg/kg.
Acute Dermal: Rabbit LD_{50} Route: Skin; Dose: 10200 mg/kg. Rat LD_{50} Route: Skin; Dose: >5600 mg/kg.

Hazard Overviews

Carcinogenicity: IARC - Not listed; NIOSH - Not listed; NTP - Not listed; ACGIH - Not listed; OSHA - Not listed; EPA - Not listed; MAK - Not listed

Environmental

Ecotoxicity: LC_{50} Rainbow trout 2.8 mg/l/96 hr LC_{50} Japanese quail (Coturnix japonica), 14 days old, oral (5 day ad libitum in diet) 993 ppm (95% confidence intervals 673-1,465 ppm)
Environmental Fate: If released to soil, it is expected to be mobile and have medium to high soil mobility. If released to water, it will be essentially nonvolatile. Aqueous base-catalyzed hydrolysis (at 25 °C has an estimated half-life of 4.5 days at pH of 8 and 45.2 days at pH of 7. Insufficient data are available to predict the relative importance or rate of biodegradation in water. If released to the atmosphere, it will exist primarily in the particulate phase. Physical removal from air can occur through wet and dry deposition.

Environmental Physical Data

Henry's Law Constant: estimated at 2.3×10^{-19}
Sorption Partition Coefficient: K_{oc} = estimated at 212
BCF: estimated at 0.38

Regulations

RCRA 40CFR: Listed Hazardous Waste No. P198 Toxic Waste
CERCLA: 40CFR 302.4: Listed per RCRA Section 3001 RQ: 1 lb (0.454 kg)
SARA 40CFR 372.65: Not listed TPQ: 500/10000 lb
SARA EHS 40CFR 355: Listed TPQ: 1 lb
TSCA: Listed

FOR4990 **CAS #: 64-18-6**

FORMIC ACID

RTECS: LQ4900000
DOT: UN1779; IMO8.0
EINECS Number: 200-579-1
Molecular Formula: CH_2O_2
Structured MF: HCO_2H
Formula Weight: 46.02

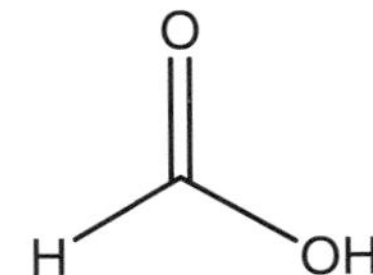

Chemical Structure

Synonyms: ACIDE FORMIQUE; ACIDO FORMICO; ADD-F; AMEISENSAEURE; AMINIC ACID; BILORIN; COLLO-BUEGLATT; COLLO-DIDAX; FORMIC ACID (85-95% IN AQUEOUS SOLUTION); FORMIRA; FORMISOTON; FORMYLIC ACID; HYDROGEN CARBOXYLIC ACID; KWAS METANIOWY; KYSELINA MRAVENCI; METHANOIC ACID; MIERENZUUR; MYRMICYL
Description: colorless liquid; pungent, penetrating odor
Use: decalcifier; wool dye reducer; depilatory for hides and tanning; rubber regeneration; electroplating; silage and grain preservation; reactive alkylating agent for alcohols; carboxylating agent for tertiary compounds; dyeing and finishing of textiles; manufacture of fumigants; insecticides; refrigerants; solvents of perfume; lacquers; acetic acid; airplane dope; allyl alcohol; cellulose formate; phenolic resins and oxalate; brewing (antiseptic); ore floatation; vinyl resin plasticizers; counterirritant; astringent; laundry and paper industries

Physical Properties

Boiling Point: 100.5 °C (213 °F)
Freezing Point: 8.4 °C (47.12 °F)
Specific Gravity: 1.22 at 20 °C/4 °C
Vapor Density: 1.59 Air=1
Saturated Vapor Density: Calculated 1.239467249 kg/m^3
Bulk Density: 10.16 lbs/gal at 20 °C
Vapor Pressure: Calculated 42.59 mm Hg at 25 °C
Water Solubility: Miscible
Other Solubilities: 95% Ethanol: >=100 mg/ml at 21 °C; Acetone: >=100 mg/ml at 21 °C; Benzene: Moderately Soluble; DMSO: >=100 mg/ml at 21 °C; Ether: Soluble; Glycerol: miscible.
Surface Tension: 38 dynes/cm at 15 °C
Odor Threshold: Detection threshold 1.50 x10^3 ppm
pH: < 7
Refraction Index: 1.3714 at 20 °C/D
Evaporation Rate: 2.1 Butyl Acetate=1
Critical Temperature: 315 °C
Ionization Potential (eV): 11.05
Flash Point: 69 °C Closed Cup
Autoignition Temperature: 540 °C
LEL: 90% Sol 18% v/v
UEL: 90% Solution 57% v/v

RTECS Toxicity Data

Acute Oral: Woman TD_{Lo} Dose: 2200 mg/kg; Toxic Effects: Lungs, Thorax, or Respiration - Sputum; Kidney, Ureter, and Bladder - Hematuria; Nutritional and gross metabolic - Metabolic acidosis. Woman LD_{Lo} Dose: 2440 ug/kg; Toxic Effects: Vascular - Shock; Blood - Other hemolysis with or without anemia; Nutritional and gross metabolic - Metabolic acidosis. Rat LD_{50} Dose: 1100 mg/kg; Toxic Effects: Behavioral - Somnolence (general depressed activity); Lungs, Thorax, or Respiration - Dyspnea.
Acute Inhalation: Rat LC_{50} Dose: 15 gm/m^3/15M; Toxic Effects: Behavioral - Somnolence (general depressed activity); Lungs, Thorax, or Respiration - Dyspnea. Man TC_{Lo} Dose: 7300 ug/m^3/8hr; Toxic Effects: Kidney, ureter, and Bladder - Other changes in urine composition.
Chronic (Multiple Dose) Oral: Rat Dose: 22680 mg/kg/9W-C; Toxic Effects: Behavioral - Food intake (animal); Nutritional and gross metabolic - Weight loss or decreased weight gain.
Chronic (Multiple Dose) Inhalation: Rat Dose: 500 ppm/6H/12D-I; Toxic Effects: Sense organs and special senses - Other; Lungs, Thorax, or Respiration - Other changes; DEATH. Rat Dose: 64 ppm/6H/13W-I; Toxic Effects: Lungs, Thorax, or Respiration - Changes in lung weight; Liver - Changes in Liver weight; Biochemical - Phosphatases. Rat Dose: 50 mg/m^3/16W-I; Toxic Effects: Brain and coverings - Recordings from specific areas of CNS; Liver - Liver function tests impaired; Kidney, Ureter, and Bladder - Proteinuria.
Irritation Eye: Rabbit Standard Draize Test Dose: 122 mg; Reaction: severe.
Irritation Skin: Rabbit Open Draize Test Dose: 610 mg open; Reaction: mild.
Mutagenic: Human Sister Chromatid Exchenge; Cell Type: lymphocyte; Dose: 10 mmol/L. Hamster Cytogenetic Analysis; Cell Type: ovary; Dose: 10 mmol/L.

Hazard Overviews

Corrosive

Flammable

Fire Diamond

Health: Corrosive to the eyes/skin/respiratory tract. Also Causes: difficulty breathing, possible bronchitis, headache, muscle weakness, kidney damage (severe poisoning).
Fire: Flammable. Can form explosive mixtures in the air. Use water spray, dry chemical, carbon dioxide, or alcohol-resistant foam.
Reactivity: Stable. Hazardous polymerization cannot occur. Incompatible with: furfuryl alcohol; hydrogen peroxide; thallium trinitrate trihydrate; nitro-methane; phosphorus pentoxide; oxidizing agents; alkaline materials; sulfuric acid. Hazardous decomposition products: acrid smoke; irritating fumes.

Carcinogenicity: IARC - Not listed; NIOSH - Not listed; NTP - Not listed; ACGIH - Not listed; OSHA - Not listed; EPA - Not listed; MAK - Not listed

Primary Target Organs:

Eyes

Skin

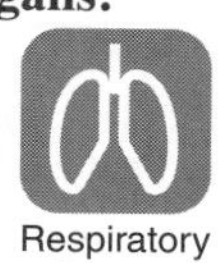
Respiratory System

Mucous Membranes

Exposure Limits

OSHA PEL: TWA: 5 ppm; 9 mg/m^3.

ACGIH TLV: TWA: 5 ppm; 9.4 mg/m^3; STEL: 10 ppm; 19 mg/m^3.

NIOSH REL: TWA: 5 ppm; 9 mg/m^3.

NIOSH IDLH: 30 ppm.

DFG MAK: TWA: 5 ppm; 9 mg/m^3.

Respirator Recommendation

Exposure Range: >5 to <30 ppm Supplied Air, Constant Flow/Pressure Demand, Half Mask

Exposure Range: 30 to unlimited ppm Self-contained Breathing Apparatus, Pressure Demand, Full Face

Note: poor warning properties

Environmental

Ecotoxicity: TLm Lepomis macrochirus (bluegill) 175 mg/l/24 hr, fresh water /Conditions of bioassay not specified TLm Daphnia 120 ppm/l/48 hr, fresh water /Conditions of bioassay not specified

Environmental Fate: If released on land, it should leach into some soils where it would probably biodegrade. In natural water it has been shown to adsorb to sediment and would probably also biodegrade. Bioconcentration in aquatic organisms is not important. In the atmosphere, it would be scavenged by rain and dissolve in cloud water where it reacts with dissolved hydroxyl radicals. It also reacts in the vapor phase with hydroxyl radicals (half-life 36 days).

Cleanup/Disposal: Guide No. 153: Eliminate all ignition sources (no smoking, flares, sparks or flames in immediate area). Do not touch damaged containers or spilled material unless wearing appropriate protective clothing. Stop leak if you can do it without risk. Prevent entry into waterways, sewers, basements or confined areas. Absorb or cover with dry earth, sand or other non-combustible material and transfer to containers. Do not get water inside containers.

Environmental Physical Data

Henry's Law Constant: 1.67 x10^{-7}

Octanol/Water Partition Coefficient: log K_{ow} = -0.54

BCF: calculated at 0.22

BOD: 2%, 5 days

Regulations

RCRA 40CFR: Listed Hazardous Waste No. U123 Toxic Waste Corrosive Waste

CERCLA: 40CFR 302.4: Listed per CWA Section 311(b)(4) per RCRA Section 3001 RQ: 5000 lb (2268 kg)

SARA 40CFR 372.65: Listed

SARA EHS 40CFR 355: Not listed

TSCA: Listed

Analytical Methods

Air: ASTM D4490

Soil: DOE OM500R

Indoor / Expired Air: NIOSH 2011

FOR6320 CAS #: 2540-82-1

FORMOTHION

RTECS: TE1050000

DOT: UN2783; UN2784; UN3017; UN3018; IMO3.2; IMO6.1

EINECS Number: 219-818-6

Molecular Formula: $C_6H_{12}NO_4PS_2$

Formula Weight: 257.29

Synonyms: ACETAMIDE,N-FORMYL-2-MERCAPTO-N-METHYL-,S-ESTER WITH O,O-DIMETHYL PHOSPHORODITHIOATE; ACETAMIDE,N-FORMYL-2-MERCAPTO-N-METHYL-,S-ESTER WITHO,O-DIMETHYL PHOSPHORODITHIOATE; AFLIX; AFLIZ; ANTHIO; ANTHIO 25; ANTIO; CP 53926; O,O-DIMETHYL DITHIOPHOSPHORYLACETIC ACID N-METHYL-N-FORMYLAMIDE; O,O-DIMETHYL DITHIOPHOSPHORYLACETIC ACIDN-METHYL-N-FORMYLAMIDE; O,O-DIMETHYL S-(N-FORMYL-N-METHYLCARBAMOYLMETHYL)PHOSPHORODITHIOATE; O,O-DIMETHYL S-(N-METHYL-N-FORMYLCARBAMOYLMETHYL)PHOSPHORODITHIOATE; O,O-DIMETHYL PHOSPHORODITHIOATE N-FORMYL-2-MERCAPTO-N-METHYLACETAMIDE S-ESTER; O,O-DIMETHYL PHOSPHORODITHIOATEN-FORMYL-2-MERCAPTO-N-METHYLACETAMIDE S-ESTER; O,O-DIMETHYL-S-(3-METHYL-2,4-DIOXO-3-AZA-BUTYL)-DITHIOFOSFAAT; O,O-DIMETHYL-S-(3-METHYL-2,4-DIOXO-3-AZA-BUTYL)-DITHIOPHOSPHAT; O,O-DIMETHYL-S-(N-METHYL-N-FORMYL-CARBAMOYLMETHYL)-DITHIOPHOSPHAT; O,O-DIMETHYLS-(N-METHYL-N-FORMYLCARBAMOYLMETHYL)PHOSPHORODITHIOATE; O,O-DIMETIL-S-(N-FORMIL-N-METIL-CARBAMOIL-METIL)-DITIOFOSFATO; ENT 27,257; S-(2-(FORMYLMETHYLAMINO)-2-OXOETHYL) O,O-DIMETHYLPHOSPHORODITHIOATE; S-(2-(FORMYLMETHYLAMINO)-2-OXOETHYL)O,O-DIMETHYLPHOSPHORODITHIOATE; N-FORMYL-N-METHYLCARBAMOYLMETHYL O,O-DIMETHYLPHOSPHORODITHIOATE; S-(N-FORMYL-N-METHYLCARBAMOYLMETHYL) O,O-DIMETHYLPHOSPHORODITHIOATE; S-(N-FORMYL-N-METHYLCARBAMOYLMETHYL) DIMETHYLPHOSPHOROTHIOLOTHIONATE; J-38; OMS-968; P 1; PHOSPHORODITHIOIC ACID,O,O-DIMETHYL ESTER,S-ESTER WITHN-FORMYL-2-MERCAPTO-N-METHYL-ACETAMIDE; PHOSPHORODITHIOIC ACID,O,O-DIMETHYL ESTER,S-ESTER WITHN-FORMYL-2-MERCAPTO-N-METHYLACETAMIDE; PHOSPHORODITHIOIC ACID,0,0-DIMETHYL ESTER,N-FORMYL-2-MERCAPTO-N-METHYLACETAMIDE S-ESTER; PHOSPHORODITHIOIC ACID,O,O-DIMETHYL ESTER,N-FORMYL-2-MERCAPTO-N-METHYLACETAMIDE S-ESTER; PHOSPHORODITHIOIC ACID,S-(2-(FORMYLMETHYLAMINO)-2-OXOETHYL) O,O-DIMETHYL ESTER; S 6900; SAN 244 I; SAN 6913 I; SAN 7107 I; SANDOZ S-6900; SPENCER S-6900; TOPROSE; VEL 4284

Description: yellowish viscous oily liquid or crystalline mass; odorless

Use: contact & systemic insecticide & acaricide; against wide range of sucking & mining insects eg, chewing insects (cydia pominella & epilachna beetles) & spider mites (non-organophosphate resistant strains only); against sap-feeding insects & mites; against aphids (non-op resistant strains only), psyllids, mealy bugs, armored scales, whiteflies, jassids, thrips, leaf miners, ermine moths, sawflies, fruit flies, olive fly, asparagus fly, mangold fly.

Physical Properties

Boiling Point: Decomposes
Freezing Point: 25 °C (77 °F) to 26 °C (78.8 °F)
Specific Gravity: 1.361 at 20 °C
Vapor Pressure: 0.113 mPa at 20 °C
Water Solubility: Slightly Soluble in Water
Other Solubilities: miscible with Ketones; miscible with Xylene; Very slightly Soluble in paraffin oil (less than 0.1%); Completely miscible with Acetone, Chloroform, Ethanol, Methanol, Xylene; slightly Soluble in hexane.
pH: Alkaline pesticide
Refraction Index: 1.5541 at 20 °C/D
Flash Point: Does not burn or burns with difficulty

RTECS Toxicity Data

Acute Oral: Rat LD_{50} Dose: 250 mg/kg. Mouse LD_{50} Dose: 83300 ug/kg; Toxic Effects: Sense organs and special senses - Miosis (pupilliary dilation); Cardiac - Pulse rate; Lungs, Thorax, or Respiration - Respiratory depression.
Acute Inhalation: Rat LC_{50} Dose: 4500 mg/m^3/4hr. Mouse LC_{50} Dose: 27 mg/m^3; Toxic Effects: Behavioral - Somnolence (general depressed activity); Gastrointestinal - Nausea or vomiting; Liver - Other changes.
Acute Dermal: Rat LD_{50} Route: Skin; Dose: 353 mg/kg. Mouse LD_{50} Route: Skin; Dose: 400 mg/kg.
Mutagenic: Rat Cytogenetic Analysis; Route: Intraperitoneal; Dose: 25 mg/kg. Rat Micronucleus Test; Route: Intraperitoneal; Dose: 250 mg/kg.

Hazard Overviews

Fire: Noncombustible.
Carcinogenicity: IARC - Not listed; NIOSH - Not listed; NTP - Not listed; ACGIH - Not listed; OSHA - Not listed; EPA - Not listed; MAK - Not listed

Environmental

Cleanup/Disposal: Guide No. 131: Fully encapsulating, vapor protective clothing should be worn for spills and leaks with no fire. Eliminate all ignition sources (no smoking, flares, sparks or flames in immediate area). All equipment used when handling the product must be grounded. Do not touch or walk through spilled material. Stop leak if you can do it without risk. Prevent entry into waterways, sewers, basements or confined areas. A vapor suppressing foam may be used to reduce vapors. Small Spills: Absorb with earth, sand or other non-combustible material and transfer to containers for later disposal. Use clean non-sparking tools to collect absorbed material. Large Spills: Dike far ahead of liquid spill for later disposal. Water spray may reduce vapor; but may not prevent ignition in closed spaces. Guide No. 152: Do not touch damaged containers or spilled material unless wearing appropriate protective clothing. Stop leak if you can do it without risk. Prevent entry into waterways, sewers, basements or confined areas. Cover with plastic sheet to prevent spreading. Absorb or cover with dry earth, sand or other non-combustible material and transfer to containers. Do not get water inside containers.

Environmental Physical Data

Regulations

RCRA 40CFR: Not listed
CERCLA: 40CFR 302.4: Not listed
SARA 40CFR 372.65: Not listed TPQ: 100 lb
SARA EHS 40CFR 355: Listed TPQ: 100 lb
TSCA: Not listed

Analytical Methods

Food: FDA 232.4, 242.1; AOAC 974.03

FOR7650 CAS #: 17702-57-7

FORMPARANATE

RTECS: FB9880000
Molecular Formula: $C_{12}H_{17}N_3O_3$
Formula Weight: 235.32
Synonyms: CARBAMIC ACID,METHYL-,ESTER WITH N'-(4-HYDROXY-O-TOLYL)-N,N-DIMETHYLFORMAMIDINE (8CI); CARBAMIC ACID,METHYL-,ESTER WITHN'-(4-HYDROXY-O-TOLYL)-N,N-DIMETHYLFORMAMIDINE; N,N-DIMETHYL-N'-(2-METHYL-4-(((METHYLAMINO)CARBONYL)OXY)PHENYL)METHANIMIDAMIDE; ENT 27,305; EPA PESTICIDE CHEMICAL CODE 359700; METHANIMIDAMIDE,N,N-DIMETHYL-N'-(2-METHYL-4-(((METHYLAMINO)CARBONYL)OXY)PHENYL)-(9CI); SCHERING 36103; UC-25074; UNION CARBIDE UC-25074

RTECS Toxicity Data

Acute Oral: Rat LD_{50} Dose: 7200 ug/kg. Mouse LD_{50} Dose: 16600 ug/kg; Toxic Effects: Sense organs and special senses - Lacrimation; Behavioral - Tremor; Gastrointestinal - Changes in structure or function of salivary glands.

Hazard Overviews

Carcinogenicity: IARC - Not listed; NIOSH - Not listed; NTP - Not listed; ACGIH - Not listed; OSHA - Not listed; EPA - Not listed; MAK - Not listed

Environmental

Regulations

RCRA 40CFR: Listed Hazardous Waste No. P197 Toxic Waste
CERCLA: 40CFR 302.4: Listed per RCRA Section 3001 RQ: 1 lb (0.454 kg)
SARA 40CFR 372.65: Not listed TPQ: 100/10000 lb
SARA EHS 40CFR 355: Listed TPQ: 1 lb
TSCA: Not listed

FOR8980 CAS #: 619-66-9

4-FORMYLBENZOIC ACID

RTECS: WZ0440000
EINECS Number: 210-607-4
Molecular Formula: $C_8H_6O_3$
Formula Weight: 150.14

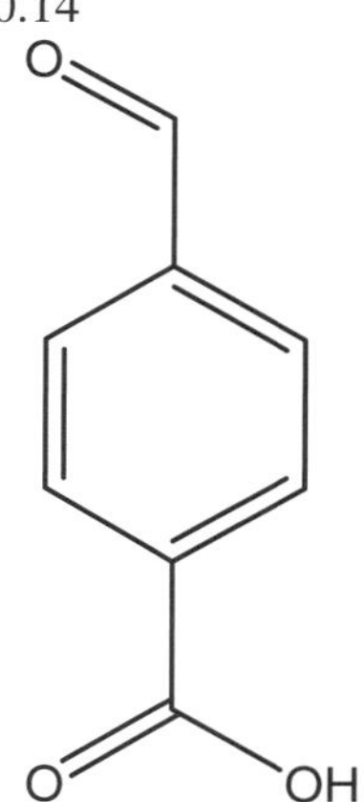

Chemical Structure

Synonyms: BENZOIC ACID,4-FORMYL-; 4-CARBOXYBENZALDEHYDE; P-CARBOXYBENZALDEHYDE; P-FORMYLBENZOIC ACID; TEREPHTHALALDEHYDIC ACID; TEREPHTHALDEHYDIC ACID

Physical Properties

Boiling Point: Sublimes
Freezing Point: 256 °C (492.8 °F)
Water Solubility: Slightly Soluble in Hot Water
Other Solubilities: Very Soluble in Alcohol; Soluble in Ether and in Chloroform

Hazard Overviews

Health: Irritating to eyes/skin/respiratory tract. Harmful. Other Acute Effects: may be harmful by inhalation, ingestion, or skin absorption. Chronic Effects: laboratory experiments have shown mutagenic effects.
Fire: Extinguishing agents: water spray; carbon dioxide, dry chemical powder or appropriate foam. Precautions: combustible liquid.
Reactivity: Incompatible with: strong oxidizing agents, strong bases. Hazardous decomposition products: toxic fumes of: carbon monoxide, carbon dioxide.
Carcinogenicity: IARC - Not listed; NIOSH - Not listed; NTP - Not listed; ACGIH - Not listed; OSHA - Not listed; EPA - Not listed; MAK - Not listed
Primary Target Organs:

Eyes

Skin

Respiratory System

Environmental

Regulations

RCRA 40CFR: Not listed
CERCLA: 40CFR 302.4: Not listed
SARA 40CFR 372.65: Not listed
SARA EHS 40CFR 355: Not listed
TSCA: Listed

FOS1000 CAS #: 25954-13-6

FOSAMINE

RTECS: BQ4112000
EINECS Number: 247-363-3
Molecular Formula: $C_3H_{11}N_2O_4P$
Formula Weight: 170.13
Synonyms: AMMONIUM ETHYL (AMINOCARBONYL)PHOSPHONATE; AMMONIUM ETHYL CARBAMOYLPHOSPHONATE; AMMONIUM ETHYL CARBAMOYLPHOSPHONATE SOLUTION; AMMONIUM-AETHYL-CARBAMOYL-PHOSPHONAT; AMMONIUM,ETHYL CARBAMOYLPHOSPHONATE; DPX-1108; DPX 1108; FOSAMINE AMMONIUM; FOSAMINE AMMONIUM SALT; FOSAMINE-AMMONIUM; KRENITE; KRENITE BRUSH CONTROL AGENT; PHOSPHONIC ACID,(AMINOCARBONYL)-,MONOETHYL ESTER,MONOAMMONIUM SALT; PHOSPHONIC ACID,CARBAMOYL-,MONOETHYL ESTER,AMMONIUMSALT
Description: colorless to white crystalline solid
Use: brush control agent & growth regulant; contact herbicide

Physical Properties

Freezing Point: 175 °C (347 °F)
Specific Gravity: 1.33
Vapor Pressure: 0.53 mPa at 25 °C
Water Solubility: 1.79 kg/kg at 25 °C
Other Solubilities: Slightly Soluble in most common organic solvents; 0.04 g/kg Chloroform at 25 °C.

RTECS Toxicity Data

Acute Oral: Rat LD_{50} Dose: 11 gm/kg. Guinea Pig LD_{50} Dose: 7380 mg/kg.
Acute Inhalation: Rat LC_{50} Dose: >57 gm/m^3/1hr.
Acute Dermal: Rabbit LD_{50} Route: Skin; Dose: >1660 mg/kg.

Hazard Overviews

Carcinogenicity: IARC - Not listed; NIOSH - Not listed; NTP - Not listed; ACGIH - Not listed; OSHA - Not listed; EPA - Not listed; MAK - Not listed

Environmental

Ecotoxicity: LD_{50} Bobwhite quail oral more than 4150 mg (as emulsifiable concentrate)/kg LC_{50} Bluegill 278 mg active ingredient (as emulsifiable concentrate)/l/96 hr /Conditions of bioassay not specified
Environmental Fate: If released to the atmosphere, it will degrade rapidly in the vapor phase by reaction with photochemically produced hydroxyl radicals (half-life of

about 7.9 hr). Particulate phase will be removed physically from air by wet and dry deposition. If released to soil or water, it can degrade through biodegradation and aqueous hydrolysis; however, hydrolysis will be important only in very acidic soils and water. Results of field, greenhouse and soil column studies have demonstrated that it is unlikely to leach under environmental conditions. Field and greenhouse persistence studies have found soil half-lives of about one week to 10 days; carbamoyl phosphonic acid has been identified as a metabolite.

Environmental Physical Data

Henry's Law Constant: estimated at 5×10^{-13}
BCF: estimated at 0.2

Regulations

RCRA 40CFR: Not listed
CERCLA: 40CFR 302.4: Not listed
SARA 40CFR 372.65: Not listed
SARA EHS 40CFR 355: Not listed
TSCA: Not listed

FOS5000 CAS #: 39148-24-8

FOSETYL-AL

RTECS: SZ9640000
DOT: NA2588; IMO6.1
EINECS Number: 254-320-2
Molecular Formula: $C_6H_{18}AlO_9P_3$
Formula Weight: 354.11
Synonyms: ALIETTE; ALIETTE EXTRA; ALUMINIUM PHOSETHYL; ALUMINIUM TRIS(ETHYL PHOSPHONATE); ALUMINUM PHOSETHYL; ALUMINUM TRIS(ETHYL PHOSPHITE); ALUMINUM TRIS(O-ETHYL PHOSPHONATE); ALUMINUM TRIS(O-ETHYLPHOSPHONATE); EFOSITE ALUMINUM; EPAL; FOSETYL AL; FOSETYL ALUMINUM; FOSETYL-ALUMINIUM; HY-COTE; HY-TONA; LS 74-783; LS 74783; LS 74 783; MIKAL; PHOSETHYL AL; PHOSETHYL ALUMINUM; PHOSPHONIC ACID MONOETHYL ESTER ALUMINUM SALT; R6 TRIPLO; RHODAX; 32545 RP; VALIANT
Description: colorless to white crystals or powder; odorless
Use: bactericide, systemic fungicide; for control of diseases caused by phycomycetes (phytophthora, plasmopara, bremia species) on lettuce, hops, strawberries, pome fruit, citrus fruit, pineapples, avocados, vines, tobacco, ornamental plants, etc; preventive & curative activity against oomycetes, alternaria & pennicillium on avocado, cacao, citrus, hops, ornamentals, pineapple, rubber, strawberries, fruit crops, tobacco, etc; suppression of bacterial pathogens: fireblight (eawinia) on pome fruit, xanthomonas & on ornamentals.

Physical Properties

Boiling Point: Decomposes > 200 °C (392 °F)
Freezing Point: > 300 °C (572 °F)
Vapor Pressure: Negligible at room temperature
Water Solubility: 120 g/L at 20 °C
Other Solubilities: Practically Insoluble (<5 mg/l) in /some/ organic solvents at 20 °C.
Flash Point: Nonflammable

RTECS Toxicity Data

Acute Oral: Rat LD_{50} Dose: 5 gm/kg. Mouse LD_{50} Dose: 3700 mg/kg.
Acute Dermal: Rabbit LD_{50} Route: Skin; Dose: >2 gm/kg. Rat LD_{50} Route: Skin; Dose: >3200 mg/kg.

Hazard Overviews

Fire: Noncombustible.
Carcinogenicity: IARC - Not listed; NIOSH - Not listed; NTP - Not listed; ACGIH - Not listed; OSHA - Not listed; EPA - Class C, Possible human carcinogen; MAK - Not listed
Exposure Limits
OSHA PEL Vacated 1989 Limits: TWA: 2 mg/m^3; as Al soluble.
ACGIH TLV: TWA: 2 mg/m^3; as Al.
NIOSH REL: TWA: 2 mg/m^3; as Al soluble salts, alkyls.

Environmental

Ecotoxicity: LC_{50} Rainbow trout 428 mg/l/96 hr /Conditions of bioassay not specified
Cleanup/Disposal: Guide No. 151: Do not touch damaged containers or spilled material unless wearing appropriate protective clothing. Stop leak if you can do it without risk. Prevent entry into waterways, sewers, basements or confined areas. Cover with plastic sheet to prevent spreading. Absorb or cover with dry earth, sand or other non-combustible material and transfer to containers. Do not get water inside containers.

Regulations

RCRA 40CFR: Not listed
CERCLA: 40CFR 302.4: Not listed
SARA 40CFR 372.65: Not listed
SARA EHS 40CFR 355: Not listed
TSCA: Not listed

FOS9000 CAS #: 21548-32-3

FOSTHIETAN

RTECS: NJ6490000
EINECS Number: 244-437-7
Molecular Formula: $C_6H_{12}NO_3PS_2$
Formula Weight: 241.26
Synonyms: AC 64,475; AC 64475; ACCONEM; CL 64,475; CL 64475; CYCLIC METHYLENE DIETHOXYPHOSPHINODITHIOIMIDOCARBONATE; CYCLIC METHYLENE(DIETHOXYPHOSPHINYL)DITHIOIMINOCARBONATE; DIETHOXYPHOSPHINYLIMINO-2 DITHIETANNE-1,3; (DIETHOXYPHOSPHINYLIMINO)-1,3-DIETHIETANE; (DIETHOXYPHOSPHINYLIMINO)-1,3-DITHIETANE; 2-(DIETHOXYPHOSPHINYLIMINO)-1,3-DITHIETANE; DIETHYL 1,3-DITHIETAN-2-YL-IDENEPHOSPHORAMIDATE; DIETHYL 1,3-DITHIETAN-2-YLIDENEPHOSPHORAMIDATE; 1,3-DITHIETAN-2-YLIDENEPHOSPHORAMIDIC ACID DIETHYL ESTER; ENT 27,873; GEOFOS; NEM-A-TAK; PHOSPHONODITHIOIMIDOCARBONIC ACID CYCLIC METHYLENEP,P-DIETHYL ESTER; PHOSPHORAMIDIC ACID,1,3-DITHIETAN-2-YLIDENE-,DIETHYLESTER

Description: pale yellow oil; mercaptan-like odor
Use: soil insecticide and nematicide

Physical Properties

Specific Gravity: 1.3 at 25 °C
Saturated Vapor Density: Technical 1.200000075 kg/m^3
Vapor Pressure: Technical 6.5 x10^{-6} mm Hg at 25 °C
Water Solubility: 50 g/kg Water at 25 °C
Other Solubilities: Soluble in Acetone, Chloroform, Methanol, Toluene.
Refraction Index: 1.5358 at 25 °C/D

RTECS Toxicity Data

Acute Oral: Rat LD_{50} Dose: 4700 ug/kg. Mouse LD_{50} Dose: 18 mg/kg.
Acute Dermal: Rabbit LD_{50} Route: Skin; Dose: 40 mg/kg.
Carcinogenicity: IARC - Not listed; NIOSH - Not listed; NTP - Not listed; ACGIH - Not listed; OSHA - Not listed; EPA - Not listed; MAK - Not listed

Environmental

Regulations

RCRA 40CFR: Not listed
CERCLA: 40CFR 302.4: Not listed
SARA 40CFR 372.65: Not listed TPQ: 500 lb
SARA EHS 40CFR 355: Listed TPQ: 500 lb
TSCA: Listed

FRE5000 **CAS #: 1420-06-0**

FRESCON

RTECS: QF2275000
EINECS Number: 215-812-2
Molecular Formula: $C_{23}H_{23}NO$
Formula Weight: 329.44
Synonyms: MORPHOLINE,4-(TRIPHENYLMETHYL)-; MORPHOLINE,4-TRITYL-; SHELL WL 8008; TRIFENMORPH; TRIPHENMORPHE; 4-(TRIPHENYLMETHYL)MORPHOLINE; N-(TRIPHENYL-METHYL)MORPHOLINE; N-TRITYLMORPHOLINE; WL 8008
Description: colorless crystals
Use: molluscicide, to kill snails which carry bilharzia flukes by application to wet pastures; effective against aquatic & semi-aquatic snails; recommended for application to irrigation, other moving water systems by drip-feed technique, & for static water; recommended for application to lymnaea truncatula habitats to control fascioliasis.

Physical Properties

Freezing Point: 174 °C (345.2 °F) to 176 °C (348.8 °F)
Saturated Vapor Density: 1.200000002 kg/m^3
Vapor Pressure: 1.4 x10^{-7} mm Hg at 20 °C
Water Solubility: 0.02 mg/L at 20 °C
Other Solubilities: at 20 °C (g/L): Carbon Tetrachloride 300; Chloroform 450; Tetrachloroethylene 255.

RTECS Toxicity Data

Acute Oral: Rat LD_{50} Dose: 83 mg/kg. Mouse LD_{50} Dose: 4809 mg/kg; Toxic Effects: Behavioral - Somnolence (general depressed activity).
Acute Dermal: Rat LD_{50} Route: Skin; Dose: >1 gm/kg.
Chronic (Multiple Dose) Oral: Rat Dose: 28440 mg/kg/17W-C; Toxic Effects: Liver - Other changes.

Hazard Overviews

Carcinogenicity: IARC - Not listed; NIOSH - Not listed; NTP - Not listed; ACGIH - Not listed; OSHA - Not listed; EPA - Not listed; MAK - Not listed

Environmental

Ecotoxicity: LC_{50} rainbow trout 0.08 mg/l/96 hr, yearlings, pH 6.9-7.0

Environmental Physical Data

Sorption Partition Coefficient: K_{oc} = estimated at 3.75 x10^4
BCF: sarotherodon mosambicus 1300

Regulations

RCRA 40CFR: Not listed
CERCLA: 40CFR 302.4: Not listed
SARA 40CFR 372.65: Not listed
SARA EHS 40CFR 355: Not listed
TSCA: Not listed

FRU5000 **CAS #: 7660-25-5**

FRUCTOSE

RTECS: LS7000000
Molecular Formula: $C_6H_{12}O_6$
Formula Weight: 180.16
Synonyms: BETA-D-FRUCTOPYRANOSE; FRUCTOPYRANOSE,BETA-D-; D-FRUCTOSE; FRUCTOSTERIL; FRUIT SUGAR; FRUTABS; HEXOSE; LAEVORAL; LAEVOSAN; LEVOLOSE; LEVUGEN; LEVULOSE
Description: colorless, clear crystals or white, crystalline or granular powder; odorless
Use: to prevent sandiness in ice cream; fluid and nutrient replenisher; vet: for bovine ketosis d-fructose; treatment of diabetic ketoacidosis; sweetening agent for foods, beverages

Physical Properties

Freezing Point: Decomposes at 103 °C (217.4 °F) to 105 °C (221 °F)
Density: 1.6 at 20 °C
Water Solubility: Freely Soluble in Water
Other Solubilities: Soluble in Pyridine; 1 g dissolves in 15 ml Alcohol, 14 ml Methanol; Slightly Soluble in cold, freely in hot Acetone; Soluble in Methylamine, Ethylamine

RTECS Toxicity Data

Acute Dermal: Rat LD_{50} Route: Subcutaneous Dose: >20 gm/kg.

Tumorigenic: Mouse Route: Subcutaneous; Dose: 5000 mg/kg; Toxic Effects: Tumorigenic - Equivocal tumorigenic agent by RTECS criteria; Tumorigenic - Tumors at site of application.

Hazard Overviews

Carcinogenicity: IARC - Not listed; NIOSH - Not listed; NTP - Not listed; ACGIH - Not listed; OSHA - Not listed; EPA - Not listed; MAK - Not listed

Environmental

Regulations

RCRA 40CFR: Not listed
CERCLA: 40CFR 302.4: Not listed
SARA 40CFR 372.65: Not listed
SARA EHS 40CFR 355: Not listed
TSCA: Not listed

FUB5000 CAS #: 3878-19-1

FUBERIDAZOLE

RTECS: DD9010000
EINECS Number: 223-404-0
Molecular Formula: $C_{11}H_8N_2O$
Formula Weight: 184.2

Chemical Structure

Synonyms: B-33172; BAY 33172; BAYER 33172; 1H-BENZIMIDAZOLE,2-(2-FURANYL)-; BENZIMIDAZOLE,2-(2-FURYL)-; FUBERIDATOL; FUBERISAZOL; FUBRIDAZOLE; 2-(2-FURANYL)-1H-BENZIMIDAZOLE; FURIDAZOL; FURIDAZOLE; 2-(2'-FURYL)-BENZIMIDAZOLE; 2-(2-FURYL)BENZIMIDAZOLE; PF 7402; VORONIT; VORONITE; W VII/117
Description: fine crystalline powder
Use: as a seed dressing protects against soil-borne fungal diseases of seedlings; for the control of vascular wilt diseases in plants

Physical Properties

Freezing Point: Decomposes at 284 °C (543.2 °F) to 288 °C (550.4 °F)
Water Solubility: 78 mg/kg in Water at room temperature
Other Solubilities: 10 g/kg in Toluene at room temperature; 1.6 g/kg in 0.1 M Hydrochloric Acid at room temperature; 2.3 g/kg in 0.1 M sodium hydroxide at room temperature; Soluble in Acetone, Ethanol, Methanol.

RTECS Toxicity Data

Acute Oral: Rat LD_{50} Dose: 500 mg/kg. Mouse LD_{50} Dose: 825 mg/kg.
Acute Inhalation: Rat LC_{50} Dose: 330 mg/m^3/4hr.
Acute Dermal: Rat LD_{50} Route: Skin; Dose: 500 mg/kg.
Mutagenic: Bacteria - S Typhimurium Mutations in Microorganisms; Dose: 100 ug/plate (+S9).

Hazard Overviews

Carcinogenicity: IARC - Not listed; NIOSH - Not listed; NTP - Not listed; ACGIH - Not listed; OSHA - Not listed; EPA - Not listed; MAK - Not listed

Environmental

Regulations

RCRA 40CFR: Not listed
CERCLA: 40CFR 302.4: Not listed
SARA 40CFR 372.65: Not listed TPQ: 100/10000 lb
SARA EHS 40CFR 355: Listed TPQ: 100 lb
TSCA: Not listed

FUE1000 CAS #: 68476-30-2

FUEL OIL NO. 2

RTECS: LS8930000
DOT: UN1202; NA1993; IMO3.3
EINECS Number: 270-671-4
Molecular Formula: Unspecified or Variable
Formula Weight: N/A
Synonyms: #2 HOME HEATING OILS; API NO. 2 FUEL OIL; GAS OIL; HOME HEATING OIL NO.2; NUMBER 2 BURNER FUEL; NUMBER 2 FUEL OIL
Description: brown slighty viscous liquid
Use: for domestic heating and power plant warm up; domestic and industrial heating; power for heavy units (ships, trucks, trains); source of synthesis gas; drilling muds; mosquito control

Physical Properties

Boiling Point: 232 °C (450 °F)
Freezing Point: -46 °C (-50.8 °F)
Specific Gravity: 0.8654 at 15 °C/15 °C
Water Solubility: Slightly Soluble in Water
Surface Tension: Estimated at 25 dynes/cm
Odor Threshold: Medium odor threshold 0.082 ppm
Flash Point: 57.778 °C Closed Cup
Autoignition Temperature: 257 °C

RTECS Toxicity Data

Acute Oral: Rat LD_{50} Dose: 12 gm/kg.
Acute Dermal: Rabbit LD; Route: Skin; Dose: >5 gm/kg.
Irritation Eye: Rabbit Standard Draize Test Dose: 100 mg/30S; Reaction: mild.
Irritation Skin: Rabbit Standard Draize Test Dose: 500 mg/24H; Reaction: moderate.

Tumorigenic: Mouse Route: Skin; Dose: 243 gm/kg/97W-I; Toxic Effects: Tumorigenic - Carcinogenic by RTECS criteria; Skin and appendages - Tumors.

Hazard Overviews

Flammable

Fire Diamond

Health: Irritating to skin/respiratory tract. Also Causes: vomiting, diarrhea, CNS depression, increased rate of respiration, rapid heart beat, cyanosis. Chronic Effects: dermatitis.

Fire: Flammable. Can form explosive mixtures in air. Use dry chemical, carbon dioxide, foam, water fog or spray. Do not use a forced water spray directly on burning oil, this scatters the fire. Use a smothering technique to extinguish fire.

Reactivity: Stable. Hazardous polymerization cannot occur. Avoid: heat and ignition sources. Incompatible with: strong oxidizing agents. Hazardous decomposition products: various hydrocarbons; hydrocarbon derivatives; partial oxidation products (carbon dioxide; carbon monoxide; sulfur dioxide).

Carcinogenicity: IARC - Group 3, Not classifiable as to carcinogenicity to humans; NIOSH - Not listed; NTP - Not listed; ACGIH - Not listed; OSHA - Not listed; EPA - Not listed; MAK - Not listed

Primary Target Organs:

Skin

Mucous Membranes

Nervous System

Cardio-vascular

Environmental

Ecotoxicity: Aquatic toxicity: 200 ppm/24 hr/juvenile American shad/TLm/fresh water 20 ppm/96 hr/rainbow trout eggs/TLm/salt water

Environmental Fate: If released to soil, it will strongly adsorb. It may biodegrade in water and soil or volatilize from water (half-life of 4.4- 4.8 hours from a model river) and moist soil surfaces, but adsorption may attenuate the rate of these processes. In water adsorption to sediment should be important. Bioconcentration in aquatic organisms may be limited for the chief components due to metabolism. If released to the atmosphere, degradation of vapor phase components by reaction with photochemically produced hydroxyl radicals (estimated half-life on the order of 1 day or less) will be important.

Cleanup/Disposal: Guide No. 128: Eliminate all ignition sources (no smoking, flares, sparks or flames in immediate area). All equipment used when handling the product must be grounded. Do not touch or walk through spilled material. Stop leak if you can do it without risk. Prevent entry into waterways, sewers, basements or confined areas. A vapor suppressing foam may be used to reduce vapors. Absorb or cover with dry earth, sand or other non-combustible material and transfer to containers. Use clean non-sparking tools to collect absorbed material. Large Spills: Dike far ahead of liquid spill for later disposal. Water spray may reduce vapor; but may not prevent ignition in closed spaces.

Environmental Physical Data

Henry's Law Constant: estimated at 29 to 68

Octanol/Water Partition Coefficient: log K_{ow} = 8.2 to 9.7

Sorption Partition Coefficient: K_{oc} = estimated at 1 $x10^4$

BCF: calculated at 1.02 $x10^4$ to 1.4 $x10^4$

Regulations

RCRA 40CFR: Not listed

CERCLA: 40CFR 302.4: Not listed

SARA 40CFR 372.65: Not listed

SARA EHS 40CFR 355: Not listed

TSCA: Listed

FUE3000 CAS #: 68476-31-3

FUEL OIL NO. 4

EINECS Number: 270-673-5

Molecular Formula: Mixture

Formula Weight: N/A

Description: moderately dark viscous liquid

Physical Properties

Freezing Point: -46 °C (-50.8 °F)

Water Solubility: Insoluble in Water

Odor Threshold: Medium odor threshold 0.5 ppm

Flash Point: 54 °C

Autoignition Temperature: 263 °C

Hazard Overviews

Flammable

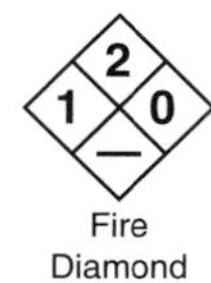

Fire Diamond

Health: Irritating to skin/respiratory tract. Also Causes: headache, dizziness, nausea, increased rate of respiration, tachycardia, stupor, convulsions, cyanosis, unconsciousness, acne, mild CNS depression. Chronic Effects: dermatitis. Possible cancer hazard. Irritating to eyes/skin/respiratory tract. Harmful. Other Acute Effects: harmful if swallowed, inhaled, or absorbed through skin; can cause CNS depression; nervous system disturbances; coughing; chest pains; difficulty in breathing; edema which may be fatal; gastrointestinal disturbances; nausea; headache; vomiting; target organ: nerves. Chronic Effects: epidemiological studies indicate persons with routine exposure may be at an increased risk to the development of benign neoplasms; digestive tract cancers; and skin cancer (melanoma); suspected neurotoxin. Possible human carcinogen.

Fire: Flammable. For small fires use dry chemical, carbon dioxide, water spray, or regular foam. For large fires use water spray, fog, or regular foam. Containers may explode in the heat of fire Flammable. Hazards: vapor may travel considerable distance to source of ignition and flash back;

container explosion may occur; emits toxic fumes. Extinguishing agents: carbon dioxide, dry chemical powder or appropriate foam. Precautions: combustible liquid.

Reactivity: Stable. Hazardous polymerization will not occur. Incompatible with: oxidizing agents, incompatible with chlorine/fluorine/magnesium perchlorate. Hazardous decomposition products: toxic fumes of: carbon monoxide, carbon dioxide.

Carcinogenicity: IARC - Not listed; NIOSH - Not listed; NTP - Not listed; ACGIH - Not listed; OSHA - Not listed; EPA - Not listed; MAK - Not listed

Primary Target Organs:

Eyes

Skin

Respiratory System

Nervous System

Environmental

Regulations

RCRA 40CFR: Not listed
CERCLA: 40CFR 302.4: Not listed
SARA 40CFR 372.65: Not listed
SARA EHS 40CFR 355: Not listed
TSCA: Not listed

FUE6000 CAS #: 68476-33-5

FUEL OIL NO. 5

RTECS: LS8960000
EINECS Number: 270-675-6
Molecular Formula: Unspecified or Variable
Formula Weight: N/A
Synonyms: RESIDUAL(HEAVY) FUEL OIL
Description: black oily liquid; asphalt-like odor

Physical Properties

Boiling Point: > 200 °C (392 °F)
Freezing Point: -29 °C (-20.2 °F) to -9 °C (15.8 °F)
Specific Gravity: 0.93 to 1.05
Vapor Density: > 6
Vapor Pressure: < 0.1 mm Hg at 20 °C
Water Solubility: Insoluble
Surface Tension: Estimated at 25 dynes/cm
Flash Point: > 60 °C Closed Cup
Autoignition Temperature: 263 °C
LEL: 1.0% v/v
UEL: 5% v/v

Hazard Overviews

Fire Diamond

Health: Irritating to skin/respiratory tract. Also Causes: vomiting, diarrhea, CNS depression, tachycardia, cyanosis, pulmonary edema, liver/renal injury. Chronic Effects: dermatitis, systemic toxicity, possible cancer hazard.

Fire: Combustible. Can form explosive mixtures in the air. Use dry chemical, carbon dioxide, foam, water fog/spray. Do not use a forced water spray directly on burning oil since this scatters the fire. Use a smothering technique to extinguish fire. Cool fire-exposed containers with water spray.

Reactivity: Stable. Hazardous polymerization cannot occur. Avoid: heat and ignition sources. Incompatible with: strong oxidizing agents. Hazardous decomposition products: low molecular weight hydrocarbons; hydrocarbon derivatives; carbon monoxide; sulfur oxides.

Carcinogenicity: IARC - Group 2B, Possibly carcinogenic to humans; NIOSH - Not listed; NTP - Not listed; ACGIH - Not listed; OSHA - Not listed; EPA - Not listed; MAK - Not listed

Primary Target Organs:

Skin

Respiratory System

Nervous System

Liver

Kidneys

Cardio-vascular

Environmental

Ecotoxicity: Aquatic toxicity: 2400 ppm/48 hr/juvenile American shad/TLm/fresh water; 2417 mg/l/48 hr/juvenile American shad/TLm/salt water

Cleanup/Disposal: Guide No. 128: Eliminate all ignition sources (no smoking, flares, sparks or flames in immediate area). All equipment used when handling the product must be grounded. Do not touch or walk through spilled material. Stop leak if you can do it without risk. Prevent entry into waterways, sewers, basements or confined areas. A vapor suppressing foam may be used to reduce vapors. Absorb or cover with dry earth, sand or other non-combustible material and transfer to containers. Use clean non-sparking tools to collect absorbed material. Large Spills: Dike far ahead of liquid spill for later disposal. Water spray may reduce vapor; but may not prevent ignition in closed spaces.

Regulations

RCRA 40CFR: Not listed
CERCLA: 40CFR 302.4: Not listed
SARA 40CFR 372.65: Not listed
SARA EHS 40CFR 355: Not listed
TSCA: Listed

FUE9000 CAS #: 68553-00-4

FUEL OIL NO. 6

RTECS: LS8940000
EINECS Number: 271-384-7
Molecular Formula: Unspecified or Variable
Formula Weight: N/A
Synonyms: NO. 6 FUEL OIL

Description: black liquid to heavy paste; petroleum odor

Physical Properties

Boiling Point: 400 °C (752 °F)
Freezing Point: -46 °C (-50.8 °F)
Specific Gravity: About 0.966
Vapor Pressure: 0.2 mm Hg at 21 °C
Water Solubility: Insoluble in Water
Odor Threshold: Medium odor threshold 0.13 ppm
Flash Point: 66 °C
Autoignition Temperature: 408 °C
LEL: 3.9% v/v
UEL: 20.1% v/v

RTECS Toxicity Data

Acute Oral: Rat LD_{50} Dose: 5300 mg/kg.
Acute Dermal: Rabbit LD_{Lo} Route: Skin; Dose: 5200 mg/kg.
Irritation Skin: Rabbit Standard Draize Test Dose: 500 mg; Reaction: mild.

Hazard Overviews

Fire Diamond

Health: Irritating to respiratory tract. Also Causes: vomiting, diarrhea, CNS depression, increased respiration rate, tachycardia, cyanosis. Chronic: dermatitis, possible systemic toxicity, possible cancer hazard.
Fire: Combustible. Can form explosive mixtures in the air. Use dry chemical, carbon dioxide, foam, water fog, or spray. Do not use a forced water spray directly on burning oil since this scatters the fire. Use a smothering technique to extinguish fire. Cool fire-exposed containers with water spray.
Reactivity: Stable. Hazardous polymerization cannot occur. Avoid: heat and ignition sources. Incompatible with: strong oxidizing agents. Hazardous decomposition products: various hydrocarbons; hydrocarbon derivatives; partial oxidation products (carbon dioxide; carbon monoxide; sulfur dioxide).
Carcinogenicity: IARC - Not listed; NIOSH - Not listed; NTP - Not listed; ACGIH - Not listed; OSHA - Not listed; EPA - Not listed; MAK - Not listed
Primary Target Organs:

Skin

Respiratory System

Nervous System

Environmental

Regulations

RCRA 40CFR: Not listed
CERCLA: 40CFR 302.4: Not listed
SARA 40CFR 372.65: Not listed
SARA EHS 40CFR 355: Not listed
TSCA: Listed

FUI5000 **CAS #: 1524-88-5**

FUIRANDRENOLIDE

RTECS: TU5024800
EINECS Number: 216-196-8
Molecular Formula: $C_{24}H_{33}FO_6$
Formula Weight: 436.52

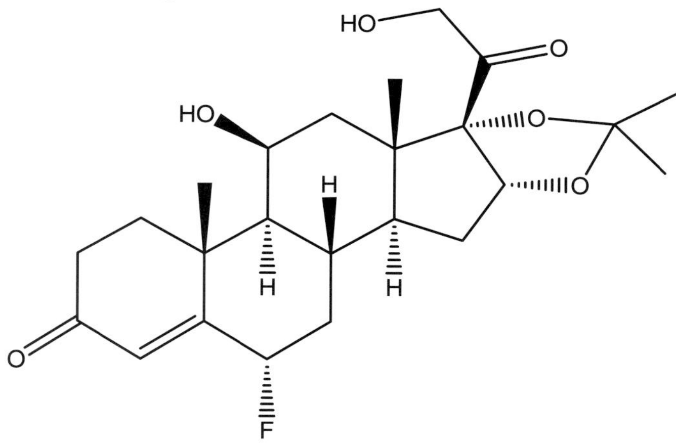

Chemical Structure

Synonyms: 33379; ACETONIDE OF 6ALPHA-FLUORO-16ALPHA-HYDROXYHYDROCORTISONE; ALONDRA-F; CORDAN; CORDRAN; DRENISON; DROCORT; FLUDROXYCORTIDE; FLUORANDRENOLONE; FLUORANDRENOLONE ACETONIDE; FLUOROANDRENOLONE ACETONIDE; 6ALPHA-FLUORO-16ALPHA-HYDROXYHYDROCORTISONE16,17-ACETONIDE; 6ALPHA-FLUORO-11BETA,16ALPHA,17,21-TETRAHYDROXYPROGESTERONE CYCLIC 16,17-ACETAL WITHACETONE; FLURANDRENOLIDE; FLURANDRENOLONE; FLURANDRENOLONE ACETONIDE; GLUCOCORTICOID; HAELAN; HALDRONE-F; L 33379; PREGN-4-ENE-3,20-DIONE,6-FLUORO-11,21-DIHYDROXY-16,17-((1-METHYLETHYLIDENE)BIS(OXY))-,(6ALPHA,11BETA,16ALPHA)-; PREGN-4-ENE-3,20-DIONE,6ALPHA-FLUORO-11BETA,16ALPHA,17,21-TETRAHYDROXY-,CYCLIC16,17-ACETAL WITH ACETONE; SERMAKA
Description: white to off-white, fluffy crystalline powder or crystals; odorless
Use: medication: glucocorticoid; anti-inflammatory agent

Physical Properties

Freezing Point: 247 °C (476.6 °F) to 255 °C (491 °F)
Water Solubility: Practically Insoluble in Water
Other Solubilities: freely Soluble in Chloroform; Soluble in Methanol; Sparingly Soluble in Alcohol; Practically Insoluble in Ether

RTECS Toxicity Data

Mutagenic: Human Other Mutation Test Systems; Route: Skin; Dose: 19 mg/kg.

Hazard Overviews

Health: Irritating to eyes/skin/respiratory tract. Other Acute Effects: may be harmful by inhalation, ingestion, or skin absorption; systemic absorption may produce reversible hypothalamic-pituitary-adrenal (HPA) axis suppression; manifestations of Cushing's syndrome; hyperglycemia; glucosuria; readily absorbed through skin; target organs: adrenal gland, skin, eyes, bones.

Fire: Hazards: emits toxic fumes. Extinguishing agents: noncombustible; use extinguishing media appropriate to surrounding fire conditions. Precautions: combustible liquid.

Reactivity: Stable. Hazardous polymerization will not occur. Incompatible with: strong oxidizing agents. Hazardous decomposition products: toxic fumes of: carbon monoxide, carbon dioxide, hydrogen fluoride.

Carcinogenicity: IARC - Not listed; NIOSH - Not listed; NTP - Not listed; ACGIH - Not listed; OSHA - Not listed; EPA - Not listed; MAK - Not listed

Primary Target Organs:

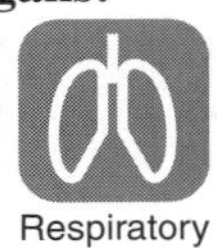

Eyes | Skin | Respiratory System | Bone | Glandular System

Environmental

Regulations

RCRA 40CFR: Not listed

CERCLA: 40CFR 302.4: Not listed

SARA 40CFR 372.65: Not listed

SARA EHS 40CFR 355: Not listed

TSCA: Not listed

FUM5000 CAS #: 110-17-8

FUMARIC ACID

RTECS: LS9625000

EINECS Number: 203-743-0

Molecular Formula: $C_4H_4O_4$

Structured MF: $HO_2CCH{=}CHCO_2H$

Formula Weight: 116.07

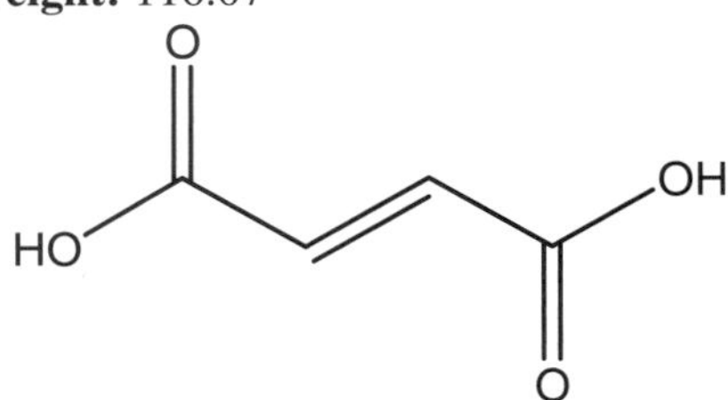

Chemical Structure

Synonyms: ALLOMALEIC ACID; BOLETIC ACID; (E)-BUTENEDIOIC ACID; 2-BUTENEDIOIC ACID (E)-; 2-BUTENEDIOIC ACID,(E)-; BUTENEDIOIC ACID; BUTENEDIOIC ACID,(E)-; TRANS-BUTENEDIOIC ACID; 1,2-ETHENEDICARBOXYLIC ACID,TRANS-; TRANS-1,2-ETHENEDICARBOXYLIC ACID; (E)-1,2-ETHYLENEDICARBOXYLIC ACID; 1,2-ETHYLENEDICARBOXYLIC ACID,(E); 1,2-ETHYLENEDICARBOXYLIC ACID,TRANS-; TRANS-1,2-ETHYLENEDICARBOXYLIC ACID; TRANS-1,2-ETHYLENEDICCARBOXYLIC ACID; KYSELINA FUMAROVA; LICHENIC ACID; NSC-2752; U-1149

Description: colorless to white crystals, powder; odorless

Use: substitute for tartaric acid in beverages and baking powders; an antioxidant; manufacture of polyhydric alcohols, synthetic resins; as mordant in dying; for polyester, alkyd, and phenolic resins; plasticizers; rosin esters and add alkyd resin coatings; upgrading natural drying oils; as acidulant and flavoring agent; organic synthesis; printing inks

Physical Properties

Boiling Point: 165 °C (329 °F) at 1.7 mm Hg

Freezing Point: 287 °C (548.6 °F)

Specific Gravity: 1.635 at 20 °C/4 °C

Bulk Density: 32.6 lbs/cu ft

Vapor Pressure: 10.1 mm Hg at 0 °C

Water Solubility: < 1 mg/mL at 22 C

Other Solubilities: Soluble in 100 g 95% Alcohol at 30 °C; 1.72 g Soluble in 100 g Acetone at 30 °C; 0.72 g Soluble in 100 g Ether at 25 °C; almost Insoluble in Olive Oil, Benzene, Xylene, molten Camphor, liquid Ammonia.

Ionization Potential (eV): 10.7 +/-0.6

Flash Point: Not available; probably combustible

Autoignition Temperature: Powder 740 °C

RTECS Toxicity Data

Acute Oral: Rat LD_{50} Dose: 9300 mg/kg. Rabbit LD_{Lo} Dose: 5 gm/kg.

Irritation Eye: Rabbit Standard Draize Test Dose: 100 mg/24H; Reaction: moderate.

Irritation Skin: Rabbit Standard Draize Test Dose: 500 mg/24H; Reaction: mild.

Mutagenic: Rat DNA Inhibition; Route: Intravenous; Dose: 40 mg/kg.

Hazard Overviews

Fire Diamond

Health: Irritating to eyes/skin/respiratory tract.

Fire: Will burn. Use water spray, dry chemical, foam, or carbon dioxide.

Reactivity: Stable. Hazardous polymerization cannot occur. Avoid: heat; ignition sources. Incompatible with: oxidizers. Hazardous decomposition products: carbon monoxide; carbon dioxide.

Carcinogenicity: IARC - Not listed; NIOSH - Not listed; NTP - Not listed; ACGIH - Not listed; OSHA - Not listed; EPA - Not listed; MAK - Not listed

Primary Target Organs:

Eyes | Skin | Respiratory System

Environmental

Environmental Fate: If released to soil or water, biodegradation is expected to be the major fate process. River die-away studies have yielded half-lives ranging from 1-15 days with degradation rates increasing as the pollution in the water source increases. If released to the atmosphere, it will exist primarily in the particulate-phase where it can be physically removed via wet and dry deposition. Vapor-phase is readily degraded by sunlight-formed hydroxyl radicals (estimated half-life of 7.3 hr).

Cleanup/Disposal: Guide No. 171: Do not touch or walk through spilled material. Stop leak if you can do it without risk. Prevent dust cloud. Avoid inhalation of asbestos dust. Small Dry Spills: With clean shovel place material into clean, dry container and cover loosely; move containers from spill area. Small Spills: Take up with sand or other noncombustible absorbent material and place into containers for later disposal. Large Spills: Dike far ahead of liquid spill for later disposal. Cover powder spill with plastic sheet or tarp to minimize spreading. Prevent entry into waterways, sewers, basements or confined areas.

Environmental Physical Data

Henry's Law Constant: estimated at 8.5 $x10^{-14}$

Octanol/Water Partition Coefficient: log K_{ow} = calculated at 0.07 to 0.56

Sorption Partition Coefficient: K_{oc} = estimated at 33.5

BCF: estimated at 4.2

BOD: 61%, 5 days

Regulations

RCRA 40CFR: Not listed

CERCLA: 40CFR 302.4: Listed per CWA Section 311(b)(4) RQ: 5000 lb (2268 kg)

SARA 40CFR 372.65: Not listed

SARA EHS 40CFR 355: Not listed

TSCA: Listed

FUR1000 CAS #: 108-31-6

2,5-FURANDIONE

RTECS: ON3675000

DOT: UN2215; IMO8.0

EINECS Number: 203-571-6

Molecular Formula: $C_4H_2O_3$

Structured MF: OCOCH=CHCO

Formula Weight: 98.06

Chemical Structure

Synonyms: ANHYDRID KYSELINY MALEINOVE; CIS-BUTENEDIOIC ANHYDRIDE; DIHYDRO-2,5-DIOXOFURAN; 2,5-FURANDION; 2,5-FURANEDIONE; LYTRON 810; LYTRON 820; MALEIC ACID ANHYDRIDE; MALEIC ANHYDRIDE; MALEINANHYDRID; TOXILIC ANHYDRIDE

Description: colorless or white solid, crystals; acrid, irritating odor

Use: manufacture of unsaturated polyester resins; manufacture of fumaric acid; pesticide manufacture; pharmaceuticals; Diels-Alder synthesis

Physical Properties

Boiling Point: 202 °C (396 °F) at 760 mm Hg

Freezing Point: 52.8 °C (127.04 °F)

Specific Gravity: 1.48

Vapor Density: 3.4 Air=1

Density: 1.314 g/mL at 60 °C

Vapor Pressure: 0.2 mm Hg

Water Solubility: Soluble in Water

Other Solubilities: 95% Ethanol: Soluble; Acetone: Soluble; Benzene: Soluble; Chloroform: Soluble; Dioxane: Soluble; DMSO: Soluble; Ether: Soluble; Ligroin: Slightly Soluble; Toluene: Soluble.

Odor Threshold: 1.84 to 1.96 mg/m^3

Ionization Potential (eV): 9.90

Flash Point: 103 °C

Autoignition Temperature: 470 °C

LEL: 1.4% v/v

UEL: 7.1% v/v

RTECS Toxicity Data

Acute Oral: Rat LD_{50} Dose: 400 mg/kg. Mouse LD_{50} Dose: 465 mg/kg.

Acute Dermal: Rabbit LD_{50} Route: Skin; Dose: 2620 mg/kg.

Chronic (Multiple Dose) Inhalation: Rat Dose: 9800 ug/m^3/6H/26W-I; Toxic Effects: Blood - Changes in spleen; Nutritional and gross metabolic - Weight loss or decreased weight gain.

Irritation Eye: Rabbit Standard Draize Test Dose: 1%; Reaction: severe.

Reproductive/Teratogenic: Rat Route: Oral; Dose: 4060 mg/kg; Duration: multigenerations; Effects on Newborn - Growth statistics.

Mutagenic: Hamster Cytogenetic Analysis; Cell Type: lung; Dose: 230 mg/L.

Tumorigenic: Rat Route: Subcutaneous; Dose: 1220 mg/kg/61W-I; Toxic Effects: Tumorigenic - Equivocal tumorigenic agent by RTECS criteria; Tumorigenic - Tumors at site of application.

Hazard Overviews

Corrosive

Fire Diamond

Health: Corrosive to eyes/skin/respiratory tract. Toxic. Also Causes: burning, irritation of the throat, asthmatic response, bronchitis, and pulmo-nary edema, irritation of eyes, corneal swelling, conjunctivitis, and burns, photophobia and double vision, bullous (blistering) dermatitis, inflammation, thermal burns. Chronic Effects: asthma bronchitis, allergies.

Fire: Will burn. May form explosive dust-air mixtures. Use alcohol foam or carbon dioxide. Dry chemicals are not recommended since they contain alkali which may cause rapid exothermic reactions. Water or foam may cause frothing.
Reactivity: Stable in properly vented tanks. Hazardous polymerization can occur when mixed with olefins and catalysts; decomposition-polymerization explosion hazard can occur in the presence of alkali or alkaline earth, metal ions, ammonium ions, or amines when heated above 302 °F (150 °C). Avoid: moisture. Incompatible with: strong oxidizing agents; alkali; alkaline earth; metal ions; ammonium ions; caustics; amines; dimethylamine, triethylamine; pyridine; quinoline. Hazardous decomposition products: irritating vapors; carbon monoxide.
Carcinogenicity: IARC - Not listed; NIOSH - Not listed; NTP - Not listed; ACGIH - Not listed; OSHA - Not listed; EPA - Not listed; MAK - Not listed
Primary Target Organs:

Eyes

Skin

Respiratory System

Mucous Membranes

Exposure Limits
OSHA PEL: TWA: 0.25 ppm; 1 mg/m^3.
ACGIH TLV: TWA: 0.25 ppm; 1 mg/m^3.
NIOSH REL: TWA: 0.25 ppm; 1 mg/m^3.
NIOSH IDLH: 10 mg/m^3.
DFG MAK: TWA: 0.1 ppm; 0.4 mg/m^3.
Respirator Recommendation
Exposure Range: >0.25 to <10 ppm Supplied Air, Constant Flow/Pressure Demand, Half Mask
Exposure Range: 10 to unlimited ppm Self-contained Breathing Apparatus, Pressure Demand, Full Face
Note: poor warning properties

Environmental

Ecotoxicity: TLm Mosquito fish 230 to 240 mg/l/24 to 96 hr /Conditions of bioassay not specified
Environmental Fate: Since it hydrolyzes rapidly in water to the acid, it would not appear in wastewater and would quickly degrade to maleic acid if spilled in water. If emitted to the atmosphere from fugitive emissions or vent gases, it will degrade in a few hours by reaction with ozone or photochemically-produced hydroxyl radicals. If spilled on land its fate is unknown but it is apt to biodegrade or hydrolyze.
Cleanup/Disposal: Guide No. 156: Eliminate all ignition sources (no smoking, flares, sparks or flames in immediate area). All equipment used when handling the product must be grounded. Do not touch damaged containers or spilled material unless wearing appropriate protective clothing. Stop leak if you can do it without risk. A vapor suppressing foam may be used to reduce vapors. For chlorosilanes, use AFFF alcohol-resistant medium expansion foam to reduce vapors. Do not get water on spilled substance or inside containers. Use water spray to reduce vapors or divert vapor cloud drift. Prevent entry into waterways, sewers, basements or confined areas. Small Spills: Cover with dry earth, dry sand, or other non-combustible material followed with plastic sheet to minimize spreading or contact with rain. Use clean non-sparking tools to collect material and place it into loosely covered plastic containers for later disposal.

Environmental Physical Data

BCF: no food chain concentration potential
BOD: 50%, 5 days

Regulations

RCRA 40CFR: Listed Hazardous Waste No. U147 Toxic Waste
CERCLA: 40CFR 302.4: Listed per CWA Section 311(b)(4) per RCRA Section 3001 RQ: 5000 lb (2268 kg)
SARA 40CFR 372.65: Listed
SARA EHS 40CFR 355: Not listed
TSCA: Listed

Analytical Methods

Soil: SW846 8270C
Indoor / Expired Air: NIOSH 3512
Plasma: EPA 29

FUR2140 CAS #: 98-01-1

FURFURAL

RTECS: LT7000000
DOT: UN1199; IMO3.3
EINECS Number: 202-627-7
Molecular Formula: $C_5H_4O_2$
Formula Weight: 96.08

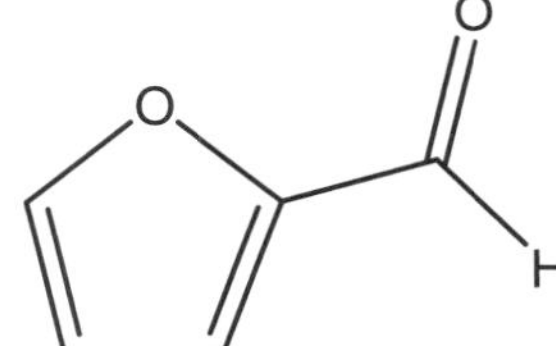

Chemical Structure

Synonyms: ANT OIL,ARTIFICIAL; ARTIFICIAL ANT OIL; ARTIFICIAL OIL OF ANTS; BRAN OIL; 2-FORMYLFURAN; 2-FORMYLOFURAN; FURAL; 2-FURALDEHYDE; ALPHA-FURALDEHYDE; FURALDEHYDE; FURALE; 2-FURANALDEHYDE; 2-FURANCARBONAL; FURANCARBONAL; 2-FURANCARBOXALDEHYDE; 2-FURANKARBALDEHYD; 2-FURFURAL; 2-FURFURALDEHYDE; FURFURALDEHYDE; FURFURALE; FURFUROL; FURFUROLE; FURFURYLALDEHYDE; 2-FURIL-METANALE; FUROL; ALPHA-FUROLE; FUROLE; 2-FURYLALDEHYDE; 2-FURYLCARBOXALDEHYDE; 2-FURYL-METHANAL; 2-FURYLMETHANAL; FURYL-METHANAL; PYROMUCIC ALDEHYDE; QUAKERAL; U1199
Description: colorless liquid
Use: flavor ingredient; solvent for wood resin, nitrated cotton, cellulose acetate lubricating oils, butadiene, rosin, nitrocellulose, shoe dyes and gums; production of phenolic plastics, thermosetting resins, refined petroleum oils,

tetrahydrofuran, furfuryl alcohol, dyes and varnishes; wetting agent in the manufacture of abrasive wheels and brake linings; manufacture of pyromucic acid, adipic acid, adiponitrile, lysine, vulcanized rubber, furan derivative insecticides, polymers and other organic chemicals; used in road construction; refining of rare earths and metals; an analytical reagent; and in fungicides herbicides and germicides

Physical Properties

Boiling Point: 161.8 °C (323 °F) at 760 mm Hg
Freezing Point: -38.7 °C (-37.66 °F)
Specific Gravity: 1.1563 at 25 °C/4 °C
Vapor Density: 3.3 Air=1
Saturated Vapor Density: 1.207124267 kg/m^3
Density: 1.159 g/mL at 20 °C
Vapor Pressure: 0.26 kPa at 20 °C
Water Solubility: 8% by weight
Other Solubilities: 95% Ethanol: >=100 mg/ml at 18 °C; Acetone: >=100 mg/ml at 18 °C; Benzene: Soluble; Chloroform: Soluble; DMSO: >=100 mg/ml at 18 °C; Ether: Very Soluble.
Surface Tension: 43.5 dynes/cm
Odor Threshold: 0.0240 to 20.0 mg/m^3
Refraction Index: 1.5261 at 20 °C/D
Critical Temperature: 396 °C
Critical Pressure: 798 psia
Ionization Potential (eV): 9.21
Flash Point: 60 °C Closed Cup
Autoignition Temperature: 316 °C
LEL: 2.1% v/v
UEL: 19.3% v/v

RTECS Toxicity Data

Acute Oral: Rat LD_{50} Dose: 65 mg/kg. Mouse LD_{50} Dose: 400 mg/kg; Toxic Effects: Tumorigenic - Active as anti-cancer agent.
Acute Inhalation: Rat LC_{50} Dose: 175 ppm/6hr; Toxic Effects: Lungs, Thorax, or Respiration - Other changes; Nutritional and gross metabolic - Weight loss or decreased weight gain. Human TC_{Lo} Dose: 310 ug/m^3.
Acute Dermal: Rabbit LD_{Lo} Route: Skin; Dose: 620 mg/kg. Rabbit LD_{Lo} Route: Subcutaneous Dose: 500 mg/kg.
Chronic (Multiple Dose) Oral: Rat Dose: 91 mg/kg/14D-I; Toxic Effects: Liver - Other changes; Kidney, Ureter, and Bladder - Other changes; Biochemical - Other enzymes. Mouse Dose: 4800 mg/kg/16D-I; Toxic Effects: DEATH. Mouse Dose: 4875 mg/kg/13W-I; Toxic Effects: Liver - Changes in liver weight.
Chronic (Multiple Dose) Inhalation: Rat Dose: 200 mg/m^3/5H/12W-I; Toxic Effects: Automatic Nervous System - Sympathomimetic; Endocrine - Other changes. Rat Dose: 10140 ug/m^3/24H/60D-C; Toxic Effects: Behavioral - Muscle contraction or spasticity; Biochemical - True cholinesterase; DEATH. Rat Dose: 500 mg/m^3/4H/4W-I; Toxic Effects: Kidney, Ureter, and Bladder - Changes in both tubules and glomeruli.
Chronic (Multiple Dose) Dermal: Rat Route: Subcutaneous; Dose: 75 mg/kg/6W-I; Toxic Effects: Kidney, Ureter, and Bladder - Chgs in tubules (inc acute renal failure, acute tubular necrosis; Blood - Pigmented or nucleated red Blood cells; Blood - Changes in erythrocite (RBC) cell count. Rat Route: Subcutaneous; Dose: 33 mg/kg/13W-I; Toxic Effects: Kidney, Ureter, and Bladder - Chgs in tubules (inc acute renal failure, acute tubular necrosis; Blood - Normocytic anemia; Blood - Leukopenia.
Irritation Eye: Rabbit Standard Draize Test Dose: 100 mg/24H; Reaction: moderate.
Irritation Skin: Rabbit Standard Draize Test Dose: 20 mg/24H; Reaction: moderate. Rabbit Standard Draize Test Dose: 500 mg/24H; Reaction: mild.
Mutagenic: Human DNA Inhibition; Cell Type: HeLa cell; Dose: 3 mmol/L. Human Sister Chromatid Exchenge; Cell Type: lymphocyte; Dose: 70 umol/L.
Tumorigenic: Rat Route: Oral; Dose: 30900 mg/kg/2Y-C; Toxic Effects: Tumorigenic - Equivocal tumorigenic agent by RTECS criteria; Liver - Tumors. Mouse Route: Oral; Dose: 90125 mg/kg/2Y-C; Toxic Effects: Tumorigenic - Carcinogenic by RTECS criteria; Liver - Tumors.

Hazard Overviews

Flammable

Fire Diamond

Health: Irritating to eyes/skin/respiratory tract. Eye contact can be damaging. Toxic. Also Causes: headache, dermatitis. Chronic Effects: eczema, allergic sensitization, photosensitization.
Fire: Flammable. Can form explosive mixtures in the air. Use dry chemical, carbon dioxide, alcohol foam, or water spray. Use a water spray to keep fire-exposed containers cool. If a spill or leak has not ignited, use a water spray to disperse the vapors and to protect anyone who may be attempting to stop the leak.
Reactivity: Stable. Hazardous polymerization can occur on contact with strong mineral acids, alkalies, elevated temperatures. Avoid: heat; ignition sources. Incompatible with: phenol; amines; urea; alkalis; oxidizing materials; strong acids; mineral oils; mineral acids; mixture with sodium hydrogen carbonate. Hazardous decomposition products: carbon monoxide; acrid smoke; fumes.
Carcinogenicity: IARC - Not listed; NIOSH - Not listed; NTP - Not listed; ACGIH - Class A3, Animal carcinogen; OSHA - Not listed; EPA - Not listed; MAK - Class B, Justifiably suspected of having carcinogenic potential
Primary Target Organs:

Eyes

Skin

Respiratory System

Exposure Limits
OSHA PEL: TWA: 5 ppm; 20 mg/m^3.
OSHA PEL Vacated 1989 Limits: TWA: 2 ppm; 8 mg/m^3.

ACGIH TLV: TWA: 2 ppm; 7.9 mg/m^3.
NIOSH IDLH: 100 ppm.
Respirator Recommendation
Exposure Range: >5 to 50 ppm Air Purifying, Negative Pressure, Half Mask
Exposure Range: >50 to <100 ppm Air Purifying, Negative Pressure, Full Face
Exposure Range: 100 to unlimited ppm Self-contained Breathing Apparatus, Pressure Demand, Full Face
Cartridge Color: black

Environmental

Ecotoxicity: LD_{50} Agelaius phoeniceus (Redwinged blackbird) oral 98.0 mg/kg
Environmental Fate: If released to soil, it is expected to display high mobility and it has the potential to leach into groundwater. Limited data suggests that it may undergo biodegradation in soil. Volatilization from the soil surface to the atmosphere may occur; however it is not expected to be a rapid process. If released to water, it is expected to undergo microbial degradation, under both aerobic and anaerobic conditions. Acclimation has been found to increase the rate of biodegradation, and high concentrations inhibit the rate. It is not expected to adsorb to sediment or suspended organic matter, nor is it expected to bioconcentrate in fish and aquatic organisms. Hydrolysis is not expected to be a significant fate process under environmental conditions. In the atmosphere, it is expected to exist predominately in the vapor phase. Destruction by the vapor phase reaction with photochemically produced hydroxyl radicals is expected to be an important process with an estimated half-life of 0.44 days. Night time destruction by the vapor phase reaction with nitrate radicals may be an important process in urban areas. Limited data suggests that direct photochemical degradation may occur in the atmosphere. Atmospheric removal by wet deposition may be a significant process.
Cleanup/Disposal: Guide No. 132: Fully encapsulating, vapor protective clothing should be worn for spills and leaks with no fire. Eliminate all ignition sources (no smoking, flares, sparks or flames in immediate area). All equipment used when handling the product must be grounded. Do not touch or walk through spilled material. Stop leak if you can do it without risk. Prevent entry into waterways, sewers, basements or confined areas. A vapor suppressing foam may be used to reduce vapors. Absorb with earth, sand or other non-combustible material and transfer to containers (except for Hydrazine). Use clean non-sparking tools to collect absorbed material. Large Spills: Dike far ahead of liquid spill for later disposal. Water spray may reduce vapor; but may not prevent ignition in closed spaces.

Environmental Physical Data

Henry's Law Constant: calculated at 3.7 x10^{-6}
Octanol/Water Partition Coefficient: log K_{ow} = 0.41
Sorption Partition Coefficient: K_{oc} = calculated at 1 to 40
BCF: calculated at 0.008 to 1.2
BOD: 0.28 to 0.77 lb/lb, 5 days

Regulations

RCRA 40CFR: Listed Hazardous Waste No. U125 Ignitable Waste
CERCLA: 40CFR 302.4: Listed per CWA Section 311(b)(4) per RCRA Section 3001 RQ: 5000 lb (2268 kg)
SARA 40CFR 372.65: Not listed
SARA EHS 40CFR 355: Not listed
TSCA: Listed

Analytical Methods

Indoor / Expired Air: NIOSH 2529, 2539
Other: EPA 1666, 1667

FUR3280 CAS #: 110-00-9

FURFURAN

RTECS: LT8524000
DOT: UN2389; IMO3.1
EINECS Number: 203-727-3
Molecular Formula: C_4H_4O
Formula Weight: 68.08

Chemical Structure

Synonyms: AXOLE; DIVINYLENE OXIDE; 1,4-EPOXY-1,3-BUTADIENE; FURAN; FURFURANE; OXACYCLOPENTADIENE; OXOLE; TETROLE
Description: colorless liquid; ethereal odor
Use: solvent for resins and in the formation of lacquers; an intermediate in organic synthesis; an intermediate in the preparation of pharmaceuticals, insecticides and fine chemicals

Physical Properties

Boiling Point: 32 °C (90 °F) at 758 mm Hg
Freezing Point: -85.65 °C (-122.17 °F)
Specific Gravity: 0.9371 at 19.4 °C/4 °C
Vapor Density: 2.35 Air=1
Saturated Vapor Density: 2.248989474 kg/m^3
Vapor Pressure: 493 mm Hg at 20 °C
Water Solubility: < 1 mg/mL at 22 C
Other Solubilities: 95% Ethanol: >=100 mg/ml at 22 °C; Acetone: >=100 mg/ml at 22 °C; Alcohol: Freely Soluble; Benzene: Soluble; DMSO: >=100 mg/ml at 22 °C; Ethanol: Soluble; Ether: Freely Soluble.
Surface Tension: 24.10 dynes/cm at 20 °C
Odor Threshold: water pollution factor 6 mg/kg
Refraction Index: 1.4214 at 20 °C/D
Critical Temperature: 214 °C
Critical Pressure: 5.32 Pa
Ionization Potential (eV): 8.88 +/-0.01
Flash Point: < 0 °C Closed Cup

LEL: 2.3% v/v
UEL: 14.3% v/v

RTECS Toxicity Data

Acute Oral: Rabbit LD_{Lo} Dose: 234 mg/kg; Toxic Effects: Gastrointestinal - Changes in structure or function of salivary glands; Gastrointestinal - Nausea or vomiting; Blood - Hemorrhage. Dog LD_{Lo} Dose: 234 mg/kg; Toxic Effects: Gastrointestinal - Changes in structure or function of salivary glands; Gastrointestinal - Nausea or vomiting; Blood - Hemorrhage.

Acute Inhalation: Rat LC_{50} Dose: 3398 ppm/1hr; Toxic Effects: Sense organs and special senses - Ptosis; Behavioral - Somnolence (general depressed activity); Lungs, Thorax, or Respiration - Dyspnea. Mouse LC_{50} Dose: 120 mg/m^3/1hr; Toxic Effects: Lungs, Thorax, or Respiration - Acute pulmonary edema; Lungs, Thorax, or Respiration - Other changes.

Chronic (Multiple Dose) Oral: Rat Dose: 240 mg/kg/16D-I; Toxic Effects: Liver - Other changes; Nutritional and gross metabolic - Weight loss or decreased weight gain. Rat Dose: 3900 mg/kg/13W-I; Toxic Effects: Liver - Hepatitis (hepatocellular necrosis), zonal; Kidney, Ureter, and Bladder - Chgs in tubules (inc acute renal failure, acute tubular necrosis; Endocrine - Changes in thymus weight.

Chronic (Multiple Dose) Inhalation: Rat Dose: 200 mg/m^3/4H/60D-I; Toxic Effects: Lungs, Thorax, or Respiration - Structural or functional change in trachea or bronchi; Liver - Other changes; DEATH. Rat Dose: 500 mg/m^3/2H/9W-I; Toxic Effects: Liver - Jaundice, other or unclassified; Nutritional and gross metabolic - Weight loss or decreased weight gain; DEATH. Rat Dose: 5 mg/m^3/4H/26W-I; Toxic Effects: Brain and coverings - Recordings from specific areas of CNS; Vascular - BP lowering not characterized in autonomic section. Rat Dose: 500 mg/m^3/2H/30D-I; Toxic Effects: Liver - Other changes; DEATH.

Reproductive/Teratogenic: Rat Route: Oral; Dose: 3900 mg/kg; Duration: male 13W prior to mating; Paternal Effects - Testes, epididymis, sperm duct; Maternal Effects - Ovaries, fallopian tubes.

Mutagenic: Mouse Mutations in Mammalian Somatic Cells; Cell Type: lymphocyte; Dose: 1139 mg/L. Hamster Cytogenetic Analysis; Cell Type: ovary; Dose: 184 mmol/L.

Tumorigenic: Rat Route: Oral; Dose: 1040 mg/kg/2Y-I; Toxic Effects: Tumorigenic - Carcinogenic by RTECS criteria; Liver - Tumors; Liver - Angiosarcoma. Mouse Route: Oral; Dose: 4160 mg/kg/2Y-I; Toxic Effects: Tumorigenic - Carcinogenic by RTECS criteria; Liver - Tumors.

Hazard Overviews

Flammable

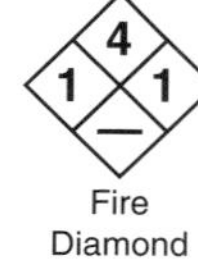

Fire Diamond

Health: Irritating to eyes/skin/respiratory tract. Also Causes: dizziness, suffocation, dryness, sensitization, fatigue, vomiting, GI and bronchial congestion, low BP, lung damage (high concentrations).

Fire: Flammable. Use dry chemical or foam. Water may be ineffective because it may spread the fire and because furan has a low flash point. Explosion hazard when exposed to heat or flame. Cool fire exposed containers with water to prevent explosion.

Reactivity: Stable. Hazardous polymerization cannot occur. Avoid: distilling to dryness. Incompatible with: acids; oxidizing agents. Hazardous decomposition products: irritating smoke; vapors.

Carcinogenicity: IARC - Group 2B, Possibly carcinogenic to humans; NIOSH - Not listed; NTP - Not listed; ACGIH - Not listed; OSHA - Not listed; EPA - Not listed; MAK - Not listed

Primary Target Organs:

Eyes

Skin

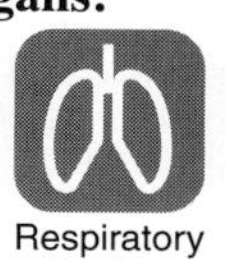
Respiratory System

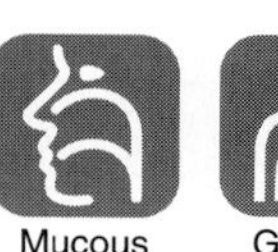
Mucous Membranes

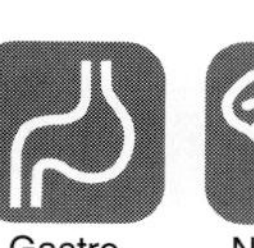
Gastro-intestinal

Nervous System

Exposure Limits
AIHA WEEL: TWA: 3 mg/m^3.

Environmental

Ecotoxicity: LC_{50} Pimephales promelas (fathead minnow) 29-31 days 61 mg/l/96 hr; Affected fish swam upside down and then laid on the bottom of the tank. Some fish were gasping at the surface. Conditions: temp = 23.2 °C; dissolved oxygen 6.7 mg/l; hardness ($CaCO_3$) = 44.5 mg/l; pH = 8.00; tank volume = 6.3 l.

Environmental Fate: If released to land, it will be susceptible to rapid volatilization and significant leaching, possibly into groundwater. Chemical hydrolysis is not expected to be an important fate process. If released to water, volatilization (t/2 of 2.5 hours from a shallow model river) and reaction with singlet oxygen (t/2 1 hour) are expected to be the dominant fate processes. This compound is not expected to chemically hydrolyze, bioaccumulate significantly in aquatic organisms, or adsorb significantly to suspended solids and sediments. If released to air, it is expected to exist almost entirely in the vapor phase. During daylight hours reaction with photochemically generated hydroxyl radicals is expected to be the dominant removal process (t/2 of 6 hours) and during night time hours reaction with nitrate radicals is expected to be the dominant removal process (t/2 of 1/2 hour). Removal from the atmosphere by reaction with ozone or physical processes is not expected to be significant.

Cleanup/Disposal: Guide No. 127: Eliminate all ignition sources (no smoking, flares, sparks or flames in immediate area). All equipment used when handling the product must be grounded. Do not touch or walk through spilled material. Stop leak if you can do it without risk. Prevent entry into waterways, sewers, basements or confined areas. A vapor suppressing foam may be used to reduce vapors. Absorb or cover with dry earth, sand or other non-combustible material and transfer to containers. Use clean non-sparking tools to collect absorbed material. Large Spills: Dike far ahead of

liquid spill for later disposal. Water spray may reduce vapor; but may not prevent ignition in closed spaces.

Environmental Physical Data

Henry's Law Constant: estimated at 5.3 x10^{-3}
Octanol/Water Partition Coefficient: log K_{ow} = 1.34
Sorption Partition Coefficient: K_{oc} = 27 to 128
BCF: estimated at 3 to 6

Regulations

RCRA 40CFR: Listed Hazardous Waste No. U124 Ignitable Waste
CERCLA: 40CFR 302.4: Listed per RCRA Section 3001 RQ: 100 lb (45.35 kg)
SARA 40CFR 372.65: Not listed TPQ: 500 lb
SARA EHS 40CFR 355: Listed TPQ: 100 lb
TSCA: Listed

FUR4420 CAS #: 98-00-0

FURFURYL ALCOHOL

RTECS: LU9100000
DOT: UN2874; IMO6.1
EINECS Number: 202-626-1
Molecular Formula: $C_5H_6O_2$
Formula Weight: 98.10

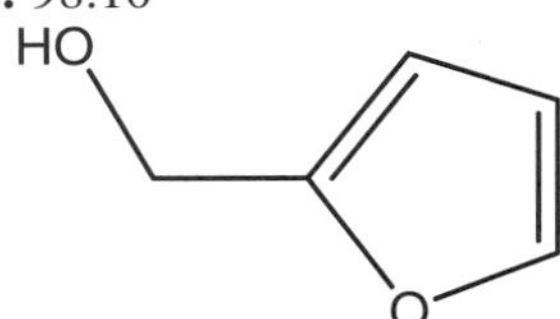

Chemical Structure

Synonyms: 2-FURANCARBINOL; 2-FURANMETHANOL; FURFURAL ALCOHOL; FURFURALCOHOL; 2-FURFURYL ALCOHOL; ALPHA-FURFURYL ALCOHOL; 2-FURFURYLALKOHOL; FURYL ALCOHOL; 2-FURYL CARBINOL; 2-FURYLCARBINOL; ALPHA-FURYLCARBINOL; FURYLCARBINOL; 2-FURYLMETHANOL; 2-(HYDROXYMETHYL)FURAN; 2-HYDROXYMETHYLFURAN; 5-HYDROXYMETHYLFURAN; METHANOL,(2-FURYL)-
Description: colorless, brown-black liquid; ether-like odor
Use: as a flavor ingredient and wetting agent, in foundry cores, polymers, sealants and cements and as a solvent for dyes and resins; a solvent for cellulose ethers and esters and ester gum; in the manufacture of abrasive wheels; as a liquid propellant, penetrant, viscosity reducer and as a solvent in textile printing; to produce THFA (tetrahydrofurfuryl alcohol)

Physical Properties

Boiling Point: 171 °C (340 °F) at 750 mm Hg
Freezing Point: -14.6 °C (5.72 °F)
Specific Gravity: 1.1296 at 20 °C/4 °C
Vapor Density: 1 Air=1
Saturated Vapor Density: 1.202291211 kg/m^3
Density: 1.13 g/mL at 20 °C
Vapor Pressure: 0.609 mm Hg at 25 °C
Water Solubility: Miscible
Other Solubilities: immiscible with petroleum hydrocarbons; miscible with most organic solvents; immiscible with most oils.
Surface Tension: 38 dynes/cm
Odor Threshold: 8 ppm
Refraction Index: 1.4868 at 20 °C/D
Flash Point: 75 °C Open Cup
Autoignition Temperature: 490 °C
LEL: 1.8% v/v
UEL: 16.3% v/v

RTECS Toxicity Data

Acute Oral: Rat LD_{50} Dose: 177 mg/kg; Toxic Effects: Behavioral - Excitment; Behavioral - Ataxia; Lungs, Thorax, or Respiration - Cyanosis. Mouse LD_{50} Dose: 160 mg/kg; Toxic Effects: Tumorigenic - Active as anti-cancer agent.
Acute Inhalation: Rat LC_{50} Dose: 233 ppm/4hr. Mouse LC_{Lo} Dose: 597 ppm/6hr.
Acute Dermal: Rabbit LD_{50} Route: Skin; Dose: 400 mg/kg; Toxic Effects: Behavioral - Convulsions or effect on seizure threshold. Rat LD_{50} Route: Skin; Dose: 3825 mg/kg. Rat LD_{50} Route: Subcutaneous Dose: 85 mg/kg; Toxic Effects: Behavioral - Convulsions or effect on seizure threshold.
Chronic (Multiple Dose) Inhalation: Rat Dose: 250 mg/m^3/2W-I; Toxic Effects: Sense organs and special senses - Other; DEATH. Rat Dose: 50 ppm/6H/16W-I; Toxic Effects: Brain and coverings - Other degenerative changes; Biochemical - Dehydrogenases; Biochemical - Other transferases.
Irritation Eye: Rabbit Standard Draize Test Dose: 100 mg/24H; Reaction: moderate.
Mutagenic: Hamster Cytogenetic Analysis; Cell Type: ovary; Dose: 2500 umol/L. Bacteria - B Subtilis DNA Repair; Dose: 2 mg/disc.

Hazard Overviews

Fire Diamond

Health: Irritating to eyes/skin/respiratory tract. Toxic. Also Causes: excitability, headache, confusion, visual disturbances, painful sensitivity to light, fatigue, nausea/vomiting, pallor, dizziness, dermatitis.
Fire: Combustible. Can form explosive mixtures in the air. For small fires use dry chemical, water spray, or regular foam. For large fires use water spray, fog, or regular foam.
Reactivity: Stable. Hazardous polymerization can occur on contact with some organic acids. Avoid: acids; strong oxidizers; heat; ignition sources. Incompatible with: acids; strong oxidizers; sulfur tetrafluoride; hydrogen peroxide. Hazardous decomposition products: toxic smoke; fumes.
Carcinogenicity: IARC - Not listed; NIOSH - Listed as carcinogen; NTP - Not listed; ACGIH - Not listed; OSHA - Not listed; EPA - Not listed; MAK - Not listed

Primary Target Organs:

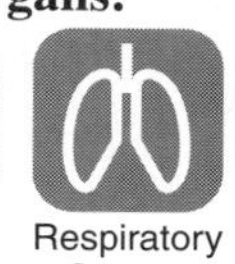

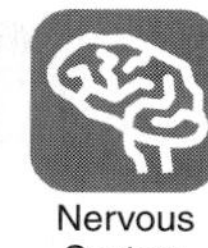

Eyes Skin Respiratory System Mucous Membranes Nervous System

Exposure Limits

OSHA PEL: TWA: 50 ppm; 200 mg/m^3.

OSHA PEL Vacated 1989 Limits: TWA: 10 ppm; 40 mg/m^3; STEL: 15 ppm; 60 mg/m^3.

ACGIH TLV: TWA: 10 ppm; 40 mg/m^3; STEL: 15 ppm; 60 mg/m^3.

NIOSH REL: TWA: 10 ppm; 40 mg/m^3. STEL: 15 ppm; 60 mg/m^3; skin.

NIOSH IDLH: 75 ppm.

DFG MAK: TWA: 10 ppm; 40 mg/m^3.

Respirator Recommendation

Exposure Range: >50 to <75 ppm Air Purifying, Negative Pressure, Half Mask

Exposure Range: 75 to unlimited ppm Self-contained Breathing Apparatus, Pressure Demand, Full Face

Cartridge Color: black

Environmental

Ecotoxicity: LC_{50} Pimephales promelas (fathead minnow) 32 mg/l/96 hr (static bioassay in reconstituted water at 18 to 22 °C) TLm Mosquito fish 44 to 24 mg/l/24 to 96 hr. /Conditions of bioassay not specified

Environmental Fate: If released to soil, it will be expected to exhibit very high mobility, based upon the reported infinite solubility of the compound in water and an estimated K_{oc} of 34. It may, therefore, leach through soil to groundwater if it does not biodegrade or otherwise decompose first. It may be subject to biodegradation in soil based upon results observed in a laboratory aqueous aerobic biodegradation screening test. It should not be subject to volatilization from moist near surface soil based upon an estimated Henry's Law constant of 7.86 x10^{-8} atm-cu m/mole calculated from the water solubility and vapor pressure. However, it may volatilize from dry near surface soil and other dry surfaces based upon its vapor pressure. If released to water, it will not be expected to adsorb to sediment or suspended particulate matter or to bioconcentrate in aquatic organisms based upon its estimated K_{oc} and BCF, respectively and its high solubility in water. It may directly photolyze in surface water based upon its absorption of light at wavelengths >290 nm. It may be subject to biodegradation in natural waters based upon results observed in a laboratory biodegradation aqueous aerobic screening test using a rigorous activated sludge inoculum. It should not be subject to volatilization from surface waters based upon the estimated Henry's Law constant. If released to the atmosphere, it can be expected to exist mainly in the vapor-phase in the ambient atmosphere based upon its vapor pressure. The estimated atmospheric half-life for vapor-phase reaction with photochemically produced hydroxyl radicals half-life is 3.7 hr at an atmospheric concentration of 5 x10^5 hydroxyl radicals per cu cm. It may be susceptible to direct photolysis in the atmosphere.

Cleanup/Disposal: Guide No. 153: Eliminate all ignition sources (no smoking, flares, sparks or flames in immediate area). Do not touch damaged containers or spilled material unless wearing appropriate protective clothing. Stop leak if you can do it without risk. Prevent entry into waterways, sewers, basements or confined areas. Absorb or cover with dry earth, sand or other non-combustible material and transfer to containers. Do not get water inside containers.

Environmental Physical Data

Henry's Law Constant: calculated at 7.86 x10^{-8}

Octanol/Water Partition Coefficient: log K_{ow} = 0.28

Sorption Partition Coefficient: K_{oc} = estimated at 34

BCF: estimated at 0.96

Regulations

RCRA 40CFR: Not listed

CERCLA: 40CFR 302.4: Not listed

SARA 40CFR 372.65: Not listed

SARA EHS 40CFR 355: Not listed

TSCA: Listed

Analytical Methods

Air: ASTM D4490

Indoor / Expired Air: NIOSH 2505

FUR5560 CAS #: 552-86-3

FUROIN

RTECS: KM5774095

EINECS Number: 209-024-8

Molecular Formula: $C_{10}H_8O_4$

Formula Weight: 192.18

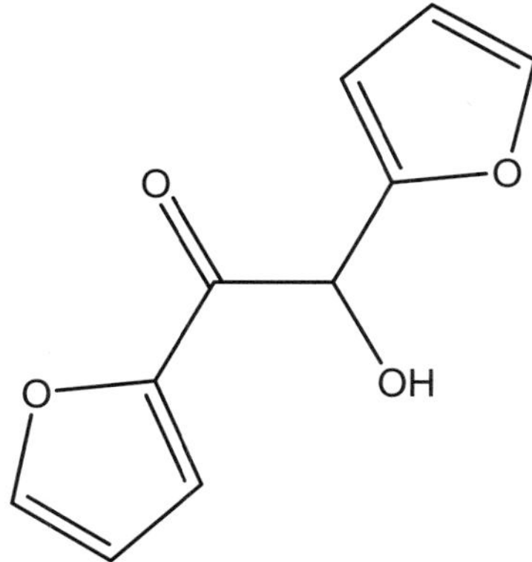

Chemical Structure

Synonyms: 1,2-DI-2-FURANYL-2-HYDROXYETHANONE; ETHANONE,1,2-DI-2-FURANYL-2-HYDROXY-; ETHANONE,1,2-DI-2-FURYL-2-HYDROXY-; ALPHA-FUROIN; FUROYLFURYLCARBINOL; KETONE,2-FURYL ALPHA-HYDROXYFURFURYL

Description: needles

Physical Properties

Freezing Point: 138 °C (280.4 °F) to 139 °C (282.2 °F)

Water Solubility: Slightly Soluble in Hot Water

Other Solubilities: Slightly Soluble in Alcohol Soluble in Ether, Methanol, hot Alcohol

Hazard Overviews

Health: Irritating to eyes/skin. Other Acute Effects: may be harmful by inhalation, ingestion, or skin absorption.
Fire: Hazards: emits toxic fumes. Extinguishing agents: water spray; carbon dioxide, dry chemical powder or appropriate foam. Precautions: combustible liquid.
Reactivity: Incompatible with: strong oxidizing agents. Hazardous decomposition products: toxic fumes of: carbon monoxide, carbon dioxide.
Carcinogenicity: IARC - Not listed; NIOSH - Not listed; NTP - Not listed; ACGIH - Not listed; OSHA - Not listed; EPA - Not listed; MAK - Not listed
Primary Target Organs:

Eyes

Skin

Environmental

Regulations

RCRA 40CFR: Not listed
CERCLA: 40CFR 302.4: Not listed
SARA 40CFR 372.65: Not listed
SARA EHS 40CFR 355: Not listed
TSCA: Listed

FUR6700 CAS #: 54-31-9

FUROSEMIDE

RTECS: CB2625000
EINECS Number: 200-203-6
Molecular Formula: $C_{12}H_{11}ClN_2O_5S$
Formula Weight: 330.77

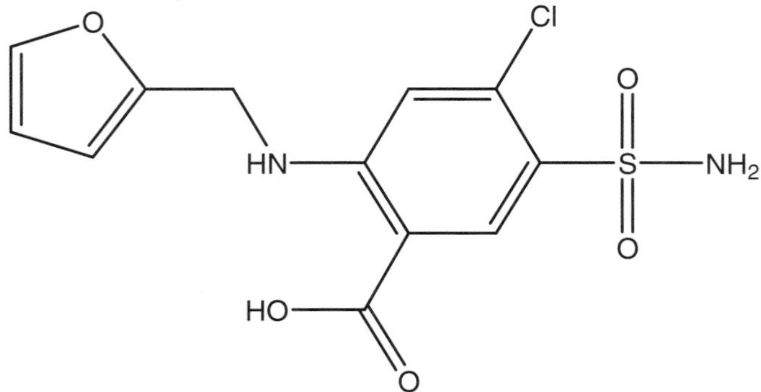

Chemical Structure

Synonyms: AISEMIDE; ALUZINE; 5-(AMINOSULFONYL)-4-CHLORO-2-((2-FURANYLMETHY)AMINO)BENZOIC ACID; 5-(AMINOSULFONYL)-4-CHLORO-2-((2-FURANYLMETHYL)AMINO)BENZOIC ACID; 5-(AMINOSULFONYL)-4-CHLORO-2-[(2-FURANYLMETHYL)AMINO]BENZOIC ACID; ANTHRANILIC ACID,4-CHLORO-N-FURFURYL-5-SULFAMOYL-; AQUASIN; ARASEMIDE; BENZOIC ACID,5-(AMINOSULFONYL)-4-CHLORO-2-((2-FURANYLMETHYL)AMINO)-; BENZOIC ACID,5-(AMINOSULFONYL)-4-CHLORO-2-((2-FURANYLMETHYL)AMINO)-(9CI); BERONALD; CHLOR-N-(2-FURYLMETHYL)-5-SULFAMYLANTHRANILSAEURE; 4-CHLORO-N-FURFURYL-5-SULFAMOYLANTHRANILIC ACID; 4-CHLORO-N-(2-FURYLMETHYL)-5-SULFAMOYLANTHRANILIC ACID; 4-CHLORO-N-(2-FURYLMETHYL)-5-SULFANOYLANTHRANILIC ACID; DESDEMIN; DISCOID; DIURAL; DIUROLASA; DRYPTAL; DURAFURID; ENDURAL; ERROLON; EUTENSIN; FRANYL; FRUSEMIDE; FRUSEMIN; FRUSETIC; FRUSID; FULSIX; FULUVAMIDE; FURANTHRIL; FURANTHRYL; FURANTRIL; FURESIS; 2-FURFURYLAMINO-4-CHLORO-5-SULFAMOYLBENZOIC ACID; FURO-PUREN; FUROSE; FUROSEDON; FUROSEMID; FUROSEMIDU; FUROSIDE; FURSEMID; FURSEMIDE; FUSID; HYDREX; HYDROLED; HYDRO-RAPID; IMPUGAN; KATLEX; LASEX; LASILETTEN; LASILIX; LASIX; LAXUR; LB 502; LOWPSTRON; MACASIROOL; MIRFAT; MOILARORIN; NEO-RENAL; NICOROL; NOVOSEMIDE; ODEMASE; ODEMEX; PREFEMIN; PROFEMIN; PROMEDES; PURESIS; RADONNA; ROSEMIDE; RUSYDE; SALIX; SEGURIL; SIGASALUR; TRANSIT; TROFURIT; UREX; UREX-M; URITOL; UROSEMIDE
Description: fine, white to slightly yellow, crystalline powder; odorless
Use: in the treatment of edema and hypertension, as a diuretic and as an antihypertensive; in veterinary medicine as a diuretic-saluretic in the treatment of acute inflammatory tissue edema in dogs and horses; in the treatment of edema (pulmonary congestion and ascites) associated with cardiac insufficiency in the dog and in treatment of physiological parturient edema of the mammary gland and associated structures in cattle

Physical Properties

Freezing Point: Decomposes at 203 °C (397.4 °F) to 205 °C (401 °F)
Water Solubility: Slightly Soluble in Water
Other Solubilities: Freely Soluble in alkali hydroxide.
Flash Point: Not available; probably combustible

RTECS Toxicity Data

Acute Oral: Woman TD_{Lo} Dose: 120 mg/kg/21W-I; Toxic Effects: Kidney, Ureter, and Bladder - Interstitial nephritis. Woman TD_{Lo} Dose: 6250 ug/kg; Toxic Effects: Behavioral - Muscle weakness; Nutritional and gross metabolic - Changes in calcium; Nutritional and gross metabolic - Changes in other metals. Rat LD_{50} Dose: 2600 mg/kg.
Chronic (Multiple Dose) Oral: Rat Dose: 1400 mg/kg/5W-I; Toxic Effects: Kidney, Ureter, and Bladder - Chgs in tubules (inc acute renal failure, acute tubular necrosis; Kidney, Ureter, and Bladder - Urine volume increased; Liver - Changes in liver weight. Rat Dose: 32200 mg/kg/14D-C; Toxic Effects: Kidney, Ureter, and Bladder - Chgs in tubules (inc acute renal failure, acute tubular necrosis; DEATH. Rat Dose: 27300 mg/kg/13W-I; Toxic Effects: Liver - Changes in liver weight; Kidney, Ureter, and Bladder - Chgs in tubules (inc acute renal failure, acute tubular necrosis. Rat Dose: 15288 mg/kg/26W-I; Toxic Effects: Kidney, Ureter, and Bladder - Urine volume increased; Kidney, Ureter, and Bladder - Changes in kidney weight; Nutritional and gross metabolic - Changes in sodium.

Reproductive/Teratogenic: Rat Route: Oral; Dose: 150 mg/kg; Duration: female 12-14D of pregnancy Specific Developmental Abnormalities - Musculoskeletal system. Rat Route: Oral; Dose: 300 mg/kg; Duration: female 16D of pregnancy; Effects on Embryo or Fetus - Extra embryonic structures; Specific Developmental Abnormalities - Musculoskeletal system. Rat Route: Oral; Dose: 122 mg/kg; Duration: male 6W prior to mating; Paternal Effects - Spermatogenesis; Testes, epididymis, sperm duct. Rat Route: Oral; Dose: 150 mg/kg; Duration: female 16D of pregnancy; Specific Developmental Abnormalities - Musculoskeletal system.

Mutagenic: Human Cytogenetic Analysis; Cell Type: leukocyte; Dose: 200 mg/L/24H. Mouse Mutations in Microorganisms; Cell Type: lymphocyte; Dose: 1 mg/L (+S9).

Tumorigenic: Rat Route: Oral; Dose: 21112 mg/kg/2Y-C; Toxic Effects: Tumorigenic - Equivocal tumorigenic agent by RTECS criteria; Brain and coverings - Tumors; Kidney, Ureter, and Bladder - Kidney tumors. Mouse Route: Oral; Dose: 15652 mg/kg/2Y-C; Toxic Effects: Tumorigenic - Carcinogenic by RTECS criteria; Skin and appendages - Tumors; Blood - Lymphomax including Hodgkin's disease. Mouse Route: Oral; Dose: 122 gm/kg/2Y-C; Toxic Effects: Tumorigenic - Carcinogenic by RTECS criteria; Skin and appendages - Tumors.

Hazard Overviews

Health: May cause irritation. Toxic. Other Acute Effects: harmful if swallowed, inhaled, or absorbed through skin. Chronic Effects: target organs: liver, kidneys, ears. Carcinogen.

Fire: Will burn. Hazards: emits toxic fumes. Extinguishing agents: water spray; carbon dioxide, dry chemical powder or appropriate foam. Precautions: combustible liquid.

Reactivity: Incompatible with: strong oxidizing agents, may discolor on exposure to light. Hazardous decomposition products: toxic fumes of: carbon monoxide, carbon dioxide, nitrogen oxides, sulfur oxides, hydrogen chloride gas.

Carcinogenicity: IARC - Group 3, Not classifiable as to carcinogenicity to humans; NIOSH - Not listed; NTP - Not listed; ACGIH - Not listed; OSHA - Not listed; EPA - Not listed; MAK - Not listed

Primary Target Organs:

Liver

Kidneys

Environmental

Regulations

RCRA 40CFR Not listed
CERCLA\\ 40CFR 302.4 Not listed
SARA 40CFR 372.65 Not listed
SARA EHS 40CFR 355 Not listed
TSCA Not listed

FUR8980 CAS #: 3688-53-7

FURYLFURAMIDE

RTECS: AS3500000
Molecular Formula: $C_{11}H_8N_2O_5$
Formula Weight: 248.21
Synonyms: ACRYLAMIDE,2-(2-FURYL)-3-(5-NITRO-2-FURYL)-; AF-2; AF 2 (PRESERVATIVE); FF; 2-FURANACETAMIDE,ALPHA-((5-NITRO-2-FURANYL)METHYLENE)-; 2-FURANACETAMIDE,ALPHA-((5-NITRO-2-FURANYL)METHYLENE)-(9CI); 2-FURANACRYLAMIDE,ALPHA-2-FURYL-5-NITRO; 2-FURANACRYLAMIDE,ALPHA-2-FURYL-5-NITRO-; 2-FURANACRYLAMIDE,ALPHA-2-FURYL-5-NITRO-(8CI); FURYLAMIDE; ALPHA-2-FURYL-5-NITRO-2-FURANACRYLAMIDE; 2-(2-FURYL)-3-(5-NITRO-2-FURYL)ACRYLAMIDE; 2-(FURYL)-3-(5-NITRO-2-FURYL)ACRYLAMIDE; 2-(2-FURYL)-3-(5-NITRO-2-FURYL)ACRYLIC ACID AMIDE; ALPHA-(FURYL)-BETA-(5-NITRO-2-FURYL)ACRYLIC AMIDE; ALPHA-((5-NITRO-2-FURANYL)METHYLENE)-2-FURANACETAMIDE; TOFURON
Description: bright orange crystals or orange solid

Physical Properties

Water Solubility: < 0.1 mg/mL at 21 C
Other Solubilities: 95% Ethanol: <1 mg/ml at 18 °C; Acetone: 10-50 mg/ml at 21 °C; DMSO: >=100 mg/ml at 21 °C.
Flash Point: Not available; probably combustible

RTECS Toxicity Data

Acute Oral: Rat LD_{50} Dose: 1554 mg/kg. Mouse LD_{50} Dose: 221 mg/kg.

Reproductive/Teratogenic: Rat Route: Oral; Dose: 6 gm/kg; Duration: female 1-20D of pregnancy; Effects on Embryo or Fetus - Fetotoxicity; Specific Developmental Abnormalities - Musculoskeletal system. Rat Route: Oral; Dose: 600 mg/kg; Duration: female 1-20D of pregnancy; Effects on Newborn - Weaning or lactation index. Rat Route: Oral; Dose: 6600 mg/kg; Duration: female 1-21D of pregnancy; Effects on Fertility - Pre-implantation mortality; Effects on Embryo or Fetus - Fetotoxicity.

Mutagenic: Human Unscheduled DNA Synthesis; Cell Type: fibroblast; Dose: 50 umol/L. Human Cytogenetic Analysis; Cell Type: embryo; Dose: 3 mg/L/2H. Human Cytogenetic Analysis; Cell Type: lymphocyte; Dose: 30 umol/L. Human Sister Chromatid Exchenge; Cell Type: lymphocyte; Dose: 500 ug/L.

Tumorigenic: Rat Route: Oral; Dose: 52 gm/kg/40W-C; Toxic Effects: Tumorigenic - Carcinogenic by RTECS criteria; Skin and appendages - Tumors. Rat Route: Oral; Dose: 25 gm/kg/78W-C; Toxic Effects: Tumorigenic - Neoplastic by RTECS criteria; Gastrointestinal - Tumors; Skin and appendages - Tumors.

Hazard Overviews

Fire: Will burn.

Carcinogenicity: IARC - Group 2B, Possibly carcinogenic to humans; NIOSH - Not listed; NTP - Not listed; ACGIH -

Not listed; OSHA - Not listed; EPA - Not listed; MAK - Not listed

Environmental

Regulations

RCRA 40CFR: Not listed
CERCLA: 40CFR 302.4: Not listed
SARA 40CFR 372.65: Not listed
SARA EHS 40CFR 355: Not listed
TSCA: Not listed

FUS3000 CAS #: 23255-69-8

FUSARENON

RTECS: YD0160000
Molecular Formula: $C_{17}H_{22}O_8$
Formula Weight: 354.13

Chemical Structure

Synonyms: 4-ACETYLOXY-12,13-EPOXY-3,7,15-TRIHYDROXY-(3-ALPHA,4-BETA,7-BETA)-TRICHOTHEC-9-EN-8-ONE; 12,13-EPOXY-3-ALPHA,4-BETA,7-BETA,15-TETRAHYDROXYTRICHOTHEC-9-EN-8-ONE 4-ACETATE; FUSARENON X; FUSARENONE X; NIVALENOL MONOACETATE; NIVALENOL-4-O-ACETATE; TRICHOTHEC-9-EN-8-ONE,4-(ACETYLOXY)-12,13-EPOXY-3,7,15-TRIHYDROXY-,(3ALPHA,4BETA,7BETA)-; TRICHOTHEC-9-EN-8-ONE,12,13-EPOXY-3ALPHA,4BETA,7BETA,15-TETRAHYDROXY-,4-ACETATE; 3ALPHA,7ALPHA,15-TRIHYDROXY-4BETA-ACETOXY-12,13-EPOXYTRICHOTHEC-9-EN-8-ONE; 3,7,15-TRIHYDROXY-4-ACETOXY-8-OXO-12,13-EPOXY-DELTA(SUP 9)-TRICHOTHECENE; 3,7,15-TRIHYDROXY-4-ACETOXY-8-OXO-12,13-EPOXY-DELTA(SUP9)-TRICHOTHECENE; 3,7,15-TRIHYDROXYSCIRP-4-ACETOXY-9-EN-8-ONE
Description: transparent crystals

Physical Properties

Freezing Point: 91 °C (195.8 °F) to 92 °C (197.6 °F)
Water Solubility: Soluble in Water
Other Solubilities: Soluble in Ethyl Acetate, Ethanol, Methanol & Chloroform; Insoluble in N-Hexane & N-Pentane

RTECS Toxicity Data

Acute Oral: Duck LD_{Lo} Dose: 5 mg/kg; Toxic Effects: Behavioral - Fluid intake; Gastrointestinal - Nausea or vomiting; Gastrointestinal - Other changes.
Acute Dermal: Cat LD_{Lo} Route: Subcutaneous Dose: 1 mg/kg; Toxic Effects: Gastrointestinal - Nausea or vomiting; Gastrointestinal - Other changes; Liver - Hepatitis (hepatocellular necrosis), diffuse.
Mutagenic: Human DNA Damage; Cell Type: HeLa cell; Dose: 32 mg/L/1H. Yeast - S Cerevisiae Mutations in Microorganisms; Dose: 250 mg/L (-S9).

Hazard Overviews

Poison

Health: Irritating to eyes/skin/respiratory tract. Poison. Other Acute Effects: may be fatal if inhaled, swallowed, or absorbed through skin; common symptoms of acute intoxication in experimental animals are diarrhea; nausea; vomiting; decreases in spontaneous movement and body temperature. The toxicological properties have not been thoroughly investigated.
Fire: Extinguishing agents: water spray; carbon dioxide, dry chemical powder or appropriate foam. Precautions: combustible liquid.
Reactivity: Stable. Hazardous polymerization will not occur. Hazardous decomposition products: toxic fumes of: carbon monoxide, carbon dioxide.
Carcinogenicity: IARC - Group 3, Not classifiable as to carcinogenicity to humans; NIOSH - Not listed; NTP - Not listed; ACGIH - Not listed; OSHA - Not listed; EPA - Not listed; MAK - Not listed
Primary Target Organs:

Eyes

Skin

Respiratory System

Environmental

Regulations

RCRA 40CFR: Not listed
CERCLA: 40CFR 302.4: Not listed
SARA 40CFR 372.65: Not listed
SARA EHS 40CFR 355: Not listed
TSCA: Not listed

FUS6000 CAS #: 536-69-6

FUSARIC ACID

RTECS: US5625000
EINECS Number: 208-643-0
Molecular Formula: $C_{10}H_{13}NO_2$

Formula Weight: 179.22

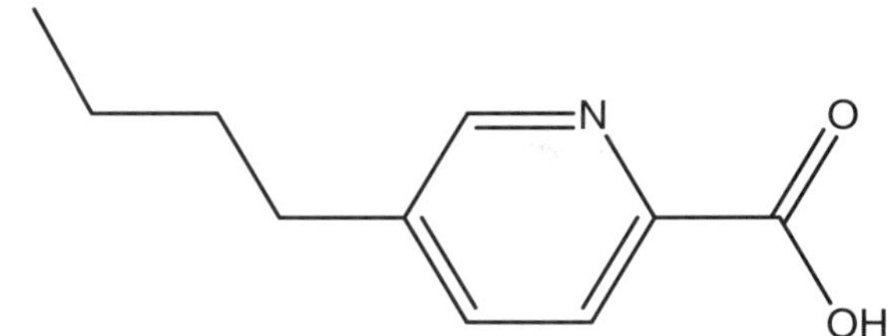

Chemical Structure

Synonyms: 5-BUTYLPICOLINIC ACID; 5-BUTYL-2-PYRIDINECARBOXYLIC ACID; 5-N-BUTYLPYRIDINE-2-CARBOXYLIC ACID; FUSARINIC ACID; PICOLINIC ACID,5-BUTYL-; 2-PYRIDINECARBOXYLIC ACID,5-BUTYL-

Description: colorless crystals

Physical Properties

Freezing Point: 96 °C (204.8 °F) to 98 °C (208.4 °F)

RTECS Toxicity Data

Acute Oral: Rat LD_{50} Dose: 480 mg/kg; Toxic Effects: Behavioral - Somnolence (general depressed activity); Vascular - BP lowering not characterized in autonomic section. Mouse LD_{50} Dose: 180 mg/kg.

Acute Dermal: Rat LD_{50} Route: Subcutaneous Dose: 300 mg/kg. Mouse LD_{50} Route: Subcutaneous Dose: 140 mg/kg.

Hazard Overviews

Health: May cause irritation to eyes/skin. Toxic. Other Acute Effects: harmful if swallowed, inhaled, or absorbed through skin.

Fire: Hazards: emits toxic fumes. Extinguishing agents: water spray; carbon dioxide, dry chemical powder or appropriate foam. Precautions: combustible liquid.

Reactivity: Incompatible with: strong oxidizing agents. Hazardous decomposition products: toxic fumes of: carbon monoxide, carbon dioxide, nitrogen oxides.

Carcinogenicity: IARC - Not listed; NIOSH - Not listed; NTP - Not listed; ACGIH - Not listed; OSHA - Not listed; EPA - Not listed; MAK - Not listed

Environmental

Regulations

RCRA 40CFR: Not listed
CERCLA: 40CFR 302.4: Not listed
SARA 40CFR 372.65: Not listed
SARA EHS 40CFR 355: Not listed
TSCA: Listed

GAL1000 **CAS #: 526-99-8**

GALACTARIC ACID

RTECS: LW5180000
EINECS Number: 208-404-0
Molecular Formula: $C_6H_{10}O_8$
Formula Weight: 210.14

Chemical Structure

Synonyms: GALACTOSACCHARIC ACID; MUCIC ACID; SACCHAROLACTIC ACID; SCHLEIMSAURE; TETRAHYDROXYADIPIC ACID; 2,3,4,5-TETRAHYDROXYHEXANEDIOIC ACID
Description: crystalline powder or prisms
Use: substitute for tartaric acid; sequesterant for metal ions (calcium, iron); retards hardening of concrete; intermediate for synthesis of heterocyclic compd (pyrroles); research chemical

Physical Properties

Boiling Point: Decomposes ~ 225 °C (437 °F)
Freezing Point: Decomposes at ~ 255 °C (491 °F)
Water Solubility: 1 parts in 300 parts Cold Water
Other Solubilities: Insoluble in Pyrimidine.

RTECS Toxicity Data

Acute Oral: Mouse LD_{50} Dose: 8 gm/kg; Toxic Effects: Tumorigenic - Active as anti-cancer agent.
Acute Dermal: Mouse LD_{50} Route: Subcutaneous Dose: 960 mg/kg.

Hazard Overviews

Health: May cause irritation to eyes/skin. Other Acute Effects: may be harmful by inhalation, ingestion, or skin absorption.
Fire: Hazards: emits toxic fumes. Extinguishing agents: water spray; carbon dioxide, dry chemical powder or appropriate foam. Precautions: combustible liquid.
Reactivity: Incompatible with: strong oxidizing agents. Hazardous decomposition products: toxic fumes of: carbon monoxide, carbon dioxide.
Carcinogenicity: IARC - Not listed; NIOSH - Not listed; NTP - Not listed; ACGIH - Not listed; OSHA - Not listed; EPA - Not listed; MAK - Not listed

Environmental

Regulations

RCRA 40CFR: Not listed
CERCLA: 40CFR 302.4: Not listed
SARA 40CFR 372.65: Not listed
SARA EHS 40CFR 355: Not listed
TSCA: Listed

GAL3000 **CAS #: 65-29-2**

GALLAMINE TRIETHIODIDE

RTECS: BS1100000
EINECS Number: 200-605-1
Molecular Formula: $C_{30}H_{60}I_3N_3O_3$
Formula Weight: 891.63

IH

Chemical Structure

Synonyms: 3.697 R.P; AMMONIUM,(V-PHENENYLTRIS(OXYETHYLENE))TRIS(TRIETHYL-,TRIIODIDE; BENZCURINE IODIDE; 2,2',2"-[1,2,3-BENZENETRIYLTRIS(OXY))TRIS(N,N,N-TRIETHYLETHANAMINIUM] TRIIODIDE; ETHANAMINIUM,2,2',2"-(1,2,3-BENZENETRIYLTRIS(OXY))TRIS(N,N,N-TRIETHYL-

,TRIIODIDE; F 2559; FLACEDIL; FLAXEDIL; FOURNEAU 2559; GALLAMIN; GALLAMIN TRIETHIODIDE; GALLAMINE; GALLAMINE IODIDE; GALLAMINE TRIIODOETHYLATE; GALLAMINE-3ETI; (V-PHENENYLTRIS(OXYETHYLENE))TRIS(TRIETHYLAMMONIUM IODIDE); (V-PHENENYLTRIS(OXYETHYLENE))TRIS(TRIETHYLAMMONIUM)IODIDE; (V-PHENENYLTRIS(OXYETHYLENE))TRIS(TRIETHYLAMMONIUMTRIIODIDE); PYROGALLOL 1,2,3-(DIETHYLAMINOETHYL ETHER) TRISETHYLIODIDE; RELAXAN; REMYOLAN; RETENSIN; RP 3697; SINCURARINE; SYNCURARINE; TRICURAN; 1,2,3-TRI(BETA-DIETHYLAMINOETHOXY)BENZENE TRIETHIODIDE; TRI(BETA-DIETHYLAMINOETHOXY)-1,2,3-BENZENETRI-IODOETHYLATE; TRIIODOETHYLATE DE GALLAMINE; TRI(IODOETHYLATE) DE TRI (BETA DIETHYLAMINOETHOXY)-1,2,3BENZENE; TRIIODOETHYLATE OF TRI(DIETHYLAMINOETHYLOXY)-1,2,3-BENZENE; TRIIODURE DE TRI(BETA-TRIETHYLAMMONIUMETHOXY)-1,2,3BENZENE; 1,2,3-TRIS(2-DIETHYLAMINOETHOXY)BENZENE TRIETHIODIDE; 1,2,3-TRIS(2-DIETHYLAMINOETHOXY)BENZENE TRIS(ETHYLIODIDE); 1,2,3-TRIS(2-TRIETHYLAMMONIUM ETHOXY)BENZENE TRIIODIDE

Description: white amorphous crystalline powder; odorless

Use: medication: skeletal muscle relaxant; (vet): skeletal muscle relaxant

Physical Properties

Freezing Point: 152 °C (305.6 °F) to 153 °C (307.4 °F)

Water Solubility: Freely Soluble in Water

Other Solubilities: freely Soluble in Alcohol, dilute Acetone Sparingly Soluble in anhydrous Acetone, Ether, Benzene, Chloroform

RTECS Toxicity Data

Acute Oral: Rat LD_{50} Dose: >1 gm/kg. Mouse LD_{50} Dose: 425 mg/kg.

Acute Dermal: Rabbit LD_{Lo} Route: Subcutaneous Dose: 2 mg/kg. Rat LD_{50} Route: Subcutaneous Dose: 28500 ug/kg.

Hazard Overviews

Health: Irritating to eyes/skin/respiratory tract. Toxic. Other Acute Effects: harmful if swallowed, inhaled, or absorbed through skin; target organs: nerves; skeletal muscle.

Fire: Hazards: emits toxic fumes. Extinguishing agents: water spray; carbon dioxide, dry chemical powder or appropriate foam. Precautions: combustible liquid.

Reactivity: Incompatible with: strong oxidizing agents, may decompose on exposure to light. Hazardous decomposition products: toxic fumes of: carbon monoxide, carbon dioxide, hydrogen iodide, nitrogen oxides.

Carcinogenicity: IARC - Not listed; NIOSH - Not listed; NTP - Not listed; ACGIH - Not listed; OSHA - Not listed; EPA - Not listed; MAK - Not listed

Primary Target Organs:

Eyes

Skin

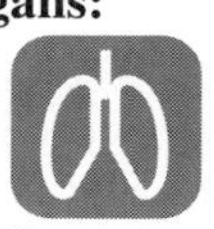
Respiratory System

Nervous System

Environmental

Regulations

RCRA 40CFR: Not listed

CERCLA: 40CFR 302.4: Not listed

SARA 40CFR 372.65: Not listed

SARA EHS 40CFR 355: Not listed

TSCA: Not listed

GAL5000 CAS #: 149-91-7

GALLIC ACID

RTECS: LW7525000

EINECS Number: 205-749-9

Molecular Formula: $C_7H_6O_5$

Structured MF: $(HO)_3C_6H_2COOH$

Formula Weight: 170.12

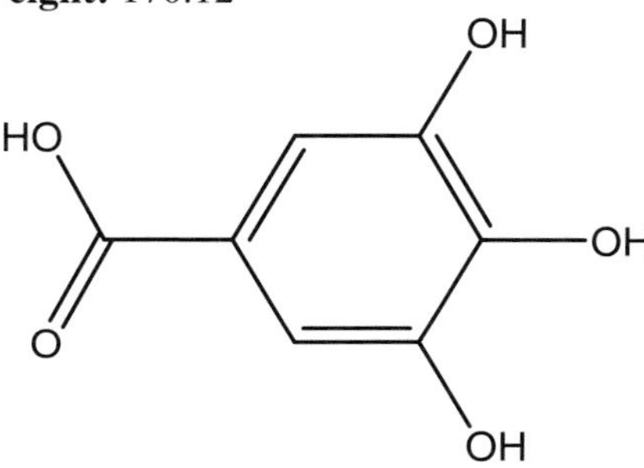

Chemical Structure

Synonyms: BENZOIC ACID,3,4,5-TRIHYDROXY-; KYSELINA GALLOVA; KYSELINA 3,4,5-TRIHYDROXYBENZOOVA; 3,4,5-TRIHYDROXYBENZOIC ACID

Description: prisms or needles; odorless

Use: photography, ink, dyeing, manufacture of pyrogallol, tanning agent, paper manufacture, pharmaceuticals, engraving and lithography, analytical reagent

Physical Properties

Boiling Point: Sublimes

Freezing Point: Decomposes at 235 °C (455 °F) to 240 °C (464 °F)

Specific Gravity: 1.694 at 6 °C/4 °C

Density: 1.69 g/mL

Water Solubility: Sparingly Soluble

Other Solubilities: 95% Ethanol: Soluble; Acetone: Soluble; Benzene: Insoluble; Ether: Sparingly Soluble.

Flash Point: Not available; probably combustible

RTECS Toxicity Data

Acute Oral: Rabbit LD_{50} Dose: 5 gm/kg; Toxic Effects: Lungs, Thorax, or Respiration - Chronic pulmonary edema; Liver - Other changes; Kidney, Ureter, and Bladder - Other changes.

Acute Dermal: Rat LD_{Lo} Route: Subcutaneous Dose: 5 gm/kg.

Reproductive/Teratogenic: Rat Route: Subcutaneous; Dose: 5 mg/kg; Duration: female 1D prior to mating; Maternal Effects - Ovaries, fallopian tubes.
Mutagenic: Mouse Cytogenetic Analysis; Route: Intraperitoneal; Dose: 100 umol/kg. Hamster Cytogenetic Analysis; Cell Type: ovary; Dose: 50 mg/L.

Hazard Overviews

Health: Irritating to eyes/skin/respiratory tract. Other Acute Effects: may be harmful by inhalation, ingestion, or skin absorption.
Fire: Will burn. Extinguishing agents: water spray; carbon dioxide, dry chemical powder or appropriate foam. Precautions: combustible liquid.
Reactivity: Incompatible with: strong oxidizing agents, strong bases, acid chlorides, acid anhydrides, may discolor on exposure to light. Hazardous decomposition products: toxic fumes of: carbon monoxide, carbon dioxide.
Carcinogenicity: IARC - Not listed; NIOSH - Not listed; NTP - Not listed; ACGIH - Not listed; OSHA - Not listed; EPA - Not listed; MAK - Not listed
Primary Target Organs:

Eyes

Skin

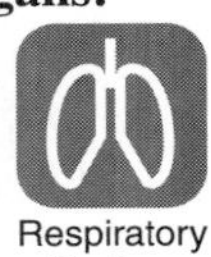
Respiratory System

Environmental

Ecotoxicity: Aquatic toxicity: 30-35 ppm/6 hr/minnow/MLD/fresh water 15-20 ppm/6 hr/minnow/MLD/fresh water

Environmental Physical Data

Octanol/Water Partition Coefficient: log K_{ow} = -0.25 to 0.15
BCF: no food chain concentration potential
BOD: 8%, 5 days

Regulations

RCRA 40CFR: Not listed
CERCLA: 40CFR 302.4: Not listed
SARA 40CFR 372.65: Not listed
SARA EHS 40CFR 355: Not listed
TSCA: Listed

Analytical Methods

Water / Groundwater: ASTM D4763

GAL7000 CAS #: 1303-00-0

GALLIUM ARSENIDE

RTECS: LW8800000
EINECS Number: 215-114-8
Molecular Formula: AsGa
Structured MF: GaAs
Formula Weight: 144.64
Synonyms: GALLIUM MONOARSENIDE; GALLIUM MONOARSENIDE (GAAS)
Description: dark gray ingots, crystals; garlic odor when moist
Use: as a semiconductor in light emitting diodes for telephone dials, injection lasers, solar cells, magneto-resistance devices, thermistors and microwave generation

Physical Properties

Freezing Point: 1238 °C (2260.4 °F)
Specific Gravity: 5.31 at 25 °C/4 °C
Density: 5.31 g/mL at 25 °C
Water Solubility: < 1 mg/mL at 20 C
Other Solubilities: 95% Ethanol: <1 mg/ml at 20 °C; Acetone: <1 mg/ml at 20 °C; DMSO: <1 mg/ml at 20 °C; Hydrochloric Acid: Soluble; Methanol: <1 mg/ml at 20 °C.
Flash Point: Not available; probably combustible

RTECS Toxicity Data

Chronic (Multiple Dose) Oral: Rat Dose: 9870 mg/kg/21D-I; Toxic Effects: Liver - Changes in liver weight; Biochemical - True cholinesterase.
Chronic (Multiple Dose) Inhalation: Rat Dose: 12 mg/m^3/17W; Toxic Effects: Brain and coverings - Recordings from specific areas of CNS; Blood - Changes in serum composition; Nutritional and gross metabolic - Weight loss or decreased weight gain. Rat Dose: 12 mg/m^3/17W; Toxic Effects: Blood - Changes in erythrocite (RBC) cell count; Nutritional and gross metabolic - Changes in calcium; Nutritional and gross metabolic - Changes in phosphorus.
Reproductive/Teratogenic: Rat Route: Inhalation; Dose: 37 mg/m^3/6H; Duration: female 4-19D of pregnancy; Maternal Effects - Other effects on females; Effects on Embryo or Fetus - Fetotoxicity. Mouse Route: Inhalation; Dose: 37 mg/m^3/6H; Duration: female 4-17D of pregnancy; Maternal Effects - Other effects on females; Effects on Embryo or Fetus - Fetotoxicity. Mouse Route: Inhalation; Dose: 75 mg/m^3/6H; Duration: female 4-17D of pregnancy; Effects on Fertility - Post-implantation mortality; Other measures of fertility; Specific Developmental Abnormalities - Craniofacial (including nose and tongue).

Hazard Overviews

Fire Diamond

Health: Toxic. Also Causes: cardiovascular changes, pulmonary edema, peripheral neuropathy, nausea, vomiting, diarrhea, dehydration, abdominal pain, muscle cramping, kidney failure, jaundice, delirium, headache, convulsions, parasthesis. Chronic Effects: skin disorders, liver damage, injury to nervous system, cancer hazard.
Fire: Will burn. Reacts with water; decomposes and gives off deadly poisonous arsine gas. Use Class D extinguisher or smother with dry sand, dry clay, or dry ground limestone. Do not use carbon dioxide, water, or halogenated compounds.

Reactivity: Stable. Hazardous polymerization cannot occur. Avoid: heat; ignition sources. Incompatible with: quartz; steam; acids; acid fumes. Hazardous decomposition products: toxic arsine; arsenic; gallium.

Carcinogenicity: IARC - Group 1, Carcinogenic to humans; NIOSH - Listed as carcinogen; NTP - Class 1, Known to be a carcinogen; ACGIH - Class A1, Confirmed human carcinogen; OSHA - Listed as a carcinogen; EPA - Class A, Human carcinogen; MAK - Class A1, Capable of inducing malignant tumors as shown by experience with humans

Primary Target Organs:

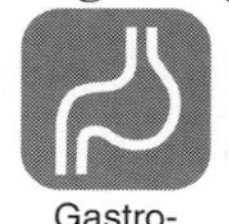
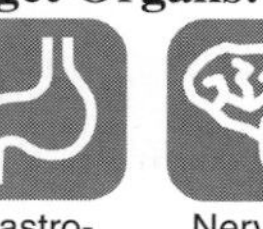

Respiratory System | Gastro-intestinal | Nervous System | Kidneys | Cardio-vascular | Blood

Exposure Limits

OSHA PEL: TWA: 0.01 mg/m^3; as As.

ACGIH TLV: TWA: 0.01 mg/m^3; as As.

NIOSH REL: STEL: 0.002 mg/m^3; ceiling (15 min) as As.

Respirator Recommendation

Exposure Range: >0.01 to <5 mg/m^3 Supplied Air, Constant Flow/Pressure Demand, Full Face

Exposure Range: 5 to unlimited mg/m^3 Self-contained Breathing Apparatus, Pressure Demand, Full Face

Note: as arsenic, inorganic compounds; refer to 29CFR 1910.1018 for more specific respirator recommendations

Environmental

Regulations

RCRA 40CFR: Not listed

CERCLA: 40CFR 302.4: Listed as Compound per CWA Section 307(a) per CAA Section 112

SARA 40CFR 372.65: Listed as Compound

SARA EHS 40CFR 355: Not listed

TSCA: Listed

GAL9000 CAS #: 13450-90-3

GALLIUM TRICHLORIDE

RTECS: LW9100000

EINECS Number: 236-610-0

Molecular Formula: Cl_3Ga

Structured MF: Cl_3Ga

Formula Weight: 176.08

Cl^-

Ga^{+++}

Cl^- Cl^-

Chemical Structure

Synonyms: GALLIUM CHLORIDE

Description: white needles

Physical Properties

Boiling Point: 201.3 °C (394 °F)

Freezing Point: 77.9 °C (172.22 °F)

Specific Gravity: 247 at 26 °C/4 deg

Water Solubility: Very Soluble in Water

Other Solubilities: Soluble in Benzene, Carbon Tetrachloride, Carbon Disulfide.

RTECS Toxicity Data

Acute Inhalation: Rat LC_{Lo} Dose: 316 mg/m^3/3hr.

Acute Dermal: Rabbit LD_{50} Route: Subcutaneous Dose: 245 mg/kg. Rat LD_{50} Route: Subcutaneous Dose: 306 mg/kg.

Mutagenic: Bacteria - B Subtilis DNA Repair; Dose: 2350 ug/disc.

Hazard Overviews

Corrosive

Health: Corrosive to eyes/skin/respiratory tract. Other Acute Effects: harmful if swallowed, inhaled, or absorbed through skin; inhalation may result in spasm, inflammation and edema of the larynx and bronchi, chemical pneumonitis and pulmonary edema; symptoms of exposure may include burning sensation; coughing; wheezing; laryngitis; shortness of breath; headache; nausea; vomiting.

Fire: Hazards: emits toxic fumes. Extinguishing agents: carbon dioxide, dry chemical powder or appropriate foam. Precautions: combustible liquid.

Reactivity: Incompatible with: strong oxidizing agents, may discolor on exposure to air, reacts violently with water. Hazardous decomposition products: hydrogen chloride gas.

Carcinogenicity: IARC - Not listed; NIOSH - Not listed; NTP - Not listed; ACGIH - Not listed; OSHA - Not listed; EPA - Not listed; MAK - Not listed

Primary Target Organs:

Eyes | Skin | Respiratory System

Environmental

Regulations

RCRA 40CFR: Not listed

CERCLA: 40CFR 302.4: Not listed

SARA 40CFR 372.65: Not listed TPQ: 500/10000 lb

SARA EHS 40CFR 355: Listed TPQ: 500 lb

TSCA: Listed

GAN5000 CAS #: 82410-32-0

GANCICLOVIR

RTECS: MF8407000
Molecular Formula: $C_9H_{13}N_5O_4$
Formula Weight: 255.23
Synonyms: 2-AMINO-1,9-((2-HYDROXY-1-(HYDROXYMETHYL)ETHOXY)METHYL)-6-H-PURIN-6-ONE; BIOLF-62; BIOLF 62; BW-759; BW-759U; BWB759U; CYMEVAN; CYMEVENE; CYTOVENE; DHPG; 9-((1,3-DIHYDROXY-2-PROPOXY)METHYL)GUANINE; 2'NDG; 2'-NOR-2'-DEOXYGUANOSINE; 6H-PURIN-6-ONE,2-AMINO-1,9-DIHYDRO-9-((2-HYDROXY-1-(HYDROXYMETHYL)ETHOXY)METHYL)-(9CI); RS-21592
Description: white to off-white solid or crystals
Use: medication: antiviral agent

Physical Properties

Freezing Point: Decomposes at 250 °C (482 °F)
Water Solubility: 4.3 mg/ml at pH 7 at 25 °C
Other Solubilities: Aqueous solubility of the drug is relatively constant over a pH range of 3.5-8.5 but increases substantially in strongly acidic or basic solutions.

RTECS Toxicity Data

Acute Oral: Mouse LD_{50} Dose: >2 gm/kg. Dog LD_{50} Dose: >1 gm/kg.
Mutagenic: Human Sister Chromatid Exchenge; Cell Type: lymphocyte; Dose: 10 mg/L.

Hazard Overviews

Carcinogenicity: IARC - Not listed; NIOSH - Not listed; NTP - Not listed; ACGIH - Not listed; OSHA - Not listed; EPA - Not listed; MAK - Not listed

Environmental

Regulations

RCRA 40CFR: Not listed
CERCLA: 40CFR 302.4: Not listed
SARA 40CFR 372.65: Not listed
SARA EHS 40CFR 355: Not listed
TSCA: Not listed

GAS5000 CAS #: 86290-81-5

GASOLINE

RTECS: DE3550000
DOT: UN1203; UN1257; IMO3.1
EINECS Number: 289-220-8
Synonyms: ANTIKNOCK GASOLINE; BENZIN; BENZINE BR-1; BENZINE BR-2; BENZYNA; CASING HEAD GASOLINE; CRACKED GASOLINE; GASOLINE BR-1; HERBICIDE ES; HIGH-OCTANE GASOLINE; LEAD-FREE GASOLINE; NATURAL GASOLINE; NEFRAS 150/200; PETROL; POLYMER GASOLINE; PYROLYSIS GASOLINE; REFORMED GASOLINE; STRAIGHT-RUN GASOLINE; WHITE GASOLINE
Description: mobile liquid; characteristic odor
Use: fuel in internal combustion engines of the spark-ignited, reciprocating type; fuel, fats, extractant or dilutant for essential oils; solvent for rubber adhesives; detergent for precision instruments; finishing agent for artificial leathers

Physical Properties

Boiling Point: 39 °C (102 °F) to 200 °C (392 °F)
Freezing Point: -95.4 °C (-139.72 °F) to -90.5 °C (-130.9 °F)
Specific Gravity: 0.7 to 0.8
Vapor Density: 3 Air=1
Water Solubility: Insoluble in Water
Other Solubilities: freely Soluble in absolute Alcohol, Ether, Chloroform, Benzene.
Refraction Index: 1.375 to 1.388
Autoignition Temperature: about 250 °C
LEL: 1.3% v/v
UEL: 7.1% v/v

RTECS Toxicity Data

Acute Oral: Rat LD_{50} Dose: 92 gm/kg. Mouse LD_{50} Dose: 60 mL/kg; Toxic Effects: Automatic Nervous System - Other (direct) parasympathomimetic; Behavioral - Convulsions or effect on seizure threshold; Lungs, Thorax, or Respiration - Respiratory stimulation.
Mutagenic: Rat Sperm Morphology; Route: Inhalation; Dose: 300 mg/m^3/10W-I.

Hazard Overviews

Poison

Fire Diamond

Health: Irritating to eyes/skin/respiratory tract. Poison. Other Acute Effects: harmful if swallowed, inhaled, or absorbed through skin; can cause CNS depression; dermatitis. Chronic Effects: may cause nervous system disturbances; inhalation studies have demonstrated the development of inflammatory and ulcerous lesions of the penis, prepuce and scrotum in animals; target organs: heart, brain, liver, kidneys, bladder.
Fire: Will burn. Hazards: emits toxic fumes; vapor may travel considerable distance to source of ignition and flash back; container explosion may occur. Extinguishing agents: water spray; carbon dioxide, dry chemical powder or appropriate foam. Precautions: combustible liquid.
Reactivity: Incompatible with: strong oxidizing agents. Hazardous decomposition products: toxic fumes of: carbon monoxide, carbon dioxide.
Carcinogenicity: IARC - Not listed; NIOSH - Not listed; NTP - Not listed; ACGIH - Not listed; OSHA - Not listed; EPA - Not listed; MAK - Not listed

Primary Target Organs:

Eyes

Skin

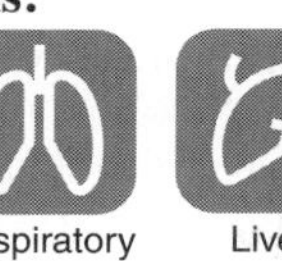
Respiratory System

Liver

Kidneys

Cardio-vascular

Environmental

Cleanup/Disposal: Guide No. 128: Eliminate all ignition sources (no smoking, flares, sparks or flames in immediate area). All equipment used when handling the product must be grounded. Do not touch or walk through spilled material. Stop leak if you can do it without risk. Prevent entry into waterways, sewers, basements or confined areas. A vapor suppressing foam may be used to reduce vapors. Absorb or cover with dry earth, sand or other non-combustible material and transfer to containers. Use clean non-sparking tools to collect absorbed material. Large Spills: Dike far ahead of liquid spill for later disposal. Water spray may reduce vapor; but may not prevent ignition in closed spaces.

Regulations

RCRA 40CFR: Not listed
CERCLA: 40CFR 302.4: Not listed
SARA 40CFR 372.65: Not listed
SARA EHS 40CFR 355: Not listed
TSCA: Not listed

GEL1000 CAS #: 9000-70-8

GELATIN

RTECS: LX8580000
EINECS Number: 232-554-6
Molecular Formula: Unknown
Formula Weight: N/A
Synonyms: ABSORBABLE GELATIN SPONGE; GELATIN FOAM; GELATINE; GELATINS; GELFOAM; GT; PHARMAGEL A; PHARMAGEL ADB; PHARMAGEL B; PURAGEL; SPONGIOFORT; VEE GEE GELATIN
Description: colorless or slightly yellow, transparent brittle, sheets, flakes, or coarse powder; odorless
Use: in mfr of inks, plastic compd, artificial silk, matches, & light filters for mercury lamps; clarifying agent; pharmaceutic aid (suspending & microencapsulating agent); in hectographic masters; sizing paper & textiles; to inhibit crystallization in bacteriology; food additive: gelling & as adjuvant protein in puddings, etc; carrier in photographic films & papers; electroplating agent, in cosmetics & ointments; med: (vet) plasma expander, hemostatic (sponge), tissue adhesive, protein source.

Physical Properties

Water Solubility: Soluble in Hot Water
Other Solubilities: Soluble in hot mixture of Glycerin & water; Insoluble in Alcohol, Chloroform, Ether, fixed & volatile oils.

RTECS Toxicity Data

Reproductive/Teratogenic: Mouse Route: Intraperitoneal; Dose: 700 mg/kg; Duration: female 7-13D of pregnancy; Specific Developmental Abnormalities - Urogenital system; Effects on Newborn - Growth statistics.

Hazard Overviews

Health: May cause irritation. Other Acute Effects: may be harmful by inhalation, ingestion, or skin absorption. The toxicological properties have not been thoroughly investigated.
Fire: Hazards: hazardous polymerization will not occur. Extinguishing agents: water spray; carbon dioxide, dry chemical powder or appropriate foam. Precautions: combustible liquid.
Reactivity: Stable. Hazardous polymerization will not occur. Hazardous decomposition products: carbon monoxide, carbon dioxide.
Carcinogenicity: IARC - Not listed; NIOSH - Not listed; NTP - Not listed; ACGIH - Not listed; OSHA - Not listed; EPA - Not listed; MAK - Not listed

Environmental

Regulations

RCRA 40CFR: Not listed
CERCLA: 40CFR 302.4: Not listed
SARA 40CFR 372.65: Not listed
SARA EHS 40CFR 355: Not listed
TSCA: Listed

GEL5000 CAS #: 509-15-9

GELSEMINE

RTECS: LX9100000
EINECS Number: 208-095-2
Molecular Formula: $C_{20}H_{22}N_2O_2$
Formula Weight: 322.44

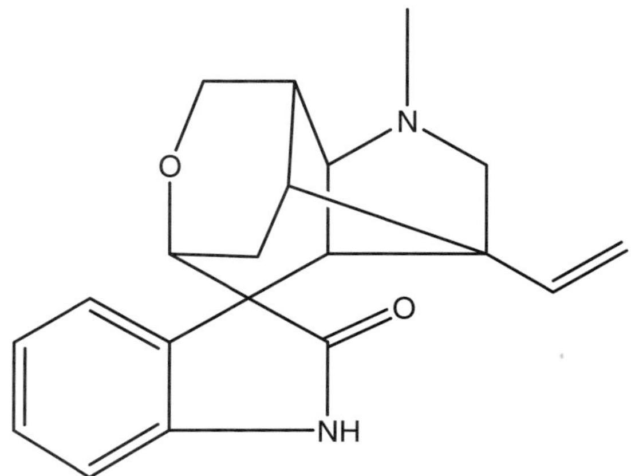

Chemical Structure

Synonyms: (3R-(3ALPHA,4A BETA,5ALPHA,8ALPHA,8A BETA,9S*,10S*))-5-ETHENYL-3,4,4A,5,6,7,8,8A-OCTAHYDRO-7-METHYLSPIRO(3,5,8-ETHANYLYLIDENE-1H-PYRANO(3,4-C)PYRIDINE-10,3'-(3H)INDOL)-2'(1',H)-ONE; GELSEMIN; SPIRO(3,5,8-ETHANYLYLIDENE-1H-PYRANO(3,4-C)PYRIDINE-10,3'-(3H)INDOL)-2'(1'H)-ONE,5-ETHENYL-3,4,4A,5,6,7,8,8A-OCTAHYDRO-7-METHYL-,(3R-(3ALPHA,4ABETA,5ALPHA,8ALPHA,8ABETA,9S*,10S*))-
Description: crystals

Physical Properties

Freezing Point: 178 °C (352.4 °F)
Water Solubility: Slightly Soluble in Water
Other Solubilities: Soluble in Alcohol, Benzene, Chloroform, Ether, Acetone, dilute acids

Hazard Overviews

Poison

Health: May cause irritation to skin. Poison. Other Acute Effects: may be fatal if inhaled, swallowed, or absorbed through skin; prolonged exposure can cause CNS stimulation; target organs: central nervous system.
Fire: Hazards: emits toxic fumes. Extinguishing agents: carbon dioxide; dry chemical powder; water spray. Precautions: combustible liquid.
Reactivity: Incompatible with: strong oxidizing agents. Hazardous decomposition products: thermal decomposition may produce carbon monoxide, carbon dioxide, and nitrogen oxides.
Carcinogenicity: IARC - Not listed; NIOSH - Not listed; NTP - Not listed; ACGIH - Not listed; OSHA - Not listed; EPA - Not listed; MAK - Not listed
Primary Target Organs:

Nervous System

Environmental

Regulations

RCRA 40CFR: Not listed
CERCLA: 40CFR 302.4: Not listed
SARA 40CFR 372.65: Not listed
SARA EHS 40CFR 355: Not listed
TSCA: Not listed

GEL9000 CAS #: 35306-33-3

GELSEMINE HYDROCHLORIDE

RTECS: LX9150000
EINECS Number: 252-504-7
Molecular Formula: $C_{20}H_{23}ClN_2O_2$
Formula Weight: 358.90
Synonyms: GELSEMINE,HYDROCHLORIDE; GELSEMINE,MONOHYDROCHLORIDE; GESEMINE HYDROCHLORIDE
Description: prisms
Use: medication: formerly as central stimulant; tool in neurological research

Physical Properties

Freezing Point: 326 °C (618.8 °F)
Water Solubility: Soluble in Water
Other Solubilities: Slightly Soluble in Alcohol

RTECS Toxicity Data

Acute Dermal: Rabbit LD_{Lo} Route: Subcutaneous Dose: 500 ug/kg; Toxic Effects: Lungs, Thorax, or Respiration - Other changes; Liver - Other changes; Kidney, Ureter, and Bladder - Other changes.

Hazard Overviews

Health: Irritating to skin. Other Acute Effects: may be fatal if swallowed or absorbed through skin; exposure can cause CNS stimulation; mydriasis; double vision; myastenia; respiratory collapse.
Fire: Hazards: emits toxic fumes. Extinguishing agents: carbon dioxide, dry chemical powder or appropriate foam. Precautions: combustible liquid.
Reactivity: Stable. Hazardous polymerization will not occur. Hazardous decomposition products: toxic fumes of: carbon monoxide, carbon dioxide, nitrogen oxides, hydrogen chloride gas.
Carcinogenicity: IARC - Not listed; NIOSH - Not listed; NTP - Not listed; ACGIH - Not listed; OSHA - Not listed; EPA - Not listed; MAK - Not listed
Primary Target Organs:

Skin

Environmental

Regulations

RCRA 40CFR: Not listed
CERCLA: 40CFR 302.4: Not listed
SARA 40CFR 372.65: Not listed
SARA EHS 40CFR 355: Not listed
TSCA: Not listed

GEN3000 CAS #: 1403-66-3

GENTAMYCIN

RTECS: LY2450000
EINECS Number: 215-765-8
Molecular Formula: Unspecified or Variable

Synonyms: CIDOMYCIN; GARAMYCIN; GARASOL; GENTACYCOL; GENTAMICIN; GENTAMYCIN-CREME; GENTAVET; GENTOCIN; REFOBACIN; REFOBACIN TM; UROMYCINE
Description: white amorphous powder; odorless
Use: medication: antibacterial agent for treating gram-neg microbial infections; medication: aminoglycoside antibiotic

Physical Properties

Freezing Point: 102 °C (215.6 °F) to 108 °C (226.4 °F)
Water Solubility: Freely Soluble in Water
Other Solubilities: Insoluble in Alcohol, Acetone, Benzene /Gentamicin C complex Sulfate/; freely Soluble in Methanol; Slightly Soluble in Ether; Practically Insoluble in other organic solvents /Gentamicin Hydrochloride/.

RTECS Toxicity Data

Acute Oral: Rat LD_{50} Dose: 6600 mg/kg; Toxic Effects: Behavioral - Tremor; Behavioral - Convulsions or effect on seizure threshold; Gastrointestinal - Other changes. Mouse LD_{50} Dose: 10 gm/kg; Toxic Effects: Behavioral - Altered sleep time (including change in righting reflex); Behavioral - Convulsions or effect on seizure threshold; Behavioral - Ataxia.
Acute Dermal: Rabbit LD_{50} Route: Subcutaneous Dose: 1230 mg/kg; Toxic Effects: Behavioral - Convulsions or effect on seizure threshold; Behavioral - Ataxia; Lungs, Thorax, or Respiration - Dyspnea. Rat LD_{50} Route: Subcutaneous Dose: 710 mg/kg.
Chronic (Multiple Dose) Dermal: Rat Route: Subcutaneous; Dose: 4200 mg/kg/4W-I; Toxic Effects: Kidney, Ureter, and Bladder - Chgs in tubules (inc acute renal failure, acute tubular necrosis. Rat Route: Subcutaneous; Dose: 120 mg/kg/6D-I; Toxic Effects: Kidney, Ureter, and Bladder - Urine volume increased; Kidney, Ureter, and Bladder - Other changes in urine composition; Biochemical - Phosphatases. Rat Route: Subcutaneous; Dose: 560 mg/kg/14D-I; Toxic Effects: Kidney, Ureter, and Bladder - Chgs in tubules (inc acute renal failure, acute tubular necrosis; Kidney, Ureter, and Bladder - Other changes in urine composition; Blood - Changes in serum composition. Rat Route: Subcutaneous; Dose: 720 mg/kg/12D-I; Toxic Effects: Immunological including allergic - Hypersensitivity delayed. Rat Route: Subcutaneous; Dose: 560 mg/kg/7D-I; Toxic Effects: Kidney, Ureter, and Bladder - Other changes in urine composition; Biochemical - Other enzymes. Rat Route: Subcutaneous; Dose: 1072 mg/kg/16D-I; Toxic Effects: Sense organs and special senses - Change in cochlear structure or function. Rat Route: Subcutaneous; Dose: 525 mg/kg/21D-I; Toxic Effects: Behavioral - Ataxia; Blood - Changes in serum composition; DEATH.
Reproductive/Teratogenic: Woman Route: Oral; Dose: 10769 units/kg; Duration: female 20W of pregnancy; Specific Developmental Abnormalities - Hepatobiliary system. Rat Route: Subcutaneous; Dose: 660 mg/kg; Duration: female 10-15D of pregnancy Effects on Newborn - Delayed effects. Rat Route: Subcutaneous; Dose: 660 mg/kg; Duration: female 15-20D of pregnancy Specific Developmental Abnormalities - Urogenital system.
Mutagenic: Human Sister Chromatid Exchenge; Cell Type: fibroblast; Dose: 75 mg/L. Mouse Cytogenetic Analysis; Cell Type: lymphocyte; Dose: 500 mg/L.

Hazard Overviews

Carcinogenicity: IARC - Not listed; NIOSH - Not listed; NTP - Not listed; ACGIH - Not listed; OSHA - Not listed; EPA - Not listed; MAK - Not listed

Environmental

Regulations

RCRA 40CFR: Not listed
CERCLA: 40CFR 302.4: Not listed
SARA 40CFR 372.65: Not listed
SARA EHS 40CFR 355: Not listed
TSCA: Not listed

GEN6000 CAS #: 72968-42-4

GENTIAN EXTRACT

EINECS Number: 277-139-0
Synonyms: BITTER ROOT; ENZIANWURZEL; PALE GENTIAN; YELLOW GENTIAN
Use: vet: bitter tonic; flavoring agent in foods, beverages, & pharmaceuticals

Hazard Overviews

Carcinogenicity: IARC - Not listed; NIOSH - Not listed; NTP - Not listed; ACGIH - Not listed; OSHA - Not listed; EPA - Not listed; MAK - Not listed

Environmental

Regulations

RCRA 40CFR: Not listed
CERCLA: 40CFR 302.4: Not listed
SARA 40CFR 372.65: Not listed
SARA EHS 40CFR 355: Not listed
TSCA: Not listed

GER1000 CAS #: 106-24-1

GERANIOL

RTECS: RG5830000
EINECS Number: 203-377-1
Molecular Formula: $C_{10}H_{18}O$
Formula Weight: 154.24

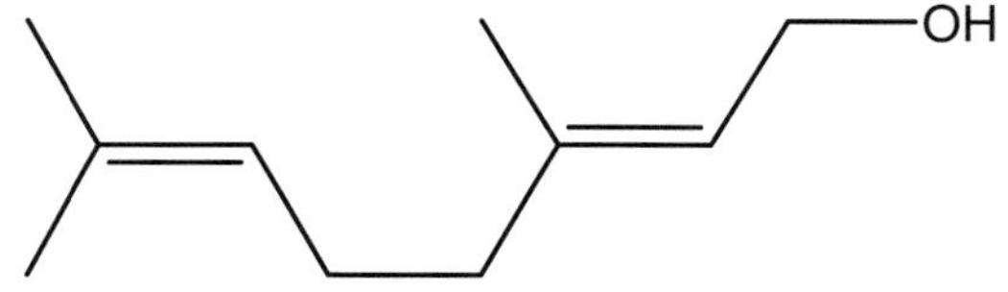

Chemical Structure

Synonyms: (E)-3,7-DIMETHYL-2,6-OCTADIEN-1-OL; 2,6-DIMETHYL-2,6-OCTADIEN-8-OL; 2,6-DIMETHYL-TRANS-2,6-OCTADIEN-8-OL; 2-TRANS-3,7-DIMETHYL-2,6-OCTADIEN-1-OL; 3,7-DIMETHYL-2,6-OCTADIEN-1-OL; 3,7-DIMETHYL-TRANS-2,6-OCTADIEN-1-OL; TRANS-GERANIOL; GERANIOL ALCOHOL; GERANIOL EXTRA; GERANYL ALCOHOL; GUANIOL; LEMONOL; 2,6-OCTADIEN-1-OL,3,7-DIMETHYL-,(E)-; 2,6-OCTADIEN-1-OL,3,7-DIMETHYL-,TRANS-

Description: colorless to pale yellow; oily liquid; sweet rose floral odor

Use: in perfumery; a synthetic flavoring substance and adjuvant

Physical Properties

Boiling Point: 230 °C (446 °F) at 760 mm Hg
Freezing Point: < -15 °C (5 °F)
Specific Gravity: 0.8894 at 20 °C
Vapor Pressure: 0.205 kPa at 75 °C
Water Solubility: Practically Insoluble in Water
Other Solubilities: Acetone: Soluble; Alcohol: Soluble; Ether: Soluble.
Refraction Index: 1.4766 at 20 °C/D
Flash Point: 101 °C

RTECS Toxicity Data

Acute Oral: Rat LD_{50} Dose: 3600 mg/kg; Toxic Effects: Behavioral - Somnolence (general depressed activity); Behavioral - Coma; Skin and appendages - Hair.

Acute Dermal: Rabbit LD_{50} Route: Skin; Dose: >5 gm/kg. Mouse LD_{50} Route: Subcutaneous Dose: 1090 mg/kg; Toxic Effects: Peripheral Nerve and sensation - Spastic parapysis with or without sensory change.

Irritation Skin: Man Standard Draize Test Dose: 16 mg/24H; Reaction: severe. Rabbit Standard Draize Test Dose: 100 mg/24H; Reaction: severe.

Hazard Overviews

Fire: Will burn.

Carcinogenicity: IARC - Not listed; NIOSH - Not listed; NTP - Not listed; ACGIH - Not listed; OSHA - Not listed; EPA - Not listed; MAK - Not listed

Environmental

Regulations

RCRA 40CFR: Not listed
CERCLA: 40CFR 302.4: Not listed
SARA 40CFR 372.65: Not listed
SARA EHS 40CFR 355: Not listed
TSCA: Listed

GER3000 CAS #: 105-87-3

GERANYL ACETATE

RTECS: RG5920000
EINECS Number: 203-341-5
Molecular Formula: $C_{12}H_{20}O_2$
Structured MF: $(CH_3)_2C{=}CHCH_2CH_2C(CH_3){=}CHCH_2OCOCH_3$
Formula Weight: 196.32

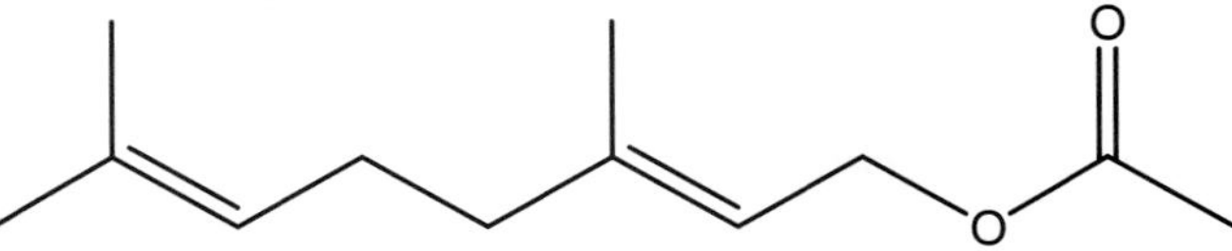

Chemical Structure

Synonyms: ACETIC ACID GERANIOL ESTER; ACETIC ACID,GERANIOL ESTER; BAY PINE (OYSTER) OIL; 2,6-DIMETHYL-2,6-OCTADIENE-8-YL ACETATE; (E)-3,7-DIMETHYL-2,6-OCTADIEN-1-OL ACETATE; TRANS-3,7-DIMETHYL-2,6-OCTADIEN-1-OL ACETATE; TRANS-3,7-DIMETHYL-2,6-OCTADIEN-1-OL,ACETATE; 3,7-DIMETHYL-2-TRANS,6-OCTADIENYL ACETATE; 3,7-DIMETHYL-2-TRANS-6-OCTADIENYL ACETATE; TRANS-3,7-DIMETHYL-2,6-OCTADIEN-1-YL ACETATE; TRANS-2,6-DIMETHYL-2,6-OCTADIEN-8-YL ETHANOATE; GERANIOL ACETATE; GERANOIL ACETATE; 2,6-OCTADIEN-1-OL,3,7-DIMETHYL-,ACETATE,(E)-; 2,6-OCTADIEN-1-OL,3,7-DIMETHYL-,ACETATE,TRANS-

Description: colorless liquid; odor of lavender

Use: in perfumery and flavoring

Physical Properties

Boiling Point: 242 °C (468 °F)
Freezing Point: Decomposes at ~ 242 °C (467.6 °F)
Specific Gravity: 0.9174 at 15 °C/15 °C
Vapor Density: 6.8 Air=1
Saturated Vapor Density: ~ 1.2001822 kg/m^3
Density: 0.907 to 0.918 g/mL at 15 °C
Vapor Pressure: ~ 0.02 mm Hg at 20 °C
Water Solubility: < 1 mg/mL at 20 C
Other Solubilities: 95% Ethanol: >=100 mg/ml at 23.5 °C; Acetone: >=100 mg/ml at 23.5 °C; Alcohol: Soluble; Corn oil: miscible; DMSO: >=100 mg/ml at 23.5 °C; Ether: Soluble; Glycerol: Insoluble.
Refraction Index: 1.4580 to 1.4640 at 20 °C
Flash Point: 104 °C

RTECS Toxicity Data

Acute Oral: Rat LD_{50} Dose: 6330 mg/kg; Toxic Effects: Behavioral - Somnolence (general depressed activity); Behavioral - Coma. Mouse LD_{Lo} Dose: 8 gm/kg; Toxic Effects: Behavioral - Somnolence (general depressed activity).

Chronic (Multiple Dose) Oral: Rat Dose: 260 gm/kg/13W-I; Toxic Effects: DEATH. Mouse Dose: 28 gm/kg/14D-I; Toxic Effects: DEATH. Mouse Dose: 130 gm/kg/13W-I; Toxic Effects: Liver - Other changes; DEATH.

Irritation Skin: Man Standard Draize Test Dose: 16 mg/48H; Reaction: mild. Rabbit Standard Draize Test Dose: 100 mg/24H; Reaction: severe.

Mutagenic: Rat Unscheduled DNA Synthesis; Route: Oral; Dose: 1 gm/kg. Mouse Mutations in Microorganisms; Cell Type: lymphocyte; Dose: 18 mg/L (+S9).

Hazard Overviews

Fire: Will burn.
Carcinogenicity: IARC - Not listed; NIOSH - Not listed; NTP - Not listed; ACGIH - Not listed; OSHA - Not listed; EPA - Not listed; MAK - Not listed

Environmental

Regulations

RCRA 40CFR: Not listed
CERCLA: 40CFR 302.4: Not listed
SARA 40CFR 372.65: Not listed
SARA EHS 40CFR 355: Not listed
TSCA: Listed

GER5000 CAS #: 7782-65-2

GERMANE

RTECS: LY4900000
DOT: UN2192
EINECS Number: 231-961-6
Molecular Formula: Ge H_4
Structured MF: GeH_4
Formula Weight: 76.63

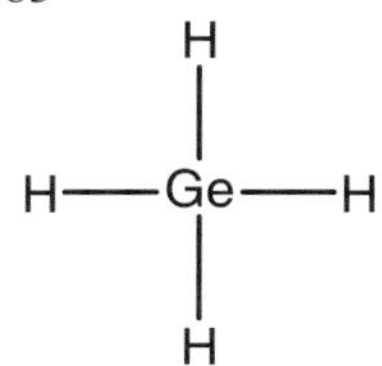

Chemical Structure

Synonyms: GERMANIUM HYDRIDE; GERMANIUM TETRAHYDRIDE; GERMANOMETHANE; MONOGERMANE
Description: colorless gas

Physical Properties

Boiling Point: -88.33 °C (-127 °F)
Freezing Point: -166.11 °C (-266.998 °F)
Vapor Density: 2.65 Air=1
Vapor Pressure: > 1 atm
Water Solubility: Insoluble
Other Solubilities: Soluble in liquid ammonia, slightly soluble in hot HCl
Ionization Potential (eV): 11.34
Flash Point: Severe explosion hazard
Autoignition Temperature: Ignites spontaneously in air

RTECS Toxicity Data

Acute Oral: Mouse LD_{50} Dose: 1250 mg/kg; Toxic Effects: Behavioral - Excitment; Behavioral - Ataxia; Behavioral - Muscle contraction or spasticity.
Acute Inhalation: Mouse LC_{50} Dose: 1380 mg/m^3; Toxic Effects: Behavioral - Excitment; Behavioral - Ataxia; Behavioral - Muscle contraction or spasticity.

Hazard Overviews

Poison

Explosive

Flammable

Fire Diamond

Health: Severely irritating to eyes/skin/respiratory tract. Poison. Also Causes: pulmonary congestion and edema. Seizure activity may be present.
Fire: Pyrophoric. Allow germane fires to burn themselves out; do not attempt to extinguish them, because combustion reduces the possibility of explosions of unstable germane/air mixtures. Treat any fire that involves it as an emergency.
Reactivity: Stable, in closed, airtight, pressurized containers. Hazardous polymerization cannot occur. Avoid: sources of ignition (open flame; unprotected heaters; lighted tobacco products; electric sparks); excessive heat; mixing without checking chemical compatibility. Incompatible with: air. Hazardous decomposition products: germanium; hydrogen gas.
Carcinogenicity: IARC - Not listed; NIOSH - Listed as carcinogen; NTP - Not listed; ACGIH - Not listed; OSHA - Not listed; EPA - Not listed; MAK - Not listed
Primary Target Organs:

Eyes

Skin

Respiratory System

Liver

Kidneys

Blood

Exposure Limits
OSHA PEL Vacated 1989 Limits: TWA: 0.2 ppm; 0.6 mg/m^3.
ACGIH TLV: TWA: 0.2 ppm; 0.63 mg/m^3.
Respirator Recommendation
Exposure Range: >0.2 to 10 ppm Supplied Air, Constant Flow/Pressure Demand, Half Mask
Exposure Range: >10 to 200 ppm Supplied Air, Constant Flow/Pressure Demand, Full Face
Exposure Range: >200 to unlimited ppm Self-contained Breathing Apparatus, Pressure Demand, Full Face
Note: odor threshold unknown

Environmental

Cleanup/Disposal: Guide No. 119: Eliminate all ignition sources (no smoking, flares, sparks or flames in immediate area). All equipment used when handling the product must be grounded. Fully encapsulating, vapor protective clothing should be worn for spills and leaks with no fire. Do not touch or walk through spilled material. Stop leak if you can do it without risk. Do not direct water at spill or source of leak. Use water spray to reduce vapors or divert vapor cloud drift. For chlorosilanes, use AFFF alcohol-resistant medium expansion foam to reduce vapors. If possible, turn leaking containers so that gas escapes rather than liquid. Prevent entry into waterways, sewers, basements or confined areas. Isolate area until gas has dispersed.

Regulations

RCRA 40CFR: Not listed

CERCLA: 40CFR 302.4: Not listed
SARA 40CFR 372.65: Not listed
SARA EHS 40CFR 355: Not listed
TSCA: Listed

GER7000 CAS #: 7440-56-4

GERMANIUM

RTECS: LY5200000
EINECS Number: 231-164-3
Molecular Formula: Ge
Structured MF: Ge
Formula Weight: 72.59

Ge

Chemical Structure

Synonyms: GERMANIUM ELEMENT
Description: grayish-white, lustrous, brittle metal
Use: in electronics: manufacture of rectifying devices (germanium diodes), transistors; in dental alloys; in prodn of glass; catalyst; in electroplating; (vet): intestinal astringent

Physical Properties

Boiling Point: 2700 °C (4892 °F)
Freezing Point: 937.2 °C (1718.96 °F)
Specific Gravity: 5.323 at 25 °C/4 °C
Water Solubility: Insoluble in Water
Other Solubilities: Soluble in hot Sulfuric acid, Aqua Regia; Insoluble in alkali.
Refraction Index: High
Ionization Potential (eV): 7.900

RTECS Toxicity Data

Acute Oral: Man TD_{Lo} Dose: 58 mg/kg/26W-I; Toxic Effects: Kidney, Ureter, and Bladder - Chgs in tubules (inc acute renal failure, acute tubular necrosis. Man TD_{Lo} Dose: 786 mg/kg/2Y-I; Toxic Effects: Gastrointestinal - Nausea or vomiting; Kidney, Ureter, and Bladder - Chgs in tubules (inc acute renal failure, acute tubular necrosis; Blood - Oxidant related (GPD deficient) anemia. Child TD_{Lo} Dose: 2 gm/kg/78W-I; Toxic Effects: Gastrointestinal - Nausea or vomiting; Kidney, Ureter, and Bladder - Chgs in tubules (inc acute renal failure, acute tubular necrosis; Kidney, Ureter, and Bladder - Changes in both tubules.
Acute Inhalation: Rat LC_{Lo} Dose: 3860 mg/m^3/4hr; Toxic Effects: Lungs, Thorax, or Respiration - Other changes.

Hazard Overviews

Health: Irritating to eyes/skin/respiratory tract. Other Acute Effects: may be harmful by inhalation, ingestion, or skin absorption; exposure can cause damage to liver, kidneys, blood; effects coughing; chest pains; difficulty in breathing.
Fire: Hazards: in powder form capable of creating a dust explosion; emits toxic fumes. Extinguishing agents: dry chemical powder. Precautions: combustible liquid.
Reactivity: Incompatible with: strong oxidizing agents, halogens. Hazardous decomposition products: nature of decomposition products not known;.
Carcinogenicity: IARC - Not listed; NIOSH - Not listed; NTP - Not listed; ACGIH - Not listed; OSHA - Not listed; EPA - Not listed; MAK - Not listed
Primary Target Organs:

Eyes

Skin

Respiratory System

Environmental

Regulations

RCRA 40CFR: Not listed
CERCLA: 40CFR 302.4: Not listed
SARA 40CFR 372.65: Not listed
SARA EHS 40CFR 355: Not listed
TSCA: Listed

Analytical Methods

Water / Groundwater: USGS E-SPEC
Other: EPA 1620

GER9000 CAS #: 1310-53-8

GERMANIUM DIOXIDE

RTECS: LY5240000
EINECS Number: 215-180-8
Molecular Formula: GeO_2
Structured MF: GeO_2
Formula Weight: 104.59

O=Ge=O

Chemical Structure

Synonyms: GERMANIA; GERMANIC ACID; GERMANIC OXIDE (CRYSTALLINE); GERMANIUM OXIDE
Description: colorless, hexagonal crystals or white powder
Use: in special glass mixtures; phosphors; transistors & diodes; in production of metallic germanium; in electroplating

Physical Properties

Freezing Point: 1111 °C (2031.8 °F) to 1119 °C (2046.2 °F)
Specific Gravity: 4.228 at 25 °C
Water Solubility: 0.447 g in 100 ml Water at 25 °C
Other Solubilities: Soluble in acid & alkali; Insoluble in Hydrofluoric & Hydrochloric Acids.
Refraction Index: 1.650

RTECS Toxicity Data

Acute Oral: Rat LD_{50} Dose: 1250 mg/kg. Mouse LD_{50} Dose: 1250 mg/kg.
Acute Inhalation: Rat LC_{50} Dose: >1420 mg/m^3/4hr; Toxic Effects: Lungs, Thorax, or Respiration - Dyspnea; Skin and

appendages - Hair; Nutritional and gross metabolic - Weight loss or decreased weight gain.

Acute Dermal: Rabbit LD_{50} Route: Subcutaneous Dose: 845 mg/kg. Rat LD_{50} Route: Subcutaneous Dose: 1910 mg/kg.

Chronic (Multiple Dose) Oral: Rat Dose: 13650 mg/kg/13W-I; Toxic Effects: Blood - Changes in serum composition; Nutritional and gross metabolic - Weight loss or decreased weight gain; Biochemical - Transaminases. Rat Dose: 10 mg/kg/14W-I; Toxic Effects: DEATH.

Chronic (Multiple Dose) Inhalation: Rat Dose: 309 mg/m^3/6H/4W-I; Toxic Effects: Liver - Changes in Liver weight; Kidney, Ureter, and Bladder - Urine volume increased; Nutritional and gross metabolic - Weight loss or decreased weight gain.

Chronic (Multiple Dose) Dermal: Rat Route: Subcutaneous; Dose: 2450 mg/kg/35D-C; Toxic Effects: Liver - Changes in Liver weight; Kidney, Ureter, and Bladder - Urine volume increased; Blood - Changes in leukocyte (WBC) cell count. Rat Route: Subcutaneous; Dose: 1125 mg/kg/90D-C; Toxic Effects: Liver - Changes in Liver weight; Kidney, Ureter, and Bladder - Other changes in urine composition; Nutritional and gross metabolic - Weight loss or decreased weight gain.

Hazard Overviews

Health: Irritating to eyes/skin. Harmful. Other Acute Effects: harmful if swallowed; may be harmful if inhaled; may be harmful if absorbed through the skin.

Fire: Extinguishing agents: noncombustible; use extinguishing media appropriate to surrounding fire conditions. Precautions: combustible liquid.

Reactivity: Incompatible with: strong oxidizing agents. Hazardous decomposition products: nature of decomposition products not known;.

Carcinogenicity: IARC - Not listed; NIOSH - Not listed; NTP - Not listed; ACGIH - Not listed; OSHA - Not listed; EPA - Not listed; MAK - Not listed

Primary Target Organs:

Eyes

Skin

Environmental

Regulations

RCRA 40CFR: Not listed
CERCLA: 40CFR 302.4: Not listed
SARA 40CFR 372.65: Not listed
SARA EHS 40CFR 355: Not listed
TSCA: Listed

GIB5000 CAS #: 77-06-5

GIBBERELLIC ACID

RTECS: LY8990000
EINECS Number: 201-001-0

Molecular Formula: $C_{19}H_{22}O_6$
Formula Weight: 346.37

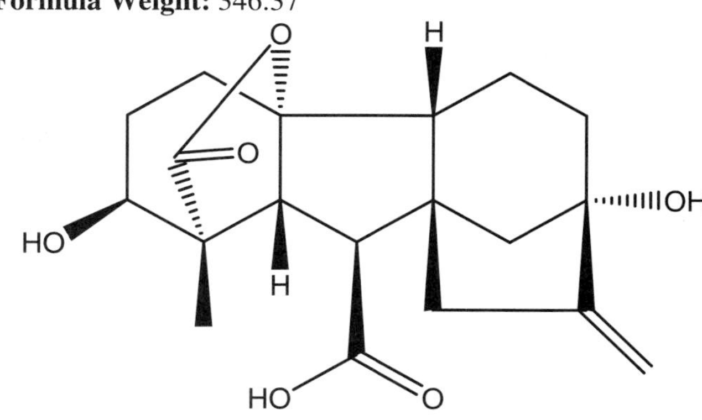

Chemical Structure

Synonyms: ACIDE GIBBERELLIQUE; ACTIVOL; ACTIVOL GA; BERELEX; BRELLIN; CEKU-GIB; CEKUGIB; (3S,3AS,4S,4AS,6S,8AS,8BS,11S)-6,11-DIHYDROXY-3-METHYL-12-METHYLENE-2-OXO-4A,6-ETHANO-3,8B-PROP-1-ENOPERHYDROINDENO[1,2-B]FURAN-4-CARBOXYLIC ACID; 3S,3AS,4S,4AS,6S,8AR,8BR,11S)-6,11-DIHYDROXY-3-METHYL-12-METHYLENE-2-OXO-4A,6-ETHANO-3,8B-PROP-L-ENOPERHYDROINDENO-(1,2-B)FURAN-4-CARBOXYLIC ACID; (3S,3AR,4S,4AS,7S,9AR,9BR,12S)-7,12-DIHYDROXY-3-METHYL-6-METHYLENE-2-OXOPERHYDRO-4A,7-METHANO-9B,3-PROPENOAZULENO[1,2-B]FURAN-4-CARBOXYLIC ACID; (3S,3AS,4S,4AS,7S,9AR,9BR,12S)-7,12-DIHYDROXY-3-METHYL-6-METHYLENE-2-OXOPERHYDRO-4A,7-METHANO-9B,3-PROPENO[1,2-B]FURAN-4-CARBOXYLIC ACID; GA; GA3; GIBB-3-ENE-1,10-DICARBOXYLIC ACID,2,4A,7-TRIHYDROXY-1-METHYL-8-METHYLENE-,1,4A-LACTONE; GIBB-3-ENE-1,10-DICARBOXYLIC ACID,2,4A,7-TRIHYDROXY-1-METHYL-8-METHYLENE-,1,4A-LACTONE,(1ALPHA,2BETA,4AALPHA,4BBETA,10BETA)-; GIBBERELLIC ACID GA3; GIBBERELLIN; GIBBERELLIN A3; GIBBERELLIN X; GIBBREL; GIBEFOL; GIBREL; GIBRESCOL; GIBRESKOL; GIB-SOL; GIB-TABS; GROCEL; LACTONE; PGR-IV; PRO-GIBB; PRO-GIBB PLUS; REGULEX; RYZUP; 2BETA,4ALPHA,7-TRIHYDROXY-1-METHYLENE-4AALPHA,4BBETA-GIBB-3-ENE-1ALPHA,10BETA-DICARBOXYLIC ACID1,4A-LACTONE; 2BETA,4A ALPHA,7-TRIHYDROXY-1BETA-METHYL-8-METHYLENE-4A ALPHA,4B BETA-GIBB-3-ENE-1ALPHA,10BETA-DICARBOXYLIC ACID 1-4A LACTONE; 2BETA,4ALPHA,7-TRIHYDROXY-1-METHYL-8-METHYLENE-4AALPHA,4BBETA-GIBB-3-ENE-1ALPHA,10BETA-DICARBOXYLIC ACID1,4A-LACTONE; 2,4A,7-TRIHYDROXY-1-METHYL-8-METHYLENEGIBB-3-ENE-1,10-CARBOXYLIC ACID 1,4-; 2,4A,7-TRIHYDROXY-1-METHYL-8-METHYLENEGIBB-3-ENE-1,10-CARBOXYLIC ACID 1-4-LACTONE; (1ALPHA,2BETA,4A ALPHA,4B BETA,10BETA)-2,4A,7-TRIHYDROXY-1-METHYL-8-METHYLENEGIBB-3-ENE-1,10-DICARBOXYLIC ACID 1,4A-LACTONE; 2,4A,7-TRIHYDROXY-1-METHYL-8-METHYLENEGIBB-3-ENE-1,10-DICARBOXYLIC ACID 1,4A-LACTONE

Description: white crystals, powder; odorless

Use: plant growth promoting hormone; malting of barley with improved enzymatic characteristics; food additive

Physical Properties

Freezing Point: 233 °C (451.4 °F) to 235 °C (455 °F)
Water Solubility: Slightly Soluble
Other Solubilities: 95% Ethanol: Soluble; Acetone: Soluble; Ether: Slightly Soluble; Ethyl Acetate: Soluble; Methanol: Soluble.
pH: Acid

Flash Point: Not available; probably combustible

RTECS Toxicity Data

Acute Oral: Rat LD_{50} Dose: 6300 mg/kg. Mouse LD_{50} Dose: 8500 mg/kg.
Acute Dermal: Rabbit LD_{50} Route: Skin; Dose: >2 gm/kg.
Mutagenic: Human Cytogenetic Analysis; Cell Type: lymphocyte; Dose: 3 mmol/L. Mammal DNA Damage; Cell Type: lymphocyte; Dose: 1 mmol/L.
Tumorigenic: Mouse Route: Oral; Dose: 142 gm/kg/78W-I; Toxic Effects: Tumorigenic - Equivocal tumorigenic agent by RTECS criteria; Lungs, Thorax, or Respiration - Tumors; Liver - Tumors. Toad Route: Oral; Dose: 13200 mg/kg/22W-I; Toxic Effects: Tumorigenic - Equivocal tumorigenic agent by RTECS criteria; Liver - Tumors.

Hazard Overviews

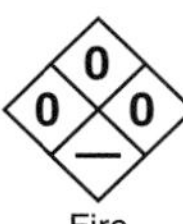

Fire Diamond

Health: Mildly irritating to eyes/skin. This material is relatively nonhazardous in routine industrial situations. It is not expected to present significant health risks to the workers who use it.
Fire: Will burn. Slight fire hazard when exposed to heat and sparks. Use water fog, dry chemical, alcohol foam, or carbon dioxide to fight fires involving gibberellic acid. Use a water spray to cool fire-exposed tanks or containers. Water or foam may cause frothing.
Reactivity: Stable. Hazardous polymerization cannot occur. Avoid: direct exposure to heat; sparks; open flame; lighted tobacco products. Incompatible with: strong oxidizers. Hazardous decomposition products: carbon monoxide;carbon dioxide.
Carcinogenicity: IARC - Not listed; NIOSH - Not listed; NTP - Not listed; ACGIH - Not listed; OSHA - Not listed; EPA - Not listed; MAK - Not listed
Primary Target Organs:

Eyes

Skin

Environmental

Environmental Physical Data

Octanol/Water Partition Coefficient: log K_{ow} = 0.24

Regulations

RCRA 40CFR: Not listed
CERCLA: 40CFR 302.4: Not listed
SARA 40CFR 372.65: Not listed
SARA EHS 40CFR 355: Not listed
TSCA: Listed

GIL5000 CAS #: 12002-43-6

GILSONITE

Molecular Formula: Unknown
Synonyms: UINTAHITE; UINTAITE
Description: black, lustrous masses or rich brown powder
Use: acid, alkali and waterproof coatings; black varnishes, lacquers, baking enamels and japans; wire-insulation compounds; linoleum and floor tile; paving; insulation; diluent in low-grade rubber compounds; a possible source of gasoline, fuel oil and metallurgical coke

Physical Properties

Freezing Point: 135 °C (275 °F) to 205 °C (401 °F)
Specific Gravity: 1.04 to 1.06 at 15/15 °C
Water Solubility: < 1 mg/mL at 21.5 C
Other Solubilities: 95% Ethanol: <1 mg/ml at 21.5 °C; Acetone: <1 mg/ml at 21.5 °C; DMSO: <1 mg/ml at 21.5 °C; Heavy lubricating petroleum: Soluble; Petroleum and chlorinated solvents: Soluble; Warm oil of turpentine: Soluble.
Flash Point: 315 °C

Hazard Overviews

Fire: Will burn.
Carcinogenicity: IARC - Not listed; NIOSH - Not listed; NTP - Not listed; ACGIH - Not listed; OSHA - Not listed; EPA - Not listed; MAK - Not listed

Environmental

Cleanup/Disposal: Guide No. 171: Do not touch or walk through spilled material. Stop leak if you can do it without risk. Prevent dust cloud. Avoid inhalation of asbestos dust. Small Dry Spills: With clean shovel place material into clean, dry container and cover loosely; move containers from spill area. Small Spills: Take up with sand or other noncombustible absorbent material and place into containers for later disposal. Large Spills: Dike far ahead of liquid spill for later disposal. Cover powder spill with plastic sheet or tarp to minimize spreading. Prevent entry into waterways, sewers, basements or confined areas.

Regulations

RCRA 40CFR: Not listed
CERCLA: 40CFR 302.4: Not listed
SARA 40CFR 372.65: Not listed
SARA EHS 40CFR 355: Not listed
TSCA: Not listed

GLE5000 CAS #: 3765-57-9

GLENBAR

RTECS: XN7525000
Molecular Formula: $C_{10}H_6Cl_4O_3S$

Formula Weight: 348.02
Synonyms: BENZOIC ACID,2,3,5,6-TETRACHLORO-4-((METHYLTHIO)CARBONYL)-,METHYLESTER; O,S-DIMETHYL TETRACHLOROTHIOTEREPHTHALATE; O,S-DIMETHYLESTER KYSELINY TETRACHLORTHIOTEREFTALOVE; O,S-DIMETHYLTETRACHLOROTHIOTEREPHTHALATE; GLENBAR; OCS-21,944; OCS 21944; TEREPHTHALIC ACID,TETRACHLOROTHIO-,O,S-DIMETHYL ESTER; TEREPHTHALIC ACID,TETRACHLOROTHIO-,O,S-DIMETHYL ESTER(8CI); THIOTEREPHTHALIC ACID,TETRACHLORO-,O,S-DIMETHYL ESTER
Description: white crystalline substance
Use: herbicide

Physical Properties

Water Solubility: Practically Insoluble in Water
Other Solubilities: Highly Soluble in most organic solvents

RTECS Toxicity Data

Acute Oral: Rat LD_{50} Dose: 3300 mg/kg.

Hazard Overviews

Carcinogenicity: IARC - Not listed; NIOSH - Not listed; NTP - Not listed; ACGIH - Not listed; OSHA - Not listed; EPA - Not listed; MAK - Not listed

Environmental

Regulations

RCRA 40CFR: Not listed
CERCLA: 40CFR 302.4: Not listed
SARA 40CFR 372.65: Not listed
SARA EHS 40CFR 355: Not listed
TSCA: Not listed

GLU1000 CAS #: 9007-92-5

GLUCAGON

RTECS: LZ3980000
EINECS Number: 232-708-2
Molecular Formula: $C_{153}H_{225}N_{43}O_{49}S$
Formula Weight: 3482.78
Synonyms: GLUKAGON NOVO; HIS-SER-GLU(NH2)-GLY-THR-PHE-THR-SER-ASP-TYR-SER-LYS-TYR-LEU-ASP-SER-ARG-ARG-ALA-GLU(NH2)-ASP-PHE-VAL-GLU(NH2)-TRP-LEU-MET-ASP(NH2)-THR
Description: white or faintly colored, fine crystalline powder; practically odorless
Use: treatment of insulin-induced hypoglycemia; in therapy of acute pancreatitis; potent stimulus of hepatocellular regeneration; ameliorated fulminant murine hepatitis; treatment for severe alcoholic hepatitis

Physical Properties

Water Solubility: Practically Insoluble in Water
Other Solubilities: Soluble in dilute alkali & acid solution; Insoluble in most organic solvents

RTECS Toxicity Data

Acute Dermal: Rat LD_{50} Route: Subcutaneous Dose: >200 mg/kg. Mouse LD_{50} Route: Subcutaneous Dose: >200 mg/kg.
Reproductive/Teratogenic: Rat Route: Subcutaneous; Dose: 3600 ug/kg; Duration: female 16-21D of pregnancy Effects on Embryo or Fetus - Fetotoxicity.
Mutagenic: Rat Unscheduled DNA Synthesis; Cell Type: liver; Dose: 75 mg/L.

Hazard Overviews

Health: May cause irritation to eyes/skin/respiratory tract. Other Acute Effects: may be harmful by inhalation, ingestion, or skin absorption; bioactive peptide; prolonged or repeated exposure may cause allergic reactions in certain sensitive individuals; exposure can cause nausea; headache; vomiting target organs: liver, heart, g.i. system, pituitary, adrenal, medulla. The chemical, physical and toxicological properties of this product have not been thoroughly investigated.
Fire: Extinguishing agents: carbon dioxide, dry chemical powder or appropriate foam. Precautions: combustible liquid.
Carcinogenicity: IARC - Not listed; NIOSH - Not listed; NTP - Not listed; ACGIH - Not listed; OSHA - Not listed; EPA - Not listed; MAK - Not listed
Primary Target Organs:

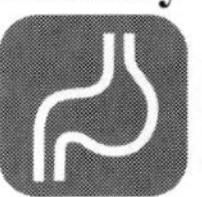
Gastro-intestinal

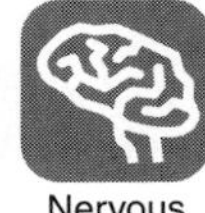
Nervous System

Liver

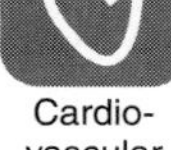
Cardio-vascular

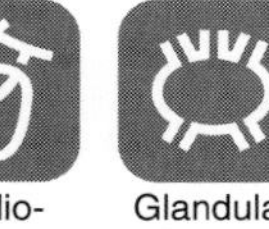
Glandular System

Environmental

Regulations

RCRA 40CFR: Not listed
CERCLA: 40CFR 302.4: Not listed
SARA 40CFR 372.65: Not listed
SARA EHS 40CFR 355: Not listed
TSCA: Not listed

GLU2140 CAS #: 526-95-4

GLUCONIC ACID

RTECS: LZ5057100
EINECS Number: 208-401-4
Molecular Formula: $C_6H_{12}O_7$
Structured MF: $C_5H_6(OH)_5COOH$
Formula Weight: 196.16

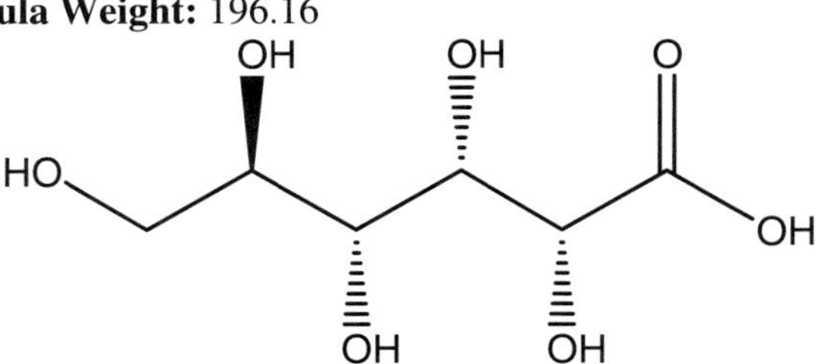

Chemical Structure

Synonyms: DEXTRONIC ACID; GLOSANTO; D-GLUCONIC ACID; GLUCONIC ACID,D-; GLYCOGENIC ACID; GLYCONIC ACID; MALTONIC ACID; PENTAHYDROXYCAPROIC ACID
Description: needles or crystals; faint odor of vinegar
Use: in dairy indust to prevent milkstone; in breweries to prevent beerstone; latent acid catalyst for acid colloid resins, particularly in textile printing component of many cleaning compounds; in pharmaceutical & food products; cleaning & pickling metals; sequestrant; paint strippers; alkaline derusters; chelating agent with utility in high alkalinity bottle washes and other cleansers; in the synthesis of water soluble cationic biguanidine skin antiseptics

Physical Properties

Freezing Point: 131 °C (267.8 °F)
Specific Gravity: 1.24 at 25 °C/4 °C
Water Solubility: Freely Soluble in Water
Other Solubilities: Slightly Soluble in Alcohol, Insoluble in Ether and most organic solvents

Hazard Overviews

Corrosive

Health: Corrosive to eyes/skin/respiratory tract. Harmful. Other Acute Effects: harmful if absorbed through skin; may be harmful if swallowed or inhaled; material is extremely destructive to tissue of the mucous membranes and upper respiratory tract, eyes and skin; inhalation may result in spasm, inflammation and edema of the larynx and bronchi, chemical pneumonitis and pulmonary edema; symptoms of exposure may include burning sensation, coughing, wheezing, laryngitis, shortness of breath, headache, nausea and vomiting; target organs: teeth, kidneys.
Fire: Hazards: emits toxic fumes. Extinguishing agents: carbon dioxide, dry chemical powder or appropriate foam. Precautions: combustible liquid.
Reactivity: Stable. Hazardous polymerization will not occur. Incompatible with: strong oxidizing agents. Hazardous decomposition products: toxic fumes of: carbon monoxide, carbon dioxide.
Carcinogenicity: IARC - Not listed; NIOSH - Not listed; NTP - Not listed; ACGIH - Not listed; OSHA - Not listed; EPA - Not listed; MAK - Not listed
Primary Target Organs:

Eyes

Skin
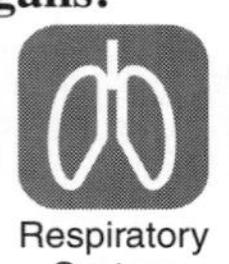
Respiratory System

Kidneys
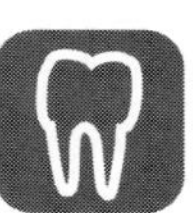
Teeth

Environmental

Environmental Physical Data

Octanol/Water Partition Coefficient: log K_{ow} = calculated at -2.57
BOD: activated > 1 lb/lb, 5 days

Regulations

RCRA 40CFR: Not listed
CERCLA: 40CFR 302.4: Not listed
SARA 40CFR 372.65: Not listed
SARA EHS 40CFR 355: Not listed
TSCA: Listed

Analytical Methods

Soil: DOE OM500R

GLU3280 CAS #: 90-80-2

GLUCONOLACTONE

RTECS: LZ5184000
EINECS Number: 202-016-5
Molecular Formula: $C_6H_{10}O_6$
Formula Weight: 178.14

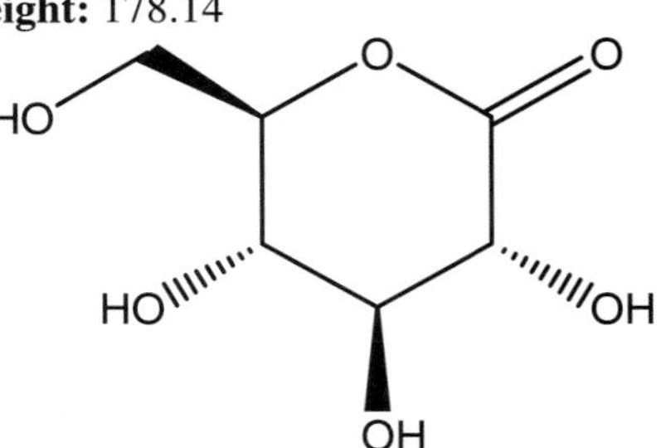

Chemical Structure

Synonyms: DELTAGLUCONOLACTONE; D-GLUCONIC ACID DELTA-LACTONE; D-GLUCONIC ACID LACTONE; GLUCONIC ACID LACTONE; D-GLUCONIC DELTA-LACTONE; GLUCONO DELTA LACTONE; D-DELTA-GLUCONOLACTONE; D-GLUCONO-1,5-LACTONE; DELTA-D-GLUCONOLACTONE; DELTA-GLUCONOLACTONE; GLUCONO-DELTA-LACTONE
Description: white crystals or needles
Use: in cleaning compounds, dairy indust, in breweries & as latent acid catalyst for acid colloid resins; acid; leavening agent; sequestering agent; food additive

Physical Properties

Boiling Point: Decomposes at 153 °C (307 °F)
Freezing Point: 153 °C (307.4 °F)
Specific Gravity: 1.61 at -5 °C
Water Solubility: 59 g Soluble in 100 ml Water
Other Solubilities: 1 g Soluble in 100 g Alcohol; Insoluble in Ether, Acetone
pH: Freshly prepared 1% aqueous solution 3.6

RTECS Toxicity Data

Mutagenic: Other Microorganisms Other Mutation Test Systems; Dose: 5 mmol/L.

Hazard Overviews

Health: May be irritating to eyes/skin/respiratory tract. Other Acute Effects: may be harmful by inhalation, ingestion, or skin absorption.

Fire: Extinguishing agents: water spray; carbon dioxide, dry chemical powder or appropriate foam. Precautions: combustible liquid.

Carcinogenicity: IARC - Not listed; NIOSH - Not listed; NTP - Not listed; ACGIH - Not listed; OSHA - Not listed; EPA - Not listed; MAK - Not listed

Environmental

Regulations

RCRA 40CFR: Not listed
CERCLA: 40CFR 302.4: Not listed
SARA 40CFR 372.65: Not listed
SARA EHS 40CFR 355: Not listed
TSCA: Listed

GLU4420 **CAS #: 50-99-7**

GLUCOSE

RTECS: LZ6600000
EINECS Number: 200-075-1
Molecular Formula: $C_6H_{12}O_6$
Formula Weight: 180.16

Chemical Structure

Synonyms: ANHYDROUS DEXTROSE; BLOOD SUGAR; CARTOSE; CERELOSE; CORN SUGAR; DEXTROPUR; DEXTROSE; DEXTROSE,ANHYDROUS; DEXTROSOL; GLUCOLIN; D(+)-GLUCOSE; D-GLUCOSE; GLUCOSE LIQUID; D-GLUCOSE,ANHYDROUS; GLUCOSE,ANHYDROUS; GRAPE SUGAR; SIRUP; SUGAR,GRAPE; TRAUBENZUCKER

Description: crystals

Use: fluid & nutrient (carbohydrate) replenisher (eg, in in pharmaceuticals); confectionary; infant foods; brewing & wine-making; intermed; caramel coloring; baking & canning; instead of lactose, in milk for infants; diluent in pilular extracts; for cocoa syrup; liq glucose; to produce methane gas; reducing agent; synthesis of amino acids; to produce sorbitol & mannitol, & oxalic acid form glucose; med:(vet) for nutrition, hypoglycemia, ketosis, to conteract hepatotoxins.

Physical Properties

Freezing Point: 146 °C (294.8 °F)
Specific Gravity: 1.56
Water Solubility: Soluble in Water
Other Solubilities: 1 g in 120 ml Ethanol at 20 °C; Very Sparingly Soluble in Absolute Alcohol, Ether, Acetone; Soluble in hot Glacial Acetic Acid, Pyridine.
pH: 0.5 molar aqueous solution 5.9
Refraction Index: 10% soluble 1.3479 at 20 °C

RTECS Toxicity Data

Acute Oral: Rat LD_{50} Dose: 25800 mg/kg; Toxic Effects: Behavioral - Coma; Lungs, Thorax, or Respiration - Cyanosis; Gastrointestinal - Hypermotility, diarrhea. Rabbit LD_{Lo} Dose: 20 gm/kg.

Reproductive/Teratogenic: Woman Route: Oral; Dose: 2 gm/kg; Duration: female 28W of pregnancy; Specific Developmental Abnormalities - Craniofacial (including nose and tongue); Other developmental abnormalities. Woman Route: Intravenous; Dose: 2 gm/kg; Duration: female 39W of pregnancy; Maternal Effects - Other effects on females; Effects on Embryo or Fetus - Other effects to embryo or fetus. Woman Route: Intravenous; Dose: 1057 ug/kg; Duration: female 39W of pregnancy; Specific Developmental Abnormalities - Hepatobiliary system. Woman Route: Intravenous; Dose: 1300 mg/kg; Duration: female 39W of pregnancy; Effects on Newborn - Biochemical and metabolic; Behavioral.

Mutagenic: Human DNA Damage; Cell Type: other cell types; Dose: 30 mmol/L. Mouse Mutations in Mammalian Somatic Cells; Cell Type: lymphocyte; Dose: 179 mmol/L.

Tumorigenic: Rat Route: Subcutaneous; Dose: 15400 gm/kg/22W-C; Toxic Effects: Tumorigenic - Equivocal tumorigenic agent by RTECS criteria; Tumorigenic - Tumors at site of application.

Hazard Overviews

Health: May be irritating to eyes/skin/respiratory tract. Other Acute Effects: may be harmful by inhalation, ingestion, or skin absorption.

Fire: Hazards: emits toxic fumes. Extinguishing agents: water spray; carbon dioxide, dry chemical powder or appropriate foam. Precautions: combustible liquid.

Reactivity: Stable. Hazardous polymerization will not occur. Incompatible with: strong oxidizing agents. Hazardous decomposition products: toxic fumes of: carbon monoxide, carbon dioxide.

Carcinogenicity: IARC - Not listed; NIOSH - Not listed; NTP - Not listed; ACGIH - Not listed; OSHA - Not listed; EPA - Not listed; MAK - Not listed

Environmental

Environmental Physical Data

BOD: sewage seed .5 to 0.78 lb/lb, 5 days

Regulations

RCRA 40CFR: Not listed
CERCLA: 40CFR 302.4: Not listed
SARA 40CFR 372.65: Not listed
SARA EHS 40CFR 355: Not listed
TSCA: Listed

GLU5560 CAS #: 138-15-8

GLUTAMIC ACID HYDROCHLORIDE

EINECS Number: 205-315-9
Molecular Formula: $C_5H_{10}ClNO_4$
Formula Weight: 183.59

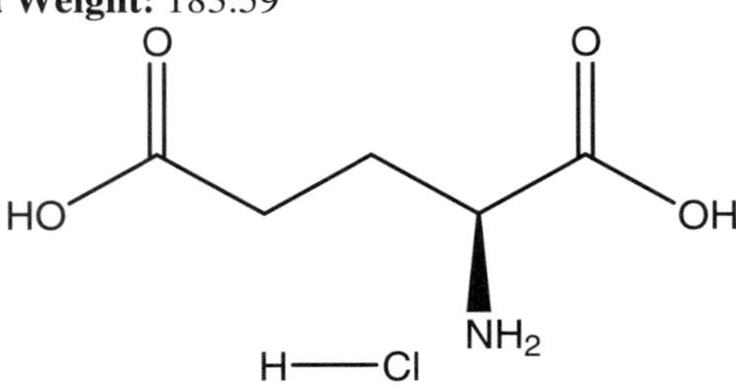

Chemical Structure

Synonyms: ACIDALIN; ACIDOGEN; ACIDORIDE; ACIDOTHYN; ACIDULEN; ACIDULIN; ACIGLUMIN; ACLOR; ACRIDOGEN; ACRIDORIDE; 2-AMINOPENTANEDIOIC ACID HYDROCHLORIDE; ANTALKA; FLAMITHIN; GASTULORIC; GLUSATIN; L-GLUTAMIC ACID HYDROCHLORIDE; GLUTAMIC ACID HYDROGEN CHLORIDE; L-GLUTAMIC ACID MONOHYDROCHLORIDE; GLUTAMIC ACID,HYDROCHLORIDE,L-; L-GLUTAMIC ACID,HYDROCHLORIDE; GLUTAMIDIN; GLUTAN HCL; GLUTAN HYDROCHLORIC; GLUTAN HYDROCHLORIDE; GLUTASIN; HYDRIONIC; HYPOCHYLIN; MURIAMIC; PEPSDOL; PEPSIDOL
Description: white crystalline powder or plates
Use: to improve taste of beer; chem int for glutamic acid and monosodium glutamate; medicinal chem int for glutamic acid and monosodium glutamate; medicinal medication: to treat achlorhydria; gastro-intestinal agent in human & vet medicine; salt substitute; flavoring agent

Physical Properties

Boiling Point: Decomposes at 214 °C (417 °F)
Freezing Point: Decomposes at 214 °C (417.2 °F)
Specific Gravity: 1.525
Water Solubility: 1 g in about 3 ml Water
Other Solubilities: Soluble in Ethanol; almost Insoluble in Ether.
pH: Solution is acid
Refraction Index: Alpha 1.546

Hazard Overviews

Health: May cause irritation to eyes/skin. Other Acute Effects: may be harmful by inhalation, ingestion, or skin absorption.
Fire: Hazards: emits toxic fumes. Extinguishing agents: water spray; carbon dioxide, dry chemical powder or appropriate foam. Precautions: combustible liquid.
Reactivity: Incompatible with: strong oxidizing agents. Hazardous decomposition products: toxic fumes of: carbon monoxide, carbon dioxide, nitrogen oxides, hydrogen chloride gas.
Carcinogenicity: IARC - Not listed; NIOSH - Not listed; NTP - Not listed; ACGIH - Not listed; OSHA - Not listed; EPA - Not listed; MAK - Not listed

Environmental

Regulations

RCRA 40CFR: Not listed
CERCLA: 40CFR 302.4: Not listed
SARA 40CFR 372.65: Not listed
SARA EHS 40CFR 355: Not listed
TSCA: Listed

GLU6700 CAS #: 111-30-8

GLUTARALDEHYDE

RTECS: MA2450000
EINECS Number: 203-856-5
Molecular Formula: $C_5H_8O_2$
Structured MF: $HCO(CH_2)_3CHO$
Formula Weight: 100.13

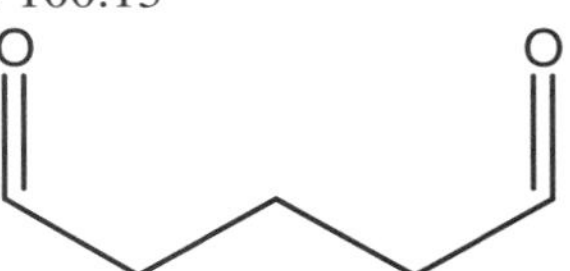

Chemical Structure

Synonyms: ALDEHYD GLUTAROWY; ALDESEN; CIDEX; 1,3-DIFORMYLPROPANE; GLUTARAL; GLUTARALDEHYD; GLUTARDIALDEHYDE; GLUTARIC ACID DIALDEHYDE; GLUTARIC ALDEHYDE; GLUTARIC DIALDEHYDE; HOSPEX; 1,5-PENTANEDIAL; PENTANEDIAL; 1,5-PENTANEDIONE; POTENTIATED ACID GLUTARALDEHYDE; SONACIDE
Description: pale yellow liquid; pungent, rotten apple odor
Use: disinfectant, as an intermediate in classical synthesis of pseudopelleterine, as a tanning agent in leather, and in the sterilization of endoscopic instruments, thermometers, rubber or plastic equipment which cannot be heat sterilized

Physical Properties

Boiling Point: Decomposes at 187 °C (369 °F) to 189 °C (372 °F)
Freezing Point: -14 °C (6.8 °F)
Specific Gravity: 0.72
Vapor Density: 3.5 Air=1
Saturated Vapor Density: 1.265837205 kg/m^3
Vapor Pressure: 17 mm Hg at 20 °C
Water Solubility: Soluble in all Proportions
Other Solubilities: 95% Ethanol: >=100 mg/ml at 22 °C; Acetone: >=100 mg/ml at 22 °C; Alcohol: Soluble; Benzene: Soluble; DMSO: >=100 mg/ml at 22 °C.
Surface Tension: Estimated at < 80 dynes/cm
Odor Threshold: 0.04 ppm
Refraction Index: 1.4338
Flash Point: Nonflammable solution

RTECS Toxicity Data

Acute Oral: Rat LD_{50} Dose: 134 mg/kg. Mouse LD_{50} Dose: 100 mg/kg.

Acute Inhalation: Rat LC_{50} Dose: 480 mg/m^3/4hr.
Acute Dermal: Rabbit LD_{50} Route: Skin; Dose: 560 uL/kg. Rat LD_{50} Route: Skin; Dose: >2500 mg/kg; Toxic Effects: Skin and appendages - Dermatitis, other. Rat LD_{50} Route: Subcutaneous Dose: 2390 mg/kg.
Chronic (Multiple Dose) Oral: Rat Dose: 12376 mg/kg/2Y-C; Toxic Effects: Kidney, Ureter, and Bladder - Other changes in urine composition; Kidney, Ureter, and Bladder - Changes in kidney weight; Nutritional and gross metabolic - Weight loss or decreased weight gain. Rat Dose: 54600 ug/kg/26W-I; Toxic Effects: Liver - Fatty liver degeneration; Nutritional and gross metabolic - Weight loss or decreased weight gain; Biochemical - True cholinesterase.
Chronic (Multiple Dose) Inhalation: Rat Dose: 5 ppm/6H/2W-I; Toxic Effects: Sense organs and special senses - Other; Lungs, Thorax, or Respiration - Other changes; DEATH. Rat Dose: 1000 ppb/6H/13W-I; Toxic Effects: Sense organs and special senses - Other; Nutritional and gross metabolic - Weight loss or decreased weight gain.
Chronic (Multiple Dose) Dermal: Rat Route: Subcutaneous; Dose: 175 mg/kg/35D-C; Toxic Effects: Liver - Changes in Liver weight; Endocrine - Changes in spleen weight; Blood - Normocytic anemia.
Irritation Eye: Rabbit Standard Draize Test Dose: 1 mg; Reaction: severe. Rabbit Standard Draize Test Dose: 250 ug/24H; Reaction: severe.
Irritation Skin: Human Standard Draize Test Dose: 6 mg/3D-I; Reaction: severe. Rabbit Standard Draize Test Dose: 2 mg/24H; Reaction: severe.
Reproductive/Teratogenic: Rat Route: Oral; Dose: 875 mg/kg; Duration: male 35D prior to mating; Paternal Effects - Testes, epididymis, sperm duct; Prostate, seminal vessicle, Cowper's gland, accessory glands. Rat Route: Oral; Dose: 4370 mg/kg; Duration: female 35D prior to mating Maternal Effects - Uterus, cervix, vagina. Rat Route: Oral; Dose: 1 gm/kg; Duration: female 6-15D of pregnancy; Effects on Embryo or Fetus - Fetotoxicity.
Mutagenic: Human DNA Damage; Cell Type: lymphocyte; Dose: 10 umol/L. Mouse Mutations in Mammalian Somatic Cells; Cell Type: lymphocyte; Dose: 8 mg/L.

Hazard Overviews

Fire Diamond

Health: Severely irritating to eyes/skin/respiratory tract. Toxic. Also Causes: headache, nosebleed, chest tightness, nausea, vomiting. Chronic Effect: dermatitis, asthma.
Fire: Noncombustible. Polymerizes when heated. Use extinguishing agents suitable for surrounding fire.
Reactivity: Stable if buffered to pH 7.5 to 8.5. Hazardous polymerization can occur with subsequent precipitation in alkaline solutions (above pH 9) or when heated. Avoid: heat; storage with alkalies. Incompatible with: heat when in contact with alkalies. Hazardous decomposition products: acrid smoke; irritating fumes.

Carcinogenicity: IARC - Not listed; NIOSH - Listed as carcinogen; NTP - Not listed; ACGIH - Not listed; OSHA - Not listed; EPA - Not listed; MAK - Not listed
Primary Target Organs:

Eyes

Skin

Respiratory System

Exposure Limits
OSHA PEL Vacated 1989 Limits: STEL: 0.2 ppm; 0.8 mg/m^3; Ceiling.
ACGIH TLV: STEL: 0.2 ppm; 0.82 mg/m^3; Ceiling.
NIOSH REL: STEL: 0.2 ppm; 0.8 mg/m^3.
DFG MAK: TWA: 0.1 ppm; 0.4 mg/m^3.
Respirator Recommendation
Exposure Range: >0.2 to 2 ppm Air Purifying, Negative Pressure, Half Mask
Exposure Range: >2 to 20 ppm Air Purifying, Negative Pressure, Full Face
Exposure Range: >20 to 200 ppm Supplied Air, Constant Flow/Pressure Demand, Full Face
Exposure Range: >200 to unlimited ppm Self-contained Breathing Apparatus, Pressure Demand, Full Face
Cartridge Color: black with dust/mist prefilter (use P100 or consult supervisor for appropriate dust/mist prefilter)

Environmental

Cleanup/Disposal: Guide No. 153: Eliminate all ignition sources (no smoking, flares, sparks or flames in immediate area). Do not touch damaged containers or spilled material unless wearing appropriate protective clothing. Stop leak if you can do it without risk. Prevent entry into waterways, sewers, basements or confined areas. Absorb or cover with dry earth, sand or other non-combustible material and transfer to containers. Do not get water inside containers.

Environmental Physical Data

BCF: no food chain concentration potential

Regulations

RCRA 40CFR: Not listed
CERCLA: 40CFR 302.4: Not listed
SARA 40CFR 372.65: Not listed
SARA EHS 40CFR 355: Not listed
TSCA: Listed

Analytical Methods

Indoor / Expired Air: NIOSH 2532

GLU7840 **CAS #: 110-94-1**

GLUTARIC ACID

RTECS: MA3740000
EINECS Number: 203-817-2
Molecular Formula: $C_5H_8O_4$
Structured MF: $HO_2C(CH_2)_3CO_2H$
Formula Weight: 132.13

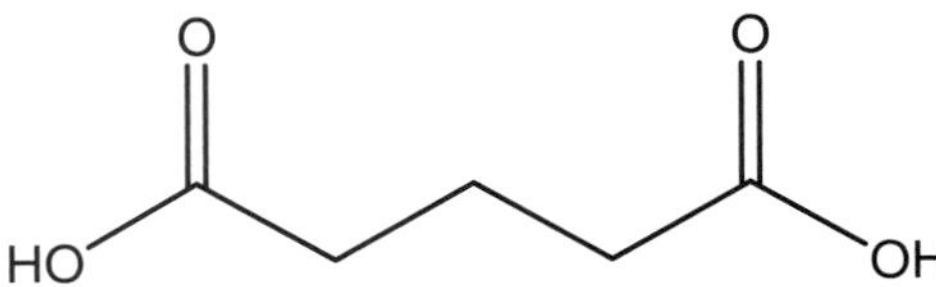

Chemical Structure

Synonyms: PENTANDIOIC ACID; 1,5-PENTANEDIOIC ACID; PENTANEDIOIC ACID; 1,3-PROPANEDICARBOXYLIC ACID; N-PYROTARTARIC ACID
Description: colorless crystals or large, monoclinic prisms
Use: in organic synthesis and biochemical research

Physical Properties

Boiling Point: 200 °C (392 °F) at 20 mm Hg
Freezing Point: 97.5 °C (207.5 °F) to 98 °C (208.4 °F)
Specific Gravity: 1.429 at 15 °C/4 °C
Vapor Pressure: 1 mm Hg at 155.5 °C
Water Solubility: 429 g/L at 0 °C
Other Solubilities: 95% Ethanol: >=100 mg/ml at 21 °C; Acetone: >=100 mg/ml at 21 °C; Alcohol: Soluble; Benzene: Soluble; Chloroform: Soluble; DMSO: >=100 mg/ml at 21 °C; Ether: Soluble; Petroleum Ether: Slightly Soluble.
Refraction Index: 1.41878 at 106 °C/D
Flash Point: Not available; probably combustible

RTECS Toxicity Data

Acute Oral: Mouse LD_{50} Dose: 6 gm/kg; Toxic Effects: Tumorigenic - Active as anti-cancer agent.

Hazard Overviews

Health: Irritating to eyes/skin. Other Acute Effects: may be harmful by inhalation, ingestion, or skin absorption.
Fire: Will burn. Hazards: in powder form capable of creating a dust explosion. Extinguishing agents: carbon dioxide, dry chemical powder or appropriate foam; water spray. Precautions: combustible liquid.
Reactivity: Incompatible with: bases, oxidizing agents, reducing agents. Hazardous decomposition products: toxic fumes of: carbon monoxide, carbon dioxide.
Carcinogenicity: IARC - Not listed; NIOSH - Not listed; NTP - Not listed; ACGIH - Not listed; OSHA - Not listed; EPA - Not listed; MAK - Not listed
Primary Target Organs:

Eyes

Skin

Environmental

Ecotoxicity: Algae: Scenedesmus no effect 1 g/l; Fish: L. macrochirus 24h LC_{50} 330 mg/l

Environmental Physical Data

Octanol/Water Partition Coefficient: log K_{ow} = -0.47 to -0.88

Regulations

RCRA 40CFR: Not listed
CERCLA: 40CFR 302.4: Not listed
SARA 40CFR 372.65: Not listed
SARA EHS 40CFR 355: Not listed
TSCA: Listed

GLU8980 CAS #: 77-21-4

GLUTETHIMIOE

RTECS: MA4725000
EINECS Number: 201-012-0
Molecular Formula: $C_{13}H_{15}NO_2$
Formula Weight: 217.26
Synonyms: ALFIMID; CC 11511; DORIDEN; DORIDEN-SED; ELRODORM; 3-ETHYL-3-PHENYL-2,6-DIKETOPIPERIDINE; 3-ETHYL-3-PHENYL-2,6-DIOXOPIPERIDINE; 2-ETHYL-2-PHENYLGLUTARIMIDE; ALPHA-ETHYL-ALPHA-PHENYLGLUTARIMIDE; 3-ETHYL-3-PHENYL-2,6-PIPERIDINEDIONE; GIMID; GLIMID; GLUTARIMIDE,2-ETHYL-2-PHENYL; GLUTATHIMID; GLUTETHIMID; GLUTETHIMIDE; GLUTETIMID; GLUTETIMIDE; GLUTETIMIDU; GLUTHETIMIDE; NOXIRON; NOXYRON; ONDASIL; PHENYL-AETHYL-GLUTARSAEUREIMID; 3-PHENYL-3-ETHYL-2,6-DIKETOPIPERIDINE; 3-PHENYL-3-ETHYL-2,6-DIOXOPIPERIDINE; 2-PHENYL-2-ETHYLGLUTARIC ACID IMIDE; ALPHA-PHENYL-ALPHA-ETHYLGLUTARIC ACID IMIDE; ALPHA-PHENYL-ALPHA-ETHYLGLUTARIMIDE; 2,6-PIPERIDINEDIONE,3-ETHYL-3-PHENYL; RIGENOX; SARODORMIN
Description: white crystalline powder
Use: medication: sedative, hypnotic

Physical Properties

Freezing Point: Crystals at 102.5 °C (216.5 °F) to 103 °C (217.4 °F)
Water Solubility: Practically Insoluble in Water
Other Solubilities: freely Soluble in Ethyl Acetate, Acetone, Ether, Chloroform; Soluble in Ethanol, Methanol.
pH: Saturated solution is slightly acid

RTECS Toxicity Data

Acute Oral: Human TD_{Lo} Dose: 171 mg/kg; Toxic Effects: Sense organs and special senses - Mydriasis (pupilliary dilation); Behavioral - Coma; Vascular - BP lowering not characterized in autonomic section. Child TD_{Lo} Dose: 25 mg/kg; Toxic Effects: Behavioral - Ataxia; Behavioral - Somnolence (general depressed activity). Monkey LD_{Lo} Dose: 300 mg/kg; Toxic Effects: Behavioral - General anesthetic; Behavioral - Muscle weakness. Rat LD_{50} Dose: 600 mg/kg.
Reproductive/Teratogenic: Rat Route: Oral; Dose: 5600 mg/kg; Duration: male 3D prior to mating; Specific Developmental Abnormalities - Musculoskeletal system; Effects on Newborn - Viability index. Rat Route: Oral; Dose: 2250 mg/kg; Duration: female 6-14D of pregnancy; Effects on Fertility - Abortion.

Hazard Overviews

Carcinogenicity: IARC - Not listed; NIOSH - Not listed; NTP - Not listed; ACGIH - Not listed; OSHA - Not listed; EPA - Not listed; MAK - Not listed

Environmental

Regulations

RCRA 40CFR: Not listed
CERCLA: 40CFR 302.4: Not listed
SARA 40CFR 372.65: Not listed
SARA EHS 40CFR 355: Not listed
TSCA: Not listed

GLY1000 CAS #: 56-81-5

GLYCERIN

RTECS: MA8050000
EINECS Number: 200-289-5
Molecular Formula: $C_3H_8O_3$
Structured MF: $HOCH_2CHOHCH_2OH$
Formula Weight: 92.09

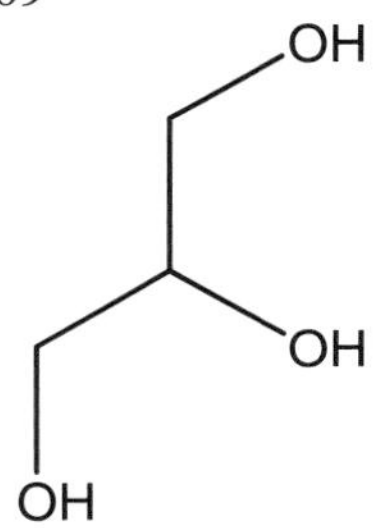

Chemical Structure

Synonyms: 90 TECHNICAL GLYCERINE; CITIFLUOR AF 2; CLYZERIN,WASSERFREI; GLYCERIN (ANHYDROUS); GLYCERIN MIST; GLYCERIN,ANHYDROUS; GLYCERINE; GLYCERIN,SYNTHETIC; GLYCERITOL; GLYCEROL; GLYCYL ALCOHOL; GLYCYLALCOHOL; GLYROL; GLYSANIN; GROCOLENE; IFP; INCORPORATION FACTOR; MOON; OPHTHALGAN; OSMOGLYN; 1,2,3-PROPANETRIOL; PROPANETRIOL; STAR; SYNTHETIC GLYCERIN; SYNTHETIC GLYCERINE; 1,2,3-TRIHYDROXYPROPANE; TRIHYDROXYPROPANE
Description: colorless oily liquid; odorless
Use: as a food additive, ingredient of some linctuses a pastilles, sweetening agent, emollient, with dried magnesium sulphate used in the treatment of septic wounds and boils, lubricating gastroscopes, preservative in some pharmaceutical preparations and in certain biological preparation and in non-alcoholic extracts and tinctures; in plasticizer manufacturing, confectionery, dynamite, nitroglycerin, antifreeze, antibiotics, rollers and hectographs, solvent, cosmetics, humectant, liquid soaps, liqueur blacking, printing and copying inks, elastic glues, lead oxide cements, to keep fabrics pliable, to preserve printing on cotton, hectographs, to keep frost from windshields, in shock absorber fluids, in fermentation of nutrients, demulcent, an oral osmotic diuretic to manage cerebral edema, reduce cerebrospinal fluid pressure and lower intraocular pressure, raw material for alkyd resins, cellophane, ester gums, perfumery, bacteriostat, penetrant, polyurethane polyols, emulsifying agent, rubber stamp inks, laxative, reducing intracranial pressure and promotes peristalsis evacuation of the lower bowel; in veterinary medicine in the treatment of bovine ketosis, pregnancy toxemia in sheep, pharmaceutical aid (humectant, solvent) emollient an emulcent

Physical Properties

Boiling Point: 290 °C (554 °F) at 760 mm Hg
Freezing Point: 20 °C (68 °F)
Specific Gravity: 1.2613 at 20 °C/4 °C
Vapor Density: 3.17 Air=1
Saturated Vapor Density: 1.200000543 kg/m^3
Density: 1.261 g/mL at 20 °C
Vapor Pressure: 1.58×10^{-4} mm Hg at 25 °C
Water Solubility: Soluble in all Proportions
Other Solubilities: 1 part dissolves in 11 parts Ethyl Acetate, in approximately 500 parts Ethyl Ether. Insoluble in Benzene, Chloroform, Carbon Tetrachloride, Carbon Disulfide, Petroleum Ether; solubility in Acetone: 1 g in 15 ml; Insoluble in fixed and volatile oils.
pH: Neutral
Refraction Index: 1.4730 at 25 °C/D
Flash Point: 177 °C Open Cup
Autoignition Temperature: 393 °C
LEL: 0.9% v/v

RTECS Toxicity Data

Acute Oral: Human TD_{Lo} Dose: 1428 mg/kg; Toxic Effects: Behavioral - Headache; Gastrointestinal - Nausea or vomiting. Rat LD_{50} Dose: 12600 mg/kg; Toxic Effects: Behavioral - General anesthetic; Behavioral - Muscle weakness; Liver - Other changes.
Acute Inhalation: Rat LC_{50} Dose: >570 mg/m^3/1hr.
Acute Dermal: Rabbit LD_{50} Route: Skin; Dose: >10 gm/kg. Rat LD_{50} Route: Subcutaneous Dose: 100 mg/kg.
Chronic (Multiple Dose) Oral: Rat Dose: 16800 mg/kg/28D-C; Toxic Effects: Endocrine - Changes in adrenal weight. Rat Dose: 96 gm/kg/30D-I; Toxic Effects: Blood - Changes in leukocyte (WBC) cell count; Blood - Changes in serum composition; Biochemical - True cholinesterase.
Irritation Eye: Rabbit Standard Draize Test Dose: 126 mg; Reaction: mild. Rabbit Standard Draize Test Dose: 500 mg/24H; Reaction: mild.
Irritation Skin: Rabbit Standard Draize Test Dose: 500 mg/24H; Reaction: mild.
Reproductive/Teratogenic: Rat Route: Oral; Dose: 100 mg/kg; Duration: male 1D prior to mating; Effects on Fertility - Post-implantation mortality. Monkey Route: Intratesticular; Dose: 119 mg/kg; Duration: male 1D prior to mating; Paternal Effects - Spermatogenesis; Testes, epididymis, sperm duct.
Mutagenic: Human DNA Inhibition; Cell Type: lymphocyte; Dose: 200 mmol/L. Rat Cytogenetic Analysis; Route: Oral; Dose: 1 gm/kg.

Hazard Overviews

Fire Diamond

Health: Irritating to eyes/skin/respiratory tract. Chronic Effects: kidney damage.

Fire: Will burn. Gives off toxic and highly irritating decomposition products. Use water as fog, dry chemical, carbon dioxide, or alcohol-resistant foam. Water or foam may cause frothing.

Reactivity: Stable. Hazardous polymerization cannot occur. Avoid: heat and ignition sources. Incompatible with: phosphorus triiodide; sodium hydride; sodium tetrahydroborate; strong oxidizers (potassium permanganate; chromium trioxide). Hazardous decomposition products: carbon oxides.

Carcinogenicity: IARC - Not listed; NIOSH - Listed as carcinogen; NTP - Not listed; ACGIH - Not listed; OSHA - Not listed; EPA - Not listed; MAK - Not listed

Primary Target Organs:

Eyes

Skin

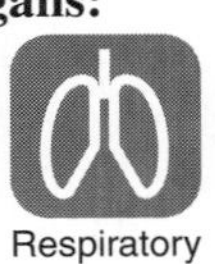
Respiratory System

Kidneys

Exposure Limits

OSHA PEL: TWA: 15 mg/m^3; total dust.

OSHA PEL Vacated 1989 Limits: TWA: 10 mg/m^3. Other Values: respirable mg/m^3; 5.

ACGIH TLV: TWA: 10 ppm.

Respirator Recommendation

Exposure Range: >5 to 50 mg/m^3 Air Purifying, Negative Pressure, Half Mask

Exposure Range: >50 to 500 mg/m^3 Air Purifying, Negative Pressure, Full Face

Exposure Range: >500 to 5000 mg/m^3 Supplied Air, Constant Flow/Pressure Demand, Full Face

Exposure Range: >5000 to unlimited mg/m^3 Self-contained Breathing Apparatus, Pressure Demand, Full Face

Cartridge Color: dust/mist filter (use P100 or consult supervisor for appropriate dust/mist filter)

Environmental

Ecotoxicity: Toxicity threshold (cell multiplication inhibition test) Algae (Microcystis aeruginosa) 2900 mg/l Toxicity threshold (cell multiplication inhibition test) Protozoa (Entosiphon sulcatum) 3200 mg/l

Environmental Fate: If released to soil, it is expected to undergo rapid biodegradation under aerobic conditions. It is expected to display very high mobility in soil and it is not expected to significantly volatilize to the atmosphere. If released to water, it is expected to rapidly degrade under aerobic conditions. Biodegradation in seawater and under anaerobic conditions is also expected. It is not expected to bioconcentrate is fish and aquatic organisms nor is it expected to adsorb to sediment and suspended organic matter. Volatilization to the atmosphere is expected to be slower then for water itself. If released to the atmosphere, it may undergo a gas-phase oxidation with photochemically produced hydroxyl radicals with a half-life of 33 hours. It may also undergo atmospheric removal by wet deposition processes.

Cleanup/Disposal: Guide No. 128: Eliminate all ignition sources (no smoking, flares, sparks or flames in immediate area). All equipment used when handling the product must be grounded. Do not touch or walk through spilled material. Stop leak if you can do it without risk. Prevent entry into waterways, sewers, basements or confined areas. A vapor suppressing foam may be used to reduce vapors. Absorb or cover with dry earth, sand or other non-combustible material and transfer to containers. Use clean non-sparking tools to collect absorbed material. Large Spills: Dike far ahead of liquid spill for later disposal. Water spray may reduce vapor; but may not prevent ignition in closed spaces.

Environmental Physical Data

Henry's Law Constant: estimated at 1.75 $x10^{11}$

Octanol/Water Partition Coefficient: log K_{ow} = -1.76

Sorption Partition Coefficient: K_{oc} = calculated at 2 to 3

BCF: estimated at 3

BOD: theoretical 20%, 5 days

Regulations

RCRA 40CFR: Not listed

CERCLA: 40CFR 302.4: Not listed

SARA 40CFR 372.65: Not listed

SARA EHS 40CFR 355: Not listed

TSCA: Listed

GLY1420 **CAS #: 25496-72-4**

GLYCEROL MONOOLEATE

RTECS: RK1300000

EINECS Number: 247-038-6

Molecular Formula: $C_{21}H_{40}O_4$

Formula Weight: 356.55

Synonyms: ADCHEM GMO; AJAX GMO; ALDO 40; ALDO MO-FG; DUR-EM 204; EMCOL O; EMERY OLEIC ACID ESTER 2221; EMRITE 6009; GLYCERIN MONOOLEATE; GLYCERINE MONOOLEATE; GLYCEROL 1-MONOOLEATE; GLYCEROL OLEATE; GLYCEROL,1-MONO (9-OCTA-DECENOATE); GLYCERYL MONOOLEATE; GLYCERYL OLEATE; GMO 8903; HAROWAX L 9; LOXIOL G 10; MONOGLYCERYL OLEATE; ALPHA-MONOOLEIN; MONOOLEIN; MONOOLEOYLGLYCEROL; 9-OCTADECENOIC ACID (Z)-,MONOESTER WITH 1,2,3-PROPANETRIOL (9CI); 9-OCTADECENOIC ACID (Z)-,MONOESTER WITH1,2,3-PROPANETRIOL; OLEIC ACID GLYCEROL MONOESTER; OLEIC ACID MONOGLYCERIDE; OLEIN,MONO-; OLEOYLGLYCEROL; OLEYLMONOGLYCERIDE; OLICINE; PECEOL; RIKEMAL O 71D; RIKEMAL OL 100; S 1096R; S 1096; S 1097; SINNOESTER OGC; SUNSOFT O 30B; SUPEOL

Description: yellow oil or soft solid; sweet odor

Use: surface-active agent in foods, pharmaceuticals and cosmetics, odorless base paints, in rust-preventive oils, textile finishing, and vinyl light stabilizers; flavoring ingredient

Physical Properties

Boiling Point: 238 °C (460 °F) to 240 °C (464 °F) at 3 mm Hg
Freezing Point: Unstable at 25 °C (77 °F)
Specific Gravity: 0.942 at 20 °C/4 °C
Water Solubility: Insoluble in Water
Other Solubilities: Soluble in Ethanol, Ether, Chloroform
Refraction Index: 1.4626 at 40 °C/D

RTECS Toxicity Data

Irritation Eye: Rabbit Standard Draize Test Dose: 100 mg; Reaction: mild.
Irritation Skin: Rabbit Standard Draize Test Dose: 500 mg; Reaction: mild.

Hazard Overviews

Carcinogenicity: IARC - Not listed; NIOSH - Not listed; NTP - Not listed; ACGIH - Not listed; OSHA - Not listed; EPA - Not listed; MAK - Not listed

Environmental

Regulations

RCRA 40CFR: Not listed
CERCLA: 40CFR 302.4: Not listed
SARA 40CFR 372.65: Not listed
SARA EHS 40CFR 355: Not listed
TSCA: Listed

GLY2260 CAS #: 93-14-1

GLYCERYL GUAIACOLATE

RTECS: TY8400000
EINECS Number: 202-222-5
Molecular Formula: $C_{10}H_{14}O_4$
Formula Weight: 198.21

O
O
OH

Chemical Structure

Synonyms: 2-G; AERONESIN; ARESOL; CALMIPAN; CORTUSSIN; CRESON; 1,2-DIHYDROXY-3-(2-METHOXYPHENOXY)PROPANE; DILYN; G 87; GAIAMAR; GGE; GGG; GLYCERIN GUAIACOLATE; GLYCERINMONOGUAIACOL ETHER; GLYCERO-GUAIACOL ETHER; GLYCEROL GUAIACOLATE; GLYCEROL ALPHA-GUAIACYL ETHER; GLYCEROL ALPHA-(2-METHOXYPHENYL) ETHER; GLYCEROL ALPHA-(O-METHOXYPHENYL)ETHER; GLYCEROL MONO(2-METHOXYPHENYL)ETHER; GLYCEROL-ALPHA-GUAJAKOLETHER; GLYCEROL-ALPHA-MONOGUAIACOL ETHER; ALPHA-GLYCERYL GUAIACOL ETHER; ALPHA-GLYCERYL GUAIACOLATE ETHER; GLYCERYL GUAIACOLATE ETHER; GLYCERYL GUAIACYL ETHER; GLYCERYLGUAIACOL; GLYCOTUSS; GNAIFENESIN; GUAIACOL GLYCEROL ETHER; GUAIACOL GLYCERYL ETHER; GUAIACOLGLICERINETERE; GUAIACURAN; GUAIACURANE; GUAIACYL GLYCERYL ETHER; GUAIAMAR; GUAIANESIN; GUAIFENESIN; GUAIPHENESIN; GUAIPHENESINE; GUAIPHESIN; GUAJACOL-GLYCERINAETHER; GUAJACOL-ALPHA-GLYCERINETHER; GUAJACURAN; GUAJAMAR; GUANAR; GUAYANESIN; GUIAPHENESIN; HUSTOSIL; METFENOSSIDIOLO; 3-O-METHOXYPHENOXYPROPANE 1:2-DIOL; 3-(2-METHOXYPHENOXY)-1,2-PROPANEDIOL; 3-(O-METHOXYPHENOXY)-1,2-PROPANEDIOL; O-METHOXYPHENYL GLYCERYL ETHER; METHOXYPROPANEDIOL; METHPHENOXYDIOL; METOSSIPROPANDIOLO; MIOCAINA; MIOCURIN; MIORELAX; MUCOSTOP; MY 301; MYOCAINE; MYORELAX; MYOSCAIN; MYOSCAINE; NEUROTON; NEUROTONE; ORESOL; ORESON; 1,2-PROPANEDIOL,3-(2-METHOXYPHENOXY)-; 1,2-PROPANEDIOL,3-(O-METHOXYPHENOXY)-; PROPANOSEDYL; REDUTON; RELAXIL G; RELAXYL-G; REORGANIN; RESIL; RESPENYL; RESPIL; RESYL; RITUSSIN; ROBITUSSIN; SIROTOL; SL-90; TENNTUSS; TRECID; TULYN; XL-90
Description: white to slightly gray, crystalline powder or prisms; slight characteristic odor
Use: expectorant in cough remedies; preanesthetic & anesthetic in veterinary medicine

Physical Properties

Boiling Point: 215 °C (419 °F) at 19 mm Hg
Freezing Point: 78.5 °C (173.3 °F) to 79 °C (174.2 °F)
Water Solubility: 1 g in 20 ml Water at 25 °C
Other Solubilities: freely Soluble in Ethanol Soluble in Chloroform, Glycerol, Propylene Glycol, N,N-Dimethylformamide; moderately Soluble in Benzene Practically Insoluble in Petroleum Ether

RTECS Toxicity Data

Acute Oral: Rat LD_{50} Dose: 1510 mg/kg; Toxic Effects: Behavioral - Altered sleep time (including change in righting reflex); Lungs, Thorax, or Respiration - Chronic pulmonary edema; Lungs, Thorax, or Respiration - Respiratory depression. Mouse LD_{50} Dose: 690 mg/kg.
Acute Dermal: Rat LD_{50} Route: Subcutaneous Dose: 2550 mg/kg. Mouse LD_{50} Route: Subcutaneous Dose: 800 mg/kg; Toxic Effects: Automatic Nervous System - Smooth muscle relaxant (mechanism undefined, spasmolytic).

Hazard Overviews

Health: Irritating to eyes/skin/respiratory tract. Harmful. Other Acute Effects: harmful if swallowed; may be harmful if inhaled or absorbed through the skin; prolonged or repeated exposure may cause allergic reactions in certain sensitive individuals; exposure can cause gastrointestinal disturbances, drowsiness.
Fire: Extinguishing agents: water spray; carbon dioxide, dry chemical powder or appropriate foam. Precautions: combustible liquid.
Carcinogenicity: IARC - Not listed; NIOSH - Not listed; NTP - Not listed; ACGIH - Not listed; OSHA - Not listed; EPA - Not listed; MAK - Not listed

Primary Target Organs:

Eyes Skin Respiratory System

Environmental

Regulations

RCRA 40CFR: Not listed
CERCLA: 40CFR 302.4: Not listed
SARA 40CFR 372.65: Not listed
SARA EHS 40CFR 355: Not listed
TSCA: Listed

GLY2680 CAS #: 122-32-7

GLYCERYL TRIOLEATE

EINECS Number: 204-534-7
Molecular Formula: $C_{57}H_{104}O_6$
Formula Weight: 885.40

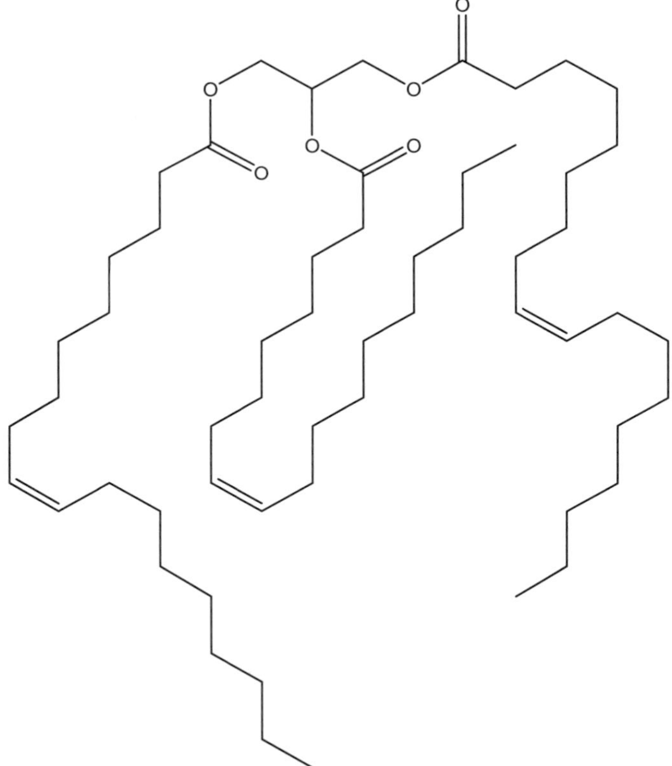

Chemical Structure

Synonyms: ALDO TO; EMERY 2423; EMERY OLEIC ACID ESTER 2230; GLYCERIN TRIOLEATE; GLYCEROL TRIOLEATE; GLYCEROL TRIOLEIN; GLYCEROL,TRI(CIS-9-OCTADECENOATE); GLYCERYL-1,2,3-TRIOLEATE; 9-OCTADECENOIC ACID (Z)-,1,2,3-PROPANETRIYL ESTER; OLEIC ACID TRIGLYCERIDE; OLEIC TRIGLYCERIDE; OLEIN; OLEIN,TRI-; OLEYL TRIGLYCERIDE; RAOLINE; TRIOLEIN; TRIOLEOYLGLYCERIDE; TRIOLEOYLGLYCEROL

Description: colorless to yellowish, oily liquid; odorless
Use: textile lubricants; lubricant for cosmetics, drugs & textiles; emulsifier for water/oil mixtures; chem int for radioactive iodine derivatives; plasticizer; in sweet almond oil for medicines & cosmetics

Physical Properties

Boiling Point: 235 °C (455 °F) to 240 °C (464 °F) at 15 mm Hg
Freezing Point: -5 °C (23 °F) to -4 °C (24.8 °F)
Specific Gravity: 0.915 at 15 °C/4 °C
Water Solubility: Practically Insoluble in Water
Other Solubilities: Soluble in Petroleum Ether.
Refraction Index: 1.4676 at 20 °C/D

Hazard Overviews

Health: May cause irritation to eyes/skin. Other Acute Effects: may be harmful by inhalation, ingestion, or skin absorption. The toxicological properties have not been thoroughly investigated.
Fire: Hazards: emits toxic fumes. Extinguishing agents: carbon dioxide, dry chemical powder or appropriate foam. Precautions: combustible liquid.
Reactivity: Stable. Hazardous polymerization will not occur. Hazardous decomposition products: toxic fumes of: carbon monoxide, carbon dioxide.
Carcinogenicity: IARC - Not listed; NIOSH - Not listed; NTP - Not listed; ACGIH - Not listed; OSHA - Not listed; EPA - Not listed; MAK - Not listed

Environmental

Regulations

RCRA 40CFR: Not listed
CERCLA: 40CFR 302.4: Not listed
SARA 40CFR 372.65: Not listed
SARA EHS 40CFR 355: Not listed
TSCA: Listed

GLY3100 CAS #: 556-52-5

GLYCIDOL

RTECS: UB4375000
EINECS Number: 209-128-3
Molecular Formula: $C_3H_6O_2$
Formula Weight: 74.08

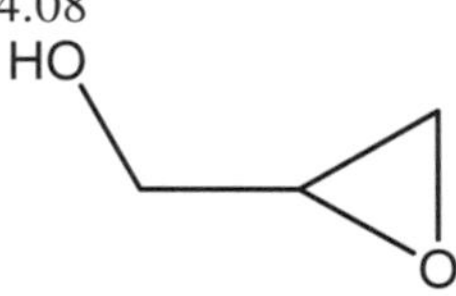

Chemical Structure

Synonyms: ALLYL ALCOHOL OXIDE; EPIHYDRIN ALCOHOL; 1,2-EPOXY-3-HYDROXY PROPANE; 1,2-EPOXY-3-HYDROXYPROPANE; 2,3-EPOXY-1-PROPANOL; 2,3-EPOXYPROPANOL; EPOXYPROPYL ALCOHOL; GLYCIDE; GLYCIDYL ALCOHOL; 1-HYDROXY-2,3-

EPOXYPROPANE; 3-HYDROXY-1,2-EPOXYPROPANE; HYDROXYMETHYL ETHYLENE OXIDE; 2-HYDROXYMETHYL OXIRAN; 2-HYDROXYMETHYLOXIRAN; 2-(HYDROXYMETHYL)OXIRANE; 3-HYDROXYPROPYLENE OXIDE; METHANOL,OXIRANYL-; OXIRANEMETHANOL; OXIRANYLMETHANOL; 1-PROPANOL,2,3-EPOXY-

Description: colorless liquid

Use: stabilizer for natural oils, demulsifier, dye-leveling agent, stabilizer for vinyl polymers and intermediate in the synthesis of glycerol, glycidyl ethers, esters and amines; in surface coatings, chemical synthesis, pharmaceuticals, sanitary chemicals and sterilizing milk of magnesia

Physical Properties

Boiling Point: 160 °C (320 °F) at 760 mm Hg
Freezing Point: -45 °C (-49 °F)
Specific Gravity: 1.115 at 20 °C/4 °C
Vapor Density: 2.15
Saturated Vapor Density: 1.202209002 kg/m^3
Density: 1.143 g/mL at 25 °C
Vapor Pressure: 0.9 mm Hg at 25 °C
Water Solubility: Soluble in Water
Other Solubilities: 95% Ethanol: >=100 mg/ml at 18 °C; Acetone: >=100 mg/ml at 18 °C; Acid: Soluble; Alcohol: Soluble; Benzene: Soluble; Chloroform: Soluble; DMSO: >=100 mg/ml at 18 °C; Ether: Soluble; Most organic solvents: Soluble; Petroleum Ether: Soluble.
Refraction Index: 1.4287
Flash Point: 72 °C

RTECS Toxicity Data

Acute Oral: Rat LD_{50} Dose: 420 mg/kg. Mouse LD_{50} Dose: 431 mg/kg.

Acute Inhalation: Rat LC_{50} Dose: 580 ppm/8hr; Toxic Effects: Lungs, Thorax, or Respiration - Emphysema; Lungs, Thorax, or Respiration - Other changes. Mouse LC_{50} Dose: 450 ppm/4hr; Toxic Effects: Lungs, Thorax, or Respiration - Emphysema; Lungs, Thorax, or Respiration - Other changes.

Acute Dermal: Rabbit LD_{50} Route: Skin; Dose: 1980 mg/kg.

Chronic (Multiple Dose) Oral: Rat Dose: 9600 mg/kg/16D-I; Toxic Effects: DEATH. Rat Dose: 36400 mg/kg/13W-I; Toxic Effects: Brain and coverings - Other degenerative changes; Kidney, Ureter, and Bladder - Chgs in tubules (inc acute renal failure, acute tubular necrosis; Endocrine - Other changes.

Chronic (Multiple Dose) Inhalation: Rat Dose: 125 mg/m^3/4H/17W-I; Toxic Effects: Brain and coverings - Recordings from specific areas of CNS; Lungs, Thorax, or Respiration - Respiratory stimulation; Nutritional and gross metabolic - Other changes. Guinea Pig Dose: 125 mg/m^3/4H/17W-I; Toxic Effects: Brain and coverings - Recordings from specific areas of CNS.

Irritation Eye: Rabbit Standard Draize Test Dose: 2 mg/24H; Reaction: severe.

Irritation Skin: Rabbit Standard Draize Test Dose: 100 mg/24H; Reaction: moderate. Rabbit Standard Draize Test Dose: 558 mg/3D; Reaction: moderate.

Reproductive/Teratogenic: Rat Route: Oral; Dose: 180 mg/kg; Duration: male 12D prior to mating; Effects on Fertility - Male fertility index. Rat Route: Oral; Dose: 18200 mg/kg; Duration: male 13W prior to mating; Paternal Effects - Spermatogenesis; Testes, epididymis, sperm duct.

Mutagenic: Human Cytogenetic Analysis; Cell Type: lymphocyte; Dose: 400 umol/L. Human Sister Chromatid Exchenge; Cell Type: lymphocyte; Dose: 50 umol/L.

Tumorigenic: Rat Route: Oral; Dose: 38625 mg/kg/2Y-C; Toxic Effects: Tumorigenic - Carcinogenic by RTECS criteria; Gastrointestinal - Tumors; Skin and appendages - Tumors. Mouse Route: Oral; Dose: 25750 mg/kg/2Y-C; Toxic Effects: Tumorigenic - Carcinogenic by RTECS criteria; Gastrointestinal - Tumors; Skin and appendages - Tumors.

Hazard Overviews

Fire Diamond

Health: Irritating to eyes/skin/respiratory tract. Toxic. Also Causes: CNS depression, unconsciousness (high concentrations), GI irritation upon ingestion.

Fire: Combustible. Use dry chemical, carbon dioxide, water spray, or regular foam.

Reactivity: Stable. Hazardous polymerization cannot occur. Avoid: heat; ignition sources. Incompatible with: strong oxidizers including nitrates; strong acids, bases, metals (copper, zinc); metal salts; (aluminum chloride, iron (III) chloride, tin (IV) chloride). Hazardous decomposition products: carbon oxide(s); acrid smoke.

Carcinogenicity: IARC - Not listed; NIOSH - Not listed; NTP - Listed; ACGIH - Class A3, Animal carcinogen; OSHA - Not listed; EPA - Not listed; MAK - Not listed

Primary Target Organs:

Eyes

Skin

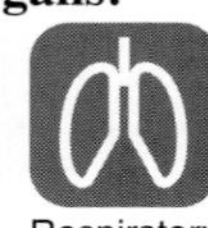
Respiratory System

Nervous System

Exposure Limits
OSHA PEL: TWA: 50 ppm; 150 mg/m^3.
OSHA PEL Vacated 1989 Limits: TWA: 25 ppm; 75 mg/m^3.
ACGIH TLV: TWA: 2 ppm; 6.1 mg/m^3.
NIOSH REL: TWA: 25 ppm; 75 mg/m^3.
NIOSH IDLH: 150 ppm.
DFG MAK: TWA: 50 ppm; 150 mg/m^3.

Respirator Recommendation
Exposure Range: >50 to <150 ppm Supplied Air, Constant Flow/Pressure Demand, Half Mask
Exposure Range: 150 to unlimited ppm Self-contained Breathing Apparatus, Pressure Demand, Full Face
Note: odor threshold unknown

Environmental

Cleanup/Disposal: Guide No. 153: Eliminate all ignition sources (no smoking, flares, sparks or flames in immediate area). Do not touch damaged containers or spilled material unless wearing appropriate protective clothing. Stop leak if

you can do it without risk. Prevent entry into waterways, sewers, basements or confined areas. Absorb or cover with dry earth, sand or other non-combustible material and transfer to containers. Do not get water inside containers.

Regulations

RCRA 40CFR: Not listed
CERCLA: 40CFR 302.4: Not listed
SARA 40CFR 372.65: Not listed
SARA EHS 40CFR 355: Not listed
TSCA: Listed

Analytical Methods

Air: ASTM D3686, D3687
Indoor / Expired Air: NIOSH 1608

GLY3520 **CAS #: 37394-13-1**

GLYCIDOXYPROPYLPENTAMETHYL DISILOXAN

Physical Properties

Specific Gravity: 0.917
Water Solubility: Insoluble

Hazard Overviews

Carcinogenicity: IARC - Not listed; NIOSH - Not listed; NTP - Not listed; ACGIH - Not listed; OSHA - Not listed; EPA - Not listed; MAK - Not listed

Environmental

Regulations

RCRA 40CFR: Not listed
CERCLA: 40CFR 302.4: Not listed
SARA 40CFR 372.65: Not listed
SARA EHS 40CFR 355: Not listed
TSCA: Not listed

GLY3940 **CAS #: 2380-83-8**

3-GLYCIDOXYPROPYLTRIMETHOXY-SILANE

Physical Properties

Boiling Point: 290 °C (554 °F)
Vapor Density: > 1 Air=1
Vapor Pressure: < 1

Hazard Overviews

Carcinogenicity: IARC - Not listed; NIOSH - Not listed; NTP - Not listed; ACGIH - Not listed; OSHA - Not listed; EPA - Not listed; MAK - Not listed

Environmental

Regulations

RCRA 40CFR: Not listed
CERCLA: 40CFR 302.4: Not listed
SARA 40CFR 372.65: Not listed
SARA EHS 40CFR 355: Not listed
TSCA: Not listed

GLY4360 **CAS #: 106-90-1**

GLYCIDYL ACRYLATE

RTECS: AS9275000
EINECS Number: 203-440-3
Molecular Formula: $C_6H_8O_3$
Formula Weight: 128.14
Synonyms: ACRYLIC ACID GLYCIDYL ESTER; ACRYLIC ACID,2,3-EPOXYPROPYL ESTER; 2,3-EPOXYPROPYL ACRYLATE; GLYCIDYL PROPENATE; GLYCIDYLESTER KYSELINY AKRYLOVE; 1-PROPANOL,2,3-EPOXY-,ACRYLATE; 2-PROPENOIC ACID,OXIRANYL METHYL ESTER (ACI); 2-PROPENOIC ACID,OXIRANYLMETHYL ESTER; 2-PROPENOIC ACID,OXIRANYLMETHYL ESTER (9CI)
Description: light amber liquid
Use: monomer for polymers used in adhesives & coatings; crosslinking agent for polymers of trialkyltin esters; epoxy resin diluent

Physical Properties

Boiling Point: 57 °C (135 °F) at 2 mm Hg
Freezing Point: -41.5 °C (-42.7 °F)
Specific Gravity: 1.1074 at 20 °C/20 °C
Density: 4.4
Water Solubility: Insoluble in Water
Other Solubilities: 95% Ethanol: >=100 mg/ml at 20.5 °C; Acetone: >=100 mg/ml at 20.5 °C; DMSO: >=100 mg/ml at 20.5 °C.
Refraction Index: 1.4490
Flash Point: 61 °C Open Cup
Autoignition Temperature: 415 °C

RTECS Toxicity Data

Acute Oral: Rat LD_{50} Dose: 210 mg/kg.
Acute Inhalation: Rat LC_{Lo} Dose: 125 ppm/4hr.
Acute Dermal: Rabbit LD_{50} Route: Skin; Dose: 400 uL/kg.
Irritation Eye: Rabbit Standard Draize Test Dose: 1 mg; Reaction: severe. Rabbit Standard Draize Test Dose: 50 ug/24H; Reaction: severe.
Irritation Skin: Rabbit Standard Draize Test Dose: 2 mg/24H; Reaction: severe.
Mutagenic: Rat Cytogenetic Analysis; Route: Intraperitoneal; Dose: 5 mg/kg. Hamster Sister Chromatid Exchenge; Cell Type: lung; Dose: 16 umol/L.

Hazard Overviews

Corrosive

Health: Corrosive to eyes/skin/respiratory tract. Toxic. Other Acute Effects: harmful if swallowed, inhaled, or absorbed through skin; inhalation may result in spasm, inflammation and edema of the larynx and bronchi, chemical pneumonitis and pulmonary edema; symptoms of exposure may include burning sensation, coughing, wheezing, laryngitis, shortness of breath, headache, nausea and vomiting.
Fire: Combustible. Extinguishing agents: water spray; carbon dioxide, dry chemical powder or appropriate foam. Precautions: combustible liquid.
Reactivity: Incompatible with: strong oxidizing agents, strong acids, strong bases, light sensitive, moisture sensitive. Hazardous decomposition products: toxic fumes of: carbon monoxide, carbon dioxide.
Carcinogenicity: IARC - Not listed; NIOSH - Not listed; NTP - Not listed; ACGIH - Not listed; OSHA - Not listed; EPA - Not listed; MAK - Not listed
Primary Target Organs:

Eyes

Skin

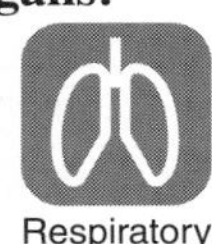
Respiratory System

Environmental

Cleanup/Disposal: Guide No. 131: Fully encapsulating, vapor protective clothing should be worn for spills and leaks with no fire. Eliminate all ignition sources (no smoking, flares, sparks or flames in immediate area). All equipment used when handling the product must be grounded. Do not touch or walk through spilled material. Stop leak if you can do it without risk. Prevent entry into waterways, sewers, basements or confined areas. A vapor suppressing foam may be used to reduce vapors. Small Spills: Absorb with earth, sand or other non-combustible material and transfer to containers for later disposal. Use clean non-sparking tools to collect absorbed material. Large Spills: Dike far ahead of liquid spill for later disposal. Water spray may reduce vapor; but may not prevent ignition in closed spaces.

Regulations

RCRA 40CFR: Not listed
CERCLA: 40CFR 302.4: Not listed
SARA 40CFR 372.65: Not listed
SARA EHS 40CFR 355: Not listed
TSCA: Listed

GLY4780 CAS #: 765-34-4

GLYCIDYLALDEHYDE

RTECS: MB3150000
DOT: UN2622; IMO3.3
EINECS Number: 212-143-8
Molecular Formula: $C_3H_4O_2$
Formula Weight: 72.07
Synonyms: EPIHYDRINALDEHYDE; EPIHYDRINE ALDEHYDE; 2,3-EPOXY-1-PROPANAL; 2,3-EPOXYPROPANAL; 2,3-EPOXYPROPIONALDEHYDE; FORMYLOXIRAN; GLYCIDAL; GLYCIDALDEHYDE; GLYCIDYL ALDEHYDE; OXIRANE-CARBOXALDEHYDE; OXIRANECARBOXALDEHYDE; PROPIONALDEHYDE,2,3-EPOXY-
Description: colorless liquid; pungent pronounced aldehyde-like odor
Use: cross-linking agent for finishing of wool, for oil tanning & fat-liquoring of leather & surgical sutures; for protein insolubilization; as vapor-phase disinfectant; research chemical

Physical Properties

Boiling Point: 112 °C (234 °F) to 113 °C (235 °F) at 760 mm Hg
Freezing Point: -62 °C (-79.6 °F)
Specific Gravity: 1.1403 at 20 °C
Vapor Density: 2.58 Air=1
Saturated Vapor Density: 1.263315245 kg/m^3
Vapor Pressure: 27 mm Hg at 25 °C
Water Solubility: Miscible with Water
Other Solubilities: miscible with all common solvents; immiscible with Petroleum Ether.
Refraction Index: 1.4225 at 22 °C/D
Flash Point: 31.111 °C Open Cup

RTECS Toxicity Data

Acute Oral: Rat LD_{Lo} Dose: 50 mg/kg.
Acute Inhalation: Rat LC_{50} Dose: 252 ppm/9hr. Human TC_{Lo} Dose: 5 ppm; Toxic Effects: Brain and coverings - Recordings from specific areas of CNS; Sense organs and special senses - Other; Behavioral - Excitement.
Acute Dermal: Rabbit LD_{50} Route: Skin; Dose: 249 mg/kg; Toxic Effects: Lungs, Thorax, or Respiration - Chronic pulmonary edema.
Chronic (Multiple Dose) Inhalation: Rat Dose: 40 ppm/4H/12W-I; Toxic Effects: Endocrine - Changes in spleen weight; Endocrine - Changes in thymus weight; Blood - Changes in bone marrow not included above.
Irritation Eye: Human Standard Draize Test Dose: 1 ppm/5M; Reaction: moderate.
Irritation Skin: Rabbit Standard Draize Test Dose: 100 mg/24H; Reaction: moderate.
Mutagenic: Mouse DNA Adduct; Route: Skin; Dose: 80 mg/kg. Mouse Mutations in Microorganisms; Cell Type: lymphocyte; Dose: 8900 ug/L (+S9).
Tumorigenic: Rat Route: Subcutaneous; Dose: 13 gm/kg/77W-I; Toxic Effects: Tumorigenic - Carcinogenic by RTECS criteria; Tumorigenic - Tumors at site of application. Rat Route: Subcutaneous; Dose: 390 mg/kg/78W-I; Toxic Effects: Tumorigenic - Equivocal tumorigenic agent by RTECS criteria; Tumorigenic - Tumors at site of application.

Hazard Overviews

Flammable

Fire: Flammable.

Carcinogenicity: IARC - Group 2B, Possibly carcinogenic to humans; NIOSH - Not listed; NTP - Not listed; ACGIH - Not listed; OSHA - Not listed; EPA - Class B2, Probable human carcinogen based on animal studies; MAK - Not listed

Environmental

Environmental Fate: If released to moist soil, it is expected to chemically hydrolyze. This compound has the potential to be quite mobile; however, hydrolysis is expected to limit the importance of this fate process. If released to dry soil, it is expected to volatilize rapidly from soil surfaces. If released to water, it is expected to chemically hydrolyze (half-life at 25 °C and neutral pH <28 days). Reaction with singlet oxygen and alkyl peroxy radicals, bioaccumulation in aquatic organisms, adsorption to suspended solids and sediments, and volatilization are not expected to be significant fate processes in water. If released to the atmosphere, the dominant removal mechanism is expected to be reaction with photochemically generated hydroxyl radicals (half-life 18 hours). Partial removal by wet deposition may also occur.

Cleanup/Disposal: Guide No. 131: Fully encapsulating, vapor protective clothing should be worn for spills and leaks with no fire. Eliminate all ignition sources (no smoking, flares, sparks or flames in immediate area). All equipment used when handling the product must be grounded. Do not touch or walk through spilled material. Stop leak if you can do it without risk. Prevent entry into waterways, sewers, basements or confined areas. A vapor suppressing foam may be used to reduce vapors. Small Spills: Absorb with earth, sand or other non-combustible material and transfer to containers for later disposal. Use clean non-sparking tools to collect absorbed material. Large Spills: Dike far ahead of liquid spill for later disposal. Water spray may reduce vapor; but may not prevent ignition in closed spaces.

Environmental Physical Data

Henry's Law Constant: estimated at 5.1 x10^{-7}

Octanol/Water Partition Coefficient: log K_{ow} = -0.73

Sorption Partition Coefficient: K_{oc} = estimated at 10

BCF: estimated at 1

Regulations

RCRA 40CFR: Listed Hazardous Waste No. U126 Toxic Waste

CERCLA: 40CFR 302.4: Listed per RCRA Section 3001 RQ: 10 lb (4.535 kg)

SARA 40CFR 372.65: Not listed

SARA EHS 40CFR 355: Not listed

TSCA: Not listed

GLY5200 CAS #: 56-40-6

GLYCINE

RTECS: MB7600000

EINECS Number: 200-272-2

Molecular Formula: $C_2H_5NO_2$

Formula Weight: 75.07

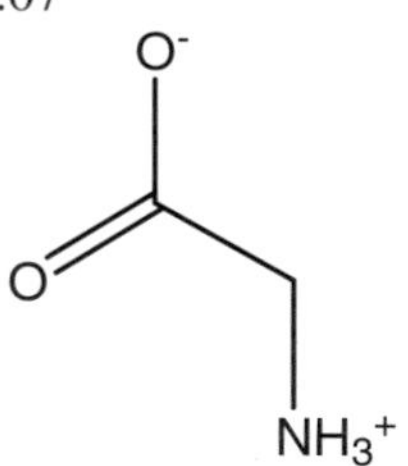

Chemical Structure

Synonyms: ACETIC ACID,AMINO-; ACIPORT; AMINO ACETIC ACID; AMINOACETIC ACID; AMINOETHANOIC ACID; GLICOAMIN; GLY; GLYCOCOLL; GLYCOLIXIR; GLYCOSTHENE; HAMPSHIRE GLYCINE; PADIL; SUCRE DE GELATINE

Description: white crystals or monoclinic or trigonal prisms; odorless

Physical Properties

Boiling Point: 289 °C (552.2 °F) to 292 °C (557.6 °F)

Freezing Point: 182 °C (359.6 °F)

Specific Gravity: 1.1607

Water Solubility: >= 100 mg/mL at 18 C

Other Solubilities: 95% Ethanol: <1 mg/ml at 18 °C; Acetone: <1 mg/ml at 18 °C; DMSO: <1 mg/ml at 18 °C; Ether: Insoluble.

pH: 0.2 molar solution in water 4

Ionization Potential (eV): 8.8 +/-1.0

Flash Point: Low, ignites at very high temperatures

RTECS Toxicity Data

Acute Oral: Rat LD_{50} Dose: 7930 mg/kg. Mouse LD_{50} Dose: 4920 mg/kg.

Acute Dermal: Rat LD_{50} Route: Subcutaneous Dose: 5200 mg/kg. Mouse LD_{50} Route: Subcutaneous Dose: 5060 mg/kg.

Mutagenic: Human Sister Chromatid Exchenge; Cell Type: lymphocyte; Dose: 100 mg/L.

Hazard Overviews

Health: May be irritating to eyes/skin/respiratory tract. Other Acute Effects: may be harmful by inhalation, ingestion, or skin absorption.

Fire: Will burn. Hazards: in powder form capable of creating a dust explosion; emits toxic fumes. Extinguishing agents: water spray; carbon dioxide, dry chemical powder or appropriate foam. Precautions: combustible liquid.

Reactivity: Stable. Hazardous polymerization will not occur. Incompatible with: strong oxidizing agents.

Carcinogenicity: IARC - Not listed; NIOSH - Not listed; NTP - Not listed; ACGIH - Not listed; OSHA - Not listed; EPA - Not listed; MAK - Not listed

Environmental

Environmental Physical Data

Octanol/Water Partition Coefficient: log K_{ow} = calculated at -3.03 to -1.7

Regulations

RCRA 40CFR: Not listed
CERCLA: 40CFR 302.4: Not listed
SARA 40CFR 372.65: Not listed
SARA EHS 40CFR 355: Not listed
TSCA: Listed

GLY5620 CAS #: 106-91-2

GLYCIOYL METHACRYLATE

RTECS: OZ4375000
EINECS Number: 203-441-9
Molecular Formula: $C_7H_{10}O_3$
Structured MF: CH_2=CH(CH_3)COOCH_2CHCH_2O
Formula Weight: 142.1

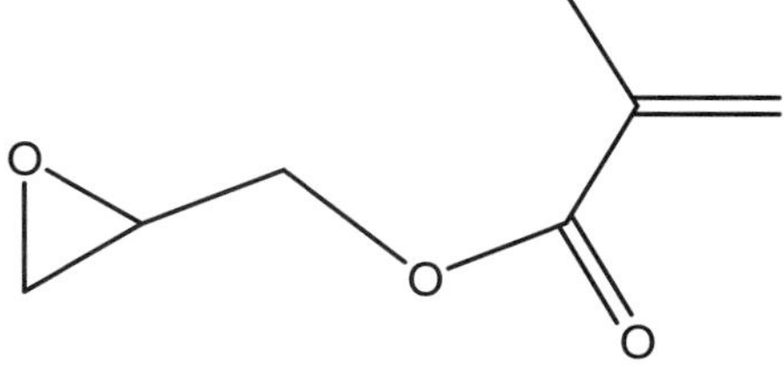

Chemical Structure

Synonyms: CP-105; CP 105; 2,3-EPOXYPROPYL METHACRYLATE; GLYCIDOL METHACRYLATE; GLYCIDYL METHACRYLATE; GLYCIDYL ALPHA-METHYL ACRYLATE; GLYCIDYL ALPHA-METHYLACRYLATE; METHACRYLIC ACID,2,3-EPOXYPROPYL ESTER; 1-PROPANOL,2,3-EPOXY-,METHACRYLATE; 2-PROPENOIC ACID,2-METHYL-,OXIRANYLMETHYL ESTER
Description: colorless liquid; fruity odor
Use: chem int for polymers; monomer & diluent in epoxy resin formulations; in hydrogel lenses; in bis-gma dental resin; co-polymerization of methyl methacrylate; produce hydrogels used for contact lenses

Physical Properties

Boiling Point: 189 °C (372 °F)
Specific Gravity: 1.042
Density: 1.073 g/cu m at 25 °C/5 °C
Water Solubility: 5 to 10 mg/mL at 20 °C
Other Solubilities: 95% Ethanol: >=100 mg/ml at 20 °C; Acetone: >=100 mg/ml at 20 °C; DMSO: >=100 mg/ml at 20 °C.
Surface Tension: Estimated at 40 dynes/cm
Refraction Index: 1.4482 at 25 °C/D
Flash Point: 84 °C Tag Open Cup

RTECS Toxicity Data

Acute Oral: Rat LD_{50} Dose: 597 mg/kg. Mouse LD_{50} Dose: 390 mg/kg.
Acute Dermal: Rabbit LD_{50} Route: Skin; Dose: 450 uL/kg.
Chronic (Multiple Dose) Oral: Rat Dose: 2100 mg/kg/30D-I; Toxic Effects: Behavioral - Food intake (animal); Gastrointestinal - Ulceration or bleeding from stomach; Kidney, Ureter, and Bladder - Changes primarily in glomeruli.
Chronic (Multiple Dose) Inhalation: Rat Dose: 206 mg/m^3/6H/26W-I; Toxic Effects: Liver - Changes in Liver weight; Endocrine - Changes in spleen weight; Biochemical - Transaminases. Rabbit Dose: 206 mg/m^3/6H/26W-I; Toxic Effects: Brain and coverings - Other degenerative changes; Cardiac - EKG changes not diagnostic of above; Blood - Changes in erythrocite (RBC) cell count.
Mutagenic: Rat Cytogenetic Analysis; Route: Intraperitoneal; Dose: 300 mg/kg. Hamster Sister Chromatid Exchenge; Cell Type: lung; Dose: 78 umol/L.

Hazard Overviews

Poison

Health: Severely irritating to eyes/skin/respiratory tract. Poison. Other Acute Effects: may be fatal if inhaled, swallowed, or absorbed through skin; high concentrations are extremely destructive to tissues of the mucous membranes and upper respiratory tract, eyes and skin; symptoms of exposure may include burning sensation, coughing, wheezing, laryngitis, shortness of breath, headache, nausea and vomiting; prolonged or repeated exposure may cause allergic reactions in certain sensitive individuals. Chronic Effects: may alter genetic material; target organs: nerves, kidneys. Carcinogen.
Fire: Combustible. Hazards: emits toxic fumes. Extinguishing agents: carbon dioxide, dry chemical powder or appropriate foam; water spray. Precautions: combustible liquid.
Reactivity: Will react with compounds containing an active hydrogen atom. Incompatible with: strong oxidizing agents, strong acids, strong bases, peroxides, may polymerize on exposure to light. Hazardous decomposition products: toxic fumes of: carbon monoxide, carbon dioxide.
Carcinogenicity: IARC - Not listed; NIOSH - Not listed; NTP - Not listed; ACGIH - Not listed; OSHA - Not listed; EPA - Not listed; MAK - Not listed
Primary Target Organs:

Eyes

Skin

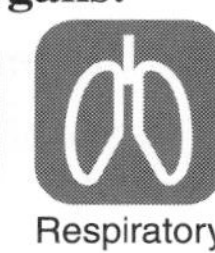
Respiratory System

Nervous System

Kidneys

Environmental

Cleanup/Disposal: Guide No. 153: Eliminate all ignition sources (no smoking, flares, sparks or flames in immediate

area). Do not touch damaged containers or spilled material unless wearing appropriate protective clothing. Stop leak if you can do it without risk. Prevent entry into waterways, sewers, basements or confined areas. Absorb or cover with dry earth, sand or other non-combustible material and transfer to containers. Do not get water inside containers.

Environmental Physical Data

BCF: no food chain concentration potential

Regulations

RCRA 40CFR: Not listed
CERCLA: 40CFR 302.4: Not listed
SARA 40CFR 372.65: Not listed
SARA EHS 40CFR 355: Not listed
TSCA: Listed

GLY6040 CAS #: 9005-65-6

GLYCOL

RTECS: WG2932500
Molecular Formula: Unknown
Formula Weight: 1309.68
Synonyms: ARMOTAN PMO-20; ATLOX 1087; ATLOX 8916TF; ATOLOX 8916TF; CAPMUL POE-O; CRILL 10; CRILL 11; CRILL S 10; CRILLET 4; DREWMULSE POE-SMO; DURFAX 80; EMSORB 6900; ETHOXYLATED SORBITAN MONOOLEATE; GLYCOSPERSE 0-20; GLYCOSPERSE 0-20X; GLYCOSPERSE O-20; GLYCOSPERSE O-20X; GLYCOSPERSE O 20; GLYCOSPERSE 0-20 VEG; GLYCOSPERSE O-20 VEG; HODAG SVO 9; LIPOSORB 0-20; LIPOSORB O-20; MO 55F; MONITAN; MONTANOX 80; NIKKOL TO; NIKKOL TO 10; OLOTHORB; POLYETHYLENE OXIDE SORBITAN MONO-OLEATE; POLYOXYETHYLENE (20) SORBITAN MONO-OLEATE; POLYOXYETHYLENE SORBITAN MONOOLEATE; POLYOXYETHYLENE SORBITAN OLEATE; POLYOXYETHYLENE(20)SORBITAN MONO-OLEATE; POLYSORBAN 80; POLYSORBATE 80 B.P.C; POLYSORBATE 80,U.S.P; POLYSORBATE 80; POLYSORBATE 81; POLYSORBATE 80 BPC; POLYSORBATE 80,USP; PROTASORB O-20; ROMULGIN O; SORBIMACROGOL OLEATE; SORBIMACROGOL OLEATE 300; SORBITAL 0 20; (X)-SORBITAN MONO-9-OCTADECENOATE POLY(OXY-1,2-ETHANEDIYL) DERIVS; SORBITAN MONO-9-OCTADECENOATE POLY(OXY-1,2-ETHANEDIYL)DERIVATIVES; SORBITAN MONO-OLEATE POLYOXYETHYLENE; SORBITAN,MONO-9-OCTADECENOATE,POLY(OXY-1,2-ETHANEDIYL)DERIVS,(Z)-; SORBITAN,MONOOLEATE POLYOXYETHYLENE DERIV; SORBITAN,MONOOLEATE,POLYOXYETHYLENE DERIVS; SORBON T 80; SORETHYTAN (20) MONO-OLEATE; SORETHYTAN (20) MONOOLEATE; SORLATE; SVO 9; TO 10; TWEEN 80; TWEEN 81; TWEEN 80 A
Description: amber colored liquid; faint odor
Use: emulsifier and dispersing agent; surfactant; foaming and defoaming agent; pharmaceuticals, shortenings and baked goods

Physical Properties

Boiling Point: > 100 °C (212 °F)
Specific Gravity: 1.06 to 1.1
Vapor Density: > 1 Air=1
Density: 1.064 g/mL
Vapor Pressure: < 1
Water Solubility: Very Soluble in Water
Other Solubilities: 95% Ethanol: >=100 mg/ml at 23 °C; Acetone: >=100 mg/ml at 23 °C; Corn oil: Soluble; Cottonseed oil: Soluble; DMSO: >=100 mg/ml at 23 °C; Ethyl Acetate: Soluble; Methanol: Soluble; Mineral oil: Insoluble; Toluene: Soluble.
pH: 5% aqueous solution 5 to 7.0
Refraction Index: 1.4756
Evaporation Rate: < 1 Butyl Acetate=1
Flash Point: > 148.889 °C

RTECS Toxicity Data

Acute Oral: Rat LD_{50} Dose: 34500 uL/kg. Mouse LD_{50} Dose: 25 gm/kg.
Irritation Eye: Rabbit Standard Draize Test Dose: 150 mg; Reaction: mild.
Reproductive/Teratogenic: Rat Route: Oral; Dose: 635 gm/kg; Duration: multigenerations; Effects on Newborn - Viability index; Weaning or lactation index. Rat Route: Oral; Dose: 1270 gm/kg; Duration: female 84D prior to mating Effects on Newborn - Viability index.
Mutagenic: Human DNA Inhibition; Cell Type: lymphocyte; Dose: 20 ppm. Mouse DNA Inhibition; Cell Type: other cell types; Dose: 20 ppm.
Tumorigenic: Rat Route: Oral; Dose: 2163 gm/kg/2Y-C; Toxic Effects: Tumorigenic - Equivocal tumorigenic agent by RTECS criteria; Endocrine - Adrenal cortex tumors. Rat Route: Subcutaneous; Dose: 10 gm/kg/27W-I; Toxic Effects: Tumorigenic - Equivocal tumorigenic agent by RTECS criteria; Tumorigenic - Tumors at site of application.

Hazard Overviews

Health: May cause irritation to eyes/skin/respiratory tract. Harmful. Other Acute Effects: may be harmful by inhalation, ingestion, or skin absorption; possible risk of irreversible effects. Chronic Effects: Possible carcinogen.
Fire: Will burn. Extinguishing agents: water spray; carbon dioxide, dry chemical powder or appropriate foam. Precautions: combustible liquid.
Carcinogenicity: IARC - Not listed; NIOSH - Not listed; NTP - Not listed; ACGIH - Not listed; OSHA - Not listed; EPA - Not listed; MAK - Not listed

Environmental

Regulations

RCRA 40CFR: Not listed
CERCLA: 40CFR 302.4: Not listed
SARA 40CFR 372.65: Not listed
SARA EHS 40CFR 355: Not listed
TSCA: Listed

GLY6460 CAS #: 107-16-4

GLYCOLONITRILE

RTECS: AM0350000
EINECS Number: 203-469-1
Molecular Formula: C_2H_3NO
Structured MF: $HOCH_2CN$
Formula Weight: 57.06

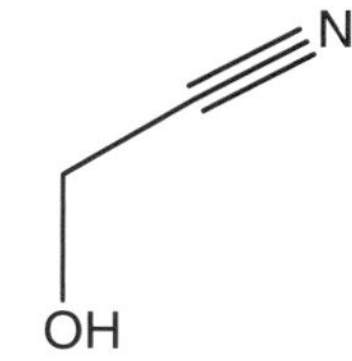

Chemical Structure

Synonyms: ACETONITRILE,HYDROXY-; CYANOMETHANOL; FORMALDEHYDE CYANOHYDRIN; FORMALDEHYDE,CYANOHYDRIN; GLYCOLIC NITRILE; GLYCOLONITRILE (8CI); GLYCONITRILE; GLYKOLONITRIL; 2-HYDROXYACETONITRILE; ALPHA-HYDROXYACETONITRILE; HYDROXYACETONITRILE; HYDROXYAZETONITRIL; ALPHA-HYDROXYMETHYLCYANIDE; HYDROXYMETHYLKYANID; 2-HYDROXYMETHYLNITRILE; HYDROXYMETHYLNITRILE; METHYLENE CYANOHYDRIN
Description: colorless to water-white oily liquid; odorless
Use: solvent and organic intermediate

Physical Properties

Boiling Point: 183 °C (361 °F)
Freezing Point: < -72 °C (-97.6 °F)
Specific Gravity: 1.1
Density: 1.97
Vapor Pressure: 1 mm Hg at 145 °F
Water Solubility: Soluble in Water
Other Solubilities: 95% Ethanol: <1 mg/ml at 20 °C; Acetone: 1-5 mg/ml at 20 °C; precipitate formed at 10 mg/ml; DMSO: 10-50 mg/ml at 20 °C; Ether: Soluble.
Refraction Index: 1.4090 at 25 °C
Flash Point: > 93.3 °C

RTECS Toxicity Data

Acute Oral: Rat LD_{50} Dose: 8 mg/kg. Mouse LD_{50} Dose: 10 mg/kg.
Acute Inhalation: Rat LC_{Lo} Dose: 27 ppm/8hr; Toxic Effects: Sense organs and special senses - Lacrimation; Behavioral - Somnolence (general depressed activity). Mouse LC_{Lo} Dose: 27 ppm/8hr; Toxic Effects: Sense organs and special senses - Lacrimation; Behavioral - Somnolence (general depressed activity).
Acute Dermal: Rabbit LD_{50} Route: Skin; Dose: 5 mg/kg; Toxic Effects: Skin and appendages - Primary irritation. Mouse LD_{Lo} Route: Subcutaneous Dose: 15 mg/kg.
Irritation Skin: Rabbit Standard Draize Test Dose: 500 mg/24H; Reaction: mild.

Hazard Overviews

Poison

Explosive

Fire Diamond

Health: Severely irritating to skin. Poison. Also Causes: rapid pulse, irregular heartbeat, pink coloration of skin, dilated pupils, nausea, dizziness, difficulty breathing, anxiety, stiffness in lower jaw, convulsions, paralysis, coma, cardiac arrhythmias, respiratory failure, dermatitis, redness, blistering, inflammation.
Fire: Explosive. Use dry chemical, carbon dioxide, water spray, fog, or regular foam.
Reactivity: Unstable, may undergo spontaneous, violent decomposition at room temperature and under normal storage and handling conditions; inhibitor (phosphoric acid) usually added for stabilization but inhibitor can be used up quickly. Hazardous polymerization can occur, regularly monitor inhibitor levels to prevent polymerization. Avoid: long term storage; decrease in inhibitor level. Incompatible with: alkalis. Hazardous decomposition products: nitrogen oxide; cyanide.
Carcinogenicity: IARC - Not listed; NIOSH - Listed as carcinogen; NTP - Not listed; ACGIH - Not listed; OSHA - Not listed; EPA - Not listed; MAK - Not listed
Primary Target Organs:

Skin

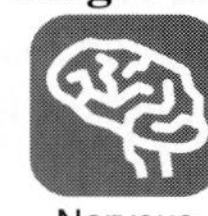
Nervous System

Cardio-vascular

Blood

Exposure Limits
NIOSH REL: STEL: 2 ppm; 5 mg/m^3; 15-minute.

Environmental

Cleanup/Disposal: Guide No. 153: Eliminate all ignition sources (no smoking, flares, sparks or flames in immediate area). Do not touch damaged containers or spilled material unless wearing appropriate protective clothing. Stop leak if you can do it without risk. Prevent entry into waterways, sewers, basements or confined areas. Absorb or cover with dry earth, sand or other non-combustible material and transfer to containers. Do not get water inside containers.

Regulations

RCRA 40CFR: Not listed
CERCLA: 40CFR 302.4: Not listed
SARA 40CFR 372.65: Not listed TPQ: 1000 lb
SARA EHS 40CFR 355: Listed TPQ: 1000 lb
TSCA: Listed

GLY6880 CAS #: 60828-78-6

GLYCOLS, POLYETHYLENE MONO(TRIMETHYLNONYL) TERGITOL TMN-10

RTECS: MD0907900
Synonyms: TERGITOL TMN-6

Physical Properties

Boiling Point: > 250 °C (482 °F)
Freezing Point: -16 °C (3.2 °F)
Specific Gravity: 1.024
Vapor Pressure: < 1
Water Solubility: Miscible
Evaporation Rate: < 1 Butyl Acetate=1

RTECS Toxicity Data

Acute Oral: Rat LD_{50} Dose: 7460 uL/kg.
Acute Dermal: Rabbit LD_{50} Route: Skin; Dose: 8480 uL/kg.
Irritation Eye: Rabbit Standard Draize Test Dose: 100 mg; Reaction: severe.

Hazard Overviews

Health: Severely irritating to eyes; irritating to skin. Other Acute Effects: may be harmful by inhalation, ingestion, or skin absorption; exposure can cause nausea; headache; vomiting; heavy or prolonged skin exposure may result in the absorption of harmful amounts.
Fire: Extinguishing agents: water spray; carbon dioxide, dry chemical powder or appropriate foam; foam and water spray are effective but may cause frothing. Precautions: combustible liquid.
Reactivity: Stable. Hazardous polymerization will not occur. Hazardous decomposition products: toxic fumes of: carbon monoxide, carbon dioxide.
Carcinogenicity: IARC - Not listed; NIOSH - Not listed; NTP - Not listed; ACGIH - Not listed; OSHA - Not listed; EPA - Not listed; MAK - Not listed
Primary Target Organs:

Eyes

Skin

Environmental

Regulations

RCRA 40CFR: Not listed
CERCLA: 40CFR 302.4: Not listed
SARA 40CFR 372.65: Not listed
SARA EHS 40CFR 355: Not listed
TSCA: Not listed

GLY7300 CAS #: 1405-86-3

GLYCYRRHIZIN

RTECS: MD2025000
EINECS Number: 215-785-7
Molecular Formula: $C_{42}H_{62}O_{16}$
Formula Weight: 822.95
Synonyms: 20BETA-CARBOXY-11-OXO-30-NOROLEAN-12-EN-3BETA-YL-2-O-BETA-D-GLUCOPYRANURONOSYL-ALPHA-D-GLUCOPYRANOSIDURONIC ACID; ALPHA-D-GLUCOPYRANOSIDURONIC ACID,(3BETA,20BETA)-20-CARBOXY-11-OXO-30-NOROLEAN-12-EN-3-YL-2-O-BETA-D-GLUCOPYRANURON OSYL-; GLYCYRON; GLYCYRRHETINIC ACID GLYCOSIDE; 18-BETA-GLYCYRRHIZIC ACID; GLYCYRRHIZIC ACID; GLYCYRRHIZIC ACID (8CI); BETA-GLYCYRRHIZIN; GLYCYRRHIZINIC ACID; LIQUORICE
Description: crystals, plates, or prisms
Use: demulcent, mild laxative; expectorant; to disguise taste of medications; foaming agent in root beer & mouthwashes; sweetener in chocolate, cocoa, & chewing gum; humectant in tobacco; flavoring in confectionary & pharmaceuticals

Physical Properties

Freezing Point: 220 °C (428 °F)
Water Solubility: Freely Soluble in Hot Water
Other Solubilities: Slightly Soluble in Ether.

RTECS Toxicity Data

Acute Oral: Human TD_{Lo} Dose: 280 mg/kg/4W; Toxic Effects: Behavioral - Somnolence (general depressed activity); Nutritional and gross metabolic - Changes in phosphorus. Man TD_{Lo} Dose: 662 mg/kg/1Y-I; Toxic Effects: Behavioral - Convulsions or effect on seizure threshold; Behavioral - Muscle weakness; Nutritional and gross metabolic - Changes in potassium.
Mutagenic: Hamster Micronucleus Test; Cell Type: ovary; Dose: 100 umol/L.

Hazard Overviews

Health: Acute Effects: may be harmful by inhalation, ingestion, or skin absorption; overexposure can cause sodium retention and potassium loss leading to hypertension, water retention and electrolyte imbalance. The toxicological properties have not been thoroughly investigated.
Fire: Extinguishing agents: carbon dioxide, dry chemical powder or appropriate foam. Precautions: combustible liquid.
Reactivity: Stable. Hazardous polymerization will not occur. Hazardous decomposition products: toxic fumes of: carbon monoxide, carbon dioxide.
Carcinogenicity: IARC - Not listed; NIOSH - Not listed; NTP - Not listed; ACGIH - Not listed; OSHA - Not listed; EPA - Not listed; MAK - Not listed

Environmental

Regulations

RCRA 40CFR: Not listed
CERCLA: 40CFR 302.4: Not listed

SARA 40CFR 372.65: Not listed
SARA EHS 40CFR 355: Not listed
TSCA: Listed

GLY7720 CAS #: 107-22-2

GLYOXAL

RTECS: MD2625000
EINECS Number: 203-474-9
Molecular Formula: $C_2H_2O_2$
Structured MF: OCHCHO
Formula Weight: 58.04

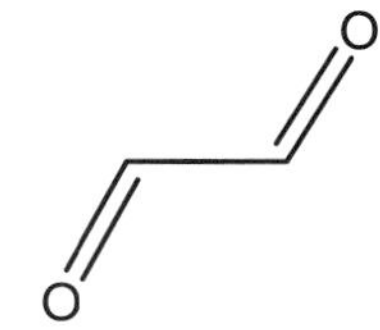

Chemical Structure

Synonyms: AEROTEX GLYOXAL 40; BIFORMAL; BIFORMYL; DIFORMAL; DIFORMYL; ETHANDIAL; ETHANEDIAL; ETHANEDIOL; 1,2-ETHANEDIONE; ETHANEDIONE; GLYOXAL ALDEHYDE; GLYOXAL,40% SOLUTION; GLYOXYLALDEHYDE; OXAL; OXALALDEHYDE
Description: yellow to white (opaque at 10 deg C) prisms or irregular pieces; green vapors; light yellow liquid; mild odor
Use: in textiles, organic synthesis, glues, biocide permanent-press fabrics, dimensional stabilization and other fibers, as an insolubilizing agent for compounds containing polyhydroxyl groups (polyvinyl alcohol, starch and cellulosic materials), in the insolubilizing of proteins (casein, gelatin and animal glue), in embalming fluids, in leather tanning, in paper coatings with hydroxyethylcellulose, as a reducing agent in dyeing textiles, and as a cosmetic ingredient and dye intermediate

Physical Properties

Boiling Point: 51 °C (124 °F) at 776 mm Hg
Freezing Point: 15 °C (59 °F)
Specific Gravity: 1.14 at 20 °C
Vapor Density: > 1 Air=1
Saturated Vapor Density: 1.228460254 kg/m^3
Density: 1.14 g/mL at 20 °C for the pure compound
Bulk Density: 10.0 lbs/gal at 20 °C
Vapor Pressure: 18.0 mm Hg at 20 °C
Water Solubility: Very Soluble in Water
Other Solubilities: 95% Ethanol: >=100 mg/ml at 22 °C; Acetone: <1 mg/ml at 22 °C; Alcohol: Soluble (pure compound); Anhydrous solvents: Soluble; DMSO: >=100 mg/ml at 22 °C; Ether: Soluble (pure compound).
pH: 40% aqueous solution 0.9
Refraction Index: 1.3826 at 20.5 °C/D
Ionization Potential (eV): 10.21 +/-0.2
Flash Point: Nonflammable solution
Autoignition Temperature: 285 °C

RTECS Toxicity Data

Acute Oral: Rat LD_{50} Dose: 200 mg/kg; Toxic Effects: Behavioral - Muscle weakness. Mouse LD_{50} Dose: 400 mg/kg; Toxic Effects: Behavioral - Muscle weakness.
Acute Dermal: Guinea Pig LD_{50} Route: Skin; Dose: 6600 mg/kg.
Chronic (Multiple Dose) Oral: Rat Dose: 26715 mg/kg/90D-C; Toxic Effects: Blood - Changes in serum composition; Nutritional and gross metabolic - Weight loss or decreased weight gain; Biochemical - Other transferases. Rat Dose: 60515 mg/kg/26W-C; Toxic Effects: Cardiac - Changes in heart weight; Liver - Changes in liver weight; Nutritional and gross metabolic - Weight loss or decreased weight gain.
Irritation Eye: Rabbit Standard Draize Test Dose: 20 mg; Reaction: severe.
Irritation Skin: Rabbit Open Draize Test Dose: 545 mg open; Reaction: mild.
Mutagenic: Human Sister Chromatid Exchenge; Cell Type: lymphocyte; Dose: 400 umol/L. Rat DNA Damage; Route: Oral; Dose: 500 mg/kg.

Hazard Overviews

Explosive

Health: Irritating to eyes/skin/respiratory tract. Other Acute Effects: may be harmful by inhalation, ingestion, or skin absorption.
Fire: Noncombustible. Hazards: emits toxic fumes. Extinguishing agents: water spray; carbon dioxide, dry chemical powder or appropriate foam. Precautions: combustible liquid.
Reactivity: Stable. Hazardous polymerization will not occur. Incompatible with: strong oxidizing agents. Hazardous decomposition products: toxic fumes of: carbon monoxide, carbon dioxide.
Carcinogenicity: IARC - Not listed; NIOSH - Not listed; NTP - Not listed; ACGIH - Not listed; OSHA - Not listed; EPA - Not listed; MAK - Not listed
Primary Target Organs:

Eyes

Skin

Respiratory System

Environmental

Cleanup/Disposal: Guide No. 171: Do not touch or walk through spilled material. Stop leak if you can do it without risk. Prevent dust cloud. Avoid inhalation of asbestos dust. Small Dry Spills: With clean shovel place material into clean, dry container and cover loosely; move containers from spill area. Small Spills: Take up with sand or other noncombustible absorbent material and place into containers for later disposal. Large Spills: Dike far ahead of liquid spill for later disposal. Cover powder spill with plastic sheet or tarp to minimize spreading. Prevent entry into waterways, sewers, basements or confined areas.

Environmental Physical Data

BCF: no food chain concentration potential

Regulations

RCRA 40CFR: Not listed
CERCLA: 40CFR 302.4: Not listed
SARA 40CFR 372.65: Not listed
SARA EHS 40CFR 355: Not listed
TSCA: Listed

GLY8140 CAS #: 298-12-4

GLYOXYLIC ACID

RTECS: MD4550000
EINECS Number: 206-058-5
Molecular Formula: $C_2H_2O_3$
Structured MF: OHCCOOH
Formula Weight: 74.04

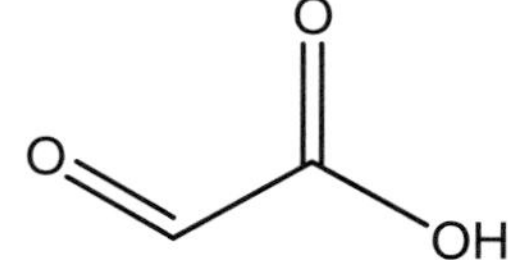

Chemical Structure

Synonyms: ACETIC ACID,OXO-; ACETIC ACID,OXO-(9CI); FORMIC ACID,FORMYL-; FORMYLFORMIC ACID; GLYOXALIC ACID; ALPHA-KETOACETIC ACID; KYSELINA GLYOXYLOVA; OXALALDEHYDIC ACID; OXOACETIC ACID; OXOETHANOIC ACID
Description: crystals; obnoxious odor
Use: intermediate for flavorings, perfumes, pharmaceuticals, dyes, plastics, agricultural chemicals; chem int for its esters-eg, methyl ester (pesticide int); chem int for vanillin

Physical Properties

Freezing Point: 98 °C (208.4 °F)
Specific Gravity: 1.42 at 20 °C/4 °C
Water Solubility: Miscible with Water
Other Solubilities: miscible with Alcohol; Insoluble in Ether, hydrocarbons
Flash Point: Not flammable

RTECS Toxicity Data

Acute Oral: Rat LD_{Lo} Dose: 3 gm/kg; Toxic Effects: Brain and coverings - Other degenerative changes; Kidney, Ureter, and Bladder - Other changes.
Mutagenic: Bacteria - S Typhimurium Mutations in Microorganisms; Dose: 200 ug/plate (+S9). Bacteria - S Typhimurium Mutations in Microorganisms; Dose: 1500 ug/plate (-S9).

Hazard Overviews

Corrosive

Health: Corrosive to eyes/skin/respiratory tract. Other Acute Effects: harmful if swallowed, inhaled, or absorbed through skin; inhalation may result in spasm; inflammation and edema of the larynx and bronchi; chemical pneumonitis and pulmonary edema; burning sensation; coughing; wheezing; laryngitis; shortness of breath; headache; nausea; vomiting; prolonged or repeated exposure may cause allergic reactions in certain sensitive individuals; possible sensitizer; possible risk of irreversible effects. Chronic Effects: laboratory experiments have shown mutagenic effects.
Fire: Noncombustible. Hazards: in powder form capable of creating a dust explosion; emits toxic fumes. Extinguishing agents: carbon dioxide, dry chemical powder or appropriate foam; water spray. Precautions: combustible liquid.
Reactivity: Incompatible with: bases, oxidizing agents, reducing agents. Hazardous decomposition products: toxic fumes of: carbon monoxide, carbon dioxide.
Carcinogenicity: IARC - Not listed; NIOSH - Not listed; NTP - Not listed; ACGIH - Not listed; OSHA - Not listed; EPA - Not listed; MAK - Not listed
Primary Target Organs:

Eyes

Skin

Respiratory System

Environmental

Regulations

RCRA 40CFR: Not listed
CERCLA: 40CFR 302.4: Not listed
SARA 40CFR 372.65: Not listed
SARA EHS 40CFR 355: Not listed
TSCA: Listed

GLY8560 CAS #: 1071-83-6

GLYPHOSATE

RTECS: MC1075000
EINECS Number: 213-997-4
Molecular Formula: $C_3H_8NO_5P$
Structured MF: $HOOCCH_2NHCH_2PO(OH)_2$
Formula Weight: 169.07

Chemical Structure

Synonyms: (CARBOXYMETHYLAMINO)METHYLPHOSPHONIC ACID; GLIALKA; GLIFONOX; GLYCEL; GLYCINE,N-(PHOSPHONOMETHYL)-; MON-0573; MON 0468; MON 0573; MON 2139; MON 6000; MUSTER; N-PHOSPHOMETHYLGLYCINE; N-(PHOSPHONOMETHYL)GLYCINE; N-PHOSPHONOMETHYLGLYCINE; PHOSPHONOMETHYLIMINOACETIC ACID; RODEO; RONDO; ROUNDUP; SONIC; SPASOR; STING; TUMBLEWEED

Description: colorless to white crystals; odorless

Use: a total pre-emergence herbicide; a broad spectrum post-emergence, translocated herbicide

Physical Properties

Freezing Point: Decomposes at 230 °C (446 °F)
Density: 0.5 g/cu cm
Vapor Pressure: Negligible
Water Solubility: 12 g/L Water at 25 °C
Other Solubilities: 95% Ethanol: <1 mg/ml at 18 °C; Acetone: <1 mg/ml at 18 °C; DMSO: <1 mg/ml at 18 °C; Most organic solvents: Insoluble.
pH: 1% solution in water 2.5
Flash Point: Not available; probably combustible

RTECS Toxicity Data

Acute Oral: Man TD_{Lo} Dose: 1214 mg/kg; Toxic Effects: Cardiac - Change in rate; Vascular - BP lowering not characterized in autonomic section; Gastrointestinal - Hypermotility, diarrhea. Man LD_{Lo} Dose: 2143 mg/kg; Toxic Effects: Vascular - BP lowering not characterized in autonomic section; Lungs, Thorax, or Respiration - Fibrosis, interstitial; Lungs, Thorax, or Respiration - Fibrosing alveolitis. Woman LD_{Lo} Dose: 4 gm/kg; Toxic Effects: Cardiac - Arrythmias (including changes in conduction); Vascular - BP lowering not characterized in autonomic section; Lungs, Thorax, or Respiration - Respiratory depression. Rat LD_{50} Dose: 4873 mg/kg; Toxic Effects: Behavioral - Convulsions or effect on seizure threshold; Lungs, Thorax, or Respiration - Respiratory stimulation; Nutritional and gross metabolic - Body temperature increase.
Acute Inhalation: Rat LC_{50} Dose: >12200 mg/m^3/4hr.
Acute Dermal: Rabbit LD_{50} Route: Skin; Dose: 7940 mg/kg.
Chronic (Multiple Dose) Oral: Rat Dose: 153 gm/kg/13W-C; Toxic Effects: Liver - Changes in liver weight; Blood - Changes in erythrocite (RBC) cell count; Biochemical - Transaminases. Rat Dose: 435 gm/kg/13W-C; Toxic Effects: Gastrointestinal - Changes in structure or function of salivary glands; Liver - Changes in liver weight; Kidney, Ureter, and Bladder - Changes in kidney weight.

Hazard Overviews

Health: Irritating to eyes/skin/respiratory tract. Harmful. Other Acute Effects: harmful if swallowed, inhaled, or absorbed through skin. Chronic Effects: may cause reproductive disorders.
Fire: Will burn. Hazards: emits toxic fumes. Extinguishing agents: water spray; carbon dioxide, dry chemical powder or appropriate foam. Precautions: combustible liquid.
Reactivity: Incompatible with: strong oxidizing agents. Hazardous decomposition products: toxic fumes of: carbon monoxide, carbon dioxide, nitrogen oxides; thermal decomposition may produce toxic fumes of phosphorus oxides and/or phosphine.
Carcinogenicity: IARC - Not listed; NIOSH - Not listed; NTP - Not listed; ACGIH - Not listed; OSHA - Not listed; EPA - Class D, Not classifiable as to human carcinogenicity; MAK - Not listed
Primary Target Organs:

Eyes

Skin

Respiratory System

Environmental

Ecotoxicity: LC_{50} Pimephales promelas (fathead minnow), wt 0.6 g, 97 mg/l/96 hr at 20 °C (95% confidence interval 79-120 mg/l) static bioassay without aeration, pH 7.2-7.5, water hardness 40-50 mg/l as calcium carbonate and alkalinity of 30-35 mg/l LC_{50} Daphnia >1000 mg/l/48 hr /Conditions of bioassay not specified LD_{50} Bobwhite quail oral >3850 mg/kg LC_{50} Quail dietary >4640 ppm/8 day
Environmental Fate: It is applied as a spray of the isopropylamine salt and is removed from the atmosphere by gravitational settling. After application to forests, fields, and other land by spraying, it is strongly adsorbed to soil, remains in the upper soil layers, and has a low propensity for leaching. It readily and completely biodegrades in soil. Its average half-life in soil is about 60 days. Biodegradation in foliage and litter is somewhat faster. In field studies, residues are often found the following year. It may enter aquatic systems through accidental spraying, spray drift, or surface runoff. It dissipates rapidly from the water column as a result of adsorption and possibly biodegradation. The half-life in water is a few days. Sediment is the primary sink. After spraying, levels in sediment rise and then decline to low levels in a few months. It does not bioconcentrate in aquatic organisms or bioaccumulate in species in higher tropic levels.

Environmental Physical Data

Sorption Partition Coefficient: K_{oc} = clay loam 76
BCF: calculated at 10 to 14

Regulations

RCRA 40CFR: Not listed
CERCLA: 40CFR 302.4: Not listed
SARA 40CFR 372.65: Not listed
SARA EHS 40CFR 355: Not listed
TSCA: Not listed

Analytical Methods

Soil: EPA PMD-GLP
Water / Groundwater: APHA 6651-B
Drinking Water: EPA 547; AOAC 991.08
Food: AOAC 983.10

GLY8980 CAS #: 38641-94-0

GLYPHOSATE ISOPROPYLAMINE SALT

RTECS: MC1080000
EINECS Number: 254-056-8
Molecular Formula: $C_6H_{17}N_2O_5P$
Formula Weight: 228.22

OH
HO P N H O⁻
O O
NH_3^+

Chemical Structure

Synonyms: BRONCO; GLIFONOX; GLIFOSATO ESTRELLA; GLYCEL; GLYCINE,N-(PHOSPHONOMETHYL)-,CMPD WITH 2-PROPANAMINE(1:1); GLYPHOSATE MONO(ISOPROPYLAMINE) SALT; ISOPROPYLAMINE SALT OF N-(PHOSPHONO-METHYL)GLYCINE; ISOPROPYLAMINE SALT OF N-(PHOSPHONOMETHYL)GLYCINE; LANDMASTER; MON-2139; MON 139; MON 39; MONO-ISOPROPYLAMMONIOVA SUL; NITOSORG; N-(PHOSPHONOMETHYL)GLYCINE ISOPROPYLAMINE SALT; N-(PHOSPHONOMETHYL)GLYCINE ISOPROPYLAMMONIUM SALT; N-(PHOSPHONOMETHYL)GLYCINE MONOISOPROPYLAMINE SALT; RATTLER; RODEO; ROUND-UP; ROUNDUP; SPASOR; STING; UTAL
Description: amber, clear, viscous solution; practically odorless to slight amine-like odor
Use: herbicide; for control of many annual & perennial grasses & broadleaf weeds plus many tree & woody brush species in cropland & noncrop sites

Physical Properties

Specific Gravity: 1.17 (Water=1)
Water Solubility: Very Soluble in Water
pH: 4.9
Flash Point: > 93.333 °C Closed Cup

RTECS Toxicity Data

Acute Oral: Rat LD_{50} Dose: 10537 mg/kg.
Acute Dermal: Rat LD_{50} Route: Skin; Dose: 7500 mg/kg.
Mutagenic: Human Sister Chromatid Exchenge; Cell Type: lymphocyte; Dose: 650 umol/L. Insects - D Melanogaster Specific Locus Test; Route: Multiple routes; Dose: 1 ppm.

Hazard Overviews

Health: Irritating to eyes/skin/respiratory tract. Toxic. Other Acute Effects: harmful if swallowed, inhaled, or absorbed through skin; readily absorbed through skin.
Fire: Will burn. Hazards: emits toxic fumes. Extinguishing agents: carbon dioxide, dry chemical powder or appropriate foam. Precautions: combustible liquid.
Reactivity: Incompatible with: strong oxidizing agents. Hazardous decomposition products: toxic fumes of: carbon monoxide, carbon dioxide, nitrogen oxides; thermal decomposition may produce toxic fumes of phosphorus oxides and/or phosphine.
Carcinogenicity: IARC - Not listed; NIOSH - Not listed; NTP - Not listed; ACGIH - Not listed; OSHA - Not listed; EPA - Not listed; MAK - Not listed
Primary Target Organs:

Eyes

Skin

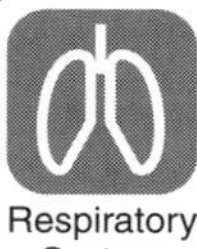
Respiratory System

Environmental

Ecotoxicity: EC_{50} Chironomus plumosus, fourth instar, 55 mg/l/96 hr at 22 °C, 95% confidence limit 31-97 mg/l, static bioassay without aeration, pH 7.2-7.5, water hardness 40-50 mg/l as calcium carbonate and alkalinity of 30-35 mg/l EC_{50} Daphnia magna, first instar, 3.0 mg/l/48 hr at 22 °C, 95% confidence limit 2.6-3.4 mg/l, static bioassay without aeration, pH 7.2-7.5, water hardness 40-50 mg/l as calcium carbonate and alkalinity of 30-35 mg/l LC_{50} Pimephals promelas (fathead minnow), wt 0.6 g, 2.3 mg/l/96 hr at 22 °C, 95% confidence limit 1.9-2.8 mg/l, static bioassay without aeration, pH 7.2-7.5, water hardness 40-50 mg/l as calcium carbonate and alkalinity of 30-35 mg/l

Regulations

RCRA 40CFR: Not listed
CERCLA: 40CFR 302.4: Not listed
SARA 40CFR 372.65: Not listed
SARA EHS 40CFR 355: Not listed
TSCA: Not listed

GOL5000 CAS #: 7440-57-5

GOLD

RTECS: MD5070000
EINECS Number: 231-165-9
Molecular Formula: Au
Structured MF: Au
Formula Weight: 196.9665

Au

Chemical Structure

Synonyms: BURNISH GOLD; C.I. 77480; C.I. PIGMENT METAL 3; COLLOIDAL GOLD; GOLD-197; GOLD FLAKE; GOLD LEAF; GOLD POWDER; MAGNESIUM GOLD PURPLE; SHELL GOLD

Description: yellow, soft metal, when finely divided may be black, ruby, or purple

Use: in mfr of jewelry; in plating other metals; alloyed with platinum, silver & copper; std of currency; antineoplastic agent, radioactive colloidal gold; in photography, dentistry; infrared reflectors; electrical contact & brazing alloys; polarographic electrodes; spinnerets; laboratory ware; in electronics; colloidal dispersions in coloring glass, nucleating agent; gold leaf in surgery; in art prodjects; plating agent in electron tubes; in fabricated bars for investment.

Physical Properties

Boiling Point: 2700 °C (4892 °F)
Freezing Point: 1064.76 °C (1948.568 °F)
Specific Gravity: 19.3
Water Solubility: Insoluble in Water
Other Solubilities: Insoluble in acid; Soluble in Aqua Regia, Potassium Cyanide, hot Sulfuric acid
Ionization Potential (eV): 9.22567

RTECS Toxicity Data

Tumorigenic: Rat Route: Implant; Dose: 200 mg/kg; Toxic Effects: Tumorigenic - Equivocal tumorigenic agent by RTECS criteria; Tumorigenic - Tumors at site of application. Rat Route: Implant; Dose: 4730 mg/kg; Toxic Effects: Tumorigenic - Equivocal tumorigenic agent by RTECS criteria; Tumorigenic - Tumors at site of application.

Hazard Overviews

Health: Irritating. Other Acute Effects: may be harmful by inhalation, ingestion, or skin absorption; exposure can cause contact dermatitis. The toxicological properties have not been thoroughly investigated.

Fire: Hazards: emits toxic fumes. Extinguishing agents: use extinguishing media appropriate to surrounding fire conditions. Precautions: combustible liquid.

Reactivity: Stable. Hazardous polymerization will not occur. Incompatible with: hydrogen peroxide, halogens, ammonia. Hazardous decomposition products: carbon monoxide, carbon dioxide.

Carcinogenicity: IARC - Not listed; NIOSH - Not listed; NTP - Not listed; ACGIH - Not listed; OSHA - Not listed; EPA - Not listed; MAK - Not listed

Primary Target Organs:

Eyes

Skin

Respiratory System

Environmental

Regulations

RCRA 40CFR: Not listed
CERCLA: 40CFR 302.4: Not listed
SARA 40CFR 372.65: Not listed
SARA EHS 40CFR 355: Not listed
TSCA: Listed

Analytical Methods

Water / Groundwater: EPA 200.0, 231.1, 231.2; APHA 3111-A, 3111-B, 3500-AU; FISON AES-0029
Other: EPA 1620

GRA5000 CAS #: 7782-42-5

GRAPHITE

RTECS: MD9659600
EINECS Number: 231-955-3
Molecular Formula: C
Structured MF: C
Formula Weight: 12.01

C

Chemical Structure

Synonyms: AERODAG G; AG 1500; AQUADAG; AS 1; AT 20; ATJ-S; ATJ-S GRAPHITE; BLACK LEAD; C.I. 77265; C.I. PIGMENT BLACK 10; CANLUB; CB 50; CEYLON BLACK LEAD; CPB 5000; DC 2; EG 0; ELECTROGRAPHITE; EXP-F; FORTAFIL 5Y; GK 2; GK 3; GP 60S; GP 60; GP 63; GRAFOIL; GRAFOIL GTA; GRAPHITE,NATURAL; GRAPHNOL N 3M; GS 2; GY 70; H 451; HITCO HMG 50; IG 11; KOROBON; MG 1; MINERAL CARBON; MPG 6; PAPYEX; PG 50; PLUMBAGO; PLUMBAGO (GRAPHITE); PYRO-CARB 406; ROCOL X 7119; S 1; S 1 (GRAPHITE); SCHUNGITE; SHUNGITE; SILVER GRAPHITE; SKLN 1; STOVE BLACK; SWEDISH BLACK LEAD; UCAR 38; VVP 66-95

Physical Properties

Boiling Point: Sublimes
Freezing Point: Sublimes at 3650 °C (6602 °F)
Specific Gravity: 2.1
Vapor Pressure: ~ 0 mm Hg
Water Solubility: Insoluble
pH: 6.6
Flash Point: Will not burn

Hazard Overviews

Reactivity: Stable. Hazardous polymerization cannot occur. Incompatible with: chlorine trifluoride; fluorine; potassium; potassium superoxide Hazardous decomposition products: carbon monoxide.

Carcinogenicity: IARC - Not listed; NIOSH - Listed as carcinogen; NTP - Not listed; ACGIH - Not listed; OSHA - Not listed; EPA - Not listed; MAK - Not listed

Exposure Limits
OSHA PEL: mppcf.
OSHA PEL Vacated 1989 Limits: TWA: 2.5 mg/m^3; respirable.
ACGIH TLV: TWA: 2 mg/m^3.
NIOSH IDLH: 1250 mg/m^3.
DFG MAK: TWA: 6 mg/m^3.

Respirator Recommendation
Exposure Range: >2.5 to 25 mg/m^3 Air Purifying, Negative Pressure, Half Mask
Exposure Range: >25 to 250 mg/m^3 Air Purifying, Negative Pressure, Full Face
Exposure Range: >250 to <1250 mg/m^3 Supplied Air, Constant Flow/Pressure Demand, Full Face
Exposure Range: 1250 to unlimited mg/m^3 Self-contained Breathing Apparatus, Pressure Demand, Full Face
Cartridge Color: dust/mist filter (use P100 or consult supervisor for appropriate dust/mist filter)

Environmental

Regulations
RCRA 40CFR: Not listed
CERCLA: 40CFR 302.4: Not listed
SARA 40CFR 372.65: Not listed
SARA EHS 40CFR 355: Not listed
TSCA: Listed

GRA9000 CAS #: 4720-09-6

GRAYANOTOXIN I

RTECS: PB9195000
Molecular Formula: $C_{22}H_{36}O_7$
Formula Weight: 412.58
Synonyms: ACETYLANDROMEDOL; ACETYLLANDROMEDOL; ANDROMEDOTOXIN; ASEBOTOXIN; G-I; GRAYANOTOXANE-3,5,6,10,14,16-HEXOL 14-ACETATE; GRAYANOTOXANE-3,5,6,10,14,16-HEXOL,14-ACETATE,(3-BETA,6-BETA,14R)-; GRAYANOTOXANE-3,5,6,10,14,16-HEXOL,14-ACETATE,(3BETA,6BETA,14R)-; 7,9A-METHANO-9AH-CYCLOPENTA(B)HEPTALENE-2,4,8,11,11A,12(1H)-HEXOL-, DODECAHYDRO-1,1,4,8-TETRAMETHYL-,12-ACETATE, (2S,3AS,4AR,7R,8R,9AS,11R,11AR,12R)-; RHODOTOXIN; RHODOTOXINE
Description: crystals
Use: physiological research

Physical Properties
Freezing Point: 258 °C (496.4 °F) to 270 °C (518 °F)
Water Solubility: Soluble in Hot Water
Other Solubilities: Soluble in Alcohol, Acetic Acid, hot Chloroform; Very Slightly Soluble in Benzene, Ether, Petroleum Ether

RTECS Toxicity Data
Acute Dermal: Rabbit LD_{Lo} Route: Subcutaneous Dose: 280 ug/kg; Toxic Effects: Behavioral - Muscle weakness; Lungs, Thorax, or Respiration - Respiratory depression. Mouse LD_{50} Route: Subcutaneous Dose: 148 ug/kg; Toxic Effects: Lungs, Thorax, or Respiration - Other changes.

Hazard Overviews
Carcinogenicity: IARC - Not listed; NIOSH - Not listed; NTP - Not listed; ACGIH - Not listed; OSHA - Not listed; EPA - Not listed; MAK - Not listed

Environmental

Regulations
RCRA 40CFR: Not listed
CERCLA: 40CFR 302.4: Not listed
SARA 40CFR 372.65: Not listed
SARA EHS 40CFR 355: Not listed
TSCA: Not listed

GRI5000 CAS #: 126-07-8

GRISEOFULVIN

RTECS: WG9800000
EINECS Number: 204-767-4
Molecular Formula: $C_{17}H_{17}ClO_6$
Formula Weight: 352.77

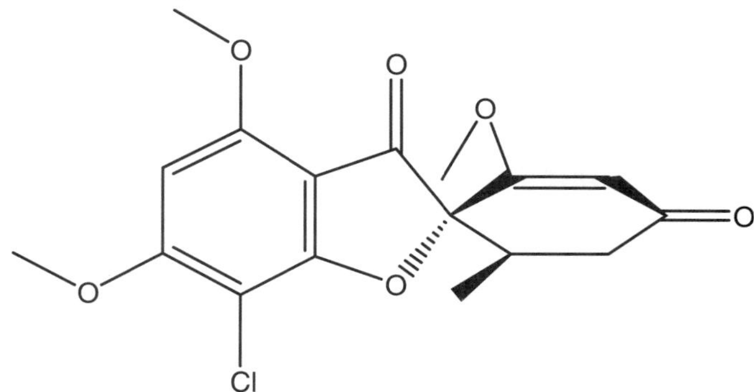

Chemical Structure

Synonyms: AMUDANE; BIOGRISIN-FP; 7-CHLORO-4,6-DIMETHOXYCOUMARAN-3-ONE-2-SPIRO-1'-(2'-METHOXY)-6'-METHYLCYCLO-; 7-CHLORO-4,6-DIMETHOXYCOUMARAN-3-ONE-2-SPIRO-1'-(2'-METHOXY-6'-METHYLCYCLOHEX-2'-EN-4'-ONE); (2S,6'R)-7-CHLORO-2',4,6-TRIMETHOXY-6'-METHYLBENZOFURAN-2-SPIRO-1'-CYCLOHEX-; 7-CHLORO-4,6,2'-TRIMETHOXY-6'-METHYLGRIS-2'-EN-3,4'-DIONE; 7-CHLORO-2',4,6-TRIMETHOXY-6'-METHYLSPIRO[BENZOFURAN-2(3H),1'-[2]CYCLOHEXENE; (2S-TRANS)-7-CHLORO-2',4,6-TRIMETHOXY-6'-METHYLSPIRO[BENZOFURAN-2(3H),1'-[2]CYCLOHEXENE]-3,4'-DIONE; 7-CHLORO-2',4,6-TRIMETHOXY-6'-BETA-SPIRO(BENZOFURAN-2(3H),1'-(2)CYCLOHEXENE; CURLING FACTOR; DELMOFULVINA; 3,4'-DIONE; 2'-ENE-3,4'-DIONE; FULCIN; FULCINE; FULVICAN GRISACTIN; FULVICIN; FULVICIN-P/G; FULVICIN-U/F; FULVINA; FULVINIL; FULVISTATIN; FUNGIVIN; GREOSIN; GRESFEED; GRICIN; GRIFULIN; GRIFULVIN; GRIFULVIN V; GRISACTIN; GRISCOFULVIN; GRISEFULINE; GRISEO; (+)-GRISEOFULVIN; GRISEOFULVIN,(+)-; GRISEOFULVIN-FORTE; GRISEOFULVINUM; GRISETIN; GRISOFULVIN; GRISOVIN; GRIS-PEG; GRIZEOFULVIN; GRYSIO; GUSERVIN; HEX-2'-EN-4'-ONE; LAMORYL; LIKUDEN; MURFULVIN; NEO-FULCIN; NSC 34533; PONCYL; SPIRO(BENZOFURAN-2(3H),1'-(2)CYCLOHEXENE)-3,4'-DIONE,7-CHLORO-2',4,6-TRIMETHOXY-6'-BETA-METHYL-; SPIRO(BENZOFURAN-2(3H),1'-(2)CYCLOHEXENE)-3,4'-DIONE,7-CHLORO-2',4,6-TRIMETHOXY-6'-METHYL-,(2S-TRANS)-; SPIROFULVIN; SPOROSTATIN
Description: white to creamy powder; odorless

Use: antifungal antibiotic; feed additive; antimitotic and treatment of fungal diseases of tomatoes, mushrooms and other crops

Physical Properties

Boiling Point: Sublimes at 210 °C (410 °F)
Freezing Point: 220 °C (428 °F)
Water Solubility: Practically Insoluble in Water
Other Solubilities: DMSO: 10-50 mg/ml at 21 °C; 95% Ethanol: <1 mg/ml at 21 °C; Acetone: 10-50 mg/ml at 21 °C; Benzene: Slightly Soluble; Chloroform: 1 in 25 at 20 °C; Mineral oil: <0.1 g/L at 25 °C; Propylene Glycol: 2 g/L at 25 °C.
Refraction Index: 1.5403 at 25 °C/D
Flash Point: Not available; probably combustible

RTECS Toxicity Data

Acute Oral: Rat LD_{50} Dose: >10 gm/kg. Mouse LD_{50} Dose: >50 gm/kg.
Acute Dermal: Mouse LD_{50} Route: Subcutaneous Dose: >12 gm/kg.
Chronic (Multiple Dose) Oral: Guinea Pig Dose: 109 gm/kg/52W-C; Toxic Effects: Nutritional and gross metabolic - Weight loss or decreased weight gain.
Reproductive/Teratogenic: Rat Route: Oral; Dose: 1250 mg/kg; Duration: female 6-15D of pregnancy; Effects on Embryo or Fetus - Fetotoxicity; Effects on Newborn - Viability index. Rat Route: Oral; Dose: 2 gm/kg; Duration: female 11-14D of pregnancy Specific Developmental Abnormalities - Central nervous system; Urogenital system. Rat Route: Oral; Dose: 12500 mg/kg; Duration: female 6-15D of pregnancy; Effects on Fertility - Post-implantation mortality; Specific Developmental Abnormalities - Musculoskeletal system. Rat Route: Oral; Dose: 500 mg/kg; Duration: female 9D of pregnancy; Effects on Embryo or Fetus - Fetal death; Specific Developmental Abnormalities - Eye, ear.
Mutagenic: Human DNA Inhibition; Cell Type: fibroblast; Dose: 20 mg/L/3D-C. Human DNA Inhibition; Cell Type: lymphocyte; Dose: 20 mg/L/3D-C. Human Cytogenetic Analysis; Cell Type: lymphocyte; Dose: 40 mg/L/3D. Human Micronucleus Test; Cell Type: lymphocyte; Dose: 5 mg/L. Human Sex Chromosome Loss; Cell Type: lymphocyte; Dose: 5 mg/L.
Tumorigenic: Rat Route: Oral; Dose: 462 gm/kg/2Y-I; Toxic Effects: Tumorigenic - Neoplastic by RTECS criteria; Blood - Lymphomax including Hodgkin's disease. Mouse Route: Oral; Dose: 440 gm/kg/52W-C; Toxic Effects: Tumorigenic - Neoplastic by RTECS criteria; Liver - Tumors. Mouse Route: Oral; Dose: 730 gm/kg/52W-C; Toxic Effects: Tumorigenic - Neoplastic by RTECS criteria; Liver - Tumors.

Hazard Overviews

Health: Irritating to eyes/skin/respiratory tract. Toxic. Other Acute Effects: toxic if swallowed, inhaled, or absorbed through skin; allergic skin reaction; photosensitivity; allergic reactions resulting in dermatologic lesions following exposure to light, from sunburnlike responses to edematous, vesiculated lesions or bullae; lupus erythematosus; anorexia; diarrhea; nausea; vomiting; cross-allergenicity with penicillin. Chronic Effects: congenital malformation in the fetus; reproductive disorders; increased toxic effects with exposure to and/or consumption of alcohol; reactions of hypersensitivity such as skin rashes, urticaria and angioneurotic edema rarely occurring; reactions of the nervous system; headache; peripheral neuritis; lethargy; mental confusion; fatigue; syncope; vertigo; blurred vision; transient macular edema and impairment of routine task performance; hematological effects; leukopenia; neutropenia; monocytosis and punctate basophilia; nausea; vomiting; oral thrush; diarrhea; epigastric distress; paresthesias of the hands and feet; gastrointestinal bleeding; menstrual irregularities and hepatotoxicity; target organs: liver, blood. Possible human carcinogen. Possibly causes reproductive effects in humans.
Fire: Will burn. Hazards: emits toxic fumes. Extinguishing agents: water spray; carbon dioxide, dry chemical powder or appropriate foam. Precautions: combustible liquid.
Reactivity: Incompatible with: strong oxidizing agents. Hazardous decomposition products: toxic fumes of: carbon monoxide, carbon dioxide, hydrogen chloride gas.
Carcinogenicity: IARC - Group 2B, Possibly carcinogenic to humans; NIOSH - Not listed; NTP - Not listed; ACGIH - Not listed; OSHA - Not listed; EPA - Not listed; MAK - Not listed
Primary Target Organs:

Eyes

Skin

Respiratory System

Liver

Blood

Environmental

Environmental Physical Data

Octanol/Water Partition Coefficient: log K_{ow} = 2.18

Regulations

RCRA 40CFR: Not listed
CERCLA: 40CFR 302.4: Not listed
SARA 40CFR 372.65: Not listed
SARA EHS 40CFR 355: Not listed
TSCA: Not listed

GUA2330 CAS #: 5051-62-7

GUANABENZ

EINECS Number: 225-750-8
Molecular Formula: $C_8H_8Cl_2N_4$
Formula Weight: 231.07

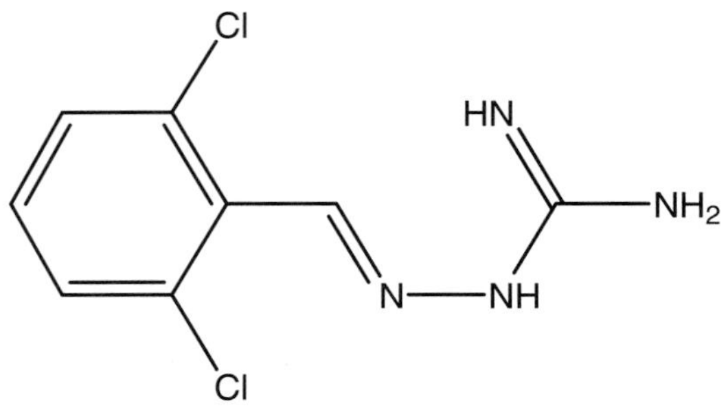

Chemical Structure

Synonyms: N-(2,6-DICHLOROBENZYLIDENE)-N'-AMIDINOHYDRAZINE; N-(2,6-DICHLOROBENZYLIDENE)AMINO]GUANIDINE; 2-[(2,6-DICHLOROPHENYL)METHYLENE]-HYDRAZINECARBOXIMIDAMIDE; NSC-68982
Description: white solid; slight odor
Use: medication: antihypertensive; fungal metabolite; potent inhibitor of human menopausal gonadotropin-coenzyme a reductase, the rate controlling enzyme in cholesterol biosynthesis

Physical Properties

Freezing Point: Decomposes at 227 °C (440.6 °F) to 229 °C (444.2 °F)
Water Solubility: 11 mg/mL
Other Solubilities: solubilities of 50 mg/ml in Alcohol, at 25 °C.

Hazard Overviews

Health: Toxic. Other Acute Effects: harmful if swallowed, inhaled, or absorbed through skin; exposure can cause nausea; dizziness; headache; sedation; drowsiness; dry mouth; weakness; constipation. Chronic Effects: possible teratogen; overexposure may cause reproductive disorder(s) based on tests with laboratory animals. The toxicological properties have not been thoroughly investigated.
Fire: Hazards: emits toxic fumes. Extinguishing agents: carbon dioxide, dry chemical powder or appropriate foam. Precautions: combustible liquid.
Reactivity: Stable. Hazardous polymerization will not occur. Hazardous decomposition products: thermal decomposition may produce carbon monoxide, carbon dioxide, and nitrogen oxides.
Carcinogenicity: IARC - Not listed; NIOSH - Not listed; NTP - Not listed; ACGIH - Not listed; OSHA - Not listed; EPA - Not listed; MAK - Not listed

Environmental

Regulations

RCRA 40CFR: Not listed
CERCLA: 40CFR 302.4: Not listed
SARA 40CFR 372.65: Not listed
SARA EHS 40CFR 355: Not listed
TSCA: Not listed

GUA3660 CAS #: 40580-59-4

GUANADREL

Molecular Formula: $C_{10}H_{19}N_3O_2$
Formula Weight: 213.28
Synonyms: (1,4-DIOXASPIRO[4.5]DEC-2-YLMETHYL)-GUANIDINE
Description: white to off-white crystalline powder or crystals
Use: medication: antihypertensive

Physical Properties

Freezing Point: Crystals at 213.5 °C (416.3 °F) to 215 °C (419 °F)
Water Solubility: 76 mg/mL at 25 °C

Hazard Overviews

Carcinogenicity: IARC - Not listed; NIOSH - Not listed; NTP - Not listed; ACGIH - Not listed; OSHA - Not listed; EPA - Not listed; MAK - Not listed

Environmental

Regulations

RCRA 40CFR: Not listed
CERCLA: 40CFR 302.4: Not listed
SARA 40CFR 372.65: Not listed
SARA EHS 40CFR 355: Not listed
TSCA: Not listed

GUA4990 CAS #: 55-65-2

GUANETHIDINE

RTECS: MF2975000
EINECS Number: 200-241-3
Molecular Formula: $C_{10}H_{22}N_4$
Formula Weight: 198.31
Synonyms: ABAPRESIN; 2-(1'-AZACYCLOOCTYL)ETHYLGUANIDINE; AZOCINE,1-((2-(AMINOIMINOMETHYL)AMINO)ETHYL)OCTAHYDRO-; AZOCINE,1-(2-GUANIDINOETHYL)OCTAHYDRO-; DOPAM; DOPOM; EUTENSOL; GUANIDINE,(2-(HEXAHYDRO-1(2H)-AZOCINYL)ETHYL)-; 2-(1-N,N-HEPTAMETHYLENEIMINO)ETHYLGUANIDINE; HEPTAMETHYLENIMINE,1-(2-GUANIDINOETHYL)-; ISMELIN; (2-(OCTAHYDRO-1-AZOCINYL)ETHYL)GUANIDINE; OCTATENSINE; OKTADIN; OKTATENSIN; OKTATENZIN; N-(2-PERHYDROAZOCIN-1-YLETHYL)GUANIDINE; SANOTENSIN; SU 5864
Description: white, crystalline powder or crystals; strong characteristic odor
Use: medication: antihypertensive; medication: in treatment of glaucoma; medication: in hyperthyroidism & thyrotoxic crisis

Physical Properties

Water Solubility: Very Soluble in Water
Other Solubilities: Slightly Soluble in Alcohol; Practically Insoluble in Chloroform.

RTECS Toxicity Data

Acute Oral: Rat LD_{50} Dose: 1050 mg/kg. Mouse LD_{50} Dose: 845 mg/kg.
Acute Dermal: Mouse LD_{50} Route: Subcutaneous Dose: 224 mg/kg.

Hazard Overviews

Carcinogenicity: IARC - Not listed; NIOSH - Not listed; NTP - Not listed; ACGIH - Not listed; OSHA - Not listed; EPA - Not listed; MAK - Not listed

Environmental

Regulations

RCRA 40CFR: Not listed
CERCLA: 40CFR 302.4: Not listed
SARA 40CFR 372.65: Not listed
SARA EHS 40CFR 355: Not listed
TSCA: Not listed

GUA6320 CAS #: 506-93-4

GUANIDINE, MONONITRATE

RTECS: MF4350000
EINECS Number: 208-060-1
Molecular Formula: $CH_6N_4O_3$
Formula Weight: 122.11

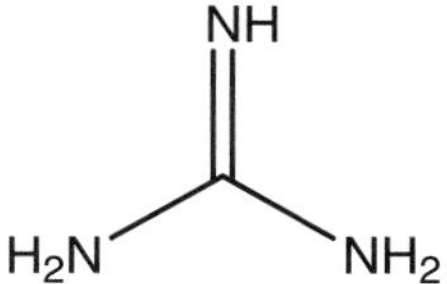

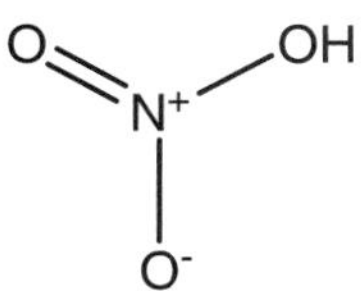

Chemical Structure

Synonyms: GUANIDINE NITRATE; GUANIDINE NITRATE (1:1); GUANIDINIUM NITRATE
Description: crystalline powder
Use: medication: striated muscle stimulant; in manufacture of explosives, disinfectants, photographic materials

Physical Properties

Boiling Point: Decomposes
Freezing Point: 214 °C (417.2 °F)
Water Solubility: 1 parts in 10 parts Water
Other Solubilities: Slightly Soluble in Acetone.
pH: Aqueous solution is neutral

RTECS Toxicity Data

Acute Oral: Rat LD_{50} Dose: 730 mg/kg; Toxic Effects: Behavioral - Altered sleep time (including change in righting reflex); Behavioral - Muscle contraction or spasticity; Behavioral - Irritability. Mouse LD_{50} Dose: 1028 mg/kg; Toxic Effects: Behavioral - Altered sleep time (including change in righting reflex); Behavioral - Excitment; Behavioral - Change in motor activity (specific assay).
Acute Dermal: Rabbit LD; Route: Skin; Dose: >2 gm/kg; Toxic Effects: Skin and appendages - Dermatitis, other.
Irritation Eye: Rabbit Standard Draize Test Dose: 92 mg; Reaction: mild.
Irritation Skin: Rabbit Standard Draize Test Dose: 500 mg; Reaction: severe.

Hazard Overviews

Health: Irritating to eyes/skin/respiratory tract. Harmful. Other Acute Effects: harmful if swallowed; may be harmful if inhaled or absorbed through the skin.
Fire: Hazards: may explode or cause fire; contact with other material may cause fire. Extinguishing agents: water spray; carbon dioxide, dry chemical powder or appropriate foam. Precautions: combustible liquid.
Reactivity: Incompatible with: strong oxidizing agents, avoid contact with metals, store away from heat and direct sunlight. Hazardous decomposition products: toxic fumes of: carbon monoxide, carbon dioxide, nitrogen oxides.
Carcinogenicity: IARC - Not listed; NIOSH - Not listed; NTP - Not listed; ACGIH - Not listed; OSHA - Not listed; EPA - Not listed; MAK - Not listed
Primary Target Organs:

Eyes

Skin

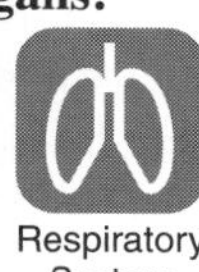
Respiratory System

Environmental

Regulations

RCRA 40CFR: Not listed
CERCLA: 40CFR 302.4: Not listed
SARA 40CFR 372.65: Not listed
SARA EHS 40CFR 355: Not listed
TSCA: Listed

GUA7650 CAS #: 73-40-5

GUANINE

RTECS: MF8260000
EINECS Number: 200-799-8
Molecular Formula: $C_5H_5N_5O$
Formula Weight: 151.13

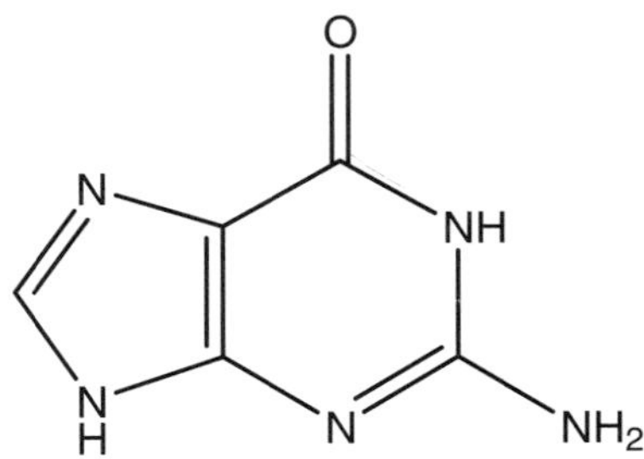

Chemical Structure

Synonyms: 2-AMINO-6-HYDROXYPURINE; 2-AMINOHYPOXANTHINE; C.I. 75170; C.I. NATURAL WHITE 1; DEW PEARL; GUANIN; GUANINE ENOL; 6-HYDROXY-2-AMINOPURINE; HYPOXANTHINE,2-AMINO-; MEARLMAID; NATURAL PEARL ESSENCE; NATURON; PATHOCIDIN; PEARL ESSENCE; 6H-PURIN-6-ONE,2-AMINO-1,7-DIHYDRO-; STELLA POLARIS

Description: usually amorphous; small rhombic crystals, needles or plates

Use: biochemical research; cosmetics; color additive

Physical Properties

Boiling Point: Sublimes

Freezing Point: Decomposes at 360 °C (680 °F)

Water Solubility: Almost Insoluble in Water

Other Solubilities: Soluble in Ammonium Hydroxide & alkali; Slightly Soluble in Ether; Insoluble in Acetic Acid, Slightly Soluble in Ether.

Ionization Potential (eV): 7.85

RTECS Toxicity Data

Mutagenic: Human Sex Chromosome Loss; Cell Type: lymphocyte; Dose: 30 umol/L. Mouse Cytogenetic Analysis; Route: Intraperitoneal; Dose: 15 mg/kg.

Tumorigenic: Rat Route: Subcutaneous; Dose: 1300 mg/kg/26W-I; Toxic Effects: Tumorigenic - Equivocal tumorigenic agent by RTECS criteria; Skin and appendages - Tumors; Tumorigenic effects - Ovarian tumors.

Hazard Overviews

Health: Irritating to eyes/skin/respiratory tract. Other Acute Effects: may be harmful by inhalation, ingestion, or skin absorption.

Fire: Hazards: emits toxic fumes. Extinguishing agents: water spray; carbon dioxide, dry chemical powder or appropriate foam. Precautions: combustible liquid.

Reactivity: Incompatible with: strong oxidizing agents. Hazardous decomposition products: toxic fumes of: carbon monoxide, carbon dioxide,nitrogen oxides.

Carcinogenicity: IARC - Not listed; NIOSH - Not listed; NTP - Not listed; ACGIH - Not listed; OSHA - Not listed; EPA - Not listed; MAK - Not listed

Primary Target Organs:

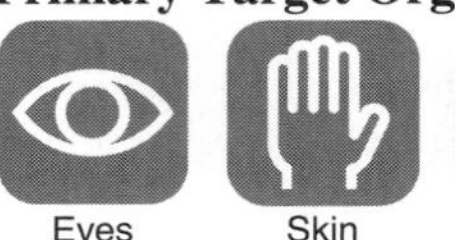

Environmental

Regulations

RCRA 40CFR: Not listed

CERCLA: 40CFR 302.4: Not listed

SARA 40CFR 372.65: Not listed

SARA EHS 40CFR 355: Not listed

TSCA: Listed

GUA8980 CAS #: 9000-30-0

GUAR GUM

RTECS: MG0185000

EINECS Number: 232-536-8

Molecular Formula: Unknown

Formula Weight: Approximately 220,000-250,000

Synonyms: 1212A; A-20D; BURTONITE V-7-E; BURTONITE V 7E; CYAMOPSIS GUM; DEALCA TP1; DEALCA TP2; DECORPA; GALACTASOL; GENDRIV 162; GUAR; GUAR FLOUR; GUARAN; GUM CYAMOPSIS; GUM GUAR; INDALCA AG; INDALCA AG-BV; INDALCA AG-HV; J 2FP; JAGUAR; JAGUAR 6000; JAGUAR A 20B; JAGUAR A 20D; JAGUAR A 40F; JAGUAR A 20 B; JAGUAR GUM A-20-D; JAGUAR NO.124; JAGUAR NO 124; JAGUAR PLUS; LYCOID DR; REGONAL; REGONOL; REIN GUARIN; SUPERCOL G.F; SUPERCOL GF; SUPERCOL U POWDER; SYNGUM D 46D; UNI-GUAR

Description: yellowish white powder; odorless

Use: reduces friction drag of water on metals; paper coating; cosmetics; pharmaceuticals; binder in tablet mixtures; interior coating of fire-hose nozzles; fracturing aid in oil wells; textiles; printing; polishing; thickener and emulsifier in food products, e.g., cheese spreads, ice cream, desserts, salad dressings, and soups; protective colloid, stabilizers, pharmaceutical jelly formulations, in suspensions, lotions, creams, toothpastes; mining industry as a flocculant; filtering agent, water treatment; as coagulant aid

Physical Properties

Bulk Density: 2 kg/cu m

Water Solubility: Completely Soluble in Water

Other Solubilities: 95% Ethanol, Acetone, and DMSO: <1 mg/ml at 19 °C; Esters: Practically Insoluble; Greases: Practically Insoluble; Hydrocarbons: Practically Insoluble; Ketones: Practically Insoluble; Oils: Practically Insoluble.

pH: 1% solution may reach 5.5 to 6.1

Flash Point: Not available; probably combustible

RTECS Toxicity Data

Acute Oral: Rat LD_{50} Dose: 6770 mg/kg. Mouse LD_{50} Dose: 8100 mg/kg.

Hazard Overviews

Health: May cause irritation to eyes/skin. Other Acute Effects: may be harmful by inhalation, ingestion, or skin absorption.
Fire: Will burn. Hazards: emits toxic fumes. Extinguishing agents: water spray; carbon dioxide, dry chemical powder or appropriate foam. Precautions: combustible liquid.
Reactivity: Incompatible with: strong oxidizing agents. Hazardous decomposition products: toxic fumes of: carbon monoxide, carbon dioxide.
Carcinogenicity: IARC - Not listed; NIOSH - Not listed; NTP - Not listed; ACGIH - Not listed; OSHA - Not listed; EPA - Not listed; MAK - Not listed

Environmental

Regulations

RCRA 40CFR: Not listed
CERCLA: 40CFR 302.4: Not listed
SARA 40CFR 372.65: Not listed
SARA EHS 40CFR 355: Not listed
TSCA: Listed

GUM1000 CAS #: 9000-01-5

GUM ARABIC

RTECS: CE5945000
EINECS Number: 232-519-5
Molecular Formula: Unknown
Formula Weight: Average about 600,000
Synonyms: ACACIA; ACACIA DEALBATA GUM; ACACIA GUM; ACACIA SENEGAL; ACACIA SYRUP; ARABIC GUM; AUSTRALIAN GUM; GUM ACACIA; GUM DRAGON; GUM OVALINE; GUM SENEGAL; INDIAN GUM; SENEGAL GUM; STARSOL NO 1; STARSOL NO. 1; WATTLE GUM
Description: colorless to light brown tears, drops, flakes, powder
Use: pharmaceuticals, adhesives, inks, textile printing, cosmetics, thickening agent and colloidal stabilizer in confectionery and food products

Physical Properties

Specific Gravity: 1.35 to 1.49
Water Solubility: Readily Soluble in Water
Other Solubilities: 95% Ethanol: <1 mg/ml at 18 °C; Acetone: <1 mg/ml at 18 °C; DMSO: <1 mg/ml at 18 °C; Glycerol: Soluble; Propylene Glycol: Soluble.
pH: Aqueous solution 4.5 to 5.0
Flash Point: Not available; probably combustible

RTECS Toxicity Data

Acute Oral: Rat LD_{50} Dose: >16 gm/kg. Mouse LD_{50} Dose: >16 gm/kg.
Irritation Eye: Rabbit Standard Draize Test Dose: 36 mg/5H; Reaction: severe.
Reproductive/Teratogenic: Rat Route: Oral; Dose: 350 gm/kg; Duration: male 10W prior to mating; Effects on Fertility - Pre-implantation mortality.
Mutagenic: Rat Dominant Lethal Test; Route: Oral; Dose: 54600 mg/kg/10W-C.

Hazard Overviews

Fire Diamond

Health: Irritating to eyes/skin/respiratory tract. Also Ccauses: shortness of breath, bronchitis, conjunctivitis, allergic reaction, diarrhea, flatulance, fecal impaction. Chronic Effects: allergic sensitization.
Fire: Use dry chemical, carbon dioxide, water spray, fog, or regular foam. Remove cylinder from fire to prevent rupture due to heat.
Reactivity: Stable. Hazardous polymerization cannot occur. Avoid: excessive heat. Incompatible with: ferric salts; borax; basic lead acetate; alcohol; sodium silicate; gelatin; ammoniated tincture of guaiac; peroxidase. Hazardous decomposition products: carbon dioxide.
Carcinogenicity: IARC - Not listed; NIOSH - Not listed; NTP - Not listed; ACGIH - Not listed; OSHA - Not listed; EPA - Not listed; MAK - Not listed
Primary Target Organs:

Eyes

Skin

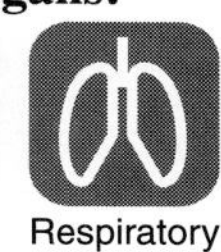
Respiratory System

Mucous Membranes

Environmental

Regulations

RCRA 40CFR: Not listed
CERCLA: 40CFR 302.4: Not listed
SARA 40CFR 372.65: Not listed
SARA EHS 40CFR 355: Not listed
TSCA: Listed

GUM3000 CAS #: 9000-05-9

GUM BENZOIN

EINECS Number: 232-523-7
Molecular Formula: Unknown
Synonyms: BENJAMIN GUM; BENZOIN GUM; BENZOIN RESIN; GUM BENJAMIN; RESIN BENJAMIN; RESIN BENZOIN; SIAM BENZOIN; SUMATRA BENZOIN
Description: reddish-brown globules; vanilla-like odor
Use: preserving ointments; preparing natural benzoic acid; for fumigating pastilles; in perfumery & cosmetics; topical protectant; vet: topically as an antiseptic & to promote healing; an inhalant for bronchitis; an expectorant tincture; expectorant; in various prepn as an antiseptic; flavor

ingredient in foods & beverages; oral protective agent for respiratory inflammations

Physical Properties

Water Solubility: Insoluble in Water
Other Solubilities: Soluble in warm Alcohol & Carbon Disulfide
pH: Alcohol solution is acid

Hazard Overviews

Carcinogenicity: IARC - Not listed; NIOSH - Not listed; NTP - Not listed; ACGIH - Not listed; OSHA - Not listed; EPA - Not listed; MAK - Not listed

Environmental

Regulations

RCRA 40CFR: Not listed
CERCLA: 40CFR 302.4: Not listed
SARA 40CFR 372.65: Not listed
SARA EHS 40CFR 355: Not listed
TSCA: Listed

GUM6000 CAS #: 9000-29-7

GUM GUAIAC

RTECS: ME6260000
EINECS Number: 232-535-2
Molecular Formula: Unknown
Formula Weight: N/A
Synonyms: GUAIAC; GUAIAC GUM; GUAIAC RESIN; GUM GUAIACUM; RESIN GUAIAC
Description: brown or greenish-brown, irregular lumps
Use: preservative in edible fats & oils; as reagent testing for occult blood; in vitro diagnostic aid (blood or hemoglobin); natural flavor ingredient & antioxidant in foods; fragrance ingredient & fixative in perfumes; varnish ingredient

Physical Properties

Freezing Point: 85 °C (185 °F) to 90 °C (194 °F)
Water Solubility: Insoluble in Water
Other Solubilities: Slightly Soluble in Benzene, Carbon Disulfide; Soluble in Creosote, solution of Chloral Hydrate, alkalies.

RTECS Toxicity Data

Acute Oral: Rat LD_{50} Dose: >5 gm/kg; Toxic Effects: Skin and appendages - Hair; Nutritional and gross metabolic - Weight loss or decreased weight gain. Mouse LD_{50} Dose: >2 gm/kg.

Hazard Overviews

Health: May cause irritation to eyes/skin/respiratory tract. Harmful. Other Acute Effects: harmful if swallowed; may be harmful if inhaled or absorbed through the skin.
Fire: Hazards: emits toxic fumes. Extinguishing agents: water spray; carbon dioxide, dry chemical powder or appropriate foam. Precautions: combustible liquid.
Reactivity: Incompatible with: strong oxidizing agents. Hazardous decomposition products: toxic fumes of: carbon monoxide, carbon dioxide.
Carcinogenicity: IARC - Not listed; NIOSH - Not listed; NTP - Not listed; ACGIH - Not listed; OSHA - Not listed; EPA - Not listed; MAK - Not listed

Environmental

Regulations

RCRA 40CFR: Not listed
CERCLA: 40CFR 302.4: Not listed
SARA 40CFR 372.65: Not listed
SARA EHS 40CFR 355: Not listed
TSCA: Not listed

GUM9000 CAS #: 39300-88-4

GUM TARA

RTECS: MG1017000
EINECS Number: 254-409-6
Molecular Formula: Unknown
Formula Weight: N/A
Synonyms: TARA GUM
Description: off-white powder; almost odorless
Use: food and cosmetic industries

Physical Properties

Water Solubility: < 1 mg/mL at 17 C
Other Solubilities: 95% Ethanol: <1 mg/ml at 17 °C; Acetone: <1 mg/ml at 17 °C; DMSO: <1 mg/ml at 17 °C.
pH: 1% aqueous solution 3.0 to 7.5
Flash Point: Not available; probably combustible

Hazard Overviews

Fire: Will burn.
Carcinogenicity: IARC - Not listed; NIOSH - Not listed; NTP - Not listed; ACGIH - Not listed; OSHA - Not listed; EPA - Not listed; MAK - Not listed

Environmental

Regulations

RCRA 40CFR: Not listed
CERCLA: 40CFR 302.4: Not listed
SARA 40CFR 372.65: Not listed
SARA EHS 40CFR 355: Not listed
TSCA: Not listed

GYP5000 CAS #: 10101-41-4

GYPSUM

RTECS: EW4150000
Molecular Formula: CaH_4O_6S
Formula Weight: 172.18

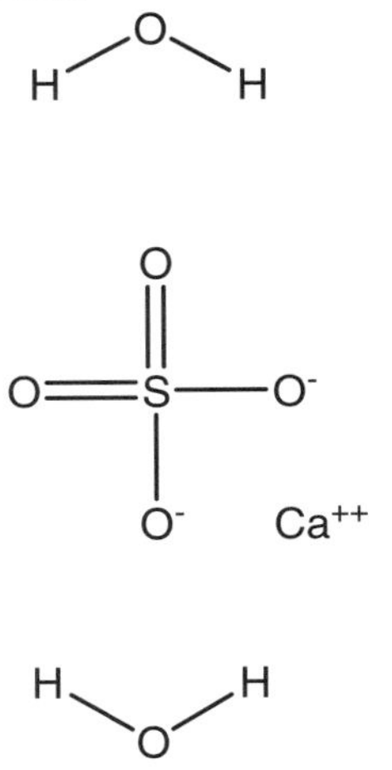

Chemical Structure

Synonyms: ANNALINE; C.I. 77231; COMPACTROL; LIGHT SPAR; MAGNESIA WHITE; MINERAL WHITE; NATIVE CALCIUM SULFATE; PRECIPITATED CALCIUM SULFATE; SATIN SPAR; SATINITE; SULFURIC ACID,CALCIUM(2+) SALT,DIHYDRATE; TERRA ALBA

Physical Properties

Freezing Point: Decomposes at 100 °C (212 °F)
Specific Gravity: 2.32
Water Solubility: 2 g/l at 20 °C
Flash Point: Noncombustible

Hazard Overviews

Health: May cause irritation to eyes/skin/respiratory tract. Other Acute Effects: may be harmful if inhaled or swallowed.
Fire: Noncombustible. Extinguishing agents: noncombustible; use extinguishing media appropriate to surrounding fire conditions. Precautions: combustible liquid.
Reactivity: Incompatible with: strong oxidizing agents, protect from moisture.
Carcinogenicity: IARC - Not listed; NIOSH - Not listed; NTP - Not listed; ACGIH - Not listed; OSHA - Not listed; EPA - Not listed; MAK - Not listed

Environmental

Regulations

RCRA 40CFR: Not listed
CERCLA: 40CFR 302.4: Not listed
SARA 40CFR 372.65: Not listed
SARA EHS 40CFR 355: Not listed
TSCA: Not listed

GYR5000 CAS #: 16568-02-8

GYROMITRIN

RTECS: LQ8500000
Molecular Formula: $C_4H_8N_2O$
Formula Weight: 100.14
Synonyms: ACETALDEHYDE N-FORMYL-N-METHYLHYDRAZONE; ACETALDEHYDE METHYLFORMYLHYDRAZONE; ACETALDEHYDE-N-FORMYL-N-METHYLHYDRAZONE; ACETALDEHYDE-N-METHYL-N-FORMYLHYDRAZONE; ACETYLALDEHYDE-N-METHYL-N-FORMYLHYDRAZONE; CETALDEHYDE METHYLFORMYLHYDRAZONE; ETHYLIDENE GYROMITRIN; FORMIC ACID,ETHYLIDENEMETHYLHYDRAZIDE; HYDRAZINE CARBOXALDEHYDE,ETHYLIDENEMETHYL-; HYDRAZINE CARBOXYALDEHYDE,ETHYLIDENEMETHYL-; HYDRAZINECARBOXALDEHYDE,ETHYLIDENEMETHYL-; N-METHYL-N-FORMYL HYDRAZONE OF ACETALDEHYDE
Description: colorless liquid

Physical Properties

Boiling Point: 143 °C (289 °F)
Freezing Point: 19.5 °C (67.1 °F) Vacuum
Specific Gravity: 1.05 at 20 °C
Water Solubility: Soluble in Water
Other Solubilities: Soluble in Acetone, Benzene, Carbon Tetrachloride, Chloroform, Diethyl Ether, Ethanol, Ethyl Acetate, Methanol, Dichloromethane.

RTECS Toxicity Data

Acute Oral: Rat LD_{50} Dose: 320 mg/kg; Toxic Effects: Behavioral - Somnolence (general depressed activity); Behavioral - Convulsions or effect on seizure threshold; Behavioral - Food intake (animal). Mouse LD_{50} Dose: 344 mg/kg.
Tumorigenic: Mouse Route: Oral; Dose: 5200 mg/kg/52W-I; Toxic Effects: Tumorigenic - Equivocal tumorigenic agent by RTECS criteria; Lungs, Thorax, or Respiration - Tumors; Endocrine - Tumors. Mouse Route: Subcutaneous; Dose: 600 mg/kg/12W-I; Toxic Effects: Tumorigenic - Carcinogenic by RTECS criteria; Lungs, Thorax, or Respiration - Tumors; Endocrine - Tumors.

Hazard Overviews

Carcinogenicity: IARC - Group 3, Not classifiable as to carcinogenicity to humans; NIOSH - Not listed; NTP - Not listed; ACGIH - Not listed; OSHA - Not listed; EPA - Not listed; MAK - Not listed

Environmental

Regulations

RCRA 40CFR: Not listed
CERCLA: 40CFR 302.4: Not listed
SARA 40CFR 372.65: Not listed
SARA EHS 40CFR 355: Not listed
TSCA: Not listed